Nutraceuticals
Efficacy, Safety, and Toxicity

Companion Web Site:

http://store.elsevier.com/product.jsp?&isbn=9780128212158

Nutraceuticals, Second Edition

Ramesh C. Gupta, Rajiv Lall, Ajay Srivastava, Editors

Available Resources:

Supplementary Files from Chapter 24: Systems Pharmacology Investigation of Mechanism of Action of Nutraceuticals

- Supplementary Table 1. Full list of epigallocatechin functional partner predictions.
- Supplementary Table 2. Full list of quercetin functional partner predictions.
- Supplementary Table 3. Full list of resveratrol functional partner predictions.
- Supplementary Table 4. Full list of TNFα functional partner predictions.
- Supplementary Table 5. Full overrepresentation analysis results of epigallocatechin.
- Supplementary Table 6. Full overrepresentation analysis results of quercetin.
- Supplementary Table 7. Full overrepresentation analysis results of resveratrol.
- Supplementary Table 8. Full overrepresentation analysis results of TNFα.
- Supplementary Figure 1. Quercetin network.
- Supplementary Figure 2. Resveratrol network.
- Supplementary Figure 3. TNFα network.

ELSEVIER

ACADEMIC PRESS

Nutraceuticals
Efficacy, Safety, and Toxicity

Second Edition

Edited by

Ramesh C. Gupta
Professor and Head, Toxicology Department
Breathitt Veterinary Center
Murray State University
Hopkinsville, KY
United States

Rajiv Lall
Founder and CEO
ProbioticSmart, LLC
and Vets Plus, Inc.
Menomonie, WI
United States

Ajay Srivastava
Chief Scientific Officer
Vets Plus, Inc.
Menomonie, WI
United States

ELSEVIER
ACADEMIC PRESS
An imprint of Elsevier

Academic Press is an imprint of Elsevier
125 London Wall, London EC2Y 5AS, United Kingdom
525 B Street, Suite 1650, San Diego, CA 92101, United States
50 Hampshire Street, 5th Floor, Cambridge, MA 02139, United States
The Boulevard, Langford Lane, Kidlington, Oxford OX5 1GB, United Kingdom

British Library Cataloguing-in-Publication Data
A catalogue record for this book is available from the British Library

Library of Congress Cataloging-in-Publication Data
A catalog record for this book is available from the Library of Congress

ISBN: 978-0-12-821038-3

For Information on all Academic Press publications
visit our website at https://www.elsevier.com/books-and-journals

Publisher: Andre Gerhard Wolff
Acquisitions Editor: Kattie Washington
Editorial Project Manager: Kristi Anderson
Production Project Manager: Kiruthika Govindaraju
Cover Designer: Matt Limbert

Typeset by MPS Limited, Chennai, India

Last digit is the print number: 10 9 8 7 6 5 4 3

Dedication

Dedicated to my wife Denise Gupta, daughter Dr. Rekha K. Gupta, and my parents,
the late Chandra and Triveni Gupta.

—Ramesh C. Gupta

Dedicated to my wife Swati Lall, sons Dr. Rohan Lall and Dr. Rishi Lall,
mother Kanak Lata Lall, and late father Professor R.B. Lall.

—Rajiv Lall

Dedicated to my parents, my wife Garima, my son Sankalp, daughter Ayana, and my friends.

—Ajay Srivastava

Contents

List of contributors

Abul H. Ahmad, G.B. Pant University of Agriculture and Technology, College of Veterinary Science, Department of Pharmacology and Toxicology, Pantnagar, India

James Akingbasote, Regulatory Department, KGK Science, London, ON, Canada

Arturo Anadón, Department of Pharmacology and Toxicology, Faculty of Veterinary Medicine, Universidad Complutense de Madrid, Madrid, Spain

Rashmi Anand, Department of Pharmacology, Vallabhbhai Patel Chest Institute, University of Delhi, Delhi, India

Vellareddy Anantharam, Parkinson's Disorder Research Laboratory, Iowa Center for Advanced Neurotoxicology, Department of Biomedical Sciences, Iowa State University, Ames, IA, United States

Csaba András, Faculty of Technical and Social Sciences, Sapientia Hungarian University of Transylvania, Miercurea Ciuc, Romania

Irma Ares, Departamento de Toxicología y Farmacología Facultad de Veterinaria, Universidad Complutense de Madrid, Madrid, Spain

Pavel V. Avdonin, Koltzov Institute of Developmental Biology, Russian Academy of Sciences, Moscow, Russia

Muhammet Ay, Parkinson's Disorder Research Laboratory, Iowa Center for Advanced Neurotoxicology, Department of Biomedical Sciences, Iowa State University, Ames, IA, United States

Aryamitra Banerjee, Chicago, IL, United States

Reid E. Barnett, Department of Biology, King University, Bristol, TN, United States

Daria A. Belinskaia, Sechenov Institute of Evolutionary Physiology and Biochemistry, Russian Academy of Sciences, Saint Petersburg, Russia

Alessia Bertero, Università degli Studi di Milano, Department of Environmental Science and Policy, Milan, Italy

Vijay K. Bharti, Nutrition and Toxicology Laboratory, DRDO-Defence Institute of High Altitude Research (DIHAR), Leh, Ladakh, India

Wan-Ping Bian, Chongqing Institute of Green and Intelligent Technology, Chinese Academy of Sciences, Chongqing, China

Karyn L. Bischoff, New York State Animal Health Diagnostic Center, Cornell Unviersity, Ithaca, NY, United States

Francesca Caloni, Department of Environmental Science and Policy, Università degli Studi di Milano, Milan, Italy

Ana Catarina Castela, Department of Life Sciences, Center for Neurosciences and Cell Biology, University of Coimbra, Coimbra, Portugal

Naresh Chand, R&D, ADRxSynergy, North Potomac, MD, United States

Ronald F. Chanderbhan, Division of Food Ingredients, Office of Food Additive Safety, Center for Food Safety and Applied Nutrition, U.S. Food and Drug Administration, FDA/CFSAN/OFAS, College Park, MD, United States

Adhithiya Charli, Parkinson's Disorder Research Laboratory, Iowa Center for Advanced Neurotoxicology, Department of Biomedical Sciences, Iowa State University, Ames, IA, United States

Andrew Charrette, Regulatory Department, KGK Science, London, ON, Canada

Si Chen, Division of Biochemical Toxicology, National Center for Toxicological Research, U.S. Food and Drug Administration, Jefferson, AR, United States

Tao Chen, Division of Genetic and Molecular Toxicology, National Center for Toxicological Research, U.S. Food and Drug Administration, Jefferson, AR, United States

Supratim Choudhuri, Division of Food Ingredients, Office of Food Additive Safety, Center for Food Safety and Applied Nutrition, U.S. Food and Drug Administration, FDA/CFSAN/OFAS, College Park, MD, United States

Soumen Choudhury, Department of Pharmacology and Toxicology, College of Veterinary Science and Animal Husbandry, U.P. Pandit Deen Dayal Upadhyaya Pashu Chikitsa Vigyan Vishwavidyalaya Evam Go-Anusandhan Sansthan (DUVASU), Mathura, India

Teresa Coccini, Laboratory of Clinical and Experimental Toxicology, Toxicology Unit, ICS Maugeri SpA-BC, IRCCS Pavia, Pavia, Italy

Rhian B. Cope, Risk Assessment Capability, Australian Pesticides and Veterinary Medicines Authority, Armidale, NSW, Australia

Robert W. Coppock, Toxicology, R. W. Coppock, DVM, Toxicologist and Assoc. Ltd, Vegreville, AB, Canada

Lucio G. Costa, University of Washington, Department of Environmental and Occupational Health Sciences, Seattle, WA, United States

Tirupapuliyur Damodaran, Department of Biological and Biomedical Sciences, North Carolina Central University, Durham, NC, United States

Altaf S. Darvesh, Department of Pharmaceutical Sciences, College of Pharmacy, Department of Psychiatry, College of Medicine, Northeast Ohio Medical University, Rootstown, OH, United States

Leena Dhruw, Center for Basic Sciences, Pt. Ravishankar Shukla University Raipur (C. G.), Raipur, India

György Dormán, Medicinal Chemistry, Targetex LLC, Kft., H-2120, Kápolna köz 4/a, Dunakeszi; Institute of Pharmaceutical Chemistry, University of Szeged, Eötvös u. 6, Szeged H-6720, Hungary

Robin B. Doss, Toxicology Department, Breathitt Veterinary Center, Murray State University, Hopkinsville, KY, United States

Dan DuBourdieu, Research Department, Vets Plus, Inc., Noble Pharma LLC, Menomonie, WI, United States

Margitta Dziwenka, Health Sciences Laboratory Animal Services (HSLAS), Faculty of Medicine and Dentistry, University of Alberta, Edmonton, AB; ToxAlta Consulting, Vegreville, AB, Canada

Joshua Eades, Tilray, San Diego, CA, United States

Harald L. Esch, University of Würzburg, Institute of Pharmacy and Food Chemistry, Würzburg, Germany

Ayhan Filazi, Ankara University Faculty of Veterinary Medicine, Department of Pharmacology and Toxicology, Ankara, Turkey

Vanessa A. Fitsanakis, Toxicology Practice Area, Robson Forensic, Inc., Charleston, SC; Department of Pharmaceutical Sciences, College of Pharmacy, Northeast Ohio Medical University, Rootstown, OH, United States

Beáta Flachner, Targetex Kft., H-2120 Dunakeszi, Kápolna köz 4/a; Institute of Enzymology, Research Centre for Natural Sciences, Hungarian Academy of Sciences, Budapest, Hungary

Paola Fossati, Università degli Studi di Milano, Dipartimento di Scienze Veterinarie per la Salute, la Produzione Animale e la Sicurezza Alimentare, Milan, Italy

Stepan P. Gambaryan, Sechenov Institute of Evolutionary Physiology and Biochemistry, Russian Academy of Sciences, Saint Petersburg, Russia

Ramesh C. Garg, Chicago, IL, United States

Satish K. Garg, Department of Pharmacology and Toxicology, College of Veterinary Science and Animal Husbandry, U.P. Pandit Deen Dayal Upadhayaya Pashu Chikitsa Vigyan Vishwavidyalaya Evam Go-Anusandhan Sansthan (DUVASU), Mathura, India

Fernando Gil, Department of Legal Medicine and Toxicology, School of Medicine, University of Granada, Granada, Spain

Nikolay V. Goncharov, Comparative Biochemistry, Sechenov Institute of Evolutionary Physiology and Biochemistry, Russian Academy of Sciences, Saint Petersburg; Toxicology, Research Institute of Hygiene, Occupational Pathology and Human Ecology (RIHOPHE), Leningrad Region, Russia

Kavita Gulati, Department of Pharmacology, Vallabhbhai Patel Chest Institute, University of Delhi, Delhi, India

Lei Guo, Division of Biochemical Toxicology, National Center for Toxicological Research, U.S. Food and Drug Administration, Jefferson, AR, United States

Xiaoqing Guo, Division of Genetic and Molecular Toxicology, National Center for Toxicological Research, U.S. Food and Drug Administration, Jefferson, AR, United States

Bhanushree Gupta, Center for Basic Sciences, Pt. Ravishankar Shukla University, Raipur, India

Ramesh C. Gupta, Toxicology Department, Breathitt Veterinary Center, Murray State University, Hopkinsville, KY, United States

Najla Guthrie, KGK Science, London, ON, Canada

Sharon M. Gwaltney-Brant, Veterinary Information Network, Mahomet, IL; Pathology, Immunology, and Laboratory Medicine, College of Medicine, University of Florida, Gainesville, FL; Comparative Biosciences, College of Veterinary Medicine, University of Illinois, Urbana, IL, United States

István Hajdú, Targetex Kft., H-2120 Dunakeszi, Kápolna köz 4/a; Institute of Enzymology, Research Centre for Natural Sciences, Hungarian Academy of Sciences, Budapest. Hungary

Joshua A. Hartsel, Delta-9 Technologies, LLC, Lake Forest, CA, United States

Trina Hazzah, Oncology Department, VCA West Los Angeles Animal Hospital, Los Angeles, CA, United States

Antonio F. Hernández, Legal Medicine and Toxicology, University of Granada, Granada, Spain

Brian Hickory, Sustainable syntheses, San Diego, CA, United States

Corey J. Hilmas, Regulatory Division, KGK Science, London, ON, Canada

Julia Hoeng, Philip Morris International R&D, Philip Morris Products S.A., Neuchâtel, Switzerland

Amul Jain, Center for Basic Sciences, Pt. Ravishankar Shukla University Raipur (C. G.), Raipur, India

Richard O. Jenkins, Leicester School of Allied Health Sciences, De Montfort University, Leicester, United Kingdom

Huajun Jin, Parkinson's Disorder Research Laboratory, Iowa Center for Advanced Neurotoxicology, Department of Biomedical Sciences, Iowa State University, Ames, IA, United States

Jagdish Chandra Joshi, Department of Pharmacology, Vallabhbhai Patel Chest Institute, University of Delhi, Delhi, India

Tatiana Kalinovsky, Dr. Rath Research Institute, Santa Clara, CA, United States

Starling Kalpana, ICAR-National Research Centre on Meat, Chengicherla, Boduppal, India

Sanyasi R. Kalidindi, Natreon Inc, New Brunswick, NJ, United States

Anumantha G. Kanthasamy, Parkinson's Disorder Research Laboratory, Iowa Center for Advanced Neurotoxicology, Department of Biomedical Sciences, Iowa State University, Ames, IA, United States

Arthi Kanthasamy, Parkinson's Disorder Research Laboratory, Iowa Center for Advanced Neurotoxicology, Department of Biomedical Sciences, Iowa State University, Ames, IA, United States

Navinchandra M. Kaore, Department of Microbiology, People's College of Medical Sciences and Research Center, Bhopal, India

Shilpa N. Kaore, Department of Pharmacology, All India Institute of Medical Sciences, Saket Nagar, Bhopal, India

Vijay Kumar Kapoor, School of Pharmaceutical Sciences, Shoolini University, Solan, India

Gurjot Kaur, Human and Environmental Toxicology, University of Konstanz, Konstanz, Germany; School of Pharmaceutical Sciences, Shoolini University, Solan, India

Renata Kolanos, College Park, MD, United States

Ekaterina A. Korf, Sechenov Institute of Evolutionary Physiology and Biochemistry, Russian Academy of Sciences, Saint Petersburg, Russia

Gopala Krishna, Toxicology, Expert consultants, Ellicott City, MD, United States

Mayur Krishna, Ellicott City, MD, United States

Kamil Kuča, Department of Chemistry, Faculty of Science, University of Hradec Kralove, Hradec Kralove; Biomedical Research Center, University Hospital Hradec Kralove, Hradec Kralove, Czech Republic

Igor V. Kudryavtsev, Institute of Experimental Medicine, Russian Academy of Medical Sciences, St. Petersburg; School of Biomedicine, Far Eastern Federal University, Vladivostok, Russia

Amit Kumar, Frederick, MD, United States

Dinesh Kumar, Division of Pharmacology and Toxicology, Indian Veterinary Research Institute, Izatnagar, India

Prafull Kumar, Vets Plus, Inc., Menomonie, WI, United States

Rajiv Lall, ProbioticSmart, LLC, and Vets Plus, Inc., Menomonie, WI, United States

Camillo La Mesa, Department of Chemistry "Stanislao Cannizzaro"—Sapienza University of Rome, Rome, Italy

Diogo A.R.S. Latino, Philip Morris International R&D, Philip Morris Products S.A., Neuchâtel, Switzerland

Leane Lehmann, Institute of Pharmacy and Food Chemistry, University of Wuerzburg, Wuerzburg, Germany

Xilin Li, Division of Genetic and Molecular Toxicology, National Center for Toxicological Research, U.S. Food and Drug Administration, Jefferson, AR, United States

Ivo F. Machado, Department of Life Sciences, Center for Neurosciences and Cell Biology, University of Coimbra, Coimbra, Portugal

Alexandros Makriyannis, Center for Drug Discovery, Northeastern University, Boston, MA, United States

Jitendra K. Malik, Division of Veterinary Pharmacology and Toxicology, Indian Veterinary Research Institute, Dehradun, India

M. Concepción Martín-Domingo, Department of Legal Medicine and Toxicology, School of Medicine, University of Granada, Granada, Spain

María-Aránzazu Martínez, Departamento de Toxicología y Farmacología Facultad de Veterinaria, Universidad Complutense de Madrid, Madrid, Spain

María-Rosa Martínez-Larrañaga, Departamento de Toxicología y Farmacología Facultad de Veterinaria, Universidad Complutense de Madrid, Madrid, Spain

Nan Mei, Division of Genetic and Molecular Toxicology, National Center for Toxicological Research, U.S. Food and Drug Administration, Jefferson, AR, United States

Dimitris E. Messinis, Philip Morris International R&D, Philip Morris Products S.A., Neuchâtel, Switzerland

Igor V. Mindukshev, Sechenov Institute of Evolutionary Physiology and Biochemistry, Russian Academy of Sciences, Saint Petersburg, Russia

Poonam Negi, School of Pharmaceutical Sciences, Shoolini University, Solan, India

Aleksandra Niedzwiecki, Dr. Rath Research Institute, Santa Clara, CA, United States

Baitang Ning, Division of Bioinformatics and Biostatistics, National Center for Toxicological Research, U.S. Food and Drug Administration, Jefferson, AR, United States

Artem V. Novozhilov, Sechenov Institute of Evolutionary Physiology and Biochemistry, Russian Academy of Sciences, Saint Petersburg, Russia

Carlos Marques Palmeira, Department of Life Sciences, Center for Neurosciences and Cell Biology, University of Coimbra, Coimbra, Portugal

Andreas M. Papas, Department of Pediatrics, East Tennessee State University, CEO, Antares Health Products, Johnson City, TN, United States

De-Sheng Pei, Chongqing Institute of Green and Intelligent Technology, Chinese Academy of Sciences, Chongqing, China

Manuel C. Peitsch, Philip Morris International, R&D, Philip Morris Products S.A., Neuchatel, Switzerland

Jeffrey D. Peterson, Director of Applied Biology, Research and Development, Perkin Elmer, Inc, Hopkinton, MA, United States

Francesca Pistollato, EU Commission Joint Research Centre (JRC), Ispra, Italy

Jason Pitt, Physical Therapy, University of Evansville, Evansville, IN, United States

Yiuka Pitt, Deaconess Henderson Hospital, Henderson, KY, United States

Pavel Pospisil, Philip Morris International R&D, Philip Morris Products S.A., Neuchâtel, Switzerland

Carine Poussin, Philip Morris International R&D, Philip Morris Products S.A., Neuchâtel, Switzerland

Aanchal Pradhan, Center for Basic Sciences, Pt. Ravishankar Shukla University Raipur (C. G.), Raipur, India

Atul Prakash, U.P. Pandit Deen Dayal Upadhyaya Pashu Chikitsa Vigyan Vishwavidyalaya Evam Go-Anusandhan Sansthan (DUVASU), College of Veterinary Science and Animal Husbandry, Department of Pharmacology and Toxicology, Mathura, India

Sahdeo Prasad, Experimental Therapeutics, UT MD Anderson Cancer Center, Hoston, TX; Department of Immunotherapeutics and Biotechnology, Texas Tech University Health Sciences Center, Lubbock, TX, United States

Anu Rahal, Animal Health, ICAR-Central Institute for Research on Goats, Mathura, India

Nishant Rai, Department of Pharmacology, Vallabhbhai Patel Chest Institute, University of Delhi, Delhi, India

Rajinder Raina, Division of Veterinary Pharmacology and Toxicology, Faculty of Veterinary Sciences and Animal Husbandry, Sher-e-Kashmir University of Agricultural Sciences and Technology Jammu, Jammu, India

Matthias Rath, Dr. Rath Research Institute, Santa Clara, CA, United States

Arunabha Ray, Department of Pharmacology, Hamdard Institute of Medical Sciences and Research (HIMSR), Hamdard University, Jamia Hamdard, New Delhi, India

Sana Rehman, Department of Pharmacology, Hamdard Institute of Medical Sciences and Research (HIMSR), Hamdard University, Jamia Hamdard, New Delhi, India

Gary Richter, Integrative Medicine, Holistic Veterinary Care, Oakland, CA, United States

Gianfranco Risuleo, Dipartimento di Biologia e Biotecnologie, Department of Biology and Biotechnologies "Charles Darwin", Sapienza Università di Roma, Rome, Italy

Ashley Roberts, Food and Nutrition, Intertek Scientific and Regulatory Consultancy, Mississauga, ON, Canada

Anabela Pinto Rolo, Department of Life Sciences, Center for Neurosciences and Cell Biology, University of Coimbra, Coimbra, Portugal

M. Waheed Roomi, Dr. Rath Research Institute, Santa Clara, CA, United States

Magdalini Sachana, Environment, Health and Safety Division, Organization for Economic Cooperation and Development (OECD), Paris, France

Tetsuo Satoh, Pharmacology and Toxicology, Chiba University, Chiba, Japan

Maria K. Serebriakova, Institute of Experimental Medicine, Russian Academy of Medical Sciences, St. Petersburg, Russia

Rahul Sharma, Department of Plant Physiology, Agri. Biochem. and Medicinal and Aromatic Plants, Indira Gandhi Krishi Vishwavidyalaya, Raipur, India

Amit Shukla, Department of Pharmacology and Toxicology, College of Veterinary Science and Animal Husbandry, U.P. Pandit Deen Dayal Upadhyaya Pashu Chikitsa Vigyan Vishwavidyalaya Evam Go-Anusandhan Sansthan (DUVASU), Mathura, India

Anita Sinha, Vets Plus, Inc., Menomonie, WI, United States

Priynak Sinha, Center for Basic Sciences, Pt. Ravishankar Shukla University Raipur (C. G.), Raipur, India

Leon J Spicer, Department of Animal and Food Sciences, Oklahoma State University, Stillwater, OK, United States

Ajay Srivastava, Vets Plus, Inc., Menomonie, WI, United States

Szabina A. Stice, College Park, MD, United States

Julia S. Sudnitsyna, Sechenov Institute of Evolutionary Physiology and Biochemistry, Russian Academy of Sciences, Saint Petersburg, Russia

Ila Taku, Training Internee, Division of Cancer Pharmacology, Indian Institute of Integrative Medicine (CSIR), Jammu, India

Jamil Talukder, Research and Development, Vets Plus, Inc., Menomonie, WI, United States

João Soeiro Teodoro, Department of Life Sciences, Center for Neurosciences and Cell Biology, University of Coimbra, Coimbra, Portugal

Suresh Kumar Thokchom, Department of Pharmacology, Vallabhbhai Patel Chest Institute, University of Delhi, Delhi, India

Andrey S. Trulioff, Institute of Experimental Medicine, Russian Academy of Medical Sciences, St. Petersburg, Russia

Anton I. Ukolov, Research Institute of Hygiene, Occupational Pathology and Human Ecology, Leningrad Region, Russia

Keizo Umegaki, Department of Food Safety and Management, Showa Women's University, Setagaya-Ku, Japan

Pankaj Verma, Department of Pharmacology, Vallabhbhai Patel Chest Institute, University of Delhi, Delhi, India

Pawan K. Verma, Division of Veterinary Pharmacology and Toxicology, Faculty of Veterinary Sciences and Animal Husbandry, Jammu, India

Dan Wan, Hunan Provincial Key Laboratory of Animal Nutritional Physiology and Metabolic Process, Key Laboratory of Agro-Ecological Processes in Subtropical Region, Institute of Subtropical Agriculture, Chinese Academy of Sciences, National Engineering Laboratory for Pollution Control and Waste Utilization in Livestock and Poultry Production, Changsha, China

Christina Wilson-Frank, College of Veterinary Medicine, Animal Disease Diagnostic Laboratory, Department of Comparative Pathobiology, Purdue University, West Lafayette, IN, United States

Moges Woldemeskel, Department of Veterinary Pathology, Tifton Veterinary Diagnostic and Investigational Laboratory, University of Georgia, College of Veterinary Medicine, Tifton, GA, United States

Qinghua Wu, College of Life Science, Yangtze University, Jingzhou, China; Department of Chemistry, Faculty of Science, University of Hradec Kralove, Hradec Kralove, Czech Republic

Mildred S. Yang, Division of Science and Technology. Food Science and Technology Programme, BNU-HKBU United International College, Zhuhai, China

Begum Yurdakok-Dikmen, Ankara University, Department of Pharmacology and Toxicology, Ankara University Faculty of Veterinary Medicine, Diskapi, Turkey

Snjezana Zaja-Milatovic, Early Product Development, Parexel International, Waltham, MA, United States

Csaba K. Zoltani, Emeritus US Army Research Lab, Aberdeen Proving Ground, MD, United States

Introduction

According to the Merriam-Webster dictionary, the term "nutraceutical" is defined as "a foodstuff (as a fortified food or dietary supplement) that provides health benefits in addition to its basic nutritional value." In 1989, Dr. Stephen DeFelice coined the term "nutraceutical" from the words nutrition and pharmaceutical and defined it as "a food (or part of a food) that provides medical or health benefits, including the prevention and/or treatment of a disease." The dietary supplement industry defines a nutraceutical as "any nontoxic food component that has scientifically proven health benefits, including treatment and prevention." Based on the definitions for nutraceuticals proposed or discussed above, a more appropriate definition would be a substance which is cultivated/produced/extracted or synthesized under optimal and reproducible conditions, which when administered orally to patients, would provide the nutrient(s) required for bringing altered body structure and function back to normal, thus improving the health and well-being of the patients. Therefore nutraceuticals fall somewhere between food nutrients and drugs.

A nutraceutical may be more than a single food or plant component(s) that has an advantage over foods and drugs; the nutraceuticals are not required to be listed in the nutrient profiles. Additionally, regulation of nutraceuticals varies widely around the world. For example, China allows nutraceutical claims to treat and prevent diseases with a formal approval from that country's regulatory body, while the United States does not allow any health claims for nutraceuticals because there is no formal review and approval process for the marketing authorization of nutraceuticals. The only major US regulation related to nutraceuticals is the 1994 passage of the Dietary Supplement Health and Education Act by the US Congress. Based on this Act, dietary supplements are classified as foods, not drugs, allowing them to be sold without proof of safety or effectiveness (US FDA, 1994).

It has been estimated that more than 150 million Americans consume either a single or multiple dietary supplements, yet it is not necessary for companies to seek Food and drug Administration (FDA) approval before marketing and manufacturing these supplements in compliance with the Dietary Supplement Health and Education Act of 1994. Nutraceuticals can be food or a part of food, but unlike foods, they all are not generally recognized as safe, nor can one assume that all nutraceuticals are safe (Gupta et al., 2018).

By the turn of the 21st century, the use of nutraceuticals became increasingly popular for both humans and animals around the world. Currently, nutraceuticals are a more than $250 billion per year industry, and the annual sale of nutraceuticals is growing at the rate of 7%–10%. While some nutraceuticals (turmeric/curcumin, *Ginkgo biloba*, St. John's wort, ginseng, glucosamine, chondroitin, and many others) are still in high demand, the hemp/cannabis industry is making headlines on a daily basis. The cannabis industry is forecast to reach $80 billion by 2030. The two main drivers of nutraceutical market growth will continue to be Baby Boomers and Millennials. Of course, senior citizens consume nutraceuticals for a number of diseases, lost muscle mass, and inactivity. The use of nutraceuticals is also becoming popular within the pediatric population. Currently, the top five trends in the nutraceutical industry include: (1) hemp/cannabis, (2) nutricosmetics, (3) supplements for man's furry friends, (4) plastic packaging problems, and (5) niche sports nutrition. The continuous growing demand of nutraceuticals for chemoprevention, antiaging, weight loss, sports and energy, and social/recreational beverages needs to be highlighted.

Most herbal supplements are classified as dietary supplements and are not subject to the regulations and safety standards applied to conventional medicine in the United States. Current European Union regulations require evidence that herbal medicinal products meet acceptable standards of quality, safety, and efficacy before a product license can be issued. As the global use of dietary supplements has increased, so have health risks to the public emerging from the safety of active and inactive components, and potential contaminants of dietary supplements, as well as likely interactions with other medications and/or dietary supplements. Consultations with physicians and pharmacists are essential for patient safety, given the potential for side effects and toxicity that may be associated with some nutraceuticals. An updated understanding of currently available scientific information for nutraceuticals and their potential side effects is therefore crucial in well-informed patient care.

Pharmacological and toxicological evaluation of phytochemicals, compared to pure synthetic pharmaceuticals, is complex due to: (1) multiple phytochemicals present in a single plant; (2) variability in phytochemical constituents due to geographic and climatic factors; (3) use of fertilizers and pesticides; and (4) diurnal variation during harvesting. Also, unlike pharmaceuticals, there is a need for more than one active component in many cases and approaches taken for extraction (standardized, normalized, or quantified extract preparation) of the active component (s). Some or all of these factors can influence the identity, purity, quality, quantity, composition, potency/strength, and safety of active component(s), thereby causing wide variability in product effectiveness from batch to batch and from one manufacturer to the other.

Herbal medicines have been used since the ancient practices of holistic healing in Chinese, Indian (Ayurveda, homeopathic, Siddha, or Unani), Middle-Eastern, Russian, and many other cultural and traditional systems for the prevention and treatment of various acute and chronic ailments. Thousands of these ancient medicines/products have been documented in Chinese, Indian, Japanese, European, and US herbal pharmacopeias. Phytochemicals have been the armamentarium of major health care since ancient human civilization. Even today, about 80% of the world's populations living in developing countries rely on traditional medicine for their primary health care. Currently, the use of nutraceuticals or complementary and alternative medicines is preferred over modern synthetic medicines for multiple reasons, such as wide margin of safety, cost-effectiveness, and lack of drug resistance.

By having antioxidative, antiinflammatory, immunomodulatory, adaptogenic, cytoprotective, apoptotic/antiapoptotic, antimicrobial, antifungal, antiparasitic, anticancer, and many other properties, nutraceuticals are used worldwide for the prevention and treatment of chronic diseases such as diabetes, arthritis, cardiovascular, respiratory, hepatic, renal, and neurodegenerative ailments, and cancer. Nutraceuticals are also used to improve general health and delay the aging process by supporting the structure and function of the body, thus contributing to an increase in life expectancy. While some nutraceuticals (e.g., curcumin, resveratrol, *Ginkgo biloba*, caffeine, anthocyanins, hyaluronate, fenugreek, ginseng, melatonin, and quercetin) have been extensively studied for their mode of action, efficacy and safety via well-characterized animal studies and clinical trials, many others are still used on an anecdotal basis.

Currently, the rapidly growing nutraceutical industry is facing many challenges, including: (1) lack of authenticity of active principles due to unavailability of reference materials or marker compounds; (2) lack of understanding of molecular interactions between bioactive phytochemicals within the same plant; (3) variability in origins of raw material (e.g., Chinese vs. American ginseng); (4) variability in processing of raw material; (5) lack of standard extraction procedures; (6) inadequacy and inconsistency in quality control standards; (7) lack of safety and toxicity data; and (8) lack of well-established and evidence-based clinical trials. An additional issue with the use of nutraceuticals is that they are poorly water soluble and highly unstable in the gastrointestinal tract, and as a result, they are inadequately absorbed with low bioavailability. To overcome this hurdle, herbal bioenhancers such as iron transport polyethylene glycol succinate have been developed. Other recent approaches for improving the solubility and bioavailability of common nutraceuticals (such as curcumin and curcuminoids) have involved complexing them with zinc, iron, and other metals. Finally, adulteration of nutraceuticals with toxic phytochemicals (such as pyrrolizidine alkaloids), metals (arsenic, lead, cadmium, and mercury), mycotoxins (aflatoxins, ochratoxins, vomitoxin, etc.), pesticides (insecticides, herbicides, fungicides, etc.), pharmaceutical drugs, and drugs of abuse can contribute to severe adverse health effects, due to a lack of national and international regulations.

Recently, herb–drug interactions have been of the utmost concern to consumers, physicians, and governmental regulatory agencies. Although the underlying mechanisms for the altered drug effects by concomitant herbal medicines have yet to be determined, both pharmacokinetic and pharmacodynamic mechanisms are considered to play a role in these interactions. Often, patients do not inform their physicians about concurrent use of nutraceuticals. This has resulted in incidences of adverse herb–-drug interactions, ranging from minor side effects to those as severe as liver or kidney failure, seizures, and even death. To address these challenges, systematic studies using integrative approaches, including in vitro and in vivo assays using animal models and human tissue/cell lines as well as human clinical trials, are warranted to better understand the bioavailability, metabolism, dose-responses, and pharmacological and toxicological actions. Additionally, scientists in the field of nutraceuticals need to develop sensitive and reliable biomarkers to determine early disease diagnosis, severity of disease and therapeutic intervention, and to validate toxicity and safety data (Gupta, 2019; Gupta et al., 2019). The partnership initiated between the United States Environmental Protection Agency, the US FDA, and the National Institutes of Health to establish a framework for mechanism-based toxicological assessment has been a great help in this direction. Above all, a thorough understanding and development of trust between the nutraceutical industry, regulators, healthcare providers, and patients is the ultimate requirement for productive and judicious use of nutraceuticals as a complementary system of medicine.

According to the WHO, traditional medicine holds great potential to improve people's health and wellness in every part of the world (Chan, 2014). There is a need to integrate traditional medicine into existing health systems, and to modernize this rich resource and cultural heritage using system biology and toxicology, "omics" technologies, bioinformatics, pharmacovigilance, and to educate consumers about what it can and cannot do. No matter the weight of historical, anecdotal data, the US FDA or European Food Safety Authority will not allow new therapeutics for human treatment without verifiable scientific evidence. Scientists in the field of nutraceuticals are applying the latest technologies in an attempt to standardize traditional treatments, especially through isolation, identification, and purification of bioactive compounds and careful analysis of their levels and activities in various herbal remedies. Both the nutraceutical industry and the US FDA acknowledge that many new products have been introduced without any scientific assessment of their efficacy and safety.

To meet the challenges of the 21st century, the nutraceutical industry needs to follow rigorous quality control, pharmacological and toxicological testing, carefully designed clinical studies, reproducibility of results, safety assessment, and proper regulations applicable to all nutraceuticals. With regards to toxicological testing, special attention needs to be paid to some toxicities that are not detected by conventional nonclinical testing, including the evaluation of acute and chronic exposure, genotoxicity, carcinogenicity, and reproductive and developmental toxicity, and a close pharmacovigilance during early drug exposure to humans (Williamson et al., 2015). In addition, the current toxicological testing paradigm embraces modern tools for predictive toxicology, molecular biology, system biology, high-throughput screening methods, computational toxicology, and bioinformatics. Good agricultural and collection practices, good laboratory practices, good manufacturing practices, and good clinical practices can help assure both practitioners and patients that a nutraceutical is effective, safe, and of high quality, meeting international standards and its acceptance in a global market.

In view of these challenges, the second edition of *Nutraceuticals: Efficacy, Safety, and Toxicity* is updated to meet the current needs of academia, industry, and regulatory agencies. The book explicitly describes the origin and historical background of many common nutraceuticals, underlying pharmacological mechanisms of action at biochemical and molecular levels (molecular pathways wherever possible), models for toxicity and safety evaluation, clinical applications, herb–drug interactions, and regulatory guidelines. There are 78 chapters, logistically arranged in seven sections. After a brief introduction, a large section covers chapters on the application of nutraceuticals in system diseases and disorders. Section II deals with various models and mechanisms involved in evaluation of the efficacy, safety, and toxicity of nutraceuticals. Thirty standalone chapters are devoted to common nutraceuticals in Section III. In Section IV, several chapters cover prebiotics, probiotics, and synbiotics. A sizeable chapter is placed in Section V dealing with veterinary nutraceuticals. The last two sections of this book deal in detail with toxicity, herb–drug interactions, and regulatory aspects of nutraceuticals from various countries and continents. Thus, this book remains the most comprehensive and authoritative text on nutraceuticals to date.

The editors remain grateful to the contributors to this book for their hard work and dedication. These contributors are international authorities in the fields of nutrition, physiology, biochemistry, neurochemistry, pharmacology, neuroscience, toxicology, molecular biology, nutraceuticals, and other areas of biomedical research. The editors express their gratitude to Robin B. Doss for her technical assistance. Finally, the editors would like to commend the tireless efforts of Kristi Anderson, Ashwathi Aravindakshan, Kattie Washington, and Kiruthika Govindaraju (Academic Press/Elsevier) for their roles in the preparation of this book.

Ramesh C. Gupta, Rajiv Lall and Ajay Srivastava

References

Chan, M., 2014. Supporting the integration and modernization of traditional medicine. Science. 346, S2.

Gupta, R.C., 2019. In: Gupta, R.C. (Ed.), Biomarkers in Toxicology, 2nd edn Academic Press/Elsevier, Amsterdam, 1222 pages.

Gupta, R.C., Srivastava, A., Lall, R., 2018. Toxicity potential of nutraceuticals. In: Nicolotti, O. (Ed.), Computational Toxicology. Methods and Protocols, Methods in Molecular Biology. Springer Nature, New York, pp. 367–394.

Gupta, R.C., Srivastava, A., Sinha, A., Lall, R., 2019. Biomarkers of foods and nutraceuticals: applications in efficacy, safety, and toxicity. In: Gupta, R.C., Srivastava, A., Lall, R. (Eds.), Nutraceuticals in Veterinary Medicine. Springer Nature, Cham, Switzerland, pp. 693–710.

US FDA, 1994. Dietary Supplement Health and Education Act of 1994. Edited by Congress, Pub. L. www.fda.gov/DietarySupplement/default.htm.

Williamson, E.M., Chan, K., Xu, Q., et al., 2015. Evaluating the safety of herbal medicines: integrated toxicological approaches. Science. 347, S47–S49.

Section I

Applications of nutraceuticals in common diseases and disorders

Chapter 1

Nutraceuticals in central nervous system diseases: potential mechanisms of neuroprotection

Ramesh C. Gupta and Lucio G. Costa

Introduction

In recent years, there has been increasing attention devoted to the possibility that several nutraceuticals may act as neuroprotective agents (Mecocci et al., 2014; Gupta et al., 2019; Singh et al., 2019). Such protective effects have often been ascribed to a direct antioxidant effect and/or to an anti-inflammatory action (Kelsey et al., 2010; Milatovic and Gupta, 2019). Some have also been shown to impact the synthesis of endogenous antioxidant molecules in cells by triggering the Nrf/ARE pathway (Dumont and Beal, 2011). However, the exact mechanisms of neuroprotection remain elusive, and various mechanisms have been proposed (Halliwell et al., 2005; Fraga et al., 2010; Gomes et al., 2018). Other chapters in this volume (e.g., Chapter 2, Prevention of Neurodegenerative Disorders and aging by Nutraceuticals, and Chapter 3, Cognitive Effects of Nutraceuticals) discuss in more detail the beneficial effects of several nutraceuticals in cognitive disorders and various neurodegenerative diseases. This chapter focuses instead on potential mechanisms of neuroprotection at the cellular, biochemical, and molecular levels (Kelsey et al., 2010; Mazzio et al., 2011; Vauzour, 2012). Polyphenols (particularly quercetin) are discussed as model nutraceuticals, although other molecules are mentioned and discussed to illustrate additional potential neuroprotective mechanisms.

Polyphenols

Several thousand molecules with a polyphenol structure have been identified in plants, and several hundred are found in edible plants. These compounds are often classified on the basis of their chemical structure, with flavonoids being one of the major classes (Manach et al., 2004; Del Rio et al., 2013; Garg et al., 2019). Among flavonoids, several subclasses can be identified, such as flavonols (e.g., quercetin), flavones, anthocyanidins, and various others. Innumerable studies support the idea that diets rich in polyphenols and/or supplementation with specific compounds are endowed with health benefits. In particular, polyphenols have been shown to exert protective actions in several pathological conditions such as cardiovascular disease, metabolic disorders, obesity, diabetes, infections, and cancer, as well as neurotoxic/neurodegenerative processes (Graf et al., 2005; Arts and Hollman, 2005; Scalbert et al., 2011; Vauzour, 2012; Del Rio et al., 2013; Bhullar and Rupasinghe, 2013; Singh et al., 2016; Chand, 2019).

Quercetin (Fig. 1.1) is found in many common fruits and vegetables such as apples, berries, onions, and capers (USDA, 2003). Its estimated dietary intake ranges from 4 to 68 mg/day, but it can increase to 200–500 mg/day in individuals who consume high quantities of fruits and vegetables rich in flavonols. Furthermore, quercetin is also sold as a dietary supplement, with a recommended dosage of 1 g/day (Harwood et al., 2007). The quercetin present in foods is not present as aglycon (i.e., without sugar groups), but rather as quercetin glycosides, which, contrary to previous belief, can be efficiently absorbed. Studies using rats and pigs have shown that quercetin distributes to several tissues, particularly the lung, kidney, colon, and liver, and lower levels in the brain (deBoer et al., 2005; Huebbe et al., 2010). Total quercetin derived from the diet is normally present in plasma in the nanomolar range (<100 nM), but it can be increased to the micromolar range after supplementation (Manach et al., 2005; Conquer et al., 1998). The half-life of quercetin ranges between 11 and 28 h, suggesting the possibility of significantly increasing plasma concentration with supplementation (Manach et al., 2004, 2005). Quercetin has an

Nutraceuticals. DOI: https://doi.org/10.1016/B978-0-12-821038-3.00001-X

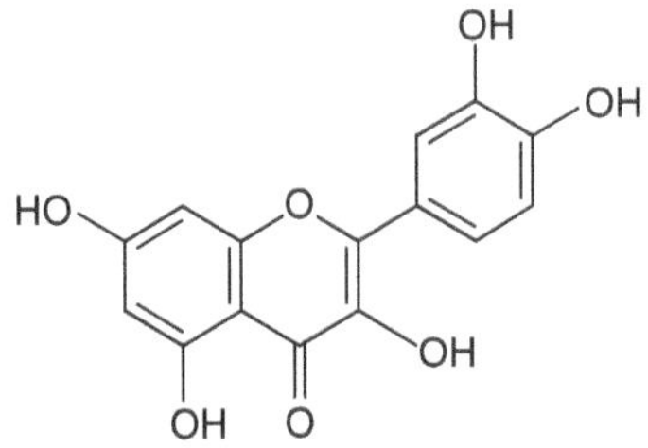

FIGURE 1.1 Structure of quercetin.

unremarkable toxicological profile, as evidenced by animal and human studies (Harwood et al., 2007; Russo et al., 2012). Similar to other polyphenols, reported beneficial effects of quercetin include effects on cardiovascular diseases, cancer, infections, inflammatory processes, gastrointestinal tract, diabetes, and nervous system disorders (reviewed in Boots et al., 2008; Kelly, 2011; Russo et al., 2012; Li et al., 2016).

Counteracting oxidative stress as a mechanism of neuroprotection

Oxidative stress is recognized as an important factor in a variety of neurodegenerative diseases, as a mediator of the adverse effects of a number of neurotoxicants, and as a mechanism for age-related degenerative processes (Halliwell, 2006; Lin and Beal, 2006; Martin and Grotewiel, 2006; Popa-Wagner et al., 2013). Oxidative stress occurs when reactive oxygen species (ROS) accumulate in cells, either from excessive production or from insufficient neutralization, causing damage to proteins, lipids, and DNA. Mitochondria are a major contributor of cellular ROS; ROS produced in the mitochondria can also target the electron transport chain (e.g., complex I), resulting in a cycle generating more ROS, followed by ATP depletion and ultimately cell death (Ott et al., 2007; Kelsey et al., 2010). Based on these premises, the identification of novel compounds that can counteract oxidative stress as potential therapeutics is a very active area of research (Linseman, 2009; Milatovic and Gupta, 2019). Natural compounds have received much attention in this regard; among these, polyphenols have been most investigated (Dajas et al., 2003; Ossola et al., 2009; Spencer, 2009; Kelsey et al., 2010).

Evidence for neuroprotection has been provided by in vitro studies showing that various polyphenols protect neuronal cells from damage due to oxidative stress, and by in vivo animal studies that have shown their ability to protect neurons against oxidative insults. Furthermore, clinical and epidemiological studies have shown that polyphenols can improve deterioration of brain function due to aging or neurodegenerative diseases (Kelsey et al., 2010; Vauzour, 2012).

Specific evidence exists regarding the neuroprotective effects of quercetin (Ossola et al., 2009). In vitro studies of neuronal cell lines and of primary neurons have shown that quercetin antagonizes cell toxicity induced by various oxidants [e.g., hydrogen peroxide (H_2O_2)] and other neurotoxic molecules believed to act by inducing oxidative stress (e.g., 6-hydroxydopamine and *N*-methyl-4-phenyl-1,2,3,6-tetrahydropyridinium) (Dajas et al., 2003; Mercer et al., 2005; Vauzour et al., 2008; Arredondo et al., 2010). Important issues for the potential use of quercetin in vivo are whether it passes the blood–brain barrier (BBB) and what concentrations of quercetin and/or its metabolites are present in brain tissue. In vitro studies with BBB models consistently indicate that quercetin enters the brain (Faria et al., 2010; Ishisaka et al., 2011; Schaffer and Halliwell, 2012). Following administration of quercetin in vivo to rats and pigs, low levels (from picomolar to nanomolar) were found in brain tissue (deBoer et al., 2005; Huebbe et al., 2010; Ishisaka et al., 2011). Of interest in this regard are the recent successful efforts to increase the bioavailability of quercetin (Russo et al., 2012). In particular, the formulation of quercetin in lipid nanoparticles significantly increases its penetration in the brain (Das et al., 2008; Dhavan et al., 2011). Several studies show that quercetin can exert neuroprotection and antagonize oxidative stress when administered in vivo. For example, oral quercetin (0.5–50 mg/kg) was shown to protect rodents from oxidative stress and neurotoxicity induced by various neurotoxic insults (Hu et al., 2008; Das et al., 2008; Barcelos et al., 2011; Ishisaka et al., 2011; Bavithra et al., 2012; Denny Joseph and Muralidhara, 2013).

Direct antioxidant action of quercetin

Quercetin is a potent scavenger of ROS, such as $O_2^{\bullet -}$, and of RNS (reactive nitrogen species), such as NO and ONOO (Boots et al., 2008). The antioxidative capacity of quercetin has been ascribed to the presence of two antioxidant pharmacophores within the molecule that have the optimal configuration for free radical scavenging, such as the catechol group in the B ring and the OH group at position 3 (Boots et al., 2008). Direct scavenging of ROS in vitro has been observed with quercetin concentrations ranging from 5 to 50 μM (Saw et al., 2014). However, it has been pointed out that the concentration of quercetin expected to be present in the brain would likely be in the nanomolar range, which is not sufficient to exert an appreciable direct antioxidant effect. In contrast, glutathione and vitamin C are present at millimolar concentrations (Schaffer and Halliwell, 2012). Thus despite its potent antioxidant capacity in vitro, it is unlikely that neuroprotective effects of quercetin observed in vivo are due to direct antioxidant action. Rather, it has been suggested that quercetin and/or its metabolites, as well as other polyphenols, may exert their neuroprotective effects by modulating the antioxidant defense mechanisms of the cell (Halliwell et al., 2005; Fraga et al., 2010; Kay, 2010). In this regard, it has been suggested that the beneficial effects of polyphenols may be due to their "prooxidant," rather than "antioxidant," properties (Halliwell, 2008, 2012). A mild degree of oxidative stress may increase the cell's own antioxidant defenses, resulting in overall cytoprotection. This important aspect of polyphenols' biological activity is discussed more in depth as it relates to the principle of hormesis.

Potential role of hormesis in neuroprotection

Hormesis is generally defined as a dose-response phenomenon characterized by low-dose stimulation and high-dose inhibition, which may be graphically represented by a J-shaped or U-shaped dose response (Mattson, 2008; Agathokleous and Calabrese, 2019). Such biphasic dose responses have been shown to occur broadly in the biomedical sciences, and are independent of biological model, end-point measure, and chemical class. Hormesis includes the phenomenon of preconditioning, whereby "exposure to a low dose of an agent that is toxic at high doses induces an adaptive, potentially beneficial effect on the cell or organism if exposed to a subsequent and more massive dose of the same or related stressor agent" (Calabrese et al., 2007; Lajqi et al., 2019). As stated, quercetin may have "prooxidant," rather than "antioxidant," properties (Halliwell, 2008, 2012; Boots et al., 2008). During its antioxidant activities, quercetin becomes oxidized into various oxidation products, including semiquinone radicals and quinones (Boots et al., 2008), which may mediate the toxicity of quercetin observed in certain conditions in what is referred to as the quercetin paradox (Boots et al., 2008; Halliwell, 2012).

Evidence is emerging to support hormetic roles for low increases in membrane oxidative stress (Calabrese et al., 2010). Lipid peroxidation generated during moderate exercise has been shown to play an important role in hormetic effects on muscles (Sachdev and Davies, 2008). Similar considerations also apply to oxidative stress in mitochondria; although high levels of oxidative stress are unquestionably detrimental to mitochondria, low levels of ROS may actually have a protective, hormetic effect, hence the term "mitohormesis" (Tapia, 2006; Calabrese et al., 2010; Yun and Finkel, 2014). There is limited evidence that neurons exposed to subtoxic levels of oxidants may be protected against a subsequent exposure to what would have otherwise been a lethal level of stress (Calabrese et al., 2007, 2010, 2012; Lajqi et al., 2019).

Some studies with the marine neurotoxin domoic acid (DomA) (Giordano et al., 2013a) provide additional support for this hypothesis. DomA is a potent human and animal neurotoxin that causes primarily apoptotic cell death of neurons as a consequence of activation of AMPA/kainate receptors (Giordano et al., 2007). DomA-induced apoptosis involves oxidative stress, is inhibited by antioxidants, and is more pronounced in neurons from transgenic mice (*Gclm*$^{-/-}$ mice), which lack the modifier subunit of glutamate cysteine ligase (GCL) and have very low glutathione (GSH) levels (Giordano et al., 2007). Prolonged exposure of mouse neurons to low, nontoxic levels of DomA (5 nM for 10 days) has been shown to protect cells from a subsequent insult of a high concentration of DomA itself and of other oxidants (Giordano et al., 2013a). The mechanism of such protection was related to the ability of chronic low DomA to increase the levels of the two subunits of GCL (GCLC and GCLM) and those of GCLholo, leading to increased GCL activity and GSH synthesis. Transcription of GCL subunits occurs through the sequence-specific binding of nuclear factor erythroid 2-related factor 2 (Nrf2) to antioxidant response elements (AREs) present in the promoters of these two genes (Moinova and Mulcahy, 1999; Wild et al., 1999). The effect observed with DomA in wild-type CGNs resembles the phenomenon of preconditioning, which is considered part of hormesis (Calabrese et al., 2007). Low levels of DomA may thus elicit a mild degree of oxidative stress, particularly in mitochondria (Giordano et al., 2007), which would lead to increased transcription of *Gclc* and *Gclm*, ultimately leading to increased GSH synthesis and neuroprotection.

Modulation of the Nrf2-ARE pathway as a mechanism of quercetin neuroprotection

Nrf2 is a master regulator of cellular defense against oxidative stress (Fig. 1.2). Under physiological conditions, Nrf2 is sequestered in the cytoplasm by the protein Keap1 (Kelch-like ECH-associated protein 1) with Cullin 3-base E3 ligase, by which Nrf2 protein is ubiquitinylated and targeted for proteasome degradation (Shih et al., 2005; Calabrese et al., 2012; Liang et al., 2013; Gan and Johnson, 2014). Keap1 has several cysteine residues that make it act as a molecular switch by responding to electrophiles and ROS with a conformational change that releases Nrf2 (Shih et al., 2005). Dissociated Nrf2 translocates into the nucleus, where it binds to small Maf proteins. The formed heterodimer binds to *cis*-acting ARE and thereby promotes the transcription of a broad range of phase II and antioxidant genes (Liang et al., 2013; Gan and Johnson, 2014).

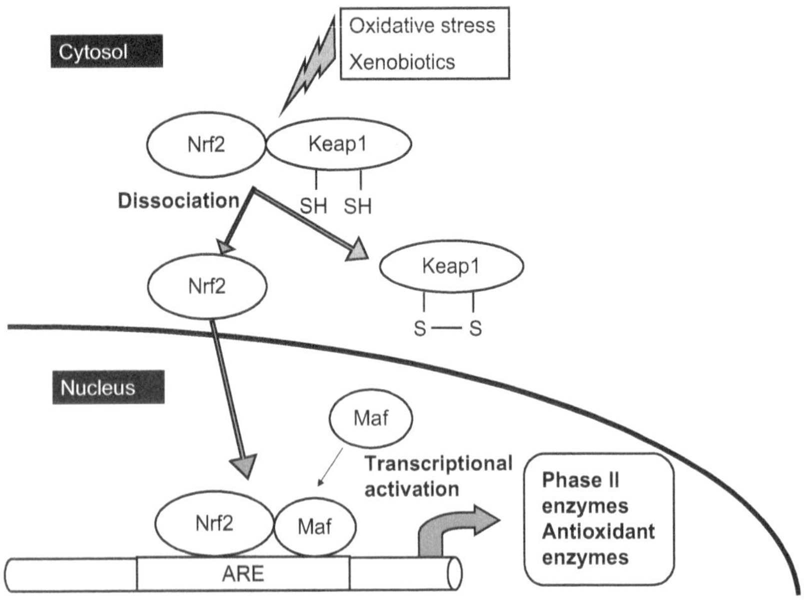

FIGURE 1.2 The Nrf2-ARE pathway. See text for details.

Proteins that are under the control of the Nrf2-ARE pathway include heme oxygenase-1, GCL, glutathione *S*-transferases, glutathione peroxidase, superoxide dismutase, catalase, sulforedoxin, thioredoxin, and several others (Shih et al., 2005; Gan and Johnson, 2014). Activation of the Nrf2-ARE pathway provides neuroprotection against oxidative damage and cell death. More recent evidence also suggests that the Nrf2-ARE pathway may modulate the formation and degradation of misfolded protein aggregates that are present in various neurodegenerative diseases, such as Parkinson, Alzheimer, and Huntington diseases, as well as amyotrophic lateral sclerosis (Gan and Johnson, 2014). For example, studies with *tert*-butylhydroquinone, a prototype Nrf2 inducer, have shown that activation of the Nrf2-ARE pathway confers protection against neurotoxicity induced by amyloid β and 3-nitropropionic acid (Shih et al., 2005; Nouhi et al., 2011).

Quercetin has been shown to counteract oxidative stress-induced cellular damage by activating the Nrf2-ARE pathway (Arredondo et al., 2010; Granado-Serrano et al., 2012; Saw et al., 2014), and similar effects have been reported for dihydroquercetin (Liang et al., 2013). Additionally, nutraceuticals, such as kaempferol and pterostilbene, have been shown to interact synergistically with quercetin (Saw et al., 2014). The Akt, extracellular signal-regulated kinase (ERK), and JNK signaling pathway may be involved in the activation of Nrf2 (Liang et al., 2013) and, in turn, they are activated by stress stimuli, including oxidative stress. Thus quercetin and many other nutraceuticals [e.g., resveratrol (RES), sulforaphane, carnosic acid, and dimethyl fumarate] may all act as neurohormetic phytochemicals (Mattson and Cheng, 2006; Calabrese et al., 2012; Li et al., 2017).

Modulation of paraoxonase 2 as a potential novel mechanism of quercetin neuroprotection

Paraoxonase 2

The paraoxonases (PONs) are a family of three genes (PON1, PON2, PON3) clustered in tandem on the long arm of human chromosome 7q21-22 and on mouse chromosome 6 (Primo-Parmo et al., 1996). Although the name of these enzymes derives from paraoxon, the active metabolite of the organophosphorus (OP) insecticide parathion, which is hydrolyzed by PON1 in vitro (Li et al., 2000), the other two PONs do not have OP esterase activity. In contrast, all three PONs are lactonases and they hydrolyze a number of acyl-homoserine lactone (acyl-HCL) molecules that mediate bacterial quorum-sensing signals and are important in regulating the expression of virulence factors and in inducing a host inflammatory response (Draganov et al., 2005; Teiber et al., 2008). PON2 also plays a significant role in atherosclerosis, as shown by studies indicating that PON2 overexpression decreases atherosclerotic lesions, although the opposite is true in PON2-deficient mice (Ng et al., 2006a, b). In contrast to PON1 and PON3, which are expressed primarily in the liver, and their protein products associated with high-density lipoproteins in the plasma, PON2 is a ubiquitously expressed intracellular enzyme not present in plasma (Ng et al., 2001; Marsillach et al., 2008; Giordano et al., 2011).

In several tissues, PON2 has been shown to exert an antioxidant effect (Ng et al., 2001; Horke et al., 2007), which is believed to play a major role in preventing the atherosclerotic process, as shown directly in PON2-knockout mice (Horke et al., 2007; Devarajan et al., 2011). Subcellular distribution studies have shown that PON2 is localized primarily in the mitochondria (Giordano et al., 2011; Devarajan et al., 2011). Mitochondria are a major source of free radical-related oxidative stress, and the preponderant localization of PON2 in mitochondria would support a role for this enzyme in protecting cells from oxidative damage. In HeLa cells, PON2 has been shown to bind to coenzyme Q_{10} that associates with complex III in mitochondria, and PON2 deficiency causes mitochondrial dysfunction (Devarajan et al., 2011). In human endothelial cells, PON2 has been shown to reduce, indirectly but specifically, the release of superoxide from the inner mitochondrial membrane, without affecting levels of other radicals such as H_2O_2 and peroxynitrite (Altenhofer et al., 2010).

PON2 in the central nervous system

PON2 mRNA has been found in mouse and human brain, and PON2 protein has been detected in mouse (Primo-Parmo et al., 1996; Ng et al., 2006a; Marsillach et al., 2008) as well as rat, human (Giordano et al., 2013b), and monkey brain (Costa et al., 2014). In a series of studies, the expression of PON2 has been characterized in mouse brain (Giordano et al., 2011, 2013b; Costa et al., 2014). The highest levels of PON2 protein were found in three dopaminergic regions, the substantia nigra, the striatum, and the nucleus accumbens, with lower levels in the cerebral cortex, cerebellum, hippocampus, and brainstem. The higher levels of PON2 in dopaminergic areas are of interest because they may be related to the higher level of oxidative stress, due to dopamine metabolism, present in these regions. The regional distribution and gender difference of PON2 were confirmed by measurements of its lactonase activity [measured by dihydrocoumarin (DHC) hydrolysis] and of PON2 mRNA levels (Giordano et al., 2011). At the cellular level, PON2 is significantly higher in astrocytes than in neurons in all brain regions, with the highest levels in cells isolated from the striatum. Striatal neurons and astrocytes isolated from female mice express higher levels of PON2 than the same cells from male animals. PON2 is also present in cortical microglia, at levels similar to those found in neurons (Giordano et al., 2011). At the subcellular level, the highest levels of PON2 are found in mitochondria, followed by membranes (microsomes), in agreement with previous observations in HeLa cells (Devarajan et al., 2011).

To provide a direct indication of whether PON2 exerts a protective effect toward oxidative stress in brain cells, as observed in other tissues and cell types, the cytotoxicity of two known oxidants, H_2O_2 and 2,3-dimethoxy-1,4-naphtoquinone (DMNQ), was investigated in neurons and astrocytes from different brain regions isolated from wild-type ($PON2^{+/+}$) and $PON2^{-/-}$ mice. In all instances, cells from mice lacking PON2 were more susceptible to the toxicity of both compounds by a factor of 5- to 11-fold. The protection afforded by PON2 to neurons and astrocytes was related to its ability to scavenge ROS on exposure to oxidants. For example, DMNQ (10 μM) increased ROS to ~400% of basal levels in cerebellar granule neurons from $PON2^{-/-}$ mice, but only to 170% in the same cells from $PON2^{+/+}$ mice (Giordano et al., 2011).

Gender differences in PON2 expression

In every brain region, PON2 levels are higher (by approximately twofold to threefold) in female mice than in male mice (Giordano et al., 2011). This may be related to a positive modulatory effect by estrogens. In striatal astrocytes from male mice, 17β-estradiol causes time-dependent and concentration-dependent increases in the levels of PON2 protein. For example, 12–24 h exposure with 200 nM estradiol increases PON2 expression in striatal astrocytes from male mice to the levels found in female striatal astrocytes (Giordano et al., 2013b). Interestingly, in female astrocytes, estradiol can further increase PON2 expression by a factor of approximately 2.5-fold. The estradiol effect is due to transcriptional activation of the PON2 gene and is mediated by activation of estrogen receptors alpha (Giordano et al., 2013b). In ovariectomized mice, PON2 levels (protein and mRNA) are significantly reduced in striatum, cerebral cortex, and liver, approaching the levels found in male mice. Given the findings of enhanced susceptibility to oxidative stress due to lack of PON2 (Giordano et al., 2011), it was of interest to ascertain whether the twofold to threefold difference in PON2 levels between genders was sufficient to confer differential susceptibility to oxidants. This was indeed the case, because striatal astrocytes and neurons from male mice were more sensitive to H_2O_2 and DMNQ-induced oxidative stress and ensuing cytotoxicity (Giordano et al., 2013b). Although gender-dependent differences in other cell defense mechanisms cannot be excluded, it is noteworthy that levels of GSH did not differ between genders.

Another important aspect is related to the lack of gender difference in susceptibility in cells from $PON2^{-/-}$ mice. Striatal astrocytes from $PON2^{-/-}$ mice of either gender are highly susceptible to oxidant-induced toxicity, as expected, but there are no significant female/male differences (Giordano et al., 2013b). Further evidence for a central role of PON2 in mediating gender differences in susceptibility to oxidative stress toxicity is provided by experiments with estradiol. In central nervous system (CNS) cells from $PON2^{+/+}$ male mice, exposure to estradiol (200 nM, 24 h) provided protection against toxicity induced by the two oxidants. This is not surprising because neuroprotective actions of estrogens are well known (Simpkins et al., 2010; Azcoitia et al., 2011). However, the protective effect of estradiol is absent in cells from PON2$^{-/-}$ mice, suggesting that a major mechanism of estrogen neuroprotection may be represented by induction of PON2 (Giordano et al., 2013b).

Modulation of PON2

The studies summarized in the previous section show that the enzyme PON2 exerts an antioxidant/antiinflammatory effect in the CNS, and that levels of expression of PON2 are an important determinant of susceptibility to neurotoxicity. Hence, modulation of PON2 by agents/factors that increase its level of expression may result in neuroprotection.

In macrophages, PON2 expression is increased by oxidative stress (Rosenblat et al., 2003), and in vascular cells an endoplasmic reticulum stress element-like sequence was found to be present in the promoter region of PON2 (Horke et al., 2007). Arachidonic acid (Rosenblat et al., 2010), the licorice phytoestrogen glabridin (Yehuda et al., 2011), and the hypocholesterolemic drug atorvastatin (Rosenblat et al., 2004) also upregulate PON2 expression in various cell types. Urokinase plasminogen activator upregulates PON2 in macrophages via NADPH oxidase and the transcription factor SREBP-2 (Fuhrman et al., 2009). Pomegranate juice was found to increase PON2 in macrophages through activation of the PPARγ and AP-1 pathways (Shiner et al., 2007), whereas extracts of Yerba mate (*Ilex paraguariensis*) have been reported to increase PON2 mRNA and lactonase activity in macrophages in vitro and after in vivo administration to healthy women (Fernandes et al., 2012).

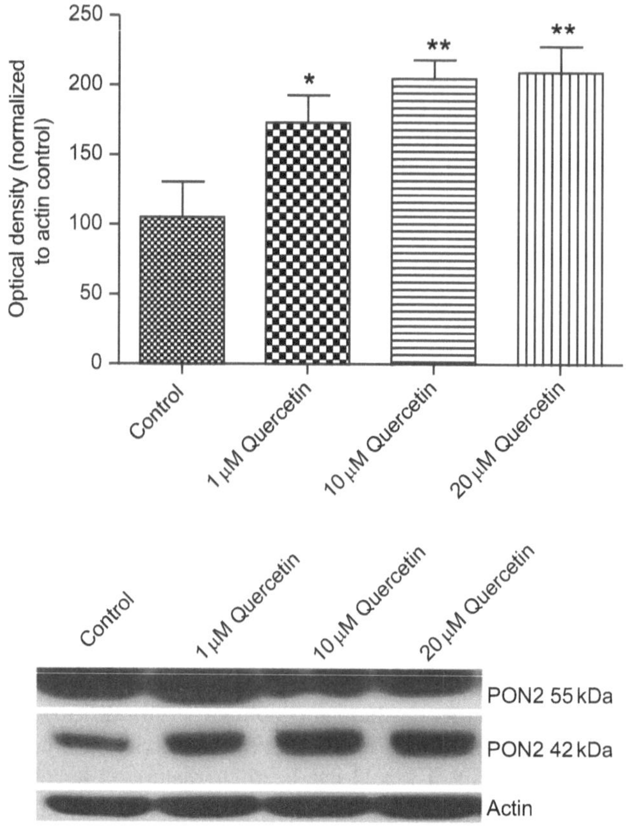

FIGURE 1.3 Induction of PON2 by quercetin. Mixed gender mouse striatal astrocytes were exposed for 24 h to quercetin (1, 10, or 20 μM). Shown are the quantification of the 42 kDa alloform (top, $n = 3$) and a representative blot (bottom). Significantly different from control: $*P < .05$, $**P < .01$. *PON2*, Paraoxonase 2. *(From Tait (2011))*

Modulation of PON2 by quercetin

Quercetin was reported to increase PON2 mRNA and protein in macrophages in vitro, although administration of 150 mg/day to human volunteers for 6 weeks had no effect (Boesch-Saadatmandi et al., 2009). In an in vitro study, Costa et al. (2013) found the induction of PON2 by quercetin. Quercetin was found to increase PON2 protein expression in mouse striatal astrocytes (mixed gender) by approximately twofold at concentrations ranging from 1 to 20 μM (Fig. 1.3). Similar results were also observed in mouse striatal neurons and in mouse RAW264.7 macrophages (Costa et al., 2013). The effect of quercetin was antagonized by SP600125, an inhibitor of the JNK/AP-1 pathway. In contrast, the PPARγ inhibitor GW9662 did not antagonize quercetin's effect on PON2, while totally abrogating the induction of PON2 by the PPARγ agonist rosiglitazone (Costa et al., 2013). Quercetin may thus induce a very low level of oxidative stress (Halliwell, 2008; Chang et al., 2009), which in turn would modulate the JNK/AP-1 pathway (Granado-Serrano et al., 2010), causing an increase in PON2 expression. Alternatively, given the effects of estradiol on PON2 expression (Giordano et al., 2013b), quercetin may induce PON2 expression by virtue of its phytoestrogen activity (Galluzzo et al., 2009; Ruotolo et al., 2014), although this would need to be investigated further.

Independent of the underlying mechanism(s), the ability of quercetin to induce PON2 may play a role in the reported neuroprotective actions of this polyphenol, which have been observed in vitro as well as in vivo (Miodini et al., 1999; Mercer et al., 2005; Boots et al., 2008; Ossola et al., 2009; Barcelos et al., 2011; Selvakumar et al., 2012). In striatal astrocytes from PON2$^{+/+}$ mice (mixed gender), exposure for 24 h to quercetin abolished the increase in ROS levels caused by H_2O_2 or DMNQ (Costa et al., 2013). This resulted in protection against the toxicity of these oxidants, as shown in Table 1.1; indeed, the IC_{50} values for cytotoxicity of H_2O_2 and DMNQ increased by 3.5- to 4-fold after treatment with quercetin, which doubled PON2 expression. This is similar to what

TABLE 1.1 Role of PON2 in quercetin protection against oxidative stress in striatal astrocytes.

	IC_{50} (μM)	
	Control	Quercetin
$PON2^{+/+}$ mice		
H_2O_2	38.9 ± 4.5	157.0 ± 8.1*
DMNQ	37.5 ± 5.6	131.3 ± 9.2*
$PON2^{-/-}$ mice		
H_2O_2	6.3 ± 1.3	11.9 ± 1.2
DMNQ	6.1 ± 1.0	8.3 ± 1.1

Notes: Striatal astrocytes from wild-type ($PON2^{+/+}$) or $PON2^{-/-}$ mice (mixed genders) were exposed for 24 h to 20 μM quercetin. After washout, cells were treated for 24 h with four to five concentrations of hydrogen peroxide (H_2O_2) or 2,3-dimethoxy-1,4-naphtoquinone (DMNQ), and cytotoxicity was measured by the MTT assay. Results indicate the IC_{50} values (μM) and are the mean (±SD) of three separate experiments. *Significantly different from wild-type control ($P < .01$). The findings indicate that exposure of wild-type cells to quercetin provides neuroprotection, as evidenced by an approximately fourfold increase in IC_{50} for the two oxidants. Control astrocytes from $PON2^{-/-}$ mice are sixfold more susceptible to the toxicity of H_2O_2 and DMNQ, likely because of the lack of PON2. Additionally, in $PON2^{-/-}$ cells, quercetin-induced protection is small (approximately 1.5-fold). *PON2*, paraoxonase 2
Source: Adapted from Costa et al. (2013)

is observed in brain cells from male and female mice, in which a twofold to threefold difference in PON2 expression resulted in a threefold to fourfold difference in susceptibility to these two oxidants (Giordano et al., 2013b). Although other neuroprotective pathways (e.g., Nrf2-ARE) may be involved in the observed neuroprotection, experiments performed in cells from $PON2^{-/-}$ mice show that modulation of PON2 expression plays an important role in the neuroprotective effect of quercetin. On quercetin exposure of $PON2^{-/-}$ cells, the IC_{50} for H_2O_2 and DMNQ were increased by only 1.9-fold and 1.4-fold, respectively (Table 1.1). This partial protection may be due to NrF2-ARE induction.

The issue of metabolites

An important issue to consider as part of the discussion on the beneficial effects of quercetin relates to the potential role played by its metabolites (Del Rio et al., 2010). Quercetin is metabolized to various conjugated metabolites: 3′-*O*-methyl-quercetin (isorhamnetin; MeQ), quercetin-3-*O*-glucuronide (Glu3Q); 3′-*O*-methylquercetin-*O*-glucuronide (Glu3MeQ); and quercetin-3′-*O*-sulfate (Sul3Q) (Day et al., 2001; Harwood et al., 2007). As stated, only limited amounts of quercetin aglycon are found after ingestion of quercetin, although there is some controversy regarding this issue (e.g., Shanely et al., 2010; Kelly, 2011), and methylated, sulfated, and glucuronide metabolites are the most prominent moieties found in plasma. Studies have shown that Glu3Q has antioxidant abilities in vitro and in vivo (Moon et al., 2001; Shirai et al., 2006; Kawai et al., 2008). Additional biological effects of methylated and sulfate metabolites have been reported (Yeh et al., 2011; Boesch-Saadatmandi et al., 2011; Ruotolo et al., 2014), although some studies have failed to observe an effect of quercetin metabolites (Cho et al., 2012). Of interest is also the observation that conjugated quercetin can enter the cell, where it is converted to its nonconjugated form (Fiorani et al., 2003). Thus quercetin metabolites should be tested to ascertain, for example, whether they induce the Nrf2-ARE pathways or modulate PON2 expression. One metabolite, Glu3Q, has recently been shown to act as an agonist at estrogen receptor α, and may thus be capable of inducing PON2 (Ruotolo et al., 2014). Furthermore, various other polyphenol catabolites have been shown to exert strong biological activity, particularly in protecting neuronal cells against DMNQ-induced oxidative stress and toxicity (Verzelloni et al., 2011).

RES and neuroprotection

Neuroprotective effects of RES have been investigated in various animal models: (1) neurodegenerative diseases and aging-related deficits (Allen et al., 2018; Gomes et al., 2018); (2) stroke and traumatic CNS injury (Gupta et al., 2001; Lopez et al., 2015); (3) hypoxic-ischemic (HI) brain injury (West et al., 2007; Jin et al., 2008; Arteaga et al., 2015; Gao et al., 2018); and (4) ischemia/reperfusion injury (Dou et al., 2019). Many molecular mechanisms and pathways have been reported to be involved in RES-mediated neuroprotection.

Huang et al. (2011) investigated the cellular effects of RES on Aβ-induced hippocampal neuronal loss and memory impairment in rats. Findings revealed that RES significantly improved spatial memory, and neuroprotective effects were attributed to reductions in the cellular levels of iNOS and lipid peroxidation and an increase in the production of HO-1. Marambaud et al. (2005) demonstrated that RES does not inhibit Aβ production, because it has no effect on the Aβ-producing enzymes (β- and γ-secretases). Instead it promotes intracellular degradation of Aβ via a mechanism that involves the proteasome. Bastianetto et al. (2015) and Jia et al. (2017) described that RES decreases the amyloidogenic cleavage of the amyloid precursor protein, enhances clearance of Aβ-peptides, and reduces Aβ aggregation. Various studies suggest that RES is capable of binding to the N-terminal of SIRT1, thereby increasing its activity, decreasing Aβ accumulation, and protecting against cell death and neurotoxicity. However, RES is not a direct activator of SIRT1 (Gomes et al., 2018). Very recently, Sanches-Melgar et al. (2019) reported that RES also targets adenosine receptors, affecting gene expression, receptor levels, and the downstream adenylyl cyclase/PKA pathway, thereby creating an opportunity for the use of RES against AD.

In a chronic study, adult male rats receiving RES (20 mg/kg/day for 4 weeks) showed RES to be restorative therapy for the impaired cognitive functions occurring along the normal aging process by preventing 5-HT, dopamine, and noradrenaline neurotransmission decline (Sarubbo et al., 2015). In a 26-week study, RES intake (200 mg/day) increased hippocampal functional connectivity and improved memory performance in association with improved glucose metabolism in older adults (Witte et al., 2014). In a clinical trial, administration of 1000 mg RES twice daily for 52 weeks in AD patients was found to be well tolerated and resulted in changes in the trajectory of some biomarkers compared to placebo (Turner et al., 2015).

Gupta et al. (2001) investigated the antioxidative and neuroprotective effects of *trans*-RES (20 and 40 mg/kg, ip) in $FeCl_3$-treated male Wistar rats. Epidural electrodes were implanted in these rats for EEG recording. RES administered 30 min prior to $FeCl_3$ treatment significantly delayed the onset of epileptiform EEG changes and reduced malondialdehyde levels. In an in vitro study, Lin et al. (2018) reported that the SH-SY5Y cells pretreated with RES lowered cellular ROS, apoptosis, and increased survival rates against rotenone treatment. Findings suggested that the neuroprotective effect of RES works in part through modulation of mitochondrial dynamics and the upregulation of autophagic flux via the mitogen-activated protein kinase/ERK signaling pathway.

Recently, Gao et al. (2018) demonstrated that RES exerted its neurotherapeutic potential by upregulating the protein expression of the Nrf2 and HO-1 signaling pathway, thereby attenuating oxidative stress and the inflammatory response in HI-induced neonatal rats.

In essence, RES has antioxidative, antiinflammatory, immunomodulatory, and neuroprotective properties, thus it can protect neurons against the insult caused by various diseases and toxicants via many mechanisms and pathways.

Ginkgo biloba and neuroprotection

Among many herbal medicines, *Ginkgo biloba* extract (GBE) has been used as a neuroprotectant in AD, PD, and other neurological conditions for some time (DeKosky et al., 2008; Vellas et al., 2012; Singh et al., 2016; Srivastav et al., 2017; Yuan et al., 2017). Standardized GBE contains 6% terpenoids (3.1% ginkgolides A, B, C, and J, and 2.9% bilobalide), 24% flavonoid glycosides (quercetin, kaempferol, isorhamnetin, etc.), and 5%–10% organic acids. Natural polyphenolic compounds, such as those present in GBE, impart their antioxidant effect by reducing free radical species, metal bonding, and/or antioxidant properties that positively contribute to their neuroprotective effects. GBE-induced neuroprotection is mediated via an apoptotic pathway and mitochondrial function. GBE appears to exert neuroprotection in AD by: (1) improvement of energy metabolism, (2) stabilization of mitochondrial membrane potential, (3) inhibition of cytochrome c release, (4) upregulation of antiapoptotic Bcl-2 protein, (5) downregulation of proapoptotic Bax protein, and (6) reduced levels of caspase 9 and caspase 3 after oxidative stress, and decrease of apoptotic cell death (reviewed in Singh et al., 2019). In essence, GBE and/or its constituents can exert antioxidative, antiinflammatory, immunomodulatory, and neuroprotective properties in AD (Watanabe et al., 2001; Yin et al., 2013), PD (Srivastav et al., 2017), brain ischemia (Calapai et al., 2000; Chandrasekaran et al., 2001), stroke (Jiang et al., 2014), amyotrophic lateral sclerosis (Ferrante et al., 2001), and other neurological conditions (Christen, 2004; Kumar et al., 2017; Singh et al., 2019). GBE and its constituents have also been shown to exert additional mechanisms, such as inhibition of monoamine oxidase activity and nitric oxide levels, thereby providing neuroprotection in neurological conditions (White et al., 1996; Bastianetto et al., 2000; Song et al., 2000).

Concluding remarks and future directions

There is great interest in the mechanisms that may underlie the neuroprotective effects of nutraceuticals. In this chapter, the focus has been on the polyphenol quercetin and on mechanisms related to its ability to counteract oxidative stress-mediated neurotoxicity. However, several other potential mechanisms have been investigated and should be considered. Because nutraceuticals comprise a large and diverse class of compounds with different chemical structures and bioavailability, multiple targets for biological activity are to be expected. For example, in a discussion on the biochemical and cellular bases for nutraceutical strategies to combat Parkinson disease, the following potential targets were indicated: α-synuclein aggregation; ubiquitin proteasome function; mTOR signaling/lysosomal-autophagy; energy failure; faulty catecholamine trafficking; dopamine oxidation; synthesis of toxic dopamine-quinones; inflammation; methylation; and oxidation of neuromelanin (Mazzio et al., 2011). Dozens of nutraceutical compounds have been identified that would affect at least one of the indicated targets (Mazzio et al., 2011). On a more general basis, neuroinflammation, which is believed to be most relevant in neurodevelopmental and neurodegenerative disorders (Skaper et al., 2014; Baune, 2015; Chitnis and Weiner, 2017; Stephenson et al., 2018; Park et al., 2019), represents an important potential target for nutraceuticals (Baune, 2015;

Nabavi et al., 2015; Li et al., 2016). Compounds that may antagonize microglia activation and reduce the release of proinflammatory cytokines would be of much relevance. Soy isoflavones daidzein and genistein are suggested to reduce microglial activation and subsequent release of proinflammatory factors (Chinta et al., 2013; Jantaratnotai et al., 2013), although further studies in this area are needed. There is also evidence in this regard that polyphenols or garlic extract may have beneficial antiinflammatory properties (Vauzour, 2012; Takechi et al., 2013; Nillert et al., 2017).

An additional field of interest with regard to the mechanisms of neuroprotection provided by nutraceuticals is that of sirtuins. These proteins (in mammals there are seven, SIRT1–SIRT7) are involved in a variety of cellular and molecular processes and pathways with distinct cellular localization and molecular targets (Dang, 2012). Of these, SIRT1 predominantly localizes in the nucleus and acts as a deacetylase for histones and other targets. SIRT1 protects cells from apoptosis and promotes differentiation of stem cells. SIRT2 is prevalent in the cytoplasm and has been found to accumulate in neurons, whereas other SIRTs localize primarily in the mitochondria (Dang, 2012). The neuroprotective effects of the polyphenol RES have been ascribed to activation of SIRT1, leading to inhibition of amyloid-β peptide, suppression of Bax-dependent apoptosis, and repression of multiple proapoptotic transcription factors (Kelsey et al., 2010).

One aspect that may deserve more attention relates to the modulation of autophagy. Autophagy (from the Greek "to eat oneself") refers to the cellular degradative pathways that involve delivery of the cytoplasmatic cargo to the lysosomes (Mariño et al., 2011; Gabryel et al., 2012; Giordano et al., 2014). Autophagy (macroautophagy) is a multistep process involving the formation of double membrane structures, the autophagosomes, which then fuse with lysosomes. The content of the resulting autophagolysosomes (misfolded proteins, cellular metabolic waste) is then degraded by hydrolytic enzymes. Autophagy is also important for removal of damaged mitochondria and normal mitochondria undergoing turnover in a process known as mitophagy. The integrity of the CNS is very dependent on normal basal autophagy because damaged organelles and misfolded proteins will accumulate in neurons unless they are successfully removed (Mariño et al., 2011). Deletion of key autophagy genes (*Atg5*, *Atg7*) causes severe neurodegeneration (Komatsu et al., 2006). Rapamycin, an inhibitor of mTOR (mammalian target of rapamycin) activity, is a potent inducer of autophagy and acts as a neuroprotector (Pan et al., 2009; Giordano et al., 2014). Stimulation of autophagy by nutraceuticals would thus lead to neuroprotection, as has been shown, for example, in the case of RES (Lin et al., 2014), a traditional Korean herbal formula (Bae et al., 2015), carnosine (Baek et al., 2014), and other compounds (Giordano et al., 2014). Because the well-being of most CNS cells is dependent on the integrity of mitochondria, these organelles represent the principal target for neuroprotective strategies, including, among several, modulation of mitophagy and oxidative stress (Perez-Pinzon et al., 2012).

Acknowledgments

The authors would like to thank Ms. Robin B. Doss for her technical assistance in the preparation of this chapter.

References

Agathokleous, E., Calabrese, E.J., 2019. Hormesis: the dose response for the 21st century: the future has arrived. Toxicology. 425, 152249.

Allen, E.N., Potdar, S., Tapias, V., et al., 2018. Resveratrol and pinostilbene confer neuroprotection against aging-related deficits through an ERK½-dependent mechanism. J. Nutr. Biochem. 54, 77–86.

Altenhofer, S., Witte, I., Teiber, J.F., et al., 2010. One enzyme, two functions. PON2 prevents mitochondrial superoxide formation and apoptosis independent from its lactonase activity. J. Biol. Chem. 285, 24398–24403.

Arredondo, F., Echeverry, C., Abin-Carriquiry, J.A., et al., 2010. After cellular internalization, quercetin causes Nrf2 nuclear translocation, increases glutathione levels, and prevents neuronal death against an oxidative insult. Free Radic. Biol. Med. 49, 738–747.

Arteaga, O., Revuelta, M., Uriguen, L., et al., 2015. Pretreatment with resveratrol prevents neuronal injury and cognitive deficits induced by perinatal hypoxia-ischemia in rats. PLoS One 10, e0142424. Available from: https://doi.org/10.1371/journal.pone.0142424.

Arts, I.C.W., Hollman, P.C.H., 2005. Polyphenols and disease risk in epidemiological studies. Am. J. Clin. Nutr. 81 (Suppl. 1), 317S–325S.

Azcoitia, I., Arevalo, M.A., De Nicola, A.F., Garcia-Segura, L.M., 2011. Neuroprotective actions of estradiol revisited. Trends Endocrinol. Metab. 22, 467–473.

Bae, N., Chung, S., Kim, H.J., Cha, J.W., Cha, J.W., Oh, H., 2015. Neuroprotective effect of modified Chungsimyeolda-tang, a traditional Korean herbal formula, via autophagy induction in models of Parkinson's disease. J. Ethnopharmacol. 159, 93–101.

Baek, S.H., Noh, A.R., Kim, K.A., Akram, M., Shin, Y.J., Kim, E.S., 2014. Modulation of mitochondrial function and autophagy mediates carnosine neuroprotection against ischemic brain damage. Stroke 45, 2438–2443.

Barcelos, G.R.M., Grotto, D., Serpeloni, J.M., Angeli, J.P., Rocha, B.A., de Oliveira Souza, V.C., 2011. Protective properties of quercetin against DNA damage and oxidative stress induced by methylmercury in rats. Arch. Toxicol. 85, 1151–1157.

Bastianetto, S., Zheng, W.H., Quirion, R., et al., 2000. The *Ginkgo biloba* extract (EGb 761) protects and rescues hippocampal cells against nitric oxide-induced toxicity: involvement of its flavonoid constituents and protein kinase C. J. Neurochem. 74, 2268–2277.

Bastianetto, S., Ménard, C., Quirion, R., 2015. Neuroprotective action of resveratrol. Biochim. Biophys. Acta 1852, 1195–1201.

Baune, B.T., 2015. Inflammation and neurodegeneration: is there still hope for therapeutic intervention? Curr. Opin. Psychiatry 28, 148–154.

Bavithra, S., Selvakumar, K., Kumari, R.P., Krishnamoorthy, G., Venkataraman, P., Arunakaran, J., 2012. Polychlorinated biphenyls (PCBs)-induced oxidative stress plays a critical role on cerebellar dopaminergic receptor expression: ameliorative role of quercetin. Neurotoxic. Res. 21, 149–159.

Bhullar, K.S., Rupasinghe, H.P.V., 2013. Polyphenols: multipotent therapeutic agents in neurodegenerative diseases. Oxid. Med. Cell. Longev. 2013, 891748.

Boesch-Saadatmandi, C., Pospissil, R.T., Graeser, A.C., Canali, R., Boomgaarden, I., Doering, F., 2009. Effect of quercetin on paraoxonase 2 levels in RAW264.7 macrophages and in human monocytes-role of quercetin metabolism. Int. J. Mol. Sci. 10, 4168–4177.

Boesch-Saadatmandi, C., Loboda, A., Wagner, A.E., Stachurska, A., Jozkowicz, A., Dulak, J., 2011. Effect of quercetin and its metabolites isorhamnetin and quercetin-3-glucuronide on inflammatory gene expression: role of miR-155. J. Nutr. Biochem. 22, 293–299.

Boots, A.W., Haenen, G.R.M.M., Bast, A., 2008. Health effects of quercetin: from antioxidant to nutraceutical. Eur. J. Pharmacol. 585, 325–337.

Calabrese, E.J., Bachmann, K.A., Bailer, A.J., Bolger, P.M., Borak, J., Cai, L., 2007. Biological stress response terminology: integrating the concepts of adaptive response and pre-conditioning stress with hormetic dose-response framework. Toxicol. Appl. Pharmacol. 222, 122–128.

Calabrese, V., Cornelius, C., Dinkova-Kostova, A.T., Calabrese, E.J., Mattson, M. P., 2010. Cellular stress responses, the hormesis paradigm, and vitagenes: novel targets for therapeutic intervantion in neurodegenerative disorders. Antioxid. Redox Signal. 13, 1763–1811.

Calabrese, V., Cornelius, C., Dinkova-Kostova, A.T., Iavicoli, I., Paola, R. Di, Koverech, A., 2012. Cellular stress response, hormetic phytochemicals and vitagenes in aging and longevity. Biochim. Biophys. Acta 1822, 753–783.

Calapai, G., Crupi, A., Firenzouli, F., et al., 2000. Neuroprotective effects of *Ginkgo biloba* extract in brain ischemia are mediated by inhibition of nitric oxide synthesis. Life Sci. 67, 2673–2683.

Chand, N., 2019. Standardized turmeric and curcumin. In: Gupta, R.C., Srivastava, A., Lall, R. (Eds.), Nutraceuticals in Veterinary Medicine. Springer Nature, Cham, Switzerland, pp. 3–23.

Chandrasekaran, K., Mehrabian, Z., Spinnewyn, B., et al., 2001. Neuroprotective effects of bilobalide, a component of the *Ginkgo biloba* extract (EGb 761), in gerbil global brain ischemia. Brain Res. 922, 282–292.

Chang, Y.F., Hsu, Y.C., Hung, H.F., et al., 2009. Quercetin induces oxidative stress and potentiates the apoptotic action of 2-methoxyestradiol in human hepatoma cells. Nutr. Cancer 61, 735–745.

Chinta, S.J., Ganesan, A., Reis-Rodrigues, P., Lithgow, G.J., Andersen, J.K., 2013. Anti-inflammatory role of isoflavone daidzein in lipopolysaccharide-stimulated microglia: implications for Parkinson's disease. Neurotox. Res. 23, 145–153.

Chitnis, T., Weiner, H.L., 2017. CNS inflammation and neurodegeneration. J. Clin. Invest. 127, 3577–3587.

Cho, J.M., Chang, S.-Y., Kim, D.B., Needs, P.W., Jo, Y.H., Kim, M.J., 2012. Effects of physiological quercetin metabolites on interleukin-1β-induced inducible NOS expression. J. Nutr. Biochem. 23, 1394–1402.

Christen, Y., 2004. *Ginkgo biloba* and neurodegenerative disorders. Front. Biosci. 9, 3091–3104.

Conquer, J.A., Maiani, G., Azzini, E., Raguzzini, A., Holub, B.J., 1998. Supplementation with quercetin markedly increases plasma quercetin concentration without effect on selected risk factors for heart disease in healthy subjects. J. Nutr. 128, 593–597.

Costa, L.G., Tait, L., de Laat, R., Dao, K., Giordano, G., Pellacani, C., 2013. Modulation of paraoxonase 2 (PON2) in mouse brain by the polyphenol quercetin: a mechanism of neuroprotection? Neurochem. Res. 38, 1809–1818.

Costa, L.G., de Laat, R., Dao, K., Pellacani, C., Cole, T.B., Furlong, C.E., 2014. Paraoxonase-2 (PON2) in brain and its potential role in neuroprotection. Neurotoxicology. 43, 3–9.

Dajas, F., Rivera, F., Blasina, F., Arredondo, F., Echeverry, C., Lafon, L., 2003. Cell culture protection and *in vivo* neuroprotective capacity of flavonoids. Neurotox. Res. 5, 425–432.

Dang, W., 2012. The controversial world of sirtuins. Drug Discov. Today 12, e9–e17.

Das, S., Mandal, A.K., Ghosh, A., Panda, S., Das, N., Sarkar, S., 2008. Nanoparticulated quercetin in combating age related cerebral oxidative injury. Curr. Aging Sci. 1, 169–174.

Day, A.J., Mellon, F., Barron, D., Sarrazin, G., Morgan, M.R., Williamson, G., 2001. Human metabolism of dietary flavonoids: identification of plasma metabolites of quercetin. Free Radic. Res. 35, 941–952.

deBoer, V.C.J., Dihal, A.A., van der Woude, H., Arts, I.C., Wolffram, S., Alink, G.M., et al., 2005. Tissue distribution of quercetin in rats and pigs. J. Nutr. 135, 1718–1725.

DeKosky, S.T., Williamson, J.D., Fitzpatrick, A.L., et al., 2008. Ginkgo biloba for prevention of dementia: a randomized controlled trial. JAMA. 300, 2253–2262.

Del Rio, D., Costa, L.G., Lean, M.E.J., Crozier, A., 2010. Polyphenols and health: what compounds are involved? Nutr. Metab. Cardiovasc. Dis. 20, 1–6.

Del Rio, D., Rodriguez-Mateos, A., Spencer, J.P.E., Tognolini, M., Borges, G., Crozier, A., 2013. Dietary (poly)phenolics in human health: structures, bioavailability, and evidence of protective effects against chronic diseases. Antioxid. Redox Signal. 18, 1818–1892.

Denny Joseph, K.M., Muralidhara, 2013. Enhanced neuroprotective effect of fish oil in combination with quercetin against 3-nitropropionic acid induced oxidative stress in rat brain. Prog. Neuropsychopharmacol. Biol. Psychiatry 40, 83–92.

Devarajan, A., Bourquard, N., Hama, S., Navab, M., Grijalva, V.R., Morvardi, S., 2011. Paraoxonase 2 deficiency alters mitochondrial function and exacerbates the development of atherosclerosis. Antioxid. Redox Signal. 14, 341–351.

Dhavan, S., Kapil, R., Singh, B., 2011. Formulation development and systematic optimization of solid lipid nanoparticles of quercetin for improved brain delivery. J. Pharm. Pharmacol. 63, 342–351.

Dou, Z., Rong, X., Zhao, E., et al., 2019. Neuroprotection of resveratrol against focal cerebral ischemia/reperfusion injury in mice through a mechanism targeting gut-brain axis. Cell Mol. Neurobiol. 39, 883–898. PMID: 31140018.

Draganov, D.I., Teiber, J.F., Speelman, A., Osawa, Y., Sunahara, R., Du, B.N. La, 2005. Human paraoxonases (PON1, PON2 and PON3) are lactonases with overlapping and distinct substrate specificities. J. Lipid Res. 46, 1239–1247.

Dumont, M., Beal, M.F., 2011. Neuroprotective strategies involving ROS in Alzheimer disease. Free Radic. Biol. Med. 51, 1014–1026.

Faria, A., Pestana, D., Teixera, D., Azevedo, J., Freitas, V. De, Mateus, N., 2010. Flavonoid transport across RBE4 cells: a blood-brain barrier model. Cell. Mol. Biol. Lett. 15, 234–241.

Fernandes, E.S., Machado Mde, O., Becker, A.M., de Andrade, F., Maraschin, M., da Silva, E.L., 2012. Yerba mate (*Ilex paraguariensis*) enhances the gene modulation and activity of paraoxonase-2: *in vitro* and *in vivo* studies. Nutrition 28, 1157–1164.

Ferrante, R.J., Klein, A.M., Dedeoglu, A., et al., 2001. Therapeutic efficacy of EGb 761 (*Ginkgo biloba* extract) in a transgenic mouse model of amyotrophic lateral sclerosis. J. Mol. Neurosci. 17, 89–96.

Fiorani, M., Accorsi, A., Cantoni, O., 2003. Human red blood cells as a natural flavonoid reservoir. Free Radic. Res. 37, 1331–1338.

Fraga, C.G., Galleano, M., Verstraeten, S.V., Oteiza, P.I., 2010. Basic biochemical mechanisms behind the health benefits of polyphenols. Mol. Asp. Med. 31, 435–445.

Fuhrman, B., Gantman, A., Khateeb, J., Volkova, N., Horke, S., Kiyan, J., 2009. Urokinase activates macrophage PON2 gene transcription via the PI3K/ROS/MEK/SREBP-2 signalling cascade mediated by the PDGFR-beta. Cardiovasc. Res. 84, 145–154.

Gabryel, B., Kost, A., Kasprowska, D., 2012. Neuronal autophagy in cerebral ischemia—a potential target for neuroprotective strategies? Pharmacol. Rep. 64, 1–15.

Galluzzo, P., Martini, C., Bulzomi, P., Leone, S., Bolli, A., Pallottini, V., 2009. Quercetin-induced apoptotic cascade in cancer cells: antioxidant versus estrogen receptor alpha-dependent mechanisms. Mol. Nutr. Food Res. 53, 699–708.

Gan, L., Johnson, J.A., 2014. Oxidative damage and the Nrf2-ARE pathway in neurodegenerative diseases. Biochim. Biophys. Acta 1842, 1208–1218.

Gao, Y., Fu, R., Wang, J., et al., 2018. Resveratrol mitigates the oxidative stress mediated by hypoxic-ischemic brain injury in neonatal rats via Nrf2/HO-1 pathway. Pharm. Biol. 56, 440–449.

Garg, S.K., Shukla, A., Choudhury, S., 2019. Polyphenols and flavonoids. In: Gupta, R.C., Srivastava, A., Lall, R. (Eds.), Nutraceuticals in Veterinary Medicine. Springer Nature, Cham, Switzerland, pp. 187–204.

Giordano, G., White, C.C., Mohar, I., Kavanagh, T.J., Costa, L.G., 2007. Glutathione levels modulate domoic acid induced apoptosis in mouse cerebellar granule cells. Toxicol. Sci. 100, 433–444.

Giordano, G., Cole, T.B., Furlong, C.E., Costa, L.G., 2011. Paraoxonase 2 (PON2) in the mouse central nervous system: a neuroprotective role? Toxicol. Appl. Pharmacol. 256, 369–378.

Giordano, G., Kavanagh, T.J., Faustman, E.M., White, C.C., Costa, L.G., 2013a. Low-level domoic acid protects cerebellar granule neurons from acute neurotoxicity: role of glutathione. Toxicol. Sci. 132, 399–408.

Giordano, G., Tait, L., Furlong, C.E., Cole, T.B., Kavanagh, T.J., Costa, L.G., 2013b. Gender differences in brain susceptibility to oxidative stress are mediated by levels of paraoxonase-2 (PON2) expression. Free Radic. Biol. Med. 58, 98–108.

Giordano, S., Darly-Usmar, V., Zhang, J., 2014. Autophagy as an essential cellular antioxidant pathway in neurodegenerative disease. Redox Biol. 2, 82–90.

Gomes, B.A.Q., Silva, J.P.B., Romeiro, C.F.R., et al., 2018. Neuroprotective mechanisms of resveratrol in Alzheimer's disease: role of SIRT1. Oxid. Med. Cell Longev. 2018, 8152373. Available from: https://doi.org/10.1155/2018/8152373.

Graf, B.A., Milbury, P.E., Blumberg, J.B., 2005. Flavonols, flavones, flavonones, and human health: epidemiological evidence. J. Med. Food 8, 281–290.

Granado-Serrano, A.B., Martin, M.A., Bravo, L., Goya, L., Ramos, S., 2010. Quercetin modulates Nf-kB and AP-1/JNK pathways to induce cell death in human hepatoma cells. Nutr. Cancer 62, 390–401.

Granado-Serrano, A.B., Martin, M.A., Bravo, L., Goya, L., Ramos, S., 2012. Quercetin modulates Nrf2 and glutathione-related defenses in HepG2 cells: involvement of p38. Chem. Biol. Interact. 195, 154–164.

Gupta, Y.K., Chaudhary, G., Sinha, K., Srivastava, A.K., 2001. Protective effect of resveratrol against intracortical $FeCl_3$-induced model of posttraumatic seizures in rats. Methods Find. Exp. Clin. Pharmacol. 23, 241–244.

Gupta, R.C., Srivastava, A., Lall, R. (Eds.), 2019. Nutraceuticals in Veterinary Medicine. Springer Nature, Cham, Switzerland, p. 877.

Halliwell, B., 2006. Oxidative stress and neurodegeneration: where are we now? J. Neurochem. 97, 1634–1658.

Halliwell, B., 2008. Are polyphenols antioxidants or pro-oxidants? What do we learn from cell culture and *in vivo* studies? Arch. Biochem. Biophys. 476, 107–112.

Halliwell, B., 2012. The antioxidant paradox: less paradoxical now? Br. J. Clin. Pharmacol. 75, 637–644.

Halliwell, B., Rafter, J., Jenner, A., 2005. Health promotion of flavonoids, tocopherols, tocotrienols, and other phenols: direct or indirect effects? Antioxidants or not? Am. J. Clin. Nutr. 81, 268S–276S.

Harwood, M., Danielewska-Nikiel, B., Borzelleca, J.F., Flamm, G.W., Williams, G.M., Lines, T.C., 2007. A critical review of the data related to the safety of quercetin and lack of evidence of in vivo toxicity, including lack of genotoxic/carcinogenic properties. Food Chem. Toxicol. 45, 2179–2205.

Horke, S., Witte, I., Wilgenbus, P., Kruger, M., Strand, D., Forstermann, U., 2007. Paraoxonase-2 reduces oxidative stress in vascular cells and decreases endoplasmic reticulum stress-induced caspase activation. Circulation 115, 2055–2064.

Hu, P., Wang, M., Chen, W.H., Liu, J., Chen, L., Yin, S.T., 2008. Quercetin relieves chronic lead exposure-induced impairment of synaptic plasticity in rat dentate gyrus *in vivo*. Naunyn-Schmiedeberg's Arch. Pharmacol. 378, 43–51.

Huang, T.C., Lu, K.T., Wo, Y.Y.P., et al., 2011. Resveratrol protects rats from Aβ-induced neurotoxicity by the reduction of iNOS expression and lipid peroxidation. PLoS One 6, e29102. Available from: https://doi.org/10.1371/journal.pone.0029102.

Huebbe, P., Wagner, A.E., Boesch-Saadatmandi, C., Sellmer, F., Wolffram, S., Rimbach, G., 2010. Effect of dietary quercetin on brain quercetin levels and the expression of antioxidant and Alzheimer's disease relevant genes in mice. Pharmacol. Res. 61, 242–246.

Ishisaka, A., Ichikawa, S., Sakakibara, H., Piskula, M.K., Nakamura, T., Kato, Y., 2011. Accumulation of orally administered quercetin in brain tissue and its antioxidative effects in rats. Free Radic. Biol. Med. 51, 329–336.

Jantaratnotai, N., Utaisincharoen, P., Sanvarinda, P., Thampithak, A., Sanvarinda, Y., 2013. Phytoestrogens mediated anti-inflammatory effect through suppression of IRF-1 and pSTAT1 expressions in lipopolysaccharide-activated microglia. Int. Immunopharmacol. 17, 483–488.

Jia, Y., Wang, N., Liu, X., 2017. Resveratrol and amyloid-beta: mechanistic insights. Nutrients 9, 1122. Available from: https://doi.org/10.3390/nu9101122.

Jiang, M., Li, J., Peng, Q., et al., 2014. Neuroprotective effects of bilobalide on cerebral ischemia and reperfusion injury are associated with inhibition of pro-inflammatory mediator production and down-regulation of JNK½ and p38 MAPK activation. J. Neuroinflamm. 11, 167.

Jin, Y.S., Min, A.S., Choi, E.J., et al., 2008. Neuroprotective effects of resveratrol via anti-apoptosis on hypoxic-ischemic brain injury in neonatal rats. Korean J. Pediatr. 51, 1102–1111.

Kawai, Y., Nishikawa, T., Shiba, Y., Saito, S., Murota, K., Shibata, N., 2008. Macrophage as a target of quercetin glucuronides in human atherosclerotic arteries. J. Biol. Chem. 283, 9424–9434.

Kay, C.D., 2010. The future of flavonoid research. Br. J. Nutr. 104, S91–S95.

Kelly, G.S., 2011. Quercetin. Monograph. Altern. Med. Rev. 16, 172–194.

Kelsey, N.A., Wilkins, H.M., Linseman, D.A., 2010. Nutraceutical antioxidants as novel neuroprotective agents. Molecules 15, 7792–7814.

Komatsu, M., Waguri, S., Chiba, T., Murata, S., Iwata, J., Tanida, I., 2006. Loss of autophagy in the central nervous system causes neurodegeneration in mice. Nature 441, 880–884.

Kumar, S.S., Barreto, E.G., Aliev, G., et al., 2017. *Ginkgo biloba* as an alternative medicine in the treatment of anxiety in dementia and other psychiatric disorders. Curr. Drug. Metab. 18, 112–119.

Lajqi, T., Stojiljkovic, M., Wetzker, R., 2019. Toxin-induced hormesis may restrain aging. Biogerontology 20, 571–581.

Li, W.F., Costa, L.G., Richter, R.J., Hagan, T., Shih, D.M., Tward, A., 2000. Catalytic efficiency determines the *in vivo* efficacy of PON1 for detoxifying organophosphates. Pharmacogenetics 10, 767–779.

Li, Y., Yao, J., Han, C., et al., 2016. Quercetin, inflammation and immunity. Nutrients 8, 167. Available from: https://doi.org/10.3390/nu8030167.

Li, Y.-R., Li, S., Lin, C.-C., 2017. Effect of resveratrol and pterostilbene on aging and longevity. BioFactors. 44, 69–82.

Liang, L.L., Gao, C., Luo, M., Wang, W., Zhao, C., Zu, Y., 2013. Dihydroquercetin (DHQ) induced HO-1 and NQO1 expression against oxidative stress through the Nrf2-dependent antioxidant pathway. J. Agric. Food Chem. 61, 2755–2761.

Lin, M.T., Beal, M.F., 2006. Mitochondrial dysfunction and oxidative stress in neurodegenerative diseases. Nature 443, 787–795.

Lin, T.K., Chen, S.D., Chuang, Y.C., Lin, H.Y., Huang, C.R., Chuang, J.H., 2014. Resveratrol partially prevents rotenone-induced neurotoxicity in dopaminergic SH-SY5Y cells through induction of heme oxygenase-1 dependent autophagy. Int. J. Mol. Sci. 15, 1625–1646.

Lin, K.L., Lin, K.J., Wang, P.M., et al., 2018. Resveratrol provides neuroprotective effects through modulation of mitochondrial dynamics and ERK½ regulated autophagy. Free Radic. Res. 52, 1371–1386.

Linseman, D.A., 2009. Targeting oxidative stress for neuroprotection. Antioxid. Redox Signal. 11, 421–424.

Lopez, M., Dempsey, R.J., Vemuganti, R., 2015. Resveratrol neuroprotection in stroke and traumatic CNS injury. Neurochem. Int. 89, 75–82.

Manach, C., Scalbert, A., Morand, C., 2004. Polyphenols: food sources and bioavailability. Am. J. Clin. Nutr. 79, 727–747.

Manach, C., Williamson, G., Morand, C., 2005. Bioavailability and bioefficacy of polyphenols in humans. I. Review of 97 bioavailability studies. Am. J. Clin. Nutr. 81, 230S–242S.

Marambaud, P., Zhao, H., Davies, P., 2005. Resveratrol promotes clearance of Alzheimer's disease amyloid-beta peptides. J. Biol. Chem. 280, 37377–37382.

Mariño, G., Madeo, F., Kroemer, G., 2011. Autophagy for tissue homeostasis and neuroprotection. Curr. Opin. Cell Biol. 23, 198–206.

Marsillach, J., Mackness, B., Mackness, M., 2008. Immunohistochemical analysis of paraoxonase-1, 2 and 3 expression in normal mouse tissues. Free Radic. Biol. Med. 45, 146–157.

Martin, I., Grotewiel, M.S., 2006. Oxidative damage and age-related functional declines. Mech. Aging Dev. 127, 411–423.

Mattson, M.P., 2008. Hormesis defined. Aging Res. Rev. 7, 1–7.

Mattson, M.P., Cheng, A., 2006. Neurohormetic phytochemicals: low dose toxins that induce adaptive stress responses. Trends Neurosci. 29, 632–639.

Mazzio, E.A., Close, F., Soliman, K.F.A., 2011. The biochemical and cellular basis for nutraceutical strategies to attenuate neurodegeneration in Parkinson's disease. Int. J. Mol. Sci. 12, 506–569.

Mecocci, P., Tinarelli, C., Schulz, R.J., Polidori, M.C., 2014. Nutraceuticals in cognitive impairment and Alzheimer's disease. Front. Pharmacol. 5, 147.

Mercer, L.D., Kelly, B.L., Horne, M.K., Beart, P.M., 2005. Dietary polyphenols protect dopamine neurons from oxidative insults and apoptosis: investigations in primary rat mesencephalic cultures. Biochem. Pharmacol. 69, 339–345.

Milatovic, D., Gupta, R.C., 2019. Antioxidants in prevention and treatment of diseases and toxicity. In: Gupta, R.C., Srivastava, A., Lall, R. (Eds.), Nutraceuticals in Veterinary Medicine. Springer Nature, Cham, Switzerland, pp. 205–213.

Miodini, P., Fioravanti, L., Fronzo, G. Di, Cappelletti, V., 1999. The two phytoestrogens genistein and quercetin exert different effects of noestrogen receptor function. Br. J. Cancer 80, 1150–1155.

Moinova, H.R., Mulcahy, R.T., 1999. Up-regulation of the human gamma glutamylcysteine synthetase regulatory subunit gene involves binding of Nrf-2 to an electrophile response element. Biochem. Biophys. Res. Commun. 261, 661–668.

Moon, J.H., Tsushida, T., Nakahara, K., Terao, J., 2001. Identification of quercetin 3-*O*-beta-d-glucuronide as an antioxidative metabolite in rat plasma after oral administration of quercetin. Free Radic. Biol. Med. 11, 1274–1285.

Nabavi, S.F., Tenore, G.C., Daglia, M., 2015. The cellular protective effects of rosmarinic acid: from bench to bedside. Curr. Neurovasc. Res. 12, 98–105.

Ng, C.J., Wadleigh, D.J., Gangopadhyyay, A., 2001. Paraoxonase-2 is a ubiquitously expressed protein with antioxidant properties and is capable of preventing cell-mediated oxidative modification of low density lipoprotein. J. Biol. Chem. 276, 44444–44449.

Ng, C.J., Bourquard, N., Grijalva, V., 2006a. Paraoxonase-2 deficiency aggravates atherosclerosis in mice despite lower apolipoprotein-B-containing lipoproteins. Antiatherogenic role for paraoxonase-2. J. Biol. Chem. 281, 29491–29500.

Ng, C.J., Hama, S.Y., Bourquard, N., 2006b. Adenovirus mediated expression of human paraoxonase 2 protects against the development of atherosclerosis in apolipoprotein E-deficient mice. Mol. Genet. Metab. 89, 368–373.

Nillert, N., Pannangrong, W., Welbat, J.U., et al., 2017. Neuroprotective effects of aged garlic extract on cognitive dysfunction and neuroinflammation induced by β-amyloid in rats. Nutrients 9, 24. Available from: https://doi.org/10.3390/nu9010024.

Nouhi, F., Tusi, S.K., Abdi, A., Khodagoli, F., 2011. Dietary supplementation with tBHQ, an Nrf2 stabilizer molecule, confers neuroprotection against apoptosis in amyloid-β-injected rat. Neurochem. Res. 36, 870–878.

Ossola, B., Kaariainen, T.M., Mannisto, P.T., 2009. The multiple faces of quercetin in neuroprotection. Expert Opin. Drug. Saf. 8, 397–409.

Ott, M., Gogvadze, V., Orrenius, S., Zhivotovsky, B., 2007. Mitochondria, oxidative stress and cell death. Apoptosis 12, 913–922.

Pan, T., Rawal, Y., Wu, Y., Xie, W., 2009. Rapamycin protects against rotenone-induced apoptosis through autophagy induction. Neuroscience 164, 541–551.

Park, J.Y., Kang, Y.W., Cho, W.G., 2019. Inflammasome-mediated inflammation in neurodegenerative diseases. Open Neurol. J. 13, 55–62.

Perez-Pinzon, M.A., Stetlen, R.A., Fiskum, G., 2012. Novel mitochondrial targets for neuroprotection. J. Cereb. Blood Flow. Metab. 32, 1362–1376.

Popa-Wagner, A., Mitran, S., Sivenesan, S., 2013. ROS and brain disease: the good, the bad, and the ugly. Oxid. Med. Cell. Longev. 2013, 963520.

Primo-Parmo, S.L., Sorenson, R.C., Teiber, J., La Du, B.N., 1996. The human serum paraoxonase/arylesterase gene (PON1) is one member of a multigene family. Genomics 33, 498–507.

Rosenblat, M., Draganov, D., Watson, C.E., 2003. Mouse macrophage paraoxonase-2 activity is increased whereas cellular paraoxonase 3 activity is decreased under oxidative stress. Arterioscler. Thromb. Vasc. Biol. 23, 468–474.

Rosenblat, M., Hayek, T., Hussein, K., Aviram, M., 2004. Decreased macrophage paraoxonase 2 expression in patients with hypercholesterolemia is the result of their increased cellular cholesterol content: effect of atorvastatin therapy. Arterioscler. Thromb. Vasc. Biol. 24, 175–180.

Rosenblat, M., Volkova, N., Roqueta-Rovera, M., 2010. Increased macrophage cholesterol biosysnthesis and decreased cellular paraoxonase 2 (PON2) expression in Δ6-desaturase knockout (6-DS-KO) mice: beneficial effects of arachidonic acid. Atherosclerosis 210, 414–421.

Ruotolo, R., Calani, L., Brighenti, F., 2014. Glucuronidation does not suppress the estrogenic activity of quercetin in yeast and human breast cancer cell model system. Arch. Biochem. Biophys. 559, 62–67.

Russo, M., Spagnuolo, C., Tedesco, I., 2012. The flavonoid quercetin in disease prevention and therapy: facts and fancies. Biochem. Pharmacol. 83, 6–15.

Sachdev, S., Davies, K.J., 2008. Production, detection, and adaptive responses to free radical in exercise. Free Radic. Biol. Med. 44, 215–223.

Sanches-Melgar, A., Albasanz, J.L., Guixa-Gonzalez, R., et al., 2019. The antioxidant resveratrol acts as a non-selective adenosine receptor agonist. Free Radic. Biol. Med. 135, 261–273.

Sarubbo, F., Ramis, M.R., Aparicio, S., et al., 2015. Improving effect of chronic resveratrol treatment on central monoamine synthesis and cognition in aged rats. Age 37, 9777.

Saw, C.L., Guo, Y., Yang, A.Y., 2014. The berry constituents quercetin, kaempferol, and pterostilbene synergistically attenuate reactive oxygen species: involvement of the Nrf2-ARE signaling pathway. Food Chem. Toxicol. 72, 303–311.

Scalbert, A., Andres-Lacueva, C., Arita, M., 2011. Databases on food phytochemicals and their health-promoting effects. J. Agric. Food Chem. 59, 4331–4348.

Schaffer, S., Halliwell, B., 2012. Do polyphenols enter the brain and does it matter? Some theoretical and practical considerations. Genes Nutr. 7, 99–109.

Selvakumar, K., Bavithra, S., Suganthi, M., 2012. Protective role of quercetin on PCBs-induced oxidative stress and apoptosis in hippocampus of adult rats. Neurochem. Res. 37, 708–721.

Shanely, R.A., Knab, A.M., Nieman, D.C., 2010. Quercetin supplementation does not alter antioxidant status in humans. Free Radic. Res. 44, 224–231.

Shih, A.Y., Imbeault, S., Barakaukas, V., 2005. Induction of the Nrf2-driven antioxidant response confers neuroprotection during mitochondrial stress *in vivo*. J. Biol. Chem. 280, 22925–22936.

Shiner, M., Fuhrman, B., Aviram, M., 2007. Macrophage paraoxonase 2 (PON2) expression is up-regulated by pomegranate juice phenolic anti-oxidants via PPARγ and AP-1 pathway activation. Atherosclerosis 195, 313–321.

Shirai, M., Kawai, Y.M., Yamanishi, R., 2006. Effect of a conjugated quercetin metabolite, quercetin 3-glucuronide, on lipid hydroperoxide-dependent formation of reactive oxygen species in differentiated PC-12 cells. Free Radic. Res. 40, 1047–1053.

Simpkins, J.W., Yi, K.D., Yang, S.H., Dykens, J.A., 2010. Mitochondrial mechanisms of estrogen neuroprotection. Biochim. Biophys. Acta 1800, 1113–1120.

Singh, S.K., Srivastav, S., Yadav, A.K., et al., 2016. Overview of Alzheimer's disease and some therapeutic approaches targeting Aβ by using several synthetic and herbal compounds. Oxid. Med. Cell Longev. 2016, 7361613.

Singh, S.K., Srivastav, S., Castellani, R.J., et al., 2019. Neuroprotective and antioxidant effect of *Ginkgo biloba* extract against AD and other neurological disorders. Neurotherapeutics 16, 666–674.

Skaper, S.D., Facci, L., Giusti, P., 2014. Neuroinflammation, microglia and mast cells in the pathophysiology of neurocognitive disorders: a review. CNS Neurol. Disord. Drug Targets 13, 1654–1666.

Song, W., Guan, H.J., Zhu, X.Z., et al., 2000. Protective effect of bilobalide against nitric oxide-induced neurotoxicity in PC12 cells. Acta Pharmacol. Sin. 21, 415–420.

Spencer, J.P.E., 2009. Flavonoids and brain health: multiple effects underpinned by common mechanisms. Genes Nutr. 4, 243–250.

Srivastav, S., Fatima, M., Mondal, A.C., 2017. Important medicinal herbs in Parkinson's disease pharmacotherapy. Biomed. Pharmacother. 92, 856–863.

Stephenson, J., Nutma, E., van der Valk, P., et al., 2018. Inflammation in CNS neurodegenerative diseases. Immunology 164, 204–219.

Tait L.J. 2011. Modulation of Paraoxonase 2 (PON2) in the CNS (M.S. thesis). University of Washington, Washington, p. 45.

Takechi, R., Pallebage-Garallage, M.M., Lam, V., 2013. Nutraceutical agents with anti-inflammatory properties prevent dietary saturated-fat induced disturbances in blood-brain barrier function in wild-tye mice. J. Neuroinflamm. 10, 842.

Tapia, P.C., 2006. Sublethal mitochondrial stress with and attendant stoichiometric augmentation of reactive oxygen species may precipitate many of the beneficial alterations in cellular physiology produced by caloric restriction, intermittent fasting, exercise and dietary phytonutrients: "mitohormesis" for health and vitality. Med. Hypotheses 66, 832–843.

Teiber, J.F., Horke, S., Haines, D.C., 2008. Dominant role of paraoxonases in inactivation of the *Pseudomonas aeruginosa* quorum-sensing signal *N*-(3-oxododecanoyl)-l-homoserine lactone. Infect. Immun. 76, 2512–2519.

Turner, R.S., Thomas, R.G., Craft, S., et al., 2015. A randomized, double-blind, placebo-controlled trial of resveratrol for Alzheimer's disease. Neurology 85, 1383–1391.

USDA (United States Department of Agriculture), 2003. USDA Database for the Flavonoid Content of Selected Foods. USDA, Beltsville Human Nutrition Research Center, Beltsville, MD, p. 77.

Vauzour, D., 2012. Dietary polyphenols as modulators of brain functions: biological actions and molecular mechanisms underpinning their beneficial effects. Oxid. Med. Cell. Longev. 2012, 914273.

Vauzour, D., Ravaioli, G., Vafeiadou, K., 2008. Peroxynitrite induced formation of the neurotoxins 5-*S*-cysteinyl-dopamine and DHBT-1: implications for Parkinson's disease and protection by polyphenols. Arch. Biochem. Biophys. 476, 145–151.

Vellas, B., Coley, N., Ousset, P.J., et al., 2012. Long-term use of standardized *Ginkgo biloba* extract for the prevention of Alzheimer's disease (GuidAge): a randomized placebo-controlled trial. Lancet Neurol. 11, 851–859.

Verzelloni, E., Pellacani, C., Tagliazucchi, D., 2011. Antiglycative and neuroprotective activity of colon-derived polyphenol catabolytes. Mol. Nutr. Food Res. 55, 1–9.

Watanabe, C.M., Wolffram, S., Ader, P., et al., 2001. The *in vivo* neuromodulatory effects of the herbal medicine *Ginkgo biloba*. Proc. Natl Acad. Sci. U.S.A. 98, 6577–6580.

West, T., Atzeva, M., Holtzman, D.M., 2007. Pomegranate polyphenols and resveratrol protect the neonatal brain against hypoxic-ischemic injury. Dev. Neurosci. 29, 363–372.

White, H.L., Scates, P.W., Cooper, B.R., 1996. Extracts of *Ginkgo biloba* leaves inhibit monoamine oxidase. Life Sci. 58, 1315–1321.

Wild, A.C., Moinova, H.R., Mulcahy, R.T., 1999. Regulation of gamma-glutamylcysteine synthetase subunit gene expression by the transcription factor Nrf2. J. Biol. Chem. 274, 33627–33636.

Witte, A.V., Kerti, L., Margulies, D.S., et al., 2014. Effects of resveratrol on memory performance, hippocampal functional connectivity, and glucose metabolism in healthy older adults. J. Neurosci. 34, 7862–7870.

Yeh, S.L., Yeh, C.L., Chan, S.T., Chuang, C.H., 2011. Plasma rich in quercetin metabolites induces G2/M arrest by upregulating PPAR-γ expression in human A549 lung cancer cells. Planta Med. 77, 992–998.

Yehuda, I., Madar, Z., Szuchman-Sapir, A., Tamir, S., 2011. Glabridin, a phytoestrogen from licorice root, up-regulates manganese superoxide dismutase, catalase and paraoxonase 2 under glucose stress. Phytother. Res. 25, 659–667.

Yin, Y., Ren, Y., Wu, W., et al., 2013. Protective effects of bilobalide on Aβ(25-35) induced learning and memory impairments in male rats. Pharmacol. Biochem. Behav. 106, 77–84.

Yuan, Q., Wang, C.W., Shi, J., et al., 2017. Effects of *Ginkgo biloba* on dementia: an overview of systematic reviews. J. Ethnopharmacol. 195, 1–9.

Yun, J., Finkel, T., 2014. Mitohormesis. Cell Metab. 19, 757–766.

Chapter 2

Prevention of neurodegenerative disorders by nutraceuticals

Francesca Pistollato and Magdalini Sachana

Introduction

Dietetic macronutrients and micronutrients play a crucial role in the control of brain physiology, and food intake is known to stimulate the activity of neurotrophic factors regulating synaptic plasticity. Among micronutrients, vitamins (e.g., B1, B6, folic acid, B12, C, D, K, and α-tocopherol) and minerals (e.g., iron, lithium, magnesium, copper, iodine, and manganese) are known to modulate central nervous system (CNS) functionality, whereas macronutrients [e.g., polyunsaturated fatty acids (PUFAs), essential amino acids, low glycemic index foods, and dietary fibers] are relevant for the maintenance of cognitive functions and the prevention of neurodegeneration (Pistollato and Battino, 2014; Pistollato et al., 2019, 2020).

The chronic persistence of neuroinflammation is currently considered the main driving force of the neurodegenerative process. Neuroinflammation is generally characterized by the activation of glia and microglia and the upregulation of inflammatory-related molecules (Morales et al., 2014), and this phenomenon seems to occur especially during the aging process (Michaud et al., 2013). Plant-derived bioactive nutrients, such as antioxidants, n-3 and n-6 PUFAs, and other antiinflammatory nutraceuticals, have been shown to prevent neuroinflammation, thus reducing the risk of neurodegeneration (Virmani et al., 2013;Nabavi et al., 2017; Omar et al., 2017). The presence of chronic neuroinflammatory status is determined by several factors, such as lifestyle and diet. Importantly, dietetic patterns characterized by high consumption of animal-derived products and a very low or null intake of plant-derived foods can lead to persistent chronic inflammation, the potential onset of metabolic syndrome (MetS)-related dysfunctions, and a high risk of cognitive impairment (Pistollato and Battino, 2014; Pistollato et al., 2019). One of the factors characteristic of both the MetS and neurodegeneration is the presence of a high plasma homocysteine level, a condition known as hyperhomocyteinemia (hHcy). Indeed, hHcy is commonly known as an independent risk factor for cardiovascular disease and stroke; however, it is also correlated with several neurodegenerative diseases (Boldyrev et al., 2013), including Alzheimer disease (AD) and vascular dementia (Troen et al., 2008). hHcy has also been found in L-DOPA-treated Parkinson disease (PD) patients (Zoccolella et al., 2005) and is correlated with neuroinflammation (Sudduth et al., 2013) and loss of central cholinergic neurons (Pirchl et al., 2010). Additionally, a high level of serum homocysteine has been found in the body fluids of autistic children (Kaluzna-Czaplinska et al., 2013). Some nutraceuticals, specifically B vitamins, are fundamental to stabilize homocysteine levels; indeed, vitamins B6 (pyridoxine), B12 (cobalamin), and B9 (folic acid) are cofactors needed to guarantee the physiological functioning of the enzymes specifically involved in homocysteine metabolism.

To counteract neuroinflammation-related processes plant-derived polyphenols, such as resveratrol, sulforaphanes, and curcumin, have been regarded as essential antioxidant factors contributing to the regulation of brain homeostasis. Besides their antioxidant capacity, these neurohormetic phytochemicals are also known to downregulate oxidative/inflammatory stress signaling pathways and to upregulate the expression of genes encoding antioxidant enzymes, phase-2 enzymes, neurotrophic factors, and cytoprotective-related signaling pathways, such as the sirtuin-Forkhead box O (FoxO) pathway, the nuclear factor-kappaB (NF-κB) pathway, the nuclear factor erythroid 2-related factor 2 (Nrf2)/antioxidant responsive element (ARE) pathway, and the cyclic AMP responsive element binding protein (CREB)-related pathway (Vauzour, 2012).

This chapter describes the molecular mechanisms underlying the beneficial effects elicited by well-described nutraceuticals in the prevention and regression of AD, PD, and autism spectrum disorders (ASDs).

Nutraceuticals. DOI: https://doi.org/10.1016/B978-0-12-821038-3.00002-1

Nutraceuticals for the prevention and amelioration of Alzheimer disease

AD is the most common form of dementia, characterized by the extracellular accumulation of beta-amyloid (Aβ) deposits and progressive microtubule disintegration, leading to dysfunctional neuronal communication and neuronal cell death. Moreover, mature long-lasting amyloid plaques appear to be less toxic than the prefibrillar aggregates, which represent their precursors. The early aggregates seem to interact with cell membranes, causing oxidative stress and an increase in free Ca^{2+} levels, eventually leading to apoptotic or necrotic cell death (Stefani and Dobson, 2003). As the disease advances, confusion, irritability, aggression, mood swings, trouble with language, and long-term memory loss often occur, with an average life expectancy upon diagnosis of approximately 7 years (Waldemar et al., 2007). Increasing evidence suggests that the pathological traits of the disease (e.g., Aβ plaques and neurofibrillary tangles formation) can be found several decades before the onset of clinical symptoms (Sperling et al., 2011; Schelke et al., 2016), which suggests that preventive intervention strategies may help reduce the risk of developing AD. Among the risk factors, hyperlipidemia, hHcy, diabetes, alcohol consumption, smoking, and obesity have been found to increase AD risk, whereas consumption of plant-based foods, enriched in bioactive phytocompounds, and also fish, unsaturated fat or *n*-3 fatty acids, a Mediterranean diet, and other emerging healthy dietary patterns, DASH (Dietary Approach to Stop Hypertension) and MIND (Mediterranean-DASH diet Intervention for Neurodegenerative Delay) (Solfrizzi et al., 2017), together with physical and social activity, have been described as protective in observational studies (Weih et al., 2007; Pistollato et al., 2019).

Lowered blood and cerebrospinal fluid levels of certain nutrients, such as docosahexaenoic acid (DHA), choline, vitamin B12, folate, vitamin C, 25-hydroxyvitamin D, and vitamin E, have been observed in AD patients (Lopes da Silva et al., 2014; de Wilde et al., 2017; Etgen et al., 2012; Shen and Ji, 2015).

Currently, given the lack of a definitive and effective treatment for AD, lifestyle changes, including plant-based nutritional interventions and natural nutraceutical supplementations, and also practicing cognitive and social activity and physical exercise are considered alternative measures to prevent AD occurrence (Mecocci et al., 2014). In particular, flavonoids, vitamins, and other natural substances have been studied in relation to AD and have been considered beneficial for the maintenance of cognitive performances (Table 2.1).

Polyphenols, which are naturally present in vegetables, fruits, herbs, and nuts, may promote prevention and regression of AD by targeting specific signaling pathways associated with protein folding and neuroinflammation (Essa et al., 2012), beneficially modulating the microbiota–gut–brain axis (Serra et al., 2020). Importantly, an effective strategy to prevent and/or reduce protein misfolding and Aβ formation and restoring cellular aggretome might be the use of amyloid-binding polyphenols. These polyphenols seem to act via different mechanisms, either inhibiting fibril formation or steering oligomer formation into unstructured, nontoxic pathways (Ngoungoure et al., 2015). Polyphenols have antioxidant and antiinflammatory effects and can modulate autophagic processes, reducing amyloid protein aggregation (Nabavi et al., 2017), and upregulating the ATP-binding cassette (ABC) transporter P-glycoprotein (ABCB1) (Sawikr et al., 2017), which actively modulates Aβ efflux across the blood–brain barrier (BBB) (Sita et al., 2017). AD and its consequential neuronal damage seem to occur as a consequence of a sustained neuroinflammatory process, which seems to be caused by a plethora of different damage signals such as trauma, infection, oxidative agents, redox-active iron, and oligomers of tau and Aβ. In this context, astrocytes and microglial cells get progressively activated, leading to overproduction of proinflammatory agents (Morales et al., 2014). Several natural antiinflammatory compounds have been tested in AD models both in vitro and in vivo and are currently regarded as preventive and coadjuvant treatments for AD. Several phytocompounds, such as curcumin, resveratrol, quercetin, piperine, genistein, and sulforaphane, have been explored as potential therapeutic and preventive agents for AD, showing promising effects in numerous AD animal models (Omar et al., 2017). In particular, curcumin, a natural phenolic compound derived from the perennial herb *Curcuma longa* (turmeric), is known to exhibit antiinflammatory and antioxidant effects (Lu et al., 2014). Curcumin has been found to activate the heat shock response, reducing oxidative damage related to AD (Calabrese et al., 2003). Moreover, curcumin treatment has been reported to attenuate cognitive impairment and stimulate neuroprotection in a mouse model of AD and to inhibit the generation of Aβ by inducing autophagy, as evidenced by analysis of the autophagy-related protein LC3, and this effect was mediated via downregulation of the phosphatidylinositol-3-kinase (PI3K)/Akt and the mammalian target of rapamycin (mTOR) signaling pathway (Wang et al., 2014). Furthermore, an in vitro study has shown that both curcumin and its analog Cur1 protect neuroblastoma SK-N-SH cells from the exogenous effects of Aβ1-42 exposure, and the protective effect is elicited via the upregulation of human telomerase reverse-transcriptase (hTERT) expression (Xiao et al., 2014).

Curcumin has also been found to reduce inflammation induced by microglia and astrocytes in primary rat mixed neuronal/glial cultures, through PPARγ (peroxisome

TABLE 2.1 Summary of nutrients and nutraceuticals and their relevance to Alzheimer disease and dementia.

Nutrient or nutraceutical	Elicited effects	References
Polyphenols, amyloid-binding polyphenols	Target-specific signaling pathways associated with protein folding and neuroinflammation Modulate the microbiota–gut–brain axis Inhibit fibril formation or steer oligomer formation Have antioxidant and antiinflammatory effects, can modulate autophagic processes, reducing amyloid protein aggregation Upregulate the ATP-binding cassette (ABC) transporter P-glycoprotein (ABCB1)	Essa et al. (2012), Serra et al. (2020), Ngoungoure et al. (2015), Nabavi et al. (2017), Sawikr et al. (2017), Omar et al. (2017)
Resveratrol	Caloric restriction mimetic Induces weight loss, lowers the decline of cerebrospinal fluid Aβ40	Turner et al. (2015)
Curcumin, Cur1, Longvida®, Biocurcumax™	Antiinflammatory and antioxidant effects Activates the heat shock response, reducing oxidative damage Attenuates cognitive impairment and stimulates neuroprotection, inhibits the generation of Aβ by inducing autophagy via downregulation PI3K/Akt and mTOR Protects from exogenous effects of Aβ1-42 via the upregulation of hTERT expression Reduces inflammation induced by microglia and astrocytes through PPARγ and inhibition of NF-κB Improves working memory tasks and sustained attention (Biocurcumax™) No significant effects on mood and general cognitive functions	Calabrese et al. (2003), Lu et al. (2014), Wang et al. (2014), Xiao et al. (2014), Liu et al. (2016), Cox et al. (2015), Rainey-Smith et al. (2016)
Isoflavones (e.g., genistein, GS-14, 25d)	Elicit neuroprotection from Aβ protein exposure, inhibits AChE, modulates estrogenic activity Antioxidative activity, inhibit Aβ aggregation and exhibit metal chelating properties; reverse scopolamine-induced memory deficit Attenuate cytotoxicity and inflammatory damage induced upon Aβ25-35 exposure by inhibiting TLR4 and NF-κB upregulation (genistein) activates PPARγ and increases the release of ApoE from primary astrocytes, improving cognitive performance in Alzheimer disease (AD) mouse model, and reducing Aβ plaques	Lee et al. (2005), Qiang et al. (2014), Shi et al. (2012), Zhou et al. (2014), Bonet-Costa et al. (2016)
Isothiocyanates (e.g., sulforaphane)	Activate Nrf2/ARE pathway, promoting the upregulation of glutathione (GSH) Antioxidant potential upon exposure to Aβ25-35, upregulate antioxidant enzymes via activation of Nrf2, prevent Aβ-mediated apoptosis Ameliorate neurobehavioral deficits (in an AD-like mouse models), promote cholinergic neuron neurogenesis and reduce aluminum load in the brain	Lee et al. (2013), Tarozzi et al. (2013), Zhang et al. (2014)

(Continued)

TABLE 2.1 (Continued)

Nutrient or nutraceutical	Elicited effects	References
B vitamins [e.g., folate (vitamin B9), vitamin B6, vitamin B12]	Prevent tau hyperphosphorylation and memory deficits induced by acute administration of homocysteine Inhibit tau hyperphosphorylation and accumulation in hippocampus and cortex; downregulate GSK-3β, CDK-5, JNK, ERK, and p38MAPK, attenuate memory deficits May increase Mini Mental State Examination (MMSE) scores and lower inflammation (i.e., higher plasmatic levels of S-adenosylmethionine and S-adenosylmethionine/S-adenosylhomocysteine), reduce the expression of Aβ40, presenilin and tumor necrosis factor-α genes Lower levels of oxidative stress-associated carbonyl proteins, increase levels of kynurenine (a metabolite of tryptophan) and kynurenine/tryptophan ratio Often ineffective in AD patients	Aisen et al. (2008), Nelson et al. (2009), Wei et al. (2011), Zhang et al. (2008), Chen et al. (2016), Rommer et al. (2016)
Fortasyn Connect™ (cocktail of docosahexaenoic acid, eicosapentaenoic acid, uridine-5′-monophosphate, choline, phospholipids, antioxidants, and B vitamins)	Designed to enhance synapse formation and functionality Improve memory performance, positively affect brain functional connectivity in drug-naïve patients with mild AD No significant effects on cognitive performance in prodromal AD, preserve hippocampal and ventricular volumes Elevate plasmatic and erythrocyte levels of uridine, choline, selenium, folate, vitamins B6 and B12, vitamin E, DHA and EPA Increase phosphatidylcholine species	Engelborghs et al. (2014a,b), Mi et al. (2013), Scheltens et al. (2010), Scheltens et al. (2012), de Waal et al. (2014), Soininen et al. (2014), Rijpma et al. (2015), Hartmann et al. (2014)
Shilajit, fulvic acid, with/without B vitamins	Contribute to AD prevention	Carrasco-Gallardo et al. (2012)
Vitamin E, vitamin C, α-lipoic acid (E/C/ALA)	Lower cerebrospinal fluid (CSF) F2-isoprostane levels, indicative of oxidative stress reduction in the brain; also associated with faster cognitive decline	Galasko et al. (2012)
Nutraceutical formulation (NF) (i.e., folate, α-tocopherol, vitamin B12, S-adenosylmethioinine, N-acetyl cysteine, and acetyl-L-carnitine)	Improves cognitive performance in MCI patients Prevents cognitive performance decline in AD patients	Remington et al. (2015), Remington et al. (2016)
Vitamin D	May enhance learning and memory performance (in vivo) No amelioration of cognitive functions in AD patients	Yu et al. (2011), Latimer et al. (2014), Landel et al. (2016)
α- and γ-Tocopherols	High α-tocopherol seems to be associated with higher amyloid load when γ-tocopherol levels were low Conversely, high α-tocopherol seems to be associated with lower amyloid levels when γ-tocopherol levels were high α-Tocopherol slows dementia progression, ameliorates cognition in people affected by MCI or AD	Morris et al. (2015), Farina et al. (2017)

(Continued)

TABLE 2.1 (Continued)

Nutrient or nutraceutical	Elicited effects	References
n-3 PUFAs	No effects in patients with mild-to-moderate AD May be more effective in patients with very mild AD May improve neurocognition in a dose–response manner	Burckhardt et al. (2016), Canhada et al. (2017), Eriksdotter et al. (2015)
S-adenosylmethionine (SAM)	Neuroprotective under conditions of reduced GST activity, diminished SAM, increased accumulation of SAH, and deprived folate	Panza et al. (2009)

proliferator-activated receptor gamma) and inhibition of NF-κB (nuclear factor kappa-light-chain-enhancer of activated B cells) signaling pathway (Liu et al., 2016).

The chronic and acute-on-chronic effects induced by supplementation with Longvida® Optimized Curcumin has been shown to improve working memory tasks and sustained attention in healthy elderlies (Cox et al., 2015). On the contrary, a 12-month randomized placebo-controlled trial of 96 healthy participants has shown no significant effects on mood and general cognitive functions upon supplementation with Biocurcumax™ (a standardized curcumin extract) (Rainey-Smith et al., 2016).

Resveratrol, a phytoestrogen and caloric restriction mimetic found in grapes, berries, and nuts has been shown to induce weight loss, and lower the decline of cerebrospinal fluid $A\beta_{40}$, which is associated with dementia progression, particularly after 52-week supplementation, although other parameters (e.g., plasma $A\beta_{42}$, cerebrospinal fluid $A\beta_{42}$, tau, and phospho-tau 181, hippocampal volume, entorhinal cortex thickness, MMSE score, etc.) were not affected (Turner et al., 2015).

Soybean isoflavones, such as genistein, have been regarded as beneficial for the prevention and regression of neurodegenerative diseases (Lee et al., 2005). In vitro treatment with a synthetic version of genistein, called GS-14, has been shown to be neuroprotective on SH-SY5Y cells previously exposed to Aβ proteins; GS-14 inhibits acetylcholinesterase (AChE) and also modulates estrogenic activity, suggesting its possible use as an effective agent for the treatment of AD (Shi et al., 2012). Analogously, some genistein derivatives, in particular one called 25d, tested both in vitro and in vivo, have been found to promote AChE inhibitory activity and to display antioxidative activity, promoting inhibition of Aβ aggregation and exhibiting metal chelating properties. The analog 25d reversed scopolamine-induced memory deficit in mice, suggesting that 25d may be a promising multifunctional agent for the treatment of AD (Qiang et al., 2014). Additionally, treatment with genistein has been found to attenuate the cytotoxicity and inflammatory damage induced by Aβ25-35 exposure in vitro by inhibiting the Toll-like receptor 4 (TLR4) and NF-κB upregulation mediated by Aβ25-35 and the DNA binding and transcriptional activities of NF-κB (Zhou et al., 2014). Moreover, genistein has been shown to activate PPARγ and increase the release of ApoE from primary astrocytes, improving cognitive performance, and reducing Aβ plaques in a mouse model of AD (Bonet-Costa et al., 2016).

Among phytochemicals, the isothiocyanate sulforaphane, derived from the hydrolysis of glucoraphanin, a glucosinolate mainly present in Brassica vegetables, has also demonstrated neuroprotective effects in several in vitro and in vivo studies. Sulforaphane, in particular, seems to activate the antioxidant Nrf2/ARE pathway, promoting the upregulation of glutathione (GSH) (Tarozzi et al., 2013). Additionally, in vitro treatment with sulforaphane exerts antioxidant potential in SH-SY5Y cells exposed to Aβ25-35, as shown by the upregulation of antioxidant enzymes [i.e., γ-glutamylcysteine ligase, NAD(P)H:quinone oxidoreductase-1, and heme oxygenase-1] via activation of Nrf2, thus preventing Aβ-mediated apoptosis (Lee et al., 2013).

Moreover, sulforaphane has been shown to ameliorate neurobehavioral deficits in an AD-like mouse models, promoting cholinergic neuron neurogenesis and reducing aluminum load in the brain (Zhang et al., 2014).

Numerous epidemiological findings, confirmed also by in vivo animal studies, show that several MetS-related factors, such as atherosclerosis, diabetes, hHcy, hypertension, and high total and LDL cholesterol levels, may play a role in the development of AD (Ehrlich and Humpel, 2012). For this reason, supplementation with folate and vitamin B12 has been found to be effective in preventing tau protein hyperphosphorylation and memory deficits induced by acute administration of homocysteine in young rats (Zhang et al., 2008). Folate and vitamin B12 supplementation has also been found to be effective in aged rats with chronic hHcy, promoting inhibition of tau hyperphosphorylation and accumulation in hippocampus and cortex, together with downregulation of glycogen synthase kinases-3β (GSK-3β), cyclin-dependent kinase-5 (CDK-5), c-Jun N-terminal kinase (JNK), extracellular signal-regulated kinase (ERK), and p38 mitogen-activated protein kinase (p38MAPK). As a consequence, a significant attenuation of memory deficits occurs (Wei et al., 2011).

Despite the promising results described in these in vivo studies, randomized controlled trials on single-compound supplementations have shown contradictory or minimal/insignificant effects on neurocognition or on reducing the risk of AD and dementia in humans (Engelborghs et al., 2014a,b; Rijpma et al., 2014; Shea and Remington, 2015; Munoz Fernandez et al., 2017).

A multicenter, randomized, double-blind, controlled clinical trial of high-dose folate, vitamin B6, and vitamin B12 supplementation in 409 (of 601 screened) individuals with mild-to-moderate AD and normal folic acid, vitamin B12, and homocysteine levels has shown that vitamin supplements effectively reduced homocysteine levels compared to the placebo group, they did not improve primary and secondary cognitive measurements, eventually leading to adverse events such as occurrence of depression (Aisen et al., 2008). For this reason, regimens with high-dose B vitamin supplements might not be advisable. Accordingly, a 2009 study examined the associations between dietary and supplementation of folate, vitamin B12, and vitamin B6 and the incidence of AD among elderly men and women (Cache County Memory, Health and Aging Study). Interestingly, the authors of that study have not observed differences in risk of AD or dementia by increasing the total intake of folate, vitamin B12, or vitamin B6 (Nelson et al., 2009).

Recently, a two-year integration with folate and vitamin B12 showed contradictory results, with higher homocysteine but also higher folate levels both associated with lower total brain volume (van der Zwaluw et al., 2016). In another study, 6-month supplementation with folate has been found to increase Mini Mental State Examination (MMSE) scores and lower inflammation, as shown by higher plasmatic levels of S-adenosylmethionine and S-adenosylmethionine/S-adenosylhomocysteine, and reduce the expression of Aβ_{40}, presenilin, and tumor necrosis factor-α genes, in newly diagnosed patients with mild-to-severe AD and treated with donepezil (Chen et al., 2016).

After a 3-month B vitamin supplementation (i.e., B1, B6, B12 and folic acid), patients affected by AD or MCI showed a trend toward a higher MMSE score, lower levels of oxidative stress-associated carbonyl proteins, higher levels of kynurenine (a metabolite of tryptophan), and a higher kynurenine/tryptophan ratio (Rommer et al., 2016). Furthermore, while in vivo studies have shown that vitamin D supplementation may enhance learning and memory performance in several AD animal models (e.g., Yu et al., 2011; Latimer et al., 2014), vitamin D supplementations in AD patients have shown no amelioration of cognitive functions (Landel et al., 2016).

Several nutritional intervention studies in patients affected by AD have been conducted in past years; some of these interventions are based on single nutrient administrations, and others are based on nutrient combinations, such as the medical food "Fortasyn Connect™," which is designed to enhance synapse formation and functionality and contains a cocktail of docosahexaenoic acid, eicosapentaenoic acid, uridine-5'-monophosphate, choline, phospholipids, antioxidants, and B vitamins (Mi et al., 2013). In two randomized controlled trials, "Fortasyn Connect™" in its nutrient formulation (Souvenaid®) has resulted in improved memory performance in mild, drug-naïve patients with AD, positively affecting brain functional connectivity (Engelborghs et al., 2014a,b; Mi et al., 2013).

Twelve- and 24-week supplementations with Souvenaid have been shown to ameliorate memory performance in drug-naïve patients with mild AD (Scheltens et al., 2010, 2012). Similarly, a 2014 study reported that Souvenaid preserves brain network organization, beneficially affecting synaptic integrity and functions in drug-naïve AD patients (de Waal et al., 2014). On the other hand, 24-month supplementation with Souvenaid has shown no significant effects on cognitive performance in prodromal AD, although higher hippocampal volume and lower ventricular volume could be observed (Soininen et al., 2017). Additionally, 12-24-week Souvenaid supplementation has been shown to elevate plasmatic and erythrocyte levels of uridine, choline, selenium, folate, vitamins B6 and B12, vitamin E, DHA, and EPA in patients with mild and mild-to-moderate AD (Rijpma et al., 2015). Along the same line, Souvenaid supplementation for 24 weeks has been shown to increase five phosphatidylcholine species in drug-naïve patients with very mild-to-moderate AD (Hartmann et al., 2014). Moreover, a clinical trial conducted on patients with mild AD has shown that both shilajit alone, a sticky tar-like substance, and its active principle, fulvic acid, as well as a combination of shilajit and B vitamins, contribute to AD prevention (Carrasco-Gallardo et al., 2012).

Another randomized clinical study has been conducted on subjects with mild-to-moderate AD and analyzed the effects elicited by a 16-week treatment intervention comparing the administration of vitamin E (α-tocopherol, 800 IU/day), vitamin C (500 mg/day), α-lipoic acid (900 mg/day), globally named the E/C/ALA group, or coenzyme Q (400 mg, three times per day), or simple placebo (Galasko et al., 2012). Importantly, the antioxidants present in the E/C/ALA treatment did not alter any cerebrospinal fluid (CSF) biomarkers related to amyloid or tau pathology, rather than lowering CSF F_2-isoprostane levels, which is indicative of oxidative stress reduction in the brain. However, the E/C/ALA treatment also induced a faster cognitive decline, which represents a possible serious concern, especially if longer term clinical trials have to be conducted (Galasko et al., 2012).

Other multicomponent nutraceutical formulations have been assessed. For instance, a nutraceutical formulation (NF) containing folate, α-tocopherol, vitamin B12, S-adenosyl methionine, N-acetyl cysteine, and acetyl-L-carnitine, supplemented for 6 months, has been shown to improve cognitive performance [i.e., higher scores in the Dementia Rating Scale (DRS) tests] in individuals affected by MCI (Remington et al., 2015). A later study from the same group has shown that 12-month supplementation with NF prevented a decline in cognitive performance in individuals with AD (Remington et al., 2016).

The correlation between α- and γ-tocopherol brain concentrations and AD neuropathology has been assessed in 115 deceased participants of the prospective Rush Memory and Aging Project by taking into account the following parameters: amyloid load and neurofibrillary tangle severity. In particular, brain concentrations of γ- and α-tocopherols seem to be associated with AD neuropathology in an interrelated mechanism. High α-tocopherol levels seem to be associated with higher amyloid load when γ-tocopherol levels were low; conversely, high α-tocopherol levels seem to be associated with lower amyloid levels when γ-tocopherol levels were high (Morris et al., 2015). Moreover, supplementation with vitamin E, in particular α-tocopherol, has not been shown to slow the progression of MCI to dementia, or to ameliorate cognition in people affected by MCI or AD (Farina et al., 2017).

With regards to *n*-3 PUFA supplementation, no evidence of beneficial effects of *n-3* PUFA supplementation for 6, 12, or 18 months in patients with mild-to-moderate AD could be found (Burckhardt et al., 2016). A 2017 systematic review has shown that *n*-3 PUFA may be more effective in patients with very mild AD, rather than in full-blown AD patients (Canhada et al., 2017). Additionally, *n*-3 PUFA supplementation for 6 months improved neurocognition in a dose–response manner (Eriksdotter et al., 2015).

A 2009 review has commented on the positive effects elicited by combined supplementation with *S*-adenosylmethionine (SAM) and PUFAs for very mild AD subgroups or mild cognitive impairment (Panza et al., 2009). In particular, the use of SAM as a neuroprotective dietary supplement for AD patients seems to be relevant because AD subjects are often characterized by reduced glutathione *S*-transferase (GST) activity, diminished SAM, and increased accumulation of the downstream metabolic product resulting from SAM-mediated transmethylation [i.e., *S*-adenosylhomocysteine (SAH)] under conditions of deprived folate (Panza et al., 2009).

These studies globally suggest that appropriately combined, rather than single, nutraceutical supplementations and synergistic dietary approaches might be more effective for the maintenance of mental health and, possibly, the reduction of AD risk (Greenwood and Parrott, 2017). Large-scale clinical trials will be essential to confirm the efficacy of nutrients and nutraceuticals for the prevention or regression of both mild cognitive impairment and AD.

Nutraceuticals for the prevention and amelioration of Parkinson disease

PD is a degenerative disorder of the CNS characterized by motor symptoms, such as shaking, rigidity, slowness of movement, and difficulty with walking and gait, possibly followed by late-onset dementia. These impairments are a consequence of a progressive degeneration of dopaminergic neurons localized in the substantia nigra. The pathology of this disease is also characterized by the accumulation of α-synuclein protein in the Lewy bodies within neurons, and the insufficient formation and activity of dopamine within the midbrain (Shulman et al., 2011).

Current treatments are effective in managing the early motor symptoms, mainly through the administration of L-DOPA and dopamine agonists; however, with the progression of the disease, these drugs eventually become ineffective and often further complications occur, such as dyskinesia, which is characterized by involuntary writhing movements (Shulman et al., 2011; de Bie et al., 2020). Besides surgery, deep brain stimulation, gene therapy, stem cell transplantations, and novel neuroprotective pharmacologic agents, supplementation with specific plant-based phytocompounds and foods and some forms of rehabilitation have been proven effective in alleviating PD-related symptoms (Hang et al., 2016). In particular, nutraceuticals play a critical role in the regulation of energy metabolism and signaling pathways that control neurotransmission and neuroinflammation (Table 2.2). Nutraceuticals have been proven to interfere with several mechanisms related to PD symptoms manifestation, such as α-synuclein aggregation, ubiquitin-proteasome function, mTOR signaling and lysosomal-mediated autophagy, energy failure, faulty trafficking of catecholamine, dopamine oxidation, hHcy, methylation patterns, neuroinflammation, and irreversible oxidation of neuromelanin (Mazzio et al., 2011). Some of the applied nutraceuticals include vitamins C, D, and E, coenzyme Q10, creatine, unsaturated fatty acids, sulfur-

TABLE 2.2 Summary of nutrients and nutraceuticals and their relevance to Parkinson disease.

Nutrient or nutraceutical	Elicited effects	References
Vitamins C, D, and E, coenzyme Q10, creatine, unsaturated fatty acids, sulfur-containing compounds, polyphenols, stilbenes, and phytoestrogens	Prevent α-synuclein aggregation, ubiquitin-proteasome function, mammalian target of rapamycin (mTOR) signaling and lysosomal-mediated autophagy, energy failure, faulty trafficking of catecholamine, dopamine oxidation, hyperhomocyteinemia (hHcy), methylation patterns, neuroinflammation, and irreversible oxidation of neuromelanin Useful for the management of Parkinson disease (PD)	Chao et al. (2012), Mazzio et al. (2011)
Sulforaphane	Reduces motor coordination deficits and rotations induced by 6-OHDA in mice, reduces apoptosis by blocking DNA fragmentation and caspase-3 activation, enhances glutathione (GSH) levels and some neuronal survival pathways (ERK1/2) Inhibits locomotor activity deficiency and dopaminergic neuronal loss in mice rotenone-induced PD model, via enhancement of the antioxidative and autophagic ability of the brain	Morroni et al. (2013), Morroni et al. (2018), Zhou et al. (2016), Huang et al. (2019)
Catechins, (−)-epigallocatechin-3-gallate, quercetin	Neutralize stress-related free radicals and inflammation Inhibit L-DOPA methylation and prevent oxidative hippocampal neurodegeneration Inhibit human liver COMT-mediated O-methylation of L-DOPA in vitro [only (+)-catechin] Inhibit L-DOPA methylation in both peripheral compartment and striatum in rats, reduce glutamate-induced oxidative cytotoxicity in vitro via inactivation of NF-κB signaling, confer neuroprotection against kainic acid in vivo Enhance mitochondrial electron transport chain complex I in rotenone-treated rats, a rodent model of PD, and increase antioxidant enzyme activity Improve locomotor activity and cognitive function by activating antioxidant and antiinflammatory pathways	Kang et al. (2010), Kang et al. (2013b), Mandel et al. (2012), Weinreb et al. (2004), Shah and Duda (2015), BituPinto et al. (2015)
Extracts of tangerine peel (rich in polymethoxylated flavones), cocoa-2 (rich in procyanidins), and red clover (rich in isoflavones)	Attenuate dopaminergic neuron loss, no protection observed in vivo	Datla et al. (2007)
Safflower flavonoid extract (SAFE)	Improves behavioral performance by suppressing α-synuclein and reactive astrogliosis in vivo and in vitro	Ren et al. (2016)
Chrysin (5,7-dihydroxyflavone)flavonoid	Reduces intracellular nitric oxide levels, regulates antioxidant pathways and increases the production of neurotrophic factors, contributing to survival of dopaminergic neurons	Angelopoulou et al. (2020)

(Continued)

TABLE 2.2 (Continued)

Nutrient or nutraceutical	Elicited effects	References
Ginkgo biloba extract 761 (EGb 761, containing 24% flavonoids and 6% terpenoids)	Antioxidant, elicits neurorecovery of damaged midbrain dopamine (DA) neurons, improves locomotion Attenuates locomotor activity and inhibits the progress of PD in A53T α-synuclein transgenic mice, due to decreased oxidative damage, and maintenance of the normal dopamine levels Ginkgolic acid promotes autophagy-dependent clearance of α-synuclein aggregates in SH-SY5Y neuroblastoma cells and rat primary cortical neurons	Rojas et al. (2012), Kuang et al. (2018), Vijayakumaran et al. (2019)
Resveratrol	Activates the SIRT1 enzyme; neuroprotective against oxidative stress Prevents H_2O_2 or 6-OHDA triggered toxicity and the toxic effects elicited by Aβ1-42 and the α-synuclein-A30P Improves mitochondrial activity by activating several metabolic sensors, resulting in the activation of PGC-1α Activates AMPK and SIRT1 and upregulates the expression of several PGC-1α target genes, resulting in enhanced mitochondrial oxidative function	Albani et al. (2009), Albani et al. (2010), Ferretta et al. (2014)
Curcumin, curcuminoids, CNB-001 (a curcumin derivative)	Prevent neuroinflammation Prevent 1-methyl-4-phenyl-1,2,3,6-tetrahydropyridine (MPTP)-mediated depletion of dopamine and tyrosine hydroxylase immunoreactivity, downregulate GFAP, iNOS proteins, proinflammatory cytokine, and total nitrite generation in the striatum of MPTP-treated mice; promote increased motor performance and gross behavioral activity Ameliorate behavioral anomalies, increase expression of monoamine transporter, and improve mitochondria functionality Attenuate motor impairments and pathological changes elicited by MPTP administration In combination with DFO elicit neuroprotection in a 6-OHDA-PD rat model, increasing the levels of PCC, SOD, and GSH	Jayaraj et al. (2014a,b), Lv et al. (2014), Ojha et al. (2012), Witkin and Li (2013)
Korean red ginseng	Improves behavioral impairment by inhibiting dopaminergic neuronal death and enhancing differentiation and proliferation of endogenous neural stem cells in the subventricular zone Restores the MPTP-induced proteomic changes in the striatum and alleviates the protein expression profile related to neuronal development and energy metabolism for the neuronal survival and neuroprotection in the same brain area Inhibits MPP^+-induced apoptosis in rat PC12 cells and decreased mRNA expressions of caspase-3 and caspase-9	Jun et al. (2015), Choi et al. (2018), Ryu et al. (2016), Ryu et al. (2018), Kim et al. (2016), Kim et al. (2018)
Vitamin D3	Stabilizes PD symptoms without triggering hypercalcemia	Suzuki et al. (2013)

(*Continued*)

TABLE 2.2 (Continued)

Nutrient or nutraceutical	Elicited effects	References
Folate, vitamin B12, vitamin B6	Only vitamin B6 decreases risk of PD	de Lau et al. (2006), Murakami et al. (2010)
Tocopherol, CoQ10, and GSH	CoQ10 and GSH in particular show a small but significant improvement in PD symptoms	Weber and Ernst (2006)
Creatine	Ameliorates PD symptoms	Gualano et al. (2010)
n3-PUFAs	Ameliorates the depression symptoms and the Unified Parkinson's Disease Rating Scale in PD patients. Omega-3 has positive effects on high-sensitivity C-reactive protein (hs-CRP), total antioxidant capacity, glutathione, and markers of insulin metabolism	Li and Song (2020), Avallone et al. (2019)
AtreMorine (E-PodoFavalin-15999)	Increases plasma noradrenaline, adrenaline, dopamine Induces neuroprotection by enhancing dopaminergic neurotransmission, antiinflammatory mechanisms and selective neurotrophic factors in dopaminergic neurons in in vitro and in vivo models	Corzo et al. (2020)

containing compounds, polyphenols, stilbenes, and phytoestrogens (Chao et al., 2012). Among plant-derived nutraceuticals, the antioxidant sulforaphane has been found to reduce motor coordination deficits and rotations induced by 6-hydroxydopamine (6-OHDA) in mice (Morroni et al., 2013); 6-OHDA is a chemical compound used in some animal models to mimic the effects of PD. Sulforaphane reduces the 6-OHDA-dependent apoptosis by blocking DNA fragmentation and caspase-3 activation and enhances GSH levels and some neuronal survival pathways, such as ERK1/2, in the murine brain (Morroni et al., 2013). Recent in vitro and in vivo studies further confirm the antioxidative and neuroprotective properties of isothiocyanate sulforaphane, derived from precursor glucosinolate present in Brassica vegetables, and its metabolite erucin (Morroni et al., 2018). In a mouse rotenone-induced PD model, chronic sulforaphane administration led to inhibition of locomotor activity deficiency and dopaminergic neuronal loss, which appear to be mediated by enhancement of the antioxidative and autophagic ability of neuronal cells (Zhou et al., 2016). These data suggest that sulforaphane might effectively slow the progression of PD by modulating oxidative stress and the apoptotic machinery (Pistollato and Battino, 2014; Huang et al., 2019).

The beneficial effects and limitations of *n*-3 PUFA supplementation on PD-related behavioral and neuropathological abnormalities in different cellular and animal models of PD and clinical patients have been recently reviewed by Li and Song (2020). The published literature emphasizes the importance of omega-3 supplements, to ameliorate the depression symptoms in PD. Supplementation with omega-3 has favorable effects not only on the PD phenotype, but also on high-sensitivity C-reactive protein (hs-CRP), total antioxidant capacity, glutathione, and markers of insulin metabolism (Avallone et al., 2019).

Moreover, natural plant polyphenols, such as the green and black flavonoid catechins present in tea, seem to neutralize stress-related free radicals and inflammation (Mandel et al., 2012; Weinreb et al., 2004). It has been shown that (−)-epigallocatechin-3-gallate, a tea polyphenol, inhibits L-DOPA methylation and prevents oxidative hippocampal neurodegeneration (Kang et al., 2010). Additionally, tea catechins [(+)-catechin and (−)-epicatechin] and quercetin have been reported to strongly inhibit human liver catechol-*O*-methyltransferase (COMT)-mediated O-methylation of L-DOPA in vitro, whereas only (+)-catechin, due to its better bioavailability in vivo, has been found to significantly inhibit L-DOPA methylation in both the peripheral compartment and striatum in rats. Quercetin has been found to enhance mitochondrial electron transport chain complex I in rotenone-treated rats, a rodent model of PD, and increase antioxidant enzyme activity (Shah and Duda, 2015). Furthermore, (+)-catechin elicits strong reduction of glutamate-induced oxidative cytotoxicity in HT22 mouse hippocampal neurons cultured in vitro, and this occurs via inactivation of the NF-κB signaling pathway. In vivo administration of (+)-catechin has been found to be neuroprotective against kainic acid-induced oxidative rat hippocampal neurodegeneration (Kang et al., 2013b). In 6-OHDA-lesioned rats, treatment with standardized extract from *Camellia sinensis*, epicatechin and (−)-epigallocatechin-3-gallate, indicated neuroprotection after improvement of locomotor activity cognitive function that was attributed to the antioxidant and antiinflammatory properties of the extract (BituPinto et al., 2015).

In the same animal disease model, the administration of standardized safflower flavonoid extract (SAFE) that mainly contained flavonoids, improved behavioral performance in PD animal models by suppressing α-synuclein and reactive astrogliosis that was additionally demonstrated in cellular models (Ren et al., 2016). Another study conducted in vivo in a 6-OHDA-induced PD rat model has reported that short-term presupplementation with plant extracts rich in flavonoids composed of extracts of tangerine peel (rich in polymethoxylated flavones), cocoa-2 (rich in procyanidins), and red clover (rich in isoflavones) significantly attenuates dopaminergic neuron loss, whereas no significant protection has been observed in animals supplemented with red and white grape seeds (rich in catechins) and cocoa-1 (rich in catechins) (Datla et al., 2007). Chrysin (5,7-dihydroxyflavone), another flavonoid that has been extensively studied, was shown to reduce intracellular nitric oxide levels, regulate antioxidant pathways, and increase the production of neurotrophic factors, contributing to survival of dopaminergic neurons (Angelopoulou et al., 2020). Combined with protocatechuic acid, chrysin inhibits neuronal loss in PD animal models, and additionally elevates dopamine levels in the striatum and improves behavioral deficits in PD animal models (Angelopoulou et al., 2020). Jung and Kim (2018) recently reviewed multiple studies demonstrating beneficial effects of flavonoids in PD treatment through their antioxidant and antiinflammatory properties and additionally by activating antiapoptotic pathways that improve mitochondrial function and induce neurotrophic factors. They concluded that flavonoids seem to be promising for the prevention of PD that could potentially be considered also as supplementary treatment for PD patients, although no clinical trials so far show restoration of the aberrant phenotypes in patients with PD (Jung and Kim, 2018).

G. biloba extract 761 (EGb 761), a patented and defined mixture of active compounds extracted from *G. biloba* leaves and containing flavonoids (24%) and terpenoids (6%), has been described as neuroprotective due to its antioxidant function (Rojas et al., 2012). In a rodent model of PD, EGb761 has been found to be neuroprotective and to elicit neurorecovery of damaged midbrain DA neurons, improving locomotion (Rojas et al., 2012). The same extract treatment in A53T α-synuclein transgenic mice attenuates locomotor activity and inhibits the progress of PD, most likely due to decrease oxidative damage, and maintenance of normal dopamine levels (Kuang et al., 2018). Ginkgolic acid, an extract from *G. biloba* leaves, has been reported to promote autophagy-dependent clearance of α-synuclein aggregates in SH-SY5Y neuroblastoma cells and rat primary cortical neurons (Vijayakumaran et al., 2019).

Resveratrol, a natural phytocompound acknowledged as a potential anticancer drug, has been shown to display neuroprotective actions by activating the sirtuin 1 (SIRT1) enzyme, one of the seven described sirtuin deacetylases involved in many physiologic and pathologic processes including apoptosis, autophagy, diabetes, cancer, cardiovascular disorders, and neurodegeneration (Albani et al., 2010). In this regard, resveratrol has been found to be neuroprotective in a neuroblastoma cell model of oxidative stress, preventing H_2O_2 or 6-OHDA triggered toxicity and also the toxic effects elicited by exposure to two aggregation-prone proteins (i.e., the Aβ1-42 and the α-synuclein-A30P) (Albani et al., 2009). Furthermore, resveratrol has been described as improving mitochondrial activity by activating several metabolic sensors, resulting in the activation of the peroxisome proliferator-activated receptor-gamma coactivator-1α (PGC-1α) (Ferretta et al., 2014). Data on primary fibroblast cultures obtained from two patients with early-onset PD linked to different Park2 mutations have revealed that resveratrol activates AMPK and SIRT1 and upregulates the expression of several PGC-1α target genes, resulting in enhanced mitochondrial oxidative function, indicative of a shift from glycolytic to oxidative metabolism (Ferretta et al., 2014).

Animal models and human-related studies have indicated curcumin as beneficial for stroke, AD, stress, and mood disorders, and also PD and its related neuroinflammation (Witkin and Li, 2013). In particular, oral administration of curcuminoids has been found to significantly prevent 1-methyl-4-phenyl-1,2,3,6-tetrahydropyridine (MPTP)-mediated depletion of dopamine and tyrosine hydroxylase immunoreactivity. Also, downregulation of glial fibrillary acidic protein (GFAP) and inducible nitric oxide synthase (iNOS) protein expression has been observed, and proinflammatory cytokine and total nitrite generation in the striatum of MPTP-treated mice have been noted to be significantly reduced by curcuminoid administration. Motor performance and gross behavioral activity have been improved, suggesting curcuminoid compounds as potential therapeutic candidate nutraceuticals for the prevention and/or management of PD (Ojha et al., 2012). Some studies have described the effects of CNB-001, a novel pyrazole derivative of curcumin known to possess various neuroprotective properties, in a subacute MPTP rodent model of PD. Pretreatment with CNB-001 ameliorates behavioral anomalies, increases expression of monoamine transporter, and improves mitochondria functionality (Jayaraj et al., 2014a). CNB-001 has also been found to significantly attenuate motor impairments and pathological changes elicited by MPTP administration (Jayaraj et al., 2014b). However, it is important to note that data related to the neuroprotective efficacy of tested agents may depend on the MPTP administration protocol adopted (i.e., acute vs. subacute administration) (Anderson et al., 2006). Combined treatment with curcumin and desferrioxamine (DFO), a bacterial siderophore produced by the actinobacteria *Streptomyces pilosus* and used as an iron chelating agent, has been found to be neuroprotective in a 6-OHDA-PD rat model, increasing the levels of propionyl-CoA carboxylase (PCC), superoxide dismutase (SOD), and GSH (Lv et al., 2014). Another combination of curcumin with niacin and adenosine $A_{2A}R$ antagonist (ZM241385) enhanced animals' behavior and restored selected biochemical and histological features in a mouse rotenone model of PD (Motawi et al., 2020).

Korean red ginseng that has been extensively studied in animal models of Parkinson disease was shown to improve the behavioral impairment by inhibiting dopaminergic neuronal death and enhancing differentiation and proliferation of endogenous neural stem cells in the subventricular zone (Jun et al., 2015; Choi et al., 2018; Ryu et al., 2018). Korean red ginseng was also found to restore the MPTP-induced proteomic changes in the striatum (Kim et al., 2016) and alleviated the protein expression profile related to neuronal development and energy metabolism for neuronal survival and neuroprotection in the same brain area (Kim et al., 2018). In a cellular model, Korean red ginseng treatment inhibited 1-methyl-4-phenylpyridinium ion ($MPP+$)-induced apoptosis on rat pheochromocytoma (PC12) cells and decreased mRNA expressions of caspase-3 and caspase-9 (Ryu et al., 2016). MohdSairazi and Sirajudeen (2020) recently also reviewed more natural derived products and their neuroprotective potentials in PD animal and cellular models, including *Withania somnifera* extracts that reduce oxidative stress and mitochondrial dysfunction, and *Uncaria rhynchophylla* and fucoidan from *Laminaria japonica* that reduce neuronal loss in dopaminergic neurons in the substantia nigra. Altogether these in vitro and in vivo animal studies suggest that plant-derived nutraceuticals may play a protective role in the prevention of PD.

There are many biological and pathophysiological links between vitamin D and PD (reviewed in Moretti et al., 2018 and Pieri et al., 2020). A 2013 study assessed whether vitamin D3 supplementation inhibits PD progression in patients with vitamin D receptor FokI CC genotype and found that vitamin D3 supplementation could stabilize PD symptoms for a short period of time without triggering hypercalcemia (Suzuki et al., 2013). Additional data derived from clinical studies showed that reduced levels of vitamin D might influence the progress of PD, but the supplementation of vitamin D did not indicate any alteration in the disease-associated outcomes, potentially because the timing of the vitamin D supplementation at the clinical trials occurred at a later stage than in animal models (Moretti et al., 2018).

Some clinical studies tried to address the efficacy of nutraceutical administration in PD-affected subjects. In the so-called Rotterdam Study, a prospective population-based cohort study including people aged 55 years and older, the association between dietary intake of folate, vitamin B12, and vitamin B6 and the incidence of PD was assessed in 5289 participants who were free of dementia and parkinsonism and underwent complete dietary assessment at baseline. The authors of this study found that a higher dietary intake of vitamin B6 is associated with a significantly decreased risk of PD, whereas no association has been observed for intake of dietary folate and vitamin B12. Thus, supplementation with vitamin B6 might decrease PD risk, probably through mechanisms not related to homocysteine metabolism (de Lau et al., 2006). The results were later confirmed by a Japanese study, where it was reported that although the intake of folate, vitamin B12, and riboflavin is not associated with the risk of PD, a low intake of vitamin B6 is related to an increased risk of PD, as observed in 249 patients with diagnosed PD and 368 controls without neurodegenerative diseases (Murakami et al., 2010).

A 2006 study revised publications on the effects of the three main antioxidants or supplements used for the prevention and/or treatment of PD: tocopherol, CoQ10, and GSH. The authors found that, besides some large observational studies focusing on tocopherol administration and one study of CoQ10 that enrolled 80 patients, other retrieved studies included less than 30 patients and were conducted for 3 months or shorter periods of time. Whereas tocopherol supplementation does not alter the course of PD, two of the CoQ10 studies and the study on GSH showed a small but significant improvement in PD symptoms (Weber and Ernst, 2006). A metaanalysis of randomized controlled trials indicated no significant difference between CoQ10 treatment when compared with placebo in terms of motor symptoms in PD patients (Zhu et al., 2017).

In the last decade, findings point to the potential use of creatine, which is implicated in energy provision through a reaction catalyzed by phosphorylcreatine kinase, for the prevention and amelioration of PD symptoms, and creatine supplementation has been proposed as a therapeutic tool for elderly subjects (Gualano et al., 2010). However, a systematic review from published randomized controlled trials has not found any association between creatine use and neuroprotection against PD and supports the need for better designed randomized controlled trials in the future (Attia et al., 2017).

AtreMorine (E-PodoFavalin-15999), which is derived from the *Vicia faba* L. plant by using nondenaturing biotechnological procedures, has been found to increase plasma catecholamines (noradrenaline, adrenaline, dopamine), but to have no effect on the levels of serotonin. The E-PodoFavalin-15999 extract is a natural source of L-DOPA that can be found also in other bioactive substances (e.g., vegetal proteins, unsaturated fatty acids, minerals, and vitamins, vegetal fiber, starch, carotenes, and phytosterols) that has neuroprotective and antiinflammatory effects on dopaminergic neurons as was demonstrated in in vitro and in animal models of PD (reviewed in Corzo et al., 2020). This neuroprotective effect is attributed not only to L-DOPA, but also to other selective neurotrophic factors of the compound. Clinical studies indicate that AtreMorine is an enhancer of dopaminergic neurotransmission in patients with PD and is proposed to be coadministrated with antiparkinsonian drugs to reduce some of their side effects and to enable their dose reduction (reviewed in Corzo et al., 2020). Further large-scale clinical studies would be required to better define the role of nutraceutical supplementations for PD prevention and amelioration of PD symptoms.

Nutraceuticals for the prevention and amelioration of autism spectrum disorders

ASD is a cluster of heterogeneous and complex neurodevelopmental disorders. The current revised definition of ASD no longer differentiates the various ASD subtypes such as autistic disorder or Asperger syndrome, but rather includes all of them under the same definition. The common features encountered in ASD are abnormal social interactions, impaired communication, and stereotyped or repetitive behaviors. There is evidence of a genetic predisposition to ASD, mainly derived from twin studies (Folstein and Rosen-Sheidley, 2001). However, it is believed that autistic phenotypes are attributed to interactions involving not only genetic factors but also environmental factors, such as chemicals, viral infections, and stress (Geier et al., 2009; Gillott and Standen, 2007; Libbey et al., 2005; Tordjman et al., 2014).

Although extensive research in the field of ASD has been performed during the past two decades, the full mechanistic understanding of ASD and the development of appropriate treatments remain limited. Currently, applied behavioral therapies and antipsychotic medications to ASD individuals are not considered effective or safe (Bobo et al., 2013). Due to these limitations, the use of nutraceuticals in ASD management is steadily increasing and a significant number of companies produce various compositions of nutraceuticals that are available over the counter (Alanazi, 2013). Despite the initial use of nutraceuticals that relied exclusively on anecdotal and empirical evidence, a substantial amount of animal and human data has been generated to provide some support and justification for the beneficial administration of nutraceuticals in ASD treatment protocols (Pistollato et al., 2020) (Table 2.3). This section aims to review the metabolic or physiological abnormalities found in ASD patients that form the basis for the development of treatments with nutraceutical origin. The presented data derive from: (1) epidemiological studies in which levels of micronutrients and specific biomarkers associated with ASD phenotype have been measured; (2) clinical trials in which nutraceuticals have been administered to ASD individuals; and (3) animal models of ASD that illustrate the mechanisms by which nutraceuticals prevent, ameliorate, or cure neurodevelopmental disorders of this spectrum.

In light of the specific nutritional and dietary deficits associated with gastrointestinal (GI) issues and gut microbiota dysbiosis that have been reported in ASD children, the integration of nutrients or specific dietary interventions have been intensively investigated as complementary management options for ASD (Trudeau et al., 2019). As reported in a recent metaanalysis, children with ASD tend to consume less protein, calcium, phosphorus, selenium, vitamin D, thiamine (vitamin B1), riboflavin (vitamin B2), cobalamin (vitamin B12), and *n*-3 PUFAs, and increase the intake of vitamin E, fruits, and vegetables compared to healthy controls (Esteban-Figuerola et al., 2019).

In general, multivitamin intake during periconception has been associated with a lower incidence of ASD in children. A 2011 study demonstrates that mothers of ASD individuals are less likely to receive vitamin supplements 3 months before and 1 month after conception than mothers with healthy children (Schmidt et al., 2011). Multivitamin and mineral supplementation to ASD patients for up to 3 months leads to improvements in sleep, reduction of autistic symptoms, and amelioration in biomarkers of energy production as well as oxidative stress, suggesting that the effectiveness of multivitamins is partially related to better mitochondrial function (Adams et al., 2011). Similarly, a case–control study comparing autistic children using conventional medication and micronutrient supplementation that included not only a multivitamin but also minerals, amino acids, and antioxidants revealed greater improvement in the micronutrient-treated group (Mehl-Madrona et al., 2010).

A 2017 large-scale Swedish study showed that nutritional supplementation with multivitamin, iron, and folic acid during pregnancy was associated with a lower risk of ASD in offspring with and without intellectual disability (DeVilbiss et al., 2017).

Besides the studies that followed the administration of cocktails containing vitamins in combination with cofactors, some other experimental approaches have focused on providing evidence through human or animal studies by using individual vitamins or nutraceuticals. The reason behind this is

TABLE 2.3 Summary of nutrients and nutraceuticals and their relevance to autism spectrum disorder.

Nutrient or nutraceutical	Elicited effects	References
Multivitamin and mineral supplements	During periconception promote lower incidence of autism spectrum disorder (ASD) children Improve sleep, reduce autistic symptoms Ameliorate biomarkers of energy production and oxidative stress, improving mitochondrial function	Adams et al. (2011), Mehl-Madrona et al. (2010), Schmidt et al. (2011), DeVilbiss et al. (2017)
Vitamin B2	Significantly elevated in children with autism	Main et al. (2015)
Vitamin B6 and magnesium	Improvement in behavioral symptoms, speech, and language skills (at appropriate doses) Possible negative effects: risk of neuropathy (from vitamin B6) and diarrhea (from magnesium) (at high doses)	Nye and Brice (2002), Mousain-Bosc et al. (2006), Li et al. (2018a,b)
Folic acid, folinic acid	Supplementation during pregnancy has been associated with higher incidence of ASD Promotes altered expression of transcription factors, imprinted genes, neurodevelopmental genes, and genes associated with ASD when administered during pregnancy in mice Folate levels in patients with ASD are lower compared to controls Impairment of folate transport across the blood–brain barrier (BBB) and into neuronal cells Folinic acid can cross the BBB, ameliorating ASD and cerebral folate deficiency (CFD) Possible negative effects: may increase the risk of ASD at high doses	Ali et al. (2011), Barua et al. (2015), Castro et al. (2014), Frye and Rossignol (2014), Wang et al. (2017), Wiens and DeSoto (2017), Levine et al. (2018), Raghavan et al. (2018), Sharman Moser et al. (2019), Frye et al. (2013), Frye et al. (2018)
Vitamin B12	When combined with glutathione (GSH) and special low-fructose and organic diet, it improves social interaction, concentration, spoken and written language, and behavior Reduces the risk of ASD (at appropriate doses), but may increase the risk of ASD (at high doses) (methyl B12) improvement in ASD symptoms, methionine metabolism, cellular methylation capacity, and GSH levels (combined with vitamin B6 and B9); reduction of urinary homocysteine; (combined with GSH and a low fructose and organic diet) improvement in concentration, social interaction, spoken and written language, and behavioral symptoms	Patel and Curtis (2007), Bertoglio et al. (2010), Hendren et al. (2016), Raghavan et al. (2018)
Vitamin D	(combined with *n*-3 PUFAs) reduces irritability and hyperactivity Increases blood levels of nerve growth factor Improves ASD core symptoms (vitamin D3) Improved in self-care and inappropriate speech, but no reduction of stereotypic behavior (vitamin D3)	Feng et al. (2017), Kerley et al. (2017), Moradi et al. (2018), Mazahery et al. (2019)
Calcium and zinc	Calcium supply during pregnancy reduces the risk of ASD Zinc supplementation improves cognitive-motor performance, increases serum levels of metallothionein, and lowers serum levels of copper	Li et al. (2018a,b), Meguid et al. (2019)
Propionic (propanoic) acid	Together with enteric short-chain fatty acids (SCFAs), it is correlated to some forms of ASD	Macfabe (2013)

(Continued)

TABLE 2.3 (Continued)

Nutrient or nutraceutical	Elicited effects	References
n-3 PUFAs	Their levels are decreased in ASD Supplementation of *n*-3 PUFAs in ASD children improves or does not affect main ASD symptoms Improve erythrocyte membrane omega-6/omega-3 ratio and some ASD symptoms (with or without vitamin D) modestly improve ASD core symptoms (docosahexaenoic acid, DHA) had no effects on ASD core symptoms	Amminger et al. (2007), Bent et al. (2011), Vancassel et al. (2001), Wilczynski-Kwaitek et al. (2009), Parellada et al. (2017), Mazahery et al. (2019), Meiri et al. (2009), Voigt et al. (2014)
Biopterin, tetrahydrobiopterin, sapropterin	Highly concentrated in urine samples and at low levels in cerebrospinal fluid (CSF) of ASD children Supplementation with tetrahydrobiopterin improves language skills, eye contact, communication, and repetitive behavior Sapropterin improves tetrahydrobiopterin metabolism in ASD individuals and ameliorates nitric oxide (NO) metabolism	Danfors et al. (2005), Frye et al. (2010), Frye et al. (2013), Tani et al. (1994)
Probiotics (e.g., *Bacteroides fragilis*)	Alleviate maternal immune activation (MIA), anxiety, and stereotyped behaviors, improve gastrointestinal (GI) function, and restore serum metabolite levels related to autism (i.e., 4-EPS and indolepyruvate); ameliorate the detoxification capability of ASD individuals and consequently provide protection from environmental chemicals Modify stool composition (with an increase of bifidobacteria and lactobacilli levels), reduce body weight, and amelioration of autistic symptoms Decrease D-arabinitol urinary levels, decrease the ratio of D-/L-arabinitol, and improve concentration ability in ASD children	Alanazi (2013), Hsiao (2014), Parracho et al. (2010), Shaaban et al. (2018), Kaluzna-Czaplinska and Blaszczyk (2012)
Flavonoids (e.g., luteolin, quercetin, and rutin, sulforaphane)	Antioxidant, antiinflammatory, and neuroprotective properties; improve ASD symptoms by targeting JAK2/STAT3 signaling Improve GI and allergy symptoms, eye contact, and social interaction in ASD children Modify urinary metabolic profile, metabolites involved in oxidative stress, amino acid and gut microbiome, hormones, neurotransmitters, and metabolism of sphingomyelin	Parker-Athill et al. (2009), Taliou et al. (2013), Theoharides et al. (2012), Tsilioni et al. (2015), Singh et al. (2014), Bent et al. (2011)
Ginseng (red and white)	Red ginseng extract improves sociability and social preference paradigms in an ASD mouse model	Kim et al. (2013)
N-acetyl-L-cysteine, methylcobolamine, folic acid, vitamin C, and other antioxidants	Positively modulate GSH levels and metabolism	Frye and Rossignol (2014), Hardan et al. (2012)

that the levels of only certain vitamins or certain biomarkers have been found to be altered in ASD children.

Vitamin B2 is significantly elevated in children with autism and their siblings compared with controls, but there is no difference in other B vitamins or homocysteine (Main et al., 2015). This study included 35 children with autism, 27 of their siblings without autism, and 25 age- and gender-matched community controls (Main et al., 2015).

Large doses of vitamin B6 and magnesium have been shown to improve behavioral symptoms, speech, and language skills in ASD children (Nye and Brice, 2002; Mousain-Bosc et al., 2006; Li et al., 2018a,b). However, attention should be paid to the administration of supratherapeutic/high doses of both supplements, which may increase the risk of neuropathy (from vitamin B6) and diarrhea (from magnesium).

Folic acid or folate (vitamin B9) supplementation during pregnancy has been accused of causing a higher incidence of ASD, whereas it is well-recognized to protect against neural tube defects. However, after reviewing the limited and contradictory studies, Castro et al. (2014) have concluded that further studies are required to determine the critical role of gestational folic acid supplementation in relation to ASD occurrence (Castro et al., 2014). It has been found that several transcription factors, imprinted genes, neurodevelopmental genes, and genes associated with ASD exhibit altered expression levels in the brain of mice pups that have been exposed to high concentrations of folic acid during gestation (Barua et al., 2015). Folate levels in patients with ASD are lower compared to controls (Ali et al., 2011; Castro et al., 2014). In addition, a number of genetic polymorphisms in enzymes involved in the folate pathway have been identified in ASD individuals (Frye and Rossignol, 2014). These abnormalities result in the impairment of folate transport across the BBB and into neuronal cells. The most studied abnormality in folate metabolism related to ASD is the formation of autoantibodies to the folate receptor α, which leads to cerebral folate deficiency (CFD). Some studies have investigated the beneficial effects of folinic acid, which is a reduced form of folate that can cross the BBB in children with ASD and CFD, revealing promising results (Castro et al., 2014; Frye and Rossignol, 2014). However, two reviews concluded that the effects of folate-enhancing interventions on the clinical symptoms of ASD have not been fully explored yet, because only one reduced form of folate has been investigated so far and only with respect to CFD (Castro et al., 2014; Frye and Rossignol, 2014).

More recently, the effects of 12-week folinic acid supplementation have been investigated in 48 ASD children versus a placebo. Children were further assessed for their FRAA and glutathione (GSH) status. Folinic acid supplementation was found to improve verbal communication, with folate receptor α (FRAA)-positive subjects showing greater effects (Frye et al., 2018). Frye et al. also assessed the effects of a 3-month supplementation with folinic acid and methylcobalamin (a form of vitamin B12) in 37 children diagnosed with ASD and with abnormal GSH and methylation metabolism. Improvements in interpersonal, play-leisure and coping social skills, expressive communication, and daily living skills (both personal and domestic) were recorded upon supplementation, and such effects were positively associated with improvement in GSH redox status (Frye et al., 2013).

Notably, high unmetabolized levels of folic acid during pregnancy have also been associated with a higher incidence of ASD (Wiens and DeSoto, 2017; Raghavan et al., 2018; Sharman Moser et al., 2019), warranting caution about oversupplementation of folate.

A 2017 metaanalysis has shown that folic acid supplementation during pregnancy could reduce the risk of ASD in children, as compared to mothers who did not undergo supplementation (Wang et al., 2017). Similarly, a 2018 study has described an inverse correlation between maternal intake of folic acid and/or multivitamin supplements before and during pregnancy and the risk of ASD in offspring (Levine et al., 2018).

A study of 1257 mother–child pairs found a U-shaped relationship between maternal supplementation of both folate and vitamin B12 during pregnancy and the risk of ASD. In particular, while middle (80th percentile) levels of both vitamins were associated with a lower ASD risk, high levels were correlated with a 2.5-fold higher risk (Raghavan et al., 2018).

The levels of vitamin B12 have been found to be decreased in a cohort of autistic children (Ali et al., 2011; Belardo et al., 2019). Improvements in social interaction, concentration, spoken and written language, and behavior have been recorded after vitamin B12 administration for 3–6 months in combination with GSH and special low-fructose and organic diets in a small group of autistic children (4–10 years old) (Patel and Curtis, 2007).

Subcutaneous injection of methyl B12 for 8 weeks in 50 ASD children has been shown to improve ASD symptoms and improve methionine metabolism and cellular methylation capacity, as shown by plasma methionine increase, S-adenosyl-L-homocysteine decrease, and a higher S-adenosylmethionine to S-adenosyl-L-homocysteine ratio (Hendren et al., 2016).

However, it has been also reported that 12-week subcutaneous injection of methyl B12 was effective only in 30% of ASD subjects, with amelioration of ASD symptoms and an increase of GSH plasma concentrations and of the GSH/GSSG (glutathione disulfide) ratio (Bertoglio et al., 2010).

It has been shown that vitamin D deficiency during periconceptional and gestational periods and during early childhood is associated with ASD risk (Cannell, 2017). Supplementation with vitamin D, *n*-3 PUFAs, or both has been shown to be beneficial for the control of irritability and hyperactivity in ASD children (Mazahery et al., 2019). Vitamin D supplementation has been shown to increase blood levels of nerve growth factor in ASD children (Moradi et al., 2018), a neurotrophin involved in neuronal cell growth and survival (Groves and Burne, 2017). Along the same line, supplementation with vitamin D3 for 3 months has been found to induce improvement of ASD symptoms, with younger children showing the greatest improvements (Feng et al., 2017).

On the contrary, supplementation with vitamin D3 for 5 months was found to be ineffective in ameliorating stereotypies in ASD children, although some improvement in self-care and inappropriate speech could be observed (Kerley et al., 2017).

Adequate calcium supply during pregnancy has been associated with a lower risk of ASD (Li et al., 2018a,b). Furthermore, zinc supplementation has been shown to be promising in improving cognitive-motor performance in a cohort of ASD children, who underwent increased serum levels of metallothionein, and lowered serum levels of copper (Meguid et al., 2019). This suggests that zinc supplementation may be considered for the treatment of ASD children.

Enteric short-chain fatty acids and, more specifically, propionic (propanoic) acid seem to be produced from ASD-related GI bacteria, and there is evidence suggesting that they may be the cause behind some forms of ASD (Macfabe, 2013). Propionic acid-exposed rats demonstrate repetitive and antisocial behaviors and similar neurochemical and neuropathological alterations as ASD patients (Macfabe, 2013). The levels of phospholipid fatty acids in the plasma of a population of autistic subjects showed a marked reduction that reflects the decrease in the levels of total *n*-3 PUFAs but not of *n*-6 PUFAs (Vancassel et al., 2001).

Some studies that have followed *n*-3 fatty acid supplementation in ASD children have revealed contradicting results. One study shows improvement in the main ASD symptoms (Amminger et al., 2007), whereas another demonstrates no beneficial effect after *n*-3 fatty acid treatment (Bent et al., 2011). However, both studies failed to include a sufficient number of ASD individuals and to provide data that can be analyzed statistically with confidence. Assessments that would include not only behavioral performance indexes but also biological indicators and, more specifically, the omega 6/3 ratios could improve the understanding of the role of fatty acids in the management of ASD (Wilczynski-Kwaitek et al., 2009).

More recently, an 8-week supplementation with omega-3 in ASD children/adolescents has been found to improve erythrocyte membrane omega-6/omega-3 ratio, with some improvements in ASD symptoms (Parellada et al., 2017). Similarly, supplementation with omega-3 (with or without vitamin D), has been shown to modestly improve ASD core symptoms in a cohort of ASD children (Mazahery et al., 2019), and improvement of ASD symptoms has been reported upon 12-week supplementation with omega-3 (Meiri et al., 2009). On the contrary, no improvement in ASD core symptoms could be observed upon 6-month supplementation with docosahexaenoic acid (DHA) in ASD children (Voigt et al., 2014).

Biopterins, which are considered important cofactors of catecholaminergic and several critical metabolic pathways, have been found to be highly concentrated in urine samples and at low levels in CSF of ASD children (Tani et al., 1994). These findings are more pronounced in ASD individuals who are younger than 6 years old. Studies targeting this age group have shown that tetrahydrobiopterin treatment can be beneficial in language skills, eye contact, communication, and repetitive behavior (Danfors et al., 2005). In these controlled and other open-label clinical trials (Frye et al., 2010, 2013), sapropterin, which is a synthetic form of tetrahydrobiopterin, has been administered to ASD children. Interestingly, sapropterin has been found not only to improve tetrahydrobiopterin metabolism in ASD individuals, but also to ameliorate nitric oxide (NO) metabolism because serum biomarkers related to NO have been suggested to have predictive value for ASD children's response to sapropterin (Frye et al., 2013).

GI symptoms, abnormal food cravings, and unique intestinal bacterial populations have been proposed to be implicated in the development and severity of ASD symptoms (Hughes et al., 2018). More specifically, certain beneficial bacteria are not present in the microbiota of ASD patients (Kang et al., 2013a). Some ASD cases are connected to maternal viral or bacterial infection during pregnancy, also known as maternal immune activation (MIA). It has been suggested that MIA leads to a decrease in gut barrier integrity and an increase in the presence of bacteria metabolites in the blood of embryos (Boulanger-Bertolus et al., 2018). It has been shown that the treatment of MIA with probiotics alleviates some of the ASD clinical symptoms in a mouse model of ASD, such as anxiety and stereotyped behaviors (Hsiao, 2014; Hsiao et al., 2013). Interestingly, these MIA offspring mice that received *Bacteroides fragilis* show improved GI function and restored the levels of serum metabolites that are considered human autism biomarkers, like 4-ethylphenylsulfate (4-EPS) and indole pyruvate (Hsiao, 2014). Regarding clinical trials assessing the beneficial role of probiotics in ASD children, a 2010 study has shown significant improvement in behavioral scores, despite the limitations of the study due to high interindividual variability and the high dropout rate of participants (Parracho et al., 2010). Moreover, probiotics improve GI function, and it has also been suggested that they ameliorate the detoxification capability of ASD individuals and consequently provide protection from environmental chemicals, such as mercurial compounds that can be more hazardous to these patients compared to typical developing children (Alanazi, 2013).

A probiotics formula containing *Lactobacillus acidophilus*, *Lactobacillus rhamnosus*, and *Bifidobacteria longum* supplemented for 12 weeks has been shown to modify stool composition (with an increase of bifidobacteria and lactobacilli levels), causing a reduction in body weight, and an amelioration of autistic symptoms (Shaaban et al., 2018). A supplementation with *Lactobacillus acidophilus* for 2 months has been shown to significantly decrease D-arabinitol urinary levels, a metabolite produced by the pathogenic *Candida* species that has been found elevated in the urine of ASD patients (Shaw et al., 1995), to decrease the ratio of D-/L-arabinitol, and to improve the concentration ability in ASD children (Kaluzna-Czaplinska and Blaszczyk, 2012).

Accumulating evidence suggests that inflammation in brain regions related to cognitive function is a hallmark of ASD. The natural flavonoid luteolin possesses antioxidant, antiinflammatory, and neuroprotective properties. Despite their antithyroid properties, which might possibly impact brain development, flavonoids have been shown to improve ASD symptoms in animal models and humans. In an MIA mouse model of ASD, luteolin inhibits autism-like behaviors by targeting Janus kinase 2 (JAK2)/Signal Transducer and Activator of Transcription 3 (STAT3) signaling (Parker-Athill et al., 2009). A mixture of flavonoids composed of luteolin, quercetin, and the quercetin glycoside rutin has been shown to improve GI and allergy symptoms, eye contact, and social interaction when administered to a group of ASD children (Theoharides et al., 2012). Accordingly, the same formulation has been given to 40 children and seems to be effective in reducing ASD symptoms without causing any major adverse effects (Taliou et al., 2013). Moreover, luteolin supplementation for 26 weeks has been found to reduce the levels of serum interleukin-6 and tumor necrosis factor, which are associated with brain inflammation, and improve ASD symptoms in ASD children (Tsilioni et al., 2015).

Twelve-week supplementation with sulforaphane has been found to ameliorate social responsiveness and behavior in 15 children with ASD, causing a beneficial modification of urinary metabolic profile (Bent et al., 2011). Along the same line, sulforaphane supplementation for 18 weeks in 29 adolescents/young men with moderate to severe ASD has been found to temporarily improve ASD symptoms (Singh et al., 2014).

Ginseng is one of the most widely used medicinal plants. Red ginseng compared to white is considered to provide superior pharmacological effects with limited adverse effects. Making use of the prenatal valproic acid-injection rat model of ASD, which produces social impairment and similar neuropathological changes monitored in ASD patients, it has been shown that the chronic administration of red ginseng extract improves sociability and social preference paradigms in a dose-dependent manner (Kim et al., 2013).

Abnormal GSH metabolism is a common finding in ASD children (Frye and Rossignol, 2014). The reduced levels of this major intracellular antioxidant lead to oxidative damage that is usually detected in the cortex of ASD individuals. A supplement that provides a precursor to GSH, named *N*-acetyl-L-cysteine, has been administered with success to ASD individuals, but the levels of GSH have not been measured (Hardan et al., 2012). Besides *N*-acetyl-L-cysteine, other novel ASD interventions that address oxidative stress are methyl cobalamin in combination with or without folic acid, vitamin C, and other antioxidants (Frye and Rossignol, 2014).

Although the etiology of ASD remains elusive, converging lines of research indicate that mitochondrial dysfunction may play a substantive role in disease pathophysiology. Some of the nutraceuticals presented here target this cellular compartment and appear to be promising ASD treatment approaches. However, without an established causal link, the generation of therapeutic targets for ASD has been relatively unsuccessful and has been focused solely on the amelioration of individual symptoms by certain nutraceuticals without providing a mechanistic understanding or reliable clinical data.

Concluding remarks and future directions

In recent years, nutraceuticals have been proven beneficial for the prevention or amelioration of cognitive impairments in degenerative diseases such as AD and PD, and manifestations of ASD. Some clinical studies have tried to address the role and molecular mechanisms underlying the effects elicited by bioactive compound supplementations in human subjects; however, the data obtained in these studies are sometimes controversial and often are obtained using small cohorts of subjects and/or for interventions with limited times. Moreover, scientific research on the role of nutraceuticals in relation to neurodegenerative diseases often has been based on observations derived from in vitro and/or animal models reproducing neurodegenerative diseases. These models have often yielded contradictory results, given the immutable interspecies differences (Jucker, 2010) and the obvious limitations of in vitro cellular models. Animal models have been useful for improving the understanding of the etiology of these neurodegenerative diseases and for assessing the effects of new treatments. Nevertheless, available animal models often do not resemble the actual neuroanatomy and pathophysiology of idiopathic diseases, such as AD (Franco and Cedazo-Minguez, 2014), PD (Potashkin et al., 2011), and ASD (Varghese et al., 2017), thus limiting data translatability into clinical practice.

Future research will be required to consider the multifaceted nature of neurodegenerative diseases and to assess, through human-based epidemiological and clinical studies, the effectiveness and utility of nutritional multivitamin, mineral, and plant-based mixtures to slow the neurodegenerative progression and their protective/preventive effects (Arab and Sabbagh, 2010). Even though human-based interventional/clinical studies remain scarce, the adoption of a "precautionary principle" in relation to food choices should be highly recommended to prevent cognitive impairment and also MetS and cancer risk. Among the "precautions," limiting or avoiding alcohol, high-fat dairy products, red meat, processed meat, and meat cooked at high temperatures and also increasing the intake of soy products and fruits and vegetables enriched in bioactive phytonutraceuticals should be advised (Gonzales et al., 2014).

References

Adams, J.B., Audhya, T., McDonough-Means, S., 2011. Effect of a vitamin/mineral supplement on children and adults with autism. BMC Pediatr. 11 (2011), 111.

Aisen, P.S., Schneider, L.S., Sano, M., 2008. High-dose B vitamin supplementation and cognitive decline in Alzheimer disease: a randomized controlled trial. JAMA. 300 (15), 1774–1783.

Alanazi, A.S., 2013. The role of nutraceuticals in the management of autism. Saudi Pharm. J. 21 (3), 233–243.

Albani, D., Polito, L., Signorini, A., 2010. Neuroprotective properties of resveratrol in different neurodegenerative disorders. Biofactors. 36 (5), 370–376.

Albani, D., Polito, L., Batelli, S., 2009. The SIRT1 activator resveratrol protects SK-N-BE cells from oxidative stress and against toxicity caused by alpha-synuclein or amyloid-beta (1—42) peptide. J. Neurochem. 110 (5), 1445–1456.

Ali, A., Waly, M.I., Al-Farsi, Y.M., 2011. Hyperhomocysteinemia among Omani autistic children: a case-control study. Acta Biochim. Pol. 58 (4), 547–551.

Amminger, G.P., Berger, G.E., Schafer, M.R., 2007. Omega-3 fatty acids supplementation in children with autism: a double-blind randomized, placebo-controlled pilot study. Biol. Psychiatry. 61 (4), 551–553.

Anderson, D.W., Bradbury, K.A., Schneider, J.S., 2006. Neuroprotection in Parkinson models varies with toxin administration protocol. Eur. J. Neurosci. 24 (11), 3174–3182.

Angelopoulou, E., Pyrgelis, E.S., Piperi, C., 2020. Neuroprotective potential of chrysin in Parkinson's disease: molecular mechanisms and clinical implications. Neurochem Int. 132, 104612.

Arab, L., Sabbagh, M.N., 2010. Are certain lifestyle habits associated with lower Alzheimer's disease risk? J. Alzheimers Dis. 20 (3), 785–794.

Attia, A.H., Gadelkarim, M., Morsi, M., et al., 2017. Meta-analysis of creatine for neuroprotection against Parkinson's disease. CNS Neurol. Disord. Drug. Targets 16 (2), 169–175.

Avallone, R., Vitale, G., Bertolotti, M., 2019. Omega-3 fatty acids and neurodegenerative diseases: new evidence in clinical trials. Int. J. Mol. Sci. 20 (17), 4256.

Barua, S., Kuizon, S., Chadman, K.K., 2015. Microarray analysis reveals higher gestational folic acid alters expression of genes in the cerebellum of mice offspring-a pilot study. Brain Sci. 5 (1), 14–31.

Belardo, A., Gevi, F., Zolla, L., 2019. The concomitant lower concentrations of vitamins B6, B9 and B12 may cause methylation deficiency in autistic children. J. Nutr. Biochem. 70, 38–46.

Bent, S., Bertoglio, K., Ashwood, P., 2011. A pilot randomized controlled trial of omega-3 fatty acids for autism spectrum disorder. J. Autism. Dev. Disord. 41 (5), 545–554.

Bertoglio, K., Jill James, S., Deprey, L., et al., 2010. Pilot study of the effect of methyl B12 treatment on behavioral and biomarker measures in children with autism. J. Altern. Complement. Med. 16 (5), 555–560.

Bitu Pinto, N., DaSilva Alexandre, B., Neves, K.R.T., et al., 2015. Neuroprotective properties of the standardized extract from Camellia sinensis (green tea) and its main bioactive components, epicatechin and epigallocatechin gallate, in the 6-OHDA model of Parkinson's disease. Evid Based Complement Altern Med. 2015, 12Article ID 161092. Available from: http://dx.doi.org/10.1155/2015/161092.

Bobo, W.V., Cooper, W.O., Stein, C.M., 2013. Antipsychotics and the risk of type 2 diabetes mellitus in children and youth. JAMA Psychiatry 70 (10), 1067–1075.

Boldyrev, A., Bryushkova, E., Mashkina, A., 2013. Why is homocysteine toxic for the nervous and immune systems? Curr. Aging Sci. 6 (1), 29–36.

Bonet-Costa, V., Herranz-Perez, V., Blanco-Gandia, M., et al., 2016. Clearing amyloid-beta through PPARgamma/ApoE activation by genistein is a treatment of experimental Alzheimer's disease. J. Alzheimers Dis. 51 (3), 701–711.

Boulanger-Bertolus, J., Pancaro, C., Mashour, G.A., 2018. Increasing role of maternal immune activation in neurodevelopmental disorders. Front. Behav. Neurosci. (12), 230.

Burckhardt, M., Herke, M., Wustmann, T., et al., 2016. Omega-3 fatty acids for the treatment of dementia. Cochrane Database Syst. Rev. (4), CD009002.

Calabrese, V., Scapagnini, G., Colombrita, C., 2003. Redox regulation of heat shock protein expression in aging and neurodegenerative disorders associated with oxidative stress: a nutritional approach. Amino Acids. 25 (3–4), 437–444.

Canhada, S., Castro, K., Perry, I.S., Luft, V.C., 2017. Omega-3 fatty acids' supplementation in Alzheimer's disease: a systematic review. Nutr. Neurosci. 1–10.

Cannell, J.J., 2017. Vitamin D and autism, what's new? Rev. Endocr. Metab. Disord. 18 (2), 183–193.

Carrasco-Gallardo, C., Farias, G.A., Fuentes, P., 2012. Can nutraceuticals prevent Alzheimer's disease? Potential therapeutic role of a formulation containing shilajit and complex B vitamins. Arch. Med. Res. 43 (8), 699–704.

Castro, K., Klein, L.D., Baronio, D., 2014. Folic acid and autism: what do we know? Nutr. Neurosci. 19 (7), 1–8.

Chao, J., Leung, Y., Wang, M., 2012. Nutraceuticals and their preventive or potential therapeutic value in Parkinson's disease. Nutr. Rev. 70 (7), 373–386.

Chen, H., Liu, S., Ji, L., et al., 2016. Folic cid supplementation mitigates Alzheimer's disease by reducing inflammation: a randomized controlled trial. Mediators Inflamm. 2016, 5912146.

Choi, J.H., Jang, M., Nah, S.-Y., et al., 2018. Multitarget effects of Korean Red Ginseng in animal model of Parkinson's disease: antiapoptosis, antioxidant, antiinflammation, and maintenance of blood-brain barrier integrity. J. Ginseng Res. 42 (3), 379–388.

Corzo, L., Fernández-Novoa, L., Carrera, I., et al., 2020. Nutrition, health, and disease: role of selected marine and vegetal nutraceuticals. Nutrients 12 (3), 747.

Cox, K.H., Pipingas, A., Scholey, A.B., 2015. Investigation of the effects of solid lipid curcumin on cognition and mood in a healthy older population. J. Psychopharmacol. 129 (5), 642–651.

Danfors, T., von Knorring, A.L., Hartvig, P., 2005. Tetrahydrobiopterin in the treatment of children with autistic disorder: a double-blind placebo-controlled crossover study. J. Clin. Psychopharmacol. 25 (5), 485–489.

Datla, K.P., Zbarsky, V., Rai, D., 2007. Short-term supplementation with plant extracts rich in flavonoids protect nigrostriatal dopaminergic neurons in a rat model of Parkinson's disease. J. Am. Coll. Nutr. 26 (4), 341–349.

de Bie, R.M.A., Clarke, C.E., Espay, A.J., et al., 2020. Initiation of pharmacological therapy in Parkinson's disease: when, why, and how. Lancet Neurol. Available from: https://doi.org/10.1016/S1474-4422(20)30036-3.

de Lau, L.M., Koudstaal, P.J., Witteman, J.C., 2006. Dietary folate, vitamin B12, and vitamin B6 and the risk of Parkinson disease. Neurology. 67 (2), 315–318.

de Waal, H., Stam, C.J., Lansbergen, M.M., et al., 2014. The effect of souvenaid on functional brain network organisation in patients with mild Alzheimer's disease: a randomised controlled study. PLoS One 9 (1), e86558.

de Wilde, M.C., Vellas, B., Girault, E., et al., 2017. Lower brain and blood nutrient status in Alzheimer's disease: results from meta-analyses. Alzheimers Dement. (N. Y.) 3 (3), 416–431.

DeVilbiss, E.A., Magnusson, C., Gardner, R.M., et al., 2017. Antenatal nutritional supplementation and autism spectrum disorders in the Stockholm youth cohort: population based cohort study. BMJ 359, j4273.

Ehrlich, D., Humpel, C., 2012. Chronic vascular risk factors (cholesterol, homocysteine, ethanol) impair spatial memory, decline cholinergic neurons and induce blood-brain barrier leakage in rats in vivo. J. Neurol. Sci. 322 (1–2), 92–95.

Engelborghs, S., Gilles, C., Ivanoiu, A., 2014a. Rationale and clinical data supporting nutritional intervention in Alzheimer's disease. Acta Clin. Belg. 69 (1), 17–24.

Engelborghs, S., Gilles, C., Ivanoiu, A., Vandewoude, M., 2014b. Rationale and clinical data supporting nutritional intervention in Alzheimer's disease. Acta Clin. Belg. 69 (1), 17–24.

Eriksdotter, M., Vedin, I., Falahati, F., et al., 2015. Plasma fatty acid profiles in relation to cognition and gender in Alzheimer's disease patients during oral omega-3 fatty acid supplementation: the omegAD study. J. Alzheimers Dis. 48 (3), 805–812.

Essa, M.M., Vijayan, R.K., Castellano-Gonzalez, G., 2012. Neuroprotective effect of natural products against Alzheimer's disease. Neurochem. Res. 37 (9), 1829–1842.

Esteban-Figuerola, P., Canals, J., Fernandez-Cao, J.C., et al., 2019. Differences in food consumption and nutritional intake between children with autism spectrum disorders and typically developing children: a meta-analysis. Autism 23 (5), 1079–1095.

Etgen, T., Sander, D., Bickel, H., et al., 2012. Vitamin D deficiency, cognitive impairment and dementia: a systematic review and meta-analysis. Dement. Geriatr. Cogn. Disord. 33 (5), 297–305.

Farina, N., Llewellyn, D., Isaac, M., et al., 2017. Vitamin E for Alzheimer's dementia and mild cognitive impairment. Cochrane Database Syst. Rev. 4, CD002854.

Feng, J., Shan, L., Du, L., et al., 2017. Clinical improvement following vitamin D3 supplementation in Autism Spectrum Disorder. Nutr. Neurosci. 20 (5), 284–290.

Ferretta, A., Gaballo, A., Tanzarella, P., 2014. Effect of resveratrol on mitochondrial function: implications in parkin-associated familiar Parkinson's disease. Biochim. Biophys. Acta. 1842 (7), 902–915.

Folstein, S.E., Rosen-Sheidley, B., 2001. Genetics of autism: complex aetiology for a heterogeneous disorder. Nat. Rev. Genet. 2 (12), 943–955.

Franco, R., Cedazo-Minguez, A., 2014. Successful therapies for Alzheimer's disease: why so many in animal models and none in humans? Front. Pharmacol. 5, 146.

Frye, R.E., Rossignol, D.A., 2014. Treatments for biomedical abnormalities associated with autism spectrum disorder. Front. Pediatr. 2, 66.

Frye, R.E., Huffman, L.C., Elliott, G.R., 2010. Tetrahydrobiopterin as a novel therapeutic intervention for autism. Neurotherapeutics. 7 (3), 241–249.

Frye, R.E., DeLatorre, R., Taylor, H.B., 2013. Metabolic effects of sapropterin treatment in autism spectrum disorder: a preliminary study. Transl. Psychiatry. 3, e237.

Frye, R.E., Slattery, J., Delhey, L., et al., 2018. Folinic acid improves verbal communication in children with autism and language impairment: a randomized double-blind placebo-controlled trial. Mol. Psychiatry 23 (2), 247–256.

Frye, R.E., Melnyk, S., Fuchs, G., et al., 2013. Effectiveness of methylcobalamin and folinic Acid treatment on adaptive behavior in children with autistic disorder is related to glutathione redox status. Autism Res. Treat. 2013, 609705.

Galasko, D.R., Peskind, E., Clark, C.M., 2012. Antioxidants for Alzheimer disease: a randomized clinical trial with cerebrospinal fluid biomarker measures. Arch. Neurol. 69 (7), 836–841.

Geier, D.A., Kern, J.K., Garver, C.R., 2009. Biomarkers of environmental toxicity and susceptibility in autism. J. Neurol. Sci. 280 (1–2), 101–108.

Gillott, A., Standen, P.J., 2007. Levels of anxiety and sources of stress in adults with autism. J. Intellect. Disabil. 11 (4), 359–370.

Gonzales, J.F., Barnard, N.D., Jenkins, D.J., 2014. Applying the precautionary principle to nutrition and cancer. J. Am. Coll. Nutr. 33 (3), 239–246.

Greenwood, C.E., Parrott, M.D., 2017. Nutrition as a component of dementia risk reduction strategies. Healthc. Manage Forum 30 (1), 40–45.

Groves, N.J., Burne, T.H.J., 2017. The impact of vitamin D deficiency on neurogenesis in the adult brain. Neural Regen. Res. 12 (3), 393–394.

Gualano, B., Artioli, G.G., Poortmans, J.R., 2010. Exploring the therapeutic role of creatine supplementation. Amino Acids. 38 (1), 31–44.

Hang, L., Basil, A.H., Lim, K.L., 2016. Nutraceuticals in Parkinson's disease. Neuromolecular Med. 18 (3), 306–321.

Hardan, A.Y., Fung, L.K., Libove, R.A., 2012. A randomized controlled pilot trial of oral N-acetylcysteine in children with autism. Biol. Psychiatry. 71 (11), 956–961.

Hartmann, T., van Wijk, N., Wurtman, R.J., et al., 2014. A nutritional approach to ameliorate altered phospholipid metabolism in Alzheimer's disease. J. Alzheimers Dis. 41 (3), 715–717.

Hendren, R.L., James, S.J., Widjaja, F., et al., 2016. Randomized, Placebo-Controlled Trial of Methyl B12 for Children with Autism. J. Child. Adolesc. Psychopharmacol. 26 (9), 774–783.

Hsiao, E.Y., 2014. Gastrointestinal issues in autism spectrum disorder. Harv. Rev. Psychiatry. 22 (2), 104–111.

Hsiao, E.Y., McBride, S.W., Hsien, S., 2013. Microbiota modulate behavioral and physiological abnormalities associated with neurodevelopmental disorders. Cell. 155 (7), 1451–1463.

Huang, C., Wu, J., Chen, D., Jin, J., Wu, Y., Chen, Z., 2019. Effects of sulforaphane in the central nervous system. Eur. J. Pharmacol. 853, 153–168.

Hughes, H.K., Rose, D., Ashwood, P., 2018. The gut microbiota and dysbiosis in Autism Spectrum Disorders. Curr. Neurol. Neurosci. Rep. 18 (11), 81.

Jayaraj, R.L., Elangovan, N., Dhanalakshmi, C., 2014b. CNB-001, a novel pyrazole derivative mitigates motor impairments associated with neurodegeneration via suppression of neuroinflammatory and apoptotic response in experimental Parkinson's disease mice. Chem. Biol. Interact. 220C (2014), 149–157.

Jayaraj, R.L., Elangovan, N., Manigandan, K., 2014a. CNB-001 a novel curcumin derivative, guards dopamine neurons in MPTP model of Parkinson's disease. Biomed. Res. Int. 2014, 236182.

Jucker, M., 2010. The benefits and limitations of animal models for translational research in neurodegenerative diseases. Nat. Med. 16 (11), 1210–1214.

Jun, Y.L., Bae, C.-H., Kim, D., et al., 2015. Korean red ginseng protects dopaminergic neurons by suppressing the cleavage of p35 to p25 in a Parkinson's disease mouse model. J. Ginseng Res. 39 (2), 148–154.

Jung, U.J., Kim, S.R., 2018. Beneficial Effects of Flavonoids Against Parkinson's Disease. J. Med. Food 21 (5), 421–432.

Kaluzna-Czaplinska, J., Zurawicz, E., Michalska, M., 2013. A focus on homocysteine in autism. Acta Biochim. Pol. 60 (2), 137–142.

Kaluzna-Czaplinska, J., Blaszczyk, S., 2012. The level of arabinitol in autistic children after probiotic therapy. Nutrition 28 (2), 124–126.

Kang, D.W., Park, J.G., Ilhan, Z.E., 2013a. Reduced incidence of Prevotella and other fermenters in intestinal microflora of autistic children. PLoS One. 8 (7), e68322.

Kang, K.S., Yamabe, N., Wen, Y., 2013b. Beneficial effects of natural phenolics on levodopa methylation and oxidative neurodegeneration. Brain Res. 1497, 1–14.

Kang, K.S., Wen, Y., Yamabe, N., 2010. Dual beneficial effects of (−)-epigallocatechin-3-gallate on levodopa methylation and hippocampal neurodegeneration: in vitro and in vivo studies. PLoS One. 5 (8), e11951.

Kerley, C.P., Power, C., Gallagher, L., Coghlan, D., 2017. Lack of effect of vitamin D(3) supplementation in autism: a 20-week, placebo-controlled RCT. Arch. Dis. Child., 102(11), 1030–1036.

Kim, D., Jeon, H., Ryu, S., et al., 2016. Proteomic analysis of the effect of Korean red ginseng in the striatum of a Parkinson's disease mouse model. PLoS One. 11 (10).

Kim, D., Kwon, S., Jeon, H., et al., 2018. Proteomic change by Korean red ginseng in the substantia nigra of a Parkinson's disease mouse model. J. Ginseng Res. 42 (4), 429–435.

Kim, P., Park, J.H., Kwon, K.J., 2013. Effects of Korean red ginseng extracts on neural tube defects and impairment of social interaction induced by prenatal exposure to valproic acid. Food Chem. Toxicol. 51, 288–296.

Kuang, S., Yang, L., Rao, Z., et al., 2018. Effects of Ginkgo Biloba Extract on A53T α-Synuclein Transgenic Mouse Models of Parkinson's Disease. Can. J. Neurol. Sci. 45 (2), 182–187.

Landel, V., Annweiler, C., Millet, P., et al., 2016. Vitamin D, cognition and Alzheimer's disease: the therapeutic benefit is in the D-tails. J. Alzheimers Dis. 53 (2), 419–444.

Latimer, C.S., Brewer, L.D., Searcy, J.L., et al., 2014. Vitamin D prevents cognitive decline and enhances hippocampal synaptic function in aging rats. Proc. Natl Acad. Sci. U. S. A. 111 (41), E4359–E4366.

Lee, C., Park, G.H., Lee, S.R., 2013. Attenuation of beta-amyloid-induced oxidative cell death by sulforaphane via activation of NF-E2-related factor 2. Oxid. Med. Cell. Longev. (2013), .

Lee, Y.B., Lee, H.J., Sohn, H.S., 2005. Soy isoflavones and cognitive function. J. Nutr. Biochem. 16 (11), 641–649.

Levine, S.Z., Kodesh, A., Viktorin, A., et al., 2018. Association of maternal use of folic acid and multivitamin supplements in the periods before and during pregnancy with the risk of Autism Spectrum Disorder in offspring. JAMA Psychiatry 75 (2), 176–184.

Li, P., Song, C., 2020. Potential treatment of Parkinson's disease with omega-3 polyunsaturated fatty acids. Nutr. Neurosci. Available from: https://doi.org/10.1080/1028415X.2020.1735143.

Li, Y.J., Li, Y.M., Xiang, D.X., 2018a. Supplement intervention associated with nutritional deficiencies in autism spectrum disorders: a systematic review. Eur. J. Nutr. 57 (7), 2571–2582.

Li, Y.M., Shen, Y.D., Li, Y.J., et al., 2018b. Maternal dietary patterns, supplements intake and autism spectrum disorders: a preliminary case-control study. Med. (Baltim.) 97 (52), e13902.

Libbey, J.E., Sweeten, T.L., McMahon, W.M., 2005. Autistic disorder and viral infections. J. Neurovirol. 11 (1), 1–10.

Liu, Z.J., Li, Z.H., Liu, L., et al., 2016. Curcumin attenuates beta-amyloid-induced neuroinflammation via activation of peroxisome proliferator-activated receptor-gamma function in a rat model of Alzheimer's disease. Front. Pharmacol. 7, 261.

Lopes da Silva, S., Vellas, B., Elemans, S., et al., 2014. Plasma nutrient status of patients with Alzheimer's disease: systematic review and meta-analysis. Alzheimers Dement. 10 (4), 485–502.

Lu, Z., Shen, Y., Wang, T., 2014. Curcumin promotes neurite outgrowth via reggie-1/flotillin-2 in cortical neurons. Neurosci. Lett. 559, 7–12.

Lv, H., Liu, J., Wang, L., 2014. Ameliorating effects of combined curcumin and desferrioxamine on 6-OHDA-induced rat mode of Parkinson's disease. Cell Biochem. Biophys. 70 (2), 1433–1438.

Macfabe, D., 2013. Autism: metabolism, mitochondria, and the microbiome. Glob. Adv. Health Med. 2 (6), 52–66.

Main, P.A.E., Thomas, P., Angley, M.T., 2015. Lack of evidence for genomic instability in autistic children as measured by the cytokinesis-block micronucleus cytome assay. Autism Res. 8, 94–104.

Mandel, S.A., Weinreb, O., Amit, T., 2012. Molecular mechanisms of the neuroprotective/neurorescue action of multi-target green tea polyphenols. Front. Biosci. (Sch. Ed.) 4, 581–598.

Mazahery, H., Conlon, C.A., Beck, K.L., et al., 2019. A randomised controlled trial of vitamin D and omega-3 long chain polyunsaturated fatty acids in the treatment of irritability and hyperactivity among children with autism spectrum disorder, J Steroid Biochem Mol Biol, 187. pp. 9–16.

Mazzio, E.A., Close, F., Soliman, K.F., 2011. The biochemical and cellular basis for nutraceutical strategies to attenuate neurodegeneration in Parkinson's disease. Int. J. Mol. Sci. 12 (1), 506–569.

Mecocci, P., Tinarelli, C., Schulz, R.J., 2014. Nutraceuticals in cognitive impairment and Alzheimer's disease. Front. Pharmacol. 5, 147.

Meguid, N.A., Bjorklund, G., Gebril, O.H., et al., 2019. The role of zinc supplementation on the metallothionein system in children with autism spectrum disorder. Acta Neurol. Belg .

Mehl-Madrona, L., Leung, B., Kennedy, C., 2010. Micronutrients versus standard medication management in autism: a naturalistic case-control study. J. Child. Adolesc. Psychopharmacol. 20 (2), 95–103.

Meiri, G., Bichovsky, Y., Belmaker, R.H., 2009. Omega 3 fatty acid treatment in autism. J. Child. Adolesc. Psychopharmacol. 19 (4), 449–451.

Mi, W., van Wijk, N., Cansev, M., 2013. Nutritional approaches in the risk reduction and management of Alzheimer's disease. Nutrition. 29 (9), 1080–1089.

Michaud, M., Balardy, L., Moulis, G., 2013. Proinflammatory cytokines, aging, and age-related diseases. J. Am. Med. Dir. Assoc. 14 (12), 877–882.

Mohd Sairazi, N.S., Sirajudeen, K.N.S., 2020. Natural products and their bioactive compounds: neuroprotective potentials against neurodegenerative diseases. Evid. Based Complement. Altern. 2020, 6565396. Available from: https://doi.org/10.1155/2020/6565396.

Moradi, H., Sohrabi, M., Taheri, H., et al., 2018. The effects of different combinations of perceptual-motor exercises, music, and vitamin D supplementation on the nerve growth factor in children with high-functioning autism. Complement. Ther. Clin. Pract. 31, 139–145.

Morales, I., Guzman-Martinez, L., Cerda-Troncoso, C., 2014. Neuroinflammation in the pathogenesis of Alzheimer's disease. A rational framework for the search of novel therapeutic approaches. Front. Cell. Neurosci. 8, 112.

Moretti, R., Morelli, M.E., Caruso, P., 2018. Vitamin D in neurological diseases: a rationale for a pathogenic impact. Int. J. Mol. Sci. 19 (8), 2245.

Morris, M.C., Schneider, J.A., Li, H., 2015. Brain tocopherols related to Alzheimer's disease neuropathology in humans. Alzheimers Dement. 11 (1), 32–39.

Morroni, F., Tarozzi, A., Sita, G., 2013. Neuroprotective effect of sulforaphane in 6-hydroxydopamine-lesioned mouse model of Parkinson's disease. Neurotoxicology. 36, 63–71.

Morroni, F., Sita, G., Djemil, A., et al., 2018. Comparison of adaptive neuroprotective mechanisms of sulforaphane and its interconversion product erucin in in vitro and in vivo models of Parkinson's disease. J. Agric. Food Chem. 66 (4), 856–865.

Motawi, T.K., Sadik, N.A.H., Hamed, M.A., et al., 2020. Potential therapeutic effects of antagonizing adenosine A2A receptor, curcumin and niacin in rotenone-induced Parkinson's disease mice model. Mol. Cell Biochem. 465 (1-2), 89–102.

Mousain-Bosc, M., Roche, M., Polge, A., et al., 2006. Improvement of neurobehavioral disorders in children supplemented with magnesium-vitamin B6. II. Pervasive developmental disorder-autism. Magnes. Res. 19 (1), 53–62.

Munoz Fernandez, S.S., Ivanauskas, T., Lima Ribeiro, S.M., 2017. Nutritional strategies in the management of Alzheimer disease: systematic review with network meta-analysis. J. Am. Med. Dir. Assoc. 18 (10), 897e813–897e830.

Murakami, K., Miyake, Y., Sasaki, S., 2010. Dietary intake of folate, vitamin B6, vitamin B12 and riboflavin and risk of Parkinson's disease: a case-control study in Japan. Br. J. Nutr. 104 (5), 757–764.

Nabavi, S.F., Sureda, A., Dehpour, A.R., et al., 2017. Regulation of autophagy by polyphenols: Paving the road for treatment of neurodegeneration. Biotechnol. Adv.

Nelson, C., Wengreen, H.J., Munger, R.G., 2009. Dietary folate, vitamin B-12, vitamin B-6 and incident Alzheimer's disease: the cache county memory, health and aging study. J. Nutr. Health Aging. 13 (10), 899–905.

Ngoungoure, V.L., Schluesener, J., Moundipa, P.F., 2015. Natural polyphenols binding to amyloid: a broad class of compounds to treat different human amyloid diseases. Mol. Nutr. Food Res. 59 (1), 8–20.

Nye, C., Brice, A., 2002. Combined vitamin B6-magnesium treatment in autism spectrum disorder. Cochrane Database Syst. Rev. (4), CD003497.

Ojha, R.P., Rastogi, M., Devi, B.P., 2012. Neuroprotective effect of curcuminoids against inflammation-mediated dopaminergic neurodegeneration in the MPTP model of Parkinson's disease. J. Neuroimmune Pharmacol. 7 (3), 609–618.

Omar, S.H., Scott, C.J., Hamlin, A.S., et al., 2017. The protective role of plant biophenols in mechanisms of Alzheimer's disease. J. Nutr. Biochem. 47, 1–20.

Panza, F., Frisardi, V., Capurso, C., 2009. Polyunsaturated fatty acid and S-adenosylmethionine supplementation in predementia syndromes and Alzheimer's disease: a review. Sci. World J. 9, 373–389.

Parellada, M., Llorente, C., Calvo, R., et al., 2017. Randomized trial of omega-3 for autism spectrum disorders: effect on cell membrane composition and behavior. Eur. Neuropsychopharmacol. 27 (12), 1319–1330.

Parker-Athill, E., Luo, D., Bailey, A., 2009. Flavonoids, a prenatal prophylaxis via targeting JAK2/STAT3 signaling to oppose IL-6/MIA associated autism. J. Neuroimmunol. 217 (1-2), 20–27.

Parracho, H.M.R.T., Gibson, G.R., Knott, F., 2010. A double-blind, placebo-controlled, crossover-designed probiotic feeding study in children diagnosed with autistic spectrum disorders. Int. J. Probiot. Prebiot. 5 (2), 69–74.

Patel, K., Curtis, L.T., 2007. A comprehensive approach to treating autism and attention-deficit hyperactivity disorder: a prepilot study. J. Altern. Complement. Med. 13 (10), 1091–1097.

Pieri, M., Morello, M., Zenobi, R., et al., 2020. The influence of vitamin D on neurodegeneration and neurological disorders: a rationale for its physio-pathological actions. Curr. Pharm. Des. Available from: https://doi.org/10.2174/1381612826666200316145725.

Pirchl, M., Ullrich, C., Humpel, C., 2010. Differential effects of short- and long-term hyperhomocysteinaemia on cholinergic neurons, spatial memory and microbleedings in vivo in rats. Eur. J. Neurosci. 32 (9), 1516–1527.

Pistollato, F., Forbes-Hernández, T.Y., Calderón Iglesias, R., et al., 2020. Pharmacological, non-pharmacological and stem cell therapies for the management of autism spectrum disorders: a focus on human studies. Pharmacol. Res. 152, 104579.

Pistollato, F., Iglesias, R.C., Ruiz, R., et al., 2019. Nutritional patterns associated with the maintenance of neurocognitive functions and the risk of dementia and Alzheimer's disease: a focus on human studies. Pharmacol. Res. 131, 32–43.

Pistollato, F., Battino, M., 2014. Role of plant-based diets in the prevention and regression of metabolic syndrome and neurodegenerative diseases. Trends Food Sci. 40 (1), 62–81.

Potashkin, J.A., Blume, S.R., Runkle, N.K., 2011. Limitations of animal models of Parkinson's disease. Parkinson's Dis. (2011).

Qiang, X., Sang, Z., Yuan, W., 2014. Design, synthesis and evaluation of genistein-O-alkylbenzylamines as potential multifunctional agents for the treatment of Alzheimer's disease. Eur. J. Med. Chem. 76, 314–331.

Raghavan, R., Riley, A.W., Volk, H., et al., 2018. Maternal multivitamin intake, plasma folate and vitamin B12 levels and autism spectrum disorder risk in offspring. Paediatr. Perinat. Epidemiol. 32 (1), 100–111.

Rainey-Smith, S.R., Brown, B.M., Sohrabi, H.R., et al., 2016. Curcumin and cognition: a randomised, placebo-controlled, double-blind study of community-dwelling older adults. Br. J. Nutr. 115 (12), 2106–2113.

Remington, R., Bechtel, C., Larsen, D., et al., 2016. Maintenance of cognitive performance and mood for individuals with Alzheimer's disease following consumption of a nutraceutical formulation: a one-year, open-label study. J. Alzheimers Dis. 51 (4), 991–995.

Remington, R., Lortie, J.J., Hoffmann, H., et al., 2015. A nutritional formulation for cognitive performance in mild cognitive impairment: a placebo-controlled trial with an open-label extension. J. Alzheimers Dis. 48 (3), 591–595.

Ren, R., Shi, C., Cao, J., et al., 2016. Neuroprotective effects of a standardized flavonoid extract of safflower against neurotoxin-induced cellular and animal models of Parkinson's disease. Sci. Rep. 6, 22135.
Rijpma, A., Meulenbroek, O., Olde Rikkert, M.G., 2014. Cholinesterase inhibitors and add-on nutritional supplements in Alzheimer's disease: a systematic review of randomized controlled trials. Ageing Res. Rev. 16, 105–112.
Rijpma, A., Meulenbroek, O., van Hees, A.M., et al., 2015. Effects of Souvenaid on plasma micronutrient levels and fatty acid profiles in mild and mild-to-moderate Alzheimer's disease. Alzheimers Res. Ther. 7 (1), 51.
Rojas, P., Montes, P., Rojas, C., 2012. Effect of a phytopharmaceutical medicine, Ginkgo biloba extract 761, in an animal model of Parkinson's disease: therapeutic perspectives. Nutrition. 28 (11–12), 1081–1088.
Rommer, P.S., Fuchs, D., Leblhuber, F., et al., 2016. Lowered levels of carbonyl proteins after vitamin B supplementation in patients with mild cognitive impairment and Alzheimer's disease. Neurodegener. Dis. 16 (3-4), 284–289.
Ryu, S., Jeon, H., Koo, S., et al., 2018. Korean red ginseng enhances neurogenesis in the subventricular zone of 1-methyl-4-phenyl-1,2,3,6-tetrahydropyridine-treated mice. Front. Aging Neurosci. 10, 355.
Ryu, S., Koo, S., Ha, K.-T., et al., 2016. Neuroprotective effect of Korea red ginseng extract on 1-methyl-4-phenylpyridinium-induced apoptosis in PC12 cells. Anim. Cell Syst. 20 (6), 363–368.
Sawikr, Y., Yarla, N.S., Peluso, I., et al., 2017. Neuroinflammation in Alzheimer's disease: the preventive and therapeutic potential of polyphenolic nutraceuticals. Adv. Protein Chem. Struct. Biol. 108, 33–57.
Schelke, M.W., Hackett, K., Chen, J.L., et al., 2016. Nutritional interventions for Alzheimer's prevention: a clinical precision medicine approach. Ann. N. Y. Acad. Sci. 1367 (1), 50–56.
Scheltens, P., Twisk, J.W., Blesa, R., et al., 2012. Efficacy of souvenaid in mild Alzheimer's disease: results from a randomized, controlled trial. J. Alzheimers Dis. 31 (1), 225–236.
Scheltens, P., Kamphuis, P.J., Verhey, F.R., et al., 2010. Efficacy of a medical food in mild Alzheimer's disease: a randomized, controlled trial. Alzheimers Dement. 6 (1), 1–10.e1.
Schmidt, R.J., Hansen, R.L., Hartiala, J., 2011. Prenatal vitamins, one-carbon metabolism gene variants, and risk for autism. Epidemiology. 22 (4), 476–485.
Serra, D., Almeida, L.M., Dinis, T.C.P., 2020. Polyphenols in the management of brain disorders: modulation of the microbiota-gut-brain axis. Adv. Food Nutr. Res. (91), 1–27.
Shaaban, S.Y., El Gendy, Y.G., Mehanna, N.S., et al., 2018. The role of probiotics in children with autism spectrum disorder: a prospective, open-label study. Nutr. Neurosci. 21 (9), 676–681.
Shah, S.P., Duda, J.E., 2015. Dietary modifications in Parkinson's disease: a neuroprotective intervention? Med. Hypotheses 85 (6), 1002–1005.
Sharman Moser, S., Davidovitch, M., Rotem, R.S., et al., 2019. High dose folic acid during pregnancy and the risk of autism; the birth order bias: a nested case-control study. Reprod. Toxicol. 89, 173–177.
Shaw, W., Kassen, E., Chaves, E., 1995. Increased urinary excretion of analogs of Krebs cycle metabolites and arabinose in two brothers with autistic features. Clin. Chem. 41 (8 Pt 1), 1094–1104.
Shea, T.B., Remington, R., 2015. Nutritional supplementation for Alzheimer's disease? Curr. Opin. Psychiatry 28 (2), 141–147.
Shen, L., Ji, H.F., 2015. Vitamin D deficiency is associated with increased risk of Alzheimer's disease and dementia: evidence from meta-analysis. Nutr. J. 14, 76.
Shi, D.H., Yan, Z.Q., Zhang, L.N., 2012. A novel 7-o-modified genistein derivative with acetylcholinesterase inhibitory effect, estrogenic activity and neuroprotective effect. Arch. Pharm. Res. 35 (9), 1645–1654.
Shulman, J.M., De Jager, P.L., Feany, M.B., 2011. Parkinson's disease: genetics and pathogenesis. Ann. Rev. Pathol. 6, 193–222.
Singh, K., Connors, S.L., Macklin, E.A., et al., 2014. Sulforaphane treatment of autism spectrum disorder (ASD). Proc. Natl Acad. Sci. U. S. A. 111 (43), 15550–15555.
Sita, G., Hrelia, P., Tarozzi, A., et al., 2017. P-glycoprotein (ABCB1) and oxidative stress: focus on Alzheimer's disease. Oxid. Med. Cell Longev. 2017, 7905486.
Soininen, B.V., Scheltens, P., 2014. A nutritional approach to ameliorate altered phospholipid metabolism in Alzheimer's disease. J. Alzheimers Dis. 41 (3), 715–717.
Soininen, H., Solomon, A., Visser, P.J., Hendrix, S.B., Blennow, K., Kivipelto, M., et al., 2017. 24-month intervention with a specific multinutrient in people with prodromal Alzheimer's disease (LipiDiDiet): a randomised, double-blind, controlled trial. Lancet Neurol. 16 (12), 965–975.
Solfrizzi, V., Custodero, C., Lozupone, M., et al., 2017. Relationships of dietary patterns, foods, and micro- and macronutrients with Alzheimer's disease and late-life cognitive disorders: a systematic review. J. Alzheimers Dis. 59 (3), 815–849.
Sperling, R.A., Aisen, P.S., Beckett, L.A., et al., 2011. Toward defining the preclinical stages of Alzheimer's disease: recommendations from the National Institute on Aging-Alzheimer's Association workgroups on diagnostic guidelines for Alzheimer's disease. Alzheimers Dement. 7 (3), 280–292.
Stefani, M., Dobson, C.M., 2003. Protein aggregation and aggregate toxicity: new insights into protein folding, misfolding diseases and biological evolution. J. Mol. Med. (Berl). 81 (11), 678–699.
Sudduth, T.L., Powell, D.K., Smith, C.D., 2013. Induction of hyperhomocysteinemia models vascular dementia by induction of cerebral microhemorrhages and neuroinflammation. J. Cereb. Blood Flow. Metab. 33 (5), 708–715.
Suzuki, M., Yoshioka, M., Hashimoto, M., 2013. Randomized, double-blind, placebo-controlled trial of vitamin D supplementation in Parkinson disease. Am. J. Clin. Nutr. 97 (5), 1004–1013.
Taliou, A., Zintzaras, E., Lykouras, L., 2013. An open-label pilot study of a formulation containing the anti-inflammatory flavonoid luteolin and its effects on behavior in children with autism spectrum disorders. Clin. Ther. 35 (5), 592–602.
Tani, Y., Fernell, E., Watanabe, Y., 1994. Decrease in 6R-5,6,7,8-tetrahydrobiopterin content in cerebrospinal fluid of autistic patients. Neurosci. Lett. 181 (1–2), 169–172.
Tarozzi, A., Angeloni, C., Malaguti, M., 2013. Sulforaphane as a potential protective phytochemical against neurodegenerative diseases. Oxid. Med. Cell. Long. 2013.
Theoharides, T.C., Asadi, S., Panagiotidou, S., 2012. A case series of a luteolin formulation (NeuroProtek(R)) in children with autism spectrum disorders. Int. J. Immunopathol. Pharmacol. 25 (2), 317–323.
Tordjman, S., Somogyi, E., Coulon, N., 2014. Gene × environment interactions in autism spectrum disorders: role of epigenetic mechanisms. Front. Psychiatry. 5, 53.
Troen, A.M., Shea-Budgell, M., Shukitt-Hale, B., 2008. B-vitamin deficiency causes hyperhomocysteinemia and vascular cognitive impairment in mice. Proc. Natl. Acad. Sci. U.S.A. 105 (34), 12474–12479.
Trudeau, M.S., Madden, R.F., Parnell, J.A., et al., 2019. Dietary and supplement-based complementary and alternative medicine Use in pediatric Autism Spectrum Disorder. Nutrients 11 (8).
Tsilioni, I., Taliou, A., Francis, K., et al., 2015. Children with autism spectrum disorders, who improved with a luteolin-containing dietary formulation, show reduced serum levels of TNF and IL-6. Transl. Psychiatry 5, e647.
Turner, R.S., Thomas, R.G., Craft, S., et al., 2015. A randomized, double-blind, placebo-controlled trial of resveratrol for Alzheimer disease. Neurology 85 (16), 1383–1391.
van der Zwaluw, N.L., Brouwer-Brolsma, E.M., van de Rest, O., et al., 2016. Folate and Vitamin B12-Related Biomarkers in Relation to Brain Volumes. Nutrients 9 (1).
Vancassel, S., Durand, G., Barthelemy, C., 2001. Plasma fatty acid levels in autistic children. Prostaglandins Leukot. Essent. Fat. Acids 65 (1), 1–7.
Varghese, M., Keshav, N., Jacot-Descombes, S., et al., 2017. Autism spectrum disorder: neuropathology and animal models. Acta Neuropathol. 134 (4), 537–566.
Vauzour, D., 2012. Dietary polyphenols as modulators of brain functions: biological actions and molecular mechanisms underpinning their beneficial effects. Oxid. Med. Cell. Long. (2012).
Vijayakumaran, S., Nakamura, Y., Henley, J.M., et al., 2019. Ginkgolic acid promotes autophagy-dependent clearance of intracellular alpha-synuclein aggregates. Mol. Cell Neurosci. 101, 103416.
Virmani, A., Pinto, L., Binienda, Z., 2013. Food, nutrigenomics, and neurodegeneration-neuroprotection by what you eat!. Mol. Neurobiol. 48 (2), 353–362.
Voigt, R.G., Mellon, M.W., Katusic, S.K., et al., 2014. Dietary docosahexaenoic acid supplementation in children with autism. J. Pediatr. Gastroenterol. Nutr. 58 (6), 715–722.
Waldemar, G., Dubois, B., Emre, M., 2007. Recommendations for the diagnosis and management of Alzheimer's disease and other disorders associated with dementia: EFNS guideline. Eur. J. Neurol. 14 (1), e1–e26.
Wang, C., Zhang, X., Teng, Z., 2014. Downregulation of PI3K/Akt/mTOR signaling pathway in curcumin-induced autophagy in APP/PS1 double transgenic mice. Eur. J. Pharmacol. 740, 312–320.

Wang, M., Li, K., Zhao, D., Li, L., 2017. The association between maternal use of folic acid supplements during pregnancy and risk of autism spectrum disorders in children: a meta-analysis. Mol. Autism 8, 51.

Weber, C.A., Ernst, M.E., 2006. Antioxidants, supplements, and Parkinson's disease. Ann. Pharmacother. 40 (5), 935–938.

Wei, W., Liu, Y.H., Zhang, C.E., 2011. Folate/vitamin-B12 prevents chronic hyperhomocysteinemia-induced tau hyperphosphorylation and memory deficits in aged rats. J. Alzheimers Dis. 27 (3), 639–650.

Weih, M., Wiltfang, J., Kornhuber, J., 2007. Non-pharmacologic prevention of Alzheimer's disease: nutritional and life-style risk factors. J. Neural. Transm. 114 (9), 1187–1197.

Weinreb, O., Mandel, S., Amit, T., 2004. Neurological mechanisms of green tea polyphenols in Alzheimer's and Parkinson's diseases. J. Nutr. Biochem. 15 (9), 506–516.

Wiens, D., DeSoto, M.C., 2017. Is high folic acid intake a risk factor for Autism?-A review. Brain Sci. 7 (11).

Wilczynski-Kwaitek, A., Singh, R.B., De Meester, F., 2009. Nutrition and behavior: the role of ω3 fatty acids. Open. Nutraceuticals J. 2, 1–10.

Witkin, J.M., Li, X., 2013. Curcumin, an active constituent of the ancient medicinal herb Curcuma longa L.: some uses and the establishment and biological basis of medical efficacy. CNS Neurol. Disord. Drug. Targets. 12 (4), 487–497.

Xiao, Z., Zhang, A., Lin, J., 2014. Telomerase: a target for therapeutic effects of curcumin and a curcumin derivative in Abeta1-42 insult in vitro. PLoS One. 9 (7), e101251.

Yu, J., Gattoni-Celli, M., Zhu, H., et al., 2011. Vitamin D3-enriched diet correlates with a decrease of amyloid plaques in the brain of AbetaPP transgenic mice. J. Alzheimers Dis. 25 (2), 295–307.

Zhang, C.E., Tian, Q., Wei, W., 2008. Homocysteine induces tau phosphorylation by inactivating protein phosphatase 2A in rat hippocampus. Neurobiol. Aging. 29 (11), 1654–1665.

Zhang, R., Zhang, J., Fang, L., et al., 2014. Neuroprotective effects of sulforaphane on cholinergic neurons in mice with Alzheimer's disease-like lesions. Int. J. Mol. Sci. 15 (8), 14396–14410.

Zhou, Q., Chen, B., Wang, X., et al., 2016. Sulforaphane protects against rotenone-induced neurotoxicity in vivo: involvement of the mTOR, Nrf2, and autophagy pathways. Sci. Rep. 6, 32206.

Zhou, X., Yuan, L., Zhao, X., 2014. Genistein antagonizes inflammatory damage induced by beta-amyloid peptide in microglia through TLR4 and NF-kappaB. Nutrition. 30 (1), 90–95.

Zhu, Z.G., Sun, M.X., Zhang, W.L., et al., 2017. The efficacy and safety of coenzyme Q10 in Parkinson's disease: a meta-analysis of randomized controlled trials. Neurol. Sci. 38 (2), 215–224.

Zoccolella, S., Lamberti, P., Armenise, E., 2005. Plasma homocysteine levels in Parkinson's disease: role of antiparkinsonian medications. Parkinsonism. Relat. Disord. 11 (2), 131–133.

Chapter 3

Cognitive effects of nutraceuticals

Jason Pitt and Yiuka Pitt

Introduction

The term nutraceutical is derived from "nutrition" and "pharmaceutical," and includes components of a food that provide health benefits beyond that of strict nutritional support (Kalra, 2003). This distinguishes nutraceuticals from typical foods that provide us with minerals and macromolecules necessary for proper metabolic support and cellular function. There are a number of nutraceuticals sold over the counter as all-natural means for boosting immune function, curing bodily ailments, and even enhancing mental function. Because plants produce a variety of biologically active small molecules, plant-based foods may have broader impacts on the body beyond nutrition.

Nutraceuticals are less regulated than pharmaceuticals. For example, in the United States, companies may market nutraceuticals as having "health" benefits that are not backed by clinical research, although no "medical" benefits may be claimed. The distinction—if there is one—between "health" and "medical" benefits is blurry. Although a discussion of the legalities of nutraceuticals is beyond the scope of this chapter, it is important to recognize that clinical research may be lacking or absent for many nutraceuticals currently marketed in the United States because of the less stringent regulatory policies. However, pharmaceuticals are validated through clinical trials before being prescribed to patients. As a result, there is much more research available on the efficacy, side effects, and mechanism of action of pharmaceuticals than nutraceuticals. This chapter describes the cognitive-enhancing effects of select nutraceuticals that have been tested in clinical trials or animal studies.

Biological basis of cognition

Overview of cognition

Cognition refers to mental processes involved in problem-solving, planning, attention, language, and memory. There are many common tests used to assess cognitive function (Table 3.1). Cognition, like all mental processes, is the result of coordinated communication between distinct brain regions that are specialized to process specific types of information. The bulk of cognitive processes are attributed to activity in the cerebral cortex—the six-layered superficial portion of the brain. The cerebral cortex is regionally specialized to process sensory information, coordinate motor output, store our memories, and integrate multiple types of information. Of particular importance in cognition is the prefrontal cortex, which may produce conscious experience by virtue of its connections with a variety of sensory, memory, and emotional circuitry (Fig. 3.1).

Long-term memories are ultimately stored in the cortex through coordinated interactions with subcortical structures, such as the amygdala, thalamus, and especially the hippocampus. Although cognition is usually distinguished from emotional or affective processes (e.g., fear, happiness, and sadness), cognition is undoubtedly affected by the activity of subcortical structures and consequent emotional states. Learning and memory impairments following lesions of the amygdala or hippocampus are well documented in both rodent models and human patients (Fig. 3.1; Scoville and Milner, 1957; Morris et al., 1982; Squire, 1992; Adolphs et al., 1994, 1997). The hippocampus is especially important for semantic memory function, as bilateral hippocampal lesions lead to complete anterograde amnesia. The connection between emotion and memory is further highlighted by the cognitive impairments in working memory, attention shifting, and verbal recall reported in subjects with depression (Austin et al., 2001).

Cellular basis of information storage

Information processing is performed by the coordinated activity of neuronal networks. Neurons are polar neural cells that are specialized for intercellular communication through specialized connections called synapses. When neurons are active, they release chemical messengers called neurotransmitters that diffuse across the synapse to

Nutraceuticals. DOI: https://doi.org/10.1016/B978-0-12-821038-3.00003-3

TABLE 3.1 A brief list of common cognitive tests.

Cognitive evaluation	Cognitive measures	Normal scores
Mini-Mental State Exam	Orientation, immediate and short-term memory, attention and calculation, language and praxis	• 0–30, with lower scores indicating greater impairment • Score ≤23 suggests cognitive impairment
Consortium to Establish a Registry for Alzheimer's Disease		
Trail-Making Test (TMT)	Sequential connection of points based on numbers (TMT-A) or alternating numbers and letters (TMT-B) estimates executive function, psychomotor speed, and visual scanning	• Completion times and number of errors, with higher scores indicating greater impairment in subjects ~70 years old • TMT-A: • Normal = 36 ± 13 s, 0.1 ± 0.4 errors • MCI = 48 ± 22 s, 0.2 ± 0.5 errors • AD = 67 ± 31 s, 0.2 ± 0.5 errors • TMT-B: • Normal = 82 ± 36 s, 0.5 ± 0.9 errors • MCI = 136 ± 78 s, 1.1 ± 1.2 errors • AD = 191 ± 82 s, 1.4 ± 1.3 errors
Clock Drawing Test	Executive function and spatial awareness	• Usually 0–10, although maximum scores vary, with lower scores indicating greater impairment • Normal = 8.7 ± 1.1 • AD = 4.9 ± 2.7
Stroop	Recognition (words or colors) and discrimination tasks (color vs word)	Number of correct responses (45 s per trial), with higher numbers indicating better cognitive performance
Expanded Memory Control Test	Attention and concentration; reproducing a list of four items, reciting the same list backwards, and performing four serial calculations	Maximum score of 24, with lower scores indicating greater impairment
Cognitive Drug Research (CDR) Testing Battery	Immediate/delayed word recall, word recognition, picture recognition, simple reaction time, digit vigilance, choice reaction time, numeric working memory, and spatial working memory	Each domain is scored separately based on speed and number of errors
Rey Auditory Verbal Learning Test	Short-term recall (Rey 1) and long-term recognition (Rey 2) of a set of 15 words presented five times	
Alzheimer's Disease Assessment Scale: Cognition	Spoken language ability and comprehension, following and recall of commands, orientation, word recall and recognition, construction drawing, and ideation praxis	• 0–70, with higher scores indicating greater impairment • 3-point decline is considered the minimal clinically relevant change in patients with early Alzheimer disease
Wechsler Memory Scale	Episodic memory, orientation, mental control, logical recall, visual reproduction, digit span (forward and backward)	Lower scores indicate greater impairment
Neuropsychiatric Inventory	Psychopathology, including delusions, hallucinations, agitation, euphoria, dysphoria, anxiety, apathy, disinhibition, irritability, aberrant motor behavior, sleep disorders, and eating disorders	Frequency, severity, and caregiver distress scores are given for each behavioral domain, with higher scores indicating greater impairment
Clinical Global Impression of Change	Judgment of change in overall impairment by a clinician following an unstructured interview	• Scale of 1–7: • 1—extreme improvement • 4—no change • 7—extreme worsening

(Continued)

TABLE 3.1 (Continued)

Cognitive evaluation	Cognitive measures	Normal scores
Quality of Life-Alzheimer's Disease	Self and carer assessment based on two 13-item scales	13–52, with higher scores indicating a better quality of life
Activities of Daily Living	Carer assessment of overall performance in bathing, dressing, going to the toilet, transferring, continence, and feeding	Qualitative assessment of A–G, with A representing complete independence and G representing complete dependence
Life Satisfaction in the Elderly Scale	Self-evaluation of eight categories of well-being: daily activities, meaning, goals, mood, self-concept, health, finances, and social contacts	• Scale of 1–5: • 1—poor • 5—optimum
Symptom Rating Scale	Self-evaluation of depression, anxiety, somatization, and inadequacy	Higher scores indicate poorer self-rating
Randt Memory Test	Age-indexed battery of tests assessing acquisition recall and delayed memory	Higher scores indicate better memory
World Health Organization Quality of Life Assessment Instruments	100-question self-reported questionnaire concerning quality of life and general health	Likert scales, with higher scores indicating better quality of life

activate or inhibit the activity of other neurons. It is the activity of neurons that is thought to account for our ability to process sensory information and execute motor commands. For example, the activity of specific cells within our hippocampus (called place cells) is controlled by our location. Based on the pattern of activity in our place cells, we are able to determine where we are at a given moment.

How might neural networks allow us to hold information in our conscious awareness (i.e., working or short-term memory)? This is likely accomplished by the activity of neurons in the prefrontal cortex (Fig. 3.1). Prefrontal cortex neurons are able to prolong their activation and maintain the activity of specific neural networks (Fuster and Alexander, 1971; Kubota and Niki, 1971). Thus by maintaining the activity of neural networks that encode the information "Durne Street" and "turn left," we can remember to turn left at Durne Street. Because the activity of neurons is critical for cognition, we should expect that any cognitive enhancer should affect the excitability of neurons or their ability to communicate across synapses.

Long-term storage of new information, another aspect of cognition, is also subserved by the activity of neurons. Information storage requires a rearrangement of neural networks, which is performed by changes in synaptic strength. For two neurons to be considered part of a network, one neuron must be able to reliably affect the activity of the other. As discussed, neuronal activity is shaped by synaptic signaling. Therefore neural networks contain strong synapses that are capable of reliably shaping the activity of connected neurons. Thus for us to connect two pieces of information, which are both encoded by the activity of distinct neural networks, we must strengthen the connections between these two networks (Fig. 3.2). This was first formally hypothesized by Canadian psychologist Donald Hebb in Hebb's postulate:

> *When an axon of cell A is near enough to excite a cell B and repeatedly or persistently takes part in firing it, some growth process or metabolic change takes place in one or both cells such that A's efficiency, as one of the cells firing B, is increased.*
>
> Hebb (1949).

Synaptic plasticity as a physical basis for memory

The rearrangement of neuronal networks is accomplished by a process called synaptic plasticity. Readers interested in synaptic plasticity and the long-term storage of memories are directed to excellent reviews by Lamprecht and LeDoux (2004) and Frankland and Bontempi (2005). Briefly, synaptic plasticity is an activity-dependent change in synaptic strength (Fig. 3.3), which is the basis of learning and memory in the synaptic plasticity and memory (SPM) hypothesis:

> *Activity-dependent synaptic plasticity is induced at appropriate synapses during memory formation, and is both necessary and sufficient for the information storage underlying the type of memory mediated by the brain area in which that plasticity is observed.*
>
> Martin et al. (2000).

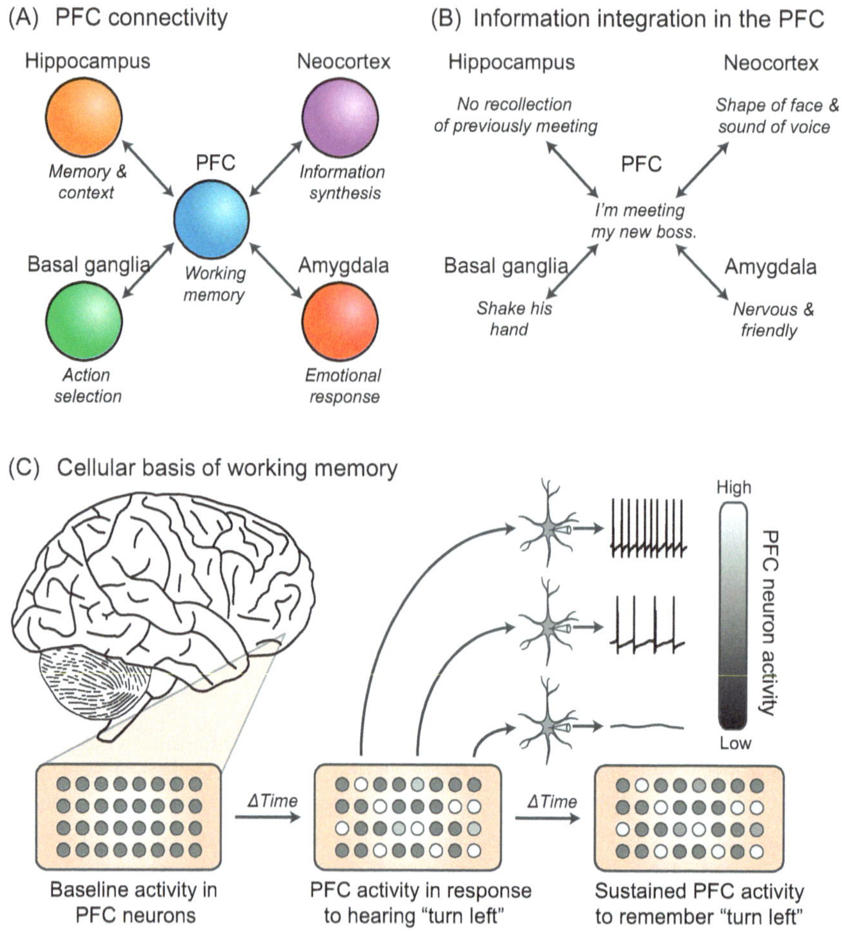

FIGURE 3.1 Cognition and memory are the products of diverse, widespread brain regions. (A) The prefrontal cortex (PFC) is connected to a variety of brain regions, allowing processing of many types of information. The hippocampus provides information about recent memories and spatial context. Other cortical regions provide sensory information. Input from the basal ganglia, particularly the ventral striatum, provides information about motivational states. Emotional information is provided from other brain regions, such as the amygdala. (B) The PFC creates our working memory (and likely consciousness awareness) by integrating a variety of information. Shown here is an example of what information the PFC might integrate when a person meets their new boss for the first time. (C) Working memory is the product of neuronal activity in the PFC. The activity pattern of neurons encodes information. Shown here is an example of how a small number of hypothetical PFC neurons might respond to the command to "turn left." Notice how some neurons become far more active (indicated as white circles), others become somewhat more active (indicated as light gray circles), and the rest do not respond (indicated as dark gray circles). Because PFC neurons can prolong their excitation, they are able to hold certain patterns of activity for extended periods of time. This allows us to hold information for a brief period of time, and thus we remember to turn left. After we have completed the immediate task, the pattern of neuronal activity can then change to encode new, immediately relevant information.

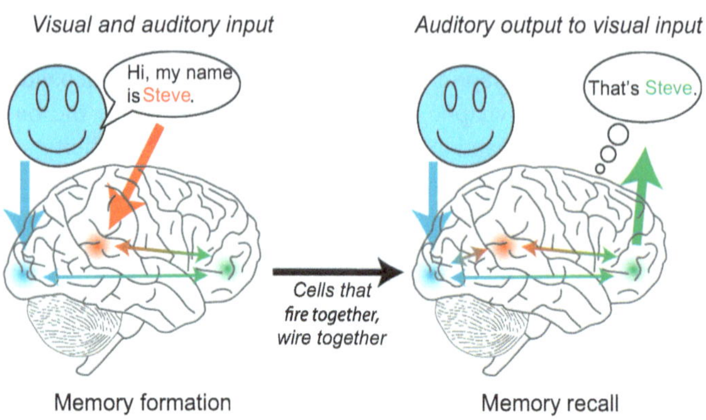

FIGURE 3.2 Hebbian plasticity allows neural network remodeling. In Hebbian plasticity, neurons that fire together, wire together. In other words, coincident activation of two or more neurons will strengthen the synaptic connections between them and allow the remodeling of neural networks. For example, we are able to associate a name with a face due to the coincident activation of neurons that encode the facial information (blue circle) and neurons that encode the person's name (red circle). Initially, the neural networks are connected through our prefrontal cortex (green circle). As we learn the person's name, we no longer rely on working memory.

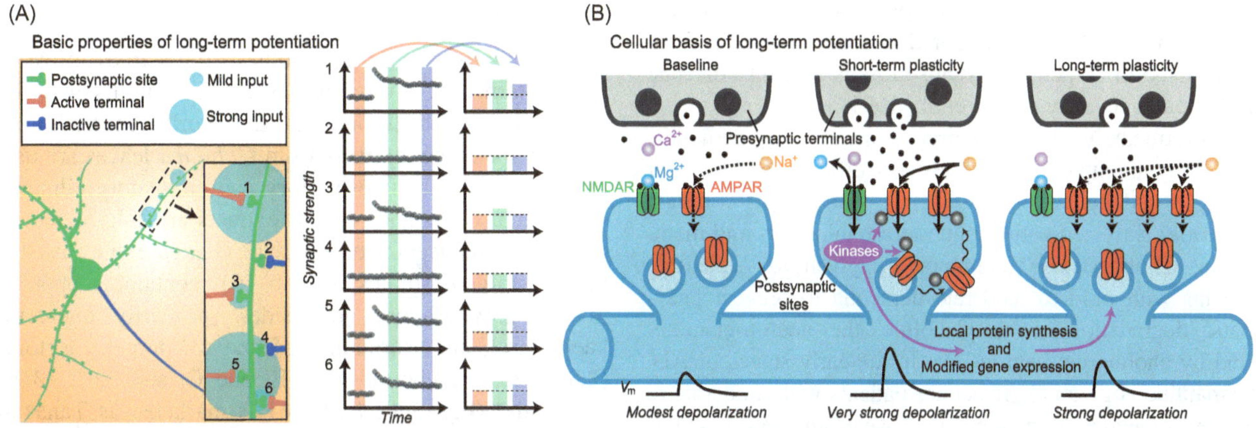

FIGURE 3.3 Synaptic plasticity is an activity-dependent change in synaptic strength. (A) long-term potentiation (LTP) is an activity-dependent change in synaptic strength that occurs following coincident activation of presynaptic and postsynaptic neurons. LTP has three basic properties: cooperativity, associativity, and specificity. LTP is cooperative, because as a stimulus it becomes more intense and involves more synapses, and the potentiation becomes longer lasting (compare synapse 1 vs synapse 3). The associativity of LTP provides a cellular analog of classical conditioning—weak input can be potentiated if active during a strong tetanus to a separate but convergent input (compare synapse 3 vs synapse 6). Finally, LTP shows specificity, because only active synapses are potentiated (compare synapse 4 vs synapse 5). (B) LTP involves short-term and long-term changes in synaptic function. LTP generally requires the activation of NMDA-type glutamate receptors. *N*-methyl-D-aspartate (NMDA) receptors are normally blocked by a Mg^{2+} ion, and thus strong synaptic input is required to depolarize the neuron sufficiently to repel the Mg^{2+} ion. When this occurs, NMDA receptors allow a substantial Ca^{2+} influx, which activates intracellular kinases that increase AMPA-type glutamate receptor conductance and trafficking to the surface. Activated kinases also modify local protein synthesis and gene expression to increase the overall level of synaptic proteins, allowing greater synaptic responses.

Indeed, the encoding of new information (i.e., learning and memory) is associated with structural alterations in neural networks, specifically at the level of synapses. Classically, synapses comprise two components: (1) the presynaptic terminal and (2) the postsynaptic site. The presynaptic terminal is almost exclusively found in axons and is the site of neurotransmitter release. The postsynaptic site can occur at a variety of locations: (1) the cell body (typically reserved for inhibitory synapses); (2) the dendrites; and (3) the dendritic spines (small protrusions from dendrites that exclusively receive excitatory input). Dendritic spines exhibit rapid, activity-dependent changes in structure, which we believe is involved in memory formation. For example, rats that are trained in a hippocampal-dependent memory task have a higher density of dendritic spines on hippocampal neurons (Moser et al., 1994; Vetere et al., 2011). Furthermore, the density of synapses decreases with age and is a reasonable indicator of memory performance—learning-impaired rats have lower synapse densities than unimpaired age-matched controls (Geinisman et al., 2004). The loss of synapse abundance is the best correlate with severity of dementia. Thus in "normal" and pathological aging, the loss of synapses reflects poorer cognitive function. It seems that memories are made and stored at synapses.

How is it that synapses adjust their strength? One cellular mechanism of synaptic plasticity is called long-term potentiation (LTP). LTP is an activity-dependent, persistent increase in synaptic strength measured by an increase in current amplitude and a decrease in failure rate. The current amplitude is based on the number of postsynaptic receptors present and their single-channel conductance (i.e., the rate at which they let ions permeate the membrane). LTP increases current amplitude through post-translational modification and increased expression of excitatory neurotransmitter receptors, such as α-amino-3-hydroxy-5-methyl-4-isoxazolepropionic acid (AMPA)-type glutamate receptors. Synaptic "failure" is a lack of neurotransmitter release when a neuron is active (i.e., when a neuron fires an action potential). LTP decreases the failure rate by increasing neurotransmitter release probability.

Biological basis of dementia

Many studies that we discuss use patients with cognitive deficits, including dementia. Individuals with dementia have profound impairments in neurocognitive function, such as memory, problem-solving, language, and judgment that compromise daily function. The use of demented patients in cognitive studies has a few advantages: (1) the outcomes of these studies have potential clinical relevance; (2) many of the cognitive tests outlined in Table 3.1 are sensitive to subnormal cognitive states, but not necessarily cognitive enhancement beyond "normal function"; and (3) starting from a lower baseline level of cognitive performance increases the odds of detecting cognitive enhancement. Although the cause of dementia is multifactorial and varies from case to case, the fundamental underlying biological mechanism of dementia is synapse loss. Therefore the loss of synapses leads to loss of memories. Although neuronal degeneration is a common theme in dementias, synapse loss is a better correlate of dementia severity (Terry et al., 1991).

Unfortunately, there is currently no cure for dementia, and many therapies have short-term efficacy at best. For example, Alzheimer disease (AD; the most common dementia) is treated with cholinesterase inhibitors and memantine. Cholinesterase inhibitors are given to Alzheimer patients with mild to moderate dementia to prevent enzymatic degradation of acetylcholine, the neurotransmitter used by cholinergic neurons lost in the early stages of AD. Memantine is given to Alzheimer patients with moderate to severe dementia. Memantine is a weak antagonist and is thought to prevent the excitotoxic loss of neurons and synapses caused by overactivation of *N*-methyl-d-aspartate (NMDA) receptors (Danysz et al., 2000; Dekundy, 2006). These two treatments temporarily slow the decline in cognitive function, but neither modifies the course of disease (Farlow and Cummings, 2007). Given the lack of effective treatments for cognitive decline, the use of natural products to enhance cognitive function is appealing.

Nutraceuticals with potential cognitive-enhancing functions

As discussed, nutraceuticals are not regulated as tightly as pharmaceuticals. As a result, research on the bioavailability, effective doses, off-target effects, and mechanism of action of nutraceuticals is not as clear as that of pharmaceuticals. Here, a brief account of some common nutraceuticals is given focusing on randomized, placebo-controlled, double-blind studies in cognitively normal and cognitively impaired human subjects. In vivo and in vitro studies are included and are summarized in the discussion of potential cellular mechanisms of action (Table 3.2).

Ginkgo biloba

Ginkgo biloba is one of the oldest living trees in evolutionary terms. This rare species grows naturally in China and is cultivated throughout the world. In traditional Chinese medicine, extracts from *G. biloba* leaves are used to treat a number of respiratory and circulatory ailments. In Western culture, *G. biloba* extracts are sold as natural cognitive enhancers. The standard *Ginkgo* extract used in the majority of studies is EGb 761, containing approximately 25% flavone glycosides (primarily quercetin, kaempferol, and isorhamnetin) and 6% terpene lactones (2.8%–3.4% ginkgolides A, B, and C, and 2.6%–3.2% bilobalide) (Fig. 3.4). EGb 761 is marketed as Tanakan, Tebonin, and Rökan. Flavone glycosides and terpene lactones ingested orally are able to cross the blood–brain barrier (BBB). In rats, a single dose of *G. biloba* extract (90 mg/kg) reaches peak levels in the brain at 1 h and remains at detectable levels for approximately 6 h (Chen et al., 2013).

Positive effects

There is little evidence that *G. biloba* enhances cognitive function in cognitively normal individuals. However, many lines of evidence suggest a small, but statistically significant, improvement in cognitive function in cognitively impaired subjects. In a randomized, double-blind, multicenter study, patients with amnestic mild cognitive impairment were given daily 240 mg doses EGb 761 for 24 weeks (Gavrilova et al., 2014). Compared to placebo, EGb 761 treatment provided small, but significant, improvements in subjective—Neuropsychiatric Inventoryand Clinical Global Impression of Change—and objective—Trail-Making Test (TMT)—measures of executive function. In another randomized, double-blind study, cognitively impaired elderly patients were given 40 or

TABLE 3.2 Nutraceuticals with potential cognitive-enhancing functions.

Nutraceutical	Effects claimed	Potential mechanism
Ginkgo biloba	Restore function in cognitively impaired individuals	Enhanced monoaminergic and cholinergic signaling
Bacopa monnieri	Restore function in cognitively impaired individuals	Facilitated cerebral blood flow, neuronal activity, and synaptic transmission
Ginseng	Improve normal cognitive function	Estrogenic activity or antistress function
Phosphatidylserine	Restore function in cognitively impaired individuals, especially Alzheimer patients	Enhanced cholinergic function
Caffeine	Acute cognitive enhancement and sparing of cognitive decline	Facilitated neuronal excitability, synaptic transmission, and synaptic plasticity through inhibition of adenosine receptors and phosphodiesterase

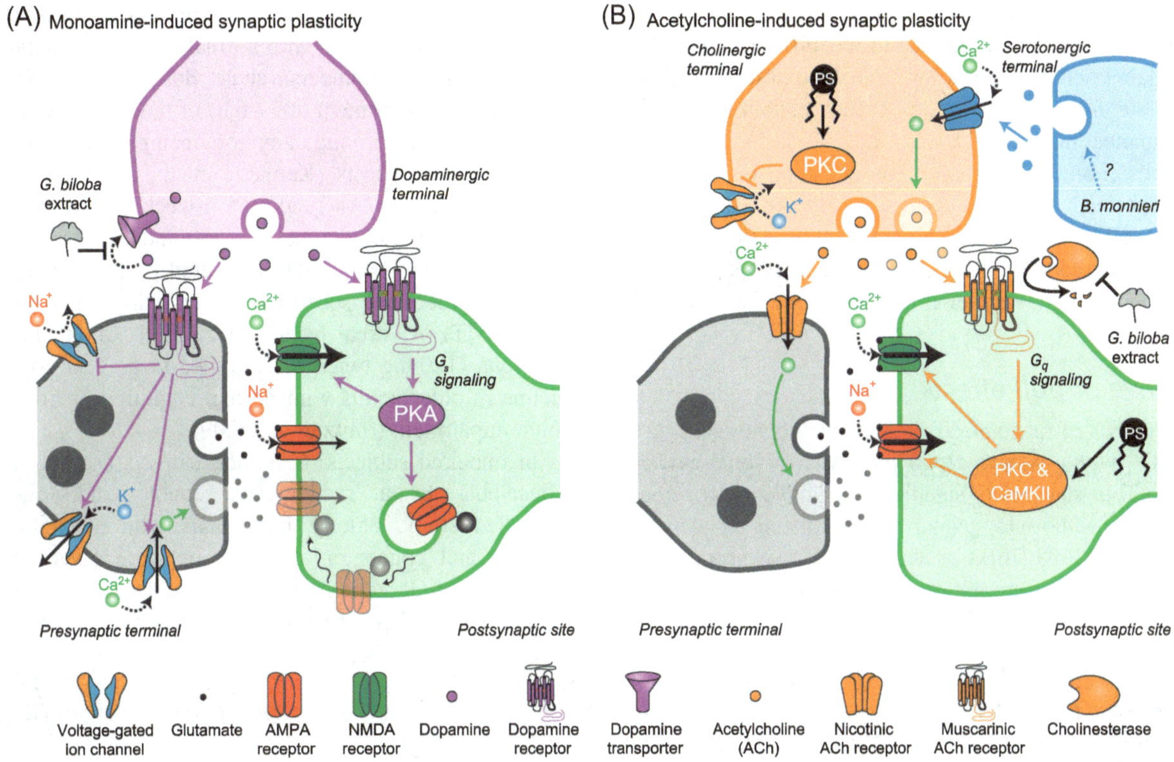

FIGURE 3.4 Monoaminergic and cholinergic activity regulate memory function. (A) Monoamines act on presynaptic and postsynaptic sites to facilitate plasticity. Postsynaptically, monoamines stimulate protein kinase A, which traffics AMPA receptors to the surface and increases *N*-methyl-D-aspartate (NMDA) receptor activity—both of these modifications make the postsynaptic site more responsive to glutamate. Presynaptically, monoamines facilitate neurotransmitter release by increasing the influx of Ca^{2+} through voltage-gated Ca^{2+} channels. Monoamines also decrease presynaptic excitability by reducing Na^+ conductance and increasing K^+ conductance. Thus it becomes more difficult for neurons to fire an action potential. However, once neurons do fire an action potential, they will release more neurotransmitter, thus producing a greater postsynaptic response with a higher probability of stimulating long-term potentiation (LTP). In this model, *G. biloba* may stimulate LTP by impairing monoamine reuptake and prolonging monoaminergic signaling. (B) Cholinergic signaling also affects presynaptic and postsynaptic sites. Presynaptically, nicotinic acetylcholine receptors allow Ca^{2+} influx, which promotes neurotransmitter release. Postsynaptically, muscarinic acetylcholine receptors stimulate protein kinase C (PKC) and Ca^{2+}/calmodulin-dependent kinase II (CaMKII), which increase the conductance of NMDA and AMPA receptors, respectively. PKC also promotes acetylcholine release by decreasing K^+ conductance in cholinergic neurons. Phosphatidylserine directly activates PKC to promote cholinergic and glutamatergic signaling. *Bacopa monnieri* may facilitate cholinergic signaling by activating serotonergic neurons. Cholinergic neurons have Ca^{2+}-permeable serotonin receptors that facilitate neurotransmitter release.

80 mg of GB-8 (*G. biloba* extract) or placebo three times daily for 3 months (Winther et al., 1998). Patients receiving 40 mg of GB-8 had increased memory function (assessed by the Wechsler Memory Scale) and decreased diastolic blood pressure after 3 months. However, dose-dependent subjects receiving 80 mg of GB-8 did not show any significant changes in memory function. Furthermore, memory function in the group receiving 40 mg of GB-8 failed to surpass that of the placebo group; observed improvements were due to differences in baseline scores.

Given the potential effectiveness of *G. biloba* in cognitively impaired subjects, it is possible that *G. biloba* extracts may offer some added benefit when used in combination with other treatments for dementia. In line with this idea, combinatorial treatment of patients with mild to moderate AD with *G. biloba* extracts and cholinesterase inhibitors offers greater improvement in some, but not all, cognitive measures compared to treatment with cholinesterase inhibitors alone (Canevelli et al., 2014).

Potential mechanisms

The *G. biloba* extract EGb 761 has a wide array of cellular effects that could enhance cognitive function (DeFeudis and Drieu, 2000). EGb 761 has antioxidant properties, which reduce the levels of reactive oxygen species and lipid peroxidation. Prevention of cellular damage is particularly important for neurons, which possess limited regenerative capabilities. EGb 761 has more specific effects on synaptic function. In particular, EGb 761 enhances monoaminergic transmission by inhibiting the uptake of monoaminergic neurotransmitters (serotonin, norepinephrine, and dopamine). Monoamines activate a number of signaling cascades that promote synaptic plasticity (Fig. 3.4).

The reported positive effects of *G. biloba* on Alzheimer patients may be due to its ability to mimic current Alzheimer therapeutics. *G. biloba* extracts have anticholinesterase activity at doses that enhance cognitive performance in mice (Das et al., 2002). EGb 761 also enhances cholinergic signaling by boosting the muscarinic acetylcholine receptors and the uptake of choline. Additionally, *G. biloba* extracts mimic the effects of memantine, reducing NMDA receptor activity and downstream signaling (Weichel et al., 1999; Li et al., 2011).

Negative or null effects

Metaanalyses of randomized, double-blind, controlled studies using *G. biloba* extracts report that there is some evidence for small but statistically significant cognitive improvements in patients with mild cognitive impairments or dementia (Oken et al., 1998; Birks et al., 2002). However, cognitive-enhancing effects were not observed in a predictable, consistent, or reliable manner (Table 3.3).

As stated, *G. biloba* extracts do not appear to offer any improvement in subjects with normal cognitive function. In young, healthy volunteers, acute doses (120, 240, and 360 mg, with assessment at 1–6 h) of the *G. biloba* extract GK 501 failed to produce any convincing change in a battery of cognitive tests (Kennedy et al., 2000). A similar lack of effect was observed in a 4-week trial with cognitively normal volunteers between 50 and 65 years of age: EGb 761 (120 mg twice daily) offered no improvement in cognitive processing speed (measured using a task similar to the TMT) or verbal recall (Cieza et al., 2003). Further, *G. biloba* (120 mg twice daily) failed to prevent cognitive decline in older adults with normal cognition or mild cognitive impairment (Snitz et al., 2009).

In impaired subjects, *G. biloba* extracts are not always efficacious. In a randomized, double-blind, placebo-controlled study, elderly individuals with a mixture of memory and concentration problems were given the *G. biloba* extract Geriaforce or a placebo control for 24 weeks. In each group, there were neither any adverse side effects

TABLE 3.3 Summary of selected studies on *Ginkgo biloba*.

Cognitive status	Improvement	Reference	Outcome measures
Normal	Yes	–	–
	No	Cieza et al. (2003)	Cognitive processing speed (Digit Connection Test, similar to TMT) and verbal recall
		Snitz et al. (2009)	MMSE and ADAS-Cog
		Kennedy et al. (2000)	Battery of cognitive measures
Mild impairment	Yes	Gavrilova et al. (2014)	NPI, CGIC, and TMT
		Winther et al. (1998)	WMS (40 mg GB-8)
	No	Snitz et al. (2009)	MMSE and ADAS-Cog
		Brautigam et al. (1998)	EMCT and AVLT
		Winther et al. (1998)	WMS (80 mg GB-8)
Dementia	Yes	Canevelli et al. (2014)	MMSE
	No	McCarney et al. (2008)	ADAS-Cog and QOL-AD
		Canevelli et al. (2014)	ADAS-Cog and ADL

ADAS-Cog, Alzheimer's Disease Assessment Scale: Cognition; *ADL*, Activities of Daily Living; *AVLT*, Rey Auditory Verbal Learning Test; *CGIC*, Clinical Global Impression of Change; *EVLT*, Expanded Mental Control Test; *MMSE*, Mini-Mental State Exam; *NPI*, Neuropsychiatric Inventor; *QOL-AD*, Quality of Life-Alzheimer's Disease; *WMS*, Wechsler Memory Scale; *TMT*, Trail-Making Test.

nor improvements in cognitive function as measured by the Expanded Mental Control Test and Rey Auditory Verbal Learning Test (AVLT) (Brautigam et al., 1998). A 6-month trial (120 mg EGb 761 and placebo control) showed a similar lack of effect on cognition (ADAS-Cog) or quality of life in patients with mild to moderate dementia (McCarney et al., 2008). *G. biloba* also proved ineffective at preventing dementia in a large randomized controlled trial of elderly subjects with normal cognition or mild cognitive impairment (DeKosky et al., 2008).

Bacopa monnieri

Bacopa monnieri is an Indian aquatic herb that has been used in traditional Ayurvedic medicine for centuries for memory decline, pain, fevers, inflammation, and epilepsy. The major ingredients in *B. monnieri* extracts (e.g., Brahmi) are steroidal saponins called bacosides. It is assumed that bacosides cross the BBB via lipid-mediated free diffusion due to their hydrophobic nature and effects on brain health and function. However, free diffusion into the brain is unlikely given the molecular weight of bacosides and their ability to form hydrogen bonds (Pardridge, 2012). Further, there is no clear evidence that bacosides are capable of crossing the BBB without chemical modification. Because of this, any cognitive-enhancing effects of *B. monnieri* may not necessarily be a consequence of direct interactions with neural cells, but rather peripheral effects, such as enhanced cerebral blood flow, or bacoside metabolites or other components of *B. monnieri* that can cross the BBB.

Positive effects

The first randomized, double-blind, placebo-controlled trial that investigated cognitive-enhancing effects of *B. monnieri* in unimpaired humans found an improvement in the speed of visual information processing and a reduction in subjective anxiety levels (Stough et al., 2001; Morgan and Stevens, 2010). Subsequent work from the same authors reported that *B. monnieri* improved spatial working memory accuracy (Stough et al., 2008). However, the effects of *B. monnieri* are small and inconsistent between these two studies. Similar subtle, but statistically significant, improvements in a select subset of cognitive outcomes were reported in other trials with healthy volunteers (Table 3.4; Roodenrys et al., 2002; Calabrese et al., 2008).

There are no randomized, placebo-controlled, double-blind studies that have examined the cognitive-enhancing functions of *B. monnieri* in impaired subjects.

Potential mechanisms

The heterogeneity of cognitive enhancement by *B. monnieri* extracts suggests diverse cellular mechanisms. This is supported by in vitro and animal studies that suggest a number of potential cognitive-enhancing mechanisms. Several studies report that *B. monnieri* is generally neuroprotective, preventing oxidative stress, lipid peroxidation, and cell death (Hosamani and Muralidhara, 2009; Thippeswamy et al., 2013; Velaga et al., 2014). Although these effects are promising, especially in disease states, the prevention of cell death does not necessarily explain any cognitive-enhancing functions that *B. monnieri* may have.

As discussed, the ability of bacosides to cross the BBB is not yet clear; therefore *B. monnieri* may act outside the central nervous system (CNS). Some of the neuroprotective and memory-restoring effects of *B. monnieri* may be due to the ability of bacosides to alter Ca^{2+} dynamics in smooth muscle cells and induce vasodilation (Kamkaew et al., 2011). Vasodilation of cerebral vasculature may facilitate cerebral blood flow, providing the brain with more oxygen and glucose to support higher levels of neuronal activity or maintain function in times of stress or injury, such as ischemia—focal blockage of cerebral blood flow. In a rat model of ischemia, *B. monnieri* reduces infarct size, attenuates signs of cell stress (e.g., free radicals and lipid peroxidation), and preserves memory function following transient occlusion of the intracarotid artery (Saraf et al., 2010).

B. monnieri appears to facilitate neuronal activity and synaptic signaling. In the hippocampus of rats fed *B. monnieri*, a variety of neurotransmitters are found at higher concentrations, including serotonin, acetylcholine, γ-aminobutyric acid, and glutamate. Given the variety of neurotransmitters involved, it is likely that *B. monnieri* increases neuronal activity rather than impairing neurotransmitter reuptake or enzymatic degradation. Elevated levels of serotonin receptors may promote information encoding by facilitating cholinergic synaptic transmission (Rajan et al., 2011; Fig. 3.4).

Negative or null effects

All of the trials discussed here that found small cognitive-enhancing effects of *B. monnieri* also reported null effects on the majority of cognitive measures. In a randomized, double-blind, placebo-controlled trial with healthy volunteers between 18 and 65 years of age, a single, acute dose of *B. monnieri* (300 mg, 2 h pretesting) did not improve performance on any aspect of a battery of cognitive evaluations (e.g., AVLT, digit span, language comprehension) (Nathan et al., 2001). In addition, *B. monnieri* treatments also commonly produce mild side effects, including nausea, increased intestinal motility, and gastrointestinal upset (Pravina et al., 2007).

Ginseng

Ginseng is a variety of plants in the genus *Panax* that contain a family of steroid glycosides called

TABLE 3.4 Summary of selected studies on *Bacopa monnieri*.

Cognitive status	Improvement	Reference	Outcome measures
Normal	Yes	Stough et al. (2001)	Inspection time
		Stough et al. (2008)	CDR (1/19 measures: working memory, % correct)
		Calabrese et al. (2008)	AVLT (# words) and Stroop Task (reaction time)
		Roodenrys et al. (2002)	Delayed recall of word pairs
		Morgan and Stevens (2010)	AVLT (some aspects)
	No	Stough et al. (2001)	AVLT, TMT, digit span
		Stough et al. (2008)	CDR (18/19 measures)
		Calabrese et al. (2008)	Other aspects of the AVLT and the Stroop Task, DAT, and WAIS
		Roodenrys et al. (2002)	Everyday memory, immediate recall of a short story, digit span (forward and backward), alphabetical speed coding, delayed recall of the short story read earlier, three trials of learning six unrelated word pairs, visual sequence recall (forward and backward), general knowledge
		Morgan and Stevens (2010)	AVLT (some aspects), TMT, Complex Figure Test
		Nathan et al. (2001)	Battery of cognitive evaluations (e.g., AVLT, digit span, language comprehension)
Mild impairment	Yes	–	–
	No	–	–
Dementia	Yes	–	–
	No	–	–

AVLT, Rey Auditory Verbal Learning Test; *DAT*, Divided Attention Task; *TMT*, Trail-Making Test; *WAIS*, Wechsler Adult Intelligence Scale.

ginsenosides—considered to be the active component of ginseng responsible for any potential beneficial effects. Ginsenosides have a four-ring, steroid-like structure with carbohydrate moieties attached. Due to their structure, ginsenosides do not readily cross the BBB and are scarcely detectable in the brain after oral administration in rats. Instead, it is the metabolites of ginsenosides that cross the BBB and act directly on neural cells.

Positive effects

Several trials in healthy subjects suggest that ginseng can improve some aspects of cognitive function, including reaction time, word recall, and mental fatigue (Table 3.5). However, these effects are generally small and are largely inconsistent between studies (negative or null effects).

There are limited data on the effects of ginseng in cognitively impaired subjects. In one study, subjects with age-associated memory impairment showed improvements in subjective assessments of psychological well-being and quality of life, as well as objective measures of memory function (Neri et al., 1995). Although we found several studies that used demented patients, they were excluded due to a lack of proper controls (i.e., nonplacebo-controlled or open-label studies). For now, high-quality evidence of its efficacy on demented patients is still lacking.

TABLE 3.5 Summary of selected studies on ginseng.

Cognitive status	Improvement	Reference	Outcome measures
Normal	Yes	D'Angelo et al. (1986)	Choice reaction time, tapping test
		Kennedy et al. (2002)	Immediate word recall
		Kennedy et al. (2001)	Simple reaction time and choice reaction time
		Reay et al. (2005)	Subjective feelings of mental fatigue
		Reay et al. (2006)	Subjective feelings of mental fatigue and some measures of cognitive demand
		Yeo et al. (2012)	Median reaction time in Vienna Test System (version IX)
	No	D'Angelo et al. (1986)	Cancellation Test, Logical Deduction Test, Digit Symbol Substitution Test
		Kennedy et al. (2002)	All measures in CDR besides immediate word recall
		Sünram-Lea et al. (2005)	Modified CDR
		Kennedy et al. (2001)	All measures in CDR besides simple reaction time and choice reaction time
		Reay et al. (2005)	Cognitive Demand Battery
		Reay et al. (2006)	Some measures of cognitive demand
		Yeo et al. (2012)	All measures except median reaction time in Vienna Test System (version IX)
		LaSala et al. (2015)	Reaction time and divided attention in a simulated driving task
Mild impairment	Yes	Neri et al. (1995)	Life satisfaction in the Elderly Scale, Symptom Rating Scale, Randt Memory Test
	No	Kim et al. (2008)	Wechsler Memory Scale, World Health Organization Quality of Life Assessment Instruments
Dementia	Yes	–	–
	No	–	–

Potential mechanism

As with other nutraceuticals, ginseng has general neuroprotective effects. Although these might be useful in disease states, they are not likely to account for cognitive enhancement in normal individuals. Instead, the positive impacts of ginseng on cognition may arise from its ability to activate estrogen receptors. Ginsenoside-Rh1 is a component of ginseng that activates estrogen receptor-dependent gene expression (Lee et al., 2003). Estrogen receptor activity is correlated with memory function. Short-term memory improves with elevated estrogen levels and declines when a decline in estrogen levels (Phillips and Sherwin, 1992a,b). At the cellular level, estrogen receptor activity promotes many memory-associated changes, including increases in dendritic spine density, NMDA receptor expression, and LTP magnitude (Smith and McMahon, 2005).

Ginseng may also boost memory function by affecting cholinergic signaling. Ginsenoside metabolites antagonize nicotinic acetylcholine receptors (nAChRs) (Tachikawa and Kudo, 2004). This could potentially enhance cognitive performance by quieting neuronal activity that may serve as a distractor, although this is as yet unfounded. Alternatively, reduced nAChR activity may suppress stress responses from the adrenal medulla. Chronic elevations in stress hormones (i.e., glucocorticoids) impede memory function by negatively impacting adult neurogenesis in the hippocampus (Cameron and Gould, 1994). Consistently, ginsenoside Rb1 improves cognitive function and the survival of newborn hippocampal neurons in adult rats (Liu et al., 2011). These data suggest that the antistress effects of ginseng enhance memory function by supporting adult hippocampal neurogenesis (Fig. 3.5).

Finally, it has been proposed that potential positive effects of ginseng may be the result of altered glucose homeostasis. In humans, a single dose of ginseng extract G115 lowers blood glucose levels while improving some aspects of cognitive function (Reay et al., 2005). The authors speculate that their reported cognitive improvements may be a consequence of increased glucose transport across the BBB, accounting for the decrease in blood glucose levels. Enhanced glucose uptake into the CNS is presumably requisite for elevated neuronal activity. However, the effects of ginseng on blood glucose levels may be independent of CNS uptake—oral administration of ginseng extracts increases hepatic glucose transporter expression in mice (Ohnishi et al., 1996). Without direct measures of CNS glucose uptake, this potential mechanism of cognitive enhancement will remain speculative.

Negative or null effects

In a randomized, double-blind, placebo-controlled trial, patients given ginseng compound HT008-1 showed no additional benefits on cognitive performance or quality of life over patients receiving placebo (Kim et al., 2008). Similar outcomes were reported in a separate trial—ginseng extract G115 had no additional benefits beyond the placebo control (Sünram-Lea et al., 2005). A single dose of ginseng offers no improvement in reaction time or divided attention in a driving task (LaSala et al., 2015).

A word of caution is offered regarding the reported positive effects of ginseng. For many of the cognitive measures, differences between baseline values between placebo and treatment groups could easily explain the small, but statistically significant, effects. Furthermore, performance was highly variable between outcome measures—ginseng improved some, diminished some, and had no effect on others. Overall, reported cognitive improvements are small, nondose-dependent, and inconsistent within and between studies.

Of note, ginseng may cause mild hypertension, coagulopathy, and hypoglycemia. In addition, the estrogenic effects of ginseng may adversely affect hormone-sensitive

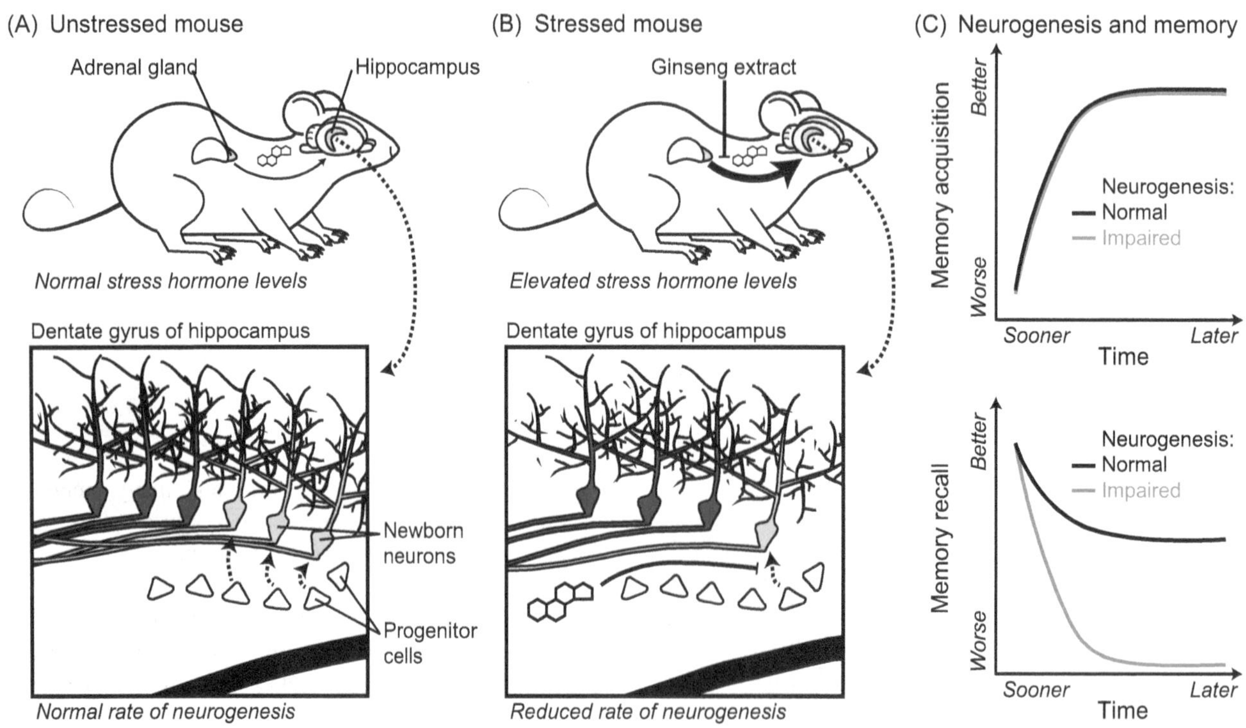

FIGURE 3.5 Chronic stress impedes memory function. (A) Adult neurogenesis occurs in discrete locations, such as the dentate gyrus of the hippocampus. The rate of neurogenesis is regulated by stress hormones (e.g., glucocorticoids) released from the adrenal gland. (B) Chronically elevated glucocorticoid levels impair neurogenesis in the dentate gyrus. The antistress effects of ginseng may promote memory function by preventing stress-induced decreases in adult neurogenesis. (C) Long-term memory requires adult neurogenesis. The learning curves (top) for mice with impaired neurogenesis (gray) are similar to those of normal mice (black). However, the long-term recall of learned information (bottom) is worsened when adult neurogenesis is impaired.

conditions such as certain types of breast cancer, endometriosis, and uterine fibroids (Coon and Ernst, 2002).

Phosphatidylserine

Phosphatidylserine (PS) is an acidic phospholipid found in the inner leaflet of the plasma membrane in animals, higher plants, and microorganisms. PS is derived from multiple sources, including bovine cortex (BC-PS), soybeans (soy-PS), and synthetic derivation. The amount of orally ingested PS that reaches the CNS is estimated to be a very small amount, with most being degraded rapidly in the intestine or by macrophages. Thus prolonged consumption of PS may be necessary to build appropriate levels for biological action.

Positive effects

There are limited data on the effects of PS on individuals with normal cognitive function (Table 3.6). In a 2-week trial with 14 college students, soy-PS (400 mg/day) reduced the correct calculation time on the Serial Subtraction Test (Parker et al., 2011). However, this difference is attributable to a single poor trial in the placebo group, likely due to small sample size, and no other differences were observed at other time points or in other outcome measures. In a larger trial with cognitively normal elderly subjects, a novel preparation of PS containing omega-3 fatty acids attached to its backbone improved a single psychometric (immediate recall on Rey AVLT) (Vakhapova et al., 2010). It remains unknown whether this small positive outcome is due to PS or the omega-3 fatty acids. Overall, the evidence for cognitive-enhancing effects of PS is dubious.

In randomized, double-blind, placebo-controlled trials, PS improves cognitive outcomes in patients with age-associated memory impairments. In a pilot study with 50- to 90-year-old patients with memory problems, 12-week treatment with soy-PS (300 mg/day) improved performance in the following cognitive parameters: memory recognition;

TABLE 3.6 Summary of selected studies on phosphatidylserine.

Cognitive status	Improvement	Reference	Outcome measures
Normal	Yes	Parker et al. (2011)	Correct calculation time on Serial Subtraction Test on a single time point
		Vakhapova et al. (2010)	Immediate recall on Rey AVLT
	No	Baumeister et al. (2008)	Stroop Color-Word Interference Test and D2 Concentration Test
		Parker et al. (2011)	Serial Subtraction Test (except correct calculation time on a single time point) and Profile of Mood States
		Vakhapova et al. (2010)	All other aspects of Rey AVLT and Rey Complex Figure Test
Mild impairment	Yes	Crook et al. (1991)	Name/face acquisition and delayed recall, facial recognition, telephone number recall, and misplaced objects recall
		Richter et al. (2013)	Memory recognition, memory recall, executive functions, mental flexibility, and Rey AVLT (total learning and immediate recall)
	No	Kato-Kataoka et al. (2010)	MMSE, Rivermead behavioral memory test, Hasegawa's dementia scale, everyday memory checklist
		Jorissen et al. (2001)	Visual Verbal Learning Test, Memory Scanning Test, Verbal Fluency Test, Stroop Color-Word Test, Signal Detection Test, Motor Choice Reaction Test, Concept Shifting Test, and Tower of London
Dementia	Yes	Engel et al. (1992)	CGIC
	No	Engel et al. (1992)	Modified Benton Test, associate learning, logical memory, TMT, Digit Symbol Substitution Test, mental arithmetic task, color-word-interference, and word fluency

AVLT, Rey Auditory Verbal Learning Test; *CGIC*, Clinical Global Impression of Change; *MMSE*, Mini-Mental State Exam; *TMT*, Trail-Making Test.

memory recall; executive functions; mental flexibility; and some components of the Rey AVLT (Richter et al., 2013). Similarly, BC-PS (100 mg three times per day) improved several psychometrics in patients with age-associated memory impairments over the course of 3, 6, and 12 weeks (Crook et al., 1991). However, the magnitude of effects in this study is difficult to determine because baseline values for each measurement were not published. PS is touted as an alternative treatment for Alzheimer patients, based on the association between aging and altered brain lipid composition. Although there are several studies that supported therapeutic benefits of PS in Alzheimer patients, there is no evidence that PS has any legitimate therapeutic efficacy in staving off dementia, aside from subjective improvements (Engel et al., 1992).

Potential mechanisms

PS may have general health benefits by improving blood lipid profile and decreasing cardiovascular risks. As noted with other nutraceuticals, PS may have antiinflammatory properties and reduce oxidative stress. Although these may stave off neural damage, these health benefits are not likely to provide cognitive enhancement per se.

The cognitive-restoring effects of PS may be due its ability to boost cholinergic function (Fig. 3.4). Treatment with BC-PS for 30 days restores ACh release in 24-month-old rats to levels observed in young rats (Pedata et al., 1985). Importantly, BC-PS does not affect release in young rats, suggesting that PS may only boost impaired cholinergic systems, such as those found in Alzheimer patients.

PS has a number of reported biochemical effects, including activation of tyrosine hydroxylase (an enzyme necessary for catecholamine synthesis), facilitated Ca^{2+} uptake, and increased glucose uptake in the CNS (cited in Pepeu et al., 1996). PS is a necessary cofactor for protein kinase C (PKC) activation: high-affinity Ca^{2+} binding to the regulatory domain requires interaction with PS, diacylglycerol, or phorbol esters (Luo et al., 1993). PKC increases the excitability of cholinergic interneurons by decreasing K^+ conductance through inactivating A-type K^+ channels (Deng et al., 2009). Because K^+ currents hyperpolarize neurons, PKC activation by PS should depolarize neurons and make it easier for them to fire action potentials in response to excitatory input.

Negative or null effects

As already mentioned, studies that reported positive effects of PS also reported null effects—in some cases the null effects vastly outnumbered the positive effects. A few other randomized, double-blind, placebo-controlled trials report no effects of PS on cognitive function. Over the course of 6 weeks, soy-PS (200 mg/day) had no effect on cognitive function in healthy subjects (Baumeister et al., 2008). In a trial involving patients with age-associated memory impairment, a 6- to 12-week treatment of soy-PS did not produce any significant improvement in performance over that of placebo control (Jorissen et al., 2001). Similarly, in a 6-month trial, treatment with soy-PS (100 mg or 300 mg/day) offered no additional improvement in cognitive function in subjects with mild cognitive impairment (Kato-Kataoka et al., 2010).

Oral administrations of PS preparations are generally considered safe. Adverse effects are rare and self-limiting, and may include excessive sweating, stomach upset, and diarrhea (Jorissen et al., 2002). A significant reduction in blood pressure has also been reported (Richter et al., 2013). Further, because of its potential anticoagulation properties, caution should be applied when taking PS together with other blood-thinning supplements or medications, such as gingko, kava, warfarin, and aspirin. Another potential serious health concern is that traditional PS preparations are made from bovine extracts, which may convey a risk of contracting prions disease—this can be avoided by using soy-based PS extracts.

Caffeine

Caffeine is a CNS stimulant found in many plants and has been consumed for thousands of years. Today, caffeine is the most widely consumed legal stimulant. Its presence is ubiquitous, from coffee and tea to products such as breath mints, lip balm, and shampoo. The per capita daily intake of caffeine in the US population has been estimated to be 3–7 mg/kg. As a reference, a 5-oz cup of coffee contains ~60–85 mg of caffeine (Wood et al., 2014).

Positive effects

Numerous double-blind, placebo-controlled, randomized trials support the beneficial effects of caffeine on acute cognitive performance (Table 3.7). Evidence on long-term cognition-enhancing effects of caffeine comes mainly from large population cohort studies. Although less convincing than randomized, controlled trial results, these cohort studies do support an association between long-term caffeine use and cognitive improvement.

Overall, results from double-blind, placebo-controlled randomized trials suggest that caffeine increases the rate of cognitive processing and information encoding. Caffeine administration enhances performance in the digit symbol substitution test and a tapping speed test. A lower dose (250 vs 500 mg) of caffeine led to performance enhancement, more favorable subjective effects (elation, pleasantness, peacefulness), and less subjective unfavorable effects (tension, nervousness, anxiety, excitement, irritability, nausea, palpitations, and restlessness)

TABLE 3.7 Summary of selected studies on caffeine.

Cognitive status	Improvement	Reference	Outcome measures
Normal	Yes	Kaplan et al. (1997)	Digit symbol substitution and tapping speed test
		Smith et al. (2013)	Reaction time and speed of encoding
		Haskell et al. (2005)	Simple reaction time, digit vigilance reaction time, numeric working memory reaction time, and sentence verification
		Van Boxtel et al. (2003)	Motor Choice Reaction Test
	No	Van Boxtel et al. (2003)	Visual Verbal Learning Test, Letter-Digit Substitution Test, Concept Shifting Test, and Stroop Color-Word Test
Mild impairment	Yes	van Gelder et al. (2007)	MMSE
		Ritchie et al. (2007)	Isaacs Set Test
		Nikić et al. (2014)	Montreal Cognitive Assessment Tool and subjective measures of vigilance and attention
		Vercambre et al. (2013)	A telephone adaptation of MMSE, immediate and delayed word recall, the East Boston Memory Test, and category fluency
	No	Ritchie et al. (2007)	MMSE and Benton Visual Retention Test
Dementia	Yes	Eskelinen and Kivipelto (2010)	Clinical examination
		Lindsay et al. (2002)	Clinical examination
	No	Laurin et al. (2004)	Clinical examination (tea drinkers)
		Dai et al. (2006)	Clinical examination (tea drinkers)
		Ritchie et al. (2007)	Clinical examination

MMSE, Mini-Mental State Exam.

(Kaplan et al., 1997). Caffeine ingestion is associated with faster encoding of new information and improved simple reaction time (Smith et al., 2013). Further, acute coffee consumption improves simple reaction time, digit vigilance reaction time, numeric working memory reaction time, and sentence verification accuracy irrespective of coffee use habits (Haskell et al., 2005).

Cohort studies suggest that consumption of caffeine, particularly coffee, may protect against cognitive decline. A large prospective cohort study investigated the association between habitual caffeine use and long-term cognitive decline in 2475 females older than 65 years of age with vascular diseases. These women were evaluated by four separate telephone cognitive assessments at 2-year intervals, assessing global cognition, verbal memory, and category fluency. Consumption of coffee, but not other caffeinated beverages, was associated with a slower rate of cognitive decline over 5 years compared to age-matched nonusers, particularly in women with stable diets (Vercambre et al., 2013). Another large prospective cohort study followed 4197 women and 2820 men recruited from three French cities at 2- and 4-year intervals. Women who consumed more than two cups of coffee per day were 33% less likely to have impairments in verbal retrieval and 22% less likely to have impairments in visuospatial memory than women who consumed one cup or less. No relation was found between caffeine intake and cognitive performance in men, or the risk of developing dementia in both genders (Ritchie et al., 2007). However, a longer 10-year longitudinal study on elderly European men (older than 75 years of age) reported that consumption of three cups of coffee per day was associated with reduced risk of cognitive decline (van Gelder et al., 2007). In a cross-sectional study of 86 hemodialysis patients, habitual caffeine consumption was associated with a lower rate (25% vs 16%) of cognitive impairment as measured by the Montreal Cognitive Assessment Tool. Caffeine users also reported subjective enhancement of vigilance and attention (Nikić et al., 2014). In contrast, a longitudinal study in the Netherlands found no association between caffeine intake and a reduced risk of cognitive decline (Van Boxtel et al., 2003). However, it should be noted that coffee and tea drinkers were grouped together in this study. Tea typically has a lower caffeine concentration than coffee, which may explain some of the differences noted between tea and coffee drinkers.

Many, but not all, studies suggest that drinking coffee may prevent the onset of AD. Adjusting for a variety of factors (i.e., age, gender, education, follow-up time, community of residence, midlife smoking, systolic blood pressure, serum total cholesterol, body mass index, and physical activity), a large cohort study in Eastern Finland concluded that drinking three to five cups of coffee in midlife was associated with an ~65% decreased risk of developing dementia or AD (Eskelinen and Kivipelto, 2010). A population-based study on Canadian seniors (older than 70 years of age) reported several lifestyle factors that are associated with a reduced risk of AD; included among these was the consumption of coffee, but not tea (Lindsay et al., 2002). Two other longitudinal studies on elderly individuals in the United States had similar conclusions: tea drinking is not associated with reduced risk of dementia or AD (Laurin et al., 2004; Dai et al., 2006). However, when the amount of caffeine in each beverage is taken into account, drinking the equivalent of three cups of coffee is not always associated with a lower risk of dementia (Ritchie et al., 2007).

In summary, drinking three to five cups of coffee per day acutely boosts cognitive performance and mood. The findings from long-term studies provide favorable support for coffee's ability to prevent cognitive decline, although perhaps not dementia—all the more reason to have your morning cup(s) of coffee.

Potential mechanisms

Caffeine is a nonselective adenosine receptor antagonist, acting on A_1 and A_{2A} receptor subtypes (Fig. 3.6). Adenosine receptors are G-protein-coupled receptors expressed in a variety of cell types. These two adenosine receptors have opposing effects on intracellular cyclic adenosine monophosphate (cAMP) levels, with A_1 receptors ($G_{i/o}$-coupled) inhibiting and A_{2A} receptors (G_s-coupled) stimulating adenylyl cyclase—an enzyme that synthesizes cAMP from adenosine triphosphate (ATP). Caffeine also affects cAMP levels by inhibiting phosphodiesterase (PDE), an enzyme that degrades cyclic nucleotides. Thus caffeine can boost cAMP levels by inhibiting either A_1 receptors or PDE. Inhibition of A_1 receptors indirectly activates adenylyl cyclase and elevates cAMP levels, whereas inhibition of PDE reduces the degradation of cAMP. Conversely, caffeine will have opposite effects on cAMP levels in cells that express higher levels of A_{2A} receptors.

cAMP is a ubiquitous secondary messenger that affects neuronal function rapidly by changing ionic conductances and over longer time periods by altering

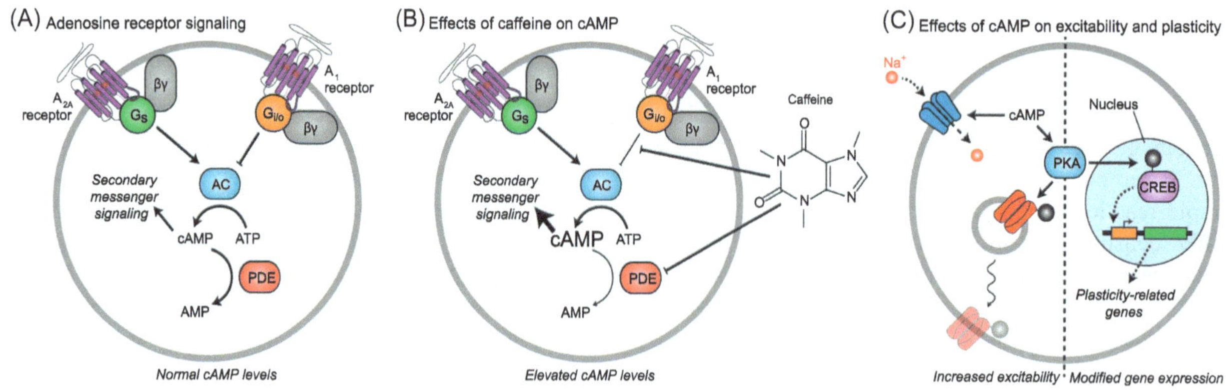

FIGURE 3.6 Neuronal activity is regulated by cyclic adenosine monophosphate (cAMP) and Ca^{2+} levels. (A) Adenosine receptors regulate intracellular cAMP levels. A_{2A} and A_1 adenosine receptors stimulate and inhibit adenylyl cyclase (AC), respectively. AC synthesizes cAMP from adenosine triphosphate. cAMP is an important secondary messenger [see (C)]; therefore its levels are tightly regulated. cAMP is degraded by phosphodiesterase (PDE) to form AMP. (B) Caffeine boosts intracellular cAMP levels by inhibiting adenosine receptors and PDE. (C) cAMP increases neuronal excitability and modifies gene expression. cAMP activates cyclic nucleotide-gated cation channels to depolarize neurons and increase their excitability. cAMP also facilitates glutamatergic signaling by directly activating protein kinase A (PKA). PKA phosphorylates intracellular AMPA receptors held in reserve and targets them to the surface. PKA also produces more long-term changes in cell function by activating the transcription factor cAMP response element binding protein, which promotes a number of plasticity-relevant changes in gene expression.

intracellular signaling cascades. Caffeine may facilitate synaptic transmission by increasing neuronal excitability and facilitating presynaptic release. Neuronal excitability is enhanced by cAMP through activation of cyclic nucleotide-gated (CNG) channels. CNG channels are cation channels that open more frequently when cAMP levels are high. Through activation of CNG channels, caffeine can depolarize neurons and facilitate the firing of action potentials. In addition to facilitating action potentials, caffeine may also promote neurotransmitter release by antagonizing A_1 receptors, which normally suppress neurotransmitter release at presynaptic sites. Consistently, inhibition of A_1 receptors facilitates synaptic plasticity (a hallmark of which is increased neurotransmitter release probability) (de Mendonca and Ribeiro, 2000). These effects on synaptic plasticity are probably a consequence of downstream signaling cascades activated by cAMP. cAMP activates protein kinase A (PKA) and the transcription factor cAMP response element binding protein (CREB). Activations of both PKA and CREB play an important role in SPM function (Frey et al., 1993; Silva et al., 1998). Briefly, PKA phosphorylates a number of synaptic proteins to boost the strength of synaptic signaling, whereas CREB alters gene expression to increase synaptic strength over long time periods.

The association of coffee consumption and reduced risk of cognitive decline may be linked to the effects of caffeine on A_{2A} receptors. Memory dysfunction has been linked to elevated A_{2A} receptor activity in astrocytes (Orr et al., 2015). Astrocytic A_{2A} receptors are upregulated in Alzheimer patients, correlating with disease severity. Interestingly, genetic downregulation of A_{2A} receptors enhances memory function in mice, whereas upregulation of astrocytic G_s-coupled signaling impairs memory function. Based on this, it may be the case that caffeine enhances memory function and prevents dementia onset by attenuating astrocytic A_{2A} receptor activity. Given their intimate association with the cerebral vasculature, astrocytes are prime targets for bloodborne agents, such as caffeine or other nutraceuticals.

Negative or null effects

Based on an extensive review of available human studies on the health effects of caffeine, caffeine consumption up to 6 mg/kg per day was considered safe, revealing no adverse effects on a number of health measures, including cardiovascular health, cancer incidence, calcium balance, and male fertility. However, there was inconclusive literature suggesting potential adverse effects on female fertility with caffeine consumption of more than 4.6 mg/kg per day (Nawrot et al., 2003). Higher doses of caffeine (>10 mg/kg per day) may lead to significant tachycardia, hypertension, and arrhythmia. Assuming normal dosing amount and time, caffeine use is not expected to produce adverse events.

Although moderate caffeine consumption is generally regarded as safe, whether caffeine has legitimate cognitive-enhancing effects is a matter of debate. The "withdrawal reversal hypothesis" suggested that caffeine does not enhance cognition or attention; instead, it reverses the negative effects of caffeine withdrawal in those who typically consume caffeine daily. Studies on this subject were inconclusive and showed conflicting results (Wood et al., 2014). Further, some studies discussed here included individuals who did not consume caffeine regularly.

Finally, not all caffeine preparations are equal. For example, most longitudinal studies find that daily consumption of coffee, but not tea, is associated with a decreased risk of developing dementia or AD. Given that tea generally has a lower concentration of caffeine than coffee, it is possible that tea drinkers do not receive an adequate dose of caffeine. However, the possibility exists that coffee contains other nutraceuticals that account for its apparent ability to stave off cognitive decline.

Centella asiatica, Ephedra sinica, and Crocus sativus

Not all nutraceuticals have received equal attention due to lack of scientific studies and/or clinical trials. For example, *Centella asiatica* (also called gotu kola) is used in traditional medicine and in the cuisine of a variety of Asian countries. There is only one clinical trial in humans studying the effect of *C. asiatica* on cognitive performance. Although the trial was suggestive of subjective mood improvement and enhanced working memory in the elderly, the trial was not randomized, blinded, or placebo-controlled, lending questionable credibility to the results (Wattanathorn et al., 2008).

Ephedra sinica has been used in Chinese medicine for thousands of years. The major active ingredients are ephedrine and pseudoephedrine, stimulants that raise blood pressure and heart rate. Ephedrine has not been studied systematically and has no proven benefit in terms of cognitive enhancement. However, ephedrine is widely used to "maintain focus," which could prevent a decline in cognitive performance during extended periods of mental activity. Although this effect of ephedrine is largely anecdotal, some peer-reviewed studies have looked at the effects of ephedrine on cognition. A single 60 mg dose of ephedrine maintains focus during a driving-related task (Moolenaar et al., 1999). Ephedrine treatment also shows some signs of improving attention in patients with Korsakoff's psychosis, although memory function was unaffected (Mair and McEntee, 1986). In a double-blind, placebo-controlled trial with normal adults, ephedrine improved performance on a spatial memory task, worsened performance on a verbal memory task, and had no effect on cognitive flexibility or problem-solving (Choi et al., 2006). However, sample sizes were small, with each group comprising only six subjects. In a separate double-blind, placebo-controlled trial, ephedrine had no effect on a variety of cognitive measures (Beversdorf et al., 1999). Given the numerous reported adverse events and ban on ephedra-containing supplements by the US Food and Drug Administration, *E. sinica* is not an attractive candidate for cognitive enhancement.

Crocus sativus stigmas, widely known as saffron (a spice), have been used in traditional medicine against AD for a long time. Two separate trials on memory function in AD patients reported improvements in memory function that were comparable to those of cholinesterase inhibitors or memantine (Akhondzadeh et al., 2010; Farokhnia et al., 2014). Indeed, saffron extract inhibits cholinesterase activity in vitro (Geromichalos et al., 2012). Saffron extract also exhibits antioxidant properties and inhibits amyloid beta-peptide (Abeta) fibrillogenesis in a concentration-dependent and time-dependent manner (Papndreou et al., 2006, 2011). A small randomized, controlled trial with patients undergoing coronary artery bypass grafting reported no effect of saffron on cognitive function or mood (Moazen-Zadeh et al., 2018).

Potential concerns of cognitive enhancement

Even in the absence of unintentional, adverse side effects, several other shortcomings must be overcome to effectively enhance cognitive function. The first shortcoming is that of spatial inaccuracy. Peripherally administered drugs are applied globally to all brain regions. Therefore drugs that target ubiquitous signaling molecules expressed throughout the entire brain are unlikely to affect select cognitive functions. This problem will require more selective drug-delivery means in the CNS or more selective targets for pharmaceutical manipulation. The second shortcoming is that of temporal inaccuracy. The timing of neuronal activity is critical for proper memory function. As already discussed, coincident activation of two or more neurons is required to induce synaptic plasticity. In fact, proper memory function requires an ordered series of molecular and cellular changes. As an example, mice that lack translation repressor 4E-BP2 have enhanced long-term synaptic plasticity (Banko et al., 2005). Despite facilitated synaptic plasticity, 4E-BP2 knockout mice have impaired memory function. In other words, one cannot simply turn on memory-promoting pathways and expect enhanced memory function.

More efficient and reliable information encoding would very likely advance the rate of human progress, particularly in any field of research and development. However, not all memories are good. An important aspect of learning and education is the forgetting of incorrect ideas as anyone learns new and correct information. In addition, there is no reason to expect memory-enhancing agents to selectively facilitate the encoding of pleasant memories. For any person with normal cognitive function, cognitive enhancers would be elective medical treatments. Given the expected out-of-pocket cost, it is far more likely that wealthier individuals will have access to cognitive-enhancing agents, further widening the gap between rich and poor.

Finally, it is important to evaluate whether the potential benefits of cognitive enhancement outweigh the risks. When specific signaling pathways are activated, cells respond by inhibiting the activated pathway to maintain homeostasis. Along these lines, neurons tightly regulate their excitability to maintain a healthy level of activity. By manipulating pharmacologically, upregulating or downregulating the specific signaling pathways in the brain, one can expect the brain to respond with the corresponding opposite effect. Therefore we may become dependent on the cognitive-enhancing agents to restore normal function. This issue is less concerning to someone with dementia, whose daily function is already compromised due to inadequate cognitive function. It is in this case when cognitive-enhancing agents have the most promise and chance for success.

Concluding remarks and future directions

Not surprisingly, the nutraceuticals covered here have diverse effects and mechanisms. *G. biloba* extracts may provide small improvements in cognitively impaired individuals, although the majority of studies suggest no effect. *G. biloba* may boost monoaminergic and serotonergic signaling. *B. monnieri* extracts may be useful in restoring some degree of cognitive function in impaired individuals, although the majority of studies suggest no effect. There is some evidence that *B. monnieri* facilitates cerebral blood flow, neuronal activity, and synaptic transmission. Ginseng may subtly improve normal cognitive function through its estrogenic or antistress functions. PS may restore cognitive function in impaired individuals, particularly patients with AD, by boosting cholinergic function. Caffeine acutely enhances cognitive function and may stave off mental decline. These protective effects were reported for individuals who drank three to five cups of coffee per day. The effects of caffeine are likely due to their inhibitory effects of adenosine A_1 and A_{2A} receptors.

Of all of the nutraceuticals covered here, caffeine appears to have the most promise in terms of consistent enhancement, safety, and ease of consumption.

References

Adolphs, R., Tranel, D., Damasio, H., 1994. Impaired recognition of emotion in facial expressions following bilateral damage to the human amygdala. Nature 372, 669–672.

Adolphs, R., Cahill, L., Schul, R., 1997. Impaired declarative memory for emotional material following bilateral amygdala damage in humans. Learn. Mem. 4, 291–300.

Akhondzadeh, S., Sabet, M.S., Harirchian, M.H., et al., 2010. A 22-week, multicenter, randomized, double-blind controlled trial of *Crocus sativus* in the treatment of mild-to-moderate Alzheimer's disease. Psychopharmacology 207, 637–643.

Austin, M.P., Mitchell, P., Goodwin, G.M., 2001. Cognitive deficits in depression: possible implications for functional neuropathology. Br. J. Psychiatry 178, 200–206.

Banko, J.L., Poulin, F., Hou, L., 2005. The translation repressor 4E-BP2 is critical for eIF4F complex formation, synaptic plasticity, and memory in the hippocampus. J. Neurosci. 25, 9581–9590.

Baumeister, J., Barthel, T., Geiss, K.R., 2008. Influence of phosphatidylserine on cognitive performance and cortical activity after induced stress. Nutr. Neurosci. 11, 103–110.

Beversdorf, D.Q., Hughes, J.D., Steinberg, B.A., 1999. Noradrenergic modulation of cognitive flexibility in problem solving. Neuroreport 10, 2763–2767.

Birks, J., Grimley, E.V., Van Dongen, M., 2002. *Ginkgo biloba* for cognitive impairment and dementia. Cochrane Database Syst. Rev. 4.

Brautigam, M.R., Blommaert, F.A., Verleye, G., 1998. Treatment of age-related memory complaints with *Ginkgo biloba* extract: a randomized double blind placebo-controlled study. Phytomedicine 5, 425–434.

Calabrese, C., Gregory, W.L., Leo, M., 2008. Effects of a standardized *Bacopa monnieri* extract on cognitive performance, anxiety, and depression in the elderly: a randomized, double-blind, placebo-controlled trial. J. Altern. Complement. Med. 14, 707–713.

Cameron, H.A., Gould, E., 1994. Adult neurogenesis is regulated by adrenal steroids in the dentate gyrus. Neuroscience 61, 203–209.

Canevelli, M., Adali, N., Kelaiditi, E., 2014. Effects of *Ginkgo biloba* supplementation in Alzheimer's disease patients receiving cholinesterase inhibitors: data from the ICTUS study. Phytomedicine 21, 888–892.

Chen, F., Li, L., Xu, F., 2013. Systemic and cerebral exposure to and pharmacokinetics of flavonols and terpene lactones after dosing standardized *Ginkgo biloba* leaf extracts to rats via different routes of administration. Br. J. Pharmacol. 170, 440–457.

Choi, Y., Novak, J.C., Hillier, A., 2006. The effect of alpha-2 adrenergic agonists on memory and cognitive flexibility. Cogn. Behav. Neurol. 19, 204–207.

Cieza, A., Maier, P., Pöppel, E., 2003. Effects of *Ginkgo biloba* on mental functioning in healthy volunteers. Arch. Med. Res. 34, 373–381.

Coon, J.T., Ernst, E., 2002. *Panax ginseng*: a systematic review of adverse effects and drug interactions. Drug. Saf. 25, 323–344.

Crook, T.H., Tinklenberg, J., Yesavage, J., 1991. Effects of phosphatidylserine in age-associated memory impairment. Neurology 41, 644–649.

D'Angelo, L., Grimaldi, R., Caravaggi, M., 1986. A double-blind, placebo-controlled clinical study on the effect of a standardized ginseng extract on psychomotor performance in healthy volunteers. J. Ethnopharmacol. 16, 15–22.

Dai, Q., Borenstein, A.R., Wu, Y., 2006. Fruit and vegetable juices and Alzheimer's disease: the Kame Project. Am. J. Med. 119, 751–759.

Danysz, W., Parsons, C.G., Möbius, H.J., 2000. Neuroprotective and symptomatological action of memantine relevant to Alzheimer's disease—a unified glutamatergic hypothesis on the mechanism of action. Neurotoxicol. Res. 2, 85–97.

Das, A., Shanker, G., Nath, C., 2002. A comparative study in rodents of standardized extracts of *Bacopa monniera* and *Ginkgo biloba*: anticholinesterase and cognitive enhancing activities. Pharmacol. Biochem. Behav. 73, 893–900.

de Mendonca, A., Ribeiro, J.A., 2000. Long-term potentiation observed upon blockade of adenosine A1 receptors in rat hippocampus is *N*-methyl-D-aspartate receptor-dependent. Neurosci. Lett. 291, 81–84.

DeFeudis, F.V., Drieu, K., 2000. *Ginkgo biloba* extract (EGb 761) and CNS functions: basic studies and clinical applications. Curr. Drug Targets 1, 25–58.

DeKosky, S.T., Williamson, J.D., Fitzpatrick, A.L., et al., 2008. *Ginkgo biloba* for prevention of dementia: a randomized controlled trial. JAMA 300, 2253–2262.

Dekundy, A., 2006. Coadministration of memantine with acetylcholinesterase inhibitors: preclinical and clinical evidence. In: Gupta, R.C. (Ed.), Toxicology of Organophosphate and Carbamate Compounds. Academic Press/Elsevier, Amsterdam, pp. 35–46.

Deng, P., Pang, Z.-P., Lei, Z., Xu, Z.C., 2009. Excitatory roles of protein kinase C in striatal cholinergic interneurons. J. Neurophysiol. 102, 2453–2461.

Engel, R.R., Satzger, W., Günther, W., 1992. Double-blind cross-over study of phosphatidylserine vs. placebo in patients with early dementia of the Alzheimer type. Eur. Neuropsychopharmacol. 2, 149–155.

Eskelinen, M.H., Kivipelto, M., 2010. Caffeine as a protective factor in dementia and Alzheimer's disease. J. Alzheimers Dis. 20 (Suppl. 1), S167–S174.

Farlow, M.R., Cummings, J.L., 2007. Effective pharmacologic management of Alzheimer's disease. Am. J. Med. 120, 388–397.

Farokhnia, M., Shafiee Sabet, M., Iranpour, N., et al., 2014. Comparing the efficacy and safety of *Crocus sativus* L. with memantine in patients with moderate to severe Alzheimer's disease: a double-blind randomized clinical trial. Hum. Psychopharmacol. 29, 351–359.

Frankland, P.W., Bontempi, B., 2005. The organization of recent and remote memories. Nat. Rev. Neurosci. 6, 119–130.

Frey, U., Huang, Y.Y., Kandel, E.R., 1993. Effects of cAMP simulate a late stage of LTP in hippocampal CA1 neurons. Science 260, 1661–1664.

Fuster, J.M., Alexander, G.E., 1971. Neuron activity related to short-term memory. Science 173, 652–654.

Gavrilova, S.I., Preuss, U.W., Wong, J.W.M., 2014. Efficacy and safety of *Ginkgo biloba* extract EGb 761 in mild cognitive impairment with neuropsychiatric symptoms: a randomized, placebo-controlled, double-blind, multi-center trial. Int. J. Geriatr. Psychiatry 29, 1087–1095.

Geinisman, Y., Ganeshina, O., Yoshida, R., et al., 2004. Aging, spatial learning, and total synapse number in the rat CA1 stratum radiatum. Neurobiol. Aging 25, 407–416.

Geromichalos, G.D., Lamari, F.N., Papndreou, M.A., 2012. Saffron as a source of novel acetylcholinesterase inhibitors: molecular docking and in vitro enzymatic studies. J. Agric. Food Chem. 60, 6131–6138.

Haskell, C.F., Kennedy, D.O., Wesnes, K.A., 2005. Cognitive and mood improvements of caffeine in habitual consumers and habitual non-consumers of caffeine. Psychopharmacology (Berl.) 179, 813–825.

Hebb, D.O., 1949. The First Stage of Perception: Growth of the Assembly. The Organization of Behavior. John Wiley and Sons, New York.

Hosamani, R., Muralidhara, 2009. Neuroprotective efficacy of *Bacopa monnieri* against rotenone induced oxidative stress and neurotoxicity in *Drosophila melanogaster*. Neurotoxicology 30, 977–985.

Jorissen, B.L., Brouns, F., Van Boxtel, M.P., 2001. The influence of soy-derived phosphatidylserine on cognition in age-associated memory impairment. Nutr. Neurosci. 4, 121–134.

Jorissen, B.L., Brouns, F., Van Boxtel, M.P.J., 2002. Safety of soy-derived phosphatidylserine in elderly people. Nutr. Neurosci. 5, 337–343.

Kalra, E.K., 2003. Nutraceutical—definition and introduction. AAPS PharmSci. 5, E25.

Kamkaew, N., Scholfield, C.N., Ingkaninan, K., 2011. *Bacopa monnieri* and its constituents is hypotensive in anaesthetized rats and vasodilator in various artery types. J. Ethnopharmacol. 137, 790–795.

Kaplan, G.B., Greenblatt, D.J., Ehrenberg, B.L., 1997. Dose-dependent pharmacokinetics and psychomotor effects of caffeine in humans. J. Clin. Pharmacol. 37, 693–703.

Kato-Kataoka, A., Sakai, M., Ebina, R., Nonaka, C., 2010. Soybean-derived phosphatidylserine improves memory function of the elderly Japanese subjects with memory complaints. J. Clin. Biochem. Nutr. 47, 246–255.

Kennedy, D.O., Scholey, A.B., Wesnes, K.A., 2000. The dose-dependent cognitive effects of acute administration of *Ginkgo biloba* to healthy young volunteers. Psychopharmacology (Berl.) 151, 416–423.

Kennedy, D.O., Scholey, A.B., Wesnes, K.A., 2001. Dose dependent changes in cognitive performance and mood following acute administration of Ginseng to healthy young volunteers. Nutr. Neurosci. 4, 295–310.

Kennedy, D.O., Scholey, A.B., Wesnes, K.A., 2002. Modulation of cognition and mood following administration of single doses of *Ginkgo biloba*, ginseng, and a ginkgo/ginseng combination to healthy young adults. Physiol. Behav. 75, 739–751.

Kim, J., Chung, S.Y., Park, S., 2008. Enhancing effect of HT008-1 on cognitive function and quality of life in cognitively declined healthy adults: a randomized, double-blind, placebo-controlled, trial. Pharmacol. Biochem. Behav. 90, 517–524.

Kubota, K., Niki, H., 1971. Prefrontal cortical unit activity and delayed alternation performance in monkeys. J. Neurophysiol. 34, 337–347.

Lamprecht, R., LeDoux, J., 2004. Structural plasticity and memory. Nat. Rev. Neurosci. 5, 45–54.

LaSala, G.S., McKeever, R.G., Patel, U., Okaneku, J., Vearrier, D., Greenberg, M. I., 2015. Effect of single-dose *Ginkgo biloba* and *Panax ginseng* on driving performance. Clin. Toxicol. (Phila.) 53, 108–112.

Laurin, D., Masaki, K.H., Foley, D.J., 2004. Midlife dietary intake of antioxidants and risk of late-life incident dementia: the Honolulu-Asia Aging Study. Am. J. Epidemiol. 159, 959–967.

Lee, Y., Jin, Y., Lim, W., 2003. A ginsenoside-Rh1, a component of ginseng saponin, activates estrogen receptor in human breast carcinoma MCF-7 cells. J. Steroid Biochem. Mol. Biol. 84, 463–468.

Li, S., Luo, J., Wang, X., 2011. Effects of *Ginkgo biloba* extracts on NMDA-activated currents in acutely isolated hippocampal neurons of the rat. Phytother. Res. 25, 137–141.

Lindsay, J., Laurin, D., Verreault, R., 2002. Risk factors for Alzheimer's disease: a prospective analysis from the Canadian Study of Health and Aging. Am. J. Epidemiol. 156, 445–453.

Liu, L., Hoang-Gia, T., Wu, H., 2011. Ginsenoside Rb1 improves spatial learning and memory by regulation of cell genesis in the hippocampal subregions of rats. Brain Res. 1382, 147–154.

Luo, J.H., Kahn, S., O'Driscoll, K., 1993. The regulatory domain of protein kinase C beta 1 contains phosphatidylserine- and phorbol ester-dependent calcium binding activity. J. Biol. Chem. 268, 3715–3719.

Mair, R.G., McEntee, W.J., 1986. Cognitive enhancement in Korsakoff's psychosis by clonidine: a comparison with l-dopa and ephedrine. Psychopharmacology (Berl.) 88, 374–380.

Martin, S.J., Grimwood, P.D., Morris, R.G., 2000. Synaptic plasticity and memory: an evaluation of the hypothesis. Annu. Rev. Neurosci. 23, 649–711.

McCarney, R., Fisher, P., Iliffe, S., 2008. *Ginkgo biloba* for mild to moderate dementia in a community setting: a pragmatic, randomised, parallel-group, double-blind, placebo-controlled trial. Int. J. Geriatr. Psychiatry 23, 1222–1230.

Moazen-Zadeh, E., Abbasi, S.H., Safi-Aghdam, H., et al., 2018. Effects of saffron on cognition, anxiety, and depression in patients undergoing coronary artery bypass grafting: a randomized double-blind placebo-controlled trial. J. Altern. Complement. Med. 24, 361–368.

Moolenaar, M., Desmond, P.A., Mascord, D.J., 1999. The effects of ephedrine on the development of fatigue in a prolonged driving-related task. Hum. Psychopharmacol. 14, 415–427.

Morgan, A., Stevens, J., 2010. Does *Bacopa monnieri* improve memory performance in older persons? Results of a randomized, placebo-controlled, double-blind trial. J. Altern. Complement. Med. 16, 753–759.

Morris, R.G., Garrud, P., Rawlins, J.N., 1982. Place navigation impaired in rats with hippocampal lesions. Nature 297, 681–683.

Moser, M.B., Trommald, M., Andersen, P., 1994. An increase in dendritic spine density on hippocampal CA1 pyramidal cells following spatial learning in adult rats suggests the formation of new synapses. Proc. Natl. Acad. Sci. U.S. A. 91, 12673–12675.

Nathan, P.J., Clarke, J., Lloyd, J., 2001. The acute effects of an extract of *Bacopa monniera* (Brahmi) on cognitive function in healthy normal subjects. Hum. Psychopharmacol. 16, 345–351.

Nawrot, P., Jordan, S., Eastwood, J., 2003. Effects of caffeine on human health. Food Addit. Contam. 20, 1–30.

Neri, M., Andermarcher, E., Pradelli, J.M., 1995. Influence of a double blind pharmacological trial on two domains of well-being in subjects with age associated memory impairment. Arch. Gerontol. Geriatr. 21, 241–252.

Nikić, P.M., Andrić, B.R., Stojimirović, B.B., 2014. Habitual coffee consumption enhances attention and vigilance in hemodialysis patients. Biomed. Res. Int. 2014, 707460.

Ohnishi, Y., Takagi, S., Miura, T., 1996. Effect of ginseng radix on GLUT2 protein content in mouse liver in normal and epinephrine-induced hyperglycemic mice. Biol. Pharm. Bull. 19, 1238–1240.

Oken, B.S., Storzbach, D.M., Kaye, J.A., 1998. The efficacy of *Ginkgo biloba* on cognitive function in Alzheimer disease. Arch. Neurol. 55, 1409–1415.

Orr, A.G., Hsiao, E.C., Wang, M.M., 2015. Astrocytic adenosine receptor A_{2A} and G_s-coupled signaling regulate memory. Nat. Neurosci. 18, 423–434.

Papndreou, M.A., Kanakis, C.D., Polissiou, M.G., 2006. Inhibitory activity on amyloid-beta aggregation and antioxidant properties of *Crocus sativus* stigmas extract and its crocin constituents. J. Agric. Food Chem. 54, 8762–8768.

Papndreou, M.A., Tsachakis, M., Efthimiopoulos, S., 2011. Memory enhancing effects of saffron in aged mice are correlated with antioxidant protection. Behav. Brain Res. 219, 197–204.

Pardridge, W.M., 2012. Drug transport across the blood-brain barrier. J. Cereb. Blood Flow Metab. 32, 1959–1972.

Parker, A.G., Gordon, J., Thornton, A., 2011. The effects of IQPLUS Focus on cognitive function, mood and endocrine response before and following acute exercise. J. Int. Soc. Sports Nutr. 8, 16.

Pedata, F., Giovannelli, L., Spignoli, G., 1985. Phosphatidylserine increases acetylcholine release from cortical slices in aged rats. Neurobiol. Aging 6, 337–339.

Pepeu, G., Pepeu, I.M., Amaducci, L., 1996. A review of phosphatidylserine pharmacological and clinical effects. Is phosphatidylserine a drug for the ageing brain? Pharmacol. Res. 33, 73–80.

Phillips, S.M., Sherwin, B.B., 1992a. Variations in memory function and sex steroid hormones across the menstrual cycle. Psychoneuroendocrinology 17, 497–506.

Phillips, S.M., Sherwin, B.B., 1992b. Effects of estrogen on memory function in surgically menopausal women. Psychoneuroendocrinology 17, 485–495.

Pravina, K., Ravindra, K.R., Goudar, K.S., et al., 2007. Safety evaluation of BacoMind in healthy volunteers: a phase I study. Phytomedicine 14, 301–308.

Rajan, K.E., Singh, H.K., Parkavi, A., 2011. Attenuation of 1-(m-chlorophenyl)-biguanide induced hippocampus-dependent memory impairment by a standardised extract of *Bacopa monniera* (BESEB CDRI-08). Neurochem. Res. 36, 2136–2144.

Reay, J.L., Kennedy, D.O., Scholey, A.B., 2005. Single doses of *Panax ginseng* (G115) reduce blood glucose levels and improve cognitive performance during sustained mental activity. J. Psychopharmacol. 19, 357–365.

Reay, J.L., Kennedy, D.O., Scholey, A.B., 2006. Effects of *Panax ginseng*, consumed with and without glucose, on blood glucose levels and cognitive performance during sustained "mentally demanding" tasks. J. Psychopharmacol. 20, 771–781.

Richter, Y., Herzog, Y., Lifshitz, Y., 2013. The effect of soybean-derived phosphatidylserine on cognitive performance in elderly with subjective memory complaints: a pilot study. Clin. Interv. Aging 8, 557–563.

Ritchie, K., Carrière, I., de Mendonca, A., 2007. The neuroprotective effects of caffeine: a prospective population study (the Three City Study). Neurology 69, 536–545.

Roodenrys, S., Booth, D., Bulzomi, S., 2002. Chronic effects of Brahmi (*Bacopa monnieri*) on human memory. Neuropsychopharmacology 27, 279–281.

Saraf, M.K., Prabhakar, S., Anand, A., 2010. Neuroprotective effect of *Bacopa monniera* on ischemia induced brain injury. Pharmacol. Biochem. Behav. 97, 192–197.

Scoville, W.B., Milner, B., 1957. Loss of recent memory after bilateral hippocampal lesions. J. Neurol. Neurosurg. Psychiatry 20, 11–21.

Silva, A.J., Kogan, J.H., Frankland, P.W., 1998. CREB and memory. Annu. Rev. Neurosci. 21, 127–148.

Smith, C.C., McMahon, L.L., 2005. Estrogen-induced increase in the magnitude of long-term potentiation occurs only when the ratio of NMDA transmission to AMPA transmission is increased. J. Neurosci. 25, 7780–7791.

Smith, A.P., Christopher, G., Sutherland, D., 2013. Acute effects of caffeine on attention: a comparison of non-consumers and withdrawn consumers. J. Psychopharmacol. 27, 77–83.

Snitz, B.E., O'Meara, E.S., Carlson, M.C., 2009. *Ginkgo biloba* for preventing cognitive decline in older adults: a randomized trial. JAMA 302, 2663–2670.

Squire, L.R., 1992. Memory and the hippocampus: a synthesis from findings with rats, monkeys, and humans. Psychol. Rev. 99, 195–231.

Stough, C., Lloyd, J., Clarke, J., 2001. The chronic effects of an extract of *Bacopa monniera* (Brahmi) on cognitive function in healthy human subjects. Psychopharmacology (Berl.) 156, 481–484.

Stough, C., Downey, L.A., Lloyd, J., et al., 2008. Examining the nootropic effects of a special extract of *Bacopa monniera* on human cognitive functioning: 90

day double-blind placebo-controlled randomized trial. Phytother. Res. 22, 1629–1634.

Sünram-Lea, S.I., Birchall, R., Wesnes, K.A., 2005. The effect of acute administration of 400 mg of *Panax ginseng* on cognitive performance and mood in healthy young volunteers. Curr. Top. Nutraceutical Res. 3, 65–74.

Tachikawa, E., Kudo, K., 2004. Proof of the mysterious efficacy of ginseng: basic and clinical trials: suppression of adrenal medullary function *in vitro* by ginseng. J. Pharmacol. Sci. 95, 140–144.

Terry, R.D., Masliah, E., Salmon, D.P., 1991. Physical basis of cognitive alterations in Alzheimer's disease: synapse loss is the major correlate of cognitive impairment. Ann. Neurol. 30, 572–580.

Thippeswamy, A.H., Rafiq, M., Viswantha, G.L.S., 2013. Evaluation of *Bacopa monniera* for its synergistic activity with rivastigmine in reversing aluminum-induced memory loss and learning deficit in rats. J. Acupunct. Meridian Stud. 6, 208–213.

Vakhapova, V., Cohen, T., Richter, Y., 2010. Phosphatidylserine containing omega-3 fatty acids may improve memory abilities in non-demented elderly with memory complaints: a double-blind placebo-controlled trial. Dement. Geriatr. Cogn. Disord. 29, 467–474.

van Boxtel, M.P.J., Schmitt, J.A.J., Bosma, H., 2003. The effects of habitual caffeine use on cognitive change: a longitudinal perspective. Pharmacol. Biochem. Behav. 75, 921–927.

van Gelder, B.M., Buijsse, B., Tijhuis, M., 2007. Coffee consumption is inversely associated with cognitive decline in elderly European men: the FINE Study. Eur. J. Clin. Nutr. 61, 226–232.

Velaga, M.K., Basuri, C.K., Robinson Taylor, K.S., 2014. Ameliorative effects of *Bacopa monniera* on lead-induced oxidative stress in different regions of rat brain. Drug. Chem. Toxicol. 37, 357–364.

Vercambre, M.-N., Berr, C., Ritchie, K., Kang, J.H., 2013. Caffeine and cognitive decline in elderly women at high vascular risk. J. Alzheimers Dis. 35, 413–421.

Vetere, G., Restivo, L., Cole, C.J., 2011. Spine growth in the anterior cingulate cortex is necessary for the consolidation of contextual fear memory. Proc. Natl. Acad. Sci. 108, 8456–8460.

Wattanathorn, J., Mator, L., Muchimapura, S., 2008. Positive modulation of cognition and mood in the healthy elderly volunteer following the administration of *Centella asiatica*. J. Ethnopharmacol. 116, 325–332.

Weichel, O., Hilgert, M., Chatterjee, S.S., 1999. Bilobalide, a constituent of *Ginkgo biloba*, inhibits NMDA-induced phospholipase A2 activation and phospholipid breakdown in rat hippocampus. Naunyn. Schmiedebergs Arch. Pharmacol. 360, 609–615.

Winther, K., Randløv, C., Rein, E., 1998. Effects of *Ginkgo biloba* extract on cognitive function and blood pressure in elderly subjects. Curr. Ther. Res. 59, 881–888.

Wood, S., Sage, J.R., Shuman, T., 2014. Psychostimulants and cognition: a continuum of behavioral and cognitive activation. Pharmacol. Rev. 66, 193–221.

Yeo, H.-B., Yoon, H.-K., Lee, H.-J., 2012. Effects of Korean red Ginseng on cognitive and motor function: a double-blind, randomized, placebo-controlled trial. J. Ginseng Res. 36, 190–197.

Chapter 4

Nutraceuticals in anxiety and stress

Ajay Srivastava and Prafulla Kumar

Introduction

Anxiety and its associated disorders are among the most common mental disorders that are responsible for poor quality of life and significant disability. Anxiety disorders are associated with a dysfunctional neurotransmitter system in the brain (Baldwin et al., 2010). The key players in the pathogenesis of anxiety disorders are γ-aminobutyric acid (GABA), serotonin, dopamine (DA), and norepinephrine (NE). In addition, neuropeptides like cholecystokinin (CCK), oxytocin, vasopressin, atrial natriuretic peptide, and substance P are also implicated in the modulation of stress-related behaviors and anxiety (Bandelow et al., 2017; MacLean et al., 2017). The role of the hypothalamic–pituitary–adrenal (HPA) axis in the pathophysiology of anxiety disorders has also been widely studied. Studies indicate that overactivation of the HPA axis leads to the stimulation of the adrenal gland to release cortisol, dehydroepiandrosterone (DHEA), adrenaline, and noradrenaline. Cortisol in humans and dogs, and corticosteroids in rodents mediate a variety of effects throughout the body (including the brain), and exert a negative feedback on the HPA axis, thereby inhibiting its own production (Herman and Cullinan, 1997).

Growing evidence suggests that herbal extracts and dietary nutrients can effectively reduce anxiety by altering both neurotransmitter levels and the HPA axis. For example, vitamins C, D, and E, omega-3 fatty acids, and the green tea amino acid L-theanine are dietary supplements known to increase the production of DA. Further, supplementation with the amino acid L-tryptophan and its precursor, 5-HTP, along with the B vitamins, vitamin D, selenium, and omega-3 fatty acids increase serotonin production. These amino acid supplements are neurotransmitter building blocks and the vitamins act as cofactors in neurotransmitter biosynthesis pathways. This dietary approach can correct the underlying neurochemistry without the side effects associated with anxiolytic drugs, which may simply mask the problem (Alramadhan et al., 2012).

This chapter describes the key neurotransmitters involved in anxiety and stress disorders, biomarkers of stress, various dietary supplements, and herbal products that can help in easing anxiety and stress levels.

Neurotransmitters, neuropeptides, and the HPA axis in stress and anxiety

γ-Aminobutyric acid

GABA is the major inhibitory neurotransmitter in the mammalian brain and it plays a central role in the maintenance of inhibitory and excitatory balance. Preclinical and clinical studies strongly suggest that impairments in GABAergic neurotransmission underpin human stress and anxiety disorders (Goddard, 2016). Hyperexcitability of the amygdala, in particular, is strongly associated with anxiety, hypervigilance, and an inability to regulate emotions. Several rodent models have highlighted the role of the GABA synthetic isoenzymes glutamic acid decarboxylase 65 (GAD65) and GAD67 in the expression of normal mammalian fear. Knockdown of the GAD67 protein in mouse amygdala impaired normal fear extinction and decreased sensitivity to diazepam (Heldt et al., 2012). Low cortical levels of GABA have been observed in panic disorders and may serve as a potential biomarker for panic and other related stress disorders (Goddard et al., 2004). Experiments in various animal models have implicated the role of GABA in separation anxiety and trait anxiety (Tasan et al., 2011; Feng et al., 2014). Additionally, chronic restraint stress can also lead to decreased expression of $GABA_A$ receptor subunits in the cortex and hippocampus, suggesting a role in HPA axis dysfunction and stress symptomatology (Wisłowska-Stanek et al., 2013). Several anxiolytic agents available on the market today mediate their calming effects by targeting the GABA receptors. Benzodiazepines (BZD) are full agonists at the BZD site of $GABA_A$ receptors and they allosterically enhance the physiological GABA responses. Preclinical work has further defined the role of

Nutraceuticals. DOI: https://doi.org/10.1016/B978-0-12-821038-3.00004-5

discrete $GABA_A$ receptor subunits in separate clinical effects of the BZDs such as anxiolysis, sedation, muscle relaxation, and anticonvulsant effects (Rudolph and Knoflach, 2011).

Recent interest in dietary supplements has led to a spike in the use of GABA as a food supplement. In Europe and the United States, GABA is considered a "food constituent" and a "dietary supplement," respectively (Boonstra et al., 2015). However, it should be noted that the ability of GABA to cross the blood–brain barrier (BBB) in humans is unclear currently and available reports are, at the very least, contradictory (Boonstra et al., 2015).

Serotonin

Serotonin or 5-hydroxytryptamine (5-HT) is a monoamine, found in the central nervous system (CNS), blood platelets, and the gastrointestinal tract. In the human brain, serotonin serves as an inhibitory neurotransmitter where it regulates anxiety, fear, and mood. In the gut, serotonin works as a hormonal messenger and regulates satiety, craving, digestion, and absorption of nutrients. Chronic stress, certain medications, heavy metals, hormonal changes, and lack of nutrition can lead to low levels of serotonin either by affecting the synthesis of serotonin or by damaging the serotonergic neurons in the brain. Genetics also plays a critical role in regulating the synthesis and metabolism of serotonin and other neurotransmitters. The principal source of serotonin release in the brain is the raphe nuclei in the brainstem. These can inhibit periaqueductal gray matter-mediated fight/flight responses from threats, and can also facilitate amygdala-mediated anxiety responses (Deakin, 2013). Animal data and genetic and neuroimaging studies in humans point to a role of the 5HT1A receptor in the neural processing of anxiety (Akimova et al., 2009). Recently, a review of the 5HT2C receptor suggested that this receptor may also play a crucial role in anxiety (Chagraoui et al., 2016). Apart from the involvement of receptors, serotonin uptake inhibition is also involved in the improvement of anxiety disorders and depression. In fact, selective serotonin reuptake inhibitors, such as fluoxetine, are the first drug of choice to treat social anxiety disorders. However, excessive serotonin in the brain may lead to unwanted side effects and may interact negatively with certain drugs. Although rare, one of the most dangerous side effects of excess serotonin is serotonin syndrome which is potentially life threatening. Therefore one should first aim to increase the brain serotonin level through proper diet and natural supplements. For example, diets rich in tryptophan and amino acids are important for the synthesis of serotonin. Certain vitamins also play a role as cofactors in the synthesis of serotonin. Studies have shown that diets rich in α-lactalbumin can improve mood and cognition in humans, presumably owing to increased serotonin (Markus et al., 2002). The banana is another good example of a food that contains serotonin. However, serotonin cannot cross the BBB. Therefore touting banana powder and banana drinks as stress relievers and mood elevators is false advertisement.

Dopamine

DA is a neurotransmitter which has been linked to pleasure and depression and is also known as an antistress molecule. Studies have suggested that DA plays a role in anxiety modulation by working on various parts of the brain, primarily the amygdala and hippocampus. DA is known to exhibit its pharmacological effects on anxiety modulation via both D1 and D2 receptors. However, in preclinical studies, there are mixed reports on anxiolytic behavior of DA receptor agonists and antagonists. For example, apomorphine, a mixed D1 and D2 receptor agonist, when injected into the amygdala of rodents exhibited an anxiolytic effect but displayed an anxiogeneic effect when injected into the hippocampus (Zarrindast et al., 2010, 2011). In clinical trials, buspirone (Buspar), which works exclusively on dopaminergic receptors, was found to be effective in the treatment of anxiety with efficacy and dosage comparable to diazepam (Duncan et al., 1985). DA can be found in many plants and foods. The banana contains a good amount of DA. However, the DA present in bananas or other foods cannot cross the BBB and thus has no effect on the brain. Certain plants such as the fava bean and velvet bean (*Mucuna pruriens*) contain a substantial amount of the DA precursor L-dopa, but there is no proven clinical evidence available on the anti-anxiety effects of these plants. Certain vitamins have also been found to modulate brain DA levels. For example, pyridoxine (vitamin B6) in malnutrition produces the loss of DA in the corpus striatum of the developing rat brain (Guilarte, 1989). However, drugs or nutraceutical herbs targeting the dopaminergic system often produce unwanted side effects and may potentially lead to harmful pharmacokinetic interactions with other medications and supplements.

Norepinephrine

NE is a monoaminergic neurotransmitter that plays a key role in regulation of the autonomic nervous system. While hypofunction is postulated for depression, hyperfunction is suggested as the reason for anxiety disorders. Stimulation of the locus coeruleus, an area containing most of the noradrenergic cell bodies of the brain, has been shown to induce anxiety and to raise the concentration of the NE metabolite, 3-methoxy-4-hydroxyphenyl glycol, in patients with panic attacks. Therefore noradrenergic activation is perceived as anxiogenic. Contrary to preclinical findings of the noradrenergic activation of anxiety, when a combination of serotonin and norepinephrine reuptake inhibitors (commonly known as SNRIs) were administered to patients with anxiety and major depressive disorder, a considerable improvement was observed. Resveratrol, a polyphenolic compound found in grapes, is known for its neuroprotective and antioxidant effects (Srivastava et al., 2002). Resveratrol inhibits 5-HT/noradrenaline uptake and MAO activity in rats and therefore may have potential as an anxiolytic. However, there is no clinical study available to support the antianxiety effect of resveratrol in human patients.

Cholecystokinin

CCK [illegible]
brain [illegible]
neuro[illegible]
and D[illegible]
import[illegible]
involv[illegible]
acts w[illegible]
system[illegible]
tides Y[illegible]
express[illegible]
anxiety[illegible]
ated vi[illegible]
by both [illegible]

Oxyto[illegible]

Oxytoci[illegible]
pituitary[illegible]
ily imp[illegible]
wealth [illegible]
an affili[illegible]
prefront[illegible]
the anxi[illegible]
humans, [illegible]
stimuli [illegible]
gene ex[illegible]
have be[illegible]
anxiety. [illegible]
understa[illegible]
might be [illegible]
in pharmacotherapy (Turcsán et al., 2017).

Vasopressin

Vasopressin (AVP) is a peptide hormone that mediates many neurophysiological and behavioral functions. AVP is an integral part of the HPA axis, which may regulate stress-related disorders such as anxiety and depression, primarily via an Avpr1 receptor-mediated mechanism in the CNS. In contrast to oxytocin's anxiolytic and antidepressive effects, AVP tends to exert anxiogenic and depressive actions (Neumann and Landgraf, 2012). Therefore AVP receptor antagonism can be a potential therapeutic target in the treatment of anxiety and depression. Preclinical findings in rodents have confirmed that intermittent alcohol-induced social anxiety-like behavior can be reversed by suppression of the AVP system (Dannenhoffer et al., 2018). Involvement of the AVP system in anxiety and depression was further supported by clinical studies wherein data suggested that administration of an Avpr1b antagonist alleviated symptoms of depression and anxiety.

HPA axis

The HPA axis is the main physiological system that mediates the body's stress response. It plays a pivotal role in regulating the synthesis and release of endocrine hormones associated with the CNS, including cortisol, a major stress hormone. Cortisol has widespread effects throughout the body. It passes through the BBB to downregulate HPA axis activity by triggering negative feedback mechanisms, thereby inhibiting its own production (Herman and Cullinan, 1997). Cortisol exerts its effects through interactions with glucocorticoid and mineralocorticoid receptors, and regulates the expression of several genes, including cytokines (Lee and Rhee, 2017). Several reports indicate that psychological stress can trigger cytokine release (Black, 2003; Zalcman and Siegel, 2006), and growing evidence has shown an important role for the immune system in regulating negative emotional states as well as personality (Maes et al., 2009; Réus et al., 2015; Takahashi et al., 2018). Activation of the HPA axis is implicated in many behavioral disorders including social and spatial restriction (Beerda et al., 1999).

Dietary supplements in anxiety and stress

Chronic anxiety status and nutrition have been demonstrated to significantly affect behavior (Bosch et al., 2007). Chronic stress and diets deficient in certain amino acids and vitamins can also affect mood and cognition, presumably by a reduction in the mood-enhancing neurotransmitter serotonin. For example, supplementation with minerals like calcium, magnesium, and selenium, and vitamins like C, D, and E are reported to reduce anxiety and improve mood in both humans and animals (Carroll et al., 2000; Benton and Cook, 1990; Hughes et al., 2011). Vitamin C, a cofactor for many enzymes, reduces anxiety by limiting the oxidative stress from metabolites and also by limiting cortisol (Hughes et al., 2011).

Betulinic acid

Betulinic acid ($C_{30}H_{48}O_3$) is a well-known triterpene present in the bark of a number of trees including white birch (*Betula pubescens*), ber tree (*Ziziphus mauritiana*), sycamore, and other members of the *Platanus* family (*Platanus occidentalis*, *Platanus orientalis*, *Platanus acerfolia*). Betulinic acid (BA) is also found in a number of small plant species, such as the herb rosemary and the pitcher plant.

An amazing array of diverse biological activity has been reported for betulinic acid and its derivatives. Betulinic acid has been identified as the key active component accounting for the antianxiety activity of the leaves and bark of *Souroubea sympetala*, and *Souroubea gilgii*, a neotropical vine native to Costa Rica and other Central American countries. The antianxiety activity of the *S. sympetala* plant extracts was demonstrated by in vivo studies in rats using a variety of accepted paradigms such as the elevated maze, fear potentiated startle, and social interaction tests. These same tests were also used to demonstrate the antianxiety activity of pure betulinic acid and a number of its derivatives. Synergism of the BA antianxiety activity by the related triterpenes α- and β-amyrin present in the extract has also been proven. It has been proposed that these compounds increase the bioavailability of betulinic acid. The mode of action of BA has been shown to involve the $GABA_A$-BZD (benzodiazepine) receptor. When animals were pretreated with the antagonist flumazenil, the antianxiety effect of both the *Souroubea* leaf extracts and the pure more soluble derivative methyl ester of betulinic acid (**2**) was extinguished. Betulinic acid and its derivatives also significantly lower levels of the stress hormone cortisol in stressed rats, dogs, piglets, and trout. The effect in dogs was shown in two efficacy trials involving a colony of more than 50 Beagles exposed to simulated thunder. The onset of activity occurred within 1 h of administration of the plant material at 1–2 mg/kg. Two safety studies of the plant material in dogs where animals received as much as 5–10 times the recommended dosage of 1 mg/kg have demonstrated a very high degree of safety. A 95% ethanol extract of a 60:40 mixture of *P. occidentalis* bark and *S. sympetala* plant material has been used in a human phase 1 safety trial, where 45 participants were divided into three groups: placebo, 0.5 mg/kg, and 1 mg/kg of BA dosages. No adverse effects were noted in any of the three groups.

Probiotics

Probiotics are bacteria that when given in adequate amounts may offer several health benefits, including, but not limited to, digestion, metabolism, and immune health. However, in recent years probiotics have received considerable attention as potential treatments for depression and anxiety. Several studies have established the role of the gut microbiome in modulating brain function and mental disorders such as depression and anxiety. Gut microbiome via the gut–brain axis can activate neural pathways and CNS signaling systems to facilitate a beneficial effect between stress response, inflammation, depression, and anxiety (Foster and McVey Neufeld, 2013). Apart from regulating the gut–brain axis, other findings in rodents suggest the gut microbiota regulates depression-like behavior in rodents via a neuroendocrine–immune–mitochondrial pathway. In this study, rats were implanted with fecal microbiota from patients with depression. The rats exhibited a decreased level of hippocampal neurotransmitters and antiinflammatory cytokines, and showed damaged mitochondria in the intestinal epithelial cells, significant endoplasmic reticulum expansion, and border aggregation of nuclear chromatin (Liu et al., 2015).

However, more preclinical and clinical studies on different probiotic strains are warranted in order to establish clarity on the mechanisms by which gut microbes modulate brain function and improve mental health. A systematic review and metaanalysis of 33 controlled clinical trials concluded that, in general, probiotics can support both antidepressant and antianxiety effects. In another systematic metaanalysis of randomized studies, Pirbaglou et al. (2016) concluded that despite methodological limitations of the included trials and the complex nature of gut–brain interactions, probiotic supplementation can have positive psychological benefits on the symptoms of anxiety and depression (Pirbaglou et al., 2016). Most recently in a randomized, double-blind, placebo-controlled study, Chong et al. (2019) reported that consumption of *Lactobacillus plantarum* DR7 strain DR7 (1×10^9 cfu/day) for 12 weeks reduced symptoms of stress, anxiety, and total psychological scores as early as 8 weeks in among stressed adults compared to the placebo group. Plasma cortisol levels and proinflammatory cytokines were also reduced in the probiotic group compared to the placebo control. Gut microbiota has been shown to improve mood and reduce anxiety in patients with chronic fatigue (Logan and Katzman, 2005; Rao et al., 2009). Similarly, studies indicate that oral intake of probiotics resulted in reduced urinary cortisol and perceived psychological stress (Messaoudi et al., 2011). Certain bacterial strains like *Lactobacillus* and *Bifidobacterium* can produce GABA in vivo and were found to be effective in increasing GABA concentrations in the enteric nervous system (Barrett et al., 2012). Each of these studies indicates that probiotics may also work by modulating the GABAergic system.

Protein hydrolysates

Studies have shown that fish hydrolysate, a natural supplement derived from fish protein, helps in reducing fear and anxiety in Beagles. In a thunderstorm model, fish hydrolysate provided reduction in the hyperactivity response to thunder (Landsberg et al., 2015). Other hydrolysates derived from milk protein have also shown efficacy in improving symptoms of anxiety in rats and dogs. A comparative study conducted in 38 dogs demonstrated that α-casozepine (a milk protein hydrolysate) and selegeline (Anipryl) were equally effective in alleviating anxiety symptoms (Beata et al., 2007a,b). In addition, in a placebo-controlled trial α-casozepine displayed an improvement in fearful behavior and anxiety in cats due to socially stressful situations. Although not clearly elucidated, some studies have shown that the antianxiety effect of α-casozepine may be due to its structural similarity to GABA and its binding to GABA receptors in the brain.

Pheromones

Pheromones are chemical substances which affect social behavior in mammals. In a triple-blinded, randomized, and placebo-controlled study in puppies newly adopted from a pet shop, the dog-appeasing pheromone (DAP) showed reductions in behaviors associated with fear of unfamiliar people and new surroundings (Mills et al., 2006). In another open clinical study, DAP was able to reduce the symptoms of anxiety in dogs that showed signs of fear in response to fireworks (Sheppard and Mills, 2003). These pheromones have also shown some efficacy in reducing anxiety and fear in puppies during puppy classes and have resulted in improved socialization (Denenberg et al., 2005). However, 11 of the 14 reports reviewed and published in the *Journal of the American Veterinary Medical Association* suggested that there is insufficient evidence to support the effectiveness of pheromones for the treatment of undesirable behavior in cats and dogs.

Omega-3 fatty acids

Omega-3 polyunsaturated fatty acids (PUFAs) are abundantly present in fish and krill oil and other vegetarian sources such as chia and flex seed. Eicosapentaenoic acid (EPA) and docosahexaenoic acid are two primary components of PUFA that are beneficial for health. EPA, in particular, has been implicated in improving mood disorders (Sublette et al., 2011). Reports indicate that omega-3 deficiency may be associated with mood and behavioral disorders (Owen et al., 2008) and dietary omega-3 fatty acids have been shown to be effective in improving mood and reducing the risk of anxiety (Appleton et al., 2008; Kiecolt-Glaser et al., 2011). In one study, the diet consisted of a mixed formula of fish proteins, rice carbohydrates, *Punica granatum*, *Valeriana officinalis*, *Rosmarinus officinalis*, *Tilia* spp., tea extract, and L-tryptophan, with an omega-3:6 (ratio of 1:0.8). When administered to canines, they showed significant improvement in clinical and behavioral symptoms (Di Cerbo et al., 2017). Su et al. (2018) performed a metaanalysis of omega-3 clinical trial in humans, in both placebo-controlled and nonplacebo-controlled designs, to assess changes in the severity of anxiety symptoms after omega-3 PUFA treatment. They found omega-3 significantly improved the symptoms of clinical anxiety over those of controls. However, this finding was only observed in subgroups with a higher dosage (at least 2000 mg/day) and not in subgroups with a lower dosage (<2000 mg/day). Overall, these studies demonstrate the positive effects of omega-3-enriched diets on neuroendocrine, behavioral, and social parameters associated with stress and anxiety.

Melatonin

Melatonin is a pineal gland hormone secreted in the circadian rhythm. The primary function of melatonin is to maintain the sleep–wakefulness cycle and it is used clinically in patients with insomnia. Additionally, in preclinical studies, melatonin has shown neuroprotective effects against posttraumatic seizures and stroke. In randomized controlled trials compared to placebo, melatonin given as premedication (tablets or sublingually) can reduce preoperative anxiety in adults (measured 50–100 min after administration). Melatonin may be equally as effective as standard treatment with midazolam in reducing preoperative anxiety in adults (measured 50–100 min after administration). The effect of melatonin on postoperative anxiety (measured 90 min and 6 h after surgery) in adults is mixed but suggests an overall attenuation of the effect compared to preoperatively.

Herbal extracts in calming

In addition to nutrients such as amino acids, minerals, and vitamins, herbs and plant products have also been shown to be effective in treating anxiety (Head and Kelly, 2009; Saeed et al., 2007). Plant extracts offer a better alternative tool in the pharmacotherapy of psychological disorders without the adverse effects associated with anxiolytic drugs. Anxiolytic drugs such as BZD display many negative side effects including, but not limited to, addiction, depression, suicide, seizures, sexual dysfunction, and headaches (Alramadhan et al., 2012). In addition, drugs to treat anxiety do not restore normal levels of neurotransmitters, instead manipulating the brain chemistry. Anxiolytic drugs often negatively interact with other medications and even with herbal and plant products and can lead to potential adverse effects. Although not as efficacious as their pharmacological counterparts, plants are safe alternatives to harmful anxiolytic drugs.

Ginkgo biloba

Ginkgo biloba is one of the oldest ornamental trees on Earth and has long been used as a traditional herbal medicine for the treatment of anxiety (Liu et al., 2015). In preclinical studies, *G. biloba* has demonstrated efficacy against various neurological disorders including dementia and anxiety. Pharmacologically, *G. biloba* extracts were shown to activate GABA pathways and act like a benzodiazepine and reduce anxiety in patients with generalized anxiety disorder (Woelk et al., 2007). In addition to its benzodiazepine-like actions, *G. biloba* also exhibits antioxidant properties, targets 1-adrenergic, 5-HT1A, and muscarinic receptors, and protects against ischemic neuronal death. Kuribara et al. (2003) found that daily administration of ginkgolide-A (1 or 2 mg/kg, po) can result in an anxiolytic-like effect in mice. In a more recent 6-week randomized controlled, pretest–posttest design study, the addition of *G. biloba* to the psycho-education group proved superior to psycho-education alone. Therefore the authors concluded that supplementing *G. biloba* may be beneficial in alleviating anxiety and fatigue in psycho-education groups (Alsmadi et al., 2017). *G. biloba* appeared to be effective in reducing stress by decreasing cortisol secretion and the secretion of corticotropic-releasing hormone.

Ashwagandha (Withania somnifera)

Ashwagandha, a perennial shrub, is also known as Indian ginseng or winter cherry, and belongs to the nightshade family *Solanaceae*. It has been used for centuries in Indian, Chinese, and Arabic traditional medicines. The root extract is recommended for medication and supplement purposes. Preclinical studies have reported that rodents treated with ashwagandha showed a reduction in anxiety behavior compared with the control treatment. This reduction matched the reduction in anxiety in these rodents when treated with several benzodiazepine drugs (Ramanathan et al., 2011; Mohan et al., 2011). It has also been shown in clinical studies to reduce anxiety in patients and was superior in action compared to the control group receiving psychotherapy (Cooley et al., 2009). Different studies reported the great beneficial impact of ashwagandha root extract on stress, anxiety, and depression and in improving general well-being. Recent clinical trials have provided ample evidence in this direction. Regular supplementation of the herb extract with B vitamins was found to be effective for mood improvement and stress reduction in women (Chai et al., 2020). Due to extensive health benefits, ashwagandha is also now available on the market through various brands. Ashwagandha root extract, with a brand KSM-66, has ~5% withanolide, is processed through a green chemistry-based approach, and does not require alcohol-based extraction. Many studies have assessed the efficiency and qualitative value of this ashwagandha through clinical trials and established its efficacy in stress and anxiety management. A landmark study conducted by Chandrasekhar et al. (2012) reported that high-concentration full-spectrum ashwagandha root extract safely and effectively improved an individual's resistance to stress by significantly reducing the serum cortisol level. Another report by Choudhary et al. (2017) suggested that a 600 mg/day dosage of ashwagandha (KSM-66) was efficacious in reducing chronic stress and body weight management.

A double-blind, placebo-controlled study carried out on patients with insomnia provided remarkable outcomes with ashwagandha root extract (Langade et al., 2019). The 10-week study implemented modern actigraphy tools to understand and evaluate sleep onset latency, total sleep time, sleep efficiency, wake after sleep onset, total time in bed (sleep log), mental alertness on rising, sleep quality, Pittsburgh Sleep Quality Index, and Hamilton Anxiety Rating Scale (HAM-A) scales for insomnia patients and controls. Significant improvement in sleep quality and reduction in anxiety were observed. Moreover, consumption of ashwagandha was reported to help in improving general well-being and quality of life (Langade et al., 2019). Recently, Salve et al. (2019) reported the adaptogenic and anxiolytic impacts of ashwagandha root extract in healthy humans. The report suggested a reduction in PSS scores and serum cortisol levels in the group consuming ashwagandha root extract compared to the placebo group participants. Improvement of sleep quality was also observed in this study for the participants receiving ashwagandha (Salve et al., 2019). The anxiolytic and stress reduction mechanisms of ashwagandha are not yet clear. However, in recent years several pharmacological mechanisms have been explored. Ashwagandha temporarily impairs the HPA axis activity. Generally, activated HPA increases the secretion of cortisol (Doom and Gunnar, 2013) and DHEA (Lennartsson et al., 2012; Morgan et al., 2004) as a physiological response to stress. Postexcitation homeostasis is attained naturally through a negative feedback system. However, depression and anxiety hinder the HPA axis normal function, and increases cortisol secretion. Ashwagandha may reduce the cortisol level and restore normal function of the HPA axis. Apart from its modulatory activity on the HPA axis, ashwagandha can also modulate GABAergic (Candelario et al., 2015) and serotoninergic activities (Tripathi et al., 1998), and therefore can aid in reducing stress and anxiety. Varying dosages are recommended while consuming ashwagandha either as a supplement or as part of a polyherbal mixture. Studies attempting to explore the impact of varying dosages of ashwagandha extract have reported dosages ranging from 240 to 600 mg/day (Choudhary et al., 2017; Salve et al., 2019). Commercially available reputable brands of ashwagandha root extract recommend 600 mg/day (Langade et al., 2019; Salve et al., 2019), or preferably 300 mg taken twice a day. Acute and subacute toxicity analysis in mouse models suggested that up to 2000 mg/kg/day for 28 days was tolerable and did not have any toxic effects (Patel et al., 2016).

Valerian (V. officinalis)

Valerian is a temperate root that is the most widely available, over-the-counter nutraceutical commonly used for the self-management of insomnia and anxiety. The anxiolytic properties of valerian root extracts are well reported in animal models and humans, suggesting its use as an natural alternative to anxiolytic drugs (Benke et al., 2009; Hattesohl et al., 2008). In a double-blind, placebo-controlled pilot study, valerian root extract showed a significant reduction in the psychic factor on the HAM-A, suggesting a potential anxiolytic effect on the psychic symptoms of anxiety. Contradictory to this report, in an Internet-based randomized, double-blind, placebo-controlled trial, valerian root extract neither improved the quality of sleep nor relieved the anxiety symptoms more than placebo (Jacobs et al., 2005). Valerian extracts have been shown to both increase GABA synthesis and decrease synaptic GABA reuptake (Ortiz et al., 1999) through the activation of glutamic acid decarboxylase, an enzyme involved in the synthesis of GABA (Awad et al., 2007). Valerinic acid is an active compound from valerian root which exerts its action through allosteric modulation of the GABA-A receptors (Benke et al., 2009).

In addition to the above-mentioned anxiolytic herbs, there are reports on the beneficial use of rosenroot, lemon balm, chamomile, and lavender oil in reducing anxiety and stress in animals and humans (Weeks, 2009; Bystritsky et al., 2008; Kennedy et al., 2006; Komiya et al., 2009). Herbal extracts contain a wide range of phytochemicals including, but not limited to, alkaloids, phenolic compounds, bioflavonoids, lignans, cinnamates, terpenes, and saponins. These phytochemicals either individually or collectively target the GABAergic, serotonergic, noradrenergic, and dopaminergic systems. Phytochemicals also modulate the HPA axis, thereby regulating the levels of cortisol and the release of proinflammatory cytokines (Farzaei et al., 2016). Taken together, these studies advocate the use of nutraceutical herbs and dietary supplements as a potential substitute to conventional anxiolytic therapies in order to improve efficacy and reduce adverse effects.

Kava (Piper methysticum)

Kava (*Piper methysticum*) has been used for a long time in its native southern Pacific islands for both therapeutic and recreational purposes (Mathews et al., 1988). Traditionally, kava is used to ameliorate anxiety and is sometimes paired with alcohol for recreational use. Beverages brewed with kava roots have also been reported to have sedative and hypnotic effects depending upon dosages (Singh, 1992). In preclinical studies, whole extract of kava showed a neuroprotective effect against ischemic injury (Backhauss and Krieglstein, 1992) and displayed a rapid anxiolytic effect in rodent model of anxiety within 10 min of administration of kavalactones. Interestingly, a higher brain tissue concentration was observed after oral ingestion compared to parenteral administration, suggesting that kavalactones are highly bioavailable via the oral route and can easily reach the brain by crossing the blood–brain barrier (Garrett et al., 2003; Keledjian et al., 1988). In addition to their antianxiety and neuroprotective properties, kava extracts have also shown clinical benefits in improving mood and cheerfulness, improving cognition in healthy volunteers, and reducing clinical anxiety. Mechanistic studies have proposed a GABA-modulating activity of kavalactones as they bind to $GABA_A$ receptors (Jussofie et al., 1994). Kavalactones have also been shown to inhibit neuronal uptake of DA and NE in prefrontal cortex (Baum et al., 1998; Seitz et al., 1997). Data from animal studies have found kava to be safe as they have a high LD_{50} of 300–400 mg/kg (Ulbricht et al., 2005). In addition, kava is also known to have sedative and hypnotic effects when given along with alcohol, and renders alcoholics more susceptible to liver damage if taking kava concurrently.

Contrary to animal safety findings, in humans there are several reports that have shown kava to be an hepatotoxic herb causing hepatitis and hepatic failure (Humberston et al., 2003; Gow et al., 2003). Many European countries, Canada, and Japan have banned the use of kava. Therefore both acute and long-term studies are warranted in order to establish the safe usage of kava (Jamieson and Duffield, 1990). In one study, kava was reported to be equally effective compared to benzodiazepine, suggesting that kava may be a safer alternative to BZD, eliminating the risk of addiction or dependence (Malsch and Kieser, 2001). In yet another 8-week randomized placebo-controlled multicentric study, kava was found to be equally effective when compared with opipramol and buspirone. In contrast, there are a few studies in which kava has failed to exert an antianxiety effect in treating generalized anxiety disorder (Connor et al., 2006). These findings suggest that there is a need for multiple large placebo-controlled randomized clinical trials to establish the efficacy of kava extracts in clinical anxiety.

In addition to the above-mentioned anxiolytic herbs, there are reports on the beneficial use of roseroot, lemon balm, chamomile, and lavender oil in reducing anxiety and stress in animals and humans (Weeks, 2009; Bystritsky et al., 2008; Kennedy et al., 2006; Komiya et al., 2009). Herbal extracts contain a wide range of phytochemicals such as alkaloids, phenolic compounds, bioflavonoids, lignans, cinnamates, terpenes, and saponins. These phytochemicals either individually or collectively target the GABAergic, serotonergic, noradrenergic, and dopaminergic systems. Phytochemicals also modulate the HPA axis, thereby regulating the levels of cortisol and release of proinflammatory cytokines (Farzaei et al., 2016). Taken together, these studies advocate the use of nutraceutical herbs and dietary supplements as a potential substitute for conventional anxiolytic therapies in order to improve efficacy and reduce adverse effects.

Concluding remarks and future directions

Anxiety and stress can be stimulated by both environmental, social, or age-related factors. The underlying pathophysiology of anxiety is diverse and may involve multiple neurotransmitters, neuroendocrine systems, including the HPA axis, neuropeptides, and a deficiency in certain dietary nutrients. While newer anxiolytics with fewer side effects and addictive potentials are available on the market, growing evidence shows that nutraceuticals are equally potent and in some cases more effective than pharmacotherapy. Additionally, while anxiolytics exert their action by manipulating the neurotransmitter systems in the brain, herbal supplements exert their action by bringing the levels of neurotransmitters to a physiological level to maintain homeostasis. This chapter describes the key neurotransmitter systems involved in the development of anxiety and stress, the various herbal and dietary supplements that are beneficial for reducing stress and anxiety, and a more meaningful insight on the usefulness of nutraceuticals for anxiety disorder. It is also important to mention here that although nutraceuticals/supplements appear to be a safer alternative to conventional pharmacological anxiolytics, there are only a handful of studies available on nutraceutical–drug interactions and long-term safety in either the healthy human population or patients with anxiety and stress. Further studies with more carefully designed protocols are warranted to investigate and validate the results presented here.

References

Alsmadi, A.M., Tawalbeh, L.I., Gammoh, O.S., Shawagfeh, M.Q., Zalloum, W., Ashour, A., The effect of *Ginkgo biloba* and psycho-education on stress, anxiety and fatigue among refugees, *Proc. Singap. Health Care* 27 (1), 2017, 26–32.

Akimova, E., Lanzenberger, R., Kasper, S., 2009. The serotonin-1A receptor in anxiety disorders. Biol. Psychiatry 66, 627–635.

Alramadhan, E., Hanna, M.S., Hanna, M.S., et al., 2012. Dietary and botanical anxiolytics. Med. Sci. Monit. 18 (4), RA40–RA48.

Appleton, K.M., Rogers, P.J., Ness, A.R., 2008. Is there a role for n-3 long-chain polyunsaturated fatty acids in the regulation of mood and behavior? A review of the evidence to date from epidemiological studies, clinical studies and intervention trials. Nutr. Res. Rev. 21 (1), 13–41.

Awad, R., Levac, D., Cybulska, P., et al., 2007. Effects of traditionally used anxiolytic botanicals on enzymes of the gamma-aminobutyric acid (GABA) system. Can. J. Physiol. Pharmacol. 85 (9), 933–942.

Backhauss, C., Krieglstein, J., 1992. Extract of kava (*Piper methysticum*) and its methysticin constituents protect brain tissue against ischemic damage in rodents. Eur. J. Pharmacol. 215 (2-3), 265–269.

Baldwin, D.S., Allgulander, C., Altamura, A.C., et al., 2010. Manifesto for a european anxiety disorders research network. Eur. Neuropsychopharmacol. 20, 426–432.

Bandelow, B., Baldwin, D., Abelli, M., et al., 2017. Biological markers for anxiety disorders, OCD and PTSD: a consensus statement. Part II: neurochemistry, neurophysiology and neurocognition. World J. Biol. Psychiatry 18 (3), 162–214.

Barrett, E., Ross, R.P., O'Toole, P.W., et al., 2012. Gamma-aminobutyric acid production by culturable bacteria from the human intestine. J. Appl. Microbiol. 113, 411–417.

Baum, S.S., Hill, R., Rommelspacher, H., 1998. Effect of kava extract and individual kavapyrones on neurotransmitter levels in the nucleus accumbens of rats. Prog. Neuropsychopharmacol. Biol. Psychiatry 22 (7), 1105–1120.

Beata, C., Beaumont-Graff, E., Coll, V., et al., 2007a. Effect of alpha-casozepine (Zylkene) on anxiety in cats. J. Vet. Behav. 2, 40–46.

Beata, C., Beaumont-Graff, E., Diaz, C., et al., 2007b. Comparison of the effect of alpha-casozepine (Zylkene) versus selegiline hydrochloride on anxiety disorders in dogs. J. Vet. Behav. 2, 175–183.

Beerda, B., Schilder, M.B., Bernadina, W., et al., 1999. Chronic stress in dogs subjected to social and spatial restriction. II. Hormonal and immunological responses. Physiol. Behav. 66 (2), 243–254.

Benke, D., Barberis, A., Kopp, S., et al., 2009. $GABA_A$ receptors as in vivo substrate for the anxiolytic action of valerenic acid, a major constituent of valerian root extracts. Neuropharmacology 56 (1), 174–181.

Benton, D., Cook, R., 1990. Selenium supplementation improves mood in a double-blind crossover trial. Psychopharmacol. (Berl.) 102 (4), 549–550.

Black, P.H., 2003. The inflammatory response is an integral part of the stress response: implications for atherosclerosis, insulin resistance, type II diabetes and metabolic syndrome X. Brain Behav. Immun. 17, 350–364.

Boonstra, E., de Kleijn, R., Colzato, L.S., et al., 2015. Neurotransmitters as food supplements: The effects of GABA on brain and behavior. Front. Psychol. 6, 1520.

Bosch, G., Beerda, B., Hendriks, W.H., et al., 2007. Impact of nutrition on canine behavior: Current status and possible mechanisms. Nutr. Res. Rev. 20, 180–194.

Bourin, M., Dailly, E., 2004. Cholecystokinin and panic disorder. Acta Neuropsychiatr. 16, 85–93.

Bystritsky, A., Kerwin, L., Feusner, J.D., 2008. A pilot study of *Rhodiola rosea* (Rhodax) for generalized anxiety disorder (GAD). J. Altern. Complement. Med. 14 (2), 175–180.

Candelario, M., Cuellar, E., Reyes-Ruiz, J.M., et al., 2015. Direct evidence for GABAergic activity of *Withania somnifera* on mammalian ionotropic $GABA_A$ and $GABA\rho$ receptors. J. Eethnopharmacol 171, 264–272.

Carroll, D., Ring, C., Suter, M., et al., 2000. The effects of an oral multivitamin combination with calcium, magnesium, and zinc on psychological well-being in healthy young male volunteers: a double-blind placebo-controlled trial. Psychopharmacol. (Berl.) 150 (2), 220–225.

Chagraoui, A., Thibaut, F., Skiba, M., et al., 2016. 5-HT2C receptors in psychiatric disorders: A review. Prog. Neuropsychopharmacol. Biol. Psychiatry 66, 120–135.

Chai, S.C., Li, I., Pacanowski, C., et al., 2020. Effects of four-week supplementation of Ashwagandha and B-Vitamins on mood and stress relief. Curr. Dev. Nutr. 4 (Suppl 2), 1195.

Chandrasekhar, K., Kapoor, J., Anishetty, S., 2012. A prospective, randomized double-blind, placebo-controlled study of safety and efficacy of a high-concentration full-spectrum extract of ashwagandha root in reducing stress and anxiety in adults. Indian J. Psychol. Med. 34 (3), 255.

Chong H.X., Yusoff, N.A.A., Hor, Y.Y., Lew, L.C., Jaafar, M.H., Choi, S.B., et al., *Lactobacillus plantarum* DR7 alleviates stress and anxiety in adults: A randomised, double-blind, placebo-controlled study, *Benef. Microbes* 10 (4) 2019, 355–373.

Choudhary, D., Bhattacharyya, S., Joshi, K., 2017. Body weight management in adults under chronic stress through treatment with Ashwagandha root extract: a double-blind, randomized, placebo-controlled trial. J. Evid-Based Complem. Altern. Med. 22 (1), 96–106.

Connor, K.M., Payne, V., Davidson, J.R., 2006. Kava in generalized anxiety disorder: Three placebo-controlled trials. Int. Clin. Psychopharmacol. 21 (5), 249–253.

Cooley, K., Szczurko, O., Perri, D., et al., 2009. Naturopathic care for anxiety: a randomized controlled trial ISRCTN78958974. PLoS One 4 (8), e6628.

Dannenhoffer, C.A., Kim, E.U., Saalfield, J., et al., 2018. Oxytocin and vasopressin modulation of social anxiety following adolescent intermittent ethanol exposure. Psychopharmacology 235 (10), 3065–3077.

Deakin, J., 2013. The origins of 5-HT and mechanisms of defense by Deakin and Graeff: a personal perspective. J. Psychopharmacol. (Oxf.) 27, 1084–1089.

Denenberg, S., Landsberg, G., Gaultier, E., 2005. Evaluation of DAP's effect on reduction of anxiety in puppies (Canis familiaris) as well as its usefulness in improving learning and socialization. In: Mills, D., Levine, E., Landsberg, G. (Eds.), Current Issues and Research in Veterinary Behavioral Medicine. Purdue Univ Press, pp. 225–228.

Di Cerbo, A., Sechi, S., Canello, S., et al., 2017. Behavioral disturbances: an innovative approach to monitor the modulatory effects of a nutraceutical diet. J. Vis. Exp. (119), e54878. Available from: https://doi.org/10.3791/54878.

Doom, J.R., Gunnar, M.R., 2013. Stress physiology and developmental psychopathology: past, present, and future. Devel Psychopathol. 25 (4 Pt 2), 1359–1373.

Duncan, P., TaylorMichael, S., Eison, L.A., et al., 1985. Pharmacological and clinical effects of buspirone. Pharmacol. Biochem. Behav. 23 (4), 687–694.

Farzaei, M.H., Bahramsoltani, R., Rahimi, R., et al., 2016. A systematic review of plant-derived natural compounds for anxiety disorders. Curr. Top. Med. Chem. 6 (17), 1924–1942.

Feng, M., Sheng, G., Li, Z., et al., 2014. Postnatal maternal separation enhances tonic GABA current of cortical layer 5 pyramidal neurons in juvenile rats and promotes genesis of GABAergic neurons in neocortical molecular layer and subventricular zone in adult rats. Behav. Brain Res. 260, 74–82.

Foster, J.A., McVey Neufeld, K.A., 2013. Gut-brain axis: how the microbiome influences anxiety and depression. Trends Neurosci. 36 (5), 305–312.

Garrett, K.M., Basmadjian, G., Khan, I.A., et al., 2003. Extracts of kava (*Piper methysticum*) induce acute anxiolytic-like behavioral changes in mice. Psychopharmacol. (Berl.) 170 (1), 33–41.

Goddard, A.W., 2016. Cortical and subcortical gamma amino acid butyric acid deficits in anxiety and stress disorders: clinical implications. World J. Psychiatr. 6 (1), 43–53.

Goddard, A.W., Mason, G.F., Rothman, D.L., et al., 2004. Family psychopathology and magnitude of reductions in occipital cortex GABA levels in panic disorder. Neuropsychopharmacology 29, 639–640.

Gow, P.J., Connelly, N.J., Hill, R.L., et al., 2003. Fatal fulminant hepatic failure induced by a natural therapy containing kava. Med. J. Aust. 178 (9), 442–443.

Guilarte, T.R., 1989. Effect of vitamin B-6 nutrition on the levels of dopamine, dopamine metabolites, dopa decarboxylase activity, tyrosine, and GABA in the developing rat corpus striatum. Neurochem. Res. 14 (6), 571–578.

Hattesohl, M., Feistel, B., Sievers, H., et al., 2008. Extracts of *Valeriana officinalis* L.s.l. show anxiolytic and antidepressant effects but neither sedative nor myorelaxant properties. Phytomedicine 15 (1–2), 2–15.

Head, K.A., Kelly, G.S., 2009. Nutrients and botanicals for treatment of stress: adrenal fatigue, neurotransmitter imbalance, anxiety, and restless sleep. Altern. Med. Rev. 14 (2), 114–140.

Heldt, S.A., Mou, L., Ressler, K.J., 2012. *In vivo* knockdown of GAD67 in the amygdala disrupts fear extinction and the anxiolytic-like effect of diazepam in mice. Transl. Psychiatr. 2, e181.

Herman, J.P., Cullinan, W.E., 1997. Neurocircuitry of stress: central control of the hypothalamo-pituitary-adrenocortical axis. Trends Neurosci. 20 (2), 78–84.

Hughes, R.N., Lowther, C.L., van Nobelen, M., 2011. Prolonged treatment with vitamins C and E separately and together decreases anxiety-related open-field behavior and acoustic startle in hooded rats. Pharmacol. Biochem. Behav. 97 (3), 494–499.

Humberston, C.L., Akhtar, J., Krenzelok, E.P., 2003. Acute hepatitis induced by kava kava. J. Toxicol. Clin. Toxicol 41 (2), 109–113.

Jacobs, B.P., Bent, S., Tice, J.A., et al., 2005. An internet-based randomized, placebo-controlled trial of kava and valerian for anxiety and insomnia. Med. (Baltim.) 84 (4), 197–207.

Jamieson, D.D., Duffield, P.H., 1990. Positive interaction of ethanol and kava resin in mice. Clin. Exp. Pharmacol. Physiol. 17 (7), 509–514.

Jussofie, A., Schmiz, A., Hiemke, C., 1994. Kavapyrone enriched extract from *Piper methysticum* as modulator of the GABA binding site in different regions of rat brain. Psychopharmacol. (Berl.) 116 (4), 469–474.

Keledjian, J., Duffield, P.H., Jamieson, D.D., et al., 1988. Uptake into mouse brain of four compounds present in the psychoactive beverage kava. J. Pharm. Sci. 77 (12), 1003–1006.

Kennedy, D.O., Little, W., Haskell, C.F., et al., 2006. Anxiolytic effects of a combination of *Melissa officinalis* and *Valeriana officinalis* during laboratory induced stress. Phytother. Res. 20 (2), 96–102.

Kiecolt-Glaser, J.K., Belury, M.A., Andridge, R., et al., 2011. Omega-3 supplementation lowers inflammation and anxiety in medical students: a randomized controlled trial. Brain Behav. Immun. 25 (8), 1725–1734.

Komiya, M., Sugiyama, A., Tanabe, K., et al., 2009. Evaluation of the effect of topical application of lavender oil on autonomic nerve activity in dogs. Am. J. Vet. Res. 70 (6), 764–769.

Kuribara, H., Weintraub, S.T., Yoshihama, T., et al., 2003. An anxiolytic-like effect of *Ginkgo biloba* extract and its constituent, ginkgolide-A, in mice. J. Nat. Prod. 66 (10), 1333–1337.

Landsberg, G.M., Mougeot, I., Kelly, S., et al., 2015. Assessment of noise-induced fear and anxiety in dogs: modification by a novel fish hydrolysate supplemented diet. J. Vet. Behav. 10 (5), 391–398.

Langade, D., Kanchi, S., Salve, J., et al., 2019. Efficacy and safety of Ashwagandha (*Withania somnifera*) root extract in insomnia and anxiety: a double-blind, randomized, placebo-controlled study. Cureus 11 (9), e5797.

Lee, S., Rhee, D.K., 2017. Effects of ginseng on stress-related depression, anxiety, and the hypothalamic-pituitary-adrenal axis. J. Ginseng Res. 41 (4), 589–594.

Lennartsson, A.K., Kushnir, M.M., Bergquist, J., et al., 2012. DHEA and DHEA-S response to acute psychosocial stress in healthy men and women. Biol. Psychol. 90 (2), 143–149.

Li, H., Ohta, H., Izumi, H., et al., 2013. Behavioral and cortical EEG evaluations confirm the roles of both CCKA and CCKB receptors in mouse CCK-induced anxiety. Behav. Brain Res. 237, 325–332.

Liu, L., Liu, C., Wan, Y., et al., 2015. Herbal medicine for anxiety, depression and insomnia current. Neuropharmacology (13), 481–493.

Logan, A.C., Katzman, M., 2005. Major depressive disorder: probiotics may be an adjuvant therapy. Med. Hypotheses 64, 533–538.

MacLean, E.L., Gesquiere, L.R., Gruen, M.E., et al., 2017. Endogenous oxytocin, vasopressin, and aggression in domestic dogs. Front. Psychol. 8, 1613.

Maes, M., Yirmyia, R., Noraberg, J., et al., 2009. The inflammatory and neurodegenerative (I&ND) hypothesis of depression: leads for future research and new drug developments in depression. Metab. Brain Dis. 24, 27–53.

Malsch, U., Kieser, M., 2001. Efficacy of kava-kava in the treatment of non-psychotic anxiety, following pretreatment with benzodiazepines. Psychopharmacol. (Berl.) 157 (3), 277–283.

Markus, C.R., Olivier, B., de Haan, E.H., 2002. Whey protein rich in alpha-lactalbumin increases the ratio of plasma tryptophan to the sum of the other large neutral amino acids and improves cognitive performance in stress-vulnerable subjects. Am. J. Clin. Nutr. 75, 1051–1056.

Mathews, J.D., Riley, M.D., Fejo, L., et al., 1988. Effects of the heavy usage of kava on physical health: summary of a pilot survey in an aboriginal community. Med. J. Aust. 148 (11), 548–555.

Messaoudi, M., Lalonde, R., Violle, N., et al., 2011. Assessment of psychotropic-like properties of a probiotic formulation (*Lactobacillus helveticus* R0052 and *Bifidobacterium longum* R0175) in rats and human subjects. Br. J. Nutr 105, 755–764.

Mills, D.S., Ramos, D., Gandia Estelles, M., et al., 2006. A triple blind placebo-controlled investigation into the assessment of the effect of dog appeasing

Pheromone (DAP) on anxiety related behavior of problem dogs in the veterinary clinic. Appl. Anim. Behav. Sci. 98, 114–126.

Mohan, L., Rao, U.S., Gopalakrishna, H.N., Nair, V., 2011. Evaluation of the anxiolytic activity of NR-ANX-C (a polyherbal formulation) in ethanol withdrawal-induced anxiety behavior in rats. Evid. Based Complement. Altern. Med. 2011, 327160.

Morgan, C.A., Southwick, S., Hazlett, G., et al., 2004. Relationships among plasma dehydroepiandrosterone sulfate and cortisollevels, symptoms of dissociation, and objective performance in humans exposedto acute stress. Arch. Gen. Psychiatr. 61 (8), 819–825.

Neumann, I.D., Landgraf, R., 2012. Balance of brain oxytocin and vasopressin: implications for anxiety, depression, and social behaviors. Trends Neurosci. 35 (11), 649–659.

Ortiz, J.G., Nieves-Natal, J., Chavez, P., 1999. Effects of *Valeriana officinalis* extraction [3H]fl unitrazepam binding, synaptosomal [3H]GABA uptake, and hippocampal [3H]GABA release. Neurochem. Res. 24 (11), 1373–1378.

Owen, C., Rees, A.M., Parker, G., 2008. The role of fatty acids in the development and treatment of mood disorders. Curr. Opin. Psychiatry 21, 19–24.

Patel, S.B., Rao, N.J., Hingorani, L.L., 2016. Safety assessment of *Withania somnifera* extract standardized for Withaferin A: acute and sub-acute toxicity study. J. Ayurveda Integr. Med. 7 (1), 30–37.

Pirbaglou, M., et al., 2016. Probiotic supplementation can positively affect anxiety and depressive symptoms: a systematic review of randomized controlled trials. Nutr. Res. 36 (9), 889–898.

Ramanathan, M., Balaji, B., Justin, A., 2011. Behavioral and neurochemical evaluation of Perment an herbal formulation in chronic unpredictable mild stress induced depressive model. Indian J. Exp. Biol. 49 (4), 269–275.

Rao, A.V., Bested, A.C., Beaulne, T.M., et al., 2009. A randomized, double-blind, placebo-controlled pilot study of a Probiotic in emotional symptoms of chronic fatigue syndrome. Gut Pathog. 1, 6. Available from: https://doi.org/10.1186/1757-4749-1-6.

Réus, G.Z., Fries, G.R., Stertz, L., et al., 2015. The role of inflammation and microglial activation in the pathophysiology of psychiatric disorders. Neuroscience 300, 141–154.

Rudolph, U., Knoflach, F., 2011. Beyond classical benzodiazepines: novel therapeutic potential of GABAA receptor subtypes. Nat. Rev. Drug. Discov. 10, 685–697.

Sabihi, S., Dong, S.M., Maurer, S.D., et al., 2017. Oxytocin in the medial prefrontal cortex attenuates anxiety: anatomical and receptor specificity and mechanism of action. Neuropharmacology 125, 1–12.

Saeed, S.A., Bloch, R.M., Antonacci, D.J., 2007. Herbal and dietary supplements for treatment of anxiety disorders. Am. Fam. Physician 76 (4), 549–556.

Salve, J., Pate, S., Debnath, K., et al., 2019. Adaptogenic and anxiolytic effects of ashwagandha root extract in healthy adults: a double-blind, randomized, placebo-controlled clinical study. Cureus 11 (12), e6466.

Seitz, U., Schüle, A., Gleitz, J., 1997. [3H]-monoamine uptake inhibition properties of kava pyrones. Planta Med. 63 (6), 548–549.

Sheppard, G., Mills, D.S., 2003. Evaluation of dog-appeasing pheromone as a potential treatment for dogs fearful of fireworks. Vet. Rec. 152, 432–436.

Singh, Y.N., 1992. Kava: an overview. J. Ethnopharmacol. 37 (1), 13–45. Available from: https://doi.org/10.1016/0378-8741(92)90003-a.

Srivastava, A.K., Gupta, S.K., Jain, S., Gupta, Y.K., 2002. Effect of melatonin and phenytoin on an intracortical ferric chloride model of posttraumatic seizures in rats. Methods Find. Exp. Clin. Pharmacol. 24 (3), 145–149.

Su, K.P., Tseng, P.T., Lin, P.Y., et al., 2018. Association of use of omega-3 polyunsaturated fatty acids with changes in severity of anxiety symptoms: a systematic review and meta-analysis. JAMA Netw. Open. 1 (5), e182327.

Sublette, M.E., Ellis, S.P., Geant, M.L., 2011. Meta-analysis of the effects of eicosapentaenoic acid (EPA) in clinical trials in depression, J Clin Psychiatr, 72. pp. 1577–1584.

Takahashi, A., Flanigan, M.E., McEwen, B.S., et al., 2018. Aggression, social stress, and the immune system in humans and animal models. Front. Behav. Neurosci. 12, 56.

Tasan, R.O., Bukovac, A., Peterschmitt, Y.N., et al., 2011. Altered GABA transmission in a mouse model of increased trait anxiety. Neuroscience 183, 71–80.

Tripathi, A.K., Dey, S., Singh, R.H., et al., 1998. Alterations in the sensitivity of 5th receptor subtypes following chronic Ashwagandha treatment in rats. Anc. Sci. Life 17 (3), 169.

Turcsán, B., Range, F., Rónai, Z., et al., 2017. Context and individual characteristics modulate the association between oxytocin receptor gene polymorphism and social behavior in border Collies. Front. Psychol. 8, 2232.

Ulbricht, C., Basch, E., Boon, H., et al., 2005. Safety review of kava (*Piper methysticum*) by the natural standard research collaboration. Expert. Opin. Drug. Saf. 4 (4), 779–794.

Weeks, B.S., 2009. Formulations of dietary supplements and herbal extracts for relaxation and anxiolytic action: Relarian. Med. Sci. Monit. 15 (11), RA256–RA262.

Wisłowska-Stanek, A., Lehner, M., Skórzewska, A., et al., 2013. Changes in the brain expression of alpha-2 subunits of the GABA-A receptor after chronic restraint stress in low- and high-anxiety rats. Behav. Brain Res. 253, 337–345.

Woelk, H., et al., 2007. *Ginkgo Biloba* special extract EGb 761 in generalized anxiety disorder and adjustment disorder with anxious mood: a randomized, double-blind, placebo-controlled trial. J. Psychiatr. Res. 41 (6), 472–480.

Zalcman, S.S., Siegel, A., 2006. The neurobiology of aggression and rage: role of cytokines. Brain Behav. Immun. 20, 507–514.

Zarrindast, M.R., Naghdi-Sedeh, N., Nasehi, M., et al., 2010. The effects of dopaminergic drugs in the ventral hippocampus of rats in the nicotine-induced anxiogenic-like response. Neurosci. Lett. 475 (3), 156–160.

Zarrindast, M.R., Mahboubi, S., Ahmadi, S., et al., 2011. Anxiolytic–like effect induced by the cannabinoid CB1 receptor agonist, arachydonilcyclopropylamide (ACPA), in the rat amygdala is mediated through the D1 and D2 dopaminergic systems. J. Psychopharmacol. 25 (1), 131–140.

Zwanzger, P., Domschke, K., Bradwejn, J., 2012. Neuronal network of panic disorder: the role of the neuropeptide cholecystokinin. Depress. Anxiety 29, 762–774.

Chapter 5

Countering cardiovascular diseases with nutraceuticals

Csaba K. Zoltani

Introduction

Nutraceuticals are primarily herbaceous plants, fruits, and berries that contain phytochemicals whose active ingredients can alleviate diseases. Plants produce a large number of secondary metabolites (Wink, 2018) which are not directly involved in the normal processes of the plant such as growth. However, these metabolites may play important role in the defense of the plant against various attacks. Numerous plant secondary metabolites are of interest (Zoltani, 2016) since they can contribute to the alleviation of cardiological diseases.

Cardiac dysfunctions, such as atheromatous vascular disease, are amenable to nutraceutical treatment. A number of plant components control oxidative stress emanating from the production of free radicals, deficiency of antioxidants, or limited reactive oxygen system (ROS) upsurge, which are beyond the endogenous antioxidant system capabilities. ROS generation impairs nitric oxide (NO). Bioavailability thereby causes endothelial dysfunction. Low NO results in vasoconstriction, platelet aggregation, and leukocyte endothelium adhesion.

Endothelial dysfunction, including the loss of NO bioavailability in vessel walls, diminishes the antiinflammatory properties of endothelial cells leading to platelet activation and increased free radical generation, elevated low-density lipoprotein (LDL), homocysteine, and changes in adhesion of platelets and leukocytes.

Bioactive plant components (phytochemicals) can supplement decreased essential micronutrients to hinder endothelial dysfunction and prevent various expressions of cardiac dysfunction that need to be addressed for normal functioning.

Nutraceuticals, through the use of epigenetic processes whereby polyphenolic compounds that possess medicinal properties such as antiinflammatory effects (Jain et al., 2018) are introduced to modify posttranslational properties, can achieve changes in their potential usefulness.

The cardioprotective effect of nutraceuticals to counter adverse cardiac effects has been clearly demonstrated for several conditions (Adegbole et al., 2017; Alves et al., 2019; Cicero et al., 2017; Cicero and Colletti, 2018; Momtazi et al., 2017; Nikhra, 2019; Poli and Visioli, 2019). Nutraceuticals have also found application in statin-intolerant patients (Banach et al., 2018).

Primary cardiovascular diseases of concern

Atherosclerosis

One of the major causes of cardiovascular diseases (CVDs) is atherosclerosis, an inflammatory disorder of the vasculature. It is primarily caused by plaques, that is, deposition of cholesterol and other substances, on medium and large arterial walls. A primary source of cholesterol in the pathogenesis of atherosclerosis is apolipoprotein B derived from plasma. Accumulation of apolipoprotein causes endothelial cell dysfunction via oxidized LDL. Eventually, foam cells are formed from the secreted chemokines and adhesion proteins that recruit monocytes. A lipid-rich necrotic core is inflammatory and capped. Inflammatory mediators cause plaque rupture, leading to thrombotic events.

Accumulation of lipids on arterial walls is associated with vascular lipid oxidation that results in atheroma buildup. Atheroma buildup by fatty deposits restricts the circulation and the buildup of plaque on the coronary artery, resulting in unstable collection of white blood cells and lipids. The ensuing endothelial dysfunction is accompanied by a loss of NO bioavailability and homocysteine elevation. This inflammatory condition leads to ischemia in extreme cases (Aquila et al., 2019; Barrios et al., 2017). Zaina et al. (2005) treated the effect of aberrant DNA methylation in atherosclerosis (Table 5.1).

Nutraceuticals. DOI: https://doi.org/10.1016/B978-0-12-821038-3.00005-7

TABLE 5.1 Effect of nutraceuticals on symptoms of atherosclerosis.

Nutraceutical	Effect	Reference
Berberine	Diminished inflammation and reduced plaque size	Chen et al. (2014)
Allicin	Reduced TC and TG	Gonen et al. (2005)
Curcumin	Antiatherogenic effect	Shin et al. (2011)
Phytosterols	Lowered LDL	Nashed et al. (2005)

LDL, low-density lipoprotein; *TC*, total cholesterol; *TG*, total triglyceride.

As noted by Madjid et al. (2007), a significant increase in inflammatory cells in the atherosclerotic coronary arteries develops during acute systemic infections. Increased inflammation in the coronary arteries can lead to the rupture of plaques. Recently, it was noted that in COVID-19 infection cases an increase in cardiac dysfunction occurs in older patients. Since atherosclerosis is present in the most endangered COVID-19 group, plaque disability is most likely a contributor. COVID-19-associated cardiac injury, based on ejection fraction decline and troponin I elevation, has also been reported (Shi et al., 2020).

Nutraceutical pleiotropic effects can modulate the proatherogenic changes taking place in the vasculature (Moss and Ramji, 2016.)

Dyslipidemia

Dyslipidemia, a major contributor to CVDs, is usually categorized based on the levels of total cholesterol (TC) as well as LDL cholesterol (LDL-C) exceeding certain levels or when high-density lipoprotein cholesterol (HDL-C) is low. Levels of TC higher than 200 mg/dL or LDL-C higher than 115 mg/dL in adults is considered abnormal. Antihyperlipidemic constituents have been described by Gao et al. (2014).

Pleiotropic effects of nutraceuticals include lipid-lowering (Cicero et al., 2017; Poli and Visioli, 2019), antioxidative, and antiinflammatory actions, which are also beneficial in diminishing sarcopenia. Primarily, they are used in conjunction with lipid-lowering drugs, and the outcome can be achieved by inhibition of cholesterol absorption from the intestine, inhibition of cholesterol synthesis, and enhancement of LDL excretion.

Plant sterols, contained in vegetable oil and nuts, and stanols, collectively referred to as PSS, reduce LDL-C via intestinal absorption (Trautwein et al., 2018). The reported dose−response relationship with 1.3−3 g/day use lowers LDL-C by up to 12%. The lipid-lowering effect of PSS is dose-dependent and achieved by soluble fibers like glucomannan and psyllium that reduce the absorption of intestinal cholesterol. Cholesterol synthesis is inhibited by a number of other nutraceuticals, like red yeast rice (RYR) obtained by fermentation of yeast in rice (*Oryza sativa*). Interestingly, the monacolin K produced is structurally similar to lovastatin.

For the management of hypercholesterolemia, proprotein convertase subtilisin/kexin type 9 (PCSK9) is the key regulator of circulating LDL-C levels (Zhang, 2017), and is mediated by modulation of LDL receptor (LDLR) density on the surface of hepatocytes as a major participant. Inhibition of PCSK9 lowers plasma LDL-C and also serves as an adjunct to statin therapy.

An LDL-C excretion inducer is berberine (BBR). This alkaloid (Bagade et al., 2017) works by downregulating LDL-C. BBR is an inhibitor of PCSK9 by modulation of LDL receptors on hepatocytes of the LDLR and promoting the cholesterol excretion into the bile. BBR with stanols inhibits cholesterol absorption. Nutraceuticals modulate PCSK9 levels by use of the sterol-responsive element binding protein (Momtazi et al., 2017).

Hypertension

Elevated blood pressure is a hallmark of CVDs. Endothelial activation with vascular smooth muscle dysfunction (hypertrophy) (Xiao et al., 2017) and lipid deposition on blood vessel walls are all indicators of hypertension. Additional markers are production of ROS that impairs antioxidant defense mechanisms.

Generation of NO in the vascular endothelium is essential for normal system functioning. NO regulates vascular tone, cellular proliferation, and leukocyte adhesion, and inhibits platelet aggregation. Impaired NO production is part of the pathogenesis of hypertension. NO also has a number of antioxidant properties, and reduces superoxide anion formation and superoxide anion in the blood. NO also upregulates heme oxygenase-1 and ferrite expression.

Hydrogen sulfide production deficiency is also correlated with expressions of hypertension. In addition, there are indications that it reduces ROS production, diminishes vascular smooth muscle cells vascular smooth muscle cells, and induces NO synthesis.

Antihypertensive effects are expressed by secondary metabolites of herbs (Borghi and Cicero, 2017). Notable among these is garlic (*Allium sativum*) which also has antioxidant, antiinflammatory, and antibacterial properties. Its hypotensive mechanisms are based on its organosulfur constituents such as allicin, diallyl disulfides, S-allylcysteine, diallyl trisulfides, and methyl thiosulfonate.

The transcription factor nuclear factor kappa B (NF-κB) induces endothelial cell dysfunction, and inflammation through the release of proinflammatory cytokines yielding derangement of vascular behavior.

The renin–angiotensin–aldosterone system (RAAS) plays an important role in regulating blood pressure. The RAAS system inhibitors increase angiotensin-converting enzyme 2, which is a negative regulator of angiotensin II receptors that the SARS-CoV-2 virus also uses to cause COVID-19 illness, and to invade cells. The connection between the COVID-19 infection and high blood pressure continues to be investigated.

Increased nervous system activity leads to hypertension (Oparil et al., 2003). Elevation of sympathetic activity induces vascular remodeling, norepinephrine level changes, and organ damage (Esler, 2000).

Vitamins have an antihypertensive effect. Vitamin C increases NO, and activates K^+ channels and endothelial function. Plasma ascorbate concentration is inversely correlated with both systolic and diastolic blood pressure. Vitamin D3 levels are also important. The likelihood of hypertension is increased when low levels of D3 are present. With adequate D3, an increase in NO and improved endothelial function are assured. A large number of nutraceuticals are excellent sources of vitamins.

Lists of nutraceuticals with clinically relevant blood pressure-lowering effects and their potential mechanisms are given by Borghi and Cicero (2017), Cicero et al. (2019), Al Disi et al. (2016), Marques et al. (2015), and Houston (2018). Anwar et al. (2016) give insights into the mechanism of plants in countering hypertension.

Herb- and plant-invoked processes

Herbs and plants, due to their nutritional and assumed medicinal value have been of interest since the beginning of time. They are the source of many medications. A large number of publications have detailed the pharmacognosy and biomolecular and clinical aspects of plants. These noteworthy sources include Benzie and Wachtel-Galor (2011), Badal and Delgoda (2017), and Craig (1999). The article by Rastogi et al. (2016) deals with cardiac disorders and traditional herbs. The potential use of ginseng and ginsenosides for CVDs was suggested by Lee and Kim (2014) and Riaz et al. (2019).

The medicinal herb, *Centella asiatica* (Razali et al., 2019) is the source of asiatic acid, a pentacyclic triterpenoid of therapeutic promise (Meeran et al., 2018). It has antioxidative, antiinflammatory, antihypertensive, and cardioprotective properties. Asiatic acid attenuates cardiomyocyte hypertrophy by inhibiting atrial natriuretic peptide and NF-κB binding activity and by reducing the activation of transforming growth factor-β1 signaling (Menezes et al., 2010; James and Dubery, 2009).

Antioxidant plants

Oxidative stress causes the development of diseases including CVDs. ROS induces oxidative damage. Plants can biosynthesize enzymatic as well as nonenzymatic antioxidants. Young and Woodside (2001) published an overview of oxidative stress and subsequent disease genesis, including aspects of atherosclerosis and consequences of LDL oxidation. Further insight was given by Pham-Huy et al. (2008) into the action of free radicals which contain unpaired electrons that make them highly reactive (formation and catabolism of radicals) in several diseases. The chain-breaking process of antioxidants and release of an electron also play an important role in oxidative stress.

Peroxisomes, chloroplasts, and mitochondria are the sites of ROS generation within plants (Kasote et al., 2015). The antioxidant defense system of plants can deal with enzymatic and nonenzymatic systems. The latter rely on carotenoids, flavonoids, and ascorbic acid as well as secondary metabolites like tannins. Carotenoids from fruits and vegetables are excellent antioxidants but their effectiveness is diminished at higher concentrations. Vitamin E (tocopherols) produced in plants is a good redox buffer system and is present in cellular lipid structures. It is an important singlet oxygen scavenger and thus provides protection against lipid peroxidation. Vitamin C is also an excellent antioxidant but the presence of catalytic metal ions in its vicinity can change it to a pro-oxidant.

Panchawat et al. (2010) list 30 antioxidant plants with short descriptions of their activity. Gupta and Sharma (2006) also detail antioxidant plants, describing the relevant parts of the plant with the desired biological activity. Sesquiterpene lactones of the Asteraceae plant family (also found in lettuce, *Lactuca sativa*) that possess strong antiinflammatory capabilities as well as a large group of secondary metabolites also mediate immune response by inhibition of NF-κB as described in Chadwick et al. (2013).

Lee and Kim (2014) claim an antioxidation, reduced platelet adhesion, and improved lipid profile potential for ginseng. Jiang et al. (2019) describe the antioxidant and antiinflammatory properties of *Salvia miltiorrhiza* (red sage), whose primary ingredient, tanshinone, has received considerable pharmacological interest.

Antiinflammatory plants

Endothelial abnormalities leading to CVDs usually derive from inflammation. This is in response to tissue injury or bacterial and viral invasion. The innate or the adaptive immune system responds as needed. The B and T cells of the adaptive immune system counter the attack, producing antibodies or specific receptors. Cytokines, prostaglandins, and leukotrienes and the inflammation-modulating transcription factor κB, regulator of immune and inflammatory response NF-κB proteins, induce transcription of inflammatory-associated genes (Jain et al, 2018; Verma, 2016).

Cyclooxygenase (COX) enzyme activity is responsible for the generation of prostaglandins involved in inflammation and platelet aggregation. Verma (2016) details close to a dozen medicinal plants known for their antiinflammatory activity. Azab et al. (2016) compiled review articles from 2005 to 2016 on the antiinflammation effects of plant products. Rahal and Kumar (2016) reviewed herb-based remedies for cardiovascular dysfunctions.

Park et al. (2017) described the mechanisms of how rubiarbonone C (a triterpenoid from *Rubia philippinensis*) exerts antioxidative and antiinflammatory effects in pathologies related to cardiac remodeling. Deng and Chow (2010) discussed hypolipidemic, antioxidative, and antiinflammatory properties of the microalgae *Spirulina*.

Lipid-lowering nutraceuticals

A large number of studies have demonstrated that nutraceuticals offer lipid-lowering activity that could alleviate hypercholesterolemia. Cicero et al. (2017) published a list of nutraceuticals that have lowering effects on LDL-C and total triglyceride, and elevating effects on HDL-C. Atherosclerotic diseases were the main lipid targets.

Lipid-lowering effects can be achieved in three ways: (1) reduction of intestinal cholesterol absorption; (2) reduction of liver cholesterol synthesis; and (3) elevation of LDL-C excretion. Selected nutraceuticals can be indicated for each action. PSS aid intestinal cholesterol absorption by the formation of solubilized micelles that move stanols from the intestinal lumen (Ras et al., 2013). Soluble fibers, including oats, pectin, and psyllium lower the levels of TC and LDL-C, however, the percentage of reduction varies widely, depending on the fiber.

Glucomannan, derived from *Amorphallus konjac*, is widely used in Asian countries. It does not bind bile acids but reduces the absorption of cholesterol in the jejunum, improves the level of LDL-C in the plasma (Behera and Ray, 2016), and improves the activity of the enzyme that converts cholesterol to bile acid.

Nutraceuticals can inhibit cholesterol synthesis. One example is RYR that results from the fermentation of *Monacus purpurens* in rice, forming monacolins which resemble lovastatin. The cholesterol-lowering mechanism is attributed to the inhibition of 5-hydroxy-3-methylglutaryl-coenzyme A reductase (HMG-CoA) reductase enzyme active in cholesterol synthesis. RYR has been effective in lowering plasma LDL-C levels.

The lipid-lowering effect of garlic is mediated by inhibiting HMG-CoA reductase as well as acetyl CoA synthetase enzymes. Reducing synthetase enzymes is needed for endogenous synthesis of cholesterol.

The fruit bergamot (*Citrus bergamia*) is rich in flavonoid content. It also inhibits HMG-CoA reductase with the ability to reduce cholesterol.

The excretion of cholesterol can be increased by BBR (an alkaloid present in a number of plants). The lipid-lowering effect of BBR comes from inhibition of proprotein convertase subtilisin/kexin type 9 protein through degradation of hepatocyte nuclear factor alpha, limiting degradation of the hepatic LDLR (Abidi et al., 2005).

Vitamin E also lowers lipid concentration by activating peroxisome proliferator-activated receptor, HMG-CoA reductase inhibitor.

There are a large number of nutraceuticals that have lipid-lowering effects. Further discussion is available in Sahebkar et al. (2016), which describes the multiple mechanisms of the lipid-lowering effect of nutraceuticals. Estrada et al. (2014) deal with *Licania cruegeriana*'s antiplatelet effects. Liwa et al. (2017) provide details of the cardiovascular effect of guggulsterone (oleogum resin), BBR, saponins, and garlic. Guggulsterone inhibits NF-κB, a regulator of inflammatory response.

C-phycocyanin, derived from *Spirulina platensis* algae (Liu et al., 2016), has antioxidative, antiinflammatory, and immune-stimulatory properties. It inhibits COX-2 enzyme, which is highly expressed in inflammation. By effectively inhibiting the activity of free radicals, C-phycocyanin regulates lipid levels. Thus, C-phycocyanin can inhibit the progress of atherosclerosis and endothelial cell apoptosis. It is also involved in autoimmune disorders.

The general consensus is that nutraceuticals cannot replace lipid-lowering therapy, but they may help to reduce CV-associated risks.

Nutraceutical modulation of immune function

Polyphenols express both immunomodulatory and antiinflammatory actions (Yahfoufi et al., 2018; Colitti et al., 2019). Basically, polyphenols regulate immunity by modulating immune cells, proinflammatory cytokine synthesis, and gene expression. In addition, they are involved in the inactivation of NF-κB, and modulation of the mitogen-activated protein kinase and arachidonic acid pathways. In addition to controlling protein synthesis, they suppress toll-like receptor and proinflammatory gene expression, inhibit enzymes that are involved in ROS generation, and upregulate endogenous antioxidant enzymes. These and several other compounds are important in the prevention and treatment of a number of diseases, including cardiovascular and other diseases in which inflammation is involved.

CVDs benefit from the effect of polyphenols due to their antihypertensive potential. The polyphenol resveratrol has several useful mechanisms including inhibition of endothelial NO synthase (eNOS) and NF-κB. A widely used nutraceutical quercetin, from *Ginkgo biloba*, is known to modulate immunity (Li et al., 2016). Quercetin is an antiinflammatory flavonol that inhibits COX activity and lipid oxidation. It is also an antioxidant, inhibits platelet aggregation, and stimulates mitochondrial biogenesis. Quercetin is an immunosuppressor of dendritic cells (Huang et al., 2010), and it integrates with incoming signals and conveys them to lymphocytes, thereby affecting the immune response.

Observable markers

Terpenes

Among the compounds synthesized by plants, terpenes play an important role. Although not crucial for normal development and reproduction, they offer protection of the plant in several ways.

Terpenoids (terpenes with an additional *O*-containing functional group), one of the major classes of secondary metabolites, constitute a large class of organic compounds resulting from the rearrangement of several isoprene molecules. Depending on the number of building blocks, terpenoids are classified into monoterpenes, diterpenes, triterpenes, and tetraterpenes. The function of terpenoids is determined by their structure–activity relationship. Thousands of terpenes that affect the cardiac system have been identified (Alves-Silva et al., 2016). Table 5.2 lists a wide range of biological activities of terpenes in relation to the CVS.

By acting directly on the cardiac muscle or indirectly through the vascular system, terpenoids can affect cardiovascular functioning. Diterpenes and triterpenes, by their homeostatic effect expressed by changes in electrophysiology, have cornered most of the attention of recent research on medical uses of herbal products. This includes myocardial enzymes, such as creatine kinase, lactate dehydrogenase, and cardiac troponin, which are also inhibited by some terpenes.

A large number of studies have investigated terpenes for their cardiovascular effects (Alves-Silva et al., 2016; Gonen et al., 2005; Menezes et al., 2010; Roy, 2019; Sharma et al., 2018; Wagner and Elmadfa, 2003). James and Dubery (2009) studied pentacyclic triterpenoids from the medicinal herb *C. asiatica* that is comprised of some 50 different species. Razali et al. (2019) reported that the triterpenes are effective in countering hypertensive, oxidative, and inflammatory conditions. However, detailed insight is needed to explore the underlying molecular pathways. Ciccone et al. (2013) list a large number of published results dealing with carotenoids. High serum levels of carotenoids/antioxidants are associated with a reduction in CVDs. Prominent mention is made of the subclinical and early stages of atherosclerosis.

Preferred terpenes to counter CVD hypertension include citronellol, linalool, and ferruginol. Catalpol (an iridoid glucoside) is used to counter atherosclerosis, while geraniol is used to counter dyslipidemia.

MicroRNA effects

MiRNAs are regulatory molecules consisting of noncoding nucleotides that posttranscriptionally regulate gene expression. These noncoding RNAs can inhibit target gene expression through binding to the 3'-untranslated regions of the target mRNA. As a result, the mRNA is degraded and translationally inhibited (Li and Sarkar, 2015).

TABLE 5.2 Selected cardiac-related terpenoids.

Plant	Terpenoid type	Effect	References
Licania cruegereriana	Lupine triterpenoid	Hypertension, platelet aggregation	Estrada et al., 2014
Centella asiatica	Pentacyclic triterpenoid	Metabolic syndrome, oxidative stress	Sharma et al. (2018); James and Dubery (2009)
Various including Semen Celosine *Trapaeolum majus* L	Eight members of terpenoid family	Aortic atherosclerosis lesion area	Liu et al. (2016)
Rubia phillippinensis	Rubiarbonone C	Inhibits VSMC	Gao et al. (2014); Park et al. (2017)
Thymus vulgaris	Thymol	Suppressed calcium and potassium channels	Magyar et al. (2002)

VSMC, vascular smooth muscle cells.

MiRNAs, posttranscriptional regulators of gene expression, can individually regulate multiple genes, while individual genes can be regulated by multiple miRNAs. This allows the regulation of cardiovascular pathologies such as hypertension and various arrhythmias.

Since miRNAs are present in body fluids including the circulation, they are useful for various therapeutic approaches. Gene regulation has a primary function in the remediation of CVD (Bauersachs and Thum, 2011).

The biogenesis of miRNAs and their use in a wide range of cardiovascular cases, including atherosclerosis, arrhythmia, and hypertension is described in Romaine et al. (2015). These authors summarize miRNAs and their targets of atherosclerosis. Their supplementary reference list provides a notable insight into our current understanding of miRNA. This includes notable miRNAs as biomarkers of acute coronary disease. The study also discusses the role of circulating miRNAs in the diagnosis of stable coronary artery disease.

Curcumin, isoflavone, green tea polyphenol (epigallocatechin gallate), and others alter the expression of miRNAs (Kura et al., 2019). This has also been exploited in cancer research (Rahman et al., 2019). Bioactive nutrients, as part of nutraceuticals, can modulate miRNA-1, miRNA-21, and miRNA-133, which are active in heart function.

Transcription factors

Recently, Zoltani (2019) stated that nutraceutical components of plants are bioactive compounds that are involved in antioxidation, signal transduction, mitochondrial integrity, and genetic changes. Induction of signaling pathways among other transcription factors include NF-κB and Nrf2. They exert antiinflammatory and antioxidative control through inhibition. NF-κB, collectively as a family of structurally related transcription factors, plays a critical role in immunity, inflammation in plasma, and controlling the gene network, especially mature B cells. Promoters and enhancers of genes hold specific NF-κB binding sites (Oeckinghaus and Ghosh, 2009). They play an important role in mediating external stimuli and thus in gene regulation. Diseases produce dysregulation of NF-κB. Inhibitory proteins, including cytoplasmic retention protein (IkB) and protein kinase, phosphorylate IkB. Posttranslational modifications modulate several of the present proteins and NF-κB.

The active ingredient of *Asteraceae* (sesquiterpenoids lactones) is of interest for the treatment of cardiovascular disease through its antiinflammatory effect and as an inhibitor of NF-κB DNA binding.

Epigenetic issues

Epigenetics refers to molecular mechanisms that affect and control gene expression without a change in DNA expression. Epigenetic mechanisms including histone modifications, DNA methylation, and gene silencing by noncoding RNAs, and play important roles in the development and regulation of aberrant behavior leading to disease. Phytochemicals modulate the epigenome by regulating proteins responsible for changes in chromatin.

Age-related changes that culminate in diseases, including autoimmune disorders, are marked by epigenetic and genetic changes (McGregor and Seo, 2016). A primary epigenetic marker is DNA methylation that consists of the addition of a methyl group to cytosine, one of the building blocks of DNA. These regions are of importance for gene silencing.

Nutraceuticals contain a large number of components, including polyphenolic compounds, with medically useful properties. Epigenetic processes, such as DNA methylation/demethylation and histone lysine acetylation/deacetylation, are examples. The epigenetic changes are carried out by enzymes that regulate posttranslational modifications. Curcumin, for example, effectively controls a large number of miRNAs. Epigenetic remodeling typically involves alteration by DNA methyl and histone modification and miRNA altering of gene expression. Epigenetic dysregulation contributes to pathogenesis. Numerous therapeutic approaches rely on correction of aberrant DNA methylation patterns.

Polyphenols, important phytochemicals contained in nutraceuticals, include flavonoids, phenolic acids, and sulforaphane produced by the hydrolysis of glucoraphanin of cruciferous vegetables such as cabbage or broccoli. They exert antiinflammatory and antioxidative properties (Khurana et al., 2013). Epigenetic modulation by several phenolic compounds is also widely used in cancer treatment.

Resveratrol (3,5,4'-trihvdroxystilbene) is a polyphenol contained in a number of plants, especially grapes. It partakes in epigenetic mechanisms that incorporate DNA methyltransferase (DNMT), histone deacetylase (HDAC), and lysine-specific demethylase-l (Fernandes et al., 2017).

The flavonoid quercetin, a plant pigment found in berries and green tea, is an antioxidant and antiinflammatory component (Cook and Samman, 1996). Synergistically used with curcumin it represses cell proliferation that was associated with downregulation of DNMT1 and HDAC1 in tumor growth. A reduction in systolic blood pressure also occurs.

Epigenetic processes also affect miRNAs that are critical regulators of cardiovascular processes. Posttranscriptional RNA modified by methylation can modulate function by creation or elimination of CpG sites (Loscalzo and Handy, 2014).

Cardioprotective effects of nutraceuticals

The effects of nutraceuticals on body processes have attracted considerable attention as noted in the literature.

These include effects on aging, cancer, neuroendocrine processes, metabolic disorders, and CVDs (McCubrey et al., 2017). In many of these treatises, miRNA regulations are prominently mentioned in relation to cardiac dysfunction.

Nutraceuticals tend to reduce circulating levels of LDL-C by modulating cholesterol synthesis in the liver, binding cholesterol within the intestines, and increasing LDL-C receptor uptake in the liver.

Alternatively, nutraceuticals reduce oxidation with antioxidants such as flavones and phenols. Fibrinolytic activity reduces artery plaques, and consequently helps to normalize blood pressure.

The following are prominent nutraceuticals for dealing with cardiac problems.

Resveratrol

This polyphenolic compound, abundant in grapes, possesses antiinflammatory and antioxidant properties. In addition, it upregulates eNOS and scavenges peroxyl radicals (Bonnefont-Rousselot, 2016).

Resveratrol (RVT) has been found useful in treating obesity, diabetes, dyslipidemia, inflammation, and hypertension. Questions of bioavailability remain unanswered. The intracellular molecular targets of this polyphenol include NF-κB and miRNAs (Fernandes et al., 2017). RVT activates the cardioprotective system silent information regulator 1 (SIRT1) that has a role in myocardial ischemia. RVT limits the phenylephrine-induced hypertrophic response. Under ischemic conditions, RVT mediates the inflammatory signaling pathways (Yang et al., 2013). Beside reducing inflammatory enzymes, it also controls molecules that mediate leukocyte adhesion including NF-κB, the mediator of inflammatory signaling.

RVT is also protective in metabolic diseases and the activation of SIRT1 regulates genes exercising metabolic control. The HDAC class of enzymes removes acetyl groups from histone and nonhistone proteins. Transcriptional repression relies on histone deacetylation. Modulating epigenetic patterns by changing the levels of S-adenosylmethionine and S-adenosylhomocyteine by RVT has also been reported.

Berberine

An isoquinoline alkaloid present in *Coptis chinensis* as well as in *Berberis vulgaris*, known to be lipid-lowering, is an inhibitor of PCSK9 (by binding to LDL receptors it reduces LDL uptake). BBR acts directly on the expression of LDL by means of upregulating the receptors that stabilize mRNA pathways (Bagade et al., 2017; also Abidi et al., 2005). It also reduces triglyceride levels and promotes cholesterol excretion into the bile. Along with plant stanols it also inhibits cholesterol absorption. It improves left ventricular function in congestive heart failure patients.

BBR is potentially useful in rendering protection against myocardial ischemia/reperfusion injury. Activation of the SIRT1 reduces oxidative damage and inflammation response (Yu et al., 2016).

Curcumin

The nutraceutical curcumin (*Curcuma longa*), a polyphenol, has found wide use in combating CVDs. As a scavenger of ROS, it primarily acts through its antioxidant and antiinflammatory mechanisms. Due to its limited bioavailability, it is usually used with piperine to increase bioavailability. It is widely used to combat metabolic syndrome that is characterized by hyperlipidemia, elevated LDL-C, triglycerides, and low values of HDL-C (Gliozzi et al., 2013). Curcumin reduces blood pressure, inflammation, and oxidative stress, as it is also a C-reactive protein-lowering agent. Hewlings and Kalman (2017) summarize the widespread use of curcumin for health preservation.

Li et al. (2020) collected the molecular targets of curcumin in CVDs. These include inhibition of p300-HAT (histone acetyltransferase) enzymes that can acetylate lysine amino acids on histone and also on nonhistone proteins. It is also protective in the sense of repressing cardiac hypertrophy and heart failure.

The nuclear factor Nrf2, a transcription factor, contributes to the maintenance of the cellular redox homeostasis. Nrf2 interacts with antioxidant response elements, and promotes transcription of a large number of genes (Houghton et al., 2016).

One of the important transcription factors, NF-κB, is indirectly regulated by curcumin. Increased NF-κB activity has been related to the development of atherosclerosis as well as to chronic inflammation.

Concluding remarks and future directions

Components of nutraceuticals are effective in ameliorating CVDs. Their identification has led to new approaches to countering these diseases. However, considerably more experimental data is needed for the verification and wide acceptance of these methods and applications. The influence of epigenetics on chromatin remodeling with the effect on cardiovascular function is now widely accepted. Quantification of the expected epigenetic alterations by various nutraceuticals on cardiovascular behavior needs further exploration.

References

Abidi, P., Zhou, Y., Jiang, J.D., et al., 2005. Extracellular signal-regulated kinase-dependent stabilization of hepatic low-density lipoprotein receptor mRNA by herbal medicine berberine. Arterioscler Thromb. Vasc. Biol. 25, 2170–2176.

Adegbole, P., Aderigbibe, I., Hammend, W., et al., 2017. Antioxidant and anti-inflammatory medicinal plants have potential role in the treatment of cardiovascular disease: a review. Am. J. Cardiovasc. Dis. 7 (2), 19–32.

Al Disi, S.S., Anwar, M.A., Eid, A.H., 2016. Anti-hypertensive herbs and their mechanism of action: part I. Front. Pharmacol. 6, 323. article.

Alves, Q.L., Camargo, S.B., Silva, D.F., 2019. Role of nutraceuticals in the prevention and treatment of hypertension and cardiovascular diseases. J. Hypertens. Manage. 5, 037. Available from: https://doi.org/10.23937/2474-3690/1510037.

Alves-Silva, J.M., Zucarte, M., Marques, C., et al., 2016. Protective effects of terpenes on the cardiovascular system: current advances and future perspectives. Curr. Med. Chem. 23, 1–42.

Anwar, M.A., Al Disi, S.S., Eid, A.H., 2016. Anti-hypertensive herbs and their mechanisms of action: part II. Front. Pharmacol. 7, Artic. 50.

Aquila, G., Marracino, L., Martino, V., et al., 2019. The use of nutraceuticals to counteract atherosclerosis: the role of the Notch pathway. Oxid. Med. Cell. Longev. 2019, Article ID 5470470.

Azab, A., Nassar, A., Azab, A.N., 2016. Anti-inflammatory activation of natural products. Molecules 21, 1321. Available from: https://doi.org/10.3390/molecules21101321.

Badal, S., Delgoda, R., 2017. Pharmacognosy, Fundamentals, Applications and Strategies. Elsevier, London.

Bagade, A., Tumbigeremutt, V., Pallavi, G., 2017. Cardiovascular effects of berberine: a review of the literature. J. Restorative Med. 6, 37–40.

Banach, M., Patti, A.M., Giglio, R.V., et al., 2018. The role of nutraceuticals in statin intolerant patients. J. Am. Coll. Cardiol. 72 (1), 96–118.

Barrios, V., Escobar, C., Cicero, A.F.G., et al., 2017. A nutraceutical approach (Armolipid Plus) to reduce total and LDL cholesterol in individuals with mild to moderate dyslipidemia: review of clinical evidence'. Atheroscler Suppl. 24, 1–15.

Bauersachs, J., Thum, T., 2011. Biogenesis and regulation of cardiovascular microRNAs. Circulation Res. 109 (3), 334–347.

Behera, S.S., Ray, R.C., 2016. Konjac glucomannan, a promising polysaccharide of *Amorphapallus konjac* K. Koch in health care. Int. J. Biol. Macromol. 92, 942–956.

Benzie, I.F.F., Wachtel-Galor, S. (Eds.), 2011. Herbal Medicine: Biomolecular and Clinical Aspects. second ed. CRC Press/Taylor & Francis, Boca Raton, FL.

Bonnefont-Rousselot, D., 2016. Resveratrol and cardiovascular diseases. Nutrients. 8, 250. Available from: https://doi.org/10.3390/nu8050250.

Borghi, C., Cicero, A.F.G., 2017. Nutraceuticals with a clinically detectable blood pressure-lowering effect: a review of available randomized clinical trials and their meta-analysis. Br. J. Clin. Pharmacol. 83, 163–171.

Chadwick, M., Trewin, H., Gawthrop, F., et al., 2013. Sesquiterpenoids lactones: benefits to plants and people. Int. J. Mol. Sci. 14, 12780–12805.

Chen, J., Cao, J., Fang, l, et al., 2014. Berberine derivatives reduce atherosclerotic plaque size and vulnerability in apoE$^{(-/-)}$ mice. J. Transl. Med. 12, 326. Article.

Ciccone, M.F., Cortese, F., Gesualdo, M., et al., 2013. Dietary intake of carotenoids and their antioxidant and anti-inflammatory effects in cardiovascular care. Mediators Inflamm. 2013, Article ID 782137.

Cicero, A.F.G., Colletti, A., 2018. Handbook of Nutraceuticals for Clinical Use. Springer, Cham.

Cicero, A.F.G., Colletti, A., Bajraktari, G., et al., 2017. Lipid lowering nutraceuticals in clinical practice: position paper from an International Lipid Expert Panel. Arch. Med. Sci. 13, 965–1005.

Cicero, A.F.G., Grassi, D., Tocci, G., et al., 2019. Nutrients and nutraceuticals for the management of high normal blood pressure: an evidence-based consensus document. High Blood Press. Cardiovasc. Prev. 26 (1), 9–25.

Colitti, M., Stefanon, B., Gabai, G., et al., 2019. Oxidative stress and nutraceuticals in the modulation of the immune function: current knowledge in animals of veterinary interest. Antioxidants 8, 28. Available from: https://doi.org/10.3390/antiox8010028.

Cook, N.C., Samman, S., 1996. Flavonoids-chemistry, metabolism, cardioprotective effects, and dietary sources. J. Nutr. Biochem. 7, 66–76.

Craig, W.J., 1999. Health-promoting properties of common herbs. Am. J. Clin. Nutr. 70 (suppl), 491S–499S.

Deng, R., Chow, T.J., 2010. Hypolipidemic, antioxidant and anti-inflammatory activities of microalgae *Spirulina*. Cardiovasc. Ther. 28 (4), e33–e45.

Esler, M., 2000. The sympathetic system and hypertension. Am. J. Hypertens. 13, 99S–105S.

Estrada, O., Contreras, W., Acha, G., et al., 2014. Chemical constituents from *Licania cruegeriana* and their cardiovascular and antiplatelet effects. Molecules 19, 21215–21225.

Fernandes, F.G.S., Silva, G.D.B., Pavan, A.R., et al., 2017. Epigenetic regulatory mechanisms induced by resveratrol. Nutrients 9, 1201.

Gao, Y., Su, Y., Huo, Y., et al., 2014. Identification of antihyperlipidemic constituents from the roots of *Rubia Yunnanensis* Diels. J. Ethnopharmacol. 155 (2), 1315–1321.

Gliozzi, M., Walker, R., Muscoli, S., et al., 2013. Bergamot polyphenolic fraction enhances rosuvastatin-induced effect on LDL-cholesterol, LOX-1 expression and protein kinase B phosphorylation in patients with hyperlipidemia. Int. J. Cardiol. 170 (2), 140–145.

Gonen, A., Harats, D., Rabinkov, A., et al., 2005. The antiatherogenic effect of allicin: possible mode of action. Pathobiology 72 (6), 325–334.

Gupta, V.K., Sharma, S.K., 2006. Plants as natural antioxidants. Nat. Prod. Radiance 5 (4), 326–334.

Hewlings, S.J., Kalman, D.S., 2017. Curcumin: a review of its' effects on human health. Foods 6, 92. Available from: https://doi.org/10.3390/foods6100092.

Houghton, C.A., Fassett, R.G., Coombes, J.S., 2016. Sulforaphane and other nutrigenomic Nrf2 activators: can the clinician's expectation be matched by the reality? Oxid. Med. Cell. Longev. 2016, Article ID 7857186.

Houston, M.C., 2018. Treatment of hypertension with nutrition and nutraceutical supplements: part 1. Alternat. Complement. Ther. 24 (6), 260–275.

Huang, R.-Y., Yu, Y.-L., Cheng, W.-C., et al., 2010. Immunosuppressive effect of quercetin on dendritic cell activation and function. J. Immunol. 184, 6815–6821.

Jain, S., Buttar, H.S., Chintameneni, M., et al., 2018. Prevention of cardiovascular diseases with anti-inflammatory and antioxidant nutraceuticals and herbal products: an overview of pre-clinical and clinical studies. Recent Pat. Inflamm. Allerergy Drug Discov. 12 (2), 145–157.

James, J.J., Dubery, I.A., 2009. Pentacyclic triterpenoids from the medicinal herb, *Centella asiatica* (L.) urban. Molecules 14, 3922–3941.

Jiang, Z., Gao, W., Huang, L., 2019. Tanshinones, critical pharmacological components in *Salvia miltiorrhiza*. Front. Pharmacol. 10, Article 202.

Kasote, D.M., Katyare, S.S., Hegde, M.V., et al., 2015. Significance of antioxidant potential of plants and its relevance to therapeutic applications. Int. J. Biol. Sci. 11, 982–991.

Khurana, S., Venkatraman, K., Hollingsworth, A., et al., 2013. Polyphenols: benefits to the cardiovascular system in health and aging. Nutrients 5 (10), 3779–3827.

Kura, B., Parikh, M., Slezak, J., et al., 2019. The influence of diet on microRNAs that impact cardiovascular disease. Molecules 24 (8), 1509–1525.

Lee, C.H., Kim, J., 2014. A review on the medical potentials of ginseng and ginsenosides on cardiovascular diseases. J. Ginseng Res. 38, 161–166.

Li, Y., Sarkar, F.H., 2015. Targeting epigenetically deregulated miRNA by nutraceuticals: focusing on cancer prevention and treatment. Curr. Pharmacol. Rep. 1, 1–10.

Li, Y., Yao, J., Han, C., et al., 2016. Quercetin, inflammation and immunity. Nutrients 8, 167. Available from: https://doi.org/10.3390/nu8030167.

Li, H., Sureda, A., Devkota, H.P., et al., 2020. Curcumin, the golden spice in treating cardiovascular diseases. Biotechnol. Adv. 38, 107343. Available from: https://doi.org/10.1016/j.biotechadv,2019.01.010.

Liu, Q., Huang, Y., Zhang, R., et al., 2016. Medical application of *Spirulina platensis* derived C- phycocyanin. Evid-Based Complement. Alternat. Med. 2016, Article ID 7803846.

Liwa, A.C., Barton, E.N., Cole, W.C., et al., 2017. Bioactive plant molecules, sources and mechanism of action in the treatment of cardiovascular disease. In: Badal, S., Delgoda, R. (Eds.), Pharmacognosy, Fundamentals, Applications and Strategies. Elsevier, London, pp. 315–328.

Loscalzo, J., Handy, D.E., 2014. Epigenetic modifications: basic mechanism and role in cardiovascular disease (2013 Grover Conference series). Pulm. Circ. 4 (2), 169–174.

Madjid, M., Vela, D., Khalili-Tabriz, H., et al., 2007. Systemic infections cause exaggerated local inflammation in atherosclerotic coronary arteries. Tex. Heart Inst. J. 34 (1), 11–18.

Magyar, J., Szentandrassy, N., Banyasz, T., et al., 2002. Effects of thymol on calcium and potassium currents in canine and human ventricular cardiomyocytes. Br. J. Pharmacol. 136 (2), 330–338.

Marques, F.Z., Booth, S.A., Charchar, F.J., 2015. The emerging role of non-coding RNA in essential hypertension and blood pressure regulation. J. Hum. Hypertens. 29, 459–467.

McCubrey, J.A., Lertpiriyapong, K., Steelman, L.S., et al., 2017. Effects of resveratrol, curcumin, berberine and other nutraceuticals on aging, cancer development, cancer stem cells and microRNAs. Aging 9 (6), 1477–1536.

McGregor, R.A., Seo, D.Y., 2016. miRNAs as nutritional targets in aging. In: Malavolta, M., Moochegiani, E. (Eds.), Molecular Basis of Nutrition and Aging. Elsevier, New York, NY, pp. 277–292.

Meeran, M.F.N., Goyal, S.N., Suchal, K., et al., 2018. Pharmacological properties, molecular mechanisms, and pharmaceutical development of asiatic acid: a pentacyclic triterpenoid of therapeutic promise. Front. Pharmacol. 8, 892. Available from: https://doi.org/10.3389/fphar.2018.00892.

Menezes, I.A.C., Barreto, C.M.N., Antoniolli, A., et al., 2010. Hypotensive activity of terpenes found in essential oils. Z. Naturforsch. C. J. Biosci. 65 (9), 562–566.

Momtazi, A.A., Banach, M., Pirro, M., et al., 2017. Regulation of PCSK9 by nutraceuticals. Pharmacol. Res. 120, 157–169.

Moss, J.W.E., Ramji, D.P., 2016. Nutraceutical therapies for atherosclerosis. Nat. Rev. Cardiol. 13, 513–532.

Nashed, B., Yeganch, B., HayGlass, K.T., et al., 2005. Antiatherogenic effects of dietary plant sterols are associated with inhibition of proinflammatory cytokine production in Apo E-KO mice. J. Nutr. 135, 2438–2444.

Nikhra, V., 2019. Nutraceuticals for improving cardiovascular health and prognosis in cardiovascular disease. EC Nutr. 14 (5), 434–448.

Oeckinghaus, A., Ghosh, S., 2009. The NF-kB family of transcription factors and its regulation. Cold Spring Harb. Perspect. Biol. 1, a000034.

Oparil, S., Zaman, M.A., Calhoun, D.A., 2003. Pathogenesis of hypertension. Ann. Intern. Med. 139, 761–776.

Panchawat, S., Rathore, K.S., Sisodia, S.S., 2010. A review on herbal antioxidants. Int. J. PharmTech. Res. 2 (1), 232–239.

Park, H.S., Quan, K.T., Han, J.H., et al., 2017. Rubiarbonone C inhibits platelet-derived growth factor-induced proliferation and migration of vascular smooth muscle cells through the focal adhesion kinase, MAPK and STAT3 Tyr^{705}signalling pathways. Br. J. Pharmacol. 174 (22), 4140–4154.

Pham-Huy, L.A., He, H., Pham-Huy, C., 2008. Free radical, antioxidants in disease and health. Int. J. Biomed. Sci. 4 (2), 89–96.

Poli, A., Visioli, F., 2019. Pharmacology of nutraceuticals with lipid lowering properties. High Blood Press. Cardiovasc. Prev. 26 (2), 113–118.

Rahal, A., Kumar, A., 2016. Cardiovascular diseases: can herb-based antioxidant and anti-inflammatory functional products be a solution? ROS 2 (6), 404–416.

Rahman, M.M., Brane, A.C., Tollesfsbol, T.O., 2019. MicroRNAs and epigenetics strategies to reverse breast cancer. Cells 8, 1214. Available from: https://doi.org/10.3390/cells8101214.

Ras, R.T., Hiemstra, H., Lin, Y., et al., 2013. Consumption of plant sterol-enriched foods and effects on plasma plant sterol concentrations-a meta-analysis of randomized controlled studies. Atherosclerosis 230, 336–346.

Rastogi, S., Pandey, M.M., Rawat, A.K.S., 2016. Traditional herbs: a remedy for cardiovascular disorders. Phytomedicine 23 (11), 1082–1089.

Razali, N.N.M., Ng, C.T., Fong, L.Y., 2019. Cardiovascular protective effects of *Centella asiatica* and its triterpenes: a review. Planta Med. 85, 1203–1215.

Riaz, M., Rahman, N.U., Zia-Ul-Haq, M., et al., 2019. Ginseng: a dietary supplement as immune-modulator in various diseases. Trends Food Sci. Technol. 83, 12–30.

Romaine, S.P.R., Tomaszewski, M., Condorelli, G., 2015. MicroRNAs in cardiovascular disease: an introduction for clinicians. Heart 101 (12), 921–928. Available from: https://doi.org/10.1136/heartjnl-2013-305402.

Roy, D.N. (Ed.), 2019. Terpenoids Against Human Diseases. CRC Press, Taylor & Francis Group, Boca Raton, FL.

Sahebkar, A., Serban, M.C., Gluba-Brzozka, A., et al., 2016. Lipid-modifying effects of nutraceuticals: an evidence-based approach. Nutrition 32 (11-12), 1179–1192.

Sharma, H., Kumar, P., Deshmukh, P.R., et al., 2018. Pentacyclic triterpenes: new tools to fight metabolic syndrome. Phytomedicine 50, 166–177.

Shi, S., Qin, M., Shen, B., et al., 2020. Association of cardiac injury with mortality in hospitalized patients with COVID-19 in Wuhan, China. JAMA Cardiol. 5, 802–810. Available from: https://doi.org/10.1001/jamacardio.2020.0950.

Shin, S.K., Ha, T.Y., McGregor, R.A., et al., 2011. Long-term curcumin administration protects against atherosclerosis via hepatic regulation of lipoprotein cholesterol metabolism. Mol. Nutr. Food Res. 55, 1829–1840.

Trautwein, E.A., Vermeer, M.A., Hiemstra, H., et al., 2018. LDL-cholesterol lowering of plant sterols and stanols——which factors influence their efficacy? Nutrients 10 (9), 1262–1277.

Verma, S., 2016. Medicinal plants with anti-inflammatory activity. J. Phytopharmacol. 5 (4), 157–159.

Wagner, K.H., Elmadfa, I., 2003. Biol, relevance of terpenoids. Ann. Nutr. Metab. 47, 95–106.

Wink, M., 2018. Plant secondary metabolites modulate insect behavior-steps toward addiction. Front. Physiol. 9, article 364.

Xiao, Y., Yang, Z., Wu, Q.Q., et al., 2017. Cucurbitacin B protects against pressure overload induced cardiac hypertrophy. J. Cell. Biochem. 118 (11), 3899–3910.

Yahfoufi, N., Alsadi, N., Jambi, M., et al., 2018. The immunomodulatory and anti-inflammatory role of polyphenols. Nutrients 10, 1618. Available from: https://doi.org/10.3390/nu10111618.

Yang, Y., Duan, W., Li, Y., et al., 2013. Novel role of silent information regulator 1 in myocardial ischemia. Circulation 128, 2232–2240.

Young, I.S., Woodside, J.V., 2001. Antioxidants in health and disease. J. Clin. Pathol 54, 176–186.

Yu, L., Li, Q., Yu, B., et al., 2016. Berberine attenuates myocardial ischemia/reperfusion injury by reducing oxidative stress and inflammation response: role of silent information regulator 1. Oxid. Med. Cell. Longev 2016, Article ID 1689602.

Zaina, S., Lindholm, M.W., Lund, G., 2005. Nutrition and aberrant DNA methylation patterns in atherosclerosis: more than just hyperhomocysteinemia? J. Nutr. 135, 5–8.

Zhang, P.Y., 2017. PCSK9 as a therapeutic target for cardiovascular diseases. Exp. Ther. Med. 13 (3), 810–814.

Zoltani, C.K., 2016. Nutraceuticals in cardiovascular diseases. In: Gupta, R.C. (Ed.), Nutraceuticals, Efficacy, Safety and Toxicity. Elsevier, Amsterdam, pp. 49–59.

Zoltani, C.K., 2019. Nutraceuticals in cardiovascular diseases. In: Gupta, R.C., Srivastava, A., Lall, R. (Eds.), Nutraceuticals in Veterinary Medicine. Springer Nature, Switzerland AG, Cham, pp. 427–435.

Chapter 6

Nutraceuticals for diabetes and glucose balance

Ramesh C. Gupta, Robin B. Doss, Ramesh C. Garg, Rajiv Lall, Ajay Srivastava and Anita Sinha

Introduction and background

Despite major scientific advances leading to the development and availability of a wide range of safe and effective oral, injectable, and inhalable pharmaceutical products for diabetes, the prevalence of and morbidity due to diabetes and metabolic syndrome is increasing around the world. Based on the Diabetes Statistics Report (Centers for Disease Control and Prevention, 2014), 29.1 million people (9.3% of the US population) have diabetes. Clinically, about 90% of diabetic patients are reported to have type II diabetes (T2DM), which is characterized by insulin resistance and β-cell dysfunction. It is well known that diabetes or prolonged hyperglycemia can affect the function of many systems and organs of the body, causing serious health complications. These complications include heart disease and stroke, nephropathies leading to renal impairment, ocular changes leading to retinal blindness, neuropathies, and peripheral gangrene leading to lower limb amputations. The diabetes-associated medical burden has resulted in a large number of hospitalizations and requires challenging attention from healthcare professionals, amounting to huge direct and indirect costs. According to the American Diabetes Association, an estimated total cost of diagnosed diabetes has risen to $327 billion in 2017 from $245 billion in 2012. Worldwide estimates for 2014 from the International Diabetes Federation (IDF Atlas 6th Edition Update, 2014) indicated that diabetes remains a major worldwide healthcare challenge. The estimates for 2014 include:

1. A total of 387 million people are living with diabetes, and this number is expected to grow to 592 million by 2035.
2. Diabetes expenditures worldwide have reached $612 billion, approximately 1 in 9 healthcare dollars are spent on diabetes.
3. Diabetes-related deaths have reached 4.9 million, approximately one in every seven deaths.
4. A total of 77% of people with diabetes live in low- and middle-income countries and the largest portion of people affected with diabetes are between the ages of 40 and 59. Even more alarming is the fact that these numbers do not include the estimated 179 million people who remain undiagnosed or are new cases. According to the World Health Organization (WHO, 2016), more than 415 million individuals with diabetes were reported in 2015, along with five million deaths in 2015. There were 79,000 estimated new cases of type I diabetes in children in 2013 alone.

The main pharmacological targets for T2DM medications currently available have largely been insulin replacement therapy, reduction of glucose absorption from the intestine, enhancement of insulin production, increasing the release of insulin from the pancreas, and sensitization of insulin receptors or pathways. Other target interventions have included affecting specific pancreatic cells directly or modulating the activity of selected peptides, such as glucagon-like peptide-1 and glucose-dependent insulinotropic polypeptide, or modulating the activity of certain enzymes like dipeptidyl peptidase-4. Some of the newer drugs for diabetes have been approved with strong cautionary statements (or a black box warning) due to the potential for adverse effects of significant impact on human health. Examples include compounds recently identified as thiazolidinediones, gliptins, sodium glucose cotransporter (SGLT), and others. The SGLTs allow the glucose filtered through the kidneys to be reabsorbed into the blood in the proximal renal tubules via SGLT1 and SGLT2 (Wright, 2001). These compounds have been discussed as a cancer risk. Although preclinical animal toxicology data did not conclude a clear cancer risk for dapagliflozin and overall tumor incidence did not increase, an increase in the numbers of female breast cancer and male bladder cancer cases was noted in preclinical animal trials without statistical significance

Nutraceuticals. DOI: https://doi.org/10.1016/B978-0-12-821038-3.00006-9

(Lin and Tseng, 2014). Because these and other potential adverse effects associated with some selected drugs for T2DM listed may require closer monitoring for safe use in patients, and because costs for some of the newer T2DM medications could be prohibitive, there is now an emerging trend for many diabetic patients to look for and use the readily available complementary and alternative medicines (CAMs). This trend is true not only for countries like India or China where alternative medicine has been practiced for centuries, but also for advanced countries like the United States and those in Europe. These approaches are gaining increasing attention because of the easy availability and the perception of fewer adverse effects with CAMs compared to prescription medications. The most commonly prescribed oral T2DM medication is "metformin," and because it originated from a plant source, *Galega officinalis* (Oubre et al., 1997), it adds to the faith in the safety and efficacy of herbal-type CAMs for T2DM.

The CAMs for T2DM have been used with varying success. Despite the skepticism and a lack of pharmaceutical-type weight of evidence for their therapeutic efficacy and safety, the use of herbal remedies increased by approximately 380% in the United States from 1990 to 1997 (Eisenberg et al., 1998). This may have been associated in part with the passage of the Dietary Supplements and Health Education Act in 1994, which does not require evidence for the efficacy and safety of CAMs prior to marketing because no medical or diagnostic claims are allowed. For overall use of CAMs, it is now estimated that 70% of the US population uses dietary supplements at least occasionally and 40% use them regularly (Becker, 2012). The consumer base does not appear to be limited only to those who are poorly informed scientifically or to those less educated, it also includes scientifically well-informed patients.

A serious gap appears to exist with respect to the availability of organized information that can provide dependable, scientifically reviewed, and well-balanced data for the efficacy and safety of selected T2DM nutraceuticals in the format as included in the US Product Insert for pharmaceutical products. This chapter intends to provide such scientific information for nutraceuticals used for managing hyperglycemia or diabetes. When possible, information relating to bioactive components in the nutraceutical, target(s) for the desired effect (mechanism of action), preclinical and clinical efficacy, and margin of safety has been summarized. This should allow the end user to make judicious use of the nutraceutical products.

Pathophysiology of diabetes mellitus

The pathophysiology of T2DM is very complex as multiple targets, mechanisms, and pathways are involved (Gupta et al., 2019). Insulin resistance in target tissues (skeletal muscle, adipose tissue, liver, etc.) is a characteristic feature and major contributing factor to T2DM (DeFronzo and Tripathy, 2009; Kang et al., 2016). Mitochondrial dysfunction and oxidative stress (Kaneto et al., 2007; Kim et al., 2008; Kangralkar et al., 2010; Raza et al., 2011), and increased ROS-impairing phosphatidylinositol 3-kinase/AKT pathway are reported to be involved in insulin resistance and T2DM (Wang et al., 2011; Naveen and Baskaran, 2018). An increase in insulin resistance leads to elevated gluconeogenesis and reduced glycogen synthesis in the liver, thereby causing decreased insulin secretion from pancreatic β cells, and hyperglycemia (Yan et al., 2018). Other important targets in T2DM include α-amylase, α-glucosidase, dipeptidyl peptidase-IV, aldolase reductase, peroxisome proliferator-activated receptor gamma, 5′-adenosine monophosphate-activated protein kinase (AMP kinase), and glucose transporter type 4 (Naveen and Baskaran, 2018). Insulin resistance and hyperglycemia can also be caused by alterations in endothelial homeostasis (due to increased proinflammatory cytokines and reduced adiponectin secretion). These conditions are associated with altered gene expression and cell signaling in the vascular endothelium, thereby affecting release of endothelium-derived factors, activation of NADPH oxidase, uncoupling of endothelial NOS, and the expression of endothelin-1 (reviewed in Gupta et al., 2019).

Nutraceuticals indicated in diabetes mellitus

The most commonly used T2DM nutraceutical compounds are discussed here; these include cinnamon, fenugreek, *Momordica charantia* (bitter melon), vitamin D, berberine (BBR), chromium, *Gymnema sylvestre* (Gur-Mar), ginger (gingerol and shogaol), and others.

Cinnamon

Cinnamon has been used as a spice in many countries for centuries. There are hundreds of species in the genus *Cinnamomum*, but only four species are used to obtain the spice cinnamon (Jayaprakasha and Rao, 2011). The commonly used varieties of cinnamon are the Ceylon or "true" cinnamon (*Cinnamomum zeylanicum*) and the Chinese Cassia cinnamon (*Cinnamomum aromaticum*). There are an abundance of in vitro and in vivo animal studies supporting the efficacy of cinnamon powder, aqueous extract fraction, cinnamaldehyde, and other active ingredients in cinnamon for use in managing hyperglycemia in animals.

Clinical studies supporting the antihyperglycemic effect of cinnamon or the active components of cinnamon in T2DM patients are limited and efficacy in humans does not seem to be well-established. There are clinical investigations that have failed to document a clearly antihyperglycemic effect in T2DM patients. Variability in clinical effectiveness could largely be related to: (1) the type of cinnamon product or formulation used in the study; (2) content of bioactive component(s); (3) dose; (4) duration of treatment; (5) differences in patient population; and (6) concomitant medications. The variability in clinical effectiveness of cinnamon is illustrated by authors urging clinicians to refrain from recommending cinnamon supplementation in place of the proven standard of care (Pham et al., 2007) and regulatory approval of cinnamon tablets (60 mg cinnamon extract isolated from 2.4 g of Chinese *C. aromaticum* and prepared by Shanghai Yitian Bio-Synthetic Co., Ltd., Shanghai, China) by the State Food and Drug Administration of China (Lu et al., 2012).

The reproducibility of some nonclinical efficacy studies is also not without questions. Because nonclinical studies for nutraceuticals are currently not performed in compliance with Good Laboratory Practice (GLP) standards, attention to reproducibility of data is not a primary focus of studies. As an example, lack of reproducibility in nonclinical data is illustrated by a marked difference in the dose of the positive control drug "glibenclamide" used at 0.6 mg/kg in one rat study (Subash Babu et al., 2007) and 200 mg/kg in another rat study (Anand et al., 2010). A positive control was needed to determine the cinnamaldehyde dose equivalent to the positive control. The cinnamaldehyde dose of 20 mg/kg in streptozotocin (STZ)-treated diabetic rats was considered almost equivalent.

With the dose of STZ for inducing experimental diabetes in two studies remaining nearly comparable (50−60 mg/kg via intraperitoneal injection), the source or stability of the "glibenclamide" test sample used in the 2010 study could have been questioned and verified if this was a GLP study. The clinical data are not abundant, with several clinical studies supporting efficacy in T2DM patients; however, some studies suggest no clear evidence of efficacy. Again, there could be several reasons for variability in efficacy in clinical studies. Additionally, authentication of the test drug with respect to content and type of bioactive component(s) used in the trials was not documented. These and other topics are discussed here in greater detail.

Bioactive components

The major bioactive components identified by different researchers include:

1. *Cinnamon aqueous extracts*: Used in a study by Ziegenfuss et al. (2006) and other investigators. Yield for extract was (1:20) Cinnulin PF used by Ziegenfuss et al. (2006).
2. *Different water-soluble fractions in cinnamon*: These have included B-type procyanidins at an early stage, methyl hydroxy chalcone polymer (MHCP) as discussed by Jarvill-Taylor et al. (2001), and the A-type doubly linked procyanidin oligomers (Anderson et al., 2004) commonly called "polyphenolic polymers."
3. *Cinnamaldehyde*: Used by several investigators, including Anand et al. (2010).
4. *Cinnamon aqueous extract approved by the Food and Drug Administration of China (CFDA) for human use (Approval Number G20110080)*: Each approved tablet contained 60 mg of cinnamon extract derived from 2.4 g of Chinese *C. aromaticum* (Lu et al., 2012).

Preclinical efficacy

The major scientific push for considering cinnamon to be a supplement of value in the management of T2DM came from the impressive work of Anderson et al. (2004). These investigators concluded that doubly linked, water-soluble A-type polymers, possibly misidentified earlier as MHCP (Jarvill-Taylor et al., 2001), are the active components. Several in vitro investigations reporting the inhibitory effect on α-amylase activity, α-glucosidase activity, increase in glucose uptake by adipocytes, effect on muscle glucose transporter-4 (GLUT4) level, and insulin receptor sensitizing effect of cinnamon were reviewed by Ranasinghe et al. (2012) (see Table 6.1). Other investigations using animals have also supported the antidiabetic effect of cinnamon. Using a rat model, Qin et al. (2003) postulated that cinnamon potentiates insulin-mediated glucose uptake by enhancing the insulin-signaling pathway in skeletal muscle. Jarvill-Taylor et al. (2001) observed that when cinnamon and insulin were combined in vitro, the effect was greater than expected, implying a synergistic relationship. Additionally, Kim et al. (2006) conducted a study using mice and suggested that cinnamon acts either by increasing sensitivity to endogenous insulin or by slowing the intestinal absorption of carbohydrates.

There are systematic reviews relating to the efficacy of *C. aromaticum* in diabetes (Baker et al., 2008), but almost no scientific reviews exploring therapeutic efficacy and safety of cinnamon for use in diabetes are available. A comprehensive scientific review about the efficacy and safety of *C. zeylanicum*, a "true" Ceylon cinnamon, has been published by Ranasinghe et al. (2012), tabulating eight in vitro and 11 in vivo preclinical investigations supporting the role of cinnamon in diabetes management.

Further, cinnamaldehyde is one of the active components of cinnamon, forming 90% of the essential oil in cinnamon bark and 2% of cinnamon powder. This is what imparts the typical cinnamon flavor. There are differing opinions about the bioactive component(s) of cinnamon

TABLE 6.1 Summary of clinical trial data for cinnamon or its components used in T2DM or related patients.

Clinical study reference/year/ country	Number of subjects randomized in study, type of diabetes	Type of cinnamon, dose, duration, information relating to trial, bioactive component	Compliance with good clinical practices	Results	Comments
				Effect on FBG and/HbA1c	
Khan et al. (2003), Pakistan (first clinical trial for cinnamon)	60; T2DM patients, continued their regular oral T2DM medications	*Cinnamon cassia powder*, 1, 3, or 6 g/day in capsules (500 mg) for 40 days +20-day washout; included three parallel placebo groups	No	Lowered the FBG by 18%–29% compared to each placebo group; no dose response	Appears to have reached the maximal effect with 1 g/day dose
Mang et al. (2006) Germany	79; T2DM patients not on insulin, continued oral T2DM medications	*Cinnamon extract* equivalent to 3 g cinnamon powder/day for 4 months	No	*HbA1c*: No effect; *mean FBG*: significantly higher reduction in the cinnamon group (10.3%) than in the placebo group (3.4%)	Cinnamon had a moderate effect in further reduction in FBG; effect took 4 months
Vanschoonbeek et al. (2006) The Netherlands	25; T2DM postmenopausal women, continued oral T2DM medicationsStable for drug effect	*Indonesian cinnamon* (*C. cassia*) 1.5 g/day or placebo (one 500 mg capsule 3 times daily), placebo-controlled, double-blind study, 6- to 7-week treatment duration	No	*HbA1c*: No effect; *FBG*: No effect	Lack of effect may be related to a shorter duration of study, particularly for *HbA1c*
Ziegenfuss et al. (2006) United States	22; Metabolic syndrome	*Cinnulin PF* 1 g/day (500 mg of cinnulin extract equivalent to 10 g of cinnamon powder) for 12 weeks, two 250-mg capsules twice daily (breakfast and dinner)	No	*HbA1c*: No data; *FBG* = lowered by 9.8 mg/dL in cinnulin group; increase of 1.1 mg/dL in the placebo group	Cinnulin PF helpful supplement for reducing FBG
Blevins et al. (2007) United States	60; T2DM patients not on insulin, continued their regular oral T2DM medications, no changes allowed in these or other related medication	*Cinnamon* (*C. cassia*) 1 g/day or placebo (wheat flour) for 3 months	No	There were no significant differences between the cinnamon and placebo groupsGroups for FBG or HbA1c	Effects of cinnamon can differ by population; not generally recommended for the US population
Crawford (2009) United States	109; T2DM patients not on insulin, continued oral T2DM medication	*Cinnamon powder* 1 g/day (Puritan Pride), 500 mg capsules for 3 months; double-blind trial, followed Title 32 Code of Federal Regulations 219	No GCPs, but followed IRB instructions	*HbA1c*: A reduction of 0.83% compared to 0.37% in placebo; *FBG*: Not monitored	Cinnamon supplementation helpful to the regular oral medications
Akilen et al. (2010) United Kingdom	58; T2DM patients not on insulin, continued oral T2DM medication	*Cinnamon powder* 2 g/day for 12 weeks in 500-mg capsule, placebo included; double-blind study; first cinnamon trial in the United Kingdom	No GCP, but good documentation for test drug	*HbA1c*: 8.22–7.86%, no effect in placebo group; *FBG*: A significant decrease	Cinnamon supplementation helpful in addition to oral medications
Lu et al. (2012) China	66; T2DM, patients not on insulin, continued on gliclazide for the entire 3-month period	*Cinnamon extract* approved by China's FDA (60 mg of extract from 2.4 g cinnamon); low (120 mg/day) and high doses 360 mg/day), and placebo for a 3-month duration	No	*HbA1c*: Average reduction of 0.92%; *FBG*: Significant decrease from 9.0 to 7.99 mmol/L	Form of cinnamon used and extraction method may affect the efficacy of cinnamon
Beejmohun et al. (2014) France	18; Healthy volunteers not on any T2DM medications	*Ceylon cinnamon extract* (CCE), a hydroalcoholic extract (10:1). 1 g of CCE dosed 30 min prior to a test meal	Yes (first cinnamon trial run as a trial for a pharmaceutical drug candidate)	For the first 60 min, the area under the blood glucose curve was 21.2% lower compared to the placebo Reached statistical significance	This confirms CCE may be working by inhibiting α-amylase activity observed in rat in this study

T2DM, Type II diabetes mellitus; FBG, fasting blood glucose; HbA1c, glycosylated hemoglobin; FDA, Food and Drug Administration; GCP, Good Clinical Practice; IRB, Institutional Review Board.
Source: Modified and updated from Crawford, P., 2009. Effectiveness of cinnamon for lowering hemoglobin A1C in patients with type 2 diabetes: a randomized, controlled trial. J. Am. Board Fam. Med. 22 (5), 507–512.

that drive the antidiabetic effect. Cinnamaldehyde has been investigated by several groups for its antidiabetic effect (Subash Babu et al., 2007; Zhang et al., 2008; Anand et al., 2010). These studies have supported the antidiabetic potential of cinnamaldehyde.

Some cinnamon varieties like *Cinnamomum cassia* contain an appreciable amount of "coumarin," but it is found only in trace amounts in *C. zeylanicum*. Because high levels of coumarin can have adverse effects (Abraham et al., 2010), several agencies have established a maximal total daily intake (TDI) of coumarin that can be consumed without any adverse effects.

Clinical efficacy

The clinical trial data for the effect of cinnamon in T2DM patients are relatively limited. A study in Pakistan by Khan et al. (2003) was the first known clinical trial performed for cinnamon, whereas studies related to the bioactive component (Anderson et al., 2004), in progress at that time, indicated that the water-soluble polyphenolic polymer found in cinnamon may be the bioactive component responsible for potentiating the insulin action and may also function as an antioxidant. Several other clinical trials (Table 6.1) have also indicated varying degrees of success, but there are few studies or analyses that did not show the glucose-lowering effect of cinnamon.

One such study was performed involving 25 postmenopausal women using cinnamon (*C. cassia* powder) at a dose of 1.5 g/day for 6 weeks (Vanschoonbeek et al., 2006). In this study, all treated and placebo T2DM patients were stable on some form of clinically established standard oral diabetic agent. Medication had not been modified for the previous 3 months. Interpretation of results as "no effect" could have been summarized as "cinnamon powder (1.5 g/day) did not show any improvement in effect in patients already established as stable for antidiabetic effect with the use of oral pharmaceutical agents." The short duration of cinnamon exposure was likely not adequate for a subject already stable on traditional oral antihyperglycemics. In another study (Baker et al., 2008), a metaanalysis of five prospective, randomized, controlled trials ($n = 282$) performed until 2007, it was concluded that cinnamon did not significantly alter hemoglobin A1c (HbA1c) and fasting blood glucose (FBG). One of the five clinical trials in this metaanalysis was a study of T2DM postmenopausal women already discussed (Vanschoonbeek et al., 2006). Another study designated as a study with negative findings (Khan et al., 2003) was the first known clinical trial conducted with cinnamon in T2DM subjects. In this study, a significant reduction (18%–29%) in serum glucose was not considered as adequate by the metaanalysis because it lacked a dose–response. The dose–response of a drug may not be an essential factor in all cases, particularly if the lowest dose in the study already shows a maximal observable effect for the test formulation. The lowest dose in this study was a relatively large dose of 1 g/day. Conversely, this study indicated that there is no clinical advantage in dosing cinnamon powder at more than 1 g/day.

For a better effect, investigators should be searching for a better formulation with a higher level of bioactive components or a mixture of two or more nutraceutical components. Several other clinical trials with cinnamon followed the first clinical trial by Khan et al. (2003). These were in different geographic regions of the world. A study by Mang et al. (2006) showed efficacy in Western patients who were also using treatments in accordance with then-current diabetes guidelines. With respect to the need for different formulations, use of an aqueous cinnamon extract led to a moderate effect, showing a significantly higher reduction in mean fasting glucose in the cinnamon group (10.3%) compared to the placebo group (3.4%). Still, further work is needed with respect to optimization of the cinnamon source, manufacturer, and formulation. Approval of a cinnamon product by the state FDA in China (Lu et al., 2012) is an indication that well-optimized and standardized formulations offer promise for better cinnamon efficacy in the future.

Acute effects of 1 g cinnamon extract administered orally in two 500-mg capsules (as a single dose) on postprandial blood glucose and insulin response in healthy male and female volunteers have been reported by Beejmohun et al. (2014). The capsules were administered approximately 30 min prior to meal time as the extract appears to work by inhibiting the α-amylase enzymes involved in carbohydrate digestion. The postmeal glycemic response reduced by approximately 20% and approximately 15% between 0 and 60 min and between 0 and 120 min postmeal absorption, respectively. This effect is shown in Fig. 6.1.

Mechanism(s) of action

Taher et al. (2006) isolated cinnamtannin B1 (a proanthocyanidin compound) from the bark of *C. zeylanicum*. Cinnamtannin B1 activated the phosphorylation of insulin receptor β-subunit on 3T3-L1 adipocytes. This effect was significantly greater than the effect produced by insulin alone (100 nm). Additionally, the enzyme "protein tyrosine phosphatase 1B," which has a role in dephosphorylating the insulin receptors, resulting in an increase in insulin resistance, is blocked by certain drugs and cinnamon improves insulin sensitivity. An investigation using the main bioactive fraction of cinnamon (0.1 *N* NH_4OH extract) was tested by Imparl-Radosevich et al. (1998) for its effect on PTP-1B activity in an in vitro experiment using rat adipose cells. The activity of enzyme PTP-1B

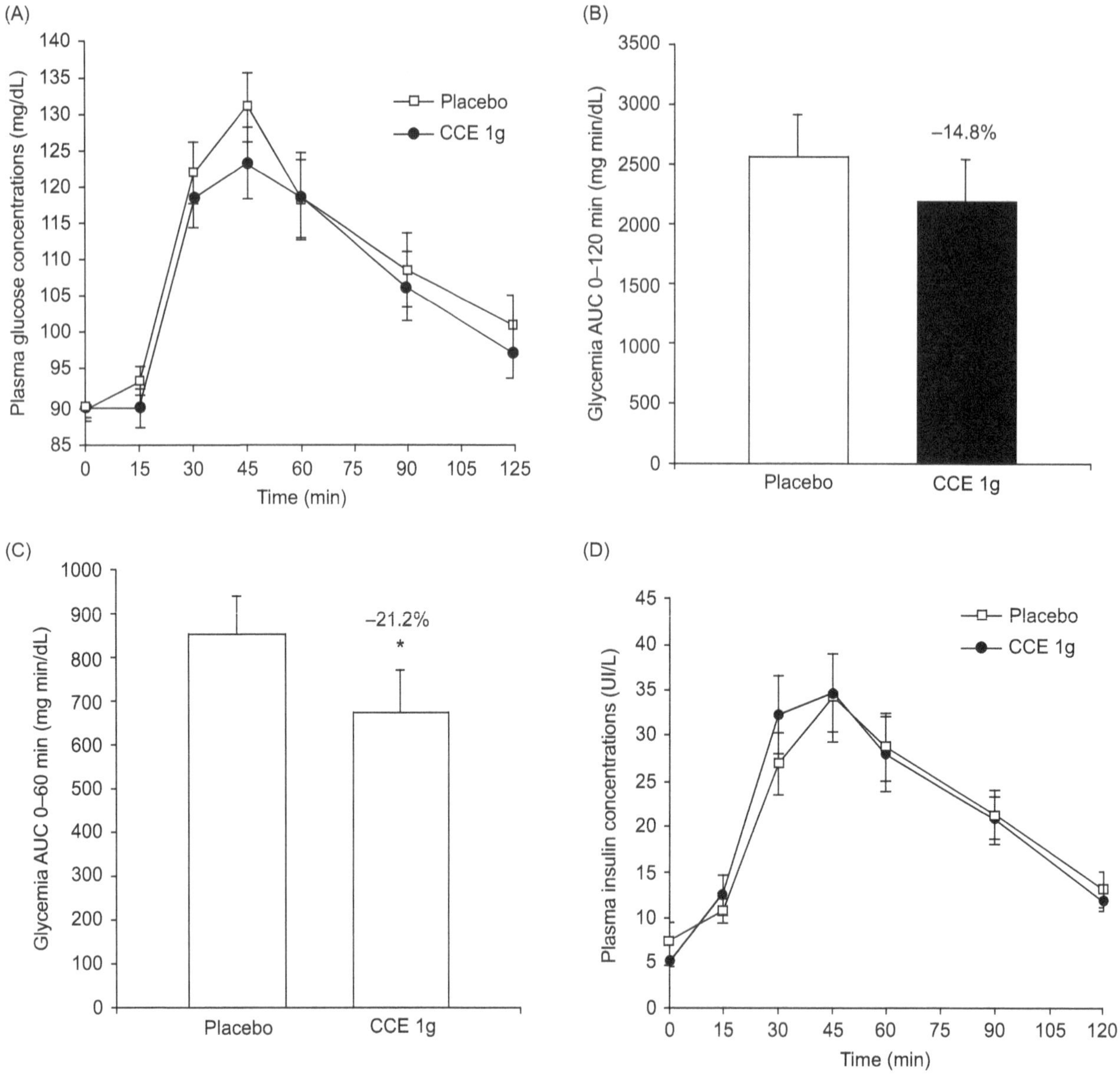

FIGURE 6.1 Effect of CCE (1 g dose) on blood glucose and insulin response after a standard meal in humans. (A) Blood glucose concentration over 120 min, (B) glycemia area under the curve (AUC) for 0–120 min, (C) glycemia AUC for 0–60 min, and (D) insulin concentrations over 120 min. Values represent mean ± SEM, $n = 16$. Percent inhibition of test meal glycemic response is indicated. Change in CCE effect from placebo, $*P < .05$. *CCE, Ceylon cinnamon extract. Modified with permission from Beejmohun, V., Peytavy-Izard, M., Mignon, C., 2014. Acute effect of Ceylon cinnamon extract on postprandial glycemia: alpha-amylase inhibition, starch tolerance test in rats, and randomized crossover clinical trial in healthy volunteers. BMC Compl. Alt. Med. 14, 351.*

was decreased by cinnamon. Based on this effect, it is believed that cinnamon helps in balancing blood glucose by lowering the insulin resistance. Cinnamon appears to lack this effect in the absence of endogenous insulin. This is supported by no efficacy in a cinnamon clinical trial in which 72 adolescents with type I diabetes were the study subjects (Altschuler et al., 2007). The observation of "no effect of cinnamon" in type I diabetics is in line with the majority of published research.

Stimulation of GLUT4 production and translocation to plasma membranes in muscles and brown adipose tissue (Shen et al., 2010) by cinnamon extract and in peripheral tissues (Anand et al., 2010) by cinnamaldehyde have been reported. Additionally, cinnamon extract has also been shown to increase the expression of GLUT1 in macrophages (Cao et al., 2008), which is also thought to be responsible for basal glucose uptake. In prediabetics, consumption of 500 mg cinnamon/day for 12 weeks may reduce oxidative stress by 14%. In addition, cinnamon may increase insulin secretion, increase insulin sensitivity, and lower blood glucose by moving glucose into cells (reviewed in Gupta et al., 2019).

In vitro, *C. zeylanicum* has demonstrated the potential for reducing postprandial intestinal glucose absorption by inhibiting the activity of enzymes like pancreatic α-amylase (Beejmohun et al., 2014) and intestinal α-glucosidase (Adisakwattana et al., 2011) involved in carbohydrate breakdown and by stimulating glucose metabolism, glucose homeostasis, and glycogen synthesis (Anand et al., 2010).

Margin of safety for clinical doses

Safety information has been discussed collectively for cinnamon powder or cinnamon extract (aqueous or alcoholic) and for cinnamaldehyde. Where no observed adverse effect level (NOAEL) information based on a repeat-dose animal toxicity study was available, an estimate of the safe dose for humans or a safety margin with respect to the highest clinical dose already used in human clinical trial(s) has been included.

Most available nonclinical data are for cinnamaldehyde. The acute oral LD_{50} of 1850 mg/kg for cinnamaldehyde has been reported in rats by Subash Babu et al. (2007). This has ranged from 2220 to 3400 mg/kg based on other data (WHO/FAO, 1981), indicating a good margin of safety following acute oral overdose. In view of the extended daily use of cinnamon anticipated for T2DM patients, data from repeat-dose animal toxicity studies are considered more relevant. Such a study was performed by Hagan et al. (1967) using cinnamaldehyde in rat diets at 1000, 2500, and 10,000 ppm for 16 weeks. In view of only minor effects (slight hepatic cell swelling and slight hyperkeratosis of the squamous portion of rat stomach) at the 10,000 ppm dose, the highest dose of 10,000 ppm can be considered the NOAEL. Assuming this NOAEL dose equates to 500 mg/kg/day in rats, it would translate to a human equivalent dose of 80.6 mg/kg/day based on comparison of the body surface area between rats and humans. The highest efficacious dose of cinnamaldehyde used in clinical trials (Table 6.1) is 20 mg/day, which equates to 0.33 mg/kg/day for a 60-kg person. Assuming 20 mg/day as the effective dose for T2DM (Table 6.1), the margin of safety for the effective cinnamaldehyde dose equates to approximately 244 (80.6 divided by 0.33), which is considered a very desirable safety margin for oral exposure.

Cinnamon powder has been used in humans up to a dose of 6 g/day (Khan et al., 2003) without any adverse effects. These human safety data, when viewed in terms of cinnamaldehyde safety data, tend to suggest that cinnamon powder could be dosed without adverse effects at even higher doses.

Coumarin is another constituent found in relatively high quantities (700–12,230 mg/kg) in certain species of cinnamon such as *C. cassia* (Abraham et al., 2010). The high exposure of coumarin can have harmful effects, such as hepatotoxicity and a delay in blood coagulation. Other potential effects of coumarin have been discussed by Abraham et al. (2010), and these relate to an increase in the incidence of liver and kidney tumors in rodents. This is not considered a risk to humans because it was observed only at high doses with a threshold effect, and human exposure to coumarin from cinnamon is anticipated to be relatively low. To determine the highest coumarin exposure without any adverse effects, a subacute and chronic toxicity study was performed using dogs (Hagan et al., 1967), which is the most sensitive species. This study indicated the dose of 10 mg/kg/day as the NOAEL. Based on this NOAEL, using a safety factor of 100, the TDI of coumarin for human was estimated to be 0.1 mg/kg body weight (EFSA, 2004).

Fenugreek

Information for fenugreek with respect to its beneficial effects in hyperglycemia or T2DM is discussed in detail in Chapter 37, Fenugreek: Multiple Health Benefits.

Bitter melon (*M. charantia*)

The *M. charantia* plant belongs to the Cucurbitaceae family and grows as a slender vine in the tropics and subtropics. In this chapter, the name *M. charantia* (MC) is used interchangeably for bitter melon. The fruit is the main part of the plant that has been used for food and medicinal purposes. It is oblong in shape, narrow at both ends, and 6–16 cm long. The outer skin of the fruit is thick and ribbed, green when raw, and orange when ripe. When ripe, it becomes soft and fleshy, opening to reveal pendulous seeds covered with red pulp. It is bitter in taste but has been used as food in many parts of the world, including India, China, East Africa, Central and South America, and the Caribbean. It is also available in the United States and the United Kingdom. For uses related to various ailments, including diabetes, the fruit has been used raw, ripe, or dried, or only the ripe seed. The fruit has shown not only potent antihyperglycemic activity in various animal models and clinical trials but also a dose-related hypoglycemic effect at higher doses in normoglycemic rats (Ojewole et al., 2006). Bitter melon should be used with great caution in diabetic patients, especially when given as an adjuvant to glucose-lowering drugs (Chan et al., 2016; Gupta et al., 2017).

Because more than 100 published reports have described the antihyperglycemic and hypolipidemic properties of the plant (Chaturvedi, 2012; Joseph and Jini, 2013), information related to the antihyperglycemic effects of MC in this chapter is largely limited to recent, comprehensive reviews only. One such publication (Alam et al., 2015) provides a detailed and complete review of biological activities and beneficial effects of MC, including its effect in hyperglycemia and T2DM conditions. In the following sections, most information is based on data summarized by Alam et al. (2015). The reader is directed to this publication for more details.

Bioactive component(s)

Bitter melon contains several biologically active chemical compounds such as glycosides, saponins, alkaloids, fixed oils, triterpenes, proteins, and steroids (Raman and Lau, 1966; Gupta et al., 2018). Although biologically active chemical constituents have been isolated from different parts of the plant, it is not clear what part of the plant or what component of the fruit is responsible for the antihyperglycemic effect.

Preclinical efficacy

Based on data summarized by Alam et al. (2015), the preclinical efficacy data can be presented in terms of the effect of various parts of bitter melon, pulp juice, different extracts, or whole bitter melon fruit on hyperglycemia in animal models of hyperglycemia or in T2DM patients. These data include: (1) aqueous extract of bitter melon lowered hyperglycemia induced by both oral and intraperitoneal glucose load without altering insulin response in normal mice; (2) aqueous and alkaline chloroform extracts also reduced hyperglycemia in diabetic mice; (3) pulp juice of bitter melon lowered FBG and glucose intolerance in T2DM model rats, whereas no effect was seen in STZ-treated type I diabetes rats; (4) in alloxan-induced diabetic rats, blood sugar level was lowered after 3 weeks of treatment with aqueous extract of bitter melon fruits; and (5) plant extract also improved glucose intolerance in STZ-treated diabetic rats and increased glycogen synthesis in the liver. The aqueous extract powder of fresh, unripe, whole fruits at a dose of 20 mg/kg body weight reduced FBG by 48%, which is comparable to the effect of glibenclamide (a well-known oral antidiabetic drug) in rats.

Even a single, high, acute oral dose of whole plant extract has led to a significant dose-related decline in glycemia not only in fasted STZ-treated diabetic rats but also in normoglycemic rats (Ojewole et al., 2006). To lower blood glucose in normoglycemic rats by 52.83%, an impractical high dose of 800 mg/kg bitter melon was needed; however, a 72.03% reduction was seen in the positive control group (chlorpropamide 250 mg/kg) compared to that achieved with the predose level. The exact mechanism for the hypoglycemic effect after a single MC dose in normoglycemic rats is not clear. However, a discussion that provides some information about the probable mechanism is provided in this chapter. Information generated by Sridhar et al. (2008) indicates bitter melon-related improvement in insulin sensitivity, glucose tolerance, and insulin signaling in rats with high-fat diet-induced insulin resistance. The hypoglycemic effect in normoglycemic and diabetic rats was not lowered significantly at doses of 10–20 mg/kg, which corresponds well with data from a clinical study (Fuangchan et al., 2011) in which a relatively high dose (200 mg/day) was needed to show any hypoglycemic effect in T2DM subjects. Even at this high dose, the effect was evident only for blood fructosamine level; the blood glucose level remained almost unchanged.

Clinical efficacy

The clinical efficacy for bitter melon has not been investigated in detail. Based on limited clinical trial information discussed by Alam et al. (2015), bitter melon is considered moderately effective as a supplement for managing glucose levels in T2DM patients. Specific summary information relating to the clinical studies providing relevant information for this chapter has been reproduced from those included by Alam et al. (2015) and summarized in Table 6.2.

Mechanisms of action

There are several mechanisms involved in the hypoglycemic effect of bitter melon or its extracts in experimental animals and in T2DM patients. Using supporting literature, many potential mechanisms have already been described by Alam et al. (2015) and are discussed in brief here:

1. *Prevention of glucose absorption in the alimentary tract*: This view is supported by a significant reduction in the Na^+/K^+-dependent absorption of glucose from the intestinal mucosa in STZ-induced diabetic rats. Intestinal inhibition of carbohydrate-metabolizing enzymes like α-amylase, α-glucosidase, and pancreatic lipase in the intestinal area by bitter melon also limits the availability and absorption of glucose through the gut wall.
2. *Enhancing glucose uptake by tissues*: Several authors have reported that the plant extract improves the glucose uptake in cells, thereby increasing glucose metabolism. Oral supplementation of the plant has increased the muscle content of facilitative GLUT4 proteins, which might be responsible for the significant improvement of oral glucose tolerance in KK-Ay mice (Miura et al., 2001), an animal model of type II diabetes with hyperinsulinemia. Similar results were also reported by other investigators. Shih et al. (2009) reported that bitter melon extract significantly increased mRNA expression and GLUT4 in skeletal

TABLE 6.2 Clinical studies of bitter melon in types I and II diabetics

Study design	Subjects	Dose and duration	Outcome	References
Case study	100 moderate type I diabetics	Aqueous, homogenized suspension of the MC vegetable pulp	Significant reduction of both fasting and postprandial serum glucose	Ahmad et al. (1999)
Case study	15 type I diabetics of either sex (age, 52–65 years)	200 mg of MC twice daily for 7 days plus half doses of metformin or glibenclamide or both daily	The extract acted in synergism with oral hypoglycemic agents and potentiated the hypoglycemia effect; most likely a noninsulin mechanism	Tongia et al. (2004)
Randomized, double-blind, placebo-controlled trial	40 patients, age 18 years or older	Two capsules of MC three times per day after meals, for 3 months	No significant effect on mean fasting blood sugar (FBS), total cholesterol, and weight or on serum creatinine, ALT, AST, sodium, and potassium	Dans et al. (2007)
Multicenter, randomized, double-blind, active-control trial	129 patients randomized to metformin or MC groups at 500, 1000, or 2000 mg/day	MC at 500 or 1000 or 2000 mg/day or metformin 1000 mg/day for 4 weeks	2000 mg/day dose led to a significant decline in fructosamine at week 4, but 500 and 1000 mg/day doses did not work	Fuangchan et al. (2011)
Open-label uncontrolled supplementation trial	21 men and 21 women with a mean age of 45.7 ± 11.4 years (23–63 years)	4.8 g lyophilized MC powder in capsules daily for 3 months	Incidence of metabolic syndrome decreased compared to baseline; waist circumference also decreased significantly	Tsai et al. (2012)
Two-arm, parallel, randomized, double-blinded, placebo-controlled trial	38 type II diabetic patients, 19 treated, 19 placebo	6 g/day of MC as fruit pulp containing 6.26 ± 0.28 mg of charantin	Significant decline of total advanced glycation end products in serum after 16 weeks of intervention	Trakoon-osot et al. (2013)

MC, *Momordica charantia* (bitter melon).
Source: Modified from Alam, M.A., Uddin, R., Subhan, N., 2015. Beneficial role of bitter melon supplementation in obesity and related complications in metabolic syndrome. J. Lipids 2015, 496169.

muscle and normalized fructose diet-induced hyperglycemia in rats. In vitro studies have also suggested that fresh juice of the plant increased the uptake of amino acids and glucose in L6 myotubes (Cummings et al., 2004). The effects of the plant on glucose uptake and adiponectin secretion were also reported in 3T3-L1 adipocytes (Kumar et al., 2009). Water-soluble components of the plant enhanced glucose uptake at suboptimal concentrations of insulin in 3T3-L1 adipocytes (Roffey et al., 2007).

3. *Enhancing insulin-like action and pancreatic β-cell stimulation*: An investigation was performed to note the effect of bitter melon fruit juice on the distribution and number of α, β, and δ cells in the pancreas of STZ-induced diabetic rats. It was found that the juice significantly increased the number of β cells in STZ-induced diabetic rats compared with untreated diabetic rats (Ahmed et al., 1998). However, α cells did not change significantly compared with untreated diabetic rats. Further, oral administration of seed extracts at a dosage of 150 mg/kg body weight for 30 days prevented degeneration of pancreatic islets and restored islets function. The acetone extract of plant fruit powder at doses of 25, 50, and 100 mg/kg body weight affected different phases of recovery of β cells of the islets of Langerhans and normalized the functioning of the relevant cells.

Studies suggested that administration of ethanolic extract of fruit pulp of the plant in neonatal STZ-induced type II diabetic rats increased islet size, total β-cell area, number of β cells, and insulin levels compared with untreated diabetic rats. It has also been suggested that bitter melon has insulin secretagogue properties. Subcutaneous administration of protein extract isolated from bitter melon fruit pulp increased plasma insulin concentrations by twofold after 4 h of administration. The fruit pulp protein extract also increased insulin secretion, but not glucagon, in perfused rat pancreas. Apart from these effects, bitter melon has also been reported to improve the sensitivity of insulin in hyperinsulinemia.

Bitter melon supplementation improved insulin resistance and lowered serum insulin and leptin in rats fed a high-fat diet. It improves insulin sensitivity in skeletal muscle by increasing skeletal muscle insulin-stimulated insulin receptor substrate-1 tyrosine phosphorylation in rats fed a high-fat diet.

Margin of safety for clinical doses

Overall, bitter melon is considered quite safe and has been used as a food for decades. Regarding the clinical trials listed in Table 6.2, no author has reported any serious side effects during the study period. In some patients, headache, dizziness, stomach pain, and bloating were reported. Because these effects are considered quite nonspecific, bitter melon may or may not be the causative agent.

Based on a very limited amount of safety information and available data dating back 30 years or more, it may be advisable to stay alert for potential adverse effects of bitter melon in pregnant women, and so until more definitive information becomes available, bitter melon should be avoided in pregnant women. Bitter melon seeds contain momorcharin, which has been shown to have antifertility effects in female mice and causes spermatogenesis inhibition in dogs (Stepka et al., 1974). A decline in fertility from 90% to 20% was seen in mice fed bitter melon juice daily. Midterm miscarriages were induced in mice given isolated momorcharins from bitter melon (Chan et al., 1984). Also, bitter melon juice (6 mL/kg) caused uterine hemorrhage and death in two pregnant rabbits, but no uterine hemorrhage occurred in nonpregnant rabbits (Sharma et al., 1960). The use of bitter melon can cause side effects, such as diarrhea, abdominal pain, headache, fever, hypoglycemia, atrial fibrillation, urinary incontinence, and chest pain (Gupta et al., 2018). Allergic reactions to bitter melon have not been reported, but in rare individuals with allergy or hypersensitivity to members of the Cucurbitaceae family (gourds and melons), allergic reactions may be seen.

In general, those who are deficient in the glucose-6-phosphate dehydrogenase gene should avoid consumption of bitter melon preparations due to the presence of vicine in the seeds (Gupta et al., 2018).

Vitamin D

Vitamin D is well known to play a significant role in maintaining calcium and phosphorus homeostasis and promoting bone mineralization. Vitamin D has been the subject of many research studies supporting its role in diabetes and several nonskeletal functions (Bikle, 2009). The increased interest in vitamin D resulted from the realization that most tissues in the body have receptors for the active form of vitamin D, "1,25-dihydroxyvitamin D" [$1,25(OH)_{2D}$] or calcitriol or vitamin D3. These receptors are appropriately named "vitamin D receptors" (VDRs), and tissues with VDR are potential target tissues. As summarized (Bikle, 2009), the preactive form of vitamin D is called vitamin D2, which is formed from 7-dehydrocholestrol in the skin under the influence of ultraviolet radiation (sunlight) and skin pigmentation. Vitamin D2 then isomerizes to vitamin D3, the most active form in a thermosensitive process. The nonclassic or nonskeletal actions of vitamin D have been categorized (Bikle, 2009) into three general effects: (1) regulation of hormonal secretion; (2) regulation of immune function; and (3) regulation of cellular proliferation and differentiation. Because the role of vitamin D in T2DM is most likely via hormonal regulation (effect on insulin), discussion related to vitamin D in this chapter is limited to such effects.

Bioactive component(s)

The bioactive form of vitamin D is "[$1,25(OH)_{2D}$]" or calcitriol or vitamin D_3.

Efficacy

In view of the many studies that have been performed during the past 25 years, information in this section is limited to the most recent review articles in which detailed information has been collected regarding whether vitamin D plays a significant role in the clinical management of diabetes. At present, there is a lack of evidence to support such interventions, and only adequately powered studies should guide clinical decisions.

In a clinical trial, Sheth et al. (2015) recruited a diabetic population from western India including 912 subjects (429 T2DM cases and 483 nondiabetic placebo controls). The enrolled study subjects were investigated for biochemical parameters, such as FBS, postprandial blood sugar, glycosylated hemoglobin (HbA1c), fasting insulin, homeostasis model of assessment-insulin resistance index (HOMA-IR), and vitamin D (25OHD) representing 25OH-D2 and 25OH-D3 levels in blood. Vitamin D deficiency was seen in 91.4% and 93.0% of T2DM cases and control subjects, respectively. There was no association between serum vitamin D deficiency and HbA1c or HOMA-IR in T2DM cases or in control subjects. Findings from this study suggest that vitamin D deficiency is prevalent both in T2DM and nondiabetic subjects; therefore the role of vitamin D in hemoglobin glycation and insulin resistance could not be established. Despite an association (inconsistent) reported between vitamin D insufficiency and incidence of T2DM in longitudinal observational studies, Pittas and Dawson-Hughes (2010) summarized that results from small underpowered trials and post hoc analyses of data from larger trials show no effect of vitamin D supplementation on glycemia in healthy adults. Vitamin D may restrict the progression to diabetes in adults with glucose intolerance. However, because these findings have not been scientifically confirmed, the hypothesis that vitamin D may modify diabetes risk needs to be confirmed in future trials specifically designed for T2DM patients. One such trial, the "SUNNY" trial, has been initiated (Krul-Poel et al., 2014) in the Netherlands, which may provide helpful information.

Mechanism of action

The mechanism of action of [1,25(OH)$_{2D}$], the active form of vitamin D, is similar to that of other steroid hormones and is mediated by its binding to VDR. VDR is a member of the superfamily of nuclear hormone receptors, including receptors for steroid and thyroid hormones and retinoic acid. Although the basis for the role of active vitamin D in stimulation of insulin secretion has been known for more than 25 years (Lee et al., 1994), the mechanism of vitamin D in diabetes has been the subject of several investigations (Tai et al., 2008; Teegarden and Donkin, 2009; Alvarez and Ashraf, 2010).

Insulin secretion is a calcium-dependent process (Milner and Hales, 1967), as reported by Bikle (2009), and both vitamin D and calbindin-D_{28k} appear to play a role in the effect of vitamin D on insulin. Calbindin-D_{28k} is a vitamin D-responsive gene in many tissues. Both VDR and calbindin-D_{28k} are found in pancreatic β cells (Morrissey et al., 1975; Clark et al., 1980), and studies using calbindin-D_{28k} null mice have suggested that calbindin-D_{28k}, by regulating intracellular calcium, can modulate depolarization-stimulated insulin release (Sooy et al., 1999). Additionally, calbindin-D_{28k} can protect against cytokine-mediated destruction of β cells (Rabinovitch et al., 2001) by buffering calcium. Several case-controlled and observational studies have suggested that vitamin D deficiency contributes to increased risk for T2DM (Pittas et al., 2007).

Margin of safety for clinical doses

Overall analysis regarding the role of vitamin D in diabetic patients has recently been reviewed by Mathieu (2015), including a discussion about the dose needed as a supplement. Based on this evaluation, the question still remains regarding whether giving more vitamin D to subjects who are already vitamin D-sufficient will help prevent diabetes. Applying international guidelines for supplementation of vitamin D, small daily doses (500–1000 IU) are considered adequate to avoid the development of deficiency. Considering the fat-soluble nature of vitamin D, high doses of vitamin D should usually be avoided, as they may contribute to hypercalcemia or may induce kidney effects such as nephrocalcinosis (Mathieu, 2015).

Chromium

Based on data compiled by the Agency for Toxic Substances and Disease Registry (1998), chromium is a naturally occurring element in rocks, animals, plants, soil, and volcanic dust and gases. Chromium occurs in the environment predominantly in one of two valence forms: trivalent chromium (Cr III), which occurs naturally and is an essential nutrient for humans, and hexavalent chromium (Cr VI), which is most commonly produced by industrial processes. Because Cr VI has been classified as a "class I carcinogen" by the International Agency for Research on Cancer (IARC, 1990), and it is associated with other toxicities and is not essential for humans, it is not discussed here as a potential chromium supplement. The most widely used chromium compound as a source for Cr III supplementation is chromium picolinate (Cr Pic) because of its solubility and increased absorption efficiency (Bailey et al., 2008). Despite the popularity of Cr Pic as a nutritional supplement for managing T2DM, the FDA has not yet listed Cr III picolinate under the category "generally recognized as safe" (Code of Federal Regulation, 2004).

The general population is exposed to chromium, generally Cr III, by eating food, drinking water, and inhaling air that contains the chemical. The average daily intake from air, water, and food is estimated to be less than 0.2–0.4, 2.0, and 60 μg, respectively. A daily dose of 120 μg Cr III is recommended for humans. The FDA allows up to 1000 μg Cr III/day. With a greater dose than this, nephrotoxicity may ensue.

Cr III is known to play an essential role in normal glucose metabolism and has been considered an essential dietary element since the 1970s. Dietary deficiency of chromium has led to disturbances in carbohydrate metabolism, increased risk of glucose intolerance, insulin resistance, hyperglycemia, and risk of developing T2DM (Jeejeebhoy et al., 1977). Based on experimental research and clinical trials, the use of Cr III for its antihyperglycemic action in T2DM patients (especially insulin resistance) is well-established (Dattilo and Miguel, 2003; Ryan et al., 2003; Biswas et al., 2010; Hua et al., 2012). Chromium has been reported to alleviate insulin resistance by antioxidant and antiinflammatory effects (Althuis et al., 2002; Rains and Jain, 2011; Hua et al., 2012). Recently, Moradi et al (2019) reviewed literature on the antiinflammatory role of chromium, using inflammatory biomarkers [TNF-α, C-reactive protein (CRP), interleukins, monocyte chemoattractant protein-1, intercellular adhesion molecule-1, and adepocytokines] in hyperglycemia and diabetes. Based on in vitro and in vivo studies, findings revealed that chromium may have potential antiinflammatory properties, but some of the studies did not support this contention. Therefore the rationale for the use of chromium as an antiinflammatory agent in hyperglycemia and diabetes remains controversial.

Although the exact mechanism of action for the beneficial role of chromium in T2DM is yet to be established, limited data indicate that chromium inhibits the enzyme phosphotyrosine phosphatase that normally cleaves phosphate from the insulin receptor, leading to an increase in insulin sensitivity (Davis et al., 1996). Also, chromium is known to activate insulin receptor kinase (Davis and Vincent, 1997). These two effects, inhibition of insulin receptor phosphatase and activation of insulin receptor kinase, would lead to increased phosphorylation of the insulin receptor, thus resulting in increased insulin sensitivity (Davis et al., 1996).

Despite a reasonable body of evidence for the potential of Cr III in the prevention and management of T2DM, the FDA, after an evidence-based review of the issue, has concluded that the relationship between Cr Pic intake and insulin resistance is highly uncertain (Trumbo and Ellwood, 2006; Moradi et al., 2019). Furthermore, Vincent (2015) reported that the movement of Cr III in the body, particularly in response to changes in insulin concentration, suggests that Cr III could act as a second messenger, thereby amplifying insulin signaling.

The use of chromium as a supplement for T2DM is extensive but somewhat controversial at this time, particularly in subjects who are not deficient in Cr III (Cefalu and Hu, 2013; Costello et al., 2016; Mauro-Martin et al., 2016). These investigators, however, have concluded that growing evidence suggests that chromium supplementation, particularly at higher doses and in the form of Cr Pic, may improve insulin sensitivity and glucose metabolism in patients with glucose intolerance and types I and II diabetes as well as in some individuals without diabetes. Most clinical studies used for efficacy of Cr III in humans have major limitations, including small size, short-term, nonrandomized design, and different doses of chromium supplementation (Abdollahi et al., 2013; Bailey, 2014; Suksomboon et al., 2014; Mauro-Martin et al., 2016; Moradi et al., 2019). Therefore more clinical trials are needed in the US population to examine the robustness of the results observed in other populations and appropriate doses.

Ginger (gingerol and shogaol)

Ginger (*Zingiber officinale*) is one of the most widely consumed spices worldwide, originating from Southeast Asia and spreading to Europe and other Western countries. Although ginger is not widely known for its effectiveness in managing diabetes, several preclinical/animal studies (Al-Amin et al., 2006; Singh et al., 2009; Rahmani et al., 2014; Araujo et al., 2018) and clinical studies (Arablou et al., 2014; Mozaffari-Khosravi et al., 2014; Mahluji et al., 2013; Daily et al., 2015; Jafarnejad et al., 2017; Araujo et al., 2018; Bahramian et al., 2018) support its potential efficacy for balancing hyperglycemia in diabetic conditions.

Al-Amin et al. (2006) studied the effect of an aqueous extract of raw ginger (500 mg/kg) administered intraperitoneally daily for a period of 7 weeks to STZ-induced diabetic rats. The results of this study showed hypoglycemic and hypolipidemic effects. These findings supported the results of previous investigators. In another study, Singh et al. (2009) also observed blood glucose-lowering, lipid-lowering, and antioxidant effects of gingerol in type II diabetic db/db mice. These investigations, added to information compiled by Rahmani et al. (2014), indicate that ginger and its constituents show significant promise as a supplement in the prevention of metabolic diseases like diabetes. Findings from recent clinical studies further increase the weight of evidence for ginger as a supplement for diabetes management. Mahluji et al. (2013) studied the effect of 2 g daily oral doses of powdered ginger in T2DM patients. After 2 months, insulin, HOMA, triglycerides (TGs), and low-density lipoprotein (LDL) decreased significantly in the ginger group compared with the placebo group, but there was no significant change in FBG, HbA1c, total cholesterol, and high-density lipoprotein levels. Later, in a double-blinded, placebo-controlled clinical trial, Arablou et al. (2014) investigated the effect of 1600 mg of ginger versus 1600 mg of wheat flour as placebo daily for 12 weeks in 70 type II diabetic patients. The ginger group had significantly lowered fasting plasma glucose, HbA1c, insulin, HOMA, TG, total cholesterol, CRP, and prostaglandin E_2 compared with the placebo group, indicating a possible effect on insulin sensitivity. Findings from this clinical study are complemented by another clinical study by Mozaffari-Khosravi et al. (2014). Results from this study indicated that intake of three 1-g capsules of ginger powder daily for 8 weeks lowered FBS and HbA1c and improved insulin resistance index as well as the quantitative insulin sensitivity check index (QUICKI).

The chemical composition of ginger (*Z. officinale*) has been well-investigated, but the bioactive component(s) of ginger responsible for an antidiabetic effect still needs to be identified (Srinivasan et al., 2019). To determine the bioactive component with antidiabetic activity, Roufogalis (2014) undertook fractionation of ginger and tested different fractions for glucose uptake using cultured L6 myotubes as the biological assay. Fractionation of the extract using a hexane and ethyl acetate mixture, with fraction F7 as a major peak, had the greatest glucose uptake activity. This fraction was shown to be concentrated (approximately 43.4%) in (*S*)-[6]-gingerol. Other fractions contained smaller amounts of gingerols, as well as flavonoids, triterpenoids, fatty acids, and lipids. Shogaols, which are structurally close to gingerols, may also have antidiabetic activity.

The mechanism of action of ginger for its effectiveness in diabetes is not yet clear, but based on several studies, ginger may modulate insulin sensitivity, cause inhibition of the oxidative process, and have an antiinflammatory effect, all of which are factors that contribute to the development of diabetes (Srinivasan et al., 2019). Also, 6-gingerol, the main bioactive component known so far, may play a role in GLUT4 translocation in L6 myotubes differently than that which results from insulin-mediated GLUT4 to undergo exocytosis and relocation to the plasma membrane, in addition to restoration of altered carbohydrate and lipid metabolism. Some additional mechanisms include inhibition of several transcriptional pathways at the molecular level, lipid peroxidation, carbohydrate-metabolizing enzymes, increase in adenosine monophosphate kinase (AMPK) phosphorylation and 3-hydroxy-3-methyl-glutaryl-coenzyme A reductase, and the activation of antioxidant enzyme capacity and LDL receptors (Li et al., 2012; Rani et al., 2011, 2012; Akash et al., 2015). Samad et al (2017) found that 4-week treatment of 6-gingerol led to a significant increase in glucose-stimulated insulin secretion and improved glucose tolerance involving the GLP-1-mediated insulin secretion pathway. All these mechanisms have been discussed in detail by Roufogalis (2014) and Srinivasan et al. (2019).

Berberine

BBR can be obtained from a number of plants of the genus *Berberis*, *Hydrastis*, *Coptis*, *Argemone*, and *Thalictrum*. BBR has been found to be effective in diabetic patients for some time. Pharmacokinetic data suggest that the oral bioavailability of BBR is <1% due to its poor aqueous solubility, first-pass elimination, interaction with P-glycoprotein, and dissolution. In the gut, BBR is usually converted to dihydroberberine, which is five times more absorbable. BBR is metabolized by a cytochrome P450 system in the liver with phase I demethylation and phase II glucuronidation. The four major metabolites of BBR are thalifendine, berberrubine, jatrorrhizine, and demethyleneberberine, and all of these are subject to glucuronidation. Following oral administration, the highest concentrations of BBR and its metabolites are found in the liver, the site for cholesterol, TGs and glucose production, metabolism, and pharmacological effects.

Several mechanisms are involved in the antidiabetic/antihyperglycemic effects of BBR (Srivastava et al., 2019), some of which are:

1. Decreases glucose absorption by inhibiting α-glucosidase and reducing glucose transport through the intestinal epithelium;
2. Activates adenosine 5'-monophosphate-activated protein kinase (AMPK);
3. Induces glucose transport by enhancing GLUT1 gene expression;
4. Increases glucose uptake from blood to target organs (skeletal muscle and adipose tissue);
5. Increases insulin receptor expression and improves insulin secretion and sensitivity, thereby reducing insulin resistance;
6. Directly inhibits gluconeogenesis in the liver; and
7. Enhances glucose metabolism by stimulation of glycolysis.

In several experimental (Wang et al., 2008; Bagade et al., 2017) and clinical studies (Li, 2007; Zhang et al., 2010a, b), BBR has been found to produce significant antihyperglycemic and antidiabetic effects. BBR has also been reported to show positive effects against diabetes associated nephropathy, neuropathy, and cardiopathy (Pang et al., 2015; Imenshahidi and Hosseinzadeh, 2015). In conclusion, BBR has been found to be highly effective in type II diabetes mellitus and diabetic complications.

In general, BBR shows very low toxicity and side effects (Zhang and Chen, 2012; Lan et al., 2015; Pang et al., 2015). Oral LD_{50} values of BBR are reported to be >29,586 and >15,000 mg/kg in mice and rats, respectively. Humans with type II diabetes have been given up to 1500 mg BBR (in three divided doses of 500 mg each) daily without any serious adverse effects. BBR has been reported to cross the placental barrier and cause harm to the developing fetus. It has also been shown to exert a uterine stimulatory effect, thus its use in pregnancy is cautioned. Further, BBR can be transferred through breast milk, and therefore caution is required for its use while breastfeeding.

Gymnema sylvestre (Gur-Mar)

G. sylvestre is a slow-growing, perennial, woody climber medicinal plant. This plant is commonly found in central and southern India and tropical Africa. In the Ayurvedic system, it is referred to as "Meshasringa" or "Gurmar" and has indications for use in glucose balance. Gymnemic acid—an active component isolated from the leaves of *G. sylvestre*—has antiobesity and antidiabetic properties; it causes a decrease in body weight and inhibits glucose absorption (Pothuraja et al., 2014). Several animal studies have been performed to investigate the previous notion concerning its role in the management of diabetes. Liu et al. (2009) reported the insulinotropic activity of an extract of *G. sylvestre* using mouse MIN6 β cells and human islets of Langerhans. Several other preclinical animal studies have been performed, as reviewed by Pothuraja et al. (2014), that support a reduction in blood glucose, but clinical trial data are limited. Al-Romaiyan et al. (2010) reported that daily oral administration of *G. sylvestre* extract (OSA 1 g/day; two capsules, 250 mg each, bid) for 60 days produced significant improvements in glycemic control, and reductions in FBG levels. No well-documented information regarding the safety of *G. sylvestre* at doses higher than 1 g/day was available.

Others

A few of the other nutraceuticals that have been reported to lower blood glucose level include *Raphanus sativus* or "radish" (Shukla et al., 2011), turmeric (Selvi et al., 2014), and Siberian ginseng (Freye and Gleske, 2013). A dose of 300 mg/kg of *R. sativus* root juice was the most effective dose for lowering blood glucose (−33.4%) in normoglycemic rats. This dose was more effective than the 2.5 mg/kg dose of glibenclamide in mildly diabetic rats (Shukla et al., 2011). Selvi et al. (2014) used turmeric (2 g) as a supplement with standard metformin therapy for 4 weeks in a group of 30 diabetic patients and compared results to another group of 30 patients who did not receive turmeric. After 1 month, fasting plasma glucose and HbA1c were further lowered in the group that received turmeric. Panax and Siberian ginseng (*Eleutherococcus senticosus*) have been investigated for antidiabetic effects. Panax ginseng has different pharmacodynamic properties when compared to Siberian ginseng. What has now been demonstrated (Freye and Gleske, 2013) is that, in contrast to Panax ginseng and placebo, Siberian ginseng is able to lower elevated blood sugar levels in patients with T2DM, both after meals and after fasting. This is also the first report demonstrating a favorable long-term effect on lipid metabolism and peripheral neuropathy. Doses taken at regular intervals were safe and effective.

Curcumin from *Curcuma longa*, stevioside from *Stevia rebaudiana*, catalpol from *Rehmannia glutinosa*, *Terminalia catappa*, diosmin, shilajit, and many other plants with potential for antihyperglycemic and antidiabetic properties are discussed in detail elsewhere (Behl and Kotwani, 2017; Hsu et al., 2017; Youssef et al., 2017; Gupta et al., 2019).

Efficacy of nutraceuticals

The preclinical efficacy of cinnamon was considered to be due to water-soluble fraction "A" type doubly linked procyanidin oligomers (Anderson et al., 2004), commonly called "polyphenolic polymers." Cinnamaldehyde content (approximately 2% in whole cinnamon powder) is also considered a bioactive component. This is because cinnamaldehyde was effective in lowering experimentally induced hyperglycemia in STZ-treated diabetic rats at a relatively low dose of 20 mg/kg and the effect was comparable to a 0.6 mg/kg dose of glibenclamide, which is a standard therapy (Subash Babu et al., 2007). Similar results have been reported for bitter melon. As summarized by Alam et al. (2015), the dose of aqueous extract of fresh, unripe, whole bitter melon fruits needed only for lowering FBG by 48% in diabetic rats was 20 mg/kg body weight. Again, this was comparable to the effect of glibenclamide, a standard oral antidiabetic drug. For lowering glucose in normoglycemic rats, a much higher oral dose was required. This was 800 mg/kg of bitter melon, which is considered too high from a patient compliance point of view. This indicates that at commonly used doses (approximately 20 mg/kg), the risk of hypoglycemia is minimal to none. Other compounds including *G. sylvestre* have also shown efficacy, as evidenced by a 2010 study (Al-Romaiyan et al., 2010) in which daily oral administration of *G. sylvestre* extract (OSA) for 60 days produced significant improvements in glycemic control. Ten of 11 patients using OSA 1 g/day (two capsules, 250 mg each, two times daily) in this study showed a reduction in FBG levels. Another nutraceutical, ginger, had to be used at higher oral doses for the antihyperglycemic effect. A clinical study reported by Mozaffari-Khosravi et al. (2014) indicated that intake of three 1-g capsules of ginger powder daily for 8 weeks lowered FBS, HbA1c, and improved insulin resistance index and QUICKI in diabetic subjects. Finally, the preclinical efficacy data for vitamin D3 and chromium are promising, but clinical trial data for efficacy remain a challenge.

Overall, nutraceuticals discussed in this chapter provide moderate beneficial effects in lowering the fasting or postprandial blood glucose based on preclinical data and some clinical trial data. The clinical trials results have varied in terms of efficacy. Translation of data from animal studies to clinical trials remains a major challenge for the establishment of good-quality clinical efficacy data. There are clinical investigations that have failed to show a clearly antihyperglycemic effect for cinnamon. As an example, cinnamon has failed to show efficacy in some clinical trials in T2DM patients (Vanschoonbeek et al., 2006), whereas a good antihyperglycemic response has been documented in animal studies (Subash Babu et al., 2007; Anand et al. (2010). The cause may be multifactorial and the following steps may be of value:

1. The bioactive component(s) should be identified in the nutraceutical product and standardized with respect to effective dose.

2. The endpoints/markers for assessing clinical efficacy should be standardized with respect to indication and should be comparable in each repeat clinical investigation.
3. In addition to adequacy of the number of subjects, proper exclusion and inclusion criteria and age range, assurance that other factors like lifestyle, timing of the medication, eating habits, amount and frequency of physical activity, and sleep pattern are comparable among the subjects as best as possible.

Considering the spiraling growth in the number of prediabetic and diabetic populations in coming decades, nutraceuticals, if used effectively with personalized treatment regimens, have a role to play. This can be achieved with a strong emphasis on managing the common risk factors such as stress, lack of exercise, and poor eating habits. Beneficial effects of nutraceuticals can be better realized if used as a supplement to regular medications prescribed by a physician. Simply being able to delay or avoid some of the potentially risky (tier 2) diabetic medications by controlling diabetes would be a huge positive contribution of nutraceuticals.

Safety of nutraceuticals

The adverse effects associated with nutraceuticals for diabetes are minimal. Cinnamon doses up to 6 g/day have been used in clinical trials without any adverse effects. For some nutraceuticals, exposure to coumarin needs to be avoided or minimized; this is true in the case of cinnamon. This can be managed either by using only "true" cinnamon (*C. aromaticum*), which contains no or only traces of coumarin, or by using the aqueous extract of cinnamon instead of the whole powder. Using aqueous extracts helps because coumarin is not extracted in an aqueous medium. The potential for adverse effects during pregnancy or for allergy has been discussed for some nutraceuticals, as applicable. Bitter melon is considered safe with respect to general toxicity because it has been used as a food for decades without any adverse effects, but adverse reproductive effects have been reported when high doses of bitter melon or bitter melon juice (6 mL/kg) were used daily for an extended duration. Use of bitter melon has lowered fertility in mice (Chan et al., 1984) and caused uterine hemorrhage in pregnant rabbits (Sharma et al., 1960). A decline in spermatogenesis has been reported in dogs (Stepka et al., 1974). Caution should be exercised in the use of high doses of bitter melon when fertility and pregnancy maintenance are important objectives of the study until good-quality reproductive data have been generated in compliance with current ICH guidelines.

Concluding remarks and future directions

Several promising nutraceuticals have been reviewed with respect to their potential for managing hyperglycemia. In most cases, declines in FBG and/or in HbA1c compared to control animals have been used to estimate efficacy in preclinical studies, whereas additional measures related to insulin sensitivity have been included in some human clinical trials.

References

Abdollahi, M., Farshchi, A., Nikfar, S., et al., 2013. Effect of chromium on glucose and lipid profiles in patients with type 2 diabetes: a meta-analysis review of randomized trials. J. Pharm. Pharm. Sci. 16, 99–114.

Abraham, K., Wohrlin, F., Lindtner, O., 2010. Toxicology and risk assessment of coumarin: focus on human data. Mol. Nutr. Food Res. 54, 228–239.

Adisakwattana, S., Lerdsuwankij, O., Poputtachai, U., 2011. Inhibitory activity of cinnamon bark species and their combination effect with acarbose against intestinal a-glucosidase and pancreatic a-amylase. Plant Foods Hum. Nutr. 66, 143–148.

Agency for Toxic Substances and Disease Registry (ATSDR), 1998. Toxicological Profile for Chromium. U.S. Public Health Service, U.S. Department of Health and Human Services, Atlanta, Georgia.

Ahmad, N., Hassan, M.R., Halder, H., 1999. Effect of *Momordica charantia* (Karolla) extracts on fasting and postprandial serum glucose levels in NIDDM patients. Bangladesh Med. Res. Counc. Bull. 25, 11–13.

Ahmed, I., Adeghate, E., Sharma, A.K., 1998. Effects of Momordica charantia fruit juice on islet morphology in the pancreas of the streptozotocin-diabetic rat. Diabetes Res. Clin. Pract. 40, 145–151.

Akash, M.S., Rehman, K., Tariq, M., et al., 2015. *Zingiber officinale* and type 2 diabetes mellitus: evidence from experimental studies. Crit. Rev. Eukaryot. Gene Expr. 25, 91–112.

Akilen, R., Tsiami, A., Devendra, D., Robinson, N., 2010. Glycated haemoglobin and blood pressure-lowering effect of cinnamon in multi-ethnic Type 2 diabetic patients in the UK: a randomized, placebo-controlled, double-blind clinical trial. Diabetic Med. 27 (10), 1159–1167.

Alam, M.A., Uddin, R., Subhan, N., 2015. Beneficial role of bitter melon supplementation in obesity and related complications in metabolic syndrome. J. Lipids 2015, 496169.

Al-Amin, Z.M., Thomson, M., Al-Qattan, K.K., Ali, M., 2006. Anti-diabetic and hypolipidaemic properties of ginger (*Zingiber officinale*) in streptozotocin-induced diabetic rats, Br. J. Nutr, 96. pp. 660–666.

Al-Romaiyan, A., Liu, B., Asare-Anane, H., 2010. A novel *Gymnema sylvestre* extract stimulates insulin secretion from human islets *in vivo* and *in vitro*. Phytother. Res. 24, 1370–1376.

Althuis, M.D., Jordan, N.E., Ludington, E.A., 2002. Glucose and insulin responses to dietary chromium supplements: a meta-analysis. Am. J. Clin. Nutr. 76, 148–155.

Altschuler, J.A., Casella, S.J., MacKenzie, T.A., 2007. The effect of cinnamon on A1c among adolescents with type 1 diabetes. Diabetes Care 30, 813–816.

Alvarez, J.A., Ashraf, A., 2010. Role of vitamin D in insulin secretion and insulin sensitivity for glucose homeostasis. Int. J. Endocrinol. 2010, 351385.

Anand, P., Murali, K.Y., Tandon, V., 2010. Insulinotropic effect of cinnamaldehyde on transcriptional regulation of pyruvate kinase, phosphoenolpyruvate carboxykinase, and GLUT4 translocation in experimental diabetic rats. Chemico-Biol. Interact. 186, 72–81.

Anderson, R.A., Broadhurst, C.L., Polansky, M.M., 2004. Isolation and characterization of polyphenol type-A polymers from cinnamon with insulin-like biological activity. J. Agric. Food Chem. 52, 65–70.

Arablou, T., Aryaeian, N., Valizadeh, M., 2014. The effect of ginger consumption on glycemic status, lipid profile and some inflammatory markers in patients with type 2 diabetes mellitus. Int. J. Food Sci. Nutr. 65, 515–520.

Araujo, A.J.S., Jesus-Lima, J.C.R., Otoch, J.P., et al., 2018. Effects of ginger (*Zingiber officinale*) supplementation on diabetes: an update. Am. J. Phytomed. Clin. Ther. 6 (3), 1–20. Available from: https://doi.org/10.21767/2321-2748.100349.

Bagade, A., Tumbigeremutt, V., Pallavi, G., 2017. Cardiovascular effects of berberine: a review of the literature. J. Retroact. Med. 6, 37–45.

Bahramian, Z., Sehhatie-Shafaie, F., Mirghafourvand, M., et al., 2018. The effect of ginger capsules on the control of blood sugar in gestational diabetes: a triple-blind randomized controlled clinical trial. Crescent J. Med. Biol. Sci. 5, 358–365.

Bailey, C.H., 2014. Improved meta-analytic methods show no effect of chromium supplements on fasting glucose. Biol. Trace Elem. Res. 157, 1–8.

Bailey, M.M., Sturdivant, J., Jernigan, P.L., 2008. Comparison of the potential for developmental toxicity of prenatal exposure to two dietary chromium supplements, chromium picolinate and $Cr_3O(O_2CCH_2CH_3)(6(H_2O)_3]+$, in mice. Birth Defects Res. B: Dev. Reprod. Toxicol. 83, 27–31.

Baker, W.L., Kluger, J., Gutierrez-Williams, G., 2008. Effect of cinnamon on glucose control and lipid parameters. Diabetes Care 31, 41–43.

Becker, M., 2012. The Blood Sugar Market Goes Boom. Neutraceutical World 15. Available from: www.nutraceuticalsworld.com/issues/2012-10/view_features/the-blood-sugar-market-goes-boom/.

Beejmohun, V., Peytavy-Izard, M., Mignon, C., 2014. Acute effect of Ceylon cinnamon extract on postprandial glycemia: alpha-amylase inhibition, starch tolerance test in rats, and randomized crossover clinical trial in healthy volunteers. BMC Compl. Alt. Med. 14, 351.

Behl, T., Kotwani, A., 2017. Proposed mechanisms of *Terminalia catappa* in hyperglycemia and associated diabetic complications. J. Pharm. Pharmacol. 69, 123–134.

Bikle, D., 2009. Nonclassic actions of vitamin D. J. Clin. Endocrinol. Metab. 94, 26–34.

Biswas, T.K., Polley, G., Pandit, S., 2010. Effects of adjunct therapy of a proprietary herbo-chromium supplement in type 2 diabetes: a randomized clinical trial. Int. J. Diabetes Dev. Countries 30, 153–161.

Blevins, S.M., Leyva, M.J., Brown, J., Wright, J., Scofield, R.H., Aston, C.E., 2007. Effect of cinnamon on glucose and lipid levels in non insulin-dependent type 2 diabetes. Diabetes Care 30 (9), 2236–2237.

Cao, H., Urban, J.F., Anderson, R.A., 2008. Cinnamon polyphenol extract affects immune responses by regulating anti- and proinflammatory and glucose transporter gene expression in mouse macrophages. J. Nutr. 138, 833–840.

Cefalu, W.T., Hu, F.B., 2013. Role of chromium in human health and in diabetes. Diabetes Care 36 (9), 2872.

Centers for Disease Control and Prevention, 2014. National Diabetes Statistics Report: Estimates of Diabetes and its Burden in the United States. U.S. Department of Health and Human Services, Atlanta, GA.

Chan, W., Tam, P., Yeung, H., 1984. The termination of early pregnancy in the mouse by beta-momorcharin. Contraception. 29, 91–100.

Chan, N., Li, S., Perez, E., 2016. Interaction between Chinese nutraceuticals and Western medicines. In: Gupta, R.C. (Ed.), Nutraceuticals: Efficacy, Safety and Toxicity. Academic Press/Elsevier, Amsterdam, pp. 875–882.

Chaturvedi, P., 2012. Antidiabetic potentials of *Momordica charantia*: multiple mechanisms behind the effects. J. Med. Food 15, 101–107.

Clark, S.A., Stumpf, W.E., Sar, M., 1980. Target cells for 1,25 dihydroxyvitamin D3 in the pancreas. Cell Tissue Res. 209, 515–520.

Code of Federal Regulation, 2004. Title 21, Part 101.9 (c) (8) (iv). US Government Printing Office, Washington, DC.

Costello, R.B., Dwyer, J.T., Bailey, R.L., 2016. Chromium supplements for glycemic control in type 2 diabetes: limited evidence of effectiveness. Nutr. Rev. 74, 455–468.

Crawford, P., 2009. Effectiveness of cinnamon for lowering hemoglobin A1C in patients with type 2 diabetes: a randomized, controlled trial. J. Am. Board Fam. Med. 22 (5), 507–512.

Cummings, E., Hundal, H.S., Wackerhage, H., 2004. *Momordica charantia* fruit juice stimulates glucose and amino acid uptakes in L6 myotubes. Mol. Cell. Biochem. 261, 99–104.

Daily, J.W., Yang, M., Kim, D.S., et al., 2015. Efficacy of ginger for treating type 2 diabetes: a systematic review and meta-analysis of randomized clinical trials. J. Ethnic Foods 2, 36–43.

Dans, A.M.L., Villarruz, M.V.C., Jimeno, C.A., 2007. The effect of *Momordica charantia* capsule preparation on glycemic control in type 2 diabetes mellitus needs further studies. J. Clin. Epidemiol. 60, 554–559.

Dattilo, A.M., Miguel, S.G., 2003. Chromium in health and disease. Nutr. Today 38, 121–133.

Davis, C.M., Vincent, J.B., 1997. Chromium oligopeptide activates insulin receptor tyrosine kinase activity. Biochemistry 6 (15), 4382–4385.

Davis, C.M., Sumrall, K.H., Vincent, J.B., 1996. A biologically active form of chromium may activate a membrane phosphotyrosine phosphatase (PTP). Biochemistry 35, 12963–12969.

DeFronzo, R.A., Tripathy, D., 2009. Skeletal muscle insulin resistance is the primary defect in type 2 diabetes. Diabetes Care 32, S157–S163.

EFSA, 2004. Opinion of the scientific panel on food additives, flavorings, processing aids and materials in contacts with food (AFC) on a request from the commission related to coumarin; Adopted on 6 October 2004. EFSA J. 104, 1–36.

Eisenberg, D.M., Davis, R.B., Ettner, S.L., 1998. Trends in alternative medicine use in the United States, 1990–1997: results of a follow-up national survey. JAMA 280, 1569–1575.

Freye, E., Gleske, J., 2013. Siberian ginseng results in beneficial effects on glucose metabolism in diabetes type 2 patients: a double blind placebo-controlled study in comparison to Panax ginseng. Int. J. Clin. Nutr. 1, 11–17.

Fuangchan, A., Sonthisombata, P., Seubnukarnb, T., 2011. Hypoglycemic effect of bitter melon compared with metformin in newly diagnosed type 2 diabetes patients. J. Ethnopharmacol. 134, 422–428.

Gupta, R.C., Chang, D., Nammi, S., et al., 2017. Interactions between antidiabetic drugs and herbs: an overview of mechanisms of action and clinical implications. Diabetol. Metab. Syndr. 9, 59.

Gupta, R.C., Lall, R., Sinha, A., Srivastava, A., 2019. Nutraceuticals for diabetes in dogs and cats. In: Gupta, R.C., Srivastava, A., Lall, R. (Eds.), Nutraceuticals in Veterinary Medicine. Springer Nature, Cham, Switzerland, pp. 523–539.

Gupta, RC, Srivastava, A, Lall, R, 2018. Toxicity potential of nutraceuticals. In: Nicolotti, O (Ed.), Computational Toxicology: Methods and Protocols, Methods in Molecular Biology. Springer Nature/Humana Press, New York, pp. 367–394.

Hagan, E.C., Hansen, W.H., Fitzhugh, O.G., 1967. Food flavorings and compounds of related structure. II. Subacute and chronic toxicity. Food Cosmet. Toxicol. 5, 141–157.

Hsu, C.C., Lin, M.H., Cheng, J.T., et al., 2017. Antihyperglycemic action of diosmin, a citrus flavonoid, is induced through endogenous- endorphin in type I-like diabetes rats. Clin. Exp. Pharmacol. Physiol. 44 (5), 549–555.

Hua, Y., Clark, S., Ren, J., 2012. Molecular mechanisms of chromium in alleviating insulin resistance. J. Nutr. Biochem. 23, 313–319.

IARC, 1990. Monograph on the Evaluation of Carcinogenic Risk to Human. Chromium, Nickel, and Welding, IARC Lyon, France, vol. 48, pp. 49–508.

IDF Update, 2014. International Diabetic Federation. IDF Diabetes Atlas, sixth ed. Available from: www.idf.org/diabetesatlas.

Imenshahidi, M., Hosseinzadeh, H., 2015. *Berberis vulgaris* and berberine: an update review. Phytother. Res. 30, 1745–1764.

Imparl-Radosevich, J., Deas, S., Polansky, M.M., 1998. Regulation of PTP-1 and insulin receptor kinase by fractions from cinnamon: implications for cinnamon regulation of insulin signalling. Horm. Res. 50, 177–182.

Jafarnejad, S., Keshavarz, S.A., Mahbubi, S., et al., 2017. Effect of ginger (*Zingiber officinale*) on blood glucose and lipid concentrations in diabetic and hyperlipidemic subjects: a meta-analysis of randomized controlled trials. J. Funct. Foods 29, 127–134.

Jarvill-Taylor, K.J., Anderson, R.A., Graves, D.J.A., 2001. A hydroxyl-chalcone derived from cinnamon functions as a mimetic for insulin in 3T3-L1 adipocytes. J. Am. Coll. Nutr. 20, 327–336.

Jayaprakasha, G., Rao, L.J.M., 2011. Chemistry, biogenesis, and biological activities of *Cinnamomum zeylanicum*. Crit. Rev. Food Sci. Nutr. 51, 547–562.

Jeejeebhoy, K.N., Chu, R.C., Marliss, E.B., 1977. Chromium deficiency, glucose intolerance, and neuropathy reversed by chromium supplementation, in a patient receiving long-term total parenteral nutrition. Am. J. Clin. Nutr. 30, 531–538.

Joseph, B., Jini, D., 2013. Antidiabetic effects of *Momordica charantia* (bitter melon) and its medicinal potency. Asian J. Trop. Dis. 3, 93–102.

Kaneto, H., Katakami, N., Kawamori, D., et al., 2007. Involvement of oxidative stress in the pathogenesis of diabetes. Antioxid. Redox Signaling 9, 355–366.

Kang, S., Tsai, L.T., Rosen, E.D., et al., 2016. Nuclear mechanisms of insulin resistance. Trends Cell Biol. 26, 341–351.

Kangralkar, V., Shivraj, A., Patil, D., 2010. Oxidative stress and diabetes: a review. Int. J. Pharm. Appl. 1, 38–45.

Khan, A., Safdar, M., Ali Khan, M.M., 2003. Cinnamon improves glucose and lipids of people with type 2 diabetes. Diabetes Care 26, 3215–3218.

Kim, S.H., Hyun, S.H., Choung, S.Y., 2006. Anti-diabetic effect of cinnamon extract on blood glucose in db/db mice. J. Ethnopharmacol. 104, 119–123.

Kim, J.A., Wei, Y., Sowers, J.R., et al., 2008. Role of mitochondrial dysfunction in insulin resistance. Circ. Res. 102, 401–414.

Krul-Poel, Y.H.M., Wijland, H.V., Stam, F.F., 2014. Study protocol: a randomized placebo-controlled clinical trial to study the effect of vitamin D supplementation on glycemic control in type 2 diabetes mellitus SUNNY trial. BMC Endocr. Disord. 14, 59.

Kumar, R., Balaji, S., Uma, T.S., 2009. Fruit extracts of *Momordica charantia* potentiate glucose uptake and up-regulate Glut-4, PPARγ and PI3K. J. Ethnopharmacol. 126, 533–537.

Lan, J., Zhao, Y., Dong, F., et al., 2015. Meta-analysis of the effect and safety of berberine in the treatment of type 2 diabetes mellitus, hyperlipidemia and hypertension. J. Ethnopharmacol. 161, 69–81.

Lee, S., Clark, S.A., Gill, R.K., 1994. 1,25-Dihydroxyvitamin D3 and pancreatic β-cell function: vitamin D receptors, gene expression, and insulin secretion. Endocrinology 134, 1602–1610.

Li X.-Y., 2007. Efficacy and safety of berberine in the treatment of diabetes with dyslipidemia. US Clinical TrialsGov.

Li, Y., Tran, V.H., Duke, C.C., et al., 2012. Gingerols of *Zingiber officinale* enhance glucose uptake by increasing cell surface GLUT4 in cultured L6 myotubes. Planta Med. 78, 1549–1555.

Lin, H.-W., Tseng, C.-H., 2014. A review on the relationship between SGLT2 inhibitors and cancer. Int. J. Endocrinol. 2014, 719578.

Liu, B., Anane, H.A., Romaiyan, A.A., 2009. Characterisation of the insulinotropic activity of an aqueous extract of *Gymnema sylvestre* in mouse β-cells and human islets of Langerhans. Cell Physiol. Biochem. 23, 125–132.

Lu, T., Sheng, H., Wu, J., 2012. Cinnamon extract improves fasting blood glucose and glycosylated hemoglobin level in Chinese patients with type 2 diabetes. Nutr. Res. 32, 408–412.

Mahluji, S., Attari, V.E., Mobasseri, M., 2013. Effects of ginger (*Zingiber officinale*) on plasma glucose level, HbA1c and insulin sensitivity in type 2 diabetic patients. Int. J. Food Sci. Nutr. 64, 682–686.

Mang, B., Wolters, M., Schmitt, B., 2006. Effects of a cinnamon extract on plasma glucose, HbA1c and serum lipids in diabetes mellitus type 2. Eur. J. Clin. Invest. 36, 340–344.

Mathieu, C., 2015. Vitamin D and diabetes: where do we stand? Diabetes Res. Clin. Pract. 108, 201–209.

Mauro-Martin, I.S., Ruiz-León, A.M., Camina-Martin, M.A., et al., 2016. Chromium supplementation in patients with type 2 diabetes and high risk of type 2 diabetes: a meta-analysis of randomized controlled trials. Nutr. Hosp. (Span.) 33, 27. Available from: https://doi.org/10.20960/nh.27.

Milner, R.D., Hales, C.N., 1967. The role of calcium and magnesium in insulin secretion from rabbit pancreas studied *in vitro*. Diabetologia 3, 47–49.

Miura, T., Itoh, C., Iwamoto, N., 2001. Hypoglycemic activity of the fruit of the *Momordica charantia* in type 2 diabetic mice. J. Nutr. Sci. Vitaminol. 47, 340–344.

Moradi, F., Maleki, V., Saleh-Ghadimi, S., et al., 2019. Potential roles of chromium on inflammatory biomarkers in diabetes: a systematic review. Clin. Exp. Pharmacol. Physiol. 46, 975–983.

Morrissey, R.L., Bucci, T.J., Richard, B., 1975. Calcium-binding protein: its cellular localization in jejunum, kidney and pancreas. Proc. Soc. Exp. Biol. Med. 149, 56–60.

Mozaffari-Khosravi, H., Talaei, B., Jalali, B.-A., 2014. The effect of ginger powder supplementation on insulin resistance and glycemic indices in patients with type 2 diabetes: a randomized, double-blind, placebo-controlled trial. Complementary Therap. Med. 22, 9–16.

Naveen, J., Baskaran, V., 2018. Antidiabetic plant-derived nutraceuticals: a critical review. Eur. J. Nutr. 57, 1275–1299.

Ojewole, J.A., Adewole, S.O., Olayiwola, G., 2006. Hypoglycaemic and hypotensive effects of *Momordica charantia* Linn (Cucurbitaceae) whole-plant aqueous extract in rats. Cardiovasc. J. S. Afr. 17, 227–232.

Oubre, A.Y., Carlson, T.J., King, S.R., 1997. From plant to patient: an ethnomedical approach to the identification of new drugs for the treatment of NIDDM. Diabetologia 40, 614–617.

Pang, B., Zhao, L.H., Zhou, Q., et al., 2015. Application of berberine on treating type 2 diabetes mellitus. Int. J. Endocrinol. 905749.

Pham, A.Q., Kourlas, H., Pham, D.Q., 2007. Cinnamon supplementation in patients with type 2 diabetes mellitus. Pharmacotherapy 27, 595–599.

Pittas, A.G., Dawson-Hughes, B., 2010. Vitamin D and diabetes. J. Steroid Biochem. Mol. Biol. 121, 425–429.

Pittas, A.G., Lau, J., Hu, F.B., 2007. The role of vitamin D and calcium in type 2 diabetes. A systematic review and meta-analysis. J. Clin. Endocrinol. Metab. 92, 2017–2029.

Pothuraja, R., Sharma, R.K., Chagalamarri, J., 2014. A systematic review of *Gymnema sylvestre* in obesity and diabetes management. J. Sci. Food Agric. 94, 834–840.

Qin, B., Nagasaki, M., Ren, M., 2003. Cinnamon extract (traditional herb) potentiates *in vivo* insulin-regulated glucose utilization via enhancing insulin signaling in rats. Diabetes Res. Clin. Pract. 62, 139–148.

Rabinovitch, A., Suarez-Pinzon, W.L., Sooy, K., 2001. Expression of calbindin-D (28k) in a pancreatic islet β-cell line protects against cytokine-induced apoptosis and necrosis. Endocrinology 142, 3649–3655.

Rahmani, A.H., Al shabrmi, F.M., Aly, S.M., 2014. Active ingredients of ginger as potential candidates in the prevention and treatment of diseases via modulation of biological activities. Int. J. Physiol. Pathophysiol. Pharmacol. 6, 125–136.

Rains, J.L., Jain, S.K., 2011. Oxidative stress, insulin signaling, and diabetes. Free Rad. Biol. Med. 50, 567–575.

Raman, A., Lau, C., 1966. Anti-diabetic properties and phytochemistry of *Momordica charantia* L. (Cucurbitaceae). Phytomedicine 2, 349–362.

Ranasinghe, P., Jayawardana, R., Galappaththy, P., 2012. Efficacy and safety of "true" cinnamon (*Cinnamomum zeylanicum*) as a pharmaceutical agent in diabetes: a systematic review and meta-analysis. Diabetic Med. 29, 1480–1492.

Rani, M.P., Padmakumari, K.P., Sankarikutty, B., et al., 2011. Inhibitory potential of ginger extracts against enzymes linked to type 2 diabetes, inflammation and induced oxidative stress. Int. J. Food Sci. Nutr. 62, 106–110.

Rani, M.P., Krishna, M.S., Padmakumari, K.P., et al., 2012. *Zingiber officinale* extract exhibits antidiabetic potential via modulating glucose uptake, protein glycation and inhibiting adipocyte differentiation: an in vitro study. J. Sci. Food Agr. 92, 1948–1955.

Raza, H., Prabu, S.K., John, A., et al., 2011. Impaired mitochondrial respiratory functions and oxidative stress in streptozotocin-induced diabetic rats. Int. J. Mol. Sci. 12, 3133–3147.

Roffey, B.W.C., Atwal, A.S., Johns, T., 2007. Water extracts from *Momordica charantia* increase glucose uptake and adiponectin secretion in 3T3-L1 adipose cells. J. Ethnopharmacol. 112, 77–84.

Roufogalis, B.D., 2014. *Zingiber officinale* (Ginger): a future outlook on its potential in prevention and treatment of diabetes and prediabetic states. N. J. Sci. 2014, 674684.

Ryan, G.J., Wanko, N.S., Redman, A.R., Cook, C.B., 2003. Chromium as adjunctive treatment for type 2 diabetes. Ann. Pharmacother. 37, 876–885.

Samad, M.B., Mohsin, M., Razu, B.A., et al., 2017. [6]-Gingerol, from *Zingiber officinale*, potentiates GLP-1 mediated glucose-stimulated insulin secretion pathway in pancreatic beta-cells and increases- RAB8/RAB10-regulated membrane presentation of GLUT4 transporters in skeletal muscle to improve hyperglycemia in Lepr (db/db) type 2 diabetic mice. BMC Complement. Altern. Med. 17, 395.

Selvi, N.M.K., Sridhar, M.G., Swaminathan, R.P., 2014. Efficacy of turmeric as adjuvant therapy in type 2 diabetic patients. Ind. J. Clin. Biochem. 30, 180–186.

Sharma, V., Sogani, R., Arora, R., 1960. Some observations on hypoglycaemic activity of *Momordica charantia*. Indian J. Med. Res. 48, 471–477.

Shen, Y., Fukushima, M., Ito, Y., 2010. Verification of the antidiabetic effects of cinnamon (*Cinnamomum zeylanicum*) using insulin-uncontrolled type 1 diabetic rats and cultured adipocytes. Biosci. Biotechnol. Biochem. 74, 2418–2425.

Sheth, J.J., Shah, A., Sheth, F.J., 2015. Does vitamin D play a significant role in type 2 diabetes? BMC Endocr. Disord. 15, 5.

Shih, C.-C., Lin, C.-H., Lin, W.-L., 2009. *Momordica charantia* extract on insulin resistance and the skeletal muscle GLUT4 protein in fructose-fed rats. J. Ethnopharmacol. 123, 82–90.

Shukla, S., Chatterji, S., Mehta, S., 2011. Antidiabetic effect of *Raphanus sativus* root juice. Pharm. Biol. 49, 32–37.

Singh, A., Akanksha, S.N., Maurya, R., 2009. Anti-hyperglycaemic, lipid lowering and anti-oxidant properties of 6-gingerol in *db/db* mice. Int. J. Med. Sci. 1, 536–544.

Sooy, K., Schermerhorn, T., Noda, M., 1999. Calbindin-D(28k) controls [Ca(2+)] (i) and insulin release. Evidence obtained from calbindin-D(28k) knockout mice and β cell lines. J. Biol. Chem. 274, 34343–34349.

Sridhar, M.G., Vinayagamoorthi, R., Suyambunathan, V.A., 2008. Bitter gourd (*Momordica charantia*) improves insulin sensitivity by increasing skeletal muscleinsulin-stimulated IRS-1 tyrosine phosphorylation in high-fatfed rats. Br. J. Nutr. 99, 806–812.

Srinivasan, K., Adhya, P., Sharma, S.S., 2019. Nutraceutical potential of ginger. In: Gupta, R.C., Srivastava, A., Lall, R. (Eds.), Nutraceuticals in Veterinary Medicine. Springer Nature, Cham, Switzerland, pp. 51–70.

Srivastava, A., Sinha, A., Lall, R., Gupta, R.C., 2019. Berberine. In: Gupta, R.C., Srivastava, A., Lall, R. (Eds.), Nutraceuticals in Veterinary Medicine. Springer, Cham, Switzerland, pp. 71–81.

Stepka, W., Wilson, K., Madge, G., 1974. Antifertility investigation on *Momordica*. Lloydia 37, 645c.

Subash Babu, P., Prabuseenivasan, S., Ignacimuthu, S., 2007. Cinnamaldehyde—a potential antidiabetic agent. Phytomedicine 14, 15–22.

Suksomboon, N., Poolsup, N., Yuwanakorn, A., 2014. Systematic review and meta-analysis of the efficacy and safety of chromium supplementation in diabetes. J. Clin. Pharm. Ther. 39, 292–306.

Taher, M., Fadzilah, A.A.M., Mohomad, R.S., 2006. A proanthocyanidin from *Cinnamomum zeylanicum* stimulates phosphorylation of insulin receptor in 3T3-L1 adipocytes. J. Teknol. 44, 53–68.

Tai, K., Need, A.G., Horowitz, M., 2008. Vitamin D, glucose, insulin, and insulin sensitivity. Nutrition 24, 279–285.

Teegarden, D., Donkin, S.S., 2009. Vitamin D: emerging new roles in insulin sensitivity. Nutr. Res. Rev. 22, 82–92.

Tongia, A., Tongia, S.K., Dave, M., 2004. Phytochemical determination and extraction of *Momordica charantia* fruit and its hypoglycemic potentiation of oral hypoglycemic drugs in diabetes mellitus (NIDDM). Ind. J. Physiol. Pharmacol. 48, 241–244.

Trakoon-osot, W., Sotanaphun, U., Phanachet, P., 2013. Pilot study: hypoglycemic and antiglycation activities of bitter melon (*Momordica charantia* L.) in type 2 diabetic patients. J. Pharm. Res. 6, 859–864.

Trumbo, P.R., Ellwood, K.C., 2006. Chromium picolinate intake and risk of type 2 diabetes: an evidence-based review by the United States Food and Drug Administration. Nutr. Rev. 64, 357–363.

Tsai, C.-H., Chen, E.C.-F., Tsay, H.-S., 2012. Wild bitter gourd improves metabolic syndrome: a preliminary dietary supplementation trial. Nutr. J. 11, 4.

Vanschoonbeek, K., Thomassen, B.J.W., Senden, J.M., 2006. Cinnamon supplementation does not improve glycemic control in postmenopausal type 2 diabetes patients. J. Nutr. 136, 977–980.

Vincent, J.B., 2015. Is the pharmacological mode of action of chromium (III) as a second messenger? Biol. Trace Elem. Res. 166, 7–12.

Wang, Z.-Q., Lu, F.R., Leng, S.-H., et al., 2008. Facilitating effects of berberine on rat pancreatic islets through modulating hepatic nuclear factor 4 alpha expression and glucokinase activity. World J. Gastroenterol. 14, 6004–6011.

Wang, R.H., Kim, H.S., Xiao, C., et al., 2011. Hepatic Sirt1 deficiency in mice impairs Torc2/Akt signaling and results in hyperglycemia, oxidative damage, and insulin resistance. J. Clin. Invest. 121, 4477–4490.

WHO (2016) Global report on diabetes. http://www.who.int/iris/bitstream/10665.

WHO/FAO, 1981. Joint Expert Committee on Food Additives (JECFA)—Monographs and Evaluations; Food Additive Series 14: Cinnamaldehyde, 471. Available from: http://www.inchem.org/pages/jecfa.html.

Wright, E.M., 2001. Renal Na (+)-glucose co-transporters. Am. J. Physiol.-Renal Physiol 280, F10–F18.

Yan, J., Wang, C., Jin, Y., et al., 2018. Catalpol ameliorates hepatic insulin resistance in type 2 diabetes through acting on AMPK/NOX4/PI3K/AKT pathway. Pharmacol. Res. 130, 466–480.

Youssef, F.S., Ashour, M.L., Ebada, S.S., et al., 2017. Antihyperglycemic activity of the methanol extract from leaves of *Eremophila maculata* (Scrophulariaceae) in streptozotocin-induced diabetic rats. J. Pharm. Pharmacol. 69, 733–742.

Zhang, W., Xu, Y.C., Guo, F.J., 2008. Anti-diabetic effects of cinnamaldehyde and berberine and their impacts on retinol-binding protein 4 expression in rats with type 2 diabetes mellitus. Chin. Med. J. (Engl.) 121, 2124–2128.

Zhang, H., Kong, W.J., Shan, Y.Q., et al., 2010a. Protein kinase D activation stimulates the transcription of the insulin receptor gene. Mol. Cell Endocrinol. 330, 25–32.

Zhang, H., Wei, J., Xue, R., et al., 2010b. Berberine lowers blood glucose in type 2 diabetes mellitus patients through increasing insulin receptor expression. Metabolism. 59, 285–292.

Zhang, M., Chen, L., 2012. Berberine in type 2 diabetes therapy: a new perspective for an old antidiarrheal drug? Acta Pharm. Sin. B 2, 379–386.

Ziegenfuss, T.N., Hofheins, J.E., Mendel, R.W., 2006. Effects of a water-soluble cinnamon extract on body composition and features of the metabolic syndrome in pre-diabetic men and women. J. Int. Soc. Sports Nutr. 3, 45–53.

Chapter 7

Role of nutraceuticals in respiratory and allied diseases

Kavita Gulati, Pankaj Verma, Nishant Rai and Arunabha Ray

Introduction

Respiratory disorders are one of the leading causes of mortality and morbidity and they impose an immense global health burden. Respiratory disorders mainly affect the airway passages, which include nasal and paranasal structures, the trachea, bronchi, bronchioles, and the lungs. Their etiopathology could be inflammatory, allergic, infectious, neoplastic, or traumatic. The most common respiratory disorders include bronchial asthma, chronic obstructive pulmonary disease (COPD), pneumonia, tuberculosis, pneumoconiosis, cystic fibrosis, lung cancer, etc., and their complex cellular and molecular mechanisms are known. For example, various factors such as allergies, toxic environmental exposure, cigarette smoking, and tobacco consumption may precipitate conditions resulting in airway obstruction, which is primarily reversible in asthma and irreversible in COPD patients. Bronchial asthma and COPD are chronic airway diseases characterized by airway obstruction, bronchial hypersensitivity, and airway inflammation. These diseases are mediated through the T helper-2 immune response followed by mast cell (MC) destabilization and eosinophilic infiltration, which results in the release of various inflammatory mediators (Chaudhary et al., 2016; Holgate et al., 2008). The expression of inflammatory mediators, IgE antibody production, and airway hyperresponsiveness contribute significantly to the occurrence of bronchial asthma (Rai et al., 2015). Various factors leading to asthma involve obesity, repeated infections, dietary factors, environmental exposure, and increased allergen exposure. Prolonged exposure to irritants and toxins results in COPD, which is characterized by chronic airway inflammation and damaged alveolar structures of the lungs, leading to chronic bronchitis, chronic bronchiolitis, and emphysema (Barnes, 2008). Further, increasing complexities in the pathophysiology of infectious diseases like pneumonia and tuberculosis, which are caused by microorganisms, have led to constant modifications in antimicrobial therapy. Some of these drugs have significant off-target toxicities that must be addressed to enhance treatment compliance. Modulation of proinflammatory molecules like chemokines, cytokines, transcription factors, enzymes due to exposure to smoking, allergens, environmental toxicants, stress, radiation, alcohol, etc., has resulted in chronic inflammation. Active participation of various transcription factors such as AP-1, Nrf2, NF-κB, STAT-3, and HIF-1, also play a vital role in the pathogenesis of chronic inflammation. All of the above cellular/molecular mechanisms may contribute to respiratory pathology and could be important targets for designing therapeutic strategies.

Several pharmacological approaches have been adopted to combat respiratory disorders which involve the use of antiinflammatories, antineoplastics, bronchodilators, antibiotics, antiallergics, and mucolytics. However, the benefit vs risk ratio of these agents is still debated. Further, some of the unwarranted adverse effects with these agents, coupled with increasing treatment costs, have prompted a search for viable alternatives from alternative/herbal sources. Cystic fibrosis continues to be an enigma and treatment strategies are still far from satisfactory. In view of the marked increase in incidences of respiratory diseases and the increasing ineffectiveness and safety concerns of conventional drug therapy, there is a pressing need for safer and potential alternative/adjunct therapies.

The growing awareness of nutrition, as well as the interest of the population in natural substances and natural supplements, has now been widely recognized. There is a growing demand for these nutritional supplements which include functional foods, beverages, and herbal products, due to safety, efficacy, wide availability, and easy consumption as well their low toxicity. These natural substances are often referred to as nutraceuticals, a term coined by Stephen De Felice in 1989, which is a

Nutraceuticals. DOI: https://doi.org/10.1016/B978-0-12-821038-3.00007-0

combination of two words, nutrient and pharmaceutical. Nutraceuticals may be considered as any substance which has physiological and therapeutic benefit and provides protection against chronic disease, extends life expectancy, or supports the structures or functions of the body. Apart from nutraceutical use in functional foods and dietary supplements, they also possess naturally bioactive ingredients, chemical compounds with therapeutic medicinal value, and help in the prevention and treatment of disease states. Therefore nutraceuticals can be used as a viable alternative for the treatment and management of respiratory disease as compared to modern pharmacological strategies. Nutraceuticals have been used traditionally and are emerging as an important class of natural bioactive compounds with scientifically supported and validated medical evidence for their safety and efficacy (Dillard and German, 2000). Nutraceuticals have also gained considerable attention due to their wide range of safety and efficacy when used in the food industry and pharmaceuticals (Das et al., 2012; Nasri et al., 2014). Moreover, the use of plant extracts and herbal molecules for the treatment and management of disease has also been widely accepted and welcomed throughout the world.

Nutraceuticals can be dietary food supplements, medicinal foods, herbal agents, or pharmaceuticals, and they can be categorized in several ways depending upon the food sources, mechanism of action, and chemical nature.

Foods and nutrients

Nutritional factors are being increasingly recognized as important determinants of chronic disease development and progression. Both preclinical and clinical studies have shown the impact of nutritional elements in the etiopathology and management of respiratory diseases. Dietary patterns, nutrient intake, and weight status at different life stages may be important modifiable risk factors for obstructive lung diseases such as asthma and COPD and may influence/modify disease management. Adequate intake of fruits and vegetables, as well as other dietary nutrients and patterns, could impact the treatment of several respiratory diseases.

Dietary fibers

Dietary fibers include hemicelluloses, cellulose, gums and pectins, lignins, resistant starches, and dextrins that are found in food material, including fruits, vegetables, oats, barley, and beans. These are digested by microflora in the gut. Dietary fibers are classified into two categories based on their solubility: (1) insoluble dietary fibers which include celluloses, some hemicelluloses, and lignins, and (2) soluble dietary fibers which include β-glucans, pectins, gums, mucilages, and hemicelluloses. Studies have shown increased immunity in humans with the consumption of soluble fibers (Watzl et al., 2005). Fruits and vegetables are rich sources of dietary fibers which may play an important role in the treatment and management of respiratory disorders. Various studies have reported an inverse association between the incidence of asthma and intake of fruits and vegetables (Nurmatov et al., 2011; Boeing et al., 2012).

Dietary fiber content affects gut microbiota, increasing the circulating levels of short-chain fatty acids, which act to drive bone marrow hematopoiesis and impair the capacity of dendritic cells to instigate Th2 cell-mediated allergic airway inflammation. The increased inflammation and mucus secretion in animals on a low-fiber diet was reported to increase airway hyperreactivity (AHR), as measured by increased airway resistance after methacholine challenge and vice versa (Trompette et al., 2014). A study suggested that high dietary fiber intake led to an improvement up to 40% in the risk of COPD in men who were exsmokers (Kaluza et al., 2018). Further, Vaughan et al. (2019) concluded that dietary fiber consumption in a balanced diet is strongly associated with reduced risk of COPD and other respiratory disorders.

Probiotics and prebiotics

Probiotics have been used for the treatment and management of necrotizing enterocolitis, acute diarrheal infections, antibiotic-associated diarrhea, infant colic, and acute respiratory tract infections (Liu et al., 2018). Probiotics are living microbial food supplements that, when administered in adequate amounts, benefit the host organism by improving the intestinal microbial balance. Most probiotics include lactic acid-producing bacteria, such as *Lactobacillus*, *Streptococcus*, *Bifidobacterium*, and *Enterococcus* speciesor nonpathogenic yeasts such as *Saccharomyces boulardii*, which are used for the treatment and management of gastroenteritis, inflammatory bowel disease, food allergies, etc. There is evidence that administration of probiotics decreases the risk of systemic conditions such as allergy, asthma, cancer, and also several other infections of the ear and urinary tract (Lenoir-Wijnkoop et al., 2007). Prebiotics are dietary ingredients that act by selectively altering the composition or metabolism of the gut microbiota. These are short-chain polysaccharides (fructose-based oligosaccharides) having unique chemical structures that are not digested by humans. The consumption of prebiotics generally promotes *Lactobacillus* and *Bifidobacterium* growth in the gut, thus aiding the metabolism (Hord, 2008). Vegetables like chicory root, banana, tomato, and allium are rich in fructo-oligosaccharides. In view of their immunomodulatory properties, probiotics, prebiotics, and synbiotics may constitute valuable tools to treat and prevent immune-related disorders such as allergy (Gourbeyre et al., 2011). However, the outcome of various experimental studies regarding the role of prebiotics and probiotics in the prevention and treatment of asthma and other manifestations of allergy are equivocal (Osborn and Sinn, 2013). In fact, the most encouraging effects have been shown in the treatment and prevention of infant eczema (Boyle et al., 2009). A recent study also showed that probiotic consumption reduced the symptoms as well as risk of respiratory tract infection (Wang et al., 2016). Probiotics have also been shown to possess antimutagenic activity, anticarcinogenic activity, and to promote T- and B-cell proliferation (Moratz et al., 2013).

Vitamins

Vitamins are essential organic nutrients, most of which are not produced in the body, or only in small amounts, and are thus mainly obtained through food. Vitamin C, vitamin E, and carotenoids are collectively known as antioxidant vitamins, and are present abundantly in fruits and vegetables. They exert their protective action by free-radical scavenging mechanisms.

A systematic review and metaanalysis by Allen et al. (2009) showed that dietary vitamin A intake was significantly lower in asthmatic patients than in healthy subjects. Supplementation of the diet with lycopene, a carotene found in tomatoes and carrots, has a protective effect against asthma development in a murine model (Hazlewood et al., 2011). It has been shown that children with vitamin A deficiency have an increased risk of developing respiratory diseases (Sommer et al., 1984). Vitamin A has been reported to play a major role in the regulation of innate and cellular immunities (Stephensen, 2001; Villamor and Fawzi, 2005). Aukrust et al. (2000) demonstrated that a deficiency of vitamin A increased the production of proinflammatory cytokines such as TNF-α and upregulated inflammatory responses, and that supplementation with vitamin A reversed these effects.

Vitamin C, better known as ascorbic acid, donates a hydrogen atom to lipid radicals, quenches singlet oxygen radicals, and removes molecular oxygen. Scavenging of aqueous radicals by the synergistic effect of ascorbic acid along with tocopherol supplementation is a well-known antioxidant mechanism (Lee et al., 2004). Administration of vitamin C in ovalbumin (OVA)-challenged mice decreased AHR, influx of inflammatory cells in bronchoalveolar lavage fluid (BALF), and attenuated lung inflammation (Jeong et al., 2010). Similarly, high-dose vitamin C supplementation significantly reduced eosinophilic infiltration in BALF and increased the Th1/Th2 cytokine secretion ratio, thus skewing the Th1/Th2 balance toward a nonallergic Th1 immune response in asthmatic mice (Chang et al., 2009). In a clinical study, vitamin C supplementation attenuated exercise-induced bronchoconstriction in asthmatic patients (Tecklenburg et al., 2007). Supplementation with vitamin C was also shown to stimulate the immune system in response to infection (Wintergerst et al., 2006).

Vitamin E is comprised of tocopherols and tocotrienols, the most important naturally occurring antioxidants. Tocopherols are found mostly in vegetable oils such as corn, soybean, sesame, nut, and cottonseed, whereas tocotrienols are plentiful in palm oil. Treatment with α-tocopherol attenuated lung edema and lipid peroxidation in animal models of acute respiratory syndrome (Demling et al., 1995; Suntres and Shek, 1996; Fan et al., 2000). The administration of α-tocopherol was found to be beneficial in the prevention of bleomycin-induced lung fibrosis in experimental animals (Kilinc et al., 1993; Suntres and Shek, 1997; Deger et al., 2007). Treatment with α-tocopherol showed a reduction in transendothelial migration of neutrophils and protected against lung injury in endotoxin- induced airway inflammation (Rocksen et al., 2003). The administration of γ-tocopherol by nebulization was shown to be effective in improving pulmonary function in sheep (Hamahata et al., 2008). Wagner et al. (2008) demonstrated that γ-tocopherol supplementation inhibited airway eosinophilia and mucus cell hyperplasia in experimental models of allergic rhinitis and asthma. γ-Tocopherol also attenuated leukotriene production by suppressing 5-LOX activity (Jiang et al., 2011). Tocotrienol was shown to inhibit the production of inflammatory mediators such as inducible nitric oxide synthase, prostaglandin E2, proinflammatory cytokines, cyclooxygenase-2, and NF-κB expression (Wu et al., 2008). A clinical study showed that tocotrienol was also important in the management of oxidative stress in patients with COPD (Rahman, 2012). Administration of vitamin E to allergen-challenged mice reduced mitochondrial dysfunction, Th2 cytokine production, allergen-specific IgE, and expression of lipid mediators in the lung, leading to alleviation of asthmatic symptoms (Mabalirajan et al., 2009). Expression of IL-5 mRNA and protein in the lung and plasma IgE levels were reduced after OVA sensitization and challenge compared to wild-type mice in vitamin E transfer protein knockout mice (Lim et al., 2008). Moreover, dietary supplementation with vitamin E provided variable degrees of protection against ozone-induced enhanced airway response in allergen-sensitized guinea pigs (Chhabra et al., 2010).

Hejazi et al. (2016) reviewed the effects of vitamin D on common respiratory diseases, such as asthma, COPD, and tuberculosis, and suggested an inverse relationship between vitamin D and asthma. Vitamin D has been shown to decrease the inflammatory cytokines IL-9, IL-5, and IL-13, all of which play an important role in the pathogenesis of bronchial asthma (Hejazi et al., 2016). A study showed that high maternal intake of vitamin D and antioxidants reduced the risk of childhood asthma and wheezing (Erkkola et al., 2009). Another study conducted among mother–infant pairs in Taiwan reported an inverse relationship between maternal vitamin D levels and IgE antibodies against allergens before age 2 (Chiu et al., 2015). A randomized, double-blind clinical trial study showed reduced risk of acute respiratory tract infections following vitamin D supplementation versus a placebo group (Martineau et al., 2017).

Polyunsaturated fatty acids

Polyunsaturated fatty acids (PUFAs) are also known as essential fatty acids as they are not formed in the body and must be taken through the diet. PUFAs have two subdivisions:

1. Omega-3 (*n*-3) fatty acids: α-linolenic acid (ALA), eicosapentaenoic acid (EPA), docosahexaenoic acid (DHA). Fatty fishes and fish oils are sources of EPA and DHA. Principal sources of ALA are mainly flaxseed, soybeans, canola, and some nuts.
2. Omega-6 (*n*-6) fatty acids: linoleic acid (LA), γ-linolenic acid, and arachidonic acid (ARA). LA occurs mainly in corn, safflower, soybean and sunflower vegetable oils. ARA is found in animal products such as meat, poultry, and eggs.

Emerging research evidence has shown the benefits of omega-3 oils in other areas of health including premature infant health, asthma, bipolar and depressive disorders, etc. (Hodge et al., 1996; Broughton et al., 1997).

PUFAs suppress T-cell activation and inhibit cytokine production from macrophages (Lo et al., 1999; Zeyda et al., 2003). Supplementation with a fish oil capsule containing EPA and docosahexanoic acid showed protective effects against the exercise-induced bronchoconstriction of asthma (Mickleborough et al., 2006). Omega-3 PUFA derived from fish oil inhibited leukotriene synthesis and release of proinflammatory cytokines (Lee et al., 1985; Endres et al., 1989). Dietary *cis*-9, *trans*-11-conjugated LA, naturally occurring in milk fat, inhibited allergic airway inflammation in a murine asthma model (Jaudszus et al., 2008). Epidemiological studies in different populations such as Americans, Australians, and Japanese have shown that supplementation with omega-3 PUFA (fish oil) was associated with lower incidences of inflammatory diseases such as asthma (Schwartz and Weiss, 1994; Peat et al., 1992; Satomi et al., 1994; Arm et al., 1989). There have been a number of clinical studies that have shown potentially protective effects of omega-3 PUFAs in childhood asthma patients. A positive association between the ratio of omega-6:omega-3 PUFAs in diet and risk for asthma has also been reported (Oddy et al., 2004). Supplementation with fish oil rich in EPA and DHA was found to be beneficial for children with bronchial asthma (Nagakura et al., 2000). According to new research, dietary intake of two fatty acids, omega-3 and omega-6, may have opposite effects on the severity of asthma in children and may also play opposite roles in modifying their response to indoor air pollution (Brigham et al., 2019). Thus the experimental and clinical evidence indicates that omega-3 PUFA might be a promising remedy for allergic diseases like asthma but the mechanism remains to be elucidated.

Another observational study evaluated increased prenatal omega-3 LC-PUFA dietary exposure in observational studies and omega-3 LC-PUFA supplementation in randomized controlled trials on outcomes of IgE-mediated allergic disease. The study concluded that prenatal omega-3 LC-PUFA consumption resulted in reduced childhood allergic disease (Best et al., 2016). Another case-control study which compared fish and PUFA intakes and plasma phospholipid PUFA between COPD and healthy subjects concluded a marginally significant correlation between total long-chain omega-3 PUFA levels in plasma phospholipids and COPD incidence (Fulton et al., 2019). A study conducted in rat pups showed increased omega-3 PUFA levels in the blood and brainstem of pups due to maternal omega-3 PUFA supplementation, which resulted in improved lung architecture and response to respiratory challenges (Tenorio-Lopes et al., 2016). Atlantis and Cochrane (2016) showed that dietary PUFA consumption improved exercise capacity and reduced inflammation in COPD patients as compared to healthy subjects. Another study evaluated associations of omega-3 PUFA biomarkers (α-linolenic acid, EPA, and DHA) with compromised pulmonary health using pulmonary function test (PFT) and showed positive associations between these biomarkers and forced expiratory volume in 1 second (FEV_1) and forced vital capacity (FVC) (Xu et al., 2018). This study reiterated the therapeutic and beneficial effects of omega-3 PUFAs on pulmonary functions.

Trace elements

Various trace elements have been shown to play beneficial roles in the etiology and pathogenesis of respiratory disorders such as bronchial asthma. Intracellular levels of calcium have been associated with muscle contraction and, hence, muscle functioning (Sahito et al., 2013). Magnesium helps in relaxing the bronchial tubes as well as the smooth muscles of the esophagus. Laboratory studies have indicated that magnesium can relax bronchial smooth muscles (Gourgoulianis et al., 2001). In some clinical trials conducted on children with mild to moderate persistent asthma, it was found that magnesium supplementation significantly improved lung function, lowered bronchial reactivity, and improved asthma control and quality of life (QOL) (Bede et al., 2003; Gontijo-Amaral et al., 2007).

Selenium is an essential nutritional antioxidant that is important for many aspects of human health. Several epidemiological studies showed reduced levels of selenium in asthmatic patients (Flatt et al., 1990; Shaw et al., 1994; Kadrabova et al., 1996). One of the studies showed a stronger selenium–asthma relationship in those exposed to cigarette smoke. About a 50% reduction in asthma prevalence was found when levels of selenium were increased in smoke-exposed individuals (Rubin et al., 2004). Another clinical study showed a significant improvement in asthma symptoms following selenium supplementation as compared to controls (Allam and Lucane, 2004).

Zinc is an antioxidant that may have antiinflammatory properties related to the pathophysiology of asthma. Lowered levels of zinc are associated with increased production of MCs, basophils, and B cells, which play an active role in the immune system and the pathophysiology of asthma. Zinc deficiency has been correlated with the onset of asthma, growth retardation, delayed wound healing, etc. Moreover, various zinc supplements have been implicated in ality of lifasthma (Ariaee et al., 2016). Several studies have shown a significant reduction in plasma, serum and hair zinc in asthmatic patients (Di Toro et al., 1987; el-Kholy et al., 1990). Zinc supplementation decreased airway hyperresponsiveness and airway inflammation in mice, indicating a potential treatment strategy for asthmatics (Morgan et al., 2011). Zinc-deficient ovalbumin-treated mice had greater eosinophilia as compared to zinc-normal ovalbumin-treated mice and zinc supplementation partially decreased eosinophil cells in the former group (Richter et al., 2003). It has been shown that serum zinc levels decrease in COPD patients, and this reduction was more prominent in more severe cases (Karadag et al., 2004). Furthermore, a reduction of zinc levels in the lungs of rats has been observed when inflammation was induced by inhalation of nitrogen dioxide (Lebedeva et al., 1992). Kirkil et al. (2008) also showed antioxidant effects of zinc supplementation in patients with COPD.

The role of iron was shown in a clinical study which reported that higher iron stores were inversely associated with asthma and lowered body iron and higher tissue iron requirements were directly related to reduced lung functions (Brigham et al., 2015). One study reported that airway manifestations of allergic asthma were suppressed by both increased levels of iron intake and systemic iron administrations in the mouse model of bronchial asthma (Maazi et al., 2011). Hypokalemia was also reported to be associated with acute episodes of bronchial asthma (Wang et al., 2016). Iron supplementation reduced airway hyperresponsiveness and eosinophilia in a murine model of allergic asthma (Maazi et al., 2011). Administration of an iron chelator significantly reduced serum IgE levels in chemically induced Th2-mediated autoimmunity in rats (Wu et al., 2004). Lactoferrin, a natural iron chelator, reduced pollen antigen-induced airway inflammation in a mouse model of allergic asthma (Kruzel et al., 2006).

Deficiency of trace elements is often associated with the incidence of infectious disease. Various trace element deficiencies, such as selenium and zinc, have been correlated with respiratory infection such as tuberculosis. A case-control study suggested that altered serum levels of trace elements (cobalt, copper, chromium, manganese, molybdenum, selenium, and zinc) are associated with mycobacterial disease (Oh et al., 2019). Regulation of T-lymphocytes is affected by zinc deficiency, which results in the development of various allergies. Cellular copper levels are regulated by IL-2, which is a T-cell-derived inflammatory mediator. Altered zinc and copper levels reduce the efficiency of the antioxidant system, which leads to hyperreactivity and inflammation in the respiratory system. Another study conducted in India showed lower levels of zinc and higher levels of copper in bronchial asthmatic patients as compared to healthy subjects (Bhaskar et al., 2019).

Medicinal plants

Traditional herbal medicines have regained importance all over the world due to extensive scientific research being carried out in this field. The cellular and molecular mechanisms of action of some of the traditionally used medicinal plant products have revealed that they may constitute a viable alternative source for therapeutic agents in respiratory disorders. In fact, many herbal drugs have provided vital therapeutic solutions in patients with resistance to synthetic drugs. Plants and plant-derived components have been major sources of health care since ancient human civilization. According to the World Health Organization (WHO), about 80% of the world's population living in developing countries rely on traditional medicine for their primary health care (World Health Organization, 2003). In every traditional system of medicine, herbal agents play an important role and are widely used due to their effectiveness, lower toxicity, and relatively low cost (Yusuf et al., 2007). The WHO also promotes the use of traditional medicines due to their safety, low cost, ease of availability, and faith of people in such remedies. Such medicinal plant-based products are commercially available in different formulations (monoherbal or polyherbal) and are used in various traditional systems of medicine throughout the world for respiratory diseases. For example, in India, the traditional systems of medicine like Ayurveda and Unani extensively use medicinal plant products either as dietary supplements or in various formulations for the prevention and treatment of a number of clinical conditions. They are not only safe and effective but also pharmacoeconomically viable alternatives to conventional therapy. Since ancient times such products have been used in the treatment of respiratory diseases and, in recent years, their use in the management of bronchial asthma, COPD, pulmonary fibrosis, and other respiratory diseases has been reported in the scientific literature.

Adhatoda vasica

Adhatoda vasica (family Acanthaceae, AV) is a shrub, used by Asian and European medical practitioners. This plant has been used in the Indian traditional system of medicine (Manjunath 1948). Numerous in vitro, in vivo, and clinical studies have acknowledged AV as an important natural agent for many medical illnesses. AV has an antiinflammatory action on the respiratory tract and is effective in respiratory tract infection. The alkaloids vasicinone and vasicine have potent bronchodilator and antiallergic activity. Owing to these activities, AV is effective in acute asthma conditions. AV has been proven for its antitussive activity as effective as codeine in irritant aerosols and citric acid-induced cough models. The leaves, roots, flowers, and bark of this plant have been used in the treatments of cough, colds, asthma, to liquefy sputum, as a bronchodilator, bronchial catarrh, bronchitis, and tuberculosis. A number of parts of the plant are commonly used in the forms of decoctions or powders. The juice from the leaves is also frequently used (Claeson et al., 2000). Vasaka syrup and vasaka liquid extract are mentioned in Pharmacopoeia of India (The Indian Pharmacopoeia), (1955). An essential oil extracted from the leaves of AV showed a relaxant effect on the airway smooth muscle in the isolated guinea pig tracheal chain (D'Cruz et al., 1979). A methanolic extract from the whole plant showed antiallergic and antiasthmatic properties in the guinea pig (Muller et al., 1993). Many herbal preparations containing AV are commercially used, such as Kada in India, Salus Tuss in Germany, and Kan Jang and Spirote in Sweden (Claeson et al., 2000). Vasicine, an alkaloid, is one of the major components of the plant and is responsible for most of its activities, including its antioxidant, antiinflammatory, and bronchodilatory qualities. Bromhexine, a derivative of vasicine, has been shown to possess mucus-liquefying/expectorant activity. A clinical trial was conducted with bromhexine in 30 patients suffering from different respiratory symptoms. A significant change in the viscosity and acid mucopolysaccharide structure was found in the bromhexine-treated group as compared to the placebo group, and this was seen in both infected and noninfected patients. Ambroxol, a widely used secretolytic agent developed from vasicine, was shown to inhibit IgE-dependent mediator secretion from human MCs and basophils—the principal effectors of allergic inflammation. As compared to vasicine, ambroxol showed a greater effect in lowering basophil IL-4 and IL-13 secretions. It was also reported to reduce IgE-dependent p38 mitogen-activated protein kinase (MAPK) phosphorylation in basophils (Rachana et al., 2011).

Albizia lebbeck

Albizia lebbeck, a flora of the family *Mimosaceae*, is a deciduous woody tree which is traditionally used for treating asthma, colds, coughs, and other allergic diseases (Zia-Ul-Haq et al., 2013). The aqueous extract of the bark of *A. lebbeck* showed antiasthmatic and antianaphylactic activity in experimental studies (Tripathi and Das, 1977). The aqueous extract of the bark also showed antiallergic and MC-stabilizing activity in experimental studies (Baruach et al., 1997). Antiallergic and MC-stabilizing activity of standardized extract of *A. lebbeck* were reported with respect to catechin as a phytomarker (Venkatesh et al., 2010). Antiallergic effects were also demonstrated for bark extract in experimental models of cutaneous anaphylaxis. Another experimental study showed that a decoction of the bark had MC-stabilizing effects in rat peritoneal MCs (comparable to cromoglycate), and antianaphylactic effects, as well as protection against histamine-induced bronchoconstriction in guinea pigs. However, there were no reductions in the weight of the thymus, adrenal gland, or spleen (Faisal et al., 2012). In a clinical study, an *A. lebbeck* stem bark decoction was administered to 81 patients with bronchial asthma and the results were assessed in terms of clinical recovery, symptomatic relief, and pulmonary function improvement. The results showed that there was a significant improvement in Peak expiratory flow rate (PEFR), total eosinophil count, and Erythrocyte Sedimentation Rate (ESR) after 6 weeks treatment and 56% of asthma patients showed a good response. These results suggest that *A. lebbeck* can be used as an effective herbal agent in bronchial asthma. In another study, 35 patients with tropical pulmonary eosinophilia who were treated with *A. lebbeck* water extract for 6 weeks showed an excellent response in 82% patients without any untoward effects (Kumar et al., 2010).

A study screened all nine active ingredients of *A. lebbeck* and concluded that lebbecacidin binding to IL-5 resulted in reduced incidence of bronchial asthma (Yim et al., 2014). Another recent study evaluated the antiasthmatic potential of *A. lebbeck* in an experimental model of bronchial asthma and suggested antiinflammatory (reduced eosinophils, neutrophils, and TNF-α) and immunomodulatory effects (reduced OVA-specific IgE, IL-4, and enhanced IFN-γ) in rats (Chaudhary et al., 2016).

Boswellia serrata

Boswellia serrata (family *Burseraceae*) is commonly found in many regions of the world, such as South Asia, Northern Africa, and the Middle East. Traditional medicine using extracts made from this plant has long been used to treat asthma. These extracts contain resin, amino acids, phenols, terpenes, polysaccharides, and β-boswellic acid, the major active antiinflammatory component. Extracts of *B. serrata* or β-boswellic acid have been reported to inhibit hypersensitivity reactions by regulating both the humoral and cellular immune systems. They decrease primary antibody synthesis, inhibit polymorphonuclear leukocyte proliferation and infiltration, enhance the phagocytotic function of macrophages, suppress the classical and alternate complement pathways, and suppress the inflammation process, one of the critical pathological features in asthma (Khajuria et al., 2008). They also inhibit histamine release from MCs challenged with G protein stimulator c48/80 in a dose-dependent manner (Pungle et al., 2003). β-Boswellic acid can downregulate the synthesis of prostaglandins by inhibiting COX-1 in intact human platelets (Siemoneit et al., 2008). It has been shown that β-boswellic acid inhibits the production of proinflammatory cytokines, including TNF-α, IL-1, IL-2, IL-6, IL-12, and IFN-γ, by suppressing the activation of NF-κB (Gayathri et al., 2007). These results suggest that *B. serrata* might be effective in controlling the inflammation process and contraction of airway smooth muscle in asthmatic conditions by inhibiting enzymes required for the production of proinflammatory mediators and bronchoconstrictors. Preliminary clinical investigation has shown that *B. serrata* has a potential therapeutic effect on asthma (Ammon et al., 1991). Clinical study results suggest that *B. serrata* extract has potential benefit for asthma patients (Gupta et al., 1998). A novel herbal composition LI13109F, containing extracts of *B. serrata* gum resin and *Aegle marmelos* fruit, improved asthma symptoms by reducing IL-4 levels and increasing IFN-γ in a double-blind controlled clinical trial (Kaluza et al., 2018).

Bryophyllum pinnatum or *Kalanchoe integra*

Kalanchoe integra (family *Crassulaceae*) has shown various pharmacological activities such as anthelmintic, immunosuppressive, wound healing, hepatoprotective, antinociceptive, antiinflammatory, antidiabetic, nephroprotective, antioxidant, antimicrobial, analgesic, anticonvulsant, and antipyretic properties. In respiratory disorders, boiled leaf extracts are useful in the management of acute and chronic bronchitis, pneumonia, bronchial asthma, and palpitation (Biswas et al., 2011). Leaves of *K. integra* contain flavonoids and tannins, which have been found to have beneficial effects in the management of bronchial asthma. The antihistaminic and antianaphylactic effects of *Kalanchoe* species have also been demonstrated by researchers (Cruz et al., 2012). Thus it may be a potential nutraceutical with antiinflammatory activity for the management of airway remodeling in bronchial asthma (Julius et al., 2014). Histopathological studies showed dose-dependent progressive bronchiolar dilatation, alveolar distention, and inflammatory cell infiltration effect of *Bryophyllum pinnatum* (500 mg/kg and 1500 mg/kg) as compared to normal control rats (Sakpa and Idodia, 2019). However, the group also suggested deleterious effects of *B. pinnatum* at comparatively higher doses.

Curcuma longa

Curcuma longa (family *Zingiberaceae*), or turmeric, is a perennial herb with pointed leaves that bears funnel-shaped yellow flowers. It grows to a height of 3–5 feet and is cultivated extensively in Asia, mostly in India, China, and other countries with a tropical climate. Curcumin is a dietary polyphenol obtained from the plant *C. longa*, and is the principal active ingredient responsible for therapeutic effects. Curcumin has been shown to have antiasthmatic effects in both in vivo and in vitro studies. In an OVA-induced asthma model in guinea pigs, curcumin administration during OVA sensitization showed significant protective effects through attenuation of bronchial constriction and hyperreactivity (Ram et al., 2003). This indicated that curcumin had both preventive and therapeutic effects on asthma that were attributed to the suppression of inducible nitric oxide synthase (iNOS) and subsequent nitric oxide (NO) production, inhibition of inflammatory cytokine synthesis, and downregulation of eosinophil recruitment to the airway. In vitro, curcumin supplementation inhibited IgE/antigen activation of MCs through the principal activation pathway mediated by FcεRI, directly inhibiting Syk kinase phosphorylation, which is crucial for gene transcription and production of proinflammatory cytokines (Lee et al., 2008). These results indicated that curcumin may attenuate asthma symptoms by inhibiting the production of cytokines related to eosinophil function and IgE synthesis, and by suppressing IgE-mediated reactions and hyperreactivity. Curcumin has also been reported to restore histone deacetylase-2 enzyme activity, the role of which is well known in airway inflammation of asthma and COPD. Thus, curcumin has potential to reverse corticosteroid resistance which is common in COPD and asthmatic patients (Meja et al., 2008). Curcumin also inhibited the formation of lipid-derived inflammatory mediators like leukotrienes (Hong et al., 2004). It was also found to be beneficial in the inhibition of lung oxidative stress and fibrosis (Zhang et al., 2011; Venkatesan et al., 1997; Punithavathi et al., 2000). It also provided protection from other acute and chronic lung diseases (Venkatesan et al., 2007). One study examined the clinical and immunologic effects of lactoferrin and curcumin (LC) in which oral supplementation in healthy children with such respiratory infections showed a reduction in infections and beneficial immune-modulatory effects (Zuccotti et al., 2009). Curcumin was also found to decrease mucus formation in the airways in a murine asthma model. Curcumin inhibited the production of proinflammatory cytokines, such as IL-1β and IL-8, suppressed inducible iNOS, and modulated steroid activity. Its effects on steroid activity may have been the result of inhibition of NF-κB through blocking IκB kinase (IKK) activity, thus providing valuable leads wherein dietary supplementation with curcumin could modulate airway inflammation in chronic diseases such as asthma in cases where corticosteroids are no longer effective (Biswas and Rahman, 2008). Evidence also showed that during allergic inflammation, curcumin may modulate both early- and late-phase responses by altering Th2 responses. Curcumin may also disrupt antigen presentation and thus may have potential therapeutic value in allergen-triggered airway inflammation (Kurup et al., 2007). One study evaluated the effects of *C. longa* and curcumin on total and differential WBC count and oxidant/antioxidant biomarkers in an experimental model of asthma. It determined specific antioxidant and antiinflammatory effects, which suggested a therapeutic potential for the plant and its constituents on asthma (Shakeri et al., 2017). Another randomized, double-blind, placebo-controlled, phase II clinical trial concluded that the administration of powdered roots of *C. longa* resulted in less frequent night-time awakenings, less frequent use of short-acting β-adrenergic agonists, and better disease control after 3 and 6 months when compared to placebo in children and adolescents (Manarin et al., 2019).

Glycyrrhiza glabra

Glycyrrhiza glabra (family *Fabaceae*), commonly known as licorice, is a herbaceous perennial that has been used as a flavoring agent in foods and medicinal remedies for thousands of years. Licorice root has been widely used around the world to treat cough since ancient times. It contains several active compounds including glycyrrhizin, glycyrrhetinic acid, flavonoids, isoflavonoids, and chalcones. Glycyrrhizin and glycyrrhetinic acid are considered to be the main active components and are potent inhibitors of cortisol metabolism, due to their steroid-like structures. The root of this plant has been used for treatment of coughs, colds, asthma, and COPD (Ram et al., 2011). Glycyrrhizin is a triterpene glycoside, a major active constituent obtained from the plant *G. glabra*. Glycyrrhizin alleviated allergic asthma in an ovalbumin-induced experimental mouse model of asthma as evidenced by increased IFNγ level, while it decreased IL-4, IL-5 levels, and eosinophil count in bronchoalveolar lavage (BAL). It also reduced OVA-specific IgE levels and also upregulated total IgG2a in serum. These results indicated that glycyrrhizin interfered with the production of IgE by decreasing the IgE-stimulating cytokines (Ram et al., 2006). It also attenuated carrageenan-induced lung inflammation and mucus production in mice (Menegazzi et al., 2008; Nishimoto et al., 2010). In an OVA-induced murine asthma model, oral administration of glycyrrhizin alleviated airway constriction, hyperreactivity, and pulmonary inflammation (Park et al., 2004). Isoliquiritigenin, a flavonoid isolated from the *G. glabra* roots, relaxed the tracheal smooth muscle of guinea pigs in vitro and in vivo (Liu et al., 2008). The effects of glycyrrhetinic acid and liquiritigenin (a flavonoid of licorice root) on asthma have been tested both in vivo and in vitro. In vitro, glycyrrhetinic acid and liquiritigenin inhibits β-hexosaminidase release from RBL-2H3 cells induced by IgE/dinitrophenyl (DNP) and from rat peritoneal MCs challenged with c48/80. In vivo, they suppressed c48/80 induced passive cutaneous anaphylactic reaction in mice. In an OVA-induced murine asthma model, glycyrrhetinic acid, but not liquiritigenin, reduced the level of IgE in serum (Shin et al., 2007). Flavonoids extracted from licorice root quenched lipopolysaccharide (LPS)-induced pulmonary inflammation by inhibiting the recruitment of neutrophils, macrophages, and lymphocytes in BALF, and by suppressing mRNA expression of TNF-α and IL-1β in LPS-challenged lung tissue in mice (Xie et al., 2009). A recent study suggested that *G. glabra* (500 mg/kg, ip) reduced levels of hydroxyproline and malondialdehyde (MDA), and increased catalase activity in a bleomycin-induced model of pulmonary fibrosis in experimental rats. This is an indicator of its potential therapeutic value in the management of respiratory disorders (Sakpa and Idodia, 2019)

Hibiscus rosa-sinensis

Hibiscus rosa-sinensis is one such medicinal plant which has been known to be traditionally used for the management of bronchial asthma. *H. rosa-sinensis* is a bushy, evergreen shrub or small tree growing 2.5–5 m tall with glossy leaves and red flowers in summer and autumn. Active constituents include β-sitosterol, stigmasterol, taraxeryl acetate, and three cyclopropane compounds and their derivatives, cyanidin glucoside, flavonoids, vitamins, thiamine, riboflavin, niacin, ascorbic acid, quercetin-3-diglucoside, 3,7-diglucoside, cyanidin-3,5-diglucoside, 6-cyanidin-3-sophoroside-5-glucoside, and kaempferol-3-xylosylglucoside. *H. rosa-sinensis* has been shown to have antiinflammatory properties in experimental models of arthritis (Verma et al., 2016). Moreover, *H. rosa-sinensis* has traditionally been used for bronchial asthma in most parts of northeastern India (Khan and Yadava, 2010). *H. rosa-sinensis* has been shown to contain vital essential nutrients such as calcium, magnesium, zinc, and potassium, which can help in the management of bronchial asthma (Sahito et al., 2013).

Hedychium spicatum

Hedychium spicatum (family *Zingiberaceae*) is a perennial rhizomatous herb called Shati in the Ayurvedic system of medicine. The rhizome of *H. spicatum* has traditionally been used to treat coughs, asthma, and other respiratory disorders, and clinically for the treatment of asthma (Chaturvedi and Sharma, 1975). One study evaluated the effects of aqueous and ethanolic extracts of the dried rhizome of *H. spicatum* for antihistaminic activities in guinea pig, antiinflammatory activities in rat, and acute toxicity in mouse. The dose-dependent antihistaminic action of both extracts was used to treat histamine-induced bronchospasm in guinea pigs. Both extracts also showed an antiinflammatory effect against carrageenan-induced paw edema in rats from up to 3 h. Neither extract showed any toxic effects such as increased motor activity, salivation, clonic convulsion, coma, or death in mice, even at the 2000 mg/kg dose (nearly 10 times the optimal effective dose), indicating their safety. The results confirm the indigenous use of this plant in respiratory disorders (Ghildiyal et al., 2012).

Ocimum sanctum

Ocimum sanctum (family Lamiaceae), commonly known as Holy Basel or Tulsi, is an annual herb that has been used in the Indian traditional system of medicine for hundreds of years. The leaves of this plant have been traditionally used in coughs, colds, asthma, and bronchitis (Ghosh, 1995). A steam-distilled extract of *O. sanctum* leaves provided a reduction in antigen-induced histamine release from peritoneal MCs (Mediratta et al., 1988).

Experimental animal studies showed the immunomodulatory potential of *O. sanctum* seed oil and an extract of leaves (Godhwani et al., 1988; Mediratta et al., 2002). Another study showed that ethanolic extract of fresh leaves and fixed oil from seeds protected against histamine and acetylcholine-induced proconvulsive dyspnea (suggestive of an antiasthmatic effect), and an antiinflammatory effect against carrageenan, serotonin, histamine, and PGE2-induced inflammation. The antiasthmatic effect was attributed to the volatile oils present in the extract. However, extracts of dried leaves did not show any protective effects in the above experimental model of bronchoconstriction, indicating the disappearance of activity during the drying process (Gupta et al., 2002). Hydroalcoholic extract of *Ocimum basilicum* showed decreased IL-4, IgE, PLA2, and total protein (TP) levels in ovalbumin-induced model of asthma in rats (Eftekhar et al., 2019). All pathological lungs indicated significantly increased interferon (IFN)-γ/IL-4 ratio and improved lung pathology, suggesting immunomodulatory and antiinflammatory effects compared to controls.

Piper nigrum

Piper nigrum (black pepper) is a flowering plant belonging to the *Piperaceae* family. It is cultivated as a fruit, dried, and used as a spice and seasoning. It grows in the southern parts of India and other tropical regions. Dried pepper has been used extensively for its flavoring properties as well as in traditional medicine. The active constituent present in *P. nigrum* is piperine. Other phytochemicals present are amides, piperidine, pyrrolidines, and trace amounts of safrole (Corinna et al., 2012). Piperine has been used as a natural bioenhancer with other drugs to potentiate their therapeutic effects. Its bioenhancer effect has been shown by either action on a metabolizing enzyme, enhancing drug transport, affecting blood supply to the gastrointestinal tract (GIT), and/or membrane fluidity (Wadhwa et al., 2014). Aqueous extract of fruits of *Sapindus mukorossi* and *P. nigrum* have been shown to have antiasthmatic properties by inhibiting acetylcholine-induced bronchoconstriction of isolated goat trachea (Parganiha et al., 2012). Another study used *P. nigrum* as one of the ingredients in a polyherbal compound and reported significant in vitro and in vivo antihistaminic and bronchodilator activity in albino rats (Amutha et al., 2015). Bhatt et al. (2013) used *P. nigrum* as one of the ingredients in a polyherbal formulation, Zeal Herbal Granules, and reported an antiasthmatic effect mediated by MC-stabilization properties (Bhatt et al., 2013).

Piper longum

Piper longum (family *Piperaceae*) is used as an important traditional medicine in Asia and the Pacific islands. *P. longum* has been shown to be a good remedy for treating tuberculosis and respiratory tract infections (Singh, 1992). The fruits and roots of this plant have been used in the treatment of childhood asthma (Dahanukar et al., 1984; Fernandez et al., 1980). The alcoholic extract of *P. longum* fruits showed immunomodulatory potential in experimental animal studies (Sunila and Kuttan, 2004). Piperine is a major alkaloid isolated from *P. longum* fruits that have been reported to inhibit the release of Th2-mediated cytokines, eosinophil infiltration, and airway hyperresponsiveness in an ovalbumin-induced asthma model (Kim and Lee, 2009).

Tylophora indica

Tylophora indica (family *Apocynaceae*) is a perennial climbing plant used in the Ayurvedic system of medicine. The leaves of the plant have been extensively used in the treatment of various inflammatory and allergic disorders including bronchial asthma, bronchitis, whooping cough, etc. (Chopra et al., 1958). The extracts of the leaves of *T. indica* have antiasthmatic and antiallergic potential (Shivpuri et al., 1972). The plant extracts also inhibited the Schultz-Dale reaction and systemic anaphylaxis in guinea pigs (Haranath and Shyamalakumari, 1975). The extract of *T. indica* leaves showed inhibition of cellular immune responses in experimental models (Ganguly and Sainis, 2001). Tylophorine is a major alkaloid obtained from the plant *T. indica*, which has been reported to have antiinflammatory, antiasthmatic, and antianaphylactic potential in experimental studies (Gopalakrishnan et al., 1980; Raina and Raina, 1980).

Solanum xanthocarpum

Solanum xanthocarpum (family *Solanaceae*) is an annual herbaceous plant commonly known as Kantkari in traditional Indian medicine. In Ayurveda, it is listed as a drug for the treatment of bronchitis and asthma. Juice of the berries is reported to be useful for a sore throat (Singh and Singh, 2010). *S. xanthocarpum* has been widely used for the treatment of respiratory diseases in the Siddha system of medicine, mainly in southern India. Treatment with the powder of the whole plant of *S. xanthocarpum* improved the different parameters of PFT in asthmatic subjects (Govindan et al., 1999, 2004). Further, aqueous standardized extract of *S. xanthocarpum* has been shown to have antiinflammatory, immunomodulatory, and antioxidant effects in an experimental model of bronchial asthma in rats (Ariaee et al., 2016).

Zingiber officinale

Zingiber officinale (family *Zingiberaceae*) is a dietary component commonly known as ginger. The rhizome of the plant has been used in the treatment of colds, asthma, and bronchitis (Ram et al., 2011). The ethanolic extracts of ginger rhizome exhibited antibacterial activity against respiratory tract pathogens, namely *Staphylococcus aureus*, *Streptococcus pyogenes*, *Streptococcus pneumoniae*, and *Haemophilus influenzae* (Akoachere et al., 2002). Fresh ginger has antiviral activity against human respiratory syncytial virus-induced plaque formation on airway epithelium (Chang et al., 2013). In a study in patients with acute respiratory distress syndrome, a diet enriched with ginger showed lower serum levels of inflammatory cytokines IL-1, IL-6, and TNF-α (Shariatpanahi et al., 2013). Gingerol, an active component of ginger, also inhibited the production of TNF-α, IL-1β, and IL-12 from LPS-stimulated macrophages (Tripathi et al., 2007).

Phytochemicals

Phytochemicals are compounds with different structures but with the same therapeutic activity isolated from various plant species and used as active components for the treatment of various diseases. Some of these active moieties are present in edible plants and their addition to the diet would be beneficial for respiratory diseases. Various phytochemicals obtained from several plant sources have been reported for the treatment of respiratory diseases.

Flavonoids

Flavonoids are polyphenolic secondary plant metabolites found abundantly in vegetables, fruits, and beverages. They play an active role in the treatment and management of various respiratory disorders such as asthma due to their antioxidant, antiallergic, and immunomodulatory activities. These activities inhibit the synthesis of Th2 cytokines such as IL-4 and IL-13, differentiation of naïve T-cells into effector T-cells, and the release of various chemical mediators (Bhatt et al., 2013).

Quercetin is an important dietary flavonoid found in various plants like onions, apples, tea, berries, and broccoli, which have antiasthmatic potential. Quercetin has been reported to have antihistaminic, antiinflammatory, and MC-stabilizing activity (Kimata et al., 2000; Kempuraj et al., 2005). Quercetin also suppressed lipopolysaccharide-induced lung inflammation in experimental studies (Takashima et al., 2014). Quercetin attenuated mitogen-activated protein kinases and NF-κB (Nanua et al., 2006). With respect to allergy, quercetin has a specific effect on MCs involved in the allergic response (Lee et al., 2010). Quercetin decreases eosinophil recruitment, reduces IL-5 and IL-4 levels, and inhibits NF-κB activation in BALF in the OVA-induced mouse model (Rogerio et al., 2010). It also regulates Th1/Th2 balance by enhancing IFN-γ and decreasing IL-4 levels in mouse asthma model (Park et al., 2009).

Red wine, another source of flavonoids, possesses antioxidant, antiinflammatory, anticancer, and immunomodulatory activities. Some flavonoids such as herbacetin, isobavachalcone, quercetin 3-β-d-glucoside, and helichrysetin were found to inhibit Middle East respiratory syndrome-coronavirus (MERS-CoV) by blocking the enzymatic activity of MERS-CoV 3C-like protease (Jo et al., 2019). Another study reported that herbacetin, rhoifolin, and pectolinarin efficiently blocked the enzymatic activity of SARS-CoV 3CLpro (Jo et al., 2019).

Alkaloids

Colchicine is an alkaloid obtained from the plant *Colchicum autumnale*, which has been used in the treatment of pulmonary fibrosis (Peters et al., 1993) and bronchial asthma (Kelly et al., 1995; Schwarz et al., 1990). The alkaloid fraction of the *Peganum harmala* plant is an effective traditional medicinal herb for the treatment of cough and asthma with potent antitussive, expectorant, and bronchodilating effects (Liu et al., 2015). Warifteine is an alkaloid obtained from the plant *Cissampelos sympodialis* that has been reported to inhibit airway hyperresponsiveness, lung remodeling, and eosinophil infiltration in a mouse model of asthma (Bezerra-Santos et al., 2012). Imperialine, an alkaloid derived from the bulbs of *Fritillaria cirrhosa*, improved pulmonary functions and structural impairment, and suppressed the inflammatory response in a COPD-like rat model (Wang et al., 2016). Synephrine and stachydrine are the alkaloids derived from *Pericarpium citri reticulata*. They have been reported to possess beneficial immunomodulatory and antiinflammatory properties, and there is evidence of their antiasthmatic potential.

Polyphenols

Resveratrol is a dietary polyphenol found in the skin and seeds of grapes. Resveratrol has been shown to have antiasthmatic potential as it modulates the Th1/Th2 balance, the expression of Th2 regulatory transcription factor, GATA-3, and decreases the levels of Th2 cytokines, IL-4, and IL-5 in bronchoalveolar lavage of sensitized animals (Lee et al., 2009). It also acts as a mucolytic agent by inhibiting mucus overproduction when administered to mice suffering from allergic airway disease. Resveratrol has been shown to inhibit airway inflammation and hyperresponsiveness in OVA-induced asthma in mice by reducing eosinophil/neutrophil infiltration, levels of IL-4 and IL-5 in plasma, and BALF. It also inhibits cytokine release in vitro by alveolar macrophages from patients with COPD, including IL-8 and granulocyte-macrophage colony-stimulating factor (GM-CSF) (Culpitt et al., 2003). Resveratrol suppressed lipopolysaccharide-induced neutrophils and inflammatory mediators in airways (Birrell et al., 2005). Resveratrol also restored glutathione synthesis by activation of Nrf2 and provide protection against cigarette smoke-induced release of reactive oxygen species in human lung epithelial cells (Kode et al., 2008). These findings suggest that resveratrol can be a very attractive compound for preventing/treating asthma.

Naringenin chalcone obtained from the tomato is a polyphenol effective in inhibiting eosinophilic recruitment around the airways in a murine model of allergic asthma. Naringenin alleviates airway inflammation and reactivity by decreasing serum total Immunoglobulin E (IgE) level and IL-4, IL-13 level in BALF, and by inhibiting NF-κB activity (Shi et al., 2009). IL-4 and IL-5 production from in vitro stimulated splenocytes in the group receiving the polyphenol was also suppressed (Iwamura et al., 2010). The efficacy of a polyphenol-rich tomato extract containing naringenin chalcone was also evaluated in subjects with perennial allergic rhinitis. Oral administration of the extract for 8 weeks to the subjects relieved sneezing and improved QOL of the subjects. Nasal discharge and obstruction also showed a decreased trend (Yoshimura et al., 2007). Shi et al. (2009) suggested the therapeutic value of naringenin as an antiasthmatic polyphenol on the basis of attenuation of airway inflammation, bronchial hyperreactivity, and airway remodeling in response to its administration in an experimental model of mice.

Concluding remarks and future directions

Respiratory disorders are by far one of the leading causes of morbidity and mortality, affecting people of all ages, and are a major global health problem. Nutraceuticals can play an important role in the prevention and treatment of various respiratory and allied disorders. Foods and nutrients play an important role in healthcare systems all over the world. They are helpful in maintaining well-being, enhancing health, modulating immunity, and in reducing the risk of various human disorders, including respiratory diseases. Medicinal plants and plant-derived products are major resources for alternative medicines and indeed many modern drugs for several human disorders, including respiratory diseases, are also derived from botanical sources. The traditional systems of medicine are found to be effective in treating respiratory diseases and their scientific validation has proven the efficacy of plants in the management of several such conditions. Medicinal herbs are potential sources of therapeutic agents and have attained a significant position in healthcare systems across the world for both animals and humans. Phytochemicals obtained from botanicals also contribute significantly as nutraceuticals for the treatment and prevention of respiratory diseases. Food-derived products/nutrients and herbal agents will continue to serve as reliable sources of alternative strategies in rapidly expanding healthcare systems due to their relative safety, efficacy, ease of availability, and low cost. Scientific validation of the safety and efficacy of these potentially useful nutraceutical substances will help in expanding their horizons as adjuncts in respiratory and allied diseases and will promote healthcare options.

References

Akoachere, J.F., Ndip, R.N., Chenwi, E.B., et al., 2002. Antibacterial effect of *Zingiber officinale* and *Garcinia kola* on respiratory tract pathogens. East Afr. Med. J. 79, 588–592.

Allam, M.F., Lucane, R.A., 2004. Selenium supplementation for asthma. Cochrane Database Syst. Rev. CD003538.

Allen, S., Britton, J.R., Leonardi-Bee, J.A., 2009. Association between antioxidant vitamins and asthma outcome measures: systematic review and meta-analysis. Thorax 64, 610–619.

Ammon, H.P., Mack, T., Singh, G.B., et al., 1991. Inhibition of leukotriene B_4 formation in rat peritoneal neutrophils by an ethanolic extract of the gum resin exudate of *Boswellia serrata*. Planta Med. 57, 203–207.

Amutha, A., Selvakumari, U., Ramachandran, M., 2015. Investigation into the studies on antihistaminic and bronchodilator activity of polyherbal compound (Siringiyathi Chooranam) in albino rats. J. Chem. Pharm. Res. 7, 777–781.

Ariaee, N., Farid, R., Shabestari, F., et al., 2016. Trace elements status in sera of patients with allergic asthma. Rep. Biochem. Mol. Biol. 5, 20–25.

Arm, J.P., Horton, C.E., Spur, B.W., et al., 1989. The effects of dietary supplementation with fish oil lipids on the airways response to inhaled allergen in bronchial asthma. Am. Rev. Respir. Dis. 139, 1395–1400.

Atlantis, E., Cochrane, B., 2016. The association of dietary intake and supplementation of specific polyunsaturated fatty acid with inflammation and functional capacity in chronic obstructive pulmonary disease: a systematic review. Int. J. Evid Based Healthc. 14, 54–63.

Aukrust, P., Muller, F., Ueland, T., et al., 2000. Decreased vitamin A levels in common variable immunodeficiency: vitamin A supplementation in vivo enhances immunoglobulin production and downregulates inflammatory responses. Eur. J. Clin. Invest. 30, 252–259.

Barnes, P.J., 2008. Immunology of asthma and chronic obstructive pulmonary disease. Nat. Rev. Immunol. 8, 183–192.

Baruach, C.C., Gupta, P.P., Patnaik, G.K., et al., 1997. Anti-allergic and mast cell stabilizing activity of *Albizzia lebbeck*. Ind. Veterinary Med. J. 21, 127–132.

Bede, O., Surányi, A., Pintér, K., et al., 2003. Urinary magnesium excretion in asthmatic children receiving magnesium supplementation: a randomized, placebo controlled, double-blind study. Magnes. Res. 16 (4), 262–270.

Best, K.P., Gold, M., Kennedy, D., et al., 2016. Omega-3 long-chain PUFA intake during pregnancy and allergic disease outcomes in the offspring: a systematic review and meta-analysis of observational studies and randomized controlled trials. Am. J. Clin. Nutr. 103, 128–143.

Bezerra-Santos, C.R., Vieira-de-Abreu, A., Vieira, G.C., et al., 2012. Effectiveness of *Cissampelos sympodialis* and its isolated alkaloid warifteine in airway hyperreactivity and lung remodeling in a mouse model of asthma. Int. Immunopharmacol. 13, 148–155.

Bhaskar, N., Mahajan, S., Bhandari, J., et al., 2019. Assessment of trace elements status in bronchial asthma. Int. J. Res. Rev. 6, 109–114.

Bhatt, S., Upadhyay, U., Upadhyay, S., et al., 2013. Evaluation of acute toxicity study and antiasthmatic activity of Zeal Herbal granules. Int. Res. J. Pharm. 4, 213–215.

Birrell, M.A., McCluskie, K., Wong, S., et al., 2005. Resveratrol, an extract of red wine, inhibits lipopolysaccharide induced airway neutrophilia and inflammatory mediators through an NF-kappa B independent mechanism. FASEB J. 19, 840–841.

Biswas, S., Rahman, I., 2008. Modulation of steroid activity in chronic inflammation: a novel anti-inflammatory role for curcumin. Mol. Nutr. Food Res. 52, 987–994.

Biswas, S.K., Chowdhury, A., Das, J., et al., 2011. Literature review on pharmacological potentials of *Kalanchoe pinnata*(Crassulaceae). Afr. J. Pharm. Pharmacol. 5, 1258–1262.

Boeing, H., Bechthold, A., Bub, A., et al., 2012. Critical review: vegetables and fruit in the prevention of chronic diseases. Eur. J. Nutr. 51, 637–663.

Boyle, R.J., Bath-Hextall, F.J., Leonardi-Bee, J., et al., 2009. Probiotics for treatment of eczema: a systematic review. Clin. Exp. Allergy 39, 1117–1127.

Brigham, E.P., McCormack, M.C., Takemoto, C.M., et al., 2015. Iron status is associated with asthma and lung function in US women. PLoS One 10, 1–12.

Brigham, E.P., Woo, H., McCormak, M.E., et al., 2019. Omega-3 and omega-6 intake modifies asthma severity and response to indoor air pollution in children. Am. J. Resp. Crit. Care Med. 199, 1478–1486.

Broughton, K.S., Johnson, C.S., Pace, B.K., et al., 1997. Reduced asthma symptoms with n-3 fatty acid ingestion are related to 5-series leukotriene production. Am. J. Clin. Nutr. 65, 1011–1017.

Chang, H.H., Chen, C.S., Lin, J.Y., 2009. High dose vitamin C supplementation increases the Th1/Th2 cytokine secretion ratio, but decreases eosinophilic infiltration in bronchoalveolar lavage fluid of ovalbumin-sensitized and challenged mice. J. Agric. Food Chem. 57, 10471–10476.

Chang, J.S., Wang, K.C., Yeh, C.F., et al., 2013. Fresh ginger (*Zingiber officinale*) has anti-viral activity against human respiratory syncytial virus in human respiratory tract cell lines. J. Ethnopharmacol. 145, 146–151.

Chaturvedi, G.N., Sharma, B.D., 1975. Clinical studies on *Hedychium spicatum* (Shati): an antiasthmatic drug. J. Res. Indian Med. 10, 6.

Chaudhary, S., Gulati, K., Rai, N., et al., 2016. Evaluation of anti-inflammatory and immunomodulatory effects of aqueous extract of *Solanum xanthocarpum* in experimental models of bronchial asthma. EC Pharmacol. Toxicol. 2, 241–250.

Chhabra, S.K., Yasir, A., Chaudhry, K., et al., 2010. Effect of ozone on response to ovalbumin & its modulation by vitamins C & E in sensitized guinea pigs. Indian J. Med. Res. 132, 87–93.

Chopra, R.N., Chopra, I.C., Handa, K.L., et al., 1958. Tylophora Asthmatica. Indigenous drugs of India, U.N. Dhar & Sons, Calcutta, p. 431.

Chiu, C.Y., Huang, S.Y., Peng, Y.C., et al., 2015. Maternal vitamin D levels are inversely related to allergic sensitization and atopic diseases in early childhood. Pediatr. Allergy Immunol. 26, 337–343.

Claeson, U.P., Malmfors, T., Wikman, G., et al., 2000. *Adhatoda vasica*: a critical review of ethnopharmacological and toxicological data. J. Ethnopharmacol. 72, 1–20.

Cruz, E., Reuter, S., Martin, H., et al., 2012. *Kalanchoe pinnata* inhibits mast cell activation and prevents allergic airway disease. Phytomedicine 19, 115–121.

Corinna, D., Andrea, H., Oliver, F., et al., 2012. Structural and sensory characterization of key pungent and tingling compounds from black pepper. J. Agric. Food Chem. 60, 2884–2895.

Culpitt, S.V., Rogers, D.F., Fenwick, P.S., et al., 2003. Inhibition by red wine extract, resveratrol, of cytokine release by alveolar macrophages in COPD. Thorax 58, 942–946.

Dahanukar, S.A., Karandikar, S.M., Desai, M., 1984. Efficacy of *Piper longum* in childhood asthma. Indian Drugs 21, 384–388.

Das, L., Bhaumik, E., Raychaudhuri, U., et al., 2012. Role of nutraceuticals in human health. J. Food Sci. Technol. 49, 173–183.

D'Cruz, J.L., Nimbkar, A.Y., Kokate, C.K., 1979. Evaluation of essential oil from leaves of *Adhatoda vasicaas* an airway smooth muscle relaxant. Indian J. Pharm. Sci. 41, 247.

Deger, Y., Yur, F., Ertekin, A., et al., 2007. Protective effect of alpha-tocopherol on oxidative stress in experimental pulmonary fibrosis in rats. Cell Biochem. Funct. 25, 633–637.

Demling, R., LaLonde, C., Ikegami, K., et al., 1995. Alpha-tocopherol attenuates lung edema and lipid peroxidation caused by acute zymosan-induced peritonitis. Surgery 117, 226–231.

Di Toro, R., Galdo Capotorti, G., Gialanella, G., et al., 1987. Zinc and copper status of allergic children. Acta Paediatr. Scand. 76, 612–617.

Dillard, C.J., German, J.B., 2000. Phytochemicals: nutraceuticals and human health. J. Sci. Food Agric. 80, 1744–1756.

Eftekhar, N., Moghimi, A., Roshan, N.M., et al., 2019. Immunomodulatory and anti-inflammatory effects of hydro-ethanolic extract of *Ocimum basilicum* leaves and its effect on lung pathological changes in an ovalbumin-induced rat model of asthma. BMC Complement. Altern. Med. 19, 349–359.

el-Kholy, M.S., Gas Allah, M.A., el-Shimi, S., et al., 1990. Zinc and copper status in children with bronchial asthma and atopic dermatitis. J. Egypt Public. Health Assoc. 65, 657–668.

Endres, S., Ghorbani, R., Kelley, V.E., et al., 1989. The effect of dietary supplementation with n-3 polyunsaturated fatty acids on the synthesis of interleukin-1 and tumor necrosis factor by mononuclear cells. N. Engl. J. Med. 320, 265–271.

Erkkola, M., Kaila, M., Nwaru, B.I., et al., 2009. Maternal vitamin D intake during pregnancy is inversely associated with asthma and allergic rhinitis in 5-year-old children. Clin. Exp. Allergy 39, 875–882.

Faisal, M., Singh, P.P., Irchhaiya, R., 2012. Review on *Albizzia lebbeck* a potent herbal drug. Int. Res. J. Pharm. 3, 63–68.

Fan, J., Shek, P.N., Suntres, Z.E., et al., 2000. Liposomal antioxidants provide prolonged protection against acute respiratory distress syndrome. Surgery 128, 332–338.

Fernandez, A., Tavares, F., Athavale, V.B., 1980. Asthma in children: a clinical controlled study of *Piper longum* in asthma. Paediatr. Clin. India 15, 45–52.

Flatt, A., Pearce, N., Thomson, C.D., et al., 1990. Reduced selenium in asthmatic subjects in New Zealand. Thorax 45, 95–99.

Fulton, A.S., Baldock, K.L., Coates, A.M., et al., 2019. Polyunsaturated fatty acid intake and lung function in a regional Australian population: a cross-sectional study with a nested case-control analysis. J. Nutr. Intermed. Metab. 18, 1–8.

Ganguly, T., Sainis, K.B., 2001. Inhibition of cellular immune responses by *Tylophora indica* in experimental models. Phytomedicine 8, 348–355.

Gayathri, B., Manjula, N., Kumar, V.K.S., et al., 2007. Pure compound from *Boswellia serrata* extract exhibits anti-inflammatory property in human PBMCs and mouse macrophages through inhibition of TNF-alpha, IL-1beta, NO and MAP kinases. Int. Immunopharmacol. 7, 473–482.

Ghildiyal, S., Gautam, M.K., Joshi, V.K., et al., 2012. Pharmacological evaluation of extracts of *Hedychium spicatum* (Ham-ex-Smith) rhizome. Anc. Sci. Life 31, 117–122.

Ghosh, G.R., 1995. Tulsi (N.O. Labiatae, Genus-*Ocimum*). N. Approaches Med. Health 3, 23–29.

Godhwani, S., Godhwani, J.L., Vyas, D.S., 1988. *Ocimum sanctum*: a preliminary study evaluating its immunoregulatory profile in albino rats. J. Ethnopharmacol. 24, 193–198.

Gontijo-Amaral, C., Ribeiro, M., Gontijo, L.S., et al., 2007. Oral magnesium supplementation in asthmatic children: a double-blind randomized placebo-controlled trial. Eur. J. Clin. Nutr. 61 (1), 54–60.

Gopalakrishnan, C., Shankaranarayanan, D., Nazimudeen, S.K., et al., 1980. Effect of tylophorine, a major alkaloid of *Tylophora indica*, on immunopathological and inflammatory reactions. Indian J. Med. Res. 71, 940–948.

Gourbeyre, P., Denery, S., Bodinier, M., 2011. Probiotics, prebiotics, and synbiotics: impact on the gut immune system and allergic reactions. J. Leukoc. Biol. 89, 685–695.

Gourgoulianis, K.I., Chatziparasidis, G., Chatziefthimiou, A., et al., 2001. Magnesium as a relaxing factor of airway smooth muscles. J. Aerosol Med. 14, 301–307.

Govindan, S., Viswanathan, S., Vijayasekaran, V., et al., 1999. A pilot study on the clinical efficacy of *Solanum xanthocarpum* and *Solanum trilobatum* in bronchial asthma. J. Ethnopharmacol. 66, 205–210.

Govindan, S., Viswanathan, S., Vijayasekaran, V., et al., 2004. Further studies on the clinical efficacy of *Solanum xanthocarpum* and *Solanum trilobatum* in bronchial asthma. Phytother. Res. 18, 805–809.

Gupta, I., Gupta, V., Parihar, A., et al., 1998. Effects of *Boswellia serrata* gum resin in patients with bronchial asthma: results of a double-blind, placebo-controlled, 6-week clinical study. Eur. J. Med. Res. 3, 511–514.

Gupta, S.K., Prakash, J., Srivastava, S., 2002. Validation of traditional claim of tulsi, *Ocimum sanctum* Linn. as a medicinal plant. Indian J. Exp. Biol. 40, 765–773.

Hamahata, A., Enkhbaatar, P., Kraft, E.R., et al., 2008. γ-Tocopherol nebulization by a lipid aerosolization device improves pulmonary function in sheep with burn and smoke inhalation injury. Free Radic. Biol. Med. 45, 425–433.

Haranath, P.S., Shyamalakumari, S., 1975. Experimental study on mode of action of *Tylophora asthmatica* in bronchial asthma. Indian J. Med. Res. 63, 661–670.

Hazlewood, L.C., Wood, L.G., Hansbro, P.M., et al., 2011. Dietary lycopene supplementation suppresses Th2 responses and lung eosinophilia in a mouse model of allergic asthma. J. Nutr. Biochem. 22, 95–100.

Hejazi, M.E., Modaressi-Ghazani, F., Entezari-Maleki, T., 2016. A review of vitamin D effects on common respiratory disease: asthma, chronic obstructive pulmonary disease and tuberculosis. J. Res. Pharm. Pract. 5, 7–15.

Hodge, L., Salome, C.M., Peak, J.K., et al., 1996. Consumption of oily fish and childhood asthma risk. Med. J. Aust. 164, 137–140.

Holgate, S.T., Davies, D.E., Puddicombe, S., et al., 2008. Epithelial mesenchymal interactions in the pathogenesis of asthma. J. Allergy Clin. Immunol. 52 (2), 45–52.

Hong, J., Bose, M., Ju, J., et al., 2004. Modulation of arachidonic acid metabolism by curcumin and related beta-diketone derivatives: effects on cytosolic phospholipase-A2, cyclooxygenases and 5-lipoxygenase. Carcinogenesis 25, 1671–1679.

Hord, N.G., 2008. Eukaryotic microbiotic crosstalk: potential mechanisms for health benefits of prebiotics and probiotics. Annu. Rev. Nutr. 28, 215–231.

Iwamura, C., Shinoda, K., Yoshimura, M., et al., 2010. Naringenin chalcone suppresses allergic asthma by inhibiting the type-2 function of CD_4 T cells. Allergol. Int. 59, 67–73.

Jaudszus, A., Krokowski, M., Mockel, P., et al., 2008. Cis-9, *trans*-11-conjugated linoleic acid inhibits allergic sensitization and airway inflammation via a PPAR-gamma related mechanism in mice. J. Nutr. 138, 1336–1342.

Jeong, Y.J., Kim, J.H., Kang, J.S., et al., 2010. Mega-dose vitamin C attenuated lung inflammation in mouse asthma model. Anat. Cell Biol. 43, 294–302.

Jiang, Z., Yin, X., Jiang, Q., 2011. Natural forms of vitamin E and 13'-carboxychromanol, a long-chain vitamin E metabolite, inhibit leukotriene generation from stimulated neutrophils by blocking calcium influx and suppressing 5-lipoxygenase activity, respectively. J. Immunol. 186, 1173–1179.

Jo, S., Kim, H., Kim, S., et al., 2019. Characteristics of flavaoids as potent MERSCoV 3C-like protease inhibitors. Chem. Biol. Drug Des. 94, 2023–2030

Julius, A.G., Charles, A., Samuel, N.E., et al., 2014. Nutraceutical with anti-inflammatory activity for the management of airway remodeling in bronchial asthma: *Kalanchoe integra* var. crenata (Andr.) cuf leaf extract. Pharmacol. Pharm. 5, 250–261.

Kadrabova, J., Mad'aric, A., Kovacikova, Z., et al., 1996. Selenium status is decreased in patients with intrinsic asthma. Biol. Trace Elem. Res. 52, 241–248.

Kaluza, J., Harris, H., Wallin, A., et al., 2018. Dietary fiber intake and risk of chronic obstructive pulmonary disease: a prospective cohort study of men. Epidemiology 29, 254–260.

Karadag, F., Cildag, O., Altinisik, M., et al., 2004. Trace elements as a component of oxidative stress in COPD. Respirology 9, 33–37.

Kelly, S.J., Uri, A.J., Freeland, H.S., et al., 1995. Effects of colchicine on IgE-mediated early and late airway reactions. Chest 107, 985–991.

Kempuraj, D., Madhappan, B., Christodoulou, S., et al., 2005. Flavonols inhibit proinflammatory mediator release, intracellular calcium ion levels and protein kinase C theta phosphorylation in human mast cells. Br. J. Pharmacol. 145, 934–944.

Khajuria, A., Gupta, A., Suden, P., et al., 2008. Immunomodulatory activity of biopolymeric fraction BOS 2000 from *Boswellia serrata*. Phytother. Res. 22, 340–348.

Khan, M.H., Yadava, P.S., 2010. Herbal remedies of asthma in Thoubal district of Manipur in North East India. Indian J. Nat. Prod. Res. 1, 80–84.

Kilinc, C., Ozcan, O., Karaoz, E., et al., 1993. Vitamin E reduces bleomycin-induced lung fibrosis in mice: biochemical and morphological studies. J. Basic Clin. Physiol. Pharmacol. 4, 249–269.

Kim, S.H., Lee, Y.C., 2009. Piperine inhibits eosinophil infiltration and airway hyperresponsiveness by suppressing T cell activity and Th2 cytokine production in the ovalbumin-induced asthma model. J. Pharm. Pharmacol. 61, 353–359.

Kimata, M., Shichijo, M., Miura, T., et al., 2000. Effects of luteolin, quercetin and baicalein on immunoglobulin E-mediated mediator release from human cultured mast cells. Clin. Exp. Allergy 30, 501–508.

Kirkil, G., Hamdi Muz, M., Seckin, D., et al., 2008. Antioxidant effect of zinc picolinate in patients with chronic obstructive pulmonary disease. Respir. Med. 102, 840–844.

Kode, A., Rajendrasozhan, S., Caito, S., et al., 2008. Resveratrol induces glutathione synthesis by activation of Nrf2 and protects against cigarette smoke-mediated oxidative stress in human lung epithelial cells. Am. J. Physiol. Lung Cell Mol. Physiol. 294, L478–L488.

Kruzel, M.L., Bacsi, A., Choudhury, B., et al., 2006. Lactoferrin decreases pollen antigen-induced allergic airway inflammation in a murine model of asthma. Immunology 119, 159–166.

Kumar, S., Bansal, P., Gupta, V., et al., 2010. The clinical effect of *Albizia lebbeck* stem bark decoction on bronchial asthma. Int. J. Pharm. Sci. Drug Res. 2, 48–50.

Kurup, V.P., Barrios, C.S., Raju, R., et al., 2007. Immune response modulation by curcumin in a latex allergy model. Clin. Mol. Allergy 5, 1.

Lebedeva, E.S., Tadzhiev, F.S., Danilov, L.N., et al., 1992. Content of trace elements and steroid hormones in blood and tissues of internal organs of rats during an inflammatory process in the bronchi. Patol. Fiziol. Eksp. Ter. 3, 3–5.

Lee, T.H., Hoover, R.L., Williams, J.D., et al., 1985. Effect of dietary enrichment with eicosapentaenoic and docosahexaenoic acids on in vitro neutrophil and monocyte leukotriene generation and neutrophil function. N. Engl. J. Med. 312, 1217–1224.

Lee, J., Koo, N., Min, D.B., 2004. Reactive oxygen species, aging and antioxidative nutraceuticals. Compr. Rev. Food Sci. Food Saf. 3, 21–33.

Lee, J.H., Kim, J.W., Ko, N.Y., et al., 2008. Curcumin, a constituent of curry, suppresses IgE-mediated allergic response and mast cell activation at the level of Syk. J. Allergy Clin. Immunol. 121, 1225–1231.

Lee, M., Kim, S., Kwon, O.K., et al., 2009. Anti-inflammatory and antiasthmatic effects of resveratrol, a polyphenolic stilbene, in a mouse model of allergic asthma. Int. Immunopharmacol. 9, 418–424.

Lee, E.J., Ji, G.E., Sung, M.K., 2010. Quercetin and kaempferol suppress immunoglobulin E-mediated allergic inflammation in RBL-2H3 and Caco-2 cells. Inflamm. Res. 59, 847–854.

Lenoir-Wijnkoop, I., Sanders, M.E., Cabana, M.D., et al., 2007. Probiotic and prebiotic influence beyond the intestinal tract. Nutr. Rev. 65, 469–489.

Lim, Y., Vasu, V.T., Valacchi, G., et al., 2008. Severe vitamin E deficiency modulates airway allergic inflammatory responses in the murine asthma model. Free Radic. Res. 42, 387–396.

Liu, B., Yang, J., Wen, Q., et al., 2008. Isoliquiritigenin, a flavonoid from licorice, relaxes guinea-pig tracheal smooth muscle in vitro and in vivo: role of cGMP/PKG pathway. Eur. J. Pharmacol. 587, 257–266.

Liu, W., Cheng, X., Wang, Y., et al., 2015. In vivo evaluation of the antitussive, expectorant and bronchodilating effects of extract and fractions from aerial parts of *Peganum harmala* Linn. J. Ethnopharmacol. 162, 79–86.

Liu, Y., Tran, D.Q., Rhoades, J.M., 2018. Probiotics in disease prevention and treatment. J. Clin. Pharmacol. 58, S164–S179.

Lo, C.J., Chiu, K.C., Fu, M., et al., 1999. Fish oil decreases macrophage tumor necrosis factor gene transcription by altering the NF-kappa B activity. J. Surg. Res. 82, 216–221.

Maazi, H., Shirinbak, S., Bloksma, N., et al., 2011. Iron administration reduces airway hyperreactivity and eosinophilia in a mouse model of allergic asthma. Clin. Exp. Immunol. 166, 80–86.

Mabalirajan, U., Aich, J., Leishangthem, G.D., et al., 2009. Effects of vitamin E on mitochondrial dysfunction and asthma features in an experimental allergic murine model. J. Appl. Physiol. 107, 1285–1292.

Manarin, G., Anderson, D., Silva, J.M.E., et al., 2019. *Curcuma longa* L. ameliorates asthma control in children and adolescents: a randomized, double-blind, controlled trial. J. Ethnopharmacol. 238, 111882.

Manjunath, B.L., 1948. The Wealth of India: A Dictionary of Indian Raw Materials and Industrial Products. CSIR, Delhi, p. 31.

Martineau, A.R., Jolliffe, D.A., Hooper, R.L., et al., 2017. Vitamin D supplement to prevent acute respiratory tract infections: systematic review and meta-analysis of individual participant data. Br. Med. J. 356, 1–14.

Mediratta, P.K., Dewan, V., Bhattacharya, S.K., et al., 1988. Effect of *Ocimum sanctum* Linn. on humoral immune responses. Indian J. Med. Res. 87, 384–386.

Mediratta, P.K., Sharma, K.K., Singh, S., 2002. Evaluation of immunomodulatory potential of *Ocimum sanctum* seed oil and its possible mechanism of action. J. Ethnopharmacol. 80, 15–20.

Meja, K.K., Rajendrasozhan, S., Adenuga, D., et al., 2008. Curcumin restores corticosteroid function in monocytes exposed to oxidants by maintaining HDAC2. Am. J. Respir. Cell Mol. Biol. 39, 312–323.

Menegazzi, M., Di Paola, R., Mazzon, E., et al., 2008. Glycyrrhizin attenuates the development of carrageenan-induced lung injury in mice. Pharmacol. Res. 58, 22–31.

Mickleborough, T.D., Lindley, M.R., Ionescu, A.A., et al., 2006. Protective effect of fish oil supplementation on exercise-induced bronchoconstriction in asthma. Chest 129, 39–49.

Moratz, E., Adcock, I.M., Folkerts, G., et al., 2013. Probiotics in the management of lung diseases. Mediators Inflam. 2013, 1–10.

Morgan, C.I., Ledford, J.R., Zhou, P., et al., 2011. Zinc supplementation alters airway inflammation and airway hyperresponsiveness to a common allergen. J. Inflamm. 8, 36.

Muller, A., Antus, S., Bittinger, M., et al., 1993. Chemistry and pharmacology of antiasthmatic *Galphimia glauca*, *Adhatoda vasica*, and *Picrorhiza kurrooa*. Planta Med. 59, A586–A587.

Nagakura, T., Matsuda, S., Shichijyo, K., et al., 2000. Dietary supplementation with fish oil rich in omega-3 polyunsaturated fatty acids in children with bronchial asthma. Eur. Respir. J. 16, 861–865.

Nanua, S., Zick, S.M., Andrade, J.E., et al., 2006. Quercetin blocks airway epithelial cell chemokine expression. Am. J. Respir. Cell Mol. Biol. 35, 602–610.

Nasri, H., Baradaran, A., Shirzad, H., et al., 2014. New concepts in nutraceuticals as alternative for pharmaceuticals. Int. J. Prev. Med. 5, 1487–1499.

Nishimoto, Y., Hisatsune, A., Katsuki, H., et al., 2010. Glycyrrhizin attenuates mucus production by inhibition of MUC_5AC mRNA expression in vivo and in vitro. J. Pharmacol. Sci. 113, 76–83.

Nurmatov, U., Devereux, G., Sheikh, A., 2011. Nutrients and foods for the primary prevention of asthma and allergy: systematic review and meta-analysis. J. Allergy Clin. Immunol. 127, 724–733.

Oddy, W.H., de Klerk, N.H., Kendall, G.E., et al., 2004. Ratio of omega-6 to omega-3 fatty acids and childhood asthma. J. Asthma 41, 319–326.

Oh, J., Shin, S.H., Choi, R., et al., 2019. Assessment of 7 trace elements in serum of patients with nontuberculous mycobacterial lung disease. J. Trace Elem. Med. Biol. 53, 84–90.

Osborn, D.A., Sinn, J.K.H., 2013. Prebiotics in infants for prevention of allergy. Cochrane Database Syst. Rev. (Issue 3), Art. No.: CD006474.

Parganiha, R., Verma, S., Wani, V., et al., 2012. In vitro anti-asthmatic activity of fruit extracts of *Sapindus mukorossi* and *Piper nigrum*. Int. J. Herbal Drug Res. 1, 12–16.

Park, H.Y., Park, S.H., Yoon, H.K., et al., 2004. Anti-allergic activity of 18 beta-glycyrrhetinic acid-3-O-beta-D-glucuronide. Arch. Pharm. Res. 27, 57–60.

Park, H.J., Lee, C.M., Jung, I.D., et al., 2009. Quercetin regulates Th1/Th2 balance in a murine model of asthma. Int. Immunopharmacol. 9, 261–267.

Peat, J.K., Salome, C.M., Woolcock, A.J., 1992. Factors associated with bronchial hyperresponsiveness in Australian adults and children. Eur. Respir. J. 5, 921–929.

Peters, S.G., McDougall, J.C., Douglas, W.W., et al., 1993. Colchicine in the treatment of pulmonary fibrosis. Chest 103, 101–104.

Pharmacopoeia of India (The Indian Pharmacopoeia). (1955) Government of India, Ministry of Health, Delhi, 741pp.

Pungle, P., Banavalikar, M., Suthar, A., et al., 2003. Immunomodulatory activity of boswellic acids of *Boswellia serrata* Roxb. Indian J. Exp. Biol. 41, 1460–1462.

Punithavathi, D., Venkatesan, N., Babu, M., 2000. Curcumin inhibition of bleomycin-induced pulmonary fibrosis in rats. Br. J. Pharmacol. 131, 169–172.

Rachana, R., Basu, S., Pant, M., et al., 2011. Review and future perspectives of using vasicine, and related compounds. Indo-Global J. Pharm. Sci. 1, 85–98.

Rahman, I., 2012. Pharmacological antioxidant strategies as therapeutic interventions for COPD. Biochim. Biophys. Acta 1822, 714–728.

Rai, N., Ray, A., Jamil, S.S., Gulati, K., 2015. Cellular and molecular mechanism of action of polyherbal preparation UNIM-352 in experimental models of bronchial asthma. Indian J. Exp. Biol. 53, 625–631.

Raina, V., Raina, S., 1980. The responsiveness of leukocyte adenyl cyclase to tylophorine in asthmatic subjects. Biochem. Biophys. Res. Commun. 94, 1074–1077.

Ram, A., Das, M., Ghosh, B., 2003. Curcumin attenuates allergen-induced airway hyperresponsiveness in sensitized guinea pigs. Biol. Pharm. Bull. 26, 1021–1024.

Ram, A., Mabalirajan, U., Das, M., et al., 2006. Glycyrrhizin alleviates experimental allergic asthma in mice. Int. Immunopharmacol. 6, 1468–1477.

Ram, A., Balachandar, S., Vijayananth, P., et al., 2011. Medicinal plants useful for treating chronic obstructive pulmonary disease (COPD): current status and future perspectives. Fitoterapia 82, 141–151.

Richter, M., Bonneau, R., Girard, M.A., et al., 2003. Zinc status modulates bronchopulmonary eosinophil infiltration in a murine model of allergic inflammation. Chest 123, 446S.

Rocksen, D., Ekstrand-Hammarstrom, B., Johansson, L., et al., 2003. Vitamin E reduces transendothelial migration of neutrophils and prevents lung injury in endotoxin-induced airway inflammation. Am. J. Respir. Cell Mol. Biol. 28, 199–207.

Rogerio, A.P., Dora, C.L., Andrade, E.L., et al., 2010. Anti-inflammatory effect of quercetin-loaded microemulsion in the airways allergic inflammatory model in mice. Pharmacol. Res. 61, 288–297.

Rubin, R.N., Navon, L., Cassano, P.A., 2004. Relationship of serum antioxidants to asthma prevalence in youth. Am. J. Respir. Crit. Care Med. 169, 393–398.

Sahito, S.B., Kazi, T.G., Jatoi, W.B., et al., 2013. Determination and evaluation of mineral constituents of medicinal plants used for the treatment of asthma and other ailments by atomic absorption spectrophotometry. Pak. J. Anal. Environ. Chem. 14, 61–67.

Sakpa, C.L., Idodia, D., 2019. Histopathological changes in lungs of Wistar rats treated with aqueous leaf extract of *Bryophyllum pinnatum*. Afr. J. Online 18, 97–106.

Satomi, H., Minowa, M., Hatano, S., et al., 1994. An epidemiological study of the preventive effect of dietary fish on the bronchial asthma. Bull. Inst. Public Health 43, 305–314.

Schwartz, J., Weiss, S.T., 1994. The relationship of dietary fish intake to level of pulmonary function in the first National Health and Nutrition Survey (NHANES I). Eur. Respir. J. 7, 1821–1824.

Schwarz, Y.A., Kivity, S., Ilfeld, D.N., et al., 1990. A clinical and immunologic study of colchicine in asthma. J. Allergy Clin. Immunol. 85, 578–582.

Shakeri, F., Soukhtanloo, M., Boskabady, M.H., 2017. The effect of hydro-ethanolic extract of *Curcuma longa* rhizome and curcumin on total and differential WBC and serum oxidant, antioxidant biomarkers in rat model of asthma. Iran. J. Basic Med. Sci. 20, 155–165.

Shariatpanahi, V.Z., Mokhtari, M., Taleban, F.A., et al., 2013. Effect of enteral feeding with ginger extract in acute respiratory distress syndrome. J. Crit. Care 28, e1–217.

Shaw, R., Woodman, K., Crane, J., et al., 1994. Risk factors for asthma symptoms in Kawerau children. N. Z. Med. J. 107, 387–391.

Shi, Y., Dai, J., Liu, H., et al., 2009. Naringenin inhibits allergen-induced airway inflammation and airway responsiveness and inhibits NF-kappaB activity in a murine model of asthma. Can. J. Physiol. Pharmacol. 87, 729–735.

Shin, Y.W., Bae, E.A., Lee, B., et al., 2007. In vitro and in vivo antiallergic effects of *Glycyrrhiza glabra* and its components. Planta Med. 73, 257–261.

Shivpuri, D.N., Singhal, S.C., Parkash, D., 1972. Treatment of asthma with an alcoholic extract of tylophora indica: a cross-over, double-blind study. Ann. Allergy 30, 407–412.

Siemoneit, U., Hofmann, B., Kather, N., et al., 2008. Identification and functional analysis of cyclooxygenase-1 as a molecular target of boswellic acids. Biochem. Pharmacol. 75, 503–513.

Singh, Y.N., 1992. Kava: an overview. J. Ethnopharmacol. 37, 13–45.

Singh, O.M., Singh, T.P., 2010. Phytochemistry of *Solanum Xanthocarpum*: an amazing traditional healer. J. Sci. Ind. Res. 69, 732–740.

Sommer, A., Katz, J., Tarwotjo, I., 1984. Increased risk of respiratory disease and diarrhea in children with preexisting mild vitamin A deficiency. Am. J. Clin. Nutr. 40, 1090–1095.

Stephensen, C.B., 2001. Vitamin A, infection, and immune function. Annu. Rev. Nutr. 21, 167–192.

Sunila, E.S., Kuttan, G., 2004. Immunomodulatory and antitumor activity of *Piper longum* Linn. and piperine. J. Ethnopharmacol. 90, 339–346.

Suntres, Z.E., Shek, P.N., 1996. Treatment of LPS-induced tissue injury: role of liposomal antioxidants. Shock 6, S57–S64.

Suntres, Z.E., Shek, P.N., 1997. Protective effect of liposomal alpha-tocopherol against bleomycin-induced lung injury. Biomed. Environ. Sci. 10, 47–59.

Takashima, K., Matsushima, M., Hashimoto, K., et al., 2014. Protective effects of intratracheally administered quercetin on lipopolysaccharide-induced acute lung injury. Respir. Res. 15, 150.

Tecklenburg, S.L., Mickleborough, T.D., Fly, A.D., et al., 2007. Ascorbic acid supplementation attenuates exercise-induced bronchoconstriction in patients with asthma. Respir. Med. 101, 1770–1778.

Tenorio-Lopes, L., Baldy, C., Jochmans-Lemione, A., et al., 2017. Consequences of maternal omega-3-polyunsaturated fatty acid supplementation on respiratory function in rat pups. J. Physiol. 595, 1637–1655.

Tripathi, R.M., Das, P.K., 1977. Studies on anti-asthmatic and antianaphylactic activity of *Albizzia lebbeck*. Indian J. Pharmacol. 9, 189–194.

Tripathi, S., Maier, K.G., Bruch, D., et al., 2007. Effect of 6-gingerol on pro-inflammatory cytokine production and costimulatory molecule expression in murine peritoneal macrophages. J. Surg. Res. 138, 209–213.

Trompette, A., Gollwitzer, E.S., Yadava, K., et al., 2014. Gut microbiota metabolism of dietary fiber influences allergic airway disease and hematopoiesis. Nat. Med. 20, 159–166.

Vaughan, A., Frazer, Z.A., Hansbro, P.M., et al., 2019. COPD and the gut-lung axis: the therapeutic potential of fibre. J. Thorac. Dis. 11, S2173–S2180.

Venkatesan, N., Punithavathi, V., Chandrakasan, G., 1997. Curcumin protects bleomycininduced lung injury in rats. Life Sci. 61, PL51–PL58.

Venkatesan, N., Punithavathi, D., Babu, M., 2007. Protection from acute and chronic lung diseases by curcumin. Adv. Exp. Med. Biol. 595, 379–405.

Venkatesh, P., Mukherjee, P.K., Kumar, N.S., et al., 2010. Anti-allergic activity of standardized extract of *Albizia lebbeck* with reference to catechin as a phytomarker. Immunopharmacol. Immunotoxicol. 32, 272–276.

Verma, P., Sharma, S., Kumar, V., 2016. Anti-inflammatory effect of hydroalcoholic extract of *Hibiscus rosa-sinensis* on acute and chronic inflammation. Int. J. Pharmacol. Toxicol. 4, 123–126.

Villamor, E., Fawzi, W.W., 2005. Effects of vitamin a supplementation on immune responses and correlation with clinical outcomes. Clin. Microbiol. Rev. 18, 446–464.

Wadhwa, S., Singhal, S., Rawat, S., 2014. Bioavailability enhancement by piperine: a review. Asian J. Biomed. Pharm. Sci. 4, 1–8.

Wagner, J.G., Jiang, Q., Harkema, J.R., et al., 2008. γ-tocopherol prevents airway eosinophilia and mucous cell hyperplasia in experimentally induced allergic rhinitis and asthma. Clin. Exp. Allergy 38, 501–511.

Wang, Y., Li, X., Ge, T., et al., 2016. Probiotics for prevention and treatment of respiratory tract infections in children: a systematic review and meta-analysis of randomized controlled trials. Medicine 95, e4509.

Watzl, B., Girrbach, S., Roller, M., 2005. Inulin, oligofructose and immunomodulation. Br. J. Nutr. 93, S49–S55.

Wintergerst, E.S., Maggini, S., Hornig, D.H., 2006. Immune-enhancing role of vitamin C and zinc and effect on clinical conditions. Ann. Nutr. Metab. 50, 85–94.

World Health Organization.(2003) Shaping the Future. The World Health Report, WHO, Geneva.

Wu, Z., Holwill, S.D., Oliveira, D.B., 2004. Desferrioxamine modulates chemically induced T helper 2-mediated autoimmunity in the rat. Clin. Exp. Immunol. 135, 194–199.

Wu, S.J., Liu, P.L., Ng, L.T., 2008. Tocotrienol-rich fraction of palm oil exhibits anti-inflammatory property by suppressing the expression of inflammatory mediators in human monocytic cells. Mol. Nutr. Food Res. 52, 921–929.

Xie, Y.C., Dong, X.W., Wu, X.M., et al., 2009. Inhibitory effects of flavonoids extracted from licorice on lipopolysaccharide-induced acute pulmonary inflammation in mice. Int. Immunopharmacol. 9, 194–200.

Xu, J., Gaddis, N.C., Bartz, T.M., et al., 2018. Omega-3 fatty acids and genome-wide interaction analyses reveal DPP1—pulmonary function association. Am. J. Resp. Crit. Care Med. 199, 631–642.

Yim, M., Sarma, B.P., Sinha, S., et al., 2014. Exploring the possible mechanism of *Albezzia lebbeck* components binding with drug targets of bronchial asthma—an insilico and clinical analysis. Int. J. Pharm. Sci. Res. 5, 5042–5051.

Yoshimura, M., Enomoto, T., Dake, Y., et al., 2007. An evaluation of the clinical efficacy of tomato extract for perennial allergic rhinitis. Allergol. Int. 56, 225–230.

Yusuf, N., Irby, C., Katiyar, S., et al., 2007. Photoprotective effects of greentea polyphenols. Photodermatol. Photoimmunol. Photomed. 23, 48–56.

Zeyda, M., Szekeres, A.B., Saemann, M.D., et al., 2003. Suppression of T cell signaling by polyunsaturated fatty acids: selectivity in inhibition of mitogen-activated protein kinase and nuclear factor activation. J. Immunol. 170, 6033–6039.

Zhang, D., Huang, C., Yang, C., et al., 2011. Antifibrotic effects of curcumin are associated with overexpression of cathepsins K and L in bleomycin treated mice and human fibroblasts. Respir. Res. 12, 154.

Zia-Ul-Haq, M., Ahmad, S., Qayum, M., et al., 2013. Compositional studies and antioxidant potential of *Albizia lebbeck* (L.) Benth.pods and seeds. Turk. J. Biol. 37, 25–32.

Zuccotti, G.V., Trabattoni, D., Morelli, M., et al., 2009. Immune modulation by lactoferrin and curcumin in children with recurrent respiratory infections. J. Biol. Regul. Homeost. Agents 23, 119–123.

Chapter 8

Nutraceuticals in hepatic diseases

Sharon M. Gwaltney-Brant

Introduction

The liver is the largest internal organ in the body, and plays vital roles in nutrient homeostasis, filtration of particulates, protein synthesis, bioactivation and detoxication, formation of bile, and biliary secretion and excretion (Jaeschke, 2008). Its extensive portal blood supply results in the liver receiving high exposures to ingested compounds that are absorbed across the gastrointestinal tract (GIT). Because of its role in processing, biotransformation, and detoxification of xenobiotics, the liver is frequently exposed to potentially injurious compounds on a regular basis. Hepatic injury due to chemicals and drugs is the most common cause of acute liver failure leading to liver transplantation in the United States, and is also the leading cause of regulatory action against drugs (Watkins and Seef, 2006).

Hepatotoxic agents may be intrinsically or extrinsically toxic to the liver (Cattley and Cullen, 2013). Intrinsic toxicants cause a high incidence of dose-dependent liver injury in one or more species, and they are generally well known as hepatotoxicants. Intrinsic hepatotoxicants include a variety of commodity chemicals (e.g., carbon tetrachloride, formaldehyde), food/feed-related compounds (e.g., mycotoxins, plant toxins), metals and minerals (e.g., iron), and pharmaceutical compounds (e.g., acetaminophen). Extrinsic hepatotoxicants cause a low incidence of toxicosis in a small number of individuals, often in a dose-independent fashion. The relationship between exposure and hepatic injury may not be recognized until a very large number of individuals have been exposed, resulting in a small population of susceptible individuals being identified. The field of pharmacogenomics has identified variations in gene expression that can make individuals more or less susceptible to the adverse effects of xenobiotics, primarily due to genetic differences in xenobiotic absorption, distribution, biotransformation, and elimination (Stephens et al., 2014; Gurley, 2012). Nongenetic factors (e.g., diet, concurrent hepatic disease) also play a role in extrinsic hepatotoxicosis but are less well understood. Because prescription pharmaceutical agents are closely monitored for adverse effects, drugs are the most common chemical entities whose extrinsic toxicities have been identified, but extrinsically toxic compounds likely exist among all classes of chemicals, both natural and anthropogenic. Nontoxicant-induced liver injury may occur due to infectious disease, immune-mediated mechanisms, metabolic derangement, disturbances in biliary formation and flow, cardiovascular and hematologic disturbances, trauma, and neoplasia (Cullen, 2007).

Liver structure and function

Hepatic anatomy

Situated in the proximal-most region of the abdominal cavity immediately distal to the diaphragm, the human liver grossly appears as four anatomically distinct lobes. The gall bladder sits immediately under, and intimately attached to the liver. Dual blood supplies ensure that the liver is a highly perfused organ. Venous blood from the GIT and spleen provides 60%–70% of the liver's blood supply via the portal vein (Crawford, 2005). Highly oxygenated blood enters through the hepatic artery, which branches from the celiac artery and supplies the remainder of the hepatic blood supply. The major blood vessels enter, and the main bile duct exits, the liver through the area known as the hilum. The bile duct, hepatic artery, and portal vein progressively branch out along portal tracts which provide the hepatic parenchyma with an extensive portal system consisting of approximately 450,000 terminal branches. Blood from the portal system filters through the parenchyma and enters terminal hepatic veins (also called central veins due to their location at the center of hepatic lobules) that feed into the hepatic vein which ultimately empties into the inferior vena cava.

The hepatic parenchyma is described by two different models: the *lobule* and the *acinus* (Fig. 8.1). The lobular model is based on the microscopic appearance of the liver

Nutraceuticals. DOI: https://doi.org/10.1016/B978-0-12-821038-3.00008-2

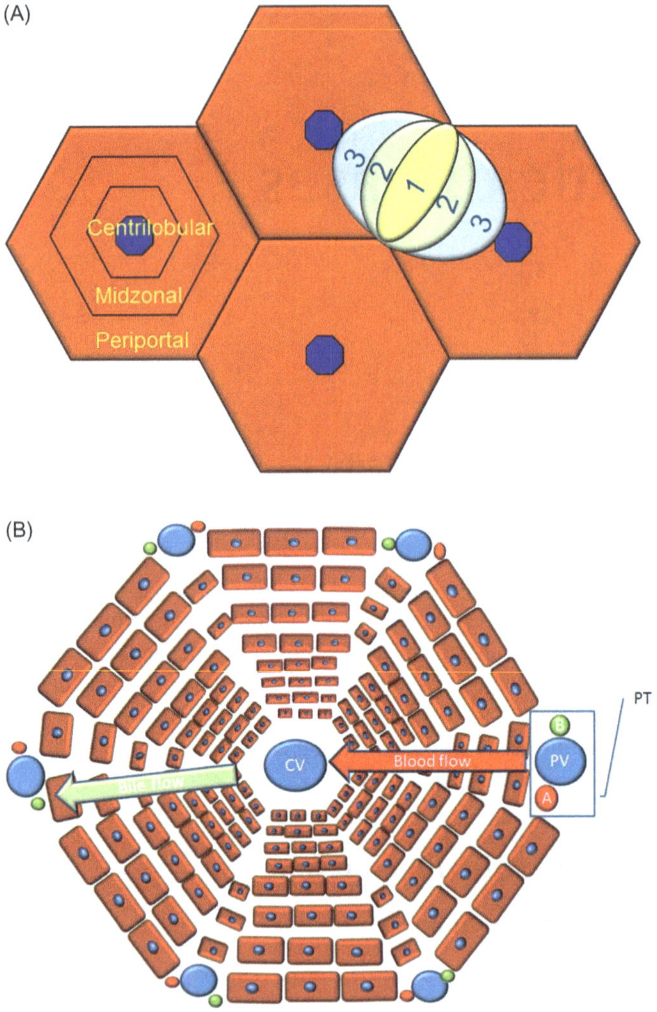

FIGURE 8.1 (A) Diagrammatic view of the liver showing lobular structure on left with centrilobular, midzonal, and periportal areas labeled. Upper right shows structure of hepatic acinus demonstrating acinar zones 1, 2, and 3. (B) Diagrammatic view of liver lobule showing portal triads (PT) each containing a hepatic artery (**A**), portal vein (PV) and bile duct (**B**). Blood flows from the portal triad through the sinusoids surrounding the hepatocytes and exits via the central vein (CV). Bile flows from zone 3 to zone 1 to empty into bile ducts within the portal triad.

as a hexagonal lobule that surrounds a central vein with a *portal triad* composed of a hepatic artery, a portal vein, and a bile duct at each of the six vertices of the hexagon. In the lobular model, blood flows from the portal triad in the *periportal region*, passes through endothelial-lined sinusoids separating anastomosing sheets of hepatocytes and enters the terminal hepatic vein (central vein) in the center of the lobule, termed the *centrilobular region*. The region between the centrilobular area and periportal area is termed the *midzonal region*. The acinar model is based on the physiologic structure of the liver, which is centered on the highly oxygenated area adjacent to the portal triads (*zone 1*). As blood filters through the sinusoids, it gives up oxygen so that blood entering the area surrounding the central vein (*zone 3*) is poorly oxygenated, containing one-third to one-half of the oxygen concentrations present in zone 1 blood (Bischoff et al., 2018). *Zone 2* is the area between zones 1 and 3 and is of intermediate oxygenation. Bile flows through the acinus via *bile canaliculi* in a direction retrograde to blood flow to empty into the terminal bile ducts at the portal triad.

Hepatic physiology

Hepatocytes play a key role in biotransformation of xenobiotics, carbohydrate metabolism and glucose regulation, lipid metabolism, production of plasma proteins, urea production, and bile formation. Not all hepatocytes perform the same functions at the same level, and the acinar model of the liver best reflects the physiologic and metabolic heterogeneity between hepatocytes in different regions (Cattley and Cullen, 2013). Hepatocytes closest to the portal regions are efficient in oxidative energy metabolism, ureagenesis from amino acids, gluconeogenesis, bile acid and bilirubin excretion, and antioxidant activity, while hepatocytes in zone 3 (centrilobular) are most efficient in glycolysis, lipogenesis, ureagenesis from ammonia, and xenobiotic biotransformation. In addition to differences in oxygenation, the different zones of the hepatic acinus are exposed to differing levels of nutrients and xenobiotics. Hepatocytes in zone 1 are first exposed to nutrients as well as a variety of xenobiotics, and are at risk of injury from toxicants that are directly injurious to cells (e.g., iron salts). Hepatocytes in zone 3 contain higher levels of metabolic enzymes, especially cytochrome P450 (CYP450) enzymes that are involved in the biotransformation of xenobiotics and endogenous intermediates. Although biotransformation is largely thought of as a means of detoxification, in some instances metabolic enzyme activity results in bioactivation of xenobiotics to more toxic compounds. Therefore hepatocytes in zone 3 are at higher risk from toxicants, such as aflatoxin B_1, that require enzymatic activation to express their harmful effects.

The liver's role in biotransformation is essential to maintaining normal homeostasis by converting potentially toxic xenobiotics and endogenous intermediates into compounds that are able to be excreted, primarily via urine and/or bile (Jaeschke, 2008). The liver has an abundant array of metabolic enzymes which drive biotransformation pathways; by convention, these pathways are termed "phase I" and "phase II" based on the type of reactions performed. Phase I reactions introduce or expose functional groups that may then undergo phase II conjugation reactions, the ultimate goal being to make the metabolite more water soluble for elimination via urine or bile. Phase I enzymes include the CYP450 oxidases which form three large families with considerable species, familial, sex, and individual heterogeneity in the relative amounts and types of enzymes present. The relative concentrations and activities of many hepatic phase I enzymes can be induced or inhibited by a variety of xenobiotics, from dietary components to pharmaceuticals. Although most phase I reactions are detoxifying in nature, a small percentage of phase I reactions result in a metabolite that is more toxic than the parent compound. For example, the metabolism of aflatoxin B_1 involves hydroxylation to form a highly electrophilic epoxide intermediate that forms covalent adducts with cellular proteins, RNA and DNA resulting in cell injury and, potentially, neoplastic transformation. Phase II reactions are conjugation reactions whereby conjugation with a variety of compounds including amino acids, sulfate, methyl groups, acetyl groups, and glucuronide enhances the water solubility and excretability of metabolites.

In addition to biotransformation, the liver plays a vital role in nutrient homeostasis, including protein synthesis and catabolism, lipid processing, carbohydrate metabolism, and bilirubin synthesis and excretion. Liver injury can result in decreased protein synthesis, resulting in cachexia, hypoalbuminemia, and edema due to loss of plasma oncotic pressure. Impaired hepatic conversion of ammonia from amino acid deamination to urea can result in elevated blood ammonia levels, which can lead to significant central nervous system dysfunction. Interference with lipid processing results in accumulation of lipid within hepatocytes (fatty change or steatosis). Hypoglycemia may occur due to disruptions in carbohydrate metabolism and decreased gluconeogenesis. Disruption of bilirubin processing can cause elevations in tissue and blood bilirubin levels, leading to jaundice and dermal photosensitization.

Hepatocytes are relatively long-lived cells that maintain cell function via degradation of senescent cellular components such as proteins and organelles using a process termed autophagy (Czaja et al., 2013). During autophagy, cytosolic components are enclosed by a double membrane which then fuses with a lysosome. Enzymes within this "autolysosome" then degrade the contents, releasing the degradation products back into the cytosol. Autophagy removes and recycles metabolic products and damaged or excess organelles (e.g., mitochondria) that otherwise would accumulate and ultimately cause injury to the cell. Autophagy plays a role in protein degradation during periods of starvation, increasing the rate of protein recycling over three times that seen during normal nutritional status (Czaja et al., 2013). Autophagy is also important in hepatocellular lipid processing, regulation of the liver immune response, and modulation of cell death.

Hepatic injury and repair

Mechanisms of injury

Hepatocellular injury can occur by a variety of mechanisms, including direct damage, hypoxia, triggering of apoptosis (programmed cell death), formation of reactive intermediates, disruption of normal cell metabolism, cytoskeletal damage, and stimulation of the immune response against cells. Direct damage to and disruption of hepatocellular membranes can occur through mechanical injury, such as lysis caused by infectious agents (e.g., bacteria, viruses, parasites), lipid peroxidation, or disruption of membrane proteins and transporters. Lipid peroxidation can be triggered by the presence of electrophiles such as reactive oxygen species (ROS) produced during inflammation or electrophilic intermediate metabolites formed during biotransformation reactions. Cellular membrane disruption results in loss of cytosolic components into the extracellular matrix which triggers an inflammatory response that may in turn cause more hepatocellular injury. Depending on the source of the hepatocellular injury, the distribution of lesions within the liver may vary. For instance, lesions due to salmonellosis tend to be multifocal in nature with no zonal distribution, whereas hepatic lesions from the viral disease yellow fever tend to occur in the midzonal regions (Stalker and Hayes, 2007; Quaresma et al., 2006).

The liver is a highly vascular organ well bathed in oxygenated blood, but there are still instances when hypoxia can result in significant hepatocellular injury, death, and fibrosis. Due to its proximity to the inferior vena cava, the hepatic vein is subject to decreased outflow when abnormal cardiovascular function, such as right heart failure, occurs. The resulting backup of blood within the liver causes pooling in the areas surrounding the central veins (zone 3) and in chronic states can result in the development of centrilobular fibrosis (Stalker and Hayes, 2007). Because zone 3 hepatocytes are already exposed to relatively oxygen-poor blood, hepatocytes in this area are also more prone to hypoxic injury in circumstances where the oxygen-carrying capacity of the blood is compromised, for example, anemia.

Apoptosis (programmed cell death) is a normal component of cell turnover, but it can also be triggered in a variety of pathologic situations, such as hepatotoxicosis, exposure to ionizing radiation, neoplasia, and infection with hepatotropic viruses (Elmore, 2007; Wang, 2014). In the latter event, apoptosis is considered to be a strategy of the host to prevent the spread of viral infection within the body (Ke and Chen, 2012). Apoptosis is a prominent feature of human hepatitis C viral infection (although the virus appears to have both anti- and proapoptotic activities), and the virus appears to sensitize host cells to apoptotic death. A large variety of xenobiotics have been associated with triggering or inhibition of apoptosis, and development of anticancer therapies using compounds that can alter apoptosis is an area of current research interest.

The generation of free radicals within hepatocytes can occur during xenobiotic biotransformation, normal metabolic processes involving oxidation/reduction reactions, inflammatory states [mediated by nitric oxide (NO)], and exposure to ionizing radiation (Bischoff et al., 2018). Free radicals have unpaired electrons, which make them highly susceptible to binding other macromolecules, including lipoproteins of cell membranes, enzymatic proteins, and DNA. Binding of free radicals to cellular macromolecules generates even more free radical production, resulting in a snowballing effect of free radical generation. Normally, cells have scavenger molecules (e.g., glutathione) to react with and detoxify free radicals, but excessive free radical production can result in depletion of scavenger stores, resulting in increased binding of free radicals to vital cellular structures. Free radical damage can cause alterations to cell membrane permeability, inactivation of membrane-associated proteins, and loss of polarity to mitochondrial membranes. Inactivation of enzymatic proteins can interfere with normal cellular metabolism, leading to cellular degeneration or necrosis. Free radical-induced DNA damage can result in interference with transcription or translation, resulting in decreased protein synthesis.

The mechanisms behind toxic hepatic injury can be divided into two categories: intrinsic and idiosyncratic. Intrinsic liver injury is a dose-dependent, predictable reaction to a toxicant and it is the most common type of liver injury in humans and animals (Bischoff et al., 2018). There is frequently a predictable latent period between exposure to the toxicant and the development of clinical signs of liver insufficiency. Intrinsic liver injury is often caused by highly reactive by-products of xenobiotic metabolism, especially free radicals and electrophiles. Idiosyncratic liver injury is an unpredictable response to a xenobiotic. Idiosyncratic reactions are rare and nondose-dependent and may be associated with extrahepatic lesions. Idiosyncratic reactions frequently occur following unremarkable initial exposure (sensitization) and manifest upon reexposure to the xenobiotic, although development of idiosyncratic hepatotoxicity after months or years of chronic exposure can also occur. With idiosyncratic drug reactions, it is not uncommon for there to be a delay of a few weeks to several months between the time of initial exposure and development of liver injury. Extrahepatic changes may include fever and dermal erythema or rash.

Within the cell, calcium is sequestered within organelles such as mitochondria and endoplasmic reticulum, which maintain a relatively low level of calcium within the cytosol. Xenobiotics that interfere with the pumps that maintain calcium balance can result in increased release of calcium into the cytosol. The released calcium ions impede normal cytoskeletal function and activate a variety of cytosolic enzymes including ATPases, phospholipases, and proteases. Subsequently, mitochondrial membrane permeability increases, which causes induction of necrosis and apoptosis (Bischoff et al., 2018).

Cytoskeletal disruption

The hepatocyte cytoskeleton aids in maintaining the shape and position of the hepatocyte. Disruption to the proteins that compose the cytoskeleton may result in loss of cell-to-cell adhesion, rounding up of hepatocytes, and release of free hepatocytes into the circulation. Phalloidin, from hepatotoxic mushrooms, and microcystin, from toxic cyanobacteria, are toxins that damage hepatocellular cytoskeletal proteins, resulting in hepatocellular dissociation and massive hepatic necrosis (Jaeschke, 2008). Leptospirosis is an infectious disease that also can result in hepatocellular dissociation (Stalker and Hayes, 2007).

Response to injury

The response of the liver to injury depends on the type of insult, duration of insult, cell populations affected, systemic response to insult, and collateral damage. Although the list of causes of liver injury is quite large (including infectious, inflammatory, circulatory, toxic, and metabolic processes), the liver is actually capable of only a limited number of responses. The clinical manifestations of these responses will vary based on the severity and extent of the hepatic injury.

Hepatocellular degeneration and death

The most basic response of hepatocytes to nonlethal insult is hepatocellular degeneration, characterized by cellular swelling (Crawford, 2005). Accumulation of substances such as copper, fat, pigments, or glycogen may occur within degenerated hepatocytes. In many instances, degeneration is a transient and reversible process, but with more severe injury degeneration can progress to cellular death via necrosis, cell lysis, or apoptosis. Influx of inflammatory cells in response to infectious organisms may result in further hepatocellular injury from ROS and other mediators produced by the inflammatory cells in order to kill invading organisms. Hepatocellular degeneration and necrosis frequently display distinct distribution patterns that may aid in narrowing down the list of potential causes. Infectious and immune-mediated hepatocellular injuries tend to produce scattered foci of affected cells, whereas many hepatotoxicants will produce characteristic zonal patterns of injury. Circulatory disturbances tend to result in hepatocellular degeneration and necrosis in the centrilobular region due to the normally hypoxic environment of this area.

Cholestasis and biliary injury

Hepatic insults that interfere with the ability of hepatocytes to excrete bile into canaliculi or with the transport of bile throughout the biliary tree can result in systemic buildup of solutes within blood and other tissues; when severe enough, a yellow-orange cast to the tissues appears which is termed icterus or jaundice. Cholestasis may be caused by metabolic derangements in hepatocytes, hepatocellular swelling, impediments to canalicular transport (e.g., presence of inflammatory cells), or obstruction of ducts of the biliary tree or gall bladder. Cholestatic disease may involve only the biliary system, or it may be a manifestation of hepatobiliary insult.

Steatosis

Steatosis, or fatty liver, is a common result of moderate to severe hepatocellular insult. In humans, the most common cause of steatosis is insulin resistance associated with central obesity and sedentary lifestyle (nonalcoholic fatty liver disease or "metabolic syndrome"), with alcohol-induced steatosis also a common occurrence (Valenti et al., 2013; Jaeschke, 2008). In steatosis, hepatocytes become distended with lipids, a condition that in early or mild stages is benign and reversible. However, steatosis may progress to steatohepatitis, which can result in chronic, progressive hepatic injury, fibrosis, and possibly hepatocellular carcinoma. Studies have shown that defects in hepatocellular autophagy may play a key role in the development of steatosis (Czaja et al., 2013).

Healing, fibrosis, and cirrhosis

The liver has a remarkable capacity for self-repair and regeneration following acute and chronic insult. Regeneration depends on the balanced interrelationship between a variety of hepatocellular factors including cell cycle regulation, metabolism, angiogenesis, cell adhesion, and cellular senescence as well as extracellular factors such as inflammation, cell recruitment, and extracellular matrix components (Duarte et al., 2015). Deposition of excessive extracellular matrix proteins most commonly begins in the periportal areas, centrilobular areas, or within the space of Disse (the space between apical hepatocyte membrane and adjacent endothelial cells), and it interferes with the exchange of nutrients and oxygen between hepatocytes and the sinusoidal blood, furthering hepatocellular stress and interfering with normal hepatocellular function (Jaeschke, 2008). Hepatic stellate cells are responsible for producing the majority of hepatic extracellular matrix proteins, and their activation is the first step in hepatic fibrosis. Hepatic stellate cells can be activated by the presence of products formed during hepatocellular injury, including ROS and lipid peroxidation products, by cytokines released by damaged sinusoidal endothelial cells or Kupffer cells within the space of Disse, or by cytokines such as platelet-derived growth factor released by activated platelets. Hepatic stellate cell activation results in changes in the nature of extracellular matrix, with the nonfibrillar basement membrane matrix being replaced by fibril-forming collagens. As collagen deposition continues, fibrous scars disrupt the hepatic architecture, resulting in islands of regenerating hepatocytes surrounded by dense fibrous connective tissue, a condition known as cirrhosis. The cirrhotic areas have limited ability to function normally and when sufficient liver tissue is affected, hepatic failure ensues. In humans, viral hepatitis is the most common cause of hepatic cirrhosis, although alcohol and nonalcoholic steatosis are increasingly important causes as well.

Neoplasia

Chronic liver injury with inflammation and/or cirrhosis may lead to the development of hepatocellular carcinoma (Jaeschke, 2008). Although direct proof of cause and effect is largely lacking, development of human hepatocellular carcinoma has been strongly associated with viral hepatitis, metabolic disorders such as hemochromatosis, nonalcoholic steatosis, and toxicants such as aflatoxin, alcohol, and exogenous androgens.

Biomarkers of hepatic injury

Clinicians use several different clinical pathology parameters to evaluate hepatic function. Because a single uniquely specific biomarker of hepatic injury does not exist, most diagnoses of liver dysfunction rely on evaluation of several parameters, including measurement of serum enzyme activities, serum bilirubin, serum albumin and protein levels, serum ammonia, serum cholesterol, coagulation parameters, and urine urobilinogen concentrations (Yang et al., 2014). In general, serum enzyme activities and bilirubin levels are the most reliable parameters measured, with the remaining parameters used to support suspicions of hepatic injury.

Alanine aminotransferase

Alanine aminotransferase (ALT) is the most widely used clinical biomarker of hepatic health (Ozer et al., 2008). As its name implies, ALT is involved in the transamination of alanine, and it is present in the liver at much higher concentrations than in other organs. Leakage of ALT from the hepatocyte into the blood occurs following hepatocellular injury, where it is cleared with a plasma half-life of approximately 42 hours. Although elevations in ALT can be highly suggestive of liver injury, ALT is not liver-specific, and extrahepatic sources of circulating ALT include areas of skeletal or cardiac muscle damage and drugs that increase ALT gene expression (e.g., fenofibrate) (Yang et al., 2014). Assays to distinguish the relative percentages of two ALT isoforms, ALT1 and ALT2, may aid in distinguishing between hepatic injury and extrahepatic injury (Rafter et al., 2012).

Aspartate aminotransferase

Aspartate aminotransferase (AST), which functions to transaminate aspartate, has been commonly used as a biomarker for hepatic injury. AST is considered a less specific marker for liver injury than ALT, due to expression from other tissues, such as brain, myocardial cells, and skeletal muscle cells (Yang et al., 2014). Some clinicians use the ratio of AST to ALT to help in distinguishing between extrahepatic and hepatic injury, with an AST/ALT ratio of 2:1 being indicative of hepatic injury (Ozer et al., 2008); others have proposed that an AST/ALT ratio of ≤ 0.4 during recovery from severe acute hepatotoxicity indicates a good prognosis for recovery (McGovern et al., 2015).

Alkaline phosphatase

Alkaline phosphatase (ALP) is another enzyme that is frequently used in evaluating liver function. Elevations in ALP occur during obstructive biliary disease; however, because there are many isoenzymes of ALP from other organs and tissues (e.g., bone), ALP should not be considered as a specific biomarker for cholestasis (Yang et al., 2014). Changes in ALP should be evaluated in tandem with other hepatic biomarkers.

Gamma-glutamyltransferase

Gamma-glutamyltransferase (GGT) is a relatively nonspecific enzyme whose activity can increase with hepatic disease, cardiac injury, and muscular injury (Yang et al., 2014). In humans, elevations in GGT do appear to be fairly consistently seen following consumption of large quantities of alcohol. GGT is approximately as predictive of biliary injury as is ALP, and it is best utilized in tandem with ALP to verify that the ALP elevations are related to biliary injury.

Total bilirubin

Total bilirubin is the sum of unconjugated bilirubin from extrahepatic sources and conjugated hepatic bilirubin. Serum bilirubin concentration reflects the ability of hepatocytes to take up, conjugate, and secrete bilirubin, so it is a functional marker rather than a marker of cellular integrity as reflected by serum transaminase levels. In the absence of biliary obstruction, total bilirubin is an insensitive measure of liver function; due to the large reserve capacity of the liver for bilirubin processing, considerable hepatic injury must occur before alterations in total bilirubin are detectable (Yang et al., 2014). This means that by the time total bilirubin is elevated, substantial hepatic injury has occurred. The combination of ALT and bilirubin is used in "Hy's law" which states that, in drug-induced liver injury, elevation of both ALT and bilirubin is more indicative of serious liver injury than is elevation in serum enzymes alone.

Nutraceuticals and the liver

Liver function is essential for the survival, and liver failure threatens the life of the individual, in some cases necessitating life-saving liver transplant. Fortunately, the liver has a large reserve capacity and a remarkable ability to adapt to alterations in its microenvironment and to regenerate following injury. Many nutraceutical compounds have been promoted to support the liver during various stages of health and disease, and these agents are roughly classified into three categories: *antihepatotoxic*

TABLE 8.1 Hepatic effects of nutraceuticals.

Compound/ source	Active constituent (s)	Hepatoprotective effects	Hepatotoxic effects	References
Andrographis paniculata	Diterpenes, lactones, flavonoids	↓ Lipid peroxidation and hepatocellular damage due to acetaminophen and carbon tetrachloride	–	Akbar (2011)
Areca catechu	Alkaloids, tannins, flavones, triterpenes, sterols, fatty acids	Antioxidant, antiinflammatory, lipid regulation	–	Peng et al. (2015)
Black cumin (*Nigella sativa*)	Thymoquinone	↑ Neoplastic cell apoptosis; antioxidant; protects from cisplatin-induced injury	–	Khan et al. (2011); Al-Malki and Sayed (2014)
Blue green algae (*Spirulina* spp.)	Carotenoids	Antioxidant	Microcystin contamination may result in hepatic injury	Madrigal-Santillan et al. (2014); Jiang et al. (2008)
Capsicum spp.	Capsaicinoids	–	Bioactivation with formation of electrophiles	Reilly and Yost (2006)
Commiphora mukul	Gum guggul	Hypolipidemic, ↓ lipid peroxidation		NTP (2005)
Luteolin	–	↓ Expression of lipogenic genes, decreased ethanol-induced steatosis	Bioactivation to reactive metabolites, glutathione depletion	Shi et al. (2015); Liu et al. (2014)
Melatonin	–	↓ Lipid peroxidation, ↓ DNA damage, ↓ necrosis, ↓ expression of inducible nitric oxide synthetase	–	Schmidt (2010); Mathes (2010)
Milk thistle (*Silybum marianum*)	Silibinin (Silybin)	Antioxidant, antiinflammatory, antifibrotic; ↓ neoplastic cell proliferation, ↑ neoplastic cell apoptosis	Exacerbates tumor promotion effects of ethanol	Loguercio and Festi (2011); Brandon-Warner et al. (2012)
	Silymarin	Antioxidant, antiinflammatory, ↑ glutathione and superoxide dismutase levels, membrane stabilization, antifibrotic, stimulates hepatocellular regeneration		Bahmani et al. (2015); Madrigal-Santillan et al. (2014)
Neem (*Azadirachta indica*)	Azadirachtin-A	↓ Necrosis from carbon tetrachloride; ↓ apoptosis from cisplatin	–	Baligar et al. (2014); Dkhil et al. (2013)
Resveratrol	–	Antioxidant, antiinflammatory, decreased hepatic triglyceride accumulation	–	Heeboll et al. (2014)

agents are said to antagonize the effects from agents causing hepatic inflammation or disease; *hepatotropic* agents provide general support to the normal liver and promote healing in liver during recovery from injury; and *hepatoprotective* agents are those that prevent liver injury from various causes (Roy et al., 2012; Parimoo et al., 2014). There can be considerable overlap between functions of these nutraceuticals, with, for instance, some hepatoprotective agents also providing hepatotrophic and/or antihepatotoxic activity as well. Although hundreds of compounds including herbals, plant extracts, and dietary supplements have been purported to provide aid in hepatic health, very few have been subject to strict scrutiny to determine their efficacy, mechanism of action, and safety (Teschke et al., 2014). Discussed below are some commonly used nutraceuticals along with their potential effects on the liver; additional products are summarized in Table 8.1.

Anthocyanins

Anthocyanins are water-soluble, phenolic flavonoid pigments that are present in many fruits, flowers, and leaves where they impart red, blue, and purple colorations (Fang, 2014). Most berries are rich in anthocyanins, especially chokeberry, elderberry, blueberries, raspberries, and blackberries (Wu et al., 2006). Anthocyanins exert hepatoprotective and hepatotropic effects in the liver through their strong antioxidant properties, decreasing injury from both ROS and lipid peroxidation (Madrigal-Santillan et al., 2014). Anthocyanins have also been shown to reduce hepatocellular lipid accumulation, and they may play a role in the prevention of nonalcoholic fatty liver disease through promotion of lipolysis, inhibition of lipogenesis, and induction of antioxidant enzymes (Valenti et al., 2013). Based on studies in mice, anthocyanins may delay aging-related degenerative changes in the liver, decreasing age-related liver injury and fibrosis (Wei et al., 2017).

Ashwagandha

Ashwagandha (*Withania somnifera*), also known as "Indian winter cherry" and "Indian ginseng," is one of the most important herbs used in traditional medicine in India, with a variety of medicinal effects attributed to its use (Singh et al., 2011). Ashwagandha contains various alkaloids, steroidal lactones, and saponins, and has been used for its purported antistress/adaptogenic, antitumor, tonic, anxiolytic, antiinflammatory, and antiarthritic properties. Studies in rabbits and rodents have shown hypolipidemic and antioxidant activity in some aqueous extracts (Shukla et al., 2014; Visavadiya and Narasimhacharya, 2011), and its antioxidant properties were shown to provide protection against radiation-induced hepatotoxicity in rats (Hosny et al., 2012) as well as iron-induced hepatotoxicity in broiler chicks (Ramakrishnan et al., 2011). Ashwagandha was found to have antioxidant and proapoptotic activity against hepatocellular carcinoma cells in vitro (Ahmed et al., 2018) Ashwagandha does not appear to induce or inhibit major CYP450 enzyme activities (CYP3A4 and CYP2D6) in human hepatic microsomes (Savai et al., 2015). Dietary supplements containing ashwagandha have recently been implicated in the development of cholestatic and mixed liver injury in humans (Bjornsson et al., 2020; LiverTox, 2019a).

Astaxanthin

Astaxanthin is a xanthophyll carotenoid present in a variety of microalgae and yeasts that has been approved in the United States and European Union as a food colorant used as a source of pigment in feed for salmon, trout, and shrimp (Ambati et al., 2014). The primary source of astaxanthin for human consumption is obtained via ingestion of seafood or extracted from the microalga *Haematococcus pluvialis* or yeast of the species *Phaffia*. Like many carotenoids, astaxanthin possesses strong antioxidant activity via neutralization of singlet oxygen, scavenging of free radicals, inhibiting lipid peroxidation, enhancing immune system function, and regulating gene expression. Astaxanthin has a protective effect against ischemia-reperfusion injury in steatotic livers (Li et al., 2017). Astaxanthin prevents activation of hepatic stellate cells, which inhibit transforming growth factor (TGF-β1)-induced profibrotic gene expression and inhibit hepatic fibrosis (Yang et al., 2015). Astaxanthin also improves glucose metabolism, decreases expression of hepatocellular endoplasmic reticulum stress proteins, and decreases hepatocellular lipid accumulation in mice fed high fat-high fructose diets (Bhuvaneswari et al., 2014). Astaxanthin has a weak inhibitory effect on human CYP2C19, but was not inhibitory on eight other CYP activities and is not considered to be a significant cause of altered CYP activity (Zheng et al., 2013).

Caffeine

Caffeine is the most widely consumed pharmacologically active substance in the world (Wang et al., 2014a). In recent years, epidemiological studies have shown apparent beneficial effects on chronic liver disease in people consuming caffeine-containing beverages. Caffeine prevents the activation of hepatic stellate cells by inhibition of the action of adenosine on hepatic stellate cell adenosine A2A receptors and by blocking the expression of TGF-β (Wang et al., 2014a; Arauz et al., 2014). As hepatic stellate cells are a major source of extracellular matrix in the liver, the inhibition of hepatic stellate cell activation by caffeine results in an overall antifibrotic effect on the liver. Caffeine also exerts antimutagenic effects against food-derived heterocyclic aromatic amines by forming heterocomplexes with the amines that reduce their bioavailability and limit their biologic effects (Woziwodzka et al., 2013). Caffeine appears to have a hepatoprotective effect against acetaminophen-induced liver injury, possibly due to improvement of mitochondrial bioenergetic function increasing cellular oxygen utilization (Goncalves et al., 2018). Rats that were exposed to caffeine in utero due to maternal exposure had increased susceptibility to nonalcoholic fatty liver disease when they matured into adults (Wang et al., 2014b). Livers of these rats showed increased lipogenesis, reduced lipid output, and enhanced gluconeogenesis, which predisposed them to the development of nonalcoholic fatty liver disease.

Coffee is a major source of daily caffeine consumption and coffee drinking has been associated with reduced incidence of nonalcoholic fatty liver disease (Chen et al., 2014). In individuals at risk for liver disease, coffee consumption was associated with decreased elevations in liver enzymes in a dose-dependent manner (Saab et al., 2014). In individuals with chronic liver disease, coffee consumption was associated with decreased progression to cirrhosis, lowered mortality from cirrhosis, and decreased rate of hepatocellular carcinoma development. Improved responses to antiviral therapy and reduced fibrosis were observed in chronic hepatitis C patients who consumed coffee (Saab et al., 2014; Rodrigues Machado et al., 2013). Many of the positive effects of coffee have been attributed to caffeine and other methylxanthines in the brew, but some studies have shown similar hepatoprotective effects in individuals who ingest decaffeinated coffee, suggesting that other coffee constituents may play a role in the beneficial effects of coffee on the liver (Xiao et al., 2014).

Curcumin

Curcumin is the major component of turmeric, a ground powder from the root of the *Curcuma longa* plant that is used widely throughout the world as a spice and coloring agent (NTP, 1993). Turmeric has also been widely used as a remedy for a variety of health disorders due to its purported antioxidant, antimicrobial, antiinflammatory, antiviral, anticarcinogenic, and antidiabetic properties (Soliman et al., 2014). Curcumin has been shown to decrease the level of ethanol-induced steatosis in rats, block carbon tetrachloride- and acetaminophen-induced hepatocellular injury, and prevent the reduction of CYP450 expression induced during inflammatory situations (Bengmark et al., 2009; Dhiman et al., 2012). Increased incidence of hepatocellular carcinomas occurred in mice exposed to dietary levels of 10,000 and 50,000 ppm tumeric oleoresin (75%–85% curcumin) in 2-year carcinogenicity studies (NTP, 1993).

Fenugreek

Fenugreek (*Trigonella foenum-graecum*) is a seed that has been used for centuries as both a culinary spice and for medicinal purposes in many traditional/folk medicine systems (Yadav and Banquer, 2014). The seed and its extracts have shown antilipidemic, antioxidant, antiinflammatory, and hepatoprotective properties. Treatment of high-fat diet-induced obese rats with fenugreek extract resulted in inhibition of hepatocellular fat accumulation and dyslipidemia, improvement of lipid metabolism, downregulation of lipogenic enzymes, improvement in serum hepatic enzyme levels, and increased antioxidant defense (Kumar et al., 2013). In studies with rats, fenugreek and its extracts were protective against oxidative hepatotoxicity induced by ethanol, cadmium, thioacetamide, and aluminum chloride (Shivashankara et al., 2012; Arafa et al., 2014; Zargar, 2014; Belaid-Nouira et al., 2013).

Garcinia cambogia

Garcinia cambogia is a fruit native to southeastern Asia and western Africa that has been promoted as a weight loss and "fat-burning" compound (Dara et al., 2008; Marquez et al., 2012). The active component of *G. cambogia* is hydroxycitric acid, a compound that inhibits the enzyme ATP-citrate lyase, which is involved in endogenous lipid biosynthesis. Hydroxycitric acid also increases hepatic glycogen synthesis, suppresses appetite, and decreases body weight gain. In rats, *G. cambogia* protected against high-fat diet-induced obesity in rats by inhibiting fatty acid synthase activity and glucose resistance and by enhancing gene expression and enzymatic activity involved in fatty acid β-oxidation (Kim et al., 2013). However, in the same study, *G. cambogia* increased hepatic collagen accumulation and lipid peroxidation and triggered expression of genes regulating oxidative stress (superoxide dismutase and glutathione peroxidase) and inflammatory responses [tumor necrosis factor-α (TNF-α) and monocyte chemoattractant protein-1]. The use of weight-loss supplements containing *G. cambogia* and extracts of other *Garcinia* species has been associated with hepatotoxicity (Dara et al., 2008; Crescioli et al., 2018; LiverTox, 2019b; Yousaf et al., 2019). Additionally, fatal liver failure following exposure to *Garcinia*-containing supplements in a patient on chronic montelukast therapy suggests the potential for adverse interactions between these supplements and pharmaceutical agents (Actis et al., 2007).

Garlic

Garlic (*Allium sativum*), a member of the onion family, has been used for centuries as a component of food as well as for its purported medicinal properties. Garlic contains more than 2000 biologically active substances including volatile, water-soluble, and oil-soluble organosulfur compounds as well as dietary fiber, sugars, flavonoids, essential oils, and pectin (Majewski, 2014). The sulfur components of garlic include alliin and scordinin A and B, as well as several cyclic sulfoxides termed garlicnins. When garlic cloves are cut or crushed, the cysteine sulfoxide alliin is rapidly broken down by alliinase into thiosulfinates (e.g., allicin), ajoenes, vinyldithiins, sulfides, and disulfides. Garlic has an effect on lipid metabolism and has been shown to aid in reduction of blood cholesterol levels in humans, possibly by altering enzymes involved in cholesterol synthesis and metabolism. Garlic essential oil was shown to protect obese mice from nonalcoholic fatty liver disease through amelioration of lipid metabolic disorders as well as decreasing oxidative stress (Lai et al., 2014). Garlic reduces hepatocellular production of low-density lipoproteins and decreases expression of inducible NO synthetase, both of which contribute to the antiatherosclerotic effects of garlic (Majewski, 2014). Some forms of garlic (e.g., aged garlic extract) are said to have a strong hepatoprotective effect due to antioxidant properties. In a study of broiler chicks, 0.1% garlic added to the diet reduced the oxidative hepatic injury associated with exposure to excessive dietary iron (Ramakrishnan et al., 2011). Garlic supplements may reduce the activity of certain CYP450 enzymes (e.g., CYP2C9), which can result in alterations in plasma concentrations of drugs that are metabolized by these enzymes (Majewski, 2014). Acute hepatic injury associated with high-dose garlic supplementation was reported in a liver transplant patient; liver parameters normalized upon cessation of the supplement (Shaikh et al., 2017)

Ginkgo biloba

Ginkgo biloba (maidenhair tree) is a popular ornamental tree that has been used for centuries for its medicinal effects. *Ginkgo* contains a variety of active constituents including flavonoid glycosides (e.g., quercetin, kaempferols), terpene lactones (e.g., ginkgolides, bilobalide), and organic acids (Gurley et al., 2012). *Ginkgo* has been shown to ameliorate insulin resistance and dyslipidemia in nonalcoholic fatty liver disorder by decreasing expression and activity of key enzymes in endogenous lipid biosynthesis (Jadeja et al., 2014). *Ginkgo* was found to be hepatoprotective against hepatocarcinogenesis in experimental studies (Madrigal-Santillan et al., 2014). *Ginkgo* extract inhibits CYP1A1, CYP1A2, CYP2C8, and CYP2C9, induces expression of CYP2B6, CYP2C19, and CYP3A4 activities in human hepatic microsomes, and alters the activity of some xenobiotic transporters (Na et al., 2011; Gurley, 2012; Gurley et al., 2012). Although the clinical significance of these effects has not been fully established, induction of CYP2C19 by a *Ginkgo* supplement was implicated in fatal breakthrough seizures and subtherapeutic drug levels in a patient on anticonvulsants metabolized via this pathway (Kupiec and Raj, 2005). In carcinogenicity studies, ingestion of *G. biloba* was associated with increased instances of hepatocellular adenomas, hepatocellular carcinomas, and hepatoblastomas in mice, and it exacerbated the metastasis of colon cancer to the liver in a mouse model (NTP, 2013; Wang et al., 2017).

Ginseng

The two most widely used ginseng species are Asian/Chinese ginseng (*Panax ginseng* CA Meyer) and American ginseng (*Panax quinquefolius* L.). Chinese ginseng has been used for almost 2000 years in traditional Chinese medicine, and ginseng is currently one of the most popular medicinal herbs in the world (Gurley et al., 2012). Ginseng contains ginsenosides which are triterpene glycosides unique to the *Panax* species. Over 40 ginsenosides have been identified, and commercially available ginseng extracts may vary considerably in the number and amount of ginsenosides they contain, frequently confounding attempts to determine levels of efficacy, drug interaction, and toxicity. The pharmacologic activity of ginsenosides is due to their bioactivation in the GIT to metabolites that are absorbed into the systemic circulation. Ginsenoside Rg1 is the primary active component of *P. ginseng* which has antiinflammatory and antiapoptotic activity within the liver via inhibition of TNF-α and has been shown to inhibit hepatic fibrosis (Zhao et al., 2014). Ginseng has been shown to have some protective effects against liver injury induced by ethanol, lipopolysaccharide, hepatitis virus, carbon tetrachloride, radiation, aflatoxin, and cadmium (Kim, 2016). Although some ginseng extracts have been shown to be weak inhibitors of some hepatic metabolic enzymes (e.g., CYP1A1, CYP2C9) and xenobiotic transporters in vitro, the levels of ginsenosides required to cause these effects are much higher than would be obtainable at normal ginseng dose ranges (Gurley et al., 2012). Carcinogenicity studies in rodents did not indicate any evidence of hepatic carcinogenicity of ginseng (NTP, 2011).

Green tea extract

After water, tea is the second most consumed beverage in the world, and green tea is the fourth most commonly used dietary supplement in the United States (Sarma et al., 2008). Because it is unfermented, green tea, unlike black tea, retains its antioxidant polyphenolic catechols, alkaloids, and other bioactive compounds. Health benefits attributed to green tea consumption include weight loss, decreased inflammation, decreased hepatic lipid accumulation, and decreased hepatic fibrosis (Heber et al., 2014; Safer et al., 2015). However, there have been numerous case reports of hepatotoxicity following the use of dietary supplements containing green tea and its extracts (Sarma et al., 2008; Krishna et al., 2011; Patel et al., 2013). Following a comprehensive review of green tea extracts, the United States Pharmacopeia has issued warnings regarding potential hepatotoxic effects; several other countries including Canada and Norway have cautionary statements about potential for hepatotoxicity associated with green tea extracts (Oketch-Rabah, et al., 2020). Although there was no evidence of carcinogenic activity of green tea extract in animal studies, it was found to be mutagenic in several in vitro studies, and administration in rats and mice was associated with increased incidences of non-neoplastic lesions in the liver (NTP, 2014).

Isoflavones

Isoflavones are phytochemicals present in many legumes including soybeans, fava beans, chickpeas, and peanuts (Qiu and Chen, 2015). The soybean-derived isoflavone genistein has been shown to attenuate alcoholic and nonalcoholic fatty liver disease through stimulation of hepatic fatty acid β-oxidation, increasing hepatic antioxidant activities, decreasing hepatic fibrosis, and reducing lipid peroxidation. Similarly, other isoflavones have demonstrated antilipogenic, antioxidant, antifibrotic, and antiinflammatory properties in the liver. Multiple chronic studies of genistein in rodents have shown no evidence of hepatic carcinogenic potential (NTP, 2007); however, recent independent studies have shown increased risk of certain types of cancers in women with breast cancer (Yu et al., 2016)

Quercetin

Quercetin is a flavonoid found in many plants including citrus, berries, leafy vegetables, herbs, spices, legumes, tea, and cocoa (NTP, 1992). Quercetin demonstrated hepatoprotective activity following acute hepatocellular and cholestatic liver injury (Lin et al., 2014; El-Shafey et al., 2014). Proposed mechanisms of hepatoprotection are via general antioxidant activity, alleviation of NO-induced changes, suppression of TGF-β1 secretion, and inhibition of collagen secretion and fibrogenesis. Multiple chronic studies of quercetin in rodents have shown no evidence of hepatic carcinogenic potential (NTP, 1992).

Concluding remarks and future directions

Hundreds of nutraceuticals have been promoted for their potential benefit to the liver, but further studies are needed for most to determine the clinical efficacy of the individual agents. Extensive research into many nutraceuticals has revealed that many do possess potential medicinal activities that can be measured at the cellular and/or molecular level, but evaluating their clinical effectiveness has been more problematic. Many reports of positive clinical effects of nutraceuticals in humans rely heavily on anecdotal evidence or individual case reports with confounding factors such as coadministered compounds (e.g., drugs, herbals, other nutraceuticals) or lack of effective control populations. Commercially available nutraceuticals can vary considerably between brands (and often between lots within a brand) in the quality and content of active constituents, and researchers may use extracts with differing proportions of active constituents, making it difficult to make comparisons between studies. In many countries, herbal and nutraceutical products are poorly regulated, with the consumer at the mercy of the manufacturer in terms of product quality and safety (Seef et al., 2015).

Many nutraceutical compounds may work well in vitro, but poor bioavailability or rapid clearance may interfere with their ability to exert their actions in vivo due to inability to reach effective plasma concentrations. Some compounds may achieve good plasma levels but be unable to distribute to the target tissue due to the presence or absence of transporters to cross physiological barriers. Others may have beneficial effects on the liver but may result in adverse effects in other tissues. For instance, silymarin shows significant hepatoprotective activity, yet it can enhance the growth of mammary tumors (Malewicz et al., 2006), while silibinin is hepatoprotective, but when coadministered with ethanol, it promotes progression of ethanol-dependent hepatocellular carcinoma (Brandon-Warner et al., 2012).

Nutraceuticals that exhibit antioxidant, antiinflammatory, and/or antifibrotic activities in the liver may aid in the reduction of liver injury by endogenous and exogenous hepatotoxic agents, and/or may aid in the prevention or progression of chronic liver disorders such as nonalcoholic fatty liver disease, alcohol-induced liver disease, hepatitis C viral infection, and hepatic toxicity and carcinogenesis (Teschke et al., 2014). However, potential adverse effects from these compounds must be studied and documented, as alteration of hepatic metabolic enzymes, decreased inflammatory/immune responses, and stimulation of hepatocellular proliferation may result in undesirable consequences such as interference with therapeutic pharmaceutical agents, decreased hepatic immune surveillance or increased risk of hepatic neoplasia. Long-term safety and carcinogenicity data are lacking for most nutraceutical agents. Contamination or intentional adulteration of some products with pesticide residues, heavy metals, mycotoxins, toxic plants, pharmaceuticals, and other potentially hazardous substances has resulted in human illness and death (Seef et al., 2015). Much research remains needed to address these issues to ensure the safety and efficacy of nutraceutical products.

References

Actis, G.C., Bugianesi, E., Ottobrelli, A., 2007. Fatal liver failure following food supplements during chronic treatment with montelukast. Dig. Liver Dis. 39, 953–955.

Ahmed, W., Mofed, D., Zekri, A.R., et al., 2018. Antioxidant activity and apoptotic induction as mechanisms of action of *Withania somnifera* (Ashwagandha) against a hepatocellular carcinoma cell line. J. Int. Med. Res. 46 (4), 1358–1369.

Akbar, S.A., 2011. *Andrographis paniculata:* a review of pharmacological activities and clinical effects. Alt. Med. Rev. 16, 66–76.

Al-Malki, A.L., Sayed, A.A.R., 2014. Thymoquinone attenuates cisplatin-induced hepatotoxicity via nuclear factor kappa-β. BMC Compl. Alt. Med. 14, 282.

Ambati, R.R., Phang, S.M., Ravi, S., 2014. Astaxanthin: sources, extraction, stability, biological activities and its commercial application—a review. Mar. Drugs 12, 128–152.

Arafa, M.H., Mohammad, N.S., Atteia, H.H., 2014. Fenugreek seed powder mitigates cadmium-induced testicular damage and hepatotoxicity in male rats. Exp. Toxicol. Pathol. 66, 293–300.

Arauz, J., Zarco, N., Segovia, J., 2014. Caffeine prevents experimental liver fibrosis by blocking the expression of TGF-β. Eur. J. Gastroenterol. Hepatol. 26, 164–173.

Bahmani, M., Shirzad, H., Rafieian, S., 2015. *Silybum marianum*: beyond hepatoprotection. J. Evid. Based Compl. Alt. Med. 20 (4), 292–301.

Baligar, N.S., Aladakatti, R.H., Ahmet, M., 2014. Hepatoprotective activity of the neem-based constituent azadirachtin-A in carbon tetrachloride intoxicated Wistar rats. Can. J. Physiol. Pharmacol. 92, 267–277.

Belaid-Nouira, Y., Bakhta, H., Haouas, Z., 2013. Fenugreek seeds, a hepatoprotector forage crop against chronic AlCl3 toxicity. Vet. Res. 9, 22.

Bengmark, S., Mesa, M.D., Gil, A., 2009. Plant-derived health: the effects of turmeric and curcuminoids. Nutr. Hosp. 24, 273–281.

Bhuvaneswari, S., Yogalakshmi, B., Sreeia, S., 2014. Astaxanthin reduces hepatic endoplasmic reticulum stress and nuclear factor kB-mediated inflammation in fructose and high fat diet-fed mice. Cell Stress Chaperones 19, 183–191.

Bischoff, K., Mukai, M., Ramaiah, S.K., 2018. Liver toxicity. In: Gupta, R.C. (Ed.), Veterinary Toxicology: Basic and Clinical Principles, third ed. Elsevier, San Diego, CA, pp. 239–258.

Bjornsson, H.K., Bjornsson, E.S., Avula, B., et al., 2020. Ashwagandha-induced liver injury: a case series from Iceland and the US Drug-Induced Liver Injury Network. Liver Int. 40, 825–829. Available from: https://doi.org/10.1111/liv.14393.

Brandon-Warner, E., Eheim, A.L., Foreau, D.M., 2012. Silibinin (milk thistle) potentiates ethanol-dependent hepatocellular carcinoma progression in male mice. Cancer Lett. 326, 88–95.

Cattley, R.C., Cullen, J.M., 2013. Liver and gall bladder. In: Hascheck, W.M., Rousseaux, C.G., Wallig, C.G. (Eds.), Haschek and Rousseaux's Handbook of Toxicologic Pathology. Academic Press—Elsevier, St. Louis, MO, pp. 1509–1566.

Chen, S., Teoh, N.C., Chitturi, G.C., et al., 2014. Coffee and non-alcoholic fatty liver disease; brewing evidence for hepatoprotection? Gastroenterol. Hepatol. 29, 435–441.

Crawford, J.M., 2005. Liver and biliary tract. In: Kumar, V., Abbas, A.K., Fausto, N. (Eds.), Robbins and Cotran Pathologic Basis of Disease, seventh ed. Saunders—Elsevier, St. Louis, MO, pp. 877–937.

Crescioli, G., Lombardi, N., Bettiol, A., 2018. Acute liver injury following *Garcinia cambogia* weight-loss supplementation: case series and literature review. Intern. Emerg. Med. 13 (6), 857–872.

Cullen, J.M., 2007. Liver, biliary system, and exocrine pancreas. In: McGavin, M.D., Zachary, J.F. (Eds.), Pathologic Basis of Veterinary Disease, fourth ed. Mosby—Elsevier, St. Louis, MO, pp. 393–461.

Czaja, M.J., Ding, W.X., Donohue, T.M., 2013. Functions of autophagy in normal and diseased liver. Autophagy 9, 1131–1158.

Dara, L., Hewett, J., Lim, J.K., 2008. Hydroxycut hepatotoxicity: a cases series and review of liver toxicity from herbal weight loss supplements. World J. Gastroentrol 14, 6999–7004.

Dhiman, A., Nanda, A., Ahmad, S., 2012. A recent update in research on the antihepatotoxic potential of medicinal plants. J. Chin. Integr. Med. 10, 117–127.

Dkhil, M.A., Al-Quraishy, S., Aref, A.M., 2013. The potential role of *Azadirachta indica* treatment in cisplatin-induced hepatotoxicity and oxidative stress in female rats. Oxid. Med. Cell Longev. 2013, 741871.

Duarte, S., Baber, J., Fujii, T., 2015. Matrix metalloproteinases in liver injury, repair and fibrosis. Matrix Biol. 44-46, 147–156.

Elmore, S., 2007. Apoptosis: a review of programmed cell death. Toxicol. Pathol. 35, 495–516.

El-Shafey, M.M., Abd-Allah, G.M., Mohamadin, A.M., 2014. Quercetin protects against acetaminophen induced hepatorenal toxicity by reducing reactive oxygen and nitrogen species. Pathophysiology 2014, S0928–S4680.

Fang, J., 2014. Bioavailability of anthocyanins. Drug. Metab. Rev. 46, 508–520.

Goncalves, D.F., deCarvalho, N.R., Leite, M.B., 2018. Caffeine and acetaminophen association: effects on mitochondrial bioenergetics. Life Sci. 193, 234–241.

Gurley, B.J., 2012. Pharmacokinetic herb–drug interactions (part 1): origins, mechanisms, and the impact of botanical dietary supplements. Planta Med. 78, 1378–1489.

Gurley, B.J., Fifer, E.K., Garnder, Z., 2012. Pharmacokinetic herb–drug interactions (part 2): drug interactions involving popular botanical dietary supplements and their clinical relevance. Planta Med. 78, 1490–1514.

Heber, D., Zhang, Y., Yang, J., 2014. Green tea, black tea and oolong tea polyphenols reduce visceral fat and inflammation in mice fed high-fat, high-sucrose obesogenic diets. J. Nutr. 144, 1393–2385.

Heeboll, S., Thomsen, K.L., Pedersen, T.B., 2014. Effects of resveratrol in experimental and clinical non-alcoholic fatty liver disease. World J. Hepatol. 6, 188–198.

Hosny, H., Mansour, H., Hafez, F., 2012. Protective effect of *Withania somnifera* against radiation-induced hepatotoxicity in rats. Ecotoxicol. Environ. Saf. 80, 14–19.

Jadeja, R., Devkar, R.V., Nammi, S., 2014. Herbal medicines for the treatment of nonalcoholic steatohepatitis: current scenario and future prospects. Evid. Based Compl. Alt. Med. 2014, 648308.

Jaeschke, H., 2008. Toxic responses of the liver. In: Klaasen, C.D. (Ed.), Casarett and Doull's Toxicology: The Basic Science of Poisons, eighth ed. McGraw-Hill Medical, New York, NY, pp. 557–582.

Jiang, Y., Xie, P., Chen, J., 2008. Detection of the hepatotoxic microcystins in 36 kinds of cyanobacteria Spirulina food products in China. Food Addit. Contam. Part A: Chem. Anal. Control. Expos. Risk Assess. 25, 885–894.

Ke, P.Y., Chen, S.S.L., 2012. Hepatitis C virus and cellular stress response: implications to molecular pathogenesis of liver disease. Viruses 4, 2251–2290.

Khan, M.A., Chen, H.C., Tania, M., 2011. Anticancer activities of *Nigella sativa* (black cumin). Afr. J. Tradit. Compl. Alt. Med. 8, 226–232.

Kim, T., 2016. Ginseng for liver injury: friend or foe? Medicines (Basel) 3 (4), 33.

Kim, Y.J., Choi, M.S., Park, Y.B., 2013. *Garcinia cambogia* attenuates diet-induced adiposity but exacerbates hepatic collagen accumulation and inflammation. World J. Gastreoenterol 19, 4689–4701.

Krishna, Y.R., Mittal, V., Grewal, P., 2011. Acute liver failure caused by "fat burners" and dietary supplements: a case report and literature review. Can. J. Gastroenterol. 25, 157–160.

Kumar, P., Bhandari, U., Jamadagni, S., 2013. Fenugreek seed extract inhibits fat accumulation and ameliorates dyslipidemia in high fat diet-induced obese rats. Med. Res. Int. 2013, 606021.

Kupiec, T., Raj, V., 2005. Fatal seizures due to potential herb-drug interactions with *Ginkgo biloba*. J. Anal. Toxicol. 29, 758–775.

Lai, Y.S., Chen, W.C., Ho, C.T., 2014. Garlic essential oil protects against obesity-triggered nonalcoholic fatty liver disease through modulation of lipid metabolism and oxidative stress. J. Agric. Food Chem. 62, 5897–5906.

Li, S., Takahara, T., Fujino, M., et al., 2017. Astaxanthin prevents ischemia-reperfusion injury in the steatotic liver in mice. PLoS One. 12 (11), e0187810.

Lin, S.Y., Wang, Y.Y., Chen, W.Y., 2014. Beneficial effect of quercetin on cholestatic liver injury. J. Nutr. Biochem. 25, 1183–1195.

Liu, G., Zhang, Y., Liu, C., 2014. Luteolin alleviates alcoholic liver disease induced by chronic and binge ethanol feeding in mice. J. Nutr. 144, 1009–1015.

LiverTox, 2019a. Clinical and Research Information on Drug-Induced Liver Injury. National Institute of Diabetes and Digestive and Kidney Diseases, Bethesda, MD; 2012 - Ashwagandha. <https://www.ncbi.nlm.nih.gov/books/NBK548536/>.

LiverTox 2019b Clinical and Research Information on Drug-Induced Liver Injury. National Institute of Diabetes and Digestive and Kidney Diseases, Bethesda, MD; 2012 - *Garcinia Cambogia*. <https://www.ncbi.nlm.nih.gov/books/NBK548087/>.

Loguercio, C., Festi, D., 2011. Silybin and the liver: from basic research to clinical practice. World J. Gastroenterol. 17, 2288–2301.

Madrigal-Santillan, E., Madrigal-Bujaidar, E., Alvarez-Gonzalez, I., 2014. Review of natural products with hepatoprotective effects. World J. Gastroenterol. 20, 14787–14804.

Majewski, M., 2014. *Allium sativum*: facts and myths regarding human health. Rocz. Panstw. Zakl. Hig. 65, 1–8.

Malewicz, B., Wang, Z., Jiang, C., 2006. Enhancement of mammary carcinogenesis in two rodent models by silymarin dietary supplements. Carcinogenesis 9, 1739–1747.

Marquez, F., Babio, N., Bullo, M., 2012. Evaluation of the safety and efficacy of hydroxycitric acid or *Garcinia cambogia* extracts in humans. Crit. Rev. Food Sci. Nutr. 52, 585–594.

Mathes, A.M., 2010. Hepatoprotective actions of melatonin: possible mediation by melatonin receptors. World J. Gastroenterol. 16, 6087–6097.

McGovern, A.J., Vltkovtsky, I.V., Jones, D.L., 2015. Can AST/ALT ratio indicate recovery after acute paracetamol poisoning. Clin. Toxicol. (Phila.) 53, 164–167.

Na, D.H., Ji, H.Y., Park, E.J., 2011. Evaluation of metabolism-mediated herb–drug interactions. Arch. Pharm. Res. 34, 1829–2842.

National Toxicology Program (NTP), 1992. NTP Technical report on the toxicology and carcinogenesis studies of quercetin (CAS no. 117-39-5) in F334/N rats (feed studies). U.S. Department of Health and Human Services, National Institutes of Health Publication No. 92-3140.

National Toxicology Program (NTP), 1993. NTP Technical report on the toxicology and carcinogenesis studies of turmeric oleoresin (CAS no. 8024-37-1) (Major component 79%–85% curcumin, CSA no. 458-37-7) in F334/N rats and B6C3F1 mice (feed studies). U.S. Department of Health and Human Services, National Institutes of Health Publication No. 93-3158.

National Toxicology Program (NTP), 2005. Gum guggl and some of its steroidal constituents: review of toxicological literature. U.S. Department of Health and Human Services, National Institutes of Health, Contract no. N01-ES-35515.

National Toxicology Program (NTP), 2007. NTP Technical report on the toxicology and carcinogenesis studies of genistein (CAS no. 446-72-0) in Sprague-Dawley rats (feed study). U.S. Department of Health and Human Services, National Institutes of Health Publication No. 08-4430.

National Toxicology Program (NTP), 2011. NTP Technical report on the toxicology and carcinogenesis studies of ginseng (CAS no. 50647-08-8) in F334/N rats and B6C3F1 mice (gavage studies). U.S. Department of Health and Human Services, National Institutes of Health Publication No. 11-5909.

National Toxicology Program (NTP), 2013. NTP Technical report on the toxicology and carcinogenesis studies of Ginkgo biloba extract (CAS no. 90045-36-6) in F334/N rats and B6C3F1 mice (gavage studies). U.S. Department of Health and Human Services, National Institutes of Health Publication No. 13-5920.

National Toxicology Program (NTP), 2014. NTP Toxicology studies of green tea extract in F334/NTac rats and B6C3F1 mice and toxicology and carcinogenesis studies of green tea extract in Wistar HAN[Crl:WI(Han)] rats and B5C3F1N mice (gavage studies). U.S. Department of Health and Human Services, National Institutes of Health Publication No. 14-5927.

Oketch-Rabah, H.A., Roe, A.L., Rider, C.V., et al., 2020. United States Pharmacopeia (USP) comprehensive review of the hepatotoxicity of green tea extracts. Toxicol. Rep. 7, 386–402.

Ozer, J., Ratner, M., Shaw, M., 2008. The current state of serum biomarkers of hepatotoxicity. Toxicology 245, 194–205.

Parimoo, H.A., Sharma, R., Patil, R.D., 2014. Hepatoprotective effect of *Ginkgo biloba* leaf extract on lantadenes-induced hepatotoxicity in guinea pigs. Toxicon 81, 1–12.

Patel, S.S., Beer, S., Kearney, D.L., 2013. Green tea extract: a potential cause of acute liver failure. World J. Gastroenterol. 19, 5174–5177.

Peng, W., Liu, Y.J., Wu, N., et al., 2015. Areca catechu L. (Arecaceae): a review of its traditional use, phytochemistry, pharmacology and toxicology. J. Ethnopharmacol. 164, 340–356.

Qiu, L.X., Chen, T., 2015. Novel insights into the mechanisms whereby isoflavones protect against fatty liver disease. World J. Gastroenterol. 21, 1099–1107.

Quaresma, J.A.S., Duarte, M.I.S., Vasconce, P.F.C., 2006. Midzonal lesions in yellow fever: a specific pattern of liver injury caused by direct virus action and in situ inflammatory response. Med. Hypothesis 67, 618–621.

Rafter, I., Graberg, T., Kotronen, A., 2012. Isoform-specific alanine aminotransferase measurement can distinguish hepatic from extrahepatic injury in humans. Int. J. Mol. Med. 30, 1241–1249.

Ramakrishnan, V., Gopala Reddy, A., Rajasekher Reddy, A., 2011. Evaluation of iron-induced oxidative stress and its amelioration by certain herbs in broilers. Toxicol. Int. 18, 54–57.

Reilly, C.A., Yost, G.S., 2006. Metabolism of capsaicinoids by P450 enzymes: a review of recent finding on reaction mechanism, bioactivation and detoxification processes. Drug. Metab. Rev. 38, 685–706.

Rodrigues Machado, S., Parise, E.R., de Carvalho, L., 2013. Coffee has hepatoprotective benefits Brazilian patients with chronic hepatitis C even in lower daily consumption than in American and European populations. Braz. J. Infect. Dis. 18, 170–176.

Roy, S.D., Das, S., Shil, D., 2012. Herbal hepatoprotective agents: a review. World J. Pharmaceut. Res. 1, 87–99.

Saab, S., Mallam, D., Cox, G.A., 2014. Impact of coffee on liver diseases: a systematic review. Liver Int. 34, 495–504.

Safer, A.M., Afzal, M., Hanafy, N., Mousa, S., 2015. Green tea extract diminishes hepatic fibrosis mediated by dual exposure to carbon tetrachloride and ethanol: a histopathological study. Exp. Ther. Med. 9, 787–794.

Sarma, D.N., Barrett, M.L., Chavez, M.L., 2008. Safety of green tea extracts: a systemic review by the US Pharmacopeia. Drug. Saf. 31, 469–484.

Savai, J., Varghese, A., Panidita, N., 2015. Investigation of CYP3A4 and CYP2D6 interactions with *Withania somnifera* and *Centella asiatica* in human liver microsomes. Phyther. Res. 29 (5), 785–790.

Schmidt, R., 2010. Hepatoprotective actions of melatonin: possible mediation by melatonin receptors. World J. Gastroenterol. 16, 6087–6097.

Seef, L.B., Bonokovsky, H.L., Navarro, V.J., 2015. Herbal products and the liver: a review of adverse effects and mechanisms. Gastroenterology 148, 517–532.

Shaikh, S.A., Tischer, S., Choi, E.K., et al., 2017. Good for the lung but bad for the liver? Garlic-induced hepatotoxicity following liver transplantation. J. Clin. Pharm. Ther. 42 (5), 646–648.

Shi, F., Zhao, P., Li, X., 2015. Cytotoxicity of luteolin in primary rat hepatocytes: the role of CYP3A-mediated ortho-benzoquinone metabolite and glutathione depletion. J. Appl. Toxicol. 35 (11), 1372–1380.

Shivashankara, A.R., Azmidah, A., Hanniadka, R., 2012. Dietary agents in the prevention of alcohol-induced hepatotoxicity: preclinical observations. Food Funct. 3, 101–109.

Shukla, K., Diskhit, P., Shukla, R., 2014. Hypolipidemic and antioxidant activity of aqueous extract of fruit of Withania coagulans (Stocks) Dunal in cholesterol-fed hyperlipidemic rabbit model. Indian. J. Exp. Biol. 52, 870–875.

Singh, N., Bhalla, M., de Jager, P., 2011. An overview on ashwagandha: a Rasayana (rejuvenator) of Ayurveda. Afr. J. Tradit. Compl. Alt. Med. 8, 208–213.

Soliman, M.M., Nassan, M.A., Ismail, T.A., 2014. Immunohistochemical and molecular study on the protective effect of curcumin against hepatic toxicity induced by paracetamol in Wistar rats. BMC Compl. Alt. Med. 14, 457–468.

Stalker, M.J., Hayes, M.A., 2007. Liver and biliary system. In: Maxie, M.G. (Ed.), Jubb, Kennedy, and Palmer's Pathology of Domestic Animals, sixth ed. Elsevier—Saunders, St. Louis, MO, pp. 297–388.

Stephens, C., Andrade, R.J., Lucena, M.I., 2014. Mechanisms of drug-induced liver injury. Curr. Opin. Allergy Clin. Immunol. 14, 286–292.

Teschke, R., Wolff, A., Frenzel, C., 2014. Herbal hepatotoxicity-an update on Chinese medicine preparations. Aliment. Pharmacol. Ther. 40, 32–50.

Valenti, L., Riso, P., Maxxocchi, A., 2013. Dietary anthocyanins as nutritional therapy for nonalcoholic fatty liver disease. Oxid. Med. Cell Longev. 2013, 145421.

Visavadiya, N.P., Narasimhacharya, A.V., 2011. Ameliorative effects of herbal combinations in hyperlipidemia. Oxid. Med. Cell Longev. 2011, 160804.

Wang, K., 2014. Molecular mechanisms of hepatic apoptosis. Cell Death Dis. 5, e996.

Wang, H., Guan, W., Yang, W., 2014a. Caffeine inhibits the activation of hepatic stellate cells induced by acetaldehyde via adenosine A2A receptor mediated by the cAMP/PKA/SRC/ERK1/2/P338 MAPK signal pathway. PLoS One. 9, e92482.

Wang, L., Shen, L., Ping, J., 2014b. Intrauterine metabolic programming alteration increased susceptibility to non-alcoholic adult fatty liver disease in prenatal caffeine-exposed rat offspring. Toxicol. Lett. 224, 311–318.

Wang, H., Wu, X., Lezmi, S., et al., 2017. Extract of *Gingko biloba* exacerbates liver metastasis in a mouse colon cancer xenograft model. BMC Compl. Altern. Med. 17 (1), 516.

Watkins, P.B., Seef, L.B., 2006. Drug-induced liver injury: summary of a single topic clinical research conference. Hepatology 43, 618–631.

Wei, J., Zhang, G., Zhang, X., et al., 2017. Anthocyanins delay ageing-related degenerative changes in the liver. Plant Foods Hum. Nutr. 72 (4), 425–431.

Woziwodzka, A., Golunski, G., Wryzykowski, D., 2013. Caffeine and other methylxanthines as interceptors of food-borne aromatic mutagens: inhibition of Trp-P-1 and Trp-P-2 mutagenic activity. Chem. Res. Toxicol. 26, 1660–1673.

Wu, X., Beecher, G.R., Holden, J.M., 2006. Concentrations of anthocyanins in common foods in the United States and estimation of normal consumption. J. Agric. Food Chem. 54, 4069–4075.

Xiao, Q., Sinha, R., Graubard, B.I., 2014. Inverse associations of total and decaffeinated coffee with liver enzyme levels in National Health and Nutrition Examination Survey 1999–2010. Hepatology 60, 2091–2098.

Yadav, U.C., Banquer, N.Z., 2014. Pharmacological effects of *Trigonella foenum-graecum* L. in health and disease. Pharm. Biol. 52, 243–254.

Yang, X., Schnackenberg, L.K., Shi, Q., 2014. Hepatic toxicity biomarkers. In: Gupta, R.C. (Ed.), Biomarkers in Toxicology. Academic Press—Elsevier, Ames, IA, pp. 241–260.

Yang, Y., Kim, B., Park, Y.K., 2015. Astaxanthin prevents TGFβ1-induced pro-fibrogenic gene expression by inhibiting Smad3 activation in hepatic stellate cells. Biochim. Biophys. Acta 1850, 178–185.

Yousaf, M.N., Chaudhary, F.S., Hodanazari, S.M., et al., 2019. Hepatotoxicity associated with *Garcinia cambogia*: a case report. World J. Hepatol. 11, 735–742.

Yu, J., Xiaojuan, B., Yu, B., et al., 2016. Isoflavones: anti-inflammatory benefit and possible caveats. Nutrients 8 (6), 361.

Zargar, S., 2014. Protective effect of *Trigonella foenum-graecum* on thioacetamide induced hepatotoxicity in rats. Saudi J. Biol. Sci. 21, 139–145.

Zhao, J., Shi, Z., Liu, S., 2014. Genosides Rg1 form *Panax ginseng:* a potential therapy for acute liver failure patients? Evid. Based Compl. Alt. Med. 2014, 538059.

Zheng, Y.F., Bae, S.H., Kwon, M.J., 2013. Inhibitory effects of astaxanthin, β-cryptoxanthin, canthaxanthin, lutein and zeaxanthin on cytochrome P450 enzyme activities. Food Chem. Toxicol. 59, 75–85.

Chapter 9

Nutraceuticals in renal diseases

Sharon M. Gwaltney-Brant

Introduction

Through maintenance of fluid balance, regulation of acid–base and electrolytes, concentration and elimination of waste products, secretion of regulatory peptides and hormones, and metabolism and elimination of endogenous compounds and xenobiotics, the kidneys play essential roles in coordinating the complex physiological processes involved in maintaining homeostasis. The human kidneys receive approximately 25% of cardiac output and convert approximately 1700 L of blood into 1 L of urine every day (Alpers, 2005). This high level of perfusion results in the kidney being exposed to blood-borne hazards such as toxicants, microorganisms, and endogenous compounds, any of which may trigger acute or chronic kidney injury. Renal injury may also occur through the bioactivation of toxicants within renal tubular epithelial cells. Renal damage can result in significant systemic derangements due to accumulation of waste products within the blood. The limited ability of the kidneys to regenerate following severe injury may result in long-term renal insufficiency following renal insult.

Chronic kidney disease is a common condition, and dialysis and renal transplantation place an enormous burden on the healthcare system (Yarnell, 2012). The prevalence of renal disease is growing, largely due to the increase in diabetes mellitus and hypertension within the population. Because of the tremendous reserve capacity of the kidneys and lack of clear tests to detect ongoing subclinical renal injury, chronic renal damage can slowly progress over years until the development of chronic renal insufficiency results in clinical evidence of disease. Research is ongoing regarding pharmaceutical and nutraceutical agents that may aid in the prevention or delay of the onset of chronic renal disease.

Kidney functional anatomy

The human kidneys are bean-shaped organs that rest retroperitoneally against the posterior muscular wall of the upper abdominal region, with the concave side of the organs facing medially. The left kidney is positioned slightly superior to the right kidney. Each kidney is surrounded by a thin, fibrous connective tissue capsule. Deep to the renal capsule, the highly vascular renal cortex forms a 1.2- to 1.5-cm-thick outer rim of renal tissue and receives approximately 90% of renal blood flow. The renal medulla is represented by seven cone-shaped renal pyramids aligned with bases facing the medulla and apices extending inward toward the center of the kidney. Each medullary pyramid connects to a small collecting tube (minor calyx), which merges with others to form three to four large major calyces, which in turn further merge to form the renal pelvis. The renal pelvis exits the kidney at the hilus and narrows to form the ureter.

The nephron is the functional unit of the kidney, and is composed of the renal corpuscle (Bowman's capsule and glomerulus), proximal tubule, loop of Henle, and distal tubule (Fig. 9.1). The ultimate number of nephrons is fixed at birth and is species-variable, with rats possessing approximately 35,000 nephrons per kidney compared to 1,000,000 in humans (Maxie and Newman, 2007). The nephron is highly dependent on its blood supply to perform its normal functions. Renal blood flow originates with the renal arteries, which enter the kidney at the hilus and bifurcate into anterior and posterior vessels at the hilus. These vessels further divide into interlobar arteries that course between lobes and give rise to arcuate arteries that arch between the cortex and medulla and divide into interlobular arteries. Afferent arterioles branch off of the interlobular arteries and progressively divide into 20–40 capillary loops to form glomerular tufts. These capillary loops reconnect to form the efferent arterioles that then form either a vascular network that surrounds the cortical tubules or straight vessels that extend deeply to vascularize the inner and outer medullary regions.

The unique vascular anatomy of the kidney has important implications in renal disease (Alpers, 2005). First, because the renal arteries are essentially end-arteries, occlusion of any one artery can lead to renal infarction

Nutraceuticals. DOI: https://doi.org/10.1016/B978-0-12-821038-3.00009-4

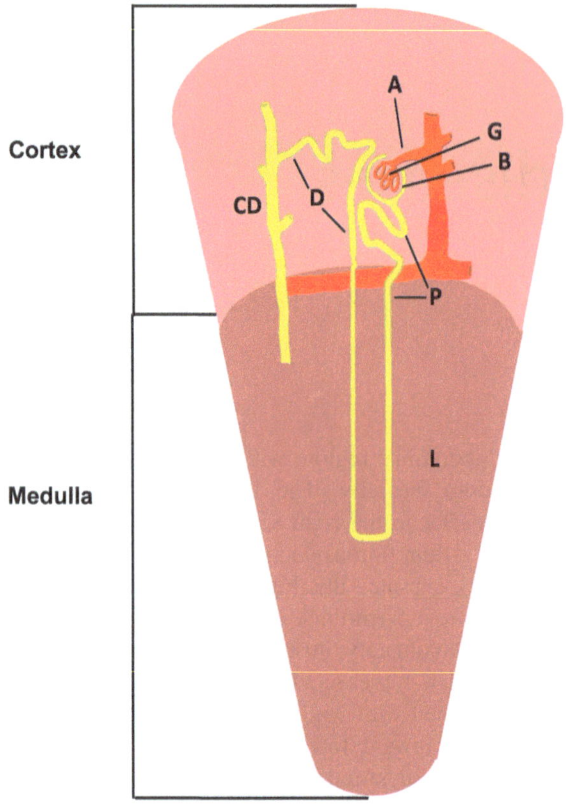

FIGURE 9.1 Expanded view of the microanatomy of the classical nephron. *A*, afferent arteriole; *B*, Bowman's capsule; *CD*, collecting duct; *D*, distal tubule; *G*, glomerulus; *L*, loop of Henle; *P*, proximal tubule.

and loss of function from the infarcted area. All blood flow beyond the level of the glomerulus depends on the flow of blood through glomerular capillaries, so any disease process that interferes with glomerular blood flow ultimately reduces the amount of blood supplied to the medulla. Because the blood within the glomerulus has relatively low oxygen content, the medullary areas tend to be oxygen-poor under normal circumstances; this makes the medulla exceptionally susceptible to ischemic injury.

The glomerulus is a tuft of branching and anastomosing capillaries supported by mesenchymal mesangial cells (Alpers, 2005), which in turn are supported on a basement membrane-like meshwork of extracellular matrix. Mesangial cells are contractile, phagocytic, capable of producing extracellular matrix and collagen, capable of proliferation, and capable of secreting a variety of biologically active mediators. Glomerular capillaries are lined by fenestrated endothelial cells resting on the basement membrane composed of a lamina rara interior, central lamina densa, and lamina rara exterior. Visceral epithelial cells (podocytes) extend interdigitating foot processes that attach to the lamina rara exterior of the basement membrane. Adjacent foot processes (pedicels) are separated by ~25-nm-wide filtration slits that are bridged by a thick diaphragm. This glomerular structure allows free passage of water and solutes through the glomerular walls and into Bowman's capsule. Other particles are filtered in a size-dependent and charge-dependent fashion, with larger and more negatively charged particles being excluded from filtration.

From Bowman's capsule, filtrate flows into the proximal renal tubule, which is divided into the proximal convoluted tubule (PCT) and the pars recta (Khan et al., 2013). The PCT has cuboidal epithelial cells with large, round central nuclei, prominent brush borders, and abundant endocytotic vesicles, lysosomes, phagolysosomes, Golgi apparatus, mitochondria, and smooth endoplasmic reticulum. The pars recta has tall brush borders and less prominent phagolysosome structures. The proximal tubule is the site of passive reabsorption of water and active reabsorption of sodium via Na^+/K^+-ATPases. Absorption of other solutes such as calcium, potassium, bicarbonate, phosphorus, glucose, amino acids, proteins, and various xenobiotics also occurs. The proximal tubule ultimately reabsorbs 60%–80% of solute and water filtered by the glomerulus, and it excretes weak organic anions and cations. A large number of cytochrome P450-dependent mixed-function oxidases mediate metabolism of a wide variety of endogenous compounds and xenobiotics. Compounds that undergo metabolic bioactivation to toxic metabolites (e.g., acetaminophen) can induce proximal renal tubular injury.

From the proximal tubule, filtrate enters the loop of Henle, which is composed of a thin-walled descending limb and a thick-walled ascending limb, both of which extend deep to the level of the outer medullary region. Some nephrons, termed long-looped nephrons, dip even further to the inner medulla. Countercurrent exchange mechanisms result in the reabsorption of ~20% of filtered water and 25% of filtered sodium and potassium (Schnellmann, 2008). Water and interstitial solutes such as electrolytes and urea are free to diffuse into the lumen of the thin descending limb. At the level of the thin ascending limb, the tubule becomes impermeable to water and urea, and electrolytes such as sodium and chloride ions are actively transported via Na^+/K^+ ATPases. Between its location in the poorly oxygenated medulla and the high oxygen demand of its Na^+/K^+-ATPases, the loop of Henle is especially susceptible to hypoxic injury (Brezis et al., 1984).

The macula densa is an area of specialized cells located between the end of the thick ascending loop of Henle and the proximal-most aspect of the distal tubule. The close proximity of this area to afferent arterioles allows communication between the macula densa and afferent arteriole of the same nephron (Khan et al., 2013). Increased intratubular solute concentration at the macula densa sends feedback signals to the afferent arteriole, triggering vasoconstriction, and reduction of the glomerular filtration rate (GFR). This feedback mechanism works to prevent massive fluid and solute loss in the face of impaired renal tubular absorption. The proximal aspect of the distal tubule functions to reabsorb sodium and potassium, with the remaining segment of the tubule responsible for elimination of excess hydrogen ions and potassium as well as absorption of additional water as needed to regulate the volume and composition of the urine. Agents that disrupt the osmotic gradient of the medullary region can disrupt the ability of the kidney to concentrate urine. Distal tubules empty into collecting ducts that progressively anastomose toward the renal papilla to ultimately empty in the renal calyces. The renal interstitium is a fibrovascular stroma that surrounds and supports the nephron. Interstitial damage can occur due to primary insult or secondarily to injury to the nephron.

The kidneys secrete a variety of regulatory peptides and hormones that maintain normal systemic homeostasis. Erythropoietin secreted by renal peritubular interstitial cells promotes red blood cell formation; chronic renal disease is frequently associated with anemia due to decreased erythropoietin secretion. Renin secreted by juxtaglomerular cells increases systemic and renal blood pressure as well as aldosterone release. A variety of renal cells produces prostaglandins, which aid in regulation of renal vascular tone, mesangial cell contractility, and processing of water and electrolytes by the renal tubules. The kidney also plays an important role in xenobiotic metabolism and/or elimination.

Renal injury and repair

The kidney can be exposed to potentially injurious agents via hematogenous spread, via ascending spread from the lower urinary tract, and via local synthesis, precipitation, or bioactivation of injurious compounds or metabolites (Newman et al., 2007). The blood can carry infectious agents such as bacteria, immunoglobins (free or complexed with antigen), thrombi or other particulates, or toxicants that can cause direct injury to renal vessels, glomeruli, or proximal tubules. Hematogenous toxicants may cause immediate injury to glomerular structures or to the PCT. Immunoglobulin complexes may adhere to glomerular structures, interfering with filtration. Other compounds with blood concentrations too low to be immediately injurious to glomerulus and proximal tubules can be concentrated to toxic levels as they progress through the tubules and cause injury further down the nephron. Some compounds may be bioactivated to toxic metabolites by tubular mesangial or epithelial cells and exert their toxic effects at the site of bioactivation. Precipitation of crystals within renal tubules, such as calcium oxalate crystals in ethylene glycol toxicosis, can cause mechanical injury to tubular epithelium. Particulates in the blood (thrombi, bacterial plaques, etc.) can lodge in arterioles or capillaries, resulting in infarction of regions supplied by those vessels. Ascending spread of bacteria or other microorganisms from the lower urinary tract can result in introduction of infection into the kidney.

Patterns of renal injury

Renal disease is often described by the region of the kidney that is initially affected: glomerular, tubular, vascular, interstitial, or tubulointerstitial. However, the interconnectedness between the nephron, vasculature, and interstitium dictates that injury to one component has adverse effects on the remaining components. With disease progression or chronicity, changes to the renal architecture tend to become more generalized until, in the end stages of renal disease, pinpointing the site of original insult is frequently impossible.

Glomerular injury may result from direct action of toxicants within the blood, entrapment of bacterial or thrombotic emboli, deposition of immune complexes on the basement membrane, or infection by microorganisms (Newman et al., 2007). Direct injury or injury secondary to inflammatory responses can result in endothelial necrosis, podocyte injury, and loss of mesangial cells and matrix. Alterations to the glomerular basement membrane due to the primary insult or secondary to cellular injury disrupt the filtration barrier of the glomerulus, allowing passage of normally excluded solutes (e.g., protein) into the renal tubules. Thickening of the glomerular basement membrane by deposition of immune complexes or other compounds also disrupts the glomerular filtration barrier, leading to membranous glomerulonephropathy.

Injury to the renal tubules is the most common manifestation of renal injury, and the proximal tubule is the most commonly affected segment of the nephron (Schnellmann, 2008). Damage to renal tubules may occur via hematogenous spread of infectious agents, ascending infections, nephrotoxicosis, and ischemia. Proximal renal tubular epithelium demonstrates a high level of endocytosis of agents that bind the brush border, and the epithelium sequesters these agents in phagolysosomes (Khan et al., 2013). When the endocytotic mechanism is overwhelmed or impeded, phagolysosome membrane integrity is lost, which results in lysosomal leakage and epithelial cell injury or death. The straight segment of the proximal tubule is the most susceptible to injury due to ischemia, transporter-associated toxicant accumulation, and bioactivation of compounds into toxic metabolites.

Interstitial fibrosis may become severe enough to interfere with normal nephron function and to contribute to a progressive loss of renal function. Inflammatory cell infiltration in the interstitium due to nephron or interstitial injury can result in edema, hemorrhage, and local injury due to cytokines or reactive oxygen species (ROS) elaborated by inflammatory cells and/or interstitial fibrosis.

Acute kidney injury

Acute renal failure is one of the most common manifestations of renal injury (Schnellmann, 2008). Acute renal failure is defined as the abrupt decrease in renal function leading to retention of nitrogenous wastes, although more recent convention is to use the term acute renal injury (AKI) to encompass the entire spectrum of renal injury, from minor alterations of serum chemistry renal parameters [i.e., blood urea nitrogen (BUN) and creatinine] to anuric renal failure (Langston, 2010).

The primary manifestation of AKI is decreased GFR, which may result from prerenal, renal, or postrenal causes, which account for 20%–80%, 10%–45%, and 5%–15%, respectively, of AKI in humans (Langston, 2010). Hypovolemia, renal vasoconstriction, and poor cardiac output are common causes of decreased GFR, whereas postrenal causes include obstruction of the ureters and lower urinary tract. Primary renal factors causing decreased GFR include renal tubular injury, glomerular injury, interstitial disease, and renal vascular compromise with ischemia–reperfusion and nephrotoxicosis causing more than 90% of primary renal disease in humans (Schnellmann, 2008). Primary renal injury from nephrotoxicants can occur due to the following: direct tubular epithelial cell injury resulting in cell necrosis, sloughing of cells, and obstruction of tubules by cellular debris; detachment of lethally injured cells from the basement membrane, resulting in leakage of filtrate across exposed basement membrane and adherence of detached cells to injured cells still attached to the basement membrane, causing tubular obstruction; damage to the glomerular filtration barrier; renal vasoconstriction causing hypoxia and ischemic necrosis of renal structures; and impairment of renal healing and repair (Counts et al., 1995).

The clinical effects of AKI or renal failure can include polydipsia, polyuria/oliguria/anuria, nausea, vomiting, lethargy, weakness, anorexia, and dehydration. More severe cases may result in halitosis, oral ulceration, renal pain, enlarged kidneys, and cardiac arrhythmias (Alpers, 2005; Langston, 2010). Clinical laboratories suggestive of AKI include elevations in BUN and serum creatinine (azotemia), hyperphosphatemia, hypokalemia or hyperkalemia, and metabolic acidosis. When azotemia is accompanied by typical clinical symptoms of AKI, the term *uremia* is used. Advanced cases of uremia may present with gastrointestinal ulceration, anemia, peripheral neuropathy, encephalopathy, and cardiac dysfunction.

The clinical course of AKI will vary with each individual circumstance, but in general most cases go through stages of initiation, maintenance, and recovery (Alpers, 2005). The initiation stage generally encompasses the first 36 h, during which time few abnormalities may be seen beyond a slight decrease in urine output or slight elevation of BUN and/or creatinine. The maintenance stage is characterized by sustained decreases in urine production (oliguria), with progressive elevations in BUN and potassium, development of metabolic acidosis, and other manifestations of uremia developing. For AKI caused by many nephrotoxicants, polyuria, not oliguria, is often the primary effect that occurs. The recovery stage is dominated by a steady increase in urine output (polyuria) accompanied by heavy losses of water and solutes. Hypokalemia now becomes an issue during this stage, and there is an increased likelihood for development of infection (Alpers, 2005).

The ability of the kidney to heal following an acute toxic insult depends on the type and severity of the insult, the amount of remaining renal function, presence and severity of secondary uremic conditions, and degree of medical intervention and supportive care provided during the acute crisis. Cases of moderate renal tubular injury with retention of tubular basement membranes have a reasonable prognosis for renal tubular regeneration provided that adequate supportive care is administered until tubules have a chance to recover. In uncomplicated acute tubular injuries, epithelial cell regeneration begins approximately 7–10 days following the renal insult, with full recovery of architecture taking 2–3 weeks or longer depending on the severity of the AKI (Maxie and Newman, 2007). Tubules with fully obstructed lumens or loss of basement membrane integrity may be unable to fully regenerate, resulting in long-term renal insufficiency that may progress to chronic renal failure.

Chronic renal failure

In chronic renal failure, decreased overall renal GFR and loss of nephrons result in hemodynamic compensation in an attempt to re-establish normal GFR levels. Blood flow and pressure to surviving nephrons increase, exacerbating pathologic conditions such as glomerular sclerosis, tubular atrophy, and interstitial fibrosis (Brenner et al., 1982). Increased intraglomerular pressure causes glomerular hypertension and hypertrophy, resulting in hyaline accumulation within glomerular capillary walls, mesangial dysfunction, abnormal mesangial matrix accumulation and microaneurysm formation, and thrombosis due to endothelial injury (Polzin, 2010). Podocytes are unable to maintain integrity of filtration slits and focal denudation of the glomerular basement membrane occurs, allowing leakage of larger solutes into the glomerular filtrate.

Chronic renal failure progresses through four stages, with some degree of overlap between the stages: diminished renal reserve; renal insufficiency; renal failure; and end-stage renal disease (Alpers, 2005). In diminished renal reserve, GFR is approximately 50% of normal, patients are asymptomatic, and BUN and serum creatinine are normal. However, individuals at this stage are at increased risk for development of azotemia with additional renal insult. Patients with renal insufficiency have GFR at 20%–50% of normal and are azotemic. Patients may also develop anemia, hypertension, polyuria, and nocturia. Sudden stress or renal insult may trigger uremia. In renal failure, GFR is less than 20%–25% of normal and kidneys have lost ability to regulate water and solute composition. Patients may develop edema, metabolic acidosis, hypocalcemia, immunosuppression, and/or uremia. Gastrointestinal (uremic gastritis, uremic enterocolitis), cardiovascular (atrial hypertension), and neurologic dysfunction (uremic encephalopathy, peripheral neuropathy) may occur (Polzin, 2010). End-stage renal disease occurs when less than 5% of renal function is available, with worsening of the signs and symptoms seen in renal failure, culminating in death.

Nutraceuticals and the kidney

The kidney has limited ability to heal following acute insult, and patients with AKI that does not fully resolve are at risk for development of progressive renal disease and, ultimately, chronic renal failure. A variety of nutraceutical products have been promoted as being able to support, protect, and/or heal renal injuries, but research to support many of these claims is lacking. Purported mechanisms by which nutraceuticals may benefit patients with chronic kidney disease include reduction of inflammation, modulation of oxidative stress, inhibition of interstitial fibrosis, promotion of renal blood flow and GFR, and stimulation of tubular regeneration. In addition to unproven efficacy, individual nutraceuticals will have their own inherent toxicity, and they may have adverse interactions with other drugs prescribed for renal disease. Discussed here are some nutraceuticals reported to have beneficial or deleterious effects on the kidney, with additional compounds summarized in Table 9.1.

Angiotensin-converting enzyme-inhibiting herbs

Pharmacologic intervention to prevent the progression of renal injury is generally restricted to the use of angiotensin-converting enzyme (ACE) inhibitors and angiotensin receptor blockers, whose antihypertensive and antifibrotic properties have been used to slow the progression of chronic kidney disease (Wojcikowski and Gobe, 2014). Several herbal/nutraceutical products have been proposed to act as ACE inhibitors, including garlic bulb (*Allium sativum*), hawthorn (*Crataegus* spp.), reishi fruit (*Ganoderma lucidum*), lespedeza (*Lespedeza capitata*), and Chinese sage (*Salvia miltiorrhiza*) (Yarnell, 2012).

Aristolochic acid

Extracts of plants of *Aristolochiaceae*, present in some Chinese herbal remedies, especially slimming agents, contain aristolochic acid, which has been implicated in "Chinese herb nephropathy" and Balkan endemic nephropathy (Ardalan et al., 2015). Aristolochic nephropathy has been characterized by small kidneys, proteinuria, tubular atrophy, and interstitial fibrosis. Some individuals have developed urothelial malignancies, Fanconi syndrome, or acute renal failure. Aristolochic acid has been shown to induce AKI, proximal renal tubular epithelial apoptosis, progressive renal interstitial fibrosis, epithelial-to-mesenchymal metaplasia, and chronic renal failure (Wojcikowski et al., 2008; Debelle et al., 2008).

Caffeine

Caffeine is the most widely consumed stimulant in the modern world, and it has been used for centuries for its stimulant and diuretic effects (Campana et al., 2014; Osswald and Schnermann, 2011). Caffeine intake has also been associated with decreased risk of kidney stones (Ferraro et al., 2014). Caffeine may enhance GFR and reduce risk of chronic kidney disease, although observational study results have been inconsistent (Cosola et al., 2018). Caffeine may induce cyst enlargement in patients with autosomal dominant polycystic kidney disease, and patients with this defect are generally counseled to limit caffeine intake (Vendramini et al., 2012). However, daily caffeine intake of ≤ 471 mg/day did not adversely affect renal cyst volume. Caffeine overdoses result in overstimulation of the central nervous and cardiovascular systems, and AKI secondary to severe rhabdomyolysis has been reported with caffeine overdosage (Campana et al., 2014).

Curcumin

Curcumin is the major component of turmeric, a ground powder from the rhizome of the *Curcuma longa* plant that is used widely throughout the world as a spice and coloring agent (NTP, 1993). Turmeric has also been widely used as a remedy for a variety of health disorders due to its purported antioxidant, antimicrobial, antiinflammatory, antiviral, anticarcinogenic, and antidiabetic properties (Soliman et al., 2014). Curcumin has been proposed as a beneficial nutraceutical in the management of AKI due to its inhibitory effect on cytokines such as transforming

TABLE 9.1 Renal effects of nutraceuticals.

Compound/source	Active constituent(s)	Renal effects	References
Astragalus membranaceus, Astragalus mongholicus	Astragalosides	Diuretic; antioxidant; antiinflammatory; attenuates complement-mediated podocyte injury; reduced proteinuria in patients with diabetic nephropathy; improved GFR	Zhong et al. (2013)
Blue green algae (*Spirulina* spp.)	Carotenoids	Antioxidant; ↓ injury due to cisplatin and gentamicin	DerMarderosian and Beutler (2014)
L-carnitine	–	Antioxidant, free radical scavenger, essential amino acid; decreased adverse effects from dialysis	Lucius (2019)
Conjugated linoleic acid	–	Worsened renal hypertrophy, glomerular and interstitial fibrosis, and inflammation in obesity-associated nephropathy	Zhan et al. (2015)
Cordyceps sinensis	H1-A	Antioxidant	Zhong et al. (2013)
Green tea and coffee	Flavonoids (catechins)	Antioxidant, antiinflammatory, anticarcinogenic; protective against diabetic nephropathy	Cosola et al. (2018)
Licorice (*Glycyrrhiza glabra*)	Glycyrrhizin	Reduced proteinuria; immunomodulation; inhibition of mesangial sclerosis and inflammation; promotes lowered potassium	Yarnell (2012)
Milk thistle (*Silybum marianum*)	Silymarin, silicristin	Antioxidant; mitigates ischemia–perfusion injury; reduced damage due to diabetic nephropathy; enhances glutathione system function; stimulates renal epithelial regeneration in vitro	Yarnell (2012)
Neem (*Azadirachta indica*)	Aqueous extract	Modulation of nitric oxide; antioxidant; reduces apoptosis of renal cells	Moneim et al. (2014)
Rhubarb (*Rheum* spp.)	Emodin, rhein	Decreased BUN, creatinine; reduced degenerative changes in glomeruli and tubules	Hamzeh et al. (2014)
Stevia (*Stevia rebaudiana*)	Steviol glycosides, primarily rebaudioside A and stevioside	Improved glomerular filtration rate, delay in cyst formation in polycystic kidney disease	Shivanna et al. (2013); Yuajit and Chatsudthipong (2016)

growth factor-β (TGF-β) (Khajehdehi, 2012). Curcumin inhibits interstitial fibroblast proliferation by reducing fibroblast proliferation and deposition of extracellular matrix by modulating TGF-β signaling (Zhou et al., 2014). Additionally, curcumin has been shown to have bimodal antioxidant activity by direct scavenging of ROS and by enhancing renal antioxidant ability via stimulation of antioxidant gene expression and increasing activity of antioxidant enzymes (Trujillo et al., 2013). Curcumin prevented apoptosis resulting from renal ischemia–reperfusion injury in animal models (Fan et al., 2017). Curcumin administration resulted in improved early graft function, decreased rate of acute rejection, and decreased neurotoxicity following renal transplantation using cadaver-donor kidneys in humans (Shoskes et al., 2005). Renoprotective effects of curcumin have been demonstrated in experimental models of chronic renal failure, ischemia–reperfusion, diabetic nephropathy, and nephrotoxicity induced by a variety of nephrotoxic compounds, including cisplatin, hexavalent chromium, gentamicin, and sodium fluoride (Chen et al., 2013; Sun et al., 2014; Trujillo et al., 2013; Ugur et al., 2015). Curcumin may be beneficial in the prevention of decline in renal function by providing antioxidant activity, improving renal blood flow and enhancing mitochondrial function (Cosola et al., 2018). In vitro studies on the effects of curcumin on clear cell renal carcinoma have shown dose-related differences in activity: at low doses, curcumin enhanced viability of renal clear cell carcinoma cells, while at high doses, curcumin induced tumor cell autophagy (Deng et al., 2018). Increased incidence of hepatocellular carcinomas occurred in mice exposed to dietary levels of 10,000 and 50,000 ppm turmeric oleoresin (75%–85% curcumin) in 2-year carcinogenicity studies (NTP, 1993).

Fenugreek

Fenugreek (*Trigonella foenum-graecum*) is a seed that has been used for centuries as both a culinary spice and for medicinal purposes in many traditional/folk medicine systems (Yadav and Banquer, 2014). The seed and its extracts have shown antioxidant and antiinflammatory properties. Fenugreek is used in folk medicine as a remedy for urolithiasis and nephrolithiasis, and fenugreek seeds reduced renal injury induced by aluminum chloride in rats (Belaïd-Nouira et al., 2013). Fenugreek seeds combined with onion showed some renoprotective activity in experimental models of diabetic nephropathy (Pradeep et al., 2019). Fenugreek was found to improve antioxidant status in kidneys exposed to the pesticide cypermethrin (Sushma and Devasena, 2010).

Quercetin

Quercetin is a flavonoid found in many plants, including citrus, berries, leafy vegetables, herbs, spices, legumes, tea, and cocoa (NTP, 1992). Quercetin is a potent scavenger of ROS, and as such has been shown to be protective against oxidative renal injury induced by ionizing radiation and valproate in rodent studies (Ozyurt et al., 2014; Chaudhary et al., 2015). Quercetin was shown to have renoprotective effects against diabetic nephropathy because it reduced oxidative stress and apoptosis in renal cells in diabetic mice (Gomes et al., 2014). Quercetin also can modulate macrophage function in the kidney, reducing inflammatory injury in the renal interstitium (Lu et al., 2018). Multiple chronic studies of quercetin in rodents have shown no evidence of hepatic carcinogenic potential (NTP, 1992).

Resveratrol

Resveratrol is a natural polyphenolic compound found in grapes, berries, wine, and peanut skins. Resveratrol is an antioxidant, and it is thought to have cytoprotective effects via antioxidant activity and activation of SIRT1, an NAD^+-dependent deacetylase that has been linked to delayed onset of age-related degeneration and prolonged life span in calorie-restricted animals (Kitada and Koya, 2013). Renoprotective effects may also come from the activity of resveratrol on the renin–angiotensin system, particularly through suppression of the angiotensin II/angiotensin II type 1 receptor axis (Jang et al., 2018). Resveratrol improved urinary protein excretion, renal dysfunction, and renal oxidative stress in diabetic rats and was renoprotective against cisplatin-induced apoptosis in murine proximal renal tubular cells. Resveratrol also reduced ischemia–reperfusion and sepsis-induced and aldosterone-induced renal injury and had antifibrotic effects in experimental models of urinary obstruction.

Concluding remarks and future directions

Further studies are needed to verify the beneficial action of select nutraceuticals on the kidney. Extensive research into nutraceuticals has revealed that many do possess potential medicinal activities that can be measured at the cellular and/or molecular level, but evaluating their clinical effectiveness has been more problematic. Many reports of positive clinical effects of nutraceuticals in humans rely heavily on anecdotal evidence or individual case reports with confounding factors such as coadministered compounds (e.g., drugs, herbals, other nutraceuticals or supplements) or lack of effective control populations. Commercially available nutraceuticals can vary considerably between brands (and often between lots within a single brand) in the quality and content of active constituents, and researchers may use extracts with differencing proportions of active constituents, making it difficult to make comparisons between studies. In many countries, herbal and nutraceutical products are poorly regulated, with the consumer at the mercy of the manufacturer in terms of product quality and safety (Seef et al., 2015; Brown, 2017a, b).

Many nutraceutical compounds may work well in vitro, but poor bioavailability or rapid clearance may interfere with their ability to exert their actions in vivo due to inability to reach effective plasma concentrations. Some compounds may achieve good plasma levels but may be unable to distribute to the target tissue due to the presence or absence of transporters to cross physiological barriers. Others may have beneficial effects on the kidney but may result in adverse effects in other tissues. For instance, silymarin shows some cytoprotective activity in the kidney, yet it can enhance the growth of mammary tumors (Yarnell, 2012; Malewicz et al., 2006).

Nutraceuticals that exhibit antioxidant, antiinflammatory, and/or antifibrotic activities in the kidney may aid in the reduction of renal injury by endogenous and exogenous nephrotoxic agents, and/or may aid in the prevention or progression of chronic renal disorders such as diabetic nephropathy, hypertension, chronic renal failure, and renal carcinogenesis. However, potential adverse effects from these compounds must be studied and documented, because alteration of renal metabolic enzymes, decreased inflammatory/immune responses, and stimulation of cellular proliferation may result in undesirable consequences such as interference with therapeutic pharmaceutical agents, decreased renal immune surveillance, or increased risk of renal neoplasia. Long-term safety and carcinogenicity data are lacking for most nutraceutical agents, and contamination or intentional adulteration of some products with pesticide residues, heavy metals, mycotoxins, toxic plants, and other potentially hazardous substances has resulted in human illness and death (Seef et al., 2015; Gupta et al., 2018). Much research remains to address these issues to ensure the safety and efficacy of nutraceutical products.

References

Alpers, C.E., 2005. The kidney. In: Kumar, V., Abbas, A.K., Fausto, N. (Eds.), Robbins and Cotran Pathologic Basis of Disease, seventh ed. Saunders: Elsevier, Philadelphia, PA, pp. 955–1021.

Ardalan, M.R., Khodaie, L., Nasri, H., et al., 2015. Risk of aristolochic acid nephropathy in Iran. Iran. J. Kidney Dis. 9, 14–17.

Belaïd-Nouira, Y., Bakhta, H., Haouas, Z., 2013. Fenugreek seeds reduce aluminum toxicity associated with renal failure in rats. Nutr. Res. Pract. 7 (6), 466–474.

Brenner, B.M., Meyer, T.H., Hotstetter, T.H., 1982. Dietary protein intake and the progressive nature of kidney disease: the role of hemodynamically mediated glomerular injury in the pathogenesis of glomerular sclerosis in angina, renal ablation and intrinsic renal disease. N. Engl. J. Med. 307, 652–659.

Brezis, M., Rosen, S., Silva, P., 1984. Transport activity modifies thick ascending limb damage in isolate perfused kidney. Kidney Int. 25, 65–72.

Brown, A.C., 2017a. Liver toxicity related to herbs and dietary supplements: online table of case reports. Part 2 of 5 series. Food Chem. Toxicol. 107, 472–501.

Brown, A.C., 2017b. Kidney toxicity related to herbs and dietary supplements: online table of case reports. Part 3 of 5 series. Food Chem. Toxicol. 107, 502–519.

Campana, C., Griffin, P.L., Simon, E.L., 2014. Caffeine overdose resulting in severe rhabdomyolysis and acute renal failure. Am. J. Emerg. Med. 32 (111), e3–e4.

Chaudhary, S., Ganjoo, P., Raiussdin, S., et al., 2015. Nephroprotective activities of quercetin with potential relevance to oxidative stress induced by valproic acid. Protoplasma 252, 208–217.

Chen, T.H., Yang, Y.C., Wang, J.C., et al., 2013. Curcumin treatment protects against renal ischemia and reperfusion injury-induced cardiac dysfunction and myocardial injury. Transplant. Proc. 45, 3546–3549.

Cosola, C., Sabatino, A., di Bari, I., et al., 2018. Nutrients, nutraceuticals, and xenobiotics affecting renal health. Nutrients 10, 808.

Counts, R.S., Nowak, G., Wyatt, R.D., et al., 1995. Nephrotoxicants inhibition of renal proximal tubule cell regeneration. Am. J. Physiol. 269, F274–F281.

Debelle, F.D., Vanherweghem, J.L., Nortier, J.L., 2008. Aristolochic acid nephropathy: a worldwide problem. Kidney Int. 74, 158–169.

Deng, Q., Liang, L., Liu, Q., et al., 2018. Autophagy is a major mechanism for the dual effects of curcumin on renal clear carcinoma cells. Eur. J. Pharmacol. 826, 24–30.

DerMarderosian, A., Beutler, J.A., 2014. The review of natural products, eighth ed. Clinical Drug Information, St. Louis, MO, pp. 1675–1679.

Fan, Y., Chen, H., Peng, H., et al., 2017. Molecular mechanisms of curcumin renoprotection in experimental acute renal injury. Front. Pharmacol. 8, 912.

Ferraro, P.M., Taylor, E.N., Gambaro, G., 2014. Caffeine intake and risk of kidney stone. Am. J. Clin. Nutr. 100, 1596–1603.

Gomes, I.B.S., Ponto, M.L., Santos, M.C.L.F.S., et al., 2014. Renoprotective, antioxidative and anti-apoptotic effects of oral low-dose quercetin in the C57BL/6J. model of diabetic nephropathy. Lipids Health Dis. 13, 184–194.

Gupta, R.C., Srivastava, A., Lall, R., 2018. Toxicity potential of nutraceuticals. In: Nicolotti, O. (Ed.), Computational Toxicology: Mehtods and Protocolas, Methods in Molecular Biology. Springer Nature, New York, pp. 367–394.

Hamzeh, S., Farokhi, F., Heydari, R., 2014. Renoprotective of hydroalcoholic extract of *Rheum ribes* root in diabetic female rats. Avicenna J. Phytomed 4, 392–401.

Jang, I.A., Kim, E.N., Lim, J.H., et al., 2018. Effects of resveratrol on the renin-angiotensin system in the aging kidney. Nutrients 10 (11).

Khajehdehi, P., 2012. Turmeric: reemerging of a neglected Asian traditional remedy. J. Nephropathol. 1, 1–22.

Khan, K.N.M., Hard, G.C., Alden, C.E., 2013. Kidney. In: Hascheck, W.M., Rousseaux, C.G., Wallig, M.A. (Eds.), Haschek and Rousseaux's Handbook of Toxicologic Pathology, third ed. Academic Press: Elsevier, Waltham, MA, pp. 1667–1773.

Kitada, M., Koya, D., 2013. Renal protective effects of resveratrol. Oxid. Med. Cell. Longev. 2013, 1–7. 568093.

Langston, C., 2010. Acute uremia. In: seventh ed. Ettinger, S.J., Feldman, E.C. (Eds.), Textbook of Veterinary Internal Medicine, vol 2. Saunders: Elsevier, St. Louis, MO, pp. 1969–1984.

Lu, H., Wu, L., Liu, L., et al., 2018. Quercetin ameliorates kidney injury and fibrosis by modulating M1/M2 macrophage polarization. Biochem. Pharmacol. 154, 203–212.

Lucius, K., 2019. Diet, botanical medicine and nutraceuticals for chronic kidney disease. Altern. Complement Ther. 25 (6), 312–319.

Malewicz, B., Wang, Z., Jiang, C., et al., 2006. Enhancement of mammary carcinogenesis in two rodent models by silymarin dietary supplements. Carcinogenesis 9, 1739–1747.

Maxie, M.G., Newman, S.J., 2007. Urinary system. In: fifth ed Maxie, M.G. (Ed.), Pathology of Domestic Animals, vol 2. Saunders: Elsevier, St. Louis, MO, pp. 425–522.

Moneim, A.E.A., Othman, M.S., Aref, A.M., 2014. *Azadirachta indica* attenuates cisplatin-induced nephrotoxicity and oxidative stress. Biomed. Res. Int. 2014, 647131.

National Toxicology ProgrAm. (NTP), 1992. NTP Technical Report on the Toxicology and Carcinogenesis Studies of Quercetin (CAS no. 117-39-5) in F334/N Rats (Feed Studies). National Institutes of Health Publication No. 92-3140. U.S. Department of Health and Human Services.

National Toxicology ProgrAm. (NTP), 1993. NTP Technical Report on the Toxicology and Carcinogenesis Studies of Turmeric Oleoresin (CAS no. 8024-37-1) (Major Component 79%85% Curcumin, CSA no. 458-37-7) in F334/N Rats and B6C3F$_1$ Mice (Feed Studies). National Institutes of Health Publication No. 93-3158. U.S. Department of Health and Human Services.

Newman, S.J., Confer, A.W., Panciera, R.J., 2007. Urinary system. In: McGavin, M.D., Zachary, J.F. (Eds.), Pathologic Basis of Veterinary Disease, fourth ed. Mosby: Elsevier, St. Louis, MO, pp. 613–691.

Osswald, H., Schnermann, J., 2011. Methylxanthines and the kidney. Handb. Exp. Pharmacol. 391–412.

Ozyurt, H., Cevik, O., Ozgen, Z., et al., 2014. Quercetin protects radiation-induced DNA damage and apoptosis in kidney and bladder tissues of rats. Free Radic. Res. 48, 1247–1255.

Polzin, D.J., 2010. Chronic kidney disease. In: seventh ed. Ettinger, S.J., Feldman, E.C. (Eds.), Textbook of Veterinary Internal Medicine, vol 2. Saunders: Elsevier, St. Louis, MO, pp. 1990–2021.

Pradeep, S.R., Barman, S., Srinivasan, K., 2019. Attenuation of diabetic nephropathy by dietary fenugreek (*Trigonella foenum-graecum*) seeds and onion (*Allium cepa*) via suppression of glucose transporters and renin-angiotensin system. Nutrition 2019, 67–68.

Schnellmann, R.G., 2008. Toxic responses of the kidney. In: Klaassen, C.D. (Ed.), Casarett and Doul's Toxicology: The Basic Science of Poisons, seventh ed. McGraw-Hill Medical, New York, pp. 583–608.

Seef, L.B., Bonokovsky, H.L., Navarro, V.J., et al., 2015. Herbal products and the liver: a review of adverse effects and mechanisms. Gastroenterology 148, 517–532.

Shivanna, N., Naika, M., Khanum, F., et al., 2013. Antioxidant, anti-diabetic and renal protective properties of *Stevia rebaudiana*. J. Diabetes Complications 27, 103–113.

Shoskes, D., Lapierre, C., Cruz-Correa, M., et al., 2005. Beneficial effects of the bioflavonoids and curcumin and quercetin on early function in cadaveric renal transplantation: a randomized placebo controlled trial. Transplantation 80, 1556–1559.

Soliman, M.M., Nassan, M.A., Ismail, T.A., 2014. Immunohistochemical and molecular study on the protective effect of curcumin against hepatic toxicity induced by paracetamol in Wistar rats. BMC Complement. Altern. Med. 14, 457–468.

Sun, L.N., Yang, Z.Y., Lv, S.S., et al., 2014. Curcumin prevents diabetic nephropathy against inflammatory responses via reversing caveolin-1 Tyr14 phosphorylation influenced TRL4 activation. Int. Immunopharmacol. 23, 236–246.

Sushma, N., Devasena, T., 2010. Aqueous extract of *Trigonella foenum graecum* (fenugreek) prevents cypermethrin-induced hepatotoxicity and nephrotoxicity. Hum. Exp. Toxicol. 29, 311–319.

Trujillo, J., Chirino, Y.I., Molina-Jijon, E., et al., 2013. Renoprotective effect of the antioxidant curcumin: recent findings. Redox Biol. 2013, 448–456.

Ugur, S., Ulu, R., Dogukan, A., et al., 2015. The renoprotective of curcumin in cisplatin-induced nephropathy. Ren. Fail. 26, 1–5.

Vendramini, L.C., Nishiura, J.L., Baxmann, A.C., et al., 2012. Caffeine intake by patients with autosomal dominant polycystic kidney disease. Braz. J. Med. Biol. Res. 45, 834–840.

Wojcikowski, K., Gobe, G.C., 2014. Use of phytochemicals as adjuncts to conventional therapies for chronic kidney disease. In: Prakash, D., Sharma, G. (Eds.), Phytochemicals of Nutraceutical Importance. CAB International, Oxfordshire, pp. 20–32.

Wojcikowski, K., Wohlmuth, H., Johnson, D.W., 2008. An in vitro investigation of herbs traditionally used for kidney and urinary system disorders: potential therapeutic and toxic effects. Nephrology 14, 70–79.

Yadav, U.C., Banquer, N.Z., 2014. Pharmacological effects of *Trigonella foenum-graecum* L. in health and disease. Pharm. Biol. 52, 243–254.

Yarnell, E., 2012. Botanical medicines used for kidney disease in the United States. Iran. J. Kidney Dis. 6, 407–418.

Yuajit, C., Chatsudthipong, V., 2016. Nutraceutical for autosomal dominant polycystic kidney disease therapy. J. Med. Assoc. Thai. 99 (suppl. 1), S97–S103.

Zhan, Y., Shi, H., Caligiuri, S.P., 2015. Trans-10, cis-12-conjugated linoleic acid worsens renal pathology and alters cyclooxygenase-derived oxylipins in obesity-associated nephropathy. J. Nutr. Biochem. 26, 130–137.

Zhong, Y., Deng, Y., Chen, Y., 2013. Therapeutic use of traditional Chinese herbal medications for chronic kidney diseases. Kidney Int. 84 (6), 1108–1118.

Zhou, X., Zhang, J., Xu, C., et al., 2014. Curcumin ameliorates renal fibrosis by inhibiting local fibroblast proliferation and extracellular matrix deposition. J. Pharmacol. Sci. 126, 344–350.

Chapter 10

Nutraceuticals in gastrointestinal disorders

Ramesh C. Gupta, Robin B. Doss, Aryamitra Banerjee, Rajiv Lall, Ajay Srivastava and Anita Sinha

Introduction

The naturally occurring bioactive compounds that bridge the gap between food products and drugs are called nutraceuticals. Nutraceutical products may provide health and medicinal benefits, including the prevention and treatment of diseases in addition to their natural nutritional value. The word "nutraceutical" was coined by Stephen L. DeFelice, founder and chairman of the Foundation of Innovation Medicine. Although it is a hybrid of the words "nutrition" and "pharmaceutical," nutraceuticals are different from pharmaceuticals and the name is applied to products ranging from dietary supplements (minerals, vitamins, and antioxidant supplements), herbal products, probiotics, processed foods, and essential oils. Although in the modern era the use of traditional medicine has been surpassed by pharmaceutical products, renewed interest in the general population for alternative medicine is apparent from the fact that the US market reached $250 billion by 2018, having stood at $35 billion in 2010.

Nutraceuticals show their biological effect by participating as substrates, cofactors, or inhibitors of enzymes in biochemical reactions, influencing the absorption and stability of nutrients as well as scavenging and eliminating toxic compounds along the gastrointestinal tract (GIT) and being selective growth factors and fermentation substrates for beneficial oral and GI bacteria as well as inhibitors of deleterious intestinal bacteria. They may act systemically as well as locally (Vitaglione et al., 2008; Dillard, 2000). This chapter describes the health benefits to the GIT and risks associated with nutraceuticals.

Pharmaceuticals versus nutraceuticals

Pharmaceuticals are synthetic molecules or peptides/biologics that are designed and manufactured to alter the function of specific enzymes or receptors known to be dysfunctional. Generally, pharmaceuticals are used to treat a disease or its symptoms. Even though pharmaceuticals are designed to interact with a particular target and provide a specific biological response, they also interact with other unintended targets and are associated with various side effects. Nutraceuticals are mostly used to prevent, rather than treat, diseases. However, there is growing evidence that targeted nutraceuticals may help prevent and/or cure diseases including osteoarthritis, metastatic cancer, and GI disorders (Akhtar and Haqqi, 2012; Romano et al., 2012; Kuppusamy et al., 2014; Li et al., 2015; Talukder, 2019). Some of the most common GI system disorders and their pharmaceutical treatments are shown in Table 10.1. These pharmaceutical options to treat various GI disorders, like chemotherapy, radiation, nonsteroidal antiinflammatory drugs (NSAIDs), antibiotics, corticosteroids, and immune-modulators are associated with severe side effects.

When the GI tract is exposed to radiation, this may lead to low blood cell counts, diarrhea, nausea, vomiting, proctosigmoiditis, and inflammation of the rectum and sigmoid colon. Chemotherapy leads to hair loss, loss of appetite, blood disorders, neurological disorders, pain, and fatigue. Chemotherapy can also produce permanent damage to the heart, lungs, liver, kidneys, central nervous system, or reproductive system. Histamine type 2 (H_2)-receptor antagonists are known for their interaction with cytochrome P450 (CYP450), leading to headache, abdominal pain, constipation, and diarrhea (Penston and Wormsley, 1986). The use of 5HT-modulators is restricted for use in GI disorders due to associated adverse effects (Gershon, 1991; Camilleri, 2009). Linaclotide, a peptide agonist of guanylate cyclase 2C, promotes bowel movement by increasing smooth muscle contraction by inhibiting activation of colonic sensory neurons and activating colonic motor neurons (Thomas and Allmond, 2013). Linaclotide is associated with diarrhea, whereas

Nutraceuticals. DOI: https://doi.org/10.1016/B978-0-12-821038-3.00010-0

TABLE 10.1 Common gastrointestinal conditions and their pharmaceutical treatments.

No.	Gastrointestinal disorders	Pharmaceutical treatments
1	Leukoplakia	Surgical removal, chemotherapy
2	Oral cancer	Radiation, chemotherapy
3	Cancer of the esophagus	Radiation, chemotherapy
4	Stomach cancer	Radiation, chemotherapy
5	Gastritis	H2-receptor blockers, antacids, corticosteroids
6	Irritable bowel disease (Crohn's disease and ulcerative colitis)	Antibiotic, NSAIDs, corticosteroids, immune-modulators
7	Peptic ulcer	H2-receptor blockers, antacid, proton pump inhibitors
8	Appendicitis	Antibiotics, NSAIDs
9	Peritonitis	Antibiotics, analgesics
10	Pancreatitis	IV fluid therapy, analgesics, enzyme supplements

NSAIDs, nonsteroidal antiinflammatory drugs.

lactulose produces gas, diarrhea, stomach upset, and stomach cramps. Proton pump inhibitors (PPI) may increase the chances of osteoporosis, pneumonia, and enteric infection, and are also known for producing drug interactions (Maffei et al., 2007). Corticosteroids affect a plethora of biological targets, and even their short-term use is associated with cutaneous effects, electrolyte abnormalities, hypertension, hyperglycemia, pancreatitis, hematologic, and immunologic systems. Their long-term use may be associated with more serious conditions like osteoporosis, aseptic joint necrosis, adrenal insufficiency, gastrointestinal, hepatic, and ophthalmologic effects, hyperlipidemia, growth suppression, and possible congenital malformations (Buchman, 2001).

By turn of the 21st century, a new wave of interest emerged dealing with research into testing naturally occurring ingredients, extracts, and formulations for treating various GIT conditions. Most nutraceuticals are very affordable in comparison to pharmaceuticals, but they cannot be patented. The most attractive feature of nutraceuticals is that, along with their therapeutic effect, they also act as a general tonic (Prakash and Sharma, 2012) (especially Ayurvedic medicines) and are not associated with severe side effects, although some people can have allergies to one or more natural ingredients.

The systemic effect of phytochemicals depends on their bioavailability and/or the activity of metabolites and biotransformation products. Their effects and mechanisms have been determined primarily based on in vitro studies using cell or tissue culture with high concentrations of these compounds, which is difficult to achieve in vivo via oral exposure or administration. Therefore extrapolating in vitro results to an in vivo system may not be appropriate, and although this has been done in the past, the limited bioavailability was not considered. For example, resveratrol demonstrated several biological effects in vitro at concentrations of 1 to 100 μM, a concentration highly unlikely to be obtained in vivo by consuming moderate amounts of red wine (containing resveratrol at 0–10 ppm) due to the rapid metabolism and urinary excretion of the molecule (Vitaglione et al., 2005; Baur and Sinclair, 2006). Fortunately, this limitation does not apply to the GIT. Hence, there are numerous studies that have been published showing a positive response to polyphenolic plant-based compounds in various conditions through GIT targets (Akhtar and Haqqi, 2012; Romano et al., 2012; Kuppusamy et al., 2014; Li et al., 2015).

Naturally derived food substances have been studied in various disease conditions as therapeutic alternates to pharmaceuticals. The roles of vitamin E, selenium, vitamin D, green tea, soy, lycopene, curcumin, and various other polyphenol and short-chain and long-chain fatty acid-containing products from plants, animals, and bacterial origins are documented as nutraceuticals and widely studied in human and animal health (Brower, 1998; Singh et al., 2012; Gupta, 2016; Gupta et al., 2019).

Nutraceutical bioenhancers in gastrointestinal health and diseases

Most nutraceuticals are unstable, poorly soluble, and absorbed in the GI tract, thereby limiting their bioavailability (Kang et al., 2009; Jhanwar and Gupta, 2014; Pandit et al., 2019). The dissolution and inability to permeate the GI wall are additional factors further restricting the bioavailability of nutrients and nutraceuticals (Yu et al., 2012; Ning et al., 2018; Li et al., 2019) leading to

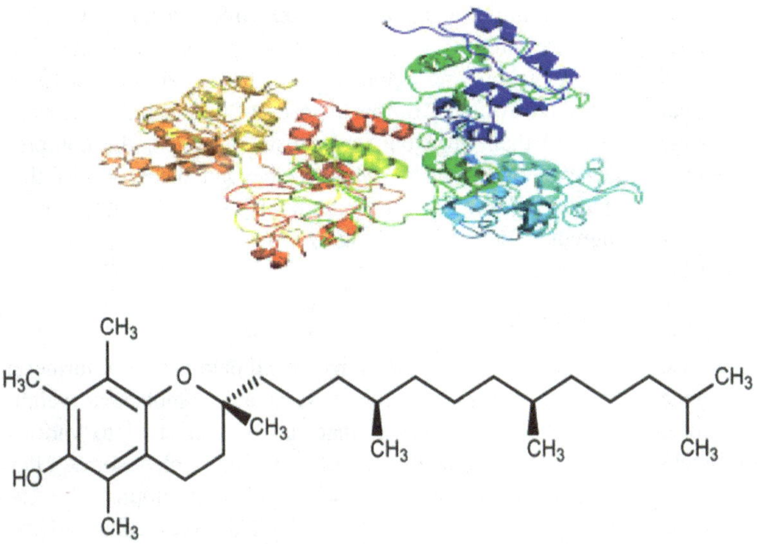

FIGURE 10.1 Structural formula of iron transport tocopheryl polyethylene glycol succinate. *Courtesy of Srivastava A., Lall R., Talukder J., DuBourdieu D., Gupta R.C. (2019). Iron transport tocopheryl polyethylene glycol succinate in animal health and diseases. Molecules, 24: 4289. doi: 10.3390/molecules24234289.*

very low concentrations of bioactive phytoconstituents at the target site(s), thereby producing minimal pharmacological and therapeutic effects. Microbiome dysbiosis in the GIT may disrupt the GI barrier system with impaired absorption of nutrients and phytomedicines, leading to ill health in humans and animals (Yu et al., 2012; Zhang et al., 2015; Redfern et al., 2017; Rowland et al., 2018; Feng et al., 2019; Hughes, 2019; Pilla and Suchodolski, 2020; Rodriguez et al., 2019). In fact, failure of the gut (abnormal colonization, infection that originated in the gut, bacterial translocation and inflammation) can not only cause disease(s) of the GIT, but may even lead to multiple organ dysfunction syndrome (Galland, 2014; Goulet, 2015; Boulangé et al., 2016; Ley et al., 2016; Tremlett et al., 2017; Sircana et al., 2019).

In the past, a number of drug-delivery systems, including liposomal formulations, nanoparticles, and micellar formulations, have been developed and investigated for enhanced absorption and bioavailability. By manipulating the formulation, enhanced absorption of drugs and herbal medicines can be achieved via micronization, nanocrystals/nanosuspensions, solid solutions, self-dispersing solid solutions with surfactants, lipid solutions, self-emulsifying drug-delivery systems, and surfactant-cosolvent systems (Gupta et al., 2013; Yurdakok-Dikmen et al., 2018).

Bioenhancers of herbal origins, such as piperine, peppermint oil, curcumin, ginseng, and many others are indicated as having greater bioavailability for therapeutic drugs and phytomedicines (Singh et al., 2009; Drabu et al., 2011; Patil et al., 2011; Keserwani and Gupta, 2013; Ajazuddin et al., 2014; Jain and Patil, 2015; Yurdakok-Dikmen et al., 2018; Pandit et al., 2019). In recent years, vitamin E polyethylene glycol succinate (TPGS) has been found to increase the bioavailability of pharmaceuticals, nutraceuticals, nanomedicines, and nutrients, which are otherwise poorly soluble and less bioavailable (Zou and Gu, 2013; Li et al., 2019). Vitamin E TPGS appears to be a powerful tool in the formulation of poorly soluble lipophilic compounds, as it enhances the absorption and bioavailability via efficient micellar uptake into enterocytes. However, the poor water solubility of vitamin E limits its absorption in the GIT. Recently, in order to achieve greater stability, bioavailability, additional biological/pharmacological effects, and better therapeutic efficacy, the TPGS molecule was further improvised by combining it with the iron transport (IT) protein of ovotransferrin (Srivastava et al., 2019). The structural formula of IT tocopheryl polyethylene glycol succinate (ITPGS) is shown in Fig. 10.1.

ITPGS is both fat- and water-soluble and it exerts multiple biological and pharmacological effects, such as antioxidative, antiinflammatory, antimicrobial, immunomodulatory, healthy gut, and overall health improvement (Srivastava et al., 2019). Like TPGS, ITPGS may exert additional properties, such as a surfactant, apoptogenic, anticancer, immunomodulator, neuroprotector, and reversal of multidrug resistance by inhibiting P-glycoprotein activity (Youlk et al., 2005; Collnot et al., 2007, 2010; Mi et al., 2012; Yang et al., 2018). In essence, ITPGS has great potential as a bioenhancer and nutraceutical, and merits further investigation for other biological and pharmacological properties supported by in vivo studies and clinical trials.

Role of nutraceuticals in cancer associated with the gastrointestinal tract

There are several pharmaceutical approaches available for cancer treatment, but there is no cure available to treat metastatic tumors, mostly due to the heterogeneity of cancer cells. In recent years, there have been numerous studies suggesting that natural extracts and substances derived from them can provide a potential breakthrough in cancer therapy. These substances with natural origins were found to interact with important pathways in cancer cell generation and propagation. Natural agents like curcumin, epigallocatechin-3-gallate (EGCG), and isoflavones alter the expression of miRNAs that are known to regulate the cellular signaling and biological behavior of cancer cells. They possess the ability to regulate DNA transcription in tumors by controlling the DNA-damaging factors. Phytoestrogens have been found to regulate IL-6 gene expression, control DNA-damaging factors in cancer cells, and regulate DNA transcription in tumors (Dijsselbloem et al., 2004; Penner et al., 2005; Sarkar, 2010).

Nutraceutical compounds and other probiotic bacteria are also known to interfere with intracellular signal transduction pathways like NF-κB and AP-1 by interacting with host cells and modulating mitogen-activated protein kinase, protein kinase C, and phosphatidylinositol 3-kinase (Penner et al., 2005; Li et al., 2010; Sarkar, 2010). These findings are also in synergy with various epidemiological and preclinical studies which suggest that nutraceuticals derived from medicinal plants such as tea, garlic, ginger, soybean, and others may act as chemopreventives via apoptotis. Flavonoids extracted from black tea induce apoptosis in human stomach cancer cells, whereas flavonoids from green tea induce apoptosis in human colon cancer cells. Stilbene from grapes, peanuts, and pines induces apoptosis in various cancer cells, including esophageal carcinoma, whereas sulfur-containing compounds form *Brassica* plants induce apoptosis in various cancer cells, including colon cancer cells (Alexander and Kuang, 2004). Curcumin, EGCG, and resveratrol from grapes have been shown to inhibit the expression of cyclooxygenase-2 (COX-2) and nitric oxide synthase (NOS) by blocking NF-κB activation (Surh et al., 2001). Various fruits containing antioxidants were found to be effective in suppressing the growth of various colon cancer cells and other intestinal tumors (Kuppusamy et al., 2014).

Carotenoids and their metabolites have been found to be very useful in the treatment of colon cancer, mostly due to their free radical scavenging activity. Carotenoids have been shown to induce anticancer effects through poly (ADP-ribose) polymerase cleavage, induction of caspase inhibitors, and nuclear condensation (Yoshida et al., 2007). Various extracts from marine and dietary supplements (omega fatty acids and vitamins) are also found to modulate the targets of importance in cancer therapy. Grape seed extract containing proanthocyanins (potent antioxidants) is known to provide protection against gastric and colon cancers. Selenium and vitamin E-containing diets, Mediterranean diets (containing antioxidants in the form of hemocyanins), and kombucha tea (free radical scavenging property) are also known to provide protection against gastric, intestinal, and colon cancer. There is an abundance of literature available on the efficacy of nutraceuticals in cancer and their probable mechanism of action, the comprehensive discussion of which is beyond the scope of this chapter. Some of the commonly studied nutraceutical ingredients in cancer are shown in Fig. 10.2.

Curcumin

Curcumin is a versatile nutraceutical obtained from turmeric powder of *Curcuma longa*. It is a low-molecular-weight hydrophobic polyphenol that can be converted to various salts. The development of curcumin phytosomes has remarkably increased its bioavailability and improved intestinal absorption. This improved GI delivery enhanced the chances for exploiting the use of curcumin in GI disorders (Anand et al., 2007). Curcumin possesses a wide range of biological properties and has been studied extensively for anticancer drug-like properties in in vitro and in vivo systems (Surh and Chun, 2007; Rajasekaran, 2011). Curcumin may slow the growth of GI cancers including esophageal, mouth, intestinal, stomach, and colon cancers, probably due to its increased bioavailability in the GI tract (Anand et al., 2007). Its anticancer properties are mainly attributed to its antioxidative (oxygen and nitrogen free radical scavenging) and antiinflammatory potential (Menon and Sudheer, 2007). Curcumin also exerts its effects by several other mechanisms. Curcumin appears to be responsible for anti-DNA damage activity. Curcumin also suppresses the production of cytokines and the activation of NF-κB, and it inhibits the inducible NOS pathways involved in GI cancer. Additionally, curcumin modulates various targets and pathways (involved in GI cancer) like Akt/mTOR, Ca^{2+}/calmodulin, COX-2, epidermal growth factor receptor, IGF-1R, c-Jun-N-terminal kinases, IL-8, and Wnt/b-catenin (Thangapazham et al., 2006; Rajasekaran, 2011).

Flavonoids

Flavonoids are a group of compounds that consist of two phenolic benzene rings linked to a heterocyclic pyran or pyrone. Flavonoids are one of the most thoroughly studied anticarcinogens, although much of the evidence has been derived from in vitro studies. The United States Department of Agriculture published a database for the flavonoid content of various foods in 2003. This lists the flavonoid values for 19 compounds from 255 foods. A comprehensive list of various food sources for flavonoids and their major ingredients can be found at http://www.nal.usda.gov/fnic/foodcomp.

The ability of flavonoids to interact with cellular and molecular pathways like cell cycle control and apoptosis

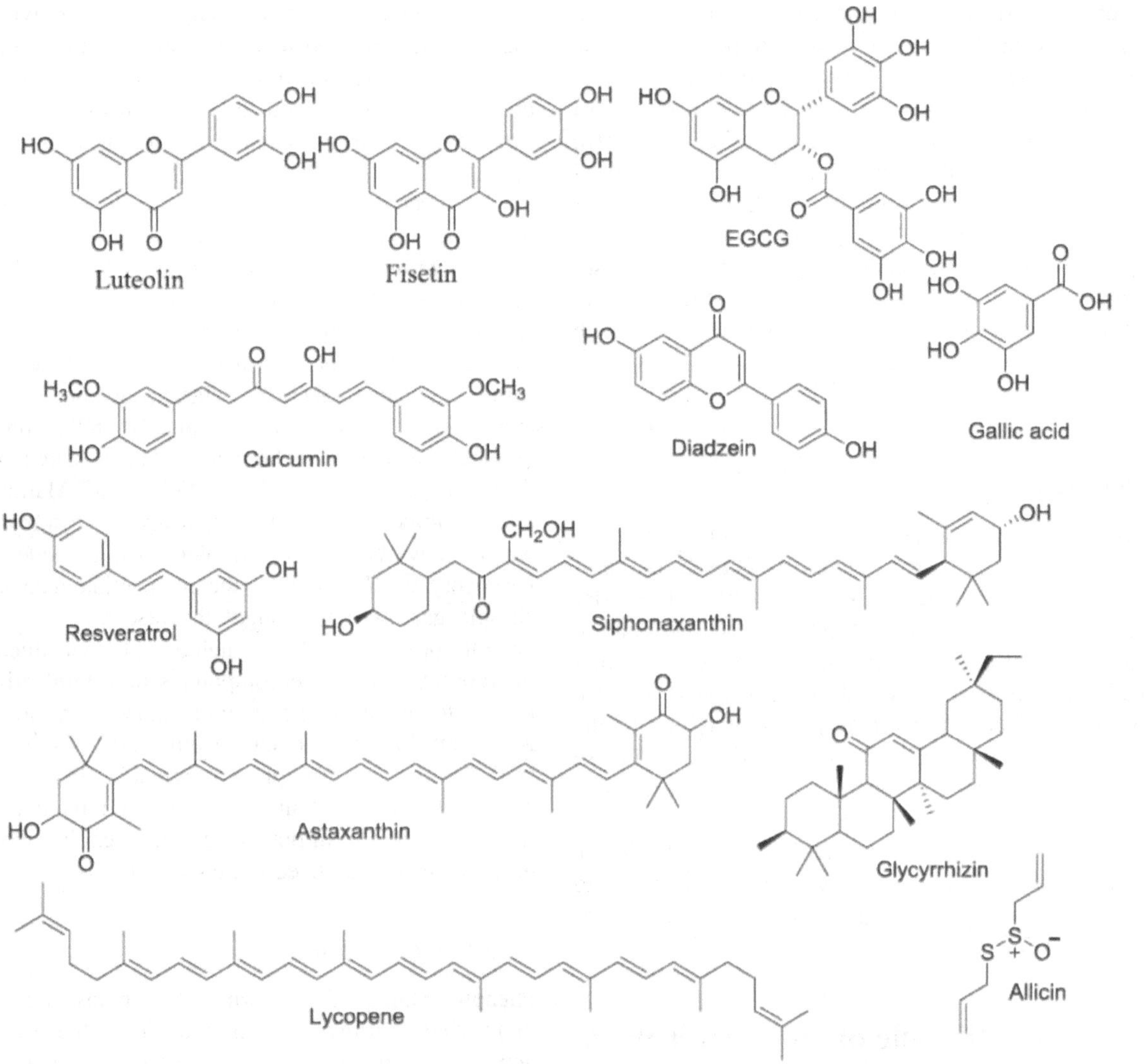

FIGURE 10.2 Clinically important nutraceuticals for treating gastrointestinal disorders.

makes them a potential candidate for GI cancer treatment. Flavonoids like apigenin, chrysin, and kaempferol are COX-2 inhibitors and can suppress COX-2 transcription by mechanisms that include activation of the peroxisome proliferator-activated receptor, gamma transcription factor, and inhibition of NF-κB expression (Baumann et al., 1980; Liang et al., 1999, 2001). Genistein suppresses the growth of colon cancer cells by attenuating activity of the PI3K/Akt pathway. Genistein provides anticolon cancer activity by blocking the expression of Wnt5a through the Wnt signaling pathway. Genistein is also known to affect the expression of estrogen receptors and tumor suppressor genes, suggesting involvement of membrane receptors and tumor suppressors in the antiproliferative effects of genistein (Qi et al., 2011). EGCG reduced tumors in a rat model of esophageal cancer in a dose-dependent manner and significantly downregulated the gene expression of both cyclin D1 and COX-2. It also inhibited DNA methyltransferase activity (Li et al., 2002). Flavopiridol has been shown to induce G1 or G2/M cell cycle arrest and apoptosis and to reduce protein levels of cyclin D1 and retinoblastoma, which are crucial regulators of the G0/G1 cell cycle checkpoint, in several GI cell lines (Carlson et al., 1996; Schrump et al., 1998).

Halocynthiaxanthin

Halocynthiaxanthin is a dietary carotenoid. In combination with peridinin, it sensitizes colon cancer cell lines to tumor necrosis factor-related apoptosis-inducing ligand. This combination regulates programmed cell death in colon cancer cells via poly(ADP-ribose) polymerase cleavage, induction of caspase inhibitors, and nuclear condensation (Yoshida et al., 2007). Halocynthiaxanthin may also induce apoptosis via inhibiting Bcl2 expression and may provide protection against various cancers, including colon cancer. Astaxanthin is another keto-carotenoid that belongs to the phytochemical class of terpenes. It is one of the most potent free radical scavengers known (Satoh, 2016), and is reported to exert protective effects against ulcer formation in murine models. Kim et al. (2016) reported that astaxanthin inhibits proliferation by interrupting cell cycle progression in KATO-III and SNU-1 gastric cancer cells, which may be due to inhibition of the phosphorylation of extracellular signal-regulated kinases and the enhanced expression of $p27^{kip-1}$. Palozza et al. (2009) demonstrated that astaxanthin inhibited the growth of HCT-116 colon cancer cells by arresting cell cycle progression and by promoting apoptosis.

Glycyrrhizin

Glycyrrhizin (GA) is the main ingredient of licorice (*Glycyrrhiza glabra*) extract. Licorice has been traditionally prescribed for treating conditions like asthma, dry cough, dyspepsia, and other "pectoral diseases," and it was also thought to be effective in preventing thirst. In a clinical trial, Raveendra et al. (2012) found licorice extract very effective against dyspepsia. It is one of the oldest and most commonly prescribed traditional Chinese medicines. Due to its apoptotic effect on tumor cells, GA has emerged as an attractive drug candidate for cancer therapy, including colon cancer (Hibasami et al., 2005). GA is also known to inhibit angiogenesis, probably due to its inhibitory activity toward reactive oxygen species (ROS) generation.

Gastritis and the role of nutraceuticals in its treatment

Gastritis is inflammation of the lining of the stomach, and it is the most common illness associated with the stomach. It occurs due to infection with *Helicobacter pylori* and/or the effect of NSAIDs. It can also be caused by irritation due to excessive alcohol, smoking, cocaine, severe illness, autoimmune problems, radiation therapy, and Crohn's disease (CD). Gastritis can be diagnosed in its acute phase, during which superficial inflammation of the stomach may cause nausea, pain, and discomfort in the upper abdomen. In the developed stage, it is termed "chronic gastritis," which may present with symptoms like pain in the upper abdomen, feeling of fullness, and loss of appetite. Gastritis due to duodenogastric bile reflux and pharmaceuticals like NSAIDs is also known as reactive or chemical gastritis, which is defined as a foveolar elongation, tortuosity, and hypercellularity of the gastric surface epithelium. It may also present with edema, vasodilatation, congestion of gastric lamina propria, and a paucity of inflammatory cells.

PPIs such as omeprazole (OPZ) and its derivatives are most commonly used to treat gastritis. OPZ is known to produce side effects and drug interaction and is also associated with gastric gland toxicity. However, this toxicity can be reversed by vitamin C administration (Kohler et al., 2010).

Antiulcer drugs are used as comedications to protect the gastric mucosa from damage due to steroidal and nonsteroidal drugs. This is especially useful during long-term pain and antiinflammatory therapies. Drugs like PPIs, histamine type 2 receptor antagonists, sucralfate, or acid neutralizers are used for neutralization of gastric acidity. The triple eradication therapy (clarithromycin, amoxicillin, and OPZ) provides approximately 80% protection, although recently its efficacy has been shown to have declined due to *H. pylori* (Wolle and Malfertheiner, 2007). Moreover, these therapies involve too many pills, which may be associated with severe side effects. Discovery of *H. pylori* as the main etiological agent of chronic gastritis has brought the research focus back to natural products, which include phytochemicals and microorganisms. These compounds of natural origins are known to modulate the immune mucosal response, H^+/K^+ pump, histamine release from mast cells, and mucus wall structure, or through prostaglandin (PG), myeloperoxidase (MPO), and nitric oxide (NO) regulation. The results obtained from published preclinical in vivo studies may also be extrapolated to human subjects.

Phenolic compounds

Phenolic compounds (known as flavonoids) are available in a variety of drinks and edible plants such as strawberry, apple, tea, red wine, beer, and others. They have been reported to exert biological properties such as antiviral, antiinflammatory, antihistaminic, antioxidant, and free radical scavenging (Kahraman et al., 2003). The gastroprotective effect of flavonoids may also be attributed to its effectiveness in increasing the endogenous PG, reduction in histamine secretion, scavenging oxygen-derived free radicals, and even to gastric mucus stimulation (Tan et al., 2002; Rocha Lapa et al., 2007). These properties may be responsible for restoring gastric mucosal integrity by a dynamic balance and homeostasis between epithelial cell renewal and cellular apoptosis.

Plant origin flavonoid substances like Sophoradin (*Sophora tonkinensis*) extract, Amaranth seed extract, grapefruit seed extract, and capsaicin (extract of the chili pepper) are reported to provide protection against gastritis by increasing gastric microcirculation to enhance the expression of constitutive NOS and release NO and neuropeptides (Zayachkivska et al., 2005). The flavonoid-rich extract of *Syngonanthus bisulcatus* was found to show gastroprotective effects via mucus and PG, which are integral parts of the GI mucosa's cytoprotective mechanisms against aggressive factors.

Quercetin, a flavonoid commonly found in a variety of vegetables, fruits, and beverages such as tea and red wines, is known for its potential to prevent gastric injuries in rats. It prevents gastric damage from the deleterious effects of activated neutrophil infiltration by inhibiting MPO activity (Kahraman et al., 2003). Quercetin can also increase superoxide dismutase (SOD) activity to reduce lipid peroxidation and protein carbonyl compounds to provide protection to the gastric mucosa from inflammation (Kahraman et al., 2003; Coşkun et al., 2004). Ethanol consumption is associated with increased erythrocyte and tissue malondialdehyde (MDA), increased serum NO concentration, and a reduction in antioxidant enzyme activities of SOD, glutathione peroxidase, and catalase. Quercetin treatment decreased the elevated MDA and NO and also increased the reduced antioxidant enzyme activities. Therefore, quercetin may protect against ethanol-induced gastric ulcers, probably decreasing oxidative stress and increasing antioxidant enzyme activity (Coşkun et al., 2004).

Wogonin (5,7-dihydroxy-8-methoxyflavone) is a strong antiinflammatory compound that targets arachidonic acid metabolism and inhibits the activation of 5-lipoxygenase and COX-2. Wogonin was found to be as efficacious as rebamipide (a well-known drug prescribed clinically for the treatment of gastritis and gastric ulcer) in the prevention of alcohol stomach injury (Park et al., 2004).

Curcumin

Curcumin-based therapy was found to be associated with clinical amelioration and a statistically significant decrease in gastric inflammation (Di Mario et al., 2007). Curcumin is also known to inhibit the growth of *H. pylori* via inhibition of the shikimate pathway, necessary for synthesis of aromatic amino acids in bacteria (Han et al., 2006).

Compounds targeting the adhesin–mucin interaction

Infection occurs when bacterial adhesins located on the outer cell wall interact with mucosal glycoproteins and epithelial mucins. This mechanism has been recently exploited to target the adhesive process of *H. pylori*. Compounds such as polysaccharides are found to interact with the bacterial adhesins before adhesin–mucin, thus avoiding the infection process (Kusters et al., 2006; Wittschier et al., 2009).

Because adhesion to epithelial cells is the most essential step of the infectious process for almost all bacterial pathogens, many efforts have been aimed at developing antiadhesin therapy. Sialyllactose (NeuAc[K2-3]Gal[L1-4]Glc), an inhibitor of the sialic acid-specific adhesin of *H. pylori*, was found to reduce the load of bacteria in monkeys. Cranberry juice has also been found to be useful in treating *H. pylori* infection via inhibition of sialic acid-specific adhesion of *H. pylori* to human gastric mucus and human erythrocytes (Burger et al., 2000). Similarly, root extracts of *Pelargonium sidoides* (*Geraniaceae*), which contain a polysaccharide fraction, EPs 7630 with antiadhesive activity against *H. pylori*, was found to be useful for treating acute respiratory infections (Wittschier et al., 2007).

G. glabra L. root is a rich source of polysaccharides composed of arabinose, galactose, glucose, and glucuronic acid and is able to interact with the outer membrane surface adhesins of *H. pylori*, avoiding its adhesion to mucus (Wittschier et al., 2007).

Nutraceuticals in irritable bowel disease

Ulcerative colitis (UC) and CD are collectively known as inflammatory bowel disease (IBD). The etiology of these conditions is currently unclear, but the hypotheses for causation of these conditions include genetic susceptibility, environmental risk factors, chronic immune responses to members of the intestinal microflora, as well as failed immune regulatory mechanisms.

The pathophysiology of IBD involves continued intestinal inflammation, presence of proinflammatory cytokines, increased ROS, and tissue injury often triggered by luminal bacteria. Avenues of natural product therapy in IBD include modulation of mediators involved in the inflammatory process, altering luminal bacteria, modifying the immune response, and rejuvenation of intestinal healing (Baumgart and Sandborn, 2007).

In terms of therapeutic agents for IBD, 5-aminosalicylic acid (5-ASA) is a well-established therapy in the management of UC, although in CD its involvement is less clear (Criscuoli et al., 2013). The NF-κB pathway is an important target. In IBD, the presence of proinflammatory cytokines such as TNF-α causes phosphorylation of Iκβ, resulting in activation of NF-κB, initiating the synthesis of more proinflammatory cytokines. 5-ASA inhibits TNF-α-induced stimulation of Iκβ phosphorylation, disrupting a critical signal transduction pathway involved in the onset and progression of chronic inflammation (Fang, 1999).

Aloe vera gel, an extract of *Aloe barbadensis*, is commonly used in the treatment of IBD (Rahimi et al., 2009; Hilsden et al., 2003). In a double-blind, randomized, controlled trial (RCT), 4 weeks of aloe vera gel administration to patients with active UC showed they had greater symptom alleviation than placebo, although this effect was not statistically significant (Langmead et al., 2004).

Andrographis paniculata has demonstrated a significant reduction in the transcriptional activity of NF-κB and decreased secretions of proinflammatory cytokines, including TNF-α and IL-6, both of which have been shown to be elevated in patients with IBD (Chiou et al., 2000; Chao et al., 2010). In a double-blind RCT ($n = 224$), greater clinical response was encountered in UC patients receiving *A. paniculata* treatment than placebo (Sandborn et al., 2013). Additionally, similar results were observed in a clinical trial comparing the efficacy of *A. paniculata* to conventional mesalazine therapy (Tang et al., 2011).

In animal studies, boswellic acid, the active component of *Boswellia serrata*, caused leukocyte recruitment and, hence, protected the intestinal mucosa from the inappropriate immune reaction in IBD (Anthoni et al., 2006; Hartmann et al., 2012). When *B. serrata* extract (BSE) was given as 400-mg capsules compared with placebo in patients with IBD, it was found that a statistically significant difference in remission rates between BSE and placebo groups in patients with collagenous colitis occurred (Madisch et al., 2007). However, in a 52-week trial, it was observed by Holtmeier et al. (2011) that remission rates did not show statistical significance between control and treatment groups ($P = .85$). BSE treatment was well tolerated and demonstrated long-term safety, as Holtmeier et al. (2011) suggested that those treated with BSE experienced less adverse events than those treated with placebo ($P = .087$).

Turmeric is predominant in the diet in Asian countries and is known to improve colonic morphology by counteracting the generation of damaging ROS produced by inflamed colon cells (Irving et al., 2011). Curcumin has also been demonstrated to inhibit the secretion of inflammatory cytokines by modulating the activation of NF-κB in patients with IBD (Atreya et al., 2008). Curcumin causes suppression of NF-κB activation by causing a blockage in the phosphorylation of inhibitory factor Iκβ kinase. This downregulates the expression of downstream proinflammatory eicosanoids (Atreya et al., 2008). Additionally, it blocks the action of proinflammatory TNF-α, primarily through the inhibition of the production of TNF transcription factors, leading to transcriptional repression (Aggarwal et al., 2013). In a pilot study of five patients with UC and five with CD, symptoms improved over a 2-month period of curcumin administration (Holt et al., 2005). In a multicenter RCT study, patients administered curcumin in conjunction with 5-ASA showed significantly improved symptoms compared to placebo with 5-ASA. Curcumin-treated patients also had a lower relapse rate 6 months after the study (Hanai et al., 2006). In a prospective phase 1 study in adults using a significantly higher dose of curcumin to treat cancer (8 g/day), no toxic effect was encountered (Cheng et al., 2001), which confirmed the safety of the dosage used to treat IBD.

Twenty-three patients with active distal UC in a double-blind RCT were given oral wheatgrass juice or placebo for up to 4 weeks (Ben-Arye et al., 2002). Patients treated with wheatgrass juice showed greater reductions in a composite clinical disease activity index, severity of rectal bleeding, and the doctor's global assessment than those in the placebo group. No side effects were reported.

Studies suggested the efficacy in UC for a germinated barley food product (GBF) consisting mainly of dietary fiber and glutamine-rich protein that are believed to act primarily as a prebiotic (Araki et al., 2001; Bamba et al., 2002; Fukuda et al., 2002; Kanauchi et al., 2002). Eleven patients given GBF for 4 weeks as an adjunctive treatment showed a greater decrease in clinical disease activity than nine patients given conventional therapy alone. In a follow-up study, 24 weeks of treatment for 21 patients with GBF together with continuing 5-ASA and steroid therapy reduced rectal bleeding and nocturnal diarrhea. Adjunctive GBF also produced a lower relapse rate over 12 months when given to 22 patients with UC in remission than did conventional therapy in 37 patients (Hanai et al., 2006). GBF was well tolerated and appeared safe in all three reports.

Polyphenols are phytochemicals present in food substances produced from plants that are believed to play a biologically active role and have been shown to be potentially immune-modulating (Shapiro et al., 2007). For IBD, polyphenols cause downregulation of inflammatory mediators and NF-κB. Several polyphenols have been identified, five (resveratrol, epigallocatechin, curcumin, quercetin, and BSE) in particular have shown proof of benefit to animals and humans with IBD.

It has been proposed that supplementation with omega-3 fatty acids may be beneficial in treating or preventing relapse of chronic inflammatory diseases. For IBD, in vitro and in vivo animal studies have shown that omega-3 fatty acids can effectively prevent and treat mouse models of colitis (Shoda et al., 1995; Arita et al., 2005; Meister and Ghosh, 2005; Whiting et al., 2005; Bassaganya-Riera and Hontecillas, 2006; Hudert et al., 2006). Previously, it was reported that the use of an enteric-coated formulation for CD found a lower relapse rate for the fish oil group than the control group (28% compared with 69%; $P < .001$) (Belluzzi et al., 1996). However, the available data are insufficient to draw conclusions about the effects of omega-3 fatty acids on clinical, endoscopic, or histologic scores or induced remission or relapse rates (Lorenz et al., 1989; O'Morain et al., 1989; Vilaseca et al., 1990; Hillier et al., 1991; Hawkey et al., 1992; Jorquera Plaza et al., 1997; Belluzzi et al., 2000; Arslan et al., 2002; Simopoulos, 2002; Razack and Seidner, 2007).

Vitamin D deficiency is common in patients with IBD, even when the disease is in remission (Bikle, 2007). Vitamin D and vitamin D receptor have been shown to be physiologic regulators of intestinal inflammation in IBD (Bikle, 2007). Vitamin D deficiency occurs frequently in both forms of IBD.

The microbial environment is very important in the establishment of IBD; therefore, targeting of the microbiota presents an option for therapeutic intervention (Fedorak and Madsen, 2004a,b; Rioux and Fedorak, 2006). Probiotics are usually considered as live microorganisms in that, when consumed in adequate quantities, they offer a health benefit to the host (Shanahan, 2004a,b, c; Sheil et al., 2004). They have been used in the treatment of numerous inflammatory conditions, including UC and CD (Madsen et al., 1999; Steidler et al., 2000; O'Mahony et al., 2001; Shibolet et al., 2002; Dieleman et al., 2003; McCarthy et al., 2003; Shiba et al., 2003; Moon et al., 2004; Gaudier et al., 2005; Herias et al., 2005; Moller et al., 2005; Rachmilewitz et al., 2004).

Based on results from the treatment of experimental colitis with VSL#3, *Lactobacillus* GG, and other strains, several clinical trials have been executed for both CD and UC. Although mixed results were observed for benefiting CD as either an induction or maintenance adjunct to standard medical therapy, probiotics have demonstrated benefits in UC for both induction and continued remission of disease (Kruis et al., 1997; Venturi et al., 1999; Rembacken et al., 1999; Guslandi et al., 2003; Ishikawa et al., 2003; Borody et al., 2003; Kruis et al., 2004; Kato et al., 2004; Furrie et al., 2005).

Nutraceuticals in peptic ulcer

Peptic ulcer is a common disease in which the gastric mucosa is damaged and perforations lead to bleeding. A number of factors are responsible for this gastric ulcer disease, among which 70%–80% are due to infection with *H. pylori*, whereas the use of NSAIDs accounts for approximately 25% of gastric ulcer cases.

Anthocyanosides, obtained from *Vaccinium myrtillus*, has demonstrated significant preventive and curative antiulcer activity by influencing the biosynthesis of mucopolysaccharides, thus improving the efficiency of the mucus barrier at the gastric level (Cristoni and Magistretti, 1987; Magistretti et al., 1988).

An ethanol extract of celery (*Apium graveolens*) has also been used as a drug against gastric ulcer in rats. At doses of 250 and 500 mg/kg body weight, the ethanolic extract of celery showed antigastric ulcer activity against indometacin, cytodestructive agents (80% ethanol, 0.2 M NaOH, and 25% NaCl), and cold-restraint stress-induced ulceration in rats by protecting the gastric mucosa and suppressing the basal gastric secretion in rats through its antioxidant potential (Al-Howiriny et al., 2010).

In animals, bilberry also decreased the incidence and severity of experimentally induced ulcers (Magistretti et al., 1988).

The roots of *Bupleurum falcatum* (a species of flowering plant in the *Apiaceae* family) have shown weak antiulcerogenic activity in a pylorus ligated ulcer model (Shibata et al., 1973). Also, the polysaccharide fraction of *B. falcatum* possesses potent antiulcer activity against HCl/ethanol-induced lesions in mice (Yamada et al., 1989).

Curcumin has shown antiinflammatory and antioxidant properties as it scavenges ROS and regulates matrix metalloproteinase activity to exert antiulcer activity (Swarnakar et al., 2005). Curcumin (diferuloylmethane) exhibits antiinflammatory activity by suppressing PG synthesis (Huang et al., 1991). Turmeric root extract has been found to relieve pain from biliary dyskinesia during a double-blinded study (Niederau and Gopfert, 1999) and improve endoscopic healing of peptic ulcers as well as symptoms of patients with nonulcer dyspepsia (Prucksunand et al., 2001).

Extensive work has been performed regarding various medicinal attributes of green tea (*Camellia sinensis*). The prophylactic actions of the tea seed-derived triterpene saponins against ethanol-induced gastric mucosal lesions (Yoshikawa et al., 2005; Morikawa et al., 2006) and black tea extract against various ulcerogens (Maity et al., 1995, 2001) have been reported in rat models. Theaflavin (TF) is a major active component of black tea. TF healed indometacin-induced gastric ulcer by its antioxidative properties, synthesis of PGE2, and enhancement of mucin secretion. TF also helps to suppress various inflammatory modulators in ulcer margin (Adhikary et al., 2011). eNOS modulation may be another possible pathway of gastric ulcer healing by TF (Adhikary et al., 2011).

Licorice-derived compounds elevate the PG levels and promote mucus secretion from the stomach, thus increasing the life span of surface cells of the stomach and presenting an antipepsin activity that ultimately leads to ulcer healing (Duke, 1985). Licorice extracts have demonstrated the ability to accelerate the healing of GI ulceration (especially in the stomach and esophagus), possibly due to their antioxidant effect. The antiulcer drug carbenoxolone, a succinate derivative of glycyrrhetinic acid, has become the preferred form of licorice used to promote healing of ulcers (Turpie et al., 1969; Dehpour et al., 1994; Bafna and Balaraman, 2005).

Saponins isolated from the fruit of *Kochia scoparia* have shown gastroprotective properties via activation of mucous membrane protective factors (Saito et al., 1977; Matsuda et al., 1998).

Lagenaria vulgaris, a popular vegetable, has shown antiulcer activity through an antioxidative pathway by modulation of glutathione level (Mehta et al., 2000).

The root of *Panax ginseng* is widely used clinically for the treatment of GI disorders as well as as an erythropoietic and tonic (Sun et al., 1992).

Phyllanthus emblica L. or *Emblica officinalis* Gaertn is one of the strongest rejuvenatives among Indian medicinal plants due to its anticancer (Khan et al., 2002; Lambertini et al., 2004), antiinflammatory (Ihantola-Vormisto et al., 1997), and antioxidative properties (Bandyopadhyay et al., 2000). A clinical study has also demonstrated its significant healing effect on gastric syndrome (Sharma, 1971). Previous reports present that the ethanolic extract of fruits of *P. emblica* showed antiulcer activity in a dose-dependent manner, mainly through its antioxidative properties by modulation of IL-10 level (Chatterjee et al., 2011). A fraction of the active component of the ethanolic extract of fruits of *P. emblica* has also shown antiulcer activity by the modulation of COX (Chatterjee et al., 2012a) or the NOS pathway (Chatterjee et al., 2012b).

Picrorhiza kurroa is a small hairy perennial herb that is a major ingredient of many indigenous medicinal preparations traditionally used for dyspepsia, bilious fever, and chronic dysentery. The most important active constituents of the roots and rhizomes of *P. kurroa* are the iridoid glycosides, picrosides I, II, and III, and kutkoside. Picroside I and kutkoside have been shown to possess potent gastroprotective properties, mainly through their antioxidative activity (Banerjee et al., 2009).

The antiulcerogenic activity of the leaf oil of *Piper betle* has been reported (Bhattacharya et al., 2007; Majumdar et al., 2002). The best antiulcer activity is demonstrated by one of its active components, allyl pyrocatechol, mainly by modulation of arginase metabolism and the shift of cytokine balance (Yadav et al., 2009).

The resinous exudate from *Pistacia lentiscus* has been used by traditional healers for the relief of upper abdominal discomfort, gastralgia, dyspepsia, and peptic ulcer (Al-Said et al., 1986). Experiments showed a significant reduction in the intensity of gastric mucosal damage induced by pyloric ligation, aspirin, phenylbutazone, reserpine, and restraint cold stress.

Pteleopsis suberosa showed protective action against ethanol-induced and indometacin-induced gastric mucosa damage due to the stimulation of PG synthesis in clinical trials (De Pasquale et al., 1995; Germano et al., 1998).

Shilajit (asphaltum or mineral pitch) has been used in traditional medicine in the treatment of bronchial asthma, diabetes, genitourinary infections, wound healing, and stomach ulcer (Goel et al., 1990). 4-Methoxy-6-carbomethoxy biphenyl from Shilajit caused changes in the mucosa and provided resistance against the effect of ulcerogens and also against shedding of mucosal cells.

Silymarin is a flavonolignan complex present in the milk thistle, *Silybum marianum*, which has been found to be effective in some types of experimentally induced gastric ulcers. The antiulcerogenic effect of silymarin could be related to its inhibitory mechanism on the lipoxygenase pathway, avoiding leukotriene synthesis (Alarcon de la Lastra et al., 1992).

Flavonoids derived from sophoradin, isolated from the root of *Sophora subprostrata*, are known to exhibit gastroprotective and ulcer-healing properties (Kyogoku et al., 1979).

Petersson et al. (2007) and Bjorne et al. (2004) showed that spinach creates a thicker layer of mucus lining of the gastric wall, in turn providing better protection from the hydrochloric acid secreted in gastric juice and reducing the risk of ulcers in rats.

Parts of the *Tectona grandis* tree cause an effect on the protein content of gastric juice, and may possibly reverse aspirin-induced changes in peptic activity, protein, and sialic acid (Goel et al., 1987).

Sanyal et al. (1961, 1963) studied the antiulcer activity of *Plantain banana* against ulcers induced by phenylbutazone, prednisolone, and restraint stress. Further, Best et al. (1984) and Elliott and Heward (1976) described the antiulcer activity of a variety of preparations of dried unripe *P. banana* against aspirin-induced ulcers in rats. Further studies indicated that banana increases the mucosal defense by promoting mucus secretion. This increases the mucoprotein content of the mucosa and decreases the shedding of cells and leakage of protein in gastric secretions in response to ulcerogenic agents, and promotes healing, as shown by the increase in thymidine uptake by gastric mucosal cells and by causing a concentration-dependent increase in the eicosanoid stimulation in the incubates of human gastric and colonic mucosa (Goel et al., 1986, 1989). It was proven endoscopically that 70% of duodenal ulcers healed after 12 weeks of treatment with banana powder as compared to approximately 16% with placebo.

Mowrey and Clayson (1982) discovered that *Zingiber officinale* is highly effective in motion sickness management and also significantly reduces GI distress in human subjects. It also has marked antiulcer activity (al-Yahya et al., 1989).

Nutraceuticals in chronic or prolonged inflammations like appendicitis

Antiinflammatory nutraceuticals have been shown to be effective in various complications/conditions resulting from chronic or prolonged inflammation, namely, rheumatoid or osteoarthritis, psoriasis, gastritis, meningitis, UC, and appendicitis. This includes omega-3 polyunsaturated fatty acids, soy isoflavonoids, plants, spices, and chitosan that work against the proinflammatory agents (Dijsselbloem et al., 2004; Goggs et al., 2005; Srinivasan, 2005; Penner et al., 2005; Cheeke et al., 2006; Li et al., 2010; Sarkar, 2010).

Nutraceuticals in other gastrointestinal conditions

Bacterial peritonitis and other infective conditions have been shown to be strongly counteracted by the use of tea polyphenols (Sajilata et al., 2008).

A pilot study was performed to evaluate the effect of oral curcumin with piperine on pain and the markers of oxidative stress in patients with tropical pancreatitis (TP). Oral curcumin with piperine could reverse lipid peroxidation in patients with TP, although this did not correspond with improvement in pain (Durgaprasad et al., 2005).

Concluding remarks and future directions

Nutraceuticals are generally considered to be safer than conventional pharmaceutical therapy, which encourages patients to turn to alternative integrative interventions to alleviate clinical symptoms and maintain remission. Further studies to establish better standardization, quality control, and process development for nutraceuticals are important if we expect these treatments to be added to standard medical strategies for the treatment of different GI ailments in the near future.

References

Adhikary, B., Yadav, S.K., Roy, K., 2011. Black tea and theaflavins assist healing of indomethacin-induced gastric ulceration in mice by antioxidative action. Evid. Based Complement. Alternat. Med. 2011 (11).

Aggarwal, B.B., Gupta, S.C., Sung, B., 2013. Curcumin: an orally bioavailable blocker of TNF and other pro-inflammatory biomarkers. Br. J. Pharmacol. 169 (8), 1672–1692.

Ajazuddin, A.A., Alexander, A., Qureshi, A., et al., 2014. Role of herbal bioactives as a potential bioavailability enhancer for active pharmaceutical ingredients. Fitoterapia 97, 1–14.

Akhtar, N., Haqqi, T.M., 2012. Current nutraceuticals in the management of osteoarthritis: a review. Ther. Adv. Musculoskelet. Dis. 4 (3), 181–207.

Alarcon de la Lastra, C., Martin, M.J., Marhuenda, E., 1992. Gastric anti-ulcer activity of silymarin, a lipoxygenase inhibitor, in rats. J. Pharm. Pharmacol. 44 (11), 929–931.

Alexander, G., Kuang, Y.C., 2004. Nutraceuticals, apoptosis, and disease prevention. Nutrition 20, 95–102.

Al-Howiriny, T., Alsheikh, A., Alqasoumi, S., 2010. Gastric antiulcer, antisecretory and cytoprotective properties of celery (*Apium graveolens*) in rats. Pharm. Biol. 48 (7), 786–793.

Al-Said, M.S., Ageel, A.M., Parmar, N.S., 1986. Evaluation of mastic, a crude drug obtained from *Pistacia lentiscus* for gastric and duodenal anti-ulcer activity. J. Ethnopharmacol. 15 (3), 271–278.

al-Yahya, M.A., Rafatullah, S., Mossa, J.S., 1989. Gastroprotective activity of ginger *Zingiber officinale* rosc., in albino rats. Am. J. Chin. Med. 17 (1-2), 51–56.

Anand, P., Kunnumakkara, A.B., Newman, R.A., et al., 2007. Bioavailability of curcumin: problems and promises. Mol. Pharm. 4, 807–818.

Anthoni, C.L.M., Rijcken, E., Vowinkel, T., 2006. Mechanisms underlying the antiinflammatory actions of boswellic acid derivatives in experimental colitis. Am. J. Physiol. Gastrointest. Liver Physiol. 290, G1131–G1137.

Araki, Y., Andoh, A., Fujiyama, Y., 2001. Germinated barley foodstuff exhibits different adsorption properties for hydrophilic versus hydrophobic bile acids. Digestion 64 (4), 248–254.

Arita, M., Yoshida, M., Hong, S., 2005. Resolvin E1, an endogenous lipid mediator derived from omega-3 eicosapentaenoic acid, protects against 2,4,6-trinitrobenzene sulfonic acid-induced colitis. Proc. Natl. Acad. Sci. U.S.A. 102 (21), 7671–7676.

Arslan, G., Brunborg, L.A., Froyland, L., 2002. Effects of duodenal seal oil administration in patients with inflammatory bowel disease. Lipids 37 (10), 935–940.

Atreya, I., Atreya, R., Neurath, M.F., 2008. NF-kappaB in inflammatory bowel disease. J. Int. Med. 263 (6), 591–596.

Bafna, P.A., Balaraman, R., 2005. Anti-ulcer and anti-oxidant activity of pepticare, a herbomineral formulation. Phytomedicine 12 (4), 264–270.

Bamba, T., Kanauchi, O., Andoh, A., 2002. A new prebiotic from germinated barley for nutraceutical treatment of ulcerative colitis. J. Gastroenterol. Hepatol. 17 (8), 818–824.

Bandyopadhyay, S.K., Pakrashi, S.C., Pakrashi, A., 2000. The role of antioxidant activity of *Phyllanthus emblica* fruits on prevention from indomethacin induced gastric ulcer. J. Ethnopharmacol. 70 (2), 171–176.

Banerjee, D., Bhattacharya, S., Bandyopadhyay, S.K., 2009. *Picrorhiza kurroa*—phytochemical and pharmacological evaluation. Rec. Progr. Med. Plants 24, 69–89.

Bassaganya-Riera, J., Hontecillas, R., 2006. CLA and n-3 PUFA differentially modulate clinical activity and colonic PPAR-responsive gene expression in a pig model of experimental IBD. Clin. Nutr. 25 (3), 454–465.

Baumann, J., von Bruchhausen, F., Wurm, G., 1980. Flavonoids and related compounds as inhibition of arachidonic acid peroxidation. Prostaglandins 20 (4), 627–639.

Baumgart, D.C., Sandborn, W.J., 2007. Inflammatory bowel disease: clinical aspects and established and evolving therapies. Lancet 369 (9573), 1641–1657.

Baur, J.A., Sinclair, D.A., 2006. Therapeutic potential of resveratrol: the *in vivo* evidence. Nat. Rev. Drug Discov. 5 (6), 493–506.

Belluzzi, A., Brignola, C., Campieri, M., 1996. Effect of an enteric-coated fish-oil preparation on relapses in Crohn's disease. N. Engl. J. Med. 334 (24), 1557–1560.

Belluzzi, A., Boschi, S., Brignola, C., 2000. Polyunsaturated fatty acids and inflammatory bowel disease. Am. J. Clin. Nutr. 71 (1 Suppl.), 339S–342SS.

Ben-Arye, E., Goldin, E., Wengrower, D., 2002. Wheat grass juice in the treatment of active distal ulcerative colitis: a randomized double-blind placebo-controlled trial. Scand. J. Gastroenterol. 37 (4), 444–449.

Best, R., Lewis, D.A., Nasser, N., 1984. The anti-ulcerogenic activity of the unripe plantain banana (*Musa* species). Br. J. Pharmacol. 82 (1), 107–116.

Bhattacharya, S., Banerjee, D., Bauri, A.K., 2007. Healing property of the *Piper betel* phenol, allylpyrocatechol against indomethacin-induced stomach ulceration and mechanism of action. World J. Gastroenterol. 13 (27), 3705–3713.

Bikle, D.D., 2007. What is new in vitamin D: 2006–2007. Curr. Opin. Rheumatol. 19 (4), 383–388.

Bjorne, H.H., Petersson, J., Phillipson, M., 2004. Nitrite in saliva increases gastric mucosal blood flow and mucus thickness. J. Clin. Invest. 113 (1), 106–114.

Borody, T.J., Warren, E.F., Leis, S., Surace, R., 2003. Treatment of ulcerative colitis using fecal bacteriotherapy. J. Clin. Gastroenterol. 37 (1), 42–47.

Boulangé, C.L., Neves, A.L., Chilloux, J., et al., 2016. Impact of gut microbiota on inflammation, obesity, and metabolic disease. Genome Med. 8, 42.

Brower, V., 1998. Nutraceuticals: poised for a healthy slice of the healthcare market? Nat. Biotechnol. 16 (8), 728–731.

Buchman, A.L., 2001. Side effects of corticosteroid therapy. J. Clin. Gastroenterol. 33 (4), 289–294.

Burger, O., Ofek, I., Tabak, M., 2000. A high molecular mass constituent of cranberry juice inhibits *Helicobacter pylori* adhesion to human gastric mucus. FEMS Immunol. Med. Microbiol. 29 (4), 295–301.

Camilleri, M., 2009. Serotonin in the gastrointestinal tract. Curr. Opin. Endocrinol. Diabetes Obes. 16 (1), 53–59.

Carlson, B.A., Dubay, M.M., Sausville, E.A., 1996. Flavopiridol induces G1 arrest with inhibition of cyclin-dependent kinase $(CDK)_2$ and CDK_4 in human breast carcinoma cells. Cancer Res. 56 (13), 2973–2978.

Chao, W.W., Kuo, Y.H., Lin, B.F., 2010. Anti-inflammatory activity of new compounds from *Andrographis paniculata* by NF-kappaB transactivation inhibition. J. Agric. Food Chem. 58 (2010), 2505–2512.

Chatterjee, A., Chattopadhyay, S., Bandyopadhyay, S.K., 2011. Biphasic effect of *Phyllanthus emblica* L. Extract on NSAID-Induced ulcer: an antioxidative trial weaved with immunomodulatory effect. Evid. Based Complement. Alternat. Med. 2011.

Chatterjee, A., Chatterjee, S., Das, S., 2012a. Ellagic acid facilitates indomethacin-induced gastric ulcer healing via COX-2 up-regulation. Acta Biochim. Biophys. Sin. (Shanghai) 44 (7), 565–576.

Chatterjee, A., Chatterjee, S., Biswas, A., 2012b. Gallic acid enriched fraction of *Phyllanthus emblica* potentiates indomethacin-induced gastric ulcer healing via e-NOS-dependent pathway. Evid. Based Complement. Alternat. Med. 2012.

Cheeke, P.R., Piacente, S., Oleszek, W., 2006. Anti-inflammatory and anti-arthritic effects of *Yucca schidigera*: a review. J. Inflamm. (Lond.) 3, 6.

Cheng, A.L., Hsu, C.H., Lin, J.K., 2001. Phase I clinical trial of curcumin, a chemopreventive agent, in patients with high-risk or pre-malignant lesions. Anticancer Res. 21 (4B), 2895–2900.

Chiou, W.F., Chen, C.F., Lin, J.J., 2000. Mechanisms of suppression of inducible nitric oxide synthase (iNOS) expression in RAW 264.7 cells by andrographolide. Br. J. Pharmacol. 129, 1553–1560.

Collnot, E.M., Baldes, C., Wempe, M.F., et al., 2007. Mechanism of inhibition of P-glycoprotein mediated efflux by vitamin E TPGS: influence of ATPase activity and membrane fluidity. Mol. Pharmacol. 4, 465–475.

Collnot, E.M., Baldes, C., Schaefer, U.F., et al., 2010. Vitamin E TPGS P-glycoprotein inhibition mechanism: influence on conformational flexibility, intracellular ATP levels, and role of time and site access. Mol. Pharm. 7, 642–651.

Coşkun, Ö., Kanter, M., Armutçu, F., Çetin, K., 2004. Protective effects of quercetin, a flavonoid antioxidant, in absolute ethanol-induced acut gastric ulcer. Eur. J. Gen. Med. 1 (3), 37–42.

Criscuoli, V.M.I., Orlando, A., Cottone, M., 2013. Mesalazine for the treatment of inflammatory bowel disease. Expert Opin. Pharmacother. 14 (12), 1669–1678.

Cristoni, A., Magistretti, M.J., 1987. Antiulcer and healing activity of *Vaccinium myrtillus* anthocyanosides. Farmaco Prat. 42 (2), 29–43.

De Pasquale, R., Germano, M.P., Keita, A., 1995. Antiulcer activity of *Pteleopsis suberosa*. J. Ethnopharmacol. 47 (1), 55–58.

Dehpour, A.R., Zolfaghari, M.E., Samadian, T., 1994. The protective effect of liquorice components and their derivatives against gastric ulcer induced by aspirin in rats. J. Pharm. Pharmacol. 46 (2), 148–149.

Di Mario, F.C.L., Nouvenne, A., Stefani, N., 2007. A curcumin-based 1-week triple therapy for eradication of *Helicobacter pylori* infection: something to learn from failure? Helicobacter 12, 238–243.

Dieleman, L.A., Goerres, M.S., Arends, A., 2003. *Lactobacillus* GG prevents recurrence of colitis in HLA-B27 transgenic rats after antibiotic treatment. Gut 52 (3), 370–376.

Dijsselbloem, N., Vanden Berghe, W., De Naeyer, A., 2004. Soy isoflavone phytopharmaceuticals in interleukin-6 affections. Multi-purpose nutraceuticals at the crossroad of hormone replacement, anti-cancer and anti-inflammatory therapy. Biochem. Pharmacol. 68 (6), 1171–1185.

Dillard, C.J.G., 2000. Phytochemicals: nutraceuticals and human health. J. Sci. Food Agric. 80, 1744–1756.

Drabu, S., Khatri, S., Babu, S., et al., 2011. Use of herbal bioenhancers to increase the bioavailability of drugs. Res. J. Pharm. Biol. Chem. Sci. 2, 107–119.

Duke, J.A., 1985. Handbook of Medicinal Herbs. CRC Press, Boca Raton, FL, 1985.

Durgaprasad, S., Pai, C.G., kumar, V., 2005. A pilot study of the antioxidant effect of curcumin in tropical pancreatitis. Indian J. Med. Res. 122 (4), 315–318.

Elliott, R.C., Heward, G.J., 1976. The influence of banana supplemented diet on gastric ulcers in mice. Pharmacol. Res. Commun. 8 (2), 167–171.

Fang, Y.P.D., 1999. Aminosalicylic acid inhibits IkB kinase alpha phosphorylation of IkBalpha in mouse intestinal epithelial cells. J. Biol. Chem. 274 (51), 36631–36636.

Fedorak, R.N., Madsen, K.L., 2004a. Probiotics and prebiotics in gastrointestinal disorders. Curr. Opin. Gastroenterol. 20 (2), 146–155.

Fedorak, R.N., Madsen, K.L., 2004b. Probiotics and the management of inflammatory bowel disease. Inflamm. Bowel Dis. 10 (3), 286–299.

Feng, W., Ao, H., Peng, C., et al., 2019. Gut microbiota, a new frontier to understand traditional Chinese medicine. Pharmacol. Res. 142, 176–191.

Fukuda, M., Kanauchi, O., Araki, Y., Andoh, A., 2002. Prebiotic treatment of experimental colitis with germinated barley foodstuff: a comparison with probiotic or antibiotic treatment. Int. J. Mol. Med. 9 (1), 65–70.

Furrie, E., Macfarlane, S., Kennedy, A., 2005. Synbiotic therapy (*Bifidobacterium longum*/Synergy 1) initiates resolution of inflammation in patients with active ulcerative colitis: a randomised controlled pilot trial. Gut 54 (2), 242–249.

Galland, L., 2014. The gut microbiome and the brain. J. Med. Food 17, 1261–1272.

Gaudier, E., Michel, C., Segain, J.P., 2005. The VSL# 3 probiotic mixture modifies microflora but does not heal chronic dextran-sodium sulfate-induced colitis or reinforce the mucus barrier in mice. J. Nutr. 135 (12), 2753–2761.

Germano, M.P., Sanogo, R., Guglielmo, M., 1998. Effects of *Pteleopsis suberosa* extracts on experimental gastric ulcers and *Helicobacter pylori* growth. J. Ethnopharmacol. 59 (3), 167–172.

Gershon, M.D., 1991. Serotonin: its role and receptors in enteric neurotransmission. Adv. Exp. Med. Biol. 294, 221–230.

Goel, R.K., Gupta, S., Shankar, R., 1986. Anti-ulcerogenic effect of banana powder (*Musa sapientum* var. *paradisiaca*) and its effect on mucosal resistance. J. Ethnopharmacol. 18 (1), 33–44.

Goel, R.K., Pathak, N.K., Biswas, M., 1987. Effect of lapachol, a naphthaquinone isolated from *Tectona grandis*, on experimental peptic ulcer and gastric secretion. J. Pharm. Pharmacol. 39 (2), 138–140.

Goel, R.K., Tavares, I.A., Bennett, A., 1989. Stimulation of gastric and colonic mucosal eicosanoid synthesis by plantain banana. J. Pharm. Pharmacol. 41 (11), 747–750.

Goel, R.K., Banerjee, R.S., Acharya, S.B., 1990. Antiulcerogenic and antiinflammatory studies with shilajit. J. Ethnopharmacol. 29 (1), 95–103.

Goggs, R., Vaughan-Thomas, A., Clegg, P.D., 2005. Nutraceutical therapies for degenerative joint diseases: a critical review. Crit. Rev. Food Sci. Nutr. 45 (3), 145–164.

Goulet, O., 2015. Potential role of the intestinal microbiota in programming health and disease. Nutr. Rev. 73 (Suppl 1), 32–40.

Gupta, R.C. (Ed.), 2016. Nutraceuticals: Efficacy, Safety and Toxicity. Academic Press/Elsevier, Amsterdam, pp. 1–1022.

Gupta, S., Kesarla, R., Omri, A., 2013. Formulation strategies to improve the bioavailability of poorly absorbed drugs with special emphasis on self-emulsifying systems. ISRN Pharm. 848043. Available from: https://doi.org/10.1155/2013/848043.

Gupta, R.C., Srivastava, A., Lall, R., 2019. In: Gupta, R.C., Srivastava, A., Lall, R. (Eds.), Nutraceuticals in Veterinary Medicine. Springer Nature, Cham, pp. 1–877.

Guslandi, M., Giollo, P., Testoni, P.A., 2003. A pilot trial of *Saccharomyces boulardii* in ulcerative colitis. Eur. J. Gastroenterol. Hepatol. 15 (6), 697–698.

Han, C.W.L., Yu, K., Chen, L., 2006. Biochemical characterization and inhibitor discovery of shikimate dehydrogenase from *Helicobacter pylori*. FEBS J. 273, 4682–4692.

Hanai, H.I.T., Takeuchi, K., Watanabe, F., 2006. Curcumin maintenance therapy for ulcerative colitis: randomized, multi-center, double-blind, placebo-controlled trial. Clin. Gastroeneterol. Hepatol. 4, 1502–1506.

Hartmann, R.M., Morgan Martins, M.I., Tieppo, J., Fillmann, H.S., 2012. Effect of *Boswellia serrate* on antioxidant status in an experimental model of colitis rats induced by acetic acid. Dig. Dis. Sci. 57, 2038–2044.

Hawkey, C.J., Mahida, Y.R., Hawthorne, A.B., 1992. Therapeutic interventions in gastrointestinal disease based on an understanding of inflammatory mediators. Agents Actions 36, C22–C26.

Herias, M.V., Koninkx, J.F., Vos, J.G., 2005. Probiotic effects of *Lactobacillus casei* on DSS-induced ulcerative colitis in mice. Int. J. Food Microbiol. 103 (2), 143–155.

Hibasami, H., Iwase, H., Yoshioka, K., 2005. Glycyrrhizin induces apoptosis in human stomach cancer KATO III and human promyelotic leukemia HL-60 cells. Int. J. Mol. Med. 16 (2), 233–236.

Hillier, K., Jewell, R., Dorrell, L., Smith, C.L., 1991. Incorporation of fatty acids from fish oil and olive oil into colonic mucosal lipids and effects upon eicosanoid synthesis in inflammatory bowel disease. Gut 32 (10), 1151–1155.

Hilsden, R.J., Verhoef, M.J., Best, A., Pocobello, G., 2003. Complementary and alternative medicine use by Canadian patients with inflammatory bowel disease: results from a national survey. Am. J. Gastroenterol. 98 (7), 1563–1568.

Holt, P.R., Katz, S., Kirshoff, R., 2005. Curcumin therapy in inflammatory bowel disease: a pilot study. Dig. Dis. Sci. 50 (11), 2191–2193.

Holtmeier, W., Zeuzem, S., Preiss, J., 2011. Randomized, placebo-controlled, double-blind trial of *Bosbellia serrata* in maintaining remission of Crohn's disease: good safety profile but lack of efficacy. Inflamm. Bowel Dis. 17 (2), 573–582.

Huang, M.-T., Lysz, T., Ferraro, T., Abidi, T.F., 1991. Inhibitory effects of curcumin on in vitro lipoxygenase and cyclooxygenase activities inmouse epidermis. Cancer Res. 51 (3), 813–819.

Hudert, C.A., Weylandt, K.H., Lu, Y., 2006. Transgenic mice rich in endogenous omega-3 fatty acids are protected from colitis. Proc. Natl. Acad. Sci. U.S.A. 103 (30), 11276–11281.

Hughes, R.L., 2019. A review of the role of the gut microbiome in personalized sports nutrition. Front. Nutr. 6, 191. Available from: https://doi.org/10.3389/fnut.2019.00191.

Ihantola-Vormisto, A., Summanen, J., Kankaanranta, H., 1997. Anti-inflammatory activity of extracts from leaves of *Phyllanthus emblica*. Planta Med. 63 (6), 518–524.

Irving, G.R., Karmokar, A., Berry, D.P., 2011. Curcumin: the potential for efficacy in gastrointestinal diseases. Best Pract. Res. Clin. Gastroenterol. 25 (4-5), 519–534.

Ishikawa, H., Akedo, I., Umesaki, Y., 2003. Randomized controlled trial of the effect of bifidobacteria-fermented milk on ulcerative colitis. J. Am. Coll. Nutr. 22 (1), 56–63.

Jain, G., Patil, U.K., 2015. Strategies for enhancement of bioavailability of medicinal agents with natural products. Int. J. Pharm. Sci. Res. 6, 5315–5324.

Jhanwar, B., Gupta, S., 2014. Biopotentiation using herbs: novel technique for poor bioavailable drugs. Int. J. PharmTech Res. 6, 443–454.

Jorquera Plaza, F., EspinelDiez, J., Olcoz Goñi, J.L., 1997. Inflammatory bowel disease: importance of nutrition today. Nutr. Hosp. 12 (6), 289–298.

Kahraman, A., Erkasap, N., Köken, T., Serteser, M., 2003. The antioxidative and antihistaminic properties of quercetin in ethanol-induced gastric lesion. Toxicology 183 (1-3), 133–142.

Kanauchi, O., Suga, T., Tochihara, M., 2002. Treatment of ulcerative colitis by feeding with germinated barley foodstuff: first report of a multicenter open control trial. J. Gastroenterol. 37 (Suppl. 14), 67–72.

Kang, M.J., Cho, J.Y., Shim, B.H., et al., 2009. Bioavailability enhancing activities of natural compounds from medicinal plants. J. Med. Plant. Res. 3, 1204–1211.

Kato, K., Mizuno, S., Umesaki, Y., 2004. Randomized placebo-controlled trial assessing the effect of bifidobacteria-fermented milk on active ulcerative colitis. Aliment. Pharmacol. Ther. 20 (10), 1133–1141.

Keserwani, K., Gupta, R., 2013. Bioavailability enhancers of herbal origin: an overview. Asian Pac. J. Trop. Biomed. 3, 253–266.

Khan, M.T., Lampronti, I., Martello, D., 2002. Identification of pyrogallol as an antiproliferative compound present in extracts from the medicinal plant *Emblica officinalis*: effects on *in vitro* cell growth of human tumor cell lines. Int. J. Oncol. 21 (1), 187–192.

Kim, et al., 2016. Reported that astaxanthin inhibits proliferation of human gastric cancer cell lines by interrupting cell cycle progression. Gut Liver 10 (3), 369–374.

Kohler, J.E., Blass, A.L., Liu, J., 2010. Antioxidant pre-treatment prevents omeprazole-induced toxicity in an in vitro model of infectious gastritis. Free Rad. Biol. Med. 49 (5), 786–791.

Kruis, W., Schutz, E., Fric, P., 1997. Double-blind comparison of an oral *Escherichia coli* preparation and mesalazine in maintaining remission of ulcerative colitis. Aliment. Pharmacol. Ther. 11 (5), 853–858.

Kruis, W., Fric, P., Pokrotnieks, J., 2004. Maintaining remission of ulcerative colitis with the probiotic *Escherichia coli* Nissle 1917 is as effective as with standard mesalazine. Gut 53 (11), 1617–1623.

Kuppusamy, P., Yusoff, M.M., Maniama, G.P., Govindana, N., 2014. Nutraceuticals as potential therapeutic agents for colon cancer: a review. Acta Pharm. Sin. B 4 (3), 173–181.

Kusters, J.G., van Vliet, A.H., Kuipers, E.J., 2006. Pathogenesis of *Helicobacter pylori* infection. Clin. Microbiol. Rev. 19 (3), 449–490.

Kyogoku, K., Hatayama, K., Yokomori, S., 1979. Anti-ulcer effect of isoprenyl flavonoids. II. Synthesis and anti-ulcer activity of new chalcones related to sophoradin. Chem. Pharm. Bull. (Tokyo). 27 (12), 2943–2953.

Lambertini, E., Piva, R., Khan, M.T., 2004. Effects of extracts from Bangladeshi medicinal plants on *in vitro* proliferation of human breast cancer cell lines and expression of estrogen receptor alpha gene. Int. J. Oncol. 24 (2), 419–423.

Langmead, L., Feakins, R.M., Goldthorpe, S., Holt, H., 2004. Randomized, double-blind, placebo-controlled trial of oral aloe vera gel for active ulcerative colitis. Aliment. Pharmacol. Ther. 19, 739–747.

Ley, R.E., Turnbaugh, P.J., Klein, D., et al., 2016. Microbial ecology: human gut microbes associated with obesity. Nature 444, 1022–1023.

Li, Z.G., Shimada, Y., Sato, F., 2002. Inhibitory effects of epigallocatechin-3-gallate on N-nitrosomethylbenzylamine-induced esophageal tumorigenesis in F344 rats. Int. J. Oncol. 21 (6), 1275–1283.

Li, Y., Kong, D., Wang, Z., 2010. Regulation of microRNAs by natural agents: an emerging field in chemoprevention and chemotherapy research. Pharm. Res. 27 (6), 1027–1041.

Li, Y., Go, V.L., Sarkar, F.H., 2015. The role of nutraceuticals in pancreatic cancer prevention and therapy: targeting cellular signaling, microRNAs, and epigenome. Pancreas 44 (1), 1–10.

Li, H., Yan, L., Tang, E.K.Y., et al., 2019. Synthesis of TPGS/curcumin nanoparticles by thin-film hydration and evaluation of their anti-colon efficacy in in vitro and in vivo. Front. Pharmacol. 10, 769.

Liang, Y.C., Huang, Y.T., Tsai, S.H., 1999. Suppression of inducible cyclooxygenase and inducible nitric oxide synthase by apigenin and related flavonoids in mouse macrophages. Carcinogenesis 20 (10), 1945–1952.

Liang, Y.C., Tsai, S.H., Tsai, D.C., 2001. Suppression of inducible cyclooxygenase and nitric oxide synthase through activation of peroxisome proliferator-activated receptor-gamma by flavonoids in mouse macrophages. FEBS Lett. 496 (1), 12–18.

Lorenz, R., Weber, P.C., Szimnau, P., 1989. Supplementation with n-3 fatty acids from fish oil in chronic inflammatory bowel disease–a randomized, placebo-controlled, double-blind cross-over trial. J. Intern. Med. Suppl. 731, 225–232.

Madisch, A., Miehlke, S., Eichele, O., 2007. *Boswellia serrata* extract for the treatment of collagenous colitis. A double-blind, randomized, placebo-controlled, multicenter trial. Int. J. Colorectal. Dis. 22 (12), 1445–1451.

Madsen, K.L., Doyle, J.S., Jewell, L.D., 1999. Lactobacillus species prevents colitis in interleukin 10 gene-deficient mice. Gastroenterology 116 (5), 1107–1114.

Maffei, M., Desmeules, J., Cereda, J.M., 2007. Side effects of proton pump inhibitors (PPIs). Rev. Med. Suisse 3 (123), 1934–1936.

Magistretti, M.J., Conti, M., Cristoni, A., 1988. Antiulcer activity of an anthocyanidin from *Vaccinium myrtillus*. Arzneimittelforschung 38 (5), 686–690.

Maity, S., Vedasiromoni, J.R., Ganguly, D.K., 1995. Anti-ulcer effect of the hot water extract of black tea (*Camellia sinensis*). J. Ethnopharmacol. 46 (3), 167–174.

Maity, S., Vedasiromoni, J.R., Chaudhuri, L., 2001. Role of reduced glutathione and nitric oxide in the black tea extract-mediated protection against ulcerogen-induced changes in motility and gastric emptying in rats. Jpn. J. Pharmacol. 85 (4), 358–364.

Majumdar, B., Chaudhuri, S.R., Ray, A., 2002. Potent antiulcerogenic activity of ethanol extract of leaf of *Piper betle* Linn by antioxidative mechanism. Indian J. Clin. Biochem. 17 (1), 49–57.

Matsuda, H., Li, Y., Murakami, T., 1998. Protective effects of oleanolic acid oligoglycosides on ethanol- or indomethacin-induced gastric mucosal lesions in rats. Life Sci. 63 (17), PL245–PL250.

McCarthy, J., O'Mahony, L., O'Callaghan, L., 2003. Double blind, placebo-controlled trial of two probiotic strains in interleukin 10 knockout mice and mechanistic link with cytokine balance. Gut 52 (7), 975–980.

Mehta, V.B., Sharma, V.J., Sadhana, S., 2000. Evaluation of antioxidant and immunomodulatory activity of *Lagenaria vulgaris*. Int. J. Res. Pharm. Sci. 2 (3), 393–398.

Meister, D., Ghosh, S., 2005. Effect of fish oil enriched enteral diet on inflammatory bowel disease tissues in organ culture: differential effects on ulcerative colitis and Crohn's disease. World J. Gastroenterol. 11 (47), 7466–7472.

Menon, V.P., Sudheer, A.R., 2007. Antioxidant and anti-inflammatory properties of curcumin. Adv. Exp. Med. Biol. 595, 105–125.

Mi, Y., Zhao, J., Feng, S.S., et al., 2012. Vitamin E TPGS prodrug micelles for hydrophilic drug delivery with neuroprotective effects. Int. J. Pharm. 438, 98–106.

Moller, P.L., Paerregaard, A., Gad, M., 2005. Colitic scid mice fed *Lactobacillus* spp. show an ameliorated gut histopathology and an altered cytokine profile by local T cells. Inflamm. Bowel Dis. 11 (9), 814–819.

Moon, G., Myung, S.J., Jeong, J.Y., 2004. Prophylactic effect of *Lactobacillus* GG in animal colitis and its effect on cytokine secretion and mucin gene expressions. Korean J. Gastroenterol. 43 (4), 234–245.

Morikawa, T., Li, N., Nagatomo, A., 2006. Triterpene saponins with gastroprotective effects from tea seed (the seeds of *Camellia sinensis*). J. Nat. Prod. 69 (2), 185–190.

Mowrey, D.B., Clayson, D.E., 1982. Motion sickness, ginger, and psychophysics. Lancet 1 (8273), 655–657.

Niederau, C., Gopfert, E., 1999. The effect of chelidonium- and turmeric root extract on upper abdominal pain due to functional disorders of the biliary system. Med. Klin. (Munich) 94 (8), 425–430.

Ning, Z., Wang, C., Liu, Y., et al., 2018. Integrating strategies of herbal metabolomics, network pharmacology, and experiment validation to investigate Frankincense processing effects. Front. Pharmacol. 9, 2018.

O'Mahony, L., Feeney, M., O'Halloran, S., 2001. Probiotic impact on microbial flora, inflammation and tumour development in IL-10 knockout mice. Aliment. Pharmacol. Ther. 15 (8), 1219–1225.

O'Morain, C., Tobin, A., Suzuki, Y., 1989. Risk factors in inflammatory bowel disease. Scand. J. Gastroenterol. Suppl. 170, 58–60.

Palozza, P., Torelli, C., Boninsegna, A., et al., 2009. Growth-inhibitory effects of the astaxanthin-rich alga *Haematococcus pluvialis* in human colon cancer cells. Cancer Lett. 283, 108–117.

Pandit, A.P., Joshi, S.R., Dalal, P.S., et al., 2019. Curcumin as a permeability enhancer enhanced the antihyperlipidemic activity of dietary green tea extract. BMC Complement. Alternat. Med. 19.

Park, S., Hahm, K.B., Oh, T.Y., 2004. Preventive effect of the flavonoid, wogonin, against ethanol-induced gastric mucosal damage in rats. Dig. Dis. Sci. 49 (3), 384–394.

Patil, U.K., Singh, A., Chakraborty, A.K., 2011. Role of piperine as a bioavailability enhancer. Int. J. Recent Adv. Pharm. Res. 4, 16–23.

Penner, R., Fedorak, R.N., Madsen, K.L., 2005. Probiotics and nutraceuticals: non-medicinal treatments of gastrointestinal diseases. Curr. Opin. Pharmacol. 5 (6), 596–603.

Penston, J., Wormsley, K.G., 1986. Adverse reactions and interactions with H2-receptor antagonists. Med. Toxicol. 1 (3), 192–216.

Petersson, J., Phillipson, M., Jansson, E.A., 2007. Dietary nitrate increases gastric mucosal blood flow and mucosal defense. Am. J. Physiol. Gastrointest. Liver Physiol. 292 (3), G718–G724.

Pilla, R., Suchodolski, J.S., 2020. The role of the canine gut microbiome and metabolome in health and gastrointestinal disease. Front. Vet. Sci. 6, 498. Available from: https://doi.org/10.3389/fvets.2019.00498.

Prakash, D.G.C., Sharma, G., 2012. Importance of phyto-chemicals in nutraceuticals. J. Chin. Med. Res. Dev. 1, 70–78.

Prucksunand, C., Indrasukhsri, B., Leethochawalit, M., 2001. Phase II clinical trial on effect of the long turmeric (*Curcuma longa* Linn) on healing of peptic ulcer. Southeast Asian J. Trop. Med. Public Health 32 (1), 208–215.

Qi, W., Weber, C.R., Wasland, K., 2011. Genistein inhibits proliferation of colon cancer cells by attenuating a negative effect of epidermal growth factor on tumor suppressor FOXO3 activity. BMC Cancer 11, 219.

Rachmilewitz, D.K.K., Karmeli, F., Hayashi, T., 2004. Toll-like receptor 9 signaling mediates the antiinflammatory effects of probiotics in murine experimental colitis. Gastroenterology 126 (2), 520–528.

Rahimi, R., Mozaffari, S., Abdollahi, M., 2009. On the use of herbal medicines in management of inflammatory bowel diseases: a systematic review of animal and human studies. Dig. Dis. Sci. 54, 471–480.

Rajasekaran, S.A., 2011. Therapeutic potential of curcumin in gastrointestinal diseases. World J. Gastrointest. Pathophysiol. 2 (1), 1–14.

Raveendra, K.R., Jayachandra, S.V., Sushma, K.R., et al., 2012. An extract of *Glycyrrhiza glabra* (Gut Gard) alleviates symptoms of functional dyspepsia: a randomized, double-blind, placebo-controlled study. Evid. Based Complement. Alternat. Med. 2012, 216970.

Razack, R., Seidner, D.L., 2007. Nutrition in inflammatory bowel disease. Curr. Opin. Gastroenterol. 23 (4), 400–405.

Redfern, A., Suchodolski, J., Jergens, A., 2017. Role of the gastrointestinal microbiota in small animal health and disease. Vet. Rec. 181Available from: https://doi.org/10.1136/vr.103826.

Rembacken, B.J., Snelling, A.M., Hawkey, P.M., 1999. Non-pathogenic *Escherichia coli* versus mesalazine for the treatment of ulcerative colitis: a randomised trial. Lancet 354 (9179), 635–639.

Rioux, K.P., Fedorak, R.N., 2006. Probiotics in the treatment of inflammatory bowel disease. J. Clin. Gastroenterol. 40 (3), 260–263.

Rocha Lapa, F.F.C., Hatsuko Baggio, C., Missau, F., 2007. Gastroprotective activity of the hydroalcoholic extract obtained from *Polygala paniculata* L. in rats. J. Pharm. Pharmacol. 59 (10), 1413–1419.

Rodriguez, D.A., Vélez, R.P., Monjaraz, E.M.T., et al., 2019. The gut microbiota: a clinically impactful factor in patient health and disease. SN Compr. Clin. Med. 1, 188–199.

Romano, M., Vitaglione, P., Sellitto, S., 2012. Nutraceuticals for protection and healing of gastrointestinal mucosa. Curr. Med. Chem. 19 (1), 109–117.

Rowland, I., Gibson, G., Heinken, A., et al., 2018. Gut microbiota function: metabolism of nutrients and other food components. Eur. J. Nutr. 57, 1–24.

Saito, H., Lee, Y.M., Takagi, K., 1977. Pharmacological studies of *Panacis Japonici* Rhizoma. I. Chem. Pharm. Bull. (Tokyo) 25 (5), 1017–1025.

Sajilata, M.G., Bajaj, P.R., Singhal, R.S., 2008. Tea polyphenols as nutraceuticals. Compr. Rev. Food Sci. Food Saf. 7 (3), 229–254.

Sandborn, W.J., Targan, S.R., Byers, V.S., 2013. Andrographis paniculata extract (HMPL-004) for active ulcerative colitis. Am. J. Gastroenterol. 108 (1), 90–98.

Sanyal, R.K., Das, P.K., Sinha, S., 1961. Banana and gastric secretion. J. Pharm. Pharmacol. 13, 318–319.

Sanyal, A.K., Gupta, K.K., Chowdhury, N.K., 1963. Banana and experimental peptic ulcer. J. Pharm. Pharmacol. 15, 283–284.

Sarkar, F.H., 2010. Current trends in the chemoprevention of cancer. Pharm. Res. 27 (6), 945–949.

Satoh, T., 2016. Astaxanthin: health benefits and toxicity. In: Gupta, R.C. (Ed.), Nutraceuticals: Efficacy, Safety and Toxicity. Academic Press/Elsevier, Amsterdam, pp. 531–539.

Schrump, D.S., Matthews, W., Chen, G.A., 1998. Flavopiridol mediates cell cycle arrest and apoptosis in esophageal cancer cells. Clin. Cancer Res. 4 (11), 2885–2890.

Shanahan, F., 2004a. Host-flora interactions in inflammatory bowel disease. Inflamm. Bowel Dis. 10 (Suppl. 1), S16–S24.

Shanahan, F., 2004b. Probiotics and the immune response: how much can we expect? J. Pediatr. Gastroenterol. Nutr. 39 (Suppl. 3), S748–S749.

Shanahan, F., 2004c. Probiotics in inflammatory bowel disease—therapeutic rationale and role. Adv. Drug Deliv. Rev. 56 (6), 809–818.

Shapiro, H., Singer, P., Halpern, Z., 2007. Polyphenols in the treatment of inflammatory bowel disease and acute pancreatitis. Gut 56 (3), 426–435.

Sharma, S.A., 1971. Acid balancing effect on digestive system: clinical studies were conducted to investigate the effect of amalaki in amlapitta (Gastritis syndrome). Indian J. Med. Res. 34, 98–99.

Sheil, B., McCarthy, J., O'Mahony, L., Bennett, M.W., 2004. Is the mucosal route of administration essential for probiotic function? Subcutaneous administration is associated with attenuation of murine colitis and arthritis. Gut 53 (5), 694–700.

Shiba, T., Aiba, Y., Ishikawa, H., 2003. The suppressive effect of bifidobacteria on *Bacteroides vulgatus*, a putative pathogenic microbe in inflammatory bowel disease. Microbiol. Immunol. 47 (6), 371–378.

Shibata, M., Yoshida, R., Motoashi, S., 1973. Pharmacological studies on *Bupleurum falcatum* L. IV. Some pharmacological effects of crude saikosides, saikogenin A and syrupy residue (author's transl). Yakugaku Zasshi. 93 (12), 1660–1667.

Shibolet, O., Karmeli, F., Eliakim, R., 2002. Variable response to probiotics in two models of experimental colitis in rats. Inflamm. Bowel Dis. 8 (6), 399–406.

Shoda, R., Matsueda, K., Yamato, S., 1995. Therapeutic efficacy of N-3 polyunsaturated fatty acid in experimental Crohn's disease. J. Gastroenterol. 8, 98–101.

Simopoulos, A.P., 2002. Omega-3 fatty acids in inflammation and autoimmune diseases. J. Am. Coll. Nutr. 21 (6), 495–505.

Singh, R., Devi, S., Patel, J.H., et al., 2009. Indian herbal bioenhancers: a review. Pharmacogn. Rev. 3, 90–92.

Singh, A., Dubey, R., Paliwal, R.T., 2012. Nutraceuticals—an emerging era in the treatment and prevention of diseases. Int. J. Pharm. Pharm. Sci. 4 (4), 39–43.

Sircana, A., De Michieli, F., Parente, R., et al., 2019. Gut microbiota, hypertension and chronic kidney disease: recent advances. Pharmacol. Res. 144, 390–408.

Srinivasan, K., 2005. Role of spices beyond food flavoring: nutraceuticals with multiple health effects. Food Rev. Int. 21 (2), 167–188.

Srivastava, A., Lall, R., Talukder, J., DuBourdieu, D., Gupta, R.C., 2019. Iron transport tocopheryl polyethylene glycol succinate in animal health and diseases. Molecules 24, 4289. Available from: https://doi.org/10.3390/molecules24234289.

Steidler, L., Hans, W., Schotte, L., 2000. Treatment of murine colitis by *Lactococcus lactis* secreting interleukin-10. Science 289 (5483), 1352–1355.

Sun, X.B., Matsumoto, T., Yamada, H., 1992. Purification of an anti-ulcer polysaccharide from the leaves of *Panax ginseng*. Planta Med. 58 (5), 445–448.

Surh, Y.J., Chun, K.S., Cha, H.H., 2001. Molecular mechanisms underlying chemopreventive activities of anti-inflammatory phytochemicals: down-regulation of COX-2 and iNOS through suppression of NF-kappa B activation. Mutat. Res. 480-481, 243–268.

Surh, Y.J., Chun, K.S., 2007. Cancer chemopreventive effects of curcumin. Adv. Exp. Med. Biol. 595, 149–172.

Swarnakar, S., Ganguly, K., Kundu, P., Banerjee, A., 2005. Curcumin regulates expression and activity of matrix metalloproteinases 9 and 2 during prevention and healing of indomethacin-induced gastric ulcer. J. Biol. Chem. 280 (10), 9409–9415.

Talukder, J., 2019. Nutraceuticals in gastrointestinal conditions. In: Gupta, R.C., Srivastava, A., Lall, R. (Eds.), Nutraceuticals in Veterinary Medicine. Springer Nature, Cham, pp. 467–479.

Tan, P.V., Nyasse, B., Dimo, T., 2002. Gastric cytoprotective anti-ulcer effects of the leaf methanol extract of *Ocimum suave* (Lamiaceae) in rats. J. Ethnopharmacol. 82 (2–3), 69–74.

Tang, T., Targan, S.R., Li, Z.S., Xu, C., 2011. Randomised clinical trial: herbal extract HMPL-004 in active ulcerative colitis—a double-blind comparison with sustained release mesalazine. Aliment. Pharmacol. Ther. 3, 194–202.

Thangapazham, R.L., Sharma, A., Maheshwari, R.K., 2006. Multiple molecular targets in cancer chemoprevention by curcumin. AAPS J. 8, E443–E449.

Thomas, R.H., Allmond, K., 2013. Linaclotide (Linzess) for irritable bowel syndrome with constipation and for chronic idiopathic constipation. P&T 38 (3), 154–160.

Tremlett, H., Bauer, K.C., Appel-Cresswell, S., et al., 2017. The gut microbiome in human neurological disease: a review. Ann. Neurol. 82, 369–382.

Turpie, A.G., Runcie, J., Thomson, T.J., 1969. Clinical trial of deglycyrrhizinized liquorice in gastric ulcer. Gut 10 (4), 299–302.

Venturi, A., Gionchetti, P., Rizzello, F., 1999. Impact on the composition of the faecal flora by a new probiotic preparation: preliminary data on maintenance treatment of patients with ulcerative colitis. Aliment. Pharmacol. Ther. 13 (8), 1103–1108.

Vilaseca, J., Salas, A., Guarner, F., 1990. Dietary fish oil reduces progression of chronic inflammatory lesions in a rat model of granulomatous colitis. Gut 31 (5), 539–544.

Vitaglione, P., Sforza, S., Galaverna, G., 2005. Bioavailability of trans-resveratrol from red wine in humans. Mol. Nutr. Food Res. 49 (5), 495–504.

Vitaglione, P., Napolitano, A., Fogliano, V., 2008. Cereal dietary fibre: a natural functional ingredient to deliver phenolic compounds into the gut. Trends Food Sci. Technol. 19, 451–463.

Whiting, C.V., Bland, P.W., Tarlton, J.F., 2005. Dietary n-3 polyunsaturated fatty acids reduce disease and colonic proinflammatory cytokines in a mouse model of colitis. Inflamm. Bowel Dis. 11 (4), 340–349.

Wittschier, N., Faller, G., Hensel, A., 2007. An extract of *Pelargonium sidoides* (EPs 7630) inhibits in situ adhesion of *Helicobacter pylori* to human stomach. Phytomed. Int. J. Phytother. Phytopharmacol. 14 (4), 285–288.

Wittschier, N., Faller, G., Hensel, A., 2009. Aqueous extracts and polysaccharides from liquorice roots (*Glycyrrhiza glabra* L.) inhibit adhesion of *Helicobacter pylori* to human gastric mucosa. J. Ethnopharmacol. 125 (2), 218–223.

Wolle, K., Malfertheiner, P., 2007. Treatment of *Helicobacter pylori*. Best Pract. Res. Clin. Gastroenterol. 21 (2), 315–324.

Yadav, S.K., Adhikary, B., Maity, B., 2009. The gastric ulcer-healing action of allylpyrocatechol is mediated by modulation of arginase metabolism and shift of cytokine balance. Eur. J. Pharmacol. 614 (1-3), 106–113.

Yamada, H., Ra, K.S., Kiyohara, H., 1989. Structural characterization of an anti-complementary pectic polysaccharide from the roots of *Bupleurum falcatum* L. Carbohydr. Res. 189, 209–226.

Yang, C., Wu, T., Qi, Y., et al., 2018. Recent advances in the application of vitamin E TPGS for drug delivery. Theranostics 8, 464–485.

Yoshida, T., Maoka, T., Das, S.K., 2007. Halocynthiaxanthin and peridinin sensitize colon cancer cell lines to tumor necrosis factor-related apoptosis-inducing ligand. Mol. Cancer Res. 5 (6), 615–625.

Yoshikawa, M., Morikawa, T., Li, N., Nagatomo, A., 2005. Bioactive saponins and glycosides. XXIII. Triterpene saponins with gastroprotective effect from the seeds of *Camellia sinensis*—theasaponins E3, E4, E5, E6, and E7. Chem. Pharm. Bull. (Tokyo) 53 (12), 1559–1564.

Youlk, H.J., Lee, E., Choi, M.K., et al., 2005. Enhanced anticancer efficacy of alpha-tocopheryl succinate by conjugation with polyethylene glycol. J. Control. Release 107, 43–52.

Yu, L.C.H., Wang, J.-T., Wei, S.-C., et al., 2012. Host-microbial interactions and regulation of intestinal epithelial barrier function: from physiology to pathology. World J. Gastrointest. Pathophysiol. 3 (1), 27–43.

Yurdakok-Dikmen, B., Turgut, Y., Filazi, A., 2018. Herbal bioenhancers in veterinary medicine. Front. Vet. Sci. 5, 249. 2018.

Zayachkivska, O.S., Konturek, S.J., Drozdowicz, D., 2005. Gastroprotective effects of flavonoids in plant extracts. J. Physiol. Pharmacol. 56 (Suppl. 1), 219–231.

Zhang, Y.-J., Li, S., Gan, R.-Y., et al., 2015. Impacts of gut bacteria on human health and diseases. Int. J. Mol. Sci. 16, 7493–7519.

Zou, T., Gu, L., 2013. TPGS emulsified zein nanoparticles enhanced oral bioavailability of daidzein: in vitro characteristics and in vivo performance. Mol. Pharm. 10, 2062.

Chapter 11

Nutraceuticals in reproductive and developmental disorders

Moges Woldemeskel

Introduction

Various diseases and abnormalities of infectious and noninfectious causes have been a threat to human health since ancient times. Plant products and their crude extracts were used in traditional medicine for the treatment of various disorders until the development of synthetic drugs in the modern era (Patel and Patel, 2017). However, despite the development of modern synthetic drugs, medicinal plants still serve as mainstream therapeutics and are central to traditional medicine (Akhtar et al., 2017) in many parts of the world. More than 40% of prescription drugs globally were mainly derived from herbal source (Patel and Patel, 2017). Although modern medicine is at the forefront in treating diseases, plant-based remedies in alternative medicine are still highly used throughout the world—mainly in developing countries (Fadus et al., 2017). Medicinal plants currently play a key role in health care, as more than 80% of the world's population relies on traditional medicine for their primary health care (Patel and Patel, 2017). This is largely because natural products are effective, relatively nontoxic, and have remedial doses far below their toxic levels (Fadus et al., 2017).

Nutraceuticals are natural products derived from various food sources that provide health benefits in addition to nutritional values (Kalra, 2003; Larussa et al., 2017). Nutraceuticals can be substances such as vitamins, minerals, amino acids, fatty acids, herbs, botanical products, or substances derived from other sources taken as dietary supplements including pyruvate, chondroitin sulfate, steroid hormone precursors, etc. (Chauhan et al., 2013).

Natural products or nutraceuticals have been shown to elicit antiaging, anticancer, and other health-enhancing effects. A key target of the effects of natural products may be the regulation of microRNA expression which results in cell death or prevents aging, diabetes, cardiovascular, and other diseases. Nutraceuticals have also been proposed to exert their effects on cancer stem cells by interacting with the expression of microRNAs (McCubrey et al., 2017). A Mediterranean-type diet was associated with a lower general cancer incidence among patients in secondary prevention of cardiovascular diseases (Hernáez and Estruch, 2019).

Various nutraceuticals have been in use for centuries to heal different types of diseases in humans and animals, including diseases that inflict the male and female reproductive systems. This chapter provides a brief account of some nutraceuticals used in managing fetal developmental disorders, and treatments for reproductive diseases in males and females.

Nutraceuticals in female reproductive abnormalities

Microorganisms, physical agents, and other etiologies are known to cause various disorders in the female reproductive system. However, due to environmental pollution, the use of synthetic chemicals in cosmetics and other items for domestic day-to-day uses, and other risk factors, currently cancers are the major diseases that affect the female reproductive tracts in women and also in pet animals. Various forms of cancer therapeutics are available. However, all the currently available cancer therapeutic options are expensive and none are safe and without side effects. On the contrary, traditional plant-derived medicines or nutraceuticals are relatively safe and are used in the treatment of diseases of the female reproductive tract. Some nutraceuticals effective in the treatment and management of female reproductive diseases—mainly cancerous diseases—are briefly highlighted below.

Effects of nutraceuticals on breast cancer

Different types of cancers affect the female reproductive tract. Breast cancer is the second leading cause of death

Nutraceuticals. DOI: https://doi.org/10.1016/B978-0-12-821038-3.00011-2

in women worldwide and more than one million women are diagnosed with breast tumor (Sayeed and Ameen, 2015; Mandal et al., 2017), with approximately 458,000 deaths each year (Sayeed and Ameen, 2015). It is also the second most frequent cancer in women in the United States, with about 246,660 new breast cancer cases and 40,450 breast cancer-related deaths estimated in 2016 (Siegel et al., 2016). Citing several authors, Mandal et al. (2017) reported that chronic inflammation is suggested to contribute to the development and progression of breast cancers. Therefore agents that inhibit chronic inflammation may be effective in the prevention and treatment of breast cancer. Several natural products, phytochemicals, and dietary agents with antiinflammatory activities have shown promise in the prevention and treatment of breast cancer (Mandal et al., 2017). The Mediterranean diet has been associated with a decrease in the risk of chronic diseases, including breast cancer (Hernáez and Estruch, 2019).

Distinct subtypes of breast tumor response differently to treatment, making breast cancer extremely intractable. Currently, surgical resection, adjuvant chemotherapy, radiotherapy, and hormone therapy represent the main treatment options for early-stage breast cancer. However, the development of drug resistance and major side effects have weakened the efficacy of these therapies. In addition, triple-negative breast cancer (TNBC) does not respond to hormone therapy. Due to the poor efficacy of modern therapies, prevention of the occurrence of breast cancer appears to be the best strategy for which a diet- and nutrition-based approach has been considered as an effective method. A group of dietary natural products have shown a potential role in the prevention and treatment of cancers (Li et al., 2017).

A recent review summarized the potential role of dietary natural products and their major bioactive components in the prevention and treatment of breast cancer. Several epidemiological and experimental studies suggested the inverse correlation between intake of vegetables and fruits, and incidence of breast cancer. Many dietary natural products such as soy, pomegranate, mangosteen, citrus fruits, apple, grape, mango, cruciferous vegetables, ginger, garlic, black cumin, edible macrofungi, and cereals could affect the development and progression of breast cancer. Experimental studies indicated that their antibreast cancer effects involve various mechanisms of action such as downregulating Estrogen-Receptor-α expression and activity; inhibiting proliferation, migration, metastasis, and angiogenesis of breast tumor cells; inducing apoptosis and cell cycle arrest; and sensitizing breast tumor cells to radiotherapy and chemotherapy. Epidemiological studies have suggested that high consumption of some dietary natural products might reduce the recurrence and increase the survival rate of breast cancer patients. Therefore the use of natural dietary substances could be a practical approach for the prevention and treatment of breast cancer (Li et al., 2017). One widely known such compound is beta-sitosterol, a plant-derived nutrient with anticancer properties against breast cancer, prostate cancer (PCa), colon cancer, lung cancer, stomach cancer, ovarian cancer, and leukemia (Sayeed and Ameen, 2015). Other natural products and plant extracts with similar effects on breast cancer are given below.

Some studies suggest an inverse association between folate and vitamin B12 status and breast cancer. A protective effect of vitamin D against human breast cancer has also been reported in some observational studies (Krusinska et al., 2019).

Berberine is a natural isoquinoline alkaloid extracted from plants such as *Berberis aquifolium*, *Berberis vulgaris*, *Berberis aristata*, and *Tinospora cordifolia* (Imanshahidi and Hosseinzadeh, 2008). The antitumor effects of berberine in various human cancer cells have been reported. Its effect was evaluated on TNBC, a cancer defined by negative expression of estrogen receptor, progesterone receptor, and human epidermal growth factor receptor 2. Because of the lack of standard molecular targets, the current viable targeted therapies are not useful for TNBC patients. TNBC is insensitive to the antihormone receptor (HR) and HER2-targeted drugs, and has a higher incidence than HR and HER2-positive patients. Berberine inhibited tumor cell proliferation and migration, and induced cellular apoptosis. It activates caspase-9/cytochrome c-mediated apoptosis to inhibit the growth of TNBC breast cancer cells in vitro and in vivo. Therefore berberine is a potential treatment agent for TNBC (Zhao et al., 2017). Furthermore, it was reported that berberine can bind to the vasodilator stimulated protein overexpressed in breast cancer cells and suppress tumor proliferation and growth (McCubrey et al., 2017).

Resveratrol, a polyphenol found in various plants, is also reported as one of the agents effective against breast cancer. Experimentally, it inhibited the proliferation of cancer stem cells isolated from breast cancer cell lines. Resveratrol reduced the presence of mammospheres in breast cancer cell lines, inhibited tumor formation of the mammospheres in mice, and also induced autophagy in the breast mammospheres (Fu et al., 2014).

Soy isoflavones have also been reported by many studies to have a protective role against breast cancer, though some adverse effects also have been reported. Whether soy isoflavones promote or inhibit the growth of breast cancer seemed to be dose dependent (Xie et al., 2013), thus the dosage and long-term safety of soy isoflavones need to be further investigated before they are recommended as supplements for breast cancer patients (Li et al., 2017).

Curcumin, a substance in turmeric, has been reported to exert antitumor development effects on breast cancer. It has been shown to have effects on drug transporter expression in breast cancer cells with cancer stem cell properties. Curcumin could increase the tumoricidal effects of mitomycin-C and also increase the sensitivity of breast cancer cell line cells to various chemotherapeutic drugs such as cisplatin, doxorubicin, and paclitaxel. The combination of curcumin and mitomycin-C together prevented the sphere-forming capacity of breast cancer cell lines (Zhou et al., 2015).

Li et al. (2017), citing various studies, indicated that fruits that normally contain a high content of polyphenols, have great antioxidant activity and may help reduce cancer risk. Fruits such as pomegranate, mangosteen, and citrus fruits, have shown inhibition activities on breast cancer cells. For example, pomegranate inhibited breast cancer growth and metastasis through various mechanisms such as inhibiting breast cancer cell growth by inducing cell cycle arrest, inducing apoptosis, exerting antiproliferative effects on rat mammary tumorigenesis, and significantly decreasing the cell viability of breast cancer cell lines. Pomegranate juice and its components blocked metastatic processes of breast cancer cells and proliferation of breast cancer cell lines in vitro, indicating a potential role of pomegranate for the prevention of estrogen-responsive breast cancers. Whole pomegranate seed oil and fermented pomegranate juice polyphenols inhibited the tumor lesion formation in a murine mammary gland organ culture suggesting a chemopreventive property and adjuvant therapeutic potential of pomegranate. Pomegranate seed oil and fermented juice concentrate suppressed preneoplastic mammary gland lesions ex vivo in a mouse mammary organ culture model and oral administration of pomegranate juice concentrate reduced the volume and weight of xenografted tumors in female athymic nude mice. Pomegranate emulsion containing most bioactive phytochemicals present in the whole fruit, exerted a significant chemopreventive activity against DMBA-initiated mammary tumorigenesis in female rats. It reduced the incidence, total burden, and average weight of mammary tumors with a concomitant inhibition of intratumor cell proliferation and induction of apoptosis (Mehta and Lansky, 2004; Banerjee et al., 2012; Li et al., 2017; Mandal et al., 2017).

Several in vitro and in vivo studies have reported antibreast cancer activities of mangosteen (*Garcinia mangostana*). Crude methanolic extract of mangosteen pericarp inhibits the proliferation and induces apoptosis on a breast cancer cell line; phenolics from the fruit pericarp produced cytotoxicities against human breast cancer cell lines; and alpha-mangostin treatment significantly increased the survival rate of mice with mammary tumors, and greatly suppressed tumor volume and lymph node metastases. Furthermore, studies report that citrus fruits such as orange, lemon, grapefruit, pomelo, and lime have an impact on the overall health of patients, including reducing the risk of developing various cancers. A number of studies pointed out an inverse association between citrus fruits intake and the risk of breast cancer. For example, polysaccharides from Korean citrus hallabong peels inhibited angiogenesis, and extracts from a citrus fruit from citrus hassaku Hort ex. Tanaka, also known as phalsak, and lemon citrus extract induced apoptosis in breast cancer stem cell lines. Furthermore, extracts of apple, grape, mango, strawberry, peach, and other citrus fruits are reported to have effects on breast cancer cells in vivo and in vitro through inducing apoptosis, inhibiting cell proliferation, prevention of breast cancer cell growth, decreasing tumor volume, and activation of oxidative stress in cancer cells. Generally, the intake of fruits is beneficial for the prevention and treatment of breast cancer, with pomegranate, mangosteen, apple, citrus fruits, grape, and mango showing the most promising effects. The antibreast cancer action of these fruits might be attributed to the presence of bioactive components such as ellagitannins in pomegranate and mangostin in mangosteen (Li et al., 2017).

Vegetables also are reported to have an effect on breast cancers. In experimental studies, several vegetables, especially cruciferous vegetables, have shown an inhibitory effect on breast cancer cells. For example intake of broccoli, cauliflower, watercress, and Brussels sprouts, was inversely associated with risk of breast cancer (Bradlow, 2008; Thomson et al., 2016). Indole-3-carbinol found in Brassica vegetables showed antibreast cancer action since it could interact directly with the estrogen receptor and inhibit its activity, or through estrogen-independent actions, such as blocking cell cycle progression, metastasis, and inducing apoptosis (Li et al., 2017). Daily intake of fresh carrot juice might also benefit patients with breast cancer (Butalla et al., 2012). Generally, isothiocyanates, indole-3-carbinol, and its metabolite 3,30-diindolylmethane found in cruciferous vegetables have shown a potential role in the prevention and treatment of breast cancer. The underlying mechanisms mainly include downregulating estrogen receptor-alpha and repressing estrogen receptor signaling, inducing apoptosis and cell cycle arrest, and inhibiting metastasis of breast cancer cells (Li et al., 2017).

Consumption of certain spices is also known to have an antibreast cancer effect. Several studies have suggested that ginger, garlic, and black cumin have shown promising antibreast cancer effects among various spices. Bioactive components in spices, such as gingerols and shogaols in ginger, and organosulfur substances including diallyl disulfide, diallyl trisulfide, S-allyl mercaptocysteine, and allicin in garlic have antibreast cancer activities and help in reducing the risks of developing breast cancer. Methanolic extract of ginger (*Zingiber officinale*) exhibited an inhibitory effect on cancer cell proliferation in breast cancer cell lines. The antibreast cancer property of ginger might be attributed to its bioactive constituents such as gingerols and shogaols (Lee et al., 2008; Joo et al., 2016). A supercritical CO_2 extract of black cumin (*Nigella sativa*) exhibited proapoptotic and antimetastatic effects on breast cancer cell lines in vitro. Thymoquinone, a major bioactive component isolated from the seeds of *Nigella sativa*, has shown potent chemopreventive and chemotherapeutic activities and an antiproliferative effect on breast cancer (Schneider-Stock et al., 2014; Li et al., 2017).

Other spices such as capsaicin, piperine isolated from black pepper (*Piper nigrum*), saffron (*Crocus sativus*), clove (*Syzygium aromaticum*), wasabi (*Wasabia japonica*), rosemary (*Rosmarinus officinalis*), and coriander are reported to have antibreast cancer activities in in vitro and in vivo experimental models. However, it should be noted that the role of capsaicin in cancer is controversial. Some studies have indicated that capsaicin itself is mutagenic and promotes tumor formation, and might increase the cancer risk in humans (Li et al., 2017), warranting more detailed studies to determine its real impact and benefit in breast cancer.

Several kinds of edible macro-fungi have shown inhibitory effects on breast cancer. Mushroom intake might be inversely associated with the risk of breast cancer. A significant inverse association between mushroom consumption and breast cancer incidence was found in postmenopausal women in Korea. Several mechanisms such as inhibiting proliferation, inducing apoptosis, and suppressing angiogenesis have been found to explain the antibreast cancer effects of edible macro-fungi. The antibreast cancer effects of edible macro-fungi are mainly attributed to the polysaccharides with different molecular weights present in the fungi. A number of studies also indicated an inverse association between cereal fiber intake and breast cancer risk. For example, sorghum suppressed tumor growth, induced cell cycle arrest, and inhibited metastasis in nude mice bearing breast cancer xenografts, young barley (the grass of the barley plant) exhibited significant antiproliferative and proapoptotic activities in a rat breast tumor model and in human breast cancer cells in vitro, and wheat (*Triticum aestivum*) flour inhibited growth of breast cancer cells and induced apoptosis in vitro. Sorghum, barley, and wheat have shown the potential to inhibit growth of breast cancer cells, mainly through inducing apoptosis and cell cycle arrest, and inhibiting metastasis (Hong et al., 2008; Li et al., 2014, 2017).

In addition to their direct effect on cancers, dietary natural products have synergistic effects with anticancer therapies. This is very important because chemotherapy and radiotherapy frequently used nowadays in cancer therapy exhibit certain toxic adverse effects, and drug resistance is a common cause of chemotherapy failure and disease recurrence. Some dietary products and their bioactive components have shown synergistic effects with chemotherapy or radiotherapy by enhancing their therapeutic effect or reducing their adverse side effects (Li et al., 2017).

Certain vitamins and minerals are also beneficial in reducing the risk of developing various cancers including breast cancer. For instance, experimental studies suggest a protective effect of B vitamins on breast cancer risk. A large prospective study also suggested dietary, supplemental, and total pyridoxine and thiamin intakes were associated with decreased breast cancer risk and a potential protective effect against breast cancer in middle-aged women (Egnell et al., 2017). Moreover, Tsuji et al. (2015), citing various experimental studies in animal models, indicated that selenium deficiency impairs immune response to infection, cancer, and other stimuli. These included reduction in CD4+ T-cell response in selenium-deficient mice challenged with a peptide/adjuvant, and increased tumor growth and spread in a mouse model of breast cancer indicating that selenium may play an important role in the prevention of breast cancer development.

Effects of nutraceuticals on cervical cancer

A number of diseases can affect the cervix in women. However, cancer is the major disease condition affecting the cervix and is globally among the most highest causes of mortality in women. Several natural products (nutraceuticals) are reported to be effective in treating and/or reducing the risk of cervical cancer in women. For example, McCubrey et al. (2017), citing several studies, indicated that berberine and curcumin may be promising agents for treatment of cervical cancer. In a study on cancer cell lines, it was reported that *Aloe vera* alone or its simultaneous use with cisplatin exhibited antineoplastic effects in breast and cervical cancers by inducing apoptosis and modulation of expression of effector molecules. These results signify that *A. vera* may be an effective antineoplastic agent to inhibit cancer cell growth and increase the therapeutic efficacy of conventional drugs like cisplatin, promoting the development of plant-derived therapeutic agents for novel cancer treatment strategies (Hussain et al., 2015). Due to the effect of various nutraceuticals on inhibiting tumor growth and reducing the risk of cancer development in general, consumption of natural plant products and other nutraceuticals undoubtedly will help in minimizing the risk of developing cervical cancer.

Effects of nutraceuticals on diseases of the uterus

Different types of diseases, including infectious diseases and neoplasms, may affect the uterus in females. A number of modern medicinal therapies have been in use to treat uterine diseases. However, nutraceuticals, as described earlier, are gaining increased preference due to the side effects that follow treatment with modern drugs and the well-documented efficacy of nutraceuticals in fighting various infections and diseases. Nutraceuticals that promote the overall health of individuals with antiinflammatory and anticancer effects described earlier certainly have an impact on the treatment and prevention of uterine diseases. For example, berberine was reported to inhibit the proliferation of human uterine leiomyoma cells (Wu et al., 2015). Some nutraceuticals have a synergetic effect, working in concert with various modern drugs and may enhance the effect of modern medicines on cancer cells. Berberine and metformin induced a lipolysis-stimulated lipoprotein receptor, which is a component of tricellular tight junctions in endometrial cancer cells, and suppressed cell migration and invasion (McCubrey et al., 2017).

Endometriosis is a common gynecological disease that affects women and other primate females. Treatment options for endometriosis include surgery and medication, mainly pain medications such as nonsteroidal antiinflammatory agents or hormone therapy including gonadotropin-releasing hormone agonists and androgens. Although such treatments can improve some symptoms, there is no cure for endometriosis. Trials have been conducted to determine the effects of some natural products on endometriosis. It was reported that *Pueraria* flower extract suppressed the adhesion of human endometriotic cells to human mesothelial cells and significantly inhibited the migration of endometriotic cells. It also significantly suppressed endometriotic lesion formation in a mouse model suggesting that *Pueraria* flower is an antiendometriotic agent for the inhibition of endometriotic cell adhesion, endometriotic cell migration, and the establishment of endometriosis-like lesions; and has potential use in the treatment and prevention of endometriosis (Kim et al., 2017).

Effects of nutraceuticals on ovarian cancer

Ovarian cancer is the most lethal gynecological malignancy in developed countries. This is due to the lack of specific symptoms that hinder early diagnosis and the high relapse rate after treatment with radical surgery and chemotherapy (Áyen et al., 2018). A number of studies have reported the effects of different nutraceuticals on ovarian cancer. Berberine suppressed proliferation of ovarian cancer cells and in concert with cisplatin it induced death of ovarian cancer cells, and augmented the effects of cisplatin in inducing cell cycle arrest in ovarian carcinoma cells (McCubrey et al., 2017). Resveratrol was also found to kill ovarian cancer stem cells. Further studies reported that bitter melon extracts contain compounds that inhibit the growth of many types of cancers, including ovarian cancer, and ameliorate resistance to common chemotherapeutic compounds (Yung et al., 2016). Bitter melon extracts inhibited ovarian cancer growth via activation of various signaling cascades. Alpha-momorcharin, a ribosome-inactivating protein present in bitter melon extracts, may in part be responsible for the anticancer effects (McCubrey et al., 2017). Dietary calcium was significantly associated with a reduced risk of ovarian cancer and increased dietary calcium intake might be inversely correlated with the risk of ovarian cancer (Song et al., 2017). Briefly, large numbers of natural products that have a significant effect on overall health of animals and humans are beneficial in minimizing the risk of developing ovarian cancer and in the management of clinical disease.

Effects of nutraceuticals on other disorders of the female reproductive system

In different parts of the world, natural products, mainly of plant origin, are used in the treatment and management of various conditions and symptoms associated with female reproductive system. For example, in India, the bark of *Adenium obesum* is used as an abortifacient agent and in the Omani ethnic community it is used for the treatment of venereal diseases, among many others (Akhtar et al., 2017).

Some nutraceuticals are also used in the treatment of a number of reproductive tract-associated ailments in females. For instance, *Agnus castus* has also been used in the treatment of many female conditions including menstrual disorders such as amenorrhea and dysmenorrhea, premenstrual dysphoric disorder, corpus luteum insufficiency, hyperprolactinemia, infertility, acne, disrupted lactation, cyclic breast pain, cyclical mastalgia, and inflammatory conditions. It was further indicated that *A. castus* and magnolia, combined with soy isoflavones plus lactobacilli were effectively and safely used in the treatment of vasomotor symptoms in postmenopausal women, mainly with poor quality of sleep. Such alternative treatments are preferred to hormone replacement therapy due to adverse side effects, poor compliance, and fear of increasing the risk of cancer or weight gain associated with hormone therapy (De Franciscis et al., 2017).

Many studies have indicated that combined supplementation with vitamin E and selenium, given a few weeks before calving, either by injection or orally, can significantly reduce the incidence of retained placenta in cattle. It was also indicated that selenium as sodium selenite injected 3 weeks prepartum, with or without oral vitamin E, reduced the incidence of metritis from 83% to 57%. Furthermore, an injection of 50 mg selenium (sodium selenite) 3 weeks before parturition reduced the incidence of cystic ovaries from 47% to 19% (Hemingway, 2003).

Nutraceuticals such as garlic that are effective against bacteria (Zardast et al., 2016) and other plant products including *Balanites aegyptiaca* and *Capparis erythrocarpos* known to be effective against fungi (Anywar et al., 2014) may help in the management and treatment of bacterial and mycotic reproductive tract infections.

Severity of polycystic ovary syndrome (PCOS), the most common female endocrine disorder characterized by chronic anovulation, hyperandrogenism, and polycystic ovaries, was reported to be associated with adherence to a Mediterranean diet. The report indicated that an unhealthy dietary pattern was associated with more severe hyperandrogenemia, inflammatory status, and insulin resistance. This association could support a therapeutic role of foods and nutrients of the Mediterranean dietary pattern in the PCOS pathogenesis, likely involving their inflammatory status, insulin resistance, and hyperandrogenemia (Barrea et al., 2019).

OleaVita, an olive leaf extract, is hypothesized to suppress placental inflammation. In human placental tissue culture, OleaVita treatment inhibited the secretion of inflammatory cytokines and NF-B p65 protein expression, has potential as a supplementation for the inhibition of NLRP3 inflammasomes, and may lead to the alleviation of various pregnancy complications (Kaneko et al., 2019).

Nutraceuticals in developmental disorders

Developmental disorders are disorders in physical and/or mental development of developing offspring and young subjects. Such disorders can be associated with nutritional deficiencies during pregnancy.

It is well accepted that a healthy pregnancy is required to ensure optimal fetal development and a healthy start to life for the developing child. Deficiencies in key dietary components during pregnancy have been shown to increase offspring risk of developing conditions such as insulin resistance, type II diabetes mellitus, and metabolic dysfunction (Hofstee et al., 2020). Nutraceuticals (vitamins and minerals) are very important for normal/proper fetal development. For example, supplementation with folic acid during preconception and early pregnancy can prevent 40%–80% of neural tube defects such as spina bifida in fetal development. Low maternal calcium intake can contribute to osteopenia, paresthesia, muscle cramps, tetanus, and tremor in the mother, as well as delayed growth, low birth weight, and poor fetal mineralization in the fetus. Iron deficiency or anemia has been associated with greater risk of preterm birth, low birth weight, or small-for-gestational-age infants, impaired maternal function, and decreased defenses against infection, as well as abnormal psychomotor development and cognitive function in infancy. Nutrients including polyunsaturated fatty acids, zinc, vitamins C and E, and B-complex vitamins have critical roles in maternal and fetal health and should be acquired through a balanced diet. Some studies suggest that thiamine (B1) deficiency may impair fetal brain development due to subclinical metabolic disturbances in the thiamine-dependent enzyme systems involved in lipid and nucleotide synthesis in the brain. Deficiencies in riboflavin and niacin have been correlated with preeclampsia, congenital heart defects, and low-birth-weight infants. Studies suggest that maternal night blindness (vitamin A deficiency) correlates with a higher risk of infant mortality and low-birth-weight infants. However, due to the teratogenicity associated with a high intake of vitamin A, intakes of vitamin A should be limited to prevent teratogenicity; an upper limit of 10,000 IU per day (3000 μg retinol) has been established and the nontoxic form (β-carotene) is preferred during pregnancy (Mousa et al., 2019).

Furthermore, increased brain ability to function and learn, indicated by the increase in nonverbal IQ, and a substantial 18-month increase in developmental ability in communication, daily living skills, and social skills is suggested by using nutritional supplements and a healthy diet. Several studies suggest that customized vitamin/mineral supplementation is beneficial for children with autism spectrum disorder. A study also reported that omega-3 fatty acid supplementation is very helpful for infants with Rett's syndrome, a disorder which often includes autistic symptoms (Adams et al., 2018). It is also reported that inadequate iodine status (insufficient iodine intake in pregnancy) affects the synthesis of the thyroid hormones and may impair brain development due to the negative impact on neurodevelopment in fetal life (Markhus et al., 2018).

Adequate selenium intake around pregnancy is important for offspring metabolic health; and selenium deficiency during pregnancy would cause small-for-gestational-age newborns. A mouse model of selenium deficiency also resulted in placental insufficiency with reduced fetal glucose concentrations and fetal growth restriction, two factors also associated with programmed disease outcomes in offspring. In addition to the potential roles of fetal hypoglycemia and altered skeletal muscle function in the development of offspring disease, selenium deficiency may have altered offspring metabolic function as a consequence of programming thyroid dysfunction. Therefore it is important that women obtain adequate selenium during pregnancy to ensure the health of the developing child (Hofstee et al., 2020).

Low-birth-weight infants have been noted to have low zinc concentrations in cord blood, and zinc deficiency in childhood is associated with reduced immunocompetence and increased infectious disease morbidity (Sazawal et al., 2001).

Generally, nutritional deficiency can have potential lifelong complications, especially when children experience these effects early on in life. For example, children who had inadequate amounts of iron in vitro show permanent cognitive deficiencies because iron affects neurotransmission and dopamine levels. When dopamine is not received properly due to iron deficiency, infants develop delayed cognition and mental growth (Young, 2012).

Nutraceuticals in male reproductive abnormalities/diseases

Abnormalities in the male reproductive system can be induced by infectious organisms, exposure to environmental toxins, and the individual's life style among other causes. Obesity is one of the conditions that have an effect on male fertility and reproductive efficiency. Subfertility associated with male obesity may be restored with very short-term micronutrient supplementation that targets the timing of the transit of sperm through the epididymis, which is the developmental window where sperm are most susceptible to oxidative damage. Improving micronutrient deficiency in obese men through oral supplementation may be a viable intervention for reducing sperm reactive oxygen species concentrations and improving fetal outcomes (McPherson et al., 2019). Male reproductive disorders such as sexual dysfunction, infertility, cryptorchidism, hypospadias, and testicular cancer are diseases of much concern currently (Bonde, 2010). Among many abnormalities and diseases of the male reproductive system, PCa is a major issue. Benign prostatic hyperplasia also affects large numbers of men above 50 years of age. It has been reported that an unhealthy dietary pattern is associated with increased risk of PCa, while a healthy dietary pattern is associated with a decreased risk of PCa (Bagheri et al., 2018). This section highlights the use of some nutraceuticals on disorders and diseases of the prostate gland.

Effects of nutraceuticals on diseases of the prostate gland

PCa is a major public health issue in the United States. Epidemiological and observational data endorse diet as a risk factor for PCa (Kumar et al., 2019). Plant-based healthy diets and life styles are reported to play an important role in reducing the risks of developing prostate diseases and cancer in men. It is reported that a Mediterranean diet is associated with lower risk of PCa (Hernáez and Estruch, 2019). A number of trials have been made to look into the effects of nutritional plants and plant derivatives on PCa. According to the World Health Organization, one-third of all cancer deaths are preventable through increased consumption of natural compounds capable of modulating key molecular signaling cascades that ultimately inhibit cancer cell proliferation and induce apoptosis (Lodi et al., 2017). Overexpression of chromatin modifier protein, metastasis-associated protein 1 (MTA1), in PCa contributes to tumor aggressiveness and metastasis (Kolhatkar et al., 2018). Gnetin C, a resveratrol dimer and dietary compound found in the melinjo plant, exerts its anticancer activity in PCa, in part, through the inhibition of cancer-promoting cooperation between MTA1 and erythroblastosis E26 transformation-specofic 2 (ETS2). Gnetin C specifically and strongly downregulates MTA1 and ETS2 expression in PCa cells, causes cytotoxicity, cell death, and a reduction of the metastatic ability of PCa cells also through MTA1-mediated mechanisms, indicating its potential for chemoprevention and therapy in PCa (Kumar et al., 2019).

A number of other dietary phytochemicals, including curcumin, ursolic acid, epigallocatechin-3-gallate, resveratrol, sulforaphane, and 6-shogaol, have shown potential chemopreventive effects in vitro and in vivo in either animal models or clinical studies on several cancers, including PCa. Ursolic acid, curcumin, and resveratrol administered to a mouse allograft model of PCa through the diet, either alone or in combination, produced synergistic effects on tumor size and weight, indicating that synergistic combinations of natural compounds may be important for cancer therapeutic interventions (Lodi et al., 2017). Curcumin has been shown to have some effects on signaling pathways implicated in PCa. It has an effect on cancer-associated fibroblasts, which are important in the regulation of epithelial to mesenchymal transition in tumor cells and in the induction of stem cell characteristics (McCubrey et al., 2017).

Other plant derivative compounds such as beta-sitosterol are reported to have anticancer properties against PCa among many others (Sayeed and Ameen, 2015). Berberine also decreased PCa growth and cellular testosterone formation. It suppressed the migration and invasion of PCa cells and decreased the expression of certain mesenchymal genes such as bone morphogenetic protein-7, nodal growth differentiation factor, and SNAIL, normally associated with shorter survival of PCa patients. Moreover, berberine inhibited hypoxia-inducible factor-1α and vascular endothelial growth factor expression in PCa cells, and increased their radiosensitivity in vitro and in vivo. Berberine also targets epidermal growth factor receptor signaling to suppress proliferation of PCa cells in vitro, and was shown to increase the induction of apoptosis (McCubrey et al., 2017).

Other plants such as pomegranate fruit have been gaining widespread reputations as dietary supplements as well as functional foods due to emerging scientific evidence on potential health benefits, including prevention and treatment of erectile dysfunction in male subjects among many other effects. Pomegranate contains phytochemicals, including flavonoids (anthocyanins and catechins), flavonols (kaempferol and quercetin), flavones (apigenin and luteolin), conjugated fatty acids, hydrolyzable tannins, and related compounds, which are thought to be responsible for various biological and pharmacological activities including cancer-preventive and therapeutic effects against many cancers including PCa (Mandal et al., 2017).

Consistent with cell and animal models, epidemiological studies have demonstrated a link between the intake of cruciferous vegetables and a reduction in the incidence or progression of PCa. Alliaceous vegetables, including garlic, onions, leeks, and shallots, are rich in a variety of bioactive compounds, such as flavonols, oligosaccharides, selenium, arginine, and organosulfur compounds. A potential antiproliferative effect of lycopene on PCa cells, with postulated mechanisms linked to cell cycle arrest and apoptosis has been demonstrated. Lycopene has been shown to exert antiproliferative effects through its ability to induce G2/M-phase cell cycle arrest and apoptotic cell death on LNCaP cells. Resveratrol in wine also appears to have a significant multimodal antitumor activity on PCa cells through increasing apoptosis, cell cycle arrest, and downregulation of signal transduction pathways. Curcumin also has a multimodal effect on PCa similar to other dietary bioactives and has been proven in vitro to inhibit the growth of PCa cells by increasing apoptosis, suppressing NF-κB activation, downregulating the expression of B-cell lymphoma 2 (Bcl-2) and B-cell lymphoma extra-large (Bcl-xL), and leading to activation of caspase-3 and -8 (Livingstone et al., 2019).

In addition to PCa, benign prostatic hyperplasia is known to commonly affect most men over age 50. Various modern drugs are prescribed for benign prostatic hyperplasia. A cure for benign prostatic hyperplasia without side effects is not available currently. Due to the adverse effects of the drugs, it is desirable to develop effective and long-term safe herbal medicines to inhibit the progress of benign prostatic hyperplasia (Chung et al., 2016). Pumpkin seed has long been used to treat micturition disorders and is found to substantially improve lower urinary tract symptoms associated with benign prostatic hyperplasia. The observed symptom relief is accompanied by a clinically significant improvement in International Prostate Symptom Score-related quality of life (Vahlensieck et al., 2015). Another study showed that garlic may have suppressing effects on benign prostatic hyperplasia. Its administration decreased the relative prostate weight ratio, suppressed the mRNA expression level of androgen receptor, dihydrotestosterone serum levels, and growth of prostatic tissue in benign prostatic hyperplasia-induced rats. Garlic administration prevented the progress of benign prostatic hyperplasia by regulating inflammatory responses through decreased levels of inflammatory proteins and apoptosis, which will be related to antioxidant properties of garlic and its components, suggesting that garlic has great potential as a treatment for benign prostatic hyperplasia (Chung et al., 2016).

All in all, many nutraceuticals with known antibacterial and antifungal properties such as garlic (Zardast et al., 2016) and potent antifungal and antibacterial nutraceutical plant species such as *B. aegyptiaca* and *C. erythrocarpos* (Anywar et al., 2014) will have potential to treat bacterial- and fungal-induced reproductive diseases in humans and animals.

Concluding remarks and future directions

Reproductive disorders in males and females, including infertility, infection-associated inflammation, cancer, developmental disorders, and many other abnormalities, are among the most common health issues in humans and animals. Environmental pollution due to immense industrial development, use of pesticides in profit-oriented modern agriculture, and chemicals for various household uses in humans and animals have immensely contributed to reproductive disorders, mainly in industrialized and developing countries. Currently, modern drugs are routinely prescribed to manage and treat reproductive abnormalities and other diseases. However, the common occurrence of side effects of modern medical treatments has encouraged patients resort to the use of traditional and alternative medicines, mainly nutraceuticals, to prevent or minimize the incidence of reproductive abnormalities and to treat clinical diseases. As a result, the use of natural products, including nutraceuticals, to tackle reproductive disorders has increased. Encouraging results supported by research undertakings and clinical trials are inspiring greater hope in the future use of nutraceuticals for the treatment of various reproductive diseases and developmental disorders in animals and humans.

References

Adams, J.B., Audhya, T., Geis, E., et al., 2018. Comprehensive nutritional and dietary intervention for autism spectrum disorder—a randomized, controlled 12-month trial. Nutrients. 10 (3), 369. Available from: https://doi.org/10.3390/nu10030369. 2018 Mar 17.

Akhtar, M.S., Hossain, M.A., Said, S.A., 2017. Isolation and characterization of antimicrobial compound from the stem-bark of the traditionally used medicinal plant, *Adenium obesum*. J. Tradit. Complement. Med. 7 (3), 296–300.

Anywar, G., Oryem-Origa, H., Kamatenesi-Mugisha, M., 2014. Antibacterial and antifungal properties of some wild nutraceutical plant species from Nebbi district, Uganda. Br. J. Pharm. Res. 4 (14), 1753–1761.

Áyen, Á., Jiménez Martínez, Y., Marchal, J.A., et al., 2018. Recent progress in gene therapy for ovarian cancer. Int. J. Mol. Sci. 19 (7), 1930. Available from: https://doi.org/10.3390/ijms19071930.

Bagheri, A., Nachvak, S.M., Rezaei, M., et al., 2018. Dietary patterns and risk of prostate cancer: a factor analysis study in a sample of Iranian men. Health Promot. Perspect. 8 (2), 133–138.

Banerjee, N., Talcott, S., Safe, S., et al., 2012. Cytotoxicity of pomegranate polyphenolics in breast cancer cells in vitro and vivo: potential role of miRNA-27a and miRNA-155 in cell survival and inflammation. Breast Cancer Res. Treat. 136, 21–34.

Barrea, L., Arnone, A., Annunziata, G., et al., 2019. Adherence to the Mediterranean diet, dietary patterns and body composition in women with polycystic ovary syndrome (PCOS). Nutrients. 11 (10), 2278. Available from: https://doi.org/10.3390/nu11102278. 2019 Sep 23.

Bonde, J.P., 2010. Male reproductive organs are at risk from environmental hazards: a review. Asian J. Androl. 12, 152–156.

Bradlow, H.L., 2008. Indole-3-carbinol as a chemoprotective agent in breast and prostate cancer. Vivo. 22, 441–445.

Butalla, A.C., Crane, T.E., Patil, B., et al., 2012. Effects of a carrot juice intervention on plasma carotenoids, oxidative stress, and inflammation in overweight breast cancer survivors. Nutr. Cancer. 64, 331–341.

Chauhan, B., Kumar, G., Kalam, N., et al., 2013. Current concepts and prospects of herbal nutraceutical: a review. J. Adv. Pharm. Technol. Res. 4, 4–8.

Chung, K.S., Shin, S.J., Lee, N.Y., et al., 2016. Anti-proliferation effects of garlic (*Allium sativum* L.) on the progression of benign prostatic hyperplasia. Phytother. Res. 30 (7), 1197–1203.

De Franciscis, P., Grauso, F., Luisi, A., et al., 2017. Adding *Agnus Castus* and Magnolia to soy isoflavones relieves sleep disturbances besides postmenopausal vasomotor symptoms long-term safety and effectiveness. Nutrients. 9, 129. Available from: https://doi.org/10.3390/nu9020129. 2017.

Egnell, M., Fassier, P., Lécuyer, L., et al., 2017. B-Vitamin intake from diet and supplements and breast cancer risk in middle-aged women: results from the prospective NutriNet-Santé cohort. Nutrients. 9, 488. Available from: https://doi.org/10.3390/nu9050488. 2017.

Fadus, M.C., Lau, C., Bikhchandani, J., et al., 2017. Curcumin: an age-old anti-inflammatory and anti-neoplastic agent. J. Tradit. Complement. Med. 7, 339–346 (2017).

Fu, Y., Chang, H., Peng, X., et al., 2014. Resveratrol inhibits breast cancer stem-like cells and induces autophagy via suppressing Wnt/β-catenin signaling pathway. PLoS One. 9, e102535. Available from: https://doi.org/10.1371/journal.pone.0102535. 2014.

Hemingway, R.G., 2003. The influences of dietary intakes and supplementation with selenium and vitamin-E on reproduction diseases and reproductive efficiency in cattle and sheep. Vet. Res. Commun. 27 (2), 159–174.

Hernáez, Á., Estruch, R., 2019. The Mediterranean diet and cancer: what do human and molecular studies have to say about It? Nutrients. 11 (9), 2155. Available from: https://doi.org/10.3390/nu11092155. 2019 Sep 9.

Hofstee, P., McKeating, D.R., Bartho, L.A., et al., 2020. Maternal selenium deficiency in mice alters offspring glucose metabolism and thyroid status in a sexually dimorphic manner. Nutrients. 12 (1), 267. 2020 Jan.

Hong, S.A., Kim, K., Nam, S.J., et al., 2008. A case-control study on the dietary intake of mushrooms and breast cancer risk among Korean women. Int. J. Cancer 122, 919–923.

Hussain, A., Sharma, C., Khan, S., et al., 2015. *Aloe vera* inhibits proliferation of human breast and cervical cancer cells and acts synergistically with cisplatin. Asian Pac. J. Cancer Prev. 16 (7), 2939–2946.

Imanshahidi, M., Hosseinzadeh, H., 2008. Pharmacological and therapeutic effects of *Berberis vulgaris* and its active constituent, berberine. Phytother. Res. 22 (8), 999–1012.

Joo, J.H., Hong, S.S., Cho, Y.R., Seo, D.W., 2016. 10-Gingerol inhibits proliferation and invasion of MDA-MB-231 breast cancer cells through suppression of Akt and p38(MAPK) activity. Oncol. Rep. 35, 779–784.

Kalra, E.K., 2003. Nutraceutical—definition and introduction. AAPS PharmSci. 5, E25. Available from: https://doi.org/10.1208/ps050325. 2003.

Kaneko, Y., Sano, M., Seno, K., et al., 2019. Olive leaf extract (OleaVita) suppresses inflammatory cytokine production and NLRP3 inflammasomes in human placenta. Nutrients. 11 (5), 970.

Kim, J.H., Woo, J.-H., Kim, H.M., et al., 2017. Anti-endometriotic effects of Pueraria flower extract in human endometriotic cells and mice. Nutrients. (9), 212. Available from: https://doi.org/10.3390/nu9030212. 2017.

Kolhatkar U., Dholakia K., Sikorska G., et al., 2018. Abstract 253: Gnetin C as a candidate for targeted chemopreventive and therapeutic measures in prostate cancer. Metrics Proceedings: AACR Annual Meeting 2018; April 14–18, 2018; Chicago, IL. DOI: 10.1158/1538-7445. AM2018-253.

Krusinska, B., Wadolowska, L., Biernacki, M., et al., 2019. Serum "vitamin-mineral" profiles: associations with postmenopausal breast cancer risk including dietary patterns and supplementation. A case-control study. Nutrients. 11 (9), 2244. Available from: https://doi.org/10.3390/nu11092244. 18.

Kumar, A., Dholakia, K., Sikorska, G., et al., 2019. MTA1-dependent anticancer activity of gnetin C in prostate cancer. Nutrients. 11 (9), 2096.

Larussa, T., Imeneo, M., Luzza, F., 2017. Potential role of nutraceutical compounds in inflammatory bowel disease. World J. Gastroenterol. 23 (14), 2483–2492.

Lee, H.S., Seo, E.Y., Kang, N.E., et al., 2008. [6]-Gingerol inhibits metastasis of MDA-MB-231 human breast cancer cells. J. Nutr. Biochem. 19, 313–319.

Li, J.Y., Zou, L., Chen, W., et al., 2014. Dietary mushroom intake may reduce the risk of breast cancer: evidence from a meta-analysis of observational studies. PLoS One 9, e93437. 2014.

Li, Y., Li, S., Meng, X., et al., 2017. Dietary natural products for prevention and treatment of breast cancer. Nutrients. 9, 728. Available from: https://doi.org/10.3390/nu9070728.

Livingstone, T.L., Beasy, G., Mills, R.D., et al., 2019. Plant bioactives and the prevention of prostate cancer: evidence from human studies. Nutrients. 11 (9), 2245. Available from: https://doi.org/10.3390/nu11092245. 18.

Lodi, A., Saha, A., Lu, X., et al., 2017. Combinatorial treatment with natural compounds in prostate cancer inhibits prostate tumor growth and leads to key modulations of cancer cell metabolism. NPJ Precis. Oncol. Available from: https://doi.org/10.1038/s41698-017-0024-z1:18.

Mandal, A., Bhatia, D., Bishayee, A., 2017. Anti-inflammatory mechanism involved in pomegranate-mediated prevention of breast cancer: the role of NF-κB and Nrf2 signaling pathways. Nutrients. 9, 436. Available from: https://doi.org/10.3390/nu9050436.

Markhus, M.W., Dahl, L., Moe, V., et al., 2018. Maternal iodine status is associated with offspring language skills in infancy and toddlerhood. Nutrients. 10 (9), 1270. Available from: https://doi.org/10.3390/nu10091270.

McCubrey, J.A., Lertpiriyapong, K., Steelman, L.S., et al., 2017. Effects of resveratrol, curcumin, berberine and other nutraceuticals on aging, cancer development, cancer stem cells and microRNAs. Aging. 9 (6).

McPherson, N.O., Shehadeh, H., Fullston, T., et al., 2019. Dietary micronutrient supplementation for 12 days in obese male mice restores sperm oxidative stress. Nutrients. 11 (9), 2196. Available from: https://doi.org/10.3390/nu11092196.

Mehta, R., Lansky, E.P., 2004. Breast cancer chemopreventive properties of pomegranate (*Punica granatum*) fruit extracts in a mouse mammary organ culture. Eur. J. Cancer Prev. 13, 345–348.

Mousa, A., Naqash, A., Lim, S., 2019. Macronutrient and micronutrient intake during pregnancy: an overview of recent evidence. Nutrients. 11 (2), 443. Available from: https://doi.org/10.3390/nu11020443.

Patel, K., Patel, D.K., 2017. Medicinal importance, pharmacological activities, and analytical aspects of hispidulin: a concise report. J. Tradit. Complement. Med. 7, 360–366 (2017).

Sayeed, B., Ameen, S.S., 2015. Beta-sitosterol: a promising but orphan nutraceutical to fight against cancer. Nutr. Cancer. 67 (8), 1214–1220.

Sazawal, S., Black, R.E., Menon, V.P., et al., 2001. Zinc supplementation in infants born small for gestational age reduces mortality: a prospective, randomized, controlled trial. Pediatrics 108 (6), 1280–1286. Available from: https://doi.org/10.1542/peds.108.6.1280.

Schneider-Stock, R., Fakhoury, I.H., Zaki, A.M., et al., 2014. Thymoquinone: fifty years of success in the battle against cancer models. Drug. Discov. Today 19, 18–30.

Siegel, R.L., Miller, K.D., Jemal, A., 2016. Cancer statistics, 2016. CA Cancer J. Clin. 66, 7–30.

Song, X., Li, Z., Ji, X., Zhang, D., 2017. Review: calcium intake and the risk of ovarian cancer: a meta-analysis. Nutrients. 679. Available from: https://doi.org/10.3390/nu9070679.

Thomson, C.A., Ho, E., Strom, M.B., 2016. Chemopreventive properties of 3,30-diindolylmethane in breast cancer: evidence from experimental and human studies. Nutr. Rev. 74, 432–443.

Tsuji, P.A., Carlson, B.A., Anderson, C.B., et al., 2015. Dietary selenium levels affect selenoprotein expression and support the interferon- and IL-6 immune response pathways in mice. Nutrients. 7, 6529–6549. Available from: https://doi.org/10.3390/nu7085297.

Vahlensieck, W., Theurer, C., Pfitzer, E., et al., 2015. Effects of pumpkin seed in men with lower urinary tract symptoms due to benign prostatic hyperplasia in the one-year, randomized, placebo-controlled GRANU study. Urol. Int. 94 (3), 286–295.

Wu, H.L., Chuang, T.Y., Al-Hendy, A., et al., 2015. Berberine inhibits the proliferation of human uterine leiomyoma cells. Fertil. Steril. 103, 1098–1106.

Xie, Q., Chen, M.L., Qin, Y., et al., 2013. Isoflavone consumption and risk of breast cancer: a dose-response meta-analysis of observational studies. Asia Pac. J. Clin. Nutr. 22, 118–127.

Young, R., 2012. Nutrient and iron deficiency: focus on fetal brain development and beyond. Int. J. Childbirth Educ 27 (4), 65–69.

Yung, M.M., Ross, F.A., Hardie, D.G., et al., 2016. Bitter melon (*Momordica charantia*) extract inhibits tumorigenicity and overcomes cisplatin-resistance in ovarian cancer cells through targeting AMPK signaling cascade. Integr. Cancer Ther. 15, 376–389.

Zardast, M., Namakin, K., Kaho, J.E., et al., 2016. Assessment of antibacterial effect of garlic in patients infected with *Helicobacter pylori*: assessment of garlic's effect using urease breath test. Avicenna J. Phytomed 6 (5), 495–501.

Zhao, Y., Jing, Z., Lv, J., et al., 2017. Berberine activates caspase-9/cytochrome c-mediated apoptosis to suppress triple-negative breast cancer cells in vitro and in vivo. Biomed. Pharmacother. 95, 18–24.

Zhou, Q., Ye, M., Lu, Y., et al., 2015. Curcumin improves the tumoricidal effect of mitomycin C by suppressing ABCG2 expression in stem cell-like breast cancer cells. PLoS One. 10, e0136694. Available from: https://doi.org/10.1371/journal.pone.0136694. 2015.

Chapter 12

Nutraceuticals in ophthalmic diseases

Szabina A. Stice and Renata Kolanos

Introduction

The number of blind people and those with visual impairment has been increasing over time due to the growth and aging of the world's population (Bourne et al., 2017). Leading causes of blindness and visual impairment are uncorrected reflective error, cataract, age-related macular degeneration (AMD), glaucoma, diabetic retinopathy (DR), corneal opacity, and trachoma (a bacterial infection) (Flaxman et al., 2017). Reasons for vision impairment vary by age, sex, region, and income level (Flaxman et al., 2017; Stevens et al., 2013). According to Flaxman et al. (2017), globally, 81.2% of people had a preventable or treatable cause of their blindness or visual impairment in 2015.

Surgery and ocular medicines can treat a wide variety of eye diseases to various degrees. Nutrition has an important role in eye health, and vitamin and essential trace element deficiencies have been associated with ophthalmic diseases and decreased ocular function (Mares et al., 2017). For example, systemic zinc depletion commonly manifests in poor dark adaptation, vitamin A deficiency may cause night blindness, specific or mixed vitamin B deficiencies are associated with nutritional amblyopia, and hypovitaminosis D has been linked to multiple ophthalmic diseases such as uveitis, AMD, and dry eye disease (DED) (Ugarte et al., 2013; Mares et al., 2017; Llop et al., 2019; Layana et al., 2017; Yang et al., 2018). On the other hand, excess intake of certain vitamins and essential trace elements also is linked to deleterious effects in ophthalmic diseases and symptoms, as well as to toxic effects. Therefore proper nutrition and/or the use of nutraceuticals may help maintain eye health and prevent, slow down, or treat certain ophthalmic diseases.

Several supplements, such as vitamins, minerals, botanicals, botanical extracts, and other products marketed for eye health and ophthalmic diseases often are referred to as nutraceuticals. The aim of this chapter is to summarize the evidence—or in many cases, the lack of it—in support of supplementation for the most common eye diseases, including cataract, glaucoma, AMD, retinopathy, retinitis pigmentosa (RP), DED, and inflammatory ophthalmic diseases. We focus mainly on summarizing the outcomes of human clinical trials evaluating the effectiveness of nutraceuticals rather than in vitro and in vivo animal studies, as beneficial effects reported in in vitro or in vivo animal studies often do not translate to humans. Unfortunately, adequate human data are lacking for many promising nutraceuticals. Additionally, this chapter briefly discusses the mode of action and safety of nutraceuticals used for the prevention, delay, or treatment of ophthalmic diseases.

Efficacy of nutraceuticals in the treatment and prevention of various ophthalmic diseases

Cataract

Cataract is a clouding of the lens of the eye (National Eye Institute, 2019). It causes vision to become blurry, foggy, or less colorful (i.e., faded), as well as double vision or trouble seeing at night. Surgery is a cost-effective and common way of treating cataracts. Globally, approximately 52.6 million people suffered from mild to moderate vision impairment and 12.6 million were blind due to cataract in 2015 (Flaxman et al., 2017). In the same year, in those aged 50 and older, cataract was the leading cause of blindness and one of the leading causes of moderate to severe visual impairment. The number of people with cataract is expected to rise in the future; therefore, reducing the risk and slowing the progression of cataracts are of utmost importance. Although protecting the eyes from direct sunlight, quitting smoking and eating a well-balanced diet rich in fruits and vegetables may help delay or reduce the risk of cataracts, more options are needed (Sella and Afshari, 2019; National Eye Institute, 2019).

Cataracts can be classified according to cause as age-related, pediatric, and secondary to other causes

Nutraceuticals. DOI: https://doi.org/10.1016/B978-0-12-821038-3.00012-4

(Liu et al., 2017). Though age is the most common cause, eye injury, diabetes, malnutrition, corticosteroid use, smoking, spending a lot of time in the sun, and having relatives with cataracts, along with other factors, increase the risk of developing cataracts. Nonenzymatic glycation of lens proteins, damage by reactive oxygen species (ROS), and aldose reductase (AR) activity resulting in intracellular accumulation of sorbitol that leads to the degeneration, collapse, and liquefaction of lens fibers may contribute to the development of lens opacity (Kaur et al., 2017). Therefore potential cataract prevention strategies include the use of antioxidants and ROS scavengers, AR inhibitors, antiglycating agents, and inhibitors of lens epithelial cell apoptosis. Vitamins (e.g., C and E), polyphenols (e.g., flavonoids), and carotenoids (e.g., lutein, zeaxanthin, and β-carotene) have the potential to prevent cataracts due to their antioxidant activities. Additionally, AR inhibitors (e.g., ellagic acid, rutin, quercetin, epicatechin, and curcumin), antiglycating agents [e.g., flavonoids, carotenoids, and polyunsaturated fatty acids (PUFAs)], and inhibitors of lens epithelial cell apoptosis [e.g., polyphenols (e.g., resveratrol), grape seed extract, and Coenzyme Q10] are also reasonable anticataract agents/nutraceuticals.

Nutraceuticals for the treatment of cataract

Numerous studies and reviews have examined the effects of various nutraceuticals for the prevention of cataracts. While some studies or reviews may find beneficial effects for a specific nutraceutical, other studies or reviews for the same substance may show no effects. For example, based on their meta-analysis of the evidence from epidemiological studies, Wang et al. (2014) concluded that greater β-carotene intakes might be inversely associated with the risk for cataract. On the other hand, a Cochrane Systematic Review found no evidence of effect for β-carotene in reducing the risk of cataract or the rates of cataract extraction (Mathew et al., 2012). The same is true for lutein and zeaxanthin; while some studies or reviews show beneficial effects for cataracts (e.g., Jiang et al., 2019; Sideri et al., 2019), others found that lutein plus zeaxanthin intake did not reduce the occurrence of cataract surgery or the progression of lens opacities (e.g., Glaser et al., 2015).

A large-scale randomized trial of middle-aged and older men, the Physicians' Health Study II (PHS II), revealed that use of a long-term daily multivitamin (Centrum Silver, Pfizer Inc.) modestly and significantly reduced the incidence of cataracts after more than 10 years of treatment and follow-up (Christen et al., 2014). While numerous studies and reviews concluded that high intakes of vitamins A, C, D, E, or K1 may be inversely associated with the risk of age-related cataract (e.g., Sella and Afshari, 2019; Liu et al., 2018a; Wei et al., 2016; Zhang et al., 2015; Wang et al., 2014), others concluded there is no evidence that many of these vitamins prevent or slow the progression of age-related cataracts (e.g., Mathew et al., 2012, vitamins C and E; Christen et al., 2015, vitamin E; and Sideri et al., 2019, vitamins A and E). A currently (as of December 2019) ongoing large-scale clinical trial, COSMOS-Eye (NCT03205202) examining whether a multivitamin supplement (Centrum Silver) reduces the risk of cataract, may provide more clarity on the role of multivitamin supplementation in cataracts. In addition, COSMOS-Eye is also examining whether a cocoa extract supplement—two capsules each day containing a total of 600 mg cocoa flavanols (polyphenols)—reduces the risk of cataract. Polyphenols derived from cocoa were shown earlier to inhibit diabetes-induced cataract formation in rats (Osakabe et al., 2004).

Sometimes, different studies of the same substance report no effects, beneficial effects, and possibly harmful effects. While in some of the abovementioned studies vitamin C or E supplementation was either found to reduce the risk of age-related cataracts or had no net effect; in a large, prospective, population-based cohort of men, those who used high-dose vitamin C or vitamin E supplements had a statistically significant increased risk of age-related cataract (Zheng Selin et al., 2013) and in a large, prospective, population-based cohort of women, vitamin C supplementation was associated with a higher risk of age-related cataract (Rautiainen et al., 2009). Of the B vitamins, B2, B6, and B12 were found to be associated with a protective effect for nuclear and/or cortical lens opacities or cataracts, while the highest intake of dietary folate (vitamin B9) was associated with an increased risk of mild posterior subcapsular lens opacity development in participants taking multivitamins (Glaser et al., 2015). In a large-scale randomized trial of women, daily supplementation with a combination of folic acid (converted to folate in the body) and vitamins B6 and B12 had no significant effect on cataracts, and may have increased the risk of cataract extraction (Christen et al., 2016). The results of a large-scale population-based prospective study suggested that the use of high-dose B vitamin supplements is associated with an increased risk for age-related cataract (Selin et al., 2017). Because many studies and reviews report conflicting results, more studies are needed to establish the long-term benefits and safety of nutraceuticals in preventing the formation of cataracts, delaying their progression, or decreasing the occurrence of cataract surgery.

Glaucoma

Glaucomas are a group of progressive optic neuropathies (i.e., damage to the optic nerve) that may lead to blindness (Weinreb et al., 2014). While intraocular pressure (IOP) is the main risk factor, oxidative stress, excitotoxicity, neuroinflammation, and impaired ocular blood flow may also play a role in the development of this ophthalmic disease (Morrone et al., 2018). Currently, reducing IOP is the most important and proven form of treatment. Nutraceuticals capable of lowering IOP, modulating excitotoxicity, reducing inflammation, increasing circulation to the optic nerve, and promoting retinal ganglion cell (a type of neuron located near the inner surface of the retina) survival may provide an additional option for reducing the risk of or slowing the progression of glaucoma, decreasing the need for conventional therapies.

Globally, in 2015, approximately 4.0 million people suffered from mild to moderate vision impairment and 2.9 million were blind due to glaucoma (Flaxman et al., 2017). In the same year, it was one of the leading causes of blindness in those aged 50 and older. The number of people with glaucoma is expected to rise in the future; therefore evaluation of the safety and efficacy of nutraceuticals with the potential to lower IOP, or with antioxidant or neuroprotective properties, is of great importance.

Nutraceuticals for the treatment of glaucoma

Numerous studies and reviews have examined whether nutritional interventions or nutraceuticals can prevent or delay the progression of glaucoma. A review of 33 trials concluded that flavonoids may have beneficial effects in glaucoma by improving ocular blood flow and potentially slowing progression of visual field loss (Loskutova et al., 2019). Additionally, the authors stated that forskolin, found in the Indian *Coleus* plant and available as a dietary supplement, has the capacity to reduce IOP beyond the levels achieved with traditional therapy alone, and therefore has a potential role in treating glaucoma. However, more evidence is needed to make any recommendations.

For most nutraceuticals, including, but not limited to, vitamins A, C, D, and E, Coenzyme Q, saffron, curcumin, ginseng, *Ginkgo biloba*, α-lipoic acid, palmitoyl ethanolamide, and citicoline, there is either a lack of consistent findings or human studies (Loskutova et al., 2019; Morrone et al., 2018; Ramdas et al., 2018). Study findings for vitamin D, for example, are inconsistent. Topical administration of vitamin D analogs (1α,25-dihydroxyvitamin D3 and its analog, 2-methylene-19-nor-(20S)-1α,25-dihydroxyvitamin D3) was found to markedly suppress IOP in nonhuman primates (Kutuzova et al., 2012). Moreover, patients with primary open-angle glaucoma had a lower mean serum 25-hydroxy-vitamin D [25(OH) D] concentration and a greater prevalence of vitamin D insufficiency than did controls (Goncalves et al., 2015). However, in healthy patients with lower 25(OH)D levels, supplementation with 20,000 IU of vitamin D3 twice a week for 6 months did not affect IOP (Krefting et al., 2014). Additionally, no associations between serum 25 (OH)D levels and IOP were revealed in this study.

Though human studies for many nutraceuticals with the potential to treat or prevent glaucoma are either lacking or conflicting, this does not mean that they are ineffective, only that additional research is needed to obtain more evidence regarding their associations with glaucoma. These research studies need to be larger, longer-term, and better designed to enable practitioners to make recommendations. A currently ongoing small study (NCT03959592) aims to determine the effect of lutein (10 mg), zeaxanthin (2 mg), and meso-zeaxanthin (10 mg) supplementation on visual performance in glaucoma patients (www.clinicaltrials.org).

Age-related macular degeneration

The macula, responsible for central, high-resolution, color vision, is located near the center of the retina. Degeneration of the macula may result in blurred or no vision in the center of the visual field. The progressive loss of this central vision can negatively impact quality of life and increases the number of falls and the risk of functional disability and cognitive impairment in the elderly (Mitchell et al., 2018). Globally, in 2015, approximately 8.4 million people suffered from mild to moderate vision impairment due to AMD, a number that was expected to increase to 8.8 million by 2020 (Flaxman et al., 2017). AMD is one of the leading causes of moderate to severe vision impairment in those aged 50 and older.

Nutraceuticals for the treatment of age-related macular degeneration

Genetic and nongenetic risk factors, such as smoking, age, and low dietary antioxidant intakes, increase one's susceptibility to AMD (Mitchell et al., 2018). Therefore, unsurprisingly, high-dose zinc, antioxidant vitamin supplements, and β-carotene (a precursor to vitamin A) have been investigated in multiple studies to determine whether they can prevent or slow the progression of AMD from early- to late-stage disease. Lutein and zeaxanthin have been investigated more thoroughly for the same purpose, since these carotenoids are present in the macula. In addition, other nutraceuticals—such as, but not limited to, vitamins B6, B9, and D, folate and calcium—have also been examined, but more research is needed.

The Rotterdam Study found that the dietary intake of both vitamin E and zinc was inversely associated with incident AMD; and that a high dietary intake of β-carotene, vitamins C and E, and zinc was associated with a substantially reduced risk of AMD in elderly persons (van Leeuwen et al., 2005). The long-term effects of a formulation of high-dose antioxidants and zinc supplement on the progression of AMD were examined in the Age-Related Eye Disease Study (AREDS) (Chew et al., 2013). In this study, participants received the following antioxidants daily, known as the AREDS formulation: vitamins C and E (500 mg and 400 IU, respectively); β-carotene (15 mg), along with zinc (80 mg as zinc oxide) and copper (2 mg as cupric oxide). Compared to the control group, the AREDS formulation demonstrated a significant odds reduction in the risk of developing advanced AMD or the development of neovascular AMD (an advanced form of AMD), in addition to a significant reduction in the development of moderate vision loss. No serious adverse effects were associated with the AREDS formulation. Unfortunately, no significant reduction was seen for central geographic atrophy, also an advanced form of AMD.

A follow-up study (AREDS2) assessed the value of replacing β-carotene with lutein (10 mg) and zeaxanthin (2 mg) due to the increased risk of lung cancer from β-carotene in former and current smokers (Chew et al., 2014) and because lutein and zeaxanthin are important components of the retina. The AREDS2 study found that, compared to β-carotene, lutein/zeaxanthin is more appropriate in AREDS-type supplements. A systematic literature review and meta-analysis to evaluate the relationship between dietary intake of lutein and zeaxanthin and AMD risk found that the dietary intake of lutein and zeaxanthin significantly reduced the risk of late, but not early, AMD; and a statistically significant inverse association was observed between lutein and zeaxanthin intake and neovascular AMD risk (Ma et al., 2012). Additionally, the Blue Mountains Eye Study (BMES), designed to assess the relationship between baseline dietary and supplement intakes of antioxidants and long-term risk of incident AMD, found that higher dietary lutein and zeaxanthin intake reduced the risk of long-term incident AMD (Tan et al., 2008).

For many nutraceuticals, the evidence is conflicting. For instance, in contrast to AREDS, in which a formulation of antioxidants, zinc, and copper showed a beneficial role in AMD, PHS II revealed that long-term daily multivitamin use had no significant effect on visually significant AMD after more than 10 years of treatment and follow-up (Christen et al., 2014). It is important to note that (1) while both the AREDS formulation and Centrum Silver used in PHS II contain vitamins C and E, zinc, and β-carotene (vitamin A: 20% as β-carotene in Centrum), the levels of these were much lower in Centrum Silver than in the AREDS formulation; and (2) the PHS II population generally had a lower risk profile than did the AREDS population. The discrepancy may or may not arise from the above two reasons or other reasons.

COSMOS-Eye is a large-scale ancillary study (NCT03205202) of the COcoa-Supplement and Multivitamins Outcomes Study (COSMOS) and aims to examine whether a multivitamin supplement (Centrum Silver) reduces the risk of AMD (www.clinicaltrial.gov). Estimated to be completed in 2020, the study may help clarify whether daily multivitamin supplementation has a beneficial effect on AMD. Despite the somewhat conflicting results, the American Academy of Ophthalmology noted that people with certain forms of AMD may benefit from a specific mix of vitamins and minerals [lutein (10 mg), zeaxanthin (2 mg), vitamins C (500 mg) and E (400 IU), zinc (80 mg), and copper (2 mg)] (www.aao.org). Interestingly, COSMOS-Eye is also examining whether a cocoa extract supplement (600 mg cocoa flavanols, including 80 mg (-)-epicatechin, and 50 mg theobromine) can reduce the risk of AMD.

Another example of conflicting results is the case of ω-3 fatty acids (FAs). While some studies show a protective role for fish and ω-3 FAs, other studies showed no net benefit of ω-3 FA supplementation for the prevention or slowing of the progression of AMD (Mitchell et al., 2018). The Macular Degeneration Omega-3 Study (MADEOSA; NCT03297515), a small study of approximately 60 participants, is in the process of assessing the therapeutic potential of ω-3 FA supplementation in dry macular degeneration, a form of AMD, and Stargardt disease, a genetic disease characterized by macular degeneration (www.clinicaltrials.gov). The study is estimated to be completed in 2020 and will help clarify the role of ω-3 FA supplementation in AMD.

Moreover, a large-scale ancillary study (NCT01782352) of the VITamin D and OmegA-3 Trial (VITAL), examining whether vitamin D3 (2,000 IU/day) and/or ω-3 FAs {840 mg of marine ω-3 FAs [465 mg of eicosapentaenoic acid (EPA) and 375 mg of docosahexaenoic acid (DHA)]}, reduce the incidence and/or progression of AMD may shed more light on the potentials of both vitamin D and ω-3 FAs to prevent or delay the onset of AMD (www.clinicaltrials.gov). This is especially important, as vitamin D deficiency has been associated with AMD, and studies assessing the effect of vitamin D supplementation in preventing the onset or progression of AMD are lacking (Layana et al., 2017). VITAL will also examine whether vitamin D3 and ω-3 FAs interact with each other, or with specific genes associated with AMD or vitamin D3 or ω-3 FA metabolic activity, to affect the risk of developing AMD. The study is estimated to be completed at the end of 2020.

Yet another example of conflicting results relates to B vitamins. A trial in a large cohort of women concluded that daily supplementation with a combination of folate and vitamins B6 and B12 may reduce the risk of AMD (Christen et al., 2009). Additionally, folate and vitamin B12 deficiencies were found to predict increased risk of incident AMD, suggesting their role in reducing AMD risk (Gopinath et al., 2013). On the other hand, the analysis of the AREDS data revealed that while high dietary folate intake is associated with a lower risk of progression to geographic atrophy in patients with a specified genotype, neither folate nor vitamin B6 or B12 were significantly associated with neovascular AMD (Merle et al., 2016).

As for many of the other eye diseases, no effects, beneficial effects, and possibly harmful effects have been reported for the same substance in different studies. For example, the authors of the BMES were not only unable to confirm the protective effect shown by the AREDS and the Rotterdam Study of a combination of high doses of zinc, β-carotene, and vitamins C and E on AMD, but also found that increasing β-carotene intake was associated with a higher risk of neovascular AMD, and high total vitamin E intake was associated with risk of late AMD or indistinct soft or reticular drusen (yellow deposits) (Tan et al., 2008). Another example relates to calcium supplementation. A secondary analysis of the AREDS study revealed that higher levels of dietary and supplementary calcium intake were associated with a lower incidence of progression to late AMD in AREDS participants, although the authors cautioned that the results may be due to chance or uncontrolled confounding, and additional studies are needed to determine whether calcium supplementation is beneficial in AMD (Tisdale et al., 2019). On the other hand, a cross-sectional study designed to investigate the association between self-reported supplementary calcium consumption and the prevalence of AMD concluded that supplementary calcium consumption is associated with increased prevalence of AMD (Kakigi et al., 2015). Therefore, as for the other ophthalmic diseases mentioned above, more research is needed to determine the efficacy and long-term safety of nutraceuticals to prevent and delay the progression of AMD.

Retinopathy

Retinopathy is damage to the retina of the eye. It can have different causes, including abnormalities in the blood vessels of the retina, a major cause of the disease. The major types of retinopathy are retinopathy of prematurity (ROP), DR, hypertensive retinopathy, and central serous retinopathy. Symptoms include blurred vision, vision loss, loss of central or color vision, and floaters (dark floating spots). DR is the most common type of retinopathy and the leading cause of blindness in US working-aged adults (Centers for Disease Control and Prevention, 2018). Globally, in 2015, approximately 2.6 million people suffered from mild to moderate vision impairment due to DR (Flaxman et al., 2017). This number may increase with the expected increase in the prevalence of diabetes. Healthy eating, reaching and maintaining a healthy weight, adequate physical activity, cessation of smoking, well-controlled blood sugar levels, and regular eye exams are recommended for those with diabetes to help decrease the risk or progression of DR. Medications, surgery, and laser treatment are often employed to treat it. Additionally, various nutraceuticals have been investigated for their ability to decrease the risk or slow the progression of DR.

Nutraceuticals for the treatment of diabetic retinopathy

Many studies investigating the possibly beneficial effects of nutraceuticals for the prevention or treatment of DR focus on carotenoids. A study by Hu et al. (2011) found that the serum concentrations of lutein and zeaxanthin were significantly lower in DR patients compared to healthy subjects. Additionally, the authors concluded that lutein (6 mg/day) and zeaxanthin (0.5 mg/day) for 3 months can improve visual acuity, contrast sensitivity, and macular edema in DR patients, showing the therapeutic potential of these two carotenoids. The ability of lutein supplementation alone to improve visual function in patients with DR was also evaluated (Zhang et al., 2017). In this study, 10 mg lutein/day for 36 weeks resulted in a slight improvement in visual acuity, marginally significant changes in contrast sensitivity, and a slight increase in glare sensitivity. The authors concluded that further studies are needed to determine if lutein supplementation could be used as an adjunct therapy in DR.

Lutein also was tested as a component of an antioxidant formulation (Garcia-Medina et al., 2011). In this study, patients received a formulation that provided daily doses of 120 mg vitamin C, 20 mg niacin (a form of vitamin B3), 20 mg α-tocopherol (a type of vitamin E), 6 mg β-carotene and 6 mg lutein (carotenoids), 27 mg of zinc, 2 mg of copper, 20 μg selenium, and 2 mg of manganese for 5 years. A retardation of DR progression was observed in the treatment group, indicating that oral antioxidant supplementation could be a beneficial adjunctive therapy. On the other hand, in the San Luis Valley Diabetes Study, not only was no protective effect observed between antioxidant nutrients such as vitamins C and E and β-carotene, but there appeared to be a potential for deleterious effects of these antioxidants in DR depending on insulin use (Mayer-Davis et al., 1998). A currently ongoing clinical trial (NCT03702374) aiming to support previous research on antioxidant therapy effects in DR outcome will also evaluate DR severity progress at the end of intervention from baseline, and may help clarify the role of antioxidants in DR treatment (www.clinicaltrials.gov). The nutraceutical being used in this study contains lutein (10 mg), astaxanthin (4 mg), zeaxanthin (1 mg), vitamin C (180 mg), vitamin E (30 mg), zinc

(zinc oxide, 20 mg), and copper (copper sulfate, 1 mg). The study is expected to be completed in 2020.

As increasing the dietary intake of PUFAs was associated with a reduced likelihood of the presence and severity of DR in well-controlled diabetes (Sasaki et al., 2015), and the intake of at least 500 mg of ω-3 FA/day at baseline had a 46% decreased risk of sight-threatening DR compared to those not meeting this target (Sala-Vila et al., 2016), ω-3 FA supplementation may be beneficial in DR. A currently (as of December 2019) ongoing trial (NCT04140201) investigating the effect of ω-3 FA (EPA) on the enhancement of visual acuity and on reduction of macular edema in patients with DR may help clarify the role of ω-3 FA in its treatment (www.clinicaltrials.gov).

The value of curcumin in the management of diabetic microangiopathy (a disease of the small blood vessels) and retinopathy was investigated in a small study (Steigerwalt et al., 2012). Treatment with 200 mg of curcumin in a highly bioavailable form for 4 weeks resulted in a significant decrease in the edema score, an improvement in retinal edema, a significant improvement in microcirculation, and an improvement in visual acuity, indicating that curcumin has potential value in the treatment of diabetic microangiopathy and retinopathy. Another study found that regular consumption of Chinese green tea every week for at least a year resulted in a DR risk reduction of approximately 50% in diabetic patients (Ma et al., 2015). The beneficial effects of green tea in diabetic retinas and in retinal cells under diabetic conditions were demonstrated in rats earlier, revealing that green tea protects the retina against neurodegeneration in DR (Silva et al., 2013). Therefore further exploring the efficacy of green tea for the treatment of DR is a desired goal.

Though melatonin has been proposed to be beneficial in DR, clinical trials are needed to confirm its efficacy in humans. An ongoing clinical trial (NCT03478306), evaluating the effect of 4 mg of melatonin on the visual function and progression of DR, slated to be completed in 2020, may shed some light on the role of melatonin in DR (www.clinicaltrials.gov). In addition to the abovementioned nutraceuticals, many others, such as, but not limited to, resveratrol, anthocyanins, saponins, quercetin, hesperidin, rutin, and catechin, were suggested to be potentially beneficial in DR based on in vitro and in vivo animal studies, but human clinical studies evaluating their efficacy and safety in DR are needed (Rossino and Casini, 2019).

Nutraceuticals for the treatment of retinopathy of prematurity

The potential role of lutein, zeaxanthin, and β-carotene, along with lycopene (also a carotenoid), was also investigated for the prevention or slowing of the progression of ROP. In a study providing 0.5 mg lutein/kg body weight (bw)/day and 0.02 mg zeaxanthin/kg bw/day, no differences were found in the incidence of ROP at any stage between treated and control groups (Romagnoli et al., 2011). While the results of a second study pointed to the protective effects of lutein on preterm retina health and maturation and showed that carotenoid supplementation (lutein, lycopene, and β-carotene) may decrease inflammation and progression to severe ROP, the total incidence of ROP was similar between the groups (Rubin et al., 2012). In two other trials, preterm infants were supplemented with 0.14 mg lutein and 0.006 or 0.0006 mg zeaxanthin with the objective of preventing ROP (Dani et al., 2012; Manzoni et al., 2013). While the supplemented groups had a slightly reduced incidence of ROP compared to the control group, in addition to a decreased progression rate from early- to higher-stage ROP, the reduced incidence rates did not reach statistical significance. As some of these studies showed beneficial effects, further studies are warranted to establish the efficacy of the preventative and therapeutic aspects of these carotenoids in ROP.

Retinitis pigmentosa

RP is a group of rare genetic disorders that involve a breakdown and loss of cells in the retina. Early symptoms include decreased night vision and loss of peripheral vision (tunnel vision). RP may lead to vision loss and blindness.

Nutraceuticals for the treatment of retinitis pigmentosa

Vitamin A showed beneficial effects for delaying vision loss in RP. In patients receiving either 15,000 IU of vitamin A, 15,000 IU of vitamin A and 400 IU of vitamin E, or 400 IU of vitamin E, vitamin A was found to slow the rate of retinal function decline, while the results for vitamin E suggested an adverse effect on the course of RP (Berson et al., 1993). Lutein supplementation (10 mg/day for 12 weeks, followed by 30 mg/day for another 12 weeks) was reported to improve the visual field and might also improve visual acuity slightly (Bahrami et al., 2006). In RP patients taking 15,000 IU of vitamin A/day, (1) lutein (12 mg/day) supplementation was observed to slow the loss of midperipheral visual field and (2) slower mean annual rates of decline were reported in distance and retinal visual acuities among those consuming a diet rich in ω-3 FAs (≥0.20 g/day), but (3) supplementation with 1200 mg DHA (an ω-3 FA) over 4 years did not, on average, slow the course of RP (Berson et al., 2004, 2010, 2012).

In a small trial, treatment with *Dunaliella bardawil* powder containing approximately 20 mg β-carotene, significantly increased retinal function was reported in patients with RP (Rotenstreich et al., 2013). As of December 2019, two clinical trials (NCT02018692 and NCT01680510) were listed on www.clinicaltrials.gov to study the substance's effects on visual functions in adolescent patients and adult patients, respectively, with RP. These two trials may shed additional light on the benefits of β-carotene in RP.

Dry eye syndrome/dry eye disease

Dry eye syndrome (DES)/DED is a condition of having dry eyes; that is, not having sufficient tears to lubricate the eyes, to wash away foreign matter, and to keep the surface of the eyes clear and smooth. This can be due to a deficiency of tear production or excessive tear evaporation. Consequently, the eyes become dry, red, irritated, scratchy, or burning, and vision may become blurry. Unfortunately, DES is a common and often chronic problem, especially in older people. The prevalence of DES is reported to vary from 5% to 34% (Messmer, 2015). Depending on the severity of the symptoms, treatment may involve artificial tear drops, medications (both oral and medicated eye drops), or plugs to occlude the efferent lacrimal ducts.

Nutraceuticals for the treatment of dry eye syndrome/dry eye disease

Once again, data regarding the efficacy of nutraceuticals to ameliorate the symptoms of DED are either conflicting or lacking. A good example of conflicting results is the case of ω-3 FAs. Based on the information obtained from the Women's Health Study, higher dietary intakes of ω-3 FAs are associated with a decreased incidence of DES (Miljanović et al., 2005). Additionally, a randomized controlled trial of ω-3 FAs in DES revealed that 65% of participants in the ω-3 FAs group (receiving 650 mg EPA and 350 mg of DHA) versus 33% of participants in the placebo group had significant improvement in symptoms at 3 months (Bhargava et al., 2013). Yet another study provided evidence that a nutraceutical formulation containing ω-3 FAs [1050 mg triglyceride bound DHA/day, 127.5 mg EPA/day, and 90 mg docosapentaenoic acid (DPA)/day]; vitamins A (400 μg/day); C (80 mg/day); and E (12 mg/day); essential trace elements [zinc (5 mg/day), copper (0.5 mg/day), magnesium (1 mg/day), and selenium (27.5 μg/day)]; and other components (tyrosine, cysteine, and glutathione) is an effective adjunctive treatment of DED in patients using artificial tears by showing a statistically significant improvement in DED after 3 months of treatment (Oleñik, 2014). A meta-analysis of five studies also suggested that ω-3 FA offers an effective therapy for DES (Liu and Ji, 2014). Moreover, a systematic literature search and analysis to assess whether ω-3 FA supplementation is more efficacious for the improvement of the symptoms and signs of DED than placebo concluded that ω-3 FA supplementation significantly improves DED symptoms and signs based on 17 randomized clinical trials involving 3363 patients (Giannaccare et al., 2019). On the other hand, in a multicenter, double-blind, placebo-controlled clinical trial, patients with DED who received supplements containing 3000 mg ω-3 FAs (2000 mg EPA/day and 1000 mg DHA/day) for a year did not have significantly better outcomes than those who were assigned to receive placebo (Dry Eye Assessment and Management Study Research Group, 2018). After a literature review and analysis of 34 randomized controlled trials involving 4314 adults, the review authors suggested that long-chain ω-3 FA supplementation may have a role in managing DED, although they found the evidence to be inconsistent and uncertain (Downie et al., 2019). Additionally, the Women's Health Study mentioned above revealed that a higher ratio of dietary ω-6 to ω-3 FA is associated with an increased risk of DES (Miljanović et al., 2005). Therefore studies clarifying the role of ω-3 FAs in DES are needed.

Fortunately, there are two ongoing clinical trials as of December 2019 that may help resolve this issue. An ancillary study (NCT 01880463) of the VITAL is examining whether vitamin D3 (2000 IU/day) and/or ω-3 FAs [840 mg of marine ω-3 FAs (465 mg of EPA and 375 mg of DHA)] reduce the incidence and/or progression of DED (www.clinicaltrials.gov). The study, with almost 26,000 participants, started in 2010 and is estimated to be completed at the end of 2021. A phase III study (NCT04181593) with approximately 300 participants designed to evaluate the safety and efficacy of OmegaD softgels (ω-3 FAs, two capsules/day for 84 days) for the treatment of DED is also ongoing (www.clinicaltrials.gov). The study was expected to be completed in mid-2020, and along with the VITAL study, may help settle the questions regarding the efficacy of ω-3 FA supplementation in DED.

Additionally, the ancillary study of the VITAL study may also shed light on the efficacy of vitamin D for alleviating symptoms of DED. It is especially important, as vitamin D deficiency has been linked to dry eye in some, but not all, studies (Yang et al., 2018). Moreover, supplemental vitamin D was reported to improve dry eye symptoms, tear quality, and ocular surface conditions (1000 IU orally; Yang et al., 2018) in addition to being effective for the treatment of patients with DES refractory to conventional treatment and with vitamin D deficiency (200,000 IU cholecalciferol via intramuscular injection; Bae et al., 2016). Additionally, vitamin D supplementation (2000 IU orally or 200,000 IU via intramuscular injection) was found to enhance the efficacy of topical treatment (hyaluronate and carbomer-based lipid-containing eye drops) and therefore could be a useful adjuvant therapy for patients with DED refectory to topical lubricants (Hwang et al., 2019). Therefore more research into vitamin D supplementation for the treatment of DED is welcomed.

While ω FAs and vitamin D appear to be the two most commonly researched nutraceuticals for the alleviation of dry eye symptoms and DES, other nutraceuticals, such as, but not limited to, ginseng, phytoestrogens, bilberry extract, and maqui berry extract were investigated for their ability to ameliorate the symptoms of DES. A small study revealed that Korean red ginseng supplementation (3 g/day for 8 weeks) may provide an additional treatment option for dry eye and for patients with glaucoma using antiglaucoma eye drops (Bae et al., 2015). A currently ongoing small clinical trial (NCT03992287) investigating the effects of daily supplementation of 2.4 g hydrolyzed red ginseng extract on dry eye for 12 weeks may help shed some more light on the potential of red ginseng extract to treat dry eye (www.clinicaltrials.gov). This trial was anticipated to be completed in mid-2020.

The effects of phytoestrogen supplementation in postmenopausal women with DES was evaluated in a small randomized clinical trial (Scuderi et al., 2012). The study authors concluded that phytoestrogen supplementation can significantly improve the signs and symptoms of DES in postmenopausal women. Based on the above study and a meta-analysis of seven randomized clinical trials that suggested sex hormone therapy plays a potentially useful role in postmenopausal women with DES (Liu et al., 2018b), phytoestrogens may be useful for the treatment of DES in this population. Bilberry extract was found to significantly improve the volume of tear secretion (Riva et al., 2017) and maqui berry extract reduced eye dryness (Yamashita et al., 2019). Unfortunately, these studies either had a small group of participants and/or were short in duration. Therefore no recommendations for the use of these and other products for DED sufferers can be made until additional larger, well-designed, multicenter randomized clinical trials are conducted to further examine their efficacy.

Inflammatory eye diseases

Inflammatory eye diseases may affect different parts or a combination of parts of the eye. *Uveitis* is a form of eye inflammation that affects the uvea (the middle layer of the three concentric layers that make up the eye wall). Uveitis symptoms include eye redness and pain, blurred and decreased vision, light sensitivity, and floaters in the field of vision. *Scleritis* is inflammation of the sclera (the white or outer layer of the eye). Symptoms of scleritis include eye redness (or having a purple hue), pain, intolerance of light, tearing, and decreased clarity of vision. *Keratitis* is inflammation of the cornea that causes eye redness, irritation and pain, light sensitivity, and blurred or decreased vision. *Conjunctivitis (pink eye)* is inflammation of the conjunctiva, the tissue lining the inside of the eyelids and covering the white of the eye (sclera). Symptoms of conjunctivitis include red, itchy, watery eyes with discharge. *Keratoconjunctivitis* is inflammation of the cornea and conjunctiva. *Blepharitis* is inflammation of the eyelids that may lead to irritated or red eyes. Other symptoms include gritty, burning, or stinging sensations in the eyes, dry eyes or excess tearing, or red, swollen, and itchy eyelids that may stick together. *Blepharokeratoconjunctivitis* (BKC) is a type of inflammation of the surface of the eye and eyelids. While inflammation of the eye can be caused by an infection, this section will discuss the use of nutraceuticals for noninfectious eye inflammations.

Nutraceuticals for the treatment of uveitis and scleritis

The results of a pilot study suggested that vitamin D (600 IU) supplementation in patients with autoimmune uveitis may have a beneficial effect (Valdez et al., 2014). In a prospective case-control study, vitamin D supplementation was associated significantly with uveitis inactivity (Chiu et al., 2019). A study conducted to determine whether an association between vitamin D and noninfectious ocular inflammation exists found that the odds of having uveitis were 1.92 times higher for patients with hypovitaminosis D compared to patients with normal vitamin D levels (Llop et al., 2019). Moreover, a secondary analysis of the results revealed that the odds of developing uveitis or scleritis were 5% lower and 4% lower, respectively, for every unit increase in vitamin D level. Therefore vitamin D supplementation potentially has a beneficial role in uveitis and scleritis, but more studies are needed to demonstrate this unambiguously.

Nutraceuticals for the treatment of blepharitis and blepharokeratoconjunctivitis

The effects of dietary supplementation with ω-3 FAs (6000 mg) on blepharitis and meibomian gland dysfunction (MGD) was evaluated in a 1-year study (Macsai, 2008). (Note: In MGD the meibomian glands do not secrete sufficient oil into the tears, causing the tears to evaporate too quickly. Therefore, MGD is associated with DES.) According to the author, this trial demonstrated an improvement in the tear film break-up time (slowing of tear evaporation) and dry eye symptoms, suggesting a potential benefit for the treatment of blepharitis. In a second trial, patients diagnosed with moderate to severe chronic blepharitis and MGD received ω-3 FA (1000 mg) for 3 months (Saif, 2017). In this study, a significant improvement in tear film break-up time was reported, indicating that ω-3 FA supplementation may be effective in the treatment of moderate to severe chronic blepharitis and MGD. The findings of a third study suggested that 2.5 mL of flaxseed oil/day (a dietary source of ω-3 FA) may be an effective antiinflammatory nutritional therapy alternative to long-term antibiotics in children with chronic BKC (Jones et al., 2007). Unfortunately, the numbers of patients enrolled in these three trials were small, therefore a need exists for larger, better-designed, and controlled clinical trials to determine the therapeutic potential of ω-3 FAs in the treatment of blepharitis and BKC.

Nutraceuticals for the treatment of keratitis, conjunctivitis, and keratoconjunctivitis

Clinical studies to evaluate the efficacy of nutraceuticals for the treatment of noninfectious keratitis, conjunctivitis, and keratoconjunctivitis are lacking; therefore this is an area of needed research. As of December 2019, a search of the clinical trials database (www.clinicaltrials.gov) did not yield any studies that are recruiting, not yet recruiting, or currently active, but not recruiting, for the treatment of any of the abovementioned ocular inflammatory diseases with oral nutraceuticals. Overall, this area of nutraceutical research seems to be relatively quiet, despite the therapeutic potentials of vitamin D and ω-3 FAs shown in the above studies to treat certain ophthalmic inflammatory diseases.

Mode of action and safety of the most commonly used nutraceuticals for ophthalmic diseases

Carotenoids

Mode of action of carotenoids

The carotenoids lutein and zeaxanthin are found in many of the human ocular structures, with the highest concentration in the macula of the retina (Mares, 2016). Macular carotenoids may protect the eye from photooxidative damage, oxidative stress, and inflammation via multiple mechanisms. As inflammation and oxidative stress play important roles in the initiation and progression of multiple ocular diseases such as, but not limited to, AMD, DR, cataract, and glaucoma; macular carotenoids with proven antioxidant and antiinflammatory properties could be beneficial in these diseases. β-Carotene, in addition to its antioxidant and antiinflammatory effects, can be converted to vitamin A, which has an important role in eye health.

Safety of carotenoids

Based on the results of several human studies, no adverse effects were reported for lutein at doses of 20 mg/day for up to 6 months, 30 mg/day for 5 months, or 40 mg/day for over 2 months (Connolly et al., 2011). The only side effect reported in humans has been carotenosis, yellowing/orange discoloration of the skin, due to excessive lutein intake. This effect is reversible and does not damage organs. Initially, the Joint FAO/WHO Expert Committee on Food Additives (JECFA) and the European Food Safety Authority (EFSA) set an acceptable daily intake of 0−2 mg lutein/kg bw/day and 1 mg/kg bw/day, respectively (JECFA, 2005; EFSA, 2010). Based on the available data, the EFSA Panel on Diabetic Products, Nutrition and Allergies concluded that intakes of 0.75 mg synthetic zeaxanthin/kg bw/day or 53 mg/day for a 70-kg adult do not raise safety concerns (EFSA, 2012a). Due to the low toxicity of lutein and zeaxanthin, JECFA raised the group ADI of lutein, lutein esters, and zeaxanthin to "not specified," and withdrew the previous group ADI of 0−2 mg/kg bw/day for lutein and zeaxanthin in 2018 (JECFA, 2018).

EFSA (2006) stated that 20 mg/day or more of supplemental β-carotene is contraindicated for use in current heavy smokers, based on evidence from human trials. As the data on the effect of dietary supplementation with β-carotene on lung cancer risk and mortality in current and former smokers is contradictory, caution is recommended regarding high-dose β-carotene supplementation in patients with lung cancer, and especially in current and former smokers (Alsharairi, 2019). For more information on the adverse effects and the modes of toxic action of β-carotene, see the subsection "Vitamin toxicity" in Chapter 69, Nutraceuticals and adverse outcome pathways.

Polyphenols

Mode of action of polyphenols

A number of polyphenols improve ocular blood flow and signal transduction and have potent antioxidant and antiinflammatory activities, along with other beneficial effects (Mares et al., 2017). They therefore have potential roles in the prevention and treatment of numerous ophthalmic diseases. Flavonoids (a group of polyphenolic compounds) are powerful antioxidants, have documented antiinflammatory properties, or increase ocular blood flow. Quercetin (a representative of the flavonoids) may protect against neuronal damage in DR by inhibiting the apoptosis of neurons and ameliorating the levels of neurotrophic factors (Ola et al., 2017). Anthocyanins (also representatives of flavonoids) have free radical scavenging activity, antioxidant capacity, and antiinflammatory activity; and thereby potentially are useful for the treatment of numerous ophthalmic diseases (Bungau et al., 2019). Resveratrol (a representative of stilbenes, a group of polyphenols) has been shown to have antiapoptotic effect, suppress the production of proinflammatory markers, and inhibit diabetes-induced stress and neuronal cell destruction, along with other potentially beneficial effects in DR and other ophthalmic diseases.

Safety of polyphenols

Dietary supplementation with polyphenols is a double-edged sword; while it may confer benefits, high doses may result in toxicity (Martin and Appel, 2009). Under certain conditions, at high doses, phenolic antioxidants can behave like prooxidants and elicit toxicity. For example, in humans, high doses of polyphenol-enriched green tea- and other tea-based dietary supplements were reported to cause hepatotoxicity. Some flavonoids may negatively affect thyroid function and exhibit antinutritional effects, along with other adverse effects.

A review of the safety aspects of quercetin used as a dietary supplement found no adequate scientific data in the published literature regarding the safety of long-term (>12 weeks) use of supplemental doses (≥ 1000 mg) of quercetin (Andres et al., 2018). Based on animal studies, very high doses of quercetin may enhance nephrotoxic effects in the predamaged kidney. Moreover, it is suspected to promote tumor development, especially in estrogen-dependent cancer, a potential adverse effect that needs further investigation. Additionally, like some other polyphenols, quercetin may interact with certain drugs, altering their bioavailability.

No tolerable upper limit[1] (UL) exists for anthocyanins. While JECFA established an acceptable daily intake of 2.5 mg/kg bw/day for anthocyanins from grape skin extracts, after reevaluation of the safety data, the EFSA (2013) concluded that the available toxicological data and information were not adequate to establish a numerical acceptable daily intake for anthocyanins.

EFSA (2016) evaluated the safety of resveratrol and stated that 1 g resveratrol/day or higher was reported to cause diarrhea or other gastrointestinal symptoms in humans. According to a second review of resveratrol safety, abdominal disorders may appear at 0.5 g resveratrol/day or higher (Cottart et al., 2014). Moreover, a metabolite of resveratrol can inhibit cytochrome P450 (CYP) enzymes and therefore, like other polyphenols, may interact with certain medicines (EFSA, 2016). Nonetheless, EFSA stated that a supplement containing up to 150 mg synthetic *trans*-resveratrol/day does not raise safety concerns for adults. The safety of long-term resveratrol use in humans needs further evaluation.

Other chapters in this book discuss the toxicity and modes of beneficial and toxic actions of polyphenols, anthocyanins, and green tea extract (high in polyphenols) in more detail. See Chapters 69 (for polyphenols), (for a and Chapter 41 (for green tea extract) for more information.

Polyunsaturated fatty acids

Mode of action of polyunsaturated fatty acids

PUFAs may have beneficial effects in ophthalmic diseases due to their antiinflammatory and antioxidant properties along with other useful properties, such as their potential to reduce IOP (Saccà et al., 2018; Mares et al., 2017). The ω-3 PUFAs EPA and DHA seem to have important roles in eye health and disease. For example, in DED, PUFA supplementation can improve the lipid layer of the lacrimal film, normalize the function of the lacrimal and meibomian glands, and reduce inflammation (Saccà et al., 2018).

Safety of polyunsaturated fatty acids

EFSA (2012b) stated that "supplemental intakes of EPA and DHA combined at doses up to 5 g/day, and supplemental intakes of EPA alone up to 1.8 g/day, do not raise safety concerns for adults," and "supplemental intakes of DHA alone up to about 1 g/day do not raise safety concerns for the general population." On the other hand, the FDA recommends not exceeding an intake of 3 g/day of ω-3 PUFAs (EPA and DHA) and the German Federal Risk Assessment Agency established a recommended upper intake level for ω-3 PUFAs of 1.5 g/day to safeguard against adverse effects, such as increased bleeding risk, seen at high intake levels (BfR, 2009; FDA, 2017). While the ratio of ω-3 to ω-6 PUFAs seems to be important in the pathogenesis, treatment, and prevention of ophthalmic and other diseases, the optimal ratio of ω-3 and ω-6 PUFAs has not been defined and may depend on multiple factors. For information on the adverse effects of fish oil (high in ω-3 PUFAs), see the subsection "Fish oil toxicity" within Chapter 69, Nutraceuticals and adverse outcome pathways, in this book.

Vitamins

Mode of action of vitamins

As mentioned earlier, deficiencies in certain vitamins, such as vitamins A and B, may increase the risk of or cause different ophthalmic diseases. Beneficial effects of

1. The maximum level of total chronic daily intake from all sources judged to be unlikely to pose a risk of adverse health effects to humans.

vitamins may be due to their antiinflammatory properties (e.g., vitamins D and B6), their ability to protect against oxidative stress and damage (e.g., vitamins B2, B6, C, and E and niacin), and/or their ability to prevent hyperhomocysteinemia[2] (B vitamins) (Mares et al., 2017).

Safety of vitamins

Both benefit and harm could result from taking vitamins for ophthalmic diseases. For some of the potential harms in ophthalmic diseases, see the section titled "Efficacy of nutraceuticals in the treatment and prevention of various ophthalmic diseases." In addition to worsening certain ophthalmic conditions, excessive intakes of vitamins may cause other health problems. For example:

1. Excessive intake of vitamin B6 is associated with neuronal damage (EFSA, 2006).
2. High-dose vitamin A was found to increase the risk of all-cause mortality and cancer mortality (Schwingshackl et al., 2017). Moreover, hypervitaminosis A may result in hepatotoxicity, skeletal problems, and teratogenesis (EFSA, 2006).
3. Excessive vitamin D intake may lead to elevated level of plasma calcium, partially due to excessive calcium mobilization from bone (EFSA, 2006). Hypervitaminosis D may also affect various organs such as the kidney.
4. Vitamin E was associated with an increased risk of all-cause mortality, and a high intake should be avoided, as it may negatively affect blood clotting (Schwingshackl et al., 2017; EFSA, 2006).
5. Due to the antagonistic interaction of vitamin K and coumarin anticoagulants, people taking these drugs should not use supplementary vitamin K without medical advice (EFSA, 2006).

Based on all the available safety data and information, EFSA (2006) established ULs for vitamins A [3000 μg retinol equivalent (RE)[3]/day], B6 (25 mg/day), D (50 μg/day), and E (300 mg/day). The Institute of Medicine (IOM), now the National Academy of Medicine, also established ULs for vitamins A (3000 μg/day), B6 (100 mg/day), D (100 μg/day), E (1000 mg/day), and folate (1000 μg/day) (Institute of Medicine, National Academy of Science, 2011a). Neither EFSA nor IOM established ULs for vitamins B1, B2, B12, and K. While EFSA did not establish a UL for vitamin C, IOM established a value of 2000 mg/day. For more information on vitamins' adverse effects and modes of toxic action, see the subsection "Vitamin toxicity" in Chapter 69, Nutraceuticals and adverse outcome pathways, in this book.

Minerals (zinc and copper)

Modes of action of zinc and copper

Outside of the prostate, the highest concentration of zinc is in the retina, implying zinc's important role in ocular physiology (Mares et al., 2017). Zinc deficiency is associated with impaired retinal function and poor dark adaptation (Mares et al., 2017; Ugarte et al., 2013). Mechanisms by which zinc deficiency induces retinal dysfunction likely involve oxidative stress, lipofuscin (considered to be one of the aging or "wear-and-tear" pigments in the retina) accumulation, and photoreceptor disruption (Ugarte et al., 2013). In vitro, zinc depletion resulted in reduced protein synthesis, disruption of growth factor signaling molecules, and cell death in retinal pigment epithelial cells. Copper also plays an important role in ocular physiology, and copper deficiency is associated with optic neuropathy (Ugarte et al., 2013). Currently, whether deficiency of certain trace elements is a cause or an effect of the retinal degeneration in AMD is unclear.

Safety of zinc and copper

In addition to retinal deficiency, retinal excess of zinc and copper is also associated with retinal dysfunction and pathology (Ugarte et al., 2013). Moreover, acute zinc toxicity is associated with gastrointestinal disturbances, and chronic zinc toxicity is associated with symptoms of copper deficiency (EFSA, 2006). Acute effects of copper toxicity are mostly observed in the gastrointestinal tract, while chronic copper toxicity has its most pronounced adverse effect on liver function. Consequently, EFSA (2006) set the UL for zinc at 25 mg/day and for copper at 5 mg/day for adults. IOM set the ULs of zinc and copper slightly higher, at 40 mg/day and 10 mg/day, respectively (Institute of Medicine, National Academy of Science, 2011b). For more information on copper and zinc toxicity, see Chapter 68, Toxic contamination of nutraceuticals and food ingredients, in this book.

Concluding remarks and future directions

While numerous nutraceuticals seem to show beneficial effects in various ophthalmic diseases, or lower their risk of development, the data for most of them are either limited or conflicting. Conflicting results may be due to various reasons, such as the different lengths of interventions,

2. Hyperhomocysteinemia is associated with ocular diseases such as retinopathy, pseudoexfoliative glaucoma maculopathy, cataract, optic atrophy, and retinal vessel atherosclerosis (Ajith, 2015).

3. One RE equals 1 μg of retinol.

the dose levels used, ages of participants at intervention, sex and numbers of participants, and stages of disease at the start of a study, among others. Nonetheless, due to the large number of people suffering from, or at risk of, developing ophthalmic diseases and the potentially beneficial effects of nutraceuticals in these diseases demonstrated in numerous studies, further research should be conducted into their benefits and safety. Unfortunately, until more data are gathered, clear recommendations for most nutraceuticals cannot be issued for most ophthalmic diseases.

References

Ajith, T.A., 2015. Homocysteine in ocular diseases. Clin. Chim. Acta 450, 316–321.

Alsharairi, N.A., 2019. The effects of dietary supplements on asthma and lung cancer risk in smokers and non-smokers: a review of the literature. Nutrients 11 (4), 725.

Andres, S., Pevny, S., Ziegenhagen, R., et al., 2018. Safety aspects of the use of quercetin as a dietary supplement. Mol. Nutr. Food Res. 62 (1), 1700447.

Bae, H.W., Kim, J.H., Kim, S., et al., 2015. Effect of Korean Red Ginseng supplementation on dry eye syndrome in glaucoma patients—a randomized, double-blind, placebo-controlled study. J. Ginseng Res. 39 (1), 7–13.

Bae, S.H., Shin, Y.J., Kim, H.K., et al., 2016. Vitamin D supplementation for patients with dry eye syndrome refractory to conventional treatment. Sci. Rep. 6, 33083.

Bahrami, H., Melia, M., Dagnelie, G., 2006. Lutein supplementation in retinitis pigmentosa: PC-based vision assessment in a randomized double-masked placebo-controlled clinical trial [NCT00029289]. BMC Ophthalmol. 6 (1), 23.

Berson, E.L., Rosner, B., Sandberg, M.A., et al., 1993. A randomized trial of vitamin A and vitamin E supplementation for retinitis pigmentosa-reply. Arch. Ophthalmol. 111 (11), 1463–1465.

Berson, E.L., Rosner, B., Sandberg, M.A., et al., 2004. Clinical trial of docosahexaenoic acid in patients with retinitis pigmentosa receiving vitamin a treatment. Arch. Ophthalmol. 122 (9), 1297–1305.

Berson, E.L., Rosner, B., Sandberg, M.A., et al., 2010. Clinical trial of lutein in patients with retinitis pigmentosa receiving vitamin A. Arch. Ophthalmol. 128 (4), 403.

Berson, E.L., Rosner, B., Sandberg, M.A., et al., 2012. ω-3 intake and visual acuity in patients with retinitis pigmentosa receiving vitamin A. Arch. Ophthalmol. 130 (6), 707–711.

BfR (Bundesinstitut fur Risikobewertung), 2009. Fur die Anreicherung von Lebensmitteln mit Omega-3-Fettsauren empfiehlt das BfR dia Festsetzung von Hochstmengen. Stellungnahme Nr. 10 pp.

Bhargava, R., Kumar, P., Kumar, M., et al., 2013. A randomized controlled trial of omega-3 fatty acids in dry eye syndrome. Int. J. Ophthalmol. 6 (6), 811.

Bourne, R.R., Flaxman, S.R., Braithwaite, T., et al., 2017. Magnitude, temporal trends, and projections of the global prevalence of blindness and distance and near vision impairment: a systematic review and meta-analysis. Lancet Glob. Health 5 (9), e888–e897.

Bungau, S., Abdel-Daim, M.M., Tit, D.M., et al., 2019. Health benefits of polyphenols and carotenoids in age-related eye diseases. Oxid. Med. Cell. Longev. 2019.

Chew, E.Y., Clemons, T.E., Agrón, E., Age-Related Eye Disease Study Research Group, et al., 2013. Long-term effects of vitamins C and E, β-carotene, and zinc on age-related macular degeneration: AREDS report no. 35. Ophthalmology 120 (8), 1604–1611.

Chew, E.Y., Clemons, T.E., SanGiovanni, J.P., et al., 2014. Secondary analyses of the effects of lutein/zeaxanthin on age-related macular degeneration progression: AREDS2 report no. 3. JAMA Ophthalmol. 132 (2), 142–149.

Chiu, Z.K., Lim, L.L., Rogers, S.L., et al., 2019. Patterns of vitamin D levels and exposures in active and inactive noninfectious uveitis patients. Ophthalmology .

Christen, W.G., Glynn, R.J., Chew, E.Y., et al., 2009. Folic acid, vitamin B6, and vitamin B12 in combination and age-related macular degeneration in a randomized trial of women. Arch. Int. Med. 169 (4), 335.

Christen, W.G., Glynn, R.J., Manson, J.E., et al., 2014. Effects of multivitamin supplement on cataract and age-related macular degeneration in a randomized trial of male physicians. Ophthalmology 121 (2), 525–534.

Christen, W.G., Glynn, R.J., Gaziano, J.M., et al., 2015. Age-related cataract in men in the selenium and vitamin e cancer prevention trial eye endpoints study: a randomized clinical trial. JAMA Ophthalmol. 133 (1), 17–24.

Christen, W.G., Glynn, R.J., Chew, E.Y., et al., 2016. Folic acid, vitamin B6, and vitamin B12 in combination and age-related cataract in a randomized trial of women. Ophthalmic Epidemiol. 23 (1), 32–39.

Connolly, E.E., Beatty, S., Loughman, J., et al., 2011. Supplementation with all three macular carotenoids: response, stability, and safety. Invest. Ophthalmol. Visual Sci. 52 (12), 9207–9217.

Cottart, C.H., Nivet-Antoine, V., Beaudeux, J.L., 2014. Review of recent data on the metabolism, biological effects, and toxicity of resveratrol in humans. Mol. Nutr. Food Res. 58 (1), 7–21.

Dani, C., Lori, I., Favelli, F., et al., 2012. Lutein and zeaxanthin supplementation in preterm infants to prevent retinopathy of prematurity: a randomized controlled study. J. Matern.-Fetal Neonat. Med. 25 (5), 523–527.

Downie, L.E., Ng, S.M., Lindsley, K.B., et al., 2019. Omega-3 and omega-6 polyunsaturated fatty acids for dry eye disease. Cochrane Database Syst. Rev. (12).

Dry Eye Assessment and Management Study Research Group, 2018. n − 3 Fatty acid supplementation for the treatment of dry eye disease. N. Engl. J. Med. 378 (18), 1681–1690.

EFSA Panel on Dietetic Products, Nutrition and Allergies (NDA), 2012a. Statement on the safety of synthetic zeaxanthin as an ingredient in food supplements. EFSA J. 10 (10), 2891.

EFSA Panel on Dietetic Products, Nutrition and Allergies (NDA), 2012b. Scientific opinion on the tolerable upper intake level of eicosapentaenoic acid (EPA), docosahexaenoic acid (DHA) and docosapentaenoic acid (DPA). EFSA J. 10 (7), 2815.

EFSA Panel on Dietetic Products, Nutrition and Allergies (NDA), 2016. Safety of synthetic trans-resveratrol as a novel food pursuant to Regulation (EC) No 258/97. EFSA J. 14 (1), 4368.

EFSA Panel on Food Additives and Nutrient Sources added to Food (ANS), 2010. Scientific opinion on the re-evaluation of lutein (E 161b) as a food additive. EFSA J. 8 (7), 1678.

EFSA Panel on Food Additives and Nutrient Sources added to Food (ANS), 2013. Scientific opinion on the re-evaluation of anthocyanins (E 163) as a food additive. EFSA J. 11 (4), 3145.

Centers for Disease Control and Prevention, 2018. Watch out for diabetic retinopathy. [online] Available at: https://www.cdc.gov/features/diabetic-retinopathy/index.html. (Accessed December 28, 2019).

EFSA Scientific Committee on Food, 2006. Tolerable upper intake levels for vitamins and minerals. European Food Safety Authority.

Flaxman, S.R., Bourne, R.R., Resnikoff, S., et al., 2017. Global causes of blindness and distance vision impairment 1990–2020: a systematic review and meta-analysis. Lancet Glob. Health 5 (12), e1221–e1234.

Food and Drug Administration (FDA), 2017. Fed. Regist 62 (108), 30751–30757.

Garcia-Medina, J.J., Pinazo-Duran, M.D., Garcia-Medina, M., et al., 2011. A 5-year follow-up of antioxidant supplementation in type 2 diabetic retinopathy. Eur. J. Ophthalmol. 21 (5), 637–643.

Giannaccare, G., Pellegrini, M., Sebastiani, S., Bernabei, F., Roda, M., Taroni, L., et al., 2019. Efficacy of omega-3 fatty acid supplementation for treatment of dry eye disease: a meta-analysis of randomized clinical trials. Cornea 38 (5), 565–573.

Glaser, T.S., Doss, L.E., Shih, G., Age-Related Eye Disease Study Research Group, et al., 2015. The association of dietary lutein plus zeaxanthin and B vitamins with cataracts in the Age-Related Eye Disease Study: AREDS report no. 37. Ophthalmology 122 (7), 1471–1479.

Goncalves, A., Milea, D., Gohier, P., et al., 2015. Serum vitamin D status is associated with the presence but not the severity of primary open angle glaucoma. Maturitas 81 (4), 470–474.

Gopinath, B., Flood, V.M., Rochtchina, E., et al., 2013. Homocysteine, folate, vitamin B-12, and 10-y incidence of age-related macular degeneration. Am. J. Clin. Nutr. 98 (1), 129–135.

Hu, B.J., Hu, Y.N., Lin, S., et al., 2011. Application of lutein and zeaxanthin in nonproliferative diabetic retinopathy. Int. J. Ophthalmol. 4 (3), 303.

Hwang, J.S., Lee, Y.P., Shin, Y.J., 2019. Vitamin D enhances the efficacy of topical artificial tears in patients with dry eye disease. Cornea 38 (3), 304–310.

Institute of Medicine, National Academy of Science, 2011a. Tolerable upper intake levels, vitamins. Available online as of 03/10/2020 at https://www.ncbi.nlm.nih.gov/books/NBK56068/table/summarytables.t7/?report = objectonly

Institute of Medicine, National Academy of Science, 2011b. Tolerable upper intake levels, elements. Available online as of 03/10/2020 at https://www.ncbi.nlm.nih.gov/books/NBK56068/table/summarytables.t8/?report = objectonly

JECFA, Joint FAO/WHO Expert Committee on Food Additives and World Health Organization, 2005. Evaluation of certain food additives: sixty-third report of the Joint FAO.

JECFA, Joint FAO/WHO Expert Committee on Food Additives, 2018. Summary and conclusions of the 86th meeting (JECFA/86/SC).

Jiang, H., Yin, Y., Wu, C.R., et al., 2019. Dietary vitamin and carotenoid intake and risk of age-related cataract. Am. J. Clin. Nutr. 109 (1), 43–54.

Jones, S.M., Weinstein, J.M., Cumberland, P., et al., 2007. Visual outcome and corneal changes in children with chronic blepharokeratoconjunctivitis. Ophthalmology 114 (12), 2271–2280.

Kakigi, C.L., Singh, K., Wang, S.Y., et al., 2015. Self-reported calcium supplementation and age-related macular degeneration. JAMA Ophthalmol. 133 (7), 746–754.

Kaur, A., Gupta, V., Christopher, A.F., et al., 2017. Nutraceuticals in prevention of cataract – an evidence-based approach. Saudi J. Ophthalmol. 31 (1), 30–37.

Krefting, E.A., Jorde, R., Christoffersen, T., Grimnes, G., 2014. Vitamin D and intraocular pressure–results from a case–control and an intervention study. Acta Ophthalmol. 92 (4), 345–349.

Kutuzova, G.D., B'Ann, T.G., Kiland, J.A., et al., 2012. 1α, 25-Dihydroxyvitamin D3 and its analog, 2-methylene-19-nor-(20S)-1α, 25-dihydroxyvitamin D3 (2MD), suppress intraocular pressure in non-human primates. Arch. Biochem. Biophys. 518 (1), 53–60.

Layana, A.G., Minnella, A.M., Garhöfer, G., et al., 2017. Vitamin D and age-related macular degeneration. Nutrients 9 (10), 1120.

Liu, A., Ji, J., 2014. Omega-3 essential fatty acids therapy for dry eye syndrome: a meta-analysis of randomized controlled studies. Med. Sci. Monit.: Int. Med. J. Exp. Clin. Res. 20, 1583.

Liu, Y.C., Wilkins, M., Kim, T., et al., 2017. Cataracts. Lancet 390 (10094), 600–612.

Liu, F., Xiong, J., Hu, J., et al., 2018a. Vitamin C and risk of age-related cataracts: a systematic review and meta-analysis. Int. J. Clin. Exp. Med. 11 (9), 8929. - + .

Liu, C., Liang, K., Jiang, Z., et al., 2018b. Sex hormone therapy's effect on dry eye syndrome in postmenopausal women: a meta-analysis of randomized controlled trials. Medicine 97 (40).

Llop, S.M., Davoudi, S., Stanwyck, L.K., et al., 2019. Association of low vitamin D levels with noninfectious uveitis and scleritis. Ocul. Immunol. Inflammation 27 (4), 602–609.

Loskutova, E., O'Brien, C., Loskutov, I., et al., 2019. Nutritional supplementation in the treatment of glaucoma: a systematic review. Surv. Ophthalmol. 64 (2), 195–216.

Ma, L., Dou, H.L., Wu, Y.Q., et al., 2012. Lutein and zeaxanthin intake and the risk of age-related macular degeneration: a systematic review and meta-analysis. Br. J. Nutr. 107 (3), 350–359.

Ma, Q., Chen, D., Sun, H.P., et al., 2015. Regular Chinese green tea consumption is protective for diabetic retinopathy: a clinic-based case-control study. J. Diabetes Res. 2015.

Macsai, M.S., 2008. The role of omega-3 dietary supplementation in blepharitis and meibomian gland dysfunction (an AOS thesis). Trans. Am. Ophthalm Soc. 106, 336.

Manzoni, P., Guardione, R., Bonetti, P., et al., 2013. Lutein and zeaxanthin supplementation in preterm very low-birth-weight neonates in neonatal intensive care units: a multicenter randomized controlled trial. Am. J. Perinatol. 30 (01), 025–032.

Mares, J., 2016. Lutein and zeaxanthin isomers in eye health and disease. Annu. Rev. Nutr. 36, 571–602.

Mares, J.A., Millen, A.E., Lawler, T.P., et al., 2017. Diet and supplements in the prevention and treatment of eye diseases. In: AnnCoulston, M., CarolBoushey, J., MarioFerruzzi, G., LindaDelahanty, M. (Eds.), Nutrition in the Prevention and Treatment of Disease. Academic Press, pp. 393–434.

Martin, K.R., Appel, C.L., 2009. Polyphenols as dietary supplements: A double-edged sword. Nutrition and Dietary Supplements 2, 1–12.

Mathew, M.C., Ervin, A.M., Tao, J., et al., 2012. Antioxidant vitamin supplementation for preventing and slowing the progression of age-related cataract. Cochrane Database Syst. Rev. (6).

Mayer-Davis, E.J., Bell, R.A., Reboussin, B.A., et al., 1998. Antioxidant nutrient intake and diabetic retinopathy: the San Luis Valley Diabetes Study. Ophthalmology 105 (12), 2264–2270.

Merle, B.M., Silver, R.E., Rosner, B., et al., 2016. Dietary folate, B vitamins, genetic susceptibility and progression to advanced nonexudative age-related macular degeneration with geographic atrophy: a prospective cohort study. Am. J. Clin. Nutr. 103 (4), 1135–1144.

Messmer, E.M., 2015. The pathophysiology, diagnosis, and treatment of dry eye disease. Dtsch. Ärztebl. Int. 112 (5), 71.

Miljanović, B., Trivedi, K.A., Dana, M.R., et al., 2005. Relation between dietary n − 3 and n − 6 fatty acids and clinically diagnosed dry eye syndrome in women. Am. J. Clin. Nutr. 82 (4), 887–893.

Mitchell, P., Liew, G., Gopinath, B., et al., 2018. Age-related macular degeneration. Lancet 392 (10153), 1147–1159.

Morrone, A.L., Rombola, L., Adornetto, A., et al., 2018. Rational basis for nutraceuticals in the treatment of glaucoma. Curr. Neuropharmacol. 16 (7), 1004–1017.

National Eye Institute, 2019. Cataracts. Available at: https://www.nei.nih.gov/learn-about-eye-health/eye-conditions-and-diseases/cataracts. (Accessed December 22, 2019).

Ola, M.S., Ahmed, M.M., Shams, S., et al., 2017. Neuroprotective effects of quercetin in diabetic rat retina. Saudi J. Biol. Sci. 24 (6), 1186–1194.

Oleñik, A., 2014. Effectiveness and tolerability of dietary supplementation with a combination of omega-3 polyunsaturated fatty acids and antioxidants in the treatment of dry eye symptoms: results of a prospective study. Clin. Ophthalmol. 8, 169.

Osakabe, N., Yamagishi, M., Natsume, M., et al., 2004. Ingestion of proanthocyanidins derived from cacao inhibits diabetes-induced cataract formation in rats. Exp. Biol. Med. 229 (1), 33–39.

Ramdas, W.D., Schouten, J.S., Webers, C.A., 2018. The effect of vitamins on glaucoma: a systematic review and meta-analysis. Nutrients 10 (3), 359.

Rautiainen, S., Lindblad, B.E., Morgenstern, R., et al., 2009. Vitamin C supplements and the risk of age-related cataract: a population-based prospective cohort study in women. Am. J. Clin. Nutr. 91 (2), 487–493.

Riva, A., Togni, S., Franceschi, F., et al., 2017. The effect of a natural, standardized bilberry extract (Mirtoselect®) in dry eye: a randomized, double blinded, placebo-controlled trial. Eur. Rev. Med. Pharmacol. Sci. 21 (10), 2518–2525.

Romagnoli, C., Giannantonio, C., Cota, F., et al., 2011. A prospective, randomized, double blind study comparing lutein to placebo for reducing occurrence and severity of retinopathy of prematurity. J. Matern.-Fetal Neonat. Med. 24 (suppl. 1), 147–150.

Rossino, M.G., Casini, G., 2019. Nutraceuticals for the treatment of diabetic retinopathy. Nutrients 11 (4), 771.

Rotenstreich, Y., Belkin, M., Sadetzki, S., et al., 2013. Treatment with 9-cis β-carotene–rich powder in patients with retinitis pigmentosa: a randomized crossover trial. JAMA Ophthalmol. 131 (8), 985–992.

Rubin, L.P., Chan, G.M., Barrett-Reis, B.M., et al., 2012. Effect of carotenoid supplementation on plasma carotenoids, inflammation and visual development in preterm infants. J. Perinatol. 32 (6), 418.

Saccà, S.C., Cutolo, C.A., Ferrari, D., et al., 2018. The eye, oxidative damage and polyunsaturated fatty acids. Nutrients 10 (6), 668.

Saif, A.T., 2017. Oral linolenic acid dietary supplementation in posterior blepharitis and meibomian gland dysfunction. Delta J. Ophthalmol. 18 (2), 51.

Sala-Vila, A., Díaz-López, A., Valls-Pedret, C., et al., 2016. Dietary marine ω-3 fatty acids and incident sight-threatening retinopathy in middle-aged and older individuals with type 2 diabetes: prospective investigation from the PREDIMED trial. JAMA Ophthalmol. 134 (10), 1142–1149.

Sasaki, M., Kawasaki, R., Rogers, S., et al., 2015. The associations of dietary intake of polyunsaturated fatty acids with diabetic retinopathy in well-controlled diabetes. Invest. Ophthalmol. Visual Sci. 56 (12), 7473–7479.

Schwingshackl, L., Boeing, H., Stelmach-Mardas, M., et al., 2017. Dietary supplements and risk of cause-specific death, cardiovascular disease, and cancer: a systematic review and meta-analysis of primary prevention trials. Adv. Nutr. 8 (1), 27–39.

Scuderi, G., Contestabile, M.T., Gagliano, C., et al., 2012. Effects of phytoestrogen supplementation in postmenopausal women with dry eye syndrome: a randomized clinical trial. Can. J. Ophthalmol. 47 (6), 489–492.

Selin, J.Z., Lindblad, B.E., Bottai, M., et al., 2017. High-dose B-vitamin supplements and risk for age-related cataract: a population-based prospective study of men and women. Br. J. Nutr. 118 (2), 154–160.

Sella, R., Afshari, N.A., 2019. Nutritional effect on age-related cataract formation and progression. Curr. Opin. Ophthalmol. 30 (1), 63–69.

Sideri, O., Tsaousis, K.T., Li, H.J., et al., 2019. The potential role of nutrition on lens pathology: A systematic review and meta-analysis. Surv. Ophthalmol .

Silva, K.C., Rosales, M.A., Hamassaki, D.E., et al., 2013. Green tea is neuroprotective in diabetic retinopathy. Invest. Ophthalmol Visual Sci. 54 (2), 1325–1336.

Steigerwalt, R., Nebbioso, M., Appendino, G., et al., 2012. Meriva®, a lecithinized curcumin delivery system, in diabetic microangiopathy and retinopathy. Panminerva Med. 54 (4), 11.

Stevens, G.A., White, R.A., Flaxman, S.R., et al., 2013. Global prevalence of vision impairment and blindness: magnitude and temporal trends, 1990–2010. Ophthalmology 120 (12), 2377–2384.

Tan, J.S., Wang, J.J., Flood, V., et al., 2008. Dietary antioxidants and the long-term incidence of age-related macular degeneration: the Blue Mountains Eye Study. Ophthalmology 115 (2), 334–341.

Tisdale, A.K., Agrón, E., Sunshine, S.B., et al., 2019. Association of dietary and supplementary calcium intake with age-related macular degeneration: Age-Related Eye Disease Study report 39. JAMA Ophthalmol. 137 (5), 543–550.

Ugarte, M., Osborne, N.N., Brown, L.A., et al., 2013. Iron, zinc, and copper in retinal physiology and disease. Surv. Ophthalmol. 58 (6), 585–609.

Valdez, M., Mirthi, A.T., Francisco, M.T., et al., 2014. Autoimmune uveitis: a pilot study of vitamin D levels and its supplementation effect. Invest. Ophthalmol. Visual Sci. 55 (13), 93.

van Leeuwen, R., Boekhoorn, S., Vingerling, J.R., et al., 2005. Dietary intake of antioxidants and risk of age-related macular degeneration. JAMA 294 (24), 3101–3107.

Wang, A., Han, J., Jiang, Y., et al., 2014. Association of vitamin A and β-carotene with risk for age-related cataract: a meta-analysis. Nutrition 30 (10), 1113–1121.

Wei, L., Liang, G., Cai, C., et al., 2016. Association of vitamin C with the risk of age-related cataract: a meta-analysis. Acta Ophthalmol. 94 (3), e170–e176.

Weinreb, R.N., Aung, T., Medeiros, F.A., 2014. The pathophysiology and treatment of glaucoma: a review. JAMA 311 (18), 1901–1911.

Yamashita, S.I., Suzuki, N., Yamamoto, K., et al., 2019. Effects of MaquiBright® on improving eye dryness and fatigue in humans: a randomized, double-blind, placebo-controlled trial. J. Tradit. Complement. Med. 9 (3), 172–178.

Yang, C.H., Albietz, J., Harkin, D.G., et al., 2018. Impact of oral vitamin D supplementation on the ocular surface in people with dry eye and/or low serum vitamin D. Cont. Lens Anterior Eye 41 (1), 69–76.

Zhang, Y., Jiang, W., Xie, Z., et al., 2015. Vitamin E and risk of age-related cataract: a meta-analysis. Public Health Nutr. 18 (15), 2804–2814.

Zhang, P.C., Wu, C.R., Wang, Z.L., et al., 2017. Effect of lutein supplementation on visual function in nonproliferative diabetic retinopathy. Asia Pac. J. Clin. Nutr. 26 (3), 406.

Zheng Selin, J., Rautiainen, S., Lindblad, B.E., et al., 2013. High-dose supplements of vitamins C and E, low-dose multivitamins, and the risk of age-related cataract: a population-based prospective cohort study of men. Am. J. Epidemiol. 177 (6), 548–555.

Chapter 13

Nutraceuticals in dermal diseases

Gurjot Kaur, Poonam Negi and Vijay Kumar Kapoor

Introduction

Skin is the largest organ of body and makes up to 16% of total body weight, with a surface area of 1.50 m^2 in adults. The most important feature of skin is to form a physical barrier to the environment, limiting the inward and outward passage of water, electrolytes, and various substances, while providing protection against microorganisms, ultraviolet radiation, toxic agents, and mechanical insults (Degim, 2006). It is a dynamic organ in a constant state of change, as cells of the outer layers are continuously shed and replaced by inner cells moving up to the surface. Although structurally consistent throughout the body, skin varies in thickness according to the anatomical site and age of the individual (Hadgraft, 2004). As the skin serves the vital functions of a physical, chemical, and bacterial barrier, it needs an efficient and robust repair mechanism. The skin repair process or wound healing repairs injuries caused by surgery or trauma and restores skin integrity (Ranjani et al., 2010). This repair process is complex, involving many cell types (local as well as systemic cells) and chemokines and accomplishes restoration of the barrier function, preventing further damage and/or infection.

Wound healing is initiated immediately after the injury by formation of a fibrin clot to restore hemostasis (Arya et al., 2014; Golebiewska and Poole, 2015). The process of wound healing may be divided into inflammation, proliferation, and epithelialization and scarring. As skin disorders are a common cause of human suffering and damaged and wrinkled skin is a sign of aging, nutraceuticals are used in maintaining skin health and to delay signs of aging.

Nutraceuticals are considered as a safe alternative to allopathic medicine and therefore are widely used for wound healing as well as part of antiaging cosmetics. In this chapter, we describe the anatomy and physiology of the skin followed by the current understanding of the wound healing process. Finally, we describe some well-known nutraceuticals/phytoconstituents that have shown wound healing potential and/or antiaging properties on the skin.

Skin anatomy and physiology

The human skin is, on an average, 0.5 mm thick (ranging from 0.05 to 2 mm) and is composed of four main layers: the stratum corneum, viable epidermis, dermis, and subcutaneous tissue. Hair, nails, sebaceous glands, sweat glands, and apocrine glands are regarded as accessories or additional features of the skin (Fig. 13.1).

Stratum corneum

The stratum corneum is the outermost layer of the skin. It is typically 10–20 μm thick and composed of 10–15 layers of corneocytes (Agache, 2004a; William, 2003). Corneocytes are nonliving cells derived from terminally differentiated keratinocytes that have originated from the deeper layers of the epidermis. Morphologically, corneocytes are flattened and elongated, measuring about 0.2 μm thick and 40–60 μm wide (Kashibuchi et al., 2002). Corneocytes have a cornified envelope in place of a plasma membrane, which is surrounded by a lipid coat. They lack nuclei and cytoplasmic organelles but are filled with keratin filaments and are interspersed in a lipid-enriched extracellular matrix containing protein/peptide components (Elias, 2012). This organization of the stratum corneum is commonly referred to as the "brick and mortar" model (Elias, 1983; Michaels et al., 1975), where the corneocytes are likened to bricks and the extracellular matrix analogous to the mortar in a brick wall. Corneocytes are connected by corneodesmosomes and are continuously shed from the skin surface via desquamation. Since diffusion across the stratum corneum is considered as the major pathway of skin permeation, the structural properties of this layer have been extensively studied to elucidate its barrier function. Significant effort has been focused on elucidating the role of stratum corneum lipids, since the lipid-enriched extracellular matrix is the only continuous domain in the stratum corneum and thus likely to be pivotal to its barrier function (Bouwstra et al., 2002). This is supported by empirical data that

Nutraceuticals. DOI: https://doi.org/10.1016/B978-0-12-821038-3.00013-6

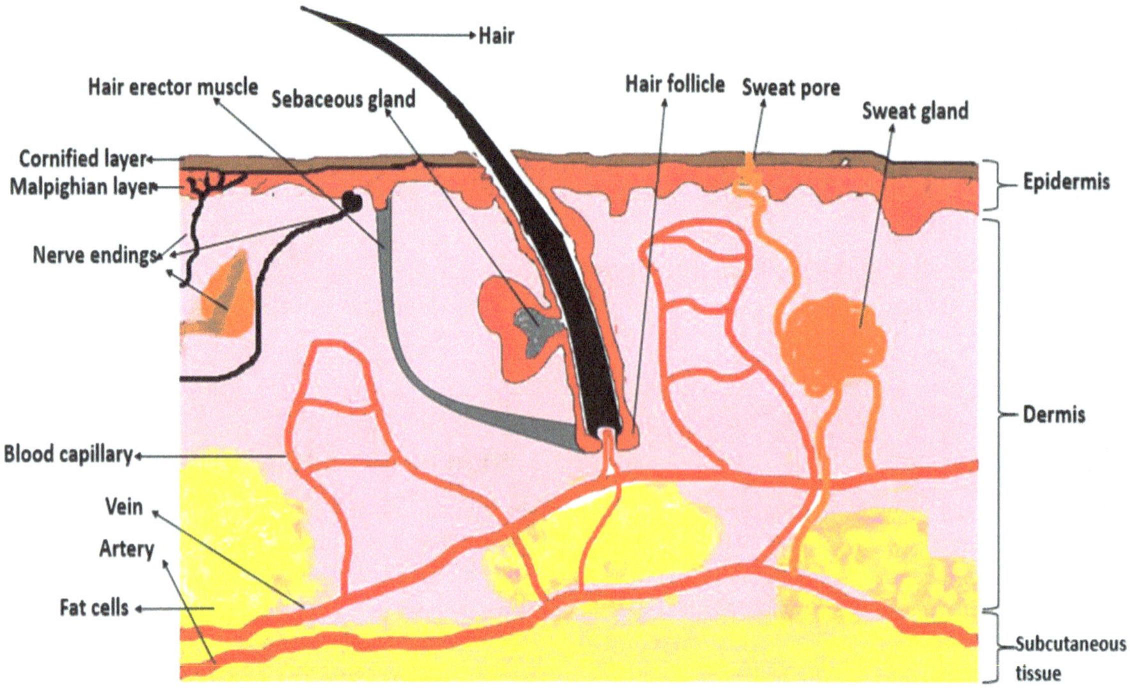

FIGURE 13.1 Structure of skin. A cross-section of skin showing epidermis, dermis, and subcutaneous tissue. The parts of each skin layer are also displayed. *Source: Adapted from Kolarsick, Kolarsick, M.A., and Goodwin, C., 2011. Anatomy and physiology of the skin. J. Dermatol. Nurses Assoc. 3, 203–213.*

demonstrate skin penetration predominantly through the extracellular matrix (Labouta et al., 2011), the marked reduction in the skin's barrier function following stratum corneum lipid extraction (Sweeney and Downing, 1970), and that chemicals that disrupt stratum corneum lipid organization enhance skin permeation (William 2003). Through freeze-fracture electron microscopy, Breathnach et al. (1973) first observed that the extracellular space in the stratum corneum was composed of a "laminated material." It is now understood that stratum corneum lipids are organized into layers (lamellae) that run parallel to the flat plane of the corneocytes. According to the widely accepted Landmann model of skin barrier lipid morphogenesis (Landmann, 1986), stratum corneum lipids are initially secreted by lamellar bodies found in keratinocytes and later fuse to form continuous lipid lamellae in the extracellular space. This model is supported by evidence of lipid membrane extrusion and fusion in the stratum corneum under electron microscopy (Madison et al., 1987). An alternative model, the membrane folding model, postulates that the extracellular lipid lamellae forms from single, coherent lipid structures through a nonfusogenic process (Norlén, 2001).

Viable epidermis

Excluding the stratum corneum, the rest of the epidermis is composed of nucleated cells and therefore is collectively referred to as the viable epidermis. The viable epidermis is typically 50–100 μm thick (Ng and Lau, 2015) and devoid of blood capillaries and sensory nerve endings. It is composed primarily of keratinocytes (95%), with the remainder being Langerhans cells, melanocytes, and Merkel cells. Keratinocytes arise from the stratum basale and undergo progressive differentiation whilst migrating toward the stratum corneum (Benson et al., 2019). Keratinocyte differentiation is characterized by increasing keratinization (formation of intracellular networks of keratin fibers), the formation of the lamellar bodies that secrete stratum corneum lipids, and the loss of intracellular organelles and nuclei. The process culminates in the formation of corneocytes in the stratum corneum. Keratinocyte differentiation serves to maintain the stratum corneum by replenishing stratum corneum lipids and corneocytes lost via desquamation (Ng and Lau, 2015).

Dermis

The dermis, typically ≥1 mm thick (Agache, 2004b; William, 2003), comprises the bulk of the skin and is responsible for its elasticity and strength. It contains polysaccharides and proteins such as glucosaminoglycan and proteoglycan and macromolecules that can help to maintain skin hydration. Aside from its remarkable capacity to retain water, the dermis also contain protein fibers such as collagen and elastin (also called extracellular matrix proteins), which gives skin its tensile strength, elasticity, and pliability (Lalan et al., 2015). Further, it also contains a range of immune cells including macrophages and dermal dendritic cells. The dermis can be subdivided into the upper papillary dermis and the lower reticular dermis, which can be distinguished microscopically from each other by the thinner and looser packing of collagen fibers in the papillary dermis. The papillary dermis contains papillae that interdigitate with the basal layer of the epidermis at the dermo–epidermal junction. The dermis contains hair follicles, sweat glands, sebaceous glands, sensory nerve endings, lymphatic vessels, and blood capillaries which extend to the dermal side of the dermo–epidermal junction. This allows nutrient and oxygen delivery to, as well as waste removal from, the avascular epidermis to occur by diffusion across the dermo–epidermal junction (Ng and Lau, 2015).

Subcutaneous tissue

Subcutaneous tissue below the dermis consists of adipose and loose connective tissue. It acts as a protective cushion and helps to insulate the body. It also houses blood vessels and nerves (Ng and Lau 2015).

Skin injury and repair: a complex interplay

As the skin serves the vital functions of a physical, chemical, and bacterial barrier, it needs an efficient and robust repair mechanism. Wound healing is the skin repair process that heals injury caused by surgery, trauma, tears, cuts, contusions, and burns and restores skin integrity (Ranjani et al., 2010). Wound healing is a complex process and involves many cell types localized at the wound site as well as participation of systemic cells. Intricate interplay between the cell types and mediators achieves restoration of the barrier function and prevents further damage and/or infection. Wound healing starts immediately after skin injury, with the three phases overlapping (and even occurring in parallel) in both time and space in the wound (Reinke and Sorg, 2012; Landen et al., 2016). Upon physical injury, a fibrin clot rapidly forms to restore hemostasis (Arya et al., 2014; Golebiewska and Poole, 2015). Platelets trigger the clotting cascade and secrete several growth factors to initiate wound healing (Enoch and Leaper, 2008). The process of wound healing may be divided into the following phases (Fig. 13.2).

Inflammation

In this phase, skin cells are exposed to acute phase signals (damage-associated molecular patterns or pathogen-specific molecular patterns) that are recognized by toll-like receptors (Landen et al., 2016; Strbo et al., 2014; Takeuchi and Akira, 2010). Inflammatory signals cause an increasing chemokine gradient coupled to transmigration of neutrophil granulocytes at the site of injury (Sinno and Prakash, 2013; Vestweber, 2015). The neutrophils migrate into the wound site engulfing foreign debris, releasing proteolytic enzymes, and killing bacteria by phagocytosis (Enoch and Leaper, 2008; Martin and Leibovich, 2005). This activates neutrophils to secrete more proinflammatory cytokines (TNF-α and IL-1), elastase, and collagenase, and amplify the inflammatory response (Eming et al., 2014; Eming et al., 2007; Werner and Grose, 2003; Lee et al., 2012). Meanwhile, blood monocytes infiltrate the injury site and differentiate into macrophages and release proteases to remove debris at the wound site (Enoch and Leaper, 2008). They secrete a mixture of bioactive molecules, including TGF-β1 and VEGF, that stimulate the migration of fibroblasts and epithelial cells (Delavary et al., 2011; Tepole and Kuhl, 2013). Also, T cells, as part of the adaptive immune system, are activated and attenuate interferon-γ production and the accumulation of proinflammatory macrophages through the epidermal growth factor receptor pathway and thus regulate tissue inflammation (Nosbaum et al., 2016).

Proliferation

Once inflammation is regulated, proliferation kicks in. The wound surface recovers through diverse activities such as reepithelialization (by keratinocytes), granulation tissue formation (by fibroblasts), and restoration of the vascular network (by endothelial cells) (Reinke and Sorg, 2012; Landen et al., 2016; Kasuya and Tokura, 2014; Fan et al., 2012). This phase usually starts 3 days after wounding.

Skin resident macrophages and macrophages differentiated from infiltrating monocytes are activated by cellular signals (Strbo et al., 2014) and differentiate into M1-type macrophages, responsible for phagocytic activity, scavenging and proinflammatory cytokines (PDGF, TGF-β, β-FGF, TNF-α, IL-1, and IL-6) production (Landen et al., 2016; Sindrilaru and Scharffetter-Kochanek, 2013; Galli et al., 2011; Mosser and Edwards, 2008). This signal results in the proliferative phase of healing. In the next step, M1 macrophages transform to M2 type macrophages and upregulate the synthesis of antiinflammatory mediators and production of extracellular matrix and collagen as an initiation of angiogenic processes by fibroblasts (Sindrilaru and Scharffetter-Kochanek, 2013; Brancato and Albina, 2011). The extracellular matrix and collagen form the granulation tissue that replaces the damaged tissue. Eventually, M2 phagocytose neutrophils through efferocytosis (bacteria and cell debris) to clean the wound

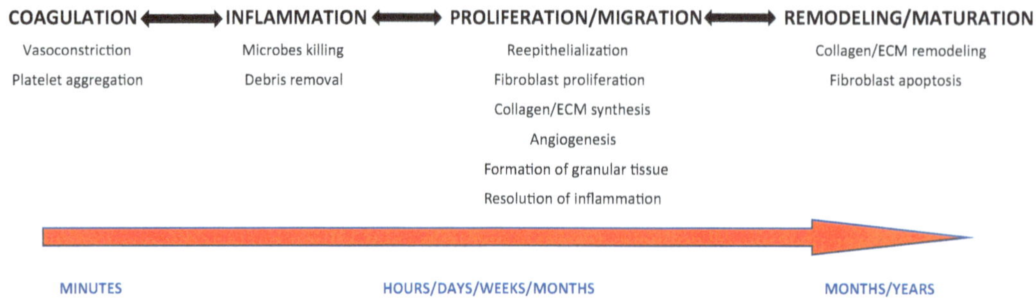

FIGURE 13.2 Timeline for wound healing showing the various overlapping phases. The three overlapping phases, inflammation, proliferation, and remodeling, are indicated. *Source: Adapted from Gonzalez, A.C.O. et al., 2016. Wound healing—a literature review. An. Bras. Dermatol. 91, 614–620.*

and avoid further damage at the wound site (Sindrilaru et al., 2011; Mirza et al., 2015; Khanna et al., 2010).

Epithelialization or epithelial remodeling

Epithelial remodeling, that is, resurfacing and wound contraction, are essential for wound closure and repair. In humans, reepithelialization is the dominant force for wound closure (Volk and Bohling, 2013). Epithelialization is highly dependent on the wound site, that is, the location, depth, size, microbial contamination, patient's health, and genetic and epigenetic factors. The thickness of the wound decides the extent of tissue rebuilding. Partial-thickness wounds involving epidermis and dermis usually heal by "primary intention" with intact skin parts, while full-thickness wounds that display complete loss of epidermis, dermis, and deeper structures require tissue granulation to replace lost regions before epithelialization can initiate (secondary intention healing). Healing by tertiary intention is utilized in complex patient cases where wounds are left intentionally open temporarily and closing carried out by sutures or surgery after regression of a highly inflammatory and life-threatening situation (Knobloch and Vogt, 2010).

Reepithelialization usually starts a few hours after injury by differentiation of cobblestone-shaped stationary keratinocytes to flat migratory keratinocytes (Rittié, 2016). The process initiates at pilosebaceous units, eccrine sweat glands (Rittié, 2016; Rittié et al., 2013), and outer root sheath of hair follicles (Rittié et al., 2016). The anatomical positioning of skin adnexa decides the distance of migration of the outgrowth before meeting another outgrowth from the opposite direction (Rittié, 2016).

The epidermal wound usually proceeds by arrangement of migratory keratinocytes in such a manner that entire rows of cells crawl over the wound to cover it (Krawczyk, 1971; Lambert et al., 1984; Gov, 2007; Omelchenko et al., 2003; Farooqui and Fenteany, 2005; Matsubayashi et al., 2011), often described by the term "leapfrogging." Leapfrogging of keratinocytes is usually accompanied by epidermal tongue, lamellipodial crawling, and shuffling (Jacinto et al., 2001; Stone et al., 2016). Epidermal tongue or extension membrane is formed adjacent to the wound site by the first rows of keratinocytes by cytoskeleton reorganization. The epidermal tongue then acts as a base for the advancement of keratinocytes to spread across the wound in a leapfrog manner, by dragging them out of blood clot-derived fibrin, fibronectin, and vitronectin and on to the wound matrix (Rittié 2016). The leading row cells do not migrate but shuffle (changing shape, loosening cell–cell contacts) and rearrange themselves in the epidermal tongue (Jacinto et al., 2001). Once keratinocyte outgrowths cover the wound (usually top to bottom for faster wound closure and to prevent fluid loss or infection) and meet in the center, the migration is stopped due to contact inhibition (Santoro and Gaudino, 2005). Cell–cell contacts are strengthened and keratinocytes revert back to their quiescent cobblestone-shaped phenotypes by epidermal stratification. Important deciding factors in appropriate keratinocyte migration are availability of extracellular matrix and underlying wound beds. Dermis, fascia, and muscle represent optimal wound beds, while adipose tissue counteracts wound coverage. In any case, granulation tissue formation is an important step in epithelial migration and all connective tissues except dermis require its formation. Granulation tissue comprises of macrophages, fibroblasts, blood vessels, and a loose matrix of type 1 collagen, glycoprotein, fibronectin, and hyaluronic acid.

The reconstitution of the cellular damage is performed by invading stem cells, that is, stem cells derived from hair follicle bulge and interfollicular epidermis niche (Blanpain and Fuchs, 2014; Donati and Watt, 2015; Pastar et al., 2014). Stem cells play important roles, such as in the recession of inflammation, cell migration, proliferation, and differentiation (Chen et al., 2016; Yang et al., 2011; Tenenhaus and Rennekampff, 2016; Kusindarta et al., 2016; Milan et al., 2016; Kong et al., 2013; Bodnar, 2015). The extent of angiogenesis also impacts the rate of wound healing and is essentially going on from the beginning to the end of wound remodeling (Demidova-Rice et al., 2012; Dulmovits and Herman, 2012). The angiogenesis and the establishing microvasculature maintain hemostasis, reduce blood loss, and prepare a wound matrix, also providing blood clot-derived cytokines and growth factors that recruit pivotal cells, crucial for healing process. These cells include immune cells to remove cell debris. In the meantime, the microvasculature provides a nutritional base and generates a high density of new capillaries. Usually disordered initially, the neovascularization usually attains a distinct pattern forming a circle with an inner ring of circularly organized vessels at the wound border followed by radially shaped vessels supplying the inner ones and connecting to normal uninjured skin (Sorg et al., 2007). Multiple factors may go wrong in chronic, nonhealing wounds, that is, unfavorable microenvironment, hyperglycemia, inflammation, and growth factor and cytokine deficiencies (Demidova-Rice et al., 2012). Recent studies have reported an important role for pericytes in wound healing due to their function in vascular development and stabilization of endothelium in newly formed blood vessels (Bodnar et al., 2016). The cells provide blood barriers and regulate capillary flow-through, and regulate immune responses. They produce adhesive substrates (VCAM-1, E-selectin, ICAM-1) and thus initiate neutrophil crawling at the endothelium and migration into the extravascular tissue (Vestweber, 2015; Proebstl et al., 2012; Voisin and Nourshargh, 2013; Ayres-Sander et al., 2013). The pericyte activation results in inhibition of antigen-specific activation of chemokine-recruited T cells (Pober and Tellides, 2012) and profibrotic activity, that is, dedifferentiation into fibroblasts producing collagen (Humphreys et al., 2010).

Scarring

Scar formation indicates the end of wound remodeling and last phase of wound healing. The original tissue is replaced by collagenous scar. Cutaneous scars have no epidermal appendages, that is, hair follicles and sebaceous glands and a collagen pattern, easily recognizable from normal unwounded skin. Collagen fibers are secreted by fibroblasts already at day 3 after wounding. As the collagen matrix forms, densely packed fibers fill the wound site.

After wounding, dermal fibroblasts at wound margins are activated by growth factors released into the wound. Further stimulated by mechanical tension and PDGF, fibroblasts turn into stress fiber-expressing protomyofibroblasts that are present in early granulation tissue. Myofibroblasts appear after day 4 after differentiation of protomyofibroblasts (Mirastschijski et al., 2010; Hinz et al., 2001). These myofibroblasts link intracellular cytoskeleton to the extracellular matrix by focal adhesion contacts. Ultimately scar contracture, that is, shortening of scar length and decreased function in the area, takes place. Although the scarring takes a comparatively short time, scar remodeling can occur for months to years. When the wound closure is finalized, capillaries regress and the redness of an early scar disappears in the next days/months, giving way to scar pigmentation. As time passes, the wound strengthens due to focal regulations by growth factors (PDGF, TGF-β3). Growth factors stimulate cellular proliferation and act as a chemoattractant for the new cells (Armour et al., 2007; Wassermann et al., 1998).

Nutraceuticals and skin

Skin disorders are a common occurrence and account for a great deal of suffering. Damaged and wrinkled skin is a sign of aging. Usually, skin disorders are a local manifestation of a generalized phenomenon of the body. The origin of the disease can be at a distant body part. In addition, skin diseases may arise due to unhygienic conditions and irregular dietary habits. [There may be different causes of skin disorders including infections (scabies, herpes, leprosy, warts, etc.); allergic (urticaria, eruptions; climatic miliaria, intertrigo, chill blains); and miscellaneous (acne, vitiligo, psoriasis, lupus, etc.). The other factors affecting skin damage are exposure to ultraviolet radiation, hormonal imbalances, and nutritional deficiencies.] Nutraceuticals, a term coined by Dr. Stephan L. F. Defelia in 1989, are the oral dietary components found in foods and believed to have a medical or health benefit. Nutraceuticals are oral dietary components that play a vital role in maintaining skin healthy and delay aging. One of the major factors responsible for skin damage is formation of free radicals and reactive oxidative species (ROS) during oxidative stress. The increased oxidative stress is a major cause of DNA damage. Since majority food items contain antioxidant components, they may act as ROS scavengers. Going by the adage "let food be thy medicine and medicine be thy food" coined by Hippocrates, all nutraceuticals act as dermal protectives. Nutraceuticals with dermal benefits include polyphenols, terpenoids, polysaturated fatty acids, and minerals. The following sections describe these common phytoconstituents, their chemistry, and occurrence.

Polyphenols

Polyphenols are the natural products which have more than one hydroxyl group attached to an aromatic ring. Polyphenols are one of the largest groups of secondary metabolites and include flavonoids and tannins. Flavonoids are polyphenolic substances occurring in a wide range of vascular plants. More than 8000 flavonoids have been reported and are considered as powerful antioxidants (Pérez-Sánchez et al., 2018). The main dietary sources of flavonoids are tea, citrus fruit, berries, red wine, apples, legumes, grapes, onions, etc. An example of an antioxidant flavonoid is kaempferol, which is found abundantly in many fruits and vegetables (Fig. 13.3; Calderon-Montano et al., 2011).

Epigallocatechin gallate, a polyphenol found abundantly in tea (*Camelia sinesis*) is used in many dietary supplements. Chemically, it is the ester of epigallocatechin (Fig. 13.4) and gallic acid (Fig. 13.5), and acts as a powerful antioxidant (Kim et al., 2018).

Besides green tea, EGCG is also found in small amounts in fruits like cranberries, avocado, etc. Curcumin, a major polyphenolic constituent of turmeric (*Curcuma longa*), is a powerful antiinflammatory and antioxidant (Fig. 13.6). It acts as an antioxidant by suppressing ROS production, scavenging free oxygen radicals, and inhibiting lipid peroxidation (Shailaja et al., 2017). Turmeric is a spice added in food items as a coloring and flavoring agent.

Pycnogenol, a standardized extract of bark of the French maritime Pine bark (*Pinus pinaster*) is rich in flavonoids [catechins and procyanidins (B1, B2, B3, B7 C1, and C2)], and phenolic acids (caffeic, ferulic, and p-hydroxybenzoic acids). Studies have proposed that this mixture formulation has a photo-protective effect against UV-light induced skin damage through a significant increase in the mRNA expression of hyaluronic acid synthase and in gene expression involved in collagen de novo synthesis. Pycnogenol supplementation alone or in

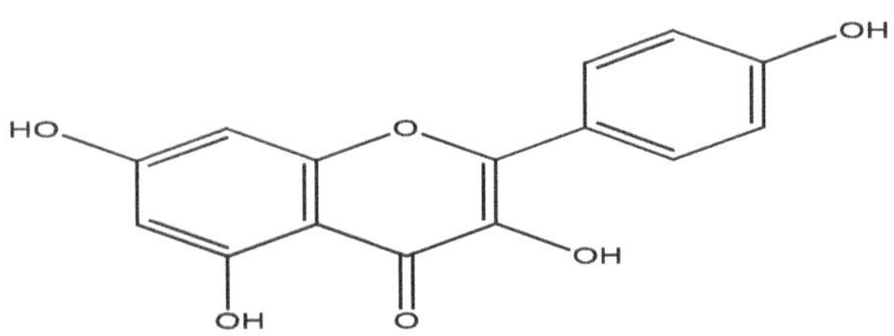

FIGURE 13.3 Kaempferol.

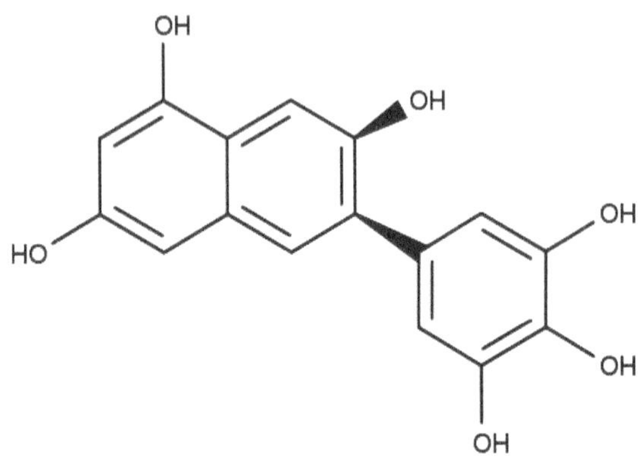

FIGURE 13.4 Epigallocatechin (EGC).

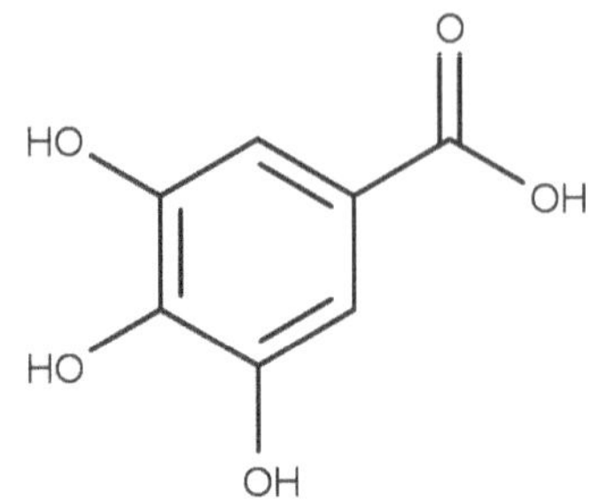

FIGURE 13.5 Gallic acid.

combination significantly improved the hydration and elasticity of skin, even in melasma, in clinical studies (Saliou et al., 2001; Marini et al., 2012; Segger and Schönlau, 2004; Ni et al., 2002). Oral administration of another French pine bark extract, Flavangenol, has been shown to be useful in photoaged facial skin (Furumura et al., 2012). NutroxSun, a formulation of citrus extract of *Citus paradise* (naringenin enriched) and rosemary extract of *Rosmarinus officinalis* (phenolic compounds and diterpenes enriched) can improve UV protection antiaging effects (Pérez-Sánchez et al., 2014). The skin photo-protective and antiaging effects have been also reported for powdered extract of red orange fruit (Red Orange Complex). The dietary supplement contained 2.8%–3.2% wt./wt. anthocyanins, 1.8%–2.2% wt./wt. hydroxycinnamic acids, 8.5%–9.5% wt./wt. flavone glycosides, and 5.5%–6.5% wt./wt. ascorbic acid (Puglia et al., 2014). *Polypodium leucotomos*/pomegranate combination (PPmix), containing different polyphenols specifically ellagitannins such as punicalagin, showed improvement in skin elasticity and hydration (Emanuele et al., 2017; Nestor et al., 2015).

Terpenoids (carotenoids)

Terpenoids form the largest group of natural products isolated from plants and consist of a definite number of isoprene (C5) units. Thus, the total number of carbon atoms in terpenoids is always a multiple of five. Carotenoids are a group of terpenoids that consist of eight isoprene units with extended conjugation. These are colored and serve as plant pigments providing red, orange, and yellow coloration. They are present in fruits, vegetable, algae, and some bacteria (Ross, 2012). They protect against excessive light. Medicinally, carotenoids are potent-reactive oxygen species (ROS) scavengers and protect the skin from oxidative stress. They have wide occurrence in fruits and vegetables which form dietary items. The four carotenoids which are of significance are β-carotene, lycopene, lutein, and zeaxanthin. A diet high in fruit and vegetables increases the carotenoid concentration in human skin and exposure to UV radiation, alcohol use,

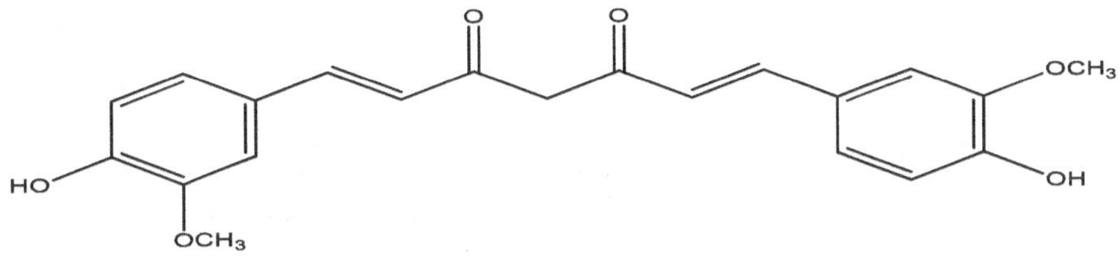

FIGURE 13.6 Curcumin.

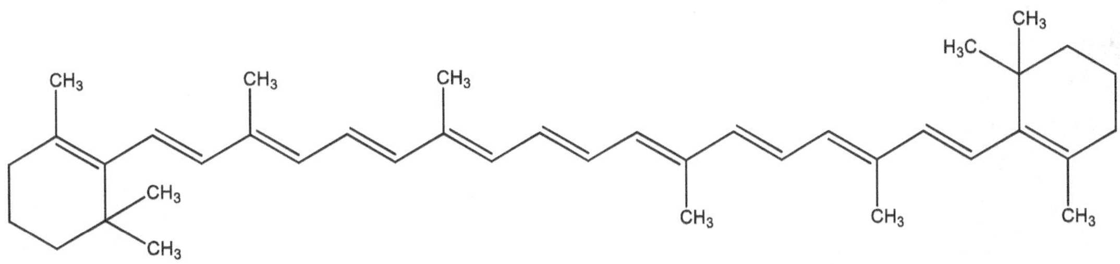

FIGURE 13.7 β-Carotene.

and smoking decrease their levels (Lademann et al., 2011). They are commonly used in nutricosmetics for skin photoprotection against UV radiation. Listed are some of the common dietary carotenoids.

β-Carotene

β-Carotene is a precursor of vitamin A. β-Carotene (Fig. 13.7) is reported to prevent wrinkle formation and decrease matrix metalloproteinase (MMP)-9 activity, protecting the extracellular matrix (ECM) from degradation in mice (Evans and Johnson, 2010).

Fruits and vegetables rich in β-carotene include carrots, green leafy vegetables, and yellow- and orange-colored fruits.

Lycopene

Lycopene rich in tomatoes is a red-colored pigment and is reported to significantly lower UV-induced erythema and decreases MMP-1 activity, an enzyme involved in the breakdown of collagen (Stahl and Sies, 2012). Besides tomatoes, pink grapefruit and watermelon are also rich in lycopene (Fig. 13.8).

Lutein and zeaxanthin

Lutein and zeaxanthin are known as xanthophyll carotenoids, and unlike β-carotene and lycopene contain hydroxyl groups. Xanthinophylls are light-gathering pigments and serve to protect humans against phototoxic damage. These are found in the macula of the human retina, and are thought to play a role in protection against age-related macular degeneration (ARMD) and age-related cataract formation (Aziz et al., 2020). Lutein (Fig. 13.9) and zeaxanthin (Fig. 13.10) are isomers differing in the location of the double bonds.

Lutein and zeaxanthin are found most commonly in leafy green vegetables, but also in other fruits and vegetables, chicken egg yolks are a rich source of both xanthinophylls. A combined oral and topical application of carotenoids lutein and zeaxanthin showed marked cutaneous hydration than either alone (Palombo et al., 2007). In addition, astaxanthin, a xanthophyll carotenoid-based supplement, improves skin fine lines/wrinkles, age spot size, skin texture, moisture content of the corneocyte layer, and skin elasticity. It stimulates collagen type I and basic fibroblast growth factor, decreases DNA damage, inflammation, and oxidative stress, and enhances the immune response, increasing natural killer cells, T cells, B cells, and IL-6 levels (Vollmer et al., 2018). Several studies have demonstrated that carotenoids prevent UV-induced skin damage in human volunteers (Bouilly-Gauthier et al., 2010; Lee et al., 2000; Gollnick et al., 1996; Postaire et al., 1997; Stahl et al., 2000; Heinrich et al., 2003).

Polyunsaturated fatty acids

Polyunsaturated fatty acids play a significant role in healthy maintenance of skin (Luigi Capella, 2011; Kendall and Nicolaou, 2013; Kowal-Bielecka et al., 2001; Ziboh et al., 2000). Depending upon the origin of these fatty acids, they are divided into omega-3 and omega-6 fatty acids. While linolenic acid (Fig. 13.11) is an omega-3 fatty acid, linoleic acid (Fig. 13.12) is an omega-6 fatty acid (Wiktorowska-Owczarek et al., 2015).

The three most common omega-3 fatty acids are α-linolenic acid (ALA), eicosapentaenoic acid (EPA), and docosahexaenoic acid (DHA) (Draelos 2010). Combinations of lipids serve as a protective barrier in the outermost layer of the skin. Here, epidermal linoleic acid

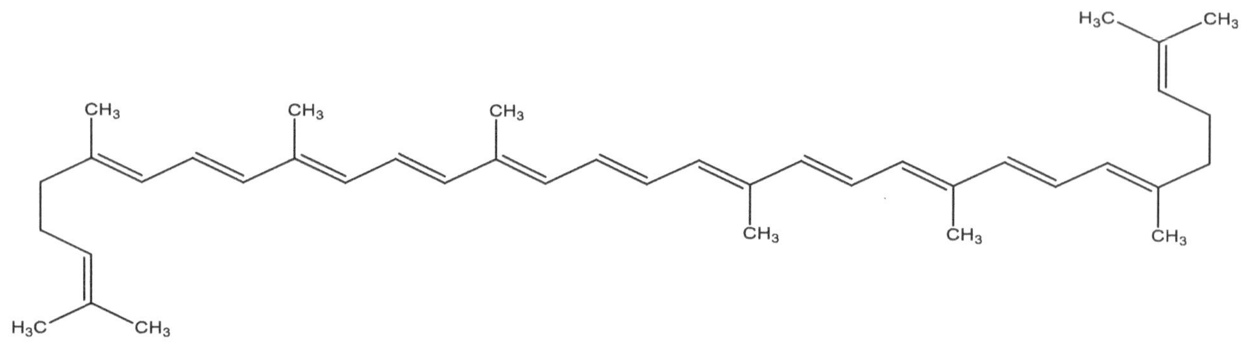

FIGURE 13.8 Lycopene.

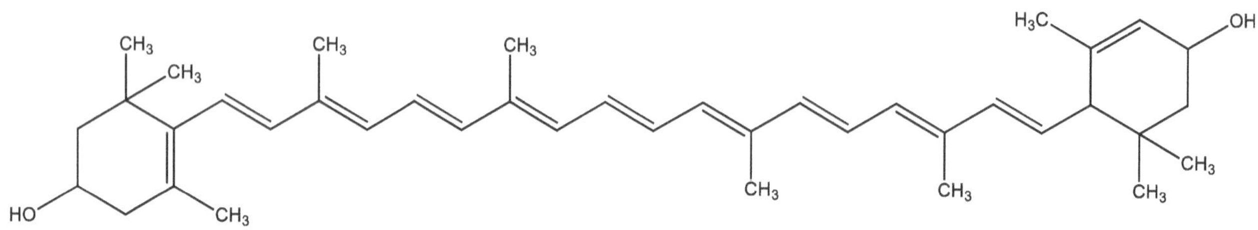

FIGURE 13.9 Lutein.

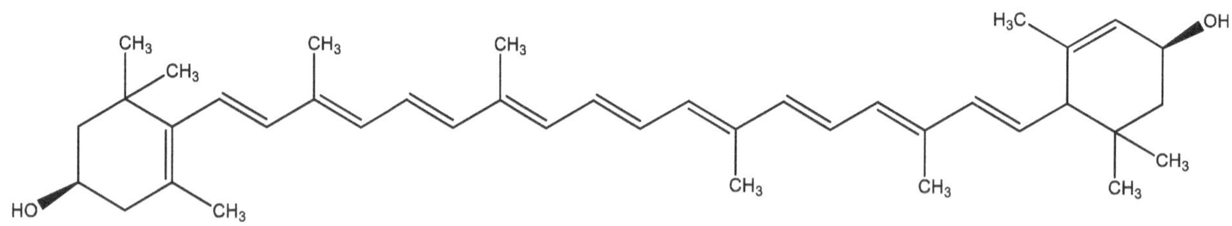

FIGURE 13.10 Zeaxanthin.

is the highest in concentration and serves as an important structural component. Polyunsaturated fatty acids activate the formation of proinflammatory cytokines such as prostaglandins, leukotrienes, tumor necrosis factor-α, interferon-γ, and interlukin-12 and 6, and regulate skin health (McCusker and Grant-Kels, 2010; Yoshida et al., 2016). Omega-3 and omega-6 fatty acids are found primarily in oils such as flaxseed oil, canola oil, hemp oil, and other foods such as fish, nuts, and seeds. The photo-protective capacity of low-dose fish oil (EPA and DHA) has been demonstrated (Rhodes et al., 2000; Orengo et al., 1992). SemoSqualene, an oral supplement rich in squalene, a polyunsaturated aliphatic hydrocarbon, reduced facial wrinkles, decreased skin reactivity to UV, increased type I procollagen gene expression, and reduced UV-induced DNA damage and apoptosis (Cho et al., 2009). Fish oil-supplemented psoriasis patients showed mild-to-moderate improvement in psoriatic lesions and overall skin condition improvement. An increase in leukotriene B5 could be linked to the reduction in inflammation (Ziboh et al., 2000; Maurice et al., 1987; Bittiner et al., 1988; Lassus et al., 1990; Wright and Burton, 1982).

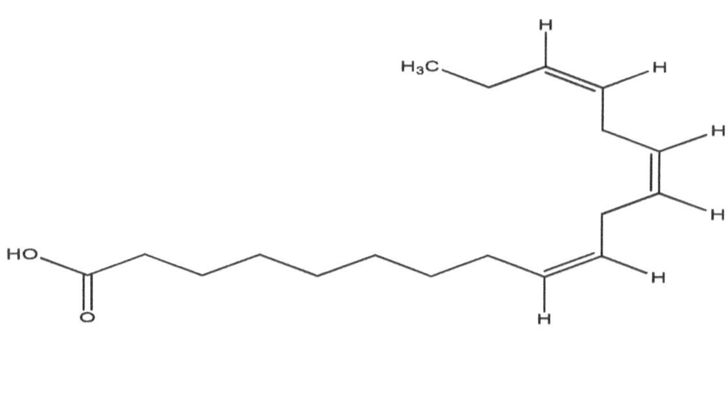

FIGURE 13.11 Linolenic acid.

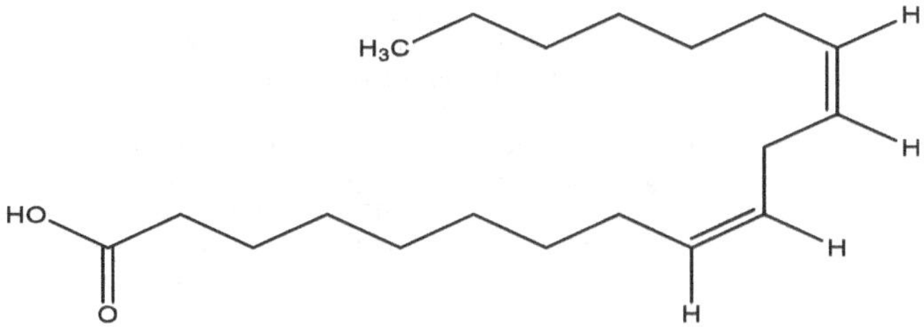

FIGURE 13.12 Linoleic acid.

Vitamins and minerals

Vitamins and minerals are accessory factors/micronutrients that keep the body healthy and also protect it from a variety of diseases. Many vitamins are "essential" because they are not synthesized in the body (except vitamin D) and therefore must be supplied by food. Inorganic minerals are present in rocks, soil, and water. Vitamins do play a role in skin health through antioxidant activity, sebum and keratinization regulation, collagen synthesis, ECM homeostasis, and photoprotection (Shapiro and Saliou, 2001). The vitamins important for dermal care are vitamin C (water-soluble) and vitamin E (fat-soluble).

Vitamin C (ascorbic acid) is obtained from raw and green peppers, oranges, grapefruits, kiwi fruit, and broccoli. Vitamin C is a cofactor in systemic biosynthesis and metabolic processes (metabolism of fats, cholesterol, and bile acids; collagen biosynthesis, biosynthesis of adrenal hormones). In addition, it speeds up wound healing and has immunostimulating and bacteriostatic properties (Szyszkowska et al., 2014). It is a powerful antioxidant and encourages intracellular and extracellular collagen production. Vitamin E (α-tocopherol) blocks the formation of reactive oxygen species and scavenges free radicals and stabilizes cell membranes. Vitamin E deficiency is associated with premature skin aging and keratoses in wound healing (Vollmer et al., 2018). Sunflower seeds, peanuts, almonds, walnuts, sesame seeds, fruits, and vegetables are sources of vitamin E.

Minerals, important for dermal care, include copper, zinc, and selenium. Copper acts as a cofactor in the enzymatic reaction for collagen cross-linking (Rucker et al., 1998) and can be obtained from nuts, seed, seafood, meat, and grains. Zinc has important roles in cell division, the metabolic process, immune system functioning, glycemic control, antioxidant activity, coagulation, and wound healing. It is supplied to the body by whole grains, red meat, and seafood. It plays an important role in the repair and maintenance of the skin as an essential cofactor for metalloenzymes, superoxide dismutase, metallothionein, and RNA polymerases (Göllner et al., 2017).

Selenium, in trace amounts, supports DNA synthesis and guards against oxidative damage, being a cofactor of glutathione peroxidases and thioredoxin reductases, and removes lipid hydroperoxides, hydrogen peroxide, and peroxynitrite (Göllner et al., 2017). It is found in two forms in nature, selenate and selenite (inorganic) and selenomethionine and selenocysteine (organic). Both forms are good dietary sources, although the organic form is absorbed considerably more than the inorganic form. Seafood and meat are good sources of selenium and absorbed selenium is stored in liver, kidneys, skeletal muscles, brain, and testis. Selenium deficiency is observed in skin cancer.

A formulation containing vitamins (*Carica papaya*: vitamin B6, folic acid), calcium, potassium, niacin, magnesium, copper, zinc, and some amino acids; and an antioxidant cocktail: *trans*-resveratrol, selenium, vitamin E, and vitamin C, showed significant improvement in skin elasticity, moisture, and antioxidant capacity (Bertuccelli et al., 2016).

Concluding remarks and future directions

Skin nutraceuticals have been proven beneficial in skin disorders and wound healing. A number of natural ingredients have been identified with potential for alleviating the signs of skin aging and skin diseases. Even though great interest in understanding the mechanism of action of these nutraceuticals is rising, experimental evidence in this direction is currently limited. More studies would not only elucidate the mechanisms but also enable the production of these nutraceuticals synthetically. In addition, current human trials in skin nutraceuticals are far from conclusive. Some report on poorly characterized single constituents or the use of mixtures. The use of insufficient numbers of subjects and/or interventions also leads to poor statistics. For sufficient clarity on the mechanism of action, better human trials coupled with in vitro skin cell models must be employed. Finally, skin nutraceuticals are expected to become very profitable in the near future as they are considered safer than allopathic drugs and encourage patients to turn to nutraceuticals to alleviate symptoms.

Acknowledgments

The authors thank Ajay Kumar Sharma and Tania Dhiman for help in preparing this chapter.

Author contributions: Poonam Negi drafted the section on skin anatomy and physiology and Vijay Kumar Kapoor and Gurjot Kaur drafted the section on nutraceutical and skin. Gurjot Kaur drafted the introduction, the section on skin injury and repair, and the concluding remarks and future directions. Gurjot Kaur edited and finalized the final version of the chapter.

References

Agache, P., 2004a. Metrology of the stratum corneum. In: Agache, P., Humbert, P. (Eds.), Measuring the Skin: Non-invasive Investigations, Physiology, Normal Constants. Springer, Berlin, pp. 101–111.

Agache, P., 2004b. Metrology of the stratum corneum. In: Agache, P., Humbert, P. (Eds.), Measuring the Skin: Non-invasive Investigations, Physiology, Normal Constants. Springer, Berlin, pp. 3–5.

Armour, A., Scott, P.G., Tredget, E.E., 2007. Cellular and molecular pathology of HTS: basis for treatment. Wound Repair Regen 15, S6–S17.

Arya, A.K., Tripathi, K., Das, P., 2014. Promising role of ANGPTL4 gene in diabetic wound healing. Int. J. Low. Extrem. Wounds 13, 58–63.

Ayres-Sander, C.E., Maier, C.L., Lauridsen, H., et al., 2013. Transendothelial migration enables subsequent transmigration of neutrophils through underlying pericytes. PloS one 8, e60025.

Aziz, E., Batool, R., Akhtar, W., et al., 2020. Xanthophyll: health benefits and therapeutic insights. Life Sci. 240, 117104.

Benson, H.A.E., Grice, J.E., Mohammed, Y., et al., 2019. Topical and transdermal drug delivery: from simple potions to smart technologies. Curr. Drug Deliv. 16, 444–460.

Bertuccelli, G., Zerbinati, N., Marcellino, M., et al., 2016. Effect of a quality-controlled fermented nutraceutical on skin aging markers: an antioxidant-control, double-blind study. Exp. Ther. Med 11, 909–916.

Bittiner, S., Tucker, W.F., Cartwright, I., et al., 1988. A double-blind, randomised, placebo-controlled trial of fish oil in psoriasis. Lancet 331, 378–380.

Blanpain, C., Fuchs, E., 2014. Plasticity of epithelial stem cells in tissue regeneration. Science 344, 1242281.

Bodnar, R.J., 2015. Chemokine regulation of angiogenesis during wound healing. Adv. Wound Care 4, 641–650.

Bodnar, R.J., Satish, L., Yates, C.C., et al., 2016. Pericytes: a newly recognized player in wound healing. Wound Repair Regen 24, 204–214.

Bouilly-Gauthier, D., Jeannes, C., Maubert, Y., et al., 2010. Clinical evidence of benefits of a dietary supplement containing probiotic and carotenoids on ultraviolet-induced skin damage. Brit. J. Dermatol 163, 536–543.

Bouwstra, J.A., Gooris, S.S., Frank, E.R., et al., 2002. Phase behavior of stratum corneum lipid mixtures based on human ceramides: the role of natural and synthetic ceramide. J. Investig. Dermatol 118, 606–617.

Brancato, S.K., Albina, J.E., 2011. Wound macrophages as key regulators of repair: origin, phenotype, and function. Am. J. Pathol 178, 19–25.

Breathnach, A., Goodman, T., Stolinski, C., et al., 1973. Freeze-fracture replication of cells of stratum corneum of human epidermis. J. Anat. 114, 65.

Calderon-Montano, J.M., Burgos-Moron, E., Perez-Guerrero, C., et al., 2011. A review on the dietary flavonoid kaempferol. Mini Rev. Med. Chem. 11, 298–344.

Chen, D., Hao, H., Fu, X., et al., 2016. Insight into reepithelialization: how do mesenchymal stem cells perform? Stem Cells Int 2016.

Cho, S., Choi, C.-W., Lee, D.H., et al., 2009. High-dose squalene ingestion increases type I procollagen and decreases ultraviolet-induced DNA damage in human skin in vivo but is associated with transient adverse effects. Clin. Exp. Dermatol 34, 500–508.

Degim, I.T., 2006. New tools and approaches for predicting skin permeability. Drug Discov. Today 11, 517–523.

Delavary, B.M., van der Veer, W.M., van Egmond, M., et al., 2011. Macrophages in skin injury and repair. Immunobiology 216, 753–762.

Demidova-Rice, T.N., Durham, J.T., Herman, I.M., 2012. Wound healing angiogenesis: innovations and challenges in acute and chronic wound healing. Adv. Wound Care 1, 17–22.

Donati, G., Watt, F.M., 2015. Stem cell heterogeneity and plasticity in epithelia. Cell stem cell 16, 465–476.

Draelos, Z.D., 2010. Nutrition and enhancing youthful-appearing skin. Clin. Dermatol. 28, 400–408.

Dulmovits, B.M., Herman, I.M., 2012. Microvascular remodeling and wound healing: a role for pericytes. Int. J. Biochem. Cell Biol. 44, 1800–1812.

Elias, P.M., 1983. Epidermal lipids, barrier function, and desquamation. J. Invest. Dermatol. 80.

Elias, P.M., 2012. Structure and function of the stratum corneum extracellular matrix. J. Invest. Dermatol 132, 2131–2133.

Emanuele, E., Bertona, M., Biagi, M., 2017. Comparative effects of a fixed *Polypodium leucotomos*/pomegranate combination versus *Polypodium leucotomos* alone on skin biophysical parameters. Neuro Endocrinol. Lett 38 (1), 38–42.

Eming, S.A., Brachvogel, B., Odorisio, T., et al., 2007. Regulation of angiogenesis: wound healing as a model. Prog. Histochem. Cytochem. 42, 115–170.

Eming, S.A., Martin, P., Tomic-Canic, M., 2014. Wound repair and regeneration: mechanisms, signaling, and translation. Sci. Transl. Med. 6 (265), 265sr6.

Enoch, S., Leaper, D.J., 2008. Basic science of wound healing. Surgery 26, 31–37.

Evans, J.A., Johnson, E.J., 2010. The role of phytonutrients in skin health. Nutrients 2, 903–928.

Fan, D., Takawale, A., Lee, J., et al., 2012. Cardiac fibroblasts, fibrosis and extracellular matrix remodeling in heart disease. Fibrogenesis Tissue Repair 5, 15.

Farooqui, R., Fenteany, G., 2005. Multiple rows of cells behind an epithelial wound edge extend cryptic lamellipodia to collectively drive cell-sheet movement. J. Cell Sci. 118, 51–63.

Furumura, M., Sato, N., Kusaba, N., et al., 2012. Oral administration of French maritime pine bark extract (Flavangenol®) improves clinical symptoms in photoaged facial skin. Clin. Interv. Aging 7, 275.

Galli, S.J., Borregaard, N., Wynn, T.A., 2011. Phenotypic and functional plasticity of cells of innate immunity: macrophages, mast cells and neutrophils. Nat. Immunol. 12, 1035.

Golebiewska, E.M., Poole, A.W., 2015. Platelet secretion: from haemostasis to wound healing and beyond. Blood Rev. 29, 153–162.

Göllner, I., Voss, W., von Hehn, U., et al., 2017. Ingestion of an oral hyaluronan solution improves skin hydration, wrinkle reduction, elasticity, and skin roughness: results of a clinical study. Evid. Based Complement. Alternat. Med 22, 816–823.

Gollnick, H.M., Hopfenmüller, W., Biesalski, H.K., et al., 1996. Systemic beta carotene plus topical UV-sunscreen are an optimal protection against harmful effects of natural UV-sunlight: results of the Berlin-Eilath study. Eur. J. Dermatol. 6, 200–205.

Gonzalez, A.C.O., et al., 2016. Wound healing—a literature review. An. Bras. Dermatol. 91, 614–620.

Gov, N.S., 2007. Collective cell migration patterns: follow the leader. Proc. Natl. Acad. Sci. U. S. A 104, 15970–15971.

Hadgraft, J., 2004. Skin deep. Eur. J. Pharm. Biopharm. 58, 291–299.

Heinrich, U., Gärtner, C., Wiebusch, M., et al., 2003. Supplementation with β-carotene or a similar amount of mixed carotenoids protects humans from UV-induced erythema. Nutr. J 133, 98–101.

Hinz, B., Mastrangelo, D., Iselin, C.E., et al., 2001. Mechanical tension controls granulation tissue contractile activity and myofibroblast differentiation. Am. J. Pathol 159, 1009–1020.

Humphreys, B.D., Lin, S.-L., Kobayashi, A., et al., 2010. Fate tracing reveals the pericyte and not epithelial origin of myofibroblasts in kidney fibrosis. Am. J. Pathol. 176, 85–97.

Jacinto, A., Martinez-Arias, A., Martin, P., 2001. Mechanisms of epithelial fusion and repair. Nat. Cell Biol. 3, E117–E123.

Kashibuchi, N., Hirai, Y., O'Goshi, K., et al., 2002. Three-dimensional analyses of individual corneocytes with atomic force microscope: morphological changes related to age, location and to the pathologic skin conditions. Skin Res. Tech 8, 203–211.

Kasuya, A., Tokura, Y., 2014. Attempts to accelerate wound healing. J. Dermatol. Sci. 76, 169–172.

Kendall, A.C., Nicolaou, A., 2013. Bioactive lipid mediators in skin inflammation and immunity. Prog. Lipid Res. 52, 141–164.

Khanna, S., Biswas, S., Shang, Y., et al., 2010. Macrophage dysfunction impairs resolution of inflammation in the wounds of diabetic mice. PloS one 5 (3), e9539.

Kim, E., Hwang, K., Lee, J., et al., 2018. Skin protective effect of epigallocatechin gallate. Int. J. Mol. Sci. 19 (1), 173.

Knobloch, K., Vogt, P., 2010. Die rekonstruktive Sequenz des 21. Jahrhunderts. Chirurg 81, 441–446.

Kolarsick, P.A.J., Kolarsick, M.A., Goodwin, C., 2011. Anatomy and physiology of the skin. J. Dermatol. Nurses Assoc. 3, 203–213.

Kong, P., Xie, X., Li, F., et al., 2013. Placenta mesenchymal stem cell accelerates wound healing by enhancing angiogenesis in diabetic Goto-Kakizaki (GK) rats. Biochem. Bioph. Res. Co 438, 410–419.

Kowal-Bielecka, O., Distler, O., Neidhart, M., et al., 2001. Evidence of 5-lipoxygenase overexpression in the skin of patients with systemic sclerosis: a newly identified pathway to skin inflammation in systemic sclerosis. Arthritis Rheumatol 44, 1865–1875.

Krawczyk, W.S., 1971. A pattern of epidermal cell migration during wound healing. J. Cell Biol 49, 247–263.

Kusindarta, D.L., Wihadmadyatami, H., Fibrianato, Y.H., et al., 2016. Human umbilical mesenchymal stem cells conditioned medium promote primary wound healing regeneration. Vet. World 9, 605.

Labouta, H.I., El-Khordagui, L.K., Kraus, T., et al., 2011. Mechanism and determinants of nanoparticle penetration through human skin. Nanoscale 3, 4989–4999.

Lademann, J., Meinke, M.C., Sterry, W., et al., 2011. Carotenoids in human skin. Exp. Dermatol 20, 377–382.

Lalan, M., Baweja, J., Misra, A., 2015. Atopic dermatitis: drug delivery approaches in disease management. Crit. Rev. Ther. Drug Carrier Syst. 32 (4), 323–361.

Lambert, W.C., Cohen, P.J., Lambert, M.W., 1984. Role of the epidermis and other epithelia in wound healing: selected concepts. Clin. Dermatol. 2, 24–33.

Landen, N.X., Li, D., Stahle, M., 2016. Transition from inflammation to proliferation: a critical step during wound healing. Cell. Mol. Life Sci. 73, 3861–3885.

Landmann, L., 1986. Epidermal permeability barrier: transformation of lamellar granule-disks into intercellular sheets by a membrane-fusion process, a freeze-fracture study. J. Invest. Dermatol. 87, 202–209.

Lassus, A., Dahlgren, A.-L., Halpern, M.J., et al., 1990. Effects of dietary supplementation with polyunsaturated ethyl ester lipids (angiosan®) in patients with psoriasis and psoriatic arthritis. J. Int. Med. Res. 18, 68–73.

Lee, J., Jiang, S., Levine, N., et al., 2000. Carotenoid supplementation reduces erythema in human skin after simulated solar radiation exposure. Proc. Soc. Exp. Biol. Med. 223, 170–174.

Lee, Y.S., Warburton, D., Wysocki, A.B., et al., 2012. Wound healing in development. Birth Defects Res. A 96, 213–222.

Luigi Capella, G., 2011. Strategies for leukotriene modulation in dermatology: even more visionary perspectives? an update. Antiinflamm. Antiallergy Agents Med. Chem 10, 407–417.

Madison, K.C., Swartzendruber, D.C., Wertz, P.W., et al., 1987. Presence of intact intercellular lipid lamellae in the upper layers of the stratum corneum. J. Invest. Dermatol. 88, 714–718.

Marini, A., Grether-Beck, S., Jaenicke, T., et al., 2012. Pycnogenol® effects on skin elasticity and hydration coincide with increased gene expressions of collagen type I and hyaluronic acid synthase in women. Skin Pharmacol. Phys 25, 86–92.

Martin, P., Leibovich, S.J., 2005. Inflammatory cells during wound repair: the good, the bad and the ugly. Trends Cell Biol 15, 599–607.

Matsubayashi, Y., Razzell, W., Martin, P., 2011. 'White wave'analysis of epithelial scratch wound healing reveals how cells mobilise back from the leading edge in a myosin-II-dependent fashion. J. Cell Sci. 124, 1017–1021.

Maurice, P.,D., Allen, B.R., Barkley, A., et al., 1987. Effects of dietary supplementation with eicosapentaenoic acid in patients with psoriasis. Adv. Prostagl, Thrombox, Leukotr Res 17, 647.

McCusker, M.M., Grant-Kels, J.M., 2010. Healing fats of the skin: the structural and immunologic roles of the omega-6 and omega-3 fatty acids. Clin. Dermatol. 28, 440–451.

Michaels, A.S., Chandrasekaran, S.K., Shaw, J.E., 1975. Drug permeation through human skin: theory and invitro experimental measurement. AIChE J. 21, 985–996.

Milan, P.B., Lotfibakhshaiesh, N., Joghtaie, M.T., et al., 2016. Accelerated wound healing in a diabetic rat model using decellularized dermal matrix and human umbilical cord perivascular cells. Acta Biomater. 45, 234–246.

Mirastschijski, U., Schnabel, R., Claes, J., et al., 2010. Matrix metalloproteinase inhibition delays wound healing and blocks latent transforming growth factor-β1-promoted myofibroblast formation and function. Wound Repair Regen 18, 223–234.

Mirza, R.E., Fang, M.M., Novak, M.L., et al., 2015. Macrophage PPARγ and impaired wound healing in type 2 diabetes. J. Pathol. 236, 433–444.

Mosser, D.M., Edwards, J.P., 2008. Exploring the full spectrum of macrophage activation. Nat. Rev. Immunol. 8, 958–969.

Nestor, M.S., Berman, B., Swenson, N., 2015. Safety and efficacy of oral Polypodium leucotomos extract in healthy adult subjects. J. Clin. Aesthet. Dermatol. 8, 19.

Ng, K., Lau, W., 2015. Skin deep: the basics of human skin structure and drug penetration. Percutaneous Penetration Enhancers Chemical Methods in Penetration Enhancement. Springer.

Ni, Z., Mu, Y., Gulati, O., 2002. Treatment of melasma with Pycnogenol®. Phytother. Res. 16, 567–571.

Norlén, L., 2001. Skin barrier structure and function: the single gel phase model. J. Invest. Dermatol. 117, 830–836.

Nosbaum, A., Prevel, N., Truong, H.-A., et al., 2016. Cutting edge: regulatory T cells facilitate cutaneous wound healing. J. Immunol 196, 2010–2014.

Omelchenko, T., Vasiliev, J.M., Gelfand, I.M., et al., 2003. Rho-dependent formation of epithelial "leader" cells during wound healing. Proc. Natl. Acad. Sci. U. S. A 100, 10788–10793.

Orengo, I., Black, H., Wolf, J., 1992. Influence of fish oil supplementation on the minimal erythema dose in humans. Arch. Dermatol. Res 284, 219–221.

Palombo, P., Fabrizi, G., Ruocco, V., et al., 2007. Beneficial long-term effects of combined oral/topical antioxidant treatment with the carotenoids lutein and zeaxanthin on human skin: a double-blind, placebo-controlled study. Skin Pharmacol. Physiol. 20, 199–210.

Pastar, I., Stojadinovic, O., Yin, N.C., et al., 2014. Epithelialization in wound healing: a comprehensive review. Adv. Wound Care 3, 445–464.

Pérez-Sánchez, A., Barrajon-Catalan, E., Caturla, N., et al., 2014. Protective effects of citrus and rosemary extracts on UV-induced damage in skin cell model and human volunteers. J. Photoch. Photobio. B 136, 12–18.

Pérez-Sánchez, A., Barrajon-Catalan, E., et al., 2018. Nutraceuticals for skin care: a comprehensive review of human clinical studies. Nutrients 10, 403.

Pober, J.S., Tellides, G., 2012. Participation of blood vessel cells in human adaptive immune responses. Trends Immunol. 33, 49–57.

Postaire, E., Jungmann, H., Bejot, M., et al., 1997. Evidence for antioxidant nutrients-induced pigmentation in skin: results of a clinical trial. Biochem Mol Biol Int 42, 1023–1033.

Proebstl, D., Voisin, M.-B., Woodfin, A., et al., 2012. Pericytes support neutrophil subendothelial cell crawling and breaching of venular walls in vivo. J. Exp. Med 209, 1219–1234.

Puglia, C., Offerta, A., Saija, A., et al., 2014. Protective effect of red orange extract supplementation against UV-induced skin damages: photoaging and solar lentigines. J. Cosmet. Dermatol. 13, 151–157.

Ranjani, M., Rajan, S., Brindha, P., 2010. Antioxidant and antibacterial potentials of Aloe vera juice extract against wound isolates. J. Pure Appl. Microbio. 4, 733–739.

Reinke, J.M., Sorg, H., 2012. Wound repair and regeneration. Eur. Surg. Res. 49, 35–43.

Rhodes, L., et al., 2000. Systemic eicosapentaenoic acid reduces UVB-induced erythema and p53 induction in skin, while increasing oxidative stress, in a double-blind randomised study. Brit. J. Dermatol. 142, 601–602.

Rittié, L., 2016. Cellular mechanisms of skin repair in humans and other mammals. J. Cell Commun. Signal 10, 103–120.

Rittié, L., Sachs, D.L., Orringer, J.S., et al., 2013. Eccrine sweat glands are major contributors to reepithelialization of human wounds. Am. J. Clin. Pathol. 182, 163–171.

Rittié, L., Farr, E.A., Orringer, J.S., et al., 2016. Reduced cell cohesiveness of outgrowths from eccrine sweat glands delays wound closure in elderly skin. Aging cell 15, 842–852.

Ross, A.C., 2012. Modern Nutrition in Health and Disease, 11th edn. Jones & Bartlett, 1616 pages.

Rucker, R.B., Kosonen, T., Clegg, M.S., et al., 1998. Copper, lysyl oxidase, and extracellular matrix protein cross-linking. Am. J. Clin. Nutr 67 (Suppl. 5), 996S–1002S.

Saliou, C., Rimbach, G., Moini, H., et al., 2001. Solar ultraviolet-induced erythema in human skin and nuclear factor-kappa-B-dependent gene expression in keratinocytes are modulated by a French maritime pine bark extract. Free Radic. Biol. Med. 30, 154–160.

Santoro, M.M., Gaudino, G., 2005. Cellular and molecular facets of keratinocyte reepithelization during wound healing. Exp. Cell Res. 304, 274–286.

Segger, D., Schönlau, F., 2004. Supplementation with Evelle® improves skin smoothness and elasticity in a double-blind, placebo-controlled study with 62 women. J. Dermatol. Treat. 15, 222–226.

Shailaja, M., Damodara Gowda, K.M., Vishakh, K., et al., 2017. Anti-aging role of curcumin by modulating the inflammatory markers in albino wistar rats. J. Natl. Med. Assoc. 109, 9–13.

Shapiro, S.S., Saliou, C., 2001. Role of vitamins in skin care. Nutrition 17, 839–844.

Sindrilaru, A., Peters, T., Wieschalka, S., et al., 2011. An unrestrained proinflammatory M1 macrophage population induced by iron impairs wound healing in humans and mice. J. Clin. Invest. 121, 985–997.

Sindrilaru, A., Scharffetter-Kochanek, K., 2013. Disclosure of the culprits: macrophages—versatile regulators of wound healing. Adv. Wound Care 2, 357–368.

Sinno, H., Prakash, S., 2013. Complements and the wound healing cascade: an updated review. Plast. Surg. Int. 2013.

Sorg, H., Krueger, C., Vollmar, B., 2007. Intravital insights in skin wound healing using the mouse dorsal skin fold chamber. J. Anat 211, 810–818.

Stahl, W., Sies, H., 2012. β-Carotene and other carotenoids in protection from sunlight. Am. J. Clin. Nutr 96, 1179S–1184S.

Stahl, W., Heinrich, U., Jungmann, H., et al., 2000. Carotenoids and carotenoids plus vitamin E protect against ultraviolet light-induced erythema in humans. Am. J. Clin. Nutr. 71, 795–798.

Stone, R.C., Pastar, I., Ojeh, N., et al., 2016. Epithelial-mesenchymal transition in tissue repair and fibrosis. Cell Tissue Res 365, 495–506.

Strbo, N., Yin, N., Stojadinovic, O., 2014. Innate and adaptive immune responses in wound epithelialization. Adv. Wound Care 3, 492–501.

Sweeney, T.M., Downing, D.T., 1970. The role of lipids in the epidermal barrier to water diffusion. J. Invest. Dermatol. 55, 135–140.

Szyszkowska, B., Lepecka-Klusek, C., Koklowicz, K., et al., 2014. The influence of selected ingredients of dietary supplements on skin condition. Adv. Dermatol. Allergol. 31, 174.

Takeuchi, O., Akira, S., 2010. Pattern recognition receptors and inflammation. Cell 140, 805–820.

Tenenhaus, M., Rennekampff, H.-O., 2016. Current concepts in tissue engineering: skin and wound. Plast. Reconstr. Surg. 138, 42S–50S.

Tepole, A.B., Kuhl, E., 2013. Systems-based approaches toward wound healing. Pediatr. Res. 73, 553–563.

Vestweber, D., 2015. How leukocytes cross the vascular endothelium. Nat. Rev. Immunol. 15, 692–704.

Voisin, M.-B., Nourshargh, S., 2013. Neutrophil transmigration: emergence of an adhesive cascade within venular walls. Innate Immun. 5, 336–347.

Volk, S.W., Bohling, M.W., 2013. Comparative wound healing—are the small animal veterinarian's clinical patients an improved translational model for human wound healing research? Wound repair and regen 21, 372–381.

Vollmer, D.L., West, V.A., Lephart, E.D., 2018. Enhancing skin health: by oral administration of natural compounds and minerals with implications to the dermal microbiome. Int. J. Mol. Sci. 19, 3059.

Wassermann, R., Polo, M., Smith, P., et al., 1998. Differential production of apoptosis-modulating proteins in patients with hypertrophic burn scar. J. Surg. Res. 75, 74–80.

Werner, S., Grose, R., 2003. Regulation of wound healing by growth factors and cytokines. Physiol. Rev. 83, 835–870.

Wiktorowska-Owczarek, A., Berezinska, M., Nowak, J.Z., 2015. PUFAs: structures, metabolism and functions. Adv. Clin. Exp. Med. 24, 931–941.

William, A., 2003. Transdermal and topical drug delivery from theory to clinical practice. Pharm. J. 37–84.

Wright, S., Burton, J., 1982. Oral evening-primrose-seed oil improves atopic eczema. The Lancet 320, 1120–1122.

Yang, M., et al., 2011. Bone marrow-derived mesenchymal stem cells transplantation accelerates tissue expansion by promoting skin regeneration during expansion. Ann. Surg. 253, 202–209.

Yoshida, S., Yasutomo, K., Watanabe, T., 2016. Treatment with DHA/EPA ameliorates atopic dermatitis-like skin disease by blocking LTB4 production. J. Med. Invest. 63, 187–191.

Ziboh, V.A., Miller, C.C., Cho, Y., 2000. Metabolism of polyunsaturated fatty acids by skin epidermal enzymes: generation of antiinflammatory and antiproliferative metabolites. Am. J. Clin. Nutr 71, 361s–366s.

Chapter 14

Nutraceuticals in arthritis

Ramesh C. Gupta, Sanyasi R. Kalidindi, Robin B. Doss, Rajiv Lall, Ajay Srivastava and Anita Sinha

Introduction

During the 20th century, the number of arthritis cases doubled. Currently, more than 55 million adults in the United States suffer from some form of arthritis, rheumatoid arthritis (RA), gout, lupus, or fibromyalgia. By the year 2040, this number will increase to 78 million adults in the United States due in part to dietary habits, obesity, and lifestyle (Hootman et al., 2016). Arthritis is also prevalent in other species, including canine and equine. Among all forms of arthritis, osteoarthritis (OA) is the most common form, afflicting nearly 31 million adults in the United States. The other common form of arthritis is RA. OA is an inflammatory degenerative joint disease (DJD) characterized by chronic and progressive cartilage degeneration, osteophyte formation, subchondral sclerosis, hypertrophy of bone at the margin, and changes in the synovial membrane. The disease can wear down cartilage in a joint to the point that bone rubs against bone, resulting in crepitus and progressive loss of cartilage. In severe cases, cartilage fragments may break off and irritate muscles that are adjacent to the bone, causing severe pain. These biochemical and morphological cascading events eventually result in decreased stability, decreased movement, loading, progressive joint destruction, deformity, stiffness of joints, lameness, pain, and loss of function (CDC, 2010; Umlauf et al., 2010; Loeser, 2013; Murdock et al., 2016; Gupta et al., 2019a).

The diagnosis of OA is usually based on physical examination and radiographic and microradiographic evidence (Neogi et al., 2011; Braun and Gold, 2012; Waung et al., 2014). Magnetic resonance imaging (MRI) findings revealed that the changes of joint and cartilage degeneration were consistent with OA (Calvo et al., 2001; Guermazi et al., 2013; Kumar et al., 2013; Eckstein et al., 2014; Runhaar et al., 2014; Holzer et al., 2018; Daugaard et al., 2020).

Both humans and animals with OA are treated with multipronged approaches involving invasive as well as noninvasive measures. The objectives in managing OA include minimizing joint pain by reducing inflammation and slowing the progression of cartilage damage, thereby increasing joint flexibility and thus improving quality of life. To achieve these goals, a variety of pharmaceuticals, nutraceuticals, disease-modifying agents, stem cell therapy, and physical therapy with acupuncture can be used (Abadie et al., 2004; Dieppe and Lohmander, 2005; Martel-Pelletier et al., 2005; Mathews and Hunter, 2011; Panahifar et al., 2014; Vilar et al., 2014; Otis et al., 2019). Cyclooxygenase (COX)-inhibiting nonsteroidal antiinflammatory drugs (NSAIDs) have most commonly been used to manage OA. Unfortunately, chronic use of some NSAIDs has been linked to adverse events, such as reduced appetite, vomiting, gastrointestinal (GI) upset and bleeding, and cardiovascular, hepatic and renal dysfunction (Muhlfeld and Floege, 2005; Raekillio et al., 2006; Coruzzi et al., 2007; Sorres et al., 2010; Fanelli et al., 2013; Tonge et al., 2014; Perkins et al., 2015; Scarpiganto et al., 2015; Breivik, 2017; Gosser et al., 2017; Sessa et al., 2017; de Salazar Alcalá et al., 2019). Therapeutic drugs, such as aspirin and ibuprofen, inhibit the activities of COX-I and COX-II as well as the production of inflammatory prostaglandins, thus providing a therapeutic effect. However, these drugs also inhibit the production of constitutive prostaglandins, causing severe side effects such as GI bleeding and renal damage (Matteson, 2000; Petersen et al., 2019). In 1998 the *American Journal of Medicine* published that, in the United States, NSAIDs cause more than 100,000 hospitalizations and 16,000 deaths among arthritis patients each year.

Alternative remedies to COX inhibitors include therapeutic diets, natural herbs, and nutraceuticals, in addition to physical exercise, massage, acupuncture, hydrotherapy, nutrition, and weight loss (Christensen et al., 2005; Messier et al., 2007; Lee et al., 2010; Gupta et al., 2019a,b; Thomas et al., 2019). Due to the increasing number of people and animals with OA and the severe side effects of COX inhibitors, there is a great need for safer remedies to alleviate the symptoms associated with arthritis.

Nutraceuticals. DOI: https://doi.org/10.1016/B978-0-12-821038-3.00014-8

Nutraceuticals such as glucosamine, chondroitin, type II cartilage, shilajit, acetyl-keto-boswellic acid, avocado/soybean unsaponifiables, curcumin, hyaluronan, and many others have gained immense popularity for their antiarthritic use in humans and animals (Goggs et al., 2005; Clegg et al., 2006; Lee et al., 2010; Wandel et al., 2010; Gupta et al., 2011; Fleck et al., 2013; Lawley et al., 2013; Murdock et al., 2016; Gupta et al., 2019a, b). This chapter describes, in brief, the efficacy, safety, and toxicity of nutraceuticals used for OA.

Pathophysiology of OA

The pathogenesis of OA is very complex because it involves multiple etiologies (aging, injury, excess or lack of exercise, nutritional deficiency, obesity, genetic predisposition, infection, etc.) and cellular and molecular mechanisms and pathways. Therefore there is an urgent need to advance understanding of the pathogenesis of OA and define new molecular pathways that may facilitate the development of novel treatments (Kim et al., 2018). OA, also known as DJD, is a functional disorder of the joint, characterized by a change in joint shape secondary to a loss of articular cartilage, osteophyte formation, subchondral sclerosis, bone marrow lesions (BMLs), and synovial proliferation, with subsequent alteration of mechanical properties that result in decreased stability, movement, and loading (Bellamy et al., 2001; CDC, 2010; Dore et al., 2010; Gupta et al., 2019a,b). It is well established that OA is a heterogeneous disease of the entire joint, encompassing multiple tissues including cartilage, bone, adipose, and skeletal muscle, thereby causing remodeling and/or failure of the joint (Yuan et al., 2014; Tonge et al., 2014; Gupta et al., 2019a). In the early stages of OA, progressive depletion of cartilage proteoglycan leads to a net loss of matrix from the cartilage (Reid and Miller, 2008; Thote et al., 2013). Normally, the joint cartilage consists of 5% chondrocytes and 95% extracellular matrix (ECM). OA is characterized by the degradation of cartilage matrix components, including cartilage-specific type II collagen and the proteoglycan aggrecan, ultimately resulting in the loss of cartilage structure and function. Breakdown and deterioration of the cartilage have been correlated with increased activities of certain enzymes, including the matrix metalloproteinases (MMPs). The increased cartilage lesions in OA have also been correlated with enhanced levels of MMP-1. There are several key enzymes in anabolic and catabolic pathways that are dysregulated in OA cartilage (Lorenz and Richter, 2006). These cascading events lead to increased friction and inflammation in the joints (Fig. 14.1).

Studies have described the details of interplay/cross-talk between the bone and the cartilage, which may be elevated in OA (Pan et al., 2012; Sharma et al., 2013; Yuan et al., 2014). Factors released from subchondral bone tissue may induce OA cartilage degradation. Conversely, subchondral bone remodeling may be a result of factors secreted from cartilage. Increased vascularization and formation of microcracks associated with abnormal bone remodeling in joints during OA facilitate molecular transport from cartilage to bone and vice versa. Procoagulant factors have also been shown to compromise subchondral vasculature, and thus may accelerate joint damage (Ghosh and Cheras, 2001). It is suggested that several critical signaling pathways and biological factors are key regulators that activate cellular and molecular processes in cross-talk among joint compartments (Yuan et al., 2014; Kim et al., 2018).

Castaneda et al. (2012) stated that bone turnover is likely to play a pivotal role in the pathogenesis of OA. OA is associated with early bone loss due to increased bone remodeling, followed by slow turnover, leading to densification of the subchondral plate and complete loss of cartilage (Burr and Gallant, 2012; Kim et al., 2018). It can be stated that elevated bone remodeling and its associated stimulated vascularity are reportedly indispensable for the progression of OA.

In the early and middle stages of OA, bone attrition (decrease or loss of bone height and contour) is observed. In late-stage OA, thickening of subchondral bone occurs, which is accompanied by decreased mineralization. Subchondral bone attrition is also associated with the severity of cartilage loss in adjacent regions and areas of BMLs detected by MRI, indicating an increased load on the bone–cartilage biomechanical unit (Roemer et al., 2010). BMLs are indicators of OA progression and are considered an important risk factor for structural deterioration. The localization of BMLs was associated with areas of severe cartilage loss in the overlying articular surface and revealed fat necrosis and localized marrow fibrosis associated with microfractures of the trabecular bone. Bone marrow edema and subchondral cysts are useful for assessing OA progression by MRI (Yuan et al., 2014).

OA cartilage degradation is due to an imbalance between the synthesis and degradation of ECM components. Oxidative/nitrosative stress and inflammation seem to be the two major mechanisms that lead to the progression of OA (Yudoh et al., 2005; Jerosch, 2011; Wan and Zhao, 2017). Nuclear factor kappa B (NF-κB) appears to be a key transcription factor that drives the expression of OA inducer genes in response to the activation of toll-like receptors and interleukin receptors by ECM fragments and IL-1β (Scanzello et al., 2008). Recently, Lepetos and Papavassiliou (2016) reported that, in arthritic tissues, the NF-κB signaling pathway can be activated by proinflammatory cytokines, mechanical stress, and ECM degradation products. This activation results in regulation of the expression of many cytokines, inflammatory mediators, transcription factors, and several matrix-degrading enzymes. Overall, NF-κB signaling affects cartilage matrix

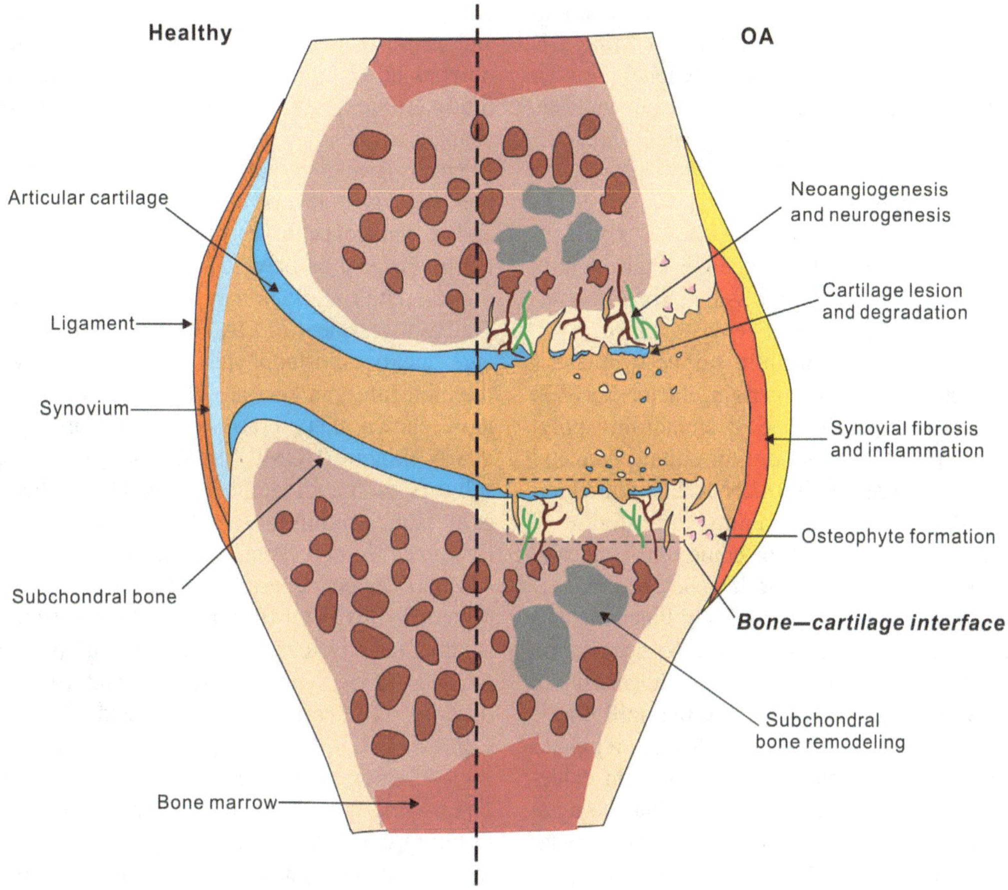

FIGURE 14.1 Differences between normal joint and OA joint. OA is considered to involve all joint tissues. The bone–cartilage interface is at the center of joint function and disease. *OA*, osteoarthritis. *Source: Courtesy of Yuan, X.L., Meng, H.Y., Wang, Y.C., 2014. Bone–cartilage interface crosstalk in osteoarthritis: potential pathways and future therapeutic strategies. Osteoarthr. Cartil. 22, 1077–1089.*

remodeling, chondrocyte apoptosis, and synovial inflammation, and has indirect stimulatory effects on downstream regulators of terminal chondrocyte differentiation. Also, the interaction between redox signaling and NF-κB transcription factors seems to play a distinctive role in OA pathogenesis (Rigoglou and Papavassiliou, 2013; Lepetos and Papavassiliou, 2016).

Proinflammatory cytokines, such as IL-1β, are produced in the OA stimulating mitogen activated protein kinase (MAPK) pathway through the extracellular signal-regulated kinases (ERK)1/2, p38 kinase, and c-Jun N terminal kinase (JNK). Elevated MAPK phosphorylation results in activation of transcription factors, which in turn upregulates the production of several molecules, such as MMPs and aggrecanases. Increased enzymatic activities of MMPs and aggrecanases are responsible for matrix degradation.

Francin et al. (2014) found that the elevated level of adiponectin in chondrocytes from OA patients that might contribute to matrix remodeling during OA. These authors provided functional evidence for a potential role of adiponectin in OA but were unable to establish a strong association with the grade of cartilage destruction.

A bone–cartilage modulatory pathway, such as transforming growth factor β (TGFβ), is required for the maintenance of metabolic homeostasis and structural integrity of healthy cartilage. TGFβ is highly expressed in normal cartilage, but is almost absent in OA cartilage (Yuan et al., 2014). Therefore loss or interruption of TGFβ signaling in cartilage results in loss of proteoglycans and cartilage degeneration. Thus inhibition of endogenous TGFβ leads to increased damage to cartilage.

Another signaling pathway, Wnt/β-catenin, plays a critical modulatory role in maintaining the bone–cartilage biochemical unit. The effects of activated Wnt signaling contribute to excessive bone remodeling and degradation of cartilage matrix (Corr, 2008). Activation or inhibition of β-catenin can result in cartilage damage, OA-like lesions, and chondrocyte apoptosis. Conversely, overexpression of Wnt signaling is deleterious to chondrocytes, leading to OA-like disease. Increased expression of Wnt signaling agonists, such as Wnt-induced signaling protein 1 (WISP-1), Wnt-16, and Wnt-28, have been reported in human OA cartilage and may contribute to cartilage degradation by upregulating MMPs and aggrecanases

(Blom et al., 2009). Ishimaru et al. (2014) determined that the structure of chondroitin sulfate (CS) in cartilage is reflected by the degree of cartilage degeneration in patients with OA of the knee and how CS biosynthesis affects cartilage degeneration. Findings revealed that CS concentration and chain length were reduced closer to the more degraded cartilage with decreasing CS glycosyltransferase gene expression. Inhibition of CS glycosyltransferase gene expression may reduce CS chain length, which may contribute to OA progression. Six glycosyltransferases have been identified as being involved in CS biosynthesis in mammals: chondroitin sulfate synthase 1 (CSS1/CHSY1); chondroitin polymerizing factor (CHPF); chondroitin sulfate synthase 3 (CSS3); chondroitin polymerizing factor 2 (CHPF2); chondroitin sulfate *N*-acetyl galactosaminyl transferase 1 (CSGALNACT1); and chondroitin sulfate *N*-acetyl galactosaminyl transferase 2 (CSGALNACT2). There may still be some other underlying pathways involved in OA that need to be explored.

Mao et al. (2017) found that miR-92a-3p functions as a negative regulator by downregulating histone deacetylase (HDAC2) in both chondrogenesis and OA pathogenesis. In another study, Si et al. (2017) have shown that miR-140 is expressed specifically in cartilage and thereby regulates ECM-degrading enzymes. Most recently, based on epidemiological studies, Valdes (2020) suggested that there are common molecular mechanisms that underlie both metabolic syndrome and OA pain. These include regulation of the endocannabinoid system, activation of the transient receptor potential cation channel vanilloid subfamily member 1 (TRPV1) channel and gut dysbiosis. These three mechanisms are interlinked and are the targets of therapeutic dietary or pharmacological interventions.

It will take decades for scientists to fully understand the pathophysiology of OA.

Biomarkers of OA

The field of OA is desperately in need of biomarkers, as there are no serum, urinary, or synovial biomarkers that are specific and validated for OA (Lafeber and van Spil, 2013; Gupta et al., 2019b). It has been recognized that the very earliest pathological changes are not captured well by plain radiography, but they are evident on ultrasound and MRI (Williams and Spector, 2008; Hall et al., 2014; Palmer et al., 2017). A biomarker detectable very early in disease that could be measured in blood or urine would enhance our ability to detect early OA and shed light on the pathophysiological mechanisms underlying its early stages. Such a biomarker would be of great value to those conducting trials for novel symptom-modifying or disease-modifying OA agents.

In a mouse model, Waung et al. (2014) reported that quantitation of subchondral bone mineral content by X-ray microradiography can be used as a rapid, sensitive, and cost-effective method to identify abnormal joint phenotypes at several ages. The Boston Osteoarthritis Knee Study Group has looked at the relationship between MRI changes at the knee and a number of markers of cartilage turnover (Hunter et al., 2007). The group hypothesized that increased levels of cartilage degradation products and/or imbalance of cartilage synthesis and degradation markers would be predictive of subsequent cartilage loss.

Molecular markers may be of immense value in quantitative assessment of OA changes, such as articular cartilage thickness, osteophytes, sclerosis, and cysts. Some of the molecular markers are discussed here in brief. Studies have shown that serum levels of cartilage oligomeric matrix protein (COMP) predict progression of disease in subjects with radiographic knee OA at baseline (Vilim et al., 2002) and in those with knee pain and normal baseline knee X-ray (Petersson et al., 1998). COMP is a 435,000-Da pentameric member of the thrombospondin protein family. COMP, initially isolated from cartilage, is synthesized by chondrocytes and is present in small amounts in the synovium, tendon, and serum but is abundant in OA cartilage (Saxne and Heinegard, 1992). Conrozier et al. (1998) reported that serum COMP seems to be a surrogate marker of OA and may be of interest for detection in patients at risk for rapidly progressing disease in hip OA. It can be used as a good biomarker for selecting patients at high risk for cartilage loss for recruitment into clinical trials, and for more rapid assessment of the efficacy of a disease-modifying agent.

Currently, accumulating evidence suggests that COMP predicts MRI cartilage loss and appears to be a useful biomarker of early OA. Changes in bone sialoprotein (BSP; a bone-specific noncollagenous matrix molecule) in serum reflect alterations in the subchondral bone turnover in OA (Conrozier et al., 1998). Petersson et al. (1998) suggested that pathological processes in cartilage and subchondral bone coincide in OA and appear to be reflected by circulating levels of COMP and BSP released from cartilage and bone. Changes in serum levels of COMP and BSP are potential tools in studies of knee joint OA in subjects with chronic knee pain. Therefore COMP and BSP both have potential as markers of OA progression. BSP also seems to be a marker of bone turnover in OA.

Measurement of joint space width using a digitized image analyzer is a sensitive way of assessing OA progression that facilitates evaluation of tissue markers in relation to anatomical changes in the joint. Radiologic evidence, patient symptoms, and OA risk factors such as age, gender, and body mass index have limited value for predicting the risk of rapid, highly predictable OA progression (Wisniewski et al., 2014). Therefore there is a need to identify early biomarkers of OA progression. Wisniewski et al. (2014) reported that TNF-stimulated gene 6 (TSG-6 or TSG-6 protein) activity is a promising independent biomarker for OA progression. Given the high NPV, this assay may be particularly suitable for identifying patients at low risk for rapid disease progression and to assist in the timing of arthroplasty. TNFIP-6, encoded by TSG-6, is a hyaluronan-binding protein associated with inflammation (Milner and Day, 2003). TSG-6 protein inhibits both osteoblast differentiation and osteoclast activation, thereby providing antiinflammatory and chondroprotective activities in OA. The presence of TSG-6 has been demonstrated in synovial fluids (SFs) or cartilage of patients with either OA or RA. Since the TSG-6 protein has not been quantified in SF or cartilage, the potential of this protein as a biomarker of OA or RA progression has yet to be explored.

Leptin (a 16-kDa protein) is a member of the adipose tissue-derived hormones that have been collectively termed "adipokines." Leptin has been shown to enhance the production of proinflammatory cytokines and destructive mediators involved in the pathogenesis of OA [such as IL-1β, IL-6, nitric oxide (NO), and prostaglandin E_2 (PGE_2)]. Leptin has also been reported to enhance the production of MMP-1, MMP-3, MMP-9, and MMP-13 in human OA cartilage (Scotece et al., 2014). SF leptin levels have been shown to correlate with the radiographic severity of OA (Ku et al., 2009) and with MMP-1 and MMP-3 levels in OA patients (Koskinen et al., 2011).

For details on biomarkers associated with OA inflammation, pain, cartilage degeneration and loss, proteomic and metabolomic analysis, and microRNAs, readers are referred to Gupta et al. (2019b). Recently, Hosnijeh et al (2019) further stated that given the paucity of robust biomarkers within the field, and the heterogeneity of the disease, enormous works are needed for the development and validation of novel and clinically applicable biomarkers to reduce the impact of this highly prevalent and debilitating condition.

Management of OA

Although great strides have been made in the recent past to further the understanding of cartilage loss, subchondral bone abnormalities, and the extraarticular inflammation of tendons and ligaments (hallmarks of OA), there are still no good animal models and no specific symptom-modifying drugs and disease-modifying agents for the management of OA (Rozendaal et al., 2005; Williams and Spector, 2008; Mevel et al., 2014; Tonge et al., 2014; Gupta et al., 2019a). Until recently, symptom-modifying drugs such as analgesics (e.g., acetaminophen) and NSAIDs were the prescriptions of choice for patients with OA. These drugs are effective, but they do not eliminate signs and symptoms of active disease. Disease-modifying drugs are not yet available.

There are three main objectives in managing or treating OA (Irsay et al., 2010; Sofat and Kuttapitiya, 2014; Gupta et al., 2019a):

1. Improvement of symptoms;
2. Stopping the progression of the disease;
3. Increased functionality.

Treatment modalities are multiple, including drugs, physiotherapy, balneotherapy, surgery, and lifestyle changes. Changes in lifestyle often include weight loss, dietary changes, and controlled exercise therapy, reducing the stress and home environment modifications to the infirmity caused by the disease (Irsay et al., 2010; Gersing et al., 2016; Van der Kraan, 2017; Daugaard et al., 2020). Although all of these methods can be very helpful, they are very difficult to put in practice due to reduced patient compliance. Physiotherapy, balneotherapy, and drug treatment all represent effective treatment modalities (Gaál et al., 2008; Sofat and Kuttapitiya, 2014; Gupta, 2016; Otis et al., 2019; Gupta et al., 2019a).

Restoration of diseased articular cartilage in patients with OA is still a challenge for researchers and clinicians. Clinical management of OA typically involves a limited combination of pharmacological and nonpharmacological interventions to reduce pain and increase tolerance for functional activity (Hunter, 2010; Sofat and Kuttapitiya, 2014; Tonge et al., 2014). Unfortunately, NSAIDs either do not modify the disease course or have been associated with serious effects in the renal, hepatic, cardiovascular, and GI systems (Mukherjee et al., 2001; Bergh and Budsberg, 2005; Coruzzi et al., 2007; Inns et al., 2010; Tonge et al., 2014; Perkins et al., 2015; Scarpiganto et al., 2015; Breivik, 2017; Gosser et al., 2017; Sessa et al., 2017; de Salazar Alcalá et al., 2019). Patients with knee OA can be given intraarticular (IA) corticosteroid injections, which can be effective in some patients in providing pain relief, but the benefit is short-lived and frequent injections can damage the articular cartilage. At one time, MMPs, a diverse family of zinc-dependent proteolytic enzymes involved in the maintenance of the ECM, were recognized as attractive drug targets for the treatment of OA. Surgery is used for improving symptoms and the sequelae of the disease with little if any change at the molecular level.

Currently, several therapeutic regenerative strategies have investigated whether autologous mesenchymal stem cells (MSCs) have significant effects on regeneration and/or maintenance of articular cartilage in OA (Oldershaw, 2012; Vilar et al., 2013, 2014; Davatchi et al., 2016; Kong et al., 2017). Autologous MSC therapy is based on the isolation of these cells from tissues, such as fat or bone marrow (Takemitsu et al., 2012), and after culture expansion they are administered back to the patient (Guercio et al., 2012) for damaged joint tissue, such as cruciate ligaments, menisci, and cartilage (Agung et al., 2006).

Because multiple mechanisms, pathways, and target sites are involved in the pathogenesis of OA, a combinatorial symptom-/disease-modifying agent could offer a holistic approach in ameliorating arthritic pain and cartilage loss. In recent years, there has been growing interest in alternative treatments for OA (Liu et al., 2018; Gupta et al., 2019a, c). Some nutraceuticals have been found to improve joint function by restoring cartilage loss, thereby easing pain. Nutraceuticals can be administered orally or intraarticularly. When administered orally, nutraceuticals are well-tolerated and safe for long-term use.

Many animal species (mouse, rat, rabbit, sheep, dog, cat, and monkey) have been used to study the effects of nutraceuticals on OA (Mevel et al., 2014; Gupta et al., 2019a). However, none of these models can perfectly mimic OA, because these animals differ physiopathologically from humans in terms of their joint anatomy, genetics, biomechanics, and disease parameters (severity, localization, and kinetics). A brief description of some common nutraceuticals used for OA is provided below.

Glucosamine and chondroitin

Currently, glucosamine [glucosamine sulfate (GS); glucosamine hydrochloride (GH)] and CS are the most commonly used supplements to ease the pain and discomfort of arthritis in humans and animals. Glucosamine and chondroitin are components of the ECM of articular cartilage. Glucosamine (2-amino-2-deoxy-D-glucose) is an amino sugar (Fig. 14.2) and a constituent of glycosaminoglycan (GAG), which plays a role in the normal growth and repair of articular cartilage. In fact, glucosamine is considered a building block of cartilage. Glucosamine is extracted from crab, lobster, and shrimp shells.

Following oral administration of either GS or GH, these salts are ionized in the stomach, making glucosamine available for absorption in the small bowel. Ninety percent of GS is absorbed, but due to an extensive first-pass metabolism, bioavailability approaches only 25% (Setnikar and Rovati, 2001). Its excretion is mainly through the kidneys, with only a small unmodified amount eliminated through the stool. For detailed pharmacokinetics of glucosamine, readers are referred to Persiani et al. (2005) and Block et al. (2010).

CS is a sulfated GAG composed of a chain of alternating sugars, *N*-acetyl-D-galactosamine and D-glucuronic acid (Fig. 14.3). CS is a normal constituent of aggrecan, the major proteoglycan of articular cartilage. It is extracted from animal cartilage, such as trachea and shark cartilage.

Due to its larger size as compared to glucosamine, approximately 30% of CS is absorbed, with 12%–13% bioavailability (Deal and Moskowitz, 1999; Huskisson, 2008). For detailed pharmacokinetics of CS, readers are referred to Jackson et al. (2010) and Jerosch (2011).

Jackson et al. (2010) studied the pharmacokinetics of glucosamine (500 mg, TID) and CS (400 mg TID) when taken separately or in combination as a single dose in normal individuals.

Studies suggest that glucosamine helps relieve pain by enhancing proteoglycan synthesis, which is impaired in osteoarthritic cartilage (Hooper, 2001). D'Abusco et al. (2007) investigated the effects of glucosamine on MMP production, MAPK phosphorylation, and activator protein (AP)-1 transcription factor activation in human chondrocytes. Findings revealed that glucosamine reduced the expression of MMPs (MMP-1, MMP-3, and MMP-13) and inhibited c-jun amino terminal kinase, p38 phosphorylation, and, consequently, c-jun binding activity. In essence, glucosamine inhibits IL-1β-stimulated MMP production in human chondrocytes by affecting MAPK phosphorylation. The glucosamine and CS combination suppresses IL-1-induced gene expression of iNOS, COX-2, mPGEs, and NF-κB in cartilage explants. This leads to reduced production of NO and PGE_2, two mediators responsible for the cell death of chondrocytes and inflammatory reactions (Chan et al., 2005).

CS provides hydration, assists in cushioning impact stress, helps create osmotic pressure within the ECM to maintain the compressive resistance of cartilage, improves function/mobility of the joint, reduces the progression of OA, and reduces joint pain (Frech and Clegg, 2007; Hochberg et al., 2008). Jackson et al. (2010) reported that pain relief perceived following ingestion of CS probably does not depend on simultaneous or prior intake of glucosamine. CS has been shown to increase hyaluronan production by human synovial cells, which has a beneficial effect on maintaining viscosity in the SF. CS stimulates

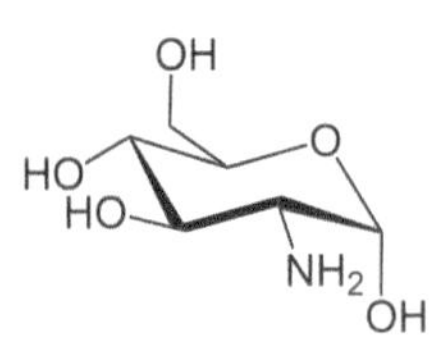

FIGURE 14.2 Structural formula of glucosamine.

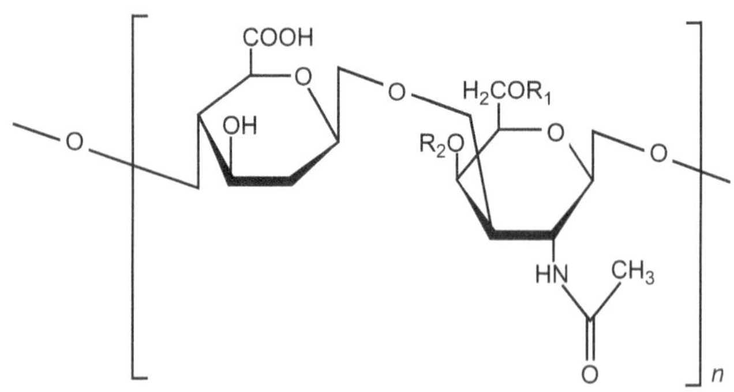

FIGURE 14.3 Structural formula of chondroitin sulfate.

chondrocyte metabolism, leading to the synthesis of collagen and proteoglycan, the basic components of new cartilage (Jerosch, 2011). CS also provides elasticity and assists in cushioning impact stress. It is suggested that CS may help the body to repair damaged cartilage and help restore joint integrity. It may also protect existing cartilage from premature breakdown. Because CS production by the body decreases with age, its supplementation may be especially helpful for older humans, dogs, and horses. Furthermore, CS inhibits the enzymes leukocyte elastase and hyaluronidase, which are found in high concentrations in the SF of patients with rheumatic diseases. It has also been hypothesized to reduce inflammation, inhibit synthesis of degradative enzymes including MMPs, increase synthesis of ECM constituents, and reduce apoptosis of articular chondrocytes (Vangness et al., 2009). Monfort et al. (2008) reviewed details of the biochemical basis of the effect of CS on OA articular tissue. At molecular levels, CS inhibits NF-κB nuclear translocation and phosphorylation of p38 MAPK, ERK1/2, and JNK (Jomphe et al., 2008).

When given in combination, these two supplements: stimulate synoviocyte and chondrocyte metabolism; inhibit enzymatic degradation and reduce the fibrin thrombi in the periarticular microcirculation; and can regulate the genetic expression and synthesis of NO and PGE_2, thereby exerting antiinflammatory properties. In addition to antiinflammatory action, GS and CS exhibit an antioxidant expression that leads to a significant reduction in iNOS expression and activity (Valvason et al., 2008), thereby reducing the otherwise NO-induced cell death of chondrocytes.

In several studies, GS and CS have been evaluated as single agents and in combination for the treatment of OA. In fact, it is a common practice for glucosamine and chondroitin to be used together because they offer a greater beneficial effect than when given alone, although they work through different mechanisms of action. However, oral supplementation with glucosamine or CS may reduce pain in knee OA, but without any additional effect using both agents in combination for the management of symptomatic knee OA. In a clinical trial, GS was found to be as effective or slightly more effective than analgesics (Lopez, 1982; Muller-Fassbender et al., 1994; Qiu et al., 1998) or NSAIDs (Towheed et al., 2001) for decreasing pain. A trial by McAlindon et al. (2004) did not show a difference between GS and placebo. In another clinical trial, GS was found to be ineffective for reducing pain in patients with severe knee OA, but it was more effective when it was used in combination with CS due to a synergistic effect in patients with moderate to severe pain (Sawitzke et al., 2008). In a recent clinical trial, Reginster et al. (2017) found that a pharmaceutical grade CS (800 mg/day) is similar to celecoxib (200 mg/day) and superior to placebo in reducing pain and improving function over 6 months in symptomatic knee OA. Wildi et al. (2011) reported that CS reduced cartilage volume loss and BMLs in knee OA patients within 6 months of treatment.

Clegg et al. (2006) reported that, in a multicenter study, 1583 patients with symptomatic OA in the knee received 1500 mg of GS and 1200 mg of CS daily. Findings revealed that the supplements alone or in combination did not reduce pain effectively in the overall treatment groups. Irsay et al. (2010) found that these chondroprotective agents (GS and CS) were effective in improving the function of patients with OA, but the radiological modifications in the knee were statistically insignificant after 12 months of monitoring. However, the findings of Narvy and Vangsness (2010) supported a role of these nutraceuticals in reducing radiographic progression of knee OA. Kozakçioğlu (2012) revealed that GS and CS both have a suppressor effect on the structural deformities of OA.

Erhan et al. (2012) showed that topical glucosamine treatment combined with physical therapy in patients with knee OA had no superiority over placebo, based on radiological findings (Kellgren–Lawrence score) and joint stiffness index (Western Ontario and McMaster Osteoarthritis, WOMAC score).

In all clinical trials, GS, GH, and CS were found to be as safe as placebo (Anderson et al., 2005; Clegg et al., 2006; Hathcock and Shao, 2007; Narvy and Vangsness, 2010; Irsay et al., 2010). The acute oral LD_{50} of glucosamine is >8 g/kg in rats and 15 g/kg in mice, which can be considered nontoxic. Oral administration of glucosamine at 2700 mg/kg for 12 months in animals produced no adverse effects. Oral administration of large doses of glucosamine in animals has no documented effects on glucose metabolism. However, Lafontaine-Lacasse et al. (2011) reported that, at beyond recommended dosages, glucosamine may damage pancreatic cells, thereby possibly increasing the risk of developing diabetes. In a few patients, hepatotoxicity (hepatitis and/or cholestasis) developed following exposure to glucosamine (Smith and Dillon, 2009). These authors established a temporal relationship between the onset of liver injury and glucosamine ingestion.

Clinical studies have not identified any significant side effects of CS, which suggests its long-term safety (Hathcock and Shao, 2007). Recently, the Task Force of the European League against Rheumatism (EULAR) committee also granted CS a level of toxicity of 6 on a 0–100 scale, confirming that CS is one of the safest remedies for OA.

Currently, glucosamine and chondroitin are not recommended by the Osteoarthritis Research Society International (OARSI) or the American College of Rheumatology (ACR) (Singh et al., 2012).

In dogs with moderate OA, daily administration of GH (2000 mg) and CS (1600 mg) for a period of 120–150 days significantly reduced pain associated with OA (Gupta et al., 2011). Gupta et al. (2009) also reported significant reduction of pain in horses with moderate OA receiving GH (5.4 g) and CS (1.8 g) daily for a period of 150 days. In both dogs and horses, GH and CS produced no side effects and were well-tolerated. Previously, in a number of in vivo and in vitro studies, GS and CS have been found to be very effective against OA in animal models (Dechant et al., 2005; Neil et al., 2005; Trumble, 2005). These studies suggested that the combination of GS and CS appears to be more effective in preventing or treating OA in animals than either product alone.

A Golden Retriever developed polyuria and polydipsia when treated with 1000 mg of glucosamine/day (Breese McCoy and Bryson, 2003). The polyuria and polydipsia resolved after the glucosamine dose was lowered to 500 mg/day. In an experimental study, glucosamine was reported to cause hyperglycemia, insulin resistance, and a diabetic-like state in rats (Giaccari et al., 1995). Beriault et al. (2011) did not confirm hyperglycemia and insulin resistance with glucosamine, but it promoted endoplasmic reticulum stress, hepatic steatosis, and accelerated atherogenesis in mice.

Hyaluronic acid

Hyaluronic acid (HA; hyaluronan, hyaluronate; Fig. 14.4) is a naturally occurring nonprotein nonsulfated GAG present in many tissues and fluids. HA is a major component of the ECM and is normally found in bone marrow, articular cartilage, and SF. With distinct physicochemical and rheological properties, HA is produced by type B synoviocytes, fibroblasts, and chondrocytes, and it contributes to cell proliferation, migration, and morphogenesis. It is an important component of articular cartilage present as a coat around each chondrocyte. HA (a polymer of disaccharides) can be 25,000 disaccharide repeats in length, with a molecular weight (MW) ranging from 5000 to 20,000,000 Da. The average MW of HA in human SF is 3–4 million Da. HA is synthesized by a class of integral membrane proteins called hyaluronan synthases (HAS), of which vertebrates have three types (HAS1, HAS2, and HAS3). These enzymes lengthen hyaluronan by repeatedly adding β-1,4-D-glucuronic acid and β-1,3-*N*-acetylglucosamine units to the nascent polysaccharide, which is extruded via the ABC transporter through the cell membrane into the extracellular space. The MW of HA in cartilage decreases with age, but the amount increases (Holmes et al., 1988). HA is degraded by a family of enzymes called hyaluronidases.

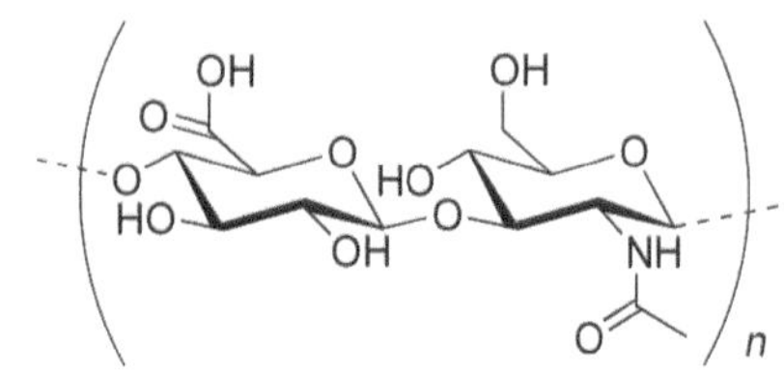

FIGURE 14.4 Chemical structure of hyaluronic acid.

HA binds to ECM molecules and cell surface receptors, thereby regulating cellular activities via control of the tissue's macro- and microenvironments. HA can bind to three main classes of cell surface receptors: (1) CD44 (a membrane glycoprotein), (2) receptor for hyaluronate-mediated motility (RHAMM), and (3) intercellular adhesion molecule 1 (ICAM-1), each of which perform different functions. HA appears to inhibit signal transduction through CD44 and RHAMM receptors. Evidently, higher- and lower-MW HA have distinct molecular and cellular mechanisms and diverse biological effects through interaction with CD44 receptors (Yang et al., 2012; Bhattacharya et al., 2017). CD44-mediated signaling affects both chondrocyte survival pathways as well as chondroptotic pathways. The other two receptors for HA binding include: (1) lymphatic vessel endothelial hyaluronan receptor (LYVE-1), and (2) HA receptor for endocytosis, also known as Stabilin-2.

By having unique rheological properties (viscoelasticity, high moisture retention capacity, high biocompatibility, and hygroscopic activity), HA at a concentration as low as 0.1%, can act as a lubricant, shock absorber, joint structure stabilizer, and water balance- and flow resistance-regulator. All these physiological and pharmacological effects of HA are MW-dependent (Gupta et al., 2019c).

HA has been used for more than four decades in the treatment of OA in humans, dogs, and horses. IA treatment with HA has been well-studied and is widely used as OA therapy (Armstrong et al., 1994; Jerosch, 2011). However, there is controversy over the efficacy of orally administered HA. Pharmacokinetic data revealed that orally administered high-MW HA also reached the joint (Balogh et al., 2008), which provides a rationale for the oral supplementation of HA (Oe et al., 2016).

HA has been used to treat OA of the knee via injection into the joint, with proven beneficial effects. In addition to its contribution to the structural properties of the ECM, HA may have a role in regulating the synthesis of proteoglycans during maturation of articular cartilage and in repair processes such as those that occur in DJD.

Ghosh and Guidolin (2002) demonstrated that HAs within the MW range of $0.5 \times 10^6 – 1.0 \times 10^6$ Da partially restore SF rheological properties and synovial fibroblast metabolism in animal models. These authors also described the interaction of HA with pain receptors and its analgesic effects. HA exerts pharmacologic actions by mitigating the activities of proinflammatory mediators and pain-producing neuropeptides released by activated synovial cells. Moreland (2003) reported that HA can reduce nerve impulses and nerve sensitivity associated with the pain of OA. In experimental studies, HA might have decreased OA-associated pain by attenuating PGE_2, bradykinin synthesis, and substance P. In clinical studies, HA has been reported to enhance several aspects of quality of life in adults with knee OA, such as improvement in physical activity and pain reduction due to decreases in nerve impulses and nerve sensitivity (Carrabba et al., 1995; Pozo et al., 1997; Kalman et al., 2008). Like glucosamine and CS, HA is a slow-acting antiarthritic agent and it can be used prophylactically or therapeutically as an antiinflammatory disease-modifying agent in OA (Aubry-Rozier, 2012).

Histological evidence suggests that HA prevents the degradation of cartilage and may promote its regeneration. Ghosh and Guidolin (2002) also provided evidence that HA treatment mitigated synovial hypertrophy and increased the numbers of synovial fibroblast-like cells while decreasing macrophages, lymphocytes, mast cells, and adipocytes. In essence, the structure-modifying effects of HA in OA joints are largely mediated by its antiinflammatory, cytoprotective, and immunomodulatory properties (Gupta et al., 2019c). However, in comparison to GS and CS, HA has been shown to exert very minor antiinflammatory and antiapoptotic effects (Campo et al., 2009). Also, HA may provide cartilage protection, not from its direct effect on chondrocytes, but rather effects secondary to the downregulation of cytokines, free radicals including NO, and proteolytic activities in the SF and synovium, which act on chondrocytes and promote matrix catabolism (Takahashi et al., 2001). In an experimental model of OA in rabbits, the administration of HA was found to decrease apoptosis and NO levels (Díaz-Gallego et al., 2005).

Currently, many formulations of HA are prepared with lower MW (500–730 kDa), intermediate MW (800–2000 kDa), and higher MW (~6000 kDa) for IA injections and are available worldwide (Berenbaum et al., 2012). The therapeutic efficacy of these products can differ based on HA origin, method of production, treatment schedule, MW, viscoelasticity and other rheological properties, half-life within the joint, and pharmacokinetics and pharmacodynamics (Gupta et al., 2019c). Some of the Food and Drug Administration (FDA)-approved HA products include Hyalgan (Fidia Pharm), Supartz (Bioventus), Orthovisc (DePuy Mitek), Euflexxa (Ferring Pharmaceuticals), Synvisc/Synvisc One (Sanofi-Aventis), and Gel-Syn (Institut Biochimique). Therapy with an IA HA formulation of a higher MW (e.g., hylan G-F 20) appears to be more effective in relieving OA-associated pain than therapies with lower MWs (e.g., sodium hyaluronate) (Pritchard et al., 1999; Santilli et al., 2016). Jerosch (2011) also reported that HA of a higher MW may be more effective than that of a lower MW. In a large number of clinical trials, HA has been found to exhibit anti-OA effects. Berenbaum et al. (2012) found that treatment with an intermediate-MW (800–1500 kDa) HA (GO-ON, Rottapharm) was significantly superior to lower-MW (500–730 kDa) HA (Hyalgan, Fidia Pharma) for knee OA symptoms.

HA is commonly used in the treatment of articular disorders (carpal and fetlock joints) in horses, especially those involved in racing competition and heavy physical labor. HA is especially used for mild to moderate levels of synovitis associated with equine OA. However, it has limitations in treating severe synovitis or OA. HA can be injected directly into an affected joint or administered intravenously for less localized disorders. During the past decade, the use of intravenous HA has become common, but there is only one licensed product made by Bayer (Legend in the United States and Hyonate elsewhere). It has been suggested that when used in combination, HA not only provides the improved beneficial effects of corticosteroids but can also placate the side effects of certain corticosteroids. It is noteworthy that, according to Canadian regulations, HA in the HY-5 preparation should not be administered to horses that are to be slaughtered for meat; however, in Europe, the same preparation does not have any such restriction.

HA is also known to produce adverse effects, including muscle pain, cramping, pain in the knee injected with HA, and swelling in the arms and legs, making movement difficult (Rutjes et al., 2012). Future studies investigating the effects of HA on OA in well-controlled clinical studies may help to confirm its therapeutic claims and to understand its side effects (Gupta et al., 2019c).

Type II collagen

It is well-established that type II collagen is the main collagen component of hyaline cartilage, which is destroyed in OA by MMP-13 (Wu et al., 2002; Xu et al., 2007). Glycosylated undenatured type II collagen (hereafter referred to as type II collagen) has been shown to be effective in ameliorating pain associated with arthritis in humans (Trentham et al., 1993; Crowley et al., 2009). In a series of investigations conducted on moderately arthritic dogs, daily administration of glycosylated undenatured type II collagen at 40 mg daily for a period of 120–150 days significantly reduced pain and improved their daily activity (DeParle et al., 2005; D'Altilio et al., 2007; Peal et al., 2007; Gupta et al., 2011). The findings of these studies demonstrated that daily treatment of OA dogs with type II collagen alone or in combination with GH and CS markedly alleviated arthritic-associated pain. Type II collagen was less effective when given in combination with GH and CS. In horses, daily administration of type II collagen (480 mg) was found to be effective in reducing arthritic pain (Gupta et al., 2009). These supplements were well-tolerated and no side effects were observed.

In OA, type II collagen can promote a reduction in inflammation because it reacts with the body's immune system to improve crippling signs and symptoms of arthritis. Type II collagen is one of the primary connective tissues of the body, providing flexibility and support to bone joints. It functions through a process called oral tolerization, which takes place in the small intestine where food is absorbed (Trentham, 1998). Oral tolerance can be induced by two major mechanisms, bystander suppression and clonal energy, depending on the dose of antigen involved. Throughout the small intestine, there are patches of gut-associated lymphoid tissue (GALT). Within the GALT, nodules called Peyer's patches are present. The Peyer's patches contain organized assemblages of T and B lymphocytes, macrophages, and dendritic cells and are the primary area within the GI tract where immune responses are generated (Weiner, 1997). Clonal energy is a mechanism by which an orally administered protein can induce downregulation of an immune response. It appears that oral tolerance of type II collagen is induced by bystander suppression. Thus type II collagen taken orally improves joint mobility and flexibility by preventing the immune system from attacking and damaging its own joint cartilage. Adverse effects have not been noted in humans (Weiner, 1997; Trentham, 1998; Marone et al., 2010), dogs, or horses (DeParle et al., 2005; D'Altilio et al., 2007; Peal et al., 2007; Gupta et al., 2009, 2011).

Methylsulfonylmethane

Methylsulfonylmethane (MSM), also known as methyl sulfonate or dimethyl sulfone, is a naturally occurring organosulfur compound and a putative methyl donor. MSM is the first oxidized metabolite of dimethyl sulfoxide (DMSO). It occurs naturally in some plants and it is added in small amounts to many foods and beverages as a dietary supplement. Currently, MSM is sold via 52 different products as a single agent in capsule, caplet, lotion, and cream forms, and also in more than 30 different products in combination with other dietary supplements.

MSM is sold as a dietary supplement and marketed with a variety of claims, often in combination with glucosamine and/or chondroitin for prevention or treatment of OA. However, there is a paucity of scientific evidence to support the use of MSM. It has been suggested that the benefits claimed for MSM far exceed the number of scientific studies. In October 2000, the US FDA warned one MSM promoter, Karl Loren, to cease and desist from making therapeutic claims for MSM.

Due to its sulfur content, MSM is used by the body to maintain normal connective tissue. MSM may have antiinflammatory activity, a chemopreventive property, prostacyclin (PGI_2) synthesis inhibition, antiatherosclerotic action, salutary effect on eicosanoid metabolism, and free radical scavenging activity (Ebisuzaki, 2003; Kim et al., 2006). MSM is an analgesic, antiinflammatory, and blood vessel dilator. In OA, MSM works by reducing inflammation and blocking the pain response in nerve fibers. However, like any other nutraceutical, MSM does not cure OA. It serves the same purpose as NSAIDs, but MSM has none of the negative outcomes associated with NSAIDs.

In a randomized, double-blind, placebo-controlled trial, Kim et al. (2006) reported that, compared to placebo, 3 g of MSM twice per day (6 g/day total) for 12 weeks produced significant decreases in WOMAC pain and physical function impairment. No notable changes were found in WOMAC stiffness and aggregated total symptoms scores. MSM ameliorated symptoms of pain and improved physical function during the intervention without major adverse events. In another randomized, double-blind, placebo-controlled clinical trial, Pagonis et al. (2014) found that 100 patients with hip and/or knee OA receiving 6 g MSM daily for 26 weeks showed significant decreases in all subscales of WOMAC with improved performance of daily living activities on the SF-36 evaluation. Patients with OA of the knee taking lower doses of MSM (1.125 g three times) for 12 weeks showed a decrease in pain and improvement in physical function (Debbi et al., 2011). In a 12-week trial, Usha and Naidu (2004) treated patients with knee OA with 1.5 g MSM (500 mg TID), 1.5 g GS, MSM plus GS, or placebo for 12 weeks. Significant decreases in the Lequesne Index were reported with MSM, GS, and their combination ($P < .05$). The authors reported a 33% decrease in pain in the MSM group; joint mobility, swelling, global evaluation, and walking time were also improved. The MSM dosage used by Usha and Naidu (2004) was lower than the recommended dosage of MSM for clinical practice. In all of these studies, the improvements were small; therefore further studies are warranted to establish their clinical significance. For further details on the effects of MSM and DMSO in OA, readers are referred to Brien et al. (2008).

Acute and subchronic toxicity studies in rats using a single dose of 2 g/kg and a daily dose of 1.5 g/kg MSM for 90 days showed no adverse events, organ pathology, or mortality (Horvath et al., 2002). These doses of MSM are considered equivalent to five to seven times the maximum dose used in humans (Kim et al., 2006). There are no clinical studies on adverse effects, changes in blood chemistry, safety monitoring data, or possible subclinical neurotoxicity symptoms. Side effects of MSM following its oral use are mild and may include digestive upset, headaches, increased blood pressure, increased hepatic enzymes, allergic reactions, and skin rashes.

Omega-3 fatty acids

Omega-3 fatty acids (ω-3 fatty acids or n-3 fatty acids) are polyunsaturated fatty acids (PUFAs), such as α-linolenic acid, eicosapentaenoic acid (EPA), and docosahexaenoic acid (DHA), which are found in fish oils, walnut, flaxseed, etc. They are known for their antiinflammatory actions and have been successfully used to treat RA and OA symptoms and progression (van der Tempel et al., 1990; Galarraga et al., 2008; Jerosch, 2011; Calder, 2012; Villalvilla et al., 2013). In vitro studies demonstrated that omega-3 fatty acids increase collagen synthesis and decrease the inflammation mediator PGE_2 (Kremer et al., 1987). Omega-3 fatty acids decrease IL-1β-induced aggrecanase and collagenase activity, reduce mRNA expression of ADAMTS-4, COX-2, IL-1α, iNOS, and TNF-α, and reduce the protein levels of MMP-3 and MMP-13 (Kremer et al., 1987; Zainal et al., 2009; Adler et al., 2018). PUFAs can modulate mammalian targets of rapamycin signaling, leading to protective activation of autophagy in chondrocytes (Villalvilla et al., 2013). Curtis et al. (2002) showed that the antiinflammatory effects of EPA and DHA abrogated OA-mediated pathologic indicators in human cartilage. In conclusion, diets supplemented with omega-3 fatty acids could prevent OA or slow its progression and improve mobility and activity in humans, dogs, cats, and horses (Knott et al., 2011; Corbee et al., 2012; Moreau et al., 2013; Hill et al., 2016; Mehler et al., 2016; Gupta et al., 2019a). In several studies, omega-3 fatty acids and resolvins (bioactive lipid mediators enzymatically biosynthesized in vivo from EPA and DHA) have also been shown to alleviate joint stiffness and pain in RA due to their antiinflammatory property (Lee et al., 2012; Norling and Perretti, 2013; Akbar et al., 2017).

Shilajit

Shilajit, also called mumie or moomiyo, is a black-brown exudate from the sedimentary rocks of the pristine Himalayan Mountains at an altitude of approximately 10,000 feet. Purified shilajit is prepared from this exudate by a proprietary extraction process (Natreon Inc., New Brunswick, NJ). The main constituents of shilajit are dibenzo-α-pyrones (DBPs), DBP-chromoproteins, and fulvic acids with a DBP core, in addition to a high phenolic content. By having multiple herbo-mineral and bioactive principles, shilajit exerts several physiological and pharmacological actions, such as cholinomimetic, antioxidative, antidyslipidemic, antiaging, antihyperglycemic, antiinflammatory, adaptogenic, antibacterial, antiulcerogenic, antiarthritic, immunomodulatory, ovogenic, spermatogenic, cardioprotective, and energetic properties. For more than 3000 years, shilajit has been used for the prevention and treatment of various human ailments, such as arthritis, diabetes, hypertension, loss of cognition, immune dysfunction, allergies, and loss of libido. Both human and animal studies support its use as a "revitalizer," enhancing physical performance and relieving fatigue with enhanced production of adenosine triphosphate (ATP). Recently, in randomized, double-blind studies, twice daily administration of purified shilajit alone (500 mg) or a supplement containing shilajit (7.5 mg) in combination with trivalent chromium (500 μg) and *Phyllanthus emblica* (amla) extract (7.5 mg) for a period of 5 months significantly ameliorated arthritis-associated pain and improved quality of life for moderately arthritic dogs without causing any untoward effects (Lawley et al., 2013; Fleck et al., 2013). Because OA pain is associated with oxidative stress and inflammation, it is highly possible that shilajit alone or in combination with chromium and amla extract attenuated arthritic pain by antioxidative and antiinflammatory mechanisms (Rege et al., 2012; Lawley et al., 2013; Fleck et al., 2013). Additional properties of shilajit, such as boosting the immune system, promoting healing of bone fractures, increasing muscle mass, enhancing endurance, and antiaging activity, might have also played roles in improvement of OA dogs. In several studies, shilajit has been found to be safe for long-term use (Velmurugan et al., 2012; Lawley et al., 2013; Fleck et al., 2013; Stoh, 2014).

Resveratrol

Resveratrol (RES) is a polyphenol phytoalexin present in the skin and seeds of grapes, red wine, plant berries (blueberries, bilberries, mulberries, cranberries, and raspberries), soybeans, pomegranates, peanuts, pistachios, rhubarb, dried roots of *Polygonum cuspidatum*, etc.

RES (*trans*-3,4′,5-trihydroxystilbene) and its structural analog (pterostilbene) are by far the two most widely researched stilbenes in terms of their beneficial bioactivities. RES has been reported to exert multiple biological (antioxidative, antiinflammatory, and immunomodulatory) activities. In addition to being antiarthritic, RES has other pharmacological effects, such as cardioprotective, neuroprotective, analgesic, antiaging, antidiabetic, antimicrobial, chemopreventive, and anticancer. Since RES has poor solubility, stability, absorption, and bioavailability, a large number of RES derivatives have been synthesized to overcome these issues in order to achieve greater pharmacological and therapeutic effects. Currently, these RES derivatives are under investigation for detailed mechanisms of action and therapeutic effects.

In a large number of in vitro and in vivo studies, RES has been found to be effective against OA and RA. RES provides anti-OA effects via multiple mechanisms, that is prevention and/reversal of pathways involved in the pathogenesis of OA. Elmali et al. (2005) and Lee et al. (2013) showed that RES provided significant protective effects against articular cartilage degradation. In human chondrocytes, RES exerted both antiapoptotic and antiinflammatory regulatory mechanisms (Lee et al., 2013). RES is a potent inhibitor of both NF-κB activation and NF-κB-dependent gene expression, as well as an inhibitor of COX-2 and iNOS expression. RES blocks TNF-α- and IL-1β-induced activation of NF-κB, thereby regulating chondrocyte apoptosis induced by NO, IL-1β, and IL-17. RES-induced activation of SIRT1 also provided protection against the IL-1β-induced inflammatory response in articular chondrocytes.

Li et al. (2015) investigated the effects of intraarticular (IA) RES administration on the expression of SIRT1, hypoxia-inducible factor-2 (HIF-2α), and catabolic factors in OA cartilage in mice. RES significantly induced the activation of SIRT1, inhibited HIF-2α, reduced the expression of iNOS and MMP-13, and retained the expression of collagen type II in OA cartilage. In a high-fat diet-induced OA in mice model, Gu et al. (2016a) found that RES appears to exert a reduction in chondrocyte apoptosis, thereby exerting anti-OA effects. Dave et al. (2008) reported that RES protects against IL-1β-induced catabolic effects and prevents chondrocyte apoptosis via inhibition of mitochondrial membrane and ATP depletion, by inhibiting COX-2-derived PGE_2 synthesis. RES also decreased the production of MMP (MMP-1, MMP-3, and MMP-13) activities. In conclusion, RES appears to provide anti-OA effects by antioxidative, antiinflammatory, and antiapoptotic activities.

Yang et al. (2018) demonstrated that RES alleviated RA by inhibiting ROS, inflammation, MAPK signaling pathways, and angiogenesis. Wang et al. (2019) provided evidence that RES ameliorated RA through activation of the Nrf2-ARE signaling pathway via SIRT1/NF-κB/miR-29a-3p/Keap1 and the SIRT1/NF-kB/miR-23a-3p/cu13 signaling pathway. Recently, Oz et al. (2019) demonstrated that RES exerted an antiinflammatory effect by inhibiting the signaling pathways such as Src kinase, STAT3, and Wnt, which are activated by collagen-induced arthritis in rats.

Thus RES appears to hold great potential for anti-OA and anti-RA effects, and clinical studies and trials are warranted.

Naturally Preferred Holistic Frozen Dog Treats

In a clinical trial, Naturally Preferred Holistic Frozen Dog Treats (a product of Henry Schein Animal Health) were given orally to client-owned moderately arthritic dogs for a period of 4 weeks to evaluate antiarthritic efficacy, safety, and tolerability (Broderick et al., 2013). Each dog in group I (<50 pounds each) received half a treat and those in group II (> 50 pounds each) received a full treat. On a weekly basis, each dog was evaluated for arthritis-associated pain (overall pain, pain upon limb manipulation, and pain after physical exertion) using different numeric scales. Dogs were also examined for physical parameters (body weight, heart rate, and body temperature), hepatic (bilirubin, ALT, and AST), renal (BUN and creatinine), heart and skeletal muscle (creatine kinase) functions, and hematological parameters. Findings of this investigation revealed that arthritic pain was significantly ameliorated in dogs of both groups within 1 week, with maximum effects after 4 weeks. The product, Naturally Preferred Holistic Frozen Dog Treats, was found to be safe and well-tolerated in moderately OA dogs. All ingredients of this product are natural, and the mechanism of action for antiarthritic effects remains proprietarily confidential at this time.

Flex Choice

In a clinical trial, Webber et al. (2020) investigated the therapeutic efficacy and safety of Flex Choice (Clinics Choice, LLC) in moderately OA dogs. Flex Choice [consisting of krill oil, HA, astaxanthin, *Boswellia serrata* extract, green-lipped mussel (GLM), and iron transport tocopheryl polyethylene glycol succinate (ITPGS)] was given as a chew twice a day for 150 days. On a monthly basis, each dog was evaluated for arthritis-associated pain (overall pain, pain upon limb manipulation, and pain after physical exertion), using the Glasgow scoring system with different numeric scales. Dogs were also examined for physical parameters (body weight, heart rate, and body temperature), hepatic (bilirubin, ALT, and AST), renal (BUN and creatinine), heart and skeletal muscle (creatine kinase) functions, and hematological parameters. Dogs showed marked reductions in overall pain (52%), pain upon limb manipulation (35%), and pain after physical exertion (40%). The active ingredients in Flex Choice exert antioxidative, antiinflammatory, immunomodulatory, analgesic, cartilage repair, and anti-OA effects. ITPGS, in addition to being a bioenhancer, exerts antioxidative and antiinflammatory effects. In conclusion, Flex Choice treatment significantly reduced OA-associated pain, and it was well tolerated by OA dogs.

Crominex-3 +

Crominex-3 + (a product of Natreon Inc.) is a complex of trivalent chromium, *P. emblica* (amla) extract, and purified shilajit. In a randomized, double-blind, placebo-controlled clinical trial, client-owned moderately OA dogs received one capsule of Crominex-3 + in the morning and one capsule in the evening for a period of 5 months. Each capsule of 25 mg Crominex-3 + contained 500 μg trivalent chromium, 7.5 mg amla extract, and 7.5 mg purified shilajit. On a monthly basis, dogs were evaluated for arthritis-associated pain, hepatic, renal, heart, and skeletal muscle functions, and hematological parameters and underwent a full physical examination. Dogs receiving Crominex-3 + showed a significant reduction in arthritic pain as early as 90 days, with a maximum effect after 5 months of treatment (Fleck et al., 2013). In a similar clinical trial, moderately OA horses received 1000 mg Crominex-3 + (20 mg trivalent chromium, 300 mg amla extract, and 300 mg purified shilajit) per day in two divided doses for a period of 5 months. Unlike dogs, horses showed significant reductions in pain within 30 days (overall pain, 32%; pain after limb manipulation, 31%; and pain after physical exertion, 32%), with maximum pain reduction after 5 months (77%, 69%, and 70%, correspondingly) (May et al., 2015). In both dogs and horses, no significant change occurred in physical parameters, serum biomarkers of liver, kidney, heart, and skeletal muscle functions, or hematological parameters, which suggested that Crominex-3 + was found to be safe and well-tolerated (Fleck et al., 2013; May et al., 2015).

All three ingredients in Crominex-3 + exert multiple pharmacological effects through multiple cellular and molecular mechanisms. Trivalent chromium is known to produce antihyperglycemic, antioxidant, and antiinflammatory effects. The amla extract, which has a high $ORAC_{FN}$ value of 47,000 μmol TE/g, exerts antioxidative and antiinflammatory effects. The antioxidative property is also due to the presence of low-MW hydrolysable tannoids (Emblicanin A, Emblicanin B, Pedunculagin, and Punigluconin) in the amla extract, which exert effects similar to those of vitamin C. Various mechanisms of action and effects of shilajit were already discussed earlier in this chapter. Thus Crominex-3 + provides significant antioxidative, antiinflammatory, and antiarthritic effects without eliciting any untoward events in moderately arthritic dogs and horses.

Terminalia chebula extract

Terminalia chebula Retzius, commonly known as Black Myrobalan or Hard (mentioned as "King of medicines" in Ayurvedic Materia Medica), is a species of *Terminalia*, native to South Asia from India and Nepal east to southwest China (Yunnan), and south to Sri Lanka, Malaysia, and Vietnam. The *T. chebula* fruit extract is known to contain several bioactive constituents (such as chebulagic acid, chebulinic acid, gallic acid, ellagic acid, tannic acid, corilagin, polyphenolic compounds, triterpenoids, and ascorbate). Chebulagic acid and chebulinic acid are the major constituents. Phytochemicals present in TCE exert a variety of pharmacological actions (Sawant et al., 2013; Upadhyay et al., 2014; Bupesh et al., 2016; Usharani et al., 2016).

In experimental and clinical studies, TCE and its components have been shown to exert antioxidant, antiinflammatory, and analgesic activities (Seo et al., 2012; Sarwar et al., 2013; Jami et al., 2014; Chandrasekhar et al., 2016; Pokuri et al., 2016), as well as an antiobesity (Rehman and Tiwari, 2017) effect, thereby exerting anti-OA (Murdock et al., 2016; Lopez et al., 2017) and anti-RA (Ramani and Pradhan, 2012; Seo et al., 2012) effects.

In a double-blind, placebo-controlled clinical trial, Murdock et al. (2016) evaluated the therapeutic efficacy and safety of TCE (AyuFlex, Natreon Inc.) in moderately OA dogs. The dogs received either 500 mg placebo or 500 mg TCE twice daily for 150 days. On a monthly basis, dogs were given a full physical exam and were evaluated for arthritic pain, inflammation (erythrocyte sedimentation rate, ESR), analysis of complete blood count, and serum biomarkers of liver (bilirubin, ALT, and AST), kidney (BUN and creatinine), and heart and skeletal muscle (CK) functions. The elbow and stifle joints were radiographed on day 0 and day 150 for evaluation of OA progression. Dogs given AyuFlex showed significant reductions in arthritic pain with maximum reductions on day 150 (overall pain, 81.2%; pain upon limb manipulation, 81.5%; and pain after physical exertion, 84.2%). A marked reduction in ESR coincided with pain reduction in AyuFlex-treated dogs, which was indicative of the antiinflammatory effect of TCE. Radiographic evidence also indicated slowed progression of OA in those joints that were examined. The study concluded that TCE, by having many active principles (chebulagic acid, chebulinic acid, corilagin, hydrolyzable tannoids, etc.), might have provided antioxidative, antiinflammatory, and antiarthritic effects in dogs without causing any side effects.

In a randomized, double-blind, placebo-controlled human study, subjects (aged 40–70) received 250 or 500 mg AyuFlex twice daily for 12 weeks (Chandrasekhar et al., 2016). Modified Western Ontario and McMaster Universities Arthritis Index (mWOMAC) and knee swelling index (KSI), as well as visual analog scale (VAS), were used to assess pain, joint stiffness, and disability. Each subject was evaluated for these measures every 4 weeks. The findings revealed that OA patients receiving 500 mg AyuFlex twice daily showed significant reductions in mWOMAC, KSI, and VAS, when compared to placebo or baseline values. In another clinical trial, using various parameters for OA and inflammation, administration of aqueous TCE (250 or 500 mg AyuFlex twice daily) for 84 days significantly improved knee and overall health and functional capacity outcomes in otherwise healthy subjects with exercise/activity-dependent knee discomfort (Lopez et al., 2017).

Ramani and Pradhan (2012) reported that acetone extract of *T. chebula* (TCE) produced a reduction of ESR values and rheumatoid factor values comparable to a dexamethasone-treated group of rats, suggesting that TCE has antiinflammatory and immunomodulatory effects in RA. Seo et al. (2012) evaluated the analgesic and anti-RA effects of NDI10218 (a standardized ethanol extract of *T. chebula*) on collagen-induced arthritis and acetic acid-induced writhing model in DBA/1J mice. The findings revealed that NDI10218 significantly reduced the arthritic index, blocked the synovial hyperplasia, decreased the number of T-cell subpopulations in the regional lymph nodes, and produced an analgesic effect in RA mice.

Boswellia serrata extract

The *B. serrata* plant is native to India, and its extract/resin/gum has been used as a traditional medicine for centuries. In the Western world, the *Boswellia* tree is known as Frankincense. The *B. serrata* extract (BSE) has boswellic acids, which provide beneficial health effects. These organic acids consist of a pentacyclic triterpene, a carboxyl group, and at least one other functional group. Alpha-boswellic acid and beta-boswellic acid both have an additional hydroxyl group; they differ only in their triterpene structure. The chemical structure of alpha-boswellic acid is shown in Fig. 14.5. One of the six boswellic acids, acetyl-keto-beta-boswellic acid (AKBA), is present at 2%–3% of the total extract and is the most important for positive health effects. The BSE extract and/or boswellic acids have been found effective not only in OA and RA (Etzel, 1996; Kimmatkar et al., 2003), but also in ulcerative colitis, chemically induced hepatic damage, bronchial asthma, and other diseases (Gupta et al., 1998; Kiela et al., 2005; Jyothi et al., 2006; Anthoni et al., 2006; Ismail et al., 2019). BSE and acetyl-boswellic acids have been reported to exhibit antiinflammatory properties by inhibiting lipoxygenase, human leukocyte elastase (HLE), and cytokines (Ammon et al., 1993; Safayhi et al., 1996; Upaganlawar and Ghule, 2009; Sengupta et al., 2011).

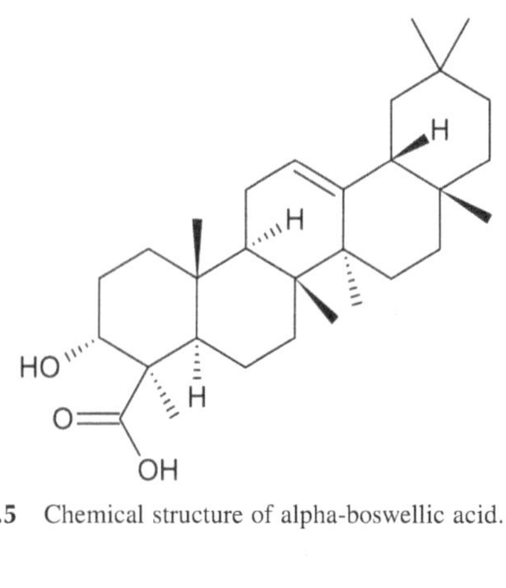

FIGURE 14.5 Chemical structure of alpha-boswellic acid.

In a clinical trial, knee OA patients receiving BSE showed a decrease in knee pain, increased knee flexion, and increased walking distance (Kimmatkar et al., 2003). The frequency of swelling in the knee joint was also decreased. The BSE extract/AKBA has been shown to exert antiinflammatory, analgesic, immunomodulatory, and antiarthritic properties. BSE has also been shown to inhibit MMP-3 and ICAM-1 activities, thereby preventing cartilage degeneration.

5-Loxin (a patented product of PLT Health Solutions, Inc.), with 30% AKBA as its active ingredient, provides improvement in joint mobility and comfort within 1 week. 5-Loxin provided significant benefits in WOMAC, Huskisson's VAS, and Lequesne's functional index (LFI) standardized testing methodologies for joint health. It inhibits MMP-3, an enzyme that breaks down cartilage, collagen, and connective tissue. 5-Loxin positively impacts biomarkers of inflammation and arthritis (such as NF-κB, TNF-α, CRP, and IL-6). BSE has also been shown to deter the activity of HLE, which has been implicated in the development of RA.

Wang et al. (2014) reported the pharmacokinetics of boswellic acids in arthritic rats and compared these with Huo Luo Xiang Ling Dan (HLXLD), which has 11 ingredients including boswellic acids. The results revealed that there were significant differences in pharmacokinetic parameters between normal and arthritic groups, and the absorptions of two boswellic acids (11-keto-β-boswellic acid and 3-*O*-acetyl-11-keto-β-boswellic acid) were significantly higher in HLXLD than in BSE alone.

In a number of studies, it has been suggested that BSE and AKBA are safe and well-tolerated in OA patients (Kimmatkar et al., 2003; Sengupta et al., 2008).

New Zealand green-lipped mussel

Perna canaliculus is a species of GLM found off the waters of New Zealand. In the 1970s and 1980s, GLM was used in the treatment of OA and RA due to its antiinflammatory properties (Heghton and McArthur, 1975; Croft, 1979; Miller and Ormrod, 1980; Caughey et al., 1983). GLM is a rich source of GAGs, omega-3 fatty acids, and eicosatetraenoic acid (ETA). GAGs exert antiinflammatory activity and lubricate joints in OA patients, and ETA and possibly some other components of GLM inhibit the activities of COX-2 and lipoxygenase (Rainsford and Whitehouse, 1980; Bierer and Bui, 2002) and the biosynthesis of prostaglandins (Miller and Wu, 1984).

In a carrageenan-induced inflammatory edema model, Miller and Ormrod (1980) found the antiinflammatory effect only when GLM crude extract was given intraperitoneally, and not after oral administration in rats. Using the same model of inflammation (footpad edema in rat), Miller et al. (2006) found that an aqueous fraction of GLM containing high-molecular-weight material (possibly a polysaccharide glycogen) produced a dose-dependent antiinflammatory effect. Mobilization of neutrophils to the site of an inflammatory stimulus was also significantly reduced.

The GLM product GlycoMarine is used extensively for the treatment of RA and OA in humans and animals because it exerts antiinflammatory and chondroprotective effects and selectively inhibits COX-2 activity. Bui and Bierer (2003) treated OA dogs with GLM powder (0.3% of a dry diet) for 6 weeks and found significant improvement in arthritic score, joint pain, and swelling, without significant improvement in crepitus and range of joint movement. In another study, Bierer and Bui (2002) treated OA dogs with GLM-containing PEDIGREE JointCare Treats (450 mg for <25 kg, 1 treat; 750 mg for 25–34 kg, 1.5 treats; 1000 mg for >34 kg, 2 treats) for 6 weeks. The findings revealed a reduction in lameness and pain associated with OA in dogs. In many other studies, GLM products have been reported to be effective against OA in dogs, cats (Pollard et al., 2006; Hielm-Björkman et al., 2009; Rialland et al., 2013; Gupta et al., 2019a), and horses (Cayzer et al., 2012). Interestingly, GLM provides antihistaminic and gastroprotective effects, in addition to the antiinflammatory and antiarthritic effects (Rainsford and Whitehouse, 1980; Kosuge et al., 1986).

Avocado and soybean unsaponifiables

Avocado and soybean unsaponifiables (ASU) are commonly used in the treatment of OA. The active ingredients of this extract are the phytosterols β-sitosterol, campesterol, and stigmasterol (Castrogiovanni et al., 2016). The synergism between the avocado and soya components and their relative ratios (one-third avocado oil and two-thirds soybean oil) appear to be very important (Maheu et al., 1998). ASU modulates OA pathogenesis by inhibiting a number of molecules and pathways implicated in OA (Christiansen et al., 2015). Their properties are generally related to potential stimulation of ECM synthesis. There is evidence that ASU can stimulate the production of collagen in cell cultures and reduce IL-1β-induced collagenolytic activity (Boumediene et al., 1999). ASU can also inhibit the stimulating action of IL-1β on MMP-3 (stromelysin), IL-6, IL-8, ERK, and PGE_2 production and IL-1β-stimulated collagenase synthesis by human articular chondrocytes (Henrotin et al., 1998). The chondroprotective effect of ASU occurs by stimulating the expression of TGF-β1 and TGF-β2 genes in chondrocytes (Boumediene et al., 1999), increasing the levels of TGF-β1 and TGF-β2 (Altinel et al., 2007), and by decreasing vascular endothelial growth factor (VEGF) in joint fluid (Christiansen et al., 2015). In addition, ASU increased the synthesis of plasminogen activator inhibitor 1 at both the mRNA and protein levels. Evidently, ASU can be useful in preventing or blocking erosion of the tissue in OA, slowing joint space narrowing, and improving the repair of cartilage (Maheu et al., 1998; Applebloom et al., 2001; Lequesne et al., 2002).

Recently, Al-Afify et al. (2018) reported that oral administration of ASU (27.5 mg/kg/day for 3 weeks) to mono-iodoacetate-induced OA rats significantly attenuated the synovium, cartilage, and subchondral degeneration. ASU also reduced the expression of TNF-α and MMP-13 in OA cartilage and the expression of iNOS in both OA cartilage and subchondral bone. Interestingly, Christiansen et al. (2015) reported that ASU inhibits cholesterol absorption and endogenous cholesterol biosynthesis, which mediate ROS pathology in chondrocytes. ASU provides structure-modifying abilities via its antioxidative and antiinflammatory properties, in addition to its ability to modulate MMP-13 activity (Christiansen et al., 2015).

In basic and clinical studies, ASU has been shown to promote the synthesis of matrix components in various cell types (dermal fibroblasts, synovial cells, and articular chondrocytes) and reduce most of the effects of IL-1β, thereby promoting matrix synthesis, reducing OA pain, and improving functional disability (Blotman et al., 1997; Henrotin et al., 1998; Christiansen et al., 2015). In a 6-month clinical trial, Maheu et al. (1998) evaluated the symptomatic efficacy of ASU in the treatment of OA of the knee and hip. Based on LFI and VAS (100 mm), ASU treatment showed significant symptomatic efficacy, including pain reduction. After 6 months of ASU treatment, LFI decreased from 9.7 ± 0.3 to 6.8 ± 0.4 and pain on VAS decreased from 56.1 ± 1.6 to 35.3 ± 2.3 mm. Overall functional disability was significantly reduced and ASU was well-tolerated.

Curcumin

Curcumin has been widely used in many human and animal diseases and poisonings for centuries. For decades, curcumin has been studied for its cellular/molecular mechanisms of action and physiological and pharmacological effects. By having strong antioxidant, antiinflammatory, and immunomodulatory properties, curcumin alone or in combination with other nutraceuticals provides antiarthritic effects in humans and animals (Zhang et al., 2016; Haroyan et al., 2018; Chand, 2019). Further details on curcumin's effects in OA and other diseases are provided in chapters on curcumin and curcuminoid metallo-complexes.

MicroRNAs

MicoRNAs (miRNAs) have been reported to be involved in the pathogenesis of OA and degradation of ECM and cartilage (iNOS expression, inflammation, autophagy, apoptosis, chondrocyte metabolism, differentiation, and homeostasis) (Miyaki and Asahara, 2012; Zhang et al., 2017; Oliviero et al., 2019). Yu et al. (2011) presented evidence of the important role played by miRNA in determining the complex gene expression patterns of OA chondrocytes and their role in the regulation of transcription and possible demethylation mechanisms that might be applicable in OA. Studies on different miRNA have revealed that they have proinflammatory and anabolic/catabolic roles in the pathophysiology of OA (Alcaraz et al., 2010). Findings indicate that the regulatory effect of miRNAs on the expression of catabolic and anabolic genes in OA may take place at upstream levels prior to their transcription. The activity of several signaling pathways, such as the NF-κB pathway (Zhang et al., 2015; Gu et al., 2016b), Wnt/beta-Catenin pathway (Cheleschi et al., 2017), SIRT1/p53 pathway (Yan et al., 2016), and the SDF1/CXCR4 pathway (Zheng et al., 2017), were found to be modulated by miRNAs in chondrocytes during the development of OA.

In an OA rat model, Wang et al. (2016) reported that miRNA-98 expression was reduced in the cartilage cells by apoptosis, and the overexpression of miRNA-98 inhibits cartilage cell apoptosis. In a similar investigation, miRNA-140 levels were significantly reduced in human OA cartilage-derived chondrocytes and SF compared with normal chondrocytes and SF. Interestingly, overexpression of miRNA-140 in primary human chondrocytes promoted collagen II synthesis and inhibited MMP-13 and ADAMST-5 expression (Si et al., 2017). IA administration of miRNA-140 alleviates OA progression by modulating ECM homeostasis in the rat and may have potential as a new therapy for OA in humans, dogs, and horses.

In essence, given the role of miRNAs in mediating the translation of target mRNAs into proteins, the identification of differentially expressed miRNAs in OA tissue and the crucial contribution that miRNAs play in the progression of OA, miRNAs may have important diagnostic and therapeutic potential, and provide a novel means of treating OA (Yu et al., 2011). Further investigations are needed to determine the specificity, sensitivity, and efficacy of miRNAs for clinical applications (Zhang et al., 2017; Oliviero et al., 2019).

Concluding remarks and future directions

Arthritis is a chronic debilitating disease that commonly afflicts millions of people around the world and has no curative treatment. Like humans, some animal species (canine and equine) are also afflicted by this crippling disease. Among the two most common forms of arthritis (OA and RA), OA occurs with a greater frequency. The pathophysiology of arthritis is very complex because of multiple causes and mechanisms; therefore treatment is not simple. Animal models and specific biomarkers for detection of early OA are warranted. Presently, among all options for treatment of OA, nutraceuticals are found to be very effective and well-tolerated, with little to no side effects. An overall recommendation to use nutraceuticals in the treatment of all patients with OA is not strongly supported by the available data. Because nutraceuticals exert antioxidant, antiinflammatory, chondroprotective, analgesic, and many other physiological and pharmacological effects, a synergic therapy involving the use of different nutraceuticals might prove to be a promising strategy for treating OA. The quest for better animal models, sensitive biomarkers for early OA detection, and more effective nutraceuticals supported by long-term scientific studies for efficacy and safety should continue.

References

Abadie, E., Ethgen, D., Avouac, B., 2004. Recommendation for the use of new methods to assess the efficacy of disease-modifying drugs in the treatment of osteoarthritis. Osteoarthr. Cartil. 12, 263–268.

Adler, N., Schoeniger, A., Fuhrmann, H., 2018. Polyunsaturated fatty acids influence inflammatory markers in a cellular model for canine osteoarthritis. J. Anim. Physiol. Anim. Nutr. 102, e623–e632.

Agung, M., Ochi, M., Yanada, S., 2006. Mobilization of bone marrow-derived mesenchymal stem cells into the injured tissues after intraarticular injection and their contribution to tissue regeneration. Knee Surg. Sports Traumatol. Arthrosc. 14, 1307–1314.

Akbar, U., Yang, M., Kurian, D., et al., 2017. Omega-3 fatty acids in rheumatic diseases. J. Clin. Rheumatol. 23, 330–339.

Al-Afify, A.S.A., El-Akabawy, G., El-Sherif, N.M., et al., 2018. Avocado soybean unsaponifiables ameliorates cartilage and subchondral bone degeneration in mono-iodoacetate-induced knee osteoarthritis in rats. Tissue Cell 52, 108–115.

Alcaraz, M.J., Megías, J., García-Arnandis, I., et al., 2010. New molecular targets for the treatment of osteoarthritis. Biochem. Pharmacol. 80, 13–21.

Altinel, L., Saritas, Z.K., Kose, K.C., 2007. Treatment with unsaponifiables extracts of avocado and soybean increases TGF-β1 and TGF-β2 levels in canine joint fluid. Tohoku J. Exp. Med. 211, 181–186.

Ammon, H.P., Safayi, H., Mack, T., Sabieraj, J., 1993. Mechanism of antiinflammatory actions of curcumin and boswellic acids. J. Ethnopharmacol. 38, 113–119.

Anderson, J.W., Nicolosi, R.J., Borzelleca, J.F., 2005. Glucosamine effects in humans: a review of effects on glucose metabolism, side effects, safety considerations and efficacy. Food Chem. Toxicol. 43, 187–201.

Anthoni, C., Laukoetter, M.G., Rijcken, E., 2006. Mechanisms underlying the antiinflammatory actions of boswellic acid derivatives in experimental colitis. Am. J. Physiol. Gastroint. Liver Physiol. 290, G1131–G1137.

Appleboom, T., Schuermans, J., Verbruggen, G., 2001. Symptoms modifying effect of avocado/soybean unsaponifiables (ASU) in knee osteoarthritis. A double blind, prospective, placebo-controlled study. Scand. J. Rheumatol. 30, 242–247.

Armstrong, S., Read, R., Ghosh, P., 1994. The effect of intra-articular hyaluronan on cartilage and subchondral bone changes in an ovine model of early osteoarthritis. J. Rheumatol. 21, 680–687.

Aubry-Rozier, B., 2012. Role of slow-acting anti-arthritic agents in osteoarthritis (chondroitin sulfate, glucosamine, hyaluronic acid). Rev. Med. Suisse 14, 571–572.

Balogh, L., Polyak, A., Mathe, D., 2008. Absorption, uptake and tissue affinity of high-molecular-weight hyaluronan after oral administration in rats and dogs. J. Agric. Food Chem. 56, 10582–10593.

Bellamy, N., Carr, A., Dougados, M., 2001. Towards a definition of "differences" in osteoarthritis. J. Rheumatol. 28, 427–430.

Berenbaum, F., Grifka, J., Cazzaniga, S., et al., 2012. A randomized, double-blind, controlled trial comparing two intraarticular hyaluronic acid preparations differing by their molecular weight in symptomatic knee osteoarthritis. Ann. Rheum. Dis. 71, 1454–1460.

Bergh, M.S., Budsberg, S.C., 2005. The coxib NSAIDs: potential clinical and pharmacologic importance in veterinary medicine. J. Vet. Intern. Med. 19, 633–643.

Beriault, D.R., Sharma, S., Shi, Y., 2011. Glucosamine-supplementation promotes endoplasmic reticulum stress, hepatic steatosis and accelerated atherogenesis in apoE$^{-/-}$ mice. Atherosclerosis 219, 134–140.

Bhattacharya, D., Svechkarev, D., Souchek, J.J., et al., 2017. Impact of structurally modifying hyaluronic acid on CD44-interaction. J. Mater. Chem. B 5, 8183–8192.

Bierer, T.L., Bui, L.M., 2002. Improvement of arthritic signs in dogs fed green-lipped mussel (*Perna canaliculus*). J. Nutr. 132, 1634s–1636s.

Block, J.A., Oegema, T.R., Sandy, J.D., Plaas, A., 2010. The effects of oral glucosamine on joint health: is a change in research approach needed? Osteoarthr. Cartil. 18, 5–11.

Blom, A.B., Brockbank, S.M., van Lent, P.L., 2009. Involvement of the Wnt signaling pathway in experimental and human osteoarthritis: prominent role of Wnt-induced signaling protein 1. Arthritis Rheumatol. 60, 501–512.

Blotman, F., Maheu, E., Wulwik, A., 1997. Efficacy and safety of avocado/soybean unsaponifiables in the treatment of symptomatic osteoarthritis of the knee and hip: a prospective, multicenter, three month, randomized, double-blind, placebo controlled trial. Rev. Rhum. Engl. Ed. 64, 825–834.

Boumediene, K., Felisaz, N., Bogdanowicz, P., 1999. Avocado/soy unsaponifiables enhance the expression of transforming growth factor β1 and β2 in cultured articular chondrocytes. Arthritis Rheumatol. 42, 148–156.

Braun, H.J., Gold, G.E., 2012. Diagnosis of osteoarthritis: imaging. Bone 51 (2), 278–288.

Breese McCoy, S.J., Bryson, J.C., 2003. High-dose glucosamine associated with polyuria and polydipsia in a dog. J. Am. Vet. Med. Assoc. 222, 431–432.

Breivik, H., 2017. NSAIDs relieve osteoarthritis (OA) pain, but cardiovascular safety in question even for diclofenac, ibuprofen, naproxen, and celecoxib: what are the alternatives? Scand. J. Pain 16, 148–149.

Brien, S., Prescott, P., Bashir, N., 2008. Systematic review of the nutritional supplements dimethyl sulfoxide (DMSO) and methylsulfonylmethane (MSM) in the treatment of osteoarthritis. Osteoarthr. Cartil. 16, 1277–1288.

Broderick, B.A., Miller, J., Goad, J.T., Gupta, R.C., 2013. Efficacy and safety of Naturally Preferred Holistic Frozen Dog Treats in moderately arthritic dogs. In: Proceedings of the Annual Meeting of Ohio Valley Chapter of the Society of Toxicology, Louisville, KY, USA, p. 20.

Bui, L.M., Bierer, T.L., 2003. Influence of green lipped mussel (*Perna canaliculus*) in alleviating signs of arthritis in dogs. Vet. Ther. 4, 397–407.

Bupesh, G., Manikandan, E., Thanigaiarul, K., et al., 2016. Enhanced antibacterial, anticancer activity from *Terminalia chebula* medicinal plant rapid extract by photosynthesis of silver nanoparticles core-shell structures. J. Nanomed. Nanotechnol. 7, 1–6. Available from: https://doi.org/10.4172/2157-7439.1000355.

Burr, D.B., Gallant, M.A., 2012. Bone remodeling in osteoarthritis. Nat. Rev. Rheumatol. 8, 665–673.

Calder, P.C., 2012. Mechanisms of action of (n-3) fatty acids. J. Nutr. 142, 592S–599S.

Calvo, E., Palacios, I., Delgado, E., 2001. High-resolution MRI detects cartilage swelling at the early stages of experimental osteoarthritis. Osteoarthr. Cartil. 9, 463–472.

Campo, G.M., Avenoso, A., Campo, S., 2009. Glycosaminoglycans modulate inflammation and apoptosis in LPS-treated chondrocytes. J. Cell. Biochem. 106, 83–92.

Carrabba, M., Paresce, E., Angelini, M., 1995. The safety and efficacy of different dose schedules of hyaluronic acid in the treatment of painful osteoarthritis of the knee with joint effusion. Eur. J. Rheumatol. Inflammation 15, 25–31.

Castaneda, S., Roman-Blas, J.A., Lago, R., Herrero-Beaumont, G., 2012. Subchondral bone as a key target for osteoarthritis treatment. Biochem. Pharmacol. 83, 315–323.

Castrogiovanni, P., Trovato, F.M., Loreto, C., et al., 2016. Nutraceutical supplements in the management and prevention of osteoarthritis. Int. J. Mol. Sci. 17, 2042. Available from: https://doi.org/10.3390/ijms17122042.

Caughey, D., Grigor, R., Caughey, E., et al., 1983. *Perna canaliculus* in the treatment of rheumatoid arthritis. Eur. J. Rheumatol. Inflammation 6, 197–200.

Cayzer, J., Hedderley, D., Gray, S., 2012. A randomized, double-blinded, placebo-controlled study on the efficacy of a unique extract of green-lipped mussel (*Perna canaliculus*) in horses with chronic fetlock lameness attributed to osteoarthritis. Equine Vet. J. 44, 393–398.

Centers for Disease Control and Prevention (CDC), 2010 Centers for Disease Control and Prevention (CDC), 2010. Arthritis-basic-definition-osteoarthritis (Retrieved 10 June 2018).

Chan, P.S., Caron, J.P., Rosa, G.J., Orth, M.W., 2005. Glucosamine and chondroitin sulfate regulate gene expression and synthesis of nitric oxide and prostaglandin E_2 in articular cartilage explants. Osteoarthr. Cartil. 13, 387–394.

Chand, N., 2019. Standardized turmeric and curcumin. In: Gupta, R.C., Srivastava, A., Lall, R. (Eds.), Nutraceuticals In Veterinary Medicine. Springer Nature, Cham, Switzerland, pp. 3–23.

Chandrasekhar, N., et al., 2016. A randomized, double-blind, placebo-controlled, parallel group clinical study to evaluate the analgesic effect of aqueous extract of *Terminalia chebula*, a proprietary chromium complex, and their combination in subjects with joint discomfort. Asian J. Pharm. Clin. Res. 9, 264–269.

Cheleschi, S., De Palma, A., Pecorelli, A., et al., 2017. Hydrostatic pressure regulates microRNA expression levels in osteoarthritic chondrocyte cultures via the Wnt/beta-catenin pathway. Int. J. Mol. Sci. 18, 133. Available from: https://doi.org/10.3390/ijms18010133.

Christensen, R., Astrup, A., Bliddal, H., 2005. Weight loss: the treatment of choice for knee osteoarthritis? A randomized trial. Osteoarthr. Cartil. 13, 20–27.

Christiansen, B.A., Bhatti, S., Goudaarzi, R., et al., 2015. Management of osteoarthritis with avocado/soybean unsaponifiables. Cartilage 6, 30–44.

Clegg, D.O., Reda, D.J., Harris, C.L., 2006. Glucosamine, chondroitin sulfate, and the two in combination for painful knee osteoarthritis. N. Engl. J. Med. 354, 795–808.

Conrozier, T., Sazne, T., Fan, C.S.S., 1998. Serum concentrations of cartilage oligomeric matrix protein and bone sialoprotein in hip osteoarthritis: a one year prospective study, Ann. Rheum. Dis., 57. pp. 527–532.

Corbee, R.J., Barnier, M.M.C., van de Lest, C.H.A., 2012. The effect of dietary long-chain omega-3 fatty acid supplementation on owner's perception of behavior and locomotion in cats with naturally occurring osteoarthritis. J. Anim. Physiol. Anim. Nutr. 97, 846–853.

Corr, M., 2008. Wnt-beta-catenin signaling in the pathogenesis of osteoarthritis. Nat. Clin. Pract. Rheumatol. 4, 550–556.

Coruzzi, G., Venturi, N., Spaggiari, S., 2007. Gastrointestinal safety of novel nonsteroidal anti-inflammatory drugs: selective COX-2 inhibitors and beyond. Acta Biomed. 78, 96–110.

Croft, J.E., 1979. Relief from arthritis. Cox and Wyman Ltd., Reading, UK.

Crowley, D.C., Lau, F.C., Sharma, P., 2009. Safety and efficacy of undenatured type II collagen in the treatment of osteoarthritis of the knee: a clinical trial. Int. J. Med. Sci. 6, 312–321.

Curtis, C.L., Rees, S.G., Little, C.B., 2002. Pathologic indicators of degradation and inflammation in human osteoarthritic cartilage are abrogated by exposure to *n*-3 fatty acids, Arthritis Rheumatol., 46. pp. 1544–1553.

D'Abusco, A.S., Calamia, V., Cicione, C., 2007. Glucosamine affects intracellular signaling through inhibition of mitogen-activated protein kinase phosphorylation in human chondrocytes. Arthritis Res. Ther. 9, R104.

D'Altilio, M., Peal, A., Alvey, M., 2007. Therapeutic efficacy and safety of undenatured type II collagen singly or in combination with glucosamine and chondroitin in arthritic dogs. Toxicol. Mech. Methods 17, 189–196.

Daugaard, C.L., Hangaard, S., Bartels, E.M., et al., 2020. The effects of weight loss on imaging outcomes in osteoarthritis of the hip or knee in people who are overweight or obese: a systematic review. Osteoarthr. Cartil. 28, 10–21.

Davatchi, F., Abdollahi, B.S., Mohyeddin, M., et al., 2016. Mesenchymal stem cell therapy for knee osteoarthritis: 5 years follow-up of three patients. Int. J. Rheum. Dis. 19, 219–225.

Dave, M., Attur, M., Palmer, G., et al., 2008. The antioxidant resveratrol protects against chondrocyte apoptosis via effects on mitochondrial polarization and ATP production. Arthritis Rheumatol. 58, 2786–2797.

de Salazar Alcalá, A.G., Gioda, L., Dehman, A., et al., 2019. Assessment of the efficacy of firocoxib (Previcox) and grapiprant (Galliprant) in an induced model of acute arthritis in dogs. BMC Vet. Res. 15 (1), 309.

Deal, C.L., Moskowitz, R.W., 1999. Nutraceuticals as therapeutic agents in osteoarthritis. The role of glucosamine, chondroitin sulfate, and collagen hydrolysate. Rheum. Dis. Clin. North Am. 25, 379–395.

Debbi, E.M., Agar, G., Fichman, G., 2011. Efficacy of methylsulfonylmethane supplementation on osteoarthritis of the knee: a randomized controlled study. BMC Complement. Altern. Med. 11, 50.

Dechant, J.E., Baxter, G.M., Frisble, D.D., 2005. Effects of glucosamine hydrochloride and chondroitin sulfate, alone and in combination, on normal and interleukin-1 conditioned equine cartilage explants metabolism. Equine Vet. J. 37, 227–231.

DeParle, L.A., Gupta, R.C., Canerdy, T.D., 2005. Efficacy and safety of glycosylated undenatured type-II collagen (UC-II) in therapy of arthritic dogs. J. Vet. Pharmacol. Ther. 28, 385–390.

Díaz-Gallego, L., Prieto, L.G., Coronel, P., et al., 2005. Apoptosis and nitric oxide in an experimental model of osteoarthritis in rabbit after hyaluronic acid treatment. J. Orthop. Res. 23, 1370–1376.

Dieppe, P., Lohmander, L.S., 2005. Pathogenesis and management of pain in osteoarthritis. Lancet 365, 965–973.

Dore, D., Martens, A., Quinn, S., 2010. Bone marrow lesions predict site-specific cartilage defect development and volume loss: a prospective study in older adults. Arthritis Res. Ther. 12, R222.

Ebisuzaki, K., 2003. Aspirin and methylsulfonylmethane (MSM): a search for common mechanisms, with implications for cancer prevention. Anticancer Res. 23, 453–458.

Eckstein, F., Guermazi, A., Gold, G., 2014. Imaging of cartilage and bone: promises and pitfalls in clinical trials of osteoarthritis. Osteoarthr. Cartil. 22, 1516–1532.

Elmali, N., Esenkaya, I., Harma, A., et al., 2005. Effect of resveratrol in experimental osteoarthritis in rabbits. Inflamm. Res. 54, 158–162.

Erhan, B., Gündüz, B., Üstünel, S.H., 2012. The efficacy of topical glucosamine sulfate-chondroitin sulfate in knee osteoarthritis treated with physical therapy: a randomized, double-blind, placebo-controlled study. Turk. J. Phys. Med. Rehab. 58, 194–198.

Etzel, R., 1996. Special extract of *Boswellia serrata* (H 15) in the treatment of rheumatoid arthritis. Phytomedicine 3, 91–94.

Fanelli, A., Romualdi, P., Vigao, R., et al., 2013. Non-selective non-steroidal anti-inflammatory drugs (NSAIDs) and cardiovascular risk. Acta Biomed. 84, 5–11.

Fleck, A., Gupta, R.C., Goad, J.T., 2013. Anti-arthritic efficacy and safety of Chrominex 3 + (trivalent chromium, *Phyllanthus emblica* extract, and shilajit) in moderately arthritic dogs. J. Vet. Sci. Anim. Husb. 1 (4e), 1–6.

Francin, P.J., Abot, A., Guillaume, C., 2014. Association between adiponectin and cartilage degradation in human osteoarthritis. Osteoarthr. Cartil. 22, 519–526.

Frech, T.M., Clegg, D.O., 2007. The utility of nutraceuticals in the treatment of osteoarthritis. Curr. Rheumatol. Rep. 9, 25–30.

Gaál, J., Varga, J., Szehanecz, Z., 2008. Balneotherapy in elderly patients: effect on pain from degenerative knee and spine conditions and on quality of life. Isr. Med. Assoc. J. 10, 365–369.

Galarraga, B., Ho, M., Youssef, H.M., 2008. Cod liver oil (*n*-3 fatty acids) as an nonsteroidal anti-inflammatory drug sparing agent in rheumatoid arthritis. Rheumatology 47, 665–669.

Gersing, A.S., Solka, M., Joseph, G.B., et al., 2016. Progression of cartilage degeneration and clinical symptoms in obese and overweight individuals is dependent on the amount of weight loss: 48-month data from the osteoarthritis initiative. Osteoarthr. Cartil. 24, 1126–1134.

Ghosh, P., Cheras, P.A., 2001. Vascular mechanisms in osteoarthritis. Best Pract. Res. Clin. Rheumatol. 15, 693–709.

Ghosh, P., Guidolin, D., 2002. Potential mechanism of action of intra-articular hyaluronan therapy in osteoarthritis: are the effects molecular weight dependent? Semin. Arthritis Rheumatol. 32, 10–37.

Giaccari, A., Morviducci, L., Zorretta, D., 1995. In vivo effects of glucosamine on insulin secretion and insulin sensitivity in the rat: possible relevance to the maladaptive responses to chronic hyperglycemia. Diabetologia 38, 518–524.

Goggs, R., Vaughan-Thomas, A., Clegg, P.D., 2005. Nutraceutical therapies for degenerative joint diseases: a critical review. Crit. Rev. Food Sci. Nutr. 45, 145–164.

Gosser, T., Ricciotti, E., FitzGerald, G.A., 2017. The cardiovascular pharmacology of nonsteroidal anti-inflammatory drugs. Trends Pharmacol. Sci. 38, 733–748.

Gu, H., Li, K., Li, X., et al., 2016a. Oral resveratrol prevents osteoarthritis progression in C57BL/6J mice fed a high fat diet. Nutrients 8, 233. Available from: https://doi.org/10.3390/nu8040233.

Gu, R., Liu, N., Luo, S., et al., 2016b. MicroRNA-9 regulates the development of knee osteoarthritis through the NF-kappaB1 pathway in chondrocytes. Medicine 95, e4315.

Guercio, A., Di Marco, P., Casella, S., 2012. Production of canine mesenchymal stem cells from adipose tissue and their application in dogs with chronic osteoarthritis of the humeroradial joints. Cell Biol. Int. 36, 189–194.

Guermazi, A., Roemer, F.W., Haugen, I.K., et al., 2013. MRI-based semiquantitative scoring of joint pathology in osteoarthritis. Nat. Rev. Rheumatol. 9, 236–251.

Gupta, I., Gupta, V., Parihar, A., 1998. Effects of *Boswellia serrata* gum resin in patients with bronchial asthma: results of a double-blind, placebo-controlled, 6-week clinical study. Eur. J. Med. Res. 3, 511–514.

Gupta, R.C., 2016. Nutraceuticals in arthritis. In: Gupta, R.C. (Ed.), Nutraceuticals: efficacy, safety and toxicity. Academic Press/Elsevier, Amsterdam, pp. 161–176.

Gupta, R.C., Canerdy, T.D., Scaggs, P., 2009. Therapeutic efficacy of type-II collagen (UC-II) in comparison of glucosamine and chondroitin in arthritic horses. J. Vet. Pharmacol. Ther. 32, 577–584.

Gupta, R.C., Canerdy, T.D., Lindley, J., 2011. Comparative therapeutic efficacy and safety of type-II collagen (UC-II), glucosamine and chondroitin in arthritic dogs: pain evaluation by ground force plate. J. Anim. Physiol. Anim. Nutr. 96, 770–777.

Gupta, R.C., Doss, R.B., Lall, R., Srivastava, A., Sinha, A., 2019a. Nutraceuticals in arthritis. Nutraceuticals in Veterinary Medicine. Springer Nature, Cham, Switzerland, pp. 365–381.

Gupta, R.C., Srivastava, A., Lall, R., Sinha, A., 2019b. Osteoarthritis biomarkers. In: Gupta, R.C. (Ed.), Biomarkers in Toxicology, 2nd edition Academic Press/Elsevier, Amsterdam, pp. 929–943.

Gupta, R.C., Lall, R., Srivastava, A., Sinha, A., 2019c. Hyaluronic acid: molecular mechanisms and therapeutic trajectory. Front. Vet. Sci. 6 (192), 1–24.

Hall, M., Doherty, S., Courtney, P., 2014. Synovial pathology detected on ultrasound correlates with the severity of radiographic knee osteoarthritis more than with symptoms. Osteoarthr. Cartil. 22, 1627–1633.

Haroyan, A., Mukuchyan, V., Mkrtchyan, N., et al., 2018. Efficacy and safety of curcumin and its combination with boswellic acid in osteoarthritis: a comparative, randomized, double-blind, placebo-controlled study. BMC Complement. Altern. Med. 18, 7. Available from: https://doi.org/10.1186/s12906-017-2062-z.

Hathcock, J.N., Shao, A., 2007. Risk assessment for glucosamine and chondroitin sulfate. Regul. Toxicol. Pharmacol. 47, 78–83.

Heghton, T., McArthur, A., 1975. Pilot study on the effect of New Zealand green mussel on rheumatoid arthritis. N. Z. Med. J. 261.

Henrotin, Y.E., Labasse, A.H., Jaspar, J.M., 1998. Effects of three avocado/soybean unsaponifiables mixtures on metalloproteinases, cytokines and prostaglandin E_2 production by human articular chondrocytes. Clin. Rheumatol. 17, 31–39.

Hielm-Björkman, A., Tulamo, R.-M., Salonen, H., et al., 2009. Evaluating complementary therapies for osteoarthritis part I: Green-lipped mussel. (*Perna canaliculus*). eCAM 6, 365–373.

Hill, C.L., March, L.M., Aitken, D., et al., 2016. Fish oil in knee osteoarthritis: a randomized clinical trial of low dose versus high dose. Ann. Rheum. Dis. 75, 23–29.

Hochberg, M.C., Zhan, M., Langenberg, P., 2008. The rate of decline of joint space width in patients with osteoarthritis of the knee: a systematic review and meta-analysis of randomized placebo-controlled trials of chondroitin sulfate. Curr. Med. Res. Opin. 24, 3029–3035.

Holmes, M.W.A., Bayliss, M.T., Muir, H., 1988. Hyaluronic acid in human articular cartilage. Biochem. J. 250, 435–441.

Holzer, N., Salvo, D., Marijnissen, A.K., et al., 2018. How to assess ankle osteoarthritis: Comparison of the Kellgren and Lawrence scale with functional outcome and digital image analysis. Orthroped. Proc. 94-B, 1–3.

Hooper, M., 2001. Is glucosamine an effective treatment for osteoarthritic pain? Clev. Clin. J. Med. 68, 494–495.

Hootman, J.M., Helmick, C.G., Barbour, K.E., et al., 2016. Updated projected prevalence of self-reported doctor-diagnosed arthritis and arthritis-attributable activity limitation among US adults, 2015-2040. Arthritis Rheumatol. 68 (7), 1582–1587.

Horvath, K., Noker, P.E., Somfai-Relle, S., 2002. Toxicity of methylsulfonylmethane in rats. Food Chem. Toxicol. 40, 1459–1462.

Hosnijeh, F.S., Bierma-Zeinstra, S.M., Bay-Jensen, A.C., 2019. Osteoarthritis year in review 2018: biomarkers (biochemical markers). Osteoarthr. Cartil. 27 (2019), 412–423.

Hunter, D.J., 2010. Pharmacologic therapy for osteoarthritis—the era of disease modification. Nat. Rev. Rheumatol. 7, 13–22.

Hunter, D.J., Li, J., LaValley, M., 2007. Cartilage markers and their association with cartilage loss on magnetic resonance imaging in knee osteoarthritis: the Boston Osteoarthritis Knee Study. Arthritis Res. Ther. 9, R108.

Huskisson, E.C., 2008. Glucosamine and chondroitin for osteoarthritis. J. Int. Med. Res. 36, 1161–1179.

Inns, J.F., Clayton, J., Lascelles, B., 2010. Review of the safety and efficacy of long-term NSAID use in the treatment of canine osteoarthritis. Vet. Rec. 166, 226–230.

Irsay, L., Borda, M.H., Nitu, A.D., 2010. Effectiveness of glucosamine and chondroitin sulfate combination in patients with primary osteoarthritis. Appl. Med. Inform. 27, 47–54.

Ishimaru, D., Sugiara, N., Akiyama, H., 2014. Alterations in the chondroitin sulfate chain in human osteoarthritic cartilage of the knee. Osteoarthr. Cartil. 22, 250–258.

Ismail, I.E., Abdelnour, S.A., Shehata, S.A., et al., 2019. Effect of dietary *Boswellia serrata* resin on growth performance, blood biochemistry, cecal microbiota of growing rabbits. Front. Vet. Sci. 6 (471), 1–8.

Jackson, C.G., Plaas, A.H., Sandy, J.D., 2010. The human pharmacokinetics of oral ingestion of glucosamine and chondroitin sulfate taken separately or in combination. Osteoarthr. Cartil. 19, 297–302.

Jami, M.S.I., Sultana, Z., Ali, M.E., et al., 2014. Evaluation of analgesic and anti-inflammatory activities of ethanolic extract of *Terminalia chebula* fruits in experimental animal models. Am. J. Plant Sci. 5, 63–69.

Jerosch, J., 2011. Effects of glucosamine and chondroitin sulfate on cartilage metabolism in OA: outlook on other nutrient partners especially omega-3 fatty acids. Int. J. Rheumatol. 2011, 1–17.

Jomphe, C.R., Gabriac, M., Hale, T.M., 2008. Chondroitin sulfate inhibits the nuclear translocation of nuclear factor-kappa B in interleukin-1beta-stimulated chondrocytes. Basic Clin. Pharmacol. Toxicol. 102, 59–65.

Jyothi, Y., Kamath, J.V., Asad, M., 2006. Effect of hexane extract of *Boswellia serrata* oleo-gum resin on chemically induced liver damage. Pak. J. Pharm. Sci. 19, 129–133.

Kalman, D.S., Heimer, M., Valdeon, A., 2008. Effect of a natural extract of chicken combs with a high content of hyaluronic acid (Hyal-joint) on pain relief and quality of life in subjects with knee osteoarthritis: a pilot randomized double-blind placebo-controlled trial. Nutr. J. 7, 3.

Kiela, P.R., Midura, A.J., Kuscuoglu, N., 2005. Effects of *Boswellia serrata* in mouse models of chemically induced colitis. Am. J. Physiol. Gastrointest. Liver Physiol. 288, G798–G808.

Kim, L.S., Axelrod, L.J., Howard, P., 2006. Efficacy of methylsulfonylmethane (MSM) in osteoarthritis pain of the knee: a pilot clinical trial. Osteoarthr. Cartil. 14, 286–294.

Kim, J.R., Yoo, J.J., Kim, H.A., 2018. Therapeutics in osteoarthritis based on an understanding of its molecular pathogenesis. Int. J. Mol. Sci. 19 (3), 674. PMID: 29495538.

Kimmatkar, N., Thawani, V., Hingorani, L., et al., 2003. Efficacy and tolerability of Boswellia serrata extract in treatment of osteoarthritis of knee—a randomized double blind placebo controlled trial. Phytomedicine 10, 3–7.

Knott, L., Avery, N.C., Hollander, A.P., et al., 2011. Regulation of osteoarthritis by omega-3 (n-3) polyunsaturated fatty acids in a naturally occurring model of disease. Osteoarthr. Cartil. 19, 1150. Available from: https://doi.org/10.1016/j.joca.2011.06.005.

Kong, L., Zheng, L.-Z., Qin, L., et al., 2017. Role of mesenchymal stem cells in osteoarthritis treatment. J. Orthop. Transl. 9, 89–103.

Koskinen, A., Vuolteenaho, K., Nieminen, R., 2011. Leptin enhances MMP-1, MMP-3, and MMP-13 production in human osteoarthritic cartilage and correlates with MMP-1 and MMP-3 in synovial fluid from OA patients. Clin. Exp. Rheumatol. 29, 57–64.

Kosuge, T., Tsuji, K., Ishida, H., et al., 1986. Isolation of an antihistamine substance in green-lipped mussel (*Perna canaliculus*). Chem. Pharmacol. Bull. 34, 4825–4828.

Kozakçioğlu, M., 2012. Osteoarthritis and glucosamine. Turk. J. Rheumatol. 24, 94–97.

Kremer, J.M., Jubiz, W., Michalek, A., 1987. Fish-oil fatty acid supplementation in active rheumatoid arthritis. A double-blinded, controlled, crossover study. Ann. Intern. Med. 106, 497–503.

Ku, J.H., Lee, C.K., Joo, B.S., 2009. Correlation of synovial fluid leptin concentrations with the severity of osteoarthritis. Clin. Rheumatol. 28, 1431–1435.

Kumar, D., Wyatt, C.R., Lee, S., 2013. Association of cartilage defects, and other MRI findings with pain and function in individuals with mild-moderate radiographic hip osteoarthritis and controls. Osteoarthr. Cartil. 21, 1685–1692.

Lafeber, F.P.J.G., van Spil, W.E., 2013. Osteoarthritis year 2013 in review: Biomarkers; reflecting before moving forward, one step at a time. Osteoarthr. Cartil. 21, 1452–1464.

Lafontaine-Lacasse, M., Dore, M., Picard, F., 2011. Hexosamine stimulate apoptosis by altering Sirt1 action and levels in pancreatic β-cells. J. Endocrinol. 208, 41–49.

Lawley, S., Gupta, R.C., Goad, J.T., 2013. Anti-inflammatory and anti-arthritic efficacy and safety of purified shilajit in moderately arthritic dogs. J. Vet. Sci. Anim. Husb. 1 (3e), 1–6.

Lee, V.H., Woo, J., Choi, S.J., 2010. Effect of glucosamine or chondroitin sulfate on the osteoarthritis progression: a meta-analysis. Rheumatol. Int. 30, 357–363.

Lee, Y.-H., Bae, S.-C., Song, G.-G., 2012. Omega-3 polyunsaturated fatty acids and the treatment of rheumatoid arthritis: a meta-analysis. Arch. Med. Res. 43, 356–362.

Lee, A.S., Ellman, M.B., Yan, D., et al., 2013. A current review of molecular mechanisms regarding osteoarthritis and pain. Gene 527, 440–447.

Lepetos, O., Papavassiliou, A.G., 2016. ROS/oxidative stress signaling in osteoarthritis. Biochim. Biophys. Acta 1862, 576–591.

Lequesne, M., Maheu, E., Cadet, C., 2002. Structural effects of avocado/soybean unsaponifiables on joint space loss in osteoarthritis of the hip. Arthritis Rheumatol. 47, 50–58.

Li, W., Cai, L., Zhang, Y., et al., 2015. Intra-articular resveratrol injection prevents osteoarthritis progression in a mouse model by activating SIRT1 and thereby silencing HIF-2α. J. Orthop. Res. 33, 1061–1070.

Liu, X., Machado, G.C., Eyles, J.P., et al., 2018. Dietary supplements for treating osteoarthritis: a systematic review and meta-analysis. Br. J. Sports Med. 52, 167–175.

Loeser, R.F., 2013. Osteoarthritis year in review 2013: biology. Osteoarthr. Cartil. 21, 1436–1442.

Lopez, V.A., 1982. Double-blind clinical evaluation of the relative efficacy of ibuprofen and glucosamine sulfate in the management of osteoarthritis of the knee in out-patients. Curr. Med. Res. Opin. 8, 145–149.

Lopez, H.L., Habowski, S.M., Sandrock, J.E., et al., 2017. Effects of dietary supplementation with a standardized aqueous extract of *Terminalia chebula* fruit (AyuFlex®) on joint mobility, comfort, and functional capacity in healthy overweight subjects: a randomized placebo-controlled clinical trial. BMC Complement. Altern. Med. 17, 475. Available from: https://doi.org/10.1186/s12906-017-1977-8.

Lorenz, H., Richter, W., 2006. Osteoarthritis: cellular and molecular changes in degenerating cartilage. Prog. Histochem. Cytochem. 40, 135–163.

Maheu, E., Mazières, B., Valat, J.-P., 1998. Symptomatic efficacy of avocado/soybean unsaponifiables in the treatment of osteoarthritis of the knee and hip. A prospective, randomized, double-blind, placebo-controlled, multicenter clinical trial with six-month treatment period and two-month follow-up demonstrating a persistent effect. Arthritis Rheumatol. 41, 81–91.

Mao, G., Zhang, Z., Huang, Z., et al., 2017. MicroRNA-92a-3p regulates the expression of cartilage-specific genes by directly targeting histone deacetylase 2 in chondrogenesis and degradation. Osteoarthr. Cartil. 25, 521–532.

Marone, P.A., Lau, F.C., Gupta, R.C., 2010. Safety and toxicological evaluation of undenatured type II collagen. Toxicol. Mech. Methods 20, 175–189.

Martel-Pelletier, J., Tardif, G., Pelletier, J., 2005. Cytokines and growth factors in the treatment of osteoarthritis: what could be the best disease modifying drugs. Curr. Med. Chem. 4, 235–249.

Mathews, G.L., Hunter, D.J., 2011. Emerging drugs for osteoarthritis. Expert Opin. Emerg. Drugs 16, 479–491.

Matteson, E.L., 2000. Current treatment strategies for rheumatoid arthritis. Mayo Clin. Proc. 75, 69–74.

May, K., Gupta, R.C., Miller, J., 2015. Therapeutic efficacy and safety evaluation of a novel chromium supplement (Crominex® + 3−) in moderately arthritic horses. J. Vet. Sci. Res. 1 (3), 014.

McAlindon, T., Formica, M., LaValley, M., 2004. Effectiveness of glucosamine for symptoms of knee osteoarthritis: results from an internet-based randomized double-blind controlled trial. Am. J. Med. 117, 643–649.

Mehler, S.J., May, L.R., King, C., et al., 2016. A prospective, randomized, double-blind, placebo-controlled evaluation of the effects of eicosapentaenoic acid and docosahexaenoic acid on the clinical signs and erythrocyte membrane polyunsaturated fatty acid concentrations in dogs with osteoarthritis. Prostaglandins Leukot. Essent. Fat. Acids 109, 1–7.

Messier, S., Mihalko, S., Loester, R., 2007. Glucosamine/chondroitin combined with exercise for the treatment of knee: a preliminary study. Osteoarthr. Cartil. 15, 1256–1266.

Mevel, E., Monfoulet, L.-E., Merceron, C., 2014. Nutraceuticals in joint health: animal models as instrumental tools. Drug Discov. Today 19, 1649–1658.

Miller, T.E., Ormrod, D.J., 1980. The anti-inflammatory activity of *Perna canaliculus* (NZ green-lipped mussel). N. Z. Med. J. 667, 187–193.

Miller, T.E., Wu, H., 1984. In vivo evidence for prostaglandin inhibitory activity in New Zealand green-lipped mussel extract. N. Z. Med. J. 97, 355–357.

Miller, T.E., Dodd, J., Ormrod, D.J., et al., 2006. Anti-inflammatory activity of glycogen extracted from *Perna canaliculus* (NZ green-lipped mussel). Agents Actions 38, 114–118.

Milner, C.M., Day, A.J., 2003. TSG-6: a multifunctional protein associated with inflammation. J. Cell Sci. 116, 1863–1873.

Miyaki, S., Asahara, H., 2012. Macro view of microRNA's function in osteoarthritis. Nat. Rev. Rheumatol. 8, 543–552.

Monfort, J., Pelletier, J.P., Garcia-Giralt, N., Martel-Pelletier, J., 2008. Biochemical basis of the effect of chondroitin sulfate on osteoarthritis articular tissues. Ann. Rheum. Dis. 67, 735–740.

Moreau, M., Troncy, E., del Castillo, J.R.E., 2013. Effects of feeding a high omega-3 fatty acids diet in dogs with naturally occurring osteoarthritis. J. Anim. Physiol. Anim. Nutr. 97, 830–837.

Moreland, L.W., 2003. Intra-articular hyaluronan (hyaluronic acid) and hylans for the treatment of osteoarthritis: mechanisms of action. Arthritis Res. Ther. 5, 54–67.

Muhlfeld, A., Floege, J., 2005. COX-2 inhibitor induced anuric renal failure in a previously healthy young woman. Clin. Nephrol. 63, 221–224.

Mukherjee, D., Nissen, S.E., Topol, E.J., 2001. Risk of cardiovascular events associated with selective COX-2 inhibitors. JAMA 286, 954–959.

Muller-Fassbender, H., Bach, G.L., Haase, W., 1994. Glucosamine sulfate compared to ibuprofen in osteoarthritis of the knee. Osteoarthr. Cartil. 2, 61–69.

Murdock, N., Gupta, R.C., Vega, N., et al., 2016. Evaluation of *Terminalia chebula* extract for anti-arthritic efficacy and safety in osteoarthritic dogs. J. Vet. Sci. Technol. 7, 1. Available from: https://doi.org/10.4172/2157-7579.1000290.

Narvy, S.J., Vangsness, C.T., 2010. Critical appraisal of the role of glucosamine and chondroitin in the management of osteoarthritis of the knee. Nutr. Diet. Suppl. 2, 13–25.

Neil, K.M., Orth, M.W., Coussens, P.M., 2005. Effects of glucosamine and chondroitin sulfate on mediators of osteoarthritis in cultured equine chondrocytes stimulated by use of recombinant equine interleukin-1 beta. Am. J. Vet. Res. 66, 1861–1869.

Neogi, T., Felson, D., Niu, J., 2011. Association between radiographic features of knee osteoarthritis and pain: results from two cohort studies. BMJ 339, b2844.

Norling, L.V., Perretti, M., 2013. The role of omega-3 derived resolvins in arthritis. Curr. Opin. Pharmacol. 13, 476–481.

Oe, M., Tashiro, T., Yoshida, H., et al., 2016. Oral hyaluronan relieves knee pain: a review. Nutr. Res. 15, 11. Available from: https://doi.org/10.1186/s12937-016-0128-2.

Oldershaw, R., 2012. Cell sources for the regeneration of articular cartilage: the past, the horizon and the future. Int. J. Exp. Pathol. 93, 389–400.

Oliviero, A., Porta, G.D., Peretti, G.M., et al., 2019. MicroRNA in osteoarthritis: physiopathology, diagnosis and therapeutic challenge. Br. Med. Bull. 130, 137–147.

Otis, C., Guillot, M., Moreau, M., et al., 2019. Sensitivity of functional targeted neuropeptide evaluation in testing pregabalin analgesic efficacy in a rat model of osteoarthritis pain. Clin. Exp. Pharmacol. Physiol. 46, 723–733.

Oz, B., Yildirim, A., Yolbas, S., et al., 2019. Resveratrol inhibits Src tyrosine kinase, STAT3, and Wnt signaling pathway in collagen induced arthritis model. BioFactors 45, 69–74.

Pagonis, T.A., Givissis, P.A., Kritis, A.C., et al., 2014. The effect of methylsulfonylmethane on osteoarthritic large joints and mobility. Int. J. Orthop. 1 (1), 19–24.

Palmer, A., Fernquest, S., Rombach, I., et al., 2017. Diagnostic and prognostic value of delayed gadolinium enhanced magnetic resonance imaging of cartilage (dGEMRIC) in early osteoarthritis of the hip. Osteoarthr. Cartil. 25, 1468–1477.

Pan, J., Wang, B., Li, W., 2012. Elevated crosstalk between subchondral bone and cartilage in osteoarthritic joints. Bone 51, 212–217.

Panahifar, A., Jaremko, J.L., Tessier, A.G., 2014. Development and reliability of a multi-modality scoring system for evaluation of disease progression in preclinical models of osteoarthritis: celecoxib may possess disease-modifying properties. Osteoarthr. Cartil. 22, 1639–1650.

Peal, A., D'Altilio, M., Simms, C., 2007. Therapeutic efficacy and safety of undenatured type-II collagen (UC-II) alone or in combination with (−)-hydroxycitric acid and chromemate in arthritic dogs. J. Vet. Pharmacol. Ther. 30, 275–278.

Perkins, J.R., Sanak, M., Canto, G., et al., 2015. Unravelling adverse reactions to NSAIDs using system biology. Trends Pharmacol. Sci. 36, 178–180.

Persiani, S., Roda, E., Rovati, L.C., 2005. Glucosamine oral bioavailability and plasma pharmacokinetics after increasing doses of crystalline glucosamine sulfate in man. Osteoarth. Cartil. 13, 1041–1049.

Petersen, J., Hansen, J.M., de Muckadell, O.B.S., et al., 2020. A model to predict the risk of aspirin/non-steroidal anti-inflammatory drugs-related upper gastrointestinal bleeding for the individual patient. Basic Clin. Pharmacol. Toxicol. 126 (5), 437–443. Available from: https://doi.org/10.1111/bcpt.13370.

Petersson, I.F., Boegard, T., Swensson, B., 1998. Changes in cartilage and bone metabolism identified by serum markers in early osteoarthritis of the knee joints. Br. J. Rheumatol. 37, 46–50.

Pokuri, V.K., Kumar, C.U., Pingali, U., 2016. A randomized, double-blind, placebo-controlled, cross-over study to evaluate analgesic activity *Terminalia chebula* in healthy human volunteers using a mechanical pain model. J. Anesth. Clin. Pharmacol. 32, 329–332.

Pollard, B., Guilford, W.G., Ankenbauer-Perkins, K.L., et al., 2006. Clinical efficacy and tolerance of an extract of green-lipped mussel (*Perna canaliculus*) in dogs presumptively diagnosed with degenerative joint disease. N. Z. Vet. J. 54, 114–118.

Pozo, M.A., Balazs, E.A., Belmonte, C., 1997. Reduction of sensory responses to passive movements of inflamed knee joints by hylan, a hyaluronan derivative. Exp. Brain Res. 116, 3–9.

Pritchard, C.H., Sripada, P., Bankers, P.F., et al., 1999. A retrospective comparison of the efficacy and tolerability of sodium hyaluronate and hylan G-F-20 in the treatment of osteoarthritis of the knee. J. Musculoskeletal Res. 6, 197–205.

Qiu, G.X., Gao, S.N., Giacovelli, G., 1998. Efficacy and safety of glucosamine sulfate versus ibuprofen in patients with knee osteoarthritis. Arzneimittelforschung 48, 469–474.

Raekillio, M., Hielm-Bjorkman, A., Kejonen, J., 2006. Evaluation of adverse effects of long-term orally administered carprofen in dogs. J. Am. Vet. Med. Assoc. 228, 876–880.

Rainsford, K.D., Whitehouse, M.W., 1980. Gastroprotective and anti-inflammatory properties of green-lipped mussel (*Perna canaliculus*) preparation. Arzneim-Frsch/Drug Res. 12, 2128–2132.

Ramani, Y.R., Pradhan, S., 2012. Antiarthritic activity of acetone extract of Terminalia chebula. WebMed Cent. Pharmacol. 17, 1–9.

Rege, A., Juvekar, P., Juvekar, A., 2012. In vitro antioxidant and anti-arthritic activities of shilajit. Int. J. Pharm. Pharm. Sci. 4, 650–653.

Reginster, J.-Y., Dudler, J., Blicharski, T., et al., 2017. Pharmaceutical-grade chondroitin sulfate is as effective as celecoxib and superior to placebo in symptomatic knee osteoarthritis: the chondroitin versus celecoxib versus placebo trial (CONCEPT). Ann. Rheum. Dis. 76, 1537–1543.

Rehman, M., Tiwari, S.B., 2017. Antiobesity effect of *Terminalia chebula* fruit extract on high fat diet induced obese animal model. IOSR J. Pharm. 7, 43–51.

Reid, D.M., Miller, G.C., 2008. Clinical Trials in Rheumatoid Arthritis and Osteoarthritis. Springer-Verlag, London Ltd., Aberdeen, UK.

Rialland, P., Bichot, S., Lussier, B., et al., 2013. Effect of a diet enriched with green-lipped mussel on pain behavior and functioning in dogs with clinical osteoarthritis. Can. J. Vet. Res. 77, 66–74.

Rigoglou, S., Papavassiliou, A.G., 2013. The NF-kappaB signaling pathway in osteoarthritis. Int. J. Biochem. Cell Biol. 45, 2580–2584.

Roemer, F.W., Neogi, T., Nevitt, M.C., 2010. Subchondral bone marrow lesions are highly associated with, and predict subchondral bone attrition longitudinally: the MOST study. Osteoarthr. Cartil. 18, 47–53.

Rozendaal, R.M., Koes, B.W., Weinans, H., 2005. The effect of glucosamine sulfate on osteoarthritis: design of a long-term randomized clinical trial (ISRCTN54513166). BMC Musculoskelet. Disord. 6, 20.

Runhaar, J., Schiphof, D., van Meer, B., 2014. How to define subregional osteoarthritis progression using semi-quantitative MRI osteoarthritis knee score. Osteoarthr. Cartil. 22, 1533–1536.

Rutjes, A.W., Jüni, P., de Costa, B.R., 2012. Viscosupplementation for osteoarthritis of the knee: a systematic review and meta-analysis. Ann. Intern. Med. 157, 180–191.

Safayhi, H., Sailer, E.R., Ammon, H.P.T., 1996. 5-Lipoxygenase inhibition by acetyl-11-keto-beta-boswellic acid (AKBA) by a novel mechanism. Phytomedicine 3, 71–72.

Santilli, V., Paoloni, M., Mangone, M., et al., 2016. Hyaluronic acid in the management of osteoarthritis: injection therapies innovations. Clin. Cases Min. Bone Metab. 13, 131–134.

Sarwar, S., Malik, A.H., Rahman, M.A., et al., 2013. Antioxidant, cytotoxic and analgesic activities of the methanolic fruit extract of *Terminalia chebula* Retz. Int. Curr. Pharm. J. 3, 219–222.

Sawant, R., Binorkar, S.V., Bhoyar, M., et al., 2013. Phyto-constituents bioefficacy and phytopharmacological activities of *Terminalia chebula*-A review. Int. J. Ayurveda Altern. Med. 1, 1–11.

Sawitzke, A.D., Shi, H., Finco, M.F., 2008. The effect of glucosamine and/or chondroitin sulfate on the progression of knee osteoarthritis: a report from the glucosamine/chondroitin arthritis intervention trial. Arthritis Rheumatol. 58, 3183–3191.

Saxne, T., Heinegard, D., 1992. Cartilage oligomeric matrix protein: a novel marker of cartilage turnover detectable in synovial fluid and blood. Br. J. Rheumatol. 31, 583–591.

Scanzello, C.R., Plaas, A., Crow, M.K., 2008. Innate immune system activation in osteoarthritis: is osteoarthritis a chronic wound? Curr. Opin. Rheumatol. 20, 565–572.

Scarpiganto, C., Lanas, A., Blandizzi, C., et al., 2015. Safe prescribing of non-steroidal anti-inflammatory drugs in patients with osteoarthritis—an expert consensus addressing benefits as well as gastrointestinal and cardiovascular risks. BMC Med. 13, 1–22. Available from: https://doi.org/10.1186/s12916-015-0285-8.

Scotece, M., Conde, J., Vuolteenaho, K., 2014. Adipokines as drug targets in joint and bone disease. Drug Discov. Today 19, 241–258.

Sengupta, K., Alluri, K.V., Satish, A.R., et al., 2008. A double-blind, randomized, placebo controlled study of the efficacy and safety of 5-Loxin® for treatment of osteoarthritis of the knee. Arthritis Res. Ther. 10 (4), R85.

Sengupta, K., Kolla, J.N., Krishnaraju, A.V., et al., 2011. Cellular and molecular mechanisms of anti-inflammatory effect of Aflapin: a novel *Boswellia serrata* extract. Mol. Cell Biochem. 354, 189–197.

Seo, J.B., Jeong, J.-Y., Park, J.Y., et al., 2012. Anti-arthritic and analgesic effect of NDI10218, a standardized extract of *Terminalia chebula*, on arthritis and pain model. Biomol. Ther. 20, 104–112.

Sessa, M., Sportiello, L., Mascolo, A., et al., 2017. Campania preventability assessment committee (Italy): a focus on the preventability of non-steroidal anti-inflammatory drug's adverse drug reactions. Front. Pharmacol. 8, 305. Available from: https://doi.org/10.3389/fphar.2017.00305.

Setnikar, J., Rovati, L.C., 2001. Absorption, distribution, metabolism and excretion of glucosamine sulfate. a review. Arzneimittelforschung 51, 699–725.

Sharma, A.R., Jagga, S., Lee, S.S., 2013. Interplay between cartilage and subchondral bone contributing to pathogenesis of osteoarthritis. Int. J. Mol. Sci. 14, 19805–19830.

Si, H.B., Zeng, Y., Liu, S.Y., et al., 2017. Intra-articular injection of microRNA-140 (miRNA-140) alleviates osteoarthritis (OA) progression by modulating extracellular matrix (ECM) homeostasis in rats. Osteoarthr. Cartil. 25, 1698–1707.

Singh, J.A., Furst, D.E., Bharat, A., 2012. Update of the 2008 American College of Rheumatology recommendation for the use of disease modifying anti-rheumatic drugs and biologic agents in the treatment of rheumatoid arthritis. Arthritis Care Res. 64, 625–639.

Smith, A., Dillon, J., 2009. Acute liver injury associated with the use of herbal preparations containing glucosamine: three case studies. BMJ Case Rep. 1603, 1–8.

Sofat, N., Kuttapitiya, A., 2014. Future directions for the management of pain in osteoarthritis. Int. J. Clin. Rheumatol. 9, 197–216.

Sorres, C., Gargallo, C.J., Arroyo, M.T., et al., 2010. Adverse effects of non-steroidal anti-inflammatory drugs (NSAIDs, aspirin and coxibs) on upper gastrointestinal tract. Best Pract. Res. Clin. Gastroienterol 24, 121–132.

Stoh, S.J., 2014. Safety and efficacy of shilajit (mumie, moomiyo). Phytother. Res. 28, 475–479.

Takahashi, K., Hashimoto, S., Kubo, T., et al., 2001. Hyaluronan suppressed nitric oxide production in the meniscus and synovium of rabbit osteoarthritis model. J. Orthop. Res. 19, 500–503.

Takemitsu, H., Zhao, D., Yamamoto, I., 2012. Comparison of bone marrow and adipose tissue-derived canine mesenchymal stem cells. BMC Vet. Res. 8, 150.

Thomas, J.E., Bhat, A.K., Rao, M., et al., 2019. Use of vitamin D supplements in osteoarthritis: an observational study in a tertiary health care facility. J. Am. Coll. Nutr. 38, 227–234.

Thote, T., Lin, A.S.P., Raji, Y., 2013. Localized 3D analysis of cartilage composition and morphology in small animal models of joint degeneration. Osteoarthr. Cartil. 21, 1132–1141.

Tonge, D.P., Pearson, M.J., Jones, S.W., 2014. The hallmarks of osteoarthritis and the potential to develop personalized disease-modifying pharmacological therapeutics, Osteoarthr. Cartil, 22. pp. 609–621.

Towheed, T.E., Anastassiades, T.P., Shea, B., 2001. Glucosamine therapy for treating osteoarthritis. Cochrane Database Syst. Rev. (2), CD002946.

Trentham, D.E., 1998. Oral tolerization as a treatment of rheumatoid arthritis. Rheum. Dis. Clin. N. Am. 24, 525–536.

Trentham, D.E., Dynesius-Trentham, R.A., Orav, E.J., 1993. Effects of oral administration of type-II collagen on rheumatoid arthritis. Science 262, 1727–1730.

Trumble, T.N., 2005. The use of nutraceuticals for osteoarthritis in horses. Vet. Clin. North. Am. Equine Pract. 21, 575–597.

Umlauf, D., Frank, S., Pap, T., 2010. Cartilage, biology, pathology, and repair. Cell. Mol. Life Sci. 67, 4197–4211.

Upadhyay, A., Agrahari, P., Singh, D.K., 2014. A review on the pharmacological aspects of *Terminalia chebula*. Int. J. Pharmacol. 10, 289–298.

Upaganlawar, A., Ghule, B., 2009. Pharmacological activities of *Boswellia serrata* RoxB-Mini review. Ethnobot. Leafl. 13, 766–774.

Usha, P., Naidu, M., 2004. Randomized, double-blind, parallel, placebo-controlled study of oral glucosamine, methylsulfonylmethane and their combination in osteoarthritis. Clin. Drug. Invest. 24, 353–363.

Usharani, P., Nutalapati, C., Pokuri, V.K., et al., 2016. A randomized, double-blind, placebo-, and positive-controlled clinical pilot study to evaluate the efficacy and tolerability of standardized aqueous extracts of *Terminalia chebula* and *Terminalia bellerica* in subjects with hyperuricemia. Clin. Pharmacol. Adv. Appl. 8, 51–59.

Valdes, A.M., 2020. Metabolic syndrome and osteoarthritis pain: common molecular mechanisms and potential therapeutic implications. Osteoarthr. Cartil. 28, 7–9.

Valvason, C., Musacchio, E., Pozzuoli, A., 2008. Influence of glucosamine sulfate on oxidative stress in human osteoarthritic chondrocytes: effects of HO-I, p22 (Phox) and iNOS expression. Rheumatology 47, 31–35.

Van der Kraan, P.M., 2017. Factors that influence outcome in experimental osteoarthritis. Osteoarthr. Cartil. 25, 369–375.

van der Tempel, H., Tulleken, J.E., Limburg, P.C., 1990. Effects of fish oil supplementation in rheumatoid arthritis. Ann. Rheum. Dis. 49, 76–80.

Vangness, C.T., Spiker, W., Erickson, J., 2009. A review of evidence-based medicine for glucosamine and chondroitin sulfate use in knee osteoarthritis, Arthroscopy, 25. pp. 86–94.

Velmurugan, C., Vivek, B., Wilson, E., 2012. Evaluation of safety profile of black shilajit after 91 days repeated administration in rats. Asian Pac. J. Trop. Biomed. 2, 210–214.

Vilar, J.M., Morales, M., Santana, A., 2013. Controlled, blinded force platform analysis of the effect of intraarticular injection of autologous adipose derived mesenchymal stem cells associated to PRGF-Endoret in osteoarthritic dogs. BMC Vet. Res. 9, 131.

Vilar, J.M., Batista, M., Moralis, M., 2014. Assessment of the effect of intra-articular injection of autologous adipose-derived mesenchymal stem cells in osteoarthritic dogs using a double blinded force platform analysis. BMC Vet. Res. 10, 143.

Vilim, V., Olejarova, M., Machacek, S., 2002. Serum levels of cartilage oligomeric matrix protein (COMP) correlate with radiographic progression of knee osteoarthritis. Osteoarthr. Cartil. 10, 707–713.

Villalvilla, A., Gómez, R., Largo, R., 2013. Lipid transport and metabolism in healthy and osteoarthritic cartilage. Int. J. Mol. Sci. 14, 20793–20808.

Wan, Z.-H., Zhao, Q., 2017. Gypenoside inhibits interleukin-1β-induced inflammatory response in human osteoarthritis. J. Biochem. Toxicol. 31 (9), PMID: 28422402.

Wandel, S., Jüni, P., Tendal, B., 2010. Effects of glucosamine, chondroitin, or placebo in patients with osteoarthritis of hip or knee: network meta-analysis. BMJ 341, c4675.

Wang, H., Zhang, C., Wu, Y., 2014. Comparative pharmacokinetic study of two boswellic acids in normal and arthritic rat plasma after oral administration of *Boswellia serrata* extract or Huo Luo Xiao Ling Dan by LC-MS. Biomed. Chromatogr. 28, 1402–1408.

Wang, G.L., Wu, Y.B., Liu, J.T., et al., 2016. Upregulation of miR-98 inhibits apoptosis in cartilage cells in osteoarthritis. Gen. Test. Mol. Biomark 20, 645–653.

Wang, G., Xie, X., Yuan, I., et al., 2019. Resveratrol ameliorates rheumatoid arthritis via activation of SIRT1-Nrf2 signaling pathway. BioFactors 43 (3), 441–453. Available from: https://doi.org/10.1002/biof.1599.

Waung, J.A., Maynard, S.A., Gopal, S., 2014. Quantitative X-ray microradiography for high-throughput phenotyping of osteoarthritis in mice. Osteoarthr. Cartil. 22, 1396–1400.

Webber, R.E., Gupta, R.C., Doss, R.B., et al., 2020. Therapeutic efficacy and safety evaluation of Flex Choice™ in moderately osteoarthritic dogs. Int. J. Vet. Health Sci. Res. 8 (1), 242–251.

Weiner, H.L., 1997. Oral tolerance: immune mechanisms and treatment of autoimmune diseases. Immunol. Today 18, 335–343.

Wildi, L.M., Raynauld, J., Martel-Pelletier, J., et al., 2011. Chondroitin sulfate reduces both cartilage volume loss and bone marrow lesions in knee osteoarthritis patients starting as early as 6 months after initiation of therapy: a randomized, double-blind, placebo-controlled pilot study using MRI. Ann. Rheum. Dis. 70, 982–989.

Williams, F.M.K., Spector, T.D., 2008. Biomarkers in osteoarthritis. Arthritis Res. Ther. 10, 1–2.

Wisniewski, H.-G., Colón, E., Liublinska, V., 2014. TSG-6 activity as a novel biomarker of progression in knee osteoarthritis. Osteoarthr. Cartil. 22, 235–241.

Wu, W., Billinghurst, R.C., Pidoux, I., 2002. Sites of collagenase cleavage and denaturation of type II collagen in aging and osteoarthritic articular cartilage and their relationship to the distribution of matrix metalloproteinase I and matrix metalloproteinase-13. Arthritis Rheumatol. 46, 2087–2094.

Xu, I., Peng, H., Glasson, S., 2007. Increased expression of the collagen receptor discoidin domain receptor 2 in articular cartilage as a key event in the pathogenesis of osteoarthritis. Arthritis Rheumatol. 56, 2663–2673.

Yan, S., Wang, M., Zhao, J., et al., 2016. MicroRNA-34a affects chondrocyte apoptosis and proliferation by targeting the SIRT1/p53 signaling pathway during the pathogenesis of osteoarthritis. Int. J. Mol. Med. 38, 201–209.

Yang, C., Cao, M., Liu, H., et al., 2012. The high and low molecular weight forms of hyaluronan have distinct effects on CD44 clustering. J. Biol. Chem. 287, 43094–43107.

Yang, G., Chang, C.-C., Yang, Y., et al., 2018. Resveratrol alleviates rheumatoid arthritis via reducing ROS and inflammation, inhibiting MAPK signaling pathways, and suppressing angiogenesis. J. Agr. Food Chem. 66, 12953–12960.

Yu, C., Chen, W.-P., Wang, X.-H., 2011. MicroRNA in osteoarthritis. J. Int. Med. Res. 39, 1–9.

Yuan, X.L., Meng, H.Y., Wang, Y.C., 2014. Bone–cartilage interface crosstalk in osteoarthritis: potential pathways and future therapeutic strategies. Osteoarthr. Cartil. 22, 1077–1089.

Yudoh, K., Nguyen, T., Nakamura, H., 2005. Potential involvement of oxidative stress in cartilage senescence and development of osteoarthritis: oxidative stress induces chondrocyte telomere instability and downregulation of chondrocyte function. Arthritis Res. Ther. 7, R380–R391.

Zainal, Z., Longman, A.J., Hurst, S., et al., 2009. Relative efficacies of omega-3 polyunsaturated fatty acids in reducing expression of key proteins in a model system for studying osteoarthritis. Osteoarthr. Cartil. 17, 896–905.

Zhang, D., Cao, X., Li, J., et al., 2015. MiR-210 inhibits NF-kappaB signaling pathway by targeting DR6 in osteoarthritis. Sci. Rep. 5, 12775. Available from: https://doi.org/10.1038/srep12775.

Zhang, Z., Leong, D.J., Xu, L., et al., 2016. Curcumin slows osteoarthritis progression and relieves osteoarthritis-associated pain symptoms in a post-traumatic osteoarthritis mouse model. Arthr Res. Ther. 18, 128. Available from: https://doi.org/10.1186/s13075-016-1025-y.

Zhang, M., Lygrisse, K., Wang, J., 2017. Role of microRNA in osteoarthritis. J. Arthr 6, 1–12. Available from: https://doi.org/10.4172/2167-7921.1000239.

Zheng, X., Zhao, F.C., Pang, Y., et al., 2017. Downregulation of miR-221-3p contributes to IL-1beta-induced cartilage degradation by directly targeting the SDF1/CXCR4 signaling pathway. J. Mol. Med. 95, 615–627.

Chapter 15

Nutraceuticals in sports activities and fatigue

Nikolay V. Goncharov, Ekaterina A. Korf, Artem V. Novozhilov, Richard O. Jenkins and Pavel V. Avdonin

Introduction and background

The boundaries of physical capacities are essentially the limits of physiological safety of healthy humans, and fatigue can be considered as a mechanism for limiting the harmful effects of exhausting muscle exercise. Fatigue during exercise is defined as the inability to maintain the required level of strength and performance (Duchateau and Hainaut, 1993). Exercise-induced muscle fatigue has to be clearly delineated from fatigability in the resting stage. Also, muscle fatigue, which can be recovered from fairly quickly, must be clearly distinguished from muscle damage, which usually requires a longer recovery period. Factors that contribute to feeling tired include neurological and non-neurological causes. Neurological causes include central fatigue [mental disorders, organic central nervous system (CNS) abnormalities] and peripheral fatigue [peripheral nervous system (PNS) or neural adaptations, and skeletal muscle diseases—contractile impairment]. Non-neurological causes include cardiac disorders, pulmonary disorders [e.g., chronic obstructive pulmonary disease (COPD)], hematologic disorders, metabolic diseases, renal problems, malignancies, overtraining syndrome, and chronic fatigue syndrome (CFS) (Finsterer, 2012). Some 60 years ago, Merton suggested that in motivated persons, fatigue is determined principally by non-neurological and peripheral causes (Merton, 1954). Nevertheless, because motivation is a subtle ingredient and there are perpetual problems with its existence, new preparations have been elaborated and extensively studied since that time, many of them affecting central and synaptic levels to the detriment of the peripheral level. For the purpose of fatigue prevention and performance enhancement, many preparations have been developed and applied, and are conventionally categorized into three groups: stimulators (amphetamines, cocaine, sidnocarb, modafinil); regulators of heart rate and tremor (β-blockers); and weight regulators (androgens, growth hormone, β_2-agonists, diuretics) (Oliynyk and Oh, 2012). New stimulators based on peptides—being as a rule fragments of hormones and having their properties—have also been developed (Van den Broek et al., 2015). However, doping agents and manipulations of any kind are undesirable companions of professional and amateur sports. Numerous adverse analytical findings and the confessions of athletes have demonstrated the variety of doping agents and methods, as well as the inventiveness, of cheating sportspersons. Emerging drugs are considered relevant for current and future doping controls due to their high potential for misuse in sports (Thevis et al., 2017).

Nutraceuticals have advantages over medicines because they avoid side effects and naturally supplement the diet (Chauhan et al., 2013). A great advantage of nutraceuticals is their safety, whereas their great disadvantage is their relatively poor effectiveness. However, this apparent disadvantage may be deceptive because of our poor understanding of the mechanistic effects of nutraceuticals and their incompetent administration. Many controlled studies have not determined the effects of nutritional supplements that are advertised as ergogenic and are commonly used by athletes (Mason and Lavallee, 2012). It should be mentioned here that the most popular stimulators or agents favoring physical or mental adaptation are essentially nutraceuticals: caffeine, nicotine, ethanol, and tetrahydrocannabinol (THC). Despite this, ethanol and THC are included in the World Anti-Doping Agency (WADA) Prohibited List; caffeine was previously included to the list but presently its level is subject to monitoring, as is the level of nicotine (Pesta et al., 2013). In this chapter experimental investigations and practical applications of some nutraceuticals are analyzed, first with regards to their capacity to enhance physical performance and reduce fatigue, but also their mechanistic properties that are important for physiological and biochemical adaptation.

Nutraceuticals. DOI: https://doi.org/10.1016/B978-0-12-821038-3.00015-X

Dietary strategies to recover from exercise-induced fatigue and muscle damage

The nutritional intake of athletes is a critical determinant of their athletic performance and ability to compete both physically and mentally. Energy intakes of >50 kcal/kg/day for male athletes who train for >90 min/day, and 45–50 kcal/kg/day for female athletes training for >90 min/day, are recommended (Economos et al., 1993). These nearly 30-year-old recommendations are still relevant today. Moreover, it was reported then and confirmed some 20 years later that with sufficient calorie intake, protein or other supplementation is not necessary, and there is no special food that will help elite athletes perform better (Desbrow et al., 2014; Lee and Lim, 2019). Athletes in training should consume 70% of total calories as carbohydrate. Athletes on low-energy diets (<2200 kcal/day) should have a diet of <25% fat, and athletes with high energy needs should consume 30% fat in their diet (Economos et al., 1993). For bodybuilding and resistance training, caloric intake should be set at a level that results in body weight losses of approximately 0.5%–1%/week to maximize muscle retention. Within this caloric intake, most but not all athletes will respond best to consuming 2.3–3.1 g/kg of lean body mass per day of protein, 15%–30% of calories from fat, and the reminder of calories from carbohydrates (Helms et al., 2014). For high-level athletes of endurance-strength sports, the protein intake should be 1.2–2.1 g/kg/day, whereas the requirements for carbohydrate and fat intakes should be >5 g/kg/day and 20%–35% of energy, respectively (Nunes et al., 2018). Vitamin and mineral supplements are not needed if adequate energy to maintain body weight is consumed from a variety of foods (Heffernan et al., 2019). However, athletes who restrict energy intake, use severe weight-loss practices, eliminate one or more food groups from their diet, or consume unbalanced diets with low micronutrient density, may require supplements. Indeed, previous studies have shown that athletes' diets are often not properly balanced, and can thus negatively affect sporting performance (Wierniuk and Włodarek, 2013; Nunes et al., 2018). Exhaustive or unaccustomed intense exercise can cause exercise-induced muscle damage (EIMD) and its undesirable consequences may decrease the ability to exercise and adhere to a training program. In this case, the combined intake of carbohydrates and proteins and the use of antioxidants and/or antiinflammatory nutrients within physiological ranges are interventions that may assist the recovery process. Again, however, although the works studying food instead of nutritional supplements are very scarce, their results seem to indicate that food might be a favorable option as a recovery strategy (Sousa et al., 2014). Nutraceuticals are foods or components of food that provide medical or health benefits including the prevention and/or treatment of a disease (the term was coined by combining "nutrition" and "pharmaceutical") (Chauhan et al., 2013). As for sports activities, enhancing aerobic and/or anaerobic performance is the principal criterion for effect (rehabilitation and recovery after muscle injury is not the theme of this chapter). Far too often, the decision to purchase and consume protein or other supplements is based on marketing claims rather than available evidence-based research. It seems that the problem of scarce and inconsistent reports on the effects of nutraceuticals lies either in the sphere of inadequate testing models or quasiscientific explanation of experimental data obtained.

Proteins, lipids, and carbohydrates: what is the main fuel during exercise?

Along with the common view that nutrition plays a key role in allowing numerous training hours to be translated into useful adaptive responses of various tissues in the individual athlete, there are original data and analytical reports on the effects of different manipulations of carbohydrate, protein, and other supplements on exercise performance. For example, a comprehensive analysis was undertaken (McLellan et al., 2014) that tested the hypothesis that protein supplements, when combined with carbohydrate, directly enhance endurance performance by sparing muscle glycogen during exercise and increasing the rate of glycogen restoration during recovery (Betts and Williams, 2010). It was concluded that protein supplements provided no ergogenic effect, regardless of the performance metric used. In addition, the data suggested that recovery of muscle glycogen stores, together with subsequent rate of utilization during exercise, is not related to the potential ergogenic effect of protein supplements. The regulatory role of muscle glycogen availability on power and endurance training capacity and performance has been emphasized in many publications (Hawley, 2013). On the other hand, carbohydrate restriction and low glycogen exercise can improve training-induced adaptation of muscle oxidative capacity, by increasing mRNA of peroxisome proliferator-activated receptor-γ coactivator-1 (PGC-1α), cytochrome c oxidase subunit I, and pyruvate dehydrogenase kinase (PDK) isozyme 4 (Psilander et al., 2013). Moreover, protein supplementation in conjunction with carbohydrate restriction further enhances anabolic skeletal muscle intracellular signaling and mitochondrial protein synthesis, which is primarily regulated by 5′AMP-activated protein kinase (AMPK) and p38 mitogen-activated protein kinase (p38 MAPK) (Margolis and Pasiakos, 2013).

Training with limited carbohydrate availability can stimulate adaptations in muscle cells to facilitate energy production via fat oxidation. Citrate synthase and β-hydroxyacyl coenzyme A dehydrogenase activity were significantly upregulated in the fasted group; moreover, only the fasted volunteers avoided the development of an exercise-induced drop in blood glucose concentration (Van Proeyen et al., 2011). Along with PGC-1α, the transcription factor/nuclear receptor PPAR-∂ is involved, binding to the carnitine palmitoyltransferase 1 (CPT1) promoter and initiating AMPK translocation in skeletal muscle in response to exercise (Philp et al., 2013). The training-induced increase in fat oxidation is due primarily to increased oxidation of nonplasma-derived fatty acids, perhaps from intramuscular triacylglycerol stores (Horowitz and Klein, 2000; Van Proeyen et al., 2011). The endurance capacity is directly correlated with fatty acid β-oxidation activity and the level of muscle peroxisome proliferator-activated receptor (PPAR) and medium-chain acyl-CoA dehydrogenase, and inversely correlated with the respiratory exchange ratio and plasma lactate levels measured immediately after exercise, suggesting that carbohydrate breakdown is suppressed by enhanced fat utilization during exercise (Haramizu et al., 2009).

PGC-1α in skeletal muscle also drives the expression of lactate dehydrogenase (LDH) B and reduces the expression of LDH A; in addition, PGC-1α enhances the transcription of monocarboxylate transporter 1 (MCT1) gene that drives lactate import into skeletal muscle (Summermatter et al., 2013). PGC-1α thereby promotes a lactate oxidizing phenotype, which is associated with improved endurance capacity and metabolic health. Fat oxidation is lower in high-intensity exercise than in moderate-intensity exercise, in part because of decreased fatty acid delivery to exercising muscles (Horowitz and Klein, 2000). Adrenaline plays a critical role in exercise-stimulated AMPKα1 and α2 activities, and in ACC (acetyl-CoA carboxylase) Ser79 phosphorylation, thus regulating lipolysis (Koh et al., 2007).

Amino acids and their derivatives

Amino acids and their derivatives are widely propagated as being effective aids to enhance physical and mental performance in many ways, even though studies have pointed out that only those individuals who are deficient are more likely to benefit from this dietary supplementation (Luckose et al., 2015). Nutritional supplements such as L-glutamine, L-arginine, and branched chain amino acids (BCAAs) are often recommended to reduce immunosuppression and excessive inflammation in sports activities (Cruzat et al., 2014). In the plasma, BCAAs compete with tryptophan (Trp) for transport into the brain, but during prolonged exercise BCAAs are used by the muscles to produce energy; also, free fatty acids (the concentration of which increases during exercise) displace albumin-bound Trp, causing an increase in free Trp levels (Boyas and Guével, 2011). After an increase in the brain's uptake of Trp, the brain's neurons synthesize serotonin, then the synaptic effects of serotoninergic neurons augment and induce an increase in the sensation of fatigue (The et al., 2014). Studies on the dietary supply of Trp and BCAA demonstrated contradictory results concerning the influence of these nutrients on the ability to sustain exercise (Van Hall et al., 1995). In contrast, a supply of carbohydrates limits the activity of Trp and prolongs cycling exercise at maximal aerobic power (Davis et al., 1992). Diets containing BCAA without Trp but with antioxidants (cashew apple juice) are reported to have potential effects on fat utilization and enhance endurance performance (Prasertsri et al., 2013).

Glutamine is a "conditionally essential amino acid" widely used in sports nutrition, especially because of its immunomodulatory role, but also due to several other biological functions, such as cell proliferation, energy production, glycogenesis, ammonia buffering, and maintenance of the acid–base balance (Coqueiro et al., 2019). Most of the studies evaluated observed that glutamine supplementation improved some fatigue markers, such as increased glycogen synthesis and reduced ammonia accumulation, but this intervention did not increase physical performance. Thus, despite improving some fatigue parameters, glutamine supplementation seems to have limited effects on performance (Coqueiro et al., 2019).

Administration of tyrosine, a precursor of catecholamines, has also not been able to improve exercise performance as it was claimed, although it has been proved to be very effective in fighting stress, and improving mood and cognitive performance, particularly in sleep-deprived subjects (Luckose et al., 2015).

Arginine (Arg) is the main precursor of nitric oxide (NO) via nitric oxide synthase (NOS) activity. Additionally, L-citrulline has been indicated to be a second NO donor in the NOS-dependent pathway, since it can be converted to Arg. The importance of L-citrulline as an ergogenic support derives from the fact that L-citrulline is not subject to presystemic elimination and, consequently, could be a more efficient way to elevate extracellular levels of Arg by itself (Sureda and Pons, 2012). L-Citrulline malate can develop beneficial effects on the elimination of NH_3 in the course of recovery from exhaustive muscular exercise and also as an effective precursor of Arg and creatine. However, dietary supplementation with L-citrulline alone does not improve exercise performance. The ergogenic response of L-citrulline or Arg supplements depends on the training status of the subjects: NO donors could improve tolerance to aerobic and anaerobic exercise if administered to untrained or moderately healthy subjects. However, when highly trained subjects were supplemented no positive effect on performance was indicated (Sureda and Pons, 2012).

N-acetylcysteine (NAC), the antioxidant derivative of L-cysteine, contains a thiol group that can interact with ROS. It easily enters cells and supports resynthesis of glutathione. There is plenty of evidence for beneficial effects of NAC on fatigue in humans during submaximal efforts (Matuszczak et al., 2005; Corn and Barstow, 2011). NAC treatment improved force capacity and fatigue properties in ex vivo extensor digitorum longus (EDL) and soleus muscles from rats submitted to an acute bout of aerobic exercise (Jannig et al., 2017), whereas in another model of peripheral arterial insufficiency, treatment with NAC improved fatigue resistance just in the soleus but not the EDL muscle (Roseguini et al., 2015). It was suggested that NAC and perhaps some other antioxidant treatments could potentially enhance fatigue resistance by counteracting any reactive oxygen species (ROS)-induced decrease in myofibrillar Ca^{2+} sensitivity (Hernández et al., 2012).

Betaine is a methyl derivative of glycine first isolated from sugar beet. Betaine consumed from food sources and through dietary supplements presents similar bioavailability and is metabolized to di-methylglycine and sarcosine in the liver. The ergogenic and clinical effects of betaine have been investigated with doses ranging from 500 to 9000 mg/day (Cholewa et al., 2014). Some studies using animal models and human subjects suggest that betaine supplementation could promote adiposity reductions and/or lean mass gains. Moreover, previous investigations report positive effects of betaine on sports performance in both endurance- and resistance-type exercise, despite some conflicting results. The mechanisms underlying these effects are poorly understood, but could involve: the stimulation of lipolysis and inhibition of lipogenesis via gene expression and subsequent activity of lipolytic-/lipogenic-related proteins; stimulation of autocrine/endocrine IGF-1 release and insulin receptor signaling pathways; stimulation of growth hormone secretion; increased creatine synthesis; increases in protein synthesis via intracellular hyperhydration; as well as exerting psychological effects such as attenuating sensations of fatigue (Cholewa et al., 2014).

Creatine is one of the most popular nutritional supplements, accounting for $400 million in sales annually (Breuner, 2014). Creatine is not an essential nutrient and is synthesized mainly in the liver and kidney from L-arginine, glycine, and L-methionine, and to a lesser extent in the pancreas. It is transported through the blood and taken up by tissues with high energy demand, such as the brain and skeletal muscle, through an active transport system (Brosnan et al., 2011). Approximately 95% of the human body's total creatine is located in skeletal muscle. The concentration of adenosine triphosphate (ATP) in skeletal muscle is usually 2–5 mM, which would result in a muscle contraction of only a few seconds. When energy demands are increased, the phosphagen (or ATP/PCr) system rapidly resynthesizes ATP from ADP with the use of phosphocreatine (PCr) through a reverse Lohmann reaction with the enzyme creatine kinase (CK). In skeletal muscle, PCr concentrations may reach 20–35 mM and even higher. In most muscles the ATP regeneration capacity of CK is very high and is not a limiting factor. Creatine has the ability to increase muscle stores of PCr, potentially increasing the muscle's ability to resynthesize ATP from ADP. The hypertrophic potential following creatine administration has been mostly linked to fluid retention in myofibers due to swelling-induced osmotic potential of high intracellular creatine (Deldicque et al., 2008). Most studies have focused on the effects of creatine monohydrate on performance. Creatine has been shown to improve performance in brief high-intensity exercises, and to increase strength and muscle morphology with heavy resistance training (Breuner, 2014). Smaller benefits are found in endurance sports of longer duration. For example, creatine supplementation attenuated the changes of inflammatory and muscle soreness markers—creatine kinase (CK), lactate dehydrogenase (LDH), prostaglandin E_2 (PGE_2) and tumor necrosis factor-α (TNF-α)—after running 30 km, where the athletes finished the race in a time equivalent to their personal best (Santos et al., 2004). Moreover, creatine-related adverse effects have been reported (Cooper et al., 2012), which include weight gain, water retention, gastrointestinal cramping, fatigue, and diarrhea.

Carnitine is a quaternary ammonium compound synthesized in the kidneys, liver, and testes from the amino acids lysine and methionine (Steiber et al., 2004). It allows the transport of long-chain fatty acids from the cytoplasm to mitochondria where they undergo β-oxidation. Carnitine is widely available as a nutritional supplement, though very scarce scientific data exist on its positive role as an exercise performance enhancer. For example, it was shown that muscle carnitine content can be increased in humans by dietary means and that carnitine plays a dual role in skeletal muscle fuel metabolism that is exercise intensity dependent (Wall et al., 2011). Specifically, increasing muscle total carnitine content reduced muscle carbohydrate use during low-intensity exercise, consistent with an increase in muscle lipid utilization. During high-intensity exercise, muscle carnitine loading results in a better matching of glycolytic, pyruvate dehydrogenase complex, and mitochondrial flux, thereby reducing muscle anaerobic energy generation. The carnitine group increased work output 11% from baseline in the performance trial, while the control group showed no change (Wall et al., 2011). In another study

(Orer and Guzel, 2014), 3 or 4 g of L-carnitine taken before physical exercise prolonged the endurance performance of footballers.

Propionyl-L-carnitine (PLC), a propionyl ester of L-carnitine, exhibits high affinity for the enzyme carnitine acetyltransferase (CAT) and, thus, is readily converted into propionyl-coenzyme A and free carnitine. PLC might reduce hydroxyl radical production via the Fenton system, by chelating the iron required for the generation of hydroxyl radicals (Vanella et al., 2000). Unfortunately, no effect on exercise performance except for patients with claudication has been shown (Brass et al., 2013). Glycine propionyl-L-carnitine (GPLC), along with carnitine supplementation, has also been suggested to increase levels of NO, although the physiological mechanisms remain to be elucidated, and again, no positive effect on performance was indicated (Bescós et al., 2012).

β-Alanine is a component of the naturally occurring peptides carnosine (β-alanyl-L-histidine) and anserine (β-alanyl-N-methylhistidine) and also of pantothenic acid (vitamin B5), which itself is a component of coenzyme A. It is formed in vivo by the degradation of dihydrouracil and carnosine, and is not used in the biosynthesis of any major proteins or enzymes. The ingestion of β-alanine, the rate-limiting precursor of carnosine, has been shown to elevate the muscle carnosine content and significantly attenuated fatigue in repeated bouts of exhaustive dynamic contractions, whereas the increase in muscle carnosine did not improve isometric endurance or 400-m race time (Derave et al., 2007). β-Alanine, taken separately or in combinations with creatine monohydrate, caffeine, and bicarbonate, was stated to render beneficial effects for high-intensity interval training (Helms et al., 2014; Sahlin, 2014). β-Hydroxy β-methylbutyric acid (HMB) is a metabolite of leucine, and has been used to increase muscle mass and decrease muscle breakdown in resistance-exercise training (Nissen et al., 1996). HMB was shown to increase sarcolemmal integrity via conversion to HMG-CoA, enhance protein synthesis via the mechanistic (or mammalian) target of rapamycin (mTOR) pathway, and slow down protein degradation through inhibition of the ubiquitin pathway (Wilson et al., 2008).

Taurine (2-aminoethanesulfonic acid, Tau) is a derivative of cysteine, and is the most abundant of the low-molecular-weight organic constituents; a 70-kg human contains up to 70 g of Tau (Huxtable, 1992). Tau has many fundamental physiological roles, such as conjugation of bile acids, antioxidation, osmoregulation, radioprotection, xenobiotic conjugation, modulation of calcium signaling, and energy storage and production. For example, in experiments with male Sprague-Dawley rats, raising Tau content above endogenous levels increased twitch and subtetanic and specific force in rat fast-twitch skeletal muscle, protected muscle function during high-frequency in vitro stimulation and the ensuing recovery period, and reduced oxidative stress during prolonged stimulation (Goodman et al., 2009). However, in another experiment with isolated mouse soleus muscle, Tau failed to elicit any significant difference in the measured parameters (acute muscle power output, time to fatigue, and recovery from fatigue) compared with the untreated controls (Tallis et al., 2014).

Polyunsaturated fatty acids

Omega-3 polyunsaturated fatty acids (ω3 PUFAs) are among the most studied nutrients (D'Antona et al., 2014). α-Linolenic acid (ALA—C18:3, ω3), which is not synthesized in the human body and therefore must be consumed with the diet, is the precursor of the two most important bioactive long-chain polyunsaturated fatty acids (ω3 LC-PUFA): eicosapentaenoic acid (EPA—C20:5 ω3) and docosahexaenoic acid (DHA—C22:6 ω3). In the endoplasmic reticulum, ALA is converted in EPA and DHA through the development of enzymatic elongation and desaturation reactions, in which Δ-6 desaturase and elongase and Δ-5 desaturase enzymes are involved. In peroxisomes, these reactions are followed by β-oxidation to produce DHA. ω3 LC-PUFAs are considered "conditionally essential" because they are not synthesized in sufficient amounts to meet human needs. This condition occurs when the dietary intake of ALA is too low in comparison with linoleic acid (LA, C18:2-ω6) and the ratio ω6/ω3 is higher than 5/1. ω3 PUFAs serve as precursors to prostaglandins such as prostaglandin E3, which are powerful hormone-like substances that reduce inflammation (Calder, 2006). They also inhibit arachidonic acid-derived 2-series prostaglandins and the 4-series leucotrienes, which are known to modulate the production of proinflammatory and immunoregulatory cytokines. Although a number of studies have assessed the efficacy of ω3 PUFAs supplementation on red blood cell (RBC) deformability, muscle damage, inflammation, and metabolism during exercise, only a few studies have evaluated the impact of ω3 PUFAs supplementation on exercise performance. For example, ω3 PUFA supplementation enhanced exercise performance in endurance-trained athletes (cyclists) (Zebrowska et al., 2014), and it was suggested that NO release may play a central role in cardiovascular adaptive mechanisms. Nevertheless, there is no consistent scientific support for the hypothesis that ω3 PUFA supplementation is effective and ergogenic. The data are inconclusive whether ω3 PUFA supplementation effectively attenuates the inflammatory and immunomodulatory response to exercise (Shei et al., 2014), though recent human studies demonstrate that they can influence not only the exercise and metabolic response of skeletal muscle, but also the functional response for a period of exercise training (Gammone et al., 2018).

Vitamins C and E

The so-called antioxidants are perhaps the most ambiguous and speculative group of dietary supplements. Here we discuss only vitamins C and E, whereas glycosides and flavonoids which sometimes are regarded as antioxidants are discussed below in a separate section. Vitamins C and E are used by many people, especially athletes. The users often believe that high dosages of supplements improve health (resistance to illness and disease) and physical performance. These assumptions are, however, generally not supported in the scientific literature. Antioxidant supplementation has recently been demonstrated to be a double-edged sword, because small to moderate doses of exogenous antioxidants are essential or beneficial, while high doses may have adverse effects and may impair exercise training adaptations (Gleeson, 2013). The adverse effects can be manifested as the antioxidants may shut down some redox-sensitive signaling in the exercised muscle fibers. High dosages of vitamins C (1000 mg/day) and E (235 mg/day) have been shown to have negative effects on adaptation to resistance exercise and training in young volunteers, but positive effects in older men (Paulsen et al., 2014). In accordance with this, vitamin C (1000 mg/day) exerted a harmful effect on the mean running speed of trained female runners (Braakhuis et al., 2014). Following vitamin C intake, certain oxidative markers were elevated: catalase at rest, protein carbonyls at rest, and superoxide dismutase post-exercise.

Some studies have reported that supplementation with vitamins C and E, other antioxidants, or antioxidant mixtures can reduce symptoms or indicators of oxidative stress as a result of exercise. However, these supplements appear to have no beneficial effect on performance. Exercise training seems to reduce the oxidative stress of exercise, such that trained athletes show less evidence of lipid peroxidation for a given bout of exercise and an enhanced defense system in relation to untrained subjects after the same bout of exercise. Whether the body's natural antioxidant defense system is sufficient to counteract the increase in ROS with exercise or whether additional exogenous supplements are needed is not known, although trained athletes who received antioxidant supplements showed evidence of reduced oxidative stress (Clarkson and Thompson, 2000).

Ubiquinone-10

Ubiquinone-10 (Q10) is a lipid-soluble antioxidant found in high concentrations in meat and fish (Martini, 2019). Despite some evidence that indicated intramuscular Q10 content had a positive relationship with exercise capacity, this was more likely a function of decreased fatigability based on the muscle properties (e.g., oxidative capacity) and not Q10 content (Hernández et al., 2012). Whereas supplementation with Q10 may assist individuals with mitochondrial disease, most investigations on healthy individuals show either no or even a deleterious effect on exercise performance (Malm et al., 1997). Therefore there is insufficient evidence to support a role for Q10 as an antioxidant having an ergogenic effect, even in combinations with vitamins C and E (Nielsen et al., 1999).

Caffeine and other alkaloids

Many beverages, foods, and supplements consumed by athletes as ergogenic aids or health enhancers contain alkaloids, that is, naturally occurring chemical compounds that contain mostly basic nitrogen atoms. There are four representative alkaloids or alkaloid sources: caffeine (including pure and plant sources, such as guarana, mate, and tea); theophylline, including its isomers theobromine and paraxanthine (from cocoa); ephedrine and pseudoephedrine (from ephedra); and gingerols, shogaols and paradols (from ginger) (Senchina et al., 2014). Caffeine functions through inhibition of phosphodiesterase (PDE) and stimulation of adenosine receptors, leading to accumulation of intracellular 3,5-cyclic-adenosine monophosphate (cAMP), which is metabolically excitatory for cells (Vaughan et al., 2014). Caffeine produces ergogenic effects at a dose as low as 200–250 mg (2.0–3.5 mg/kg), and may be effective for sports involving short bursts of activity, such as tennis and team sports, combat contesting, and single and multiple sprint sports (Glaister et al., 2008; Spriet, 2014). The ergogenic effect of low caffeine doses appears to result from alterations in the central nervous system (Spriet, 2014). Abstaining from caffeine for at least 7 days before use will give the greatest chance of optimizing the ergogenic effect (Ganio et al., 2009).

Prospective alkaloid sources (Goji berry, Noni berry, and bloodroot) were recently under investigation for the purposes of sports activities, though without definite conclusions on their effectiveness (Senchina et al., 2014; Sellami et al., 2018).

Glycosides, saponins, and flavonoids

Glycoside is a compound formed from a simple sugar and another compound by replacement of a hydroxyl group in the sugar molecule. Glycosides found in plants include some pharmacologically important products. Saponins are amphipathic glycosides grouped phenomenologically by the soap-like foaming they produce when shaken in aqueous solutions, and structurally by having one or more hydrophilic glycoside moieties combined with a lipophilic triterpene derivative. Flavonoids are the largest group of plant polyphenols. Chemically, they have the general structure of a 15-carbon skeleton, which consists of two phenyl rings (A and B) and a heterocyclic ring (C) (McNaught and Wilkinson, 1997). We have compiled these different groups in one section because of their frequent occurrence in one plant or its extract. Of course, caffeine and alkaloids can also be found in these plants (and tea is the best example), but we tried to select those experimental data in which caffeine content was controlled or minimized.

Herbal stimulators or adaptogens are derived as extracts from many plants, for example, *Panax ginseng* (Bahrke et al., 2009), *Astragalus membranaceus* (Yeh et al., 2014), *Rhaponticum carthamoides* (Wilborn et al., 2006), *Ginkgo biloba* (Bing and Zhaobao, 2010), *Tribulus terrestris* (Pokrywka et al., 2014), *Polygonatum alte-lobatum* (Horng et al., 2014), *Camellia sinensis* (Hodgson et al., 2013), and some others. Most herbal supplement effects are likely due to activation of the central nervous system via stimulation of catecholamines (Sellami et al., 2018). Ginseng is one of the most popular herbal supplements in the world. Athletes use ginseng for its alleged performance-enhancing attributes. However, many studies examining the pharmacological effects of ginseng on physical performance have not employed sound scientific design and methodology (Bahrke et al., 2009). As a component of proprietary blends, ginseng saponins increased physical performance and delayed fatigue. For example, in exhaustive cycling test of humans, time to exhaustion was significantly greater, the plasma lactate concentrations were significantly lower, and nonesterified fatty acid levels were significantly higher than those in the placebo treatment (Yeh et al., 2011). Recently, antifatigue properties of an American ginseng (*Panax quinquefolium*) were demonstrated using a forced swimming test (Qi et al., 2014): the swimming time was found to be longer, blood lactate and serum urea nitrogen levels decreased, and hepatic glycogen level increased in mice treated with American ginseng extract for 28 days, as compared to the control group. Additionally, the American ginseng lowered malonic dialdehyde (MDA) content and increased the levels of glutathione peroxidase (GPX) and superoxide dismutase (SOD).

Saponins, originated from leuzea or maral root (*Rhaponticum carthamoides*) were first declared to be a new class of anabolic substances in the Soviet Union (Syrov and Kurmukov, 1976), as this plant inhabits the subalpine zone of southern Siberia, Kazakhstan, the Altay region, and western Sayani. The tested preparations taken orally diminished fat content and elevated muscle mass under conditions of daily aerobic-anaerobic training, and elevated the magnitude of "total work" estimated per 1 kg of body weight (Gadzhieva et al., 1995). Phytoecdysteroids are structural analogs of the insect molting hormone ecdysone, they have a number of proven beneficial effects on mammals but the hormonal effects of ecdysteroids have been proven only in arthropods. Ecdysteroids were believed not to bind to the cytosolic steroid receptors, instead, they were likely to influence signal transduction pathways via membrane-bound receptors (Báthori et al., 2008). However, recently the phytoectysteroid ecdysterone was reported to stimulate protein synthesis and enhance physical performance through estrogen receptor beta (ERβ) (Parr et al., 2014). It is worth noting in connection to this a representative study with 45 resistance-trained males, which indicated that ecdysterone (20-hydroxyecdysone, 20E) supplementation did not affect body composition or training adaptations, nor did it influence the anabolic/catabolic hormone status or general markers of catabolism (Taylor et al., 2006).

Astragalus membranaceus is a popular "Qi-tonifying" herb with a long history of use as a traditional Chinese medicine with multiple biological functions. In experiments with male ICR (Imprinting Control Region) strain mice, exercise training combined with *Astragalus* supplementation increased endurance exercise capacity and increased hepatic and muscle glycogen content, as well as reduced exercise-induced accumulation of the byproducts blood lactate and ammonia with acute exercise challenge (Yeh et al., 2014).

An extract of an exotic plant *Tribulus terrestris* is a component of several supplements that are available over-the-counter and widely recommended, generally as enhancers of human vitality. This plant extract is touted as a testosterone booster and remedy for impaired erectile function; therefore it is targeted at physically active men, including male athletes. Nevertheless, a *Tribulus terrestris* extract used alone without additional components does not improve androgenic status or physical performance among athletes (Pokrywka et al., 2014).

Polygonatum alte-lobatum Hayata, a rhizomatous perennial herb with high levels of polyphenols, flavonoids, and especially polysaccharides, belongs to the Liliaceae family and is endemic to Taiwan. In experiments with Sprague-Dawley rats, extracts of *Polygonatum alte-lobatum* dose-dependently increased endurance running time to exhaustion, SOD activity, and total antioxidant ability of blood, while they dose-dependently decreased serum urea nitrogen and MDA levels after exercise. Hepatic glycogen content, an important energy source for exercise, was significantly increased with the treatment (Horng et al., 2014).

Ginkgo biloba extract can increase the body's endurance exercise capacity in mice and delay fatigue (Bing and Zhaobao, 2010). The swimming time to exhaustion was significantly prolonged in the exercised drug-treated group as compared with the exercised control group. In addition, *Ginkgo biloba* extract can help to increase the activity of the antioxidant enzymes in liver tissue, reduce the lipid peroxidation injury in liver tissue caused by free radicals, improve athletic ability, and promote the recovery process after exercise in mice. The combined herbal supplement of *Rhodiola* and *Gingko* could also improve the endurance performance by increasing oxygen consumption and protecting against fatigue in healthy male volunteers (Zhang et al., 2009). However, inconsistent results were obtained with *Ginkgo biloba* extract in studies on the incidence and severity of acute mountain sickness following rapid ascent of human volunteers to high altitude (Leadbetter et al., 2009). It was concluded that the source and composition of *Ginkgo biloba* products may determine their effectiveness for prophylaxis of acute mountain sickness.

Rhodiola crenulata and *Cordyceps sinensis* are commonly used herbs that promote health in traditional Chinese medicine. Despite these two herbs exhibiting antiinflammation and antioxidant functions, their combined supplementation (20 mg/kg/day) faintly enhanced endurance training-induced positive adaptations in body composition in young sedentary individuals, whereas the blood lipid profile and systemic oxidative stress states were not altered after such intervention (Liao et al., 2019).

Green tea (*Camellia sinensis*) extract (GTE) and its decaffeinated preparations are widely used as antioxidants and fat burners, and their antidiabetic, anticarcinogenic, and other properties are also under investigation. Pleiotropic effects of GTE consist of stimulation of lipid metabolism and modulation of gluconeogenesis, influence upon respiration and antioxidative effects, cytoskeletal rearrangements, signaling, reparation of DNA and processing of mRNA, iron deposition and chaperon expression; the effects of GTE are associated with AMP-activated protein kinase (AMPK) and increased levels of ROS (Hwang et al., 2009). Short-term effects of GTE are implemented through activation of, inter alia, catechol-*O*-methyl transferase (COMT), which inactivates catecholamines (dopamine, epinephrine, norepinephrine). Long-term effects include intensification of fat metabolism in skeletal muscles and impairment of hepatic adipogenesis via PGC-1α and different isoforms of PPAR, key regulators of energetic metabolism (Hodgson et al., 2013). Activation of PPARα is associated with adaptive thermogenesis and is observed under the influence of low temperatures contributing to the formation of slow-twitch oxidative fibers (Liang and Ward, 2006). Activating the PPARβ/δ, GTE prevents expression of inducible NO-synthase in cardiomyocytes (Danesi et al., 2009). Activation of PPARγ in endothelial cells enhances their viability through modulation of the expression of serine/threonine kinases (Liu et al., 2013). GTE elevates total antioxidative status and the level of catalase, simultaneously reducing the MDA level and expression of angiotensin II receptor in renal and hepatic tissues (Thomson et al., 2012). Not only epigallocatechin gallate, the major polyphenol of GTE, but also other catechins and/or components of GTE exert physiological effects supplementing or amplifying the final effects: these are modulation of expression of the gluconeogenesis enzymes in hepatic tissue (glucose-6-phosphatase and phosphoenolpyruvate carboxykinase), and reduced expression of the liponeogenetic enzymes (fatty acid synthase, hydroxymethylglutaryl-CoA reductase, acetyl-CoA carboxylase α). It is associated with reduced expression of Srebf1 and/or Srebf2 (sterol regulatory element-binding transcription factors), which bind to the promoter and suppress transcription of the above-mentioned enzymes (Yasui et al., 2012). GTE regulates the level of testosterone due to influence upon UDP-glucuronosyltransferase (Jenkinson et al., 2013), and inhibits glycerol-3-phosphate dehydrogenase (Kao et al., 2010). Positive effects of GTE were demonstrated in chronic fatigue syndrome (Sachdeva et al., 2010) and Duchenne muscular dystrophy (Dorchies et al., 2009).

The influence of GTE on endurance capacity and metabolism was investigated in various models of swimming or running workloads (Call et al., 2008; Novozhilov et al., 2013, 2015). Ambiguous and inconsistent data were obtained, obviously dependent upon the manufacturing company, features of experimental model, regimen, and duration of GTE consumption. In our experiments, we compared decaffeinated GTE of several companies and found a decaffeinated GTE of the Sunphenon series to be the most effective. In a model of exhaustive running of rats, the GTE resulted in preventing oxidative stress in RBC, decreasing acid hemolysis of RBC, and reducing reticulocytosis (Novozhilov et al., 2013). In a forced swimming test up to exhaustion, GTE administered to rats twice a day for 2 weeks provided a statistically significant increase in the swimming duration after the first and more so after the second week, as compared with control animals that consumed water (Novozhilov et al., 2015). It was also established that GTE increased endurance due to involvement of slow-twitch muscles whose adaptation is associated with enhanced expression of the genes responsible for the regulation of Ca^{2+} balance (Korf et al., 2017).

Water and acid–base balance

In summing up conclusions on the causes of peripheral fatigue, the following intracellular changes emerge: (1) accumulation of lactate and hydrogen ions, which is partly buffered by bicarbonate; (2) accumulation of phosphate and ammonia; (3) accumulation of heat, which induces greater sweat secretion. The water loss associated with this last phenomenon could lead to dehydration, and adversely affect aerobic exercise, potentially more so at higher exercise intensities in warm environments (Hackney et al., 2012). Thermal hypohydration redistributes water largely from the intracellular (41%) and extracellular (59%) spaces of muscle (40%) and skin (30%), but neither the brain nor liver loses significant quantities of water (Nose et al., 1983). Because dehydration limits the appropriate delivery of oxygen and substrates to the working muscle, one can readily infer that water and sodium replacement during prolonged exercise in the heat may be critically important to maintaining fluid and electrolyte balance and muscle contractility (Anastasiou et al., 2009). Nevertheless, sodium supplements had no effect on time-trial performance of well-trained cyclists during exercise of approximately 3 h duration (Cosgrove and Black, 2013). This is not surprising since dehydration during exercising and environmental heat stress reduces water volume more than electrolyte content, causing hyperosmotic hypohydration (Cotter et al., 2014).

In the prominent work of Carr et al. (2011), a meta-analysis was presented on the effects of acute ingestion of three agents—sodium bicarbonate (SB), sodium citrate (SC), and ammonium chloride (AC)—on performance and related physiological variables (blood bicarbonate, pH, and lactate). It was found that SB produced a possibly moderate performance enhancement of 1.7% (± 2.0%) with a typical dose of ~0.3 g/kg in a single 1-min sprint; SC with a typical dose of ~0.5 g/kg had an unclear effect on performance of 0.0% (± 1.3%); and finally, AC produced a moderate performance impairment of 1.6% (± 1.9%) with a typical dose of ~0.3 g/kg. SB is a buffering agent that is suggested to improve performance by promoting the efflux of lactate and hydrogen ions from working cells and tissues. Research surrounding its efficacy as an ergogenic aid is conflicting, making it difficult to draw conclusions as to its effectiveness for training and competition (Hadzic et al., 2019). In general, the overall effect size for the influence of SB on performance is significantly lower for specifically trained as opposed to recreationally trained participants (Zabala et al., 2011; Peart et al., 2012). Gastrointestinal side effects are the main problem with SB use in sports, and to avoid these new SB loading regimens were proposed, either by decreasing the dose to 100 mg/kg (Durkalec-Michalski et al., 2018), or by administering gelatine capsules with a higher dose of SB (600 mg/kg) over 19.5 h (Marcus et al., 2019). In both cases gastrointestinal symptoms were eliminated, but in the former case the doses could have been too small to elicit additional improvements in anaerobic power, while in the latter case there was no improving effectiveness described in the paper.

Prior studies have found that the administration of high levels of AC (0.3 g/kg) to induce metabolic acidosis has a negative effect on endurance (Hollidge-Horvat et al., 1999; Carr et al., 2011). In our research, surprisingly, the administration of low doses of AC (10 mg/kg) 5 min before the swimming exercise during normalized and exhaustive loading can actually enhance endurance; furthermore, an additional benefit was observed when AC was administered in combination with GTE (Goncharov et al., 2018). A mechanism was proposed for optimizing the oxygen transport and shuttle function of erythrocytes by AC under conditions of intense physical activity. A higher efficacy of CA at a dose of 20 mg/kg as compared with a dose of 10 mg/kg was then established (Novozhilov et al., 2020). Morphometric studies of muscle fibers of rats suggested that the effectiveness of preparations was mainly due to adaptive changes in T-tubules and mitochondria of fast-twitch muscles, but it increases with the participation of slow-twitch muscle fibers in the structural and functional adaptation to the exhaustive physical load (Baidyuk et al., 2020).

Some other supplements

Besides water and minerals, enhanced levels of NO are associated with an increase in blood flow to working skeletal muscle, which in turn is associated with greater delivery of oxygen and nutrients. NO synthesis occurs not only via NOS-dependent pathways, but also via NOS-independent when nitrate and nitrite are reduced in vivo to NO in conditions of low oxygen availability, as may occur in skeletal muscle during exercise (Jones et al., 2013). Several studies using NO donors have assessed this hypothesis in healthy, trained participants. However, the conclusions from these studies showed several discrepancies. Studies involving untrained or moderately trained healthy subjects showed that NO donors could improve tolerance to aerobic and anaerobic exercise. However, when highly trained subjects were supplemented, no positive effect on performance was indicated (Bescós et al., 2012).

Iron is an essential micronutrient involved in oxidative metabolism and is critical to exercise performance. Iron status can be assessed using a number of indicators; however clinical cut-offs for active populations remain controversial. Randomized, placebo-controlled supplementation trials of iron-depleted female athletes have shown that oral iron supplementation in doses of 100 mg $FeSO_4$/day (approximately 20 mg elemental iron) improves iron status and may improve measures of physical performance (DellaValle, 2013).

Succinate-containing preparations have been studied in Russia and some are claimed to exert actoprotective and antihypoxic effects and promote physical fitness (Kosinets et al., 2012), although there are convincing data that succinate and other tricarboxylic-acid-cycle intermediates do not improve physical performance (Brown et al., 2004).

Multiingredient preworkout supplements

In recent years, a new class of dietary supplements called multiingredient preworkout supplements (MIPS) has increased in popularity. These supplements are intended to be taken prior to exercise and typically contain a blend of ingredients, the combination of which may elicit a synergistic effect on acute exercise performance and subsequent training adaptations compared to single ingredients alone (Harty et al., 2018). The relative prevalence and average amounts of the top ingredients were: β-alanine (87%; 2.0 ± 0.8 g), caffeine (86%; 254.0 ± 79.5 mg), citrulline (71%; 4.0 ± 2.5 g), tyrosine (63%; 348.0 ± 305.7 mg), taurine (51%; 1.3 ± 0.6 g), and creatine (49%; 2.1 ± 1.0 g) (Jagim et al., 2019a). Nearly half of all ingredients are included as part of a proprietary blend with undisclosed amounts of each ingredient, therefore direct comparisons between formulations of MIPS or between a MIPS and a single ingredient are challenging and often impossible. Nevertheless, a substantial body of evidence suggests that the acute pre-exercise consumption of MIPS may positively influence muscular endurance and subjective mood, though mixed results have been reported regarding the acute effect of MIPS on force and power production. The chronic consumption of MIPS in conjunction with a periodized resistance training program appears to augment beneficial changes in body composition through increased lean mass accretion (Harty et al., 2018). However, the impact of acute and even long-term MIPS supplementation on force production, muscular endurance, aerobic performance, and subjective measures is much less clear (Harty et al., 2018; Lutsch et al., 2019; Blake et al., 2020).

MIPS ingestion appears to be relatively safe, though most studies that have assessed the safety of MIPS are relatively short (less than 8 weeks) and thus more information is needed regarding the safety of long-term supplementation. As with any dietary supplement, the use of MIPS carries implications for the athlete, as many formulations may intentionally contain banned substances as ingredients or unintentionally as contaminants (Harty et al., 2018; Goncharov et al., 2019). Importantly, over half (54%) of 1045 individuals that responded experienced side effects following MIPS use, including skin reactions, heart abnormalities, and nausea (Jagim et al., 2019b). Females were more likely than males to experience these side effects, despite being less likely to consume two or more serving sizes per dose.

Thus, it appears that MIPS have promise as an ergogenic aid for active individuals, though further information is required regarding long-term efficacy and safety in a wider variety of populations.

Concluding remarks and future directions

It is impossible within a chapter to give an exhaustive description of all the nutraceuticals that are produced and represented at the market of sport physiology and medicine. Lots of preparations are advertised as actoprotectors or ergogenic remedies—different vitamins, ω3 PUFA, creatine, glutamine, choline, Zn/Mg-aspartate, curcumin, resveratrol, quercetin, etc.—but scientific evidence is either not available or rather ambiguous to support the use of these supplements for performance enhancement (Mason and Lavallee, 2012; Heaton et al., 2017). Very often conclusions from the studies conducted are deeply hampered by methodological problems including difficulties in statistical power. Limitations in study designs together with the large variability in surrogate markers of muscle damage reduce the strength of the evidence base (Pasiakos et al., 2014). Indeed, even for those investigations where reliability seems to be beyond doubt there is a need for additional evidence from either experimental or clinical laboratories and from sports competitions. To demonstrate measureable gains in strength and performance through exercise training and any supplementation, many of the studies recruited untrained participants. Since skeletal muscle responses to exercise and supplementation differ between trained and untrained individuals, findings are not easily generalized for all consumers (Pasiakos et al., 2015). Future human studies should assess the effectiveness of nutraceuticals on exercise performance and fatigue in multisport athletes and military personnel who typically engage in more than one bout of exercise per day. At the same time, scientists and physicians should together carry the burden of responsibility for the health and life of athletes. If in cases of incurable disease there is often no choice for patients and they lose nothing by taking a new remedy, healthy and qualified and well-paid athletes run the risk of losing all these achievements due to incompetent advice from coaches and/or physicians.

Many nutraceuticals, especially herbal extracts, have multiple molecular and functional targets in humans and animals, and at least two consequences follow from this fact: (1) experimental animal and in vitro studies can hardly be overestimated; (2) studies of nutraceuticals should be based on the multitarget drug approach, according to which focus should be placed on selecting molecular targets, judging their relevance and estimating safety concerns (Koeberle and Werz, 2014; Goncharov et al., 2019). For example, many natural products are agonists of PPARγ, and the bioactivity pattern of the active compounds and plant extracts warrants future research regarding their therapeutic (ergogenic in our case) potential and the possibility to modulate PPARγ activation by dietary interventions or food supplementations (Wang et al., 2014). Also, inappropriate nutritional status and muscle wasting can be caused by gut microbiota, which is an environmental factor involved in energy sparing from the diet, and in the regulation of host immunity and metabolism (Bindels and Delzenne, 2013; Pane et al., 2018). Although athletes may have elevated physiological protein requirements and they may benefit from dietary supplements, the evidence regarding the role of dietary protein and supplements in the nutrition of recreational sportspeople and sedentary populations is somewhat complex and contradictory. In high-protein diets, more undigested protein-derived constituents end up in the large intestine compared to moderate- or low-protein diets, and hence, more bacterial amino acid metabolism takes place in the colon, having both positive and negative systemic and metabolic effects on the host (Kårlund et al., 2019). For example, a protein supplement (whey isolate and beef hydrolysate) had no impact on several important fermentation-derived parameters and metabolites (fecal pH, water content, ammonia, and short-chain fatty acid concentrations); however, it increased the abundance of the Bacteroidetes phylum and decreased the presence of health-related taxa including *Roseburia*, *Blautia*, and *Bifidobacterium longum*, indicating that a long-term protein supplementation may have a negative impact on gut microbiota (Moreno-Pérez et al., 2018). On the other hand, selected beneficial bacteria given with probiotics could positively affect athletes undergoing periods of intense training and may assist in performance recovery (Pane et al., 2018).

Combinations of nutraceuticals are possible, if they exert individual efficiency and keeping in mind the above-mentioned safety concerns. For example, acute pre-exercise coingestion of beetroot juice (providing 8 mmol of nitrate) with caffeine has additive effects on exercise performance compared with either beetroot or caffeine alone (Handzlik and Gleeson, 2013). On the other hand, carbohydrate and caffeine are known to independently improve certain aspects of athletic performance, but the addition of caffeine to carbohydrate did not improve repeated sprint performance with short rest intervals or agility (Lee et al., 2014). Also, a MIPS containing creatine monohydrate, β-alanine, taurine, leucine, and caffeine (according to different studies, all of these ingredients can improve either resistance or endurance performance), combined with 8 days of training, did not significantly improve muscular strength, anaerobic power, or performance (Outlaw et al., 2014).

One possible contributor to discrepancies within the literature regarding efficacy of individual ingredients is experimental variability. Multiple confounds exist in energy expenditure measurements such as duration of measurement, type of analytical and measurement software, duration of measurement after ingestion of stimulant, and the amount of purported metabolic stimulator consumed (Vaughan et al., 2014). Proprietary blends found within dietary supplements have numerous issues that are not encountered when testing pharmaceutical-grade single chemicals (such as caffeine). For example, dietary supplements may vary in content from one batch to the next, suggesting that some batches function better than others. Another interesting and significant variable worth discussing is the matter of funding for research on dietary supplements. Multiple articles disclose that funding is contributed by the companies which market the product for retail purposes (Vaughan et al., 2014). It is not surprising that the results of consuming various compounds and their mixtures advertised to be helpful in reducing fatigue and to enhance physical performance are at least ambiguous. At the same time, there are lots of examples described of side effects and delayed pathologies (e.g., insomnia, irritability, lethargy, acne, gynecomasty, acromegaly, cardiovascular diseases, toxic hepatitis, etc.) emerging in the case of their long-term consumption (Timcheh-Hariri et al., 2012; Goncharov et al., 2019). To finalize, we state two simple but urgent requirements: (1) for approbation of nutraceuticals preventing fatigue and enhancing sports performance, while being actually effective and safe, we need to develop and consistently apply adequate experimental models to elaborate methodologies for diagnostics and monitoring of fatigue and muscle damage; (2) the effectiveness and safety of nutraceuticals must be confirmed in preclinical and clinical experiments according to demands of GLP, GMP, and evidence-based medicine.

Acknowledgments

In memory of Vladimir Morozov, our colleague, tutor, and a great man. This work was supported by the State Program AAAA-A18-118012290142-9.

References

Anastasiou, C.A., Kavouras, S.A., Arnaoutis, G., et al., 2009. Sodium replacement and plasma sodium drop during exercise in the heat when fluid intake matches fluid loss. J. Athl. Train. 44 (2), 117–123.

Bahrke, M.S., Morgan, W.P., Stegner, A., 2009. Is ginseng an ergogenic aid? Int. J. Sport. Nutr. Exerc. Metab. 19 (3), 298–322.

Báthori, M., Tóth, N., Hunyadi, A., et al., 2008. Phytoecdysteroids and anabolic-androgenic steroids structure and effects on humans. Curr. Med. Chem. 15 (1), 75–91.

Baidyuk, E.V., Sobolev, V.E., Korf, E.A., et al., 2020. Ultrastructural characteristics of slow and fast twitch muscle fibers of rats after a forced swimming cycle against a background of introduction of green tea extract and ammonium salts. Cell Tissue Biol. 14 (1), 65–73. Available from: https://doi.org/10.1134/S1990519X20010022.

Bescós, R., Sureda, A., Tur, J.A., et al., 2012. The effect of nitric-oxide-related supplements on human performance. Sports Med. 42 (2), 99–117.

Betts, J.A., Williams, C., 2010. Short-term recovery from prolonged exercise: exploring the potential for protein ingestion to accentuate the benefits of carbohydrate supplements. Sports Med. 40 (11), 941–959.

Bindels, L.B., Delzenne, N.M., 2013. Muscle wasting: the gut microbiota as a new therapeutic target? Int. J. Biochem. Cell Biol. 45 (10), 2186–2190.

Bing, Y., Zhaobao, W., 2010. Effects of Ginkgo biloba extract on free radical metabolism of liver in mice during endurance exercise. Afr. J. Tradit. Complement. Altern. Med. 7 (4), 291–295.

Blake, M.S., Johnson, N.R., Trautman, K.A., et al., 2020. Neither a multi-ingredient pre-workout supplement nor caffeine were effective at improving. Int. J. Exerc. Sci. 13 (2), 167–182.

Boyas, S., Guével, A., 2011. Neuromuscular fatigue in healthy muscle: underlying factors and adaptation mechanisms. Ann. Phys. Rehabil. Med. 54 (2), 88–108.

Braakhuis, A.J., Hopkins, W.G., Lowe, T.E., 2014. Effects of dietary antioxidants on training and performance in female runners. Eur. J. Sport. Sci. 14 (2), 160–168.

Brass, E.P., Koster, D., Hiatt, W.R., et al., 2013. A systematic review and meta-analysis of propionyl-L-carnitine effects on exercise performance in patients with claudication. Vasc. Med. 18 (1), 3–12.

Breuner, C.C., 2014. Performance-enhancing substances. Adolesc. Med. State Art. Rev. 25 (1), 113–125.

Brosnan, J.T., da Silva, R.P., Brosnan, M.E., 2011. The metabolic burden of creatine synthesis. Amino Acids 40 (5), 1325–1331.

Brown, A.C., Macrae, H.S., Turner, N.S., 2004. Tricarboxylic-acid-cycle intermediates and cycle endurance capacity. Int. J. Sport. Nutr. Exerc. Metab. 14 (6), 720–729.

Calder, P.C., 2006. n-3 Polyunsaturated fatty acids, inflammation, and inflammatory diseases. Am. J. Clin. Nutr. 83 (6), 1505S–1519S.

Call, J.A., Voelker, K.A., Wolff, A.V., et al., 2008. Endurance capacity in maturing mdx mice is markedly enhanced by combined voluntary wheel running and green tea extract. J. Appl. Physiol. 105 (3), 923–932.

Carr, A.J., Hopkins, W.G., Gore, C.J., 2011. Effects of acute alkalosis and acidosis on performance: a meta-analysis. Sports Med. 41 (10), 801–814.

Chauhan, B., Kumar, G., Kalam, N., et al., 2013. Current concepts and prospects of herbal nutraceutical: a review. J. Adv. Pharm. Technol. Res. 4 (1), 4–8.

Cholewa, J.M., Guimarães-Ferreira, L., Zanchi, N.E., 2014. Effects of betaine on performance and body composition: a review of recent findings and potential mechanisms. Amino Acids 46 (8), 1785–1793.

Clarkson, P.M., Thompson, H.S., 2000. Antioxidants: what role do they play in physical activity and health? Am. J. Clin. Nutr. 72 (2), 637S–646S.

Cooper, R., Naclerio, F., Allgrove, J., et al., 2012. Creatine supplementation with specific view to exercise/sports performance: an update. J. Int. Soc. Sports Nutr. 9 (1), 33.

Coqueiro, A., Rogero, M.M., Tirapegui, J., 2019. Glutamine as an anti-fatigue amino acid in sports nutrition. Nutrients. 11 (4), 863. Available from: https://doi.org/10.3390/nu11040863. pii: E863.

Corn, S.D., Barstow, T.J., 2011. Effects of oral *N*-acetylcysteine on fatigue, critical power, and W' in exercising humans. Respir. Physiol. Neurobiol. 178 (2), 261–268.

Cosgrove, S.D., Black, K.E., 2013. Sodium supplementation has no effect on endurance performance during a cycling time-trial in cool conditions: a randomized cross-over trial. J. Int. Soc. Sports Nutr. 10, 30.

Cotter, J.D., Thornton, S.N., Lee, J.K., et al., 2014. Are we being drowned in hydration advice? thirsty for more? Extrem. Physiol. Med. 3, 18.

Cruzat, V.F., Krause, M., Newsholme, P., 2014. Amino acid supplementation and impact on immune function in the context of exercise. J. Int. Soc. Sports Nutr. 11 (1), 61.

Danesi, F., Di Nunzio, M., Boschetti, E., et al., 2009. Green tea extract selectively activates peroxisome proliferator-activated receptor beta/delta in cultured cardiomyocytes. Br. J. Nutr. 101 (12), 1736–1739.

D'Antona, G., Nabavi, S.M., Micheletti, P., et al., 2014. Creatine, L-carnitine, and ω3 polyunsaturated fatty acid supplementation from healthy to diseased skeletal muscle. Biomed. Res. Int. 2014, Article ID 613890.

Davis, J.M., Bailey, S.P., Woods, J.A., et al., 1992. Effects of carbohydrate feedings on plasma free tryptophan and branched-chain amino acids during prolonged cycling. Eur. J. Appl. Physiol. Occup. Physiol 65, 513–519.

Deldicque, L., Atherton, P., Patel, R., et al., 2008. Effects of resistance exercise with and without creatine supplementation on gene expression and cell signaling in human skeletal muscle. J. Appl. Physiol. 104 (2), 371–378.

DellaValle, D.M., 2013. Iron supplementation for female athletes: effects on iron status and performance outcomes. Curr. Sports Med. Rep. 12 (4), 234–239.

Derave, W., Ozdemir, M.S., Harris, R.C., et al., 2007. beta-Alanine supplementation augments muscle carnosine content and attenuates fatigue during repeated isokinetic contraction bouts in trained sprinters. J. Appl. Physiol. 103 (5), 1736–1743.

Desbrow, B., McCormack, J., Burke, L.M., et al., 2014. Sports dietitians Australia position statement: sports nutrition for the adolescent athlete. Int. J. Sport. Nutr. Exerc. Metab. 24 (5), 570–584.

Dorchies, O.M., Wagner, S., Buetler, T.M., et al., 2009. Protection of dystrophic muscle cells with polyphenols from green tea correlates with improved glutathione balance and increased expression of 67LR, a receptor for (−)-epigallocatechin gallate. Biofactors 35 (3), 279–294.

Duchateau, J., Hainaut, K., 1993. Behavior of short and long latency reflexes in fatigued human muscles. J. Physiol. 471, 787–799.

Durkalec-Michalski, K., Zawieja, E.E., Podgórski, T., et al., 2018. The effect of a new sodium bicarbonate loading regimen on anaerobic capacity and wrestling performance. Nutrients. 10 (6), 697. Available from: https://doi.org/10.3390/nu10060697. pii: E697.

Economos, C.D., Bortz, S.S., Nelson, M.E., 1993. Nutritional practices of elite athletes: practical recommendations. Sports Med. 16 (6), 381–399.

Finsterer, J., 2012. Biomarkers of peripheral muscle fatigue during exercise. BMC Musculoskelet. Disord. 13, 218.

Gadzhieva, R.M., Portugalov, S.N., Paniushkin, V.V., et al., 1995. A comparative study of the anabolic action of ecdysten, leveton and Prime Plus, preparations of plant origin. Eksp. Klin. Farmakol. 58 (5), 46–48 [Article in Russian].

Gammone, M.A., Riccioni, G., Parrinello, G., et al., 2018. Omega-3 polyunsaturated fatty acids: benefits and endpoints in sport. Nutrients. 11 (1), 46. pii: E46.

Ganio, M.S., Klau, J.F., Casa, D.J., et al., 2009. Effect of caffeine on sport-specific endurance performance: a systematic review. J. Strength. Cond. Res. 23 (1), 315–324.

Glaister, M., Howatson, G., Abraham, C.S., et al., 2008. Caffeine supplementation and multiple sprint running performance. Med. Sci. Sports Exerc. 40 (10), 1835–1840.

Gleeson, M., 2013. Nutritional support to maintain proper immune status during intense training. Nestle Nutr. Inst. Workshop Ser. 75, 85–97.

Goncharov, N.V., Mindukshev, I.V., Novozhilov, A.V., et al., 2018. Adaptive biochemical changes of rat erythrocytes after the cycle of forced swimming with the use of green tea extract and ammonium preconditioning. Ross. Fiziol. Zh Im. I M. Sechenova 104 (12), 1428–1441 [Article in Russian].

Goncharov, N., Sobolev, V., Terpilowski, M., et al., 2019. Safety and toxicity evaluation of nutraceuticals in animal models. In: Gupta, R., Srivastava, A., Lall, R. (Eds.), Nutraceuticals in Veterinary Medicine. Springer Nature, Cham, Switzerland, pp. 675–684.

Goodman, C.A., Horvath, D., Stathis, C., et al., 2009. Taurine supplementation increases skeletal muscle force production and protects muscle function during and after high-frequency in vitro stimulation. J. Appl. Physiol. 107 (1), 144–154.

Hackney, K.J., Cook, S.B., Fairchild, T.J., et al., 2012. Skeletal muscle volume following dehydration induced by exercise in heat. Extrem. Physiol. Med. 1 (1), 3.

Hadzic, M., Eckstein, M.L., Schugardt, M., 2019. The impact of sodium bicarbonate on performance in response to exercise duration in athletes: a systematic review. J. Sports Sci. Med. 18 (2), 271–281.

Handzlik, M.K., Gleeson, M., 2013. Likely additive ergogenic effects of combined preexercise dietary nitrate and caffeine ingestion in trained cyclists. ISRN Nutr. 2013, Article ID 396581.

Haramizu, S., Nagasawa, A., Ota, N., et al., 2009. Different contribution of muscle and liver lipid metabolism to endurance capacity and obesity susceptibility of mice. J. Appl. Physiol. 106 (3), 871–879.

Hawley, J.A., 2013. Nutritional strategies to modulate the adaptive response to endurance training. Nestle Nutr. Inst. Workshop Ser. 75, 1–14.

Harty, P.S., Zabriskie, H.A., Erickson, J.L., et al., 2018. Multi-ingredient pre-workout supplements, safety implications, and performance outcomes: a brief review. J. Int. Soc. Sports Nutr. 15 (1), 41. Available from: https://doi.org/10.1186/s12970-018-0247-6.

Heaton, L.E., Davis, J.K., Rawson, E.S., et al., 2017. Selected in-season nutritional strategies to enhance recovery for team sport athletes: a practical overview. Sports Med. 47 (11), 2201–2218.

Heffernan, S.M., Horner, K., De Vito, G., et al., 2019. The role of mineral and trace element supplementation in exercise and athletic performance: a systematic review. Nutrients. 11 (3), 696.

Helms, E.R., Aragon, A.A., Fitschen, P.J., 2014. Evidence-based recommendations for natural bodybuilding contest preparation: nutrition and supplementation. J. Int. Soc. Sports Nutr. 11, 20.

Hernández, A., Cheng, A., Westerblad, H., 2012. Antioxidants and skeletal muscle performance: "common knowledge" vs. experimental evidence. Front. Physiol. 3, 46. Available from: https://doi.org/10.3389/fphys.2012.00046.

Hodgson, A.B., Randell, R.K., Jeukendrup, A.E., 2013. The effect of green tea extract on fat oxidation at rest and during exercise: evidence of efficacy and proposed mechanisms. Adv. Nutr. 4 (2), 129–140.

Hollidge-Horvat, M.G., Parolin, M.L., Wong, D., et al., 1999. Effect of induced metabolic acidosis on human skeletal muscle metabolism during exercise. Am. J. Physiol. 277 (4), E647–E658.

Horng, C.T., Huang, J.K., Wang, H.Y., et al., 2014. Antioxidant and antifatigue activities of *Polygonatum alte-lobatum* Hayata rhizomes in rats. Nutrients 6 (11), 5327–5337.

Horowitz, J.F., Klein, S., 2000. Lipid metabolism during endurance exercise. Am. J. Clin. Nutr. 72 (2), 558S–563S.

Huxtable, R.J., 1992. Physiological actions of taurine. Physiol. Rev. 72 (1), 101–163.

Hwang, J.T., Kwon, D.Y., Yoon, S.H., 2009. AMP-activated protein kinase: a potential target for the diseases prevention by natural occurring polyphenols. N. Biotechnol. 26 (1-2), 17–22.

Jagim, A.R., Harty, P.S., Camic, C.L., 2019a. Common ingredient profiles of multi-ingredient pre-workout supplements. Nutrients. 11 (2), 254. Available from: https://doi.org/10.3390/nu11020254.

Jagim, A.R., Camic, C.L., Harty, P.S., 2019b. Common habits, adverse events, and opinions regarding pre-workout supplement use among regular consumers. Nutrients. 11 (4), 855.

Jannig, P.R., Alves, C.R.R., Voltarelli, V.A., et al., 2017. Effects of N-acetylcysteine on isolated skeletal muscle contractile properties after an acute bout of aerobic exercise. Life Sci. 191, 46–51.

Jenkinson, C., Petroczi, A., Naughton, D.P., 2013. Effects of dietary components on testosterone metabolism via UDP-Glucuronosyltransferase. Front. Endocrinol. (Lausanne) 4, 80.

Jones, A.M., Vanhatalo, A., Bailey, S.J., 2013. Influence of dietary nitrate supplementation on exercise tolerance and performance. Nestle Nutr. Inst. Workshop Ser. 75, 27–40.

Kao, C.C., Wu, B.T., Tsuei, Y.W., et al., 2010. Green tea catechins: inhibitors of glycerol-3-phosphate dehydrogenase. Planta Med. 76 (7), 694–696.

Kårlund, A., Gómez-Gallego, C., Turpeinen, A.M., et al., 2019. Protein supplements and their relation with nutrition, microbiota composition and health: is more protein always better for sports people? Nutrients. 11 (4). Available from: https://doi.org/10.3390/nu11040829.

Koeberle, A., Werz, O., 2014. Multi-target approach for natural products in inflammation. Drug. Discov. Today 19 (12), 1871–1882.

Koh, H.J., Hirshman, M.F., He, H., et al., 2007. Adrenaline is a critical mediator of acute exercise-induced AMP-activated protein kinase activation in adipocytes. Biochem. J. 403 (3), 473–481.

Korf, E.A., Kubasov, I.V., Vonsky, M.S., et al., 2017. Green tea extract increases the expression of genes responsible for regulation of calcium balance in rat slow-twitch muscles under conditions of exhausting exercise. Bull. Exp. Biol. Med. 164 (1), 6–9. Available from: https://doi.org/10.1007/s10517-017-3913-9.

Kosinets, V.A., Stolbitski, V.V., Shturich, I.P., 2012. The use of cytoflavin in sports nutrition. Klin. Med. (Mosk.) 90 (7), 56–58 [Article in Russian].

Leadbetter, G., Keyes, L.E., Maakestad, K.M., et al., 2009. *Ginkgo biloba* does and does not prevent acute mountain sickness. Wilderness Environ. Med. 20 (1), 66–71.

Lee, S., Lim, H., 2019. Development of an evidence-based nutritional intervention protocol for adolescent athletes. J. Exerc. Nutr. Biochem. 23 (3), 29–38.

Lee, C.L., Cheng, C.F., Astorino, T.A., et al., 2014. Effects of carbohydrate combined with caffeine on repeated sprint cycling and agility performance in female athletes. J. Int. Soc. Sports Nutr. 11, 17.

Liang, H., Ward, W.F., 2006. PGC-1alpha: a key regulator of energy metabolism. Adv. Physiol. Educ. 30 (4), 145–151.

Liao, Y.H., Chao, Y.C., Sim, B.Y., et al., 2019. Rhodiola/cordyceps-based herbal supplement promotes endurance training-improved body composition but not oxidative stress and metabolic biomarkers: a preliminary randomized controlled study. Nutrients. 11 (10).

Liu, Y., Zhao, B., Mao, G., et al., 2013. Epigallocatechin-3-O-gallate, a green tea polyphenol, induces expression of pim-1 kinase via PPARγ in human vascular endothelial cells. Cardiovasc. Toxicol. 13 (4), 391–395.

Luckose, F., Pandey, M.C., Radhakrishna, K., 2015. Effects of amino acid derivatives on physical, mental and physiological activities. Crit. Rev. Food Sci. Nutr. 55 (13), 1793–1807.

Lutsch, D.J., Camic, C.L., Jagim, A.R., et al., 2019. Acute effects of a multi-ingredient pre-workout supplement on 5-KM running performance in recreationally-trained athletes. Int. J. Exerc. Sci. 12 (2), 1045–1056.

Malm, C., Svensson, M., Ekblom, B., et al., 1997. Effects of ubiquinone-10 supplementation and high intensity training on physical performance in humans. Acta Physiol. Scand. 161 (3), 379–384.

Marcus, A., Rossi, A., Cornwell, A., et al., 2019. The effects of a novel bicarbonate loading protocol on serum bicarbonate concentration: a randomized controlled trial. J. Int. Soc. Sports Nutr. 16 (1), 41. Available from: https://doi.org/10.1186/s12970-019-0309-4.

Margolis, L.M., Pasiakos, S.M., 2013. Optimizing intramuscular adaptations to aerobic exercise: effects of carbohydrate restriction and protein supplementation on mitochondrial biogenesis. Adv. Nutr. 4 (6), 657–664.

Martini, N., 2019. Potion or poison? coenzyme Q10. J. Prim. Health Care 11 (3), 290–291.

Mason, B.C., Lavallee, M.E., 2012. Emerging supplements in sports. Sports Health 4 (2), 142–146.

Matuszczak, Y., Farid, M., Jones, J., et al., 2005. Effects of N-acetylcysteine on glutathione oxidation and fatigue during handgrip exercise. Muscle Nerve. 32 (5), 633–638.

McLellan, T.M., Pasiakos, S.M., Lieberman, H.R., 2014. Effects of protein in combination with carbohydrate supplements on acute or repeat endurance exercise performance: a systematic review. Sports Med. 44 (4), 535–550.

McNaught, A.D., Wilkinson, A., 1997. IUPAC Compendium of Chemical Terminology, 2 ed. Blackwell Scientific, Oxford.

Merton, P.A., 1954. Voluntary strength and fatigue. J. Physiol. 123 (3), 553–564.

Moreno-Pérez, D., Bressa, C., Bailén, M., et al., 2018. Effect of a protein supplement on the gut microbiota of endurance athletes: a randomized, controlled, double-blind pilot study. Nutrients. 10 (3). Available from: https://doi.org/10.3390/nu10030337.

Nielsen, A.N., Mizuno, M., Ratkevicius, A., et al., 1999. No effect of antioxidant supplementation in triathletes on maximal oxygen uptake, 31P-NMRS detected muscle energy metabolism and muscle fatigue. Int. J. Sports Med. 20 (3), 154–158.

Nissen, S., Sharp, R., Ray, M., et al., 1996. Effect of leucine metabolite beta-hydroxy-beta-methylbutyrate on muscle metabolism during resistance-exercise training. J. Appl. Physiol. 81 (5), 2095–2104.

Nose, H., Morimoto, T., Ogura, K., 1983. Distribution of water losses among fluid compartments of tissues under thermal dehydration in the rat. Jpn. J. Physiol. 33 (6), 1019–1029.

Novozhilov, A.V., Tavrovskaya, T.V., Voitenko, N.G., et al., 2013. The effect of antioxidants on erythrocytes in rats during an exhaustive run. Ross. Fiziol. Zh Im. I M. Sechenova 99 (10), 1223–1232 [Article in Russian].

Novozhilov, A.V., Tavrovskaya, T.V., Voitenko, N.G., et al., 2015. Efficacy of green tea extract in two exercise models. Bull. Exp. Biol. Med. 158 (3), 342–345.

Novozhilov, A.V., Mindukshev, I.V., Korf, E.A., et al., 2020. Ammonium salts promote the functional adaptation of rat erythrocytes in the model of forced swimming. Bull. Exp. Biol. Med. 168 (4), 444–448. Available from: https://doi.org/10.1007/s10517-020-04728-6.

Nunes, C.L., Matias, C.N., Santos, D.A., et al., 2018. Characterization and comparison of nutritional intake between preparatory and competitive phase of highly trained athletes. Medicina (Kaunas.) 54 (3), 41.

Oliynyk, S., Oh, S., 2012. The pharmacology of actoprotectors: practical application for improvement of mental and physical performance. Biomol. Ther. (Seoul.) 20 (5), 446–456.

Orer, G.E., Guzel, N.A., 2014. The effects of acute L-carnitine supplementation on endurance performance of athletes. J. Strength. Cond. Res. 28 (2), 514–519.

Outlaw, J.J., Wilborn, C.D., Smith-Ryan, A.E., et al., 2014. Acute effects of a commercially-available pre-workout supplement on markers of training: a double-blind study. J. Int. Soc. Sports Nutr. 11, 40.

Pane, M., Amoruso, A., Deidda, F., et al., 2018. Gut microbiota, probiotics, and sport: from clinical evidence to agonistic performance. J. Clin. Gastroenterol. 52 (Suppl 1). Available from: https://doi.org/10.1097/MCG.0000000000001058 Proceedings from the 9th Probiotics, Prebiotics and New Foods, Nutraceuticals and Botanicals for Nutrition & Human and Microbiota Health Meeting, held in Rome, Italy from September 10 to 12, 2017:S46-S49.

Parr, M.K., Zhao, P., Haupt, O., et al., 2014. Estrogen receptor beta is involved in skeletal muscle hypertrophy induced by the phytoecdysteroid ecdysterone. Mol. Nutr. Food Res. 58 (9), 1861–1872.

Pasiakos, S.M., Lieberman, H.R., McLellan, T.M., 2014. Effects of protein supplements on muscle damage, soreness and recovery of muscle function and physical performance: a systematic review. Sports Med. 44 (5), 655–670.

Pasiakos, S.M., McLellan, T.M., Lieberman, H.R., 2015. The effects of protein supplements on muscle mass, strength, and aerobic and anaerobic power in healthy adults: a systematic review. Sports Med. 45 (1), 111–131. Available from: https://doi.org/10.1007/s40279-014-0242-2.

Paulsen, G., Cumming, K.T., Hamarsland, H., et al., 2014. Can supplementation with vitamin C and E alter physiological adaptations to strength training? BMC Sports Sci. Med. Rehabil. 6, 28.

Peart, D.J., Siegler, J.C., Vince, R.V., 2012. Practical recommendations for coaches and athletes: a meta-analysis of sodium bicarbonate use for athletic performance. J. Strength. Cond. Res. 26 (7), 1975–1983.

Pesta, D.H., Angadi, S.S., Burtscher, M., et al., 2013. The effects of caffeine, nicotine, ethanol, and tetrahydrocannabinol on exercise performance. Nutr. Metab. (Lond.) 10 (1), 71.

Philp, A., MacKenzie, M.G., Belew, M.Y., et al., 2013. Glycogen content regulates peroxisome proliferator activated receptor-∂ (PPAR-∂) activity in rat skeletal muscle. PLoS One 8 (10), e77200.

Pokrywka, A., Obmiński, Z., Malczewska-Lenczowska, J., et al., 2014. Insights into supplements with tribulus terrestris used by athletes. J. Hum. Kinet. 41, 99–105.

Prasertsri, P., Roengrit, T., Kanpetta, Y., et al., 2013. Cashew apple juice supplementation enhanced fat utilization during high-intensity exercise in trained and untrained men. J. Int. Soc. Sports Nutr. 10 (1), 13.

Psilander, N., Frank, P., Flockhart, M., et al., 2013. Exercise with low glycogen increases PGC-1α gene expression in human skeletal muscle. Eur. J. Appl. Physiol. 113 (4), 951–963.

Qi, B., Liu, L., Zhang, H., et al., 2014. Anti-fatigue effects of proteins isolated from Panax quinquefolium. J. Ethnopharmacol. 153 (2), 430–434.

Roseguini, B.T., Silva, L.M., Polotow, T.G., et al., 2015. Effects of N-acetylcysteine on skeletal muscle structure and function in a mouse model of peripheral arterial insufficiency. J. Vasc. Surg. 61 (3), 777–786.

Sachdeva, A.K., Kuhad, A., Tiwari, V., et al., 2010. Protective effect of epigallocatechin gallate in murine water-immersion stress model of chronic fatigue syndrome. Basic. Clin. Pharmacol. Toxicol. 106 (6), 490–496.

Sahlin, K., 2014. Muscle energetics during explosive activities and potential effects of nutrition and training. Sports Med. 44 (2), 167–173.

Santos, R.V., Bassit, R.A., Caperuto, E.C., et al., 2004. The effect of creatine supplementation upon inflammatory and muscle soreness markers after a 30km race. Life Sci. 75 (16), 1917–1924.

Sellami, M., Slimeni, O., Pokrywka, A., et al., 2018. Herbal medicine for sports: a review. J. Int. Soc. Sports Nutr. 15, 14. Available from: https://doi.org/10.1186/s12970-018-0218-y.

Senchina, D.S., Hallam, J.E., Kohut, M.L., et al., 2014. Alkaloids and athlete immune function: caffeine, theophylline, gingerol, ephedrine, and their congeners. Exerc. Immunol. Rev. 20, 68–93.

Shei, R.J., Lindley, M.R., Mickleborough, T.D., 2014. Omega-3 polyunsaturated fatty acids in the optimization of physical performance. Mil. Med. 179 (11), 144–156.

Sousa, M., Teixeira, V.H., Soares, J., 2014. Dietary strategies to recover from exercise-induced muscle damage. Int. J. Food Sci. Nutr. 65 (2), 151–163.

Spriet, L.L., 2014. Exercise and sport performance with low doses of caffeine. Sports Med. 44 (2), S175–S184.

Steiber, A., Kerner, J., Hoppel, C., 2004. Carnitine: a nutritional, biosynthetic, and functional perspective. Mol. Asp. Med. 25 (5–6), 455–457.

Summermatter, S., Santos, G., Pérez-Schindler, J., et al., 2013. Skeletal muscle PGC-1α controls whole-body lactate homeostasis through estrogen-related receptor α-dependent activation of LDH B and repression of LDH A. Proc. Natl Acad. Sci. USA 110 (21), 8738–8743.

Sureda, A., Pons, A., 2012. Arginine and citrulline supplementation in sports and exercise: ergogenic nutrients? Med. Sport. Sci. 59, 18–28.

Syrov, V.N., Kurmukov, A.G., 1976. Anabolic activity of phytoecdysone-ecdysterone isolated from Rhaponticum carthamoides (Willd.) Iljin. Farmakol. Toksikol. 39 (6), 690–693 [Article in Russian].

Tallis, J., Higgins, M.F., Cox, V.M., et al., 2014. Does a physiological concentration of taurine increase acute muscle power output, time to fatigue, and recovery in isolated mouse soleus (slow) muscle with or without the presence of caffeine? Can. J. Physiol. Pharmacol. 92 (1), 42–49.

Taylor, L.W., Campbell, B.I., Kerksick, C., et al., 2006. Effects of methoxyisoflavone, ecdysterone, and sulfo-polysaccharide supplementation on training adaptations in resistance-trained males. J. Int. Soc. Sports Nutr. 3, 19–27.

The, G.K., Verkes, R.J., Fekkes, D., et al., 2014. Tryptophan depletion in chronic fatigue syndrome, a pilot cross-over study. BMC Res. Notes 7, 650.

Thevis, M., Kuuranne, T., Geyer, H., Schänzer, W., 2017. Annual banned-substance review: analytical approaches in human sports drug testing. Drug. Test. Anal. 9 (1), 6–29.

Thomson, M., Al-Qattan, K., Mansour, M.H., et al., 2012. Green tea attenuates oxidative stress and downregulates the expression of angiotensin II AT(1) receptor in renal and hepatic tissues of streptozotocin-induced diabetic rats. Evid. Based Complement. Altern. Med. 2012, Article ID 409047.

Timcheh-Hariri, A., Balali-Mood, M., Aryan, E., et al., 2012. Toxic hepatitis in a group of 20 male body-builders taking dietary supplements. Food Chem. Toxicol. 50 (10), 3826–3832.

Van den Broek, I., Blokland, M., Nessen, M.A., et al., 2015. Current trends in mass spectrometry of peptides and proteins: application to veterinary and sports-doping control. Mass. Spectrom. Rev. 34 (6), 571–594. Available from: https://doi.org/10.1002/mas.21419.

Van Hall, G., Raaymakers, J.S., Saris, W.H., et al., 1995. Ingestion of branched-chain amino acids and tryptophan during sustained exercise in man: failure to affect performance. J. Physiol. 486 (3), 789–794.

Van Proeyen, K., Szlufcik, K., Nielens, H., et al., 2011. Beneficial metabolic adaptations due to endurance exercise training in the fasted state. J. Appl. Physiol. 110 (1), 236–245.

Vanella, A., Russo, A., Acquaviva, R., et al., 2000. L-propionyl-carnitine as superoxide scavenger, antioxidant, and DNA cleavage protector. Cell Biol. Toxicol. 16 (2), 99–104.

Vaughan, R.A., Conn, C.A., Mermier, C.M., 2014. Effects of commercially available dietary supplements on resting energy expenditure: a brief report. ISRN Nutr. 2014, Article ID 650264.

Wall, B.T., Stephens, F.B., Constantin-Teodosiu, D., et al., 2011. Chronic oral ingestion of L-carnitine and carbohydrate increases muscle carnitine content and alters muscle fuel metabolism during exercise in humans. J. Physiol. 589 (4), 963–973.

Wang, L., Waltenberger, B., Pferschy-Wenzig, E.M., et al., 2014. Natural product agonists of peroxisome proliferator-activated receptor gamma (PPARγ): a review. Biochem. Pharmacol. 92 (1), 73–89.

Wierniuk, A., Włodarek, D., 2013. Estimation of energy and nutritional intake of young men practicing aerobic sports. Rocz. Panstw. Zakl. Hig. 64 (2), 143–148.

Wilborn, C.D., Taylor, L.W., Campbell, B.I., et al., 2006. Effects of methoxyisoflavone, ecdysterone, and sulfo-polysaccharide supplementation on training adaptations in resistance-trained males. J. Int. Soc. Sports Nutr. 3, 19–27.

Wilson, G.J., Wilson, J.M., Manninen, A.H., 2008. Effects of beta-hydroxy-beta-methylbutyrate (HMB) on exercise performance and body composition across varying levels of age, sex, and training experience: a review. Nutr. Metab. (Lond.) 5, 1.

Yasui, K., Paeng, N., Miyoshi, N., et al., 2012. Effects of a catechin-free fraction derived from green tea on gene expression of enzymes related to lipid metabolism in the mouse liver. Biomed. Res. 33 (1), 9–13.

Yeh, T.S., Chan, K.H., Hsu, M.C., et al., 2011. Supplementation with soybean peptides, taurine, pueraria isoflavone, and ginseng saponin complex improves endurance exercise capacity in humans. J. Med. Food 14 (3), 219–225.

Yeh, T.S., Chuang, H.L., Huang, W.C., et al., 2014. *Astragalus membranaceus* improves exercise performance and ameliorates exercise-induced fatigue in trained mice. Molecules 19 (3), 2793–2807.

Zabala, M., Peinado, A.B., Calderón, F.J., et al., 2011. Bicarbonate ingestion has no ergogenic effect on consecutive all out sprint tests in BMX elite cyclists. Eur. J. Appl. Physiol. 111 (12), 3127–3134.

Zebrowska, A., Mizia-Stec, K., Mizia, M., et al., 2014. Omega-3 fatty acids supplementation improves endothelial function and maximal oxygen uptake in endurance-trained athletes. Eur. J. Sport. Sci. 1, 1–10.

Zhang, Z.J., Tong, Y., Zou, J., et al., 2009. Dietary supplement with a combination of Rhodiola crenulata and Ginkgo biloba enhances the endurance performance in healthy volunteers. Chin. J. Integr. Med. 15 (3), 177–183.

Chapter 16

Role of nutraceuticals as adaptogens

Arunabha Ray, Kavita Gulati, Sana Rehman, Nishant Rai and Rashmi Anand

Introduction

Living organisms are constantly exposed to emotionally and environmentally challenging situations, and the unique ability to cope with such aversive stimuli has been a distinctive feature of both animals and plants. In today's world, modified human lifestyles constantly exposed to stressful situations and maladaptation to such inputs can result in a plethora of disease states. Adaptive or coping processes help living organisms combat occupational stress, physical and emotional trauma, bacterial infections, industrial pollution, sleep deprivation, etc., which in turn enhances resistance to disease and promotes good health. Any agent that prevents aging, bolsters immunity, and reduces oxidant effects would be expected to protect against stress. Acting as adjuvants, they could actually help in positively modifying the course or pattern of disease. They supplement the body's ability to restore the disrupted dynamic equilibrium or homeostasis or resist damaging effects of stress. Integration of traditional wisdom regarding diet, nutritional supplements, and medicinal products with modern research and concepts can help us overcome stress, improve energy levels, prevent aging, and stall disease processes.

Stress, health, and disease

Stress is an aversive stimulus, external or internal, which challenges the physiological homeostasis, and physiological and behavioral adaptive responses come into play. These coordinated adaptive responses enhance the probability of survival. Such stress responses are mediated by the stress system and orchestrated by the brain (CNS) in conjunction with peripheral organs. Failure to adapt to stressors results in dysregulation of the stress system and may lead to psychiatric, endocrine/metabolic, and/or autoimmune diseases, or increase vulnerability to such diseases (Chrousos, 2009; Yaribeygi et al., 2017). Stress and stress-related disorders account for over 70% of global illnesses. Scientific evidence suggests that stress is the causative factor in the development, maintenance, or exacerbation of several mental and physical health conditions, including asthma, rheumatoid arthritis, anxiety disorders, depression, cardiovascular disease, chronic pain, human immunodeficiency virus/AIDS, stroke, and certain types of cancer. Stress has also been implicated in accelerated biological aging and premature mortality (Slavich, 2016). The pathophysiological aspects of stress have been a subject of research for a long time and Walter Cannon in 1929 first suggested that any aversive emotional stimulus is capable of causing physical damage to the body and can produce disease states like anxiety and depression, hypertension, immunosuppression, endocrine disorders, diabetes mellitus, and peptic ulcer (Cannon, 1929). In 1936, Hans Selye introduced the concept of stress in biology and medicine and defined it as a nonspecific response of the body to any environmental demand placed upon it. He described a response triad, such as adrenal hypertrophy, gastrointestinal ulceration, and thymolymphatic involution/atrophy, which should be elicited by any stressor and suggested the hypothalamic–pituitary–adrenal (HPA) axis as the principal regulator of the stress response. He considered the adrenal cortex to be the organ of integration during stress. Selye also introduced the term general adaptation syndrome with its three successive phases: the alarm reaction, stage of resistance, and exhaustion stage (Selye, 1950). First, an initial alarm reaction, analogous to Cannon's "fight-or-flight" response, second, a stage of adaptation associated with resistance to the stressor, and eventually a stage of exhaustion and the organism's death. However, his idea of nonspecificity of stress response was replaced by one of specificity and accordingly the concept of stress was altered. Stress and stress research has evolved considerably over the years and it is now increasingly being recognized as a highly interactive phenomenon and terms like stress syndrome, stress system, and allostasis have emerged. Allostasis is the process of adaptation of the body upon exposure to various stressors, wherein mediators like cortisol or adrenaline are released, which in turn

Nutraceuticals. DOI: https://doi.org/10.1016/B978-0-12-821038-3.00016-1

promote adaptation. However, when these stress mediators are not turned on adequately during stress, or when they are not shut down after the stressor, or when there is overuse, it leads to "allostatic overload." It is thus implied that the outcome of long-term physical or mental stress would depend on adequate regulation of such allostatic load (McEwen and Gianaros, 2010). More recent studies have indicated that these complex central nervous system (CNS)-mediated pathways in stress may involve both epigenetic and nongenomic mechanisms, and structural remodeling of neural networks may be crucial for adaptive processes (McEwen et al., 2015).

Stress research has revealed that acute stress effects are quite different from those of chronic stress. Whereas acute stress promotes adaptation and survival via responses of neural, cardiovascular, autonomic, immune, and metabolic systems, chronic stress results in exacerbations in pathophysiology due to dysregulation of the stress system. Chronic stress also predisposes to altered and detrimental lifestyle changes like smoking, eating too much, drinking, poor-quality sleep—which is also labeled as allostatic overload. Brain regions such as the hippocampus, prefrontal cortex, and amygdala respond to acute and chronic stress and show changes in morphology and chemistry that are largely reversible if the chronic stress lasts for weeks. However, it is not clear whether prolonged stress for many months or years may have irreversible effects on the brain. The adaptive plasticity of chronic stress involves many mediators, including glucocorticoids, excitatory amino acids, and endogenous factors such as brain neurotrophic factor (McEwen, 2008).

It is now well known that the CNS is central to stress effects/outcomes and its neural networks are also vulnerable to emotional and environmental stressors which could act as a trigger and orchestrate the genesis of many neuropsychiatric/neurodegenerative, cardiovascular, gastrointestinal, and immunological disorders. Complex neurochemical pathways/interactions have been proposed to explain stress responses and resultant pathophysiological states and brain amines, amino acids, and neuropeptides have been implicated. The activation of the hypothalamic–pituitary–adrenocortical axis and renin–angiotensin system plays a major role in the regulation of the stress response. Neuroendocrine components activated by stressors include the increased secretion of epinephrine and norepinephrine from the sympathetic nervous system and adrenal medulla. Further, the release of corticotropin-releasing factor (CRF) induces the secretion of pituitary adrenocorticotropin (ACTH), leading to secretion of glucocorticoids by the adrenal gland. CRF coordinates the endocrine, autonomic, behavioral, and immune responses to stress and also acts as a neurotransmitter or neuromodulator in the amygdala, dorsal raphe nucleus, hippocampus, and locus coeruleus, to integrate brain multisystem responses to stress (Kvetnansky et al., 2009).

The response of the cardiovascular system to stress has been ascribed mainly to excessive stimulation of catecholamine and involves increased cardiac output and vascular resistance, lipid mobilization, and stimulation of platelet aggregation (Chrousos, 1998). When exaggerated, these effects can cause hypertension, atherosclerosis, and ischemic heart disease. The gastrointestinal and immune systems are particularly susceptible to stressful inputs and result in complex disease states. More recently, it has been proposed that the CNS and its neural circuitry could regulate peripheral stress effects on the gastrointestinal and immune systems and concepts like the CNS–gut and CNS–immune axis have emerged (Ray et al., 1992, 1993).

The brain, which is the key initiator/regulator of the stress response, is highly susceptible to the detrimental effects of reactive oxygen species due to their high metabolic rate, rich composition of fatty acids prone to peroxidation, high intracellular concentrations of transition metals capable of catalyzing the formation of reactive hydroxyl radicals, low levels of antioxidants, and reduced capability to regenerate. Generated free radicals can attack proteins, polysaccharides, lipid bilayers, and DNA, causing cellular oxidative damage. Damage of nervous tissue by free radicals may contribute to neuronal loss in cerebral ischemia and may also be involved in normal aging and neurodegenerative diseases, for example, epilepsy, schizophrenia, Parkinson, Alzheimer, and other diseases (Wang and Michaelis, 2010). More recently, it has been suggested that the gasotransmitter, nitric oxide (NO) and its complex signaling pathways could regulate stress effects and NO-dependent physiological variables and pharmacological responses were identified (Ray et al., 2007, 2009; Gulati et al., 2007, 2009; Anand et al., 2012).

Classically, antistress agents are those which help to combat or cope with stressful situations and include both typical and atypical anxiolytic agents like benzodiazepines, β-blockers (propranolol), and 5-HT1A partial agonist (Buspirone), which have different pharmacological profiles. For example, the prototype antistress agents of the benzodiazepine class (diazepam, chlordiazepoxide, alprazolam, etc.) facilitate GABAergic neurotransmission and also have sedative/hypnotic, anticonvulsant, and muscle-relaxant effects. Further, long-term treatment with benzodiazepines incurs the risk of cognitive dysfunction, dependence, and induces withdrawal signs upon discontinuation.

Stress also results in a profound effect on the balance of intestinal microflora. In a pioneering study, Moore et al. (1978) concluded that the changes in diet did not result in significant shifts in intestinal flora composition, but anger or fear stress resulted in statistically significant shifts in the proportions of some species. Lizko et al. (1984) investigated the microbial load during preparation for and participation in space flight and found a significant decrease in the numbers of *Bifidobacterium* and lactobacilli and an increase in the numbers of *E. coli* and enterobacteria. The effect of nervous emotional stress caused by being launched into space resulted in worsening of these imbalances, that is, altered balance of beneficial and pathogenic organisms. After the flight, the numbers of pathogenic enterobacteria and clostridia were also substantially increased, while the number of lactobacilli decreased, thus suggesting the disruption of microbial balance by the physiological strain of space flight.

Although modern medicine has developed several strategies, including pharmacological interventions, to attenuate/cope with stress, several gray areas still exist and the prevention and management of stress and stress-related disorders remain a problem. Hence, balancing safety and efficacy as well as identification of the right target are essential for developing drugs in stress and stress-related disorders. In addition, there is a dire need to develop a battery of tests, which need to be comprehensive and confirmatory to screen new agents with antistress profiles.

Adaptogens

Adaptogens are agents that improve the nonspecific response to stress and promote recovery from the damaging effects of stress. The term adaptogen was coined by I.I. Brekhman and, according to him, an adaptogen has four general properties: (1) it is harmless to the host; (2) it has a general, rather nonspecific effect; (3) it increases the resistance of the recipient to a variety of physical, chemical, or biological stressors; and (4) it acts as a general stabilizer/normalizer. In the 1950s, Russian researchers identified that many plants, especially from the *Araliaceae* family, have adaptogenic properties. Some of these adaptogens included *Schisandra chinensis*, *Panax ginseng*, *Eleutherococcus senticosus* (ES), *Rhodiola rosea*, etc., which are known to exert adaptogenic effects primarily on the adrenal glands; whereas, *Rhodiola* is known to exert adaptogenic effects primarily on the CNS. Most recent studies have further elaborated on the concept of adaptogens and suggested that these plant-based products could counteract the effects of physical, chemical, and biological stress, for example, environmental population, radiation, infectious disease, psychological trauma in a nonspecific manner, and by being nontoxic to the host. They have also been labeled as bioregulators and metabolic regulators, which could improve the body's ability to recognize, respond, recover, and restore or regenerate. Adaptogens have also been categorized as primary and secondary, based on clinical experience, and their ability to impact the HPA axis (Ray et al., 2016; Liao et al., 2018).

Around 2000 year ago, Hippocrates said "Let food be your medicine and medicine be your food," thus blurring the gap between food and medicine. In 1989, the term nutraceutical was coined by Stephen DeFelice, founder and chairman of the Foundation for Innovation in Medicine, an American organization which encourages medical health. According to him "a nutraceutical is any substance that is a food or a part of food and provides medical or health benefits, including the prevention and treatment of disease."

Nutraceuticals are derived from food sources to provide extra health benefits, in addition to the basic nutritional value found in foods. Depending on the jurisdiction, products may claim to prevent chronic diseases, improve health, delay the aging process, increase life expectancy, or support the structure or function of the body (Boothe, 1997). Such products may range from isolated nutrients, dietary supplements and specific diets to genetically engineered designer foods and herbal products. In the United States, "nutraceuticals" are commonly used, without any regulatory definition. The Health Ministry of Canada defined a nutraceutical as "a product which is isolated or purified from the food, generally sold in medicinal form not associated with food and demonstrated to have a physiological benefit. It also provides benefit against chronic disease."

The term "functional food" is also gaining importance and in the United Kingdom, the Ministry of Agriculture, Fisheries, and Food defined functional food as "a food that has a component incorporated into it to give it a specific medical or physiological benefit, other than purely nutritional benefit." Although both terms seem to intermingle yet there is a slight difference between the functional foods and nutraceuticals (Boothe, 1997, 1998). Functional food helps to provide the body with the required amount of vitamins, fats, proteins, carbohydrates, etc. needed for its healthy survival and maintenance. When a functional food is used in the prevention or treatment of diseased states, it is called a nutraceutical (e.g., fortified dairy products, vitamin D addition to milk). Functional foods are defined as enriched foods close to their natural state, that have been either enriched or fortified, by a process called nutrification. This restores the nutrient content in a food back to similar levels from before the food was processed. However, when a dietary supplement, nutraceutical, or other feed is intended to be used for the treatment or prevention of disease, it "becomes" a drug (Dzanis, 1998). Drugs have to undergo a regulatory approval process prior to marketing to demonstrate their safety and efficacy. On the other hand, nutraceuticals do not usually undergo an approval process, and thus are different from classical drugs (Boothe, 1997). This shows that nutraceuticals fall somewhere in between foods and drugs. They have many advantages over either food or drugs since they are not required to list nutrient profiles as required by feeds, and in many cases are intended to treat or prevent disease without first undergoing proper drug approval. Thus, if the substance contributes only to the maintenance of healthy tissues and organs, it may be considered to be a food ingredient. If, however, it can be shown to have a modifying effect on one or more of the body's physiological processes, it is likely to be considered to be a medicinal substance.

Commonly used nutraceuticals could be prebiotics, dietary fibers, omega 3 fatty acids, or antioxidants. A prebiotic is a specialized plant fiber that nourishes the good bacteria already present in the large bowel or colon to promote the growth of many of the good bacteria in the gut. The healthy bacteria that live in the colon strengthen the bowel wall, improve mineral absorption, and aid in the regulation of hormone production. Prebiotics fertilize these good bacteria which attenuate the production of disease-causing microbes, but are also associated with some untoward effects. Prebiotics do not contain dangerous or untested chemicals, and are suitable for vegetarian diets. Prebiotics are known to enhance the immune system, increase calcium absorption and bone density, and help in weight management control and correction of harmful colon bacteria, leaky gut, and endotoxemia. Synbiotics are a combination of probiotics and prebiotics administered together.

Dietary fibers are of two types, insoluble and soluble, and both are delivered unchanged to the colon. Insoluble fiber is water insoluble and is not fermented or broken down by colon bacteria, but retains water, thereby helping in increasing the bulk and softness of the stool. On the other hand, soluble fiber is broken down and fermented by colon bacteria. Insoluble fiber could be further classified into two types: fermentable and nonfermentable. Nonfermentable insoluble fiber is a bulking agent, and consuming adequate insoluble fiber keeps people regular, while fermentable insoluble fiber—such as resistant starch—produces the same healthy gases and acids in the large intestine that soluble fiber does. One important difference between the two types of fibers is that soluble fiber tends to slow digestion while insoluble fiber enhances it.

Omega-3 fatty acids are polyunsaturated fatty acids with a double bond ($C = C$) at the third carbon atom from the end of the carbon chain. They are commonly found in plant oils, for example, walnut, edible seeds, flaxseed oil, and hemp oil. Animal sources of omega-3 fatty acids include fish oils, egg oil, squid oils, and krill oil. They are required for normal metabolism and humans have a limited ability to synthesize omega-3 fats (Zheng et al., 2013). The ability to make the longer-chain omega-3 fatty acids from ALA is impaired with advancing age. In foods exposed to air, unsaturated fatty acids are vulnerable to oxidation and rancidity. Omega-3 fatty acids have been shown to have antiinflammatory properties (Miles and Calder, 2012), and play a role in breast cancer (Zheng et al., 2013), mental health (Hegarty and Parker, 2013), and cognitive function (Mazereeuw et al., 2012).

Antioxidants are molecules that inhibit the oxidation of other molecules. Oxidation is a chemical reaction involving the loss of electrons or an increase in the oxidation state, and such reactions can generate free radicals. In turn, these free radicals can start chain reactions which could cause cell damage and death. Antioxidants terminate these chain reactions by removing free radical intermediates, and inhibit other oxidation reactions. They themselves get oxidized in the bargain—thus behaving like reducing agents, for example, thiols, ascorbic acid, or polyphenols (Sies, 1997). Oxidative stress seems to play a significant role in many human diseases, including cardiovascular, respiratory, cancers, etc. The use of antioxidants in therapeutics has been extensively studied, particularly in cardiovascular diseases (e.g., heart attack and stroke) and neurodegenerative diseases (e.g., Alzheimer, Parkinson, ALS). Antioxidants also have many industrial uses, such as preservatives in food and cosmetics and to prevent the degradation of rubber and gasoline. Cells are protected against oxidative damage by an interacting network of antioxidant enzymes like superoxide dismutase, catalase and peroxiredoxins, thioredoxin, and glutathione systems (Di Matteo and Esposito, 2003; Baillie et al., 2009).

Nutraceutical products cover all therapeutic areas such as antiarthritic, pain killers, cold and cough, sleeping disorders, digestion, prevention of certain cancers, osteoporosis, blood pressure, cholesterol, depression, and diabetes. Powerful market forces are fueling the interest in nutraceuticals as healthcare costs are soaring, and there is an aging population which is health conscious. There have been also rapid advances in scientific knowledge supporting the vital role of diet in health and disease prevention and the regulatory environment is changing, thus allowing the development of health-promoting foods.

The question of safety of nutraceuticals has always been a contentious issue. Like many other substances, nutraceuticals, though efficacious, may cause problems due to direct toxic effects or a delay in more appropriate treatments (Dzanis, 1998). Safety of a nutraceutical product is often easier to establish than efficacy. Studies that test doses of nutraceuticals several fold greater than the intended (recommended) dose help to establish both acute and chronic toxicity data. However, a lack of reported toxicity problems with any nutraceutical should not be taken as evidence of safety (Boothe, 1998). Nevertheless, regulatory guidelines in this matter vary from country to country depending on the nature of substances.

The pharmaceutical, pharmacokinetic, and pharmacodynamic studies of a compound provide evidence of the efficacy of that compound (Boothe, 1998). Pharmaceutical data provide an evaluation of the quality of manufacturing, purity of product, and labeling accuracy. Pharmacokinetic data consist of tracking the compound through the animal's body and provide information about absorption, distribution, metabolism, and excretion. The most difficult question to answer for nutraceuticals is the pharmacodynamics, that is, how the animal responds to the compound, since most of these compounds are involved in a cascade of different reactions throughout the body.

Stress, nutraceuticals, and adaptogens

Herbal agents

In recent years, several therapeutic drugs have been rationally designed and synthesized based on the novel knowledge gained from investigating the actions of biologically active chemicals derived from foods, plants, and medicinal herbs. One of the major advantages of these naturalistic chemicals is their ability to interact with multiple targets in the body resulting in a combined beneficial effect. Adaptogens are such plant-derived agents, which are both nonspecific and nontoxic, and which help in coping with stress and stress-related problems. Initial studies in the area of adaptogens led to the pioneering observations of Russian scientists who identified and mostly focused on three herbs, *S. chinensisat*, *ES*, and *R. rosea*, as they fitted well into the antistress models being tested at that time and qualified as adaptogens by being nonspecific, nontoxic, and general revitalizers. Their effects were extensively studied against a variety of toxic substances, pollutants, and microbes, and were shown to protect the main targets of stress, including the CNS, the neuroendocrine axis, and the immune system. Additionally, differing degrees of cardioprotection, hepatoprotection, and gastroprotection were also reported with these agents (Panossian and Wikman, 2010).

More recently, plants like *P. ginseng*, *Acanthopanax senticosus*, *Lepidium meyenii*, etc. were categorized as true adaptogens, with the Korean plant, *P. ginseng*, being universally labeled as the prototype. In addition to the adaptogenic herbs initially proposed, several other medicinal plants have been documented in Indian traditional medicine as sources of antistress agents or adaptogens, for example, *Bacopa monniera*, *Centella asiatica*, *Passiflora incarnata*, *Valeriana officinalis*, *Withania somnifera* (WS), *Humulus lupulus*, *Matricaria chamomilla*, *Galphimia glauca*, *Melissa officinalis*, *Piper methysticum*, *Scutellaria lateriflora*, and *Ziziphus jujuba*. Several others, like *Ocimum sanctum*, *Tinospora cordifolia*, *Asparagus racemosus*, *Azadirachta indica*, etc., have been used extensively in the Ayurveda and Unani systems of medicine, and many of their effects have been scientifically validated. They act by complex mechanisms and by neutralizing one or a combination of stress mechanisms. The net outcome is a benefit in stress and stress-related disorders, which is of great relevance in today's changing life style. The proposed mechanisms of action include modulation of the HPA axis, neurotransmitter mechanisms/interactions, immune mechanisms, and cellular and molecular biomarkers of stress, for example, heat shock proteins, oxidative/nitrosative stress markers, etc. (Gulati et al., 2015; Ray et al., 2016). A brief description of some of these phytomedicinal adaptogenic agents is given next.

Eleuthrococcus senticosus

This is commonly known as Siberian ginseng or touch me not, and the whole plant is used. Its chemical constituents include glycosides called eleutherosides A–M; saponins, ciwujianosides; eleutherans, β-sitosterol (BSS), isofraxidin, syringin, chlorogenic acid, sesamin, and triterpene friedelin. *ES*, along with *R. rosea* and *S. chinensis* are among the most extensively studied adaptogens. Both extracts and phytochemicals have been shown to have adaptogenic effects. Adaptogens induced a state of nonspecific resistance to highly toxic chemicals (e.g., chlorophos, phosphorus, cyclophosphane, strychnine, aniline, sodium nitrite, narcotics like sodium barbital, hexenal, chloralhydrate, benzene, acetone, ether, etc.) and microbes demonstrated in many pharmacological/toxicological studies. Several clinical studies with *ES* in healthy human subjects were performed in normal and stressful conditions (e.g., high-temperature environment, forced work periods, loud noise conditions, motion sickness, varying degrees of deafness, heavy physical burden, hypertension, mountain rescuers under forced conditions, athletes, deep sea divers, intense mental work and physical work, factory workers under extreme working conditions). It was observed that there was an improvement in the physical and mental work capacities in all cases. Several other studies were conducted in sick patients with atherosclerosis, acute pyelonephritis, diabetes, hypertension, trauma, neuroses, rheumatic heart disease, chronic bronchitis, insomnia, cancer, and several other ailments. In most cases, a moderate improvement relative to the initial conditions was observed. The most convincing evidence of the efficacy of adaptogens was found in studies related to their neuroprotective effects, effects on cognitive functions and mental performance in fatigue, and their efficacy in asthenia and depression. The beneficial stress-protective effect of adaptogens is related to the regulation of homeostasis via several mechanisms of action, which are associated with the HPA axis and the regulation of key mediators of the stress response, such as molecular chaperons (e.g., Hsp70), stress-activated c-Jun N-terminal protein kinase (JNK1), cortisol, and NO. It was suggested that adaptogens have potential use in age- and stress-related disorders, such as neurodegenerative diseases and cardiovascular diseases. Elderly people may be able to maintain their health status on a normal level, improve their quality of life, and may increase longevity. ES was also shown to be safe in acute and subacute toxicity studies (Panossian and Wikman, 2010). Further, immuno-enhancing effects of ES extract as evidenced by an increase in the antibody levels in mouse blood serum and stimulated cellular system have been documented (Drozd et al., 2002), including against bovine serum albumin or ovalbumin (Hwang et al., 2003). ES is known to attenuate LPS-induced iNOS expression but not COX-2 expression (Jung et al., 2007). It was further proposed that attenuations in stress-induced impairments and in stress-related disorders were due to effects on the brain and neuroendocrine–immune axis.

Panax ginseng (family Araliaceae)

There are at least three varieties of *Panax* spp., with the Korean variety (*P. ginseng*) being the most popular. The other varieties are Chinese and American (Kim et al., 2018). The Indian adaptogen, WS, is often referred to as the Indian ginseng. *P. ginseng* is widely recognized as the prototype adaptogen or revitalizing agent. The name is derived from the Chinese word "Jensheng" meaning man-herb due to the humanoid shape of the root. The part of the plant used is the root, and it contains triterpene glycosides, or saponins, commonly called ginsenosides: protopanaxadiol (Rb1, Rb2, Rc, and Rd) and protopanaxatriol (Re, Rf, Rg1, and Rg2). Another botanical species, although commonly called Siberian ginseng (*ES*), is not a true ginseng. As an adaptogen, it enhances physical performance (including sexual), promotes vitality and increases resistance to stress and aging (Ray et al., 2016). The adaptogenic properties of ginseng are believed to be due to its effects on the HPA axis. Yue et al. (2007) have suggested antitumor, angiomodulating, and steroid-like activities of ginsenosides.

Volate et al. (2005) have also reported that dietary herbal supplements (quercetin, curcumin, silymarin, ginseng, and rutin) can modulate aberrant crypt foci and influence carcinogenesis, apoptosis, angiogenesis, and metastasis. Hofseth and Wargovich (2007) have suggested that *Panax* affects multiple targets within the inflammatory cascade, including inhibition of COX-2, iNOS, and NF-κB, and plays an antiinflammatory role in carcinogenesis. Studies also showed inhibitory effects of the ginsenoside Rg3 on phorbol ester-induced COX-2 expression, NF-κB activation, and tumor promotion (Keum et al., 2003). A clinical study compared the effects of *P. ginseng* on HIV-1-infected patients and showed that the herbal agent slowed the depletion of CD4 T cells in these patients (Sung et al., 2005). Another randomized controlled clinical trial examined the effects of *P. ginseng* in type II diabetes, showing that, when compared to placebo, the ginseng group experienced better mood, improved physical performance, and decreased fasting blood glucose levels as compared to the placebo group (Sotaniemi et al., 1995). Reay et al. (2005), in a double-blind, cross-over randomized clinical trial (RCT), concluded that a single dose of *P. ginseng* reduces blood glucose levels and improves cognitive performance during sustained mental activity by regulating glucose metabolism. A clinical study by Vuksan et al. (2008) showed that *P. ginseng* improved glucose and insulin regulation in type II diabetes.

P. ginseng and its components have also been shown to prevent neurodegeneration, and the beneficial effects in neurodegenerative diseases have been proposed to be due to the antioxidative and immunomodulatory activities of its ginsenoside components. Studies with ginsenosides revealed that the neuroprotection was due to the antioxidant effects as well as by modulations of intracellular neuronal signaling and metabolism, cell survival/death genes, and mitochondrial function (Kim et al., 2018). When used appropriately, ginseng appears to be safe. However, the various reported side effects include hypertension, diarrhea, restlessness, mastalgia, and vaginal bleeding. A review by Seely et al. (2008), however, suggested cautious use of *P. ginseng* in pregnancy and lactation. Chan et al. (2003) reported that various ginsenosides had differential embryotoxic effects.

Valeriana officinalis

V. officinalis (valerian) essential oils are reported for their sedative and anxiolytic activity, while its valepotriates are known to exert a regulatory effect on the autonomic nervous system. More than 150 constituents have been identified but none have been claimed to be solely responsible for valerian's effects, suggesting that its compounds act synergistically. Valerian is known to cause a dose-dependent release of neurotransmitters such as gamma-aminobutyric acid (GABA) and also inhibits the enzyme-induced breakdown of GABA in the brain. However, valerian lignan hydroxypinoresinol has also been reported to bind to benzodiazepine receptors (Bodesheim and Holzl, 1997). Valerian acts more as a nervous system depressant than as a muscle relaxant to cause its sedative effect (Hattesohl et al., 2008). In comparison to diazepam (2.5 mg three times daily), a valerian preparation (50 mg three times daily, standardized to 80% dihydrovaltrate) resulted in a significant reduction in symptoms of anxiety measured on HAM-A after 4 weeks (Andreatini et al., 2002). In another study, either 600 mg/day valerian extract or placebo were given for 4 weeks and assessed. After 14 days, valerian was rated better than placebo on the Clinical Global Impression Scale (CGIS). On day 28, 66% of patients rated valerian effective for sleep as compared to 26% who were taking placebo (Trevena, 2004). In a study using polysomnographic recordings and questionnaires,

Donath et al. (2000) found that sleep latency was significantly reduced in 16 insomnia patients treated with valerian as compared to placebo. Valerian (600 mg daily) was compared to the benzodiazepine oxazepam (10 mg daily) in 202 patients for 6 weeks with positive effects on sleep quality, measured by the Sleep Questionnaire, CGIS, and Global Assessment of Efficacy. Mild-to-moderate adverse effects were reported in 36% of patients taking oxazepam as compared with 28% in the valerian group (Ziegler et al., 2002). A trial of valerian 100 mg three times daily for 2 weeks after benzodiazepine withdrawal resulted in improved sleep quality (Poyares et al., 2002). In a study of patients complaining of insufficient sleep, significant improvement in sleep was noted after 2 weeks using 410 mg of valerian at bedtime (Dominguez et al., 2000). An animal study comparing valerian with a combination of valerian, Rhodiola, and L-theanine found significant and comparable shortening of sleep latency in both groups (Tokunaga et al., 2007).

Withania somnifera

WS (dunal; winter cherry; Indian ginseng) is a herb which is extensively grown in the Indian subcontinent and has the characteristic odor of a sweaty horse. In Indian traditional medicine (Ayurveda) it is classified as a Rasayana, a group of plant-derived drugs reputed to promote physical and mental health, augment resistance of the body against disease and many adverse environmental factors, revitalize the body in debilitated conditions, and increase longevity. These properties closely resemble the Korean adaptogen, *P. ginseng*, which is referred to as the gold standard in adaptogens. However, studies have shown that WS has a more favorable efficacy and safety profile as compared to *Panax*. Both the roots and aerial parts (leaves) of the plant are used for therapeutic purposes. Over 35 chemical constituents have been isolated: alkaloids like isopelletierine, anaferine, ashwagandhin, cuscohygrine, anahygrine, topine, somniferin saponins, sitoindosides, withanolides, withaferin A, withasomnidieone, withanone, etc. The herb is commonly used in Indian traditional medicine (Ayurveda and Unani) to promote vigor and vitality (including as an aphrodisiac), enhance muscle strength and endurance, and improve overall health (Kulkarni and Dhir, 2008; Singh et al., 2011; Ray et al., 2016). Several clinical and experimental studies have shown its beneficial effects. It was shown to have health restorative properties by counteracting chronic fatigue, weakness, impotence, sterility, nervous exhaustion, senility, and premature aging (Kulkarni and Dhir, 2008). The mental and physical effects of ashwagandha have been investigated in several research studies, with identified positive effects in reducing stress and anxiety (Pratte et al., 2014) and increasing memory and cognition (Choudhary et al., 2017). A study showed that WS influenced ethanol-induced anxiety modulation in mice (Gupta and Rana, 2008). It was suggested that WS induced a state of nonspecific increase in resistance during stress (Singh et al., 2011). In a recent meta-analysis comprising four clinical trials, it was concluded that ashwagandha supplementation was associated with significant increases in sperm concentration, semen volume, and sperm motility in oligospermic males. Increases in serum testosterone and luteinizing hormone levels were also identified (Durg et al., 2018).

Immunomodulatory effects of WS extracts were also reported as evidenced by a significant increase in the cell-mediated immunity of mice. WS root extract enhances the levels of IFN-γ, IL-2, and GM-CSF in mice, suggestive of an immunopotentiating and myeloprotective effect. WS also enhanced nitric oxide synthase (NOS) activity of the macrophages, which in turn increased the microbial killing power of these immune cells. In fact, both immunostimulatory and immunosuppressive properties have been reported with WS (Tiwari et al., 2014). It has also been demonstrated in several animal studies that ashwagandha can influence the hypothalamic–pituitary–gonadal hormonal axis and increase testosterone concentrations (Azgomi et al., 2018; Sengupta et al., 2018). The ability of WS extracts to influence oxidative stress parameters has also been widely reported (Ray et al., 2016).

In a study on chronically stressed adults, ashwagandha supplementation for 60 days was associated with reductions in anxiety (as measured by the Hamilton Anxiety Scale) and increases in serum DHEA-S. A study reported that WS supplementation is associated with significant increases in muscle mass and strength and it may also be useful in conjunction with a resistance training program (Wankhede et al., 2015). Another study demonstrated that supplementation with a WS extract was associated with significant improvements in salivary DHEA-S (neurosteroid) and testosterone, but not cortisol and estradiol, in healthy males aged between 40 and 70 years (Lopresti et al., 2019). The antistress property of neurosteroids has been demonstrated to be NO dependent (Chakraborty et al., 2011). All the above observations strongly support an adaptogenic profile of the herb and are in concurrence with the reported traditional use of the medicinal plant and its phytochemicals.

Bacopa monnieri

Bacopa monnieri (Brahmi) belongs to the family *Scrophularaciae* and the plant part used is the leaf. The pharmacological effects of *B. monnieri* are attributed to the presence of a number of biologically active compounds, including alkaloids like brahmine, nicotinine, and herpestine; saponins like herasapononin; potassium salts and sterols. The compounds responsible for the memory-enhancing effects of *Bacopa* are attributed to triterpenoid saponins called bacosides. Apart from memory- and learning-facilitating properties, Limpeanchob et al. (2008) have demonstrated a neuroprotective effect of *B. monnieri* on β-amyloid-induced cell death but not glutamate-induced excitotoxicity in primary cortical culture. This neuroprotection may possibly be due to its ability to suppress cellular acetylcholinesterase activity but not the inhibition of glutamate-mediated toxicity. Antioxidant properties of *B. monnieri* were also shown, which prolonged the life span of cultured neurons. Another experimental study reported antioxidant properties including metal ion reduction, free radical scavenging, and lipid peroxidation inhibition could have potential in the treatment of asthma, epilepsy, insanity, inflammation, and cardiovascular system diseases (Channa et al., 2006). Both animal and clinical research data support the traditional (Indian) Ayurvedic use of *B. monnieri* for stress and anxiety. A study in a rat model of anxiety demonstrated a *Bacopa* extract exerted anxiolytic activity comparable to lorazepam, a widely used benzodiazepine anxiolytic drug. However, the *Bacopa* extract did not induce amnesia, as lorazepam did, but instead had a memory-enhancing effect (Bhattacharya and Ghosal, 1998). Another experimental study conducted in rats showed that bacosides of *B. monnieri* had the potential to modulate the activities of Hsp70, P450, and SOD, thereby possibly allowing the brain to be prepared to act under adverse conditions such as stress (Chowdhuri et al., 2002).

In a clinical study, 35 patients with anxiety neurosis were administered Brahmi syrup, resulting in a significant decrease in anxiety symptoms, level of anxiety, level of disability, and mental fatigue, and an increase in immediate memory span. In another RCT, *Bacopa*'s effect on cognitive function was assessed and it was found that its administration resulted in significant improvement in anxiety after 12 weeks (Stough et al., 2001). Further, Calabrese et al. (2008) conducted a study on the mental and emotional effects of *Bacopa* in the elderly, and the herb-treated group experienced significant improvement in anxiety, cognitive performance, and depression scores as assessed by combined state plus trait anxiety scores as compared to placebo (Table 16.1).

Nutritional supplements

Despite the number of nutritional supplements advertised and publicized in the electronic media to treat anxiety, only inositol (part of the vitamin B complex, B8) and an intracellular second messenger, has shown evidence suggesting superiority to placebo and even comparability with the selective serotonin reuptake inhibitor (SSRI) fluvoxamine (Palatnik et al., 2001). Inositol is a carbohydrate (though not a classical sugar) and showed a positive effect on patients with panic disorder; however, its effect on patients with obsessive compulsive disorder is less clear. However, there are limited data to date which suggest partial responses with a side-effect profile that may be comparable with that of SSRIs.

Neurotransmitter or hormonal precursors

The vast majority of neurotransmitter or hormonal precursors that claim to be useful for treating anxiety disorders have no evidence supporting clinical utility. Only 5-hydroxytryptophan (5-HTP) appeared to show clinical effectiveness among the precursor preparations. Although there is some indication that 5-HTP can reduce anxiety symptoms among patients with anxiety disorders, the evidence is inconclusive and unproven (Graeff et al., 1996). Also, it has been known to cause eosinophilia-myalgia syndrome, a significantly dangerous side effect. Therefore, the risk/benefit ratio does not favor use in patients.

Cortisol modulators

Phosphatidylserine

In a crossover RCT, significant decreases in peak cortisol and area under the curve for cortisol as compared to placebo have been reported in 10 healthy males given 600 mg phosphatidylserine (PS) for 10 days (Starks et al., 2008). In another study, blunting of serum ACTH and serum salivary cortisol and decreased emotional responses to Trier Social Stress Test-induced stress has been reported when four groups of 20 subjects each were given a phosphatidic acid complex and PS at a dose of 400, 600, 800 mg, or placebo for 3 weeks (Hellhammer et al., 2004).

Plant sterols and sterolins

Plant sterols and sterolins are described as phytochemicals or plant "fats," chemically similar to cholesterol, and have biological "adaptogenic" activity. In a double-blind clinical trial involving marathon runners, Bouic et al. (1999) studied the effects of BSS and β-sitosterol glucoside mixture on selected immune parameters of marathon runners and reported inhibition of post marathon immune suppression and inflammation as compared to the placebo group. Cortisol levels increased in marathon runners receiving the placebo; however, cortisol levels remained constant in the sterol/sterolin treatment group, indicating a reduction in the adrenal stress response to the event. Also, it was observed that CD3 and CD4 lymphocyte subsets increased in the sterol/sterolin group and declined in the placebo group. However, IL-6 (an inflammatory cytokine) decreased in the sterol/sterolin treatment group but increased in the placebo group.

TABLE 16.1 Some medicinal plants and their active phytochemicals with adaptogenic potential.

Medicinal plant	Active phytochemical	Adaptogenic mechanisms
Panax ginseng (Korean ginseng)	Ginsenosides: protopanaxadiol, protopanaxatriol	Immunomodulator, antioxidant, COX-2 inhibition, NOS inhibition
Withania somnifera (ashwagandha, Indian ginseng)	Alkaloids, saponins, steroidal lactones (withaferin A, other withanolides), sitoindosides	GABA mimetic, antioxidant, antiinflammatory, immunomodulatory, neuroprotective
Asparagus racemosus (shatavari)	Steroidal saponins (shatavarin I–IV), isoflavones, asparagamine	By interacting with the serotonin and GABA systems, the levels of all monoamines in the HPA axis are increased
Ocimum sanctum (holy basil, tulsi)	Essential oil, eugenol, euginal, urosolic acid, carvacrol, linaloo, limatrol, caryophyllen, methyl carvicol	Inhibition of neurotransmitter reuptake (NA, 5-HT, DA). Modifies GABAergic and glutamatergic transmission; antioxidant, neuroprotective, antibacterial
Azadirachta indica (neem)	Azadirachtin, nimbin, nimbidin, nimbidol, nimbolinin, salanin, quercitin	Antiinflammatory, antioxidant immunomodulator, DA mimetic
Bacopa monnieri (brahmi)	Saponins (bacosides), betulinic acid, D-mannitol, β-stigmasterol, and stigmastanol	Mediates calcium ion influx; anxiolytic effects not completely understood
C. asiatica (mandukparni, gotu kola)	Saponins (asiaticosides), madecassoside, flavonoids, bramhinoside, tannins	Stimulates conversion of glutamic acid to GABA
L-Theanine (from *Camellia sinensis*)	Epigallocatechin, epicatechin, gallocatechin, and catechin	Increases alpha-brain wave activity; increases dopamine, serotonin, and glycine
Matricaria chamomilla (German chamomile)	Essential oil and chamazulene	Interacts with benzodiazepine receptors
Melissa officinalis (lemon balm)	Essential oil (citral) and flavonoids	Inhibits GABA degradation
P. incarnata (passionflower)	Benzoflavone compound (BZF) and other flavonoids	Interacts with opioid and GABA-benzodiazepine receptors
Piper methysticum (kava kava)	Dihydrokawain, yangonin, demethoxyangonin, kawain, dihydromethysticinor methysticin, flavokawain A, and yangonin	Binds to GABA receptors; inhibits norepinephrine uptake
Scutellaria lateriflora (blue American skullcap)	Phenols (scutellarin, baicalin, baicalein, calamenene), tannin, ascorbic acid	Binds to GABA-benzodiazepine receptors; contains GABA
Valeriana officinalis (valerian)	Pyridine alkaloids, some organic acids and terpenes, especially the so-called valepotriates, esterified iridoid monoterpenes	Increases GABA release and inhibits GABA breakdown; binds to benzodiazepine receptors

Lipoic acid

Lipoic acid, an antioxidant, known for its unique ability to neutralize both intracellular and extracellular free radicals, has also been shown to prevent accumulation of catecholamines in cardiac tissue during stress and enhance the elimination of catecholamine degradation products. Thus, it can be said that α-lipoic acid results in indirect benefit when cortisol levels are high as it can partially restore hydrocortisone-induced suppression of helper T-cell activity (Fomichev and Pchelintsev, 1993).

L-theanine

Tea is often used as a relaxing beverage. L-theanine is an amino acid extracted from green or black tea. In the brain, L-theanine increases dopamine, serotonin, and the inhibitory neurotransmitter glycine (Yamada et al., 2009). Kakuda et al. (2000) have reported inhibitory effects of higher doses of theanine on caffeine stimulation evaluated by electroencephalogram (EEG) in the rat. However, when given by itself, lower doses of theanine resulted in excitatory effects, suggesting a dual activity of theanine depending on the dose. Nobre et al. (2008) have reported that even lower doses of L-theanine can induce alpha-wave production as observed in EEG tracings from 54 healthy participants at baseline and 45, 60, 75, 90, and 105 min after 50 mg L-theanine or placebo. Statistical analysis of data showed a significantly greater increase in alpha-wave production in the theanine group than the placebo group. Kimura et al. (2007) have reported that the acute stress response elicited by a math test was attenuated by 200 mg theanine—assessed by heart rate and salivary IgA.

Neurotransmitters and their precursors

Gamma aminobutyric acid

GABA is an inhibitory neurotransmitter widely distributed throughout the CNS. Too much excitation can lead to irritability, restlessness, insomnia, seizures, and movement disorders, so it must be balanced with inhibition which is provided by GABA acting like a "brake" during times of stress. Medications for anxiety, such as benzodiazepines, stimulate GABA receptors and induce relaxation. Either low GABA levels or decreased GABA function in the brain is associated with several psychiatric and neurological disorders, including anxiety, depression, insomnia, and epilepsy. Studies indicate that GABA can improve relaxation and enhance sleep. Because of the association between low GABA levels and these conditions, many antianxiety and sleep-enhancing drugs have been developed that interact primarily with GABA receptors. These include the benzodiazepine drugs—alprazolam, diazepam, flurazepam, quazepam, temazepam, and triazolam—and zolpidem tartrate and baclofen. GABA mediates presynaptic inhibition of primary afferent fibers in the motor system. It regulates brain excitability via $GABA_A$ receptors, which are classified into three major groups (alpha, beta, and gamma) with subunits that determine its pharmacological activity. GABA A receptors are highly expressed in the thalamus, a region of the brain involved with sleep processes (Orser, 2006). Bazil et al. (2005) have reported that gabapentin (structurally similar to GABA; increases brain GABA levels) has been found to be effective for panic disorder and improves sleep disturbances associated with alcohol consumption. Roth et al. (2006) showed that GABA-agonist drugs are sedatives that can be used to treat insomnia. Natural therapies that produce relaxation also act, at least in part, by enhancing GABA levels. Khom et al. (2007) have reported that valerenic acid potentiates and inhibits $GABA_A$ receptors. Geuze et al. (2008) have reported reduced $GABA_A$ receptor binding in veterans with posttraumatic stress disorder as demonstrated by positron emission tomography scan. On EEG, alpha waves are generated in a relaxed state, whereas beta waves are seen in stressful situations that make mental concentration difficult. Therefore, the ratio of alpha-to-beta waves is used as an indication of relaxation and better concentration. In general, the greater the alpha-to-beta ratio, the more relaxed and alert the person is. Abdou et al. (2006) have reported relaxation and immunity enhancement effects of GABA administration in humans and it was seen that GABA produced significant effects on both increasing alpha waves and decreasing beta waves, resulting in a significant increase in the alpha-to-beta wave ratio. Due to its relaxation effects, GABA may be considered to be a sleep aid.

L-tryptophan/5-hydroxytryptophan

L-tryptophan, a large neutral amino acid essential to human metabolism, is the metabolic precursor of serotonin (a neurotransmitter), melatonin (a neurohormone), and niacin (vitamin B3). Tryptophan hydroxylase is the rate-limiting enzyme for serotonin production and involves the conversion of tryptophan to 5-HTP. This enzyme can be inhibited by stress, insulin resistance, magnesium or vitamin B6 deficiency, or increasing age. Maron et al. (2004) have demonstrated the inhibitory effect of 5-HTP on cholecystokinin-4-induced panic attacks in healthy volunteers. Another study conducted by Schruers et al. (2002) examined that acute L-5-hydroxytryptophan administration inhibits carbon dioxide-induced panic attacks in panic disorder patients as displayed by reduced panic symptom scores, and number of panic attacks in comparison to placebo. Bruni et al. (2004) have reported that L-5-HT treatment helps in reduction of sleep terrors in children because of its enhancement of serotonin and then melatonin as compared to the untreated group.

Melatonin

Melatonin, the primary hormone of the pineal gland, acts as a powerful "chronobiotic," maintaining normal circadian rhythms. In patients with sleep disorders and altered circadian rhythms, such as occur in jet lag, nightshift work, and various neuropsychiatric disorders, oral administration of melatonin can provide the necessary resynchronization of those cycles. The primary physiological role identified for melatonin is its ability to influence circadian rhythms. Because the hours of highest melatonin secretion correlate to normal hours of sleep, it has been investigated for use in sleep disorders. Attenburrow et al. (1996) demonstrated that patients with insomnia have decreased nocturnal melatonin secretion. In a placebo-controlled trial of eight subjects with delayed sleep-phase insomnia, Dahlitz et al. (1991) found melatonin acts as a "phase-setter" for sleep–wake cycles. Similar results were obtained by another group of researchers who administered melatonin to six subjects with delayed sleep-phase insomnia at night. The onset of sleep started an average of 115 min earlier when taking melatonin as compared to no melatonin administration. Recent experimental studies have also shown that melatonin attenuated stress-induced anxiety and had antiseizure effects, which could be attributable to the ability of this pineal hormone to alter the balance between oxidative and nitrosative stress in the brain (Gulati and Ray, 2006; Gulati et al., 2009).

L-tyrosine

Findings from several studies suggest that supplementation with the amino acid tyrosine reduces the acute effects of stress and fatigue on task performance. During stressful episodes the brain reserves of the catecholamine neurotransmitters norepinephrine and dopamine are depleted in animals, which results in a decline in stress-induced performance. Tyrosine, which is the precursor for catecholamine synthesis, antagonizes the stress-induced decline in performance resulting from brain catecholamine depletion. To test the effect of tyrosine under these circumstances, Neri et al. (1995) studied a battery of cognitive performance tasks and mood scales during an extended wakefulness test and showed that tyrosine (150 mg/kg) was able to neutralize declines in performance/vigilance when compared to placebo. Deijen and Orlebeke (1994) investigated the effects of tyrosine administration on cadets during a military combat training course in which one group received a protein-rich drink containing 2 g tyrosine and another group received a carbohydrate-rich drink with the same amount of calories. The results showed that the group with the tyrosine-rich drink performed better on tasks involving memory and tracking and also decreased systolic blood pressure as compared to the carbohydrate-rich drink group. In a within-subject RCT, individuals ingesting 300 mg/kg tyrosine prior to cold emersion showed better resistance to stress as compared to placebo. It was observed that tyrosine supplementation attenuated cold exposure-induced reductions in working memory. In another study it was observed that tyrosine (100 mg/kg) supplementation also enhanced cardiovascular and cognitive performance in subjects exposed to stress-inducing low negative-pressure sessions (−50 mm Hg) for a maximum of 30 min.

Nitric oxide

The discovery and evolution of NO has been one of the most exciting areas of medicine and biology. The journey of NO from being an environmental pollutant, to a vasodilator (EDRF) and now a gasotransmitter/signaling molecule, has revolutionized medical sciences. Its ubiquitous distribution and multifunctional role in physiology and pathophysiology has been recognized and its therapeutic potential acknowledged. Modulations of NO levels by drugs are now proposed as mechanisms of action of some landmark therapeutic agents. In the CNS (neurons), as well as in the periphery (nitrergic nerves), NO is known to play a neurotransmitter role and various disease states have been explained on the basis of deficiency or excess of this molecule. The role of NO in stress and stress-related diseases has now been speculated. NO is well known to modulate the HPA axis. It is an intracellular messenger in the brain, and plays an important role in various physiological and pathological processes. It is generated from L-arginine via NOS. NO and stress physiology share many common pathways and molecular mechanisms, and NO plays a role in many stress-related diseases. NO is widely distributed in the CNS and recent studies have shown that free radicals and in particular NO may play a crucial role in the regulation of acute and chronic stress effects. In fact, NO has been implicated in behavioral disorders like anxiety, seizures, Alzheimer disease, etc. and also in the effects of neurotherapeutic agents like opioids, benzodiazepines, neurosteroids, etc. (Ray et al., 2007; Chakraborty et al., 2011; Gądek-Michalska et al., 2013; Gulati et al., 2015; Dubey et al., 2018).

Classically, NO is generated from L-arginine via the help of NOS and various cofactors endogenously. Although supplemental arginine may augment NO activity in NO-deficient situations, in patients with endothelial dysfunction having an impaired ability to convert l-arginine to NO, such L-arginine supplementation is not beneficial. Currently, it is known that NO synthesis occurs via at least two physiological pathways: NOS dependent and NOS independent. In the former, the precursor amino acid, L-arginine is oxidized to NO and citrulline by the action of the NOS enzymes (NOS1, NOS2, and NOS3). Additionally, L-citrulline has been indicated to be a secondary NO donor in the NOS-dependent pathway, since it can recycle back to L-arginine. Alternatively, dietary nitrates and nitrites are the main substrates to produce NO via the NOS-independent pathway. These anions can be reduced in vivo to NO and other bioactive nitrogen oxides. Other molecules, such as the dietary supplement glycine propionyl-L-carnitine, have also been suggested to increase levels of NO, although the physiological mechanisms remain to be elucidated (Bescos et al., 2012).

Thus, dietary supplements can be a source for NO in NO-deficient states and pathophysiological states. There are two ways by which such NO activity can be increased:

1. L-arginine supplementation by food rich in L-arginine such as turkey, chicken, pumpkin seeds, soybeans, peanuts, spirulina, dairy products, chickpeas, lentils; and
2. Nitrate-rich food supplements, for example, celery, cress, chervil, lettuce, red beetroot, spinach, which are converted into nitrites by the salivary bacteria (Kurtz et al., 2018; Hord et al., 2009).

Vitamins as neurotransmitters or cofactors

Thiamine (vitamin B1)

Studies have shown that thiamine is effective as a nutrient in protecting the adrenal gland from functional exhaustion secondary to surgery. Intramuscular injections of thiamine at a dose of 120 mg/day, starting several days prior to surgery and 1.5–2.0 h immediately prior to surgery, reduced the cortisol reaction, both prior to and at the height of the surgery. Continued administration of thiamine post-surgery prevented the usual post-surgery reduction in blood cortisol levels (Vinogradov et al., 1981).

Niacinamide (vitamin B3)

Vitamin B3 is found in blood in the form of niacinamide (also known as nicotinamide). It can enter the BBB easily, is nonaddictive, and is devoid of the flush effect of the niacin form of vitamin B3. Tryptophan can either be converted to niacin, or serotonin, and exogenous niacinamide may signal via feedback inhibition to decrease the activity of tryptophan pyrrolase (the enzyme that converts tryptophan to niacin). This allowed more tryptophan to be converted to 5-HTP and then to serotonin. In 1979, Hoffman La Roche, the Swiss drug company which manufactured Valium, described niacinamide as "a brain constituent that has BZD-like actions." Niacinamide helps in sleep enhancement as evidenced in a 3-week study of six subjects with normal sleep patterns and two with insomnia using EEGs, electromyograms, and electrooculograms to evaluate sleep patterns at baseline and after niacinamide treatment. There was a significant increase in rapid eye movement (REM) sleep in all normal-sleeping subjects, while the two subjects with moderate-to-severe insomnia experienced significant increases in REM sleep by the third week; awake time was also significantly decreased (Robinson et al., 1977).

Pantethine/pantothenic acid (vitamin B5)

Evidence indicates that adrenal cortex function is compromised in the event of a deficiency of vitamin B5 derivatives and metabolites. On the other hand, administration of pantethine (active vitamin B5) in several experimental animal models appeared to enhance adrenal cortex function. Administration of pantethine to humans with a variety of clinical conditions buffered the rise in urinary cortisol metabolites expected to occur secondary to a loading dose of ACTH, suggesting pantethine can downregulate hypersecretion of cortisol secondary to high-stress conditions (Onuki and Hoshino, 1970; Onuki and Suzawa, 1970).

Pyridoxal 5′-phosphate (active vitamin B6)

Pyridoxal 5′-phosphate (the active form of vitamin B6) is a necessary cofactor for the formation of several important neurotransmitters associated with stress. Within the brain, glutamic acid is converted to GABA via the enzyme glutamate decarboxylase and its cofactor pyridoxal 5′-phosphate. GABA is metabolized by gamma-aminobutyrate transaminase, also a P5P-dependent enzyme, forming an intermediate metabolite succinate semialdehyde. P5P is a cofactor in the conversion of 5-HTP to serotonin. Furthermore, conversion of L-tryptophan to 5-HTP, the rate-limiting step in serotonin synthesis, can be inhibited by stress, insulin resistance, magnesium or vitamin B6 deficiency, or increasing age. The decarboxylation of 5-HTP to serotonin is dependent on the presence of pyridoxal 5′-phosphate. P5P is also a cofactor in the synthesis of dopa to dopamine in the pathway converting tyrosine to epinephrine and norepinephrine (van Praag and Lemus, 1986).

Methylcobalamin (vitamin B12)

Stress disturbs the circadian rhythmic secretion of cortisol. To correct this disturbance in cortisol secretion an effective method is a combination of bright-light exposure and methylcobalamin. Methylcobalamin enhances the light sensitivity of the circadian clock (Honma et al., 1992; Hashimoto et al., 1996) and also improves the quality of sleep (Uchiyama et al., 1995). Perhaps the greatest advantage of methylcobalamin's effect on resetting circadian rhythms secondary to stress is its impact on cortisol. It is thought that methylcobalamin does not impact total levels of cortisol, rather it helps shift the cortisol secretion peak, helping place the cortisol clock back on schedule (Tomoda et al., 1995).

5-Methyltetrahydrofolate (active folate)

Tetrahydrobiopterin (BH4) is necessary for the formation of the monoamine neurotransmitters serotonin, dopamine, norepinephrine, and epinephrine and, in the process, it gets oxidized. Folate helps in regenerating tetrahydrobiopterin. BH4 acts as a rate-limiting enzyme cofactor to the hydroxylase enzymes that metabolize tryptophan to 5-HTP, phenylalanine to tyrosine, and tyrosine to dopa (Anderson et al., 1992).

Ascorbic acid (vitamin C)

Ascorbic acid is involved in monoamine neurotransmitter synthesis and acts both as a cofactor at the enzyme level and a stabilizer of BH4, which prevents oxidation of BH4 and increases BH4 levels. Therefore intracellular BH4 levels are dependent on cellular levels of ascorbate (Heller et al., 2001). Ascorbic acid in levels significantly greater than the RDA can support adrenal function and decrease high cortisol levels (Nakai et al., 2003). Administration of ascorbic acid improved the capacity of the adrenals to adapt to surgical stress by normalizing cortisol and ACTH in patients with lung cancer (Gromova et al., 1990). Antistress and immunomodulatory effects of ascorbic (alone and/or in combination with tocopherol) have been shown in different experimental models in which stress-induced anxiety and immune suppression have been attenuated by the compound which were attributable to its antioxidant properties (Pal et al., 2006, 2009, 2011). A combination of ascorbic acid (300 mg three times daily) and vitamins B1 and B6 administered intravenously improved glucocorticoid function of the adrenal glands and simultaneously normalized the rhythmic activity of the gland.

Lycopene

Lycopene is a bright red carotene and carotenoid pigment and phytochemical found in tomatoes and other red fruits and vegetables, such as red carrots, watermelons, and papayas, but not in strawberries, red bell peppers, or cherries. Although lycopene is chemically a carotene, yet it has no vitamin A activity. Carotenoids like lycopene are important pigments found in photosynthetic pigment–protein complexes in plants, photosynthetic bacteria, fungi, and algae. Structurally, lycopene is a tetraterpene and assembled from eight isoprene units that are composed entirely of carbon and hydrogen. It is insoluble in water. Carotenoids act as antioxidants in oxidative stress damage (Basu and Imrhan, 2007). Preliminary research has shown that people who consume tomatoes may have a lower cancer risk, possibly due to lycopene affecting mechanisms of prostate cancer although data are insufficient (Vance Terrence et al., 2013; Khan et al., 2008). Lycopene is nontoxic and is commonly found in the diet, but cases of excessive carotenoid intake have been reported. It causes discoloration of the skin, known as lycopenodermia, and is nontoxic. There are also cases of intolerance or allergic reaction to dietary lycopene, which may cause diarrhea, nausea, stomach pain or cramps, gas, vomiting, and loss of appetite.

Nutraceuticals for life style diseases: the future

The rapidly evolving life style and its problems have drawn attention toward improving people's fitness and health to face these challenges. Spurred by media coverage, it is prompting a majority of people to approach for healthier lifestyles. The expanding nutraceutical market indicates that people are avoiding highly processed food (e.g., junk food) and are seeking items with extra nutritional benefits and organoleptic value. The unprecedented growth of the nutraceutical industry has had a telling impact on the food, pharmaceutical, healthcare, and agricultural sectors and it seems destined to expand much more in the years to come. As for the future, scientists currently believe that enzymes and fermentation technology using microbes represent another exciting domain to create new food products in nutraceuticals. This global trend of people seeking nutrified products in search of a healthy lifestyle is in the greater interest of humanity. Entrepreneurs are now taking the initiative by investing heavily and strategically in relevant science, product development, marketing, and consumer education. The advent of rational use of herbal preparations and phytopharmaceuticals has further facilitated this issue. The use of medicinal plants and other food supplements as adaptogens for the prevention and cure of stress-related pathophysiological states in the traditional as well as modern systems of medicine is encouraging and likely to have a great impact on healthcare costs in lifestyle diseases.

Concluding remarks and future directions

The biological system is constantly exposed to aversive stimuli like stress which disrupt the physiological homeostasis if coping mechanisms are inadequate. Several pathophysiological conditions can be precipitated by chronic stress, which involves complexly interactive pathways and body systems, such as the cardiovascular, neuroendocrine, immune, CNS, gastrointestinal, systems. The neuroendocrine–immune axis plays a key role in stress response/effect regulation. Stress-related diseases are lifestyle disorders and certain changes have to be made in the daily routine to enhance coping abilities. For example, good diet, exercise, meditation, positive approach, etc. play a major role in combating stress-related health abnormalities. A number of nutrients and botanicals can provide support for stress-related conditions. Such support requires a five-pronged approach: (1) use of adaptogens for support of the adrenals, (2) normalization of cortisol levels by usage of dietary nutrients, (3) use of anxiolytic herbs for correction of sleep disorders and anxiety, (4) balancing neurotransmitters with amino acid precursors, and (5) providing necessary dietary nutrient cofactors. Adaptogens are agents which protect against the harmful effects of stress and modify stress-related disease states. Nutraceuticals from diverse sources, which are sometimes products associated with our daily lives, could prevent the occurrence and reduce the impact of stress effects on the biological system. These agents have significant potential in the promotion of human health and alleviating disease and thus, scientists, medical professionals, and nutritionists can collaborate for planning strategies to provide the ultimate health and therapeutic benefit. In recent years, there has been an increasing body of evidence to suggest that nutraceuticals like food supplements, antioxidants, vitamins, and herbal preparations can play a major role in such situations. However, more randomized clinical trials are required to scientifically validate the use of such nutraceuticals in various stress-related conditions—both for prophylaxis and treatment.

References

Abdou, A.M., Higashiguchi, S., Horie, K., et al., 2006. Relaxation and immunity enhancement effects of gamma aminobutyric acid (GABA) administration in humans. Biofactors 26, 201–208.

Anand, R., Gulati, K., Ray, A., 2012. Pharmacological evidence for the role of nitric oxide in the modulation of stress-induced anxiety by morphine in rats. Eur. J. Pharmacol. 676, 71–74.

Anderson, D.N., Abou-Saleh, M.T., Collins, J., et al., 1992. Pterin metabolism in depression: an extension of the amine hypothesis and possible marker of response to ECT. Psychol. Med. 22, 863–869.

Andreatini, R., Sartori, V.A., Seabra, M.L., et al., 2002. Effect of valepotriates (valerian extract) in generalized anxiety disorder: a randomized placebo-controlled pilot study. Phytother. Res. 16, 650–654.

Attenburrow, M.E., Dowling, B.A., Sharpley, A.L., et al., 1996. Case-control study of evening melatonin concentration in primary insomnia. BMJ 312, 1263–1264.

Azgomi, R.N.D., Nazemiyeh, H., Sadeghi Bazargani, H.S., et al., 2018. Comparative evaluation of the effects of *Withania somnifera* with pentoxifylline on the sperm parameters in idiopathic male infertility: A triple-blind randomized clinical trial. Andrologia 50 (7), e13041. Available from: https://doi.org/10.1111/and.13041.

Baillie, J.K., Thompson, A.A.R., Irving, J.B., et al., 2009. Oral antioxidant supplementation does not prevent acute mountain sickness: double blind, randomized placebo-controlled trial. QJM 102 (5), 341–348.

Basu, A., Imrhan, V., 2007. Tomatoes versus lycopene in oxidative stress and carcinogenesis: conclusions from clinical trials. Eur. J. Clin. Nutr. 61 (3), 295–303.

Bazil, C.W., Battista, J., Basner, R.C., 2005. Gabapentin improves sleep in the presence of alcohol. J. Clin. Sleep Med. 1, 284–287.

Bescos, R., Sureda, A., Tur, J.A., et al., 2012. The effect of nitric oxide-related supplements on human performance. Sports Med. 42 (3), 1–19.

Bhattacharya, S.K., Ghosal, S., 1998. Anxiolytic activity of a standardized extract of *Bacopa monniera* in an experimental study. Phytomedicine 5, 77–82.

Bodesheim, U., Holzl, J., 1997. Isolation and receptor binding properties of alkaloids and lignans from *Valeriana officialis* L. Pharmazie 52, 386–391.

Boothe, D.M., 1997. Nutraceuticals in veterinary medicine. Part I. Definitions and regulations. Comp. Cont. 19 (11), 1248–1255.

Boothe, D.M., 1998. Nutraceuticals in veterinary medicine. Part II. Safety and efficacy. Comp. Cont. 20 (1), 15–21.

Bouic, P.J., Clark, A., Lamprecht, J., et al., 1999. The effects of s-sitosterol (BSS) and s-sitosterol glucoside (BSSG) mixture on selected immune parameters of marathon runners: inhibition of post marathon immune suppression and inflammation. Int. J. Sports Med. 20, 258–262.

Bruni, O., Ferri, R., Miano, S., et al., 2004. L-5- hydroxytryptophan treatment of sleep terrors in children. Eur. J. Pediatr. 163, 402–407.

Calabrese, C., Gregory, W.L., Leo, M., et al., 2008. Effects of a standardized *Bacopa monnieri* extract on cognitive performance, anxiety, and depression in the elderly: a randomized, double-blind, placebo-controlled trial. J. Altern. Complementary Med. 14, 707–713.

Cannon, W.B., 1929. Organization for physiological homeostasis. Physiol. Rev. 9, 399–431.

Chakraborty, A., Gulati, K., Ray, A., 2011. Involvement of nitric oxide in the protective effects of dehydroepiandrosterone sulphate on stress induced neurobehavioral suppression and brain oxidative injury in rats. Eur. J. Pharmacol. 652, 55–59.

Chan, L.Y., Chiu, P.Y., Lau, T.K., 2003. An in vitro study of ginsenoside Rb1-induced teratogenicity using a whole rat embryo culture model. Hum. Reprod. 18, 2166–2168.

Channa, S., Dar, A.S., Yaqoob, M., et al., 2006. Anti-inflammatory activity of *Bacopa monnieri* in rodents. J. Ethnopharmacol. 104, 286–289.

Choudhary, D., Bhattacharyya, S., Bose, S., 2017. Efficacy and safety of ashwagandha (*Withania somnifera* (L.) Dunal) root extract in improving memory and cognitive functions. J. Diet. Suppl. 14 (6), 599–612. Available from: https://doi.org/10.1080/19390211.2017.1284970.

Chowdhuri, D.K., Parmar, D., Kakkar, P., et al., 2002. Anti-stress effects of bacosides of *Bacopa monnieri*: modulation of Hsp70 expression, superoxide dismutase and cytochrome P450 activity in rat brain. Phytother. Res. 16 (7), 639–645.

Chrousos, G.P., 1998. Stressors, stress and neuroendocrine integration of the adaptive response, The 1997 Hans Selye Memorial Lecture. Ann. N. Y. Acad. Sci. 851, 311–335.

Chrousos, G.P., 2009. Stress and disorders of the stress system. Nat. Rev. Endocrinol. 5, 374–381.

Dahlitz, M., Alvarez, B., Vignau, J., et al., 1991. Delayed sleep phase syndrome response to melatonin. Lancet 337, 1121–1124.

Deijen, J.B., Orlebeke, J.F., 1994. Effect of tyrosine on cognitive function and blood pressure under stress. Brain Res. Bull. 33, 319–323.

Di Matteo, V., Esposito, E., 2003. Biochemical and therapeutic effects of antioxidants in the treatment of Alzheimer's disease, Parkinson's disease, and amyotrophic lateral sclerosis. Curr. Drug. Targets CNS Neurol. Disord. 2 (2), 95–107.

Dominguez, R.A., Bravo-Valverde, R.L., Kaplowitz, B.R., et al., 2000. Valerian as a hypnotic for Hispanic patients. Cult. Divers. Ethnic Minor. Psychol. 6, 84–92.

Donath, F., Quispe, S., Diefenbach, K., et al., 2000. Critical evaluation of the effect of valerian extract on sleep structure and sleep quality. Pharmacopsychiatry 33, 47–53.

Drozd, J., Sawicka, T., Prosinska, J., 2002. Estimation of humoral activity of *Eleutherococcus senticosus*. Acta Pol. Pharm. Drug. Res. 59, 395–401.

Dubey, H., Gulati, K., Ray, A., 2018. Effects of s-nitrosoglutathione (NO donor) on intracerebroventricular streptozotocin induced neurobehavioral and biochemical changes in rats. NeuroToxicology 66, 58–65.

Durg, S., Shivaram, S.B., Bavage, S., 2018. *Withania somnifera* (Indian ginseng) in male infertility: an evidence-based systematic review and meta-analysis. Phytomedicine 50, 247–256. Available from: https://doi.org/10.1016/j.phymed.2017.11.011.

Dzanis, D.A., 1998. Nutraceuticals: Food or Drug? TNAVC Proceedings, pp. 430–431.

Fomichev, V.I., Pchelintsev, V.P., 1993. The neurohumoral systems of patients with ischemic heart disease and under emotional-pain stress: the means for their pharmacological regulation. Kardiologiia 33, 15–18.

Gądek-Michalska, A., Tadeusz, J., Rachwalska, P., et al., 2013. Cytokines, prostaglandins and nitric oxide in the regulation of stress-response systems. Pharmacol. Rep. 65, 1655–1666.

Geuze, E., van Berckel, B.N., Lammertsma, A.A., et al., 2008. Reduced $GABA_A$ benzodiazepine receptor binding in veterans with post-traumatic stress disorder. Mol. Psychiatr. 13, 74–83.

Graeff, F.G., Guimarães, F.S., De Andrade, T.G., et al., 1996. Role of 5-HT in stress, anxiety, and depression. Pharmacol. Biochem. Behav. 54 (1), 129–141.

Gromova, E.G., Sviridova, S.P., Kushlinskii, N.E., et al., 1990. Regulation of the indices of neuroendocrine status in surgical patients with lung cancer using optimal doses of ascorbic acid. Anesteziol. Reanimatol. 5, 71–74.

Gulati, K., Chakrabort, A., Ray, A., 2007. Modulation of stress induced neurobehavioral changes and brain oxidative injury by nitric oxide (NO) mimetics in rats. Behav. Brain Res. 183, 226–230.

Gulati, K., Joshi, J.C., Ray, A., 2015. Recent advances in stress research: Focus on nitric oxide. Eur. J. Pharmacol. 765, 406–414.

Gulati, K., Chakraborti, A., Ray, A., 2009. Differential role of nitric oxide (NO) in acute and chronic stress induced neurobehavioral modulation and oxidative injury in rats. Pharmacol. Biochem. Behav. 92, 272–276.

Gulati, K., Ray, A., 2006. Studies on the mechanisms of anxiolytic and neuropeotective effects of melatonin in rats. Front. Neuroendocrinol. 27, 129–130.

Gupta, G.L., Rana, A.C., 2008. Effect of *Withania somnifera* Dunal in ethanol-induced anxiolysis and withdrawal anxiety in rats. Indian J. Exp. Biol. 46, 470–475.

Hashimoto, S., Kohsaka, M., Morita, N., et al., 1996. Vitamin B12 enhances the phase-response of circadian melatonin rhythm to a single bright light exposure in humans. Neurosci. Lett. 220, 129–132.

Hattesohl, M., Feistel, B., Sievers, H., et al., 2008. Extracts of *Valeriana officinalis* L. show anxiolytic and antidepressant effects but neither sedative nor myorelaxant properties. Phytomedicine 15, 2–15.

Hegarty, B., Parker, G., 2013. Fish oil as a management component for mood disorders—an evolving signal. Curr. Opin. Psychiat 26 (1), 33–40.

Heller, R., Unbehaun, A., Schellenberg, B., et al., 2001. L-ascorbic acid potentiates endothelial nitric oxide synthesis via a chemical stabilization of tetrahydrobiopterin. J. Biol. Chem. 276, 40–47.

Hellhammer, J., Fries, E., Buss, C., et al., 2004. Effects of soy lecithin phosphatidic acid and phosphatidylserine complex (PAS) on the endocrine and psychological responses to mental stress. Stress 7, 119–126.

Hofseth, L.J., Wargovich, M.J., 2007. Inflammation, cancer, and targets of ginseng. J. Nutr. 137, 183S–185S.

Honma, K., Kohsaka, M., Fukuda, N., et al., 1992. Effects of vitamin B12 on plasma melatonin rhythm in humans: increased light sensitivity phase-advances the circadian clock? Experientia 48, 716–720.

Hord, N.G., Tang, Y., Bryan, N.S., 2009. Food sources of nitrates and nitrites: the physiologic context for potential health benefits. Am. J. Clin. Nutr. 90, 1–10.

Hwang, S.H., Ha, E.S., Yu, K.W., et al., 2003. Effect of the crude polysaccharides fraction from *Eleutherococcus senicosus* as a immunoadjuvant to soluble antigens (BSA and OVA). Yakhak Hoeji 47, 167–175.

Jung, C.H., Jung, H., Shin, Y.C., et al., 2007. *Eleutherococcus senticosus* extract attenuates LPS-induced iNOS expression through the inhibition of Akt and JNK pathways in murine macrophage. J. Ethnopharmacol. 113, 183–187.

Kakuda, T., Nozawa, A., Unno, T., et al., 2000. Inhibiting effects of theanine on caffeine stimulation evaluated by EEG in the rat. Biosci. Biotechnol. Biochem. 64, 287–293.

Keum, Y.S., Han, S.S., Chun, K.S., 2003. Inhibitory effects of the ginsenoside Rg3 on phorbol ester-induced cyclooxygenase-2 expression, NF-kappaB activation and tumor promotion. Mutat. Res. 523-524, 75–85.

Khan, N., Afaq, F., Mukhtar, H., 2008. Cancer chemoprevention through dietary antioxidants: progress and promise. Antioxid. Redox Signal. 10 (3), 475–510.

Khom, S., Baburin, I., Timin, E., et al., 2007. Valerenic acid potentiates and inhibits GABA(A) receptors: molecular mechanism and subunit specificity. Neuropharmacology 53, 178–187.

Kim, K.H., Lee, D., Lee, H.L., et al., 2018. Beneficial effects of *Panax ginseng* for the treatment and prevention of neurodegenerative diseases: past findings and future directions. J. Ginseng Res. 42, 239–247.

Kimura, K., Ozeki, M., Juneja, L.R., et al., 2007. L-theanine reduces psychological and physiological stress responses. Biol. Psychol. 74, 39–45.

Kulkarni, S.K., Dhir, A., 2008. *Withania somnifera*: an Indian ginseng. Prog. Neuropsychopharmacol. Biol. Psychiat. 32 (5), 1093–1105.

Kurtz, T.W., DiCarlo, S.E., Pravenec, M., et al., 2018. Functional foods for augmenting nitric oxide activity and reducing the risk for salt-induced hypertension and cardiovascular disease in Japan. J. Cardiol. 72, 42–49.

Kvetnansky, R., Sabban, E.L., Palkovits, M., 2009. Catecholaminergic systems in stress: structural and molecular genetic approaches. Physiol. Rev. 89, 535–606.

Liao, L., He, Y., Li, L., et al., 2018. A preliminary review of studies on adaptogens: comparison of their bioactivity in TCM with that of ginseng-like herbs used worldwide. Chin. Med. 13, 57–68. Available from: https://doi.org/10.1186/s13020-018-0214-9.

Limpeanchob, N., Jaipan, S., Rattanakaruna, S., et al., 2008. Neuroprotective effect of *Bacopa monnieri* on beta-amyloid-induced cell death in primary cortical culture. J. Ethnopharmacol. 120, 112–117.

Lizko, N.N., Silov, V.M., Syrych, G.D., 1984. Events in the development of dysbacteriosis of the intestines in man under extreme conditions. Nahrung 28, 599–605.

Lopresti, A.L., Drummond, P.D., Smith, S.J., 2019. A randomized, double-blind, placebo-controlled, crossover study examining the hormonal and vitality effects of ashwagandha (*Withania somnifera*) in aging, overweight males. Am. J. Men's Health 13, 1–15. Available from: https://doi.org/10.1177/1557988319835985.

Maron, E., Toru, I., Vasar, V., et al., 2004. The effect of 5-hydroxytryptophan on cholecystokinin-4-induced panic attacks in healthy volunteers. J. Psychopharmacol. 18, 194–199.

Mazereeuw, G., Lanctôt, K.L., Chau, S.A., et al., 2012. Effects of omega-3 fatty acids on cognitive performance: a meta-analysis. Neurobiol. Aging 33 (7), e17–29.

McEwen, B.S., 2008. Central effects of stress hormones in health and disease: understanding the protective and damaging effects of stress and stress mediators. Eur. J. Pharmacol. 583, 174–185.

McEwen, B.S., Gianaros, P.J., 2010. Central role of the brain in stress and adaptation: links to socio-economic status, health, and disease. Ann. N. Y. Acad. Sci. 1186, 190–222. Available from: https://doi.org/10.1111/j.1749-6632.2009.05331.x.

McEwen, B.S., Bowles, N.P., Gray, J.D., et al., 2015. Mechanisms of stress in the brain. Nat. Neurosci. 18, 1353–1363.

Miles, E.A., Calder, P.C., 2012. Influence of marine n-3 polyunsaturated fatty acids on immune function and a systematic review of their effects on clinical outcomes in rheumatoid arthritis. Br. J. Nutr. 107 (Suppl 2), S171–S184.

Moore, W.E., Cato, E.P., Holdeman, L.V., 1978. Some current concepts in intestinal bacteriology. Am. J. Clin. Nutr. 31, S33–S42.

Nakai, K., Urushihara, M., Kubota, Y., et al., 2003. Ascorbate enhances iNOS activity by increasing tetrahydrobiopterin in *RAW* 264.7 cells. Free Radic. Biol. Med. 35, 929–937.

Neri, D.F., Wiegmann, D., Stanny, R.R., et al., 1995. The effects of tyrosine on cognitive performance during extended wakefulness. Aviat. Space Environ. Med. 66, 313–319.

Nobre, A.C., Rao, A., Owen, G.N., 2008. L-theanine, a natural constituent in tea, and its effect on mental state. Asia Pac. J. Clin. Nutr. 17, 167–168.

Onuki, M., Hoshino, H., 1970. Effect of pantetheine on the adrenocortical function. 1. Experimental results using rabbits. Horumon Rinsho 18, 601–605.

Onuki, M., Suzawa, A., 1970. Effect of pantethine on the function of the adrenal cortex. Clinical experience using pantethine cases steroid hormone treatment. Horumon Rinsho 18, 937–940.

Orser, B.A., 2006. Extrasynaptic GABAA receptors are critical targets for sedative-hypnotic drugs. J. Clin. Sleep. Med. 2, S12–S18.

Pal, R., Gulati, K., Banerji, B.D., et al., 2006. Role of free radicals in stress-induced neurobehavioral changes in rats. Indian J. Exp. Biol. 44, 809–815.

Pal, R., Ahmed, T., Kumar, V., et al., 2009. Protective effects of different antioxidants against enosulfan induced oxidative stress and immunotoxicity in albino rats. Indian J. Exp. Biol. 47, 723–729.

Pal, R., Gulati, K., Banerji, B.D., Ray, A., 2011. Possible involvement of free radicals in modulation of stress-induced immunological and biochemical changes in rats. Int. Immunopharmacol. 11, 1680–1684.

Palatnik, A., Frolov, K., Fux, M., et al., 2001. Double-blind, controlled, crossover trial of inositol versus fluvoxamine for the treatment of panic disorder. J. Clin. Psychopharmacol. 21 (3), 335–339.

Panossian, A., Wikman, G., 2010. Effects of adaptogens on the central nervous system and the molecular mechanisms associated with their stress-protective activity. Pharmaceuticals 3, 188–224. Available from: https://doi.org/10.3390/ph3010188.

Poyares, D.R., Guilleminault, C., Ohayon, M.M., et al., 2002. Can valerian improve the sleep of insomniacs after benzodiazepine withdrawal? Prog. Neuropsychopharmacol. Biol. Psychiatry 26, 539–545.

Pratte, M.A., Kaushal, B.B.S., et al., 2014. An alternative treatment for anxiety: a systematic review of human trial results reported for the ayurvedic herb ashwagandha (*Withania somnifera*). J. Altern. Comp. Med. 20, 901–908.

Ray, A., Puri, S., Chakravarty, A.K., et al., 1992. Central histaminergic involvement during stress in rats. Indian J. Exp. Biol. 30, 724–728.

Ray, A., Henke, P.G., Gulati, K., et al., 1993. The amygdaloid complex, corticotropin releasing factor and stress-induced gastric ulcerogenesis in rats. Brain Res. 624, 286–290.

Ray, A., Chakraborti, A., Gulati, K., 2007. Current trends in nitric oxide research. Cell Mol. Biol. 53, 3–14.

Ray, A., Chakraborti, A., Gulati, K., 2009. Stress induced anxiogenesis and brain oxidative injury in rats: influence of age, sex and emotionality. J. Cerebr. Blood Flow. Metab. 29, S455–S466.

Ray, A., Gulati, K., Anand, R., 2016. Stress, adaptogens and their evaluation: an Overview. J. Pharma Rep. 1 (2), 1–10.

Reay, J.L., Kennedy, D.O., Scholey, A.B., 2005. Single doses of *Panax ginseng* (G115) reduce. Psychopharmacology 19, 357–365.

Robinson, C.R., Pegram, G.V., Hyde, P.R., et al., 1977. The effects of nicotinamide upon sleep in humans. Biol. Psychiatry 12, 139–143.

Roth, T., Soubrane, C., Titeux, L., et al., 2006. Efficacy and safety of zolpidem-MR: a double-blind, placebo-controlled study in adults with primary insomnia. Sleep Med. 7, 397–406.

Schruers, K., van Diest, R., Overbeek, T., et al., 2002. Acute L-5-hydroxytryptophan administration inhibits carbon dioxide-induced panic in panic disorder patients. Psychiatry Res. 113, 237–243.

Seely, D., Dugoua, J.J., Perri, D., Mills, E., et al., 2008. Safety and efficacy of *Panax ginseng* during pregnancy and lactation. Can. J. Clin. Pharmacol. 15, e87–e94.

Selye, H., 1950. Stress and the general adaptation syndrome. Br. Med. J. 4667, 1383–1392.

Sengupta, P., Agarwal, A., Pogrebetskaya, M., et al., 2018. Role of *Withania somnifera* (Ashwagandha) in the management of male infertility. Reprod. Biomed. Online 36 (3), 311–326. Available from: https://doi.org/10.1016/j.rbmo.2017.11.007.

Sies, H., 1997. Oxidative stress: oxidants and antioxidants. Exp. Physiol. 82 (2), 291–295.

Singh, N., Bhalla, M., de Jager, P., et al., 2011. An overview on ashwagandha: a rasayana (rejuvenator) of ayurveda. Afr. J. Tradit. Complementary Altern. Med. 8 (S), 208–213.

Slavich, G.M., 2016. Life stress and health: a review of conceptual issues and recent findings. Teach. Psychol. 43 (4), 346–355. Available from: https://doi.org/10.1177/0098628316662768.

Sotaniemi, E.A., Haapakoski, E., Rautio, A., 1995. Ginseng therapy in non-insulin-dependent diabetic patients. Diabetes Care 18, 1373–1375.

Starks, M.A., Starks, S.L., Kingsley, M., et al., 2008. The effects of phosphatidylserine on endocrine response to moderate intensity exercise. J. Int. Soc. Sports Nutr. 5, 11.

Stough, C., Lloyd, J., Clarke, J., et al., 2001. The chronic effects of an extract of *Bacopa monnieri* (Brahmi) on cognitive function in healthy human subjects. Psychopharmacol. (Berl.) 156, 481–484.

Sung, H., Kang, S.M., Lee, M.S., 2005. Korean red ginseng slows depletion of CD4 T cells in human immunodeficiency virus type 1-infected patients. Clin. Diagn. Lab. Immunol. 12, 497–501.

Tiwari, R., Chakraborty, S., Saminathan, M., et al., 2014. Ashwagandha (*Withania somnifera*): role in safeguarding health, immunomodulatory effects, combating infections and therapeutic applications: A review. J. Biol. Sci. 14, 77–94.

Tokunaga, S., Takeda, Y., Niimoto, T., et al., 2007. Effect of valerian extract preparation (BIM) on the sleep-wake cycle in rats. Biol. Pharm. Bull. 30, 363–366.

Tomoda, A., Miike, T., Matsukura, M., 1995. Circadian rhythm abnormalities in adrenoleukodystrophy and methyl B12 treatment. Brain Dev. 17, 428–431.

Trevena, L., 2004. Sleepless in Sydney—is valerian an effective alternative to benzodiazepines in the treatment of insomnia? ACP J. Club 141, A14–A16.

Uchiyama, M., Mayer, G., Okawa, M., et al., 1995. Effects of vitamin B12 on human circadian body temperature rhythm. Neurosci. Lett. 192, 1–4.

van Praag, H.M., Lemus, C., 1986. Monoamine precursors in the treatment of psychiatric disorders. In: Wurtman, R.J., Wurtman, J.J. (Eds.), Nutrition and the Brain. Raven Press, New York, NY, pp. 89–139.

Vance Terrence, M., Su, J., Fontham, E.T.H., et al., 2013. Dietary antioxidants and prostate cancer: a review. Nutr. Cancer 65 (6), 793–801.

Vinogradov, V.V., Tarasov, I.A., Tishin, V.S., et al., 1981. Thiamine prevention of the corticosteroid reaction after surgery. Probl. Endokrinol. (Mosk.) 27, 11–16.

Volate, S.R., Davenport, D.M., Muga, S.J., et al., 2005. Modulation of aberrant crypt foci and apoptosis by dietary herbal supplements (quercetin, curcumin, silymarin, ginseng and rutin). Carcinogenesis (26), 1450–1456.

Vuksan, V., Sung, M.K., Sievenpiper, J.L., 2008. Korean red ginseng (*Panax ginseng*) improves glucose and insulin regulation in well-controlled, type 2 diabetes: results of a randomized, double-blind, placebo-controlled study of efficacy and safety. Nutr. Metab. Cardiovasc. Dis. 18, 46–56.

Wang, X., Michaelis, E.K., 2010. Selective neuronal vulnerability to oxidative stress in the brain. Front. Aging Neurosci. 2, 1–13.

Wankhede, S., Langade, D., Joshi, K., et al., 2015. Examining the effect of *Withania somnifera* supplementation on muscle strength and recovery: a randomized controlled trial. J. Intern. Soc. Sports Nutr. 12, 1–11. Available from: https://doi.org/10.1186/s12970-015-0104-9.

Yamada, T., Terashima, T., Kawano, S., et al., 2009. Theanine, gamma-glutamylethylamide, a unique amino acid in tea leaves, modulates neurotransmitter concentrations in the brain striatum interstitium in conscious rats. Amino Acids 36, 21–27.

Yaribeygi, H., Panahi, Y., Sahraei, H., et al., 2017. The impact of stress on body function: a review. EXCLI J. 16, 1057–1072.

Yue, P.Y., Mak, N.K., Cheng, Y.K., 2007. Pharmacogenomics and the yin/yang actions of ginseng: anti-tumor, angiomodulating and steroid-like activities of ginsenosides. Chin. Med. 2, 6.

Zheng, J.S., Hu, X.J., Zhao, Y.M., et al., 2013. Intake of fish and marine n-3 polyunsaturated fatty acids and risk of breast cancer: meta-analysis of data from 21 independent prospective cohort studies. BMJ 346 (5), f3706.

Ziegler, G., Ploch, M., Miettinen-Baumann, A., et al., 2002. Efficacy and tolerability of valerian extract LI 156 compared with oxazepam in the treatment of non-organic insomnia—a randomized, double-blind, comparative clinical study. Eur. J. Med. Res. 7, 480–486.

Chapter 17

Nutraceuticals for weight loss

Rhian B. Cope

Introduction

In this chapter, the term nutraceuticals will be considered separately from the functional foods concept. A nutraceutical is defined as a pharmaceutical-grade and standardized nutrient (or dietary supplement or food additive), whereas a functional food is a food given an additional function by the existence or addition of a specific food component. Foods can, of course, be fortified with a specific nutraceutical and such products would then be considered a functional food; however, for reasons of expediency and simplicity this chapter will focus on specific plants from which various claimed antiobesity nutraceuticals have been extracted. Common foods which are claimed to have antiobesity effects such as peanuts, citrus fruits, and pomegranates are not covered by this chapter. This chapter is not intended as a review of the pathophysiology of obesity and metabolic syndrome(s). The currently proposed nutraceuticals are mostly those with claimed effects on dietary fat bioavailability, antioxidant effects, antiinflammatory effects, modification of lipogenesis and preadipocyte maturation, increased lipid metabolism, and effects on satiety. However, many critical modes of action remain incompletely understood.

Curcumin [(1E,6E)-1,7-bis(4-hydroxy-3-methoxyphenyl)-1,6-heptadiene-3,5-dione]

Curcumin is the major curcuminoid (linear diarylheptanoid) found in turmeric. Human phase I and II clinical trials have demonstrated that this food coloring and flavoring agent is relatively well tolerated (Soleimani et al., 2018). Some trial subjects developed diarrhea, headache, skin rashes, and yellow stools. Overall, oral curcumin dosing at 6 g/day for 4–7 weeks was reasonably well tolerated in humans. Oral bioavailable curcumin formulations were reasonably well tolerated following dosing at 500 mg BID for 30 days; however, there is more limited information on curcumin nanoformulations. Curcumin is not genotoxic in classical assays, was not a reproductive toxin in animals following oral dosing studies, in humans did not show toxic effects, and was safe at the dose of 6 g/day orally for 4–7 weeks. While curcumin has multiple pharmacological actions, it has poor oral bioavailability and is rapidly excreted. Phospholipidation and nanoparticle dose forms codosing with piperine have been used to increase oral bioavailability. Curcumin inhibits several CYP450 enzymes, glutathione-S-transferases, and UDP glucuronosyltransferases. Thus, its use may be associated with drug interactional effects.

Because of its antiinflammatory actions, curcumin treatment has been applied to a wide range of inflammatory conditions, including obesity and metabolic disorders. Within the inflammation signaling milieu curcumin interacts directly with cyclooxygenase 2, DNA polymerase, lipoxygenase, glycogen synthase kinase 3b, and tumor necrosis factor alpha (TNF-α). It also modulates the actions of a number of signal transducers (AP-1, β-catenin, STAT, and PPARc).

The antioxidant and antiinflammatory effects of curcumin have been evaluated in several small human clinical trials. In a human double-blind placebo-controlled trial, the effect of curcumin and phospholipidated curcumin on antibody titers to heat shock protein 27 (a biomarker of inflammation and oxidant stress) was evaluated in patients with metabolic syndrome (Mohammadi et al., 2018). Dosing with 1 g/day for 6 weeks with curcumin or phospholipidated curcumin had no significant ($P > .6$) effect on heat shock protein 27 titers. However, oral dosing of patients with metabolic syndrome with curcumin (1 g/day; with or without 10 mg/day of piperine) for 6–8 weeks resulted in a significant ($P < .05$) increase in serum prooxidant–antioxidant balance (Ghazimoradi et al., 2017; Panahi et al., 2015); however curcumin dosing had no beneficial effects on circulating levels of vitamin E (Mohammadi et al., 2017). A human double-blind placebo-controlled trial that evaluated the effect of curcumin or phospholipidated curcumin on serum copper, zinc, and zinc/copper ratio demonstrated that, compared with

Nutraceuticals. DOI: https://doi.org/10.1016/B978-0-12-821038-3.00017-3

placebo, both curcumin and phospholipidated curcumin complex, given at a dose of 1 g/day for 6 weeks, were associated with a significant ($P < .05$) increase in serum zinc and serum zinc/copper ratio in patients with metabolic syndrome (Safarian et al., 2018). In this trial, phospholipidated curcumin (1 g = 200 mg of curcumin) was significantly ($P < .05$) more efficacious in this effect than curcumin. This is likely due to the higher bioavailability of the phospholipidated dose form. Based on the results of a double-blind randomized control trial, the effects of curcumin (1 g/day) combined with piperine (10 mg/day) resulted in increased serum adiponectin, decreased serum leptin, and decreased leptin:adiponectin ratio (a biomarker of atherosclerosis) (Panahi et al., 2016, 2017). These changes were claimed to reflect a decrease in the level of systemic inflammation as measured by circulating TNF-α levels.

A limited number of human clinical trials have also examined the effects of curcumin on obesity and dyslipidemias. A small, preliminary human randomized clinical trial demonstrated patients who did not lose body weight (< 2% reduction) after 30 days of diet and lifestyle intervention had increased weight loss (about 2%–5%) and increased reductions in BMI (about 2%–6%) compared with controls after daily oral dosing with 800 mg/day of curcumin complexed with sunflower phospholipids and 8 mg of piperine. The plasma lipid-modifying effects of a bioavailable curcumin complex (1 g/day with 10 mg of piperidine; dosing for 8 weeks) was evaluated in patients who had been diagnosed with metabolic syndrome (Panahi et al., 2014). In this trial, curcumin was more effective than placebo in reducing serum LDL-C, non-HDL-C, total cholesterol, triglycerides, and Lp(a), and elevating HDL-C concentrations. Similar effects were observed in a second trial which utilized TID dosing at 630 mg/day (Yang et al., 2014).

Overall, while there is reasonable quality preliminary, small-scale human clinical trial evidence that bioavailable forms of curcumin may have beneficial effects on metabolic syndrome, obesity, and dyslipidemias, evidence of overall long-term beneficial outcomes in patients with metabolic syndrome and obesity is lacking.

Lagenaria siceraria (bottle gourd)

Bottle gourd, a plant native to India, has a long history of ethnopharmacological use. Traditional medicinal uses of the plant include: as an aliuretic, purgative, and laxative; as a diuretic for cardiogenic ascites; for the relief of headache and toothache; as an anticonvulsant; as a treatment for boils, tetanus, fever, wounds, and rheumatism; as a treatment for asthma and coughs; as an anticancer agent; as a treatment for alopecia; and as a poison antidote. The plant is reputed to have cardioprotective, antihyperlipidemic, antioxidant, antihyperglycemic, analgesic, antiinflammatory, immunomodulatory, and hepatoprotective effects in humans. Fresh bottle gourd juice is considered an effective treatment for obesity in India.

Small controlled studies have claimed that consuming freshly prepared bottle gourd fruit extract once per day for 90 days reduced plasma total cholesterol (marginal), plasma triglycerides, plasma low-density lipoproteins, and BMI, as well as having antioxidant properties in humans (Kaur et al., 2015; Katare et al., 2014). These findings are supported by studies in rats where oral dosing of bottle gourd methanol extracts at 100–400 mg/kg body weight/day resulted in dose-dependent reductions in plasma total cholesterol, triglycerides, low-density lipoproteins level, and increased plasma high-density lipoproteins (Ghule et al., 2006, 2009). Based on the results of this small study, fresh bottle gourd juice may have some limited beneficial effects on dyslipidemias in humans.

Overall, while there is reasonable preliminary exploratory data on the possible beneficial effects of bottle gourd products on dyslipidemia in humans, evidence of long-term beneficial outcomes in patients is lacking.

Trigonella foenum-graecum (fenugreek)

One of the world's oldest medicinal plants, fenugreek, originated in India and North Africa (Kaur et al., 2015). There are three groups of medicinal compounds found in this plant: steroidal saponins, galactomannans, and isoleucine. Fenugreek is an important source of the drug precursor diosgenin, which is used in the synthesis of steroidal drugs.

While fenugreek is generally considered a safe product, patients with allergies to the plant or allergies to chick peas (cross reactive with fenugreek) should avoid it (Kaur et al., 2015). Consumption of fenugreek is known to cause dizziness, transient diarrhea, and flatulence. Hypoglycemia and reduced T3 levels have been reported in some patients. Fenugreek contains coumarin and consumption may result in increased prothrombin time and a bleeding tendency. Fenugreek should be avoided during pregnancy because it stimulates uterine contractions in animal studies. Because fenugreek is rich in dietary fiber it may decrease the oral bioavailability of drugs. Since fenugreek induces hypoglycemia, concurrent use with other hypoglycemic agents may exacerbate their pharmacological effects.

The ability of fenugreek preparations to attenuate postprandial glycemia and insulinemia is reputed to be due to its soluble dietary fiber effects (Repin et al., 2017). These effects are claimed to be due to modification of ingesta viscosity and delaying of gastric emptying. The effects of fenugreek soluble dietary fiber on these types of parameters are about the same as other sources of soluble dietary fiber.

The fenugreek steroidal saponins are claimed to have antihypercholesterolemia and antihyperlipidemia effects in humans (Kassaian et al., 2009). The mechanism of these effects is incompletely understood, but may be a manifestation of increased soluble dietary fiber intake.

Galactomannans are polymers of galactose and mannose present in the seeds of fenugreek. They are hydrophilic and form viscous aqueous solutions at low concentrations (Kaur et al., 2015) and have been used as gastric "fillers" in an attempt to stimulate satiety and decrease caloric intake. Isoleucine is an amino acid precursor which regulates insulin secretion.

Emblica officinalis (syn. *Phyllanthus emblica*; Indian gooseberry)

Indian gooseberry is an important dietary source of vitamin C and minerals in many areas of India and has a long history of use as a traditional medicine (treatment of fever, antiinflammatory, diuretic, hair and liver tonic, and as a digestive remedy) (Kaur et al., 2015). The fruit also has religious significance and has also been used to make ink due to its high tannin content.

Preparations of Indian gooseberry have been demonstrated to decrease serum cholesterol, low-density lipoprotein, and triglycerides in humans and are PPAR-α agonists (Kaur et al., 2015; Akhtar et al., 2011). They are also claimed to exert some of their effects due to their antioxidant and free radical scavenging properties (D'souza et al., 2014).

Murraya koenigii (curry tree)

The leaves of this plant are commonly used in Indian cooking (typically in curries). The plant also has a long history of use as a traditional medicine and is known for having antioxidant, antitumor, antiinflammatory, antihyperglycemic, hypoglycemic, and hypolipidemic effects (Kaur et al., 2015). Curry tree products act as hypoglycemics by increasing hepatic glycogen storage and decreasing gluconeogenesis. Part of the nutraceutical effects of the curry tree may be due to the actions of the carbazole alkaloid mahanimbine. In mice, mahanimbine inhibits the obesogenic effects of a high-fat diet, including reducing body weight gain, preventing hyperlipidemia, fat accumulation in adipose tissue, and hepatic steatosis, and improving glucose clearance and insulin responsiveness (Jagtap et al., 2017).

Vigna sp. (black gram, black lentil, cowpea, black-eyed pea, adzuki bean, mung bean)

Black lentils (*Vigna mungo*) have been used as a safe, high-protein traditional food source in Nepal and India for millennia. The claimed benefits of consuming black lentils are (Kaur et al., 2015):

- The high fiber, low glycemic index properties of this food are supposedly able to modulate lipid homeostasis in people with a high-saturated-fat diet.
- The high fiber, low glycemic index properties of this food reputedly help to main blood glucose control in people with diabetes mellitus.
- Reputedly, diets high in black lentils have some benefit in controlling body weight since they are claimed to have satiety effects, thus limiting overall food consumption.
- Black lentils inhibit α-amylase. α-Amylases are known to delay carbohydrate absorption and to reduce peak postprandial plasma glucose concentration.

Other members of the *Vigna* genus are reputed to have similar effects. In a human randomized double-blind cross-over trial, consumption of *V. unguiculata* protein reduced total cholesterol (12%), LDL cholesterol (18.9%), non-HDL cholesterol (16%), and apoB (14%), and increased HDL cholesterol (+2.7%); however, no effects on the serum inflammasome were noted (Frota Kde et al., 2015). Adzuki bean consumption by mice with high-fat diet-induced nonalcoholic liver disease resulted in reduced fasting blood and hepatic triglyceride and cholesterol levels with evidence of effects on hepatic lipogenesis and decreased inflammation (Kim et al., 2015).

Camellia sinensis (tea)

Claims regarding the health benefits of the three main types of tea—black, oolong, and green—extend back in history for thousands of years. More research effort has been devoted to *Camellia sinensis* than any other potential nutraceutical. However, despite this substantial research effort, a definitive consensus on the health benefits and appropriate uses of tea as nutraceuticals and complimentary medicines has not been reached (Jurgens et al., 2012).

There are three main categories of claimed beneficial nutraceutical compounds in tea:

- Xanthine bases and methyl xanthines (caffeine, theophylline, and theobromine);
- Essential oils;
- Polyphenolic compounds (notably epigallocatechin, epigallocatechin gallate, epicatechin gallate, and gallocatechin gallate).

Various studies have demonstrated that tea phytochemicals appear to have beneficial effects on obesity and metabolic syndrome in high-fat diet animal models. The following studies are representative of the findings in rodents. Green tea phytochemicals administered in drinking water to Wistar rats at concentrations up to 3.2 g/L over a period of 26 weeks reduced visceral adipose tissue accumulation by about 40% compared with controls (no dose response) (Tian et al., 2013). The effect was claimed to be due to modulation of the erk1/2-PPARγ-adiponectin pathway. Oral dosing of mice fed a high-fat diet with 400 mg/kg body weight/day of green tea extract (containing 15 μg/mg epigallocatechin, 95 μg/mg epigallocatechin gallate, 20.8 μg/mg epicatechin gallate, and 4.9 μg/mg gallocatechin gallate) for 8 weeks was associated with increased hormone-sensitive lipase and perilipin in mesenteric adipose tissue, thus reputedly increasing lipolysis and reducing low-grade inflammation (as measured by TNF-α levels and reductions in TLR4, MYD88, and TRAF6 proinflammatory signaling), and reducing body weight by about 16% compared with controls (Cunha et al., 2013). The antiobesity effects of tea are not only associated with leaf and leaf extract products. Ethanolic extracts of the fruit peel of the tea plant, which is normally regarded as an agricultural waste product, have been demonstrated to have antiobesity effects in rats (Chaudhary et al., 2014). In this study, administration of 100 mg/kg body weight/day of an ethanolic green tea fruit peel extract as part of a high-fat diet fed to rats for 50 days resulted in about a 25% reduction in body weight and a 74% reduction in adipose tissue fat pad weights compared with controls.

Suggested modes of action of tea extracts on obesity and metabolic syndromes include inhibition of pancreatic lipase, appetite-repression activity, downregulation of adipogenesis, thermogenesis, alterations in lipid metabolism including increased lipolysis and decreased lipogenesis, increased antioxidant defenses, changes in gut microflora, increased whole-body fat utilization, increased fat oxidation, and antiinflammatory effects (Grove et al., 2012; Yuda et al., 2012; Moon et al., 2007; Wolfram et al., 2006; Lu et al., 2012; Axling et al., 2012; Basu et al., 2010; Park et al., 2011).

The presumed major chemicals involved in the antiobesity effects of tea products in animals lie within the polyphenol fraction and include the epigallocatechin 3-gallate, the major catechin-like flavonoid.

The safety properties of tea have been best studied for green tea preparations. The consumption of moderate amounts of green tea as a beverage is currently regarded as safe. Concentrated great tea extracts have been reputedly associated with hepatoxicity in a small number of people; however, the US National Center for Complementary and Integrative Health has concluded that in the case of green tea "definite conclusions cannot yet be reached on whether green tea is helpful for most of the purposes for which it is used" (National Center for Complementary and Integrative Health, 2016).

Hibiscus sabdariffa

This well-studied plant is commonly prepared as a soft drink and is used to reduce inflammation, hypertension, and liver disorders. In the monosodium glutamate-induced obese mouse model, *Hibiscus sabdariffa* calyces aqueous extract administered at 120 mg/kg body weight/day for 60 days resulted in an approximately 22% reduction in body weight gain compared with control and reduced liver steatosis (Alarcon-Aguilar et al., 2007). Reductions in body weight and adipose tissue weight compared with controls have also been demonstrated in obese rats treated with extracts from this plant (Gamboa-Gómez et al., 2015). Similar results have been achieved in other animal models. Promising results have also been detected in human clinical trials where consumption of *H. sabdariffa* extract for 12 weeks resulted in body weight loss, reduced BMI, reduced body fat, reduced waist-to-hip ratio, and reduced liver steatosis in subjects with a BMI $\geq$ 27 and between the ages of 16 and 65 years (Chang et al., 2014).

The critical phytochemicals identified in this plant are phenolic and flavonoid compounds (Peng et al., 2011). Multiple putative modes of action relating to the plant's antiobesity, antimetabolic disorder, and hepatoprotective effects have been proposed including downregulation of the genes coding for free fatty acid synthase and sterol regulatory element binding proteins, inhibition of lipid droplet accumulation in adipocytes and adipocyte hypertrophy, antioxidant effects, antiinflammatory effects, and downregulation of pancreatic lipases (Gamboa-Gómez et al., 2015; Kim et al., 2003).

Hypericum perforatum (St. John's wort)

A limited database of studies suggest that St. John's wort may have antiobesity effects in addition to its well-known uses as a sedative, anxiolytic, and antidepressant. Oral dosing of high-fat-diet-induced obese rats with a St. John's wort extract at 100 and 200 mg/kg body weight/day for 15 days resulted in 12% and 15% reductions in body weight, respectively, compared with controls, as well as reducing blood total cholesterol and low-density cholesterol in normal rats (Husain et al., 2011). The same extract normalized dyslipidemia and improved insulin sensitivity in fructose-fed rats. Similar results with extracts of other *Hypericum* sp. have been reported by other investigators (García-de la Cruz et al., 2013).

Possible relevant modes of action for *Hypericum* sp. extracts have included modulation of the serotonin system which, in turn, may reduce food intake, suppress appetite, and reduce anxiety (Ganji et al., 2017). Reduced oxidative stress, inhibition of inflammation, inhibition of adipogenesis, attenuation of insulin-sensitive glucose uptake by adipocytes, and limiting the differentiation of preadipocytes have also been associated with the antiobesity effects (Richard et al., 2012).

The specific antiobesogenic compounds in St. John's wort have not been fully described but may include hyperforin, hypericin, flavonoids, and condensed tannins. The safety and drug interactional properties of St. John's wort extracts are well known, with the most import ones being serotonin syndrome-like effects and drug–drug interactional effects due to induction of CYP3A4 and CYP1A2.

Persea americana (avocado)

Antiobesity effects associated with avocados have been reported for both fruit, leaf, and oil extract products (Brai et al., 2007). Eight weeks of dosing of rats with hypercholesterolemia induced by a high-saturated-fat diet with 10 mg/kg body weight/day of aqueous and methanolic extracts of avocado leaves resulted in a 25% reduction in body weight gain compared with controls. Likewise, rats fed defatted avocado fruit pulp (100 g/kg of diet) for 28 days had a 26% reduced body weight gain and reduced total hepatic fat levels compared with controls; however, when this diet and cholesterol were co-fed, higher serum cholesterol occurred (Naveh et al., 2002). In the case of avocado pulp, the antiobesity effects were attributed to reduced food consumption. Similar effects have been reported for alcoholic fruit extracts (100 mg/kg body weight/day for 14 weeks) and hydroalcoholic extracts (Padmanabhan and Arumugam, 2014; Monika and Geetha, 2016). These effects were associated with antioxidant effects, changes in adiponectin and PPARγ, alterations in key lipid metabolic enzymes, and lower blood LDL and lipid peroxides.

The key antiobesity compounds in avocado products have not yet been defined.

Avocados are a safe food in humans; however, they are toxic to a variety of domestic animals and birds.

Capsicum annuum (capsicum)

Capsicum extracts have been reported to be antiadipogenic in animal models. A dietary 13.35% capsicoside G fraction (dosed at 10 and 100 mg/kg body weight/day) isolated from capsicum seeds modulated the obesogenic effects of a high-fat diet in C57BL/6J mice (Sung et al., 2016). Key effects included reduced epididymal adipose tissue weight, reduced adipocyte hypertrophy, reduced hepatic steatosis, and reduced expression of adipocyte differentiation regulators (PPARγ, CCATα, and sterol regulatory element binding protein 1c). Likewise, C57BL/6 mice fed a high-fat diet with a green pepper juice supplement (10 mL/kg body weight/day) had lower weight gain, decreased serum triglycerides, reduced serum cholesterol and LDL, and lower abdominal fat (subcutaneous and visceral) volume compared with controls (Kim and Park, 2015).

Antiobesogenic effects of capsaicinoids from *Capsicum* sp. have also been demonstrated in a limited human clinical trial. A dose of 6 mg/day for 12 weeks resulted in a small reduction in abdominal adiposity compared with placebo.

Various modes of action have been described for the antiobesity effects of capsicum including inhibition of adipogenesis, changes in gut microflora with a resulting reduction in chronic low-grade inflammation, antioxidant effects, modulation of inflammation, and alteration in lipoprotein metabolism (Kim et al., 2018; Kang et al., 2017; Marrelli et al., 2016; Yu et al., 2012).

Rosmarinus officinalis (rosemary)

Rosemary is a well-known medicinal spice with a long history of historical use as a herbal abortifacient, however there is now an expanding body of animal data demonstrating potential efficacy in obesity and metabolic disorders. Dosing of high-fat diet-induced obese mice with 200 mg/kg body weight/day of a rosemary leaf extract resulted in reduced weight gain by about 64% accompanied by an approximate 57% reduction in fat mass gain (Harach et al., 2010). This was accompanied by increased fecal lipid excretion and an approximate 39% reduction in hepatic triglyceride levels. Similar results were obtained when the rosemary leaf extract was standardized to a 20% carnosic acid content and dosed at 500 mg/kg body weight/day of the extract (Ibarra et al., 2011). Several investigators have now replicated these findings (Zhao et al., 2015).

The best investigated active antiobesity compound in rosemary is carnosic acid, which has a number of potentially relevant modes of action including inhibition of gastric lipase, modulation of genes involved in metabolism (SIRT1, PPARγ coactivator 1α, glucose-6-phosphatase, ACC, and low-density lipoprotein receptor), modulation of PPARγ, activation of the AMPK and PPAR pathways, increased hepatic glycolysis and fatty acid oxidation, decreasing inflammation, decreasing leptin, increasing adiponectin, modifying the cecal microbiome, inhibition of cecal β-glucosidase, increasing fecal short-chain fatty acid and fiber excretion, stimulation of muscle glucose uptake via modulation of the PME-1/PP2A/PKB signaling axis, modulation of TLR4-MyD88 signaling in adipocytes, reducing hepatic lipogenesis while increasing hepatic lipolysis (Romo Vaquero et al., 2012; Tu et al., 2013; Romo-Vaquero et al., 2014a,b; Park and Mun, 2014). Carnosic acid inhibits TLR4-MyD88 signaling pathway in LPS-stimulated 3T3-L1 adipocytes (Park and Sung, 2015).

The safety properties of rosemary and rosemary extract have been well-documented. The key effects are on male fertility, establishment/maintenance of pregnancy, and changes in estrogen metabolism.

Concluding remarks and future directions

While obesity and metabolic disorders continue to be an ever-increasing, worldwide public health problem, a market will exist for nutraceutical products that target these diseases. Accordingly, molecule mining in the nutraceutical area is likely to continue in the foreseeable future. While there is an increasing body of proof of concept type literature in various animal models, there remain relatively few large, high-quality clinical trials. Mechanistic data are also often of a preliminary, explorative nature. Given the current status quo, substantially more translational research is required to fully establish the therapeutic credibility of many of these materials.

References

Akhtar, M., Ramzan, A., Ali, A., et al., 2011. Effect of Amla fruit (*Emblica officinalis* Gaertn.) on blood glucose and lipid profile of normal subjects and type 2 diabetic patients. Int. J. Food Sci. Nutr. 62 (6), 609–616.

Alarcon-Aguilar, F., Zamilpa, A., Perez-Garcia, M., et al., 2007. Effect of *Hibiscus sabdariffa* on obesity in MSG mice. J. Ethnopharmacol. 114 (1), 66–71.

Axling, U., Olsson, C., Xu, J., et al., 2012. Green tea powder and *Lactobacillus plantarum* affect gut microbiota, lipid metabolism and inflammation in high-fat fed C57BL/6J mice. Nutr. Metab. 9 (1), 105.

Basu, A., Sanchez, K., Leyva, M., et al., 2010. Green tea supplementation affects body weight, lipids, and lipid peroxidation in obese subjects with metabolic syndrome. J. Am. Coll. Nutr. 29 (1), 31–40.

Brai, B., Odetla, A., Agomo, P., 2007. Effects of *Persea americana* leaf extracts on body weight and liver lipids in rats fed hyperlipidaemic diet. Afr. J. Biotechnol. 6, 1007–1011.

Chang, H., Peng, C., Yeh, D., et al., 2014. *Hibiscus sabdariffa* extract inhibits obesity and fat accumulation, and improves liver steatosis in humans. Food Funct. 5 (4), 734–739.

Chaudhary, N., Bhardwaj, J., Seo, H., et al., 2014. *Camellia sinensis* fruit peel extract inhibits angiogenesis and ameliorates obesity induced by high-fat diet in rats. J. Funct. Foods 7, 479–486.

Cunha, C., Lira, F., Rosa Neto, J., et al., 2013. Green tea extract supplementation induces the lipolytic pathway, attenuates obesity, and reduces low-grade inflammation in mice fed a high-fat diet. Mediators Inflammation 2013, 635470.

D'souza, J., D'souza, P., Fazal, F., et al., 2014. Anti-diabetic effects of the Indian indigenous fruit *Emblica officinalis* Gaertn: active constituents and modes of action. Food Funct. 5 (4), 635–644.

Frota Kde, M., dos Santos Filho, R., Ribeiro, V., et al., 2015. Cowpea protein reduces LDL-cholesterol and apolipoprotein B concentrations, but does not improve biomarkers of inflammation or endothelial dysfunction in adults with moderate hypercholesterolemia. Nutr. Hosp. 31 (4), 1611–1619.

Gamboa-Gómez, C., Rocha-Guzmán, N., Gallegos-Infante, J., et al., 2015. Plants with potential use on obesity and its complications. EXCLI J. 14, 809–831.

Ganji, A., Salehi, I., Sarihi, A., et al., 2017. Effects of *Hypericum scabrum* extract on anxiety and oxidative stress biomarkers in rats fed a long-term high-fat diet. Metab. Brain Dis. 32 (2), 503–511.

García-de la Cruz, L., Galvan-Goiz, Y., Caballero-Caballero, S., et al., 2013. *Hypericum silenoides* Juss. and *Hypericum philonotis* Cham. and Schlecht. extracts: in vivo hypolipidaemic and weight-reducing effects in obese rats. J. Pharm. Pharmacol. 65 (4), 591–603.

Ghazimoradi, M., Saberi-Karimian, M., Mohammadi, F., et al., 2017. The effects of curcumin and curcumin-phospholipid complex on the serum pro-oxidant-antioxidant balance in subjects with metabolic syndrome. Phytother. Res. 31 (11), 1715–1721.

Ghule, B., Ghante, M., Saoji, A., et al., 2006. Hypolipidemic and antihyperlipidemic effects of *Lagenaria siceraria* (Mol.) fruit extracts. Indian J. Exp. Biol. 44 (11), 905–909.

Ghule, B., Ghante, M., Saoji, A., et al., 2009. Antihyperlipidemic effect of the methanolic extract from *Lagenaria siceraria* Stand. fruit in hyperlipidemic rats. J. Ethnopharmacol. 124 (2), 333–337.

Grove, K., Sae-Tan, S., Kennett, M., et al., 2012. Epigallocatechin-3-gallate inhibits pancreatic lipase and reduces body weight gain in high fat-fed obese mice. Obesity 20, 2311–2313.

Harach, T., Aprikian, O., Monnard, I., et al., 2010. Rosemary (*Rosmarinus officinalis* L.) leaf extract limits weight gain and liver steatosis in mice fed a high-fat diet. Planta Med. 76 (6), 566–571.

Husain, G., Chatterjee, S., Singh, P., et al., 2011. Hypolipidemic and antiobesity-like activity of standardised extract of *Hypericum perforatum* L. in rats. ISRN Pharmacol. 2011, 505247.

Ibarra, A., Cases, J., Roller, M., et al., 2011. Carnosic acid-rich rosemary (*Rosmarinus officinalis* L.) leaf extract limits weight gain and improves cholesterol levels and glycaemia in mice on a high-fat diet. Br. J. Nutr. 106 (8), 1182–1189.

Jagtap, S., Khare, P., Mangal, P., et al., 2017. Effect of mahanimbine, an alkaloid from curry leaves, on high-fat diet-induced adiposity, insulin resistance, and inflammatory alterations. Biofactors 43 (2), 220–231.

Jurgens, T., Whelan, A., Killian, L., et al., 2012. Green tea for weight loss and weight maintenance in overweight or obese adults. Cochrane Database Syst. Rev. 1465–1858. Available from: https://doi.org/10.1002/14651858.CD008650.pub2. Retrieved from.

Kang, C., Wang, B., Kaliannan, K., et al., 2017. Gut microbiota mediates the protective effects of dietary capsaicin against chronic low-grade inflammation and associated obesity induced by high-fat diet. mBio 8 (4).

Kassaian, N., Azadbakht, L., Forghani, B., et al., 2009. Effect of fenugreek seeds on blood glucose and lipid profiles in type 2 diabetic patients. Int. J. Vitam. Nutr. Res. 79 (1), 34–39.

Katare, C., Saxena, S., Agrawal, S., et al., 2014. Lipid-lowering and antioxidant functions of Bottle gourd (*Lagenaria siceraria*) extract in human dyslipidemia. J. Evid. Based Complement. Alternat. Med. 19 (2), 112–118.

Kaur, G., Mukundan, S., Wani, V., et al., 2015. Nutraceuticals in the management and prevention of metabolic syndrome. Austin J. Pharmacol. Ther. 3 (1), 1–6.

Kim, N., Park, S., 2015. Evaluation of green pepper (*Capsicum annuum* L.) juice on the weight gain and changes in lipid profile in C57BL/6 mice fed a high-fat diet. J. Sci. Food Agric. 95 (1), 79–87.

Kim, M., Kim, J., Kim, H., et al., 2003. Hibiscus extract inhibits the lipid droplet accumulation and adipogenic transcription factors expression of 3T3-L1 preadipocytes. J. Altern. Complement. Med. 9 (4), 499–504.

Kim, S., Hong, J., Jeon, R., et al., 2015. Adzuki bean ameliorates hepatic lipogenesis and proinflammatory mediator expression in mice fed a high-cholesterol and high-fat diet to induce nonalcoholic fatty liver disease. Nutr. Res. 36 (1), 90–100.

Kim, H., You, M., Wang, Z., et al., 2018. Red pepper seed inhibits differentiation of 3T3-L1 cells during the early phase of adipogenesis via the activation of AMPK. Am. J. Chin. Med. 46 (1), 107–118.

Lu, C., Zhu, W., Shen, C., et al., 2012. Green tea polyphenols reduce body weight in rats by modulating obesity-related genes. PLoS One 7, e38332.

Marrelli, M., Menichini, F., Conforti, F., 2016.). Hypolipidemic and antioxidant properties of hot pepper flower (*Capsicum annuum* L.). Plant Foods Hum. Nutr. 71 (3), 301–306.

Mohammadi, A., Sadeghnia, H., Saberi-Karimian, M., et al., 2017. Effects of curcumin on serum vitamin E concentrations in individuals with metabolic syndrome. Phytother. Res. 31 (4), 657–662.

Mohammadi, F., Ghazi-Moradi, M., Ghayour-Mobarhan, M., et al., 2018. The effects of curcumin on serum heat shock protein 27 antibody titers in patients with metabolic syndrome. J. Diet. Suppl. 29, 1–10.

Monika, P., Geetha, A., 2016. Effect of hydroalcoholic fruit extract of *Persea americana* Mill. on high fat diet induced obesity: a dose response study in rats. Indian J. Exp. Biol. 54 (6), 370–378.

Moon, H., Chung, C., Lee, H., et al., 2007. Inhibitory effect of (-)epigallocatechin-3-gallate on lipid accumulation of 3T3-L1 cells. Obeisty 15, 2571–2582.

National Center for Complementary and Integrative Health, 2016, September. Green tea. Retrieved from Health Topics: https://nccih.nih.gov/health/greentea.

Naveh, E., Werman, M., Sabo, E., et al., 2002. Defatted avocado pulp reduces body weight and total hepatic fat but increases plasma cholesterol in male rats fed diets with cholesterol. J. Nutr 132 (7), 2015–2018.

Padmanabhan, M., Arumugam, G., 2014. Effect of *Persea americana* (avocado) fruit extract on the level of expression of adiponectin and PPAR-γ in rats subjected to experimental hyperlipidemia and obesity. J. Complementary Integr. Med. 11 (2), 107–119.

Panahi, Y., Khalili, N., Hosseini, M., et al., 2014. Lipid-modifying effects of adjunctive therapy with curcuminoids-piperine combination in patients with metabolic syndrome: results of a randomized controlled trial. Complement. Ther. Med. 22 (5), 851–857.

Panahi, Y., Hosseini, M., Khalili, N., et al., 2015. Antioxidant and anti-inflammatory effects of curcuminoid-piperine combination in subjects with metabolic syndrome: a randomized controlled trial and an updated meta-analysis. Clin. Nutr. 34 (6), 1101–1108.

Panahi, Y., Hosseini, M., Khalili, N., et al., 2016. Effects of supplementation with curcumin on serum adipokine concentrations: a randomized controlled trial. Nutrition 32 (10), 1116–1122.

Panahi, Y., Khalili, N., Sahebi, E., et al., 2017. Curcuminoids plus piperine modulate adipokines in type 2 diabetes mellitus. Curr. Clin. Pharmacol. 12 (4), 253–258.

Park, M., Mun, S., 2014. Carnosic acid inhibits TLR4-MyD88 signaling pathway in LPS-stimulated 3T3-L1 adipocytes. Nutr. Res. Pract. 8 (5), 516–520.

Park, M., Sung, M., 2015. Carnosic acid attenuates obesity-induced glucose intolerance and hepatic fat accumulation by modulating genes of lipid metabolism in C57BL/6Job/ob mice. J. Sci. Food Agric. 95 (4), 828–835.

Park, H., DiNatale, D., Chung, M., et al., 2011. Green tea extract attenuates hepatic steatosis by decreasing adipose lipogenesis and enhancing hepatic antioxidant defenses in ob/ob mice. J. Nutr. Biochem. 22 (4), 393–400.

Peng, C., Chyau, C., Chan, K., et al., 2011. *Hibiscus sabdariffa* polyphenolic extract inhibits hyperglycemia, hyperlipidemia, and glycation-oxidative stress while improving insulin resistance. J. Agric. Food Chem. 59 (18), 9901–9909.

Repin, N., Kay, B., Cui, S.W., 2017. Investigation of mechanisms involved in postprandial glycemia and insulinemia attenuation with dietary fibre consumption. Food Funct. 8 (6), 2142–2154.

Richard, A., Amini, Z., Ribnicky, D., et al., 2012. St. John's Wort inhibits insulin signalling in murine and human adipocytes. Biochim. Biophys. Acta 1822 (4), 557–563.

Romo Vaquero, M., Yáñez-Gascón, M., García Villalba, R., et al., 2012. Inhibition of gastric lipase as a mechanism for body weight and plasma lipids reduction in Zucker rats fed a rosemary extract rich in carnosic acid. PLoS One. 7 (6), e39773.

Romo-Vaquero, M., Larrosa, M., Yáñez-Gascón, M., et al., 2014a. A rosemary extract enriched in carnosic acid improves circulating adipocytokines and modulates key metabolic sensors in lean Zucker rats: critical and contrasting differences in the obese genotype. Mol. Nutr. Food Res. 58 (5), 942–953.

Romo-Vaquero, M., Selma, M., Larrosa, M., et al., 2014b. A rosemary extract rich in carnosic acid selectively modulates caecum microbiota and inhibits β-glucosidase activity, altering fiber and short chain fatty acids fecal excretion in lean and obese female rats. PLoS One 9 (4), e94687.

Safarian, H., Parizadeh, S., Saberi-Karimain, M., et al., 2018. The effect of curcumin on serum copper and zinc and Zn/Cu ratio in individuals with metabolic syndrome: a double-blind clinical trial. J. Diet. Suppl. 18, 0. Available from: https://doi.org/10.1080/19390211.2018.1472711.

Soleimani, V., Sahebkar, A., Hosseinzadeh, H., 2018. Turmeric (*Curcuma longa*) and its major constituent (curcumin) as nontoxic and safe substances: review. Phytother. Res. 32 (6), 985–995.

Sung, J., Jeong, H., Lee, J., 2016. Effect of the capsicoside G-rich fraction from pepper (*Capsicum annuum* L.) seeds on high-fat diet-induced obesity in mice. Phytother. Res. 30 (11), 1848–1855.

Tian, C., Ye, X., Zhang, R., et al., 2013. Green tea polyphenols reduced fat deposits in high fat-fed rats via erk1/2-PPARγ-adiponectin pathway. PLoS One 8 (1), e53796.

Tu, Z., Moss-Pierce, T., Ford, P., et al., 2013. Rosemary (*Rosmarinus officinalis* L.) extract regulates glucose and lipid metabolism by activating AMPK and PPAR pathways in HepG2 cells. J. Agric. Food Chem. 61 (11), 2803–2810.

Wolfram, S., Wang, Y., Thielecke, F., 2006. Anhti-obesity effects of green tea: from bedside to bench. Mol. Nutr. Food Res. 50, 176–187.

Yang, Y., Su, Y., Yang, H., et al., 2014. Lipid-lowering effects of curcumin in patients with metabolic syndrome: a randomized, double-blind, placebo-controlled trial. Phytother. Res. 28 (12), 1770–1777.

Yu, Q., Wang, Y., Yu, Y., et al., 2012. Expression of TRPV1 in rabbits and consuming hot pepper affects its body weight. Mol. Biol. Rep. 39 (7), 7583–7589.

Yuda, N., Tanaka, M., Suzuki, M., et al., 2012. Polyphenols extracted from black tea (*Camellia sinensis*) residue by hot-compressed water and their inhibitory effectg on pancreatic lipase in vitro. J. Food Sci. 77, H254–H261.

Zhao, Y., Sedighi, R., Wang, P., et al., 2015. Carnosic acid as a major bioactive component in rosemary extract ameliorates high-fat-diet-induced obesity and metabolic syndrome in mice. J. Agric. Food Chem. 63 (19), 4843–4852.

Chapter 18

Nutraceuticals in cancer prevention

M. Waheed Roomi, Tatiana Kalinovsky, Matthias Rath and Aleksandra Niedzwiecki

Introduction: targeting hallmarks of cancer

Because 90% of cancer deaths occur secondary to metastasis, any successful anticancer treatment has to target this stage of cancer development. Rath and Pauling (1992) proposed using micronutrients (NM) to target common pathomechanisms involved in growth and invasion of all types of cancer. In this aspect, measures assuring optimum collagen stability were proposed as the most effective and universal approaches in controlling tumor growth and invasiveness (Rath and Pauling, 1992). These include controlling matrix proteolytic degradation and enhancing extracellular matrix (ECM) structure and integrity (Rath and Pauling, 1992). Studies by Almholt et al. (2005) have demonstrated limited metastasis in a uPA-deficient mouse model. Also, highly metastatic cancer cells secrete higher amounts of matrix metalloproteinases (MMPs) than do poorly metastatic cells, demonstrating that the invasive and metastatic abilities of these cancer cells correlate with an ability of these enzymes to degrade basement membranes (Stetler-Stevenson, 2001; Duffy, 1992). Because this process is involved not only in metastasis but also in angiogenesis and tumor growth, control of the proteolytic activity of ECM provides an opportunity to modulate key common aspects of malignancy.

A major structural protein for ECM and basement membrane is type IV collagen. Therefore research has focused on type IV collagenases MMP-2 (72-kDa gelatinase A) and MMP-9 (92-kDa gelatinase B), especially because elevated MMP-2 and MMP-9 expression has been associated with cancer cell aggression in many experimental and clinical studies and is elevated in a variety of malignancies (Chambers and Matrisian, 1997; Kleiner and Stetler-Stevenson, 1999; Liotta et al., 1980; Stetler-Stevenson et al., 1996).

Nutrient synergy targets pathomechanisms of cancer

Rath and Pauling (1992) proposed the use of nutritional components, such as vitamin C, lysine, and lysine analogs, to target plasmin-mediated connective tissue degradation as a universal approach to controlling these critical pathomechanisms of cancer. Lysine interferes with the activation of plasminogen into plasmin by tissue plasminogen activator (tPA) by binding to plasminogen active sites, thereby affecting the plasmin-induced MMP activation cascade (Rath and Pauling, 1992). Our subsequent studies confirmed this approach and resulted in identifying a novel formulation composed of lysine, ascorbic acid, proline, and green tea extract and other NM that have shown significant anticancer activity against a large number ($\sim$40) of cancer cell lines, blocking cancer growth, tissue invasion, and MMP expression both in vitro and in vivo (Niedzwiecki et al., 2010).

Defining anticancer activities of NM components

Consumption of a plant-based diet has been shown to have preventive effects on cancer development and its progression, which has been attributed to its high antioxidant content (Aldercreutz, 1990; Miller, 1990). The nutrients selected in NM act primarily on critical physiological targets in cancer progression and metastasis, which is the stability and integrity of the ECM. The anticancer effects of the individual constituents of NM have been reported in both clinical and experimental studies. Adequate supplies of ascorbic acid and the amino acids lysine and proline are essential for optimal ECM formation and structure because these nutrients insure proper synthesis and hydroxylation of collagen fibers. Manganese and copper are also essential for collagen formation. Lysine contributes to ECM stability as a natural inhibitor of plasmin-induced proteolysis (Rath and Pauling, 1992; Sun et al., 2002). Green tea extract has been shown to control cancer cell growth, metastasis, angiogenesis, and other aspects of cancer progression (Kemberling et al., 2003; Sato and Matsushima, 2003; Valcic et al., 1996; Mukhtar and Ahmed, 2000; Yang et al., 1998; Taniguchi et al., 1992;

Nutraceuticals. DOI: https://doi.org/10.1016/B978-0-12-821038-3.00018-5

Hara, 2001). *N*-acetyl cysteine and selenium have been observed to inhibit MMP-9 and the invasive activities of tumor cells, as well as migration of endothelial cells through ECM (Kawakami et al., 2001; Morini et al., 1999; Yoon et al., 2001). Ascorbic acid has been reported to exert cytotoxic and antimetastatic actions on malignant cell lines (Naidu et al., 2003; Anthony and Schorah, 1982; Maramag et al., 1997; Koh et al., 1998; Chen et al., 2005); in addition, low levels of ascorbic acid have been reported in cancer patients (Nunez et al., 1995; Kurbacher et al., 1996). Arginine is a precursor of nitric oxide (NO); any deficiency of arginine can limit the production of NO, which has been shown to predominantly act as an inducer of apoptosis, such as in breast cancer cells (Cooke and Dzau, 1997).

Combining these NM expands metabolic targets synergistically, maximizing the biological impact with lower doses of components. Furthermore, effective lower doses of components in nutrient synergy also assure overcoming absorption barriers characteristic for high vitamin doses.

Nutrient mixture inhibits tumor growth and cancer cell proliferation

Inhibition of tumor growth (xenografts) and proliferation marker Ki67

NM have been shown to modulate xenograft tumor growth in a variety of cancer cell lines. Table 18.1 presents results obtained with some of these cell lines. For example, female athymic mice at 6 weeks of age ($n = 12$) were inoculated subcutaneously with 3×10^6 HeLa cells in 0.2 mL phosphate buffered saline (PBS) and 0.1 mL Matrigel and randomly divided into two groups; the control group mice were fed regular Purina mouse chow and the NM group mice were fed the regular diet supplemented with 0.5% NM (w/w). After 4 weeks, the mice were sacrificed and their tumors were excised and processed for histology. Dimensions (length and width) of tumors were measured using a digital caliper. Mean weight of mice at initiation of study and termination of study did not differ significantly between the groups. The HeLa xenograft study showed that NM supplementation inhibited tumor weight by 59% ($p = 0.001$) compared with the control diet (Roomi et al., 2014a). Furthermore, immunohistochemical analysis of the proliferation marker Ki67 showed less intensity and frequency of staining in the tumors from the NM group as compared to the control group (Roomi et al., 2014a). Tumors from the control group showed more than 60%–70% of cells positive for Ki67, with positive cells dispersed uniformly through the tumor amid Ki67-negative cells, whereas NM-treated tumor sections showed large areas of Ki67-negative nucleated cells amid areas of positive cells for greatly reduced Ki67 expression (not shown).

Decreased cancer cell proliferation in vitro in the presence of nutrient synergy

We also evaluated the effect of NM on cancer cell proliferation in vitro utilizing the MTT assay, a colorimetric assay based on the ability of viable cells to reduce a soluble yellow tetrazolium salt [3-(4,5-dimethylthiazol-2-yl) 2,5-diphenyl tetrazolium bromide] (MTT) to a blue formazan crystal by mitochondrial succinate dehydrogenase activity of viable cells. This test is a good index of mitochondrial activity and thus of cell viability. Inhibition of cancer cell proliferation by NM was confirmed in a variety of cancer cell lines, as shown in Table 18.2, with the highest inhibition (80%) at NM 1000 μg/mL for prostate LnCap cells.

NM inhibits chemically induced carcinogenesis

We have found that NM also significantly inhibited chemically induced carcinogenesis in the following models: *N*-methyl-*N*-nitrosourea (NMU)-induced mammary

TABLE 18.1 Effect of NM diet on tumor weight in xenograft studies.

Cancer cell line	Reduction of tumor weight by NM diet over control diet
Hepatocellular carcinoma SK-Hep-1	42% ($P = .09$)
Colon HCT-116	63% ($P = .0002$)
Prostate PC-3	47% ($P < .0001$)
Lung A-549	44% ($P = .001$)
Melanoma A2058	57% ($P < .0001$)
Cervical HeLa	59% ($P = .001$)
Head and neck squamous cell carcinoma OHSU-974	47% ($P = .0009$)

TABLE 18.2 Effect of NM on cancer cell proliferation.

Cancer cell line	% Inhibition of cell growth at NM 1000 μg/mL	Cancer cell line	% Inhibition of cell growth at NM 1000 μg/mL
Prostate LNCaP	80	Chondrosarcoma SW-1353	39
Melanoma A2058	64	Pancreas MIA PACA-2	38
Liposarcoma SW-872	61	Cervical cancer HeLa	37
Osteosarcoma U2OS	60	Fibrosarcoma HT1080	37
Glioma A-172	50	Hepatocellular carcinoma SK-Hep-1	33
Breast MBA-MB-231	50	Synovial carcinoma	28
Prostate DU-145	47	Rhabdomyosarcoma	20
Cervical cancer DoTc2451	45	Lung carcinoma A-549	20
Prostate PC-3	40	Osteosarcoma MNNG	10

carcinogenesis in Sprague Dawley rats (Roomi et al., 2005a), urethane-induced lung carcinogenesis in A/J mice (Roomi et al., 2009c), and dimethylbenzoanthracene (DMBA)-induced skin carcinogenesis in SENCAR mice (Roomi et al., 2008). See Table 18.3 for a summary of these studies.

Induction of cancer cell apoptosis in vivo and in vitro by NM

Pro-apoptotic effects of nutrient synergy in various types of tumors were investigated using the following tests: TUNEL and Bcl-2 in vivo and live-green caspases in vitro. Immunohistochemical analysis of HeLa xenograft tumors from mice on NM 0.5% supplemented diet compared to control diet demonstrated increased apoptosis occurring in the NM-treated group (Roomi et al., 2014a). The control group tumors showed regions of cells not undergoing apoptosis within larger regions of cells undergoing apoptosis for a heterogeneous pattern of normal cells interspersed with apoptotic cells, whereas NM-treated tumors showed uniform areas of homogeneous apoptotic cells with peripheral areas of nonapoptotic cells (not shown). Some apoptosis was detected in cells of the fibrous capsule of NM-treated tumors. High staining for Bcl-2, a pro-survival, antiapoptotic protein, was observed in untreated tumors and was greatly reduced Bcl-2 in NM-treated tumors (Roomi et al., 2014a). Untreated tumors had extensive regions of positive cells inside the tumors adjacent to and infiltrating regions of cells without Bcl-2, whereas NM-treated tumors had vast regions of Bcl-2-free cells that were clearly nucleated with much less Bcl-2 expression (not shown).

In vitro studies of the effect of NM on apoptosis in a variety of cancer cells used live-green caspases. To study the effect of NM on apoptosis of cancer cells, cells were grown to confluence, challenged with NM dissolved in media at 0, 50, 100, 250, 500, and 1000 μg/mL, and incubated for 24 h. The cell culture was washed with PBS and treated with the caspase reagent as specified in the manufacturer's protocol (Molecular Probes Image-IT Live Green Poly Caspases Detection Kit 135104; Invitrogen). The cells were photographed under a fluorescence microscope and counted. Green cells represent viable cells, while, yellow, and orange represent early apoptosis, and red represents late apoptosis. Dose-dependent induction of apoptosis in a variety of cancer cell lines was confirmed with the NM challenge. The effect of NM on apoptosis in HeLa cells is shown in Fig. 18.1. At 100 μg/mL NM, 30% of cells were viable (22% in early apoptosis and 48% in late apoptosis); at 500 μg/mL, NM 9% of cells were viable (8% in early apoptosis and 83% in late apoptosis); at 1000 μg/mL NM, 1% were viable (8% in early apoptosis and 91% in late apoptosis). Studies by Harakeh et al. (2006) have shown that NM induces apoptosis of adult T-cell leukemia cells and acute leukemia cells by the modulation of p53, p21, Bcl-2, and Bax protein levels.

TABLE 18.3 Effect of NM on carcinogen induced tumors.

Carcinogen used for induction of cancer	NM treatment	Tumor incidence and multiplicity	Tumor weight and burden
N-methyl-*N*-nitrosourea (NMU) single dose 50 mg/kg, IP, into Sprague Dawley rats ($n = 20$)	2 weeks post NMU injection, rats fed control or NM 0.5% supplemented diet for 24 weeks	NM inhibited **mammary** tumor incidence and multiplicity by 68.4%	NM inhibited **mammary** tumor weight by 78% ($P = .0001$) and tumor burden by 60.5% ($P < .0001$)
Urethane Single dose 1 mg/g, IP, into A/J mice ($n = 20$)	Mice fed control or NM 0.5% supplemented diet for 20 weeks	NM inhibited **lung** tumor incidence by 49% ($P < .0001$)	NM inhibited **lung** tumor weight by 18% ($P = .003$)
7,12-dimethlbenzanthracene (DMBA) Topical application of 100 nM DMBA in 0.2 mL acetone to SENCAR mice 2 × per week for 4 weeks: groups 1 ($n = 10$), 3 ($n = 20$), 4 ($n = 20$) (group 2 treated with acetone)	Groups 1 and 2 were fed control diet Group 3A was fed NM 0.5% supplemented diet Group 3B was fed control diet and topical application of NM 75 mg in 0.4 mL acetone/water (1:1) prior to DMBA application Group 4 was fed NM 0.5% diet 2 weeks prior to DMBA; then, 4A continued on NM diet and 4B was fed control diet for 10 weeks	NM inhibited **skin** tumor multiplicity by 59% (group 3A), 62% (group 3B), 68% (group 4A), and 86% (group 4B)	N/A

Nutrient mixture modulates various parameters critical to metastasis: secretion of MMPs, uPA, and TIMPs, cancer cell migration, and Matrigel invasion

Proteases play a key role in tumor cell invasion and metastasis due to their ability to digest basement membrane and ECM components. Two families of proteases: serine (uPA) and MMP-2 and MMP-9 are necessary for invasive and metastatic potential. Strong clinical and experimental evidence shows that elevated levels of uPA and MMPs are associated with tumor growth, cancer progression, metastasis, and shortened survival in patients. MMP activities are regulated by specific tissue inhibitors of metalloproteinases (TIMPs). Elevated levels of MMP-9, a type IV collagenase, have been shown to be associated with cancer progression, metastasis, and shortened patient survival, because they play a key role in tumor cell invasion and metastasis by digesting the basement membrane and ECM components (Kleiner and Stetler-Stevenson, 1999; Yurchenko and Schitny, 1990; Barsky et al., 1983; Liotta et al., 1980; Stetler-Stevenson et al., 1996; Stetler-Stevenson, 2001). In addition to proteolysis, MMP-9, specifically the MMP-9 dimer, has been shown to play an important role in cell migration (Dufour et al., 2010; Bjorland and Koivunen, 2005). MMP-9 is unique in its ability to be secreted in both the monomeric and a disulfide-bonded dimeric form. The MMP-9 dimer (present usually as 10%–15% of the MMP-9 population), not the monomer, appears necessary for MMP-9 enhanced cell migration (Dufour et al., 2010).

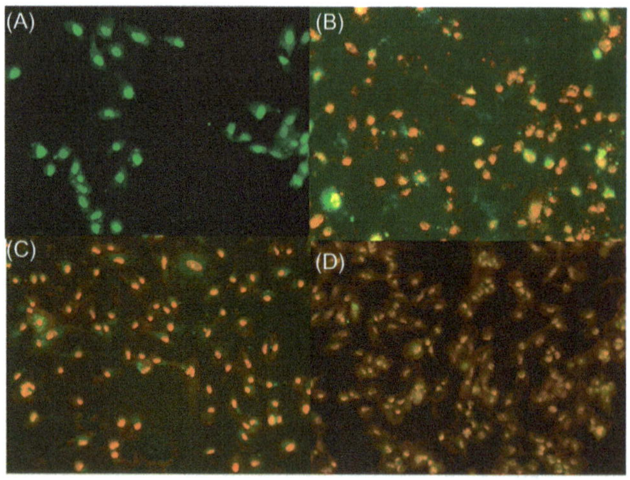

FIGURE 18.1 Effect of NM on cervical HeLa cell apoptosis: live green caspases. (A) Control. (B) NM 100 μg/mL. (C) NM 500 μg/mL. (D) NM 1000 μg/mL. *NM*, micronutrients.

Nutrient mixture inhibits MMP-2, MMP-9, and MMP-9 dimer secretion

In vivo results demonstrated modulation of MMP-2 and MMP-9 secretion with NM treatment. Immunohistochemical analysis of tumor tissue from HeLa xenograft studies demonstrated reduction of MMP-2 and MMP-9 secretion when female athymic mice were fed a diet supplemented with NM 0.5% (Roomi et al., 2014a). Tumors from the control group showed intense uniform staining of MMP-2 in and around each cell, whereas NM-treated tumors exhibited greatly reduced central expression/secretion of MMP-2 with a ring of peripheral cells positive for MMP-2 (not shown). NM-treated tumors showed much less MMP-9 staining than did control group tumors.

Using gelatinase zymography, MMP-2, MMP-9, and MMP-9 dimer secretion in vitro in many cancer cell lines were evaluated with some expressing MMP-2 or MMP-9 and some expressing both. Gelatinase zymography was performed in 10% Novex Pre-Cast SDS Polyacrylamide Gel (Invitrogen Corporation) in the presence of 0.1% gelatin under nonreducing conditions. Culture media (20 μL) were mixed with sample buffer and loaded for sodium dodecyl sulfate-polyacrylamide gel electrophoresis (SDS-PAGE) with Tris glycine SDS buffer, as suggested by the manufacturer (Novex). Samples were not boiled before electrophoresis. Following electrophoresis, the gels were washed twice in 2.5% Triton X-100 for 30 min at room temperature to remove SDS. The gels were then incubated at 37°C overnight in substrate buffer containing 50 mM Tris−HCl and 10 mM $CaCl_2$ at pH 8.0 and stained with 0.5% Coomassie Blue R250 in 50% methanol and 10% glacial acetic acid for 30 min and destained. Upon renaturation of the enzyme, the gelatinases digested the gelatin in the gel, producing clear bands against an intensely stained background. Protein standards were run concurrently and approximate molecular weights were determined by plotting the relative mobilities of known proteins. Gelatinase zymograms were scanned using a CanoScan 9950F scanner at 300 dpi.

NM inhibited MMPs in a dose-dependent manner in all cancer cells tested. Zymograms of a representative cancer cell line expressing both MMP-2 and MMP-9 monomer and dimer, chondrosarcoma SW-1353, are shown in Fig. 18.2. Gelatinase zymography showed bands corresponding to MMP-2 and MMP-9 and strongly induced MMP-9 and MMP-9 dimer with phorbol 12-myristate 13-acetate (PMA) (100 ng/mL) treatment. NM inhibited their secretions in a dose-dependent manner with total block of MMP-9 dimer at 250 μg/mL (linear trend $R^2 = 0.843$), MMP-9 monomer at 500 μg/mL (linear trend $R^2 = 0.729$), and MMP-2 at 500 μg/mL (linear trend $R^2 = 0.866$). Secretion of MMP-9 dimer, which has been shown to be required for cell migration, was found to be

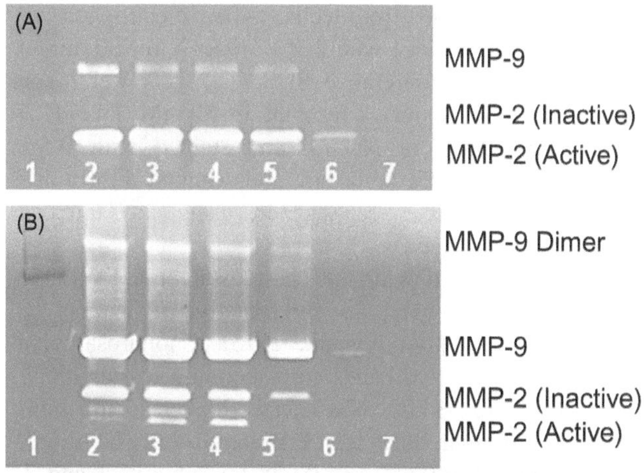

FIGURE 18.2 Effect of NM on chondrosarcoma SW-1353 MMP-2 and MMP-9 expression. Gelatinase zymograms of normal SW-1353 (A) and PMA-treated SW-1353 (B) MMP-2 and MMP-9 expression. Legend: (1) markers; (2) control; (3−7) NM 10, 50, 100, 500, and 1000 μg/mL. *NM*, micronutrients.

25% of that of MMP-9, indicating the aggressiveness of chondrosarcoma, because MMP-9 dimer is usually present as 10%−15% of the MMP-9 level (Roomi et al., 2014b).

Of interest, in evaluating dimer activity, sarcomas had the highest levels of MMP-9 monomer and dimer combined with and without PMA among cancer cell lines from various organ malignancies, with osteosarcoma showing the highest MMP-9 dimer to MMP-9 ratio, indicating a very aggressive cancer. Cervical, uterine, and male breast cancer cell lines showed the next highest levels of combined MMP-9, followed by breast cancer cell lines. Melanoma, renal, lung, head, and neck and hepatic carcinoma showed lower levels, and prostate, glioblastoma, bladder, and leukemia cell lines showed the lowest levels. The NM showed dose-dependent inhibition of MMP-9 monomer and dimer in all cell lines tested (Roomi et al., 2014b).

NM decreases uPA and increases TIMP-2 activities

Studies of several cancer cell lines demonstrated downregulated activity of uPA and upregulated activity of TIMPs. Cells were treated with NM at 0, 10, 50, 100, 500, and 1000 μg/mL. Fibrin zymography was used to analyze uPA activity on 10% SDS-PAGE gels containing fibrinogen (5.5 mg/mL) and plasminogen (50 μg/mL). After electrophoresis, the gels were washed twice with 2.5% Triton X-100 for 30 min. The gels were then incubated overnight at 37°C with 0.1% glycine buffer pH 7.5 and then stained with 0.5% Coomassie Brilliant Blue R250 and destained. TIMPs were analyzed by reverse zymography on 15% SDS gels containing serum-free conditioned medium from cells. After electrophoresis, gels were washed twice with 2.5% Triton-X for 30 min at room temperature to remove SDS. The gels were then incubated at 37 degrees overnight in 50 mM Tris–HCl and 10 mM $CaCl_2$ at pH 7.6 and stained with 0.5% Coomassie Blue R25, destained, and scanned. Fibrin and reverse zymograms were scanned using a CanoScan 9950F Canon Scanner.

Results obtained with the chondrosarcoma SW-1353 cell line, presented here, are representative of findings for other cell lines (Roomi et al., 2013). Chondrosarcoma SW-1353 showed two bands corresponding to subunits 1 and 2 at 55 and 33 kDa. NM exerted dose-response inhibition with virtual block of uPA activity at 100 μg/mL (linear trends $R^2 = 0.690$ and 0.698 for subunits 1 and 2, respectively), as shown in Fig. 18.3. Chondrosarcoma activity of TIMP-2 was upregulated by NM in a dose-dependent manner, as shown in Fig. 18.4.

Nutrient synergy has inhibitory effect on cancer cell migration

To study cell migration, a 2-mm wide single uninterrupted scratch was made from the top to the bottom of culture plates of cancer cells grown to confluence. Culture plates were washed with PBS and incubated with NM in medium and tested at 0, 10, 50, 100, 500, and 1000 μg/mL in triplicate at each dose for 24 h. Cells were washed with PBS and fixed and stained with H&E, and photomicrographs were taken. A representative example, breast 4T1 cell line (Roomi et al., 2014c), demonstrated reduced breast cancer 4T1 cell migration by NM treatment in a dose-dependent fashion, with total inhibition at 1000 μg/mL.

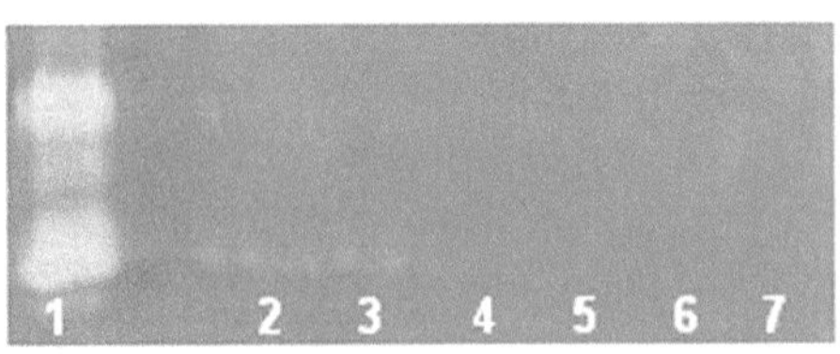

FIGURE 18.3 Effect of NM on chondrosarcoma SW-1353 uPA expression. Fibrin zymogram of SW-1353 uPA expression. Legend: (1) uPA; (2) markers; (3) control; (4–8) NM 50, 100, 250, 500, and 1000 μg/mL. *NM*, micronutrients.

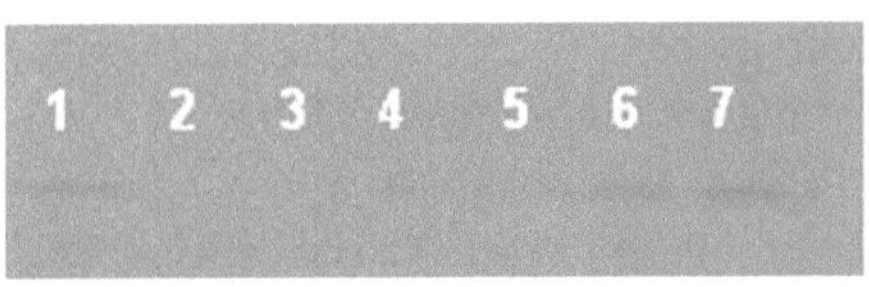

FIGURE 18.4 Effect of NM on chondrosarcoma SW-1353 TIMP-2 expression. Reverse zymogram of SW-1353 TIMP-2 expression. Legend: (1) markers; (2) control; (3–7) NM 50, 100, 250, 500, and 1000 μg/mL. *NM*, micronutrients.

Nutrient synergy inhibits Matrigel invasion by cancer cells

Invasion studies were conducted on many cancer cell lines using Matrigel (Becton Dickinson) inserts in 24-well plates. Table 18.4 summarizes the concentration of NM required to completely block Matrigel invasion in a variety of cancer cell lines. Suspended in medium, cancer cells were supplemented with nutrients, as specified in the design of the experiments, and seeded on the insert in the well. Thus both the medium on the insert and in the well contained the same supplements. The plates with the inserts were then incubated in a culture incubator equilibrated with 95% air and 5% CO_2 for 24 h. After incubation, the media from the wells were withdrawn. The cells on the upper surface of the inserts were gently scrubbed away with cotton swabs. The cells that had penetrated the Matrigel membrane and migrated onto the lower surface of the Matrigel were stained with H&E and visually counted under the microscope.

As a representative example, the effect of NM on breast cancer 4T1 is presented here. NM significantly inhibited 4T1 invasion through Matrigel in a dose-dependent manner, with total block at 250 μg/mL (Fig. 18.5) (Roomi et al., 2014c).

Nutrient mixture inhibits angiogenesis in vivo and in vitro

Immunohistochemical analysis of tumor tissue from the HeLa xenograft study demonstrated modulation of vascular endothelial growth factor (VEGF), a pro-angiogenic factor that is critical for both primary tumor growth and metastasis with NM treatment (Roomi et al., 2014a). The majority of the cross-sectional areas of control group tumors exhibited VEGF expression in a uniform distribution pattern. NM-treated tumors showed dramatically reduced VEGF staining with large central areas of cells having no VEGF expression (not shown).

NM also inhibited angiogenesis in an in vivo model of a chick chorioallantoic membrane assay in a chicken embryo, inhibited b-FGF-induced vessel growth in a mouse Matrigel plug assay, inhibited blood vessel formation in tumors in a xenograft mouse model, and also inhibited capillary tube formation (Roomi et al., 2005b). At a biochemical level, NM decreased in vitro expression of pro-angiogenic factors such as VEGF, angiopoietin-2, basic fibroblast growth factor, PDGD, and TDGβ-1 by U2OS cells (Roomi et al., 2005b).

Nutrient synergy controls metastasis

Experimental results have confirmed the effectiveness of the NM mixture in controlling metastasis, supporting our results with its effect on invasive parameters in vitro (Table 18.5).

TABLE 18.4 Effect of NM on cancer cell invasion through Matrigel.

Cancer cell line	NM concentration to achieve 100% block of ECM invasion
Breast MBA-MB-231 Breast MCF-7 (estrogen-sensitive) Osteosarcoma MNNG & U2OS	100 μg/mL
Cervical cancer HeLa Lung carcinoma A-549 Pancreas MIA PACA-2 Prostate LNCaP Colon HCT 116 Rhabdomyosarcoma Prostate DU-145 Thyroid SW 579	500 μg/mL
Hepatocellular carcinoma SK-Hep-1 Bladder cancer T-24 Fibrosarcoma HT1080 Cervical cancer DoTc2451 Ovarian SK-OV-3 Prostate PC-3 Renal carcinoma 786-0 Synovial carcinoma Melanoma A2058 Glioma A-172 Liposarcoma SW-872	1000 μg/mL

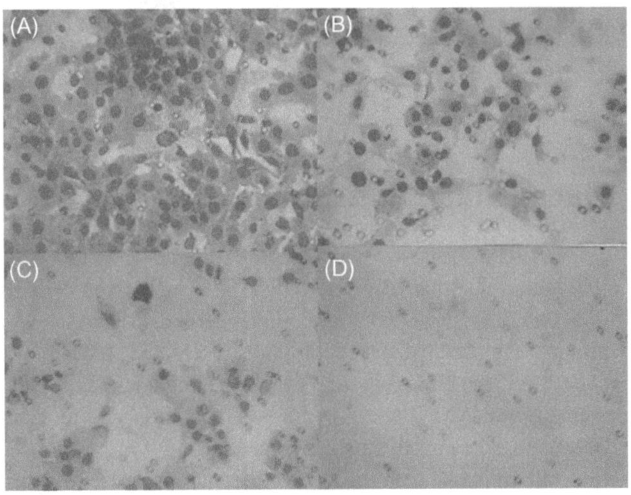

FIGURE 18.5 Effect of NM on breast cancer 4T1 invasion through Matrigel and photomicrographs. (A) Control. (B) NM 50 μg/mL. (C) NM 100 μg/mL. (D) NM 250 μg/mL. *NM*, micronutrients.

Metastasis from breast

We investigated the effect of NM in vivo on murine 4T1 breast cancer, a representative model for metastatic breast cancer. After 1 week of isolation, 5- to 6-week-old female Balb/C mice were inoculated with 5×10^5 4T1 cells into the mammary pad and randomly divided into two groups. The control group was fed a regular diet and the NM group was fed a regular diet supplemented with 0.5% NM. After 4 weeks, the mice were sacrificed and their tumors, lungs, livers, kidneys, hearts, and spleens were excised and processed for histology. Dimensions (length and width) of tumors were measured using a digital caliper, and the tumor burden was calculated using the following formula: 0.5 × length × width. NM inhibited tumor weight and burden of 4T1 tumors by 50% ($P = .02$) and 53.4% ($P \leq .0001$), respectively (Roomi et al., 2014b). Lung metastasis was profoundly inhibited by NM supplementation. The mean number of colonies was reduced by 87% ($P < .0001$) and the mean weight of lungs was reduced by 60% ($P = .0001$) compared to control mice (Roomi et al., 2014b). Metastasis to liver, spleen, kidney, and heart was significantly reduced with NM supplementation (Roomi et al., 2014b) (Table 18.5).

Pulmonary metastasis from vein

Pulmonary metastasis was studied in 6- to 7-week-old female C57BL/6 mice, which were divided into six groups of six mice each. Groups 1–4 were injected via the tail vein with B16FO melanoma cells (5×10^4/mouse). Group 5 was injected in the tail vein with the same number of B16FO melanoma cells pretreated with NM (500 μg/mL for 18 h). Group 6 (untreated control) was injected with saline. Subsequently, the mice were placed on the following diet regimens and administration routes for NM: groups 1 and 5 were fed Purina mouse

TABLE 18.5 Effect of NM on metastasis.

Metastasis study	Inhibition of tumor growth by NM compared to control	Inhibition of metastasis by NM compared to control
Lung, liver, spleen, kidney, heart metastasis Breast cancer 4T1 injection into mammary pad of female Balb/C mice NM by diet	Mean tumor weight by 50% ($P = .02$) Mean tumor burden by 53.4% ($P \leq .0001$)	Lung metastasis: number of colonies inhibited by 87% ($P < .0001$) and lung weight inhibited by 60% ($P = .0001$) Metastasis to liver, spleen, kidney, and heart was significantly reduced
Lung metastasis B16FO melanoma cells injection into tail vein of female C57BL/six mice NM by diet, IP and IV	N/A	Pulmonary colonization was inhibited by 63% ($P < .0001$) in mice receiving NM by diet and by 86% ($P < .0001$) in mice receiving NM by IP and IV injections; the lungs from mice injected with viable melanoma cells pretreated with NM were free from any metastasis
Hepatic metastasis B16FO melanoma cells injection into the spleen of athymic nude mice NM by diet	Tumor growth (mean spleen weight) was reduced by 64% ($P = .001$)	Hepatic metastasis in NM-supplemented mice was reduced by 55% ($P = .006$); no metastasis to the kidneys and lungs was evident in either group
Peritoneal and lung metastasis B16FO melanoma cells injection into right testis of male athymic mice NM by diet	N/A	Peritoneal metastasis extensive in all control mice; in contrast, in 50% of the NM group there was no evidence of peritoneal metastasis and mild metastasis in the remaining 50% Severe lung metastasis in two of six mice in the control group and mild metastasis in two of six mice in the NM group No metastasis to the liver, kidney, or spleen was evident in either group

chow, the control diet; group 2 mice were given NM intraperitoneally (IP) (4 mg/mouse, three times per week for 2 weeks) and fed the control diet; and group 4 was given NM intravenously (IV) (4 mg/mouse, three times per week for 2 weeks) and fed the control diet. Group 6 mice were fed the control diet. Two weeks later, the mice were sacrificed, and their lungs were excised, weighed, and processed for histopathologic examination. The weights of the lungs were obtained and the number of metastases and number of cells in each metastatic lesion were counted. Lungs isolated from mice injected with 5×10^4 B16FO cells and fed NM 0.5% diet (group 2) had fewer metastatic colonies (by 63%; $P < .0001$) than did the lungs of mice fed the control diet (group 1) (Roomi et al., 2006). Pulmonary colonization was inhibited by 86% ($P < .0001$) in mice receiving NM by IP and IV injections (groups 3 and 4) (Roomi et al., 2006). The lungs from mice injected with viable melanoma cells pretreated with NM (group 5) were free from any metastasis (Roomi et al., 2006). Thus exposing melanoma cells to NM inhibited their ability to metastasize without the need for dietary supplementation, suggesting profound changes on a cellular level.

Hepatic metastasis from spleen

Hepatic metastasis was studied in 10- to 12-week-old athymic nude mice that received 10^6 B16FO melanoma cells by injection into the spleen and then were divided into two groups. The control group of mice received Purina mouse chow and the NM group received the regular diet supplemented with NM 0.5%. After 2 weeks, animals were sacrificed and spleens, livers, kidneys, and lungs were excised from all animals, examined, weighed, and processed for histology. The control group mice developed large black spleens and livers, indicating growth in the spleen and metastasis to the liver, whereas mice supplemented with NM not only showed less tumor growth in the spleen but also drastically reduced metastasis to the liver (Roomi et al., 2009a). Intrasplenic tumor growth (mean spleen weight) was reduced significantly (64%; $P = .001$) in the NM-supplemented group compared to the control group (Roomi et al., 2009a). Hepatic metastasis in NM-supplemented mice was reduced by 55% ($P = .006$) compared to the control group, based on mean liver weights of the groups (Roomi et al., 2009a). In all groups, no metastasis to the kidneys and lungs was evident.

Peritoneal and lung metastasis from testicle

A major cause of mortality from testicular cancer is metastasis to the lung. We examined the effect of NM on metastasis from intratesticular inoculation of B16FO cells into nude mice (Roomi et al., 2012). Male 10- to 12-week-old athymic mice ($n = 12$) were inoculated with 5 $\times$ 10^5 B16FO melanoma cells in 100 μL of PBS into the right testis and the left testis was left untreated. Following inoculation, the mice were randomly divided into two groups. The control group ($n = 6$) was fed a regular mouse chow diet and the NM 1% group ($n = 6$) was fed the same diet but supplemented with 1% NM. Four weeks later, the mice were sacrificed and the abdominal cavity was opened. Mice in the control group exhibited extensive metastasis in the peritoneal cavity and severely enlarged right testes and necrotic seminiferous tubules. By contrast, in the NM 1% group there was no evidence of peritoneal metastasis in 50% of the animals and mild metastasis in the remaining 50%. The right testes were enlarged and seminiferous tubules in the area of invasion showed evidence of degeneration. No metastasis to the liver, kidney, or spleen were evident in either group. However, severe lung metastasis was observed in two of six mice in the control group and mild metastasis was observed in two of six mice in the NM 1% group.

Nutrient synergy modulates inflammatory proteins Cox-2 and iNOS expression in cancer cells

Inflammatory proteins such as Cox-2 and iNOS contribute to carcinogenesis of human tumors and have been found to be elevated in cancer patients. Immunohistochemical analysis of Cox-2 and iNOS in HeLa xenograft tumors from female nude mice demonstrated the inhibition of Cox-2 and iNOS staining with NM 0.5% supplementation (Roomi et al., 2014a). Tumors from the control group exhibited large dispersed regions of central Cox-2 enzymes, whereas NM-treated tumors showed greatly decreased Cox-2 (not shown). Tumors from the control group had extensive diffuse cytoplasmic as well as intense punctate iNOS staining within the tumor, with the majority of the tumor positive for iNOS; in contrast, in the NM-treated group, tumors exhibited large areas devoid of iNOS (not shown).

In vitro, NM effects on Cox-2 and iNOS protein expression in cancer cells were determined by Western blot analysis using polyclonal antibodies against Cox-2. Briefly, the cells were lysed for 30 min in ice cold radio-immunoprecipitation buffer (50 mM Tris–HCl, pH 7.5, 10% Triton X-100, 1.0% Na-deoxycholate, 150 mM NaCl, 2 mM EDTA) with 1 mM activated orthovandate and 1 $\times$ Complete Mini Protease Inhibitor Cocktail tablet. Clarified cell lysates were denatured and equal proteins were loaded and fractionated by SDS-polyacrylamide gel electrophoresis. After electrophoresis, the proteins were transferred to nitrocellulose membrane. The membranes were then incubated with the respective antibodies and developed by the enhanced chemiluminescence system. Roomi et al. (2009b) found Cox-2 and iNOS expression by bladder cancer T-24 cells exposed to NM to be reduced in a dose-dependent fashion. A study of the effect of another NM mixture including ascorbic acid, quercetin, naringenin, hesperitin, tea catechins, lysine, proline, arginine, and *N*-acetyl cysteine on inflammatory gene expression found that this mixture inhibited Cox-2 enzymatic activity and downregulated pro-inflammatory cytokine protein expression at the transcription level complementing a blockade in NFkβ activation (Ivanov et al., 2008).

Safety of NM

In contrast to the toxic side effects of current treatments, the NM mixture has been shown to be a safe nutraceutical. In a previous in vivo study addressing safety issues, authors of this chapter found that gavaging adult female ODS rats (weighing 250–300 g) with the nutrient mixture (at 30, 90, or 150 mg per day for 7 days) did not have adverse effects on vital organs (heart, liver, and kidney) on the associated functional serum enzymes, indicating that this mixture is safe to use even at these high doses that far exceed the normal equivalent dosage of the nutrient (Roomi et al., 2003).

Concluding remarks and future directions

In conclusion, the synergistic NM mixture has been shown to be effective against multiple targets critical to development and progression of many types of cancer. These results indicate that this nontoxic comprehensive approach, with its pleiotropic effects, should be considered in cancer therapy as a safe and economical approach.

References

Aldercreutz, H., 1990. Western diet and Western disease: some hormonal and biochemical mechanisms and associations. Scand. J. Clin. Lab. Invest. Suppl. 201, 3–23.

Almholt, K., Lundt, L.R., Rygaard, J., Nielsen, B.S., Danø, K., Rømer, J., et al., 2005. Reduced metastasis of transgenic mammary cancer in urokinase-deficient mice. Int. J. Cancer 113 (4), 525–532.

Anthony, H.M., Schorah, C.J., 1982. Severe hypovitaminosis C in lung-cancer patients: the utilization of vitamin C in surgical repair and lymphocyte-related host resistance. Br. J. Cancer 46, 354–367.

Barsky, S.H., Siegel, G.P., Jannotta, F., 1983. Loss of basement membrane components by invasive tumors but not by their benign counterparts. Lab. Invest. 49, 140–147.

Bjorland, M., Koivunen, E., 2005. Gelatinase-mediated migration and invasion of cancer cells. Biochim. Biophys. Acta 1755, 37–69.

Chambers, A.F., Matrisian, L.M., 1997. Changing views on the role of matrix metalloproteinases in metastasis. J. Natl. Cancer Inst. 89 (17), 1260–1270.

Chen, Q., Espey, M.G., Krishna, M.C., 2005. Pharmacologic ascorbic acid concentrations selectively kill cancer cells: action as a pro-drug to deliver hydrogen peroxide to tissues. Proc. Natl. Acad. Sci. U.S.A 102 (38), 13604–13609.

Cooke, J.P., Dzau, V.J., 1997. Nitric oxide synthase: role in the genesis of vascular disease. Annu. Rev. Med. 48, 489–509.

Duffy, M.J., 1992. The role of proteolytic enzymes in cancer invasion and metastasis. Clin. Exp. Metastasis 10, 145–155.

Dufour, A., Zucker, S., Sampson, N.S., Kuscu, C., Cao, J., 2010. Role of matrix metalloproteinase-9 dimers in cell migration: design of inhibitory pepides. J. Biol. Chem. 285 (46), 35944–35956.

Hara, Y., 2001. Green Tea: Health Benefits and Applications. Marcel Dekker, New York, Basel.

Harakeh, S., Diab-Assaf, M., Niedzwiecki, A., Khalife, J., Abu-El-Ardat, K., Rath, M., 2006. Apoptosis induction by Epican Forte in HTLV-1 positive and negative malignant T-cells. Leuk. Res. 30 (7), 869–881.

Ivanov, V., Cha, J., Ivanova, S., Kalinovsky, T., Roomi, M.W., Rath, M., et al., 2008. Essential nutrients suppress inflammation by modulating key inflammatory gene expression. Int. J. Mol. Med. 22 (6), 731–742.

Kawakami, S., Kageyama, Y., Fujii, Y., Kihara, K., Oshima, H., 2001. Inhibitory effects of *N*-acetyl cysteine on invasion and MMP 9 production of T24 human bladder cancer cells. Anticancer Res. 21, 213–219.

Kemberling, J.K., Hampton, J.A., Keck, R.W., Gomez, M.A., Selman, S.H., 2003. Inhibition of bladder tumor growth by the green tea derivative epigallocatechin-3-gallate. J. Urol. 170 (3), 773–776.

Kleiner, D.L., Stetler-Stevenson, W.G., 1999. Matrix metalloproteinases and metastasis. Cancer Chemother. Pharmacol. 43 (Suppl), 42s–51s.

Koh, W.S., Lee, S.J., Lee, H., Park, C., Park, M.H., Kim, W.S., et al., 1998. Differential effects and transport kinetics of ascorbate derivatives in leukemic cell lines. Anticancer Res. 8, 2487–2493.

Kurbacher, C.M., Wagner, U., Kolster, B., Andreotti, P.E., Krebs, D., Bruckner, H. W., 1996. Ascorbic acid (vitamin C) improves the antineoplastic activity doxorubicin, cisplatin and paclitaxel in human breast carcinoma cells *in vitro*. Cancer Lett. 103 (2), 183–189.

Liotta, L.A., Tryggvason, K., Garbisa, S., Hart, I., Foltz, C.M., Shafie, S., 1980. Metastatic potential correlates with enzymatic degradation of basement membrane collagen. Nature 284, 67–68.

Maramag, C., Menon, M., Balaji, K.C., Reddy, P.G., Laxmanan, S., 1997. Effect of vitamin C on prostate cancer cells in vitro: effect on cell number, viability and DNA synthesis. Prostate 32, 188–195.

Miller, A.B., 1990. Diet and cancer: a review. Acta Oncol. 29 (1), 87–95.

Morini, M., Cai, T., Aluigi, M.G., Noonan, D.M., Masiello, L., De Flora, S., et al., 1999. The role of the thiol *N*-acetyl cysteine in the prevention of tumor invasion and angiogenesis. Int. J. Biol. Markers 14, 268–271.

Mukhtar, H., Ahmed, N., 2000. Tea polyphenols: prevention of cancer and optimizing health. Am. J. Clin. Nutr. 71, 1698s–1702s.

Naidu, K.A., Karl, R.C., Coppola, D., 2003. Antiproliferative and proapoptotic effect of ascorbyl stearate in human pancreatic cancer cells: association with decreased expression of insulin-like growth factor 1 receptor. Dig. Dis. Sci. 48 (1), 230–237.

Niedzwiecki, A., Roomi, M.W., Kalinovsky, T., Rath, M., 2010. Micronutrient synergy—a new tool in effective control of metastasis and other key mechanisms of cancer. Cancer Metastasis Rev. 29 (3), 529–543.

Nunez, C., Ortiz de Apodaca, Y., Ruiz, A., 1995. Ascorbic acid in the plasma and blood cells of women with breast cancer. The effect of consumption of food with an elevated content of this vitamin. Nutr. Hosp. 10, 68–372.

Rath, M., Pauling, L., 1992. Plasmin-induced proteolysis and the role of apoprotein(a), lysine and synthetic analogs. J. Orthomol. Med. 7, 17–23.

Roomi, M.W., Ivanov, V., Netke, S.P., 2003. Serum markers of the liver, heart, and kidney and lipid profile and histopathology in ODS rats treated with nutrient synergy. J. Am. Coll. Nutr. 22, 477.

Roomi, M.W., Roomi, N.W., Ivanov, V., 2005a. Modulation of *N*-methyl–*N*-nitrosourea-induced mammary tumors in Sprague-Dawley rats by combination of lysine, proline, arginine, ascorbic acid and green tea extract. Breast Cancer Res. 7 (3), R291–R295.

Roomi, M.W., Roomi, N.W., Ivanov, V., 2005b. Inhibitory effect of a mixture containing ascorbic acid, lysine, proline, and green tea extract on critical parameters in angiogenesis. Oncol. Rep. 14 (4), 807–815.

Roomi, M.W., Ivanov, V., Kalinovsky, T., 2006. Inhibition of pulmonary metastasis of melanoma B16FO cells in C57BL/6 mice by a nutrient mixture consisting of ascorbic acid, lysine, proline, arginine, and green tea extract. Exp. Lung Res. 32 (10), 517–530.

Roomi, M.W., Roomi, N.W., Kalinovsky, T., 2008. Inhibition of 7, 12-dimethylbenzathracene-induced skin tumors by a nutrient mixture. Med. Oncol. 25 (3), 330–340.

Roomi, M.W., Kalinovsky, T., Roomi, N.W., 2009a. A nutrient mixture suppresses hepatic metastasis in athymic nude mice injected with murine B16FO melanoma cells. BioFactors 33, 181–189.

Roomi, M.W., Kalinovsky, T., Niedzwiecki, A., Nilsson, W.E., 2009b. Pleiotropic effects of a micronutrient mixture on critical parameters of bladder cancer. In: Nilsson, W.E. (Ed.), Bladder Cancer Etymology, Diagnosis and Treatments. Nova Science Publishers, Inc., New York.

Roomi, M.W., Roomi, N.W., Kalinovsky, T., 2009c. Chemopreventive effect of a novel nutrient mixture on lung tumorigenesis induced by urethane in male A/J mice. Tumori 95, 508–513.

Roomi, M.W., Kalinovsky, T., Roomi, N.W., 2012. Suppression of metastasis of intratesticular inoculation of B16FO melanoma cells by a novel nutrient mixture in male athymic mice. Exp. Ther. Med. 4, 775–780.

Roomi, M.W., Kalinovsky, T., Niedzwiecki, A., 2013. Modulation of u-PA, MMPs and their inhibitors by a novel nutrient mixture in adult human sarcoma cell lines. Int. J. Oncol. 43, 39–49.

Roomi, M.W., Kalinovsky, T., Cha, J., 2014a. Effect of NM on immunohistochemical cancer markers in human cervical cancer HeLa cell tumor xenograft in female nude mice. Exp. Ther. Med. 9 (2), 294–302.

Roomi, M.W., Kalinovsky, T., Rath, M., 2014b. Effect of a nutrient mixture on matrix metalloproteinase-9 dimers in various human cancer cell lines. Int. J. Oncol. 44 (3), 986–992.

Roomi, M.W., Kalinovsky, T., Roomi, N.M., 2014c. *In vitro* and *in vivo* effects of a nutrient mixture on breast cancer progression. Int. J. Oncol. 44 (6), 1933–1944.

Sato, D., Matsushima, M., 2003. Preventive effects of urinary bladder tumors induced by *N*-butyl-*N*-(4-hydroxybutyl)-nitrosamine in rat by green tea leaves. Int. J. Urol. 10 (3), 160–166.

Stetler-Stevenson, W.G., 2001. The role of matrix metalloproteinases in tumor invasion, metastasis and angiogenesis. Surg. Oncol. Clin. N. Am. 10, 383–392.

Stetler-Stevenson, W.G., Hewitt, R., Corcoran, M., 1996. Matrix metalloproteinases and tumor invasion from correlation and causality to the clinic. Semin. Cancer Biol. 7, 147–154.

Sun, Z., Chen, Y.H., Wang, P., 2002. The blockage of high-affinity lysine binding sites of plasminogen by EACA significantly inhibits prourokinase-induced plasminogen activation. Biochem. Biophys. Acta 1596, 182–192.

Taniguchi, S., Fujiki, H., Kobayashi, H., 1992. Effect of (−) epigallocatechin gallate, the main constituent of green tea, on lung metastasis with mouse B16 melanoma cell lines. Cancer Lett. 65, 51–54.

Valcic, S., Timmermann, B.N., Alberts, D.S., 1996. Inhibitory effect of six green tea catechins and caffeine on the growth of four selected human tumor cell lines. Anticancer. Drugs 7, 461–468.

Yang, G.Y., Liao, J., Kim, K., 1998. Inhibition of growth and induction of apoptosis in human cancer cell lines by tea polyphenols. Carcinogenesis 19, 611–616.

Yoon, S.O., Kim, M.M., Chung, A.S., 2001. Inhibitory effects of selenite on invasion of HT 1080 tumor cells. J. Biol. Chem. 276, 20085–20092.

Yurchenko, P.D., Schitny, J.C., 1990. Molecular architecture of basement membranes. FASEB J. 4, 1577–1590.

Chapter 19

Application of "nano" nutraceuticals in medicine

Alessia Bertero, Paola Fossati, Teresa Coccini, Leon J Spicer and Francesca Caloni

Introduction

The use of nutraceuticals to improve health and prevent or even cure some disorders is a practice dating back to ancient times (Ruchi et al., 2017). Since then, continuous efforts have been made to address the main limitations concerning the application of these compounds, namely their generally low bioavailability and unstable nature, contextually trying to enhance their targetability toward specific sites of action (Zaki, 2014). In this context nanotechnologies may offer new perspectives, as they have done in many other scientific and technological fields, thanks to their versatility and rapid development. Indeed, the application of this technique to nutritional medicine seems to be promising, leading to the production of nanonutraceuticals that have often shown better efficacy in comparison with traditional formulations (Jampilek et al., 2019). The great potentialities shown by nanotechnologies rely on their dimensions; nanoparticles are objects with at least one external diameter ranging from 1 to 100 nm; this characteristic strongly influences the physical, chemical, and biological properties of the material (i.e., causing the alteration of the surface/volume ratio and surface energy), leading to a different behavior in comparison with the bulk material (Phan and Haes, 2019). There are many advantages related to the application of nanoparticles for nutraceutical delivery; for instance, better bioavailability, solubility, and stability may be obtained, and this is particularly beneficial for the oral route, which is preferred for nutraceutical administration but, on the other hand, is often characterized by slow/poor/unpredictable absorption of the active substance also because of degradation processes that may occur in the gastrointestinal tract (GIT) (enzymatic and chemical degradation due to the digestive activity) and the potential poor solubility of the compound. Other benefits related to the application of nanoformulations may include the possibility to attain a targeted delivery and controlled release into the GIT. Given that, a smaller amount of the active substance can be administrated, with a reduction of the undesired side effects (Jampilek et al., 2019). On the other hand, the application of nanotechnology to nutraceutical delivery has also some potential disadvantages and risks that need to be considered: among them, the costs of certain nanoformulations that require expensive and complex technology for production, and the unpredictable effects on health and the environment of this powerful and relatively new technology are not completely known yet.

Nanoparticles as delivery systems for nutraceuticals

Nanoliposomes

Among all the studied nanocarriers for bioactive compounds, nanoliposomes are considered one of the most promising (Su et al., 2018; Javeri, 2016). These nanostructures are colloidal systems of closed lipid bilayer vesicles with a hydrophilic core; one or more bilayers, made mainly of natural nontoxic phospholipids and segregated by aqueous interphases, can be identified (Khorasani et al., 2018; Danaei et al., 2018), with the polar head groups oriented toward the aqueous domains. The dispersion of phospholipids in the aqueous phase is obtained through the application of a source of energy (Mozafari, 2010). Nanoliposomes, having both lipophilic and hydrophilic domains (lipid bilayer, bilayer interface, aqueous core), are able to carry molecules with different physicochemical characteristics (i.e., polar as well as nonpolar compounds) (Bozzuto and Molinari, 2015; Akbarzadeh et al., 2013; Su et al., 2018), even allowing incompatible ingredients to be encapsulated into the same carrier system. Other advantages related to the application of nanoliposomes as nutraceutical carriers lie in the ability to improve the efficacy of the bioactive agents by enhancing

Nutraceuticals. DOI: https://doi.org/10.1016/B978-0-12-821038-3.00019-7

their solubility, bioavailability, and stability, and also avoiding undesirable interactions with other substances (Mozafari, 2010) and enclosing unpalatable tastes. Nanoliposomes are also being applied for targeted delivery, a technique that allows the transport of the active compound into the organ/tissue of action while minimizing its concentration/effects in other tissues, optimizing the treatment efficacy, and potentially avoiding undesirable side effects (Zarrabi et al., 2020).

Nanoliposomes may also include other substances in their structure such as sterols, proteins, carbohydrates, and polymers, which are usually added to obtain increased stability, better protection against degradation, or targeted delivery of the encapsulated substances (Maherani et al., 2011). Moreover, since their bilayered structure is close to those of cellular membranes (Javeri, 2016), these nanocarriers are characterized by great biocompatibility (Su et al., 2018; Danaei et al., 2018).

Nanoliposome encapsulation technology has been applied on many nutraceuticals to evaluate the benefits brought by this delivery system. Omega-3 polyunsaturated fatty acids (PUFAs), like eicosapentaenoic acid (EPA) and docosahexaenoic acid (DHA), have shown many beneficial effects on health, cardiovascular diseases (i.e., atrial fibrillation, thrombosis, atherosclerosis), and other disorders including mental illnesses and depression, cognitive decline, rheumatoid arthritis, and inflammatory conditions (Shahidi and Ambigaipalan, 2018). The main dietary source of omega-3 PUFAs, and specifically of long-chain omega-3 PUFAs, is fish oil, but its use as a nutraceutical compound is difficult to implement because of two main issues: poor organoleptic properties and a high oxidation potential. To overcome these hurdles, fish oil has been encapsulated in nanoliposomes and added to fortified bread (Ojagh and Hasani, 2018). After 25 days of refrigerated storage, physical properties, including oxidative stability, were assessed. This formulation demonstrated good nanoencapsulation efficiency (around 90%) and was effective in providing protection against lipid oxidation during storage. Moreover, the addition of nanoliposome-encapsulated fish oil did not alter the organoleptic characteristics of bread, rather it produced positive effects on technological features (higher loaf volume). Nanoliposome-encapsulated fish oil was also added to fortified yogurt (Ghorbanzade et al., 2017). As in the case of the previous work, the physicochemical characteristics of the obtained product were investigated during refrigerated storage (21 days). The results showed higher DHA and EPA contents in the yogurt fortified with nanoencapsulated fish oil in comparison with that with only free fish oil added. A good nanoencapsulation efficiency (around 92%) and stability, with a decrease of acidity, syneresis (released whey) and peroxide value were observed. As for the sensory compliance, the organoleptic properties were closer to those of the nonfortified product, in comparison with the yogurt added with free fish oil, thus proving to be a suitable formulation for the fortification of this particular food. Fish oil has been also added to milk. In a work by Rasti et al. (2017), nanoliposome-encapsulated fish oil was used for food enrichment and the results in terms of stability and sensory characteristics were compared to those obtained with microencapsulated and unencapsulated fish oil. They observed that the nanoformulation, in contrast to the micro- and unencapsulated forms, did not show any detectable difference in comparison with the unfortified milk. Furthermore, a higher percentage of omega-3 PUFAs, together with lower peroxide and anisidine values, was detected in the nanoformulation in comparison with microencapsulated and unencapsulated fish oil. In the same work, similar results were obtained with bread, findings that are in accordance with those obtained by Ojagh and Hasani (2018), thus suggesting good reproducibility for this technique.

Vitamins are another class of nutraceuticals that have been formulated in nanoliposomes. Vitamin C is an essential micronutrient involved in many biochemical pathways, such as the synthesis of amino acids, some peptide hormones, carnitine, catecholamines, cholesterol, and collagen. Moreover, vitamin C is also involved in iron absorption, has antioxidant properties, and sustains immune system function (Carr and Maggini, 2017; Grosso et al., 2013). Vitamin C deficiency can lead to tiredness, depression, connective tissue alteration, gingivitis, petechiae, and internal hemorrhages, while in children it can lead to bone growth impairment (Fain, 2004). Furthermore, potential effects of vitamin C on the prevention of cancer and cardiovascular diseases, together with actions on nervous and chronic disorders have been pointed out (Grosso et al., 2013). In a study by Amiri et al. (2019), a new nanoliposome formulation for the encapsulation of vitamin C was produced, using different sonication times and concentrations of milk phospholipids, cholesterol, and phytosterols, finding that positive effects were obtained by replacing cholesterol with campesterol in terms of encapsulation efficiency, stability, and controlled release of the vitamin C, while a decreased sonication time caused an increase in the controlled release (quantity of vitamin C released from the liposomes after 7 days of refrigerated storage). These results illustrate the great flexibility of this technology, showing that by acting on parameters such as composition, sonication energy, etc., different nanoliposomes with particular characteristics can be produced, with distinct release patterns, allowing the production of specific formulations for a wide range of bioactive molecules.

Other vitamins have also been successfully encapsulated in nanoliposomes, for instance vitamin D, which is crucial for calcium and phosphorus homeostasis and for the correct development of bone, and cartilaginous and dental tissues, but is also thought to offer protection from certain diseases (i.e., cardiovascular and immune diseases) (Skaaby et al., 2017; Sassi et al., 2018), and vitamin K, which is a cofactor for the synthesis of blood clotting factors and regulates calcium binding in tissues (Shearer et al., 2012). A review by van Ballegooijen et al. (2017) provided indications that the efficacy of joint supplementation of vitamins D and K may be higher (in terms of bone strength, cardiovascular health, and immune system activity) than those reached via oral administration and, in a recent work by Dalmoro et al. (2019), vitamins K2 and D3 have been encapsulated, using a new high-yield technique, in nanoliposomes for joint administration. They produced two types of nanoliposomes—uncoated and coated with chitosan—with the latter showing better performances in terms of load capacity, stability, and mucoadhesiveness. Moreover, they succeeded in obtaining massive production, a necessary condition for the commercial feasibility of delivery systems for nutraceutical applications.

It is worth mentioning a recently developed bioactive carrier called "tocosome" (Zarrabi et al., 2020). This is a colloidal system characterized by the presence of vesicular bioactive carriers made of tocopheryl phosphates (a phosphoric acid ester of α-tocopherol), biological compounds that have been isolated from both animal and plant tissues (Gianello et al., 2005). It has been shown that these molecules may exert antiinflammatory, antiatherosclerotic, antioxidant, and cardioprotective actions (Ghayour-Mobarhan et al., 2015; Libinaki et al., 2010), thus these nanocarriers (their structure can include sterols, polymers, and proteins) seem to possess good opportunities to be exploited for nutraceutical purposes (Zarrabi et al., 2020).

Solid lipid nanoparticles

Solid lipid nanoparticles (SLNs) are gaining increasing attention nowadays, showing good potential for targeted and controlled delivery of bioactive substances, especially antioxidant molecules (Mishra et al., 2018), as the solid state of the lipids forming these nanosystems is able to provide remarkable protection against degradation to labile encapsulated compounds (Hsu et al., 2019). Indeed SLNs are made up of lipids that are in a solid state at room and body temperature, thus SLN dispersions are comparable to oil-in-water emulsions in which the lipid phase is completely crystallized (Salminen et al., 2016). For the production of SLNs, natural and synthetic solid lipids can be used, for instance, cholesterol, fatty acids, wax, triglycerides, glyceryl monostearate, etc. The preparation usually involves the application of a hot homogenization technique but cold homogenization can also be used for the encapsulation of heat-sensitive substances (Hsu et al., 2019).

SLN technology has been applied to prepare oral formulations for micronutrients such as carotenoids. For example, astaxanthin (ATX), a xanthophyll carotenoid that is endowed with neurovascular, cardiovascular, and skin-protective actions, together with antioxidant, antiinflammatory, anticancer, antidiabetic, and immune-enhancing properties (Davinelli et al., 2018), has been encapsulated in SLNs made of soybean oil, stearic acid, and Tween 20 as an emulsifier (Li et al., 2016). Indeed, despite the great potential of ATX for applications as a nutraceutical compound, the real use has been limited so far due to its scarce water solubility and bioavailability as well as its sensitivity to degradation induced by environmental factors such as heat, light, and O_2. Encapsulation in SLNs provided a formulation characterized by good long-term stability of the particle size (at 4°C and 25°C), improved chemical stability of ATX, and prolonged release of the active compound in a simulated gastric and intestinal environment, demonstrating that the above-mentioned drawbacks could be overcome with the application of this nanotechnology.

Phenolics represent another class of natural and synthetic compounds that have attracted interest over recent years mainly because of their antioxidant properties and several other potential health benefits. Naturally occurring phenolics are found in plants (mainly fruits and leaves) and beverages like tea and wine, and they are responsible for many of the sensory and functional properties of these products (Dimitrios, 2006). It has been proven that diets including abundant vegetable and fruit intakes are associated with a significant reduction of diabetes, cancer, and cardiovascular risks, all oxidative stress-connected disorders (Kumar and Goel, 2019). Specifically, phenolic acids (one of the main polyphenols synthesized by plants) are known to exert direct as well as indirect antioxidant actions stimulating endogenous antioxidants and protective enzymes (Tan et al., 2018), to affect glucose and insulin receptors acting as antidiabetic compounds (Ong et al., 2012) and to decrease cancer initiation via different mechanisms, such as operating a regulation of carcinogen-detoxifying enzymes (Kou et al., 2013) and mutagen-transforming enzymes (Słoczyńska et al., 2014). Because of their multiple beneficial effects, the use of these compounds for nutraceutical applications has been widely studied, and the necessity to address their poor bioavailability (due to scarce absorption, rapid elimination, and metabolization) and high reactivity (leading to protein complexation and browning reactions) highlighted. For these reasons, SLN encapsulation may offer interesting improvements in terms of nutraceutical potential, acting on chemical stability, release, transport to the target sites, uptake efficiency, and bioavailability of these substances (Nunes et al., 2017). For example, rosmarinic acid and herbal extracts of sage and savory have been encapsulated in SLNs made of Witepsol and Carnauba added with 10% mannitol to stabilize the system (Campos et al., 2017). The evaluation of storage shelf-life and stability, carried out over a period of 365 days, showed that SLNs in a liquid system maintained their characteristics until 3 months of storage in plastic bottles at 4°C, whereas SLNs in solid state, stored at room temperature, maintained their stability during the whole storage period only if packaged in a protective atmosphere (N_2) in glass bottles. Other phenolic antioxidants have been loaded into SLNs, namely ferulic acid (FA) and tocopherol (Toc). A good encapsulation efficiency was found for these compounds, even for the hydrophilic one (the FA). In general, the SLN system proved to be quite stable; no free Toc was detected during the whole period of storage (15 weeks, room temperature), while AF was subjected to a partial release, showing a final encapsulation rate of 55%–45%. Nevertheless, good radical scavenging activity was conserved (Oehlke et al., 2017). These results seem to point out a possible commercial feasibility for these nutraceutical formulations.

Nanoemulsions

Nanoemulsions are stable colloidal systems in which, by means of a surfactant agent, two immiscible liquids are melded to form a single phase characterized by a nanodimensional droplet size (Jaiswal et al., 2015), through energy application (high- or low-energy methods are available). More specifically, there are several different shapes that the drops in the continuous phase can assume, not just spherical but also more complex structures (Gupta, 2019). These formulations, and in particular oil-in-water nanoemulsions, are notably convenient for the encapsulation of lipophilic nutraceuticals, being able to form stable systems generally characterized by high oral bioavailability (Guttoff et al., 2015). Moreover, nanoemulsions are flexible tools since, acting on lipid composition, droplet size, surfactant, and method of preparation, very different carriers can be obtained, with specific features, to vehiculate a large variety of active compounds, both hydrophobic and hydrophilic, also for targeted drug delivery (Ostróżka-Cieślik and Sarecka-Hujar, 2017).

For instance, several nanoemulsions have been used as curcumin delivery systems. Curcumin is a natural polyphenol, extracted from the rhizome of *Curcuma longa*, which exhibits a wide array of biological activities: antioxidant, antibacterial, antiinflammatory, and even anticancer actions have been described (Tsuda, 2018). Despite the promising potential of this molecule as a nutraceutical compound, effective applications have been limited due to its scarce water solubility and oral bioavailability (Gupta et al., 2013).

Curcumin-loaded nanoemulsions have been studied for potential application in the beverage industry, after the discovery that, when added to milk, a significant reduction of the lipid oxidation was obtained (Joung et al., 2016). Curcumin-loaded nanoemulsions have also been produced with lysophosphatidylcholine, a phospholipid known for its good emulsification ability, obtaining a stable formulation with good oral bioavailability in mice (Chavez-Zamudio et al., 2017), and with medium-chain triglyceride oil droplets using whey protein concentrate-70 and Tween-80 as emulsifiers. The latter formulation was able to protect curcumin against the simulated gastric environment (90% of curcumin maintained the emulsified state) whereas during the in vitro intestinal digestion, 77% of curcumin was released within 2 h, with a slower release of the remaining percentage afterward. It is thought that this fractioned intestinal liberation may lead to increased bioavailability, together with the enhancement provided by this nanoformulation in terms of increased hydrophilicity and stability (Sari et al., 2015). Furthermore, to improve the physical stability in challenging environmental conditions (pH, temperature, etc.), a multilayered nanosystem has been developed by covering curcumin nanoemulsions made of medium-chain triglycerides (oil phase) and sodium dodecyl sulfate as emulsifier with chitosan and alginate layers. Higher curcumin bioaccessibility was observed with uncoated nanoemulsions, whereas multilayer nanoemulsions caused a decrease in the lipid digestion rate and of the fraction of absorbed lipids showing a delayed lipid digestion together with increased protection against the challenging digestive tract environment, leading to better preservation of curcumin antioxidant properties (Silva et al., 2018).

Other nutraceuticals that have been loaded in nanoemulsions are tea polyphenols (catechins); this formulation was prepared in corn oil and polysorbate 80, obtaining a stable system with no particle aggregation after storage (4°C–40°C for 20 days) (Peng et al., 2018). Nanoemulsified tea polyphenols proved to be more stable after mouth and gastric in vitro simulated digestion than those in aqueous solution, whereas after simulated small intestinal digestion, only one catechin showed a recovery rate greater than those in aqueous solution; however, it should be noted that the tea catechins are known to be subjected to auto-oxidation in neutral or alkaline environments. The same formulation was also tested in vivo in rats, showing that the nanoemulsion increased the oral bioavailability of the two major catechins of the tea extract, and finding a kinetic consistent with controlled release of the polyphenols (Peng et al., 2018). Nanoemulsions have also been investigated as carriers for liposoluble vitamins, such as vitamin D, which has been successfully loaded in this nanosystem using the cheap and easy technique of spontaneous emulsification. This formulation maintained its stability during storage (1 month) at room temperature, thus showing a possible effective application in functional food production (Guttoff et al., 2015). In another work (Golfomitsou et al., 2018), whole-fat milk was fortified with vitamin D (cholecalciferol) encapsulated in nanoemulsions. Polysorbate 20 and soybean lecithin were used as emulsifiers, whereas the dispersed oil phase was made of soybean oil and mixtures with cocoa butter, with the application of high-pressure homogenization to generate the nanoemulsion. This product was able to maintain its stability for 10 days with no signs of particle growth or gravitational separation; these results seem to indicate a great potential for food-grade nanoemulsions as carriers of nutraceuticals in the food and beverage industry (Golfomitsou et al., 2018).

Nanoemulsification has also been used for the encapsulation of vitamin E (Saberi et al., 2013, 2014), an antioxidant compound reported to be involved in the prevention of heart diseases (Saremi and Arora, 2010) and carcinogenesis (Jiang, 2017), and of Kenaf seed oil (*Hibiscus cannabinus L.*), a source of vitamin E and phytosterols (Cheong et al., 2018), opening the possibility for broader applications of this nanotechnology.

Polymeric nanoparticles

Among the nanoformulations, polymeric nanoparticles are colloidal systems that have captured attention because of their great stability, uniformity, loading efficiency, and targeted gastrointestinal delivery of nutraceuticals, providing protection from gastrointestinal degradation of the active substances (Ernest et al., 2018).

Polymeric nanoparticles are usually characterized by a spherical or irregular shape and can be produced using a huge range of natural and synthetic polymers.

Nanoparticles of chitosan, a polysaccharide characterized by the capability to adhere to the gastrointestinal mucosa and cross the epithelial barrier, have been studied for the encapsulation of nanonutraceuticals (both hydrophobic and hydrophilic) to achieve protection against degradation and, when associated with other polysaccharide coatings, targeted delivery (small or large intestine). *Withania coagulans*, a solanaceae showing anticancer, antidiabetic, and antihyperlipemic effects, has been encapsulated in chitosan nanoparticles, then other polysaccharides, characterized by different enzymatic susceptibility, have been applied as a second coating to allow the release of the chitosan-active substance complex into the small or large intestine (Sampathkumar and Loo, 2018). The efficacy of this delivery system has been evaluated in vitro using simulated gastrointestinal fluid, finding that the outer polysaccharide coating composed of retrograded soluble starch or retrograded high amylose corn starch (Hylon VII) was able to protect the nanochitosan from the simulated gastrointestinal fluid, allowing its release into the small or large intestine, respectively (Sampathkumar and Loo, 2018). Moreover, using an in vitro intestinal model (Caco-2 cells) it has been demonstrated that chitosan nanoparticles are able to be taken up by the intestinal cells (Sampathkumar and Loo, 2018). All these results together show that this delivery system could be a promising tool for nutraceutical delivery, allowing the protection of labile compounds from degradation, targeting the small or large intestine, and facilitating also the intestinal barrier penetration. Another polymeric nanoparticle developed to achieve controlled release of nutraceutical compounds and preserve them from degradation has been produced with poly-ε-caprolactone and alginate. This carrier has been used to encapsulate white tea extract, whose beneficial health effects, linked to its antioxidant properties, are due to the presence of polyphenolic constituents (catechins), which are characterized by great chemical instability (pH, light, temperature, and oxygen are involved factors) and poor bioavailability (Sanna et al., 2015). These issues, that have limited the application of catechins as nutraceuticals, may be overcome through encapsulation in nanocarriers. Using this technique, controlled release of the white tea extract into simulated gastrointestinal fluids (20% of the encapsulated polyphenols were released into the simulated gastric fluid, 80% into the simulated intestinal fluid within 5 h) was achieved and the antioxidant properties and stability were maintained, also in the case of long-term storage (30 days) (Sanna et al., 2015). All these results taken together reveal good potential for polymeric nanoparticles as nutraceutical delivery systems.

Regulatory concerns with nanonutraceuticals

Despite all the potentialities and possibilities offered by nanotechnologies in many fields, because of the particular and not yet fully known properties and behavior of these materials, safety for humans and the environment has not been conclusively proved and further studies are needed to obtain specific data in order to perform a proper risk evaluation, also with regulatory purposes (Javeri, 2016).

In the United States, the FDA regulates a great variety of products (i.e., drugs, foods, cosmetics, etc.) that may contain nanomaterials and its attitude toward legislation is to "establish a flexible, product-focused, science-based approach to regulation of products that involve nanomaterials or applications of nanotechnology" (FDA, 2018). With this aim, an FDA Nanotechnology Task Force was created in August 2006, to delineate regulatory approaches based on scientific knowledge and to monitor possible adverse health effects connected to regulated nanoproducts.

In 2014, guidance for industry entitled "Considering Whether an FDA-Regulated Product Involves the Application of Nanotechnology" (FDA, 2014a), including dietary supplements, was published, finalizing the draft guidance issued in June 2011. Since the FDA has not established regulatory definitions of "nanotechnology" and "nanomaterial," this guide illustrates criteria for determining whether or not an FDA-regulated product implies the application of nanotechnology, and thus if potential implications have to be considered in terms of regulatory status, efficacy, and safety.

In "Dietary Supplements: New Dietary Ingredient Notifications and Related Issues: Draft Guidance for Industry" (FDA, 2016), issued by the FDA in August 2016, it is stated that manufacturers or distributors of dietary supplements containing a new dietary ingredient (NDI), namely a product that has not been present in the food supply as an article used for food "in a form in which it has not been chemically altered" must submit a premarket safety notification before introducing it into the market. In this category are comprised products subjected to manufacturing processes intended to produce an ingredient with particles in the 1–100 nm nanoscale range, since nanodimensions may alter the chemical characteristics of the product. In this regard, considering that a limited amount of scientific literature concerning the safety of nanomaterials in dietary supplements is available, the FDA recommends that manufacturers contact the FDA before the submission of an NDI notification for a product containing nanomaterials or based on nanotechnology. The guidance also draws attention to the fact that the different chemical, physical, and biological properties that can be acquired by a material reduced into the nanodimensional range, may question the applicability of the traditional safety tests applied on conventional materials bringing up the necessity to use tests validated for these specific materials.

As for products already in the market, in 2014, "Guidance for Industry – Assessing the Effects of Significant Manufacturing Process Changes, Including Emerging Technologies, on the Safety and Regulatory Status of Food Ingredients and Food Contact Substances, Including Food Ingredients that are Color Additives" (FDA, 2014b) was issued with the aim of providing help to manufacturers of food ingredients in order to establish if a change in the manufacturing process may affect the identity, safety, and regulatory status of a food substance, as in the case of the application of new formulation techniques that imply nanotechnologies.

In the EU, nanotechnologies are addressed by a regulatory framework based on horizontal and sector-specific legislation, to ensure the safety of products on the market (biocidal products, cosmetics, medical devices, and novel foods), together with other regulations that cover nanomaterials implicitly, such as REACH (EC 1907/2006). Proper labeling (for transparency reasons, nanomaterials must be indicated on the label in the list of ingredients, with the word "nano" in brackets following the name of the substance; labeling of novel food and food additives containing nanomaterials is required under the Regulation on the Provision of Food Information to Consumers 1169/2011) and safety assessments are also required before marketing authorization. The regulation on novel foods, which applies to food consisting of engineered nanomaterials, and thus to nanonutraceuticals, is the (EU) 2015/2283 of the European Parliament and of the Council of 25 November 2015, which states that novel foods are authorized only if safe (according to the given criteria), otherwise the precautionary principle (i.e., the general principle of food law that food shall not be placed on the market if it is unsafe under Article 14(1) of Regulation 178/2002 may be applied). Together with this regulation, to better address the needs raised by the widespread diffusion of nanotechnologies, guidelines, best practices, and test methods for safety assessment are being implemented, also within the Organisation for Economic Co-operation and Development (OECD) test guidelines.

Powerful emerging technologies like nanotechnologies always bring, together with their positive potential, concerns about risks and how to assess them properly. More research is needed to address regulatory questions for the implementation of robust and complete regulatory guidance based on scientific evidence in order to ensure a conscious approach to the development, use, and diffusion of this technology.

Concluding remarks and future directions

The application of nanotechnology to nutraceuticals as a novel formulation strategy has shown many potentialities and benefits in terms of enhanced stability, pharmacokinetics, and pharmacodynamics of the active compound, with a considerable improvement of its efficacy. Nowadays many nanoformulations are available and feasible with industrial-scale production, but further studies are necessary to better investigate the in vitro and in vivo efficacy and effects, also performing complete risk evaluations from a global perspective (humans, animals, and the environment). A lack of knowledge regarding nanotechnology impact on the environment and health needs to be addressed before claiming this technique as safe and in order to fulfill the challenging task for the regulatory authorities throughout the world of creating robust and uniform legislation on this complex topic.

References

Akbarzadeh, A., Rezaei-Sadabady, R., Davaran, S., et al., 2013. Liposome: classification, preparation, and applications. Nanoscale Res. Lett. 8, 102.

Amiri, S., Rezazadeh-Bari, M., Alizadeh-Khaledabad, M., et al., 2019. New formulation of vitamin C encapsulation by nanoliposomes: production and evaluation of particle size, stability and control release. Food Sci. Biotechnol. 28, 423–432.

Bozzuto, G., Molinari, A., 2015. Liposomes as nanomedical devices. Int. J. Nanomed. 10, 975–999.

Campos, D.A., Madureira, A.R., Sarmento, B., et al., 2017. Technological stability of solid lipid nanoparticles loaded with phenolic compounds: drying process and stability along storage. J. Food Eng. 196, 1–10.

Carr, A.C., Maggini, S., 2017. Vitamin C and immune function. Nutrients 9, 1211.

Chavez-Zamudio, R., Ochoa-Flores, A.A., Soto-Rodriguez, I., et al., 2017. Preparation, characterization and bioavailability by oral administration of O/W curcumin nanoemulsions stabilized with lysophosphatidylcholine. Food Funct. 8, 3346–3354.

Cheong, A.M., Tan, C.P., Nyam, K.L., 2018. Stability of bioactive compounds and antioxidant activities of kenaf seed oil-in-water nanoemulsions under different storage temperatures. J. Food Sci. 83, 2457–2465.

Dalmoro, A., Bochicchio, S., Lamberti, G., et al., 2019. Micronutrients encapsulation in enhanced nanoliposomal carriers by a novel preparative technology. RSC Adv. 9, 19800–19812.

Danaei, M., Kalantari, M., Raji, M., et al., 2018. Probing nanoliposomes using single particle analytical techniques: effect of excipients, solvents, phase transition and zeta potential. Heliyon 4, e01088.

Davinelli, S., Nielsen, M.E., Scapagnini, G., 2018. Astaxanthin in skin health, repair, and disease: a comprehensive review. Nutrients 10, 522.

Dimitrios, B., 2006. Sources of natural phenolic antioxidants. Trends Food Sci. Technol. 17, 505–512.

Ernest, U., Chen, H.Y., Xu, M.J., et al., 2018. Anti-cancerous potential of polyphenol-loaded polymeric nanotherapeutics. Molecules 23.

Fain, O., 2004. Carences en vitamine C. Rev. Méd. Interne 25, 872–880.

FDA. 2014a. Considering Whether an FDA-Regulated Product Involves the Application of Nanotechnology—Guidance for Industry [Online]. Available: https://www.fda.gov/regulatory-information/search-fda-guidance-documents/considering-whether-fda-regulated-product-involves-application-nanotechnology [Accessed].

FDA. 2014b. Guidance for Industry: Assessing the Effects of Significant Manufacturing Process Changes, Including Emerging Technologies, on the Safety and Regulatory Status of Food Ingredients and Food Contact Substances, Including Food Ingredients that Are Color Additives [Online]. Available: https://www.fda.gov/regulatory-information/search-fda-guidance-documents/guidance-industry-assessing-effects-significant-manufacturing-process-changes-including-emerging [Accessed].

FDA. 2016. Draft Guidance for Industry: New Dietary Ingredient Notifications and Related Issues [Online]. Available: https://www.fda.gov/regulatory-information/search-fda-guidance-documents/draft-guidance-industry-new-dietary-ingredient-notifications-and-related-issues [Accessed].

FDA. 2018. Nanotechnology Programs at FDA [Online]. Available: https://www.fda.gov/science-research/science-and-research-special-topics/nanotechnology-programs-fda [Accessed].

Ghayour-Mobarhan, M., Saghiri, Z., Ferns, G., et al., 2015. alpha-Tocopheryl phosphate as a bioactive derivative of vitamin E: a review of the literature. J. Diet. Suppl. 12, 359–372.

Ghorbanzade, T., Jafari, S.M., Akhavan, S., et al., 2017. Nano-encapsulation of fish oil in nano-liposomes and its application in fortification of yogurt. Food Chem. 216, 146–152.

Gianello, R., Libinaki, R., Azzi, A., et al., 2005. Alpha-tocopheryl phosphate: a novel, natural form of vitamin E. Free Radical Biol. Med. 39, 970–976.

Golfomitsou, I., Mitsou, E., Xenakis, A., et al., 2018. Development of food grade O/W nanoemulsions as carriers of vitamin D for the fortification of emulsion based food matrices: a structural and activity study. J. Mol. Liq. 268, 734–742.

Grosso, G., Bei, R., Mistretta, A., et al., 2013. Effects of vitamin C on health: a review of evidence. Front. Biosci. 18, 1017–1029.

Gupta, P.K., 2019. In: Bhandari, N. (Ed.), An Update on Nanoemulsions Using Nanosized Liquid in Liquid Colloidal Systems. IntechOpen, Rijeka.

Gupta, S.C., Kismali, G., Aggarwal, B.B., 2013. Curcumin, a component of turmeric: from farm to pharmacy. BioFactors 39, 2–13.

Guttoff, M., Saberi, A.H., Mcclements, D.J., 2015. Formation of vitamin D nanoemulsion-based delivery systems by spontaneous emulsification: factors affecting particle size and stability. Food Chem. 171, 117–122.

Hsu, C.Y., Wang, P.W., Alalaiwe, A., et al., 2019. Use of lipid nanocarriers to improve oral delivery of vitamins. Nutrients 11.

Jaiswal, M., Dudhe, R., Sharma, P.K., 2015. Nanoemulsion: an advanced mode of drug delivery system. 3 Biotech 5, 123–127.

Jampilek, J., Kos, J., Kralova, K., 2019. Potential of nanomaterial applications in dietary supplements and foods for special medical purposes. Nanomaterials 9.

Javeri, I., 2016. Application of "nano" nutraceuticals in medicine. In: Gupta, R.C. (Ed.), Nutraceuticals: Efficacy, Safety and Toxicity. Academic Press/Elsevier, Amsterdam, pp. 189–192.

Jiang, Q., 2017. Natural forms of vitamin E as effective agents for cancer prevention and therapy. Adv. Nutr. 8, 850–867.

Joung, H.J., Choi, M.J., Kim, J.T., et al., 2016. Development of food-grade curcumin nanoemulsion and its potential application to food beverage system: antioxidant property and in vitro digestion. J. Food Sci. 81, N745–N753.

Khorasani, S., Danaei, M., Mozafari, M.R., 2018. Nanoliposome technology for the food and nutraceutical industries. Trends Food Sci. Technol. 79, 106–115.

Kou, X., Kirberger, M., Yang, Y., et al., 2013. Natural products for cancer prevention associated with Nrf2–ARE pathway. Food Sci. Hum. Wellness 2, 22–28.

Kumar, N., Goel, N., 2019. Phenolic acids: natural versatile molecules with promising therapeutic applications. Biotechnol. Rep. 24, e00370-e00370.

Li, M., Zahi, M.R., Yuan, Q., et al., 2016. Preparation and stability of astaxanthin solid lipid nanoparticles based on stearic acid. Eur. J. Lipid Sci. Technol. 118, 592–602.

Libinaki, R., Tesanovic, S., Heal, A., et al., 2010. Effect of tocopheryl phosphate on key biomarkers of inflammation: implication in the reduction of atherosclerosis progression in a hypercholesterolaemic rabbit model. Clin. Exp. Pharmacol. Physiol. 37, 587–592.

Maherani, B., Arab-Tehrany, E., Mozafari, M.R., et al., 2011. Liposomes: a review of manufacturing techniques and targeting strategies. Curr. Nanosci. 7, 436–452.

Mishra, V., Bansal, K.K., Verma, A., et al., 2018. Solid lipid nanoparticles: emerging colloidal nano drug delivery systems. Pharmaceutics 10.

Mozafari, M.R., 2010. Nanoliposomes: preparation and analysis. Methods Mol. Biol. 605, 29–50.

Nunes, S., Madureira, A.R., Campos, D., et al., 2017. Solid lipid nanoparticles as oral delivery systems of phenolic compounds: overcoming pharmacokinetic limitations for nutraceutical applications. Crit. Rev. Food Sci. Nutr. 57, 1863–1873.

Oehlke, K., Behsnilian, D., Mayer-Miebach, E., et al., 2017. Edible solid lipid nanoparticles (SLN) as carrier system for antioxidants of different lipophilicity. PLoS One 12, e0171662.

Ojagh, S.M., Hasani, S., 2018. Characteristics and oxidative stability of fish oil nano-liposomes and its application in functional bread. J. Food Meas. Charact. 12, 1084–1092.

Ong, K.W., Hsu, A., Tan, B.K.H., 2012. Chlorogenic acid stimulates glucose transport in skeletal muscle via AMPK activation: a contributor to the beneficial effects of coffee on diabetes. PLoS One 7, e32718.

Ostróżka-Cieślik, A., Sarecka-Hujar, B., 2017. Chapter 7—the use of nanotechnology in modern pharmacotherapy. In: Grumezescu, A.M. (Ed.), Multifunctional Systems for Combined Delivery, Biosensing and Diagnostics. Elsevier, Amsterdam.

Peng, Y., Meng, Q., Zhou, J., et al., 2018. Nanoemulsion delivery system of tea polyphenols enhanced the bioavailability of catechins in rats. Food Chem. 242, 527–532.

Phan, H.T., Haes, A.J., 2019. What does nanoparticle stability mean? J. Phys. Chem. C Nanomater. Interfaces 123, 16495–16507.

Rasti, B., Erfanian, A., Selamat, J., 2017. Novel nanoliposomal encapsulated omega-3 fatty acids and their applications in food. Food Chem. 230, 690–696.

Regulation (EC) No 178/2002 of the European Parliament and of the Council of 28 January 2002 Laying Down the General Principles and Requirements of Food Law, Establishing the European Food Safety Authority and Laying Down Procedures in Matters of Food Safety.

Regulation (EC) No 1169/2011 of the European Parliament and of the Council of 25 October 2011 on the Provision of Food Information to Consumers, Amending Regulations (EC) No 1924/2006 and (EC) No 1925/2006 of the European Parliament and of the Council, and Repealing Commission Directive 87/250/EEC, Council Directive 90/496/EEC, Commission Directive

1999/10/EC, Directive 2000/13/EC of the European Parliament and of the Council, Commission Directives 2002/67/EC and 2008/5/EC and Commission Regulation (EC) No 608/2004 Text with EEA relevance.

Regulation (EC) No 1907/2006 of the European Parliament and of the Council of 18 December 2006 Concerning the Registration, Evaluation, Authorisation and Restriction of Chemicals (REACH), Establishing a European Chemicals Agency, Amending Directive 1999/45/EC and Repealing Council Regulation (EEC) No 793/93 and Commission Regulation (EC) No 1488/94 as well as Council Directive 76/769/EEC and Commission Directives 91/155/EEC, 93/67/EEC, 93/105/EC and 2000/21/EC (Text with EEA relevance).

Regulation (EC) No 2015/2283 of the European Parliament and of the Council of 25 November 2015 on Novel Foods, Amending Regulation (EU) No 1169/2011 of the European Parliament and of the Council and Repealing Regulation (EC) No 258/97 of the European Parliament and of the Council and Commission Regulation (EC) No 1852/2001.

Ruchi, S., Amanjot, K., Sourav, T., et al., 2017. Role of nutraceuticals in health care: a review. Int. J. Green Pharm. 11.

Saberi, A.H., Fang, Y., Mcclements, D.J., 2013. Fabrication of vitamin E-enriched nanoemulsions: factors affecting particle size using spontaneous emulsification. J. Colloid Interface Sci. 391, 95–102.

Saberi, A.H., Fang, Y., Mcclements, D.J., 2014. Stabilization of vitamin E-enriched nanoemulsions: influence of post-homogenization cosurfactant addition. J. Agric. Food Chem. 62, 1625–1633.

Salminen, H., Gömmel, C., Leuenberger, B.H., et al., 2016. Influence of encapsulated functional lipids on crystal structure and chemical stability in solid lipid nanoparticles: towards bioactive-based design of delivery systems. Food Chem. 190, 928–937.

Sampathkumar, K., Loo, S.C.J., 2018. Targeted gastrointestinal delivery of nutraceuticals with polysaccharide-based coatings. Macromol. Biosci. 18, e1700363.

Sanna, V., Lubinu, G., Madau, P., et al., 2015. Polymeric nanoparticles encapsulating white tea extract for nutraceutical application. J. Agric. Food Chem. 63, 2026–2032.

Saremi, A., Arora, R., 2010. Vitamin E and cardiovascular disease. Am. J. Ther. 17, e56–e65.

Sari, T.P., Mann, B., Kumar, R., et al., 2015. Preparation and characterization of nanoemulsion encapsulating curcumin. Food Hydrocoll. 43, 540–546.

Sassi, F., Tamone, C., D'amelio, P., 2018. Vitamin D: nutrient, hormone, and immunomodulator. Nutrients 10.

Shahidi, F., Ambigaipalan, P., 2018. Omega-3 polyunsaturated fatty acids and their health benefits. Annu. Rev. Food Sci. Technol. 9, 345–381.

Shearer, M.J., Fu, X., Booth, S.L., 2012. Vitamin K nutrition, metabolism, and requirements: current concepts and future research. Adv. Nutr 3, 182–195.

Silva, H.D., Poejo, J., Pinheiro, A.C., et al., 2018. Evaluating the behaviour of curcumin nanoemulsions and multilayer nanoemulsions during dynamic in vitro digestion. J. Funct. Foods 48, 605–613.

Skaaby, T., Thuesen, B.H., Linneberg, A., 2017. Vitamin D, cardiovascular disease and risk factors. Adv. Exp. Med. Biol. 996, 221–230.

Słoczyńska, K., Powroźnik, B., Pękala, E., et al., 2014. Antimutagenic compounds and their possible mechanisms of action. J. Appl. Genet. 55, 273–285.

Su, C., Liu, Y., He, Y., et al., 2018. Analytical methods for investigating in vivo fate of nanoliposomes: a review. J. Pharm. Anal. 8, 219–225.

Tan, B.L., Norhaizan, M.E., Liew, W.P., et al., 2018. Antioxidant and oxidative stress: a mutual interplay in age-related diseases. Front. Pharmacol. 9, 1162.

Tsuda, T., 2018. Curcumin as a functional food-derived factor: degradation products, metabolites, bioactivity, and future perspectives. Food Funct. 9, 705–714.

van Ballegooijen, A.J., Pilz, S., Tomaschitz, A., et al., 2017. The synergistic interplay between vitamins D and K for bone and cardiovascular health: a narrative review. Int. J. Endocrinol 2017, 7454376.

Zaki, N.M., 2014. Progress and problems in nutraceuticals delivery. J. Bioequiv. Availab. 6.

Zarrabi, A., Alipoor Amro Abadi, M., Khorasani, S., et al., 2020. Nanoliposomes and tocosomes as multifunctional nanocarriers for the encapsulation of nutraceutical and dietary molecules. Molecules 25.

Section II

Models and mechanisms in evaluation of nutraceuticals

Chapter 20

The biology of nutrients: genetic and molecular principles*

Supratim Choudhuri and Ronald F. Chanderbhan

Introduction

Every nutritional process utilizes a plethora of proteins, such as the digestive enzymes, nutrient transporters, and various enzymes in metabolic pathways. Proper expression of these proteins underlies the nutrient homeostasis in the cell. The nutrients themselves can also modulate cell signaling and gene expression. Consequently, studies on the expression of proteins involved in the absorption, distribution, metabolism, and excretion of nutrients as well as nutrient-mediated expression of genes, form an integral part of the molecular basis of nutrition. The ability to study the effects of nutrients on gene expression has been greatly refined and expanded by high-throughput techniques to study global gene expression profiles. Advances in DNA sequencing technology have also aided in understanding the genetic basis of interindividual differences. Therefore it is reasonable to hope that such knowledge can be utilized to understand the science of nutritional genomics and the molecular basis of human nutrition as well as individual genetic differences that affect it. Personalized medicine and personalized nutrition hold the greatest hopes for human health going forward.

Nutritional implications of interindividual genetic differences

The genomes of organisms in a population contain many polymorphisms. In other words, the genomes are full of sites at which the organisms in a population differ from one another by having different bases caused by mutations. Single nucleotide polymorphisms (SNPs; pronounced *snips*) and copy number variations (CNVs) represent two common types of heritable variations that are responsible for many apparent interindividual genetic differences in a population. SNPs not only include variants that affect a single base, but also include single-base insertions and deletions. Originally, a point mutation with at least a 1% frequency in the population was regarded as a SNP, but currently almost all point mutations are referred to as SNPs. In the human genome, two-thirds of all SNPs involve C→T mutations (transitions). SNPs can occur in both coding and noncoding regions of the genome. Many SNPs have no effect on cell function, but scientists believe other SNPs could predispose people to disease or influence their response to drugs and xenobiotics. The human genome has SNPs at a frequency of about one per 1000 bases, but SNPs that make up about 90% of all human genetic variations, occur every 100–300 bases in the genome (Choudhuri, 2009a).

A set of linked SNPs that tend to be inherited together as a unit is referred to as a *SNP haplotype*. The term haplotype indicates a specific combination of linked loci that are usually inherited together as a unit. The extent of coinheritance of SNPs as a haplotype is an indication of how closely associated they are, and *linkage disequilibrium* (LD) is a quantitative measure of such association. Higher LD values indicate greater association. Thus, SNPs that are in strong LD with disease-causing genes can be used as markers to identify the gene. The interindividual differences in the ability to produce functional enzymes, transporters, and other proteins that aid in the digestion, transport, and assimilation of nutrients are important determinants of personalized nutrition (Choudhuri, 2009a).

There are interindividual genetic differences in taste preference, food tolerance, metabolism, food ingredient absorption, and other nutrition-related characteristics in the human population (Lampe, 2009). For example, two

*The opinions expressed in this chapter are the authors' personal opinions and do not necessarily reflect those of the FDA, DHHS, or the Federal Government.

Nutraceuticals. DOI: https://doi.org/10.1016/B978-0-12-821038-3.00020-3

polymorphisms in the methylenetetrahydrofolate reductase gene (C677T→Ala222Val and A1298C→Glu429Ala) are associated with higher homocysteine levels in the blood. Carriers of these mutations could be at higher risk of cardiovascular disease. Such individuals should consult their physicians to determine whether they should increase their intake of folate in order to reduce blood homocysteine levels. Another example is glutathione *S*-transferase M1 (GSTM1) deficiency. Mutations suppressing GSTM1 expression are associated with more rapid excretion of sulforaphane, an isothiocyanate. Because sulforaphane is obtained from cruciferous vegetables, such as broccoli, Brussels sprout, and cabbage, individuals with this mutation may not be able to derive the full benefit of consuming cruciferous vegetables. Likewise, polymorphisms in the genes encoding UDP-glucuronosyltransferase and sulfotransferase may contribute to variability in phytochemical (e.g., flavonoid) clearance and efficacy (Lampe, 2009; Lampe et al., 2013). Human milk oligosaccharides modulate the microbiota of breast-fed infants (Wang et al., 2015). However, some mothers are deficient in glycosyl transferase enzymes involved in the synthesis of specific oligosaccharides. Thus, the interindividual genetic differences among mothers can determine a baby's exposure/lack of exposure to specific oligosaccharides from the mother's milk (Muñoz et al., 2013). Other examples include glucose-6-phosphate dehydrogenase deficiency and favism, iron overload and hemochromatosis, copper malabsorption and Menkes disease, etc.

Just as the genotype could determine the ability to utilize food ingredients/nutrients, food habits could also drive the evolution of traits necessary for nutritional adaptation. An example of how gene CNV may play a role in nutritional adaptation comes from the study by Perry et al. (2007). There is a direct correlation between the α-amylase (*AMY1*) gene copy number and salivary amylase content. When seven different populations were examined for their salivary amylase content and the diploid *AMY1* gene copy number, it was found that the high starch-consuming populations (e.g., European Americans, Japanese, and Hazda hunter-gatherers) had a higher diploid *AMY1* gene copy number (median 7) compared to the low starch-consuming populations (e.g., Biaka and Mbuti rainforest hunter-gatherers, Datog pastoralists, and Yakut pastoralist and fishing society) (median 5). Further comparison of chimpanzee, Bonobo, and human *AMY1* genes suggest (1) that the expansion of *AMY1* gene copy number took place in human lineage, suggesting that a shift toward a high-starch diet resulted in positive selection of the high copy number variants of the *AMY1* gene in the population, and (2) the selection appears to be ongoing even after the evolution of *Homo sapiens*.

The concept of "nutritional phenotype" provides a good tool in the development of personalized nutrition. Nutritional phenotype has been defined as "an integrated set of genetic, proteomic, metabolomic, functional, and behavioral factors that, when measured, form the basis for assessment of human nutritional status" (Zeisel et al., 2005).

Membrane transporters

Following digestion, the absorption, distribution, and excretion of nutrients are facilitated by transporters localized strategically in the cellular plasma membrane, such as the apical versus basolateral side. Hydrophilic molecules are transported by specific transporters. Even hydrophobic molecules need to be transported by specific transporters in order to maintain a tightly regulated cellular homeostasis. The concept of regulated transmembrane transport of hydrophobic molecules is contrary to the traditional knowledge that assumed that hydrophobic molecules could simply diffuse through the plasma membrane. Once inside the cell, the nutrients exert a multitude of effects from providing energy and being part of the metabolic network, to regulating specific gene expression and cellular homeostasis and the survival/death of the cell. The narrative below focuses on how nutrients are absorbed and transported at the molecular level, and discusses some important transporters. An overview of these transporters is presented in Table 20.1. Readers are referred to http://slc.bioparadigms.org/ for the detailed SLC classification of transporters.

Monosaccharide transporters (SGLT and GLUT)

Glucose is a key fuel in tissues. It is obtained directly from the diet by digestion of ingested carbohydrates, and also by synthesis in the cell from other substrates through gluconeogenesis, which occurs primarily in the liver.

Glucose is transported by two structurally and functionally distinct types of transporters: (1) Na^+-glucose cotransporter (SGLT; gene *SLC5A*), a symporter; (2) Na^+-independent glucose transporter (GLUT; gene *SLC2A*), a facilitative uniporter. The solute carrier (*SLC*) designation of the genes is given according to the Human Genome Organization (HUGO) Gene Nomenclature Committee (HGNC) (Joost and Thorens, 2001).

SGLTs transport glucose and galactose through a secondary (2°) active transport mechanism (see Box 20.1). SGLT couples the transport of glucose and Na^+ in the same direction (hence symporter). There are at least six SGLT transporters (SGLT1–6) identified to date, of which SGLT1–3 are well characterized. SGLTs are expressed in all major tissues but each member exhibits different substrate specificity, kinetics, and expression profiles. This facilitates tissue-specific adaptation of glucose uptake through differential regulation of these

TABLE 20.1 An overview of some nutritionally relevant transporters.

Transporter	*SLC* (gene) family designation	Known substrates	2° active (Na^+/H^+) or facilitative (F)
Sugar transporters			
SGLT	*SLC5A*	Glucose, galactose,	Na^+
GLUT	*SLC2A*	Glucose, fructose, myo-inositol, glucosamine, dehydroascorbate	F
Amino acid transporters			
AA transporters	SLC 1A, 6A, 7A, 16A, 36A, 38A, 43A	Amino acids, monocarboxylates (e.g., lactate, pyruvate)	Na^+/H^+/F (depending on the transporter)
Peptide transporters			
PEPT	*SLC15A*	Dipeptides, tripeptides	H^+
Fatty acid transporters			
MCT	*SLC16A*	Short-chain fatty acids (monocarboxylates; e.g., acetate, lactate, pyruvate, butyrate, propionate, succinate)	H^+
FATP	*SLC27A*	Long-chain fatty acids	
Nucleoside transporters			
CNT	*SLC28A*	Nucleosides	Na^+
ENT	*SLC29A*	Nucleosides	F
Water-soluble vitamin transporters			
SVCT	*SLC23A*	Vitamin C (L-ascorbate)	Na^+
SMVT	*SLC5A6*	Biotin, pantothenate, α-lipoate	Na^+
PCFT	*SLC46A1*	Folate, antifolate (e.g., methotrexate)	H^+
RFC1	*SLC19A1*	Folate, antifolate (e.g., methotrexate)	Organic phosphate
THTR	*SLC19A2* and *-3*	Thiamine	F
Fat-soluble vitamin transporters			
SR-B1		Vitamin D, vitamin E, lutein, lycopene, cholesterol, cholesteryl esters	
NPC1L1		Vitamin D, vitamin E, cholesterol	
Other transporters			
OCTN, CT2	*SLC22A*	L-carnitine	Na^+/F (depending on the transporter)
TAUT	*SLC6A6*	Taurine	
ZIP	*SLC39*	Metals (e.g., Zn)	
ZnT	*SLC30*	Metals (e.g., Zn)	
Nramp (Nramp1; Nramp2 also known as DCT1/ DMT1)	*SLC11A*	Divalent metals and Iron	H^+

BOX 20.1 Primary and secondary active transport, facilitative transport, symporter, antiporter, uniporter.

In *primary active transport*, ATP directly provides the energy needed to move a molecule against its concentration gradient, that is, from low to high concentration (uphill movement). The Na^+/K^+ pump is an example of a primary active transport system that maintains a gradient of about 140 mM Na^+ outside of the cell and 5 mM Na^+ inside the cell. In *secondary active transport*, the energy from ATP is not directly utilized. Instead, the electrochemical gradient of an ion, already established through primary active transport, is utilized to move a molecule across the membrane, usually against its concentration gradient. The ion that is moved down its electrochemical gradient is referred to as the *driving ion* (usually Na^+ or H^+), and the downhill movement of the driving ion is coupled with the uphill movement of the molecule. In other words, the energy-expending diffusion of the driving ion powers the energy-absorbing uphill movement of the driven molecule. Secondary active transport can be of two types: *cotransport* (or *symport*) and *exchange* (or *antiport*); thus a transporter can be a *cotransporter* (*symporter*) or *exchanger* (*antiporter*). In cotransport (symport), Na^+ or H^+ and the transported molecule are transported in the same direction across the plasma membrane, whereas in exchange (antiport), Na^+ or H^+ and the transported molecule are transported in the opposite direction. In contrast to active transport, there is *facilitative (facilitated) or passive transport*. In facilitative transport, transporters passively transport substrates down the concentration gradient, that is, from high to low concentration and does not require energy from ATP. Because the transport process is driven by concentration gradient, it is unidirectional; hence the facilitative transporter is a *uniporter*.

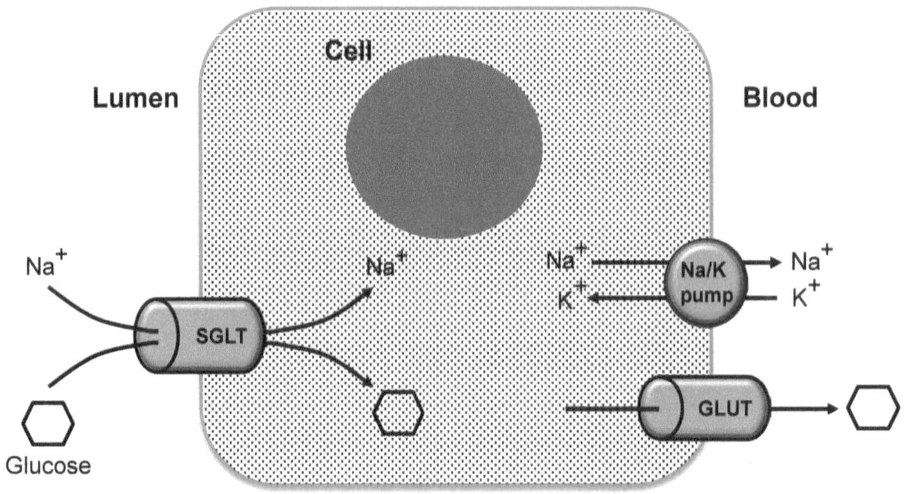

FIGURE 20.1 Cotransport (symport) by SGLT and facilitative transport by GLUT. SGLT is a Na^+-dependent cotransporter that transports Na^+ and glucose molecules into the cell. The higher glucose concentration gradient inside the cell triggers the facilitative transport of glucose by GLUT into the blood.

transporters. SGLT1 was the first member of this group to be cloned and characterized; it has 14 transmembrane domains (TMDs) and is strongly expressed in the small intestine but at lower levels in the kidneys, liver, and lung. SGLT2 is predominantly expressed in the proximal convoluted tubules in the kidney. *SGLT1 is a high-affinity but low-capacity glucose transporter that also transports galactose. In contrast, SGLT2 is a low-affinity but high-capacity glucose transporter and it does not transport galactose* (Wright 2001; Wright et al., 2011). SGLT2 transports the bulk of plasma glucose (low-affinity, high-capacity) from the glomerular filtrate into the cell, and any remaining glucose is recovered by SGLT1 (high affinity, low capacity).

GLUT2 is the major glucose transporter across the basolateral membrane. Concerted action of SGLT and GLUT thus prevents glucose loss in the urine (Wood and Trayhurn, 2003; Wright et al., 2011). Fig. 20.1 shows the Na^+-glucose cotransport. The regulation of SGLT1 has a significant translational or posttranslational component because infusion of D-glucose in the intestine increases *SGLT1* mRNA by about 2-fold, while the transporter mass and activity increases 60–90-fold. *SGLT1* mRNA expression in rat intestine shows circadian rhythm, being sevenfold higher in the morning (10–11 a.m.) than in the afternoon (4–5 p.m.) (Rhoads et al., 1998). SGLT3 is not a transporter in some species. In humans, SGLT3 is a glucosensor expressed in the enteric nervous system ("brain of the gut") and muscles, and colocalizes with the acetylcholine receptor. SGLT3 uses glucose as the signaling molecule and responds to changes in extracellular glucose concentrations, such as stimulating pancreatic and

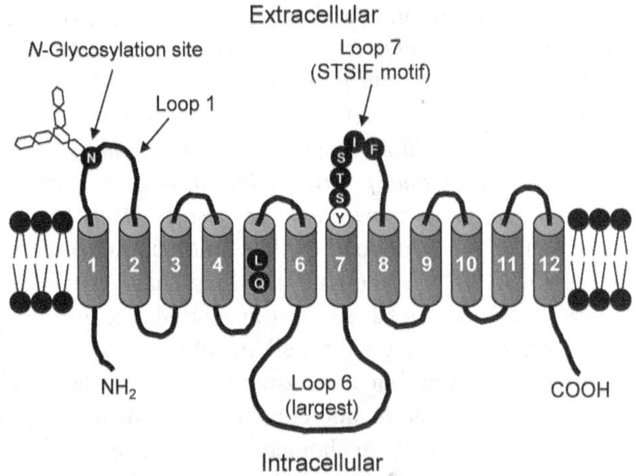

FIGURE 20.2 Membrane topology of class I GLUT transporters. Residues that appear specific for class I are a glutamine (QL motif) in transmembrane helix 5 corresponding to Q161 in GLUT-1 and the STSIF motif in loop 7 which is extracellular.

intestinal secretions through changes in the membrane potential. SGLT3 does not bind galactose.

GLUTs are facilitative transporters (uniporters) (see Box 20.1) and exhibit different substrate specificities, kinetic properties, and tissue expression profiles (Wood and Trayhurn, 2003). The prototype of this family is GLUT1 (Fig. 20.2). GLUT transporters contain *sugar transporter signatures* that consist of numerous conserved glycine, tryptophan, and tyrosine residues, which are regarded as essential for general facilitative transporter function (Joost and Thorens, 2001). There are 14 isoforms (GLUT1–12 and -14, genes *SLC2A1–12* and *-14*; and H^+-coupled myo-inositol transporter or HMIT, gene *SLC2A13*). The functions of GLUT1–5 and their tissue distribution are well known. All 14 isoforms share common structural features, such as 12 TMDs, N- and C-termini facing the cytoplasm of the cell, and an *N*-glycosylation side either within the first (for class I and II) or fifth (for class III) extracellular loop (Augustin, 2010).

Based on sequence similarity, GLUT transporters are divided into three classes: classes I, II, and III. Class I comprises GLUT1–4 and GLUT14, which is the product of GLUT3 gene duplication. Residues that appear specific for class I are a glutamine (QL motif) in TMD 5 corresponding to Q161 in GLUT1 and the STSIF motif in extracellular loop 7 (Augustin, 2010; Cura and Carruthers, 2012). Class II includes the "odd transporters," that is, the high-affinity fructose-specific transporter GLUT5 and three other related transporters: GLUT7, -9, and -11. They are all characterized by the lack of a tryptophan residue in intracellular loop 10 that corresponds to tryptophan 388 in GLUT1. Class III includes the "even transporters," that is, GLUT6, -8, -10, -12, and the myo-inositol transporter HMIT1 (GLUT13) (Augustin, 2010). Class III GLUTs are characterized by a shorter extracellular loop 1 that lacks a glycosylation site, and by the presence of such a site in the larger loop 9 (Joost and Thorens, 2001).

All GLUT transporters mediate monosaccharide transport following concentration gradient. GLUT1 is ubiquitously expressed and mediates glucose transfer across the blood–brain barrier. GLUT2 is the major transporter of glucose across the basolateral membrane in both the liver and kidney (Wright et al., 2011). Translocation of GLUT4 from the intracellular stores to the plasma membrane in response to insulin represents the rate-limiting step in insulin-controlled glucose uptake of skeletal and heart muscle as well as adipose tissue. *Insulin-induced translocation of GLUT4 from the intracellular stores to the plasma membrane results in an immediate 10–20-fold increase in glucose transport* (Bryant et al., 2002). GLUT2 functions as the glucose sensor of the insulin-secreting β-cells in the pancreas (Augustin, 2010). GLUT1 also transports galactose, mannose, and glucosamine. *GLUT5 is the high-affinity fructose transporter* that is primarily expressed in the small intestine in human, rat, and mouse. The expression of GLUT5 is regulated by consumption of a high-sucrose and high-fructose diet. GLUT5 expression in adult rats shows a diurnal rhythm similar to that of SGLT1.

Amino acid transporters

Amino acid (AA) transporters accept groups of AAs rather than individual AAs, such as small neutral AAs, large neutral AAs, anionic AAs, and cationic AAs. Some transporters are Na^+-dependent, while others are Na^+-independent. Initially, AA transporters were classified as "systems," such as *System L*, *System A*, *System ASC*, *System* X^-_{AG}, *System* y^+, *System* $b^{0,+}$, and many others. In this nomenclature, names in uppercase mean Na^+-dependent transporters whereas names in lowercase mean Na^+-independent transporters. AA transporters can be symporter, antiporter, or uniporter. Neutral and anionic AAs are taken up into intestinal epithelial cells by Na^+-coupled cotransport, whereas cationic/dibasic AAs are transported largely by an AA exchange mechanism (Poncet and Taylor, 2013).

The modern classification of AA transporters in mammalian cells, based on similarity between transporter gene sequences, has largely replaced the classical "systems"-based classification (Taylor, 2014). There are seven families of AA transporters in the *SLC* gene superfamily, such as *SLC1A* (high-affinity glutamate and neutral AA transporters), *SLC6A* (Na^+-dependent neurotransmitter transporters), *SLC7A* (cationic and neutral AA transporters), *SLC16A* (monocarboxylate, such as lactate, pyruvate, etc. and aromatic AA transporters; could be H^+-dependent or facilitative), *SLC36A* (H^+-dependent AA transporters),

SLC38A (Na^+-dependent neutral AA transporters), and *SLC43A* (facilitative branched-chain AA and amino alcohol transporters). Each family has many isoforms. The monocarboxylate transporters (MCT1–9; *SLC16A1–9*) are important in fatty acid transport as well (discussed later). The protein names of the AA transporters frequently reflect the substrates they transport. For example, ASCT1 (*SLC1A4*) transports A (Ala), S (Ser), C (Cys), T (Thr). AA transporters usually contain 10–12 TMDs.

Peptide transporters

Uptake transporters for di- and tripeptides belong to the SLC15 family in mammals. There are two main forms of peptide transporters, PEPT1 (gene *SLC15A1*) and PEPT2 (gene *SLC15A2*), which are proton (H^+)-peptide cotransporters. These transporters have 12 TMDs with their N- and C-termini in the cytoplasm and a large extracellular loop between TMD 9 and 10 (Rubio-Aliaga and Daniel, 2008). Di- and tripeptides taken up in the cells are rapidly hydrolyzed to release AAs. In contrast to individual AA transporters that are selective, di- and tripeptide transporters are less selective. Therefore the presence of di- and tripeptide transporters, in addition to the AA transporter, most likely facilitates the efficiency of transport, specifically during the time of increased need.

PEPT1 shows higher expression in the apical membrane of intestinal epithelial cells and appears to be responsible for the intestinal uptake of the major bulk of di- and tripeptides. PEPT2 is highly expressed in proximal tubular cells in the kidney and is the dominant transporter for the reabsorption of di- and tripeptides and its pharmacological substrates. PEPT1 and PEPT2 also transport various peptidomimetics and peptidase inhibitors. *The oral bioavailability of these drugs is determined by PEPT1-mediated transport from the intestine while the plasma half-life is increased by PEPT2-mediated reabsorption from the glomerular filtrate* (Rubio-Aliaga and Daniel, 2008). The human PEPT2 has two main genetic variants, PEPT2*1 and PEPT2*2, found in all ethnic groups, but having different affinities for specific substrates (Pinsonneault et al., 2004).

Fatty acid transporters

Short-chain fatty acids (SCFAs), such as acetate, lactate, pyruvate, butyrate, propionate, and succinate are monocarboxylates that are metabolically important molecules. They are generated in large quantities in the colon by microbial fermentation of fibers. Monocarboxylate transporter 1 (MCT1) plays an important role in the colonic transport of these monocarboxylates (Stein et al., 2000). In humans, nine MCTs have been reported, MCT1–9 (gene *SLC16A1–9*). *MCT1 is expressed ubiquitously in all mammalian tissues with very strong expression in heart and red muscle. Its expression is upregulated in response to exercise and increased physical activity, suggesting its important role in lactate transport* (Glatz et al., 2010). *In white muscle, MCT4 is a major transporter of lactate and pyruvate, but it is a low-affinity transporter of these substrates compared to MCT1.* MCTs are proton (H^+) – monocarboxylate anion cotransporters.

Previously it was thought that long-chain fatty acids (LCFAs) did not need to be transported because they could permeate through the cell membrane due to their hydrophobic nature, but it is now realized that such nonregulated movement is not conducive to maintaining a tightly regulated cellular homeostasis. The presence of various fatty acid transport mechanisms has been demonstrated and many proteins involved in such transport have been cloned. However, the often-used designation "fatty acid transporter" (also used in the following narrative) is a loosely used term; it does not necessarily imply a classic transmembrane transporter, rather the proteins share the feature of facilitating the transmembrane translocation of LCFAs (Glatz et al., 2010).

LCFAs released following digestion are bound by circulating albumin; the complex is dissociated during fatty acid uptake by tissues. Various transport proteins that bind and transport LCFAs include the peripheral membrane fatty acid binding protein ($FABP_{pm}$), cytosolic fatty acid binding protein ($FABP_c$), transmembrane fatty acid translocase (FAT/CD36), and fatty acid transporter proteins 1–6 (FATP1–6; gene *SLC27A1–6*). $FABP_{pm}$, $FABP_c$, and FAT/CD36 are all expressed in multiple tissues (Glatz et al., 2010), whereas FATPs have predominant expressions in specific tissues where they are involved in LCFA uptake. All mammalian FATPs can increase fatty acid uptake upon overexpression (Gimeno, 2007). The precise mechanism of LCFA uptake by FATPs is still debatable (Anderson and Stahl, 2013). Inside the cell, fatty acids are bound to $FABP_c$, which functions as a sink for incoming fatty acids (Glatz et al., 2010).

FATP1 is highly expressed in the skeletal muscle, heart, and adipose tissue, while FATP2 is expressed primarily in the liver and kidney. In mouse, FATP3 is highly expressed in the adrenal gland, testis, ovary, and lung. *FATP4 is the primary fatty acid transporter in the intestine.* It is also expressed in the sebaceous glands in the adult. *FATP5 is expressed exclusively in the liver*, whereas FATP6 is expressed primarily in the heart. The location of FATPs is not static; they can translocate between plasma membrane and cytoplasm. Such ability of translocation may be reflective of their bifunctional nature. The bifunctional nature of the FATPs also supports the notion that fatty acid transport and activation are coupled (Anderson and Stahl, 2013).

The transport proteins are up- or downregulated depending on the diet. For example, the ingestion of a fat-rich diet increases $FABP_{pm}$, $FABP_c$, and FAT/CD36 proteins in human skeletal muscle but a carbohydrate-rich diet has the opposite effect (Roepstorff et al., 2004).

Nucleoside transporters

There are two types of nucleoside transporters: concentrative nucleoside transporter 1–3 (CNT1–3; gene *SLC28A1–3*); and equilibrative nucleoside transporter 1–4 (ENT1–4; gene *SLC29A1–4*). CNTs are Na^+-nucleoside cotransporters and are predicted to possess 13 TMDs, with an intracellular N-terminus and an extracellular C-terminus. The region encompassing TMD 7–9 is responsible for substrate selectivity (Choi and Berdis, 2012). CNT1 transports pyrimidine nucleosides and adenosine; CNT2 transports purine nucleosides and uridine; and CNT3 transports both pyrimidine and purine nucleosides. CNT3 is unique among the CNTs in the sense that it couples the cotransport of both Na^+ and H^+ with nucleoside translocation (Young et al., 2013).

ENTs are facilitative transporters that can transport bidirectionally across the membrane depending upon the concentration gradient. ENTs are more permissive regarding their substrates compared to CNTs. However, ENT1 does not transport cytosine (Choi and Berdis, 2012). ENTs are predicted to possess 11 TMDs with an intracellular N-terminus and an extracellular C-terminus. In mammalian ENTs, the region spanning TMD 3–6 is implicated in solute recognition (Hyde et al., 2001). ENT2, and to a lesser extent ENT1, transport nucleobases as well. ENT3 functions predominantly intracellularly, including in lysosomal and mitochondrial membranes. ENT3 possesses a very long (51 AA residues), hydrophilic N-terminus that contains a dileucine motif characteristic of endosomal/lysosomal targeting sequences (Choi and Berdis, 2012). The basolateral location of ENT1 and the apical location of CNT1 suggest that both transporters operate in tandem to achieve optimal transepithelial nucleoside transport (Choi and Berdis, 2012). Nucleoside transporters also transport various nucleoside analog drugs; thus they are clinically relevant transporters.

Vitamin transporters

Water-soluble vitamins represent a group of structurally and functionally unrelated compounds. Vitamins are transported by specific transporters with the exception of vitamin B_{12} (cobalamin), which is transported by a receptor-mediated mechanism. Some of the transporters that have been identified are discussed below. Vitamin C exists in the diet in two forms; the reduced form, L-ascorbic acid, and the oxidized form, dehydroascorbic acid (DHAA). The reduced form L-ascorbic acid is transported by two distinct isoforms of Na^+-dependent vitamin C transporters1 and 2 (SVCT1 and 2; gene *SLC23A1–2*). *Both SVCT1 and -2 transport L-ascorbic acid but SVCT1 is the main system in the gut.* DHAA is transported by GLUT1, GLUT3, and GLUT4 (Said and Mohammed, 2006). *Vitamin C (as ascorbic acid) concentrations in the brain exceed those in blood by 10-fold.*

Biotin (vitamin B7) transport is mediated by Na^+-dependent multivitamin transporter (SMVT; gene *SLC5A6*), which also transports pantothenic acid (vitamin B5) and α-lipoic acid (a metabolic antioxidant) (Prasad and Ganapathy, 2000). SMVT is a very high-affinity transporter of α-lipoic acid, which can inhibit the transport of both biotin and pantothenic acid with an inhibition constant (K_i) of about 5 μM. SMVT is expressed in the colon, blood–brain barrier, heart, and placenta. SMVT is predicted to have 12 TMDs.

Folate (vitamin B9) is obtained through the diet. Dietary folate exists mainly in polyglutamate form that is converted by folate hydrolase to monoglutamate form, which is transported by folate-uptake transporters. There are two main folate uptake transporters: reduced folate carrier1 (RFC1; gene *SLC19A1*) and proton-coupled folate transporter (PCFT; gene *SLC46A1*). In the intestine, both RFC and PCFT are expressed across the apical brush-border membrane in polarized enterocytes. However, PCFT is highly expressed in the proximal small intestine and functions in acidic pH. Thus, PCFT plays a major role in folate absorption in the proximal small intestine, whereas RFC plays a major role in their transport in the distal small intestine. RFC is also the major transport system for folates and classical antifolates in other mammalian cells and tissues. RFC functions as an anion exchanger with organic phosphates in which the downhill flow of organic phosphates is linked to the uphill transport of folate (Said and Mohammed, 2006; Zhao and Goldman, 2013). Both RFC and PCFT also transport antifolate therapeutics with antitumor activity, such as methotrexate and raltitrexed (Tomudex).

Thiamine (vitamin B1) is transported by thiamine transporter 1 (THTR1; gene *SLC19A2*) and THTR2 (gene *SLC19A3*), which are Na^+-independent transporters and close relatives of the folate transporter RFC1. It is widely expressed in human tissues. In the gut, THTR1 is expressed at both the apical and basolateral membrane of polarized human enterocytes, whereas THTR2 expression is restricted to the apical membrane. The transporter contains 12 putative TMDs (Hediger et al., 2013).

Even fat-soluble vitamins can be transported by specific transporters. Traditionally, it was thought that the fat-soluble vitamins, such as vitamins A (or its precursor β-carotene), D, E, and K were absorbed by a passive diffusion process, just like fat. However, interindividual differences in the absorption of these vitamins were also known. A passive diffusion process does not explain such interindividual differences. Additionally, a passive diffusion process is not conducive to a tightly regulated cellular homeostasis. Recent advances show that there are a number of transporters that transport fat-soluble vitamins. Some examples are scavenger receptor class B type1 (SR-B1), Niemann–Pick C1-like1 (NPC1L1), and ATP-Binding Cassette A1 (ABCA1); these proteins display broad substrate specificity. In enterocytes, SR-BI and NPC1L1 are apically located, whereas ABCA1 is basolaterally located. The substrate spectrum of these transporters overlaps. For example, vitamins D and E are transported by both SR-B1 and NPC1L1. *SR-B1 can serve as both uptake and efflux transporter and it transports lutein and lycopene*. Many other such transporters remain to be identified (Reboul and Borel, 2011).

Other transporters

In addition to the transporters discussed above, there are various other nutritionally relevant transporters, such as organic anion and cation transporters, metal transporters. Examples include organic cation transporters OCTN1 (Na^+-dependent carnitine cotransporter; gene *SLC22A4*), OCTN2 (Na^+-independent transporter; gene *SLC22A5*), and testis-specific CT2 (gene *SLC22A6*); taurine transporter (TAUT; gene *SLC6A6*); metal uptake ZRT-IRT-like-Protein transporters (ZIP; gene *SLC39*) and efflux cation diffusion facilitator/zinc transporter (CDF/ZnT; gene *SLC30*); H^+-dependent divalent cation transporter that also transports iron (DCT1, also called divalent metal transporter or DMT1 and natural resistance-associated macrophage protein or Nramp; gene *SLC11A*). ZIP transporters are predicted to have eight TMDs and they increase cytoplasmic zinc (Zn) concentrations by Zn uptake when cellular Zn is depleted. ZnT transporters are predicted to have six TMDs and they efflux Zn to the extracellular space or sequester cytoplasmic Zn into intracellular compartments when cellular Zn levels are elevated (Picard et al., 2000; Huang and Tepaamorndech, 2013).

Nutritional regulation of cell signaling, cell growth, proliferation, and survival

Cellular growth, proliferation, and survival are intimately coupled with the energy status of the cell and the availability of nutrients. Nutrient signaling is also tightly coupled to longevity. In conditions of nutrient overabundance organisms develop, grow, and age quickly. In contrast, during nutritional restriction, as happens with dietary restriction, growth and development decline, stress response pathways become induced and organisms live longer. Some of the important pathways involved in nutrient signaling regulating cell survival, growth, and proliferation are insulin/insulin-like growth factor-1 (IGF-1), AMP kinase (AMPK), protein kinase A, and the target of rapamycin (TOR, mTOR in mammals) pathway (Stanfel et al., 2009). Recent evidence suggests that these are functionally interrelated, with the TOR signaling pathway playing a crucial role. TOR was so named because it is the target of a molecule named rapamycin (sirolimus), which is a macrolide produced by the bacteria *Streptomyces hygroscopicus*. The following discussion will focus on TOR.

TOR is an evolutionarily conserved serine/threonine protein kinase that belongs to the phosphoinositide 3-kinase (PI3K)-related kinase family. The TOR signaling pathway integrates both intracellular and extracellular signals and serves as a central regulator of cell metabolism, growth, proliferation, and survival by regulating translation. TOR integrates signals from nutritionally relevant inputs, such as nutrients, growth factors, cellular energy status, and stress, among others. Overstimulation of TOR signaling through excess energy and nutrient intake may underlie obesity-induced diseases, including diabetes and even tumor development. In contrast, impairing TOR signaling through dietary restriction can extend lifespan. Consequently, the TOR pathway is activated during many normal and disease-related cellular processes, such as nutrient abundance, adipogenesis, T-lymphocyte activation, insulin resistance, type II diabetes, tumor formation, and angiogenesis (Laplante and Sabatini, 2009).

The TOR protein complex (TORC) exists in two distinct forms, TORC1 and TORC2. TORC1 positively regulates cell growth and proliferation by promoting many anabolic processes, such as biosynthesis of proteins, lipids, and organelles, and by limiting catabolic processes, such as autophagy. TORC1 directly phosphorylates the eukaryotic translation initiation factor eIF4E-binding protein (4E-BP1) and S6 kinase1 (S6K1), which, in turn, promote protein synthesis. The phosphorylation of 4E-BP1 prevents its binding to eIF4E, enabling eIF4E to promote cap-dependent translation (Laplante and Sabatini, 2012). The signal inputs that are integrated by TORC1 may come from growth factors, energy status, oxygen, AAs, genotoxic stress, inflammation, and others. TORC2 plays important roles in cell survival, metabolism, proliferation, and cytoskeleton organization. However, in contrast to TORC1, for which many upstream signals and cellular functions are known, the biology of TORC2 is yet to be understood well (Laplante and Sabatini, 2009).

AAs positively regulate TOR signaling, although the degree of dependency on external AAs varies between cell types. Leucine, a branched-chain AA, is the most effective in most cell types (Laplante and Sabatini, 2012). Inhibition of TOR function leads to the inhibition of translation, resulting in the accumulation of cells in G1 phase of the cell cycle and potential apoptosis. Thus, TOR inhibition in cancer cells mimics nutrient and growth factor deprivation, limiting cell growth and proliferation (Huang and Houghton, 2003). The data from various invertebrate species demonstrate that TOR signaling plays an important role in the aging process. A growing body of evidence shows that the same may be true for mammals. Studies in invertebrates indicate that reduced TOR signaling is likely to account for the lifespan extension associated with dietary restrictions. However, pharmacological inhibition of the TORC1 complex may not produce consistent results in mammals (Stanfel et al., 2009).

AMPK regulates metabolic energy balance at the whole-body level. The sensing of the cellular energy status by TOR is thought to be AMPK-mediated. AMPK is activated under states of low cellular energy (high AMP: ATP ratio) and it acts to reduce ATP expenditure and also to increase the ATP supply by activating pathways producing ATP (Hardie and Hawley, 2001). Besides AMP accumulation, AMPK can be activated by different insults such as oxidative stress and DNA damage. AMPK also plays a major role in glucose homeostasis by regulating GLUT4 translocation to the cell surface. One of the downstream targets of AMPK is TOR. Activated AMPK phosphorylates tuberous sclerosis complex2 (TSC2), which is in complex with TSC1 and suppresses TORC1 activation and signaling (Liu and Qian, 2011).

In the insulin signaling cascade, the binding of insulin or IGF-1 to their receptors triggers the PI3K pathways, leading to the phosphorylation and activation of Akt (a serine/threonine protein kinase). Activated Akt then phosphorylates and suppresses the TSC2, which is in complex with TSC1 and acts as a negative regulator of TORC1 signaling (Liu and Qian, 2011). Thus, the net effect of insulin/IGF-1 signaling is TSC1/TSC2 suppression, TORC1 activation, and increased translation. The insulin signaling also impinges on the downstream TOR signaling pathway, such as S6K1 and 4E-BP1. Phosphorylation of 4E-BP1 releases the translation initiation factor eIF4E, resulting in translational activation (Gingras et al., 2001). In contrast, withdrawal of AAs leads to rapid dephosphorylation of S6K1 and 4E-BP1, which mimics rapamycin treatment. Furthermore, rapamycin treatment renders the phosphorylation of S6K1 refractory to AA stimulation (Hara et al., 1998). Fig. 20.3 shows the stimulatory and inhibitory signals for TORC1.

Studies on the molecular basis of nutrient-mediated modulation of cell signaling, cell proliferation, and programmed cell death are likely to yield useful information on the biochemical pathways and intermediates that could serve as potential molecular targets for therapy.

Effects of nutrients on gene expression

One of the first examples of nutrient-mediated modulation of gene expression came from the discovery of the lactose operon in *E. coli*. Growing *E. coli* in lactose instead of glucose led to the de novo transcription and expression of three enzymes, β-galactosidase, *trans*-acetylase, and permease, that are all involved in the uptake and utilization of lactose as the carbon source in *E. coli*. Likewise, in higher eukaryote nutrients, such as glucose, AAs, fatty acids, as well as vitamins and minerals are known to regulate gene expression.

Carbohydrates and gene expression

Many enzymes involved in carbohydrate metabolism are induced in the liver and other tissues by a high-carbohydrate diet. Consumption of a high-carbohydrate meal leads to a rise in plasma glucose that triggers insulin secretion from the pancreas. Insulin acts on its target tissues and orchestrates a series of anabolic effects that include inhibition of gluconeogenesis in the liver, increase

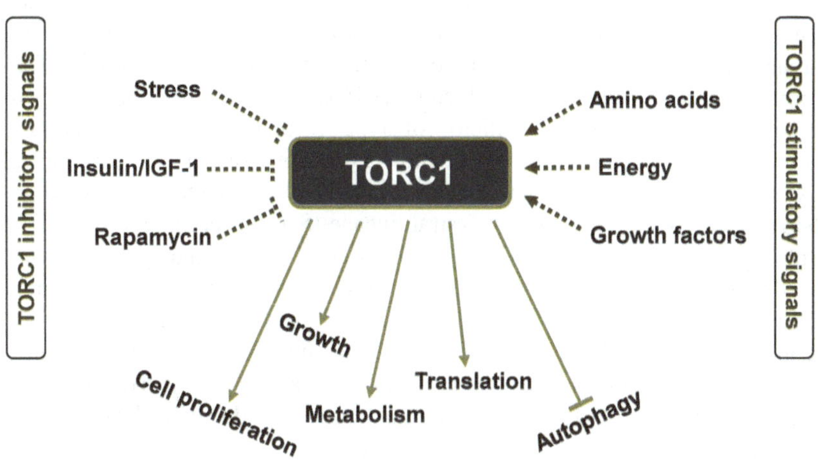

FIGURE 20.3 Some important stimulatory and inhibitory signals of TORC1. The dashed lines indicate that the effects of the signals involve other molecular players that are not shown in the diagram.

in glucose uptake and oxidation in skeletal muscle, suppression of lipolysis, and increasing lipogenesis in adipose tissues. The lipogenic actions of insulin are mediated via transcriptional activation of target genes controlled by the insulin-responsive transcription factor sterol regulatory element binging protein (SREBP)-1c (Foufelle and Ferré, 2002; Shankar et al., 2010). Glucose itself regulates many targets genes via the carbohydrate-response element ChoRE (also known as glucose-insulin response element or GIRE) and ChoRE-binding proteins (ChoRE-BP). High-carbohydrate diets modulate the expression of genes regulating glucose transport, glycolysis, fatty acid and triglyceride biosynthesis, fatty acid desaturation and elongation, adipogenesis, and adipokines (Shankar et al., 2010).

Glucose induces the expression of many enzymes involved in its utilization. For example, the levels of glucokinase mRNA are low during fasting but increase more than 50-fold above the fasting level within 6 h of an oral glucose load. The mRNA level returns toward basal values by 18 h (Iynedjian et al., 1987). It is now established that glucose also induces the expression of many other hepatic genes, such as those of glucose transporters, lipogenic enzymes (e.g., acetyl CoA carboxylase, fatty acid synthase), and represses the expression of genes of the gluconeogenetic pathway, such as, that of phosphoenolpyruvate carboxykinase (PEPCK). Both insulin and glucose can independently repress *PEPCK* gene expression. Thus, for most genes regulated by a carbohydrate-rich diet, the up- or downregulation of their mRNA is substantial and relatively rapid and is also reversible. High glucose has also been shown to induce a proinflammatory response, increase reactive oxygen species, and activate proinflammatory transcription factors, including activator protein 1, early growth response 1 (Egr-1), and nuclear factor κB (NF-κB) (Aljada et al., 2004).

Mourrieras et al. (1997) proposed that glucose-6-phosphate is the most likely signal metabolite responsible for the glucose-induced transcription of this group of genes (Fig. 20.4). Insulin signaling may also play an important role in this context. Insulin is required along with glucose to induce the expression of a number of glucose-inducible genes, such as those of L-type pyruvate kinase, fatty acid synthase, acetyl CoA carboxylase, and stearoyl-CoA desaturase. Fig. 20.4 shows some major pathways for transcriptional regulation of target genes in the liver, mediated by glucose and insulin. Consumption of high-sucrose diets also alters the expression of genes involved in cell adhesion, cell cycle control, signaling, and proliferation. It was proposed by Wang et al. (2009) that the promotion of intestinal tumorigenesis by high sucrose consumption in mice may in part be mediated by increased IGF-I signaling in intestinal cells.

Studies in humans with high- and low-carbohydrate diets showed differential gene expression in subcutaneous fat as well. Kallio et al. (2007) showed that, in humans, feeding a rye-pasta (low insulin response) or an oat-wheat-potato (high insulin response) diet for 12 weeks resulted in a marked difference in gene expression in subcutaneous fat. In individuals in the low-insulin-response group, 71 genes showed decreased expression and none showed increased expression. Several downregulated genes have links to insulin-signaling pathways. In contrast, in individuals from the high-insulin-response group, 62 genes showed increased expression and none showed decreased expression. Two of the noteworthy downregulated genes were the hormone-sensitive lipase (*HSL*) gene and the transcription factor *TCF7L2* gene. Downregulation of HSL, which is a key enzyme in the release of fatty acids from adipose tissue, is associated with resistance to genetic or diet-induced obesity. The transcription factor TCF7L2 is the strongest known genetic predictor of type II diabetes. In humans, the stable but differential expression pattern of many genes involved in carbohydrate metabolism may be rooted in geographic ancestral lineage (Schisler et al., 2009).

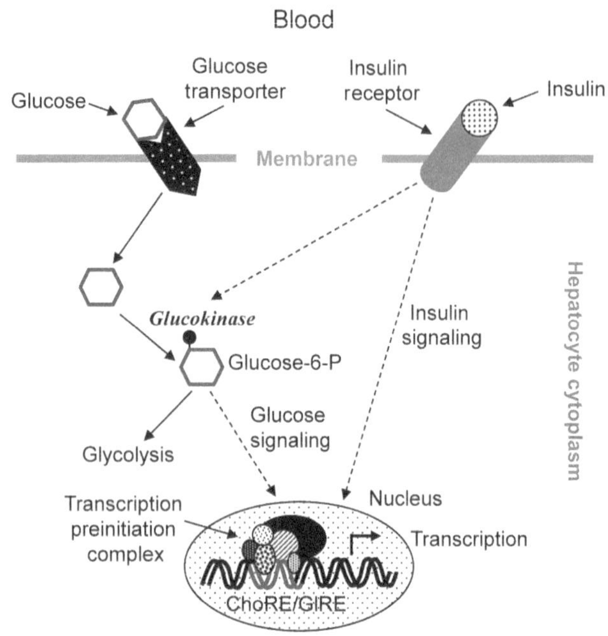

FIGURE 20.4 Transcriptional regulation of target genes in the liver by glucose and insulin signaling. Formation of glucose-6-phophate is apparently an important step in the glucose-signaling pathway. Insulin signaling pathway may involve insulin-responsive transcription factor SREBP-1c.

Changes in the levels of glucose and glucagon can also regulate the expression of protooncogenes, which could play a role in liver cell growth control. Corcos et al. (1987) reported that in rats, short-time physiological fasting results in a dramatic decrease of c-myc expression in the liver, and administration of glucagon leads to an increase in *c-myc* mRNA.

AAs and gene expression

AA availability regulates the increased mRNA expression of many genes both in vivo and in vitro. Some examples of genes that are upregulated by AA starvation include those encoding enzymes of the AA synthesis pathway (asparagine synthetase or ASNS, glutamine synthetase), enzymes of the urea cycle (argininosuccinate synthetase), the rate-controlling enzyme in polyamine biosynthesis pathway [ornithine decarboxylase (*ODC*)], transporters, (e.g., sodium-coupled neutral AA transporter 2 or SNAT 2), transcriptional activator proteins (e.g., CCAAT/enhancer-binding protein or C/EBP, C/EBP homologous protein or CHOP), insulin-like growth factor-binding protein-1, and proto-oncogenes (jun and myc) (Brasse-Lagnel et al., 2009; Kilberg et al., 2012).

Initial studies were performed to explore the molecular mechanisms of the inhibitory effect of asparagine and histidine on *ASNS* and that of leucine on *CHOP* (also known as *GADD 153*) gene expression (Brasse-Lagnel et al., 2009). Transcription of the *ASNS* gene is increased in response to both AA and glucose starvation. CHOP is a nuclear protein related to the C/EBP family of transcription factors. *CHOP* gene expression is strongly induced by cellular stress, such as oxidative stress, DNA damage, as well as AA starvation. For example, leucine limitation in human cell lines results in the induction of *CHOP* mRNA and protein (Brasse-Lagnel et al., 2009; Kilberg et al., 2012). Several response elements regulating transcription have been identified in the promoter regions of *ASNS* and *CHOP* genes. For example, the *ASNS* gene promoter has three GC boxes (GC-I through III) that bind the transcription factors, specificity protein1 and 3 (Sp1 and Sp3), and two nutrient-sensing response elements (NSRE-1 and NSRE-2). For the GC boxes, the hierarchy of importance was found to be GC-III > GC-II > GC-I (Jousse et al., 2004). The amino acid response (AAR) element within the *CHOP* gene promoter binds the activating transcription factors 2 and 4 (ATF2 and ATF4) to activate transcription (Brasse-Lagnel et al., 2009; Kilberg et al., 2012). The *ASNS* and *CHOP* genes are also induced by the unfolded protein response, which is characterized by the accumulation of unfolded proteins in the endoplasmic reticulum (Brasse-Lagnel et al., 2009; Kilberg et al., 2012).

The majority of the mRNA increase by AA starvation or supplementation is through an increase in de novo transcription, but mRNA stabilization also plays an important role. Similarly, a high asparagine concentration induces *ODC* mRNA expression primarily by posttranscriptional stabilization of the *ODC* mRNA, and also by stimulating the synthesis and suppressing the degradation of the ODC protein. During AA starvation, the translation of mRNAs of some of these AA-regulated genes (e.g., arginine/lysine transporter *Cat-1* mRNA) is initiated through internal ribosomal entry sites located in the 5′-untranslated region of the mRNAs (Jousse et al., 2004). These mRNAs may be part of the cellular "survival" mRNAs that are selectively translated during conditions of significant stress.

A wide spectrum of adaptive mechanisms has evolved to sense and respond to fluctuations in dietary nutrients. For example, general control nonderepressible 2 (GCN2) is a serine/threonine kinase that acts as a sensor of AA deficiency. The AAR involves a number of signaling pathways that include many proteins, but the central pathway of GCN2−eIF2−ATF4 appears to be involved in all tissues. Many of the ATF4-activated genes, such as those of the transcription factors C/EBP, ATF3, and CHOP act as feedback repressors of the ATF4 signal. Investigation of the signaling pathways and gene expression programs triggered by the AAR has shown that the bZIP superfamily of transcription factors plays the central role during these cellular stress responses. Formation of homo- and heterodimers among the ATF, C/EBP, and FOS/JUN families of proteins constitutes an integrated transcription factor network that determines the initiation, magnitude, and length of the cellular response to AA limitation (Kilberg et al., 2012).

Fatty acids and gene expression

After uptake, fatty acids are bound by FABPs. Fatty acids can undergo a number of metabolic fates but more importantly they can serve as signaling molecules that can interact with intra- and extracellular receptor sensor systems, either directly or after conversion to specific fatty acid derivatives. An example of these lipid sensors is the nuclear receptors that mediate the activation of gene transcription by a variety of hydrophobic compounds, such as retinoic acid, steroid hormones, oxysterols, and bile acids. The peroxisome proliferator-activated receptor (PPAR) is probably the best sensor system for fatty acids. PPAR was shown to cause proliferation of peroxisomes and be activated by a variety of fibrate hypolipidemic drugs and non-genotoxic rodent hepatocarcinogens that are collectively termed peroxisome proliferators (Issemann et al., 1993).

A great majority of the reported actions of fatty acids on gene expression have come from studies on LCFAs. Analyses have focused on several candidate mechanisms, including PPAR- and prostanoid-regulated processes (Jump and Clarke, 1999) (see Box 20.2). PPARs are ligand-dependent transcription factors that regulate target gene expression by binding to specific peroxisome proliferator response elements (PPREs) in the enhancers of PPAR-target genes. Each receptor binds to its PPRE as a heterodimer with the retinoid X receptor (RXR). PPAR-RXR heterodimer, upon binding PPRE, recruits transcriptional coactivators. The result is an increase in gene

BOX 20.2 Cyclooxygenase and lipoxygenase pathways.

Fatty acid derivatives in tiny concentrations can exert profound effects on cellular physiological or pathophysiological reactions. These derivatives are classified according to the enzymatic pathways of their formation. The cyclooxygenase (COX) pathway products include prostaglandins, prostacyclins, and thromboxanes, collectively referred to as *prostanoids*. The lipoxygenase pathway products include leukotrienes and lipoxins. The term *eicosanoids* is a collective name for molecules derived from 20-carbon fatty acids. Thus, there can be prostaglandin and leukotriene forms of eicosanoids. Fatty acids of the linoleic acid family are the main source of eicosanoids, ARA (an ω-6 fatty acid) being the major precursor. PUFAs (ω-3 and ω-6), released from the plasma membrane phospholipids by the action of phospholipase A2, are also metabolized by COX-1 and COX-2 to form prostaglandins and thromboxanes.

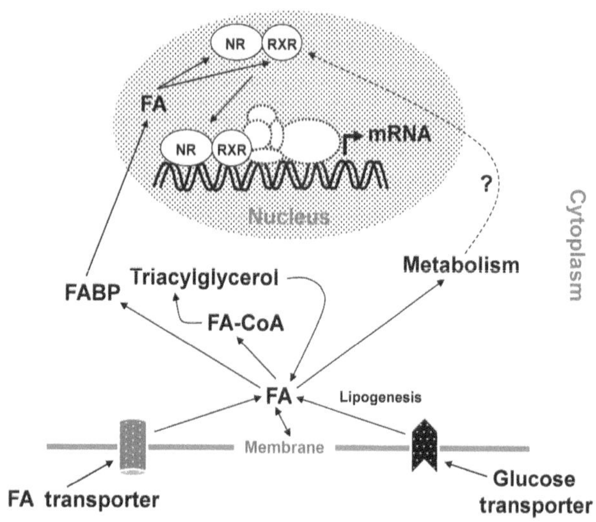

FIGURE 20.5 Transcriptional regulation of target genes by fatty acids. Transport of fatty acids inside the nucleus results in the activation of nuclear receptors, such as PPAR, LXR, FXR, and PXR. The target receptor dimerizes with retinoid X receptor (RXR). The receptor dimer binds to the appropriate response elements in the promoter of the target genes and activates transcription. *PPAR*, Peroxisome proliferator-activated receptor; *LXR*, liver X receptor; *FXR*, farnesoid X receptor; *PXR*, pregnane X receptor.

transcription. Fig. 20.5 summarizes some pathways of fatty acid regulation of gene transcription. Three different PPAR subtypes have been cloned: PPARα, PPARβ/δ, and PPARγ. PPARα is predominant in brown adipose tissue, cardiac muscle, skeletal muscle, and liver. PPARβ/δ is found in many cell types, whereas PPARγ expression is more restricted in adipocytes, macrophages, and tumor cells. *All three PPARs are able to bind fatty acids with a general preference for polyunsaturated fatty acids (PUFAs), such as arachidonic acid (ARA) and docosoahexanoic acid (DHA). PPARα mediates the effects of dietary fatty acids on hepatic gene expression*, including many genes involved in fatty acid catabolism (Georgiadi and Kersten, 2012).

PPARα activation results in increased transcription of genes and expression of proteins involved in intracellular fatty acid transport (e.g., FABP), mitochondrial and peroxisomal β-oxidation (e.g., acyl-CoA oxidase and acyl-CoA-dehydrogenase), and peroxisomal ω-oxidation (e.g., cytochrome P4504A1). The effects of PPARγ on lipid metabolism are mainly mediated by genes expressed in adipocytes. PPARγ activation results in the upregulation of genes involved in fatty acid transport (e.g., FABP, FATP), lipoprotein hydrolysis (e.g., lipoprotein lipase), lipogenesis (e.g., acyl-CoA synthetase), thermogenesis (e.g., mitochondrial uncoupling proteins), proinflammatory cytokines (e.g., TNF-α), regulator of food intake hormone (leptin), etc. (Berger and Moller, 2002).

In addition to PPARs, other transcription factors that are activated by fatty acids include SREBP-1c, nuclear receptors in addition to RXR (e.g., liver X receptor or LXR, farnesoid X receptor or FXR, also known as bile acid receptor or BAR), hepatic nuclear factor 4α, and NF-κB (Georgiadi and Kersten, 2012).

FXR is highly expressed in liver and intestine and is activated by bile acids. Its ligands include PUFAs, like arachidonic, linolenic, or docosahexaenoic acid, as well as intermediates of the bile acid synthetic pathways. FXR plays an important role in glucose and lipid metabolism (Claudel et al., 2005). The two forms of SREBP, SREBP-1c and SREBP-2, together activate the expression of more than 30 genes involved in the synthesis and uptake of cholesterol, fatty acids, triglycerides, and phospholipids. Although PUFAs inhibit both SREBP-1c and SREBP-2, SREBP-1c is seemingly more important in the downregulation of gene expression by PUFAs. PUFAs downregulate *SREBP-1c* mRNA either by stimulating the mRNA decay or by antagonizing enhanced gene transcription by LXR. PUFAs also reduce the expression of the glycolytic gene for pyruvate kinase, probably by inhibiting the nuclear translocation of the ChoRE-BP, which induces the expression of genes involved in glycolysis and lipogenesis. Interestingly, the ω-3 long-chain PUFAs (e.g., EPA and DHA), and the ω-6 PUFA (e.g., ARA) have far more inhibitory capacity on SREBP than do the shorter chain unsaturated fatty acids (e.g., 18:1 oleic acid, 18:2 linoleic acid), and saturated fatty acids have almost no effect on SREBP (Deckelbaum et al., 2006; Georgiadi and Kersten, 2012).

Omega-3 PUFAs (DHA and EPA) inhibits endothelial cell production of the acute-phase cytokine interleukin-6 (Khalfoun et al., 1997) and this may involve the inhibition of NF-κB, which is a pleiotropic transcriptional activator for the proinflammatory cytokines. Diets containing ω-3 and ω-6 PUFAs inhibit hepatic de novo lipogenesis by downregulating the expression of genes for enzymes involved in both lipogenesis and glucose metabolism, including acetyl-CoA carboxylase, fatty acid synthase, L-type pyruvate kinase, stearoyl-coenzyme A desaturase-1, glucose-6-phosphate dehydrogenase, ATP citrate-lyase, malic enzyme, small acidic S14 protein, Δ-5 and Δ-6 desaturase, etc. (Duplus et al., 2000; Jump and Clarke, 1999). The cycloxygenase products of another ω-6 PUFA linoleic acid, hydroperoxy-octadecadienoic acid, induce *c-fos*, *c-jun*, and *c-myc* mRNA expression in aortic smooth muscle cells (Rao et al., 1995). In Swiss 3T3 cells, ARA conversion to prostaglandin E_2 stimulates DNA synthesis and mitosis by activating the expression of c-fos and Egr-1 (Danesch et al., 1994).

Among SCFAs, the effect of butyrate on gene expression has been studied for a long time by many laboratories. The ability of butyrate in increasing histone acetylation, which is a transcriptional activating chromatin modification, was reported by Riggs et al. (1977). This finding provided the molecular mechanism of action of butyrate in gene regulation. Butyrate was later found to increase histone acetylation and gene expression by inhibiting histone deacetylase activity.

Other nutrients and gene expression

Like the macronutrients (protein, carbohydrate, and fat), micronutrients (vitamins and minerals) can also modulate gene expression. The following discussion will briefly mention a couple of examples each of vitamin- and mineral-regulated gene expression.

Vitamin A-mediated regulation of gene expression is a well-characterized example of the direct involvement of micronutrients in gene expression. The bioactive form of vitamin A, retinoic acid, is a ligand for the nuclear retinoic acid receptor (RAR). Ligand-activated RAR heterodimerizes with RXR; the RAR–RXR complex binds to the cognate retinoic acid response element in the regulatory region of the target genes, recruits coregulatory proteins, and modulates the transcriptional output. RAR–RXR also downregulates genes by antagonizing certain other transcription factors. One of the target genes of vitamin A is *PEPCK* that encodes phosphoenolpyruvate carboxykinase, which is highly induced under reduced blood glucose as happens during food deprivation (McGrane, 2007). Dietary vitamin A levels can also modulate the expression of *IGF-I*, *IGF-IR*, and *IR* genes in many different tissues. Serum IGF-I concentrations and tissue *IGF-I* mRNA are downregulated by vitamin A deficiency.

The model of vitamin D (1,25-dihydroxyvitamin D_3)-mediated gene expression involves similar steps of events as for vitamin A, that is, vitamin D binding-mediated activation of vitamin D receptor (VDR), formation of VDR–RXR complex, its binding to VDRE, and modification of transcriptional output. Genome-wide mapping of VDR binding revealed that vitamin D could influence over 200 genes, highlighting links to disease states (Ramagopalan et al., 2010).

Zinc is an essential cofactor for many enzymes as well as Zn-finger transcription factors involved in gene regulation. For example, Metal Responsive element Factor-I is a Zn-dependent transcriptional activator that regulates metallothionein (MT) I and II gene expression through a metal response element. MT proteins are involved in metal homeostasis in the cell (Klaassen et al., 1999). Cells contain a plethora of Zn-finger transcription factors that are important for regulating gene expression at all stages of life, and under various physiological conditions. Large-scale gene expression analyses show that Zn-deficient conditions also trigger the expression of many genes. Recently, Zn-finger nucleases have been engineered for genome-editing purposes. The Zn-fingers in these proteins guide the nuclease to a specific site in the genome where the nuclease exerts its editing function by cutting the DNA strands, which is followed by nonhomologous end joining and mutation.

Iron is another nutrient that controls gene expression depending on the iron status in the cell. Many proteins involved in iron metabolism are regulated by the iron response element (IRE) in the mRNA and its interaction with the IRE-binding protein (IRE-BP). Ferritin is an intracellular iron-binding protein, whereas transferrin is a major iron transport protein in the plasma. When intracellular iron levels are high IRE-BP binds iron (Fe), and the IRE-BP–Fe complex cannot bind to IRE in the 5′-UTR of *ferritin* mRNA. This allows *ferritin* mRNA to be translated to ferritin, which then sequesters excess intracellular iron. When intracellular iron levels are low, the IRE-BP does not bind iron; hence it can bind to the IRE. As a result, the translation of the *ferritin* mRNA is prevented. In contrast to ferritin, transferrin binds to transferrin receptor located on the plasma membrane. When iron levels are low, *transferrin receptor* mRNAs are stabilized by the binding of IRE-BP with the IREs in its 3′-UTR so that more transferrin receptors can be translated and cells can take up more iron from transferrin. Unlike *ferritin* mRNA which has a single IRE in 5′-UTR, *transferrin receptor* mRNA has multiple IREs in 3′-UTR. When iron levels are high, IRE-BP binds iron and cannot interact with IREs; as a result, *transferrin receptor* mRNA is degraded at a faster rate to prevent over-uptake

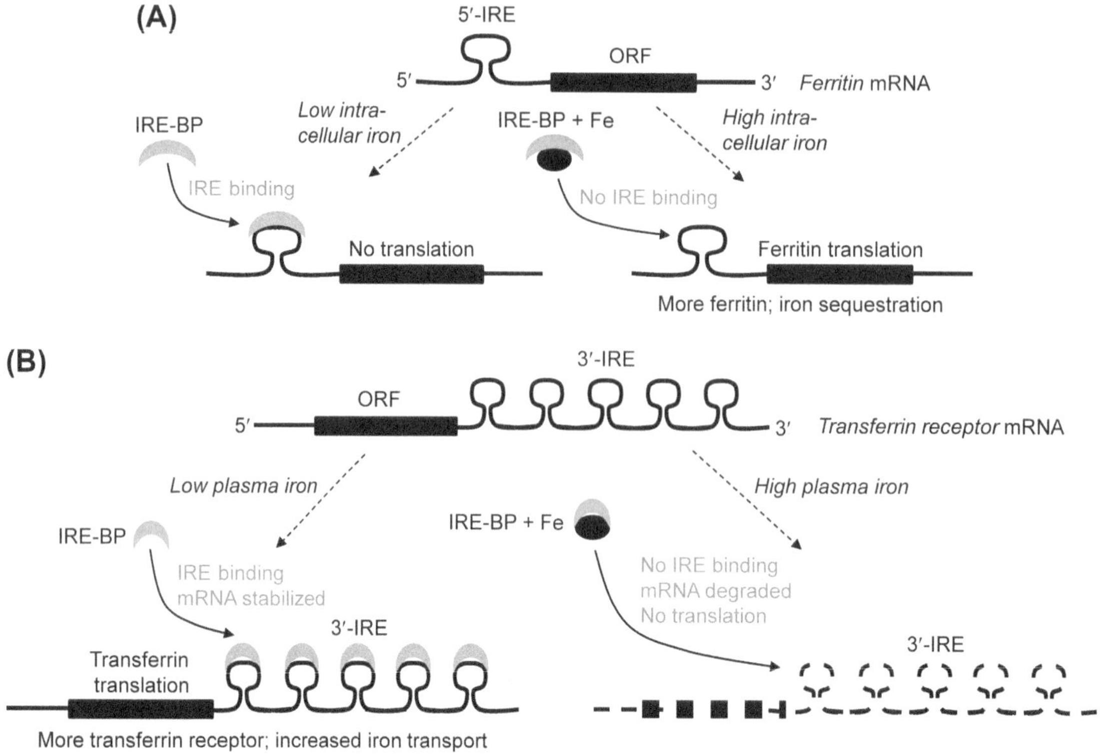

FIGURE 20.6 Interactions between the IRE and the IRE-BP result in different translational outcomes depending on whether the IREs are in the 5′-UTR or the 3′-UTR. (A) The IRE–IRE-BP interaction in the 5′-UTR of *ferritin* mRNA downregulates translation, resulting in lower availability of the ferritin protein. (B) The same interaction in the 3′-UTR of *transferrin receptor* mRNA directly increases the half-life of *transferrin receptor* mRNA by stabilizing it, resulting in its increased translation and increased availability of the transferrin receptor protein. *IRE*, Iron response element; *IRE-BP*, IRE-binding protein.

(thus, over-accumulation) of iron in the cell (Fig. 20.6) (Choudhuri, 2009b).

Concluding remarks and future directions

Investigations into the molecular mechanism of action of nutrients have helped in understanding the complexity of genomic responses to nutrition while opening up the possibilities of nutritional intervention, both clinical and population-based. Although a long-term goal, basic research using genomic and proteomic techniques is expected to elucidate the molecular basis of nutrient-mediated effects on genome stability, imprinting, and expression. The full complement of genes that are up- or downregulated in response to macro- and micronutrients is yet to be identified, as are the regulatory networks involving and controlling these genes. Advances in genomic technologies, transcriptional network analysis, and systems biology are expected to greatly help in understanding how to utilize the scientific knowledge in developing guidance for optimal health and a healthy lifestyle. Correlating interindividual genetic differences with differential responses to nutrients may provide a tool to understand and underscore the molecular basis of "individualized nutrition."

References

Aljada, A., Ghanim, H., Mohanty, P., et al., 2004. Glucose intake induces an increase in activator protein 1 and early growth response 1 binding activities, in the expression of tissue factor and matrix metalloproteinase in mononuclear cells, and in plasma tissue factor and matrix metalloproteinase concentrations. Am. J. Clin. Nutr. 80, 51–57.

Anderson, C.M., Stahl, A., 2013. SLC27 fatty acid transport proteins. Mol. Asp. Med. 34, 516–528.

Augustin, R., 2010. The protein family of glucose transport facilitators: it's not only about glucose after all. IUBMB Life 62, 315–333.

Berger, J., Moller, D.E., 2002. The mechanisms of action of PPARs. Annu. Rev. Med. 53, 409–435.

Bryant, N.J., Govers, R., James, D.E., 2002. Regulated transport of the glucose transporter GLUT4. Nat. Rev. Mol. Cell Biol. 3, 267–277.

Brasse-Lagnel, C., Lavoinne, A., Husson, A., 2009. Control of mammalian gene expression by amino acids, especially glutamine. FEBS J. 276, 1826–1844.

Choi, J.S., Berdis, A.J., 2012. Nucleoside transporters: biological insights and therapeutic applications. Future Med. Chem. 4, 1461–1478.

Choudhuri, S., 2009a. The eukaryotic genome: a brief overview of structure, function and evolution. In: Choudhuri, S., Carlson, D.B. (Eds.), Genomics: Fundamentals and Applications. Informa., New York, pp. 49–99.

Choudhuri, S., 2009b. Fundamentals of structure-function analysis of eukaryotic protein-coding genes. In: Choudhuri, S., Carlson, D.B. (Eds.), Genomics: Fundamentals and Applications. Informa., New York, pp. 3–48.

Claudel, T., Staels, B., Kuipers, F., 2005. The farnesoid X receptor: a molecular link between bile acid and lipid and glucose metabolism. Arterioscler. Thromb. Vasc. Biol. 25, 2020–2031.

Corcos, D., Vaulont, S., Denis, N., et al., 1987. Expression of c-myc is under dietary control in rat liver. Oncogene Res. 1, 193–199.

Cura, A.J., Carruthers, A., 2012. The role of monosaccharide transport proteins in carbohydrate assimilation, distribution, metabolism and homeostasis. Comp. Physiol. 2, 863–914.

Danesch, U., Weber, P.C., Sellmayer, A., 1994. Arachidonic acid increases c-fos and Egr-1 mRNA in 3T3 fibroblasts by formation of prostaglandin E2 and activation of protein kinase C. J. Biol. Chem. 269, 27258–27263.

Deckelbaum, R.J., Worgall, T.S., Seo, T., 2006. n-3 Fatty acids and gene expression. Am. J. Clin. Nutr. 83 (Suppl), 1520S–1525S.

Duplus, E., Glorian, M., Forest, C., 2000. Fatty acid regulation of gene transcription. J. Biol. Chem. 275, 30749–30752.

Foufelle, F., Ferré, P., 2002. New perspectives in the regulation of hepatic glycolytic and lipogenic genes by insulin and glucose: a role for the transcription factor sterol regulatory element binding protein-1c. Biochem. J. 366, 377–391.

Georgiadi, A., Kersten, S., 2012. Mechanisms of gene regulation by fatty acids. Adv. Nutr. 3, 127–134.

Gimeno, R.E., 2007. Fatty acid transport proteins. Curr. Opin. Lipidol. 18, 271–276.

Gingras, A.C., Raught, B., Sonenberg, N., 2001. Regulation of translation initiation by FRAP/mTOR. Genes. Dev. 15, 807–826.

Glatz, J.F., Luiken, J.J., Bonen, A., 2010. Membrane fatty acid transporters as regulators of lipid metabolism: implications for metabolic disease. Physiol. Rev. 90, 367–417.

Hara, K., Yonezawa, K., Weng, Q.P., et al., 1998. Amino acid sufficiency and mTOR regulate p70 S6 kinase and eIF-4E BP1 through a common effector mechanism. J. Biol. Chem. 273, 14484–14494.

Hardie, D.G., Hawley, S.A., 2001. AMP-activated protein kinase: the energy charge hypothesis revisited. BioEssays 23, 1112–1119.

Hediger, M.A., Clémencon, B., Burrier, R.E., et al., 2013. The ABCs of membrane transporters in health and disease (SLC series): introduction. Mol. Asp. Med. 34, 95–107.

Huang, S., Houghton, P.J., 2003. Targeting mTOR signaling for cancer therapy. Curr. Opin. Pharmacol. 3, 371–377.

Huang, L., Tepaamorndech, 2013. The SLC30 family of zinc transporters—a review of current understanding of their biological and pathophysiological roles. Mol. Asp. Med. 34, 548–560.

Hyde, R.J., Cass, C.E., Young, J.D., et al., 2001. The ENT family of eukaryote nucleoside and nucleobase transporters: recent advances in the investigation of structure/function relationships and the identification of novel isoforms. Mol. Membr. Biol. 18, 53–63.

Issemann, I., Prince, R.A., Tugwood, J.D., 1993. The peroxisome proliferator-activated receptor:retinoid X receptor heterodimer is activated by fatty acids and fibrate hypolipidaemic drugs. J. Mol. Endocrinol. 11, 37–47.

Iynedjian, P.B., Ucla, C., Mach, B., 1987. Molecular cloning of glucokinase cDNA. Developmental and dietary regulation of glucokinase mRNA in rat liver. J. Biol. Chem. 262, 6032–6038.

Joost, H.G., Thorens, B., 2001. The extended GLUT-family of sugar/polyol transport facilitators: nomenclature, sequence characteristics, and potential function of its novel members. Mol. Membr. Biol. 18, 247–256.

Jousse, C., Averous, J., Bruhat, A., et al., 2004. Amino acids as regulators of gene expression: molecular mechanisms. Biochem. Biophys. Res. Commun. 313, 447–452.

Jump, D.B., Clarke, S.D., 1999. Regulation of gene expression by dietary fat. Annu. Rev. Nutr. 19, 63–90.

Kallio, P., Kolehmainen, M., Laaksonen, D.E., et al., 2007. Dietary carbohydrate modification induces alterations in gene expression in abdominal subcutaneous adipose tissue in persons with the metabolic syndrome: the FUNGENUT Study. Am. J. Clin. Nutr. 85, 1417–1427.

Khalfoun, B., Thibault, F., Watier, H., et al., 1997. Docosahexaenoic and eicosapentaenoic acids inhibit in vitro human endothelial cell production of interleukin-6. Adv. Exp. Med. Biol. 400B, 589–597.

Kilberg, M.S., Balasubramanian, M., Fu, L., 2012. The transcription factor network associated with the amino acid response in mammalian cells. Adv. Nutr. 3, 295–306.

Klaassen, C.D., Liu, J., Choudhuri, S., 1999. Metallothionein: an intracellular protein to protect against cadmium toxicity. Annu. Rev. Pharmacol. Toxicol. 39, 267–294.

Lampe, J.W., 2009. Interindividual differences in response to plant-based diets: implications for cancer risk. Am. J. Clin. Nutr. 89 (Suppl), 1553S–1557S.

Lampe, J.W., Navarro, S.L., Hullar, M.A., et al., 2013. Inter-individual differences in response to dietary intervention: integrating omics platforms towards personalised dietary recommendations. Proc. Nutr. Soc. 72, 207–218.

Laplante, M., Sabatini, D.M., 2009. mTOR signaling at a glance. J. Cell Sci. 122, 3589–3594.

Laplante, M., Sabatini, D.M., 2012. mTOR signaling in growth control and disease. Cell 149, 274–293.

Liu, B., Qian, S.B., 2011. Translational regulation in nutrigenomics. Adv. Nutr. 2, 511–519.

McGrane, M.M., 2007. Vitamin A regulation of gene expression: molecular mechanism of a prototype gene. J. Nutr. Biochem. 18, 497–508.

Mourrieras, F., Foufelle, F., Foretz, M., et al., 1997. Induction of fatty acid synthase and S14 gene expression by glucose, xylitol and dihydroxyacetone in cultured rat hepatocytes is closely correlated with glucose 6-phosphate concentrations. Biochem. J. 326, 345–349.

Muñoz, E.C., Martin, M.J., Prieto, P.A., 2013. 2′-fucosyllactose: an abundant, genetically determined soluble glycan present in human milk. Nutr. Rev. 71, 773–789.

Perry, G.H., Dominy, N.J., Claw, K.G., et al., 2007. Diet and the evolution of human amylase gene copy number variation. Nat. Genet. 39, 1256–1260.

Picard, V., Govoni, G., Jabado, N., et al., 2000. Nramp 2 (DCT1/DMT1) expressed at the plasma membrane transports iron and other divalent cations into a calcein-accessible cytoplasmic pool. J. Biol. Chem. 275, 35738–35745.

Pinsonneault, J., Nielsen, C.U., Sadée, W., 2004. Genetic variants of the human H + /dipeptide transporter PEPT2: analysis of haplotype functions. J. Pharmacol. Exp. Ther. 311, 1088–1096.

Poncet, N., Taylor, P.M., 2013. The role of amino acid transporters in nutrition. Curr. Opin. Clin. Nutr. Metab. Care. 16, 57–65.

Prasad, P.D., Ganapathy, V., 2000. Structure and function of mammalian sodium-dependent multivitamin transporter. Curr. Opin. Clin. Nutr. Metab. Care 3, 263–266.

Ramagopalan, S.V., Heger, A., Berlanga, A.J., et al., 2010. A ChIP-seq defined genome-wide map of vitamin D receptor binding: associations with disease and evolution. Genome Res. 20, 1352–1360.

Rao, G.N., Alexander, R.W., Runge, M.S., 1995. Linoleic acid and its metabolites, hydroperoxyoctadecadienoic acids, stimulate c-*Fos*, c-*Jun*, and c-*Myc* mRNA expression, mitogen-activated protein kinase activation, and growth in rat aortic smooth muscle cells. J. Clin. Invest. 96, 842–847.

Reboul, E., Borel, P., 2011. Proteins involved in uptake, intracellular transport and basolateral secretion of fat-soluble vitamins and carotenoids by mammalian enterocytes. Prog. Lipid Res. 50, 388–402.

Rhoads, D.B., Rosenbaum, D.H., Unsal, H., et al., 1998. Circadian periodicity of intestinal Na + /glucose cotransporter 1 mRNA levels is transcriptionally regulated. J. Biol. Chem. 273, 9510–9516.

Riggs, M.G., Whittaker, R.G., Neumann, J.R., et al., 1977. n-Butyrate causes histone modification in HeLa and Friend erythroleukaemia cells. Nature 268, 462–464.

Roepstorff, C., Wulff-Helge, J., Vistisen, B., et al., 2004. Studies of plasma membrane fatty acid-binding protein and other lipid-binding proteins in human skeletal muscle. Proc. Nutr. Soc. 63, 239–244.

Rubio-Aliaga, I., Daniel, H., 2008. Peptide transporters and their roles in physiological processes and drug disposition. Xenobiotica 38, 1022–1042.

Said, H.M., Mohammed, Z.M., 2006. Intestinal absorption of water-soluble vitamins: an update. Curr. Opin. Gastroenterol. 22, 140–146.

Schisler, J.C., Charles, P.C., Parker, J.S., et al., 2009. Stable patterns of gene expression regulating carbohydrate metabolism determined by geographic ancestry. PLoS One 4, e8183.

Shankar, K., Harrell, A., Kang, P., et al., 2010. Carbohydrate-responsive gene expression in the adipose tissue of rats. Endocrinology 151, 153–164.

Stanfel, M.N., Shamieh, L.S., Kaeberlein, M., et al., 2009. The TOR pathway comes of age. Biochim. Biophys. Acta 1790, 1067–1074.

Stein, J., Zores, M., Schroder, O., 2000. Short-chain fatty acid (SCFA) uptake into Caco-2 cells by a pH-dependent and carrier mediated transport mechanism. Eur. J. Nutr. 39, 121–125.

Taylor, P.M., 2014. Role of amino acid transporters in amino acid sensing. Am. J. Clin. Nutr. 99 (Suppl), 223S–230S.

Wang, B., Bobe, G., LaPres, J.J., et al., 2009. Dietary carbohydrate source alters gene expression profile of intestinal epithelium in mice. Nutr. Cancer 61, 146–155.

Wang, M., Li, M., Wu, S., et al., 2015. Fecal microbiota composition of breast-fed infants is correlated with human milk oligosaccharides consumed. J. Pediatr. Gastroenterol. Nutr. 60 (6), 825–833.

Wood, I.S., Trayhurn, P., 2003. Glucose transporters (GLUT and SGLT): expanded families of sugar transport proteins. Br. J. Nutr. 89, 3–9.

Wright, E.M., 2001. Renal Na(+)-glucose cotransporters. Am. J. Physiol. Ren. Physiol. 280, F10–F18.

Wright, E.M., Loo, D.D., Hirayama, B.A., 2011. Biology of human sodium glucose transporters. Physiol. Rev. 91, 733–794.

Young, J.D., Yao, S.Y.M., Baldwin, J.M., et al., 2013. The human concentrative and equilibrative nucleoside transporter families, SLC28 and SLC29. Mol. Asp. Med. 34, 529–547.

Zeisel, S.H., Freake, H.C., Bauman, D.E., et al., 2005. The nutritional phenotype in the age of metabolomics. J. Nutr. 135, 1613–1616.

Zhao, R., Goldman, I.D., 2013. Folate and thiamine transporters mediated by facilitative carriers (SLC19A1-3 and SLC46A1) and folate receptors. Mol. Asp. Med. 34, 373–385.

Chapter 21

Genomic and proteomic mechanisms and models in toxicity and safety evaluation of nutraceuticals

Christina Wilson-Frank

Introduction

The term "nutraceutical" is used to describe foods or food-derived products that can be used nutritionally or medicinally to improve health. Although there has been a long history of using nutraceuticals to support health and disease prevention, the role they play in human health continues to evolve as an important component of the diet–health paradigm. For example, a wide variety of nutraceuticals have been recognized for their ability to be used as preventative or supportive treatments for diseases such as hepatic disease, renal disease, reproductive or gastrointestinal disorders, CNS diseases, diabetes, and cancer (Farag et al., 2016; Abenavoli et al., 2018; Sun et al., 2018; Ruscica et al., 2019; Saeed et al., 2019; Calvani et al., 2020; Xiang et al., 2020; Lupo et al., 2020). Historically, evaluation of these health-promoting benefits and functional properties of nutraceuticals has been established using animal and cultured cell models (Baker, 2008; Botham, 2015). These investigations have ameliorated our understanding of how nutraceuticals influence cell physiology and have given us insight into the physiological responses to those compounds when associated with disease. Although these approaches have been helpful in elucidating their primary mechanisms of action, these studies often involve a more targeted model system measuring changes in single genes or proteins. Nutraceuticals can have multiple functions and impact more than one disease state, which emphasizes their potential to effect complex, biochemical processes that could include hundreds of genes and proteins. Therefore targeting single genes or proteins may not be sufficient to provide a complete understanding of the complex interrelationships among genes and proteins affected by nutraceuticals with respect to their health-promoting benefits. Recent advances in rapid, high-throughput genomics and proteomics technologies have provided a more comprehensive platform for quantifying changes and comparing patterns of groups of genes and proteins that are expressed at any given time in a cell, tissue, or organism. These "omics" technologies are being used to obtain a comprehensive view of the global changes in gene and protein expression with respect to the beneficial and adverse effects of nutraceuticals on human health. This chapter describes the genomic and proteomic investigations that have been conducted in order to gain a better understanding of the underlying mechanisms involved in the efficacy, safety, and toxicity of nutraceuticals.

Nutrigenomics

Due to the completion of the Human Genome Project in 2003 (International Human Sequencing Consortium, 2004) and rapid advances in genomics technologies, measuring the impact of nutraceuticals on gene expression and their effects on the molecular processes in organisms has led to the inception of a new frontier in nutrition science termed "nutrigenomics" (Peregrin, 2001; Zhang et al., 2015). Nutrigenomics is an emerging field used to investigate changes in gene expression, at the molecular level, in response to nutraceuticals or diet as applied to individuals or population groups (Sales et al., 2014). This global strategy of measuring changes in gene expression in the whole genome of an organism is crucial in understanding the multifactorial complex, biochemical processes that are impacted by nutraceuticals. Several genomics technologies have been employed to investigate how nutraceuticals affect health through altering gene expression. The most common genomics technologies used are real-time quantitative PCR, RNA-Seq, and microarrays.

Nutraceuticals. DOI: https://doi.org/10.1016/B978-0-12-821038-3.00021-5

Application of high-throughput technologies

Real-time quantitative polymerase chain reaction

Real-time quantitative polymerase chain reaction (qRT-PCR) is a sensitive, high-throughput technique used to rapidly detect and quantify gene expression in real time. qRT-PCR is a polymerase-dependent, repetitive, thermal reaction that generates copies of a specific target DNA segment that spans two primers (Raso and Biassoni, 2014). These amplified DNA segments (amplicons) are detected and quantified using intercalating dyes or fluorescent probes such that at the end of each cycle, the fluorescence emitted by the PCR amplicons and the signal intensity is proportional to the concentration of the target DNA in the sample (Carey et al., 2007; Raso and Biassoni, 2014). Due to its large dynamic range, this highly sensitive assay can detect and quantify changes in gene expression of approximately 100–400 genes at a time with a precision of 5%–10% (Schmittgen et al., 2008; Singh and Roy-Chowdhuri, 2016). qRT-PCR is commonly used as a validation tool for confirming gene expression results obtained from other genomic technologies such as RNA-Seq and microarrays and is considered the "gold standard" in gene expression detection and quantification (Rajkumar et al., 2015).

qRT-PCR has been applied to quantifying changes in gene expression, as well as identifying the genes involved in the complex interrelationships between nutraceuticals, genes, and disease.

Microarrays

The human genome contains approximately 3 billion bases and 20,000–25,000 protein-coding genes (International Human Sequencing Consortium, 2004). Given that disease states often involve a multitude of biochemical pathways, a variety of genes could be up- or downregulated at any given time. Although the other genomics techniques have their advantages, microarrays are superior with respect to throughput, being capable of measuring the entire transcriptome or genome simultaneously (Khelurkar et al., 2017). Microarrays, which are commonly referred to as DNA microchips or microchips, are comprised of a solid support (such as a microscope slide or "chip") with thousands of microscopic spots in defined positions that contain picomole amounts of a known gene or DNA sequence (also called probes or reporters) (Khelurkar et al., 2017). These probes are used to hybridize a cRNA or cDNA sample (referred to as the target) and the probe–target hybridization can be quantified and detected using a fluorophore, chemiluminescent-label, enzyme detection system, laser scanning, or autoradiography (Khelurkar et al., 2017). Ultimately, the signal intensity from each spot correlates with the amount of target sample binding to the probes. The use of microarray technology has advanced nutrition research, providing a genome-wide view of the impacts of nutraceuticals at the molecular level. Although microarrays are highly advantageous with regard to the number of genes that can be profiled simultaneously, the technique can be costly and often qRT-PCR has to be used as a validation tool for confirming gene expression results obtained from microarrays due to their limited sensitivity and specificity (Morey et al., 2006; Schmittgen et al., 2008; Rajkumar et al., 2015).

RNA-Seq

High-throughput transcriptome sequencing (RNA-Seq) is a genomics technique used to identify and quantify differentially expressed genes (DEGs). This technique is emerging as the preferred technique to measure expression levels as it can be performed without prior knowledge of the reference sequence and allows for "de novo" reconstruction of the transcript without prior knowledge of the reference sequence (Cokus et al., 2008; Montazavi et al., 2008). RNA-Seq is a genomic method that involves isolating RNA, sequencing the RNA with a high-throughput sequencer, and then subsequently reannealing the RNA (Korf, 2013). The process first involves fragmenting RNA isolated from samples into complementary DNA sequences and sequencing the complementary DNA using a high-throughput sequencer (Wang et al., 2009; Oshlack et al., 2010). The sequences generated are mapped to a transcriptome or genome and the expression levels for each isoform or gene (Wang et al., 2009; Oshlack et al., 2010; Costa-Silva et al., 2017). The mapped data are then normalized using bioinformatics tools and the DEGs are identified (Wang et al., 2009; Oshlack et al., 2010; Costa-Silva et al., 2017). RNA-Seq has some advantages over DNA microarrays in that it is a less expensive pooling strategy, it has a high level of data reproducibility, and allows for the identification and quantification of the expression of isoforms and unknown transcripts and is evolving as a popular tool for comparing and cataloging gene expression (Agrawal et al., 2010; Korf, 2013).

Nutriproteomics

Proteomics technologies were developed in order to provide a platform for isolating and identifying proteins and peptides in complex biological samples that change with disease or in response to external stimuli. Measuring changes in global protein expression has remained a challenge in proteomics due to the heterogeneity of protein expression and large dynamic range of proteins in complex biological samples. For example, the human serum proteome can contain approximately 20,000 proteins at any given time, highlighting the need for high-throughput platforms that can adequately isolate and detect proteins of interest (Anderson and Anderson, 2002). In order to overcome this challenge, proteomics studies first begin with global separation of the complex mixtures of proteins or peptides using multidimensional, chromatographic, or gel-based separation platforms. Chromatographic separation platforms such as high-performance liquid chromatography (HPLC) and ultra-high performance liquid chromatography (UHPLC) and gel-based separation platforms such as one-dimensional gel electrophoresis, two-dimensional gel electrophoresis (2-DE), and two-dimensional in-gel electrophoresis (2D-DIGE) have evolved considerably, offering increased resolution and improved isolation of proteins or peptides from complex samples (Isaaq et al., 2005; Meleady, 2018; Shahali et al., 2019; Sheflin et al., 2019). Once separation is achieved, protein sequencing and identification are accomplished using mass spectrometry in combination with a variety of bioinformatics tools. The most common mass spectrometers used for proteomics studies include soft ionization techniques such as matrix-assisted laser desorption/time-of flight mass spectrometry (MALDI-TOF MS) and electrospray ionization/tandem mass spectrometry (ESI/MS/MS).

Historically, nutrition research utilized traditional in vivo and in vitro methods in order to reveal the molecular mechanisms that link nutraceuticals with host protein expression and health outcomes. Although the results of these investigations have been rewarding, these traditional approaches limit the ability to elucidate the global changes in protein expression due to the inherent complexity of the host proteome and the nutraceuticals themselves. In order to overcome this limitation, nutriproteomics has emerged as a novel tool in nutrition research that exploits the use of these proteomics platforms to assess changes in global protein expression influenced by nutraceuticals.

Application of high-throughput technologies

Gel-based separation technologies

Two-dimensional gel electrophoresis (2-DE and 2D-DIGE) remains the most common, protein separation technique used in proteomics research (Rabilloud, 2002). Although these gel-based separation techniques have been used for decades, they still provide the highest resolution of proteins from complex samples and are capable of separating up to 10,000 proteins per gel (Daniel, 2002; Saraswathy and Ramalingam, 2011a). 2-DE is a high-throughput separation technique that separates proteins in two dimensions. The first dimension separates proteins based on their isoelectric point and the second dimension separates them based on their molecular weight in a single gel. The proteins are then detected using a variety of staining techniques and image analysis software to visualize the protein spots on the gel. Coomassie Brilliant Blue and SYBRO Ruby dyes are more commonly used in proteomics because they are MS compatible (Saraswathy and Ramalingam, 2011a). The visualized protein spots are then excised and digested into peptides for identification using mass spectrometry. Despite the fact that the 2-DE technique has the highest resolution of proteins from samples, gel-to-gel variation can occur and reproducibility is a limitation (Minden, 2012). In order to overcome this limitation, 2D-DIGE was developed. Two-dimensional difference gel electrophoresis is similar to 2-DE; however, it allows for comparison of two to three protein samples in the same gel. In 2D-DIGE, proteins in two or three different samples are labeled with different colored, fluorescent dyes such as Cy3-NHS, Cy5-NHS, Cy3-Mal, and Cy5-Mal (Minden, 2012). The samples are labeled with the fluorescent tags, mixed together, then subjected to 2-DE separation on one gel. The proteins are visualized by imaging the gel on a fluorescent gel imager at the Cy dye wavelengths. The ratio of fluorescence of the proteins from the samples analyzed will be constant for proteins that are common in those samples; whereas differentially expressed proteins will have a higher fluorescent signal of one fluorophore (e.g., Cy3 vs Cy5) versus the other (Minden, 2012). Like 2-DE, the visualized proteins are excised and enzymatically digested for mass spectrometric identification. Protein identification from gel-based separation techniques is often accomplished using MALDI-TOF MS. This technique can detect as low as 0.2 femtomole of protein and has a linear dynamic range of 10,000-fold concentration (Minden, 2012).

Chromatography-based separation techniques

Due to its reproducibility, high analyte resolution, and compatibility with ESI/MS/MS, liquid chromatography (LC) has evolved as a successful, alternative separation technique. One advantage of using HPLC or UHPLC is that there are a variety of chromatographic modalities that can be used to separate proteins and peptides based on their ability to adsorb and desorb to the stationary phase through mobile phase manipulation. These chromatographic modalities can be used alone or in a multidimensional format and include: hydrophobic or hydrophilic interaction chromatography, affinity chromatography, capillary electrophoresis chromatography, ion exchange chromatography, size exclusion chromatography, strong cation exchange chromatography, or reversed-phase chromatography (Shi et al., 2004; Coskun, 2016). The most common multidimensional approach used in proteomics is 2D-LC-MS, where the two-dimensional LC separations are accomplished using ion-exchange chromatographic separation followed by reversed-phase chromatography (Shi, et al., 2004). Ion exchange chromatography separates peptides or proteins in the first dimension based on their charge (positive or negative) in the presence of an anionic or cationic stationary phase. In the second dimension, reversed-phase chromatography uses a hydrophobic stationary phase to separate proteins and peptides based on hydrophobicity. In this multidimensional approach, the peak capacity can be 5000 or greater and could yield anywhere from 800 to more than 2000 protein identifications when combined with mass spectrometry (Opiteck et al., 1997; Hoffman et al., 2007). Although LC can be used to separate intact proteins for proteomics experiments, 2-DE is superior when separating intact proteins. Because peptides have narrow isoelectric points and lower molecular weights, 2-DE is not a suitable technique to use for separating peptides in proteomics studies. However, LC is recognized as the optimal proteomics tool for separating peptides from proteolytic digests due to its resolving power and high-throughput capacity. Another advantage of LC is that it can be directly interfaced with ESI/MS/MS instrumentation, allowing for high-throughput peptide identification.

Mass spectrometry

Over the past three decades, significant advancements in mass spectrometry instrumentation have improved the sensitivity, resolving power, high-throughput automation, and mass range of detection of peptides and proteins. It was not until the introduction of MALDI-TOF/MS and ESI/MS that proteomics emerged as an indispensable tool for detecting and quantifying changes in protein expression. Depending on the accuracy of the peptide mass of comparison, these mass spectrometers have a mass accuracy of approximately 50 ppm, can detect peptides in the femtomolar range, and have a mass range of detection of up to 200 kDa (Saraswathy and Ramalingam, 2011b). MALDI-TOF/MS is frequently used to detect peptides generated from proteolytic digests of proteins from gel-based separation techniques; whereas ESI/MS/MS is frequently used as the mass spectrometer of choice when using chromatography-based separations. MALDI-TOF/MS and ESI/MS/MS experiments first detect the molecular weights of the peptides in the sample followed by peptide mass with subsequent fragmentation of the peptides in the tandem MS (Saraswathy and Ramalingam, 2011b). Protein identification is achieved by comparing the experimental, molecular weights of the peptide fragments to theoretical, peptide masses in protein sequence databases, known as peptide mass fingerprinting. In addition to being indispensable for protein identification, MS instrumentation is also essential for quantitating changes in global protein expression. Quantitating changes in protein expression can include chemical or proteolytic labeling of peptides with isotope-coded mass tags prior to chromatography separation. This process involves labeling the peptides from a control sample with a light isotope (label with a lower molecular weight) and labeling peptides from an experimental sample with a heavy isotope tag (label with a higher molecular weight). After labeling the peptides from each sample group, the samples are mixed together and peptides separated using LC. When introduced into the mass spectrometer, the mass spectrometer can detect the difference between the molecular weights of the isotope-coded peptide samples. Quantitating changes in these peptides is achieved by calculating the difference between the ion intensities of the light isotope labeled control sample and the heavy isotope labeled experimental sample. The differences between the ion intensities indicate which peptides are increased or decreased with treatment. The isotope-coded reagents frequently used in quantitative proteomics include ICAT, GIST, $^{14}N/^{15}N$-labeling, $^{16}O/^{18}O$-labeling, and iTRAQ (Goshe and Smith, 2003; Zieske, 2006).

Genomics and proteomics applications in nutraceutical research

Determining molecular mechanisms and toxicity of nutraceuticals

Green tea

Green tea (*Camellia sinensis*) is a nutraceutical that has long been recognized for its health benefits and its ability to have anticancer, antidiabetic, and antimicrobial properties (Ahmad et al., 2014). Polyphenolic catechins, such as epigallocatechin-3-gallate, epigallocatechin, epicatechin-3-gallate, and epicatechin in green tea are the powerful antioxidants responsible for these health benefits and comprise approximately 42% of the dry weight of green tea extract (Ahmad et al., 2014). Most of the proteomics and genomics applications in nutraceutical research with respect to green tea polyphenols have been used to elucidate the molecular mechanisms responsible for their antimicrobial and anticancer benefits. In one study, the cellular responses and proteomic analysis of *Escherichia coli* exposed to green tea polyphenols were investigated using proteomics analysis. Using 2-DE and MALDI-TOF MS, the study concluded that there were nine proteins upregulated in *E. coli* with green tea extract exposure and eight proteins that were downregulated (Cho et al., 2007).

The upregulated proteins included those involved in cellular defense, such as EmrK, GyrA, RpoS, and SodC. GroEL was another protein that was upregulated, which is a bacterial chaperonin protein important for protein folding. The proteins that were downregulated proteins in *E. coli* with green tea polyphenol exposure included Eno, SdhA, and UgpQ (proteins involved in carbon and energy metabolism) and BltK and TyrB (proteins involved in amino-acid biosynthesis). This study provided some insight into the polyphenol-induced protein changes in *E. coli* that cause cytotoxicity, and it also revealed that these polyphenols have a dose-dependent, bactericidal effect on *E. coli*.

Nutraceutical researchers are also using proteomics and genomics tools to help reveal the molecular mechanisms responsible for the anticancer benefits of green tea. Notably, 2-DE and LC/MS/MS and qRT-PCR have been implemented to understand the anticancer mechanisms of green tea extract on lung cancer using lung cancer A549 cell lines. In one study, 2-DE combined with LC/MS/MS and qRT-PCR were used to analyze changes in proteins and genes from A549 cells before and after green tea extract exposure at 20 and 40 μg/mL concentrations (Lu et al., 2009). Fourteen tumor-associated proteins were identified that increased greater than twofold after green tea extract treatment which included fascin and lamin A/C (associated with cytoskeletin and motility), annexin I and II (calcium-binding proteins) and other proteins involved in metabolism, detoxification, and gene regulation. Upregulation of these proteins may explain the multiple, antitumor pathways of green tea extract in lung cancer (Lu et al., 2009). Previous "omics" investigations by these researchers infer that annexin-I may serve as a molecular target associated with the anticancer activity of green tea extract (Lu et al., 2007). Through the use of proteomics and genomics techniques, annexin-I has also been identified as a potential target for the anticancer activity of green tea in bladder cancer (Xiao et al., 2007).

Gingko biloba

Extracts from the leaves and seeds of the *Gingko biloba* tree have long been used as a nutraceutical in Chinese medicine. *G. biloba* extracts have been reported to have many health benefits which include anticancer activity and cardioprotective effects; however, its purported ability to provide neuroprotection against neurovascular diseases such as ischemic brain injury has become of great interest to researchers (Mahadevan and Park, 2008). In order to understand these neuroprotective mechanisms, proteomics applications have been used in order to determine changes in proteins involved in the neuroprotective mechanisms of *G. biloba* extracts. For example, a proteomics approach was used to investigate protein changes that occurred when adult, male rats were dosed with 100 mg/kg of a *G. biloba* extract (EGb 761) prior to middle cerebral artery occlusion (Koh, 2011, 2012). From the cerebral cortex tissues collected after occlusion, it was determined that EGb 761 pretreatment prevented ischemic injury-induced reduction of PEA-15 and hippocalcin expression (Koh, 2011, 2012). In astocytes, decreased expression of PEA-15 and hippocalcin cause apoptotic cell death; thereby explaining the plausible neuroprotective mechanisms of *G. biloba* extract.

In addition to using proteomics tools to understand the neuroprotective mechanisms of *G. biloba* extract, genomics and proteomics techniques have been employed to identify changes in gene and protein expression associated with *G. biloba* extract-induced hepatotoxicity. These studies highlighted that *G. biloba* extract can induce and inhibit expression of hepatic enzymes important for the metabolism of xenobiotics, suggesting a potential for herb–drug interactions. In one study, gene expression changes were determined following dosing of B6C3F1 mice with *G. biloba* extract (Guo et al., 2010a). Of the 31,802 genes analyzed, 33 genes involved in phase I metabolism and 18 genes involved in phase II metabolism were significantly altered. The genes that were significantly overexpressed are in the CYP_{450} superfamily of genes which included CYP1A1, CYP1A2, CYP2B10, CYP2C55, CES2, and GSTA2. These studies were also able to show that *G. biloba* extract inhibits FMO3, a flavin-containing monooxygenase that plays a major role in phase I oxidative metabolism of xenobiotics (Guo et al., 2010a). This suggests that *G. biloba* extract may alter pharmacokinetics due to herb–drug interactions, potentially causing an increase in bioactivation of tumorigenic xenobiotics through inducing genes in the CYP_{450} superfamily or decreasing the efficacy of drugs through inhibition of FMAO3, thereby contributing to the *G. biloba*'s hepatotoxic potential.

Kava

Kava is a nutraceutical that is extracted from the roots of the plant *Piper methysticum.* The presence of kavalactones and flavakavains in kava have been reported to have anxiolytic and muscle-relaxing properties; therefore this nutraceutical has been used to alleviate generalized anxiety disorder, insomnia, fatigue, and muscle strain (Smith and Leiras, 2019). Since its increased use as a plausible alternative to antianxiety drugs, there have been some reports of hepatotoxicity occurring with kava (Campo et al., 2002). In a study similar to the one conducted for *G. biloba* extract, B6C3F1 male mice were treated with varying doses of kava and the liver gene expression changes quantified (Guo et al., 2010b). At the 2.0 g/kg dose, 95 genes associated with drug-metabolizing enzymes changed significantly, 28 of those genes being associated with phase I and 29 genes associated with phase II enzymes. Pathway analysis revealed that the genes upregulated by kava treatment include the xenobiotic metabolism pathways and signaling pathways involved in oxidative stress response. Genes noted that were significantly upregulated included the CYP_{450} superfamily and those associated with the NRF2 oxidative stress pathway including Nqo1, Gstms, Gstas, and Gstp1. This study suggested that kava-induced hepatotoxicity is likely due to herb–drug interactions through modulation of metabolizing enzymes that could alter drug metabolism (Guo et al., 2010b).

Determining molecular mechanisms of nutraceuticals and disease

Cancer and inflammation

The genomics techniques have been used to investigate the use of nutraceuticals in the prevention of disease including their impact on gene changes associated with cancer and inflammatory immune responses. For example, isoliquiritigenin, a flavonoid that is found in licorice, has been recognized as a nutraceutical that has antitumor potential (DeQiang et al., 2019). Using qRT-PCR, investigators evaluated whether isoliquiritigenin controlled cell proliferation and apoptosis through regulating forkhead box G1 (a transcription factor repressor) expression via miR-21 (a cancer-promoting microRNA) in tongue squamous cell carcinoma cells (DeQiang et al., 2019). The results of this study revealed that isoliquiritigenin decreased miR-21 levels and promoted forkhead box G1 expression, thereby inhibiting proliferation and inducing apoptosis of tongue squamous cell carcinoma cells. This study provided the genomic data to suggest that isoliquiritigenin could serve as a novel nutraceutical for the treatment of tongue squamous cell carcinoma cell. *Melicope ptelefolia* is a well-known herb in Asian culture that has been touted for its medicinal benefits and anticancer activity (Kabir et al., 2017). Using microarray gene expression profiling, leaf extracts from this plant were applied to colorectal (HCT116) and hepatocellular (HepG2) carcinoma cell lines to elucidate transcriptome profiles that may be associated with anticancer activity (Kabir et al., 2018). The microarray data indicated that leaf extracts induced differential expression of approximately 1290 genes in the HCT116 cell line and 1325 genes in the HepG2 cell line. The leaf extract was shown to modulate gene expression of several genes supporting antiproliferative activity including upregulation of genes that promote apoptosis (FFIT3, BBC3, JUN), cell cycle arrest (CDKN1A, CDKN2B), growth inhibition (TP53, GADD45A), and metastasis suppression (NDRG1); whereas other genes, such as those involved with promoting cell cycle progression, DNA replication, and cell proliferation were downregulated, including BIRC5, CCNA2, CCNB1, CCNB2, CCNE2, CDK1/2/6, GINS2, HELLS, RRM2, SKP2, PLK1, MCM2, MCM3, MCM7, MCM10. These gene expression changes revealed that using microarray technology further supported the anticancer potential of the nutraceutical *M. ptelefolia*.

In addition to antitumor activity, some nutraceuticals have been shown to affect the immune system either by enhancing immune activity or suppressing inflammatory immune responses. Egg yolk phosvitin is a highly phosphorylated glycoprotein in egg yolk and comprises approximately 7% of the total yolk proteins (Jung et al., 2012). It is a nutraceutical that has been touted to have strong metal-chelating ability and immune-enhancing potential (Lee et al., 2017). qRT-PCR was used to determine the effect of phosvitin on proinflammatory mediators in RAW 264.7 macrophages (Lee et al., 2017). The results from this study revealed that phosvitin significantly increased the mRNA expression for inducible nitric oxide synthase, TNF-α, and IL-1β at approximately 46, 9, and 85 times more (respectively) when compared to controls. This demonstrated that phosvitin enhanced the immune functions of RAW 264.7 macrophages through increasing production of immune mediators and phagocytic activity and provided evidence that phosvitin could serve as an immune-boosting nutraceutical.

Nutraceuticals that suppress the immune system can also be beneficial to human health. For example, *Lactobacillus casei* (*L. casei*) is a nutraceutical that has been shown to suppress the inflammatory immune responses in models of rheumatoid arthritis (So et al., 2008). In one study, qRT-PCR was used to evaluate the expression of inflammatory biomolecules in an experimental rat model for osteoarthritis given oral doses of *L. casei* alone versus *L. casei* given in combination with type II collagen and glucosamine (So et al., 2011). The hallmark findings in this study revealed that the coadministration of *L. casei*, type II collagen, and glucosamine decreased expression of proinflammatory cytokines (IL-β, IL-2, IL-6, IL-12, IL-17), TNF-α, metalloproteinases (MMP1, MMP3 and MMP13), and IFN-γ and also upregulated expression of antiinflammatory cytokines IL-4 and IL-10. Overall, this study concluded that the combined administration of *L. casei* with type II collagen and glucosamine was more effective in reducing pain, lymphocyte filtration, and cartilage destruction.

Metabolic syndromes

Undaria pinnatifida, also known as wakame, is an edible brown seaweed that is rich in nutrients and is a common staple in the Japanese diet (Yoshinaga et al., 2018).

Wakame consumption has been shown to have antihypertensive effects and improve glucose and lipid metabolism; however, the molecular mechanisms involved were not been entirely understood (Yoshinago et al., 2018). In order to characterize the underlying mechanisms involved, DNA microarray technology was used to evaluate hepatic gene expression changes in Sprague-Dawley rats fed experimental diets containing dried wakame powder for 28 days (Yoshinago et al., 2018). Approximately 22 genes were differentially expressed that were related to energy metabolism in the rat liver after consumption of a diet containing 1% wakame. Of this subset of genes, there were two genes that changed significantly in this study. Consumption of 1% wakame in the diet caused downregulated SREBF-1 (sterol regulatory element binding protein-1), thereby suppressing the lipogenic pathway. Conversely, 1% wakame in the diet promoted upregulation of the PPAR (peroxisome proliferator-activated receptor) signaling pathway leading to reduced accumulation of cholesterol, increased biosynthesis of bile acid, and promotion of gluconeogenesis. Using microarray technology, the investigators were able to show that wakame ingestion has the potential to affect lipid and glucose metabolism by altering the expression of PPAR and SREBF-1, implicating wakame as a potential nutraceutical to prevent metabolic syndromes in humans.

Determining nutraceutical quality

Some examples of how RNA-Seq, a genomics technique, has been used in the field of nutrigenomics involve identifying transcriptome alterations in order to ascertain nutraceutical quality and characteristics. In one study, de novo characterization of the *Lycium ruthenicum* transcriptome and DGE profiles were analyzed during different stages of fruit development and ripening (Peng et al., 2017). *Ly. ruthenicum* is a perennial, thorny shrub that produces black, goji berries and has been shown to be an important nutraceutical for the treatment and prevention of a wide range of diseases (Wang et al., 2017). These investigators compared gene expression profiles of fruits from this plant at different ripening stages in order to characterize the transcriptome responsible for the biosynthesis of anthocyanins, which are flavonoids responsible for the plant's nutraceutical qualities (Yong et al., 2017). There were a total of 458 DEGs related to flavonoid biosynthesis, phenylpropanoid biosynthesis, and starch and sucrose metabolism detected at the ripening stages of the fruit. They found that almost all of the regulatory genes involved in the anthocyanin pathway were upregulated during the fully ripe stage and that genes responsible for starch and sucrose metabolism decreased during fruit development. Overall, this study was able to compare gene expression changes at different stages of fruit development and was able to highlight at what stage during the ripening process the optimal, nutraceutical quality was provided.

In the nutraceutical industry, DNA barcoding can be used to authenticate raw natural health product botanical ingredients (Faller et al., 2019). In order to determine if botanical processing could affect DNA integrity and interfere with DNA barcoding of product, qRT-PCR was used as a tool to determine the DNA quality and quantity in *C. sinensis* (green tea) plants. This study investigated whether DNA would degrade or be variable when comparing green tea being processed at a Chinese tea farm versus that being processed at a natural health product facility (Faller et al., 2019). The DNA from the end-production of green tea was sequenced in order to authenticate the green tea as the "botanical of interest." The study revealed that there was a 41.1% decrease in extractable DNA from the Chinese tea farm processing facility and a 99.7% decrease in extractable DNA from the natural health product processing facility. However, the researchers noted that there was sufficient intact DNA at the end of processing from each method to sequence and authenticate.

Concluding remarks and future directions

Integrating high-throughput genomics and proteomics technologies in nutraceutical research has provided more comprehensive platforms for understanding the complex interrelationships between nutraceuticals and disease. Although most of the "omics" research conducted has focused on understanding the molecular mechanisms of action of nutraceuticals as they relate to their health-promoting benefits, these technologies are gaining ground with respect to elucidating the safety, toxicity, and quality of nutraceuticals. Some challenges that may be encountered in future "omics" studies in nutraceutical research may include the recognition that host responses to nutraceuticals may vary with the individual. Therefore investigating the safety, efficacy, and toxicity of some of these nutrient-rich products may be difficult to standardize, but may also provide opportunities for personalized medicine.

References

Abenavoli, L., Izzo, A.A., Milic, N., et al., 2018. Milk thistle (*Dilybum marianum*): a concise review on its chemistry, pharmacological, and nutraceutical uses in liver diseases. Phytother. Res. 32 (11), 2202–2213.

Agrawal, A., Koopstein, D., Rozowsky, J., et al., 2010. Comparison and calibration of transcriptome data from RNA-Seq and tilling arrays. BMS Genomics 11 (1), 383. Available from: https://doi.org/10.1186/1471-2164-11-383.

Ahmed, M., Baba, W.N., Shah, U., et al., 2014. Nutraceutical properties of the green tea polyphenols. J. Food Proc. Technol. 5, 11. Available from: https://doi.org/10.4172/2157-7110.1000390.

Anderson, N.L., Anderson, N.G., 2002. The human plasma proteome. Mol. Cell Proteom. 1, 311–326.

Baker, D.H., 2008. Animal models in nutrition research. Symposium: animal models in nutrition research. J. Nutr. 138, 391–396.

Botham, K.M., 2015. Cellular models in nutrition research. In: Lovegrove, J.A., Hodson, L., Sharma, S., Lanham-New, S.A. (Eds.), Nutrition Research Methodologies. John Wiley & Sons, Chichester, West Sussex, pp. 278–306.

Calvani, M., Pasha, A., Favre, C., 2020. Nutraceutical boom in cancer: inside the labyrinth of reactive oxygen species. Mol. Sci. 21 (6), 1936. Available from: https://doi.org/10.3390/ijms21061936. pii: E1936.

Campo, J.V., McNabb, J., Perel, J.M., et al., 2002. Kava-induced fulminant hepatic failure. J. Am. Acad. Child Adolesc. Psychiatry 41, 631–632.

Carey, C.M., Kirk, J.L., Ojha, S., et al., 2007. Current and future uses of real-time polymerase chain reaction and microarrays in the study of intestinal microbiota and probiotic used and effectiveness. Can. J. Microbiol. 53, 537–550.

Cho, Y.S., Schiller, N.L., Kahng, H.Y., et al., 2007. Cellular responses and proteomic analysis of *Escherichia coli* exposed to green tea polyphenols. Curr. Microbiol. 55 (6), 501–506.

Cokus, S.J., Feng, S., Zhang, X., et al., 2008. Shotgun bisulphite sequencing of the *Arabidopsis* genome reveals DNA methylation patterning. Nature 452 (7184), 215–219.

Coskun, O., 2016. Separation techniques: chromatography. North Clin. Istanb. 3 (2), 156–160.

Costa-Silva, J., Domingues, D., Lopes, F.M., 2017. RNA-Seq differential expression analysis: an extended review and a software tool. PLoS One 12 (12), e0190152. Available from: https://doi.org/10.1371/journal.pone.01090152.

Daniel, H., 2002. Genomics and proteomics: importance for the future of nutrition research. Br. J. Nutr. 87, S305–S311.

DeQiang, H., YuFeng, G., Ning, B., et al., 2019. Isoliquiritigenin inhibits proliferation and induces apoptosis in tongue squamous cell carcinoma by modulating miR21/FOXG1 pathway. Curr. Top. Nutraceut. Sci. 17 (4), 463–469.

Faller, A.C., Subramanyam, R., Shanmughanandhan, D., et al., 2019. DNA quality and quantity analysis of *Camellia sinensis* through processing from fresh leaves to a green tea extract. J. AOAC Int 102 (6), 1798–1807.

Farag, M.R., Alagawany, M., El-hack, M.E.A., et al., 2016. Nutritional and healthical aspects of Spirulina (*Arthrospira*) for poultry, animals and human. Intl J. Pharmacol. 12 (1), 36–51.

Goshe, M.B., Smith, R.D., 2003. Stable isotope-coded proteomic mass spectrometry. Curr. Opin. Biotechnol. 14, 101–109.

Guo, L., Mei, N., Liao, W., et al., 2010a. *Ginkgo biloba* extract induces gene expression changes in xenobiotics metabolism and the myc-centered network. OMICS 14, 75–90.

Guo, L., Shi, Q., Dial, S., Xia, Q., Mei, N., Li, Q., et al., 2010b. Gene expression profiling in male B6C3F1 mouse livers exposed to kava identifies changes in drug metabolizing genes and potential mechanisms linked to kava toxicity. Food Chem. Toxicolol 48, 686–696.

Hoffman, S.A., Joo, W.A., Echan, L.A., et al., 2007. Higher dimensional (Hi-D) separation strategies dramatically improve the potential for cancer biomarker detection in serum and plasma. J. Chromatogr. B: Anal. Technol. Biomed. Life Sci. 849, 43–52.

International Human Genome Sequencing Consortium, 2004. Finishing the euchromatic sequence of the human genome. Nature 431 (7011), 931–945.

Issaq, H.J., Chan, K.C., Janini, G.M., et al., 2005. Multidimensional separation of peptides for effective proteomic analysis. J. Chromatogr. B 817, 35–47.

Jung, S., Kim, H.D., Son, J.H., et al., 2012. The functional property of egg yolk phosvitin as a melanogenesis inhibitor. Food Chem. 135 (3), 993–998.

Kabir, M.F., Mohd, A.J., Abolmaesoomi, M., et al., 2017. *Melicope ptelefolia* leaf extracts exhibit antioxidant activity and exert anti-proliferative effects with apoptosis induction on four different cancer cell lines. BMC Complem. Altern. Med. 17 (1), 1761–1769.

Kabir, M.F., Mohd, A.J., Hashim, O.H., 2018. Microarray gene expression profiling in colorectal (HCT116) and hepatocellular (HepG2) carcinoma cell lines treated with *Melicope ptelefolia* leaf extract reveals trancriptome profiles exhibiting anti-cancer activity. Peer J. 6, e5203. Available from: https://doi.org/10.7717/peerj.5203.

Khelurkar, V.C., Ingle, K.P., Padole, D., 2017. DNA microarray: basic principle and its applications. Trends Biosci. 10 (2), 488–490.

Koh, P., 2011. *Gingko biloba* extract (EGb 761) attenuates the focal cerebral ischemic injury-induced decrease in astrocytic phosphoprotein PEA-15 levels. Am. J. Chin. Med. 39 (5), 971–979.

Koh, P., 2012. *Gingko biloba* extract (EGb 761) attenuates ischemic brain injury-induced reduction in Ca^{+2} sensor protein hippocalcin. Lab. Anim. Res. 28 (3), 199–204.

Korf, I., 2013. Genomics: the state of the art in RNA-Seq analysis. Nat. Methods 10 (12), 1165–1166.

Lee, J.H., Moon, S.H., Kim, H.S., et al., 2017. Immune-enhancing activity of phosvitin by stimulating the production of pro-inflammatory mediator. Poult. Sci. 96 (11), 3872–3878.

Lu, Q.Y., Jin, Y.S., Zhang, Z.F., et al., 2007. Green tea induces annexin-I expression in human lung adenocarcinoma A549 cells: involvement of annexin-I in actin remodeling. Lab. Invest. 87 (5), 456–465.

Lu, Q.Y., Yang, Y.A., Jin, Y.S., et al., 2009. Effects of green tea extract on lung cancer A549 cells: proteomic identification of proteins associated with cell migration. Proteomics 9 (3), 757–767.

Lupo, M.G., Biancorosso, N., Brilli, E., et al., 2020. Cholesterol-lowering action of a novel nutraceutical combination in uremic rats: insights into the molecular mechanism in a hepatoma cell line. Nutrients 12 (2), 436. Available from: https://doi.org/10.3390/nu12020436.

Mahadevan, S., Park, Y., 2008. Multifaceted therapeutic benefits of *Ginkgo biloba* L.: chemistry, efficacy, safety, and uses. J. Food Sci. 73, R14–R19.

Meleady, P., 2018. Two-dimensional gel electrophoresis and 2-D DIGE. Methods Mol. Biol. 1664, 3–14.

Minden, J.S., 2012. Two-dimensional difference gel electrophoresis (2D DIGE). Methods Cell Biol. 112, 111–141.

Montazavi, A., Williams, B.A., McCue, K., et al., 2008. Mapping and quantifying mammalian transcriptomes by RNA-Seq. Nat. Methods 5 (7), 621–628.

Opiteck, G.J., Lewis, K.C., Jorgenson, R.J., et al., 1997. Comprehensive on-line LC/LC/MS of proteins. Anal. Chem. 69 (8), 1518–1524.

Oshlack, A., Robinson, M.D., Young, M.D., 2010. From RNA-Seq reads to differential expression results. Genome Biol. 11 (12), 220. Available from: https://doi.org/10.1186/gb-2010-11-12-220.

Peng, Y., Ma, H., Chen, S., 2017. *De novo* characterization of the *Lycium ruthenicum* transcriptome and analysis of its digital gene expression profiles during fruit development and ripening. Arch. Biol. Sci. 69 (1), 181–190.

Raso, A., Biassoni, R., 2014. Twenty years of qPCR: a mature technology? In: Biassoni, R., Raso, A. (Eds.), Quantitative Real-Time PCR, Methods and Protocols. Springer, New York, pp. 1–3.

Peregrin, T., 2001. The new frontier of nutrition science: nutrigenomics. J. Am. Dietet. Assoc. 101 (11), 1306.

Rabilloud, T., 2002. Two-dimensional gel electrophoresis in proteomics: old, old fashioned, but it still climbs up the mountains. Proteomics 2, 3–10.

Rajkumar, A.P., Qvist, P., Lazarus, R., et al., 2015. Experimental validation of methods for differential gene expression analysis and sample pooling in RNA-seq. BMC Genomics 16 (1), 548. Available from: https://doi.org/10.1186/s12864-015-1767-y.

Ruscica, M., Pavanello, C., Gandini, S., et al., 2019. Nutraceutical approach for the management of cardiovascular risk—a combination containing the probiotic *Bifidobacterium longum* BB536 and red yeast extract: results from a randomized, double-blind, placebo-controlled study. Nutr. J. 18, 13. Available from: https://doi.org/10.1186/s12937-019-0438-2.

Saeed, M., Naveed, M., Bibi, J., et al., 2019. Potential nutraceutical and food additive properties and risks of coffee: a comprehensive review. Crit. Rev. Food Sci. Nutr. 59 (20), 3239–3319.

Saraswathy, N., Ramalingam, P., 2011a. Two-dimensional gel electrophoresis of proteins. In: Saraswathy, N., Ramalingam, P. (Eds.), Concepts and Techniques in Genomics and Proteomics. Woodhead Publishing Series Biomedicine, Sawston, Cambridge, UK, pp. 159–170.

Saraswathy, N., Ramalingam, P., 2011b. Protein identification by peptide mass fingerprinting (PMF). In: Saraswathy, N., Ramalingam, P. (Eds.), Concepts and Techniques in Genomics and Proteomics. Woodhead Publishing Series Biomedicine, Sawston, Cambridge, UK, pp. 185–192.

Sales, N.M.R., Pelegrini, P.B., Goerch, M.C., 2014. Nutrigenomics: definitions and advances of this new science. J. Nutr. Metab. 2014, 202759. Available from: https://doi.org/10.1155/2014/202759.

Schmittgen, T.D., Lee, E.J., Jiang, J., 2008. High-throughput real-time PCR. In: Marx, A., Seitz, O. (Eds.), Molecular Beacons: Signaling Nucleic Acid Probes, Methods and Protocols, Methods in Molecular Biology, Vol. 429. Humana Press Inc, Totowa, NJ, pp. 89–98.

Shahali, Y., Senechal, H., Poncet, P., 2019. Double one-dimensional electrophoresis (D1-DE) adapted for immunoproteomics. Methods Mol. Biol. 1871, 133–141.

Sheflin, A., Kirkwood, J.S., Wolfe, L.M., et al., 2019. High-throughput quantitative analysis of phytohormones in sorghum leaf and root tissue by ultra-performance liquid chromatography-mass spectrometry. Anal. Bioanal. Chem. 411 (19), 4839–4848.

Shi, Y., Xiang, R., Horvath, C., et al., 2004. The role of liquid chromatography in proteomics. J. Chromatogr. A 1053, 27–36.

Singh, C., Roy-Chowdhuri, S., 2016. Quantitative real-time PCR: recent advances. Clin. Appl. PCR 1392, 161–176.

Smith, K., Leira, C., 2019. The effectiveness and safety of Kava for treating anxiety symptoms: a systematic review and analsysis of randomized clinical trials. Complem. Ther. Clin. Pract. 33, 107–117.

So, J.S., Lee, C.G., Kwon, H.K., et al., 2008. *Lactobacillus casei* potentiates induction of oral tolerance in experimental arthritis. Mol. Immunol. 46 (1), 172–180.

So, J.S., Song, M.K., Kwon, H.K., et al., 2011. *Lactobacillus casei* enhances type II collagen/glucosamine-mediated suppression of inflammatory responses in experimental osteoarthritis. Life Sci. 88 (7/8), 358–366.

Sun, G.Y., Simonyi, A., Fritsche, K.L., et al., 2018. Docosahexanoic acid (DHA): an essential nutrient and a nutraceutical for brain health and diseases. Prostaglandids Leukot. Essent. Fat. Acids 136, 3–16.

Wang, Z., Gerstein, M., Snyder, M., 2009. RNA-Seq: a revolutionary tool for transcriptomics. Nat. Rev. Genet. 10 (1), 57–63.

Wang, H., Li, J., Tao, W., et al., 2017. *Lycium ruthenicum* studies: molecular biology, phyochemistry and pharmacology. Food Chem. 240, 759–766.

Xiang, G., JingWen, L., Li, L., et al., 2020. A brief review of nutraceutical ingredients in gastrointestinal disorders: evidence and suggestions. Int. J. Mol. Sci. 21 (5), 1822. Available from: https://doi.org/10.3390/ijms21051822.

Xiao, G.S., Jin, Y.S., Lu, Q.Y., et al., 2007. Annexin-I as a potential target for green tea extract induced actin remodeling. Intl J. Cancer 120 (1), 111–120.

Yong, Peng, HuiQin, Ma, ShangWu, Chen, et al., 2017. *De novo* characterization of the *Lycium ruthenicum* transcriptome and analysis of its digital gene expression profiles during fruit development and ripening. Archives of Biological Sciences 69 (1), 181–190.

Yoshinaga, K., Nakai, Y., Izumi, H., et al., 2018. Oral administration of edible seaweed *Undaria pinnatifida* (wakame) modifies glucose and lipid metabolism in rats: a DNA microarray analysis. Mol. Nutr. Food Res. 62, e1700828. Available from: https://doi.org/10.1002/mnfr.201700828.

Zhang, X., You, L., Wang, W., et al., 2015. Novel omics technologies in food nutrition. In: Bagchi, D., Swaroop, A., Bagchi, M. (Eds.), Genomics, Proteomics and Metabolomics in Nutraceuticals and Functional Foods, 2nd edition John Wiley & Sons, Chichester, West Sussex, pp. 1–12.

Zieske, L.R., 2006. A perspective on the use of iTRAQ™ reagent technology for protein complex and profiling studies. J. Exp. Bot. 57 (7), 1501–1508.

Chapter 22

Transcriptomic profiling for safety and toxicity evaluation of nutraceuticals

Nan Mei, Tao Chen, Baitang Ning and Lei Guo

Introduction

In 1994 the US Congress passed the Dietary Supplement Health and Education Act. Based on this act, dietary supplements are classified as foods, not drugs, allowing them to be sold without proof of effectiveness or safety (Public Law, 1994). Since then the use of dietary supplements has greatly increased in popularity in the United States with increased numbers of dietary supplement products from 4000 in 1994 to more than 85,000 currently (Dwyer et al., 2018). It has been estimated that more than 18% of Americans use dietary supplements (Clarke et al., 2015), with out-of-pocket expenditures of $30 billion (Nahin et al., 2016). Nutraceuticals, defined as a fortified food or dietary supplement, have medical or health benefits, but are not essential to the diet (Aronson, 2017). Therefore some herbal supplements are also called nutraceuticals, a term commonly used in marketing, but having no regulatory definition. As the global use and commercial success of dietary supplements (or nutraceuticals) have grown, so have concerns about public health risks regarding the concentration, composition, and individual contaminants of dietary supplements, as well as their interactions with drugs or other dietary supplements (Petroczi et al., 2011; Ronis et al., 2018). It has been reported that about 23,000 emergency room visits in the United States every year are attributable to adverse events related to dietary supplements (Geller et al., 2015), and a majority of dietary supplement-related adverse events ($\sim$50,000 per year as estimated) are believed to remain unreported (Cohen, 2009). Since little has been done to determine the potential risks associated with prolonged or high-dose use of dietary supplements, toxicological profiles of many dietary supplements that are currently on the market need to be scientifically evaluated (Fu et al., 2009).

Dietary supplements can be found in many forms, including capsules, gel caps, liquids, powders, soft gels, and tablets. The US Food and Drug Administration (FDA) can take action against the sale of any unsafe dietary supplement after it reaches the market (Public Law, 1994). The FDA issued more than 200 recalls between 2004 and 2012 for dietary supplements containing unapproved substances or impurities (Geller et al., 2015). To assess the potential risks of dietary supplements, the FDA and other government agencies have also nominated several dietary supplements and their active ingredients to the US National Toxicology Program (NTP) for toxicological evaluation. In 1998, the NTP held an international workshop to evaluate research needs on the use and safety of medicinal herbs (Matthews et al., 1999). Subsequently, the NTP initiated a number of studies to evaluate the toxicity of various dietary supplements and their constituents. The NTP studies are mainly conducted in mice and rats with multiple doses for 2 years to draw conclusions regarding carcinogenic activity based on a four-level scale (clear, some, equivocal, and no evidence) (NTP, 2020a). Several completed studies have indicated that *Aloe vera*, *Ginkgo biloba*, goldenseal, and kava show different levels of evidence of carcinogenic activity in rodents, while ginseng, green tea extract, milk thistle, and *Senna* show no evidence (NTP, 2016). Studies with black cohosh, dong quai, *Echinacea purpurea*, *Garcinia cambogia*, gum guggul, *Usnea* lichen, valerian root, and vinpocetine are ongoing (NTP, 2020b).

Traditionally, toxicology studies employ various in vivo and in vitro models to investigate the potential adverse effects of chemicals or mixtures. Since the introduction of DNA microarray technology in the mid-1990s, toxicogenomics has become an important subdiscipline in the field of toxicology (Nuwaysir et al., 1999), resulting from the natural convergence of the field of conventional toxicological research and the emergent field of functional genomics (Mei et al., 2010a). Technology evolution has made toxicogenomics approaches increasingly

Nutraceuticals. DOI: https://doi.org/10.1016/B978-0-12-821038-3.00022-7

powerful and cost effective, and has also extended the impact and reach of toxicogenomics (Liu et al., 2019). Microarray-based gene expression profiling has been identified as a powerful approach to address biological questions and is extensively used in toxicological research. Differentially expressed genes (DEGs) may be used to elucidate molecular mechanisms that can provide an improved understanding of toxicity and create gene signatures that can identify classes of chemicals (Gatzidou et al., 2007). Due to concern regarding the reliability and utility of microarray technology, the FDA-led, community-wide MicroArray Quality Control (MAQC) project provided evidence for intraplatform consistency across test sites and a high level of interplatform concordance in terms of DEGs (Shi et al., 2006). To validate and extend the findings of this study, a rat toxicogenomic study of a dietary supplement (comfrey) and two herbal constituents [riddelliine and aristolochic acid (AA)] was also conducted, which showed high concordance in intersite and cross-platform comparisons (Guo et al., 2006b). The MAQC consortium also suggested that a straightforward approach of fold change selection plus a nonstringent *P* cutoff is useful in identifying reproducible gene lists (Guo et al., 2006b; Shi et al., 2006).

In 2007, next-generation sequencing (NGS) technologies were developed, resulting in a diversity of genomics applications, including whole-genome sequencing, chromatin immunoprecipitation sequencing, and transcriptome sequencing (RNA-seq) (Chi, 2008). A comprehensive assessment of RNA-seq accuracy, reproducibility, and information content was conducted by the Sequencing Quality Control project, which was coordinated by the FDA (SEQC et al., 2014). The major difference between RNA-seq and microarrays is their sensitivity for genes with low expression. Furthermore, RNA-seq has advantages over microarrays (Wang et al., 2014). In a comprehensive study of rat RNA-seq transcriptomic BodyMap data across multiple organs and developmental stages, a large number of genes show organ-specific, age-dependent, or sex-specific differential expression patterns (Yu et al., 2014). RNA-seq has been found to have good overall agreement with other established technologies, such as microarray and qPCR for differential expression analysis; and it has been increasingly used in toxicogenomic studies.

Transcription is believed to be an intermediate step in response to xenobiotic exposure and subsequently represents a possible precursor to the ultimate phenotypic outcome. Therefore the identification of unique gene expression patterns produced by a chemical or dietary supplement may facilitate the elucidation of the mechanisms of action leading to toxicity. During the last decade we have performed several toxicogenomic studies to evaluate a number of herbal dietary supplements and their constituents (e.g., comfrey, kava, *G. biloba*, riddelliine, AA, and usnic acid) (Chen et al., 2006b, 2017b; Guo et al., 2007, 2009, 2010a; Mei et al., 2006b, 2007), drugs (e.g., pioglitazone, rosiglitazone, and troglitazone) (Guo et al., 2006a,c), and chemicals in food (e.g., acrylamide) (Mei et al., 2008) for safety and toxicity. In this chapter we provide examples of the application of transcriptomic profiling in evaluating the safety and toxicity of dietary supplements.

Ginkgo biloba

Therapeutic use of *Ginkgo biloba*

G. biloba (ginkgo) is one of the world's oldest living tree species that can live for over 1000 years, and ginkgo trees have become popular around the world as ornamental trees in city streets, parks, and gardens (IARC, 2016; Mei et al., 2017; Wang et al., 2020). Ginkgo seeds have been used as a source of traditional Chinese medicine for many thousands of years against cough, asthma, enuresis, alcohol misuse, pyogenic skin infections, and worm infections in the intestinal tract (van Beek and Montoro, 2009). The seeds are also occasionally cooked and eaten in restaurants. Ginkgo leaves have also been adopted for medicinal uses in treating memory loss, cognitive disorders, arrhythmias, ischemic heart disease, cancer, diabetes, and thrombosis (EGb, 2003), and recently several clinical studies have been conducted with mixed results. Ginkgo extracts in different forms can be purchased in many countries as dietary supplements, and the total worldwide sales of ginkgo products were $1.26 billion U.S. dollars in 2012 (IARC, 2016). According to the report of 2018 herbal supplement sales in the United States, ginkgo was one of the best-selling products with $16 million in mainstream multi-outlet (or mass market) stores and $4.6 million in nature and health food stores (Smith et al., 2019).

Toxicity of *Ginkgo biloba*

In 1998, *G. biloba* leaf extract was nominated by the National Cancer Institute to the NTP for a 2-year chronic carcinogenicity bioassay, because (1) ginkgo is a well-defined product and it or its active ingredients have demonstrated biological activities, (2) it can be consumed in rather large doses for an extended period of time, and (3) some of its components are mutagens (NTP, 1998). Although no clinical findings were observed during the 2-year gavage study, after 18 months of ginkgo administration the body weights of mice in the middle- and high-dose groups decreased gradually by about 17% and 25%, respectively, compared to the control group (Guo et al., 2010a). The NTP technical report concluded there was clear evidence of carcinogenic activity of *G. biloba* leaf extract in male and female B6C3F1/N mice based on increased incidences of hepatocellular carcinoma (HCC) and hepatoblastoma, and there was some evidence of carcinogenic activity of *G. biloba* leaf extract in male and female F344/N rats based on increased incidences of thyroid gland follicular cell neoplasms (NTP, 2013). Using a bacterial mutagenicity test, *G. biloba* extract was demonstrated to induce frameshift and base substitution mutations in *Salmonella typhimurium* strains TA98 and TA100 and in *Escherichia coli* strain WP2 *uvrA*/pKM101, with and without metabolic activation (NTP, 2013). In addition, the mutagenic activities induced by two ginkgo constituents, quercetin and kaempferol, were documented using the Ames test, chromosomal aberration analysis, micronucleus assay, and the mouse lymphoma assay (Lin et al., 2014). The ginkgo leaf extract used in the NTP study, one commercial ginkgo product, and three major flavonoids (quercetin, kaempferol, and isorhamnetin) also induced DNA damage in cultured human hepatic cells (Zhang et al., 2015).

Gene expression profiling of *Ginkgo biloba*

It is a challenge to determine the mechanisms of toxicity induced by mixtures containing many chemical components, such as herbal dietary supplements and other non-nutraceutical public health concerns such as cigarette smoke. Therefore we first evaluated *G. biloba* leaf extract in male B6C3F1 mice and utilized microarrays to analyze liver gene expression changes and determine affected pathways and mechanisms associated with ginkgo treatments (Guo et al., 2010a). Male B6C3F1 mice (8–9 weeks old) were administered *G. biloba* leaf extract in corn oil by oral gavage at 0, 200, 600, or 2000 mg/kg body weight, 5 days a week for 104 weeks. After all treatments a section of the left lateral lobe of the liver was collected randomly from four animals in each group (Guo et al., 2010a). A total of 16 RNA samples were isolated and three technical replicate arrays were performed for each RNA sample. Thus, 48 microarrays were used in this study. It is worth noting that in order to reduce research costs, technical replicate arrays are not used in most labs. Hierarchical cluster analysis (HCA) showed that the three technical replicates for the same RNA sample were tightly grouped, showing close similarity (i.e., high reproducibility) (Fig. 22.1). Ginkgo-treated samples were well separated from the control samples into high-dose, middle-dose, and low-dose groups. Criteria for differential expression were set as a twofold change in gene expression compared with controls and a P-value less than 0.01. As a result, totals of 129, 289, and 2011 DEGs were identified from the 200, 600, and 2000 mg/kg treatment groups, respectively (Guo et al., 2010a).

Hepatotoxicity has been reported as one of the major adverse effects caused by herbal dietary supplements.

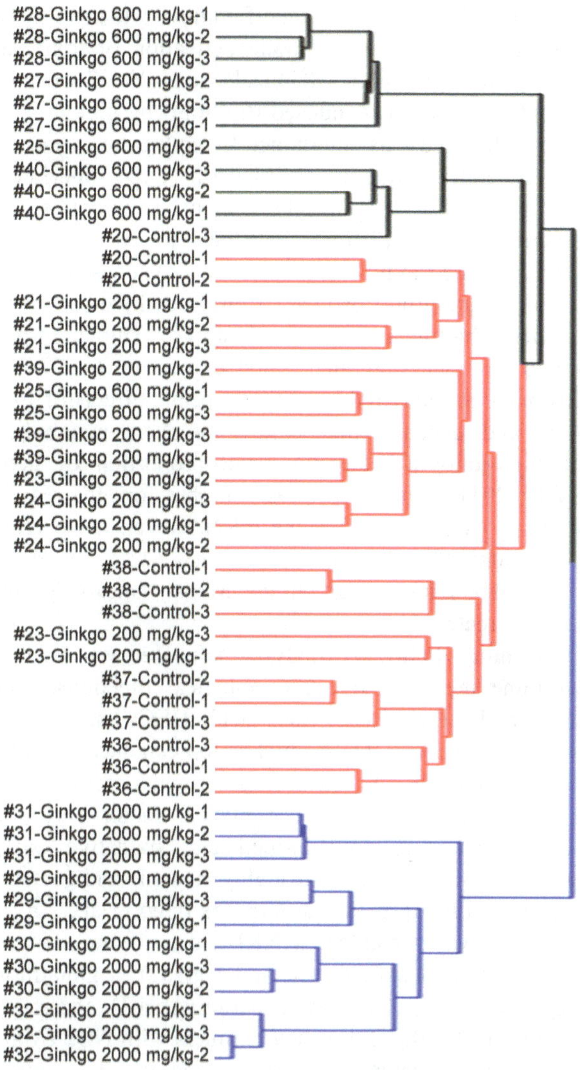

FIGURE 22.1 Hierarchical cluster analysis (HCA) of gene expression profiles from the livers of control and ginkgo-treated mice. The $\log_2$ intensity data of the entire gene set from all 48 microarrays (16 RNA samples, each with three technical replicates) were hierarchically clustered using the distance metric of 1-*R* and average linkage. The first cluster consists of samples from the control, low- (200 mg/kg) and middle-dose (600 mg/kg) ginkgo treatment groups, and the three groups (control and low-dose groups in red and middle-dose group in black) are separated well with few exceptions. The second cluster (in blue) including the samples from the high-dose (2000 mg/kg) ginkgo treatment group is clearly separated from other treatment groups.

Alterations of drug-metabolizing genes are suspected to contribute to the chemical-induced liver toxicity. After ginkgo treatments, alterations in the expression of a large number of drug-metabolizing genes were observed. For example, 2000 mg/kg ginkgo treatment significantly altered the expression of 68 drug-metabolizing genes, including 33 phase I metabolizing enzymes (~21%), 18 phase II metabolizing enzymes (~21%), and 17 phase III transporters (~24%) (Guo et al., 2010a). Nineteen of the 33 differentially expressed phase I genes belong to the

Cyp1, Cyp2, and Cyp3 superfamilies, which play an important role in the metabolic activation of toxic and carcinogenic xenobiotics (Gonzalez and Yu, 2006). For example, Cyp1a1 was induced (6.8-fold) by ginkgo treatment in this study, as determined by both gene expression profiling and TaqMan assays. Since the Cyp1 superfamily is important in the metabolism of xenobiotics, the induction of Cyp1a1 may contribute to the toxicity of *G. biloba*. The flavin-containing monooxygenase (Fmo) is another major group of phase I enzymes involved in the oxidative biotransformation of drugs. The high-dose ginkgo treatment inhibited Fmo3 expression by 77-fold. This reduction may reduce the therapeutic effects of some drugs and enhance the toxicity of their metabolites. Although only a relatively small proportion of genes (6.7%, 2011 genes of 29,922 probes on the Mouse OneArray) were altered at the mRNA level, the percentage of altered drug-metabolizing genes (22%, 68 of 313 drug-metabolizing genes on the Mouse OneArray) was notably greater.

Ingenuity Pathway Analysis software was used to determine the most relevant biological functions, pathways, and networks of the 2011 DEGs induced by 2000 mg/kg ginkgo treatment. The top canonical pathway was "metabolism of xenobiotics by cytochrome P450" with 36 genes involved. The nuclear factor erythroid-related factor 2 (NRF2)-mediated oxidative stress response pathway was also highly ranked, and included 23 genes with expression differences (Guo et al., 2010a). In the pathway regulated by NRF2, detoxifying enzymes (Nqo1, Gstms, Gstas, Gstp1, and Ugt1a2), a glutathione homeostasis gene (Gsr), a transporter (Abcb1), and a transcription factor that forms heterodimers with Nrf2 (Maf2) were all upregulated compared to controls. In contrast, PI3k and Sod3 were downregulated, indicating lower phosphorylation levels of Nrf2 and lower superoxide defense, which can increase intracellular stress. The middle- and high-dose ginkgo treatments produced pronounced changes in the expression of Nqo1, resulting in 7.5- and 11-fold induction, respectively (Guo et al., 2010a). Keap1-Nrf2-ARE signaling is involved in the antioxidative response, detoxification of xenobiotics, and proteome maintenance, and thus plays a critical role in protecting cells from endogenous and exogenous stresses (Kensler et al., 2007).

After ginkgo treatment, 87 networks were significantly changed, each of which was associated with specific genes and involved different functions. The network containing the largest number of genes (34 genes) was "cell cycle, cellular movement, and cancer." Centered in this network is the proto-oncogene Myc, which was upregulated by about sixfold. In addition, 21 genes (15 upregulated and 6 downregulated) were directly associated with Myc gene activities in the cell cycle, cell growth/proliferation, and cell death processes (Guo et al., 2010a). Cidec (cell-death-inducing DFFA-like effectors c), which is involved in energy metabolism and apoptosis, displayed a 47-fold induction. Cell division cycle associated 8 (Cdca8) and cell division cycle 45-like (Cdc45L), which have a critical role in DNA replication, were also upregulated. The growth arrest-specific gene, Gas1, which inhibits DNA synthesis as a tumor suppressor gene, was downregulated 20-fold after ginkgo treatment. It has been reported that Myc is overexpressed during hepatocarcinogenesis induced by several factors including a choline-devoid diet, arsenite, 5-diethoxycarbonyl-1,4-dihydrocollidine, and carbon tetrachloride (Beer et al., 2008; Chandar et al., 1989; Chen et al., 2001). Therefore stimulation of Myc pathways is suggested to cause liver damage and even liver tumors.

When the gene expression profiles between ginkgo-induced HCC and spontaneous HCC were compared using the Affymetrix Mouse Genome GeneChip arrays, there was significant separation of the two groups. Although ginkgo-induced HCC were not histologically distinguishable from spontaneous HCC, there were significant differences in global gene expression between the two. Ginkgo-induced HCC indicated overpresentation of genes associated with pathways related to cancer signaling, HCC development, and xenobiotic/oxidative stress (Hoenerhoff et al., 2013). The results from gene expression profiling provided important clues for the study of the mechanisms associated with the putative toxicity of *G. biloba*. Using biochemical and molecular approaches, we evaluated the in vitro genotoxicity of *G. biloba* leaf extract and seven of its major chemical constituents, and demonstrated that the DNA damage and carcinogenicity of ginkgo extract in animals may be related to topoisomerase II inhibition (Zhang et al., 2015).

Aristolochic acid

Therapeutic use of aristolochic acid

AA is an active component of herbal drugs derived from the plants in *Aristolochiaceae* family, which is comprised of more than 450 species primarily in *Aristolochia* and *Asarum* genera (Li et al., 2020; Luciano and Perazella, 2015). AA-containing herbal drugs were used worldwide as antiinflammatory and diuretic agents or for other therapeutic purposes for many years until their toxicity were known. AA can inhibit inflammation induced by immune complexes as well as other immunological and nonimmunological agents. Its activity mainly results from its inhibitory effects on phospholipase A_2 (PLA_2) in regulating the synthesis of arachidonic acid, an intermediate in the production of prostaglandins (Moreno, 1993; Rosenthal et al., 1989). Prostaglandins have a wide variety of effects, such as causing fever, constriction or dilation in vascular smooth muscle cells, aggregation or disaggregation of platelets, regulating inflammation, calcium movement and hormones, and increasing the glomerular filtration rate. AA has been used in the treatment of arthritis, gout, rheumatism, and festering wounds, as well as in obstetrics (IARC, 2002). In addition, it has also been used to treat snake bites because it can interact with PLA_2 found in snake venom and inhibit its edema-inducing activity.

Toxicity of aristolochic acid

Plants containing AA are nephrotoxic, genotoxic, and carcinogenic in humans. In 1992, an outbreak of rapidly progressive renal fibrosis in Belgium involved at least 100 patients who were undergoing a weight-loss regimen that included the use of AA-containing Chinese herbs. About half of these patients required renal replacement therapy (Vanherweghem et al., 1993; Vanherweghem, 1998), while another half developed urothelial malignancies (Cosyns et al., 1994, 1999; Nortier et al., 2000). AA is also related to the etiology of Balkan endemic nephropathy, a chronic tubulointerstitial disease associated with a high frequency of urothelial atypia, occasionally culminating in tumors of the renal pelvis and urethra (Arlt et al., 2007). Specific AA mutation signature (A:T → T:A) in critical genes for carcinogenesis, such as the *TP53* gene, was found in the Belgium patients, Balkan endemic nephropathy, and other cases worldwide, particularly in Taiwan, suggesting AA exposure is a general public health issue (Jelakovic et al., 2019). A ban on AA-containing Chinese herbal products in Taiwan might have reduced the incidence of cancer in the human renal and urinary organs (Jhuang et al., 2019).

In addition to renal and urinary organs, recent studies have shown that AA exposure has been statistically associated with human liver cancers. It was reported that there was significant correlation between consumption of AA-containing herbs and HCC in patients with hepatitis B virus infection (Chen et al., 2018). AA-induced DNA adducts and mutation signature were found in a large subset of human HCCs (Lu et al., 2019; Nault and Letouze, 2019).

AA also induces tumors in rabbits, rats, and mice (Cosyns et al., 2001; Mengs, 1988; Mengs et al., 1982). Rabbits treated with AA develop severe hypocellular interstitial fibrosis, urothelial dysplasia, and tumors of the urinary tract (Cosyns et al., 2001). AA also induces squamous cell carcinomas in the forestomach as well as malignant tumors in the kidney, urinary tract, ovary, and other tissues in mice and rats (Kwak et al., 2014; Mengs, 1988; Mengs et al., 1982). Extensive investigations have been conducted to explore the possible mechanisms underlying the carcinogenicity of AA. It has been found that AA induces DNA adducts in the kidney, ureter, bladder, liver, lung, and spleen of patients with AA nephropathy (Lord et al., 2004; Schmeiser et al., 1996). Four major AA-DNA adducts have been found, among which the dA-AA I adduct is the most mutagenic. AA also induces mutations in target tissues of humans, mice, and rats (Chen et al., 2006a; Kohara et al., 2002; McDaniel et al., 2012; Mei et al., 2006a). AA was positive in many different *S. typhimurium* strains and induced mutations in the *Hprt* and *Tk* genes in cell lines. The major type of mutation found in rodents and humans is A:T → T:A transversions. Mutations have been detected in the *p53* and *ras* genes in vivo and in vitro. AA-induced tumors are associated with activation of the H-*ras* oncogene and inactivation of the *p53* tumor suppressor gene by AA-specific A:T → T:A mutations (Lord et al., 2004; Wang et al., 2011). These evidences suggest a genotoxic mechanism for AA-induced carcinogenesis.

Based on the toxicological evidence, many countries including Australia, Canada, Germany, and the United Kingdom have banned the use of herbs containing AA (Kessler, 2000). In 2001, the FDA issued a Consumer Advisory warning consumers against using dietary supplements and herbs containing AA (FDA, 2001). The International Agency for Research on Cancer has classified herbal remedies containing plant species of the genus *Aristolochia* as human carcinogens (Group 1) (Grosse et al., 2009).

Gene and microRNA expression profiling of aristolochic acid

Gene expression

Gene expression profiling has been used to identify the potential mechanism underlying the toxicity of AA (Arlt et al., 2011; Chen et al., 2006b; Li et al., 2009; Simoes et al., 2008). Chen et al. (2006b) demonstrated gene expression changes in response to AA exposure. Rats treated with AA at carcinogenic doses developed hyperplasia, adenomas, and adenocarcinomas in the kidney. HCA and principal component analysis (PCA) clearly showed that kidney or liver samples of rats exposed to AA grouped together by their expression profiles according to the treatments, suggesting that AA exposure can be identified based on gene expression profiles. A total of 2172 and 2225 DEGs were identified in the kidney and liver, respectively. Most DEGs identified in the kidney, a target tissue of AA, were different from those observed in the liver, a nontumor target tissue in rats. Functional analysis indicated that more DEGs involved in cancer-related pathways were altered by AA treatment in the kidney when compared to the liver (Table 22.1). In addition, fold change ranking showed that there were more DEGs with higher fold changes in kidney than in liver, indicating that

TABLE 22.1 Numbers of differentially expressed genes (> 1.5-fold and $P < .01$) involved in cancer-related pathways altered by aristolochic acid treatment in the kidney and liver of rats.

Functions	Kidney	Liver
Cancer	92	25
Cell death	54	2
Cell-to-cell and signaling and interaction	41	8
Cellular growth and proliferation	37	8
Cell morphology	36	15
Cell cycle	26	13
Tumor morphology	19	8
Cellular development	18	7
Genetic disorder	17	6
Cellular function and maintenance	16	5
Drug metabolism	14	10
DNA replication, recombination, and repair	2	4

Source: Adapted data from Chen, T., et al. (2006) and the same genes may exist in different functions.

AA treatment in rats resulted in greater changes in gene expression in kidney than in liver. This difference might indicate an important role in the differential carcinogenicity and toxicity of AA in kidney and liver.

The large number of DEGs in kidney and liver mainly resulted from different properties of AA in the tissues, including the toxicological effects of this chemical. Functional analysis of these DEGs demonstrated that there were many more DEGs involved in cancer-related pathways in the kidney than in the liver. These genes were involved in functions associated with different stages of carcinogenesis, including AA metabolism, DNA repair, cell cycle, cellular development, cell signaling, cellular growth and proliferation, cell morphology, cell death, and tumor morphology. By comparing the alterations in biological processes and functions in liver, very different biological alterations were found. Most of the altered biological processes predicted by mRNA changes in liver were related to lipid metabolism, including steroid and other lipid metabolism pathways. Most of these genes were upregulated, implicating the pharmacological effects of AA in liver as an inhibitor of PLA_2 that can hydrolyze phospholipids to fatty acids and lysophospholipids. In summary, gene expression profiling has shown that AA treatment can produce significantly different alterations of gene expression in different tissues, which reflects the tissue-specific mechanisms of AA carcinogenicity. There were many more DEGs associated with carcinogenesis in kidney than in liver after AA exposure. Significant changes in biological processes related to defense response, apoptosis, and immune response were found in kidney, while alterations in the expression of genes related to lipid metabolism occurred in liver.

Since the *P53* gene in AA-induced tumors is frequently mutated, a research group conducted a study using $Trp53^{(+/+)}$, $Trp53^{(+/-)}$, and $Trp53^{(-/-)}$ mouse models to investigate effects of P53 on gene expression profiles from the mice treated with AA. The microarray analysis of the mouse kidneys after AA treatment showed that gene expression was differentially altered by the treatment for the different *Trp53* genotype mice. The treatment resulted in changes in apoptotic pathways for the $Trp53^{(+/+)}$ genotype, but alterations in oncogenic and prosurvival pathways for the $Trp53^{(+/-)}$ and $Trp53^{(-/-)}$, respectively. Alterations of biological processes by AA in mouse kidneys suggest the mechanisms by which loss of the P53 can lead to cell proliferation without apoptosis and promotion of cell growth (Sborchia et al., 2019).

MicroRNA expression

MicroRNAs (miRNAs) are a class of small nucleic acids that directly regulate gene expressions. They regulate gene expression by binding to mRNA to inhibit protein translation or destabilize target transcripts (Ambros, 2003). miRNA expression profiles have a predictive value for chemical carcinogenicity (Chen, 2010) and specific miRNAs could represent attractive molecules as informative biomarkers for a specific carcinogen or a type of carcinogens. Our research group conducted a microarray analysis of miRNA expression in tissues from rats treated with AA. Global analysis of miRNA expression for kidney and liver samples of rats treated with AA found that 19 miRNAs were significantly dysregulated in the kidney, with most of the miRNAs related to carcinogenesis (Table 22.2). Only one miRNA, miR-34a (a tumor suppressor), was differentially expressed in the liver. The results indicate that AA treatment of rats produced dysregulation of a large number of miRNAs in the tumor target tissue, and that the upregulation of miR-21 correlated with the carcinogenicity of AA, while the upregulation of miR-34a correlated with its mutagenicity in liver and kidney (Meng et al., 2014).

We also examined effects of AA treatment on the whole process of gene expression including mRNA, miRNA, and protein expressions in kidney samples from the rats treated with 10 mg/kg AA and vehicle control for 12 weeks. The miRNA and mRNA expressions were examined using deep sequencing, and the protein expression was determined using proteomics. AA treatment resulted in significant differential expression of miRNAs, mRNAs, and proteins. Specifically, 63 miRNAs, 6794

TABLE 22.2 Differentially expressed miRNAs (> twofold and $P < .05$) altered by AA treatment in the kidney and liver of rats.

miRNA	Liver	Kidney
rno-miR-21-5p		9.3
rno-miR-34a-5p	5.2	9.1
rno-miR-10b-5p		5.5
rno-miR-182-5p		4.1
rno-miR-30e-3p		3.2
rno-miR-183-5p		2.6
rno-miR-10a-5p		2.6
rno-miR-15b-5p		2.5
rno-miR-206-3p		−2.0
rno-miR-409a-3p		−2.1
rno-miR-140-3p		−2.2
rno-miR-103-3p		−2.2
rno-miR-107-3p		−2.2
rno-miR-192-5p		−2.3
rno-miR-320-3p		−2.7
rno-miR-194-5p		−3.0
rno-miR-122-5p		−3.0
rno-miR-296-5p		−3.0
rno-miR-378a-3p		−3.0

Source: Adapted data from Meng, F., Li, Z., Yan, J., et al., 2014.

mRNAs, and 800 proteins were significantly altered by AA treatment. Ingenuity Pathways Analysis showed that cancer was the top network and disease associated with those dysregulated miRNAs, mRNAs, and proteins, suggesting that AA is a kidney carcinogen. Further analysis of miRNAs' effects on kidney mRNA and protein expression showed that in addition to translational repression and transcriptional destabilization, miRNAs and their target genes were expressed in the same direction at levels of transcription (169) or translation (227). Furthermore, upregulation of 13 oncogenic miRNAs was associated with translational activation of 45 of 54 cancer-related target genes. Thus, these findings suggest that dysregulated miRNA expression in rat kidney plays an important role in AA-induced carcinogenesis (Li et al., 2015).

To search for early biomarkers of AA-induced kidney injury, Pu and his colleagues performed profiling miRNA expression in rat kidney and plasma and found that 21 miRNAs were significantly dysregulated in kidney of rats treated with AA. miR-21-3p was identified as an upregulated miRNA in both the kidney and plasma, but not in nontarget tissues, and increased as early as 2 days after treatment. The elevation of plasma miR-21-3p even preceded the traditional indicators of renal tubular injury, such as increase in blood urea nitrogen and creatinine. The findings suggest that plasma miR-21-3p may be able to serve as a new potential biomarker for early diagnosis of AA-induced kidney injury in rats, and possibly in humans (Pu et al., 2017).

Kava

Therapeutic use of kava

Prepared from the rhizome of a tropical shrub (*Piper methysticum* Forst F.), kava has been used for hundreds of years in beverages for social and recreational events in the South Pacific (IARC, 2016). In Europe, kava has been used as a diuretic, as well as a treatment for gonorrhea and nervous disorders. In recent years kava has been used as a dietary supplement for therapeutic purposes in Europe and the United States to relieve tension, anxiety, stress, and restlessness. Kava-containing products are available on the US market in a variety of forms, such as the crude herb, extracts, and tinctures (Fu et al., 2008). It is believed that kavalactones are the active components that contribute to the pharmacological activities of kava. Standardized kava preparations contain more than 30% kavalactones. The six major kavalactone constituents, including kawain, 7,8-dihydrokawain, 5,6-dehydrokawain, methysticin, dihydromethysticin, and yangonin, have been reported to be responsible for 95% of the total therapeutic activity of kava (Jiang et al., 2007). According to the report of 2018 herbal supplement sales in the United States, kava was the 24th top-selling dietary supplement in nature and health food stores with $3.6 million (Smith et al., 2019). The total sales of kava in the United States was about $20 million during 2005–2011 (IARC, 2016).

Toxicity of kava

Hepatotoxicity associated with the use of kava has been reported (Campo et al., 2002; Clough et al., 2003; Humberston et al., 2003). The use of kava in patients can result in acute liver failure and even death (Pantano et al., 2016). For example, a female Brazilian needed a liver transplantation due to kava-induced acute liver failure after 52-day use of kava at standard dosages (Becker et al., 2019). Clinically, both hepatic necrosis and cholestatic hepatitis have been observed (Stickel et al., 2003). The reports of acute liver failure in humans have prompted regulatory agencies in many countries to take action, from warning consumers to banning kava products from the markets. In 2002, the FDA warned consumers about the potential risk of liver toxicity associated with the intake of kava-containing products and issued a Consumer Advisory (FDA, 2002).

Because of its use as a dietary supplement in the United States and reports of liver toxicity among humans, kava extract was nominated to the NTP for toxicity evaluation. F344/N rats and B6C3F1 mice were administered kava extract by oral gavage for 2 weeks, 3 months, or 2 years; after which, general toxicity and tumorigenicity were evaluated (NTP, 2011). A notable increase in hepatoblastomas in male mice and in HCC in female mice was observed in the 2-year chronic study, therefore the NTP concluded that kava extract caused cancers of the liver in male and female mice (NTP, 2011). Besides liver tumors, increased incidence of testicular tumors in male rats was also reported. In addition, kava also caused increased incidence of lesions in many other tissues, including liver, forestomach, kidney, eye, and pancreas of rats, and liver and forestomach of mice (NTP, 2011).

The mechanism of kava-associated carcinogenicity in rodents is not clear. It is possibly mediated through nongenotoxic mechanisms because kava is not mutagenic based on the results of genotoxicity evaluations using bacterial mutagenicity tests, in vivo micronucleus tests (Behl et al., 2011; NTP, 2011), and mouse lymphoma assays (Whittaker et al., 2008). The activation of drug-metabolizing enzymes may be involved in the carcinogenicity of kava, because toxic metabolites may be accumulated. The activation of genes involved in the oxidative stress pathway in the livers of mice exposed to kava may also contribute to its hepatocarcinogenicity (Guo et al., 2010c).

The mechanism of kava-associated hepatotoxicity has also not been clearly defined. In vitro data suggest that some kava constituents, such as pipermethystine and flavokavain B, may play roles in kava-associated toxicity. The formation of activated metabolites is another proposed mechanism of kava-induced liver toxicity. By profiling the metabolic pathway of kavain (accounting 6% of kava extract) in mouse liver, urine, and feces, 28 kavain metabolites were identified and its reactive metabolites could induce kavain–GSH adducts (Wang et al., 2019). Electrophilic quinone metabolites of kava have been detected by incubating kava extract with liver microsomes; glucuronic acid and sulfate conjugates of some of these metabolites have been detected in urine from a human volunteer who ingested a single dose of a kava-containing dietary supplement (Johnson et al., 2003). The most studied mechanism of kava-related toxicity is herb–drug interaction via the induction or inhibition of drug-metabolizing genes, which in turn influence the metabolism of concomitantly taken drugs, thus potentiating kava-associated toxicity. Herb–drug interaction is widely accepted as a major mechanism for herbal dietary supplement-associated toxicity because of the complex chemical constituents of herbal dietary supplements. In vitro studies using human hepatocytes or microsomes have revealed that kava and its constituents can modulate a number of drug-metabolizing enzymes, including CYP1A1, 1A2, 2C9, 2E1, and 3A4 (Fu et al., 2008; Li et al., 2011; Unger et al., 2002).

Gene expression profiling of kava

The NTP conducted a 3-month subchronic study to characterize potential target tissues for the toxicity of kava extract in rats and mice (NTP, 2011). Groups of 10 male and 10 female Fischer 344 rats and B6C3F1 mice were orally administered kava extract at doses of 0, 0.125, 0.25, 0.5, 1.0, or 2.0 g/kg/day. RNA samples were isolated from rat or mouse livers and gene expression profiles were determined by genome-wide gene expression microarray analysis (Guo et al., 2009, 2010c). Using the criteria of a >1.5-fold change (up or down) and a <.05 *P*-value, 14, 41, 110, 386, and 916 genes in the rat study were identified as DEGs of 22,226 probes (Guo et al., 2009), and 349, 353, 880, 1339, and 1674 genes in the mouse study were identified as DEGs among the 45,281 probed transcripts (Guo et al., 2010c), in the 0.125, 0.25, 0.5, 1.0, and 2.0 g/kg kava-treatment groups, respectively. In both the rat and mouse studies, the numbers of DEGs showed a dose–response relationship in kava-treated animals. Since herb–drug interaction is considered as an important mechanism of kava-associated toxicity, gene expression changes of drug-metabolizing enzymes, including phase I and phase II enzymes and phase III transporters, were investigated in detail.

Gene expression analysis revealed that there were 72 and 95 drug-metabolizing genes significantly altered in the 2.0 g/kg kava-treatment group of rats and mice, respectively. Among the 72 drug-metabolizing genes identified in the rat study, 19, 21, and 32 genes were classified in the phase I, phase II, and phase III metabolism groups, respectively (Guo et al., 2009). Of the 95 drug-metabolizing genes detected in the mouse study, 28, 29, and 38 genes are classified in the phase I, phase II, and phase III metabolism groups (Guo et al., 2010c). There were 20 drug-metabolizing genes that were altered by kava treatment both in mice and rats (Table 22.3) with five genes in phase I, nine in phase II, and six in phase III metabolism groups. Only three of these genes were regulated in the opposite direction between species (e.g., downregulation in rats and upregulation in mice). Although drug metabolism is complicated, these results indicate that, at least in part, both mice and rats share a common transcriptional response of drug metabolism genes after kava treatment.

The general approach for studying drug–drug or herb–drug interactions is to investigate metabolic

TABLE 22.3 Differentially expressed genes (≥1.5-fold and $P < .05$) involved in drug metabolism altered by kava treatment of 2 g/kg in both rat and mouse livers.

	Gene symbol	Fold change	
		Rat[a]	Mouse[b]
Phase I metabolism	Aldh1a1	2.8	2.3
	Cyp1a2	6.4	1.8
	Cyp2c37	2.1	2.8
	Cyp2c55	4.0	16.6
	Cyp4a12	1.5	1.6
Phase II metabolism	Ephx1	3.6	1.7
	Gnmt	−2.1	−3.1
	Gsta2	2.9	24.9
	Gsta4	2.0	1.9
	Gstm1	1.8	4.1
	Gstm2	1.6	3.7
	Gstm3	*−1.6*	*2.9*
	Gstp1	1.6	1.9
	Nqo1	11.4	1.5
Phase III metabolism	Abcc2	1.7	2.0
	Abcc3	15.9	1.5
	Slc2a2	*−1.6*	*1.6*
	Slc6a9	1.5	1.9
	Slc25a17	1.5	1.7
	Slc38a4	*−1.7*	*2.3*

Negative values represent downregulation compared to controls. Fold changes in italics denote genes were regulated in opposite directions between mice and rats.
[a]*Data from Guo et al. (2009).*
[b]*Data from Guo et al. (2010c).*

inhibitory or inductive effects on one or several drug-metabolizing enzymes. However, in whole-genome expression profiling, relative mRNA expression levels for the full spectrum of drug-metabolizing enzymes can be monitored simultaneously. This rapid and efficient approach is useful in predicting potential herb–drug interactions; however, well-designed biochemical experiments are needed to confirm the hypotheses generated from gene expression studies and to elucidate the mechanism. For example, CYP1A1 is involved in metabolizing the polycyclic aromatic hydrocarbon (PAH) class of chemicals to the reactive metabolites responsible for the carcinogenicity of PAHs (Whitlock, 1999). When CYP1A1 activity is elevated, the production of reactive metabolites of PAHs can be increased. Therefore any agent that induces CYP1A1 activity may enhance the toxicity of PAHs, and thus may raise safety concerns. Using microarray analysis, a significant induction of *CYP1A1* gene expression was observed in the livers of rats treated with kava extract (Guo et al., 2009). To expand this work, the underlying mechanism of CYP1A1 induction by kava extract was studied using molecular and biochemical approaches in cultured cells, and the active kavalactones contributing to CYP1A1 induction were also examined (Li et al., 2011). In mouse hepatoma Hepa1c1c7 cells, kava extract and the major kavalactones all showed induction of CYP1A1 at transcriptional, protein, and enzymatic activity levels. Among the six major kavalactones examined, methysticin and 7,8-dihydromethysticin were the most potent CYP1A1 inducers. Interestingly, in silico molecular docking results demonstrated that methysticin and 7,8-dihydromethysticin were the two molecules with the greatest potential to induce CYP1A1. While it is difficult to single out which components in herbal dietary supplements contribute to their biological or toxicological effects, by combining biological observations with in silico predictions, this study has generated a refined hypothesis to be further tested (Li et al., 2011).

Comfrey

Therapeutic use of comfrey

Comfrey belongs to the plant family *Boraginaceae* and there are at least five species referred to as comfrey, including *Symphytum officinale* L. (the major comfrey species), *Symphytum asperum* Lepech, *Symphytum x uplandicum* Nyman, *Symphytum tuberosum* L., and *Symphytum caucasicum* Bieb (Mei et al., 2010b). Comfrey has been used both internally and externally as a herbal medicine for more than 2000 years in many countries (Rode, 2002). Comfrey is applied externally in the form of extracts, ointments, or compress pastes of comfrey leaves and roots to treat joint inflammatory disorders, wounds, bone fractures, gout, hematomas, and thrombophlebitis. Several randomized, controlled trials have demonstrated the efficacy of topical comfrey preparations in the treatment of muscle and joint pain (Staiger, 2012). A herbal practitioner survey of the external use of comfrey in the United Kingdom indicated that comfrey is rated most effective for fractures, wounds, post-surgery healing, and problems with tendon, ligament, and muscle; while it is less effective for treating boils, hemorrhoids, and varicose veins (Frost et al., 2014). Humans have widely consumed comfrey as a vegetable and taken comfrey formulations of tea or tablets. For internal applications, comfrey has been used as infusions and extracts for the treatment of gastritis, gastroduodenal ulcers, and lung congestion (Roeder, 1995).

Toxicity of comfrey

The major hepatotoxic manifestation is hepatic veno-occlusive disease (VOD), both in humans (Bach et al., 1989; Ridker et al., 1985; Weston et al., 1987; Yeong et al., 1990) and in experimental animals (Yeong et al., 1991). A 13-year-old boy who drank comfrey tea containing unknown quantities of comfrey leaves for the treatment of Crohn disease was diagnosed with hepatic VOD by liver biopsy (Weston et al., 1987). A 23-year-old man in New Zealand who ate a vegetarian diet which included steamed young comfrey leaves presented with hepatic VOD and severe portal hypertension, and later died due to liver failure (Yeong et al., 1990). A 47-year-old woman who consumed comfrey tea (≤ 10 cups per day) and comfrey pills for more than 1 year developed ascites 8 years after beginning the consumption of comfrey and dense fibrosis of portal tracts containing proliferating bile ductules (Bach et al., 1989). A 49-year-old woman who consumed comfrey during a 6-month period had portal hypertension associated with obliteration of the smaller hepatic venules (Ridker et al., 1985). A 66-year-old woman with chronic diseases who drank herbal tea containing comfrey (1–1.5 L/day) for months prior to hospitalization was diagnosed with severe pulmonary hypertension because of progressive dyspnea (Gyorik and Stricker, 2009). In addition to essential nutrients, comfrey contains at least nine pyrrolizidine alkaloids (PAs) that are hepatotoxic and carcinogenic in humans and animals. Comfrey roots contain higher concentrations of PAs than comfrey leaves. A person who consumed two capsules per meal for 6 months received 162 or 1740 mg PAs from the herbal preparation of comfrey leaves or roots, respectively (Huxtable et al., 1986).

Gene expression profiling of comfrey

Rats fed diets containing up to 33% (weight/weight) comfrey leaves or 8% comfrey roots developed liver tumors, and comfrey-fed rats also had higher mutant frequencies in the liver with G:C→T:A transversions as the major type of mutation (Mei et al., 2005). In order to determine gene expression changes associated with comfrey exposure, total RNA was isolated from the livers of rats fed 8% comfrey diets and control diets. A microarray containing 26,857 verified rat genes was employed. PCA of the intensities of the whole rat genome expression data revealed a clear separation between controls and comfrey-treated groups, indicating a clear comfrey-treatment effect on liver gene expression (Mei et al., 2006b). Using a twofold change (up or down) in gene expression compared to controls and a P-value less than .01, a total of 2726 genes was selected as DEGs, with about half being upregulated and half being downregulated in response to comfrey treatment.

PAs contained in comfrey need to be metabolically activated to alkylating pyrroles by mixed-function oxidases. The liver is the major organ for biotransformation of xenobiotics and drugs. Comfrey exposure resulted in expression changes of many genes, including phase I, II, and III drug-metabolizing genes (Mei et al., 2006b). For example, Cyp2c12, Cyp7a1, Cyp26, Gsta3, and Abcc3 were increased 6- to 21-fold, whereas Cyp2c, Cyp39a1, Gstm3, Abcc8, and others were reduced in the comfrey-treated livers. The altered expression of phase I and II enzymes as well as phase III transporters could contribute to the increased susceptibility of rats exposed to comfrey.

PAs first target sinusoidal endothelial cells in the liver, which then results in VOD (DeLeve et al., 1999). Gene expression profiling showed that many genes associated with apoptosis, cell death, adhesion, and cell movement of endothelial cells were differentially expressed in comfrey-treated livers (Mei et al., 2006b), including the endothelial cell markers plasminogen activator inhibitor type 1 (PAI-1, also known as Serpine1 and Serpine2), tissue plasminogen activator (Plat), cytokine tumor necrosis factors (Tnfrsf6, Tnfrsf12a, and Tnfsf10), and transforming growth factors (Tgfb1 and Tgfb2). Endothelin 1 (Edn1), urokinase plasminogen activator receptor (Plaur), collagen type IV alpha 2 (Col4a2), matrix metalloproteinase 2 (Mmp2), mitogen-activated protein kinase 9 (Mapk9), and secreted phosphoprotein 1 (Spp1) are involved in the injury of endothelial cells. Endothelial injury is the initiating event in the cascade of events leading to the hepatic changes and clinical manifestation of VOD (Kumar et al., 2003).

Alterations in the expression of genes involved in liver cell death and growth were also observed in response to comfrey treatment, because PAs in comfrey can reach the hepatocytes via the sinusoidal blood and their toxic metabolites can damage hepatocytes (Prakash et al., 1999). These genes included tumor necrosis factors, transforming growth factor β, chemokine receptor CCR2 gene (Ccr2), heme oxygenase 1 (Hmox1), immediate early response 3 (Ier3), cyclin-dependent kinase inhibitor (Cdkn1a), inhibin beta A (Inhba), and tissue inhibitor of metalloproteinase 1 (Timp1). Hmox1 induction is considered to be a protective mechanism against the oxidative stress associated with liver injury (Bauer and Bauer, 2002).

Comfrey treatment also altered the expression of 16 genes that were involved in liver fibrosis. Liver fibrosis is characterized by cell proliferation and accumulation of extracellular matrix components. Upregulation of cytokines Hgf and Tnfrsf6 and downregulation of Igf1 and Lif play an important role in the pathogenesis of liver injury and fibrosis. Reduced serum Igf1 levels provide a useful index of hepatocellular dysfunction and impaired nutritional status, and elevated serum Hgf levels appear to limit liver fibrosis (Sanz et al., 2005). Agt and Agtr1a,

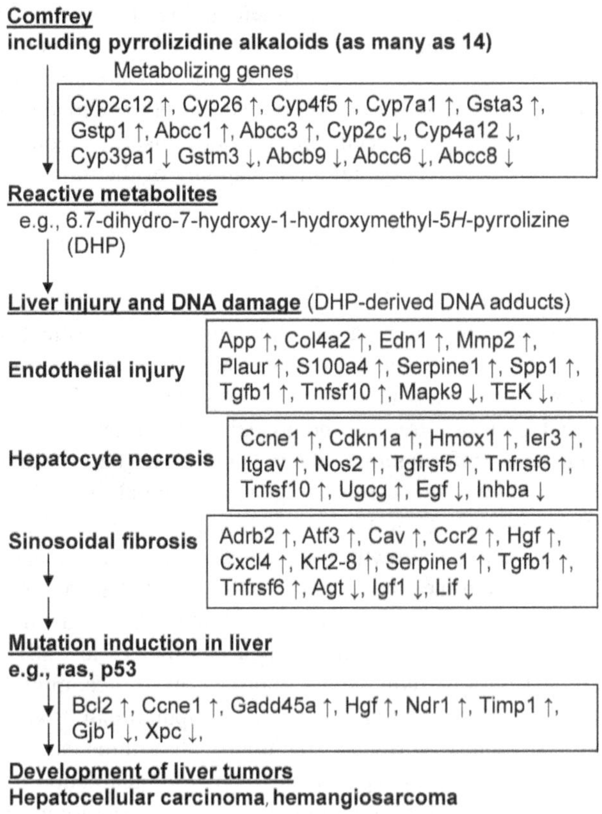

FIGURE 22.2 Integration of gene expression with the proposed mechanistic pathway after comfrey treatment. *Modified with permission from our previous work (Mei et al., 2006b).*

cytokines with vasoactive properties, also regulate liver fibrogenesis. Chemokines in the liver (Cxcl4 and Cxcl12) may modulate the progression of liver fibrosis through their actions on hepatic stellate cells.

The course of tumorigenesis via mutation induction following comfrey treatment was revealed by the integration of gene expression changes with a mechanistic pathway analysis, as shown in Fig. 22.2. Active metabolites of PAs contained in comfrey can target endothelial and hepatocyte DNA, leading to damage to hepatic endothelial cells and hepatocytes. Such damage, if not repaired, may result in liver fibrosis and genetic mutations. These mutations can be associated with the development of HCC and hemangiosarcoma. Gene expression profiling is a powerful approach for identifying DEGs and discovering affected pathways, and this approach provides further insight into the mechanisms involved in the development of VOD and tumorigenesis after exposure to comfrey.

Concluding remarks and future directions

Recently, the US government developed two federal dietary supplement databases, the Dietary Supplement Ingredient Database and the Dietary Supplement Label Database, that are required to address public health issues involving dietary supplements (Dwyer and Coates, 2018). Although many dietary supplements have been reported to exhibit health benefits in humans, several important safety-related issues are major concerns that should be addressed. First, most dietary supplements are a mixture of chemicals in which the components exhibiting pharmacological effects or toxicity are largely unknown (Ernst, 2000; Guo et al., 2010b). Second, a given dietary supplement may be contaminated by other herbs, pesticides, metals, or other chemicals (Guo et al., 2010b), and may even contain a different plant species than the manufacturer's claim. Third, for any dietary supplement, the safe daily dose and appropriate pharmacokinetic information need to be evaluated (Cohen, 2012). Fourth, dietary supplement/herb–drug interactions can cause unexpected side effects. Fifth, dietary supplements can interact with gut microbiota to have an impact on the safety of dietary supplements. Finally, in addition to transcriptomic profiling, interrogation of epigenetic alternations, such as the expression of noncoding RNAs, DNA methylation, and/or histone modification status and patterns, can be well adopted for the assessment of nutraceutical safety. To address these challenges, systematic studies, using integrative approaches, including in vitro and in vivo assays using animal models and human tissue samples, cell lines or single cells, as well as human clinical trials, are warranted to better understand the bioavailability, metabolism, dose responsiveness, pharmacological action, and toxicity of dietary supplements.

The complex chemical nature of dietary supplements makes it difficult to evaluate their efficacy and safety. However, the integration of traditional toxicological approaches with high-throughput molecular profiling technologies, such as functional genomics, transcriptomics, proteomics, metabonomics, toxicogenomics, and bioinformatics, has been providing new approaches to investigate alterations in gene/protein activities and biological pathways/functions responding to toxic substances in animals and humans. Whereas traditional toxicological approaches rely on conventional in vitro and in vivo assays, toxicogenomics has advantages in the collection and interpretation of high-throughput and high-dimensional data, making it possible to overcome challenges that are difficult to address using conventional toxicology methods. Among these new technologies, as reviewed in this chapter, transcriptomic profiling has been used broadly for mechanistic studies, since it can simultaneously and inexpensively evaluate the expression profiles of thousands of genes to interrogate biological and toxicological responses at the transcriptional level.

Advances in NGS technologies enable researchers to detect gene expression changes, mutations, and epigenetic alterations in toxicogenomics studies in ways that are more rapid, accurate, precise, and economical. These high-throughput sequencing technologies could provide tremendous potential and opportunities to revolutionize biomedical science and clinical medicine (Collins and Hamburg, 2013). Indeed, NGS has much higher resolution for gene expression analysis by offering details of transcriptional boundaries, differential expression, mutations, rare transcripts, and alternative splicing (Kaur et al., 2013). However, certain technical challenges may present hurdles for NGS technologies to overcome. Although the cost of NGS instruments and reagents is relatively low (NIH, 2019), the massive amounts of data that result from NGS studies present unique data storage, management, analysis, and interpretation difficulties. It is worth noting that the single-base resolution data "could be compromised by biases, artifacts, or systematic errors derived from NGS-related protocols, platforms, sequencing depth, data analysis pipelines, and reference genomes or transcriptomes, resulting in both high rates of false positive and false negative results" (Ning et al., 2014).

In addition to gene expression profiles, epigenetic changes, such as the expression of noncoding RNAs (ncRNAs) and DNA methylation profiles, have drawn increasing attention in toxicological research. The human genome consists of a large portion (~98%) of noncoding genes, producing various of types of noncoding transcripts, including long noncoding RNA, miRNA, small interfering RNA (siRNA), and other types of ncRNAs. These ncRNAs do not only have biological functions in gene regulation, but also can serve as biomarkers to monitor hosts' responses to environmental stimuli (Koturbash et al., 2015; Li et al., 2019). For example, miRNAs are a class of small noncoding RNA molecules that function as regulators for posttranscriptional gene regulation and are involved in almost every aspect of biology and pathology. MiRNAs exist in many biofluids including plasma, saliva, and urine; miRNAs are biochemically stable (resistant to degradation); some miRNAs are tissue/organ expression specific, or disease-state expression specific, or tissue-injury-specific; some miRNAs are early responders towards stimuli; and miRNAs are genetically conservative among species. These characteristics of miRNAs support the idea that miRNAs could serve as biomarkers for drug efficacy and safety (Koturbash et al., 2015; Li et al., 2019). On the other hand, DNA methylation and histone modification mediate chromatin conformation changes leading to gene expression alteration. Quantitative measurement and/or pattern characterization of DNA methylation and/or histone modification can be utilized to assess the safety of nutraceuticals (Rasoulpour et al., 2011). The principle for detection of DNA methylation is to distinguish unmethylated cytosine with 5-methylcytosine. Currently, several sequencing-based approaches have been widely used to analyze methylation status of genomic DNA samples, such as whole-genome bisulfite sequencing, methylated DNA immunoprecipitation-sequencing, methyl-binding domain capture sequencing (MBDCap-Seq), reduced representation bisulfite sequencing, and so on. By quantitative mapping of the amount of 5-methylcytosine in DNA sequences, methylation preference can be calculated at single-base resolution level (Feng and Lou, 2019).

Advances in technologies have been promoting data generation and computation. Emerging as a new scientific field, big data technology systematically manages, analyzes, and utilizes an extremely large amount of data to reveal patterns, trends, and associations among multidimensional information, which "could not be captured, managed and processed by general computers with an acceptable scope" in the past (Chen et al., 2017a). In the big data era, the growth of chemical data sets is exponential: for example, the largest toxicological database now contains data on approximately 10,000 chemicals from 800,000 toxicological studies (Luechtefeld et al., 2018). To deal with the complexity and quantity of accumulated data, artificial intelligence-based computational toxicology is rapidly evolving (including algorithms such as machine learning, deep learning, convolutional neural network, etc.) and is revamping systems toxicology and methodologies in risk assessment.

To better understand the safety and toxicity of dietary supplements, not only gene expression data but also in vitro and in vivo phenotype data, clinical trial data, lifestyle records, healthcare records, and other public health information are needed. The generation, analysis, integration, and interpretation of these comprehensive data are critical steps for translating the knowledge related to the safety and toxicity of dietary supplements into regulatory decisions and other administrative actions that promote public health.

Acknowledgments

The authors thank Drs. Page McKinzie and Zhichao Liu for their critical review of this chapter. The information in this chapter is not a formal dissemination of information by the US Food and Drug Administration (FDA) and does not represent agency position or policy.

References

Ambros, V., 2003. MicroRNA pathways in flies and worms: growth, death, fat, stress, and timing. Cell 113, 673–676.

Arlt, V.M., Stiborova, M., vom Brocke, J., et al., 2007. Aristolochic acid mutagenesis: molecular clues to the aetiology of Balkan endemic nephropathy-associated urothelial cancer. Carcinogenesis 28, 2253–2261.

Arlt, V.M., Zuo, J., Trenz, K., et al., 2011. Gene expression changes induced by the human carcinogen aristolochic acid I in renal and hepatic tissue of mice. Int. J. Cancer 128, 21–32.

Aronson, J.K., 2017. Defining 'nutraceuticals': neither nutritious nor pharmaceutical. Br. J. Clin. Pharmacol. 83, 8–19.

Bach, N., Thung, S.N., Schaffner, F., 1989. Comfrey herb tea-induced hepatic veno-occlusive disease. Am. J. Med. 87, 97–99.

Bauer, M., Bauer, I., 2002. Heme oxygenase-1: redox regulation and role in the hepatic response to oxidative stress. Antioxid. Redox Signal. 4, 749–758.

Becker, M.W., Lourencone, E.M.S., De Mello, A.F., et al., 2019. Liver transplantation and the use of kava: case report. Phytomedicine 56, 21–26.

Beer, S., Komatsubara, K., Bellovin, D.I., et al., 2008. Hepatotoxin-induced changes in the adult murine liver promote MYC-induced tumorigenesis. PLoS One 3, e2493.

Behl, M., Nyska, A., Chhabra, R.S., et al., 2011. Liver toxicity and carcinogenicity in F344/N rats and B6C3F1 mice exposed to kava kava. Food Chem. Toxicol. 49, 2820–2829.

Campo, J.V., McNabb, J., Perel, J.M., et al., 2002. Kava-induced fulminant hepatic failure. J. Am. Acad. Child. Adolesc. Psychiatry 41, 631–632.

Chandar, N., Lombardi, B., Locker, J., 1989. c-myc gene amplification during hepatocarcinogenesis by a choline-devoid diet. Proc. Natl. Acad. Sci. U S A 86, 2703–2707.

Chen, T., 2010. The role of MicroRNA in chemical carcinogenesis. J. Environ. Sci. Health C Environ. Carcinog. Ecotoxicol. Rev. 28, 89–124.

Chen, H., Liu, J., Zhao, C.Q., et al., 2001. Association of c-myc overexpression and hyperproliferation with arsenite-induced malignant transformation. Toxicol. Appl. Pharmacol. 175, 260–268.

Chen, L., Mei, N., Yao, L., et al., 2006a. Mutations induced by carcinogenic doses of aristolochic acid in kidney of Big Blue transgenic rats. Toxicol. Lett. 165, 250–256.

Chen, T., Guo, L., Zhang, L., et al., 2006b. Gene expression profiles distinguish the carcinogenic effects of aristolochic acid in target (kidney) and non-target (liver) tissues in rats. BMC Bioinformatics. 7 (Suppl. 2), S20.

Chen, M., Mao, S., Liu, Y., 2017a. Big data: a survey. Mob. Netw. Appl. 19, 171–209.

Chen, S., Zhang, Z., Qing, T., et al., 2017b. Activation of the Nrf2 signaling pathway in usnic acid-induced toxicity in HepG2 cells. Arch. Toxicol. 91, 1293–1307.

Chen, C.J., Yang, Y.H., Lin, M.H., et al., 2018. Herbal medicine containing aristolochic acid and the risk of hepatocellular carcinoma in patients with hepatitis B virus infection. Int. J. Cancer 143, 1578–1587.

Chi, K.R., 2008. The year of sequencing. Nat. Methods 5, 11–14.

Clarke, T.C., Black, L.I., Stussman, B.J., et al., 2015. Trends in the use of complementary health approaches among adults: United States, 2002-2012. Natl. Health Stat. Rep. 1–16.

Clough, A.R., Bailie, R.S., Currie, B., 2003. Liver function test abnormalities in users of aqueous kava extracts. J. Toxicol. Clin. Toxicol. 41, 821–829.

Cohen, P.A., 2009. American roulette—contaminated dietary supplements. N. Engl. J. Med. 361, 1523–1525.

Cohen, P.A., 2012. Assessing supplement safety—the FDA's controversial proposal. N. Engl. J. Med. 366, 389–391.

Collins, F.S., Hamburg, M.A., 2013. First FDA authorization for next-generation sequencer. N. Engl. J. Med. 369, 2369–2371.

Cosyns, J.P., Jadoul, M., Squifflet, J.P., et al., 1994. Urothelial malignancy in nephropathy due to Chinese herbs. Lancet 344, 188.

Cosyns, J.P., Jadoul, M., Squifflet, J.P., et al., 1999. Urothelial lesions in Chinese-herb nephropathy. Am. J. Kidney Dis. 33, 1011–1017.

Cosyns, J.P., Dehoux, J.P., Guiot, Y., et al., 2001. Chronic aristolochic acid toxicity in rabbits: a model of Chinese herbs nephropathy? Kidney Int. 59, 2164–2173.

DeLeve, L.D., McCuskey, R.S., Wang, X., et al., 1999. Characterization of a reproducible rat model of hepatic veno-occlusive disease. Hepatology 29, 1779–1791.

Dwyer, J.T., Coates, P.M., 2018. Why Americans need information on dietary supplements. J. Nutr. 148, 1401S–1405S.

Dwyer, J.T., Coates, P.M., Smith, M.J., 2018. Dietary supplements: regulatory challenges and research resources. Nutrients 10, 41.

EGb, 2003. EGb 761: *Ginkgo biloba* extract, Ginkor. Drugs R D 4, 188–193.

Ernst, E., 2000. Herbal medicines: where is the evidence? BMJ 321, 395–396.

FDA (2001). FDA concerned about botanical products, including dietary supplements, containing aristolochic acid. https://wayback.archive-it.org/7993/20170722144008/https://www.fda.gov/Food/RecallsOutbreaksEmergencies/SafetyAlertsAdvisories/ucm095272.htm. (Accessed February 26, 2020).

FDA (2002). Consumer advisory: kava-containing dietary supplements may be associated with severe liver injury. https://ods.od.nih.gov/Health_Information/kava.aspx. (Accessed February 26, 2020).

Feng, L., Lou, J., 2019. DNA methylation analysis. Methods Mol. Biol. 1894, 181–227.

Frost, R., O'Meara, S., MacPherson, H., 2014. The external use of comfrey: a practitioner survey. Complement. Ther. Clin. Pract. 20, 347–355.

Fu, P.P., Xia, Q., Guo, L., et al., 2008. Toxicity of kava kava. J. Environ. Sci. Health C Environ. Carcinog. Ecotoxicol. Rev. 26, 89–112.

Fu, P.P., Chiang, H.M., Xia, Q., et al., 2009. Quality assurance and safety of herbal dietary supplements. J. Environ. Sci. Health C Environ. Carcinog. Ecotoxicol. Rev. 27, 91–119.

Gatzidou, E.T., Zira, A.N., Theocharis, S.E., 2007. Toxicogenomics: a pivotal piece in the puzzle of toxicological research. J. Appl. Toxicol. 27, 302–309.

Geller, A.I., Shehab, N., Weidle, N.J., et al., 2015. Emergency department visits for adverse events related to dietary supplements. N. Engl. J. Med. 373, 1531–1540.

Gonzalez, F.J., Yu, A.M., 2006. Cytochrome P450 and xenobiotic receptor humanized mice. Annu. Rev. Pharmacol. Toxicol. 46, 41–64.

Grosse, Y., Baan, R., Straif, K., et al., 2009. A review of human carcinogens-part A: pharmaceuticals. Lancet Oncol. 10, 13–14.

Guo, L., Fang, H., Collins, J., et al., 2006a. Differential gene expression in mouse primary hepatocytes exposed to the peroxisome proliferator-activated receptor alpha agonists. BMC Bioinformatics 7 (Suppl. 2), S18.

Guo, L., Lobenhofer, E.K., Wang, C., et al., 2006b. Rat toxicogenomic study reveals analytical consistency across microarray platforms. Nat. Biotechnol. 24, 1162–1169.

Guo, L., Zhang, L., Sun, Y., et al., 2006c. Differences in hepatotoxicity and gene expression profiles by anti-diabetic PPAR gamma agonists on rat primary hepatocytes and human HepG2 cells. Mol. Diversity. 10, 349–360.

Guo, L., Mei, N., Dial, S., et al., 2007. Comparison of gene expression profiles altered by comfrey and riddelliine in rat liver. BMC Bioinformatics 8 (Suppl. 7), S22.

Guo, L., Li, Q., Xia, Q., et al., 2009. Analysis of gene expression changes of drug metabolizing enzymes in the livers of F344 rats following oral treatment with kava extract. Food Chem. Toxicol. 47, 433–442.

Guo, L., Mei, N., Liao, W., et al., 2010a. *Ginkgo biloba* extract induces gene expression changes in xenobiotics metabolism and the Myc-centered network. OMICS 14, 75–90.

Guo, L., Mei, N., Xia, Q., et al., 2010b. Gene expression profiling as an initial approach for mechanistic studies of toxicity and tumorigenicity of herbal plants and herbal dietary supplements. J. Environ. Sci. Health C Environ. Carcinog. Ecotoxicol. Rev. 28, 60–87.

Guo, L., Shi, Q., Dial, S., et al., 2010c. Gene expression profiling in male B6C3F1 mouse livers exposed to kava identifies—changes in drug metabolizing genes and potential mechanisms linked to kava toxicity. Food Chem. Toxicol. 48, 686–696.

Gyorik, S., Stricker, H., 2009. Severe pulmonary hypertension possibly due to pyrrolizidine alkaloids in polyphytotherapy. Swiss Med. Wkly. 139, 210–211.

Hoenerhoff, M.J., Pandiri, A.R., Snyder, S.A., et al., 2013. Hepatocellular carcinomas in B6C3F1 mice treated with *Ginkgo biloba* extract for two years differ from spontaneous liver tumors in cancer gene mutations and genomic pathways. Toxicol. Pathol. 41, 826–841.

Humberston, C.L., Akhtar, J., Krenzelok, E.P., 2003. Acute hepatitis induced by kava kava. J. Toxicol. Clin. Toxicol. 41, 109–113.

Huxtable, R.J., Luthy, J., Zweifel, U., 1986. Toxicity of comfrey-pepsin preparations. N. Engl. J. Med. 315, 1095.

IARC (2002). Some traditional herbal medicines, some mycotoxins, nephthalene and styrene. IARC Monographs on the Evaluation of Carcinogenic Risks to Humans 82, Lyon, France. http://monographs.iarc.fr/ENG/Monographs/vol82/mono82.pdf. (Accessed February 26, 2020).

IARC (2016). Some drugs and herbal products. IARC Working Group on the Evaluation of Carcinogenic Risks to Humans 108. http://monographs.iarc.fr/ENG/Monographs/vol108/index.php. (Accessed January 23, 2020).

Jelakovic, B., Dika, Z., Arlt, V.M., et al., 2019. Balkan endemic nephropathy and the causative role of aristolochic acid. Semin. Nephrol. 39, 284–296.

Jhuang, J.R., Chiang, C.J., Su, S.Y., et al., 2019. Reduction in the incidence of urological cancers after the ban on Chinese herbal products containing aristolochic acid: an interrupted time-series analysis. Sci. Rep. 9, 19860.

Jiang, T.F., Wang, Y.H., Lv, Z.H., et al., 2007. Determination of kava lactones and flavonoid glycoside in *Scorzonera austriaca* by capillary zone electrophoresis. J. Pharm. Biomed. Anal. 43, 854–858.

Johnson, B.M., Qiu, S.X., Zhang, S., et al., 2003. Identification of novel electrophilic metabolites of piper methysticum Forst (kava). Chem. Res. Toxicol. 16, 733–740.

Kaur, H., Li, J.J., Bay, B.H., et al., 2013. Investigating the antiproliferative activity of high affinity DNA aptamer on cancer cells. PLoS One 8, e50964.

Kensler, T.W., Wakabayashi, N., Biswal, S., 2007. Cell survival responses to environmental stresses via the Keap1-Nrf2-ARE pathway. Annu. Rev. Pharmacol. Toxicol. 47, 89–116.

Kessler, D.A., 2000. Cancer and herbs. N. Engl. J. Med. 342, 1742–1743.

Kohara, A., Suzuki, T., Honma, M., et al., 2002. Mutagenicity of aristolochic acid in the lambda/lacZ transgenic mouse (MutaMouse). Mutat. Res. 515, 63–72.

Koturbash, I., Tolleson, W.H., Guo, L., et al., 2015. MicroRNAs as pharmacogenomic biomarkers for drug efficacy and drug safety assessment. Biomarkers Med. 9, 1153–1176.

Kumar, S., DeLeve, L.D., Kamath, P.S., et al., 2003. Hepatic veno-occlusive disease (sinusoidal obstruction syndrome) after hematopoietic stem cell transplantation. Mayo Clin. Proc. 78, 589–598.

Kwak, D.H., Park, J.H., Lee, H.S., et al., 2014. Aristolochic acid I induces ovarian toxicity by inhibition of akt phosphorylation. Chem. Res. Toxicol. 27, 2128–2135.

Li, Z., Su, Z., Wen, Z., et al., 2009. Microarray platform consistency is revealed by biologically functional analysis of gene expression profiles. BMC Bioinformatics. 10 (Suppl. 11), S12.

Li, Y., Mei, H., Wu, Q., et al., 2011. Methysticin and 7,8-dihydromethysticin are two major kavalactones in kava extract to induce CYP1A1. Toxicol. Sci. 124, 388–399.

Li, Z., Qin, T., Wang, K., et al., 2015. Integrated microRNA, mRNA, and protein expression profiling reveals microRNA regulatory networks in rat kidney treated with a carcinogenic dose of aristolochic acid. BMC Genomics 16, 365.

Li, D., Tolleson, W.H., Yu, D., et al., 2019. Regulation of cytochrome P450 expression by microRNAs and long noncoding RNAs: epigenetic mechanisms in environmental toxicology and carcinogenesis. J. Environ. Sci. Health C Environ. Carcinog. Ecotoxicol. Rev. 37, 180–214.

Li, X., Guo, X., Wang, H.R., et al., 2020. Aristolochic acid-induced genotoxicity and toxicogenomic changes in rodents. World J. Tradit. Chin. Med. 6, 12–25.

Lin, H., Guo, X., Zhang, S., et al., 2014. Mechanistic evaluation of *Ginkgo biloba* leaf extract-induced genotoxicity in L5178Y cells. Toxicol. Sci. 139, 338–349.

Liu, Z., Huang, R., Roberts, R., et al., 2019. Toxicogenomics: a 2020 vision. Trends Pharmacol. Sci. 40, 92–103.

Lord, G.M., Hollstein, M., Arlt, V.M., et al., 2004. DNA adducts and p53 mutations in a patient with aristolochic acid-associated nephropathy. Am. J. Kidney Dis. 43, e11–e17.

Lu, Z.N., Luo, Q., Zhao, L.N., et al., 2019. The mutational features of aristolochic acid-induced mouse and human liver cancers. Hepatology 71 (3), 929–942.

Luciano, R.L., Perazella, M.A., 2015. Aristolochic acid nephropathy: epidemiology, clinical presentation, and treatment. Drug Saf. 38, 55–64.

Luechtefeld, T., Rowlands, C., Hartung, T., 2018. Big-data and machine learning to revamp computational toxicology and its use in risk assessment. Toxicol. Res. 7, 732–744.

Matthews, H.B., Lucier, G.W., Fisher, K.D., 1999. Medicinal herbs in the United States: research needs. Environ. Health Perspect. 107, 773–778.

McDaniel, L.P., Elander, E.R., Guo, X., et al., 2012. Mutagenicity and DNA adduct formation by aristolochic acid in the spleen of Big Blue(R) rats. Environ. Mol. Mutagen. 53, 358–368.

Mei, N., Guo, L., Fu, P.P., et al., 2005. Mutagenicity of comfrey (*Symphytum officinale*) in rat liver. Br. J. Cancer 92, 873–875.

Mei, N., Arlt, V.M., Phillips, D.H., et al., 2006a. DNA adduct formation and mutation induction by aristolochic acid in rat kidney and liver. Mutat. Res. 602, 83–91.

Mei, N., Guo, L., Zhang, L., et al., 2006b. Analysis of gene expression changes in relation to toxicity and tumorigenesis in the livers of Big Blue transgenic rats fed comfrey (*Symphytum officinale*). BMC Bioinformatics. 7 (Suppl. 2), S16.

Mei, N., Guo, L., Liu, R., et al., 2007. Gene expression changes induced by the tumorigenic pyrrolizidine alkaloid riddelliine in liver of Big Blue rats. BMC Bioinformatics. 8 (Suppl. 7), S4.

Mei, N., Guo, L., Tseng, J., et al., 2008. Gene expression changes associated with xenobiotic metabolism pathways in mice exposed to acrylamide. Environ. Mol. Mutagen. 49, 741–745.

Mei, N., Fuscoe, J.C., Lobenhofer, E.K., et al., 2010a. Application of microarray-based analysis of gene expression in the field of toxicogenomics. Methods Mol. Biol. 597, 227–241.

Mei, N., Guo, L., Fu, P.P., et al., 2010b. Metabolism, genotoxicity, and carcinogenicity of comfrey. J. Toxicol. Environ. Health B Crit. Rev. 13, 509–526.

Mei, N., Guo, X., Ren, Z., et al., 2017. Review of *Ginkgo biloba*-induced toxicity, from experimental studies to human case reports. J. Environ. Sci. Health C Environ. Carcinog. Ecotoxicol. Rev. 35, 1–28.

Meng, F., Li, Z., Yan, J., et al., 2014. Tissue-specific microRNA responses in rats treated with mutagenic and carcinogenic doses of aristolochic acid. Mutagenesis 29, 357–365.

Mengs, U., 1988. Tumour induction in mice following exposure to aristolochic acid. Arch. Toxicol. 61, 504–505.

Mengs, U., Lang, W., Poch, J.A., 1982. The carcinogenic action of aristolochic acid in rats. Arch. Toxicol. 51, 107–119.

Moreno, J.J., 1993. Effect of aristolochic acid on arachidonic acid cascade and in vivo models of inflammation. Immunopharmacology 26, 1–9.

Nahin, R.L., Barnes, P.M., Stussman, B.J., 2016. Expenditures on complementary health approaches: United States, 2012. Natl. Health Stat. Rep. 1–11.

Nault, J.C., Letouze, E., 2019. Mutational processes in hepatocellular carcinoma: the story of aristolochic acid. Semin. Liver Dis. 39, 334–340.

NIH (2019). DNA sequencing costs: data. https://www.genome.gov/about-genomics/fact-sheets/DNA-Sequencing-Costs-Data. (Accessed January 29, 2020).

Ning, B., Su, Z., Mei, N., et al., 2014. Toxicogenomics and cancer susceptibility: advances with next-generation sequencing. J. Environ. Sci. Health C Environ. Carcinog. Ecotoxicol. Rev. 32, 121–158.

Nortier, J.L., Martinez, M.C., Schmeiser, H.H., et al., 2000. Urothelial carcinoma associated with the use of a Chinese herb (*Aristolochia fangchi*). N. Engl. J. Med. 342, 1686–1692.

NTP (1998). Nomination of *Ginkgo biloba* extract. http://ntp.niehs.nih.gov/ntp/htdocs/chem_background/exsumpdf/ginkgo_gbe_508.pdf. (Accessed February 26, 2020).

NTP (2011). NTP technical report on toxicology and carcinogenesis studies of kava kava extract in F344/N rats and B6C3F1 mice. NTP TR 571, NIH Publication No. 11-5913.

NTP (2013). Toxicology and carcinogenesis studies of *Ginkgo biloba* extract (CAS No. 90045-36-6) in F344/N rats and B6C3F1/N mice (Gavage studies). Natl Toxicol Program Tech Rep Ser, 1-183.

NTP (2016). NTP botanical dietary supplements program. https://www.niehs.nih.gov/health/materials/botanical_dietary_supplements_program_ntp_508.pdf. (Accessed January 27, 2020).

NTP (2020a). Cancer evaluation criteria. https://ntp.niehs.nih.gov/whatwestudy/testpgm/cartox/criteria/index.html. (Accessed January 22, 2020).

NTP (2020b). Botanical dietary supplements. https://ntp.niehs.nih.gov/whatwestudy/topics/botanical/index.html?utm_source = direct&utm_medium = prod&utm_campaign = ntpgolinks&utm_term = botanical. (Accessed January 22, 2020).

Nuwaysir, E.F., Bittner, M., Trent, J., et al., 1999. Microarrays and toxicology: the advent of toxicogenomics. Mol. Carcinog. 24, 153–159.

Pantano, F., Tittarelli, R., Mannocchi, G., et al., 2016. Hepatotoxicity induced by "the 3Ks": kava, kratom and khat. Int. J. Mol. Sci. 17, 580.

Petroczi, A., Taylor, G., Naughton, D.P., 2011. Mission impossible? Regulatory and enforcement issues to ensure safety of dietary supplements. Food Chem. Toxicol. 49, 393–402.

Prakash, A.S., Pereira, T.N., Reilly, P.E., et al., 1999. Pyrrolizidine alkaloids in human diet. Mutat. Res. 443, 53–67.

Public Law (1994). Dietary Supplement Health and Education Act of 1994. Public Law 103–417. https://ods.od.nih.gov/About/DSHEA_Wording.aspx. (Accessed January 22, 2020).

Pu, X.Y., Shen, J.Y., Deng, Z.P., et al., 2017. Plasma-specific microRNA response induced by acute exposure to aristolochic acid I in rats. Arch. Toxicol. 91, 1473–1483.

Rasoulpour, R.J., LeBaron, M.J., Ellis-Hutchings, R.G., et al., 2011. Epigenetic screening in product safety assessment: are we there yet? Toxicol. Mech. Methods 21, 298–311.

Ridker, P.M., Ohkuma, S., McDermott, W.V., et al., 1985. Hepatic venocclusive disease associated with the consumption of pyrrolizidine-containing dietary supplements. Gastroenterology 88, 1050–1054.

Rode, D., 2002. Comfrey toxicity revisited. Trends Pharmacol. Sci. 23, 497–499.

Roeder, E., 1995. Medicinal plants in Europe containing pyrrolizidine alkaloids. Pharmazie 50, 83–98.

Ronis, M.J.J., Pedersen, K.B., Watt, J., 2018. Adverse effects of nutraceuticals and dietary supplements. Annu. Rev. Pharmacol. Toxicol. 58, 583–601.

Rosenthal, M.D., Vishwanath, B.S., Franson, R.C., 1989. Effects of aristolochic acid on phospholipase A2 activity and arachidonate metabolism of human neutrophils. Biochim. Biophys. Acta 1001, 1–8.

Sanz, S., Pucilowska, J.B., Liu, S., et al., 2005. Expression of insulin-like growth factor I by activated hepatic stellate cells reduces fibrogenesis and enhances regeneration after liver injury. Gut 54, 134–141.

Sborchia, M., Keun, H.C., Phillips, D.H., et al., 2019. The Impact of p53 on aristolochic acid I-induced gene expression in vivo. Int. J. Mol. Sci. 20.

Schmeiser, H.H., Bieler, C.A., Wiessler, M., et al., 1996. Detection of DNA adducts formed by aristolochic acid in renal tissue from patients with Chinese herbs nephropathy. Cancer Res. 56, 2025–2028.

SEQC/MAQC-III Consortium, 2014. A comprehensive assessment of RNA-seq accuracy, reproducibility and information content by the Sequencing Quality Control Consortium. Nat. Biotechnol. 32, 903–914.

Shi, L., Reid, L.H., Jones, W.D., et al., 2006. The MicroArray Quality Control (MAQC) project shows inter- and intraplatform reproducibility of gene expression measurements. Nat. Biotechnol. 24, 1151–1161.

Simoes, M.L., Hockley, S.L., Schwerdtle, T., et al., 2008. Gene expression profiles modulated by the human carcinogen aristolochic acid I in human cancer cells and their dependence on TP53. Toxicol. Appl. Pharmacol. 232, 86–98.

Smith, T., Gillespie, M., Eckl, V., et al., 2019. Herbal supplement sales in US increased 9.4% in 2018. HerbalGram 123, 62–73.

Staiger, C., 2012. Comfrey: a clinical overview. Phytother. Res. 26, 1441–1448.

Stickel, F., Baumuller, H.M., Seitz, K., et al., 2003. Hepatitis induced by kava (*Piper methysticum* rhizoma). J. Hepatol. 39, 62–67.

Unger, M., Holzgrabe, U., Jacobsen, W., et al., 2002. Inhibition of cytochrome P450 3A4 by extracts and kavalactones of *Piper methysticum* (kava-kava). Planta Med. 68, 1055–1058.

van Beek, T.A., Montoro, P., 2009. Chemical analysis and quality control of *Ginkgo biloba* leaves, extracts, and phytopharmaceuticals. J. Chromatogr. A 1216, 2002–2032.

Vanherweghem, L.J., 1998. Misuse of herbal remedies: the case of an outbreak of terminal renal failure in Belgium (Chinese herbs nephropathy). J. Altern. Complement. Med. 4, 9–13.

Vanherweghem, J.L., Depierreux, M., Tielemans, C., et al., 1993. Rapidly progressive interstitial renal fibrosis in young women: association with slimming regimen including Chinese herbs. Lancet 341, 387–391.

Wang, Y., Meng, F., Arlt, V.M., et al., 2011. Aristolochic acid-induced carcinogenesis examined by ACB-PCR quantification of H-Ras and K-Ras mutant fraction. Mutagenesis 26, 619–628.

Wang, C., Gong, B., Bushel, P.R., et al., 2014. The concordance between RNA-seq and microarray data depends on chemical treatment and transcript abundance. Natl. Biotechnol. 32, 926–932.

Wang, P., Zhu, J., Shehu, A.I., et al., 2019. Enzymes and pathways of kavain bioactivation and biotransformation. Chem. Res. Toxicol. 32, 1335–1342.

Wang, L., Cui, J., Jin, B., et al., 2020. Multifeature analyses of vascular cambial cells reveal longevity mechanisms in old *Ginkgo biloba* trees. Proc. Natl. Acad. Sci. U S A 117, 2201–2210.

Weston, C.F., Cooper, B.T., Davies, J.D., et al., 1987. Veno-occlusive disease of the liver secondary to ingestion of comfrey. Br. Med. J. 295, 183.

Whitlock Jr., J.P., 1999. Induction of cytochrome P4501A1. Annu. Rev. Pharmacol. Toxicol. 39, 103–125.

Whittaker, P., Clarke, J.J., San, R.H., et al., 2008. Evaluation of commercial kava extracts and kavalactone standards for mutagenicity and toxicity using the mammalian cell gene mutation assay in L5178Y mouse lymphoma cells. Food Chem. Toxicol. 46, 168–174.

Yeong, M.L., Swinburn, B., Kennedy, M., et al., 1990. Hepatic veno-occlusive disease associated with comfrey ingestion. J. Gastroenterol. Hepatol. 5, 211–214.

Yeong, M.L., Clark, S.P., Waring, J.M., et al., 1991. The effects of comfrey derived pyrrolizidine alkaloids on rat liver. Pathology 23, 35–38.

Yu, Y., Fuscoe, J.C., Zhao, C., et al., 2014. A rat RNA-Seq transcriptomic BodyMap across 11 organs and 4 developmental stages. Nat. Commun. 3230.

Zhang, Z., Chen, S., Mei, H., et al., 2015. *Ginkgo biloba* leaf extract induces DNA damage by inhibiting topoisomerase II activity in human hepatic cells. Sci. Rep. 5, 14633.

Chapter 23

Target identification and polypharmacology of nutraceuticals

György Dormán, Beáta Flachner, István Hajdú and Csaba András

Introduction

For many years, the dominant paradigm in drug discovery was the concept of designing maximally selective ligands to act on individual drug targets, thus minimizing the adverse side effects. This is, in fact, the direct extension of the old lock-and-key concept, thus selectivity is considered a centuries-old dogma.

Structure-based and rational drug design reflects the "one gene, one drug, one disease" paradigm, however, much evidence supports the fact that most diseases are "polygenic" or multifactorial based on a complicated network "system." The completion of the human genome project significantly contributed to this paradigm shift, leading the way for examining the global function of the genome through systems biology and computational approaches. Furthermore, many agents that target a specific biomolecule for the treatment of complex diseases are found ineffective while several recently developed drugs have many targets ("one drug, multiple targets," called polypharmacology). Although drug promiscuity ["dirty" drugs (Frantz, 2005)] is an old observation, polypharmacology is a relatively young concept (Reddy and Zhang, 2013) and it has only recently become widely accepted by investigators searching for more effective drug therapies (Proschak et al., 2019). This concept includes drugs acting on multiple "networked" targets of unique single or multiple disease pathways. Polypharmacology has gained enormous attention in the past 5 years due to the recognition of the proven therapeutic benefits. In the meantime, several multitarget drugs reached the market (Bolognesi, 2019).

In general, current drug discovery programs reflect a more holistic view (looking at disease pathways with their multiple targets) instead of the reductionistic approach dominating the 1970s–1990s drug discovery efforts (Medina-Franco et al., 2013).

Multitarget compounds, polypharmacology

Adapting the old lock-and-key analogy to polypharmacology means that the active components might have multiple keys that could bind to a large diversity of ligand-binding domains (Granier and Kobilka, 2012) or with different domains of a single protein leading to various activities (Fig. 23.1). Previously the pioneers who investigated the polypharmacology of existing drugs estimated that a drug might interact on an average with $\sim 3-6$ targets and about half of all drugs might exhibit activity against more than five targets ($<10\ \mu$mol) (Jalencas and Mestres, 2013). The exact degree of promiscuity (polypharmacology) of the small-molecule natural products and nutraceuticals can only be predicted due to the current insufficient data availability for activity profiling (Hu et al., 2014). On the other hand, the development of virtual screening protocols has accelerated the identification of multitarget therapeutics (Jenwitheesuk et al., 2008). Multitarget drugs became particularly important in anticancer therapy (Giordano and Petrelli, 2008). Oncogenesis is predominantly multigenic, and most cancers are driven by four to seven independent mutations. There is an obvious shift in treatment from single-targeted therapies to combination usage or to multitargeted therapies (Le Tourneau et al., 2008).

Multitarget therapeutics can be more efficacious and less vulnerable to adaptive resistance (Zimmermann et al., 2007). There are the following variations: (1) single chemical entities act simultaneously at multiple molecular targets, and (2) combination of two or more chemical entities that either modulate separate targets/pathways to create a simple combination effect or one component alters the ability of another compound to reach its target by altering its metabolism or blocking the efflux pumps. Alternatively, the components bind separate sites on the same target.

Nutraceuticals. DOI: https://doi.org/10.1016/B978-0-12-821038-3.00023-9

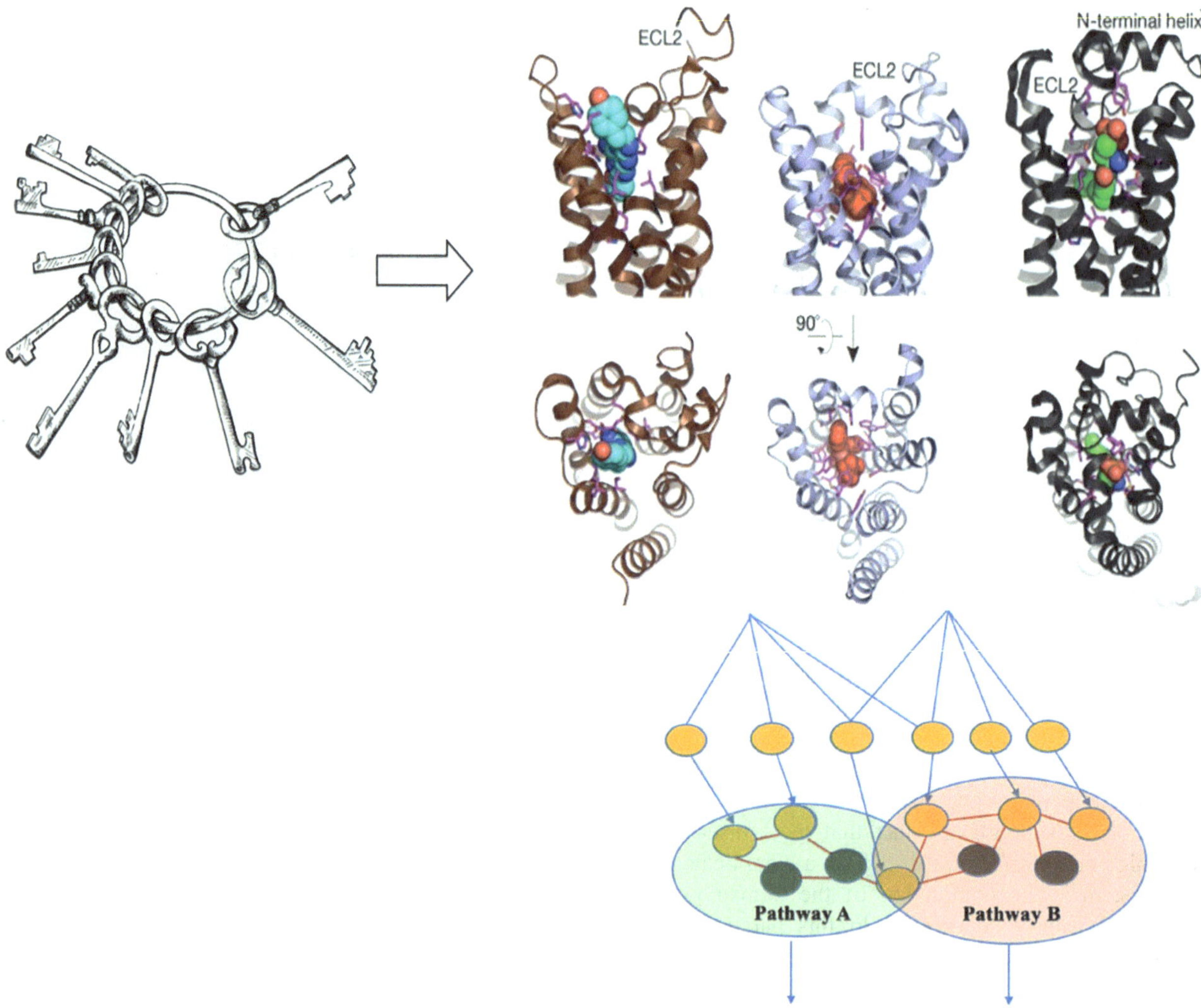

FIGURE 23.1 Molecular entities with multiple "keys" might interact with various network proteins as well as different domains of various proteins. *Partly adapted from Granier, S., Kobilka, B. 2012. A new era of GPCR structural and chemical biology. Nat. Chem. Biol. 8, 670–673; Chamberlin, S.R., Blucher, A., Wu, G., et al., 2019. Natural product target network reveals potential for cancer combination therapies. Front. Pharmacol. 10, 557.*

Drug combinations (mixtures of monotherapies) are also a promising strategy to overcome the compensatory mechanisms and unwanted off-target effects that limit the utility of many potential drugs (Lehár et al., 2009). Nature-guided drug discovery led to the discovery of natural product-based combination drugs that exploit the many known pharmacokinetic and pharmacodynamic benefits (Isgut et al., 2018).

The major limitation of polypharmacology and the multitarget approach is that the pathways/mechanisms of many diseases are only partially understood at the molecular levels.

The advent of network pharmacology

As a consequence of large-scale data accumulation, systems biology appeared as a multidisciplinary approach and considered the biological systems as a whole. Its objective is to elucidate biological mechanisms and to understand the mechanisms of action of active substances using computational and experimental methods (Peitsch and De Graaf, 2013).

The traditional drug design at the single molecular target level is "blind" to other processes that are connected through complex networks (Maggiora, 2011). Therefore, network-based approaches are needed to elucidate the molecular mechanisms of human disease (Arrell and Terzic, 2010). Network pharmacology emerged as another recent approach to target disease-causing networks instead of single genes/targets (Hopkins, 2008). Network pharmacology identifies the combination of nodes in a biological network whose perturbation results in a desired therapeutic outcome and identifies discovering agents with the desired polypharmacology profile to perturb those nodes. To cure complex, multifactorial, polygenic diseases that have extensive biological networks such as cancer, diabetes, or Alzheimer disease single target intervention appears ineffective.

Network pharmacology-based drug design integrates systems biology, metabolomics, network analysis, and connectivity. This new paradigm enables the drugs to target several different proteins or networks implicated in a disease. Nutraceuticals target whole networks, allowing the components to act in an additive or synergistic manner, as well as possibly leading to improved pharmacokinetics (Poornima et al., 2016).

Csermely et al. (2005) states that drugs with multiple targets might have a better chance of affecting the complex equilibrium of whole cellular networks than drugs that act on a single target. Low-affinity, multitarget drugs might have further advantages, namely weak links that are able to stabilize complex networks (Csermely et al., 2005). This is particularly valid for nutraceuticals present in herbal compositions: since the majority of medicinal plants contain broad-spectrum compounds with rather weak or moderate bioactivity.

Polypharmacology and synergy in the area of nutraceuticals

Chaudhari et al. describe various computational approaches that have been developed to study the polypharmacology profiles of small molecules (Chaudhari et al., 2017).

Herbal extracts containing nutraceutical components are often more active than the single isolated compound (s), since a potentiating, additive, or synergistic interaction between the different natural phytochemicals can be assumed (Wagner and Ulrich-Merzenich, 2009). "Synergistic multitarget effects" in the herbal area means that the single constituents of a mono-extract or a multi-extract combination affect not only one single target, but several targets, and therefore cooperate in an agonistic, synergistic way (Fig. 23.2).

The thousand years' old traditional Chinese medicine (TCM) represents a holistic treatment for diseases. The molecular mechanisms responsible for their therapeutic effectiveness have begun to be elucidated. The development of network-based systems biology opened new avenues for these efforts. In TCM theory, disease status is considered as an imbalance of the whole body system, and the combination of many natural species helps to restore this balance.

The TCM formulae contain multicomponent and multitarget agents, acting in the same way as the combination therapy of multicomponent drugs. Network-based systems biology allows studying TCM pharmacology to understand the targets of the active ingredients and their interactions within the molecular networks (Zhao et al., 2010). It often reveals synergy between the components.

Nutraceuticals or phytochemicals offer novel treatment opportunities through a synergistic mode of action. As part of the biological defense function, plant secondary metabolites act by targeting and disrupting the cell membrane. By binding and inhibiting specific proteins they adhere to or intercalate into the RNA or DNA (Efferth and Koch, 2011). Furthermore, plants have learned to cope with resistance during evolution by combinations of multitargeted compound complexes. Fairly or weakly active broad-spectrum nutraceuticals are usually compounds from the classes of flavonoids, coumarins, tannins, terpenoids (monoterpens, triterpenes, sesquiterpenes, saponines, iridoids, lignans), and anthracenes. The success of using many different herbal mixtures in European phytomedicine, TCM, or Ayurveda might be due to synergistic effects rather than the pharmacological activity of the single active components that have a high and specific mode of action (Nahrstedt and Butterweck, 2010). For example, genistein (GEN), baicalein, hesperetin, naringenin, and quercetin act synergistically to inhibit the growth of cancer cells. Consequently, Li et al. (2011) proposed the "network target"-based paradigm for studying the mechanism of TCM. In their definition "network

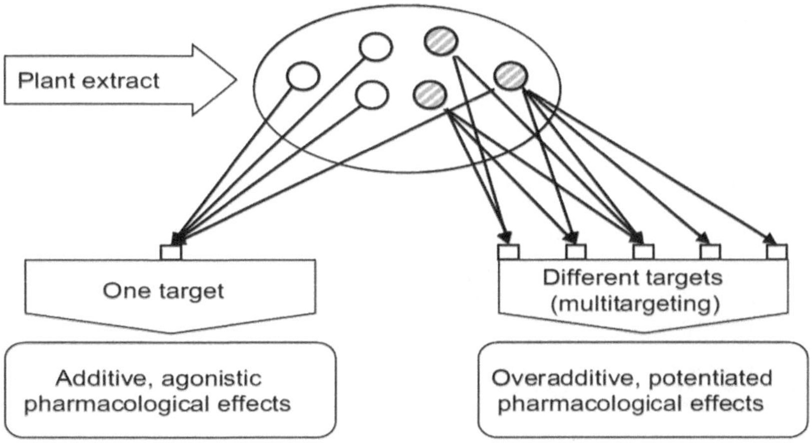

FIGURE 23.2 Various nutraceutical components might interact with various cellular targets. *Adapted from Wagner, H., Ulrich-Merzenich, G., 2009. Synergy research: approaching a new generation of phytopharmaceuticals. Phytomedicine 16, 97–110.*

target" is a therapeutic target which is derived from systematic interventions of the biological network.

Recent advances and developments of systems pharmacology and computational (in silico) approaches provide powerful tools for exploring the polypharmacological profiles of natural products. Fang et al. (2018) introduced recent progress in and advances of computational tools as well as systems pharmacology approaches for identifying drug targets of natural products that may be involved in targeted cancer therapy. The authors investigated the polypharmacological and systems immunology profiles of five representative natural products [curcumin, epigallocatechin-3-gallate (EGCG), quercetin, resveratrol, and berberine] (Fang et al., 2018) (Fig. 23.3). Apart from direct targeting of various oncogenic pathways, the selected natural products affected the tumor microenvironments (TMEs) for cancer chemoprevention as well as demonstrated immunomodulatory or inflammatory "side" effects.

Target identification and deconvolution techniques

Target fishing, target identification, or ligand profiling is used to investigate and validate biological targets that interact directly with small organic molecules. However, in network pharmacology-driven drug discovery the active ingredients (nutraceuticals) could directly and indirectly affect various molecular targets and signaling pathways at various stages.

The interaction between the molecular targets and the effector small molecules could be either direct through binding to a particular cavity of the protein, or indirect when the direct, initial signal is transmitted through various direct protein–protein interactions and enzymatic transformations (e.g., phosphorylation) followed by conformational changes (downstream effects on targets). The identification of the direct and indirect molecular targets is a critical issue in elucidating the mechanism of action (MoA) of the nutraceuticals and requires different approaches. While the direct molecular binding partners can be deconvoluted by affinity-based methods, the downstream (indirect) targets require various functional genetics and molecular biology approaches including cell-phenotypic assays, exploring the phosphorylation pattern, gene-expression profiling, etc.

In order to reflect all these aspects, a new term "efficacy target" was recently introduced, which collects all the proteins in various pathways that are influenced directly or indirectly by the small molecule upon treatment of the cells. Various approaches and techniques should be combined in order to uncover the total number

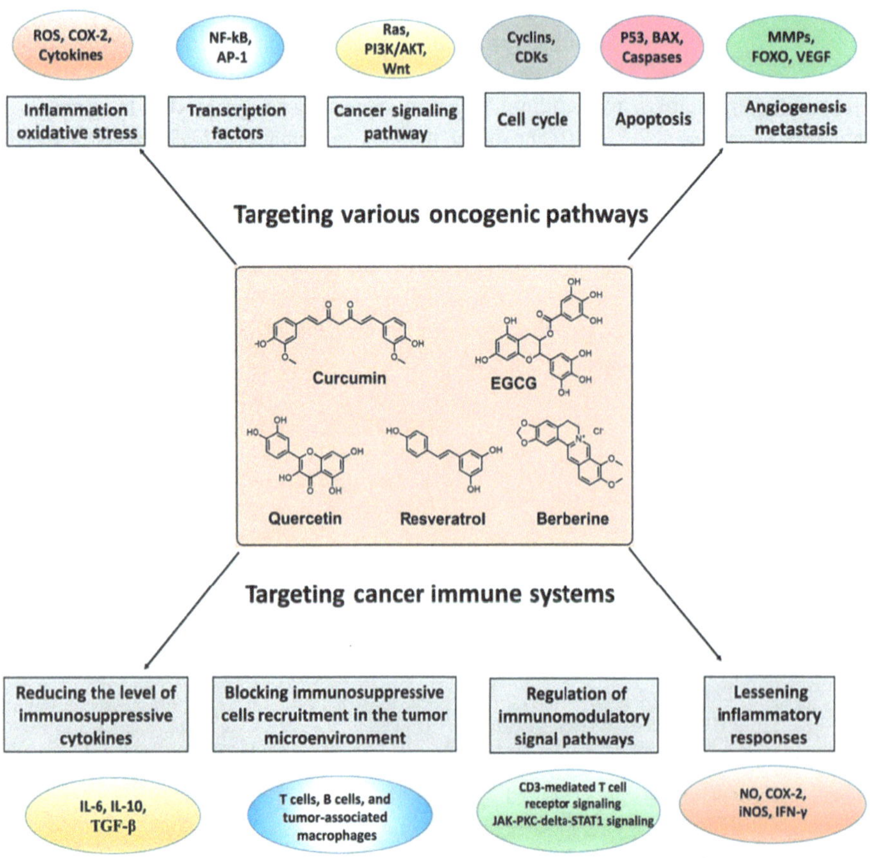

FIGURE 23.3 Mechanism of action of anticancer effects for five selected natural products. *COX-2*, cyclooxygenase 2; *ROS*, reactive oxygen species. *Adapted from Fang, J., Liu, C., Wang, Q., et al., 2018. In silico polypharmacology of natural products. Briefings Bioinf. 19, 1153–1171.*

of efficacy targets including affinity-based, functional genomics, cellular profiling, and knowledge-based approaches (Schirle and Jenkins, 2016) (Fig. 23.4).

In silico target and mechanism of action prediction

In silico identification of potential biological targets for chemical compounds offers a cheap alternative for the exploration of ligand–target interactions. Various methods for computational target prediction have been developed with the help of advances in molecular descriptors and data-mining algorithms. The contemporary approaches can be divided into four classes: chemical similarity searching, data mining/machine learning, panel docking, and the analysis of bioactivity spectra. Chemical similarity is based on the well-established medicinal chemistry concept that structurally similar compounds have similar physicochemical properties and possibly similar biological profiles (Bender et al., 2007). Computational target fishing mines drug–target databases and then maps compound structures into chemogenomical space in order to predict the biological targets. There are various commercial and public databases (PubChem, Chem-Bank, ChEMBL, etc.), as recently reviewed by Chen et al. (2016).

A recent review summarizes the various ligand- and target-based as well as hybrid approaches for computational target prediction and is shown in Fig. 23.5 (Sydow et al., 2019).

Proteochemometrics can be used to extend the single target models to multiple targets. If there are close sequence similarities between proteins, data can be extrapolated from one protein to a related protein target. Structure-based target prediction includes binding site comparison, inverse docking (screening), reverse pharmacophore screening, and interaction fingerprint comparison. In inverse docking one query ligand is docked to a set of target proteins to identify possible targets. In reverse pharmacophore screening a query ligand-associated ligand-based pharmacophore is investigated against a panel of pharmacophore models, coming from protein – ligand complexes. Binding site or target comparison is performed between similar proteins or between their particular binding sites that bind similar small molecules. In interaction fingerprint comparison, interaction fingerprints are generated encoding the interaction between ligands and the possible target. Such fingerprints encode hydrogen bonds, hydrophobic, charged, aromatic, and metal-binding interactions (Salentin et al., 2014). Bioactivity spectra are the response of a compound to a series of cell lines, DNA microarrays, or proteins. If two compounds target the same signaling pathway or protein, they generate similar bioactivity spectra. They induce similar patterns of either gene expression or phenotypic responses. The bioactivity spectra-based computational method predicts potential targets of a chemical by profiling similarity measurements and hierarchical clustering.

The rapid development of the computational power and the novel simulation and model building approaches led to data analysis via a new generation of machine-learning applications such as "deep-learning." Such novel methods use molecular graphs and deep-learning-based algorithmic frameworks (Wen et al., 2017; Byrne and Schneider, 2019).

In silico target prediction for nutraceuticals as herbal components

Herb ingredients' targets (HIT) is linking a single herbal ingredient to its potential protein targets derived from

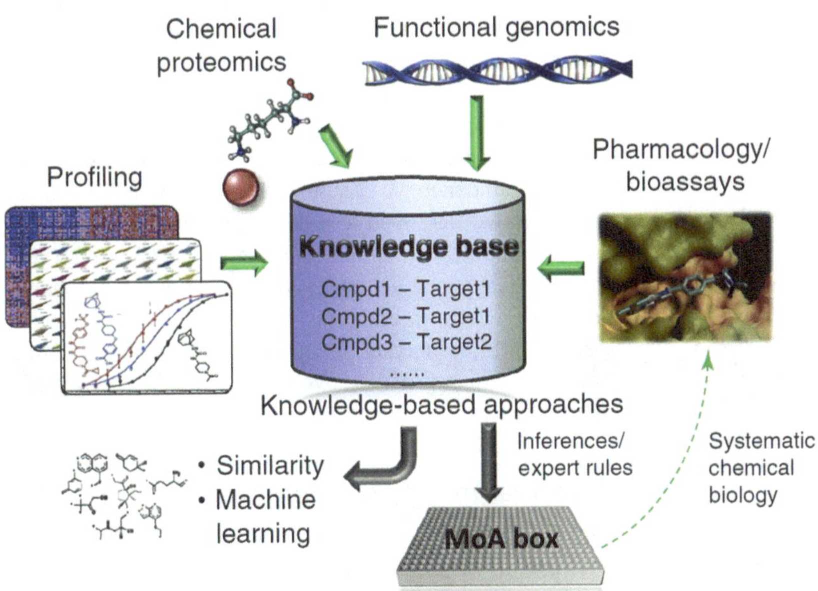

FIGURE 23.4 State-of-the art technologies for target deconvolution and mechanism of action studies. *MoA*, mechanism of action. *Adapted from Schirle, M., Jenkins, J.L., 2016. Identifying compound efficacy targets in phenotypic drug discovery. Drug Discov. Today 21, 82–89.*

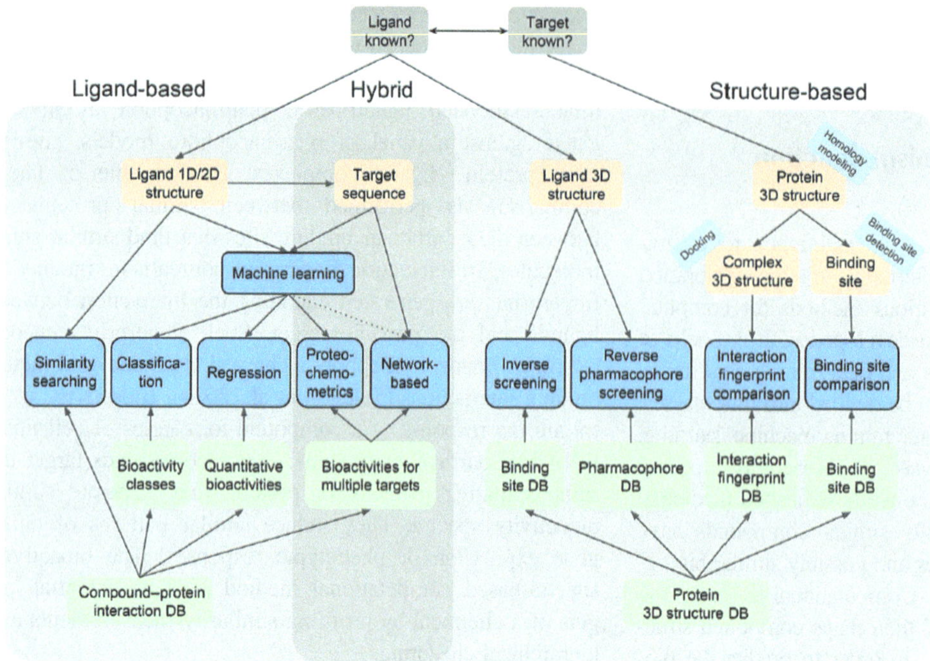

FIGURE 23.5 Overview of ligand- and structure-based as well as hybrid methods for target prediction. *Adapted from Sydow, D., Burggraaff, L., Szengel, A., et al., 2019. Advances and challenges in computational target prediction. J. Chem. Inf. Model. 59 (5), 1728–1742.*

experimental results (Ye et al., 2011). The methods applied include: (1) text-mining technologies applied to PubMed abstracts in order to collect related literature; (2) retrieving relevant information about protein target; and (3) providing clues to the potential mechanism (indirect targets are also collected).

Yu et al. (2012) has developed a simple, in sil*ico* target identification approach on the basis of random forest (RF) and support vector machine (SVM) methods to predict the putative targets for TCM treatment. RF and SVM techniques integrate chemical, genomic, and pharmacological information for drug targeting.

In TCM, a widely accepted theory ("multiple herbal drugs for one disease") proposes that if different active ingredients treat the same disease they share common targets. Recent advances in systems biology and medicine have allowed the application of new profiling technologies in the study of the potential targets of Chinese herbs (Wang et al., 2012). This model requires a large pharmacological database (drugs and targets). The authors aimed at identifying the molecular targets of Chinese herbs that are administered for cardiovascular diseases (CVD). They performed a multiple docking analysis to select the ligand–protein interactions with binding free energies of ≤ -5.0 kcal/mol, and generated a ligand–potential target network. The ligand–candidate target and ligand–potential target networks were correlated with the target–disease network and target–pathway network. As a result, the authors identified 54 protein targets for the investigated Chinese herbs, of which 29 are common targets in CVD (52.7%) (Li et al., 2012). Recently, Huang and co-workers used artificial intelligence technology to identify active component–target interactions, which promotes the development of anticancer agents from Chinese medicinal herbs (Huang et al., 2018).

In vitro target identification approaches

Current methods for target identification of small molecules are described by various reviews (Ares et al., 2013; Ziegler et al., 2013; Schirle and Jenkins, 2016). There are direct, indirect, and label-free approaches (Tashiro and Imoto, 2012). Yoshida (2019) reviewed the state-of-the-art methods for the elucidation of the mechanism of how bioactive natural products act within cells to induce biological activity. There are physical and genetic interaction-based methods. Sometimes direct detection of the binding proteins is rather difficult due to low affinity or low abundance. A genetic interaction between a ligand and its target may lead to a change in the cell's sensitivity to the compound, thus the function of the protein target is altered. Chemical genomics offers a novel approach to identify genetic interactions between the ligands and all genes and the analysis of the gathered genomic information (gene disruption or overexpression) helps to predict the pathways and targets affected.

Direct (affinity-based) approaches

Direct approaches comprise the typical affinity-based methods (Sato et al., 2010), while indirect approaches are based on various bioinformatics, molecular biology, or genetic methods. Apart from exploring the functional interactions with the molecular targets (on-targets) the additional binding proteins ("off-targets") could also be identified, which may lead to undesired biological activity and toxicity. However, the recent concept of polypharmacology raised the importance of the direct target deconvolution.

Affinity/activity-based proteomics ("pull-down") is one of the most widely applied methodologies to identify the targets of biologically active compounds (Fig. 23.6). The pull-down probe is immobilized on a solid matrix and exposed to a protein extract (e.g., cell lysate) to bind the target protein(s). The proteins that bind nonspecifically to the probe and the matrix are removed by successive washing prior to release of the bound proteins by elution. The target proteins are identified by SDS–PAGE (sodium dodecyl sulfate–polyacrylamide gel electrophoresis), tryptic digestion, and identification of the resulting peptides by MS/MS (mass spectrometry/mass spectrometry) analysis. Structure–activity relationship (SAR) data help to determine the site for attachment of the linker which tolerates substitutions without the loss of activity. The pull-down probe (small molecule) can be covalently immobilized to an activated matrix (beads) or noncovalently make use of the strong affinity between biotin and streptavidin. To avoid steric interference between the matrix and the target proteins, the length of the linker can be properly adjusted. Recent mass-spectrometric techniques allow the identification of putative targets directly from complex mixtures immediately after release from the affinity resin (Wiśniewski et al., 2009). In order to confirm the specificity to the putative binding protein, an excess of the bioactive compound is added to the protein mixture before pulldown. It binds to the target and thus prevents binding of the target protein to the immobilized affinity probe.

Simple biotin-affinity tagging also belongs in this category (Galat et al., 1992). A combination of an affinity tag (biotin) with a fluorophore has been successfully applied for target isolation in several cases.

The major trend recently shifted from noncovalent affinity-based methods toward covalent techniques. Activity-based protein profiling (ABPP) (Evans and Cravatt, 2006) relies on small molecules with a so-called "warhead," which react with residues in the active site of enzymes in an irreversible manner. The small molecules are attached via a linker to a functionality, such as biotin for affinity purification, or to a fluorophore for visualization. As the small molecules react with their respective target proteins to form a covalent bond, no additional cross-linking is required for further affinity purification. ABPP enables activity assignment to certain proteins, not only within the entire proteome, but also within a protein family, thereby creating activity-based subclasses. Since only a certain class of proteins could react with the reactive group of ABPPs, photoactivatable units could also be attached to the pharmacophore portion of the small molecule (Dormán and Prestwich, 2000), providing a general covalent labeling opportunity.

Chemical proteomics is a systematic affinity-based target profiling which provides a very powerful technique to identify kinases in the signaling pathways as well as the promiscuity of the kinase inhibitors (Bantscheff et al., 2009). This approach employed chemical probes specifically engineered to capture and identify protein targets, or classes of structurally or functionally related proteins, often referred to as subproteomes. Bantscheff et al. (2007) applied such techniques to study the multitarget features

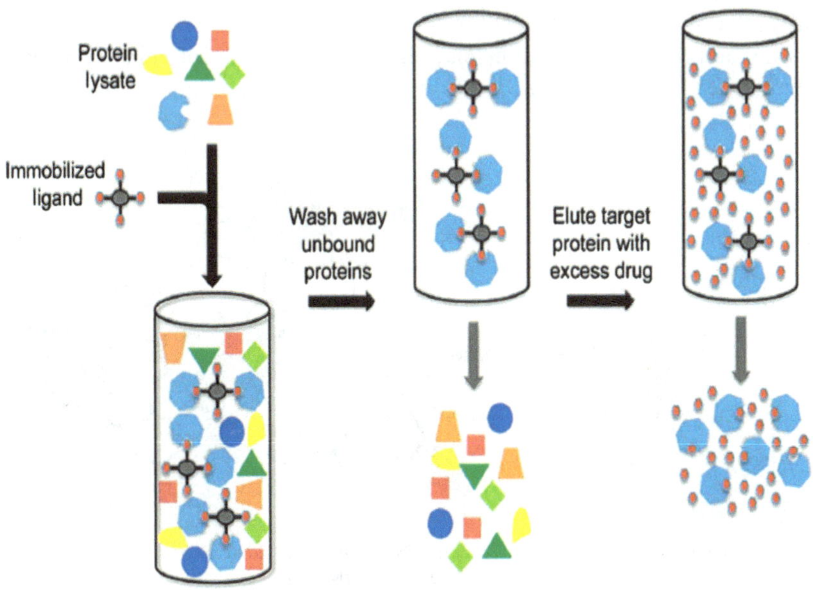

FIGURE 23.6 Conventional affinity-based methods ("pull-down" assay) to identify direct molecular targets. *Adapted from Lomenick, B., Olsen, R.W., Huang, J., 2010. Identification of direct protein targets of small molecules. ACS Chem. Biol. 6, 34–46.*

of the small molecules that interact with the kinome (the set of protein kinases in the genome).

In affinity-based methods, the small molecule should be modified by introducing a linker for immobilization. Kanoh et al. developed a photo-cross-linked small-molecule affinity matrix, which does not require such structural modifications (Kanoh, 2016). It utilizes the diverse reactivity pattern of the photogenerated carbene to immobilize various structurally different bioactive small molecules in a chemo- and site-nonselective manner (Fig. 23.7).

For example, aromatic rings react with the photogenerated carbene at various positions and with different mechanisms. Hydroxyl group or carboxylic acid functionalities are also compatible with the photoconjugation approach to producing C–H and O–H insertion derivatives, while thiols and amines lead to primarily N–H or S–H insertion products (Fig. 23.8).

The target proteins of xanthohumol (XN, Fig. 23.9) were investigated using the photoconjugated affinity matrix. XN is a prenylated chalcone found in hop and is also an ingredient of beer. XN modulates autophagy, blocking the maturation of the autophagosome. However, to identify the exact MoA, the direct protein targets should be elucidated. The authors applied XN photochemically generated affinity beads for pull-down assays and identified several potential protein targets in human epidermoid carcinoma A431 cells: (1) valosin-containing protein (VCP), (2) voltage-dependent anion channel, (3) prohibitin, and (4) prohibitin-2 (PHB2). Competition experiments revealed VCP as the specific binding partner (Kanoh, 2016).

Indirect approaches

Affinity chromatography has a certain limitation since it requires a SAR study in order to identify the most suitable site for modification without loss of activity. SAR studies are time-consuming and require extensive medicinal chemistry expertise. The indirect and label-free methods were reviewed by Park et al. (2019). The drug affinity responsive target stability (DARTS) method was developed as a universal strategy to identify drug-binding proteins (Fig. 23.10). DARTS is based on the finding that a target protein bound to a small molecule might be less susceptible to proteolysis than its unbound state. A major advantage of DARTS is the use of unmodified compounds for target identification. Small-molecule binding to a target protein stabilizes protein conformation and reduces accessibility to proteolysis (Lomenick et al., 2009). In other words, compound binding may mask proteolytic cleavage sites. A comparison of affinity chromatography and DARTS is reported by Lomenick et al. (2010). An example for resveratrol is discussed under section "Drug affinity responsive target stability."

Target deconvolution by comparison

One possibility to profile known compounds is the screening for cytotoxicity or growth inhibition of cancer-cell-line panels. The National Cancer Institute established a panel of 60 human cancer cell lines (NCI60) derived from various organs (Sharma et al., 2010). In a COMPARE analysis, the pattern of toxicity of small molecules to the NCI60 cell line panel is compared. Small molecules that have similar patterns of toxicity are correlated by the COMPARE algorithm, and the resulting pattern suggests that the small molecules may have a similar MoA (Tashiro and Imoto, 2012). In addition, the connectivity Map database was also used which collects gene expression data of cells treated with small molecules. The transcript profile data of a compound of interest are compared to analogous data that have been collected on hundreds of small molecules with known molecular targets.

Phage display

In phage display the genes encoding proteins of interest are cloned into bacteriophages to form fusion proteins

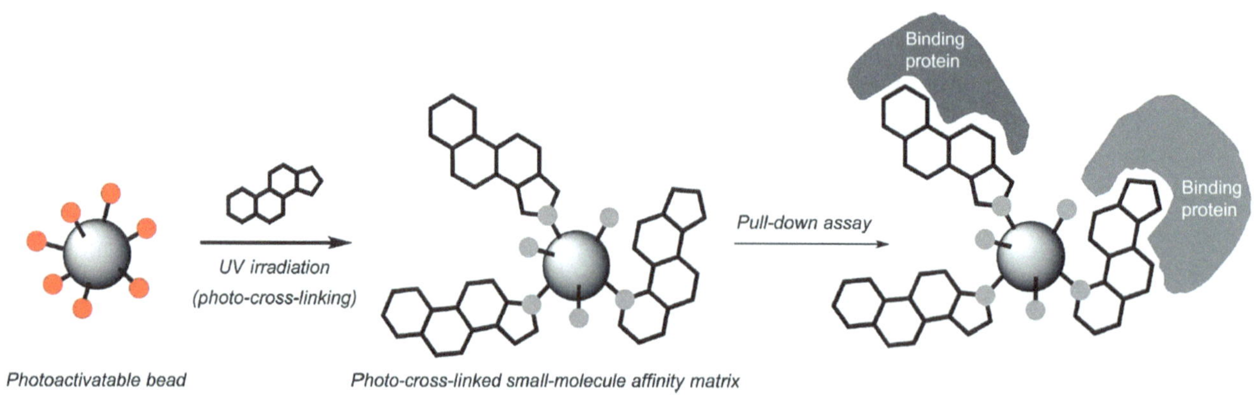

FIGURE 23.7 Application of photo-cross-linked small-molecule affinity matrices. *Adapted partly from Kanoh, N., 2016. Photo-cross-linked small-molecule affinity matrix as a tool for target identification of bioactive small molecules. Nat. Prod. Rep. 33, 709–718.*

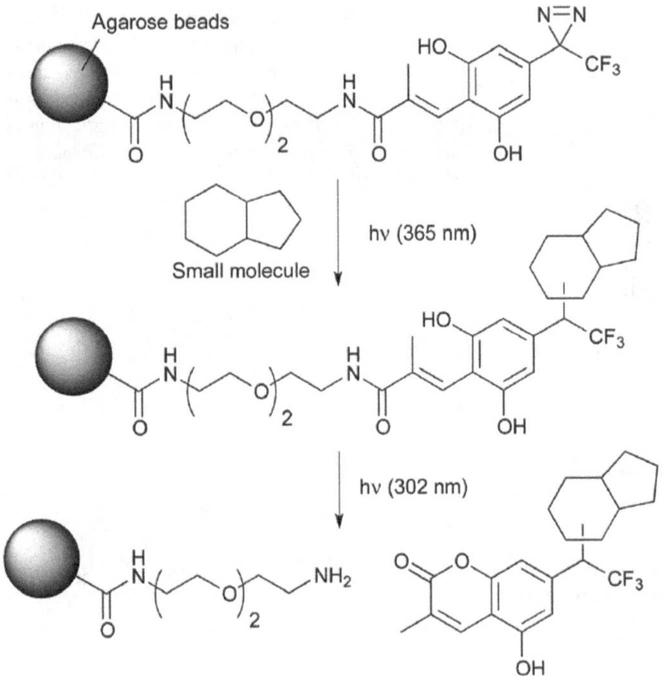

FIGURE 23.8 Reactivity pattern of the photogenerated carbens. *Adapted partly from Kanoh, N., 2016. Photo-cross-linked small-molecule affinity matrix as a tool for target identification of bioactive small molecules. Nat. Prod. Rep. 33, 709–718.*

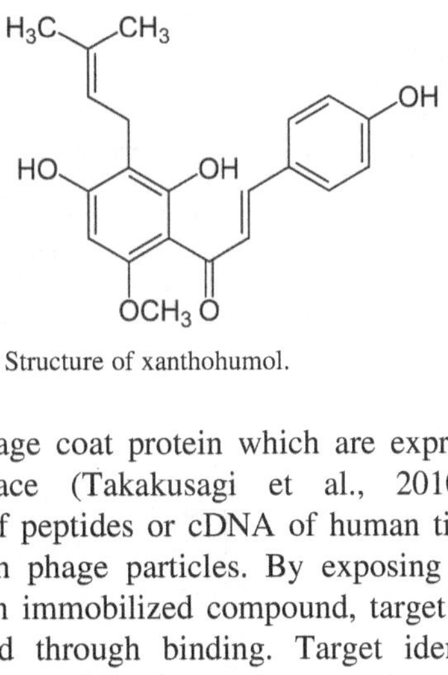

FIGURE 23.9 Structure of xanthohumol.

with the phage coat protein which are expressed on the phage surface (Takakusagi et al., 2010). Random sequences of peptides or cDNA of human tissues can be displayed on phage particles. By exposing and binding phages to an immobilized compound, target proteins can be identified through binding. Target identification is achieved by amplification and sequencing of the phage DNA (Fig. 23.11).

Cellular thermal shift assay

Thermal shift assay was extended to investigate target interactions of drug molecules in cells and tissues. Intact cells were heated to various temperatures, in the absence or presence of drug molecules. After lysis of the cells, the soluble fraction was examined by Western blot analysis, which revealed that the proteins were denatured in a temperature-dependent manner. The melting curve of certain proteins is shifted upon interaction with the drug molecules in living cells. The only disadvantage of cellular thermal shift assay (CETSA) is that it requires various antibodies, thus this method cannot be used for the identification of unknown target proteins (Jafari et al., 2014) (Fig. 23.12).

The principle of CETSA has been expanded to proteome-wide investigations. Thermal proteome profiling uses quantitative mass spectrometry with isobaric mass tags (Franken et al., 2015). Live cells were heated to 10 different temperatures in the absence or presence of drug molecules followed by digestion of the peptides and conjugation with 10 different tandem mass tags (TMTs). The TMT-conjugated peptides were then pooled and separated by liquid chromatography. Tandem mass spectrometry was used to identify and quantify the peptides at each temperature. The melting curve and melting temperature (Tm) of each protein were compared in the absence and presence of drug molecules. Significant Tm shift was characteristic of those proteins that interacted with the drug molecules.

Stability of protein from rates of oxidation

Stability of protein from rates of oxidation is based on the principle of thermodynamic changes in protein folding (Strickland et al., 2013) (Fig. 23.13). Cell lysates (in the absence or presence of a drug molecule) are treated with various concentrations of a protein-denaturing agent (guanidinium chloride). Upon denaturing, hydrogen peroxide is used to oxidize the methionine residues. After digestion of the proteins, the oxidized methionine residues in the peptide fragments are analyzed by MS. The drug-binding targets are elucidated by comparing the shifts of the curves, which show the amount of oxidized methionine

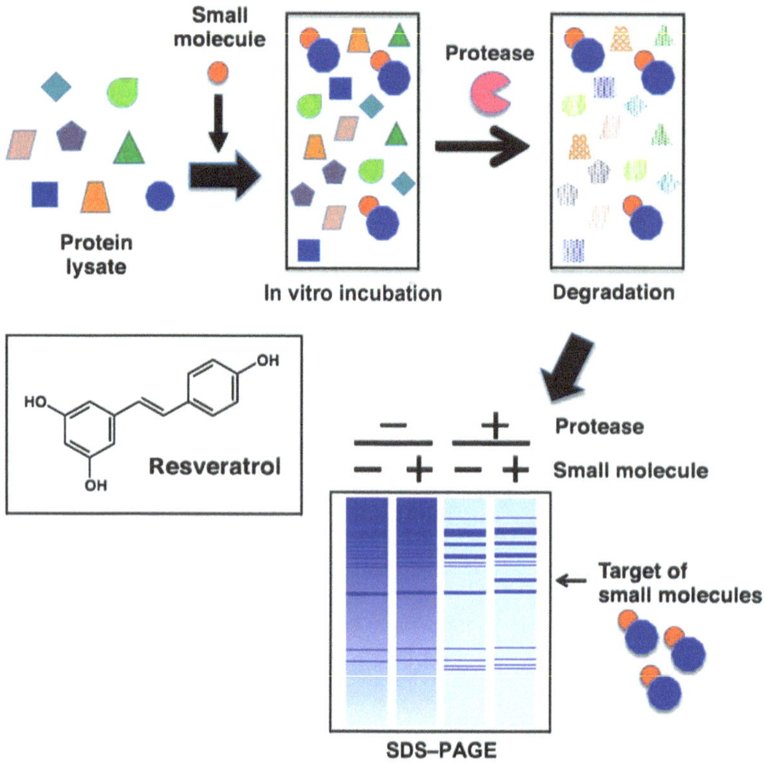

FIGURE 23.10 Identification of the direct molecular target for resveratrol applying drug affinity responsive target stability. *SDS–PAGE*, sodium dodecyl sulfate–polyacrylamide gel electrophoresis. *Adapted from Tashiro, E., Imoto, M., 2012. Target identification of bioactive compounds. Bioorg. Med. Chem. 20, 1910–1921.*

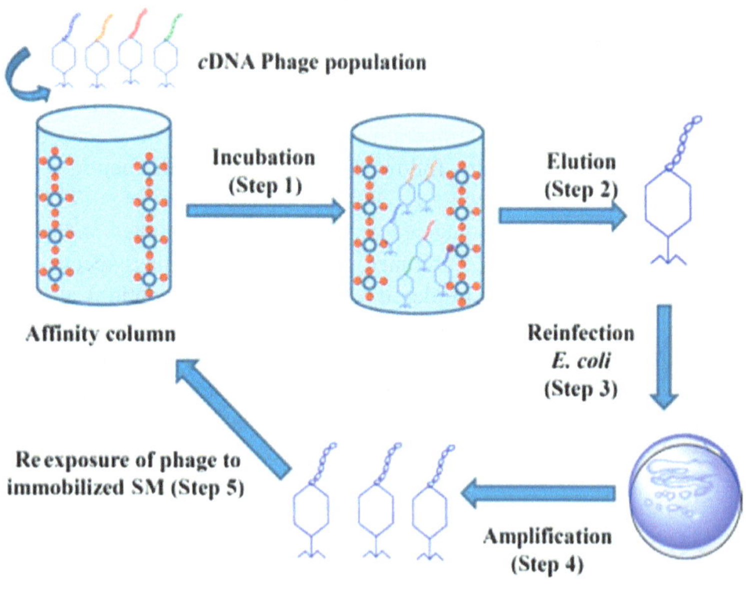

FIGURE 23.11 The principle of the phage display approach. *E. coli, Escherichia coli. Adapted from Ares, J. M.B., Durán-Peña, M.J., Hernández-Galán, R. et al., 2013. Chemical genetics strategies for identification of molecular targets. Phytochem. Rev. 12, 895–914.*

and the concentration of guanidinium chloride. However, there is a serious limitation to this concept. Since the occurrence of methionine in proteins is rather low; proteins that lack methionine cannot be identified.

Natural products containing a chemically reactive substructure (such as Michael acceptor enones) can create covalent linkages with their protein target. This unique feature allows rapid target identification and mapping combined with ABPP-based chemoproteomic strategies (Nomura and Maimone, 2018).

Quantitative proteomics approaches

Affinity chromatography approaches coupled with mass spectrometry are able to systematically identify small molecule–protein interactions. On the other hand,

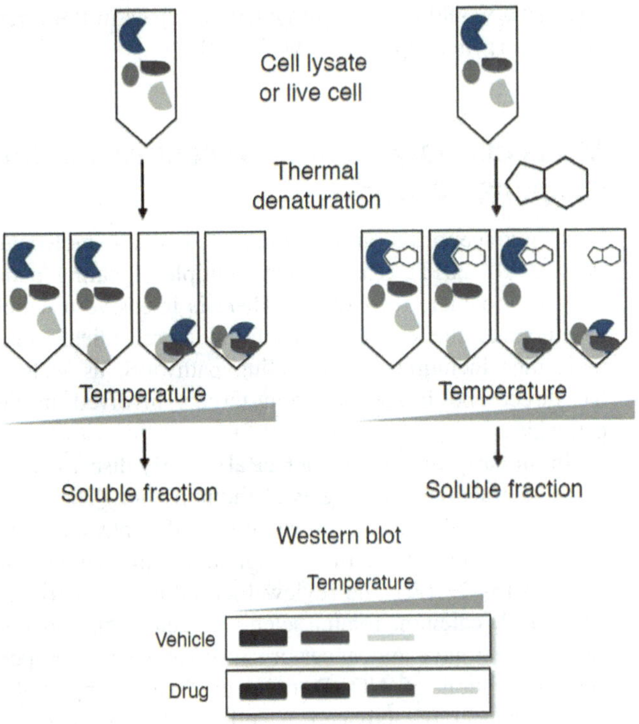

FIGURE 23.12 Major steps of cellular thermal shift assay. *Adapted from Park, H., Ha, J., Park, S.B., 2019. Label-free target identification in drug discovery via phenotypic screening. Curr. Opin. Chem. Biol. 50, 66–72.*

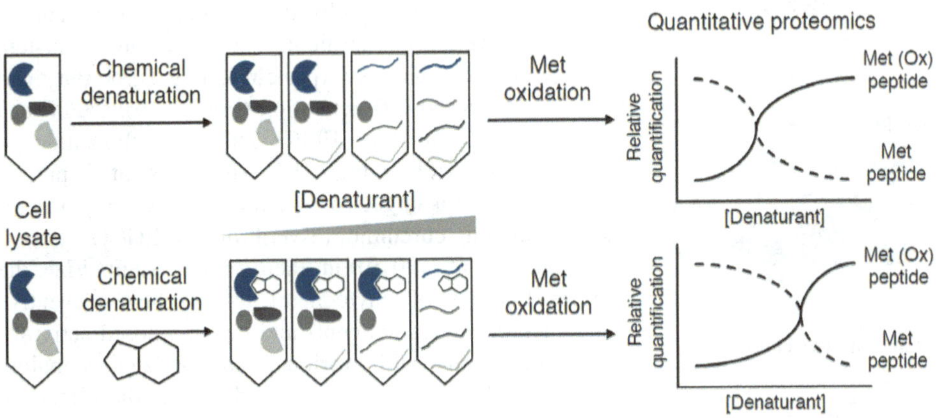

FIGURE 23.13 The principle of stability of protein from rates of oxidation. *Adapted from Park, H., Ha, J., Park, S.B., 2019. Label-free target identification in drug discovery via phenotypic screening. Curr. Opin. Chem. Biol. 50, 66–72.*

nonspecific (NS) binding remains a critical issue, which often provides false results. Metabolic or chemical labeling, or label-free approaches, could overcome these limitations (Wang et al., 2016).

Stable isotope labeling by amino acids in cell culture

In order to identify the potential binding partners of curcumin, it was added to a cultured medium containing live cells. Dimethyl sulfoxide (DMSO) was used as a negative control. Curcumin binds to its protein targets and then the cells are allowed to grow in "light"-labeled medium. The treated cells are then incubated with both JW-RF-010 (a cyclic hypervalent iodine reagent) and iodoacetamide cysteine probes. In the next step, the cells are lysed to release the proteome. The proteins in the "light" and "heavy" cells are combined and subjected to click chemistry with biotin that contains either the azide or the alkyne group. For enrichment, the protein mixture is bound to streptavidin agarose beads and washed, removing unbound proteins. The mixture is then subjected to trypsin digestion and washed again to separate unbound peptides. Finally, photocleavage is used to release the bound peptides. The eluted peptides are subjected to LC-MS/MS. Those proteins that have similar "light" and "heavy" intensities are considered to be NS binders of curcumin, while those that show lower "light" intensity compared to the "heavy" intensity are considered to be specific (S) binding proteins (Fig. 23.14). Principally, pretreatment of curcumin blocked the cysteine residues within the specific binding proteins, which prevented binding to the cysteine specific probes, thus its intensities were reduced (Wang et al., 2016).

Chemotype-specific resistance for drug target identification

"Chemotype-specific resistance" is the result of a mutation in the protein target that suppresses inhibitor binding. Normally, such mutations are "silent" and do not alter protein function to cause a phenotype. In this genetic approach, the mutant forms of the tentative targets are generated and tested as to whether such mutations could suppress the inhibitory activity of the drugs in in vitro cell-based assays. This approach provides a validation of the drug's direct targets in human cells. On the other hand, this approach has certain limitations. While inhibitor-resistant human cells can be generated, it is difficult to separate mutations that cause resistance from those that are unrelated genes. The human genome is rather complex, therefore genetically tractable organisms such as yeast can be used. Sometimes it is not completely clear whether the frequently chemotype-specific resistance would come from genetic changes in the direct protein targets (Kapoor and Miller, 2017).

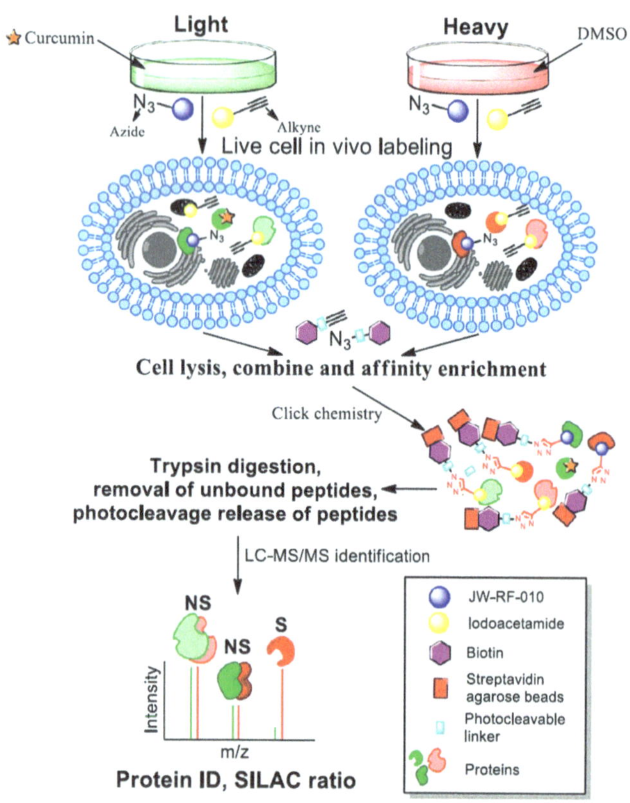

FIGURE 23.14 Identification of curcumine-binding proteins by stable isotope labeling by amino acids in cell culture. *SILAC*, stable isotope labeling by amino acids in cell culture; *NS*, nonspecific; *S*, specific. *Adapted from Wang, J., Gao, L., Lee, Y.M., et al., 2016. Target identification of natural and traditional medicines with quantitative chemical proteomics approaches. Pharmacol. Ther. 162, 10–22.*

Molecular targets of nutraceuticals in the signaling networks

Most nutraceuticals are multitargeted small molecules (Lee et al., 2011). They affect multiple signaling pathways to modulate their effect. There is intensive research to elucidate the exact MoA of the nutraceutical small molecules including the signaling pathways, as well as the direct and indirect protein targets involved in the pathways.

In an early review, Milner et al. (2001) discussed the major interactions and targets of the nutrients. (The structures of the major nutraceuticals involved in physiological processes and intracellular signaling are shown in Figs. 23.15–23.18.) The review focused on six nutrients, vitamin D, calcium, folate, selenium, genistein, and resveratrol, that have anticancer or cancer-preventive properties. Activation of NF-κB (nuclear factor kappa-light-chain-enhancer of activated B cells) and AP1 (activator protein-1) (Surh, 2003) is one of the best studied cancer preventive signaling pathways affected by curcumin, EGCG, resveratrol, capsaicin, and genistein. Genistein specifically inhibits AKT (protein kinase B) activity and AKT-mediated NF-κB activation (Fig. 23.19). EGCG can block the activities of PI3K (phosphatidylinositide 3-kinase) and AKT, while IKK (inhibitor of kappa B kinase) is also a target for chemopreventive phytochemicals, including curcumin, resveratrol, and EGCG.

Singh et al. (2019) in their review summarize the recent findings of phytochemicals as chemopreventive and chemotherapeutic agents and their proposed activities and interactions with various cancer-related signaling pathways. Another recent review discusses the effects of phytochemicals on cellular signaling (Pham et al., 2019).

Another important chemopreventive pathway affected by nutraceuticals is the activation of NRF2 [nuclear factor-erythroid 2p45 (NF-E2)-related factors] (Lee and Surh, 2005) (Fig. 23.20). NRF2 is a transcription factor that regulates the expression of many detoxification or antioxidant enzymes. Curcumin and caffeic acid phenethyl ester (CAPE) disrupt the NRF2–KEAP1 (Kelch-like-ECH-associated protein 1) complex, leading to increased NRF2 binding to ARE. Sulforaphane (SFN) directly interacts with KEAP1 by covalent binding to its thiol groups. 6-(Methylsulfinyl)hexyl isothiocyanate (6-HITC)—a sulforaphane analog from Japanese horseradish—stimulates nuclear translocation of NRF2, which subsequently activates ARE (antioxidant-responsive element). The recent developments in this research were summarized by Kwak and Kensler (2010).

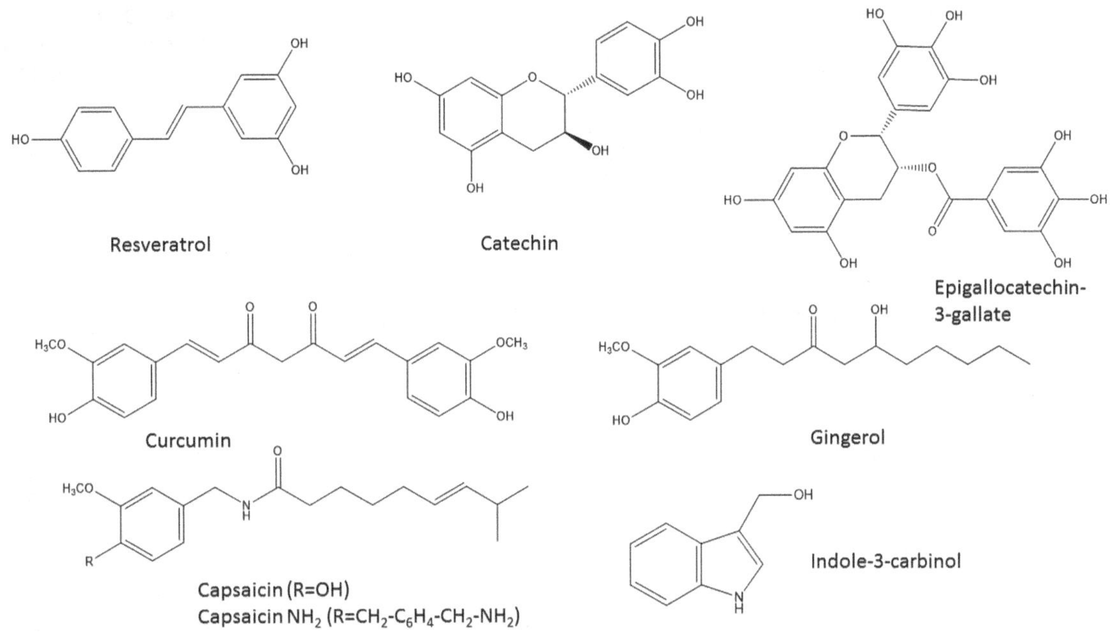

FIGURE 23.15 Various nutraceuticals that interact with protein targets in intracellular signaling.

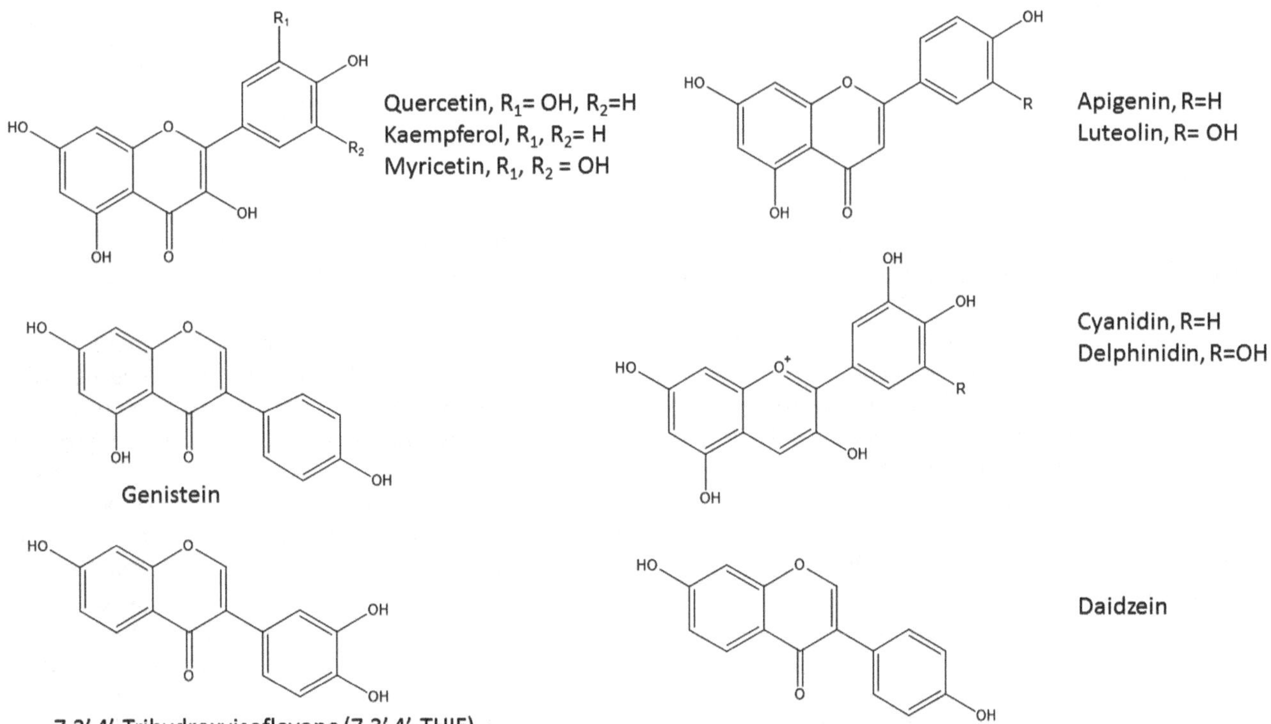

FIGURE 23.16 Flavone and isoflavone nutraceuticals that interact with protein targets in intracellular signaling.

A recent study revealed that sulforaphane attenuates tumor necrosis factor-α (TNF-α)-induced expression of intracellular adhesion molecule-1 (ICAM-1) in ECV 304 cells. Pretreatment of ECV 304 cells with SFN dose-dependently inhibited the secretion of proinflammatory cytokines, such as interleukin (IL)-1b, IL-6, and IL-8. SFN inhibited TNF-α-induced NF-κB DNA-binding activity. Furthermore, SFN decreased TNF-α-mediated

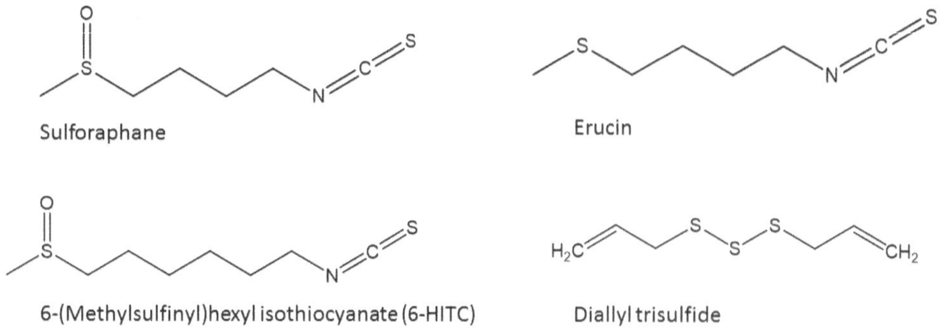

FIGURE 23.17 Sulfur-containing nutraceuticals that interact with protein targets in intracellular signaling.

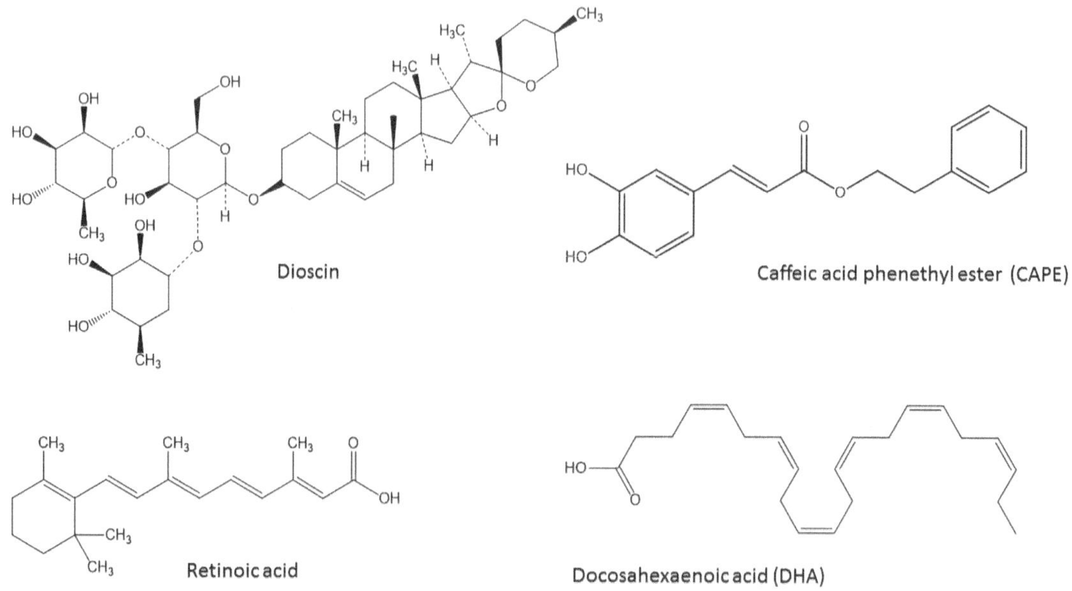

FIGURE 23.18 Additional nutraceuticals involved in physiological processes.

phosphorylation of IKK, Rho A, ROCK, ERK1/2, and PAI-1 (plasminogen activator inhibitor-1) levels. As an overall effect, SFN inhibited NF-κB DNA-binding activity and downregulated TNF-α-mediated induction of ICAM-1 in endothelial cells by inhibiting the Rho A/ROCK/NF-κB signaling pathway (Hung et al., 2014).

Another chemopreventive pathway is based on β-catenin signaling (Fig. 23.21). β-Catenin is a multifunctional protein which binds to the cytosolic tail of E-cadherin and connects actin filaments through α-catenin to form the cytoskeleton. This route is associated with Wnt signaling and is involved in tumorigenesis.

Curcumin downregulates β-catenin through caspase-mediated degradation of the protein, resulting in decreased DNA-promoter-binding activity of the β-catenin−TCF/LEF (T-cell factor/lymphoid enhancer factor) complex and reduced levels of c-MYC protein. CAPE and resveratrol also attenuate expression of β-catenin. EGCG inhibits β-catenin−TCF4 reporter activity and reduces β-catenin protein levels. Indole-3-carbinol shifts the pattern of β-catenin mutations, hindering their nuclear translocation. Further studies confirmed that Wnt and Hedgehog (Hh) signaling pathways are critical in the activation of cancer stem cells and are also associated with cancer progression and metastasis. The latest knowledge on targeted inactivation of Wnt and Hh signaling pathways for the prevention and treatment of tumor progression by nutraceuticals have been reviewed (Sarkar et al., 2010). The curcumin-affected molecular targets and pathways have been summarized by Anand et al. (2008) and show how curcumin mediates its anticancer and chemopreventive activities. The latest results on spice-derived nutraceuticals affecting various pathways and biological targets through the modulation of various transcription/growth factors, protein kinases, and inflammatory mediators were reviewed by Sung et al. (2012) (Fig. 23.22).

Another study reports that nutraceuticals could influence cancer development and tumor cell behavior through modulating the *p53* tumor-suppressor gene pathway (Fig. 23.23) (Smeltzer and Kim, 2005).

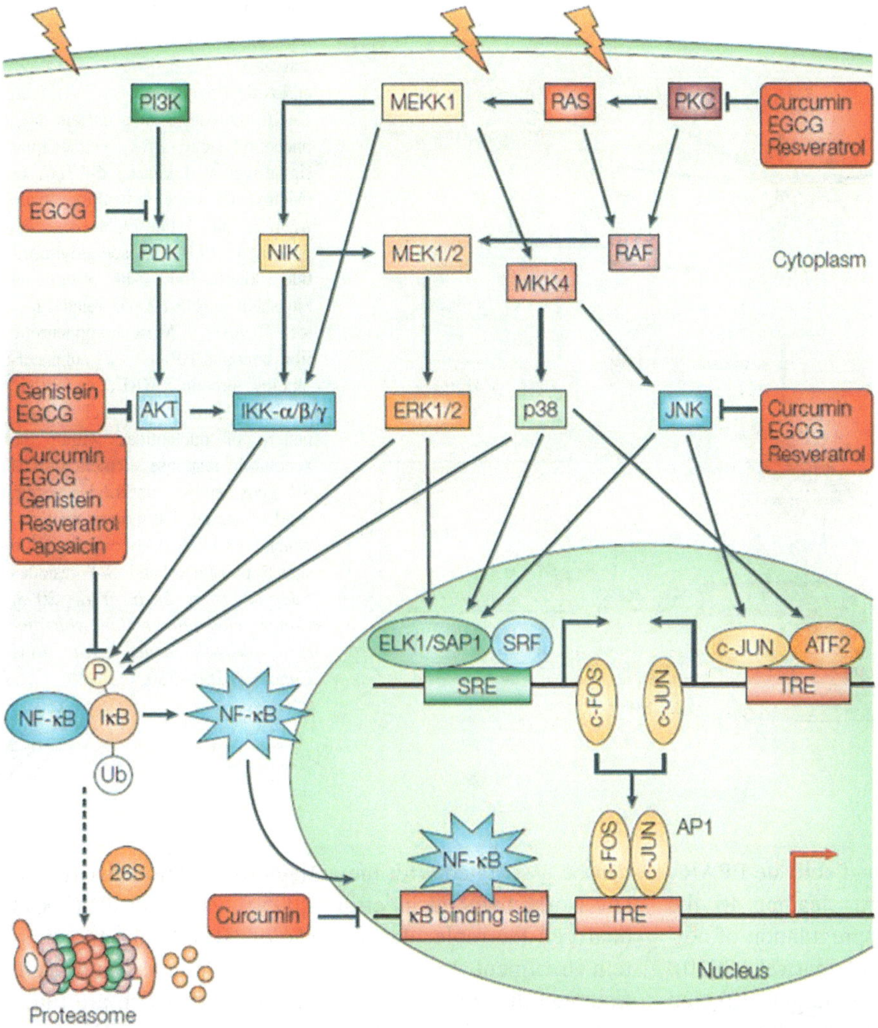

FIGURE 23.19 Interaction of various nutraceuticals with the NF-κB and AP1 activation pathway. *AP1*, activator protein-1. *JNK*, c-Jun N-terminal kinase;*SRE, serum response element;SRF, serum response factor; RAF, rapidly accelerated fibrosarcoma; TRE, TPA-responsive element; TPA, 12-O-tetradecanoyl-phorbol-13-acetate. Adapted from Surh, Y.J., 2003. Cancer chemoprevention with dietary phytochemicals. Nat. Rev. Cancer 3, 768–780.*

Anthocyanins, ellagitannins, and hydrolyzable tannins in pomegranate fruit extract possess antitumor effects (Afaq et al., 2005), via the inhibition of TPA-induced phosphorylation of ERK1/2, p38 and JNK1/2, as well as through activation of NF-κB and IKKα and phosphorylation and degradation of IκBα. Bioflavonoids were also examined for their antiinflammatory effects through their interaction with the protein targets involved in the arachidonic acid cascade (Yoon and Baek, 2005). Catechin and EGCG inhibited cyclooxygenase 2 (COX-2) expression. Green tea catechin and EGCG displayed COX inhibition activity. EGCG and resveratrol were particularly effective COX-1 inhibitors. Genistein downregulates COX-2 promoter activity in colon cancer cells transfected with a COX-2 reporter gene system.

Aggarwal and Shishodia (2006) extensively investigated the MoA of a number of nutraceuticals for various diseases. They provided evidence that numerous agents identified from fruits and vegetables can interfere with several cell-signaling pathways and act on multiple molecular targets. They claimed that curcumin is one of the broadest spectrum nutraceuticals. Cell-signaling pathways inhibited by curcumin alone include NF-κB, AP-1, STAT3, Akt, Bcl-2, Bcl-X(L), caspases, PARP, IKK, EGFR, HER2, JNK, MAPK, COX2, and 5-LOX. Curcumin has been shown to interfere with signaling pathways, including cell cycle (cyclin D1 and cyclin E), apoptosis (activation of caspases and downregulation of antiapoptotic gene products), proliferation (HER-2, EGFR, and AP-1), survival (PI3K/AKT pathway), invasion (MMP-9 and adhesion molecules), angiogenesis (VEGF), metastasis (CXCR-4), and inflammation (NF-jB, TNF, IL-6, IL-1, COX-2, and 5-LOX). The activity of curcumin reported against various cancers reflects its ability to affect multiple targets (Fig. 23.24).

Curcumin and related nutraceuticals are also beneficial for inflammation-induced obesity and metabolic diseases. They suppress the proinflammatory transcription factor

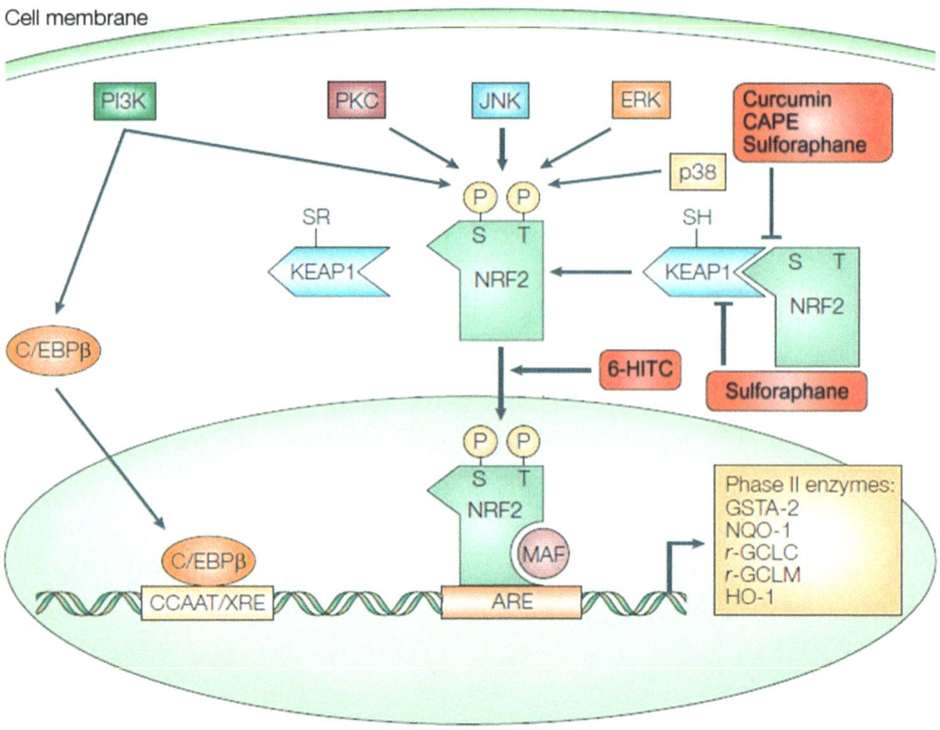

FIGURE 23.20 Interaction of various nutraceuticals with the NRF2 activation pathway. *ARE*, antioxidant-responsive element; *CAPE*, curcumin and caffeic acid phenethyl ester; *ERK*, extracellular signal-regulated kinase; *6-HITC*, 6-(Methylsulfinyl)hexyl isothiocyanate; *KEAP1*, Kelch-like-ECH-associated protein 1; *PI3K*, phosphatidylinositide 3-kinase. JNK: c-Jun N-terminal kinaseNRF2: NF-E2 p45-related factor 2MAF: Musculoaponeurotic fibrosarcomaEBPβ: enhancer-binding protein βXRE: xenobiotic response element. CCAAT distinct pattern of nucleotides within the xenobiotic response element (XRE) SR covalently modified cystein residueS serine, T threonine residues, which provide O-phosphorylation sites,P phosphorylated S/T residues *Adapted from Surh, Y.J., 2003. Cancer chemoprevention with dietary phytochemicals. Nat. Rev. Cancer 3, 768–780.*

NF-κB, STAT-3, and Wnt/ β-catenin, and activate PPARγ and Nrf2 cell-signaling pathways, thus leading to the downregulation of adipokines and the upregulation of adiponectin and related gene products (Aggarwal, 2010). Potential treatment and the pathways and targets affected by nutraceuticals in chronic inflammation have been reviewed (Gupta et al., 2010). A review on curcumin summarized the knowledge of the potential molecular target of this nutraceutical (Esatbeyoglu et al., 2012). Kunnumakkara et al. (2017) reported curcumin to be a multitargeting agent for multiple chronic diseases.

Studying molecular targets by expression proteomics

In expression proteomic studies of quercetin (Wenzel et al., 2004) and genistein (Fuchs et al., 2005a), the changes in the protein expression level revealed which downstream proteins are up- or downregulated. Homocysteine (Hcy) caused dysregulation of gene expression. Cells were exposed for 24 h to 25 mM Hcy alone and in combination with 2.5 mM genistein. Protein expression was analyzed by two-dimensional gel electrophoresis. The altered spot intensities were identified by peptide mass fingerprinting. Genistein reversed Hcy-induced changes of the proteins involved in metabolism, detoxification, and gene regulation. Some of those effects can be associated with the antiatherosclerotic activities of soy isoflavone (Fuchs et al., 2005b). Zhang et al. (2007) analyzed the molecular response of HL-60 cells to genistein treatment by the gel-based proteomics approach. Fourteen differentially expressed proteins which are functionally involved in metabolism, cell signaling, RNA processing, cell proliferation and motility, and chaperones were identified.

In another study, Chinese medicines such as saponins that express cancer prevention and chemotherapeutic agents were studied to give an understanding of the molecular basis of their activities. Wang et al. (2007) used expression proteomics to examine the cytotoxic effect of dioscin, a glucoside saponin, on human myeloblast leukemia HL-60 cells. Dioscin induces apoptosis in HL-60 cells in a time-dependent manner. Protein profiling of the microsomal fraction with enriched plasma membrane proteins isolated from HL-60 cells revealed that proteins act as chaperones and/or mediators of protein folding and were substantially altered in expression cells upon dioscin stimuli.

Erucin, a recently identified dietary isothiocyanate, is present in cruciferous vegetables and has recently been considered a promising cancer chemopreventive nutraceutical. It affects the molecular pathways involved in apoptosis and cell cycle arrest, such as PARP-1 cleavage, p53, and p21 protein expression (Melchini et al., 2009).

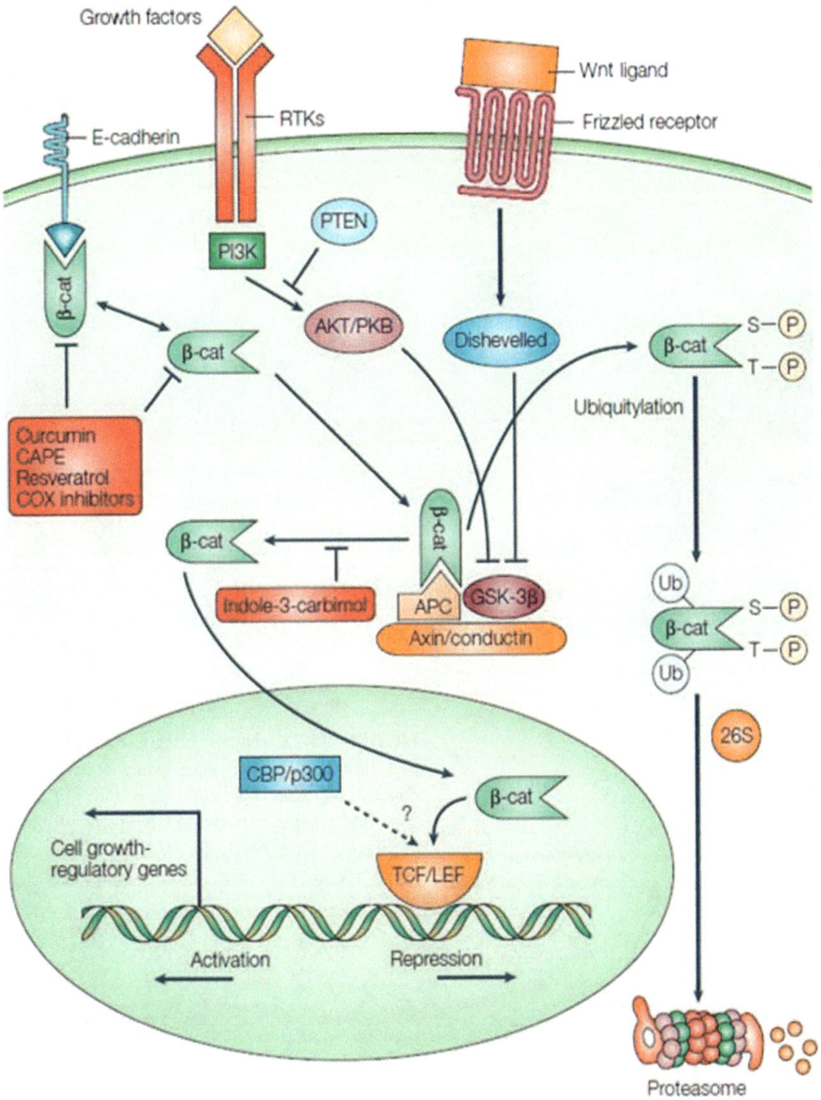

FIGURE 23.21 Interaction of various nutraceuticals with the β-catenin activation pathway. *LEF*, lymphoid enhancer factor; *PI3K* (phosphatidylinositide 3-kinase); *TCF*, T-cell factor. PTEN: Phosphatase and tensin homologAPC: adenomatous polyposis coli geneRTK: Receptor tyrosine kinasePKB: Protein kinase BAKT: AKR mouse strain developing thymic ymphoma (AKT is a synonym for PKB)CBP: CREB-binding protein (CREB cAMP-response-element-binding protein) GSK: Glycogen synthase kinaseUb: Ubiquitinated sitesS serine, T threonine residues, which provide phosphorylation sites,P phosphorylated S/T residues. *Adapted from Surh, Y.J., 2003. Cancer chemoprevention with dietary phytochemicals. Nat. Rev. Cancer 3, 768–780.*

Italian researchers revealed that the molecular mechanisms of action of erucin include modulation of phase I (Cyp450), phase II [quinone reductase (QR) glutatione transferase (GST)], and phase III detoxification, regulation of cell growth by induction of apoptosis and cell cycle arrest, induction of reactive oxygen species (ROS) mechanisms, and modulation of tumor suppressor proteins (p53, p21) (Melchini and Traka, 2010). The related compound, sulforaphane, present in broccoli, resulted in dose-dependent apoptosis and G2/M cell cycle arrest, which was associated with the activation of caspases (3, 8, and 9), inactivation of PARP, p53-independent upregulation of p21(CIP1/WAF1), and inhibition of the Cdc2/Cyclin B1 complex. Interestingly, sulforaphane also inhibited the AKT and mTOR survival pathways. However, the direct molecular target has not been identified (Suppipat et al., 2012).

Identification of the molecular targets of nutraceuticals

Application of affinity-based methods

Pull-down assays

Resveratrol

Various direct resveratrol targeting proteins (RTPs) were identified by affinity-based approaches in the last decade, including cytosolic protein quinone reductase 2 (NQO2) and glutathione sulfotransferase (GSTP1) as chemical detoxification enzymes and estrogen receptor ERβ, 12-o-tetradecanoylphorbol-13-acetate (TPA)-induced IjB kinase, and COX-2 (Hsieh and Wu, 2010).

Resveratrol was immobilized on epoxy-activated agarose forming a resveratrol affinity column, which was

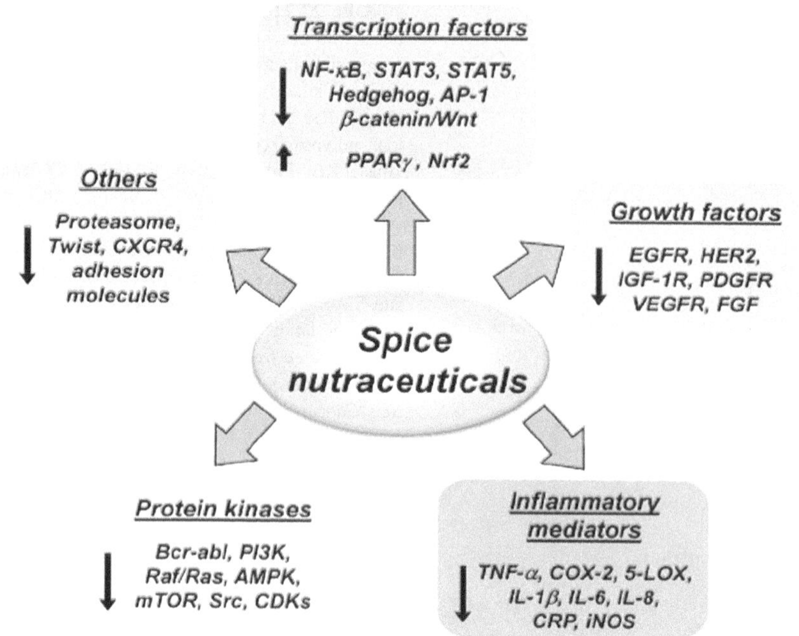

FIGURE 23.22 Putative molecular targets of various anticancer nutraceuticals. *Adapted from Sung, B., Prasad, S., Yadav, V.R., et al., 2012. Cancer cell signaling pathways targeted by spice-derived nutraceuticals. Nutr. Cancer 64, 173–197.*

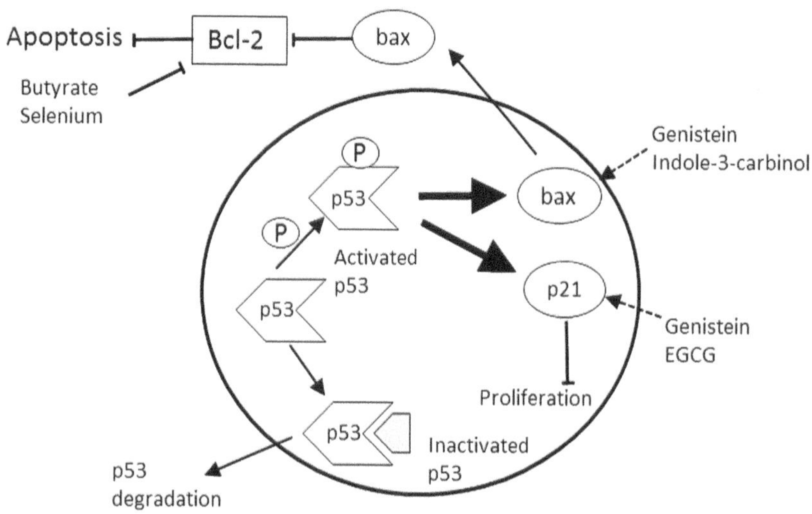

FIGURE 23.23 Nutraceuticals that affect the *p53* tumor-suppressor gene pathway. *Based on Smeltzer, A., Kim, Y.S., 2005. The effects of bioactive food components on p53 pathway in cancer prevention. Nutr. Today 40, 50–53.*

used to detect and isolate RTPs. A 22-kDa polypeptide, RTP-22, eluted with resveratrol was identified by MALDI-TOF MS. Cloning and expression of the protein in *Escherichia coli* revealed that it is a dihydronicotinamide riboside quinone reductase 2 (NQO2). The utility of a resveratrol affinity column was additionally explored with extracts derived from different-stage prostate cancer cells (Wang et al., 2004).

Genistein

In order to identify proteins that are targets of GEN, Belenky et al. (2003) used a GEN-agarose–affinity matrix. 2-Carboxygenistein derivative was first synthesized and then attached to the resin. The eluted proteins in the various fractions were subjected to SDS–PAGE analysis. Affinity elution resulted in a p38 protein from MCF-7 cells, which belongs to the DING proteins. However, there were no matching peptide sequences in the available protein databases. Previously, a similar N-terminal sequence was linked to a p205 synovial T-cell stimulatory protein.

Epigallocatechin-3-gallate

Ermakova et al. (2005) identified a novel molecular target for EGCG using affinity chromatography, two-dimensional electrophoresis, and mass spectrometry.

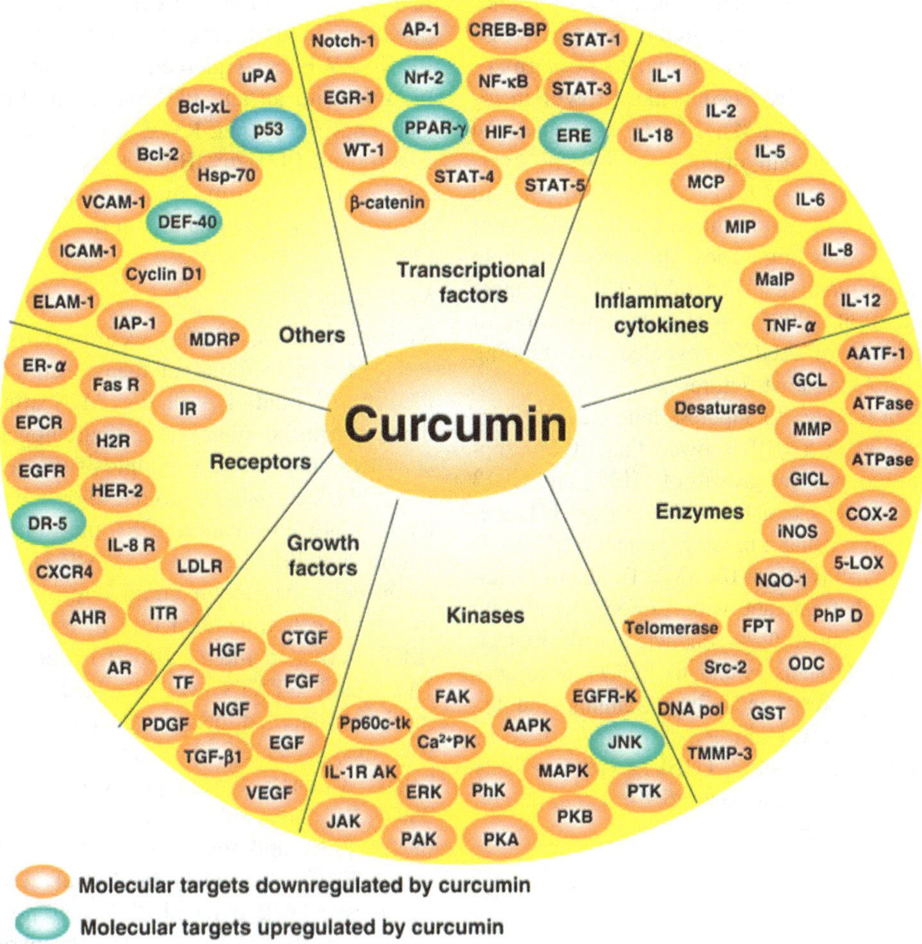

FIGURE 23.24 Molecular targets that are affected by curcumin treatment. NF-kB, nuclear factor-kappa B; AP-1, activating protein1; STAT, signal transducers and activators of transcription; Nrf-2, nuclear factor 2-related factor; Egr-1, early growth response gene-1; PPAR-c, peroxisome proliferator-activated receptor-gamma; CBP, CREB-binding protein; EpRE; CTGF, connective tissue growth factor; EGF, epidermal growth factor; EGFRK, epidermal growth factor receptor-kinase; FGF, fibroblast growth factor; HGF, hepatocyte growth factor; NGF, nerve growth factor; PDGF, platelet-derived growth factor; TGF-b1, transforming growth factor-b1; VEGF, vascular endothelial growth factor; AR, androgen receptor; Arh-R, aryl hydrocarbon receptor; DR-5, death receptor-5; EGF-R, epidermal growth factor-receptor; EPC-R, endothelial protein C-receptor; ER-a, estrogen receptor-alpha; Fas-R, Fas receptor; H2-R, histamine (2)-receptor; InsP3-R, inositol 1,4,5-triphosphate receptor; IR, integrin receptor; IL-8-R, interleukin 8-receptor; LDL-R, low density lipoprotein–receptor; MMP, matrix metalloproteinase; TIMP, tissue inhibitor of metalloproteinase-3; iNOS, inducible nitric oxide oxidase; COX-2, cyclooxygenase-2; LOX, lipoxygenase; Gcl, glutamate-cysteine ligase; NAT, arylamine N-acetyltransferases; IAP, inhibitory apoptosis protein; HSP-70, heat-shock protein 70; TNF-a, tumor necrosis factor alpha; IL, interleukin; MCP, monocyte chemoattractant protein; MIF, migration inhibition protein; MIP, macrophage inflammatory protein; ERK, extracellular receptor kinase; IARK, IL-1 receptor-associated kinase; cAK, autophosphorylation-activated protein kinase; CDPK, Ca2 + -dependent protein kinase; cPK, protamine kinase; JAK, janus kinase; JNK, c-jun N-terminal kinase; MAPK, mitogen-activated protein kinase; TK, protein tyrosine kinase; FAK, focal adhesion kinase; PhK, phosphorylase kinase; pp60c-src, pp60c-src tyrosine kinase; PKA, protein kinase A; PKB, protein kinase B; PKC, protein kinase C; FPTase, farnesyl protein transferase; GST, glutathione S-transferase; HO, hemeoxygenase; ICAM-1, intracellular adhesion molecule-1; VCAM-1, vascular cell adhesion molecule-1; ELAM-1, endothelial leukocyte adhesion molecule-1; SHP-2, Src homology 2 domain-containing tyrosine phosphatase 2, uPA, urokinase-type plasminogen activator *Adapted from Anand, P., Sundaram, C., Jhurani, S., et al., 2008. Curcumin and cancer: an "old-age" disease with an "age-old" solution. Cancer Lett. 267, 133–164.*

Spots of interest were identified as vimentin, which is one of the type III intermediate filament (IF) proteins that is expressed during development in a wide range of cells. IFs are essential for the structure and mechanical integration of the cellular space and a variety of cellular functions such as mitosis, locomotion, and organizational cell architecture. Identification was confirmed by Western blot analysis using an antivimentin antibody (Ermakova et al., 2005). The binding of EGCG to vimentin has a K_d of 3.3 nM. EGCG inhibited phosphorylation of vimentin by cyclin-dependent kinase 2 (CDK2) and cAMP-dependent protein kinase at serines 50 and 55, proving that EGCG specifically inhibits cell proliferation by binding to vimentin. In a follow-up study, using EGCG-Sepharose 4B affinity, EGCG was found to be a small-molecule inhibitor of IGF-IR (insulin-like growth factor receptor) activity ($IC_{50} = 14\ \mu M$). EGCG abolished the anchorage-independent growth induced by IGF-IR overexpression and also prevented human breast and cervical cancer cell phenotype expression through the inhibition of IGF-IR downstream signaling. These findings may help explain the chemopreventive effect of EGCG on cancer development (Li et al., 2007). Molecular modeling of IGF-IR and docking of EGCG revealed the binding pocket of the IGF-IR crystal structure.

EGCG has been reported to suppress lung cancer (Shim et al., 2010). Shim et al. (2010) showed that EGCG suppressed anchorage-independent growth of H1299 and CL13 lung cancer cells. EGCG interacted with the Ras–GTPase-activating protein SH3 domain-binding protein 1 (G3BP1) with a high binding affinity ($K_d = 0.4\ \mu M$). To identify putative target proteins that interact with EGCG, an affinity chromatography experiment was conducted with EGCG-Sepharose 4B beads. Proteins bound to EGCG were subjected to 2D electrophoresis and silver staining, and the respective spots were digested with trypsin and identified by MALDI-TOF mass spectrometry. G3BP1 was positively identified by 13 matched peptides and 46.9% sequence coverage using mass spectrometry.

Quercetin

Quercetin inhibited 12-O-tetradecanoylphorbol-13-acetate (TPA)-induced transformation of JB6 promotion-sensitive mouse skin epidermal (JB6 P+) cells (Lee et al., 2008). The activation of AP1 and NF-κB induced by TPA was dose-dependently inhibited by red wine extract (RWE) or quercetin treatment. Western blot and kinase assay data revealed that RWE or quercetin inhibited MEK1 [mitogen-activated protein kinase/extracellular signal-regulated kinase (ERK)] and Raf1 kinase activities. Quercetin inhibited MEK1 completely at 5 μM, and Raf1 activity was inhibited by 24% at 10 μM. In comparison, resveratrol did not affect either MEK1 or Raf1 kinase activity. Pull-down assays revealed that RWE or quercetin (but not resveratrol) directly bound either MEK1 or Raf1.

The possible mechanism of red wine polyphenols was further investigated for their inhibitory effects on TNF-α-induced upregulation of MMP-9 and on the migratory phenotype of JB6 P+ mouse epidermal (JB6 P+) cells (Hwang et al., 2009a). RWE and quercetin inhibited the TNFα-induced upregulation of MMP-9 and cell migration, whereas resveratrol did not have significant inhibitory effects. The inhibitory effects of RWE and quercetin were mediated by suppression of the phosphorylation of Akt and the transactivation of AP1 and nuclear factor-αB. Kinase assay and pull-down assay revealed that RWE and quercetin bind directly to PI3K. The direct binding of PI3K–RWE or –quercetin was confirmed by immunoblotting with an anti-PI3K antibody. Furthermore, it was also confirmed that RWE and quercetin bind to TNF-α-activated PI3K in JB6 P+ cells. Molecular targets of quercetin with antiinflammatory properties were recently discussed (Karuppagounder et al., 2016).

Myricetin

Kumamoto et al. (2009) examined myricetin, a typical flavonol that is found in many fruits and vegetables, to identify its binding proteins. Myricetin acts by antioxidant, antiinflammatory, anticarcinogen, and antiviral activities and induces apoptosis in several cancer cell lines. Myricetin also inhibited thioredoxin reductase activity in A549 lung cancer cells and the enzyme activity of matrix metalloproteinase 2 (MMP-2) in colorectal carcinoma cells. Myricetin could interact with JAK1/STAT3 to inhibit cell transformation in epidermal growth factor (EGF)-activated mouse JB6 P+ cells. Direct binding of JAK1 and STAT3 was detected with Sepharose 4B-myricetin-coupled beads. As a positive control, MEK1 was also detected in Sepharose 4B-myricetin-coupled beads, which had been previously reported (Lee et al., 2007). Recently, pull-down assay revealed the direct binding of PI3K and myricetin in mouse skin lysates (Jung et al., 2010). Previously Jung et al. found that myricetin suppressed UVB-induced PI3K activity and subsequently attenuated the UVB-induced phosphorylation of Akt/$p70^{S6K}$ in mouse skin lysates.

Kaempferol

Kaempferol (KF), a natural dietary flavonoid, is a potent chemopreventive agent for cancer (Lee et al., 2010b,c). The activation of AP1 and NF-κB induced by EGF was also prevented by KF. The EGF-induced phosphorylation of Akt was completely suppressed by KF, although p38, JNK, and p90 ribosomal S6 kinase were unaffected by KF. Kinase profiling assay revealed that KF bound directly to PI3K. To investigate the direct binding interaction, the authors performed a KF pull-down assay, which revealed that PI3K specifically bound to the KF-Sepharose 4B beads. In order to confirm the specific inhibition of PI3K by KF, the pull-down assay was negative for ERK, p38, or JNK kinase. These data indicate that KF binds to PI3K and subsequently inhibits PI3K activation and downstream signaling.

The same investigators found that kaempferol suppressed UVB-induced COX-2 protein expression in mouse skin epidermal JB6 P+ cells and attenuated the UVB-induced transcriptional activities of Cox-2 and AP-1 through several kinases. However, it had no effect on the phosphorylation of the upstream MAPK regulator Src. On the other hand, other studies confirmed that kaempferol blocked Src kinase activity. A pull-down assay revealed that kaempferol competes with ATP for direct binding to Src and subsequent docking data suggest that kaempferol docks easily into the ATP-binding site of Src (Lee et al., 2010b,c). Therefore it can be considered as its direct molecular target.

7,3′,4′-Trihydroxyisoflavone

7,3',4'-Trihydroxyisoflavone (7,3',4'-THIF) is one of the metabolites of daidzein, which is a isoflavone from the soy bean. Lee et al. (2010a) investigated the molecular mechanisms of its chemopreventive activity. 7,3',4'-THIF prevented EGF-induced neoplastic transformation and proliferation of JB6 P+ mouse epidermal cells. 7,3',4'-THIF suppressed the phosphorylation of retinoblastoma protein at Ser-795 and Ser-807/Ser-811, which are the specific sites of phosphorylation by cyclin-dependent kinase (CDK) 4. It also inhibited the expression of G1 phase-regulatory proteins, including cyclin D1, CDK4, cyclin E, and CDK2. It also bound to PI3K, strongly inhibiting its kinase activity and thereby suppressing the Akt/GSK-3α/AP-1 pathway (Fig. 23.25). For direct binding studies, recombinant cyclin D1-CDK4, cyclin E-CDK2 complex, or PI3K was incubated with 7,3',4'-THIF-conjugated Sepharose 4B. 7,3',4'-THIF specifically binds CDK4 and CDK2 but does not compete with ATP for CDK4 and CDK2 binding. Similarly, 7,3',4'-THIF can bind directly to PI3K to inhibit its activation and downstream signaling.

Delphinidin

Delphinidin (3,3',4',5,5',7-hexahydroxyflavylium) is a dominant anthocyanidin component in red wine and berries. Delphinidin strongly inhibited TNF-α-induced COX-2 expression, and such effects are attributed to the inhibition of Fyn kinase activity (Hwang et al., 2009b). Kinase and pull-down assay data revealed that delphinidin inhibited Fyn kinase activity and directly bound to Fyn kinase noncompetitively with ATP. In contrast, resveratrol and gallic acid did not show significant inhibitory effects. Delphinidin inhibited the TNF-α-induced phosphorylations of JNK, p38 MAP kinase, Akt, p90RSK, MSK1, and ERK, and subsequently blocked the activation of the eukaryotic transcription factors AP-1 and NF-κB as a downstream effect.

In another study, Kwon et al. (2009) investigated the mechanism of the strong inhibition of MAPKK-4 and PI3K by delphinidin. A pull-down assay using delphinidin−Sepharose beads revealed that delphinidin binds directly to MAPKK4 or PI3K in an ATP-competitive manner. In vivo investigations using mouse skin revealed that the upregulation of COX-2 expression, MAPKK4 activity, and PI3K activity induced by UVB was eliminated with delphinidin treatment. Collectively, their results demonstrated that delphinidin targets MAPKK4 and PI3K directly to suppress COX-2

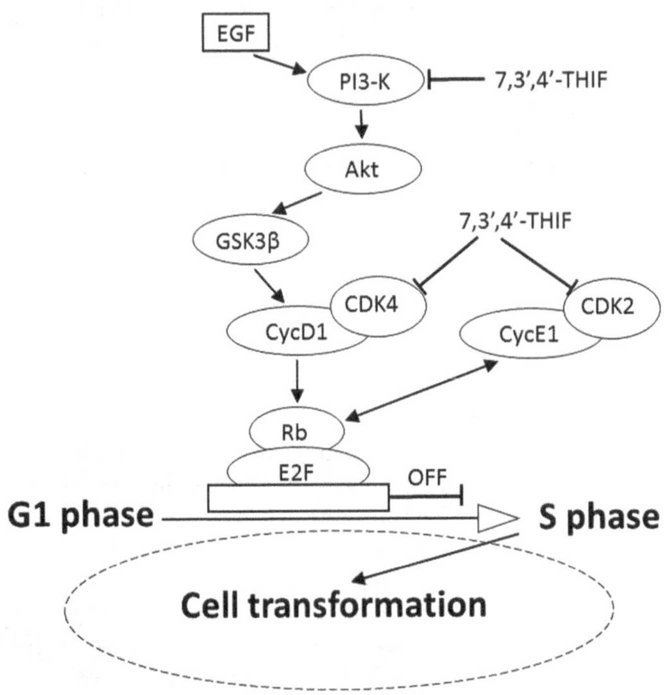

FIGURE 23.25 Hypothetical scheme of the mechanism of the chemopreventive action of 7,3',4'-THIF. *EGF*, epidermal growth factor; *7,3',4'-THIF*, 7,3',4'-Trihydroxyisoflavone. EGF: epidermal growth factorPI3-K: phosphoinositide 3-kinaseAkt: AKR mouse strain developing thymic lymphomaCycD1: G1/S-specific cyclin-D1CDK4: Cyclin-dependent kinase 4CycE1: G1/S-specific cyclin-E1CDK2 Cyclin-dependent kinase 2Rb: ribosomeE2F: Transcription activator that binds E2 recognition siteGSK3b: Glycogen synthase kinase 3b. *Based on Lee, D.E., Lee, K.W., Song, N.R., et al., 2010a. 7, 3′, 4′-Trihydroxyisoflavone inhibits epidermal growth factor-induced proliferation and transformation of JB6 P+ mouse epidermal cells by suppressing cyclin-dependent kinases and phosphatidylinositol 3-kinase. J. Biol. Chem. 285, 21458–21466.*

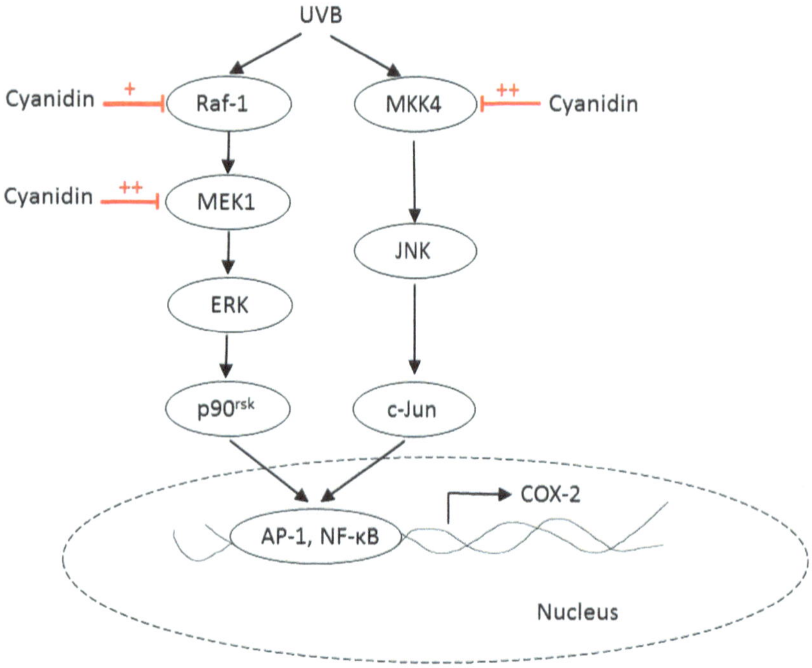

FIGURE 23.26 Simplified model of the effects of cyanidin on UVB-induced COX-2 expression. *COX-2*, cyclooxygenase 2. UVB: ultraviolet B irradiationRaf-1: Rapidly Accelerated Fibrosarcoma 1MEK1: Mitogen-activated protein kinase kinaseERK: extracellular receptor kinasep90rsk: 90 kDa ribosomal protein S6 kinase 1AP-1: activating protein1NF-kB: nuclear factor-kappa BCOX-2: cyclooxygenase-2c-Jun: protein encoded by the JUN geneJNK: c-Jun N-terminal kinaseMKK4: Mitogen-activated protein kinase kinase 4. *Based on Kim, J.E., Kwon, J.Y., Seo, S. K., et al., 2010. Cyanidin suppresses ultraviolet B-induced COX-2 expression in epidermal cells by targeting MKK4, MEK1, and Raf-1. Biochem. Pharmacol. 79, 1473–1482.*

overexpression, confirming a potential protective role against UVB-mediated skin carcinogenesis.

Cyanidin

Cyanidin, a plant pigment anthocyanidin, is another potent chemopreventive agent. The direct molecular targets that could reveal the molecular mechanism were investigated by Kim et al. (2010). Cyanidin inhibited UVB-induced COX-2 expression and prostaglandin E2 secretion in an epidermal skin cell line by suppressing the transactivation of nuclear factor-κB and AP1. Cyanidin inhibited the phosphorylation of JNK1/2, ERK1/2, and MEK1/2 by MKK4 and Raf-1. Using cyanidin-Sepharose 4B affinity matrix beads it was confirmed that cyanidin inhibited the activities of MKK4, MEK1, and Raf-1 through direct binding (Fig. 23.26). ATP competed with cyanidin for MKK4 binding but not for MEK1 or Raf-1 binding. These results indicate that cyanidin binds to the ATP-binding pocket of MKK4 and binds to Raf-1 and MEK1 at allosteric sites.

Apigenin

Several polyphenols, such as apigenin, have been reported to overcome resistance to Apo2L/TRAIL by upregulating death receptor 5 (DR5) in malignant tumor cells. However, the detailed mechanisms through which polyphenols increase Apo2L/TRAIL sensitivity remain unknown. Oishi et al. (2013) identified adenine nucleotide translocase-2 (ANT2) as a binding protein of apigenin which upregulated DR5 using magnetic fiberglass (FG) beads (Fig. 23.27). Knockdown of ANT2 enhanced Apo2L/TRAIL-induced apoptosis by upregulating DR5, similar to treatment with apigenin. Furthermore, knockdown of ANT2 attenuated the enhancement of DR5 expression and Apo2L/TRAIL-induced apoptosis by apigenin. These results propose that apigenin binds and inhibits ANT2, which induces DR5 expression and potentiates Apo2L/TRAIL sensitivity.

For immobilization of apigenin and genistein onto the beads, the magnetic FG beads with an epoxy linker were mixed with apigenin in DMF containing potassium carbonate at 37°C or 60°C for 24 h. The binding proteins of apigenin and genistein were purified from the human prostate cancer DU145 cell extracts and were identified by MALDI-TOF MS analysis. ANT2 was identified as a binding protein of apigenin, but not genistein. Therefore apigenin is responsible for inducing DR5 expression (Fig. 23.28).

Gingerol

A combination of 3D modeling (Fig. 23.29) and pull-down assay revealed the direct molecular target of gingerol (Jeong et al., 2009). [6]-Gingerol, a major component of ginger, exhibits antiinflammatory and antitumorigenic activities. Despite its potential efficacy in cancer, the mechanism by which [6]-gingerol shows such effects has still not been explored. The leukotriene A_4 hydrolase (LTA_4H) protein is considered an important target for cancer therapy. Jeong et al. (2009) used a reverse-docking approach that suggested LTA_4H to be a potential target (Fig. 23.29). Such prediction was supported by showing that [6]-gingerol suppresses anchorage-independent

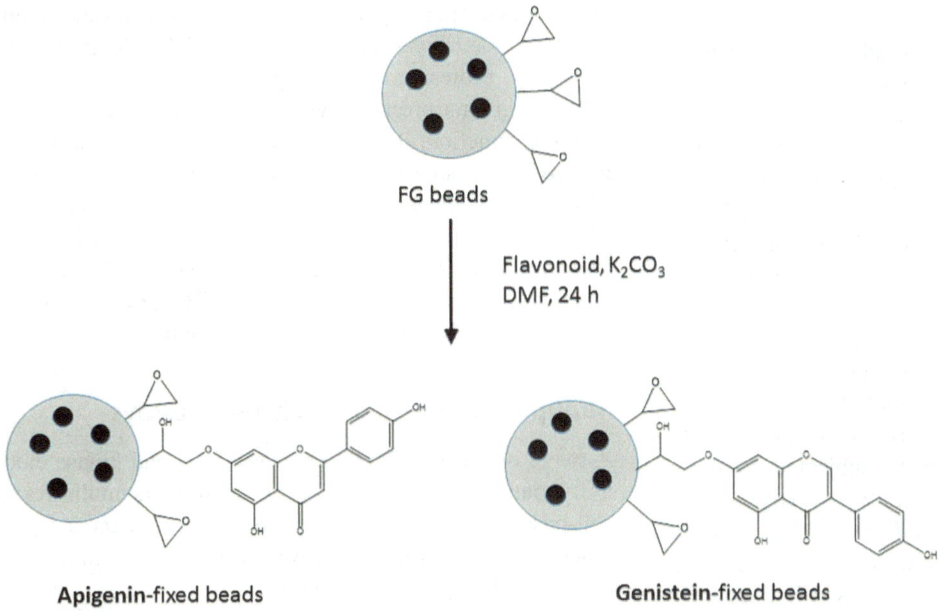

FIGURE 23.27 The immobilization of apigenin and genistein onto the magnetic FG beads. *DMF*, dimethyl formamide; *FG*, fiberglass. *Based on Oishi, M., Iizumi, Y., Taniguchi, T., et al., 2013. Apigenin sensitizes prostate cancer cells to Apo2L/TRAIL by targeting adenine nucleotide translocase-2. PLoS One 8, e55922.*

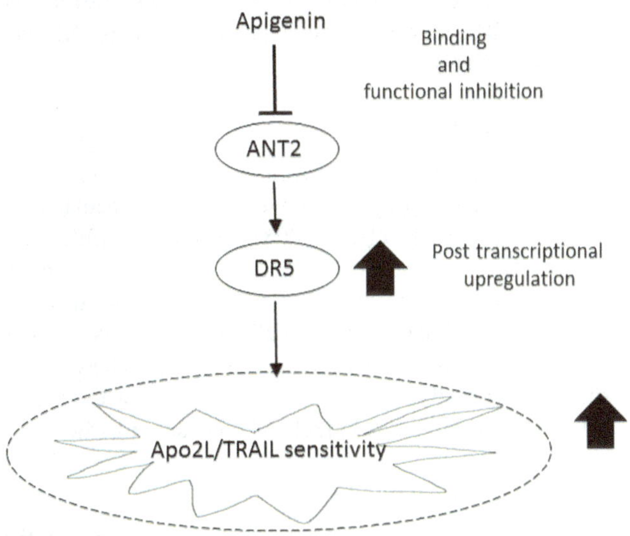

FIGURE 23.28 Schematic model of the enhancement of Apo2L/TRAIL sensitivity by apigenin. *Apo2L/TRAIL;* TNF-related apoptosis-inducing ligand; *ANT2*, adenine nucleotide translocase-2; *DR5*, death receptor 5. *Based on Oishi, M., Iizumi, Y., Taniguchi, T., et al., 2013. Apigenin sensitizes prostate cancer cells to Apo2L/TRAIL by targeting adenine nucleotide translocase-2. PLoS One 8, e55922.*

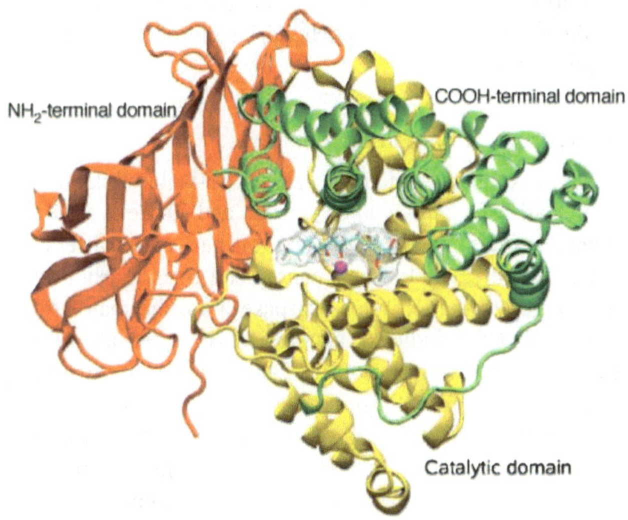

FIGURE 23.29 3D modeling of the binding of gingerol to leukotriene A_4 hydrolase. *Adapted from Jeong, C.H., Bode, A.M., Pugliese, A., et al., 2009. [6]-Gingerol suppresses colon cancer growth by targeting leukotriene A4 hydrolase. Cancer Res. 69, 5584–5591.*

cancer cell growth by inhibiting LTA_4H activity in HCT116 colorectal cancer cells. Direct binding was confirmed by pull-down assay with [6]-gingerol-Sepharose 4B beads.

In another study, Oi et al. (2010) investigated the direct binding of resveratrol or quercetin to LTA_4H with conjugated Sepharose 4B beads (Oi et al., 2010). The data showed that recombinant LTA_4H ($rLTA_4H$) bound to resveratrol- or quercetin-conjugated Sepharose 4B beads.

Furthermore, resveratrol-conjugated Sepharose 4B beads were bound to endogenous LTA_4H, whereas RSVL3- or quercetin-conjugated beads were not. The above results suggest that resveratrol might be more effective against LTA_4H than quercetin and that its hydroxyl groups are critical for the binding. In additional studies, resveratrol suppressed proliferation and anchorage-independent growth in PANC-1 pancreatic cancer cells.

Capsaicin

Capsaicin is widely used as a food additive and analgesic agent. These activities are mediated by vanilloid receptor 1 expressed in ganglion neurons. Capsaicin has also been considered a potential anticancer agent, since it inhibits cell proliferation and induces apoptosis. Kuramori et al. (2009) identified a new molecular target of capsaicin from human myeloid leukemia cells by using capsaicin-conjugated magnetic beads. The affinity beads were prepared by incubating magnetic nanocarriers with a linker possessing a carboxyl group and were incubated with Cap-NH_2 in DMF using standard coupling reagents. After affinity separation, the bound proteins were separated by SDS–PAGE and visualized by silver or Coomassie Blue staining. The selected bands were subjected to trypsin digestion, and the resultant peptides were analyzed by Q-TOF MS. Two proteins (30 and 37 kDa) were isolated on capsaicin-conjugated beads after elution with human myeloid leukemia NB4 cells, and were identified as PHB1 (prohibitin 1) and PHB2, and confirmed by Western blotting using anti-PHB1 and PHB2 antibodies. PHB2, localized to the inner mitochondrial membrane, was implicated in the maintenance of mitochondrial morphology and the control of apoptosis.

2′-Hydroxycinnamaldehyde

2'-Hydroxycinnamaldehyde (HCA) is isolated from the stem bark of *Cinnamomum cassia*, and its derivative 2'-benzoyloxycinnamaldehyde (BCA) has antiproliferative activities on several human cancer cell lines. Earlier studies suggested that ROS and caspase-3 are the major regulators of HCA-induced apoptosis. Hong et al. (2007) demonstrated a novel molecular target using in vitro pull-down assay by biotin-labeled HCA (biotinyl-HCA; Fig. 23.30) in SW620 cells. A biotinylated HCA was synthesized to serve as an affinity chromatography reagent. An in vitro pull-down assay was performed from the whole-cell lysates of SW620 by biotin-HCA and DMSO vehicle, respectively. Avidin affinity column, 2D GE, and MALDI-TOF analyses revealed that 5 of 11 differentially pull-downed spots by biotin-HCA are proteasome subunits. The identification of the major HCA-binding proteins as proteasome subunits suggested that HCA mediates its antiproliferative effect via inhibition of proteasome function. DNA microarray showed that HCA-induced heat shock family and endoplasmic reticulum (ER) stress-responsive genes, which reflects the accumulation of misfolded proteins by proteasome inhibition. Western blot analysis confirmed that HCA induces glucose-regulated protein, 78 kDa (GRP78), and some representative ER stress-responsive proteins.

Multiple binding and mechanistic assays

A range of assays (Cerep, www.cerep.fr) and kinase electivity panel (Wan et al., 2004), as well as multiplexed cell-based assays were introduced in recent years and provided a good starting point for the estimation of molecular targets for many natural components.

Studies with human cancer cell lines demonstrated that EGCG, inhibits MAP kinases, CDK kinases, growth factor-related cell signaling, AP1, NF-κB, topoisomerase I, and MMPs, as well as other potential targets (Lambert and Yang, 2003).

Diallyl trisulfide

Diallyl trisulfide, derived from garlic, was investigated for its anticancer property as well as the molecular basis of its activity (Seki et al., 2008). Diallyl trisulfide binds to specific cysteine residues in β-tubulin to form S-allylmercaptocystein and the authors suggest that this could be the primary cause of cell cycle arrest and successive apoptosis with activation of caspase-3. These results suggest that diallyl trisulfide is responsible for the proven anticancer effect in individuals who consume garlic.

Apigenin and luteolin

Chin et al. (2013) investigated various flavonoids that were assessed for their Fms-like tyrosine kinase 3 (FLT3) inhibitory activities using homogeneous time-resolved

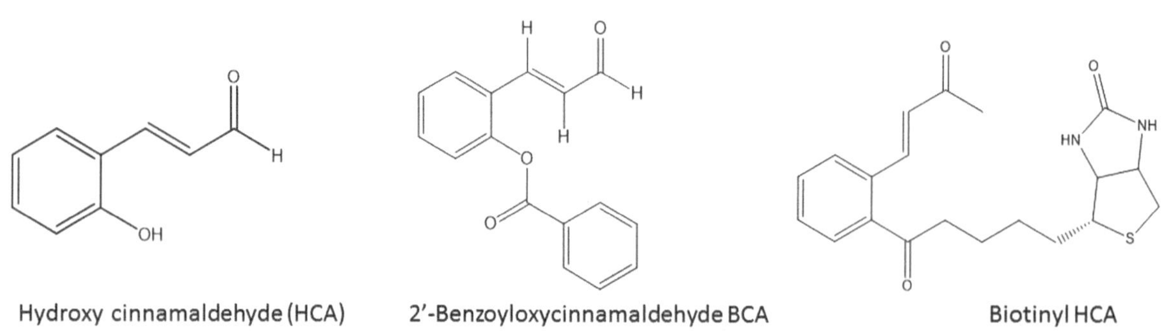

FIGURE 23.30 The structure of hydroxycinnamaldehyde and its biotinylated analog.

fluorescence kinase assay. Apigenin and luteolin inhibited FLT3 enzyme activities with $IC_{50} = 1.45$ and 0.83 μM, respectively. Kaempferol ($IC_{50} = 2.32$ μM), quercetin ($IC_{50} = 0.59$ μM), and myricetin ($IC_{50} = 1.68$ μM) also demonstrated their inhibitory activities against the FLT3 enzyme. FLT3, expressed at high levels in the majority of acute myeloid leukemia patients, is considered a potential therapeutic target.

Several flavones exhibit strong anticancer activities on various cell systems and animal models. Pyruvate kinase isoenzyme type PKM2 is an important target in anticancer research. Inhibitors and activators of this enzyme are promising molecules for use in treating cancer. Aslan and Adem (2015) tested the in vitro effects of some flavones on recombinant human PKM2. Apigenin, flavone, 3-hydroxyflavone, 5-hydroxyflavone, 6-hydroxyflavone, and 7-hydroxyflavone effectively inhibited PKM2, with IC_{50} in the range of 0.99−2.12 μM. PKM2 activity was assayed according to a described method by a continuous assay coupled to lactate dehydrogenase (LDH).

Synergy investigations

Apigenin

Apigenin induces apoptosis and cell death in lung epithelium cancer (A549) cells with an IC_{50} value of 93.7 μM for 48 h treatment (Choudhury et al., 2013). Target identification investigations using A549 cells and also in cell-free systems, demonstrated that apigenin depolymerized microtubules and inhibited reassembly of cold depolymerized microtubules of A549 cells. Apigenin induces a mitochondrial death cascade by inhibition of VEGF and HIF-1 expression via pI3K/AKT/p70S6K1 and HDM2/p53 pathways. Apigenin inhibited polymerization of purified tubulin with an IC_{50} value of 79.8 μM. Choudhury et al. (2013) demonstrated that apigenin and curcumin showed anticancer activity synergistically at lower concentrations through binding to tubulin at different sites. Apigenin and curcumin synergistically induced cell death and apoptosis and also blocked cell cycle progression in the G2/M phase of A549 cells. In silico modeling suggested that apigenin is bound at the interphase of α/β-subunit of tubulin. The binding site is 19 Å in distance from the previously predicted curcumin-binding site. Apigenin and curcumin mixtures could potentially be used clinically for anticancer therapy.

Genetic methods

Epigallocatechin-3-gallate

Tachibana et al. (2004) identified a receptor that mediates the anticancer activity of EGCG. Expression of the metastasis-associated 67-kDa laminin receptor (67 LR) confers EGCG responsiveness to cancer cells at physiologically relevant concentrations. Experiments using surface plasmon resonance demonstrated binding of EGCG to the 67 LR with a nanomolar K_d value (Tachibana et al., 2004). It was demonstrated that all-*trans* retinoic acid (ATRA) enhances the binding of EGCG to the cell surface of cancer cells. To identify target candidates through which EGCG inhibits cell growth, the authors used a subtraction cloning strategy involving cDNA libraries constructed from cells treated or untreated with ATRA. In this study, a single target was identified that allows EGCG to bind to the cell surface and analysis of the DNA sequence confirmed it as the 67 LR. In fact, the expression of this 67 LR was enhanced by ATRA treatment. The 67 LR is expressed in a variety of tumor cells, and the expression level of this protein strongly correlates with the risk of tumor invasion and metastasis.

Drug affinity responsive target stability

DARTS was used to identify a molecular target of resveratrol (Lomenick et al., 2009), known for various health benefits including lifespan extension. Yeast cell lysates or human HeLa cell lysates were treated with resveratrol in vitro, followed by thermolysin digestion and silver staining after separation by SDS−PAGE. Bands of resveratrol-treated lysate protected from thermolysin digestion were analyzed by MS, and eIF4A (eukaryotic translation initiation factor 4A) was identified. This protein is required for the binding of mRNA to 40S ribosomal subunits. Similarly, purified wild-type eIF4A was protected by resveratrol from proteolysis, which suggested that resveratrol directly bound to eIF4A. This confirms that resveratrol inhibits the biological function of eIF4A in vivo. Resveratrol normally extends the lifespan of wild-type N2 worms. Since this longevity effect was lost in eIF4A knockdown worms, it suggested that eIF4A is indeed a physiological target of resveratrol.

Phage display

HBC (4-{3,5-bis-[2-(4-hydroxy-3-methoxy-phenyl)-ethyl]-4,5-dihydro-pyrazol-1-yl}-benzoic acid) is a recently developed curcumin derivative which showed antiproliferative effects on various tumor cell lines. Shim et al. (2004) investigated and identified Ca^{2+}/calmodulin (Ca^{2+}/CaM) as a direct target protein of HBC using phage display biopanning. Ca^{2+}/CaM-expressing phages specifically bound to the immobilized HBC, and the binding was Ca^{2+} dependent.

Biotinyl-HBC (Fig. 23.31) was immobilized onto streptavidin-coated wells, and T7 phages encoding human cDNA libraries were used for affinity selection of HBC-binding proteins. An increased number of phage particles were bound to HBC in proportion to each biopanning

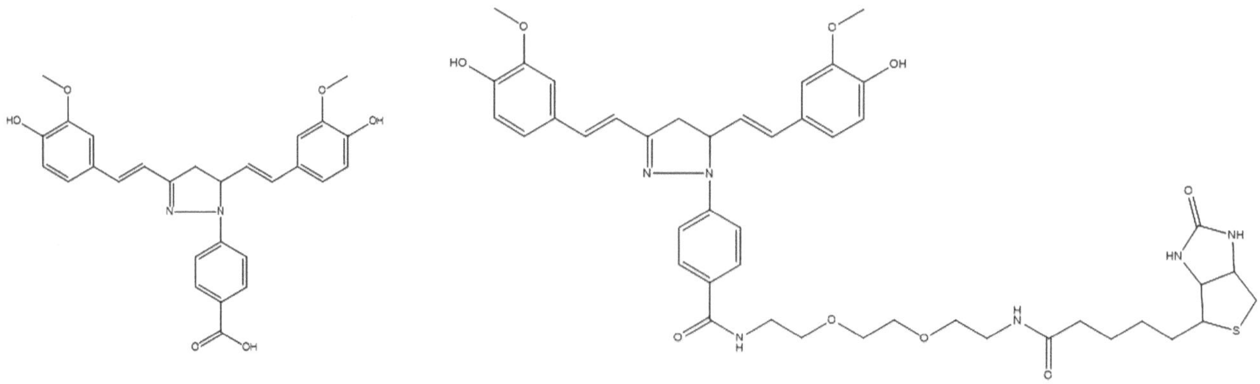

HBC (curcumin derivative) Biotinylated HBC (curcumin derivative)

FIGURE 23.31 The structure of a modified curcumin derivative and its biotinylated analog.

round, indicating that specific HBC-binding phages were enriched. After the fourth round of biopanning, phage particles that specifically bound to HBC were isolated using a plaque-forming assay and then analyzed by DNA sequencing. After sequence analysis, the resulting data were subjected to sequence homology analysis, which confirmed that the coding sequences of the CaM phages are 100% identical to the C-terminal fragment of human CaM.

In a similar study (Feng et al., 2006) resveratrol-binding proteins were identified by biotinylated resveratrol immobilized into streptavidin-coated microplate wells and with subsequent phage display analysis. A random library of phage-displayed peptides was screened for binding to immobilized resveratrol to isolate resveratrol-binding proteins. Several peptides were found to bind to resveratrol specifically, which was proven by enzyme-linked immunosorbent assay. The results from sequence comparison revealed high homology with several proteins, including death-associated transcription factor, immunoglobulin E, F-box and leucine-rich repeat protein, protein kinase C and CKII, retinoblastoma-binding protein, breast cancer-associated antigen BRCAA1, oncostatin M receptor, etc. Some of these proteins have been associated with the anticancer activity of resveratrol.

In silico target identification

In silico target identification was discussed previously in detail (see section "In silico target and mechanism of action prediction"). Therefore here only some specific applications are addressed. It is well known that several food-related compounds have gene regulatory action on inflammation. However, the direct protein target in the cascade is not known. The direct inhibitory binding of nutraceuticals to proinflammatory enzymes, such as COX-2, was investigated by in silico approaches such as 3D molecular docking. Curcumin and ATRA showed maximal binding affinities (9.3 kcal/mol), but genistein, apigenin, cyanidin, kaempferol, and docosahexaenoic acid, were also close to this value. These results suggest that the antiinflammatory properties of some food-derived molecules could be the result of their direct binding capabilities to COX-2, and this process can be modeled using protein–ligand docking methodologies (Maldonado-Rojas and Olivero-Verbel, 2011).

Plasmodium falciparum is a protozoan parasite, one of the species of *Plasmodium* that cause malaria in humans (Nag et al., 2013). Protein kinases present in *P. falciparum* catalyze the phosphorylation reaction to control growth and differentiation of the parasite throughout its life cycle. Protein kinase inhibitors are found to kill the parasite, but their cellular target enzymes are unknown. Potential protein kinase inhibitors, including nutraceuticals, were evaluated in in silico docking studies using *P. falciparum* RIO-2 kinase (right open reading frame-2 protein kinase) as the target enzyme. Most of the protein kinase inhibitors showed appropriate docking within the ATP-binding domain of the PfRIO-2 kinase. Docking and substrate competition studies identified, among others, apigenin and kaempferol against the native/active form of the PfRIO-2 kinase.

Sommer et al. (2017) reported the identification of the beer component hordenine as a food-derived dopamine D2 receptor (D2R) agonist by virtual screening (VS) using a 3D compound database. D2R is involved in food reward and compulsive food intake. A VS method was developed to identify nutraceutical components, which could have a role in modulating D2R signaling. Altogether, 17 compounds were selected by VS and then tested in radioligand assays. Seven compounds were identified as HITS (hit rate = 41%) upon their D2R binding affinities. D2R-induced G-protein activation was detected for one small molecule, hordenine, which is a natural

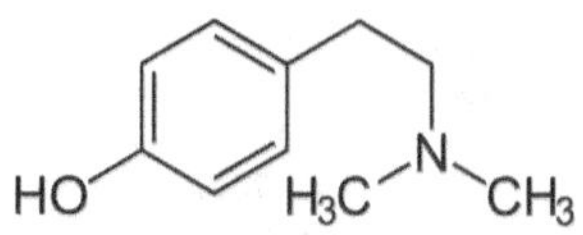

FIGURE 23.32 Structure of hordenine.

ingredient of barley and beer. The ligand efficacy is close to dopamine (76%). Thus, the authors suggest that hordenine (Fig. 23.32) contributes to the mood-increasing effects of beer (Sommer et al., 2017).

Concluding remarks and future directions

Nutraceuticals are mostly studied as extracts or isolated components in cell-based assays, as well as in genetic assays, which reveal several targets that are affected in particular signaling pathways. There are many molecular targets that are downregulated or upregulated upon incubation of the cell with the particular nutraceutical agent. Of course, based on the pathway analysis and modeling one could make a prediction of which molecular targets are affected. Consequently, it would involve in silico docking studies which would predict potential binding interaction that can be confirmed in in vitro assays. Furthermore, pathway analysis would predict which target families could be investigated. For example, kinase screening panels are typically used. On the other hand, identifying the direct binding partners with affinity-based methods seems to be the most widely used technique due its simplicity. The identification of the direct partner more precisely elucidates or provides further evidence for the particular signaling pathways affected. The recently developed magnetic beads further popularize this technique. Additional approaches, such as DARTS or phage display selection, provide additional tools for practitioners. In essence, the contemporary target identification methods of nutraceuticals combines network pharmacology, in silico chemoinformatics, genetics, and simple affinity-based methods in an innovative and pragmatic variation. In the future, this integration would be more apparent with some focus shifted toward the system biology and network-based approaches.

References

Afaq, F., Saleem, M., Krueger, C.G., et al., 2005. Anthocyanin-and hydrolyzable tannin-rich pomegranate fruit extract modulates MAPK and NF-κB pathways and inhibits skin tumorigenesis in CD-1 mice. Int. J. Cancer 113, 423–433.

Aggarwal, B.B., 2010. Targeting inflammation-induced obesity and metabolic diseases by curcumin and other nutraceuticals. Annu. Rev. Nutr. 30, 173–199.

Aggarwal, B.B., Shishodia, S., 2006. Molecular targets of dietary agents for prevention and therapy of cancer. Biochem. Pharmacol. 71, 1397–1421.

Anand, P., Sundaram, C., Jhurani, S., et al., 2008. Curcumin and cancer: an "old-age" disease with an "age-old" solution. Cancer Lett. 267, 133–164.

Ares, J.M.B., Durán-Peña, M.J., Hernández-Galán, R., et al., 2013. Chemical genetics strategies for identification of molecular targets. Phytochem. Rev. 12, 895–914.

Arrell, D.K., Terzic, A., 2010. Network systems biology for drug discovery. Clin. Pharmacol. Ther. 88, 120–125.

Aslan, E., Adem, S., 2015. In vitro effects of some flavones on human pyruvate kinase isoenzyme M2. J. Biochem. Mol. Toxicol. 29, 109–M13.

Bantscheff, M., Eberhard, D., Abraham, Y., et al., 2007. Quantitative chemical proteomics reveals mechanisms of action of clinical ABL kinase inhibitors. Nat. Biotechnol. 25, 1035–1044.

Bantscheff, M., Scholten, A., Heck, A.J., 2009. Revealing promiscuous drug–target interactions by chemical proteomics. Drug Discov. Today 14, 1021–1029.

Belenky, M., Prasain, J., Kim, H., et al., 2003. DING, a genistein target in human breast cancer: a protein without a gene. J. Nutr. 133, 2497S–2501S.

Bender, A., Young, D.W., Jenkins, J.L., et al., 2007. Chemogenomic data analysis: prediction of small-molecule targets and the advent of biological fingerprints. Comb. Chem. High Throughput Screening. 10, 719–731.

Bolognesi, M.L., 2019. Harnessing polypharmacology with medicinal chemistry. ACS Med. Chem. Lett. 10, 273–275.

Byrne, R., Schneider, G., 2019. In silico target prediction for small molecules. Methods Mol. Biol. 1888, 73–309.

Chamberlin, S.R., Blucher, A., Wu, G., et al., 2019. Natural product target network reveals potential for cancer combination therapies. Front. Pharmacol. 10, 557.

Chaudhari, R., Tan, Z., Huang, B., et al., 2017. Computational polypharmacology: a new paradigm for drug discovery. Expert Opin. Drug Discov. 12, 279–291.

Chen, X., Yan, C.C., Zhang, X., et al., 2016. Drug–target interaction prediction: databases, web servers and computational models. Briefings Bioinf. 17 (4), 696–712.

Chin, Y.W., Kong, J.Y., Han, S.Y., 2013. Flavonoids as receptor tyrosine kinase FLT3 inhibitors. Bioorg. Med. Chem. Lett. 23, 1768–1770.

Choudhury, D., Ganguli, A., Dastidar, D.G., et al., 2013. Apigenin shows synergistic anticancer activity with curcumin by binding at different sites of tubulin. Biochimie 95, 1297–1309.

Csermely, P., Agoston, V., Pongor, S., 2005. The efficiency of multi-target drugs: the network approach might help drug design. Trends Pharmacol. Sci. 26, 178–182.

Dormán, G., Prestwich, G.D., 2000. Using photolabile ligands in drug discovery and development. Trends Biotechnol. 18, 64–77.

Efferth, T., Koch, E., 2011. Complex interactions between phytochemicals. The multi-target therapeutic concept of phytotherapy. Curr. Drug Targets 12, 122–132.

Ermakova, S., Choi, B.Y., Choi, H.S., et al., 2005. The intermediate filament protein vimentin is a new target for epigallocatechin gallate. J. Biol. Chem. 280, 16882–16890.

Esatbeyoglu, T., Huebbe, P., Ernst, I., et al., 2012. Curcumin—from molecule to biological function. Angew. Chem. Int. Ed. 51, 5308–5332.

Evans, M.J., Cravatt, B.F., 2006. Mechanism-based profiling of enzyme families. Chem. Rev. 106, 3279–3301.

Fang, J., Liu, C., Wang, Q., et al., 2018. In silico polypharmacology of natural products. Briefings Bioinf. 19, 1153–1171.

Feng, L., Jin, J., Zhang, L.F., et al., 2006. Analysis of the resveratrol-binding protein using phage-displayed random peptide library. Acta Biochim. Biophys. Sin. 38, 342–348.

Franken, H., Mathieson, T., Childs, D., et al., 2015. Thermal proteome profiling for unbiased identification of direct and indirect drug targets using multiplexed quantitative mass spectrometry. Nat. Protoc. 10, 1567–1593.

Frantz, S., 2005. Drug discovery: playing dirty. Nature 437, 942–943.

Fuchs, D., de Pascual-Teresa, S., Rimbach, G., et al., 2005a. Proteome analysis for identification of target proteins of genistein in primary human endothelial cells stressed with oxidized LDL or homocysteine. Eur. J. Nutr. 44, 95–104.

Fuchs, D., Erhard, P., Rimbach, G., et al., 2005b. Genistein blocks homocysteine-induced alterations in the proteome of human endothelial cells. Proteomics 5, 2808–2818.

Galat, A., Lane, W.S., Standaert, R.F., et al., 1992. A rapamycin-selective 25-kDa immunophilin. Biochemistry 31, 2427–2434.

Giordano, S., Petrelli, A., 2008. From single-to multi-target drugs in cancer therapy: when aspecificity becomes an advantage. Curr. Med. Chem. 15, 422–432.

Granier, S., Kobilka, B., 2012. A new era of GPCR structural and chemical biology. Nat. Chem. Biol. 8, 670–673.

Gupta, S.C., Kim, J.H., Prasad, S., et al., 2010. Regulation of survival, proliferation, invasion, angiogenesis, and metastasis of tumor cells through modulation of inflammatory pathways by nutraceuticals. Cancer Metastasis Rev. 29, 405–434.

Hong, S.H., Kim, J., Kim, J.M., et al., 2007. Apoptosis induction of 2′-hydroxycinnamaldehyde as a proteasome inhibitor is associated with ER stress and mitochondrial perturbation in cancer cells. Biochem. Pharmacol. 74, 557–565.

Hopkins, A.L., 2008. Network pharmacology: the next paradigm in drug discovery. Nat. Chem. Biol. 4, 682–690.

Hsieh, T.C., Wu, J.M., 2010. Resveratrol: Biological and pharmaceutical properties as anticancer molecule. Biofactors 36, 360–369.

Hu, Y., Gupta-Ostermann, D., Bajorath, J., 2014. Exploring compound promiscuity patterns and multi-target activity spaces. Comput. Struct. Biotechnol. J. 9, 1–11.

Huang, M.Y., Zhang, L.L., Ding, J., et al., 2018. Anticancer drug discovery from Chinese medicinal herbs. Chin. Med. 13, 35.

Hung, C.N., Huang, H.P., Wang, C.J., et al., 2014. Sulforaphane inhibits TNF-α-induced adhesion molecule expression through the Rho A/ROCK/NF-κB signaling pathway. J. Med. Food 17, 1095–1102.

Hwang, M.K., Kang, N.J., Heo, Y.S., et al., 2009a. Fyn kinase is a direct molecular target of delphinidin for the inhibition of cyclooxygenase-2 expression induced by tumor necrosis factor-α. Biochem. Pharmacol. 77, 1213–1222.

Hwang, M.K., Song, N.R., Kang, N.J., et al., 2009b. Activation of phosphatidylinositol 3-kinase is required for tumor necrosis factor-α-induced upregulation of matrix metalloproteinase-9: its direct inhibition by quercetin. Int. J. Biochem. Cell Biol. 41, 1592–1600.

Isgut, M., Rao, M., Yang, C., et al., 2018. Application of combination high-throughput phenotypic screening and target identification methods for the discovery of natural product-based combination drugs. Med. Res. Rev. 38, 504–524.

Jafari, R., Almqvist, H., Axelsson, H., et al., 2014. The cellular thermal shift assay for evaluating drug target interactions in cells. Nat. Protoc. 9, 2100–2122.

Jalencas, X., Mestres, J., 2013. On the origins of drug polypharmacology. MedChemComm 4, 80–87.

Jenwitheesuk, E., Horst, J.A., Rivas, K.L., et al., 2008. Novel paradigms for drug discovery: computational multitarget screening. Trends Pharmacol. Sci. 29, 62–71.

Jeong, C.H., Bode, A.M., Pugliese, A., et al., 2009. [6]-Gingerol suppresses colon cancer growth by targeting leukotriene A4 hydrolase. Cancer Res. 69, 5584–5591.

Jung, S.K., Lee, K.W., Byun, S., et al., 2010. Myricetin inhibits UVB-induced angiogenesis by regulating PI-3 kinase in vivo. Carcinogenesis 31, 911–917.

Kanoh, N., 2016. Photo-cross-linked small-molecule affinity matrix as a tool for target identification of bioactive small molecules. Nat. Prod. Rep. 33, 709–718.

Kapoor, T.M., Miller, R.M., 2017. Leveraging chemotype-specific resistance for drug target identification and chemical biology. Trends Pharmacol. Sci. 38, 1100–1109.

Karuppagounder, V., Arumugam, S., Thandavarayan, R.A., et al., 2016. Molecular targets of quercetin with anti-inflammatory properties in atopic dermatitis. Drug Discov. Today 21, 632–639.

Kim, J.E., Kwon, J.Y., Seo, S.K., et al., 2010. Cyanidin suppresses ultraviolet B-induced COX-2 expression in epidermal cells by targeting MKK4, MEK1, and Raf-1. Biochem. Pharmacol. 79, 1473–1482.

Kumamoto, T., Fujii, M., Hou, D.X., 2009. Myricetin directly targets JAK1 to inhibit cell transformation. Cancer Lett. 275, 17–26.

Kunnumakkara, A.B., Bordoloi, D., Padmavathi, G., et al., 2017. Curcumin, the golden nutraceutical: multitargeting for multiple chronic diseases. Br. J. Pharmacol. 174, 1325–1348.

Kuramori, C., Azuma, M., Kume, K., et al., 2009. Capsaicin binds to prohibitin 2 and displaces it from the mitochondria to the nucleus. Biochem. Biophys. Res. Commun. 379, 519–525.

Kwak, M.K., Kensler, T.W., 2010. Targeting NRF2 signaling for cancer chemoprevention. Toxicol. Appl. Pharmacol. 244, 66–76.

Kwon, J.Y., Lee, K.W., Kim, J.E., et al., 2009. Delphinidin suppresses ultraviolet B-induced cyclooxygenases-2 expression through inhibition of MAPKK4 and PI-3 kinase. Carcinogenesis 30, 1932–1940.

Lambert, J.D., Yang, C.S., 2003. Mechanisms of cancer prevention by tea constituents. J. Nutr. 133, 3262S–3267S.

Le Tourneau, C., Faivre, S., Raymond, E., 2008. New developments in multitargeted therapy for patients with solid tumours. Cancer Treat. Rev. 34, 37–48.

Lee, J.S., Surh, Y.J., 2005. Nrf2 as a novel molecular target for chemoprevention. Cancer Lett. 224, 171–184.

Lee, K.W., Kang, N.J., Rogozin, E.A., et al., 2007. Myricetin is a novel natural inhibitor of neoplastic cell transformation and MEK1. Carcinogenesis 28, 1918–1927.

Lee, K.W., Kang, N.J., Heo, Y.S., et al., 2008. Raf and MEK protein kinases are direct molecular targets for the chemopreventive effect of quercetin, a major flavonol in red wine. Cancer Res. 68, 946–955.

Lee, D.E., Lee, K.W., Song, N.R., et al., 2010a. 7, 3′, 4′-Trihydroxyisoflavone inhibits epidermal growth factor-induced proliferation and transformation of JB6 P+ mouse epidermal cells by suppressing cyclin-dependent kinases and phosphatidylinositol 3-kinase. J. Biol. Chem. 285, 21458–21466.

Lee, K.M., Lee, D.E., Seo, S.K., et al., 2010b. Phosphatidylinositol 3-kinase, a novel target molecule for the inhibitory effects of kaempferol on neoplastic cell transformation. Carcinogenesis 31, 1338–1343.

Lee, K.M., Lee, K.W., Jung, S.K., et al., 2010c. Kaempferol inhibits UVB-induced COX-2 expression by suppressing Src kinase activity. Biochem. Pharmacol. 80, 2042–2049.

Lee, K.W., Bode, A.M., Dong, Z., 2011. Molecular targets of phytochemicals for cancer prevention. Nat. Rev. Cancer 11, 211–218.

Lehár, J., Krueger, A.S., Avery, W., et al., 2009. Synergistic drug combinations tend to improve therapeutically relevant selectivity. Nat. Biotechnol. 27, 659–666.

Li, M., He, Z., Ermakova, S., et al., 2007. Direct inhibition of insulin-like growth factor-I receptor kinase activity by (−) − epigallocatechin-3-gallate regulates cell transformation. Cancer Epidemiol. Biomarkers Prev. 16, 598–605.

Li, S., Zhang, B., Zhang, N., 2011. Network target for screening synergistic drug combinations with application to traditional Chinese medicine. BMC Syst. Biol. 5 (Suppl. 1), S10.

Li, X., Xu, X., Wang, J., et al., 2012. A system-level investigation into the mechanisms of Chinese traditional medicine: Compound Danshen Formula for cardiovascular disease treatment. PLoS One 7, e43918.

Lomenick, B., Hao, R., Jonai, N., et al., 2009. Target identification using drug affinity responsive target stability (DARTS). Proc. Natl. Acad. Sci. USA 106, 21984–21989.

Lomenick, B., Olsen, R.W., Huang, J., 2010. Identification of direct protein targets of small molecules. ACS Chem. Biol. 6, 34–46.

Maggiora, G.M., 2011. The reductionist paradox: are the laws of chemistry and physics sufficient for the discovery of new drugs? J. Comput. Aided Mol. Des. 25, 699–708.

Maldonado-Rojas, W., Olivero-Verbel, J., 2011. Potential interaction of natural dietary bioactive compounds with COX-2. J. Mol. Graph. Model. 30, 157–166.

Medina-Franco, J.L., Giulianotti, M.A., Welmaker, G.S., et al., 2013. Shifting from the single to the multitarget paradigm in drug discovery. Drug Discov. Today 18, 495–501.

Melchini, A., Costa, C., Traka, M., et al., 2009. Erucin, a new promising cancer chemopreventive agent from rocket salads, shows anti-proliferative activity on human lung carcinoma A549 cells. Food Chem. Toxicol. 47, 1430–1436.

Melchini, A., Traka, M.H., 2010. Biological profile of erucin: a new promising anticancer agent from cruciferous vegetables. Toxins 2, 593–612.

Milner, J.A., McDonald, S.S., Anderson, D.E., et al., 2001. Molecular targets for nutrients involved with cancer prevention. Nutr. Cancer 41, 1–16.

Nag, S., Prasad, K.M.N., Bhowmick, A., et al., 2013. PfRIO-2 kinase is a potential therapeutic target of antimalarial protein kinase inhibitors. Curr. Drug Discov. Technol. 10, 85–91.

Nahrstedt, A., Butterweck, V., 2010. Lessons learned from herbal medicinal products: the example of St. John's wort. J. Nat. Prod. 73, 1015–1021.

Nomura, D.K., Maimone, T.J., 2018. Target identification of bioactive covalently acting natural products. *Activity-Based Protein Profiling*. Springer, Cham, pp. 351–374.

Oi, N., Jeong, C.H., Nadas, J., et al., 2010. Resveratrol, a red wine polyphenol, suppresses pancreatic cancer by inhibiting leukotriene a4 hydrolase. Cancer Res. 70, 9755–9764.

Oishi, M., Iizumi, Y., Taniguchi, T., et al., 2013. Apigenin sensitizes prostate cancer cells to Apo2L/TRAIL by targeting adenine nucleotide translocase-2. PLoS One 8, e55922.

Park, H., Ha, J., Park, S.B., 2019. Label-free target identification in drug discovery via phenotypic screening. Curr. Opin. Chem. Biol. 50, 66–72.

Peitsch, M.C., De Graaf, D., 2013. A decade of systems biology: where are we and where are we going to? Drug Discov. Today 19, 105–107.

Pham, D.C., Shibu, M.A., Mahalakshmi, B., et al., 2019. Effects of phytochemicals on cellular signaling: reviewing their recent usage approaches. Crit. Rev. Food Sci. Nutr. 1–25.

Poornima, P., Kumar, J.D., Zhao, Q., et al., 2016. Network pharmacology of cancer: from understanding of complex interactomes to the design of multi-target specific therapeutics from nature. Pharm. Res. 111, 290–302.

Proschak, E., Stark, H., Merk, D., 2019. Polypharmacology by design: a medicinal chemist's perspective on multitargeting compounds. J. Med. Chem. 62 (2), 420–444.

Reddy, A.S., Zhang, S., 2013. Polypharmacology: drug discovery for the future. Expert Rev. Clin. Pharmacol. 6, 41–47.

Salentin, S., Haupt, V.J., Daminelli, S., et al., 2014. Polypharmacology rescored: protein-ligand interaction profiles for remote binding site similarity assessment. Prog. Biophys. Mol. Biol. 116, 174–186.

Sarkar, F.H., Li, Y., Wang, Z., et al., 2010. The role of nutraceuticals in the regulation of Wnt and Hedgehog signaling in cancer. Cancer Metastasis Rev. 29, 383–394.

Sato, S.I., Murata, A., Shirakawa, T., et al., 2010. Biochemical target isolation for novices: affinity-based strategies. Chem. Biol. 17, 616–623.

Schirle, M., Jenkins, J.L., 2016. Identifying compound efficacy targets in phenotypic drug discovery. Drug Discov. Today 21, 82–89.

Seki, T., Hosono, T., Hosono-Fukao, T., et al., 2008. Anticancer effects of diallyl trisulfide derived from garlic. Asia Pac. J. Clin. Nutr. 17 (Suppl. 1), 249–252.

Sharma, S.V., Haber, D.A., Settleman, J., 2010. Cell line-based platforms to evaluate the therapeutic efficacy of candidate anticancer agents. Nat. Rev. Cancer 10, 241–253.

Shim, J.S., Lee, J., Park, H.J., et al., 2004. A new curcumin derivative, HBC, interferes with the cell cycle progression of colon cancer cells via antagonization of the Ca^{2+}/calmodulin function. Chem. Biol. 11, 1455–1463.

Shim, J.H., Su, Z.Y., Chae, J.I., et al., 2010. Epigallocatechin gallate suppresses lung cancer cell growth through Ras–GTPase-activating protein SH3 domain-binding protein 1. Cancer Prev. Res. 3, 670–679.

Singh, V.K., Arora, D., Ansari, M.I., et al., 2019. Phytochemicals based chemopreventive and chemotherapeutic strategies and modern technologies to overcome limitations for better clinical applications. Phytother. Res. 33, 3064–3089.

Smeltzer, A., Kim, Y.S., 2005. The effects of bioactive food components on p53 pathway in cancer prevention. Nutr. Today 40, 50–53.

Sommer, T., Hübner, H., El Kerdawy, A., et al., 2017. Identification of the beer component hordenine as food-derived dopamine D2 receptor agonist by virtual screening a 3D compound database. Sci. Rep. 7, 44201.

Strickland, E.C., Geer, M.A., Tran, D.T., et al., 2013. Thermodynamic analysis of protein-ligand binding interactions in complex biological mixtures using the stability of proteins from rates of oxidation. Nat. Protoc. 8, 148–161.

Sung, B., Prasad, S., Yadav, V.R., et al., 2012. Cancer cell signaling pathways targeted by spice-derived nutraceuticals. Nutr. Cancer 64, 173–197.

Suppipat, K., Park, C.S., Shen, Y., et al., 2012. Sulforaphane induces cell cycle arrest and apoptosis in acute lymphoblastic leukemia cells. PLoS One 7, e51251.

Surh, Y.J., 2003. Cancer chemoprevention with dietary phytochemicals. Nat. Rev. Cancer 3, 768–780.

Sydow, D., Burggraaff, L., Szengel, A., et al., 2019. Advances and challenges in computational target prediction. J. Chem. Inf. Model. 59 (5), 1728–1742.

Tachibana, H., Koga, K., Fujimura, Y., et al., 2004. A receptor for green tea polyphenol EGCG. Nat. Struct. Mol. Biol. 11, 380–381.

Takakusagi, Y., Takakusagi, K., Sugawara, F., et al., 2010. Use of phage display technology for the determination of the targets for small-molecule therapeutics. Expert Opin. Drug. Discov. 5, 361–389.

Tashiro, E., Imoto, M., 2012. Target identification of bioactive compounds. Bioorg. Med. Chem. 20, 1910–1921.

Wagner, H., Ulrich-Merzenich, G., 2009. Synergy research: approaching a new generation of phytopharmaceuticals. Phytomedicine 16, 97–110.

Wan, Y., Hur, W., Cho, C.Y., et al., 2004. Synthesis and target identification of hymenialdisine analogs. Chem. Biol. 11, 247–259.

Wang, Z., Hsieh, T.C., Zhang, Z., et al., 2004. Identification and purification of resveratrol targeting proteins using immobilized resveratrol affinity chromatography. Biochem. Biophys. Res. Commun. 323, 743–749.

Wang, Y., Che, C.M., Chiu, J.F., et al., 2007. Dioscin (saponin)-induced generation of reactive oxygen species through mitochondria dysfunction: a proteomic-based study. J. Proteome Res. 6, 4703–4710.

Wang, X., Xu, X., Tao, W., et al., 2012. A systems biology approach to uncovering pharmacological synergy in herbal medicines with applications to cardiovascular disease. Evid. Based Complement. Alternat. Med. 2012, 519031.

Wang, J., Gao, L., Lee, Y.M., et al., 2016. Target identification of natural and traditional medicines with quantitative chemical proteomics approaches. Pharmacol. Ther. 162, 10–22.

Wen, M., Zhang, Z., Niu, S., et al., 2017. Deep-learning-based drug–target interaction prediction. J. Proteome Res. 16, 1401–1409.

Wenzel, U., Herzog, A., Kuntz, S., et al., 2004. Protein expression profiling identifies molecular targets of quercetin as a major dietary flavonoid in human colon cancer cells. Proteomics 4, 2160–2174.

Wiśniewski, J.R., Zougman, A., Nagaraj, N., et al., 2009. Universal sample preparation method for proteome analysis. Nat. Methods 6, 359–362.

Ye, H., Ye, L., Kang, H., et al., 2011. HIT: linking herbal active ingredients to targets. Nucleic Acids Res. 39 (Suppl. 1), D1055–D1059.

Yoon, J.H., Baek, S.J., 2005. Molecular targets of dietary polyphenols with anti-inflammatory properties. Yonsei Med. J. 46, 585–596.

Yoshida, M., 2019. Recent advances in target identification of bioactive natural products. Biosci. Biotechnol. Biochem. 83, 1–9.

Yu, H., Chen, J., Xu, X., et al., 2012. A systematic prediction of multiple drug-target interactions from chemical, genomic, and pharmacological data. PLoS One 7, e37608.

Zhang, D., Tai, Y.C., Wong, C.H.S., et al., 2007. Molecular response of leukemia HL-60 cells to genistein treatment, a proteomics study. Leuk. Res. 31, 75–82.

Zhao, J., Jiang, P., Zhang, W., 2010. Molecular networks for the study of TCM pharmacology. Briefings Bioinf. 11, 417–430.

Ziegler, S., Pries, V., Hedberg, C., et al., 2013. Target identification for small bioactive molecules: finding the needle in the haystack. Angew. Chem. Int. Ed. 52, 2744–2792.

Zimmermann, G.R., Lehar, J., Keith, C.T., 2007. Multi-target therapeutics: when the whole is greater than the sum of the parts. Drug Discov. Today 12, 34–42.

Chapter 24

Systems pharmacology investigation of mechanism of action of nutraceuticals

Dimitris E. Messinis, Diogo A.R.S. Latino, Pavel Pospisil, Carine Poussin, Manuel C. Peitsch and Julia Hoeng

Introduction to systems-based mechanism of action discovery

Natural products and nutraceuticals

The term "nutraceuticals" covers a range of naturally occurring products that are believed to positively influence human health (Andrew and Izzo, 2017). Many natural products have been reportedly used as nutraceuticals since 2600 BC (Cragg and Newman, 2013) and exhibit various therapeutic antiinflammatory, analgesic, anticancer, antimalarial, and antimicrobial effects (Thomford et al., 2018). Their properties render them a good source of potential drug leads in pharmaceutical research (Dias et al., 2012).

Natural products are also termed secondary metabolites, to distinguish them from the primary metabolites that are essential to all living organisms. Secondary metabolites are not necessarily essential for an organism's development, and they are mainly produced for two reasons: either in adaptation to the surrounding environment or as a defense mechanism against predators. Therefore the mechanism of production of a natural product is often unique to an organism (Dewick, 1997), adding to the complexity of the mechanism of action (MoA) of nutraceuticals.

MoA of nutraceuticals

The MoA of a compound typically refers to the sequence of specific molecular interactions through which the compound mediates its pharmacological effect on the organism. Elucidating the MoA of plant-derived natural compounds can be challenging because these compounds can bind multiple protein targets with unrelated structures (Breinbauer et al., 2002). Drug discovery from natural products requires a nonreductionist strategy for understanding their complex mechanisms of action at the molecular level (Thomford et al., 2018). However, the lack of a full understanding of the MoA of a compound has not been necessarily a barrier to commercialization. An example of this is acetaminophen, which is one of the most widely used over-the-counter analgesic drugs worldwide, although its MoA is still unclear. The development of an experimental strategy that will holistically cover the biological systems, combined with computational approaches for integrating large-scale data, will support understanding of the MoAs of existing compounds for their therapeutic effects and accelerate the development of new drugs.

The discovery of the MoA of a compound encompasses identification of its direct and indirect biological targets. A direct target generally refers to the native protein (e.g., receptor, enzyme, ion channel, or nuclear hormone) or nucleic acid to which the compound physically binds, changing its conformation. This event subsequently activates or inhibits downstream molecules—seen as indirect targets or functional partners—participating in the causal chain of events resulting in the pharmacological effect of the compound. The terms "target" and "functional partners," when used in this chapter, refer to both direct and indirect targets unless otherwise specified.

The systems biology approach

Systems biology is an emerging field encompassing tools such as transcriptomics, proteomics, and metabolomics, with a goal to characterize all gene and cell products (e.g., mRNA, proteins, and metabolites) in a comprehensive manner (Dias et al., 2012). A systems approach is necessary for integrating as many factors involved in a physiological function in response to a perturbation as possible. A systems biology perspective combines structural analysis and evaluation of regulatory networks to elucidate the mechanisms of pathophysiology and drug

Nutraceuticals. DOI: https://doi.org/10.1016/B978-0-12-821038-3.00024-0

action (Boran and Iyengar, 2010). The plethora of systems biology computational methods developed raises several issues, including which approaches are the most useful in a specific problem, how to select the top-performing algorithms, and, maybe most importantly, how to integrate knowledge from the results of multiple methods. Crowdsourcing is a concept employed for solving the issue of the extended range of different methods by bringing together experts from multiple fields to solve a given problem. Examples of such methods can be seen in the "Systems Biology Verification combined with Industrial Methodology for Process Verification in Research" (sbv IMPROVER) Species Translation Challenge for assessing the limits of using rat models to predict human biology (Bilal et al., 2015; Poussin et al., 2014) and in the various "Dialogue on Reverse Engineering Assessment and Methods" challenges (Prill et al., 2011). Regarding the need for a consensus result, there are other methods used for integrating knowledge—for example, consensus networks (Bilal et al., 2015; Prill et al., 2011) and consensus scored lists (Szklarczyk et al., 2016).

Multiomics data generation

Technologies such as mass spectrometry and microarrays enable scientists to simultaneously measure thousands of molecules in a single biological sample. Genomics, transcriptomics, proteomics, and metabolomics data obtained in a single experiment can provide a system-level view of the mechanisms involved. The ability to produce extended profiles of cellular perturbations has led to the construction of reference databases of such profiles, which enable scientists to compare perturbations among test compounds and develop more informed hypotheses from the similarities and differences. Since the first functional discovery via a compendium of expression profiles in 2000 (Hughes et al., 2000), reference databases such as Connectivity Map (Lamb et al., 2006) and the L1000 data set (Subramanian et al., 2017) have been compiled. The high cost of commercial gene expression microarrays inspired the team behind the L1000 method to develop a new approach to gene expression profiling by using the Luminex technology. The technology uses capture antibodies coupled to beads that target a protein residue of interest. Similar to the classic sandwich ELISA, the assay uses a detection antibody coupled with biotin and a reporter molecule such as phycoerythrin coupled with streptavidin, which has affinity to the biotin on the detection antibody. The advantage of the technique is that a collection of beads that target analytes can be multiplexed, allowing measurement of multiple analytes in a single sample. However, the multiplexability of beads requires many cross-reactivity tests and quality control checks, which are conducted by the vendors of these assays. Although the Luminex scanner can detect only 500 bead colors, the National Institutes of Health Library of Integrated Network-Based Cellular Signatures team developed a method for measuring 978 genes that is sufficient to understand 82% of the full transcriptome (Subramanian et al., 2017), after analyzing 12,031 expression profiles from Gene Expression Omnibus (GEO).

Having access to big data and sophisticated assays enables researchers to compare their results in a robust manner and leverage the accumulating wealth of biological information to identify mechanisms of action.

Chemoinformatics for target identification

Chemoinformatics is a scientific field that considers the chemical structure of a substance as the primary source of information and relates it to the molecular phenotype, which refers to the physicochemical and biological properties of the substance. In the human body, substances interact with thousands of macromolecules such as proteins and genes via chemical bonds. A high degree of specificity and selectivity in these interactions can indicate that the substance–target pair is the key to the MoA.

Several databases of compound targets are available currently. Database screening techniques presuppose that compounds that are structurally similar might have the same target and that the pharmacophore of a compound might fit the known active site of the target.

The SwissTargetPrediction tool predicts macromolecular targets by using the two- and three-dimensional similarity of small compounds, assumed to be bioactive, from a library of 370,000 compounds known to be active against more than 3000 proteins from three species (Gfeller et al., 2014; Gfeller et al., 2013). This approach combines several measures of chemical similarity that are based both on chemical structure and molecular shape. This is especially efficient when no ligand with the same scaffold is known, and the target is, therefore, matched by structure and shape. SwissTargetPrediction can be used online: the input is the chemical structure, and the output is a list of potential therapeutic targets ranked by a specific match score.

ChEMBL is a database that contains manually curated data on bioactive molecules with drug-like properties that are extracted regularly from the literature. ChEMBL is a large-scale repository of chemical, bioactivity, and genomic information that allows researchers to derive relevant knowledge for drug development. The database contains data extracted from more than 72,000 scientific articles and other publications to date and 54 deposited data sets from other databases. More than 1.8 million compounds, 15.5 million activity value entries, 12,482 targets, and 4277 mechanisms of action are accessible in ChEMBL (Gaulton et al., 2017). Some ChEMBL database applications include identification of lead compounds for potential therapeutic targets, extraction of data on a specific target for developing quantitative structure–activity relationship models and structure–activity relationship studies, and information on mechanisms of action.

Network-based methods for the mechanism of action discovery

For gaining a systems-level perspective of many different perturbed molecules, a network view depicting them as nodes and their interactions as edges is often helpful.

Methods for constructing such a network are typically classified as being data- or topology-driven. Data-driven approaches are generally based on regression analysis such as Bayesian models or principal component analysis. Topology-driven approaches require prior knowledge of the network topology and use various signal-propagation assumptions to construct the experiment-specific network, including ordinary differential equations, Boolean models, fuzzy models, and integer linear programming formulations (Melas et al., 2011). Here, we focus on topology-driven methods for network construction.

There are several resources of signaling network data as a priori biological knowledge, including the Kyoto Encyclopedia of Genes and Genomes (KEGG) (Kanehisa et al., 2017) and Reactome (Fabregat et al., 2018). Wu and colleagues gathered functional interaction knowledge in a resource termed "Reactome Functional Interactions" (Reactome FI), which pools information from Reactome, Panther, CellMap, NCI Nature, NCI BioCarta, KEGG, and Transcriptional Regulatory Element Database (TRED) (Wu et al., 2010). This resource is updated annually and can be downloaded from the Reactome website (https://reactome.org/).

Several computational methods exist for analyzing cellular signal transduction information from these networks and cross-referencing it with experimental data. Melas and colleagues developed an approach that uses integer linear programming to encode constraints posed by the network topology on the behavior of the nodes, such as upregulation and downregulation (Melas et al., 2013). Philip Morris International has developed computational methods which leverage causal biological network models representative of biological processes/mechanisms (Boue et al., 2015) and gene expression data (Hoeng et al., 2012; Martin et al., 2014; Thomson et al., 2013); these methods enable investigation of the underlying mechanisms and prediction of the effects of exposure to a stimulus (e.g. compound, natural extract, toxicants). The methodology translates gene expression fold changes into differential values for each network node and summarizes this at the network level to provide quantitative assessment of the degree of perturbation in the network (Hoeng et al., 2012; Martin et al., 2014). Combining multiple relevant network models allows calculation of the biological impact factor, the overall biological effect of a perturbing agent, by aggregating individual network perturbation amplitude scores (Hoeng et al., 2012; Thomson et al., 2013).

A more recently developed pipeline (Liu et al., 2019) is the "CAusal Reasoning for Network identification using Integer VALue" program (CARNIVAL), which identifies upstream regulatory signaling networks explained by downstream gene expression changes. CARNIVAL selects a subset of interactions from a prior knowledge network that links known or potential targets of perturbation and active transcription factors. CARNIVAL users adjust the settings and data types to customize the default inferences of transcription factor activities from gene expression data by using the algorithm "Discriminant Regulon Expression Analysis" (DoRothEA) (Garcia-Alonso et al., 2018). Along with these inferences, a protein–protein interaction network and the targets of the perturbation yield a series of linear constraints for integer linear programming problems (Mitsos et al., 2009). Then, an integer linear programming solver such as IBM ILOG CPLEX is used to identify the subnetwork topology.

In Table 24.1, we summarize the resources mentioned in this chapter, which are also integrated into our computational pipeline presented here.

Integrated systems-based approaches for target identification and the MoA discovery

In this chapter, our goal is to understand the MoA of compounds such as nutraceuticals by using systems biology approaches. More specifically, we present a selection of the most up-to-date tools in systems pharmacology, proposing a pipeline for integrating knowledge from several methods and creating ensemble predictions that empower further hypothesis development and experimental design.

To illustrate the relevance and value of our integrated approach, we provide use cases on three extensively studied nutraceuticals, epigallocatechin-3-gallate (hereinafter called epigallocatechin), quercetin, and resveratrol, as well as on a prototypical proinflammatory stimulus, tumor necrosis factor alpha (TNF-α).

Overview of deployed computational tools and strategy

The computational approaches deployed in this chapter are depicted in Fig. 24.1.

TABLE 24.1 Summary of tools and knowledge resources mentioned in this chapter.

	Short name	Full name	References
Tools	CARNIVAL	CAusal Reasoning for Network identification using Integer VALue programming	Liu et al. (2019)
	DataWarrior	DataWarrior	Sander et al. (2015)
	ILP formulations	Integer Linear Programming formulations	Mitsos et al. (2009)
	NPA and BIF	Network Perturbation Amplitude and Biological Impact Factor	Hoeng et al. (2012), Martin et al. (2014), (Thomson et al., 2013)
	SwissTargetPrediction	SwissTargetPrediction	Gfeller et al. (2014)
Knowledge resources	ChEMBL	Ch European Molecular Biology Laboratory	Gaulton et al. (2017)
	DoRothEA	Discriminant Regulon Expression Analysis	Garcia-Alonso et al. (2018)
	KEGG	Kyoto Encyclopedia of Genes and Genomes	Kanehisa et al. (2017)
	LINCS L1000 dataset	Library of Integrated Network-Based Cellular Signatures L1000 dataset	Subramanian et al. (2017)
	OmniPath	OmniPath	Turei et al. (2016)
	Reactome FI	Reactome Functional Interactions	Wu et al. (2010)
	STITCH	Search Tool for InteracTions of CHemicals	Szklarczyk et al. (2016)
	STRING	STRING	Szklarczyk et al. (2019)

The resources are divided into knowledge resources and tools. Their short name, full name, and reference publication(s) are listed.

Our aim is to identify potential direct and indirect targets and generate hypotheses on the MoA of a compound that can be tested in a new cycle of experiments. This pipeline consists of complementary computational systems biology and chemoinformatics-based approaches that integrate results from omics data and different profile- and network-based computational approaches that leverage chemical structures and omics data.

With regard to profile comparison methods, we compared our gene expression data with the wide range of perturbations in the L1000 data set (Subramanian et al., 2017). For perturbations with a high degree of similarity to our compounds, we retrieved predicted functional partners of chemicals and proteins from the databases "Search Tool for InteracTions of CHemicals" (STITCH) (Szklarczyk et al., 2016) and STRING (Szklarczyk et al., 2019), respectively.

Employing network-based reverse engineering, we used the CARNIVAL pipeline (Liu et al., 2019) to infer transcription factor activities from the gene expression data using the algorithm DoRothEA (Garcia-Alonso et al., 2018). We then derived appropriate linear constraints to generate an integer linear programming problem to solve with the IBM ILOG CPLEX software package and predict possible compound targets and functional partners from the signaling network topology.

On the chemoinformatics front, we compared the chemical structures of the tested compounds with those of approximately 1.87 million compounds from the ChEMBL database (Gaulton et al., 2017) using the DataWarrior software (Sander et al., 2015) and obtained a list of potential targets from SwissTargetPrediction (Gfeller et al., 2014).

Finally, the predictions from all these sources were integrated in a Bayesian fashion to develop hypotheses for the functional partners of the compounds. By performing overrepresentation analysis (ORA) of various gene lists throughout the pipeline, we were able to identify most likely associated biological processes/pathways.

Transcriptomics data set generation

For the purpose of the case studies, we have generated transcriptomics data for three well-studied nutraceuticals—epigallocatechin, quercetin, and resveratrol—as described previously (Gonzalez-Suarez et al., 2016). Briefly, cryopreserved human hepatocytes were exposed to various concentrations of epigallocatechin (1, 5, and 10 μM), quercetin (5, 10, and 25 μM), resveratrol (10, 25, and 100 μM), or vehicle controls for 24, 48, or 72 h. The concentrations of the nutraceuticals were selected so that no more than a 30% decrease in cell count was observed after 72 h of exposure. The transcriptome of the hepatocytes was analyzed using

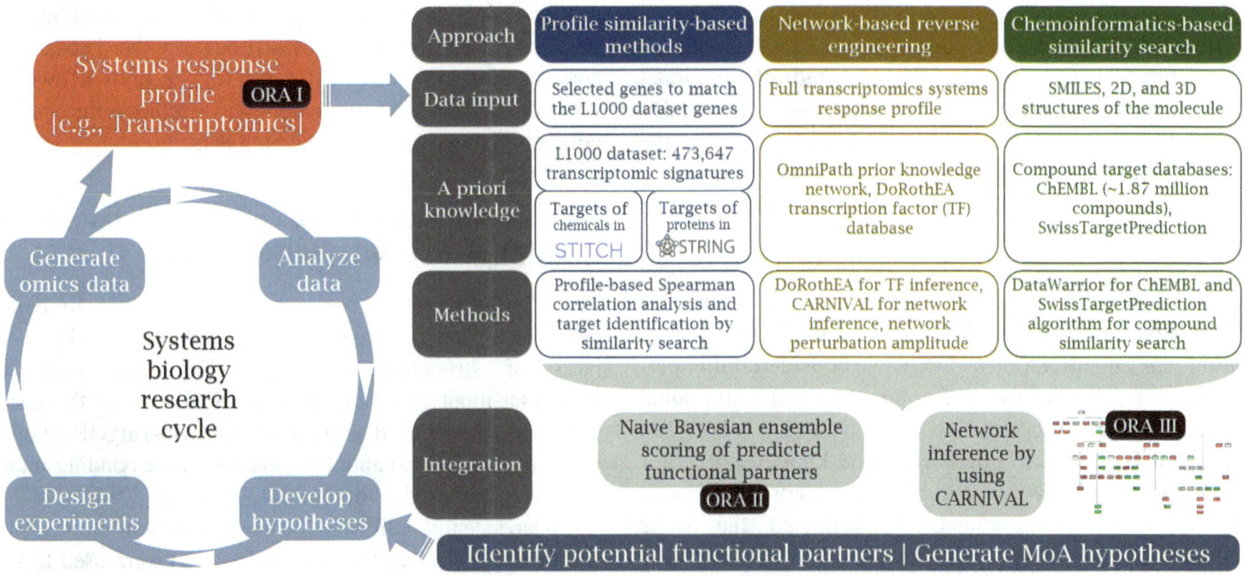

FIGURE 24.1 Integrated systems-based approach for target prediction and MoA discovery. Starting with the gene expression data of the compound at hand, we use different approaches to integrate knowledge and investigate the MoA of the compound. On the profile comparison front, we compare its transcriptomic signature to the 473,647 signatures of L1000 data set perturbations and acquire the similar compounds' predicted functional partners from the STITCH or STRING database. On the network-based reverse engineering front, we deploy the CARNIVAL pipeline to predict possible compound functional partners from the signaling network topology. On the chemoinformatics front, we compare the 2D and 3D structures of the molecule with those of approximately 1.87 million compounds. Predictions from all these sources are integrated in a Bayesian fashion to develop hypotheses for the next experimental design iteration. ORA allows the identification of most likely associated biological processes and pathways. ILP, integer linear programming; MoA, mechanism of action; ORA, overrepresentation analysis.

Affymetrix microarrays. Raw data were quality checked, normalized, and processed as systems response profiles (SRP). A SRP corresponds to a pairwise comparison, computed by comparing each concentration treatment with its respective vehicle control. The data set for each nutraceutical consisted of nine SRPs.

To demonstrate the adaptability of our integrated approach to other stimuli types, we applied our pipeline on a transcriptomics dataset previously generated after stimulation of primary human coronary artery endothelial cells with TNF-α (0.05, 0.1, 0.5, 1, or 10 ng/mL) for 4 and 16 h (Poussin et al., 2019), resulting in 10 SRPs as described above.

SRP similarity search based on the L1000 data set

To compare our gene expression data with publicly available data of other compounds, we used the Level 5 LINCS L1000 data set, which consists of 473,647 differential gene expression signatures. The data can be accessed with GEO accession number GSE92742 and file name "GSE92742_Broad_LINCS_Level5_COMPZ.MODZ_n473647x12328.gctx." The data set provides expression values for 12,328 genes inferred by the 978 landmark genes measured with the L1000 technology as described above. We used only the 978 landmark genes for our case in order to remain close to measured rather than computationally inferred data. The overlap between the 978 genes from the LINCS L1000 dataset and our measured genes was 938, which were the genes to be used in the comparison.

To compare our 37 SRPs with the 473,647 L1000 signatures, we selected the 938 overlapping genes as explained above, creating 37 transcriptomic signatures. We performed Spearman's correlation analysis, because it is rank-based and, therefore, more appropriate when the data are not derived from the same experiment and might have been computationally processed in different ways. For each of the 37 transcriptomic signatures, we retained the highest Spearman score of each L1000 perturbation and discarded negative Spearman correlation scores. In parallel, we ran 100 random transcriptomic signatures against the 473,647 L1000 signatures. For each of our compounds, we calculated the *P* value of the Spearman correlation coefficient values of the nine observations and 100 random signatures and discarded any observation with a *P* value greater than .001. The resulting table consisted of L1000 perturbations and their Spearman correlation scores, to which we added the type of perturbation as noted in the L1000 dataset (chemical or genetic reagents). Chemical perturbations are small-molecule compounds, while genetic perturbations can be CRISPR/Cas9 constructs, shRNAs, and open reading frames used to edit, knockdown, or overexpress genes, respectively.

We identified the predicted functional partners of each chemical perturbation using STITCH (Szklarczyk et al.,

2016) and computationally matched the PubChem chemical identifier of each compound with its PubChem identifier in the STITCH database. We acquired all functional partners and their associated STITCH scores and multiplied the Spearman correlation coefficient with the STITCH functional partner score, because both scores range from 0 to 1. We followed a similar approach for all L1000 perturbations in the ligand category, which are proteins such as TNF. We then identified the functional partners and relevant scores from the STRING database (Szklarczyk et al., 2019). When a matching L1000 perturbation was an shRNA or a cDNA, we considered this perturbation a potential molecular target of our compound, using its Spearman correlation coefficient as a score. The L1000 data set lists a known MoA and known targets for 1902 compounds. When our compound correlated with one of these perturbations, we followed the same approach as that with the chemicals and STITCH database described above, setting the known targets of the perturbation at a score that STITCH would list if the predicted target was certain. The process is depicted in Fig. 24.1.

Prior knowledge-driven network construction for investigating functional partners

To infer the potential functional partners of a compound through network construction, we used the CARNIVAL pipeline (Liu et al., 2019), which, unlike previous integer linear programming-based biological network optimization efforts, can hypothesize the molecular targets of a perturbation from the available data; this approach is termed "inverted CARNIVAL." As a first step, we ran DoRothEA (Garcia-Alonso et al., 2018) to infer transcription factor activities from our gene expression data. The only other input needed for the inverted CARNIVAL approach is the prior knowledge network on which the algorithm will fit the experimental data. For this, we used OmniPath (Turei et al., 2016), a network curated from 41,237 references, which supports each of its interactions by an average of 2.88 peer-reviewed publications. We then generated the integer linear programming problem and solved it with IBM ILOG CPLEX. The network solution included a weight ranging from 0 to 100 for every node of the network, corresponding to its predicted activation. Additionally, having run the algorithm for all transcriptomic signatures at hand, we extracted a metric of how many times each node appears across the number of algorithm runs, because this frequency might be indicative of the importance of the specific node in explaining the MoA of the compound. We normalized the scores and multiplied them to yield a score from 0 to 1, which included information from both metrics. Of note, the integer linear programming solver "trims" prior knowledge network edges to minimize the error between the experimental data and prior knowledge network. Therefore the resulting network might be missing nodes and edges, either because the corresponding protein is not a part of the prior knowledge network, or because it was optimal to remove the specific connection to explain the remaining data.

Chemoinformatics-based target prediction using ChEMBL and SwissTargetPrediction

SwissTargetPrediction tool was used in order to derive potential molecular targets of the tested compounds on the basis of structural similarities. For that, simplified molecular-input line-entry system (SMILES) of the tested compounds were used as input in the SwissTargetPrediction website. A list of potential targets and corresponding match score was collected as output (Gfeller et al., 2014).

Independently, ChEMBL database (Gaulton et al., 2017) and DataWarrior tool (Sander et al., 2015) were used to perform a target search based on the similarity comparison between our compounds and those in the ChEMBL database. Similarity descriptors such as Fragment Fingerprints (FragFP), SkeletonSpheres (SkelSpheres), and Flexophores can be calculated to represent the compounds and compare ChEMBL and query compounds. For the present case studies, only FragFP was used as a descriptor. FragFp is a substructure fragment dictionary-based binary fingerprint similar to MDL keys (Xue and Bajorath, 2000); it contains 512 predefined structure fragments. The FragFp descriptor contains 1 bit for each fragment in the dictionary. A bit is set to 1 if the corresponding fragment is present in the molecule at least one time. We used a similarity threshold value of 0.9 for the comparison, meaning that only compounds with similarity higher than or equal to 0.9 were retrieved in the target search.

Consolidating and ranking approaches

There is a need for a consensus result when multiple computational approaches are used. The STRING (Szklarczyk et al., 2019) and STITCH (Szklarczyk et al., 2016) databases both have to consolidate information from various sources—such as experiments, literature, biochemical data, and other databases—and reach a consensus on a single score that represents the likelihood that a molecule is a functional partner. To accomplish this, these databases combine scores from various sources in a naïve Bayesian fashion:

$$\text{Ensemble score} = 1 - (1 - a) \times (1 - b) \times (1 - c),$$

where a, b, and c are scores from several sources, normalized to a value ranging from 0 to 1.

To integrate and prioritize the targets predicted by the different approaches described earlier (Fig. 24.1), we created an ensemble score for each molecular target (gene name) similarly to the above-mentioned approach. More specifically, our ensemble score described in Fig. 24.2 includes predicted targets resulting from: (1) analysis of nutraceutical-derived SRP comparison with L1000-derived transcriptomic signatures, including targets from STITCH (L1000 chemicals) and STRING (L1000 proteins), shRNA and cDNA (L1000 shRNA and cDNA), and known MoAs (L1000 known MoA); (2) inference of a network by using the CARNIVAL pipeline; and (3) chemoinformatics-based similarity search in ChEMBL by using DataWarrior and the SwissTargetPrediction database and tool.

Predicted molecular targets and network visualization

The CARNIVAL pipeline that was used in the network-based predictions described above was run again, using as an input the top 10 targets of each compound as predicted by our computational pipeline and gene expression data (Fig. 24.2). However, we selected genes that had a positive or negative t statistic across all transcriptomic signatures for each compound, calculated the average t statistic of the nine signatures, and selected the top 100 genes with the highest absolute average t statistic (either positive or negative). This procedure allowed to select genes that were consistently and most significantly perturbed across all concentrations and time points.

The resulting signaling networks were visualized using Cytoscape software (Shannon et al., 2003). Here, we present as an example the resulting signaling network of epigallocatechin in Fig. 24.4. The signaling networks of the other tested compounds are presented in Supplementary Figs. 24.1–24.3.

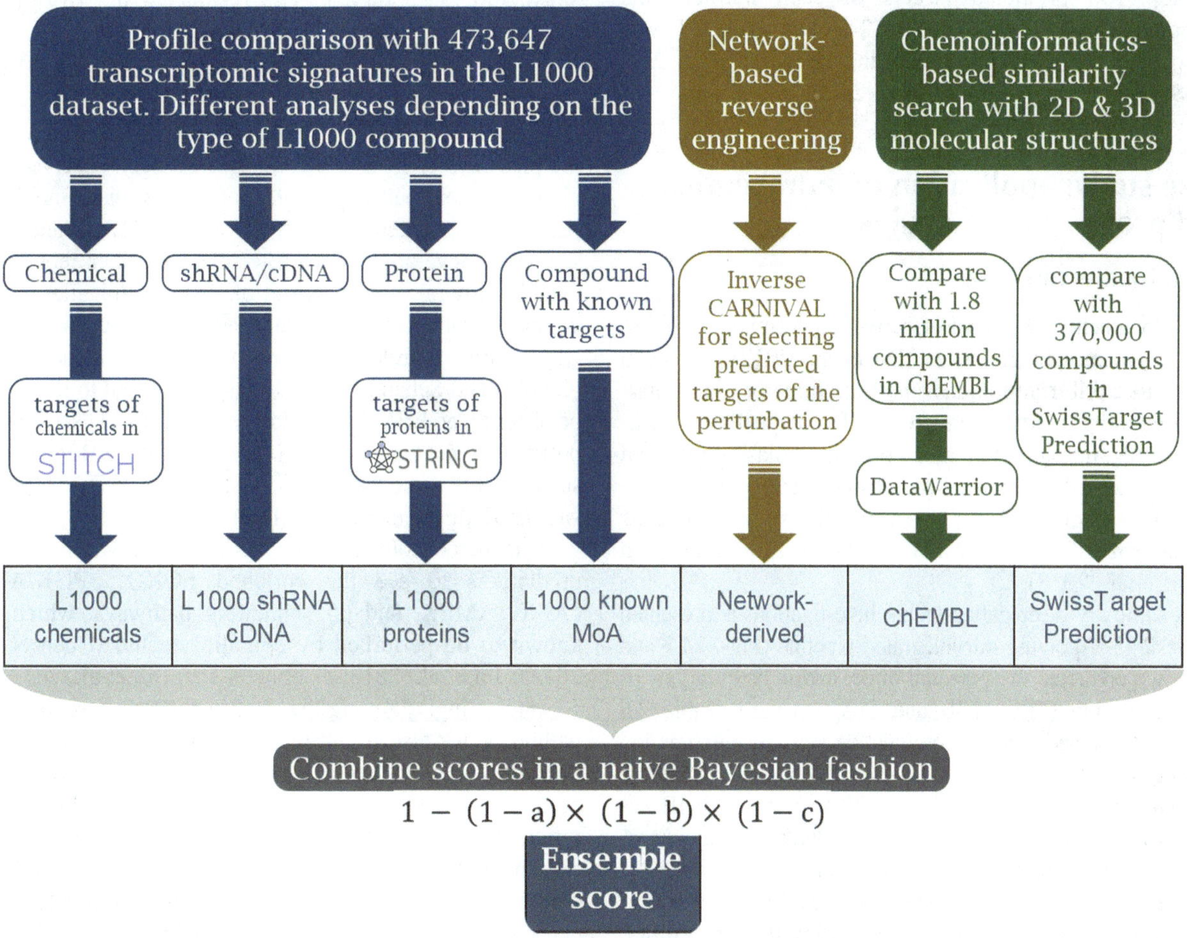

FIGURE 24.2 Ensemble scoring for integrating and prioritizing potential direct and indirect targets predicted using our computational pipeline. To create an ensemble score for each molecular target (gene name), we consolidated our sources: (1) targets from STITCH, (2) targets from STRING, (3) targets from L1000 compounds with known MoA (corresponding to L1000 perturbations whose transcriptomic signatures correlate with the signature of our compound), (4) L1000 perturbations derived from shRNA or cDNA that correlated with the signature of our compound, (5) network-derived potential targets from the CARNIVAL pipeline, (6) structural similarity-derived compounds from ChEMBL, and (7) structural similarity-derived compounds from SwissTargetPrediction.

Pathway overrepresentation analysis for biological insights

In order to support the biological interpretation of gene expression data as well as direct and indirect targets identified and prioritized by our pipeline, ORAs were performed using the KEGG database (Kanehisa et al., 2017) as source of a priori knowledge including 209 gene sets representative of canonical pathways. For ORAs, Fischer's exact test was used setting the size of gene universe to the maximal number of genes that could be measured using the Affymetrix microarray. ORAs were conducted at three stages of the computational pipeline as annotated ORA I, II, and III in Fig. 24.1. ORA I was based on the genes that were consistently and most significantly perturbed across all concentrations and time points of the experiments, selected as described above in the section "Predicted molecular targets and network visualization." ORA II was based on the top 100 predicted targets of each nutraceutical (Supplementary Tables 24.A1–24.A4). ORA III was based on the nodes of the resulting networks (Fig. 24.4 and Supplementary Figures 24.1–24.3).

Case study: application of nutraceuticals and prototypical stimulus

Epigallocatechin

Epigallocatechin, the most effective cancer chemopreventive polyphenol in green tea (Du et al., 2012), has been shown to exhibit antioxidant, antiinflammatory, and antiatherogenic properties via growth factor-mediated pathways, the mitogen-activated protein kinase-dependent pathway, and the ubiquitin/proteasome degradation pathway. The compound has been widely studied, but its mechanisms of action are not yet completely understood (Negri et al., 2018).

A summary of epigallocatechin investigations is presented in Fig. 24.3. Applying our integrated pipeline (Figs. 24.1 and 24.2), we predicted the potential direct and indirect targets of epigallocatechin (Fig. 24.3B and Supplementary Table 24.A1) and leveraged them in combination with gene expression data to build a network (Fig. 24.4). ORA was performed at different stages of the analysis (ORA I, II, and III, as described above) to gain biological insights into the MoA (Fig. 24.3C and Supplementary Table 24.A5).

We performed a literature review to evaluate the biological significance of our results, first going through the top 15 predicted functional partners of epigallocatechin, shown in Fig. 24.3C. Epigallocatechin is a known NRF2/NFE2L2 activator (Sun et al., 2017). One of the top predicted functional partners, DNA cytosine-5-methyltransferase 1, is a direct molecular target of epigallocatechin (Khan et al., 2015). Epigallocatechin is an ATP-competitive inhibitor of phosphoinositide-3-kinase (PI3K) and a mechanistic target of rapamycin, with Ki values of ~ 300 nM (Van Aller et al., 2011). It has also been shown to target the heat shock protein HSP90α (HSP90AA1) (Moses et al., 2015), inhibit phosphorylation of the Jun protooncogene (JUN) (Yang et al., 2000), and inhibit the EP300 acetyltransferase (Choi et al., 2009). As for MAPK1/ERK2, epigallocatechin has been shown to inhibit PC3 prostate cancer cell proliferation via MEK-independent ERK1/2 activation (Albrecht et al., 2008). Epigallocatechin inhibits activation of receptor tyrosine-protein kinase erbB-2 (ERBB2) and downstream signaling networks in human head and neck and breast carcinoma cells (Masuda et al., 2003). There are conflicting findings regarding JAK1, with reports demonstrating that epigallocatechin pretreatment reduces phosphorylated JAK1 levels (Liu et al., 2014) or suggesting that the action of epigallocatechin on p-STAT1—which is also present among the top predicted functional partners is mediated by downregulation of JAK2 but not JAK1 (Hamed et al., 2018). It has been shown that epigallocatechin promotes p53 accumulation and activity via inhibition of MDM2-mediated p53 ubiquitination (Jin et al., 2013).

Among the top 15 significantly over-represented gene set pathways (Fig. 24.3B), we interestingly observe three viral infection pathways, relevant to Epstein–Barr virus (EBV), the oncogenic human retrovirus HTLV-I, and measles virus. Several green tea catechins have been reported to exhibit antiviral effects (Xu et al., 2017). Epigallocatechin competes with heparan sulfate and sialic acid and, in this way, inhibits unrelated viruses by a common mechanism (Colpitts and Schang, 2014). It has been found to hinder the persistence of EBV latency by reducing the DNA-binding potency of nuclear antigen 1 (Chen et al., 2012b) and to suppress the oncogenic human retrovirus HTLV-I pX as well as *Tax* gene expression (Li et al., 2000).

It is also encouraging for the biological validity of our predictions to see the enriched FOXO1, PI3K/AKT, cAMP, AMPK, and p53 signaling pathways, which are known to be perturbed by epigallocatechin (Chen et al., 2012a; Jin et al., 2013; Li et al., 2019; Liu et al., 2013).

Interestingly, among the significantly over-represented pathways, we also found representatives of two key functions of epigallocatechin: regulation of the cell cycle (Han et al., 2011) and anticancer effects (Du et al., 2012) (significant over-represented pathways: proteoglycans in cancer and pancreatic cancer). Proliferation is an important part of cancer development, expressed by the activity of cell cycle-related proteins. In general, nutraceuticals have been shown to possess the potential to prevent cancer by halting cell cycle progression by targeting one or more steps of the cell cycle; specifically, most nutraceuticals prevent the transition of cancer cells from the G1 to S phase (Gupta et al., 2010).

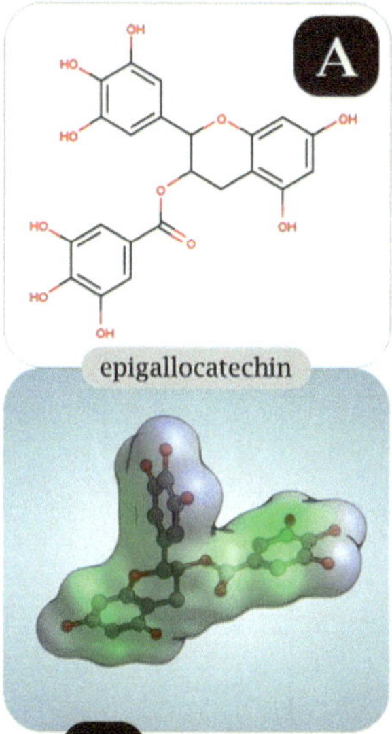

C

Gene set name	ORA I	ORA II	ORA III	Mean rank
Chemokine signaling pathway	8.50E-02	5.00E-19	4.33E-16	11.33
HTLV-I infection	3.37E-01	7.14E-20	2.13E-14	13.67
FoxO signaling pathway	4.40E-01	6.46E-19	1.10E-12	16.83
MicroRNAs in cancer	4.70E-01	5.45E-24	7.32E-11	16.83
Cell cycle	8.70E-03	1.10E-10	1.13E-15	23.17
PI3K-Akt signaling pathway	9.15E-01	1.61E-30	1.96E-14	23.33
Proteoglycans in cancer	9.46E-01	3.71E-28	9.35E-17	23.83
Measles	7.20E-01	1.70E-14	2.30E-12	28.00
cAMP signaling pathway	7.20E-01	5.04E-12	2.46E-14	29.67
Pancreatic cancer	9.77E-01	1.45E-25	3.89E-11	31.17
Rap1 signaling pathway	7.20E-01	2.02E-23	4.20E-07	32.33
Epstein-Barr virus infection	9.97E-01	2.73E-20	3.72E-13	34.33
Prolactin signaling pathway	9.97E-01	8.38E-18	1.07E-13	36.67
AMPK signaling pathway	2.97E-01	9.68E-12	2.16E-08	37.83
p53 signaling pathway	3.12E-02	6.50E-09	1.53E-09	37.83

B

Gene name	Ensemble score	L1000 chemicals	L1000 shRNA cDNA	L1000 proteins	L1000 known MoA	Network-derived	ChEMBL	SwissTarget Prediction
MAPK1	0.742	0.444	0.162	0.168	0.000	0.187	0.180	0.000
JUN	0.741	0.441	0.211	0.000	0.000	0.412	0.000	0.000
DNMT1	0.721	0.415	0.168	0.000	0.172	0.000	0.180	0.156
TP53	0.715	0.431	0.000	0.000	0.157	0.404	0.000	0.000
MTOR	0.710	0.438	0.198	0.171	0.225	0.000	0.000	0.000
NFE2L2	0.682	0.381	0.167	0.000	0.000	0.247	0.180	0.000
HSP90AA1	0.681	0.433	0.151	0.171	0.200	0.000	0.000	0.000
EP300	0.671	0.437	0.231	0.000	0.000	0.075	0.180	0.000
JAK1	0.669	0.429	0.156	0.170	0.166	0.009	0.000	0.000
PIK3CB	0.668	0.387	0.156	0.171	0.225	0.000	0.000	0.000
ERBB2	0.661	0.394	0.164	0.183	0.000	0.000	0.180	0.000
IKBKB	0.657	0.432	0.205	0.000	0.236	0.006	0.000	0.000
MAPK3	0.656	0.450	0.169	0.170	0.000	0.091	0.000	0.000
MEN1	0.652	0.403	0.151	0.163	0.000	0.000	0.180	0.000
STAT1	0.650	0.264	0.000	0.169	0.000	0.173	0.180	0.156

FIGURE 24.3 Summary of epigallocatechin investigations. (A) 2D (top) and 3D (bottom) structural representations of epigallocatechin. The 3D representation depicts the molecular surface, where the green area represents the lipophilic portions of the molecule, and light blue represents the polar portions. (B) Top predicted functional partners of epigallocatechin with their corresponding ensemble scores, calculated from the seven integrated sources of computational prediction. Extended table: Supplementary Table 24.A1. (C) Summary of the ORA results. Three different sets of genes (ORA I, II, and III) were tested as depicted in Fig. 24.1 and described in the section "Enrichment and predicted MoA." The table is sorted by descending average rank of *P* values across the three ORA groups. Extended table: Supplementary Table 24.A5. MoA, mechanism of action; ORA, overrepresentation analysis.

We did not find papers reporting links between epigallocatechin and signaling pathways of RAP1, and prolactin or the protein Menin (MEN1), part of the list of predicted targets in Fig. 24.3; The name of this protein is derived from the autosomal dominant disorder "multiple endocrine neoplasia type 1" (MEN1) characterized by the combined occurrence of parathyroid and adrenocortical tumors and neuroendocrine tumors of the pancreas and pituitary. Interestingly, MEN1 is a nuclear tumor suppressor that has been reported to interact in vitro with proteins involved in cell cycle control (Schnepp et al., 2006). These potential links could be worth further investigations.

Finally, as detailed in the section "Predicted molecular targets and network visualization," we created—on the basis of the predicted functional partners and gene expression data at hand—a visualization of what the predicted signaling network of epigallocatechin could looks like (Fig. 24.4).

Quercetin

Quercetin—found in fruits, vegetables, nuts, and berries (Tutel'ian and Lashneva, 2013)—has been reported to exhibit anticarcinogenic, antiinflammatory, and antiviral

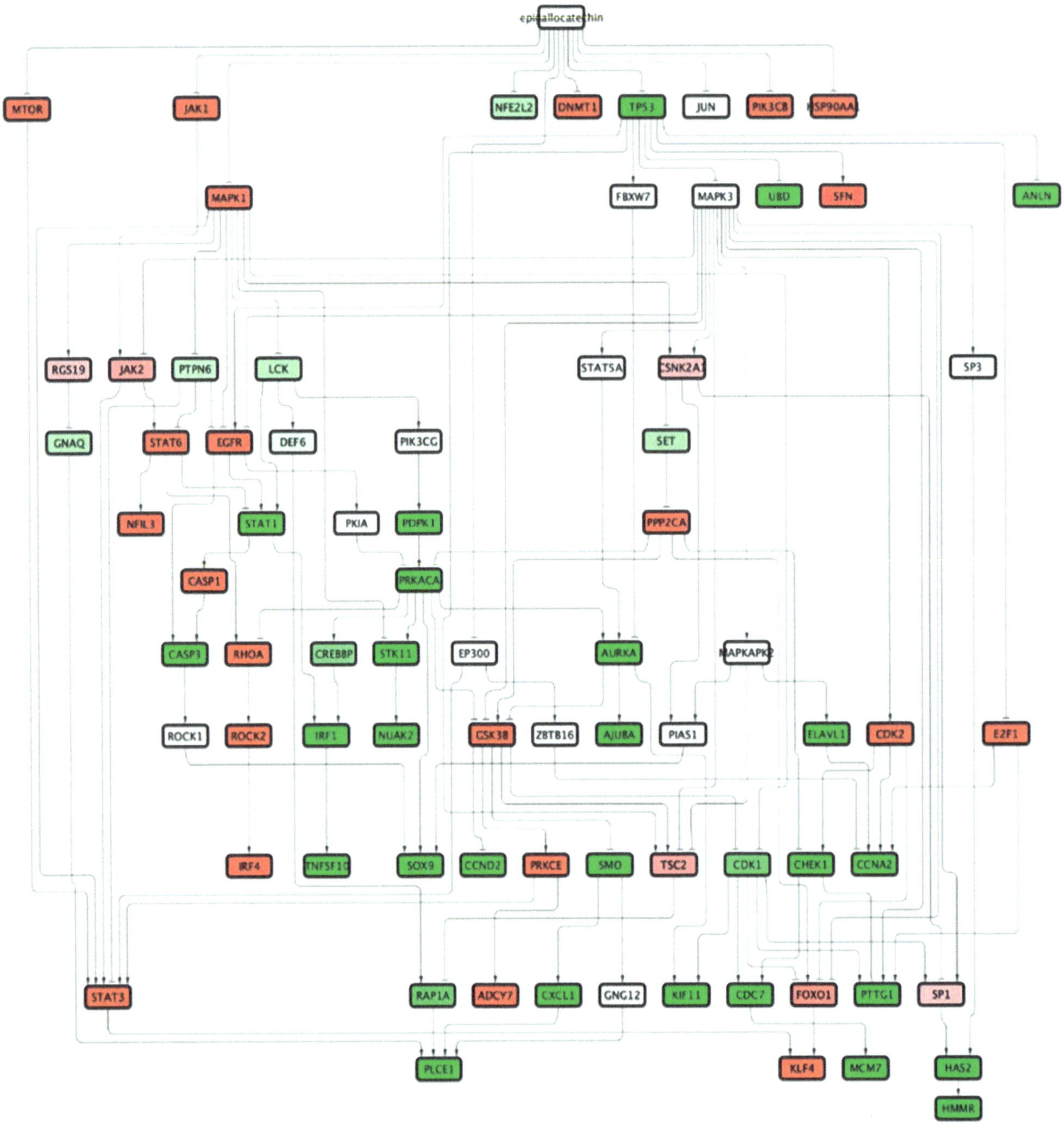

FIGURE 24.4 Epigallocatechin network. Signaling network illustrating the hypothesized MoA of epigallocatechin. Nodes that are predicted to be up/downregulated are highlighted in green/red, respectively. MoA, mechanism of action.

properties as well as reduce lipid peroxidation, platelet aggregation, and capillary permeability (Li et al., 2016).

A summary of quercetin investigations is presented in Fig. 24.5. Applying our integrated pipeline (Figs. 24.1 and 24.2), we predicted the potential direct and indirect targets of quercetin (Fig. 24.5B and Supplementary Table 24.A2) and leveraged them in combination with gene expression data to build a network (Supplementary Fig. 24.2). ORA was performed at different stages of the analysis (ORA I, II, and III, as described above) to gain biological insights into the MoA (Fig. 24.5C and Supplementary Table 24.A6).

We performed a literature search to evaluate the biological relevance of our results, first going through the top 15 predicted functional partners of quercetin, shown in Fig. 24.5C.

It is interesting to find at the top of the list the transcription factor SP1, which has been strongly suggested to be a novel molecular target of quercetin (Chae et al.,

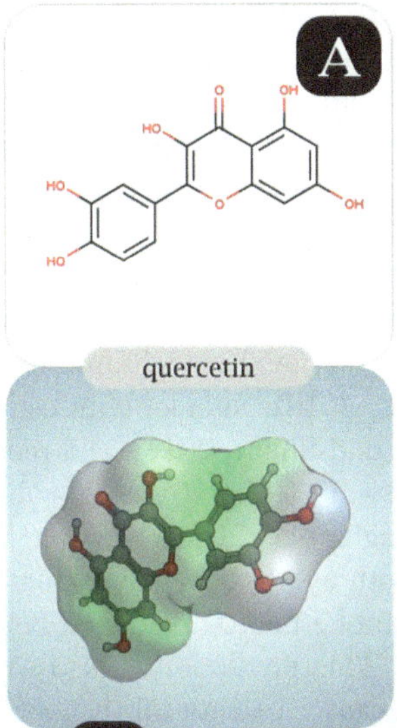

C

Gene set name	ORA I	ORA II	ORA III	Mean rank
Osteoclast differentiation	6.91E-01	6.88E-18	1.57E-17	18.33
Pathways in cancer	1.00E+00	8.66E-56	1.53E-28	36.00
Prostate cancer	1.00E+00	4.61E-44	7.28E-23	37.33
Proteoglycans in cancer	1.00E+00	2.01E-34	2.55E-23	37.50
PI3K-Akt signaling pathway	1.00E+00	9.39E-38	6.42E-22	38.00
Hepatitis B	1.00E+00	3.60E-26	2.55E-23	38.17
Focal adhesion	1.00E+00	1.40E-24	5.81E-20	40.67
Thyroid hormone signaling pathway	1.00E+00	1.53E-22	5.52E-21	41.67
ErbB signaling pathway	1.00E+00	4.98E-23	3.00E-19	41.83
Rap1 signaling pathway	1.00E+00	2.86E-23	1.06E-18	42.67
Hepatitis C	1.00E+00	1.12E-22	4.76E-19	43.00
Pancreatic cancer	1.00E+00	6.03E-22	3.12E-19	44.00
Colorectal cancer	1.00E+00	4.86E-22	6.28E-17	45.67
FoxO signaling pathway	1.00E+00	6.55E-22	1.13E-15	47.67
Endometrial cancer	1.00E+00	5.03E-19	3.00E-19	48.83

B

Gene name	Ensemble score	L1000 chemicals	L1000 shRNA cDNA	L1000 proteins	L1000 known MoA	Network-derived	ChEMBL	SwissTarget Prediction
SP1	0.825	0.289	0.000	0.000	0.000	0.754	0.000	0.000
GSK3B	0.798	0.298	0.000	0.139	0.155	0.396	0.180	0.200
STAT3	0.772	0.336	0.159	0.154	0.000	0.519	0.000	0.000
FOXO1	0.723	0.330	0.000	0.138	0.000	0.520	0.000	0.000
PPARG	0.716	0.335	0.172	0.142	0.000	0.400	0.000	0.000
JUN	0.694	0.276	0.187	0.000	0.000	0.480	0.000	0.000
KDR	0.692	0.319	0.000	0.154	0.184	0.000	0.180	0.200
MAPK1	0.679	0.284	0.000	0.143	0.000	0.361	0.180	0.000
CTNNB1	0.670	0.331	0.000	0.151	0.000	0.419	0.000	0.000
PTK2	0.668	0.294	0.156	0.151	0.000	0.000	0.180	0.200
ERBB2	0.647	0.392	0.157	0.163	0.178	0.000	0.000	0.000
FLT3	0.647	0.340	0.000	0.000	0.184	0.000	0.180	0.200
HIF1A	0.644	0.000	0.000	0.154	0.000	0.488	0.180	0.000
RUNX1	0.638	0.064	0.164	0.000	0.000	0.437	0.180	0.000
MMP9	0.634	0.335	0.000	0.161	0.000	0.000	0.180	0.200

FIGURE 24.5 Summary of quercetin investigations. (A) 2D (up) and 3D (down) structural representations of quercetin. The 3D representation depicts the molecular surface, where the green area represents the lipophilic portions of the molecule, and light blue represents the polar portions. (B) Top predicted functional partners of quercetin with their corresponding ensemble scores, calculated from the seven integrated sources of computational prediction. Extended table: Supplementary Table 24.A2. (C) Summary of the ORA results. Three different sets of genes (ORA I, II, and III) were tested as depicted in Fig. 24.1 and described in the section "Enrichment and predicted MoA." The table is sorted by descending average rank of *P* values across the three ORA groups. Extended table: Supplementary Table 24.A6. MoA, mechanism of action; ORA, overrepresentation analysis.

2012). Likewise, transcription factors STAT3, JUN, FOXO1, and HIF1A have also been experimentally shown to be perturbed by quercetin (Huang et al., 2013; Michaud-Levesque et al., 2012; Wang et al., 2014; Wilson and Poellinger, 2002). Inactivation of GSK3B via the PI3K/Akt pathway negatively regulates quercetin-induced apoptosis by inhibiting JNK activity (Lee and Yoo, 2013). Quercetin activates PPARG (peroxisome proliferator-activated receptor-gamma) (Beekmann et al., 2015), targets VEGFR2/KDR (Li et al., 2015), and inhibits migration through a MAPK-dependent mechanism (Nam et al., 2008). As part of its anticancer properties, quercetin has been shown to regulate β-catenin (CTNNB1) signaling (Srinivasan et al., 2016). Quercetin can also suppress the activity of protein tyrosine kinase 2 (PTK2) (Huang et al., 2005), has been shown to induce ubiquitination and downregulation of ERBB2 (Jeong et al., 2008), inhibit strongly FMS-like tyrosine kinase 3 (Boly et al., 2011), and downregulate MMP9 (Vijayababu et al., 2006).

Among the top 15 over-represented gene set pathways (Fig. 24.5B), "osteoclast differentiation" popped up in first position. Quercetin has been shown to decrease osteoclastic differentiation via a mechanism involving transcription factors NF-κB and AP-1 (Wattel et al., 2004). ORAs also highlighted significant over-represention of gene sets representative of PI3K-Akt, ErbB, thyroid hormone, and FoxO signaling pathways, which have all been shown to be perturbed by quercetin (Firdous et al., 2014; Giuliani et al., 2014; Granato et al., 2017; Huang et al., 2013).

The antiviral activity of quercetin is reflected by the significant over-representation of Hepatitis B gene set. Indeed, quercetin has been shown to significantly reduce the levels of hepatitis B antigens, reduce the secretion and genomic DNA levels of hepatitis B virus (Cheng et al., 2015), and possess antihepatitis C virus properties (Rojas et al., 2016).

Over-representation of gene set representative of focal adhesion pathway is also biologically relevant, because we have already identified the focal adhesion kinase PTK2 among the perturbed functional partners (Fig. 24.5C), and quercetin has been shown to impair the formation of focal adhesions and actin bundles by suppressing the production of focal adhesion kinase and paxillin (Cui et al., 2019).

Finally, six of the top 15 over-represented gene set pathways were related to cancer, in line with known anticancer effects of quercetin. Indeed, quercetin has been shown to inhibit the proliferation of a wide range of cancers, including prostate, cervical, lung, breast, and colon cancers (Jeong et al., 2009).

Resveratrol

Resveratrol—found in grapes, peanuts, soy, and berries (Breuss et al., 2019)—has been shown to exert antiinflammatory and antioxidative effects, although clinical trials must contend with its poor bioavailability (Berman et al., 2017).

A summary of resveratrol investigations is presented in Fig. 24.6. Applying our integrated pipeline (Figs. 24.1 and 24.2), we predicted the potential direct and indirect targets of resveratrol (Fig. 24.6B and Supplementary Table 24.A3) and leveraged them in combination with gene expression data to build a network (Supplementary Fig. 24.3). ORA was performed at different stages of the analysis (ORA I, II, and III, as described above) to gain biological insights into the MoA (Fig. 24.6C and Supplementary Table 24.A7).

We investigated the literature to evaluate the biological relevance of our results, first going through the top 15 predicted functional partners of quercetin, shown in Fig. 24.6C.

Resveratrol has been reported to inhibit directly the transcription factor SP1 (Lee et al., 2012) and to modulatethe expression of transcription factors JUN, STAT3, SREBF1 (sterol regulatory element-binding transcription factor 1), and CEBPB (Baek et al., 2016; Santos et al., 2014; Wang et al., 2019; Wang et al., 2009). Resveratrol also enhances the expression of CREB1 (cAMP-response element-binding protein) (Lei et al., 2017). Activation of TP53 is one of the mechanisms by which resveratrol exerts its anticancer effects, and it has also been reported that resveratrol directly interacts with TP53 (Muniz De Queiroz and Prives, 2018).

Resveratrol has been experimentally shown to inhibit retinoblastoma protein (RB1) (Choi et al., 2019), carbonic anhydrase 2 (Senturk et al., 2011), and cytochrome P450 3A4 (CYP3A4) (Mohos et al., 2018); interact with estrogen receptor 2 (ESR2) (Robb and Stuart, 2011); and modulate the expression of PTGS2 (prostaglandin-endoperoxide synthase 2) (Singh et al., 2011).

Interestingly, we found both cyclin-dependent kinases 1 and 2 (CDK1 and CDK2) among the top predicted functional partners of resveratrol. This prediction is also supported by the literature, because resveratrol has been experimentally shown to increase the levels of CDK1 and CDK2 and promotes CDK1 phosphorylation (Delmas et al., 2002).

Finally, we analyze the top 15 over-represented gene sets (Fig. 24.6B). Interestingly, similar to quercetin, resveratrol prevents osteoclast differentiation (He et al., 2010), a biological process highlighted in our analysis. It has been reported that resveratrol has cardioprotective effects that could improve viral myocarditis (Riba et al., 2017), induces apoptosis in HTLV-I infected cell lines (Hayashibara et al., 2002), selectively inhibits acute myeloid leukemia cells (Gautam et al., 2000), has dose-dependent antileishmanial activity (Ferreira et al., 2014), and alleviates *Mycobacterium tuberculosis* infection through antiinflammatory mechanisms (Yang et al., 2019).

Estrogen signaling is part of the top 15 over-represented pathways and the connection between resveratrol and ESR2, also part of the list of the predicted functional partners has been discussed earlier (Fig. 24.6C). The signaling pathways of Ras, ErbB, MAPK, and Fc epsilon RI also highlighted in our analysis have been shown to be perturbed by resveratrol treatment previoulsy (Gu et al., 2018; Jin et al., 2017; Koo et al., 2006; Sui et al., 2014).

Interestingly, "platelet activation" process is also part of the top 15 over-represented pathways, and it was recently shown that resveratrol has the dual ability to reduce unwanted platelet activation during storage, while preserving critical hemostatic function (Lannan et al., 2016).

The mechanisms of action underlying the prevention of prion protein-mediated toxicity by resveratrol are under investigations, however, with autophagy being revealed as one of the MoAs (Jeong et al., 2012).

No information connectingresveratrol and RUNX1 or the Rap1 signaling pathway was reported in the literature.

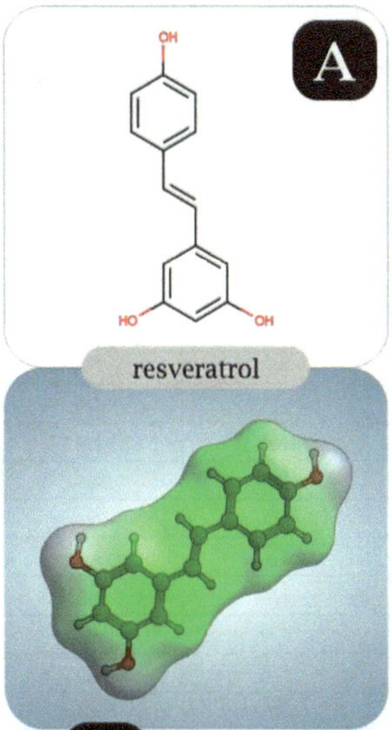

C

Gene set name	ORA I	ORA II	ORA III	Mean rank
Chemokine signaling pathway	8.50E-01	3.31E-17	6.13E-07	17.67
Rap1 signaling pathway	6.15E-01	2.55E-10	1.46E-06	19.17
Osteoclast differentiation	6.15E-01	2.50E-09	5.39E-08	19.50
HTLV-I infection	8.91E-01	2.59E-14	1.16E-08	28.50
Acute myeloid leukemia	8.91E-01	3.31E-17	6.51E-06	29.17
Estrogen signaling pathway	9.52E-01	1.30E-17	1.29E-07	30.33
Tuberculosis	8.50E-01	3.48E-08	3.63E-06	33.83
Ras signaling pathway	9.45E-01	1.81E-13	1.82E-07	34.33
ErbB signaling pathway	9.52E-01	1.23E-17	3.38E-05	36.67
Leishmaniasis	8.51E-01	1.21E-08	2.30E-06	40.00
MAPK signaling pathway	9.52E-01	2.33E-09	5.36E-07	44.33
Fc epsilon RI signaling pathway	8.91E-01	8.31E-08	6.95E-06	45.67
Viral myocarditis	8.50E-01	2.48E-05	8.75E-06	45.83
Platelet activation	8.50E-01	4.46E-08	1.31E-03	47.33
Prion diseases	8.50E-01	1.10E-05	1.63E-04	48.83

B

Gene name	Ensemble score	L1000 chemicals	L1000 shRNA cDNA	L1000 proteins	L1000 known MoA	Network-derived	ChEMBL	SwissTarget Prediction
CREB1	0.825	0.287	0.164	0.150	0.000	0.656	0.000	0.000
TP53	0.792	0.310	0.000	0.000	0.000	0.633	0.180	0.000
SP1	0.766	0.246	0.000	0.000	0.000	0.690	0.000	0.000
JUN	0.695	0.266	0.000	0.000	0.000	0.584	0.000	0.000
STAT3	0.692	0.303	0.000	0.000	0.000	0.558	0.000	0.000
SREBF1	0.641	0.287	0.000	0.000	0.000	0.497	0.000	0.000
RUNX1	0.630	0.060	0.158	0.000	0.000	0.429	0.180	0.000
RB1	0.601	0.246	0.170	0.000	0.000	0.363	0.000	0.000
CEBPB	0.591	0.289	0.000	0.000	0.000	0.424	0.000	0.000
ESR2	0.565	0.350	0.000	0.000	0.176	0.000	0.180	0.010
PTGS2	0.541	0.300	0.000	0.000	0.000	0.000	0.180	0.200
CA2	0.534	0.289	0.000	0.000	0.000	0.000	0.180	0.200
CYP3A4	0.523	0.272	0.000	0.000	0.000	0.000	0.180	0.200
CDK2	0.513	0.357	0.000	0.000	0.195	0.051	0.000	0.010
CDK1	0.509	0.351	0.000	0.000	0.195	0.052	0.000	0.010

FIGURE 24.6 Summary of resveratrol investigations. (A) 2D (up) and 3D (down) structural representations of resveratrol. The 3D representation depicts the molecular surface, where the green area represents the lipophilic portions of the molecule, and light blue represents the polar portions. (B) Top predicted functional partners of resveratrol with their corresponding ensemble scores, calculated from the seven integrated sources of computational prediction. Extended table: Supplementary Table 24.A3. (C) Summary of the ORA results. Three different sets of genes (ORA I, II, and III) were tested as depicted in Fig. 24.1 and described in the section "Enrichment and predicted MoA." The table is sorted by descending average rank of *P* values across the three ORA groups. Extended table: Supplementary Table 24.A7. MoA, mechanism of action; ORA, overrepresentation analysis.

TNF-α

We used TNF-α, a prototypical inflammatory stimulus, as an additional positive control to demonstrate the relevance of the methodology for exploring mechanisms of action and to highlight the applicability of the approach to other types of molecules such as protein ligands.

A summary of TNF-α investigations is presented in Fig. 24.7. Applying our integrated pipeline (Figs. 24.1 and 24.2), we predicted the potential direct and indirect targets of TNF-α (Fig. 24.7A and Supplementary Table 24.A4) and leveraged them in combination with gene expression data to build a network (Supplementary Fig. 24.3). ORA was performed at different stages of the analysis (ORA I, II, and III, as described above) to gain biological insights into the MoA (Fig. 24.7B and Supplementary Table 24.A8). Because TNF-α is a protein, we could not perform 2D and 3D structure similarity comparison using SwissTargetPrediction or ChEMBL.

Of our top 15 predicted functional partners, 10 are key components of the TNF pathway (Aggarwal, 2003): TRAF2, IKBKB, JUN, NFKBIA, MAPK11, BIRC3, CXCL8, and the TNF receptor family members LTBR, CD40, and, of course, TNFRSF1A, the main direct target of TNF-α. TNF-α stimulation has also been associated with tyrosine phosphorylation of STAT1 and STAT3 (Guo et al., 1998), while PPARG activity is regulated by TNF-α

A

Gene name	Ensemble score	L1000 chemicals	L1000 shRNA cDNA	L1000 proteins	L1000 known MoA	Network-derived	ChEMBL	SwissTarget Prediction
STAT3	0.875	0.300	0.000	0.741	0.310	0.332	NA	NA
IKBKB	0.865	0.220	0.000	0.758	0.284	0.000	NA	NA
PPARG	0.859	0.255	0.000	0.745	0.258	0.000	NA	NA
TRAF2	0.833	0.061	0.265	0.758	0.000	0.000	NA	NA
TNFRSF1A	0.825	0.000	0.276	0.758	0.000	0.000	NA	NA
PSMB1	0.824	0.253	0.000	0.684	0.256	0.000	NA	NA
LTBR	0.818	0.062	0.259	0.739	0.000	0.000	NA	NA
JUN	0.812	0.254	0.000	0.748	0.000	0.143	NA	NA
NFKBIA	0.812	0.239	0.000	0.753	0.000	0.000	NA	NA
CD40	0.812	0.042	0.335	0.704	0.000	0.000	NA	NA
STAT1	0.810	0.000	0.277	0.738	0.000	0.669	NA	NA
IL10	0.809	0.213	0.000	0.757	0.000	0.000	NA	NA
MAPK11	0.805	0.275	0.000	0.732	0.000	0.000	NA	NA
BIRC3	0.802	0.219	0.000	0.747	0.000	0.000	NA	NA
CXCL8	0.802	0.244	0.000	0.739	0.000	0.000	NA	NA

B

Gene set name	ORA I	ORA II	ORA III	Mean rank
Toxoplasmosis	3.81E-03	3.70E-35	7.33E-22	3.50
NF-kappa B signaling pathway	6.87E-07	5.49E-55	2.12E-18	4.33
Toll-like receptor signaling pathway	1.88E-02	4.63E-35	2.84E-20	8.33
Hepatitis B	1.13E-02	2.83E-23	4.45E-21	9.33
Leishmaniasis	3.78E-03	4.51E-29	2.78E-14	10.67
Apoptosis	1.49E-02	2.33E-25	6.78E-18	12.00
Pathways in cancer	4.85E-02	5.96E-29	4.63E-21	12.50
Osteoclast differentiation	3.35E-02	1.26E-30	2.64E-14	15.50
Chagas disease (American trypanosomiasis)	8.31E-02	7.41E-25	4.89E-19	15.67
Herpes simplex infection	4.11E-03	4.51E-29	7.35E-11	15.83
Cytokine-cytokine receptor interaction	3.84E-10	2.18E-23	1.23E-10	16.00
Tuberculosis	1.88E-02	4.78E-23	1.51E-15	16.17
Influenza A	1.13E-03	2.15E-12	1.97E-16	16.33
Hepatitis C	8.80E-02	1.60E-19	2.58E-25	17.00
TNF signaling pathway	3.17E-01	2.20E-32	3.61E-17	18.00

FIGURE 24.7 Summary of TNF-α investigations. (A) Top predicted functional partners of TNF-α with their corresponding ensemble scores, calculated from the seven integrated sources of computational prediction. Extended table: Supplementary Table 24.A4. (B) Summary of the ORA results. Three different sets of genes (ORA I, II, and III) were tested as depicted in Fig. 24.1 and described in the section "Enrichment and predicted MoA." The table is sorted by descending average rank of *P* values across the three ORA groups. Extended table: Supplementary Table 24.A8. MoA, mechanism of action; ORA, overrepresentation analysis.

at pre- and post-translational levels (Ye, 2008). Interestingly, among the predicted functional partners of TNF-α, we found two molecules that have been shown to be connected to TNF-α levels, although it is not yet completely understood how. Gene expression and protein levels of TNF-α are increased by silencing PSMB1 (proteasome subunit beta type 1) (Wu et al., 2019), while TNF-α blockade induces IL10 expression (Evans et al., 2014).

Among the over-represented pathways, we identified several known TNF-α-related pathways, namely cytokine–cytokine receptor interaction, apoptosis, Toll-like receptor signaling pathway, NF-κB signaling pathway, and TNF signaling pathway.

These results are in line with the reported effects of TNF-α, a potent proinflammatory cytokine involved in the inflammatory response to infections or tissue damage.

Concluding remarks and future directions

In this chapter, we have presented a pipeline that integrates complementary systems biology and chemoinformatics-based approaches with the goal of identifying potential direct and indirect targets and generating hypotheses for the MoA of a compound that can be tested in a new cycle of experiments.

Because of the high importance of—but also multiple difficulties involved in—elucidating the MoA of nutraceuticals, we have provided use cases of this methodology on three extensively studied nutraceuticals—epigallocatechin, quercetin, and resveratrol. To further validate the model, we have also applied the same pipeline on a prototypical proinflammatory stimulus, TNF-α.

In the era of big data and an exploding range of available computational algorithms, integration of knowledge from different data and methods is crucial for better experimental design and hypothesis development. Therefore we focus how we can bring together knowledge from multiple sources and integrate them into an ensemble score. Additionally, the integration of results from pathway over-representation analysis at various stages of the computational pipeline provide more robust biological insights supporting decision-making for the next iteration of the systems biology research cycle (Fig. 24.1).

Our use cases show that out of a total of 60 predicted functional partners, 57 had been already reported in the literature to be perturbed by or have a connection with the corresponding compounds.

The results obtained with TNF-α demonstrate the relevance of the approach and its applicability to protein ligands as well.

Among the evaluated nutraceuticals, we interestingly observed that multiple cancer pathways were enriched, highlighting the importance of nutraceuticals in cancer treatment. Epigallocatechin, quercetin, and resveratrol have all been studied for cancer treatment, either alone or in combination with conventional therapeutics, because of their proapoptotic, antioxidant, and antiproliferative effects (Brito et al., 2015; Ko et al., 2017; Min and Kwon, 2014).

Another important shared pattern among the nutraceuticals is their antiviral and antibacterial properties, highlighted by multiple enriched pathways involved in diseases caused by viruses and bacteria.

Additional use cases will further strengthen the validation of the methodology proposed in this chapter to investigate MoA of nutraceuticals. The approach can be extended by integrating new complementary tools selected on the basis of benchmarking procedures.

In conclusion, this work shows how different knowledge sources, computational tools and omics data can be integrated to investigate the MoA of nutraceuticals and enable the development of more informed hypotheses that can be verified in the next experimental design iteration of the systems biology research cycle.

References

Aggarwal, B.B., 2003. Signalling pathways of the TNF superfamily: a double-edged sword. Nat. Rev. Immunol. 3, 745–756.

Albrecht, D.S., Clubbs, E.A., Ferruzzi, M., et al., 2008. Epigallocatechin-3-gallate (EGCG) inhibits PC-3 prostate cancer cell proliferation via MEK-independent ERK1/2 activation. Chem. Biol. Interact. 171, 89–95.

Andrew, R., Izzo, A.A., 2017. Principles of pharmacological research of nutraceuticals. Br. J. Pharmacol. 174, 1177–1194.

Baek, S.H., Ko, J.H., Lee, H., et al., 2016. Resveratrol inhibits STAT3 signaling pathway through the induction of SOCS-1: role in apoptosis induction and radiosensitization in head and neck tumor cells. Phytomedicine 23, 566–577.

Beekmann, K., Rubio, L., De Haan, L.H., et al., 2015. The effect of quercetin and kaempferol aglycones and glucuronides on peroxisome proliferator-activated receptor-gamma (PPAR-gamma). Food Funct. 6, 1098–1107.

Berman, A.Y., Motechin, R.A., Wiesenfeld, M.Y., et al., 2017. The therapeutic potential of resveratrol: a review of clinical trials. NPJ Precis. Oncol. 1, 35.

Bilal, E., Sakellaropoulos, T., Melas, I.N., et al., 2015. A crowd-sourcing approach for the construction of species-specific cell signaling networks. Bioinformatics 31, 484–491.

Boly, R., Gras, T., Lamkami, T., et al., 2011. Quercetin inhibits a large panel of kinases implicated in cancer cell biology. Int. J. Oncol. 38, 833–842.

Boran, A.D., Iyengar, R., 2010. Systems pharmacology. Mt. Sinai J. Med. 77, 333–344.

Boue, S., Talikka, M., Westra, J.W., et al., 2015. Causal biological network database: a comprehensive platform of causal biological network models focused on the pulmonary and vascular systems. Database (Oxf.) 2015, bav030.

Breinbauer, R., Vetter, I.R., Waldmann, H., 2002. From protein domains to drug candidates-natural products as guiding principles in the design and synthesis of compound libraries. Angew. Chem. Int. Ed. Engl. 41, 2879–2890.

Breuss, J.M., Atanasov, A.G., Uhrin, P., 2019. Resveratrol and its effects on the vascular system. Int. J. Mol. Sci. 20, 1523.

Brito, A.F., Ribeiro, M., Abrantes, A.M., et al., 2015. Quercetin in cancer treatment, alone or in combination with conventional therapeutics? Curr. Med. Chem. 22, 3025–3039.

Chae, J.I., Cho, J.H., Lee, K.A., et al., 2012. Role of transcription factor Spl in the quercetin-mediated inhibitory effect on human malignant pleural mesothelioma. Int. J. Mol. Med. 30, 835–841.

Chen, D., Pamu, S., Cui, Q., et al., 2012a. Novel epigallocatechin gallate (EGCG) analogs activate AMP-activated protein kinase pathway and target cancer stem cells. Bioorg. Med. Chem. 20, 3031–3037.

Chen, Y.L., Tsai, H.L., Peng, C.W., 2012b. EGCG debilitates the persistence of EBV latency by reducing the DNA binding potency of nuclear antigen 1. Biochem. Biophys. Res. Commun. 417, 1093–1099.

Cheng, Z., Sun, G., Guo, W., et al., 2015. Inhibition of hepatitis B virus replication by quercetin in human hepatoma cell lines. Virol. Sin. 30, 261–268.

Choi, K.C., Jung, M.G., Lee, Y.H., et al., 2009. Epigallocatechin-3-gallate, a histone acetyltransferase inhibitor, inhibits EBV-induced B lymphocyte transformation via suppression of RelA acetylation. Cancer Res. 69, 583–592.

Choi, S., Sa, M., Cho, N., et al., 2019. Rbfox2 dissociation from stress granules suppresses cancer progression. Exp. Mol. Med. 51, 1–12.

Colpitts, C.C., Schang, L.M., 2014. A small molecule inhibits virion attachment to heparan sulfate- or sialic acid-containing glycans. J. Virol. 88, 7806–7817.

Cragg, G.M., Newman, D.J., 2013. Natural products: a continuing source of novel drug leads. Biochim. Biophys. Acta 1830, 3670–3695.

Cui, S., Wu, Q., Wang, J., et al., 2019. Quercetin inhibits LPS-induced macrophage migration by suppressing the iNOS/FAK/paxillin pathway and modulating the cytoskeleton. Cell Adh. Migr. 13, 1–12.

Delmas, D., Passilly-Degrace, P., Jannin, B., et al., 2002. Resveratrol, a chemopreventive agent, disrupts the cell cycle control of human SW480 colorectal tumor cells. Int. J. Mol. Med. 10, 193–199.

Dewick, P.M., 1997. Medicinal Natural Products: A Biosynthetic Approach. John Wiley & Sons, Chichester.

Dias, D.A., Urban, S., Roessner, U., 2012. A historical overview of natural products in drug discovery. Metabolites 2, 303–336.

Du, G.J., Zhang, Z., Wen, X.D., et al., 2012. Epigallocatechin Gallate (EGCG) is the most effective cancer chemopreventive polyphenol in green tea. Nutrients 4, 1679–1691.

Evans, H.G., Roostalu, U., Walter, G.J., et al., 2014. TNF-alpha blockade induces IL-10 expression in human CD4 + T cells. Nat. Commun. 5, 3199.

Fabregat, A., Jupe, S., Matthews, L., et al., 2018. The reactome pathway knowledge base. Nucleic Acids Res. 46, D649–D655.

Ferreira, C., Soares, D.C., Nascimento, M.T., et al., 2014. Resveratrol is active against Leishmania amazonensis: in vitro effect of its association with Amphotericin B. Antimicrob. Agents Chemother. 58, 6197–6208.

Firdous, A.B., Sharmila, G., Balakrishnan, S., et al., 2014. Quercetin, a natural dietary flavonoid, acts as a chemopreventive agent against prostate cancer in an in vivo model by inhibiting the EGFR signaling pathway. Food Funct. 5, 2632–2645.

Garcia-Alonso, L., Iorio, F., Matchan, A., et al., 2018. Transcription factor activities enhance markers of drug sensitivity in cancer. Cancer Res. 78, 769–780.

Gaulton, A., Hersey, A., Nowotka, M., et al., 2017. The ChEMBL database in 2017. Nucleic Acids Res. 45, D945–D954.

Gautam, S.C., Xu, Y.X., Dumaguin, M., et al., 2000. Resveratrol selectively inhibits leukemia cells: a prospective agent for ex vivo bone marrow purging. Bone Marrow Transpl. 25, 639–645.

Gfeller, D., Michielin, O., Zoete, V., 2013. Shaping the interaction landscape of bioactive molecules. Bioinformatics 29, 3073–3079.

Gfeller, D., Grosdidier, A., Wirth, M., et al., 2014. SwissTargetPrediction: a web server for target prediction of bioactive small molecules. Nucleic Acids Res. 42, W32–W38.

Giuliani, C., Bucci, I., Di Santo, S., et al., 2014. The flavonoid quercetin inhibits thyroid-restricted genes expression and thyroid function. Food Chem. Toxicol. 66, 23–29.

Gonzalez-Suarez, I., Martin, F., Hoeng, J., Peitsch, M.C., 2016. Chapter 22—Mechanistic network models in safety and toxicity evaluation of nutraceuticals. In: Gupta, R.C. (Ed.), Nutraceuticals: Efficacy, Safety and Toxicity. Academic Press/Elsevier, Amsterdam, pp. 287–304.

Granato, M., Rizzello, C., Gilardini Montani, M.S., et al., 2017. Quercetin induces apoptosis and autophagy in primary effusion lymphoma cells by inhibiting PI3K/AKT/mTOR and STAT3 signaling pathways. J. Nutr. Biochem. 41, 124–136.

Gu, X., Cai, Z., Cai, M., et al., 2018. AMPK/SIRT1/p38 MAPK signaling pathway regulates alcoholinduced neurodegeneration by resveratrol. Mol. Med. Rep. 17, 5402–5408.

Guo, D., Dunbar, J.D., Yang, C.H., et al., 1998. Induction of Jak/STAT signaling by activation of the type 1 TNF receptor. J. Immunol. 160, 2742–2750.

Gupta, S.C., Kim, J.H., Prasad, S., et al., 2010. Regulation of survival, proliferation, invasion, angiogenesis, and metastasis of tumor cells through modulation of inflammatory pathways by nutraceuticals. Cancer Metastasis Rev. 29, 405–434.

Hamed, F.N., Mcdonagh, A.J.G., Almaghrabi, S., et al., 2018. Epigallocatechin-3 gallate inhibits STAT-1/JAK2/IRF-1/HLA-DR/HLA-B and reduces CD8 MKG2D lymphocytes of alopecia areata patients. Int. J. Environ. Res. Public Health 15, 2882.

Han, D.W., Lee, M.H., Kim, H.H., et al., 2011. Epigallocatechin-3-gallate regulates cell growth, cell cycle and phosphorylated nuclear factor-kappaB in human dermal fibroblasts. Acta Pharmacol. Sin. 32, 637–646.

Hayashibara, T., Yamada, Y., Nakayama, S., et al., 2002. Resveratrol induces downregulation in survivin expression and apoptosis in HTLV-1-infected cell lines: a prospective agent for adult T cell leukemia chemotherapy. Nutr. Cancer 44, 193–201.

He, X., Andersson, G., Lindgren, U., et al., 2010. Resveratrol prevents RANKL-induced osteoclast differentiation of murine osteoclast progenitor RAW 264.7 cells through inhibition of ROS production. Biochem. Biophys. Res. Commun. 401, 356–562.

Hoeng, J., Deehan, R., Pratt, D., et al., 2012. A network-based approach to quantifying the impact of biologically active substances. Drug Discov. Today 17, 413–418.

Huang, Y.T., Lee, L.T., Lee, P.P., et al., 2005. Targeting of focal adhesion kinase by flavonoids and small-interfering RNAs reduces tumor cell migration ability. Anticancer Res. 25, 2017–2025.

Huang, C.Y., Chan, C.Y., Chou, I.T., et al., 2013. Quercetin induces growth arrest through activation of FOXO1 transcription factor in EGFR-overexpressing oral cancer cells. J. Nutr. Biochem. 24, 1596–1603.

Hughes, T.R., Marton, M.J., Jones, A.R., et al., 2000. Functional discovery via a compendium of expression profiles. Cell 102, 109–126.

Jeong, J.H., An, J.Y., Kwon, Y.T., et al., 2008. Quercetin-induced ubiquitination and down-regulation of Her-2/neu. J. Cell Biochem. 105, 585–595.

Jeong, J.H., An, J.Y., Kwon, Y.T., et al., 2009. Effects of low dose quercetin: cancer cell-specific inhibition of cell cycle progression. J. Cell Biochem. 106, 73–82.

Jeong, J.K., Moon, M.H., Bae, B.C., et al., 2012. Autophagy induced by resveratrol prevents human prion protein-mediated neurotoxicity. Neurosci. Res. 73, 99–105.

Jin, L., Li, C., Xu, Y., et al., 2013. Epigallocatechin gallate promotes p53 accumulation and activity via the inhibition of MDM2-mediated p53 ubiquitination in human lung cancer cells. Oncol. Rep. 29, 1983–1990.

Jin, Z., Feng, W., Ji, Y., et al., 2017. Resveratrol mediates cell cycle arrest and cell death in human esophageal squamous cell carcinoma by directly targeting the EGFR signaling pathway. Oncol. Lett. 13, 347–355.

Kanehisa, M., Furumichi, M., Tanabe, M., et al., 2017. KEGG: new perspectives on genomes, pathways, diseases and drugs. Nucleic Acids Res. 45, D353–D361.

Khan, M.A., Hussain, A., Sundaram, M.K., et al., 2015. (-)-Epigallocatechin-3-gallate reverses the expression of various tumor-suppressor genes by inhibiting DNA methyltransferases and histone deacetylases in human cervical cancer cells. Oncol. Rep. 33, 1976–1984.

Ko, J.H., Sethi, G., Um, J.Y., et al., 2017. The role of resveratrol in cancer therapy. Int. J. Mol. Sci. 18, 2589.

Koo, N., Cho, D., Kim, Y., et al., 2006. Effects of resveratrol on mast cell degranulation and tyrosine phosphorylation of the signaling components of the IgE receptor. Planta Med. 72, 659–661.

Lamb, J., Crawford, E.D., Peck, D., et al., 2006. The Connectivity Map: using gene-expression signatures to connect small molecules, genes, and disease. Science 313, 1929–1935.

Lannan, K.L., Refaai, M.A., Ture, S.K., et al., 2016. Resveratrol preserves the function of human platelets stored for transfusion. Br. J. Haematol. 172, 794–806.

Lee, K.A., Lee, Y.J., Ban, J.O., et al., 2012. The flavonoid resveratrol suppresses growth of human malignant pleural mesothelioma cells through direct inhibition of specificity protein 1. Int. J. Mol. Med. 30, 21–27.

Lee, K.H., Yoo, C.G., 2013. Simultaneous inactivation of GSK-3beta suppresses quercetin-induced apoptosis by inhibiting the JNK pathway. Am. J. Physiol. Lung Cell Mol. Physiol. 304, L782–L789.

Lei, M.J., Dong, Y., Sun, C.X., et al., 2017. Resveratrol inhibits proliferation, promotes differentiation and melanogenesis in HT-144 melanoma cells through inhibition of MEK/ERK kinase pathway. Microb. Pathog. 111, 410–413.

Li, H.C., Yashiki, S., Sonoda, J., et al., 2000. Green tea polyphenols induce apoptosis in vitro in peripheral blood T lymphocytes of adult T-cell leukemia patients. Jpn. J. Cancer Res. 91, 34–40.

Li, F., Bai, Y., Zhao, M., et al., 2015. Quercetin inhibits vascular endothelial growth factor-induced choroidal and retinal angiogenesis in vitro. Ophthalmic Res. 53, 109–116.

Li, Y., Yao, J., Han, C., et al., 2016. Quercetin, inflammation and immunity. Nutrients 8, 167.

Li, X., Chen, Y., Shen, J.Z., et al., 2019. Epigallocatechin gallate inhibits hepatic glucose production in primary hepatocytes via downregulating PKA signaling pathways and transcriptional factor FoxO1. J. Agric. Food Chem. 67, 3651–3661.

Liu, S., Wang, X.J., Liu, Y., Cui, Y.F., 2013. PI3K/AKT/mTOR signaling is involved in (-)-epigallocatechin-3-gallate-induced apoptosis of human pancreatic carcinoma cells. Am. J. Chin. Med. 41, 629–642.

Liu, W., Dong, M., Bo, L., et al., 2014. Epigallocatechin-3-gallate ameliorates seawater aspiration-induced acute lung injury via regulating inflammatory cytokines and inhibiting JAK/STAT1 pathway in rats. Mediators Inflamm. 2014, 612593.

Liu, A., Trairatphisan, P., Gjerga, E., et al., 2019. From expression footprints to causal pathways: contextualizing large signaling networks with CARNIVAL. NPJ Syst. Biol. Appl. 5, 40.

Martin, F., Sewer, A., Talikka, M., et al., 2014. Quantification of biological network perturbations for mechanistic insight and diagnostics using two-layer causal models. BMC Bioinforma. 15, 238.

Masuda, M., Suzui, M., Lim, J.T., et al., 2003. Epigallocatechin-3-gallate inhibits activation of HER-2/neu and downstream signaling pathways in human head and neck and breast carcinoma cells. Clin. Cancer Res. 9, 3486–3491.

Melas, I.N., Mitsos, A., Messinis, D.E., et al., 2011. Combined logical and data-driven models for linking signalling pathways to cellular response. BMC Syst. Biol. 5, 107.

Melas, I.N., Samaga, R., Alexopoulos, L.G., et al., 2013. Detecting and removing inconsistencies between experimental data and signaling network topologies using integer linear programming on interaction graphs. PLoS Comput. Biol. 9, e1003204.

Michaud-Levesque, J., Bousquet-Gagnon, N., Beliveau, R., 2012. Quercetin abrogates IL-6/STAT3 signaling and inhibits glioblastoma cell line growth and migration. Exp. Cell Res. 318, 925–935.

Min, K.J., Kwon, T.K., 2014. Anticancer effects and molecular mechanisms of epigallocatechin-3-gallate. Integr. Med. Res. 3, 16–24.

Mitsos, A., Melas, I.N., Siminelakis, P., et al., 2009. Identifying drug effects via pathway alterations using an integer linear programming optimization formulation on phosphoproteomic data. PLoS Comput. Biol. 5, e1000591.

Mohos, V., Bencsik, T., Boda, G., et al., 2018. Interactions of casticin, ipriflavone, and resveratrol with serum albumin and their inhibitory effects on CYP2C9 and CYP3A4 enzymes. Biomed. Pharmacother. 107, 777–784.

Moses, M.A., Henry, E.C., Ricke, W.A., et al., 2015. The heat shock protein 90 inhibitor, (-)-epigallocatechin gallate, has anticancer activity in a novel human prostate cancer progression model. Cancer Prev. Res. (Phila.) 8, 249–257.

Muniz De Queiroz, R., Prives, C., 2018. Unfolding the roles of resveratrol in p53 regulation. Oncotarget 9, 34455–34456.

Nam, T.W., Yoo, C.I., Kim, H.T., et al., 2008. The flavonoid quercetin induces apoptosis and inhibits migration through a MAPK-dependent mechanism in osteoblasts. J. Bone Min. Metab. 26, 551–560.

Negri, A., Naponelli, V., Rizzi, F., et al., 2018. Molecular targets of epigallocatechin-gallate (EGCG): a special focus on signal transduction and cancer. Nutrients 10, 1936.

Poussin, C., Kramer, B., Lanz, H.L., et al., 2019. 3D human microvessel-on-a-chip model for studying monocyte-to-endothelium adhesion under flow—application in systems toxicology. ALTEX 37 (1), 47–63.

Poussin, C., Mathis, C., Alexopoulos, L.G., et al., 2014. The species translation challenge-a systems biology perspective on human and rat bronchial epithelial cells. Sci. Data 1, 140009.

Prill, R.J., Saez-Rodriguez, J., Alexopoulos, L.G., et al., 2011. Crowdsourcing network inference: the DREAM predictive signaling network challenge. Sci. Signal. 4, mr7.

Riba, A., Deres, L., Sumegi, B., et al., 2017. Cardioprotective effect of resveratrol in a postinfarction heart failure model. Oxid. Med. Cell Longev. 2017, 6819281.

Robb, E.L., Stuart, J.A., 2011. Resveratrol interacts with estrogen receptor-beta to inhibit cell replicative growth and enhance stress resistance by upregulating mitochondrial superoxide dismutase. Free Radic. Biol. Med. 50, 821–831.

Rojas, A., Del Campo, J.A., Clement, S., et al., 2016. Effect of quercetin on hepatitis C virus life cycle: from viral to host targets. Sci. Rep. 6, 31777.

Sander, T., Freyss, J., Von Korff, M., et al., 2015. DataWarrior: an open-source program for chemistry aware data visualization and analysis. J. Chem. Inf. Model. 55, 460–473.

Santos, J.C., Gotardo, E.M., Brianti, M.T., et al., 2014. Effects of yerba mate, a plant extract formulation ("YGD") and resveratrol in 3T3-L1 adipogenesis. Molecules 19, 16909–16924.

Schnepp, R., Chen, Y.-X., Wang, H., et al., 2006. Mutation of tumor suppressor gene Men1 acutely enhances proliferation of pancreatic islet cells. Cancer Res 66 (11), 5707–5715.

Senturk, M., Gulcin, I., Beydemir, S., et al., 2011. In Vitro inhibition of human carbonic anhydrase I and II isozymes with natural phenolic compounds. Chem. Biol. Drug Des. 77, 494–499.

Shannon, P., Markiel, A., Ozier, O., et al., 2003. Cytoscape: a software environment for integrated models of biomolecular interaction networks. Genome Res. 13, 2498–2504.

Singh, M., Parent, S., Leblanc, V., et al., 2011. Resveratrol modulates the expression of PTGS2 and cellular proliferation in the normal rat endometrium in an AKT-dependent manner. Biol. Reprod. 84, 1045–1052.

Srinivasan, A., Thangavel, C., Liu, Y., et al., 2016. Quercetin regulates beta-catenin signaling and reduces the migration of triple negative breast cancer. Mol. Carcinog. 55, 743–756.

Subramanian, A., Narayan, R., Corsello, S.M., et al., 2017. A next generation connectivity map: L1000 platform and the first 1,000,000 profiles. Cell 171, 1437–1452 e17.

Sui, T., Ma, L., Bai, X., et al., 2014. Resveratrol inhibits the phosphatidylinositide 3-kinase/protein kinase B/mammalian target of rapamycin signaling pathway in the human chronic myeloid leukemia K562 cell line. Oncol. Lett. 7, 2093–2098.

Sun, W., Liu, X., Zhang, H., et al., 2017. Epigallocatechin gallate upregulates NRF2 to prevent diabetic nephropathy via disabling KEAP1. Free Radic. Biol. Med. 108, 840–857.

Szklarczyk, D., Santos, A., Von Mering, C., et al., 2016. STITCH 5: augmenting protein-chemical interaction networks with tissue and affinity data. Nucleic Acids Res. 44, D380–D384.

Szklarczyk, D., Gable, A.L., Lyon, D., et al., 2019. STRING v11: protein-protein association networks with increased coverage, supporting functional discovery in genome-wide experimental datasets. Nucleic Acids Res. 47, D607–D613.

Thomford, N.E., Senthebane, D.A., Rowe, A., et al., 2018. Natural products for drug discovery in the 21st century: innovations for novel drug discovery. Int. J. Mol. Sci. 19, 1578.

Thomson, T.M., Sewer, A., Martin, F., et al., 2013. Quantitative assessment of biological impact using transcriptomic data and mechanistic network models. Toxicol. Appl. Pharmacol. 272 (3), 863–878. https://doi.org/10.1016/j.taap.2013.07.007. 23933166.

Turei, D., Korcsmaros, T., Saez-Rodriguez, J., 2016. OmniPath: guidelines and gateway for literature-curated signaling pathway resources. Nat. Methods 13, 966–967.

Tutel'ian, V.A., Lashneva, N.V., 2013. Biologically active substances of plant origin. Flavonols and flavones: prevalence, dietary sourses and consumption. Vopr. Pitan. 82, 4–22.

Van Aller, G.S., Carson, J.D., Tang, W., et al., 2011. Epigallocatechin gallate (EGCG), a major component of green tea, is a dual phosphoinositide-3-kinase/mTOR inhibitor. Biochem. Biophys. Res. Commun. 406, 194–199.

Vijayababu, M.R., Arunkumar, A., Kanagaraj, P., et al., 2006. Quercetin downregulates matrix metalloproteinases 2 and 9 proteins expression in prostate cancer cells (PC-3). Mol. Cell Biochem. 287, 109–116.

Wang, G.L., Fu, Y.C., Xu, W.C., et al., 2009. Resveratrol inhibits the expression of SREBP1 in cell model of steatosis via Sirt1-FOXO1 signaling pathway. Biochem. Biophys. Res. Commun 380, 644–649.

Wang, L., Cheng, X., Li, H., et al., 2014. Quercetin reduces oxidative stress and inhibits activation of cJun Nterminal kinase/activator protein1 signaling in an experimental mouse model of abdominal aortic aneurysm. Mol. Med. Rep. 9, 435–442.

Wang, F.M., Hu, Z., Liu, X., et al., 2019. Resveratrol represses tumor necrosis factor alpha/c-Jun N-terminal kinase signaling via autophagy in human dental pulp stem cells. Arch. Oral Biol. 97, 116–121.

Wattel, A., Kamel, S., Prouillet, C., et al., 2004. Flavonoid quercetin decreases osteoclastic differentiation induced by RANKL via a mechanism involving NF kappa B and AP-1. J. Cell Biochem. 92, 285–295.

Wilson, W.J., Poellinger, L., 2002. The dietary flavonoid quercetin modulates HIF-1 alpha activity in endothelial cells. Biochem. Biophys. Res. Commun. 293, 446–450.

Wu, G., Feng, X., Stein, L., 2010. A human functional protein interaction network and its application to cancer data analysis. Genome Biol. 11, R53.

Wu, F., Niu, Z., Zhou, B., et al., 2019. PSMB1 negatively regulates the innate antiviral immunity by facilitating degradation of IKK-epsilon. Viruses 11, 99.

Xu, J., Xu, Z., Zheng, W., 2017. A review of the antiviral role of green tea catechins. Molecules 22, 1337.

Xue, L., Bajorath, J., 2000. Molecular descriptors in chemoinformatics, computational combinatorial chemistry, and virtual screening. Comb. Chem. High Throughput Screen. 3, 363–372.

Yang, G.Y., Liao, J., Li, C., et al., 2000. Effect of black and green tea polyphenols on c-jun phosphorylation and H(2)O(2) production in transformed and non-transformed human bronchial cell lines: possible mechanisms of cell growth inhibition and apoptosis induction. Carcinogenesis 21, 2035–2039.

Yang, H., Hu, J., Chen, Y.J., et al., 2019. Role of Sirt1 in innate immune mechanisms against *Mycobacterium tuberculosis* via the inhibition of TAK1 activation. Arch. Biochem. Biophys 667, 49–58.

Ye, J., 2008. Regulation of PPARgamma function by TNF-alpha. Biochem. Biophys. Res. Commun. 374, 405–408.

Chapter 25

Noninvasive in vivo optical imaging models for safety and toxicity testing

Jeffrey D. Peterson

Introduction and background

The global pharmaceutical market exceeds \$1.2 trillion, expected to grow to \$1.5 trillion by 2023, and the global nutraceutical market was estimated at ~\$241 billion for 2019 and is expected to grow from \$240 billion to over \$350 billion. Though the majority of people in developed countries depend on pharmaceutical remedies, it is estimated that more than 50%–70% also use some sort of alternative nutraceutical therapies, with herbal/botanical supplements accounting for a significant proportion. This use of herbal remedies has been increasing at a steady pace over the past decade, in large part because these products are receiving increasing exposure through the media, in popular journals, and even in scientific research articles. Interest in herbal medicine has also been facilitated by criticism of the pharmaceutical industry based on perceptions that chemistry-based medications are expensive, dangerous, and overprescribed. Alternatively, herbal supplements are generally seen as "natural" (and therefore safe), with some epidemiological studies supporting that consumption of plant-derived foods provides a range of health benefits. Indeed, a variety of natural products have shown some efficacy across a wide variety of diseases and conditions, including cancer, Alzheimer disease, epilepsy, depression, arthritis, autoimmunity, diabetes, viral infection, and cardiovascular disease, including curcumin derived from ginger (Javeri and Chand, 2016), quercetin (Ay et al., 2016) and resveratrol from fruits (Risuleo, 2016), arginine and citrulline (Kaore and Kaore, 2016), thymoquinone from *Negilla sativa* plants (Gupta et al., 2016), natural organosulfur compounds (Goncharov et al., 2016), algae-derived spirulina (Wan et al., 2016), green tea extract (Coppock and Dziwenka, 2016), Chinese ginseng (Yang and Wu, 2016), and cannabinoids derived from *Cannabis sativa* or hemp (Hartsel et al., 2016). Overall, however, the amount of efficacy data and knowledge of specific bioactive ingredients vary tremendously among different nutraceuticals, and their increasing use warrants the generation of more scientific evidence for their perceived harmlessness.

Despite the standards of rigorous testing of industrial chemicals, pesticides, and pharmaceuticals, at the present time nutraceuticals exist largely unscrutinized and unrestrained by the toxicology testing guidelines used in the pharmaceutical industry. They are classified as conventional foods, dietary supplements, or medical foods, with no requirement for toxicology testing if the manufacturer/vendor can confirm that all ingredients are generally regarded as safe. Consequently, very few supplements have been studied thoroughly for efficacy, safety, and toxicity (Ronis et al., 2018). Questions of purity and standardization (Gil et al., 2016; Nounou et al., 2018), drug interactions (Dasgupta, 2002; Kupiec and Raj, 2005; Saxena et al., 2008; Anadon et al., 2016; Chan et al., 2016), and the potential for direct organ/tissue toxicity are now coming to the forefront as doctors are reporting an increased incidence of patient health issues related to the prevalence of nutraceutical use (Haller et al., 2002; Gabardi et al., 2007; Debelle et al., 2008; Parvez and Rishi, 2019). Consistent findings that some commercial nutraceuticals may not actually contain the ingredients that are claimed (Nounou et al., 2018) further warrants better testing, quality control, and oversight. Without rigorous control of purity it is impossible to predict potential health risks presented by unknown or uncharacterized ingredients. Even in the case of high-quality nutraceuticals it is important to note that any bioactive molecule, whether a synthetic drug or a natural product, has the potential for both positive and negative effects depending on dose, bioactive metabolites, tissue accumulation, chronic cumulative effects, and the health/treatment status of the patient.

Typically, drug safety/toxicology is assessed using a variety of in vitro, in vivo, and ex vivo assessments of cellular, physiologic, and tissue effects, respectively, and

Nutraceuticals. DOI: https://doi.org/10.1016/B978-0-12-821038-3.00025-2

these tests can be readily applied to the assessment of nutraceutical safety. Large panels of in vitro assays determine potential reactivity with enzymes, proteases, kinases, and receptors. In vivo testing is generally an iterative process including single-dose (acute) toxicology testing, repeat dosing, and full toxicokinetic and pharmacokinetic studies. The predominant reliance, however, is on labor-intensive histologic assessment of tissues and blood samples generated with chronic dose-ranging in rats, requiring large numbers of animals, extensive tissue preparation, and the educated eye of a pathologist. In addition, a variety of safety pharmacology studies are performed to assess cardiovascular, respiratory, and central nervous system (CNS) effects, as well as genotoxicity, carcinogenicity, teratogenicity, and immunogenicity. These assessments generally rely mostly on measurements of physiologic changes. If toxic effects are found, then the window between toxicity and efficacy must be established to determine whether the candidate has the potential for development as a clinical therapy. For the most part, these toxicology studies occur late in drug discovery, after potential candidates for clinical testing have been identified and moved into development. However, the failure rate of such advanced drug development candidates due to toxicity findings (predominantly liver and kidney) is encouraging adoption of earlier measures of potential adverse effects prior to selection of development candidates (Ahuja and Sharma, 2014).

Of course, full toxicology profiling of early drug discovery molecules is expensive and impractical, so appropriate quick screening approaches, in vitro and in vivo, are seeing more use. In this case, the major utility of this would not be comprehensive toxicology assessment; rather, the focus would be quick screening of key organ system effects for risk assessment and prioritization of early-stage bioactive therapeutics. This chapter highlights advances in a relatively new approach to safety assessment, building upon recent advances in optical imaging, near-infrared (NIR) agents, and imaging strategies (Weissleder and Ntziachristos, 2003; Korideck and Peterson, 2009; Krautz-Peterson et al., 2009; Kossodo et al., 2010; Peterson et al., 2010; Bao et al., 2012; Lin et al., 2012; Zhang et al., 2012; Eaton et al., 2013; Daghighi et al., 2014; Tseng et al., 2017; Peterson, 2019). Noninvasive optical imaging technology has grown dramatically with the increased interest in biological readouts, rather than anatomical readouts, and the possibility of alternatives to the use of radioactive ligands for imaging. High-throughput epifluorescence imaging approaches have been further augmented by recent advances in three-dimensional fluorescent tomographic imaging. Such technology has opened the door not only to deep tissue detection and localization but also to the new capability of absolute quantification of fluorescence at any site within the body. Although imaging of preclinical models of safety/toxicology has not been pursued extensively to date, the ability to use fluorescent imaging agents that detect and quantify a variety of biological activities has the potential to expand the horizons of early safety screening in drug and nutraceutical development. In addition, the general ability to use fewer animals per study, and to perform longitudinal noninvasive assessments in the same animals, allows a significant decrease in animal use, a potentially major benefit in support of current worldwide initiatives in the 3Rs (reduce, refine, and replace) of animal welfare (Balls, 2015). We at PerkinElmer have continued to aggressively test the hypotheses that fluorescence imaging (FLI) technology, paired with NIR fluorescent probes to important biomarkers, can provide useful quantitative readouts for assessing the induction of organ/tissue injury.

Noninvasive optical imaging

Conventional preclinical toxicology biomarkers include a variety of blood or urine molecules, including DNA, proteins, multiple enzymes, bile acids/pigments, and urea nitrogen, among others, and generally these are accepted as systemic markers of local tissue injury. In some instances, because these markers are collected far from the site of tissue damage, this can lead to some doubt regarding the precise tissue origin of the circulating biomarker. The transient nature of many of these biomarkers can also raise doubts regarding the magnitude of the toxicity and can raise the possibility of false-negative results. In contrast, various modalities of biomedical imaging can offer some useful options for directly assessing sites of tissue injury. In preclinical research, nuclear imaging techniques, such as positron-emission tomography (PET) and single-photon emission computed tomography, can be used to monitor drug pharmacokinetics and localization or regions of increased metabolic activity. Other imaging modalities, like ultrasound, μCT, and MRI, can often provide valuable morphological and physiological information to characterize the safety of new therapeutics (Beckmann et al., 2004), with MRI being the most versatile and widely used technique across indications like liver, cardiovascular, CNS, and kidney safety assessment (Wang and Yan, 2008). Unfortunately, these anatomical modalities often face limitations in sensitivity, spatial resolution, and/or specificity and are generally focused on specific types of toxicity, like reproductive or developmental toxicology. A general need for CT and MR contrast agents, which themselves often have toxicity issues (Hasebroock and Serkova, 2009), can further limit the utility of anatomical imaging modalities for routine and general toxicology screening.

FLI, however, has the potential to detect in situ dynamic changes in tissue biology through the use of highly stable, nonradioactive vascular contrast agents as well as targeted or protease-activated imaging probes. Imaging probes are designed to detect tissue biomarkers and accumulate directly at the local site of tissue injury rather than detecting the released/circulating biomarker fraction. Other advantages include a favorable safety profile, ability to distinguish between different soft tissues and organ systems, the potential for multiplexing disparate biological readouts, and the ease of use and short acquisition time for imaging (2–15 min). Most of the precedents in preclinical toxicology optical imaging to date have been generated using surgically invasive approaches, like intravital multiphoton microscopy, highlighting the potential to detect and quantify biological changes in living tissue (Goetz et al., 2008; He et al., 2010; Peti-Peterdi et al., 2012; Hall et al., 2013; Thorling et al., 2014; Schuh et al., 2016) with minimal animal morbidity. Such an approach can provide rich detail at the cellular level; however, this technique faces challenges with regard to potential sampling error (due to small fields of view), technical challenges in longitudinal applications, and a focus on limited perfusion data with optical reporters or small-molecule fluorescent dyes, like fluorescein. In addition, low-throughput surgical imaging by intravital microscopy is challenging to apply robustly in preclinical animal safety/toxicology screening, which may require testing of large numbers of compounds in large numbers of animals.

Noninvasive in vivo FLI of live animals can potentially provide a tool for noninvasive, repeated assessments (either longitudinal or kinetic) of in situ biological processes, and there are several preclinical imaging systems currently used in academic and industry research. The majority of these systems use basic epifluorescence (2D) bioluminescence imaging or FLI capabilities, but the IVIS Lumina III, Lumina Kinetic, and Lumina S5/X5 (PerkinElmer, Hopkinton, MA) also offer advanced modalities such as spectral unmixing, dynamic contrast enhancement (DyCE), kinetic imaging, and/or paired X-ray acquisition. Recent advances in optical imaging technology have yielded some unique optical imaging systems (FMT 4000 and IVIS Spectrum CT; PerkinElmer) that can also offer deep-tissue tomographic (3D) imaging capabilities for localization and quantification of signals in entire animals or organ systems. The critical pairing of FLI systems with appropriate NIR imaging probes further allows the detection and quantification of important biological processes, such as cellular protease activity, vascular leak, and receptor upregulation. NIR fluorescent probes have been effectively used to image a variety of disease conditions, including cancer (Weissleder and Ntziachristos, 2003; Montet et al., 2005; Kossodo et al., 2010; Bao et al., 2012; Tseng et al., 2017), atherosclerosis (Chen et al., 2002; Deguchi et al., 2006; Lin et al., 2012), cardiac pathologies (Nahrendorf et al., 2007; Sosnovik et al., 2007), arthritis (Wunder et al., 2004; Peterson et al., 2010), CNS demyelination (Eaton et al., 2013), and osteoarthritis (Vermeij et al., 2014). Because optical imaging of toxicology is a relatively new research area, there is not yet a concerted effort to align imaging probes with current assays for circulating markers. Rather, many fluorescent imaging probes, those validated for the detection of important biomarkers associated with disease, are currently being examined for their potential in toxicology imaging (Table 25.1). This can be a valuable approach because many of the same biological/physiological processes and cellular players are involved in disease-related and bioactive molecule-induced toxicity.

In vivo preclinical NIR FLI in liver safety/toxicology

The liver is often a target for drug-induced injury due to its significant role in drug metabolism (Grattagliano et al., 2009; Corsini and Bortolini, 2013). Liver toxicity does not appear to be limited to pharmaceuticals (Alempijevic et al., 2017), although the greater focus of toxicology studies is on drug development. There are a variety of studies that clearly reveal a number of herbs and herbal treatments to be toxic to the liver, including chaparral, germander, pennyroyal, glue thistle, Jin Bu Huan, mistletoe, and kava, among others (Ramachandran and Kakar, 2009). The injury-inducing component can be the parent drug/nutraceutical itself, or often, metabolites (Wen and Gorycki, 2019), and outcomes of improper self-treatment with certain herbal products can range from idiosyncratic and variable liver function tests (Teschke and Xuan, 2019), to short-term hepatitis or outright hepatic failure (Haller et al., 2002). Herb toxicity can also occur through the actions of reactive metabolites (e.g., pennyroyal oil) (Anderson et al., 1996) or through the induction of downstream pathways (Haller et al., 2002). Whether induced by drugs or natural herbal products, there are a variety of toxicity mechanisms, including hepatocellular, cholestatic, steatosis, hepatitis, and mixed forms (Ramachandran and Kakar, 2009). The histological phenotypes, that generally focus on late-stage presentation, are diverse, and it is expected that early stages of the induction of liver injury, prior to overt histologic presentation, are likely diverse in underlying biological changes as well.

It is important to understand immediate cellular stress and death pathways as well as downstream biological changes resulting from chemical insult to the liver, when developing approaches to detect and understand drug-induced tissue injury (Kuijper et al., 2017). Oxidative

TABLE 25.1 Toxicology tissue changes and NIR imaging agents.

Toxicity/tissue	Biological tissue response	Potential imaging agents	Source
General markers	Fibrosis	MMPSense 680	PerkinElmer
		MMPSense 750 FAST	
	Inflammation	MMPSense 680	PerkinElmer
		MMPSense 750 FAST	
		ProSense 680	
		ProSense 750EX	
		ProSense 750 FAST	
		Neutrophil Elastase 680 FAST	
	Apoptosis/necrosis	Annexin-Vivo 750	PerkinElmer
		DyLight Polycaspase Probe	Thermo Scientific
		Hoechst-IR	Georgia Institute of Technology
	Vascular damage	AngioSense 680EX	PerkinElmer
		AngioSense 750EX	
		Qtracker Probes	Life Technologies
	Vascular protease	MMPSense 750 FAST	
	Oxidative stress	CF-SPN 775	Stanford University
		P-ONOO	Jiangxi Key Laboratory for Nano-Biomaterials, Institute of Advanced Materials (IAM), East China Jiaotong University
Liver	Iron metabolism	Transferrin-Vivo 750	PerkinElmer
		LCY7 Fe^{2+} detection	Shandong University
	Enzyme	Cl2-BDCM-ALP Alkaline Phosphatase	Chongqing University
	Enzyme	Leucine aminopeptidase (LAP)	South China University of Technology
Kidney	Glomerular filtration	GFR-Vivo 680	PerkinElmer
		CDIR2	Nanyang Technological University Tianjin University
	Renin–angiotensin system	ReninSense 680 FAST	PerkinElmer
	Oxidative stress	Lyso-SiR-2S (HOCl)	Hunan University
		Naph-O2 (-) NIR-OS (-) (superoxide)	
Stomach	Gastric emptying	GastroSense 680	PerkinElmer
Soft tissue calcification	Calcium deposition	OsteoSense 750	PerkinElmer
		IRDye 800CW BoneTag	LI-COR
	Cathepsin K activity	Cat K 680 FAST	PerkinElmer

Notes: Underlined probes indicate those that are commercially available with some early validation data for use in preclinical drug-induced tissue injury applications. *NIR*, near-infrared.

stress pathways can be activated almost immediately in the liver upon chemical insult, inducing a number of responses to directly remove the threat or minimize damage. This can play a pivotal role in necrotic or regulated apoptotic cell death, designed to minimize damage to surrounding unaffected cells. The rapid effect of certain drugs on cellular redox in the liver makes the detection of relevant reactive oxygen species (ROS) and reactive nitrogen species (RNS) attractive targets for early detection of drug-induced liver injury (DILI). Researchers are now pursuing optical imaging strategies to exploit these very early biological changes in early detection of DILI. Recently, a dual-emitting phosphorescent probe (P-ONOO) was developed that can detect increases in peroxynitrite, ONOO(-) in vivo, and in both acetaminophen (APAP)- and ketoconazole-induced liver injury increases could be observed and quantified in living animals (Chen et al., 2020), both drugs known to cause centrilobular hepatic necrosis (Hinson et al., 2010; Fukami et al., 2016). Shuhendler et al. (2014) used a semiconducting polymer-based probe (CF-SPN 775) designed to detect RNS and ROS and found it to be highly effective at imaging levels of tissue nitrosative/oxidative stress associated with APAP-induced toxicity in female nu/nu mice. CF-SPN 775 could detect these changes in the liver within 20–60 min using noninvasive epifluorescence and chemiluminescence imaging on the IVIS Spectrum (PerkinElmer). Toxic effects were imaged and quantified at 300 mg/kg, and the toxicity-related signal was reduced by glutathione therapy, as expected. Similar results were seen with isoniazid, a widely used treatment for tuberculosis that has been shown to cause oxidative covalent protein modification and cellular damage in the liver (Metushi et al., 2011).

Downstream of the immediate death and cellular stress pathways, depending on the drug or bioactive molecule being tested, a variety of different responses can occur in tissue, from subtle changes in cell surface biomarkers, to changes in metabolism, to overt changes in physiology and microvascular flow, to indirect tissue destruction. Many of these mechanisms can be easily imaged by current optical imaging technology using available imaging probes and multiplex imaging to simultaneously detect and quantify different aspects of biological change. In the liver, Kupffer cell oxidative burst (Hinson et al., 2010) can cause an influx of neutrophils that can yield concomitant biomarker changes in protease activity (Jaeschke and Smith, 1997). Additional liver changes in many instances can be measured, for example, as progressive damage to the microvasculature (DeLeve, 2007) or changes in iron metabolism (Izawa et al., 2014), among other things. Ultimately, such changes can culminate in cellular apoptosis and/or necrosis. Many NIR imaging probes exist for the assessment of these tissue injury biomarkers; however, few of these mechanisms have been fully explored for their utility in liver injury imaging. The increasing interest in applying optical imaging to safety screening and the basic research currently growing in various drug-discovery organizations are likely to yield numerous practical preclinical screening approaches.

A couple of optical imaging approaches have been recently explored based upon known enzymes involved in liver responses to drugs or xenobiotics. Jie et al. (2019) adapted an established serum marker of liver injury, alkaline phosphatase (ALP), for the purpose of in vivo imaging. An activatable NIR fluorescent probe (Cl2-BDCM-ALP) was used for time-resolved imaging of the variations in ALP activity and for the first time ALP activity in situ was successfully used to detect acute DILI in mice. More novel imaging targets are also in the early stages of exploration for DILI imaging, including the use of an iron-detecting fluorescent probe to measure iron accumulation and homeostasis in injured livers (Wu et al., 2020). Leucine aminopeptidase, a liver enzyme important in detoxification, has also been effectively targeted in a preliminary study by using an activatable optoacoustic probe for imaging DILI (Huang et al., 2019b).

In 2012 a simple but elegant noninvasive optical imaging approach was used that explored in vivo changes in liver vasculature by imaging the accumulation of indocyanine green (ICG), a fluorescent dye used clinically for ophthalmic angiography, following induction of liver injury with carbon tetrachloride (CCl_4) (Amoozegar et al., 2012). High-speed DyCE imaging of ICG was used to assess vascular flow/clearance dynamics in multiple organ regions simultaneously, before and after CCl_4 administration. ICG uptake in the liver was decreased in mice showing onset of liver injury, with a return to relatively normal rates seen as mice recovered 2 days later. Results were attributed to decreases in active uptake and processing of albumin-bound ICG rather than simple perfusion changes; however, the kinetic differences within the first seconds would suggest that the majority of the effects were due to alterations in liver microperfusion. This is supported by the known vascular changes as detected by microscopy (Hennenberg et al., 2009) and the functional changes in leukocyte velocity and microvascular capacity (Vollmar et al., 1998). Other drugs and/or types of liver injury remain to be explored with this specific in vivo imaging technique, but liver dynamic changes could be expected to differ (i.e., either increase or decrease) dependent on the toxicology mechanism and the degree of vascular damage.

Similar research within PerkinElmer R&D has broadened the approach to liver injury imaging by exploring a panel of NIR fluorescent imaging probes that detect different biological aspects of DILI (Vasquez and Peterson, 2017). The imaging procedure itself is very simple, involving the placement of up to five animals in the imaging field for a 2D epifluorescence acquisition that takes less than 2 min. We first established the use of NIR-labeled Annexin V (Annexin-Vivo 750 [AV750]) to detect cell surface phosphatidylserine present on apoptotic and early necrotic cells and found it to be an effective strategy for detecting hepatocellular injury induced by single bolus administration of agents like APAP and thioacetamide (TAA) (Vasquez and Peterson, 2017). Other agents, rifampicin (RMP) and chlorpromazine (CPZ) that induce nonhepatocellular forms of liver injury, such as cholestasis or mixed forms of injury, however, were not found to be as effectively imaged by AV750, likely because these mechanisms do not involve much direct induction of cellular death following a single bolus injection. We identified two additional probes that we demonstrated to be more useful for cholestasis and cholestasis/hepatocellular toxicity, Transferrin-Vivo 750 (TfV750) and MMPSense 645 (MMP645), that detect key changes in iron metabolism and matrix metalloprotease (MMP) activity, respectively, that are associated with DILI (Bansal et al., 2005; Calabro et al., 2014; Izawa et al., 2014). To simplify this method for universal detection of all forms of liver injury, we developed an imaging strategy using a cocktail of all three probes (AV750, MMP750, and TfV750) in the same spectral channel (AMT750 cocktail). This approach facilitated the testing of uncharacterized compounds, making no assumptions regarding the mechanism of potential liver toxicity. Further, by implementing early (2 h) and late (18–24 h) times for probe injection, it also allowed the detection of early vs late biological changes, and the strategy prevented the possibility of missing significant biological responses that resolved by 24 h. Bolus dosing of RMP, CPZ, (cholestasis- and cholestasis/hepatocellular-inducing drugs), as well as APAP, and TAA yielded clear, liver region fluorescence by 2D epifluorescence imaging and fivefold to sevenfold increases in signal over controls (Fig. 25.1). These results in the AMT750 cocktail effectively and equivalently detecting hepatocellular, cholestatic, and mixed liver injury in mice. It is important to note that both RMP and CPZ mice also showed increased kidney signal, which was sufficiently bright to contribute to the liver region imaging results, further indicating the potential of AMT750 to detect not only DILI but also injury to other tissues.

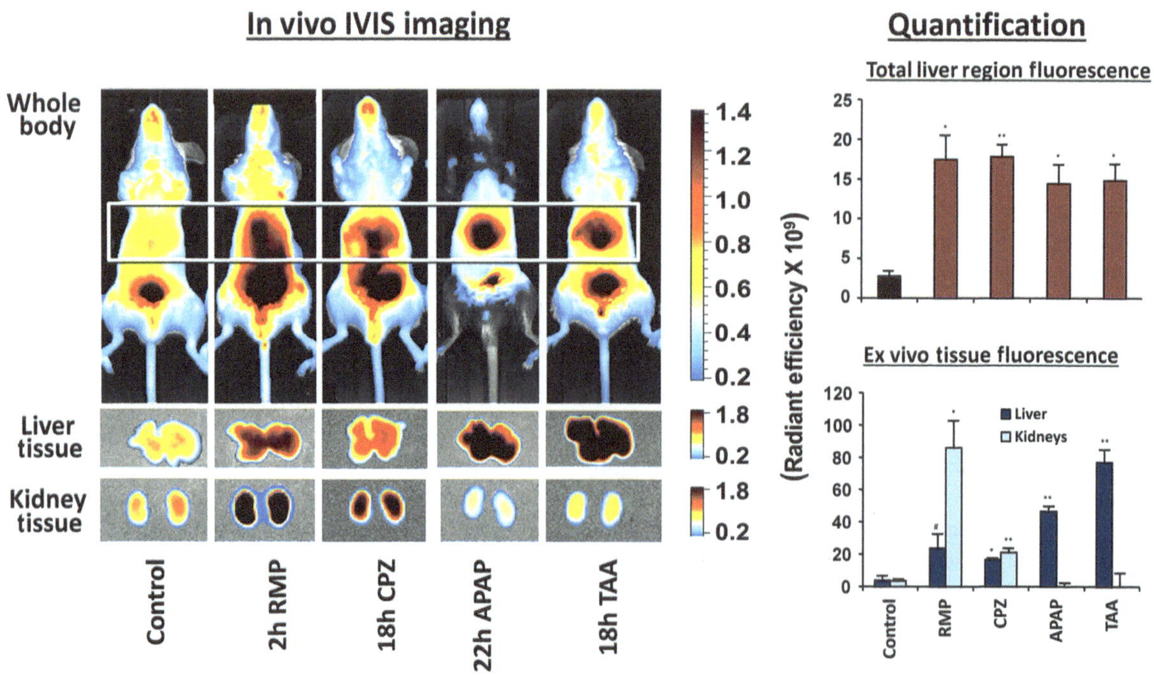

FIGURE 25.1 Validation of a three-probe fluorescent cocktail for general detection of DILI. A cocktail (called AMT750) of three imaging probes (AV750, MMP750, and TfV750) was optimized for 24 h imaging in mice treated with four different liver injury-inducing agents. Agent-dosed mice were injected with the imaging cocktail at two different times post-treatment (18 h for TAA and APAP; 2 h for CPZ and RMP), and all mice were imaged 24 h post-AMT750 for both noninvasive epifluorescence liver assessment (upper left panel) and ex vivo tissues (lower left panel). Quantification of liver signal from noninvasive imaging (upper right panel) and ex vivo liver and kidney signal (lower right panel) were determined by placement of analytical boxes to capture regions of interest (ROI) for the entire liver or individual tissues, and results were measured and represented as total liver fluorescent signal ± SE. Statistical significance was assessed by ANOVA with Dunnett's post-test (# $P < .05$, * $P < .01$, ** $P < .001$, $n = 3$). *TAA*, thioacetamide; *APAP*, acetaminophen; *CPZ*, chlorpromazine; *RMP*, rifampicin.

Though little has been routinely studied regarding acute drug effects on vascular leak in preclinical animal models, in these same studies we assessed the vascular imaging probe, AngioSense 680EX (AS680), to discover drug-dependent effects on liver vascular damage as well as a pattern of diverse organ effects that differed between drugs. As this vascular leak could also have an effect on the accumulation of AMT750 within the liver, we adopted a co-injection protocol in which all animals received an IV injection of AMT750/AS680. This allowed independent assessment of vascular damage as well as correcting AMT750 data for the contribution of the vascular leak (Fig. 25.2). Using pixel-by-pixel ratioing of the images, a more accurate representation of AMT750 localization quantification was revealed (Fig. 25.3), clarifying the refractory response with RMP (early onset with later recovery) and the opposite late-onset kinetics of TAA. Results were confirmed by ex vivo imaging of livers as

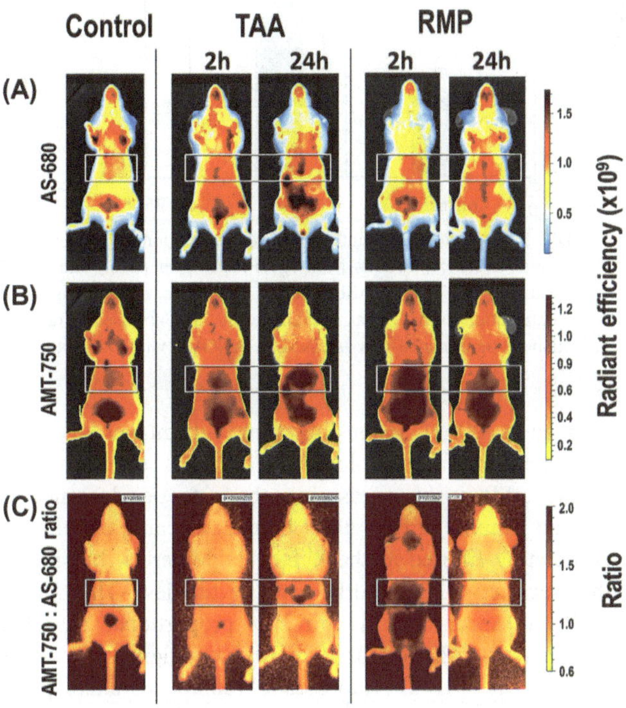

FIGURE 25.2 Ratioing of AMT750 imaging to AS680 imaging allows correction for nonspecific accumulation of AMT750 in tissues due to vascular leak. Mice that were pretreated with TAA or RMP were co-injected with the AMT750 cocktail and AS680, a well-established vascular leak imaging probe, to allow normalization of AMT750 data. Whole-mouse epifluorescence imaging was used to detect accumulation of the 680 and 750 nm probes in the liver regions of different cohorts of mice post-treatment. (A) Epifluorescence images of AS680 and (B) AMT750 (middle panel) signal in controls and in treated mice at two times post-treatment. (C) Normalized ratio images (AMT750:AS680) were generated on the IVIS SpectrumCT Living Image 4.5 software. White boxes indicate the liver regions analyzed for fluorescent signal. Results are representative of three studies using probe cocktail imaging. *TAA*, thioacetamide.

well as a variety of other tissues, showing excellent agreement with published drug effects on nonhepatic tissues as well. Our imaging probes, and the strategy of taking into account vascular drug effects, may be critical for accurately interpreting results with biomarker-targeted probes in cases where there is accompanying vascular damage.

In vivo preclinical NIR imaging in kidney safety/toxicology

Like the liver, the kidneys serve an important role as central detoxification organs (Frazier et al., 2019), leading to very high exposure of renal tissue to drugs, reactive metabolites, environmental compounds, or other bioactive molecules. This exposure can lead to significant cellular damage, and drug-induced kidney injury is reported to be the second-leading cause of drug attrition, after liver toxicity. Ironically, there are a number of herbal supplements suggested to be useful for kidney health, yet some herbs have been found to cause kidney damage. One of the biggest culprits is naturally occurring aristolochic acid, contained at high levels in some herbal supplements (Vanhaelen et al., 1994; Quan et al., 2020). In addition, many traditional medicines and foods from the tropical regions of Africa and Asia contain renal toxic plants (Jha and Chugh, 2003). Of course, undetected contaminants like pesticides, microbial contaminants, chemical toxins, and heavy metals can also be damaging to the kidneys (Wojcikowski et al., 2004; Posadzki et al., 2013).

It can be difficult to accurately assess kidney toxicity in preclinical studies using traditional markers, like blood urea nitrogen (BUN) and serum creatinine, which are relatively insensitive markers in mice and rats; increases in serum concentrations of these biomarkers occur only after significant renal injury, generally after losing two-thirds of kidney function (Pfaller and Gstraunthaler, 1998). Thus there is a need for more sensitive biomarkers and/or assessment approaches to be applied to kidney injury measurement. Additional measurements of kidney health are certainly possible, based on the multiple biological changes that can occur in the kidney in response to toxic insult. Subtle changes in cell surface biomarkers, overt changes in physiology and microvascular flow, changes in overall filtration function, and cellular damage all can occur as a result of toxic kidney injury. Therefore one possible method for early and sensitive detection of kidney injury involves using different biomarkers tailored to specific mechanisms of toxicity of each drug or drug class. Recent research has been pursuing mechanisms involving oxidative stress in the kidneys. Wang et al. (2019) developed a highly selective lysosome-targeting NIR fluorescent probe (Lyso-SiR-2S) that senses HOCl fluctuations driven by inflammatory cell myeloperoxidase, and they used it in

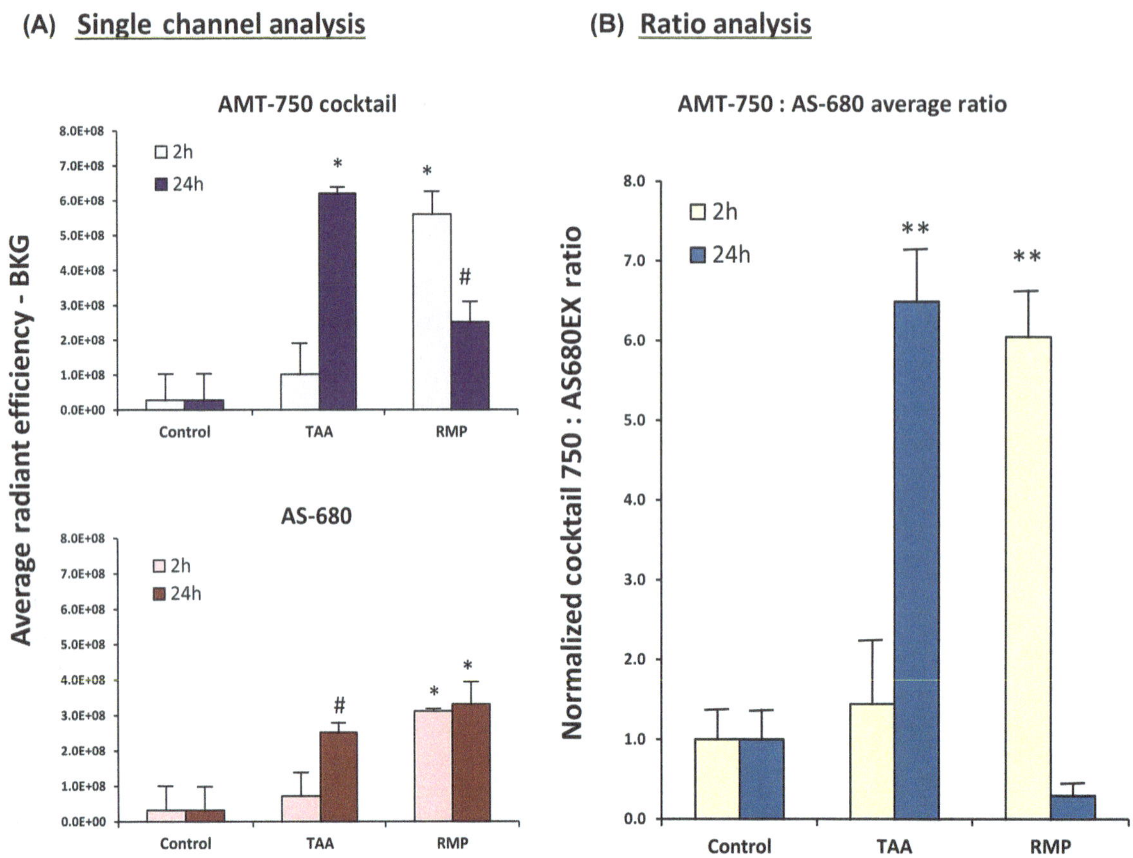

FIGURE 25.3 Quantification of AMT750 to AS680 ratios for quantification of DILI. Single-channel quantification of liver region fluorescence signal from noninvasive imaging was determined by ROI placement to capture the entire liver. (A) Results from AMT750 (upper panel) and AS680 (lower panel) signals are represented as radiant efficiency without correction. (B) Ratio of AMT750 to AS680, using Living Image 4.5 image math functions was performed to correct AMT750 results for nonspecific vascular leak contribution to the overall signal. Results are expressed as ratios of AMT750 to AS680 subsequently normalized to set average control ratios to 1. Statistical significance was assessed by ANOVA with Dunnett's post-test (#$P < .05$, *$P < .01$, **$P < .001$, $n = 3$).

preliminary studies to visualize kidney changes resulting from gentamycin-induced injury. Lv et al. (2018) used novel mitochondria-targeted fluorescent probes (Naph-O2 (-) and NIR-O2 (-)) to detect superoxide anion (O2 (-)) associated with cisplatin-induced nephrotoxicity. Although the imaging was performed ex vivo by 2-photon microscopy in this preliminary study, such probes have enormous potential for in vivo use.

Specific biological approaches for imaging drug-induced kidney injury can be difficult when the mechanism of toxicity is not known, and the range of available in vivo probes is currently limited. Alternatively, direct functional assessment of the kidney, for example, measurement of glomerular filtration rate (GFR), offers a general readout with high clinical relevance and independence of precise toxic mechanism. GFR is considered the gold standard in kidney function assessment and is used preclinically (as a blood assay) to determine the progression of kidney disease and drug-induced kidney toxicity. Large changes in GFR can occur in the absence of significant changes in serum creatinine or BUN (Waikar and Bonventre, 2009), suggesting some advantages over current common methodologies. The most direct way to assess GFR is by measuring the rate of clearance of labeled inulin from the blood, because inulin is completely filtered at the kidneys' glomeruli (but is neither secreted nor reabsorbed by the tubules). This rate of inulin disappearance is directly proportional to GFR. Huang et al. (2019a) used real-time imaging of a long-wavelength NIR-II fluorescent renal clearance probe (CDIR2) to measure kidney dysfunction in living mice. This probe showed relatively slow blood clearance as compared to inulin, and serial blood collection was required to determine actual GFR changes. However, they were able to detect clear cisplatin-induced changes in probe accumulation in the kidneys and bladder when the probe was injected 72 h post-drug treatment, correlating with histologic renal tubule damage.

For direct measurement of GFR, noninvasively, PerkinElmer developed an NIR fluorescent-labeled form of inulin (GFR-Vivo 680) with spectral characteristics (ex/em = 670/685 nm) optimal for in vivo application

(Bao et al., 2019). Blood levels of GFR-Vivo 680 (GFR680) were measured noninvasively by using a unique, rapid fluorescence molecular tomography imaging system (FMT 4000; PerkinElmer) specifically adapted for longitudinal quantification of the fluorescence of heart blood. Bodily blood flow showed a unique pattern with an obvious 3D heart region that was analyzed by careful quantification over time.

Following an intravenous bolus of this agent in cyclosporin A (CsA)-treated, or control SKH-1E mice, FMT images were acquired as 2-min tomographic scans at 1, 5, 15, 30, 45, and 60 min after injection (Fig. 25.4A). Blood clearance rates were calculated for each individual animal using a two-compartment model (*PK* Solver 2.0, an Excel Add-In), yielding average rates of 270 ± 6 mL/min in normal mice. Modification of kidney function by treating mice for 14 days with 80 mg/kg/day of CsA significantly slowed blood clearance rates by 47% (Fig. 25.4B and C) and caused accompanying increases in plasma creatinine and BUN (Fig. 25.4D). GFR680, in combination with FMT heart imaging, provides a noninvasive fluorescent imaging approach to generate consistent GFR measurements with as few as five mice per group, whereas similar approaches with blood collection would require nearly four or five times more mice. Results are robust and consistent enough to allow determination of multiple pharmacokinetic parameters

In vivo preclinical NIR imaging in gastric emptying toxicology

Gastric emptying is a complex process involving excitatory and inhibitory nerves, which may contribute to both

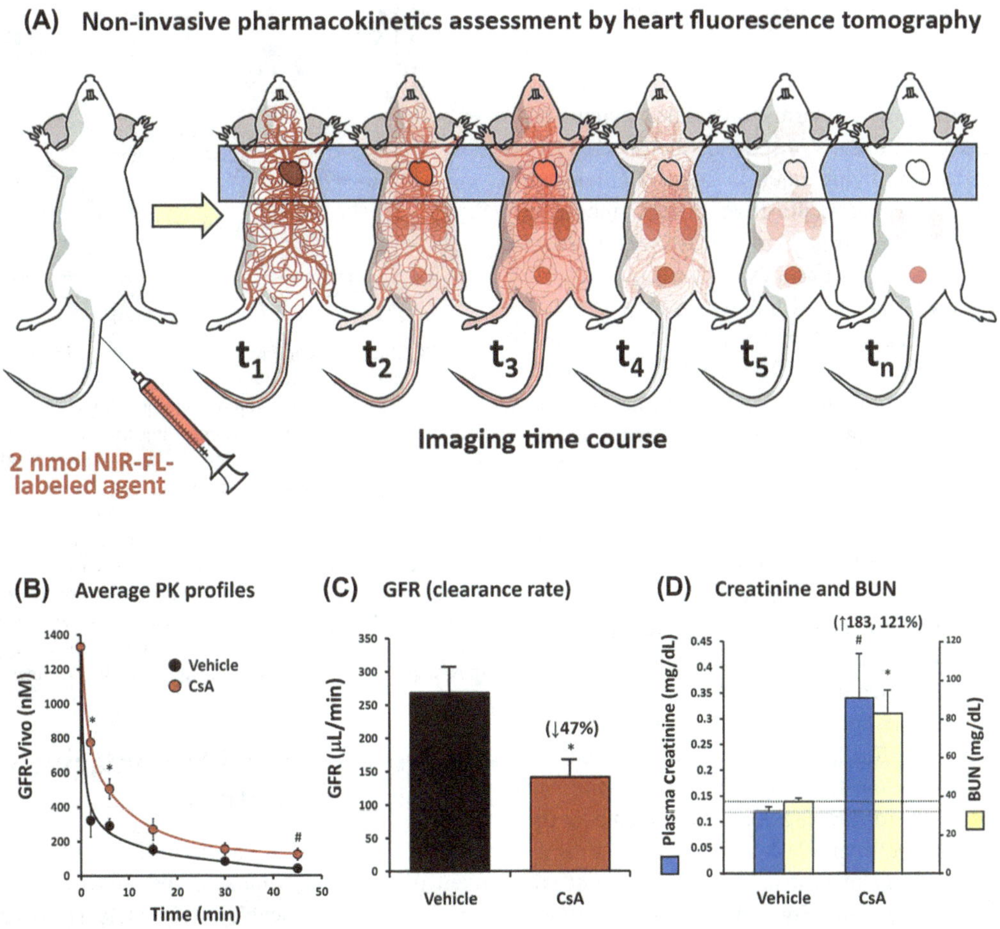

FIGURE 25.4 Kinetic FMT 4000 tomographic imaging of the heart region of mice after GFR-Vivo 680 IV injection offers a noninvasive approach to quantifying circulating fluorescence that can only be achieved by this laser-based system. (A) A diagrammatic representation of kinetic heart imaging in mice to establish blood pharmacokinetic profiles of fluorescent-labeled molecules in vivo. (B) Mice were treated with cyclosporine A (CsA; 80 mg/kg/day) or vehicle (olive oil) for 14 days. Heart fluorescence was identified and quantified at the indicated imaging times after intravenous injection with 2 nmol of GFR-Vivo 680. (C) GFR values were calculated from heart imaging data with GFR-Vivo 680 based on calculated pharmacokinetic parameters using PK Solver 2.0. The results revealed a 47% decrease in GFR in animals with CsA treatment. (D) Plasma creatinine and blood urea nitrogen (BUN) were assessed by microplate assay, revealing CsA-induced increases. Asterisks indicate statistically significant differences between control and CsA groups (#$P < .05$; *$P < .005$). *GFR*, glomerular filtration rate.

acceleration and delay of the emptying process. Food that remains in the stomach too long can lead to either obstructions or fermentation, which can then allow growth of bacteria. Overly rapid gastric emptying can lead to issues in maintaining body weight and proper blood glucose levels. Some medications such as narcotic pain relievers and anticholinergic medications (Osinski et al., 2002; Inada et al., 2004) can delay gastric emptying, whereas other medications, like metoclopramide and erythromycin, can cause rapid emptying of the stomach (MacLaren et al., 2008). Many natural products have long been known to be beneficial in treating either constipation (e.g., flax seed and senna) or diarrhea (e.g., fennel, chamomile, and green tea). Unfortunately, herbs intended for particular indications may have off-target side effects that affect gastric emptying or gastrointestinal motility, like black cohosh and some of the herbal "fat burners." These side effects can create an unwanted and unintended impact on prescription drug absorption and efficacy.

The accurate quantification of gastric emptying in preclinical models is crucial for understanding the physiological, pathological, and pharmacological mechanisms underlying gastric motility. Current methods used for determining gastric emptying rates in small animals are either terminal or involve radioactive tracers, incurring high financial and time costs. An alternative precedent was set in 2004 for using FLI as a tool for tracking gastric emptying using Cy5.5-labeled Tentagel (Gremlich et al., 2004). PerkinElmer has developed a novel NIR fluorescent imaging agent, GastroSense 750 (GS750), to monitor and quantify gastric emptying rates in murine models in vivo, noninvasively, and in real time (Morin et al., 2011). The imaging agent comprises an acid-stable NIR fluorophore conjugated to a nonabsorbable pharmacokinetic modifier designed to be formulated in liquid or incorporated into a solid meal. Using female 8-week-old BALB/c mice recipients, 0.25 nmol GS750 was fed in egg yolk as an experimental approach designed to detect changes in gastric emptying. In vivo 3D fluorescent imaging (Fig. 25.5A) and quantification using the FMT 4000 system at multiple time points revealed a gastric half-life ($t_{1/2}$) of 63 $\pm$ 18 min for the egg formulation. Clonidine (an α2 adrenergic agonist, 1 mg/kg) further inhibited gastric emptying time for the egg formulation, halting the process completely between 2 and 4 h and ultimately delaying the completion of gastric emptying to approximately more than 6 or 7 h (Fig. 25.5B). GS750 signal (either in solid food or liquid administration) can also be quantified as it transits into the intestines (not shown), ultimately decreasing in intensity and volume as it is excreted. These findings reveal that GS750 and FMT 3D fluorescent in vivo imaging can be used successfully to quantify gastrointestinal transit of both liquid and solid diets in small animals and the effect of drug treatment.

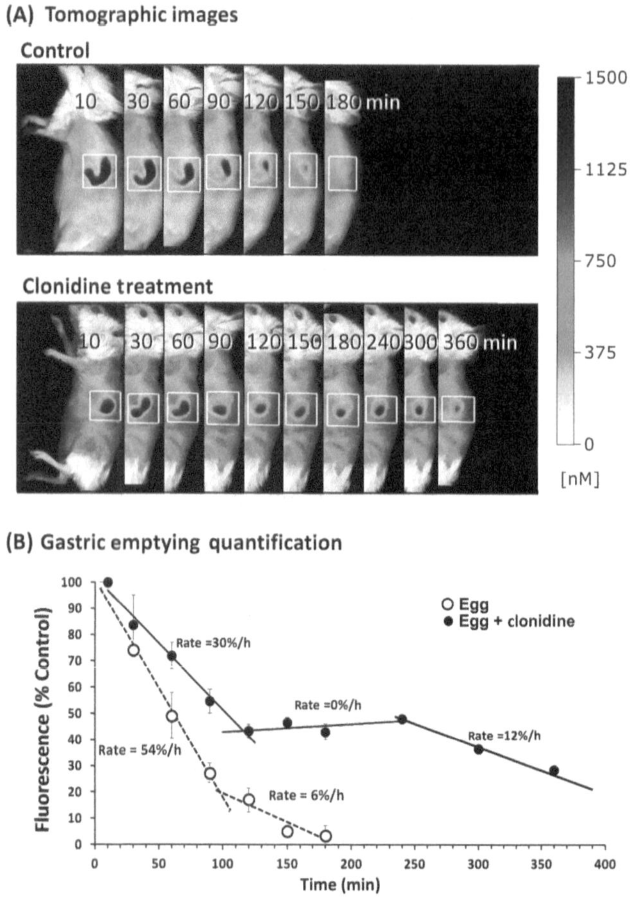

FIGURE 25.5 Kinetic FMT 4000 tomographic imaging of the stomach region of depilated BALB/c mice after GastroSense 680 oral administration offers a noninvasive approach to quantifying gastric emptying. (A) Mice were treated by intraperitoneal injection with clonidine (1 mg/kg), a gastric emptying inhibitor. Fifteen minutes after treatment, Gastrosense, formulated in egg yolk, was fed to mice, and animals ($n = 6$) were imaged over time. Fluorescence brightness is represented in an inverse scale, with black indicating maximum signal. (B) Fluorescence tomographic data sets were analyzed by placing three-dimensional regions of interest to capture the stomach region. For each gastric clearance phase, the linear rate of clearance (% clearance per hour) was also determined; complete stoppage of gastric emptying was shown between 120 and 240 min with clonidine treatment.

In vivo preclinical NIR imaging of vitamin D toxicity

Eating a normal, well-balanced diet that provides an assortment of nutrients, vitamins, and minerals will not lead to toxicity in normal healthy individuals. However, more than 60,000 instances of vitamin toxicity are reported annually to US poison control centers (Mowry et al., 2013) because supplements contain megadoses of vitamins, allowing the potential for overdose. In particular, fat-soluble vitamins A, D, E, and K may lead to toxicity because these vitamins can be stored in the body for long periods of time. For example, vitamin A is highly teratogenic in pregnancy, especially during the first 8 weeks with daily intake more than 10,000 IU. Megadoses of supplemental vitamin E may pose a hazard to people using blood-thinning medications and those using statin drugs. Excessive amounts of vitamin K can cause liver damage and the breakdown of red blood cells and can alter blood clotting times.

A fluorescent bioimaging approach tailored to specific, expected biological changes can be used to profile vitamin toxicity, as can be illustrated by the imaging of vitamin D toxicity. Vitamin D plays an important role in mineral homeostasis and skeletal health maintenance through the regulation of metabolism, calcium, and phosphorus absorption. Very high doses of vitamin D supplements coupled with large amounts of fortified foods, health issues impacting liver or kidney function, or steroid use may cause accumulation in the liver and produce signs of poisoning. Signs of vitamin D toxicity include excess calcium in the blood, decreased bone calcium, soft tissue calcification, slowed mental and physical growth, decreased appetite, nausea, and vomiting (Alshahrani and Aljohani, 2013). Hypercalcemia is responsible for producing most of these overt symptoms.

There are a few possible imaging strategies for detecting the effects of vitamin D-induced hypercalcemia, but first and foremost would be the detection of bone mineral loss within the proximal tibia as well as accumulation of calcium in soft tissue. This can be achieved effectively by using an imaging probe designed to detect calcium loss/deposition using the bisphosphonate-based OsteoSense 750 (OS750) (Zilberman et al., 2008; Hoff et al., 2012; Tower et al., 2014) or osteoclastic protease activity using Cat K 680 FAST™ (CK680) (Scales et al., 2016). Bisphosphonates are a class of drugs that prevent the loss of bone mass by targeting osteoclasts via accumulation on exposed hydroxyapatite within the bone matrix. OS750 detected twofold to threefold increased signal in the knee regions of mice treated for 3 days with a high dose (2.5 μg/kg) of vitamin D3, indicating increased exposure of hydroxyapatite due to bone loss in this region (Fig. 25.6A). Consistent with this bone loss, the CK680 signal increased approximately sevenfold in the knee regions, indicating that vitamin D3 increased osteoclast-driven bone degradation. OS750 was also able to detect vitamin D3-induced calcium deposition in the kidneys, showing >10-fold fluorescent increases within the kidneys of treated mice, correlating with assays for kidney calcium and supported by the significantly increased plasma levels of calcium in these mice (Fig. 25.6B). This data set shows that imaging approaches using appropriate fluorescent probes can further be used as a tool to better understand mechanistic toxicity by direct in situ assessment of soft tissue calcification.

In vivo preclinical NIR imaging of vascular toxicity

Drug-induced vascular injury (DIVI) can be a significant concern in the development of novel therapeutics and is a

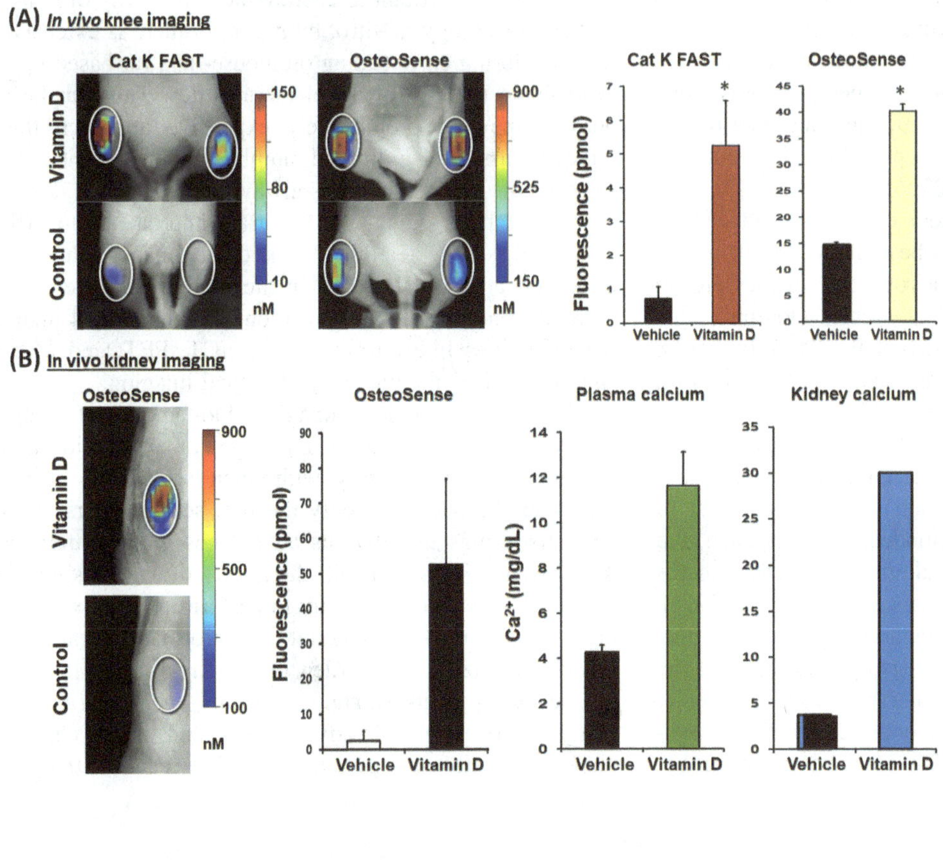

FIGURE 25.6 SKH-1E mice were injected daily (3 days) with 0.06 mg of vitamin D3, a vitamin known to cause bone loss by increasing calcium mobilization and reabsorption. (A) Mice were imaged by fluorescent tomography on the FMT 4000 and showed clear changes in the knees as detected by increased bone turnover detected by a bisphosphonate-based NIR fluorescent imaging probe, OsteoSense 750, and the cathepsin K bone resorption probe, Cat K 680 FAST. (B) Soft tissue calcification was imaged and quantified in the kidneys by OsteoSense, and plasma kidney calcium levels were confirmed by calcium microplate assay. *NIR*, near-infrared.

significant cause of termination of candidate drugs during preclinical safety assessment. Vascular injury has been associated with exposure to vasoactive drugs like minoxidil, theobromine, and digoxin (Clemo et al., 2003), and even herbal medicines containing amfepramone, black cohosh, or passiflora extract (Smith et al., 1993; Ingraffea et al., 2007; Halbsguth et al., 2009). Yet vascular injury often occurs in animals that have exhibited no clinical signs or changes in routine clinical pathology parameters. To further complicate this issue, the clinical relevance of DIVI in preclinical studies is unclear, with some drugs showing significant preclinical vascular injury with little or no evidence of this in clinical populations at therapeutic doses (Johansson, 1981; Collins et al., 1988; Sobota, 1989; National Toxicology Program, 1998). The absence of qualified biomarkers to detect and monitor DIVI in animals during drug development further limits options for detection in the clinical setting and hinders our ability to understand the ramifications of DIVI with regard to potential occurrence in humans.

Multiple mechanisms are implicated in DIVI, including direct chemical interactions of therapeutics with the vascular endothelium, induced variations in blood pressure (vasoconstriction/vasodilation), altered endothelial signaling, direct or indirect mechanical damage to vessel walls, hemorrhage, necrosis, secondary inflammation, and vascular smooth muscle cell (VSMC) proliferation (Mikaelian et al., 2014). These putative biological changes offer opportunities for applying optical imaging techniques to quantify DIVI, including the use of vascular probes to detect changes in vessel integrity, $\alpha v\beta 3$ integrin-targeted probes for vascular remodeling, and protease activity for inflammation. In particular, it has been reported that MMPs, such as MMP3 and MMP9, can contribute in disease situations involving vascular pathology and remodeling (Johnson, 2007; Newby, 2008; Monach et al., 2011).

The histopathologic time course of DIVI induced by vasoactive compounds has been best characterized in rats, with the ability to distinguish across drug classes and to predict DIVI in larger preclinical species (Joseph et al., 1996; Albassam et al., 2001; Losco et al., 2004). In the rat, because DIVI primarily affects muscular resistance arteries of the mesentery, evaluations have historically relied on histopathological examination of the mesenteric arteries. This requires the sacrifice of large numbers of animals to visualize and perform the semiquantitative assessment of changes in vascular integrity, remodeling, hemorrhage, and inflammatory infiltration. As an alternative approach that is quicker and requires fewer animals, researchers at Merck & Co. have applied NIR molecular imaging probes that have the potential to be effective regardless of the initiating insult to reveal common physiologic responses to vascular injury (Lin et al., 2015; Gonzalez et al., 2017). They evaluated three commercially available optical imaging probes from PerkinElmer to assess vascular permeability, AngioSense 750EX (AS750), $\alpha v\beta 3$ integrin expression, IntegriSense 750 (IS750), and MMP activity, MMPSense 750 FAST (MMP750). These three imaging probes would be expected to detect biomechanical vessel damage, integrin upregulation due to VSMC proliferation, and MMP activity associated with vascular remodeling. Rats were treated with a proprietary soluble guanylate cyclase activator (sGCa) known to damage arteries through the nitric oxide-soluble guanylate cyclase-cyclic guanosine monophosphate signaling pathway (Groneberg et al., 2010; Homer et al., 2014). The NIR probes were used for ex vivo quantitative measurements of DIVI in rat mesenteric arteries after oral dosing of the sGCa, with threefold to fourfold increases in mesenteric artery signal observed with AS750 and MMP750, and twofold increases observed with IS750. Histopathology, immunohistochemistry, serum biomarker assessment, and flow cytometry confirmed the mechanistic upregulation and cellular localization of MMP activity. Their collective findings validated a strategy of using fluorescent imaging in conjunction with circulating biomarker data for monitoring DIVI in preclinical studies (Lin et al., 2015; Gonzalez et al., 2017).

Concluding remarks and future directions

The goal of toxicology testing is to understand and predict adverse effects of drugs and other xenobiotics. To do this, researchers use a variety of in vitro and in vivo assessments. As it is difficult to capture the complexity of mammalian physiology in vitro, by necessity there is extensive reliance during early testing on mouse- and rat-based tests to evaluate toxicity of novel therapeutics. This can be a burdensome and prohibitively expensive approach that typically assesses a limited number of end points. To improve the process, there is always a need for the development of new, quick, and accurate approaches to efficacy and safety testing. Biomedical imaging has an important and growing role in these areas, but there are currently limited safety applications for biomedical imaging modalities like ultrasound and μCT, PET, and MRI, and even less in the modality of optical imaging.

Research in optical imaging biomarkers of drug-induced toxicity is at the early stages of exploration and adoption in drug discovery, with much of this work occurring over the course of only the past several years. This modality brings the obvious advantages of the safety of the imaging approach itself, the ability to collect longitudinal data sets, and the ability to perform multiplex imaging, maximizing the amount of information gathered while minimizing the burden of animal usage. Many of the necessary tools (imaging systems and a variety of imaging probes) are already in place, and in this chapter there are several new advances in safety readouts that are highlighted, including liver, kidney, gastric, and vascular injury, as well as soft tissue calcification. However, there is still a considerable distance to go before optical imaging fully realizes its potential. More safety applications need to be explored, existing imaging probes must be validated fully in comparison to a variety of current metrics, and a focused effort on developing toxicity-specific probes is needed. Additional research can be performed to translate some of the more useful nonimaging ex vivo markers for the purpose of noninvasive optical imaging.

It is important to note that optical imaging will not provide a solution for every toxicology mechanism of interest, and there are many current validated approaches that cover these important needs. There may be opportunities for optical imaging in some instances to provide superior readouts, or perhaps equivalent readouts with improved ease of performance, or even simpler readouts with their main advantages being efficiency and consistency. Regardless, the focus for optical imaging will likely be in surrogate early risk assessment screening in mice or rats, limiting the amounts of test materials required and facilitating internal decision-making for discovery projects. Ultimately, the final assessment of drug-induced toxicity will continue to be performed by more conventional means in late-stage drug development candidates in rats and larger species. Over the next several years, however, the advances in optical imaging technology and probes will likely yield some valuable, validated toxicology readouts with the potential for regulatory acceptance.

Competing Interests

Jeffrey D. Peterson is an employee of PerkinElmer, Inc.

References

Ahuja, V., Sharma, S., 2014. Drug safety testing paradigm, current progress and future challenges: an overview. J. Appl. Toxicol. 34, 576–594.

Albassam, M.A., Metz, A.L., Potoczak, R.E., Gallagher, K.P., Haleen, S., Hallak, H., et al., 2001. Studies on coronary arteriopathy in dogs following administration of CI-1020, an endothelin A receptor antagonist. Toxicol. Pathol. 29, 277–284.

Alempijevic, T., Zec, S., Milosavljevic, T., 2017. Drug-induced liver injury: do we know everything? World J. Hepatol. 9, 491–502.

Alshahrani, F., Aljohani, N., 2013. Vitamin D: deficiency, sufficiency and toxicity. Nutrients 5, 3605–3616.

Amoozegar, C.B., Wang, T., Bouchard, M.B., et al., 2012. Dynamic contrast-enhanced optical imaging of in vivo organ function. J. Biomed. Opt. 17 (9), 96003-1.

Anadon, A., Martinez-Larranaga, M.R., Ares, I., Martinez, M.A., 2016. Interactions between nutraceuticals/nutrients and therapeutic drugs. In: Gupta, R.C. (Ed.), Nutraceuticals: Efficacy, Safety and Toxicity. Academic Press/Elsevier, Amsterdam, pp. 855–874.

Anderson, I.B., Mullen, W.H., Meeker, J.E., et al., 1996. Pennyroyal toxicity: measurement of toxic metabolite levels in two cases and review of the literature. Ann. Intern. Med. 124, 726–734.

Ay, M., Charli, A., Jin, H., et al., 2016. Quercetin. In: Gupta, R.C. (Ed.), Nutraceuticals: Efficacy, Safety and Toxicity. Academic Press/Elsevier, Amsterdam, pp. 447–452.

Balls, M., 2015. The three Rs: the way forward. Altern. Lab. Anim. 43, P25–P27.

Bansal, M.B., Kovalovich, K., Gupta, R., et al., 2005. Interleukin-6 protects hepatocytes from CCl4-mediated necrosis and apoptosis in mice by reducing MMP-2 expression. J. Hepatol. 42, 548–556.

Bao, B., Groves, K., Zhang, J., et al., 2012. In vivo imaging and quantification of carbonic anhydrase IX expression as an endogenous biomarker of tumor hypoxia. PLoS One 7, e50860.

Bao, B., Vasquez, K.O., Ho, G., Zhang, J., Delaney, J., Rajopadhye, M., et al., 2019. Blood pharmacokinetics imaging by noninvasive heart fluorescence tomography and application to kidney glomerular filtration rate assessment. J. Pharmacol. Exp. Ther. 370, 288–298.

Beckmann, N., Laurent, D., Tigani, B., et al., 2004. Magnetic resonance imaging in drug discovery: lessons from disease areas. Drug Discov. Today 9, 35–42.

Calabro, S.R., Maczurek, A.E., Morgan, A.J., et al., 2014. Hepatocyte produced matrix metalloproteinases are regulated by CD147 in liver fibrogenesis. PLoS One 9, e90571.

Chan, N., Li, S., Perez, E., 2016. Interactions between Chinese nutraceuticals and western medicines. In: Gupta, R.C. (Ed.), Nutraceuticals: Efficacy, Safety and Toxicity. Academic Press, London, pp. 875–882.

Chen, J., Tung, C.H., Mahmood, U., et al., 2002. In vivo imaging of proteolytic activity in atherosclerosis. Circulation 105, 2766–2771.

Chen, Z., Meng, X., Zou, L., et al., 2020. A dual-emissive phosphorescent polymeric probe for exploring drug-induced liver injury via imaging of peroxynitrite elevation in vivo. ACS Appl. Mater. Interfaces 12, 12383–12394.

Clemo, H.R., Keniston, L., Meredith, M.A., 2003. A comparison of the distribution of GABA-ergic neurons in cortices representing different sensory modalities. J. Chem. Neuroanat. 26, 51–63.

Collins, J.J., Elwell, M.R., Lamb, J.C.T., et al., 1988. Subchronic toxicity of orally administered (gavage and dosed-feed) theophylline in Fischer 344 rats and B6C3F1 mice. Fundam. Appl. Toxicol. 11, 472–484.

Coppock, R.W., Dziwenka, M., 2016. Green tea extract. In: Gupta, R.C. (Ed.), Nutraceuticals: Efficacy, Safety and Toxicity. Academic Press, London, pp. 633–652.

Corsini, A., Bortolini, M., 2013. Drug-induced liver injury: the role of drug metabolism and transport. J. Clin. Pharmacol. 53, 463–474.

Daghighi, S., Sjollema, J., Dijkstra, R.J., et al., 2014. Real-time quantification of matrix metalloproteinase and integrin alphavbeta3 expression during biomaterial-associated infection in a murine model. Eur. Cell Mater. 27, 26–37. discussion 37-28.

Dasgupta, A., 2002. Endogenous and exogenous digoxin-like immunoreactive substances: impact on therapeutic drug monitoring of digoxin. Am. J. Clin. Pathol. 118, 132–140.

Debelle, F.D., Vanherweghem, J.L., Nortier, J.L., 2008. Aristolochic acid nephropathy: a worldwide problem. Kidney Int. 74, 158–169.

Deguchi, J.O., Aikawa, M., Tung, C.H., et al., 2006. Inflammation in atherosclerosis: visualizing matrix metalloproteinase action in macrophages in vivo. Circulation 114, 55–62.

DeLeve, L.D., 2007. Hepatic microvasculature in liver injury. Semin. Liver Dis. 27, 390–400.

Eaton, V.L., Vasquez, K.O., Goings, G.E., Hunter, Z.N., Peterson, J.D., Miller, S. D., 2013. Optical tomographic imaging of near infrared imaging agents quantifies disease severity and immunomodulation of experimental autoimmune encephalomyelitis in vivo. J. Neuroinflammation 10, 138.

Frazier, K.S., Ryan, A.M., Peterson, R.A., et al., 2019. Kidney pathology and investigative nephrotoxicology strategies across species. Semin. Nephrol. 39, 190–201.

Fukami, T., Iida, A., Konishi, K., et al., 2016. Human arylacetamide deacetylase hydrolyzes ketoconazole to trigger hepatocellular toxicity. Biochem. Pharmacol. 116, 153–161.

Gabardi, S., Munz, K., Ulbricht, C., 2007. A review of dietary supplement-induced renal dysfunction. Clin. J. Am. Soc. Nephrol. 2, 757–765.

Gil, F., Hernandez, A.F., Martin-Domingo, M.C., 2016. Toxic contamination of nutraceuticals and food ingredients. In: Gupta, R.C. (Ed.), Nutraceuticals: Efficacy, Safety and Toxicity. Academic Press, London, pp. 825–838.

Goetz, M., Thomas, S., Heimann, A., et al., 2008. Dynamic in vivo imaging of microvasculature and perfusion by miniaturized confocal laser microscopy. Eur. Surg. Res. 41, 290–297.

Goncharov, N., Orekhov, A.N., Voitenko, N., et al., 2016. Organosulfur compounds as nutraceuticals. In: Gupta, R.C. (Ed.), Nutraceuticals: Efficacy, Safety and Toxicity. Academic Press, London, pp. 555–568.

Gonzalez, R.J., Lin, S.A., Bednar, B., et al., 2017. Vascular imaging of matrix metalloproteinase activity as an informative preclinical biomarker of drug-induced vascular injury. Toxicol. Pathol. 45, 633–648.

Grattagliano, I., Bonfrate, L., Diogo, C.V., et al., 2009. Biochemical mechanisms in drug-induced liver injury: certainties and doubts. World J. Gastroenterol. 15, 4865–4876.

Gremlich, H.U., Martinez, V., Kneuer, R., et al., 2004. Noninvasive assessment of gastric emptying by near-infrared fluorescence reflectance imaging in mice: pharmacological validation with tegaserod, cisapride, and clonidine. Mol. Imaging 3, 303–311.

Groneberg, D., Konig, P., Wirth, A., et al., 2010. Smooth muscle-specific deletion of nitric oxide-sensitive guanylyl cyclase is sufficient to induce hypertension in mice. Circulation 121, 401–409.

Gupta, B., Ghosh, K.K., Gupta, R.C., 2016. Thymoquinone. In: Gupta, R.C. (Ed.), Nutraceuticals: Efficacy, Safety and Toxicity. Academic Press/Elsevier, Amsterdam, pp. 541–550.

Halbsguth, U., Schwanda, S., Lehmann, T., et al., 2009. Necrotising vasculitis of the skin associated with an herbal medicine containing amfepramone. Eur. J. Clin. Pharmacol. 65, 647–648.

Hall, A.M., Rhodes, G.J., Sandoval, R.M., et al., 2013. In vivo multiphoton imaging of mitochondrial structure and function during acute kidney injury. Kidney Int. 83, 72–83.

Haller, C.A., Anderson, I.B., Kim, S.Y., et al., 2002. An evaluation of selected herbal reference texts and comparison to published reports of adverse herbal events. Adverse Drug React. Toxicol. Rev. 21, 143–150.

Hartsel, J.A., Eades, J., Hickory, B., Makriyannis, A., 2016. *Cannabis sativa* and hemp. In: Gupta, R.C. (Ed.), Nutraceuticals: Efficacy, Safety and Toxicity. Academic Press/Elsevier, Amsterdam, pp. 735–756.

Hasebroock, K.M., Serkova, N.J., 2009. Toxicity of MRI and CT contrast agents. Expert Opin. Drug Metab. Toxicol. 5, 403–416.

He, Y., Kang, C.H., Xu, S., et al., 2010. Toward surface quantification of liver fibrosis progression. J. Biomed. Opt. 15, 056007.

Hennenberg, M., Trebicka, J., Kohistani, A.Z., et al., 2009. Vascular hyporesponsiveness to angiotensin II in rats with CCl(4)-induced liver cirrhosis. Eur. J. Clin. Invest. 39, 906–913.

Hinson, J.A., Roberts, D.W., James, L.P., 2010. Mechanisms of acetaminophen-induced liver necrosis. Handb. Exp. Pharmacol. 369–405.

Hoff, B.A., Chughtai, K., Jeon, Y.H., et al., 2012. Multimodality imaging of tumor and bone response in a mouse model of bony metastasis. Transl. Oncol. 5, 415–421.

Homer, B.L., Morton, D., Bagi, C.M., et al., 2014. Oral administration of soluble guanylate cyclase agonists to rats results in osteoclastic bone resorption and remodeling with new bone formation in the appendicular and axial skeleton. Toxicol. Pathol. 43 (3), 411–423.

Huang, J., Xie, C., Zhang, X., et al., 2019a. Renal-clearable molecular semiconductor for second near-infrared fluorescence imaging of kidney dysfunction. Angew. Chem. Int. Ed. Engl. 58, 15120–15127.

Huang, Y., Qi, Y., Zhan, C., et al., 2019b. Diagnosing drug-induced liver injury by multispectral optoacoustic tomography and fluorescence imaging using a leucine-aminopeptidase-activated probe. Anal. Chem. 91, 8085–8092.

Inada, T., Asai, T., Yamada, M., et al., 2004. A new method using flow cytometry to measure the effects of drugs on gastric emptying and gastrointestinal transit in mice. Arzneimittelforschung 54, 557–562.

Ingraffea, A., Donohue, K., Wilkel, C., et al., 2007. Cutaneous vasculitis in two patients taking an herbal supplement containing black cohosh. J. Am. Acad. Dermatol. 56, S124–S126.

Izawa, T., Murakami, H., Wijesundera, K.K., et al., 2014. Inflammatory regulation of iron metabolism during thioacetamide-induced acute liver injury in rats. Exp. Toxicol. Pathol. 66, 155–162.

Jaeschke, H., Smith, C.W., 1997. Mechanisms of neutrophil-induced parenchymal cell injury. J. Leukoc. Biol. 61, 647–653.

Javeri, I., Chand, N., 2016. Curcumin. In: Gupta, R.C. (Ed.), Nutraceuticals: Efficacy, Safety and Toxicity. Academic Press/Elsevier, Amsterdam, pp. 435–446.

Jha, V., Chugh, K.S., 2003. Nephropathy associated with animal, plant, and chemical toxins in the tropics. Semin. Nephrol. 23, 49–65.

Jie, X., Wu, M., Yang, H., et al., 2019. Red-near-infrared fluorescent probe for time-resolved in vivo alkaline phosphatase detection with the assistance of a photoresponsive nanocontainer. Anal. Chem. 91, 13174–13182.

Johansson, S., 1981. Cardiovascular lesions in Sprague-Dawley rats induced by long-term treatment with caffeine. Acta Pathol. Microbiol. Scand. A 89, 185–191.

Johnson, J.L., 2007. Matrix metalloproteinases: influence on smooth muscle cells and atherosclerotic plaque stability. Expert Rev. Cardiovasc. Ther. 5, 265–282.

Joseph, E.C., Rees, J.A., Dayan, A.D., 1996. Mesenteric arteriopathy in the rat induced by phosphodiesterase III inhibitors: an investigation of morphological, ultrastructural, and hemodynamic changes. Toxicol. Pathol. 24, 436–450.

Kaore, S.N., Kaore, N.M., 2016. Arginine and citrulline as nutraceuticals: efficacy and safety in diseases. In: Gupta, R.C. (Ed.), Nutraceuticals: Efficacy, Safety and Toxicity. Academic Press, London, pp. 511–530.

Korideck, H., Peterson, J.D., 2009. Noninvasive quantitative tomography of the therapeutic response to dexamethasone in ovalbumin-induced murine asthma. J. Pharmacol. Exp. Ther. 329, 882–889.

Kossodo, S., Pickarski, M., Lin, S.A., et al., 2010. Dual in vivo quantification of integrin-targeted and protease-activated agents in cancer using fluorescence molecular tomography (FMT). Mol. Imaging Biol. 12, 488–499.

Krautz-Peterson, G., Ndegwa, D., Vasquez, K., Korideck, H., Zhang, J., Peterson, J.D., et al., 2009. Imaging schistosomes in vivo. FASEB J. 23, 2673–2680.

Kuijper, I.A., Yang, H., Van De Water, B., et al., 2017. Unraveling cellular pathways contributing to drug-induced liver injury by dynamical modeling. Expert Opin. Drug Metab. Toxicol. 13, 5–17.

Kupiec, T., Raj, V., 2005. Fatal seizures due to potential herb-drug interactions with *Ginkgo biloba*. J. Anal. Toxicol. 29, 755–758.

Lin, S.A., Patel, M., Suresch, D., et al., 2012. Quantitative longitudinal imaging of vascular inflammation and treatment by ezetimibe in apoE mice by FMT using new optical imaging biomarkers of cathepsin activity and alpha(v)beta (3) integrin. Int. J. Mol. Imaging 2012, 189254.

Lin, S.A., Suresch, D.L., Connolly, B., et al., 2015. Optical imaging biomarkers of drug-induced vascular injury. Mol. Imaging 14.

Losco, P.E., Evans, E.W., Barat, S.A., et al., 2004. The toxicity of SCH 351591, a novel phosphodiesterase-4 inhibitor, in Cynomolgus monkeys. Toxicol. Pathol. 32, 295–308.

Lv, Y., Dan, C., Dongdong, S., et al., 2018. Visualization of oxidative injury in the mouse kidney using selective superoxide anion fluorescent probes. Chem. Sci. 9, 7606–7613.

MacLaren, R., Kiser, T.H., Fish, D.N., et al., 2008. Erythromycin vs metoclopramide for facilitating gastric emptying and tolerance to intragastric nutrition in critically ill patients. J. Parenter. Enter. Nutr. 32, 412–419.

Metushi, I.G., Cai, P., Zhu, X., et al., 2011. A fresh look at the mechanism of isoniazid-induced hepatotoxicity. Clin. Pharmacol. Ther. 89, 911–914.

Mikaelian, I., Cameron, M., Dalmas, D.A., The Vascular Injury Working Group of the Predictive Safety Consortium, et al., 2014. Nonclinical safety biomarkers of drug-induced vascular injury: current status and blueprint for the future. Toxicol. Pathol. 42, 635–657.

Monach, P.A., Tomasson, G., Specks, U., et al., 2011. Circulating markers of vascular injury and angiogenesis in antineutrophil cytoplasmic antibody-associated vasculitis. Arthritis Rheum. 63, 3988–3997.

Montet, X., Ntziachristos, V., Grimm, J., et al., 2005. Tomographic fluorescence mapping of tumor targets. Cancer Res. 65, 6330–6336.

Morin J., Kossodo S., Peterson J.D., 2011. Imaging gastric emptying and intestinal transit using oral administration of a near infrared imaging agent, in *World Molecular Imaging Congress*, San Diego, California.

Mowry, J.B., Spyker, D.A., Cantilena Jr., L.R., et al., 2013. 2012 Annual report of the American Association of Poison Control Centers' National Poison Data System (NPDS): 30th annual report. Clin. Toxicol. (Phila.) 51, 949–1229.

Nahrendorf, M., Swirski, F.K., Aikawa, E., et al., 2007. The healing myocardium sequentially mobilizes two monocyte subsets with divergent and complementary functions. J. Exp. Med. 204, 3037–3047.

National Toxicology Program, 1998. NTP toxicology and carcinogenesis studies of theophylline (CAS No. 58-55-9) in F344/N rats and B6C3F1 mice (feed and gavage studies). Natl. Toxicol. Program Tech. Rep. Ser. 473, 1–326.

Newby, A.C., 2008. Metalloproteinase expression in monocytes and macrophages and its relationship to atherosclerotic plaque instability. Arterioscler. Thromb. Vasc. Biol. 28, 2108–2114.

Nounou, M.I., Ko, Y., Helal, N.A., et al., 2018. Adulteration and counterfeiting of online nutraceutical formulations in the United States: time for intervention? J. Diet. Suppl. 15, 789–804.

Osinski, M.A., Seifert, T.R., Cox, B.F., et al., 2002. An improved method of evaluation of drug-evoked changes in gastric emptying in mice. J. Pharmacol. Toxicol. Methods 47, 115–120.

Parvez, M.K., Rishi, V., 2019. Herb-drug interactions and hepatotoxicity. Curr. Drug Metab. 20, 275–282.

Peterson, J.D., 2019. Paradigms in fluorescence molecular imaging: maximizing measurement of biological changes in disease, therapeutic efficacy, and toxicology/safety. Mol. Imaging Biol. 21, 599–611.

Peterson, J.D., Labranche, T.P., Vasquez, K.O., et al., 2010. Optical tomographic imaging discriminates between disease-modifying anti-rheumatic drug (DMARD) and non-DMARD efficacy in collagen antibody-induced arthritis. Arthritis Res. Ther. 12, R105.

Peti-Peterdi, J., Burford, J.L., Hackl, M.J., 2012. The first decade of using multiphoton microscopy for high-power kidney imaging. Am. J. Physiol. Ren. Physiol. 302, F227–F233.

Pfaller, W., Gstraunthaler, G., 1998. Nephrotoxicity testing in vitro—what we know and what we need to know. Env. Health Perspect. 106 (Suppl 2), 559–569.

Posadzki, P., Watson, L., Ernst, E., 2013. Contamination and adulteration of herbal medicinal products (HMPs): an overview of systematic reviews. Eur. J. Clin. Pharmacol. 69, 295–307.

Quan, Y., Jin, L., Luo, K., et al., 2020. Assessment of nephrotoxicity of herbal medicine containing aristolochic acid in mice. Korean J. Intern. Med. 35, 400–407.

Ramachandran, R., Kakar, S., 2009. Histological patterns in drug-induced liver disease. J. Clin. Pathol. 62, 481–492.

Risuleo, G., 2016. Resveratrol: multiple activities on the biological functionality of the cell. In: Gupta, R.C. (Ed.), Nutraceuticals: Efficacy, Safety and Toxicity. Academic Press/Elsevier, Amsterdam, pp. 453–464.

Ronis, M.J.J., Pedersen, K.B., Watt, J., 2018. Adverse effects of nutraceuticals and dietary supplements. Annu. Rev. Pharmacol. Toxicol. 58, 583–601.

Saxena, A., Tripathi, K.P., Roy, S., et al., 2008. Pharmacovigilance: effects of herbal components on human drugs interactions involving cytochrome P450. Bioinformation 3, 198–204.

Scales, H.E., Ierna, M., Smith, K.M., et al., 2016. Assessment of murine collagen-induced arthritis by longitudinal non-invasive duplexed molecular optical imaging. Rheumatology 55, 564–572.

Schuh, C.D., Haenni, D., Craigie, E., et al., 2016. Long wavelength multiphoton excitation is advantageous for intravital kidney imaging. Kidney Int. 89, 712–719.

Shuhendler, A.J., Pu, K., Cui, L., et al., 2014. Real-time imaging of oxidative and nitrosative stress in the liver of live animals for drug-toxicity testing. Nat. Biotechnol. 32, 373–380.

Smith, G.W., Chalmers, T.M., Nuki, G., 1993. Vasculitis associated with herbal preparation containing Passiflora extract. Br. J. Rheumatol. 32, 87–88.

Sobota, J.T., 1989. Review of cardiovascular findings in humans treated with minoxidil. Toxicol. Pathol. 17, 193–202.

Sosnovik, D.E., Nahrendorf, M., Weissleder, R., 2007. Molecular magnetic resonance imaging in cardiovascular medicine. Circulation 115, 2076–2086.

Teschke, R., Xuan, T.D., 2019. How can green tea polyphenols affect drug metabolism and should we be concerned? Expert Opin. Drug Metab. Toxicol. 15, 989–991.

Thorling, C.A., Crawford, D., Burczynski, F.J., et al., 2014. Multiphoton microscopy in defining liver function. J. Biomed. Opt. 19, 90901.

Tower, R.J., Campbell, G.M., Muller, M., et al., 2014. Binding kinetics of a fluorescently labeled bisphosphonate as a tool for dynamic monitoring of bone mineral deposition in vivo. J. Bone Miner. Res. 29, 1993–2003.

Tseng, J.C., Narayanan, N., Ho, G., et al., 2017. Fluorescence imaging of bombesin and transferrin receptor expression is comparable to 18F-FDG PET in early detection of sorafenib-induced changes in tumor metabolism. PLoS One 12, e0182689.

Vanhaelen, M., Vanhaelen-Fastre, R., But, P., et al., 1994. Identification of aristolochic acid in Chinese herbs. Lancet 343, 174.

Vasquez, K.O., Peterson, J.D., 2017. Early detection of acute drug-induced liver injury in mice by noninvasive near-infrared fluorescence imaging. J. Pharmacol. Exp. Ther. 361, 87–98.

Vermeij, E.A., Koenders, M.I., Blom, A.B., et al., 2014. In vivo molecular imaging of cathepsin and matrix metalloproteinase activity discriminates between arthritic and osteoarthritic processes in mice. Mol. Imaging 13, 1–10.

Vollmar, B., Siegmund, S., Menger, M.D., 1998. An intravital fluorescence microscopic study of hepatic microvascular and cellular derangements in developing cirrhosis in rats. Hepatology 27, 1544–1553.

Waikar, S.S., Bonventre, J.V., 2009. Creatinine kinetics and the definition of acute kidney injury. J. Am. Soc. Nephrol. 20, 672–679.

Wan, D., Wu, Q., Kuca, K., 2016. Spirulina. In: Gupta, R.C. (Ed.), Nutraceuticals: Efficacy, Safety and Toxicity. Academic Press, London, pp. 569–584.

Wang, Y.X., Yan, S.X., 2008. Biomedical imaging in the safety evaluation of new drugs. Lab. Anim. 42, 433–441.

Wang, J., Cheng, D., Zhu, L., et al., 2019. Engineering dithiobenzoic acid lactone-decorated Si-rhodamine as a highly selective near-infrared HOCl fluorescent probe for imaging drug-induced acute nephrotoxicity. Chem. Commun. (Camb.) 55, 10916–10919.

Weissleder, R., Ntziachristos, V., 2003. Shedding light onto live molecular targets. Nat. Med. 9, 123–128.

Wen, B., Gorycki, P., 2019. Bioactivation of herbal constituents: mechanisms and toxicological relevance. Drug. Metab. Rev. 51, 453–497.

Wojcikowski, K., Johnson, D.W., Gobe, G., 2004. Medicinal herbal extracts—renal friend or foe? Part two: herbal extracts with potential renal benefits. Nephrology (Carlton) 9, 400–405.

Wu, L., Ding, Q., Wang, X., Li, P., et al., 2020. Visualization of dynamic changes in labile iron(II) pools in endoplasmic reticulum stress-mediated drug-induced liver injury. Anal. Chem. 92, 1245–1251.

Wunder, A., Tung, C.H., Muller-Ladner, U., et al., 2004. In vivo imaging of protease activity in arthritis: a novel approach for monitoring treatment response. Arthritis Rheum. 50, 2459–2465.

Yang, M.S., Wu, M.Y., 2016. Chinese ginseng. In: Gupta, R.C. (Ed.), Nutraceuticals: Efficacy, Safety and Toxicity. Academic Press/Elsevier, Amsterdam, pp. 693–706.

Zhang, J., Preda, D.V., Vasquez, K.O., et al., 2012. A fluorogenic near-infrared imaging agent for quantifying plasma and local tissue renin activity in vivo and ex vivo. Am. J. Physiol. Ren. Physiol 303, F593–F603.

Zilberman, Y., Kallai, I., Gafni, Y., et al., 2008. Fluorescence molecular tomography enables in vivo visualization and quantification of nonunion fracture repair induced by genetically engineered mesenchymal stem cells. J. Orthop. Res. 26, 522–530.

Chapter 26

Flow cytometry and light-scattering techniques in evaluation of nutraceuticals

Igor V. Mindukshev, Igor V. Kudryavtsev, Maria K. Serebriakova, Andrey S. Trulioff, Stepan P. Gambaryan, Julia S. Sudnitsyna, Pavel V. Avdonin, Richard O. Jenkins and Nikolay V. Goncharov

Introduction and background

Modern automated particle- and cell-counting instruments use optical methods (light scatter), impedance-based methods based on the Coulter principle (changes in electrical current induced by blood cells flowing through an electrically charged opening), or a combination of both (Green and Wachsmann-Hogiu, 2015). They are used for a variety of applications including routine cell culture, hematological analysis, and industrial controls (Dittami et al., 2012). A limitation with existing Coulter technologies is the lack of metrics on the overall health of cell samples. Consequently, additional techniques must often be used in conjunction with Coulter counting to assess cell viability. This extends the experimental setup time and cost since the traditional methods of viability assessment require cell staining and use of more elaborated and expensive equipment such as a flow cytometer. In contrast to light scattering, which cannot distinguish damaged cells from debris, fluorescent dyes used in membrane integrity and mitochondrial polarization assays are capable of labeling and discriminating all cells in the suspension (Reardon et al., 2014). Nevertheless, there is an advantage, besides low cost, to the light-scattering technique: the capability for online recording of the dynamic changes of suspended viable cells in response to receptor agonists, toxic agents, pharmaceuticals, etc. At the turn of the century, the authors of this chapter developed a method for simultaneous monitoring of changes in shape and aggregation of platelets (Mindukshev et al., 2005a). The signal of light-scattering alterations at angles below six degrees was shown to be caused by platelet aggregation dynamics (aggregation, disaggregation, coagulation), while over a range of larger angles (6–15 degrees) cell shape changes also contributed to the cell spherization and pseudopodia formation. Aggregation kinetics in saline solution under turbulent flow showed second-order kinetics in relation to the initial cell concentration. The rate constant depended on stirring conditions and the calcium concentration in the medium. We subsequently applied the low-angle light-scattering technique to studies of tumor cells and erythrocytes (Zinchenko et al., 2006; Mindukshev et al., 2007).

Numerous investigations of toxic agents and pharmaceuticals have employed flow cytometry or light-scattering techniques (Mindukshev et al., 2005b, 2006; Goncharov et al., 2009; Radilov et al., 2009; Kudryavtsev et al., 2014; Geisler et al., 2014; Prado et al., 2015), though there are not many examples of combined applications of both technical approaches. Also, compared with pharmaceuticals and toxic agents, the toxic and mechanistic properties of nutraceuticals have not been as extensively studied. For example, as of April 1, 2020 the PubMed database contained nearly two times more papers with keywords "flow cytometry toxic agents" (912) than with keywords "flow cytometry nutraceuticals" (470). Nevertheless, there are many common points and correlations between the techniques, so that a convergence seems to be inevitable. For example, changes of mitochondrial membrane potential lead to morphological and volume changes of the cells (Zinchenko et al., 2006). There are continuing concerns over the toxicity and purity of many plant extracts used as nutraceuticals or supplements, as well as over the simple methods used to evaluate them (Goncharov et al., 2019). Cells affected with nutraceuticals/supplements can be easily analyzed using flow cytometry and light scattering—and sometimes this is made together with electron microscopy, Western blotting, and immunohistochemical analysis—for cell size and mitochondrial membrane potential, as well as cell cycle profile (LeBlanc et al., 2007; Buenz, 2008; Nielsen et al., 2008; Attilio et al., 2010; Yang et al., 2012).

In this chapter, we briefly describe the methodologies of low-angle light scattering and flow cytometry in relation to mechanisms of cell death and survival, and present an

Nutraceuticals. DOI: https://doi.org/10.1016/B978-0-12-821038-3.00026-4

overview of some cellular studies of nutraceuticals involving light scattering and/or flow cytometry techniques.

Methodology of research

Mechanisms of cell death

Studies of various cellular death pathways can aid our understanding of the mechanistic properties of toxic or therapeutic effects of chemical agents, and are important in the development of therapeutic strategies. There are three major types of morphologically distinct cell death: apoptosis (type I), autophagic cell death (type II), and necrosis (type III) (Green and Llambi, 2015). Initially they were thought to constitute mutually exclusive cellular states, though a balanced interplay between them was revealed at the turn of the century. DNA fragmentation is one of the principal characteristics of apoptosis, which is used in flow cytometry for identification of apoptotic cells. The fragmentation is a result of activation of caspase-3, -6, and -7, which initiates several mechanisms of fragmentation (Fan et al., 2005). The first one is bound to impairment of DNA reparation due to inactivation by caspase-3 of poly[adenosine diphosphate (ADP)-ribose] polymerase-1 (PARP-1) and topoisomerase I, which both participate in DNA recombination. Another mechanism is activation of DNases. A further substrate of caspases is histone H1, which blocks boarding of endonucleases on DNA fragments between separate nucleosomes (Kitazumi and Tsukahara, 2011). Mitochondrial damage is a universal sign of apoptosis specific to eukaryotic organisms (Abdelwahid et al., 2011). As a result, a series of proapoptotic factors are released to the cytoplasm: cytochrome C, AIF (apoptosis inducing factor), Smac/DIABLO, endonuclease G, and procaspases 2, 3, and 9 (Orrenius et al., 2007). This is coupled with changes in the membrane potential ($\Delta\Psi m$). Finally, the principal sign of apoptosis is phosphatidylserine exposure to the extracellular space due to impairment of ATP-dependent maintenance of the cell membrane lipid distribution (Verhoven et al., 1995). Also, exposure of phosphatidylserine can be initiated by scramblase, which is activated by Ca^{2+} at early stages of apoptosis (Kalienkova et al., 2019).

Necrosis is characterized by the loss of plasma membrane integrity followed by cytoplasmic leakage, and can occur as a consequence of either extensive damage (high or low temperatures, pH, etc.) or in a regulated receptor interacting protein (RIP) kinase-dependent and caspase-independent manner (necroptosis) (Green and Llambi, 2015). Several death initiator and effector molecules, signaling pathways, and subcellular sites have been identified as key mediators in both processes, either by constituting common modules or by functioning as a switch allowing cells to decide which route to take.

As for autophagy, this is a predominantly cytoprotective process that has been linked to both types of cell death, serving either a prosurvival or prodeath function (Nikoletopoulou et al., 2013). Autophagy is a central mechanism mediating the lysosomal degradation of cytoplasmic components. The central autophagic organelle is the autophagosome, a double membrane-bound vacuole that sequesters the cytoplasmic material destined for disposal. The ultimate destiny of the autophagosome is to fuse with a lysosome, resulting in the degradation of the autophagic cargo (Dupont et al., 2014). Dysregulation in autophagy is associated with a diverse range of pathologies. Both natural and synthetic compounds regulate the interplay between apoptosis, autophagy, and necroptosis, stimulating common molecular mediators and sharing common organelles (Radogna et al., 2015).

Recently, ferroptosis was described as a nonapoptotic form of cell death that may facilitate the selective elimination of some tumor cells or be activated in specific pathological states (Dixon et al., 2012; Conrad et al., 2018). Ferroptosis is dependent upon intracellular iron, but not other metals, and is morphologically, biochemically, and genetically distinct from apoptosis, necrosis, and autophagy. Cells transformed with oncogenic Ras have increased iron content relative to their normal cell counterparts through upregulation of transferrin receptor 1 and downregulation of ferritin heavy chain 1 and ferritin light chain (Yang and Stockwell, 2008). Ferrostatin-1 and deferoxamine are inhibitors of ferroptosis in cancer cells and glutamate-induced cell death in organotypic rat brain slices, suggesting similarities between these two processes. Erastin, as well as glutamate, inhibits cystine uptake by the cystine/glutamate antiporter, ultimately leading to iron-dependent oxidative death (Dixon et al., 2012). Numerous overlapping mechanisms using thiol compounds exist to decompose hydrogen peroxide to water in order to bypass the generation of hydroxyl radicals. Hydrogen peroxide is now recognized as a signaling molecule whose main regulator is the peroxiredoxin/sulfiredoxin systems, which are at least partially under the control of the Nrf2/Keap1 system. Ironically, the persistent activation of antioxidant systems via genetic alterations in Nrf2 and Keap1 contributes to carcinogenesis. Some cancers hijack the Nrf2/Keap1 system with mutation in these genes, which persistently activate antioxidant systems in the cancer cells; overexpression of CD44v (8–11) stabilizes the cystine/glutamate antiporter to increase cysteine and GSH levels in cancer stem cells (Toyokuni, 2014).

Pyroptosis—a highly pyrogenic inflammatory form of cell death—develops due to formation of inflammasomes, which are cytosolic multiprotein platforms assembled in response to invading pathogens and other danger signals. Typically, inflammasome complexes contain a sensor protein, an adaptor protein and a zymogen—procaspase-1. The assembly results in processing of inactive procaspase-1 into an active cysteine-protease enzyme (caspase-1), which subsequently activates the proinflammatory cytokines, interleukins IL-1β and IL-18 (Vanaja et al., 2015). Although these processes are intended to protect the body from insults, prolonged or exacerbated inflammatory responses associated with inflammasome activation are related to a growing number of diseases. Inflammasome activation and autophagy were shown to be linked and to mutually influence each other (Martins et al., 2015).

Eryptosis, the suicidal erythrocyte death characterized by cell shrinkage and cell membrane scrambling, is stimulated by Ca^{2+} entry through Ca^{2+}-permeable PGE_2-activated cation channels, by ceramide, caspases, calpain, complement, hyperosmotic shock, energy depletion, oxidative stress, and deranged activity of several kinases (e.g., AMPK, GK, PAK2, CK1α, JAK3, PKC, p38-MAPK). Eryptosis is triggered by intoxication, malignancy, hepatic failure, diabetes, chronic renal insufficiency, hemolytic uremic syndrome, phosphate depletion, malaria, iron deficiency, sickle cell anemia, thalassemia, glucose 6-phosphate dehydrogenase deficiency, and many other factors or pathological states. Eryptosis may precede and protect against hemolysis, but by the same token result in anemia and deranged microcirculation (Lang and Lang, 2015).

Flow cytometry in assessment of cellular viability

In this section, we briefly outline commonly used methods in flow cytometry that are based on the assessment of phosphatidylserine exposure, mitochondrial transmembrane potential, and DNA accessibility and/or fragmentation. It should be noted here that methods of revealing the DNA fragmentation are rather laborious, greatly depend upon the professional skills of personnel, and require costly reagents. Therefore other methods based upon staining of DNA have received wide application. As for the apoptotic cells, they bear an incomplete or hypodiploid number of DNA (less than 2 N) forming the so-called "sub-G1 peak." A distribution of cells according to fluorescence intensity after staining of DNA is caused by the stoichiometric character of the staining, and this kind of binding is a distinguishing feature for all the probes of this group. Besides apoptotic studies, these probes are widely used to investigate proliferative activity of various cells. Moreover, within the frame of apoptotic studies, they are used together with other probes oriented at revealing asymmetry of the cell membrane, changes of mitochondrial membrane potential, activity of caspases, etc. (Zhao et al., 2009; Wlodkowic et al., 2012).

YO-PRO-1 and *PO-PRO-1* are DNA-binding probes, which are used together with PI to stoichiometrically bind to nucleic acids, that is, RNA in the cytoplasm and DNA in the nucleus (Idziorek et al., 1995). Being excited at 488 nm, YO-PRO-1 emits fluorescence in the green region (near 509 nm) and PI in the red region (near 617 nm) of the spectrum. But the principal difference between them is the method of penetration into the cells. YO-PRO-1 enters cells through purinergic receptors, P2RX7, which are ligand-dependent ionic channels (Glisic-Milosavljevic et al., 2005; Stokes et al., 2006). In the living cells, YO-PRO-1 would not accumulate because of almost inactive channels and very low transporting ability of the membrane. The channels become activated at early stages of apoptosis development, together with disturbance of plasma membrane asymmetry. To reveal the later stages, cells should be stained with DNA-binding PI, which does not need special transporters and is able to penetrate into cells and nuclei only via damaged/fragmented membranes, the process taking place at late stages of apoptosis or necrotic cell death. Thus, the living cells would not be stained at all, the cells at early stages of apoptosis would be only YO-PRO-1 positive, and the cells at later stages would be effectively stained by both probes.

The *SYTO family* of dyes, capable of measuring apoptosis with the flow cytometry technique, were first described in 1995 (Frey, 1995). Presently there are four main groups of cyanide stains distinguished by excitation wavelengths. Dyes of the SYTO blue group (SYTO 40–45) have an excitation maximum at 419–452 nm, with fluorescent emission maximum at 445–484 nm. Dyes of the SYTO green group (SYTO 11–16 and 20–25) have an excitation maximum at 483–521 nm, and fluorescent emission maximum at 500–556 nm. The two other groups are SYTO orange (SYTO 80–86) and SYTO red (SYTO 17 and 59–64), with corresponding excitation maxima at 528–567 and 598–654 nm, and fluorescent emission maxima at 544–583 and 620–680 nm, respectively. A distinguishing feature of SYTO dyes is their ability to spontaneously penetrate the lipid bilayers and bind with nucleic acids (DNA and RNA) in the cytoplasm, nucleus, and mitochondria. This interaction is accompanied with multiple (up to 40 times) enhancement of fluorescence. Another advantage is an absence of visible toxic effects, which was demonstrated in long-term experiments (up to 72 h) with cells (Wlodkowic et al., 2007). The exact mechanism of interaction of SYTO dyes with nucleic acids is not known (Wlodkowic et al., 2012), but their principal difference from other DNA dyes is a low interaction and decrease of fluorescence in apoptotic cells. Thus, live cells are designated as $SYTO^{bright}$, while dying cells are $SYTO^{dim}$. Interestingly, it may be that there is a correlation between the fluorescence intensity of SYTO dyes and lipophilic cationic mitochondrial probes (e.g., JC-1 or TMRM) (Wlodkowic et al., 2008), which is indicative of an analogy in their mechanism of interaction due to positive charged groups within molecules and dependence of interaction and fluorescence upon $\Delta\Psi m$.

Methods of studying *mitochondrial membrane potential $\Delta\Psi m$* with flow cytometry are based on use of cationic lipophilic dyes, which are also known as "mitochondrial probes." These dyes spontaneously penetrate through lipid bilayers and accumulate in areas with a high concentration of protons, that is, under inner mitochondrial membrane. This is accompanied by changes of fluorescence. If proton concentration is low—decrease of $\Delta\Psi m$—then accumulation of dye and its fluorescence is also low. Most probes are related to three principal groups: (1) rhodamine derivatives [rhodamine 123 (Rh123), ethyl (TMRE) and methyl (TMRM) ethers of tetramethylrhodamine]; (2) carbocyanine derivatives [5,5'-6,6'-tetrachloro-1,1',3,3'-tetraethylbenzimidazolylcarbocyanine iodide (JC-1), 3,3'-dimethyl-α-naphthoxacarbocyanine iodide (JC-9), 3,3'-dihexiloxacarbocyanine iodide $DiOC_6(3)$]; and (3) MitoTracker dyes (Cottet-Rousselle et al., 2011).

In order to investigate the asymmetry of the apoptotic cell membranes, namely to reveal *phosphatidylserine*, conjugates of *annexin V with fluorochromes (usually FITS or PE)* are used. In the presence of Ca^{2+} this protein can conjugate with negatively charged phospholipids, with the greatest affinity to phosphatidylserine ($K_D = 5 \times 10^{-10}$ M). The changes in mitochondrial membrane potential and activation of caspase-3 precede the onset of phosphatidylserine in the outer layer, and the latter precedes reorganization of cytoskeleton and cell membrane integrity. Therefore there are many combinations of cell staining with Annexin V and other probes for revealing both early and late phases of apoptosis (Wlodkowic et al., 2011). The cell membrane of healthy cells is impermeable to DNA-binding stains, so they are negative to both Annexin-FITC and propidium iodide ($AnnV\text{-}FITC^-PI^-$). At early apoptosis, phosphatidylserine is exposed on the cell surface, making it possible to conjugate with AnnV-FITC in the presence of Ca^{2+} ions. This exposure coincides with chromatin condensation and DNA fragmentation, though the membrane is yet not permeable to DNA-binding stains. On a histogram, the early apoptotic cells are AnnV-FITC positive and PI negative ($AnnV\text{-}FITC^+PI^-$); dead cells or those at late stages of apoptosis would be AnnV-FITC and PI positive ($AnnV\text{-}FITC^+PI^+$). The same characteristics of fluorescence would also have necrotic cells, though these cells could be AnnV-FITC negative ($AnnV\text{-}FITC^-PI^+$).

Light-scattering technique in studies of platelets and red blood cells

Laser particle size analyzers available from commercial sources (for example, Malvern, Beckman, Horiba, Fritch, and others) are widely used for size distribution analysis of natural and artificial particles. However, up until now these devices have not been routinely applied to biological systems. Based on the technology used in these analyzers we created a new device (LaSca, Biomedsystems Ltd, Russia) and developed original software for the registration of the kinetics of changes in cell volume, morphology, and aggregation (Mindukshev et al., 2005b, 2010, 2012). A quantitative characterization of cell volume can only be achieved under single scatter conditions. For this, the cell density must be adjusted to yield a transmittance of at least 50% of the transmittance of the blank cuvette (Van de Hulst, 1981).

Cell volume and cell shape undergo dynamic changes in many physiological and pathological processes. Active volume regulation to protect cells from lysis and apoptosis and to maintain optimal concentration of intracellular components is one of the most basic and essential cellular functions. Several methods are applied for measurement of cell volume changes evoked by different stimuli. Changes in volume of attached cells can be accurately measured by the 3D reconstruction of cell shape based on cell images acquired from two perpendicular directions (Boudreault and Grygorczyk, 2004). A laser light-scattering method was introduced and used for volume changes registration of attached cells (McManus et al., 1993) and cells in suspension (Faggio et al., 2011). However, no quantitative or time-resolved data of cell volume have yet been presented by these methods. For analysis of cell volume in suspension, which is especially important for blood cells, the Cell Lab Quanta SC Flow analyzer was introduced by Beckman and has been used for analysis of cellular apoptosis (Bortner et al., 2007). However, flow cytometry methods are not suitable for online registration of fast changes of cell volumes.

For demonstrating the feasibility of a method for characterization of cell volume regulation we used osmotic stress-induced volume changes in HL-60 cells (Fig. 26.1). The cell density range applicable for HL-60 cells corresponds to 120–130 cells/μL. To measure cell volume distribution, we monitored light scattering from 28 channels in a range from 1 to 12 degrees. For simplification of the graph, changes of light-scattering intensity (LSI) are presented only for 1 and 12 degrees (Fig. 26.1A). In hypertonic conditions, LSI decreased at 1 degree but increased at 12 degrees. In hypotonic conditions the changes in LSI are inversed for 1 and 12 degrees. It should be noted that cells are highly tolerable to phase changes of osmolarity, while acute changes from iso- to hypoosmotic (45 mOsm)

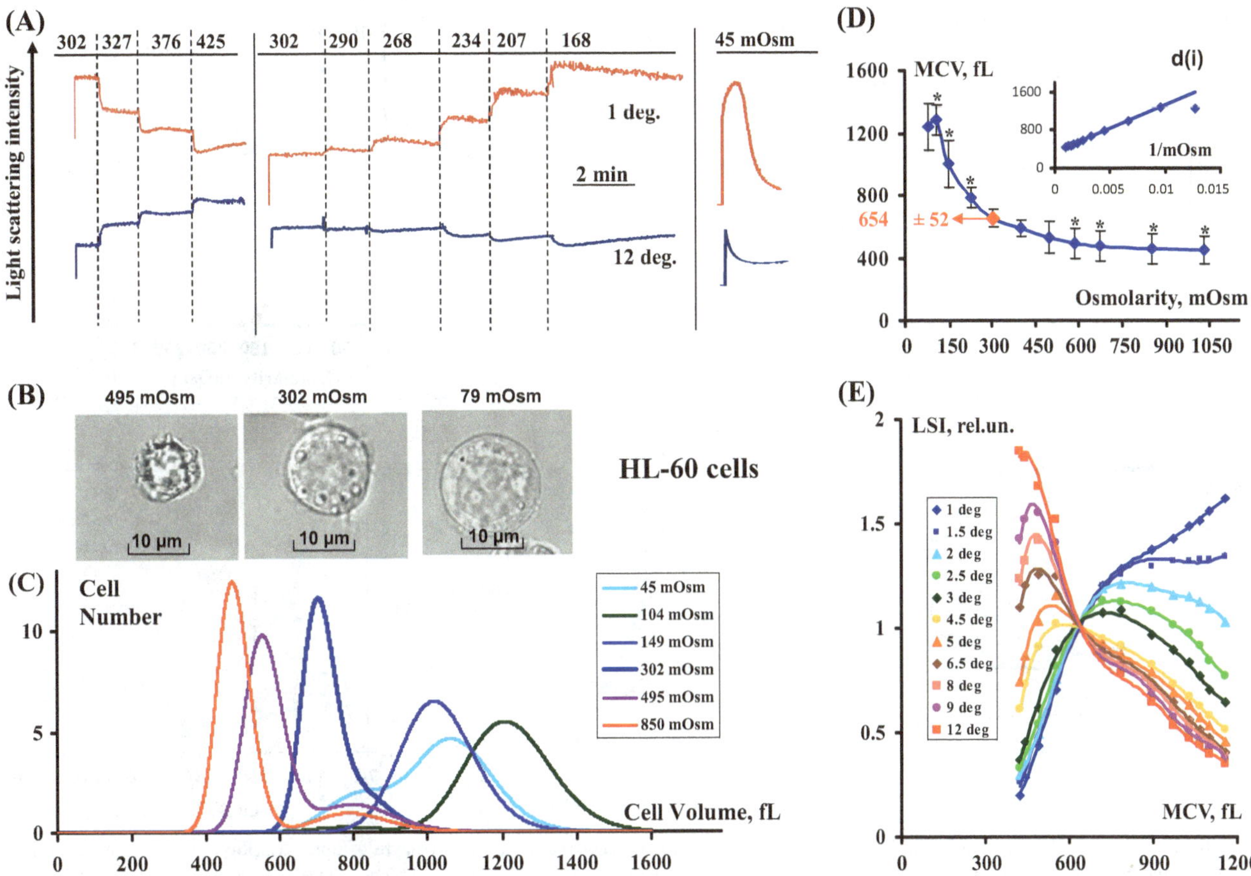

FIGURE 26.1 **Application of the low-angle light-scattering method for registration of HL-60 cell volume changes.** (A) Original recordings (representative for four independent experiments) of light-scattering intensity (LSI) measured at 1 and 12 degrees of HL-60 cells incubated in media with different osmolarity. (B) Light microscopical images of HL-60 cells incubated for 5 min in media with the indicated osmolarity. The mean diameter calculated from 60 individual images of control cells (302 mOsm) was 11.2 ± 0.6 μm, in 495 mOsm was 8.1 ± 0.3 μm, in 79 mOsm was 14. 5 ± 0.5 μm, which corresponds to the calculated mean cell volume (MCV, data from part D) 654 ± 52 fL, 531 ± 96 fL, and 1240 ± 93 fL, respectively. (C) Histogram of cell volume distribution in media of the osmolarity indicated. (D) Calculated relations of MCV with the osmolarity of the media. (E) Dependence of relative LSI (normalized to LSI in isotonic media) in indicated LSI angles on changes of MCV.

solution lead to immediate cell lysis, which is reflected in the strong decrease in LSI (Fig. 26.1A). The registered scatter intensity correlates with changes in cell size observed by light microscopy (Fig. 26.1B). Cell volume distribution in hyper-, iso-, and hypoosmotic solutions was constructed according to the changes of LSI (Fig. 26.1C). Based on data of cell volume distribution, the mean cell volume (MCV) was calculated (Fig. 26.1D). Because the relation between LSI and cell volume is nonlinear (Fig. 26.1E), all data registered at 28 channels are necessary for calculation of cell volume distribution and MCV. The method shows very high sensitivity. We can measure and calculate acute MCV changes of as little as 1%, making this method especially applicable for receptor-mediated kinetic analysis of fast and reversible cell volume changes.

By this method, we can also demonstrate the registration of erythrocyte osmotic fragility (Fig. 26.2). To maintain single scattering conditions for erythrocytes, the cell density was adjusted to 500–600 cells/μL. Under hypoosmotic conditions the LSI at 1 degree increased, but decreased at 12 degrees (Fig. 26.2A). Cell volume distribution for these conditions was calculated according to the changes in LSI (Fig. 26.2B). The maximal increase in erythrocyte MCV was achieved at 110 mOsm from 84 ± 6 fL (control, isotonic conditions) to 142 ± 8 fL. Lower osmolarity caused a fast and irreversible lysis reaction (Fig. 26.2C). Oxidative stress induced by *tert*-butylhydroperoxide (tBH) and calcium increase induced by the calcium ionophore (A23187) causes erythrocyte transformations and changes in osmotic fragility (Fig. 26.2D,E). Strong oxidative stress (2 mM tBH) and increased intracellular calcium induce cell rigidity. A variety of methods, including light-scattering analysis (Turcu et al., 2006), are used for investigation of erythrocyte transformations under different conditions. The fast and rather easy registration of erythrocyte transformations is one of the advantages of the method.

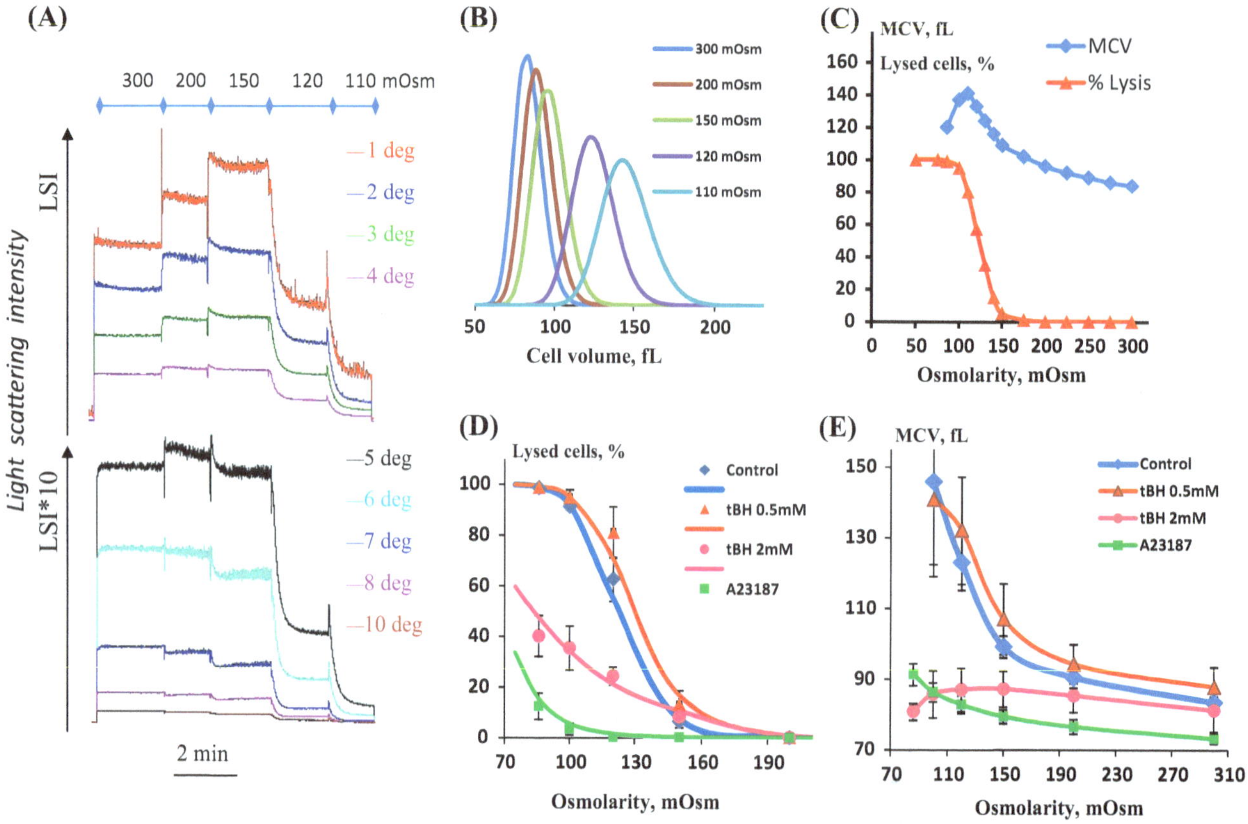

FIGURE 26.2 Application of low-angle light-scattering method to the registration of erythrocyte osmotic fragility. (A) Original recordings (representative for four independent experiments) of light-scattering intensity (LSI) measured at 1 and 10 degrees of erythrocytes incubated in iso- and hypoosmotic media. (B) Histogram of cell volume distribution in media of the osmolarity indicated. (C) Calculated relation of MCV and erythrocytes hemolysis to the osmolarity of the media. Relation of lysed cells (D) and MCV (E) to the media osmolarity of control erythrocytes and cells incubated for 40 min with either tBH (0.5 and 2 mM) or A23187 (1 uM).

The ammonia/ammonium influx into erythrocytes is mediated by surface glycoprotein RhAG that forms a structural complex with anion exchanger 1 (AE1, band 3). In contrast to other blood cells, erythrocytes exposed to isotonic ammonium buffer (where Na^+ is replaced by NH_4^+) swell and upon reaching critical volumes undergo hemolysis. In the NH_4^+ loading experiments, the cell volume increase was reflected by an increase in the light-scattering intensity in the angle range of 0–6 degrees (Fig. 26.3A) and a decrease in the light-scattering intensity in the angle range of 6–12 degrees. During hemolysis, the light-scattering intensity was declining at all the scattering angles. Inhibition of AE1 function by DIDS (300 μM) completely prevents hemolysis in ammonium media (Fig. 26.3B,D). In this case pH_i reaches its maximum values, but the volume increases only up to 30%. There is no secondary acidification and recovery of pH_i (Fig. 26.3C). Therefore inhibition of AE1 function plays an important role in NH_3/NH_4^+ accumulation by human erythrocytes. Our data support the functional interaction between the transporters and a crucial role of the anion transporter in the intracellular pH maintenance.

Next we applied this method to registration of blood platelet shape change and aggregation. During the aggregation process platelets first undergo a change from an oblate spheroidal shape to a spherical shape, and then to a complex shape characterized by cell protrusions (shape change) linking the cells to a microaggregate and in a later step to a macroscopic clot. The widely used Born aggregometry method, which is based on multiple scattering, registers changes in the transmission of the whole sample, thus providing relatively low sensitivity and limited information about net changes in cell shape or volume; this is overcome in our technique. The theoretical basis of single scattering is more elaborate than for multiple scattering and allows the combined analysis of cell size, volume, and distribution, together with shape change and aggregation. Also, the sensitivity of our method is considerably increased due to the fact that the scattered light is observed against a virtually dark background. It is widely recognized that particularly mild platelet function disorders demand a highly sensitive analysis not provided by conventional aggregometry methods (Watson et al., 2010). To establish and verify this novel method, we

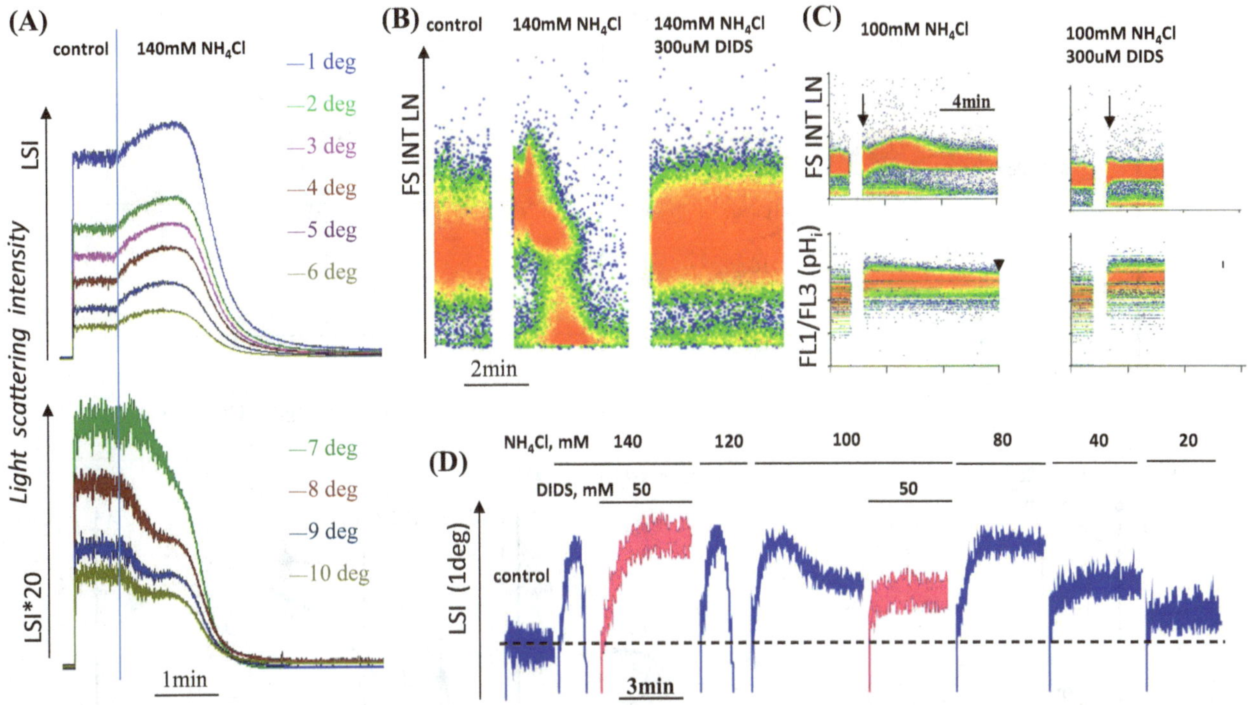

FIGURE 26.3 Low-angle light-scattering (A,D) and flow cytometry (B,C) analysis of the correlation between erythrocyte volume changes and pH$_i$ under conditions of ammonium loading and inhibition of AE1. (A) Original recordings (representative for five independent experiments) of light-scattering intensity (LSI) measured at 1 and 10 degrees of erythrocytes incubated in the ammonium media. (B) Dot plot of the erythrocyte volume distribution (forward scattering FS INT LIN) under ammonium loading (140 mM NH_4Cl) and in the presence of AE1 inhibitor DIDS (300 μM). (C) Dynamics of pH$_i$ and volume increase in 100 mM NH_4^+ containing media. The lack of secondary acidification in the presence of DIDS (original FACS experiment). Erythrocytes loaded with BCECF-AM (5 μM). (D) Control erythrocytes (10^6/mL) were incubated in isotonic buffer (pH 7.4, 37°C), and other aliquots were incubated in NH_4^+ containing media with or without DIDS (50 μM, 30 min).

monitored ADP-induced platelet shape change and aggregation. The maximum platelet density allowing for single light scattering was determined to be 10,000 platelets/μL, therefore platelet-rich plasma (PRP) used in the experiments was diluted with buffer accordingly. For verification of the underlying theory a model for the platelet aggregation processes was established. The model predictions were experimentally proven for the aggregation of human platelet suspensions. Angle-dependent scatter intensities were measured in a range from 1 to 12 degrees using human platelet suspensions stimulated with 500 nM ADP. The experimental data were in good accordance with the calculated scatter curves. As predicted by the model, the intensity of the scattered light increased between 1 and 5 degrees and decreased between 9 and 12 degrees (Fig. 26.4A) during ADP-stimulated platelet aggregation. In the range between 5 and 8 degrees the changes in scatter intensity are less pronounced and are thus not suited for aggregation monitoring. From the scatter curves it is evident that changes in scatter intensity are maximal at 1 degree, making this angle particularly suitable for aggregation measurement. Platelet shape change is reflected by an increase in scatter intensity at 8 to 12 degrees with a maximal increase at 12 degrees (Fig. 26.4B). Consequently, a scatter angle of 1 degree was chosen for the detection of platelet aggregation and 12 degrees for monitoring of platelet shape change. Maximal shape change was induced by 90 nM ADP, a concentration not sufficient to induce platelet aggregation (Fig. 26.4B). For quantitative characterization of platelet aggregation response different ADP concentrations were applied, showing a dose-dependent increase in aggregation velocity and amplitude with a maximum response at 1 μM of ADP (Fig. 26.4C). Verification of the observations with the light-scattering method was accomplished with measurement of the established platelet activation markers P-selectin expression and activated integrin αIIbβ3 in flow cytometry (Fig. 26.4D). These data show a good correlation of light scatter intensity changes with common activation markers (Fig. 26.4D). Platelet aggregation is characterized by the initial velocity (V_{ia}) of aggregation, which was determined from the first 60 s interval after stimulation and the maximal amplitude [ΔI(1)], calculated as differences between the maximal scatter intensity and the initial scatter intensity prior to stimulation. The aggregation velocity was used as an important parameter for pharmacological experiments. The maximal aggregation velocity (V_{max}) was determined for each experiment.

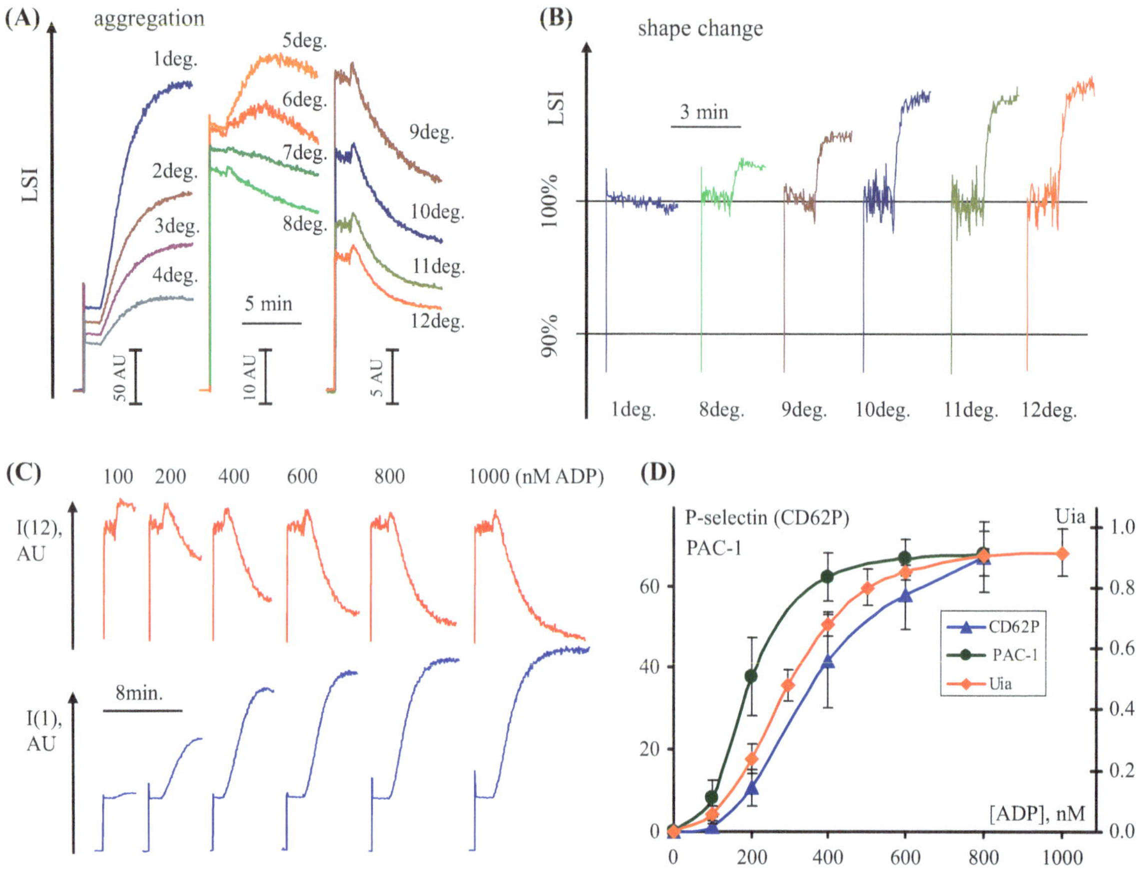

FIGURE 26.4 Application of low-angle light-scattering method for measuring platelet aggregation and shape change. (A) Light-scattering intensity (LSI) measured at 12 different scatter angles of ADP (500 nM) stimulated human platelet suspension. For better resolution the ordinate scale is enlarged five times for the 5–8 degree curves and 10 times for the 9–12 degree curves. According to the predicted changes in LSI during aggregation, the LSI increases in the range of 1–5 degrees and decreases in the range of 8–12 degrees. (B) Characterization of platelet shape change by LSI. LSI presented as percent increase from basal (100%) measured at seven different scatter angles with ADP (90 nM) stimulated human platelet suspension. The maximal increase in LSI is observed at 12 degrees. (C) Original recording of LSI measured at 1 and 12 degrees of ADP stimulated human platelets with the concentrations indicated. I(1) and I(12) correspond to LSI at 1 or 12 degrees, respectively. (D) Dependence of platelet activation [flow cytometry analysis of platelet P-selectin—CD62P—surface exposure and α2bβ3 integrin activation (PAC-1 binding)] and initial velocity (Uia) of aggregation from ADP concentrations. LSI is given in arbitrary units of the intensity of the scattered light. Flow cytometry data presented as mean fluorescence. For flow cytometry analysis platelets were taken directly from the LaSca cuvette 2 min after stimulation.

For a direct comparison of different experiments, the relative initial aggregation velocity—designated as U_{ia}—was calculated as $U_{ia} = V_{ia}/V_{max}$, and the normalized maximal aggregation velocity (U_{max}) was also determined. The normalized initial aggregation velocity U_{ia} according to the Hill equation shows the expected sigmoid dose dependence on ADP concentration (Fig. 26.4D). The parameters for this equation are obtained from a Schild plot. The value for the Hill coefficient (2.53 ± 0.29) reflects the observed cooperativity of $P2Y_1$ and $P2Y_{12}$ receptor activation in platelets (Gachet, 2006). ADP concentrations in routinely used clinical platelet aggregation tests are in the range of 5–10 μM, which gives a maximal aggregation response. The EC50 for ADP-induced platelet aggregation calculated from our data was 307 ± 17 nM, with a maximum response to 1 μM.

As nucleotide-induced platelet activation is mediated by three different purinergic receptors (Geiger et al., 1998), we carried out detailed determination of the individual contribution of these receptors with established specific inhibitors (Fig. 26.5). Inhibition of the P2Y1 receptor with MRS significantly increased the EC50 for ADP, but did not affect U_{max} (Fig. 26.5A). P2Y12 receptor inhibition with ARC significantly decreased U_{max}, leaving the EC50 unchanged (Fig. 26.5B). These data underline the hypothesized function of the P2Y1 receptor, being necessary for the initiation of platelet aggregation, and the P2Y12 receptor for completion of ADP-induced aggregation (Gachet, 2006). However, as was described for P2Y1 (Léon et al., 1999) and P2Y12 (Cattaneo et al., 2003) knockout mouse models, lack of either receptor does still allow for a small aggregation response. We

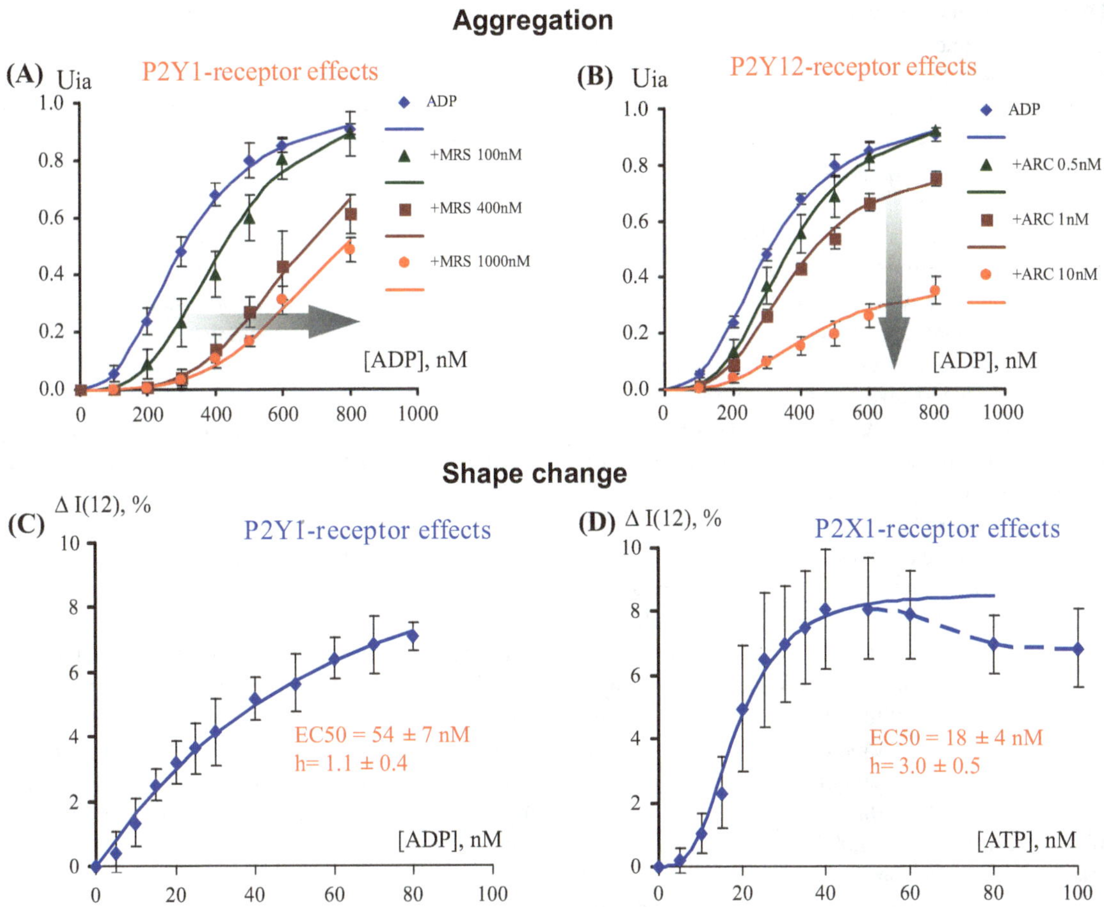

FIGURE 26.5 Quantitative characterization of P2-receptor-induced platelet aggregation and shape change. Diluted human platelet suspension was analyzed by light-scattering method for ADP-induced aggregation (A, B), ADP- (C), and ATP-induced shape change (D). Dependence of initial velocity (Uia) of aggregation from ADP concentrations in the absence (ADP) or presence of different concentrations of P2Y1 receptor antagonist (+MRS2179) (a) and P2Y12 receptor antagonist (+ARC69931) (b). Inhibition of P2Y1 receptor significantly increases EC_{50}, without changes of U_{max}, while inhibition of P2Y12 receptor significantly decreases U_{max}. (C, D) Characterization of platelet shape changes induced by different ADP (C) and ATP (D) concentrations calculated as percent increase of LSI at 12 degrees from the basal signal level (**Δ**I(12), %). The P2Y1 receptor can be described by the Langmuir equation ($h = 1.1 \pm 0.3$), while P2X1 receptor can be described as the sigmoid shape of curves in normal axes and Hill coefficient higher than unity ($h = 2.6 \pm 0.5$).

obtained similar results in light scatter experiments (Mindukshev et al., 2012). For platelet shape change, EC50 for ADP was 47 ± 7 nM with a maximal response at 90 nM. This ADP response could be completely blocked by a specific P2Y1 receptor inhibitor, without any effects of P2Y12 and P2X1 receptor inhibitors (Fig. 26.5C). ADP stimulation of shape change can be described by the Langmuir equation (Hill coefficient 1.1 ± 0.3). The P2X1 receptor is a well characterized nonselective ligand gated cation channel activated by extracellular ATP. It was shown that the P2X1 receptor in platelets is activated at high concentrations (10 μM and more) of nonhydrolyzed ATP analogs (Rolf and Mahaut-Smith, 2002), while the EC50 for this receptor expressed in oocytes is in the nanomolar range (Rettinger and Schmalzing, 2003). We can directly demonstrate the P2X1 receptor-mediated activation of platelet shape change by ATP (Fig. 26.5D). Our technique allows also to quantitatively characterize P2Y1- and P2X1-mediated platelet shape changes. This method could be used for sensitive detection of receptor defects or drug action. The small sample volumes make this method ideally suited for investigation of mouse platelets, and for transgenic mouse models in particular (Mindukshev et al., 2012).

Studies of nutraceuticals with the help of light scattering and flow cytometry

Flow cytometry and/or light scattering has been applied to the study of: the effects of omega-3 fatty acids supplementation on platelet and endothelial activation (Mackay et al., 2012); effects of conjugated linoleic acid or combination omega-3 on metabolic characteristics in muscle cells (Vaughan et al., 2012); interrelationships among the hypothalamus–pituitary–adrenal stress axis and T-cell maturation in both zinc deficiency and responses during zinc repletion (Blewett and Taylor, 2012); protective effects of gallic acid on endothelial cell death and the mechanisms involved (Kam et al., 2014); ability of quercetin and apigenin to modulate platelet activation and aggregation (Del Turco et al., 2015); the potential of epigallocatechin gallate to suppress cell proliferation of SSC-4 human oral squamous cell carcinoma (Irimie et al., 2015) and human nonsmall-cell lung cancer A549 cells (Sonoda et al., 2014); curcumin-induced apoptotic cell death of NCCIT human embryonic carcinoma cells (Yun et al., 2015); toxic effects of garlic (*Allium sativum*) upon pathogenic and beneficial bacteria (Booyens and Thantsha, 2014); effects of sodium 2-propenylthiosulfate, allicin, and diallyl trisulfide (organosulfane sulfur compound from garlic) in cancer cell cytotoxicity (Sabelli et al., 2008; Chandra-Kuntal et al., 2013; Tao et al., 2013); and many other investigations.

A part of our work not long ago was aimed at the development of a methodology for cytophysiological screening of antidote substances and/or those enhancing the therapeutic effect of antineoplastic and other drugs. Flow cytometry, together with low-angle light-scattering technique and some other methods of cell viability assessment were valuable instruments for this research (Mindukshev et al., 2005a, 2006, 2007; Zinchenko et al., 2006, 2007; Radilov et al., 2009; Kudryavtsev et al., 2014; Prokofieva and Goncharov, 2014; Goncharov et al., 2009, 2015a). The development of cancer is dependent on different cells (cancer and stem cancer cells, endothelial cells, blood platelets, red blood cells, neutrophiles, and stromal cells) contributing to tumor growth and forming a vicious cycle (Szabo et al., 2010; Qu et al., 2013). To break this cycle, an effective therapeutic complex should be multitargeted, that is, be simultaneously directed at different molecular and cellular targets. Nutraceuticals seem to be an essential part of the complex, being applied as adjuvant and/or neoadjuvant therapy. The essence of an anticancer screening algorithm involves treating tumor and nontumor cells (endothelial, platelets, red blood cells, etc.) in vitro with various concentrations of camptothecin (positive control) or other cytotoxic drugs with known molecular targets (rothenon, iodoacetate, etc.), then with neutraceuticals, such as garlic oil, diallyl disulfide (DADS), green tea extract, and epigallocatechin gallate, followed by testing of their combinations to reveal if there was an amplification, attenuation, or no combined effect at all. This is a rather simple and effective approach, giving an opportunity to reveal or to suggest a "personal" response of the cells participating in the tumor growth. Further analysis of the possible impact on the cells relating to effective concentrations of the substances tested, renders it possible to suggest an effective combination to test it in vivo. Such an algorithm could be developed with nutraceuticals, after the first positive results were obtained with fluoroacetate applied in combination with cyclophosphamide that proved to be no less and even slightly more effective than the combination of metformin with cyclophosphamide (Anikin et al., 2013, 2014). Figs. 26.6–26.9 show some primary results obtained with flow cytometry, and those obtained with low-angle light-scattering technique within the frame of complex anticancer screening experiments are shown in Fig. 26.10.

Concluding remarks and future directions

The search for drugs with low toxicity, together with the development of technical and methodological approaches for their testing in vitro is an urgent task for experimental biology and medicine. Discriminating between the toxic, nontoxic, and protective effects, and understanding of how cell death or cell viability is regulated upon exposure

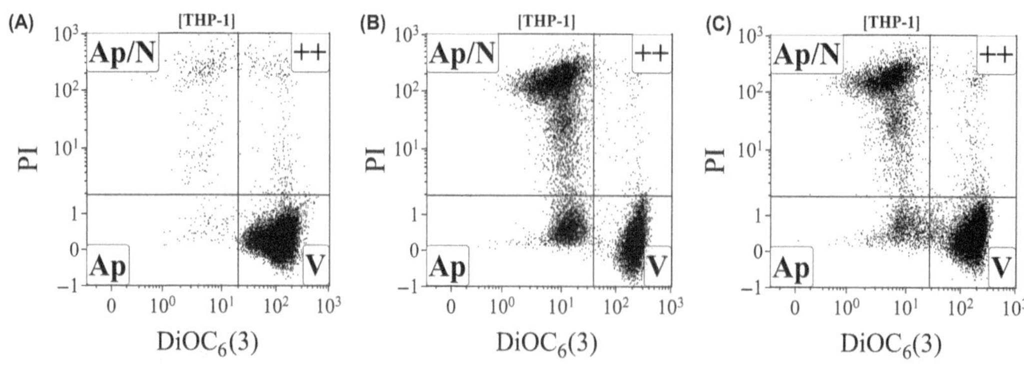

FIGURE 26.6 Dissipation of mitochondrial transmembrane potential and plasma membrane permeability assessment. Analysis by concurrent staining with lipophilic cationic carboxycyanine dye $DiOC_6(3)$ and propidium iodide (PI). THP-1 cells (human monocytic leukemia cell line) were either unexposed (A) or exposed to camptothecin, 5 μM (B) or to DADS, 100 μM (C). Viable cells (*V*) accumulate $DiOC_6(3)$ in mitochondria and are impermeable to PI. THP-1 cells during early apoptosis (*Ap*) have an impaired mitochondrial transmembrane potential and the impermeable cell membrane [decreased intensity of $DIOC_6(3)$ and PI negative]. Ap/N cells exhibit loss of mitochondrial transmembrane potential [low $DiOC_6(3)$ fluorescence] and become permeable to PI.

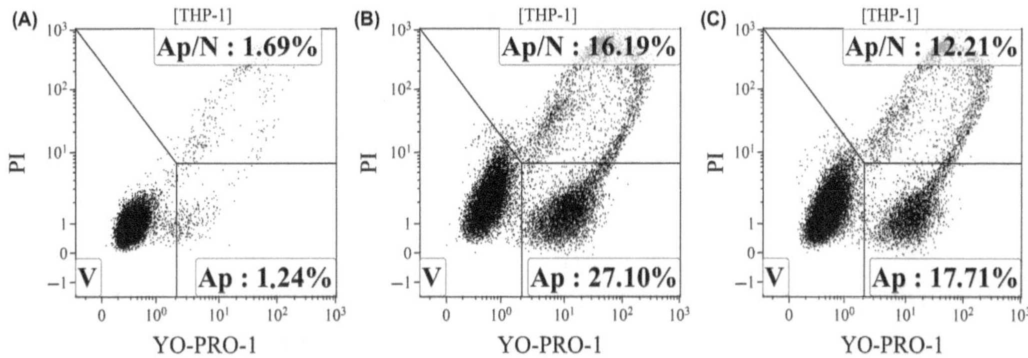

FIGURE 26.7 Analysis of plasma membrane permeability with cyanine dyes. THP-1 cells either unexposed (A) or exposed to DADS, 100 μM (B), or DADS, 33 μM (C). Live cells (*V*) remain impermeable to both dyes (YO-PRO-1neg/PIneg). Early apoptotic cells (Ap) are characterized by cell membrane permeability for YO-PRO-1 due to activation of P2X7 channels (YO-PRO-1pos/PIneg). THP-1 cells in late stages of apoptosis and necrotic cells (Ap/N) lose cell membrane integrity and become permeable to both YO-PRO-1 and PI.

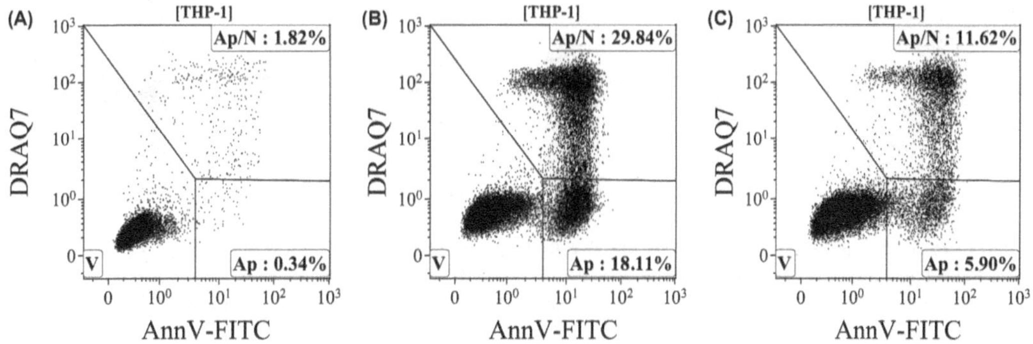

FIGURE 26.8 Detection of early and late stages of apoptosis by concurrent staining with annexinV-FITC and DRAQ7. Unexposed (A), exposed to camptothecin (B), or DADS (33 μM) (C) THP-1 cells were treated with annexin V-FITC conjugate and DRAQ7 and their green and far red fluorescences were measured by flow cytometry. Viable (*V*) cells make up a double-negative population in the lower left corner. At an early stage of apoptosis (*Ap*) annexin V-FITC interacts with phosphatidylserine molecules at the outer lipid bilayer of cell membranes, but the membrane does not lose its integrity, so the cells are DRAQ7 negative. At a late stage of apoptosis (*Ap/N*), the cells bind annexin V-FITC and are stained with DRAQ7 that binds readily to nuclear DNA and reports about cell death by strong far-red fluorescence.

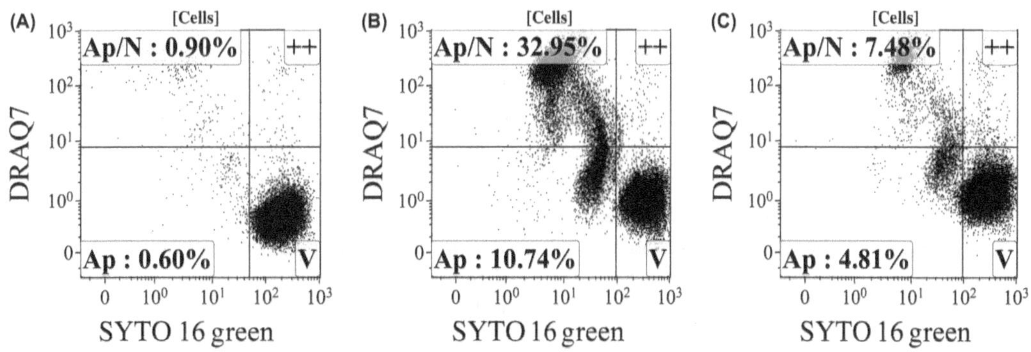

FIGURE 26.9 Detection of apoptosis by SYTO 16 dye and DNA-binding dye DRAQ7. The cells were treated with both dyes for 15 min in final concentrations 250 nM and 3 μM, respectively, then washed with PBS supplemented with 2% of fetal calf serum (FCS), centrifuged for 7 min (330 g), resuspended in fresh PBS and analyzed on flow cytometer. THP-1 cells were either unexposed (a) or exposed to camptothecin, 2.5 μM (b) or to DADS, 33 μM (c) for 24 h. Live cells (*V*) are SYTO 16^{bright} and DRAQ7 negative. Early apoptotic cells (*Ap*) are SYTO 16^{dim} but still DRAQ7 negative. Late apoptotic/necrotic cells (*Ap/N*) are SYTO 16^{low} and $DRAQ7^{bright}$.

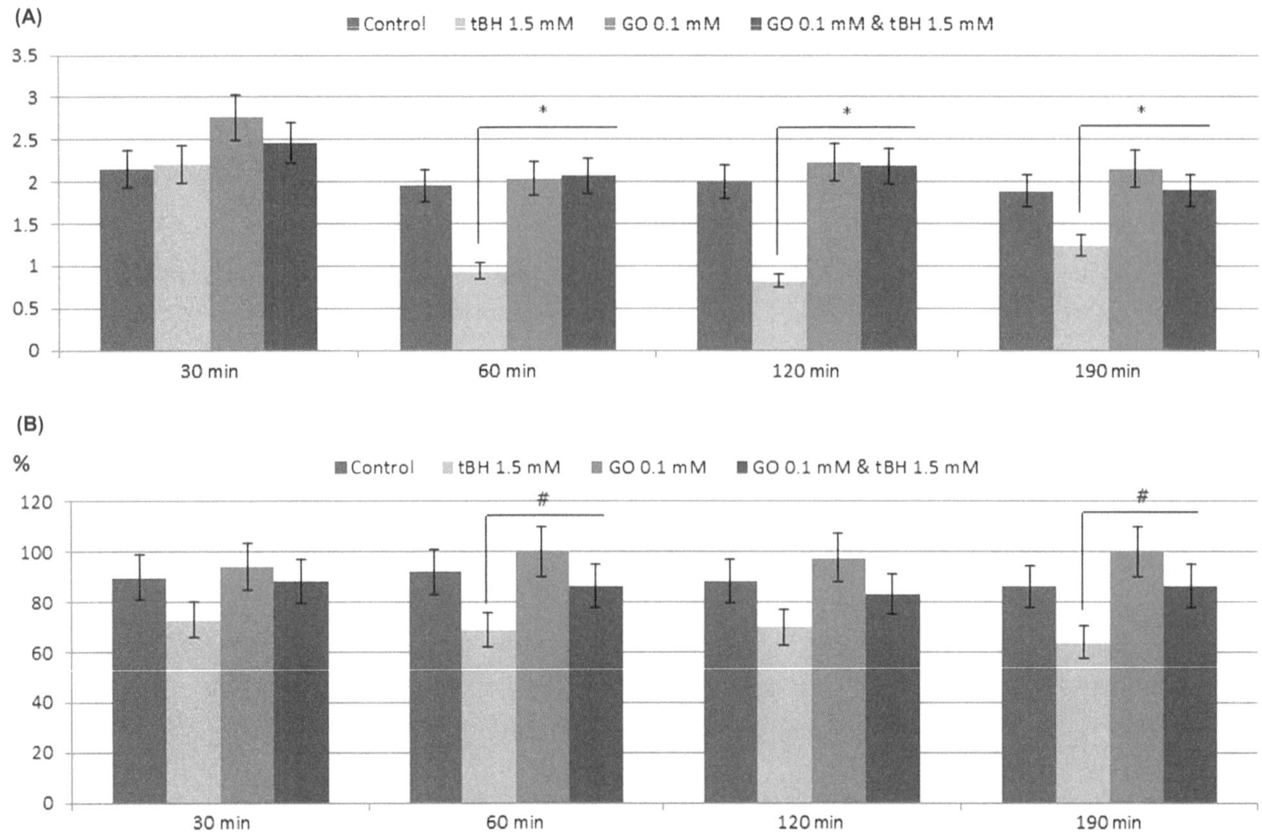

FIGURE 26.10 Parameters of hemolysis of human red blood cells measured by low-angle light-scattering technique at an angle of 1.5 degrees. (A) Maximal rate (velocity) of hemolysis. (B) Quantity of hemolyzed cells (%). Red blood cells taken from venous blood of a healthy individual were washed with saline solution #1 (see below) and treated with tert-butyl-hydroperoxide (tBH), garlic oil (GO), or GO followed by tBH (with 1.5 min interval) for 30, 60, 120, 180, or 1500 minutes. Two saline solutions were used in the experiments (mM): (1) NaCl 140, KCl 2, HEPES 10, $MgCl_2$ 1, glucose 5, EGTA 2; and (2) NH_4Cl 140, KCl 2, HEPES 10, glucose 5. RBCs were washed and incubated in saline 1, and at each time point hemolysis was initiated with saline 2 (140 mM NH_4Cl). Decrease of maximum velocity of hemolysis (A) and decreased percent of hemolysis (B) after exposure to tBH indicates enhanced rigidity of erythrocytes upon development of eryptosis. A total of 0.1 mM GO shows a protective effect, if added to the cells 1.5 min before 1.5 mM tBH. * *P* 0.05; # *P* 0.3.

to cytotoxic agents and/or nutraceuticals may provide an insight into how these compounds can be combined and manipulated in the treatment of disease. Most exogenous compounds—including toxic agents, pharma compounds, and more so nutraceuticals (having a low specificity)—have multiple cellular and molecular targets. For example, some thiols and disulfides are used as nutraceuticals in complex cardiovascular or cancer therapy (Cerella et al., 2011; Ried and Fakler, 2014), and at certain concentrations they can be toxic to other cells (Munday, 2012; Prokofieva and Goncharov, 2014; Alzoubi et al., 2015; Goncharov et al., 2019), with the immune system and/or ROS being the most likely mediators of their toxic effects (Rao et al., 2020). On the other hand, there are different cellular and molecular sources of ROS with respect to their participation in the pathogenesis of diseases (Goncharov et al., 2015b), and to reveal the impact of each of them at different stages of pathogenesis is a major challenge. Therefore the methodologies of experimental and preclinical studies should be revised and expanded to bring into play more in vitro models and fewer in vivo or invasive experiments; even more so, ethical and financial considerations favor these scientific endeavors. Intuitively, there should be a strong correlation between cellular responses in vitro and the final efficacy of the therapy, though this correlation has yet to be proved. Thus, the development of tools for interventions aimed at improving human health require a comprehensive knowledge of the cellular reactions to positive or negative stimuli, correlating these in vitro and/or ex vivo obtained data with final in vivo therapeutic or toxic effects. In spite of many studies on separate cell lines with flow cytometry or light scattering, there are very few examples of successful development of such a complex methodology. Dr. F. Lang and collaborators have made much progress in establishing mechanisms of eryptosis evoked by different agents including nutraceuticals, through flow cytometry and light-scattering studies (based just on forward scatter) (Alzoubi et al., 2015; Peter et al., 2015). Importantly, they also applied some other biochemical techniques to reveal the molecular mechanisms (Lang et al., 2015; Lang and Lang, 2015).

Low-angle light-scattering analysis of erythrocytes, platelets, and other cells allows very sensitive and simultaneous quantitative online registration of changes in cell volume, shape change, and cell–cell interactions. This ensures detection of subtle changes in cell volume and shape changes when compared to other techniques such as Born aggregometry. On the other hand, flow cytometry has greater potential to enhance sensitivity for measuring small particles of submicrometer size, primarily due to fluorescence triggering. For example, about 50 × more Annexin-5 (Anx5)-positive extracellular vesicles can be detected by flow cytometry, with the limit of detection of the fluorescence triggering method estimated at about 1000–2500 Anx5 molecules (Arraud et al., 2015). Nevertheless, label-free and rapid classification of cells can have a wide range of applications in experimental biology. A new robust method of polarization diffraction imaging flow cytometry (p-DIFC) has been developed for achieving this goal (Feng et al., 2014). This method has the capacity to discriminate cells of high similarity in their morphology with "fingerprint" features extracted from the diffraction images, which may be attributed to subtle but statistically significant differences in the nucleus-to-cell volume ratio in some cells.

It was noted recently that future directions of in vitro instrumentation for cell screening research should include the addition of new parameters and the development of point-of-care instrumentation (Green and Wachsmann-Hogiu, 2015). The low-angle light-scattering technique is a simple though unique supplement to the methodology of investigating natural supplements, that is, nutraceuticals. We believe that these supplements, taken together, would greatly add to human health and knowledge about it.

Acknowledgment

This work was supported by the State Program AAAA-A18-118012290142-9.

References

Abdelwahid, E., Rolland, S., Teng, X., et al., 2011. Mitochondrial involvement in cell death of non-mammalian eukaryotes. Biochim. Biophys. Acta. 1813, 597–607.

Alzoubi, K., Calabrò, S., Faggio, C., et al., 2015. Stimulation of Suicidal Erythrocyte Death by Sulforaphane. Basic. Clin. Pharmacol. Toxicol. 116 (3), 229–235.

Anikin, I.V., Goncharov, N.V., Tyndyk, M.L., et al., 2013. Effect of sodium fluoroacetate on growth of Ehrlich solid tumor and autochthonous mouse sarcoma. Vopr. Oncol. 59 (6), 777–780 [Article in Russian].

Anikin, I.V., Goncharov, N.V., Tyndyk, M.L., et al., 2014. Influence of sodium fluoroacetate and metformin on antitumor effect of cyclophosphamide in autochthonous mouse sarcoma model. Vopr. Oncol. 60 (4), 514–516 [Article in Russian].

Arraud, N., Gounou, C., Linares, R., et al., 2015. A simple flow cytometry method improves the detection of phosphatidylserine-exposing extracellular vesicles. J. Thromb. Haemost. 13 (2), 237–247.

Attilio, P., Merritt, C., Sims, J., et al., 2010. The effect of ellagic acid on platelet activation as measured by the quantification of P-selectin using flow cytometry. AANAJ 78 (6), 453–459.

Blewett, H.J., Taylor, C.G., 2012. Dietary zinc deficiency in rodents: effects on T-cell development, maturation and phenotypes. Nutrients. 4 (6), 449–466.

Booyens, J., Thantsha, M.S., 2014. Fourier transform infra-red spectroscopy and flow cytometric assessment of the antibacterial mechanism of action of aqueous extract of garlic (Allium sativum) against selected probiotic Bifidobacterium strains. BMC Complement. Altern. Med. 14, 289.

Bortner, C.D., Sifre, M.I., Cidlowski, J.A., 2007. New approaches for determining apoptotic volume decrease in cells. Methods Enzymol. 428, 161–181.

Boudreault, F., Grygorczyk, R., 2004. Evaluation of rapid volume changes of substrate-adherent cells by conventional microscopy 3D imaging. J. Microsc. 215 (3), 302–312.

Buenz, E.J., 2008. Aloin induces apoptosis in Jurkat cells. Toxicol. Vitro 22 (2), 422–429.

Cattaneo, M., Zighetti, M.L., Lombardi, R., et al., 2003. Molecular bases of defective signal transduction in the platelet P2Y12 receptor of a patient with congenital bleeding. Proc. Natl. Acad. Sci. USA. 100 (4), 1978–1983.

Cerella, C., Dicato, M., Jacob, C., et al., 2011. Chemical properties and mechanisms determining the anti-cancer action of garlic-derived organic sulfur compounds. Anticancer. Agents Med. Chem. 11 (3), 267–271.

Chandra-Kuntal, K., Lee, J., Singh, S.V., 2013. Critical role for reactive oxygen species in apoptosis induction and cell migration inhibition by diallyl trisulfide, a cancer chemopreventive component of garlic. Breast Cancer Res. Treat. 138 (1), 69–79.

Conrad, M., Kagan, V.E., Bayir, H., et al., 2018. Regulation of lipid peroxidation and ferroptosis in diverse species. Genes. Dev. 32 (9-10), 602–619.

Cottet-Rousselle, C., Ronot, X., Leverve, X., et al., 2011. Cytometric assessment of mitochondria using fluorescent probes. Cytometry A 79A, 405–425.

Del Turco, S., Sartini, S., Cigni, G., et al., 2015. Synthetic analogues of flavonoids with improved activity against platelet activation and aggregation as novel prototypes of food supplements. Food Chem. 175, 494–499.

Dittami, G.M., Sethi, M., Rabbitt, R.D., et al., 2012. Determination of mammalian cell counts, cell size and cell health using the Moxi Z mini automated cell counter. J. Vis. Exp. 64. Available from: https://doi.org/10.3791/3842.

Dixon, S.J., Lemberg, K.M., Lamprecht, M.R., et al., 2012. Ferroptosis: an iron-dependent form of nonapoptotic cell death. Cell. 149 (5), 1060–1072.

Dupont, N., Orhon, I., Bauvy, C., et al., 2014. Autophagy and autophagic flux in tumor cells. Methods Enzymol. 543, 73–88.

Faggio, C., Torre, A., Pelle, E., et al., 2011. Cell volume regulation following hypotonic shock in hepatocytes isolated from Sparus aurata. Comp. Biochem. Physiol. A Mol. Integr. Physiol. 158 (1), 143–149.

Fan, T, Han, L., Cong, R., et al., 2005. Caspase family proteases and apoptosis. Acta Biochim. Biophys. Sin. 37, 719–727.

Feng, Y., Zhang, N., Jacobs, K.M., et al., 2014. Polarization imaging and classification of Jurkat T and Ramos B cells using a flow cytometer. Cytometry A 85 (9), 817–826.

Frey, T., 1995. Nucleic acid dyes for detection of apoptosis in live cells. Cytometry. 21 (3), 265–274.

Gachet, C., 2006. Regulation of platelet functions by P2 receptors. Annu. Rev. Pharmacol. Toxicol. 46, 277–300.

Geiger, J., Hönig-Liedl, P., Schanzenbächer, P., et al., 1998. Ligand specificity and ticlopidine effects distinguish three human platelet ADP receptors. Eur. J. Pharmacol. 351 (2), 235–246.

Geisler, K., Reischer, A., Kroeger, I., et al., 2014. Nilotinib combined with interleukin-2 mediates antitumor and immunological effects in a B16 melanoma model. Oncol. Rep. 31 (5), 2015–2020.

Glisic-Milosavljevic, S., Waukau, J., Jana, S., et al., 2005. Comparison of apoptosis and mortality measurements in peripheral blood mononuclear cells (PBMCs) using multiple methods. Cell Prolif. 38 (5), 301–311.

Goncharov, N., Glashkina, L., Savelieva, E., et al., 2009. Fluoroacetate. In: Gupta, R.C. (Ed.), Handbook of the Toxicology of Chemical Warfare Agents. Elsevier Inc, Oxford, pp. 177–198.

Goncharov, N., Savelieva, E., Zinchenko, V., et al., 2015a. Fluoroacetate. In: Gupta, R.C. (Ed.), Handbook of Toxicology of Chemical Warfare Agents, 2nd edition Academic Press, pp. 193–214.

Goncharov, N.V., Avdonin, P.V., Nadeev, A.D., et al., 2015b. Reactive oxygen species in pathogenesis of atherosclerosis. Curr. Pharm. Des. 21 (9), 1134–1146.

Goncharov, N., Sobolev, V., Terpilowski, M., Korf, E., Jenkins, R., 2019. Safety and Toxicity Evaluation of Nutraceuticals in Animal Models. In: Gupta, R., Srivastava, A., Lall, R. (Eds.), Nutraceuticals in Veterinary Medicine. Springer, Cham, pp. 675–684.

Green, D.R., Llambi, F., 2015. Cell Death Signaling. Cold Spring Harb. Perspect. Biol. 7 (12), . Available from: https://doi.org/10.1101/cshperspect.a006080pii: a006080.

Green, R., Wachsmann-Hogiu, S., 2015. Development, history, and future of automated cell counters. Clin. Lab. Med. 35 (1), 1–10.

Idziorek, T., Estaquier, J., DeBels, F., et al., 1995. YOPRO-1 permits cytofluorometric analysis of programmed cell death (apoptosis) without interfering with cell viability. J. Immunol. Methods. 185 (2), 249–258.

Irimie, A.I., Braicu, C., Zanoaga, O., et al., 2015. Epigallocatechin-3-gallate suppresses cell proliferation and promotes apoptosis and autophagy in oral cancer SSC-4 cells. Onco Targets Ther. 8, 461–470.

Kalienkova, V., Clerico Mosina, V., Bryner, L., et al., 2019. Stepwise activation mechanism of the scramblase nhTMEM16 revealed by cryo-EM. Elife. 8, . Available from: https://doi.org/10.7554/eLife.44364pii: e44364.

Kam, A., Li, K.M., Razmovski-Naumovski, V., et al., 2014. Gallic acid protects against endothelial injury by restoring the depletion of DNA methyltransferase 1 and inhibiting proteasome activities. Int. J. Cardiol. 171 (2), 231–242.

Kitazumi, I., Tsukahara, M., 2011. Regulation of DNA fragmentation: the role of caspases and phosphorylation. FEBS J. 278, 427–441.

Kudryavtsev, I.V., Garnyuk, V.V., Nadeev, A.D., et al., 2014. Hydrogen peroxide modulates expression of surface antigens by human umbilical vein endothelial cells in vitro. Biochem. (Mosc.) Supp. S. A: Membr. Cell Biol. 8 (1), 97–102.

Lang, E., Bissinger, R., Gulbins, E., et al., 2015. Ceramide in the regulation of eryptosis, the suicidal erythrocyte death. Apoptosis. 20 (5), 758–767.

Lang, E., Lang, F., 2015. Mechanisms and pathophysiological significance of eryptosis, the suicidal erythrocyte death. Semin. Cell Dev. Biol. 39, 35–42.

LeBlanc, C.J., Dietrich, M.A., Horohov, D.W., et al., 2007. Effects of dietary fish oil and vitamin E supplementation on canine lymphocyte proliferation evaluated using a flow cytometric technique. Vet. Immunol. Immunopathol. 119 (3-4), 180–188.

Léon, C., Hechler, B., Freund, M., et al., 1999. Defective platelet aggregation and increased resistance to thrombosis in purinergic P2Y(1) receptor-null mice. J. Clin. Invest. 104 (12), 1731–1737.

Mackay, I., Ford, I., Thies, F., et al., 2012. Effect of Omega-3 fatty acid supplementation on markers of platelet and endothelial function in patients with peripheral arterial disease. Atherosclerosis. 221 (2), 514–520.

Martins, J.D., Liberal, J., Silva, A., et al., 2015. Autophagy and inflammasome interplay. DNA Cell Biol. 34 (4), 274–281.

McManus, M., Fischbarg, J., Sun, A., et al., 1993. Laser light-scattering system for studying cell volume regulation and membrane transport processes. Am. J. Physiol. 265 (2), C562–C570.

Mindukshev, I.V., Jahatspanian, I.E., Goncharov, N.V., et al., 2005a. A new method for studying platelets, based upon the low-angle light scattering technique. 1. Theoretical and experimental foundations of the method. Spectrosc. Int. J. 19 (5-6), 235–246.

Mindukshev, I.V., Ermolaeva, E.E., Vivulanets, E.V., et al., 2005b. A new method for studying platelets, based upon the low-angle light scattering technique. 2. Application of the method in experimental toxicology and clinical pathology. Spectrosc. Int. J. 19 (5-6), 247–257.

Mindukshev, I.V., Goncharov, N.V., Shabanova, E.Yu, et al., 2006. A new method for studying platelets, based upon the low-angle light scattering technique. 3. Aggregation hypersensitivity of platelets (ADP agonist) and search for corrective agents. Spectrosc. Int. J. 20 (2), 57–66.

Mindukshev, I.V., Krivoshlyk, V.V., Ermolaeva, E.E., et al., 2007. Necrotic and apoptotic volume changes of red blood cells investigated by low-angle light scattering technique. Spectrosc. Int. J. 21 (2), 105–120.

Mindukshev, I.V., Senchenkova, E.Y., Goncharov, N.V., et al., 2010. New methods for studying platelets and red blood cells, based upon the low-angle light scattering technique. In: Columbus, F., Nova, N.Y. (Eds.), Hemorheology: Blood Flow, Disturbance and Prognosis. Science Publishers, pp. 87–124.

Mindukshev, I., Gambaryan, S., Kehrer, L., et al., 2012. Low angle light scattering analysis: a novel quantitative method for functional characterization of human and murine platelet receptors. Clin. Chem. Lab. Med. 50, 1253–1262.

Munday, R., 2012. Harmful and beneficial effects of organic monosulfides, disulfides, and polysulfides in animals and humans. Chem. Res. Toxicol. 25 (1), 47–60.

Nielsen, H.G., Skjønsberg, O.H., Lyberg, T., 2008. Effect of antioxidant supplementation on leucocyte expression of reactive oxygen species in athletes. Scand. J. Clin. Lab. Invest. 68 (7), 526–533.

Nikoletopoulou, V., Markaki, M., Palikaras, K., et al., 2013. Crosstalk between apoptosis, necrosis and autophagy. Biochim. Biophys. Acta. 1833 (12), 3448–3459.

Orrenius, S., Gogvadze, V., Zhivotovsky, B., 2007. Mitochondrial oxidative stress: implications for cell death. Ann. Rev. Pharmacol. Toxicol. 47, 143–183.

Peter, T., Bissinger, R., Enkel, S., et al., 2015. Programmed erythrocyte death following in vitro treosulfan treatment. Cell Physiol. Biochem. 35 (4), 1372–1380.

Prado, R., García, R., Rioboo, C., et al., 2015. Suitability of cytotoxicity endpoints and test microalgal species to disclose the toxic effect of common aquatic pollutants. Ecotoxicol. Environ. Saf. 114, 117–125.

Prokofieva, D.S., Goncharov, N.G., 2014. Effects of biogenic and abiogenic disulphides upon endothelial cells in culture: comparison of three methods of viability assessment. Tsitologiya 56 (6), 410–418.

Qu, Y., Zhang, L., Rong, Z., et al., 2013. Number of glioma polyploid giant cancer cells (PGCCs) associated with vasculogenic mimicry formation and tumor grade in human glioma. J. Exp. Clin. Cancer Res. 32 (1), 75.

Radilov, A., Rembovskiy, V., Rybalchenko, I., et al., 2009. Russian VX. In: Gupta, R.C. (Ed.), Handbook of the Toxicology of Chemical Warfare Agents. Elsevier Inc, Oxford, pp. 69–91. , 2009.

Radogna, F., Dicato, M., Diederich, M., 2015. Cancer-type-specific crosstalk between autophagy, necroptosis and apoptosis as a pharmacological target. Biochem. Pharmacol. 94 (1), 1–11.

Rao, T., Liu, Y.T., Zeng, X.C., Li, C.P., Ou-Yang, D.S., 2020. The hepatotoxicity of *Polygonum multiflorum*: The emerging role of the immune-mediated liver injury. Acta Pharmacol. Sin. Available from: https://doi.org/10.1038/s41401-020-0360-3 [Epub ahead of print].

Reardon, A.J., Elliott, J.A., McGann, L.E., 2014. Fluorescence as an alternative to light-scatter gating strategies to identify frozen-thawed cells with flow cytometry. Cryobiology. 69 (1), 91–99.

Rettinger, J., Schmalzing, G., 2003. Activation and desensitization of the recombinant P2X1 receptor at nanomolar ATP concentrations. J. Gen. Physiol. 121 (5), 451–461.

Ried, K., Fakler, P., 2014. Potential of garlic (*Allium sativum*) in lowering high blood pressure: mechanisms of action and clinical relevance. Integr. Blood Press. Control. 7, 71–82.

Rolf, M.G., Mahaut-Smith, M.P., 2002. Effects of enhanced P2X1 receptor Ca2 + influx on functional responses in human platelets. Thromb. Haemost. 88 (3), 495–502.

Sabelli, R., Iorio, E., De Martino, A., et al., 2008. Rhodanese-thioredoxin system and allyl sulfur compounds. FEBS J. 275 (15), 3884–3899.

Sonoda, J.I., Ikeda, R., Baba, Y., et al., 2014. Green tea catechin, epigallocatechin-3-gallate, attenuates the cell viability of human non-small-cell lung cancer A549 cells via reducing Bcl-xL expression. Exp. Ther. Med. 8 (1), 59–63.

Stokes, L., Jiang, L.H., Alcaraz, L., et al., 2006. Characterization of a selective and potent antagonist of human P2X(7) receptors, AZ11645373. Br. J. Pharmacol. 149 (7), 880–887.

Szabo, E., Rampalli, S., Risueño, R.M., et al., 2010. Direct conversion of human fibroblasts to multilineage blood progenitors. Nature. 468 (7323), 521–526.

Tao, M., Gao, L., Pan, J., et al., 2013. Study on the inhibitory effect of allicin on human gastric cancer cell line SGC-7901 and its mechanism. Afr. J. Tradit. Complement. Altern. Med. 11 (1), 176–179.

Toyokuni, S., 2014. Iron and thiols as two major players in carcinogenesis: friends or foes? Front. Pharmacol. 5, 200.

Turcu, I., Pop, C.V., Neamtu, S., 2006. High-resolution angle-resolved measurements of light scattered at small angles by red blood cells in suspension. Appl. Opt. 45 (9), 1964–1971.

Van de Hulst, H.C., 1981. Light Scattering by Small Particles. Dover Publications, New York, p. 478.

Vanaja, S.K., Rathinam, V.A., Fitzgerald, K.A., 2015. Mechanisms of inflammasome activation: recent advances and novel insights. Trends Cell Biol. 25 (5), 308–315.

Vaughan, R.A., Garcia-Smith, R., Bisoffi, M., et al., 2012. Conjugated linoleic acid or omega 3 fatty acids increase mitochondrial biosynthesis and metabolism in skeletal muscle cells. Lipids Health Dis. 11, 142.

Verhoven, B., Schlegel, R.A., Williamson, P., 1995. Mechanisms of phosphatidylserine exposure, a phagocyte recognition signal, on apoptotic T lymphocytes. J. Exp. Med. 182, 1597–1601.

Watson, S., Daly, M., Dawood, B., et al., 2010. Phenotypic approaches to gene mapping in platelet function disorders -identification of new variant of P2Y12, TxA2 and GPVI receptors. Hamostaseologie. 30 (1), 29–38.

Wlodkowic, D., Skommer, J., Pelkonen, J., 2007. Towards an understanding of apoptosis detection by SYTO dyes. Cytometry A. 71 (2), 61–72.

Wlodkowic, D., Skommer, J., Darzynkiewicz, Z., 2008. SYTO probes in the cytometry of tumor cell death. Cytometry A. 73 (6), 496–507.

Wlodkowic, D., Telford, W., Skommer, J., et al., 2011. Apoptosis and beyond: cytometry in studies of programmed cell death. Methods Cell Biol. 103, 55–98.

Wlodkowic, D., Skommer, J., Darzynkiewicz, Z., 2012. Cytometry of apoptosis. Historical perspective and new advances. Exp. Oncol. 34 (3), 255–262.

Yang, W.S., Stockwell, B.R., 2008. Synthetic lethal screening identifies compounds activating iron-dependent, nonapoptotic cell death in oncogenic-RAS-harboring cancer cells. Chem. Biol. 15 (3), 234–245.

Yang, X.R., Wang, Y.Y., La, K.K., et al., 2012. Inhibitory effects of cocoa tea (Camellia ptilophylla) in human hepatocellular carcinoma HepG2 in vitro and in vivo through apoptosis. Nutr. Biochem. 23 (9), 1051–1057.

Yun, J.H., Park, Y.G., Lee, K., et al., 2015. Curcumin induces apoptotic cell death via Oct4 inhibition and GSK-3β activation in NCCIT cells. Mol. Nutr. Food Res. 59 (6), 1053–1062.

Zhao, H., Albino, A.P., Jorgensen, E., et al., 2009. DNA damage response induced by tobacco smoke in normal human bronchial epithelial and A549 pulmonary adenocarcinoma cells assessed by laser scanning cytometry. Cytometry A. 75 (10), 840–847.

Zinchenko, V.P., Lee, V.V., Berezhnov, A.V., et al., 2006. Application of a low-angle light scattering technique to cell volume and cell signaling studies on Ehrlich ascite tumor cells. Spectrosc. Int. J. 20 (2), 45–55.

Zinchenko, V.P., Goncharov, N.V., Teplova, V.V., et al., 2007. Polarographic and spectroscopic studies of the effects of fluoroacetate/fluorocitrate on cells and mitochondria. Spectrosc. Int. J. 21 (2), 121–134.

Chapter 27

Zebrafish as a model system to evaluate the safety and toxicity of nutraceuticals

Wan-Ping Bian and De-Sheng Pei

Introduction

Compared to conventional mammalian animals, the zebrafish (*Danio rerio*) is a favorite vertebrate model for evaluating the toxicity and safety of chemical compounds and for assessing their effects on human health. The zebrafish not only has a similar physiology, development, and metabolic pathway to nonhuman mammals, but it also has certain unique advantages, such as its small size, visible embryo development, and short reproductive cycle (Dai et al., 2014; Pei and Strauss, 2013). Nutraceuticals commonly have positive physiological effects on the human body and bridge the gap between food and medicine. However, nutraceuticals supplied with different concentrations and durations can cause adverse effects or toxicity (Goncharov et al., 2019; Gupta et al., 2018). This chapter describes the zebrafish as an excellent in vivo model system for safety and toxicity assessments of common nutraceuticals.

Zebrafish model for safety and toxicity assay of nutraceuticals

Zebrafish embryonic safety and toxicity

Zebrafish embryo toxicity assays include the fish acute toxicity test (OPPTS 850.1075) and the fish early-life stage toxicity test (OPPTS 850.1400) (He et al., 2014). Due to the presence of unfertilized eggs and the time required to select healthy embryos, the treatment time of drugs and chemicals is usually 6 h postfertilization (hpf). The earliest stage for distinguishing healthy, synchronized embryos is at the 4 hpf stage, leaving a 2 h window for selecting embryos before treatment. Embryos at 4 hpf with intact chorion membranes are selected and reexamined 2 h later. The fertilized embryos are pooled at random for subsequent treatment. Generally, six-well microplates are used for treating 30 embryos simultaneously with optimal drug and chemical solutions at 28°C to detect the lethal dose 50 (LD_{50}) end-point (Lammer et al., 2009; Njiwa et al., 2010; Truong et al., 2011; Vaughan and van Egmond, 2010).

Zebrafish developmental and teratogenic safety and toxicity

Compared to other animal models, the visual embryonic development of zebrafish is very rapid, which is helpful to assess the developmental and teratogenic toxicity of nutraceuticals in vivo. The developmental processes in zebrafish are highly conserved and the genome of zebrafish is well characterized (Pei et al., 2013). Importantly, zebrafish embryos can be cultured for organogenesis (Brannen et al., 2010; Busch et al., 2011). So far, various official test guidelines of toxicity have been published (e.g., DIN 38415-6, OECD 203, 210, 215, 229, 230, 234) using the zebrafish model (He et al., 2014). Teratogenicity can be evaluated by comparing nutraceutical concentrations that cause lethality and developmental defects. The teratogenic index (TI) is defined as the LC_{50}/EC_{50} ratio for nutraceuticals. A nutraceutical is teratogenic if the TI is >1, indicating that the concentration at which 50% of the zebrafish exhibits developmental abnormalities is lower than the concentration that is observed. TI <1 is considered nonteratogenic. If TI $= 1$, then a toxicity assessment must be performed at the exact LC_{50} concentration to determine whether zebrafish exhibit malformations at that concentration (Weigt et al., 2011).

Zebrafish reproductive safety and toxicity

The zebrafish is a simpler and more attractive animal model to investigate the reproductive safety and toxicity of nutraceuticals due to fertilization and embryogenesis in vitro (Deng et al., 2010; Lawrence, 2011; Uusi-

Nutraceuticals. DOI: https://doi.org/10.1016/B978-0-12-821038-3.00027-6

Heikkilä et al., 2018). The zebrafish has a short life cycle (3 months from eggs to mature adults), and the assessment of reproductive safety and toxicity for nutraceuticals can shorten the test period, reduce cost, and increase throughput. The short-term reproduction assay of zebrafish can be performed according to OECD 229 (Organization for Economic Co-operation and Development). Several factors related to reproduction may be evaluated, such as the relative gonad index, reproduction ability, the expression pattern of the hypothalamus–pituitary–gonad axis, and gonad histology (He et al., 2014; Ma et al., 2018).

Assessment of common nutraceuticals using zebrafish

Zebrafish embryos, larvae, and adults have been widely used to assess the safety and toxicity of common nutraceuticals. We have summarized recent reports about common nutraceuticals (Table 27.1) and zebrafish tissue that the compounds targeted (Fig. 27.1), including caffeine, curcumin, quercetin, etc. More novel nutraceuticals will be defined and assessed using the zebrafish model in the future.

Caffeine

Caffeine is a common trimethylxanthine found in food and beverages, such as coffee, tea, colas, and chocolate. Lots of works have been carried out using the zebrafish model to evaluate its toxicity and safety. In 2011, Capiotti et al. evaluated the effect of caffeine on zebrafish larvae. After exposure to 100 μM caffeine, the expression pattern of cellular target genes of the larvae was affected, but morphological alterations and tactile sensibility were not detected (Capiotti et al., 2011). Another study with 300 ppm caffeine was performed on zebrafish embryos. Real-time polymerase chain reaction (qPCR) results showed that the angiogenesis of embryos was disturbed. However, qPCR did not show any defects in vasculogenesis via a green fluorescent vascular endothelium transgenic line (Yeh et al., 2012). The effect of caffeine on heartbeat, survival, and cell damage-related gene expression was reported by Abdelkader et al. (2013). Increasing exposure time to caffeine may trigger apoptosis, but neither significant mortality nor morphological changes were detected. Moreover, heart rate was increased, and embryo hatching was delayed (Abdelkader et al., 2013). Interestingly, with exposure to a high concentration of caffeine, the zebrafish body shape, notochord, and heart rate were significantly malformed, compared to control (Rah et al., 2017). Wu et al. (2015) exposed zebrafish embryos to both 2-aminothoxyldiphenyl borate (2-APB) and caffeine, and found no malformed phenotypes, which was different from those treated with 2-APB alone. These results indicated that caffeine was important for maintaining the balance of cytosolic Ca^{2+} levels (Wu et al., 2015). Khor et al. (2013) investigated the potential function of caffeine in protecting the brain during early development in male zebrafish, and they found that caffeine could protect zebrafish from dexamethasone-induced anxiety. Caffeine had another role in the suppression of hepatic lipid accumulation by downregulation of genes associated with lipogenesis, ER stress, inflammatory response, and enhancement of lipid oxidation (Zheng et al., 2015). In zebrafish, caffeine reduced the lipopolysaccharide (LPS)-stimulated elevation of the inflammatory mediator nitric oxide (NO), implying that caffeine inhibited LPS-stimulated NO production in vivo (Hwang et al., 2016). In recent years, more and more studies have investigated the role of caffeine in the behavior of adult and larval zebrafish. Different concentrations of caffeine may alter vertical swimming, freezing, and erratic movements in both wild-type and leopard zebrafish (Rosa et al., 2018).

A significant anxiety-like behavior effect was elicited by caffeine treatment, and de Carvalho et al. (2019) found that α-tocopherol could protect against this behavior. Interestingly, Gutierrez et al. (2019) screened caffeine as a drug to reduce the aggression of zebrafish. They identified caffeine as a drug that consistently decreased aggression with less of an effect upon locomotion. Through a series of reward experiments, Ruiz-Oliveira et al. (2019) found that low-dose caffeine exposure could increase zebrafish performance in a multicue discrimination task. No abnormalities in locomotor behavior were observed.

Curcumin

Curcumin is a polyphenol derived from the plant *Curcuma longa*, and its toxicity and teratogenic effects were tested using zebrafish embryos. Significant decreases in survival and hatching rates were observed when embryos were treated with 7.5 μM or higher of curcumin. All embryos died within 2 days when treated with 15 μM curcumin. These results show that curcumin is lethal to vertebrates in the micromolar range (Wu et al., 2007). A similar result was reported by Shiau et al. (2011). They found that no zebrafish embryos or larvae survived for 3 days after administration of 7.5 μM curcumin. Further, all embryos died within 2 days of exposure to a high concentration (12.5 μM) (Shiau et al., 2011). After exposure to curcumin nanoparticles for 5 days, Krausz et al. (2015) found no statistical differences for mortality, development, larval morphology, and behavioral endpoints. A different kind of nanoparticle, chitosan–curcumin nanoparticle (CSCur NP), was used because of its biocompatibility and nontoxicity. The exposure experiment indicated that CSCur NPs did not elicit

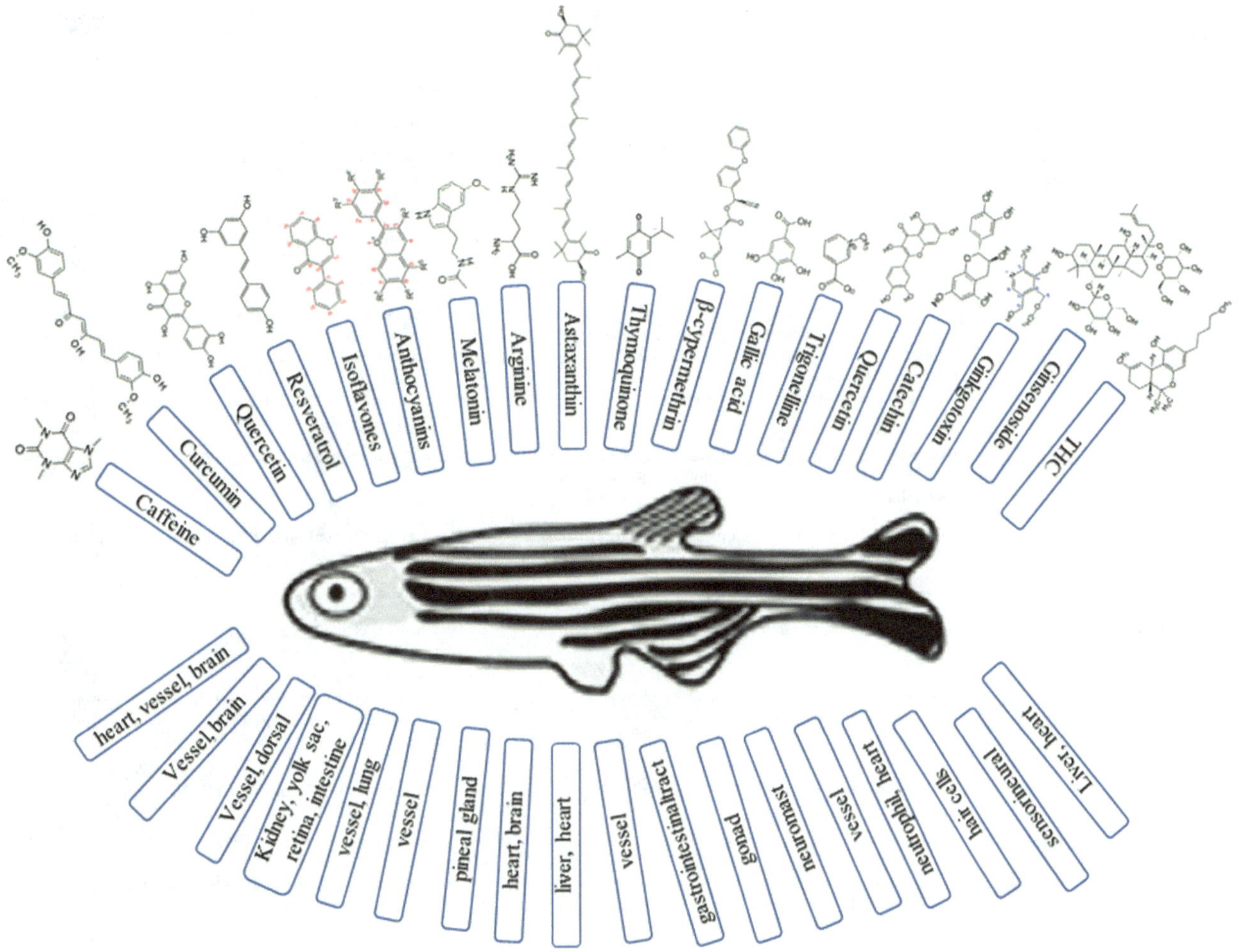

FIGURE 27.1 The structural formula of the main compounds in the common nutraceuticals (upper) and the corresponding tissues that the compounds target (lower).

TABLE 27.1 Assessment of common nutraceuticals using the zebrafish model.

Nutraceutical	Zebrafish	Biomarker/end-point	References
Caffeine	Embryos larvae adults	Angiogenesis, vasculogenesis, vascular development, myofibril alignment, brain function, heartbeat, survival, cell damage genes expression, vascular endothelial growth factor (VEGF) expression, phosphoprotein, and neurotrophic factor expression, morphological alterations, tactile sensibility, anxiety, lipogenesis, inflammatory response, vertical swimming, freezing, and erratic movements	Capiotti et al. (2011), Yeh et al. (2012), Abdelkader et al. (2013), Khor et al. (2013), Wu et al. (2015), Rah et al. (2017), Zheng et al. (2015), Hwang et al. (2016), Rosa et al. (2018), de Carvalho et al. (2019), Gutierrez et al. (2019), Ruiz-Oliveira et al. (2019)
Curcumin	Embryos larvae	Survival rates, LD_{50} values, morphology, tumor growth, antioxidant, cognitive function	Shiau et al. (2011), Wu et al. (2007), Pal et al. (2018), Krausz et al. (2015), Yadav et al. (2018), Xu et al. (2018), Harishkumar et al. (2019), Muthuraman et al. (2019)
Quercetin	Embryos larvae adults	Hair cell damage, blood vessel, neurodegenerative diseases, neuroprotective, angiogenesis, tumor-induced angiogenesis, tumor growth and tumor metastasis, lipid accumulation, curved dorsal, swimming velocity, immunity, neurocognitive	Lee et al. (2015a,b,c,d), Zhao et al. (2014), Richetti et al. (2011), Zhang et al. (2011, 2017), Wu et al. (2013), Fan et al. (2019), Zhao et al. (2014), Hirose et al. (2016), Seo et al. (2015), Wang et al. (2020), Rishitha and Muthuraman (2018)
Resveratrol	Embryos, larvae, adults	Nephroprotective effect, kidney and red blood cell circulation, fat consumption, yolk sac fat content, AMPKα-Sirt1-PPARγ pathway, pronephric cysts, body curvature, invasion of the gastrointestinal tract, retinal damage, mitochondrial DNA repair, pigment production, VEGF, angiogenesis, osteoblast formation, bone mineralization, osteogenic differentiation	Ding et al. (2015), Pardal et al. (2014), Nishimura et al. (2015), Ran et al. (2017), Wu et al. (2016), Euba et al. (2017), Al Azzaz et al. (2018), Schirmer et al. (2012), Sheng et al. (2018), Wang et al. (2019), Eo and Kim (2019), Hu et al. (2019a,b), Decui et al. (2019), Luo et al. (2019a,b)

(*Continued*)

TABLE 27.1 (Continued)

Nutraceutical	Zebrafish	Biomarker/end-point	References
Isoflavones	Embryos, larvae, adults	Angiogenesis, tumor angiogenesis, lung metastasis, antiinflammatory, tumor growth, VEGF-VEGFR2, and MAPK signaling pathways	Tang et al. (2010), Li et al. (2011, 2012, 2015), Yang et al. (2014), Luo et al. (2019a)
Anthocyanins	Embryos, larvae, adults	Antigenotoxic, DNA damage, angiogenesis, melanogenesis, intersegmental vessel	Rocco et al. (2015), Sun et al. (2017), Bucci et al. (2018), Karunarathne et al. (2019)
Melatonin	Embryos, larvae, adults	Pineal gland, daily rhythm, circadian clock, behavioral asymmetry, cell proliferation, immune response, neutrophil recruitment, parkinsonian phenotype, neurotoxicity, sleep/wake behaviors, depressive symptoms, neurobiological traits, cardiovascular, sleep-waking behavior	Vuilleumier et al. (2006), Dekens and Whitmore (2008), Budaev and Andrew (2009), Lima-Cabello et al. (2014), Ren et al. (2015a, b), Ren et al. (2017), Gandhi et al. (2015), Diaz-Casado et al. (2016, 2018), Han et al. (2017), Mi et al. (2019), Tang et al. (2019), Genario et al. (2020), Li et al. (2019a), Montalbano et al. (2018), Oh et al. (2017)
Arginine	Embryos. larvae	Locomotor activity, morphological alterations, extracellular nucleotide hydrolysis, body size, peripheral microcirculation, neutrophils	Capiotti et al. (2013), Lee et al. (2017b)
Astaxanthin	Embryos, larvae, adults	Liver steatosis, adiposity level, cardiac defects, Wnt signaling, oxidative stress	Her et al. (2013), Lutfi et al. (2017), Takemoto et al. (2018), Cao et al. (2020)
Thymoquinone	Embryos, larvae	Angiogenic activity, melanin content, melanogenesis	Paramasivam et al. (2012), Jeong et al. (2020)
Spirulina	Embryos, larvae, adults	Reproduction, egg production, immune stimulatory, immunotoxic, survival rate	Geffroy and Simon (2013), Zhang et al. (2018), Rajasekar et al. (2019), Lo Cascio et al. (2017, 2018), Edirisinghe et al. (2019), Hedayatirad et al. (2019)
Neem extract	Embryos, larvae, adults	Anxiety-like effects, learning and memory, melanin, gonad, reproductive	Bernardi et al. (2013), Kumar et al. (2013), Techer et al. (2017), Singulani et al. (2017)
Fenugreek	Embryos, larvae, adults	Inflammatory cytokine IL-1β and TNF-α expression, tumor growth, hair cell, neuromast damage, antiinflammatory, antidiabetic	Teyssier et al. (2017), Thakur and Ahirwar (2019), Castaneda et al. (2017), Oyelaja-Akinsipo et al. (2020)
St. John's wort	Embryos, larvae, adults	Angiogenesis, embryos vascular	Lin et al. (2012)
Green tea extract	Embryos, larvae, adults	Adiposity, lipid catabolism genes, body fat accumulation, body fat volume, visceral adipose tissue volume, plasma triglyceride, total cholesterol levels, WNT/β-catenin and AMPK signaling pathways, obesity, neutrophil accumulation, travel speed and distance, expression of IL-1β and TNF-α, embryotoxicity, mitochondria dysfunction, survival rates, cardiogenesis impairment	Hasumura et al. (2012), Meguro et al. (2015), Zang et al. (2019), Wang et al. (2009), Nguyen et al. (2019), Pyun et al. (2011), Shin et al. (2014), Li et al. (2019b), Lombo et al. (2019)
Gingko biloba	Embryos, larvae	Seizure-like behavior, zebrafish lateral line hair cells, antithrombotic, antiinflammatory	Lee et al. (2012), Neveux et al. (2017), Hu et al. (2019a)
Chinese ginseng	Embryos, larvae, adults	Monitor rest/wake behavior, melanogenesis, sensorineural damage, melanin content, tyrosinase activity, protein expression	Lee et al. (2015a,b,c, 2017a, 2018), Liu et al. (2019), Nam et al. (2019)
Shilajit	Adults	X-ray-induced oxidative stress and DNA damage	Musthafa et al. (2017)
Ashwagandha	Adults	Antiinflammatory, tissue regeneration, reduced congestion in blood vessels	Mohanty et al. (2016), Rajamohamed and Siddharthan (2019)
Cannabis sativa	Embryos, larvae, adults	Behavioral analysis, yolk sac, pericardial edema, bent tail/body axis, curved axis, eye/snout/jaw/trunk/fin deformities, swim bladder distention, swimming speed and distance, yolk lipid mobilization, hepatosteatosis, heart rates, axial malformations and shorter trunks, hatch delayed, birth defects in the eyes, Shh signaling	Akhtar et al. (2013), Carty et al. (2019), Stewart and Kalueff (2014), Achenbach et al. (2018), Jensen et al. (2018), Silvestri et al. (2015), Ahmed et al. (2018), Valim Brigante et al. (2018), Samarut et al. (2019), Fish et al. (2019), Janssens et al. (2018), Serra et al. (2019)

any toxic responses (Yadav et al., 2018). A surface-passivated carbon dot (CDP) from curcumin was synthesized and used in zebrafish embryos. The result showed that the survivability of the embryo treated with CDP at 0.5 μg/mL was about 52%, and the initial low concentrations had a nonsignificant impact on the zebrafish embryo (Pal et al., 2018). Liposomal curcumin (LC) that encapsulated the curcumin with the liposomal delivery system was used to expose the zebrafish. The result indicated that exposure to 80 μM LC for 6 h did not cause any toxic effects on the zebrafish embryos, and LC dramatically suppressed the growth of tumors in a concentration-dependent manner (Xu et al., 2018). By comparing the toxicity of three compounds, the authors found that curcumin had more toxic effects on antioxidant parameters, biochemical tests, and morphological deformation (Harishkumar et al., 2019). A recent study demonstrated that curcumin possesses potential ameliorative effects in cognitive function against cigarette smoke extract exposure in zebrafish (Muthuraman et al., 2019).

Quercetin

Quercetin, a traditional Chinese medicinal herb, is associated with anticancer activity. Quercetin may be used as an effective therapeutic agent to prevent and treat neurodegenerative diseases, such as Parkinson's or Alzheimer's (Richetti et al., 2011; Zhang et al., 2011). Wu et al. (2013) found that quercetin has strong inhibitory effects on embryonic angiogenesis, tumor-induced angiogenesis, tumor growth, and tumor metastasis. Therefore it may serve as a potential drug for cancer therapy (Wu et al., 2013). In 2019, Fan et al. (2019) reported that quercetin could decrease the metastasis of tumor cells by inhibiting Src/Stat3/S100A7 signaling. Zhao et al. (2014) found that the formation of intersegmented vessels, the dorsal aorta, and the posterior cardinal vein was disrupted by quercetin, indicating that it had a potent antiangiogenic activity. Lee et al. (2015d) evaluated the protective effects of quercetin on cisplatin-induced hair cell damage in transgenic zebrafish embryos, and found protective activity against apoptotic cell death (Lee et al., 2015d). Another study suggested that quercetin could be a new tool in the prevention of progressive hearing loss (Hirose et al., 2016). Quercetin inhibited lipid accumulation and NO generation in a dose-dependent fashion (Seo et al., 2015). In 2015, Zhang et al. (2017) showed that quercetin could induce an upward curved dorsal and reduce the swimming velocity of larvae. Further, it altered the muscle-related gene myostatin A and myogenic differentiation, which could be the reason for the locomotion defect and deformities. Another work evaluated the toxicity of quercetin with a pentylenetetrazole-induced neurocognitive impairment in zebrafish. It indicated that the solid lipid nanoparticle of quercetin produced potential neuroprotective and cognitive improvement in the model. Thus it might be a future medicine for the treatment of neurodegenerative disorders like memory disorders (Rishitha and Muthuraman, 2018). A comprehensive study was performed on immunity, antioxidant indices, and disease resistance in adult zebrafish after a 2-month chronic exposure to quercetin (Wang et al., 2020). The authors showed that 1 μg/L quercetin would be the best concentration among seven different levels, and quercetin could promote immunostimulant properties, antioxidant indices, and disease resistance.

Resveratrol

Resveratrol is a polyphenol that has been reported to possess antioxidant, anticancer, and antiinflammatory properties. Pardal et al. (2014) evaluated the metabolism of resveratrol and its ability to consume fat in zebrafish larvae. Their results showed that resveratrol could reduce the fat content of the yolk sac. Further, Ding et al. (2015) used two transgenic zebrafish lines to evaluate the nephron-protective effects of resveratrol by observing changes in the kidney and in red blood cell circulation. Their results revealed that resveratrol protects kidneys from malformations and improves blood circulation. Through transcriptome sequencing and qPCR, Nishimura et al. (2015) found that the inhibition of EP300 might be a common therapeutic mechanism of resveratrol in the adipose tissues of obesity. Ran et al. (2017) showed that resveratrol may regulate lipid metabolism through the AMP-activated protein kinase α (AMPKα)-Sirt1-PPARγ pathway in zebrafish that overfed. After exposure to 150 μM resveratrol, pronephric cysts and body curvature at 3 dpf were shown in *pkd2* knockdown zebrafish. The authors found that cyst formation was inhibited significantly and expression levels of complement factor, tumor necrosis factor α (TNF-α), and monocyte chemotactic protein 1 (MCP-1) were downregulated (Wu et al., 2016). Euba et al. (2017) used adult zebrafish as a model to evaluate the function of resveratrol in treating nontypeable *Haemophilus influenzae* (NTHi) infestation. The result showed that zebrafish as a preclinical sepsis model system was useful for in vivo therapeutic evaluation against NTHi infection (Euba et al., 2017). Another report showed that *trans*-resveratrol treatment was efficient for inducing autophagy and limiting invasion of the gastrointestinal tract by pathogens (Al Azzaz et al., 2018). Two previous studies indicated that resveratrol offered protection against N-Methyl-D-aspartic acid-induced retinal damage and mitochondrial DNA repair by regulating the expression of related genes (Schirmer et al., 2012; Sheng et al., 2018). Wang et al. (2019) found that resveratrol could promote mitochondrial function by regulating the activity of Ampk/Sirt1/Pgc1α and Akt/mTOR pathways, suggesting that resveratrol might be able to prevent age-related oculopathy. Resveratrol potently inhibited pigment production via the extracellular signal-regulated kinase (ERK)1/2 and PI-3K/Akt pathways in zebrafish (Eo and Kim, 2019). Hu et al. (2019a,b) found that resveratrol had a suppression activity on vascular endothelial growth factor (VEGF) and induced angiogenesis by inhibiting receptor signaling (Hu et al., 2019a,b). Using the zebrafish seizure model, an anticonvulsant effect was identified from micronized resveratrol, which became a safe candidate drug for neurodevelopment (Decui et al., 2019). Resveratrol effectively promoted the formation of osteoblasts, bone mineralization, and osteogenic differentiation and also maintained a balance between bone formation and resorption (Luo et al., 2019b).

Isoflavones

Isoflavones are types of polyphenols found in legumes, including soybeans, chickpeas, fava beans, pistachios, peanuts, and other fruits and nuts. One of the isoflavones, calycosin, was reported to induce angiogenesis in zebrafish. Study results indicated that calycosin had selective affinity for estrogen receptors ERα and ERβ, and that it promoted angiogenesis via the activation of the VEGF-VEGFR2 and mitogen-activated protein kinase (MAPK) signaling pathways (Tang et al., 2010). Li et al. (2011) performed transcriptomics and qPCR analysis to demonstrate the proangiogenic effects of calycosin and found that it regulated VEGF, fibroblast growth factor, and ErbB signaling pathways (Li et al., 2011). Li et al. (2011) also validated that barbigerone, an isoflavone, inhibited tumor angiogenesis and tumor growth through the VEGFR2 signaling pathway. Another work reported that barbigerone could inhibit tumor angiogenesis, tumor growth, and lung metastasis via the downregulation of the MEK3/6/p38 MAPK signaling pathway (Yang et al., 2014). Another isoflavone, formononetin (MBHS), induced endothelial cell migration, stress fiber formation, and angiogenesis through an ERα-enhanced Rho-associated protein kinases (ROCK) pathway by forming an ERα/ROCK-II activated complex (Li et al., 2015). Luo et al. (2019a) reported that two isoflavones, MBHS and its glycosylated derivative ononin, exerted potent antiinflammatory effects on LPS-induced zebrafish models through the MyD88, Toll/IL-1R-domain-containing adaptor-inducing IFN-β MAPK/ERK, and MAPK/Jun N-terminal kinases pathways. The results provided new insights into the antiinflammatory mechanisms of isoflavones (Luo et al., 2019a).

Anthocyanins

Anthocyanins are members of the flavonoid group of phytochemicals, a group predominant in teas, honey, wines, fruits, vegetables, nuts, olive oil, cocoa, and cereals. Rocco et al. (2015) demonstrated that anthocyanin can activate the natural defense systems of the cell, and that anthocyanin possesses an antigenotoxic action, reducing DNA damage, and providing stimulus in the field of genotoxicity. Several works have been performed with the extracts of red wine, blueberry, and the petals of *Hibiscus syriacus* L., and the main compound of all extracts have been anthocyanins. Red wine has the benefit of inhibiting angiogenesis in embryos, which had a lower number of intersegmental vessels than those of the control (Sun et al., 2017). Blueberry was effective in protecting the skin from photodamage (Bucci et al., 2018). The petals extract of *H. syriacus* L. may inhibit melanogenesis by activating the ERK signaling pathway (Karunarathne et al., 2019).

Melatonin

Melatonin is synthesized from tryptophan by alkylamine *N*-acetyltransferase (*aanat1* and *aanat2*), mostly in the pineal and retina cells. Melatonin plays a key homeostatic role in zebrafish physiology, including sleep, development, nutrition, reproduction, learning, and memory (Lima-Cabello et al., 2014). Studies in zebrafish embryos revealed that melatonin could induce cell proliferation. Many studies found that light could initiate molecular oscillations in zebrafish and influence the timing of the cell cycle (Budaev and Andrew, 2009; Dekens and Whitmore, 2008; Vuilleumier et al., 2006). In the immunological processes, the immune response also exhibited circadian rhythmicity. A low dosage of melatonin significantly regulated the recruitment of rhythmic neutrophils toward the injured site, but a higher exogenous dose dramatically inhibited the migration of injury-induced neutrophils (Ren et al., 2015a, b). Ren et al. (2017) found that endogenous melatonin promoted rhythmic migration of neutrophils through cytokines. The result from another zebrafish mutant line, *aanat2*$^{-/-}$, indicated that melatonin was required for circadian regulation of sleep (Gandhi et al., 2015). Using zebrafish as a Parkinson disease model, melatonin was thought to be a useful drug in the treatment of the disease, because it could prevent and recover damage induced by 1-methyl-4-phenyl-1,2,3,6-tetrahydropyridine. Further, the efficacy of melatonin was found to restore parkinsonian phenotypes in animals (Diaz-Casado et al., 2016, 2018). Recently, more works have been performed to study the behavior of zebrafish after exposure to melatonin. Melatonin inhibited fenvalerate-mediated oxidative stress and protected against neurotoxicity (Han et al., 2017). Mi et al. (2019) evaluated the protection of melatonin against fluorene-9-bisphenol (BHPF)-induced mortality in zebrafish. Their results showed that melatonin could correct an adverse influence on the development and sleep/wake behaviors caused by

BHPF exposure. Tang et al. (2019) evaluated the function of venlafaxine combined with melatonin in depression-like behavior in zebrafish induced by reserpine. The results showed that venlafaxine combined with melatonin played a more effective role in the remedy of depressive symptoms in zebrafish. After exposure to melatonin, there was sex specificity in pharmacological responses, which demonstrated the potential of zebrafish for studying sex differences (Genario et al., 2020). Li et al. (2019a) demonstrated that exposure to melatonin and sodium tanshinone IIA sulfonate could protect cardiovascular function and sleep-waking behavior (Li et al., 2019a). Montalbano et al. (2018) showed antiobesity protective effects using the zebrafish model with diet-induced obesity after exposure to melatonin. After exposure to melatonin, the injury of neomycin was alleviated and apoptosis was reduced, suggesting that melatonin had the potential to reduce aminoglycosides-induced apoptosis (Oh et al., 2017).

Arginine

Capiotti et al. (2013) reported that exposure to arginine decreased body size and induced adenosine triphosphate , adenosine diphosphate, and adenosine monophosphate (AMP) hydrolysis. These results demonstrated that arginine affected morphological parameters and ectonucleotidase activity in zebrafish larvae. The homocysteine-treated zebrafish exhibited peripheral microcirculation dysfunction after treatment with L-arginine and the peripheral microcirculation was attenuated (Lee et al., 2017b).

Astaxanthin

Astaxanthin (ATX), one of the dominant carotenoids, has shown both a strong quenching effect against singlet oxygen and a strong scavenging effect against free radicals. Using a transgenic zebrafish model with hepatic lipid accumulation (GY), Her et al. (2013) found that ATX-treated larvae showed no liver steatosis via inhibition of PPAR-γ (Her et al., 2013). Interestingly, another report showed that ATX did not modify the adiposity level of zebrafish (Lutfi et al., 2017). The nanoemulsion of ATX protected against neomycin-induced hair-cell damage in zebrafish, but not the ATX due to the stability and absorbency (Takemoto et al., 2018). A previous study showed that ATX was able to correct cardiac defects by downregulating oxidative stress and inducing Wingless-INT (WNT) signaling pathway (Cao et al., 2020).

Thymoquinone

Thymoquinone is a natural product derived from black cumin (*Nigella sativa*), which can exert antiinflammatory, antioxidant, and antineoplastic effects both in vitro and in vivo. Thymoquinone has a potential effect on antiangiogenesis due to downregulation of the expression of angiogenesis-related genes (Paramasivam et al., 2012). The zebrafish showed a dose-dependent response to thymoquinone for decreasing the content of melanin (Jeong et al., 2020).

Spirulina

Geffroy and Simon (2013) investigated the effects of spirulina when fed to female zebrafish as their sole food source. Compared to a commercial diet, the production of zebrafish eggs was lower, but the survival rate increased and larvae developed more quickly in the group fed with *Spirulina platensis*. There were no differences in weight or size (Geffroy and Simon, 2013). In the reproductive system of male zebrafish exposed to β-cypermethrin (β-CYP), selenium-enriched spirulina can effectively prevent oxidative damage to the reproductive system during spermatocyte differentiation (Zhang et al., 2018). Rajasekar et al. (2019) demonstrated that sulfated polysaccharide), isolated from *S. platensis*, enhanced the growth and reproductive performances of adult zebrafish. After fasting and refeeding of spirulina, zebrafish can overcome feed deprivation periods and will compensate by increasing gastrointestinal protein PepT1 and CCK8 expression. A compensatory mechanism may be induced by fasting and refeeding a higher protein spirulina diet. Microalgae are an excellent candidate for animal breeding and the human diet (Lo Cascio et al., 2017, 2018). Pectin isolated from *Spirulina maxima* may enhance the early hatching ratio of zebrafish embryos, and it had a beneficial immune-stimulatory effect (Edirisinghe et al., 2019). After exposure to atrazine (AZ, endocrine disruptor), dietary supplementation of spirulina could assist female zebrafish in overcoming the immunotoxic effects of AZ. Females with spirulina added to their diet produced eggs with higher fertilization and hatching success (Hedayatirad et al., 2019).

Neem extract

Bernardi et al. (2013) examined the neurotoxic effects of neem in adult zebrafish. They assessed general activity, anxiety-like effects, and learning and memory in a passive avoidance task after exposure to 20 or 40 μL/L neem extract. Neem extract decreased the general activity of zebrafish and increased anxiety-like behavior. However, learning and memory were not affected (Bernardi et al., 2013). Kumar et al. (2013) used gallic acid to expose zebrafish larvae, and found that melanin production was significantly inhibited by the downregulation of microphthalmia-associated transcription factor (MITF), tyrosinase, and their regulatory proteins. Techer et al.

(2017) found that glutathione peroxidase activity was increased and the level of 11 ketotestosterone was decreased. However, no alterations were found in the gonads and reproductive performance. Interestingly, no toxic effects on zebrafish embryos were found after gallic acid treatment (Singulani et al., 2017).

Fenugreek

The commercial Plant defense stimulators Stifenia (FEN560) that is composed of crushed fenugreek seeds was studied using the zebrafish model. The authors observed a dose-dependent effect of Stifenia on the expression levels of inflammatory cytokine Interleukin-1 beta (IL-1β) and TNF-α (Teyssier et al., 2017). The ethyl iso-allocholate isolated from *Trigonella foenum-graecum* L. (fenugreek) seeds, inhibited tumor growth and reduced the migration of tumor cells into the liver (Thakur and Ahirwar, 2019). Castaneda et al. (2017) found that trigonelline significantly reduced hair cell loss and neuromast damage in diabetic zebrafish (Castaneda et al., 2017). The results described by Oyelaja-Akinsipo et al. (2020) indicated that diosgenin was an antiinflammatory, antidiabetic, and neuroprotective agent.

St. John's wort

St. John's wort is a perennial plant with the common name, *Hypericum perforatum*. Zebrafish were exposed to a flavonoid, quercetin-49-O-b-D-glucopyranoside, isolated from *Hypericum attenuatum*, for the study of health safety. The results indicated that angiogenesis was inhibited via restraint of the phosphorylation of VEGFR2 tyrosine kinase and suppression of the VEGFR2-mediated signaling pathway (Lin et al., 2012).

Green tea extract

Zebrafish can be used as an obesity model for studying the effect of green tea extract (GTE). Hasumura et al. (2012) used a diet-induced obesity zebrafish model and found that GTE exposure significantly decreased the visceral volume and increased the expression of lipid catabolism genes. An excellent model of vertebrate lipid metabolism was established by high-fat diets. Exposure to GTE significantly suppressed body fat accumulation and body fat volume (Meguro et al., 2015). Another work showed that GTE reduced visceral adipose tissue volume, plasma triglyceride, and total cholesterol levels. RNA-Seq data revealed the reduction effects are attributed to the activation of the WNT/β-catenin and AMPK signaling pathways (Zang et al., 2019), indicating that daily consumption of green tea may be beneficial for the prevention and treatment of obesity. GTE had the function of protecting against ultraviolet B exposure (Wang et al., 2009). Epigallocatechin gallate (EGCG), a major catechin in green tea, could reduce neutrophil accumulation and downregulate the expression of IL-1β, TNF-α, and related signaling pathways (Nguyen et al., 2019). Epicatechin (EC), a minor component of GTE could protect against radiation-induced toxicity. Pyun's work showed that EC markedly attenuated embryotoxicity and protected against the loss of auditory neuromasts by inhibiting changes in the mitochondrial membrane potential and MAPK signals (Pyun et al., 2011). Shin et al. (2014) showed that EC inhibited the generation of radiation-induced reactive oxygen species and mitochondrial dysfunction, and attenuated embryotoxicity. After exposure to a green tea–water solution, the length, and size of the head and eyes were decreased in zebrafish. Moreover, embryo development was influenced and the survival rates of embryos and larvae were decreased in a dose-dependent manner (Li et al., 2019b). Lombo et al. (2019) used EGCG to neutralize cardiogenesis impairment, and found that the EGCG treatment was able to recover phenotype and reverse all deleterious effects.

Ginkgo biloba

Ginkgotoxin, a natural neurotoxin in *Ginkgo biloba*, was hypothesized to be an inhibitor of the formation of neurotransmitter gamma-aminobutyric acid. Lee et al. (2012) found that seizure-like behavior was induced in zebrafish after 2 h of exposure to ginkgotoxin. Data suggested that ginkgotoxin-induced seizures in zebrafish larvae could serve as an in vivo model for epileptic seizure research (Lee et al., 2012). Neveux et al. (2017) found that flavonoids from the *G. biloba* plant showed significant ototoxicity on the lateral hair cells of zebrafish, and that it could kill up to 30% of hair cells (Neveux et al., 2017). A thrombosis model induced by phenylhydrazine was used to evaluate the antithrombotic effect of the major active components of *G. biloba*. The results indicated that different proportions of the extracts significantly increased the intensity of heart red blood cells and reduced thrombosis (Lu et al., 2018). Ginkgetin and resveratrol combined synergistically exerted antiinflammatory effects, and these phytochemicals show potential for the treatment of inflammation-related diseases, such as rheumatoid arthritis (Hu et al., 2019a).

Chinese ginseng

Lee and his colleagues performed a series of experiments to elucidate the inhibition of *Panax ginseng* on zebrafish melanogenesis. Zebrafish were exposed to different compounds from *P. ginseng*, including ginsenoside Rh6, vina-ginsenoside R4, vina-ginsenoside R13, ginsenoside Rb2, ginsenoside Rh23, floral ginsenoside A, and picrionoside A for melanogenic analysis. The authors found that all these compounds with varying contents in *P. ginseng* could decrease melanin content and body pigmentation (Lee et al., 2015a,b, 2018). Furthermore, the tyrosinase activity of zebrafish was reduced after treatment with ginsenoside Rb2 and floral ginsenoside A (Lee et al., 2015c, 2017a). Ginsenoside compound-Y, a metabolite of *Panax ginseng*, could downregulate protein expression levels of MITF, tyrosinase-related protein 1 (TRP-1), TRP-2, and tyrosinase (Liu et al., 2019). The efficacy of Korean red ginseng (RG, *P. ginseng*) extract was evaluated in alloxan-induced diabetic neuromast (AIDN) zebrafish. The results showed that RG recovered sensorineural damage in AIDN zebrafish (Nam et al., 2019).

Shilajit

Shilajit is a natural mineral originating in India and widely used in Indian traditional medicine. Using *Chironomus* larvae bioencapsulated with shilajit, Musthafa et al. (2017) fed zebrafish for 2 months and then subjected them to irradiation with a single acute dose of X-ray. The results indicated that pretreatment with shilajit inhibited the oxidative stress and DNA damage induced by irradiation.

Ashwagandha

Ashwagandha (*Withania somnifera*), one of the most powerful medicinal plants, plays a role in the regulation of thyroid function. Sivamani et al. (2014) reported that zebrafish treated with *W. somnifera* extract (via stainless steel implants), exhibited an antiinflammatory effect by inhibiting the expression of TNF-α. The authors indicated that the effects might be due to the rich phenolic acids and flavonoids. Zebrafish treated with B[a]P showed a reversal in scototaxis (anxiety-like) behavior after exposure to *W. somnifera* leaf extract (WSLE). Mohanty et al. (2016) found that the WSLE could restore normal scototaxis due to its neuroprotective role. Another study using zebrafish as an infection model was established by infection with *Candida albicans*. The results revealed that a biofilm which formed in the intestine was disseminated, when treated with *Amukkara choornam* (AC, *W. somnifera*). In addition, the treatment of the fish with AC also promoted tissue regeneration and reduced congestion in the blood vessels (Rajamohamed and Siddharthan, 2019).

Cannabis sativa

Cannabis sativa is a remarkable plant containing many valuable natural components, and Δ^9-tetrahydrocannabinol (THC) and cannabidiol (CBD) make up the majority of compounds. Behavioral analysis after THC treatment was performed in zebrafish larvae using an automated video-tracking system. After exposure, zebrafish larvae showed stimulation of the locomotor activity at low concentrations and suppression at high doses. The larvae exhibited a range of malformation, yolk sac and pericardial edema, and bent tail/body axis (Akhtar et al., 2013). Carty et al. (2019) showed that larvae exposed to THC and CBD showed edema, curved axis, eye/snout/jaw/trunk/fin deformities, swim bladder distention, and behavioral abnormalities. Acute exposure to THC was performed in adult zebrafish. The results indicated that THC produced an anxiogenic-like reduction of top swimming, paralleled with slower and continuous bottom swimming (Stewart and Kalueff, 2014). Further, both THC and CBD were used for individual and combined exposure. The authors found that THC decreased but CBD increased enzyme activity. However, the combination exposure showed that activity was enhanced following the increased THC concentration (Achenbach et al., 2018). Interestingly, Jensen et al. found that CBD can decrease the swimming speed and distance of adult zebrafish. Moreover, six immune-related genes were changed significantly after CBD exposure (Jensen et al., 2018). Following exposure to Δ^9-tetrahydrocannabivarin (THCV) and CBD, the lipid mobilization of zebrafish yolk was increased, but the development of hepatosteatosis was inhibited in larvae (Silvestri et al., 2015). Both Δ^9-THC and CBD caused reduced heart rates, axial malformations, and shorter trunks. The synaptic activity at neuromuscular junctions was altered and the C-star escape responses to sound were drastically reduced (Ahmed et al., 2018).

Valim Brigante et al. (2018) evaluated the toxicity of 300 μg/L CBD to zebrafish embryos and found that no malformations were observed, but 20% of the embryos had delayed hatching. THC and CBD had a synergistic effect when the ratio was 1:1, which could be a benefit for reducing side effects when cannabinoids are used as therapeutics (Samarut et al., 2019). Combined exposure to THC and alcohol caused a greater incidence of birth defects in the eyes than that with individual THC or alcohol, possibly because cannabinoids reduced Shh signaling by inhibiting Smoothened (Fish et al., 2019). THCV had effects on genes drTRPV5/6 in a strong and concentration-dependent manner. Janssens et al. (2018) showed that TRPV5 and TRPV6, as novel targets of phytocannabinoids, modulated the epithelial calcium transport (Janssens et al., 2018). In a *tsc2* mutant zebrafish, CBD did not affect larvae motility and survival ratio, but possibly reduced the number of phosphorylated rpS6 positive cells (Serra et al., 2019).

Transparent zebrafish and fluorescent transgenic zebrafish for safety and toxicity testing of nutraceuticals

The optical transparency of the embryo is an advantage for investigating the early stages of zebrafish development. However, common zebrafish adults are opaque. The transparent zebrafish line, *casper*, was developed by White et al. using both mutant *nacre* and *roy orbison* as founders (Lister et al., 1999; White et al., 2008). The *casper* zebrafish lacks both pigmented melanocytes and reflective iridophores for its entire life. Thus the organs of *casper*, including the heart, intestinal tube, liver, and gallbladder, can be observed using a standard stereomicroscope (White et al., 2008). The transparent zebrafish will be a useful model for safety and toxicity testing of nutraceuticals. After exposure to different nutraceuticals, one can easily check the development of vasculature, kidney, or other organs, and quickly determine tissue-specific toxicology.

The transgenic zebrafish lines can be constructed with tissue-specific promoters linked to different fluorescent reporter genes (Pei et al., 2011), which opens a novel window for monitoring the safety and toxicity of nutraceuticals. Transgenic zebrafish lines for labeling vasculature *Tg(fli1:EGFP*(enhanced green fluorescent protein)*)* (Lawson and Weinstein, 2002), epithelium *Tg(krt8:GFP* (green fluorescent protein)*)* (Gong et al., 2002), kidney *Tg(wt1b:EGFP)* (Perner et al., 2007), pancreas *Tg(pdx1:GFP)* (Huang et al., 2001), and liver *Tg(lfabp:GFP)* (Her et al., 2003) have been developed. However, the best observation time of fluorescence is limited to 5 days because of the presence of pigments. If fluorescent transgenic *casper* zebrafish can be developed with specific fluorescent proteins expressed in different organs and tissues, both acute and chronic toxicity analyses of nutraceuticals can be performed with real-time monitoring. Moss et al. (2013) developed a transparent transgenic zebrafish expressing both mCherry and GFP to evaluate morphologies from embryo to 2-week larvae. They indicated that these zebrafish can be utilized as a vertebrate metabolic model for generating new insights for safety and toxicity evaluation (Moss et al., 2013).

Genome-editing technique for safety and toxicity testing of nutraceuticals in zebrafish

Zinc finger nucleases (ZFNs), transcription activator-like effector nucleases (TALENs), and clustered regularly interspaced short palindromic repeats (CRISPRs) are three commonly used biotechnologies in genome editing of mammalian animals, fish, and other species. ZFNs were obtained by fusing DNA-binding domains of eukaryotic transcription factors with the cleavage domain of FokI restriction endonuclease (Urnov et al., 2010). TALENs were artificial restriction enzymes generated by fusing a TAL effector DNA-binding domain to the cleavage domain of FokI (Boch et al., 2009). Currently, the CRISPR/Cas system is the most promising genome-editing tool, because it is simple, inexpensive, easily programmed, and extremely efficient when compared to ZFNs and TALENs (Jinek et al., 2012). All three of these techniques first introduce double-stranded breaks (DSBs) in the genome, then repair with NHEJ (nonhomologous end joining), and then leave insertion or deletion of DNA fragments. The CRISPR/Cas system has been successfully introduced in the zebrafish model (Cong et al., 2013; Hwang et al., 2013; Jao et al., 2013). Recently, other CRISPR/Cas systems have been developed and widely used, such as Cas12a (Cpf1) (Fernandez et al., 2018), Cas13 (Cox et al., 2017), and Cas14 (Harrington et al., 2018). Based on the CRISPR/Cas system, more base-editing systems have been developed for editing a single base, such as the cytidine base editor and the adenine base editor (Gaudelli et al., 2017; Komor et al., 2016). Thus more sensitive zebrafish lines for assessing the safety and toxicity of nutraceuticals can be developed by knocking out key genes that sensitively respond to different nutraceuticals (Fig. 27.2).

Concluding remarks and future directions

In conclusion, nutraceuticals are becoming more popular in our daily lives. Assessing the safety and toxicity of nutraceuticals is vital to human health. This chapter describes the zebrafish as an excellent in vivo model system for safety and toxicity assessments and highlighted the common nutraceuticals assessments. Fluorescent transgenic *casper* may be utilized to generate new insights into the function of nutraceuticals. A novel genome-editing technique, such as the CRISPR/Cas system, would provide more information about safety and toxicity testing of nutraceuticals in the future.

Acknowledgments

W.P.B (the first author) and D.S.P. (corresponding author, peids@-cigit.ac.cn) contributed equally to this manuscript. The authors are grateful for support from the CAS Team Project of the Belt and Road (to D.S.P.), Chongqing Research Program of Basic Research and Frontier Technology (no. cstc2019jcyj-msxm0700 to W.P.B), Chongqing Key Program of Basic Research and Advanced Exploration Project (no. cstc2019jcyj-zdxmX0035 to D.S.P.), Basic Research and Frontier Exploration of Chongqing Science and Technology Commission and Beibei District, Chongqing (no. 2019-04 to W.P.B.), Three Hundred Leading Talents in Scientific and Technological Innovation Program of Chongqing (no.

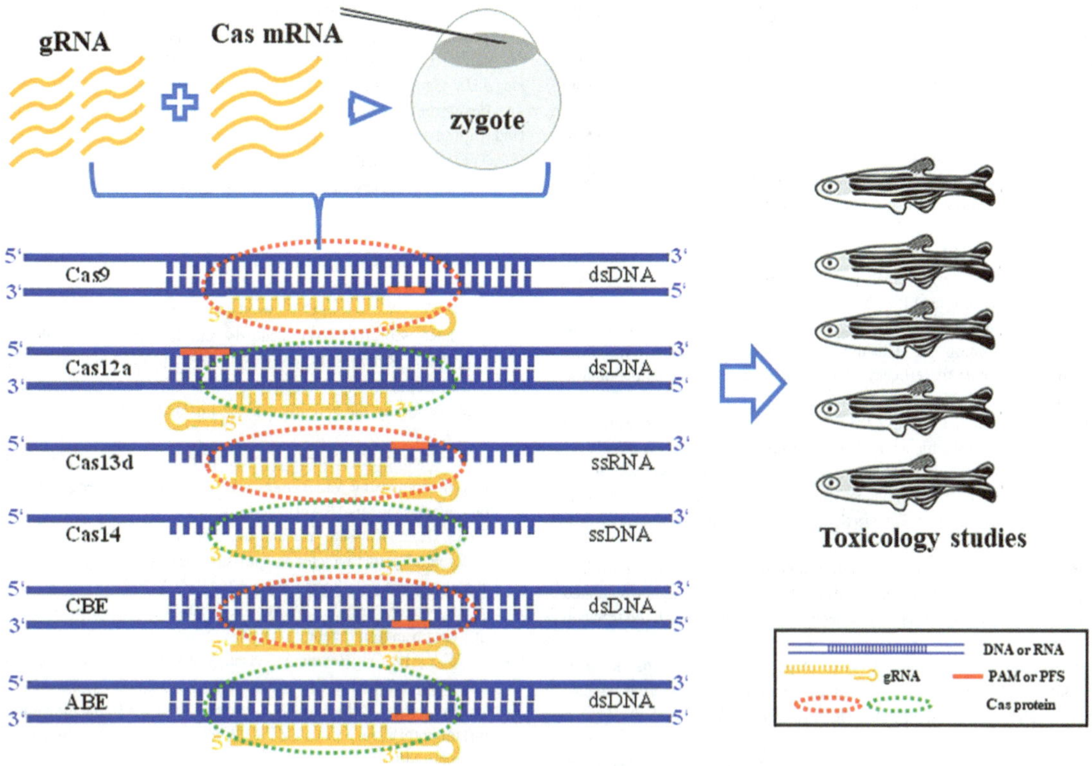

FIGURE 27.2 A schematic diagram of safety and toxicity testing of nutraceuticals by using different CRISPR/Cas systems in zebrafish. *ABE*, adenine base editor; *CBE*, cytidine base editor; *CRISPs*, clustered regularly interspaced short palindromic repeats.

CSTCCXLJRC201714 to D.S.P.), and Program of China-Sri Lanka Joint Research and Demonstration Center for Water Technology and China-Sri Lanka Joint Center for Education and Research by Chinese Academy of Sciences, China.

Conflict of interest

The authors declare no conflict of interest.

References

Abdelkader, T.S., Chang, S.N., Kim, T.H., et al., 2013. Exposure time to caffeine affects heartbeat and cell damage-related gene expression of zebrafish *Danio rerio* embryos at early developmental stages. J. Appl. Toxicol. 33, 1277–1283.

Achenbach, J.C., Hill, J., Hui, J.P.M., et al., 2018. Analysis of the uptake, metabolism, and behavioral effects of cannabinoids on zebrafish larvae. Zebrafish 15, 349–360.

Ahmed, K.T., Amin, M.R., Shah, P., et al., 2018. Motor neuron development in zebrafish is altered by brief (5-hr) exposures to THC ((9)-tetrahydrocannabinol) or CBD (cannabidiol) during gastrulation. Sci. Rep. 8, 10518.

Akhtar, M.T., Ali, S., Rashidi, H., et al., 2013. Developmental effects of cannabinoids on zebrafish larvae. Zebrafish 10, 283–293.

Al Azzaz, J., Rieu, A., Aires, V., et al., 2018. Resveratrol-Induced xenophagy promotes intracellular bacteria clearance in intestinal epithelial cells and macrophages. Front. Immunol. 9, 3149.

Bernardi, M.M., Dias, S.G., Barbosa, V.E., 2013. Neurotoxicity of neem commercial formulation (*Azadirachta indica* A. Juss) in adult zebrafish (*Danio rerio*). Environ. Toxicol. Pharmacol. 36, 1276–1282.

Boch, J., Scholze, H., Schornack, S., et al., 2009. Breaking the code of DNA binding specificity of TAL-type III effectors. Science 326, 1509–1512.

Brannen, K.C., Panzica-Kelly, J.M., Danberry, T.L., et al., 2010. Development of a zebrafish embryo teratogenicity assay and quantitative prediction model. Birth Defects Res. B Dev. Reprod. Toxicol. 89, 66–77.

Bucci, P., Prieto, M.J., Milla, L., et al., 2018. Skin penetration and UV-damage prevention by nanoberries. J. Cosmet. Dermatol. 17, 889–899.

Budaev, S., Andrew, R.J., 2009. Patterns of early embryonic light exposure determine behavioural asymmetries in zebrafish: a habenular hypothesis. Behav. Brain Res. 200, 91–94.

Busch, W., Duis, K., Fenske, M., et al., 2011. The zebrafish embryo model in toxicology and teratology, September 2-3, 2010, Karlsruhe, Germany. Reprod. Toxicol. 31, 585–588.

Cao, Z., Huang, Y., Xiao, J., et al., 2020. Exposure to diclofop-methyl induces cardiac developmental toxicity in zebrafish embryos. Environ. Pollut. 259, 113926.

Capiotti, K.M., Menezes, F.P., Nazario, L.R., et al., 2011. Early exposure to caffeine affects gene expression of adenosine receptors, DARPP-32 and BDNF without affecting sensibility and morphology of developing zebrafish (*Danio rerio*). Neurotoxicol. Teratol. 33, 680–685.

Capiotti, K.M., Fazenda, L., Nazario, L.R., et al., 2013. Arginine exposure alters ectonucleotidase activities and morphology of zebrafish larvae (*Danio rerio*). Int. J. Dev. Neurosci. Off. J. Int. Soc. Dev. Neurosci 31, 75–81.

Carty, D.R., Miller, Z.S., Thornton, C., et al., 2019. Multigenerational consequences of early-life cannabinoid exposure in zebrafish. Toxicol. Appl. Pharmacol. 364, 133–143.

Castaneda, R., Rodriguez, I., Nam, Y.H., et al., 2017. Trigonelline promotes auditory function through nerve growth factor signaling on diabetic animal models. Phytomedicine 36, 128–136.

Cong, L., Ran, F.A., Cox, D., et al., 2013. Multiplex genome engineering using CRISPR/Cas systems. Science 339, 819–823.

Cox, D.B.T., Gootenberg, J.S., Abudayyeh, O.O., et al., 2017. RNA editing with CRISPR-Cas13. Science 358, 1019–1027.

Dai, Y.J., Jia, Y.F., Chen, N., et al., 2014. Zebrafish as a model system to study toxicology. Environ. Toxicol. Chem. SETAC 33, 11–17.

de Carvalho, T.S., Cardoso, P.B., Santos-Silva, M., et al., 2019. Oxidative stress mediates anxiety-like behavior induced by high caffeine intake in zebrafish:

protective effect of alpha-tocopherol. Oxid. Med. Cell Longev. 2019, 8419810.

Decui, L., Garbinato, C.L.L., Schneider, S.E., et al., 2019. Micronized resveratrol shows promising effects in a seizure model in zebrafish and signalizes an important advance in epilepsy treatment. Epilepsy Res. 159, 106243.

Dekens, M.P., Whitmore, D., 2008. Autonomous onset of the circadian clock in the zebrafish embryo. EMBO J. 27, 2757–2765.

Deng, J., Liu, C., Yu, L., et al., 2010. Chronic exposure to environmental levels of tribromophenol impairs zebrafish reproduction. Toxicol. Appl. Pharmacol. 243, 87–95.

Diaz-Casado, M.E., Lima, E., Garcia, J.A., et al., 2016. Melatonin rescues zebrafish embryos from the parkinsonian phenotype restoring the parkin/PINK1/DJ-1/MUL1 network. J. Pineal Res. 61, 96–107.

Diaz-Casado, M.E., Rusanova, I., Aranda, P., et al., 2018. In vivo determination of mitochondrial respiration in 1-methyl-4-phenyl-1,2,3,6-tetrahydropyridine-treated zebrafish reveals the efficacy of melatonin in restoring mitochondrial normalcy. Zebrafish 15, 15–26.

Ding, Y.J., Sun, C.Y., Wen, C.C., et al., 2015. Nephroprotective role of resveratrol and ursolic acid in aristolochic acid intoxicated zebrafish. Toxins 7, 97–109.

Edirisinghe, S.L., Dananjaya, S.H.S., Nikapitiya, C., et al., 2019. Novel pectin isolated from *Spirulina maxima* enhances the disease resistance and immune responses in zebrafish against *Edwardsiella piscicida* and *Aeromonas hydrophila*. Fish. Shellfish Immunol. 94, 558–565.

Eo, S.H., Kim, S.J., 2019. Resveratrol-mediated inhibition of cyclooxygenase-2 in melanocytes suppresses melanogenesis through extracellular signal-regulated kinase 1/2 and phosphoinositide 3-kinase/Akt signalling. Eur. J. Pharmacol. 860, 172586.

Euba, B., Lopez-Lopez, N., Rodriguez-Arce, I., et al., 2017. Resveratrol therapeutics combines both antimicrobial and immunomodulatory properties against respiratory infection by nontypeable *Haemophilus influenzae*. Sci. Rep. 7, 12860.

Fan, J.J., Hsu, W.H., Lee, K.H., et al., 2019. Dietary flavonoids luteolin and quercetin inhibit migration and invasion of squamous carcinoma through reduction of Src/Stat3/S100A7 signaling. Antioxidants (Basel) 8, 557.

Fernandez, J.P., Vejnar, C.E., Giraldez, A.J., et al., 2018. Optimized CRISPR-Cpf1 system for genome editing in zebrafish. Methods 150, 11–18.

Fish, E.W., Murdaugh, L.B., Zhang, C., et al., 2019. Cannabinoids exacerbate alcohol teratogenesis by a CB1-hedgehog interaction. Sci. Rep. 9, 16057.

Gandhi, A.V., Mosser, E.A., Oikonomou, G., et al., 2015. Melatonin is required for the circadian regulation of sleep. Neuron 85, 1193–1199.

Gaudelli, N.M., Komor, A.C., Rees, H.A., et al., 2017. Programmable base editing of A*T to G*C in genomic DNA without DNA cleavage. Nature 551, 464–471.

Geffroy, B., Simon, O., 2013. Effects of a *Spirulina platensis*-based diet on zebrafish female reproductive performance and larval survival rate. Cybium 37, 31–38.

Genario, R., Giacomini, A., de Abreu, M.S., et al., 2020. Sex differences in adult zebrafish anxiolytic-like responses to diazepam and melatonin. Neurosci. Lett. 714, 134548.

Goncharov, N., Sobolev, V., Terpilowski, M., et al., 2019. Safety and toxicity evaluation of nutraceuticals in animal models. In: Gupta, R.C., Srivastava, A., Lall, R. (Eds.), Nutraceuticals in Veterinary Medicine. Springer Nature, Cham, pp. 675–684.

Gong, Z., Ju, B., Wang, X., et al., 2002. Green fluorescent protein expression in germ-line transmitted transgenic zebrafish under a stratified epithelial promoter from keratin8. Dev. Dyn. Off. Publ. Am. Assoc. Anatomists 223, 204–215.

Gupta, R.C., Srivastava, A., Lall, R., 2018. Toxicity potential of nutraceuticals. In: Nicolotti, O. (Ed.), Computational Toxicology. Methods and Protocoals. Humana Press, Springer Protocols, New York, pp. 367–394.

Gutierrez, H.C., Vacca, I., Schoenmacker, G., et al., 2019. Screening for drugs to reduce zebrafish aggression identifies caffeine and sildenafil. Eur. Neuropsychopharmacol. 30, 17–29.

Han, J., Ji, C., Guo, Y., et al., 2017. Mechanisms underlying melatonin-mediated prevention of fenvalerate-induced behavioral and oxidative toxicity in zebrafish. J. Toxicol. Environ. Health A 80, 1331–1341.

Harishkumar, R., Reddy, L.P.K., Karadkar, S.H., et al., 2019. Toxicity and selective biochemical assessment of quercetin, gallic acid, and curcumin in zebrafish. Biol. Pharm. Bull. 42, 1969–1976.

Harrington, L.B., Burstein, D., Chen, J.S., et al., 2018. Programmed DNA destruction by miniature CRISPR-Cas14 enzymes. Science 362, 839–842.

Hasumura, T., Shimada, Y., Kuroyanagi, J., et al., 2012. Green tea extract suppresses adiposity and affects the expression of lipid metabolism genes in diet-induced obese zebrafish. Nutr. Metab. 9, 73.

He, J.H., Gao, J.M., Huang, C.J., et al., 2014. Zebrafish models for assessing developmental and reproductive toxicity. Neurotoxicol. Teratol. 42, 35–42.

Hedayatirad, M., Mirvaghefi, A., Nematollahi, M.A., et al., 2019. Transgenerational disrupting impacts of atrazine in zebrafish: beneficial effects of dietary spirulina. Comp. Biochem. Physiol. Toxicol. Pharmacol. CBP 230, 108685.

Her, G.M., Chiang, C.C., Chen, W.Y., et al., 2003. In vivo studies of liver-type fatty acid binding protein (L-FABP) gene expression in liver of transgenic zebrafish (*Danio rerio*). FEBS Lett. 538, 125–133.

Her, G.M., Pai, W.Y., Lai, C.Y., et al., 2013. Ubiquitous transcription factor YY1 promotes zebrafish liver steatosis and lipotoxicity by inhibiting CHOP-10 expression. Biochim. Biophys. Acta 1831, 1037–1051.

Hirose, Y., Sugahara, K., Kanagawa, E., et al., 2016. Quercetin protects against hair cell loss in the zebrafish lateral line and guinea pig cochlea. Hearing Res. 342, 80–85.

Hu, W.H., Chan, G.K., Duan, R., et al., 2019a. Synergy of ginkgetin and resveratrol in suppressing VEGF-induced angiogenesis: a therapy in treating colorectal cancer. Cancers (Basel) 11, 1828.

Hu, W.H., Duan, R., Xia, Y.T., et al., 2019b. Binding of resveratrol to vascular endothelial growth factor suppresses angiogenesis by inhibiting the receptor signaling. J. Agric. Food Chem. 67, 1127–1137.

Huang, H., Vogel, S.S., Liu, N., et al., 2001. Analysis of pancreatic development in living transgenic zebrafish embryos. Mol. Cell Endocrinol. 177, 117–124.

Hwang, W.Y., Fu, Y., Reyon, D., et al., 2013. Efficient genome editing in zebrafish using a CRISPR-Cas system. Nat. Biotechnol. 31, 227–229.

Hwang, J.H., Kim, K.J., Ryu, S.J., et al., 2016. Caffeine prevents LPS-induced inflammatory responses in RAW264.7 cells and zebrafish. Chem. Biol. Interact. 248, 1–7.

Janssens, A., Silvestri, C., Martella, A., et al., 2018. Delta(9)-tetrahydrocannabivarin impairs epithelial calcium transport through inhibition of TRPV5 and TRPV6. Pharmacol. Res. 136, 83–89.

Jao, L.E., Wente, S.R., Chen, W., 2013. Efficient multiplex biallelic zebrafish genome editing using a CRISPR nuclease system. Proc. Nat. Acad. Sci. USA 110, 13904–13909.

Jensen, H.M., Korbut, R., Kania, P.W., et al., 2018. Cannabidiol effects on behaviour and immune gene expression in zebrafish (*Danio rerio*). PLoS One 13, e0200016.

Jeong, H., Yu, S.M., Kim, S.J., 2020. Inhibitory effects on melanogenesis by thymoquinone are mediated through the betacatenin pathway in B16F10 mouse melanoma cells. Int. J. Oncol. 56, 379–389.

Jinek, M., Chylinski, K., Fonfara, I., et al., 2012. A programmable dual-RNA-guided DNA endonuclease in adaptive bacterial immunity. Science 337, 816–821.

Karunarathne, W., Molagoda, I.M.N., Park, S.R., et al., 2019. Anthocyanins from *Hibiscus syriacus* L. inhibit melanogenesis by activating the ERK signaling pathway. Biomolecules 9, 645.

Khor, Y.M., Soga, T., Parhar, I.S., 2013. Caffeine neuroprotects against dexamethasone-induced anxiety-like behaviour in the Zebrafish (*Danio rerio*). Gen. Comp. Endocrinol. 181, 310–315.

Komor, A.C., Kim, Y.B., Packer, M.S., et al., 2016. Programmable editing of a target base in genomic DNA without double-stranded DNA cleavage. Nature 533, 420–424.

Krausz, A.E., Adler, B.L., Cabral, V., et al., 2015. Curcumin-encapsulated nanoparticles as innovative antimicrobial and wound healing agent. Nanomedicine 11, 195–206.

Kumar, K.J., Vani, M.G., Wang, S.Y., et al., 2013. In vitro and in vivo studies disclosed the depigmenting effects of gallic acid: a novel skin lightening agent for hyperpigmentary skin diseases. Biofactors 39, 259–270.

Lammer, E., Carr, G.J., Wendler, K., et al., 2009. Is the fish embryo toxicity test (FET) with the zebrafish (*Danio rerio*) a potential alternative for the fish acute toxicity test? Comp. Biochem. Physiol. Toxicol. Pharmacol: CBP 149, 196–209.

Lawrence, C., 2011. The reproductive biology and spawning of zebrafish in laboratory settings. In: McGrath, P. (Ed.), Zebrafish. pp. 1–13.

Lawson, N.D., Weinstein, B.M., 2002. In vivo imaging of embryonic vascular development using transgenic zebrafish. Dev. Biol. 248, 307–318.

Lee, G.H., Sung, S.Y., Chang, W.N., et al., 2012. Zebrafish larvae exposed to ginkgotoxin exhibit seizure-like behavior that is relieved by pyridoxal-5'-phosphate, GABA and anti-epileptic drugs. Dis. Model. Mechan 5, 785–795.

Lee, D.Y., Cha, B.J., Lee, Y.S., et al., 2015a. The potential of minor ginsenosides isolated from the leaves of *Panax ginseng* as inhibitors of melanogenesis. Int. J. Mol. Sci. 16, 1677–1690.

Lee, D.Y., Jeong, S.C., Jeong, Y.T., et al., 2015b. Antimelanogenic effects of picrionoside a isolated from the leaves of Korean Ginseng. Biol. Pharm. Bull. 38, 1663–1667.

Lee, D.Y., Jeong, Y.T., Jeong, S.C., et al., 2015c. Melanin biosynthesis inhibition effects of ginsenoside Rb2 isolated from *Panax ginseng* Berry. J. Microbiol. Biotechnol. 25, 2011–2015.

Lee, S.K., Oh, K.H., Chung, A.Y., et al., 2015d. Protective role of quercetin against cisplatin-induced hair cell damage in zebrafish embryos. Hum. Exp. Toxicol. 34, 1043–1052.

Lee, D.Y., Lee, J., Jeong, Y.T., et al., 2017a. Melanogenesis inhibition activity of floralginsenoside A from *Panax ginseng* berry. J. Ginseng Res. 41, 602–607.

Lee, S.J., Park, S.H., Chung, J.F., et al., 2017b. Homocysteine-induced peripheral microcirculation dysfunction in zebrafish and its attenuation by L-arginine. Oncotarget 8, 58264–58271.

Lee, D.Y., Kim, H.G., Lee, Y.G., et al., 2018. Isolation and quantification of ginsenoside Rh23, a new anti-melanogenic compound from the leaves of *Panax ginseng*. Molecules 23, 267.

Li, S., Lou, S., Lei, B.U., et al., 2011. Transcriptional profiling of angiogenesis activities of calycosin in zebrafish. Mol. BioSyst. 7, 3112–3121.

Li, X., Wang, X., Ye, H., et al., 2012. Barbigerone, an isoflavone, inhibits tumor angiogenesis and human non-small-cell lung cancer xenografts growth through VEGFR2 signaling pathways. Cancer Chemother. Pharmacol. 70, 425–437.

Li, S., Dang, Y., Zhou, X., et al., 2015. Formononetin promotes angiogenesis through the estrogen receptor alpha-enhanced ROCK pathway. Sci. Res. 5, 16815.

Li, M., Zhang, T., Jia, Y., et al., 2019a. Combined treatment of melatonin and sodium tanshinone IIA sulfonate reduced the neurological and cardiovascular toxicity induced by deltamethrin in zebrafish. Chemosphere 243, 125373.

Li, X., Zhang, Q., Peng, X., et al., 2019b. The effects of five types of tea solutions on epiboly process, neural and cardiovascular development, and locomotor capacity of zebrafish. Cell Biol. Toxicol. 35, 205–217.

Lima-Cabello, E., Diaz-Casado, M.E., Guerrero, J.A., et al., 2014. A review of the melatonin functions in zebrafish physiology. J. Pineal Res. 57, 1–9.

Lin, C., Wu, M., Dong, J., 2012. Quercetin-4'-O-beta-D-glucopyranoside (QODG) inhibits angiogenesis by suppressing VEGFR2-mediated signaling in zebrafish and endothelial cells. PLoS One 7, e31708.

Lister, J.A., Robertson, C.P., Lepage, T., et al., 1999. Nacre encodes a zebrafish microphthalmia-related protein that regulates neural-crest-derived pigment cell fate. Development 126, 3757–3767.

Liu, X.Y., Xiao, Y.K., Hwang, E., et al., 2019. Antiphotoaging and antimelanogenesis properties of ginsenoside C-Y, a ginsenoside Rb2 metabolite from American ginseng PDD-ginsenoside. Photochem. Photobiol. 95, 1412–1423.

Lo Cascio, P., Calabro, C., Bertuccio, C., et al., 2017. Effects of fasting and refeeding on the digestive tract of zebrafish (*Danio rerio*) fed with Spirulina (*Arthrospira platensis*), a high protein feed source. Nat. Prod. Res. 31, 1478–1485.

Lo Cascio, P., Calabro, C., Bertuccio, C., et al., 2018. Immunohistochemical characterization of PepT1 and ghrelin in gastrointestinal tract of zebrafish: effects of Spirulina vegetarian diet on the neuroendocrine system cells after alimentary stress. Front. Physiol. 9, 614.

Lombo, M., Gonzalez-Rojo, S., Fernandez-Diez, C., et al., 2019. Cardiogenesis impairment promoted by bisphenol A exposure is successfully counteracted by epigallocatechin gallate. Environ. Pollut. 246, 1008–1019.

Lu, X., Li, C., Liu, T., et al., 2018. Chemical analysis, pharmacological activity and process optimization of the proportion of bilobalide and ginkgolides in *Ginkgo biloba* extract. J. Pharm. Biomed. Anal. 160, 46–54.

Luo, L., Zhou, J., Zhao, H., et al., 2019a. The anti-inflammatory effects of formononetin and ononin on lipopolysaccharide-induced zebrafish models based on lipidomics and targeted transcriptomics. Metabolomics 15, 153.

Luo, Q., Liu, S., Xie, L., et al., 2019b. Resveratrol ameliorates glucocorticoid-induced bone damage in a zebrafish model. Front. Pharmacol. 10, 195.

Lutfi, E., Babin, P.J., Gutierrez, J., et al., 2017. Caffeic acid and hydroxytyrosol have anti-obesogenic properties in zebrafish and rainbow trout models. PLoS One 12, e0178833.

Ma, Y.B., Jia, P.P., Junaid, M., et al., 2018. Reproductive effects linked to DNA methylation in male zebrafish chronically exposed to environmentally relevant concentrations of di-(2-ethylhexyl) phthalate. Environ. Pollut. 237, 1050–1061.

Meguro, S., Hasumura, T., Hase, T., 2015. Body fat accumulation in zebrafish is induced by a diet rich in fat and reduced by supplementation with green tea extract. PLoS One 10, e0120142.

Mi, P., Zhang, Q.P., Li, S.B., et al., 2019. Melatonin protects embryonic development and maintains sleep/wake behaviors from the deleterious effects of fluorene-9-bisphenol in zebrafish (*Danio rerio*). J. Pineal Res. 66, e12530.

Mohanty, R., Das, S.K., Singh, N.R., et al., 2016. *Withania somnifera* Leaf extract ameliorates benzo[a]pyrene-induced behavioral and neuromorphological alterations by improving brain antioxidant status in zebrafish (*Danio rerio*). Zebrafish 13, 188–196.

Montalbano, G., Mania, M., Abbate, F., et al., 2018. Melatonin treatment suppresses appetite genes and improves adipose tissue plasticity in diet-induced obese zebrafish. Endocrine 62, 381–393.

Moss, L.G., Caplan, T.V., Moss, J.B., 2013. Imaging beta cell regeneration and interactions with islet vasculature in transparent adult zebrafish. Zebrafish 10, 249–257.

Musthafa, M.S., Athaullah, A., Anbumani, S., et al., 2017. Ameliorative efficacy of bioencapsulated *Chironomous* larvae with shilajit on zebrafish (*Danio rerio*) exposed to ionizing radiation. Appl. Radiat. Isot. 128, 108–113.

Muthuraman, A., Thilagavathi, L., Jabeen, S., et al., 2019. Curcumin prevents cigarette smoke extract induced cognitive impairment. Front. Biosci. (Elite Ed.) 11, 109–120.

Nam, Y.H., Moon, H.W., Lee, Y.R., et al., 2019. *Panax ginseng* (Korea Red Ginseng) repairs diabetic sensorineural damage through promotion of the nerve growth factor pathway in diabetic zebrafish. J. Ginseng Res. 43, 272–281.

Neveux, S., Smith, N.K., Roche, A., et al., 2017. Natural compounds as cccult ototoxins? *Ginkgo biloba* flavonoids moderately damage lateral line hair cells. J. Assoc. Res. Otolaryngol. 18, 275–289.

Nguyen, T., Payan, B., Zambrano, A., et al., 2019. Epigallocatechin-3-gallate suppresses neutrophil migration speed in a transgenic zebrafish model accompanied by reduced inflammatory mediators. J. Inflamm. Res. 12, 231–239.

Nishimura, Y., Sasagawa, S., Ariyoshi, M., et al., 2015. Systems pharmacology of adiposity reveals inhibition of EP300 as a common therapeutic mechanism of caloric restriction and resveratrol for obesity. Front. Pharmacol. 6, 199.

Njiwa, J.R.K., Suter, M.J.F., Eggen, R.I., 2010. Zebrafish embryo toxicity assay, combining molecular and integrative endpoints at various developmental stages. In: Timmis, K.N. (Ed.), Handbook of Hydrocarbon and Lipid Microbiology. Springer, Berlin, Heidelberg, pp. 4481–4489. <https://doi.org/10.1007/978-3-540-77587-4_351>.

Oh, K.H., Rah, Y.C., Hwang, K.H., et al., 2017. Melatonin mitigates neomycin-induced hair cell injury in zebrafish. Drug Chem. Toxicol. 40, 390–396.

Oyelaja-Akinsipo, O.B., Dare, E.O., Katare, D.P., 2020. Protective role of diosgenin against hyperglycaemia-mediated cerebral ischemic brain injury in zebrafish model of type II diabetes mellitus. Heliyon 6, e03296.

Pal, T., Mohiyuddin, S., Packirisamy, G., 2018. Facile and green synthesis of multicolor fluorescence carbon dots from curcumin: *In Vitro* and *in Vivo* bioimaging and other applications. ACS Omega 3, 831–843.

Paramasivam, A., Kalaimangai, M., Sambantham, S., et al., 2012. Anti-angiogenic activity of thymoquinone by the down-regulation of VEGF using zebrafish (*Danio rerio*) model. Biomed. Prev. Nutr. 2, 169–173.

Pardal, D., Caro, M., Tueros, I., et al., 2014. Resveratrol and piceid metabolites and their fat-reduction effects in zebrafish larvae. Zebrafish 11, 32–40.

Pei, D.S., Strauss, P.R., 2013. Zebrafish as a model system to study DNA damage and repair. Mut. Res. 743-744, 151–159.

Pei, D.S., Yang, X.J., Liu, W., et al., 2011. A novel regulatory circuit in base excision repair involving AP endonuclease 1, Creb1 and DNA polymerase beta. Nucleic Acids Res. 39, 3156–3165.

Pei, D., Luther, W., Wang, W., et al., 2013. Distinct neuroblastoma-associated alterations of PHOX2B impair sympathetic neuronal differentiation in zebrafish models. PLoS Genet. 9, e1003533.

Perner, B., Englert, C., Bollig, F., 2007. The Wilms tumor genes wt1a and wt1b control different steps during formation of the zebrafish pronephros. Dev. Biol. 309, 87–96.

Pyun, J.H., Kang, S.U., Hwang, H.S., et al., 2011. Epicatechin inhibits radiation-induced auditory cell death by suppression of reactive oxygen species generation. Neuroscience 199, 410–420.

Rah, Y.C., Yoo, M.H., Choi, J., et al., 2017. In vivo assessment of hair cell damage and developmental toxicity caused by gestational caffeine exposure using zebrafish (*Danio rerio*) models. Neurotoxicol. Teratol. 64, 1–7.

Rajamohamed, B.S., Siddharthan, S., 2019. Modulatory effects of Amukkara Choornam on *Candida albicans* biofilm: in vitro and in vivo study. Mol. Biol. Rep. 46, 2961–2969.

Rajasekar, P., Palanisamy, S., Anjali, R., et al., 2019. Isolation and structural characterization of sulfated polysaccharide from *Spirulina platensis* and its bioactive potential: in vitro antioxidant, antibacterial activity and zebrafish growth and reproductive performance. Int. J. Biol. Macromol. 141, 809–821.

Ran, G., Ying, L., Li, L., et al., 2017. Resveratrol ameliorates diet-induced dysregulation of lipid metabolism in zebrafish (*Danio rerio*). PLoS One 12, e0180865.

Ren, D.L., Li, Y.J., Hu, B.B., et al., 2015a. Melatonin regulates the rhythmic migration of neutrophils in live zebrafish. J. Pineal Res. 58, 452–460.

Ren, D.L., Sun, A.A., Li, Y.J., et al., 2015b. Exogenous melatonin inhibits neutrophil migration through suppression of ERK activation. J. Endocrinol. 227, 49–60.

Ren, D.L., Ji, C., Wang, X.B., et al., 2017. Endogenous melatonin promotes rhythmic recruitment of neutrophils toward an injury in zebrafish. Sci. Rep. 7, 4696.

Richetti, S.K., Blank, M., Capiotti, K.M., et al., 2011. Quercetin and rutin prevent scopolamine-induced memory impairment in zebrafish. Behav. Brain Res. 217, 10–15.

Rishitha, N., Muthuraman, A., 2018. Therapeutic evaluation of solid lipid nanoparticle of quercetin in pentylenetetrazole induced cognitive impairment of zebrafish. Life Sci. 199, 80–87.

Rocco, L., Mottola, F., Santonastaso, M., et al., 2015. Anti-genotoxic ability of alpha-tocopherol and Anthocyanin to counteract fish DNA damage induced by musk xylene. Ecotoxicology 24, 2026–2035.

Rosa, L.V., Ardais, A.P., Costa, F.V., et al., 2018. Different effects of caffeine on behavioral neurophenotypes of two zebrafish populations. Pharmacol. Biochem. Behav. 165, 1–8.

Ruiz-Oliveira, J., Silva, P.F., Luchiari, A.C., 2019. Coffee time: low caffeine dose promotes attention and focus in zebrafish. Learn. Behav. 47, 227–233.

Samarut, E., Nixon, J., Kundap, U.P., et al., 2019. Single and synergistic effects of cannabidiol and delta-9-Tetrahydrocannabinol on zebrafish models of neuro-hyperactivity. Front. Pharmacol. 10, 226.

Schirmer, H., Pereira, T.C., Rico, E.P., et al., 2012. Modulatory effect of resveratrol on SIRT1, SIRT3, SIRT4, PGC1alpha and NAMPT gene expression profiles in wild-type adult zebrafish liver. Mol. Biol. Rep. 39, 3281–3289.

Seo, M.J., Lee, Y.J., Hwang, J.H., et al., 2015. The inhibitory effects of quercetin on obesity and obesity-induced inflammation by regulation of MAPK signaling. J. Nutr. Biochem. 26, 1308–1316.

Serra, I., Scheldeman, C., Bazelot, M., et al., 2019. Cannabidiol modulates phosphorylated rpS6 signalling in a zebrafish model of Tuberous Sclerosis Complex. Behav. Brain Res. 363, 135–144.

Sheng, W., Lu, Y., Mei, F., et al., 2018. Effect of resveratrol on sirtuins, OPA1, and Fis1 expression in adult zebrafish retina. Invest. Ophthalmol. Vis. Sci. 59, 4542–4551.

Shiau, R.J., Shih, P.C., Wen, Y.D., 2011. Effect of silymarin on curcumin-induced mortality in zebrafish (*Danio rerio*) embryos and larvae. Ind. J. Exp. Biol. 49, 491–497.

Shin, H.A., Shin, Y.S., Kang, S.U., et al., 2014. Radioprotective effect of epicatechin in cultured human fibroblasts and zebrafish. J. Radiat. Res. 55, 32–40.

Silvestri, C., Paris, D., Martella, A., et al., 2015. Two non-psychoactive cannabinoids reduce intracellular lipid levels and inhibit hepatosteatosis. J. Hepatol. 62, 1382–1390.

Singulani, J.L., Scorzoni, L., Gomes, P.C., et al., 2017. Activity of gallic acid and its ester derivatives in *Caenorhabditis elegans* and zebrafish (*Danio rerio*) models. Future Med. Chem. 9, 1863–1872.

Sivamani, S., Joseph, B., Kar, B., 2014. Anti-inflammatory activity of *Withania somnifera* leaf extract in stainless steel implant induced inflammation in adult zebrafish. J. Genet. Eng. Biotechnol. 12, 1–6.

Stewart, A.M., Kalueff, A.V., 2014. The behavioral effects of acute Delta(9)-tetrahydrocannabinol and heroin (diacetylmorphine) exposure in adult zebrafish. Brain Res. 1543, 109–119.

Sun, H., Zhang, Y., Shen, Y., et al., 2017. Inhibitory effects of red wine on lipid oxidation in fish oil emulsion and angiogenesis in zebrafish embryo. J. Food Sci. 82, 781–786.

Takemoto, Y., Hirose, Y., Sugahara, K., et al., 2018. Protective effect of an astaxanthin nanoemulsion against neomycin-induced hair-cell damage in zebrafish. Auris Nasus Larynx 45, 20–25.

Tang, J.Y., Li, S., Li, Z.H., et al., 2010. Calycosin promotes angiogenesis involving estrogen receptor and mitogen-activated protein kinase (MAPK) signaling pathway in zebrafish and HUVEC. PLoS One 5, e11822.

Tang, Y.Q., Li, Z.R., Zhang, S.Z., et al., 2019. Venlafaxine plus melatonin ameliorate reserpine-induced depression-like behavior in zebrafish. Neurotoxicol. Teratol. 76, 106835.

Techer, D., Milla, S., Fontaine, P., et al., 2017. Influence of waterborne gallic and pelargonic acid exposures on biochemical and reproductive parameters in the zebrafish (*Danio rerio*). Environ. Toxicol. 32, 227–240.

Teyssier, L., Colussi, J., Delemasure, S., et al., 2017. Inflammatory effects of the plant protection product stifenia (FEN560) on vertebrates. Front. Public Health 5, 74.

Thakur, R.S., Ahirwar, B., 2019. A steroidal derivative from *Trigonella foenum graecum* L. that induces apoptosis in vitro and in vivo. J. Food Drug Anal. 27, 231–239.

Truong, L., Harper, S.L., Tanguay, R.L., 2011. Evaluation of embryotoxicity using the zebrafish model. Methods Mol. Biol. 691, 271–279.

Urnov, F.D., Rebar, E.J., Holmes, M.C., et al., 2010. Genome editing with engineered zinc finger nucleases. Nat. Rev. Genet. 11, 636–646.

Uusi-Heikkilä, S., Bierbach, D., Alós, J., et al., 2018. Relatively large males lower reproductive success in female zebrafish. Environ. Biol. Fishes 101, 1625–1638.

Valim Brigante, T.A., Abe, F.R., Zuardi, A.W., et al., 2018. Cannabidiol did not induce teratogenicity or neurotoxicity in exposed zebrafish embryos. Chem. Biol. Interact. 291, 81–86.

Vaughan, M., van Egmond, R., 2010. The use of the zebrafish (*Danio rerio*) embryo for the acute toxicity testing of surfactants, as a possible alternative to the acute fish test. Altern. Lab. Anim. ATLA 38, 231–238.

Vuilleumier, R., Besseau, L., Boeuf, G., et al., 2006. Starting the zebrafish pineal circadian clock with a single photic transition. Endocrinology 147, 2273–2279.

Wang, Y.H., Wen, C.C., Yang, Z.S., et al., 2009. Development of a whole-organism model to screen new compounds for sun protection. Mar. Biotechnol. 11, 419–429.

Wang, N., Luo, Z., Jin, M., et al., 2019. Exploration of age-related mitochondrial dysfunction and the anti-aging effects of resveratrol in zebrafish retina. Aging (Albany, NY.) 11, 3117–3137.

Wang, J., Zhang, C., Zhang, J., Xie, J., et al., 2020. The effects of quercetin on immunity, antioxidant indices, and disease resistance in zebrafish (*Danio rerio*). Fish Physiol. Biochem. 46, 759–770.

Weigt, S., Huebler, N., Strecker, R., et al., 2011. Zebrafish (*Danio rerio*) embryos as a model for testing proteratogens. Toxicology 281, 25–36.

White, R.M., Sessa, A., Burke, C., et al., 2008. Transparent adult zebrafish as a tool for in vivo transplantation analysis. Cell Stem Cell 2, 183–189.

Wu, J.Y., Lin, C.Y., Lin, T.W., et al., 2007. Curcumin affects development of zebrafish embryo. Biol. Pharm. Bull. 30, 1336–1339.

Wu, Q., Deng, S., Li, L., et al., 2013. Biodegradable polymeric micelle-encapsulated quercetin suppresses tumor growth and metastasis in both transgenic zebrafish and mouse models. Nanoscale 5, 12480–12493.

Wu, H.J., Fong, T.H., Chen, S.L., et al., 2015. Perturbation of cytosolic calcium by 2-aminoethoxydiphenyl borate and caffeine affects zebrafish myofibril alignment. J. Appl. Toxicol. 35, 287–294.

Wu, M., Gu, J., Mei, S., et al., 2016. Resveratrol delays polycystic kidney disease progression through attenuation of nuclear factor kappaB-induced inflammation. Nephrol. Dial. Transplant. 31, 1826–1834.

Xu, H., Gong, Z., Zhou, S., et al., 2018. Liposomal curcumin targeting endometrial cancer through the NF-kappaB Pathway. Cell Physiol. Biochem. 48, 569–582.

Yadav, P., Bandyopadhyay, A., Chakraborty, A., et al., 2018. Enhancement of anticancer activity and drug delivery of chitosan-curcumin nanoparticle via molecular docking and simulation analysis. Carbohydr. Polym. 182, 188–198.

Yang, J.H., Hu, J., Wan, L., et al., 2014. Barbigerone inhibits tumor angiogenesis, growth and metastasis in melanoma. Asian Pac. J. Cancer Prev. 15, 167–174.

Yeh, C.H., Liao, Y.F., Chang, C.Y., et al., 2012. Caffeine treatment disturbs the angiogenesis of zebrafish embryos. Drug Chem. Toxicol. 35, 361–365.

Zang, L., Shimada, Y., Nakayama, H., et al., 2019. RNA-seq based transcriptome analysis of the anti-obesity effect of green tea extract using zebrafish obesity models. Molecules 24, 3256.

Zhang, Z.J., Cheang, L.C., Wang, M.W., et al., 2011. Quercetin exerts a neuroprotective effect through inhibition of the iNOS/NO system and pro-inflammation gene expression in PC12 cells and in zebrafish. Int. J. Mol. Med. 27, 195–203.

Zhang, J.L., Laurence Souders 2nd, C., Denslow, N.D., et al., 2017. Quercetin, a natural product supplement, impairs mitochondrial bioenergetics and locomotor behavior in larval zebrafish (*Danio rerio*). Toxicol. Appl. Pharmacol. 327, 30–38.

Zhang, Y., Zhou, Y., Tang, Q., et al., 2018. The protective effects of selenium-enriched spirulina on the reproductive system of male zebrafish (*Danio rerio*) exposed to beta-cypermethrin. Food Funct. 9, 5791–5804.

Zhao, D., Qin, C., Fan, X., et al., 2014. Inhibitory effects of quercetin on angiogenesis in larval zebrafish and human umbilical vein endothelial cells. Eur. J. Pharmacol. 723, 360–367.

Zheng, X., Dai, W., Chen, X., et al., 2015. Caffeine reduces hepatic lipid accumulation through regulation of lipogenesis and ER stress in zebrafish larvae. J. Biomed. Sci. 22, 105.

Chapter 28

Caenorhabditis elegans: an elegant model organism for evaluating the neuroprotective and neurotherapeutic potential of nutraceuticals

Altaf S. Darvesh, Reid E. Barnett and Vanessa A. Fitsanakis

Introduction

The term nutraceuticals encompasses a wide range of natural products. These can include, among others, fruits, vegetables, nuts, spices, vitamins, minerals, and herbs. The medicinal properties of these products have been used for millennia in various cultures across the world to prevent and cure disease. In recent decades, nutraceuticals, as isolated compounds and nutrients, as well as botanical extracts, have gained tremendous popularity. Nutraceuticals are often consumed as part of the diet, but also taken as dietary supplements that are available over-the-counter (OTC) (Chauhan et al., 2013; Nasri et al., 2014). The term nutraceutical was coined by Dr. Stephen DeFelice in the 1980s as a portmanteau word of "nutrition" and "pharmaceutical." Thus, a nutraceutical is generally recognized as a substance that may be considered food, or part of food, provides medicinal or health benefits, and can be useful for prevention and treatment of illness. It was Hippocrates, recognized as the Father of Medicine, who famously said "*Let food be thy medicine and medicine be thy food.*" In doing so, he emphasized the health benefits and curative potential of foods, and the relationship between good health and healthy eating (DeFelice, 1992; Andlauer and Furst, 2002).

The prophylactic and therapeutic benefits of nutraceuticals have been critically evaluated in recent years. As the sale and consumption of nutraceuticals and dietary supplements has increased, there is considerable debate about their potential usefulness, as well as their ability to cause harm (Aronson, 2017). There is lack of clarity and information regarding the associated health claims of preventing and curing disease, appropriate dosing, manufacturing standards, safety and adulteration information, and regulations (Santini et al., 2018). As the use of dietary supplements and nutraceuticals has grown into a multibillion dollar market, their reported adverse and toxic effects, as well as drug interactions, are on the rise. Lack of significant regulation further compounds the confusion surrounding the proper use of dietary supplements and nutraceuticals (Young and Bass, 1995; Jiang, 2009; Frankos et al, 2010; Schmitz et al., 2014; Ronis et al., 2018).

In view of the aforementioned concerns, there exists a critical need to systematically, and rigorously, evaluate the prophylactic and therapeutic properties of dietary supplements and nutraceuticals. In order to benefit from the use of nutraceuticals and prevent harm, it is imperative to conduct systematic preclinical and clinical evaluations. Since clinical trials require considerable time, expense, and effort, conducting mechanistic pharmacological and toxicological evaluation of nutraceuticals in model organisms before proceeding with clinical investigation is imperative. This approach is supported by various governmental agencies that fund research for dietary supplements and nutraceuticals, such as the National Center for Complementary and Integrative Health (https://www.nccih.nih.gov/) and the Office of Dietary Supplements (https://ods.od.nih.gov/). Preclinical investigation of dietary supplements and nutraceuticals utilizes both in vitro and in vivo models. Although use of traditional rodent and nonrodent mammalian models is essential for regulatory approval, early studies can utilize high-throughput techniques to elucidate mechanisms of action, as well as the pharmacological and toxicological profile of

Nutraceuticals. DOI: https://doi.org/10.1016/B978-0-12-821038-3.00028-8

nutraceutical molecules. In vitro models are cost- and time-effective, and often utilize cell culture, and isolated proteins, peptides, lipids, and nucleic acids to provide relevant mechanistic information. They are unable, however, to provide the big picture that in vivo models can due to the presence and integration of multiple organ systems (Nasri et al., 2014; Dwyer et al., 2018; Kellogg et al., 2019). The nematode *Caenorhabditis elegans* (*C. elegans*) is an excellent in vivo screening tool, and it provides a more efficient and economical means to screen nutraceuticals than traditional animal models (O'Reilly et al., 2014; Carretero et al., 2017; Hunt, 2017). This chapter highlights the usefulness of *C. elegans* as a useful model to evaluate the neurotherapeutic and neuroprotective potential of dietary supplements and nutraceuticals.

Caenorhabditis elegans as a model organism

Model organisms are those used in research settings as surrogates for humans. Inbred rodents, such as rats and mice, have been widely used for years in pharmaceutical, toxicological, and nutraceutical testing. They can be expensive to maintain, however, and they require special facilities for housing (Committee, 2011). Members of both the scientific and animal welfare communities have encouraged researchers to reduce the number of vertebrates in scientific research. In fact, the goal of finding ways to reduce, refine, or replace (known as the 3Rs) laboratory animals is encouraged by many professional and scientific organizations, such as the Society of Toxicology (Society of Toxicology, 2016). In addition to ethical concerns regarding animal testing, it is quite difficult and expensive to use mammals for medium- or high-throughput screening of new compounds (Giacomotto and Segalat, 2010). The push to screen large numbers of compounds, collect "omics" data, and identify novel genes or pathways important in diseases and their therapies have increased the use of alternative model organisms (Juntti, 2019).

Studies with bacteria (Burby and Simmons, 2020) and primary or immortalized cell cultures (Hirsch and Schildknecht, 2019) have long supplemented in vivo studies. These in vitro model systems, however, lack the complexity of eukaryotic organisms. In recognition of the need to combine simplicity with a robust in vivo system, Noble Laureate Sydney Brenner proposed using the nematode *C. elegans* as a model organism for basic cellular and developmental biology research (Brenner, 1974). He reasoned that its small size (1 mm), rapid reproduction cycle (Byerly et al., 1976), transparent nature, short life span of around 21 days (Geanacopoulos, 2004), and ability to feed on the ubiquitous bacteria, *Escherichia coli* (*E. coli*), fulfilled many of the characteristics required of a successful model organism (Bolker, 2019). *C. elegans* also have two sexes—male and hermaphrodite. This means that a single hermaphrodite *C. elegans* can give rise to an entire population, an advantage when generating mutants through random mutagenesis (Corsi et al., 2015).

Scientists moved rapidly to further characterize this worm, which enhanced its usefulness as a model organism. The fate of each cell from fertilization to adult, which is highly conserved, was published in the early 1980s (Sulston et al., 1983). This cell fate mapping was facilitated by the fact that each hermaphrodite has 959 cells, and each male, due to additional reproductive cells, has a total of 1031. In 1986, the specific neuronal connections (connectome) of the entire nervous system were mapped and published (White et al., 1986). Later, the publication of the complete genome of *C. elegans* (Elegans Sequencing Consortium, 1998) enabled scientists to identify homologs and orthologs of important human genes (Shaye and Greenwald, 2011). Once the genome was published, numerous research groups began comparing the nervous system anatomy and physiology of this worm to that of humans. This included signal transduction (Bargmann and Kaplan, 1998), neuronal gene expression (Von Stetina et al., 2007), neuronal control of behavior (Singh and Aballay, 2019), and growth and development of dendritic spines (Cuentas-Condori et al., 2019).

The tractability of the worm's genome has also enabled the scientific community to make numerous mutants, many of which are available through the *Caenorhabditis* Genetics Center for a nominal fee to researchers (https://cgc.umn.edu/). Many strains originally had green fluorescent protein (GFP) incorporated into their genome to enable visualization of discrete neuronal populations (Boulin et al., 2006). Later, other fluorescent proteins (red, yellow, etc.) were integrated into various genes so that multiple proteins can be differentially labeled in vivo (Toyoshima et al., 2020; Feng et al., 2019). Strains were also developed with inserts of human disease-associated genes (Table 28.1), including genes from Parkinson disease (PD) and Alzheimer disease (AD), to better model the disease and investigate relevant pathogenic pathways (Bornhorst et al., 2014; Springer et al., 2005; Mukherjee et al., 2017; Sutphin et al., 2017; Wang et al., 2017). Finally, research in *C. elegans* has been instrumental in characterizing gene-editing techniques, including RNA interference (RNAi) (Apfeld and Alper, 2018; Simpson et al., 2012; Billi et al., 2014) and clustered regularly interspaced short palindromic repeats (CRISPR)-CRISPR-associated protein 9 (CAS 9) mechanisms (Kim and Colaiacovo, 2019).

TABLE 28.1 Worm strains with disease-related mutations or insertions.

Disease models	Strain name	Description
Alzheimer disease	BR5270	Pan-neuronal overexpression of the F3 pro-aggregation fragment of the human tau with K280 deleted
	BR5271	Control strain for BR5270 (above). Contains two isoleucines to proline substitutions in the hexapeptide motifs of the repeat region
	CL4176	This strain produces low levels of Aβ(1-42) peptide
	GMC101	Expression of full-length human Aβ(1-42) peptide in body-wall muscle cells that aggregate in vivo
Parkinson disease	PE871	YFP expression associated with α-synuclein in body-wall muscle cells
	PE872	YFP expression associated with α-synuclein in body-wall muscle cells
	RB2547	Knockout of PINK1
	VC1024	Knockout of *pdr-1*, the ortholog of human *parkin*
	WLZ1	Insert of normal human *LRRK2*
	WLZ3	Insert of human *LRRK2* with the G2019S mutation

Aβ, Amyloid beta; *LRRK2*, leucine-rich repeat kinase 2; *PDR-1*, Parkinson disease related; *PINK1*, PTEN-induced putative kinase; *YFP*, yellow fluorescent protein.
From the *Caenorhabditis* Genetics Center (CGC) at https://cgc.umn.edu/.

The *Caenorhabditis elegans* nervous system

One of the first systems described and characterized in *C. elegans* was the nervous system (Sulston et al., 1983; White et al., 1986). Compared to other organisms, *C. elegans* has very few neurons: the hermaphrodite has 302, and the male has 385 (Hobert, 2010). Although initially controversial, it is now generally accepted that *C. elegans* also have glial cells that provide support for the associated neurons (Stout et al., 2014; Singhvi and Shaham, 2019). While the pathways and connections among neurons in *C. elegans* typically are much simpler than those in mammals, many characteristics of the human nervous system are highly conserved. For example, *C. elegans* have both ionotropic (Brockie and Maricq, 2006) and metabotropic (Chun et al., 2015) receptors, neuropeptide (Clark et al., 2018a,b) and classic (Gasnier, 2000) neurotransmitters, as well as neurotransmitter synthesis, storage, and release proteins (Hobert, 2013).

These similarities between humans and *C. elegans* have facilitated research into complex physiology and pathophysiology processes. For example, by inserting fluorescent tags and/or human genes into the worm's genome, researchers have gained a better understanding of mechanisms associated with neurodegeneration (Smith et al., 2019; Gaeta et al., 2019), neurotoxicity (Peres et al., 2019; Montgomery et al., 2018), and neurodevelopment (Wong et al., 2019). In the case of fluorescence, individual neuronal populations can be tagged (Hutter, 2004) so that consequences of toxic or therapeutic chemical exposure can be visualized (Negga et al., 2012). Many of these fluorescent-reporter strains are also available through the *Caenorhabditis* Genetics Center (Table 28.2). Furthermore, multiple proteins can be tagged simultaneously to assess neuron-specific processes. Some labs have used this approach to follow mitochondrial dynamics (Laker et al., 2014) or assess oxidative stress in discrete neuron populations (Young et al., 2019).

Important target pathways in *Caenorhabditis elegans*

In addition to early studies involving the nervous system, *C. elegans* were also used to evaluate and characterize processes instrumental to cell and developmental biology. As such, many important metabolic and regulatory pathways were quickly delineated. Among these were the transforming growth factor-β (TGF-β) and the insulin/insulin-like growth factor-1 (IGF-1) pathways (Gumienny and Savage-Dunn, 2013; Murphy and Hu, 2013). Damage to or interference with either pathway can lead to or hasten disease progression. On the other hand, molecules in both pathways are often targeted by pharmaceuticals or nutraceuticals with the intent of inhibiting or ameliorating disease progression. While the pathways delineated in *C. elegans* typically are much simpler than those in humans, in these cases both are highly evolutionarily conserved, and similar to those in humans (Gumienny and Savage-Dunn, 2013; Murphy and Hu, 2013). Although it is outside the scope of this chapter to fully delineate the TGF-β or insulin/IGF-1

TABLE 28.2 Worm strains with fluorescent protein-labeled neuronal populations.

Neuron populations	Strain name	Type of tag
Pan-neuronal	NW1229	GFP expression in all neuron populations
	OH441	GFP expression in all neuron populations
Acetylcholine	OH11319	GFP expression in DA/DB/VA/VB neurons
GABA	BN544	GFP expression in all GABAergic neurons following heat shock
	EG1285	GFP expression in all GABAergic neurons
	ZM7648	GFP expression in all GABAergic neurons; YFP expression in glutamate receptor (glr-1) containing interneurons
Dopamine	BZ555	GFP expression in dopaminergic neurons
	BN617	GFP expression in all dopaminergic neurons following heat shock
	OH7547	GFP expression in all dopaminergic neurons; RFP expression in all other neurons
	XE1311	GFP expression in all dopaminergic neurons
Serotonin	BN499	GFP expression in all serotonergic neurons following heat shock

DA, Dorsal A motor neurons located in the ventral cord; *DB*, dorsal B motor and proprioceptive neurons located in the ventral cord; *GABA*, γ-aminobutyric acid; *GFP*, green fluorescent protein; *glr-1*, glutamate receptor-1; *RFP*, red fluorescent protein; *VA*, ventral A motor neurons located in the ventral cord; *VB*, ventral B motor and proprioceptive neurons located in the ventral cord; *YFP*, yellow fluorescent protein.
From the *Caenorhabditis* Genetics Center (CGC) at https://cgc.umn.edu/.

signaling (IIS), their primary ligands or transcription factors related to relevant nutraceuticals are presented.

Transforming growth factor-β pathway in Caenorhabditis elegans

In the TGF-β pathway (Fig. 28.1), there are five known peptide ligands in *C. elegans*: ***D**rosophila* deca**p**enta**p**legic (Dpp)/vertebrate **b**one morphogenetic proteins (BMPs)-**l**ike (DBL-1); abnormal **da**uer **f**ormation-7 protein (DAF-7), **unc**oordinated-129 (UNC-129); and **t**ransforming **g**rowth factor-1 and -2 (TIG-1 and TIG-2) proteins (Table 28.3). Since the roles of DBL-1, DAF-7, and UNC-129 are fairly well established, only these pathways will be further discussed.

In addition to regulating the size of *C. elegans*, the DBL-1 ligand also modulates innate immunity, reproductive aging, and longevity (Lakdawala et al., 2019; Clark et al., 2018a,b). In general, DBL-1 binding results in activation of the SMA-2, -3, and -4 pathways, leading to nuclear translocation of these transcription factors. DAF-7 also regulates the response of *C. elegans* to environmental changes, which leads to the regulation of fat storage, alterations in foraging behavior, and modification of feeding rates (Gumienny and Savage-Dunn, 2013). Through DAF-7, TGF-β and IIS pathways complement each other to provide a level of functional redundancy (Shaw et al., 2007). This pathway can also affect neurodevelopment through UNC-129. In this case, *unc-129* expression facilitates the long-range migration of motor neurons during development (MacNeil et al., 2009). Thus, nutraceuticals that function as UNC-129 agonists may lead to improper axon guidance cues and adversely affect nervous system plasticity or development.

Insulin/IGF-1 signaling pathway in Caenorhabditis elegans

The ligands for the IIS pathway (Fig. 28.2) are classified as **i**nsulin-**l**ike **p**eptides (ILPs). These ILPs activate DAF-2/IGF transmembrane receptors, the only identified insulin receptor family in *C. elegans* (Kimura et al., 1997). Ultimately, this activation results in the nuclear translocation of the transcription factors DAF-16/**fo**rkhead bo**x** (FoxO), **h**eat **s**hock transcription **f**actor-1 (HSF-1), and **skin**head-1 (SKN-1)/**n**ose **r**esistant to **f**luoxetine (Kimura et al., 2011). Similar to the TGF-β pathway, the IIS pathway also plays a seminal role in dauer formation, fat storage, longevity, and development (Murphy and Hu, 2013). Unlike that pathway, however, over 40 ILPs have been identified in *C. elegans* (Table 28.2). This enables the system to respond to numerous environmental and extracellular cues that require metabolic changes to promote survivability. Within the organism, it is well-documented that some of the ILPs act in agonist/antagonist pairs (Chen et al., 2013). The target molecule HSF-1, as the name implies, is activated during times of stress, particularly stress related to extreme changes in temperature (Morley and Morimoto, 2004). The result is the transcription of multiple

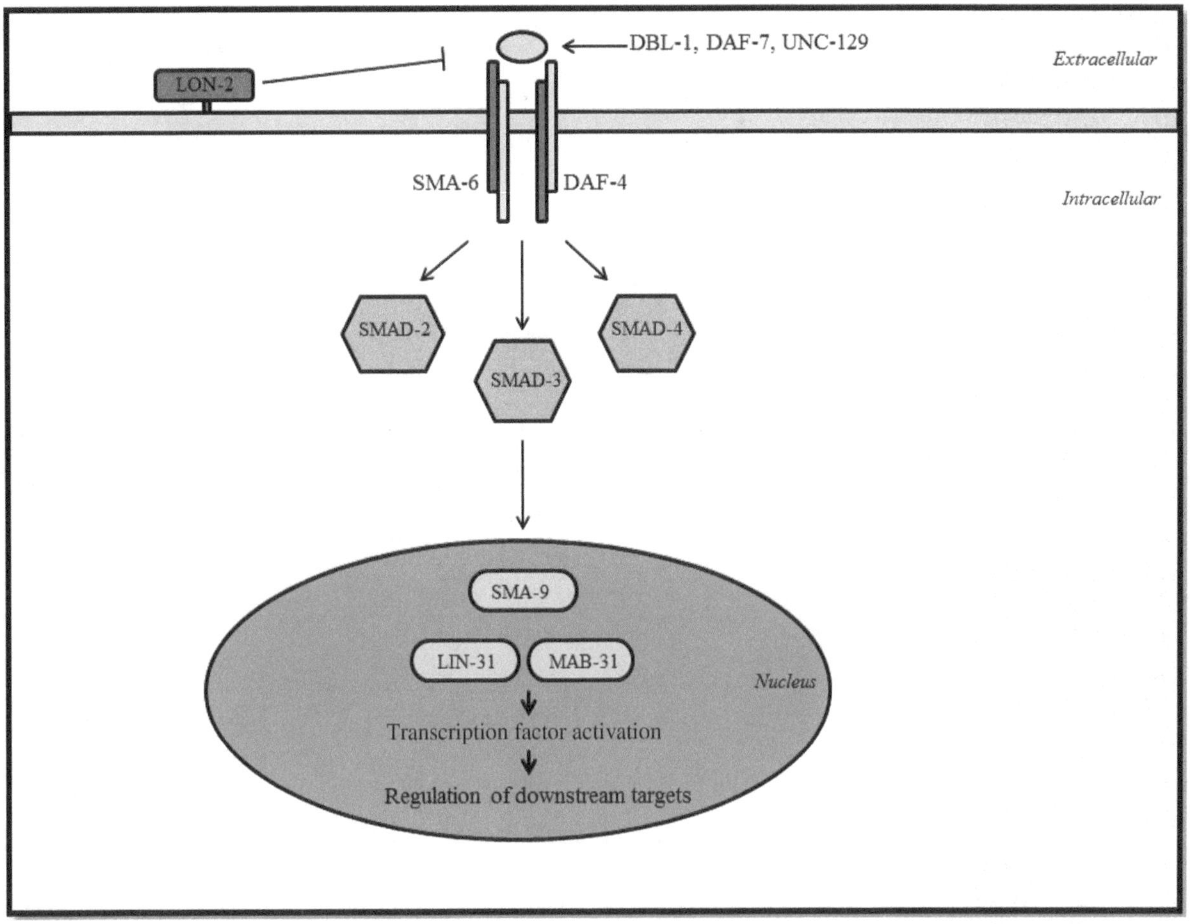

FIGURE 28.1 Transforming growth factor-β (TGF-β) pathway. DBL-1 ligand binding regulates innate immunity, body size, male tail morphogenesis and longevity. The DBL-1 signal is transduced via SMA-6 and DAF-4 receptors allowing phosphorylation of the various Smads.
Abbrevations: *DAF*, Abnormal dauer formation; *DBL*, Drosophila decapentaplegic (Dpp)/vertebrate bone morphogenetic proteins (BMPs)-like; *LIN*, Abnormal cell lineage; *LON*, Long body size (phenotype); *MAB*, Male abnormal (phenotype); *SMA*, Small body size (phenotype); *SMAD*, Caenorhabditis elegans; *UNC*, Uncoordinated.

heat shock chaperone proteins. Similarly, SKN-1/Nrf is important in upregulating the transcription of antioxidant proteins (Blackwell et al., 2015). Although HSF-1 and SKN-1/Nrf are important in providing protection, the major transcription factor regulated in this pathway is DAF-16/FoxO. Many of the products of the target genes of DAF-16/FoxO play pivotal roles in the immune response, cell proliferation, and metabolism (Tissenbaum, 2018).

When DAF-16/FoxO transcription factors are activated, they upregulate the expression of protein products involved in stress responses. For example, superoxide dismutase-3 (*sod-3*) and catalase-1 (*ctl-1*) are targets, as are metallothionein-1 (*mtl-1*) and various heat shock proteins (Lin et al., 2018). DAF-16/FoxO also transcriptionally regulates enzymes involved in biotransformation. These targets include alcohol dehydrogenases (*sodh-1*), and various P450 subtypes (*cyp-13A4*, *cyp-35B1*, *cyp-34A9*), which detoxicate numerous xenobiotic insults (Oliveira et al., 2009).

Screening strategies for *Caenorhabditis elegans*

Using *C. elegans* in high-throughput screening can be an extremely valuable tool for nutraceutical researchers focused on finding novel bioactive phytochemicals. Some characteristics of these worms that are particularly easy to exploit in screening platforms include their transparency, cost effectiveness for generating and maintaining large populations, and the limited laboratory equipment required for worm husbandry (Stiernagle, 2006). Once a lab is established, mutant strains of *C. elegans*, or RNAi and CRISPR-Cas9 gene editing can be combined with optical (typically descriptive) or biochemical (typically mechanistic) techniques (Cipriani and Piano, 2011; Palikaras and Tavernarakis, 2017) to quickly and efficiently evaluate mechanisms and pathways that typically require expensive and time-consuming modeling. Initial studies in *C. elegans* can provide direction for future assays that refine endpoints and/or reduce the number of vertebrates in future, more

TABLE 28.3 Transforming growth factor-β pathway comparison between *Caenorhabditis elegans* and humans.

Component	*C. elegans* gene	Human gene	Molecule or family member
Ligands	*dbl-1*	BMP5	Transforming growth factor-β
	daf-7	GDF11	
	unc-129	BMP3	
Receptors	*sma-6*	BMPRIB	Serine/threonine kinase receptors
	daf-1	TGF-βRI	
	daf-4	ACTRIIB	
Signal transduction proteins	*sma-2*	Smad1	Smad
	sma-3	Smad5	
	daf-8	Smad8	
	daf-14	Smad2	
	sma-4	Smad4	
	daf-3	Smad4	
	tag-68	Smad6	
Transcription factors	*sma-9*	Shn1/Shn2/Shn3	Zn-finger protein
	lin-31	FoxB1	HNF3/forkhead family
	daf-5	Sno/Ski	Proline-rich protein
	daf-12	Vitamin D receptor	Nuclear hormone receptor

complex studies. This is particularly important in instances where traditional techniques or model organisms will be required to provide translational data, or if protein(s) or pathways of interest do not behave linearly in expensive luminescent/phosphorescent assays. In these cases, using nematodes first can provide necessary preliminary indications of how to structure mammalian studies.

In addition to genetic manipulation, spectroscopic techniques for *C. elegans* are valuable preliminary tools for finding novel developmental disruptors, anthelminthic agents, neurotransmitter agonists/antagonists, and proteasome regulators. For example, Gerstbrein et al. (2005) developed spectrofluorimetric protocols to quickly and cheaply assess age-related autofluorescing pigments in vivo. It is important, however, that high-throughput studies using *C. elegans* use standardized isolation, feeding, and assay protocols to minimize idiosyncratic results across labs (Boyd et al., 2012; Scanlan et al., 2018; Koopman et al., 2016). As with other model organisms, interactions between test substances and feeding substrates need to be optimized so results across labs are comparable (Hoss et al., 2012) (Table 28.4).

Such high-throughput and multipronged approaches in an early in vivo model could aid researchers focusing on development and testing of novel nutraceuticals. This could be particularly helpful for diseases involving the nervous system. For example, current treatment options for neurodegenerative diseases such as AD, PD, Huntington disease (HD), multiple sclerosis (MS), and amyotrophic lateral sclerosis (ALS) are severely limited. Furthermore, these treatments generally focus only on symptom management. Thus, there exists a critical need to develop novel therapeutic strategies for the pharmacotherapy of these diseases. In recent years, a multitude of dietary supplements and nutraceuticals, such as polyphenols, have been investigated for their neuroprotective and neurotherapeutic potential (Darvesh et al., 2012; Tarazi et al., 2014; Devi and Chamoli, 2020; Mann et al., 2020). It is likely that using *C. elegans* to better characterize the therapeutic and toxic mechanisms of these compounds could push research forward much faster and in a more economical manner than what can be done by completing much of the preliminary mechanistic research in mammals.

Nutraceutical compounds and *Caenorhabditis elegans*

C. elegans have been used to study a number of nutraceuticals in the context of aging, stress, neurotoxicity, and neurodegeneration. In many instances, the focus of the research is on limiting, preventing, or reversing the effects

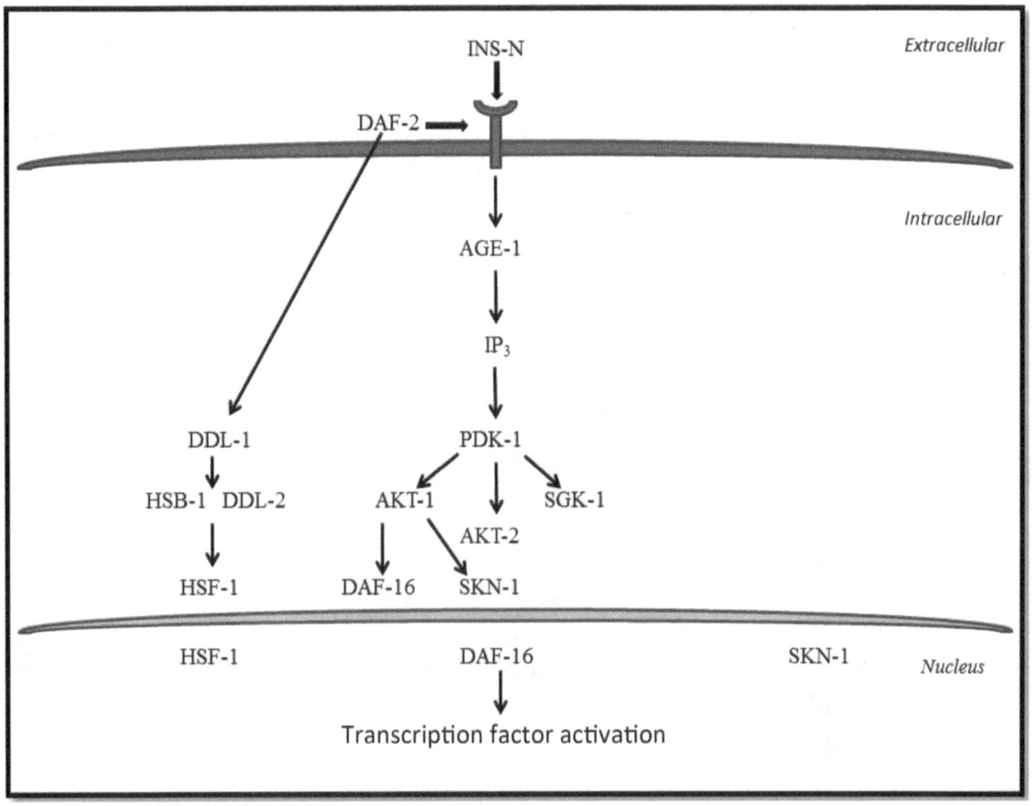

FIGURE 28.2 ILP binding regulates fat storage, longevity, and dauer formation. The ILP signal is transduced via DAF-2 allowing various proteins in the pathway to be phosphorylated. Phosphorylation leads to nuclear translocation of transcription factors where various antioxidant and longevity genes are then transcribed to counteract environmental stressors.
Abbrevations: *AGE*; Ageing alteration protein; *AKT*; Also known as PKB for protein kinase; *BSGK*; Serum- and glucocorticoid-inducible kinase homolog; *DAF*; Abnormal dauer formation; *DDL*; Daf-16-dependent longevity; *HSB*; Heat shock factor binding protein; *HSF*; Heat shock factor; *INS-N*; Insulin related protein, N-terminus; *IP3*; Inositol trisphosphate; *PDK*; Phosphoinositide-dependent kinase; *SKN*; Skinhead mutation.

of these characteristics. While some studies use only one strain of *C. elegans*, others incorporate various mutant strains into the study design in order to assess the mechanism of action of the nutraceuticals. The reader is encouraged to refer to Table 28.5 for a summary of some of the information presented in the following sections.

Catechins

Catechins, phytochemicals found in especially high concentrations in teas and cocoa (Lee et al., 2003), are known to provide protection to *C. elegans* from oxidative and thermal stressors, particularly when used as a pretreatment (Zheng et al., 2009; González-Manzano et al., 2012). Work by Sucro-Laos et al. (2012) also showed that pretreatment with catechins increased survival when oxidative stress was induced by juglone (79% in controls vs 89%–98% in pretreated) when assayed on the first day of adulthood. This increase persisted for up to 6 days post-juglone exposure (56% in controls vs 84%–99% in pretreated), suggesting potential long-term changes in the oxidative stress response. Furthermore, catechin pretreatment protection extends to other forms of stress as well. Worms exposed to thermal stress following being pretreated with catechins had a survivability rate of 87%–99%, compared to only 83% in the control worms (Sucro-Laos et al., 2012). The protection also appeared to be chemical-specific and age-specific. This was borne out in studies showing that methylated catechins may have a greater protective effect in older worms compared to younger ones.

Along these lines, Abbas and Wink (2009) demonstrated that galloate moieties contributed to the stress-resistant properties of catechins in *C. elegans*. If life span, and not survival, was used as an endpoint, only the methylated catechins were effective, but marginally (6%–12%, compared to controls). These studies also demonstrate that *C. elegans* can be used to inform structure–function studies when moving to more complex model organisms.

Saul et al. (2009) provided evidence that catechin treatment (200 μM) resulted in a significantly enhanced life span, adding approximately 10 days to their normal life span of around 21 days. These data were followed up

TABLE 28.4 Insulin/IGF-1 signaling pathway comparison between *Caenorhabditis elegans* and humans.

Component	*C. elegans* gene	Human gene	Molecule or family member
Ligands[a]	*daf-28*	Insulin	Insulin-like peptides (*ins-7* and *ins-18* expression can also be regulated through the TGF-β pathway)
	ins-1	Insulin	
	ins-6	β-Type insulin	
	ins-7	Insulin/IGF-1	
	ins-18	Insulin/IGF-1	
Receptor[b]	*daf-2*	IGFRs	Tyrosine kinase receptor
Signal transduction proteins	*age-1*	PI3K	Various kinases and phosphatases
	ist-1	IRS	
	daf-18	PTEN	
	akt-1, akt-2	Akt/PKB	
	pdk-1	PDK	
	pptr-1	PP2A	
	sgk-1	SGK	
Transcription factors	*daf-16*	FoxO	Forkhead family
	hsf-1	HSF-1	Heat shock factor
	skn-1	NFE2L2	Basic leucine zipper protein
Selected target proteins	*sod-3*	SOD2	Fe/Mn superoxide dismutase
	ctl-1	CAT	Catalase
	mtl-1	MT-3	Metallothionein
	hsp-16	HSP-B1	Heat shock protein
	cyp-13A4	CYP3A7	NADPH-dependent monooxygenase
	gcs-1	GCLC	Glutamate-cysteine ligase

[a]*Over 40 insulin-like proteins have been identified in* C. elegans. *Only major ligands are listed.*
[b]*There is only one known insulin receptor family member in* C. elegans.

with oxidative stress tests using hydrogen peroxide to assess survival rate independent of longevity. As before, worms treated with catechin showed a dose-dependent protective response to the stress, with increasing survival rates of 9%, 15%, and 28% (100 μM, 200 μM, and 300 μM, respectively). When worms were co-exposed to catechins, rather than being pretreated, protection was no longer afforded. Rather, co-administration resulted in shorter body length and increased pharyngeal pumping rates compared to control worms grown in standard media (Saul et al., 2009).

It has been suggested that the disposable soma hypothesis (Saul, et al., 2009) may help explain the positive effects of catechins. This hypothesis states that energy normally devoted to increasing body length is diverted to maintenance. This leads to an increased life span, but shorter body length, which is often observed following catechin exposure. This was tested (Saul et al., 2009) using various short-lived mutants (*age-1*, *daf-16*, *jnk-1*, *osr-1*, *sek-1*, *sir-2.1*, *unc-43*, *akt-2*, *daf-2*, abnormal **me**thyl **v**iologen sensivity-1 (*mev-1*), and *nhr-8*; see Table 28.3). This study design also helped rule out the possibility that catechins were acting as direct antioxidants capable of detoxicating or inhibiting the production of reactive oxygen species (ROS).

These studies were confirmed using *mev-1* mutant strains (Saul, et al., 2009). *mev-1* is a succinate dehydrogenase cytochrome b560 subunit ortholog that plays a fundamental role in mitochondrial oxidative phosphorylation. Generally, *mev-1* mutants show decreased energy metabolism coupled with an increase in vulnerability to oxidative stress (Ishii et al., 1990, 1998). Concomitant with abnormal energy production, this mutant strain also displays a shorter life span comparted to control worms. Since treatment with catechins was unable to rescue the shortened life span in these worms, the data

TABLE 28.5 **Screening of nutraceuticals in *Caenorhabditis elegans* models of aging, stress, neurotoxicity, and neurodegeneration.**

Nutraceutical	Pharmacological effect	References
Acai berry (*Euterpe precatoria*) Anthocyanin-rich extract	Reduced ROS levels, and increased SOD gene expression under juglone-induced oxidative stress (Strains: N2, TK-22, TJ375, CF1553, TJ356, CF1038, GR1307, BA17)	Peixoto et al. (2016)
Apple extract (*Malus domestica*)	Increased life span, improved motility, and reduced lipofuscin accumulation under thermal stress, radiation stress, and oxidative stress (Toxin: Paraquat) [Strains: N2 (*var.* Bristol), DAF-16, AGE-1]	Vayndorf et al. (2013)
Blueberry extract (*Vaccinium corymbosum*)	Increased survival and expression of antioxidant genes under thermal stress, radiation stress, and oxidative stress (Toxin: Paraquat) (Strain: N2)	Wang et al. (2018)
Chinese currant (*Ribes fasciculatum*)	Increased life span, reduced ROS levels, and increased SOD expression (Toxin: Paraquat) [Strains: N2 (var. Bristol), CF1553, GR1307, VC199, DR1572, TJ1052, FK171]	Jeon and Cha (2016)
Honey bee propolis (*Apis mellifera*) CAPE	Increased survival, and reduced ROS levels under H_2O_2-induced oxidative stress through modulation of the DAF-16 signaling pathway [Strains: N2 (*var.* Bristol), LD001, CF1038]	Havermann et al. (2014)
Mongolian milkvetch (*Astragalus propinquus*) Calycosin	Increased survival through insulin signaling pathway (Toxin: Paraquat) [Strains: N2, daf-16, daf-2, eat-2, unc-43, and skn-1(zu169)]	Lu et al. (2017)
Olive oil (*Olea europaea*) Hydroxytyrosol Oleuropein aglycone	Increased survival after thermal stress, increased life span in unstressed conditions, and improved locomotion in aged worms (Strain: N2)	Brunetti et al. (2020)
Olive oil (*O. europaea*) Hydroxytyrosol Oleuropein aglycone	Enhanced locomotion (thrashing rate, body wave) (Toxin: Rotenone) (Strain: N2)	Brunetti et al. (2020)
Raspberry extract (*Rubus idaeus*)	Increased life span, improved health span under thermal stress and radiation stress (Strains: N2, DAF-16)	Song et al. (2020)
Roseroot extract (*Rhodiola rosea*)	Increased survival and HSP16.2 expression under thermal stress and oxidative stress (Toxin: Paraquat) [Strains: N2 (var. Bristol), TJ356]	Wiegant et al. (2009)
Siberian ginseng (*Eleutherococcus senticosus*)	Increased survival and HSP16.2 expression under thermal stress, and oxidative stress (Toxin: Paraquat) [Strains: N2 (var. Bristol), TJ356]	Wiegant et al. (2009)
Tea (green) (*Camellia sinensis*) EGCG	Increased life span under thermal stress, and under juglone-induced oxidative stress (Strains: N2, BA17, TJ375, TK22)	Abbas and Wink (2009)
Tea (green) (*C. sinensis*) Catechin	Increased life span and reduced body length under heat stress, and under H_2O_2-induced oxidative stress (Strains: N2, AE501, AM1, AU1, DR1567, DR20, EU1, GR1307, MT2605, TJ1052, TK22, VC199, VC204, VC8)	Saul et al. (2009)
Tea (green) (*C. sinensis*) EGCG	Increased life span under thermal stress, upregulated levels of SOD and HSP16.2 (Strain: N2)	Zheng et al. (2009)
Tea (green) (*C. sinensis*) Epicatechin	Reduced levels of ROS, and increased production of glutathione (GSH) under thermal stress, and under juglone-induced oxidative stress (Strain: N2)	González-Manzano et al. (2012)

(Continued)

TABLE 28.5 (Continued)

Nutraceutical	Pharmacological effect	References
Tea (green) (*C. sinensis*) Catechin Epicatechin 3′-O-Methylepicatechin 4′-O-Methylepicatechin	Increased survival under thermal stress and under juglone-induced oxidative stress (Strain: N2)	Sucro-Laos et al. (2012)
Tea (green) (*C. sinensis*) L-Theanine	Increased survival under thermal stress and under juglone-induced oxidative stress (Strain: N2)	Gong et al. (2012)
Tea (green) (*C. sinensis*) L-Theanine	Increased survival (Toxin: Paraquat) [Strain: N2 (var. Bristol)]	Zarse et al. (2012)
Turmeric (*Curcuma longa*) Curcumin	Increased survival, decreased lipofuscin accumulation, and reduced body length under juglone-induced oxidative stress (Strain: N2)	Liao et al. (2011)
Turmeric (*C. longa*) Curcumin	Increased survival, decreased ROS levels, and increased expression of GST-4 and HSP16.2 under juglone-induced oxidative stress [Strains: N2 (var. Bristol), TK22, CB1370, TJ1052, GR1310, GR1318, GR1307, AM1, MT2605, AU1, EU1, VC199, FK171, CF1553, CL2166, CL2070]	Yu et al. (2014)

CAPE, Caffeic acid phenethylester; EGCG, epigallocatechin-3-gallate; *GSH*, glutathione; *GST-4*, glutathione-S-transferase 4; H_2O_2, hydrogen peroxide; *HSP16.2*, heat shock protein 16.2; *ROS*, reactive oxygen species; *SOD*, superoxide dismutase.

suggest that catechins must exert their protective effect via other, non-*mev-1*-related pathways (Saul et al., 2009). Rather, it is likely that the observed life-prolonging effects of catechin are more dependent on the IIS pathway (Fig. 28.3), particularly through DAF-2 and AKT-2, to mediate stress resistance and cellular repair (Zarse et al., 2012).

Theanine

The nutraceutical compound theanine, also known as L-γ-glutamylethylamide, is extracted from green tea (*Camelia sinensis*). Zarse et al. (2012) showed that is can protect *C. elegans* from oxidative stress. As with studies using catechins, worms were pretreated with theanine (100 nM, 1 μM, and 10 μM) followed by paraquat, an oxidative stress inducer. To address claims supporting green tea consumption, both survival and longevity were assayed as endpoints. Worms were exposed to theanine added to the media plates, and then treated with paraquat. Although 1 μM theanine significantly increased nematode resistance to paraquat, maximal protection (survival) was noted at 100 nM theanine. At this concentration, however, mean life span was only increased by 0.8 days, while maximum life span was only increased by 1.1 days (Zarse et al., 2012).

A similar protection from oxidative stress has also been observed by others (Gong et al., 2012). In studies using juglone to initiate oxidative stress, *C. elegans* exposed to 50 μg/mL, 100 μg/mL, or 200 μg/mL of theanine added to nematode growth medium (NGM) showed increased survivability rates of 21% compared to non-theanine-treated controls. If worms were stressed by both juglone and heat shock (35°C), theanine-treated populations showed only a 13% increased survival rate compared to controls. To elucidate the potential protective mechanism involved following heat shock, a GFP strain (CL2070) was also used. The worms have a *gfp*-tag integrated into the *hsp-16-2* promoter, providing a strong and reproducible induction of GFP expression following exposure to high temperatures (WormBase Community, 2000). In these studies, theanine-treated CL2070 worms, in the absence of heat shock, demonstrated upregulation of GFP expression (Gong et al., 2012), strongly suggesting that theanine induces transcription of *hsp-16.2*. This finding was confirmed by Western blot analysis of HSP-16.2, and likely explains the mechanism by which animals are partially protected from heat shock.

Eleuthrococcus senticosus and *Rhodiola rosea*

Herbs or herbal extracts that are thought to improve adrenal health through activation of the sympatho-adrenal system (SAS) are commonly referred to as adaptogens. Two commonly used adaptogens, *Eleutherococcus* and *Rhodiola*, have been demonstrated to increase stress resistance in a multitude of model organisms (Wiegant et al.,

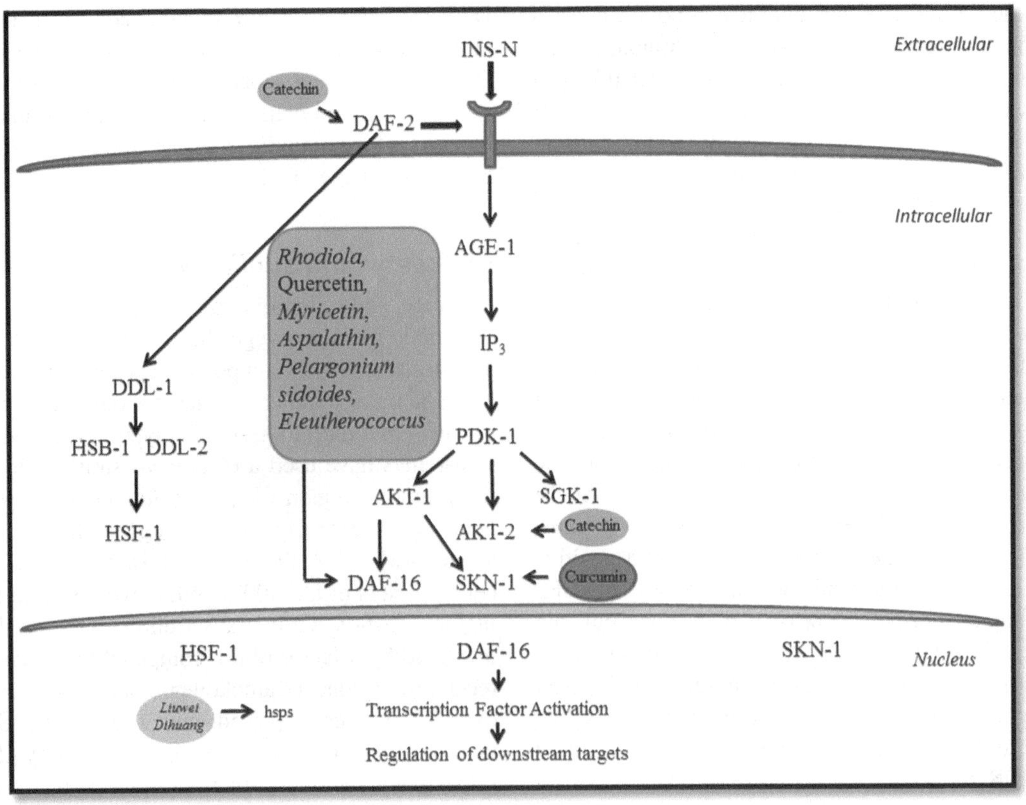

FIGURE 28.3 Putative nutraceutical intervention in the insulin/IGF-1 signaling pathway. *Nutraceutical compounds may interact at various locations along the insulin/IGF-1 signaling pathway resulting in the promotion or inhibition of the ability of transcriptions factors (predominantly DAF-16) to translocate to the nucleus. Abbrevations: DAF, Abnormal dauer formation; DBL, Drosophila decapentaplegic (Dpp)/vertebrate bone morphogenetic proteins (BMPs)-like; LIN, Abnormal cell lineage;LON, Long body size (phenotype); MAB, Male abnormal (phenotype); SMA, Small body size (phenotype); SMAD, Caenorhabditis elegans; UNC, Uncoordinated.*

2009). Although these ethnobotanical agents are toxic to *C. elegans* at higher doses, they appear to be protective at lower concentrations. For example, nematodes treated with either 250 μg/mL *Eleutherococcus* or 10 μg/mL *Rhodiola* show an increase of 2.5–3 days compared to control worms. This positive effect seems to be independent of when worms are exposed to the adaptogens since this finding holds even if worms were allowed to mature to day 8 of adulthood before they were treated (Wiegant et al., 2009). While the effect was not as pronounced when worms were treated well into adulthood, where life span was only increased by an average of 1.5 days, the data suggest that administration of adaptogens later in life still maintain their positive effects.

Similar to other nutraceuticals, worms treated with either of these adaptogens had increased resistance to thermal stress, regardless of whether the duration of the heat shock was short or chronic. *Eleutherococcus*-250 and *Rhodiola*-25 increased protection of *C. elegans* by 69% and 74%, respectively, following 3 hours of exposure to 35°C (Wiegant et al., 2009). In addition to heat stress, these herbal extracts also protected *C. elegans* from UV-induced stress. When control worms (nonadaptogen treated) were exposed to UV, their life spans were decreased to fewer than 10 days (Wiegant et al., 2009). Worms incubated with *Eleutherococcus*-250 prior to and during UV exposure, however, lived for a total of 11 days. Although this is less than the average life span of these worms of 18 days at 20°C, it is indicative of protection compared to no pretreatment with *Eleutherococcus*-250.

Studies evaluating whether *Eleutherococcus*-250 or *Rhodiola*-25 ameliorated oxidative stress also have been conducted. Worms pretreated with the nutraceuticals, followed by paraquat, showed that *Eleutherococcus*-250 increased overall survival by 21%, while *Rhodiola*-25 increased survival by 17% compared to nonpretreated worms. If worms were co-incubated with paraquat and *Eleutherococcus* or *Rhodiola* extracts for 4 days (chronic exposure for worms), *Eleutherococcus*-250, but not *Rhodiola*-25, provided significant protection from ROS (30% vs 24%, respectively) compared to control worms that did not receive co-treatment. As mentioned previously, *mev-1* mutants typically have a reduced life span.

In order to determine if either herbal compound exerted its protection through this pathway, *mev-1* mutants were pretreated with each adaptogen to assess whether life span could be rescued. *Eleutherococcus* extended the *mev-1* mutant life span by 9.2% (Wiegant et al., 2009). In contrast, *Rhodiola* did not have a significant protective effect.

Whether these nutraceuticals also affected DAF-16 localization was also investigated. Since the IIS pathway predominantly translocates DAF-16 to the nucleus, it plays a significant role in the dauer formation and longevity. In order to test this hypothesis, a DAF-16-GFP tagged strain was used to monitor DAF-16 movement throughout the cell (Hsu et al., 2003). In the presence of heat or ROS, DAF-16A/B moves from the cytosol to the nucleus. In the absence of additional exposure to other toxicants, *Eleutherococcus* and *Rhodiola* both induced DAF-16 translocation in a dose-dependent manner in *C. elegans*. To verify these results, heat shock was used as a positive control. As expected, transgenic worms showed no GFP expression under normal (control) conditions, but did express increased GFP levels when exposed to either botanical extracts, albeit to a lesser degree than by heat stress (Wiegant et al., 2009). This strongly suggests that these compounds are exerting their effect at least partially through the IIS pathway.

Curcumin

Curcumin, the primary constituent of turmeric (*Cucuma longa*) extract, participates in a number of signaling pathways, including that of Nrf2 (see information related to IIS). When *C. elegans* were grown on media containing 20 mM curcumin, their maximum life span increased by 3 days (Liao et al., 2011). Intracellular ROS levels were also significantly lower in 4-day-old worms versus 1-day-old worms. Consistent with a mechanism associated with Nrf2, the data suggested that curcumin may delay aging and cellular damage through activation of antioxidant genes. Curcumin also mitigated lipofuscin accumulation in the intestine, which is thought to positively correlate with age in *C. elegans* (Gerstbrein et al., 2005). Lipofuscin accumulation was decreased by 40% and 48% in curcumin-treated worms at both 4 and 8 days post-hatching, respectively. Furthermore, worms treated with this compound displayed decreased pharyngeal pumping rates, significantly reduced body lengths (disposable soma hypothesis), and resistance to both heat shock and juglone-induced oxidative stress (Liao et al., 2011).

It is possible that through reduction of pharyngeal pumping, curcumin may indirectly limit food consumption in *C. elegans*, which is known to prolong life in these worms (Kenyon, 2010), as well as other organisms. Further studies involving multiple mutant worm strains to address the mechanism indicated that curcumin likely acts independently of DAF-16, although *mev-1* mutants confirmed that curcumin protected against mild to severe oxidative stress (Liao et al., 2011). Additionally, since curcumin does not extend the life spanspan in *skn-1* mutants, this suggests that SKN-1 is required for curcumin-mediated protection and increased longevity (Liao et al., 2011).

Quercetin and similar compounds

Quercetin and rutin are two polyphenols common in numerous edible fruits and vegetables. Consistent with their chemical structures, both compounds appear to interact with free radicals in a manner similar to other antioxidant compounds. In order to test this hypothesis, however, many researchers have used a *C. elegans* strain with *gfp* fused to the promoter region of *gst-4* (*glutathione-S-transferase 4*). When *gst-4* transcription is upregulated, often in response to increased ROS, the worms show a strong green fluorescence (Leiers et al., 2003). When worms were treated with juglone, *gst-4p::*GFP transcription and translation were increased by a factor of 30, compared to control worms not receiving juglone (Kampkötter et al., 2007, 2008). When experiments were repeated, and worms were also exposed to quercetin or rutin, GFP levels decreased by 15% or 50%, respectively, demonstrating the ROS-scavenging potential and protection provided by both compounds. Furthermore, in the absence of juglone, *C. elegans* treated with either compound also showed increased *gst-4* promotor induction, suggesting a second mechanism of protection for these flavonols (Kampkötter et al.,2008). Similar results were obtained by these researchers when intracellular ROS production was assayed using the fluorescent probe 2',7'-Dichlorodihydrofluorescein diacetate (H_2DCFDA).

Further confirmation of the ability of quercetin to protect from oxidative stress was observed in studies involving worms with *gfp* fused to the *sod-3* promoter region (WormBase Community, 2004). This construct results in GFP production following increased transcription and translation of *sod-3*. In these experiments, addition of quercetin reduced the fluorescent signal by 85% versus control (Kampkötter et al., 2008), suggesting that quercetin either acted directly as an antioxidant, decreasing the need for *sod-3* transcription, or that quercetin suppressed *sod-3* induction. In order to delineate the mechanism(s) of action, Kampkötter et al. (2008) also used a *daf-16-gfp* transgenic strain. Their data indicated that quercetin could also induce DAF-16 localization to the nucleus in the absence of any stressors. On the other hand, translocation was increased threefold by thermal or oxidative stress, suggesting that quercetin may also work through a DAF-16 pathway.

When heat (37°C) was used as the stressor, quercetin, but not rutin, increased the survivability of *C. elegans* by 12% when compared to control worms (Kampkötter et al., 2008). When *daf-16-gfp* transgenic worms were assessed, quercetin or rutin treatment studies revealed that quercetin, but not rutin, significantly increased DAF-16 translocation. Consistent with other nutraceutical studies, following exposure of *C. elegans* to quercetin, lipofuscin accumulation decreased by more than 40% in the intestinal tracts of treated versus controls. Saul et al. (2009) expanded the mechanistic information related to the effect of quercetin on the DAF-16 pathway by using a *daf-16* deletion strain. This mutant showed increased longevity similar to that of control worms, indicating that non-DAF-16-mediated pathways also play a role in quercetin-mediated protection and longevity.

Whole-plant extracts and *Caenorhabditis elegans*

In addition to purified phytochemicals, various whole fruit/plant extracts have been investigated. For example, the effects of either blueberry (BB) polyphenolic extract or crude BB extract both prolong life in *C. elegans*. Although the polyphenolic and crude extract are sometimes presented as a homogeneous mixture, in reality they are mixtures of several chemicals. Of the primary constituent chemicals in BB extracts, anthocyanins (ATC) and chlorogenic acid (CA) have negative effects on nematode life span, whereas the proanthocyanidin (PAC) fraction increases *C. elegans* life span similarly to complete or crude BB extracts (Wilson et al., 2006). When worms were treated with 200 μg/mL BB extracts, and not specific constituents, life span was increased by 28% over control worms, in the absence of any additional stressors. Pharyngeal pumping, a measure of nematode vitality, was also preserved in older worms treated with BB extracts over nontreated worms. When studies were repeated with heat stress (Wilson et al., 2006), exposure to BB extracts increased thermotolerance by 250%. As before, this protection was not afforded to worms treated with ATC-enriched BB extract or CA. These data indicate that the primary bioactivity of BB extract is dependent upon its PAC content.

While BB seem to increase longevity in wild-type worms, this does not hold true in the mutant *mev-1* strain of *C. elegans*, where phenotype rescue was not achievable (Wilson et al., 2006). These results strongly suggest that the life-prolonging effect of BB is independent of the *mev-1* pathway. Furthermore, the antioxidant capacity attributed to BB extracts does not provide protection against mild to severe oxidative stress in this strain. Taken together, these data indicate that BB polyphenols positively affect *C. elegans* life span and thermoresistance. Such studies demonstrate that, while nutraceuticals may possess strong antioxidant potential in vitro, translation of these results to in vivo systems, in this case *C. elegans*, may differ. Thus, in vitro assessment of antioxidant potential is not necessarily a reliable predictor of bioactivity in vivo (Wilson et al., 2006).

Nutraceutical testing in a *Caenorhabditis elegans* model of neurodegenerative diseases

C. elegans is extensively used as a model system to study the mechanisms involved in neurodegeneration, neurotoxicity, aging, and neurodegenerative diseases at the cellular, molecular, signaling, and circuit levels (Silverman et al., 2009; Wolozin et al., 2011; Therrien and Parker, 2014; Van Damme et al., 2017). The etiological factors, pathophysiological mechanisms, and manifestations of neurodegeneration are complex and varied. The structural and/or functional loss of neurons can arise from a multitude of reasons such as hypoxia, oxidative stress, inflammation, and/or mitochondrial dysfunction. *C. elegans* provides an efficient means to investigate many of the aforementioned nutraceuticals in the context of neurodegeneration (Kim and Jin, 2015; De Araujo Boleti et al., 2020). To facilitate this research, it is possible to make sure the large number of *C. elegans* strains have been developed that express proteins implicated in neurodegeneration (Table 28.1). The *C. elegans* transgenic models provide an extremely powerful tool to elucidate the pathways involved in neurodegeneration, neurotoxicity, and neurodegenerative diseases (Gama Sosa et al., 2012; Li and Le, 2013; Wang et al., 2017). As such, it is possible to incorporate the lessons learned from these studies with those involving nutraceuticals. Indeed, *C. elegans* has already shown its utility in the development of therapeutic agents for neurodegenerative diseases such as AD, HD, PD, and ALS (Ma et al., 2018; Pohl and Lin, 2018).

The *C. elegans* model has been immensely useful in high-throughput toxicological and nutraceutical evaluation. Toxicity screens in *C. elegans* have been consistently shown to be predictive of toxicity data obtained in rodents (Hunt, 2017), and there is little reason at this time to suggest that data would be different if nutraceuticals were screened in a similar manner for their therapeutic effects. Furthermore, such high-throughput screens need not be limited to a single endpoint. For example, Tang et al. (2019) used *C. elegans* to assess the toxicity of three metals individually as well as in binary and tertiary mixtures. They examined the effect these treatments had on growth rate, brood size, feeding and feeding behaviors, and dopaminergic neurodegeneration. Since many nutraceuticals are taken in combination with food or other therapeutic agents, this study highlights the ability to inexpensively test multiple chemicals while assessing numerous endpoints across various time points and life stages. The short life span of *C. elegans* would also enable similar studies to determine what effects nutraceutical supplementation would have if consumed chronically. If transgenic worms were also incorporated into a similar study design, transgenic *C. elegans* would add much needed mechanistic information that could be used to inform safety studies. This is already being done with toxic compounds (Meyer and Williams, 2014; Helmcke et al., 2010), and could be easily adapted to nutraceuticals. Since many children are also given nutraceuticals, the nematode model could be used to address whether development is adversely affected as well (Ruszkiewicz et al., 2018).

Alzheimer's disease and nutraceutical testing in *Caenorhabditis elegans*

AD, which is the most common form of age-related dementia, has shown an alarming increase worldwide. The neuronal degeneration in AD is characterized by senile plaques (SPs) and neurofibrillary tangles (NFTs). The major component of SPs is β-amyloid (Aβ) peptide (primarily Aβ_{1-42}), and the NFTs contain hyperphosphorylated and aggregated forms of tau protein. Aβ peptide is formed by the amyloidogenic pathway—the sequential cleavage of the β-amyloid precursor protein (APP) by the β-secretase and γ-secretase enzyme systems (Huang and Mucke, 2012). Mutations in APP, as well as two components of the γ-secretase enzymatic complex, presenilin-1 (PS1) and presenilin-2 (PS2), have been implicated in the Aβ hypothesis of AD pathogenesis (Sherrington et al., 1996; Bertram et al., 2010). While mutations in the tau protein likely do not cause AD, they have been implicated in another neurodegenerative disease called frontotemporal dementia with parkinsonian chromosome 17 type (FTDP-17) (Rademakers et al., 2012). Extremely well-characterized transgenic *C. elegans* Aβ and tau models are commercially available (Table 28.1). These provide efficient, rapid, and complete in vivo systems for mechanistic screening of potential therapeutic compounds for AD, as well as in pathophysiological investigations of AD and related neurodegenerative diseases (Lublin and Link, 2013; Alexander et al., 2014; Griffen et al., 2017). Current pharmacotherapy, however, primarily addresses symptoms without slowing or stopping the progression of the disease. Despite major efforts in drug development, there has been little clinical success, suggesting a critical need to utilize cost-effective, efficient, and rapid high-throughput screening systems to enhance the drug-discovery process (Goedert and Spillantini, 2006; Geldenhuys and Darvesh, 2015). Many examples of using *C. elegans* to screen for potential therapeutics are outlined in Table 28.6.

Parkinson's disease and nutraceutical testing in *Caenorhabditis elegans*

PD, the second most common neurodegenerative disorder after AD, affects about 1% of the population at 60 years or older. PD is a progressive and chronic movement disorder characterized by bradykinesia, resting tremors, imbalance, and muscle rigidity (Beitz, 2014; Kalia and Lang, 2015). PD occurs due to a gradual loss of dopaminergic neurons in the substantia nigra (*pars compacta*), and the development of Lewy bodies, which are mainly composed of the protein α-synuclein (α-syn) that is found primarily in nuclei and presynaptic terminals (Rocha et al., 2018; Mor et al., 2019). Besides mutations in the *PARK1 (α-syn) (SNCA)* gene, mutations in eight additional genes have been implicated in the pathogenesis of PD: *PARK2 (parkin, PRKN), PARK5 (UCHL-1), PARK6 (PTEN-induced putative kinase 1, PINK1), PARK7 (DJ-1), PARK8 (leucine rich repeat kinase 2, LRRK2), PARK9 (ATP13A2), PARK11 (GIGYF2)*, and *PARK13 (HTRA2)*. Several well-characterized transgenic *C. elegans* models for the aforementioned PD genes (Table 28.1) have been utilized in rapid high-throughput drug-discovery programs for PD therapeutics (Harrington et al., 2010; Vistbakka et al., 2012). Besides the transgenic models, toxin-based *C. elegans* models have also been developed to enable comparison to mammalian models. As such, *C. elegans* have been treated with 6-hydroxydopamine (6-OHDA), 1-methyl-1,2,3,6-tetrahydropyridine (MPTP), rotenone, paraquat, or manganese to produce PD pathology in *C. elegans* (Maulik et al., 2017). Along with the neurochemical models, behavioral models of PD also have been developed using *C. elegans*. The behavioral assays in the nematode utilize behavioral phenotypes such as food sensing, area-restricted searching, and mechanosensory responses (Maulik et al., 2017). The in vivo nematode models are also utilized in molecular investigations of PD pathogenesis (Cooper and Van Raamsdonk, 2018; Gaeta et al, 2019). The contribution of aging, oxidative stress, inflammation, and mitochondrial dysfunction in the pathogenesis of PD can also be studied in the *C. elegans* model (Chakraborty et al., 2013; Maglioni and Ventura, 2016). Using the plethora of genetic and behavioral tools available to interrogate PD physiology in *C. elegans*, it is perhaps not surprising that many researchers have used these worms to investigate the potential therapeutic role nutraceuticals may have in this disease (Table 28.7).

Concluding remarks and future directions

The current pharmacotherapy for neurodegenerative diseases presents limited treatment options other than addressing symptom management. Thus, there is a critical need to investigate alternative treatment options for neurodegenerative diseases such as AD and PD. In the past two decades, there has been a substantial increase in reports in the scientific literature demonstrating the neurotherapeutic potential of molecules obtained from nature. These include nutraceuticals from varied sources such as fruits,

TABLE 28.6 Screening of nutraceuticals in *Caenorhabditis elegans* models of Alzheimer disease.

Nutraceutical	Pharmacological effect	References
Barquillo (Cocoa by product) (*Theobroma cacao)* Peptide 13L	Reduced Aβ aggregation, oxidative stress, and paralysis (Strain: CL4176)	Martorell et al. (2013)
Cranberry extract (*Vaccinium macrocarpon*)	Increased life span, delayed worm paralysis, improved proteostasis, decreased Aβ aggregation, and attenuated Aβ-induced toxicity [Strains: N2 (var. Bristol), CL2006, and hsf-1]	Guo et al. (2016)
Ginkgo biloba leaf extract (EGb 761) Kaempferol Quercetin	Attenuated elevated levels of ROS (Strain: CL2006)	Smith and Luo (2003)
Ginkgo biloba extract (EGb 761) Ginkgolide A	Reduced paralysis and hypersensitivity, and inhibited Aβ oligomerization and deposits [Strains: N2 (var. Bristol), CL1175, CL2006, CL2122, CL2179, CL2355, CL4176]	Wu et al. (2006)
Indian snakeroot Sarpagandha (*Rauvolfia serpentina*) Reserpine	Extended life span, caused synaptic ACh release, and ameliorated Aβ-induced toxicity [Strains: N2 (var. Bristol), CD1111, CB112, MT7988, NM440, CB246, CF1038, NM1968, PR1152, CB933, DH404, RB918]	Saharia et al. (2012)
Jujube Chinese date (*Ziziphus jujube*) 6‴-Feruloylspinosin	Prevented Aβ-induced toxicity, decreased paralysis, promoted mitophagy, provided protection under thermal stress (Strain: GMC101)	Yang et al. (2020)
Liuwei Dihuang TCM (six herbs)	Increased survival, delayed paralysis, and reduced ROS levels under juglone-induced oxidative stress [Strains: N2 (var. Bristol), TJ375, CF1553, CL4176, CL2006]	Sangha et al. (2012)
Maple syrup (*Acer saccharum*)	Increased survival, reduced Aβ aggregation, Aβ-induced toxicity, and paralysis (Strain: CL4176)	Ma et al. (2016)
Olive (*Olea europaea*) Oleuropein aglycone	Reduced Aβ aggregation, Aβ plaque deposition, Aβ oligomer levels, paralysis, and increased life span (Strains: CL2006, CL4176)	Diomede et al. (2013)
Quercetin (dietary polyphenol)	Reduced Aβ aggregation, proteosomal activity, and worm paralysis (Strain: CL2006)	Regitz et al. (2014)
Resveratrol (berries, grapes, peanuts, red wine)	Reduced worm paralysis, Aβ aggregation, Aβ-induced toxicity, and lysosomes (Strain: CL2006)	Regitz et al. (2016)
Rose essential oil (*Rosa rugosa*) β-Citronellol Geraniol	Inhibited worm paralysis and hypersensitivity, suppressed Aβ deposits, and reduced Aβ oligomers (Strains: CL2006, CL2122, CL2166, Cl2355, CL4176, LG333, TJ356)	Zhu et al. (2017)
Tea (green) (*Camellia sinensis*) EGCG	Reduced Aβ deposits, and Aβ oligomerization, and increased HSP16.1, HSP16.2 expression under juglone-induced oxidative stress (Strains: N2, CL2006, TJ356)	Abbas and Wink (2010)
Tea (green) Aroma fraction (*C. sinensis*)	Increased life span, reduced paralysis, and provided antioxidant activity (Strain: CL4176, CL802)	Takahashi et al. (2014)
Yerba mate (*Ilex paraguariensis*)	Delayed paralysis, reduced AChE activity, Aβ mRNA, induced HSP16.2 expression (Strains: N2, CL2006)	Machado et al. (2019)

Aβ, β-amyloid; *ACh*, acetylcholine; *AChE*, acetylcholinesterase; *EGCG*, epigallocatechin-3-gallate; *HSP16.2*, heat shock protein 16.2; *ROS*, reactive oxygen species; *TCM*, traditional Chinese medicine.

TABLE 28.7 Screening of nutraceuticals in *Caenorhabditis elegans* models of Parkinson disease.

Nutraceutical	Pharmacological effect	References
Acetylcorynoline (*Corydalis bungeana*) TCM	Prolonged life span, prevented α-syn aggregation, protected DA neurons, and restored food-sensing behavior (Toxin: 6-OHDA) (Strains: BZ555, OW13)	Fu et al. (2014a)
Female ginseng Dong quai (*Angelica sinensis*) n-Butylidenephthalide TCM	Prolonged life span, attenuated degeneration of DA neurons, decreased α-syn accumulation, restored DA levels, enhanced somatic proteasome activity, and recovered food-sensing behavior (Toxin: 6-OHDA) (Strains: BZ555, OW13)	Fu et al. (2014b)
Gastrodin (*Gastrodia elata*) TCM	Attenuated the accumulation of α-syn, prevented toxicity of DA neurons, improved chemotaxis. Neuroprotection through DAF-2 insulin-like pathway (Toxin: 6-OHDA) (Strains: N2, BZ555, OW13, CB1370, GR1309, CF1038, TJ356)	Yan et al. (2019)
Indian pennywort Brahmi (*Bacopa monnieri*)	Reduced α-syn accumulation, and prevented DA neurodegeneration (Toxin: 6-OHDA) (Strains: BZ555, NL5901)	Jadiya et al. (2011)
Indian thorny ash Chinese coriander (*Zanthoxylum armatum*) Tambulin	Enhanced oxidative stress tolerance, reduced ROS levels, enhanced locomotor behavior, reduced α-syn aggregation, increased DA levels (Strains: N2, TJ356, GR1307, CF1553, NL5901)	Pandey et al. (2019)
Olive oil (extra virgin) (*Olea europaea*) Tyrosol	Extended life span, decreased α-syn inclusion, improved motility, delayed degeneration of DA neurons, decreased ROS levels, upregulated chaperone and antioxidant enzymes (Strains: N2, NL5901, UA44)	Garcia-Moreno et al. (2019)
Olive oil (*O. europaea*) Hydroxytyrosol Oleuropein aglycone	Enhanced locomotion (thrashing rate, body wave), reduced α-syn accumulation, prevented neurodegeneration in DA neurons (Strains: OW13, UA44)	Brunetti et al. (2020)
Sea cucumber Sandfish (*Holothuria scabra*)	Prolonged life span, reduced α-syn accumulation, prevented neurodegeneration in DA neurons, and improved food-sensing behavior (Toxin: 6-OHDA) (Strains: N2, BZ555, NL5901)	Chalorak et al. (2018)
Sea cucumber Sandfish (*Holothuria scabra*)	Prolonged life span, decreased ROS levels, and showed antioxidant effects under thermal and oxidative stress Toxin: Paraquat (Strains: N2, VC199, AM1, VC8, TJ1052, TJ356, CF1553, CL2070)	Jattujan et al. (2018)
Sea cucumber (black) (*Holothuria leucospilota*)	Increased longevity, prevented neurodegeneration of DA neurons, suppressed α-syn accumulation, improved food-sensing behavior, restored lipid deposition (Toxin: 6-OHDA) (Strains: N2, BZ555, NL5901)	Malaiwong et al. (2019)
Spermidine (whole grains, legumes, cheese)	Reduced α-syn-induced toxicity of DA neurons (Strains: BZ555, UA44, N2Ex, baln11)	Büttner et al. (2014)
Velvet bean Seed extract (*Mucuna pruriens*)	Increased median and maximum life span (Toxin: MPP +) (Strain: N2)	Johnson et al. (2018)

α-syn, α-synuclein; *DA*, dopamine; *MPP*$^+$, 1-methyl-4-phenylpyridinium; *6-OHDA*, 6-hydroxydopamine; *ROS*, reactive oxygen species; *TCM*, traditional Chinese medicine.

vegetables, herbs, botanical and zoological sources, and traditional medicine. Studies utilizing isolated compounds as well as complex extracts have demonstrated their antioxidant and antiinflammatory properties, as well as their ability to modulate the molecular pathway implicated in neurodegeneration. Both preclinical and clinical investigations are critical to ensure the efficient use and application of these molecules provided by nature. The nematode model *C. elegans* presented in this chapter demonstrates an extremely efficient screening system that can parallel in vitro cell-based models in terms of cost and speed while generating in vivo data much more rapidly and economically than mammalian models. This nematode, and its nervous system, are exceedingly well characterized, and the wide array of transgenic worms makes it possible for researchers to investigate distinct molecular pathways of neurodegeneration through descriptive behavioral and biochemical assays, as well as mechanistic pharmacological and neuroprotective pathways. A review of the studies presented in the chapter shows the amazing flexibility of the model in its ability to screen both isolated compounds and extracts from a wide variety of sources. The *C. elegans* model thus provides a unique opportunity to rapidly and efficiently screen nutraceuticals to evaluate their therapeutic properties in age-related neurodegenerative diseases, and can be effective in the preclinical use of these agents.

Abbreviations

MPTP	1-Methyl-1,2,3,6-tetrahydropyridine
H_2DCFDA	2',7'-Dichlorodihydrofluorescein diacetate
6-OHDA	6-Hydroxydopamine
AD	Alzheimer disease
ALS	Amyotrophic lateral sclerosis
ATC	Anthocyanins
Aβ	β-amyloid
APP	β-amyloid precursor protein
C. elegans	*Caenorhabditis elegans*
ctl-1	Catalase-1
CA	Chlorogenic acid
CRISPR	Clustered regularly interspaced short palindromic repeats
CAS 9	CRISPR-associated protein 9
DAF-7	Dauer formation-7 protein
Dpp	*Drosophila* decapentaplegic
DBL-1	*Drosophila* decapentaplegic (Dpp)/vertebrate bone morphogenetic proteins (BMPs)-like-1
E. coli	*Escherichia coli*
FoxO	Forkhead box
FTDP-17	Frontotemporal dementia with parkinsonian chromosome 17 type
GFP	Green fluorescent protein
hsp-16	Heat shock protein
HSF-1	Heat shock transcription factor-1
HD	Huntington disease
IGF-1	Insulin/insulin-like growth factor-1
IIS	Insulin/IGF-1 signaling
ILPs	Insulin-like peptides
mtl-1	Metallothionein-1
mev-1	Abnormal methyl viologen sensivity-1
MS	Multiple sclerosis
NFTs	Neurofibrillary tangles
Nrf	Nose resistant to fluoxetine
OTC	Over-the-counter
PD	Parkinson disease
PS1	Presenilin-1
PS2	Presenilin-2
PAC	Proanthocyanidin
RNAi	RNA interference
SPs	Senile plaques
SKN-1	Skinhead-1
sod-3	Superoxide dismutase-3
α-syn	α-Synuclein
SAS	Sympatho-adrenal system
TGF-1	Transforming growth factor-1
TGF-2	Transforming growth factor-2
TGF-β	Transforming growth factor-β
BMPs	Vertebrate bone morphogenetic proteins
UNC-129	Uncoordinated-129

References

Abbas, S., Wink, M., 2009. Epigallocatechin gallate from green tea (*Camellia sinensis*) increases lifespan and stress resistance in *Caenorhabditis elegans*. Planta Med. 75, 216–221.

Abbas, S., Wink, M., 2010. Epigallocatechin gallate inhibits beta amyloid oligomerization in *Caenorhabditis elegans* and affects the daf-2/insulin-like signaling pathway. Phytomedicine 17, 902–909.

Alexander, A.G., Marfil, V., Li, C., 2014. Use of *Caenorhabditis elegans* as a model to study Alzheimer's disease and other neurodegenerative diseases. Front. Genet. 5 (279), 1–21.

Andlauer, W., Furst, P., 2002. Nutraceuticals: a piece of history, present status and outlook. Food Res. Int. 35, 171–176.

Apfeld, J., Alper, S., 2018. What can we learn about human disease from the nematode *C. elegans?* Methods Mol. Biol. 1706, 53–75.

Aronson, J.K., 2017. Defining 'nutraceuticals': neither nutritious nor pharmaceutical. Br. J. Clin. Pharmacol. 83, 8–19.

Bargmann, C.I., Kaplan, J.M., 1998. Signal transduction in the *Caenorhabditis elegans* nervous system. Annu. Rev. Neurosci. 21, 279–308.

Beitz, J.M., 2014. Parkinson's disease: a review. Front. Biosci. 6, 65–74.

Bertram, L., Lill, C.M., Tanzi, R.E., 2010. The genetics of Alzheimer's disease: back to the future. Neuron 68, 270–281.

Billi, A.C., Fischer, S.E.J., Kim, J.K., 2014. Endogenous RNAi pathways in *C. elegans*. In: Ahringer, J. (Ed.), WormBook. WormBook.

Blackwell, T.K., Steinbaugh, M.J., Hourihan, J.M., et al., 2015. SKN-1/Nrf, stress responses, and aging in *Caenorhabditis elegans*. Free Radic. Biol. Med. 88 (Pt B), 290–301.

Bolker, J.A., 2019. Selection of models: evolution and the choice of species for translational research. Brain Behav. Evol. 93, 82–91.

Bornhorst, J., Chakraborty, S., Meyer, S., et al., 2014. The effects of *pdr1*, *djr1.1* and *pink1* loss in manganese-induced toxicity and the role of alpha-synuclein in *C. elegans*. Metallomics 6, 476–490.

Boulin, T., Etchberger, J.F., Hobert, O., 2006. Reporter gene fusions. WormBook.

Boyd, W.A., Smith, M.V., Freedman, J.H., 2012. *Caenorhabditis elegans* as a model in developmental toxicology. Methods Mol. Biol. 889, 15–24.

Brenner, S., 1974. The genetics of *Caenorhabditis elegans*. Genetics 77, 71–94.

Brockie, P.J., Maricq, A.V., 2006. Ionotropic glutamate receptors: genetics, behavior and electrophysiology. WormBook 1–16.

Brunetti, G., Di Rosa, G., Scuto, M., et al., 2020. Health span maintenance and prevention of Parkinson's-like phenotypes with hydroxytyrosol and oleuropein aglycone in *C. elegans*. Int. J. Mol. Sci. 21 (7), E2588.

Burby, P.E., Simmons, L.A., 2020. Regulation of cell division in bacteria by monitoring genome integrity and DNA replication status. J. Bacteriol. 202.

Büttner, S., Broeskamp, F., Sommer, C., et al., 2014. Spermidine protects against α-synuclein neurotoxicity. Cell Cycle 13, 3903–3908.

Byerly, L., Cassada, R.C., Russell, R.L., 1976. The life cycle of the nematode *Caenorhabditis elegans*. I. Wild-type growth and reproduction. Dev. Biol. 51, 23–33.

Carretero, M., Solis, G.M., Petrascheck, M., 2017. *C. elegans* as model for drug discovery. Curr. Top. Med. Chem. 17, 2067–2076.

C. Elegans Sequencing Consortium, 1998. Genome sequence of the nematode *C. elegans*: a platform for investigating biology. Science 282, 2012–2018.

Chakraborty, S., Bornhorst, J., Nguyen, T.T., et al., 2013. Oxidative stress mechanisms underlying Parkinson's disease-associated neurodegeneration in *C. elegans*. Int. J. Mol. Sci. 14, 23103–23128.

Chalorak, P., Jattujan, P., Nobsathian, S., et al., 2018. *Holothuria scabra* extracts exhibit anti-Parkinson potential in *C. elegans*: a model for anti-Parkinson testing. Nutr. Neurosci. 21, 427–438.

Chauhan, B., Kumar, B., Kalam, N., et al., 2013. Current concepts and prospects of herbal nutraceutical: a review. J. Adv. Pharm. Technol. Res. 4, 4–8.

Chen, Z., Hendricks, M., Cornils, A., et al., 2013. Two insulin-like peptides antagonistically regulate aversive olfactory learning in *C. elegans*. Neuron. 77, 572–585.

Chun, L., Gong, J., Yuan, F., et al., 2015. Metabotropic GABA signalling modulates longevity in *C. elegans*. Nat. Commun. 6, 8828.

Cipriani, P.G., Piano, F., 2011. RNAi methods and screening: RNAi based high-throughput genetic interaction screening. Methods Cell Biol. 106, 89–111.

Clark, J.F., Meade, M., Ranepura, G., et al., 2018a. *Caenorhabditis elegans* DBL-1/BMP Regulates Lipid Accumulation via Interaction with Insulin Signaling. G3 (Bethesda) 8, 343–351.

Clark, T., Hapiak, V., Oakes, M., et al., 2018b. Monoamines differentially modulate neuropeptide release from distinct sites within a single neuron pair. PLoS One 13, e0196954.

Committee for the Update of the Guide for the Care and Use of Laboratory Animals, Institute for Laboratory Animal Research, Division on Earth and Life Studies & National Research Council of the National Academies, 2011. Guide for the Care and Use of Laboratory Animals, Washington, DC, National Academies Press.

Cooper, J.F., Van Raamsdonk, M., 2018. Modeling Parkinson's disease in *C. elegans*. J. Parkinson's Dis. 8, 17–32.

Corsi, A.K., Wightman, B., Chalfie, M., 2015. A transparent window into biology: a primer on *Caenorhabditis elegans*. Genetics 200, 387–407.

Cuentas-Condori, A., Mulcahy, B., He, S., et al., 2019. *C. elegans* neurons have functional dendritic spines. eLife 8.

Darvesh, A.S., Carroll, R.T., Bishayee, A., et al., 2012. Curcumin and neurodegenerative diseases: a perspective. Expert Opin. Investig. Drugs 21, 1123–1140.

De Araujo Boleti, A.P., de Oliveira Flores, T.M., Moreno, S.E., et al., 2020. Neuroinflammation: an overview of neurodegenerative and metabolic diseases and of biotechnological studies. Neurochem. Int. 9, 104714.

DeFelice, S.L., 1992. The nutraceutical initiative: a recommendation for U.S. economic and regulatory reforms. Genet. Eng. News 12, 13–15.

Devi, S.A., Chamoli, A., 2020. Polyphenols as an effective therapeutic intervention against cognitive decline during normal and pathological brain aging. Adv. Exp. Med. Biol. 1260, 159–174.

Diomede, L., Ragicci, S., Romeo, M., et al., 2013. Oleuropein aglycone protects transgenic *C. elegans* strains expressing Aβ42 by reducing plaque load and motor deficit. PLoS One 8, e58893.

Dwyer, J.T., Coates, P.M., Smith, M.J., 2018. Dietary supplements: regulatory challenges and research resources. Nutrients 10, E41.

Feng, S., Varshney, A., Coto Villa, D., et al., 2019. Bright split red fluorescent proteins for the visualization of endogenous proteins and synapses. Commun. Biol. 2, 344.

Frankos, V.H., Street, D.H., O'Neill, R.K., 2010. FDA regulation of dietary supplements and requirements regarding adverse event reporting. Clin. Pharmacol. Ther. 87, 239–244.

Fu, R.H., Wang, Y.C., Chen, C.S., et al., 2014a. Acetylcorynoline attenuates dopaminergic neuron degeneration and α-synuclein aggregation in animal models of Parkinson's disease. Neuropharmacology 82, 108–120.

Fu, R.H., Harn, H.J., Liu, S.P., et al., 2014b. N-butylidenephthalide protects against dopaminergic neuron degeneration and α-synuclein accumulation in *Caenorhabditis elegans* models of Parkinson's disease. PLoS One 9, e85305.

Gaeta, A.L., Caldwell, K.A., Caldwell, G.A., 2019. Found in translation: the utility of *C. elegans* alpha-synuclein models of Parkinson's disease. Brain Sci. 9.

Gama Sosa, M.A., De Gasperi, R., Elder, G.A., 2012. Modeling human neurodegenerative diseases in transgenic systems. Hum. Genet. 131, 535–563.

Garcia-Moreno, J.C., Porta de la Riva, M., Martínez-Lara, E., et al., 2019. Tyrosol, a simple phenol from EVOO, targets multiple pathogenic mechanisms of neurodegeneration in a *C. elegans* model of Parkinson's disease. Neurobiol. Aging 82, 60–68.

Gasnier, B., 2000. The loading of neurotransmitters into synaptic vesicles. Biochimie 82, 327–337.

Geanacopoulos, M., 2004. The determinants of lifespan in the nematode *Caenorhabditis elegans*: a short primer. Sci. Prog. 87, 227–247.

Geldenhuys, W.G., Darvesh, A.S., 2015. Pharmacotherapy of Alzheimer's disease: current and future trends. Expert Rev. Neurother. 15, 3–5.

Gerstbrein, B., Stamatas, G., Kollias, N., et al., 2005. *In vivo* spectrofluorimetry reveals endogenous biomarkers that report health span and dietary restriction in *Caenorhabditis elegans*. Aging Cell 4, 127–137.

Giacomotto, J., Segalat, L., 2010. High-throughput screening and small animal models, where are we? Br. J. Pharmacol. 160, 204–216.

Goedert, M., Spillantini, M.G., 2006. A century of Alzheimer's disease. Science 314, 777–781.

Gong, Y., Luo, Y., Huang, J.A., et al., 2012. Theanine improves stress resistance in *Caenorhabditis elegans*. J. Funct. Foods 4, 988–993.

González-Manzano, S., González-Paramás, A.M., Delgado, L., et al., 2012. Oxidative status of stressed *Caenorhabditis elegans* treated with epicatechin. J. Agric. Food Chem. 60, 8911–8916.

Griffen, E.F., Caldwell, K.A., Caldwell, G.A., 2017. Genetic and pharmacological discovery for Alzheimer's disease using *Caenorhabditis elegans*. ACS Chem. Neurosci. 8, 2596–2606.

Gumienny, T.L., Savage-Dunn, C., 2013. TGF-β signaling in *C. elegans*. WormBook. The C. elegans Research Community, WormBook http://doi.org/10.1895/wormbook.1.22.2. Available from: http://www.wormbook.org.

Guo, H., Cao, M., Zou, S., et al., 2016. Cranberry extract standardized for proanthocyanidins alleviates β-Amyloid peptide toxicity by improving proteostasis through HSF-1 in *Caenorhabditis elegans* model of Alzheimer's disease. J. Gerontol. A Biol. Sci. Med. Sci. 71, 1564–1573.

Harrington, A.J., Hamamichi, S., Caldwell, G.A., et al., 2010. *C. elegans* as a model organism to investigate molecular pathways involved with Parkinson's disease. Dev. Dyn. 239, 1282–1295.

Havermann, S., Chovolou, Y., Humpf, H.U., et al., 2014. Caffeic acid phenethylester increases stress resistance and enhances lifespan in *Caenorhabditis elegans* by modulation of the insulin-like DAF-16 signalling pathway. PLoS One 9, e100256.

Helmcke, K.J., Avila, D.S., Aschner, M., 2010. Utility of *Caenorhabditis elegans* in high throughput neurotoxicological research. Neurotoxicol. Teratol. 32, 62–67.

Hirsch, C., Schildknecht, S., 2019. *In vitro* research reproducibility: keeping up high standards. Front. Pharmacol. 10, 1484.

Hobert, O., 2010. Neurogenesis in the nematode *Caenorhabditis elegans*. *WormBook*.

Hobert, O., 2013. The neuronal genome of *Caenorhabditis elegans*. *WormBook*, pp. 1–106.

Hoss, S., Ahlf, W., Bergtold, M., et al., 2012. Interlaboratory comparison of a standardized toxicity test using the nematode *Caenorhabditis elegans* (ISO 10872). Environ. Toxicol. Chem. 31, 1525–1535.

Hsu, A.L., Murphy, C.T., Kenyon, C., 2003. Regulation of agining and age-related disease by DAF-16 and heat shock factor. Science 300 (5622), 1142–1145.

Huang, Y., Mucke, L., 2012. Alzheimer's mechanisms and therapeutic strategies. Cell 148, 1204–1222.

Hunt, P.R., 2017. The *C. elegans* models in toxicity testing. J. Appl. Toxicol. 37, 50–59.

Hutter, H., 2004. Five-colour in vivo imaging of neurons in *Caenorhabditis elegans*. J. Microsc. 215, 213–218.

Ishii, N., Takahashi, K., Tomita, S., et al., 1990. A methyl viologen-sensitive mutant of the nematode *Caenorhabditis elegans*. Mut. Res. 237, 165–171.

Ishii, N., Fujii, M., Hartman, P.S., et al., 1998. A mutation in succinate dehydrogenase cytochrome b causes oxidative stress and ageing in nematodes. Nature 394, 694–697.

Jadiya, P., Khan, A., Sammi, S.R., et al., 2011. Anti-Parkinsonian effects of *Bacopa monnieri*: insights from transgenic and pharmacological *Caenorhabditis elegans* models of Parkinson's disease. Biochem. Biophys. Res. Commun. 413, 605–610.

Jattujan, P., Chalorak, P., Siangcham, T., et al., 2018. Holothuria scabra extracts possess anti-oxidant activity and promote stress resistance and lifespan extension in *Caenorhabditis elegans*. Exp. Gerontol. 110, 158–171.

Jeon, H., Cha, D.S., 2016. Anti-aging properties of *Ribes fasciculatum* in *Caenorhabditis elegans*. Chin. J. Nat. Med. 14, 335–342.

Jiang, T., 2009. Re-thinking the dietary supplement laws and regulations 14 years after the Dietary Supplement Health and Education Act implementation. Int. J. Food Sci. Nutr. 60, 293–301.

Johnson, S.L., Park, H.Y., DaSilva, N.A., et al., 2018. Levodopa-reduced *Mucuna pruriens* seed extract shows neuroprotective effects against

Parkinson's disease in murine microglia and human neuroblastoma cells, *Caenorhabditis elegans*, and *Drosophila melanogastmer*. Nutrients 10, E1139.

Juntti, S., 2019. The future of gene-guided neuroscience research in non-traditional model organisms. Brain Behav. Evol. 93, 108–121.

Kalia, L.V., Lang, A.E., 2015. Parkinson's disease. Lancet 386, 896–912.

Kampkötter, A., Nkwonkam, C.G., Zurawski, R.F., et al., 2008. Increase of stress resistance and lifespan of *C. elegans* by quercetin. Comp. Biochem. Physiol. B Biochem Mol. Biol. 149 (2), 314–323.

Kampkötter, A., Nkwonkam, C.G., Zurawski, R.F., Timpel, C., Chovolou, Y., Wätjen, W., et al., 2007. Ivestigations of Protective Effects of the Flavonoids Quercetin and Rutin on Stress Resistance in the Model Organism C.elegans. Toxicology 234, 113–123.

Kellogg, J.J., Paine, M.F., McCune, J.S., et al., 2019. Selection and characterization of botanical natural products for research studies: a NaPDI center recommended approach. Nat. Prod. Rep. 36, 1196–1221.

Kenyon, C., 2010. A pathway that links reproductive status to lifespan in *Caenorhabditis elegans*. Ann. N. Y. Acad. Sci. 1204 (1), 156–162.

Kim, K.W., Jin, Y., 2015. Neuronal responses to stress and injury in *C. elegans*. FEBS Lett. 589, 1644–1652.

Kim, H.M., Colaiacovo, M.P., 2019. CRISPR-Cas9-guided genome engineering in *Caenorhabditis elegans*. Curr. Protoc. Mol. Biol. 129, e106.

Kimura, K., Riddle, D., Ruvkun, G., 2011. The C. elegans DAF-2 insulin-like receptor is abundantly expressed in the nervous system and regulated by nutritional status. Cold Spring Harbor Symposium on Quantitative Biology 76, 113–120.

Kimura, K., Tissenbaum, H., Liu, Y., et al., 1997. Daf-2, an insulin receptor-like gene that regulates longevity and diapause in *Caenorhabditis elegans*. Science 277, 942–946.

Koopman, M., Michels, H., Dancy, B.M., et al., 2016. A screening-based platform for the assessment of cellular respiration in *Caenorhabditis elegans*. Nat. Protoc. 11, 1798–1816.

Lakdawala, M.F., Madhu, B., Faure, L., et al., 2019. Genetic interactions between the DBL-1/BMP-like pathway and dpy body size-associated genes in *Caenorhabditis elegans*. Mol. Biol. Cell. 30, 3151–3160.

Laker, R.C., Xu, P., Ryall, K.A., et al., 2014. A novel MitoTimer reporter gene for mitochondrial content, structure, stress, and damage *in vivo*. J. Biol. Chem. 289, 12005–12015.

Lee, K.W., Kim, Y.J., Lee, H.J., et al., 2003. Cocoa has more phenolic phytochemicals and higher antioxidant capacity than teas and red wine. J. Agric. Food Chem. 51 (25), 7292–7295.

Leiers, B., Kampkotter, A., Grevelding, C.G., et al., 2003. A stress-responsive glutathione *S*-transferase confers resistance to oxidative stress in *Caenorhabditis elegans*. Free Radic. Biol. Med. 34, 1405–1415.

Li, J., Le, W., 2013. Modeling neurodegenerative diseases in *Caenorhabditis elegans*. Exp. Neurol. 250, 94–103.

Liao, V.H., Yu, C.W., Chu, Y.J., et al., 2011. Curcumin-mediated lifespan extension in *Caenorhabditis elegans*. Mech. Aging Dev. 132, 480–487.

Lin, X.X., Sen, I., Janssens, G.E., et al., 2018. DAF-16/FOXO and HLH-30/TFEB function as combinatorial transcription factors to promote stress resistance and longevity. Nat. Commun. 9, 4400.

Lu, L., Zhao, X., Zhang, J., et al., 2017. Calycosin promotes lifespan in *Caenorhabditis elegans* through insulin signaling pathway via daf-16, age-1 and daf-2. J. Biosci. Bioeng. 124, 1–7.

Lublin, A.L., Link, C.D., 2013. Alzheimer's disease drug discovery: *in vivo* screening using *Caenorhabditis elegans* as a model for β-amyloid peptide-induced toxicity. Drug Discov. Today Technol. 10, e115–e119.

Ma, H., DaSilva, N.A., Liu, W., et al., 2016. Effects of a standardized phenolic-enriched maple syrup extract on β-amyloid aggregation, neuroinflammation in microglial and neuronal cells, and β-amyloid induced neurotoxicity in *Caenorhabditis elegans*. Neurochem. Res. 41, 2836–2847.

Ma, L., Zhao, Y., Chen, Y., et al., 2018. *Caenorhabditis elegans* as a model system for target identification and drug screening against neurodegenerative diseases. Eur. J. Pharmacol. 819, 169–180.

Machado, M.L., Arantes, L.P., da Silveira, T.L., et al., 2019. Ilex paraguariensis extract provides increased resistance against oxidative stress and protection against Amyloid beta-induced toxicity compared to caffeine in *Caenorhabditis elegans*. Nutr. Neurosci. 9, 1–13.

MacNeil, L., Hardy, W., Pawson, T., et al., 2009. UNC-129 regulates the balance between UNC-40 dependent and independent UNC-5 signaling pathways. Nat. Neurosci. 12, 150–155.

Maglioni, S., Ventura, N., 2016. *C. elegans* as a model organism for human mitochondrial associated disorders. Mitochondrion 30, 117–125.

Malaiwong, N., Chalorak, P., Jattujan, P., et al., 2019. Anti-Parkinson activity of bioactive substances extracted from *Holothuria leucospilota*. Biomed. Pharmacother. 109, 1967–1977.

Mann, G., Sidkar, B., Kumar, A., et al., 2020. Role of flavonoids in neurodegenerative diseases: limitations and future perspectives. Curr. Top. Med. Chem. Available from: http://doi.org/10.2174/1568026620666200416085330.

Martorell, P., Bataller, E., Llopis, S., et al., 2013. A cocoa peptide protects *Caenorhabditis elegans* from oxidative stress and β-amyloid peptide toxicity. PLoS One 8 (5), e63283.

Maulik, M., Mitra, S., Bult-Ito, A., et al., 2017. Behavioral phenotyping and pathological indicators of Parkinson's disease in *C. elegans* models. Front. Genet. 8 (77), 1–21.

Meyer, D., Williams, P.L., 2014. Toxicity testing of neurotoxic pesticides in *Caenorhabditis elegans*. J. Toxicol. Environ. Health B Crit. Rev. 17, 284–306.

Montgomery, K., Corona, C., Frye, R., et al., 2018. Transport of a manganese/zinc ethylene-*bis*-dithiocarbamate fungicide may involve pre-synaptic dopaminergic transporters. Neurotoxicol. Teratol. 68, 66–71.

Mor, D.E., Daniels, M.J., Ischiropoulos, H., 2019. The usual suspects, dopamine and alpha-synuclein, conspire to cause neurodegeneration. Mov. Disord. 34, 167–179.

Morley, J.F., Morimoto, R.I., 2004. Regulation of longevity in *Caenorhabditis elegans* by heat shock factor and molecular chaperones. Mol. Biol. Cell 15, 657–664.

Mukherjee, S., Russell, J.C., Carr, D.T., et al., 2017. Systems biology approach to late-onset Alzheimer's disease genome-wide association study identifies novel candidate genes validated using brain expression data and *Caenorhabditis elegans* experiments. Alzheimers Dement. 13, 1133–1142.

Murphy, C.T., Hu, P.J., 2013. Insulin/insulin-like growth factor signaling in C. elegans. WormBook. The C. elegans Research Community, WormBook. Available from: http://www.wormbook.org/chapters/www_insulingrowthsignal/insulingrowthsignal.html.

Nasri, H., Baradaran, A., Shirzad, H., et al., 2014. New concepts in nutraceuticals as alternative for pharmaceuticals. Int. J. Prev. Med. 5, 1487–1499.

Negga, R., Stuart, J.A., Machen, M.L., et al., 2012. Exposure to glyphosate- and/or Mn/Zn-ethylene-*bis*-dithiocarbamate-containing pesticides leads to degeneration of gamma-aminobutyric acid and dopamine neurons in *Caenorhabditis elegans*. Neurotox. Res. 21, 281–290.

O'Reilly, L.P., Luke, C.J., Perlmutter, D.H., et al., 2014. *C. elegans* in high-throughput drug discovery. Adv. Drug Deliv. Rev. 69-70, 247–253.

Oliveira, R.P., Abate, J.P., Dilks, K., et al., 2009. Condition-adapted stress and longevity gene regulation by *Caenorhabditis elegans* SKN-1/Nrf. Aging Cell 8, 524–541.

Palikaras, K., Tavernarakis, N., 2017. *In vivo* mitophagy monitoring in *Caenorhabditis elegans* to determine mitochondrial homeostasis. Bio Protoc. 7.

Pandey, T., Sammi, S.R., Nooreen, Z., et al., 2019. Anti-ageing and anti-Parkinsonian effects of natural flavonol, tambulin from *Zanthoxyllum aramatum* promotes longevity in *Caenorhabditis elegans*. Exp. Gerontol. 120, 50–61.

Peixoto, H., Roxo, M., Krstin, S., et al., 2016. An anthocyanin-rich extract of Acai (*Euterpe precatoria* Mart.) increases stress resistance and retards aging-related markers in *Caenorhabditis elegans*. J. Agric. Food Chem. 64, 1283–1290.

Peres, T.V., Horning, K.J., Bornhorst, J., et al., 2019. Small molecule modifiers of *in vitro* manganese transport alter toxicity *in vivo*. Biol. Trace Elem. Res. 188, 127–134.

Pohl, F., Lin, P.K.T., 2018. The potential use of plant natural products and plant extracts with antioxidant properties for the prevention/treatment of neurodegenerative diseases: *In vitro, in vivo* and clinical trials. Molecules 23 (3823), 1–30.

Rademakers, R., Neumann, M., Mackenzie, I.R., 2012. Advances in understanding the molecular basis of frontotemporal dementia. Nat. Rev. Neurol. 8, 423–434.

Regitz, C., Dußling, L.M., Wenzel, U., 2014. Amyloid-beta ($A\beta_{1-42}$)-induced paralysis in *Caenorhabditis elegans* is inhibited by the polyphenol quercetin through activation of protein degradation pathways. Mol. Nutr. Food. Res. 58, 1931–1940.

Regitz, C., Fitzenberger, E., Mahn, F.L., et al., 2016. Resveratrol reduces amyloid-beta ($A\beta_{1-42}$)-induced paralysis through targeting proteostasis in an Alzheimer model of *Caenorhabditis elegans*. Eur. J. Nutr. 55, 741–747.

Rocha, E.M., De Miranda, B., Sanders, L.H., 2018. Alpha-synuclein: pathology, mitochondrial dysfunction and neuroinflammation in Parkinson's disease. Neurobiol. Dis. 109, 249–257.

Ronis, M.J.J., Pedersen, K.B., Watt, J., 2018. Adverse effects of nutraceuticals and dietary supplements. Annu. Rev. Pharmacol. Toxicol. 58, 583–601.

Ruszkiewicz, J.A., Pinkas, A., Miah, M.R., et al., 2018. *C. elegans* as a model in developmental neurotoxicology. Toxicol. Appl. Pharm. 354, 126–135.

Saharia, K., Arya, U., Kumar, R., et al., 2012. Reserpine modulates neurotransmitter release to extend lifespan and alleviate age-dependent Aβ proteotoxicity in *Caenorhabditis elegans*. Exp. Gerontol. 47, 188–197.

Sangha, J.S., Sun, X., Wally, O.S., et al., 2012. Liuwei Dihuang (LWDH), a traditional Chinese medicinal formula, protects against β-amyloid toxicity in transgenic *Caenorhabditis elegans*. PLoS One 7, e43990.

Santini, A., Cammarata, S.M., Capone, G., et al., 2018. Nutraceuticals: opening the debate for a regulatory framework. Br. J. Clin. Pharmacol. 84, 659–672.

Saul, N., Pietsch, K., Menzel, R., et al., 2009. Catechin induced longevity in *C. elegans*: from key regulator genes to disposable soma. Mech. Aging Dev. 130, 477–486.

Scanlan, L.D., Lund, S.P., Coskun, S.H., et al., 2018. Counting *Caenorhabditis elegans*: protocol optimization and applications for population growth and toxicity studies in liquid medium. Sci. Rep. 8, 904.

Schmitz, S.M., Lopez, H.L., MacKay, D., 2014. Nutravigilance: principles and practices to enhance adverse event reporting in the dietary supplement and natural products industry. Int. J. Food Sci. Nutr. 65, 129–134.

Shaw, W., Luo, S., Landis, J., et al., 2007. The *C. elegans* TGF-Beta dauer pathway regulates longevity via insulin signaling. Curr. Biol. 17, 1635–1645.

Shaye, D.D., Greenwald, I., 2011. OrthoList: a compendium of *C. elegans* genes with human orthologs. PLoS One 6, e20085.

Sherrington, R., Froelich, S., Sorbi, S., et al., 1996. Alzheimer's disease associated with mutations in presenilin 2 is rare and variably penetrant. Hum. Mol. Genet. 5, 985–988.

Silverman, G.A., Luke, C.J., Bhatia, R.S., et al., 2009. Modeling molecular and cellular aspects of human disease using the nematode *Caenorhabditis elegans*. Pediatr. Res. 65, 10–18.

Simpson, K.J., Davis, G.M., Boag, P.R., 2012. Comparative high-throughput RNAi screening methodologies in *C. elegans* and mammalian cells. N. Biotechnol. 29, 459–470.

Singh, J., Aballay, A., 2019. Neural control of behavioral and molecular defenses in C. elegans. Curr. Opin. Neurobiol. 62, 34–40.

Singhvi, A., Shaham, S., 2019. Glia-neuron interactions in *Caenorhabditis elegans*. Annu. Rev. Neurosci. 42, 149–168.

Smith, J.V., Luo, Y., 2003. Elevation of oxidative free radicals in Alzheimer's disease models can be attenuated by *Ginkgo biloba* Extract EGb 761. J. Alzheimer's Dis. 5, 287–300.

Smith, L.L., Ryde, I.T., Hartman, J.H., et al., 2019. Strengths and limitations of morphological and behavioral analyses in detecting dopaminergic deficiency in *Caenorhabditis elegans*. Neurotoxicology 74, 209–220.

Society of Toxicology, 2016. Guiding Principles in the Use of Animals in Toxicology. www.toxicology.org: Society of Toxicology.

Song, B., Zheng, B., Li, T., et al., 2020. Raspberry extract promoted longevity and stress tolerance via the insulin/IGF signaling pathway and DAF-16 in *Caenorhabditis elegans*. Food Funct. 11, 3598–3609.

Springer, W., Hoppe, T., Schmidt, E., et al., 2005. A *Caenorhabditis elegans* parkin mutant with altered solubility couples alpha-synuclein aggregation to proteotoxic stress. Hum. Mol. Genet. 14, 3407–3423.

Stiernagle, T., 2006. Maintenance of *C. elegans*. WormBook. The *C. elegans* Research Community, WormBook.

Stout Jr., R.F., Verkhratsky, A., Parpura, V., 2014. *Caenorhabditis elegans* glia modulate neuronal activity and behavior. Front. Cell Neurosci. 8, 67.

Sucro-Laos, F., Dueñas, M., González-Manzano, S., et al., 2012. Influence of catechins and their methylated metabolites on lifespan and resistance to oxidative and thermal stress of *Caenorhabditis elegans* and epicatechin uptake. Food Res. Int. 46, 514–521.

Sulston, J.E., Schierenberg, E., White, J.G., et al., 1983. The embryonic cell lineage of the nematode *Caenorhabditis elegans*. Dev. Biol. 100, 64–119.

Sutphin, G.L., Backer, G., Sheehan, S., Aging Research in Genomic Epidemiology Consortium, Gene Expression Working Group, et al., 2017. *Caenorhabditis elegans* orthologs of human genes differentially expressed with age are enriched for determinants of longevity. Aging Cell 16, 672–682.

Takahashi, A., Watanabe, T., Fujita, T., et al., 2014. Green tea aroma fraction reduces β-amyloid peptide-induced toxicity in *Caenorhabditis elegans* transfected with human β-amyloid minigene. Biosci. Biotechnol. Biochem. 78, 1206–1211.

Tang, B., Tong, P., Xue, K.S., et al., 2019. High-throughput assessment of toxic effects of metal mixtures of cadmium (Cd), lead (Pb), and manganese (Mn) in nematode *Caenorhabditis elegans*. Chemosphere. 234, 232–241.

Tarazi, F.I., Sahli, Z.T., Wolny, M., et al., 2014. Emerging therapies for Parkinson's disease: from bench to bedside. Pharmacol. Ther. 144, 123–133.

Therrien, M., Parker, J.A., 2014. Worming forward: amyotropic lateral sclerosis toxicity mechanisms and genetic interactions in *Caenorhabditis elegans*. Front. Genet. 5, 85.

Tissenbaum, H.A., 2018. DAF-16: FOXO in the context of *C. elegans*. Curr. Top. Dev. Biol. 127, 1–21.

Toyoshima, Y., Wu, S., Kanamori, M., et al., 2020. Neuron ID dataset facilitates neuronal annotation for whole-brain activity imaging of *C. elegans*. BMC Biol. 18, 30.

Van Damme, P., Robberecht, W., Van Den Bosch, L., 2017. Modelling amyotropic lateral sclerosis: progress and possibilities. Dis. Model. Mech. 10, 537–549.

Vayndorf, E.M., Lee, S.S., Liu, R.H., 2013. Whole apple extracts increase lifespan, healthspan and resistance to stress in *Caenorhabditis elegans*. J. Funct. Foods. 5, 1236–1243.

Vistbakka, J., VanDuyn, N., Wong, G., et al., 2012. *C. elegans* as a genetic model system to identify Parkinson's disease-associated therapeutic targets. CNS Neurol. Disord. Drug. Targets 11, 957–964.

Von Stetina, S.E., Watson, J.D., Fox, R.M., et al., 2007. Cell-specific microarray profiling experiments reveal a comprehensive picture of gene expression in the *C. elegans* nervous system. Genome Biol. 8, R135.

Wang, Y.A., Kammenga, J.E., Harvey, S.C., 2017. Genetic variation in neurodegenerative diseases and its accessibility in the model organism *Caenorhabditis elegans*. Hum. Genomics 11, 12.

Wang, H., Liu, J., Li, T., et al., 2018. Blueberry extract promotes longevity and stress tolerance via DAF-16 in *Caenorhabditis elegans*. Food Funct. 9, 5273–5282.

White, J.G., Southgate, E., Thomson, J.N., et al., 1986. The structure of the nervous system of the nematode *Caenorhabditis elegans*. Philos. Trans. R Soc. Lond. B Biol. Sci. 314, 1–340.

Wiegant, F.A., Surinova, S., Ytsma, E., et al., 2009. Plant adaptogens increase lifespan and stress resistance in *C. elegans*. Biogerontology 10, 27–42.

Wilson, M.A., Shukitt-Hale, B., Kalt, W., et al., 2006. Blueberry polyphenols increase lifespan and thermotolerance in *Caenorhabditis elegans*. Aging Cell 5, 59–68.

Wolozin, B., Gabel, C., Ferree, A., et al., 2011. Watching worms whither: modeling neurodegeneration in *C. elegans*. Prog. Mol. Biol. Transl. Sci. 100, 499–514.

Wong, W.R., Brugman, K.I., Maher, S., et al., 2019. Autism-associated missense genetic variants impact locomotion and neurodevelopment in *Caenorhabditis elegans*. Hum. Mol. Genet. 28, 2271–2281.

WormBase Community, 2000. *Strain CL2070*. Retrieved February 8, 2015, from WormBase: http://www.wormbase.org/species/c_elegans/strain/CL2070#02–10WormBase Community. (2002, August 9). *TJ356 Strain*. Retrieved February 10, 2015, from WormBase: http://www.wormbase.org/species/c_elegans/strain/TJ356#02--10.

WormBase Community, 2004. *CF1553 STrain*. Retrieved February 10, 2015, from WormBase: http://www.wormbase.org/species/c_elegans/strain/CF1553#02--10.

Wu, Y., Wu, Z., Butko, P., et al., 2006. Amyloid-beta-induced pathological behaviors are suppressed by *Ginkgo biloba* extract EGb 761 and ginkgolides in transgenic *Caenorhabditis elegans*. J. Neurosci. 26, 13102–13113.

Yan, J., Yang, Z., Zhao, N., et al., 2019. Gastrodin protects dopaminergic neurons via insulin-like pathway in a Parkinson's disease model. BMC Neurosci. 20, 31.

Yang, T., Zhao, X., Zhang, Y., et al., 2020. 6‴-Feruloylspinosin alleviated beta-amyloid induced toxicity by promoting mitophagy in *Caenorhabditis elegans* (GMC101) and PC12 cells. Sci. Total Environ. 715, 136953.

Young, A.L., Bass, I.S., 1995. The Dietary Supplement Health and Education Act. Food Drug Law J. 50, 285–292.

Young, L.E.A., Shoben, C., Ricci, K., et al., 2019. Genetic analysis of KillerRed in *C. elegans* identifies a shared role of calcium genes in ROS-mediated neurodegeneration. J. Neurogenet. 33, 1–9.

Yu, C.W., Wei, C.C., Liao, V.H., 2014. Curcumin-mediated oxidative stress resistance in *Caenorhabditis elegans* is modulated by *age-1*, *akt-1*, *pdk-1*, *osr-1*, *unc-43*, *sek-1*, *skn-1*, *sir-2.1*, and *mev-1*. Free Radic. Res. 48, 371–379.

Zarse, K., Jabin, S., Ristow, M., 2012. *L*-Theanine extends lifespan of adult *Caenorhabditis elegans*. Eur. J. Nutr. 51, 765–768.

Zheng, L., Jie, G., Zhang, J., et al., 2009. Significant longevity-extending effects of EGCG on *Caenorhabditis elegans* under stress. Free Radic. Biol. Med. 46, 414–421.

Zhu, S., Li, H., Dong, J., et al., 2017. Rose essential oil delayed Alzheimer's disease-like symptoms by SKN-1 pathway in *C. elegans*. J. Agric. Food Chem. 65, 8855–8865.

Chapter 29

Alternative in vitro models for safety and toxicity evaluation of nutraceuticals

Gopala Krishna and Mayur Krishna

Introduction

In recent years, the subject of nutraceuticals or "bioceuticals" as a pharmaceutical alternative has gained enormous publicity due to the physiological benefits of such products. The term "nutraceutical" was coined in 1979 by Stephen DeFelice and is defined as "a food or part of food, which provides medical or health benefits, including the prevention and treatment of disease" (Biesalski, 2001). Further, these products are variously called vitamins, dietary supplements, functional foods, "nutraceuticals," phytochemicals, biochemopreventatives, and designer foods. These terms vary in meaning from country to country, as does regulation of these products (Zeisel, 1999). Two examples of nutraceuticals are shown in Fig. 29.1.

A brief summary of classification of nutraceuticals is listed in Table 29.1 (Mestrovic, 2015).

The general position of regulatory agencies regarding the formal definition of nutraceuticals is included in the following box.

In this chapter, nutraceuticals are defined in a broad sense as those dietary supplements that deliver a concentrated form of a presumed bioactive agent from a food, presented in a nonfood matrix, and used to enhance health at dosages that exceed those that are obtained from traditional foods. A good example is genistein purified from soybeans and delivered as pills (Fig. 29.1). "Functional foods" are consumed as part of a normal diet but are designed to deliver one or more active ingredients that have physiological or health-promoting effects within the matrix of traditional foods, for example, a bread or breakfast cereal with added high-dose folic acid. However, the dosage to be consumed is not included as part of the definition.

Nutraceuticals in general are natural, bioactive, chemical compounds that have health-promoting, disease-preventing, or medicinal properties and range from isolated nutrients, herbal products, dietary supplements, and diets to genetically engineered foods and processed products such as cereals, soups, and beverages (Andlauer and

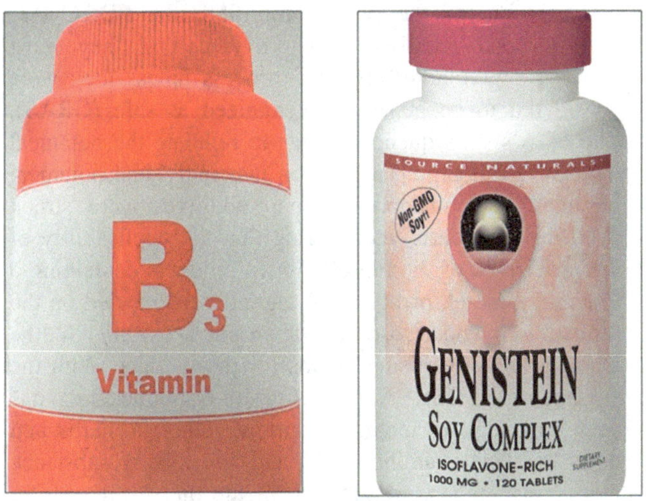

FIGURE 29.1 As an example to illustrate the benefits of nutraceuticals, a study conducted at the University of Sydney found that a form of vitamin B3 (nicotinamide) has the potential to lower the risk of developing skin cancer in people who are at higher risk for the disease (Damian, 2015). Similarly, Genistein, an isoflavone phytonutrient derived from soybeans, can bind to the same receptor sites as estrogen-causing effects in the body similar to those caused by the hormone estrogen (Dang, 2009).

Nutraceuticals. DOI: https://doi.org/10.1016/B978-0-12-821038-3.00029-X

TABLE 29.1 General definition and classification of nutraceuticals.

Nutraceuticals

- Certain form of drugs with a focus on nutrition and health to prevent disease
- Definition depends on the source of nutraceuticals: natural, pharmaceutical condition, or chemical constituents that include:
  - Products derived from food source
  - Have added health benefits
  - Serve as nonspecific biological therapy to:
  - Promote general well-being
  - Control disease symptoms
  - Prevent malignant processes

Classification of nutraceuticals

- Dietary supplements:
  - Nutrients derived from food products (could be liquid, capsules, powder, or pill)
  - FDA regulates these as foods, but pathway differs from drugs or other foods
- Functional foods:
  - Whole foods, fortified or enriched, known to reduce the risk of chronic disease
  - Provide health benefits beyond the traditional nutrition
- Medical foods:
  - Formulated to be administered with medical supervision
  - Aimed at management of certain diseases or conditions
- Farmaceuticals:
  - Medically valuable components produced from genetically modified organisms (GMOs)
  - These could be from the farm, such as agricultural crops or animals that are used as nutraceutical factories, so they are cost-effective

Potential health benefits: nutritional, safety, and therapeutic

- Antioxidant defenses
- Cell proliferation
- Gene expression
- Safeguarding of mitochondrial integrity

BOX 29.1 Formal definition of nutraceuticals

The term dietary supplements was formally defined for US government offices in 1994 as a product (other than tobacco) intended to supplement the diet to enhance health that bears or contains one or more of the following dietary ingredients: a vitamin; a mineral; an amino acid; a herb or other botanical; a dietary substance for use to supplement the diet by increasing the total dietary intake; or a concentrate, metabolite, constituent, extract, or combination of any ingredient described in this list. Dietary supplements are further defined as products that are labeled as dietary supplements and are intended for ingestion in the form of a capsule, powder, soft gel, or gel cap, and not represented as a conventional food or as a sole item of a meal or the diet. The Dietary Supplement Health and Education Act (DSHEA) of 1994 defined both of the terms "dietary ingredient" and "new dietary ingredient (NDI)" as components of dietary supplements.

The Food and Drug Administration (FDA) approach and/or thought process to the term "nutraceuticals" is as follows. Terms such as "functional foods" or "nutraceuticals" are widely used in the marketplace. Such foods are regulated by FDA under the authority of the Federal Food, Drug, and Cosmetic (FD&C) Act, even though they are not specifically defined by law (FDA, 2015a).

Furst, 2002; Bagchi, 2006). A large percentage of the US population uses a botanical or nutritional or dietary supplement on a daily basis as an adjunct to an improved quality of life or for their purported medical benefits. The regulation of botanicals by the FDA is governed by the provisions of the Federal FD&C Act. However, the FDA's attempts to regulate dietary supplements have historically been met with resistance by industry as well as by segments of the public. In 1958, the FD&C Act defined food additives and provided that they must undergo a premarket approval process unless they were considered to be generally recognized as safe (GRAS). The FDA's subsequent attempts to regulate the botanical industry by assuming a position that individual botanical products were unapproved food additives were struck down by the courts, recognizing that the applicability of the provisions of the FD&C Act to products containing a vitamin, mineral, or botanical ingredient depended on the product's intended use, as determined usually by the labeling and advertising claims for the product. With the passage of the DSHEA, the definition of nutraceutical has been expanded to include vitamins, minerals, herbs and other botanicals, amino acids, and any dietary substance for use by humans to supplement the diet by increasing total dietary intake (Whitman, 2001; Stauffer, 1999). An example of drug versus nondrug products that include nutraceuticals and their pathways for classification is shown in Fig. 29.2.

The concepts of nutraceuticals, functional foods, or dietary supplements are confusing and often used interchangeably. As with traditional foods, when used as recommended, many dietary supplements may be safe. However, the broad spectrum of products that qualify as dietary supplements per DSHEA coupled with their ever-increasing use makes assessment of risk to the health of the consumer an arduous task. In addition to the limitations imposed by DSHEA, the FDA determines what is unsafe without requiring specific information on safety to be presented by manufacturers prior to marketing, or that manufacturers submit to the FDA any reports they have received on serious adverse events associated with dietary supplement (e.g., nutraceutical) use serve to make the safety regulation of dietary supplements a sizeable challenge.

For NDIs (those not marketed in the United States prior to passage of DSHEA in 1994), manufacturers or distributors must notify the FDA at least 75 days before introducing a dietary supplement ingredient and must provide the FDA with the information that is the basis on which the manufacturer has concluded that dietary supplement or ingredient will be expected to be reasonably safe. Because the regulatory agencies have not established any specific guidance for the safety evaluation of nutraceuticals per se, this chapter refers to the guidance applicable to the safety assessment of food ingredients and/or dietary supplements interchangeably (FDA Redbook, 2007; EFSA, 2009). DSHEA established the first comprehensive definition of dietary supplements as legally equivalent to foods. Most importantly, DSHEA established a regulatory framework for dietary supplements that defined the FDA's authority over these products. An "NDI" is one that meets the aforementioned definition for a "dietary ingredient" and was not sold in the United States as a dietary supplement before October 15, 1994.

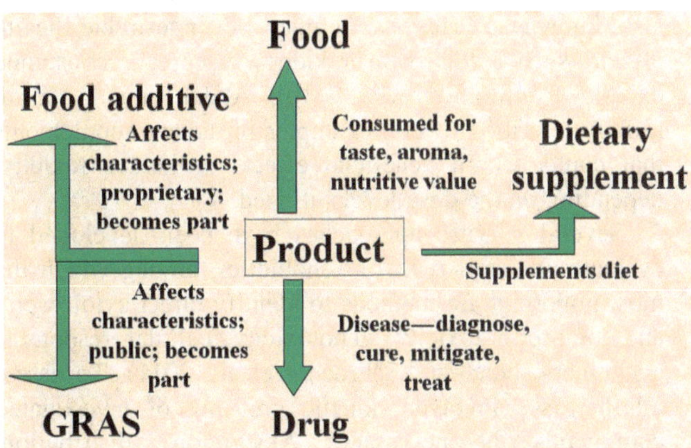

FIGURE 29.2 Categories of products intended for oral consumption; nutraceuticals are generally included under dietary supplements (Soni et al., 2010).

Scope of this chapter

This chapter provides a brief overview of available in vitro toxicity testing alternatives for nutraceutical safety assessment. Several publications have described in vitro experiments on a variety of dietary supplement ingredients, some of which were specifically designed to evaluate safety endpoints. Because a battery of tests is not mandatory for dietary supplement ingredients, results from such safety assessments have limited prevalence in the literature, unlike that for new pharmaceutical or chemical entities. In this chapter, the types of in vitro studies available for safety assessment, their potential utility, general considerations for their interpretation, the types of endpoints examined, and some of the commonly used model systems are discussed. Also included is a listing and brief description of select validated in vitro assays that are generally accepted for use in predicting effects on whole organisms and in humans.

General concepts for safety assessment of nutraceuticals

Prior to DSHEA (1994), dietary supplements were subject to the same requirements as other components of food under the FD&C Act and the same regulatory scrutiny as food additives (FDA, 2015b). DSHEA amended FD&C, providing regulations specific for dietary supplements. Presently, manufacturers do not need to register their products with the FDA before producing or marketing them, nor are they required to furnish information regarding their safety or efficacy, with the exception of "NDI." Further, manufacturers also have no legal obligation under DSHEA to report data about safety and purported benefits of products to the FDA or to the public, in contrast to the requirements for food additives and drugs. Under DSHEA, the FDA has the legal burden to determine whether a dietary supplement (nutraceutical) is unsafe before it can restrict production or sale in the

United States. A generalized regulatory safety assessment approach for food-related products is shown in Fig. 29.3.

Rao (2017) recommended the safety testing of nutraceuticals in simple terms as follows: "There is no single thumb rule. However, testing needs depends upon how novel the product and the ingredient(s) are. Based on various considerations of the product content, one has to decide on the extent of testing. However, as a general rule, typically, a 4-week to 13-week study along with mutagenicity (Ames and CHO) should be sufficient. If the product is intended for women of child bearing age and for extended long-term use, a teratology and reproduction studies may be required." Similarly, Mendes et al. (2019) talked about generally suggested safety assessment for herbal food ingredients and nutraceuticals, including compounds generally GRAS. However, in this chapter, the role of in vitro testing is described so potential safety can be assessed without the use of an excessive number of animals.

Overview of in vitro experimental systems

The in vitro systems play a significant role in the safety assessment of nutraceuticals. In vitro experimental systems including isolated organs or cells, microorganisms, subcellular organelles, and molecular entities (e.g., receptors, transport proteins) are commonly used to gain insight into the risk of adverse effects of xenobiotics. Such studies enable close scrutiny of effects on target processes or structures in isolation from confounding factors, allowing insight into its mechanism of action and adverse outcome pathways. In addition, they are generally of high throughput, more rapid, and less expensive compared to in vivo studies. Therefore, in vitro studies are much more likely to be utilized for assessment of nutraceutical safety. An in vitro assay will have the most direct predictive value when the measured effects are known to correspond to a specific functional change characteristic of an adverse health effect. It may often be appropriate to use in vitro data as hypothesis generators, as potential indicators of harmful health effects, or as information about biological plausibility or mechanism, rather than as a standalone demonstrated indicator that in itself indicates risk. However, some types of in vitro assays should be considered validated as predictors of potential toxicity and, when carefully conducted and interpreted, will provide valuable information beyond simply reinforcing observations from other categories of data or generating hypotheses. In vitro assays are appropriate to use as standalone indicators of risk to human health when a correlation is established between the results of an in vitro assay and the actual animal and/or human outcome, validating the predictive value of the assay.

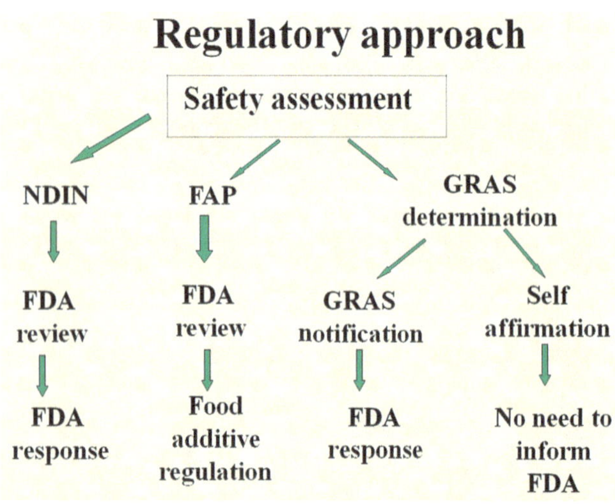

FIGURE 29.3 Regulatory considerations for classifying various food categories regarding safety assessment (Soni et al., 2010; FDA, 2015a, b). *FAP*, food additive petition; *GRAS*, generally recognized as safe; *NDIN*, new dietary ingredient notification.

Although the regulatory situation is such that completion of these or other in vitro tests is not required for dietary supplement ingredients, the scientific value of these assays in predicting adverse effects in humans is valuable in assessing the safety of dietary supplements. In vitro assays such as those contained in the FDA Redbook (2007) should be considered, among others, as important validated tools for the prediction of possible adverse effects in humans. Additionally, although it is not necessary to establish a rational mechanism of toxicity to conclude that an ingredient poses an unreasonable risk, it is nonetheless valuable to identify potential mechanisms and causal relationships. Assays that warrant the most attention are those that predict biological effects/outcomes associated with toxicological manifestations or effects that could be reasonably hypothesized to lead to serious risks. To quote the Paracelsus statement: "All substances are poisons: there is none which is not a poison. The right dose differentiates a poison and a remedy" (SOT, 2015). Vitamin C, a nutraceutical that is generally known to provide beneficial health effects at recommended levels, is also known to cause a number of undesirable health effects, such as diarrhea or kidney disease, if consumed excessively (Soni et al., 2010). Abdel-Rahman et al. (2010) described several examples of dietary supplements that could accord beneficial effects or induce toxicity, depending on the dose levels utilized.

Several in vitro approaches have been developed to evaluate potential risks of xenobiotics, among which the most important are methods to identify direct cytotoxicity and long-term toxicity, genotoxicity, cellular responses, and kinetic behavior (Schroeder et al., 2011). To assess all the risks involved, kinetic profiling of investigated compounds is necessary. Absorption, distribution, metabolism, excretion (ADME) studies are performed to determine the bioaccessibility and biotransformation as well as accumulation of bioactive substances (Fernandez-Garcia et al., 2009). It is essential to relate toxicodynamic information from in vitro systems to real-life situations by translating concentration–effect relationships to dose–effect relationships, which can be achieved by physiologically based toxicokinetic modeling (Adler et al., 2011).

While planning in vitro experiments, it is important to consider the degree to which a given in vitro system replicates the biology of the human target cells and their responses to xenobiotics. For example, these procedures may not yield relevant information if performed at substrate concentrations that are not comparable to target tissue concentrations or if the substrate is not metabolized similarly in vitro as in vivo. Knowledge of a dietary supplement ingredient's pharmacokinetics and in vivo metabolism, if available from animal or human studies, will allow the most appropriate interpretation of the relevance of compound concentrations used in in vitro experiments to amounts consumed by humans.

General types of in vitro assays

As described, in vitro assays are valuable because they allow scientists to answer questions that may not otherwise be answerable due to various constraints with whole-animal and human studies. It is not possible to describe the complete realm of in vitro assays in this chapter, but consideration of the value provided by those described here gives perspective to the value of in vitro assays as a scientific tool for predicting safety as measured by risk versus benefit.

Assay types by effect observed

Effects on cellular receptors

A number of different assays focus on examining the effects of supplement ingredients on cellular receptors. Understanding the types of receptors that are activated or inhibited, considered together with the general knowledge of their role in physiological processes, can enhance the understanding of how dietary supplement ingredients will affect humans. General types of assays for examining receptor activity are:

- Receptor-binding assays—used to measure how strongly a substrate binds to various receptor types and to determine whether the substance is likely to activate or inhibit cellular receptor activity;
- Cell function assays—used to assess agonist (a substance that initiates a physiological response when combined with a receptor in the same fashion as the endogenous ligand for the receptor) or antagonistic (a substance that interferes with or inhibits the physiological action of another) activity of receptor activity in the intact cell by examining effects that are downstream of the receptor.

Receptor activation is often a critical event in biological pathways that leads to a functional effect. Many receptors are characterized well enough to reasonably predict whether their activation or inhibition in target tissues would lead to a pharmacological or toxicological process that presents a health risk or benefit. Thus, whether cellular receptor effects should be considered as indicative of a risk depends on what is known about the downstream effects of receptor activation or inhibition if an agent in question reaches the receptor at sufficient concentrations.

Effects on ion channel activity and electrophysiology

A number of assays can predict possible effects on the electrophysiology of the cells, either by directly measuring the membrane potential of the cells or by using other methods to examine effects on ion channels, the cellular proteins that control the cell's ion distribution.

Effects on some ion channels are associated with specific adverse physiological outcomes. For example, some drugs can induce "long QT syndrome," a potentially fatal cardiac disorder that may trigger arrhythmias. This syndrome is the result of a compound interaction with the ether-a-go-go-related gene (hERG), which codes for the potassium channel, changing the action potential duration in cardiac cells. Well-characterized in vitro assays are presently used to screen compounds (such as nutraceuticals), which may exert this adverse effect on potassium channels (Roche et al., 2002).

Effects on enzyme activity

At the molecular level, enzyme activity assays measure a substance's ability to inhibit or induce enzyme activity, either by affecting the enzyme activity itself or by changing the enzyme content in the cell (e.g., affecting translation or enzyme stability). For example, effects on cytochrome P450 (CYP450) enzymes may suggest risk for pharmacokinetic interaction with other ingested substances.

Effects on genes and nucleic acids

A number of techniques are capable of identifying genetic alterations that may be responsible for disease processes, alterations such as nucleic acid mutations, or effects on chromosomes. These assays are of growing importance for risk assessment because they measure events that could lead to delayed or latent adverse health effects, including carcinogenicity.

Assay types by experimental system

The previous sections organized in vitro assays by the effects assessed: effects on enzymes; ion channels; and nucleic acids. These and other in vitro tests may be conducted in a variety of experimental systems, such as isolated tissues or in isolated cells, each of which has its own considerations.

Isolated tissues

In some experiments, animal tissues are isolated and exposed to test substances (e.g., dietary supplement ingredients). Tissues in many ways resemble an in vivo system because they contain a variety of cell types organized in their native structure, and because the protein expression is relatively similar to that in vivo. Nevertheless, the toxicological value of these effects can be limited by the inability of an isolated tissue to react with other reciprocating organ-body systems.

Botanical extracts often contain polyphenolic compounds (e.g., tannins) that can reversibly or irreversibly bind to subcellular components, such as enzymes, signal transduction factors, and receptors. When administered orally to humans, however, these compounds may bind to food components or gastrointestinal cells, or they can be metabolized by gastrointestinal mucosal or bacterial enzymes and, therefore, do not exert the same effects in vivo as seen with isolated tissue (Bravo, 1998; Yang et al., 2001).

In summary, if the impact of excluding effects of other organ systems is considered or compensated for, isolated tissue can be a valuable tool for studying physiological effects of dietary supplement ingredients.

Isolated cells

Functional cell models have an essential role in nutraceutical risk assessment and toxicology as an alternative to animal studies. These models offer a fast and reliable way to characterize new functional foods and obtain mechanistic insights at the cellular or molecular level. Despite the fact that such models lack the complexity of animal models, they have several advantages such as reproducibility of results, controlled environment, and a reductionist approach enabling mechanistic insights.

Suspensions of isolated cells or cultures of cells derived from animal tissue or from continuously replicating cell lines offer numerous advantages for studying xenobiotic toxicity. Due to the growing demand for natural compounds and new protective/probiotic strains as components of functional foods, cytotoxicity assays constitute useful tools for quickly identifying potential toxic effects. The test substance can be examined in direct contact with the cell type of interest, the concentration of the test substance can be rigorously controlled, and the secondary effects of such extracellular influences as metabolic factors, matrix, and cell-to-cell contact can be manipulated or controlled. A broad classification of cell-based in vitro model systems and their potential applications in toxicology including evaluation of cytotoxicity, elucidating mechanism of action, and kinetics are shown in Fig. 29.4.

In vitro models of normal human tissues are gaining importance due to their relevance and wide applicability. Because in vitro models do not replicate the complex biology of the human body, the model system chosen should relate to specific organ systems that are mimicked by utilizing different cell culture models, such as models of the gastrointestinal tract or liver in the case of risk assessment of nutraceuticals.

Endpoint-based in vitro toxicity assessment of nutraceuticals

In the following, some of the commonly used endpoint assays such as genotoxicity, teratogenicity, and major organ systems are discussed.

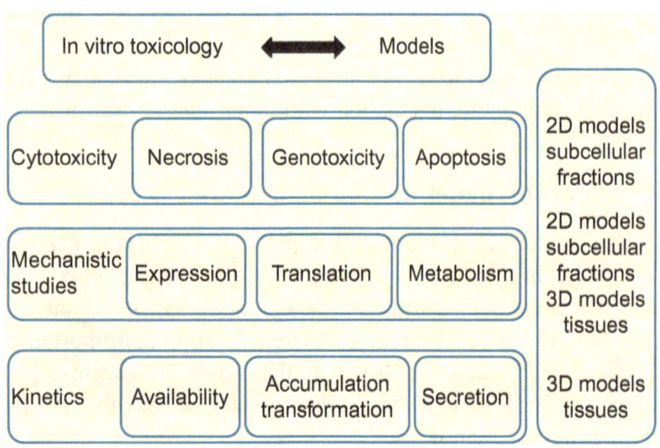

FIGURE 29.4 Functional cell models used in practicing in vitro toxicology. Monolayer models are suitable for quick determination of cytotoxicity or to monitor specific markers where cell differentiation is not as important. Two-dimensional (2D) models are suitable for high-throughput screening, enabling a fast initial screening of numerous compounds and concentrations. However, 3D models with fully differentiated cell lines are more appropriate for mechanistic and kinetic studies.

Genotoxicity assessment

Genetic toxicology is a laboratory-based science that identifies agents with toxicity directed toward the hereditary material of living systems. Genotoxic substances cause DNA and/or nuclear damage ranging from alterations or substitutions of single nucleic acids and/or nucleotides to chromosomal deletions and duplications involving thousands of DNA base pairs to losses or duplication of entire chromosomes. These adverse reactions to genetic material can be analyzed utilizing a variety of short-term test systems. In vitro, microbial and mammalian cells are used in gene mutation assays; human lymphocytes or rodent primary cells or cell lines are analyzed for chromosomal affects. These effects may include structural changes (membrane, organelles) or functional changes (ion transport, synthesis of macromolecules). Genetic toxicology tests are used to assess chemicals for the potential to cause permanent, heritable changes in test organisms (Gollapudi and Krishna, 2000; Krishna, 2010). Generally, for nutraceutical safety assessment, these assays are routinely used.

Standard in vitro test battery

In vitro, typically, a battery consisting of bacterial mutagenicity tests and mammalian micronucleus and/or chromosome aberration tests either in primary cells or in cell lines is used. The standard test battery is compliant with ICH (International Conference on Harmonization) guidelines and each test is conducted under relevant Good Laboratory Practice guidelines according to Organization for Economic Cooperation and Development (OECD) protocol guidelines. These protocols have specific requirements that must be met for an assay to be acceptable by the regulatory agencies. Specific requirements include, but are not limited to, the following criteria:

- Must be tested up to levels of cytotoxicity, limitations of solubility, or upper limit dose defined for each assay;
- Must be conducted with all relevant bacterial strains or acceptable cell systems;
- Must use appropriate metabolic conditions, treatment times, and tissue/sample collection times;
- Must use the appropriate number of replicates for each assay and repeat tests as needed to obtain conclusive results;
- Must be performed with appropriate positive and negative controls and each must respond within historical control ranges.

Metabolic transformation

All in vitro tests are conducted by exposing cells to the intended nutraceutical in culture in the presence and absence of an exogenous metabolic activation system. The metabolic system is typically prepared from the livers of rats that are pretreated with Aroclor 1254, a mixture of polychlorinated biphenyls that promotes the induction of P450 isozymes. This family of enzymes is responsible for various oxidative reactions that can biotransform a promutagen (indirect mutagen) into a mutagen. Hepatocytes are particularly rich in the P450 enzymes required for metabolic activation; however, these enzymes are also found in lower amounts in numerous other tissues.

Bacterial mutation assays (Ames test)

This test is also known as the "Ames" test after Dr. Bruce Ames, the original developer of the tester strains (Ames et al., 1975; Maron and Ames, 1983). In this assay, four *Salmonella typhimurium* and one *Escherichia coli* strain are used as the standard strains for detection of point mutations. Because bacteria do not have chromosomes, chromosomal damage is not detected in this assay. The *Salmonella* strains each have specific mutations at the histidine locus (his^-) that allow us to detect specific types of point mutations. Two of the strains, TA1535 and TA100, detect substitution-type mutations and the other two (TA1537 and TA98) detect frameshift mutations that arise through the addition or deletion of one or more nucleotides. The *E. coli* strain has a unique mutation at the tryptophan (trp^-) locus that is particularly rich in adenosine and thymidine repeats (ATATAT), making it susceptible to mutations that prefer these targets. Mutations are detected as reversions to the wild-type ($his^- \rightarrow his^+$ or $trp^- \rightarrow trp^+$) phenotype; thus this assay is considered to be a reverse mutation assay. By using a selective agar that is deficient in these nutrients, only mutated bacteria can survive and form colonies. Because bacteria divide rapidly (once every 20–30 min), colonies can be detected and counted visually or quantified with the aid of an automated machine within 48–72 h after treatment (Fig. 29.5).

This test is typically conducted earliest in the battery because of its high sensitivity, minimal compound requirements (1 g), and relatively rapid turnaround time (2–4 weeks). In addition to the mutations at the histidine or tryptophan loci that are used as a basis for mutant selection, all of the strains have engineered mutations that make their cell walls more permeable to test chemicals. Some of the strains (TA98, TA100, and *E. coli*) also contain plasmids that render them more susceptible to error-prone DNA repair, further increasing the sensitivity of this assay.

Data interpretation

When possible, test concentrations are selected to achieve dose-related cytotoxicity to the bacteria. This

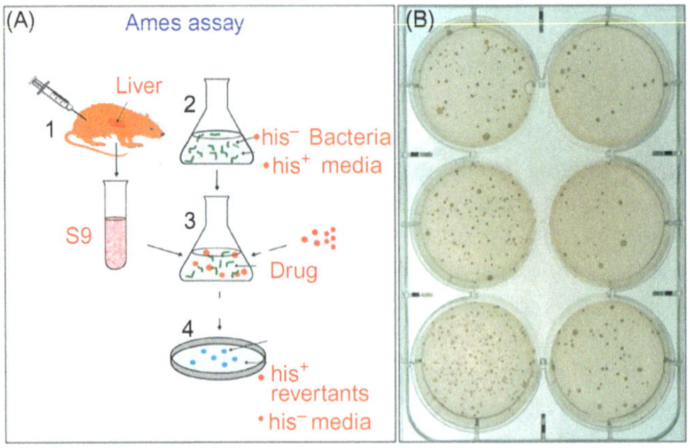

FIGURE 29.5 The in vitro Ames assay in specially constructed strains of bacteria is used to evaluate the mutagenicity of nutraceuticals, drugs, or chemicals. (A) Principle 1: S9 liver microsomal fraction (metabolic transformation) is prepared from rats pretreated with a liver enzyme inducer; principle 2: histidine mutant bacteria are grown in the presence of histidine media; principle 3: S9, bacteria, and the test substance are mixed in top agar containing trace amounts of histidine to allow several DNA replications and poured onto agar plates with minimal media without the histidine and incubated; and principle 4: if the test substance and/or its metabolite(s) cause mutations, then it would reverse the his^- gene into the his^+ gene, which can grow in minimal media and form colonies and thus is considered a mutagen. (B) The results of an Ames test are shown in a six-well plate that is usually used in high-throughput screening. From the top left each well represents: vehicle control, test article doses 1, 2, 3, and 4, and positive control showing an increasing number of colonies with dose, suggesting the test article as a positive Ames test mutagen.

demonstrates that sufficiently high enough concentrations were tested to assure detection of potential mutagens. However, in the case of antibiotics (or other compounds that are relatively more toxic to bacteria than mammalian cells), toxicity may mask mutagenic activity that may be induced at higher concentrations in mammalian eukaryotic cells. Therefore excessive cytotoxicity in bacterial test systems may drive the need to expedite testing in mammalian cells (see "Mammalian Cell Mutation Assays").

A reproducible concentration-dependent increase in mutant (revertant) colonies more than threefold over background in one or more strains of bacteria is considered to be indicative of a positive mutagenic response. A positive response seen only in the presence of metabolic activation implies that an active metabolite has been generated. Because each of the test strains has unique sensitivity and specificity to different classes of mutagens, a response in only one bacterial strain is not weighted any differently than if a response was seen with all strains. Weak, equivocal (two- to threefold) increases in mutants are confirmed in an independent repeat test. A reproducible weak increase may be indicative of a low-level, mutagenic impurity. Additional analytical testing and mutagenicity testing with different lots may be warranted to establish the significance of this response.

Mammalian cell mutation assays

These assays are the mammalian counterparts to the "Ames" bacterial mutation tests. Because these assays are conducted with mammalian cells that have a much longer cell division time (12–24 h compared to 20–30 min for bacteria), they require a proportionately longer assay time compared to bacterial tests, generally 4–8 weeks. Furthermore, because culture times are greater, the risks for culture contamination and subsequent repeat testing are also increased. Although there are several different mammalian systems that are used for assessing induction of *point mutations* in mammalian cells, the current systems of choice are the Chinese hamster ovary/hypoxanthine guanine phosphoribosyl transferase (CHO/HGPRT) assay system and mouse lymphoma assay (MLA) system.

Chinese hamster ovary/hypoxanthine guanine phosphoribosyl transferase assay system

The HGPRT gene is used as the basis for mutant selection for detecting forward mutations (normal to mutant). This gene is located on the X-chromosome that is functionally hemizygous in ovary cells (female). Therefore only one chromosome (gene copy) needs to be mutated to observe the mutant phenotype (gene expression). In normal cells, the HGPRT gene is used for purine salvage for the formation of adenosine triphosphate and guanosine triphosphate. In the absence of a functional HGPRT gene, the cell can still produce these nucleotides de novo. Therefore a mutation at this locus alone is generally not lethal. As a means for selecting mutant cells, cells are cultured in the presence of the antimetabolite 6-thioguanine (6-TG). A normal "wild-type" cell with a functional HGPRT gene will incorporate the toxic antimetabolite in place of guanine, resulting in the death of the normal cell. Conversely, a cell with a mutant HGPRT gene will not take up the antimetabolite and will survive through de novo synthesis of purines.

Because mutagenicity is typically correlated with cytotoxicity, it is necessary to perform preliminary cytotoxicity tests to fully assess the dose–response curve prior to dose selection. To assess induction of mammalian mutations, duplicate cultures of CHO cells are cultured for 4 h with several concentrations of xenobiotic in the presence or absence of *metabolic activation*. After treatment, known numbers of cells are cloned to determine nutraceutical-induced cytotoxicity, whereas a subset of cells is carried in culture for 7–9 days to express the mutant phenotype (clear out endogenous levels of hypoxanthine and guanine). Mutants are then selected by culturing a million cells at each concentration in the presence of the antimetabolite (6-TG). After an additional week in culture, each mutant cell will form a colony that is visible to the naked eye.

Data interpretation

Increased numbers of mutant colonies per million cells compared to spontaneous numbers of mutants in the control cultures are indicative of a "positive" induction of mutations. This assay is less sensitive than the *bacterial mutation assays* because it is conducted with a mammalian cell line that has some mechanisms for DNA repair and because mammalian DNA is afforded some protection from chromosomal proteins that are not present in bacteria. In addition, this test is not sensitive to larger *frameshift mutations* and *chromosomal aberrations* that usually affect other essential (needed for survival) genes downstream from the HGPRT gene. Therefore the absence of mammalian gene mutations in this assay does not negate or reduce risk if effects are seen in other assays. Conversely, a positive response in this assay in addition to a positive response in one or more other in vitro assays has a high risk for further product development.

Mouse lymphoma assay

MLA can detect a variety of mutations, including point mutations, and is performed similar to HGPRT assay. In contrast to the in vitro micronucleus test (described in this chapter), MLA also detects translocations (Liechty et al., 1998). Most of the substances that are positive in this mammalian gene mutation test also induce clastogenic effects (Kirkland et al., 2011).

In vitro cytogenetic assays

A variety of cytogenetic assays have been developed and validated for genotoxicity assessment (Krishna et al., 1984, 1985, 1986).

In vitro chromosome aberration assay

This assay is used to measure the ability of a xenobiotic to induce chromosomal damage (clastogenicity). It is also performed early in product development because it is a very sensitive test for detecting chromosome-type damage that cannot be detected in bacterial assays. These assays must be conducted with eukaryotic cells that have DNA packaged with chromosomal proteins. Cells are most susceptible to chemically induced chromosomal damage during replicative DNA synthesis. Thus any primary cell or continuous cell line that is actively dividing can be used. This assay is typically performed using cell lines (such as CHO) or with cultured human lymphocytes from normal (prescreened) donors that are not currently ill or using medication. Using human cells and generating data may be closer to and more relevant to human safety assessment and is used as an example here. Because circulating lymphocytes in the peripheral blood do not normally divide, DNA synthesis and cell division are stimulated with the mitogen, such as phytohemagglutinin. Furthermore, because a large proportion of cells with chromosome damage die and fragments of chromosomes do not migrate into the nucleus of daughter cells, it is necessary to assess chromosomes at the first metaphase after treatment.

Preliminary testing is performed to assess the cytotoxicity of the nutraceutical being tested on cell division as the basis for selecting doses for definitive tests. When possible, the highest concentrations are expected to produce a minimum 50% suppression in numbers of metaphase cells (mitotic index). All tests are conducted with duplicate cultures using varied treatment conditions. Direct testing is conducted without metabolic activation using a 3-h and a continuous 24-h treatment with collection and fixation (harvest) of the cells at 24 h after the onset of treatment.

Testing with metabolic activation is conducted for 3 h with the harvest at 24 h because, after longer treatments, the metabolic system begins to break-down and becomes cytotoxic to the cells. At least five concentrations of the nutraceutical are tested over a range expected to achieve a dose-related reduction in the mitotic index in treated cultures compared to controls. After preparing the chromosomes on slides and staining, 100 metaphase cells/slide are examined from each culture for chromosome damage. Damaged cells are further classified regarding their type of damage. This classification is useful when interpreting the significance of small increases in chromosome damage. This test generally takes 4–6 weeks to complete.

Data interpretation

A reproducible concentration-dependent increase in cells with one or more chromosome aberrations is considered indicative of a positive clastogenic (chromosome breaking) response. In general, a small reproducible increase (<10% aberrant cells) may not be as likely to be a risk if only simple breaks are observed. More significance is placed on higher increases (>10%) in numbers of aberrant cells or if the damage includes rearrangements that involve multiple breakage events on more than one chromosome/cell. Regardless, any positive increase must be noted in the labeling for the xenobiotic and may indicate a higher risk for adverse effects such as carcinogenesis.

Significance of gaps

Gaps are defined as achromatic regions (nonstaining) along the width of a chromosome/chromatid that is smaller than the width of a chromosome/chromatid and aligned with the rest of the chromosome/chromatid. Some gaps are truly breaks in DNA, whereas some may result from staining artifact or xenobiotic interactions with chromosomal proteins. The biological significance of an increase in gaps alone is unclear unless additional testing is performed. The significance of gaps can be verified through tests for reproducibility and in vitro testing for micronuclei (fragments of DNA that are not contained with the main nucleus) because a gap that is truly representative of a break would appear as a micronucleus in an interphase cell.

Significance of ploidy changes

Polyploidy is defined as an increased number of chromosomes seen as an even multiple of the normal chromosome number for a given cell type. In the case of human lymphocytes, the normal 2N chromosome number is 46; therefore, increases of two times, three times, or four times would be indicative of polyploidy. These cells are quantified while examining cells for chromosome aberrations. Increases in polyploidy are typically associated with increased nongenotoxic forms of cytotoxicity that may affect the structures and processes involved in cell division. The consequence of polyploidy in most cell types is death, and because it is associated with relatively high xenobiotic concentrations in vitro, these effects generally do not occur at efficacious plasma concentrations in vivo. However, increases in polyploidy at higher concentrations can sometimes be associated with aneuploidy at lower (noncytotoxic) concentrations.

Aneuploidy is defined as the gain or loss of one or more chromosomes from the normal 2N chromosome number (Nath and Krishna, 1997). For example, children born with Down syndrome typically have 47 chromosomes (2N +1, with three copies of chromosome 21 instead of two). Aneuploidy is usually associated with adverse effects on the cell (tumorigenesis) or individual (in the case where all cells are affected). Unlike chromosome damage, numerical changes occur as a result of adverse effects to the spindle apparatus involved in cell division rather than direct DNA damage. Therefore a threshold level and "no-effect level" can usually be defined using special assays designed to visualize the processes of cell division. As long as the threshold level is well above clinical plasma concentrations, aneuploidy should not be a concern in vivo. The in vivo micronucleus test can also be used as an in vivo assay for assessing the potential of a nutraceutical to induce aneuploidy on a case-by-case basis.

In vitro micronucleus assay

This assay detects chromosome damage in vitro using cell lines or human lymphocytes treated with the test substance both in the presence and absence of metabolic activation (Fig. 29.6). Chromosome damage is assessed as an increase in numbers of micronucleated cells in the first interphase cell after exposure (Krishna et al., 1989). This population of cells is identified as binucleated (two nuclei) cells when cytochalasin B is used during testing. Cytochalasin B is a compound that has been shown to block cytokinesis (cytoplasmic division) without affecting nuclear division. Relative changes in the proportion of

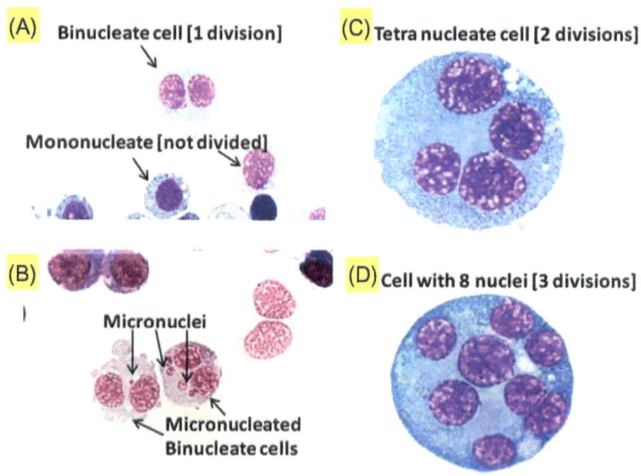

FIGURE 29.6 The in vitro and/or in vivo/in vitro micronucleus assay determines the cytotoxic and clastogenic effects of chemicals on the chromosomes. This can be evaluated in vitro using primary human peripheral blood lymphocytes, animal cells, and/or cell lines cultured in the presence of, for example, a nutraceutical. Cells were cultured in the presence of a cytokinesis-blocking agent, cytochalasin B, stained with a Giemsa stain (Diff-Quick), and prepared for light microscopic evaluation. (A) Normal mononucleate cells (not divided in culture) and binucleate cells that have undergone one division. (B) Micronuclei in binucleate cells. (C) Tetranucleate cell that has undergone two divisions. (D) Cell with eight nuclei that has undergone three divisions. This assay is used to evaluate clastogenicity (chromosome breakage), aneugenicity (chromosome lagging due to spindle dysfunction), and cytotoxicity (delay in cell cycling) with appropriate staining techniques (Krishna et al., 1989).

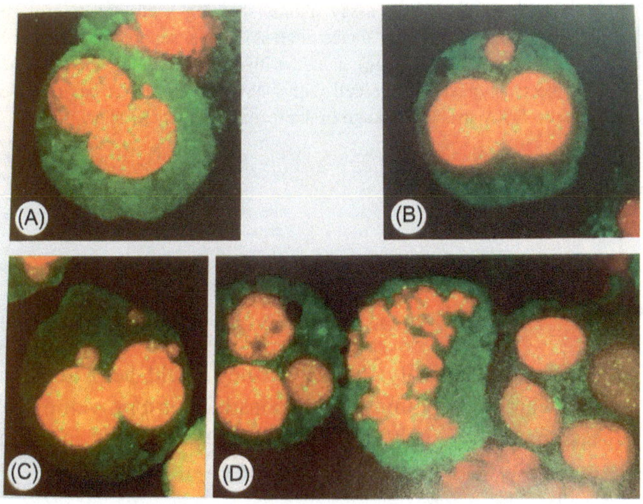

FIGURE 29.7 This is the same in vitro micronucleus assay described previously in Fig. 27.6, but the slides were stained with an antikinetochore antibody and counter-stained with acridine orange and prepared for fluorescent microscopic observation so cytotoxicity, clastogenicity, and anuegenicity are evaluated concurrently. (A) A binucleate cell with a kinetochore-negative micronucleus. (B) A binucleate cell with a kinetochore-positive micronucleus. (C) A binucleate cell with multiple kinetochore-positive micronuclei. (D) A binucleate cell with a large micronucleus showing multiple kinetochore signals, a metaphase/anaphase cell showing kinetochore-positive centromeres, and a tetranucleate cell (Krishna et al., 1992).

binucleated cells of treated cultures compared to control cultures can be used as an indicator of xenobiotic-induced cytotoxicity. Micronuclei (small nuclei) are formed from acentric fragments of chromosomes (clastogenic mechanism) or whole chromosomes (aneugenic mechanism) that are not included in the main nucleus of the interphase cell (Krishna et al., 1992). Variations of this methodology have also been used for test article-induced mechanisms (Krishna et al., 1995a,b). An immunostaining technique is used to differentiate micronuclei containing kinetochore (s), thus assessing the aneugenic potential of a test substance (shown in Fig. 29.7).

To assess the induction of micronuclei, cells are cultured in six-well tissue culture plates and then treated with the xenobiotic in the presence of cytochalasin B (a compound that blocks cytoplasmic division resulting in binucleated cells). Cells are harvested and placed directly on the microscope slides. Binucleated cells (treated population) are evaluated in interphase for the proportion containing micronuclei. Micronuclei can be quantified faster, more objectively, and with minimal technical training compared to the assessment of chromosome damage in metaphase cells. Because a substantially larger portion of cells is in interphase compared to metaphase, more cells can be scored, further increasing sensitivity.

Data interpretation

A treatment-related increase in micronucleated or binucleated cells is indicative of a positive response. Because micronuclei can originate from either chromosome fragments (clastogens) or whole chromosomes (aneugens), additional staining procedures may be used to determine the mechanism (Fig. 29.7). A no-effect level can only be assessed for aneugenic mechanisms because they are related to cytotoxicity to the organelles involved in cell division rather than DNA damage. Test compounds that induce dose-related apoptosis (cytotoxic programmed cell death) cannot be evaluated in this test because the fragmentation of the nuclei, characteristic of apoptosis, can produce artifactual increases in nuclear bodies that resemble micronuclei. Compounds that compete with cytochalasin B, or in themselves interfere with cell division and cytokinesis, also produce inconclusive results in this test.

In vitro comet assay (single-cell gel electrophoresis assay)

This is a well-established, highly sensitive, rapid, and simple genotoxicity test. Isolated cells embedded in agarose are lysed, washed to remove membranes and proteins, briefly electrophoresed, stained, and examined under epifluorescence microscopy; strand breaks, coming from either strand breakage or excision repair, result in DNA extending toward the anode in a structure resembling a "comet" (Singh et al., 1988; Tice et al., 2000). Depending on experimental conditions, the migrating DNA (comet tail or derived parameters) reflects the amount of single-strand or double-strand breaks, alkali-labile sites, including incomplete excision repair sites, and also of DNA–DNA and DNA–protein cross-links (Duez et al., 2003). A broad spectrum of DNA damage can then be detected either by visual classification of comet morphologies ("visual scoring") (Ramos et al., 2001) or from morphological parameters obtained by image analysis and integration of intensity profiles using in-house or commercially available systems. There are only few limitations of the comet assay with regard to its application and interpretation. Short-lived primary DNA lesions such as single-strand breaks, which may undergo rapid DNA repair, could be missed when using inadequate sampling times. Another limitation is that indirect mechanisms related to cytotoxicity (e.g., DNA fragmentation in apoptosis) can lead to positive effects (Speit et al.,

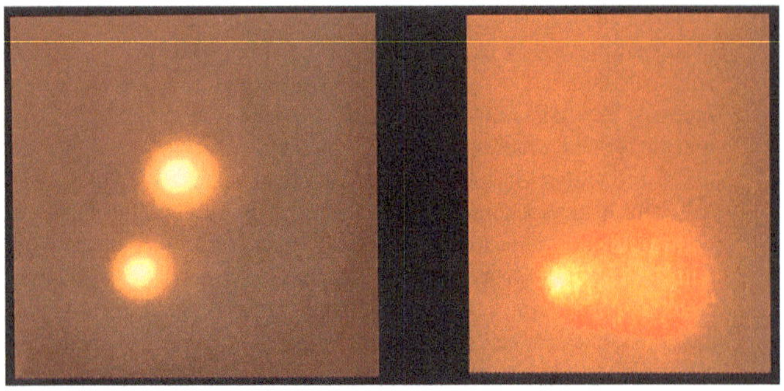

FIGURE 29.8 Comet assay photos showing a normal control cell (left; no DNA damage, intact nucleus) and a xenobiotic-treated cell (right; with DNA damage) stained with acridine orange and subjected to electrophoresis. In the cell with DNA damage, DNA has migrated due to DNA strand breakage in the form of comet (Krishna, 2010).

2005). An example of such an assay is shown in Fig. 29.8.

Analytical methods

Analytical methods with suitable detection limits have been developed for the characterization and quantification of many known molecules or molecular functionalities associated with genotoxicity; they are now being implemented in official pharmacopeias. Such methods are based on spectrophotometry (the Ehrlich reagent for pyrrolizidine alkaloids), thin-layer chromatography, gas chromatography/mass spectroscopy (GC/MS), and liquid chromatography/mass spectroscopy (LC/MS and LC/MS/MS) (Zenga et al., 2007; Napoli et al., 2010).

In silico methods

Computer models used for genotoxicity prediction fall into three principal categories (Votano et al., 2004): (1) rule-based expert systems such as DEREK that estimates the presence of a DNA-reactive moiety in a given molecule (Greene, 2002); (2) quantitative structure–activity relationship models (QSAR) such as TOPKAT that uses "electro-topological" descriptors (atom-type, bond-type, and group-type E-state) rather than chemical structure to predict mutagenic reactivity with DNA (Votano et al., 2004; Snyder and Smith, 2005); and (3) three-dimensional (3D) computational DNA-docking model to identify molecules capable of noncovalent DNA interaction (Snyder et al., 2004; Snyder and Smith, 2005). In silico prediction systems are cheaper, rapid, have higher reproducibility, have low compound synthesis requirements, can undergo constant optimization, and have potential to reduce or replace the use of animals (Valerio-Jr, 2009). Their limitations are the lack of available toxicity data, inappropriate (simplistic) modeling of some endpoints, and poor domain applicability of models (Cronin, 2002). The application of in silico methods to complex mixtures such as botanical dietary supplements is evidently limited to the detection of known phytochemical compounds bearing known or new structural alerts for genotoxicity activity; however, they could help to elucidate which compounds are responsible for a proven effect. A summary of such predictive models available in literature is shown in Fig. 29.9 (Krishna et al., 2014).

"Omics" technologies

The field of toxicogenomics is a relatively new technology that deals with the interaction between the structure and activity of the genome and the adverse biological effects of exogenous agents (Bishop et al., 2001). This discipline is based on the concept that the toxic effects of xenobiotics on biological systems are generally reflected at the cellular level by their impact on the expression of genes (transcriptomics) and proteins (proteomics) and on the production of metabolites (metabonomics) (Heijne et al., 2005; Borner et al., 2011). The usefulness of such technologies and models is shown schematically in Fig. 29.10.

Transcriptomics

Transcriptomics analyze the expression level of genes by measuring the transcriptome and the genome-wide mRNA expression (Davies, 2010). Transcriptomics uses high-density and/or high-throughput methods of assessing mRNA expression.

Microarrays

Microarrays ("DNA arrays," "DNA chips," "biochips," "gene chips"), the most common approach used for gene expression profiling (Davies, 2010), are generated by immobilizing a high number of oligonucleotides on a very small surface (up to 200,000 spots/cm^2). Based on the target sequences, significant changes of mRNA can be estimated for thousands of genes (Oberemm et al., 2005; Ulrich-Merzenich et al., 2007a). Specialized subsets of gene expression changes and quantitative real-time reverse-transcriptase polymerase chain reaction (qRT-PCR)-based approaches that focus on specific genes have also been developed.

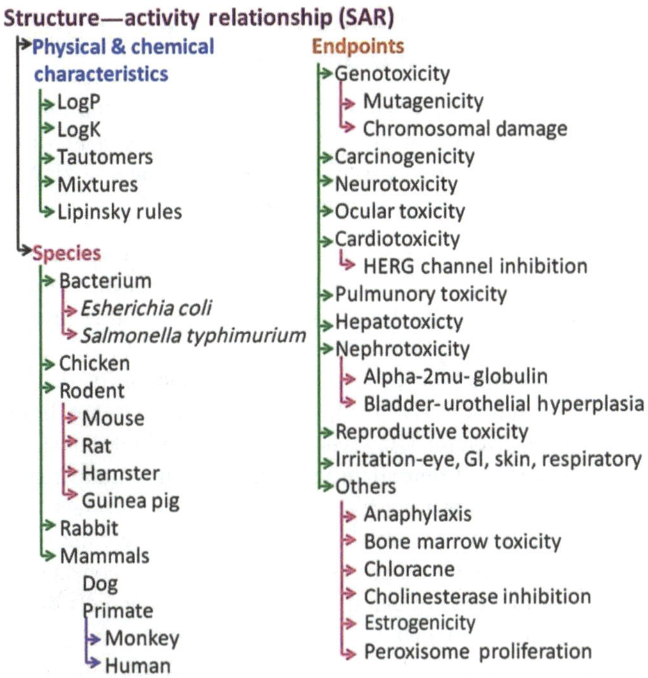

FIGURE 29.9 In silico endpoints for detection of structural alerts for a variety of adverse effect potential of nutraceuticals in a variety of toxicity models.

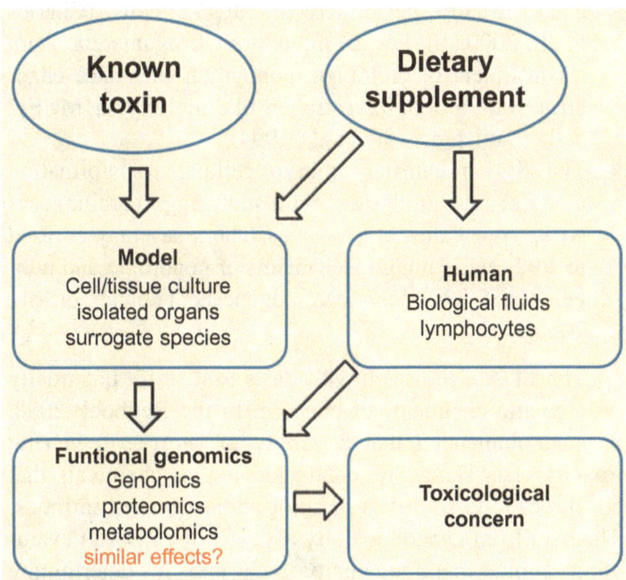

FIGURE 29.10 Testing strategies for "omics" methods. An in vitro and/or in vivo model is exposed to a chemically defined toxicant with known properties ("prototoxicant"); the response of the model is evaluated by an "omics" method and possible/likely biomarkers of exposure are examined. When the same model is exposed to an unknown compound mixture or a botanical extract, a similar pattern of response indicates probable toxicological concern.

Proteomics

Proteomics is the study of a broad spectrum of proteins within cell or tissue, including protein expression, structural status, functional states, and their interactions with other cellular components (Kussmann et al., 2006). Key technologies rely on two-dimensional (2D) gel electrophoresis coupled to MS (Ulrich-Merzenich et al., 2007a), on antibody microarrays, and on LC-MS MS of proteins fragments, a technique known as "shotgun proteomics" (Wolters et al., 2001) and its specific platform variations, called isotope coding affinity tags and MuDPIT (multidimensional protein identification technology) (Yates, 2000). Affinity chromatography, fluorescence resonance energy transfer, and surface plasmon resonance are also used to identify protein–protein or protein–DNA interactions. X-ray tomography is used to determine the location of proteins or protein complexes in labeled cells (Ulrich-Merzenich et al., 2007a).

Techniques are available but a large database of proteomics "fingerprints" is still needed for compounds of known toxicity; once available for a series of known carcinogens, it will be possible to identify changes in biochemical pathways and to assess the toxicity of unknown compounds. The products of the genes identified so far in transcriptomics studies are probably promising candidates as proteomics markers.

Metabonomics

Metabolomics is defined as the "systematic study of the unique chemical fingerprints that specific cellular processes leave behind." The study of their small-molecule metabolite profiles and metabonomics relates to its application to studies of human nutrition and responses to xenobiotics or disease. It usually compares profiles without identifying individual compounds. This implies the quantitative and qualitative study of a wide range of low-molecular-weight molecules produced as the net result of cellular functions (Van Ravenzwaay et al., 2007; Youns et al., 2010) by nuclear magnetic resonance spectroscopy, Fourier transform infrared and near-infrared spectroscopy, or MS. So far, metabonomics has been scarcely applied to genotoxicity studies; major applications concern the urinary profiling of damaged bases excreted on DNA excision repair and the search for activated metabolites of precarcinogens (Kirkland et al., 2005).

Applications in herbal products

As genotoxicity testing aims to yield information on all types of genetic damage, including gene mutations, structural chromosome aberrations (clastogenicity), and numerical chromosome aberrations (aneugenicity) as described previously, it is quite a challenge to apply these tests (which have been developed for profiling individual compounds) to complex mixtures like botanicals (ICH, 2008; WHO, 2007; EMEA, 2007, 2008). Nonetheless, the Ames test has been widely used to assess the mutagenicity of herbal products, including extracts from *Calendula officinalis* L., *Echinodorus macrophyllus* (Kunth) Micheli, *Mouriri pusa* Gardner, *Phyllantus orbicularis* Kunth, *Punica granatum* L., *Parthenium hysterophorus* L., and a green tea catechins preparation (Sanchez-Lamar et al., 2002; Santos et al., 2002; Nohynek et al., 2004), to cite a few examples. Among these extracts, *P. granatum* whole fruit extract and the enriched fractions of flavonoids and tannins from *M. pusa* were found to be positive in genotoxicity tests. The genotoxicity of copaiba oil (*Copaifera langsdorffii* Desf.) was demonstrated by in vitro micronucleus and comet assays (Rao et al., 2006). Metabonomics techniques have already been used in the toxicity studies of guan mutong (*Aristolochia manshuriensis* Kom.) (Zhao et al., 2006) and its toxic component, aristolochic acid (Jia et al., 2006); aristolochic acid I was also suggested to possess genotoxic potency by QSAR modeling (Arvidson et al., 2008; Chen et al., 2011). In the in vitro comet assay, artesunate, a semisynthetic derivative from artemisinin (*Artemisia annua* L.), induced DNA breakage in a dose-dependent manner (Efferth et al., 2008). The transgenic rodent mutation assay has been used to assess the genotoxicity of phytochemicals such as riddelliine (pyrrolizidine alkaloid) (Ulrich-Merzenich et al., 2007b). There is certainly a need for the development of validated methods to rapidly pinpoint indicators of genotoxicity that yield warning signals and indicate which products need further assessment through a complete test battery.

In vitro tests validated for use in regulatory toxicology contexts

In addition to genotoxicity tests, there is the opportunity to examine agents of interest in a battery of other in vitro tests that measure cytotoxicity, developmental toxicity, estrogenicity, and cell transformation.

- Cytotoxicity assays evaluate the ability of test materials to inhibit cellular proliferation in cells in culture. Often, CHO-K1-B4 cells are used in this assay (Morita et al., 1992).
- Several methods are available for detection of developmental toxicity in vitro. One of these is the limb bud micromass assay where fetal mouse limb buds are allowed to differentiate via incubation in culture for 5 days with a test material. The test material's effects on cytotoxicity and limb bud differentiation are compared to determine if differentiation is inhibited at concentrations that are not cytotoxic, a sign of a potentially teratogenic compound (Flint and Orton, 1984).
- Estrogenic activity of test materials in vitro may be assessed relative to the activity of 17-β-estradiol. One such assay is an estrogen receptor transactivation assay usually performed in MCF-7 cells (Charles et al., 2000a,b). For example, norhydroguaiaretic acid, a constituent of prototype monograph substance chaparral, was tested for estrogen-like activity in MCF-7 cells (Sathyamoorthy et al., 1994).
- The Syrian hamster embryo cellular transformation assay shares similarities with multistage carcinogenesis in vivo (Isfort et al., 1996). The assay may be used to study underlying mechanisms of genotoxic and nongenotoxic chemical carcinogenesis (Zhang et al., 2000).

The FDA also uses in vitro tests to identify potentially problematic chemicals of concern. In the Redbook 2000, it is recommended that a variety of short-term genetic toxicity tests should be conducted for all chemicals that are direct food additives or color additives used in foods. The data from genetic toxicity assays assist FDA in evaluating animal carcinogenicity data and in determining heritable effects of chemicals.

Use of S9 metabolic activation revolutionized in vitro testing

Correlations between the mutagenic and carcinogenic properties of chemicals were weak prior to 1960, but they have strengthened as a result of the discovery that many chemicals undergo various metabolic changes before they become biologically active (Ames et al., 1975). Despite the great diversity of chemical structures, most carcinogens/mutagens appear to give rise to electrophilic derivatives (compounds with electron-deficient atoms), which then react with nucleophilic centers (sites with atoms rich in electrons) that are present in cellular constituents. The growing experimental evidence that links the carcinogenic activity of numerous chemicals to their capacity to be converted into electrophilic derivatives, which can react with DNA and other cellular macromolecules and therefore exert a mutagenic effect, has led to the suggestion that there may be a formal relationship between carcinogenesis and mutagenesis.

Methods for teratogenicity assessment

"Teratogens" refer to toxins that can lead to fetal death or developmental abnormality. The baseline birth defects rate is approximately 6% (Christianson et al., 2006); it is not known if this baseline rate is a normal frequency of errors in the highly complex developmental process or the consequence of any yet to be identified environmental or dietary factors. Major identified teratogenic mechanisms include folate antagonism, neural crest cell disruption, endocrine disruption, oxidative stress, vascular disruption, and specific receptor-mediated or enzyme-mediated teratogenesis (Van et al., 2010). Studies have shown that plant alkaloids like indolizidine (Lather et al., 2011), pyrrolizidine (Prakash et al., 1999), piperidine, triterpene (Gaffield and Keeler, 1994), cyclopamine, veratramine, veratrosine, and germine or their metabolites exert teratogenic effects in animals. Ginsenoside Rb1(triterpenic saponin), extracted from ginseng (a commonly used botanical), exerts direct teratogenic effects on rat embryos (Chan et al., 2003). However, most dietary supplements including botanicals have never been assessed for their teratogenic potential.

The Organization for Economic Cooperation and Development (OECD, 2015) and the European Centre for the Validation of Alternative Methods (ECVAM, 2015) have largely investigated the validation of teratogenicity tests; the FP6 European project "Reprotect" has investigated possible strategies to cover the entire mammalian reproductive cycle and has published a series of articles (ReproTect, 2015).

In vitro methods for teratogenicity assessment in stem cells and embryos

Embryonic stem cell tests

Pluripotency of embryonic stem cells (ESCs) provides an important tool to assess adverse effects of xenobiotics on their differentiation, effects that might be relevant for in vivo embryo toxicity. Taking into account that general cytotoxicity is a critical parameter for identifying embryotoxicants, ESC tests are based on three endpoints, that is, differentiation inhibition of ESCs and cytotoxicity of both ESCs and fibroblasts (Stummann et al., 2008). The limitations of ESC tests are the relatively long culture duration, the subjective and laborious endpoint scoring, and the limited characterization of its applicability domain and predictability (Marx-Stoelting et al., 2009).

Mammalian micromass test

The mammalian micromass test (MM test) is based on the ability of a xenobiotic to inhibit differentiation of embryonic tissue cells in the absence of cell death; the test is supposed to reflect the in vivo teratogenic potential (Hansen and Abbott, 2009). The test based on limb buds is an attractive teratogenic model because it reproduces cartilage histogenesis, a fundamental step in skeleton morphogenesis, and the protocol has been validated by ECVAM (Genschow et al., 2002).

Mammalian whole-embryo culture test

For the postimplantation whole-embryo culture (WEC) test, mammalian (rat or mouse) embryos are maintained in culture for short periods throughout the phase from fertilization to the end of organogenesis. This allows probing major aspects of organogenesis, including heart development, closure of the neural tube, and development of ear, eye, bronchial arch, and limb buds; disturbance during this period may lead to general restriction of growth and development or to specific malformations in one or several organs (Spielmann, 2005). Xenobiotic metabolizing enzymes can be incorporated and concentrations of tested compounds and metabolites can easily be monitored in the culture medium and embryonic tissues (Bechter et al., 1991). Mouse or rat embryos with two to five somites are explanted from the uterus, cultured (Piersma et al., 2004), and exposed to various concentrations of the tested compounds. The morphology of the embryos is carefully assessed and a series of endpoints are scored, including the yolk sac diameter, crown–rump length, and head length. The protocol has been validated by ECVAM (Genschow et al., 2002). Studies of 20 compounds in a test panel reported that the WEC test showed the best concordance between in vivo classification and in vitro test results, with 80% accurate classifications compared to 78% for the ESC test and 70% for the MM test (Genschow et al., 2002). The MM test is much faster, but includes only differentiation of chondrocytes as the toxicological endpoint (Stummann et al., 2008). The ESC test is presently the only in vitro test based on a cell line that has undergone a formal validation trial (Stummann et al., 2008). However, all tests require the sacrifice of pregnant mammals for harvesting early embryos.

Lower vertebrate animal models (e.g., anuran amphibian) are increasingly proposed as alternatives to mammals to assess teratogenicity of xenobiotics because embryogenesis is highly conserved across amphibians and mammals (Di Renzo et al., 2011). Furthermore, the use of lower-order species is in line with the current trend of applying the 3Rs concept in minimizing the use of animals for safety assessment (Van den Bulck et al., 2011).

The frog embryo teratogenesis assay Xenopus

Xenopus laevis is one of the most used models for vertebrate experimental embryology due to its aquatic development, large eggs with easily manipulated embryos, low cost, reliability, and reproducibility (Richards and Cole, 2006). It has been suggested as an alternative screening model for developmental toxicity purposes (Mouche et al., 2011). Harvested embryos are cultured, exposed to xenobiotics, and evaluated for mortality (absence of heartbeat). Surviving embryos are anesthetized, fixed in formalin, and examined under a dissecting microscope for measurement of body length and assessment of malformations number and types (Di Renzo et al., 2011). This test has been evaluated by the US Interagency Coordinating Committee on the validation of alternative methods. However, they concluded that frog embryo teratogenesis assay *xenopus* was not sufficiently validated or optimized for regulatory applications (ICCVAM). The assay is limited by the aqueous solubility of test substances; its overall accuracy in predicting teratogenic potential has been claimed to be 79%–83% (Selderslaghs et al., 2009).

Fish embryo toxicity test

Most of the information to support the use and interpretation of fish embryo toxicity test has been generated using zebrafish (Selderslaghs et al., 2009). Zebrafish eggs remain transparent from fertilization up to and beyond pharyngulation, which enables observation of morphological changes during earlier developmental stages. Moreover, zebrafish embryos can activate proteratogenic/procarcinogenic substances without any addition of exogenous metabolic activation systems because they present phase I enzyme activities at very early stages of development (Weigt et al., 2011). Endpoints used for assessing the effects of the tested compound include embryo mortality, yolk sac edema, tail malformation, changes in hatching rates, and deficiencies in gastrulation, somite formation, head formation, heartbeat, and pigmentation (Van den Bulck et al., 2011). The zebrafish embryo toxicity test presents a 75% success rate in identifying nonteratogenic compounds and a 100% success rate in identifying teratogenic compounds (Ton et al., 2006).

Although in vitro models provide valuable teratogenicity data, none of these in vitro methods can currently be regarded as a full replacement for existing in vivo studies; however, they can be considered as a prescreen strategy to prioritize xenobiotics for in vivo assessment, reducing and optimizing animal use in teratogenicity evaluation.

To the best of our knowledge, teratogenicity of dietary supplement botanicals has not been assessed with the help of "omics" technologies. Conventional in vitro and in vivo methods have been used to evaluate the teratogenicity potency of herbal products. The zebrafish embryo toxicity test has been used to evaluate the teratogenicity of several natural products, including various flavonoids (Jones et al., 1964), delta-9-tetrahydrocannabinol, the major psychoactive constituent of marijuana (Thomas, 1975), and arecoline, the major alkaloid in betel nuts (Chang et al., 2004).

In vitro models for safety assessment and toxicology studies of the gastrointestinal tract

The most widely utilized cell cultures in food toxicology are cancer-derived or transformed cell lines like CaCo-2 and HT-29. Despite their human origin, they have a phenotype and glycosylation pattern distinct from normal gut epithelia and thus are unlikely to serve as an ideal model system (Cencic and Langerholc, 2010; Nissen et al., 2009). Therefore close consideration of the phenotype and proper characterization of cell lines are of utmost importance to obtain the relevant results. Generally, use of untransformed cell lines produces more relevant results, as demonstrated by Trapecar and Cencic (2012) in a recent study for risk assessment of *Bacillus cereus*, to validate a model that could easily differentiate toxigenic strains from the ones with probiotic properties. Likewise, Marques et al. (2011) successfully deployed a 3D model of the small intestine to assess toxicity of heavy metals (as food pollutants) as well as to predict their kinetic behavior, demonstrating that such models can be easily translated into nutraceutical toxicity testing. When designing an in vitro toxicological study and choosing the right cell model, it is essential to address the following major concerns:

- Which cell line has the most relevant phenotype to address the study objectives?
- Which other organ systems have an influence in vivo on the system used and how can we integrate these into the study?
- Is the model validated and, if not, which validated model could be used as a comparative control?

Cell lines differing (due to either transformation or inappropriate origin) from the targeting organ may have altered metabolic and morphological properties. It is therefore necessary to choose cell lines with phenotypes as close as possible to the in vivo target. At the same time, cells may behave differently when cultured in an isolated environment missing cell–cell interaction and chemical cross-talk (Langerholc et al., 2011). Cultivating cell lines in 3D cocultures or 3D spheres can largely circumvent this problem.

Primary cell culture models offer an advantage of freshness in cell integrity, avoiding the influence of transformation, and the data obtained can be better related to data from cell sample studies in humans (Krishna et al., 1984, 1985, 1986). An essential part of developing cell culture models is their validation and demonstration that their use provides equally viable results as animal tests. Several national and supranational institutions like the ECVAM, Japanese Center for the validation of alternative methods, and OECD validate and keep track of new models (Daneshian et al., 2011). Depending on the culturing technique, functional cell models can be divided into monolayer or 2D models, multilayer models, and 3D models (Fig. 29.11).

Monolayer models are suitable, especially for quick determination of cytotoxicity or to monitor specific markers where cell differentiation is not as important, because the objects of observation are individually expressed processes. However, single cell-based assays are not able to adequately represent the complex interplay between different cell types in a specific organ (Adler et al., 2011). Also, although cytotoxicity studies expose a direct, radical impact on cell survival, they do not offer much insight into the underlying mechanisms or long-term effects of exposure.

Two-dimensional models are usually cultivated on plastic surfaces that do not allow a full differentiation of the cell line (Cencic and Langerholc, 2010; Peracaula et al., 2008). Such models are suitable for high-throughput screening (HTS), enabling a fast initial screening of hundreds of compounds and concentrations. However, 3D models with fully differentiated cell lines are more appropriate for mechanistic and kinetic studies. Regarding risk assessment and toxicology studies of nutraceuticals, the most interesting alternative is the human 3D coculture cell model of the gut (Fig. 29.11). The 3D gut model mimicking the intestinal barrier offers immense possibilities from mechanistic to bioaccessibility studies. Moreover, it can be improved with an additional culture seeded in the basal compartment such as monocytes/macrophages, skeletal muscle cells, adipocytes, or hepatocytes simulating complex systems. Typically, objects of investigation are apical and basal supernatants as well as cell lysates to monitor, for example, signaling molecules, activation factors, compound concentrations, and other parameters. Basal supernatants can, after the appropriate incubation period, be further transferred on different 2D models for HTS. Toxicological studies with compounds that have passed the simulated intestinal barrier offer much more relevant results than those investigated directly.

In light of recent advances in the science of stem cells, intestinal stem cells (ISCs) are becoming more and more utilized in food and nutraceutical research. The ISCs are able to differentiate into absorptive enterocyte, goblet, Paneth, and enteroendocrine lineages, but require extracellular signals such as Wnt to avoid loss of differentiation (Ootani et al., 2009). As demonstrated, ISC can be cultured in a 3D matrix environment, developing in vivo-like properties and structures. In a model such as this they could be sustained in culture for more than 30 days. The ISC in combination with organ-shaped biomaterials as well as incorporation of microfluidics will probably shape the future of in vitro food and nutraceutical research once the model has been validated. Untransformed cell lines like H4-1 (human small intestinal epithelial cell line) and PSI-1 (pig small intestinal epithelial cell line) can be cultivated on membranes reaching high transepithelial resistance and polarization. Epithelial cell lines could be cocultured with human and pig monocyte/macrophage cell lines mimicking the gut and gut-associated lymphoid tissue. This and similar models have been successfully used and documented before (Marques et al., 2011; Nissen et al., 2009).

Several new methods are being currently developed, such as integration of biochips, rotation systems, and long-exposure static or microfluidic systems (Seidle and Stephens, 2009). These recent advances in microengineering redefine our conception of 3D models because they enable true organ-like shapes. We therefore distinguish 3D coculture models using membranes and 3D sphere models that use supportive biomaterials that allow cells to form a lumen. Elamin et al. (2012) used a base membrane matrix that promoted the development of hollow multicellular spheroid structures by intestinal epithelial cells. In this model, cells were able to polarize, form tight junctions, and even microvilli.

In vitro chip technology: organ, human, or lab on chips

A step further and what is called the future of preclinical research is "organs-on-chips" that integrate microfluidics technologies with living cells cultured within 3D devices created with microfabrication techniques. The base of this technology is lithography and microcontact printing that form defined shapes on a micrometer scale. Different types of cells are allowed to align to predefined shapes, and a microfluidic system can be integrated to mimic the flow of body fluids (Huh et al., 2011). By extending the

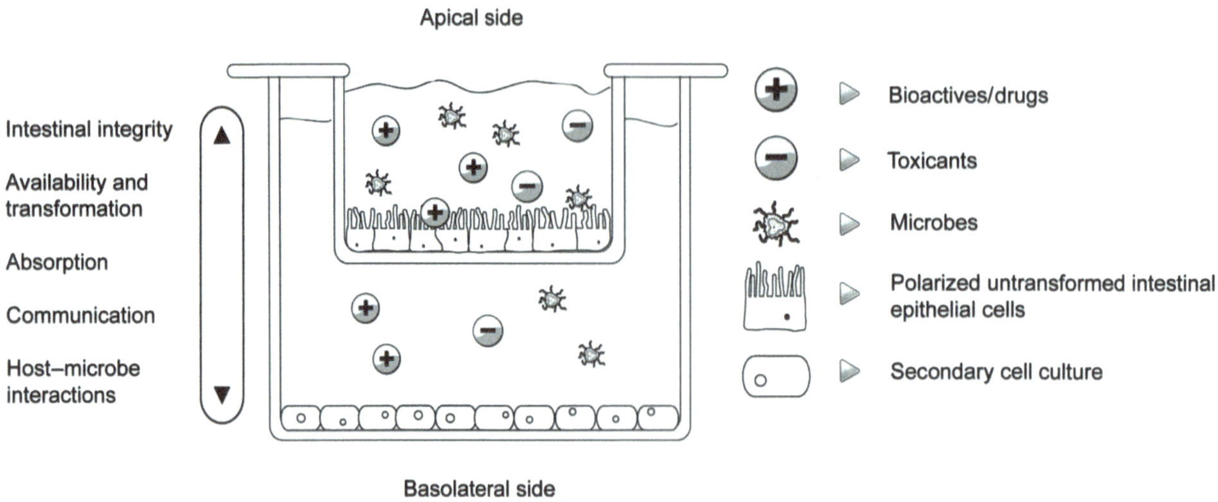

FIGURE 29.11 An example of the 3D functional cell model of the gut. *Adapted from Trapecar, M., Cencic, A., 2012. Application of gut cell models for toxicological and bioactivity studies of functional and novel foods. Foods. 1, 40–51.*

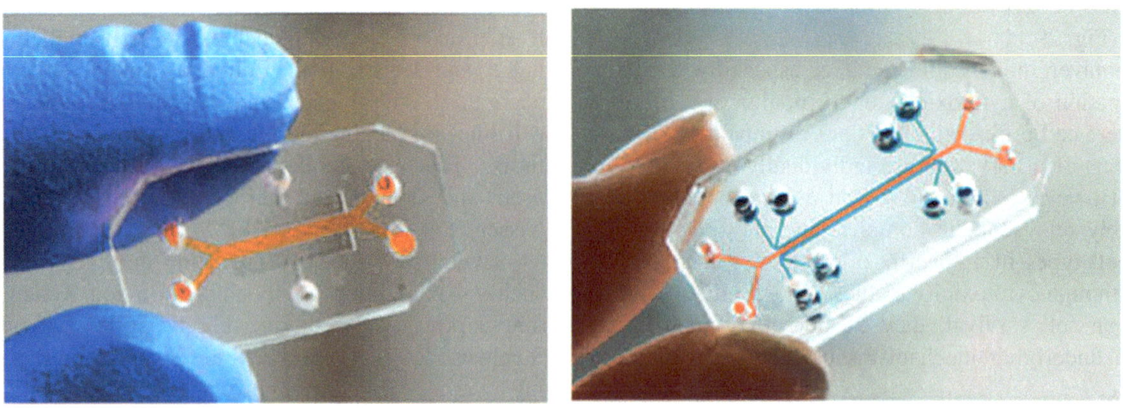

FIGURE 29.12 An example of human lung-on-a-chip (left) and human gut-on-a-chip (right).

microchip architecture "human on a chip" and "lab on a chip," concepts can be developed whereby different organ systems and laboratory techniques can be interconnected. Their potential to predict responses in humans will have profound effects on preclinical in vitro testing.

Chip technology, as it is applicable to evaluating the safety of test articles for human use, has been advancing at a fast pace (Krishna et al., 2014). Dr. Don Ingber from Harvard University pointed out that the product development costs have been increasing astronomically (Ingber, 2015). Further, cells cultured in vitro do not function like human bodies, animal studies take years to complete, and, more often than not, the results of an animal study do not accurately predict what happens in humans. To fix this, Ingber is researching engineering microchips containing living human cells to replicate the way human organs actually function, not just tissues and cultured cells. These human organs-on-chips are composed of a clear, flexible polymer approximately the size of a computer memory stick and contain hollow microfluidic channels lined by living human cells, thus allowing researchers to recreate the physiological and mechanical functions of the organ, and to observe what happens in real time (Fig. 29.12). The goal is to provide more predictive and useful measures of the efficacy and safety of potential new products in humans, which could represent an important step in reducing the need for traditional animal testing. It is hoped that these technologies will be available in a validated form soon. It is very likely that human-on-a-chip technologies will be the core of future pharmaceutical and food research; although they have not been validated and are far from being commercially available.

Therefore, it is believed that, currently, the most valuable tools are 3D coculture models that allow complete polarization, differentiation, and crosstalk between different cell types. They can be used to measure bioavailability, biotransformation, immune responses, and several other important toxicological parameters and at the same time can be linked with HTS 2D systems.

In vitro models for safety assessment and toxicology studies of the liver

In vitro liver toxicity testing comprises a major portion of any in vitro toxicology because liver is a major source of metabolism and xenobiotic biotransformation. High-throughput in vitro liver models would enable the shift to greater use of the alternative toxicity testing methods.

Immortalized cell lines

HepaRG, which has been established from a liver tumor associated with chronic hepatitis C, is a human bipotent progenitor cell line capable of differentiation toward two different cell phenotypes (i.e., biliary-like and hepatocyte-like cells). This cell line is particularly useful to evaluate xenobiotics and perform xenobiotic metabolism studies, because they express many detoxifying enzymes. It is also an interesting tool to study some aspects of progenitor biology (e.g., differentiation process), carcinogenesis, and the infection by some pathogens for which the cell line is permissive (Marion et al., 2010). HepaRG retains the expression of many liver-specific functions as well as many CYP450s, nuclear receptors, membrane transporters, and phase II enzymes (Aninat et al., 2006). Additionally, HepaRG cells have a stable karyotype, can differentiate into either hepatocyte or biliary lineages, have high proliferative capacity, and have been shown to produce data that are both reproducible and consistent among experiments (Guguen-Guillouzo and Guillouzo, 2010; Marion et al., 2010; Lubberstedt et al., 2011). Although HepaRG cells express liver-specific functions, the expression levels are much lower than that of primary hepatocytes (Lubberstedt et al., 2011).

Primary hepatocyte suspensions

Suspensions of hepatocytes are an easy to use method for moderately high-throughput toxicity studies. Griffin and Houston (2005) demonstrated that suspensions of hepatocytes provided a more accurate estimate of internal clearance rates when compared to conventional monolayer cultures. Suspensions also retain high levels of functionality, allowing for a more accurate correlation to in vivo toxicity than cultured cells (LeCluyse, 2001; Hewitt et al., 2007). However, isolation protocols involving collagenase digestion can induce a loss of cytochrome enzyme activity after isolation (Duval et al., 1995; Wang et al., 1998; Tirmenstein et al., 2000), making this a less desirable model because hepatotoxicity is often a process that occurs over several hours.

Primary hepatocyte cultures

Cultures of primary hepatocytes have been the gold standard for in vitro testing because they can maintain functional activities for 24–72 h, can be used for enzyme induction and inhibition studies, allow for medium-throughput screening of compounds, and are ideal for examining interspecies and interindividual differences in metabolism (LeCluyse, 2001; Hewitt et al., 2007). Primary hepatocytes are often used with the expectation that substances will affect or be affected by an isolated cell in the same manner that would occur in the whole organ and, in this way, will be a predictive model (DelRaso, 1993). Unfortunately, primary hepatocyte cultures are limited as in vitro systems as well.

Traditional primary hepatocyte cell culture involves the method of plating cells on a rigid substratum in which they proceed to form a monolayer across the bottom of the culture plate well. Under these conditions, primary hepatocytes undergo changes in cell morphology, structure, polarity, gene expression, and liver-specific functions (LeCluyse et al., 1996; Nelson et al., 1982; Dunn et al., 1989), a process referred to as dedifferentiation. Hepatocyte dedifferentiation is an important limitation of this in vitro system. Because the isolated cells have lost their normal microenvironment structure, cell-to-cell interactions, and cell membrane structures, responses to chemical exposures can be different than those occurring in vivo. Finally, one of the most important disadvantages of the use of primary hepatocytes for toxicological research is the drastic decrease in cell functionality over time. Liver-specific functions such as albumin production and CYP450 expression decline rapidly over the first 24–48 h of culture as the cells begin to lose their differentiated status (LeCluyse et al., 1996; Nelson et al., 1982).

A simple sandwich configuration in which hepatocytes are placed between two layers of matrix (traditionally collagen or Matrigel®47) maintains liver-specific functionality over longer periods of time (Dunn et al., 1989; Bader et al., 1992; LeCluyse et al., 2000; Liu et al., 1999; Mingoia et al., 2007). It has been suggested that the sandwich culture model is most useful for mechanistic studies of hepatobiliary toxicity (Meng, 2010). Although a number of investigators have found certain approaches to benefit hepatocyte culture in their own laboratories, difficulties in comparison of experimental results occur and continue to hinder the identification and adoption of optimal culture conditions. In addition, choices of species, strain, and sex of animal vary greatly across experiments. Conditions of cell isolation are also variable, as are the methods for cell purification, conditions of cell culture, and endpoint assay methods. Yields and viabilities of final cell preparations can differ depending on the types of collagenase and perfusion methods used. These factors must be taken into consideration when using primary hepatocyte cultures for toxicity testing and developing the proper procedures in one's own laboratory.

Primary hepatocyte cultures combined with inflammatory mediators

Traditional 2D hepatocyte cultures are not particularly well suited for HTS applications. Despite this, the availability of robotics-equipped facilities and relatively low per-sample cost of data generation using these systems have prompted exploration of novel culture formats. For example, a well-established role for inflammation in toxicity has been demonstrated with idiosyncratic drug toxicity (Shaw et al., 2010). An in vitro drug/cytokine/inflammatory mediator co-treatment approach was used to reproduce human drug hepatotoxicity signatures, particularly for idiosyncratic drugs, in cultured primary human and rat hepatocytes, as well as HepG2 cells (Cosgrove et al., 2007). Using this approach, drug–cytokine hepatotoxicity synergies were shown to be detectable for multiple idiosyncratic hepatotoxicants.

Novel liver-derived cell culture systems

Three-dimensional culture systems

Three-dimensional cultures range in complexity from monolayer sandwich culture and spheroids to more advanced systems involving porous materials, packed-bed reactors, hollow fibers, and perfusion flow.

The microenvironment of the hepatocyte is critical in maintaining its normal function, including responses to endogenous and exogenous substrates. Traditional 2D culture systems exhibit reduced expression of genes involved in xenobiotic metabolism within the first 48 h and such culture systems have been shown to have low sensitivity for drug hepatotoxicity detection (Meng, 2010). The use of 3D liver cultures can overcome deficits of the 2D culture system by providing models that reestablish cellular polarity and create more complex local environments.

Rat hepatocyte spheroids maintain a more differentiated state compared to the monolayer culture (Wu et al., 1996). Although Cyp1a2 expression and Cyp1a1 expression decline rapidly in traditional 2D monolayer culture of mouse hepatocytes, their levels remain high over 5 days in 3D spheroids (Nemoto and Sakurai, 1992, 1993). Spheroid cultures of mouse hepatocytes also maintain expression of Cyp2b9 and Cyp2b10 for several days at levels similar to that in vivo (Nemoto et al., 1995). However, due to limitations with oxygen and nutrient diffusion, spheroid culture is restricted in its utility as a liver model or for long-term culture.

Gelled materials have been devised to combine the complexity of a Matrigel®-type extracellular matrix with spheroid formation. Similar to the spheroids mentioned, a general limitation of the gel culture systems is impairment of oxygen and nutrient transfer to cells. In addition, these systems tend to be labor-intensive and low-throughput and require relatively large samples.

An extension of this line of research was the development of models that use scaffolding to provide structural support. The scaffolding increased cell adhesion, exhibited lower lactate dehydrogenase (LDH) release, and maintained biochemical functions such as albumin secretion, urea synthesis, and glycogen synthesis. Stabilization of albumin and urea levels after an initial decline was noted in a functional 20-day rat cell culture (Miyoshi et al., 1998). A simple micro-hollow fiber reactor system for rat hepatocytes had superior functionality (in terms of urea synthesis and albumin secretion) compared to simple monolayer culture. Such a system also improved the metabolic capability of cells, as indicated by greater sensitivity to chemical insult than seen in cells in monolayer culture. Nevertheless, these models are difficult to produce in large quantities and also lack flow to facilitate oxygen and nutrient delivery to the cultured cells.

Incorporating fluid flow into 3D culture systems was an important step in tissue engineering. Packed-bead, hollow fiber, or other types of "reactors" integrate microcirculation of medium to recreate the fluid flow within the liver. Systems incorporating shear flow have been shown to promote round cell aggregates that are more similar to in vivo morphology and increase and maintain liver-specific functions (Kang et al., 2004). Incorporating flow into the culture system has been shown to increase and maintain a higher level of urea synthesis (Taguchi et al., 1996) and to increase albumin, lactate, and glucose secretion (Kim et al., 1998). Physiologic oxygen gradients can be established in these microfluidic systems, which can assist in elucidating the physiological mechanisms involved in hepatotoxicity (Kim et al., 1998).

There continue to be efforts to engineer more advanced 3D culture systems that would constitute a fully functional liver culture model that can be used for toxicological and pharmacological research.

HμREL biochip

Chao et al. (2009) published an evaluation of a novel microfluidic device, an improvement compared to the HμREL biochip (an earlier model of this system), previously described by Sin et al. (2004). In this model, four biochips are enclosed in polycarbonate housing and connected by tubing lines to a fluid reservoir and peristaltic pump. Each biochip has one or more separate compartments in which different cells can be housed. The compartments are microfluidically connected in a linear path to allow for interaction between various cell types. This novel design allows for separate microfluidic experiments to be run in parallel on a larger set of housing plates. Evaluation of the system using primary human hepatocytes indicated that the system preserved cell viability and metabolic competency at similar or higher levels compared to traditional static culture conditions (Chao et al., 2009; Novik et al., 2010). It was also demonstrated that data from this model system were more correlative with in vivo data than those derived from the static hepatocyte cultures, and this correlation further improved when cocultures were used (Novik et al., 2010).

Single-well and multiwell perfused bioreactor

Sivaraman et al. (2005) described a 3D bioreactor system in which spheroids of rat hepatocytes were attached within channels of a silicon scaffold. Perfusion of the medium was directed over the top of the channels as well as through the tissue in the channels. A high-resistance filter controlled the perfusion through the channels so that the flow was uniform through all channels of the scaffold. Spheroids seeded in the bioreactor system had increased and sustained functionality when compared to spheroids in either static culture or single cells seeded into the bioreactor system. This bioreactor system can be scaled up easily to incorporate more channels and up to 12 different reactors controlled by the same unit while still maintaining cell viability and albumin production for 7 days (Domansky et al., 2010). The throughput and ease of handling of this model system are superior to other 3D systems. Toxicity testing relevant uses of this system include studies showing that clearance rates of compounds with known liver metabolism are comparable to those obtained in vivo and showing promising results with known inflammatory-mediated idiosyncratic toxicants (Dash et al., 2009). However, this cell culture system still lacks high-throughput capabilities and does not accommodate a large number of cells. It also needs to be evaluated with hepatocytes from multiple species.

Although much progress has been made over the traditional 2D static culture system, many of the available models have yet to incorporate and sustain hepatocytes from different species. Applications of these methods in toxicity testing have thus far been limited. Further research is needed to establish whether these models would provide more relevant information compared to traditional culture systems.

Coculture systems

The reconstruction of functional hepatic tissue is dependent on the ability to control factors that influence the cell environment, including cell–matrix interactions, soluble stimuli, and cell–cell interactions (Cho et al., 2010). Cell–cell interactions play a critical role in tissue morphogenesis, embryogenesis, and organ development. In vivo, the liver is a structurally and functionally heterocellular construct composed of primary hepatocytes, endothelial cells, Küpffer cells, stellate cells, and fibroblasts (Michalopoulos and DeFrances, 1997). In an effort to reconstruct liver tissue in vitro, several studies have demonstrated that coculturing primary hepatocytes with nonparenchymal cells such as fibroblasts or endothelial cells maintains hepatocyte viability and function, whereas hepatocytes cultured alone rapidly lose their function (Corlu et al., 1997; Mesnil et al., 1987). Hepatocyte cocultures are usually composed of hepatocytes with one other cell type, often other liver cells (Ries et al., 2000; Tukov et al., 2006; Krause et al., 2009), nonliver epithelial or endothelial cells (Talamini et al., 1998; Gebhardt et al., 1998), fibroblasts, or other cell lines (Kang et al., 2004). Coculturing hepatocytes with other cell types has been shown to be one of the most successful techniques for maintaining hepatocyte function under in vitro conditions (Guguen-Guillouzo and Guillouzo, 1983; Begue et al., 1984). Human hepatocytes cocultured with nonparenchymal cells in a system with flow were shown to have greater xenobiotic metabolism capabilities and better maintenance of metabolic capacity than cultures composed solely of hepatocytes (Novik et al., 2010).

Stem cells

Current procedures used to isolate primary hepatocytes tend to be labor-intensive and results vary greatly between laboratories and protocols. Although use of cryopreserved hepatocytes has become more popular, cell quality upon thawing may not always remain consistent. A more robust and reproducible source of hepatocytes would greatly benefit the field of in vitro toxicity testing and assist in standardizing research. Theoretically, stem cells would represent a renewable source of cells and would potentially provide large numbers of functionally equivalent cells that could be stored for later use. A toxicity model that incorporates a relatively unlimited supply of human hepatocytes with defined phenotypes would facilitate more accurate prediction of responses to test articles while accounting for the inherent genetic diversity within the human population.

Induced pluripotent stem cells

Induced pluripotent stem cells (iPSCs) transform into hepatocyte-like cells by following the same differentiation processes as stem cells; however, they originate from somatic cell types that must first be "reprogrammed." The iPSCs retain great capacity for cell division and differentiation (Shafritz et al., 2009). Because these cells do not come from an embryo, the use of these cells is not as ethically controversial. Fully functional hepatocytes derived from induced pluripotent stem cells would be a valuable tool for product development or evaluation of the contribution of genetic variation to variable responses. However, due to current disadvantages and technical limitations, neither induced pluripotent nor ESC are a widely accepted option for toxicological and pharmacological studies.

Much progress has been made toward improving traditional in vitro models that sustain liver-specific functions and can accurately predict in vivo responses to toxicants. These models, however, still lack the ability to provide long-term culture without cell necrosis or loss of differentiation. Although each traditional in vitro system has advantages and disadvantages, the use of 3D systems with dynamic flow for primary cell and/or stem cell cultures appears to be the most promising for toxicological studies. In addition, the ability to provide large biomasses of human disease-relevant cells could significantly reduce the need for animal use in toxicological studies. The use of cocultures could enable close simulation of in vivo-like characteristics and improve predictability. Because the need for toxicity testing of xenobiotics is unlikely to diminish in the future, the development of the models for reliable, quick, and economical toxicity testing is of major importance.

In vitro models for safety assessment and toxicology studies of the cardiovascular system

Evaluation of cardiovascular safety is an integral part of product development. As an initial screen, multiple in vitro assays have been used. A few routinely used assays are briefly described here.

hERG assay

hERG is a gene that codes for a protein known as Kv11.1, the alpha subunit of a potassium ion channel. This ion channel, sometimes simply denoted as "hERG" (Trudeau et al., 1995) is best known for its contribution to the electrical activity of the heart. hERG forms the major portion of one of the ion channel proteins [the "rapid" delayed rectifier current (IKr)] that conducts potassium (K^+) ions out of cardiomyocytes, and this current is critical in correctly timing the return of the cell membrane to its resting state (repolarization) during the cardiac action potential (Sanguinetti and Tristani-Firouzi, 2006). When this channel's ability to conduct electrical current across the cell membrane is inhibited or compromised, usually by a test substance, it can result in a potentially fatal disorder called long QT syndrome, which has made hERG inhibition an important antitarget in product development. Due to the potential risk of certain xenobiotics inducing long QT syndrome, regulatory authorities have issued recommendations for the establishment of cardiac safety during preclinical drug development; ICH S7B covers nonclinical evaluation of the potential for delayed ventricular repolarization (ICH S7B, 2005). The hERG assay is used to determine safety margins and the proarrhythmic potential of xenobiotics under development.

A number of assays have been developed to gain a picture of compound effects on the cardiac action potential and, in particular, on repolarization effects. These assays include hERG ion channel assays using (but not limited to) fluorescence, binding, atomic absorption, or electrophysiological techniques measuring interaction with or the function of the hERG ion channel, with different throughput inherent to these techniques. In addition, assays have been developed that assess the effects of compounds on other cardiac ion channels. These data may add important information and draw a more complete picture on cardiac effects than is possible with hERG channel data alone, because effects on other cardiac ion channels may potentiate or reduce hERG channel effects. Also, assays that could potentially deliver more physiologically relevant data than assays relying on only one ion channel type are used. In past years, techniques based on stem cell-derived cardiomyocytes particularly gained significant interest (Hamdam et al., 2013). Such assays integrate effects on several ion channels beyond the hERG channel. Therefore effects on several channels and interactions of such effects might be detected in assays using these preparations. The manual patch clamp hERG assay is generally considered the "gold standard" for assessment of xenobiotic interactions with the hERG potassium channel.

It is possible to identify cardiac safety issues caused by hERG ion channel blockade early in product development utilizing high-quality medium-throughput automated electrophysiology screening using cells expressing the hERG ion channel. The physiological significance of these recordings can be improved, at reduced throughput, by performing recordings at or near physiological temperatures (Polonchuk, 2012). In addition to screening of the particularly important hERG ion channel, assessing the effects of compounds on other cardiac ion channels may be required. Recently, automated patch clamp screening of compound effects on the action potential in stem cell-derived cardiomyocytes has become possible (Stoelzle et al., 2011; Milligan and Möller, 2013). These assays provide information to further assess the cardiac safety of compounds and are briefly discussed in the following paragraphs. However, currently, more validation of stem cell-based assays is required before these can be routinely used for cardiac safety investigations.

In vitro screening for cardiotoxicity using stem cell-derived cardiomyocytes

Human-induced pluripotent stem cell (hiPSC)-derived cardiomyocytes show promise for screening during early drug development. Sirenko et al. (2013) recently demonstrated that multiparameter screening of beating profiles allows for cardiotoxicity risk assessment and identification of specific patterns defining mechanism-specific effects. They tested the hypothesis that in vitro assessment of multiple cardiomyocyte physiological parameters will enable predictive and mechanistically interpretable evaluation of cardiotoxicity in a high-throughput format. With this objective, they treated human iPSC-derived cardiomyocytes for 30 min or 24 h with 131 compounds in up to six concentrations (3 nM to 30 μM). Fast kinetic imaging was used to monitor changes in cardiomyocyte function using intracellular Ca^{2+} flux readouts synchronous with beating and cell viability. A number of physiological parameters of cardiomyocyte beating, such as beat rate, peak shape (amplitude, width, raise, decay, etc.), and regularity, were collected using automated data analysis. Concentration–response profiles were evaluated using logistic modeling to derive a benchmark concentration (BMC) point-of-departure value based on one standard deviation departure from the estimated baseline in vehicle-treated (0.3% dimethylsulfoxide) cells. BMC values were used for cardiotoxicity classification and ranking of compounds. They found that beat rate and several peak shape parameters were good predictors for cardiotoxicity.

Some of the other commonly used in vitro cardiovascular toxicology models and their advantages and disadvantages are summarized in Table 29.2.

In vitro models for safety assessment and toxicology studies of the kidney

Factors contributing to the nephrotoxicity of herbal dietary supplements may be related to intrinsic toxic plant constituents or contamination of the product with nephrotoxic agents. Kidneys are particularly vulnerable to xenobiotic toxicity because they are highly vascularized and play a major role in the metabolism and elimination of toxicants. Because proximal tubular cells (PTC) are a major target for toxic drug effects in the kidney, this is the preferred cell type for in vitro models, although models with other renal cell types have also been developed (Rodriguez-Barbero et al., 2000; Sakairi et al., 2010; Takeuchi et al., 2010). PTC actively transports many drugs, organic compounds, and xenobiotics from the bloodstream into the glomerular filtrate and also metabolizes many of these compounds. They express a broad variety of drug transporters and drug-metabolizing enzymes (Lohr et al., 1998; Morrissey et al., 2013), and toxic effects depend on these specific expression patterns.

In the following paragraphs, some of the in vitro methods that are used for the assessment of xenobiotic-induced renal toxicity are discussed. In silico methods have also been applied to nephrotoxicity studies (Ursem et al., 2009), and some structural alerts such as aristolochic acids (Fang et al., 2011), ochratoxin A (Sieber et al., 2009), and diterpen triptolide (Xia et al., 2009) have been well recognized.

Isolated perfused kidney

Isolated perfused kidney from rat or mouse (de Broe and Porter, 2008) is ideal for the assessment of renal pathophysiology, especially for evaluating vascular, tubular, and glomerular functions. It can be also used to evaluate test articles for their potential nephrotoxicity (Gandolfi and Brendel, 1990). Because the isolated perfused kidney is an intact organ and maintains structural integrity, it can reproduce the in vivo toxicity for many acute toxicants (Shanley et al., 1986). However, this system represents a difficult technique, suffers from limited viability, and is time-consuming.

Renal cell culture models (de Broe and Porter, 2008) have been used for a long time. Cell lines offer several advantages (easier to culture, unlimited life span, no isolation procedure, and better stability) but are probably more distant from in vivo situations than primary cells. Several renal cell lines reproducing different renal cell types are available: cell lines with characteristics of the proximal tubule such as normal rat kidney cells (NRK52E), Lewis lung carcinoma porcine kidney (LLC-PK1), opossum kidney, or human kidney-2 (HK-2 cells) generated by transduction of human primary PTC with human papilloma virus; and cell lines with characteristics of the distal tubule or the collecting duct such as Madin Darby canine kidney cells or Madin Darby bovine kidney cells. With respect to human cells, two cell lines have been frequently used: human embryonic kidney-293 (HEK293) (and derivatives) and HK-2. HEK293 cells express various neuronal markers and are probably not kidney epithelial cells, as thought for a long time (Shaw et al., 2002). Primary PTCs share more similarities with in vivo cells. However, the isolation procedure and the limited access to these cells, particularly those from a human source, hamper their utilization. Also, primary cell lines have a limited life span and their phenotype can change rapidly due to different culture conditions. Renal cell culture

TABLE 29.2 **Some commonly used in vitro cardiovascular models.**

Method	Advantages	Disadvantages
Perfused:	Maintains structural organization	Small number of replicates
Langendorff	High correlation with drugs associated with TdP in clinical setting	Time constraints and preparation
Vascular rings	Pressure–volume relationship can be maintained	Compounds may alter coronary vasculature, impacting rate, output, and contraction
	Continuous infusion	
Organ culture:	Cell–cell/cell–substratum interactions	Small number of replicates
Isolated vessels	Effects of endothelium can be removed	Long-term effects cannot be monitored
Fetal hearts	Simple, cheap, and robust	No metabolism
Frog or mouse whole hearts	Preserve tissue architecture	Phenotypic changes in VSMCs cultured with FBS
Isolated muscle:	Examination of tension development	Oxygenation depends on diffusion
Atrial	Plasma membrane can be removed	Stability of contraction/relaxation reliability short-lived
Ventricular	Isotonic, isometric force development can be examined	Xenobiotic effects not accounted for
Papillary	Confounding effects of drugs on heart or pulmonary circulation removed	
Vascular beds		
Tissue slices:	Direct interspecies comparison	Need to co-culture with liver or intestinal slices to observe biotransformation of xenobiotics
	Easy preparation and standardization	
	Multiple cellular functions studied at same time	
	Cell architecture and heterogeneity maintained	
Single cell suspension:	Different cell types can be studied alone	Variants such as temperature, enzymes or pH/osmolarity can critically affect experiments
Embryonic	Use of voltage clamp possible	Reduction of surface receptor-binding sites can occur with isolation
Adult		
HERG channel assay:	Reliable preclinical model of drug-induced QT prolongation	No activators of ERG available
Isolated cardiac tissue	Assay for APD prolongation or shortening	Isolation of cells difficult
Heterologous hERG	In vitro results easily integrated with in vivo	Cloned ion channels lack the native conditions and structural architecture
Purkinje fibers	Experimental manipulations difficult to achieve in vivo can be changed (i.e., stimulation rate, K^+ concentration) Identification of proarrhythmic surrogates	Time-consuming
Papillary muscle		
Transmural wedge		
Cell culture:	Primary cultures can be established	Fibroblast contamination can be problematic

(Continued)

TABLE 29.2 (Continued)

Method	Advantages	Disadvantages
Endothelial	Readily proliferate under appropriate conditions	Interactions between different cell types not easily determined
Myocyte	Effect of different ECM components can be examined	Tissue architecture not maintained
Smooth muscle		
Cell fractionation:	Low-abundant proteins/complexes can be enriched	Pure fractions can be tedious and difficult to obtain
	Complexity of sample can be reduced	Quality of fractions time-consuming
		Possibility of fraction cross-contamination

ECM, Extracellular matrix; *FBS*, fetal bovine serum; *VSMC*, vascular smooth muscle cells.

models offer a better way to understand biochemical and cellular mechanisms of cytotoxicity and are inexpensive, with quick turnaround compared to animal studies.

Pluripotent stem cell-derived cells

Another approach to developing an unlimited cell source is to use stem cell-derived renal cells. A protocol has been developed for the differentiation of hiPSC into podocyte-like cells (Song et al., 2012). A method for the generation of stem cell-derived human PTC (HPTC)-like cells in vitro has been published (Narayanan et al., 2013).

Microfluidic and 3D models

Generally, it is assumed that 2D and static cultivation conditions are artificial; therefore results could be improved by using 3D and microfluidic models. Two different 3D models based on hydrogel cultures have been used for nephrotoxicology (Astashkina et al., 2012a,b). One model used freshly isolated murine PTC embedded in a modified hyaluronic acid-based hydrogel (Astashkina et al., 2012a,b). Cell viability decreased approximately twofold within the first 2 weeks of culture. This murine model would require the use of animals, and results are expected to be affected by interspecies variability. Due to the requirement for fresh tissue, it will be difficult to adapt it to human material. The model was characterized by immunostaining, reverse-transcriptase polymerase chain reaction, and assays for enzymatic activities and gluconeogenesis.

Another 3D model (DesRochers et al., 2013) used NKi-2 cells, which are HPTC with human telomerase reverse-transcriptase immortalized human renal cortical cells. These cells displayed some features of PTC. However, NKi-2 cells responded to antidiuretic hormone, were not responsive to parathyroid hormone, and expressed very low levels of gamma-glutamyl transferase, features not typical for PTC. After embedment into a mixture of matrigel and rat tail collagen, tubule-like structures were formed by branching morphogenesis. A clear disadvantage of current microfluidic and 3D models is their incompatibility with HTS. In addition, 3D gels compromise the controlled application of compounds and probably also the detection of cellular responses (DesRochers et al., 2013).

Renal slice models

Contrasting with hepatotoxicity studies, only a few renal slice models have been used to study nephrotoxicity related to iodinated contrast compounds such as cisplatin (Vickers and Fisher, 2004; Vickers et al., 2004; Harmon et al., 2009). Many nephrotoxic chemicals induce selective injury within the kidney, affecting either renal proximal straight (PST) or proximal convoluted tubules (PCT). Selective injury has also been observed following in vitro nephrotoxicant exposure to precision-cut renal slices and isolated PCT and PST segments (Ruegg, 1994). This in vitro model provides a means of comparing and contrasting basic mechanistic differences, thus rendering these segments innately susceptible to nephrotoxicant injury. Renal cortical slices are generally prepared from kidneys of untreated rats and incubated with tested compound or vehicle for 2 h. At the end of the 2-h incubation, tissue gluconeogenesis capacity (pyruvate-stimulated gluconeogenesis) and LDH release are determined as measures of cellular function and cytotoxicity (Hong et al., 2002). This model provides interaction between various cells within the nephron and has proven to be a useful in vitro system for biotransformation and toxicity studies (Gandolfi et al., 1995; Minigh and Valentovic, 2003); however, it is not useful to assess chronic renal toxicity.

Renal slice models have also been used in investigating aristolochic acid and tripterygium toxicity (Dickman et al., 2011).

No study has directly compared these different methods of renal toxicity screening. In addition to the classical new kidney injury biomarkers, several other urinary markers have been identified with "omics"-based methods. However, they were rarely compared to other toxicological screening approaches and they were not prospectively validated. A review published in 2005 (Gibbs, 2005) evaluated the relative sensitivity and specificity of proteomics and metabonomics techniques compared with traditional techniques for assessing xenobiotic-induced nephrotoxicity. The authors concluded that there was no consistent evidence that the novel methodologies were any more sensitive than the traditional methods for assessing nephrotoxicity. Newer "omics"-based techniques, although helpful in identifying new biomarkers of renal injury, lack validation and should be considered complementary tools to detect renal injury in toxicology screening programs (Rouse et al., 2011).

Many methods described here have been used to assess the nephrotoxicity of several herbal products. However, they were not applied for screening purposes and no prospective validation is available. Based on the current knowledge, renal cell culture models have been mainly used in aristolochic acids as models. A series of aristolochic acid derivatives isolated from *Aristolochia* species were analyzed for their nephrotoxic potential using the neutral red dye exclusion assay in cultures of renal epithelial cells (Balachandran et al., 2005). An example of *Aristolochia* species that causes nephrotoxic syndrome is shown in Fig. 29.13.

In addition, a recent study has investigated the renal toxicity of 47 herbs traditionally used for kidney and urinary disorders in two cell line models (Wojcikowski et al., 2009).

In conclusion, several in vitro approaches have been developed to assess the renal toxicity of xenobiotics. The safety evaluation needs to be focused on the material effectively taken by a subject. It should be recognized that there may be significant differences in pharmacokinetic behavior, such as the rate or extent of absorption and pharmacological/toxic potency between a whole botanical and the equivalent amount of an isolated active principle. In most instances, there may be no available data regarding metabolism and ADME study for a given botanical product. Despite several limitations, animal models remain the gold standard to assess the nephrotoxicity of drugs.

Toxicity testing in the twenty-first century and the human toxome project

Traditional toxicity testing typically involves studying adverse health outcomes in animals subjected to high doses of toxicants with subsequent extrapolation to expected human responses at lower doses. This approach makes testing of any single chemical expensive, time-consuming, and exorbitant in the use of animals (Hartung and Rovida, 2009; Hartung, 2011). In 2003, the US Environmental Protection Agency (EPA) report (EPA, 2003) noted that a computational toxicology program would have several advantages, primarily in prioritizing chemicals for testing and in developing predictive models for quantitative risk assessment. At the request of the EPA and the National Institute of Environmental Health Sciences, the US National Research Council (NRC) conducted a review of toxicity testing methods, producing the NRC Report on Toxicity Testing in the 21st century: A Vision and a Strategy (TT21c) (NRC, 2007). The TT21c encompasses a long-range strategic plan for transforming

FIGURE 29.13 An example of the *Aristolochia* species that causes nephrotoxic syndrome is shown.

toxicity testing. The major components of the plan include the use of predictive, high-throughput cell-based assays (of human origin) to evaluate perturbations in key toxicity pathways and to conduct targeted testing against those pathways. This approach will greatly accelerate our ability to test the vast "storehouses" of chemical compounds and nutraceuticals using a rational, risk-based approach to prioritization and provide test results that are

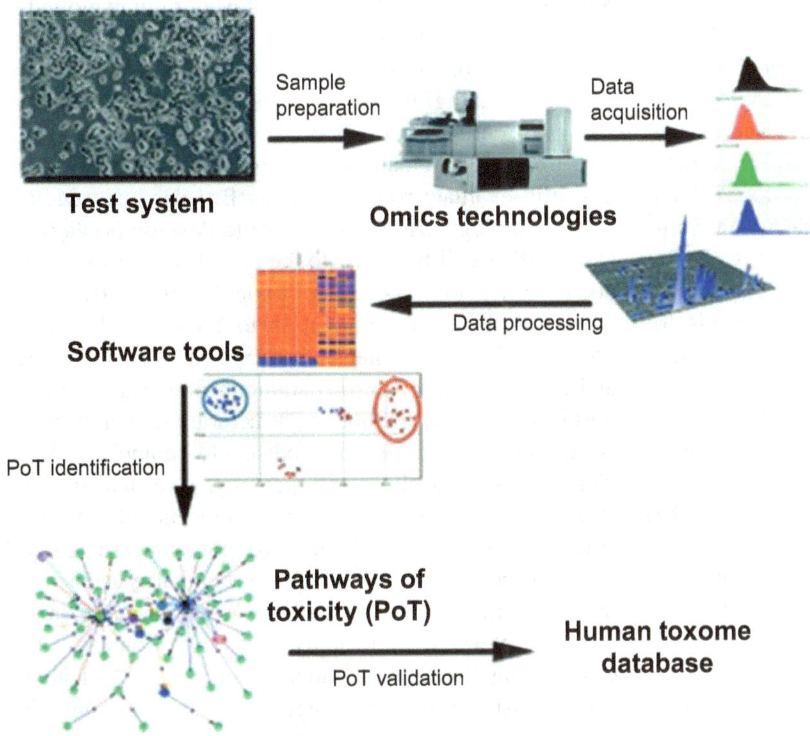

FIGURE 29.14 Components of the project "Mapping the Human Toxome by Systems Toxicology". *Adapted from Bouhifd, M., Andersen, M.E., Baghdikian, C., 2015. The human toxome project. ALTEX. 32 (2), 112–124.*

TABLE 29.3 Vision for in vitro toxicity testing for the 21st century.

Tools	Applications
High-throughput screens (HTS)	Efficiently identify critical toxicity pathway perturbations across a range of doses and molecular and cellular targets
Stem cell biology	Develop in vitro toxicity pathway assays using human cells produced from directed stem cell differentiation
Functional genomics	Identify the structure of cellular circuits involved in toxicity pathway responses to assist computational dose–response modeling
Bioinformatics	Interpret complex multivariable data from HTS and genomic assays in relation to target identification and effects of sustained perturbations on organs and tissues
Systems biology	Organize information from multiple cellular response pathways to understand integrated cellular and tissue responses
Computational systems biology	Describe dose–response relationships based on perturbations of cell circuitry underlying toxicity pathway responses giving rise to thresholds, dose-dependent transitions, and other dose-related biological behaviors
PBPK models	Identify human exposure situations likely to provide tissue concentrations equivalent to in vitro activation of toxicity pathways
Structure–activity relationships	Predict toxicological responses and metabolic pathways based on the chemical properties of environmental agents and comparison to other active structures
Biomarkers	Establish biomarkers of biological change representing critical toxicity pathway perturbations

PBPK, Physiologically based pharmacokinetic modeling.
Source: Modified from Andersen, M.E., Krewski, D., 2010. The vision of toxicity testing in the 21st century: moving from discussion to action. Toxicol. Sci. 117 (1), 17–24.

hopefully far more predictive of human toxicity than current methods.

To this end, although some toxicity pathways have already been identified, most are only partially known and no common annotation exists. Mapping the entirety of these pathways (i.e., the Human Toxome) will be a large-scale effort, perhaps on the order of the Human Genome Project. The TT21c report has prompted a number of activities for modernizing regulatory toxicology. TT21c called for embracing new technologies and basing assessments on toxicological mechanisms. In alignment with TT21c, the Human Toxome Project is focused on developing the concepts and the means for deducing, validating, and sharing molecular pathways of toxicity (PoT). The bioinformatics tools for PoT deduction constitute a core deliverable of the project. The project intends to address a number of challenges regarding the quality and standardization of cell systems, omics technologies, and bioinformatics. In parallel, concepts for annotation, validation, and sharing of PoT information (as well as their link to adverse outcomes) are planned to be developed. A reasonably comprehensive public database of PoT, the Human Toxome Knowledge base, could become a point of reference for toxicological research and regulatory test strategies. The components of the Human Toxome Project are represented in Fig. 29.14.

A summary of newer in vitro approaches to toxicology testing envisioned in TT21c is shown in Table 29.3. Once these technologies have been sufficiently validated, it is hoped that the data generated by using such technologies would form a sound basis for safety and toxicity assessment of test compounds, including nutraceuticals, as we advance into the 21st century (Andersen and Krewski, 2010).

Concluding remarks and future directions

The practicality of using alternative in vitro tests for safety and toxicity assessment of nutraceuticals, for the most part, is aligned with the DSHEA act and the general guidelines described in the FDA Redbook, with relevant advancements in technology and science. The DSHEA authorizes the FDA to intervene only when a dietary supplement ingredient, including nutraceuticals, presents a significant and unreasonable risk of illness, injury, or toxicity under conditions of use recommended in the labeling or when the product is adulterated. Although the regulatory situation is such that completion of in vitro and/or in vivo tests is not required for dietary supplement ingredients prior to their commercialization, the scientific value of these assays in predicting adverse effects in humans is immense in the safety assessment of nutraceuticals. In vitro assays such as those contained in the Redbook should be considered, among others, as important validated tools for the prediction of possible adverse effects in humans. In this chapter, important in vitro methods that are publicly available for safety assessment are briefly described, along with newer technologies. Despite many promising attributes of in vitro models, their potential has still not been fully exploited.

The future direction for in vitro toxicology depends on validation of the newer in vitro assays, so a well thought-out and meaningful risk assessment of nutraceuticals can be made. Further, it is important to establish comprehensive toxicological databases linked with biomarkers and critically evaluate rooted dogmas. Broader integration of systems biology will be necessary to develop organotypic systems that will be closer to the complexity of the human body, especially regarding toxicokinetics. The main objective of risk assessment is to predict and evaluate potential risks to humans; therefore appropriate models and relevant results are of paramount importance. In this postgenome era, a plethora of structural alerts and in silico, in vitro, and in vivo methods could be applied to predict the genotoxicity, teratogenicity, and major organ toxicity of botanicals and/or nutraceutical products. However, these technologies are not standardized yet. This will be the main challenge for regulatory agencies to overcome in establishing uniform and transparent methods to interpret and apply available "omics" data. Considering the aforementioned issues, it is critical that toxicologists in industry, regulatory agencies, and academic institutions develop rigorous "omics" methods to reach a consensus about the reliability and interpretation of endpoints. It will also be important to regulate the puzzle of conventional methods of genotoxicity assessments and their relation with new "omics" and chip technologies. It is anticipated that the integration of in vitro toxicity assessment along with the Human Toxome research project would yield more meaningful safety assessment of a variety of test compounds (Hartung, 2010; Hartung et al., 2013; Bouhifd et al., 2015). Furthermore, investigations must be undertaken to discover novel structural alerts of toxicity from nutraceuticals, which will certainly facilitate an analytical assessment of the toxicity of traditional medicinal plants at early stages of testing and development.

References

Abdel-Rahman, A., Anyangwe, A., Carlacci, L., 2010. FORUM: safety of vitamins and minerals: controversies and perspective. Toxicol. Sci. 118 (2), 348–355.

Adler, S., Basketter, D., Creton, S., 2011. Alternative (non-animal) methods for cosmetics testing: current status and future prospects-2010. Arch. Toxicol. 85, 367–485.

Ames, B.N., McCann, J., Yamasaki, E., 1975. Methods for detecting carcinogens and mutagens with the Salmonella/mammalian-microsome mutagenicity test. Mutat. Res. 31, 347–364.

Andersen, M.E., Krewski, D., 2010. The vision of toxicity testing in the 21st century: moving from discussion to action. Toxicol. Sci. 117 (1), 17–24.

Andlauer, W., Furst, P., 2002. Nutraceuticals: a piece of history, present status and outlook. Food Res. Int. 35, 171–176.

Aninat, C., Piton, A., Glaise, D., Le Charpentier, T., 2006. Expression of cytochromes P450, conjugating enzymes and nuclear receptors in human hepatoma HepaRG cells. Drug Metab. Dispos. 34, 75–83.

Arvidson, K.B., Valerio, L.G., Diaz, M., 2008. *In silico* toxicological screening of natural products. Toxicol. Mechan. Methods 18, 229–242.

Astashkina, A.I., Mann, B.K., Prestwich, G.D., 2012a. A 3-D organoid kidney culture model engineered for high-throughput nephrotoxicity assays. Biomaterials 33 (18), 4700–4711.

Astashkina, A.I., Mann, B.K., Prestwich, G.D., 2012b. Comparing predictive drug nephrotoxicity biomarkers in kidney 3-D primary organoid culture and immortalized cell lines. Biomaterials 33 (18), 4712–4721.

Bader, A., Borel Rinkes, I.H., Closs, E.I., 1992. A stable long-term hepatocyte culture system for studies of physiologic processes: cytokine stimulation of the acute phase response in rat and human hepatocytes. Biotechnol. Progr. 8, 219–225.

Bagchi, D., 2006. Nutraceuticals and functional foods regulations in the United States and around the world. Toxicology 221, 1–3.

Balachandran, P., Wei, F., Lin, R.-C., 2005. Structure activity relationships of aristolochic acid analogues: toxicity in cultured renal epithelial cells. Kidney Int. 67, 1797–1805.

Bechter, R., Terlouw, G.D., Lee, Q.P., 1991. Effects of QA 208–199 and its metabolite 209–668 on embryonic development *in vitro* after microinjection into the exocoelomic space or into the amniotic cavity of cultured rat conceptuses. Teratogen. Carcinogen. Mutagen. 11, 185–194.

Begue, J.M., Guguen-Guillouzo, C., Pasdeloup, N., 1984. Prolonged maintenance of active cytochrome P-450 in adult rat hepatocytes co-cultured with another liver cell type. Hepatology 4, 839–842.

Biesalski, H.K., 2001. Nutraceuticals: the link between nutrition and medicine. In: Kramer, K., Hoppe, P.P., Packer, L. (Eds.), Nutraceuticals in Health and Disease Prevention. Marcel Dekker Inc, New York, pp. 1–26.

Bishop, W.E., Clarke, D.P., Travis, C.C., 2001. The genomic revolution: what does it mean for risk assessment? Risk Anal. 21, 983–987.

Borner, F.U., Schutz, H., Wiedemann, P., 2011. The fragility of omics risk and benefit perceptions. Toxicol. Lett. 201, 249–257.

Bouhifd, M., Andersen, M.E., Baghdikian, C., 2015. The human toxome project. ALTEX. 32 (2), 112–124.

Bravo, L., 1998. Polyphenols: chemistry, dietary sources, metabolism, and nutritional significance. Nutr. Rev. 56, 317–333.

de Broe, M.E., Porter, G.A., 2008. Clinical Nephrotoxins: Renal Injury from Drugs and Chemicals, third ed Springer, New York, p. 990, Springer, New York 2008.

Cencic, A., Langerholc, T., 2010. Functional cell models of the gut and their applications in food microbiology—a review. Int. J. Food Microbiol. 141, S4–S14.

Chan, L.Y., Chiu, P.Y., Lau, T.K., 2003. An *in vitro* study of ginsenoside Rb1-induced teratogenicity using a whole rat embryo culture model. Hum. Reprod. 18, 2166–2168.

Chang, B.E., Liao, M.H., Kuo, M.Y., 2004. Developmental toxicity of arecoline, the major alkaloid in betel nuts, in zebrafish embryos. Birth Def. Res. A Clin. Mol. Teratol. 70, 28–36.

Chao, P., Maguire, T., Novik, E., 2009. Evaluation of a microfluidic based cell culture platform with primary human hepatocytes for the prediction of hepatic clearance in human. Biochem. Pharmacol. 78, 625–632.

Charles, G.D., Bartels, M.J., Zacharewski, T.R., 2000a. Activity of benzo[a]pyrene and its hydroxylated metabolites in an estrogen receptor-alpha reporter gene assay. Toxicol. Sci. 55, 320–326.

Charles, G.D., Gennings, C., Clemons, J., 2000b. Optimization of an estrogen receptor-alpha transcriptional activation assay for testing a diverse spectrum of chemicals. J. Appl. Toxicol. 20, 449–454.

Chen, X.W., Serag, E.S., Sneed, K.B., 2011. Herbal bioactivation, molecular targets and the toxicity relevance. Chemico-Biol. Interact. 192, 161–176.

Cho, C.H., Park, J., Tilles, A.W., 2010. Layered patterning of hepatocytes in co-culture systems using microfabricated stencils. Biotechniques 48, 47–52.

Christianson, A., Howson, C.P., Modell, B., 2006. Executive Summary—March of Dimes Report on Birth Defects. Available from: <http://www.neonatology.org/pdf/MODBDExecutiveSummary.pdf/>.

Corlu, A., Ilyin, G., Cariou, S., 1997. The co-culture: a system for studying the regulation of liver differentiation/proliferation activity and its control. Cell Biol. Toxicol. 13, 235–242.

Cosgrove, B.D., King, B.M., Hasan, M.A., 2007. Synergistic drug-cytokine induction of hepatocellular death as an *in vitro* approach for the study of inflammation-associated idiosyncratic drug hepatotoxicity. Toxicol. Appl. Pharmacol. 237, 317–330.

Cronin, M.T.D., 2002. The current status and future applicability of quantitative structure-activity relationships (QSARs) in predicting toxicity. Altern. Lab. Anim. ATLA 30, 81–84.

Damian, D., 2015. B vitamin may prevent recurrence of skin cancers. DDDmag, May 2015. Available from: <http://www.dddmag.com/articles/2015/05/b-vitamin-may-prevent-recurrence-skin-cancers?et_cid = 4570746&et_rid = 454958281&location = top/> (Accessed 14 May 2015).

Daneshian, M., Akbarsha, M.A., Blaauboer, B., 2011. A framework program for the teaching of alternative methods (replacement, reduction, refinement) to animal experimentation. ALTEX. 28, 341–352.

Dang, Z.C., 2009. Dose-dependent effects of soy phyto-oestrogen genistein on adipocytes: mechanisms of action. Obes. Rev. 10 (3), 342–349.

Dash, A., Inman, W., Hoffmaster, K., 2009. Liver tissue engineering in the evaluation of drug safety. Expert Opin. Drug Metabol. Toxicol. 5, 1159–1174.

Davies, H., 2010. A role for "omics" technologies in food safety assessment. Food Control 21, 1601–1610.

DelRaso, N.J., 1993. *In vitro* methodologies for enhanced toxicity testing. Toxicol. Lett. 68, 91–99.

DesRochers, T.M., Suter, L., Roth, A., 2013. Bioengineered 3D human kidney tissue, a platform for the determination of nephrotoxicity. PLoS One 8 (3), e59219.

Di Renzo, F., Bacchetta, R., Bizzo, A., 2011. Is the amphibian *X. laevis* WEC a good alternative method to rodent WEC teratogenicity assay? The example of the three triazole derivative fungicides Triadimefon, Tebuconazole, Cyproconazole. Reprod. Toxicol. 32, 220–226.

Dickman, K.G., Sweet, D.H., Bonala, R., 2011. Physiological and molecular characterization of aristolochic acid transport by the kidney. J. Pharmacol. Exp. Ther. 338, 588–597.

Domansky, K., Inman, W., Serdy, J., 2010. Perfused multiwell plate for 3D liver tissue engineering. Lab. Chip. 10, 51–58.

DSHEA, 1994. Dietary Supplement Health and Education Act of 1994.

Duez, P., Dehon, G., Kumps, A., 2003. Statistics of the comet assay: a key to discriminate between genotoxic effects. Mutagenesis 18, 159–166.

Dunn, J.C., Yarmush, M.L., Koebe, H.G., 1989. Hepatocyte function and extracellular matrix geometry: long-term culture in a sandwich configuration. FASEB J. 3, 174–177.

Duval, D.L., Sieg, D.J., Billings, R.E., 1995. Regulation of hepatic nitric oxide synthase by reactive oxygen intermediates and glutathione. Arch. Biochem. Biophys. 316, 699–706.

ECVAM, 2015. <https://eurl-ecvam.jrc.ec.europa.eu/> (Accessed 14 May 15).

Efferth, T., Li, P.C.H., Lam, E., 2008. Artesunate derived from traditional Chinese medicine induces DNA damage and repair. Cancer Res. 68, 4347–4351.

EFSA, 2009. Guidance on safety assessment of botanicals and botanical preparations intended for use as ingredients in food supplements. EFSA J. 7 (9), 1249.

Elamin, E., Jonkers, D., Juuti-Uusitalo, K., 2012. Effects of ethanol and acetaldehyde on tight junction integrity: *in vitro* study in a three dimensional intestinal epithelial cell culture model. PLoS One 7, e35008.

EMEA, 2007. Guideline on the Assessment of Genotoxic Constituents in Herbal Substances/Preparations. European Medicines Agency: London [EMEA/HMPC/107079/2007]. Avaialble from: <http://www.ema.europa.eu/ema/pages/includes/document/open_document.jsp?webContentId = WC500003569>.

EMEA, 2008. Overview of Comments Received on Draft 'Guideline on the Assessment of Genotoxic Constituents in Herbal Substances/Preparations' (emea/hmpc/107079/2007). European Medicines Agency: London [EMEA/HMPC/232888/2008]. Available from: <www.ema.europa.eu/docs/en_GB/document/WC500003570.pdf>.

EPA, 2003. A Framework for a Computational Toxicology Research Program Within ORD. EPA/600/R-03/065. <http://epa.gov/comptox/download_files/basic_information/comptoxframework06_02_04.pdf/>.

Fang, Z.-Z., Zhang, Y.-Y., Wang, X.-L., 2011. Bioactivation of herbal constituents: simple alerts in the complex system. Expert Opin. Drug Metab. Toxicol. 7, 1–19.

FDA, 2015a. Ingredients, Packaging & Labeling: Labeling & Nutrition. US Food and Drug Administration. Available from: <http://www.fda.gov/Food/IngredientsPackagingLabeling/LabelingNutrition/default.htm>.

FDA, 2015b. Dietary Supplements: Questions and Answers on Dietary Supplements. US Food and Drug Administration. Available at: <http://www.fda.gov/Food/DietarySupplements/QADietarySupplements/default.htm>.

FDA Redbook, 2007. Guidance for Industry and Other Stakeholders Toxicological Principles for the Safety Assessment of Food Ingredients—Redbook, 2000, revised 2007.

Fernandez-Garcia, E., Carvajal-Lerida, I., Perez-Galvez, A., 2009. *In vitro* bioaccessibility assessment as a prediction tool of nutritional efficiency. Nutr. Res. 29, 751–760.

Flint, O.P., Orton, T.C., 1984. An *in vitro* assay for teratogens with cultures of rat embryo midbrain and limb bud cells. Toxicol. Appl. Pharmacol. 76, 383–395.

Gaffield, W., Keeler, R.F., 1994. Structure-activity relations of teratogenic natural products. Pure Appl. Chem. 66, 2407–2410.

Gandolfi, A.J., Brendel, K., 1990. *In vitro* systems for nephrotoxicity studies. Toxicol. In Vitro 4, 337–345.

Gandolfi, A.J., Brendel, K., Fisher, R.L., 1995. Use of tissue slices in chemical mixture toxicology and interspecies investigations. Toxicology 105, 285–290.

Gebhardt, R., Schrode, W., Eisenmann-Tappe, I., 1998. Cellular characteristics of epithelial cell lines from juvenile rat liver: selective induction of glutamine synthetase by dexamethasone. Cell Biol. Toxicol. 14, 55–67.

Genschow, E., Spielmann, H., Scholz, G., 2002. The ECVAM international validation study on *in vitro* embryotoxicity tests: results of the definitive phase and evaluation of prediction models. European centre for the validation of alternative methods. Altern. Lab. Anim. 30, 151–176.

Gibbs, A., 2005. Comparison of the specificity and sensitivity of traditional methods for assessment of nephrotoxicity in the rat with metabonomic and proteomic methodologies. J. Appl. Toxicol. 25, 277–295.

Gollapudi, B., Krishna, G., 2000. Practical aspects of mutagenicity testing strategy: an industrial perspective. Mutat. Res. 455, 21–28.

Greene, N., 2002. Computer systems for the prediction of toxicity: an update. Adv. Drug. Deliv. Rev. 54, 417–431.

Griffin, S.J., Houston, J.B., 2005. Prediction of *in vitro* intrinsic clearance from hepatocytes: comparison of suspensions and monolayer cultures. Drug Metab. Dispos. 33, 115–120.

Guguen-Guillouzo, C., Guillouzo, A., 1983. Modulation of functional activities in cultured rat hepatocytes. Mol. Cell Biochem. 53-54, 35–56.

Guguen-Guillouzo, C., Guillouzo, A., 2010. General review on *in vitro* hepatocyte models and their applications. Methods Mol. Biol. 640, 1–40.

Hamdam, J., Sethu, S., Smith, T., 2013. Safety pharmacology–current and emerging concepts. Toxicol. Appl. Pharmacol. 273, 229–241.

Hansen, D.K., Abbott, B.D., 2009. Developmental toxicology. In: Hayes, A.W., Thomas, J.A., Gardner, D.E. (Eds.), Target Organ Toxicology Series. Informa Health-Care, New York, p. 384.

Harmon, R.C., Duffy, S.P., Terneus, M.V., 2009. Characterization of a novel model for investigation of radiocontrast nephrotoxicity. Nephrol. Dial. Transpl. 24, 763–768.

Hartung, T., 2010. Comparative analysis of the revised Directive 2010/63/EU for the protection of laboratory animals with its predecessor 86/609/EEC—a t4 report. ALTEX 27, 285–303.

Hartung, T., 2011. From alternative methods to a new toxicology. Eur. J. Pharm. Biopharm. 77, 338–349.

Hartung, T., Rovida, C., 2009. Chemical regulators have overreached. Nature 460, 1080–1081.

Hartung, T., Luechtefeld, T., Maetens, A., 2013. Integrated testing strategies for safety assessments. ALTEX 30, 3–18.

Heijne, W.H., Kienhuis, A.S., van Ommen, B., 2005. Systems toxicology: applications of toxicogenomics, transcriptomics, proteomics and metabolomics in toxicology. Expert Rev. Proteomics 2, 767–780.

Hewitt, N.J., Lechon, M.J., Houston, J.B., 2007. Primary hepatocytes: current understanding of the regulation of metabolic enzymes and transporter proteins, and pharmaceutical practice for the use of hepatocytes in metabolism, enzyme induction, transporter, clearance, and hepatotoxicity studies. Drug Metab. Rev. 39, 159–234.

Hong, S.K., Anestis, D.K., Ball, J.G., 2002. *In vitro* nephrotoxicity induced by chloronitrobenzenes in renal cortical slices from Fischer 344 rats. Toxicol. Lett. 129, 133–141.

Huh, D., Hamilton, G.A., Ingber, D.E., 2011. From 3D cell culture to organs-on-chips. Trends Cell Biol. 21, 745–754.

ICH, 2008. Guidance on Genotoxicity Testing and Data Interpretation for Pharmaceuticals Intended for Human Use. Draft January 28th 2008. Version 5.3.

ICH S7B, 2005. The Non-Clinical Evaluation of the Potential for Delayed Ventricular Repolarization (QT Interval Prolongation) by Human Pharmaceuticals. Current *Step* 4 version dated 12 May 2005.

Ingber, D., 2015. Reducing animal testing one chip at a time. <http://www.dddmag.com/articles/2015/02/reducing-animal-testing-one-chip-time?et_cid = 4408194&et_rid = 45522067&type = cta/> (Accessed 10 February 2015).

Isfort, R.J., Kerckaert, G.A., LeBoeuf, R.A., 1996. Comparison of the standard and reduced pH Syrian Hamster Embryo (SHE) cell *in vitro* transformation assays in predicting the carcinogenic potential of chemicals. Mutat. Res. 1, 11–63.

Jia, W., Chen, M.J., Su, M.M., 2006. Metabonomic study of aristolochic acid-induced nephrotoxicity in rats. J. Proteom. Res. 5, 995–1002.

Jones, R.W., Stout, M.G., Reich, H., 1964. Cytotoxic activities of certain flavonoids against zebra-fish embryos. Cancer Chemother. Rep. 34, 19–20.

Kang, I.K., Kim, G.J., Kwon, O.H., 2004. Co-culture of hepatocytes and fibroblasts by micropatterned immobilization of [beta]-galactose derivatives. Biomaterials 25, 4225–4232.

Kim, S.S., Utsunomiya, H., Koski, J.A., 1998. Survival and function of hepatocytes on a novel three-dimensional synthetic biodegradable polymer scaffold with an intrinsic network of channels. Ann. Surg. 228, 8–13.

Kirkland, D., Aardema, M., Henderson, L., Muller, L., 2005. Evaluation of the ability of a battery of three *in vitro* genotoxicity tests to discriminate rodent carcinogens and non-carcinogens I. Sensitivity, specificity and relative predictivity. Mut. Res. 584, 1–256.

Kirkland, D., Reeve, L., Gatehouse, D., 2011. A core *in vitro* genotoxicity battery comprising the Ames test plus the *in vitro* micronucleus test is sufficient to detect rodent carcinogens and *in vivo* genotoxins. Mut. Res. 721, 27–73.

Krause, P., Saghatolislam, F., Koenig, S., 2009. Maintaining hepatocyte differentiation *in vitro* through co-culture with hepatic stellate cells. In Vitro Cell. Dev. Biol. Anim. 45, 205–212.

Krishna, G., 2010. Alternatives to animal experimentation in research. In: Jagadeesh (Ed.), Biomedical Research—From Ideation to Publication. Wolters Kluwer/Lippincott, Williams & Wilkins, pp. 155–177.

Krishna, G., Nath, J., Ong, T., 1984. Preparation of mouse bone marrow primary cultures for sister chromatid exchange and chromosomal aberration studies. J. Tissue Cult. Methods 9, 193–198.

Krishna, G., Nath, J., Ong, T., 1985. Preparation of Chinese hamster bone marrow and spleen primary cell cultures for sister chromatid exchange and chromosomal aberration studies. J. Tissue Cult. Methods 9, 199–203.

Krishna, G., Nath, J., Ong, T., 1986. Murine bone marrow culture system for cytogenetic analysis. Mut. Res. 164, 91–99.

Krishna, G., Kropko, M., Theiss, J.C., 1989. The use of cytokinesis-block method for the analysis of micronuclei in V79 Chinese hamster lung cell line: results on mitomycin C and cyclophosphamide. Mut. Res. 222, 63–69.

Krishna, G., Fiedler, R., Theiss, J.C., 1992. Simultaneous analysis of chromosome damage and aneuploidy in cytokinesis-blocked V79 Chinese hamster lung cells using an antikinetochore antibody. Mut. Res. 282, 79–88.

Krishna, G., Urda, G., Lalwani, N.D., 1995a. Immunofluorescent and confocal laser cytometric analyses of centromeres in V79 cells. Mutat. Res. 328, 1–9.

Krishna, G., Urda, G., Tefera, W., 1995b. Simultaneous evaluation of dexamethasone-induced apoptosis and micronuclei in rat primary spleen cell cultures. Mutat. Res. 332, 1–8.

Krishna, G., Goel, S., Krishna, K., Gupta, R.C., 2014. Alternative animal toxicity testing and biomarkers. In: Gupta, R.C. (Ed.), Biomarkers in Toxicology. Academic Press/Elsevier, pp. 129–147.

Kussmann, M., Raymond, F., Affolter, M., 2006. OMICS-driven biomarker discovery in nutrition and health. J. Biotechnol. 124, 758–787.

Langerholc, T., Maragkoudakis, P.A., Wollgast, J., 2011. Novel and established intestinal cell line models—an indispensable tool in food science and nutrition. Trends Food Sci. Technol. 22, S11–S20.

Lather, A., Valecha, R., Sharma, K., 2011. Worldwide potential of plants causing teratogenicityan-overview. Spatula DD 1, 101–106.

LeCluyse, E.L., 2001. Human hepatocyte culture systems for the *in vitro* evaluation of cytochrome P450 expression and regulation. Eur. J. Pharmaceut. Sci. 13, 343–368.

LeCluyse, E.L., Bullock, P.L., Parkinson, A., 1996. Strategies for restoration and maintenance of normal hepatic structure and function in long-term cultures of rat hepatocytes. Adv. Drug Delivery Rev. 22, 133–186.

LeCluyse, E.L., Ahlgren-Beckendorf, J.A., Carroll, K., 2000. Regulation of glutathione S-transferase enzymes in primary cultures of rat hepatocytes maintained under various matrix configurations. Toxicol. In Vitro 14, 101–115.

Liechty, M.C., Scalzi, J.M., Sims, K.R., 1998. Analysis of largeand small colony L5178Y tk − / − mouse lymphoma mutants by loss of heterozygosity (LOH) and by whole chromosome 11 painting: detection of recombination. Mutagenesis 13, 461–474.

Liu, X., Chism, J.P., LeCluyse, E.L., 1999. Correlation of biliary excretion in sandwich-cultured rat hepatocytes and *in vivo* in rats. Drug Metab. Dispos. 27, 637–644.

Lohr, J.W., Willsky, G.R., Acara, M.A., 1998. Renal drug metabolism. Pharmacol. Rev. 50 (1), 107–141.

Lubberstedt, M., Muller-Vieira, U., Mayer, M., 2011. HepaRG human hepatic cell line utility as a surrogate for primary human hepatocytes in drug metabolism assessment *in vitro*. J. Pharmacol. Toxicol. Methods 63, 59–68.

Marion, M.J., Hantz, O., Durantel, D., 2010. The HepaRG cell line: biological properties and relevance as a tool for cell biology, drug metabolism, and virology studies. Methods Mol. Biol. 640, 261–272.

Maron, D.M., Ames, B.N., 1983. Revised methods for the Salmonella mutagenicity test. Mutat. Res. 113, 173–215.

Marques, A., Lourenço, H.M., Nunes, M.L., 2011. New tools to assess toxicity, bioaccessibility and uptake of chemical contaminants in meat and seafood. Food Res. Int. 44, 510–522.

Marx-Stoelting, P., Adriaens, E., Ahr, H.J., 2009. A review of the implementation of the embryonic stem cell test (EST). The report and recommendations of an ECVAM/ReProTect workshop. Altern. Lab. Anim. 37, 313–328.

Mendes, O., Chen, J., Gowda, R., 2019. Chapter 9—Safety assessment of herbal food ingredients and nutraceuticals, Nutraceutical and Functional Food Regulations in the United States and Around the World, third ed. Academic Press, pp. 107–120.

Meng, Q., 2010. Three-dimensional culture of hepatocytes for prediction of drug-induced hepatotoxicity. Expert Opin. Drug Metab. Toxicol. 6, 733–746.

Mesnil, M., Fraslin, J.M., Piccoli, C., 1987. Cell contact but not junctional communication (dye coupling) with biliary epithelial cells is required for hepatocytes to maintain differentiated functions. Exp. Cell Res. 173, 524–533.

Mestrovic T., 2015. What are nutraceuticals? <http://www.news-medical.net/health/What-are-Nutraceuticals.aspx/> (Accessed 01 May 2015).

Michalopoulos, G.K., DeFrances, M.C., 1997. Liver regeneration. Science 276, 60–66.

Milligan, C.J., Möller, C., 2013. Automated planar patch-clamp. Methods Mol. Biol. 998, 171–187.

Mingoia, R.T., Nabb, D.L., Yang, C.H., 2007. Primary culture of rat hepatocytes in 96-well plates: effects of extracellular matrix configuration on cytochromeP450 enzyme activity and inducibility, and its application in *in vitro* cytotoxicity screening. Toxicol. In Vitro 21, 165–173.

Minigh, J.L., Valentovic, M.A., 2003. Characterization of myoglobin toxicity in renal cortical slices from Fischer 344 rats. Toxicology 184, 113–123.

Miyoshi, H., Ookawa, K., Ohshima, N., 1998. Hepatocyte culture utilizing porous polyvinyl formal resin maintains long-term stable albumin secretion activity. J. Biomat. Sci. Polym. Ed. 9, 227–237.

Morita, T., Nagaki, T., Fukuda, I., 1992. Clastogenicity of low pH to various cultured mammalian cells. Mutat. Res. 268, 297–305.

Morrissey, K.M., Stocker, S.L., Wittwer, M.B., 2013. Renal transporters in drug development. Annu. Rev. Pharmacol. Toxicol. 53, 503–529.

Mouche, I., Malesic, L., Gillardeaux, O., 2011. FETAX assay for evaluation of developmental toxicity. Methods Mol. Biol. 691, 257–269.

Napoli, E.M., Curcuruto, G., Ruberto, G., 2010. Screening the essential oil composition of wild Sicilian fennel. Biochem. Syst. Ecol. 38, 213–223.

Narayanan, K., Schumacher, K.M., Tasnim, F., 2013. Human embryonic stem cells differentiate into functional renal proximal tubular-like cells. Kidney Int. 83 (4), 593–603.

Nath, J., Krishna, G., 1997. Fundamental and applied genetic toxicology. In: Craig, C.R., Stitzel, R.E. (Eds.), Modern pharmacology with clinical applications. Little, Brown and Company, pp. 69–78.

Nelson, K.F., Acosta, D., Bruckner, J.V., 1982. Long-term maintenance and induction of cytochrome P-450 in primary cultures of rat hepatocytes. Biochem. Pharm. 31, 2211–2214.

Nemoto, N., Sakurai, J., 1992. Differences in regulation of gene expression between Cyp1a-1 and Cyp1a-2 in adult mouse hepatocytes in primary culture. Carcinogenesis 13, 2249–2254.

Nemoto, N., Sakurai, J., 1993. Activation of Cyp1a1 and Cyp1a2 genes in adult mouse hepatocytes in primary culture. Jpn. J. Cancer Res. 84, 272–278.

Nemoto, N., Sakurai, J., Funae, Y., 1995. Maintenance of phenobarbital-inducible Cyp2b gene expression in C57BL/6 mouse hepatocytes in primary culture as spheroids. Arch. Biochem. Biophys. 316, 362–369.

Nissen, L., Chingwaru, W., Sgorbati, B., 2009. Gut health promoting activity of new putative probiotic/protective *Lactobacillus* spp. strains: a functional study in the small intestinal cell model. Int. J. Food Microbiol. 135, 288–294.

Nohynek, G.J., Kirkland, D., Marzin, D., 2004. An assessment of the genotoxicity and human health risk of topical use of kojic acid [5-hydroxy-2-(hydroxymethyl)-4H-pyran-4-one]. Food Chem. Toxicol. 42, 93–105.

Novik, E., Maguire, T.J., Chao, P., 2010. A microfluidic hepatic coculture platform for cell-based drug metabolism studies. Biochem. Pharmacol. 79, 1036–1044.

NRC, 2007. Toxicity Testing in the 21st Century—A Vision and a Strategy. The National Academies Press, Washington, DC.

Oberemm, A., Onyon, L., Gundert-Remy, U., 2005. How can toxicogenomics inform risk assessment? Toxicol. Appl. Pharmacol. 207, 592–598.

OECD, 2015. <http://www.oecd-ilibrary.org/environment/oecd-guidelines-for-the-testing-of-chemicals-section-4-health-effects_20745788/> (Accessed 13 May 2015).

Ootani, A., Li, X., Sangiorgi, E., 2009. Sustained *in vitro* intestinal epithelial culture within a Wnt-dependent stem cell niche. Nat. Med. 15, 701–706.

Peracaula, R., Barrabes, S., Sarrats, A., 2008. Altered glycosylation in tumours focused to cancer diagnosis. Dis. Markers 25, 207–218.

Piersma, A.H., Genschow, E., Verhoef, A., 2004. Validation of the post-implantation rat whole-embryo culture test in the international ECVAM validation study on three *in vitro* embryotoxicity tests. Altern. Lab. Anim. 32, 275–307.

Polonchuk, L., 2012. Toward a new gold standard for early safety: automated temperature-controlled hERG test on the patchliner. Front. Pharmacol. 3, 3.

Prakash, A.S., Pereira, T.N., Reilly, P.E., 1999. Pyrrolizidine alkaloids in human diet. Mut. Res. 443, 53–67.

Ramos, A., Rivero, R., Victoria, M.C., 2001. Assessment of mutagenicity in *Parthenium hysterophorus* L. J. Ethnopharmacol. 77, 25–30.

Rao, K.S., 2017. Safety assessment of nutraceuticals. BEMS Rep. 3 (2), 70–72.

Rao, V.S.N., Cavalcanti, B.C., Costa-lotufo, L.V., 2006. Genotoxicity evaluation of kaurenoic acid, a bioactive diterpenoid present in Copaiba oil. Food Chem. Toxicol. 44, 388–392.

ReproTect, 2015. <http://www.reprotect.eu/index.php?id = 7000/> (Accessed 25 May 15).

Richards, S.M., Cole, S.E., 2006. A toxicity and hazard assessment of fourteen pharmaceuticals to *Xenopus laevis* larvae. Ecotoxicology 15, 647–656.

Ries, K., Krause, P., Solsbacher, M., 2000. Elevated expression of hormone-regulated rat hepatocyte functions in a new serum-free hepatocyte-stromal cell coculture model. In Vitro Cell. Dev. Biol. Anim. 36, 502–512.

Roche, O., Trube, G., Zuegge, J., 2002. A virtual screening method for prediction of the hERG potassium channel liability of compound libraries. Chembiochem 3, 455–459.

Rodriguez-Barbero, A., L'Azou, B., Cambar, J., 2000. Potential use of isolated glomeruli and cultured mesangial cells as *in vitro* models to assess nephrotoxicity. Cell Biol. Toxicol. 16 (3), 145–153.

Rouse, R.L., Zhang, J., Stewart, S.R., 2011. Comparative profile of commercially available urinary biomarkers in pre-clinical drug-induced kidney injury and recovery in rats. Kidney Int. 79, 1186–1197.

Ruegg, C.E., 1994. Preparation of precision-cut renal slices and renal proximal tubular fragments for evaluating segment-specific nephrotoxicity. J. Pharmacol. Toxicol. Methods 31, 125–133.

Sakairi, T., Abe, Y., Kajiyama, H., 2010. Conditionally immortalized human podocyte cell lines established from urine. Am. J. Physiol. 298 (3), F557–F567.

Sanchez-Lamar, A., Fuentes, J.L., Fonseca, G., 2002. Assessment of the potential genotoxic risk of Phyllantus orbicularis HBK aqueous extract using *in vitro* and *in vivo* assays. Toxicol. Lett. 136, 87–96.

Sanguinetti, M.C., Tristani-Firouzi, M., 2006. hERG potassium channels and cardiac arrhythmia. Nature 440 (7083), 463–469.

Santos, F.V., Tubaldini, F.R., Cólus, I.M.S., 2002. Mutagenicity of *Mouriri pusa* gardner and *Mouriri elliptica* Martius. Toxicol. Lett. 136, 87–96.

Sathyamoorthy, N., Wang, T.T., Phang, J.M., 1994. Stimulation of pS2 expression by diet-derived compounds. Cancer Res. 54, 957–961.

Schroeder, K., Bremm, K.D., Alépée, N., 2011. Report from the EPAA workshop: *in vitro* ADME in safety testing used by EPAA industry sectors. Toxicol. In Vitro 25, 589–604.

Seidle, T., Stephens, M.L., 2009. Bringing toxicology into the 21st century: a global call to action. Toxicol. In Vitro 23, 1576–1579.

Selderslaghs, I.W., Van Rompay, A.R., De Coen, W., 2009. Development of a screening assay to identify teratogenic and embryotoxic chemicals using the zebrafish embryo. Reprod. Toxicol. 28, 308–320.

Shafritz, D.A., Oertel, M., Dabeva, M.D., 2009. The liver: biology and pathobiology. In: Arias, H.J.A.I.M., Boyer, J.L., Cohen, D.E., Fausto, N., Shafritz, D.

A., Wolkoff, A.W. (Eds.), The Liver: Biology and Pathobiology. John Wiley & Sons, pp. 577–595.

Shanley, P.F., Brezis, M., Spokes, K., 1986. Hypoxic injury in the proximal tubule of the isolated perfused rat kidney. Kidney Int. 29, 1021–1032.

Shaw, G., Morse, S., Ararat, M., 2002. Preferential transformation of human neuronal cells by human adenoviruses and the origin of HEK 293 cells. FASEB J. 16 (8), 869–871.

Shaw, P.J., Ganey, P.E., Roth, R.A., 2010. Idiosyncratic drug-induced liver injury and the role of inflammatory stress with an emphasis on an animal model of trovafloxacin hepatotoxicity. Toxicol. Sci. 118, 7–18.

Sieber, M., Wagner, S., Rached, E., 2009. Metabonomic study of ochratoxin a toxicity in rats after repeated administration: phenotypic anchoring enhances the ability for biomarker discovery. Chem. Res. Toxicol. 22, 1221–1231.

Sin, A., Chin, K.C., Jamil, M.F., 2004. The design and fabrication of three-chamber microscale cell culture analog devices with integrated dissolved oxygen sensors. Biotechnol. Progr. 20, 338–345.

Singh, N.P., McCoy, M.T., Tice, R.R., 1988. A simple technique for quantitation of low levels of DNA damage in individual cells. Exp. Cell Res. 175, 184–191.

Sirenko, O., Cromwell, E.F., Crittenden, C., 2013. Assessment of beating parameters in human induced pluripotent stem cells enables quantitative *in vitro* screening for cardiotoxicity. Toxicol. Appl. Pharmacol. 273 (3), 500–507.

Sivaraman, A., Leach, J.K., Townsend, S., 2005. A microscale *in vitro* physiological model of the liver: predictive screens for drug metabolism and enzyme induction. Curr. Drug Metab. 6, 569–591.

Snyder, R.D., Smith, M.D., 2005. Computational prediction of genotoxicity: room for improvement. Drug Discov. Today 10, 1119–1124.

Snyder, R.D., Ewing, D.E., Hendry, L.B., 2004. Evaluation of DNA intercalation potential of pharmaceuticals and other chemicals by cell-based and three-dimensional computational approaches. Environ. Mol. Mutagen. 44, 163–173.

Song, B., Smink, A.M., Jones, C.V., 2012. The directed differentiation of human iPS cells into kidney podocytes. PLoS One 7 (9), e46453.

Soni, M.G., Thurmond, S.T., Miller, E.R., 2010. FORUM safety of vitamins and minerals: controversies and perspective. Toxicol. Sci. 118 (2), 348–355.

SOT, 2015. Paracelsus statement. <https://www.toxicology.org/AI/EO/principl.asp/> (Accessed 15 May 15).

Speit, G., Brendler-Schwaab, S., Hartmann, A., 2005. The *in vivo* comet assay: use and status in genotoxicity testing. Mutagenesis 20, 245–254.

Spielmann, H., 2005. Predicting the risk of developmental toxicity from *in vitro* assays. Toxicol. Appl. Pharmacol. 207, 375–380.

Stauffer, J.E., 1999. Nutraceuticals. Cereal Foods World 44, 115–117.

Stoelzle, S., Haythornthwaite, A., Kettenhofen, R., 2011. Automated patch clamp on mESC-derived cardiomyocytes for cardiotoxicity prediction. J. Biomol. Screen. 16, 910–916.

Stummann, T.C., Hareng, L., Bremer, S., 2008. Embryotoxicity hazard assessment of cadmium and arsenic compounds using embryonic stem cells. Toxicology 252, 118–122.

Taguchi, K., Matsushita, M., Takahashi, M., 1996. Development of a bioartificial liver with sandwiched-cultured hepatocytes between two collagen gel layers. Artif. Organs 20, 178–185.

Takeuchi, S., Hiromura, K., Tomioka, M., 2010. The immunosuppressive drug mizoribine directly prevents podocyte injuryin puromycin aminonucleoside nephrosis. Nephron Exp. Nephrol. 116 (1), e3–e10.

Talamini, M.A., McCluskey, M.P., Buchman, T.G., 1998. Expression of alpha2-macroglobulin by the interaction between hepatocytes and endothelial cells in coculture. Am. J. Physiol. 275, R203–R211.

Thomas, R.J., 1975. The toxicologic and teratologic effects of delta-9-tetrahydrocannabinol in the zebrafish embryo. Toxicol. Appl. Pharmacol. 32, 184–190.

Tice, R.R., Agurell, E., Anderson, D., Burlinson, B., 2000. Single cell gel/comet assay: guidelines for *in vitro* and *in vivo* genetic toxicology testing. Environ. Mol. Mutagen 35, 206–221.

Tirmenstein, M.A., Nicholls-Grzemski, F.A., Schmittgen, T.D., 2000. Characterization of nitric oxide production following isolation of rat hepatocytes. Toxicol. Sci. 53, 56–62.

Ton, C., Lin, Y., Willett, C., 2006. Zebrafish as a model for developmental neurotoxicity testing. Birth Defects Res. A Clin. Mol. Teratol. 76, 553–567.

Trapecar, M., Cencic, A., 2012. Application of gut cell models for toxicological and bioactivity studies of functional and novel foods. Foods 1, 40–51.

Trudeau, M.C., Warmke, J.W., Ganetzky, B., 1995. HERG, a human inward rectifier in the voltage-gated potassium channel family. Science 269 (5220), 92–95.

Tukov, F.F., Maddox, J.F., Amacher, D.E., 2006. Modeling inflammation-drug interactions *in vitro*: a rat Kupffer cell-hepatocyte coculture system. Toxicol. In Vitro 20, 1488–1499.

Ulrich-Merzenich, G., Zeitler, H., Jobst, D., 2007a. Application of the "-Omic-" technologies in phytomedicine. Phytomedicine 14, 70–82.

Ulrich-Merzenich, G., Jobst, D., Zeitler, H., 2007b. Measurement of synergistic effects of a phytopharmaceutical by microarray-analysis. Planta Med. 73, SL026.

Ursem, C.J., Kruhlak, N.L., Contrera, J.F., 2009. Identification of structure-activity relationships for adverse effects of pharmaceuticals in humans. Part A: use of FDA post-market reports to create a database of hepatobiliary and urinary tract toxicities. Regul. Toxicol. Pharmacol. 54, 1–22.

Valerio-Jr, L.G., 2009. *In silico* toxicology for the pharmaceutical sciences. Toxicol. Appl. Pharmacol. 241, 356–370.

Van, G.M.M., Van, R.I.A., Miller, R.K., 2010. Teratogenic mechanisms of medical drugs. Hum. Reprod. Update 16, 378–394.

Van den Bulck, K., Hill, A., Mesens, N., 2011. Zebrafish developmental toxicity assay: a fishy solution to reproductive toxicity screening, or just a red herring? Reprod. Toxicol. 32, 213–219.

Van Ravenzwaay, B., Cunha, G.C.P., Leibold, E., 2007. The use of metabolomics for the discovery of new biomarkers of effect. Toxicol. Lett. 172, 21–28.

Vickers, A.E., Fisher, R.L., 2004. Organ slices for the evaluation of human drug toxicity. Chemico-Biol. Interact. 150, 87–96.

Vickers, A.E., Rose, K., Fisher, R., 2004. Kidney slices of human and rat to characterize cisplatin-induced injury on cellular pathways and morphology. Toxicol. Pathol. 32, 577–590.

Votano, J.R., Parham, M., Hall, L.H., 2004. Three new consensus QSAR models for the prediction of Ames genotoxicity. Mutagenesis 19, 365–377.

Wang, H., Gao, X., Fukumoto, S., Tademoto, S., 1998. Post-isolation inducible nitric oxide synthase gene expression due to collagenase buffer perfusion and characterization of the gene regulation in primary cultured murine hepatocytes. J. Biochem. 124, 892–899.

Weigt, S., Huebler, N., Strecker, R., 2011. Zebrafish (*Danio rerio*) embryos as a model for testing proteratogens. Toxicology 281, 25–36.

Whitman, M., 2001. Understanding the perceived need for complementary and alternative nutraceuticals: lifestyle issues. Clin. J. Oncol. Nurs. 5, 190–194.

WHO, 2007. WHO Guidelines for Assessing Quality of Herbal Medicines With Reference to Contaminants and Residues. World Health Organization; Geneva. Available from: <http://apps.who.int/medicinedocs/documents/s14878e/s14878e.pdf/>.

Wojcikowski, K., Wohlmuth, H., Johnson, D.W., 2009. An *in vitro* investigation of herbs traditionally used for kidney and urinary system disorders: potential therapeutic and toxic effects. Nephrology 14, 70–79.

Wolters, D.A., Washburn, M.P., Yates, J.R., 2001. An automated multidimensional protein identification technology for shotgun proteomics. Anal. Chem. 73, 5683–5690.

Wu, F.J., Friend, J.R., Hsiao, C.C., 1996. Efficient assembly of rat hepatocyte spheroids for tissue engineering applications. Biotechnol. Bioeng. 50, 404–415.

Xia, S., Liu, H., Zhu, H., 2009. NMR-based metabonomic study on rat's urinary metabolic response to dosage of triptolide. Chin. J. Chem. 27, 751–758.

Yang, C.S., Landau, J.M., Huang, M.T., 2001. Inhibition of carcinogenesis by dietary polyphenolic compounds. Annu. Rev. Nutr. 21, 381–406.

Yates, J.R., 2000. Mass spectrometry. From genomics to proteomics. Trends Genet. 16, 5–8.

Youns, M., Hoheisel, J.D., Efferth, T., 2010. Toxicogenomics for the prediction of toxicity related to herbs from traditional Chinese medicine. Planta Med. 76, 2019–2025.

Zeisel, S.H., 1999. Regulation of nutraceuticals. Science 285, 1853–1855.

Zenga, Y.-X., Zhao, C.-X., Liang, Y.-Z., 2007. Comparative analysis of volatile components from *Clematis* species growing in China. Anal. Chim. Acta 595, 328–339.

Zhang, H., Xu, Y., Kamendulis, L.M., 2000. Morphological transformation by 8-hydroxy-2′-deoxyguanosine in Syrian hamster embryo (SHE) cells. Toxicol. Sci. 56, 303–312.

Zhao, J.Y., Yan, X.Z., Peng, S.Q., 2006. Metabonomics study on nephrotoxicity of *Aristolochia manshuriensis*. Chin. Traditional Herbal Drugs 37, 725–730.

Chapter 30

Mitochondria as a target for safety and toxicity evaluation of nutraceuticals

João Soeiro Teodoro, Ivo F. Machado, Ana Catarina Castela, Anabela Pinto Rolo and Carlos Marques Palmeira

Introduction

Since ancient times, plants, herbs, and other natural products have been used as healing agents. Advances in organic chemistry since the early 19th century have enabled the preparation of numerous synthetic medicines. However, the majority of the medicinal substances available today have their origin in natural compounds. The best known example is aspirin (acetylsalicylic acid), originally derived from the bark of the white willow tree (Mahdi et al., 2006). Other examples include the immunosuppressive cyclosporines (Borel et al., 1990), and the β-Hydroxy β-methylglutaryl-CoA reductase inhibitors, commonly known as statins (Endo, 2004). As such, there is still immense potential for future medicinal compounds to be used, and research on various types of molecules has been conducted for some time now. We focus on various compounds which we believe have tremendous potential for introduction into the market in the near future, and discuss the possible applications of these compounds.

Mitochondria are the eukaryotic cells' powerhouses, as over 90% of all the cells' adenosine triphosphate (ATP) is produced here. However, they also play many other vital roles in cellular function, from calcium homeostasis to biochemical reactions compartmentalization, and to cellular death by programmed apoptotic pathways (for a complete review on mitochondria, see Palmeira et al., 2014). Due to these numerous functions and the relative ease of obtaining large amounts of these organelles, isolated mitochondrial function assessment (and also the study of mitochondrial function while inside the cells) has been, for several decades now, a tested and proved method of toxicological evaluation. But this is just the tip of the iceberg in terms of utility of mitochondria. The functionality of mitochondria is compromised in an enormous array of diseases, and the restoration of mitochondrial function and integrity is a common step toward improvement/cure of these pathologies (Palmeira et al., 2014). As such, it comes as no surprise that the evaluation of mitochondrial function and parameters is a staple of thousands of studies. In the following pages we will discuss various nutraceutical compounds and, where information is available, relate their effects to mitochondrial function.

List of nutraceuticals

Berberine

Berberine (BBR) is isoquinoline flavonoid present in the root, bark, and other structures of plants typically found in traditional Chinese and East Asian medicine (Teodoro et al., 2013). Due to its structure (Fig. 30.1), BBR has strong fluorescence capability, when stimulated with UV light. The first recorded use of BBR-containing plants in a medical setting goes back over 2000 years, with the first use estimated to be over 5000 years ago, and is thought to be due to its antimicrobial activity. Along with medical applications, BBR-enriched plants are also, to this day, used for ornamentation and as a fabric dye, due to its strong yellow color.

Traditional Chinese medicine used BBR as a fungicide and against bacteria and other unicellular pathogenic agents. However, the use of BBR in diarrhea-suffering obese patients led to the discovery of its antiobesity effects. In fact, BBR has been extensively studied for various

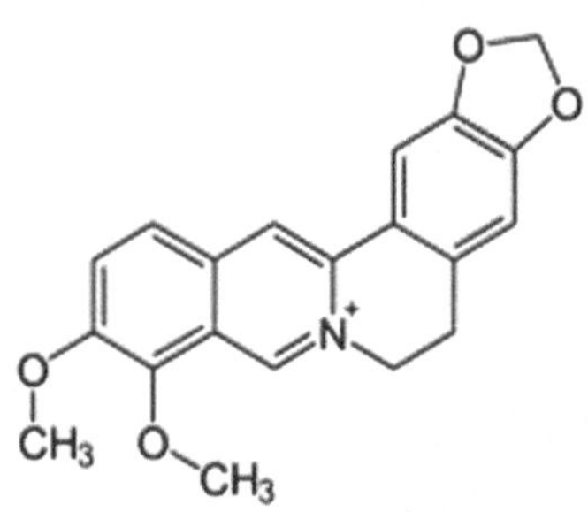

FIGURE 30.1 Structural formula of berberine.

Nutraceuticals. DOI: https://doi.org/10.1016/B978-0-12-821038-3.00030-6

decades, with powerful antiprotozoal, antihypertensive, antibacterial, antiinflammatory, anticholinergic, antiarrhythmic, and even anti-HIV activities (Hwang et al., 2002), to name a few. With such a wide and efficient array of effects, it comes as no surprise that BBR is one of the most interesting and studied nutraceutical compounds in the world. In the next sections, we focus on the effects of BBR on diabetes and obesity, cancer, and neurodegenerative disorders.

Berberine in diabetes and obesity

As mentioned above, BBR's hypoglycemic effects were accidently discovered (Yin et al., 2002), leading to growing interest in this molecule. In fact, BBR was found to be as potent as metformin at regulating glucose levels in obese, high fat-fed rats (HFD) (Gao et al., 1997). An elegant study demonstrated that BBR was capable of increasing liver hepatoma HepG2 cells' glucose intake in an insulin-independent fashion (Yin et al., 2002). Not only that, BBR also decreased hyperglycemia in animal (Teodoro et al., 2013; Wang et al., 2011) and human diabetic patients (Zhang et al., 2008, 2010a). BBR's effects appear to be multifactorial, since BBR is capable of increasing glycolytic rates (Yin et al., 2008), decreasing insulin resistance (Gomes et al., 2012; Teodoro et al., 2013; Kong et al., 2009; Liu et al., 2010b), promoting incretin-like action (increased release of insulin, decreased release of glucagon) (Lu et al., 2009), normalizing insulin levels (Teodoro et al., 2013), and increasing insulin receptor (IR) expression (Zhang et al., 2010a). As a result, the general marker of a (pre)diabetic condition (elevated glycated hemoglobin) is also reduced by BBR (Teodoro et al., 2013). These insulin-sensitizing capacities of BBR have been partially attributed to a heightened insulin pathway activation (Liu et al., 2010c), coupled with increased insulin release by up-expression of HNF-4α and increased glycolytic activity in pancreatic ß-cells (Wang et al., 2008). On the other hand, BBR also decreases intestinal absorption of sugars, through a variety of mechanisms (Liu et al., 2010a; Al-Masri et al., 2009). It is also interesting to note that BBR is also efficient in treating metabolic alterations such as dysglycemia in combination with other nutraceuticals, however, the significance of this combination is complicated to fully understand given the lack of proper controls in the study (Derosa et al., 2020).

As with polysaccharides, excess availability of lipids is also a major contributing factor to obesity and diabetes. As such, it comes as no surprise that BBR's antiobesity and antidiabetic effects also heavily circulate around altered lipid metabolism. In fact, the levels of free fatty acids (FFAs) in circulation in diabetic patients are reduced by BBR (Gu et al., 2010). We also have previously demonstrated that BBR not only reduces overall adiposity, but also decreases circulating triglycerides (TG), cholesterol, phospholipids, the levels of leptin, and changes the overall composition of deposited fatty acids, which are clearly shifted from a majority of saturated to unsaturated FFA (Teodoro et al., 2013). Low-density lipoprotein (LDL) cholesterol is also actively reduced by BBR supplementation, as it can lower intestinal cholesterol absorption (Wang et al., 2010a) and promotes transcription and stabilization of the mRNA of the LDL receptor (Li et al., 2009; Abidi et al., 2005). Obviously, the reduction in circulating nutrients causes implications for circulatory and heart disease prevention (Derosa et al., 2012). In fact, BBR supplementation leads to the activation of the AMP-activated protein kinase, AMPK (Turner et al., 2008), whose activation is a well-known step in improvement of heart disease (Shirwany and Zou, 2010).

In our studies, we have also demonstrated that mitochondria from the liver (Teodoro et al., 2013) and skeletal muscle (Gomes et al., 2012) of HFD rats have severely compromised oxidative and respiration capacities, which BBR was capable of recovering. This was accompanied by a normalization of various mitochondrial parameters (ATP levels, membrane potential, individual oxidative phosphorylation complex activities, amongst others), in a mechanism that involves the activation of various members of the family of Nicotinamide adenine dinucleotide, oxidized (NAD^+)-dependent deacetylases known as sirtuins (SirT). These enzymes are powerful metabolic sensors, whose increased activity has been shown to be beneficial in a wide range of pathologies (Guarente, 2007). Accordingly, we have shown that the actions of BBR on improving mitochondrial function relied on increased sirtuin 1 (SirT1) (Gomes et al., 2012) and SirT3 (Teodoro et al., 2013) activities. This was possible due to the increased NAD^+/Nicotinamide adenine dinucleotide, reduced ratio generated by BBR, coupled with increased NAD^+ generation (by elevated levels of the rate-limiting enzyme of the NAD^+ biosynthesis pathway, nicotinamide phosphoribosyl transferase) and as above, increased activity of AMP-activated protein kinase (AMPK), another metabolic sensor whose function is also well studied (Gomes et al., 2012). This improvement of mitochondrial function can also be aided by the inhibition of the forkhead box protein O1 (FOXO1) by BBR, which, amongst other functions, reduces gluconeogenesis and glycogenolysis, and links the insulin pathway to mitochondrial function. In fact, the removal of FOXO1 on insulin-resistant mice leads to normalized mitochondrial function and levels (Cheng et al., 2009). Conversely, the addition of berberine directly to isolated rat liver mitochondria demonstrated a strikingly different pattern, as BBR (in a dose-dependent fashion) caused a severe compromise to mitochondrial function, an effect which seems to be based on the respiratory chain's complex I activity inhibition (Pereira et al., 2007). As such, it is still not clear if the effects of BBR directly target mitochondria, or if this duality of effects reported above is dependent on BBR's activity on extramitochondrial effectors. A recent work has hypothesized that BBR might cause mitochondrial function compromise, thus reducing ATP levels and causing the activation of AMPK, which will then lead to improved metabolic status and, consequently, reverting mitochondrial activity back to normal (or even improving it) (Yin et al., 2012).

BBR is also a powerful inhibitor of hepatic injury, preventing damage caused by known hepatic toxicants (Domitrović et al., 2011; Feng et al., 2010) and steatosis and steatohepatitis (Liu et al., 2013). Not just the liver, BBR is also a cardioprotective compound with experimental and clinical proof (Zeng et al., 2003).

It is also interesting to note that some of BBR's effects might hinge on the modulation of the intestinal microbiome's activity, and some studies have already shown that BBR has a potent effect of metabolism that relies on microbiome alterations, such as improving glycometabolism (Li et al., 2020; Pan et al., 2019) or decreasing inflammation and altering the microbiome, leading to gene expression alterations and improved metabolic health (Tian et al., 2019; Zhang et al., 2012, 2019).

Berberine in cancer

BBR is also a powerful antineoplastic agent (Sun et al., 2009). Experiments using a wide array of cell lines and tumors have shown that BBR is able to arrest growth, carcinogenesis, and metastization in virtually all types of cancers (Ortiz et al., 2014). These effects range from radiosensitization of modified cells, but not of normal ones, by modulating the levels of anti- and proapoptotic proteins, amongst others (Khan et al., 2010). As such, BBR has powerful effects in crucial pathways for cancer development, such as the mitogen-activated protein kinases (MAPK) and the NF-κB pathways (Lin et al., 2004). Surprisingly, BBR can also protect normal cells from the nefarious side effects of radiotherapy (Liu et al., 2008).

BBR appears to bind to molecular DNA and several other proteins involved in cancerous processes, such as telomerase, topoisomerase I, p53, amongst others (Ortiz et al., 2014). All this results in DNA strand breaks, cell cycle arrest, and production of proapoptotic factors. Curiously, these proapoptotic effects of BBR appear to be p53 dependent, as its removal hinders BBR's anticancerous effects (Zhang et al., 2010b). As a result, BBR, in particular when associated with photodynamic therapy, promoted an autophagic response and cellular apoptosis due to reactive oxygen species (ROS) generation (Lopes et al., 2020). BBR has been reported to induce cell cycle arrest at both G_0/G_1, G_2, and also at G_2/M phases, depending on cell type and dose used (Ortiz et al., 2014). In fact, the differences in BBR being protective and apoptosis-inducing might depend on the used dose. A similar effect is produced in mitochondrial function, as low doses of BBR cause improved mitochondrial function (Gomes et al., 2012; Teodoro et al., 2013), while higher doses cause mitochondrial complex I dysfunction (Diogo et al., 2011), which might also explain some of the anticancerous properties (Pereira et al., 2007). Currently, these types of dose-dependent effects on mitochondrial function are known as mitohormetic (Palmeira et al., 2019). In fact, BBR has been identified as a mitohormetic compound (Zhu et al., 2020). The levels of proteins involved in cell cycle regulation, DNA integrity validation, oxidative stress, autophagy, and tumor growth and metastization are also altered by BBR exposure. Such is the case with DNA damage-inducible transcript 3, cyclooxygenases (COX-1 and 2), myeloid cell leukemia (Mcl-1), STAT3, nucleophosmin/B23, telomerase, Wnt and ß-catenin, LC3, the death-domain-associated protein, AMPK, amongst many others (reviewed in Ortiz et al., 2014). On a related note, BBR has also been recently demonstrated to interact with microRNAs (miR), such as miR-21, contributing to diminished cellular proliferation and increased apoptosis (Lo et al., 2013). Finally, BBR is also an effective potentiator of typically used chemo- and radiotherapeutic agents, significantly increasing their cytotoxic effect to cancerous cells (Ortiz et al., 2014). In conclusion, BBR and derivatives could provide a powerful tool in combating virtually all types of cancer.

Berberine in neurodegenerative diseases

BBR has also been studied as a potential antidepressant (Kulkarni and Dhir, 2009) and as a therapeutic agent for neurodegenerative disorders (Kulkarni and Dhir, 2010). A common model of neurodegeneration (exposure to aluminum) had its complications (memory loss, hippocampal cell loss, oxidative stress, neurotransmitter levels decrease) were prevented by BBR (Zhang et al., 2009). On a similar way, BBR prevented stroke damage both in in vivo and in vitro models (Zhou et al., 2008). Even the unpurified plant extracts (though the main component was BBR) of the Giloy plant (*Tinospora cordifolia*) were powerful enough to counter an established in vivo model of Parkinson disease (Kosaraju et al., 2014). Interestingly, these extracts were able to not only normalize dopamine levels, but they also reduced oxidative stress markers and increased mitochondrial activity. In particular, complex I activity was markedly elevated upon extract exposure, once more illustrating the subtleties of utilized dosages and the care required for data analysis.

BBR is also a potential therapeutic agent for Alzheimer disease (Ji and Shen, 2011). This appears to rest on the various effects of BBR on the brain, ranging from decreasing oxidative stress, inhibiting acetylcholinesterase, butyrylcholinesterase (Jung et al., 2009), and monoamine oxidases A (Kong et al., 2001) and B (Castillo et al., 2005), while reducing amyloid-ß (Aß) peptide levels (Asai et al., 2007) and cholesterol (Simons et al., 2001; Teodoro et al., 2013), all actions that have been shown to be beneficial in preventing the onset and progression of Alzheimer's. Other factors might also be in play when considering how BBR can be a potent therapeutic agent for Alzheimer's (Ji and Shen, 2011).

We have recently conducted a study on the effects of BBR as a preventer and reverser of Parkinson disease (unpublished data). Our data show that, in both an in vivo and in vitro model of Parkinson's [using the neurotoxic agent 6-hydroxydopamine, (6-OHDA)], there is a significant protection of dopaminergic neurons against 6-OHDA toxicity. In fact, animals pretreated with BBR had significantly higher test scores on behavioral, morphological, and neurochemical assays when compared with control 6-OHDA littermates. The assays on cultured differentiated dopaminergic neurons point to the effects of BBR on mitochondrial biogenesis, potentiating the number of functional mitochondria, thus preventing one of the most important hallmarks of parkinsonism. Unfortunately, these promising results were not reproducible in animals with already established Parkinson's, which is expectable since mitochondria from the neurons of these animals had a damaged mitochondrial population, which renders the production of new mitochondria (from the templates of damaged old mitochondria) an ineffective process. Fig. 30.2 summarizes the discussed BBR effects on several mitochondrial-related pathologies.

Resveratrol

Resveratrol (RES) is a highly studied stilbenoid phenol, produced by several plants (most predominant in the skin of berries such as grapes, blueberries, and raspberries) with promising and potent effects on various aspects of human health (Frémont, 2000) (Fig. 30.3). It is found in both *cis* and *trans* conformations, of which the latter is the most common.

For over 20 years, RES's presence in wine and grapes was acknowledged as a potential (at least partial) explanation for the "French paradox," that is, a low incidence of cardiovascular disorders, obesity, and diabetes despite a diet rich in saturated fats (Novelle et al., 2015). Since then, countless studies have implicated RES in protection against oxidative stress, inflammation, and cancer, and as a caloric restriction mimetic, with the associated benefits of reduced obesity and associated conditions and extended lifespan (Baur and Sinclair, 2006). It also shows promising effects in some skin diseases and in modulating the effects of exercise (reviewed in Novelle et al., 2015). In the following sections, we focus on the effects of RES on lifespan, obesity and diabetes, cancer, and neurodegeneration.

Resveratrol and lifespan extension

RES has been extensively and definitively demonstrated to extend the lifespan of various organisms, from yeast (*Saccharomyces cerevisiae*) to the roundworm (*Caenorhabditis elegans*), and to the fruit fly (*Drosophila melanogaster*) (Agarwal and Baur, 2011). However, in higher organisms, no lifespan extension effects were found in standard diet ad libitum-fed animals, despite beneficial effects on age-related diseases, such as type II diabetes, cardiovascular, and neurodegenerative disorders, and even cancer (Pearson et al., 2008; Strong et al., 2013). However, this appears not to be an absolute truth, since mice fed an HFD supplemented with RES had enhanced lifespan (Baur et al., 2006). As such, it appears that RES effects might be dependent on the status of the organism (e.g., its metabolic condition) in order for effects to take place, which means that RES might not be a "magic" treatment for everyone. Calorie restriction, which is considered as a process that extends lifespan in humans, is yet to be scientifically proven. However, some data indicate that this might be true, as humans practicing long-term calorie restriction without malnutrition have improved markers of aging, such as cardiovascular alterations, skeletal muscle efficiency, and reduced core body temperature (reviewed in Rizza et al., 2014), indicating that perhaps calorie restriction mimetics (such as RES) might contribute to extended lifespan in healthy humans. In accordance, RES use triggers a similar pattern of gene expression with calorie restriction, which might account for some of the effects noticed in some studies (Novelle et al., 2015).

Resveratrol in obesity and diabetes

In HFD-fed mice, RES supplementation increased insulin sensitivity (Baur et al., 2006). However, as with aging, there is a dual behavior of RES in regard to weight loss, as low doses of RES cause HFD-fed mice to gain weight (Pearson et al., 2008), but high doses cause weight loss (Lagouge et al., 2006), once more demonstrating that "one size fits all" treatment with RES is not adequate and in fact might be harmful. Accordingly, insulin sensitivity was also only improved in diabetic patients (Liu et al., 2014a). Nevertheless, RES increases fatty acid oxidation and reduces blood pressure and cardiac hypertrophy (Dolinsky et al., 2013). Similarly, lower glucose levels, improved ß-cell function and increased insulin sensitivity were achieved with RES in nonhuman primates (Jimenez-Gomez et al., 2013) and in diabetic humans (Elliott et al.,

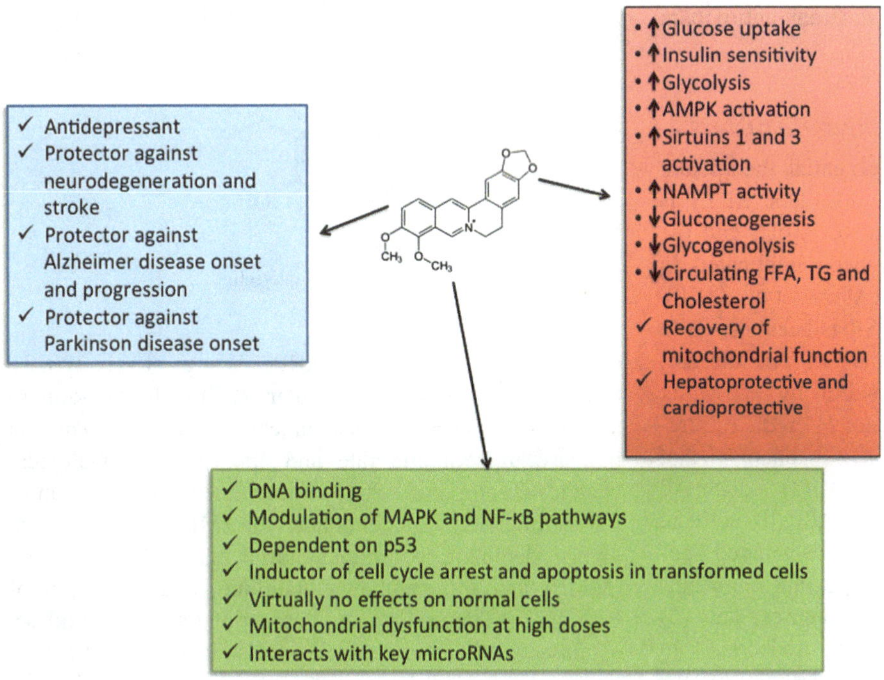

FIGURE 30.2 The effects of berberine on neurodegeneration, cancer, obesity, and diabetes.

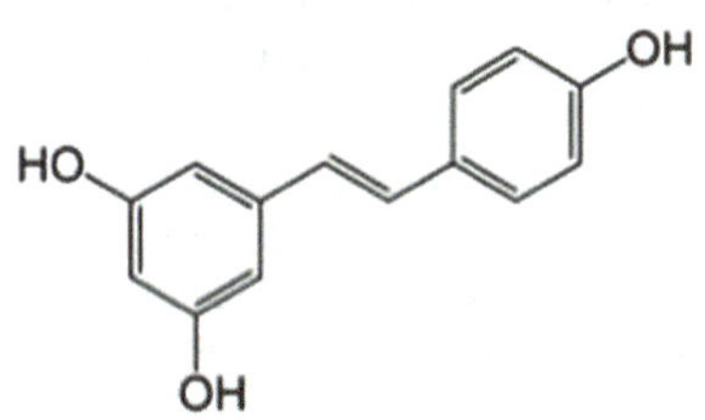

FIGURE 30.3 Structural formula of resveratrol.

2009). These effects occur synergistically with current antidiabetic treatments (Novelle et al., 2015). Furthermore, activation of SirT1 with RES causes activation of the peroxisome proliferation-activated receptor γ (PPARγ) co-activator 1α (PGC-1α), a known master regulator of mitochondrial homeostasis and biogenesis, which is accompanied by a reduction in ROS generation (Lagouge et al., 2006).

Typically associated with obesity is low-grade inflammation, in part due to elevated proinflammatory cytokine production by lipid-engorged adipocytes. RES supplementation was able to reduce adipocyte size in HFD-fed animals, while decreasing several classes of inflammatory markers (Jimenez-Gomez et al., 2013; Novelle et al., 2015). Hepatocellular damage in a condition commonly associated with obesity [nonalcoholic fatty liver disease (NAFLD)] was reduced by RES supplementation (Faghihzadeh et al., 2014), although some other works failed to obtain the same results (Chachay et al., 2014), a phenomenon once more associated with the metabolic status of the patients (Novelle et al., 2015), illustrating the need for careful analysis of the literature and conclusive studies on the matter.

Resveratrol in cancer

Since the late 1990s, RES has been on the forefront of cancer research, as some published works demonstrated a quasimiraculous cure for many cancer types, particularly in rodent animal models. From low doses of RES (obtainable through the dietary intake of RES-enriched foods) to pharmacologically obtainable high doses, the effects on cancer growth and proliferation were simply astonishing (Baur et al., 2006). RES can cause such effects through an inhibition of COX1 and 2 and other related proteins involved in cancer progression (Khanduja et al., 2004), consequently lowering angiogenic and metastization rates (Tseng et al., 2004). It can also lower the levels of drug-metabolizing enzymes, decreasing the processing of carcinogens by the cells and thus allowing the body to dispose of them instead of trying to cope with them (Baur et al., 2006). As with BBR, RES also has a cell cycle arrest and apoptotic-inducing capacity. A seminal compilation of data on RES and tumor cell lines and in vivo tumor models (Aggarwal et al., 2004) clearly illustrates that RES anticancer effects rely heavily on the downregulation of cell cycle proteins and the induction of apoptosis. However, it remains unclear whether these effects occur through direct RES action or due to sensitization of cancer cells to other proapoptotic effectors (Baur et al., 2006). Finally, RES is also capable of altering the antioxidant capacity of cells, either by modulating the levels of antioxidant enzymes or through its inherent antioxidant activity (Sengottuvelan et al., 2006).

Resveratrol and neurodegenerative disorders

RES has also been suggested as a potential therapeutic agent for various neurodegenerative disorders. SirT1 is activated by RES treatment (Price et al., 2012) and the dysfunction of this deacetylase has been implicated in various neurodegenerative situations (Albani et al., 2010). In the case of Alzheimer disease, RES has been shown to counter Aß toxicity in cellular models of this disease, either by its antioxidant abilities and/or by activating SirT1. In fact, RES has been shown to: (1) reduce intracellular and secreted levels of Aß (in a mechanism that involves proteasomal degradation of Aß); (2) reduce Aß fibril formation by directly binding to Aß; (3) activate protein kinase C, which protects against Aß-induced apoptosis; (4) increase glutathione production, thus contributing to the refreshment of antioxidant defenses; and (5) reduce NF-κB pathway activity, causing a decrease in inflammation and neuronal cell loss (Albani et al., 2010).

In the case of Parkinson disease, RES has been shown to exert protective, SirT1-dependent effects on α-synuclein or 6-OHDA-induced neuronal dysfunction (Albani et al., 2009). RES contribution to the prevention of other neurodegenerative disorders also has been extensively studied. For example, RES pretreatment alleviated some stroke characteristics through SirT1 activation, in a curious mechanism that involved the downregulation of the levels of the mitochondrial uncoupling protein 2 and heme oxygenase upregulation (Della-Morte et al., 2009). Also, the use of RES for the treatment of Huntington disease has also been showing some promising in vivo and in vitro results (Parker et al., 2005). RES can also be helpful for the treatment of amyotrophic lateral sclerosis (Kim et al., 2007) and epilepsy (Wu et al., 2009). Finally, a recent report has shown that RES has potential to reduce neurotoxicity in an in vivo model of prionic disease (Bizat et al., 2010).

Quercetin

Quercetin is one of the most widely distributed flavonoids. The name has been used since 1857, and is derived from quercetum (oak forest), after the *Quercus* genus. It is a naturally occurring polar auxin transport inhibitor, being widely present in the vegetables, fruits, nuts, and beverages derived from them (Lin et al., 2014) (Fig. 30.4).

It is well known as a strong antioxidant and free radical scavenger. It has been documented to possess antiinflammatory, antiischemic, and neuroprotective properties (Qu et al., 2014), exerting also some protective effects against cadmium-induced toxicity by inhibiting apoptosis and oxidative damage, both in vitro and in vivo

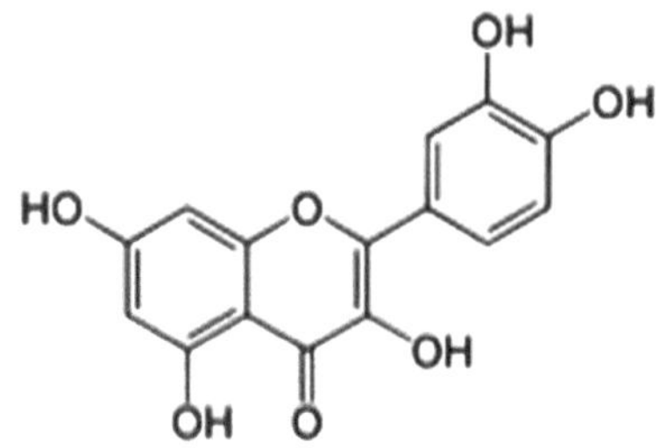

FIGURE 30.4 Structural formula of quercetin.

(Renugadevi and Prabu, 2010; Wang et al., 2013). Exposure of rat proximal tubular (rPT) cells to cadmium acetate induced a decrease in cell viability, causing an increase in apoptotic rate and apoptotic morphological changes (Wang et al., 2013). Simultaneously, elevation of intracellular ROS, malondialdehyde (MDA), and calcium levels, depletion of mitochondrial membrane potential (MMP) and intracellular glutathione, and inhibition of Na^+, K^+-ATPase, Ca^{2+}-ATPase, glutathione peroxidase, catalase (CAT), and superoxide dismutase (SOD) activities were revealed during cadmium exposure of rPT cells. However, simultaneous supplementation with quercetin protected rPT cells against cadmium-induced cytotoxicity through inhibition of apoptosis, attenuation of lipid peroxidation, maintenance of mitochondrial function, and elevation of the intracellular antioxidants levels (nonenzymatic and enzymatic). Quercetin has therefore been appointed as a widely distributed dietary antioxidant that could potentially contribute to prevent cadmium-induced cytotoxicity in kidney (Wang et al., 2013).

The protective role of quercetin, particularly when acting on mitochondria, has also been revealed when defending brain mitochondria from poisoning with endosulfan in the rat (Lakroun et al., 2015). Oxidative stress in mitochondria, namely reduction of CAT and SOD activity and reduction of glutathione level, and associated neurological damage has been inhibited by quercetin when given to rats then exposed to endosulfan by oral gavage. Particularly at the subcellular level, in the brain, mitochondrial swelling was inhibited, as well as lipid peroxidation being diminished and the activities of antioxidant enzymes were reverted by quercetin to non-poisoned levels.

Quercetin has already been reported as a metal chelator and free radical scavenger, displaying significant protection against cisplatin-induced oxidative damage in the kidney of rats (Vlachodimitropoulou et al., 2011). Moreover, when used in Sprague-Dawley rats to prevent acute kidney injury induced by mercury, quercetin proved to be useful to protect from nephrotoxicity (Shin et al., 2015). Quercetin markedly decreased the accumulation of mercury in the kidney, and also decreased the urinary excretion of biomarkers associated with kidney injury, while the excretion of mercury was significantly elevated.

Furthermore, a histological examination showed that quercetin protected against mercury-induced proximal tubular damage in the kidney (Shin et al., 2015).

Quercetin has also been found to protect gastric mucosa against ischemia/reperfusion, ethanol, indometacin, cold restraint stress, and *H. pylori* in different animal models (Arsic et al., 2010; Yan et al., 2011; Hu et al., 2015). Given that epithelial injury caused by ROS plays a critical role in the pathogenesis of gastric disorders, pharmacological intervention targeting ROS elimination has highly clinical values in therapy of gastric diseases. Using human gastric epithelial GES-1 cells, pretreated with quercetin and then challenged with hydrogen peroxide, it was found that quercetin can significantly diminish H_2O_2-induced cell death, decrease intracellular ROS and calcium influx, and restore H_2O_2-induced $\Delta\Psi_m$ dissipation (Hu et al., 2015). It also upregulates peroxisome proliferator-activated receptor-γ coactivator (PGC-1α) expression under the state of oxidative stress, and the downstream cell apoptosis significantly decreased. In vivo, chemiluminescence imaging has shown that quercetin attenuates ROS production and gastric damage in acute gastric mucosal injury. Thus quercetin can be gastroprotective due to its inhibition of oxidative stress, regulation of mitochondrial dysfunction, initiation of antioxidant defense, and inhibition of apoptosis (Hu et al., 2015).

Several viral infections, including human immunodeficiency virus (HIV), hepatitis B and C, rhinovirus, and influenza, have been shown to induce ROS formation. In this context, quercetin has also been proved to be beneficial. In respiratory viral infections, resveratrol and quercetin treatment had a protective effect against influenza infection (Uchide and Toyoda, 2011), while quercetin protected against rhinovirus both in vitro and in vivo (Ganesan et al., 2012).

Human metapneumovirus (hMPV) is a major cause of respiratory tract infections in children, the elderly, and immunocompromised hosts, for which no vaccine or treatment are currently available. Oxidative stress and inflammatory responses represent important pathogenic mechanism(s) of hMPV infection. Treatment of airway epithelial cells with quercetin during hMPV infection significantly reduced cellular oxidative damage, inflammatory mediator secretion, and viral replication (Komaravelli et al., 2015), supporting the use of dietary antioxidants such as quercetin as an effective treatment approach for modulating hMPV-induced lung oxidative damage and inflammation (Komaravelli et al., 2015).

Regarding the nervous system, quercetin has already been shown to be able of improving learning and memory ability, and to be a potential novel therapeutic strategy against Alzheimer disease (Phachonpai et al., 2010). Using rat hippocampal cultures, some authors have been already capable of delivering encapsulated quercetin, protecting neuronal cultures from oxidative stress and then hypothesizing its use against Alzheimer disease or other neurodegenerative disorders (Nday et al., 2015). Particularly in Alzheimer's, amyloid protein and/or amyloid fragments are responsible for the pathophysiology, involving mitochondrial dysfunction. Quercetin and other polyphenols, such as resveratrol, were used in a study to protect brain mitochondria from amyloid fragment fibrillogenesis (Ghobeb et al., 2014). Quercetin was able to destabilize amyloid aggregates, therefore inhibiting the fibrillation process and rescuing the mitochondrial function, providing some optimistic insights toward the prevention of Alzheimer disease. Another work, studying myelination failure associated with perinatal cerebral hypoxia-ischemia (PHI) and induced brain injury in premature infants, demonstrated that quercetin significantly improved cognitive deficits and the behavior of PHI rats in Morris water maze and open field tasks (Qu et al., 2014). After administration of quercetin, the number of neogenetic oligodendrocyte progenitor cells was evidently increased in the subventricular zone. Additionally, in corpus callosum (CC), the expression of myelin basic protein was increased, and the myelin sheaths reached normal level at 30 days with more compact while less damaged myelin sheaths and more mature oligodendrocytes repopulating the CC compared with PHI groups (Qu et al., 2014). These findings indicate that quercetin could remarkably improve both cognition performance and myelination in the context of PHI-induced brain injury.

Acute hypobaric hypoxia (HH), which causes persistent cognitive impairment, affects memory function specifically. In the brain, mitochondrial dysfunction and synaptic morphological changes are the main pathological features of HH. Quercetin has also been reported to prevent ischemia induced by neuronal injury (Liu et al., 2015). Quercetin significantly reduced HH-induced memory decline, particularly ameliorating hippocampus mitochondrial and synaptic lesions. Furthermore, quercetin regulated the expression of Sirt1, PGC-1α, and the proteins related with mitochondrial biogenesis and dynamics, suggesting that quercetin could be used as a prophylactic in HH-induced memory impairment, which is associated with mitochondrial and neuronal adaptation in hippocampus.

Additionally, quercetin has been reported to stimulate glucose uptake in cultured C2C12 skeletal muscle through an insulin-independent mechanism involving AMPK, a key regulator of the whole body—energy homeostasis. Treatment with quercetin stimulated AMPK and increased GLUT4 translocation and protein content in cultured rat skeletal muscle cells, as well as hepatic AMPK activation in hepatocytes and there was a mild tendency to increase the activity of glycogen synthase, the rate-limiting enzyme of glycogen synthesis, in HepG2 hepatocytes (Eid et al., 2015). This demonstrated that quercetin positively influences glucose metabolism in the liver and skeletal muscle, and therefore appears to be a promising therapeutic candidate in type II diabetes. In this context of metabolic disorders, quercetin was also shown to increase insulin sensitivity and improve glucose tolerance in mice (Henagan et al., 2015). Specifically, dietary supplementation with red onion extract containing quercetin attenuates HFD-induced obesity and insulin resistance by increasing energy expenditure through a mechanism involving skeletal muscle mitochondrial adaptations. Quercetin supplementation decreased HFD-induced fat mass accumulation and insulin resistance, and also increased energy expenditure. Moreover, quercetin increased skeletal muscle mitochondrial number and decreased incomplete beta oxidation. Quercetin-driven improvements in adiposity, insulin resistance, and energy expenditure occur through differential mechanisms, namely improved skeletal muscle mitochondrial number and function, as well as increased mtDNA-encoded transcript levels (Henagan et al., 2015).

Metabolic disorders have also been associated with endoplasmic reticulum (ER) stress, a condition that affects protein folding homeostasis in the cell. Emerging clinical data support the notion that pharmacological modulators of ER stress have therapeutic potential against metabolic diseases. ER is in physical contact with mitochondria, and there is strong cross-talk between these organelles at a functional level. Particularly, tunicamycin-induced ER stress causes changes in MMP, mitochondrial biogenesis, ROS production, and adiponectin secretion in 3T3-L1 adipocytes (Nisha et al., 2014). Interestingly, quercetin ameliorated these mitochondrial alterations, protecting mitochondria during ER stress and reestablishing adiponectin secretion. The main effects of quercetin, including those exerted on mitochondria, are depicted in Fig. 30.5.

Bitter kola

Garcinia kola (Guttiferae) is a widely cultivated tree that is highly valued in West and Central Africa for its edible nuts. The extracts of the seed, commonly known as "bitter kola" (Fig. 30.6), have been employed in African herbal medicine for the treatment of illnesses such as laryngitis, liver diseases, cough, hoarseness of voice, and diabetes and its associated complications (Adedara et al., 2014; Akinmoladun et al., 2015). The seed is eaten raw as a stimulant because of its bitter astringent taste, which is followed by a slight sweetness. *G. kola* seed is recommended traditionally for the treatment of diabetes (Adaramoye and Adeyemi, 2006). It is also a potent remedy for liver disorders, hepatitis, as well as an being aphrodisiac and fertility-enhancing substance (Yakubu and Quadri, 2012).

Traditionally, it is believed that regular consumption of *G. kola* seeds lowers blood glucose levels and improves the complications of diabetes mellitus. Indeed, seed administration significantly ameliorates hyperglycemia-mediated damage by decreasing the blood glucose level, enhancing the antioxidant system, inhibiting lipid peroxidation, and improving the architecture of the kidney, liver, and testes in streptozotocin (STZ)-induced diabetic rats (Adedara et al., 2014). In addition, *G. kola* seed administration restored kidney and liver function biomarkers, sperm characteristics, as well as the plasma levels of luteinizing hormone, follicle-stimulating hormone, testosterone, triiodothyronine, and thyroxine in STZ-induced diabetic rats. Thus there are nowadays several studies that provide persuasive scientific support for the traditional use of *G. kola* seed in the treatment of diabetes and its associated complications (Adedara et al., 2014).

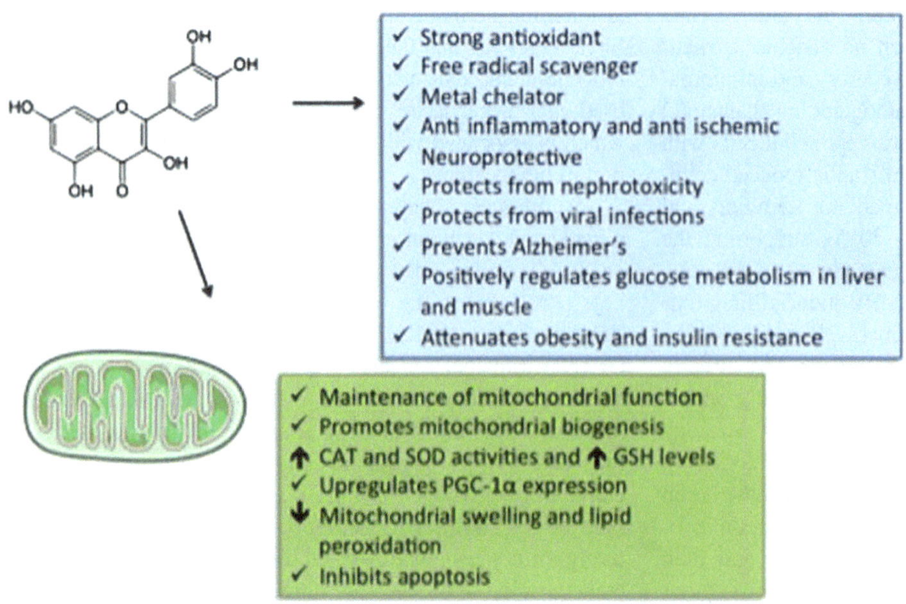

FIGURE 30.5 Summary of the main effects of quercetin, including the effects being exerted on mitochondria.

Kolaviron is a fraction of the defatted ethanol extract, containing *Garcinia* biflavonoids GB1, GB2, and kolaflavanone (Fig. 30.7). Several pharmacological properties of kolaviron have been reported in animal models including hepatoprotection (Farombi et al., 2005), nephroprotection (Adedara et al., 2015), gonadoprotection (Adaramoye et al., 2013), and cardioprotection (Adaramoye and Lawal, 2015). In particular, a recent study evaluated the cardioprotective effects of kolaviron in rats intoxicated with isoproterenol chloride (ISO) (Adaramoye and Lawal, 2015). Administration of ISO caused significant elevation of serum creatine phosphokinase, lactate dehydrogenase, alkaline phosphatase, alanine aminotransferase (ALT), and aspartate aminotransferase, with a concomitant decrease in the cardiac activities of these enzymes. Administration of ISO led to a significant decrease in the levels of cardiac SOD, catalase, glutathione *S*-transferase, reduced glutathione, and an increase in MDA. Also, ISO-treated rats had significantly higher values of serum total cholesterol, TG, and low-density lipoprotein cholesterol relative to controls. Supplementation with kolaviron caused significant elevation of cardiac antioxidant enzymes, normalized the marker enzymes, and reduced MDA levels. Kolaviron (KV) protects against ISO-induced cardiotoxicity in vivo, suggesting its usefulness as a possible chemoprophylactic agent against cardiotoxic drugs.

In the kidney, benzo(a)pyrene (B[a]P), a polycyclic aromatic hydrocarbon generally formed from incomplete combustion of organic matter, reportedly causes renal injury and elicits a nephropathic response. Kolaviron has already been used in Wistar rats to modulate this response (Adedara et al., 2015). Administration of B[a]P alone resulted in a significant increase in plasma levels of urea and creatinine in the treated rats. Moreover, B[a]P exposure significantly decreased the activities of SOD, catalase, glutathione peroxidase, and glutathione S-transferase, as well as the glutathione level in the kidneys of treated rats. Conversely, myeloperoxidase activity, hydrogen peroxide, and MDA levels were markedly elevated in kidneys of B[a]P-treated rats. Further, B[a]P exposure significantly decreased the circulatory concentrations of triiodothyronine (T3) and T3/T4 ratio without affecting thyroxine (T4). Light microscopy revealed a tubular lumen with numerous protein casts in kidneys of rats exposed to B[a]P alone. However, kolaviron administration significantly improved the renal antioxidant status, thyroid gland function, and restored the renal histology, thus demonstrating the protective effect of kolaviron. Dietary inclusion of kolaviron could exert protective effects against renal toxicity resulting from exposure to toxicants.

Another particular study investigated the protective property of kolaviron in a global model of stroke in rats (Akinmoladun et al., 2015). The protection afforded by pretreatment with kolaviron in rats subjected to bilateral common carotid artery occlusion ischemia/reperfusion injury was assessed, and there was a remarkable ablation of I/R-induced oxidative stress, neurochemical aberrations, and brain edema in animals pretreated with kolaviron. The results suggested that the protection afforded by kolaviron probably involved regulation of redox and electrolyte homeostasis as well as antiinflammatory and antiexcitotoxic mechanisms.

Furthermore, kolaviron already exhibited potent antiinflammatory effects (Abarikwu, 2014), as well as high antimalarial activities in infected mice, in addition to its known antioxidant properties (Oluwatosin et al., 2014). In a recent study, some authors used macrophages, pretreated with kolaviron, and activated them with lipopolysaccharide (LPS) (Abarikwu, 2014). The results showed that treatment with kolaviron blocked phosphorylation of p38 and p-c-JUN but not IκBα degradation, and also phosphorylation of nuclear factor-kappa B (NF-κB) (p65), extracellular signal-regulated kinases (ERK1/2), and IκBα were blocked. Furthermore, kolaviron suppressed

FIGURE 30.6 Seeds of *Garcinia kola*.

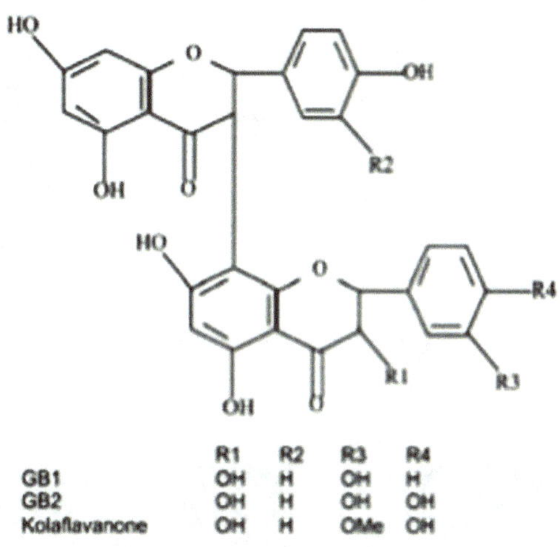

FIGURE 30.7 Structural formula of kolaviron.

LPS-induced increase in the expression of IL-18 gene and LPS-induced decrease in the mRNA expression of IP-10 but it had no effect on the LPS-induced decrease in the gene expression levels of IL-1α, IL-33, IL-1β, and IFNβ1-1. When macrophages pretreated with kolaviron were activated with LPS, phosphorylation of Akt, ERK1/2, IκBα, and NFκB (p65) but not of cAMP response element-binding protein (CREB) was blocked. Summarizing, kolaviron interferes with LPS signaling by reducing the activation of several inflammatory transcription factors, establishing an antiinflammatory potential of kolaviron via inhibition of IL-6 secretion in macrophages.

The antimalarial potential of kolaviron has also been studied, through *Plasmodium berghei* infection in Swiss albino mice (Oluwatosin et al., 2014). Administration of kolaviron KV1 and KV2 significantly suppressed *P. berghei*-infection in the mice. Also, KV2 significantly increased the mean survival time of the infected mice by 175%. The biflavonoid displayed high efficacy in ameliorating anemia, and also kolaviron significantly ameliorated the *P. berghei*-induced decrease in antioxidant status of the infected mice. Thus kolaviron was shown to have high antimalarial activity in *P. berghei*-infected mice, in addition to its known antioxidant properties (Oluwatosin et al., 2014).

In the brain, kolaviron could act as an acetylcholinesterase inhibitor (Ijomone and Obi, 2013), unveiling the development of a herbal-based natural product with therapeutic potential in the management of neurodegenerative disorders associated with disturbed cholinergic neurotransmitter systems. Explicitly, the effects of kolaviron on acetylcholinesterase activities in the hippocampus and striatum of adult Wistar rats were assessed. The results showed that kolaviron produced no neurodegenerative changes in the hippocampus and striatum, and did not significantly alter neuronal density in these brain regions. Additionally, kolaviron significantly reduced acetylcholinesterase staining intensity, suggesting a likely inhibiting effect on this enzyme (Ijomone and Obi, 2013). The main effects of "bitter kola" and kolaviron are depicted in Fig. 30.8.

Nexrutine

Regarding the phytochemicals, a considerable effort has also been made to search for naturally occurring molecules that could prove useful in the treatment of cancer (Kwon et al., 2007). Nexrutine (NX), a commercially available herbal extract from *Philodendron amurense* (commonly called the Amur cork tree), is widely used for the treatment of inflammation, gastroenteritis, abdominal pain, and diarrhea. Further, it has been shown to exhibit minimal toxicity to normal tissues (Kumar et al., 2007; Alam et al., 2015).

Active components of NX are isoquinoline alkaloids, phenolic compounds, and flavone glycosides. A recent study revealed that NX inhibited the proliferation of prostate and lung cancer cells through the modulation of Akt and CREB-mediated signaling pathways, and that its antiproliferative effects are comparable to those of berberine (James et al., 2011). Other findings also established NX to be effective against early-stage prostate tumor development as well as tumor progression in the transgenic adenocarcinoma of the mouse prostate model.

Additionally, it was recently shown that NX inhibited the promotion of skin tumorigenesis in the two-stage mouse skin tumorigenesis model (Kumar et al., 2012). The same group further reported an antitumor result in a rat liver tumorigenesis model (Alam et al., 2015). The mechanisms involved inhibition of cell proliferation and induction of apoptosis, mediated by modulation of p53 and cyclin-dependent kinase inhibitor levels, alternations

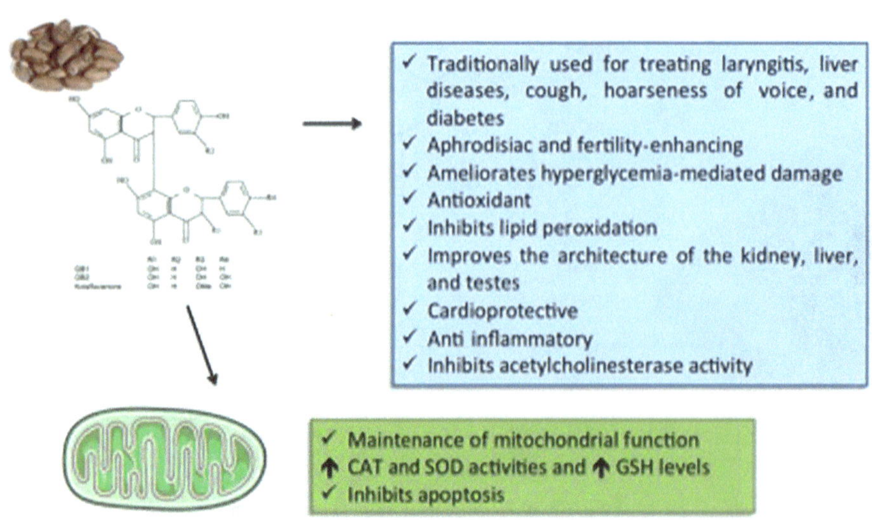

FIGURE 30.8 Summary of the main effects of "bitter kola" and kolaviron, including the effects being exerted on mitochondria.

in MAPKs signaling, and activation of caspases 3 and 9 (Alam et al., 2015).

The mechanisms of chemopreventive/chemotherapeutic efficacy of NX against skin cancer have been already explored. Single applications of NX in mice, prior to 12-O-tetradecanoylphorbol 13-acetate (TPA) application significantly inhibited TPA-induced skin edema, hyperplasia, thymidine incorporation, and ornithine decarboxylase activity (Kumar et al., 2012). Moreover, expression of cyclooxygenase-2 (COX-2) and inducible nitric oxide synthase were inhibited, as well as phosphorylation of ERK 1/2, p38, and c-jun N-terminal kinase MAPKs; and activation of I kappa B kinase, IκBα, and NF-κB in mouse skin. In a two-stage mouse skin tumorigenesis model, it was found that twice-weekly treatment of NX prior to TPA application showed reduced tumor incidence, lower tumor body burden, and significant delay in latency period compared with 7,12-dimethylbenz[α] anthracene-initiated and TPA-promoted animals (Kumar et al., 2012). Furthermore, the therapeutic efficacy of NX was assessed against human squamous carcinoma (A431) and human melanoma (A375) cells. A431 and A375 cells treated with NX for 48 h showed a decrease in viability and enhanced cell cycle arrest at the G0/G1 phase and apoptosis; however, NX had minimal cytotoxic effect on HaCaT cells and primary murine keratinocytes, suggesting its high therapeutic index. In addition, NX treatment also modulates the levels of Bax and Bcl-2 proteins along with cytochrome *c* release, cleavage and enhanced expression of poly(adenosine diphosphate-ribose) polymerase, as well as catalytic activities of caspases 3 and 9 in both A431 and A375 cells. Based on these findings, NX also could be useful in the management (chemoprevention as well as chemotherapy) of skin cancer.

Curcumin

Curcumin (CUR) is a polyphenol curcuminoid derived from the tuberous rhizomes of turmeric (*Curcuma longa*). This compound is characterized by its bright yellow color, being commonly used as a colorant in the food and textile industries or as an ingredient of curries. In addition, it has been extensively used for about 4000 years in Ayurvedic medicine due to its therapeutic properties. CUR's structure (diferuloylmethane, Fig. 30.9) confers it with the capacity to interact with several targets of distinct molecular pathways, and thus it has many biological activities (Sharma et al., 2005). It is described as possessing antioxidant, antiinflammatory, anticancer, and antidiabetic functions (Gonzales and Orlando, 2008; Jiménez-Flores et al., 2014; Ringman et al., 2005). Additionally, when administered up to 8 g/day in human patients it presented no toxicity (Cheng et al., 2001). However, studies have shown that CUR is rapidly metabolized, poorly absorbed in the gastrointestinal tract, and quickly excreted when orally administered (Ireson et al., 2001; Wahlstrom and Blennow, 1978), thus presenting a low oral bioavailability.

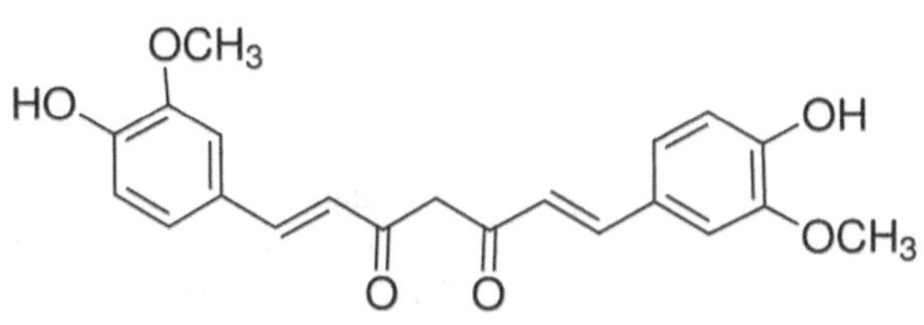

FIGURE 30.9 Structural formula of curcumin.

Curcumin in diabetes

As mentioned above, CUR has a relevant role in diabetes and its related diseases. In fact, the first report that demonstrated the involvement of CUR in diabetes dates back to 1972 (Srinivasan, 1972). Since then several studies reporting the effects of CUR in such pathological conditions have been published. Many reports have demonstrated the antihyperglycemic effect of CUR in STZ-induced diabetic rats (Gutierres et al., 2012, 2019) and in diabetic rats fed with a HFD (El-Moselhy et al., 2011; Kaur and Meena, 2012; Na et al., 2011). Diabetes is typically characterized by an elevated glucose concentration in the blood and insulin resistance. In fact, it is thought that CUR is able to decrease the glucose concentration of the plasma and blood of diabetic animal models (El-Moselhy et al., 2011; Seo et al., 2008; Weisberg et al., 2008). However, some reports contradict the previous evidence by claiming that CUR does not affect blood glucose levels (Majithiya and Balaraman, 2005; Nishizono et al., 2000). Nevertheless, CUR was also reported to improve insulin sensitivity in HFD-fed rats (El-Moselhy et al., 2011) and to reduce the insulin resistance of diabetic animals (Jang et al., 2008; Na et al., 2011; Seo et al., 2008). CUR's function regulating insulin signaling has been partly attributed to its role on reducing proinflammatory cytokine tumor necrosis factor-α (TNF-α) levels (El-Moselhy et al., 2011), and on inhibiting the hepatic protein-tyrosine phosphatase 1B (PTP1B), which consequently led to positive regulation of the insulin signaling pathway, increasing insulin receptor substrate 1 (IRS1) and IR (Li et al., 2010).

Diabetes is also often associated with an excessive accumulation of lipids, which per se further aggravates the disease. Actually, CUR was found to improve hyperlipidemia in diabetic animals by decreasing plasma FFAs and TG in liver (Jang et al., 2008; Karthikesan et al., 2010). This antihyperlipidemic effect is presumably related with CUR's ability to decrease lipogenesis, stimulate β-oxidation, and decrease the activity of hepatic lipid metabolism-related enzymes, such as fatty acid synthase, carnitine palmitoyl transferase, and acyl-CoA cholesterol acyltransferase (Jang et al., 2008; Seo et al., 2008). Such events might be intrinsically related with the upregulation of transcription factors, such as protein peroxisome proliferator-activated receptor α (PPARα) (Um et al., 2013), and of transcription factors involved in hepatic lipogenesis, such as the sterol regulatory element-binding protein 1c and the carbohydrate response element-binding protein (Shao et al., 2012). CUR can also negatively interfere with gluconeogenesis. Indeed, it was found that CUR decreases the expression and activity of enzymes involved in this metabolic pathway, such as glucose-6-phosphatase and phosphoenolpyruvate carboxykinase (Kim et al., 2009a; Seo et al., 2008). Regulation of lipid metabolism and suppression of gluconeogenesis by CUR seem to be mediated in part through activation of the energy sensor enzyme, AMPK, with it being reported that CUR leads to the phosphorylation of AMPK (Kim et al., 2009b; Liu et al., 2017; Um et al., 2013).

AMPK in known as a central regulator of mitochondrial signaling, being involved in the stimulation of mitochondrial biogenesis mainly through the positive regulation of PPARγ coactivator-1α (PGC-1α). Despite mitochondria's extensive functions in cell respiration and cellular signaling, they are also one of the main sources of ROS, that when in excessive levels contribute to the aggravation of diabetes (Brownlee, 2001; Rolo and Palmeira, 2006). In fact, CUR was reported to protect against oxidative stress in diabetic animal models by decreasing lipid peroxidation and normalizing the activity of antioxidant enzymes such as SOD and CAT (Seo et al., 2008). Both CUR's actions on AMPK and ROS signaling are suggestive of a role in mitochondrial signaling (de Oliveira et al., 2016; Trujillo et al., 2014). Indeed, CUR was shown to increase the expression of PGC-1α through AMPK, which then led to the stimulation of mitochondrial biogenesis in adipocytes (Lone et al., 2016). Additionally, mitochondrial biogenesis modulators, such as nuclear respiratory factor 1 and mitochondrial transcription factor A, and mitochondrial DNA were found to be increased by CUR pretreatment in cerebral ischemia reperfusion (I/R) injury (Liu et al., 2014b). CUR treatment, alongside an exercise regimen, resulted in elevated mitochondrial biogenesis and the increase in mitochondrial function-related proteins such as cytochrome *c* oxidase subunit IV (COX-IV) and other oxidative phosphorylation protein complexes. This was shown to be mediated through AMPK and NAD^+-dependent deacetylase SIRT1 signaling (Ray Hamidie et al., 2015). Finally, mitochondrial function impaired by hyperglycemic stress was also proven to be restored by CUR in diabetic mice (Soto-Urquieta et al., 2014).

Curcumin in cancer

Among CUR's many roles it has also been found to have interesting therapeutic potential as a chemotherapeutic agent against most types of cancer (Perrone et al., 2015; Wojcik et al., 2018). CUR is known to interfere with several tumor processes such as inflammatory processes, tumor proliferation, or apoptotic mechanisms. In fact, CUR treatment powerfully inhibits cancer cell proliferation, either by interfering with the cell cycle or by stimulating apoptosis. For instance, in human breast cancer cells it hampered proliferation by provoking cell cycle arrest at the G1 phase (Hua et al., 2010), and it also interfered with microtubule assembly dynamics (Banerjee et al., 2010). CUR can also interfere with tumor cell proliferation by forcing them to undergo programmed cell death in a process called apoptosis. This was achieved through stimulation of the proapoptotic proteins Bcl-2-associated x protein (Bax), p53, caspase 3 and caspase 8, and through downregulation of antiapoptotic proteins such as Mcl-1 and Bcl-2 (Jiang et al., 2015; Li et al., 2013). These effects were attributed to CUR's ability to inhibit the nuclear factor κ-light-chain-enhancer of activated B cells (NF-κB) and p38 MAPK signaling pathways observed in human melanoma cells (Jiang et al., 2015). It is unsurprising that CUR also stimulated mitochondria-related apoptosis by promoting cytochrome c release in human nonsmall-cell lung cancer cells (Li et al., 2013). NF-κB is a proinflammatory transcription factor known to be activated through the proinflammatory cytokine TNF-α and, indeed, CUR was also shown to downregulate TNF-α expression (Das and Vinayak, 2014). In fact, this nutraceutical is known to interfere with a large array of inflammatory cytokines, such as the interleukin IL-6 (Das and Vinayak, 2014) or COX-2 (Lev-Ari et al., 2006). Besides CUR's antiproliferative and proapoptotic effects in tumor cells, it was also observed that it is capable of affecting tumor migration and invasion, possibly impairing tumor metastatic capacity (Tsang et al., 2012).

Cancer development can be positively affected by epigenetic mechanisms deregulation. In fact, tumor cells suffer extensive epigenetic reprogramming. Interestingly, CUR has been associated with cancer epigenetic modulation through the regulation of miRs that are small noncoding sequences (~22 nucleotides) able to interfere with mRNA transcription. It was shown that CUR upregulated tumor-suppressive miR-15 and miR-16 (Yang et al., 2010) and suppressed oncogenic miR-19a and mir-19b (Li et al., 2014).

CUR was also shown to play a major role in sensitizing tumor cells to treatment, therefore proving to have chemosensitizing and radiosensitizing properties. In fact, it was shown to enhance and improve the efficacy of chemotherapeutics, and to potentiate the effects of several chemotherapeutical agents (Kumar et al., 2016). Additionally, it was observed that CUR was able to protect normal cells from the toxicity of radiotherapy (Goel and Aggarwal, 2010).

Curcumin in neurodegenerative diseases

As shown above, CUR regulates several molecular targets that are crucial in diabetic and cancer pathologies. The many functions of CUR, allied to the fact that it is able to cross the blood–brain barrier (Priyadarsini, 2014), confer it with the capacity to successfully treat several neurological conditions, such as Alzheimer and Parkinson diseases, and it was shown to improve the cognitive function of animals (Kumar et al., 2011; Mandal et al., 2020). Alzheimer disease results from the accumulation of amyloid β (Aβ) protein plaques between neurons, accumulation of neurofibrillary tangles of tau proteins inside neurons, and chronic inflammation, that consequently lead to the disruption of neuronal function (Bloom, 2014). In this context, CUR was found to prevent the accumulation of Aβ peptides and formation of Aβ fibrils (Hotsumi et al., 2019; Ono et al., 2004; Shytle et al., 2009). The accumulation of Aβ plaques leads to neurotoxicity, oxidative damage, and inflammation. Aβ-induced toxicity was shown to be counteracted by CUR treatment in PC12 cells (Park et al., 2008). This neuroprotective effect was achieved due to CUR inherent antioxidant activity, decreased Aβ-induced oxidative damage, and its ability to inhibit hyperphosphorylation of tau proteins (Park et al., 2008). Additionally, due to its antiinflammatory function, CUR proved to have a role of the utmost importance in the repression of cytokine and chemokine generation, such as TNF-α, IL-1β, and IL-8, induced by Aβ (Giri et al., 2004).

CUR also plays a major role as a potential therapeutic agent in Parkinson disease. This disease is typically characterized by the gradual loss of dopaminergic neurons in great part due to events promoted by abnormal accumulation and aggregation of α-synuclein in nerve cells, mitochondrial dysfunctionality, apoptotic events, and oxidative stress (Lotharius and Brundin, 2002). Indeed, it was demonstrated that CUR acted as an inhibitor of the aggregation of α-synuclein (Pandey et al., 2008; Sharma and Nehru, 2018). Additionally, this compound was able to protect cells against 6-OHDA neurotoxicity, as it was found to restore MMP and suppress excessive ROS production and increase SOD levels, consequently protecting against 6-ODHA-induced cell death (Wang et al., 2009). In fact, in Parkinson disease models, CUR was demonstrated to have antiapoptotic functions by decreasing Bax protein, increasing Bcl2, decreasing phosphorylation of p53 (Jaisin et al., 2011), and impairing caspase 3 and caspase 9 activities (Sharma and Nehru, 2018; Wang et al., 2010b). Recently, CUR was found to improve the motor function of mice injected with LPS and to protect dopaminergic neurons against inflammation by impairing inflammatory events, such as NF-κB signaling, and by inhibiting the generation of proinflammatory cytokines, such as TNF-α and IL-1β (Sharma et al., 2017).

Astaxanthin

Astaxanthin (ATX) is a xanthophyll carotenoid characterized by its red coloration, and is extremely abundant in microorganisms such as algae and yeast, and in marine animals such as salmon, trout, shrimp, and crayfish (Higuera-Ciapara et al., 2006). Commercial ATX can be extracted from *Phaffia* yeast and *Haematococcus pluvialis* algae, or it can be chemically synthesized. Regarding human consumption, the largest source of ATX comes from *H. pluvialis*, which under stress conditions accumulates ATX (Sarada et al., 2002; Ravi et al., 2012; Lorenz, 1999; EFSA, 2005).

The structure of this carotenoid (Fig. 30.10) explains some of its characteristics, such as its more polar nature, high antioxidant capacity, and esterification capacity (Liu and Osawa, 2007). In addition, this red pigment has the particularity of being soluble in fat (incorporated in cell membranes), while being unable to be converted into vitamin A in the human body (Hussein et al., 2006).

The European Commission considers natural ATX as a food coloring, whereas for the United States Food and Drug Administration ATX has been approved as a food coloring (color additive) in animal feed (Administration U.F. and D., 2016). With regard to human consumption, this carotenoid has been approved as a food supplement (Visioli and Artaria, 2017), and its use is growing as a food supplement in foods, nutraceuticals, medicines, and diets.

Due to its antioxidant capacity (Park et al., 2010; McNulty et al., 2007; Kamath et al., 2008; Wolf et al., 2010; Kim and Kim, 2018), ATX exhibits a variety of

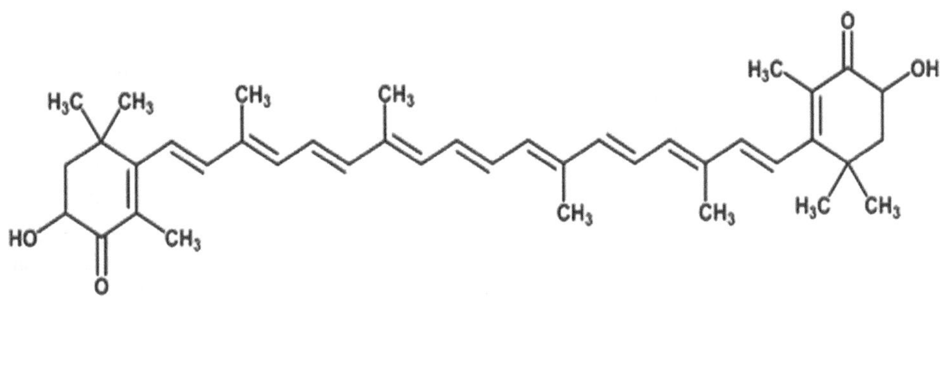

FIGURE 30.10 Structural formula of astaxanthin.

biological and beneficial effects. Thus it has been gaining relevance due to its potential effects in the prevention and treatment of diseases such as diabetes (Arunkumar et al., 2012; Mashhadi et al., 2018), cancer (Rao et al., 2013; Wolf, 1992; Chew et al., 1999), inflammatory diseases (Park et al., 2010), aging diseases, neurodegenerative and cardiovascular diseases, and hepatic, ocular, and skin diseases (Yuan et al., 2011; Hama et al., 2012; Santos et al., 2012; Barros et al., 2014; Rao et al., 2013).

ATX crosses the double layer cell membrane, and detects radicals in the conjugated polyene chain and in the terminal ring portion. This molecule has a characteristic that gives it greater protection, as it is exposed both inside and outside the cell (Yamashita, 2013).

Mitochondrial homeostasis and diabetes

Skeletal muscle is an essential tissue in the metabolism of the body, and it plays an important role in the metabolism and whole-body homeostasis, being responsible for much of the insulin-induced glucose uptake (Thiebaud et al., 1982). GLUT4 (type 4 glucose transporter) is a glucose transporter responsible for the uptake of glucose in muscle tissue and adipocytes. The amount of this transporter in type II diabetes mellitus and insulin resistance is reduced (Schreiber et al., 2017) and its translocation is impaired (Jaldin-Fincati et al., 2017). ATX accumulates in skeletal muscle leading to the activation of IRS-1, to the phosphorylation of Akt and to GLUT4 translocation in skeletal muscle, thus causing an increase in secretion and sensitivity to insulin and a decrease in blood glucose levels (Sztretye et al., 2019).

Oxidative stress, as a result of increased ROS production or a reduced antioxidant enzyme system, leads to disease. Often, mitochondrial dysfunction is involved in the onset of diseases associated with oxidative stress. ATX is a powerful antioxidant that maintains the metabolic efficiency of mitochondria; therefore it is a potential therapeutic agent to prevent or slow the progression of the disease.

Wolf and colleagues found that ATX maintains mitochondrial function, protecting mitochondrial redox balance, significantly reducing oxidative stress, and keeping mitochondria in a reduced state. It also prevents the loss of MMP and increases the consumption of mitochondrial oxygen (Wolf et al., 2010). In addition, it can prevent mitochondrial dysfunction (Zhang et al., 2016; Kuroki et al., 2013). It also suppresses ROS generation induced by 1-methyl-4-phenylpyridinium +, increasing SOD and catalase activities, thus preventing a decrease in the opening of MMP and mitochondrial permeability transition pore (Lee et al., 2011).

ATX reduces ROS levels and increases MMP in an in vitro model of inflammatory preeclampsia (Xuan et al., 2016). In another study, in vivo oxidative damage was mitigated, compromised mitochondrial function was restored, and mitochondrial content, ATP production, and complex respiratory chain activity were increased, thus suggesting that ATX increases mitochondrial efficiency and prevents aging (Park et al., 2013).

The effect of ATX on mitochondrial dysfunction in ischemic mice was investigated. The results revealed that the mitochondria isolated from the ischemic myocardium of mice show higher levels of mtROS production and mitochondrial depolarization. Treatment with ATX reduces the production of mtROS and depolarization and swelling of the mitochondria (Pongkan et al., 2017).

A study revealed that ATX inhibits the release of cytochrome c resulting from the permeabilization of mitochondria, and thus prevents mitochondria-mediated apoptotic death (Kim and Kim, 2018).

Diabetes is a serious disease that affects millions of people worldwide. Hyperglycemia and other metabolic syndromes that increase the risk of diabetes stimulate the production of ROS in mitochondria (Ballinger, 2005). Oxidative stress leads to a chronic inflammatory state associated with increased production of ROS, causing cell dysfunction and apoptosis (Ballinger, 2005; Murillo and Fernandez, 2016). ATX reduces the production of ROS and reactive nitrogen species (RNS) induced by hyperglycemia, especially in mitochondria (Kim et al., 2009a). It also inhibits inflammatory signaling and apoptosis that occurs in diabetes (Otton et al., 2010), reduces oxidative stress, and improves the metabolic profile of glucose and fatty acids, preventing damage and cellular dysfunction in the organs (Naito et al., 2004; Xu et al., 2015; Yeh et al., 2016).

It also ameliorates diabetic complications, such as neuropathy, retinopathy, and nephropathy (Naito et al., 2004; Xu et al., 2015; Yeh et al., 2016; Kim and Kim, 2018). Animal studies have also shown that ATX reduces hyperglycemia and increases insulin secretion and sensitivity by improving glucose metabolism and β-cell dysfunction by GLUT4 regulation (Ishiki et al., 2013; Bhuvaneswari and Anuradha, 2012). ATX in mice has been shown to have antihyperglycemic effects, preserves the function of β cells, and reduces glycotoxicity (Naito et al., 2004; Uchiyama et al., 2002).

In human mesenchymal cells (NHMCs) stimulated with high glucose levels, ATX reduces the production of mtROS and inhibits high glucose-induced NF-κB activation. It decreases the levels of 4-hydroxy-2,3-nonenal protein adducts, markers of lipid oxidation, in the mitochondria of NHMCs (Manabe et al., 2008).

Often, patients with diabetes mellitus have high levels of oxidative stress. Studies have shown that ATX, in diabetes, improves serum glucose and insulin levels, reduces oxidative stress (Uchiyama et al., 2002), and protects pancreatic β cells against glycotoxicity (Otton et al., 2010).

In a clinical trial the effects of supplementation with ATX in people with type II diabetes were examined. They observed that there was a reduction in blood pressure, body fat, and blood TG, and an improvement in glucose metabolism (Mashhadi et al., 2018). In addition, in the same study, an inverse relationship between adiponectin and cholesterol TG and very-low-density lipoprotein (VLDL) was also observed. Enhanced VLDL catabolism through increased lipoprotein lipase and VLDL receptor expression related to improved insulin resistance is considered to be the mechanism underlying the reduction in serum TG concentration (Hussein et al., 2007; Yoshida et al., 2010).

Liver and nonalcoholic fatty liver disease

The liver is a complex organ of great importance because it has several functions such as detoxifying contaminants, oxidizing lipids, and destroying viruses, pathogenic bacteria, and dead red blood cells. All of these functions can lead to the release of free radicals. In this context, ATX appears to have a protective role in mitochondria in the liver cells of rats against lipid peroxidation (Kurashige et al., 1990).

Currently, with diets increasingly rich in saturated fats and sugars, NAFLD has increased in prevalence worldwide. This disease is generally associated with insulin resistance and can be considered a component of the metabolic syndrome. Fat accumulation and oxidative stress impair the function of mitochondria by morphological changes, increased membrane peroxidation, ROS production, beta oxidation, mitochondrial permeability, and defective mitochondrial breathing (McCarty, 2011).

Studies with ATX in rats have been shown to prevent oxidative damage to the liver, improve the metabolic profile, and reduce liver inflammation. These studies suggest that ATX may be an effective therapy for the treatment of liver diseases mediated by oxidative stress (Curek et al., 2010; Bhuvaneswari et al., 2010; Islam et al., 2017). In another study with mice, the authors investigated the level of ALT, a specific marker for liver function, under treatment with ATX and found that ATX significantly reduced plasma levels of ALT (Yang et al., 2014).

In a study with a dietary model of fatty liver disease associated with IR, which mimics human NAFLD, it was found that ATX inhibited weight gain, provided insulin sensitivity, and reduced liver disease (Bhuvaneswari et al., 2010). Another study revealed that ATX in NAFLD significantly attenuates inflammation, and decreases the accumulation of lipids and serum markers of liver damage (Liu et al., 2018).

In a study in mice fed with a high-fat diet, it was found that ATX reduced the levels of liver and blood TG and cholesterol (Ikeuchi et al., 2007). Another study in hepatic hyperlipidemia reported that ATX significantly reduced the accumulation of lipids in hepatocytes (Jia et al., 2012).

In humans, treatment with ATX significantly reduced plasma lipid peroxidation markers, including 12-hydroxy and 15-hydroxy fatty acids (Karppi et al., 2007). In addition, another study revealed a significant reduction in oxidative stress markers, such as MDA, isoprostanes, and antioxidant markers, such as SOD and total antioxidant capacity (Choi et al., 2011).

Astaxanthin and the heart

The increase in oxidative stress resulting from ROS/RNS and inflammation are involved in several manifestations of cardiovascular disease. Arterial hypertension is a considerable cardiovascular risk factor, therefore assessing blood pressure with ATX supplementation can play an important role in the assessment of the cardiovascular system. Studies in spontaneously hypertensive rats administered with ATX (orally) showed a significant reduction in blood pressure (Hussein et al., 2005; Preuss et al., 2011)

In vivo and in vitro studies (in animals and humans) showed that at the cardiac level, ATX decreases the formation of oxidative damage products, ROS and RNS levels, reduces lipid peroxidation, suppresses inflammatory signaling, and increases enzyme antioxidant activity (Pashkow et al., 2008; Nakao et al., 2010; Abdelzaher et al., 2016; Alam et al., 2018).

In a study of ischemia and cardiac reperfusion, treatment with ATX led to a significant reduction in the size of the infarction, mitigating apoptosis and mitochondrial dysfunction (Pongkan et al., 2017). In a homocysteine-induced cardiotoxicity model, it suppressed cardiotoxicity, and inhibited mitochondrial dysfunction and oxidative damage. In addition, ATX stopped ROS overproduction, fragmentation of mitochondria, and loss of MMP, and normalized the expression of Bcl-2 family proteins, suppressing mediators of mitochondrial apoptosis (poly-ADP-ribose polymerase and caspases) (Fan et al., 2017).

In essence, ATX appears to be promising, possibly with the potential to be considered a therapeutic agent to treat several diseases associated with mitochondrial dysfunction and oxidative stress. These diseases include NAFLD, cardiovascular diseases, hyperglycemia, neurodegenerative diseases, metabolic syndrome, and diabetes.

However, there remain some gaps in knowledge with regard to its effect on health, as well as the mechanisms involved in its action in various diseases.

Concluding remarks and future directions

The virtually untapped potential of nutraceuticals is undeniable. Not only do these molecules possess interesting and often powerful biological activities, they can also be the scaffold for synthetic derivatives, significantly reducing the time and cost of development of new molecules with therapeutic potential. Mitochondria play a key role in mediating or at least participating in the vast majority of the effects of nutraceutical compounds. As such, the evaluation of mitochondrial effects (either beneficial or toxic) is a vital initial step in screening for the effects of nutraceuticals. In fact, the ease of obtaining and abundance of mitochondria allow for sensitive, cost-effective, and fast initial screenings of nutraceutical effects. Some of the nutraceuticals highlighted in this chapter may be the most interesting for therapeutic use in the near future. Some nutraceuticals possess almost "miraculous" effects on a tremendously broad range of pathological conditions, and appear to be evident candidates for drug trials and pilot studies for obesity, diabetes, Alzheimer's and other neurological disorders, and even cancer. It will also be extremely important to understand if it is possible to develop synthetic analogs to these molecules that possess higher levels of activity and, when applicable, lower toxicity. The development of these molecules is also attractive from a commercial standpoint, as it will allow for the establishment of commercial patents, further bolstering the interest in the development of these analogs.

References

Abarikwu, S.O., 2014. Kolaviron, a natural flavonoid from the seeds of *Garcinia kola*, reduces LPS-induced inflammation in macrophages by combined inhibition of IL-6 secretion, and inflammatory transcription factors, ERK1/2, NF-κB, p38, Akt, p-c-JUN and JNK. Biochim. Biophys. Acta 1840 (7), 2373–2381.

Abdelzaher, L.A., Imaizumi, T., Suzuki, T., et al., 2016. Astaxanthin alleviates oxidative stress insults-related derangements in human vascular endothelial cells exposed to glucose fluctuations. Life Sci. 150 (1), 24–31.

Abidi, P., Zhou, Y., Jiang, J.D., Liu, J., 2005. Extracellular signal-regulated kinase– dependent stabilization of hepatic low-density lipoprotein receptor mRNA by herbal medicine berberine. Arterioscler. Thromb. Vasc. Biol. 25 (10), 2170–2176.

Adaramoye, O.A., Adeyemi, E.O., 2006. Hypoglyaemic and hypolipidemic effects of fractions from kolaviron, a biflavonoid complex from *Garcinia Kola* in streptozotocin-induced diabetes mellitus rats. J. Pharm. Pharmacol. 58 (1), 121–128.

Adaramoye, O.A., Lawal, S.O., 2015. Kolaviron, a biflavonoid fraction from *Garcinia kola*, protects against isoproterenol-induced injury by mitigating cardiac dysfunction and oxidative stress in rats. J. Basic Clin. Physiol. Pharmacol. 26 (1), 65–72.

Adaramoye, O.A., Akanni, O.O., Farombi, E.O., 2013. Nevirapine induces testicular toxicity in Wistar rats: reversal effect of kolaviron (biflavonoid from *Garcinia kola* seeds). J. Basic Clin. Physiol. Pharmacol. 24 (4), 313–320.

Adedara, I.A., Awogbindin, I.O., Anamelechi, J.P., et al., 2014. *Garcinia kola* seed ameliorates renal, hepatic, and testicular oxidative damage in streptozotocin-induced diabetic rats. Pharm. Biol. 22, 1–10.

Adedara, I.A., Daramola, Y.M., Dagunduro, J.O., et al., 2015. Renoprotection of Kolaviron against benzo (A) pyrene-induced renal toxicity in rats. Ren. Fail. 37 (3), 497–504.

Administration U.F. and D. 2016. Summary of Color Additives for Use in the United States in Foods, Drugs, Cosmetics, and Medical Devices. 1–20.

Agarwal, B., Baur, J.A., 2011. Resveratrol and life extension. Ann. N. Y. Acad. Sci. 1215, 138–143.

Aggarwal, B.B., Bhardwaj, A., Aggarwal, R.S., et al., 2004. Role of resveratrol in prevention and therapy of cancer: preclinical and clinical studies. Anticancer Res. 24 (5A), 2783–2840.

Akinmoladun, A.C., Akinrinola, B.L., Olaleye, M.T., et al., 2015. Kolaviron, a *Garcinia kola* biflavonoid complex, protects against ischemia/reperfusion injury: pertinent mechanistic insights from biochemical and physical evaluations in rat brain. Neurochem. Res. 40 (4), 777–787.

Alam, S., Yadav, R.S., Pal, A., et al., 2015. Dietary administration of Nexrutine inhibits rat liver tumorigenesis and induces apoptotic cell death in human hepatocellular carcinoma cells. Toxicol. Rep. 2, 1–11.

Alam, M.N., Hossain, M.M., Rahman, M.M., et al., 2018. Astaxanthin prevented oxidative stress in heart and kidneys of isoproterenol-administered aged rats. J. Diet. Suppl. 15 (1), 42–54.

Albani, D., Polito, L., Batelli, S., et al., 2009. The SIRT1 activator resveratrol protects SK-N-BE cells from oxidative stress and against toxicity caused by alpha-synuclein or amyloid-beta (1-42) peptide. J. Neurochem. 110 (5), 1445–1456.

Albani, D., Polito, L., Signorini, A., et al., 2010. Neuroprotective properties of resveratrol in different neurodegenerative disorders. Biofactors 36 (5), 370–376.

Al-Masri, I.M., Mohammad, M.K., Tahaa, M.O., 2009. Inhibition of dipeptidyl peptidase IV (DPP IV) is one of the mechanisms explaining the hypoglycemic effect of berberine. J. Enzyme Inhib. Med. Chem. 24 (5), 1061–1066.

Arsic, I., Zugic, A., Antic, D.R., et al., 2010. *Hypericum perforatum* L. Hypericaceae/Guttiferae sunflower, olive and palm oil extracts attenuate cold restraint stress–induced gastric lesions. Molecules 15 (10), 6688–6698.

Arunkumar, E., Bhuvaneswari, S., Anuradha, C.V., 2012. An intervention study in obese mice with astaxanthin, a marine carotenoid—effects on insulin signaling and pro-inflammatory cytokines. Food Funct. 3 (2), 120–126.

Asai, M., Iwata, N., Yoshikawa, A., et al., 2007. Berberine alters the processing of Alzheimer's amyloid precursor protein to decrease Abeta secretion. Biochem. Biophys. Res. Commun. 352 (2), 498–502.

Ballinger, S.W., 2005. Mitochondrial dysfunction in cardiovascular disease. Free Rad. Biol. Med. 38 (10), 1278–1295.

Banerjee, M., Singh, P., Panda, D., 2010. Curcumin suppresses the dynamic instability of microtubules, activates the mitotic checkpoint and induces apoptosis in MCF-7 cells. FEBS J. 277, 3437–3448.

Barros, M., Poppe, S., Bondan, E., 2014. Neuroprotective properties of the marine carotenoid astaxanthin and omega-3 fatty acids, and perspectives for the natural combination of both in krill oil. Nutrients 6 (3), 1293–1317.

Baur, J.A., Pearson, K.J., Price, N.L., et al., 2006. Resveratrol improves health and survival of mice on a high calorie diet. Nature 444 (7117), 337–342.

Baur, J.A., Sinclair, D.A., 2006. Therapeutic potential of resveratrol: the in vivo evidence. Nat. Rev. Drug Discov. 5 (6), 493–506.

Bhuvaneswari, S., Anuradha, C.V., 2012. Astaxanthin prevents loss of insulin signaling and improves glucose metabolism in liver of insulin resistant mice. Can. J. Physiol. Pharmacol. 90 (11), 1544–1552.

Bhuvaneswari, S., Arunkumar, E., Viswanathan, P., et al., 2010. Astaxanthin restricts weight gain, promotes insulin sensitivity and curtails fatty liver disease in mice fed a obesity-promoting diet. Process Biochem. 45 (8), 1406–1414.

Bizat, N., Peyrin, J.M., Haik, S., et al., 2010. Neuron dysfunction is induced by prion protein with an insertional mutation via a Fyn kinase and reversed by sirtuin activation in *Caenorhabditis elegans*. J. Neurosci. 30 (15), 5394–5403.

Bloom, G.S., 2014. Amyloid-β and tau: the trigger and bullet in Alzheimer disease pathogenesis. JAMA Neurol. 71, 505–508.

Borel, J.F., Di Padova, F., Mason, J., et al., 1990. Pharmacology of cyclosporine (sandimmune). I. Introduction. Pharmacol. Rev. 41 (3), 239–242.

Brownlee, M., 2001. Biochemistry and molecular cell biology of diabetic complications. Nature 414, 813–820.

Castillo, J., Hung, J., Rodriguez, M., et al., 2005. LED fluorescence spectroscopy for direct determination of monoamine oxidase B inactivation. Anal. Biochem. 343 (2), 293–298.

Chachay, V.S., Macdonald, G.A., Martin, J.H., et al., 2014. Resveratrol does not benefit patients with nonalcoholic fatty liver disease. Clin. Gastroenterol. Hepatol. 12 (12), 2092–2103. e1–6.

Cheng, A.-L., Hsu, C.-H., Lin, J.-K., et al., 2001. Phase I clinical trial of curcumin, a chemopreventive agent, in patients with high-risk or pre-malignant lesions. Anticancer Res. 21, 2895–2900.

Cheng, Z., Guo, S., Copps, K., et al., 2009. Foxo1 integrates insulin signaling with mitochondrial function in the liver. Nat. Med. 15 (11), 1307–1311.

Chew, B.P., Park, J.S., Wong, M.W., et al., 1999. A comparison of the anticancer activities of dietary β-carotene, canthaxanthin and astaxanthin in mice in vivo. Anticancer Res. 19 (3A), 1849–1853.

Choi, H.D., Kim, J.H., Chang, M.J., et al., 2011. Effects of astaxanthin on oxidative stress in overweight and obese adults. Phytother. Res. 25 (12), 1813–1818.

Curek, G.D., Cort, A., Yucel, G., et al., 2010. Effect of astaxanthin on hepatocellular injury following ischemia/reperfusion. Toxicology 267 (1–3), 147–153.

Das, L., Vinayak, M., 2014. Long-term effect of curcumin down-regulates expression of tumor necrosis factor-α and interleukin-6 via modulation of E26 transformation-specific protein and nuclear factor-κB transcription factors in livers of lymphoma bearing mice. Leuk. Lymphoma 55, 2627–2636.

de Oliveira, M.R., Jardim, F.R., Setzer, W.N., et al., 2016. Curcumin, mitochondrial biogenesis, and mitophagy: exploring recent data and indicating future needs. Biotechnol. Adv. 34, 813–826.

Della-Morte, D., Dave, K.R., DeFazio, R.A., et al., 2009. Resveratrol pretreatment protects rat brain from cerebral ischemic damage via a sirtuin 1–uncoupling protein 2 pathway. Neuroscience 159 (3), 993–1002.

Derosa, G., Maffioli, P., Cicero, A.F., 2012. Berberine on metabolic and cardiovascular risk factors: an analysis from preclinical evidences to clinical trials. Expert Opin. Biol. Ther. 12 (8), 1113–1124.

Derosa, G., D'Angelo, A., Vanelli, A., Maffioli, P., 2020. An evaluation of a nutraceutical with berberine, curcumin, inositol, banaba and chromium picolinate in patients with fasting dysglycemia. Diabetes Metab. Syndr. Obes. 13, 653–661.

Diogo, C.V., Machado, N.G., Barbosa, I.A., et al., 2011. Berberine as a promising safe anti-cancer agent—is there a role for mitochondria? Curr. Drug. Targets 12 (6), 850–859.

Dolinsky, V.W., Charkrabarti, S., Pereira, T.J., et al., 2013. Resveratrol prevents hypertension and cardiac hypertrophy in hypertensive rats and mice. Biochim. Biophys. Acta 1832 (10), 1723–1733.

Domitrović, R., Jakovac, H., Blagojević, G., 2011. Hepatoprotective activity of berberine is mediated by inhibition of TNF-α, COX-2, and iNOS expression in CCl(4)-intoxicated mice. Toxicology 280 (1-2), 33–43.

Eid, H.M., Nachar, A., Thing, F., et al., 2015. The molecular basis of the antidiabetic action of quercetin in cultured skeletal muscle cells and hepatocytes. Pharmacogn. Mag. 11 (41), 74–81.

Elliott, P.J., Walpole, S., Morelli, L., et al., 2009. Resveratrol/SRT-501. Drugs Future 34 (4), 291.

El-Moselhy, M.A., Taye, A., Sharkawi, S.S., et al., 2011. The antihyperglycemic effect of curcumin in high fat diet fed rats. Role of TNF-α and free fatty acids. Food Chem. Toxicol. 49, 1129–1140.

Endo, A., 2004. The origin of the statins. 2004. Atheroscler. Suppl. 5 (3), 125–130.

European Food Safety Authority (EFSA), 2005. Opinion of the scientific panel on additives and products or substances used in animal feed (FEEDAP) on the updating of the criteria used in the assessment of bacteria for resistance to antibiotics of human or veterinary importance. EFSA J. 223 (1), 1–12.

Faghihzadeh, F., Adibi, P., Rafiei, R., et al., 2014. Resveratrol supplementation improves inflammatory biomarkers in patients with nonalcoholic fatty liver disease. Nutr. Res. 34 (10), 837–843.

Fan, C.D., Sun, J.Y., Fu, X.T., et al., 2017. Astaxanthin attenuates homocysteine-induced cardiotoxicity in vitro and in vivo by inhibiting mitochondrial dysfunction and oxidative damage. Front. Physiol. 8 (1041), 1–10.

Farombi, E.O., Adepoju, B.F., Ola-Davies, O.E., et al., 2005. Chemoprevention of aflatoxin B1-induced genotoxicity and hepatic oxidative damage in rats by kolaviron, a natural bioflavonoid of *Garcinia kola* seeds. Eur. J. Cancer Prev. 14 (3), 207–214.

Feng, Y., Siu, K.Y., Wang, N., et al., 2010. Hepatoprotective effects of berberine on carbon tetrachloride-induced acute hepatotoxicity in rats. Chin. Med. 5, 33.

Frémont, L., 2000. Biological effects of resveratrol. Life Sci. 66 (8), 663–673.

Ganesan, S., Faris, A.N., Comstock, A.T., et al., 2012. Quercetin inhibits rhinovirus replication in vitro and in vivo. Antivir. Res. 94 (3), 258–271.

Gao, C., Zhang, J., Huang, Q., 1997. Experimental study on the insulin-sensitivity increasing effect of berberine in insulin resistant rat models. Zhongguo Zhong Xi Yi Jie He Za Zhi 17 (3), 162–164.

Ghobeh, M., Ahmadian, S., Meratan, A.A., et al., 2014. Interaction of Aβ(25-35) fibrillation products with mitochondria: effect of small-molecule natural products. Biopolymers 102 (6), 473–486.

Giri, R.K., Rajagopal, V., Kalra, V.K., 2004. Curcumin, the active constituent of turmeric, inhibits amyloid peptide-induced cytochemokine gene expression and CCR5-mediated chemotaxis of THP-1 monocytes by modulating early growth response-1 transcription factor. J. Neurochem. 91, 1199–1210.

Goel, A., Aggarwal, B.B., 2010. Curcumin, the golden spice from Indian saffron, is a chemosensitizer and radiosensitizer for tumors and chemoprotector and radioprotector for normal organs. Nutr. Cancer 62, 919–930.

Gomes, A.P., Duarte, F.V., Nunes, P., et al., 2012. Berberine protects against high fat diet-induced dysfunction in muscle mitochondria by inducing SIRT1-dependent mitochondrial biogenesis. Biochim. Biophys. Acta 1822 (2), 185–195.

Gonzales, A.M., Orlando, R.A., 2008. Curcumin and resveratrol inhibit nuclear factor kappaB-mediated cytokine expression in adipocytes. Nutr. Metab. 5, 17.

Gu, Y., Zhang, Y., Shi, X., et al., 2010. Effect of traditional Chinese medicine berberine on type 2 diabetes based on comprehensive metabonomics. Talanta 81 (3), 766–772.

Guarente, L., 2007. Sirtuins in aging and disease. Cold Spring Harb. Symp. Quant. Biol. 72, 483–488.

Gutierres, V.O., Pinheiro, C.M., Assis, R.P., et al., 2012. Curcumin-supplemented yoghurt improves physiological and biochemical markers of experimental diabetes. Br. J. Nutr. 108, 440–448.

Gutierres, V.O., Assis, R.P., Arcaro, C.A., et al., 2019. Curcumin improves the effect of a reduced insulin dose on glycemic control and oxidative stress in streptozotocin-diabetic rats. Phytother. Res. 33, 976–988.

Hama, S., Takahashi, K., Inai, Y., et al., 2012. Protective effects of topical application of a poorly soluble antioxidant astaxanthin liposomal formulation on ultraviolet-induced skin damage. J. Pharm. Sci. 101 (8), 2909–2916.

Henagan, T.M., Cefalu, W.T., Ribnicky, D.M., et al., 2015. In vivo effects of dietary quercetin and quercetin-rich red onion extract on skeletal muscle mitochondria, metabolism, and insulin sensitivity. Genes Nutr. 10 (1), 451.

Higuera-Ciapara, I., Felix-Valenzuela, L., Goycoolea, F.M., 2006. Astaxanthin: a review of its chemistry and applications. Crit. Rev. Food Sci. Nutr. 46 (2), 185–196.

Hotsumi, M., Tajiri, M., Nikaido, Y., et al., 2019. Design, synthesis, and evaluation of a water soluble C5-monoketone type curcumin analogue as a potent amyloid β aggregation inhibitor. Bioorganic Med. Chem. Lett. 29, 2157–2161.

Hu, X.T., Ding, C., Zhou, N., et al., 2015. Quercetin protects gastric epithelial cell from oxidative damage in vitro and in vivo. Eur. J. Pharmacol. 5, 115–124.

Hua, W.-F., Fu, Y.-S., Liao, Y.-J., et al., 2010. Curcumin induces down-regulation of EZH2 expression through the MAPK pathway in MDA-MB-435 human breast cancer cells. Eur. J. Pharmacol. 637, 16–21.

Hussein, G., Nakamura, M., Zhao, Q., et al., 2005. Antihypertensive and neuroprotective effects of astaxanthin in experimental animals. Biol. Pharm. Bull. 28 (1), 47–52.

Hussein, G., Sankawa, U., Goto, H., et al., 2006. Astaxanthin, a carotenoid with potential in human health and nutrition. J. Nat. Prod. 69 (3), 443–449.

Hussein, G., Nakagawa, T., Goto, H., et al., 2007. Astaxanthin ameliorates features of metabolic syndrome in SHR/NDmcr-cp. Life Sci. 80 (6), 522–529.

Hwang, J.M., Wang, C.J., Chou, F.P., et al., 2002. Inhibitory effect of berberine on tert-butyl hydroperoxide-induced oxidative damage in rat liver. Arch. Toxicol. 76 (11), 664–670.

Ijomone, O.M., Obi, A.U., 2013. Kolaviron, isolated from *Garcinia kola*, inhibits acetylcholinesterase activities in the hippocampus and striatum of wistar rats. Ann. Neurosci. 20 (2), 42–46.

Ikeuchi, M., Koyama, T., Takahashi, J., et al., 2007. Effects of astaxanthin in obese mice fed a high-fat diet. Biosci. Biotechnol. Biochem. 71 (4), 893–899.

Ireson, C., Orr, S., Jones, D.J.L., et al., 2001. Characterization of metabolites of the chemopreventive agent curcumin in human and rat hepatocytes and in the rat in vivo, and evaluation of their ability to inhibit phorbol ester-induced prostaglandin E2 production. Cancer Res. 61, 1058–1064.

Ishiki, M., Nishida, Y., Ishibashi, H., et al., 2013. Impact of divergent effects of astaxanthin on insulin signaling in L6 cells. Endocrinology 154 (8), 2600–2612.

Islam, M.A., Al Mamun, M.A., Faruk, M., et al., 2017. Astaxanthin ameliorates hepatic damage and oxidative stress in carbon tetrachloride-administered rats. Pharmacognosy Res. 9 (suppl. 1), S84–S91.

Jaisin, Y., Thampithak, A., Meesarapee, B., et al., 2011. Curcumin I protects the dopaminergic cell line SH-SY5Y from 6-hydroxydopamine-induced neurotoxicity through attenuation of p53- mediated apoptosis. Neurosci. Lett. 489, 192–196.

Jaldin-Fincati, J.R., Pavarotti, M., Frendo-Cumbo, S., et al., 2017. Update on GLUT4 vesicle traffic: a cornerstone of insulin action. Trends Endocrinol. Metab. 28 (8), 597–611.

James, M.A., Fu, H., Liu, Y., et al., 2011. Dietary administration of berberine or Phellodendron amurense extract inhibits cell cycle progression and lung tumorigenesis. Mol. Carcinog. 50 (1), 1–7.

Jang, E.-M., Choi, M.-S., Jung, U.J., et al., 2008. Beneficial effects of curcumin on hyperlipidemia and insulin resistance in high-fat-fed hamsters. Metabolism 57, 1576–1583.

Ji, H.F., Shen, L., 2011. Berberine: a potential multipotent natural product to combat Alzheimer's disease. Molecules 16 (8), 6732–6740.

Jia, Y., Kim, J.Y., Jun, H.J., et al., 2012. The natural carotenoid astaxanthin, a PPAR-α agonist and PPAR-γ antagonist, reduces hepatic lipid accumulation by rewiring the transcriptome in lipid-loaded hepatocytes. Mol. Nutr. Food Res. 56 (6), 878–888.

Jiang, A.-J., Jiang, G., Li, L.-T., et al., 2015. Curcumin induces apoptosis through mitochondrial pathway and caspases activation in human melanoma cells. Mol. Biol. Rep. 42, 267–275.

Jimenez-Gomez, Y., Mattison, J.A., Pearson, K.J., et al., 2013. Resveratrol improves adipose insulin signaling and reduces the inflammatory response in adipose tissue of rhesus monkeys on high-fat, high-sugar diet. Cell Metab. 18 (4), 533–545.

Jiménez-Flores, L.M., López-Briones, S., Macías-Cervantes, M.H., et al., 2014. A PPARγ, NF-κB and AMPK-dependent mechanism may be involved in the beneficial effects of curcumin in the diabetic db/db mice liver. Molecules 19, 8289–8302.

Jung, H.A., Min, B.S., Yokozawa, T., et al., 2009. Anti- Alzheimer and antioxidant activities of *Coptidis Rhizoma* alkaloids. Biol. Pharm. Bull. 32 (8), 1433–1438.

Kamath, B.S., Srikanta, B.M., Dharmesh, S.M., et al., 2008. Ulcer preventive and antioxidative properties of astaxanthin from *Haematococcus pluvialis*. Eur. J. Pharmacol. 590 (1–3), 387–395.

Karppi, J., Rissanen, T.H., Nyyssönen, K., et al., 2007. Effects of astaxanthin supplementation on lipid peroxidation. Int. J. Vitam. Nutr. Res. 77 (1), 3–11.

Karthikesan, K., Pari, L., Menon, V.P., 2010. Antihyperlipidemic effect of chlorogenic acid and tetrahydrocurcumin in rats subjected to diabetogenic agents. Chem. Biol. Interact. 188, 643–650.

Kaur, G., Meena, C., 2012. Amelioration of obesity, glucose intolerance, and oxidative stress in high-fat diet and low-dose streptozotocin-induced diabetic rats by combination consisting of "curcumin with piperine and quercetin.". ISRN Pharmacol. 2012, 957283.

Khan, M., Giessrigi, B., Vonach, C., et al., 2010. Berberine and a Berberis lycium extract inactivate Cdc25A and induce α-tubulin acetylation that correlate with HL-60 cell cycle inhibition and apoptosis. Mutat. Res. 683 (1-2), 123–130.

Khanduja, K.L., Bhardwaj, A., Kaushik, G., 2004. Resveratrol inhibits N-nitrosodiethylamine- induced ornithine decarboxylase and cyclooxygenase in mice. J. Nutr. Sci. Vitaminol. (Tokyo) 50 (1), 61–65.

Kim, S.H., Kim, H., 2018. Inhibitory effect of astaxanthin on oxidative stress-induced mitochondrial dysfunction-a mini-review. Nutrients 10 (9), pii: E1137.

Kim, D., Nguyen, M.D., Dobbin, M.M., et al., 2007. SIRT1 deacetylase protects against neurodegeneration in models for Alzheimer's disease and amyotrophic lateral sclerosis. EMBO J. 26 (13), 3169–3179.

Kim, T., Davis, J., Zhang, A.J., et al., 2009a. Curcumin activates AMPK and suppresses gluconeogenic gene expression in hepatoma cells. Biochem. Biophys. Res. Commun. 388, 377–382.

Kim, Y.J., Kim, Y.A., Yokozawa, T., 2009b. Protection against oxidative stress, inflammation, and apoptosis of high-glucose-exposed proximal tubular epithelial cells by astaxanthin. J. Agric. Food Chem. 57 (19), 8793–8797.

Komaravelli, N., Kelly, J.P., Garofalo, M.P., et al., 2015. Role of dietary antioxidants in human metapneumovirus infection. Virus Res. 200C, 19–23.

Kong, L.D., Cheng, C.H., Tan, R.X., 2001. Monoamine oxidase inhibitors from rhizome of *Coptis chinensis*. Planta Med. 67 (1), 74–76.

Kong, W.J., Zhang, H., Song, D.Q., et al., 2009. Berberine reduces insulin resistance through protein kinase C-dependent upregulation of insulin receptor expression. Metabolism 58 (1), 109–119.

Kosaraju, J., Chinni, S., Roy, P.D., et al., 2014. Neuroprotective effect of *Tinospora cordifolia* ethanol extract on 6-hydroxy dopamine induced Parkinsonism. Indian J. Pharmacol. 46 (2), 176–180.

Kulkarni, S.K., Dhir, A., 2009. Current investigational drugs for major depression. Expert Opin. Investig. Drugs 18 (6), 767–788.

Kulkarni, S.K., Dhir, A., 2010. Berberine: a plant alkaloid with therapeutic potential for central nervous system disorders. Phytother. Res. 24 (3), 317–324.

Kumar, A.P., Bhaskaran, S., Ganapathy, M., et al., 2007. Akt/cAMP-responsive element binding protein/cyclin D1 network: a novel target for prostate cancer inhibition in transgenic adenocarcinoma of mouse prostate model mediated by Nexrutine, a *Philodendron amurense* bark extract. Clin. Cancer Res. 13 (9), 2784–2794.

Kumar, A., Prakash, A., Dogra, S., 2011. Protective effect of curcumin (*Curcuma longa*) against D-galactose-induced senescence in mice. J. Asian Nat. Prod. Res. 13, 42–55.

Kumar, R., Das, M., Ansari, K.M., 2012. Nexrutine(R) inhibits tumorigenesis in mouse skin and induces apoptotic cell death in human squamous carcinoma A431 and human melanoma A375 cells. Carcinogenesis 33 (10), 1909–1918.

Kumar, G., Mittal, S., Sak, K., Tuli, H.S., 2016. Molecular mechanisms underlying chemopreventive potential of curcumin: current challenges and future perspectives. Life Sci. 148, 313–328.

Kurashige, M., Okimasu, E., Inoue, M., et al., 1990. Inhibition of oxidative injury of biological membranes by astaxanthin. Physiol. Chem. Phys. Med. NMR 22 (1), 27–38.

Kuroki, T., Ikeda, S., Okada, T., et al., 2013. Astaxanthin ameliorates heat stress-induced impairment of blastocyst development in vitro: astaxanthin colocalization with and action on mitochondria. J. Assist. Reprod. Genet. 30 (5), 623–631.

Kwon, K.H., Barve, A., Yu, S., et al., 2007. Cancer chemoprevention by phytochemicals: potential molecular targets, biomarkers and animal models. Acta Pharmacol. Sin. 28 (9), 1409–1421.

Lagouge, M., Argmann, C., Gerhart-Hines, Z., et al., 2006. Resveratrol improves mitochondrial function and protects against metabolic disease by activating SIRT1 and PGC-1alpha. Cell 127 (6), 1109–1122.

Lakroun, Z., Kebieche, M., Lahouel, A., et al., 2015. Oxidative stress and brain mitochondria swelling induced by endosulfan and protective role of quercetin in rat. Environ. Sci. Pollut. Res. Int. 22 (10), 7776–7781.

Lee, D.H., Kim, C.S., Lee, Y.J., 2011. Astaxanthin protects against MPTP/MPP + induced mitochondrial dysfunction and ROS production in vivo and in vitro. Food Chem. Toxicol. 49 (1), 271–280.

Lev-Ari, S., Starr, A., Vexler, A., et al., 2006. Inhibition of pancreatic and lung adenocarcinoma cell survival by curcumin is associated with increased apoptosis, down-regulation of COX-2 and EGFR and inhibition of Erk1/2 activity. Anticancer Res. 26, 4423–4430.

Li, H., Dong, B., Park, S.W., et al., 2009. Hepatocyte nuclear factor 1alpha plays a critical role in PCSK9 gene transcription and regulation by the natural hypocholesterolemic compound berberine. J. Biol. Chem. 284 (42), 28885–28895.

Li, J.-M., Li, Y.-C., Kong, L.-D., et al., 2010. Curcumin inhibits hepatic protein tyrosine phosphatase 1B and prevents hypertriglyceridemia and hepatic steatosis in fructose-fed rats. Hepatology 51, 1555–1566.

Li, Y., Zhang, S., Geng, J., et al., 2013. Curcumin inhibits human non-small cell lung cancer A549 cell proliferation through regulation of Bcl-2/Bax and cytochrome c. Asian Pac. J. Cancer Prev. 14, 4599–4602.

Li, X., Xie, W., Xie, C., et al., 2014. Curcumin modulates miR-19/PTEN/AKT/p53 axis to suppress bisphenol A-induced MCF-7 breast cancer cell proliferation. Phytother. Res. 28, 1553–1560.

Li, C.-N., Wang, X., Lei, L., et al., 2020. Berberine combined with stachyose induces better glycometabolism than berberine alone through modulating gut microbiota and fecal metabolomics in diabetic mice. Phytother Res. 34, 1166–1174 (Epub ahead of print).

Lin, S., Tsai, S.C., Lee, C.C., et al., 2004. Berberine inhibits HIF-1alpha expression via enhanced proteolysis. Mol. Pharmacol. 66 (3), 612–619.

Lin, Q., Zhang, L., Yang, D., et al., 2014. Contribution of phenolics and essential oils to the antioxidant and antimicrobial properties of *Disporopsis pernyi* (Hua) Diels. J. Med. Food 17 (6), 714–722.

Liu, X., Osawa, T., 2007. Cis astaxanthin and especially 9-cis astaxanthin exhibits a higher antioxidant activity in vitro compared to the all-trans isomer. Biochem. Biophys. Res. Commun. 357 (1), 187–193.

Liu, Y., Yu, H., Zhang, C., et al., 2008. Protective effects of berberine on radiation-induced lung injury via intercellular adhesion molecular 1 and transforming growth factor-beta-1 in patients with lung cancer. Eur. J. Cancer 44 (16), 2425–2432.

Liu, L., Yu, L., Yang, J.S., et al., 2010a. Berberine suppresses intestinal disaccharidases with beneficial metabolic effects in diabetic states, evidences from in vivo and in vitro study. Naunyn- Schmiedeberg's Arch. Pharmacol. 381 (4), 371–381.

Liu, L.Z., Cheung, S.C., Lan, L.L., et al., 2010b. Berberine modulates insulin signaling transduction in insulin-resistant cells. Mol. Cell Endocrinol. 317 (1-2), 148–153.

Liu, X., Li, G., Zhu, H., et al., 2010c. Beneficial effect of berberine on hepatic insulin resistance in diabetic hamsters possibly involves in SREBPs, LXRα and PPARα transcriptional programs. Endocr. J. 57 (10), 881–893.

Liu, Y., Zhang, L., Song, H., et al., 2013. Update on berberine in nonalcoholic fatty liver disease. Evid. Based Complement. Altern. Med. 2013, 308134.

Liu, K., Zhou, R., Wang, B., et al., 2014a. Effect of resveratrol on glucose control and insulin sensitivity: a meta-analysis of 11 randomized controlled trials. Am. J. Clin. Nutr. 99 (6), 1510–1519.

Liu, L., Zhang, W., Wang, L., et al., 2014b. Curcumin prevents cerebral ischemia reperfusion injury via increase of mitochondrial biogenesis. Neurochem. Res. 39, 1322–1331.

Liu, P., Zou, D., Yi, L., et al., 2015. Quercetin ameliorates hypobaric hypoxia-induced memory impairment through mitochondrial and neuron function adaptation via the PGC 1α pathway. Restor. Neurol. Neurosci. 33 (2), 143–157.

Liu, Z., Cui, C., Xu, P., et al., 2017. Curcumin activates AMPK pathway and regulates lipid metabolism in rats following prolonged clozapine exposure. Front. Neurosci. 11, 558.

Liu, H., Liu, M., Fu, X., et al., 2018. Astaxanthin prevents alcoholic fatty liver disease by modulating mouse gut microbiota. Nutrients 10 (9), E1298.

Lo, T.F., Tsai, W.C., Chen, S.T., 2013. MicroRNA-21-3p, a berberine-induced miRNA, directly down-regulates human methionine adenosyl transferases 2A and 2B and inhibits hepatoma cell growth. PLoS One 8 (9), e75628.

Lone, J., Choi, J.H., Kim, S.W., et al., 2016. Curcumin induces brown fat-like phenotype in 3T3-L1 and primary white adipocytes. J. Nutr. Biochem. 27, 193–202.

Lopes, T.Z., de Moraes, F.B., Tedesco, A.C., et al., 2020. Berberine associated photodynamic therapy promotes autophagy and apoptosis via ROS generation in renal carcinoma cells. Biomed. Pharacother. 123, 109794.

Lorenz, R., 1999. A technical review of *Haematococcus* algae. NatuRose Tech. Bull. 60, 1–12.

Lotharius, J., Brundin, P., 2002. Pathogenesis of Parkinson's disease: dopamine, vesicles and α-synuclein. Nat. Rev. Neurosci. 3, 932–942.

Lu, S.S., Yu, Y.L., Zhu, H.J., et al., 2009. Berberine promotes glucagon-like peptide-1 (7-36) amide secretion in streptozotocin-induced diabetic rats. J. Endocrinol. 200 (2), 159–165.

Mahdi, J.G., Mahdi, A.J., Mahdi, A.J., et al., 2006. The historical analysis of aspirin discovery, its relation to the willow tree and antiproliferative and anticancer potential. Cell Prolif. 39 (2), 147–155.

Majithiya, J.B., Balaraman, R., 2005. Time-dependent changes in antioxidant enzymes and vascular reactivity of aorta in streptozotocin-induced diabetic rats treated with curcumin. J. Cardiovasc. Pharmacol. 46, 697–705.

Manabe, E., Handa, O., Naito, Y., et al., 2008. Astaxanthin protects mesangial cells from hyperglycemia-induced oxidative signaling. J. Cell Biochem. 103 (6), 1925–1937.

Mandal, M., Jaiswal, P., Mishra, A., 2020. Role of curcumin and its nanoformulations in neurotherapeutics: a comprehensive review. J. Biochem. Mol. Toxicol. 34, e22478.

Mashhadi, N.S., Zakerkish, M., Mohammadiasl, J., et al., 2018. Astaxanthin improves glucose metabolism and reduces blood pressure in patients with type 2 diabetes mellitus. Asia Pac. J. Clin. Nutr. 27 (2), 341–346.

McCarty, M.F., 2011. Full-spectrum antioxidant therapy featuring astaxanthin coupled with lipoprivic strategies and salsalate for management of non-alcoholic fatty liver disease. Med. Hypotheses 77 (4), 550–556.

McNulty, H.P., Byun, J., Lockwood, S.F., et al., 2007. Differential effects of carotenoids on lipid peroxidation due to membrane interactions: X-ray diffraction analysis. Biochim. Biophys. Acta Biomemb 1768 (1), 167–174.

Murillo, A.G., Fernandez, M.L., 2016. Potential of dietary non-provitamin a carotenoids in the prevention and treatment of diabetic microvascular complications. Adv. Nutr. 7 (1), 14–24.

Na, L.-X., Zhang, Y.-L., Li, Y., Liu, L.-Y., Li, R., Kong, T., et al., 2011. Curcumin improves insulin resistance in skeletal muscle of rats. Nutr. Metab. Cardiovasc. Dis. 21, 526–533.

Naito, Y., Uchiyama, K., Aoi, W., et al., 2004. Prevention of diabetic nephropathy by treatment with astaxanthin in diabetic db/db mice. BioFactors 20 (1), 49–59.

Nakao, R., Nelson, O.L., Park, J.S., et al., 2010. Effect of astaxanthin supplementation on inflammation and cardiac function in BALB/c mice. Anticancer Res. 30 (7), 2721–2725.

Nday, C.M., Halevas, E., Jackson, G.E., et al., 2015. Quercetin encapsulation in modified silica nanoparticles: potential use against Cu(II)- induced oxidative stress in neurodegeneration. J. Inorg. Biochem. 145C, 51–64.

Nisha, V.M., Anusree, S.S., Priyanka, A., et al., 2014. Apigenin and quercetin ameliorate mitochondrial alterations by tunicamycin-induced ER stress in 3T3-L1 adipocytes. Appl. Biochem. Biotechnol. 174 (4), 1365–1375.

Nishizono, S., Hayami, T., Ikeda, I., et al., 2000. Protection against the diabetogenic effect of feeding tert-butylhydroquinone to rats prior to the administration of streptozotocin. Biosci. Biotechnol. Biochem. 64, 1153–1158.

Novelle, M.G., Wahl, D., Diéguez, C., Bernier, M., de Cabo, R., 2015. Resveratrol supplementation: where are we now and where should we go? Ageing Res. Rev. 21C, 1–15.

Oluwatosin, A., Tolulope, A., Ayokulehin, K., et al., 2014. Antimalarial potential of kolaviron, a biflavonoid from *Garcinia kola* seeds, against *Plasmodium berghei* infection in Swiss albino mice. Asian Pac. J. Trop. Med. 7 (2), 97–104.

Ono, K., Hasegawa, K., Naiki, H., et al., 2004. Curcumin has potent antiamyloidogenic effects for Alzheimer's β-amyloid fibrils in vitro. J. Neurosci. Res. 75, 742–750.

Ortiz, L., Lombardi, P., Tillhon, M., et al., 2014. Berberine, an epiphany against cancer. Molecules. 19 (8), 12349–12367.

Otton, R., Marin, D.P., Bolin, A.P., et al., 2010. Astaxanthin ameliorates the redox imbalance in lymphocytes of experimental diabetic rats. Chem. Biol. Interact. 186 (3), 306–315.

Palmeira, C.M., Duarte, F.V., Teodoro, J.S., Varela, A.T., Rolo, A.P., 2014. Biomarkers of mitochondrial dysfunction and toxicity. In: Gupta, R. (Ed.), Biomarkers Toxicology, 55. pp. 847–862.

Palmeira, C.M., Teodoro, J.S., Amorim, J.Á., et al., 2019. Mitohormesis and metabolic health: the interplay between ROS, cAMP and sirtuins. Free Radical Biol. Med. 141, 483–491.

Pan, H., Li, Z., Xie, J., Liu, D., et al., 2019. Berberine influences blood glucose via modulating the gut microbiome in grass carp. Front. Microbiol. 10, 1066.

Pandey, N., Strider, J., Nolan, W.C., et al., 2008. Curcumin inhibits aggregation of α-synuclein. Acta Neuropathol. 115, 479–489.

Park, S.-Y., Kim, H.-S., Cho, E.-K., et al., 2008. Curcumin protected PC12 cells against beta-amyloid-induced toxicity through the inhibition of oxidative damage and tau hyperphosphorylation. Food Chem. Toxicol. 46, 2881–2887.

Park, J.S., Chyun, J.H., Kim, Y.K., et al., 2010. Astaxanthin decreased oxidative stress and inflammation and enhanced immune response in humans. Nutr. Metab. (Lond.) 7, 18.

Park, J.S., Mathison, B.D., Hayek, M.G., et al., 2013. Astaxanthin modulates age-associated mitochondrial dysfunction in healthy dogs. J. Anim. Sci. 91 (1), 268–275.

Parker, J.A., Arango, M., Abderrahmane, S., et al., 2005. Resveratrol rescues mutant polyglutamine cytotoxicity in nematode and mammalian neurons. Nat. Genet. 37 (4), 349–350.

Pashkow, F.J., Watumull, D.G., Campbell, C.L., 2008. Astaxanthin: a novel potential treatment for oxidative stress and inflammation in cardiovascular disease. Am. J. Cardiol. 101 (10A), 58–68.

Pearson, K.J., Baur, J.A., Lewis, K.N., et al., 2008. Resveratrol delays age-related deterioration and mimics transcriptional aspects of dietary restriction without extending life span. Cell Metab. 8 (2), 157–168.

Pereira, G.C., Branco, A.F., Matos, J.A., et al., 2007. Mitochondrially targeted effects of berberine [Natural Yellow 18, 5, 6-dihydro- 9, 10 dimethoxybenzo (g)-1, 3-benzodioxolo (5, 6-a) quinolizinium] on K1735-M2 mouse melanoma cells: comparison with direct effects on isolated mitochondrial fractions. J. Pharmacol. Exp. Ther. 323 (2), 636–649.

Perrone, D., Ardito, F., Giannatempo, G., et al., 2015. Biological and therapeutic activities, and anticancer properties of curcumin. Exp. Ther. Med. 10, 1615–1623.

Phachonpai, W., Wattanathorn, J., Muchimapura, S., et al., 2010. Neuroprotective effect of quercetin encapsulated liposomes: a novel therapeutic strategy against Alzheimer's disease. Am. J. Appl. Sci. 7 (4), 480–485.

Pongkan, W., Takatori, O., Ni, Y., et al., 2017. β-Cryptoxanthin exerts greater cardioprotective effects on cardiac ischemia-reperfusion injury than astaxanthin by attenuating mitochondrial dysfunction in mice. Mol. Nutr. Food Res. 61 (10).

Preuss, H.G., Echard, B., Yamashita, E., et al., 2011. High dose astaxanthin lowers blood pressure and increases insulin sensitivity in rats: are these effects interdependent? Int. J. Med. Sci. 8 (2), 126–138.

Price, N.L., Gomes, A.P., Ling, A.J., et al., 2012. SIRT1 is required for AMPK activation and the beneficial effects of resveratrol on mitochondrial function. Cell Metab. 15 (5), 675–690.

Priyadarsini, K.I., 2014. The chemistry of curcumin: from extraction to therapeutic agent. Molecules 19 (12), 20091–20112.

Qu, X., Qi, D., Dong, F., et al., 2014. Quercetin improves hypoxia-ischemia induced cognitive deficits via promoting remyelination in neonatal rat. Brain Res. 1553, 31–40.

Rao, A.R., Sindhuja, H.N., Dharmesh, S.M., et al., 2013. Effective inhibition of skin cancer, tyrosinase, and antioxidative properties by astaxanthin and astaxanthin esters from the green alga *Haematococcus pluvialis*. J. Agric. Food Chem. 61 (16), 3842–3851.

Ravi, S., Ambati, R.R., Kamath, S.B., et al., 2012. Influence of different culture conditions on yield of biomass and value added products in microalgae. Dyn. Biochem. Process Biotechnol. Mol. Biol. 6 (2), 77–85.

Ray Hamidie, R.D., Yamada, T., Ishizawa, R., et al., 2015. Curcumin treatment enhances the effect of exercise on mitochondrial biogenesis in skeletal muscle by increasing cAMP levels. Metabolism 64, 1334–1347.

Renugadevi, J., Prabu, S.M., 2010. Quercetin protects against oxidative stress related renal dysfunction by cadmium in rats. Exp. Toxicol. Pathol. 62 (5), 471–481.

Ringman, J.M., Frautschy, S.A., Cole, G.M., et al., 2005. A potential role of the curry spice curcumin in Alzheimer's disease. Curr. Alzheimer Res. 2, 131–136.

Rizza, W., Veronese, N., Fontana, L., 2014. What are the roles of calorie restriction and diet quality in promoting healthy longevity? Ageing Res. Rev. 13, 38–45.

Rolo, A.P., Palmeira, C.M., 2006. Diabetes and mitochondrial function: role of hyperglycemia and oxidative stress. Toxicol. Appl. Pharmacol. 212, 167–178.

Santos, S.D., Cahú, T.B., Firmino, G.O., et al., 2012. Shrimp waste extract and astaxanthin: rat alveolar macrophage, oxidative stress and inflammation. J. Food Sci. 77 (7), 141–146.

Sarada, R., Tripathi, U., Ravishankar, G.A., 2002. Influence of stress on astaxanthin production in *Haematococcus pluvialis* grown under different culture conditions. Process. Biochem. 37 (6), 623–627.

Schreiber, I., Dorpholz, G., Ott, C.E., et al., 2017. BMPs as new insulin sensitizers: enhanced glucose uptake in mature 3T3-L1 adipocytes via PPARγ and GLUT4 upregulation. Sci. Rep. 7 (1), 17192.

Sengottuvelan, M., Viswanathan, P., Nalini, N., 2006. Chemopreventive effect of trans-resveratrol—a phytoalexin against colonic aberrant crypt foci and cell proliferation in 1,2-dimethylhydrazine induced colon carcinogenesis. Carcinogenesis 27 (5), 1038–1046.

Seo, K.-I., Choi, M.-S., Jung, U.J., et al., 2008. Effect of curcumin supplementation on blood glucose, plasma insulin, and glucose homeostasis related enzyme activities in diabetic db/db mice. Mol. Nutr. Food Res. 52, 995–1004.

Shao, W., Yu, Z., Chiang, Y., et al., 2012. Curcumin prevents high fat diet induced insulin resistance and obesity via attenuating lipogenesis in liver and inflammatory pathway in adipocytes. PLoS One 7, e28784.

Sharma, N., Nehru, B., 2018. Curcumin affords neuroprotection and inhibits α-synuclein aggregation in lipopolysaccharide-induced Parkinson's disease model. Inflammopharmacology 26, 349–360.

Sharma, R.A., Gescher, A.J., Steward, W.P., 2005. Curcumin: the story so far. Eur. J. Cancer 41, 1955–1968.

Sharma, N., Sharma, S., Nehru, B., 2017. Curcumin protects dopaminergic neurons against inflammation-mediated damage and improves motor dysfunction induced by single intranigral lipopolysaccharide injection. Inflammopharmacology 25, 351–368.

Shin, Y.J., Kim, J.J., Kim, Y.J., et al., 2015. Protective effects of quercetin against HgCl2-induced nephrotoxicity in Sprague-Dawley Rats. J. Med. Food 18 (5), 524–534.

Shirwany, N.A., Zou, M.H., 2010. AMPK in cardiovascular health and disease. Acta Pharmacol. Sin. 31 (9), 1075–1084.

Shytle, R.D., Bickford, P.C., Rezai-zadeh, K., et al., 2009. Optimized turmeric extracts have potent anti-amyloidogenic effects. Curr. Alzheimer Res. 6, 564–571.

Simons, M., Keller, P., Dichgans, J., et al., 2001. Cholesterol and Alzheimer's disease: is there a link? Neurology 57 (6), 1089–1093.

Soto-Urquieta, M.G., López-Briones, S., Pérez-Vázquez, V., et al., 2014. Curcumin restores mitochondrial functions and decreases lipid peroxidation in liver and kidneys of diabetic db/db mice. Biol. Res. 47, 74.

Srinivasan, M., 1972. Effect of curcumin on blood sugar as seen in a diabetic subject. Indian J. Med. Sci. 26, 269–270.

Strong, R., Miller, R.A., Astle, C.M., et al., 2013. Evaluation of resveratrol, green tea extract, curcumin, oxaloacetic acid, and medium-chain triglyceride oil on life span of genetically heterogeneous mice. J. Gerontol. A Biol. Sci. Med. Sci 68 (1), 6–16.

Sun, Y., Xun, K., Wang, Y., et al., 2009. A systematic review of the anticancer properties of berberine, a natural product from Chinese herbs. Anticancer Drugs 20 (9), 757–769.

Sztretye, M., Dienes, B., Gonczi, M., et al., 2019. Astaxanthin: a potential mitochondrial-targeted antioxidant treatment in diseases and with aging. Oxid. Med. Cell Longev. 2019, 3849692.

Teodoro, J.S., Duarte, F.V., Gomes, A.P., et al., 2013. Berberine reverts hepatic mitochondrial dysfunction in high-fat fed rats: a possible role for SirT3 activation. Mitochondrion 13 (6), 637–646.

Thiebaud, D., Jacot, E., Maeder, E., et al., 1982. The effect of graded doses of insulin on total glucose uptake, glucose oxidation, and glucose storage in man. Diabetes 31 (11), 957–963.

Tian, Y., Cai, J., Gui, W., et al., 2019. Berberine directly affects the gut microbiota to promote intestinal Farnesoid X Receptor activation. Drug Metab. Dispos. 47 (2), 86–93.

Trujillo, J., Granados-Castro, L.F., Zazueta, C., et al., 2014. Mitochondria as a target in the therapeutic properties of curcumin. Arch. Pharm. (Weinh.) 347, 873–884.

Tsang, R.K.-Y., Tang, W.W.-Y., Gao, W., et al., 2012. Curcumin inhibits tongue carcinoma cells migration and invasion through downregulation of matrix metallopeptidase 10. Cancer Invest. 30, 503–512.

Tseng, S.H., Lin, S.M., Chen, J.C., et al., 2004. Resveratrol suppresses the angiogenesis and tumor growth of gliomas in rats. Clin. Cancer Res. 10 (6), 2190–2202.

Turner, N., Li, J.Y., Gosby, A., et al., 2008. Berberine and its more biologically available derivative, dihydroberberine, inhibit mitochondrial respiratory complex I: a mechanism for the action of berberine to activate AMP activated protein kinase and improve insulin action. Diabetes 57 (5), 1414–1418.

Uchide, N., Toyoda, H., 2011. Antioxidant therapy as a potential approach to severe influenza-associated complications. Molecules 16 (3), 2032–2052.

Uchiyama, K., Naito, Y., Hasegawa, G., et al., 2002. Astaxanthin protects β-cells against glucose toxicity in diabetic db/db mice. Redox Rep. 7 (5), 290–293.

Um, M.Y., Hwang, K.H., Ahn, J., et al., 2013. Curcumin attenuates diet-induced hepatic steatosis by activating AMP-activated protein kinase. Basic Clin. Pharmacol. Toxicol. 113, 152–157.

Visioli, F., Artaria, C., 2017. Astaxanthin in cardiovascular health and disease: mechanisms of action, therapeutic merits, and knowledge gaps. Food Funct. 8 (1), 39–63.

Vlachodimitropoulou, E., Sharp, P.A., Naftalin, R.J., 2011. Quercetin-iron chelates are transported via glucose transporters. Free Radic. Biol. Med. 50 (8), 934–944.

Wahlstrom, B., Blennow, G., 1978. A study on the fate of curcumin in the rat. Acta Pharmacol. Toxicol. (Copenh) 43, 86–92.

Wang, Z.Q., Lu, F.E., Leng, S.H., et al., 2008. Facilitating effects of berberine on rat pancreatic islets through modulating hepatic nuclear factor 4 alpha expression and glucokinase activity. World J. Gastroenterol. 14 (39), 6004–6011.

Wang, J., Du, X.-X., Jiang, H., et al., 2009. Curcumin attenuates 6-hydroxydopamine-induced cytotoxicity by anti-oxidation and nuclear factor kappaB modulation in MES23.5 cells. Biochem. Pharmacol. 78, 178–183.

Wang, M.S., Boddapati, S., Emadi, S., et al., 2010a. Curcumin reduces α- synuclein induced cytotoxicity in Parkinson's disease cell model. BMC Neurosci. 11, 57.

Wang, Y., Jia, X., Ghanam, K., et al., 2010b. Berberine and plant stanols synergistically inhibit cholesterol absorption in hamsters. Atherosclerosis 209 (1), 111–117.

Wang, Y., Campbell, T., Perry, B., et al., 2011. Hypoglycemic and insulin-sensitizing effects of berberine in high-fat diet- and streptozotocin induced diabetic rats. Metabolism 60 (2), 298–305.

Wang, L., Lin, S.Q., He, Y.L., et al., 2013. Protective effects of quercetin on cadmium-induced cytotoxicity in primary cultures of rat proximal tubular cells. Biomed. Environ. Sci. 26 (4), 258–267.

Weisberg, S.P., Leibel, R., Tortoriello, D.V., 2008. Dietary curcumin significantly improves obesity-associated inflammation and diabetes in mouse models of diabesity. Endocrinology 149, 3549–3558.

Wojcik, M., Krawczyk, M., Wojcik, P., et al., 2018. Molecular mechanisms underlying curcumin-mediated therapeutic effects in type 2 diabetes and cancer. Oxid. Med. Cell Longev. 2018, 9698258.

Wolf, G., 1992. Retinoids and carotenoids as inhibitors of carcinogenesis and inducers of cell-cell communication. Nutr. Rev. 50 (9), 270–274.

Wolf, A.M., Asoh, S., Hiranuma, H., et al., 2010. Astaxanthin protects mitochondrial redox state and functional integrity against oxidative stress. J. Nutr. Biochem. 21 (5), 381–389.

Wu, Z., Xu, Q., Zhang, L., et al., 2009. Protective effect of resveratrol against kainate-induced temporal lobe epilepsy in rats. Neurochem. Res. 34 (8), 1393–1400.

Xu, L., Zhu, J., Yin, W., et al., 2015. Astaxanthin improves cognitive deficits from oxidative stress, nitric oxide synthase and inflammation through upregulation of PI3K/Akt in diabetes rat. Int. J. Clin. Exp. Pathol. 8 (6), 6083–6094.

Xuan, R.R., Niu, T.T., Chen, H.M., 2016. Astaxanthin blocks preeclampsia progression by suppressing oxidative stress and inflammation. Mol. Med. Rep. 14 (3), 2697–2704.

Yakubu, M.T., Quadri, A.L., 2012. *Garcinia kola* seeds: is the aqueous extract a true aphrodisiac in male Wistar rats? Afr. J. Tradit. Complement. Altern. Med. 9 (4), 530–535.

Yamashita, E., 2013. Astaxanthin as a medical food. Funct. Foods Health Dis. 3 (7), 254–258.

Yan, X.M., Joo, M.J., Lim, J.C., et al., 2011. The effect of quercetin-3-O-β-D-glucuronopyranoside on indomethacin-induced gastric damage in rats via induction of mucus secretion and down-regulation of ICAM-1 expression. Arch. Pharm. Res. 34 (9), 1527–1534.

Yang, J., Cao, Y., Sun, J., et al., 2010. Curcumin reduces the expression of Bcl- 2 by upregulating miR-15a and miR-16 in MCF-7 cells. Med. Oncol. 27, 1114–1118.

Yang, Y., Pham, T.X., Wegner, C.J., et al., 2014. Astaxanthin lowers plasma TAG concentrations and increases hepatic antioxidant gene expression in diet-induced obesity mice. Br. J. Nutr. 112 (11), 1797–1804.

Yeh, P.T., Huang, H.W., Yang, C.M., et al., 2016. Astaxanthin inhibits expression of retinal oxidative stress and inflammatory mediators in streptozotocin-induced diabetic rats. PLoS One 11 (1), e0146438.

Yin, J., Hu, R., Chen, M., et al., 2002. Effects of berberine on glucose metabolism in vitro. Metabolism 51 (11), 1439–1443.

Yin, J., Gao, Z., Liu, Z., et al., 2008. Berberine improves glucose metabolism through induction of glycolysis. Am. J. Physiol. Endocrinol. Metab. 294 (1), E148–E156.

Yin, J., Ye, J., Jia, W., 2012. Effects and mechanisms of berberine in diabetes treatment. Acta Pharm. Sin. B. 2 (4), 327–334.

Yoshida, H., Yanai, H., Ito, K., et al., 2010. Administration of natural astaxanthin increases serum HDL-cholesterol and adiponectin in subjects with mild hyperlipidemia. Atherosclerosis 209 (2), 520–523.

Yuan, J.P., Peng, J., Yin, K., et al., 2011. Potential health-promoting effects of astaxanthin: a high-value carotenoid mostly from microalgae. Mol. Nutr. Food Res. 55 (1), 150–165.

Zeng, X.H., Zeng, X.J., Li, Y.Y., 2003. Efficacy and safety of berberine for congestive heart failure secondary to ischemic or idiopathic dilated cardiomyopathy. Am. J. Cardiol. 92 (2), 173–176.

Zhang, Y., Li, X., Zou, D., et al., 2008. Treatment of type 2 diabetes and dyslipidemia with the natural plant alkaloid berberine. J. Clin. Endocrinol. Metab. 93 (7), 2559–2565.

Zhang, J., Yang, J.Q., He, B.C., et al., 2009. Berberine and total base from rhizoma coptis chinensis attenuate brain injury in an aluminum-induced rat model of neurodegenerative disease. Saudi Med. J. 30 (6), 760–766.

Zhang, H., Wei, J., Xue, R., et al., 2010a. Berberine lowers blood glucose in type 2 diabetes mellitus patients through increasing insulin receptor expression. Metabolism 59 (2), 285–292.

Zhang, X., Gu, L., Li, J., et al., 2010b. Degradation of MDM2 by the interaction between berberine and DAXX leads to potent apoptosis in MDM2-overexpressing cancer cells. Cancer Res. 70 (23), 9895-904.

Zhang, X., Zhao, Y., Zhang, M., et al., 2012. Structural changes of gut microbiota during berberine-mediated prevention of obesity and insulin resistance in high-fat diet-fed rats. PLoS One 7 (8), e42529.

Zhang, Z.W., Xu, X.C., Liu, T., et al., 2016. Mitochondrion-permeable antioxidants to treat ROS-burst-mediated acute diseases. Oxid. Med. Cell Longev. 2016 (1), 6859523.

Zhang, W., Xu, J.-H., Yu, T., et al., 2019. Effects of berberine and metformin on intestinal inflammation and gut microbiome composition in db/db mice. Biomed. Pharmacother. 118, 109131.

Zhou, X.Q., Zeng, X.N., Kong, H., et al., 2008. Neuroprotective effects of berberine on stroke models in vitro and in vivo. Neurosci. Lett. 447 (1), 31–36.

Zhu, X., Wei, Y., Yang, B., et al., 2020. The mitohormetic response as part of the cytoprotection mechanism of berberine: berberine induces mitohormesis and mechanisms. Mol. Med. 26 (1), 10.

Chapter 31

Oxidative stress and excitotoxicity: antioxidants from nutraceuticals

Snjezana Zaja-Milatovic and Ramesh C. Gupta

Introduction

Oxidative stress is defined as the imbalance between the generation of reactive oxygen/nitrogen species (ROS/RNS) and the ability of the cells to neutralize them by antioxidant defense. Mitochondrial oxidative phosphorylation generates the majority of free radicals in the cell (Quinlan et al., 2012a, b; Federico et al., 2012; Trewin et al., 2018; Sas et al., 2018). ROS are produced mainly as superoxide when electrons leak onto molecular oxygen inside reactions from prosthetic groups or coenzymes involved in these redox reactions. It has been estimated that as much as 1% of total mitochondrial O_2 consumption is used to produce superoxide (Quinlan et al., 2012a,b; Handy and Loscalzo, 2012). Superoxide further serves as a precursor to the formation of the hydroxyl radical (OH^-), the most reactive radical. The hydroxyl radical is produced by the Fenton reaction and the Haber-Weiss reaction from hydrogen peroxide (H_2O_2) and metal species (iron, copper) (Dix and Aikens, 1993; Guéraud et al., 2010). Superoxide easily reacts with nitric oxide (NO) and forms peroxynitrite ($ONOO^-$), an RNS with very high reactivity (Vatassery, 2004). Peroxynitrite is a powerful oxidant exhibiting a wide array of tissue-damaging effects ranging from lipid peroxidation, inactivation of enzymes, and ion channels via protein oxidation and nitration, to inhibition of mitochondrial respiration (Virag et al., 2003). Peroxynitrite, which dissipates during oxidation (Wang et al., 2003), has also been found to nitrate as well as oxidize adenine, guanine, and xanthine nucleosides (Sodum and Fiala, 2001). Low concentrations of peroxynitrite trigger apoptotic cell death, whereas higher concentrations induce necrosis with cellular energetics (adenosine triphosphate, ATP, and nicotinamide adenine dinucleotide, NAD) serving as a switch between the modes of cell death. Superoxide is also involved in the production of nitrogen dioxide radicals and hydrogen peroxide, which can be further transformed into lipid peroxidation-initiating species, namely, peroxyl and alkoxyl radicals ($ROO^{\cdot}$ and $RO^{\cdot}$) (Dix and Aikens, 1993; Guéraud et al., 2010).

ROS and RNS are also normally generated by tightly regulated enzymes or enzyme systems located in or associated with cellular membranes or organelles in both phagocytic and nonphagocytic cells. One major source of intracellular ROS is nicotinamide adenine dinucleotide phosphate (NADPH) oxidases. NADPH oxidases catalyze the production of superoxide from O_2 and NADPH. These enzymes were originally described in phagocytes, where they were shown to kill engulfed pathogens by creating locally high levels of oxidative stress (Babior, 1999; Sullivan and Chandel, 2014). The presence of enzymes that specifically produce ROS validates the model that ROS serve a controlled function in the cell rather than simply acting as toxic by-products (Babior, 1999). NO is produced from the substrate L-arginine via the enzyme NO synthase (NOS) involving NADPH and oxygen. There are three major isoforms of NOS: (1) neuronal NOS produced by neurons; (2) endothelial NOS expressed mainly by endothelial cells; and (3) inducible NOS induced in immune cells, astrocytes, microglia, and neurons. The roles performed by NO vary based on its synthesis from NOS isoforms and the tissues in which it is produced (Knowles and Moncada, 1994; Alderton et al., 2001; Puttachary et al., 2015). NOS and NADPH oxidase isoforms that produce ROS/RNS are also involved in cellular signaling, synaptic plasticity in the central nervous system (CNS), neuronal transmission, reactions to stress and various agents, and the induction of mitogenic and apoptotic responses (Beal, 2000; Valko et al., 2007; Margialasche et al., 2009; Forrester et al., 2018).

ROS and RNS, produced under various mechanisms, can react with lipids, sugars, proteins, and nucleic acids, and inhibit normal cell functions (Deavall et al., 2012; Zhang et al., 2019; Kowaltowski, 2019). This lipid peroxidation initiating species-induced damage can compromise cell viability or induce cellular response, leading to cell death by necrosis or apoptosis (Valko et al., 2007; Margialasche et al., 2009).

Nutraceuticals. DOI: https://doi.org/10.1016/B978-0-12-821038-3.00031-8

Lipid peroxidation and markers of oxidative stress

Due to the high concentration of substrate polyunsaturated fatty acids (PUFAs) in cells, lipid peroxidation is a major outcome of free radical-mediated injury (Montine et al., 2002a,b). Lipid peroxidation is the mechanism by which lipids are attacked by chemical species that have sufficient reactivity to abstract a hydrogen atom from a methylene carbon in their chain. Lipid peroxidation in vivo, through a free radical pathway, requires a PUFA and a reactant oxidant inducer that together form a free radical intermediate. The free radical intermediate subsequently reacts with oxygen to generate a peroxyl radical, which with unpaired electrons may additionally abstract a hydrogen atom from another PUFA. The greater number of double bonds in the molecule and the higher instability of the hydrogen atom adjacent to the double bond explain why unsaturated lipids are particularly susceptible to peroxidation (Pratico et al., 2004; Gao et al., 2006).

A critically important aspect of lipid peroxidation is that it will proceed until the oxidizable substrate is consumed or termination occurs, making this fundamentally different from many other forms of free radical injury in that the self-sustaining nature of the process may entail extensive tissue damage (Porter et al., 1995). Decreased membrane fluidity following lipid peroxidation makes it easier for phospholipids to exchange between the two halves of the bilayer, increase the leakiness of the membrane to substances that do not normally cross it other than through specific channels (e.g., K^+, Ca^{2+}), and damage membrane proteins, inactivating receptors, enzymes, and ion channels (Halliwell, 2007). Increases in Ca^{2+} induced by oxidative stress can activate phospholipase A_2, which then releases arachidonic acid (AA) from membrane phospholipids. The free AA can then undergo lipid peroxidation (Farooqui et al., 2001) and act as a substrate for eicosanoid synthesis. Increased prostaglandin (PG) synthesis is immediately linked to lipid peroxidation, because low levels of peroxides accelerate cyclooxygenase (COX) action on PUFA (Smith, 2005). Phospholipase A_2 can also cleave oxidized AA residues from membranes.

The use of reactive products of lipid peroxidation as in vivo biomarkers is limited because of their chemical instability and rapid and extensive metabolism (Gutteridge and Halliwell, 1990; Moore and Roberts, 1998). For these reasons, other more stable lipid products of oxidative damage have generated intense interest in recent years as in vivo markers of oxidative damage (De Zwart et al., 1999). These compounds include the F_2-isoprostanes (F_2-IsoPs), F_3-Isops, isofurans (IsoFs), and F_4-neuroprostanes (F_4-NeuroPs) (Morrow et al., 1990; Fessel et al., 2002; Yin et al., 2007; Janicka et al., 2010; Milne, 2017).

Prostaglandin-like compounds as in vivo markers of oxidative stress

F_2-IsoPs are PG-like compounds that are produced by a non-COX free radical-catalyzed mechanism involving the peroxidation of the PUFA, AA (C20:4, omega-6). Formation of these compounds initially involves the generation of four positional peroxyl radical isomers of arachidonate, which undergo endocyclization to PGG_2-like compounds. These intermediates are reduced to form four F_2-IsoP regioisomers, each of which can consist of eight racemic diastereomers (Morrow et al., 1990). In contrast to COX-derived PGs, nonenzymatic generation of F_2-IsoPs favors the formation of compounds in which the stereochemistry of the side chains is oriented *cis* in relation to the prostane ring. A second important difference between F_2-IsoPs and PGs is that F_2-IsoPs are formed primarily in situ, esterified to phospholipids, and subsequently released by phospholipases (Gao et al., 2006; Famm and Morrow, 2003), whereas PGs are generated only from free AA (Morrow et al., 1990).

Over the course of the past 25 years, F_2-IsoPs have been shown to be a reliable biomarker of endogenous lipid peroxidation because they are ubiquitous in the body and are chemically stable in biological fluids when stored correctly (Milne et al., 2015; Milne, 2017). An NIH-sponsored study, called the biomarkers of oxidative stress study (BOSS), found that plasma F_2-IsoPs are increased in a time-dependent and dose-dependent manner in rats administered carbon tetrachloride, a toxic agent known to systematically generate free radicals in the liver (Milne et al., 2015; Kadiiska et al., 2005).

F_2-IsoP analogs may be formed by peroxidation of docosahexaenoic acid (DHA; C22:6, omega-3), which generates F_4-IsoPs. These compounds are also termed "neuroprostanes" (F_4-NeuroPs) due to the high levels of their precursor in the brain (Roberts et al., 1998). In contrast to AA, which is evenly distributed in all cell types in all tissues, DHA is highly concentrated in neuronal membranes (Salem et al., 1986). DHA is obtained mainly through dietary means because the human body can only minimally synthesize this fatty acid. DHA deficiency has been linked to slow mental development (Connor et al., 1992), whereas DHA supplementation has been linked to a variety of health benefits including decreased rates of neurodegenerative diseases (Schaefer et al., 2006). DHA can undergo oxidation both in vitro and in vivo, and increased units of unsaturation suggest its higher susceptibility to lipid peroxidation compared to AA. Thus, although the measurement of F_2-IsoPs provides an index of global oxidative damage in the brain, integrating data from both glial and neuronal cells, determination of F_4-NeuroPs permits the specific quantification of oxidative damage to neuronal membranes in vivo. In fact,

F_4-NeuroPs are the only quantitative in vivo marker of oxidative damage selective for neurons. This is particularly important because of the implication of oxidative damage and lipid peroxidation being causative factors in numerous neurodegenerative diseases (Montine et al., 2002b, 2004; Milatovic et al., 2005b).

Another F_2-IsoPs analog may be formed by peroxidation of eicosapentaenoic acid (EPA; C20:5, omega-3) that leads to the production of F_3-IsoPs. Levels of F_3-IsoPs can significantly exceed those of F_2-IsoPs generated from AA, perhaps because EPA contains more double bonds and is more easily oxidizable (Gao et al., 2006). It has also been shown that in the presence of increased oxygen tension in the microenvironment in which lipid peroxidation occurs, an additional oxygen insertion step may take place (Fessel et al., 2002). This step diverts the IsoP pathway to form tetrahydrofuran ring-containing compounds termed IsoFs, which are functional markers of lipid peroxidation under conditions of increased oxygen tension. Measurements of IsoFs represent a much more robust indicator of hyperoxia-induced lung injury than measurements of F_2-IsoPs. Like IsoPs, the IsoFs are chemically and metabolically stable, so they are well-suited to act as in vivo biomarkers of oxidative damage.

Aldehydes as lipid peroxidation products

The enzymatic and free-radical peroxidation of PUFAs, which contain at least three double bonds, like AA and DHA, could lead to malondialdehyde (MDA). This product can be generated by thromboxane synthase, but a report from the BOSS showed that peripheral levels of MDA derive primarily from nonenzymatic peroxidative degradation of unsaturated lipids (Kadiiska et al., 2005). Because MDA reacts with thiobarbituric acid (TBA), MDA-TBA adducts are used to spectrophotometrically measure the levels of oxidative stress and consequent lipid peroxidation (Spickett et al., 2010; Fang et al., 2017). MDA and MDA-TBA complex assessment can also be measured by high-performance liquid chromatography or fas chromatography (GC)—mass spectrometry (MS) (Maboudou et al., 2002). Although more specific, this approach does not deduce all the limitations of this biomarker (Sultana et al., 2006; Devasagayam et al., 2003). In a recent report, Kadiiska et al. (2015) confirmed their previous findings in different rat models that measurement of antioxidants in plasma is not useful for the assessment of oxidative damage in vivo.

4-Hydroxy-2-nonenal (HNE) is a reactive aldehyde arising from peroxidation of omega-6 fatty acid (Guéraud et al., 2010; Uchida, 2003; Barrera et al., 2018). HNE is formed under various conditions like autooxidation and stimulated microsomal lipid peroxidation, and plays a critical role in the pathophysiology of various human pathologies (Neely et al., 2005; Sonowal and Ramana, 2019). HNE is found in food (Gasc et al., 2007) and detected in vivo (Neely et al., 2005). MDA and HNE are able to covalently modify proteins and alter their functions (Butterfield et al., 2006). In addition to protein modification, these lipid peroxidation products can also interfere with the synthesis of DNA and RNA, alter cell metabolism and signaling, and mediate brain-induced oxidative damage.

Another highly reactive aldehydic product, acrolein, derived from the metal-catalyzed oxidation of PUFA, including AA and DHA, can also promote the formation of protein adducts, and thus promote neuronal damage (LoPachin et al., 2008). Studies suggest that acrolein can modify DNA, promote lipid peroxidation, inflammatory processes, induce atherothrombotic effects, tau oligomerization, and promote the formation of neurofibrillary tangles (Kuhla et al., 2007; Uchida et al., 1998; Lovell et al., 2000; Zirak et al., 2019).

Reactivity of lipid peroxidation products

Metabolization of secondary lipid peroxidation products, such as HNE, in most cells and tissues is rapid and complete. Glutathione (GSH) conjugation seems to be the primary and major step. GSH maintains protein sulfhydryl groups in the reduced form and has been implicated in the regulation of cytoskeletal function. SH groups of cystein constitute a main protein-associated target of HNE (Esterbauer et al., 1998). Cellular detoxification of HNE is mainly achieved through conjugation to cystein GSH. This reaction is catalyzed by GSH S-transferase and leads to a transient decrease in cellular GSH (Cheng et al., 1999; Hayes and Pulford, 1995; Johnson et al., 1993; Kinter and Roberts, 1996). Additionally, the cellular target of HNE is also tubulin, the core protein of microtubules containing abundant cystein (Neely et al., 2005). Studies demonstrated that exposure of Neuro2A cells to HNE results in the inhibition of cytosolic taxol-induced tubulin polymerization, thus supporting the hypothesis that HNE adduction to tubulin is the primary mechanism involved in HNE-induced loss of the highly dynamic neuronal microtubule network (Neely et al., 2005).

ROS and RNS can react with the DNA molecule and induce purine or pyrimidine base or sugar lesions, nitration and deaminations of purines, DNA—DNA or DNA—protein cross-links (Dizdaroglu et al., 2002). These processes lead to mutations and impaired transcriptional and posttranscriptional processes and compromise protein synthesis (Colurso et al., 2003; Poetsch, 2020). In addition, DNA damage, oxidative phosphorylation, and altered cell metabolism may lead to apoptosis and promote neuronal death (Fishel et al., 2007; Becker and Bonni, 2004; Poetsch, 2020).

Although ROS and RNS can attack any amino acid, sulfur-containing and aromatic amino acids are the most susceptible (Stadtman and Levine, 2003). The oxidation of amino acids leads to the formation of carbonyl derivatives produced by fragmentation (due to an attack on several amino acid side chains) and by formation of adducts between some amino acid residues and products of lipid peroxidation. Protein carbonyls can be detected with 2,4-dinitrophenylhydrazine and used as biomarkers of oxidative stress (Dalle-Donne et al., 2003). Oxidation of protein also leads to protein fragmentation and protein cross-linking. In addition, peroxynitrite and a hydroxyl radical can react with tyrosine and form other indexes of protein oxidation, 3-nitrotyrosine and ortho-tyrosine, respectively. These products of oxidative/nitrosative modification of proteins are relatively stable with sensitive assays available for their detection (Chakravarti and Chakravarti, 2007; Ore and Akinloye, 2019; Chen et al., 2020).

Oxidative stress and excitotoxicity

Lipid peroxidation, mitochondrial dysfunction, reduced neuronal energy levels, and reduced cytochrome *c* oxidase activity support the contention that excitotoxicity and neuronal injury are associated with excessive formation of ROS and RNS (Yang and Dettbarn, 1998; Gupta et al., 2001a,b; Milatovic et al., 2000a,b, 2001, 2005a).

The hypothesis concerning excitotoxicity-induced neuronal oxidative injury is that overstimulation of glutamatergic receptors results in sustained epileptic activity in the hippocampus and neuropathologic changes predominantly in the pyramidal neurons. Cell damage is thought to result from intense transient influx of calcium leading to mitochondrial functional impairments, characterized by activation of the permeability transition pores in the inner mitochondrial membrane, cytochrome *c* release, depletion of ATP, and simultaneous formation of ROS (Cadenas and Davies, 2000; Patel, 2002; Nicholls et al., 2003). In addition, an increase in cytoplasmic calcium ions triggers intracellular cascades through the stimulation of enzymes, including proteases, phospholipase A_2, and NOS, which also lead to increased levels of free radical species and oxidative stress (Lafon-Cazal et al., 1993; Farooqui et al., 2001).

Because free radicals are direct inhibitors of the mitochondrial respiratory chain, ROS generation perpetuates a reinforcing cycle, leading to extensive lipid peroxidation and oxidative cell damage (Cock et al., 2002; Cadenas and Davies, 2000). Previous studies have supported a role of oxidative stress and excessive generation of ROS and RNS in anticholinesterase-induced neurotoxicity (Dettbarn et al., 2001; Gupta et al., 2001a,b, 2007; Milatovic et al., 2005a, 2006; Zaja-Milatovic et al., 2009).

Previous studies have also investigated the role of glutamatergic excitation, oxidative injury, and neurodegeneration in the model of kainic acid (KA) excitotoxicity. Kainate is a rigid analog of glutamate, the principal excitatory neurotransmitter in the CNS, and it is a very potent stimulant of a subset of the ligand-gated ion channel, called KA receptors (Milatovic et al., 2005b). Activation of the KA subtype of ionotropic glutamate receptors results in sustained epileptic activity in the hippocampus, followed by a selective pattern of neuropathology that is similar to human temporal lobe epilepsy (Ben-Ari and Cossart, 2000; Schwob et al., 1980). Animals used in the study were exposed to intracerebroventricular (ICV) injection of KA, known as an experimental model for investigation of cerebral vulnerability, particularly during acute brain disorders and status epilepticus (Ben-Ari and Cossart, 2000; Schwob et al., 1980). Study results showed that ICV KA induced an early increase in biomarkers of oxidative damage, F_2-IsoPs, and F_4-NeuroPs. The earliest time point evaluated was 30 min because seizures start immediately after the KA injection (Table 31.1).

Elevated levels of these in vivo markers of oxidative damage are in agreement with previous findings (Montine et al., 2002a; Milatovic et al., 2005b; Gupta et al., 2007; Patel et al., 2001) and indicate that KA injection leads to profound cerebral and neuronal oxidative damage in

TABLE 31.1 Cerebral concentrations of F_2-IsoPs and F_4-NeuroPs and dendritic degeneration of hippocampal pyramidal neurons following kainic acid-induced seizures in mice.

	F_2-IsoPs (ng/g)	F_4-NeuroPs (ng/g)	Dendritic length (μm)	Spine density (number/100 μm dendrite)
Control	3.07 ± 0.05	13.89 ± 0.58	1032.10 ± 61.41	16.45 ± 0.55
KA 30 min	4.81 ± 0.19[a]	34.27 ± 2.7[a]	363.44 ± 20.78[a]	8.81 ± 0.55[a]
KA 60 min	3.40 ± 0.18	18.55 ± 1.26	425.71 ± 23.04[a]	7.44 ± 0.56[a]

Data from kainic acid-exposed mice were collected 30 or 60 min after injection. *F_2-IsoPs*, F_2-isoprostanes; *F_4-NeuroPs*, F_4-neuroprostanes; *KA*, kainic acid.
[a]*One-way ANOVA showed P < .0001 for each end-point. Bonferroni's multiple comparison test showed significant difference (P < .001) compared to vehicle-injected control.*

mice. The study was also designed to investigate whether F_2-IsoP and F_4-NeuroP formation was correlated with the vulnerability of pyramidal neurons in the CA1 hippocampal area following KA-induced excitotoxicity (Fig. 31.1).

Dendritic spines represent the structural sites of contact for the majority of excitatory, glutamatergic synaptic inputs onto neurons and are strongly implicated in mechanisms of synaptic plasticity and learning. NMDA and other glutamate receptor subtypes are clustered in dendritic spines (Rao and Craig, 1997; O'Brien et al., 1998), which serve as integrative units in synaptic circuitry and participate in synaptic plasticity (Yuste and Denk, 1995). The accumulation of glutamate receptor clusters in spines is governed by excitatory synaptic activity and increases when activity is suppressed (Rao and Craig, 1997; O'Brien et al., 1998). Conversely, excitotoxicity produces a rapid and profound loss of dendritic spines in cultured neurons (Halpain et al., 1998), mimicking the loss in dendritic spine synapses in neurological conditions including epilepsy, aging, and schizophrenia (Jeffrey et al., 1997; Jiang et al., 1998). This suggests that receptor localization at synapses might be critical to excitotoxicity and govern neuronal vulnerability to excitotoxicity.

Study results also showed that the transient increase in F_2-IsoPs and F_4-NeuroPs is accompanied by rapid evolution of dendritic abnormalities, apparent in the significant decrease in dendritic length and spine density of pyramidal neurons as early as 30 min after KA injection. However, the recovery in both oxidative damage biomarkers at 60 min after injection was not paralleled by the rescue of damaged neurons from the CA1 hippocampal area. Extended seizure activity (60 min) induced the same level of dendritic length and spine density decrease when compared to 30 min after KA injection (Table 31.1). Together, these data suggest that both oxidative stress and neurodegeneration occur as an early response to seizures but do not establish whether oxidative stress is a cause or effect of seizure-induced CA1 cell damage. Neuronal damage processes triggered by sustained seizure activity may occur as a continuum, last longer than formation of oxidative lipids, and, although not evident by markers, may already be in progress when the peak increases in F_2-IsoPs and F_4-NeuroPs occur. Thus, described studies investigated dynamic changes in lipid peroxidation and dendritic structures immediately after seizures, but future studies over a longer period should be able to determine the long-term time course of spine and dendritic changes. It is very likely that the spine loss seen in our study is the initial phase of more chronic spine loss and progressive neurodegeneration reported in other studies (Muller et al., 1993; Jiang et al., 1998; Zeng et al., 2007).

In vivo data have also established that KA induced a significant increase (more than twofold) in citrulline concentrations 30 min after the injection (Zaja-Milatovic et al., 2008). Although the study was not designed to determine whether increased citrulline originated from a combination of NOS isozymes or from one in particular, in vivo data are in agreement with results from the models of anticholinesterase toxicity and activated innate immunity (Gupta et al., 2007; Milatovic et al., 2003, 2004) and indicate that a subset of NOS activity also contributes to cerebral oxidative damage in the model of KA-induced excitotoxicity. Together, these studies showed that seizure-induced cerebral oxidative damage in adult animals is accompanied by alterations in integrity of the hippocampal CA1 dendritic system (Gupta et al., 2007; Zaja-Milatovic et al., 2008, 2009; Milatovic et al., 2010).

Antioxidant defense and suppression of oxidative injury

Living systems have developed several mechanisms to control the harmful effects of ROS and RNS. These mechanisms are mainly based on the presence of antioxidants (enzymatic and nonenzymatic) and the repair or removal of injured molecules/systems. Antioxidants are natural or synthetic molecules preventing excessive formation of ROS or inhibiting their reaction with biological structures and/or molecules. Antioxidant defense involves a variety of strategies, both enzymatic, such as superoxide dismutase (SOD), and nonenzymatic, involving tocopherols, carotenes, ascorbate, and GSH (Halliwell, 2007, 2012; Day, 2014). Different systems/mechanisms cooperate in the regulation of ROS/RNS production and neutralization, which is essential for avoiding their detrimental

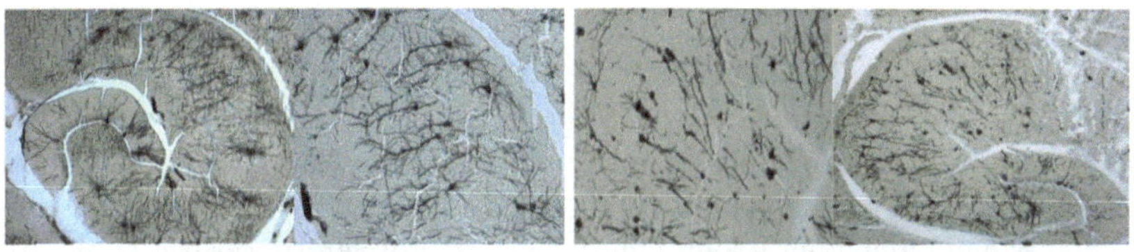

Saline exposure　　Kainic acid exposure

FIGURE 31.1 Photomicrographs of mouse hippocampi with pyramidal neurons from the CA1 hippocampal area of brains 1 h after saline (control) and KA (1 nmol/5 μL, intracerebroventricular) injections. Treatment with KA induced degeneration of the hippocampal dendritic system and a decrease in total length of the dendrite and spine density of hippocampal pyramidal neurons. *KA*, kainic acid.

effects and preserving equilibrium. This is termed redox homeostasis (Droge, 2002; He et al., 2017).

A large body of literature supports the notion that the antioxidant system plays an important role in preventing many human diseases (e.g., cancer, atherosclerosis, stroke, rheumatoid arthritis, and neurodegeneration) (Fang et al., 2002; Lorenzon Dos Santos et al, 2020; Zhou and Toan (2020)). Antioxidant enzymes, such as SOD, catalase, GSH peroxidase, and reductase, and antioxidants exert synergistic actions in scavenging free radicals. SOD is a metalloprotein, an important element of the antioxidant system considered to be the first line of defense against free radical formation. It catalyzes dismutation of superoxide radicals into oxygen and hydrogen peroxide. Superoxide radicals, if not scavenged effectively, may directly inactivate several enzymes, like catalase and GSH peroxidase, which are needed to eliminate hydrogen peroxide from intracellular media (Johnson and Giulivi, 2005; Tymko et al., 2019; Sultan et al., 2018). An important role in antioxidant defense also belongs to catalase. Catalase, located in peroxisomes (80%) and cytosol (20%), decomposes hydrogen peroxide to water and oxygen without the production of free radicals (Jones et al., 1981). Catalase does not show significant activity under physiological conditions due to its lower affinity for H_2O_2 as compared to GSH peroxidase, but it becomes an important enzyme in disease states where the concentration of H_2O_2 is elevated (Chance et al., 1979; Trieb et al., 2019).

GSH peroxidase and reductase are GSH-dependent enzymes located in cytoplasm, mitochondria, and the nucleus (Gamble et al., 1999). GSH peroxidase metabolizes hydrogen peroxide to water using reduced GSH as a hydrogen donor and is recycled back to GSH reductase by cofactor NADPH. It plays an important role in the defense mechanism of erythrocytes against lipid peroxidation damage (Sen, 2000; Shah et al., 2014).

GSH (l-γ-glutamyl-l-cysteinyl glycine), an important element of the nonenzymatic antioxidant system, is required for many critical cellular processes and plays a particularly important role in the maintenance and regulation of the thiol-redox status of the cell. In healthy cells and tissues, more than 90% of the total GSH pool is in reduced form (glutathione, GSH) and less than 10% exists in the oxidized form (glutathione disulfide) (Schafer and Buettner, 2001; Prasai et al., 2018; Gould and Pazdro, 2019).

Vitamin E has been recognized as one of the most important antioxidants. Vitamin E is a lipid-soluble vitamin found in cell membranes and circulating lipoproteins. Vitamin E inhibits the ROS-inducing generation of lipid peroxyl radicals, thereby protecting cells from peroxidation of PUFA in membrane phospholipids, from oxidative damage of cellular proteins and DNA, and from membrane degeneration (Topinka et al., 1989; Azlina et al., 2018; Simioni et al., 2018). Vitamin E mainly acts as a chain-breaking antioxidant and radical scavenger, protecting cell membranes against oxidative damage (Van Acker et al., 1993). In addition, vitamin E regulates ROS production (Chow et al., 1999), maintains oxidative phosphorylation in mitochondria, and accelerates restitution of high-energy metabolites (Punz et al., 1998). Decreased levels of vitamin E in response to hyperoxia or treatment with convulsants reported in studies (Mori et al., 2004; Onodera et al., 2003; Rauca et al., 2004) suggest that vitamin E in the brain is consumed to prevent oxidative damage. Vitamin E also prevented changes in lipase activity and lipid peroxidation in the brain and spinal cord of rats that were induced by metasystox (organophosphate insecticide) (Tayyaba and Hasan, 1985). Vitamin E was also found to exert its anticonvulsive effects by upregulating catalase activity in pilocarpine rodent models of epilepsy (Xavier et al., 2007; Barros et al., 2007; Santos et al., 2008). During pilocarpine-induced seizures, vitamin E concentrations were found to decrease in brain cortex (Tejada et al., 2007). In kindling rat models of epilepsy, the antioxidant treatment (vitamin E and GSH) reduced neuronal death and lipid peroxidation; however, it did not prevent development of recurrent seizures (Frantseva et al., 2000; Puttachary et al., 2015).

Vitamin C is the major hydrophilic antioxidant and a powerful inhibitor of lipid peroxidation. In membranes, this molecule rapidly reduces α-tocopheroxyl radicals and inhibits propagation of free radicals (Huang et al., 2002; Righi et al., 2020). Vitamin C also increased the latency to the onset of seizures after status epilepticus, while reducing the mortality rates in rat models (Santos et al., 2008; Puttachary et al., 2015). Recent evidence suggests that maintaining healthy vitamin C levels can have a protective function against age-related cognitive decline and Alzheimer disease, but avoiding vitamin C deficiency is likely to be more beneficial than taking supplements on top of a normal, healthy diet (Harrison, 2012).

Vitamin A belongs to a group of unsaturated hydrocarbons, with multiple functions in growth and development, maintenance of the immune system, and good vision. It can be found in two main lipid-soluble forms, retinol or retinyl esters and carotenes. The most well-known β-carotene is a potent antioxidant able to quench singlet oxygen (Di Mascio et al., 1991) and reduce lipid peroxidation (Upritchard et al., 2003); its depletion, as well as that of α-carotene, β-cryptoxanthin, lycopene, and lutein/zeaxanthin, has been associated with a variety of diseases including atherosclerosis (Prince et al., 1988) and cardiovascular diseases (Gaziano et al., 1995). More generally, vitamin C, vitamin E, and carotenoids have been shown to synergistically counteract lipid peroxidation (Niki et al., 1995).

In addition to vitamins, certain herbal extracts and other nutraceuticals have been shown to exert antioxidant activity. Due to chapter limitations, not all nutraceuticals with antioxidant properties will be discussed in this chapter; instead, a few examples are given here to support the context. Among many phytochemicals, polyphenols (with subclass flavonoids), commonly present in berries, fruits, and nuts, seem to possess strong antioxidant activity. These antioxidants have proven applications in cardiovascular, respiratory, renal, hepatic, and neurodegenerative diseases, diabetes, and cancer. Flavonoids have been reported to be active for the treatment of disorders of peripheral circulation and flavonoid-based herbal medicines are used as antiinflammatory, antispasmodic, antiallergic, and antiviral remedies. The underlying mechanism appears to be free radical (including peroxynitrite and hydroxyl) scavenging, preventing lipid peroxidation, metal chelation, inhibition of calcium influx, regulation of cell signaling, and modification of gene expression (Dikalov and Harrison, 2006; Boots et al., 2008; Lee et al., 2010; Rahal et al., 2014; Valentová et al., 2014; Serreli and Deiana, 2020). Of course, there are no prospective studies of flavonoid treatments that have shown benefit in any large clinical trials. It needs to be emphasized that human intestinal microbiota can significantly influence (activate or deactivate) the activity of dietary polyphenols (Duda-Chodak et al., 2015).

Details of some other polyphenols (such as resveratrol and curcumin), including their applications in various human ailments due to their antioxidant activity, are presented in standalone chapters in this book. Recent literature supports that curcumin also protects and attenuates oxidative stress against metal diseases (Zoheb et al., 2014; Srivastava et al., 2014; García-Niño and Pedraza-Chaverrí, 2014), diabetes (Abo-Salem et al., 2014; Rashid and Sil, 2015), and neurodegenerative diseases (Hamaguchi et al., 2010; Ji and Shen, 2014). Recently, Poodineh et al. (2015) reported that *Caralluma tuberculata* at appropriate doses (100 and 200 mg/kg body weight, oral, for 45 days) provided both antioxidant and antihyperglycemic activities in streptozotocin (60 mg/kg body weight, intraperitoneal) diabetic rats.

Lycopene (Ψ,Ψ-carotene), the major carotenoid in tomato and tomato products, exerts antioxidant effects by excellent singlet oxygen-quenching activity and free radical-scavenging activity (Maoka et al., 2015). Lycopene has been shown to exert anticarcinogenic effects. Recently, Maoka et al. (2015) identified several oxidative metabolites of lycopene with a unique five-member ring end group.

Many other nutraceuticals, such as fenugreek, thymoquinone, astaxanthin, and Spirulina, produce beneficial health effects by having antioxidant activity, and standalone chapters on these topics are present in this book.

Interestingly, Joseph et al. (2014) reported that the postprandial period is a time of active oxidative metabolism and formation of ROS. Excess energy intake stimulates overproduction of ROS, resulting in metabolic oxidative stress and cellular redox imbalance. This change in a cell's redox status activates redox-sensitive signaling molecules, such as NF-κB, JNK, and other stress-signaling molecules, ultimately resulting in increased expression of genes, many of which are inflammation mediators. Hence, elevation in inflammatory molecules in the postprandial state is often accompanied, and perhaps preceded, by increases in oxidative markers. Regular consumption of berries, which are rich in flavonoids (particularly anthocyanins or some other nutraceuticals with antioxidant properties), may prevent and attenuate such oxidative stress and inflammation states.

Melatonin, a mammalian hormone synthesized from serotonin, has also shown antioxidant properties (Nishida, 2005). Melatonin exerts its antioxidant capacity by stimulating the expression and activity of GSH peroxidase and SOD and inhibiting that of NOS (Pieri et al., 1995; Dato et al., 2013; Singh and Jadhav, 2014; Mantzarlis et al., 2017). Details of melatonin's application and mechanism of action can be found in the chapter on melatonin. A synthetic spin trapping agent such as phenyl-*N-tert*-butylnitrone (PBN) is also capable of scavenging many types of free radicals. This compound is widely used to trap ROS in a variety of physical, chemical, and biological studies using electron paramagnetic resonance spectrometry. PBN is known to be concentrated in mitochondria, where it reacts with ROS and forms stable adducts, thereby maintaining normal levels of energy metabolites. Numerous in vitro and in vivo experiments have shown the beneficial effects of PBN on the prevention of neuronal degeneration. Protective effects are described in experimental models of brain ischemia/reperfusion (Carney and Floyd, 1991; Fetcher et al., 1997; Gido et al., 1997), excitotoxicity (Lancelot et al., 1997; Milatovic et al., 2002), and inhibition of NOS induction (Miyajima and Kotake, 1995), and in different models of seizures (He et al., 1997; Thomas et al., 1997). Additional findings also corroborate that PBN effectively prevents neurodegeneration in Parkinson disease (PD) (Frederiksson et al., 1997; Sack et al., 1996), Alzheimer disease, and anticholinesterase neurotoxicity (Sack et al., 1996; Gupta et al., 2001a,b). Thus, PBN has been proven to rescue neurons in multiple experimental injury models. Other pharmacological properties of spin trapping agents have been described that could influence the outcome of oxidant injury. These have been described for PBN as reversible Ca^{2+} channel blockade in vascular muscle causing vasodilatation (Anderson et al., 1993), direct effect on striatal function, including inhibition of excitation–contraction coupling (Andersen et al., 1996), and induction of hypothermia (Pazos et al., 1999).

Previous studies have shown that antioxidant pretreatment suppressed anticholinesterase-induced alterations in high-energy phosphates (HEPs), their metabolites, and citrulline levels, supporting the possibility that increased generation of ROS/RNS contributes to depletion of energy phosphates (Gupta et al., 2001a,b). PBN or vitamin E treatment alone did not alter the levels of HEPs, their major metabolites, or citrulline levels in any of the brain regions. Vitamin E pretreatment suppressed the depletion of HEPs and their metabolites and increased citrulline levels without preventing seizures (Gupta et al., 2001a). The protective efficacy provided by vitamin E against organophosphate-induced or carbamate-induced changes in energy metabolites was of varying degrees in different brain regions and could partly be due to pharmacokinetic variables involved in attaining different levels of vitamin E in different brain regions. However, PBN pretreatment, 1 h before anticholinesterase agent, protected mitochondria, maintained the cellular level of high-energy metabolites, and also prevented anticholinesterase-induced convulsions and seizures (Gupta et al., 2001a,b).

The efficacy of the spin trapping agent PBN and of the antioxidant vitamin E were also tested to suppress an increase in NO and lipid peroxidation and prevent neurodegeneration of pyramidal neurons in the CA1 hippocampal area in the model of KA-induced excitotoxicity (Zaja-Milatovic et al., 2008). Vitamin E or PBN alone did not alter basal citrulline and F_4-NeuroP levels or dendritic arborization. However, vitamin E and PBN suppressed KA-induced increases in citrulline, cerebral, and neuronal markers of oxidative damage, F_2-IsoPs, and F_4-NeuroPs, respectively (Figs. 31.2 and 31.3).

Importantly, vitamin E and PBN completely suppressed both a reduction in dendrite length and a reduction in spine density of pyramidal neurons from the CA hippocampal area in KA-exposed mice (Fig. 31.4).

Close concordance was found between these results, showing that protection of the cerebrum from neuronal oxidative damage also protected hippocampal CA1 pyramidal neurons from dendritic degeneration. These agents did not alter kainate-induced seizure severity, indicating that the protective effects of vitamin E and PBN are most likely mediated by scavenging ROS, preventing lipid

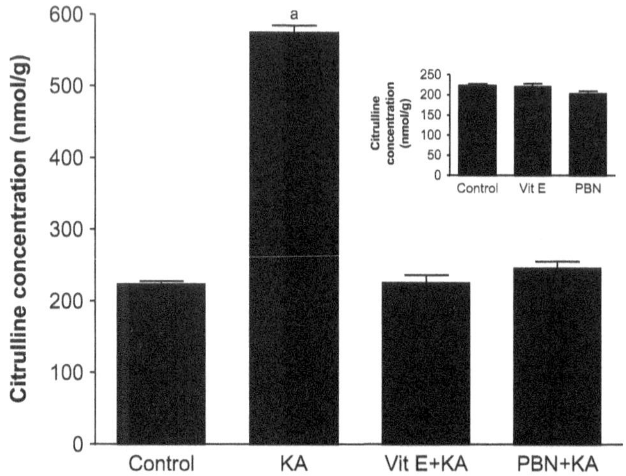

FIGURE 31.3 Ipsilateral cerebral citrulline concentrations following intracerebroventricular KA with or without Vit E or PBN pretreatment. Brains from mice exposed to KA were collected 30 min after injections ($n \geq 5$ for each group). One-way ANOVA yielded $P < .001$ with Bonferroni's multiple comparison tests significant for KA versus control, Vit E + KA, or PBN + KA treatment. *KA*, kainic acid; *PBN*, *N-tert*-butyl-α-phenylnitrone; *Vit E*, vitamin E.

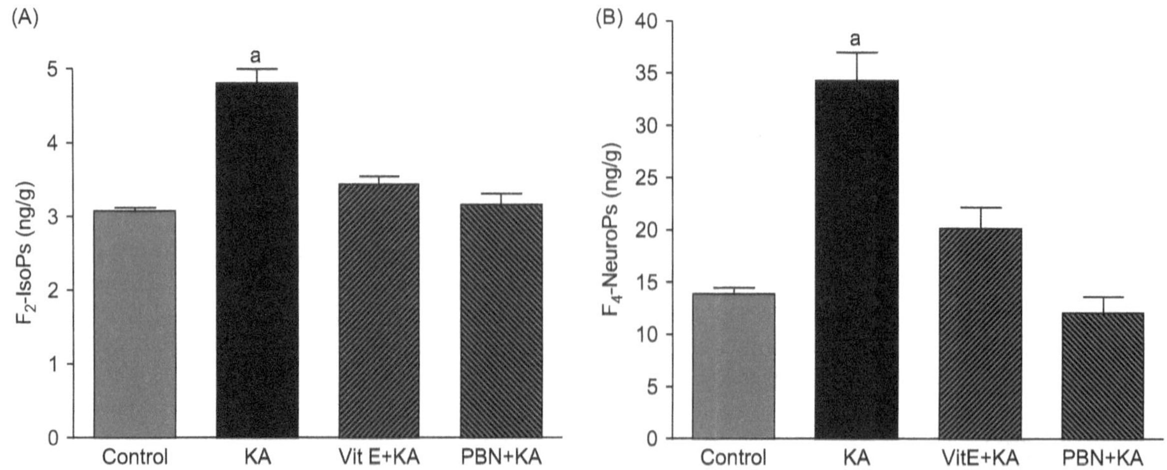

FIGURE 31.2 Ipsilateral cerebral F_2-IsoP (A) and F_4-NeuroP (B) concentrations following intracerebroventricular KA with or without Vit E or PBN pretreatment. Brains from mice exposed to KA were collected 30 min after injections ($n \geq 5$ for each group). One-way ANOVA yielded $P < .0001$ with Bonferroni's multiple comparison tests significant for KA versus control, Vit E + KA, or PBN + KA treatment. *F_2-IsoP*, F_2-isoprostanes; *F_4-NeuroP*, F_4-neuroprostanes; *KA*, kainic acid; *PBN*, *N-tert*-butyl-α-phenylnitrone; *Vit E*, vitamin E.

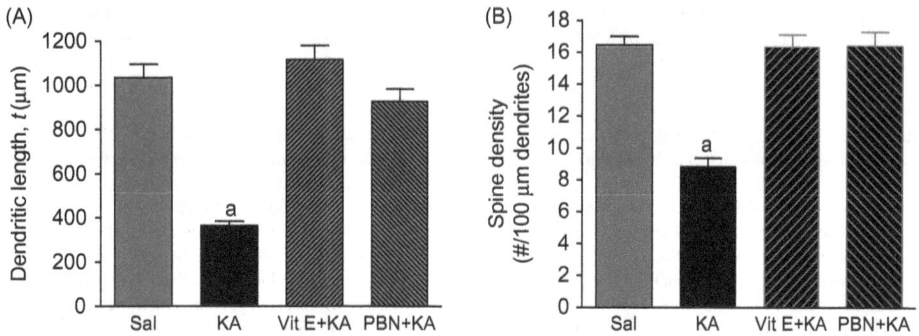

FIGURE 31.4 Dendritic length (A) and spine density (B) of pyramidal neurons from the CA1 hippocampal area of mice following intracerebroventricular KA with or without Vit E or PBN pretreatment. Brains from mice exposed to KA were collected 30 min after injections ($n \geq 5$ for each group). One-way ANOVA yielded $P < .001$ with Bonferroni's multiple comparison tests significant for KA versus control, Vit E + KA, or PBN + KA treatment. *KA*, kainic acid; *PBN*, *N-tert*-butyl-α-phenylnitrone; *Vit E*, vitamin E.

peroxidation and consequent neuronal damage, and not by a specific effect on seizures per se. Furthermore, because antioxidants minimize lipid peroxidation following an increase in α-tocopherol and PBN, a parallel reduction in neuronal damage provides strong evidence that oxidative stress and lipid peroxidation mediate seizures and corresponding injury. One limitation to potential therapeutic application is that drugs, to be effective, need to be administered prophylactically before the onset of seizures. Future research should address the efficacy of these agents when administered at higher concentrations, either during or possibly even after seizures, in preventing seizure-induced oxidative and dendritic changes and potentially reducing resultant neurocognitive deficits.

Concluding remarks and future directions

It is becoming increasingly evident that oxidative stress is a significant element in the progression of brain diseases. Oxidative stress, in combination with lowered antioxidant defense, creates a harmful combination that could disturb functions and damage organelles such as mitochondria. This would eventually lead to loss of synapses, neuronal damage, and cell death, thus giving rise to clinical symptoms associated with brain disease. Results from multiple studies support an association between oxidative stress and seizures induced by KA and suggested that ROS generation is underlying mechanisms in excitotoxicity-induced degeneration of pyramidal dendritic systems in hippocampus. Future studies should be directed at deciphering the mechanisms of protection and address attenuation of neurotoxicity via radical-scavenging mechanisms. Complementary studies should also guide the development of selective and efficacious antioxidant therapies that target neurotoxic mechanisms while maintaining neuroprotective actions. Antioxidant therapies, including natural antioxidants provided by nutrition, hold great promise.

Acknowledgments

The authors thank Ms Robin B. Doss for her technical assistance in the preparation of this chapter.

References

Abo-Salem, O.M., Harisa, G.I., Ali, T.M., 2014. Curcumin ameliorates streptozotocin-induced heart injury in rats. J. Biochem. Mol. Toxicol. 28, 263–270.

Alderton, W.K., Cooper, C.E., Knowles, R.G., 2001. Nitric oxide synthases: structure, function and inhibition. Biochem. J. 357 (3), 593–615.

Anderson, D.E., Yuan, X.J., Tseng, C.M., 1993. Nitrone spin traps block calcium channels and induce pulmonary artery relaxation independent of free radicals. Biochem. Biophys. Res. Commun. 193, 878–885.

Andersen, K.A., Diaz, P.T., Wright, V.P., 1996. *N-tert*-butyl-alpha-phenylnitrone: a free radical trap with unanticipated effects on diaphragm function. J. Appl. Physiol. 80, 862–868.

Azlina, M.F.N., Kamisah, Y., Qodriyah, M.S., 2018. Tocopherol and tocotrienol: therapeutic potential in animal models of stress. Curr. Drug Targets 19 (12), 1456–1462.

Babior, B.M., 1999. NADPH oxidase: an update. Blood. 93 (5), 1464–1476.

Barrera, G., Pizzimenti, S., Daga, M., et al., 2018. Lipid peroxidation-derived aldehydes, 4-hydroxynonenal and malondialdehyde in aging-related disorders. Antioxidants 7 (8), 102.

Barros, D.O., Xavier, S.M.L., Barbosa, C.O., 2007. Effects of the vitamin E in catalase activities in hippocampus after status epilepticus induced by pilocarpine in Wistar rats. Neurosci. Lett. 416 (3), 227–230.

Beal, M.F., 2000. Oxidative metabolism. Ann. NY Acad. Sci. 924, 164–169.

Becker, E.B., Bonni, A., 2004. Cell cycle regulation of neuronal apoptosis in development and disease. Prog. Neurobiol. 1, 1–25.

Ben-Ari, Y., Cossart, R., 2000. Kainate, a double agent that generates seizures: two decades of progress. Trends Neurosci. 23, 580–587.

Boots, A.W., Haenen, G.R.M.M., Bast, A., 2008. Health effects of quercetin: from antioxidant to nutraceutical. Eur. J. Pharmacol. 585, 325–337.

Butterfield, D.A., Reed, T., Perluigi, M., 2006. Elevated protein-bound levels of the lipid peroxidation product, 4-hydroxy-2-nonenal, in brain from persons with mild cognitive impairment. Neurosci. Lett. 3, 170–173.

Cadenas, E., Davies, K.J., 2000. Mitochondrial free radical generation, oxidative stress, and aging. Free Radic. Biol. Med. 29, 222–230.

Carney, J.M., Floyd, R.A., 1991. Protection against oxidative damage to CNS by alpha-phenyl-*tert*-butylnitrone (PBN) and other spin-trapping agents: a novel series of nonlipid free radical scavengers. J. Mol. Neurosci. 3, 47–57.

Chakravarti, B., Chakravarti, D.N., 2007. Oxidative modification of proteins: age-related changes. Gerontology. 3, 128–139.

Chance, B., Sies, H., Boveris, A., 1979. Hydroperoxide metabolism in mammalian organs. Physiol. Rev. 59, 527–605.

Chen, Z., Tian, R., She, Z., et al., 2020. Role of oxidative stress in the pathogenesis of nonalcoholic fatty liver disease. Free Radic. Biol. Med. 152, 116–141.

Cheng, J.Z., Singhal, S.S., Saini, M., 1999. Effects of mGST A4 transfection on 4-hydroxynonenal-mediated apoptosis and differentiation of K562 human erythroleukemia cells. Arch. Biochem. Biophys. 372, 29–36.

Chow, C.K., Ibrahim, W., Wei, Z., 1999. Vitamin E regulates mitochondrial hydrogen peroxide generation. Free Radic. Biol. Med. 27, 580–587.

Cock, H.R., Tong, X., Hargreaves, I.P., 2002. Mitochondrial dysfunction associated with neuronal death following status epilepticus in rat. Epilepsy Res. 48, 157–168.

Colurso, G.J., Nilson, J.E., Vervoort, L.G., 2003. Quantitative assessment of DNA fragmentation and beta-amyloid deposition in insular cortex and midfrontal gyrus from patients with Alzheimer's disease. Life Sci. 14, 1795–1803.

Connor, W.E., Neuringer, M., Reisbick, S., 1992. Essential fatty acids: the importance of n-3 fatty acids in the retina and brain. Nutr. Rev. 50, 21–29.

Dalle-Donne, I., Rossi, R., Giustarini, D., 2003. Protein carbonyl groups as biomarkers of oxidative stress. Clin. Chim. Acta. 329, 23–38.

Dato, S., Crocco, P., D'Aguila, P., 2013. Exploring the role of genetic variability and lifestyle in oxidative stress response for healthy aging and longevity. Int. J. Mol. Sci. 14 (8), 16443–16472.

Day, B.J., 2014. Antioxidant therapeutics; pandora's box. Free Radic. Biol. Med. 66, 58–64.

Deavall, D.G., Martin, E.A., Horner, J.M., 2012. Drug-induced oxidative stress and toxicity. J. Toxicol. 2012, 645–660.

Dettbarn W.-D., D. Milatovic, M. Zivin, R.C. Gupta. 2001 Oxidative stress, acetylcholine and excitoxicity international conference on antioxidants. In: J. Marwah, A. Kanthasamy, International Conference on Antioxidants, Prominant Press, Scottsdale, AZ 183–211.

Devasagayam, T.P., Boloor, K.K., Ramasarma, T., 2003. Methods for estimating lipid peroxidation: an analysis of merits and demerits. Indian J. Biochem. Biophys. 40, 300–308.

De Zwart, L.L., Meerman, J.H.N., Commandeur, J.N.M., 1999. Biomarkers of free radical damage applications in experimental animals and in humans. Free Radic. Biol. Med. 26, 202–226.

Dikalov, S., Harrison, D.G., 2006. Pharmacological compounds with antioxidant activity. Antioxid. Cardiovasc. Dis. 258, 167–194.

Di Mascio, P., Kaiser, S.P., Devasagayam, T.P., 1991. Biological significance of active oxygen species: *in vitro* studies on singlet oxygen-induced DNA damage and on the singlet oxygen quenching ability of carotenoids, tocopherols and thiols. Adv. Exp. Med. Biol. 283, 71–77.

Dix, T.A., Aikens, J., 1993. Mechanisms and biological relevance of lipid peroxidation initiation. Chem. Res. Toxicol. 6, 2–18.

Dizdaroglu, M., Jaruga, P., Birinciogl, M., 2002. Free radical-induced damage to DNA: mechanisms and measurement. Free Radic. Biol. Med. 11, 1102–1115.

Droge, W., 2002. Aging-related changes in the thiol/disulfide redox state: implications for the use of thiol antioxidants. Exp. Gerontol. 37, 1333–1345.

Duda-Chodak, A., Tarko, T., Satora, P., 2015. Interaction of dietary compounds, especially polyphenols, with the intestinal microbiota. A review. Eur. J. Nutr. 54, 325–341.

Esterbauer, H., Schaur, R.J., Zolner, H., 1998. Chemistry and biochemistry of 4-hydroxynonenal, malonaldehyde and related aldehydes. Free Radic. Biol. Med. 11, 81–128.

Famm, S.S., Morrow, J.D., 2003. The isoprostanes: unique products of arachidonic acid oxidation—a review. Curr. Med. Chem. 10, 1723–1740.

Fang, Y.Z., Yang, S., Guoyao, W., 2002. Free radicals, antioxidant, and nutricion. Nutrition. 18, 872–879.

Fang, C., Gu, L., Smerin, D., et al., 2017. The interrelation between reactive oxygen species and autophagy in neurological disorders. Oxid. Med. Cell. Longev. 1–16.

Farooqui, A.A., Ong, Yi, Lu, X.R., 2001. Neurochemical consequences of kainate-induced toxicity in brain: involvement of arachidonic acid release and prevention of toxicity by phospholipase A(2) inhibitors. Brain Res. Rev. 38, 61–78.

Federico, A., Cardaioli, E., Pozzo, P.D., 2012. Mitochondria, oxidative stress and neurodegeneration. J. Neurol. Sci. 322, 254–262.

Fessel, J.P., Porter, N.A., Moore, K.P., 2002. Discovery of lipid peroxidation products formed *in vivo* with a substituted tetrahydrofuran ring (isofurans) that are favored by increased oxygen tension. Proc. Natl. Acad. Sci. USA. 99, 16713–16718.

Fetcher, L.D., Liu, Y., Pearce, T.A., 1997. Cochlear protection from carbon monoxide exposure by free radical blockers in the guinea pig. Toxicol. Appl. Pharmacol. 142, 47–55.

Fishel, M.L., Vasko, M.R., Kelley, M.R., 2007. DNA repair in neurons: so if they don't divide what's to repair? Mutat. Res. 1–2, 24–36.

Forrester, S.J., Kikuchi, D.S., Hernandes, M.S., et al., 2018. Reactive oxygen species in metabolic and inflammatory signaling. Circ. Res. 122, 877–902.

Frantseva, M.V., Velazquez, J.L.P., Tsoraklidis, G., 2000. Oxidative stress is involved in seizure-induced neurodegeneration in the kindling model of epilepsy. Neuroscience. 97 (3), 431–435.

Frederiksson, A., Eriksson, P., Archer, T., 1997. MPTP-induced deficits in motor activity: neuroprotective effects of the spin trapping agents, alpha-phenyl-*tert*-butylnitrone (PBN). J. Neural Transm. 104, 579–592.

Gamble, S.C., Wiseman, A., Goldfarb, P.S., 1999. Selenium-dependent glutathione peroxidase and other selenoproteins: their synthesis and biochemical roles. J. Chem. Technol. Biotechnol. 68, 123–134.

Gao, L., Yin, H., Milne, G.L., 2006. Formation of F-ring isoprostane-like compounds (F_3-isoprostanes) *in vivo* from eicosapentaenoic acid. J. Biol. Chem. 281, 14092–14099.

García-Niño, W.R., Pedraza-Chaverrí, J., 2014. Protective effect of curcumin against heavy metals-induced liver damage. Food Chem. Toxicol. 69, 182–201.

Gasc, N., Taché, S., Rathahao, E., 2007. 4-Hydroxynonenal in foodstuffs: heme concentration, fatty acid composition and freeze-drying are determining factors. Redox Rep. 12, 40–44.

Gaziano, J.M., Manson, J.E., Branch, L.G., 1995. A prospective study of consumption of carotenoids in fruits and vegetables and decreased cardiovascular mortality in the elderly. Ann. Epidemiol. 5, 255–260.

Gido, G., Kristian, T., Siesjo, B.K., 1997. Extracellular potassium in a neocortical cone area after transient focal ischemia. Stroke. 28, 206–210.

Gould, R.L., Pazdro, R., 2019. Impact of supplementary amino acids, micronutrients, and overall diet on glutathione homeostasis. Nutrients 11 (5), 1–21.

Guéraud, F., Atalay, M., Bresgen, N., 2010. Advances in methods for the determination of biologically relevant lipid peroxidation products. Free Radic. Res. 44, 1098–1124.

Gupta, R.C., Milatovic, D., Dettbarn, W.-D., 2001a. Depletion of energy metabolites following acetylcholinesterase inhibitor-induced status epilepticus: protection by antioxidants. Neurotoxicol 22, 271–282.

Gupta, R.C., Milatovic, D., Dettbarn, W.-D., 2001b. Nitric oxide modulates high-energy phosphates in brain regions of rats intoxicated with diisopropylphosphorofluoridate orcarbofuran: prevention by *N-tert*-butyl-α-phenylnitrone or vitamin E. Arch. Toxicol. 75, 346–356.

Gupta, R.C., Milatovic, S., Dettbarn, W.-D., Aschner, M., Milatovic, D., 2007. Neuronal oxidative injury and dendritic damage induced by carbofuran: protection by memantine. Toxicol. Appl. Pharmacol. 219, 97–105.

Gutteridge, J.M., Halliwell, B., 1990. The measurement and mechanism of lipid peroxidation in biological systems. Trends Biochem. Sci. 15, 129–135.

Halliwell, B., 2007. Oxidative stress and neurodegeneration: where are we now? J. Neurochem. 97, 1634–1658.

Halliwell, B., 2012. Free radicals and antioxidants: updating a personal view. Nutr. Rev. 70, 257–265.

Halpain, S., Hipolito, A., Saffer, L., 1998. Regulation of F-actin stability in dendritic spines by glutamate receptors and calcineurin. J. Neurosci. 18, 9835–9844.

Hamaguchi, T., Ono, K., Yamada, M., 2010. Curcumin and Alzheimer's disease. CNS Neurosci. Ther. 16, 285–297.

Handy, D.E., Loscalzo, J., 2012. Redox regulation of mitochondrial function. Antioxid. Redox Signal. 16 (11), 1323–1367.

Harrison, F.E., 2012. A critical review of vitamin C for the prevention of age-related cognitive decline and Alzheimer's disease. J. Alzheimers Dis. 29, 711–726.

Hayes, J.D., Pulford, D.J., 1995. The glutathione *S*-transferase supergene family: regulation of GST and the contribution of the isoenzymes to cancer chemoprotection and drug resistance. Crit. Rev. Biochem. Mol. Biol. 30, 445–600.

He, Q.P., Smith, M.L., Li, P.A., 1997. Necrosis of the substantia nigra, pars reticulata, influorothyl-induced status epilepticus is ameliorated by the spin trap alpha-phenyl-*N-tert*-butylnitrone. Free Radic. Biol. Med. 22, 917–922.

He, L., He, T., Farrar, S., et al., 2017. Antioxidants maintain cellular redox homeostasis by elimination of reactive oxygen species. Cell. Physiol. Biochem. 44, 532–553.

Huang, H.Y., Appel, L.J., Croft, K.D., 2002. Effects of vitamin C and vitamin E on in vivo lipid peroxidation: results of a randomized controlled trial. Am. J. Clin. Nutr. 76, 549–555.

Janicka, M., Kot-Wasik, A., Kot, J., 2010. Isoprostanes-biomarkers of lipid peroxidation: their utility in evaluating oxidative stress and analysis. Int. J. Mol. Sci. 11, 4631–4659.

Jeffrey, M., Goodsir, C.M., Bruce, M.E., 1997. In vivo toxicity of prion protein in murine scrapie: ultrastructural and immunogold studies. Neuropathol. Appl. Neurobiol. 23, 93–100.

Ji, H.-F., Shen, L., 2014. The multiple pharmaceutical potential of curcumin in Parkinson's disease. CNS Neurol. Disord. 13, 369–373.

Jiang, M., Lee, C.L., Smith, K.L., 1998. Spine loss and other persistent alterations of hippocampal pyramidal cell dendrites in a model of early-onset epilepsy. J. Neurosci. 18, 8356–8368.

Johnson, F., Giulivi, C., 2005. Superoxide dismutases and their impact upon human health. Mol. Aspects Med. 26, 340–352.

Johnson, J.A., Barbary, A., Kornguth, S.E., 1993. Glutathione S-transferase isoenzymes in rat brain neurons and glia. J. Neurosci. 13, 2013–2023.

Jones, D.P., Eklow, L., Thor, H., 1981. Metabolism of hydrogen peroxide in isolated hepatocytes: relative contributions of catalase and glutathione peroxidase in decomposition of endogenously generated H_2O_2. Arch. Biochem. Biophys. 210, 505–516.

Joseph, S.V., Edirishinghe, I., Burton-Freeman, B.M., 2014. Berries: anti-inflammatory effects in humans. J. Agric. Food Chem. 62, 3886–3903.

Kadiiska, M.B., Gladen, B.C., Baird, D.D., 2005. Biomarkers of oxidative stress study II: are oxidation products of lipids, proteins, and DNA markers of CCl_4 poisoning? Free Radic. Bio. Med. 38, 698–710.

Kadiiska, M.B., Peddada, S., Herbert, R.A., 2015. Biomarkers of oxidative stress study VI. Endogenous plasma antioxidants fail as useful biomarkers of endotoxin-induced oxidative stress. Free Radic. Biol. Med. 81, 100–106.

Kinter, M., Roberts, R.J., 1996. Glutathione consumption and glutathione peroxidase inactivation in fibroblast cell lines by 4-hydroxy-2-nonenal. Free Radic. Biol. Med. 21, 457–462.

Kowaltowski, A.J., 2019. Strategies to detect mitochondrial oxidants. Redox Biol. 21, 1–6.

Knowles, R.G., Moncada, S., 1994. Nitric oxide synthases in mammals. Biochem. J. 298 (2), 249–258.

Kuhla, B., Haase, C., Flach, K., 2007. Effect of pseudophosphorylation and cross-linking by lipid peroxidation and advanced glycation end product precursors on tau aggregation and filament formation. J. Biol. Chem. 82, 6984–6991.

Lafon-Cazal, M., Pietri, S., Culcasi, M., 1993. NMDA-dependent superoxide production and neurotoxicity. Nature. 364, 535–537.

Lancelot, E., Revaud, M.L., Boulee, R.G., 1997. Alpha-N-tert-butylnitone attenuates excitotoxicity in rat striatum by preventing hydroxyl radical accumulation. Free Radic. Biol. Med. 23, 1031–1034.

Lee, J., Hahm, E.R., Singh, S.V., 2010. Withaferin A inhibits activation of signal transducer and activator of transcription 3 in human breast cancer cells. Carcinogenesis. 31 (11), 1991–1998.

LoPachin, R.M., Barber, D.S., Gavin, T., 2008. Molecular mechanisms of the conjugated alpha, beta unsaturated carbonyl derivatives: relevance to neurotoxicity and neurodegenerative diseases. Toxicol. Sci. 2, 235–249.

Lorenzon Dos Santos, J., Quadros, A.S., Weschenfelder, C., Garofallo, S.B., Marcadenti, A., 2020. Oxidative stress biomarkers, nut-related antioxidants, and cardiovascular disease. Nutrients 12 (3), 1–17.

Lovell, M.A., Xie, C., Markesbery, W.R., 2000. Acrolein, a product of lipid peroxidation, inhibits glucose and glutamate uptake in primary neuronal cultures. Free Radic. Biol. Med. 9, 714–720.

Maboudou, P., Mathieu, D., Bachelet, H., 2002. Detection of oxidative stress. Interest of GC-MS for malondialdehyde and formaldehyde monitoring. Biomed. Chromatogr. 16, 199–202.

Maoka, T., Yamano, Y., Wada, A., 2015. Oxidative metabolites of lycopene and γ-carotene in Gac (Momordica cochinchinensis). J. Agric. Food Chem. 63, 1622–1630.

Mantzarlis K., V. Tsolaki, E. Zakynthionoc, Role of oxidative stress and mitochondrial dysfunction in sepsis and potential therapies. Oxid. Med. Cell. Longev. ID 5985209. Published online 2017 Aug 20.

Margialasche, F., Polidori, M.C., Monastero, R., 2009. Biomarkers of oxidative and nitrosative damage in Alzheimer's disease and mild cognitive impairment. Ageing Res. Rev. 8, 285–305.

Milatovic, D., Radic, Z., Zivin, M., Dettbarn, W.-D., 2000a. Atypical effect of some spin trapping agents: reversible inhibition of acetylcholinesterase. Free Radic. Biol. Med. 28, 597–603.

Milatovic, D., Zivin, M., Dettbarn, W.-D., 2000b. The spin trapping agent phenyl-N-tert-butyl-nitrone (PBN) prevents excitotoxicity in skeletal muscle. Neurosci. Lett. 278, 25–28.

Milatovic, D., Zivin, M., Gupta, R.C., Dettbarn, W.-D., 2001. Alterations in cytochrome-c-oxidase and energy metabolites in response to kainic acid-induced status epilepticus. Brain Res. 912, 67–78.

Milatovic, D., Gupta, R.C., Dettbarn, W.-D., 2002. Involvement of nitric oxide in kainic acid-induced excitotoxicity in rat brain. Brain Res. 957, 330–337.

Milatovic, D., Zaja-Milatovic, S., Montine, K.S., 2003. Pharmacologic suppression of neuronal oxidative damage and dendritic degeneration following direct activation of glial innate immunity in mouse cerebrum. J. Neurochem. 87, 1518–1526.

Milatovic, D., Milatovic, S., Montine, K., 2004. Neuronal oxidative damage and dendritic degeneration following activation of CD14-dependent innate immunity response in vivo. J. Neuroinflamm. 1, 20.

Milatovic, D., Gupta, R.C., Dekundy, A., 2005a. Carbofuran-induced oxidative stress in slow and fast skeletal muscles: prevention by memantine. Toxicology. 208, 13–24.

Milatovic, D., VanRollins, M., Li, K., 2005b. Suppression of murine cerebral F_2-isoprostanes and F_4-neuroprostanes from excitotoxicity and innate immune response in vivo by alpha- or gamma-tocopherol. J. Chromatogr. B Analyt. Technol. Biomed. Life Sci. 827, 88–93.

Milatovic, D., Gupta, R.C., Aschner, M., 2006. Anticholinesterase toxicity and oxidative stress. Sci. World J. 6, 295–310.

Milatovic, D., Montine, T.J., Zaja-Milatovic, S., Madison, J.D., 2010. Morphometric analysis in neurodegenerative disease. Curr. Protoc. Toxicol. Unit. 12 (16), 1–14.

Milne, G.L., 2017. Classifying oxidative stress by F_2-Isoprostane levels in human disease: the re-imagining of a biomarker. Redox Biol. 12, 897–898.

Milne, G.L., Dai, Q., Roberts, J.L., 2015. The isoprostanes—25 years later. Biochim. Biophys. Acta 1851 (4), 433–445.

Miyajima, T., Kotake, Y., 1995. Spin trap phenyl-N-tert-butylnitone (PBN) inhibits induction of nitric oxide synthase in endotoxin-induction in mice. Free Radic. Biol. Med. 22, 463–470.

Montine, T.J., Milatovic, D., Gupta, R.C., 2002a. Neuronal oxidative damage from activated innate immunity is EP2 receptor-dependent. J. Neurochem. 83, 463–470.

Montine, T.J., Quinn, L.F., Milatovic, D., 2002b. Peripheral F_2-isoprostanes and F_4-neuroprostanes are not increased in patients with Alzheimer's disease or in animal model of cerebral oxidative damage. Ann. Neurol. 52, 175–179.

Montine, K.S., Quinn, J.F., Zhang, J., 2004. Isoprostanes and related products of lipid peroxidation in neurodegenerative diseases. Chem. Phys. Lipids. 128, 117–718.

Moore, K., Roberts, L.J., 1998. Measurement of lipid peroxidation. Free Radic. Res. 28, 659–671.

Mori, A., Yokoi, I., Noda, Y., 2004. Natural antioxidants may prevent posttraumatic epilepsy: a proposal based on experimental animal studies. Acta Med. Okayama. 58, 111–118.

Morrow, J.D., Hill, K.E., Burk, R.F., 1990. A series of prostaglandin F_2-like compounds are produced in vivo in humans by a non-cyclooxygenase, free radical-catalyzed mechanism. Proc. Natl. Acad. Sci. USA. 87, 9383–9387.

Muller, M., Gahwiler, B.H., Rietschin, L., 1993. Reversible loss of dendritic spines and altered excitability after chronic epilepsy in hippocampal slice cultures. Proc. Natl. Acad. Sci. USA. 90, 257–261.

Neely, M.D., Boutte, A., Milatovic, D., 2005. Mechanisms of 4-hydroxynonenal-induced neuronal microtubule dysfunction. Brain Res. 1037, 90–98.

Nicholls, D.G., Vesce, S., Kirk, L., 2003. Interactions between mitochondrial bioenergetics and cytoplasmic calcium in cultured cerebellar granule cells. Cell Calcium. 34, 407–424.

Niki, E., Noguchi, N., Tsuchihashi, H., 1995. Interaction among vitamin C, vitamin E, and beta-carotene. Am. J. Clin. Nutr. 62, 1322S–1326S.

Nishida, S., 2005. Metabolic effects of melatonin on oxidative stress and diabetes mellitus. Endocrine. 27, 131–136.

O'Brien, R.J., Kamboj, S., Ehlers, M.D., 1998. Activity-dependent modulation of synaptic AMPA receptor accumulation. Neuron. 21, 1067–1078.

Onodera, K., Omoi, N.O., Fukui, K., 2003. Oxidative damage of rat cerebral cortex and hippocampus, and changes in antioxidative defense systems caused by hyperoxia. Free Radic. Res. 37, 367–372.

Ore, A., Akinloye, O.A., 2019. Oxidative stress and antioxidant biomarkers in clinical and experimental models of non-alcoholic fatty liver disease. Medicina. 55 (26), 1–13.

Patel, M.N., 2002. Oxidative stress, mitochondrial dysfunction, and epilepsy. Free Radic. Res. 36, 1139–1146.

Patel, M.N., Liang, L.P., Roberts, L.J., 2001. Enhanced hippocampal F_2-isoprostane formation following kainate-induced seizures. J. Neurochem. 79, 1065–1069.

Pazos, A.J., Green, E.J., Busto, R., 1999. Effects of combined postischemic hypothermia and delayed *N-tert*-butyl-alpha-phenylnitrone (PBN) administration on histopatological and behavioral deficits associated with transient global ischemia. Brain Res. 846, 186–195.

Pieri, C., Marra, M., Moroni, F., 1995. The modulation of intracellular glutathione level modulates the mitochondrial response in proliferating rat splenocytes. Arch. Gerontol. Geriatr. 21, 115–125.

Poetsch, A.R., 2020. The genomics of oxidative DNA damage, repair, and resulting mutagenesis. Comput. Struct. Biotechnol. J. 18, 207–219.

Poodineh, J., Feizabad, A.K., Nakhaee, A., 2015. Antioxidant activities of *Caralluma tuberculata* on streptozotocin-induced diabetic rats. Drug Dev. Res. 76, 40–47.

Porter, N.A., Caldwell, S.E., Mills, K.A., 1995. Mechanisms of free radical oxidation of unsaturated lipids. Lipids. 30, 277–290.

Prasai, P.K., Shrestha, B., Orr, A.W., Pattillo, C.B., 2018. Decreases in GSH: GSSG activate vascular endothelial growth factor receptor 2 (VEGFR2) in human aortic endothelial cells. Redox Biol. 19, 22–27.

Pratico, D., Rokach, J., Lawson, J., 2004. F2-isoprostanes as indices of lipid peroxidation in inflammatory diseases. Chem. Phys. Lipids. 128, 165–171.

Prince, M.R., LaMuraglia, G.M., MacNichol, E.F., 1988. Increased preferential absorption in human atherosclerotic plaque with oral beta carotene. Implications for laser endarterectomy. Circulation 78, 338–344.

Punz, A., Nanobashvili, N., Feugl, A., 1998. Effect of α-tocopherol pretreatment on high energy metabolites in rabbit skeletal muscle after ischemia-reperfusion. Clin. Nutr. 17, 85–87.

Puttachary, S., Sharma, S., Stark, S., 2015. Seizure-induced oxidative stress in temporal lobe epilepsy. Biomed. Res. Int. 2015, 745613.

Quinlan, C.L., Orr, A.L., Perevoshchikova, I.V., 2012a. Mitochondrial complex II can generate reactive oxygen species at high rates in both the forward and reverse reactions. J. Biol. Chem. 287, 27255–27264.

Quinlan, C.L., Treberg, J.R., Perevoshchikova, I.V., 2012b. Native rates of superoxide production from multiple sites in isolated mitochondria measured using endodegous reporters. Free Radic. Biol. Med. 53 (9), 1807–1817.

Rahal, A., Kumar, A., Singh, V., 2014. Oxidative stress, prooxidants, and antioxidants: the interplay. Biomed. Res. Int. 2014.

Rao, A., Craig, A.M., 1997. Activity regulates the synaptic localization of the NMDA receptor in hippocampal neurons. Neuron. 19, 801–812.

Rashid, K., Sil, R.C., 2015. Curcumin enhances recovery of pancreatic islets from cellular stress induced inflammation and apoptosis in diabetic rats. Toxicol. Appl. Pharmacol. 282, 297–310.

Rauca, C., Wiswedel, I., Zerbe, R., 2004. The role of superoxide dismutase and alpha-tocopherol in the development of seizures and kindling induced by pentylenetetrazol—influence of the radical scavenger alpha-phenyl-N-*tert*-butyl nitrone. Brain Res. 1009, 203–212.

Righi, N.C., Schuch, F.B., et al., 2020. Effects of vitamin C on oxidative stress, inflammation, muscle soreness, and strength following acute exercise: meta-analyses of randomized clinical trials. Eur. J. Nutr. 59 (2).

Roberts, L.J., Montine, T.J., Markesbery, W.R., 1998. Formation of isoprostane-like compounds (neuroprostanes) in vivo from docosahexaenoic acid. J. Biol. Chem. 273, 13605–13612.

Sack, C.A., Socci, D.J., Crandall, B.M., 1996. Antioxidant treatment with phenyl-alpha-tert-butylnitrone (PBN) improves the cognitive performance and survival of aging rats. Neurosci. Lett. 205, 181–184.

Salem, N., Kim, H.Y., Lyergey, J.A., Martin, R.E., 1986. Docosahexaenoic acid: membrane function and metabolism. In: Martin, R.E. (Ed.), Health Effects of Polyunsaturated Acids in Seafoods. Academic Press, New York, NY, pp. 263–317.

Santos, L.F.L., Freitas, R.L.M., Xavier, S.M.L., 2008. Neuroprotective actions of vitamin C related to decreased lipid peroxidation and increased catalase activity in adult rats after pilocarpine-induced seizures. Pharmacol. Biochem. Behav. 89 (1), 1–5.

Sas, K., Szabó, E., Vécsei, L., 2018. Mitochondria, oxidative stress and the kynurenine system, with a focus on ageing and neuroprotection. Molecules 23 (191), 1–28.

Schaefer, E.J., Bongard, V., Beiser, A.S., 2006. Plasma phosphatidylcholine docosahexaenoic acid content and risk of dementia and Alzheimer disease: the Framingham Heart Study. Arch. Neurol. 63, 1545–1550.

Schafer, F.Q., Buettner, G.R., 2001. Redox environment of the cell as viewed through the redox state of the glutathione disulfide/glutathione couple. Free Radic. Biol. Med. 30, 1191–1212.

Schwob, J.E., Fuller, T., Price, J.L., 1980. Widespread patterns of neuronal damage following systemic or intracerebral injections of kainic acid: a histological study. Neuroscience. 5, 991–1014.

Sen, C.K., 2000. Cellular thiols and redox-regulated signal transduction. Curr. Top. Cell. Regul. 36, 1–30.

Serreli, G., Deiana, M., 2020. Extra virgin olive oil polyphenols: modulation of cellular pathways related to oxidant species and inflammation in aging. Cells 9 (2), 478 (1-21).

Shah, D., Mahajan, N., Paudyal, B., 2014. Oxidative stress and its biomarkers in systemic lupus erythematosus. J. Biomed. Sci. 21 (1), 23.

Simioni, C., Zauli, G., Meartelli, A.M., et al., 2018. Oxidative stress: role of physical exercise and antioxidant nutraceuticals in adulthood and aging. Oncotarget 9 (24), 17181–17198.

Singh, M., Jadhav, H.R., 2014. Melatonin: functions and ligands. Drug. Discov. Today. 19 (9), 1410–1418.

Smith, W.L., 2005. Cyclooxygenases, peroxide tone and the allure of fish oil. Curr. Opin. Cell Biol. 17, 174–182.

Sodum, R.S., Fiala, E.S., 2001. Analysis of peroxynitrite reactions with guanine, xanthine, and adenine nucleosides by high-pressure liquid chromatography with electrochemical detection: C8-nitration and oxidation. Chem. Res. Toxicol. 14, 438–450.

Sonowal, H., Ramana, K.V., 2019. 4-Hydroxy-*trans*-2-nonenal in the regulation of anti-oxidative and pro-inflammatory signaling pathways. Oxid. Med. Cell. Longev. 2019, 1–17.

Spickett, C.M., Wiswedel, I., Siems, W., et al., 2010. Advances in methods for the determination of biologically relevant lipid peroxidation products. Free Radic. Res 44, 1172–1202.

Srivastava, P., Yadav, R.S., Chandravanshi, L.P., 2014. Unraveling the mechanism of neuroprotection of curcumin in arsenic induced cholinergic dysfunction in rats. Toxicol. Appl. Pharmacol. 279, 428–440.

Stadtman, E.R., Levine, R.L., 2003. Freeradical-mediated oxidation of free amino acids and amino acids residue in proteins. Amino Acids. 3–4, 207–218.

Sullivan, L.B., Chandel, N.S., 2014. Mitochondrial reactive oxygen species and cancer. Cancer Metab. 2, 17.

Sultana, R., Perluigi, M., Butterfield, D.A., 2006. Protein oxidation and lipid peroxidation in brain of subjects with Alzheimer's disease: insights into mechanism of neurodegeneration from redox proteomics. Antioxid. Redox Signal. 8, 2021–2037.

Sultan, C.S., Saackel, A., Stank, A., et al., 2018. Impact of carbonylation on glutathione peroxidase-1 activity in human hyperglycemic endothelial cells. Redox Biol. 16, 113–122.

Tayyaba, K., Hasan, M., 1985. Vitamin E protects against metasystox-induced adverse effect on lipid metabolism in the rat brain and spinal cord. Acta Pharmacol. Toxicol. 57, 190–196.

Tejada, S., Sureda, A., Roca, C., 2007. Antioxidant response and oxidative damage in brain cortex after high dose of pilocarpine. Brain Res. Bull. 71 (4), 372–375.

Thomas, C.E., Ohlweiler, D.F., Taylor, V.L., 1997. Radical trapping and inhibition of iron-dependent CNS damage by cyclic nitrone spin traps. J. Neurochem. 68, 1173–1182.

Topinka, J., Bincova, B., Sram, R.J., 1989. The influence of α-tocopherol and pyritinol on oxidative DNA damage and lipid peroxidation in human lymphocytes. Mutat. Res. 225, 131–136.

Trewin, A.J., Berry, B.J., Wojtovich, A.P., 2018. Exercise and mitochondrial dynamics: keeping in shape with ROS and AMPK. Antioxidants 7 (1), 1–8.

Trieb, M., Kornej, J., Knuplez, E., et al., 2019. Atrial fibrillation is associated with alterations in HDL function, metabolism, and particle number. Basic Res. Cardiol. 114 (4), 1–10.

Tymko, M.M., Tremblay, J.C., Bailey, D.M., et al., 2019. The impact of hypoxaemia on vascular function in lowlanders and high-altitude indigenous populations. J. Physiol. 597, 5759–5776.

Uchida, K., 2003. 4-hyroxy-2-nonenal: a product and mediator of oxidative stress. Prog. Lipid Res. 42, 318–343.

Uchida, K., Kanematsu, M., Morimitsu, Y., 1998. Acrolein is a product of lipid peroxidation reaction. Formation of free acrolein and its conjugate with lysine residues in oxidized low density lipoproteins. J. Biol. Chem. 273, 16058–16066.

Upritchard, J.E., Schuurman, C.R., Wiersma, A., 2003. Spread supplemented with moderate doses of vitamin E and carotenoids reduces lipid peroxidation in healthy, nonsmoking adults. Am. J. Clin. Nutr. 78, 985–992.

Valentová, K., Vrba, J., Bancířová, M., 2014. Isoquercitrin: pharmacology, toxicology, and metabolism. Food Chem. Toxicol. 68, 267–282.

Valko, M., Leibfritz, D., Moncol, J., 2007. Free radicals and antioxidants in normal physiological functions and human disease. Int. J. Biochem. Cell Biol. 39, 44–84.

Van Acker, S.A.B.E., Koymans, L.M.H., Bast, A., 1993. Molecular pharmacology of vitamin E: structural aspects of antioxidant activity. Free Radic. Bio. Med. 15, 311–328.

Vatassery, G.T., 2004. Impairment of brain mitochondrial oxidative phosphorylation accompanying vitamin E oxidation induced by iron or reactive nitrogen species: a selective review. Neurochem. Res. 29, 1951–1959.

Virag, L., Szabo, E., Gergely, P., 2003. Peroxynitrite-induced citotoxicity: mechanism and opportunity for intervention. Toxicol. Lett. 140-141, 113–124.

Wang, J.Y., Shum, A.Y., Ho, Y.J., 2003. Oxidative neurotoxicity in rat cerebral cortex neurons: synergistic effects of H_2O_2 and NO on apoptosis involving activation of p38 mitogen-activated protein kinase and caspase-3. J. Neurosci. Res. 72, 508–519.

Xavier, S.M., Barbosa, C.O., Barros, D.O., 2007. Vitamin C antioxidant effects in hippocampus of adult Wistar rats after seizures and status epilepticus induced by pilocarpine. Neurosci. Lett. 420 (1), 76–79.

Yang, Z.P., Dettbarn, W.-D., 1998. Lipid peroxidation and changes of cytochrome c oxidase and xanthine oxidase in organophosphorous anticholinesterase induced myopathy. Xth International Symposium on Cholinergic Mechanisms. J. Physiol. 92, 157–162.

Yin, H., Gao, L., Tai, H.H., 2007. Urinary prostaglandin F2alpha is generated from the isoprostane pathway and not the cyclooxygenase in humans. J. Biol. Chem. 282, 329–336.

Yuste, R., Denk, W., 1995. Dendritic spines as basic functional units of neuronal integration. Nature. 375, 682–684.

Zaja-Milatovic, S., Gupta, R.C., Aschner, M., 2008. Pharmacologic suppression of oxidative damage and dendritic degeneration following kainic acid-induced excitotoxicity in mouse cerebrum. Neurotoxicology. 29, 621–627.

Zaja-Milatovic, S., Gupta, R.C., Aschner, M., Milatovic, D., 2009. Protection of DFP-induced oxidative damage and neurodegeneration by antioxidants and NMDA receptor antagonist. Toxicol. Appl. Pharmacol. 240, 124–131.

Zeng, L.H., Xu, L., Rensing, N.R., 2007. Kainate seizures cause acute dendritic injury and actin depolymerization in vivo. J. Neurosci. 27, 11604–11613.

Zhang, L., Wang, X., Cueto, R., et al., 2019. Biochemical basis and metabolic interplay of redox regulation. Redox Biol. 26 (1-16), 101284.

Zhou, H., Toan, S., 2020. Pathological roles of mitochondrial oxidative stress and mitochondrial dynamics in cardiac microvascular ischemia/reperfusion injury. Biomolecules 10 (1), 85 (1-21).

Zoheb, S.M., Prakash, A., Rahal, A., 2014. Curcumin attenuates oxidative stress-induced altered histoarchitecture of testes in experimentally exposed rats to metal mixture (lead, arsenic, cadmium, mercury, iron, and copper) for 28 days. Toxicol. Environ. Chem. 96, 660–679.

Zirak, M.R., Mehri, S., Karimani, S., et al., 2019. Mechanisms behind the atherothrombotic effects of acrolein, a review. Food Chem. Toxicol. 129, 38–53.

Section III

Common nutraceuticals

Chapter 32

Caffeine: an evaluation of the safety database

Ashley Roberts

Introduction and background

Human exposure to caffeine: current trends

Caffeine is the one of the most widely consumed food ingredients in the world. Every day, millions of people around the world enjoy beverages and other foods containing caffeine including coffee, tea, soft drinks, and chocolate. It has been estimated that in the order of 2.25 billion cups of coffee, the major source of dietary caffeine, are served every day, and that more than 80% of the world's population consume at least one caffeine-containing beverage on a daily basis (Heckman et al., 2010). Global consumption of coffee for the year 2018–19 was estimated in the order of 165.35 million bags (each bag weights approximately 60 kg) or approximately 10 million tons, making coffee the world's most consumed substance (ICA 2018–2019). As a food ingredient, caffeine also has one of the longest histories of human consumption due to its natural occurrence in the beans, leaves, or fruit of more than 60 different plants including coffee, tea, guarana, cola nuts, and cocoa pods.

The principal natural dietary sources of caffeine are coffee and tea, accounting for up to 90% of total caffeine consumption in some geographies. Coffee, which is extracted from roasted beans of the coffee plant, contains the highest levels of caffeine per unit weight among natural sources. However, because of the high volume of other types of caffeinated beverages consumed per capita in some geographies, coffee may not be the major source of caffeine consumption in some regions. For instance, tea beverages are the major source of caffeine is some regions of Asia, while the popularity of mate beverages in areas of South America makes Yerba Mate a major source of caffeine in that region. For comparison, an eight ounce (237 mL) serving of plain brewed coffee is reported to contain about 133 mg of caffeine (16.6 mg/ounce or 0.56 mg/mL) while the same volume of tea (green or black) or Yerba mate contains two- to threefold less caffeine at 47 mg (5.9 mg/ounce or 0.20 mg/mL) and 78 mg (9.8 mg/ounce or 0.33 mg/mL), respectively. Other natural dietary sources of caffeine include cocoa and chocolate; however, their relative contribution to the overall daily caffeine intake is considered negligible in most countries when compared to the levels of caffeine in coffee and tea (Heckman et al., 2010).

In addition to its natural occurrence in many plants, caffeine is also a food ingredient with a long history of safe use. As far back as the late 1800s, caffeine has been added to soft drinks, contributing its bitterness to the overall beverage flavor. The levels of caffeine added to soft drinks, however, are far less than those contained in an equivalent volume of brewed coffee or tea. For instance, a typical 12 ounce (355 mL) carbonated soft drink contains approximately 34 mg of caffeine, while the same volume of plain brewed coffee and tea contain 200 and 80 mg of caffeine, respectively (Heckman et al., 2010). The market and popularity for caffeine-containing beverages has grown extensively to the point that caffeinated beverages represent a major dietary source of caffeine (following coffee and tea) in countries such as the United States and in Europe (EFSA, 2015; Mitchell et al., 2014). Table 32.1 lists a compilation of some of the most commonly consumed beverages and their "typical" caffeine content in comparison to solid foods containing caffeine. The Center for Science in the Public Interest (CSPI; http://www.cspinet.org/new/cafchart.htm#table_energy_-drinks) also reports the caffeine content of a number of foods, beverages, and medication products in the US marketplace. High street specialty coffees including an Espresso shot contain caffeine at levels several-fold higher than regular brewed coffee at values up to a maximum of 398 mg per 20 ounce serving. Regular energy drink products typically contain caffeine at levels of 80 mg per 8 ounces, however products do exist in the marketplace at

Nutraceuticals. DOI: https://doi.org/10.1016/B978-0-12-821038-3.00032-X

TABLE 32.1 Caffeine content of commonly consumed beverages and foods.

Dietary source	Caffeine (mg/fluid ounce)	Caffeine (mg/mL)	Caffeine (mg/8 ounce)	References
Beverages				
Specialty coffee (Espresso)	46.7–62.8	1.6–2.1	373.6–502.4	Mitchell et al. (2014)
Brewed coffee (Regular)	11.9	0.40	95.2	Mitchell et al. (2014)
Energy drinks (Regular)	10.0	0.34	80.0	Mitchell et al. (2014)
Black tea	5.9	0.20	47.2	Mitchell et al. (2014)
Green tea	3.1	0.10	24.8	Mitchell et al. (2014)
Carbonated soft drinks (cola type)	3.0	0.10	24.0	Mitchell et al. (2014)
Chocolate milk or drink	0.2–2.0	0.0068–0.068	1.6–16	Mitchell et al. (2014)
Solid foods				
Chocolate-covered coffee beans	338–355 mg/serving[a]	–	–	Dietitians of Canada (2014)
Dark chocolate	27 mg/bar	–	–	Dietitians of Canada (2014)
Milk chocolate	8–12 mg/bar	–	–	Dietitians of Canada (2014)
Chocolate brownie	1–4 mg/brownie	–	–	Dietitians of Canada (2014)

[a]*A serving size is assumed to be 60 mL or a quarter cup.*

levels of 160 mg per 8 ounce serving. In contrast, carbonated soft drinks contain slightly lower but comparable levels of caffeine, similar to those present in some high street teas. Notably, a similar amount of caffeine in an 8-ounce carbonated soft drink may be ingested through the consumption of a dark chocolate bar. Over-the-counter pills can also include between 100 and 150 mg per capsule/tablet. In contrast to the food, beverage, and over-the-counter products permitted above, the FDA in 2018 released guidance for industry on highly concentrated caffeine in liquid and powdered dietary supplements, in order to help them determine when a product is considered adulterated and illegal by the FDA.

Caffeine is considered a mild stimulant, and as such, is commonly ingested to enhance wakefulness, improve mood and energy levels, and even enhance athletic performance (Burns et al., 2014). This stimulatory property of caffeine has opened a market for so-called "functional beverages" that includes energy drinks and energy shots, which contain a caffeine dose in a smaller volume of liquid (usually 50 mL) as compared to other beverages and an array of high street caffeine-concentrated specialty coffees and even purified caffeine tablets.

New uses of caffeine continue to be introduced into the marketplace, including dietary supplements, waffles, ice cream, jelly-beans, pancake syrup, flavored waters, and chewing gums. Moderate levels of caffeine are also present in some medicines including pain relievers, diuretics, cold remedies, and weight control preparations (Heckman et al., 2010; US FDA and 21 CFR 340.50). The use of caffeine in over-the-counter stimulant products to restore mental alertness or wakefulness during fatigue or drowsiness is also considered acceptable for adults and for children 12 years of age (i.e., adolescents) in amounts of 100–200 mg to be consumed not more often than every 3–4 hours (21 CFR 340.50; U.S. FDA, 2018b). Caffeine also has a long history of safe use for pediatric treatment of apnea and attention deficit disorder in children and infants with caffeine levels of 5–10 mg/kg body weight having been reported as being safe within the first 10 days of life for a median duration of 37 days (Erenberg et al., 2000). Caffeine for the treatment of apnea did not affect general health, motor function, behavior, cognition, or other developmental measures (deafness, blindness) during a 5-year follow-up period (Francart et al., 2013; Schmidt et al. (2006, 2007, 2012)). Furthermore, the results of a systematic review and meta-analysis indicated that consumption of caffeine at dosages ranging from 20 to 80 mg/kg body weight by infants less than 3 days old with maintenance doses of 5–10 mg/kg body weight/day for an unreported duration of time was without adverse effects (Kua and Lee, 2017). Following another systematic review and meta-analysis, Chen et al. (2018) concluded that maintenance doses of 10–20 mg/kg body weight/day were also safe.

Although there is no evidence of moderate caffeine intake being associated with adverse health effects, the increasing number of consumer products containing caffeine has triggered calls for regulatory agencies such as the United States Food and Drug Administration (US

FDA) and the European Food Safety Authority (EFSA) to reexamine both the safety profile and dietary intake of caffeine. Years of scientific research have shown that moderate consumption of caffeine is not associated with adverse health effects (Heckman et al., 2010). On the contrary, numerous health-protective benefits associated with caffeine consumption have been reported (Heckman et al., 2010). In this chapter, the safety and dietary intake information for caffeine are reviewed to affirm its safety.

Regulatory position of caffeine in the United States and Europe

Currently, there are no official recommendations on the total daily caffeine intake in the United States. The US FDA provided guidance in 2012 on the recommended total caffeine intake for the US population based primarily on a comprehensive review endorsed by Health Canada (Nawrot et al., 2003). The agency stated in a letter that caffeine intake up to 400 mg/day in healthy adults is *not* associated with adverse health effects (U.S. FDA, 2012). As a food ingredient, caffeine has been listed since 1959 as a generally recognized as safe (GRAS) substance with tolerance set at 0.02% for addition to cola-type beverages (21 CFR, Section 182.1180). The US FDA has expressed concern that under the GRAS umbrella food manufacturers may be adding caffeine indiscriminately, raising concerns about excessive exposure and safety (McGuire, 2014). Further complicating the issue is the fact that the new and existing uses of caffeine as a food ingredient, and as a dietary supplement can fall under different regulatory frameworks.

In the European Union (EU), the European Food Safety Authority (EFSA) conducted a safety review of caffeine at the request of the EU Commission, taking into account intakes from all dietary sources as well as addressing possible interactions between caffeine and other constituents within energy drinks. The request was made in response to concerns raised by Member States including Belgium and France regarding potential health effects from a potential increased caffeine intake in relation to the findings of the Scientific Committee of Food in 1999 (SCF Scientific Committee on Food, 1999). Following their review, the EFSA Panel on Dietetic Products Nutrition and Allergies concluded that habitual caffeine consumption of up to 400 mg per day does not give rise to safety concerns for nonpregnant adults and that habitual consumption up to 200 mg per day by pregnant women does not give rise to safety concerns for the fetus. Also, the lack of concern of acute intakes in adults (3 mg/kg bw/day) was also considered to serve as the basis for no concern in both children and adolescents. The EFSA also concluded that other constituents of energy drinks would not adversely interact with caffeine on its effects on the cardiovascular system, the central nervous system, or hydration status, and thus the safety of single doses of caffeine up to 200 mg was confirmed (EFSA, 2015).

Molecular characterization of caffeine

As a naturally occurring substance with a long and widespread history of safe human consumption and approved food use, the caffeine molecule has been thoroughly characterized. Importantly, the caffeine molecule is the same regardless of the method of preparation or purification from natural products (e.g., coffee) as long as the purity and the grade specifications are the same. As a molecule, caffeine is a member of the methylxanthine family of compounds which are biosynthesized from a purine ring backbone. Fig. 32.1 shows the molecular structure of caffeine as the fully N-methylated form of xanthine (i.e., trimethylxanthine) along with the official assigned IUPAC name and CAS registry number (IPCS, 1998). When isolated or synthetically produced, caffeine is a white odorless powder with a bitter taste. It is a relatively lipophilic and water-soluble molecule as indicated by its octanol:water partition coefficient (log P_{ow}) of −0.07. Its water solubility has been estimated at about 2 g/100 mL of water. Sources of caffeine approved for food use include extraction and purification from coffee or tea, and one-step synthesis from precursor compounds such as theobromine. The food-grade quality of caffeine has been specified by the major regulatory sources including the Food Chemicals Codex (FCC), the United States Pharmacopoeia (USP), and the European Pharmacopoeia (EP).

The pharmacokinetics of caffeine

By definition, the term pharmacokinetics covers how an ingredient is handled in the body following administration, including absorption, distribution, metabolism, and excretion (ADME). Pharmacokinetic analysis is critical within the overall safety evaluation of an ingredient, since it is the blood or more specifically the target tissue levels of a substance that ultimately determine its potential

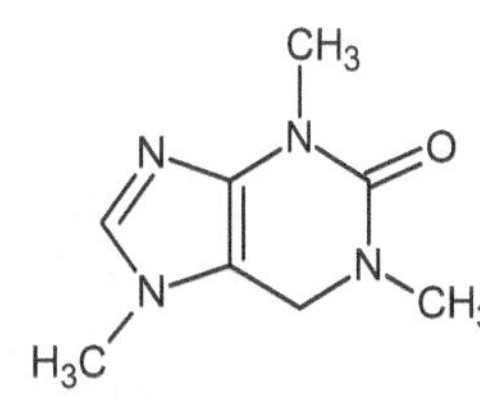

Common Name: Caffeine
IUPAC name: 1,3,7-Trimethylpurine-2,6-dione
CAS Registry number: 58-08-2

FIGURE 32.1 Chemical structure and designation for caffeine.

biological effects. The potential biological effect of a substance can also be informed by the degree of ADME in the body. Thus, a lower safety concern can be expected for a substance that is minimally absorbed and/or quickly eliminated versus one that bioaccumulates and is slowly eliminated from the body (Barton et al., 2006).

Absorption

As one of the most researched substances in the world, the pharmacokinetics of caffeine has been well-characterized following oral administration at different dose levels and in multiple species including humans. Reported information indicates that when ingested, the absorption of caffeine from the gastrointestinal tract is rapid and essentially complete. A study by Blanchard and Sawers (1983) reported that when healthy human subjects (age range of 18.8–30 years old) were administered caffeine at 5 mg/kg body weight (bw) either orally as an aqueous solution or as an intravenous (IV) infusion, peak plasma levels of caffeine were reached rapidly within 30 minutes. When the resultant area under the plasma–time concentration curve (AUC) via the oral route was compared to that obtained via the IV route, the estimated oral bioavailability was reported as 108.3% ± 3.6%, indicative of complete absorption from the gastrointestinal tract (Blanchard and Sawers, 1983). The rapid and extensive absorption of caffeine into the systemic circulation of humans is also outlined within multiple studies in different species, further indicating a very limited first-pass metabolism (Burg, 1975; Lachance, 1982). Burg (1975) reported the results of an oral administration of 25 mg/kg ^{14}C-radiolabeled caffeine to different adult (nonpregnant) animals species including rats, hamsters, rabbits, and monkeys. The results indicate marked consistency in absorption, peak plasma levels, and elimination across all the species tested.

A notable feature of the pharmacokinetics of caffeine is its linear behavior. Increasing administered oral doses of caffeine results in proportional increases in blood levels of caffeine. In a study by Newton et al. (1981), a group of six healthy subjects were administered oral caffeine solutions at increasing dose levels of 50, 300, 500, and 750 mg. When the resulting peak plasma concentrations were analyzed against administered doses, there was a proportional relationship, indicating that caffeine exhibits linear pharmacokinetic behavior at the dose levels tested. Furthermore, when plasma AUCs were normalized to administered dose per unit body weight, the plasma concentration–time curves were superimposable, indicative of a very consistent pharmacokinetic behavior across different human subjects despite intrinsic individual differences (Newton et al., 1981).

Distribution

Once absorbed into the systemic circulation, a fraction of caffeine (estimated at 10%–30%) binds to plasma albumin and the rest distributes throughout the body with *no* evidence of tissue sequestration of the parent chemical or any of its metabolites (Bonati et al., 1982, 1984). There is no evidence of physiological barriers limiting the distribution of caffeine as it has been reported to readily cross the blood–brain and blood–placenta barriers and has also

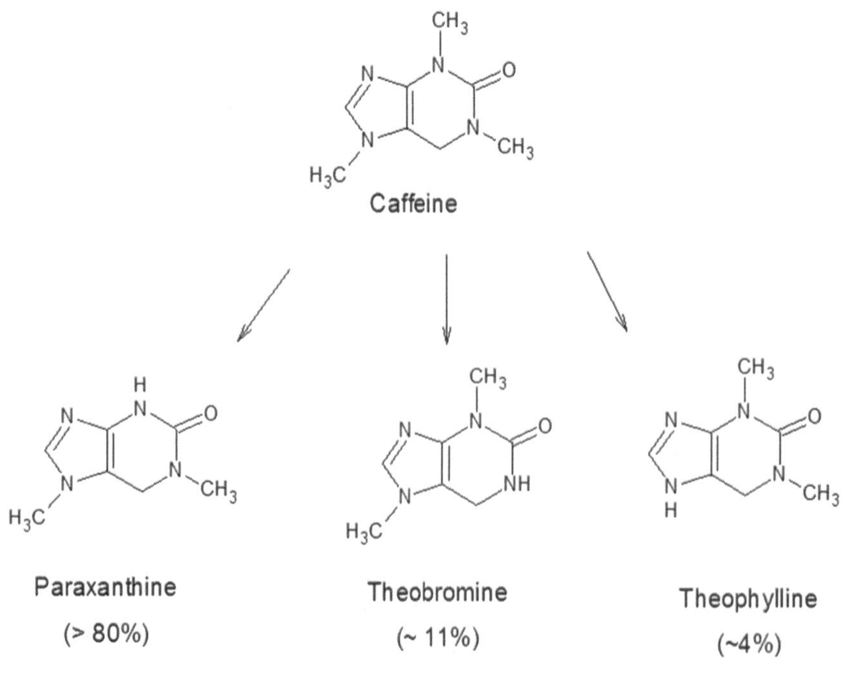

FIGURE 32.2 N-demethylation of caffeine by CYP1A2.

been found in bodily fluids such as amniotic fluid and human breast milk (Nawrot et al., 2003). This distribution pattern is consistent with its significant lipophilicity and water solubility, as indicated by its log P_{ow} of −0.07 (IPCS, 1998). The mean volume of distribution of caffeine has been estimated in humans and other species to be in the order of 0.8 L/kg bw, a value that is also consistent with distribution past the plasma compartment in the absence of any tissue sequestration (Bonati et al., 1984).

Metabolism

Caffeine is extensively metabolized in the liver by the cytochrome P450 enzyme system, particularly the CYP1A2 isozyme, and represents a clearance pathway for caffeine in the body. The metabolism consists of oxidative N-demethylation, resulting in the primary mono-demethylated metabolites: paraxanthine, theobromine, and theophylline. Fig. 32.2 shows paraxanthine as the major human metabolite produced at more than 80% of a given oral dose of caffeine, followed by theobromine at about 11%, and theophylline at about 4% (Tang-Liu et al., 1983). Other metabolites have been reported as a result of further demethylation and oxidation; however, these account for less than 6% of total caffeine metabolites. Overall, for a given ingested dose of caffeine, less than 3% of the parent caffeine compound can be expected to be excreted in urine, the primary excretion route for caffeine and its metabolites (Tang-Liu et al., 1983). There is no evidence that caffeine or any of its metabolites are bioactivated to reactive intermediates capable of causing toxicity.

Given that CYP1A2 is a critical enzyme in the metabolism of caffeine, any age-related, genetic, or environmental factors that affect the activity of this enzyme are likely to also affect the metabolism of caffeine. For instance, polycyclic aromatic hydrocarbons in cigarette smoke are known to be CYP1A2 inducers, and as a result smokers are likely to be rapid metabolizers of caffeine (Cook et al., 1996). Similarly, reductions of CYP1A2 activity during pregnancy have been associated with increases in the plasma half-life of caffeine during this time. Brazier et al. (1983) reported on pregnancy-related increases in the plasma half-life of caffeine which disappeared postnatally and were closely associated with CYP1A2 activity (Brazier et al., 1983). The immature expression of hepatic metabolic systems in the neonate has also been identified as the primary cause for the much longer half-life in this age group. The half-life of caffeine in infants approaches that of an adult by 5−6 months of age when CYP1A2 activity increases (Arnaud, 2011). The excretion of unchanged caffeine by the kidneys is minimal throughout mammalian species, for example, only 1% of a 1 g oral dose of caffeine was recovered in the urine of three healthy subjects over a 48-hour period (Burg, 1975).

Excretion

Urinary excretion is the primary route of excretion for caffeine and its metabolites. Following oral administration of caffeine at 7.5 mg/kg body weight to a group of six healthy subjects (one male, five females; 24−32 years of age; 61−80 kg in bw), 70% of the administered dose was detected in the urine (Tang-Liu et al., 1983). In another study, Callahan et al. (1982) reported that 85.8% of a ^{14}C-radiolabeled caffeine dose of 5 mg/kg body weight could be recovered in the urine from 12 healthy male volunteers (21−39 years of age) within 48 hours. Recovery of the administered dose in feces and expired air was 2.0% and 1.3% respectively (Callahan et al., 1982).

In summary, the extensive information on the pharmacokinetics of caffeine does not raise a safety concern for moderate consumption of caffeine because, although absorption is rapid and complete when ingested, there is *no* evidence of tissue sequestration of caffeine or any of its metabolites. Moreover, metabolism is extensive, serves as a clearance pathway, does not yield reactive species capable of causing toxicity and human hepatic metabolic capacity is achieved early in life, at approximately 5−6 months of age. Lastly, the body eliminates caffeine rapidly with a plasma half-life of about 5 hours, with urine being the primary route of excretion of caffeine and its metabolites.

Pharmacokinetic outcomes following caffeine administration

The pharmacokinetics of caffeine are similar between men and women and the comparative pharmacokinetics in young adults and the elderly shows no significant difference in half-life (Nawrot et al., 2003; Arnaud, 2011). The consistency and predictability of the pharmacokinetic behavior of caffeine have allowed for the development of pharmacokinetic models which are increasingly being used for extrapolating across different species, dose levels, and other situations for which there is no available information. In a recent report by Burns et al. (2014), the pharmacokinetics of orally administered caffeine was analyzed using linear one- and two-compartment pharmacokinetic models. The results showed that the population-level pharmacokinetic behavior of caffeine is best described by a one-compartment model with first-order absorption and elimination (Burns et al., 2014). The best fit to a simple one-compartment pharmacokinetic model is consistent with the observed pharmacokinetic profile of caffeine that includes rapid and complete absorption, distribution of caffeine throughout the body without any significant barrier or tissue sequestration, and relatively fast elimination. Using statistical analysis and plasma caffeine data obtained from 12 healthy subjects between the ages of 20 to 39 years old (11 males and 1 female), characterized as being moderate caffeine consumers (two to four cups of coffee consumed daily), Burns et al. (2014) estimated a population-level plasma half-life of 4.3 hours with a 95% confidence interval of 3.5–6.1 hours. The human plasma half-life of 3.5–6.1 hours is comparable to those reported for rats, hamsters, rabbits, and monkeys (Burg, 1975).

The pharmacokinetics of caffeine was also determined to be independent of the food/beverage matrix and whether present in its natural form or as an additive. In one study by Hodgson et al. (2013), a comparison of the plasma caffeine concentrations resulting from the consumption of caffeine present within coffee, to an equivalent dose of anhydrous caffeine dissolved in water was reported. Eight trained cyclists ($n = 8$; aged 41.7 ± 1 years; 78.9 ± 4.1 kg) were administered each of the test preparations (i.e., coffee, anhydrous caffeine + water, decaffeinated coffee, water) in a randomized cross-over fashion. The caffeinated test drinks of equal volumes provided 5 mg/kg bw of caffeine to each subject and all drinks were served hot ($94°C \pm 2°C$) and consumed within 15 minutes. No differences in the plasma concentration–time profiles were observed between groups. Similar observations were demonstrated in a study comparing the pharmacokinetics of caffeine following an equivalent dose via hot and cold coffee and energy drinks, consumed quickly within 2 minutes or slowly within 20 minutes (White et al., 2016). Twenty-four healthy (12 men, 12 women; aged 18–30 years), noncaffeine-naïve subjects (habitual consumption of one to three caffeinated beverages per day) consumed hot coffee over 20 minutes, cold coffee over 2 minutes, cold coffee over 20 minutes, a sugar-free energy drink over 2 minutes, and a sugar-free energy drink over 20 minutes, in a cross-over study, each providing 160 mg caffeine, in random order. The mean caffeine plasma concentration–time profiles for each test condition were similar and no significant differences between the five test conditions were reported for caffeine area under the plasma concentration–time curve (AUC) from zero to infinity, time to maximum plasma concentration, mean residence time, mean absorption time, half-life, or clearance divided by assumed bioavailability. The results of these pharmacokinetic studies confirm that the caffeine source (natural or additive), the type of beverage (coffee, energy drink), temperature, and time taken to ingest an equivalent dose, do not affect the absorption and therefore exposure (AUC) to caffeine.

The biological activity of caffeine

The major mechanism of action of caffeine in the body is postulated to be antagonism of the adenosine receptor subtypes, namely A_1, A_{2A}, A_{2B}, and A_3 receptors (Rivera-Oliver, 2014). Adenosine acts as an inhibitory neurotransmitter that suppresses activity in the central nervous system (CNS) and other tissues by its ligand (agonist) action on its receptors. For instance, in the brain, adenosine slows down metabolic activity by a combination of actions including reduction in synaptic vesicle release of glutamate and dopamine, while in the heart, it reduces the heart rate by suppressing the electrical impulses pacemaker function. The caffeine molecule is structurally similar to adenosine (Fig. 32.3), and is capable of binding to adenosine receptors on the surface of cells without activating them (i.e., acting as an antagonist) (Layland et al., 2014). In essence, caffeine opposes the effects of adenosine, including drowsiness. As the activation of adenosine receptors in the CNS promotes drowsiness, when an individual is awake and alert, little adenosine is present in the CNS neurons. However, over time in a continued wakeful state, adenosine begins to accumulate in the neuronal synapse, binding to and activating its receptors. When activated, these receptors produce a cellular response that ultimately increases drowsiness. When caffeine is consumed, it prevents adenosine from binding and activating its receptors, temporarily relieving drowsiness and restoring alertness (Lazarus et al., 2011). It has been reported using gene knockout technology that caffeine-induced wakefulness is at least in significant part due to its actions on the adenosine A_{2A} receptor (Huang et al., 2005).

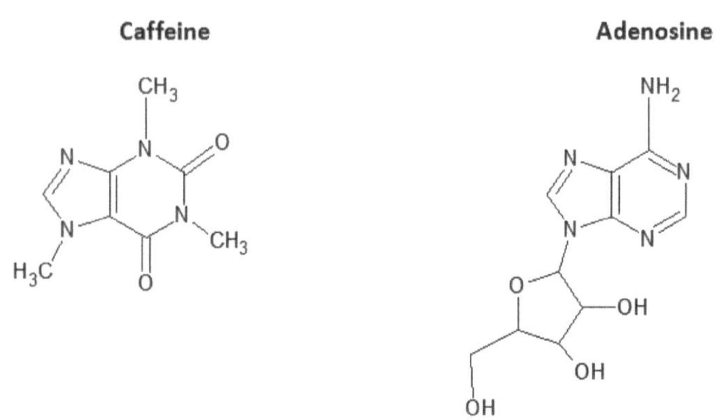

FIGURE 32.3 Chemical structure of caffeine in comparison to adenosine.

Other effects resulting from the antagonism of the adenosine receptors that may occur following high levels of caffeine exposure include stimulation of the vagal nucleus, reduction in heart rate; the vasomotor center, constricting blood vessels, and the respiratory center, increasing respiratory rate (Layland et al., 2014). The release of monoamine neurotransmitters and acetylcholine has also been associated with adenosine receptor antagonists including caffeine. Some of the findings from safety studies can be explained by the adenosine opposing actions of caffeine (described below).

The safety data available for caffeine

The safety of caffeine (naturally occurring or synthetic) has been the subject of extensive investigation over the course of many years. An overwhelming body of evidence from animal toxicity and human studies supports the safety of caffeine when consumed in moderation. In fact, caffeine may actually be protective of certain ailments such as Parkinson disease and some cancers (Heckman et al., 2010). Caffeine has been tested in animal toxicity studies over different durations reflective of acute, subchronic, and chronic/lifetime exposure scenarios. In all cases, caffeine has been consistently demonstrated to be well-tolerated at high doses (summarized below).

Safety information from animal toxicity studies

The acute toxicity of caffeine has been evaluated, following oral gavage in order to achieve high-intake dose levels in a single dosing event that is unlikely to be achieved via dietary intake (i.e., solid food or drinking water). When evaluated for acute toxicity via oral gavage, caffeine has exhibited relatively low acute oral toxicity in several species tested including rats, mice, and rabbits. Reported effects at a high dose of 185 mg/kg bw consisted of depressed activity for about 2 days, followed by repetitive movements such as running backwards. Also reported were secondary effects on the adrenal and thymus glands consisting of increases in relative organ weights. Estimated LD_{50} values were relatively high: 127 and 246 mg/kg bw for mice and rabbits, respectively (IARC, 1991), while Adamson (2016) calculated the mean LD_{50} value for male albino rats from six identified studies to be 367 ± 37 mg/kg body weight.

The US National Toxicology Program (NTP) investigated the safety of caffeine by performing a 90-day study in F344 rats in which caffeine was administered in the drinking water at concentrations of 0, 188, 375, 750, 1500, or 3000 ppm, corresponding to dose levels of 0, 19.7, 42, 85.4, 151, and 272 mg/kg bw/day for males, and 0, 23, 51, 104, 174, and 287 mg/kg bw/day for females, respectively. The major effect reported from the study was a decrease in body weight gain at the highest concentration. Water consumption also decreased by approximately 10% in the group consuming the highest caffeine concentration. No effects were noted based on clinical signs (up to 1500 ppm; clinical signs at 3000 ppm, if any, were not described), clinical chemistry, or for macroscopic lesions. Cell enlargement of the salivary gland was noted at all exposure levels but was considered to be an adaptive and reversible response to caffeine exposure. A conservative no-observed adverse effect level (NOAEL) was set at 1500 ppm (or 151 and 174 mg/kg bw/day for males and females, respectively) (OECD-SIDS, 2002).

Several 2-year rodent chronic toxicity/carcinogenicity studies have been carried out to investigate the chronic toxicity and potential carcinogenicity of caffeine from lifetime dietary exposures. One study consisted of the administration of powdered freeze-dried coffee in the diets of Sprague-Dawley rats. The highest concentration tested of 6% (corresponding to 60,000 ppm) was not associated with any significant toxicity in the animals (Würzner et al., 1977a, b). The estimated caffeine intake in the rats was estimated at approximately 100 mg/kg bw/day. The effects reported were largely limited to a 20% reduction in body weight and a corresponding increase in feed intake in the animals. Such changes were not considered adverse and were attributed to thermogenic effects of caffeine and a related increase in motor activity in the treated animals. For animals consuming the freeze-dried coffee preparation, a protective effect was observed in the form of a negative correlation between tumor incidence and coffee consumption in exposed animals as compared to controls. Similar findings of no toxicity were reported in another chronic study where Sprague-Dawley rats were exposed to caffeine in the drinking water at levels of 200, 430, 930, or 2000 mg/L for 2 years (Mohr et al., 1984). Furthermore, there were no unusual tumors or sites of origin for neoplastic growth found in any animal as compared to the control group. A NOAEL of 102 mg/kg bw/day, the highest achieved dose for male rats, was established based on reduced weight gain, an effect that was also attributed to caffeine-induced thermogenic effects and related motor activity increases.

The reproductive and developmental effects of caffeine have also been extensively evaluated in multiple animal species including rats, mice, rabbits, and nonhuman primates (Tarka, 1982; IARC, 1991; Christian and Brent, 2001; Nawrot et al., 2003; Brent et al., 2011). In rodent studies at relatively high caffeine dose levels, the most common finding has been delayed ossification, an effect that is reversible shortly after birth (Collins et al., 1987). At higher doses that resulted in maternal toxicity, caffeine produced reduced fetal body weight, resorption, and an increased incidence of malformations of the limbs and palate (Christian and Brent, 2001). The reported

NOAEL for the developmental effects in rodents was between 80 and 120 mg/kg bw/day (Brent et al., 2011).

There is no evidence that caffeine is mutagenic when tested in the Ames assay using *Salmonella* tester strains (IARC, 1991). Genotoxicity assays based on in vitro mammalian systems lacking intrinsic metabolic capability have yielded equivocal results (i.e., both positive and negative findings) as have ex vivo genotoxicity studies in mice and rats (IARC, 1991). Despite the equivocal observations from in vitro/ex vivo systems, caffeine has been consistently reported to be noncarcinogenic in rats and mice. Moreover, there is no evidence from large prospective human cohort studies of any association between coffee/caffeine intake and tumor incidences (described below) (Nawrot et al., 2003; Nkondjock, 2009). The World Health Organization's International Agency for Research on Cancer (WHO IARC) also has concluded that there is inadequate evidence to confirm that caffeine is carcinogenic (IARC, 1991).

Caffeine, as a nonselective antagonist of the adenosine receptors, can produce seizures in animals at high doses and can induce proconvulsive effects in vitro (Chrościńska-Krawczyk et al., 2011). Although caffeine is reported to promote seizures in animal models of epilepsy, the convulsive threshold in experimental models of epilepsy is induced in healthy rats at toxic doses (i.e., >400 mg/kg bw) and the effective dose eliciting a 50% convulsant response in mice is 400 mg/kg body weight (Czuczwar et al., 1990).

In summary, the evidence from animal toxicity studies suggests that caffeine is well tolerated at high dose levels based on subchronic and chronic dietary exposures (administered in drinking water or solid food). Besides mild effects on body weight or body weight gain, there is no evidence of systemic or direct organ-specific toxicity. Moreover, the effects on body weight are attributed to a thermogenic effect characteristic of caffeine and related motor activity increases. No increases in rodent tumor incidences of any type have been observed. In fact, a consistent negative correlation between tumor incidences and caffeine consumption has been frequently observed in chronic toxicity studies, an effect that also has been reported in prospective cohort studies in humans (described below). Developmental effects of concern such as reduced fetal weight, resorption, and malformation are only observed at high dose levels and in the presence of maternal toxicity, indicative of no increased susceptibility of the embryo/fetus.

Safety information from human studies

Potential health concerns associated with caffeine consumption have long been expressed with regards to pregnancy outcomes, the cardiovascular system, and the central nervous system. Numerous controlled human studies have evaluated the effects of caffeine and/or coffee on various physiological parameters such as blood pressure, heart rate, arrhythmia, glucose tolerance, and calcium balance. Further, caffeine has been used safely and effectively at levels greater than 7.9 mg/kg bw/day to treat apnea of prematurity in young infants (Erenberg et al., 2000; Francart et al., 2013). Epidemiological investigations, including retrospective and prospective cohort studies have also evaluated the consumption of caffeine, and/or coffee, and its association with various health outcomes (e.g., cancer, cardiovascular disease, diabetes, and pregnancy outcomes). A summary of the effects/associations between caffeine and various health outcomes is described below, including several cases of reported caffeine intoxication.

Reports of caffeine intoxication are relatively rare, but when it happens, the clinical signs include restlessness, nervousness, excitement, insomnia, facial flushing, diuresis, and gastrointestinal complaints. Plasma concentrations associated with these effects are reported to be in excess of 30 μg/mL (equivalent to 150 μmol/L) (Sawynok, 1995). Fatal reactions to acute exposures to caffeine are also rare but generally have been associated with blood caffeine concentrations in excess of 80 μg/mL (Moffat and Widdop, 2004). The fatal acute oral toxic dose of caffeine in adults is estimated to be in the area of 10 g (Nawrot et al., 2003), while the U.S. FDA (2018a) reported a potentially lethal dose range of 10–14 g. Reports of death from consumption of 6.5 g of caffeine have been reported and survival also has been reported for individuals consuming 24–30 g of caffeine (James, 1990; Moffat and Widdop, 2004). In children, 3 g of caffeine (183 mg caffeine/kg body weight) has been shown to be fatal (Dimaio and Garriott, 1974). In a review of 51 caffeine-related fatalities along with their postmortem blood concentrations, the median concentrations in males (161 mg/L) were not significantly different from those of females (182 mg/L). It was reported that within these individuals, caffeine had been ingested in the form of tablets or powder and that "It does not seem likely that toxic concentrations of caffeine can be achieved from over-consumption of caffeinated beverages alone" (Jones, 2017).

Caffeine is known as a mild stimulant that can produce transient increases in heart rate and promotes vasoconstriction through antagonism of the adenosine receptor (Gronroos and Alonso, 2010). Findings from controlled human studies conducted in coffee consumers and abstainers suggest that caffeine consumption at acute intakes exceeding 250 mg/person may produce mild, temporary increases in systolic and/or diastolic blood pressure (BP) of 5–15 mmHg and 5–10 mmHg, respectively (Josse et al., 2012; Nawrot et al., 2003). Within a review published by Wikoff et al. (2017), the effects of caffeine on BP were found to be inconsistent, with consumption of <100 mg caffeine reported to cause small, but significant increases in BP in some studies, while consumption of ≥400 mg/day caffeine in other studies was not associated with a significant increase in BP. Similarly, inconsistent results were also reported for the effects on heart rate, which led to the conclusion that while intakes <400 mg/day demonstrated small increases in BP in some individuals, consumption of 400 mg caffeine/day by healthy adults was concluded to not be associated with adverse effects on BP or heart rate. Slight decreases in heart rate of 1–3 beats per minute also have been reported in adolescent subjects consuming caffeine (Temple et al., 2010). EFSA also noted that the results of randomized controlled trials suggest that there is no significant effect of caffeine dosages of ≤400 mg on fasting BP after caffeine habituation has occurred (i.e., after repeated consumption for ≥7 days) (EFSA, 2015). Such transient increases in BP are attenuated following regular use of caffeine (Robertson et al., 1981; Myers and Reeves, 1991; Corti et al., 2002; Noordzij et al., 2005) with tolerance to caffeine's pressor effects seen to develop within 1–3 days (Nawrot et al., 2003). However, tolerance may be partially lost after abstinence for as little as 12 hours (Nawrot et al., 2003). The caffeine withdrawal syndrome is probably one of the most common complaints associated with regular consumption of caffeine (Budney et al., 2013). When a regular consumer of coffee suddenly stops, he/she experiences symptoms such as headache, fatigue, decreased energy and alertness, drowsiness, and irritability. These withdrawal symptoms experienced by regular consumers of caffeine are relatively mild and normally subside in a short time.

The effects of caffeine on cardiac arrhythmia have also been evaluated in controlled studies and using prospective cohorts of caffeine consumers (Conen et al., 2010; Myers et al., 1987; Sutherland et al., 1985). Reviews of the available data have concluded that caffeine consumption has minimal effects on ECG tracings and that moderate consumption is well-tolerated, even with persons with known arrhythmias (Pelchovitz and Goldberger, 2011). Another review conducted by (Cavalcante et al. (2000) concluded similarly that a large body of evidence supports the conclusion that moderate caffeine consumption does not increase the risk for atrial fibrillation (Cavalcante et al., 2000). In Cheng et al. (2014), a meta-analysis was conducted that included a dose–response evaluation, caffeine intake, and the incidence of atrial fibrillation. The meta-analysis included six prospective cohort studies and 228,465 human subjects. The overall analysis indicated that caffeine exposure was weakly associated with a reduced risk for atrial fibrillation (Cheng et al., 2014). EFSA likewise evaluated two systematic reviews and meta-analyses of prospective cohort studies and noted that there was no significant association between atrial fibrillation and habitual caffeine intakes (ranging up to 1000 mg/day) in primary or subgroup analyses, considering age, sex, race, study quality, caffeine from coffee versus caffeine from all sources, caffeine dose, or duration of follow-up (EFSA, 2015). EFSA further noted that in no single prospective cohort study was caffeine intake associated with an increased risk of atrial fibrillation (EFSA, 2015).

The potential effects of caffeine on reproduction and development have been the subject of several systematic and comprehensive reviews (Brent et al., 2011; Nawrot et al., 2003; Peck et al., 2010; Wikoff et al., 2017). In 2003, Health Canada indicated that caffeine intake at dose levels of ≥300 mg/day may have adverse effects on some reproductive/developmental parameters and they suggested that reproductive-aged women (i.e., women who plan to become pregnant, pregnant women, and nursing mothers) should consume ≤300 mg caffeine/day. Peck et al. (2010) reviewed human studies published between January 2000 and December 2009 and that measured (1) fertility, (2) spontaneous abortion, (3) fetal death, (4) preterm birth, (5) congenital malformations, and/or (6) fetal growth restriction. The results of the analysis did not support a positive relationship between caffeine consumption and adverse reproductive or perinatal outcomes. The review identified significant methodological weaknesses common to studies of spontaneous abortion, fetal death, and fetal growth restriction, which limit confidence in causal interpretation (Peck et al., 2010). After reviewing the 2000–10 scientific epidemiological literature concerning the reproductive and developmental toxicology risks of caffeine, Brent et al. (2011) concluded that confounding phenomena in caffeine studies make it unclear whether the increased risk estimates for some developmental and reproductive effects at higher exposures are due to caffeine or confounding factors. Their conclusion was that the dietary exposures of caffeine are not teratogenic (Peck et al., 2010). In addition, based upon a weight of evidence approach, Wikoff et al. (2017) concluded that "An intake of 300 mg caffeine/day in pregnant women and 400 mg/day in healthy adults are generally without significant concern regarding overt adverse effects on fecundability, fertility, male reproductive endpoints, spontaneous abortion, recurrent miscarriage, stillbirth, preterm birth and gestational age, birth defects, and childhood behavior." Overall, while there is a general consensus in the literature and expert scientific community that regular caffeine intakes ≤300 mg/day are safe for reproductive-aged women during pregnancy (Nawrot et al., 2003; Peck et al., 2010; Brent et al., 2011; Wikoff et al., 2017), a more conservative, lower limit of 200 mg caffeine/day during pregnancy was recommended by the EFSA's NDA Panel (EFSA, 2015).

IARC evaluated the carcinogenic potential of both caffeine and coffee in 1991 (IARC, 1991). The agency concluded that for coffee there was limited evidence from epidemiological studies of carcinogenic effects in the urinary bladder, and therefore classified coffee as possibly carcinogenic to the urinary bladder. However, for caffeine specifically, IARC (1991) classified caffeine as a group 3 agent, meaning that it is not classifiable as to its carcinogenicity to humans. Upon reevaluation in 2016, the IARC concluded that there is evidence supporting a lack of carcinogenicity from coffee to the prostate, pancreas, and female breast; evidence supporting an inverse association for cancers of the uterine endometrium and liver; and inadequate evidence for cancers of the lung, bladder, colon, rectum, esophagus, stomach, oral cavity, brain, ovary, and childhood leukemia (IARC, 2018). Overall, coffee drinking was classified under group 3 (not classifiable as to its carcinogenicity to humans) and the classification of caffeine as a group 3 agent was maintained.

A meta-analysis of 37 published studies (involving 966,263 individuals) conducted to evaluate the effect of coffee or caffeine on breast cancer identified a weak negative association between coffee or caffeine intake and breast cancer risk (Jiang et al., 2013). Using data from a large prospective US cohort of 968,432 men and women who were cancer free at the time of enrollment, Hildebrand et al. (2013) reported that caffeinated coffee consumption was inversely associated with oropharyngeal cancer mortality (Hildebrand et al., 2013). In a 2009 review of epidemiological studies conducted by Nkondjock (2009), coffee consumption was not associated with cancers of the prostate, ovary, and pancreas, and found to actually reduce the risk of cancers of the liver and kidney and to a lesser extent of the breast and colon (Nkondjock, 2009). Grosso et al. (2017) likewise reported a significant inverse association between coffee consumption and the risk of endometrial, liver, prostate, pancreatic, colon, skin, and oral cancers, and colorectal adenoma. Incremental increases in coffee consumption from one cup to four cups/day were reported to significantly decrease the risk of liver, breast, endometrial, skin (melanoma), colon, and prostate cancers. Similarly, Malmir and Esmaillzadeh (2017) investigated 14 current publications and demonstrated that consumption of coffee and tea is associated with a reduced risk of glioma and a recent epidemiological and meta-analyses conducted by Romualdo et al. (2019) reported that coffee intake is associated with a decreased risk of oropharyngeal and liver cancer. Overall, the evidence indicates that caffeine is unlikely to be carcinogenic to humans (IARC, 1991, 2018), while some studies indicate that caffeine, as present in coffee, may have anticancer properties (Nkondjock, 2009; Grosso et al., 2017; Malmir and Esmaillzadeh, 2017; Romualdo et al., 2019). These findings are consistent with observations in carcinogenicity studies in animals, which consistently demonstrate that dietary caffeine does not represent a carcinogenic risk.

Anecdotal findings in patients with epilepsy have implicated caffeine in lowering the seizure threshold in these individuals and under certain circumstances may interfere with anticonvulsant therapy (Boison, 2011). Other reports have reviewed several case studies documenting reports of tonic clonic seizures in energy drink consumers (Babu et al., 2011; Dikici et al., 2013; Trabulo et al., 2011). The majority of these cases were associated with individuals sensitive to caffeine and who have significant predisposing conditions (e.g., alcohol use, cerebral ischemia, or drug use) and are considered to be likely associated with individuals who are sensitive to caffeine and have underlying preexisting conditions. A recent prospective study by Dworetzky et al. (2010) did not identify an association between caffeine consumption and seizure incidence in the Nurses' Health Study cohort, a population of 105,941 participants, disease-free at enrollment, for a total of 1,440,850 person-years of follow up (Dworetzky et al., 2010). It is noted that there is neither epidemiological evidence, nor anecdotal evidence, to suggest that dietary caffeine represents an increased risk for seizure in otherwise healthy individuals, even among heavy frequent consumers of caffeinated products.

Summary of human safety data

In summary, the information from human studies suggests that, as a mild stimulant, caffeine can produce effects consistent with those actions including increases in blood pressure and heart rate; however, such effects are transient, relatively mild, and can decrease with habitual consumption of caffeine. There is no evidence that moderate consumption of caffeine increases the risk of cardiac arrhythmias or atrial fibrillation or that it is associated with adverse reproductive or perinatal outcomes. Coffee and caffeine are not considered to be carcinogenic and in the case of some cancers, there seems to be a protective effect, as indicated by reported inverse correlations between caffeine consumption and cancers of the breast, prostate, liver, and colorectal adenoma.

Since caffeine can cross the placenta into the fetus, which has an immature metabolic system (i.e., CYP1A2) and as caffeine half-life increases during pregnancy, concern might be raised for pregnancy to be a sensitive lifestage for caffeine consumption. However, epidemiological studies do not support any association between caffeine/coffee consumption and developmental effects. Further, caffeine has been used safely and effectively at levels greater than 7.9 mg/kg/day in the treatment of apnea of prematurity in young infants 28–32 weeks postconception (Erenberg et al., 2000; Francart et al., 2013).

Dietary intake studies of caffeine

Estimating the dietary intake levels of caffeine across the population is key to assessing the level of safety concern for current caffeine exposures. Caffeinated food products (primarily beverages) are consumed by virtually all segments of the population including children, adolescents, and adults. However, there is significant variation in not only the types of caffeinated beverages, but also in the daily volume (i.e., the amount of caffeine) consumed. For example, coffee has long been reported as the primary source of caffeine for adults, while carbonated soft drinks are a major contributor of caffeine in adolescents (Mitchell et al., 2014). Additionally, the caffeine content present in coffee, tea, cocoa, and other sources is highly variable depending on the type of coffee and method of preparation. A wide range of specialty high street coffees have risen in popularity and it is challenging to assign caffeine values for these products since, unlike commercially available bottled coffees or caffeinated beverages, there is no label specifying the amount of caffeine. With the increasing number of caffeinated consumer food and beverage products including energy drinks, it is imperative to use the latest and most accurate consumer and product information available, including the type of beverage consumed, the levels of caffeine, and the frequency of consumption for estimating caffeine intake.

The level of caffeine consumption within the diet including the possibility of a "substitution effect," wherein one caffeine source is substituted for another, has been evaluated in a number of publications. A population-based caffeine dietary intake study in the United States was reported by Mitchell et al. (2014). In collaboration with the Kantar World Panel, a company that specializes in documenting and analyzing trends in consumer behaviors, a population-based survey (2010–2011) was used that included new caffeinated beverages such as energy drinks and energy shots. A total of 42,000 respondents representative of the US population were recruited for the study. The collected consumption information was coupled with a comprehensive compilation of beverage caffeine values from various sources including the USDA Food and Nutrient Database for Dietary Studies, the USDA Food and Nutrient Database for Standard Reference, the Nutrition Data System for Research, and food and beverage industry websites.

The results of the study showed that 85% of the respondents consumed at least one caffeinated beverage per day and using brand-specific data on the caffeine content from all caffeinated beverages (i.e., coffee, carbonated soft drinks, tea, energy drinks, coca and chocolate-containing beverages, and energy shots) the estimated consumer only mean and 90th percentile caffeine intakes were 164.5 and 379.5 mg/day, respectively, across all age groups. Several previously reported trends were noted, including an increase in caffeine consumption (mg/day or mg/kg body weight) with age that appears to plateau after 64 years of age. Furthermore, nearly all the caffeine consumed came from either coffee, tea, or carbonated soft drinks. Fig. 32.3 shows the caffeine intake distribution for the different age groups evaluated. Adults in the age group 50–64 consumed the highest levels of caffeine, primarily through coffee. The contribution of carbonated soft drinks was comparable to that of tea beverages for the adult age groups (35–49, 50–64, and ≥65 years old), while having a slightly greater contribution for the younger age groups, except for the youngest age group (2–5 years old), where tea seems to contribute slightly more to the total caffeine daily intake. Overall, the estimated mean and 90th percentile caffeine intake for adults (50–64 years old) consuming the highest levels of caffeine were 225 and 467.4 mg/day, respectively. The corresponding mean daily caffeine intake for the younger age groups of 6–12 and 13–17 years old was much less at approximately 16.2 and 37% of adult values. An estimated 43% of children 2–5 years old consumed some type of caffeinated beverage, while essentially 100% of adults over 65 years old consumed caffeine. An important

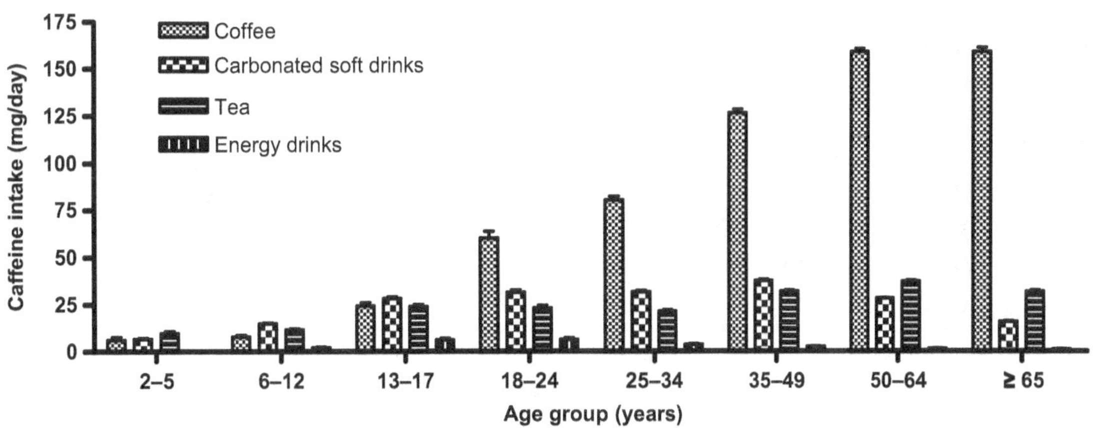

FIGURE 32.4 Caffeine daily intake distribution for the US population. Source*: Adapted from Mitchell et al. (2014).*

observation from the study is that energy drinks were found to contribute very little to the overall beverage caffeine daily intake for all age groups (Fig. 32.4).

In Europe, EFSA calculated caffeine intake from food and beverages using the Comprehensive European Food Consumption Database incorporating 39 surveys from 22 different European countries including a total of 66,531 participants. As outlined previously, while the EFSA concluded that caffeine intakes from all sources up to 400 mg per day provide no safety concerns for healthy adults in the general population, the results outlined in their opinion indicated that this level was exceeded at the 95th percentile in 7 out of 13 countries (EFSA, 2015). In these countries, the estimated proportion of the adult population exceeding a daily intake of 400 mg ranged up to almost one-third (33%), with coffee being the main source of caffeine exposure. Within the 16 Member States, for adults the mean and 95th percentile caffeine daily intakes ranged from 37 to 319 mg and from 109 to 742 mg, respectively, with the value reaching up to 776 mg of caffeine from coffee in consumers of coffee. The highest 95th percentile of caffeine intake on a single day in adults was determined to be 809 mg (10.8 mg/kg body weight) and on only coffee consumption days, the highest 95th percentile of caffeine intake from coffee increased to 890 mg. The EFSA intake analyses therefore reported daily caffeine exposures in coffee drinkers at the 95th percentile to be more than twice the level the Committee considered provides "no safety concerns."

In addition to the calculated daily caffeine exposures above, a number of studies have examined temporal trends in caffeine consumption. Fulgoni et al. (2015) examined temporal trends in caffeine consumption in adults 19 years of age or older from 2000 to 2010 using consumption data from the United States Department of Agriculture (USDA) National Health and Nutrition Examination Survey (NHANES) and caffeine content information compiled by Food & Nutrition Database Research, Inc. (Fulgoni et al., 2015). NHANES is a continuous dietary intake study designed to collect nationally representative information on health and nutrition status of the US general population. The consumption information is based on 24-hour recall information on two nonconsecutive days that is released in 2-year increments. Thus, for the time period of 2001–2010, the individual 2-year periods evaluated by Fulgoni and co-workers were 2001–02, 2003–04, 2005–06, 2007–08, and 2009–10 (Fulgoni et al., 2015). The most important result from the analysis is that caffeine consumption has remained remarkably similar for US adults (19 years of age and older) during the 10-year time period analyzed (i.e., 2001–2010), irrespective of the age or sex group. Fig. 32.5 shows the individual results for each NHANES 2-year individual consumption data set (genders combined). Only actual caffeine consumers were included in the analysis. Men were found to have a higher rate (20–40%) of daily caffeine consumption than women, but the same general trend was noted for both genders (Fulgoni et al., 2015). Such observation strongly supports the notion that caffeine consumption has not significantly increased in US adults and adds support to a product substitution behavior by consumers when new caffeinated products are introduced into the US market.

Caffeine daily intake was expressed as on a per capita (i.e., nonconsumers were also included) basis as well as on actual consumers of caffeine. The results indicate very comparable qualitative and quantitative findings as those reported by Mitchell and co-workers (Mitchell et al., 2014). The main drivers of caffeine intake reported were caffeinated beverages, particularly coffee. The mean

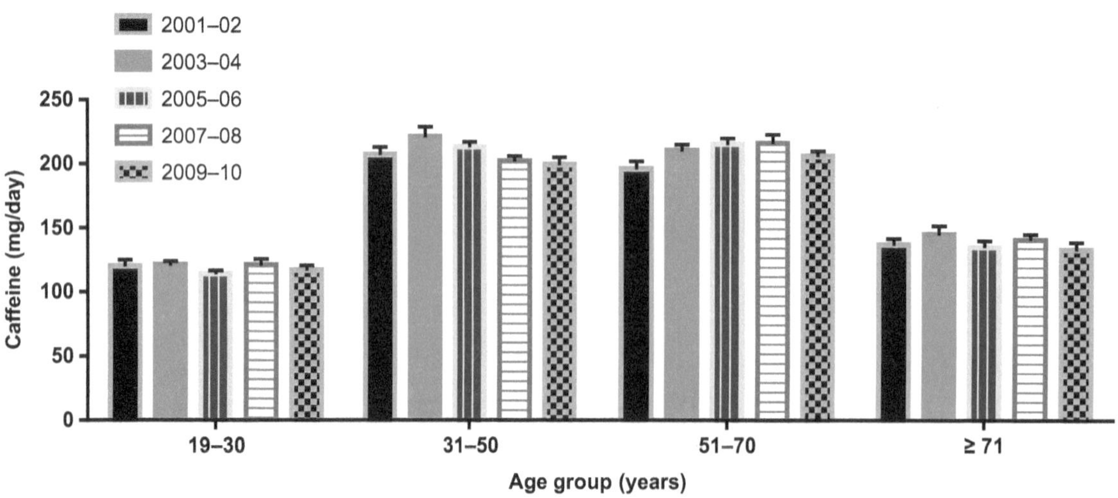

FIGURE 32.5 Caffeine daily intake for different age groups based on consumption information from NHANES 2001–2010. Source: *Adapted from Fulgoni et al. (2015).*

caffeine intake on a per capita basis (i.e., caffeine nonusers were included) was estimated at 186 mg/day for adults 19 years or older, whereas for actual caffeine consumers of the same age range, the estimated intake value was 211 mg/day (Fulgoni et al., 2015). The comparable caffeine intake on a per capita and actual consumption basis suggests that a very high percentage of the US population consumes caffeine on a daily basis. The percentage estimated by Fulgoni et al. (2015) was at about 89%, which is comparable with the 85% value estimated by Mitchell and co-workers described above (Fulgoni et al., 2015; Mitchell et al., 2014). The mean caffeine intake was highest at 275 mg/day for men in the age group 51–70 years old with a 90th percentile of 541 mg/day. Similar results were obtained by Mitchell et al. (2014), where the estimated highest mean and 90th percentile caffeine intake was for adults 50–64 years old at 225 and 467.4 mg/day, respectively. Although only adults were evaluated by Fulgoni and co-workers, there are striking similarities in the results from both studies in the prevalence of caffeine consumption in the United States, caffeine intake estimates, the age group with the highest caffeine consumption, and the main driver of caffeine consumption being primarily coffee.

In 2010, the US FDA published a caffeine intake study representative of the US population. The study consisted of coupling the caffeine content of foods from the National Nutrient Database for Standard Reference (NBD) with either the United States Department of Agriculture (USDA) National Health and Nutrition Examination Survey (NHANES) or the NPD's Group Food Consumption Survey. The two surveys have strengths and limitations that need to be considered when deriving conclusions about caffeine dietary intake. For example, while the NHANES data set is based on a greater number of respondents and thus provides a more robust population distribution of daily consumption, it is not specific for caffeine (more than 60 food components were included) and does not provide information on the specific sources of caffeine. Further, NHANES data are based on 24-hour recall information on two nonconsecutive days that is released in 2-year increments. Therefore analysis based on the NHANES data may not necessarily be reflective of current consumption patterns since the information was collected at least 2 years beforehand. The US FDA study was based on the 2005–06 NHANES survey information, the most recent data set available at the time the caffeine study was performed in 2009. Nonetheless, a major strength of the NHANES data is that historical trends in consumer information can be obtained by comparing the previous 2-year profiles as performed by Fulgoni and co-workers (see above) (Fulgoni et al., 2015).

The NPD group's survey was compiled in 2008 from 14 consecutive days of 24-h recall food diaries, conducted in sequence throughout the year with each 2-week reporting period sequence completed by 60 households. In contrast to the NHANES data, the NPD data show the specific source and quantity of caffeine and overall consumption by age and gender.

The US FDA-sponsored study is robust in that the two data sources used (NHANES and NPD) were validated and adjusted if necessary using market information provided by the National Coffee Association, the US Tea Association, and the American Beverage Association.

The results of the 2005–06 NHANES portion of the study are shown in Fig. 32.6. In contrast to the Mitchell et al. (2014) study, the NHANES data provide gender-specific information in caffeine intake but not the contribution of the individual sources of caffeine. In general, the same trends are supported as the Mitchell study, namely that caffeine consumption increases with age,

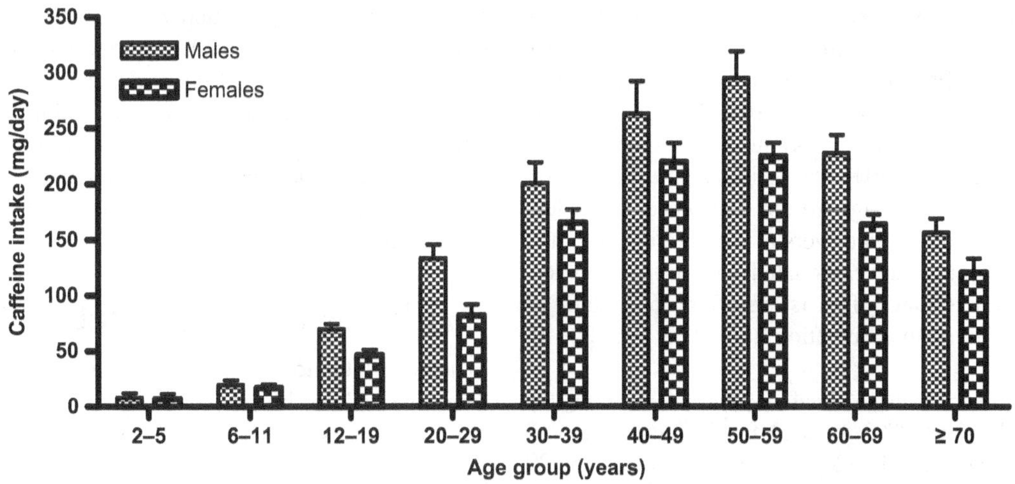

FIGURE 32.6 Caffeine daily intake from the FDA study based on the 2005–06 NHANES data. Source: *Adapted from US FDA and Oakridge National Laboratory (2010).*

with the 50–59 age group (or 50–64 age group in the Mitchell et al., 2014 study) exhibiting the highest caffeine daily intake. No major differences were noted between males and females.

Although the data obtained from the published reports herein described (Mitchell et al., 2014; Fulgoni et al., 2015, and FFDA-2005-06 NHANES) are not directly comparable (Figs. 32.3, 32.4, and 32.5) because of mismatched age groups and other factors, the total caffeine daily intake estimated from the individual studies can still be compared. Thus, gender-specific values from the NHANES datasets (2001–10 and 2005–06) were averaged so that they could be compared to the total caffeine intake estimates from the Mitchell et al. (2014) report in which genders were combined. Fig. 32.6 shows the average results for the most closely matched age groups in the three studies (coupled with brackets). Although the three datasets not only represent different time periods (Mitchell: 2013, NHANES: 2001–10, and NHANES: 2005–06) and different methodologies (Mitchell: 7-day prospective survey and NHANES: 24-hour recall information on two nonconsecutive days), this analysis indicates remarkably similar caffeine daily intakes for the different age groups evaluated in all three studies. The only major difference noted was the caffeine daily intake for the 2–5-year age group which was more than twofold higher based on the Mitchell report, but nonetheless both datasets reported the lowest daily intake by this particular age group. This analysis strongly supports the conclusion that while there has been variation in the sources of caffeine among the different age groups of the US population, the overall daily intake has not dramatically increased during the time periods (2001–10 and 2013) evaluated.

The NPD's survey data compiled in 2008 provides the specific source of caffeine as well as the quantity consumed daily by participants. The results for the age groups assessed are shown in Fig. 32.7. Overall, the caffeine daily intake estimated from the NPD's data are less than half of those estimated from the other studies (NHANES and Mitchell et al., 2014). The reason for this discrepancy was attributed to the estimated low caffeine content of several caffeinated products. Although quantitatively different, the NPD's data still support the general trends that the younger age groups consumed much less caffeine because the main drivers of caffeine intake in the younger age groups are carbonated soft drinks and tea. Also supported is that coffee is the main contributor of caffeine in the adult population (noted as $\geq$ 22) in Fig. 32.8.

Likewise, the percentage of caffeine consumers during the time period between 2003 to 2012 for adolescents 13–17 years old, young adults 18–24 years old, and adults 25–29 years old were found to remain relatively stable (Tran et al., 2016). A similar trend was reported for groups aged 2–22 years during 2003–12 (Branum et al., 2014) and for adults during 2001–10 (Fulgoni et al., 2015). While the levels of intake and the numbers of consumers have remained constant, a change in the source of dietary caffeine intakes has however occurred over time. While intakes of caffeine from coffee and tea during 1999–2012 were reported to be relatively stable in adults, caffeine from soda decreased. During this period, a significant increase in caffeine intake from energy drinks was reported in adults; however, an overall decrease in caffeine intake was observed due to higher decreases from soda consumption (Fulgoni et al., 2015; Drewnowski and Rehm, 2016). Surveys obtained by NHANES during 1999–2010 also showed decreased consumption of soda and increased consumption of coffee and energy drinks, as well as a steady consumption of tea in groups aged 2–22 years (Branum et al., 2014). Similar results were reported by Tran et al. (2016) using NHANES survey data from 2003 to 2012 that showed decreased consumption of soda and increased consumption of coffee and tea in teenagers, young adults, and adults. In contrast, using data from the 2003–16 NHANES, Vercammen et al. (2019) reported that the prevalence of energy drink consumption was increased from 2003 to 2016 in adolescents from 0.2% to 1.45%, in young adults 0.5% to 5.5%, and middle-aged adults from 0.0% to 1.2%. In general, despite the indication of decreased caffeine intake from soda and increased caffeine intake from coffee, tea, and energy drinks, the overall caffeine intakes were shown to be stable.

Based upon the introduction of products such as energy drinks and energy shots in to the marketplace and the knowledge that overall caffeine intakes have remained stable over time it would appear that consumers substitute one caffeine source for another when they switch to a new caffeinated product introduced into the market. The reality of a substitution effect occurring amongst consumers was outlined by Tran et al. (2016), who reported a significant negative relationship between caffeine intake occurring from energy drinks and that occurring from coffee, tea, and soda in the upper 50th percentile of caffeine consumers. Tran et al. (2016) further demonstrated that the intake of an energy drink is decreased if an individual has consumed another caffeine-containing product in comparison to if an individual is consuming an energy drink alone. The data also demonstrated that self-regulation of caffeine intake by consumers can occur regardless of source.

Concluding remarks and future directions

Whether synthetically derived or naturally occurring, caffeine has a long history of safe and widespread human consumption, with reports of adverse health outcomes being relatively rare. The animal toxicity information supports the notion that caffeine is well-tolerated at relatively high intake dose levels. Part of this chapter reports on the

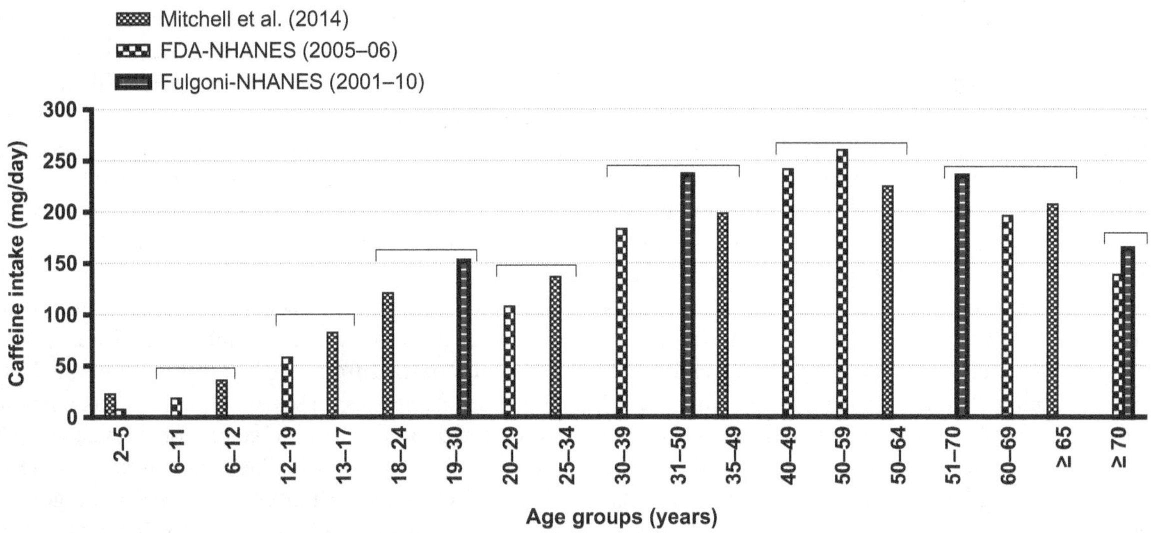

FIGURE 32.7 Caffeine daily intake from the 2005–06 FDA-NHANES, Fulgoni-NHANES 2001–10, and Mitchell et al. (2014) reports. Source: *Adapted from Fulgoni et al. (2015), Mitchell et al. (2014), US FDA and Oakridge National Laboratory (2010).*

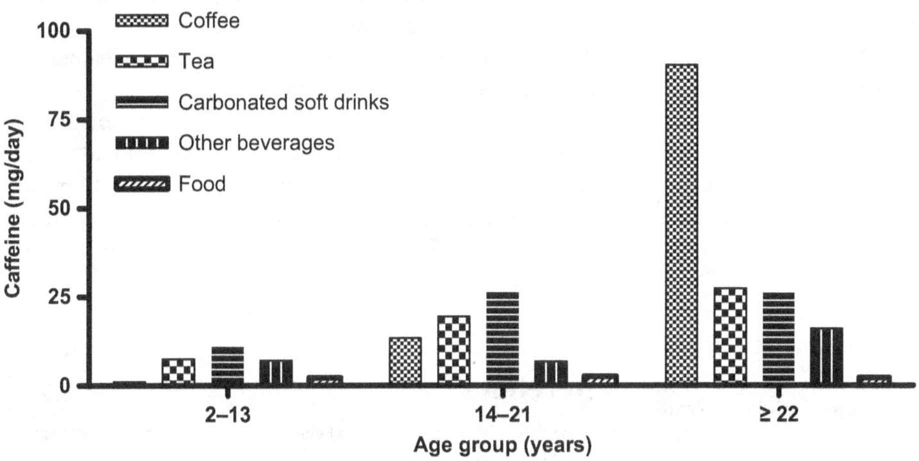

FIGURE 32.8 Caffeine daily intake. Source: *Adapted from the NPD's 2008 data set.*

effects from chronic exposures to caffeine being generally limited to reduced body weight gain and increased motor activity, which are not adverse in nature and characteristic of the thermogenic effects of caffeine. The most common finding from developmental toxicity studies conducted at high dose levels is delayed ossification, which was shown to be completely reversed following birth. At higher doses, reduced fetal body weight, resorption, and an increased incidence of malformation of the limbs and palate are observed. However, these more serious developmental effects are only seen in the presence of maternal toxicity, suggesting that there is no increased susceptibility exhibited by the fetus. Brent et al. (2011) reported a rough estimate of the necessary human consumption of coffee to approximate the animal doses associated with developmental effects at more than 40 cups per day.

There is no evidence that caffeine represents a carcinogenic risk to humans. The International Agency for Research on Cancer (IARC) concluded in 1991 and 2018 that there is inadequate evidence that caffeine (as present in coffee) is carcinogenic (IARC, 1991, 2018). On the contrary, the results of epidemiological studies suggest that regular coffee consumption actually reduces the risk of some cancers including endometrial, liver, breast, prostate, pancreatic, colon, skin, oral cancers, and colorectal adenoma. Incremental increases in coffee consumption from one cup to four cups/day were reported to significantly decrease the risk of liver, breast, endometrial, skin (melanoma), colon, and prostate cancers (Nkondjock, 2009; Grosso et al., 2017; Malmir and Esmaillzadeh, 2017; Romualdo et al., 2019). In a prospective cohort study by Loftfield et al. (2018), coffee consumption up to eight cups per day was reported to be inversely associated with overall mortality regardless of genetic influences on caffeine metabolism, although the association was weak for cardiovascular mortality.

The daily caffeine intake of 400 mg/day of caffeine, which has become the standard guidance offered by regulatory agencies, was originally based on a 2003 review of the caffeine literature by Nawrot et al. (2003) and reported as "not being associated with adverse health effects." It should be noted that such value was based on the precautionary principle and does not represent a reference dose or threshold value above which adverse effects can be expected. EFSA published its Scientific Opinion on the Safety of Caffeine whereby the 400 mg/day daily caffeine intake was again referenced as "not giving rise to safety concerns" for adults (excluding pregnant and lactating women), noting that the 95th percentile in 7 of 13 countries and as much as one-third of the adult population in these seven countries exceeded the 400 mg/day value (EFSA, 2015). In fact, EFSA estimated caffeine consumption per day in coffee consumers to be 776 mg of caffeine from coffee, with the highest 95th percentile of caffeine intake on a single day in adults determined to be 809 mg (10.8 mg/kg body weight) and when considering only coffee consumption days, the highest 95th percentile of caffeine intake from coffee on a single day rose to 890 mg. These values are up to more than double the 400 mg/day as that "not giving rise to safety concerns." Likewise, in the United States, Mitchell et al. (2014), reported that the 90th percentile caffeine intake of 541 mg/day in the age group 51–70 years old is also above the recommended 400 mg/day level and supports the IOM opinion that consumption of caffeine by military personnel up to 600 mg would improve performance without producing adverse health effects (IOM, 2001). Similarly, a value of 200 mg/day was referenced by EFSA for pregnant and lactating women on the basis of either reported associations in prospective cohort studies, or results from a dose-range study, respectively. No information was reported on the percentiles that may exceed this value in the countries evaluated. In both cases, the scarcity of adequate caffeine data to characterize risk in pregnant and lactating women was noted. For children and adolescent, the same value of 400 mg/day as adults was referenced, noting that there are no data indicating higher sensitivity (EFSA, 2015).

The overall safety information reviewed in this evaluation suggests that the safety level of concern for dietary exposures to caffeine is relatively low. In general, daily dietary exposures to caffeine from various foods and beverages are relatively low, with repeat exposures usually separated by several hours. Therefore it is highly unlikely that plasma levels of caffeine resulting from dietary exposures will build up to levels associated with any type of toxicity, particularly given the rapid rate of plasma clearance (pharmacokinetics) exhibited by caffeine when taken orally in the diet. The data from human pharmacokinetic studies indicating similar caffeine pharmacokinetics from different beverage sources, whether consumed hot or cold (speed of consumption) and whether synthetic or natural along with the understanding of metabolic development in human infants and children, coupled with the safe use of relatively large doses of caffeine for apnea in premature infants further demonstrates the safety of caffeine for all life stages. Interestingly, even though significant changes have occurred in the marketplace with the rise in consumption of designer coffees with higher caffeine levels and the increase in consumption of energy drinks and other caffeinated snack foods, the overall consumption level of caffeine and the numbers of consumers has remained constant from 2001 to 2015. While intakes have remained stable, decreases in caffeine intake from soda were noted along with a stable or increased caffeine intake from coffee and tea, and increased caffeine intake from energy drinks (Branum et al., 2014; Fulgoni et al., 2015; Drewnowski and Rehm, 2016; Tran et al., 2016). The basis for the stable exposure level over time has been demonstrated to occur as a result of a "substitution effect," wherein one caffeine source is being substituted for another by the consumer. The data also demonstrated that self-regulation of caffeine intake by consumers is likely to occur regardless of source.

In conclusion, taking into account the long history of consumption, coupled with the wealth of critical safety data in both animals and humans and detailed dietary intake information over a considerable period of time, suggests that when consumed at current dietary levels, caffeine is safe for all ages.

References

Adamson, R.H., 2016. The acute lethal dose 50 (LD50) of caffeine in albino rats. Regul. Toxicol. Pharmacol. 80, 274–276.

Arnaud, M.J., 2011. Pharmacokinetics and metabolism of natural methylxanthines in animal and man. Handb. Exp. Pharmacol. 33–91.

Babu, K.M., Zuckerman, M.D., Cherkes, J.K., et al., 2011. First-onsetseizure after use of an energy drink [corrected]. Pediat. Emerg. Care 27, 539–540.

Barton, H.A., Pastoor, T.P., Baetcke, K., et al., 2006. The acquisition and application of absorption, distribution, metabolism, and excretion (adme) data in agricultural chemical safety assessments. Crit. Rev. Toxicol. 36, 9–35.

Blanchard, J., Sawers, S.J., 1983. The absolute bioavailability of caffeine in man. Eur. J. Clin. Pharmacol. 24, 93–98.

Boison, D., 2011. Methylxanthines, seizures, and excitotoxicity. Handb. Exp. Pharmacol. 251–266.

Bonati, M., Latini, R., Galletti, F., et al., 1982. Caffeine disposition after oral doses. Clin. Pharmacol. Ther. 32, 98–106.

Bonati, M., Latini, R., Tognoni, G., et al., 1984. Interspecies comparison of in vivo caffeine pharmacokinetics in man, monkey, rabbit, rat, and mouse. Drug. Metab. Rev. 15, 1355–1383.

Branum, A.M., Rossen, L.M., Schoendorf, K.C., 2014. Trends in caffeine intake among US children and adolescents. Pediatrics 133 (3), 386–393.

Brazier, J.L., Ritter, J., Berland, M., et al., 1983. Pharmacokinetics of caffeine during and after pregnancy. Dev. Pharmacol. Ther. 6, 315–322.

Brent, R.L., Christian, M.S., Diener, R.M., 2011. Evaluation of the reproductive and developmental risks of caffeine. Birth Defects Res. B Dev Reprod. Toxicol. 92, 152–187.

Budney, A.J., Brown, P.C., Griffiths, R.R., et al., 2013. Caffeine withdrawal and dependence: a convenience survey among addiction professionals. J. Caffeine Res. 3, 67–71.

Burg, A.W., 1975. Physiological disposition of caffeine. Drug. Metab. Rev. 4, 199–228.

Burns, R., Chaturvedula, A., Turner, D.C., et al., 2014. Population pharmacokinetic pharmacodynamic modeling of caffeine using visual analogue scales. Pharmacol. Pharm. 5, 444–454.

Callahan, M.M., Robertson, R.S., Arnaud, M.J., et al., 1982. Human metabolism of [1-methyl-14c]- and [2-14c]caffeine after oral administration. Drug. Metab. Dispos. 10, 417–423.

Cavalcante, J.W., Santos Jr., P.R., Menezes, M.G., et al., 2000. Influence of caffeine on blood pressure and platelet aggregation. Arq. Bras. Cardiol. 75, 97–105.

Chen, J., Jin, L., Chen, X., 2018. Efficacy and safety of different maintenance doses of caffeine citrate for treatment of apnea in premature infants: a systematic review and meta-analysis. Biomed. Res. Int. 2018, 11. Article ID 9061234.

Cheng, M., Hu, Z., Lu, X., et al., 2014. Caffeine intake and atrial fibrillation incidence: dose response meta-analysis of prospective cohort studies. Can. J. Cardiol. 30, 448–454.

Christian, M.S., Brent, R.L., 2001. Teratogen update: evaluation of the reproductive and developmental risks of caffeine. Teratology 64, 51–78.

Chrościńska-Krawczyk, M., Jargiełło-Baszak, M., Wałek, M., et al., 2011. Caffeine and the anticonvulsant potency of antiepileptic drugs:experimental and clinical data. Pharmacol. Rep. 63 (1), 12–18.

Collins, T., Welsh, J., Black, T., et al., 1987. Potential reversibility of skeletal effects in rats exposed in utero to caffeine. Food Chem. Toxicol. 25, 647–662.

Conen, D., Chiuve, S.E., Everett, B.M., et al., 2010. Caffeine consumption and incident atrial fibrillation in women. Am. J. Clin. Nutr. 92, 509–514.

Cook, D.G., Peacock, J.L., Feyerabend, C., et al., 1996. Relation of caffeine intake and blood caffeine concentrations during pregnancy to fetal growth: prospective population based study. BMJ 313, 1358–1362.

Corti, R., Binggeli, C., Sudano, I., et al., 2002. Coffee acutely increases sympathetic nerve activity and blood pressure independently of caffeine content: role of habitual versus nonhabitual drinking. Circulation 106 (23), 2935–2940.

Czuczwar, S.J., Gasior, M., Janusz, W., et al., 1990. Influence of differentmethylxanthines on the anticonvulsant action of common antiepilepticdrugs in mice. Epilepsia 31 (3), 318–323.

Dietitians of Canada. 2014. Food sources of caffeine. Available: http://www.dietitians.ca/Your-Health/Nutrition-A-Z/Caffeine/Food-Sources-of-Caffeine.aspx [accessed 06/03/2015].

Dikici, S., Saritas, A., Besir, F.H., et al., 2013. Do energy drinks cause epileptic seizure and ischemic stroke? Am. J. Emerg. Med. 31 (274), e271–e274.

Dimaio, V.J., Garriott, J.C., 1974. Lethal caffeine poisoning in a child. Forens. Sci. 3, 275–278.

Drewnowski, A., Rehm, C.D., 2016. Sources of caffeine in diets of US children and adults: trends by beverage type and purchase location. Nutrients 8 (3), 154.

Dworetzky, B.A., Bromfield, E.B., Townsend, M.K., et al., 2010. A prospective study of smoking, caffeine, and alcohol as risk factors for seizures or epilepsy in young adult women: data from the nurses' health study ii. Epilepsia 51, 198–205.

EFSA NDA Panel (EFSA Panel on Dietetic Products, Nutrition and Allergies/NDA), 2015. Scientific opinion on the safety of caffeine. EFSA J. 13 (5), 4102.

Erenberg, A., Leff, R., Haack, D., et al., 2000. Caffeine citrate for the treatment of apnea of prematurity: a double-blind, placebo-controlled study. Pharmacotherapy 20, 644–652.

Francart, S.J., Allen, M.K., Stegall-Zanation, J., 2013. Apnea of prematurity: caffeine dose optimization. J. Pediatr. Pharmacol. Ther. 18, 45–52.

Fulgoni 3rd, V.L., Keast, D.R., Lieberman, H.R., 2015. Trends in intake and sources of caffeine in the diets of us adults: 2001-2010. Am. J. Clin. Nutr. 101, 1081–1087.

Gronroos, N.N., Alonso, A., 2010. Diet and risk of atrial fibrillation—epidemiologic and clinical evidence. Circ. J. 74, 2029–2038.

Grosso, G., Godos, J., Galvano, F., et al., 2017. Coffee, caffeine, and health outcomes: an umbrella review. Annu. Rev. Nutr. 37, 131–156.

Heckman, M.A., Weil, J., Gonzalez de Mejia, E., 2010. Caffeine (1, 3, 7-trimethylxanthine) in foods: a comprehensive review on consumption, functionality, safety, and regulatory matters. J. Food Sci. 75, R77–R87.

Hildebrand, J.S., Patel, A.V., McCullough, M.L., et al., 2013. Coffee, tea, and fatal oral/pharyngeal cancer in a large prospective us cohort. Am. J. Epidemiol. 177, 50–58.

Hodgson, A.B., Randell, R.K., Jeukendrup, A.E., 2013. The metabolic and performance effects of caffeine compared to coffee during endurance exercise. PLoS One 8(4):e59561 [10pp] .

Huang, Z.L., Qu, W.M., Eguchi, N., et al., 2005. Adenosine a2a, but not a1, receptors mediate the arousal effect of caffeine. Nat. Neurosci. 8, 858–859.

IARC (1991). Caffeine. In: Coffee, Tea, Mate, Methylxanthines and Methylglyoxal. IARC Working Group, Feb. 27-Mar. 6, 1990, Lyon, France. (IARC Monographs on the Evaluation of Carcinogenic Risks to Humans, vol 51). World Health Organization (WHO), Lyon, France: International Agency for Research on Cancer (IARC), pp. 291-390.

IARC (2018). Drinking Coffee, Mate, and Very Hot Beverages. Expert Opinions of IARC Working Group on the Evaluation of Carcinogenic Risks to Humans, May 24-31, 2016, Lyon, France. (IARC Monographs on the Evaluation of Carcinogenic Risks to Humans, Vol. 116). Lyon, France: International Agency for Research on Cancer (IARC).

IOM, 2001. Caffeine for the sustainment of mental task performance: fomulations for military operations. National Academy of Sciences, Committee on Military Nutrition Research, Food and Nutrition Board, Institute of Medicine (IOM). National Academy Press (NAP), Washington (DC).

IPCS. 1998. Caffeine.

James, J.E., 1990. The influence of user status and anxious disposition on the hypertensive effects of caffeine. Int. J. Psychophysiol. 10, 171–179.

Jiang, W., Wu, Y., Jiang, X., 2013. Coffee and caffeine intake and breast cancer risk: an updated dose-response meta-analysis of 37 published studies. Gynecol. Oncol. 129, 620–629.

Jones, A.W., 2017. Review of caffeine-related fatalities along with postmortem blood concentrations in 51 poisoning deaths. J. Anal. Toxicol. 41 (3), 167–172.

Josse, A.R., Da Costa, L.A., Campos, H., et al., 2012. Associations between polymorphisms in the ahr and cyp1a1-cyp1a2 gene regions and habitual caffeine consumption. Am. J. Clin. Nutr. 96, 665–671.

Kua, K.P., Lee, S.W., 2017. Systematic review and meta-analysis of clinical outcomes of early caffeine therapy in preterm neonates. Br. J. Clin. Pharmacol. 83 (1), 180–191 [plus supplementary data].

Lachance, M., 1982. The pharmacology and toxicology of caffeine. J. Food Saf. 4, 71–112.

Layland, J., Carrick, D., Lee, M., et al., 2014. Adenosine: physiology, pharmacology, and clinical applications. JACC Cardiovasc. Interv. 7, 581–591.

Lazarus, M., Shen, H.Y., Cherasse, Y., et al., 2011. Arousal effect of caffeine depends on adenosine a2a receptors in the shell of the nucleus accumbens. J. Neurosci. 31, 10067–10075.

Loftfield, E., Cornelis, M.C., Caporaso, N., et al., 2018. Association of coffee drinking with mortality by genetic variation in caffeine metabolism: findings from the UK biobank. JAMA Intern. Med. 178 (8), 1086–1097.

Malmir, H., Esmaillzadeh, A., 2017. The relationship between tea and coffee consumption and glioma: a systematic review. J. Babol Univ. Med. Sci. 19 (10), 69–75.

McGuire, S., 2014. Institute of Medicine, 2014. Caffeine in food anddietary supplements: examining safety-workshop summary. Washington, DC: The National Academies Press, 2014. Adv. Nutr. 5, 585–586.

Mitchell, D.C., Knight, C.A., Hockenberry, J., et al., 2014. Beverage caffeine intakes in the US. Food Chem. Toxicol. 63, 136–142.

Moffat, A.C.O.M., Widdop, B., 2004. Clarke's analysis of drugs and poisons. Pharmaceutical Press, London, England.

Mohr, U., Althoff, J., Ketkar, M., et al., 1984. The influence of caffeine on tumour incidence in sprague-dawley rats. Food Chem. Toxicol. 22, 377–382.

Myers, M.G., Harris, L., Leenen, F.H., et al., 1987. Caffeine as a possible cause of ventricular arrhythmias during the healing phase of acute myocardial infarction. Am. J. Cardiol. 59, 1024–1028.

Myers, M.G., Reeves, R.A., 1991. The effect of caffeine on daytime ambulatory blood pressure. Am. J. Hypertens. 4(5), Part. 1, 427–431.

Nawrot, P., Jordan, S., Eastwood, J., et al., 2003. Effects of caffeine on human health. Food Add. Contam. 20, 1–30.

Newton, R., Broughton, L.J., Lind, M.J., et al., 1981. Plasma and salivary pharmacokinetics of caffeine in man. Eur. J. Clin. Pharmacol. 21, 45–52.

Nkondjock, A., 2009. Coffee consumption and the risk of cancer: an overview. Cancer Lett. 277, 121–125.

Noordzij, M., Uiterwaal, C.S., Arends, L.R., et al., 2005. Blood pressure response to chronic intake of coffee and caffeine: a meta-analysis of randomized controlled trials. J. Hypertens. 23 (5), 921–928.

OECD-SIDS. 2002. Organisation for economic co-operation and development screening information dataset (OECD SIDS). Initial assessment profile: Caffeine. Available at www.chemunepch/irptc/sids/OECDSIDS/CAFFEINE.pdf.

Peck, J.D., Leviton, A., Cowan, L.D., 2010. A review of the epidemiologic evidence concerning the reproductive health effects of caffeine consumption: a 2000-2009 update. Food Chem. Toxicol. 48, 2549–2576.

Pelchovitz, D.J., Goldberger, J.J., 2011. Caffeine and cardiac arrhythmias: a review of the evidence. Am. J. Med. 124, 284–289.

Rivera-Oliver, M.D.-R.M., 2014. Using caffeine and other adenosine receptor antagonists and agonists as therapeutic tools against neurodegenerative diseases: a review. Life Sci. 101, 1–9.

Robertson, D., Wade, D., Workman, R., et al., 1981. Tolerance to the humoral and hemodynamic effects of caffeine in man. J. Clin. Invest. 67, 1111–1117.

Romualdo, G.R., Rocha, A.B., Vinken, M., et al., 2019. Drinking for protection? Epidemiological and experimental evidence on the beneficial effects of coffee or major coffee compounds against gastrointestinal and liver carcinogenesis. Food Res. Int. 123, 567–589.

Sawynok, J., 1995. Pharmacological rationale for the clinical use of caffeine. Drugs 49, 37–50.

SCF (Scientific Committee on Food), 1999. Opinion on Caffeine, Taurine and D-Glucurono -γ-Lactone as constituents of so-called "energy" drinks. 15 pp.

Schmidt, B., Anderson, P.J., Doyle, L.W., et al., 2012. Survival without disability to age 5 years after neonatal caffeine therapy for apnea of prematurity. JAMA 307 (3), 275–282.

Schmidt, B., Roberts, R.S., Davis, P., et al., 2006. Caffeine therapy for apnea of prematurity. N. Engl. J. Med. 354 (20), 2112–2121.

Schmidt, B., Roberts, R.S., Davis, P., et al., 2007. Long-term effects of caffeine therapy for apnea of prematurity. N. Engl. J. Med. 357 (19), 1893–1902.

Sutherland, D.J., McPherson, D.D., Renton, K.W., et al., 1985. The effect of caffeine on cardiac rate, rhythm, and ventricular repolarization. Analysis of 18 normal subjects and 18 patients with primary ventricular dysrhythmia. Chest 87, 319–324.

Tang-Liu, D.D., Williams, R.L., Riegelman, S., 1983. Disposition of caffeine and its metabolites in man. J. Pharmacol. Exp. Ther. 224, 180–185.

Tarka Jr., S.M., 1982. The toxicology of cocoa and methylxanthines: are view of the literature. Crit. Rev. Toxicol. 9 (4), 275–312.

Temple, J.L., Dewey, A.M., Briatico, L.N., 2010. Effects of acute caffeine administration on adolescents. Exp. Clin. Psychopharmacol. 18, 510–520.

Trabulo, D., Marques, S., Pedroso, E., 2011. Caffeinated energy drink intoxication. BMJ Case Rep. 2011.

Tran, N.L., Barraj, L.M., Bi, X., et al., 2016. Trends and patterns of caffeine consumption among US teenagers and young adults, NHANES 2003-2012. Food Chem. Toxicol. 94, 227–242.

US FDA (2018a). FDA Warns Companies to Stop Selling Dangerous and Illegal Pure and Highly Concentrated Caffeine Products. Silver Spring (MD): U.S. Food and Drug Administration (U.S. FDA).

US FDA, 2018b. Part 340—Stimulant drug products for over-the-counter human use. §340.50—Labeling of stimulant drug products. U.S. Code of Federal Regulations (CFR). Title 21: Food and Drugs. (U.S. Food and Drug Administration). U.S. Food and Drug Administration (U.S. FDA), U.S. Government Printing Office (GPO), Washington (DC). Available at. Available from: https://www.govinfo.gov/app/collection/cfr/.

US FDA, 2012 FDA, 2012. Letter from Jeanne Ireland, Assistant Commissioner for Legislation to the Honorable Richard J. Durbin, Senator dated August 10, 2012 [Re: Letter of April 3, 2012 Expressing Concern About Potential Safety Issues Associated With the Consumption of so-called "Energy Drinks"]. Available from: http://www.durbin.senate.gov/imo/media/doc/09112012FDA.pdf.

US FDA, 21 CFR 340.50. Part 340—stimulant drug products for over-the-counter human use. http://www.accessdata.fda.gov/scripts/cdrh/cfdocs/cfcfr/CFRSearch.cfm?fr = 340.50.

US FDA, Oakridge National Laboratory. 2010. Caffeine intake by the U.S population.

Vercammen, K.A., Koma, J.W., Bleich, S.N., 2019. Trends in energy drink consumption among U.S. adolescents and adults, 2003-2016. Am. J. Prev. Med. 56 (6), 827–833.

White, J.R., Padowski, J.M., Zhong, Y., et al., 2016. Pharmacokinetic analysis and comparison of caffeine administered rapidly or slowly in coffee chilled or hot versus chilled energy drink in healthy young adults. Clin. Toxicol. 54 (4), 308–312.

Wikoff, D., Welsh, B.T., Henderson, R., et al., 2017. Systematic review of the potential adverse effects of caffeine consumption in healthy adults, pregnant women, adolescents, and children. Food Chem. Toxicol. 109 (1), 585–648.

Würzner, H.-P., Lindström, E., Vuataz, L., et al., 1977a. A 2-year feeding study of instant coffees in rats. II. Incidence types neoplasms. Food Cosmet. Toxicol. 15, 289–296.

Würzner, H.-P., Lindström, E., Vuataz, L., et al., 1977b. A 2-year feeding study of instant coffees in rats. I. Body weight, food consumption, haematological parameters and plasma chemistry. Food Cosm. Toxicol. 15, 7–16.

Chapter 33

Biological activities and potential nanotechnological delivery of resveratrol

Gianfranco Risuleo and Camillo La Mesa

Introduction

To initiate this chapter about resveratrol (RV) and its potential usage in advanced nanotechnologies in biomedicine we would like to adopt an analogous consideration that was published in the previous edition of this book. Is a new edition of the book needed after such a short interval has lapsed from the second edition? To answer this question, we will again report on the scientific literature data. A PubMed search using "natural products" as keywords, scored, as of May 2020, an astounding 33,344 entries in English, during the last 10 years. In the first edition of this book, the number of review articles published from the 1960s to 2014 on the general subject "natural substances and their biological activities" was about 1650 (Risuleo, 2016). This makes about a 20-fold increase in scientific production in just a few years. The interest in this topic remains a central theme for research in a number of different biotechnologies, including the search for alternative commodities, foodstuffs, and drugs aimed at therapeutic usages. In this work, we also discuss methods for transferring exogenous materials and drugs into biological tissues. The focus will be on matrices such as micelles, vesicles, and carbon nanotubes (CNT). An ensemble of physical forces play a fundamental role in drug dispersion. The combination of these forces is responsible for drug uptake in various matrices and for their release in tissues. Uptake of exogenous either macro- or small molecules into cargo particles and their transfer to recipient cells is the result of complex processes. These processes are concomitant to drug partition among supramolecular aggregates. Similar conclusions apply to drug release as far as the kinetic features are concerned. Therefore the adsorption of nutraceuticals and their release within target organs are of particular relevance. We first consider the forces responsible for the formation of such supramolecular entities on physicochemical grounds as well as drug uptake. Finally, we review the possibility of transfecting cargo-mediated aggregates of nanoparticle/drug complexes to cells or tissues of interest and their bioactivity upon release within the cell matrix.

The number of bioactive compounds is increasing at an incredible pace. In their works from 1997–2014, Newman and Cragg (2016) produced a classification and catalog of the different drugs of natural origin. These can be divided into several different categories. the ones we deal within this chapter fall within types 2 and 3 of their classification. Type 2 are natural products, unmodified in structure, but semi- or totally synthetic; natural products or "botanical drugs" and type 3 compounds are derived from a natural product and semisynthetically modified. In addition to RV, we also review the bioproperties of other known and widely used natural substances. In any case, this chapter is far from exhaustive but should serve as stimulation for the reader to deepen their understanding of this expanding field of multidisciplinary research.

In conclusion, the phrase cited in the previous edition of this text, "the research on natural compounds and nutraceuticals is in continuous and ever-growing expansion," is not over-used, abused or trivial, but rather timely and logical.

A concise overview of natural substances

Mankind has found in nature sources of nutrition and protection (fuel, clothing) from often hostile environmental conditions. Natural products have thus become a very useful tool in ritual and archeo-medicinal procedures. Many fossil artifacts indicate, for instance, that the plant *Papaver somniferum* and the opium obtained from its flower, were known in Central and Southern Asia already during the Neolithic and Bronze Ages. During that time, it was used in magic and religious rituals. Plant products were the basis of traditional medicine systems and the

Nutraceuticals. DOI: https://doi.org/10.1016/B978-0-12-821038-3.00033-1

earliest records date from around 2600 BC. They document the use of approximately 1000 plant-derived substances in Mesopotamia. These include oils of cedar, cypress, licorice, myrrh, garlic, and the above-mentioned poppy flowers. Most, if not all, of these preparations are still used today for the treatment of a variety of conditions ranging from the common cold to inflammation and parasitic infections (Newman and Cragg, 2016;). The curative properties of garlic, for instance, are found in the Egyptian Ebers Papyrus, dating to about 1550 BC, where this herb is indicated to combat a number of diverse human conditions. Within Greek and Roman civilizations, two different concepts of natural medicaments were established: (1) *pharmakon*, an active principle alleviating a series of maladies, and (2) *toxikon*, a lethal poison. With respect to this, Hippocrates, as early as 400 BC, produced the aphorism (better known in its Latin form) "*primum non nocere*," that is, "In the first place do not inflict detriment" (even though scholars doubt Hippocrates's ownership of this aphorism). However, the quest for an equilibrium between life and death, benefit and damage, can be considered as a sort of primordial risk management. Antioxidant substances certainly still play a primary role in human well-being and the management of many afflictions. A comprehensive review of the history of medicine can be accessed at the US NIH website: www.nlm.nih.gov/hmd/medieval/arabic.html.

Some commonly known antioxidants

Antioxidants are compounds which inhibit oxidation. They play a crucial role in the maintenance of cell homeostasis, and their principal "assignment" involves the detoxification of active oxygen radicals. Oxidation processes, from which these radicals are generated, normally comprise a series of chemical reactions producing free radicals, which are detrimental to cell survival. Natural antioxidants such as thiols or ascorbic acid (vitamin C) suppress these chain reactions. Therefore to control oxidative stress, plants and animals maintain a complex redox system in which overlapping antioxidants play a crucial role. These can be either biomacromolecule like catalase and superoxide dismutase (such as SOD-2), or small molecules like glutathione. They may also be endogenous or derived from a dietary supplement. In any case, the main role of antioxidants consists of scavenging and eliminating Reactive Oxygen Species (ROS) produced by normal physiological/biochemical pathways. These ROS scavengers may be endogenous or supplemented exogenously with food. Vegetables and fruits are very rich in natural antioxidants, but synthetic antioxidant molecules are also available. Polyphenols are possibly one of the most important antioxidant compounds. The activity of these molecules is multifaceted since they may also play a role as anticarcinogens and antimutagens. Their role in controlling allergies and their antiaging capacity have also been reported. The most common and important sources of natural antioxidants are vegetables and fruits. However, antioxidant molecules from plants not commonly consumed in agricultural waste should be taken into account. Table 33.1 lists some common fruits and vegetables with their antioxidant content (Anwar et al., 2018). An important compendium on antioxidants in food and their biological significance was recently published (Shalaby, 2018).

A review of the bioproperties of RV, with special reference to its antioxidant power, cannot ignore the existence of other natural molecules of similar characteristics widely used in human and animal fields. Here, a brief description of commonly used natural products and their mode of action is provided.

Curcumin

Curcumin is the principal curcuminoid present in turmeric, a commonly used spice in Oriental cuisine. It is also widely present as an ingredient in cosmetics, as a herbal food supplement, or for flavoring and coloring in food preparations (Majeed, 2015). The commercially available yellow powder is obtained from the roots of the gourd *Curcuma longa* belonging to the *Zingiberaceae* family, which is the same family as ginger (Fig. 33.1). The anti-inflammatory action exhibited by curcumin is exerted by transcriptional activation of NF-κB and AP-1 (Maheshwari et al., 2006; Sarvjeet and Ashok, 2006). Its antioxidant and cytoprotective properties have also been assessed. In addition, it has been suggested that curcumin has a potential role in the prevention of neurodegenerative diseases since it can bind the β-amyloid protein. This protein is involved in the pathogenesis of Alzheimer disease (AD), since it inhibits the formation of the typical senile plaques. Curcumin is also able to cross the blood–brain barrier due to its lipophilic character. It can therefore localize at brain level, thus destabilizing the polymers of β-amyloid polypeptides (Frautschy et al., 2001; Lim et al., 2001). The reader is directed to Jiang et al. (2020), Mazumder et al. (2020), Moradi et al. (2020), and references therein for recent reviews on the bioproperties and nanoformulations of curcumin.

Forskolin

Forskolin (Forsk) is a diterpene compound belonging to the Labdane family. It is extracted from the roots of *Coleus forskohlii*, a common ornamental plant. The plant has long been used in traditional oriental medicine, but interest in this plant has expanded in an enormous way as recent studies have evidenced its inhibitory action toward

TABLE 33.1 Species and concentrations of antioxidants in various fruits and vegetables. ORAC scale is the Oxygen Radical Absorbance Capacity. The ORAC test allows measurement of the inhibitory capacity of an antioxidant on free radicals. However, since no physiological proof in vivo existed in support of the free-radical theory or that ORAC provided information relevant to biological antioxidant potential, it was withdrawn in 2012. For details see: https://en.wikipedia.org/wiki/Oxygen_radical_absorbance_capacity.

Name	Antioxidant contents	Concentration (ORAC value)
Pears	Ascorbic acid, flavonoids (quercetin, isorhamnetin, myricetin, kaempferol, and luteolin), betalains, taurine, total carotenoids, and total phenolics	135 (mmol TE/g)
Pomegranate	Vitamin C and polyphenols	1245 (mmol TE/g)
Papaya	Quercetin and β-sitosterol	300 (mmol TE/g)
Water melon	Lycopene, β-carotene, vitamin C	100
Apple	Proanthocyanidins, flavonoids; phenolic acids; hydroxychalcones; isoprenoid glycosides. Flavanols, flavonols, dihydrochalcones, hydroxycinnamates	16.78 ± 0.25 (mmol TE/g), 1508 ± 44 (μmol/100 g)
Plum	Proanthocyanidins, flavonoids; phenolic acids; hydroxychalcones; isoprenoid glycosides	14.55 ± 0.21 (mmol TE/g), 94.8 (mmol TE/g)
Guava	β-Carotene, lycopene, vitamin C, ellagic acid, anthocyanin	
Beet root	Betalains, vitamins C and E, carotenoids, flavonoids, and thiol compounds	4100 (dry extract), 115 (μmol TE/g)
Pea	Vitamins C and E, carotenoids, flavonoids, and thiol compounds	19 (μmol TE/g)
Carrot	Vitamins C and E, carotenoids, flavonoids, and thiol compounds	60 (μmol TE/g)
White cabbage	Vitamins C and E, carotenoids, flavonoids, and thiol compounds	61 (μmol TE/g)
Tomato	Vitamins C and E, carotenoids, flavonoids, and thiol compounds	67 (μmol TE/g)
White onion	Vitamins C and E, carotenoids, flavonoids, and thiol compounds	85 (μmol TE/g)
Cauliflower	Vitamins C and E, carotenoids, flavonoids, and thiol compounds	102 (μmol TE/g)
Spinach	Vitamins C and E, carotenoids, flavonoids, and thiol compounds	152 (μmol TE/g)

Source: This table is redrawn and simplified from Anwar H., Hussain G., Mustafa I., 2018. Antioxidants from natural sources. In: Shalaby E., (Ed.) Antioxidants in Food and Its Applications. Chapter 1 eBook. IntechOpen Limited, London. doi: 10.5772/intechopen.72008.

the growth of colorectal cancer cells. The compound even shows a synergistic activity with the anti-inflammatory drug Rolipram (McEwan et al., 2007). This activity was observed in the CA1 hippocampal cells where parallel administration of the two drugs seemed to exert a potentiated long-term action (Otmakhov et al., 2004). Commercial interest has been increasing in the last few years since many pharmaceutical industries are introducing to the market drugs based on Forsk. This is essentially due to its possible role in the reduction of fatty mass and an increment of both testosterone and bone mass in overweight or obese people. In this regard, Forsk has been envisaged as a treatment for obesity, especially in men (Godard et al., 2005).

The increase of the intracellular concentration of cAMP is an activity of Forsk. It has been demonstrated in "classical" physiology studies that this action is actually exerted via activation of adenylate-cyclase. This enzyme specifically mediates the conversion of ATP into cAMP (Seamon et al., 1981). It is also an integral part of the cell plasma membrane. Studies from our laboratory have demonstrated that the cell membrane is one of the main targets of other natural products such as resveratrol and neem oil (see below). Therefore it is plausible to assume that these natural products have similar mechanisms of action and cellular targets.

Neem oil

The neem tree, *Azadirachta indica*, is an evergreen plant belonging to the mahogany family of *Meliaceae*. It thrives in subtropical semiarid to arid areas of the globe. It was originally native of Myanmar, but is now also found in India, Africa, Central America, the Caribbean, and the

Philippines. It has also spread to warm/temperate areas such as southern Italy. The plant seeds contain high concentrations of azadirachtin. The best characterized principle consists of a tetra-nor-triterpenoid and is used for pest and insect control in agriculture. Whole neem oil is a natural mélange of a few hundred different organic compounds such as steroids (campesterol, β-sitosterol, stigmasterol) and triterpenoids. The azadirachtin content ranges from 300 to over 2000 ppm, depending on the origin and quality of the crushed seeds. The oil prepared from the seeds has been extensively used in Ayurveda, Unani, and homeopathic medicines for centuries (Brahmachari, 2004). The Sanskrit name *sarva roga nivarini* means "universal healer of all illnesses" and the common Indian name for the tree is "the village pharmacy" (Gupta et al., 2017; http://www.natureneem.com/index.htm).

In general, neem products represent good means for the biological control of pests. In addition to being affordable, they are ideal multipurpose tools in developing countries (Koul and Wahab, 2004).

The bioactive properties of a partially purified derivative of whole neem oil were extensively characterized in our laboratory. This product, dubbed H-MeX, exhibited a number of biological activities such as: (1) differential cytotoxicity on tumor cells as compared to normal fibroblasts in culture; (2) antiviral action exerted on de novo viral DNA synthesis; (3) increase in membrane fluidity and control of the differential entry of exogenous material; (4) effect on cell survival/proliferation; and (5) control of apoptosis. However, an unfortunate conclusion from the works on neem products is that potential applications in advanced biotechnology and medicine seem to be very few. The use of nanocomposites, and nanotechnologies in general, to deliver exogenous agents within a cell is largely absent from neem-related literature, even though this is a field in extremely rapid expansion, especially in biomedically oriented applications (Muzi et al., 2016a,b; Risuleo and La Mesa, 2019a,b). This may be due to the fact that no single pure compound utilizable in advanced biomedicine has been as yet identified. The data briefly summarized above were extensively presented and critically discussed in the book, *Nuts and Seeds in Health and Disease Prevention* (1st ed.), and recently updated in the second edition. For details, the reader is referred to the following works (Aiello et al., 2011; Risuleo, 2020).

Klamath algae

The *Aphanizomenon flos-aquae* (AFA) is a ubiquitous species of cyanobacteria, or blue-green algae, that grows in brackish and freshwater environments. It is also known as Klamath algae from the region of Klamath Lake (North America) and its surroundings. The aboriginal inhabitants of these North American areas were known as the Klamath people, but the Modoc population also lived in this area (principally, the Lower Klamath and Tule lakes). For some reason, the Modoc people never achieved the same degree of visibility, at least as far as the nutritional usage of this algae is concerned. However, these green-blue algae are found in different ecosystems and latitudes from the Baltic Sea in northern Europe to the Great Lakes areas in the United States. Phycocyanin is the molecule that gives the green-blue color to the algae. Depending on the source, the amount of phycocyanin can be as high as 15% of the dry weight of the organism. Phycocyanin has antioxidant and anti-inflammatory effects. One study evaluated the ability of a novel natural extract of AFA, enriched with phycocyanin, to protect normal human erythrocytes and blood plasma samples from oxidative damage in vitro (Benedetti et al., 2004). This species of cyanobacteria is also commercially available as a dietary supplement. Benefits have been indicated from its consumption. However, potential toxicity has been shown in some strains of the AFA species since they contain toxins such as cylindrospermopsin and axitoxins as well as microcystins. Therefore some sources recommend extreme caution when using AFA and its derivatives (Saker et al., 2005; Preußel et al., 2006). The microorganism AFA contains vitamins, minerals, chlorophyll (at very high concentrations), and various other antioxidant molecules. For example, polyunsaturated fatty acids, which are very important in maintaining membrane fluidity, account for up to 10% of AFA's dry weight. Evidence exists that AFA raises the blood levels of eicosapentaenoic acid and docosahexaenoic acid. These two compounds are known to improve the functions of the nervous system among other organs (Kushak et al., 2000). It is worth noting that a phycocyanin water extract inhibited the growth of tumor cell lines in vitro; therefore an antitumor property of this antioxidant molecule has also been suggested. To conclude, blue-green algae in general contain a significant amount of carotenoids, lycopene, and lutein that confer strong antioxidant properties via quenching action on reactive oxygen species (ROS, see also above) and carry intrinsic antiinflammatory properties (Shytle et al., 2010, and references therein). Despite their controversial effects, cyanobacteria can represent either a beneficial nutritional principle or produce toxic effects on human health. However, they have been a staple food in many cultures for centuries. They were also an object of commerce and food for indigenous peoples in various continents. Testimonies of their use are found in Africa, Asia, and the United States. The Chinese, Aztecs, and Mayans also consumed this food supplement (Carmichael et al., 2000; Barsanti and Gualtieri, 2006).

The four natural products discussed above present an intrinsic difference. The first two, forskolin and curcumin,

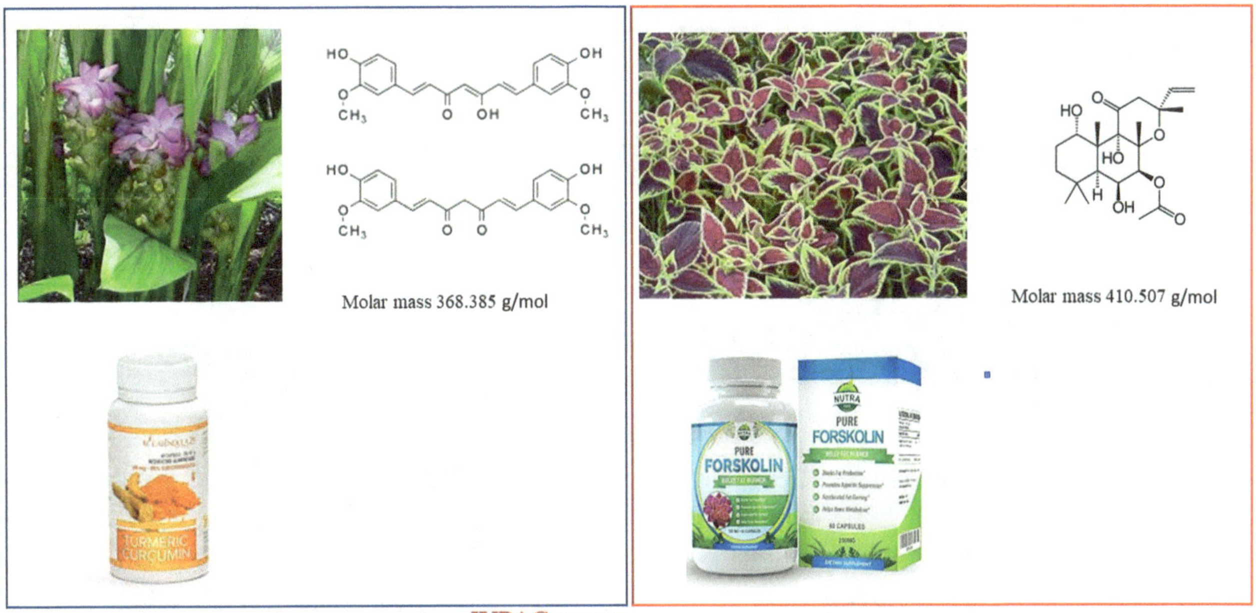

IUPAC names

(1E,6E)-1,7-Bis(4-hydroxy-3-methoxyphenyl)hepta-1,6-diene-3,5-dione

(3R,4aR,5S,6S,6aS,10S,10aR,10bS)-6,10,10b-Trihydroxy-3,4a,7,7,10a-pentamethyl-1-oxo-3-vinyldodecahydro-1H-benzo[f]chromen-5-yl acetate

FIGURE 33.1 **Curcumin and forskolin.** Curcumin: left. The formulas show curcumin in the eno- and keto- form (upper and lower formulas, respectively). Forskolin: right. The images are self-explanatory. The containers are commercially available preparations of the two compounds.

are pure molecules. Neem oil and Klamath algae are, on the contrary, complex coacervates of different components. This renders the characterization of their biological properties and the evaluation of their actions extremely difficult in a biochemically sound manner. The characteristics of these natural products are illustrated in Figs. 33.1 and 33.2.

Phytoalexins and resveratrol

Resveratrol (RV) belongs to the chemical family of phytoalexins, which also includes terpenoids, glycosteroids, and alkaloids. The term phytoalexin is applied to all types of phytochemical molecules involved in plant defense, and therefore they may be figuratively compared to the animal immune system. These natural compounds show antimicrobial and, often, very potent antioxidative actions. They are synthesized de novo in plants under a variety of stress conditions, such as pathogenic attacks and/or climate adversities, such as drought. Pathogens may cause a rapid accumulation of plant-toxic substances at the site of infection. This is where phytoalexins show their character as broad-spectrum inhibitors of infection progression. If the biosynthesis of phytoalexin is inhibited, susceptibility of the plant tissue to infection increases. Accordingly, mutant plants incapable of producing these natural "defenders" are more vulnerable to pathogen proliferation. Conversely, pathogens able to degrade phytoalexins are more virulent toward the plant (Glazebrook and Ausbel, 1994; Thomma et al., 1999). In conclusion, phytoalexins are natural compounds secreted by plants endowed with antimicrobial and antioxidative actions.

Polyphenols, flavonoids, and chemically related substances play a key role in the defense against fungal and other pathogens. A few examples of these "plant-protective" molecules include: (1) danielone (phytoalexin), which is found in the papaya fruit (*Carica papaya*) and shows a potent antifungal activity against the infectious pathogenic fungus *Colletotrichum gloesporioides* (Echeverri et al., 1997); (2) sakuranetin (flavanone), which is found in rice and *Polymnia fruticosa*, an exotic plant originating from the Andean regions. Sakuranectin and centaureidin (flavones), also present in *P. fruticosa* are known to inhibit the spore germination of *Magnaporthe grisea*, an ascomycete fungus pathogenic toward several species of cereals such as corn and rice (Molnár et al., 2010; Cho and Lee, 2015); and (3) stilbenes, which are produced in *Eucalyptus sideroxylon* following pathogens attacks. Such compounds can be implied in the hypersensitive response of plants. One of the reasons why some woods show a natural preservation against rot may reside in the presence of high levels of polyphenols as reported by the seminal work of Hart and Hillis (1974). The importance of these molecules is reflected in their enormous commercial significance. Phytoalexins are very important in counteracting plant

FIGURE 33.2 Neem and Klamat. Neem: left. Native of Chhattisgarh with leaves of the neem tree (*Azadirachta indica*) heading for Hareli Festival. Neem flowers are known with the various names of the tree being: Margosa, Neeb, Nimtree, Nimba, Vepu, Vempu, Vepa, Bevu, Veppam, Aarya Veppu, or Indian lilac. *Courtesy of Pankaj Oudhia, J.M. Garg, Hayavadhan. Right: Cells of* Aphanizomenon flos-aquae *in their aquatic environment. At the bottom of each are shown commercial preparations of these two natural products.*

disease and in pest control. Reader are referred to very recent reviews dealing specifically with this matter (Goufo et al., 2020; Yessenkyzy et al., 2020; Di Fonzo et al., 2020).

Resveratrol: in detail

As mentioned at the beginning of this section, RV belongs to the chemical family of phytoalexins. In particular, *trans*-RV, is produced in the grapes of *Vitis vinifera*, to combat infection and proliferation of fungal pathogens such as *Botrytis cinerea* (Favaron et al., 2009). Fig. 33.3 shows the structures of some common phytoalexins with different chemical features. However, RV is a phytoalexin of natural origin found in a wide evolutionary panorama ranging from bryophytes to higher plants, where it responds to different stimuli. RV is a stilbenoid with a low molecular weight (MW = 228.25 Dal). In plants, the synthesis of RV is mediated by the enzyme RV synthase (Schröder et al., 1988). The drug is normally found in two geometric isomer forms: *cis*- (Z) and *trans*- (E) RV. Both isomers can exist in a free or glucose-bound form. The *cis*–*trans* isomer transition occurs via photoisomerization, a process mediated by UV light (Mattivi et al., 1995; Lamuela-Raventos et al., 1995).

RV has relatively high solubility in water, but its solubility increases in ethanol. Its melting point is 253°C. In addition, *trans*-RV powder is stable at 40°C in the presence of air and in a relatively dry atmosphere (75% humidity). This isomer is also stabilized by the presence of transport proteins (Prokop et al., 2006). The commercially available preparation consists of a white powder with a yellowish-orange hue. RV is synthesized in *V. vinifera*, where it was originally isolated. In this plant RV is synthesized following the metabolic pathway of the phenyl-propanoids (Dixon and Paiva, 1995), where the starting compound is phenylalanine (Jeandet et al., 2002; Cichewicz and Kouzi, 2002 and references therein). Red wine contains a significant amount of RV, but the compound can be isolated from all intensely pigmented fruits (Fig. 33.4). However, the drug can be also obtained from sources as diverse as peanuts, pistachios, dark chocolate, and peanuts. RV was also obtained from the desiccated roots of the Japanese knotweed (*Polygonum cuspidatum*). It is present in this plant in a crude form, at a 400-fold higher concentration as compared to grapes or red wine. RV has been known for centuries and was used, along with other natural products, in popular traditional medicine (Risuleo, 2016). However, the presence of RV in red wine has suggested the possibility that this beverage may play a role in cardio-protection and other circulatory diseases. This peculiar property attributed to red wine is also known as the French paradox (Ferrieres, 2004; de Lorgeril and Salen, 2005). Although the authors of this

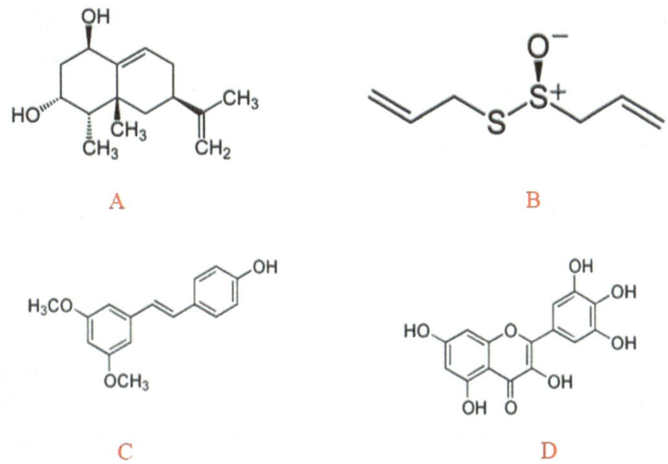

FIGURE 33.3 Structures of some phytoalexins. Phytoalexins' response to infection usually involves two chronological steps: a short-term one which produces free radicals near the site of infection. A second long-term response which ends in the synthesis of hormones and the increase of enzymes stimulating the production of phytoalexins such as capsidiol (see below). Phytoalexins are often antioxidative substances synthesized de novo. Phytoalexins tend to fall into several classes including terpenoids, glycosteroids, and alkaloids. For the sake of convenience, the definition includes all phytochemicals that are part of the defensive counter-action strategy. (A) Capsidiol is a low-molecular-weight plant secondary metabolite produced during fungal or bacterial plant infection. It is accumulated in tobacco (*Nicotiana tabacum*) and chili pepper (*Capsicum annuum*) in response to a fungal infection. (B) Pterostilbene is a stilbenoid chemically related to resveratrol. In plants it has a defensive phytoalexin role. This molecule is found in almonds, blueberries, grape leaves, and vines. However, pterostilbene is considered a corrosive substance. It is an eye irritant and is environmentally toxic, especially to aquatic life. (C) Allicin is present in garlic (*Allium sativum*) and consists of an oily, slightly yellow liquid. Allicin is released upon disruption of the berry, thus releasing the typical scent, some may not appreciate the aroma and it can remain with those who consume it. From a chemical point of view, allicin is a thioester of sulfenic acid and is also known as allyl-thiosulfinate. Its biological activity is attributed to its antioxidant property and its reaction with thiol-containing proteins. (D) Myricetin is a polyphenolic compound of the flavonoid class with antioxidant properties. It is commonly consumed in the diet, including tomatoes, oranges, nuts, berries, tea, and red wine. Myricetin is structurally similar to members of the flavonol class of flavonoids (fisetin, luteolin, and quercetin) with which it shares similar functions.

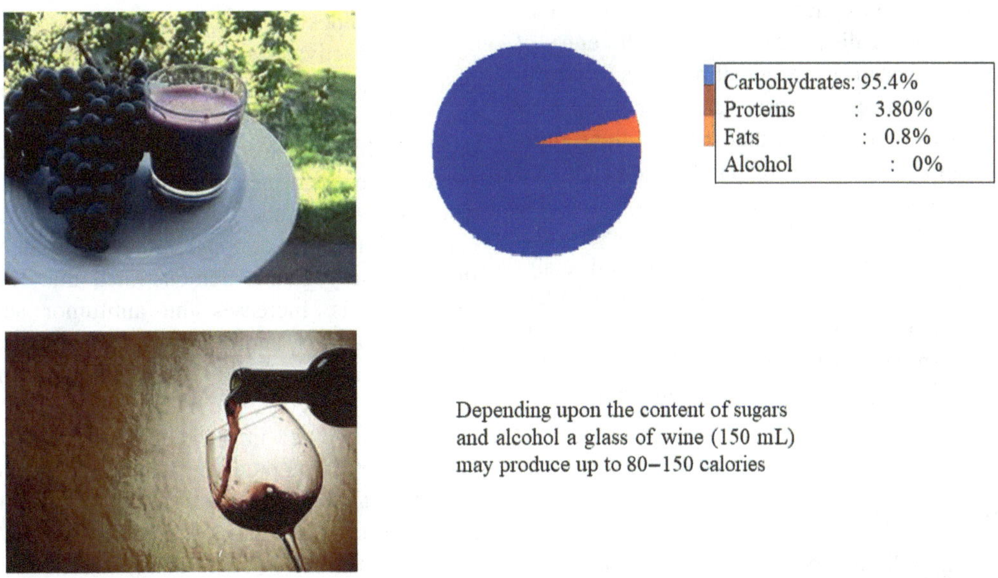

FIGURE 33.4 Resveratrol in red wine and dietary caloric contribution of wine. (Upper row, left) A red grape bundle and a glass of unfermented must. (Right) Compounds present in must. Since this is an unfermented product, the alcohol content is zero. (Lower row, left) A glass of red wine. (Right) Average amount of calories in a standard glass of wine. These depend very much upon the alcohol, sugar content, and type of wine grape. White wine tends to be more calorific, even if generally less alcoholic than red wine. This is due to the higher amount of unfermented sugar present in the final product. The presence of resveratrol in red wine has given rise to its alleged benefits for human health, especially as far as protection from cardiac conditions is concerned (the so-called French paradox) (Ferrieres, 2004; De Lorgeril and Salen, 2005).

work regularly drink red wine, neither would sponsor an uncontrolled consumption of this drink. It is true, however, that dietary habits in the Mediterranean areas are considered healthier than other regimes adopted elsewhere. In particular, this has given rise to the interesting, though less known, Albanian paradox (Gjonça and Bobak, 1997).

Finally, RV acts at the molecular level as a regulator of the human SIRT-1 gene (see also section "Pharmacological aspects: pharmacodynamics and pharmacokinetics.") This gene encodes a protein involved in aging and is considered as a potential factor of longevity (Signorelli and Ghidoni, 2005; Singh et al., 2011). However, evidence that RV has a life-prolonging property in humans remains to be demonstrated.

Biological actions

Resveratrol and the cell membrane

RV exerts its actions after entering the cell medium. This gives rise to a couple of questions. How does RV enter the target cell? How does it bypass the highly selective filter constituted by the cell membrane? Albumin, for instance, could act as a good transporter since this protein is able to bind amphiphilic molecules and, in addition to RV, it interacts with other natural substances such as genistein and curcumin (Bourassa, 2010). In addition, studies demonstrate that RV is able to bind one of the major human plasma proteins. However, the puzzling result is the amount of bound drug is inversely proportional to its plasma concentration, which in purely biochemical terms is rather difficult to explain. The entry of RV has been suggested to occur through a clathrin-independent process of endocytosis. However, lipid rafts seem to play a role since agents that damage their functionality also inhibit penetration of the drug into the cell (Jiang, 2008; Colin et al., 2011). The authors investigated the mechanism of RV entry into the cell by electrorotation: a biophysical approach allowing a single-cell analysis of the membrane function (Bonincontro and Risuleo, 2015). Our data show prompt uptake of the drug by the cell population, whereupon no relevant block in the G1 phase occurs (Risuleo, 2016; Risuleo and La Mesa, 2019a,b, and references therein).

Antioxidant properties

Tobacco smoking, pollution, hypercholesterolemia, hypertension, as well as other conditions (e.g., diabetes and obesity) either directly or indirectly lead to cardiac/coronary conditions. Atherosclerosis and chronic artery inflammation depend on lifestyle and environmental conditions. The complex genesis of atherosclerosis ends in damage to the vasal endothelium, deposit of the low-density lipid fraction (LDL) on the blood vessel walls and, finally, in the aggregation of platelets. Due to its very high antioxidant character, RV may play a protective role in this phenomenon. For example, treatment of bovine aorta cells with RV delays the formation of the atheroma and may act synergistically with PDGF, a mitosis-stimulating factor. Lipid peroxidation of unsaturated fatty acids or their esters is a direct consequence of the formation of ROS. This oxidation is a primary event in the development of atherosclerosis. RV inhibits oxidation of the LDL fraction circulating in the blood, thus limiting the formation of ROS. As a consequence, the cytotoxicity induced by LDL peroxidation, as well as the accumulation of intracellular calcium, is reduced by the drug. The final result is the inhibition of caspase-3, an enzyme that plays a crucial role in the execution of apoptosis (Araim et al., 2002; Matos et al., 2012; Shalini et al., 2015).

Antitumor activity in mammals

The main feature of a transformed tumor cell consists of unregulated proliferation that determines a significant increase in growth rate. This may originate from loss of cell cycle control and/or evasion from the apoptotic mechanisms. These two phenomena essentially keep in check cell differentiation and magnitude of the population. Many factors are implied in this metabolic dysregulation. Cyclins are a family of protein factors that regulate the correct execution and completion of the cell cycle. If a block in the G1 phase takes place, the apoptotic pathway is activated (Ahmad et al., 2001). Data have also demonstrated that the hindrance of progression towards the S phase is a determining trigger of apoptosis (Liao et al., 2010). RV has been shown to modulate the cdk-cyclin-dependent mechanisms. The molecule has been also involved in the development of human colorectal cancer. Studies in a mouse model of epidermal tumor in the mammary glands demonstrated that RV counteracts tumorigenesis, increases the antitumor activity of the mTOR inhibitors, and represses the activation pathway of tumor-stimulating factors in mammary tumor cell lines (Delmas et al., 2011; He et al., 2011). The gene product NF-κB is involved in the regulation of a vast number of protein factors modulating very complex biological phenomena such as inflammation, cell proliferation, and carcinogenesis. It has been suggested that RV inhibits the action of this factor (Singh et al., 2011).

Resveratrol and cell death

The strong antioxidant activity of RV, with consequent protection from cell damage, is common knowledge.

However, the drug may also paradoxically induce apoptosis and autophagy. These effects, monitored in a time- and concentration-dependent mode, were demonstrated in different models of cell cultures. Cytofluorimetric analyses also showed activation of the apoptotic marker factors and the occurrence of typical apoptotic morphofunctional alterations (Zhang et al., 2012). The preferential target of RV seems to be represented by the transformed cells (Berardi et al., 2009; Hasima and Ozpolat, 2014; Zhang et al., 2012; Risuleo, 2016). As far as the cell cycle is concerned, RV can induce an arrest in the S phase in murine preadipocytes and a significant increase in the intracellular concentration of lactic dehydrogenase, a hallmark of apoptosis. However, the paradoxical effects of the drug, which can protect the cell or induce its death and its synergistic effects with diverse antioxidants, are a clear sign of the complex interactions of the drug with the target cells. A number of studies have demonstrated that this behavior is shared by other natural products (De la Lastra and Villegas, 2007; Ricci et al., 2009; Chen et al., 2011; Ullah et al., 2012; Wang et al., 2011).

Antiviral activity

The first evidence of the antiviral action of RV is about two decades old (Docherty et al., 1999) and demonstrated that the drug was able to inhibit of the replication of herpes simplex virus. Subsequently, its beneficial action was shown in a mouse model system of vaginal infection (Docherty et al., 2005). The same authors later demonstrated RV-dependent inhibition of varicella-zoster DNA replication in human fibroblasts. This action was exerted in the early phases of the infection, mainly at the level of viral mRNA transcription (Docherty et al., 2006). However, this indirect antiviral action was ascribable to the inhibition by RV of the formation of the NF-κB dimeric complex, an essential prerequisite DNA synthesis. A similar observation was made in the case of influenza A virus, where the proteins necessary for the assembly and maturation of the viral capsid after cytoplasmic synthesis are not efficiently transported to the nucleus (Gregory et al., 2004; Palamara et al., 2005). A possible role of RV in the control of the complications after infection by the H1N-virus due to RV-mediated neutralization of the superoxide anion produced by macrophages has also been suggested. RV also seems to negatively influence the vitality of lymphoma cells infected with the Epstein-Barr virus (De Leo et al., 2012), which is involved in different human pathologies (Tooze, 1981). RV may also have a role in the control of HIV infection in humans since it potentiates the action of antiviral agents used in the treatment of AIDS. However, due to the extremely complicated pathological picture of the syndrome, this claim deserves further and deeper evaluation. RV has shown strong antiviral properties. It inhibits de novo viral DNA synthesis in murine fibroblasts and in human promyelocytic leukemia cells. The inhibition occurs with noncytotoxic doses of the drug. At these dosages, viral adsorption and virion entry into the cell are seemingly not affected. These observations purport the idea that the cell membrane, not the virus itself, is the target of the drug (Ricci et al., 2009; Aiello et al., 2011; Risuleo and La Mesa, 2019a,b, and references therein) (Tables 33.2 and 33.3) (Risuleo and La Mesa, 2019a,b).

Pharmacological aspects: pharmacodynamics and pharmacokinetics

The pharmacological aspects of RV are indeed complicated by the very high number of potential intracellular targets of this drug. RV has been identified as a pan-assay interference compound which produces positive results in many different laboratory assays. Therefore some caution is necessary when analyzing the ample spectrum of actions exhibited by RV. In relation to the multitarget role of RV, the authors recall that chemical compounds giving false-positive results are defined as pan-assay interference compounds and are known by the acronym PAINS. They often react nonspecifically with biological targets other than those desired, and therefore they tend to react with numerous nonspecific biological targets within the cell (Baell, 2016). Common PAINS, in addition to RV, include other closely related chemical compounds such as toxoflavin, isothiazolones, hydroxyphenyl hydrazones, curcumin, phenol-sulfonamides, rhodanines, enones, quinones, and catechols. One way to rationalize the multitarget effect of RV points to its ability to interact with the cell membrane. This could elicit a cascade of effects in other districts of the cell not directly involved with the membrane. These have been identified as hormone receptors, apoptotic factors, and cell cycle regulators (Ingólfsson et al., 2014; Bonincontro and Risuleo, 2015; Vang, 2015). The cell membrane allows the cytoplasm matrix to communicate with the outer world and acts as a highly selective filter. Biological membranes participate in many transport mechanisms, such as passive osmosis and diffusion. Therefore their integrity plays a fundamental role in numerous cell functions such as maintenance of basal homeostasis and cell differentiation as well as development via "filtering" paracrine signals. With respect to this, permeability of the cell membrane is of crucial importance in the recognition of apoptogenic external and/or intrinsic death signals. Therefore any agent affecting the efficiency and function of the cell membrane could play a binary role: help cell survival and correct proliferation or trigger the damage/death mechanisms (Mattetti and Risuleo, 2014).

TABLE 33.2 Resveratrol in various wine grapes and cultivars. The top four types are aboriginal Italian cultivars.

Grape cultivar	*Trans*-resveratrol (mg/L)
Aglianico	2.1
Piedirosso	2.4
Nerello Mascalese	2.5
Recioto[a]	0.1
Chardonnay	0.2
Bordeaux	0.05–0.08
Cabernet Sauvignon	0.09–2
Beaujolaes	3.2–3.6
Pinot Noir	0.4–2

[a]*Technically speaking, Recioto is not a cultivar but a grape-harvesting method. The berries are partially dried to concentrate the sugar (alcohol) level. Also, particular organoleptic properties of the wine are thus obtained.*

TABLE 33.3 Biological activities of resveratrol.

Action/role	System(s)
Aging	AMS, HS
Autophagy/cell death	CCS, AMS
Brain	HS, AMS

AMS, animal model systems; *CCS*, cell culture systems; *HS*, humans. Source: Redrawn and simplified from Risuleo G., La Mesa C., 2019a. Nanoparticles and molecular delivery: state of the art and future perspectives. In: Nutraceuticals in Veterinary Medicine. Gupta R., Srivastava A., Lall R. (Eds.) Springer Nature, Cham, pp. 737–747; Risuleo, G., La Mesa, C., 2019b. Resveratrol: biological activities and potential use in human and veterinary medicine. In: Nutraceuticals in Veterinary Medicine. Gupta R., Srivastava A., Lall R. (Eds.) Springer Nature, Cham, pp. 215–225.

Other studies conducted in vitro indicate that RV activates the sirtuin genes. These proteins influence a plethora of cellular processes such as aging, transcription, apoptosis, inflammation, and stress resistance, as well as energy efficiency and alertness in situations requiring an enhanced energy output. RV seems to also stimulate the production of SOD-2 and GPER activity, features that are coherent with its antioxidant activity and its interaction with the cell membrane. As a matter of fact, G proteins are guanine-binding proteins. They constitute a protein family that acts as molecular switches and are involved in the transmission of signal from a variety of stimuli from the outer environment to the interior of a cell. Interestingly, in humans, GPER interacts with the estrogen receptor-1 GPER and is activated by the female sex hormone. It has also been shown that RV acts in vitro as an agonist of peroxisome proliferator-activated receptor gamma, a nuclear receptor with pharmacologically interesting potentials as an agent for the treatment of type II diabetes. Buccal delivery without swallowing is one of the most common routes of RV administration. This involves direct absorption through the mucous membranes of the mouth. One milligram of RV in solution (50% alcohol/water) retained in the mouth for 1 minute can be promptly monitored (2 minutes later) in the plasma (37 ng/mL of free drug). Native RV is found at the same concentration after administration in a 250 mg tablet. However, the solubility of RV in water limits the amount of compound that can be absorbed through the buccal mucosa. Therefore the viability of the buccal delivery method is questionable and the efficacy of RV buccal delivery remains to be evaluated. About 70% of RV is absorbed after oral administration. However, its oral bioavailability is approximately 0.5% due to extensive chemical modifications undergone in the liver, which implies glucuronidation and sulfation of the drug. RV is extensively metabolized in the liver and lungs. The drug was detected also in the human cerebrospinal fluid after a study involving oral administration of 500 mg over a period of 13 weeks. The drug is able to cross the blood–cerebrospinal fluid barrier, even though the consequences of this transport have yet to be fully understood. It should be added that the production of the RV proprietary formulation SRT501 was discontinued by GlaxoSmith and Kline since "the company decided to terminate the Phase 2 trial of SRT501 in multiple myeloma and halt development of the drug as a potential myeloma treatment. The SRT501 formulation of RV may only offer minimal efficacy, while increasing the chances of kidney failure" (https://myelomabeacon.org/news/2010/11/30/glaxosmithkline-halts-all-further-development-of-resveratrol-drug-srt501/).

Adverse effects

Resveratrol is generally well-tolerated, although the number of studies on its effects in humans is rather limited. However, taking a 1000 mg dose per day provoked a rash that was resolved after discontinuation. Blood pressure seemed to be affected but this side effect also disappeared after discontinuing administration of the drug. In four published trials, discomfort in bowel movements and loose stools in the first month of treatment were reported. These problems seem to be confirmed after a 1-year-long phase II trial in AD patients. The most frequent adverse effects were diarrhea, weight loss, and nausea. These effects do seem to be very important within the general potential benefits derived from a controlled treatment with RV. For general aspects, action, and possible applications of resveratrol discussed in the present and previous section, the reader is referred to https://en.wikipedia.org/wiki/Resveratrol, which also contains detailed and relatively updated literature references.

After this discussion about the biomolecular and physiological properties of resveratrol, questions remain to be answered. Can the multifaceted "therapeutic" efficacy be improved and mediated by binding the drug to molecular carriers? How can penetration and intracellular survival be extended? Finally, is it possible to construct nanomolecular "cargo" vehicles able to deliver RV to the right cell district, thus improving its efficacy? Tentative answers will be given in the next section.

Nanomolecular cargos for the delivery of biological molecules

We focus mainly on three different kinds of nanoparticle for the delivery of exogenous material within a living cell:

- Cationic liposomes;
- Cat-anionic vesicles;
- Single-walled CNT.

It should be pointed out that the number, nature, and sophistication of molecular carriers are very diversified and may serve different purposes. Figs. 33.5 and 33.6 summarize some structural features of these supramolecular aggregates.

Liposomes

Due to their biocompatibility, liposomes are among the most studied drug-delivery systems. The possibility of targeting them in a selective manner to different tissues and the favorable cost/production ratio play an important role in their application in advanced medicine. Several physicochemical parameters determine the validity of liposomes as cargo particles for the transport of specific molecules. Since they are endowed with a net positive charge, they are able to interact with negatively charged cell membranes and nucleic acids. Liposomes have been shown to act as valid vehicles of macromolecules such as DNA, RNA, and proteins, but they can also function as cargo carriers of small molecules like hormones, natural compounds, and/or drugs in general. Their ability as carriers of nucleic acids and vaccine carrier/adjuvants has

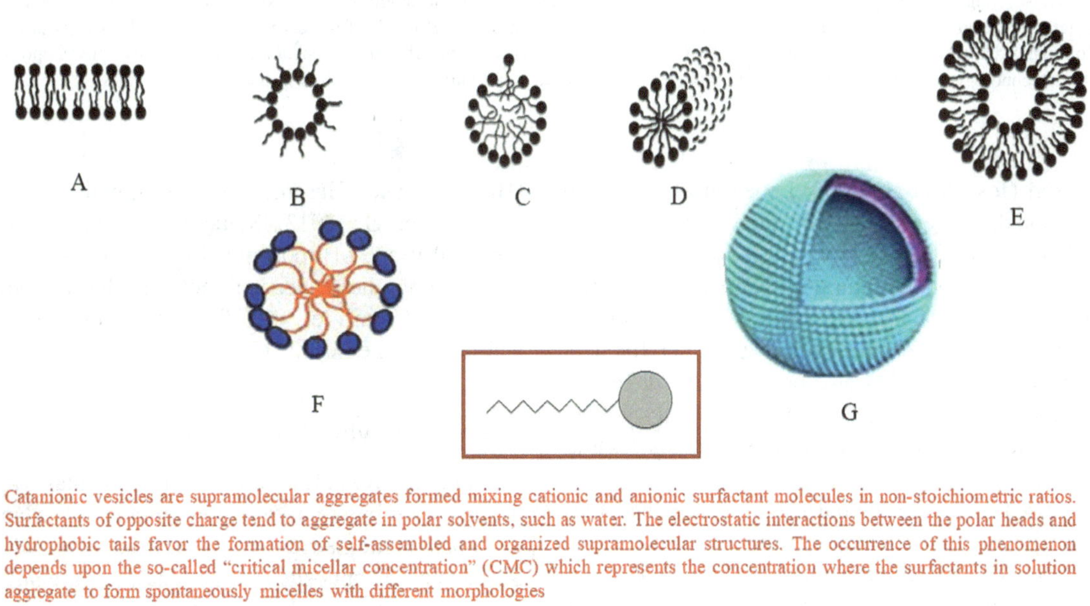

FIGURE 33.5 Phospholipid bilayers. These molecular aggregates form a flat polar membrane made of two layers of lipid molecules. The thickness of the bilayer is only a few nanometers range. Bilayers are not permeable to ions, therefore the internal salt concentration and pH are finely regulated by transporting ions across the membranes by proteins contributing to the establishment of the ion pumps. The cytoplasmic, nuclear, and endoplasmic reticulum membrane are made of a lipid bilayer (but also viruses, described as enveloped, may also be surrounded by a membrane). The lipid bilayer is the barrier for filtering macro- and small molecules. Transit from the outer environment to the inside of a cell is dictated by precise and metabolically regulated processes which allow accurate maintenance of cell homeostasis. (Upper row) Different shapes of self-assembled aggregates: (A) double layer, (B) spherical micelles, (C) wormlike, (D) vesicles, (E) "inverted" vesicles, (F) scheme of a surfactant aggregate: note the polar heads are located on the outer face of the aggregate, (G) pictorial representation of a vesicle suited for molecular biology and nanobiotechtechnology. The inner aqueous lumen may be loaded with molecules for delivery. Inset: Schematic structure of a typical surfactant molecule. The zig-zag-like line represents the hydrophobic nonpolar tail, while the circle is the polar head.

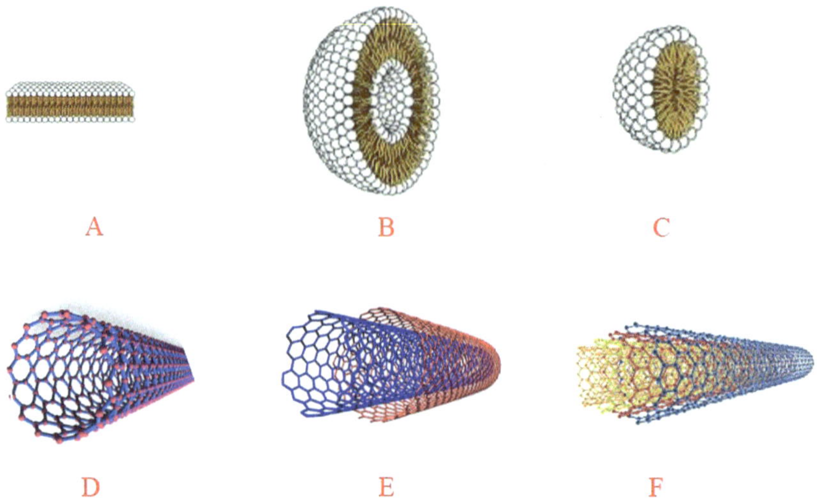

FIGURE 33.6 Liposomes, micelles and nanotubes. (A) Phospholipids exposed to water aggregate spontaneously, forming a two-layered sheet. The hydrophobic tails points to the internal face, away from the aqueous surrounding. The center of this bilayer contains neither water nor water-soluble molecules. The assembly process is driven by interactions between hydrophobic molecules and includes noncovalent interactions such as van der Waals forces, electrostatic and, to a lesser extent, hydrogen bonds. (B) Liposomes are spherical vesicles formed by one or more lipid bilayer(s). They are usually made of phospholipids, phosphatidylcholine being a major component, and for instance phosphatidylethanolamine, a natural lipid present in eggs: the essential prerequisite for a lipid to be part of a liposome is that it must be able to interact with the lipid bilayer structure. Liposomes have a very wide range of uses, from targeted drug delivery to dietary food supplements and cosmetics. (C) Micelles are formed by supramolecular aggregations of surfactants (see inset in Fig. 33.5) dispersed in a colloid form. Micelles have a *quasi*-spherical shape. Micellization, that is the formation of micelles, occurs when the concentration of surfactant is greater than the critical micelle concentration (CMC). When surfactants are present above the CMC, they can act as emulsifiers. This will cause the solubilization of a compound that normally would be insoluble in a particular solvent. (D–F) Carbon nanotubes (also known by the acronym CNTs) are tubes made of carbon with nanometric diameter. They can be either single- or multiwalled structures consisting of nested tubes. Since carbon is invariably present in these superstructures, their acronyms are often rationalized as NT, SWNT and MWNT (D–F, single-, two-layer and three-layer nanotubes, respectively). Carbon nanotubes show very good electrical conductivity and have remarkable tensile strength, essentially due to their nanosize as well as the length and strength of the carbon–carbon bonds. They can be chemically modified following relatively simple chemical protocols. These features render them superb tools in technological areas: electronics, optics, composite materials, nanotechnology, and other applications of materials science. Their importance in biomedicine is also of primary interest.

been assessed (Joseph et al., 2006; Lonez et al., 2008). It should be noted, however, that the type and amount of lipid in their formulation is crucial. Cationic lipids, for instance, are not found in naturally occurring liposomes, but they may be chemically synthesized. The possibility of using this type of liposome as a drug carrier for neoplastic and other diseases has also been examined. In particular, cationic liposomes are able to specifically deliver their payload to embryologically different tissues such as lung, liver, and gastrointestinal tract tumors, as well as to neoplastic diseases of endothelial origin. Since they may trespass the gastric environment, they have become a very attractive tool for oral administration in cancer therapy. The interaction of this type of liposome does not interfere with the morphofunctional features of the target cells as observed by optical and SFM imaging, as well as in NMR-metabolomic studies (Piccioni et al., 2007).

Finally, the antibacterial action of cationic liposomes has been also examined in Gram-negative or antibiotic-resistant pathogenic microorganisms. Liposomes may in fact increase bacterial membrane permeability, which consequently causes higher susceptibility to drug uptake (Hamblin and Hasan, 2004; Bombelli et al., 2008; Cosimati et al., 2013; Stefanutti et al., 2014). For the potential usage of liposomes in antiproliferative (anticancer) therapy, the reader is directed to the following recent reviews: Heidarli et al. (2017), Olusanya et al. (2018), and Deshantri et al. (2018).

Cat-anionic vesicles

Cat-anionic vesicles are supramolecular aggregates formed by mixing cationic and anionic surfactant species in nonstoichiometric ratios (Letizia et al., 2007). Surfactants of opposite charge tend to aggregate in aqueous polar solvents. The electrostatic interactions between the polar heads and hydrophobic tails favor the formation of self-assembled and organized supramolecular structures. An extensive analysis of the parameters governing formation and morphology can be found in the pioneering work by Israelachvili and collaborators (Israelachvili et al., 1976; Israelachvili et al., 1977). Vesicles may easily interact with DNA, RNA, or other biopolymers. Their behavior with respect to this particular aspect recalls the above-discussed liposomes. In this case, the complexes of vesicle/macromolecule, commonly defined as lipoplexes, assume fundamental importance in biotechnological and biomedical applications. Their intrinsic structural and functional properties grant them the capacity to deliver genetic material across the cell membrane via plasma–membrane fusion and/or endocytosis. Work from our laboratory showed that it is possible to form complexes between DNA and vesicles with an excess positive charge at the surface, which could be potentially delivered within the cell (Bonincontro et al., 2007; Bonincontro et al., 2008). However, due to their biocompatibility, some *caveats* are to be considered. Vesicles may show a cytotoxic effect, which is directly related to time of exposure and dose of administration as well as to the nature of the cat-anionic moieties (Kuo et al., 2005; Lozano et al., 2011). In particular, work from our laboratory showed that SDS-DDAB are far more cytotoxic than those formed with SDS and CTAB as a hydrophobic counterion. One should also always evaluate the toxicity of the separate components as compared to the assembled particle (Russo et al., 2013). Furthermore, the SDS-CTAB cat-anionic complexes are able to "hit" tumor cells more efficiently than normal mouse fibroblasts, which show a higher survival/proliferation rate as compared to the previous cell population (Aiello et al., 2010). However, despite the intrnsic cytotoxicity of cat-anionic vesicles it is possible to adjust the experimental conditions such as time and dose to allow successful vesicle-mediated transfection with subsequent expression of exogenous genetic material (Loftsson et al., 2005; Muzzalupo et al., 2017). In this respect, one main interest in the use of nanoparticles as molecular carriers of genetic material consists in ascertaining the fate of this material, be it DNA or RNA, upon entry in the recipient cell, that is, is the transfected macromolecule stable? Does it maintain its functional features? Can the genetic features encoded within the macromolecule be transformed into the mature functional product? The authors addressed these questions by transfecting messenger RNA into recipient cells and measuring the level of the encoded gene product by standard CAT assays. The sharp and clear-cut result is that the transfected cat-anionic/RNA lipoplex is efficiently metabolically translated into protein which exhibits functional epitopes, promptly recognized by the cognate antibody as shown by ELISA assays (for a study of the effects of nanoparticle-mediated gene transfer on the overall cell metabolism, see: Aiello et al., 2010; Russo et al., 2013;

TABLE 33.4 Acronyms indicating the species used to date to prepare cat-anionic vesicles.

Cationic	
Acronym	*Chemical name*
CTAB	Cetyltrimethylammonium bromide
TTAB	Tetradecyltrimethylammonium bromide
DTAB	Dodecyltrimethylammonium bromide
DDAB	Didodecildimethylammonium bromide
DODAB	Dioctyldimethylammonium bromide
AABS14	1,2-dimyristoyl-rac-glycero-3-O-(NR-acetyl-L-arginine) HCl
AABS12	1,2-dilauroyl-rac-glycero-3-O-(NR-acetyl-L-arginine) HCl
Anionic	
Acronym	*Chemical name*
SDS	Sodium dodecylsulfate
SOS	Sodium octylsulfate
AOT	Sodium bis-2-ethylhexylsulfosuccinate
SL	Sodium laurate
KPFH	Potassium perfluorohexanoate
TEAPFOS	Tetraethylammonium perfluorooctansulfonate
DPPA	Sodium 1,2-dipalmitoyl-sn-glycero-3-phosphate
STDC	Sodium taurodeoxycholate

Cosimati et al., 2013; Stefanutti et al., 2014) Finally, pilot studies indicate that vesicles may be utilized in anticancer therapy (Wright, 2008; van der Pol et al., 2012; Freire et al., 2015). Common surfactant molecules used in the preparation of vesicles are given in Table 33.4.

Nanotubes

Carbon NanoTubes (CNT) are rolled graphene sheets that form cylinders or quasicylindrical structures. These are found in single- or multiwalled concentric CNT shapes (Fig. 33.6). Nanotubes provide numerous potential applications in biochemistry, nanomedicine, pharmacology, and industry, such as avionics and space engineering (Prato et al., 2008; Ménard-Moyon et al., 2010; Venkatesan et al., 2014; Shtansky et al., 2018; Lo et al., 2018; Krishna et al., 2018; Mohajeri et al., 2018). As with other nanoparticles, CNT biocompatibility must be carefully assessed both in cell culture systems and animal models. The ascertainment that they are not toxic and do not elicit immune responses is very important in this case (Bianco et al., 2005; Bianco, 2013). CNT are poorly dispersible in aqueous solution. They establish van der Waals interactions leading to aggregation in large bundles which may be detrimental to cell membrane integrity and/or may hinder their passage across it. In the cell this represents the barrier encountered by the nanotube upon cell entry (Holt et al., 2010). Several dispersant molecules improve dispersibility, albumins in particular. Recent investigations conducted in our laboratories demonstrated that noncovalent carbon nanotube/BSA complexes have little if any effect on cell viability. These nanocomposites were administered to different cell lines such as murine fibroblasts (NH3T6), human embryonic kidney cells (HEK-293), or murine macrophages (RAW-294). No significant changes in cell survival were noted in 3T6 and HEK-293 cells after exposure to different concentrations of pristine or BSA-CNTs. A cytotoxic effect, on the contrary, was observed in the case of murine macrophages treated with both pristine and BSA-stabilized CNTs. Toxicity was diminished when CNTs were coated with BSA. This underscores the biocompatibility-enhancing property of the protein/nanotube complex. The good dispersion quality of these BSA-CNT complexes caused an efficient internalization in both phagocytic (macrophages) and nonphagocytic (fibroblasts) cell types. These complexes are able to enter the cells by passive diffusion or endocytosis, the latter being an active energy-dependent mechanism of assumption (Muzi et al., 2015; Muzi et al., 2016a). Furthermore, crossing of the plasma membrane by nanotube complexes causes no alterations of the dielectric parameters of the cell membrane as measured by electrorotation (Bonincontro and Risuleo, 2015).

Graphene

In the context of nanotube technology, the role of graphene cannot be ignored. Graphene is a carbon allotrope at the basis of nanotube technology. It consists of a one-atom-thick planar sheet of carbon atoms densely packed in a honeycomb crystal lattice. Graphene can display different forms which are conventionally combined under the so-called graphene family materials and include few-layer graphene (FLG), graphene oxide, reduced graphene oxide, graphene nanosheets, ultrafine graphite, graphene ribbons, and graphene dots (Fig. 33.7). FLG is made up of 2–10 graphene layers (Bianco et al., 2013) and it shares the basic structural element of other carbon allotropes, including CNT and fullerenes, but its history is quite recent. However, since its discovery, graphene has been the object of exciting interest in many fields of basic science and industry, mainly due to its extraordinary features.

Today, graphene-based nanotechnology represents an area of scientific research and industrial applications in rapid expansion (Liang and Chen, 2010). Graphene was first exploited in the material sciences but recently has become a very good tool in electronics, photonics, composite materials, energy generation, energy storage, and sensors, as well as in biological applications. FLG is easily produced at high yield and relatively low cost (Novoselov et al., 2012). A new caveat is that the rapid increase in production and applications of FLG implies its potential release into the environment, especially into the aquatic areas that are generally considered as a fragile and sensitive macrosphere. With respect to this, the release of FLG could occur accidentally from inadequate or uncontrolled disposal of such commercial products and the chemical modification and degradation following the waste. Possible ecosystem risks induced by FLG have been examined (Hu and Zhou, 2013). A recent review also discusses the toxicity of GFMs toward the aquatic environment including bacteria, crustaceans, nematodes, and fish (Zhao et al., 2014). A generally accepted idea is that the toxicity of FLG against aquatic organisms may not be as high as carbon nanomaterials such as fullerenes and CNT. However, the results are controversial. In fact, the biological response often depends on the intrinsic structural nature of graphene, such as the number of layers, rigidity, hydrophobicity, dose, and purity of the material under scrutiny, as well as the use of diverse cell/organism models (Bianco, 2013). Results from our laboratory on *Xenopus laevis* larvae indicate that FLG is substantially nontoxic (Muzi et al., 2016b). This seems to apply also to organisms belonging to diverse evolutionary levels, like *Pseudomonas aeruginosa*, *Caenorhabditis elegans*, *Vibrio fischeri*, and *Artemia salina* (Gollavelli and Ling, 2012; Zanni et al., 2012; Guo et al., 2013; Pretti

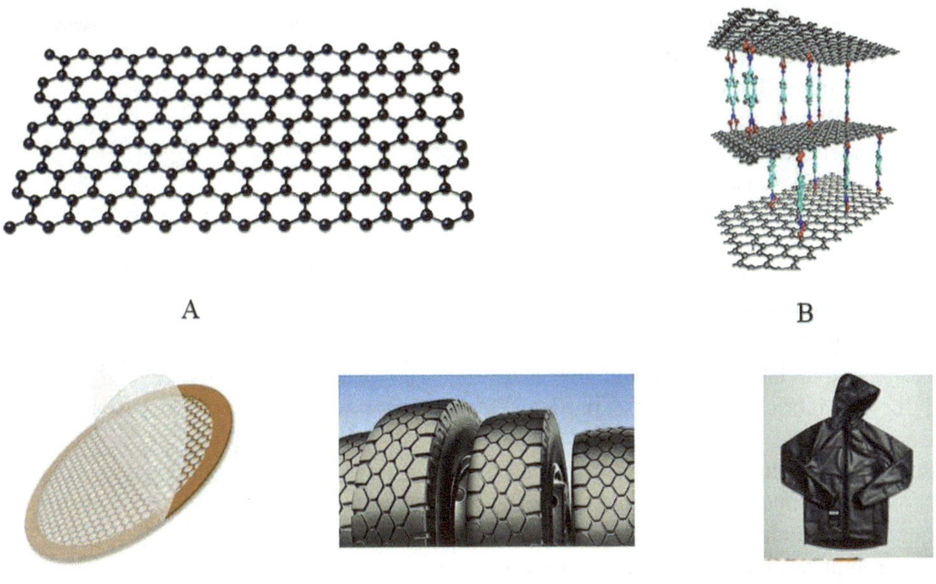

FIGURE 33.7 Graphene sheets and some of their potential uses in science and everyday life. Graphene is one of carbon's allotropic forms. Its simplest structure consists of a single layer of atoms in a two-dimensional hexagonal lattice, also known as chicken mesh. It has particular properties which make it different from other carbon allotropes. For example, it is about 100 times stronger than a steel sheet of the same thickness, but its density is significantly lower, with a surface mass of 0.763 mg/m^2. It is almost transparent and is a very good heat and electricity conductor. This may be attributable to the very short carbon–carbon bond length (about 0.140 nm). It may form single (A) or multiple layers (B). The range of technological applications is extremely vast and goes from the constructions of grids for transmission electron microscopy (C), from vehicle tires to clothing (D, E), including armored gear for security and military usages, such as bullet-proof vests and eye-glasses as well as aeronautic and drone contraptions. A somewhat extravagant potential application was the production of extremely thin, highly resistant, and impenetrable condoms. This project was funded by the Bill and Melinda Gates Foundation, although the actual feasibility of making condoms from pure graphene is disputed (https://www.extremetech.com/extreme/171417-bill-gates-funds-creation-of-thin-light-impenetrable-graphene-condoms).

et al., 2014). In particular, it has been evidenced that in a crustacean FLG, it did not induce oxidative stress. In *Xenopus* larvae, the growth rate seemed to be affected by the material to some extent. Larval size was decreased by about 40% only upon exposure to the highest concentrations of FLG, which recalls the effect on monitored larvae of the same organism exposed to CNT (Muzi et al., 2016a,b). This phenomenon has been attributed to digestive and respiratory obstructions due to the uptake of the carbon-based nanomaterial (Mouchet et al., 2011).

Concluding remarks and future directions

A very relevant aspect of the works discussed here is in regards to the interdisciplinary character of the nanotechnological approach. A number of diversified areas of expertise are required: molecular biology, physical chemistry, biophysics, and consequences of biotechnological strategies, as well as marketing. Hence, socioeconomic aspects cannot be disregarded even though this does not directly fall within the scope of this work. To this end, we have tried to give a glimpse into the potential of natural substances, with particular reference to resveratrol, and the possibility of delivering these drugs via supramolecular aggregates. The features of different nanocomposites with different physicochemical features, such as liposomes, vesicles, nanotubes, and graphene sheets, have been illustrated. As far as exposure to RV is concerned, control of cell death and senescence seem to be the most relevant effects. This action is attributable to RV's antioxidant properties that are exerted essentially through its interaction with the cell membrane. Paradoxical effects have been observed and therefore its use as a multitarget therapeutic means should be carefully evaluated. Resveratrol passes through the membrane without causing serious consequences for cell viability and survival. However, its solubility in aqueous media and bioavailability are limited. This may reduce the use of this natural substance for therapeutic purposes. Therefore new methods of delivering the drug have become a primary goal in order to control any possible adverse effects. Delivery mediated by nanoparticles such as liposomes, vesicles, or nanotubes may represent the new method for safe and efficient dispatch of molecules endowed with therapeutic capacities. In this respect, the administration of pristine and chemically modified nanocomposites was examined in cultured cells. In the case of graphene sheets, effects also were investigated in *X. laevis* larvae. These nanoparticles can be considered as a safe and efficient tool. For the time being their application is only experimental, but their future large-scale application seems to be realistic. Nanoparticles are internalized within the target cells where they deliver their molecular cargo. Little, if any, permanent damage is inflicted to the cell membrane, which is the first barrier encountered by a nanocomplex upon cell entry. Therefore it is reasonable to assume that the overall cell metabolism and survival suffer little if any significant damage. This renders nanoparticles very attractive in advanced biomedicine. The variegated field of nanotechnologies is continuously growing, and it is no easy task to predict what its future developments will be. Despite the relatively simple laboratory techniques and low cost required for the fabrication of these particles, two questions arise. Will it be possible to apply these technologies in disadvantaged situations, such as in developing countries? Will personnel with adequate backgrounds be found in those unfavorable conditions? These questions deserve a prompt answer and when required a remedy, which as the authors envision, is unfortunately of a sociopolitical nature.

References

Ahmad, N., Adhami, V.M., Afaq, F., et al., 2001. Resveratrol causes WAF-1/p21-mediated G(1)-phase arrest of cell cycle and induction of apoptosis in human epidermoid carcinoma A431 cells. Clin. Cancer Res. 7, 1466–1473.

Aiello, C., Andreozzi, P., La Mesa, C., Risuleo, G., 2010. Biological activity of SDS-CTAB cat-anionic vesicles in cultured cells and assessment of their cytotoxicity ending in apoptosis. Colloids Surf. B Biointerfaces 78, 149–154.

Aiello, C., Berardi, V., Ricci, F., Risuleo, G., 2011. Chapter 96-Biological properties of a methanolic extract of neem oil, a natural oil from the seeds of the Neem Tree (Azadirachta indica var. A. Juss). In: Preedy, V.R., Watson, R.R., Patel, V.V. (Eds.), Nuts and Seeds in Health and Disease Prevention. Academic Press/Elsevier, London, ISBN: 978-0-12-375688-6, pp. 813-821.

Anwar, H., Hussain, G., Mustafa, I., 2018. Antioxidants from natural sources. In: Shalaby, E. (Ed.), Antioxidants in Food and Its Applications. IntechOpen Limited, London, Chapter 1 eBook. doi: 10.5772/intechopen.72008.

Araim, O., Ballantyne, J., Waterhouse, A.L., et al., 2002. Inhibition of vascular smooth muscle cell proliferation with red wine and red wine polyphenols. J. Vasc. Surg. 35 (6), 1226–1232.

Baell, J.B., 2016. Feeling nature's pains: natural products, natural product drugs, and pan assay interference compounds (PAINS). J. Nat. Prod. 79 (3), 616–628.

Barsanti, L., Gualtieri, P., 2006. Algae: Anatomy, Biochemistry, and Biotechnology. CRC Press, Boca Raton, ISBN 978-0-8493-1467-4.

Benedetti, S., Benvenuti, F., Pagliarani, S., et al., 2004. Antioxidant properties of a novel phycocyanin extract from the blue-green alga *Aphanizomenon flos-aquae*. Life Sci. 75 (19), 2353–2362.

Berardi, V., Ricci, F., Castelli, M., Galati, G., Risuleo, G., 2009. Resveratrol exhibits a strong cytotoxic activity in cultured cells and has an antiviral action against polyomavirus: potential clinical use. J. Exp. Clin. Cancer Res. 28, 96–105.

Bianco, A., 2013. Graphene: safe or toxic? The two faces of the medal. Angew. Chem. Int. Ed. Engl. 52, 4986–4997.

Bianco, A., Kostarelos, K., Prato, M., 2005. Applications of carbon nanotubes in drug delivery. Curr. Opin. Chem. Biol. 9, 674–679.

Bianco, A., Cheng, H.-M., Enoki, T., et al., 2013. All in the graphene family—a recommended nomenclature for two-dimensional carbon materials. Carbon 65, 1–6.

Bombelli, C., Bordi, F., Ferro, L.S., et al., 2008. New cationic liposomes as vehicles of m-tetrahydroxyphenylchlorine in photodynamic therapy of infectious diseases. Mol. Pharm. 5, 672–679.

Bonincontro, A., Risuleo, G., 2015. Electrorotation: a spectroscopic imaging approach to study the alterations of the cytoplasmic membrane. Adv. Mol. Imag. 5, 1–15.

Bonincontro, A., La Mesa, C., Proietti, C., Risuleo, G., 2007. A biophysical investigation on the binding and controlled DNA release in a cetyltrimethylammonium bromide-sodium octyl sulfate cat-anionic vesicle system. Biomacromolecules 8, 1824–1829.

Bonincontro, A., Falivene, M., La Mesa, C., Risuleo, G., 2008. Dynamics of DNA adsorption on and release from SDS-DDAB cat-anionic vesicles: a multitechnique study. Langmuir 24, 1973–1978.

Bourassa, P., 2010. Resveratrol, genistein, and curcumin bind bovine serum albumin. J. Phys. Chem. B 114, 3348–3354.

Brahmachari, G., 2004. Neem—an omnipotent plant: a retrospection. Chembiochem 5 (4), 408–421.

Carmichael, W.W., Drapeau, C., Anderson, D.M., 2000. Harvesting of *Aphanizomenon flos-aquae* Ralfs ex Born. & Flah. var. *flos-aquae* (Cyanobacteria) from Klamath lake for human dietary use. J. Appl. Phycol. 12 (6), 585–595.

Chen, S., Xiao, X., Feng, X., et al., 2011. Resveratrol induces Sirt1-dependent apoptosis in 3T3-L1 preadipocytes by activating AMPK and suppressing AKT activity and survivin expression. J. Nutr. Biochem. 9.

Cho, M.H., Lee, S.W., 2015. Phenolic phytoalexins in rice: biological functions and biosynthesis. Int. J. Mol. Sci. 16 (12), 29120–29133.

Cichewicz, R.H., Kouzi, S.A., 2002. Resveratrol oligomers: structure, chemistry, and biological activity. In: Atta-ur-Rahman (Ed.), Studies in Natural Products Chemistry, 26. Elsevier Science B.V, pp. 507–557.

Colin, Dr, Limagne, E., Jeanningros-Arnaud, S., et al., 2011. Endocytosis of resveratrol via lipid rafts and activation of downstream signaling pathways in cancer cells. Cancer Prev. Res 4, 1065–1106. Available from: https://doi.org/10.1158/1940-6207.CAPR-10-0274.

Cosimati, R., Milardi, G.L., Bombelli, C., Bonincontro, A., Bordi, F., Mancini, G., et al., 2013. Interactions of DMPC and DMPC/gemini liposomes with the cell membrane investigated by electrorotation. Biochim. Biophys. Acta 1828, 352–356.

De la Lastra, C.A., Villegas, I., 2007. Resveratrol as an antioxidant and pro-oxidant agent: mechanisms and clinical implications. Biochem. Soc. Trans. 35 (Pt 5), 1156–1160. Nov.

De Leo, A., Arena, G., Lacanna, E., et al., 2012. Resveratrol inhibits Epstein Barr virus lytic cycle in Burkitt's lymphoma cells by affecting multiple molecular targets. Antiviral Res. 96 (2), 196–202.

de Lorgeril, M., Salen, P., 2005. Dietary prevention of coronary heart disease: the Lyon diet heart study and after. World Rev. Nutr. Diet. 95, 103–114.

Delmas, D., Solary, E., Latruffe, N., 2011. Resveratrol, a phytochemical inducer of multiple cell death pathways: apoptosis, autophagy and mitotic catastrophe. Curr. Med. Chem. 18 (8), 1100–1121.

Deshantri, A.K., Varela Moreira, A., et al., 2018. Nanomedicines for the treatment of hematological malignancies. J. Control. Release 287, 194–215.

Di Fonzo, G., Troilo, M., Squeo, G., et al., 2020. Functional compounds from olive pomace to obtain high-added value foods—a review. J. Sci. Food Agric. Available from: https://doi.org/10.1002/jsfa.104789.

Dixon, R.A., Paiva, N.L., 1995. Stress-induced phenylpropanoid metabolism. Plant Cell 7, 1085–1097.

Docherty, J.J., Fu, M.M., Stiffler, B.S., et al., 1999. Resveratrol inhibition of *Herpes simplex* virus replication. Antiviral Res. 43, 145–155.

Docherty, J.J., Fu, M.M., Hah, J.M., et al., 2005. Effect of resveratrol on herpes simplex virus vaginal infection in the mouse. Antiviral Res. 67, 155–162.

Docherty, J.J., Sweet, T.J., Bailey, E., et al., 2006. Resveratrol inhibition of varicella-zoster virus replication *in vitro*. Antiviral Res. 72, 171–177.

Echeverri, F., Torres, F., Quinones, W., et al., 1997. Danielone, a phytoalexin from papaya fruit. Phytochemistry 44 (2), 255–256.

Favaron, F., Lucchetta, M., et al., 2009. The role of grape polyphenols on trans-resveratrol activity against *Botrytis cinerea* and of fungal laccase on the solubility of putative grape PR proteins. J. Plant Pathol. 91, 579–588.

Ferrieres, J., 2004. The French paradox: lessons for other countries. Heart 90, 107–111.

Frautschy, S.A., Hu, W., Kim, P., et al., 2001. Phenolic anti-inflammatory antioxidant reversal of Abeta-induced cognitive deficits and neuropathology. Neurobiol. Aging 22, 993–1005.

Freire, J.M., Gaspar, D., Veiga, A.S., et al., 2015. Shifting gear in antimicrobial and anticancer peptides biophysical studies: from vesicles to cells. J. Pept. Sci. 21, 178–185.

Gjonça, A., Bobak, M., 1997. Albanian paradox, another example of protective effect of Mediterranean lifestyle? Lancet 350, 1815–1817.

Glazebrook, J., Ausbel, F.M., 1994. Isolation of phytoalexin-deficient mutants of *Arabidopsis thaliana* and characterization of their interactions with bacterial pathogens. PNAS 91, 8955–8959.

Godard, M.P., Johnson, B.A., Richmond, S.R., 2005. Body composition and hormonal adaptations associated with forskolin consumption in overweight and obese men. Obes. Res. 13, 1335–1343.

Gollavelli, G., Ling, Y.C., 2012. Multi-functional graphene as an in vitro and in vivo imaging probe. Biomaterials 33, 2532–2545.

Goufo, P., Singh, R.K., Cortez, I., 2020. A reference list of phenolic compounds (including stilbenes) in grapevine (*Vitis vinifera* L.) roots, woods, canes, stems, and leaves. Antioxidants (Basel) 9 (5), . Available from: https://doi.org/10.3390/antiox9050398pii: E398.

Gregory, D., Hargett, D., Holmes, D., et al., 2004. Efficient replication by herpes simplex virus type 1 involves activation of the IFkappaB kinase-IkappaB-p65 pathway. J. Virol. 78 (24), 13582–13590.

Guo, X., Dong, S., Petersen, E.J., et al., 2013. Biological uptake and depuration of radio-labeled graphene by *Daphnia magna*. Environ. Sci. Technol. 47, 12524–12531.

Gupta, S.C., Prasad, S., Tyagi, A.K., et al., 2017. Neem (*Azadirachta indica*): an Indian traditional panacea with modern molecular basis. Phytomedicine 34, 14–20.

Hamblin, M.R., Hasan, T., 2004. Photodynamic therapy: a new antimicrobial approach to infectious disease? Photochem. Photobiol. Sci. 3, 436–450.

Hart, J.H., Hillis, W.E., 1974. Inhibition of wood-rotting fungi by stilbenes and other polyphenols in *Eucalyptus sideroxylon*. Phytopathology 64, 939–948. Available from: https://doi.org/10.1094/Phyto-64-939.

Hasima, N., Ozpolat, B., 2014. Regulation of autophagy by polyphenolic compounds as a potential therapeutic strategy for cancer. Cell Death Dis. 5, e1509. Available from: https://doi.org/10.1038/cddis.2014.467.

He, X., Wang, Y., Zhu, J., et al., 2011. Resveratrol enhances the anti-tumor activity of the mTOR inhibitor rapamycin in multiple breast cancer cell lines mainly by suppressing rapamycin-induced AKT signaling. Cancer Lett. 301 (2), 168–176.

Heidarli, E., Dadashzadeh, S., Haeri, A., 2017. State of the art of stimuli-responsive liposomes for cancer therapy. Iran. J. Pharm. Res. 16, 1273–1304.

Holt, B.D., Short, P.A., Rape, A.D., et al., 2010. Carbon nanotubes reorganize actin structures in cells and ex vivo. ACS Nano 4, 4872–4878.

Hu, X., Zhou, Q., 2013. Health and ecosystem risks of graphene. Chem. Rev. 113, 3815–3835.

Ingólfsson, H.I., Thakur, P., Herold, K.F., et al., 2014. Phytochemicals perturb membranes and promiscuously alter protein function. ACS Chem. Biol. 9, 1788–1798.

Israelachvili, J., Mitchell, D.J., Ninham, B.W.J., 1976. Theory of self-assembly of hydrocarbon amphiphiles into micelles and bilayers. J. Chem. Soc., Faraday Trans. 2 Mol. Chem. Phys. 72, 1525–1568.

Israelachvili, J.N., Mitchell, D.J., Ninham, B.W., 1977. Theory of self-assembly of lipid bilayers and vesicles. Biochim. Biophys. Acta 470, 185–201.

Jeandet, P., Douillet-Breuil, A.C., Bessis, R., et al., 2002. Phytoalexins from the *Vitaceae*: biosynthesis, phytoalexin gene expression in transgenic plants, antifungal activity, and metabolism. J. Agric. Food Chem. 50, 2731–2741.

Jiang, J., 2008. Design, synthesis and spectroscopic studies of resveratrol aliphatic acid ligands of human serum albumin. Bioorg. Med. Chem. 16, 6406–6414.

Jiang, T., Liao, W., Charcosset, C., 2020. Recent advances in encapsulation of curcumin in nanoemulsions: a review of encapsulation technologies, bioaccessibility and applications. Food Res. Int. 132, 109035. Available from: https://doi.org/10.1016/j.foodres.2020.109035.

Joseph, A., Itskovitz-Copper, N., Samira, S., et al., 2006. A new intranasal influenza vaccine based on a novel polycationic lipid—ceramide carbamoyl-spermine (CCS): I. Immunogenicity and efficacy studies in mice. Vaccine 24, 3990–4006.

Koul, O., Wahab, S., 2004. Neem: today and in new millenniumIn: Koul, O., Wahab, S. (Eds.), Kluwer Academic Publishers, Dordrecht.

Krishna, V.D., Wu, K., Su, D., et al., 2018. Nanotechnology: of concepts and potential application of sensing platforms in food safety. Food Microbiol. 75, 47–54.

Kuo, J.H., Jan, M.S., Chang, C.H., et al., 2005. Cytotoxicity characterization of catanionic vesicles in RAW 264.7 murine macrophage-like cells. Colloids Surf. B Biointerfaces 41, 189–196.

Kushak, R.I., Drapeau, C., Van Cott, E.M., et al., 2000. Favorable effects of blue-green algae (*Aphanizomenon flos-aquae*) on rat plasma lipids (PDF). J. Am. Nutraceut. Assoc. 2 (3), 59–65.

Lamuela-Raventos, R.M., Romero-Perez, A.I., Waterhouse, A.L., et al., 1995. Direct HPLC analysis of *cis*- and *trans*-resveratrol and piceid isomers in Spanish Red *Vitis vinifera* wines. J. Agr. Food Chem. 43 (2), 281–283.

Letizia, C., Andreozzi, P., Scipioni, A., et al., 2007. Protein binding onto surfactant-based synthetic vesicles. J. Phys. Chem. B. 111, 898–908.

Liang, F., Chen, B., 2010. A review on biomedical applications of single-walled carbon nanotubes. Curr. Med. Chem. 17, 10–24.

Liao, P.C., Ng, L.T., Lin, L.T., et al., 2010. Resveratrol arrests cell cycle and induces apoptosis in human hepatocellular carcinoma Huh-7 cells. J. Med. Food. 13 (6), 1.

Lim, G.P., Chu, T., Yang, F., et al., 2001. The curry spice curcumin reduces oxidative damage and amyloid pathology in an Alzheimer transgenic mouse. J. Neurosci. 21 (21), 8370–8377.

Lo, C.T., Jahn, A., Locascio, L.E., et al., 2018. Controlled self-assembly of monodisperse niosomes by microfluidic hydrodynamic focusing. Langmuir 26, 8559–8566.

Loftsson, T., Jarho, P., Másson, M., et al., 2005. Cyclodextrins in drug delivery. Expert Opin. Drug Delivery 2, 335–351.

Lonez, C., Vandenbranden, M., Ruysschaert, J.-M., 2008. Cationic liposomal lipids: from gene carriers to cell signaling. Prog. Lipid Res. 47, 340–347.

Lozano, N., Perez, L., Pons, R., et al., 2011. Diacyl glycerol arginine-based surfactants: biological and physicochemical properties of catanionic formulations. Amino Acids 40, 721–729.

Maheshwari, R.K., et al., 2006. Multiple biological activities of curcumin: a short review. Life Sci. 78, 2081–2087.

Majeed, S., 2015. The state of the curcumin market. Natural products insider. 28 December <https://www.naturalproductsinsider.com/herbs-botanicals/state-curcumin-market>.

Matos, R.S., Baroncini, L.A., Précoma, L.B., et al., 2012. Resveratrol causes antiatherogenic effects in an animal model of atherosclerosis. Arq. Bras. Cardiol. 98, 136–142. pii: S0066-782X2012005000006.

Mattetti, A., Risuleo, G., 2014. Apoptosis: a mode Cell death. Biochem. Mol. Biol. 2, 34–39.

Mattivi, F., Reniero, F., Korhammer, S., 1995. Isolation, characterization, and evolution in red wine vinification of resveratrol monomers. J. Agric. Food Chem. 43 (7), 1820–1823.

Mazumder, K., Biswas, B., Raja, I.M., et al., 2020. A review of cytotoxic plants of the Indian subcontinent and a broad-spectrum analysis of their bioactive compounds. Molecules 25 (8), 1904. Available from: https://doi.org/10.3390/molecules25081904.

McEwan, D.G., Brunton, V.G., Baillie, G.S., et al., 2007. Chemoresistant KM12C colon cancer cells are addicted to low cyclic AMP levels in a phosphodiesterase 4-regulated compartment via effects on phosphoinositide 3-kinase. Cancer Res. 67, 5248–5257.

Ménard-Moyon, C., Kostarelos, K., Prato, M., et al., 2010. Functionalized carbon nanotubes for probing and modulating molecular functions. Chem. Biol. 17, 107–115.

Mohajeri, M., Behnam, B., Sahebkar, A., 2018. Biomedical applications of carbon-nanomaterials: drug and gene delivery potentials. J. Cell Physiol. 234, 298–319. Available from: https://doi.org/10.1002/jcp.26899.

Molnár, J., Engi, H., Hohmann, J., et al., 2010. Reversal of multidrug resitance by natural substances from plants. Curr. Top. Med. Chem. 10 (17), 1757–1768.

Moradi, S.Z., Momtaz, S., Bayrami, Z., et al., 2020. Nanoformulations of herbal extracts in treatment of neurodegenerative disorders. Front. Bioeng. Biotechnol. 8, 238. Available from: https://doi.org/10.3389/fbioe.2020.00238.

Mouchet, F., Landois, P., Datsyuk, V., et al., 2011. International amphibian micronucleus standardized procedure (ISO 21427-1) for *in vivo* evaluation of double-walled carbon nanotubes toxicity and genotoxicity in water. Environ. Toxicol. 26, 136–145.

Muzi, L., Ménard-Moyon, C., Russier, J., Ang, W.H., Pastorin, G., Risuleo, G., et al., 2015. A comparative study on the anticancer efficacy of two types of functionalized multi-walled carbon nanotubes filled with a cisplatin prodrug. Nanoscale 7, 5383–5394.

Muzi, L., Tardani, F., La Mesa, C., Bonincontro, A., Bianco, A., Risuleo, G., 2016a. Interactions and effects of BSA-functionalized single-wall carbon nanotubes on different cell lines. Nanotechnology 27 (15), 155704. Available from: https://doi.org/10.1088/0957-4484/27/15/155704.

Muzi, L., Mouchet, F., Cadarsi, S., et al., 2016b. Examining the impact of multilayer graphene using cellular and amphibian models. 2D Mater. 3, 1–10.

Muzzalupo, R., Pérez, L., Pinazo, A., et al., 2017. Pharmaceutical versatility of cationic noises derived from amino acid-based surfac-tants: skin penetration behavior and controlled drug release. Int. J. Pharm. 29, 245–252.

Newman, D.J., Cragg, G.M., 2016. Natural products as sources of new drugs from 1981 to 2014. J. Nat. Prod. 79 (3), 629–661.

Novoselov, K.S., Fal'ko, V.I., Colombo, L., et al., 2012. A roadmap for graphene. Nature 490, 192–200.

Olusanya, T.O.B., Haj Ahmad, R.R., Ibegbu, D.M., et al., 2018. Liposomal drug delivery systems and anticancer drugs. Molecules 23, 907–911.

Otmakhov, N., Khibnik, L., Otmakhova, N., et al., 2004. Forskolin-induced LTP in the CA1 hippocampal region is NMDA receptor dependent. J. Neurophysiol. 91, 1955–1962.

Palamara, A.T., Nencioni, L., Aquilano, K., et al., 2005. Inhibition of influenza A virus replication by resveratrol. J. Infect. Dis. 191, 1719–1729.

Piccioni, F., Borioni, A., Delfini, M., Rosaria, M., et al., 2007. Metabolic alterations in cultured mouse fibroblasts induced by an inhibitor of the tyrosine kinase receptors fibroblast growth factor receptor 1. Anal. Biochem. 367, 111–121.

Prato, M., Kostarelos, K., Bianco, A., 2008. Functionalized carbon nanotubes in drug design and discovery. Acc. Chem. Res 41, 60–68.

Pretti, C., Oliva, M., Di Pietro, R., et al., 2014. Ecotoxicity of pristine graphene to marine organisms. Ecotoxicol. Environ. Saf. 101, 138–145.

Preußel, K., Stüken, A., Wiedner, C., et al., 2006. First report on cylindrospermopsin producing *Aphanizomenon flos-aquae* (Cyanobacteria) isolated from two German lakes. Toxicon 47 (2), 156–162.

Prokop, J., Abrman, P., Seligson, A.L., et al., 2006. Resveratrol and its glycon piceid are stable polyphenols. J. Med. Food 9 (1), 11–14.

Ricci, F., Berardi, V., Risuleo, G., 2009. Differential cytotoxicity of MEX: a component of neem oil whose action is exerted at the cell membrane level. Molecules 14, 122–132.

Risuleo, G., 2016. Resveratrol: multiple activities on the biological functionality of the cell. In: Gupta, R.C. (Ed.), Nutraceuticals: Efficacy, Safety and Toxicity. Academic Press/Elsevier, Amsterdam, pp. 453–464. , Chapter 33.

Risuleo, G., 2020. Biological properties of a partially purified component of neem oil: an updated and revised work. In: Preedy, V.R., Watson, R.R. (Eds.), Nuts and Seeds in Health and Disease Prevention. Academic Press/Elsevier, San Diego, pp. 67–72. , Chapter 6.

Risuleo, G., La Mesa, C., 2019a. Nanoparticles and molecular delivery: state of the art and future perspectives. In: Gupta, R., Srivastava, A., Lall, R. (Eds.), Nutraceuticals in Veterinary Medicine. Springer Nature, Cham, pp. 737–747.

Risuleo, G., La Mesa, C., 2019b. Resveratrol: biological activities and potential use in human and veterinary medicine. In: Gupta, R., Srivastava, A., Lall, R. (Eds.), Nutraceuticals in Veterinary Medicine. Springer Nature, Cham, pp. 215–225.

Russo, L., Berardi, V., Tardani, F., Risuleo, G., 2013. Delivery of RNA and its intracellular translation into protein mediated by SDS-CTAB vesicles: potential use in nanobiotechnology. Biomed. Res. Int. 2013, 734596.

Saker, M.L., Jungblut, A.-D., Neilan, B.A., et al., 2005. Detection of microcystin synthetase genes in health food supplements containing the freshwater cyanobacterium *Aphanizomenon flos-aquae*. Toxicon 46, 555–562.

Sarvjeet, S., Ashok, K., 2006. Biological effects of curcumin and its role in cancer chemoprevention and therapy. Anticancer Agents Med. Chem. 6, 259–270.

Schröder, G., Brown, J.W., Schröder, J., et al., 1988. Molecular analysis of resveratrol synthase. cDNA, genomic clones and relationship with chalcone synthase. Eur. J. Biochem. 172 (1), 161–169.

Seamon, K.B., Padgett, W., et al., 1981. Forskolin: unique diterpene activator of adenylate cyclase in membranes and intact cells. Proc. Natl. Acad. Sci. USA 78, 3363–3367.

Shalaby, E., 2018. Antioxidants in Food and Its Applications. IntechOpen Limited, London. Available from: http://doi.org/10.5772/intechopen.72008.

Shalini, S., Dorstyn, L., Dawar, S., et al., 2015. Old, new and emerging functions of caspases. Cell Death Differ. 22, 526–539.

Shtansky, D.V., Firestein, K.L., Golberg, D.V., 2018. Fabrication and application of BN nanoparticles, nanosheets and their nanohybrids. Nanoscale 10, 17477–17493.

Shytle, D.R., Tan, J., Ehrhart, J., et al., 2010. Effects of blue-green algae extracts on the proliferation of human adult stem cells in vitro: a preliminary study. Med. Sci. Monit. 16 (1), BR1–5. PMID 20037479.

Signorelli, P., Ghidoni, R., 2005. Resveratrol as an anticancer nutrient: molecular basis, open questions and promises. J. Nutr. Biochem. 16 (8), 449–466.

Singh, N.P., Singh, U.P., Hegde, V.L., et al., 2011. Resveratrol (trans-3,5,4'-trihydroxystilbene) suppresses EL4 tumor growth by induction of apoptosis involving reciprocal regulation of SIRT1 and NF-κB. Mol. Nutr. Food Res. 55 (8), 1207–1218.

Stefanutti, E., Papacci, F., Sennato, S., et al., 2014. Cationic liposomes formulated with DMPC and agemini surfactant traverse the cell membrane without causing a significant bio-damage. Biochim. Biophys. Acta 1838, 2646–2655.

Thomma, B.P., Nelissen, I., Eggermont, K., et al., 1999. Deficiency in phytoalexin production causes enhanced susceptibility of *Arabidopsis thaliana* to the fungus *Alternaria brassicicola*. Plant. J. 19 (2), 163–171.

Tooze J. (Ed.)., 1981. Molecular Biology of Tumor Viruses: DNA Tumor Viruses, Part 2. Cold Spring Harbor Laboratory Press, Cold Spring Harbor, New York.

Ullah, M.F., Bhat, S.H., Hussain, E., et al., 2012. Ascorbic acid in cancer chemoprevention: translational perspectives and efficacy. Curr. Drug Targets 13 (14), 1757–1771.

van der Pol, E., Anita, N., Böing, P.H., et al., 2012. Classification, functions, and clinical relevance of extracellular vesicles. Pharmacol. Rev. 64, 676–705.

Vang, O., 2015. Resveratrol: challenges in analyzing its biological effects. Ann. N. Y. Acad. Sci. 1348, 161–170.

Venkatesan, J., Pallela, R., Kim, S.K., 2014. Applications of carbon nanomaterials in bone tissue engineering. J. Biomed. Nanotechnol. 10, 3105–3123.

Wang, Z., Li, W., Meng, X., et al., 2011. Resveratrol induces gastric cancer cell apoptosis via ROS, but independent of sirtuin1. Clin. Exp. Pharmacol. Physiol. 39, 227–232.

Wright, P.K., 2008. Targeting vesicle trafficking: an important approach to cancer chemotherapy. Recent Pat. Anticancer Drug Discov. 3 (2), 137–147.

Yessenkyzy, A., Saliev, T., Zhanaliyeva, M., et al., 2020. Polyphenols as caloric-restriction mimetics and autophagy inducers in aging research. Nutrients 12 (5), 1344. Available from: https://doi.org/10.3390/nu12051344.

Zanni, E., De Bellis, G., Bracciale, M.P., et al., 2012. Graphite nanoplatelets and *Caenorhabditis elegans*: insights from an *in vivo* model. Nano Lett. 12, 2740–2744.

Zhang, W., Wang, X., Chen, T., 2012. Resveratrol induces apoptosis via a Bak-mediated intrinsic pathway in human lung adenocarcinoma cells. Cell. Signal.

Zhao, J., Wang, Z., White, J.C., et al., 2014. Graphene in the aquatic environment: adsorption, dispersion, toxicity and transformation. Environ. Sci. Technol. 48, 9995–10009.

Newman, D.J., Cragg, G.M., 2020. Natural products as sources of new drugs over the nearly four decades from 01/1981 to 09/2019. J. Nat. Prod. 83 (3), 770–803. Available from: https://doi.org/10.1021/acs.jnatprod.9b01285.

Chapter 34

Melatonin: a safe nutraceutical and clinical agent

Arturo Anadón, Irma Ares, María-Rosa Martínez-Larrañaga and María-Aránzazu Martínez

Introduction

Melatonin (*N*-acethyl-5-methoxytrytamine) hormone is a methoxyindole produced (synthesized/secreted) in the pineal gland by pinealocytes during the hours of dark conditions (i.e., at night under normal light/dark conditions), and by enterochromaffin cells of the digestive mucosa. Its peak levels at night are associated with age as well as various illnesses. Melatonin was isolated from the pineal gland of bovines (Lerner et al., 1958), more commonly known as the sleep hormone, and it regulates the circadian rhythm. Melatonin is often termed the *hormone of darkness* because the timing of its production is influenced by light perception of the retina and the endogenous rhythmicity of neurons within the suprachiasmatic nucleus (SCN), a tiny region of the brain in the hypothalamus, situated directly above the optic chiasm. Lerner et al. (1958) found that melatonin had no acute pharmacological effects in the systems studied apart from its effect on sleep mechanisms.

The chemical structure of melatonin, along with its molecular formula, molecular weight, and Chemical Abstracts Service (CAS) number are shown in Fig. 34.1. Melatonin is a small lipophilic molecule, slightly off-white in color crystalline powder, which is soluble in tetrahydrofuran and methanol. It is only slightly soluble in ethyl acetate and insoluble in water. Melatonin is commonly given orally, intravenously, intranasally, or via the transbuccal route.

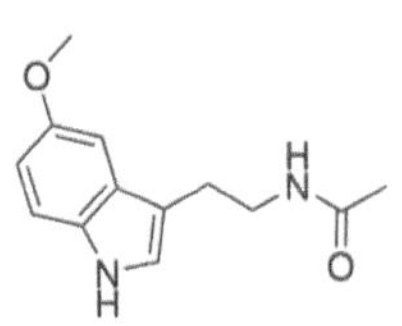

FIGURE 34.1 Chemical structure of melatonin (*N*-[2-(5-methoxy-1*H*-indol-3-yl) ethyl]acetamide), molecular formula $C_{13}H_{16}N_2O_2$, molecular weight 232.28 g/mol, and CAS number 73-31-4.

In humans, the most important circadian rhythm is the *sleep—wake cycle*. The *sleep—wake cycle* may be pathologically affected in different ways and the disturbances of this cycle are called *circadian rhythm* disorders. These disorders include jet lag syndrome, shift work sleep disorder, advanced sleep phase syndrome, and non-24 hour sleep—wake syndrome (EMEA, 2007).

Synthesis of melatonin

Melatonin is a derivative of tryptophan that was discovered by Aron B. Lerner in 1958 (Lerner et al.,1958). It is synthesized mainly in the pineal gland by parenchymatous cells in response to light information received through the retinohypothalamic pathways. Further light information reaches the SCN where the circadian clock exists. This synthesis in the pineal gland involves: (1) L-tryptophan, which is taken up from the cerebral vessels and converted to serotonin, (2) serotonin is subsequently metabolized by the rate-limiting arylalkylamine *N*-acetyltransferase (AANAT) to *N*-acetyl-5-hydroxytryptamine, and (3) *N*-acetyl-5-hydroxytryptamine is converted to melatonin by the enzyme, hydroxyindole-o-methyltransferase (Ekmekcioglu, 2006). Tryptophan is the starting molecule for melatonin production in cells. The conversion occurs through hydroxylation, decarboxylation, acetylation, and methylation (Zhao et al., 2019). The sequence of enzymatic steps that convert tryptophan to melatonin varies among species. A reduction of tryptophan leads to the marked lowering of melatonin production in animals; tryptophan decarboxylation is the initial step in serotonin production and animals first hydroxylate tryptophan using tryptophan hydroxylase (TPH) to form 5-hydroxytryptophan. 5-Hydroxytryptophan is then decarboxylated by the aromatic amino acid decarboxylase (AADC) to form serotonin (Zhao et al., 2019). The melatonin synthesis process involves the enzyme TPH and the AADC to produce the key intermediate serotonin (Fig. 34.2). Serotonin is catalyzed by

Nutraceuticals. DOI: https://doi.org/10.1016/B978-0-12-821038-3.00034-3

serotonin-N-acetyl-transferase (formerly known as hydroxy indole-O-methyltransferase, HIOMT), which is the limiting enzyme for the synthesis of melatonin. The biosynthetic process continues by involving N-acetyl-methoxy tryptamine to form the final product, melatonin. The mRNAs encoding enzymes are expressed with a day/night rhythm in the pineal gland (Claustrat and Lestonb, 2015; Zhao et al., 2019).

Regulation of melatonin

The melatonin secretion rate is controlled by an internal clock located in the suprachiasmatic nuclei of the hypothalamus. The SCN controls the pineal gland via neural signals in a multisynaptic pathway that involves the superior cervical ganglia and adrenergic neurons innervating the pineal gland. This control system allows the duration and timing of melatonin secretion to change with the seasons such that the duration of melatonin secretion is longer during the short days of winter than the long days of summer in a wide range of animal species (Kennaway, 2015). The alternation between light/dark is the main factor regulating melatonin secretion. The reduction in light perception causes the release of noradrenaline, which interacts with the postsynaptic β_1 and α_1 adrenergic receptors of the pinealocytes. As a consequence, cAMP synthesis is augmented, which in turn increases serotonin *N-acetyltransferase* activity and leads to the synthesis of melatonin in the pinealocytes. High melatonin production is maintained during the dark phase of the nychthemeron (i.e., a period of 24 consecutive hours), as long as the subject is not exposed to light. Indeed, in the presence of light, noradrenaline secretion is inhibited (Claustrat, 2009; Cipolla-Neto et al., 2014; Singh and Jadhav, 2014).

Characterization and endogenous secretion of melatonin

Melatonin has an indole structure that is neutral, and lipophilic at physiological pH (Claustrat, 2009). It is synthesized by the body in two steps from serotonin resulting from the transformation of tryptophan (see Fig. 34.2) (Welford et al., 2016). The mRNA encoding the enzymes arylamine-*N*-acetyl-transferase and HIOMT are expressed according to the circadian rhythm in the pineal gland (Claustrat et al., 2005). Adults produce approximately 20–60 μg of melatonin endogenously over a period of 8–10 hours each day (Lane and Moss, 1985). Most melatonin tablet preparations contain from 2 to 5 mg. Following oral administration of *fast-release* preparations, melatonin is rapidly absorbed and metabolized on first pass through the liver, with an elimination half-life ($t_{1/23}$) of 30–40 minutes and a bioavailability (F) of 1%–37% (Fourtillan et al., 2000).

Melatonin is mainly secreted by the pinealocytes of the pineal gland, and incidentally by the retina, intestine, skin, platelets, and bone marrow (Singh and Jadhav, 2014). This secretion occurs preferentially at night and over a period of around 10 hours. The maximum concentration is reached on average at around 3:00 to 4:00 in the morning. In adults, diurnal and nocturnal blood concentrations of melatonin are about 10 and 60 pg/mL, respectively (Comai and Gobbi, 2014). Diurnal concentrations are low, and even undetectable, in blood (Claustrat, 2009). The plasma profile of melatonin has considerable interindividual variability. In this respect, it is highly reproducible from one day to another in the same individual. The exogenous administration of a dose as low as 1 mg leads to supraphysiological concentrations (> 100 pg/mL) (Claustrat, 2009). Generally, the concentration of melatonin in the gastrointestinal tract (GIT) is higher than in the blood (by 10–100 times) or pineal gland.

Mechanisms of action of melatonin

Melatonin's four mechanisms of action in mammalian species have been described and include: (1) binding to plasma membrane localized melatonin receptors; (2) binding to intracellular proteins such as calmodulin; (3) binding to nuclear receptors of the orphan family; and (4) antioxidative effects (Macchi and Bruce, 2004).

Melatonin receptors and physiological function

In the mechanism of action of melatonin, distinct classes of receptors are associated as follows: (1) melatonin receptor type MT_1 (on cell membrane receptors) (formerly designated Mel_{1a}); (2) melatonin receptor type MT_2 (on cell membrane receptors) (formerly designated Mel_{1b}); (3) melatonin receptor type 1c (found in fish,

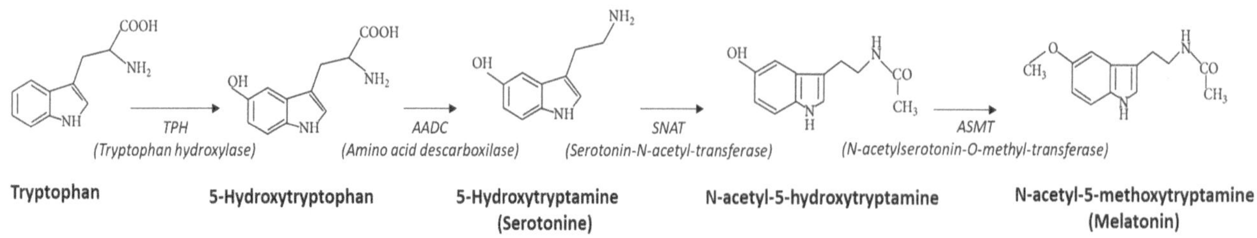

FIGURE 34.2 Pathway of melatonin biosynthesis.

amphibians, and birds); (4) quinone reductase 2 enzyme (MT_3 receptor, a detoxification enzyme). This enzyme, found in different tissues and structurally unrelated to the MT_1 and MT_2 receptors, may play a detoxification role and could participate in the regulation of intraocular pressure (Emet et al., 2016); (5) retinoid Z receptor (RZR)/related orphan receptor alpha (RORα): retinoid-related orphan nuclear hormone receptor (through this receptor, melatonin binds to the transcription factors in nucleus); and (6) GPR50: X-linked melatonin-related orphan receptor (this is effective in binding of melatonin to MT_1) (Allegra et al., 2003; Emet et al., 2016).

In humans, melatonin exerts its actions by the interaction of MT_1 and MT_2 receptors, which are transmembrane G protein-coupled receptors. They are expressed alone or co-expressed in the tissues, particularly in the CNS (SCN, retina, hippocampus, cerebellum, and central dopaminergic system). Melatonin inhibits dopamine release in the hypothalamus and retina. Activation of the MT_1 receptors inhibits the neuronal activity of the suprachiasmatic nuclei (i.e., a tiny region of the brain in the hypothalamus, located directly above the optic chiasm that is responsible for controlling circadian rhythms). Among diurnal species such as humans, this inhibition reduces the biological clock's activator effects on vigilance, which increases drowsiness and the tendency to sleep (Ekmekcioglu, 2006). Activation of the MT_1 receptors is therefore responsible for the hypnotic effects of melatonin. At the peripheral level, the MT_1 receptors are expressed in the ovaries, testes, mammary glands, gall bladder, liver, kidneys, lungs, thyroid gland, skin, adipose tissue, skeletal muscle, bone marrow, and the cardiovascular and immune systems. The MT_2 receptors are expressed in the retina, thymus, lungs, ovaries, stomach, duodenum, colon, and pancreas (Ekmekcioglu, 2006). In contrast, the activation of the MT_2 receptors has a dominant role in phase-shifting the circadian rhythm of the neuronal activity of the suprachiasmatic nuclei (Dubocovich et al., 2010). The MT_1 and MT_2 receptors may have opposing functions on different tissues or organs and complementary functions from a physiological point of view. For example, the MT_1 and MT_2 receptors act differently on the vascular system, producing, respectively, vasoconstriction and vasodilation (Comai and Gobbi, 2014). The lowering of body temperature induced by melatonin, which may contribute to enhancing sleepiness, may be a consequence of its vasoconstrictor effect from the activation of MT_1 receptors.

In addition to its effects on the biological clock, melatonin has other physiological properties such as modulation of mood, sexual behavior (studied extensively in animals because of its activity on reproduction), immune system, and regulation of body temperature or blood pressure. These effects are based on (1) a broad distribution of MT_1 and MT_2 receptors, and (2) on the pharmacological effects of melatonin, independent from the MT_1 and MT_2 receptors and involving the binding site, MT_3 (Singh and Jadhav, 2014).

Overall, the locations of melatonin receptors are as follows: brain, retina, cardiovascular system, cardiac ventricular wall, aorta, coronary and cerebral arteries, liver and gallbladder, duodenal enterocytes, colon, cecum and appendix vermiformis, skin, parotid gland, exocrine pancreas, kidneys, cells of the immune system, platelets, brown and white adipocytes, epithelial cells of the prostate and breast, ovary/granulosa cells, myometrium, placenta, and fetal kidney. In the gastrointestinal (GI) system there is another large source of endogenous melatonin, melatonin receptors that are most commonly found in jejunal and colonic mucosa (Emet et al., 2016).

Melatonin agonist drugs

There are several melatonin agonist drugs (Emet et al., 2016):

1. *Ramelteon*: This is a nonselective (MT_1/MT_2) melatonin receptor agonist. It was approved by the US FDA in 2005 and is used as a sedative/hypnotic for the treatment of insomnia. It helps in regulating the "sleep–wake cycle." Among melatonergic agonist drugs, ramelteon has a relatively higher affinity for both receptor subtypes, but it has no serotoninergic antagonistic activity. Ramelteon is rapidly absorbed from the gastrointestinal system at a rate of 84% and its elimination half-life is 1–2 hours. In elderly individuals with primary chronic insomnia, the effect of ramelteon has been found to be highly variable in maintaining sleep. Ramelteon is metabolized basically by CYP1A2, CYP2C9, and CYP3A4. When it is used, special attention must be given to patients that may also take fluvosamine. Moreover, ramelteon may also cause a severe allergic reaction (Fourman and Robert Meyer, 2013).
2. *Agomelatine* (N-[2-(7- methoxynaphthalen-1-yl) ethyl] acetamide): This drug is a synthetic analog of melatonin, approved by the European Medicines Agency (EMA) in 2008, as a melatonergic drug for the treatment of major depression in adults. It is an oral nonselective melatonergic (MT_1/MT_2) agonist and serotonergic ($5HT_{2C}$) antagonist. It also has a role as a serotonin antagonist ($5HT_{2c}$), giving rise to a direct antidepressant effect and anxiolytic potential. Agomelatine is the first reported melatonergic drug having anxiolytic and antidepressant effects. In major depression, it is used at a daily dose of 25 mg. In seasonal affective disorder, it is used at doses of 0.225–0.3 mg/day, without affecting sleep. The elimination half-life of agomelatine is 1–2 hours. The

advantage of agomelatine is not its greater antidepressant effect, but its combination of improving sleep, together with its antidepressant effects. Agomelatine is metabolized basically by CYP1A1, CYP1A2, and CYP2C9. Agomelatine normalizes the timing and continuity of sleep in patients with depression and improves daytime alertness (Singh and Jadhav, 2014).

3. *TIK-301(LY-156735)* (β-methyl-6-chloromelatonin): This is an agonist for melatonin receptors MT_1 and MT_2 and a serotonergic antagonist drug for the treatment of insomnia and other sleep disorders. Its agonist action on MT_1 and MT_2 receptors in the SCN in the brain enables its action as a chronobiotic. It belongs to the same class of melatonin receptor agonists as ramelteon and tasimelteon. The US FDA approved its use in sleep disorders of visually impaired individuals. TIK-301 is a more powerful antagonist of serotonergic 5-HT(2C) and 5-HT(2B) receptors when compared to agomelatine. The elimination half-life for TIK-301 is 1 hour.
4. *Tasimelteon N*-[[(1*R*,2*R*)-2-(2,3-dihydro-1-benzofuran-4 yl)cyclopropyl]methyl] propanamide: This drug was approved by the US FDA as a melatonin receptor agonist with high affinity for human MT_1 and MT_2 receptors, for the treatment of non-24-hour sleep–wake disorder and by the EMA for the treatment of non-24-hour sleep–wake rhythm disorder in totally blind adults. Its elimination half-life is approximately 2 hours. It is basically metabolized by CYP1A1, CYP1A2, CYP2D6, and CYP2C9. In some preclinical studies, it was reported that tasimelteon has similar phase-shifting properties to melatonin, but with fewer vasoconstrictive effects (Singh and Jadhav, 2014).
5. *Hormesis*: The hormesis induced by stress components [e.g., reactive chemical species (RCS), and hormones] by coordination or cross-talk with melatonin may suggest that melatonin may act within a hormetic framework (Calabrese, 2001). Circadian rhythms are of adaptive significance and display a high plasticity for adjusting the endogenous clock with the environment, such as changing day/night duration. It was hypothesized that circadian rhythms may also be mediated via a hormetic-type dose-dependent mechanism. Some patients with circadian rhythm sleep disorders exhibit familial inheritance. Animals and plants have shown a similar multifunctional role of melatonin. It became of interest to assess the nature of the dose–response relationship in organisms exposed to increasing doses of melatonin and whether or not they may follow a similar biological strategy in response to diverse sets of environmental Stressors (Agathokleousa et al., 2019). A review study on stress-related neuropsychiatric disorders suggested that the stress hormones, glucocorticoids, induce a biphasic inverted U-shape response, where low doses enhance and high doses inhibit neurotransmission, synaptic growth, and mitochondrial functions, thus providing a dual action on learning and memory. The hormetic dose- and duration-dependent responses to stress and glucocorticoids display consistent circadian and ultradian modes (Jeanneteau and Arango-Lievano, 2016). It is now recognized that RCS, key components of stress biology, display a dual action, with moderate elevation being necessary for normal functioning and acclimation to the environment (Drzeżdżon et al., 2018). Melatonin has the capacity to protect animals against oxidative stress, particularly against inflammation, as it is involved in the organismic adaptive response via hormetic processes which protect against environmental/health threats (Agathokleousa et al., 2019).

Pharmacokinetics of melatonin

Several pharmacokinetics studies have been performed with melatonin in different drug forms and administration routes.

Absorption and bioavailability of melatonin after oral infusion administration

One review was conducted to investigate the pharmacokinetics of exogenous melatonin in humans to provide recommendations for clinical use. The pharmacokinetics of Circadin (medicinal product containing 2 mg of melatonin gradually released), showed that the absorption of ingested melatonin is complete in adults but may be decreased by up to 50% in the elderly. The human bioavailability (F) of orally administered melatonin is estimated to be 15% due to a significant hepatic first-pass effect. It is known that up to 90% of blood melatonin is cleared by the liver in a single passage, thereby its elimination half-life ($t_{1/2\beta}$) is very short (30–60 minutes) (Bartoli et al., 2012). Following oral administration of Circadin 10 mg, a mean C_{max} of 3970 pg/mL, mean t_{max} of 1.37 hours, mean $AUC_{0-\infty}$ of 13,300 pg.h/mL, and elimination $t_{½}$ of 4.44 hours were observed (EMA, 2014). Peak blood concentrations (1020 pg/mL) were reached 3 hours after oral administration. For oral drug formulation without controlled release, this time varies from 15 to 90 minutes, depending on the dose studied (2 mg and 25 mg, respectively). The bioavailability of oral melatonin ranged from 9% to 33% for doses between 0.3 mg and 100 mg. Overall, pharmacokinetic parameters were affected by age, caffeine, smoking, oral contraceptives, feeding status, and use of fluvoxamine. For example, critically ill patients displayed accelerated absorption and compromised elimination (Harpsøe et al., 2015). The bioavailability of a new food-grade liquid emulsion formulated in a spray form was studied and compared with the administration of an oral tablet of melatonin. The spray form assured an additional increase in the contact surface and the possibility of efficient mucosal absorption. In a single-dose, open-label, cross-over study, eight subjects were randomly assigned to receive 5 mg of oral spray or oral tablet melatonin. After a washout of 1 week, plasma was collected at 6 hours postdosing and assayed by a validated liquid chromatography tandem mass spectrometry method. Data obtained in this study showed that the extent of melatonin absorption after oral spray delivery was 1.8 times that observed after administration of the standard oral tablet; the peak concentration was also significantly higher, 1.5 times the corresponding oral tablet value. The absorption rate, expressed as T_{max} and K_a, was comparable between the two products. After administration of melatonin in emulsion spray, concentrations of melatonin at 5, 10, and 20 minutes were 4.6, 3.5, and 3.0 times higher than the corresponding concentrations of melatonin after tablet administration, respectively. Thereafter, the concentrations of melatonin after administration of melatonin in emulsion spray were 1.5–2.4 times greater than the corresponding concentrations of melatonin after tablet administration (Bartoli et al., 2012).

It is noted that melatonin possesses potential efficacy in perinatal brain injuries, and it has been proposed as an adjunctive pharmacological therapy in combination with hypothermia in the clinical setting. In this regard, an oral (intragastric administration) pharmacokinetics study was performed in preterm neonates (Carloni et al., 2017). Preterm newborns were enrolled 24–72 hours after birth and randomly assigned to three groups receiving a single bolus of 0.5 mg/kg melatonin, or three boluses of 1 or 5 mg/kg of melatonin at 24-hour intervals. The elimination half-life ($t_{1/2\beta}$) of melatonin in plasma ranged from 7.98 to 10.94 hours, and the area under the curve (AUC) from 10.48 to 118.17 μg/mL/hours. These results indicate a different pharmacokinetic profile in premature newborns compared to adults and experimental animals. The high peak plasma concentrations and the long $t_{1/2\beta}$ indicate that in the neonatal clinical setting, it is possible to obtain and maintain neuroprotective high serum concentrations using a single administration of melatonin repeated every 12–24 hours. Finally, a pharmacokinetics and safety study in which melatonin was administered enterally to neonates with hypoxic–ischemic encephalopathy undergoing hypothermia was carried out (Balduini et al., 2019). Melatonin was infused at 0.5 mg/kg in five neonates with hypoxic–ischemic encephalopathy undergoing hypothermia. Infusion started 1 hour after the neonates reached the target temperature of 33.5°C. Blood samples were collected before and at selective times after melatonin infusion. The peak plasma concentration reached 0.25 μg/mL. The AUC in 24 hours was 4.35 μg/mL/hours Melatonin elimination half-life and clearance were prolonged, and the distribution volume decreased compared to adults.

Distribution

Melatonin is a lipophilic hormone that is widely distributed throughout the human body. After passing into the bloodstream, 50%–60% of melatonin is bound to plasma proteins (albumin and α_1-acid glycoprotein and high-density lipoproteins) (Cardinali et al., 1972; Morin et al., 1997). Melatonin crosses the blood–brain barrier to exert its effect on the CNS (Claustrat, 2009). From the bloodstream, it is also distributed in the tissues, cellular compartments, and biological fluids, and significant concentrations of melatonin are achieved in a number of fluids such as saliva, bile, synovial fluid, urine, cerebrospinal fluid (CSF), seminal fluid, amniotic fluid, and breast milk (Claustrat et al., 2005; Acuña-Castroviejo et al., 2014). The binding of ^{3}H-melatonin to human serum proteins was investigated by equilibrium dialysis at 37°C and pH 7.4 (Morin et al., 1997). The binding to serum was moderate (53%) for physiological melatonin concentrations below 1 nmol/L. α_1-Acid glycoprotein and albumin-bound melatonin had a high of 27 and low of 1.5 (mmol/L) affinity, respectively. Serum binding was characterized by a saturable and a nonsaturable component. The saturable component resulted from the high-affinity binding to α_1-acid glycoprotein and the nonsaturable component resulted from the low-affinity binding to albumin. The number of binding sites was 0.36/molecule of α_1-acid glycoprotein, with either pure α_1-acid glycoprotein or serum, indicating that only a fraction of α_1-acid glycoprotein bound melatonin. The high-affinity melatonin binding to α_1-acid glycoprotein might result from a potentiation of the binding interaction of albumin and the amount of melatonin bound in plasma which might vary according to the α_1-acid glycoprotein concentration (Morin et al., 1997).

A cross-over pharmacokinetics study in human volunteers treated with an intravenous bolus of 10 mg and 100 mg of melatonin was conducted (Andersen et al., 2016a). Blood samples were collected at 60, 120, 180, 240, 300, 360, and 420 minutes after administration of the intravenous bolus. Pharmacokinetic parameters were estimated by a compartmental pharmacokinetic analysis. In this study, the possible adverse effects were also recorded (i.e., sedation and other clinical symptoms). Sedation was measured at 120, 180, 300, and 420 minutes after the intravenous bolus. Median C_{max} after the bolus injections of 10 mg and 100 mg of melatonin were 221 and 500.0 pg/mL, and 1 251 and 500.0 pg/mL, respectively; mean elimination half-life ($t_{1/2\beta}$) was 42 and 46 minutes; mean volume of distribution (Vd) was 1.6 L/kg and 2.0 L/kg; mean clearance (CL) was 0.025 L/minute · kg and 0.030 L/minute · kg; and median

AUC_0 values varied extensively between the volunteers in both doses (up to a factor-7 and factor-5 difference, respectively). High-dose intravenous melatonin did not induce sedation. No adverse effects were reported in the study (Andersen et al., 2016b).

In an overview of studies investigating the pharmacokinetics of exogenous melatonin in humans, methodological differences were observed between the included studies. The review indicated that melatonin dosages varied between 0.3 and 100 mg and were administered either orally or intravenously. C_{max} ranged from 72.1 (10 mL/hour; 0.02 mg, IV) to 101.163 pg/mL (100 mg, oral). T_{max} ranged from 15 (2 mg) to 210 minutes (10 mg). $T_{1/2}$ ranged from 28 (0.005 mg, IV) to 126 minutes (4 mg, oral), whereas AUC ranged from 5400 (0.005 mg, IV) to 6.56×1010 pg/mL $\times$ minutes (1 mg, oral). CL ranged from 0.97 (0.005 mg, IV) to 132.50 L/minutes (6 mg, oral), whereas V_d ranged from 35 (0.005 mg, IV) to 1602 L (4 mg, oral). The bioavailability of oral melatonin ranged from 9% to 33% (Harpsøe et al., 2015).

Metabolism and excretion

Melatonin is metabolized primarily in the liver (90% of circulating melatonin), via hydroxylation at 6-hydroxymelatonin by CYP1A1, CYP1A2, CYP1B1, or CYP2C19 (Ma et al., 2005). After hydroxylation, melatonin is excreted as a sulfate conjugate, although glucuronide conjugation can occur. The 6-hydroxymelatonin is then mainly sulfo-conjugated to a metabolite that is fully excreted in the urine within 12 hours of ingestion. Around 1%–2% of melatonin is excreted unchanged directly through the urine (Claustrat et al., 2005; Singh and Jadhav, 2014).

Melatonin is deacetylated in the pineal gland and retina by melatonin deacetylase. In the brain, melatonin is oxidized to N_1-acetyl-N_2-formyl-5-methoxykynuramine (AFMK). This compound is derived from the reaction of melatonin with reactive oxygen species (ROS) and it has in vitro antioxidant properties (Singh and Jadhav, 2014). In other cells, it can be metabolized by free radicals and converted to cyclic 3- and 6-hydroxymelatonin (Tan et al., 1998). AFMK and N1-acetyl-5-methoxykinuramine (AMK) are two important melatonin metabolites that have excellent radical-scavenging activity. AMK is predicted to be better than melatonin in scavenging hydroxyl radicals and their scavenging efficiency depends on the radical with which they are reacting (Cavanaugh and Witt-Enderby, 2010).

In a study, ^{3}H-labeled melatonin was not metabolized when incubated with rat venous blood in vitro and 72%–82% of the added radioactivity was recovered in the plasma fractions. The binding of ^{3}H-melatonin to rat and human plasma proteins demonstrated there were no significant differences between the binding capacities of human and rat plasma, and the proportions bound (77% and 60% at 4oC and 37oC, respectively) were independent of melatonin concentration up to 1.5 HIM, indicating a high-capacity binder protein. The in vitro binding of melatonin to plasma proteins was not modified by the presence of other indole derivatives. Equilibrium dialysis of ^{3}H-melatonin with purified plasma protein fractions and electrophoresis of plasma preincubated with melatonin revealed that albumin was the only melatonin-binding protein detectable in plasma. No significant association was observed between melatonin and macromolecules in fresh CSF from goats unless the protein concentration of the CSF was increased by ultrafiltration (Cardinali et al., 1972) (Fig. 34.3).

Melatonin can be used in a series of disorders where melatonin may have clinical effectiveness, as described next.

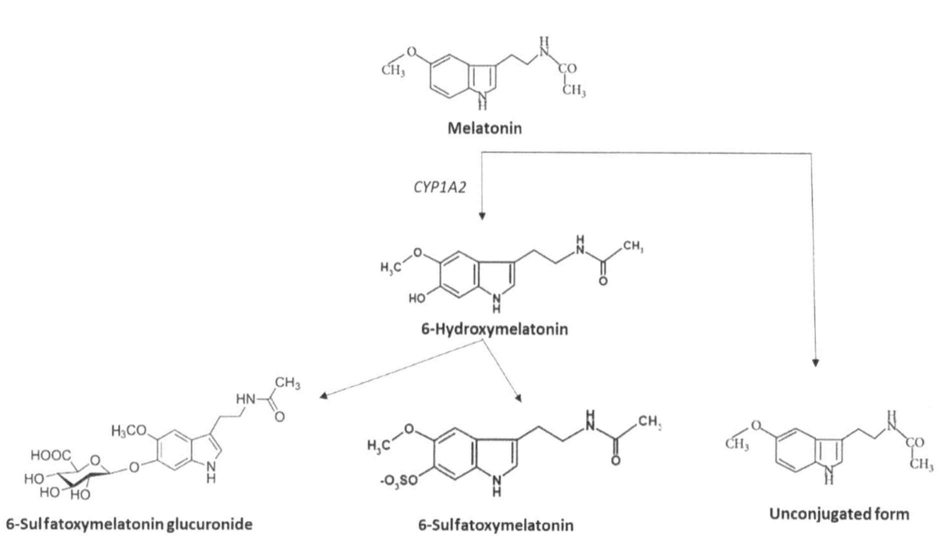

FIGURE 34.3 Pathway of melatonin metabolism. The liver is the primary site for melatonin metabolism. Melatonin is hydroxylated by CYP1A2 to 6-hydroxymelatonin. This hydroxylated metabolite is largely excreted as the sulfate conjugate 6-sulfametoximelatonin and to a lesser extent as the glucuronide metabolite 6-sulfatoxymelatonin glucuronide and as the unconjugated form in a very small proportion (Skene et al., 2001; Zhao et al., 2019).

Melatonin and circadian rhythms

Circadian rhythms are of adaptive significance and display a high plasticity for adjusting the endogenous clock with the environment (e.g., changing day/night duration). Given the critical role of melatonin in circadian rhythms and hormesis, it seems that circadian rhythms may also be mediated via a hormetic-type dose-dependent mechanism (Agathokleousa et al., 2019). The primary physiological function of melatonin, whose secretion adjusts to night length, is to convey information concerning the daily cycle of light and darkness to body structures. This information is used for the organization of functions, which respond to changes in the photoperiod such as the seasonal rhythms. Seasonal rhythmicity of physiological functions in humans related to possible alteration of the melatonin message remains, however, of limited evidence in temperate areas under field conditions. In addition, daily melatonin secretion, which is a very robust biochemical signal of night, can be used for the organization of circadian rhythms. Although functions of melatonin in humans are mainly based on correlations between clinical observations and melatonin secretion, there is some evidence that melatonin stabilizes and strengthens coupling of circadian rhythms, especially of core temperature and sleep–wake rhythms. The circadian organization of other physiological functions depends also on the melatonin signal, that is, immunity, antioxidant defenses, hemostasis, and glucose regulation. The difference between the physiological and pharmacological effects of melatonin is not always clear but is based upon consideration of dose and not of duration of the hormone message. It has been shown that a physiological dose provides plasma melatonin levels in the same order of magnitude as a nocturnal peak. The regulating system of melatonin secretion is complex, following central and autonomic pathways, and there are many pathophysiological situations where melatonin secretion can be disturbed. The resulting alteration could increase predisposition to disease, add to the severity of clinical symptoms, or modify the course and outcome of the disorder. Since melatonin receptors display a very wide distribution in the body, putative therapeutic indications of this compound are numerous. Great advances in this field could be achieved by developing multicenter trials in a large series of patients, in order to establish the efficacy of melatonin and absence of long-term toxicity (Claustrat and Lestonb, 2015)

The effects of melatonin on sleep result from the activation of receptors in the SCN in the hypothalamus. The endogenous rhythm of secretion is generated by the SCN and entrained to the light/dark cycle. Light is able to either suppress or synchronize melatonin production according to the light schedule. The nyctohemeral rhythm of this hormone can be evaluated by repeated measurement of plasma or saliva melatonin or urine sulfatoxymelatonin, the main hepatic metabolite.

When melatonin is administered in the afternoon or evening, a *phase advance* is observed (the peak of melatonin secretion occurs earlier), with a sedative effect causing an increase in feelings of tiredness, a slower reaction time, and a decrease in sleep latency (Claustrat, 2009). In contrast, administration in the morning or at noon leads to a *phase delay*, which is smaller or may even be absent. The *phase delay* effect is less pronounced than the *phase advance* effect. The effect changes direction at around 15:00 (*phase delay* effect for administration before 15:00, *phase advance* effect after 15:00). This time marker is modulated by the subject's chronotype.

The most commonly proposed mechanism for melatonin to induce sleepiness relates to its effects on the circadian clock (i.e., it *opens the sleep gate*) probably via thermoregulatory mechanisms and slightly reduced body temperature to promote the sleep. Melatonin has these effects using a wide range of dose levels [from physiological (250 μg) to pharmacological (1–10 mg)] (Kennaway and Wright, 2002).

Melatonin and oxidative stress

Melatonin shows strong antioxidant activity and possesses protective properties against oxidative stress in several circumstances that are for the most part characterized by an excessive generation of ROS. Other actions of melatonin that have evolved and are related to its antioxidant activity include retarding some age-related processes, antiinflammatory activity, resisting neurodegenerative changes, prevention of apoptosis in normal cells, and preservation of mitochondrial and chloroplast physiology (Nabavi et al., 2018) (Fig. 34.2). These functions are associated, at least in part, with melatonin's ability to neutralize free radicals.

The actions of melatonin in different species have clearly diverged during the differentiation of major animal and plant taxa. These functions show a close relationship with the characteristics of the specific taxa. In mammals, melatonin is a molecule with hormonal properties (Nabavi et al., 2018). These properties are apparent in the regulation of seasonal reproductive activity, facilitation of sleep physiology, promotion of immunoresponsiveness, suppression of carcinogenesis, promotion of stem cell proliferation, antiinflammation, and in the modulation of aging (Zhao et al., 2019). Some of these actions are surely mandated by the interaction of melatonin with cell membrane receptors, that is, MT1 and MT2, and/or perhaps with nuclear binding sites and are, therefore considered hormonal (see below). For example, melatonin's ability to constrain cancer cell proliferation often involves membrane receptors (Zhao et al., 2019). There is also evidence, however, that receptor-independent actions such as free radical-generating capacity ("pro-oxidant"), an action possibly unique to cancer cells, also kills tumor cells (Bizzarri et al., 2013). Thus in mammals, melatonin is not a typical hormone and functions via receptor-dependent and receptor-independent means. On the evolutionary scale, the free radical-scavenging properties, which continue to exist in mammalian cells, preceded the evolution of receptors for this indoleamine; thus the initial actions of melatonin were receptor-independent (Zhao et al., 2019). Melatonin also has antioxidant, antiinflammatory, antiapoptotic, and many other crucial properties (Tan et al., 2015). Melatonin is a tryptophan-derived molecule with pleiotropic activities. For example, the tryptophan to melatonin pathway differs between plants and animals. Melatonin metabolism is highly complex with these enzymatic processes having evolved from cytochrome C. In addition to its enzymatic degradation, melatonin is metabolized via pseudo enzymatic and free radical interactive processes. The metabolic products of these processes overlap and it is often difficult to determine which process is dominant (Tan et al., 2015).

Its intrinsic properties, including high cell permeability, the ability to easily cross both the blood–brain and placenta morphophysiological barriers with ease, and its role as an endogenous reservoir of free radical scavengers (with indirect extra activities), confer on it beneficial uses as an adjuvant in the biomedical field. Melatonin can exert its effects by acting through specific cellular receptors on the plasma membrane, similar to other hormones, or through receptor-independent mechanisms that involve complex molecular cross-talk with other players. There is increasing evidence regarding the extraordinary beneficial effects of melatonin also via exogenous administration. In this chapter, we summarize molecular pathways in which melatonin is considered a master regulator, with attention to cell death and inflammation mechanisms from basic, translational, and clinical points of view in the context of newborn care (Tarocco et al., 2019).

Melatonin is a *powerful scavenger* of a number of ROS and RNS (reactive nitrogen species) both in vitro and in vivo. It has long been known to play a role in the direct scavenging of ROS and RNS, as well as inducing antioxidant defense mechanisms and ameliorating inflammation. It is well documented that ROS/RNS are powerful inducers of redox-sensitive transcription factor Nrf2. Melatonin directly scavenges a number of oxygen and nitrogen-based radicals, and they reduce the damage due to toxic chemicals which are proven to be harmful to cells. These receptor-independent actions of melatonin are similar to those of other intracellular nonenzymatic antioxidants such as glutathione (GSH). Melatonin also blocks proinflammatory genes by interfering with the genomic binding of NF-κB. However, melatonin possesses genomic actions and regulates the expression of several genes including those for superoxide dismutase (SOD) and glutathione peroxide (GSX-Px) (Choi et al., 2011). In conclusion, melatonin can reduce the index of oxidative stress in various clinical pathological states.

Melatonin in neuroprotection and central nervous system disorders

Nrf2-antioxidative stress signaling has the appearance of being ideal to prevent or treat nitroxidative stress-related diseases. Our findings indicate that melatonin is a potential neuroprotective agent in cyfluthrin intoxication. It was speculated that the potential of melatonin as a new palliative and neuroprotective agent was because melatonin impairs the activation of important signaling pathways that are responsible for neuronal damage. Reduction of inflammatory mediators, apoptosis, and unbalance of redox state are involved in the neuroprotection induced by melatonin. Furthermore, disruption of the melatoninergic system in diverse neurodegenerative diseases reinforces the potential neuroprotective effects of melatonin against the toxicity induced by pyrethroids (Martínez et al., 2019).

Radical formation is the key reaction in many diseases involving neurodegeneration, and immune, inflammatory, and mitochondrial diseases. The electrophiles and radicals generated and their metabolic products do not only cause damage to cellular components, but can also result in mutations leading to alterations in structural proteins and signaling pathways (Siah et al., 2014). Melatonin also stimulates the synthesis of glutathione, which is another antioxidant that reduces electron leakage from the mitochondrial electron chain. Melatonin is metabolized by free radicals and is converted to cyclic 3-hydroxymelatonin, providing antioxidant potential (Siah et al., 2014). Melatonin is a powerful scavenger of hydroxyl and peroxyl radicals and stimulates superoxide dismutase, glutathione peroxidase, and catalase, which have antioxidant activity (Roberts-Thomson et al., 1988). Decreased melatonin production and altered nocturnal melatonin secretion have been linked to various CNS disorders, such as stroke, obsessive-compulsive disorder, mood, and schizophrenia (Papavasiliou et al., 1972). The human brain represents only 2% of the total body weight but consumes 20% of the total oxygen intake of the body. It also generates more ROS compared with other body tissues. The brain produces polyunsaturated fatty acids and ascorbic acid, which are susceptible to damage by free radicals because of inefficient enzymatic oxidative defenses. Melatonin has a neuroprotective effect in many disorders, such as epilepsy, Parkinson disease, amyotrophic lateral sclerosis, Alzheimer disease, ischemic injury, and head injury. In many of these disorders, the loss of neurons and mitochondrial dysfunction involving free radicals and glutamate over-activity have been linked to disease progression (Siah et al., 2014).

Melatonin and inflammation

Chronic inflammation characterizes several common diseases. It has been shown that the use of melatonin is beneficial in several disorders where inflammation is implicated. Therefore melatonin is considered a potent and widespread antiinflammatory agent. Several studies have shown that melatonin can regulate the activation of the immune system, reducing chronic and acute inflammation. Experimental and clinical data suggest that melatonin exerts its antiinflammatory effects by modulating both pro- and antiinflammatory cytokines in various pathophysiological situations (Mauriz et al., 2013). Since different cytokines are associated with inflammatory diseases, wherein the balance between proinflammatory and antiinflammatory molecules determines the clinical outcome to some degree, melatonin may modulate serum inflammatory parameters. In addition, melatonin inhibits the expression of cyclooxygenase (COX) and inducible nitric oxide (NO) synthase (iNOS), and limits the production of excessive amounts of prostanoids, leukotrienes, and NO, as well as other mediators of the inflammatory process, such as chemokines and adhesion molecules (Tarocco et al., 2019). Melatonin is mainly reported to possess antiinflammatory properties by inhibiting inflammasome activation, thus inhibiting caspase-1 activation, cytokines release, and pyroptosis. In this respect, melatonin can also inhibit the expression of COX and iNOS by inhibiting nuclear NF-κB translocation (Tarocco et al., 2019). It is shown that aging status is associated with an increasingly high level of inflammatory circumstances and melatonin use can be of much interest.

Melatonin and immune responses

Currently there is accumulated evidence showing that the pineal gland itself is able to play an important role modulating the immune response (i.e., melatonin function is mediated through modulating the synthesis and release of cytokines). Melatonin can interact with specific membrane-binding sites in cells from lymphoid organs. Furthermore, interactions between the pineal gland and the immune system are bidirectional since interleukins and cytokines (interferon gamma) affect melatonin synthesis and release. Melatonin is known to have a relationship with the immune system because it is a positive regulator of the immune response and can activate both innate and adaptive immune responses at the time of inflammation (Claustrat and Lestonb, 2015). Melatonin regulates the immune system by exerting its action at different levels, through the stimulation of its receptors found on the immunocompetent cells, the maintenance of cell proliferation, and the modulation of cytokine production (Rivara et al., 2015). It is important to note that MT_1 receptors are expressed in the thymus and spleen as well as in the CD-4, CD-8, and B cells of the immune system, whereas MT_2 receptors are not expressed in any of these cells. MT_2 receptors help in preventing leukocyte rolling and MT_3 receptors regulate the adhesion of leukocytes to the vascular endothelium (Siah et al., 2014). In human lymphocytes, chronic administration of melatonin increases T-helper cell activity and IL-2 production. Furthermore, melatonin activates human monocytes by inducing cellular cytotoxic agents (interleukins and TNF-α) (Ekmekcioglu, 2006). Melatonin scavenging of NO or free radicals in lymphoid cells has also been demonstrated (Claustrat and Lestonb, 2015).

Melatonin and analgesia and other modulatory roles

Melatonin could be effective for reducing chronic pain associated with some clinical diseases (antinociceptive action). The molecular pathways underlying the analgesic action of melatonin have not been completely addressed, and an abundance of mechanisms have been proposed (e.g., endorphin release from the pituitary gland, modulation of the central GABAergic system, additional interaction with multiple receptor types). Those ligand–target bindings would include benzodiazepine, opioid, α_1 and α_2-adrenergic, serotonergic and cholinergic receptors, and particularly the antinociceptive role of melatonin mediated by MT_1 and MT_2 receptors (Dubocovich et al., 2010; Tarocco et al., 2019). Studies have postulated that this may be ascribed to the reduction in the excitability of pain-transmitting dorsal horn neurons via hyperpolarization due to $MT_{1/2}$–melatonin binding (Laurido et al., 2002). Among the effects provided by melatonin administration, the modulation of N-methyl-D-aspartate (NMDA) receptors and the conservation of myelin sheets in the CNS are probably the least treated. NMDA receptors are sensitive to glutamate binding, the major excitatory neurotransmitter in the brain whose extracellular levels increase abnormally during insults such as ischemia. Excessive release of glutamate and its binding with NMDA, α-amino-3-hydroxy-5-methyl-4-isoxazolepropionic acid (AMPA), or kainite receptors promotes excitotoxicity cell injury with a consequent activation of many molecular pathways, resulting in intracellular calcium (Ca^{2+}) overload, ROS production, mitochondrial permeability transition, and cell swelling with brain edema (Tarocco et al., 2019). Patch-clamp experiments demonstrated how melatonin is able to modulate NMDA receptor activity by the drastic attenuation of their currents in neurons located in the spinal cord dorsal horn. Focused experiments conducted in rat striatum synaptosome preparations have reported that melatonin inhibits the excitatory response in a partially Ca^{2+}-dependent manner in which the direct effect on membrane hyperpolarization promoted intracellular Ca^{2+} influx reduction. The involvement of melatonin is dose-dependent, and at physiological concentrations, it inhibited NMDA-induced current. At higher dosages, it also participated in the modulation of AMPA-glutamate binding. It follows that melatonin exerts powerful protective mechanisms against oxidative damage and excitotoxicity mediated by glutamate receptors (Kim and Kwon, 1999)

Melatonin as a neuroprotective agent was also recommended in those pathologies involving white matter damage. Indeed, researchers have reported a plethora of positive effects with different exposures of melatonin following CNS trauma, such as axonal regrowth and sprouting, conservation of a given thickness of myelin sheath in oligodendrocytes, and improved nerve maturation (Olivier et al., 2009; Tarocco et al., 2019).

Melatonin plays an important role in gastrointestinal physiology as well, which includes regulation of gastrointestinal motility, local antiinflammatory response, and moderation of visceral sensation. Melatonin is a potential candidate for the treatment of irritable bowel syndrome (IBS), since melatonin has analgesic effects, regulates gastrointestinal motility, and is a promoter of the sensation to sleep. Its analgesic effect may be due to the interaction among melatonin receptors and two endogenous opioids, Met-enkephalin and β-endorphin. Several well-designed placebo-controlled trials in IBS patients consistently showed improvement of abdominal pain when taking a 3 mg dose of melatonin with no serious side effects (Siah et al., 2014). Melatonin plays several pivotal local intestinal functions such as: (1) regulation of GI motility (excitatory and inhibitory effects on gut smooth muscles); (2) antiinflammatory reaction (increases natural killer cell activity and Th2 cell-mediated immune responses); and (3) moderation of visceral sensation (Siah et al., 2014). Melatonin might be a candidate for IBS treatment in light of its: (1) analgesic effects; (2) regulatory effects on GIT motility and sensation; (3) sleep-promoting effect (IBS patients can have sleep disturbance); and (4) mood regulation and antistress effects (Siah et al., 2014). Melatonin has shown clinical applicability from its action on specific receptors to its action on depressive and anxiety mood disorders.

Toxicity and safety of melatonin

To date, no toxicology data based on regulatory requirements has been undertaken on melatonin, particularly for dietary supplements. In contrast, medicines containing melatonin are available. In subjects taking melatonin, no evidence of toxicity has been reported (i.e., no deaths or serious accidents). However, secondary effects (mild or severe) which occurred after administration of melatonin or its metabolites (i.e., linked to its pharmacological effects), have been reported on the cardiovascular system, gastrointestinal system, glucose metabolism, and in cancer and immunology (Guardiola-Lemaître, 1997). Melatonin treatment has an extremely wide margin of safety though it may cause minor adverse effects, such as headaches, rashes, and nightmares.

While the current understanding of endogenous melatonin is concrete, especially regarding melatonin's involvement in the circadian timing system, its putative role in regulating sleep and how this role is mediated remains unclear (EMEA, 2007). In vivo studies in animals essentially looked at sleep induction effects, but the results are difficult to interpret and extrapolate to humans (EMEA, 2007).

The following toxicological data are available for melatonin:

1. *Acute toxicity*. Melatonin has low toxicity after a single administration. The median lethal dose (LD_{50}) value is 1250 mg/kg p.c. in mice and more than 3200 mg/kg p.c. in rats (EMEA, 2007). The intravenous LD_{50} value is 180 mg/kg to 472 mg/kg in mice and 356 mg/kg in rats. The main effects observed at high doses were sedation, lethargy, and vasodilation. The higher doses led to impairment of righting, placing and flexor reflexes, marked reduction in body temperature, and respiratory distress preceding death (EMEA, 2007).
2. *Toxicity after repeated administration*. The toxicity of melatonin after repeated administration was evaluated in the rat and dog. After repeated administration of melatonin in rats (13 weeks, doses of 0.3, 1.2, and 6 mg/kg/day; or 26 weeks, doses of 15, 75, and 150 mg/kg/day) and dogs (6 months, doses of 0.4, 1.5, and 8 mg/kg), toxic effects were observed on the liver (hypertrophy) and the genital tract of the male rats and female dogs (adenomyosis). These effects were observed at 75 mg/kg/day or higher in rats. These effects were observed in humans at doses greater than the therapeutic dose (EMEA, 2007).
3. *Genotoxicity and carcinogenicity*. The full battery of genotoxicity tests according to the International Conference for Harmonisation standard have been performed (EMEA, 2007). The Ames test, in vitro gene mutations in mouse lymphoma cells, in vitro chromosome aberration in human lymphocytes, and in vivo mouse micronucleus were all negative. Both tests showed that melatonin has no mutagenic, clastogenic, or aneugenic effects. Further

literature data report that melatonin and two related compounds, 6-hydroxymelatonin (main metabolite) and 5-methoxyindoleacetric acid (5-MIAA), were screened for relevant information associating these chemicals with respect to mutagenic or carcinogenic effects (in silico DEREK system). No structural alerts were identified in silico. Overall, it is concluded that melatonin does not present any genotoxic potential (EMEA, 2007).

The carcinogenic potential of melatonin was evaluated in one long-term rat study and a short-term mouse study (US, National Toxicology Program (NTP) study). In the combined rat chronic toxicity and oncogenicity study (104 weeks), an increased incidence of pituitary adenomas in high-dose males was reported with statistical significance. An increase in thyroid tumors was observed at the highest doses in the study. Since no genotoxic properties have been identified for melatonin, and as the animal exposure was in large excess to that expected in the clinic, considering also that the treatment is not proposed to be continuous for long periods, and provided that this well-established mechanism of thyroid tumorigenesis in rodents (this type of tumor is classified as common in the rat) can be clearly proven by the applicant, the risk to humans appears to be minimal (EMEA, 2007). Given that no genotoxic properties have been identified for melatonin, that the exposure doses in rats were significantly higher than those expected in humans, and that taking melatonin for long periods is not recommended, the risk of carcinogenicity to humans is minimal (EMEA, 2007). Since 2007, numerous in vivo and in vitro experimental data have been published on the oncostatic effect of melatonin on different tumors (breast, prostate, colorectal cancer, etc.) (Reiter et al., 2017).

4. *Reproductive and developmental toxicity*. In reproductive studies, such as fertility and early embryonic development, in the rat treated orally by gavage with 0, 15, 55, or 200 mg/kg/day of melatonin; embryo/fetal development in rats, melatonin was administered by gavage to 25 timed-mated Sprague Dawley (CD^R) female rats on gestation days 6 to 19, at doses of 50, 100, and 200 mg/kg/day, and in New Zealand White (NZW) rabbits (oral administration of melatonin at 15, 50, and 150 mg/kg/day from days 7 to 19 of gestation); and pre- and postnatal developmental study in rats (treated with 0, 15, 55, and 200 mg/kg/day of melatonin from day 6 of gestation to day 21 postpartum, inclusive). Findings revealed a toxic effect of melatonin on the embryo/fetal development in rabbits and on postnatal development in rats.

In addition, melatonin is present in human breast milk and may communicate time of day information to breastfed infants. Exogenously ingested melatonin may also reach breast milk and reflect blood concentrations of melatonin with a short delay. Although there is no evidence that melatonin has an effect on breastfed babies, treatment in breastfeeding mothers is not recommended, as reflected in the summary of product characteristics (SPC) ("there are data in animal models including rodents, sheep, bovine and primates that indicate maternal transfer of melatonin to the foetus via the placenta or in the milk"). Therefore the use of Circadin is not recommended during pregnancy and lactation. The EMEA advises against melatonin intake by pregnant and breastfeeding women. Since Circadin is proposed only for use in patients 55 years of age or over, reproduction toxicity and milk transfer are of little relevance to this age group (EMEA, 2007). The new data available in the literature on animals do not alter the conclusions of the EMA (Singh et al., 2013).

Adverse effects in humans

Between the establishment of nutrivigilance (scheme national toxicovigilance network) in 2009 and May 2017, the French agency ANSES received 90 reports of adverse effects likely to be associated with the consumption of food supplements mainly containing melatonin. A literature search was conducted for adverse effects reported in nutrivigilance (ANSES, 2018) in order to observe their frequency of occurrence and the melatonin doses at which they appeared. These different adverse effects have been reported in the scientific literature, mainly during clinical trials, to assess the efficacy and safety of melatonin. The most frequently reported effects were general symptoms [headaches (Buscemi et al., 2005, 2006), dizziness, drowsiness], neurological (tremors, migraine), digestive (nausea, vomiting, abdominal pain), and psychological (nightmares, irritability) disorders.

The occurrence of headaches has been reported where melatonin doses were between 2 and 10 mg/day (Smits et al., 2001; Campos et al., 2004; EMEA, 2007; Hoebert et al., 2009; Costello et al., 2014; Xu et al., 2015). The vasodilatory action of melatonin has been offered as the mechanism responsible for this effect (Claustrat, 2009).

Many other publications dealing with the efficacy and safety of melatonin administered over several days, weeks, or months at doses between 0.05 and 10 mg/day, have listed sleepiness as an adverse effect felt by the participants (Arendt, 1997; Force et al., 1997; Buscemi et al., 2005, 2006; EMEA, 2007; Miano et al., 2008; Hoebert et al., 2009; Gringras et al., 2012; Costello et al., 2014; Fallah et al., 2015; Xu et al., 2015). Based on this finding, Suhner et al. (1998) studied the impact of melatonin administration on road-driving performance. Since subjective sleepiness increased after the administration of this sleep hormone, the authors recommended that caution should be exercised when driving under the influence of melatonin. The SPC for Circadin states that "Circadin may cause drowsiness, therefore the product should be used with caution if the effects of drowsiness are likely to be associated with a risk to safety" (EMA, 2014).

The occurrence of sleepiness following the administration of melatonin is explained by its hypnotic properties. While the cases of sleepiness were mostly reported for melatonin doses higher than those found in food supplements, people practicing activities where a reduction in vigilance could have implications on their personal safety or that of others should not consume food supplements containing melatonin.

Two cases of abdominal pain and vomiting, where causality was deemed likely, were reported under the nutrivigilance scheme. Gastrointestinal disorders are adverse effects frequently reported following the administration of melatonin (Claustrat, 2009). Studies have revealed the existence of melatonin receptors in the GIT. The activation of these receptors could explain certain digestive disorders observed via the effect on intestinal motricity (Guardiola-Lemaître, 1997).

Several cases of onset or reactivation of inflammatory diseases (Crohn disease, ulcerative colitis, and rheumatoid arthritis) and autoimmune diseases (hepatitis) after the consumption of melatonin have been described (Fourman and Robert Meyer, 2013). The consumption of food supplements containing melatonin should be avoided in people suffering from inflammatory or autoimmune diseases (ANSES, 2018).

Adverse events with a possible or probable relationship to the administration of 2 and 10 mg Circadin included fatigue (in 7 out of 16 subjects after 2 mg; 8 out of 16 after 10 mg), sensation of heaviness (1 out of 16 after 2 mg), somnolence (3 out of 16 after 2 mg; 2 out of 16 after 10 mg), falling asleep (2 out of 16 after both 2 and 10 mg), headache (1 out of 16 after 2 mg; 3 out of 16 after 10 mg), and nausea (1 out of 16 after 10 mg). All adverse events were transient and mild. Nausea, fatigue, and headache were more frequently reported following Circadin 10 mg compared with Circadin 2 mg (EMA, 2014).

Drug interactions

It is becoming increasingly clear that the interactions which occur when drugs are given concurrently are often the result, not of a single mechanism, but of two or more mechanisms acting in concert.

Drug–metabolism interactions

Many adult and pediatric patients being treated with melatonin for a sleep disorder are receiving other prescription drugs (and perhaps complementary and alternate medicines too) where the clinical relevance of the findings is known. The primary pathway of melatonin metabolism involves CYP1A2 (and to a lesser extent CYP2C19), and as these enzymes are involved in the metabolism of many drugs, there is the opportunity for important drug interactions (Kennaway, 2014). The metabolism of melatonin to 6-sulfatoxymelatonin and *N*-acetylserotonin is catalyzed by cytochrome-P450 (CYP) isozymes CYP1A2 and CYP2C19, respectively. Melatonin may modify the plasma concentration of drugs metabolized by the isoenzymes CYP1A1, CYP1A2, and CYP1B1 or CYP2C19 because these isoenzymes are involved in the metabolism of melatonin (Ma et al., 2005). Because liver CYP1A2 levels are at 20%–25% of adult levels in children aged 3–12 months and 50%–55% of adult levels in children from 1 to 9 years of age (Sonnier and Cresteil, 1998), children who are administered melatonin are clearly being exposed to extremely high levels of the hormone compared with adults receiving the same dose. Interactions are therefore possible between melatonin and other active substances that induce or inhibit the microsomal and other enzymes and/or are metabolized by them. In patients treated with 5- or 8-methoxypsoralen (CYP1A1 inhibitor), cimetidine (CYP2D6 inhibitor), or quinolones (CYP1A2 inhibitors), or those on estrogen (CYP1A2 inhibitor) therapy (e.g., contraceptives or hormone replacement therapy), an increase in plasma concentrations of melatonin may be observed as a result of a decrease in its metabolism. Fluvoxamine [antidepressant inhibitor of the active reuptake (transport) of serotonin (5-hydroxytryptamine, 5-HT)] inhibits CYP1A2, but also the reuptake of serotonin, a precursor of melatonin. An increase in the plasma concentration of melatonin administered with fluvoxamine has been observed because fluvoxamine is an inhibitor of CYP1A2 and CYP2C19, as is caffeine (Härtter et al., 2003) and presumably also theophylline. Therefore these combinations must be avoided. Inhibitors of CYP2C19, like the antidepressant citalopram, also altered the metabolism of orally administered melatonin (Huuhka et al., 2006). It has also been stated that melatonin must not be used with benzodiazepines and related hypnotic drugs because this increases the risk of cognitive and psychomotor adverse effects, and there are no efficacy data for this combination (Australian Government, 2009). Moreover, caution should be exercised in patients on cimetidine (a CYP2D inhibitor, which increases plasma melatonin levels, by inhibiting its metabolism), and in patients on estrogens (e.g., contraceptive or hormone replacement therapy), which increase melatonin levels by inhibiting its metabolism by CYP1A1 and CYP1A2. CYP1A2 inhibitors such as quinolones may give rise to increased melatonin exposure and CYP1A2 inducers such as carbamazepine and rifampicin may give rise to reduced plasma concentrations of melatonin (EMA, 2014).

The in vitro induction of CYP3A by melatonin has also been shown. Although the clinical relevance of this enzyme induction is not known, a decrease in the plasma concentration of medicinal products metabolized by this isoform in the event of co-administration with melatonin is possible. Given the strong binding of melatonin to albumin, α_1-acid glycoprotein, and high-density lipoprotein, interactions with drugs binding to the same plasma macromolecules may occur.

Drug pharmacodynamic interactions

Melatonin may amplify the sedative properties of novel benzodiazepine-receptor agonists whose chemical structures do not resemble those of benzodiazepines. This class includes zolpidem, zolpicone, and zaleplon (pyrazolopyrimidines are hypnotic agents that act at the benzodiazepine recognition site of $GABA_A$ (gamma-aminobutyric acid) receptors, and exhibit similar patterns of efficacy among the different receptor subtypes). Zaleplon and zolpidem are effective in relieving sleep onset insomnia. The concomitant administration of melatonin and zolpidem resulted in increased impairment of attention, memory, and coordination, compared to zolpidem used alone. Similarly, studies have shown that melatonin, when administered concomitantly with imipramine (tertiary-amine-tricyclic antidepressant) and thioridazine (antipsychotic drug), increased difficulty in performing tasks, compared to imipramine used alone, and caused more pronounced dizziness compared to thioridazine used alone (ANSES, 2018).

Lusardi et al. (2000) showed that the concomitant use of melatonin with nifedipine (calcium channel antagonist used to manage angina, high blood pressure, Raynaud syndrome, and premature labor) reduced the latter's antihypertensive effect and could lead to a loss of control of hypertension. However, it is also possible that melatonin may potentiate the effects of antihypertensive drugs (Sajith and Clarke, 2007). These effects could be related to melatonin's effects on vasomotricity. Melatonin has an inhibitory effect on platelet aggregation, and may potentiate the effects of other platelet inhibitors such as aspirin or anticoagulants such as warfarin. In addition, the immunomodulatory effects of melatonin could counteract the effects of antiinflammatory agents such as corticosteroids (Sajith and Clarke, 2007).

A pharmacodynamic interaction is possible between melatonin and platelet inhibitors, anticoagulants, antiinflammatory agents, and substances acting on the CNS, particularly hypnotic and antiepileptic agents.

In general, the consumption of food supplements containing melatonin in combination with drug therapy should be discussed with a physician or pharmacist (ANSES, 2018).

Vulnerable populations and at-risk situations

Pregnancy and breastfeeding

It has been demonstrated that endogenous melatonin plays an essential role during pregnancy and in fetal development, especially in the brain. For this, melatonin is proposed as a possible treatment for certain conditions in pregnancy (preeclampsia, threat of premature birth, intrauterine growth restriction) and brain damage in the perinatal period. Clinical studies are currently in progress (Sagrillo-Fagundes et al., 2016). However, no clinical data are available on the effect of melatonin administration to pregnant women experiencing normal pregnancy. Endogenously produced melatonin has been detected in breast milk (Illnerova et al., 1993; ANSES, 2018). This observation suggests that melatonin from an exogenous source could pass into breast milk (Vidal, 2017). Amer et al. (2015) believe that the melatonin concentration in breast milk could increase from 0.4 to 1 μg/L for 1 mg of exogenous melatonin consumed by the mother. They conclude that long-term consumption of this substance could alter the natural development of the child's circadian rhythm and their sleep cycle.

In light of the results of reproductive toxicity studies conducted in animals and in the absence of clinical data, the consumption of food supplements containing melatonin is not recommended during pregnancy (Andersen et al., 2016a,b). Because exogenous melatonin passes into breast milk, breastfeeding women should not consume it as a supplement (ANSES, 2018).

Children and adolescents

There is emerging evidence that the use of medications for pediatric sleep problems is common practice in the United States. Over 50% of pediatricians recommended nonprescription medications for sleep problems in children, and 15% of respondents recommended either melatonin or herbal preparations. Melatonin is more appealing than hypnotics because of its sleep-promoting effect for sleep problems without alteration of sleep function (Owens et al., 2005). Data are available in the literature on the effectiveness and safety of melatonin in the treatment of sleep disorders in children and adolescents with mental and neurological disorders (Sánchez-Barceló et al., 2011). It is important to note that the main outcomes of the efficacy of sleep disorders in children must take into consideration: (1) sleep onset latency (the time between laying down to sleep and the onset of sleep), and (2) sleep efficiency (the amount of time spent asleep as a percentage of the total time spent in bed) (Buscemi and Witmans, 2006). Studies suggest that melatonin may help to reduce sleep onset latency in children with primary and secondary sleep disorders; however, the clinical significance of these findings is unclear. Sleep disturbance is often multifactorial, and differences in age and comorbid conditions may explain the range in findings among studies with respect to the efficacy of melatonin (Buscemi and Witmans, 2006).

It has been speculated that long-term administration of melatonin may influence sexual maturation and large-scale randomized controlled trials are definitely needed before long-term safety can be established in children and adolescents. In France, Circadin benefits from a recommendation for temporary use for children over 6 years of age, for disturbances of the sleep–wake cycle related to developmental disorders and neurogenic diseases. This treatment is subject to a strict monitoring protocol. In the absence of sufficient safety data, the consumption of food supplements containing melatonin by children and adolescents is not recommended (ANSES, 2018).

Other populations

Other sensitive populations or at-risk situations were identified following the analysis of the literature and nutrivigilance cases. People with epilepsy, asthma, or suffering from mood, behavior, or personality disorders should not consume food supplements containing melatonin without seeking the advice of their general practitioner (ANSES, 2018).

Uses of melatonin

Melatonin is commonly given orally. In the United States, melatonin is available as an over-the-counter nonprescription drug. Moreover, melatonin is considered a dietary supplement and is not regulated by the US FDA. As a consequence, it is freely available in pharmacies and supermarkets. Consultations with the focal points of the European Food Safety Authority (EFSA) revealed the wide variability in the regulatory status of melatonin in the European Union. In Spain and Italy, melatonin is authorized for use in food supplements up to 1 mg/day. In France and Latvia in particular, regulations authorize the marketing of food supplements that provide less than 2 mg of melatonin per day. It needs to be emphasized that products containing melatonin (<2 mg), legally marketed in some European Member States, may be placed on other European country markets as food supplements, with a "principle of mutual recognition" and complying with the notification of their placing on the market to the competent national authorities. In some countries, including Denmark and the Czech Republic, melatonin is prohibited in food supplements.

The EFSA has released two opinions regarding melatonin as a food constituent: (1) The first opinion (EFSA, 2010) addresses the scientific substantiation of health claims in relation to melatonin and alleviation of subjective feelings of jet lag, reduction of sleep onset latency, and contribution to sleep quality. The claimed effect is sleep–wake cycle regulation. The target population is assumed to be the general population. In the context of the proposed wordings, the Panel assumes that sleep–wake cycle regulation encompasses subjective feelings of jet lag. The Panel considers that alleviation of subjective feelings of jet lag might be a beneficial physiological effect. On the basis of the data presented, the Panel concludes that: The food constituent melatonin, that is the subject of the health claim, is sufficiently characterized: (a) *Alleviation of subjective feelings of jet lag* (ID 1953): (i) the claimed effect is *sleep–wake cycle regulation*, (ii) the target population is assumed to be the general population, (iii) alleviation of subjective feelings of jet lag might be a beneficial physiological effect, (iv) a cause–effect relationship has been established between the consumption of melatonin and alleviation of subjective feelings of jet lag, (v) the following wording reflects the scientific evidence: "Melatonin contributes to the alleviation of subjective feelings of jet lag", (vi) in order to bear the claim, the melatonin dose should be between 0.5 and 5 mg and should be taken close to bedtime on the first day (and any subsequent day) of travel and on the following few days after arrival at the destination. (b) Reduction of sleep onset latency and improvement of sleep quality (ID 1953): (i) the claimed effect is "sleep–wake cycle regulation." The target population is assumed to be the general population, (ii) Reduction of sleep onset latency and improvement of sleep quality might be beneficial physiological effects. (iii) A cause and effect relationship has not been established between the consumption of melatonin and reduction of sleep onset latency or improvement of sleep quality. (2) The second opinion (EFSA, 2011) addresses the scientific substantiation of a health claim in relation to melatonin and reduction of sleep onset latency. The food constituent that is the subject of the health claim is melatonin. The NDA-EFSA Panel considers that the food constituent, melatonin, which is the subject of the health claim, is sufficiently characterized. The claimed effects are *sleep–wake cycle regulation*, *relaxation*, and *sleep patterns*. The NDA-EFSA Panel assumes that the target population is the general population. In the context of the proposed wordings and the clarifications provided from Member States and the references provided, the Dietetic Products, Nutrition and Allergies (NDA) Panel assumes that the claimed effects refer to the reduction of sleep onset latency (time taken to fall asleep). Sleep onset latency can be measured by validated methods. On the basis of the data presented, the NDA Panel concludes that: (a) the food constituent, melatonin, which is the subject of the health claim, is sufficiently characterized; (b) the claimed effects are *sleep–wake cycle regulation*, *relaxation*, and *sleep patterns*. The target population is assumed to be the general population. Reduction of sleep onset latency might be a beneficial physiological effect, (c) a cause and effect relationship has been established between the consumption of melatonin and reduction of sleep onset latency, (d) the following wording reflects the scientific evidence: "Melatonin helps to reduce the time to fall asleep," (e) in order to obtain the claimed effect, 1 mg of melatonin should be consumed close to bedtime (EFSA, 2011). In conclusion, EFSA has issued a favorable opinion for two claims relating to the presence of melatonin in foodstuffs (EU, 2012): (1) melatonin contributes to the alleviation of subjective feelings of jet lag. This claim may be used only for food which contains at least 0.5 mg of melatonin per quantified portion. In order to bear the claim, information shall be given to the consumer that the beneficial effect is obtained with a minimum intake of 0.5 mg to be taken close to bedtime on the first day of travel and on the following few days after arrival at the destination, and (2) melatonin contributes to the reduction of time taken to fall asleep. This claim may be used only for food which contains 1 mg of melatonin per quantified portion. In order to bear the claim, information shall be given to the consumer that the beneficial effect is obtained by consuming 1 mg of melatonin close to bedtime.

In Australia and many European Union countries, melatonin remains a prescription drug and can only be obtained through compounding pharmacies as medicinal products (Kenaway et al., 2014). In Belgium and Germany, products providing 0.3 mg or more of melatonin per day are considered medicinal products by function, in light of their pharmacological activity. Table 34.1 lists the health claims and clinical indications for melatonin in humans in the EU (EFSA, 2010, 2011). Melatonin as a prescription drug has received from the EMA two authorizations (Circadin and Slenyto). The first human medicine was Circadin (melatonin controlled-release tablets, presented in 2 mg tablets), that received approval from EMA for short-term treatment (up to 3 weeks) of primary insomnia characterized by poor quality of sleep in people aged $\geq$ 55 years (EMEA, 2007). The claimed therapeutic indication for Circadin was: "The relief of primary insomnia characterized by poor quality of sleep in patients aged 55 or over," and the indication was: "Circadin is indicated as monotherapy for the short-term treatment of primary insomnia characterized by poor quality of sleep in patients who are aged 55 or over." The second human medicine was Slenyto (containing the pharmacologically active substance melatonin, tablets 1 and 5 mg) was approved as a medicine for treating insomnia in children and adolescents (2–18 years old) who have: (1) autism spectrum disorder, a range of conditions that affects the patient's social interactions; and (2)

TABLE 34.1 Use of melatonin as dietary supplement and/or medicinal drug in the EU.

Dietary supplement (health claim)	"Melatonin contributes to the alleviation of subjective feelings of jet lag." The dose of melatonin should be between 0.5 and 5 mg, should be taken close to bedtime on the first day (and any subsequent day) of travel and on the following few days after arrival at the destination
Dietary supplement (health claim)	"Melatonin contributes to the reduction of time taken to fall asleep." This claim may be used only for food which contains 1 mg of melatonin per quantified portion
Medicinal drug (Circadin, melatonin controlled-release, 2 mg tablets)	"For short-term treatment (up to 3 weeks) of primary insomnia characterized by poor quality of sleep in people aged $\geq$ 55 years." Therapeutic indication: "the relief of primary insomnia characterized by poor quality of sleep in patients aged 55 or over." The recommended dose of Circadin is one tablet a day, taken 1–2 h before bedtime and after food. This dose can be continued for up to 13 weeks. Single morning doses of 2 and 10 mg Circadin mini-tablets administration in children with neurodevelopmental disorders with sleep disturbances over the age of 2 and under the age of 18 years
Medicinal drug (Slenyto, melatonin, 1–5 mg tablets)	"For treating insomnia in children and adolescents (2–18 years old) who have: (1) autism spectrum disorder, a range of conditions that affects the patient's social interactions, and (2) Smith–Magenis syndrome, a condition that can lead to learning difficulties." Slenyto has been developed specifically for children and is available in an age-appropriate form (small tablets). The normal dose is 2 mg taken 30 min to 1 h before bedtime. The dose can be increased up to a maximum of 10 mg if the medicine is not working well enough. The physician should review treatment regularly, at least every 6 months, and only continue treatment if the patient benefits from it.

Smith–Magenis syndrome, a condition that can lead to learning difficulties. Slenyto was developed specifically for children and is available in an age-appropriate form (small tablets) (EMA, 2018). In a main study of 125 patients, those given Slenyto over 13 weeks slept on average 51 extra minutes a night compared with 19 extra minutes for those given placebo. In addition, children who took Slenyto fell asleep around 38 minutes earlier than normal, while those taking placebo fell asleep only 13 minutes earlier. All patients had previously tried other measures, such as keeping to a regular sleeping routine, which did not work. The most common side effects with Slenyto (which may affect up to 1 in 10 people) are sleepiness, tiredness, mood swings, headache, irritability, aggression, and feeling hungover (EMA, 2018). Slenyto has been shown to be effective and safe in this population. Clinical trial data and the scientific literature suggest that the medicine is associated with a significant increase in total sleep time, shortened sleep latency, and longer duration of uninterrupted sleep. The main side effects observed in the clinical trial were somnolence, headache, and fatigue.

Because melatonin can reset rhythms, it has been tested as a treatment for initial insomnia and jet lag. However, it is not recommended, because morning exposure to bright light is more effective that melatonin at resetting rhythms for healthy individuals.

Acknowledgments

This work was supported by Project Ref. RTA2015-00010-C03-03 from Ministerio de Economía, Industria y Competitividad, Spain.

References

Acuña-Castroviejo, D., Escames, G., Venegas, C., et al., 2014. Extrapineal melatonin: sources, regulation, and potential functions. Cell Mol. Life Sci. 71 (16), 2997–3025.

Agathokleousa, E., Kitaob, M., Calabrese, E.J., 2019. New insights into the role of melatonin in plants and animal. Chem. Biol. Interact. 299, 163–167.

Allegra, M., Reiter, R.J., Tan, D.X., et al., 2003. The chemistry of melatonin's interaction with reactive species. J. Pineal Res. 34, 1–10.

Amer, M.R., Cipriano, G.C., Venci, J.V., et al., 2015. Safety of popular herbal supplements in lactating women. J. Hum. Lact. 31 (3), 348–353.

Andersen, L.P.H., Werner, M.U., Rosenkilde, M.M., et al., 2016a. Pharmacokinetics of high-dose intravenous melatoninin humans. Br. J. Clin. Pharmacol. 56 (3), 324–329.

Andersen, L.P.H., Gögenur, I., Rosenberg, J., et al., 2016b. The safety of melatonin in humans. Clin. Drug Invest. 36 (3), 169–175.

ANSES, 2018. Opinion of the French Agency for Food, Environmental and Occupational Health & Safety on the Risks Associated With the Consumption of Food Supplements Containing Melatonin. Opinion Request No. 2016-SA-0209, Maisons-Alfort, 23 February 2018.

Arendt, J., 1997. Safety of melatonin in long-term use. J. Biol. Rhythms 12 (6), 673–681.

Australian Government, 2009. Department of Health and Ageing Therapeutic Goods Administration. Australian Public Assessment Report for Melatonin. Submission No. PM-2008-2125-1- 2009.

Balduini, W., Weiss, M.D., Carloni, S., et al., 2019. Melatonin pharmacokinetics and dose extrapolation after enteral infusion in neonates subjected to hypothermia. J. Pineal Res. 66, e12565.

Bartoli, A.N., De Gregori, S., Molinaro, M., et al., 2012. Bioavailability of a new oral spray melatonin emulsion compared with a standard oral formulation in healthy volunteers. J. Bioequiv. Bioavailability 4 (7), 96–99.

Bizzarri, M., Proietti, S., Cucina, A., et al., 2013. Molecular mechanisms of the pro-apoptotic actions of melatonin in cancer: a review. Expert Opin. Ther. Targets 17 (12), 1483–1496.

Buscemi, N., Witmans, M., 2006. What is the role of melatonin in the management of sleep disorders in children? Paediatr. Child. Health 11 (8), 517–519.

Buscemi, N., Vandermeer, B., Hooton, N., et al., 2005. The efficacy and safety of exogenous melatonin for primary sleep disorders: a meta-analysis. J. Gen. Intern. Med. 20 (12), 1151–1158.

Buscemi, N., Vandermeer, B., Hooton, N., et al., 2006. Efficacy and safety of exogenous melatonin for secondary sleep disorders and sleep disorders accompanying sleep restriction: meta-analysis. BMJ 332 (7538), 385–388.

Calabrese, E.J., 2001. Apoptois: biphasic dose responses. Crit. Rev. Toxicol. 31, 607–613.

Campos, F.L., Da Silva-Júnior, F.P., De Bruin, V.M.S., et al., 2004. Melatonin improves sleep in asthma: a randomized, double-blind, placebo-controlled study. Am. J. Respir. Crit. Care Med. 170 (9), 947–951.

Cardinali, D.P., Lynch, H.J., Wurtman, R.J., 1972. Binding of melatonin to human and rat plasma proteins. Endocrinology 91 (5), 1213–1218.

Carloni, S., Proietti, F., Rocchi, M., 2017. Melatonin pharmacokinetics following oral administration in preterm neonates. Molecules 22, 2115. Available from: https://doi.org/10.3390/molecules22122115.

Cavanaugh, J.E., Witt-Enderby, P.A., 2010. CNS melatonin receptors and signaling: focus on ageing related disease and future perspective. Open Neuroendocrinol. J. 3, 96–104.

Choi, E.Y., Jin, J.Y., Lee, J.Y., et al., 2011. Melatonin inhibits *Prevotella intermedia* lipopolysaccharide-induced production of nitric oxide and interleukin-6 in murine macrophages by suppressing NF-κB and STAT-1 activity. J. Pineal Res. 50, 197–206.

Cipolla-Neto, J., Amaral, F.G., Afeche, S.C., et al., 2014. Melatonin, energy metabolism, and obesity: a review. J. Pineal Res. 56 (4), 371–381.

Claustrat, B., 2009. Mélatonine et troubles du rythme veille-sommeil. Médecine du. Sommeil 6 (1), 12–24.

Claustrat, B., Lestonb, J., 2015. Melatonin: physiological effects in humans. Neurochirurgie 61, 77–84.

Claustrat, B., Brun, J., Chazot, G., 2005. The basic physiology and pathophysiology of melatonin. Sleep Med. Rev. 9 (1), 11–24.

Comai, S., Gobbi, G., 2014. Unveiling the role of melatonin MT2 receptors in sleep, anxiety and other neuropsychiatric diseases: a novel target in psychopharmacology. J. Psychiatry Neurosci. 39 (1), 6–21.

Costello, R.B., Lentino, C.V., Boyd, C.C., et al., 2014. The effectiveness of melatonin for promoting healthy sleep: a rapid evidence assessment of the literature. Nutr. J. 13, 106.

Drzeżdżon, J., Jacewicz, D., Chmurzyński, L., 2018. The impact of environmental contamination on the generation of reactive oxygen and nitrogen species—consequences for plants and humans. Environ. Int. 119, 133–151.

Dubocovich, M.L., Delagrange, P., Krause, D.N., et al., 2010. International union of basic and clinical pharmacology. LXXV. Nomenclature, classification, and pharmacology of G protein-coupled melatonin receptors. Pharmacol. Rev. 62 (3), 343–380.

EFSA, 2010. Scientific opinion on the substantiation of health claims related to melatonin and alleviation of subjective feelings of jet lag (ID 1953), and reduction of sleep onset latency, and improvement of sleep quality (ID 1953) pursuant to Article 13(1) of Regulation (EC) No 1924/2006. EFSA J. 8 (2), 1467. 18 pp. EFSA Panel on Dietetic Products Nutrition and Allergies (NDA).

EFSA, 2011. Scientific opinion on the substantiation of a health claim related to melatonin and reduction of sleep onset latency (ID 1698, 1780, 4080) pursuant to Article 13(1) of Regulation (EC) No 1924/2006. EFSA J. 9 (6), 2241. EFSA Panel on Dietetic Products, Nutrition and Allergies (NDA).

Ekmekcioglu, C., 2006. Melatonin receptors in humans: biological role and clinical relevance. Biomed. Pharmacother. 60 (3), 97–108.

EMA, 2014. Circadin (Melatonin). Procedure No. EMA/H/C/695/P46/019 Marketing Authorisation Holder (MAH): RAD Neurim Pharmaceuticals EEC Ltd. 18 December 2014, EMA/85025/2015, CHMP.

EMA, 2018. Slenyto (Melatonin). Procedure No. EMEA/H/C/004425/0000. Assessment Report 16 July 2018, EMA/556280/2018, CHMP.

EMEA, 2007. Assessment Report for Circadin. International Nonproprietary Name: Melatonin. Procedure No. EMEA/H/C/695.

Emet, M., Ozcan, H., Ozel, L., et al., 2016. A review of melatonin, its receptors and drugs. Eurasian J. Med. 48 (2), 135–141.

EU, 2012. Commission Regulation (EU) No 432/2012 of 16 May 2012 Establishing a List of Permitted Health Claims Made on Foods, Other Than Those Referring to the Reduction of Disease Risk and to Children's Development and Health. OJ of EU L 136/1, 25.5.2012.

Fallah, R., Shoroki, F.F., Ferdosian, F., 2015. Safety and efficacy of melatonin in pediatric migraine prophylaxis. Curr. Drug Saf. 10 (2), 132–135.

Force, R.W., Hansen, L., Bedell, M., 1997. Psychotic episode after melatonin. Ann. Pharmacother. 31 (11), 1408.

Fourtillan, J.B., Brisson, A.M., Gobin, P., et al., 2000. Bioavailability of melatonin in humans after day-time administration of D7 melatonin. Biopharm. Drug Dispos. 21, 15–22.

Fourman, L.T., Robert Meyer, B., 2013. Autoimmune hepatitis in association with ramelteon. J. Clin. Gastroenterol 47 (7), 651–654.

Gringras, P., Gamble, C., Jones, A.P., et al., 2012. Melatonin for sleep problems in children with neurodevelopmental disorders: randomised double masked placebo controlled trial. BMJ 345 (7882). Available from: https://doi.org/10.1136/bmj.e6664.

Guardiola-Lemaître, B., 1997. Toxicology of melatonin. J. Biol. Rhythms 12 (6), 697–706.

Harpsøe, N.G., Andersen, L.P., Gögenur, I., et al., 2015. Clinical pharmacokinetics of melatonin: a systematic review. Eur. J. Clin. Pharmacol. 71 (8), 901–909.

Härtter, S., Nordmark, A., Rose, D.-M., et al., 2003. Effects of caffeine intake on the pharmacokinetics of melatonin, a probe drug for CYP1A2 activity. Br. J. Clin. Pharmacol. 56, 679–682.

Hoebert, M., Van Der Heijden, K.B., Van Geijlswijk, I.M., et al., 2009. Long-term follow-up of melatonin treatment in children with ADHD and chronic sleep onset insomnia. J. Pineal Res. 47 (1), 1–7.

Huuhka, K., Riutta, A., Haataja, R., et al., 2006. The effect of CYP2C19 substrate on the metabolism of melatonin in the elderly: a randomized, double-blind, placebo-controlled study. Methods Find. Exp. Clin. Pharmacol. 28, 447–450.

Illnerova, H., Buresova, M., Presl, J., 1993. Melatonin rhythm in human milk. J. Clin. Endocrinol. Metab. 77 (3), 838–841.

Jeanneteau, F., Arango-Lievano, M., 2016. Linking mitochondria to synapses: new insights for stress-related neuropsychiatric disorders. Neural Plast. 2016, 3985063.

Kennaway, D.J., Wright, H., 2002. Melatonin and circadian rhythms. Curr. Top. Med. Chem. 2, 199–209.

Kennaway, D.J., 2014. Light at night, melatonin and breast cancer. Chronobiol Int. 31(2), 297–298.

Kennaway, D.J., 2015. Potential safety issues in the use of the hormone melatonin in paediatrics. J. Paediatr. Child Health 51(6), 584–589.

Kim, H.J., Kwon, J.S., 1999. Effects of placing micro-implants of melatonin in striatum on oxidative stress and neuronal damage mediated by N-methyl-D-aspartate (NMDA) and non-NMDA receptors. Arch. Pharm. Res. 22, 35–43.

Lane, E.A., Moss, H.B., 1985. Pharmacokinetics of melatonin in man: first pass hepatic metabolism. J. Clin. Endocrinol. Metab. 61, 1214–1216.

Laurido, C., Pelissier, T., Soto-Moyano, R., et al., 2002. Effect of melatonin on rat spinal cord nociceptive transmission. Neuroreport 13 (1), 89–91.

Lerner, A.B., Case, J.D., Takahashi, Y., et al., 1958. Isolation of melatonin, the pineal gland factor that lightens melanocyte S1. J. Am. Chem. Soc. 80, 2587.

Lusardi, P., Piazza, E., Fogari, R., 2000. Cardiovascular effects of melatonin in hypertensive patients well controlled by nifedipine: a 24-hour study. Br. J. Clin. Pharmacol. 49 (5), 423–427.

Ma, X., Idle, J.R., Krausz, K.W., et al., 2005. Metabolism of melatonin by human cytochromes P450. Drug Metab. Dispos. 33 (4), 489–494.

Macchi, M.M., Bruce, J.N., 2004. Human pineal physiology and functional significance of melatonin. Front. Neuroendocrinol. 25, 177–195.

Martínez, M.A., Rodriguez, J.L., Lopez-Torres, et al., 2019. Oxidative stress and related gene expression effects of cyfluthrin in human neuroblastoma SH-SY5Y cells: protective effect of melatonin. Environ. Res. 177, 108579.

Mauriz, J.L., Collado, P.S., Veneroso, C., et al., 2013. Review of the molecular aspects of melatonin's anti-inflammatory actions: recent insights and new perspectives. J. Pineal Res. 54, 1–14.

Miano, S., Parisi, P., Pelliccia, A., et al., 2008. Melatonin to prevent migraine or tension-type headache in children. Neurol. Sci. 29 (4), 285–287.

Morin, D., Simon, N., Depres-Brummer, P., et al., 1997. Melatonin high-affinity binding to alpha-1-acid glycoprotein in human serum. Pharmacology 54 (5), 271–275.

Nabavi, S.F., Nabavi, S.F., Sureda, A., et al., 2018. Antiinflammatory effects of melatonin: a mechanistic review. Crit. Rev. Food Sci. Nutr. 14, 1–13.

Olivier, P., Fontaine, R.H., Loron, G., et al., 2009. Melatonin promotes oligodendroglial maturation of injured white matter in neonatal rats. PLoS One 4 (9), e7128.

Owens, J.A., Babcock, D., Blumer, J., et al., 2005. The use of pharmacotherapy in the treatment of pediatric insomnia in primary care: rational approaches. A consensus meeting summary. J. Clin. Sleep Med. 1 (1), 49–59.

Papavasiliou, P.S., Cotzias, G.C., Düby, S.E., et al., 1972. Melatonin and parkinsonism. JAMA 221, 88–89.

Reiter, R.J., Rosales-Corral, S.A., Tan, T.X., et al., 2017. Melatonin, a full service anti-cancer agent: inhibition of initiation, progression and metastasis. Int. J. Mol. Sci. 18 (4), E843.

Rivara, S., Pala, D., Bedini, A., et al., 2015. Therapeutic uses of melatonin and melatonin derivatives: a patent review (2012–2014). Expert Opin. Ther. Pat. 25 (4), 425–441.

Roberts-Thomson, I.C., Knight, R.E., Kennaway, D.J., et al., 1988. Circadian rhythms in patients with abdominal pain syndromes. Aust. N. Z. J. Med. 18, 569–574.

Sagrillo-Fagundes, L., Salustiano, M.A., Yen, P.W., et al., 2016. Melatonin in pregnancy: effects on brain development and CNS programming disorders. Curr. Pharm. Des. 22 (8), 978–986.

Sajith, S.G., Clarke, D., 2007. Melatonin and sleep disorders associated with intellectual disability: a clinical review. J. Intellect. Disabil. Res. 51 (1), 2–13.

Sánchez-Barceló, E.J., Mediavilla, M.D., Reiter, R.J., 2011. Clinical uses of melatonin in pediatrics. Int. J. Pediat. 2011, 892624. Available from: https://doi.org/10.1155/2011/892624.

Siah, K.T.H., Wong, R.K.M., Ho, K.Y., 2014. Melatonin for the treatment of irritable bowel syndrome. World J. Gastroenterol. 20 (10), 2492–2498.

Singh, M., Jadhav, H.R., 2014. Melatonin: functions and ligands. Drug Discov. Today 19 (9), 1410–1418.

Singh, H.J., Saleh, H.I., Gupalo, S., et al., 2013. Effect of melatonin supplementation on pregnancy outcome in Wistar-Kyoto and Sprague-Dawley rats. Acta Physiol. Sin. 65 (2), 149–157.

Skene, D.J., Papagiannidou, E., Hashemi, E., et al., 2001. Contribution of CYP1A2 in the hepatic metabolism of melatonin: studies with isolated microsomal preparations and liver slices. Contribution of CYP1A2 in the hepatic metabolism of melatonin: studies with isolated microsomal preparations and liver slices. J. Pineal Res. 31, 333–342.

Smits, M.G., Nagtegaal, E.E., van der Heijden, J., et al., 2001. Melatonin for chronic sleep onset insomnia in children: a randomized placebo-controlled trial. J. Child Neurol. 16 (2), 86–92.

Sonnier, M., Cresteil, T., 1998. Delayed ontogenesis of CYP1A2 in the human liver. Eur. J. Biochem. 251, 893–898.

Suhner, A., Schlagenhauf, P., Tschopp, A., et al., 1998. Impact of melatonin on driving performance. J. Travel Med. 5 (1), 7–13.

Tan, D.X., Manchester, L.C., Reiter, R.J., et al., 1998. A novel melatonin metabolite, cyclic 3-hydroxymelatonin: a bio marker of in vivo hydroxyl radical generation. Biochem. Biophys. Res. Commun. 253, 614–620.

Tan, D.X., Manchester, L.C., Esteban-Zubero, E., et al., 2015. Melatonin as a potent and inducible endogenous antioxidant: synthesis and metabolism. Molecules 20, 18886–18906.

Tarocco, A., Caroccia, N., Morciano, G., et al., 2019. Melatonin as a master regulator of cell death and inflammation: molecular mechanisms and clinical implications for newborn care. Cell Death Dis. 10 (4), 317.

Vidal 2017: Le Dictionaire. 92e édition.

Welford, R.W., Vercauteren, M., Trebaul, A., et al., 2016. Serotonin biosynthesis as a predictive marker of serotonin pharmacodynamics and disease-induced dysregulation. Sci. Rep. 6, 30059.

Xu, J., Wang, L.L., Dammer, E.B., et al., 2015. Melatonin for sleep disorders and cognition in dementia: a meta-analysis of randomized controlled trials. Am. J. Alzheimers Dis. Other Demen 30 (5), 439–447.

Zhao, D., Yu, Y., Shen, Y., et al., 2019. Melatonin synthesis and function: evolutionary history in animals and plants. Front. Endocrinol. 10, 249.

Chapter 35

Standardized turmeric and curcumin

Naresh Chand and Amit Kumar

Introduction

The turmeric plant is native to the Indian subcontinent and Southeast Asia. It is a herbaceous perennial plant belonging to the ginger family, Zingiberaceae. Turmeric (pulverized turmeric rhizome, root) is an ancient Ayurvedic herb that has been traditionally used for cosmetic, religious, and festivity-related reasons as well as for enhancing the aroma, flavor, and color of foods in traditional cooking since ancient times in turmeric-consuming nations. *Curcuma longa* (turmeric) has been traditionally used in Asian countries for its health benefits, such as abdominal pain, amenorrhea, congestive heart failure, constipation, and stomach pain. The curcumin extracted from *Curcuma* rhizome is an extensively studied natural phytochemical. Studies conducted in animals suggest that turmeric or its major ingredients (curcumin or curcuminoids, turmerones, and turmerosaccharides) may have some therapeutic potential, including antiinflammatory, analgesic, immunomodulatory, neuroprotective, hepatoprotective, nephroprotective, chemopreventive, anticancer, antiproliferative, antimicrobial, antifungal, antiviral, carminative, diuretic, antiarthritic, antidiabetic, antidiarrheal, and antithrombotic activities, among others (Aggarwal et al., 2013; Dosoky and Setzer, 2018; Xu et al., 2018; Chand, 2019).

More than 800 dietary supplements (DS) contain turmeric or standardized turmeric (ST) or its extracts or novel formulations in the United States. The clinical efficacy and safety of these combination products (DS) and novel formulations of curcumin and curcuminoids, including ADRxSynergy, need to be explored using well-designed large clinical trials in people living with early disease states of the eyes, joints, and brain, etc.

The chemical structure of curcuminoids, curcumin [diferuloylmethane; 1,7-bis(4-hydroxy-3-methoxyphenyl)-1,6-heptadiene-3,5-dione], demethoxycurcumin, and bisdemethoxycurcumin are shown in Fig. 35.1 (Xu et al., 2018).

There are about 93–100 different species of turmeric (*C. longa*). *Curcuma aromatica* (wild turmeric), *Curcuma wenyujin* (China), and *Curcuma domestica* (Thailand) are among the most popular species of turmeric plant. Turmeric contains curcuminoids (5%–10%), and curcuminoid-free ingredients (>90%). The most active curcuminoid-free pharmacological ingredients include essential oils (turmerones) and turmerosaccharides. The chemical composition of turmeric root and plant may vary considerably. The cultivation practices, fertilizer application, stress during growth or maturity, harvesting time, stage of maturity, genotype, variety, differential geography (climate, season), extraction, and analysis methods may result in significant differences in the amount of active ingredients in turmeric (Dosoky and Setzer, 2018; Chand, 2019).

Native turmeric has >235 complex phytochemicals naturally assembled in the roots. Cooking may activate complex curcuminoids (89), fiber, essential oils, and turmerosaccharides. In the process of cooking turmeric in oil, the activated phytochemicals and fiber may exert additive or synergistic beneficial health effects in turmeric-consuming nations in Asia and other parts of the world (Dosoky and Setzer, 2018; Chand, 2019).

Turmeric has been described as reducing aches, pain, flu, cold, boils in the skin, and even concussion (Dosoky and Setzer, 2018; Chand, 2019; Meng et al., 2018). These broad spectra of health benefits may be attributed to the placebo effect, from a scientific perspective. However, many professionals working with turmeric strongly believe that >235 complex phytochemicals may act in an additive or synergistic fashion to reduce chronic dysregulated immune inflammation in many diseases. The health benefits of turmeric and its extracts and novel formulations may be more apparent in early disease states. The mild laxative effects of turmeric and its interaction with gut microbiota are considered significant in Ayurvedic medicine. Turmeric is also known as the "poor man's aspirin" as it is available in every home and affordable as food in Asia, the Middle East, and Africa.

The scientific community, including global formulation experts, is highly successful in improving the bioavailability of curcumin or curcuminoids (Table 35.2). In

Nutraceuticals. DOI: https://doi.org/10.1016/B978-0-12-821038-3.00035-5

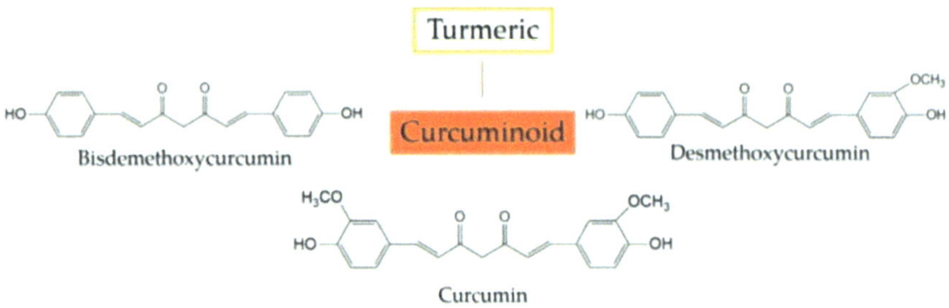

FIGURE 35.1 The chemical structure of principal curcuminoids: curcumin, demethoxycurcumin, and bisdemethoxycurcumin. *Modified after Xu, X.Y., Meng, X., Li, S., et al., 2018. Bioactivity, health benefits, and related molecular mechanisms of curcumin: current progress, challenges, and perspectives. Nutrients, 10.*

TABLE 35.1 Composition of turmeric and three major curcuminoids.

Turmeric constituents		Curcuminoids (%)	
%(w/w)			
Curcuminoids	1–6	Curcumin	60–70
Essential oils (TEO)	3–7	Demethoxycurcumin	20–27
Fiber	2–7	Bismethoxycurcumin	10–15
Minerals	3–7		
Fat	5–10		
Protein	6–8	Turmerosaccharide	10
Carbohydrate	60–70		
Moisture	6–13		

TEO, turmeric essential oil.
Source: Modified from Nelson, K.M., Dahlin, J.L., Bisson, J., et al., 2017. The essential medicinal chemistry of curcumin. J. Med. Chem., 60, 1620–1637.

addition to improvement in curcumin's pharmacokinetics (PK), high doses of curcumin were utilized in most clinical trials conducted so far. Neither improvement of bioavailability nor increased dose improved curcumin's efficacy in Alzheimer disease (AD) (Mazzanti and Di Giacomo, 2016). Turmeric and curcumin are often considered as safe. Many healthy people in the nonturmeric-consuming nations (such as the United States) use too much turmeric and its extracts as food or as DS. They may not be benefit from using higher doses, and in fact, some may face adverse consequences, especially seniors taking several medications.

Myth or reality?

Nelson et al. (2017) suggested that curcumin is an unstable, reactive, and nonbioavailable compound, and therefore that it is not a lead compound for R&D. This concept may be a myth or a fact depending on how one looks at the scientific published literature or unpublished observations. The astronomical doses of turmeric or curcumin used in in vitro studies, animals, or healthy people or patients may validate the concept proposed by Nelson and her associates (Nelson et al., 2017). In this chapter, we take a different perspective on this subject of great economic and healthcare importance for aging populations around the world. The goal of this chapter is to summarize the effects of turmeric or low doses of curcumin in animals and humans (Begum et al., 2008; Lim et al., 2001; Teter et al., 2019; Chand, 2019; Moghaddam et al., 2019; Receno et al., 2019). We will offer preclinical scientific evidence that low doses of turmeric or its extracts may exert modulation of immune inflammation in many disease states and thus may reduce the need for opiates, NSAIDs, steroids, and anticancer agents, etc. In other words, low doses of turmeric or ST, or novel formulations, may act as an adjunct therapy in many diseases with underlying persistent chronic dysregulated immune inflammation.

Composition of turmeric

Nelson et al. (2017) summarized the composition of turmeric (Table 35.1). The chemical structures of curcuminoids are depicted in Fig. 35.1. Turmeric powder contains three major curcuminoids (curcumin, demethoxycurcumin, and bismethoxycurcuminoid of 89 curcuminoids) (Jia et al., 2017), essential oils (aromatic-turmerone, α-turmerone, and β-turmerone, and many others), curcumin-free components (Aggarwal et al., 2013), and turmerosaccharides (Bethapudi et al., 2017). These complex phytochemicals in native turmeric or ST and their potential metabolites may act in an additive or synergistic fashion to exert diverse pharmacologic activities in animal models of diseases. Turmeric or its extracts exert protective activity in inflammatory diseases such as inflammatory bowel disease (IBD), chronic pain, cancer, neurodegenerative diseases, depression, diabetes, obesity, and atherosclerosis in animal models (Bethapudi et al., 2017; Chand, 2019; Gupta et al., 2013).

Aggarwal et al. (2013) reviewed the antiinflammatory and anticancer activities of curcumin-free components of turmeric (CFT). Turmerin, turmerone, elemene, furanodiene, curdione, bisacurone, cyclocurcumin, calebin A, and germacrone are some of the ingredients in CFT. Elemene is an approved anticancer agent in China. The curcumin-free turmeric (CFT) components exert activities such as antiinflammatory, anticancer, and antidiabetic activities similar to turmeric. CFT and its active ingredients need more extensive pharmacological evaluations in natural disease states in animals and humans (Aggarwal et al., 2013).

The crude turmeric essential oil (TEO) is more effective in blocking nonarticular inflammation than purified TEO (Funk et al., 2010). This observation suggests that the additional components present only in the crude extract are also bioactive with independent and/or synergistic antiinflammatory effects. Dosoky and Setzer (2018) have summarized the volatile components of various *Curcuma* species, their pharmacological activities, and potential safety concerns in an outstanding review.

Low doses of ADRxSynergy (>325 heat-activated essential oils) and other active ingredients in turmeric standardized to curcumin ($4 + 0.8\%$) may have additive or synergistic immunomodulatory effects, suppressing the generation and secretion of tumor necrosis factor-α (TNF-α) and interleukin IL-1β, IL-6, and improving antiinflammatory IL-10 secretion. It may act via multiple modes of action.

Consumption

Prasad and Aggarwal (2011) summarized the history, cultivation, uses, and processing of turmeric in detail. It is used as a preservative and antimicrobial agent, and in making pickles (turmeric, mango, limes, lemons, and others), savory and sweet dishes, and is widely used in Eastern culinary specialties. The consumption of turmeric in Asian countries is in the range of 200–1000 mg/day or 160–440 g/person/year. Intake in urban areas is lower (200 mg/day) than in rural areas (600 mg/day/person). The estimated average daily use of curcuminoids may be 10–100 mg daily in India and other turmeric-consuming nations (personal observation). The old-fashioned way to formulate turmeric (Haldi) was to heat one teaspoon of coconut oil (or other oils, ghee, or butter) in a cooking pan, then add an equal amount of turmeric powder with slow mixing for 60–120 seconds until the color became brownish. Then, grains, beans, or vegetables were added to the cooking pan. This preparation (soup, dal, beans) was often shared by 5–10 people in the family. The estimated amount of turmeric consumed daily is about 100 mg twice or three times a day. Higher doses may act as a proinflammatory and prooxidant. The estimated cost of turmeric (160–440 g/person/year) may vary from the place of cultivation, origin, and international trade (import and export, etc). For example, it is less than one dollar in India or China, and \$3–100 in the United States (Chand, 2019).

Methods to improve the bioactivity of turmeric

In traditional Indian and turmeric-consuming nations, turmeric is always used for enhancing the color, flavor, aroma, and presentation of the dishes. It has been reported that cooking doubles the antioxidative activity of turmeric (Tiwari et al., 2006) and curry paste retains its antioxidant activity even after heating (Romson et al., 2011), therefore it could be valuable as a functional food. Furthermore, Choi et al. (2019) discovered that puffing (a cooking process using heat and pressure) is a promising and simple method for enhancing the antioxidant and antiinflammatory activities of turmeric. Sun et al. (2019) reviewed the effect of heating on the biological activity of curcuminoids. They observed that boiled curcuminoids (mimicking the cooking scenario) possess relatively stronger antioxidant activity than the roasted and fried forms. Heating increased the solubility of boiled curcuminoids in phosphate-buffered saline. They concluded that the cooked curcuminoids, including boiled and roasted forms, still exert antioxidant and neuroprotective activities. Cooking may also soften the fiber content of turmeric, so it is a palatable food (prebiotic) for the gut microbiota. The products of this pathway may have some health benefits in the body (Sun et al., 2019). It is important to note that excessive and prolonged heat treatment of turmeric may degrade curcuminoids and reduce antioxidant activity (Park et al., 2019). The effects of cooking on curcuminoid-free components of turmeric (turmerones—the essential oils and turmerosaccharides, and others) are not yet known.

Safety: potential risks and adverse side effects

Curcuminoids are labeled as generally recognized as safe by the US Food and Drug Administration (FDA). Curcumin has good tolerability and safety profiles even at oral doses between 4000 and 8000 mg/day, and of doses up to 12,000 mg/day of 95% concentration of three curcuminoids—curcumin, bisdemethoxycurcumin, and demethoxycurcumin (Hewlings and Kalman, 2017). Soleimani et al. (2018) reviewed the safety of turmeric and curcumin and concluded that turmeric and curcumin are nontoxic to humans after oral administration. In addition, oral bioavailable formulations of curcumin at a dose of 500 mg twice a day for 30 days were also safe. In fact, curcumin after metabolic activation may exert prooxidant effects, similar to numerous other antioxidants (Kawanishi et al., 2005). Synthetic (VEAMIN 99) or extracted curcumin may be used as a food additive in various food categories at levels of 0.5–100 mg/100 g, depending upon the specific food category. The synthetic analogs of curcumin have not been studied in an animal model or clinical studies. Therefore the usefulness of synthetic curcumin as an R&D candidate remains unknown (Kumar et al., 2017).

Curcumin and other ingredients interact with drug-metabolizing enzymes, such as the cytochrome P450 (CYP) isozymes CYP1A2, CYP2A6, CYP2C6, CYP2C9, and CYP2DS (Concetta Scuto et al., 2019). Therefore the safety profile for novel formulations (Table 35.2) and ST (ADRxSynergy) still needs to be explored. Caution must be exercised when using high doses of DS (curcuminoids, curcumin, ST, and others) especially among seniors who are taking multiple medications every day. The interactions between the DS, medications (drugs), and diseases in aging populations are still an underexplored area in human clinical safety (Jamwal, 2018).

A case study

A male (age >60, had some arthritis in one knee) and his doctor who believes in holistic medicine suggested using turmeric extract. He bought turmeric with 95% curcuminoids [C3 complex 1500 mg; plus, black pepper extract (Bioperine 5 mg)] from Costco ($65 for 90 capsules: with a dose of two tablets every day). C3 represents curcumin, demethoxycurcumin, and bisdemethoxycurcumin. He was supposed to take this formulation for at least 3 months (one capsule twice a day), however, within a few weeks, he started experiencing the following symptoms:

1. Stiffness in the left leg and thigh;
2. Frequent urination;
3. numbness, pain, tingling, and burning sensations starting in the toes and fingers then continuing up the legs or arms.

The diagnostic blood work demonstrated that his blood level of vitamin B6 was high (72.0; approximately 4–36 times higher when compared to a normal B6 blood level of 2.1–21.7). His physician recommended stopping the use of this dietary supplement. His symptoms resolved 4–6 weeks after discontinuation of the high dosage of this formulation. The gut–microbiota–DS interaction may be part of this increased vitamin B6 synthesis. The exact mechanism or prevalence of this neuropathy remains unknown. (personal communication).

Formulations

A small quantity of powdered turmeric is utilized two or three times a day as a flavoring and coloring spice in cooking in the Indian subcontinent and Southeast Asia (Prasad and Aggarwal, 2011). Turmeric powder alone (often in combination with other spices) is cooked in heated vegetable oil or ghee/butter for 1–2 minutes before adding to foods (soaked beans, grains, rice, and vegetables). This ancient herbal formulation in Indian food often consists of more than 1000 complex phytochemicals from many other herbs (spices). The essential oils and other active ingredients in this ancient formulation of spices in Asian cooking may improve the bioavailability of phytochemicals. For example, piperine (an active component of black pepper), when combined with curcumin, increases curcumin's bioavailability by 2000% (Hewlings and Kalman, 2017). This may have profound effects on the PK and pharmacodynamics of turmeric, ST, and curcuminoids.

Many R&D experts over the past 30 years have reported that low water solubility, instability, and short half-life of curcumin in plasma poses significant challenges in R&D (Jamwal, 2018; Zhang et al., 2019b; Nelson et al., 2017). The 11 bioavailable curcumin/curcuminoid formulations exhibit an improved pharmacokinetic profile (Table 35.2). NovaSol, CurcuWin, and LongVida exhibit over 100-fold higher bioavailability relative to unformulated curcumin. Jamwal suggested that the variability in the amounts of curcumin/curcuminoids in these 11 formulations does not permit a direct comparison of PK parameters [area under curve (AUC), T_{max}, and C_{max}]. Still, the PK observations demonstrate that the curcumin formulations achieve higher AUC and C_{max} when compared to unformulated curcumin (Jamwal, 2018). The relative tissue distribution and dose-response of these formulations in disease states remain unexplored.

The advances in fabrication and therapeutic applications of curcumin-encapsulated liposomes, micelles, nanoemulsions, nanosuspensions, nanofibers, nanocomposites, and some nanoparticles including solid lipid nanoparticles, nanostructured lipid carriers, protein nanoparticles, and polysaccharide nanoparticles thus improve bioavailability and overcome inherent challenges as reviewed recently (Zhang et al., 2019a). These nanopreparations may maintain the pharmacological effect of herbal extracts and allow drug (active ingredient) delivery to the site of action in the disease state over time. These innovative nanopreparations of curcumin may be applied

TABLE 35.2 Pharmacokinetic parameters of curcumin from the different curcumin-based formulations and unformulated curcumin.

Formulation	Intervention	Dose	C_{max} (ng/mL)	T_{max} (h)	AUC_{0-t} (ng/mL · h)	$t_{1/2}$ (h)	RB curcumin
Meriva	Formulation	297 mg curcumin	50.3 ± 12.7	3.8 ± 0.6	538.0 ± 130.7[1]	NR	48
	Reference	1295 mg curcumin	9.0 ± 2.8	6.9 ± 2.2	122.5 ± 29.3[1]	NR	
LongVida	Formulation	650 mg curcuminoids	22.4 ± 1.9	2.4 ± 0.4	95.3 ± 4.6[1]	7.5 ± 2.4	100
	Reference	650 mg curcuminoids	<1	ND	ND	ND	
CurQfen	Formulation	600 mg curcumin	0.4 ± 0.2 (μg/g)	1	8100 ± 287[2] (μg/g · h)	NR	15.8
	Reference	1000 mg curcumin	0.02 ± 0.01 (μg/g)	0.5	510 ± 123[2] (μg/g · h)	NR	
MicroActive curcumin	Formulation	500 mg curcumin	NR	4*	887.5 ± 549.9[3]	NR	9.7
	Reference	500 mg curcumin	NR	NR	91.8 ± 50.0[3]	NR	
Micronised curcumin	Formulation	410 mg curcumin	15.3 ± 8.9	8.8 ± 6.4	214.6 ± 106.4[3]	NR	9
	Reference	410 mg curcumin	2.6 ± 4.9	7.5 ± 8.2	24.1 ± 42.6[3]	NR	
NovaSol	Formulation	410 mg curcumin	1189.1 ± 518.7	1.1 ± 0.4	4474.7 ± 1675.2[3]	NR	185
	Reference	410 mg curcumin	2.6 ± 4.9	7.5 ± 8.2	24.1 ± 42.6[3]	NR	
CurcuWin	Formulation	376 mg curcuminoids	27.3 ± 6.4	1.4 ± 0.5	307.6 ± 44.6[3]	NR	136.3
	Reference	1800 mg total curcuminoids	2.3 ± 0.3	7.4 ± 1.0	10.8 ± 1.7[3]	NR	
Biocurcumax (BCM-95)	Formulation	2000 mg curcuminoids	456.9* (μg/g)	3.44*	3201.3[4]* (μg/g · h)	4.96*	6.9
	Reference	2000 mg curcuminoids	149.8* (μg/g)	2*	461.9[4]* (μg/g · h)	2.63*	
Curcumin C3 complex + bioperine	Formulation	2000 mg curcumin with bioperine	180 ± 30	0.69 ± 0.07	80 ± 10[5]	0.11 ± 0.02	20
	Reference	2000 mg curcumin	6 ± 5	1*	4[5]*	NR	
Cavacurmin	Formulation	376 mg curcuminoids	73.2 ± 17.5	1*	327.7 ± 58.1[3]	NR	85
	Reference	1800 mg total curcuminoids	ND	12*	3.9 ± 0.5[3]	NR	
Theracurmin	Formulation	30 mg curcumin	29.5 ± 12.9	1*	113 ± 61[2]	NR	15.9
	Reference	30 mg curcumin	1.8 ± 2.0	6*	4.1 ± 7[2]	NR	

Source: Adapted from Jamwal, R., 2018. Bioavailable curcumin formulations: a review of pharmacokinetic studies in healthy volunteers. J. Integr. Med., 16, 367–374.

for the prevention and treatment of various diseases, especially cancer and inflammatory disorders.

The R&D experts need to establish dose-responses (ADME, PK/PD), and the safety and efficacy of these novel formulations (Table 35.2), ST, and nanopreparations, especially in age-related disease states (cancer, tumors, MCI, dementia, IBD, ulcerative colitis (UC), sarcopenia, osteoporosis, and osteoarthritis (OA), etc.).

Ayurveda is an ancient art and science of restoring homeostasis under early disease states. Ayurveda proposes that the amount of spice does not follow a perfect relationship to effectiveness (PK/PD modeling)—meaning that more is not always better. The opposite situation may happen at higher doses, meaning higher doses may act as a prooxidant or proinflammatory and may negate its health benefits (Calabrese et al., 2019). One of Ayurvedic medicine's basic principles is that efficacy cannot be related to plasma levels of a significant ingredient in a spice or herb, and the whole herb is often more efficacious than its single chemical parts (Meng et al., 2018; Chand, 2019). Preclinical studies support this concept that low doses are more effective than higher doses of curcumin (Begum et al., 2008; Lim et al., 2001; Teter et al., 2019; Bastaki et al., 2016; Moghaddam et al., 2019).

The Ayurvedic principle should be kept in mind when using natural phytochemicals (novel formulations) for restoring homeostasis in the early disease states. It suggests that higher doses, and higher bioavailability, meaning the higher plasma levels (AUC or C_{max}) of curcumin, may not lead to improved efficacy. It may negate its delivery to the site of immune inflammation or disease states, such as brain diseases (Begum et al., 2008; Lim et al., 2001; Teter et al., 2019). Therefore investigators should consider using a wide range of doses of these novel formulations, ST, and nanopreparations in future clinical trials.

Aging pets (dogs, cats, horses) and zoo animals offer a golden opportunity for testing the efficacy and safety of phytochemicals. The dose-response (ADME, PK/PD, the safety and efficacy) of these formulations, ST, and nanopreparations in age-related disease states (cancer, tumors, MCI, dementia, IBD, UC, sarcopenia, osteoporosis, and OA, etc.) need to be evaluated.

Central nervous system diseases

Alzheimer disease

AD is a growing cause of cognitive deficits in the aging population worldwide. Dysregulated immune inflammation, oxidative stress, mitochondrial damage or autophagy, and aging may play a significant role in AD pathogenesis. Curcumin has received increased interest due to its unique molecular structure that targets multiple pathways as well as amyloid aggregation—one of the significant hallmarks of AD in animal models. Voulgaropoulou et al. (2019) conducted a systematic review of preclinical and clinical studies and reported the effect of curcumin on cognition in people living with AD and healthy aging. They concluded that the preclinical animal studies are very promising, meaning curcumin may slow AD pathology and progression (Voulgaropoulou et al., 2019; Chand, 2019). However, human clinical studies are limited, and results are inconsistent, thus making any interpretation very challenging. These inconsistencies may be related to differences in methodology, doses, state of AD and formulations, etc. The measurement of AD biomarkers, disease state, optimal dosages of curcumin, or its novel formulations, ST, interactions with food and medications in aging populations, and the duration and onset of treatment would increase the understanding of curcumin's effects on cognitive deficits (Voulgaropoulou et al., 2019; Chand, 2019).

Low levels of brain-derived neurotrophic factor (BDNF) in serum are associated with cognitive, depressive, bipolar disorder, schizophrenia, and anxiety-related disorders. In a recent review analyzing the impact of curcumin supplementation on serum BDNF levels, Sarraf et al. (2019) found that curcumin supplementation (200–1820 mg/day and 8–12 weeks, respectively) increased the serum level of BDNF in four (total seven) randomized control trials with 139 participants. Interestingly, they observed a nonlinear dose-response association between curcumin intake (dose) and the mean difference in serum BDNF levels. Based on prior publications and their observations, they concluded that increased BDNF levels might improve cognition, learning, and memory by modulating immune inflammation, and restoration of the redox state (Sarraf et al., 2019). Low doses of turmeric (ST) or its extracts (curcumin, curcuminoids, and others) may help in slowing the progression of neurodegenerative diseases and comorbidities by improving the level of BDNF in the brain and serum. Consistent with this, Teter et al. (2019) demonstrated that a low dose of curcumin restores innate immune gene expression in an AD mouse model. In their experiment, mice were fed either chow containing a low (160 ppm) or high (5000 ppm) dose of curcumin, or no drug for 6 months before euthanasia. The curcumin treatment reduced proinflammatory microglia biomarkers (CD11b, iNOS, COX-2, IL1β), CD33 expression, while increasing TREM2 and TYROBP expression and activated phagocytosis, and increasing CD68 and Arg1 expression (Teter et al., 2019). These results demonstrate a novel mode of action for curcumin. Based on these findings they concluded that curcumin is an immunomodulatory agent that acts via stimulating phagocytic clearance of amyloid while restoring neuroinflammatory networks implicated in neurodegenerative diseases.

Preclinical studies tend to suggest that curcumin may have therapeutic potential for ischemic stroke (Liu et al., 2017). Microglia activation is involved in many neurologic disorders such as traumatic brain injury, Parkinson's, AD, and stroke. Hucklenbroich et al. (2014) summarized the importance of microglia activation, which is a hallmark of an innate immune-inflammatory response of the central nervous system (CNS). Aromatic (ar-) turmerone, a significant bioactive compound in turmeric, inhibited microglial activation, a property that may be useful in treating brain disease. Besides some promising studies, evidence-based medicine has demonstrated the lack of any therapeutic effect of curcumin in the onset or progression of neurodegeneration and related diseases (Hucklenbroich et al., 2014). Finally, the curcumin safety profile imposes a careful analysis of the risk–benefit balance before proposing chronic supplementation with curcumin (Concetta Scuto et al., 2019). Yavarpour-Bali and colleagues reviewed the effects of curcumin nanoparticles (CN) and their possible mechanisms of neuroprotective action in various CNS-related diseases including Parkinson's, Huntington's, Alzheimer's, multiple sclerosis, epilepsy, and amyotrophic lateral sclerosis (Yavarpour-Bali et al., 2019).

Depression

Zhu et al. (2019) reviewed preclinical studies and meta-analyses on curcumin's effect on depression. The animal studies suggested possible antidepressant effects. However, curcumin did not significantly influence self-reported measures of depressive symptoms in human trials. More clinical well-designed long-term studies are needed using a novel formulation of curcumin and ST (ADRxSynergy) in people living with the symptoms of depression and comorbidities (cognitive deficits, anxiety, schizophrenia, etc.).

Chronic pain

The prolonged use of opiates for the treatment of chronic pain leads to opioid-induced hyperalgesia (OIH). This is a major health problem in chronic pain management and fuels the opiate epidemic. Grover and Samson (2016) suggested that cuisines with the appropriate antioxidants (spice and herb combinations) may be more economical and more natural to incorporate into the lifestyle to slow or prevent the progression of OA and chronic pain. Orally administered PLGA-curcumin in mice prevented the development of opioid (morphine) tolerance and dependence in a dose-dependent manner, with ED50 values of 3.9 and 3.2 mg/kg, respectively. Furthermore, PLGA-curcumin dose-dependently attenuated already-established opioid tolerance (ED50 = 12.6 mg/kg p.o.) and dependence (ED50 = 3.1 mg/kg p.o.) in mice (Hu et al., 2015). In a separate study, PLGA-curcumin administered intrathecally or orally significantly attenuates morphine-induced hyperalgesia in mice. This formulation seems to act by suppressing chronic morphine-induced CaMKIIα activation in the superficial laminae of the spinal dorsal horn (Hu et al., 2016), suggesting that PLGA-curcumin may reverse OIH, possibly by inhibiting CaMKIIα and its downstream signaling. These studies suggest that novel formulations of curcumin, curcuminoids, and ST may reduce opiate abuse liability and exert opiate-sparing effects in patients living with chronic pain.

Eye diseases

The underlying causes of blindness in the aging population include diabetic retinopathy, age-related macular degeneration, corneal neovascularization, uveitis, glaucoma, and cataract. Curcumin's potential utility in preventing and treating many eye diseases has been supported by numerous preclinical and clinical investigations (Radomska-Lesniewska et al., 2019). In one of these studies, Pradhan et al. (2015) showed that the topical delivery of CN enhanced retention of curcumin in the cornea and prevented corneal neovascularization as graded clinically and by histopathology. This treatment suppressed the expression of VEGF, inflammatory cytokines, and MMP, suggesting the antiinflammatory mechanism for the prevention of corneal neovascularization. More recently, Mendes et al. (2020) demonstrated that intravitreal injection of curcumin (0.3 mg for 14 days) in rabbit eyes reduces retinal neovascularization. The lower dose (0.1 mg) was partially effective in reducing persistent vascular tortuosity. These investigators suggested that a higher dose of intravitreal curcumin may be effective in decreasing retinal neovascularization and restoring ocular microvasculature (Mendes et al., 2020). Curcumin also has the potential for treating ocular conditions associated with increased intraocular pressure (Zhang et al., 2015). In a retinal ischemia–reperfusion injury (RIRI) model of rats, curcumin was delivered by intraperitoneal injection (20 mg/kg/day and 100 mg/kg/day) at 30 minutes before RIRI and once daily after RIRI. Curcumin treatment inhibited IL-23 expression peaking at 24 hour and IL-17 expressions peaking at 72 hour in the retina after RIRI in a dose-dependent manner and reduced the ocular pressure. In another study, Davis et al. (2018) discovered that the topical application of CN twice daily for 3 weeks significantly reduced retinal ganglion cell loss in the rat and curcumin-loaded nanocarriers protected retinal cell lines against glutamate and hypoxia-induced injury. These findings suggest that a novel topical CN formulation may exert neuroprotective effects in glaucoma and other ocular diseases with neuronal pathology. The novel formulations of ST and novel curcuminoids and curcumin (Table 35.2) may help in preventing or treating uveitis, diabetic retinopathy, age-related macular degeneration, corneal neovascularization, glaucoma, cataract, corneal wound healing, dry eye disease, conjunctivitis, and pterygium. Double-blind, long-term clinical trials are warranted to conclusively support their preventive and treatment uses in clinics.

Colitis

IBD includes Crohn disease (CD) and UC. IBD is a chronic, idiopathic, and repeated inflammatory disease.

UC affects the mucosal layer of the colon and is becoming increasingly prevalent worldwide. Patients living with IBD may develop cancers known as colitis-associated colorectal cancer.

Bastaki et al. (2016) evaluated the effect of turmeric on colon histology, body weight, ulcer, IL-23, MPO, and glutathione in acetic acid-induced IBD in rats. Turmeric powder (1, 10, and 100 mg/kg/day) was administered orally for 3 days before or 30 minutes after the induction of IBD. This treatment was found to reduce macro- and microscopic ulcers, and also reduce IL-23, myeloperoxidase, and GSH (reduced glutathione peroxidase). The lowest dose of turmeric (1 mg/kg) caused a significant decrease in mean macroscopic ulcer and score (MaUS) after day 7 when compared to untreated groups. Interestingly, a high dose (100 mg/kg) also caused a significant ($\sim$50%) reduction in MaUS after 2 days of IBD. The high dose did not have a significant effect on MaUS after 4 and 7 days of IBD. The rats gained body weight and had a reduction in oxidative stress related to colitis (Bastaki et al., 2016). The 10 mg/kg dose appeared to exert the optimum benefit and was suggested to be the ideal dose. These investigators suggested a possibility of developing turmeric as a safe and potent antiinflammatory and antioxidant herbal remedy in the management of IBD. In another study, curcumin improved micro- and macroscopic lesions, prevented apoptosis of intestinal cells, and also induced the restoration of the mitogen-activated protein kinase immune reaction (Cunha Neto et al., 2019). A combination of mesalamine and curcumin therapy is associated with roughly threefold better odds of a clinical response compared to placebo, with minimal side effects (Chandan et al., 2020). Administration of curcumin complexed with essential turmeric oils (ETO) enhanced curcumin's bioavailability in the circulation by 7–10-fold as compared to standard curcumin, thus leading to significantly improved bioactivity (Shishu, 2010; Antony et al., 2008). In addition, it produces markedly increased antiinflammatory activity as compared to standard curcumin in a mouse model of dextran sodium sulfate (DSS)-induced colitis. ETO-curcumin acts by stimulating the T_{reg} cell production and subsequently upregulating IL-10, IL-11, and FOXP3 (Toden et al., 2017).

Recurrence of Crohn disease (CD) after surgery is a significant clinical concern. In a randomized controlled trial of patients following surgery for CD and receiving thiopurine treatment, Bommelaer and colleagues discovered that curcumin (3 g/day for 6 months) was no more effective than a placebo in preventing CD recurrence. There are no significant differences between groups in quality of life or severe adverse events (Bommelaer et al., 2019).

Curcumin, curcuminoid, and novel formulations including ADRxSynergy may exert medication-sparing effects in IBD and CRC. The use of turmeric/curcumin/curcuminoids and their novel formulations is affordable, efficient, and often free from any serious adverse side effects. These products may become an alternative adjunct treatment for IBD (Cunha Neto et al., 2019; Chand, 2019). The potential of curcumin's (its novel formulations, including ADRxSynergy) use in colitis treatment needs appropriately sized long-term randomized clinical trials (RCT) studies and confirmation of the safety of their use before their recommended use for IBD (Holleran et al., 2020).

Liver diseases

Alcohol, drugs, heavy metals, viral infections, environmental pollutants, and dietary components may cause excessive oxidative stress, which is a significant causative factor in the progression of liver injury, nonalcoholic steatohepatitis, liver fibrosis, cirrhosis, etc. Farzaei et al. (2018) reviewed the potential uses of curcumin in preventing oxidative stress-associated liver diseases. Curcumin's broad pharmacological profile and multiple molecular mechanisms including free radical scavenger activity, as well as few clinical studies, suggest that curcumin may be a lead compound in the prevention and treatment of oxidative stress-associated liver diseases. In the future, large well-controlled clinical studies are needed to confirm curcumin's potential for the prevention and treatment of liver pathologies.

Cardiometabolic diseases

Jin et al. (2018) reviewed the therapeutic potential of curcumin in cardiometabolic diseases—obesity and diabetes. The preclinical investigations demonstrated that curcumin reduces body weight gain, improves insulin sensitivity, and prevents diabetes progression in rodent models and prediabetic subjects. These beneficial cardiometabolic effects of curcumin are linked to changes in the gut microbiota. Though promising, these effects have not been well established because of very few small trials in human subjects (Jin et al., 2018). Further clinical research is needed to provide insights into curcumin's (of turmeric or curcuminoids) ability to modulate cardiometabolic disease progression and metabolic regulation that may ultimately translate into effective treatments for obesity and diabetes.

Musculoskeletal system

Inflammation and oxidative stress exacerbate dystrophic muscle pathology (Woodman et al., 2016). The nutraceuticals such as turmeric or curcuminoids and others may never be a cure for Duchenne muscular dystrophy (Woodman et al., 2016), but the stress can be reduced by curcumin via multiple pathways (Alamdari et al., 2009) and thus help reduce the need for steroids in treating dystrophic muscle-wasting diseases (Woodman et al., 2016). Detailed studies on muscle-sparing effects of ST, curcumin, or their novel formulations are warranted to confirm their role in the prevention and treatment of muscle wasting (Alamdari et al., 2009). Sarcopenia is the decline in muscle mass and functional quality with advancing age. It afflicts 5%–13% of younger seniors (60–70 years) and 11%–50% of older seniors (>80 years of age). Recently, Receno et al. (2019) demonstrated that curcumin in diets (0.2%) for 4 months reduced sarcopenia in aging rats. This effect of low dose (77 μg daily for 120 days) may be mediated via increasing the expression of nuclear factor erythroid-2 related-factor-2 (Nrf2) in the aging skeletal muscles (Receno et al., 2019). Siddiqui et al. (2009) developed a mouse model of progressive muscle wasting by implanting cachexia inducing colon tumors. The low doses of C3 complex (100 mg/kg) prevented weight loss, whereas higher doses (250 mg/kg) produced ~25% weight gain as compared with placebo-treated mice. The gains in the weight of the gastrocnemius muscle and the increased size of the muscle fibers demonstrate that this treatment prevents and/or reverses cachexia. This effect is due to curcumin-mediated inhibition of proteasome complex activity and reduced expression of muscle-specific ubiquitin ligases: atrogin-1/muscle atrophy F-box (MAFbx) and muscle RING finger 1 (MURF-1). These findings suggest that curcumin c3 may be an effective adjuvant therapy against cancer-induced cachexia (Siddiqui et al., 2009).

ADRxSynergy is a novel formulation of turmeric standardized to curcumin 4 + 0.8%. It is blended and activated in heated (boiling) coconut oil for 2 minutes. The coconut oil contains 16% myristic acid, a potent sigma$_1$ receptor agonist. It is being developed for dystrophic muscle diseases and the immune-inflammatory conditions in the body including the joints and brain. Recent personal observation. a young man (age 60+; HM) was having major difficulties in standing and walking due to limb-girdle muscular dystrophy, genetic muscle wasting, and weakness in his upper thighs. He topically applied ADRxSynergy (yellow turmeric oil extract about 5–7 drops massaged on his affected thighs). He noticed a remarkable benefit the next morning. These observations need to be repeated in a large number of subjects living with genetic weakness limbs (genetic muscular dystrophy states) and joint-related illnesses (personal communication). R&D efforts are underway to develop a transdermal patch and as a rapidly dissolving sublingual drop and film of ADRxSynergy. It is expected to prevent and treat dysregulated persistent chronic immune-inflammatory diseases of the eyes, muscles, and brain or regenerate muscles and nerves in the affected legs, shoulders, and arms of people living with genetic muscle wasting diseases. The estimated dose is 100–150 mg of heat-activated ADRxSynergy.

Arthritis

Arthritis (OA) is characterized by chronic inflammation in one or more joints that usually results in constant pain. OA incidence is rising due to the increasing aging population and obesity, and is a leading cause of disability (Johnson and Hunter, 2014).

In an animal model of the destabilization of the medial meniscus (DMM), Zhang et al. (2016) observed that oral administration of curcumin (50 mg/kg daily) for 8 weeks reduced OA disease progression but showed no significant pain relief. On the other hand, when applied topically in the form of CN (0.07 mg of 10 μg curcumin/1 mg nanoparticles) on the skin, directly above the DMM-operated knee, once daily for 8 weeks, it reduced tactile hypersensitivity and improved locomotor behavior. In a more recent animal study (Nicoliche et al., 2020), OA was induced in the right knee of rats by administering an intra-articular injection of 1 mg of zymosan. Fourteen days after induction, 50 mg/kg curcumin was administered by gavage daily for 60 days. This treatment increased IHH, Col2, and SOX-5 expression and the number of chondrocytes, and decreased the thicknesses of the middle and deep layers. In an in vitro cartilage explant assay, turmeric extracts at a relatively high concentration reduced LPS-induced release of glycosaminoglycan (a measure of cartilage breakdown), PGE2, NO, and calcein (Pearson and Kott, 2019). To understand the molecular mechanism in detail, Jiang et al. used a temporomandibular joint OA (TMJ OA) mouse model and human TMJ cartilage tissue. Curcumin inhibited the expression of the inflammation mediators, IL-6, iNOS, and COX-2, and of the matrix-degrading proteinases, MMP-1, MMP-3, MMP-9, MMP-13, ADAMTS-4, and ADAMTS-5, and upregulated the mRNA levels of the cartilage anabolic factors, COL2A1 and ACAN. Curcumin activated the Nrf2/ARE pathway in TMJ inflammatory chondrocytes (Jiang et al., 2020). Kang et al. (2020) utilized a mouse model of monoidoacetic acid-induced knee OA. They injected acid-activatable curcumin polymer micelles intramuscularly at 2.5 mg/kg and 5 mg/kg every 3 days for 28 days. This treatment protected the articular structures from arthritis by suppressing the release of TNF-α and interleukin 1β. These studies demonstrate that curcumin treatment inhibits inflammation, oxidative stress, and matrix degradation and protects the degeneration of cartilage and bones (Kang et al., 2020). Curcuma-containing products exerted effects similar to nonsteroidal antiinflammatory drugs and glucosamine (Perkins et al., 2017). A combination of glucosamine or NSAIDs with turmeric or curcuminoids may exert an additive or synergistic efficacy. A metaanalysis of RCT that studied the effects of turmeric or curcumin in arthritis found three of eight RTCs reported a reduction of pain visual analog score with turmeric/curcumin treatment, while no significant difference was found in the rest of the five RTCs in comparison with the placebo (Daily et al., 2016). A metaanalysis of four studies showed a decrease in Western Ontario and McMaster Universities Osteoarthritis Index with turmeric/curcumin treatment. Eight RCTs included in the evaluation exhibited low to moderate risk of bias. Based on these analyses, the reviewers concluded that the number of RCTs included in the metaanalysis, the sample size, and the methodological quality of clinical studies was not sufficient to draw definitive conclusions about the efficacy of turmeric/curcumin (Daily et al., 2016; Perkins et al., 2017). However, in another clinical study, Hashemzadeh et al. (2019) administered 40 mg of nanocurcumin capsule twice a day for 6 weeks in people living with OA and reported that it improved the pain, stiffness, and physical activity scores.

Henrotin et al. (2019) investigated the clinical efficacy of two doses of bio-optimized *C. longa* extract (BCL) in patients living with symptomatic OA of the knees. The OA patients were treated with BCL at a low dose (93.3 mg) and a high dose (140.1 mg BID) in capsules for 3 months. The efficacy was assessed by the Patient Global Assessment of Disease Activity (PGADA) and serum levels of Coll2-1 (sColl2-1), a biomarker of cartilage degradation. Both doses of BCL, used as an adjuvant treatment to paracetamol, produced similar efficacy—reduction in pain and PGADA from the first month of treatment. NSAID consumption decreased in the low-dose group over time, indicating that a low dose of BCL decreases pain and NSAID consumption. A low dose exerted no effect on Coll2-1 (sColl2-1) level, and a high dose slightly reduced sColl2-1 ($P = 0.031$). They concluded that daily intake of a low dose of BCL in patients with symptomatic knee OA reduces pain better than placebo. The number of adverse (gastrointestinal) effects was higher in the high-dose BCL group as compared to the placebo group ($P = 0.012$). They suggested conducting a larger phase III clinical trial including the large sample size estimate, the choice of dose, and the selection of primary outcomes. In the future, more rigorous and more extensive clinical studies are needed to confirm the therapeutic efficacy of curcumin for arthritis (Henrotin et al., 2019).

Recently, a 73-year-old lady was having an OA attack after getting up from the toilet. She topically applied ADRxSynergy (yellow warm turmeric oil) on her affected knee. She experienced tremendous relief in pain and swelling the next morning (personal communication). Similarly, a young man (age 60 +) was having difficulties in standing and walking due to some genetic weakness in his upper thigh. He topically applied ADRxSynergy (yellow warm turmeric oil) on his affected legs. He noticed a remarkable benefit the next morning. These observations suggest that this extract containing >235 phytochemicals may be developed as a knee patch to avoid coloring of the clothes. Such a clinical study needs to be confirmed in many people living with OA or genetic weakness limbs (muscular dystrophy; personal observation). More rigorous and large clinical studies including the large sample size estimate, the choice of the dose, and the selection of primary outcomes are needed to confirm the therapeutic efficacy of curcumin for arthritis.

Wound healing

The usual low aqueous solubility, poor tissue absorption, rapid metabolism, and short plasma half-life makes curcumin unsuitable for systemic administration for wound healing. In recent years, efforts have been made for various topical formulations of curcumin such as films, fibers, emulsion, hydrogels, and different nanoformulations that can be used for the targeted delivery of curcumin at wound sites in animal models (Mohanty and Sahoo, 2017). Seyhan (2020) demonstrated that curcumin application enhances burn wound healing in rats. In addition, curcumin gel has been shown to be effective in enhancing wound healing and reducing tissue edema and pain perception and normalizing tissue color in patients, and may be used as an alternative in periodontal dressing in chronic periodontitis patients (Meghana et al., 2020). Oral submucous fibrosis is a chronic, slow-progressing (insidious) disease and is associated with an increased risk of mouth cancer (malignancy). Curcumin is found to be effective in improving mouth opening and decreasing burning sensation in OSMF patients (Saran et al., 2018).

Cancer

Cancer is one of the most common causes of death worldwide. There appears to be a link between cancer and diet. Various dietary components such as turmeric (curcumin), resveratrol, and isothiocyanate are often utilized in cancer prevention and treatment. The dietary modulation of gut microbiota may also play an important role in cancer development and prevalence. Besides providing essential nutrients, various dietary components and fibers also serve as probiotics and alter gut microbiota and associated transcriptome and metabolome. This interaction of functional foods, herbs, and fiber and dietary supplement modulate the vital pathways involved in metastasis, invasion, apoptosis, tumor growth, and cell proliferation (Riaz Rajoka et al., 2018). In a DSS-dimethylhydrazine-induced inflammation-associated model of CRC, a combined treatment of curcumin and ETO-turmerone reduced adenoma multiplicity by 73% (Murakami et al., 2013). Curcumin could enhance the effect of radiation therapy, inhibit angiogenesis and cell proliferation by suppressing NF-κB and its target genes in colon cancer cells, and inhibit cell growth by modulating the Akt/mTOR pathways via the downregulation of EGFR (Giordano and Tommonaro, 2019; Johnson et al., 2009; Kunnumakkara et al., 2008). The curcumin affects several miRNA in pancreatic cells and the upregulation of miR-22 led to the suppression of ESR-1 (estrogen receptor 1) and SP1 transcription factors (Sun et al., 2008) (Sun et al., 2008). Curcumin is capable of triggering autophagy in several cancer cells lines (Shakeri et al., 2019). The synthetic monocarbonyl analogs of curcumin act via the generation of reactive oxygen species (ROS). One such analog CA10 inhibits cell growth by causing G2/M cell cycle arrest and induction of apoptotic death. CA10 induced ROS, activated the ER stress pathway, and inhibited STAT3 phosphorylation and gastric xenograft tumor growth in mice (Rajamanickam et al., 2017). A recent review summarized the chemopreventive and chemotherapeutic potential of curcumin in esophageal cancer in both preclinical and clinical settings (Hesari et al., 2019). Besides GI, curcumin may help prevent other cancers including breast cancer, papillary thyroid (Yang et al., 2020), and glioblastoma (Zhao et al., 2018). In breast cancer, it appears to act via a complex molecular signaling network, involving proliferation, apoptosis, ER, and human epidermal growth factor receptor 2 (HER2) pathways (Song et al., 2019). Curcumin can inhibit angiogenesis, thus depriving the cancer cells of essential nutrients. This effect is mediated by the modulation of the FAK/P-38 MAPK signaling pathway (Hosseini et al., 2019). Curcumin also sensitizes cancer cells to chemotherapeutic drugs. This chemosensitization may help in targeting drug resistance in cancer therapy (Bordoloi et al., 2016).

Glioma is the most aggressive, lethal, and prevalent primary brain tumor. Glioblastoma (glioblastoma multiforme) is a malignant glioma that is almost impossible to cure because of poor drug transportation across the blood–brain barrier and the existence of glioma stem cells. Recently, Zhao et al. (2018) discovered that curcumin-loaded RDP-modified liposome (RCL) inhibited glioma cell proliferation and tumor growth in an intracranial glioma mice model. RCL prolonged the survival time of the glioma-bearing mice from 23 to 33 days. This study provides evidence that nano-delivery of curcumin has potential for the treatment of human malignant gliomas. Recently, a few studies have summarized the anticancer activity of curcumin and its derivatives in several cancer cell lines, animal models, and human subjects (Tomeh et al., 2019; Kunnumakkara et al., 2017). They covered the medicinal chemistry and pharmacology of curcumin and its derivatives, and discussed the advances in curcumin-delivery systems to cancer cells and its multiple mechanisms of action targeting various tissues. In summary, a combination of curcuminoids, turmerones (essential oils), tumerosaccharides, and fiber in ADRxSynergy (a novel formulation) may improve the PK and PD (efficacy and safety) in IBD, UC, CRC, and many aging-related chronic disease states in animals and humans. ST and novel formulations of curcuminoids or curcumin may exert cancer medicine-sparing and analgesic-sparing effects. In clinical terms, it may mean that low doses of turmeric, ST, and novel formulations of curcuminoids or curcumin may act as a chemosensitizer and thus improve the safety and efficacy of FDA-approved anticancer medicines. Lifetime exposure to low doses of turmeric may reduce cancer risk and may also reduce the doses of anticancer medications and radiation required by those undergoing cancer therapy.

Mechanisms of action

Turmeric and the various complex physiochemicals act via different signaling pathways to exert their effect in various tissues. Interleukin-10 (IL_{10}) is a pleiotropic and effective antiinflammatory and immunosuppressive interleukin. It is produced by innate and adaptive immunity cells including TH1, TH2, and TH17 and regulatory T cells, dendritic cells, macrophages, mast cells, natural killer cells, eosinophils, neutrophils, B cells, etc. The decline in IL_{10} secretion or function (dysregulation) may play a role in the pathophysiology of several diseases including neurodegenerative diseases, neuropathic pain, OA, rheumatoid arthritis, psoriasis, systemic lupus erythematosus, type I diabetes, allergy, IBD, cancer, etc. Curcumin induces the expression and production of IL-10 and enhances its action in the body (Mollazadeh et al., 2019). In vitro and preclinical models suggest that curcumin may modulate the pathophysiology of pain and neurodegenerative diseases, IBD, allergy, infections, and cancer by enhancing IL_{10} secretion or its function (Mollazadeh et al., 2019). Clinical studies are needed to establish this mode of action.

In the DMM mouse model of OA, Sun et al. (2017) demonstrated that curcumin inhibited the release of inflammasome NLRP3, leading to the downregulation of inflammatory cytokines interleukin ($IL_{1\beta}$, IL_{17}, Il_{18}) and TNF-α at both RNA and protein levels (Sun et al., 2017). The molecular mechanisms for neuroprotective effects of curcumin and its nanoformulations have been described (Mandal et al., 2020; Teter et al., 2019). In breast cancer, curcumin affects molecular signaling networks, involving proliferation, apoptosis, ER, and HER2 pathways (Song et al., 2019). Curcumin uses the FAK/P-38 MAPK signaling pathway to suppress angiogenesis (Hosseini et al., 2019). Curcumin may restore mitochondrial functioning in the rat heart and thus reduce the cardiotoxic effects of bevacizumab by reducing its oxidative stress and mitochondrial dysfunction (Sabet et al., 2020). Curcumin exerts a promising protective effect against cisplatin-induced nephrotoxicity (Abd El-Kader and Taha, 2020).

Turmeric, ST (ADRxSynergy) and its extracts (curcuminoids, turmerosaccharides, essential oils, noncurcuminoids, and fiber) modulate persistent chronic dysregulated immune inflammation and may help to prevent or reverse immune inflammation and associated complex progression of the pathogenesis of diverse chronic diseases by multiple modes of actions (Sun et al., 2017; Tang and Taghibiglou, 2017; Tomeh et al., 2019; Mollazadeh et al., 2019; Chand, 2019). Lifetime exposure to an ultralow dose of hundreds of active phytochemicals in herbs and spices, including turmeric in cooking, is a part of the food exposure in turmeric-consuming nations. This traditional culinary use may act in an additive or synergistic fashion in the prevention of many aging-related diseases such as uveitis, sarcopenia, osteopenia, muscle wasting, dementia, dysbiosis (constipation), and others. However, it is not a panacea.

Concluding remarks and future directions

Many Ayurvedic and modern DS (> 800) contain turmeric or its extracts. Low doses of turmeric and its novel formulations of curcumin and ST (ADRxSynergy) may modulate persistent dysregulated chronic immune inflammation in the body. More than 235 active ingredients, including curcuminoids (> 89), and noncurcuminoids (turmerosaccharides, essential oils-tumerones, and fiber) in turmeric/ST and their potential metabolites may act in an additive or synergistic fashion as a modulator of aging-related chronic immune inflammation. The preclinical pharmacological studies in animal models of various diseases suggest that low doses of turmeric or its active ingredients (curcumin, curcuminoids, turmerosaccharides, turmerones) may have potential utility in the prevention and management of immune-inflammatory diseases. ST (such as ADRxSynergy) contains a predetermined (specific) amount of curcumin while still retaining most of the phytochemicals and fibers of the native turmeric. ST and the novel formulations of curcumin, curcuminoids, tumerones, and turmerosaccharides may reduce the need for medications for immune inflammation (steroids), chronic pain (opiates, NSAIDs), depression, and anticancer medicines.

The safety and efficacy of lower doses of novel formulations of curcumin, curcuminoids, and ST, such as ADRxSynergy, need to be explored in aging pets, such as dogs, cats, pigs, and horses living with cancer and aging-related diseases OA, chronic pain, and dementia and others). These animals may serve as realistic and the most appropriate disease models for discovering the efficacy and safety of complex phytochemicals. Veterinarians and their patients can help in finding cures for treating age-related diseases of the eyes, brain, and digestive system and cancer among aging pets and people. The mechanism of selective delivery of curcumin, TEO, and other active phytochemicals in ADRxSynergy to the target sites in disease states also need to be explored in these aging pets living with cancer and aging-related diseases OA, chronic pain, and dementia, etc.).

The topical applications of novel formulations of curcumin, curcuminoids, and ST may offer potential prevention and cure for diseases of the eyes, muscles and joints, and skin, etc. The medication-sparing effect of novel

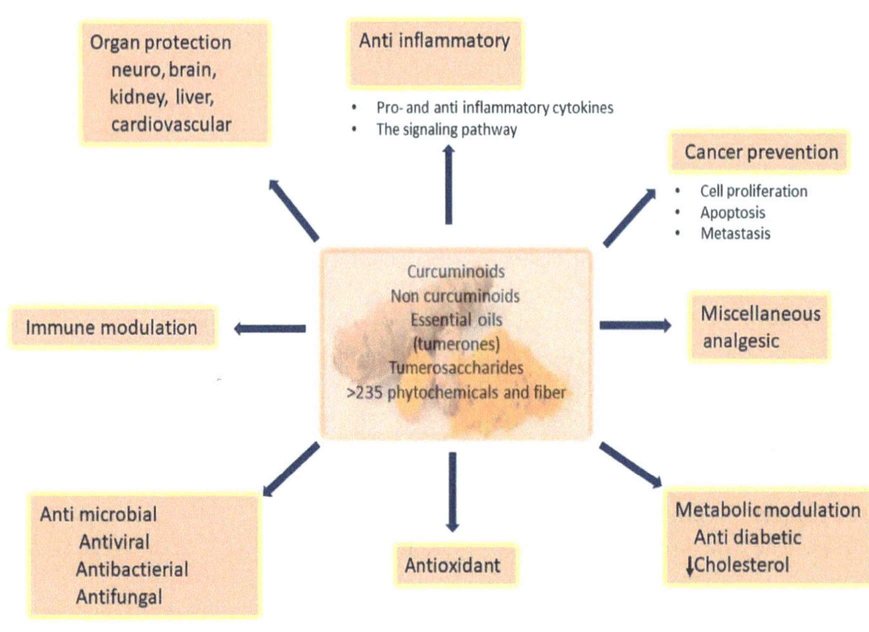

FIGURE 35.2 The biological effects of turmeric and its major ingredients (curcumin, curcuminoids, turmerones, and turmerosaccharides) in in vitro and animal models of diseases.

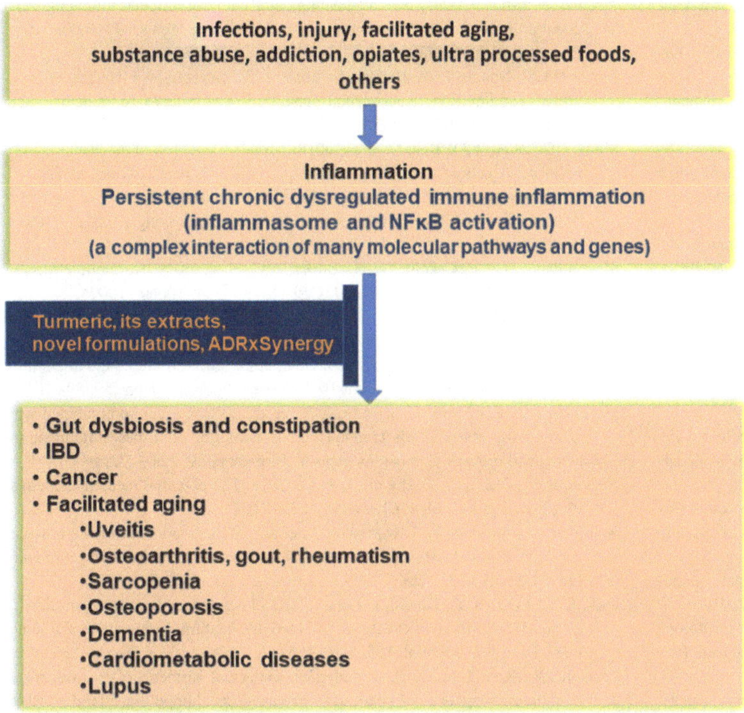

FIGURE 35.3 The pathogenesis of chronic diseases. Low doses of spices/herbs in cooking including turmeric, ST (ADRxSynergy), its extracts (curcuminoids, turmerosaccharides, essential oils, noncurcuminoids, and fiber), and novel formulations may modulate persistent chronic dysregulated immune inflammation and may prevent or reverse immune inflammation and associated complex progression of the pathogenesis of diverse chronic diseases.

formulations of curcumin, curcuminoids, and ST needs to be evaluated in older people and pets. This activity would improve the safety and efficacy of medications currently used for the treatment of cancer, diseases of the eyes, joints, digestive system and brain, and rare diseases, etc. The transdermal patch, sublingual film, or drops of novel formulations and ADRxSynergy may exert steroid-, methotrexate-, and opiate-sparing effects (uveitis, fasciitis, OA, RA, and neurodegenerative illnesses involving immune inflammation such as AD, PD, MS, HD, and ALS) and may help in reducing the ongoing cancer and opiate epidemic (Figs. 35.2 and 35.3).

References

Abd El-Kader, M., Taha, R.I., 2020. Comparative nephroprotective effects of curcumin and etoricoxib against cisplatin-induced acute kidney injury in rats. Acta Histochem. 122, 151534.

Aggarwal, B.B., Yuan, W., Li, S., et al., 2013. Curcumin-free turmeric exhibits anti-inflammatory and anticancer activities: identification of novel components of turmeric. Mol. Nutr. Food Res. 57, 1529–1542.

Alamdari, N., O'neal, P., Hasselgren, P.O., 2009. Curcumin and muscle wasting: a new role for an old drug? Nutrition 25, 125–129.

Antony, B., Merina, B., Iyer, V.S., et al., 2008. A pilot cross-over study to evaluate human oral bioavailability of BCM-95CG (Biocurcumax), a novel bioenhanced preparation of curcumin. Indian J. Pharm. Sci. 70, 445–449.

Bastaki, S.M.A., Al Ahmed, M.M., Al Zaabi, A., et al., 2016. Effect of turmeric on colon histology, body weight, ulcer, IL-23, MPO and glutathione in acetic-acid-induced inflammatory bowel disease in rats. BMC Complement. Altern. Med. 16, 72.

Begum, A.N., Jones, M.R., Lim, G.P., et al., 2008. Curcumin structure-function, bioavailability, and efficacy in models of neuroinflammation and Alzheimer's disease. J. Pharmacol. Exp. Ther. 326, 196–208.

Bethapudi, B., Murugan, S., Illuri, R., et al., 2017. Bioactive turmerosaccharides from *Curcuma longa* extract (NR-INF-02): potential ameliorating effect on osteoarthritis pain. Pharmacogn. Mag. 13, S623–s627.

Bommelaer, G., Laharie, D., Nancey, S., et al., 2019. Oral curcumin no more effective than placebo in preventing recurrence of Crohn's disease after surgery in a randomized controlled trial. Clin. Gastroenterol. Hepatol. 18, 1553–1560.

Bordoloi, D., Roy, N.K., Monisha, J., et al., 2016. Multi-targeted agents in cancer cell chemosensitization: what we learnt from curcumin thus far. Recent. Pat. Anticancer Drug Discov. 11, 67–97.

Calabrese, E.J., Dhawan, G., Kapoor, R., et al., 2019. Curcumin and hormesis with particular emphasis on neural cells. Food Chem. Toxicol. 129, 399–404.

Chand, N., 2019. Standardized turmeric and curcumin. In: Gupta, R., Srivastava, A., Lall, R. (Eds.), Nutraceuticals in Veterinary Medicine. Springer, Cham, pp. 3–23.

Chandan, S., Mohan, B.P., Chandan, O.C., et al., 2020. Curcumin use in ulcerative colitis: is it ready for prime time? A systematic review and meta-analysis of clinical trials. Ann. Gastroenterol. 33, 53–58.

Choi, Y., Ban, I., Lee, H., et al., 2019. Puffing as a novel process to enhance the antioxidant and anti-inflammatory properties of *Curcuma longa* L. (Turmeric). Antioxid. (Basel) 8, 506.

Concetta Scuto, M., Mancuso, C., Tomasello, B., et al., 2019. Curcumin, hormesis and the nervous system. Nutrients 11.

Cunha Neto, F., Marton, L.T., et al., 2019. Curcuminoids from *Curcuma longa*: new adjuvants for the treatment of Crohn's disease and ulcerative colitis? Crit. Rev. Food Sci. Nutr. 59, 2136–2143.

Daily, J.W., Yang, M., Park, S., 2016. Efficacy of turmeric extracts and curcumin for alleviating the symptoms of joint arthritis: a systematic review and meta-analysis of randomized clinical trials. J. Med. Food 19, 717–729.

Davis, B.M., Pahlitzsch, M., Guo, L., et al., 2018. Topical curcumin nanocarriers are neuroprotective in eye disease. Sci. Rep. 8, 11066.

Dosoky, N.S., Setzer, W.N., 2018. Chemical composition and biological activities of essential oils of *Curcuma* species. Nutrients 10, 1196.

Farzaei, M.H., Zobeiri, M., Parvizi, F., et al., 2018. Curcumin in liver diseases: a systematic review of the cellular mechanisms of oxidative stress and clinical perspective. Nutrients 10, 855.

Funk, J.L., Frye, J.B., Oyarzo, J.N., et al., 2010. Anti-arthritic effects and toxicity of the essential oils of turmeric (*Curcuma longa* L.). J. Agric. Food Chem. 58, 842–849.

Giordano, A., Tommonaro, G., 2019. Curcumin and cancer. Nutrients 11, 2376.

Grover, A.K., Samson, S.E., 2016. Benefits of antioxidant supplements for knee osteoarthritis: rationale and reality. Nutr. J. 15, 1.

Gupta, S.C., Kismali, G., Aggarwal, B.B., 2013. Curcumin, a component of turmeric: from farm to pharmacy. Biofactors 39, 2–13.

Hashemzadeh, K., Davoudian, N., Jaafari, M.R., et al., 2019. The effect of nanocurcumin on the improvement symptoms of knee osteoarthritis: a randomized clinical trial. Curr. Rheumatol. Rev.

Henrotin, Y., Malaise, M., Wittoek, R., et al., 2019. Bio-optimized *Curcuma longa* extract is efficient on knee osteoarthritis pain: a double-blind multicenter randomized placebo controlled three-arm study. Arthritis Res. Ther. 21, 179.

Hesari, A., Azizian, M., Sheikhi, A., et al., 2019. Chemopreventive and therapeutic potential of curcumin in esophageal cancer: current and future status. Int. J. Cancer 144, 1215–1226.

Hewlings, S.J., Kalman, D.S., 2017. Curcumin: a review of its' effects on human health. Foods 6, 92.

Holleran, G., Scaldaferri, F., Gasbarrini, A., et al., 2020. Herbal medicinal products for inflammatory bowel disease: a focus on those assessed in double-blind randomised controlled trials. Phytother. Res. 34, 77–93.

Hosseini, A., Rasmi, Y., Rahbarghazi, R., et al., 2019. Curcumin modulates the angiogenic potential of human endothelial cells via FAK/P-38 MAPK signaling pathway. Gene 688, 7–12.

Hu, X., Huang, F., Szymusiak, M., et al., 2015. Curcumin attenuates opioid tolerance and dependence by inhibiting Ca2 + /calmodulin-dependent protein kinase II alpha activity. J. Pharmacol. Exp. Ther. 352, 420–428.

Hu, X., Huang, F., Szymusiak, M., et al., 2016. PLGA-Curcumin attenuates opioid-induced hyperalgesia and inhibits spinal CaMKIIalpha. PLoS One 11, e0146393.

Hucklenbroich, J., Klein, R., Neumaier, B., et al., 2014. Aromatic-turmerone induces neural stem cell proliferation in vitro and in vivo. Stem Cell Res. Ther. 5, 100.

Jamwal, R., 2018. Bioavailable curcumin formulations: a review of pharmacokinetic studies in healthy volunteers. J. Integr. Med. 16, 367–374.

Jia, S., Du, Z., Song, C., et al., 2017. Identification and characterization of curcuminoids in turmeric using ultra-high performance liquid chromatography-quadrupole time of flight tandem mass spectrometry. J. Chromatogr. A 1521, 110–122.

Jiang, C., Luo, P., Li, X., et al., 2020. Nrf2/ARE is a key pathway for curcumin-mediated protection of TMJ chondrocytes from oxidative stress and inflammation. Cell Stress Chaperones 25, 395–406.

Jin, T., Song, Z., Weng, J., et al., 2018. Curcumin and other dietary polyphenols: potential mechanisms of metabolic actions and therapy for diabetes and obesity. Am. J. Physiol. Endocrinol. Metab. 314, E201–E205.

Johnson, S.M., Gulhati, P., Arrieta, I., et al., 2009. Curcumin inhibits proliferation of colorectal carcinoma by modulating Akt/mTOR signaling. Anticancer Res. 29, 3185–3190.

Johnson, V.L., Hunter, D.J., 2014. The epidemiology of osteoarthritis. Best Pract. Res. Clin. Rheumatol. 28, 5–15.

Kang, C., Jung, E., Hyeon, H., et al., 2020. Acid-activatable polymeric curcumin nanoparticles as therapeutic agents for osteoarthritis. Nanomedicine 23, 102104.

Kawanishi, S., Oikawa, S., Murata, M., 2005. Evaluation for safety of antioxidant chemopreventive agents. Antioxid. Redox Signal. 7, 1728–1739.

Kumar, B., Singh, V., Shankar, R., et al., 2017. Synthetic and medicinal prospective of structurally modified curcumins. Curr. Top. Med. Chem. 17, 148–161.

Kunnumakkara, A.B., Ichikawa, H., Anand, P., et al., 2008. Coronarin D, a labdane diterpene, inhibits both constitutive and inducible nuclear factor-kappa B pathway activation, leading to potentiation of apoptosis, inhibition of invasion, and suppression of osteoclastogenesis. Mol. Cancer Ther. 7, 3306–3317.

Kunnumakkara, A.B., Bordoloi, D., Harsha, C., et al., 2017. Curcumin mediates anticancer effects by modulating multiple cell signaling pathways. Clin. Sci. (Lond.) 131, 1781–1799.

Lim, G.P., Chu, T., Yang, F., et al., 2001. The curry spice curcumin reduces oxidative damage and amyloid pathology in an Alzheimer transgenic mouse. J. Neurosci. 21, 8370–8377.

Liu, Z., Ran, Y., Huang, S., et al., 2017. Curcumin protects against ischemic stroke by titrating microglia/macrophage polarization. Front. Aging Neurosci. 9, 233.

Mandal, M., Jaiswal, P., Mishra, A., 2020. Role of curcumin and its nanoformulations in neurotherapeutics: a comprehensive review. J. Biochem. Mol. Toxicol. 34, e22478.

Mazzanti, G., Di Giacomo, S., 2016. Curcumin and resveratrol in the management of cognitive disorders: what is the clinical evidence? Molecules 21, 1243.

Meghana, M.V.S., Deshmukh, J., Devarathanamma, M.V., et al., 2020. Comparison of effect of curcumin gel and noneugenol periodontal dressing in tissue response, early wound healing, and pain assessment following periodontal flap surgery in chronic periodontitis patients. J. Indian Soc. Periodontol. 24, 54–59.

Mendes, T.S., Novais, E.A., Badaro, E., et al., 2020. Antiangiogenic effect of intravitreal curcumin in experimental model of proliferative retinopathy. Acta Ophthalmol. 98, e132–e133.

Meng, F.C., Zhou, Y.Q., Ren, D., et al., 2018. Turmeric: a review of its chemical composition, quality control, bioactivity, and pharmaceutical application. In: Grumezescu, A.M., Holban, A.M. (Eds.), Natural and Artificial Flavoring Agents and Food Dyes: Handbook of Food Bioengineering. Academic Press, pp. 299–350.

Moghaddam, N.S.A., Oskouie, M.N., Butler, A.E., et al., 2019. Hormetic effects of curcumin: what is the evidence? J. Cell Physiol. 234, 10060–10071.

Mohanty, C., Sahoo, S.K., 2017. Curcumin and its topical formulations for wound healing applications. Drug Discov. Today 22, 1582–1592.

Mollazadeh, H., Cicero, A.F.G., Blesso, C.N., et al., 2019. Immune modulation by curcumin: the role of interleukin-10. Crit. Rev. Food Sci. Nutr. 59, 89–101.

Murakami, A., Furukawa, I., Miyamoto, S., et al., 2013. Curcumin combined with turmerones, essential oil components of turmeric, abolishes inflammation-associated mouse colon carcinogenesis. Biofactors 39, 221–232.

Nelson, K.M., Dahlin, J.L., Bisson, J., et al., 2017. The essential medicinal chemistry of curcumin. J. Med. Chem. 60, 1620–1637.

Nicoliche, T., Maldonado, D.C., Faber, J., et al., 2020. Evaluation of the articular cartilage in the knees of rats with induced arthritis treated with curcumin. PLoS One 15, e0230228.

Park, C.Y., Leea, K.Y., Khalid, G., et al., 2019. Phenolics and antioxidant activity of aqueous turmeric extracts as affected by heating temperature and time. LWT Food Sci. Technol. 105, 149–155.

Pearson, W., Kott, L.S., 2019. A biological extract of turmeric (*Curcuma longa*) modulates response of cartilage explants to lipopolysaccharide. BMC Complement. Altern. Med. 19, 252.

Perkins, K., Sahy, W., Beckett, R.D., 2017. Efficacy of curcuma for treatment of osteoarthritis. J. Evid. Based Complement. Altern. Med. 22, 156–165.

Pradhan, N., Guha, R., Chowdhury, S., et al., 2015. Curcumin nanoparticles inhibit corneal neovascularization. J. Mol. Med. (Berl.) 93, 1095–1106.

Prasad, S., Aggarwal, B.B., 2011. Turmeric, the golden spice: from traditional medicine to modern medicine. In: Benzie, I.F.F., Wachtel-Galor, S. (Eds.), Herbal Medicine: Biomolecular and Clinical Aspects. CRC Press/Taylor & Francis, Llc, Boca Raton, FL, ND.

Radomska-Lesniewska, D.M., Osiecka-Iwan, A., Hyc, A., et al., 2019. Therapeutic potential of curcumin in eye diseases. Cent. Eur. J. Immunol. 44, 181–189.

Rajamanickam, V., Zhu, H., Feng, C., et al., 2017. Novel allylated monocarbonyl analogs of curcumin induce mitotic arrest and apoptosis by reactive oxygen species-mediated endoplasmic reticulum stress and inhibition of STAT3. Oncotarget 8, 101112–101129.

Receno, C.N., Liang, C., Korol, D.L., et al., 2019. Effects of prolonged dietary curcumin exposure on skeletal muscle biochemical and functional responses of aged male rats. Int. J. Mol. Sci. 20, 1178.

Riaz Rajoka, M.S., Jin, M., Haobin, Z., et al., 2018. Impact of dietary compounds on cancer-related gut microbiota and microRNA. Appl. Microbiol. Biotechnol. 102, 4291–4303.

Romson, S., Sunisa, S., Worapong, U., 2011. Stability of antioxidant and antibacterial properties in heated turmeric-chili paste and its ingredients. Int. Food Res. J. 18, 397–404.

Sabet, N.S., Atashbar, S., Khanlou, E.M., et al., 2020. Curcumin attenuates bevacizumab-induced toxicity via suppressing oxidative stress and preventing mitochondrial dysfunction in heart mitochondria. Naunyn Schmiedebergs Arch. Pharmacol. 393, 1447–1457.

Saran, G., Umapathy, D., Misra, N., et al., 2018. A comparative study to evaluate the efficacy of lycopene and curcumin in oral submucous fibrosis patients: a randomized clinical trial. Indian J. Dent. Res. 29, 303–312.

Sarraf, P., Parohan, M., Javanbakht, M.H., et al., 2019. Short-term curcumin supplementation enhances serum brain-derived neurotrophic factor in adult men and women: a systematic review and dose-response meta-analysis of randomized controlled trials. Nutr. Res. 69, 1–8.

Seyhan, N., 2020. Evaluation of the healing effects of hypericum perforatum and curcumin on burn wounds in rats. Evid. Based Complement. Altern. Med. 2020, 6462956.

Shakeri, A., Cicero, A.F.G., Panahi, Y., et al., 2019. Curcumin: a naturally occurring autophagy modulator. J. Cell. Physiol. 234, 5643–5654.

Shishu, M.M., 2010. Comparative bioavailability of curcumin, turmeric and Biocurcumax in traditional vehicles using non-everted rat intestinal sac model. J. Funct. Foods 2, 60–65.

Siddiqui, R.A., Hassan, S., Harvey, K.A., et al., 2009. Attenuation of proteolysis and muscle wasting by curcumin c3 complex in MAC16 colon tumour-bearing mice. Br. J. Nutr. 102, 967–975.

Soleimani, V., Sahebkar, A., Hosseinzadeh, H., 2018. Turmeric (*Curcuma longa*) and its major constituent (curcumin) as nontoxic and safe substances: review. Phytother. Res. 32, 985–995.

Song, X., Zhang, M., Dai, E., et al., 2019. Molecular targets of curcumin in breast cancer (Review). Mol. Med. Rep. 19, 23–29.

Sun, M., Estrov, Z., Ji, Y., et al., 2008. Curcumin (diferuloylmethane) alters the expression profiles of microRNAs in human pancreatic cancer cells. Mol. Cancer Ther. 7, 464–473.

Sun, Y., Liu, W., Zhang, H., et al., 2017. Curcumin prevents osteoarthritis by inhibiting the activation of inflammasome NLRP3. J. Interferon Cytokine Res. 37, 449–455.

Sun, J.L., Ji, H.F., Shen, L., 2019. Impact of cooking on the antioxidant activity of spice turmeric. Food Nutr. Res. 63.

Tang, M., Taghibiglou, C., 2017. The mechanisms of action of curcumin in Alzheimer's disease. J. Alzheimers Dis. 58, 1003–1016.

Teter, B., Morihara, T., Lim, G.P., et al., 2019. Curcumin restores innate immune Alzheimer's disease risk gene expression to ameliorate Alzheimer pathogenesis. Neurobiol. Dis. 127, 432–448.

Tiwari, V., Shanker, R., Srivastava, J., et al., 2006. Change in antioxidant activity of spices turmeric and ginger on heat treatment. Electron. J. Environ. Agric. Food Chem. 5.

Toden, S., Theiss, A.L., Wang, X., et al., 2017. Essential turmeric oils enhance anti-inflammatory efficacy of curcumin in dextran sulfate sodium-induced colitis. Sci. Rep. 7, 814.

Tomeh, M.A., Hadianamrei, R., Zhao, X., 2019. A review of curcumin and its derivatives as anticancer agents. Int. J. Mol. Sci. 20, 1033.

Voulgaropoulou, S.D., Van Amelsvoort, T., Prickaerts, J., et al., 2019. The effect of curcumin on cognition in Alzheimer's disease and healthy aging: a systematic review of pre-clinical and clinical studies. Brain Res. 1725, 146476.

Woodman, K.G., Coles, C.A., Lamande, S.R., et al., 2016. Nutraceuticals and their potential to treat Duchenne Muscular Dystrophy: separating the credible from the conjecture. Nutrients 8, 713.

Xu, X.Y., Meng, X., Li, S., et al., 2018. Bioactivity, health benefits, and related molecular mechanisms of curcumin: current progress, challenges, and perspectives. Nutrients 10, 1553.

Yang, Y., Chen, Q., Yu, W.Y., et al., 2020. Herbal active ingredients: an emerging potential for the prevention and treatment of papillary thyroid carcinoma. Biomed. Res. Int. 2020, 1340153.

Yavarpour-Bali, H., Ghasemi-Kasman, M., Pirzadeh, M., 2019. Curcumin-loaded nanoparticles: a novel therapeutic strategy in treatment of central nervous system disorders. Int. J. Nanomed. 14, 4449–4460.

Zhang, H.J., Xing, Y.Q., Jin, W., et al., 2015. Effects of curcumin on interleukin-23 and interleukin-17 expression in rat retina after retinal ischemia-reperfusion injury. Int. J. Clin. Exp. Pathol. 8, 9223–9231.

Zhang, Z., Leong, D.J., Xu, L., et al., 2016. Curcumin slows osteoarthritis progression and relieves osteoarthritis-associated pain symptoms in a post-traumatic osteoarthritis mouse model. Arthritis Res. Ther 18, 128.

Zhang, H., Jiang, X., Song, W., et al., 2019a. Review-novel synthetic curcuminoids: not merely an anticancer agent. Pak. J. Pharm. Sci. 32, 1723–1747.

Zhang, Y., Rauf Khan, A., Fu, M., et al., 2019b. Advances in curcumin-loaded nanopreparations: improving bioavailability and overcoming inherent drawbacks. J. Drug Target 27, 917–931.

Zhao, M., Zhao, M., Fu, C., et al., 2018. Targeted therapy of intracranial glioma model mice with curcumin nanoliposomes. Int. J. Nanomed. 13, 1601–1610.

Zhu, L.N., Mei, X., Zhang, Z.G., et al., 2019. Curcumin intervention for cognitive function in different types of people: a systematic review and meta-analysis. Phytother. Res. 33, 524–533.

Chapter 36

Curcuminoid–metal complexes for oxidative stress

Dan DuBourdieu, Sahdeo Prasad and Rajiv Lall

Introduction

Many health issues can be improved by supporting and having a healthy immune system. On the whole, the immune system does a remarkable job of defending against disease-causing microorganisms and other sources of stress on the immune system. Bacteria and flu viruses are sources of disease, but less obvious sources of stress on the immune system include simply being old (Montecino-Rodriguez et al., 2013), when the immune system has a characteristic change termed immunosenescence, which can lead to increased disease incidence and severity (Weinberger et al., 2008). Simple things like traveling or having emotional issues (Segerstrom and Miller, 2004) can also be as a source of stress. As good as the immune system is, it sometimes fails and a germ invades successfully and sickness occurs. There are some common underlying aspects as to how all these sources of stress affect the body: this would be oxidative stress from reactive oxygen species (ROS). Various infective agents such as viruses can cause the production of free radicals and oxidative stress (Zhang et al., 2019) that, in turn, cause all sorts of problems for cells, and this damage can go throughout the body. Oxygen free radicals are very reactive molecules which can react with every cellular component to damage cell membranes, proteins, and DNA. To combat the free radicals are substances called antioxidants, which donate an electron to the free radical of ROS. Antioxidants prevent free radicals from taking electrons and causing cellular damage. Antioxidants are able to give an electron to a free radical without becoming destabilized themselves because they are inherently stable without having a complete electron valence shell. As such, antioxidants can stop free radical chain reactions and limit damage to the body. The body has some inherent antioxidants such as glutathione but obtaining antioxidants from the diet or supplements can greatly help reduce oxidative stress and aid the immune system. Once the balance between the accumulation of ROS and the removal of oxides is broken, oxidative stress is established. Therefore the ultimate consequences of this oxidative stress include tissue damage, inflammation responses, and cell death. A rational strategy to improve general health would be a strategy that helps reduce oxidative stress and also helps the immune system in other more specific ways. The challenge is then to find natural antioxidants that can also boost the immune system. There are numerous nutraceuticals that act as antioxidants. However, curcumin is one that has the ability to be chemically modified readily to improve its natural activities in the body.

Curcumin

Curcumin [1,7-bis(4-hydroxy-3-methoxyphenyl)-1,6-heptadiene-3,5-dione], also called diferuloylmethane, is the main natural polyphenol found in the rhizome of turmeric (*Curcuma longa*). Turmeric has been used in several Asian countries as a flavoring agent in various foods and as a traditional medicine against various ailments (Prasad and Aggarwal, 2011). Curcumin is a deep yellow, poorly water-soluble substance. In the last few decades, curcumin has been shown to exhibit a variety of effects including antioxidant, antiinflammatory, antimicrobial, antiviral, antifungal, antiallergic, anticancer, antidiabetic, cardioprotective, nephroprotective, and neuroprotective properties (Prasad et al., 2014). Curcumin can also boost β-interferon levels and reduce oxidative stress during influenza (Dai et al., 2018). However, the poor systemic bioavailability of curcumin has been noted, which creates a hindrance to its biological applications. It has been found that curcumin metabolized to various hydrocurcumins, such as dihydrocurcumin (DHC), tetrahydrocurcumin, hexahydrocurcumin, octahydrocurcumin (OHC), as well as curcumin-O-glucuronide and curcumin-O-sulfate,

Nutraceuticals. DOI: https://doi.org/10.1016/B978-0-12-821038-3.00036-7

after oral administration. Among these hydrocurcumins, tetrahydrocurcumin has been characterized as the primary metabolite and exhibits diverse biological effects (Zhongfa et al., 2012).

Tetrahydrocurcumin

Curcumin is yellow in color, whereas tetrahydrocurcumin is colorless or white. Curcumin metabolizes into tetrahydrocurcumin by bacterial enzyme NADPH-dependent curcumin reductase in the intestine. This bacterial (*Escherichia coli*) enzyme catalyzes the curcumin to DHC as an intermediate product, followed by further reduction to tetrahydrocurcumin as an end product (Hassaninasab et al., 2011). Tetrahydrocurcumin has also reported to be produced in vivo through phase I metabolism in the liver by hepatic reductases. Besides these, it can be synthesized in the laboratory by hydrogenation of curcumin, yielding a colorless substance (Novaes et al., 2017). Compared with curcumin, tetrahydrocurcumin has certain superior properties. For example, tetrahydrocurcumin is more soluble in water, and is also more stable under physiological pH and temperature conditions than curcumin (Aggarwal et al., 2014). These properties of tetrahydrocurcumin make it biologically more potent than curcumin and easier to formulate into an acceptable dosage.

Tetrahydrocurcumin has been shown to exhibit antioxidant, antiinflammatory, antiviral, antibacterial, anticancer, antidiabetic, cardioprotective, and neuroprotective effects (Aggarwal et al., 2014). In a study, it exerted a potent neuroprotective effect against glutamate-induced neuronal cell death by inhibiting the accumulation of oxidative stress and phosphorylation of mitogen-activated protein kinases (Park et al., 2019). In a diabetic mice model, oral administration of tetrahydrocurcumin (120 mg/kg per day) significantly improved the cardiac function and ameliorated myocardial fibrosis and cardiac hypertrophy, accompanied by reduced ROS generation, indicating its cardioprotective effect (Li et al., 2019). Tetrahydrocurcumin also has potential for prevention of HIV-1 infection. In silico screening studies using homology model of gp120-CD4 binding showed that tetrahydrocurcumin inhibits gp120-CD4 binding activity due to its stable hydrophobic interactions with residues Asp368 and Trp427 deeper in the Phe43 cavity of CD4 receptor (Mirani et al., 2019). Tetrahydrocurcumin has also shown a nephroprotective property. In a study, tetrahydrocurcumin ameliorated the tacrolimus-induced renal damage via an antioxidant effect and apoptosis inhibition in LLC-PK1 porcine kidney cells (Park et al., 2018). The anticancer effect of tetrahydrocurcumin has been also determined using in vitro and in vivo models. It exhibited anticancer and antiangiogenic effects in multiple cancer types including hepatocellular (Yoysungnoen et al., 2008), breast (Han et al., 2016), colon (Plyduang et al., 2014), and cervical cancer (Yoysungnoen et al., 2016). Besides these potent biological properties of tetrahydrocurcumin, it is preclinically safe. Tetrahydrocurcumin at up to 400 mg/kg per day in rats did not shown any toxic effects and thus may be considered as having no observed adverse effect (Majeed et al., 2019) and has been shown to be safe for topical skin applications (Majeed et al., 2010).

Tetrahydrocurcumin exhibits many physiologic and pharmacological activities similar to curcumin but, in some systems, tetrahydrocurcumin exerts greater antioxidant activity than curcumin. The higher antioxidant activity of tetrahydrocurcumin is thought to be due to the lack of α,β-unsaturated carbonyl moiety in tetrahydrocurcumin (Aggarwal et al., 2014). In addition, unlike curcumin, tetrahydrocurcumin is quite stable in the intestine and retains antioxidant activities at neutral or basic pH. However, tetrahydrocurcumin is unable to form Michael adducts with intracellular proteins, phase II enzymes which are responsible for inducing heme oxygenase-1 (HO-1), and NF-κB suppression in cells that causes less antitumor and antiinflammatory activity than curcumin (Aggarwal et al., 2014; Anand et al., 2008). Besides these, tetrahydrocurcumin has been shown to be useful as a safe whitening agent, antiaging, skin rejuvenator, wound healing, treating postmenopausal, and osteoarthritis symptom-relieving properties (Trivedi et al., 2017).

Besides tetrahydrocurcumin, other metabolites of curcumin have been also reported to exert biological activities. DHC has shown antioxidative property by reducing the levels of cellular nitric oxide (NO) and ROS via Nrf2 signaling pathways (Yu et al., 2018). OHC exerted appreciable antiinflammatory properties in a lipopolysaccharide-challenged RAW 264.7 murine macrophages model (Zhao et al., 2015). OHC also exhibited favorable hepatoprotective effect through restoring antioxidant status, inhibiting CYP2E1, and activating the Keap1-Nrf2 pathway (Luo et al., 2019). Tetrahydrocurcumin treatment alleviated brain damage and protected the blood–brain barrier (BBB) from cerebral ischemia/reperfusion (I/R) injury by regulating tight junction proteins, attenuating neutrophil infiltration, and reducing brain edema formation in a rat model (Wicha et al., 2020). Moreover, tetrahydrocurcumin has been shown to attenuate oxidative stress and inflammation by decreasing malondialdehyde, NO, NF-κB, and cyclooxygenase-2, and by increasing Nrf2, HO-1, and antioxidative enzymes in I/R rats (Wicha et al., 2017).

Curcumin and bioavailability

The use of curcumin has been shown to be effective in treating any number of human diseases, as discussed earlier in this chapter. However, one of the problems of using curcumin for treating diseases is its low bioavailability. This low bioavailability is primarily due to poor solubility in water, and low absorption in the digestive tract, and then rapid metabolism. All of these issues and problems would suggest that curcumin is not a very promising molecule to be developed into any kind of drug or lead drug candidate for further development by the pharmaceutical industry. As such, it is curious how curcumin can be effective in treating many diseases because it has very low water solubility and bioavailability. In addition, curcumin is not very stable under various conditions such as aqueous phosphate buffer or serum-free medium. It is quite possible that metabolites or curcumin's degradation products into different bioactive compounds may be responsible for its biological activities rather than curcumin itself (Jankun et al., 2016). This situation has enabled curcumin or remain in the nutraceutical realm and to be marketed as a nutraceutical, with its promising efficacy in many human diseases.

One of the issues of curcumin having low bioavailability relates to dosing of curcumin or turmeric on a daily basis. In Asian countries that commonly use turmeric/curcumin daily, the typical daily intake of turmeric might average 15 mg/day from uses in cooking or other forms of intake (Chand, 2019). In the United States and other countries that do not use turmeric regularly, the amount of turmeric in a typical nutraceutical supplement dose is 1–5 g. This large difference of intake might actually be detrimental for health, as relatively higher amounts might have proinflammatory activity (Chand, 2019). Furthermore, many clinical trials have used gram amounts of curcumin on a daily basis without necessarily seeing increased benefits in cancer trials (Chand, 2019). As such, it would be beneficial to be able to get the benefits of curcumin in lower doses, especially in human nutraceutical supplements.

There are a number of methods that have been employed to increase the stability and bioavailability of curcumin. Methods that have been employed include binding curcumin to lipids (Shulga, 2014; Sokolik, 2015), as well as creating phospholipid complexes of curcumin, albumin, and other molecules (Gupta et al., 2012; Tiyaboonchai et al., 2007; Li et al., 2005). Surfactants have also been used to create curcumin micellar solutions in water (Priyadarsini, 2009) as a method to increase solubility. Another approach to increase its stability, solubility, and bioavailability has been the use of metals that are complexed to curcumin.

Curcumin–metal complexes

The idea of complexing metals to curcumin goes back to the 1980s when it was noticed that curcumin has three sites on the molecular structure that can be potentially reacted with other substances to create a new molecular structure. These reactive sites include two phenol groups at the ends of the molecule and one diketone group in the central part of the molecule (Fig. 36.1), as well other potential sites.

When curcumin is in a solution it will exist as *cis* and *trans* isomers. It should be noted that curcumin is practically insoluble in water, but is quite soluble in polar solvents such as methanol, ethanol, DMSO, acetonitrile, chloroform, and ethyl acetate (Kaniuk, 2016). The solubility of curcumin in these various solvents has been taken advantage of when complexing to metals and the various formats that metals themselves can have solubility in solvents like water or other organic solvents. For example, a metal like zinc chloride is quite soluble in methanol, whereas zinc acetate dihydrate is poorly soluble in methanol but much more soluble in water. While the phenolic rings at the ends of the molecular structure of curcumin can be sites to modify, the diketal portion of curcumin in the middle of the molecule presumably is the reactive site that could be modified with metals by chelation bonding. These diketo groups possess keto-enol tautomerism (see Fig. 36.2).

Both keto and enol tautomers of curcuminoids are present in varied ratios that will depend on the particular medium used. While the keto form predominates in acidic pH, the enol form predominates at neutral or alkaline pH. The enol form appears to be more energetically stable in a solid phase. Depending on the solvent, the enol form will predominate, as when in ethanol, but if water is added, the keto form will start to appear (Manolova et al., 2014). It is thought that the keto tautomer is more responsible for the therapeutic activities of curcumin, while the enol tautomer has a tendency to degrade (Rege et al., 2019). This also makes the enol tautomer more ideally suited to chelating to metals.

The use of chelation is a preferred medical treatment for eliminating toxic effects of metals. When a chelating agent binds to a toxic metal and forms a complex, the complexed structure can be excreted and thus eliminated from the body (Flora and Pachauri, 2010). The term chelation was first used in the 1920s as the term for bringing two associating units to be fastened to a central atom (Morgan et al., 1920). Chelation is a form of chemical

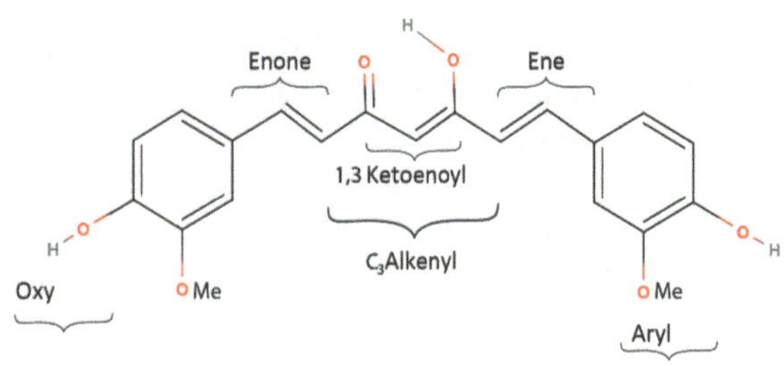

FIGURE 36.1 Potential reactive sites of curcumin.

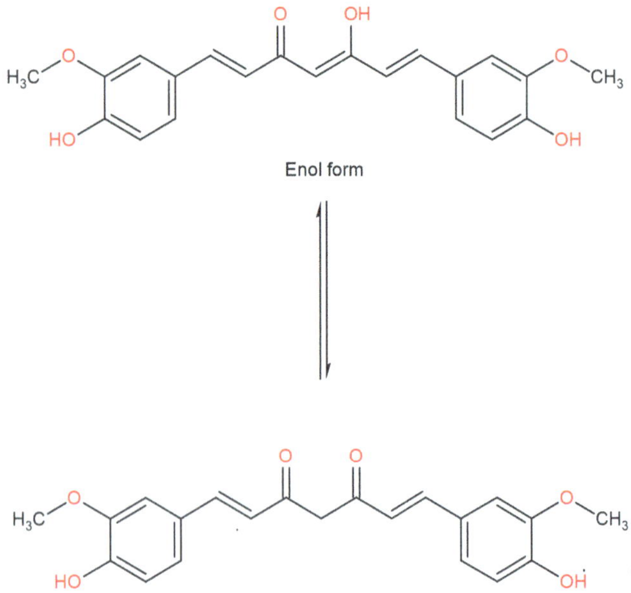

FIGURE 36.2 Keto:enol tautomers of curcumin.

bonding of metal ions that involves the formation of a central metal ion to two or more separate molecular ligands. It involves a coordinate covalent bond (also known as a dative covalent bond) where a shared pair of electrons both come from the same atom instead of from both atoms. When metal ions bond to ligands, the bond uses the coordinate covalent bonding type of interaction. In curcumin a coordinate bond with metal ions is formed because of the enol group where the enol proton is replaced by metal ion. The O-methoxy-phenol part of curcumin in complexes remains unchanged.

One of the first published papers involving a metal being complexed to curcumin was published in 1987 (Sharma et al., 1987). This involved complexing gold to curcumin by simply mixing gold chloride with curcumin using ethanol at a 1:2 molar ratio and subsequently using this complex for antiarthritis studies in rats. These initial trials have spawned extensive research into the use of curcumin–metal complexes for medical uses. Besides gold, curcumin binds to a wide range of metals including S, Se, and B (Ramanjaneyulu et al., 2007; Wanninger et al., 2014; Refat, 2013; Sagnou et al., 2011). Curcumin binds to both divalent metal ions such as Zn^{2+}, Ni^{2+}, PD $^{2+}$, Mn^{2+}, Cu^{2+}, and Co^{2+} (see Figs. 36.3–36.6) and trivalent ions such as Al^{3+}, Fe $^{3+}$, and Cr^{3+}, and a number of rare earth metals that can result in various configurations.

The ionic form of the metal is critical to the success of chelation to the ligand, as divalent ions and trivalent ions do not always bind in the same manner. Indeed, it is known that divalent metal ions will not readily bind to demethoxycurcumin, whereas trivalent metal ions are more successful (Ostrowski et al., 2013).

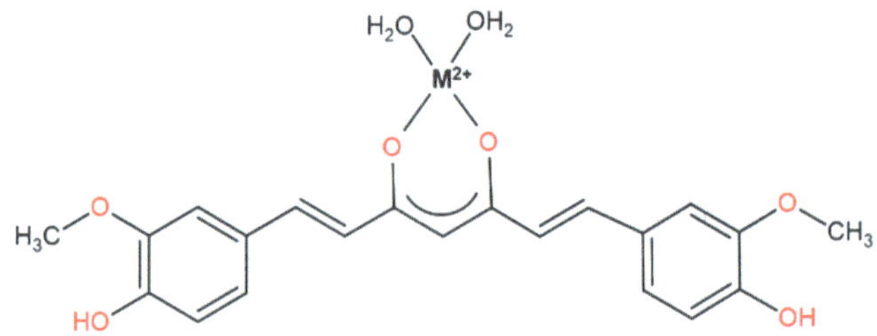

FIGURE 36.3 Divalent metal–curcumin complex.

The complexation of curcumin to metals will result in stable structures for 1:1 and 2:1 ligand:divalent metal ions and 3:1 ligand:trivalent metal ions. The use of organic solvents typically has been used to solubilize curcumin and the metals used in the complexation. These solvents tend to be volatile such as ethanol, or methanol, and other organic solvents that are of a polar nature. However, manufacturing methods using volatile solvents to solubilize both the curcumin and the particular metal format used have drawbacks of using potentially dangerous solvents. However, there are other solvating approaches that do not use volatile solvents and are potentially less dangerous. This can include using water and glycerol (Zebib et al., 2010) or water and propylene glycol. Various metal formats can be dissolved in water, such as zinc sulfate, while curcumin has the ability to be dissolved in glycerol or propylene glycol and these solvents are also compatible with each other, to allow complexes to be formed of the metal with curcumin.

To analyze curcumin–metal complexes nuclear magnetic resonance spectroscopy (NMR), UV, fourier-transform infrared spectroscopy (FTIR), and mass

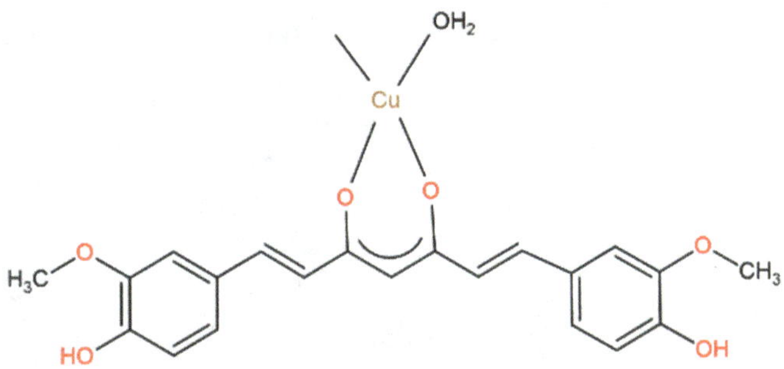

FIGURE 36.4 Copper–curcumin complex.

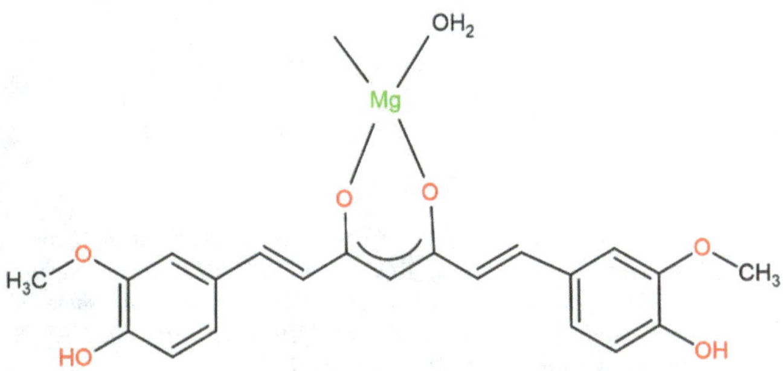

FIGURE 36.5 Magnesium–curcumin complex.

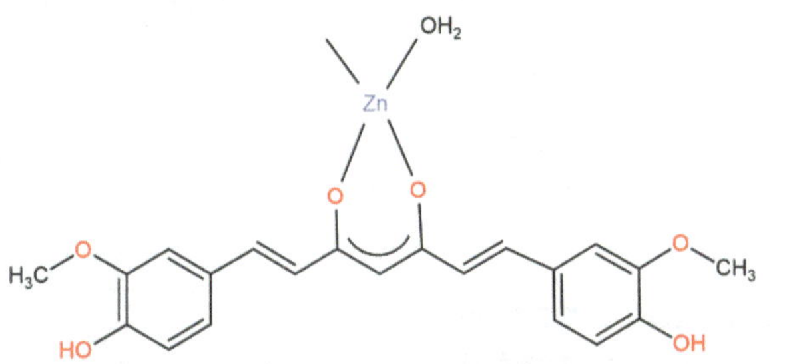

FIGURE 36.6 Zinc–curcumin complex.

spectroscopy (MS) methods have been employed. In addition, elemental analysis has been employed to look for the particular metal used when making the complex. There are shifts in UV spectra that will occur when metals bind to oxygen of the enol group in the region of 425 nm. Characteristic shifts in FTIR profile can occur in carbonyl bands and decreases in the intensity of the free hydroxyl (OH) group of curcumin can also be observed (Zebib et al., 2010). NMR data can show a shift in the curcumin when it interacts with metal ions (Priyadarsini, 2014).

The oral bioavailability of curcumin is low because of low absorption by the small intestine along with metabolism in the liver and elimination by the gall bladder. In addition, poor bioavailability is also exacerbated by curcumin binding to enterocyte proteins that can change its structure (Heger et al., 2013). Despite its poor bioavailability, curcumin does cross the BBB, presumably because of its lipophilic nature. As such curcumin is able to reach the brain in effective concentrations and promote neuroprotection. Methods that have been employed to increase bioavailability of curcumin have included the use of combinations with other compounds such as piperine or lecithin (Dei Cas and Ghidoni, 2019). Curcumin–metal complexes are also be taken up by cells and cross membranes. These complexes can be visualized inside cells using confocal microscopy. This is due to their fluorescence capability (Xing et al., 2014; Baum and Ng, 2004). The complexes get internalized by receptor-mediated endocytosis. Once inside cells, the complexes have the ability to perform various activities that curcumin or tetrahydrocurcumin carry out but in potentially a better manner to treat diseases or even cancer. For example, the curcumin–metal complexes can bind to DNA (Fig. 36.7).

While free curcumin typically binds to DNA via a groove-binding mechanism, a curcumin–metal complex can bind with additional bonds with DNA through electrostatic interaction and intercalation (Zebib et al., 2010; Vellampatti et al., 2018).

Uses of curcuminoid–metal complexes

Antioxidant activity of curcumin–metal complexes

These complexes have multiple capabilities and can have so-called "multi-anti" actions (Refat, 2013) by having antimicrobial and antioxidant actions better than those of free curcumin and free metal ions. Curcumin and

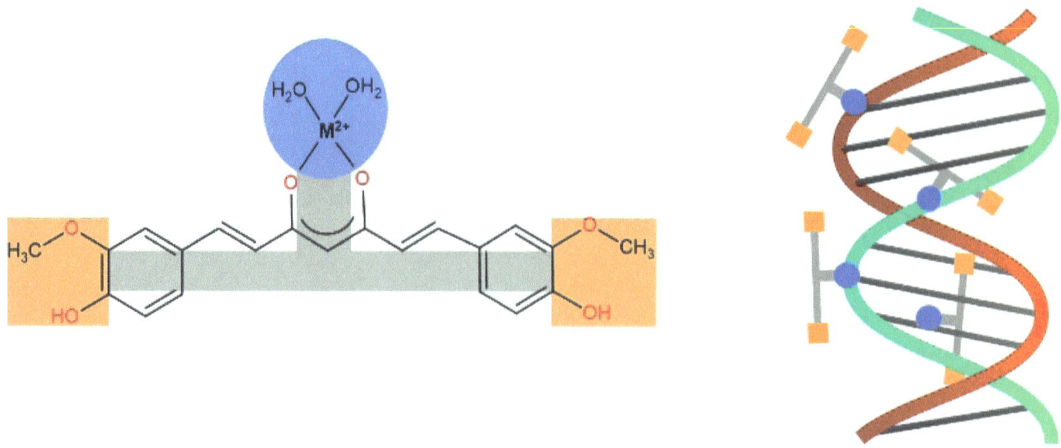

FIGURE 36.7 Divalent metal–curcumin complexes binding to DNA.

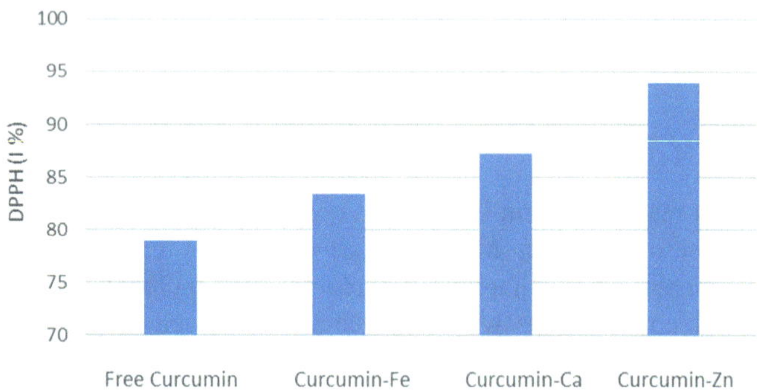

FIGURE 36.8 Antioxidant capability of curcumin–metal complexes. *Adapted from Hieu, T.Q., Thao D.T., 2019. Enhancing the solubility of curcumin metal complexes and investigating some of their biological activities. J. Chem. 2019, Article ID 8082195. 8 pages. https://doi.org/10.1155/2019/8082195.*

tetrahydrocurcumin have antioxidant capabilities that have been taken advantage of for various health conditions where oxidative stress is an underlying issue as in various diseases or due to the aging process. The concept that curcumin–metal complexes and tetrahydrocurcumin–metal complexes may be a step forward in antioxidant research is starting to emerge. Indeed, it is known that curcumin–metal complexes can show better antioxidative capability than free curcumin alone (Takam and Saewan, 2011; Hieu and Thao, 2019). Curcumin complexes with Cu(II), Zn(II), Mn(II), Mg(II), and Fe(II) have better 2,2-diphenyl-1-picrylhydrazyl (DPPH) radical-scavenging and ferrous-reducing power activities than free curcumin at the same dosage. As an example, see Fig. 36.8. The reason why curcumin–metal complexes are better antioxidants is apparently due to a stabilization of the curcumin by the metal (Barik et al., 2007) when a superoxide radical binds to it. Normally, free curcumin becomes easily oxidized and changes its structure when free radicals bind. But the when curcumin is bound by a metal the physical geometry of the structure is not changed as much and thus allows a curcumin–metal complex to remain intact and allow for many redox cycles and is also a better antioxidant.

Antiinflammatory activity of curcumin–metal complexes

It is expected that curcumin–metal complexes have better antiinflammatory activity than just curcumin alone. It is known that free curcumin can inhibit signal pathways for inflammation that involve a multitude of receptors and enzymes. Some of these actions include the protection of *S*-nitrosylation on IKKβ; binding with myeloid differentiation proteins; inhibition of nuclear factor kinase subunit β, actions on Toll-like receptors, and actions on many other proteins, receptors, and enzymes (Xu et al., 2018). While inflammation is very complicated and involves numerous pathways, many inflammation issues have an underlying characteristic of free radicals that antioxidants can help treat. As such, curcumin–zinc complexes are suggested to be promising molecules to treat inflammatory bowel disease (Sareen et al., 2015). Other metals can also be used for inflammation reduction. The use of curcumin–ferrous complexes has been observed to have significant antiinflammatory activity. In a rat paw edema model, curcumin alone reduced edema by 51.32% ($P < .05$), while a curcumin–ferrous complex was able to reduce edema by 59.76% ($P < .05$). This reduction is close to the standard drug regimen with indometacin with a 61.76% reduction (Sandhya and Renu, 2015). These results give additional credence to the idea that complexing metals to curcumin is a step in the right direction for the use of curcumin–metal complexes as a nutraceutical to help fight inflammation.

Antimicrobial activity of curcumin–metal complexes

The antimicrobial activity of curcumin–metal complexes has been studied. It has been shown (Hieu and Thao, 2019) that metal complexes are more efficient than curcumin alone in killing different types of bacteria and certain metal complexes are more efficient than others. For example, a Cur–Ca (II) worked best against *Salmonella*, followed by the Cur–Fe (III) complex, while the Cur–Zn (II) complex only worked as well as free curcumin. However, in killing *Staphylococcus aureus*, the Cur–Zn (II) complex killed better than Cur–Fe (III), although the Cur–Ca (II) remained the most efficient. However, some bacteria, like *E. coli*, may still prove somewhat difficult to kill using free curcumin or curcumin–metal complexes. It is possible that using a higher ligand:metal ratio such as 3:1 may prove to be more efficient in killing certain types of bacteria (Bagchi et al., 2015). It is possible with higher amounts of curcumin present in a curcumin–metal complex, that greater killing efficiency may occur.

Alzheimer disease use

The use of curcumin–metal complexes has the ability to bind to various cellular components. This has made them of interest in Alzheimer disease research. The use of Al^{3+}–curcumin in varying ratios of ligand:metal has been examined in relation to Alzheimer disease (Khalil et al., 2013). The use of copper–curcumin and zinc–curcumin complexes has been shown to have antifibrillogenic properties. These complexes have been studied on the aggregation of amyloid-β peptides involved in Alzheimer disease (Banerjee, 2014). Other research has used copper and iron to form complexes with curcumin for mechanisms of action in an Alzheimer disease model animal (Baum and Ng, 2004). In addition, the use of curcumin–copper or curcumin–zinc complexes shows significant neuroprotective effects in neuronal PC12 cells (Yan et al., 2018). These results show strong potential for these curcumin–metal complexes for Alzheimer disease treatments (Shakeri et al., 2019).

Cancer use

The use of curcumin–metal complexes has been shown to be better as antitumor agents than free curcumin (Valentini et al., 2009; John et al., 2002). Examples of curcumin bound to ruthenium that may have use in cancer therapies (Ma et al., 2019) are starting to appear to have efficacy.

Gastroprotective

Complexes of curcumin and zinc can have gastroprotective effects. It was shown in rodent models of gastric ulcers that a curcumin–zinc complex significantly was more effective as compared to just free curcumin (Mei et al., 2011).

Tetrahydrocurcumin–metal complexes

Tetrahydrocurcumin has the ability to chelate metals in a similar manner to curcumin as the same diketal structure is present. This has been taken advantage of in binding divalent ions of Cd. Cadmium toxicity can result in hypertension, vascular remodeling, increased arterial stiffness, and oxidative stress. Tetrahydrocurcumin can complex with cadmium and reduce the adverse effects of this divalent ion (Sangartit et al., 2014). The use of tetrahydrocurcumin as an antioxidant activity against Cd intoxication results directly and/or indirectly through free radical scavenging, metal chelation, and other modes of action including regulation of inflammatory enzymes (Kukongviriyapan et al., 2016). The use of tetrahydrocurcumin's chelation ability to metals has been used beyond its use as a medical means to remove cadmium. Tetrahydrocurcumin has been successfully complexed to metals in a manner similar to curcumin that can be put to additional medical and nutraceutical uses. For example, iridium has been complexed to tetrahydrocurcumin and been shown to have a lot of potential in cancer treatment (Colombo et al., 2019). These tetrahydrocurcumin–iridium complexes are readily taken up in human tumor cells and can induce rapid apoptosis in these cells after being irradiated with visible light.

Zinc as a metal to use in complexes

One of the metals that is known to bind curcumin is zinc. Making zinc complexes with curcumin will make it more soluble for transportation by the bloodstream. As zinc is an essential trace element it also has many beneficial health roles in the body. This makes zinc an interesting metal for the possibility of using curcumin as a delivery system for getting zinc into the body that also allows taking advantage of the benefits of curcumin. Zinc plays important roles in many physiological functions and is an essential trace element. It has biochemical functions in the body as a cofactor for enzymes; cellular functions for cell growth, cell membrane integrity, tissue growth, wound healing; immunological function in neutrophils, T cells, B cells, and HK cells; endocrine function for spermatogenesis, thyroid function, pancreatic function of insulin storage and release, prolactin secretion, thymopoietin synthesis; neurological functions for cognition, memory, taste acuity, and vision; hematological function in coagulation factors; and in skeletal functions for bone formation.

Zinc has a key function within the immune system (Khanam, 2018). It is necessary for the functioning and development of immune cells. Its importance can be seen as zinc is required in over 300 enzymes and more than 1000 transcription factors for their activities (Prasad et al., 2014). Because zinc homeostasis in immune cells is finely controlled, any changes in its regulation can result in impairment of its usual functions. As such, these impairments can be seen in all sorts of diseases, such as infections, autoimmune diseases, and cancers. Severe zinc deficiency in humans can include a whole host of conditions including bullous pustular dermatitis, diarrhea, alopecia, emotional disorder, intercurrent infections due to cell-mediated immune dysfunctions, weight loss, hypogonadism in males, neurosensory disorders, and issues with healing of ulcers. Some of these issues can become fatal (Prasad, 2008). Zinc deficiencies also compromise the immune system at various levels from the molecular to the systemic level (Gammoh and Rink, 2019). Some 1.1 billion people are deficient in zinc globally (Kumssa et al., 2015), especially children and the elderly (Roohani et al., 2013).

Zinc is critical for normal development and function of cells in the immune system. If zinc is deficient, then the growth and function of T cells, B cells, neutrophils, and NK cells will be affected (Shankar and Prasad, 1998). Phagocytosis, intracellular killing, and cytokine production by macrophages also are affected by zinc deficiency. Zinc also functions as an antioxidant and helps stabilize membranes. This suggests that zinc can play a role in preventing free radical-induced problems during inflammation. It is known that supplementing with zinc can be effective in decreasing oxidative stress and producing inflammatory cytokines including TNF-α and IL-1β in elderly people or patients with sickle cell disease (Prasad, 2009). These are among the reasons why supplementation with zinc has become a large part of the nutraceutical industry.

Since low levels of zinc can lead to impaired cell-mediated immunity, the role of zinc in the thymus gland becomes of great interest. The thymus has control of cell-mediated immunity through its effects on thymus-derived hormones such as thymosin, thymopoietin, and serum thymic factor that regulate many immune functions. As such, the health of the immune system is dependent on the health of the thymus gland, which in turn is dependent on zinc availability (Wong et al., 2009; Hojyo and Fukada, 2016). Low levels of zinc can increase risks for infection and autoimmune diseases along with inflammation but these kinds of problems can be reversed by zinc supplementation. It has been shown (Haase and Rink, 2009) that oral zinc supplementation can improve immunity and help reduce chronic inflammatory responses in the elderly.

Zinc's role as an antiviral agent

Besides the role of zinc in the immune system, it also has a role in preventing infections. There is considerable evidence that zinc has antiviral activity against a wide range of viruses using numerous mechanisms (Read et al., 2019). Exactly how zinc works as an antiviral is not clear, but it does appear that zinc can inhibit viral proteases and various polymerase processes, as well as through physical means of preventing viral physical entry into cells (Fig. 36.9) and subsequent replication and spreading to other organs with both RNA and DNA viruses (Ishida, 2019). As such the use of zinc for immune-boosting purposes and as an antiviral is well founded for nutraceutical supplements. However, the amount of the daily dose of

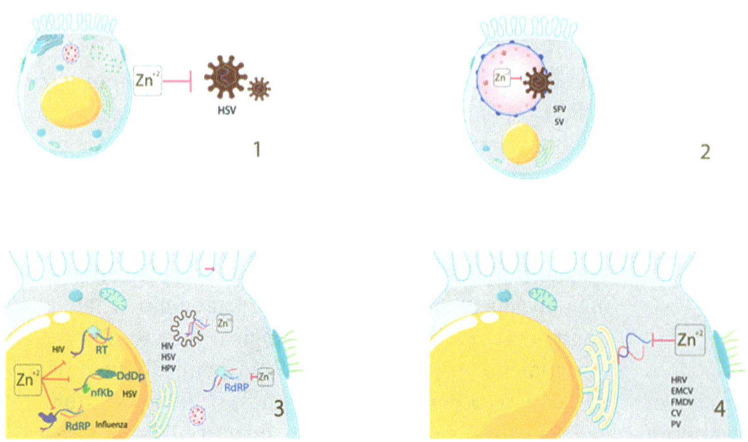

FIGURE 36.9 Antiviral actions of zinc modes of action in cells. Panel 1: free virus inactivation. Panel 2: inhibition of viral uncoating. Panel 3: viral genome transcription. Panel 4: viral protein translation and polyprotein processing. *DdDp*, DNA-dependent DNA polymerase; *EMCV*, encephalomyocarditis virus; *FMDV*, foot and mouth disease virus; *HCV*, hepatitis C virus; *HIV*, human immunodeficiency virus; HPV, human papilloma virus; *HRV*, human rhinovirus; *HSV*, herpes simplex virus; *PV*, polio virus; *RdR*p, RNA-dependent RNA polymerase; *RT*, reverse transcriptase; *SARS*, severe acute respiratory syndrome coronavirus; *SFV*, Semliki Forest virus; *SV*, sindbis virus; *VZV*, varicella-zoster virus.

zinc can be important. There are recommended levels (NIH, 2019) for adult males of 11 mg daily, and excess amounts in the 150 mg daily range can cause various problems such as vomiting, nausea, loss of appetite, and other issues that may result in hospitalization under certain circumstances (Johnson et al., 2017). As such, it would be best to get the beneficial effects of zinc in as small a dose as possible.

Curcumin as a platform for novel complexes

While the use of metals to bind only curcumin has been explored in some detail, the idea of using this platform to further modify this complex is just beginning. Another approach to cancer research has involved the use of curcumin analogs to improve the activity and stability of curcumin. This research has shown some of the endless possibilities that can occur when modifying the structure of curcumin at its reactive groups on the phenol ends of the structure and the diketal central structure in the center of curcumin. By taking advantage of the diketal structure and using various metals, a number of different molecules can be complexed to curcumin (Vyasa et al., 2013), as seen in Fig. 36.10.

Tetrahydrocurcumin–Zn–curcumin complexes

A new approach for creating novel curcuminoid–metal complexes is to use curcumin as a platform to bind useful molecules like tetrahydrocurcumin via zinc as the central atom. This type of complex will have the potential to have the antioxidant capabilities of tetrahydrocurcumin along with the antiinflammatory capabilities of curcumin, while providing zinc to the body. This approach has been carried out on an experimental basis. Taking advantage of the ligand to metal ratio of 2:1, a characteristic change in the UV profile occurs when zinc binds to both curcumin and tetrahydrocurcumin (Fig. 36.11).

This UV change coincides with a shift in the yellow color of the mixture curcumin/tetrahydrocurcumin to a red color when zinc complexes with curcumin and tetrahydrocurcumin. It is also consistent with a shift in the peak UV wavelength from a lower wavelength and the appearance of a new side peak at a higher wavelength. Changes in FTIR profiles (Fig. 36.12) are also noted with wavelength shifts occurring that would be consistent with changes in the curcumin and tetrahydrocurcumin molecular structure as zinc binds to these and forms complexes.

A proposed molecular structure is seen in Fig. 36.13 for a tetrahydrocurcumin–metal–curcumin complex.

The tetrahydrocurcumin–zinc–curcumin complex has been commercialized as TurmiZn-OS (probioticsmart.com). This complex has the ability to enter cells and the capability to perform the kinds of activities that curcumin and tetrahydrocurcumin have. This may include antiinflammatory actions as a result of downregulation or upregulation of various signal pathways, as in Fig. 36.14. As zinc is present in this complex, the zinc will be able to participate in the kinds of activities that zinc is known to do in the body.

Concluding remarks and future directions

Innumerable curcumin analogs and curcumin–metal complexes have been synthesized in the past decades (Nasir et al., 2013). There is continuing progress with the creation of the first Si-based metal–curcumin complex being reported (Lorenz et al., 2019) and examples of curcumin bound to ruthenium that may have use in cancer therapies (Ma et al., 2019). These new analogs and novel metal complexes will continue to help solve the poor bioavailability of curcumin and help reap the benefits of curcumin in the future. Using the diketal portion of curcumin and metals like zinc as a platform to add on additional molecules such as tetrahydrocurcumin, is an approach that will help increase the efficacies and bioavailability of curcumin, tetrahydrocurcumin, and the zinc itself. This novel complex may help in reducing oxidative stress in the body reacting to diseases such as the flu that can cause the production of free radicals. The approach of complexing curcumin to tetrahydrocurcumin via zinc may also help stimulate the immune system in times of stress or from aging. In addition, using the other reactive sites on curcumin or tetrahydrocurcumin to create novel complexes with yet other molecules will open doors to areas that have not been examined previously. The use of curcumin and tetrahydrocurcumin that are complexed together by metals may have some advantages of delivery of useful trace metals like zinc to the body while potentially being delivered in smaller safer overall doses than in their free formats. This complex can be delivered in a variety of formats including capsules, foods, and liquids. It will be very interesting to see how these complexes will help reduce oxidative stress, help the immune system, and improve general health in the future.

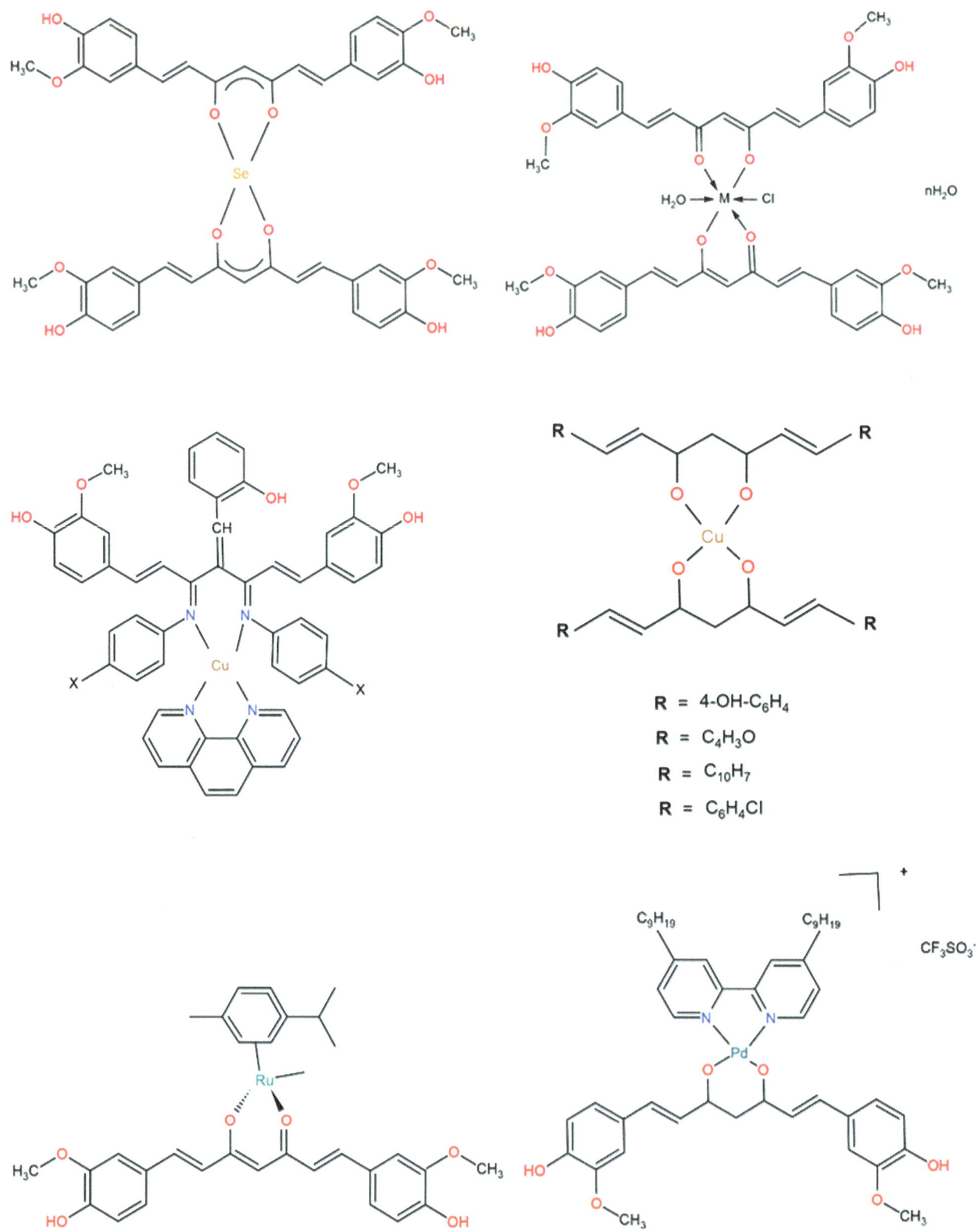

FIGURE 36.10 Curcumin–metal complexed to various molecules.

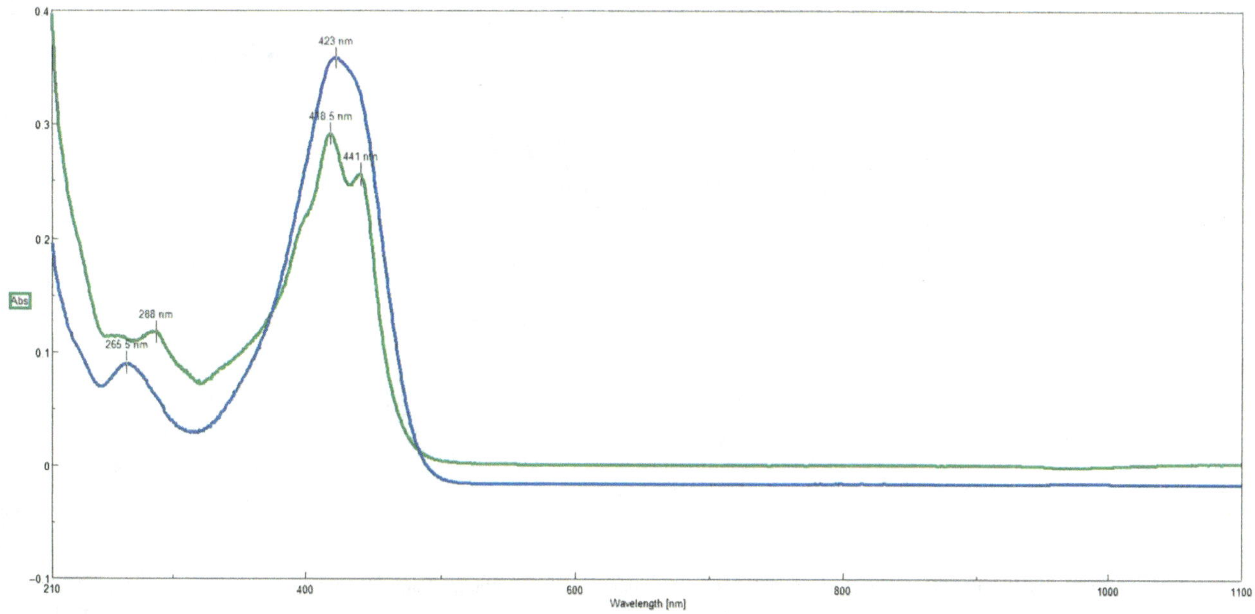

FIGURE 36.11 UV profiles of a curcumin and tetrahydrocurcumin mixture in the absence of zinc (blue trace) and curcumin–Zn–tetrahydrocurcumin complex (green trace).

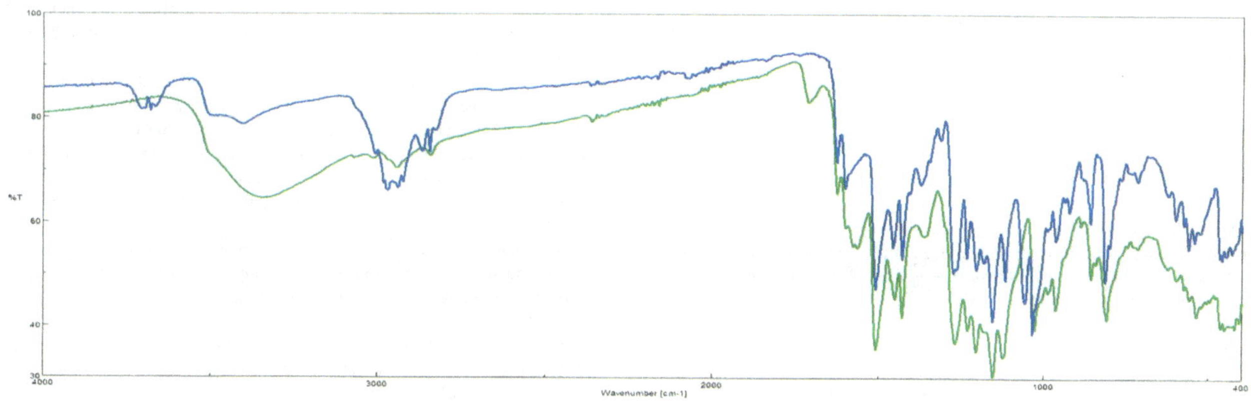

FIGURE 36.12 Frontier-transform infrared spectroscopy (FTIR) profiles of tetrahydrocurcumin/curcumin without zinc (blue trace) and tetrahydrocurcumin–Zn–curcumin complex (green trace).

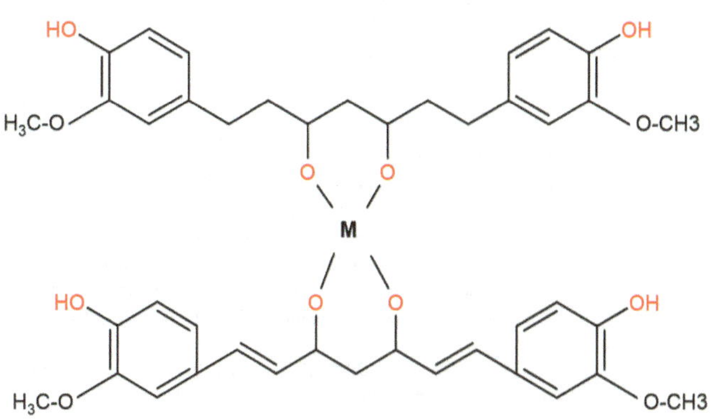

FIGURE 36.13 Proposed structure of a tetrahydrocurcumin–metal–curcumin complex. The metal could be zinc, copper, iron, or another.

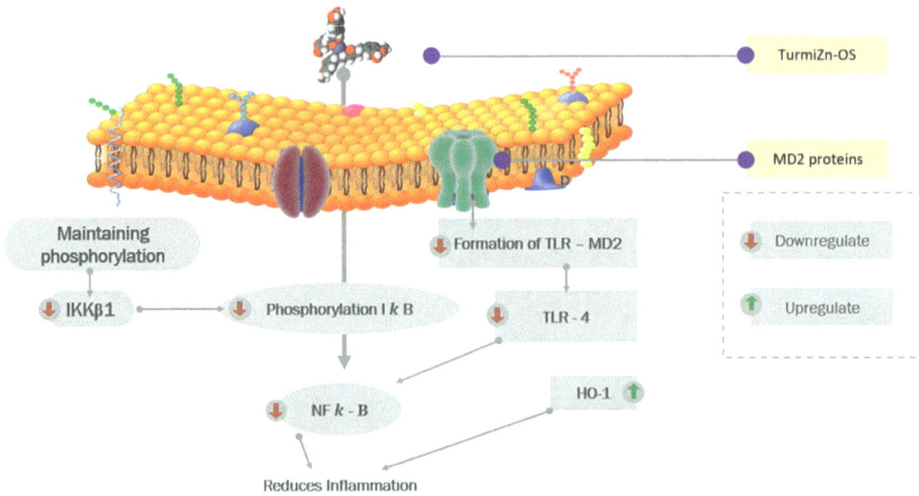

FIGURE 36.14 Tetrahydrocurcumin–zinc–curcumin complexes and cellular uptake with potential antiinflammatory mechanisms.

Acknowledgments

The authors wish to thank Sarath Nalla, Jamil Talukder, and Prafulla Kumar for their technical assistance in preparing this manuscript.

References

Aggarwal, B.B., Deb, L., Prasad, S., 2014. Curcumin differs from tetrahydrocurcumin for molecular targets, signaling pathways and cellular responses. Molecules. 20 (1), 185–205.

Anand, P., Thomas, S.G., Kunnumakkara, A.B., et al., 2008. Biological activities of curcumin and its analogues (congeners) made by man and mother nature. Biochem. Pharmacol. 76, 1590–1611.

Bagchi, A., Mukherjee, P., Bhowmick, S., et al., 2015. Synthesis, characterization and antibacterial activity of a novel curcumin metal complex. Int. J. Drug Dev. Res. 7 (2), 11–14.

Banerjee, R., 2014. Inhibitory effect of curcumin-Cu(II) and curcumin-Zn(II) complexes on amyloid-beta peptide fibrillation. Bioinorg. Chem. Appl. 2014, 325873.

Barik, A., Mishra, B., Kunwar, A., 2007. Comparative study of copper(II)–curcumin complexes as superoxide dismutase mimics and free radical scavengers. Eur. J. Med. Chem. 42 (4), 431–439.

Baum, L., Ng, A., 2004. Curcumin interaction with copper and iron suggests one possible mechanism oaction in Alzheimer's disease animal models. J. Alzheimers Dis. 6 (4), 367–377.

Chand, N., 2019. Standardized turmeric and curcumin. In: Gupta, R.C., Srivastava, A., Lall, R. (Eds.), Nutraceuticals in Veterinary Medicine. Springer Nature, Cham, pp. 3–23.

Colombo, A., Fontani, M., Dragonetti, C., et al., 2019. Dietary curcumin: correlation between bioavailability and health potential. Nutrients. 11 (9), 2147.

Dai, J., Gu, L., Su, Y., et al., 2018. Inhibition of curcumin on influenza A virus infection and influenzal pneumonia via oxidative stress, TLR2/4, p38/JNK MAPK and NF-κB pathways. Int. Immunopharmacol. 54, 177–187.

Dei Cas, M., Ghidoni, R., 2019. Dietary curcumin: correlation between bioavailability and health Potential. Nutrients. 11 (9), 2147.

Flora, S.J.S., Pachauri, V., 2010. Chelation in metal intoxication. Int. J. Environ. Res. Public Health 7 (7), 2745–2788.

Gammoh, N.Z., Rink, L., 2019. Zinc and the immune system. In: Mahmoudi, M., Rezaei, N. (Eds.), Nutrition and Immunity. Springer Nature, Cham, Switzerland.

Gupta, S.C., Patchva, S., Koh, W., et al., 2012. Discovery of curcumin, a component of the golden spice, and its miraculous biological activities. Clin. Exp. Pharmacol. Physiol. 39 (3), 283–299.

Haase, H., Rink, L., 2009. The immune system and the impact of zinc during aging. Immun. Ageing 6, 9.

Han, X., Deng, S., Wang, N., et al., 2016. Inhibitory effects and molecular mechanisms of tetrahydrocurcumin against human breast cancer MCF-7 cells. Food Nutr. Res. 60, 30616.

Hassaninasab, A., Hashimoto, Y., Tomita-Yokotani, K., et al., 2011. Discovery of the curcumin metabolic pathway involving a unique enzyme in an intestinal microorganism. Proc. Natl. Acad. Sci. USA. 108, 6615–6620.

Heger, M., van Golen, R.F., Broekgaarden, M., et al., 2013. The molecular basis for the pharmacokinetics and pharmacodynamics of curcumin and its metabolites in relation to cancer. Pharmacol. Rev. Available from: 10.1124/pr.110.004044.

Hieu, T.Q., Thao, D.T., 2019. Enhancing the solubility of curcumin metal complexes and investigating some of their biological activities. J. Chem. 2019, Article ID 8082195. 8 pages. Available from: https://doi.org/10.1155/2019/8082195.

Hojyo, S., Fukada, T., 2016. Roles of zinc signaling in the immune system. J. Immunol. Res. 2016, Article ID 6762343. 21 pages. Available from: https://doi.org/10.1155/2016/6762343.

Ishida, T., 2019. Anti-viral vaccine activity of zinc(II) for viral prevention, entry, replication, and spreading during pathogenesis process. Curr. Trends Biomed. Eng. Biosc. 19, 83–87.

Jankun, J., Wyganowskia-Swiatokowska, M., Dettlaffk, K., Jelinkska, A., et al., 2016. Determining whether curcumin degradation/condensation is actually bioactivation (Review). Int. J. Mol. Med. 37, 1151–1158.

John, V.D., Kuttan, G., Krishnankutty, K., 2002. Anti-tumor studies of metal chelates of synthetic curcuminoids. J. Exp. Clin. Cancer Res. 21, 219–224.

Johnson, A.R., Munoz, A., Gottlieb, J.L., et al., 2017. High dose zinc increases hospital admissions due to genitourinary complications. J. Urol. 177, 639–643.

Kaniuk, M.I., 2016. Prospects of curcumin use in nanobiotechnology. Biotechnol. Acta 9, 23–36.

Khalil, M.I., Al-Zahem, A.M., Al-Qunaibit, M.H., 2013. Synthesis, characterization, Mössbauer parameters and anti-tumor activity of Fe(III) curcumin complex. Bioinorg. Chem. Appl. 2013, Article ID 982423.

Khanam, S., 2018. Impact of zinc on immune response. Immunol. Curr. Res. 2, 1.

Kukongviriyapan, U., Apaijit, K., Kukongviriyapan, V., 2016. Oxidative stress and cardiovascular dysfunction associated with cadmium exposure: beneficial effects of curcumin and tetrahydrocurcumin. Tohoku J. Exp. Med. 239, 25–38.

Kumssa, D., Joy, E., Ander, E., et al., 2015. Dietary calcium and zinc deficiency risks are decreasing but remain prevalent. Sci. Rep. 5, 10974.

Li, K., Zhai, M., Jiang, L., et al., 2019. Tetrahydrocurcumin ameliorates diabetic cardiomyopathy by attenuating high glucose-induced oxidative stress and fibrosis via activating the SIRT1 pathway. Oxid. Med. Cell Longev. 2019, Article ID 6746907. Available from: https://doi.org/10.1155/2019/6746907.

Li, L., Braiteh, F.S., Kurzrock, R., 2005. Liposome encapsulated curcumin: in vitro and *in vivo* effects on proliferation, apoptosis, signaling, and angiogenesis. Cancer. 104, 1322–1331.

Lorenz, V., Liebing, P., Suta, M., et al., 2019. Synthesis, structure, complexation, and luminescence properties of the first metal-organic curcumin compound Bis(4-triphenylsiloxy)curcumin. J. Luminesc. 211, 243–250.

Luo, D.D., Chen, J.F., Liu, J.J., et al., 2019. Tetrahydrocurcumin and octahydrocurcumin, the primary and final hydrogenated metabolites of curcumin, possess superior hepatic-protective effect against acetaminophen-induced liver injury: role of CYP2E1 and Keap1-Nrf2 pathway. Food Chem. Toxicol. 123, 349–362.

Ma, D.L., Wu, C., Cheng, S.S., et al., 2019. Development of natural product-conjugated metal complexes as cancer therapies. Int. J. Mol. Sci. 20, 341.

Majeed, M., Pineda, M., Chan, G., et al., 2010. A randomized, double-blind, placebo-controlled, comparative study; The safety and efficacy of 0.25% tetrahydrocurcumin (turmeric) cream as depigment agent against 4% hydroquinone cream. Household Personal Care Today n 3, 44–46.

Majeed, M., Natarajan, S., Pandey, A., et al., 2019. Subchronic and reproductive/developmental toxicity studies of tetrahydrocurcumin in rats. Toxicol. Res. 35 (1), 65–74.

Manolova, Y., Deneva, V., Antonov, L., 2014. The effect of the water on the curcumin tautomerism: a quantitative approach. Spectrochimica Acta A Mol. Biomol. Spectrosc. 132C, 815–820.

Mei, X., Xu, D., Xu, S., et al., 2011. Gastroprotective and antidepressant effects of a new zinc(II)-curcumin complex in rodent models of gastric ulcer and depression induced by stresses. Pharmacol. Biochem. Behav. 99, 66–74.

Mirani, A., Kundaikar, H., Velhal, S., et al., 2019. Tetrahydrocurcumin-loaded vaginal nanomicrobicide for prophylaxis of HIV/AIDS: in silico study, formulation development, and in vitro evaluation. Drug. Deliv. Transl. Res. 9 (4), 828–847.

Montecino-Rodriguez, E., Berent-Maoz, B., Dorshkind, K., 2013. Causes, consequences, and reversal of immune system aging. J. Clin. Invest. 123 (3), 958–965.

Morgan, T., Gilbert, T., Drew, Harry, D.K., 1920. CLXII.—Researches on residual affinity and co-ordination. Part II acetylacetones of selenium and tellurium. J. Chem. Soc. 117, 1456–1465.

Nasir, S., Bukhari, A., Jantan, I.B., et al., 2013. Synthesis and biological evaluation of curcumin analogues. J. Med. Sci. 13, 501–513.

NIH, 2019. https://ods.od.nih.gov/factsheets/Zinc-HealthProfessional.

Novaes, J.T., Lillico, R., Sayre, C.L., et al., 2017. Disposition, metabolism and histone deacetylase and acetyltransferase inhibition activity of tetrahydrocurcumin and other curcuminoids. Pharmaceutics. 9 (4), 45.

Ostrowski, W., Uniecikowska, L., Hoffmann, M., et al., 2013. Demethoxycurcumin-metal complexes: fragmentation and comparison with curcumin-metal complexes, as studied by ESI-MS/MS. J Spectrosc. 2013, Article ID 749641, 8 pages.

Park, C.S., Jang, H.J., Lee, J.H., et al., 2018. Tetrahydrocurcumin ameliorates tacrolimus-induced nephrotoxicity via inhibiting apoptosis. Transpl. Proc. 50 (9), 2854–2859.

Park, C.H., Song, J.H., Kim, S.N., et al., 2019. Neuroprotective effects of tetrahydrocurcumin against glutamate-induced oxidative stress in hippocampal HT22 cells. Molecules. 25 (1), 144. 30.

Plyduang, T., Lomlim, L., Yuenyongsawad, S., et al., 2014. Carboxymethylcellulose-tetrahydrocurcumin conjugates for colon-specific delivery of a novel anti-cancer agent, 4-amino tetrahydrocurcumin. Eur. J. Pharm. Biopharm. 88 (2), 351–360.

Prasad, A.S., 2008. Zinc in human health: effect of zinc on immune cells. Mol. Med. 14 (5-6), 353–357.

Prasad, A.S., 2009. Zinc: role in immunity, oxidative stress and chronic inflammation. Curr. Opin. Clin. Nutr. Metab. Care. 12 (6), 646–652.

Prasad, S., Aggarwal, B.B., 2011. Turmeric, the golden spice: from traditional medicine to modern medicine. In: Benzie, I.F.F., Wachtel-Galor, S. (Eds.), Herbal Medicine: Biomolecular and Clinical Aspects, 2nd edition CRC Press/Taylor & Francis, Boca Raton, Chapter 13.

Prasad, S., Gupta, S.C., Tyagi, A.K., et al., 2014. Curcumin, a component of golden spice: from bedside to bench and back. Biotechnol. Adv. 32 (6), 1053–1064.

Priyadarsini, K.I., 2009. Photophysics, photochemistry and photobiology of curcumin: studies from organic solutions, biomimetics and living cells. J. Photochem. Photobiol. C. Photochem. 10 (2), 81–96.

Priyadarsini, K.I., 2014. The chemistry of curcumin: from extraction to therapeutic agent. Molecules. 19 (12), 20091–20112.

Ramanjaneyulu, P.S., Sayi, Y.S., Raman, V.A., et al., 2007. Spectrophotometric determination of boron in nuclear grade uranium compounds with curcumin and studies on effect of HNO_3. J. Rad. Anal. Nucl. Chem. 274 (1), 109–114.

Read, S.A., Obeid, S., Ahlenstiel, C., et al., 2019. The role of Zinc in antiviral immunity. Adv. Nutr. 10 (4), 696–710.

Refat, M.S., 2013. Synthesis and characterization of ligational behavior of curcumin drug towards some transition metal ions: chelation effect on their thermal stability and biological activity. Spectrochim. Acta. A Mol. Biomol. Spectrosc. 15 (105), 326–337.

Rege, S.A., Arya, M., Momin, S.A., 2019. Mini review on Keto-Enol ratio of curcuminoids. Ukr. J. Food Sci. 7 (1), 27–32.

Roohani, N., Hurrell, R., Kelishadi, R., et al., 2013. Zinc and its importance for human health: an integrative review. J. Res. Med. Sci. 18 (2), 144–157.

Sagnou, M., Benaki, D., Triantis, C., et al., 2011. Curcumin as the OO bidentate ligand in "2 + 1" complexes with the [M(CO)3] + (M = Re, 99m Tc) tricarbonyl core for radiodiagnostic applications. Inorg. Chem. 50 (4), 1295–1303.

Sandhya, P., Renu, K., 2015. Stabilization of curcumin with ferrous ion and evaluation of its pharmacological property. Int. J. Pharmacogn. Phytochem. Res. 7 (5), 943–947.

Sangartit, W., Kukongviriyapan, U., Donpunha, W., et al., 2014. Tetrahydrocurcumin protects against cadmium-induced hypertension, raised arterial stiffness and vascular remodeling in mice. PLoS One 9 (12), e114908. Available from: 10.1371/journal.pone.0114908.

Sareen, R., Jain, N., Dhar, K.L., 2015. Curcumin–Zn(II) complex for enhanced solubility and stability:an approach for improved delivery and pharmacodynamic effects. Pharm. Dev. Technol 21 (5), 630–635. Available from: 10.3109/10837450.2015.1041042.

Segerstrom, S.C., Miller, G.E., 2004. Psychological stress and the human immune system: a meta-analytic study of 30 years of inquiry. Psychol. Bull. 130 (4), 601–630.

Shankar, A.H., Prasad, A.S., 1998. Zinc and immune function: the biological basis of altered resistance to infection. Am. J. Clin. Nutr. 68 (suppl. 2), 447S–463S.

Shakeri, A., Panahi, Y., Johnston, T.P., et al., 2019. Biological properties of metal complexes of curcumin. Biofactors 45 (3), 304–317.

Sharma, K.K., Chandra, S., Basu, D.K., 1987. Synthesis and antiarthritic study of a new orally active diferuloyl methane (curcumin) gold complex. Inorg. Chim. Acta 135 (1), 47–48.

Shulga, S.M., 2014. Obtaining and characteristic of curcumin liposomal form. Biotechnol. Acta 7 (5), 55–61.

Sokolik, V.V., 2015. Effect of curcumin liposomal form on angiotensin converting activity, cytokines and cognitive characteristics of the rats with Alzheimer's disease model. Biotechnol. Acta 8 (6), 48–55.

Takam, A., Saewan, N., 2011. Antioxidant activities of curcumin-metal complexes. Thai J. Agric. Sci. 44, 188–193.

Tiyaboonchai, W., Tungpradit, W., Plianbangchang, P., 2007. Formulation and characterization of curcuminoids loaded solid lipid nanoparticles. Int. J. Pharmacol. 337 (1–2), 299–306.

Trivedi, M.K., Gangwar, M., Mondal, S.C., et al., 2017. Protective effects of tetrahydrocurcumin (THC) on fibroblast and melanoma cell lines in vitro: it's implication for wound healing. J. Food Sci. Technol. 54 (5), 1137–1145.

Valentini, A., Conforti, F., Crispini, A., et al., 2009. Synthesis, oxidant properties, and antitumoral effects of a heteroleptic palladium (II) complex of curcumin on human prostate cancer cells. J. Med. Chem. 52 (2), 484–491.

Vellampatti, S., Chandrasekaran, G., Mitta, S.B., et al., 2018. Metallo-curcumin-conjugated DNA complexes induces preferential prostate cancer cells cytotoxicity and pause growth of bacterial cells. Sci. Rep. 8, 14929. Available from: https://doi.org/10.1038/s41598-018-33369-z.

Vyasa, A., Dandawatea, P., Padhyea, S., et al., 2013. Perspectives on new synthetic curcumin analogs and their potential anticancer properties. Curr. Pharm. Des. 19 (11), 2047–2069.

Wanninger, S., Lorenz, V., Subhan, A., et al., 2014. Metal complexes of curcumin —synthetic strategies, structures and medicinal applications. Chem. Soc. Rev. 44 (15), 4986–5002.

Weinberger, B., Herndler-Brandstetter, D., Schwanninger, A., et al., 2008. Biology of immune responses to vaccines in elderly persons. Vaccines. 46, 1078–1084.

Wicha, P., Tocharus, J., Janyou, A., et al., 2017. Hexahydrocurcumin protects against cerebral ischemia/reperfusion injury, attenuates inflammation, and improves antioxidant defenses in a rat stroke model. PLoS One. 12 (12), e0189211. 2017.

Wicha, P., Tocharus, J., Janyou, A., et al., 2020. Hexahydrocurcumin alleviated blood-brain barrier dysfunction in cerebral ischemia/reperfusion rats. Pharmacol. Rep. 72. Available from: https://doi.org/10.1007/s43440-019-00050-9.

Wong, C.P., Yang, S., et al., 2009. Zinc supplementation increases Zinc status and thymopoiesis in aged mice. J. Nutr. 139 (7), 1393–1397.

Xing, Z.H., Wei, J.H., Cheang, T.Y., et al., 2014. Bifunctional pH-sensitive Zn (II)–curcumin nanoparticles/ siRNA effectively inhibit growth of human bladder cancer cells in vitro and in vivo. J. Mater. Chem. B 2, 2714–2724.

Xu, X.-Y., Meng, X., Li, S., et al., 2018. Bioactivity, health benefits, and related molecular mechanisms of curcumin: current progress, challenges, and perspectives. Nutrients. 10 (10), 1553.

Yan, F.S., Sun, J.L., Xie, W.H., et al., 2018. Neuroprotective effects and mechanisms of curcumin–Cu(II) and –Zn(II) complexes systems and their pharmacological implications. Nutrients 10 (1), 28.

Yoysungnoen, P., Wirachwong, P., Changtam, C., et al., 2008. Anti-cancer and anti-angiogenic effects of curcumin and tetrahydrocurcumin on implanted hepatocellular carcinoma in nude mice. World J. Gastroenterol. 14 (13), 2003–2009.

Yoysungnoen, B., Bhattarakosol, P., Changtam, C., et al., 2016. Effects of tetrahydrocurcumin on tumor growth and cellular signaling in cervical cancer xenografts in nude mice. Biomed. Res. Int. 2016, 1781208.

Yu, Q., Liu, Y., Wu, Y., et al., 2018. Dihydrocurcumin ameliorates the lipid accumulation, oxidative stress and insulin resistance in oleic acid-induced L02 and HepG2 cells. Biomed. Pharmacother. 103, 1327–1336.

Zebib B., Mouloungui Z., Noirot V. 2010. Stabilization of curcumin by complexation with divalent cations in glycerol/water system. Bioinorg Chem Appl. 2010, Article ID 292760, 8 https://doi.org/10.1155/2010/292760.

Zhang, Z., Rong, L., Ping, Y., 2019. Flaviviridae viruses and oxidative stress: implications for viral pathogenesis. Redox Biology of Infection and Consequent Disease. Oxid. Med. Cell Longev. 2019, 17Article ID 1409582. 17 pages. Available from: https://doi.org/10.1155/2019/1409582.

Zhao, F., Gong, Y., Hu, Y., et al., 2015. Curcumin and its major metabolites inhibit the inflammatory response induced by lipopolysaccharide: translocation of nuclear factor-κB as potential target. Mol. Med. Rep. 11 (4), 3087–3093.

Zhongfa, L., Chiu, M., Wang, J., et al., 2012. Enhancement of curcumin oral absorption and pharmacokinetics of curcuminoids and curcumin metabolites in mice. Cancer Chemother. Pharmacol. 69, 679–689.

Chapter 37

Fenugreek: multiple health benefits

Ramesh C. Gupta, Robin B. Doss, Ramesh C. Garg, Ajay Srivastava, Rajiv Lall and Anita Sinha

Introduction and background

Fenugreek [*Trigonella foenum-graecum* (TFG)] is one of the oldest medicinal plants. It belongs to the family Leguminoseae and is cultivated widely in the Mediterranean, North African regions, India, and other countries. In many regions of the world, it is commonly used as a condiment and seasoning in food preparations, and it is assumed to possess nutritive and restorative properties. The plant grows to 1–2 ft in height and resembles sweet clover and lucerne (alfalfa) plants. In addition to using the seeds of fenugreek, many people use fresh leaves from the fenugreek plant as part of several food preparations. The seeds are yellowish and complete their ripening in sickle-shaped pods while still on the plants.

A large amount of information relating to fenugreek has been published in scientific literature during the past several decades, which has supported a wide range of medical benefits. Because the published information varies significantly in consistency of scientific standards and quality, this chapter is intended to summarize and present the available scientific information in a format as is expected and needed for a pharmaceutical drug product. The main emphasis of the information is to provide a good analysis of the scientific data, identification of therapeutic targets, discussion of the weight of evidence for the therapeutic effects, estimation of the effective human dose, assessment of practicality of the effective dose, and estimation of safety margins and risks. The data needed for assessing the precautions and/or contraindications, even if applicable only to select group(s) of the population, have also been analyzed and summarized. The overall information is intended to be in an easy-to-read format, allowing the average reader to make a balanced interpretation of the medical benefits and safety.

Major active components of fenugreek

Fenugreek contains several types of compounds. These include proteins, amino acids, flavonoids, steroidal saponins, coumarin, lipids, vitamins, minerals, galactomannan fiber, and alkaloids, such as trigonelline. The defatted part of the fenugreek seed, called subfraction "a" which contains the seed parts like testa (seed coat) and endosperm (a source of stored food), has been reported to be the portion associated with the antihyperglycemic effect in diabetic dogs (Ribes et al., 1986).

The main bioactive components, particularly important for the antihyperglycemic effect, are related to the nonlipid, defatted, or alcoholic extract fraction of fenugreek seeds. This is because the lipid fraction is not associated with the antihyperglycemic effect (Ribes et al., 1986).

Steroidal saponin—diosgenin

One of the saponins of known therapeutic value is "diosgenin." This is a spirostanol saponin found in the seed and has been designated as the most bioactive component of fenugreek (Raju and Rao, 2011; Kanchan et al., 2016). It has a chemical formula of $C_{27}H_{42}O_3$ and a molecular weight of 414.630. Its structural formula is shown below in Fig. 37.1.

Diosgenin is used as a starting material for more than 60% of the commercial synthesis of cortisone, pregnenolone, progesterone, and other steroids (Raju and Rao, 2011). Saponins have been reported to play a role in lipid and glucose balance (Gupta et al., 2001; Sharma and Raghuram, 1990; Sharma et al., 1990; Neveen et al., 2007; Kassaian et al., 2009; Uemura et al., 2010), in prevention of fat deposition in adipocytes (Kumar et al., 2014; Vijayakumar et al., 2010), and in stimulation of

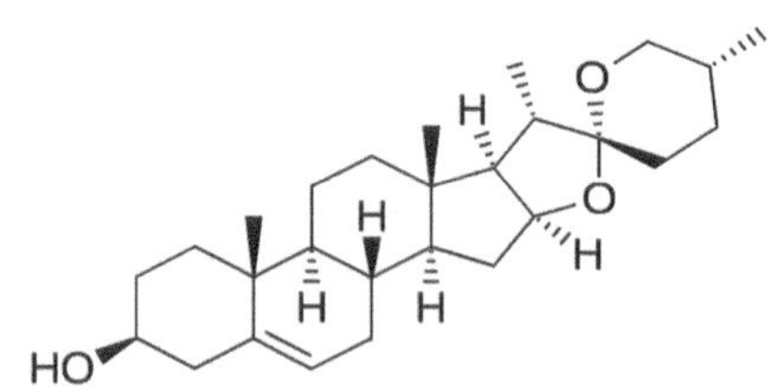

FIGURE 37.1 Structural formula of diosgenin.

Nutraceuticals. DOI: https://doi.org/10.1016/B978-0-12-821038-3.00037-9

growth hormone (GH) release from the pituitary (Shim et al., 2008). Details of these potential therapeutic targets are discussed within this chapter.

The alkaloid—trigonelline

Trigonelline is a plant hormone that has diverse regulatory functions with respect to plant cell cycle regulation, nodulation, and oxidative stress, and in helping survival and growth of the plant. It is found in significant quantities in many plants including coffee beans and fenugreek seeds. Because it was first isolated from the fenugreek seeds TFG, the name assigned to it was "trigonelline." The chemical formula for trigonelline is $C_7H_7NO_2$, with a molecular weight of 137.138. Its structural formula is shown in Fig. 37.2.

Trigonelline is a methylation product of niacin (vitamin B3), and thus is also known as "methylated niacin." At higher temperatures, trigonelline breaks down to niacin. In addition to trigonelline, fenugreek seeds contain other alkaloids such as gentianine and carpaine.

Trigonelline is present in several other plants such as coffee beans, garden peas, hemp seed, and oats. Plants containing trigonelline at levels more than 1000 ppm include coffee and fenugreek (Tice, 1997). Coffee consumption has been reported to lower the risk of colon cancer (Lee et al., 2007), type II diabetes (Campos and Baylin, 2007), and Alzheimer disease (AD) (Barranco Quintana et al., 2007; Arendash and Cao, 2010; Prasanthi et al., 2010). Trigonelline has been claimed to have other therapeutic benefits, such as anticarcinogenic (Tice, 1997) and hypocholesterolemic effects (Abdel-Mawla and Osman, 2011). Caution may be needed for use in females because phytoestrogenic activity has also been reported for trigonelline (Allred et al., 2009).

The amino acid—4-hydroxyisoleucine

Fowden et al. (1973) isolated and identified the amino acid, 4-hydroxyisoleucine, in fenugreek seed for the first time. 4-hydroxyisoleucine has been shown to cause an increase in glucose-dependent insulin secretion (Sauvaire et al., 1998; Jette et al., 2009) and to lower elevated plasma triglycerides and total cholesterol levels (Narender et al., 2006). Abdel-Mawla and Osman (2011) used cell suspension cultures to raise the concentration of 4-hydroxyisoleucine content in fenugreek seed, which increased serum insulin levels and showed improvement in lowering the fasting blood glucose level in streptozotocin (STZ)-treated diabetic rats. Haeri et al. (2012) further demonstrated that 4-hydroxyisoleucine administered orally (50 mg/kg per day) for 4 weeks to STZ-treated (type I diabetes model) rats not only lowered the blood glucose level from 500 to 330 mg/dL but also significantly lowered the elevated lipid [total cholesterol, triglycerides (TGs), and LDL (low-density lipoprotein)] levels in STZ-treated diabetic rats, but not to the control levels. Further discussions of these activities are included in appropriate sections herein.

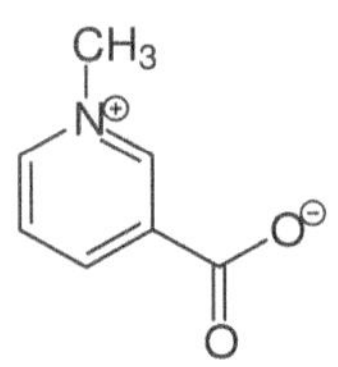

FIGURE 37.2 Structural formula of trigonelline.

Galactomannan

Galactomannan is a carbohydrate (soluble fiber) fraction representing 45%–60% of the fenugreek seed. The galactomannan derivative and guar gum have been reported to lower postprandial excursion in the glucose response (Fairchild et al., 1996; Trask et al., 2014). This and an in vitro investigation using galactomannan for its role in lowering glucose uptake from the small intestine are discussed herein.

Another bioactive fraction present in fenugreek, which is not associated with the listed beneficial effects including glucose or lipid level management, is coumarin. Potential undesirable effects that may be associated with a high content of coumarin are summarized in this chapter.

Coumarin

Coumarin is a component of many plants and plant products. It is present in certain varieties of cinnamon, and is also used for glucose balance. When present in an appreciable amount, coumarin may have an aspirin-like effect of prolonging the blood coagulation time (Egan et al., 1990) and may play a role in photosensitization reactions (Lewis, 1994; McNeely and Goa, 1998). Products containing coumarin need to be used with caution in patients using anticoagulant therapy.

Potential for an antihyperglycemic effect in diabetic conditions

This therapeutic target is based on a number of investigations extending from in vitro studies to data generated using animal models, in vivo studies in various animal species, and human clinical studies. The human data (both clinical and epidemiological) are limited and, in some cases, of poor quality (Basch et al., 2003). Only the selected studies based on relevance to the target and quality of information are discussed here. Additionally, considerations related to the effect size for normalizing the blood glucose level in diabetic animals, extrapolation of the human dose from animal data, practicality of the effective clinical dose, and safety margins are discussed. The potential for the antihyperglycemic effect has been discussed based on:

- In vitro data;
- In vivo/animal data;
- Human/clinical studies as evidence for the antihyperglycemic effect;
- Likely mechanism(s) responsible for the antihyperglycemic effect.

In vitro data supporting an antihyperglycemic effect

Several early investigations related to fenugreek or its components support either insulin mimetic and/or secretagogue effects on the pancreas in cases of elevated blood glucose level or diabetes or inhibition of intestinal absorption of glucose. Sauvaire et al. (1998) demonstrated in an in vitro study that 4-hydroxyisoleucine from fenugreek seeds increased glucose-induced insulin secretion in human and rat pancreatic islet cells. This amino acid appears to act only on pancreatic beta cells, because the levels of somatostatin and glucagon were not altered. Further, the effect of 4-hydroxyisoleucine occurred only when the glucose level was elevated to a supraphysiological range. The effect on the pancreas appears to be comparable to the mechanism exerted by compounds of the "sulfonylurea" class of drugs, but 4-hydroxyisoleucine provides the advantage that the effect occurs only when the blood glucose level is elevated, thus mitigating the unnecessary and potential risk for hypoglycemia seen with sulfonylureas.

Srichamroen et al. (2009) investigated the effect of the fenugreek component, galactomannan, on uptake of glucose from the jejunum and ileum segments of the intestine in both lean and obese rats. Based on findings presented in this chapter, with increasing concentrations of galactomannan from 0.1% to 0.35%, there was a progressive decrease in glucose uptake (estimated as nmol/100 mg dry weight per minute), which reached more than threefold at a glucose concentration of 2 mM. Increasing the galactomannan concentration more than 0.35% did not inhibit the glucose uptake further. The mechanism for this effect was considered to be possibly due to an increase in the viscous property of galactomannan. The other components of fenugreek, diosgenin (saponin) and trigonelline, have also been reported to inhibit glucose uptake in vitro (Al-Habori et al., 2001). Other investigators (Mohammad et al., 2006) clarified the mechanism by documenting that *Trigonella* seed extract modulates the expression of glucose transporter 4 (GLUT4), which is present in skeletal muscle, but may be operating at the intestinal level as well and influencing the glucose uptake. This is because another transporter, GLUT2, which is a low-affinity (high *Km*) glucose transporter, has also been evidenced in the intestinal epithelial cells (Watson and Pessin, 2001). Also, supporting fenugreek's ability to lower intestinal absorption of glucose, Amin et al. (1987) reported that the aqueous extract of fenugreek could inhibit intestinal α-amylase and sucrase enzymes.

In vivo/animal data supporting the antihyperglycemic effect

There is a large amount of in vivo data available for the antihyperglycemic effect of fenugreek that have been generated mostly during the past 30 years. In view of these extensive data, only selected studies in which fenugreek has been used as a seed powder (Raju et al., 2001; Mohammad et al., 2006; Kumar et al., 2005; Kumar et al., 2011), as a fiber-rich extract, as an aqueous or solvent extract, or as an active component, are discussed.

Kumar et al. (2005) found that mucilage from fenugreek seed in rat diet (approximately 10 mL water extract of 25 g of fenugreek seeds/rat/day for 5 weeks) could lower the fasting blood glucose by 26% and urinary sugar elimination by 30% in STZ-treated diabetic rats. Similarly, using the fenugreek seed powder without any extraction, Kumar et al. (2011) demonstrated that seed powder as 5% in rat diet (5 g mixed with 95 g of feed) lowered the elevated blood glucose level from 432 mg/dL in alloxan-induced diabetic rats to 129 mg/dL in 21 days. This was comparable with the insulin-treated group (103 mg/dL), but still approximately 50% higher than the glucose level in the control group (87 mg/dL). The 5% level of fenugreek in the diet has also been used in previous studies (Raju et al., 2001). The effect of *Trigonella* seed powder (TSP) without any extraction on key enzymes of glycolysis and gluconeogenesis and on GLUT4, the skeletal muscle glucose transporter (Mohammad et al., 2006), has been investigated in alloxan-induced diabetic rats (with a glucose level >300 mmol/L). TSP treatment resulted in a marked decrease in the plasma glucose levels by normalizing the altered activities of the referenced enzymes and the transporter process.

In addition to evidence for the antihyperglycemic effect afforded by the whole-seed powder, different types of fenugreek extracts have also been investigated. Oral doses (0.5 and 1.0 g/kg) of aqueous extract of fenugreek seed lowered the blood glucose level by approximately 33% and 51%, respectively, at 4-h intervals in normal mice, whereas the methanolic extract needed the higher dose of 1.0 g/kg for lowering the blood glucose level by nearly 16% (Zia et al., 2001). The evidence of antihyperglycemic activity with aqueous or methanolic extract indicates that extracts retain the bioactive component and are mainly responsible for the antihyperglycemic effect of fenugreek seed that may be polar in nature to be extracted in the aqueous extract. Extracts responsible for the antihyperglycemic effect are further supported by decreases in serum glucose, total cholesterol, and triacylglycerol and increase the serum insulin in diabetic rats at an oral dose of 0.5 g/kg for fenugreek extract (Eidi et al., 2007). The effect size in this study was comparable to the antidiabetic effect provided by glibenclamide at an oral dose of 600 μg/kg per day.

This has been further investigated by determining what part of the fenugreek seed gets extracted. The fiber-rich extract of fenugreek seeds, called subfraction "a" (testa and endosperm), at a dose of 1.145 g/kg has been demonstrated to reverse the hyperglycemia and glycosuria induced in alloxan-treated diabetic dogs (Ribes et al., 1986). Later, Petit et al. (1993) reported that chronic dosing of fenugreek seed extract obtained from the cotyledons and axis parts of the seed also raised the plasma insulin levels in normal rats. From the extract of cotyledons and axes, Sauvaire et al. (1998) isolated a water-soluble substance called 4-hydroxyisoleucine. This substance stimulated glucose-induced insulin release in a concentration range of 100 μmol/L to 1 mmol/L by direct effect on isolated islets of Langerhans from both rats and humans.

Both the preventive and curative effects of fenugreek (hydroalcoholic extract of seed) at a dose of 2 g/day were investigated in a male C57BL/6J mouse model of diabetes with a high-fat diet (HFD) when treatment was started at the time HFD was initiated or 17 weeks after development of diabetes for a duration of 18–20 weeks (Hamza et al., 2012). Fenugreek extract demonstrated both preventive and curative effects.

Effectiveness of diosgenin has been evaluated in obese diabetic KK-Ay mice by Uemura et al. (2010). They evaluated increased mRNA expression for the markers of adipocyte differentiation and suppression of mRNA for active macrophage-derived proinflammatory factor TNF-α, which is responsible for adipocyte inflammation and induction of insulin resistance. The KK-Ay mice were fed either a HFD and designated the control group or a HFD containing 0.5% or 2% fenugreek and designated as the 0.5F group or the 2.0F group, respectively, for 4 weeks. Plasma glucose was lowered from 386.5 to 226.9 mg/dL in the high-dose 2.0F group. These researchers confirmed that it was the hydrolyzed saponin, diosgenin, in fenugreek that was responsible for differentiating the adipocytes and for lowering the glucose level. This was most likely via a decrease in inflammation in the white adipose tissue (WAT), thus reducing insulin resistance.

Human/clinical studies as evidence for the antihyperglycemic effect

Selected fenugreek clinical studies have been summarized by Basch et al. (2003) and more recently by Neelakantan et al. (2014). Several of the 10 clinical trials identified and summarized by Neelakantan et al. (2014) support beneficial effects of fenugreek seed on glycemic control in diabetic subjects (Table 37.1). These workers also commented that higher quality clinical trials using a well-characterized fenugreek preparation are needed to provide a more conclusive outcome. Trials have generally been conducted in adult diabetic patients, where glycemic effects in terms of changes in fasting blood glucose, 2-h postload glucose, and HbA1c have been evaluated. Overall quality of the five clinical trials tabulated by Basch et al. (2003) was also evaluated as poor. Additionally, the clinical trials that could be identified by these two groups were small, with less than 40 subjects in most trials and one trial (Lu et al., 2008) with 69 subjects included.

Both acute dosing (single dose) and chronic dosing of fenugreek have been effective for lowering glucose and/or glycated hemoglobin (HbA1c) in diabetic patients. Sharma (1986) evidenced the effect of acute administration of fenugreek in humans to be associated with a reduction in postprandial glucose and insulin levels, whereas others (Sharma and Raghuram, 1990; Gupta et al., 2001) have used chronic dosing in clinical trials to show reductions in fasting and postprandial glucose levels and in HbA1c. This effect did not appear to be consistent because a comparable effect has not been observed in some of the other clinical trials. Fenugreek fiber up to 8 g/day in addition to a regular breakfast given to 18 subjects had no effect on serum glucose and insulin response (Mathern et al., 2009). Also, fenugreek seed extract up to 1176 mg/day for 14 days administered to healthy young males did not affect the serum glucose and insulin levels but did lower the fat intake (Chevassus et al., 2009). This could be evidence of safety, supporting the fact that fenugreek does not cause hypoglycemia in subjects with glucose levels in the normal physiological range.

The variability in antihyperglycemic response noted in clinical trials could be related to many factors, such as healthy versus diabetic patients, duration of treatment, active component(s) of fenugreek included in the test product, and, most importantly, the dose used. With respect to dose, high daily doses in the range of 5–25 g/day were required for the fenugreek seed powder for any notable efficacy and doses of 2 g/day or less were not effective (Table 37.1). At the lowest dose of 1 g/day used for 2 months in only one of the clinical trials (Gupta et al., 2001) listed in Table 37.1, there was no effect on the fasting blood glucose or on the 2-h postdose glucose.

TABLE 37.1 Characteristics of the 10 clinical trials included in the meta-analysis of the effects of fenugreek seeds on glycemia.

Lead author, year, and country	Population, medication	Design	Age (y)	Male (%)	Sample size (*N*)	Duration (days)	Fenugreek, preparation	Daily dose (g)	Control	Outcomes	Study quality		
											RCT	Blinding	Dropout
Bordia et al. (1997), India	Mild T2DM, NR	PL	NR	NR	40	30	Capsule, PS	5	Placebo	FBG, 2h glucose	NR	NR	NR
Bordia et al. (1997), India	Severe T2DM, NR	PL	NR	NR	40	30	Capsule, PS	5	Placebo	FBG, 2h glucose	NR	NR	NR
Chevassus et al. (2010), France	Overweight, NR	PL	38.0	100	40	42	Capsule, H	1.176	Placebo	FBG, FSI	Yes	DB	5%
Gupta et al. (2001), India	T2DM, SU, BI	PL	51.0	76	25	56	Capsule, H	1	Placebo	FBG, 2h glucose, HbA1c, FSI	Yes	DB	4%
Lu et al. (2008), China	T2DM, SU	PL	54.4	55	69	84	Capsule, PS	6.3	Placebo	FBG, 2h glucose, HbA1c	Yes	DB	NR
Alamdari et al., (2009), Iran	T2DM, Diet, OAD	CO	43.1	100	12	56	PS	8	Unspecified	FBG, 2h glucose, HbA1c	Yes	NR	25%
Chevassus et al. (2009), France	Healthy, NR	CO	22.0	100	12	14	Capsule, H	1.176	Placebo	FBG, FSI	Yes	DB	0%
Raghuram et al. (1994), India	T2DM, BI	CO	46.6	NR	10	15	Chapati[a], PS	25	Chapati	FBG, 1h glucose, HbA1c	Yes	NR	NR
Sharma et al. (1990), India	T1DM, Insulin therapy	CO	22.7	70	10	10	Chapati[a], DPS	100	Chapati	FBG, 2h glucose, FSI	Yes	NR	NR
Sharma et al. (1990), India	T2DM, BI/ metformin	CO	46.0	67	15	10	Chapati[a], DPS	100	Chapati	FBG, 2h glucose, FSI	Yes	NR	NR
Sharma et al. (1990), India	T2DM, NR	CO	42.0	NR	5	20	Chapati[a], DPS	100	Chapati	FBG, 2h glucose	Yes	NR	NR

NR, not reported; *T1DM*, type I diabetes mellitus; *T2DM*, type 2 diabetes mellitus; *SU*, sulfonylurea; *BI*, biguanides; *OAD*, oral antidiabetic drug; *PL*, parallel; *CO*, crossover trials; Daily dose *g*, fenugreek total daily dose in grams; *PS*, powdered fenugreek seeds; *H*, hydroalcoholic extract of fenugreek seeds; *DPS*, debitterized fenugreek seed powder; *FBG*, fasting blood glucose; *HbA1c* %, glycosylated hemoglobin; *FSI*, fasting serum insulin; *RCT*, randomized control trial; *DB*, double blinded.
[a]*Fenugreek incorporated into chapati (flat bread).*

The areas under the curve for blood glucose (2375 ± 574 vs 27,597 ± 274) and for insulin (2492 ± 2536 vs 5631 ± 2428), however, were significantly lower ($P < .001$) when the fenugreek-treated group was compared to the control diabetic group.

Likely mechanisms for the antihyperglycemic effect

Several studies (in vitro and in vivo) have been performed to illustrate different mechanisms for the antihyperglycemic effect of fenugreek or its active components. References to eight such studies have been included in Fig. 37.3 to illustrate the four likely mechanisms for the antihyperglycemic effect. These mechanisms are:

- Inhibition of glucose uptake or absorption from the intestine or gastrointestinal tract;
- Effect on the pancreas, exerting an insulin mimetic or an insulin secretagogue effect;
- Enhancement of peripheral blood glucose uptake, normalization of select processes, and increase in insulin receptor density;
- Suppression of proinflammatory factors, thus reducing insulin resistance.

Other fenugreek effects and limitations related to normalization of hyperglycemia

Ameliorating effect on some of the diabetic complications

Fenugreek helps in the normalization of hyperglycemia in diabetics and also has beneficial effects in selected conditions considered to be diabetes-related complications.

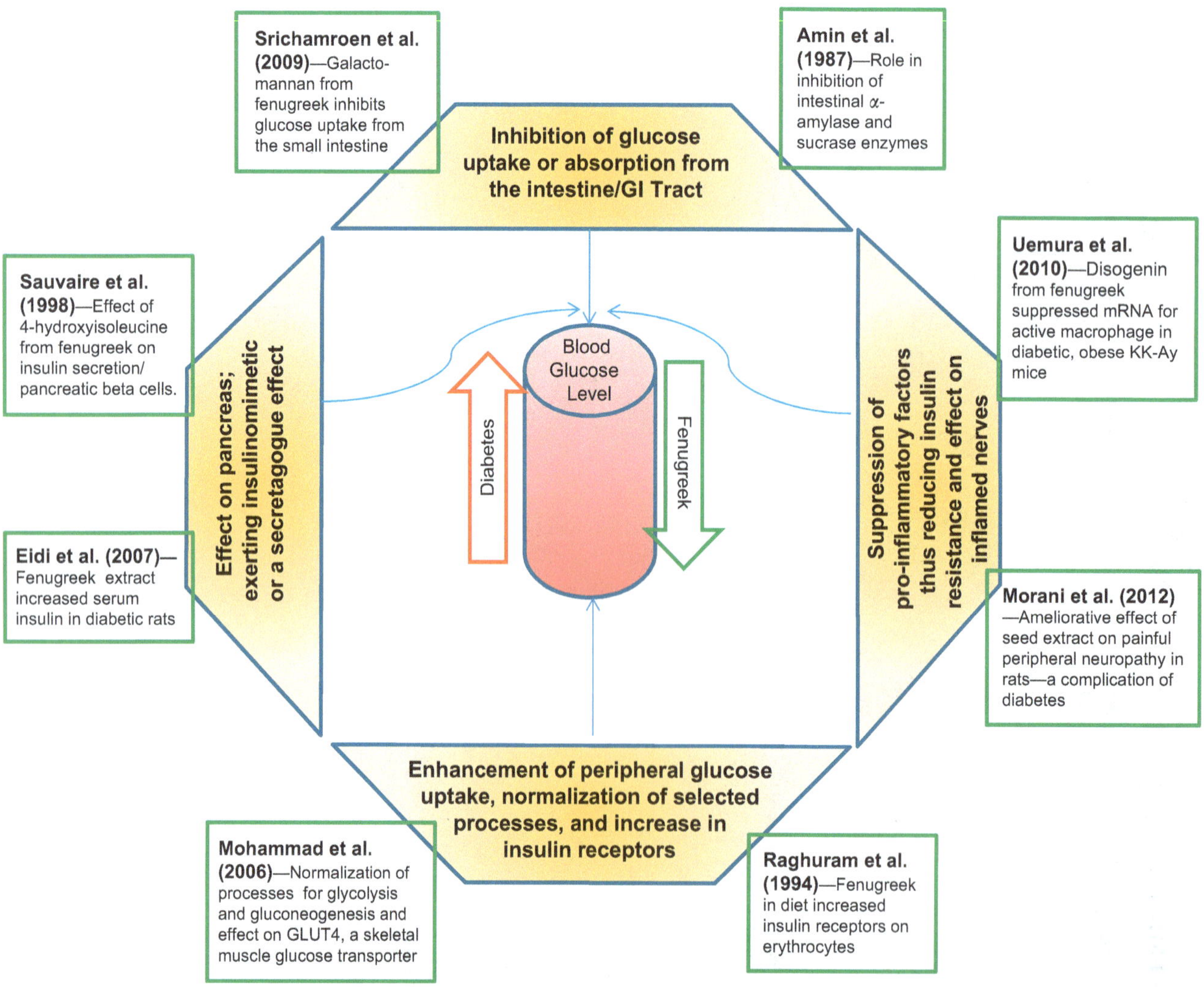

FIGURE 37.3 Potential mechanisms for the hypoglycemic effect of fenugreek.

In vivo animal data for fenugreek's benefit in diabetic neuropathy

Some of the active components of fenugreek have demonstrated neuroprotective and nephron-protective effects, which are important for addressing diabetes-related complications. Preet et al. (2005) showed that feeding fenugreek seed powder to diabetic rats (5% w/w in feed) not only lowered their blood glucose level to nearly normal ranges but also protected the sciatic nerves against structural abnormalities observed in the diabetic untreated group of rats. Recently, Morani et al. (2012) also demonstrated the ameliorative effects of *Trigonella* seed extract on painful peripheral neuropathy in rats. A fenugreek seed-derived fraction, named IND01, has been purified and standardized by high-performance liquid chromatography to a marker compound trigonelline. Daily oral administration of IND01 at $\geq$ 100 mg/kg for 15 days restored motor nerve conduction velocity in rats with experimental sciatic nerve crush injury to nearly normal ranges. The results from this investigation suggested a neuroprotective role for IND01 or trigonelline in painful peripheral neuropathy conditions commonly observed in diabetes (Morani et al., 2012).

In vivo animal data for fenugreek's benefit in diabetic nephropathy

With respect to nephropathic complications related to prolonged diabetes, Xue et al. (2011) studied the effects of low (440 mg/kg), medium (870 mg/kg), and high (1740 mg/kg) oral doses of fenugreek extract for 6 weeks in rats with diabetic nephropathy induced by STZ treatment. All of the ultramorphologic abnormalities in the kidney of diabetic rats, including the uneven thickening of the glomerular base membrane, were markedly ameliorated. This suggested that fenugreek extract offered significant value in restoring the kidney structures to normal, but meaningful benefit occurred only at the high dose (1740 mg/kg).

Limitations of fenugreek's use for the antihyperglycemic effect

Animal doses attaining the antihyperglycemic effect in animals extrapolate to an unusually high human equivalent dose

The need for a relatively high dose for the antihyperglycemic effect is supported by animal data summarized in Table 37.2. The lowest oral dose considered effective in animal experiments has been extrapolated to human equivalent dose (HED) based on body surface area. Table 37.2 also summarizes the safety of the extrapolated HED with respect to a no observed adverse effect level (NOAEL) dose in a 90-day safety/toxicity study for fenugreek in an animal species.

TABLE 37.2 Extrapolation of human equivalent dose (HED) from the lowest effective dose in diabetic animals for hypoglycemic effect and estimation of the margin of safety.

In vivo animal study/ reference	Animal species/ treatment duration	Fenugreek formulation	Lowest fenugreek dose that was effective in animals	HED[a] for the lowest effective animal dose (mg/kg per day)	Safety margin[b] for HED
Ribes et al. (1986)	Diabetic dogs/ for 21 days	Subfraction "a" (testa + endosperm)	1.145 g/kg per day	636	12.6
Kumar et al. (2011)	Diabetic rats/for 21 days	Mixed at 5% in diet	1 g/day[c] = 4 g/kg per day	645	12.4
Zia et al. (2001)	Normal mice/ value at 4 h	Methanolic extract 1 g/kg	1 g/kg per day	83	96.4
Hamza et al. (2012)	Diabetic C57BL/ 6J mice/for 18 weeks	Hydroalcoholic extract	2 g/kg per day	166	48.2
Uemura et al. (2010)	KK-Ay mice/for 4 weeks	Mixed at 2% in high-fat diet (HFD)	90 mg/day[d] = 3.6 g/kg per day	580	13.8

[a]*HED, human equivalent dose for a 60-kg person.*
[b]*Safety margin: With respect to no observed adverse effect level (NOAEL) in a 90-day toxicity study in rats (Muralidhara et al., 1999). Estimated as 2 g/day per rat translated from daily intake of approximately 20 g/day (U.S. EPA, 1988) of rat diet with 10% fenugreek. This equates to 8 g/kg per day based on average rat body weight of 250 g.*
[c]*Calculated for 5% fenugreek in rat diet using average daily feed consumption of 20 g/day for rat (U.S. EPA, 1988).*
[d]*Calculated for 2% fenugreek in mouse diet using average daily feed consumption of 4.5 g/day for mouse (U.S. EPA, 1988).*

Poor practicality of the high human doses needed for the antihyperglycemic effect

The HED for a reasonably predictable antihyperglycemic effect averages 422 mg/kg per day based on five animal studies listed in Table 37.2. The estimated HED is reasonably safe with respect to NOAEL in a 90-day toxicity study in rats (Muralidhara et al., 1999). However, the dose equates to approximately 25 g/day for a 60-kg person. The dose of 25 g/day or higher has been used in some of the clinical trials listed in Table 37.1, but only as part of the main food (Indian flat bread). This is not a practical way for dosing a nutraceutical or a pharmaceutical product. The commonly used oral dosage forms for better compliance are tablets or capsules. Too many tablets or capsules would be needed per day for a dose of 25 g/day. A lower dosage of 5 g/day or less used in some of the clinical trials (Table 37.1) seems to have contributed to a less than optimal, inconsistent, and variable hypoglycemic effect. This is an area that needs improvement and refinement for fenugreek to be a consistently effective and successful nutraceutical and for dependable management of hyperglycemia in diabetic patients. A few proposals that may help overcome this gap and possibly lower the clinically effective dose are:

- Including a detailed characterization of fenugreek test material as a fenugreek nutraceutical drug product (FNDP) used in clinical trials. All aspects of a certificate of analysis for an FNDP, such as source, purity, active components, and stability, allowing transparency for reconstruction of the study (when needed) should be provided.
- Optimization of the most effective levels of major active components like 4-hydroxyisoleucine, trigonelline, diosgenin, and others would be needed in the description of the FNDP. This may even require spiking of one or more of the active components in the FNDP.
- Performing clinical trials as double-blinded studies supervised by trained and experienced medical professionals, enrolling approximately 50 diabetic patients or more per arm with a study duration of at least 12 weeks, and using both the untreated and the positive control arms.
- Establishing and meeting the predetermined and acceptable criteria for efficacy as performed for new oral diabetic pharmaceutical products.
- This should include all applicable exclusion and inclusion criteria, as used for the diabetes clinical studies, to allow evaluation of the effect of different variables, such as diet, exercise, age, and other factors.

Adequacy of safety margin at the high human dose needed for the antihyperglycemic effect

Fenugreek's safety with respect to hypersensitivity and allergenic potential, effects on the developing fetus, and reproductive effects is discussed in a separate section of this chapter. The safety discussion here is only in regard to the high HED needed for clinical effectiveness (Table 37.3). The high HED does not pose a safety concern based on acceptable safety margins derived from animal toxicity data (Table 37.3). This estimation is also supported by the human safety data (Sharma et al., 1996). This group evaluated the long-term safety of fenugreek in humans. Sixty diabetic patients were given fenugreek seed powder at an oral dose of 25 g/day for 24 weeks. None of the participants had any evidence of clinical, hematological, hepatic, or renal abnormalities at the end of 24 weeks.

Normalization of lipid levels and effect on fat accumulation

The WAT in the body is now known to play a central role in the regulation of energy balance and to act as an endocrine organ mediating the number of physiological and pathological processes (Gregoire, 2001). Because changes in adipocyte size and/or in number can cause alterations in WAT mass, and because fenugreek or its active components can modulate the activity of the differentiated or undifferentiated adipocytes (Vijayakumar et al., 2010), the potential for fenugreek to affect lipid function and/or to affect fat accumulation is of significant interest.

In vitro investigations related to fenugreek's effect on adipocyte function and accumulation of lipids

The hypolipidemic potential of fenugreek has been investigated by using the differentiating and differentiated mouse embryo 3T3-L1 cells (representing preadipocytes) and the HepG2 cells (Vijayakumar et al., 2010; Luang et al., 2018). The mouse embryo 3T3-L1 cells are a widely used cell model for studying adipocyte differentiation and the HepG2 cells are used as a model system for studying liver metabolism and the toxicity of xenobiotics. When stimulated to differentiate, the 3T3-L1 cells assume the role of adipocytes capable of lipid accumulation function. The thermostable extract of fenugreek seeds (TEFS) used by Vijayakumar et al. (2010) inhibited accumulation of fat in both the differentiating and differentiated 3T3-L1 cells. This was most likely via a decrease in the expression of adipogenic factors. The TEFS effect on HepG2 cells indicated a decrease in concentrations of cellular triglycerides and cholesterol. The TEFS effect in HepG2 cells may be mediated via upregulation of the low-density lipoprotein receptor expression, thereby enhancing the LDL uptake.

At approximately the same time, Uemura et al. (2011) also investigated the effect of 15 hydrolyzed saponin

TABLE 37.3 Safety information related to general toxicity following single or repeated fenugreek dosing.

Type of toxicity or adverse effect	Fenugreek product/active component				
	Fenugreek seed or powder/defatted seed powder	Nonlipid, aqueous or alcoholic extract	Trigonelline (an active component)	Diosgenin (an active component)	4-hydroxy-isoleucine (an active component)
Acute oral toxicity	LD_{50} > 5 g/kg rat[a] LD_{50} > 5 g/kg rat[b] LD_{50} > 2 g/kg mice[b]	Rat[c]: NETD = 1.5 g/kg LETD = 2 g/kg	LD_{50} > 5 g/kg mice[e]	Rat[f]: NTED = 0.5 g/kg LD_{10} = 2 g/kg	No data
Acute dermal toxicity	>2 g/kg rabbit[a]	No data	No data	No data	No data
Subchronic toxicity	90-Day exposure: NOAEL = 10% in rat diet = approximately 2 g/day[a] NOAEL = 20% in rat diet = approximately 4 g/day[d]	No data	• Mice[g]: Feeding 50 mg/kg daily for 21 days: No major effects category[g] • When fed 3500 mg for 62 to 70 days: No major effects	No data	No data
Chronic toxicity (exposure over 90 days)	25 mg/day in 60 diabetes patients for 24 weeks—no adverse effects[e]	No data	No data	No data	No data

LD_{50}, dose that is expected to be lethal in 50% of test subjects; *NETD*, no effect toxic dose; *LETD*, low effect toxic dose.
[a]*Opdyke (1978).*
[b]*Muralidhara et al. (1999).*
[c]*Dande and Patil (2012).*
[d]*Rao et al. (1996).*
[e]*Aswar et al. (2009).*
[f]*Hooker (2004).*
[g]*Tice (1997).*

fractions of fenugreek on TG accumulation in HepG2 cells to further clarify which fenugreek component has the most active lipid-lowering effect. These investigators found that it was fraction 11 that inhibited accumulation of TG in HepG2 cells and the active substance contained in fraction 11 was diosgenin. Diosgenin at a concentration of 5–10 mmol/L inhibited accumulation of TG in HepG2 cells and expression of lipogenic genes in HepG2 cells. Additionally, they concluded that diosgenin inhibited transactivation of liver-X-receptor-alpha, a critical regulator of cholesterol homeostasis in hepatocytes, thereby inhibiting overall fat accumulation.

In vivo investigations related to the hypolipidemic effect of fenugreek

Well-documented reports relating to the hypocholesterolemic effect of fenugreek have appeared in the literature during the past few decades. Early work by Sharma (1986) evidenced that hypercholesterolemia could be induced in rats with a high-cholesterol diet (HCD), and this could partly be reversed with the addition of fenugreek in the diet. When fenugreek was used as a whole-seed powder or as a defatted fraction (26.1% and 30%, respectively) of the HCD or as a lipid fraction (2.1%) or trigonelline (0.12%) for 4 weeks, the hypercholesterolemia (serum cholesterol, 423 mg/dL) observed in the control rats could be managed by trigonelline or lipid extract. This is because the serum cholesterol values in these groups at the end of the 4-week period were comparable to that of the control group. The most effective fenugreek fraction was the defatted fraction, lowering the serum cholesterol to 214 mg/dL. The defatted fraction formed 93% of the seed powder after removal of the 7% hexane-separated lipid fraction. When the effective dose (daily intake) of the defatted fraction was calculated based on the average daily feed consumption of approximately 20 g/day for rats (U.S. EPA, 1988), the defatted fenugreek fraction of 30% in the rat diet translated to approximately 6 g/day per rat of fenugreek intake or nearly 24 g/kg per day based on 250 g body weight. The HED for this rat dose equated to nearly 3.8 g/kg per day. This is an extremely high dose and highly impractical for a 60-kg person.

Xue et al. (2007) studied the effect of fenugreek extract on blood lipids using relatively lower doses of 0.44, 0.87, and 1.74 g/kg per day administered orally to experimental diabetic rats for 6 weeks in combination with metformin HCl (0.175 g/kg per day). Compared to the untreated experimental diabetic rats (control group), the fenugreek extract-treated diabetic rats had a decrease in TGs and total cholesterol and an increase in high-density lipoprotein (HDL) cholesterol in a dose-dependent manner ($P < .05$). The dose of 1.74 g/kg in rats equated to a human dose of approximately 0.28 g/kg per day. This is a dose of approximately 16.8 g/day for a 60-kg person and is still too high for good patient compliance. More recently, Kumar et al. (2014) investigated the inhibitory effect of aqueous extract of fenugreek seeds (AqE-TFG) on fat accumulation and dyslipidemia induced in rats fed the HFD. Groups of female Wistar rats were fed HFD ad libitum for 28 days. Two of the HFD groups received solutions of AqE-TFG (0.5 and 1.0 g/kg) and one HFD group received orlistat (10 mg/kg) orally in addition to HFD from day 8 to day 28.

Treatment with AqE-TFG produced a significant reduction in body weight gain, body mass index, weight of WAT, blood glucose, serum insulin, lipids, leptin, lipase, and apolipoprotein B levels and caused an elevation in adiponectin levels. The elevation in lipid levels when compared to the HFD positive control group was significantly lower in the 1.0 g/kg per day AqE-TFG group but was still 10%–30% higher (depending on lipid parameter) compared to the control group fed a normal diet. AqE-TFG also reversed serum levels of aspartate aminotransferase (AST), alanine aminotransferase (ALT), and lactate dehydrogenase (LDH). AqE-TFG treatment also reduced the hepatic and cardiac thiobarbituric acid reactive substances and elevated the antioxidant enzyme (glutathione, GSH), superoxide dismutase, and catalase levels. In addition, liver and uterine WAT and lipogenic enzyme (fatty acid synthetase) and glucose-6-phosphate dehydrogenase activities were also restored to normal levels. These findings demonstrated a preventive effect of AqE-TFG on fat accumulation in tissues and for management of dyslipidemia. The estimated comparable HED was approximately 9.6 g/day.

These experiments indicate that using the extracts instead of the whole-seed powder may allow for a reduction in effective dose. Additional reduction may be possible by using the appropriate fenugreek product (or formulation) for a much longer duration, but this needs to be well documented. Experimental studies of longer duration (90 plus least) could be helpful to document such an effect.

Clinical investigations

The effect of 100 g fenugreek (defatted seed powder), divided into two equal doses daily for 10 days, was evaluated on serum lipids of 10 insulin-dependent (type I) diabetic patients between 12 and 37 years of age who were on insulin therapy (Sharma et al., 1990). Total serum cholesterol, very low density lipoprotein cholesterol, and LDL cholesterol were significantly reduced after fenugreek diet compared to the controls. The HDL cholesterol remained unchanged.

Well-controlled, scientifically evaluated mechanistic studies on the efficacy of fenugreek for the prevention and management of dyslipidemia and obesity-related complications are not yet available.

Maintenance of the growth hormone level helping in the normalization of metabolic syndrome and related altered functions

Fenugreek has been claimed to modulate the level of GH (Shim et al., 2008) as demonstrated in vitro by stimulation of release of the rat growth hormone (rGH) in the pituitary cells. Further evaluation of this observation involved testing of different fractions of methyl alcohol (MeOH) extract of fenugreek seeds. Results of this testing revealed that two of the several fenugreek fractions, saponin I and dioscin, in MeOH had the highest activity, causing approximately 12.5-fold and 17.7-fold stimulation of rGH release, respectively, compared to the control group, assumed as the 100% effect. Other fractions did have some activity but were moderate to low at best.

Metabolic syndrome is a term that has been used for decades. It is used to indicate the association of multiple risk factors such as obesity, dyslipoproteinemia, hepatic steatosis, insulin resistance, and hypertension with type II (noninsulin-dependent) diabetes mellitus and may include myocardial infarction. There are striking similarities between the conditions of metabolic syndrome and untreated GH deficiency in adults. The central findings in both of these syndromes are abdominal/visceral obesity and insulin resistance (Johannsson and Bengtsson, 1999). It was concluded that GH replacement in men with abdominal obesity can diminish the negative consequences of visceral obesity.

Many other reports regarding the use of GH have appeared since this report approximately 20 years ago. Liu et al. (2007) reported a detailed meta-analysis of findings from 220 healthy elderly subjects exposed to GH for 2 weeks or more. They concluded that data do suggest that GH is associated with changes in body composition, but adverse events are also observed with exogenous GH use. Based on this evidence, exogenous use of GH cannot be recommended as an antiaging therapy in otherwise healthy elderly subjects until more safety data are available.

At this time, there are no in vivo or clinical investigations that support or negate the effect of fenugreek extract/fractions on GH as reported by Shim et al. (2008). In view of adverse findings for exogenous GH supplementation in humans, fenugreek merits further investigation to determine if endogenously stimulated release of rGH in rats would be safe and indicative of such an effect in humans.

Potential for antiinflammatory and anticancer benefits

Antiinflammatory effect

The antiinflammatory effect was documented as early as 2001 when Ahmadiani et al. (2001) evaluated the antiinflammatory and antipyretic effects of the TFG leaf extract in rats using the formalin-induced hind paw edema model for inflammation and intraperitoneal (ip) injection of aqueous suspension of brewer's yeast as a model for hyperthermia. Sodium salicylate (SS) was used as a positive control. Both TFG and SS significantly reduced formalin-induced edema following a single dose of TFG (1000 or 2000 mg/kg ip or 2000 mg/kg oral) and of SS (300 mg/kg ip). Chronic doses were 1000 mg/kg ip for TFG and 300 mg/kg ip for SS. The antiinflammatory effect for single-dose SS (300 mg/kg ip) and TFG (1,000 mg/kg ip) was comparable and the effect size was higher for the TFG (2000 mg/kg ip) dose. The effect size via the oral route (2000 mg/kg) was relatively lower. TFG and SS also significantly reduced the hyperthermia effect induced by ip brewer's yeast injections. The results indicate that the TFG leaf extract possesses antiinflammatory as well as antipyretic properties. The antiinflammatory effect of mucilage from fenugreek was also evaluated using the adjuvant-induced rat arthritis model (Sindhu et al., 2012).

Fenugreek mucilage daily at an oral dose of 75 mg/kg produced more inhibition of paw edema compared to inhibition produced by an oral 3 mg/kg dose of indometacin (81% vs 71%) on day 21 of adjuvant arthritis. This is a highly practical dose for arthritis in humans, with the HED being approximately 725 mg/day. Fenugreek may also afford a preventive benefit for arthritic inflammation that, if validated by well-managed clinical trials, can be of immense medical benefit to the aging human population.

For exploring the mechanism of antiinflammatory effects of fenugreek, Liu et al. (2012) investigated the effect of fenugreek seed extracts (hexane, ethyl acetate, methanolic, and water extracts) on lipid peroxidation (LPO) and cyclooxygenase (COX) enzyme-inhibitory activities using MTT, LPO, COX-1, and COX-2 enzyme-inhibitory assays. The extracts inhibited LPO by 55%–95%, COX-1 by 6%–87%, and COX-2 by 36%–70%, respectively, at a level of 250 μg/mL. This is the first report relating to the mechanism for the antiinflammatory effect of fenugreek. No data from clinical studies documenting the antiinflammatory effect have been found.

When the antiinflammatory potential of fenugreek is fully established, and if fenugreek or its active component can be detected in the cerebrospinal fluid, fenugreek may be of benefit not only for arthritic inflammation, but also for avoiding the risk of developing neurodegenerative diseases like AD, which in many cases may initiate from an neuroinflammatory process. Näqqa et al. (2014) indicated that inflammation, and not the amyloid or tau pathology, is an independent underlying mechanism in the pathology of AD.

Anticancer potential

Inflammation appears to be a trigger for tumor promotion in some cases. This is because inflammatory cells, such as neutrophils, macrophages, and eosinophils, are an important endogenous source of oxygen radicals. Stimulation of these cells by tumor promoters or by foreign bodies (parasites, bacteria, etc.) causes the release of reactive oxygen species (Rosin et al., 1994). These researchers indicated that there are laboratory studies supporting genetic damage and neoplastic transformation occurring in vitro in cells co-cultured with activated inflammatory cells.

The anticancer potential of fenugreek seed extract (FE) has recently been investigated by Shabbeer et al. (2009). These authors have demonstrated the cytotoxic activity of FE in vitro on a panel of cancer cells with no effect on normal cells. Treatment with 10–15 μg/mL of FE for 72 h was inhibitory to growth in breast, pancreatic, and prostate cancer cell lines (PCa). When tested at higher doses (15–20 μg/mL), FE continued to be inhibitory to growth in PCa cell lines but not in either primary prostate or human telomerase reverse-transcriptase (htert) immortalized prostate cells. At least part of the growth inhibition is due to the induction of cell death, as seen by incorporation of ethidium bromide III into cancer cells exposed to fenugreek extracts. The surprising finding of this investigation was that death of cancer cells occurred despite growth-stimulatory pathways being simultaneously upregulated (phosphorylated) by FE.

Additionally, crude fenugreek extracts have been found to contain an active component with respect to activity against selected neoplastic targets. This active agent is a sapogenin called diosgenin. The anticarcinogenicity potential of diosgenin has been summarized in detail using more than 15 preclinical cancer models/cells with different molecular endpoints and targets (Raju and Rao, 2011). The reader is directed to this review for detailed information. The current chapter provides only significant anticancer findings after 2012. One such investigation relates to the transcription factor, nuclear factor E2-related factor 2 (Nrf2). This is considered to play a key role in cancer development and chemoresistance. Trigonelline (an active component of fenugreek and the coffee bean) has been evaluated using pancreatic carcinoma cell lines (Panc1, Colo357, and MiaPaca2) and H6c7 pancreatic duct cells. The study included analysis of the Nrf2-inhibitory effect and its impact on Nrf2-dependent proteasome activity (Arlt et al., 2013). The results of the study indicated that trigonelline caused sufficient Nrf2 inhibition, capable of blocking Nrf2-dependent proteasome activity, thereby promoting apoptosis (programmed cell death) in pancreatic cancer cells. This is considered a significant effect and should be investigated for diosgenin, also. Currently, it can be suggested that it is not only the fenugreek extract but also the active components, diosgenin and trigonelline, that seem to support anticancer activity at the preclinical level.

Enhancing milk production in new mothers

Anecdotal reports relating to successful use of fenugreek as a galactogogue have been documented as far back as 1945 (Gabay, 2002). The mechanism for this effect is not known and this observation is not supported by all studies. In a double-blind placebo-controlled study involving 26 mothers of preterm infants born at less than 31 weeks of gestation (Reader et al., 2013), the effect of fenugreek on milk production was evaluated. Starting on postpartum day 5, each mother took three fenugreek capsules (575 mg each), three times/day for 21 days. Data revealed no effect on milk production or on prolactin levels evaluated on postpartum days 5, 10, and 15.

Potential for increasing male libido, testosterone production, or testosterone maintenance in blood

Many fenugreek products are promoted by nutritional supplement companies for potential benefits related to male libido and increasing serum testosterone (T) levels. At this time, only a limited amount of published information is available on this topic.

The effects of a standardized TFG extract and mineral formulation on the physiological aspects of male libido have been evaluated by Steels et al. (2011) in 60 healthy males aged 25 and 52. Subjects were given two 600-mg Maxibed/Testofen tablets/day for 6 weeks in the treated group. The primary measure was a score on the Derogatis Interview for Sexual Functioning Self-Report (DISF-SR) scale. The treatment had a positive effect on the physiological aspects of libido and on the quality of life, along with self-reported satisfaction with respect to muscle strength, energy, and well-being. There was no difference in the serum T level in the treated and untreated subjects. This suggests that the effect on libido may be related to an unknown mechanism other than an increase in serum T levels usually needed for male libido. Additionally, in a clinical trial in college-aged men (Wilborn et al., 2010), there was only a marginal increase in serum T (total, 6.57%; bioavailable, 12.26%), which remained within the normal physiological range. This increase may have had a statistical significance in the study but does not seem to be adequate for a biological significance. This is supported by comments that: (1) serum dihydrotestosterone or estradiol (E2) levels were not affected, and (2) there was no effect on 5α-reductase or aromatase enzyme activities. Furthermore, the marginal increase in T level was probably not adequate for increasing the androgen receptor expression, as interpreted from no change in the fat-free body mass, which is a function expected to be dependent on bioavailable T. Some studies have suggested that the enhanced male libido may be due to improved organ blood flow speculated from inhibition of coagulation associated with coumarin content present in the unextracted fenugreek seed powder. This is a far-fetched speculation because if prevention of clot formation would have improved male libido, then the 81-mg aspirin/day (commonly prescribed to many elderly patients) or warfarin used in some patients would have been associated with improved male libido. Conversely, fenugreek at an extremely high dose has been reported to be toxic to the testes and to adversely affect the serum T level and spermatogenesis (Kassem et al., 2006). This observation, however, is not supported by a more recent publication in which fenugreek has been documented to provide protection not only against carbofuran-inhibited spermatogenesis and decease in serum T level, but also with testes weight and histopathological abnormalities induced by carbofuran toxicity (Sakr and Shalaby, 2014).

Hepatoprotective effect

The safety of fenugreek with respect to the liver and other target organs and systems has been discussed elsewhere in this chapter. In addition to evaluation of any acute effects, chronic administration for up to 24 weeks at 25 g/day to diabetic patients had no adverse effect (Sharma et al., 1996). This indicates a good overall safety profile for fenugreek. Fenugreek is not only safe at higher doses but also appears to have a protective effect on the liver with respect to liver damage caused by other potential toxicants. Kaviarasan and Anuradha (2007) investigated the hepatoprotective effect of fenugreek seed polyphenolic extract (FPEt) against ethanol-induced hepatic injury and apoptosis in rats. Chronic ethanol administration (6 g/kg per day for 60 days) caused liver damage, as evidenced by increases in AST, ALT, alkaline phosphatase, LDH, bilirubin, and gamma-glutamyl transferase in plasma and reduction in liver glycogen. The effects on alcohol-metabolizing enzymes such as alcohol dehydrogenase and aldehyde dehydrogenase were also found to be altered in the alcohol-treated group. Ethanol administration resulted in adaptive induction of the activities of cytochrome P450 and cytochrome-b5 and reduction in cytochrome-c-reductase and a phase II enzyme, glutathione-*S*-transferase.

Treatment with FPEt restored the levels of markers of liver injury and reversed the changes in alcohol-metabolizing and detoxification enzymes as well as the electron transport component and cytochrome-c reductase. Increased hepatocyte viability and reduced apoptotic nuclei were observed in FPEt-treated rats. These findings demonstrate that FPEt acts as a protective agent against ethanol-induced abnormalities in the liver. The effects of FPEt are comparable with those of the known hepatoprotective agent silymarin.

The liver-protective effect has also been observed when fenugreek seed powder mixed in rat diet (5%) was fed for 2 months to rats with severe liver toxicity induced by aluminum chloride ($AlCl_3$). Fenugreek (5% in diet) for 2 months resulted in near complete recovery based on biochemical and histopathological parameters (Belaïd-Nouira et al., 2013).

Safety and toxicity profile for fenugreek and its active components

The toxicity profile for either fenugreek seed powder or fenugreek extracts or for the active components like diosgenin and trigonelline is quite low with respect to any risk related to accidental overdose. This is based on the high fenugreek single dose needed for lethality in 50% of test subjects (LD_{50} value) via the oral route (Table 37.3). These dosages are higher than 5 g/kg in many studies. This places fenugreek in the category of being virtually nontoxic for acute exposure. Repeated daily oral doses of fenugreek up to 25 g/day have also been used without any adverse effects in adult diabetic patients.

General safety and toxicity profile

Single dose (acute) toxicity

Findings of single acute toxicity studies are summarized in Table 37.3. In an animal study, the acute oral LD_{50} of fenugreek was found to be higher than 5 g/kg in rats and the acute dermal LD_{50} was found to be more than 2 g/kg in rabbits (Opdyke, 1978). Another study (Muralidhara et al., 1999) also concluded that the oral LD_{50} for debitterized fenugreek powder was higher than 2 and 5 g/kg in mice and rats, respectively. Even for the extracts of diosgenin or trigonelline, the acute toxicity hazard is quite low (Table 37.3), indicating reasonable safety in cases of overdose irrespective of the type of fenugreek product (whole-seed powder or extract) or active component, such as diosgenin or trigonelline. No acute toxicity data were available for 4-hydroxyisoleucine.

Repeated dose (subchronic and chronic) toxicity

In a 90-day subchronic study in weanling rats, DFG was fed at doses of up to 10% in the diet with no effect on either the daily food intake or growth or any other parameter indicating any concern with respect to safety (Muralidhara et al., 1999). Comparable data with respect to the safety of fenugreek were reported from another study in which weanling rats were fed up to 20% fenugreek seed powder for 90 days (Rao et al., 1996). With respect to safety of trigonelline, Tice (1997) summarized that feeding mice 50 mg/kg of trigonelline daily for 21 days did not alter the weights of the liver, kidney, thymus, thyroid, adrenals, uterus, or ovaries. No visible effects were noted when cats were fed trigonelline at 3500 mg/day for 62–70 days.

With respect to fenugreek safety following chronic use in humans, data from a clinical study in which 60 diabetic patients took powdered fenugreek seeds at a dose of 25 g/day for 24 weeks showed no evidence of clinical, hematological, hepatic, or renal abnormalities (Sharma et al., 1996). This indicates that high oral doses of fenugreek can be well tolerated by adult humans as predicted by animal studies. No chronic toxicity data were available for the extracts or active components.

Effect on pregnancy and embryo/fetus or pup development

This information is summarized in Table 37.4.

Effect on pregnancy and fetal development

The effect of fenugreek on pregnancy and fetal development remains a concern in view of divergent published

TABLE 37.4 Safety information related to reproduction, fertility, pregnancy, embryo-fetal development, and postnatal pup development.

Type of toxicity or adverse effect	Fenugreek product/formulation/active component				
	Species/fenugreek seed or powder/defatted seed powder	Nonlipid, aqueous, or alcoholic extract	Trigonelline (an active component)	Diosgenin (an active component)	4-Hydroxy-isoleucine (an active component)
Male or female fertility	*Guinea pig, stimulation of uterus*[a] *Rabbit, female*, 30% in diet = approximately 36 g/day—stimulated uterus, reduced placental weight, reduced the litter size, caused changes in endometrial glands[b] *Rabbit, male* = approximately 36 g/day—adverse effects on testes, plasma testosterone lowered (50%), sperm concentration lowered by 50%[b]	• *Rats, male*, effect of aqueous extract 0.4 g/kg, 3 times weekly, oral, for 6 weeks on carbofuran-induced testicular toxicity including inhibition of spermatogenesis[f] • Fenugreek ameliorated the testicular toxicity caused by carbofuran[f]	• *Nonpregnant rats, females*, at 75 mg/kg oral dose[e]: No estrogenic activity based on vaginal cornification • No effect on estrous cycle	No data	No data
Effect on pregnancy and embryo-fetal development (segment II)	No data	*Mice, female*, on aqueous extract—doses up to 1 g/kg/day for the entire gestation period Fetal mortality, decrease in litter size and fetal weights, fetal abnormalities[c]	• *Rat, female*, 75 mg/kg/day for GD 1–7[e]; no effect on implantation • No abortions or teratogenic effects	No data	No data
Postnatal fetal development (segment III)	No data	*Mice, female*, on aqueous extract doses up to 1 g/kg per day for the entire gestation period—adverse effects locomotion, motor coordination, and T-maze[d]	No data	No data	No data

GD, gestation days. Rabbit feed intake assumed to be approximately 120 g/day (Lebas et al., 1997) translates to fenugreek intake of 36 g/day (30% level) in feed. This is 12 g/kg/day for a 3-kg rabbit. The human equivalent dose (HED) for the rabbit daily intake would estimate to approximately 4 g/kg/day or 240 g/day for a 60-kg person.
[a]*Abdo and al-Kafawi (1969).*
[b]*Kassem et al. (2006).*
[c]*Khalki et al. (2010).*
[d]*Khalki et al. (2012).*
[e]*Aswar et al. (2009).*
[f]*Sakr and Shalaby (2014).*

information and limited quality of studies. Diosgenin containing extract from wild yam (*Dioscorea villosa*) has been reviewed by the Cosmetic Ingredient Review Panel and evaluated for estrogen- or progesterone-like activities (Hooker, 2004). The specified extract (containing diosgenin up to 3.5%) did not exhibit any estrogenic activity in the juvenile rat uterotrophic assay at the dose of 500 mg/kg per day. Comparable findings were found in the work of Aswar et al. (2009), in which trigonelline, the active alkaloid in fenugreek, at a dose of 75 mg/kg orally in rats had no effect on estrous cycling (estrogenic activity), implantation, and abortifacient or teratogenic activity.

However, fenugreek has been reported to have a stimulatory effect on the gravid uterus in guinea pigs (Abdo and al-Kafawi, 1969), which may cause reductions in placental weights and a reduction in litter size, and also proliferative changes in the endometrial glands of rabbits (Kassem et al., 2006). The uterine effects identified in rabbits were found after an extremely high dose not practical for consideration in humans. This is because 30% fenugreek used in the rabbit diet equated to a human daily dose of 240 g/day (Table 37.4). Dande and Patil (2012) used a much lower oral dose (200 mg/kg) for the saponin extract of fenugreek (approximate HED of 1.9 g/day) and still observed the antiimplantation and abortifacient effects in rats. The saponin extract in this study also exhibited a weak estrogenic activity in immature ovariectomized rats.

The effect on the developing fetus has also been investigated in a study reported by Khalki et al. (2010), in which lyophilized aqueous extract from fenugreek seeds was administered to mated female mice for the entire duration of pregnancy at doses of 0, 500, and 1000 mg/kg per day. Developmental toxicity in offspring included an increase in fetal deaths, decrease in litter size, and a reduction in fetal body weight. In addition, there was an increase in the incidence of morphological fetal abnormalities.

Effect on postnatal pup development

Khalki et al. (2012) evaluated the effect of an aqueous extract of fenugreek at oral doses of 0, 500, or 1000 mg/kg per day administered to mice for the duration of pregnancy on postnatal pup development. There was a decrease in locomotor activity, motor coordination, and T-maze activity observed only at the 1000 mg/kg per day dose, indicating decreased neurobehavioral development of pups at the higher dose. Although the rat is the more accepted and commonly used rodent species for evaluating postnatal developmental effects, no reason has been discussed for selecting the mouse as the toxicology species for this postnatal investigation.

Effect on fertility and other reproductive effects

With respect to the effect on male fertility, Kassem et al. (2006) reported a reduction in testes weight with evidence for damage to the seminiferous tubules and interstitial tissues of rabbits fed a diet containing 30% fenugreek seeds. In addition, the plasma concentration of T and sperm concentrations were nearly 50% lower in the treated group of rabbits. This risk may have been overly identified because it was only observed at an extremely high dose (as discussed in the section "Effect on pregnancy and fetal development"), which may not be a realistic risk. This is also because a protective effect has been reported for fenugreek not only in the case of carbofuran-inhibited spermatogenesis but also in testes weight and histopathological abnormalities induced by carbofuran exposure (Sakr and Shalaby, 2014). Additionally, a double-blind, clinical study involving healthy, college-aged men given fenugreek at a dose of 500 mg once daily for 8 weeks showed a significant effect in lowering body fat mass and did not lower bioavailable and total serum T at week 8 compared to the baseline reading (Wilborn et al., 2010).

With respect to females, ethanol and water extracts of TFG seeds were investigated separately to determine their effect on implantation in female albino rats and estrogenic/antiestrogenic activity in Swiss mice. Among these, the ethanol extract dose needed for a significant adverse effect on implantation was 500 mg/kg per day from gestation days 1 to 10 in rats and for days 1 to 7 in 21-day-old bilaterally ovariectomized mice (Ahirwar et al., 2010).

Genotoxicity and carcinogenicity

Fenugreek seed extract containing a minimum of 40% of 4-hydroxyisoleucine was evaluated using the standard battery of genotoxicity tests (reverse mutation assay; mouse lymphoma forward mutation assay; mouse micronucleus assay) recommended by the US Food and Drug Administration for food ingredients (Flammang et al., 2004). Fenugreek seed extract was found not to be genotoxic under the conditions of the tests for high-diosgenin content (up to 3.5%). The wild yam extract was negative in both the bacterial and mammalian assay systems for mutagenicity, except that the Ames test strain TA1537 was positive at one of the dose levels in the plate incorporation method, but not in the preincubation method (Hooker, 2004).

Trigonelline was also found not to be mutagenic in either the Ames assay or the L5178Y $TK^{+/-}$ mouse lymphoma mutation assay, with and without metabolic activation (Tice, 1997).

No carcinogenicity data have been found for fenugreek, but several preclinical anticancer studies have been performed that indicate an anticancer promise. This aspect has been discussed in the section "Potential for Antiinflammatory and Anticancer Benefits."

Allergy and related effects

Dermal irritation, sensitization, and phototoxicity

Opdyke (1978) summarized this information in *Fragrance Raw Material Monograph*. Fenugreek absolute, when applied undiluted to the back of hairless mice and swine, was nonirritating. When applied full strength to intact or abraded rabbit skin for 24 h under occlusion, it was moderately irritating. When tested as 2% in petrolatum, it produced no irritation after a 48-h closed-patch test on human subjects, and no sensitization reaction was observed. This suggests that dermal exposure to fenugreek has no or negligible potential for any adverse skin reaction. Neither has any phototoxic risk been reported in this summary.

Potential for allergy and antigenicity

Botanically, fenugreek belongs to the family *Papilionideae*, which includes the peanut, chick pea, and soy bean under the main family, Leguminoseae (Patil et al., 1977). As peanut allergies are observed in some humans, the potential for a similar allergy to fenugreek cannot be excluded due to the potential for cross-reactivity. The respiratory symptoms of allergy in reported cases of reaction to fenugreek were predominately after oral or dermal exposure (Patil et al., 1977). Cross-reactivity to a peanut allergy is believed to be the cause (Faeste et al., 2009). The potential for cross-reactivity with a peanut allergy was also documented using a mouse model that represented a test model for human allergy (Vinje et al., 2011).

Concluding remarks and future directions

Based on the scientific information presented in this chapter, fenugreek presents a unique and rare nutraceutical opportunity for providing multiple health benefits. Fenugreek can supplement conventional medical therapy for chronic illnesses like type II diabetes, dyslipidemia, joint and gastrointestinal inflammatory conditions, and modulation of fat accumulation in adipocytes. It may also prevent or delay onset of neurodegenerative diseases that have inflammation as the earliest trigger for pathogenesis, such as AD. A lack of safety concerns following acute and chronic dosing is an added beneficial feature of fenugreek. Limitations and caution need to be followed with respect to risks related to pregnancy, fetal development, and allergies, such as peanut-type allergies.

There is a large amount of preclinical data for fenugreek, but clinical trials have been of marginal quality. These studies have not been conducted using a fully characterized test drug product. The reproducibility of results has not been established and there is no evidence of a good effect size and practicality of the high dose needed for a consistent and reproducible therapeutic benefit. With respect to the practicality of the effective dose, based on preclinical data, the use of fenugreek appears to be more promising for the management of inflammation. This is because fenugreek doses as low as 75 mg/kg per day in rats have been effective in experimental cases of adjuvant-induced arthritis (Sindhu et al., 2012), which equates to a highly practical human dose for arthritic inflammation, with the HED being approximately 725 mg per day. Based on this target, fenugreek may afford a benefit for the prevention and control of arthritic inflammation if validated in a randomized, double-blind, clinical trial. Further, if fenugreek can be determined to cross the blood–brain barrier, it may be of value in managing cerebral inflammation, which then may aid in the management of early stages of AD (Näqqa et al., 2014).

Fenugreek is a very well-tolerated nutraceutical even at extremely high doses following acute and chronic exposures. Doses of up to 25 g/day for 24 weeks in humans have been tolerated without any health concerns. Caution is strongly recommended for women of child-bearing age with respect to the effect on pregnancy and development of the fetus. Risk related to fertility and embryo/fetal development can only be confirmed after the good laboratory practice-compliant reproductive toxicity, embryo/fetal developmental toxicity (segment II), and fertility (segment I) studies have been performed in accordance with ICH regulations. Caution is also recommended with respect to allergic potential, especially in individuals with peanut allergies. Fenugreek is negative in repeated genotoxicity studies and shows promise in anticancer preclinical studies. It is not expected to pose any carcinogenicity risk.

The future for fenugreek can be even brighter if additional clinical studies are undertaken to increase the weight of evidence with respect to efficacy in humans using the same quality and standards as are used for pharmaceuticals.

Other points with respect to future directions are proposed:

- After having the effective dose lowered in animal studies, in addition to the example of type II diabetes, perform well-controlled clinical trials for dyslipidemia and for the prevention of fat accumulation in adipocytes in accordance with regulations applicable to pharmaceutical development.
- Due to the promising effect of controlling inflammation in animal studies at relatively low doses, there appears to be an area with a large scope for fenugreek use extending to conditions like osteoarthritis, gastrointestinal inflammatory conditions, and others.
- Perform pharmacokinetic investigations to evaluate fenugreek levels in the cerebrospinal fluid. If the results are promising, then this can allow opportunities for evaluating the use of fenugreek in neurodegenerative diseases like AD that are initially triggered by a neuroinflammatory process.

References

Abdel-Mawla, A.M.A., Osman, H.E.H., 2011. Elicitation of trigonelline and 4-hydroxyisoleucine with hypoglycemic activity in cell suspension cultures of *Trigonella foenum-graecum L.* Open Conf. Proc. 2, 80–87.

Abdo, M.S., al-Kafawi, A.A., 1969. Experimental studies on the effect of *Trigonella foenum-graecum*. Planta Med. 17, 14–18.

Ahirwar, D., Ahirwar, B., Kharya, M.D., 2010. Evaluation of antifertility activity of *Trigonella foenum-graecum* seeds. Der. Pharm. Sin. 1, 33–39.

Ahmadiani, A., Javan, M., Semnanian, S., 2001. Anti-inflammatory and antipyretic effects of *Trigonella foenum-graecum* leaves extract in the rat. J. Ethnopharmacol. 75, 283–286.

Alamdari, K.A., Choobineh, S., Jadidi, J.P., 2009. Antidiabetic effects of exercise and fenugreek supplementation in males with NIDDM. Med. Sport 62, 315–324.

Al-Habori, M., Raman, A., Lawrence, M.J., Skett, P., 2001. *In vitro* effect of fenugreek extracts on intestinal sodium-dependent glucose uptake and hepatic glycogen phosphorylase A. Int. J. Exp. Diab. Res. 2, 91–99.

Allred, K.F., Yackley, K.M., Vanamala, J., 2009. Trigonelline is a novel phytoestrogen in coffee beans. J. Nutr. 139, 1833–1838.

Amin, R., Abdul-Ghani, A.-S., Suleiman, M.S., 1987. Effect of *Trigonella feonum graecum* on intestinal absorption. Diabetes 36 (Suppl. 1).

Arendash, G.W., Cao, C., 2010. Caffeine and coffee as therapeutics against Alzheimer's disease. J. Alzheimers Dis. Child. 54, 946–949.

Arlt, A., Sebens, S., Krebs, S., 2013. Inhibition of the Nrf2 transcription factor by the alkaloid trigonelline renders pancreatic cancer cells more susceptible to apoptosis through decreased proteasomal gene expression and proteasome activity. Oncogene. 32, 4825–4835.

Aswar, U., Mohan, V., Bodhankar, S.L., 2009. Effect of trigonelline on fertility in female rats. Int. J. Green Pharm. 3, 220–223.

Barranco Quintana, J.L., Allam, M.F., et al., 2007. Alzheimer's disease and coffee: a quantitative review. Neurol. Res. 29, 91–95.

Basch, E., Ulbricht, C., Kuo, G., 2003. Therapeutic applications of fenugreek. Altern. Med. Rev. 8, 20–27.

Belaïd-Nouira, Y., Bakhta, H., Haouas, Z., 2013. Fenugreek seeds, a hepatoprotector forage crop against chronic AlCl3 toxicity. BMC Vet. Res. 9, 22.

Bordia, A., Verma, S.K., Srivastava, K.C., 1997. Effect of ginger (*Zingiber officinale Rosc.*) and fenugreek (*Trigonella foenum-graecum L.*) on blood lipids, blood sugar and platelet aggregation in patients with coronary artery disease. Prostaglandins Leukot. Essent. Fatty Acids. 56, 379–384.

Campos, H., Baylin, A., 2007. Coffee consumption and risk of type 2 diabetes and heart disease. Nutr. Res. 65, 173–179.

Chevassus, H., Molinier, N., Costa, F., 2009. A fenugreek seed extract selectively reduces spontaneous fat consumption in healthy volunteers. Eur. J. Clin. Pharmacol. 65, 1175–1178.

Chevassus, H., Gaillard, J.B., Farret, A., 2010. A fenugreek seed extract selectively reduces spontaneous fat intake in overweight subjects. Eur. J. Clin. Pharmacol. 66, 449–455.

Dande, P., Patil, S., 2012. Evaluation of saponins from *trigonella foenum-graecum seeds* for its antifertility activity. Asian J. Pharmaceut. Clin. Res. 5 (Suppl. 3), 154–157.

Egan, D., O'Kennady, R., Moran, E., 1990. The pharmacology, metabolism, analysis, and applications of coumarin and coumarin-related compounds. Drug Metab. Rev. 22, 503–529.

Eidi, A., Eidi, M., Sokhteh, M., 2007. Effect of fenugreek (*Trigonella foenum-graecum* L) seeds on serum parameters in normal and streptozotocin-induced diabetic rats. Nutr. Res. 27, 728–733.

Faeste, K.E., Namork, E., Lindnik, H., 2009. Allergenicity and antigenicity of fenugreek (*Trigonella foenum-graecum*) proteins in food. J. Allergy Clin. Immunol. 123, 187–194.

Fairchild, R.M., Ellis, P.R., Byrne, A.J., 1996. A new breakfast cereal containing guar gum reduces postprandial plasma glucose and insulin concentrations in normal-weight human subjects. Br. J. Nutr. 76, 63–73.

Flammang, A.M., Cifone, M.A., Erexson, G.L., 2004. Genotoxicity testing of a fenugreek extract. Food Chem. Toxicol. 42, 1769–1775.

Fowden, L., Pratt, H.M., Smith, A., 1973. 4-Hydroxyisoleucine from seed of *Trigonella foenum-graecum.* Phytochemistry 12, 1707–1711.

Gabay, M.P., 2002. Galactogogues: medications that induce lactation. J. Hum. Lact. 18 (3), 274–279.

Gregoire, F.M., 2001. Adipocyte differentiation: from fibroblast to endocrine cell. Soc. Exp. Biol. Med 226, 997–1002.

Gupta, A., Gupta, R., Lal, B., 2001. Effect of hydroalcoholic extracts of fenugreek seeds (*Trigonella foenum-graecum*) on insulin resistance and glycemic control in type 2 diabetes. J. Assoc. Physc. India. 49, 1057–1061.

Haeri, M.R., Limaki, H.M., White, C.J.B., 2012. Non-insulin dependent antidiabetic activity of (2S, 3R, 4S) 4-hydroxyisoleucine of fenugreek (*Trigonella foenum greacum*) in stretozotocin-induced type I diabetic rats. Phytomedicine 19, 571–574.

Hamza, N., Berke, B., Cheze, C., 2012. Preventive and curative effect of *Trigonella foenum-graecum L.* seeds in C57BL/6J models of type-2 diabetes induced by high-fat diet. J. Ethnopharmacol. 142, 516–522.

Hooker, E., 2004. Final report of the amended safety assessment of *Dioscorea villosa* (Wild Yam) root extract. Int. J. Toxicol. 23 (Suppl.), 49–54.

Jette, L., Harvey, L., Eugeni, K., 2009. 4-hydroxyisoleucine: a plant-derived treatment for metabolic syndrome. Curr. Opin. Investig. Drugs. 10, 353–358.

Johannsson, G., Bengtsson, B.A., 1999. Growth hormone and the metabolic syndrome. J. Endocrinol. Invest. 22, 41–46.

Kanchan, D.M., Somani, G.S., Peshattiwar, V.V., et al., 2016. Renoprotective effect of diosgenin in streptozotocin induced diabetic rats. Pharmacol. Rep. 68 (2), 370–377.

Kassaian, N., Azadbakht, L., Forghani, B., 2009. Effect of fenugreek seeds on blood glucose and lipid profiles in type 2 diabetic patients. Int. J. Vitam. Nutr. Res. 79, 34–39.

Kassem, A., Al-Aghbari, A., AL-Habori, M., 2006. Evaluation of the potential antifertility effect of fenugreek seeds in male and female rabbits. Contraception. 73, 301–306.

Kaviarasan, S., Anuradha, C.V., 2007. Fenugreek (*Trigonella foenum-graecum*) seed polyphenols protect liver from alcohol toxicity: a role on hepatic detoxification system and apoptosis. Pharmazie. 62, 299–304.

Khalki, L., M'hamed, S.B., Bennis, M., 2010. Evaluation of the developmental toxicity of the aqueous extract from *Trigonella foenum-graecum (L.)* in mice. J. Ethnopharmacol. 131, 321–325.

Khalki, L., Bennis, M., Sokar, Z., 2012. The developmental neurobehavioral effects of fenugreek seeds on prenatally exposed mice. J. Ethnopharmacol. 139, 672–677.

Kumar, G.S., Shetty, A.K., Sambaiah, K., 2005. Antidiabetic property of fenugreek seed mucilage and spent turmeric in streptozotocin-induced diabetic rats. Nutr. Res. 25, 1021–1028.

Kumar, P., Kale, R.K., Mukherjee, S., 2011. Antidiabetic effects of *Trigonella foenum-graecum* seed powder in a rat model. Toxicol. Environ. Chem. 93, 2085–2097.

Kumar, P., Bhandari, U., Jamadagni, S., 2014. Fenugreek seed extract inhibit fat accumulation and ameliorates dyslipidemia in high fat diet-induced obese rats. Biomed. Res. Int. 2014, 1–11.

Lebas, F., Coudert, P., de Rochambrau, H. et al., 1997. Table 17. In: The Rabbit: Husbandry, Health and Production (new revised version). FAO Animal Production and Health Series No 21.

Lee, K.J., Inoue, M., Otani, T., 2007. Coffee consumption and risk of colorectal cancer in a population-based prospective cohort of Japanese men and women. Int. J. Cancer. 121, 1312–1318.

Lewis, H.M., 1994. Therapeutic progress II: treatment of psoriasis, J. Clin. Pharm. Therap., 19. pp. 223–232.

Liu, H., Bravata, D.M., Olkin, I., 2007. Systematic review: the safety and efficacy of growth hormone in the healthy elderly. Ann. Intern. Med. 146, 104–115.

Liu, Y., Kakani, R., Nair, M.G., 2012. Compounds in functional food fenugreek spice exhibit anti-inflammatory and antioxidant activities. Food Chem. 131, 1187–1192.

Lu, F., Shen, L., Qin, Y., 2008. Clinical observation on *Trigonella foenum-graecum L.* total Saponins in combination with sulfonylureas in the treatment of type 2 diabetes mellitus. Chin. J. Integr. Med. 14, 56–60.

Luang, G., Wang, Y., Wang, Z., et al., 2018. Flavonoid glycosides from fenugreek seeds regulate glycolipid metabolism by improving mitochondrial function in 3T3-L1 adipocytes *in vitro.* J. Agric. Food Chem. 66, 3169–3178.

Mathern, J.R., Raatz, S.K., Thomas, W., 2009. Effect of fenugreek fiber on satiety, blood glucose and insulin response and energy intake in obese subjects. Phytother. Res. 23, 1543–1548.

McNeely, W., Goa, K., 1998. 5-Methoxypsoralen – a review of its effects in psoriasis and vitiligo. Drugs. 56, 667–690.

Mohammad, S., Taha Akhtar, A.K., Bamezai, R.N., 2006. In vivo effect of *Trigonella foenum-graecum* on the expression of pyruvate kinase, phosphoenolpyruvate carboxykinase and distribution of glucose transporter (GLUT4) in alloxan diabetic rats. Can. J. Physiol. Pharmacol. 84, 647–654.

Morani, A.S., Bodhankar, S.L., Mohan, V., 2012. Ameliorative effects of standardized extract from *Trigonella foenum-graecum L.* seeds on painful peripheral neuropathy in rats. Asian Pac. J. Trop. Med. 12, 385–390.

Muralidhara, N.K., Viswanatha, S., Ramesh, B., 1999. Acute and subchronic toxicity assessment of debitterized fenugreek powder in the mouse and rat. Food Chem. Toxicol. 37, 831–838.

Näqqa, K., Wattmo, C., Zhang, Y., 2014. Cerebral inflammation is an underlying mechanism of early death in Alzheimer's disease: a 13-year cause-specific multivariate mortality study. Alzheimers Res. Ther. 6 (4), 41.

Narender, T., Puri, A., Khaliq, T., 2006. 4-hydroxyisoleucine an unusual amino acid as antidyslipedemic and antihyperglycemic agent. J. Bioorg. Med. Chem. Lett. 16, 293–296.

Neelakantan, N., Narayanan, M., de Souza, R.J., 2014. Effect of fenugreek (*Trigonella foenum-greacum L.*) intake on glycemia: a meta-analysis of clinical trials. Nutr. J. 13, 7.

Neveen, H., El-Soud, A., Khalil, M.Y., Hussein, J.S., 2007. Antidiabetic effects of fenugreek alkaloid extract in streptozotocin-induced hyperglycemic rats. J. Appl. Sci. Res. 3, 1073–1083.

Opdyke, D.L.J., 1978. Fenugreek absolute. Food Cosmet. Toxicol. 16, S755–S756.

Patil, S.P., Niphadkar, P.V., Bapat, M.M., 1977. Allergy to fenugreek (*Trigonella foenum-graecum*). Ann. Allergy Asthma Immunol. 78, 297–300.

Petit, P., Sauvaire, Y., Ponsin, G., 1993. Effects of a fenugreek seed extracts on feeding behaviour in the rat: metabolic-endocrine correlates. Pharmacol. Biochem. Behav. 45, 369–374.

Prasanthi, J.R.P., Dasari, B., Marwarha, G., 2010. Caffeine protects against oxidative stress and Alzheimer's disease-like pathology in rabbit hippocampus induced by cholesterol-enriched diet. Free. Rad. Biol. Med. 49, 1212–1220.

Preet, A., Gupta, B.L., Siddiqui, M.R., 2005. Restoration of ultrastructural and biochemical changes in alloxan-induced diabetic rat sciatic nerve on treatment with Na3VO4 and *Trigonella* – a promising antidiabetic agent. Mol. Cell. Biochem. 278, 21–31.

Raju, J., Rao, C.V., 2011. Diosgenin, a steroid saponin constituent of yams and Fenugreek: emerging evidence for applications in medicine. In: Rasooli, I. (Ed.), Bioactive Compounds in Phytomedicine. InTech, Rijeka, Croatia.

Raju, J., Gupta, D., Rao, A.R., 2001. *Trigonella foenum-graecum* (fenugreek) seed powder improves glucose homeostasis in alloxan diabetic rat tissues by reversing the altered glycolytic, gluconeogenic and lipogenic enzymes. Mol. Cell. Biochem. 224, 45–51.

Rao, P.U., Sesikeran, B., Rao, P.S., 1996. Short term nutritional and safety evaluation of fenugreek. Nutr. Res. 16, 1495–1505.

Reader, C., LeGrand, A., O'Conner-Von, S.K., 2013. The effect of fenugreek on milk production and prolactin level in mothers of preterm infants. Clin. Lactat 4, 159–164.

Ribes, G., Sauvaire, Y., Costa, C. Da, 1986. Antidiabetic effects of subfractions from fenugreek seeds in diabetic dogs. Proc. Soc. Exp. Biol. Med. 182, 159–166.

Rosin, M.P., Anwar, W.A., Ward, A.J., 1994. Inflammation, chromosomal instability, and cancer: the schistosomiasis model. Cancer Res. 54 (Suppl. 7), 1929s–1933s.

Sakr, S.A., Shalaby, S.Y., 2014. Effect of fenugreek seed extract on carbofuran-inhibited spermatogenesis and induced apoptosis in albino rats. J. Infert. Reprod. Biol. 2, 36–42.

Sauvaire, Y., Petit, P., Broca, C., 1998. 4-hydroxyisoleucine: a novel amino acid potentiator of insulin secretion. Diabetes 47, 206–210.

Shabbeer, S., Sobolewski, M., Kachhap, S., 2009. Fenugreek: a naturally occurring edible spice as an anticancer agent. Cancer Biol. Ther. 8, 272–278.

Sharma, R.D., 1986. An evaluation of hypocholestrolemic factors of fenugreek seed (*T. foenum-graecum*) in rats. Nutr. Rep. Int 33, 669–677.

Sharma, R.D., Raghuram, T.C., 1990. Hypoglycemic effect of fenugreek seeds in non-insulin dependent diabetic subjects. Nutr. Res. 10, 731–739.

Sharma, R.D., Raghuram, T.C., Rao, N.S., 1990. Effect of fenugreek seeds on blood glucose and serum lipids in type 1 diabetes. Eur. J. Clin. Nutr. 44, 301–306.

Sharma, R.D., Sarkar, A., Hazra, D.K., 1996. Toxicological evaluation of fenugreek seeds: a long term feeding experiment in diabetic patients. Phytother. Res. 10, 519–520.

Shim, S.H., Lee, E.J., Kim, J.S., 2008. Rat growth-hormone release stimulators from fenugreek seeds. Chem. Biodivers 5, 1753–1761.

Sindhu, G., Ratheesh, M., Shyni, G.L., 2012. Anti-inflammatory and antioxidative effects of mucilage of *Trigonella foenum-graecum* (Fenugreek) on adjuvant induced arthritic rats. Int. Immunopharmacol. 12, 205–211.

Srichamroen, A., Thomson, A.B.R., Field, C.J., 2009. *In vitro* intestinal glucose uptake is inhibited by galactomannan from Canadian fenugreek seed (*Trigonella foenum-graecum L*) in genetically lean and obese rats, Nutr. Res., 29. pp. 49–54.

Steels, E., Rao, A., Vitetta, L., 2011. Physiological aspects of male libido enhanced by standardized *Trigonella foenum-greacum* extract and mineral formulation. Phyther. Res. 25, 1294–1300.

Tice, R., 1997. Trigonelline (535-83-1): review of toxicological literature. A report submitted to National Institute of Environmental Health Sciences (Contract No. N01-ES-65402).

Trask, L.E., Chaidarun, S.S., Platt, D., 2014. Treatment with novel galactomannan derivative reduces 2-hour postprandial glucose excursions in individuals with type 2 diabetes treated with oral medications and/or insulin. J. Diabetes Sci. Technol. 8, 1018–1022.

U.S. EPA, 1988. Recommendations for and Documentation of Biological Values for Use in Risk Assessment. EPA/600/6-87/008.

Uemura, T., Hirai, S., Mizoguchi, N., 2010. Diosgenin, present in fenugreek improves glucose metabolism by promoting adipocyte differentiation and inhibiting inflammation in adipose tissues. Mol. Nutr. Food Res. 54, 1596–1608.

Uemura, T., Goto, T., Kang, M., 2011. Diosgenin, the main aglycon of fenugreek, inhibits LXRa activity in HepG2 cells and decreases plasma and hepatic triglycerides in obese diabetic mice. J. Nutr. 141, 17–23.

Vijayakumar, M.V., Pandey, V., Mishra, G.C., 2010. Hypolipidemic effect of fenugreek seeds is mediated through inhibition of fat accumulation and upregulation of LDL receptor. Obesity. 18, 667–674.

Vinje, N.E., Namork, E., Lovik, M., 2011. Anaphylactic reactions in mice with fenugreek allergy. Scand. J. Immunol. 74, 342–353.

Watson, R.T., Pessin, J.E., 2001. Intracellular organization of insulin signaling and GLUT4 translocation. Recent. Prog. Horm. Res. 56, 175–193.

Wilborn, C., Taylor, L., Poole, C., 2010. Effects of a purported aromatase and 5α-reductase inhibitor on hormone profiles in college-age men. Int. J. Sport Nutr. Exer. Metab. 20, 457–465.

Xue, W.L., Li, X.S., Zhang, J., 2007. Effect of *Trigonella foenum-graecum* (fenugreek) extract on blood glucose, blood lipids, hemorheological properties of streptozotocin-induced diabetic rats. Asia Pac. J. Clin. Nutr. 16, 422–426.

Xue, W., Lei, J., Li, X., 2011. *Trigonella foenum-graecum* seed extract protects kidney function and morphology in diabetic rats via its antioxidant activity. Nutr. Res. 31, 555–562.

Zia, T., Hasnain, S.N., Hasan, S.K., 2001. Evaluation of the oral hypoglycemic effect of *Trigonella foenum-graecum L.* (methi) in normal mice. J. Ethnopharmacol. 75, 191–195.

Chapter 38

Cannabis sativa: an overview

Gary Richter, Trina Hazzah, Joshua A. Hartsel, Joshua Eades, Brian Hickory and Alexandros Makriyannis

Introduction

Cannabis sativa is a remarkable plant containing many valuable natural components. It has been cultivated throughout the world and history for use as a food, fuel source, nutritional supplement, body care product, source of paper, building material, medicine, and in textiles (Small and Marcus, 2002). In 1938, *Popular Mechanics* magazine noted the economic value of hemp, one member of the *C. sativa* taxon, as the new billion-dollar crop of the era. The article reported that 30,000 products could be made from only one component of the hemp plant, the fiber, and the hurd of the stalk (Mechanics, 1938). One key aspect of *C. sativa* that *Popular Mechanics* did not mention was the flower, which produces cannabinoids, terpenes, and seeds with a healthy balance of fatty acids. This chapter describes the scientific data as they relate to *C. sativa* flowers, biology, and chemistry, and how these components contribute to its recognized therapeutic value.

Cannabis sativa taxonomy

Among all genera belonging to the *Cannabaceae* family, *C. sativa* has attracted special interest from the public because of its psychoactive effects. The genetic plasticity of cannabis has made it difficult to catalog, and there is still an ongoing discussion about its proper classification. In 1737, Carl Linnaeus, the father of modern taxonomy, described *C. sativa* as a genus composed of a single species, *C. sativa*. Linnaeus was unaware of drug-type cultivars that were prevalent in Asia and classified cannabis as a single species from his experience with fiber-type crops that were common in Europe (Linnaeus, 1800). It is uncertain how he would have cataloged the new Indian and Asian cannabis cultivars had he been aware of them during his era. Comparative analyses between the Indian and the European hemp varieties, based on size, shape, leaf structure, and psychoactive effects, inspired Jean-Baptiste Lamark to classify the Indian cultivars as a separate species, *Cannabis indica* (Lamarck, 1811). The Soviet botanist Janischevsky revisited the polytypic (multispecies) view when he recognized that local Russian plants did not fit the characteristics of *C. sativa* or *C. indica*, but fell within the *Cannabis* taxa. He named the short, wild Russian autoflowering plants *Cannabis ruderalis* (Small, 1975) and divided the genus into three distinct species, *C. sativa* L., *C. indica* Lam., and *C. ruderalis* Janisch. According to the *American Herbal Pharmacopeia*, *C. sativa* L. was historically bred to be tall and is used mainly for fiber and seed (Fig. 38.1; Upton et al., 2014).

Cannabis sativa indica Lam. is characterized by a short, densely branched structure and potent level of Δ^9-tetrahydrocannabinol (Δ^9-THC), which is the main psychoactive ingredient (Upton et al., 2014). To date, scholars still disagree on how to correctly catalog cannabis species. The two competing schools of cannabis taxonomy are divided between a monotypic (single-species) and a polytypic perspective. The current debate centers on one question: Are all cannabis cultivars *C. sativa*? *Sativa* and *indica* cannabis types are frequently crossbred to produce hybrid phenotypes with desired characteristics. The viability of interbred *sativa* and *indica* cannabis types supports the cataloging of all varieties as subspecies of *C. sativa* (i.e., *C. sativa*, *C. sativa indica*, and *Cannabis sativa ruderalis*) (Anderson, 1980; Schultes et al., 1974). To simplify the text, the monotypic subspecies nomenclature is used throughout this chapter.

European approved hemp cultivars and medical *Cannabis sativa*

Unlike medical cannabis abundant in the psychoactive ingredient Δ^9-THC, modern hemp has been selectively bred to produce low levels of Δ^9-THC and high levels of fiber, seed, and, more recently, cannabidiol (CBD). Most of the European Union and Canada have recognized the value of hemp and have defined a legal limit of 0.3% Δ^9-THC in the dry plant material (Small and Marcus, 2003), with the exception of Italy (0.2% Δ^9-THC) (Cappelletto et al., 2001). Many hemp cultivars are rich in cellulose

Nutraceuticals. DOI: https://doi.org/10.1016/B978-0-12-821038-3.00038-0

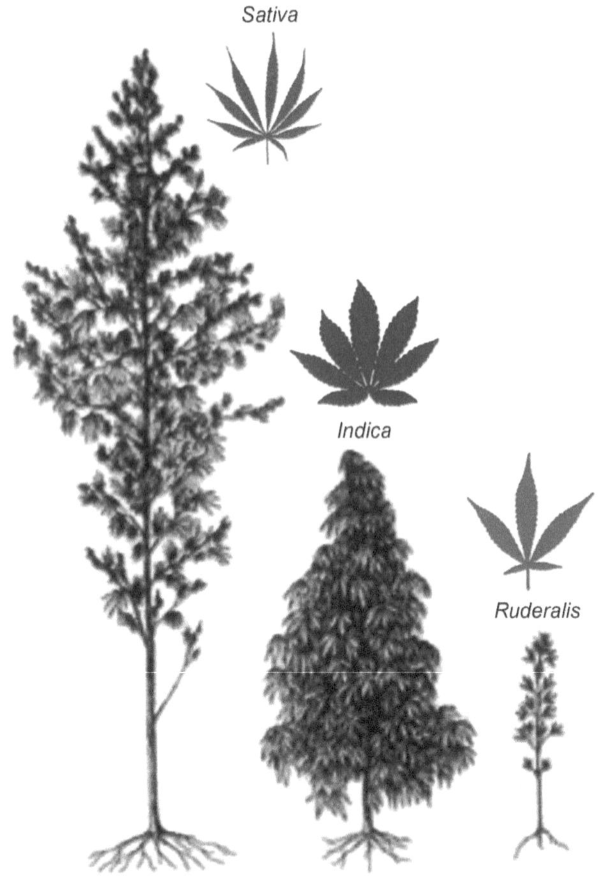

FIGURE 38.1 Subspecies of *Cannabis sativa* include *Cannabis sativa*, *Cannabis sativa indica*, and *Cannabis sativa ruderalis*. *wikicommons; public domain.*

used for biofuel, primary fibers for pulping, fine fiber textiles, dense flowering varieties that offer abundant seed, or high CBD-to-Δ^9-THC ratios. Hemp varieties resistant to drought, high soil salinity, cold, heat, humidity, and common pests and diseases have been selected to improve yields (Watson and Clarke, 2014). To date, 70 hemp cultivars have been approved for commercial use by the European Union (EU plant variety database, https://ec.europa.eu/food/plant/plant_propagation_material/plant_variety_catalogues_databases/search/public/index.cfm?event = SearchVariety&ctl_type = A&species_id = 240&variety_name = &listed_in = 0&show_current = on&show_deleted = on). These registered varieties originate from high-latitude European nations, which are not acclimated for equatorial regions of the world. Little is known about the viability of the European cultivars in the United States, and it is currently being examined by agronomists.

Chemotaxonomy of *Cannabis sativa*

Chemical phenotypes (chemotypes) can be useful to classify *C. sativa* as drug- or fiber-type varieties. The United Nations Office on Drugs and Crime categorizes *C. sativa* into three chemotypes based on the proportion of THC and CBN relative to CBD (Eq. (38.1): classification of cannabis by chemotype) (Drugs, 2009). Chemotype I (drug-type) cultivars are characterized by X values greater than 1, whereas values lower than 1 are representative of chemotype III cultivars (fiber-type).

$$X = \frac{[\mathrm{THC}] + [\mathrm{CBN}]}{[\mathrm{CBD}]} \tag{38.1}$$

The evidence suggests that drug-type or chemotype I *C. sativa* cultivars with high levels of Δ^9-THC originated from below the 30°N latitude (Hillig and Mahlberg, 2004). CBD-rich cultivars containing low levels of THC are regarded as fiber-type or chemotype III cultivars. Chemotype II cultivars are characterized by equivalent levels of Δ^9-THC and CBD and, along with chemotype III cultivars, are typically found above the 30°N latitude (Hillig and Mahlberg, 2004). Hillig and Mahlberg (2004) used a statistical approach to define chemotaxonomic trends in *C. sativa* and noticed that most cultivars did not fall within the arbitrary values set by the United Nations Office on Drugs and Crime. Instead, most cultivars clustered into chemotype I ($X > 10$), chemotype II ($0.2 < X < 10$), or chemotype III ($X < 0.2$). The relative cannabinoid levels in *C. sativa* remain constant from the seedling stage throughout the plant lifecycle (Broséus et al., 2010), making it possible to determine the chemotype at early development stages prior to flowering (Barni-Comparini et al., 1984; Vogelmann et al., 1988).

Hemp and drug-type cultivars are members of both *C. sativa* and *C. sativa indica* subspecies. Chemotaxonomic analysis is useful to differentiate hemp from drug-type *C. sativa* based on acceptable levels of Δ^9-THC established by regulating bodies. Cannabinoid detection methods have not been useful to distinguish *C. sativa* from *C. sativa indica*. Incremental progress has been made distinguishing the subspecies using terpene fingerprinting techniques, but has not provided the practical level of significance for routine application (Hillig, 2004). In addition, the draft genome of *C. sativa* has been published, opening new opportunities to identify genetic markers unique to *C. sativa* subspecies (van Bakel et al., 2011).

Potency trends

Historically, *C. sativa indica* originating from Southeast Asia has been known for its high levels of Δ^9-THC. Until recently, it was unclear whether the high Δ^9-THC cultivars were present prior to human intervention. Genetic evidence suggests that the high B_T allele frequency, responsible for Δ^9-THC production, was present in *C. sativa indica* prior to human domestication (Hillig and Mahlberg, 2004). These data indicate that natural evolution played a prominent role in *C. sativa indica* cultivars with high levels of Δ^9-THC and predated selective breeding practices. Within recent decades breeders have brought the Δ^9-THC content to unprecedented levels not observed for a single component in a plant species. A potency monitoring report put forth in collaboration between The University of Mississippi and the National Institute on Drug Abuse (NIDA) identified levels of up to 37% Δ^9-THC in 45,603 cannabis inflorescences confiscated internationally from 1993 to 2008 (Mehmedic et al., 2010). The High Times Cannabis Cup is a competitive event for breeders to showcase their superlative cannabis products. The organization submits competitive strains for independent testing and reports that the highest strains are approximately 25% (Hellerman, 2013). The high Δ^9-THC levels achieved over recent decades have largely been attributed to selective breeding, indoor cultivation, and worldwide access to potent seed varieties (Brenneisen, 2007). From 1980 to 2008, the average Δ^9-THC content in confiscated cannabis increased from 1.5% to 8.8% and was the direct result of human intervention (ElSohly et al., 2000; Mehmedic et al., 2010).

Cannabinoid levels other than Δ^9-THC have remained relatively constant from 1993 to 2008. For example, average CBD levels went from 0.2% in 1993 to 0.4% in 2008 (Mehmedic et al., 2010). However, high-CBD medical cultivars with up to 18% CBD have recently been developed. High CBD cultivars include Harlequin, ACDC, Cannatonic, and an Israeli strain, Avidekel. Despite the high CBD levels in these cultivars, they fail to meet the 0.3% Δ^9-THC threshold to be defined as hemp. Agricultural hemp cultivars producing up to 5% CBD have been described in the literature. However, the high-CBD hemp varieties also tested above the legal limit of 0.3% Δ^9-THC and could not be cultivated as agricultural hemp (Virovets, 1996). An analysis of more than 100 Italian fiber-type hemp samples indicates that CBD values range from 0.07% to 2.3%, whereas the Δ^9-THC values ranged from 0.01% to 0.44% (Rustichelli et al., 1998). These data suggest that the 5% CBD hemp cultivars were outliers and selected for high-CBD cultivar breeding projects. Alternatively, other hemp cultivars have been developed that produce little to no detectable cannabinoids (Virovets, 1996).

Cannabis sativa chemistry

C. sativa produces more than 750 natural compounds of different chemical classes (Upton et al., 2014). Cannabinoids are grouped in one class composed of over 100 terpenophenolic secondary metabolites identified in *C. sativa* (Pertwee, 2014). Another important class of compounds produced by *C. sativa* includes over 200 terpenes that have been identified to date (Russo, 2011). In addition to the fragrance that the terpenes contribute, there is also evidence to suggest that they are a critical component in modulating the strain-specific physiological effects (Russo, 2011). Other compound classes include 50 identified hydrocarbons, 34 sugars and related compounds, 27 nitrogenous compounds, 25 noncannabinoid phenols, 23 fatty acids, 23 flavonoids, 20 simple acids, 13 simple ketones, 13 simple esters and lactones, 12 simple aldehydes, 11 proteins, 11 steroids, 9 elements, 3 vitamins, and 2 pigments that are summarized in Table 38.1 (Brenneisen, 2007; Callaway, 2004; ElSohly and Slade, 2005; Turner et al., 1980).

It is worth noting that no free flavonoids have been identified in *C. sativa*, but they are found in their corresponding glycosides (Turner et al., 1980). Covering each component of *C. sativa* is beyond the scope of this chapter. However, an excellent review of *C. sativa* chemistry can be found in the body of work compiled by ElSohly and coworkers (Brenneisen, 2007; Callaway, 2004; ElSohly and Slade, 2005; Turner et al., 1980).

Cannabinoids

Of the 113 cannabinoids (Hartsel et al., 2019) identified in *C. sativa*, the majority can be categorized as analogs of

TABLE 38.1 Chemical constituents identified in *Cannabis sativa* by chemical class.

Chemical class	Identified compounds
Terpenes	140
Cannabinoids	86
Hydrocarbons	50
Sugars and related compounds	34
Nitrogenous compounds	27
Noncannabinoid phenols	25
Fatty acids	23
Flavonoids	23
Simple acids	20
Simple ketones	13
Simple esters and lactones	13
Simple aldehydes	12
Proteins, enzymes, and glycoproteins	11
Steroids	11
Elements	9
Simple alcohols	7
Vitamins	3
Pigments	2

Δ^9-THC, CBD, cannabichromene (CBC), cannabigerol (CBG), cannabinol (CBN), cannabicyclol (CBL), cannabielsoin (CBE), and cannabitriol (CBT). There are seven known analogs of CBD in *C. sativa*, and Fig. 38.2 illustrates common structural trends that are found within cannabinoid classes (ElSohly and Slade, 2005). Although the general public is most familiar with neutral cannabinoids such as CBD and Δ^9-THC, cannabinoids exist in an acid form, meaning they contain a carboxylic acid pendant to the aromatic ring [e.g., cannabidiolic acid (CBDA), R = H, with pendant COOH]. The length of the alkyl side chain is a common site of structural variability and can range from one to five carbons. The most abundant cannabinoids in *C. sativa* contain *n*-pentyl side chains, but *n*-butyl, *n*-propyl, ethyl, and methyl side chains also have been identified in lower abundances (ElSohly and Slade, 2005; Turner et al., 1980). Cannabinoids bearing an *n*-propyl side chain are referred to as cannabivarins, indicated by the suffix -varin. Tetrahydrocannabivarin (THCV), the *n*-propyl side chain analog of THC, is often found in samples collected from *C. sativa indica* subspecies (Hillig and Mahlberg, 2004). Although not common, the phenol portion of the cannabinoid can be protected as monomethyl ether, as is observed in analogs of CBD (R = H, Fig. 38.2). In addition, there are 30 miscellaneous cannabinoids present in low concentrations, but they will be excluded from this text based on the limited understanding of their therapeutic potential.

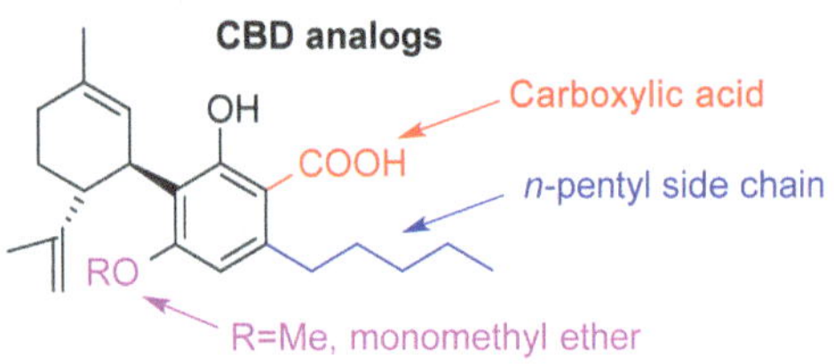

FIGURE 38.2 Typical cannabinoid analogs (e.g., CBD, R = H) differ with respect to the carboxylic acid, alkyl side chain, or methyl ether group. *CBD*, cannabidiol.

It is important to note that *C. sativa* does not produce Δ^9-THC, CBD, CGG, or CBC, but rather carboxylic acid-containing precursors referred to as Δ^9-tetrahydrocannabinoic acid (Δ^9-THCA), CBDA, cannabigerolic acid (CBGA), and cannabichromenic acid (CBCA), respectively. *C. sativa* extracts are nonpsychoactive until sufficient heat is supplied to cause a chemical reaction known as decarboxylation. Decarboxylation of cannabinoic acids occurs rapidly above 105°C, which is a condition obtained during the smoking or baking process (Fig. 38.3; Veress et al., 1990). Decarboxylation occurs slowly under ambient conditions, but the rate increases with temperature. High levels of decarboxylated cannabinoids in flowers can indicate that a sample has been stored improperly or is aging. Δ^9-THC can undergo oxidation to CBN and is also a chemical indicator of poor or lengthy storage conditions. Carbon dioxide is released during the decarboxylation process.

C. sativa produces four types of cannabinoic acids, CBGA-, CBDA-, Δ^9-THCA-, and CBCA-type, through the biosynthetic pathways shown in Fig. 38.4 (Fellermeier et al., 2001; Sirikantaramas et al., 2007; Taura et al., 2007). CBGA acts as substrate for *C. sativa*-specific oxidoreductases CBDA (Taura et al., 1996), THCA (Shoyama et al., 2005; Sirikantaramas et al., 2004), or CBCA synthase (Morimoto et al., 1997) and is the convergent intermediate in cannabinoid production. Prior to the discovery of THCA synthase, Δ^9-THC and its acid form were believed to be produced through a ring closure of CBD or CBDA, respectively (Shoyama et al., 1975). It was later shown that Δ^9-THCA was produced through independent enzyme-catalyzed conversion of CBGA, and then decarboxylated to form Δ^9-THC (Shoyama et al., 2005; Sirikantaramas et al., 2004).

The CBGA carboxyl group is critical for cannabinoid synthase substrate recognition. CBG, without the pendant carboxyl group, did not undergo enzyme catalysis through the cannabinoid synthase pathways (Taura et al., 2007). Both Δ^9-THCA and CBDA are produced with greater than 95%

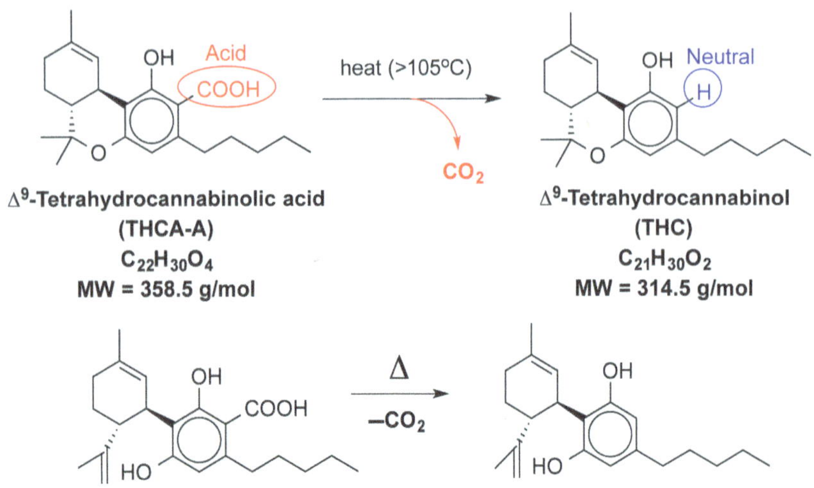

FIGURE 38.3 Conditions required to decarboxylate native cannabinoic acids to their neutral form.

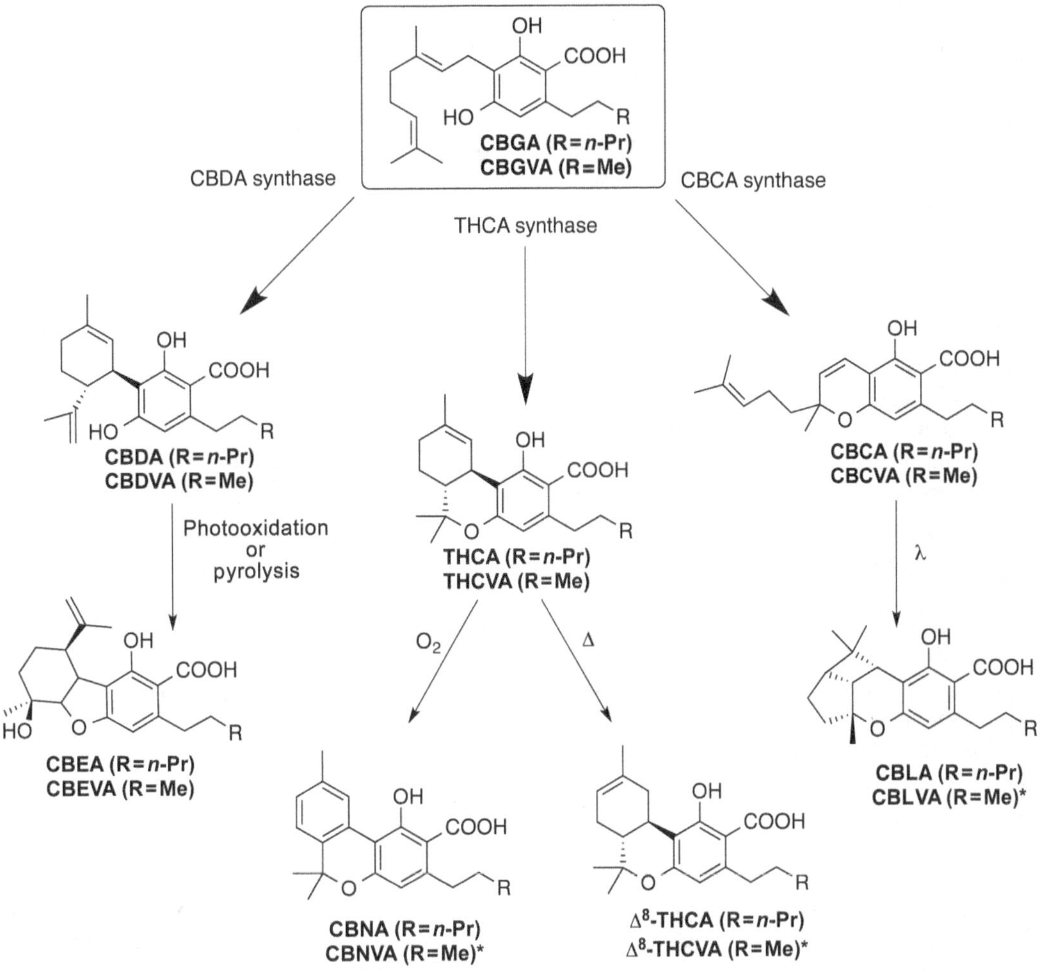

FIGURE 38.4 Biosynthesis of CBDA, THCA, and CBCA from CBGA through the cannabinoid synthase pathways. Degradation of the CBD-, THC-, and CBC-type cannabinoids leads to CBE-, CBN-, THC-, and CBL-type cannabinoids. *This compound was not identified in cannabis. *CBCA*, cannabichromenic acid; *CBDA*, cannabidiolic acid; *CBGA*, cannabigerolic acid; *CBGVA*, cannabigerovarinic acid; *THCA*, Δ9-tetrahydrocannabinoic acid.

optical purity. CBCA, however, is a 5:1 ratio of enantiomers (Morimoto et al., 1997). The evidence suggests that CBCA synthase releases the substrate before the reaction is complete, resulting in lower enantiomeric purity compared to Δ^9-THCA and CBDA (Taura et al., 2007). THCA synthase is FAD- and oxygen-dependent and generates hydrogen peroxide as a by-product. The release of hydrogen peroxide could act as a component of the plant disease defense mechanism (Taura et al., 2007). In contrast to THCA synthase, CBDA synthase does not require molecular oxygen for the conversion of CBGA to CBDA. A comparison of THCA synthase with other plant enzymes revealed that it shares high homology with the alkaloid-producing berberine-bridge enzyme from *Eschscholzia californica*, or California poppy (Dittrich and Kutchan, 1991; Taura et al., 2007).

Cannabinoids other than Δ^9-THCA, CBDA, and CBCA (primary cannabinoids) are generated through non-enzymatic degradation pathways. The primary cannabinoids either can be decarboxylated to the neutral form or converted to CBEs, CBNs, Δ^8-THCs, or CBLs via exposure to light, heat, and oxygen as shown in Fig. 38.4 (ElSohly and Slade, 2005; Turner et al., 1980). CBDs can undergo photo-oxidation or pyrolysis to CBEs. Alternatively, THCs can be degraded to CBNs in the presence of oxygen, or to more thermodynamically stable Δ^8-THCs when exposed to heat. CBCs are converted into CBL-type cannabinoids in the presence of light. The shorter side-chain cannabinoid analogs, such as three-carbon-containing cannabivarins, are generated through identical cannabinoid synthase pathways from cannabigerovarinic acid (CBGVA), the homologous propyl-containing CBGA precursor (Shoyama et al., 1984).

Cannabinoids from non-*Cannabis sativa* species

As a complement to *C. sativa*, South African *Helichrysum umbraculigerum* has been shown to contain CBGA-like,

CBG-like compounds, and other prenylated dibenzyls similar to cannabinoids in *C. sativa*. The cannabinoid-like molecules derived from *Helichrysum* and other *Asteracea* genera generally have a phenethyl group pendant at the *n*-pentyl region of cannabinoids and have been used in traditional medicine to treat a host of inflammations and infections (Appendino et al., 2011; Pollastro et al., 2017). *Echinacea* (Woelkart et al., 2005) and liverworts (Appendino et al., 2011) are known to produce cannabinoid-like molecules as well. In particular, *Echinacea* roots produce endocannabinoid-like molecules that have been shown to bind to cannabinoid receptors in rodents (Woelkart et al., 2005).

Cannabinoid-like compounds can arise from nonplant sources as well. It has long been understood that omega fatty acids have antiinflammatory properties. One of the mechanisms appears to be by interacting with the ECS. Ingested EPA and DHA found in fish and marine-based oils are converted to omega-3 endocannabinoid epoxides via crosstalk between endocannabinoid and cytochrome P450 epoxygenase metabolic pathways. These omega-3 polyunsaturated fatty acid lipid metabolites are found at concentrations similar to other endocannabinoids in research models and are expected to play a critical role in regulating inflammation (McDougle et al., 2017).

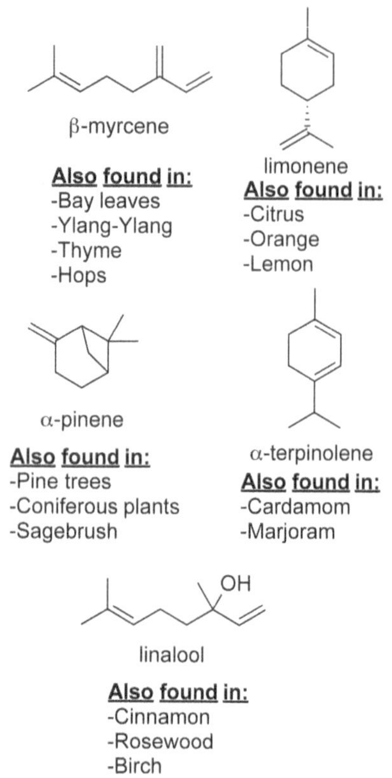

FIGURE 38.5 Common monoterpenes found in cannabis and other plants.

Cannabis sativa terpenes

The combination of more than 200 terpenes identified in *C. sativa* contributes to its unique aroma (Brenneisen, 2007). The pungent inflorescences can be described as skunky, fruity, or piney, among others, depending on the aromatic components present. Terpenes are a group of volatile acyclic, monocyclic, and polycyclic hydrocarbons that contribute to the complex aroma profile of *C. sativa* and many other terpene-containing plants (Fig. 38.5; Brenneisen, 2007).

The volatility of terpenes imparts aroma but these molecules evaporate upon standing. In one study, approximately 55% of the volatile terpene constituents were lost after being stored 3 months under drying conditions (Ross and ElSohly, 1996). Terpenes isolated through steam distillation from indoor drug-type *C. sativa* constituted a low percentage (0.29% w/w) of the total predried biomass. The main fraction of the distillate comprised monoterpenes (92% of total terpenes), with a small amount of sesquiterpenes (7%) and other simple ketones and esters (Ahmed et al., 2008; Ross and ElSohly, 1996). Of the monoterpenes present in the fresh flowers, myrcene (67%) was the most abundant component, followed by D-limonene (16%) and linalool (3%). Monoterpene content in *C. sativa* cultivated outdoors ranged from 48% to 92% of total terpenes present (Mediavilla and Steinemann, 1997). The sesquiterpene content was lower, ranging from 5% to 49%. Higher levels of monoterpenes α-pinene, *trans*-ocimene, and α-terpinolene were observed in *C. sativa* cultivated outdoors (Mediavilla and Steinemann, 1997) compared to indoor varieties (Ross and ElSohly, 1996). α-Terpinolene and α-pinene were the most abundant terpenes found in European cultivars, which contained significantly higher amounts of α-humulene (6%–9% of total terpenes) compared to the varieties cultivated in the United States (Novak et al., 2001). Sabinene, α-terpinene, eucalyptol, pulegone, γ-terpinene, terpineol-4-ol, bornyl acetate, α-copaene, alloaromadendrene, viridiflorene, α-bisabolene, γ-cadinene, trans-nerolidol, and β-bisabolol were terpenes found in trace amounts (McPartland and Russo, 2001; Mediavilla and Steinemann, 1997; Ross and ElSohly, 1996). Characteristic levels of particular terpenes have been linked to geographic origin (Brenneisen and ElSohly, 1988).

The terpene composition of 10 hemp varieties was investigated over two growing seasons and ranged from 0.11% to 0.25% (w/w) of the fresh plant material. The most abundant monoterpenes present were myrcene (8%–45%), terpinolene (0.12%–22%), α-pinene (3%–20%), *E*-ocimene (1%–10%), and β-pinene (1%–8%). The main sesquiterpenes present were β-caryophyllene (7%–28%), α-humulene (3%–12%), and caryophyllene oxide (0.12%–22%). The dioecious cultivars tended to have higher myrcene and terpinolene contents, whereas high α-pinene and β-pinene contents tended to be more characteristic of the monoecious cultivars (Bertoli et al., 2010). The publicly available literature on terpene composition in *C. sativa* was compiled for drug- and fiber-type varieties, which are summarized in Table 38.2 (Bertoli et al., 2010; Fischedick et al., 2010; Mediavilla and Steinemann, 1997; Novak et al., 2001; Ross and ElSohly, 1996).

Growing techniques, environmental factors, and genetics contribute to strain-specific aromas. Although the olfactory response is subjective, high sesquiterpene content was considered by test subjects to have a poor aroma compared to "pleasant" cultivars high in monoterpene content (Mediavilla and Steinemann, 1997). The unique skunk smell emitted by *C. sativa* may not be a traditional terpene. For example, skunks produce a group of nonterpene thiols that contribute the pungent aroma (Wood et al., 2002). It is possible that the skunk smell in *C. sativa* is also a thiol that does not fit within the terpene class; the constituent responsible for the skunky aroma of cannabis remains to be elucidated.

Phytocannabinoid and terpene biochemistry

Terpenes, phytocannabinoids, flavonoids, lignans, coumarins, polyketides, and lipids are produced through shared biosynthetic pathways in *C. sativa* (Fig. 38.6). Phosphoenolpyruvate is a key intermediate in the biosynthesis of many plant compounds and is produced through glycolysis. One pathway exploited by plants is the conversion of phosphoenolpyruvate to shikimic acid. Shikimic acid is the precursor to flavonoids, lignans, and coumarins. Alternatively, phosphoenolpyruvate can be catabolized to acetyl coenzyme A, which is a building block for botanical lipids and polyketides.

Cannabis has specially adapted type III polyketide synthase and olivetolic acid cyclase enzymes to convert acetyl coenzyme A to olivetolic acid (Gagne et al., 2012). Acetyl coenzyme A is also consumed in a competing biosynthetic pathway leading to geranyl diphosphate. Geranyl diphosphate and olivetolic acid are both downstream products derived from acetyl coenzyme A. Interestingly, geranyl diphosphate and olivetolic acid can be enzymatically coupled by olivetolate geranyl transferase in CBGA biosynthesis (Marks et al., 2009). Geranyl diphosphate also acts as a substrate in monoterpene (Taura et al., 2007) and sesquiterpene (Sell, 2009) biosynthesis. The abundance of each cannabinoid and terpene is modulated by the relative expression and efficiency of the competing enzymatic pathways. Terpenes are produced in low levels compared to cannabinoids in *C. sativa*.

Trichomes

The biosynthesis of terpenes and cannabinoids occurs within the extracellular secretory cavity, referred to as a

TABLE 38.2 **Terpene composition in freshly harvested cannabis inflorescences by percent of total steam-distillate I drug-type and fiber-type cannabis.**

Monoterpenes				
Compound	Drug type (%)	Average ± SD (*N*)	Fiber type (%)	Average ± SEM (*N*)
β-Myrcene	0.4–67	21.9 ± 5.1 (17)	8.2–66	30.7 ± 2.8 (39)
α-Pinene	ND–31.0	5.2 ± 1.9 (17)	2.3–20	8.4 ± 0.7 (39)
Terpineolene	ND–15.3	3.2 ± 1.1 (17)	tr–24	10.1 ± 1.0 (39)
trans-Ocimene	0.3–5.9	3.0 ± 1.2 (5)	0.4–10.3	4.9 ± 0.5 (39)
D-Limonene	ND–16.4	2.7 ± 1.0 (17)	0–6.4	1.9 ± 0.3 (39)
β-Pinene	tr–7.8	1.7 ± 0.5 (17)	0.6–8.0	3.1 ± 0.3 (39)
cis-Ocimene	ND–3.9	0.6 ± 0.2 (17)	0–1.0	0.3 ± 0.0 (39)
Linalool	ND–2.8	0.4 ± 0.2 (12)	ND–2.7	0.5 ± 0.1 (20)
Sesquiterpenes				
β-Caryophyllene	0.5–31	5.8 ± 2.1 (17)	7.3–37.5	19.0 ± 1.0 (39)
Humulene	ND–1.2	1.0 ± 0.3 (17)	1.3–12.6	5.5 ± 0.5 (39)
Caryophyllene oxide	tr	0.4 ± 0.2 (6)	0.3–11.3	2.6 ± 0.3 (39)
β-Eudesmol	ND–0.4	0.04 ± 0.03 (12)	ND–1.6	0.2 ± 0.1 (20)
trans-Nerolidol	tr	*N* = 1	0.1–1.7	0.9 ± 0.1 (20)

ND, not detected (<LOD); *tr*, trace levels detected (<LOQ). Drug type: Ross and ElSohly (1996), Fischedick et al. (2010), Mediavilla and Steinemann (1997). Hemp varieties: Bertoli et al. (2010); Mediavilla and Steinemann (1997); Novak et al. (2001).

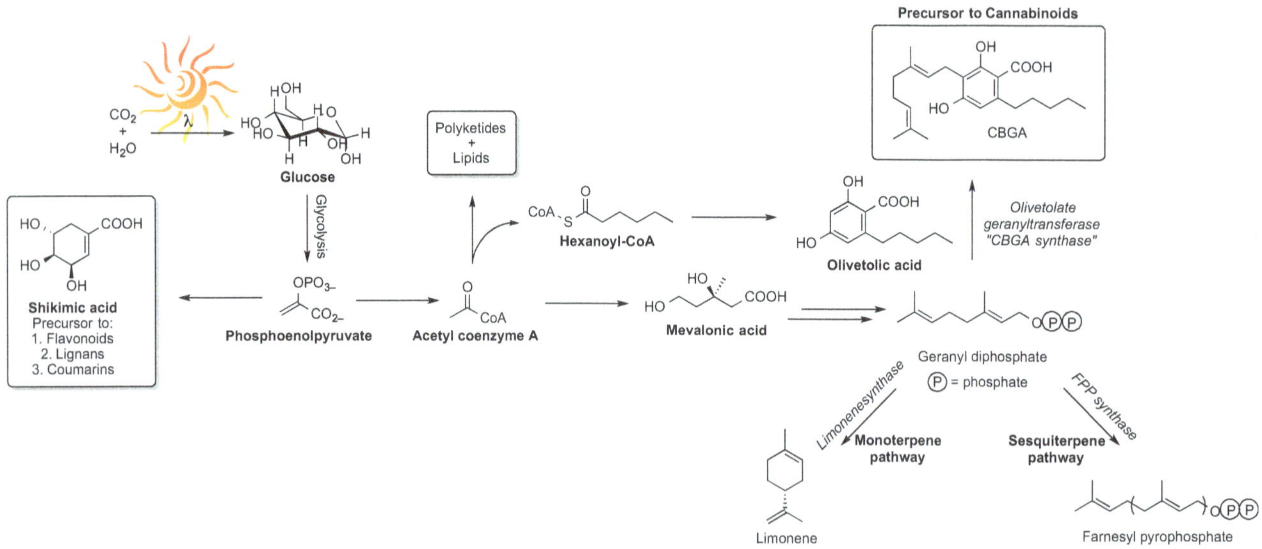

FIGURE 38.6 Plants use sunlight, carbon dioxide, and water to generate many useful botanical ingredients. Terpenes and cannabinoids are generated biosynthetically through convergent intermediate geranyl diphosphate. Geranyl diphosphate can be diverted through several competing enzymatic pathways to produce a variety of useful botanical compounds including CBGA, monoterpenes, and sesquiterpenes. *CBCVA*, cannabigerovarinic acid; *CBDVA*, cannabidivarinic acid; *CBEA*, cannabielsoic acid; *CBEVA*, cannabivarinselsoinic acid; *CBGA*, cannabigerolic acid; *CBLA*, cannabicyclolic acid; *CBLVA*, cannabicyclolvarinic acid; *CBNA*, cannabinolic acid; *CBNVA*, cannabivarinic acid; *FPP*, farnesyl pyrophosphate; *THCVA*; tetrahydrocannabivarinic acid.

trichome. Trichomes are present in more than 300 plants (Dayanandan and Kaufman, 1976), and over 300 morphologically distinct trichomes have been reported (Wagner, 1991). The TrichOME project (www.planttrichome.org) is a database of EST sequences, microarray hybridizations, mass spectrometry-based metabolite profiles, and literature compiled with funding from the NSF and the Samuel Roberts Nobel Foundation. Trichomes serve several functions in plants, including keeping the frost from the surface cells, reducing transpiration, and increasing light reflectance in desert species. Specialized glandular trichomes produce and emit compounds that interact with insects, deterring pests or promoting pollination (Wagner, 1991). The trichomes act as a first line of defense to protect the plant from the environment (Taura et al., 2007). Interestingly, some of the very cannabinoids (THCA and CBGA) produced within the *C. sativa* trichome have been found to be toxic to *C. sativa* cells (Sirikantaramas et al., 2005). Cannabinoids are secreted into the trichome to prevent cellular damage to the plant. THCA and CBGA have been shown to be toxic to certain insect species as well (Taura et al., 2007).

The term trichome is derived from the Greek word meaning hair, because of the hair-like appearance. Trichomes found on *C. sativa* have been divided into two types, glandular and nonglandular. Nonglandular *C. sativa* trichomes cover the shoot from early seedling stages to the end of flowering and are highly silicified unicellular hair-like appendages (Dayanandan and Kaufman, 1976). Cystolithic nonglandular trichomes are short hairs ranging from 150 to 220 μm in height and containing basal deposits of calcium carbonate. In contrast, noncystolithic and nonglandular are generally longer (340–500 μm), with a more slender structure, and do not contain a basal deposit of calcium carbonate (Upton et al., 2014). The nonglandular trichomes cover the majority of the plant's surface, including the stems, petioles, stipules, leaf blades, bracts, and both surfaces of the tepals (Dayanandan and Kaufman, 1976).

The cannabinoids are produced in two types of glandular trichomes, capitate-stalked and capitate-sessile. The capitate-stalked glandular trichome heads are composed of eight cells and reside on the bracts of the female plants and anthers of the male plants. They are easily recognized by the mushroom-like stalk and head appearance (Fig. 38.7C; Dayanandan and Kaufman, 1976). Capitate-stalked trichomes are the primary source of cannabinoid and terpene resin production in flowering *C. sativa*. Dayanandan and Kaufman (1976) have shown that male *C. sativa* plants produce equal quantities of glandular trichomes compared to females and bear similar trichome types. It has also been demonstrated that some resinous cannabinoid-producing glands develop prior to flower formation. For example, capitate-sessile trichomes produce cannabinoids throughout the plant life cycle, but at lower levels compared to the capitate-stalked type. Capitate-sessile trichome heads are composed of between two and four cells. In contrast, the bulbous glandular trichomes are found throughout the plant and do not produce cannabinoids or terpenes (Fig. 38.7B). The head of the trichome is composed of one, two, or four cells with stalks either one or two cells high.

The cannabinoid precursors, olivetolic acid and geranyl diphosphate, are transported through the stalk of

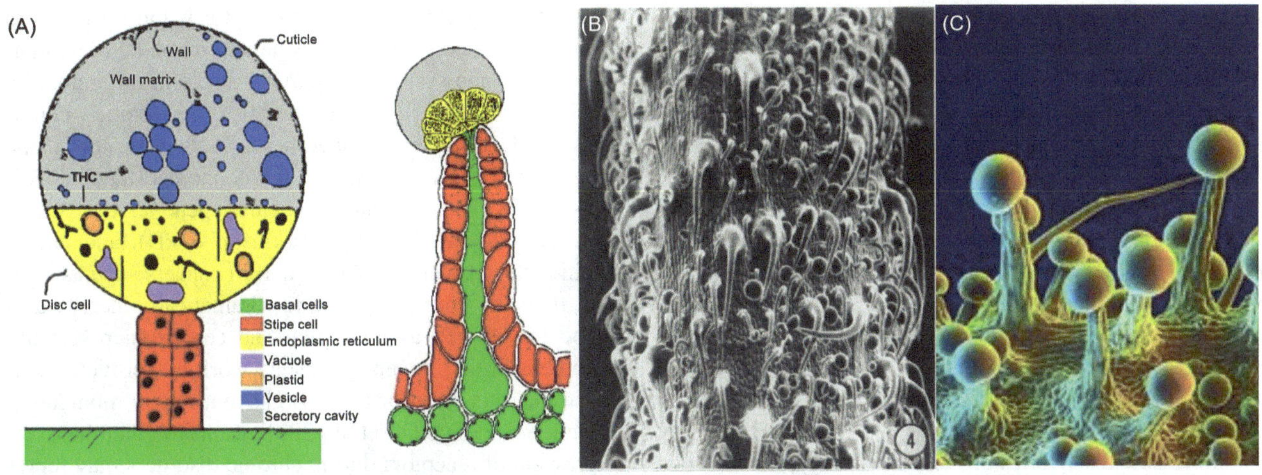

FIGURE 38.7 Structure of glandular and cystolith hair trichomes. (A) A cartoon depicts the components of the glandular capitate-stalked trichome. (B) Scanning electron microscope image of alternative bulbous and capitate-sessile glandular, and cystolithic hair-type trichomes. (C) Scanning electron microscope image of the glandular capitate-stalked trichomes. *(A) Photo credit: http://medicalmarijuana.com/experts/expert/title.cfm?artID = 140; (C) Photo credit: http://medicalmarijuana.com/experts/expert/title.cfm?artID = 140.*

glandular trichomes via vacuoles and plastids, respectively (Fig. 38.7A). The cannabinoid precursors are released into the secretory cavity, where they are transformed biosynthetically through resident cannabinoid synthase enzymes in the extracellular space (shown in Fig. 38.4). Monoterpenes are formed in the plastids via the enzymatic conversion of geranyl diphosphate. Alternatively, geranyl diphosphate can couple with isopentenyl diphosphate in the cytoplasm to form the parent compound to the sesquiterpenoids, farnesyl pyrophosphate (Russo, 2011). THCA synthase fused to green fluorescent protein helped visually demonstrate that this enzyme is accumulated in the secretory cavity (Sirikantaramas et al., 2005) located in the head of glandular capitate-stalked and capitate-sessile trichomes (Fig. 38.7C).

The pharmacology of cannabis

The therapeutic properties of cannabis have been appreciated over several millennia and its pharmacological properties have been studied since the mid-1900s. This effort was enhanced in the 1960s with the isolation, characterization, and subsequent synthesis of its major psychoactive ingredient (−)-Δ^9-THC, and much attention was given to developing synthetic compounds with more potent and targeted therapeutic effects. Concurrent approaches by individual research laboratories and the pharmaceutical industry produced a plethora of structurally related new compounds, which were collectively called cannabinoids. One of these drug development projects by Eli Lilly led to the first synthetic cannabinoid, Nabilone (Cesamet), which was used to treat nausea, pain, and reduced appetite associated with cancer chemotherapy (Makriyannis, 2014; Pertwee, 2008). Notwithstanding these major efforts by the industry, the enthusiasm to develop additional medications waned principally because of lack of knowledge of their pharmacological mechanism of action. There was also accumulating evidence that the therapeutic effects of smoked cannabis in certain instances provided superior relief when compared to the synthetic compounds.

The major breakthrough in understanding the mechanism of action of cannabis, and more specifically Δ^9-THC, came with the discovery of the first biological targets for this compound. Following a very successful research meeting in 1987 by the NIDA in Bethesda, Maryland (Makriyannis, 2014), collaborative work among the participating laboratories led to the discovery of a new G-protein-coupled receptor (GPCR) that was named CB1, to be followed 2 years later by a second THC-related GPCR named CB2. The next discovery was a group of compounds found in mammalian tissues, which were named endocannabinoids, the most representative of which were arachidonyl ethanolamide or anandamide (AEA), a long-chain fatty acid amide and 2-arachidonoyl glycerol (2AG) (Fig. 38.8), the respective ester (Pertwee, 2000). These new substances are capable of activating

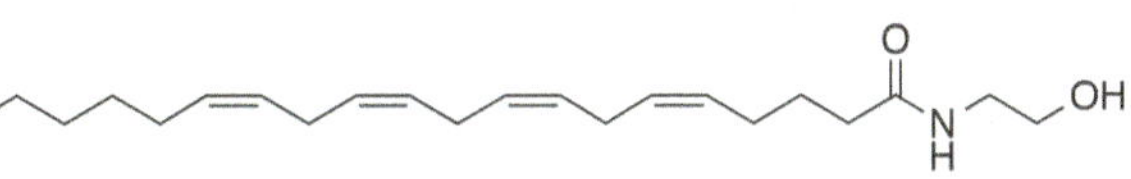

Anandamide (AEA)

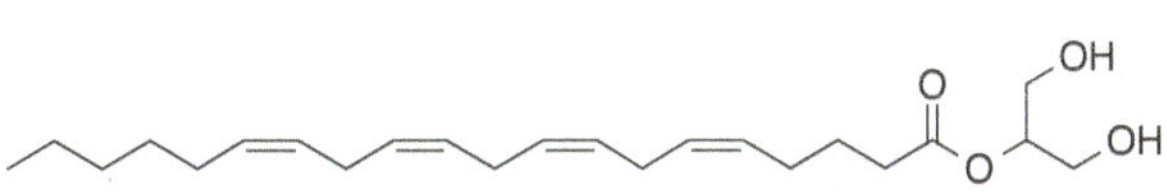

2-arachidonoyl glycerol (2AG)

FIGURE 38.8 Naturally occurring endocannabinoids that bind to the endogenous endocannabinoid receptors in mammals.

both CB1 and CB2 receptors and, when tested in animals, produce biological effects paralleling those of Δ^9-THC.

Endocannabinoids are produced on demand by a series of enzymes that are present within the cell membrane and are activated by elevated levels of calcium ions. Endocannabinoid levels also referred to as the "endocannabinoid tone" are tissue-dependent and regulated by another set of enzymes, the most prominent of which are fatty acid amide hydrolase (FAAH) and monoacylglycerol lipase, whose role is to deactivate AEA and 2AG and their congeners, respectively. An additional feature in the regulation of endocannabinoid tone is a transport mechanism that is involved in transporting released endocannabinoids into the cell (Fig. 38.9; Vemuri and Makriyannis, 2015).

These new discoveries opened the door for characterizing the endocannabinoid system (ECS) that encompasses two cannabinoid receptors, the proteins that modulate their function, and the endocannabinoid family of compounds. The ECS plays a major role in the regulation of many aspects of human physiology. It is well established by now that the CB1 cannabinoid receptor is the most abundant GPCR in the brain, but it is also present in many other organs such as the heart, blood vessels, liver, lungs, and the digestive system, as well as fat and sperm cells. However, the CB2 cannabinoid receptor is mostly found in immunity-related cells, and it is largely involved in regulating the immune system and playing a role in inflammatory conditions. Although under normal "homeostatic" conditions the CB2 receptors have a very low presence in the central nervous system (CNS), in "nonhomeostatic" conditions such as inflammation, neurodegenerative diseases such as Alzheimer's (AD), Parkinson's (PD), and amyotrophic lateral sclerosis (ALS), and in cancers such as gliomas, their presence in the brain is dramatically increased in astrocytes as well as microglial and cerebromicrovascular endothelial cells. Recent published work has provided evidence that this receptor may also have a role in modulating addictive disorders (Xi et al., 2011). Consider that the overexpression of receptors during chronic disorders may represent a protective mechanism in that the body is attempting to balance itself.

It has been shown that many, although not all, of the effects of cannabis can be linked to the ubiquitous nature of the endocannabinoid system. This can explain the multitude of physiological and potentially therapeutic effects associated with the use of cannabis as medicine (Hill et al., 2012; Izzo et al., 2009; McPartland et al., 2015; Pertwee, 2008; Pertwee et al., 2007).

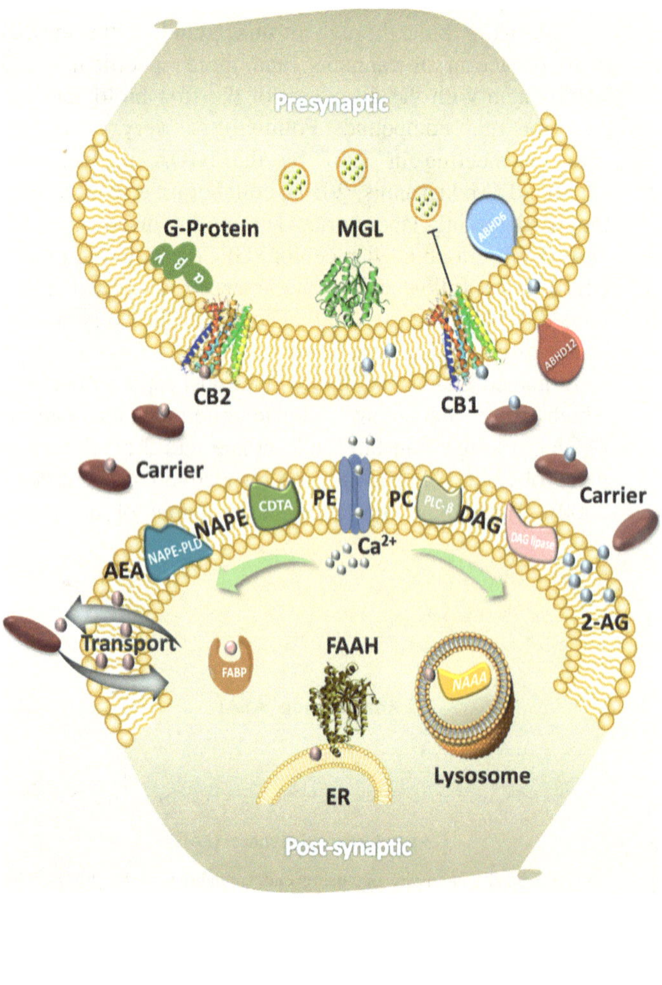

FIGURE 38.9 Schematic representation of the endocannabinoid signaling system. *CB1*, cannabinoid receptor 1; *CB2*, cannabinoid receptor 2; *FAAH*, fatty acid amide hydrolase; *MGL*, monoacylglycerol lipase; *ABHD6*, α-β-hydrolase domain-containing protein 6; *ABHD12*, α-β-hydrolase domain-containing protein 12; *NAPE*, N-arachidonoyl phosphatidylethanolamine; *PE*, phosphatidylethanolamine; *PC*, phospholipase C; *PD*, phospholipase D; *DGL*, diacylglycerol lipase; *FABP*, fatty acid–binding protein; *AEA*, arachidonoylethanolamide; *2-AG*, 2-arachidonoylglycerol; *ER*, endoplasmic reticulum. *Adapted from Vemuri V.K., Makriyannis, A., 2015. The medicinal chemistry of cannabinoids. Clin. Pharmacol. Therap. 97 (6), 553–558.*

The medicinal benefits of cannabis are derived from a variety of compounds produced naturally by the plant. These compounds can be grouped into several distinct compound classes based on structure, function, and pharmacological effect. The compound groups with the most significant body of evidence supporting therapeutic benefit are the cannabinoids and terpenes. These compounds may represent more than 20%–30% of the dry cannabis flowers by weight. Another group of compounds with demonstrated therapeutic value is the flavonoids, which are generally present as a more minor fraction of cannabis (on average between 1% and 5% by weight).

Because the cannabis varieties have different phytocannabinoid and terpene compositions, as reflected in their distinct pharmacological profiles, it can be argued that overall modulation of the ratios of specific phytocannabinoids and terpenes may offer therapeutic potential dependent on the nature of the target disease.

The various cannabinoids, terpenes, and flavonoids found in *C. sativa* have physiological effects individually as well as potential synergies between one another. This synergistic effect between two or more cannabis phytochemicals is often referred to as the "entourage effect." Specifically, the interactions between cannabinoids and terpenoids appear to lead to effects that are greater than any one component or the summation of the components. Potential beneficial medical benefits including control of pain, inflammation, depression, anxiety, addiction, epilepsy, cancer, and fungal/bacterial infections are under study (Russo, 2011).

A brief description of the pharmacological profiles of the phytocannabinoids, terpenes, and other beneficial compounds is provided, with a particular emphasis on those that have undergone or are currently undergoing clinical trials and/or advanced preclinical evaluation with in vivo animal models of CNS diseases.

Phytocannabinoids

(−)Δ^9-Tetrahydrocannabinol

Δ^9-THC is the key psychoactive phytocannabinoid and is known to have a broad spectrum of pharmacological properties. Within the ECS, Δ^9-THC is capable of activating both CB1 and CB2 receptors. THC has the highest potency at both CB1 and CB2 receptors of all the phytocannabinoids. However, it behaves as a partial agonist toward both receptors because it cannot induce their full activation to produce a maximal response. When compared to the endogenous cannabinoids, the CB1 affinities and potencies of Δ^9-THC are less than those of both AEA and 2AG. Δ^9-THC also resembles AEA in its functional properties (both are partial agonists) and differs from 2AG, which, as a full agonist, is capable of inducing a full response from both receptors. When tested in healthy mice or rats, Δ^9-THC is shown to suppress locomotor activity, produce a decrease in body temperature (hypothermia), induce catalepsy (immobility), and show analgesic properties, all of which are considered CB1-related effects. Additionally, Δ^9-THC has immunomodulating and antiinflammatory properties that are produced through the activation of the CB2 receptor. Δ^9-THC is also effective in the treatment of certain disorders such as anorexia as well as nausea and emesis, and for the relief of neuropathic pain that accompanies multiple sclerosis and cancer (Hill et al., 2012; Pertwee, 2008). Currently, an orally administered medication of pure synthetic Δ^9-THC (dronabinol) is used in cancer chemotherapy-induced nausea and vomiting as well as for appetite stimulation in HIV/AIDS patients. More recently, a medication that predominantly contains plant-extracted Δ^9-THC and CBD (Sativex) is used for the relief of neuropathic pain in patients with multiple sclerosis and cancer (Hill et al., 2012; Pertwee, 2008; Russo, 2011). Sativex has also been shown to have antineoplastic effects in animal glioma models, particularly when used in combination with temozolamide (López-Valero et al., 2018).

Other forms of THC similar to Δ^9-THC have been isolated and found to have different effects. Δ^9-Tetrahydrocannabiphorol (THCP) is naturally occurring and possesses a seven-term alkyl side chain rather than the five-carbon chain found on Δ^9-THC. Unlike Δ^9-THC, THCP acts as a full CB1 agonist and thus has more profound psychoactive and physiologic effects. The presence of THCP is postulated to be responsible for the pharmacological effects of certain strains of cannabis that cannot be solely explained by their Δ^9-THC content (Citti et al., 2019).

(−)-Cannabidiol

Except for Δ^9-THC, CBD is the cannabis component that has received the most attention. Unlike Δ^9-THC, CBD is nonintoxicating, although it is still considered to have psychoactivity. This explains CBD's utility as an anxiolytic. Furthermore, it displays low affinities for both CB1 and CB2 receptors' orthostatic site and, for this reason, its cannabinoid receptor-related effects, until recently, had not received much attention. However, more recent studies of its functional properties revealed that it behaves as a CB1 receptor antagonist when tested in mouse brain membranes and other tissues with CB1 receptors (Hill et al., 2012; McPartland et al., 2015; Pertwee, 2008). Importantly, CBD was found to block several of the THC effects in mice, rats, rabbits, and humans and was also shown to antagonize the effects of known potent synthetic CB1 agonists. With regard to its effects on the CB2 receptor, CBD appears to behave as an inverse agonist. It was also found to inhibit immune cell migration when tested in murine macrophage preparations (Hill et al., 2012; Pertwee, 2008). These in vitro observations may account for some of the antiinflammatory effects observed in human studies.

Although the nonintoxicating properties of CBD are consistent with its effects on the two cannabinoid receptors, it has also become clear that other targets contribute to the compound's overall pharmacological profile. Among those to be considered are the various transient receptor potential (TRP) channels, most notably TRPV1, TRPA1, and TRPM8, in addition to serotonin, GPR55, and PPAR receptors. The specific CBD contributions to these not yet fully characterized receptors are being studied.

Beyond the receptor-based biological properties of CBD, an important aspect of its potential therapeutic effects can be attributed to its strong antioxidant properties and its ability to interact with what are known as reactive oxygen species (ROS). These very reactive products of the oxygenation cycle interact with and damage several cellular components and contribute to inflammation. It can be argued that CBD's role in neutralizing ROS species is an important contribution to its well-known antiinflammatory and antioxidant effects.

The antiepileptic effects of cannabis have been known for many centuries, having been recognized as Ayurvedic medicine for the treatment of epilepsy. A more detailed examination of cannabis constituents starting from the 1970s identified CBD's anticonvulsant properties in rats (Consroe and Wolkin, 1977) and performed initial testing in humans in small groups (Carlini and Cunha, 1981). This was followed by a larger human study using 900–1200 mg/kg daily, which showed that CBD was effective in treating seizure states (Trembly and Sherman, 1990). In 2018, the FDA approved Epidiolex, a cannabis-based CBD pharmaceutical for the treatment of children with refractory seizures caused by Lennox-Gastaut or Dravet syndrome. Categorized as a Schedule V controlled substance, this marked the first cannabis-derived drug approved by the FDA.

The efficacy of CBD for the treatment of seizures is significantly impacted by the entourage effect. Specifically, full-spectrum CBD-rich extracts containing other cannabinoids, terpenes, and flavonoids have been shown to be more effective at reducing seizure frequency (71% of patients reporting reduction) compared to CBD isolates (46%). Additionally, patients using CBD-rich extracts report lower average effective doses (6 mg/kg per day) compared to CBD isolates (25.3 mg/kg per day). Incidence of side effects is also mitigated with full-spectrum extracts (mild 33%, severe 7%) versus CBD isolates (mild 75%, severe 26%) (Pamplona et al, 2018).

When tested in animal models, CBD was shown to have anxiolytic properties (Zuardi and Karniol, 1983), as confirmed in a number of studies involving humans. In an early study using healthy human volunteers subjected to a stressful public speaking test (SPST), a 300-mg dose reduced the volunteers' subjective anxiety to a level comparable to the standard anxiolytic diazepam (Zuardi et al., 1993). Neuroimaging confirmed these effects in follow-up studies (Bergamaschi et al., 2011; Crippa et al., 2009). CBD has also been shown to ameliorate unwanted effects of THC and may be responsible for the improved properties of smoked cannabis compared to THC as an isolate.

Ligresti et al. (2006) examined the antitumor effects of a variety of cannabinoids, including both neutral and acidic forms. Excluding THC, CBD was determined to be the most potent antitumor cannabinoid tested, CBDA was the least. The activity of CBD was due to its capability of inducing apoptosis via direct or indirect activation of cannabinoid CB2 and potential vanilloid type 1 receptors as well as inducing reactive oxygen species.

One potentially useful feature of the cannabinoids is the antimicrobial properties that could protect the plant and humans from disease. The prenyl-containing cannabinoids are highly effective against meticillin-resistant *Staphylococcus aureus* (MRSA) strains relevant to the clinical setting, with CBD being the most potent (Appendino et al., 2008). CBDA (Schultz and Haffner, 1958), CBC (Turner and Elsohly, 1981), and CBG (ElSohly et al., 1982) are powerful plant antibiotics and likely play a role in maintaining plant health.

(−) Δ^9-Tetrahydrocannabivarin

THCV is the propyl analog of THC in which the five-carbon side chain is shortened by two methylene units. Its pharmacological properties have been studied since the early 1970s, when it was recognized that it behaves as a significantly weaker agonist (approximately fivefold) compared to THC.

Thus, as with THC, THCV produces analgesic but also cataleptic effects in animals (Hill et al., 2012; Pertwee, 2008). These effects can be antagonized by synthetic CB1 antagonists such as SR1716A or AM251. This property of being a weaker agonist than THC is consistent with extensive medicinal chemistry work showing the decreased potency of the cannabivarin family relative to the longer side-chain analogs. Maximum potency is observed with side chains seven to eight carbons long (Makriyannis, 2014).

Although THCV can behave as a CB1 agonist in vitro, it can also act as a CB1 antagonist capable of blocking the effects of more potent synthetic CB1 agonists as well as THC and the endocannabinoids AEA and 2AG. For this reason, Δ^9-THCV can act as an in vivo CB1 antagonist.

At the CB2 receptor, Δ^9-THCV behaves as a partial agonist and is capable of activating this receptor, although only to a limited extent. Because CB2 receptor activation has been shown to be associated with attenuation of inflammation, the antiepileptic effects of this compound are currently being tested in humans.

THCV has been shown to be beneficial for type II diabetes by improving glucose tolerance and increasing insulin sensitivity in mouse models (Wargent et al., 2013). GW Pharmaceuticals ran a pilot study (https://clinicaltrials.gov/ct2/show/NCT01217112) that evaluated the use of THCV in patients with type II diabetes and found a variety of desirable antidiabetic effects including reduced fasting plasma glucose levels, reduced systolic blood pressure, and reduced serum IL-6 levels.

Cannabigerol

CBG is the principal precursor of all phytocannabinoids in its acid form. This compound is nonpsychoactive and preliminary studies showed that CBG may be a partial agonist at CB1 and CB2 (Gauson et al., 2007). However, subsequent findings revealed that CBG binds weakly to both CB1 and CB2 receptors. CBG is a partial agonist at CB2, but the underlying molecular mechanism on its effect on CB1 remain uncertain (Navarro et al., 2018). Because of its low cannabinoid receptor potency, it can functionally antagonize the CB1 effects of THC. CBG has also been shown to effect other noncannabinoid receptors, it is an agonist of TRPA1, TRPV1, TRPV2, and an antagonist of TRPM8 channels, as well as being an inhibitor of AEA cell uptake (De Petrocellis et al., 2011).

It has been shown to relieve intraocular pressure, which is potentially useful in the treatment of glaucoma. Additionally, its antioxidant and antiinflammatory properties make it a potential candidate for inflammatory bowel disease. Recent evidence identifies CBG as a potential candidate for treatment of colon cancer (Ligresti et al., 2006) and mouse skin melanoma cells. Lastly, a preclinical (unpublished) study suggests CBG is more effective then CBGA on stomach and bone cancers (Baek et al., 1996; https://www.hempgrower.com/article/study-suggests-cbg-more-effective-than-cbga-on-stomach-bone-cancer/).

Cannabichromene

CBC is a nonpsychotropic phytocannabinoid shown to be a potent TRPA1 agonist and to have antiinflammatory and antimicrobial properties. A preclinical (unpublished) study demonstrated antitumor effect of both CBG and CBC against gastrointestinal cancer cells. Both cannabinoids were shown to induce significantly higher rates of cancer cell death compared to other cannabinoids and suggest the need for further investigation into synergistic efficacy (https://www.hempgrower.com/article/cbg-and-cbc-kill-gastrointestinal-cancer-cells-in-preliminary-study/).

Tetrahydrocannabinoic acid and cannabidiolic acid

There is some information about Δ^9-THCA and CBDA, the two naturally nonpsychotropic precursors of Δ^9-THC and Δ^9-CBD, that shows that their therapeutic value is derived from mechanisms other than classical CB1/CB2 receptor binding (Ahmed et al., 2008). THCA in vitro is capable of modulating the functions of two TRP channel receptors acting as a potent TRPA1 agonist and TRPM8 antagonist and inhibits both cyclooxygenase enzymes, COX1 and COX2. When tested in rat models of nausea, THCA appears to be a better alternative for treating nausea and vomiting associated with cancer chemotherapy (Rock et al., 2014; Rock and Parker, 2013). THCA has also been shown to reduce levels of TNF-α in vitro, suggesting a mechanism for immune modulation (Ligresti et al., 2006). Also, THCA has been found to possess anticonvulsant activity in preliminary preclinical investigations, and is devoid of adverse psychoactive effects (Sulak et al., 2017).

CBDA was shown to be a selective inhibitor of COX2 (Takeda et al., 2008), implying antiinflammatory activity. Unlike its congener, CBDA effectively reduced anticipatory nausea and may be useful against acute nausea induced by chemotherapy (Rock and Parker, 2013).

Cannabinol

CBN is a product of Δ^9-THC oxidation and was the first phytocannabinoid to be isolated in pure form. CBN can activate both CB1 and CB2 receptors with a potency approximately 10-fold lower than Δ^9-THC. It can thus be viewed as a weak Δ^9-THC relative (Izzo et al., 2009).

Cannabis terpenes

An evaluation of the many terpenes found in cannabis and their complex synergies with the cannabinoids is an important consideration although a comprehensive discussion is beyond the scope of this chapter. A summary of the most abundant terpenes in *C. sativa*, as well as their effects on biological systems, and potential cannabinoid interactions is presented.

Myrcene

Myrcene, better known as the active sedating principle of hops and lemongrass, is also found in basil, mangos, and its namesake, *Myrcia sphaerocarpa*, a medicinal shrub from Brazil traditionally used to treat diabetes, diarrhea, dysentery, and hypertension (Ulbricht, 2011). In culinary and perfume use, myrcene's aroma is earthy, fruity, and clove-like; it is pungent in higher concentrations.

Myrcene synergizes the activities of terpenes and other compounds in a variety of ways. One mechanism that would be of particular note in cannabis is its claimed effect on the permeability of cell membranes, particularly the blood–brain barrier (BBB), increasing transport of cannabinoids into the brain; however, perusal of claimed references in the popular literature shows a lack of hard data regarding brain transport. Myrcene has been shown to enhance transdermal absorption (Schmitt et al., 2009). It has a significant analgesic effect, which is blocked by the action of naloxone, an opioid antagonist, suggesting a mechanism of action through the opioid receptor (Rao et al., 1990). However, myrcene's lack of affinity for opioid receptors points to α-2-adrenoceptor-stimulated release of endogenous opiates. In contrast to morphine, no tolerance was observed after repeated dosing in rats (Lorenzetti et al., 1991). At very high doses, myrcene in mice was a sedative comparable to phenobarbital (Gurgel do Vale et al, 2002); the effect was increased by simultaneous administration of citral, a mixture of other terpenes. Al-Omari (2007) demonstrated that myrcene improved glucose tolerance in alloxan diabetic rats comparable to metformin, without an effect on glucose levels in normal rats. Myrcene also showed powerful antiinflammatory and anticatabolic effects in a human chondrocyte model of osteoarthritis (Rufino et al., 2015). With inflammation underlying numerous diseases, myrcene is the subject of a broad array of current research.

β-Caryophyllene

β-Caryophyllene has the distinction of being the first known "dietary cannabinoid," a common component of food that has GRAS (generally recognized as safe) status and is approved by the FDA for food use. β-Caryophyllene is the primary sesquiterpene contributing to the spiciness of black pepper; it is also a major constituent of cloves, hops, rosemary, copaiba, and cannabis. It was one of the first cannabis-derived compounds other than THC, CBD, and CBN shown to bind directly to endocannabinoid receptors (Gertsch, 2008). In fact, it was one of the first cannabis-derived compounds with a fundamentally different structure from the classical cannabinoids that interacts with the endocannabinoid system in humans. β-Caryophyllene has been reported to selectively bind to the CB2 receptor, although recent research has brought this into question (Gertsch et al., 2010; Santiago et al., 2019). Due to the proposed CB2 binding, β-caryophyllene is sometimes classed as an atypical cannabinoid. As described elsewhere in this chapter, CB1 is responsible for the psychoactive effects associated with certain cannabinoids such as THC. However, CB2, particularly in peripheral tissues in the body, is a therapeutic target for treatment of inflammation, pain, atherosclerosis, and osteoporosis (Gertsch, 2008; Gertsch et al., 2008). β-Caryophyllene has now been shown to be directly beneficial for colitis (Bento et al., 2011), osteoarthritis (Rufino et al., 2015), diabetes (Basha and Sankaranarayanan, 2014), cerebral ischemia (Chang et al., 2013), anxiety and depression (Bahi et al., 2014), liver fibrosis (Calleja et al., 2013; Mahmoud et al., 2014), and Alzheimer-like disease types (Cheng et al., 2014). In cancer studies, β-caryophyllene demonstrated synergy with the chemotherapy drug Paclitaxel on human tumor cell lines, and alone it stimulates apoptosis and suppresses tumor growth (Legault and Pichette, 2007). In a *Caenorhabditis elegans* model, β-caryophyllene modulated stress-related genes and extended the lifespan of the organism (Pant et al., 2014). Importantly, it has been shown to be orally bioavailable; therefore it would provide an important medicinal benefit to oral cannabis preparations.

Limonene

Limonene is one of the most abundant terpenes in cannabis, and it may be found in concentrations as high as 16% of the essential oil fraction. Ubiquitous in citrus rind, limonene is a monoterpene commonly used in perfumes, household cleaners, foods, and medicines. Limonene has numerous medicinal benefits demonstrated in human and animal studies. Limonene is among a number of plant essential oils that have been identified as having antioxidant and anticancer properties. Limonene has therefore been suggested as an excellent dietary source for cancer prevention (Aggarwal and Shishodia, 2006). Anxiolytic effects in a mouse maze model were comparable to diazepam, but not antagonized by flumazenil, implying a nonbenzodiazepine mechanism (Lima et al., 2013). This is in contrast to previous results that also demonstrated antidepressant activity via the 5-HT_{1A} receptor pathway (Komiya et al., 2006). Limonene has antiinflammatory effects in models of osteoarthritis (Rufino et al., 2015) and asthma (Hirota et al., 2010, 2012). Multiple modes of anticancer activity were observed, including chemoprevention (Crowell and Gould, 1994). Evidence from a phase I clinical trial shows a partial response in a patient with breast cancer and stable disease for more than 6 months in three patients with colorectal cancer (Vigushin et al., 1998). Limonene is metabolized into perillyl alcohol, which is also a subject of numerous cancer-related studies (Thomas Prates Ong et al., 2012).

Humulene

Humulene is the characteristic terpene of hops, *Humulus lupulus*, but it is also found in cannabis, sage, and ginseng. Humulene, also known as α-caryophyllene, is a ring-opened isomer of β-caryophyllene, which is notably lacking in CB2 activity. Nonetheless, it also possesses powerful antiinflammatory activity equal to dexamethasone in an animal model (Fernandes et al., 2007). Humulene possesses both topical and systemic antiinflammatory properties (Chaves et al., 2008) and is an effective analgesic when taken topically, orally, or by aerosol (Rogerio et al., 2009). Humulene can produce ROS, leading to an antineoplastic effect by inducing apoptosis; however, at moderate levels, ROS can increase tumor growth. β-Caryophyllene provided synergistic activity along with humulene and paclitaxel as anticancer agents on human tumor cell lines (Legault and Pichette, 2007). Interestingly, humulene was shown to increase secretion of IL-8, a chemokine with various functions, including promoting angiogenesis which is helpful in wound healing (Satsu et al., 2004).

Cannabinoid–terpene interaction

The entourage effect does not solely refer to cannabis, rather it applies to the vast majority of medicinal botanical plants and extracts. As with other botanicals, the entourage effect describes the synergistic interactions between any/all phytochemicals present. Physiologic responses resulting from the interaction, or co-administration, of phytochemicals have been described by four basic mechanisms: multitarget effects, pharmacokinetic effects including improvements in solubility and/or bioavailability, interactions that affect bacterial resistance, and diminished adverse events (Wagner and Ulrich-Merzenich, 2009). With respect to cannabis, the most attention has been directed toward cannabinoid–terpene synergy (Russo, 2011). These types of synergies are the subject of ongoing debate, although definitive examples of entourage effects such as those noted earlier with regard to CBD and seizures are known to be clinically relevant and significant. The level of scrutiny being directed toward cannabis by the research community, and consideration of cannabis as a plant containing many biologically active compounds, some of which interact synergistically, has created a challenge for much of the scientific community, which tends to be reductionist in nature.

Flavonoids

Whole-cannabis extracts contain approximately 20 flavonoids that contribute useful activity (McPartland and Russo, 2001; Turner et al., 1980). Volatile and lipophilic, many flavonoids retain activity in smoke (Sauer et al., 1983). Apigenin inhibits TNF-α (Gerritsen et al., 1995), a potentially disease-modifying treatment target in multiple sclerosis, rheumatoid arthritis, psoriasis, and other autoimmune diseases, and is also a potent inhibitor of CYP2C9 (Si et al., 2009), a liver enzyme that metabolizes numerous drugs, most relevantly THC. Also found in chamomile (*Matricaria recutita*), apigenin has high affinity for central benzodiazepine receptors, showing powerful anxiolytic activity comparable to diazepam but without the negative effect on memory and learning (Salgueiro et al., 1997). Cannflavin A, a flavone unique to cannabis, inhibits PGE_2 with 30 times the activity of aspirin (Barrett et al., 1986). An isomer of cannaflavin B has been shown to decrease the survival of two pancreatic cancer cell models as well as pancreatic cells treated with radiotherapy. Meanwhile, in vivo, results demonstrate that cannaflavin B has therapeutic efficacy in delaying both local and metastatic tumor progression, as well as increased survival in animals with pancreatic cancer (Moreau et al., 2019).

Phytosterols

β-Sitosterol, a phytosterol found in cannabis, improves urinary symptoms and flow measures in benign prostatic hyperplasia (Wilt et al., 1999). It was also active in a model of skin diseases, reducing topical inflammation and edema (Gómez et al., 1999).

Cannabis sativa as a nutraceutical

The term "nutraceutical" was introduced to define ingredients that contain both nutritional and pharmaceutical properties; it was defined by Dr. Stephen DeFelice of the Foundation for Innovation in Medicine as, "Food, or parts of food, that provide medical or health benefits, including the prevention and treatment of disease." This broad view encompasses isolated nutrients, dietary supplements, botanicals, functional foods including prebiotics, and medicinal products thereof. However, nutraceuticals are clearly distinct from dietary supplements in that they not only supplement the diet but also aid in the prevention and/or treatment of disease (Kalra, 2003). *C. sativa* is unique in that many of its constituents can be classified as a pharmaceutical ingredient, nutrient, dietary supplement, or a herbal product.

Cannabis as food

Cannabis seed, especially its oil, is particularly nutritious and is often consumed whole or used in food preparations. Whole hemp seed contains approximately 20%–25% protein, 20%–30% carbohydrates, and 10%–15% insoluble fiber (Callaway, 2004; Deferne and Pate, 1996). In addition, it contains a mixture of the saturated fatty acids palmitic and stearic acid as well as oleic acid (Callaway, 2004; Leizer et al., 2000). Hemp seed oil is an extremely rich source of unsaturated fatty acids, especially the essential fatty acids linoleic acid (LA) and α-linolenic acid (LNA) (Callaway, 2004; Leizer et al., 2000). Essential fatty acids (EFAs) cannot be produced naturally by the human body and must be sourced from the diet—LA and LNA are omega-6 and omega-3 essential fatty acids, respectively, that are well known for their general health benefits. Hempseed oil also tends to contain high amounts of gamma-linolenic acid (GLA) and stearidonic acid (SDA), which are metabolites of LA and LNA (Callaway, 2004). Because these metabolites are produced by breakdown of dietary LA and LNA, they are not considered EFAs. However, supplementation in the diet can be extremely beneficial. Many chronic diseases of modern society, including cancer, diabetes, heart disease, arthritis, and AD, have an inflammatory component (Kapoor and Huang, 2006). Diets enriched in GLA have been shown to reduce inflammation (Tate et al., 1989); therefore the nutritional value of GLA from hempseed oil is clear.

A common claim in popular literature is that cannabis seed oil has the proper ratio (2–3:1) of omega-6 to omega-3 acids for optimal human health. Although many benefits of supplementation have been observed, it seems a simple conclusion that consumption would lead to general health benefits. Supplementation levels are not high enough to correct an otherwise high omega-6 diet (cf. fish oil chapter). However, this claim is based on the hypothesized diet of Paleolithic humans and the reality is more complicated, with health effects depending on the individual and specific health concern in question. For example, a ratio of 4:1 was associated with a 70% decrease in total mortality from cardiovascular disease, whereas in colorectal cancer a decrease in rectal cell proliferation was only observed when the ratio decreased to 2.5:1. A ratio of 5:1 was beneficial in asthma (Simopoulos, 2002).

Historical/culture uses

Throughout history, cannabis has been consumed in various ways that have provided therapeutic benefits. Cannabis preparations have been used for home remedies, as medicine, as a functional food, and as a source of nutrition.

In India, a number of oral preparations of cannabis are prevalent—these may be known as "bhang," "ghota," "thandai," or "pang" (Chopra and Chopra, 1957). The simplest bhang beverages are prepared by pounding cannabis leaves with black pepper, sugar, and water. Other added ingredients may include milk, ghee, almonds, and other herbs and spices. Although bhang is often consumed for its intoxicating properties, it has also been used for medicinal purposes. Bhang beverages have often been used for treatment of a variety of gastrointestinal and rheumatic issues, which may range from simple digestive improvement and appetite stimulation to gonorrhea and dysuria and to diarrhea and dysentery. Other medicinal uses of bhang as a household remedy include pain relief, prevention of malaria, and improvement of nervous diseases and/or epileptic disorders (Chopra and Chopra, 1957; Touw, 1981). Although many oral preparations of bhang are prevalent, such as beverages, curry, syrups, and even ice creams, other modes of application of bhang are described, for instance, as a poultice, tincture, or alcoholic extraction.

The use of cannabis in Ayurvedic medicine dates back many centuries. Early Indian texts describe cannabis as one of five sacred plants used for "freedom from distress" (Aldrich and Mathre, 1997). Cannabis has numerous known uses in Ayurvedic medicine, including as a sleep aid, appetite stimulant, analgesic, excitant, aphrodisiac, antiphlegmatic, and intoxicant (Aldrich and Mathre, 1997; Touw, 1981). Cannabis is described in one of the earliest Indian medical texts, the *Sushruta Samhita*, with its use characterized as an antiphlegmatic (Touw, 1981). Its medicinal benefit would thus include its value as an expectorant and diuretic; however, in this cultural context, phlegmatic tendencies also included sluggishness of energy and thought, and so the use of cannabis to "enliven" the spirit was considered another important therapeutic benefit. Although the *Sushruta Samhita* is the first documented evidence of medicinal cannabis use in India, dating back to 400–600 BCE, through time its use in Indian (Ayurvedic) and Islamic (Unani) medicine has spread (Touw, 1981). Both agreed that cannabis increased digestion and appetite, and both used it as an antispasmodic and anticonvulsive. There are more cannabis-related preparations and uses in common Ayurvedic practice than are found in the Ayurvedic pharmacopeia (Dwarakanath, 1965; Touw, 1981). It has been described as the "penicillin of Ayurvedic medicine" (Touw, 1981).

Cannabis was a staple crop in ancient China, where it was valued as food, fiber, and also as a medicine (Aldrich and Mathre, 1997; Li, 1973; Touw, 1981). As a food crop, cannabis seed was one of the major grains cultivated to produce cooking oil in ancient China (Li, 1973). Along with millet, soybean, rice, and barley, hempseed was considered one of the five major grains in Chinese culture. The medicinal properties of cannabis were also well known in the early Chinese pharmacopeia. Commonly described as the "Father of Chinese Medicine," Shen Nong is thought to have used cannabis as medicine as early as 2700 BCE. Some of the earliest pharmacopeia in the world listed more than 100 ailments treated by cannabis, including gout, rheumatism, malaria, constipation, and diarrhea. Other historical Chinese medical texts described its use as an antibiotic and anthelmintic, and for the treatment of leprosy (Li, 1973). Although less prevalent, there are also references to the medicinal value of *C. sativa* in the Egyptian, Greek, Roman Persian, and Indian cultures (Butrica, 2002; Aziz, 2016; Abbasi et al., 2013).

Juicing of cannabis leaves and flowers has been used historically and is gaining in popularity as a method of preparation and consumption. The cold extraction of raw cannabis juice results in preparations with much higher proportions of naturally occurring acidic cannabinoids such as THCA and CBDA. Cold extraction precludes the decarboxylation of these acidic forms to their neutral counterparts, a reaction that typically occurs at temperatures required for combustion (smoking), vaporization, or hot extraction. Because THCA has a significantly lower affinity for the endogenous cannabinoid receptor CB1, such preparations will typically not have the same psychoactivity as preparations containing neutral THC. The ingestion of unheated cannabis preparations would therefore allow one to administer significantly higher doses of acidic cannabinoids without the concomitant psychoactive effects of THC. Despite these limitations, cold extracts of raw cannabis (i.e., cannabis juice) would have the advantage of retaining a larger percentage of terpenes and other volatile compounds, which also have important medicinal properties, as described herein, and which are typically lost when using extraction or preparation methods involving heat. It is therefore clear that different cannabis preparations may have different medicinal benefits as a nutraceutical.

General health benefits

Compounds found in cannabis are known to provide a number of general health benefits that are not necessarily disease-specific. Consumption of cannabis as a nutraceutical should therefore provide general improvement of health. With regard to the treatment of specific disease conditions however, definitive research is lacking in many/most cases. Despite intense interest in both the scientific community and the public about the potential benefits of cannabis as medicine, legal hurdles continue to severely limit the type and quantity of studies being done.

In 2017, the National Academy of Science, Engineering, and Medicine evaluated current cannabis research and published a summary of their evaluation of the health effects of cannabis and cannabinoid use. The investigators concluded there is "conclusive or substantial evidence" cannabis is effective in the treatment of chronic pain in adults, as an antiemetic for chemotherapy-induced nausea, and for spasticity associated with multiple sclerosis. They found "moderate evidence" cannabis improves short-term sleep outcomes for patients with obstructive sleep apnea syndrome, fibromyalgia, chronic pain, and multiple sclerosis. All other conditions evaluated were found to have "limited evidence of efficacy" or "limited evidence of inefficacy" (National Academies Press, 2017). The researchers who compiled these outcomes recognized their conclusions were affected by both a very high standard of evidence to recognize efficacy and a deficiency of quality clinical trials due to the reasons previously mentioned.

While the list of conditions for which cannabis is an effective therapy is significantly shorter than the list for which it is reported to be effective (via anecdotal reports, in vitro study, and preclinical trials), there continues to be great promise and attention paid expanding the clinical use profile of this versatile plant. For instance, certain cannabinoids and terpenes have been shown to have antioxidant and/or neuroprotective properties. They have also been shown to be effective against inflammation, which is an underlying factor in many types of disease. Such compounds may also provide beneficial effects on metabolic pathways, which may be useful for conditions related to cancer, diabetes, metabolic syndrome, obesity, and others.

Inflammation

Inflammation is a highly prevalent condition that contributes to the development and progression of many diseases and conditions. Compounds found in cannabis that reduce inflammation are abundant and diverse. Cannabinoids have strong antiinflammatory properties though multiple mechanisms including induction of apoptosis, regulation of T cells, modulation of NF-κB, reduction of proinflammatory cytokines (IL-1, IL-6, TNF-α), upregulation of antiinflammatory cytokines (IL-4, IL-10), and PPAR activation (Nagarkatti et al., 2018). Additionally, some of the cannabinoids, CBDA, CBG, CBGA, and THCA in particular show significant COX inhibition (Ruhaak et al., 2011). Among the terpenes, α-pinene, β-myrcene, and β-caryophyllene appear to act through prostaglandin receptors (PGE1 and/or PGE2) and to have an antiinflammatory effect as well as modulate cytokine activity (Russo, 2011, deCassia da Silveira et al., 2013).

Cancer

Over the past 20 years there has been an emerging body of research, mainly using in vitro models of different cancers, to further elucidate the antineoplastic activity of cannabinoid receptor agonists (like THC, anandamide, 2-AG or synthetic compounds) and other cannabinoids and terpenes (Hinz and Ramer, 2018).

Cannabinoids can induce autophagy, apoptosis, cell cycle arrest, reduce angiogenesis, tissue invasion and metastasis, and tumor proliferation without affecting normal cells. Also, cannabinoids may elicit an antitumor immune response via multiple mechanisms, enabling a more effective action of immune cells to combat cancer or by favoring conditions that result in a local reduction of inflammation in the tumor microenvironment. It has been previously documented that chronic inflammation is associated with the development of neoplasia; therefore reducing inflammation may, to some extent, contribute to the prevention of carcinogenesis. Phytocannabinoids or terpenes with an affinity for CB2 receptors, such as THC and β-caryophyllene respectively, exhibit modulatory effects on both cellular and humoral immunity. THC action was linked to the inhibition of IFN-γ production and suppression of T-cell proliferation.

Despite studies supporting the hypothesis that cannabinoids may enhance immune responses against the progressive growth and spread of tumors, there are some reports that have found drastically different results and in fact showed cannabinoids having immunosuppressive action via activation of CB2 receptors as well as causing upregulation of the regulatory T cells (Tregs), which leads to the inhibition of the immune response. Due to the potential suppression of the host immune response against specific cancers, it is crucial that the legal environment changes to allow for additional research and well-prepared clinical trials.

Terpenes that have been proven to have antineoplastic activity include D-limonene, α-pinenes (also moderate cytotoxicity seen with β-pinene), perillyl acid (and its metabolite perillic acid), myrcene, α-caryophyllene, β-caryophyllene, and P-cymene (Jung et al., 2015; Kyoung et al., 2017).

Neuroprotection

Certain cannabinoids and terpenes have been demonstrated as neuroprotectants. The neurotransmitter glutamate is released during periods of ischemia and other traumatic brain events. In excess, glutamate itself is toxic and can lead to neuronal cell death in a process known as excitotoxic stress. Compounds with antioxidant properties are often neuroprotective, for instance, through reduction of toxic ROS produced during ischemic metabolism. Both THC and CBD have been shown to have antioxidant properties (Hampson et al., 1998). Hampson et al. (1998) also demonstrated that THC and CBD are both able to prevent glutamate-induced neurotoxicity. Interestingly, the neuroprotective effect of these compounds was found to be independent of their CB receptor binding activity. THC and CBD were both found to reduce ROS in vitro, with similar potency to known antioxidants such as ascorbate and butylated hydroxytoluene. CBD has been shown to protect against cerebral ischemic injury (Hayakawa et al., 2008), and also attenuates AD-related neuroinflammation in animal models (Esposito et al., 2007).

Cannabis also shows some promise in the treatment of ALS. In canines with degenerative myelopathy, a research model for ALS, an upregulation of CB2 receptors has been noted in affected astrocytes, suggesting a potential compensatory effect by the ECS to mitigate neuronal inflammation and degeneration (Fernández-Trapero et al., 2017). In a mouse model of ALS, a CBG derivative given intraperitoneally led to reductions in weight loss, lessening of anomalies of neurological parameters, preserved spinal cholinergic neurons, and reduced astroglial activity (Rodríguez-Cueto et al., 2018).

Certain terpenes also have strong antioxidant properties and may be useful neuroprotectants. In a mouse study of cerebral ischemia, Ciftci et al. (2011) demonstrated that β-myrcene protected against oxidative stress and histological damage induced by ischemia–reperfusion, and is thus an effective neuroprotectant. The compound is suggested as a good candidate for treatment of ischemic stroke.

Obesity

Certain cannabinoids have demonstrated effects that may be useful for obesity treatment and prevention. The cannabinoid THCV has been shown to produce weight loss and decreased body fat and serum leptin concentrations in obese mice (Riedel et al., 2009; Wargent et al., 2013). While seemingly paradoxical, Le Foll et al. (2013) found that the prevalence of obesity is lower in regular cannabis users compared to nonusers, even after adjusting for important variables such as age, sex, and tobacco smoking status. Additional support for this idea comes from recent work by Silvestri et al. (2015) in which THCV and CBD were both shown to reduce accumulated lipid levels in adipocytes and in a model of hepatosteatosis.

Antioxidant

Plant antioxidants are important for human health and include compounds such as ascorbic acid, tocopherols, polyphenolic compounds, and terpenes. Numerous mono- and sesquiterpenes have antioxidant properties, and the essential oils of a multitude of plant species have been tested for their antioxidant properties—this includes many plant species that are rich in terpenes, which are abundant in cannabis. Terpenes in cannabis with demonstrated antioxidant properties include β-caryophyllene (Calleja et al., 2013), limonene, and β-myrcene (Ciftci et al., 2011). Several cannabinoids also have demonstrated antioxidant properties, including THC and CBD (Hampson et al., 1998), as well as CBG (Borrelli et al., 2013).

Concluding remarks and future directions

C. sativa is the quintessential nutraceutical product—its seeds provide rich nutritional value and are a source of essential omega-3 and omega-6 fatty acids. Of course, dried flower products and other extracts derived from them have been consumed both medicinally and recreationally throughout history. The plant is held sacred by numerous cultures and people throughout the world, which speaks to its value to humankind, who will continue to use this nutraceutical to its full advantage.

Over recent decades, empirical evidence has provided tremendous insight into the machinery used by *C. sativa* to produce these compounds, as well as their application in disease therapy. Although there is much to learn about the complex ECS, it is clear that the components of *C. sativa* exhibit a wide range of actions. This chapter summarizes the literature available to date and provides a reference for cannabis scholars and enthusiasts working in the era of legalized cannabis.

References

Abbasi, A., Khan, S., Ahmad, M., Khan, M., Quave, C., Pieroni, A., 2013. Botanical ethnoveterinary therapies in three districts of the Lesser Himalayas of Pakistan. J. Ethnobiol. Ethnomed. 9 (1), 84. Available from: https://doi.org/10.1186/1746-4269-9-84.

Abdel-Aziz, S.M., Aeron, A., Kahil, T.A., 2016. Health benefits and possible risks of herbal medicine. Microbes in Food and Health 97–116. Available from: https://doi.org/10.1007/978-3-319-25277-3_6.

Aggarwal, B.B., Shishodia, S., 2006. Molecular targets of dietary agents for prevention and therapy of cancer. Biochem. Pharmacol. 71 (10), 1397–1421.

Ahmed, S.A., Ross, S.A., Slade, D., 2008. Cannabinoid ester constituents from high-potency *Cannabis sativa*. J. Nat. Prod. 71 (4), 536–542.

Aldrich, M., Mathre, M., 1997. History of Therapeutic Cannabis. Cannabis in Medical Practice. McFarland, Jefferson, NC.

Al-Omari S.M. 2007. The Effect of Thujone and Myrcene on Diabetes Mellitus in Albino Rats. (MS), University of Jordan.

Anderson, L.C., 1980. Leaf variation among *Cannabis* species from a controlled garden. Bot. Mus. Leaf. Harv. Univ. 28 (1), 61–69.

Appendino, G., Gibbons, S., Giana, A., 2008. Antibacterial cannabinoids from *Cannabis sativa*: a structure – activity study. J. Nat. Prod. 71 (8), 1427–1430.

Appendino, G., Chianese, G., Taglialatela-Scafati, O., 2011. Cannabinoids: occurrence and medicinal chemistry. Curr. Med. Chem. 18 (7), 1085–1099.

Baek, S., Du Han, S., Yook, C.N., et al., 1996. Synthesis and antitumor activity of cannabigerol. Arch. Pharm. Res. 19, 228–230. Available from: https://doi.org/10.1007/BF02976895.

Bahi, A., Mansouri, S. Al, Al Memari, E., 2014. β-Caryophyllene, a CB 2 receptor agonist produces multiple behavioral changes relevant to anxiety and depression in mice. Physiol. Behav. 135, 119–124.

Barni-Comparini, I., Ferri, S., Centini, F., 1984. Cannabinoid level in the leaves as a tool for the early discrimination of Cannabis chemiovariants. Forensic Sci. Int. 24 (1), 37–42.

Barrett, M.L., Scutt, A.M., Evans, F.J., 1986. Cannflavin A and B, prenylated flavones from *Cannabis sativa* L. Experientia. 42 (4), 452–453.

Basha, R.H., Sankaranarayanan, C., 2014. β-Caryophyllene, a natural sesquiterpene, modulates carbohydrate metabolism in streptozotocin-induced diabetic rats. Acta Histochem. 116 (8), 1469–1479.

Bento, A.F., Marcon, R., Dutra, R.C., 2011. β-Caryophyllene inhibits dextran sulfate sodium-induced colitis in mice through CB2 receptor activation and PPARγ pathway. Am. J. Pathol. 178 (3), 1153–1166.

Bergamaschi, M.M., Queiroz, R.H.C., Chagas, M.H.N., 2011. Cannabidiol reduces the anxiety induced by simulated public speaking in treatment-naive social phobia patients. Neuropsychopharmacology. 36 (6), 1219–1226.

Bertoli, A., Tozzi, S., Pistelli, L., 2010. Fibre hemp inflorescences: from crop-residues to essential oil production. Ind. Crop. Prod. 32 (3), 329–337.

Borrelli, F., Fasolino, I., Romano, B., 2013. Beneficial effect of the non-psychotropic plant cannabinoid cannabigerol on experimental inflammatory bowel disease. Biochem. Pharmacol. 85 (9), 1306–1316.

Brenneisen, R., 2007. Chemistry and Analysis of Phytocannabinoids and Other Cannabis Constituents Marijuana and the Cannabinoids. Springer.

Brenneisen, R., ElSohly, M.A., 1988. Chromatographic and spectroscopic profiles of Cannabis of different origins: part I. J. Forensic Sci. 33 (6), 1385–1404.

Broséus, J., Anglada, F., Esseiva, P., 2010. The differentiation of fibre- and drug type Cannabis seedlings by gas chromatography/mass spectrometry and chemometric tools. Forensic Sci. Int. 200 (1–3), 87–92.

Butrica, J.L., 2002. The medical use of Cannabis among the Greeks and Romans. J. Cannabis Ther. 2 (2), 51–70. Available from: https://doi.org/10.1300/j175v02n02_04.

Callaway, J.C., 2004. Hempseed as a nutritional resource: an overview. Euphytica. 140 (1–2), 65–72.

Calleja, M.A., Vieites, J.M., Montero-Meterdez, T., 2013. The antioxidant effect of β-caryophyllene protects rat liver from carbon tetrachloride-induced fibrosis by inhibiting hepatic stellate cell activation. Br. J. Nutr. 109 (03), 394–401.

Cappelletto, P., Brizzi, M., Mongardini, F., 2001. Italy-grown hemp: yield, composition and cannabinoid content. Ind. Crops Prod. 13 (2), 101–113.

Carlini, E.A., Cunha, J.M., 1981. Hypnotic and antiepileptic effects of cannabidiol. J. Clin. Pharmacol. 21 (S1), 417S–427S.

Chang, H.-J., Kim, J.-M., Lee, J.-C., 2013. Protective effect of β-caryophyllene, a natural bicyclic sesquiterpene, against cerebral ischemic injury. J. Med. Food. 16 (6), 471–480.

Chaves, J.S., Leal, P.C., Pianowisky, L., 2008. Pharmacokinetics and tissue distribution of the sesquiterpene alpha-humulene in mice. Planta Med. 74 (14), 1678–1683.

Cheng, Y., Dong, Z., Liu, S., 2014. β-Caryophyllene ameliorates the Alzheimer-like phenotype in APP/PS1 mice through CB2 receptor activation and the PPARγ pathway. Pharmacology. 94 (1–2), 1–12.

Chopra, I.C., Chopra, R.N., 1957. The use of the cannabis drugs in India. Bull. Narcot. 9 (1), 4–29.

Ciftci, O., Ozdemir, I., Tanyildizi, S., 2011. Antioxidative effects of curcumin, β-myrcene and 1, 8-cineole against 2, 3, 7, 8-tetrachlorodibenzo-p-dioxin-induced oxidative stress in rats liver. Toxicol. Ind. Health. 27 (5), 447–453.

Citti, C., Linciano, P., Russo, F., Luongo, L., Iannotta, M., Maione, S., et al., 2019. A novel phytocannabinoid isolated from *Cannabis sativa* L. with an in vivo cannabimimetic activity higher than Δ9-tetrahydrocannabinol: Δ9-Tetrahydrocannabiphorol. Sci. Rep. 9 (1). Available from: 10.1038/s41598-019-56785-1.

Consroe, P., Wolkin, A., 1977. Cannabidiol – antiepileptic drug comparisons and interactions in experimentally induced seizures in rats. J. Pharmacol. Exp. Therap. 201 (1), 26–32.

Crippa, J.A., Zuardi, A.W., Martín-Santos, R., 2009. Cannabis and anxiety: a critical review of the evidence. Hum. Psychopharmacol. 24 (7), 515–523.

Crowell, P.L., Gould, M.N., 1994. Chemoprevention and therapy of cancer by d-limonene. Crit. Rev. Oncog. 5 (1), 1–22.

Dayanandan, P., Kaufman, P.B., 1976. Trichomes of *Cannabis sativa L.* (Cannabaceae). Am. J. Bot. 63 (5), 578–591.

De Cássia da Silveira e Sá, R., Andrade, L.N., de Sousa, D.P., 2013. A review on anti-inflammatory activity of monoterpenes. Molecules 18 (1), 1227–1254. Available from: https://doi.org/10.3390/molecules18011227. Published 2013 Jan 18.

De Petrocellis, L., Ligresti, A., Moriello, A.S., Allarà, M., Bisogno, T., Petrosino, S., et al., 2011. Effects of cannabinoids and cannabinoid-enriched *Cannabis* extracts on TRP channels and endocannabinoid metabolic enzymes. Br. J. Pharmacol. 163, 1479–1494. Available from: https://doi.org/10.1111/j.1476-5381.2010.01166.x.

Deferne, J.-L., Pate, D.W., 1996. Hemp seed oil: a source of valuable essential fatty acids. J. Int. Hemp. Assoc. 3 (1), 4–7.

Dittrich, H., Kutchan, T.M., 1991. Molecular cloning, expression, and induction of berberine bridge enzyme, an enzyme essential to the formation of benzophenanthridine alkaloids in the response of plants to pathogenic attack. Proc. Natl. Acad. Sci. USA. 88 (22), 9969–9973.

Drugs, U.N.O.O., 2009. Recommended Methods for the Identification and Analysis of Cannabis and Cannabis Products. United Nations Publications.

Dwarakanath, C., 1965. Use of opium and cannabis in the traditional systems of medicine in India. Bull. Narcot. 17 (1), 15–19.

ElSohly, M.A., Slade, D., 2005. Chemical constituents of marijuana: the complex mixture of natural cannabinoids. Life Sci. 78 (5), 539–548.

ElSohly, H.N., Turner, C.E., Clark, A.M., 1982. Synthesis and antimicrobial activities of certain cannabichromene and cannabigerol related compounds. J. Pharmaceut. Sci. 71 (12), 1319–1323.

ElSohly, M.A., Ross, S.A., Mehmedic, Z., 2000. Potency trends of delta9-THC and other cannabinoids in confiscated marijuana from 1980–1997. J. Forensic Sci. 45 (1), 24–30.

Esposito, G., Scuderi, C., Savani, C., 2007. Cannabidiol *in vivo* blunts β-amyloid induced neuroinflammation by suppressing IL-1β and iNOS expression. Br. J. Pharmacol. 151 (8), 1272–1279.

Fellermeier, M., Eisenreich, W., Bacher, A., 2001. Biosynthesis of cannabinoids. Eur. J. Biochem. 268 (6), 1596–1604.

Fernandes, E.S., Passos, G.F., Medeiros, R., 2007. Anti-inflammatory effects of compounds alpha-humulene and (−)-trans-caryophyllene isolated from the essential oil of *Cordia verbenacea*. Eur. J. Pharmacol. 569 (3), 228–236.

Fernández-Trapero, M., Espejo-Porras, F., Rodríguez-Cueto, C., Coates, J.R., Pérez-Díaz, C., Lago, E.D., et al., 2017. Upregulation of CB2 receptors in reactive astrocytes in canine degenerative myelopathy, a disease model of amyotrophic lateral sclerosis. Dis. Models Mech. 10 (5), 551–558. Available from: https://doi.org/10.1242/dmm.028373.

Fischedick, J.T., Hazekamp, A., Erkelens, T., 2010. Metabolic fingerprinting of *Cannabis sativa* L., cannabinoids and terpenoids for chemotaxonomic and drug standardization purposes. Phytochemistry. 71 (17–18), 2058–2073.

Gagne, S.J., Stout, J.M., Liu, E., 2012. Identification of olivetolic acid cyclase from *Cannabis sativa* reveals a unique catalytic route to plant polyketides. Proc. Natl. Acad. Sci. USA. 109 (31), 12811–12816.

Gauson L.A., Stevenson L.A., Thomas A., Baillie G.L., Ross R.A., Pertwee R.G., 2007. Cannabigerol behaves as a partial agonist at both CB1 and CB2 receptors. In: Proceedings of the 17th Annual Symposium on the Cannabinoids, Burlington, VT: International Cannabinoid Research Society, 206.

Gerritsen, M.E., Carley, W.W., Ranges, G.E., 1995. Flavonoids inhibit cytokine-induced endothelial cell adhesion protein gene expression. Am. J. Pathol. 147 (2), 278–292.

Gertsch, J., 2008. Anti-inflammatory cannabinoids in diet – towards a better understanding of CB2 receptor action? Towards a better understanding of CB2 receptor action? Commun. Integr. Biol. 1 (1), 26–28.

Gertsch, J., Leonti, M., Raduner, S., 2008. Beta-caryophyllene is a dietary cannabinoid. Proc. Natl. Acad. Sci. USA. 105 (26), 9099–9104.

Gertsch, J., Pertwee, R.G., Marzo, V. Di, 2010. Phytocannabinoids beyond the Cannabis plant–do they exist? Br. J. Pharmacol. 160 (3), 523–529.

Gómez, M.T.S., García, M.D., Fernández, M.A., 1999. Study of the topical anti-inflammatory activity of *Achillea ageratum* on chronic and acute inflammation models. Z für Naturforschung C. 54c (11), 937–941.

Gurgel do Vale, T., Couto Furtado, E., Santos, J.G., 2002. Central effects of citral, myrcene and limonene, constituents of essential oil chemotypes from *Lippia alba* (Mill.) N.E. Brown. Phytomedicine. 9 (8), 709–714.

Hampson, A., Grimaldi, M., Axelrod, J., 1998. Cannabidiol and (−)Δ9-tetrahydrocannabinol are neuroprotective antioxidants. Proc. Natl. Acad. Sci. USA. 95 (14), 8268–8273.

Hartsel, J., Boyar, K., Pham, A., et al., 2019. Cannabis in veterinary medicine: cannabinoid therapies for animals. In: Gupta, R.C., Srivastava, A., Lall, R. (Eds.), Nutraceuticals in Veterinary Medicine. Springer Nature, Cham, Switzerland, pp. 121–155.

Hayakawa, K., Mishima, K., Irie, K., 2008. Cannabidiol prevents a post-ischemic injury progressively induced by cerebral ischemia via a high-mobility group box1-inhibiting mechanism. Neuropharmacology 55 (8), 1280–1286.

Hellerman, C., 2013. Is super weed, super bad? CNN news. Available from: http://www.cnn.com/2013/08/09/health/weed-potency-levels/.

Hill, A.J., Williams, C.M., Whalley, B.J., 2012. Phytocannabinoids as novel therapeutic agents in CNS disorders. Pharmacol. Ther. 133 (1), 79–97.

Hillig, K.W., 2004. A chemotaxonomic analysis of terpenoid variation in Cannabis. Biochem. Systemat. Ecol. 32 (10), 875–891.

Hillig, K.W., Mahlberg, P.G., 2004. A chemotaxonomic analysis of cannabinoid variation in *Cannabis* (Cannabaceae). Am. J. Bot. 91 (6), 966–975.

Hinz, B., Ramer, R., 2018. Anti-tumour actions of cannabinoids. Br. J. Pharmacol. 10, 1384.

Hirota, R., Roger, N.N., Nakamura, H., 2010. Anti-inflammatory effects of limonene from Yuzu (citrus junos tanaka) essential oil on eosinophils. J. Food Sci. 75 (3), H87–H92.

Hirota, R., Nakamura, H., Bhatti, S.A., 2012. Limonene inhalation reduces allergic airway inflammation in *Dermatophagoides farinae*-treated mice. Inhalation Toxicol. 24 (6), 373–381.

Izzo, A.A., Borrelli, F., Capasso, R., 2009. Non-psychotropic plant cannabinoids: new therapeutic opportunities from an ancient herb. Trends Pharmacol. Sci. 30 (10), 515–527.

Jung, J.I., Kim, E.J., Kwon, G.T., Jung, Y.J., Park, T., Kim, Y., et al., 2015. β-Caryophyllene potently inhibits solid tumor growth and lymph node metastasis of B16F10 melanoma cells in high-fat diet-induced obese C57BL/6N mice. Carcinogenesis. 36, 1028–1039.

Kalra, E., 2003. Nutraceutical-definition and introduction. AAPS Pharm. Sci. 5 (3), 27–28.

Kapoor, R., Huang, Y.-S., 2006. Gamma linolenic acid: an anti-inflammatory omega-6 fatty acid. Curr. Pharm. Biotechnol. 7 (6), 531–534.

Komiya, M., Takeuchi, T., Harada, E., 2006. Lemon oil vapor causes an anti-stress effect via modulating the 5-HT and DA activities in mice. Behav. Brain Res. 172 (2), 240–249.

Kyoung, C.S., Young-Ran, L., Kyungho, L., Jaeseok, L., Ho, L.J., Im-Soon, L., 2017. Terpenes from forests and human health. Toxicol. Res. 33 (2), 97–106.

Lamarck J.-B., 1811. Encyclopédie méthodique, botanique: Suppl. 2: Panckoucke.

Le Foll, B., Trigo, J.M., Sharkey, K.A., 2013. Cannabis and Δ9-tetrahydrocannabinol (THC) for weight loss? Med. Hypotheses. 80 (5), 564–567.

Legault, J., Pichette, A., 2007. Potentiating effect of β-caryophyllene on anticancer activity of α-humulene, isocaryophyllene and paclitaxel. J. Pharm. Pharmacol. 59 (12), 1643–1647.

Leizer, C., Ribnicky, D., Poulev, 2000. The composition of hemp seed oil and its potential as an important source of nutrition. J. Nutraceut. Funct. Med. Foods. 2 (4), 35–54.

Li, H.-L., 1973. An archaeological and historical account of cannabis in China. Econ. Bot. 28 (4), 437–448.

Ligresti, A., Moriello, A.S., Starowicz, K., 2006. Antitumor activity of plant cannabinoids with emphasis on the effect of cannabidiol on human breast carcinoma. J. Pharmacol. Exp. Ther. 318 (3), 1375–1387.

Lima, N.G.P.B., De Sousa, D.P., Pimenta, F.C.F., 2013. Anxiolytic-like activity and GC–MS analysis of (R)-(+)-limonene fragrance, a natural compound found in foods and plants. Pharmacol. Biochem. Behav. 103 (3), 450–454.

Linnaeus C. Impensis GC Nauk, 1800.

López-Valero, I., Torres, S., Salazar-Roa, M., García-Taboada, E., Hernández-Tiedra, S., Guzmán, M., et al., 2018. Optimization of a preclinical therapy of cannabinoids in combination with temozolomide against glioma. Biochem. Pharmacol. 157, 275–284. Available from: https://doi.org/10.1016/j.bcp.2018.08.023.

Lorenzetti, B.B., Souza, G.E.P., Sarti, S.J., 1991. Myrcene mimics the peripheral analgesic activity of lemongrass tea. J. Ethnopharmacol. 34 (1), 43–48.

Mahmoud, M.F., Swefy, S. El, Hasan, R.A., 2014. Role of cannabinoid receptors in hepatic fibrosis and apoptosis associated with bile duct ligation in rats. Eur. J. Pharmacol. 742, 118–124.

Makriyannis, A., 2014. 2012 Division of medicinal chemistry award address. Tracking the cannabinoid road: a personal perspective. J. Med. Chem. 57 (10), 3891–3911.

Marks, M.D., Tian, L., Wenger, J.P., 2009. Identification of candidate genes affecting Δ9-tetrahydrocannabinol biosynthesis in *Cannabis sativa*. J. Exp. Bot. erp210.

McDougle, D.R., Watson, J.E., Abdeen, A.A., Adili, R., Caputo, M.P., Krapf, J.E., et al., 2017. Anti-inflammatory ω-3 endocannabinoid epoxides. Proc. Natl. Acad. Sci. 201610325. Available from: https://doi.org/10.1073/pnas.1610325114.

McPartland, J.M., Russo, E.B., 2001. Cannabis and cannabis extracts: greater than the sum of their parts? J. Cannabis Ther. 1 (3–4), 103–132.

McPartland, J.M., Duncan, M., Di Marzo, V., 2015. Are cannabidiol and Δ9-tetrahydrocannabivarin negative modulators of the endocannabinoid system? A systematic review. Br. J. Pharmacol. 172 (3), 737–753.

Mechanics P., 1938. New billion-dollar crop. Popular Mechanics.

Mediavilla, V., Steinemann, S., 1997. Essential oil of *Cannabis sativa* L. strains. J. Int. Hemp Assoc. 4, 80–82.

Mehmedic, Z., Chandra, S., Slade, D., 2010. Potency trends of Δ9-THC and other cannabinoids in confiscated cannabis preparations from 1993 to 2008. J. Forensic Sci. 55 (5), 1209–1217.

Moreau, M., Ibeh, U., Decosmo, K., Bih, N., Yasmin-Karim, S., Toyang, N., et al., 2019. Flavonoid derivative of Cannabis demonstrates therapeutic potential in preclinical models of metastatic pancreatic cancer. Front. Oncol. 9, 660. Available from: https://doi.org/10.3389/fonc.2019.00660.

Morimoto, S., Komatsu, K., Taura, F., 1997. Enzymological evidence for cannabichromenic acid biosynthesis. J. Nat. Prod. 60 (8), 854–857.

Nagarkatti, P., et al., 2018. Cannabinoids as novel anti-inflammatory drugs. Future Med. Chem. 1 (7), 1333–1349. *PMC*. Web. 5 Oct. 2009.

Navarro, G., Varani, K., Reyes-Resina, I., et al., 2018. Cannabigerol action at cannabinoid CB_1 and CB_2 receptors and at CB_1-CB_2 heteroreceptor complexes. Front. Pharmacol. 9, 632. Available from: https://doi.org/10.3389/fphar.2018.00632. Published 2018 Jun 21.

Novak, J., Zitterl-Eglseer, K., Deans, S.G., 2001. Essential oils of different cultivars of *Cannabis sativa* L. and their antimicrobial activity. Flavour Fragr. J. 16 (4), 259–262.

Pamplona, F.A., Silva, L.R.D., Coan, A.C., 2018. Potential clinical benefits of CBD-rich cannabis extracts over purified CBD in treatment-resistant epilepsy: observational data meta-analysis. Front. Neurol. 9. Available from: https://doi.org/10.3389/fneur.2018.00759.

Pant, A., Saikia, S.K., Shukla, V., 2014. Beta-caryophyllene modulates expression of stress response genes and mediates longevity in *Caenorhabditis elegans*. Exp. Gerontol. 57, 81–95.

Pertwee, R.G., 2000. Cannabinoid receptor ligands: clinical and neuropharmacological considerations, relevant to future drug discovery and development. Exp. Opin. Investig. Drugs. 9 (7), 1553–1571.

Pertwee, R.G., 2008. The diverse CB1 and CB2 receptor pharmacology of three plant cannabinoids: Δ9-tetrahydrocannabinol, cannabidiol and Δ9-tetrahydrocannabivarin. Br. J. Pharmacol. 153 (2), 199–215.

Pertwee, R.G., 2014. Handbook of Cannabis. Oxford University Press, Oxford, pp. 3–22.

Pertwee, R.G., Thomas, A., Stevenson, L.A., 2007. The psychoactive plant cannabinoid, Δ9-tetrahydrocannabinol, is antagonized by Δ8- and Δ9-tetrahydrocannabivarin in mice *in vivo*. Br. J. Pharmacol. 150 (5), 586–594.

Pollastro, F., Petrocellis, L.D., Schiano-Moriello, A., Chianese, G., Heyman, H., Appendino, G., et al., 2017. Amorfrutin-type phytocannabinoids from *Helichrysum umbraculigerum*. Fitoterapia 123, 13–17. Available from: https://doi.org/10.1016/j.fitote.2017.09.010.

Rao, V.S.N., Menezes, A.M.S., Viana, G.S.B., 1990. Effect of myrcene on nociception in mice. J. Pharm. Pharmacol. 42 (12), 877–878.

Riedel, G., Fadda, P., McKillop-Smith, S., 2009. Synthetic and plant-derived cannabinoid receptor antagonists show hypophagic properties in fasted and non-fasted mice. Br. J. Pharmacol. 156 (7), 1154–1166.

Rock, E.M., Parker, L.A., 2013. Effect of low doses of cannabidiolic acid and ondansetron on LiCl-induced conditioned gaping (a model of nausea-induced behaviour) in rats. Br. J. Pharmacol. 169 (3), 685–692.

Rock, E.M., Limebeer, C.L., Navaratnam, R., 2014. A comparison of cannabidiolic acid with other treatments for anticipatory nausea using a rat model of contextually elicited conditioned gaping. Psychopharmacology. 231 (16), 3207–3215.

Rodríguez-Cueto, C., Santos-García, I., García-Toscano, L., Espejo-Porras, F., Bellido, M., Fernández-Ruiz, J., et al., 2018. Neuroprotective effects of the cannabigerol quinone derivative VCE-003.2 in SOD1G93A transgenic mice, an experimental model of amyotrophic lateral sclerosis. Biochem. Pharmacol. 157, 217–226. Available from: https://doi.org/10.1016/j.bcp.2018.07.049.

Rogerio, A.P., Andrade, E.L., Leite, D.F., 2009. Preventive and therapeutic anti-inflammatory properties of the sesquiterpene α-humulene in experimental airways allergic inflammation. Br. J. Pharmacol. 158 (4), 1074–1087.

Ross, S.A., ElSohly, M.A., 1996. The volatile oil composition of fresh and air-dried buds of *Cannabis sativa*. J. Nat. Prod. 59 (1), 49–51.

Rufino, A.T., Ribeiro, M., Sousa, C., 2015. Evaluation of the anti-inflammatory, anti-catabolic and pro-anabolic effects of E-caryophyllene, myrcene and limonene in a cell model of osteoarthritis. Eur. J. Pharmacol. 750 (0), 141–150.

Ruhaak, L.R., Felth, J., Karlsson, P.C., Rafter, J.J., Verpoorte, R., Bohlin, L., 2011. Evaluation of the cyclooxygenase inhibiting effects of six major cannabinoids isolated from *Cannabis sativa*. Biol. Pharm. Bull. 34 (5), 774–778. Available from: 10.1248/bpb.34.774.

Russo, E.B., 2011. Taming THC: potential cannabis synergy and phytocannabinoid-terpenoid entourage effects. Br. J. Pharmacol. 163 (7), 1344–1364.

Rustichelli, C., Ferioli, V., Baraldi, M., 1998. Analysis of cannabinoids in fiber hemp plant varieties (*Cannabis sativa* L.) by high-performance liquid chromatography. Chromatographia. 48 (3-4), 215–222.

Salgueiro, J.B., Ardenghi, P., Dias, M., 1997. Anxiolytic natural and synthetic flavonoid ligands of the central benzodiazepine receptor have no effect on memory tasks in rats. Pharmacol. Biochem. Behav. 58 (4), 887–891.

Santiago, M., Sachdev, S., Arnold, J.C., Mcgregor, I.S., Connor, M., 2019. Absence of entourage: terpenoids commonly found in *Cannabis sativa* do not modulate the functional activity of Δ9-THC at human CB1and CB2 receptors. Cannabis Cannabinoid Res. 4 (3), 165–176. Available from: https://doi.org/10.1089/can.2019.0016.

Satsu, H., Matsuda, T., Toshimitsu, T., 2004. Regulation of interleukin-8 secretion in human intestinal epithelial Caco-2 cells by α-humulene. Biofactors. 21 (1–4), 137–139.

Sauer, M.A., Rifka, S.M., Hawks, R.L., 1983. Marijuana: interaction with the estrogen receptor. J. Pharmacol. Exp. Ther. 224 (2), 404–407.

Schmitt, S., Schaefer, U.F., Doebler, L., 2009. Cooperative interaction of monoterpenes and phenylpropanoids on the *in vitro* human skin permeation of complex composed essential oils. Planta Med. 75 (13), 1381–1385.

Schultes, R.E., Klein, W.M., Plowman, T., 1974. Cannabis: An Example of Taxonomic Neglect. Botanical Museum Leaflets. *Harvard University*.

Schultz, O.E., Haffner, G.A., 1958. A sedative and antibacterial principle in German hemp (*Cannabis sativa*). Z. Naturforsch B 14b, 98–100.

Sell, C., 2009. Handbook of Essential Oils: Science, Technology, and Applications. CRC Press, New York, NY.

Shoyama, Y., Yagi, M., Nishioka, I., 1975. Biosynthesis of cannabinoid acids. Phytochemistry. 14 (10), 2189–2192.

Shoyama, Y., Hirano, H., Nishioka, I., 1984. Biosynthesis of propyl cannabinoid acid and its biosynthetic relationship with pentyl and methyl cannabinoid acids. Phytochemistry. 23 (9), 1909–1912.

Shoyama, Y., Takeuchi, A., Taura, F., 2005. Crystallization of 1-tetrahydrocannabinolic acid (THCA) synthase from *Cannabis sativa*. Acta Crystallogr. Sect. F: Struct. Biol. Cryst. Commun. 61 (8), 799–801.

Si, D., Wang, Y., Zhou, Y.-H., Guo, Y., 2009. Mechanism of CYP2C9 inhibition by flavones and flavonols. Drug Metab. Dispos. 37 (3), 629–634.

Silvestri, C., Paris, D., Martella, A., 2015. Two non-psychoactive cannabinoids reduce intra-cellular lipid levels and inhibit hepatosteatosis. J. Hepatol.

Simopoulos, A.P., 2002. The importance of the ratio of omega-6/omega-3 essential fatty acids. Biomed. Pharmacother. 56 (8), 365–379.

Sirikantaramas, S., Morimoto, S., Shoyama, Y., 2004. The gene controlling marijuana psychoactivity. Molecular cloning andheterologous expression of Δ1-tetrahydrocannabinolic acid synthase from *Cannabis sativa* L. J. Biol. Chem. 279 (38), 39767–39774.

Sirikantaramas, S., Taura, F., Tanaka, Y., 2005. Tetrahydro-cannabinolic acid synthase, the enzyme controlling marijuana psychoactivity, is secreted into the storage cavity of the glandular trichomes. Plant Cell Physiol. 46 (9), 1578–1582.

Sirikantaramas, S., Taura, F., Morimoto, S., 2007. Recent advances in *Cannabis sativa* research: biosynthetic studies and its potential in biotechnology. Curr. Pharm. Biotechnol. 8 (4), 237–243.

Small, E., 1975. American law and the species problem in *Cannabis*: science and semantics. Bull. Narcot. 27 (3), 1–20.

Small, E., Marcus, D., 2002. Hemp: a new crop with new uses for North America. Trends in New Crops and New Uses. pp. 284–326.

Small, E., Marcus, D., 2003. Tetrahydrocannabinol levels in hemp (*Cannabis sativa*) germplasm resources. Econ. Bot. 57 (4), 545–558.

Sulak, D., Saneto, R., Goldstein, B., 2017. The current status of artisanal cannabis for the treatment of epilepsy in the United States. Epilepsy Behav. 70, 328–333. May.

Takeda, S., Misawa, K., Yamamoto, I., 2008. Cannabidiolic acid as a selective cyclooxygenase-2 inhibitory component in cannabis. Drug Metab. Dispos. 36 (9), 1917–1921.

Tate, G., Mandell, B., Laposata, M., 1989. Suppression of acute and chronic inflammation by dietary gamma linolenic acid. J. Rheumatol. 16 (6), 729–734.

Taura, F., Morimoto, S., Shoyama, Y., 1996. Purification and characterization of cannabidiolic-acid synthase from *Cannabis sativa* L. Biochemical analysis of a novel enzyme that catalyzes the oxidocyclization of cannabigerolic acid to cannabidiolic acid. J. Biol. Chem. 271 (29), 17411–17416.

Taura, F., Sirikantaramas, S., Shoyama, Y., 2007. Phytocannabinoids in *Cannabis sativa*: recent studies on biosynthetic enzymes. Chem. Biodiver. 4 (8), 1649–1663.

The National Academies Press. (2017). The health effects of cannabis and cannabinoids: the current state of evidence and recommendations for research. Washington, DC.

Thomas Prates Ong, M.T.C., de Conti, A., Moreno, F.S., 2012. Chemoprevention of hepatocarcinogenesis with dietary isoprenic derivatives: cellular and molecular aspects. Curr. Cancer Drug Targets. 12 (9), 1173–1190.

Touw, M., 1981. The religious and medicinal uses of Cannabis in China, India and Tibet. J. Psychoact. Drugs. 13 (1), 23–34.

Trembly, B., Sherman, M., 1990. Double-blind clinical study of cannabidiol as a secondary anticonvulsant. Paper Presented at the Marijuana'90 International Conference on Cannabis and Cannabinoids. Kolympari, Crete, July 8.

Turner, C.E., Elsohly, M.A., 1981. Biological activity of cannabichromene, its homologs and isomers. J. Clin. Pharmacol. 21 (S1), 283S–291S.

Turner, C.E., Elsohly, M.A., Boeren, E.G., 1980. Constituents of *Cannabis sativa* L. XVII. A review of the natural constituents. J. Nat. Prod. 43 (2), 169–234.

Ulbricht, C., 2011. Focus: diabetes. J. Diet. Suppl. 8 (3), 239–256.

Upton, R., Craker, L., ElSohly, M., et al., 2014. Cannabis Inflorescence: Standards of Identity, Analysis, and Quality Control. Scott's Valley, CA.

van Bakel, H., Stout, J.M., Cote, A.G., 2011. The draft genome and transcriptome of *Cannabis sativa*. Genome Biol. 12 (10), R102.

Vemuri, V.K., Makriyannis, A., 2015. The medicinal chemistry of cannabinoids. Clin. Pharmacol. Therap. 97 (6), 553–558.

Veress, T., Szanto, J., Leisztner, L., 1990. Determination of cannabinoid acids by high-performance liquid chromatography of their neutral derivatives formed by thermal decarboxylation: I. Study of the decarboxylation process in open reactors. J. Chromatogr. A. 520, 339–347.

Vigushin, D.M., Poon, G.K., Boddy, A., Cancer Research Campaign Phase I/II Clinical Trials Committee, et al., 1998. Phase I and pharmacokinetic study of d-limonene in patients with advanced cancer. Cancer Chemother. Pharmacol. 42, 111–117.

Virovets, V., 1996. Selection for non-psychoactive hemp varieties (*Cannabis sativa* L.) in the CIS (former USSR). J. Int. Hemp. Assoc. 3, 13–15.

Vogelmann, A.F., Turner, J.C., Mahlberg, P.G., 1988. Cannabinoid composition in seedlings compared to adult plants of *Cannabis sativa*. J. Nat. Prod. 51 (6), 1075–1079.

Wagner, G.J., 1991. Secreting glandular trichomes: more than just hairs. Plant Physiol. 96 (3), 675–679.

Wagner, H., Ulrich-Merzenich, G., 2009. Synergy research: approaching a new generation of phytopharmaceuticals. Phytomedicine 16 (2-3), 97–110. Available from: https://doi.org/10.1016/j.phymed.2008.12.018.

Wargent, E., Zaibi, M., Silvestri, C., Hislop, D.C., Stocker, C.J., Stott, C.G., et al., 2013. The cannabinoid Δ9-tetrahydrocannabivarin (THCV) ameliorates insulin sensitivity in two mouse models of obesity. Nutr. Diabetes. 3 (5), e68. Available from: https://doi.org/10.1038/nutd.2013.9.

Watson, D.P., Clarke, R.C., 2014. The genetic future of hemp. Int. Hemp. Assoc.

Wilt, T.J., Macdonald, R., Ishani, A., 1999. β-sitosterol for the treatment of benign prostatic hyperplasia. A systamic review. BJU Int. 83 (9), 976–983.

Woelkart, K., Xu, W., Pei, Y., 2005. The endocannabinoid system as a target for alkamides from *Echinacea angustifolia* roots. Planta Med. 71 (08), 701–705.

Wood, W.F., Sollers, B.G., Dragoo, G.A., 2002. Volatile components in defensive spray of the hooded skunk, *Mephitis macroura*. J. Chem. Ecol. 28 (9), 1865–1870.

Xi, Z.-X., Peng, X.-Q., Li, X., 2011. Brain cannabinoid CB2 receptors modulate cocaine's actions in mice. Nat. Neurosci. 14 (9), 1160–1166.

Zuardi, A., Karniol, I., 1983. Effects on variable-interval performance in rats of delta 9-tetrahydrocannabinol and cannabidiol, separately and in combination. Brazil. J. Med. Biol. Res. 16 (2), 141–146.

Zuardi, A., Guimaraes, F., Moreira, A., 1993. Effect of cannabidiol on plasma prolactin, growth hormone and cortisol in human volunteers. Brazil. J. Med. Biol. Res. 26 (2), 213–217.

Chapter 39

Cannabidiol safety

Andrew Charrette, James Akingbasote, Najla Guthrie and Corey J. Hilmas

Introduction

The regulatory status of cannabis and its constituents is changing globally. Canada made bold moves to create a regulatory framework for cannabis-containing compounds with their Cannabis Act. On October 17, 2018, Canada lifted the curtain after nine decades of recreational cannabis prohibition and allowed the legal sales of adult-use marijuana. When the Cannabis Act was passed, only dried flower, cannabis oil, and sublingual sprays received a green light in the first wave. A second wave of regulations allowed for the sales of "derivative" products, including edibles, infused beverages (nonalcoholic), topicals, concentrates, and vapes. In June 2019, Health Canada released a proposed regulatory framework for stakeholder comment on Cannabis Health Products, the initial step toward establishing the world's first federally regulated legal market for over-the-counter cannabis health products.

The regulatory landscape is changing in the United States for hemp-derived nutraceuticals. While cannabis is a Drug Enforcement Agency (DEA) schedule I drug, defined as "no medicinal use with high abuse potential," hemp and hemp-derived cannabinoid products are no longer considered schedule I with passage of the Agriculture Improvement Act of 2018, known as the 2018 Farm Bill. Hemp is now defined as cannabis that contains $<0.3\%$ Δ-9-tetrahydrocannabinol (Δ-9-THC). While the Farm Bill eliminated regulatory risk for cannabidiol (CBD) companies by removing DEA enforcement, FDA still has jurisdiction over finished hemp-derived cannabinoid nutraceutical products in the United States. The Agency has opted for a period of enforcement discretion on nutraceuticals containing hemp-derived dietary ingredients, and the US market has grown to over 1500 products. In May 2019, several months after the passage of the Farm Bill, the US FDA held a public hearing with stakeholders to ask questions about the safety of the CBD-containing products being sold in the United States. The following month, the US House of Representatives passed report language in their 2020 Agriculture Appropriations Bill directing FDA to look for a regulatory pathway forward and lift the exclusion ban over CBD in dietary supplements. The Senate Appropriations held a similar debate later in 2019. The FDA reported back to the House Committee on Appropriations (Hahn, 2020) that the FDA is actively considering potential pathways for certain CBD products to be marketed as dietary supplements. The number one area of focus in the FDA's report was safety of CBD.

While the FDA is actively evaluating what and how much data are required to support a conclusion that CBD can safely be allowed in dietary supplements, we review the safety data regarding CBD here. We start with an overview of the constituents of hemp and focus on the safety of CBD, its interactions with a variety of receptor systems in the body, relationship to the endocannabinoid system (ECS), and pharmacokinetics. The safety of CBD is evaluated next from available genotoxicity data, in vitro toxicity studies, preclinical oral toxicity (acute and repeat-dose), reproductive toxicity, developmental toxicity, and finally a critical review of recent literature on CBD and liver toxicity in mice. We finally look at the safety evidence of CBD from its use in human clinical trials. We wind up our safety review of CBD with a look at its effects on specific cytochrome P450 (CYP) liver enzymes and draw comparisons to other compounds that act on the same set of liver enzymes. It is our hope that this concise safety review of CBD will provide regulatory agencies and industry with the information necessary to develop safe serving levels, safety recommendations to the public, and a regulatory path forward for the intended use of CBD in nutraceutical products.

Constituents of hemp

Cannabis sativa L. and *C. indica* L. are dioicous plants of the family *Cannabaceae* (Hartsel et al., 2016). Widely distributed all over the globe, cannabis and hemp have been used as a psychoactive drug, ingredient in folk

Nutraceuticals. DOI: https://doi.org/10.1016/B978-0-12-821038-3.00039-2

medicine, a cash crop in the American colonies, and a source of textile fiber for the ancient world. The hemp plant is making an economic impact in the US economy due to its unique pharmacology and fiber properties. Cannabis and hemp extracts, isolated constituents of the plant, and synthetic cannabinoids have enjoyed a resurgence of interest in new drug therapeutics, nutraceuticals, topical cosmetics, recreational edibles in finished packaged form, and inhalational vapes. Classification has been difficult for taxonomists due to its genetic variability and the ease of crossbreeding between the cultivar species to generate hybrids. *C. sativa* is subdivided into five chemotypes based upon cannabinoid profile and fiber content.

Cannabis plants are characterized by their complex chemical composition, consisting of terpenes (cyclic and acyclic), carbohydrates, fatty acids and esters, amides, amines, phytosterols, phenolic compounds, and the class of compounds referred to in modern vernacular as cannabinoids. Phytocannabinoids are alicyclic meroterpenoids (typically C21 or C22 terpenophenolic compounds). They accumulate in the secretory cavities of glandular trichomes, predominantly in female flower buds as well as other aerial plant parts. The *C. sativa* plant produces over 421 chemical compounds, including over 100 cannabinoids, which have been isolated, characterized, and divided into 11 chemical classes (Hanuš et al., 2016; Thomas and ElSohly, 2015). The most abundant cannabinoid present in drug-type (chemotype I) cannabis plants is Δ^9-tetrahydrocannabinolic acid and Δ^9-THC. Fiber-type plants (hemp) should contain very little Δ^9-THC but more of the cannabinoic acids, such as cannabidiolic acid and cannabigerolic acid, as well as their decarboxylated forms, namely CBD and cannabigerol. There are other minor cannabinoids, including cannabichromenic acid, cannabichromene, cannabinolic acid, and cannabinol (CBN); CBN is typically found in aged cannabis as a result of oxidative degradation of Δ^9-THC. The best characterized cannabinoid constituents of cannabis and hemp are Δ^9-THC and CBD. The THC content of recreational cannabis has increased from 3% to over 20% through traditional breeding (Andre et al., 2016; ElSohly et al., 2016; Jikomes and Zoorob, 2018; Mehmedic et al., 2010). For example, the median total THC content for chemotype I flower products in the state of Washington varies between 17.7%–23.2% by dry weight. In contrast, textile hemp and chemotype II and III flower has a THC content <0.3%. Total CBD content by dry weight in these chemotypes constitutes 9%–12.8% in some states (Jikomes and Zoorob, 2018). While these two cannabinoids have very similar structures (see Fig. 39.1), their slight differences impart unique pharmacological and physiological profiles in humans.

Tetrahydrocannabinol

Δ^9-THC and CBD are structurally similar but demonstrate different pharmacological activities. Δ^9-THC and CBD are both 21-carbon terpenophenolic compounds, but the subtle differences in their structures lead to a vastly different array of molecular targets and activity in the body. Δ^9-THC (herein referred to as THC) is the primary psychotropic ingredient of cannabis. THC was first isolated in 1964 by Gaoni and Mechoulam (1964). THC is the most psychotropic component of cannabis constituent and binds specific G-protein-coupled receptors, named cannabinoid 1 and 2 (CB_1 and CB_2), located in the central and peripheral nervous systems. These are known as the endocannabinoid receptors. THC is a partial agonist at CB_1 and CB_2 receptors. It also activates PPAR-γ and TRPA1. It is therapeutically used as an antiemetic and to boost appetite in AIDS patients. A cannabis-based extract

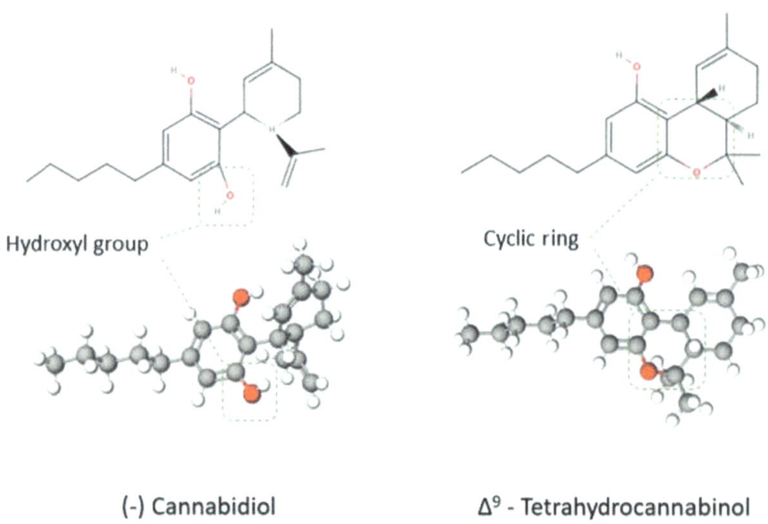

FIGURE 39.1 The differences in structure between cannabidiol and Δ^9-tetrahydrocannabinol.

with a 1:1 ratio of THC and CBD (Sativex) is marketed in Canada for the symptomatic relief of neuropathic pain in adults with multiple sclerosis and as an adjunctive analgesic treatment for adult patients with advanced cancer (Russo et al., 2007).

The discovery of cell membrane receptors for THC was followed by isolation and identification of endogenous ligands, synthesized by humans, termed endocannabinoids. The two main endocannabinoids are anandamide, which is metabolized by fatty acid amide hydrolase, and 2-arachidonoylglycerol, which is enzymatically inactivated by monoglyceride lipase (Di Marzo, 2008; Kunos et al., 2009). Cannabinoid receptors CB_1 and CB_2, the endogenous ligands that activate them, and the mechanisms for endocannabinoid biosynthesis and inactivation constitute the "endocannabinoid system." It is this ECS and other molecular targets that provide potential strategies for ameliorating pain perception and inflammation by constituents of cannabis.

Cannabidiol

CBD was first isolated by Adams et al. (1940) and its structure was identified 23 years later (Mechoulam and Shvo, 1963). Since those discoveries, a considerable wealth of published articles has been reported on the chemistry, biochemistry, pharmacology, and toxicology of CBD. Over the past 20 years, the principal focus for CBD research centered around its antiepileptic, sedative, antioxidative, anxiolytic, and antipsychotic activities (Cunha et al., 1980; Zuardi et al., 2006a) with the past decade focused on uncovering the antiinflammatory, neuroprotective, and antinociceptive roles of CBD.

There are two major stereoisomeric forms of CBD since the molecule contains a chiral center. CBD is normally taken to refer to the naturally occurring (−) enantiomer. (+) CBD ($K_i = 0.84$ μM at CB1) has been shown to have slight affinity at CB1 and CB2 receptors (Leite et al., 1982; WHO, 2018a), unlike the naturally occurring, cannabis-derived (−) CBD which displays little binding affinity. CBD is typically extracted from the cannabis plant, but CBD can also be produced synthetically. CBD's mechanism of action is thought to be mediated by the ECS by reducing neurohyperexcitability through the following ligand−receptor interactions: (1) modulation of intracellular calcium via the orphan G-protein-coupled receptor (GPR55); (2) binding to and agonist activity at the transient receptor potential vanilloid type 1 and type 2 channels (TRPV1 and TRPV2); (3) agonist activity at serotonin 5-HT_{1A} receptors (Izzo et al., 2009); and (4) modulation of adenosine-mediated signaling. Hemp-derived CBD does not activate the cannabinoid type 1 (CB1) receptor, which is typically associated with psychoactivity (Devinsky et al., 2014). Table 39.1 summarizes molecular targets for CBD, some of which are involved in modulating inflammation and pain management. CBD is probably the most promising of the cannabinoids for neurology because of its affinity for molecular targets in the body involved in chronic pain of diabetic neuropathy, inflammation, acute pain/analgesia, ischemia (cause of heart pain), and neurodegenerative diseases (see Table 39.1) (Mechoulam et al., 2007; Pertwee, 2008, 2004; Zuardi, 2008).

Endocannabinoid system

ECS is largely thought to underlie both the pharmacological activity and toxicity of endocannabinoids made endogenously and phytocannabinoids absorbed after oral ingestion or dermal penetration. The ECS is a complex cell-signaling system, which comprises cannabinoid receptors (CBRs), endocannabinoid molecules, and their metabolic enzymes, in a wide variety of organ systems. The ECS was discovered by researchers investigating the effects of Δ-9-THC (Gaoni and Mechoulam, 1964), a potent, centrally active cannabinoid found in *C. sativa* and *indica* cultivars.

See Fig. 39.2 for a description and illustration of the individual components of the ECS. CBRs can be subdivided into types 1 and 2 (CB1 and CB2) (Pertwee et al., 2010). These receptors are activated by a host of ligands which range from endogenous compounds or endocannabinoids [e.g., arachidonoyl ethanolamide or anandamide and 2-arachidonoyl glycerol (2-AG)], phytocannabinoids (e.g., Δ^9-THC and select cannabinoids) and synthetic compounds (Table 39.2).

CBRs have a differential expression throughout the human body (see Fig. 39.3). CB1 receptors are mainly localized in the central and peripheral nervous systems where they modulate the release of neurotransmitters at the presynaptic portion of neurons (Callén et al., 2012). CB1 receptors have been linked to the psychotropic activities of THC. On the other hand, CB2 receptors are mainly located on peripheral organs and on such immune cells as monocytes, macrophages, basophils, lymphocytes, and dendritic cells where they are involved in immunological responses and in the control of inflammation (Cabral et al., 2015). Activation of cannabinoid receptors by phyto- and synthetic cannabinoids has been shown to be involved in cellular homeostasis, and the pathophysiology of organ/systems such as the liver (Avraham et al., 2008; Dibba et al., 2018; Jeong et al., 2008; Mallat et al., 2011) and cardiovascular system (Defer et al., 2009; Montecucco et al., 2009; Sierra et al., 2018), among others. Interestingly, while most cannabinoids have been found to be agonists of CB1 and/or CB2 receptors, CBD does not have agonistic affinity for either CB1 or CB2 receptors; however, modulation of the ECS does not only

TABLE 39.1 Molecular targets of cannabidiol.

Molecular targets for cannabidiol	Ligand-binding effect	Description of molecular target
Cannabinoid (CBD) receptors	Inverse agonist of human CB2	CBD has very low binding affinity for CB1 and CB2 receptors. It displays inverse agonism in Chinese hamster ovary (CHO) cells transfected with human CB2 receptors. It is a high-potency antagonist in vitro of CB1 and CB2 receptors in mouse brain (Thomas et al., 2007)
	High-potency antagonist of CB1 and CB2 receptor agonists	
Human transient receptor potential (TRP)	Agonist	TRP is a superfamily of nonselective, ligand-gated cation channels. They can be divided into six subfamilies: (1) TRPC (Canonical), (2) TRPV (Vanilloid), (3) TRPM (Melastatin), (4) TRPP (Polycystin), (5) TRPML (Mucolipin), and (6) TRPA (Ankryrin) groups. Several TRPs (TRPV1, TRPV2, TRPV3, TRPV4, TRPM8, and TRPA1) have been shown to be expressed in primary afferent pain fiber receptors (nociceptors), where they act as transducers for thermal, chemical, and mechanical stimuli. CBD has affinity for subtypes within the vanilloid subfamily (TRPV1 and TRPV2) (Bisogno et al., 2001)
Adenosine uptake	Antagonist	Uptake of adenosine into neurons is a primary method of terminating adenosine signaling. Adenosine A_{2A} receptors are an important regulator of inflammation. CBD is a competitive inhibitor of adenosine uptake, causing reduction in inflammation
Orphan G-protein-coupled receptors (GPR55)	Antagonist	GPR55 is subtype of orphan G-protein-coupled receptor, activated by plant- and synthetically derived cannabinoids, as well as anandamide-related acylethanolamides. CBD acts as an antagonist of GPR55 (Ryberg et al., 2007). CBD also interacts with GPR3, GPR6, and GPR12 (Laun et al., 2019)
Fatty acid amide hydrolase (FAAH)	Antagonist	FAAH is the enzyme responsible for metabolizing endocannabinoids in the body. CBD inhibits anandamide hydrolysis through the action of FAAH (see Table 39.2) (Bisogno et al., 2001)
Peroxisome proliferator-activated receptors (PPARs)	Agonist	PPARs belong to a family of nucleus-residing receptors involved in the attenuation of neurodegenerative and inflammatory processes. CBD is an agonist of the $PPAR_{\gamma}$ subtype
Lipoxygenase	Antagonist	Lipoxygenases are nonheme iron-containing enzymes that play a role in atherosclerosis and inflammation. CBD is an inhibitor of lipoxygenases
Mu (μ)-subtype opioid receptors	Allosteric modulator	These receptors are involved in communicating and mitigating painful stimuli experienced from the environment to the brain. CBD may enhance the effects of opiates at μ-receptors (Kathmann et al., 2006)
T-type voltage-gated Ca^{2+} channels	Antagonist	These are low voltage-gated calcium ion channels found in cardiac muscle, smooth muscle, and neurons. CBD is an inhibitor of T-type Ca^{2+} channels
Sodium channels	Antagonist	CBD inhibits persistent and resurgent sodium current from WT-Nav1.6 sodium channels (Estacion et al., 2010)
Serotonin or 5-hydroxytryptamine (5-HT) receptors	Partial agonist	The 5-HT1A receptor is a subtype of 5-HT receptors and is involved in anxiety, mood, depression, headaches, ischemia, and immunological regulation. CBD acts as a partial $5\text{-}HT_{1A}$ agonist (Russo et al., 2005)
Glycine receptors	Allosteric modulator	Glycine receptors are an inhibitory superfamily of ligand-gated ion channels and the predominant inhibitor receptor of the brainstem and spinal cord. CBD is a positive, allosteric modulating ligand at α1 and α1β subtype glycine receptors

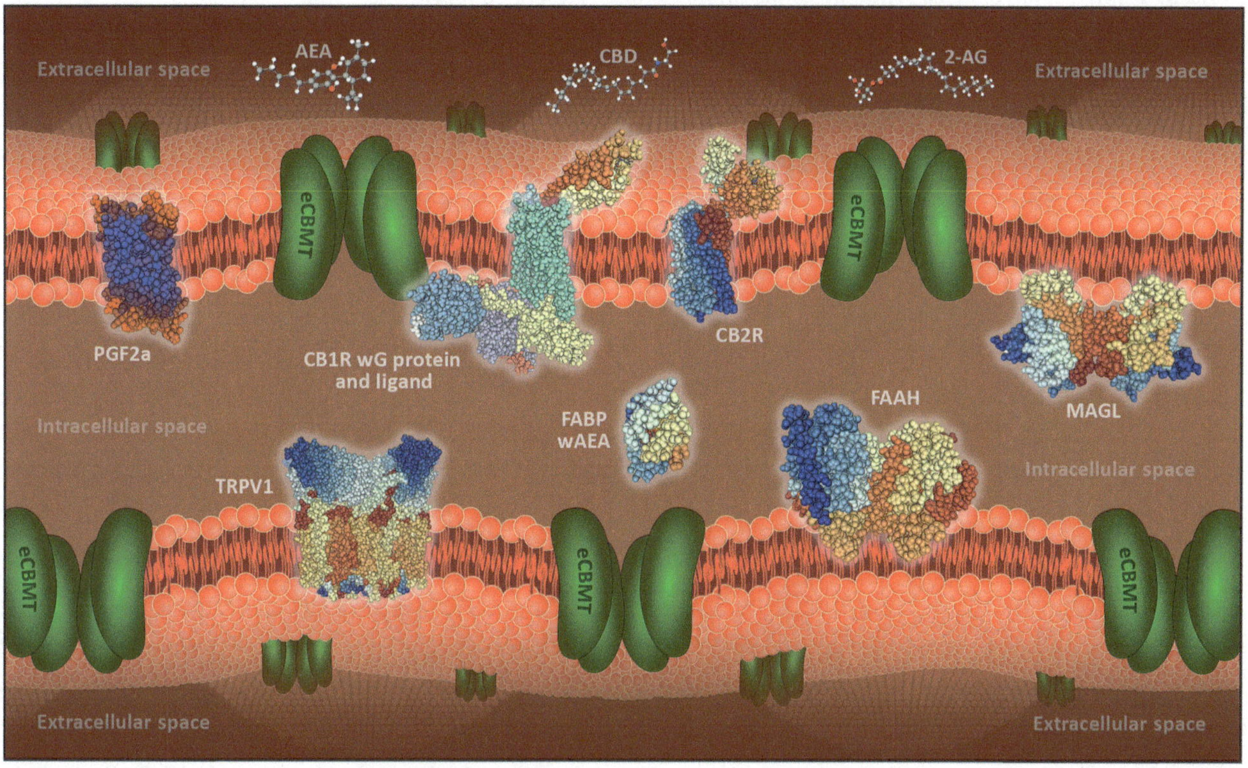

FIGURE 39.2 Cellular view of the endocannabinoid system. *AEA*, anandamide; *CBD*, cannabidiol; *2-AG*, 2-arachidonoyl glycerol; prostaglandin $F_{2\alpha}$, *PGF-2*$_\alpha$; *eCBMT*, endocannabinoid membrane transporter (proposed); *CB1R*, cannabinoid 1 receptor; *CB2R*, cannabinoid 2 receptor; *MAGL*, monoacylglycerol lipase; *FAAH*, fatty acid amide hydrolase; *FABP*, fatty-acid-binding protein; *TRPV1*, transient receptor potential cation channel subfamily V member 1. Image created by Ken Whittaker of KGK Science Inc.

occur through agonist activation of CBRs. In contrast to its reportedly low binding affinity to CB1 and CB2 receptors, it is an inverse agonist of human CB2 receptors as well as a high-potency antagonist of CB1 and CB2 receptor agonists in vitro (Thomas et al., 2007).

Pharmacokinetics of cannabidiol

The safety evaluation of any chemical compound is incomplete without an evaluation of its toxicokinetics. The biological effect of any substance depends to a large extent on the route of administration, rapidity of metabolism, organ-system distribution to gain access to the tissues and cells, and interaction with relevant cellular targets—receptors, ion channels, carrier molecules, and enzymes—to bring about downstream signaling that can be measured in terms of physiological effects (Ritter et al., 2011). In addition, the potential for metabolic bioactivation and/or inactivation also has a profound impact on the ability of a compound to exert its intended pharmacological effect on the biological system or to generate potentially toxic metabolic derivatives (Buratti and Lavine, 2002).

The toxicokinetics of CBD varies across different populations as shown in clinical and preclinical studies (FDA, 2018a,b). Tables 39.3 and 39.4 describe pharmacokinetic data derived from animal and human studies, respectively. In human subjects, CBD can be detected in the plasma when given as a single oral (Ohlsson et al., 1986) or multiple doses (FDA, 2018b), and when administered at low doses of 5.4 mg (Nadulski et al., 2005) and very high doses of 4500 mg (FDA, 2018b). In general, the plasma concentration of CBD is dose-dependent, although this can be nonlinear, especially at high doses (FDA, 2018b). The C_{max} of CBD can range between 0.5 ng/mL following a dose of 5.4 mg (Nadulski et al., 2005) and 782 ng/mL following a very high dose of 6000 mg (FDA, 2018b). More pronounced is the effect of food on the plasma concentration of CBD, whereby ingestion of CBD simultaneously with a high-fat meal can increase its plasma C_{max} by almost fivefold (FDA, 2018b; Taylor et al., 2018).

CBD is metabolized by different CYP isoforms, mainly in the liver. The resulting metabolites depend on the CYP isoform involved; CYP2C19 converts CBD to 7-COOH-CBD which is then oxidized to form its carboxylic acid derivatives. CYP 2C19 and 3A4 could hydroxylate CBD on the aromatic ring to form 6α-OH-CBD. Also, in vitro studies in hepatocytes and liver microsomes of rat, dog, and human have yielded 7-OH-CBD, 7-COOH-

TABLE 39.2 Phytocannabinoids, synthetic cannabinoids, and endocannabinoids.

Phytocannabinoids				
Δ^9-Tetrahydrocannabinol	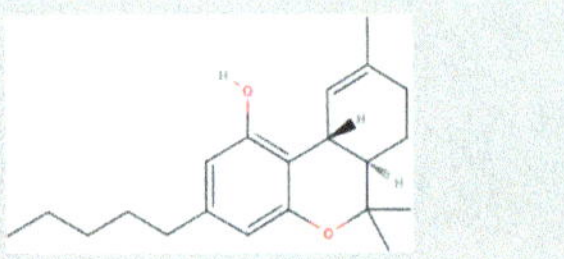	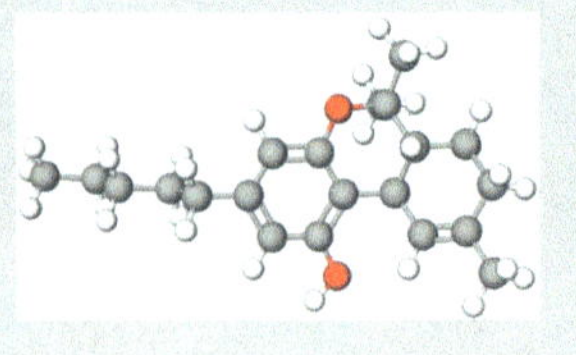	$CB_1 \approx CB_2$ agonist	Gaoni and Mechoulam (1964), Felder et al. (1995), Showalter et al. (1996), Rinaldi-Carmona et al. (1994), Rhee et al. (1997)
(−)Cannabidiol	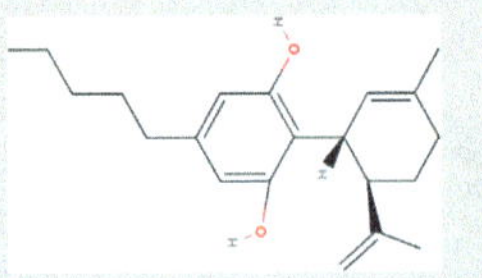	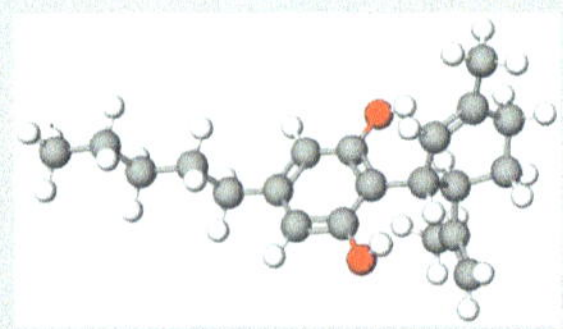	No activity at CB_1, inverse agonist at CB_2, high-potency antagonism of CB_1 and CB_2 agonists, modulator of α_1-adrenoreceptor, inhibition of AEA uptake and metabolism, TRP1A agonist	Adams et al. (1940), Mechoulam and Shvo (1963), Showalter et al. (1996), Jàrai et al. (1999), Pertwee et al. (2002), Bisogno et al. (2001), Thomas et al. (2007), De Petrocellis et al. (2011)
Cannabivarin (cannabivarol)	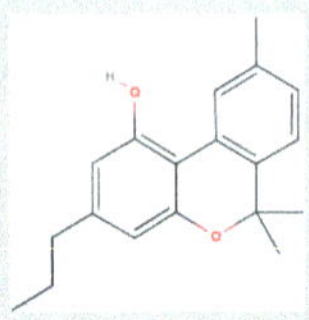	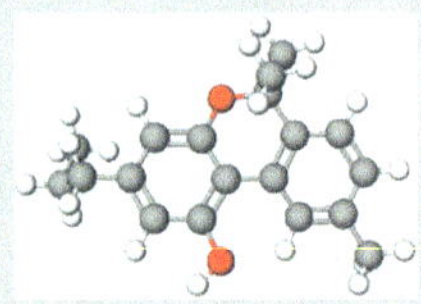	$CB_1 \approx CB_2$ antagonist	Thomas et al. (2005)
Cannabinol	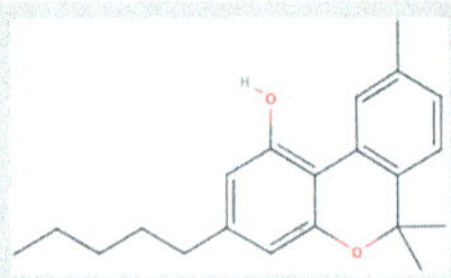	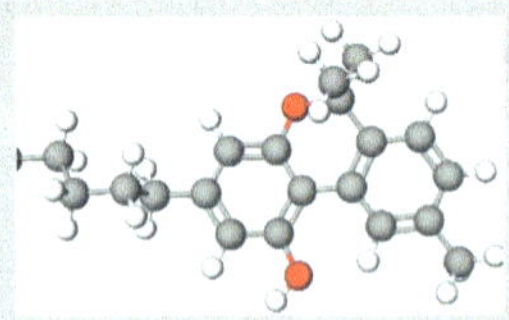	$CB_1 \approx CB_2$ agonist, TRP1A agonist, Transient receptor potential channel type M8 (TPRM8) antagonist	Rhee et al. (1997), McPartland et al. (2007), De Petrocellis et al. (2011)
Cannabigerol	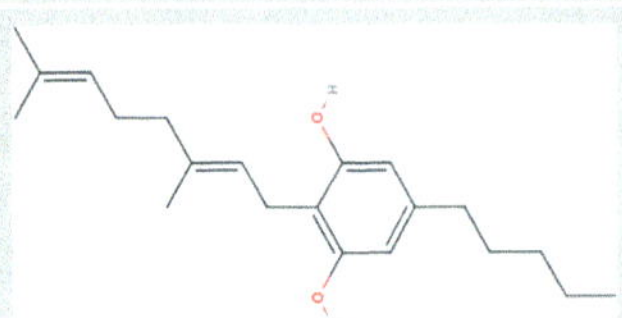	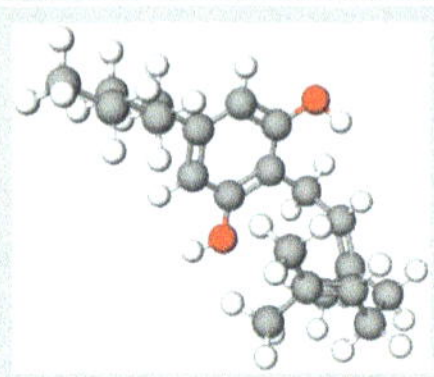	$CB_2 > CB_1$ partial agonist	Rosenthaler et al. (2014), Navarro et al. (2018)
Cannabichromene	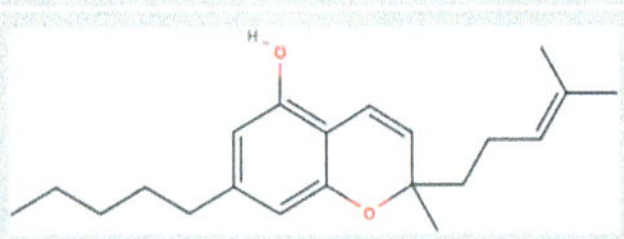	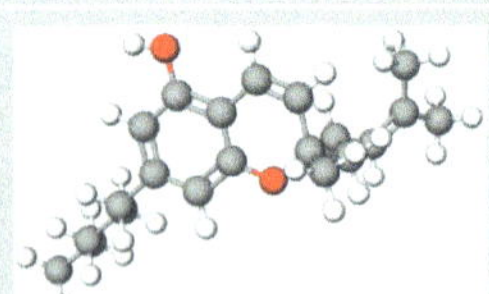	CB_2 selective agonist, TRP1A agonist	Udoh et al. (2019), De Petrocellis et al. (2011)
Δ^9-Tetrahydrocannabivarin	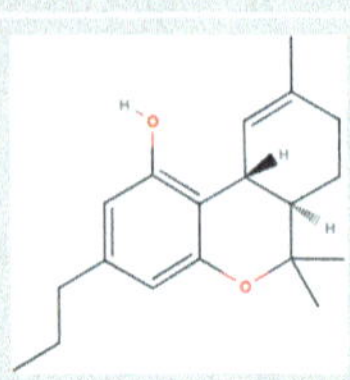	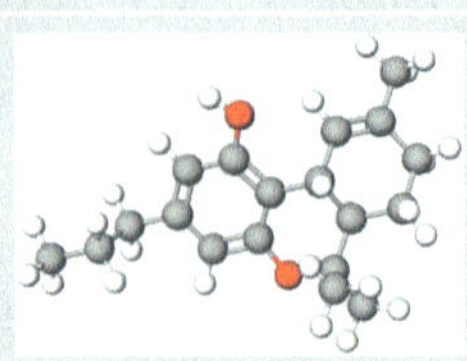	CB_1 antagonist, CB_2 partial agonist, TRP1A agonist	Pertwee (2008), McPartland et al. (2015), De Petrocellis et al. (2011)
Synthetic cannabinoids				
Ajulemic acid (CT-3, IP-751)	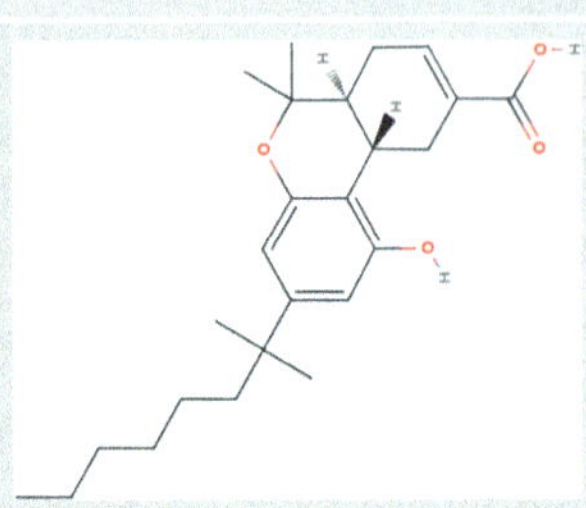	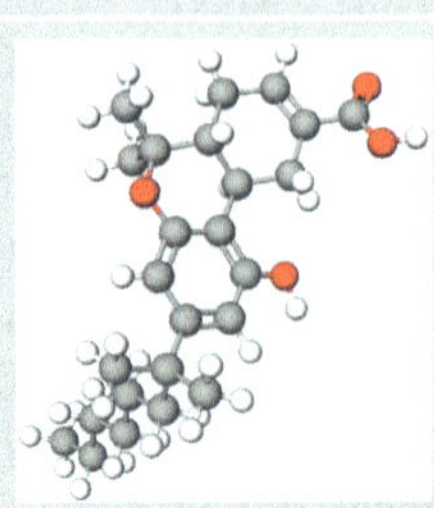	$CB_1 \approx CB_2$ agonist	Dyson et al. (2005)

(Continued)

TABLE 39.2 (Continued)

Phytocannabinoids				
(−) HU-210	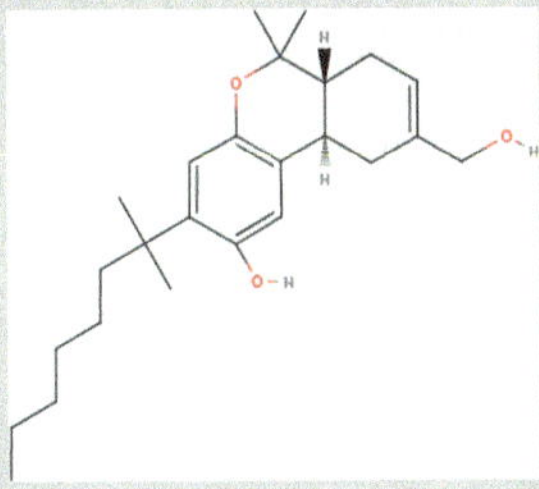	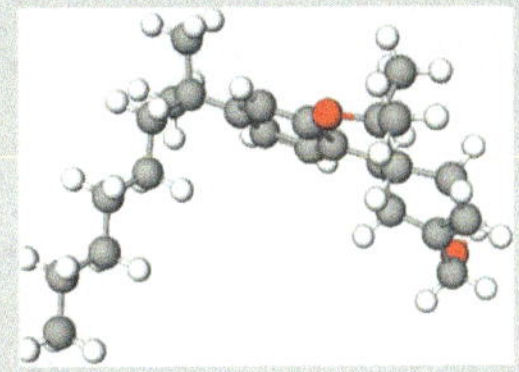	$CB_1 \approx CB_2$, agonist	Felder et al. (1995), Showalter et al. (1996)
(+) HU-211 (dexanabinol)	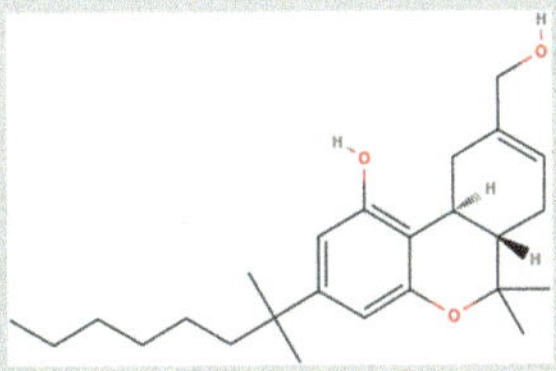	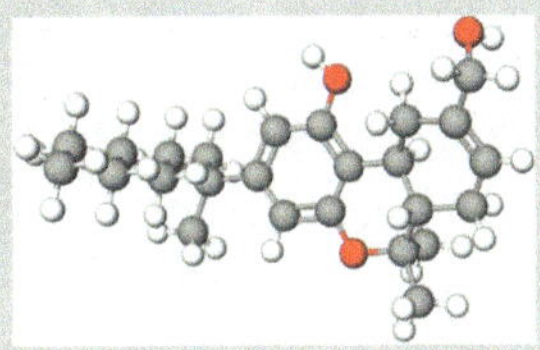	No activity at CB_1 or CB_2, noncompetitive N-Methyl-D-aspartate receptor (NMDA) antagonist	Showalter et al. (1996), Bar-Joseph et al. (1994)
WIN 55, 212-2	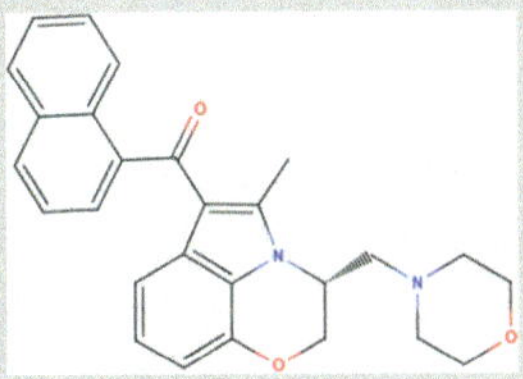	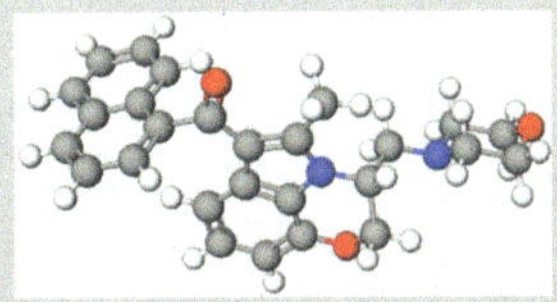	$CB_1 \approx CB_2$ agonist	Rinaldi-Carmona et al. (1994), Hillard et al. (1999), Felder et al. (1995)
(−) CP-55940	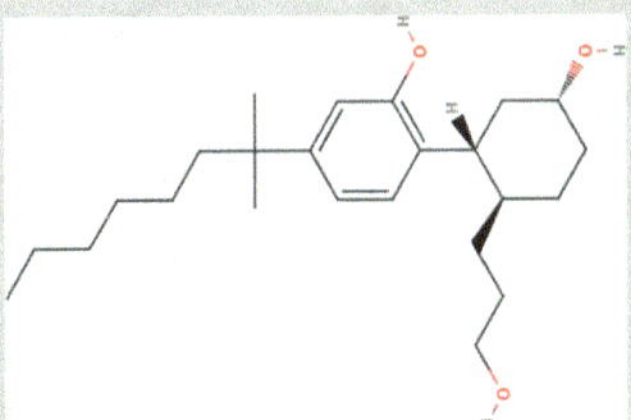	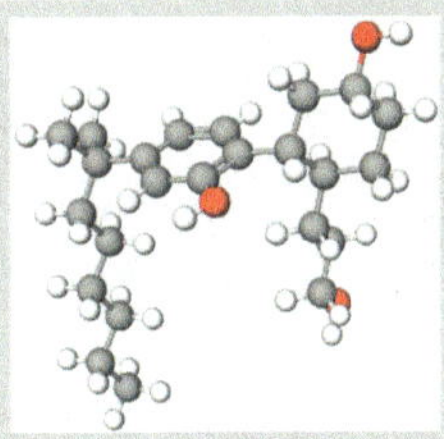	$CB_1 \approx CB_2$ agonist	Rinaldi-Carmona et al. (1994), Hillard et al. (1999), Felder et al. (1995), Ross et al. (1999)
Nabilone	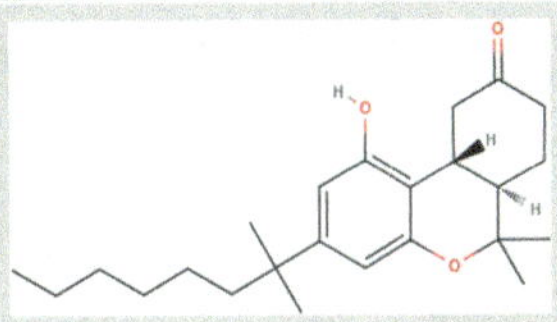	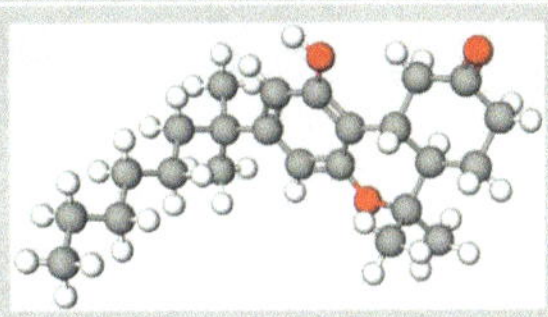	$CB_1 \approx CB_2$, agonist	Gareau et al. (1996)
SR141716A (Rimonabant)	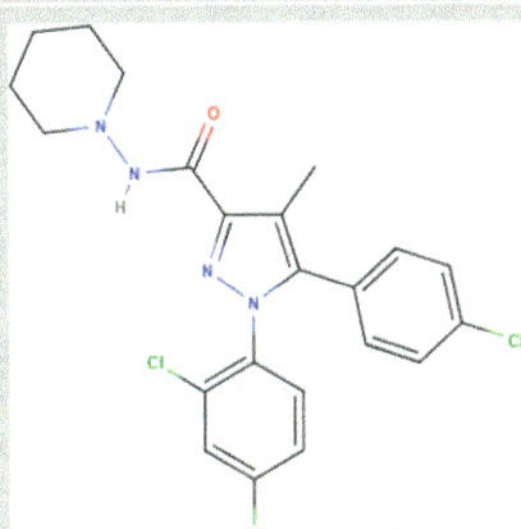	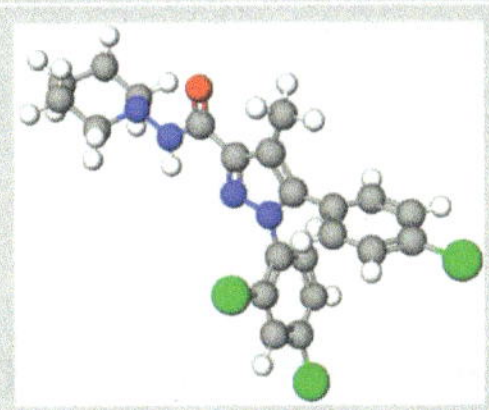	CB_1 selective antagonist	Rinaldi-Carmona et al. (1994)
AKB-48	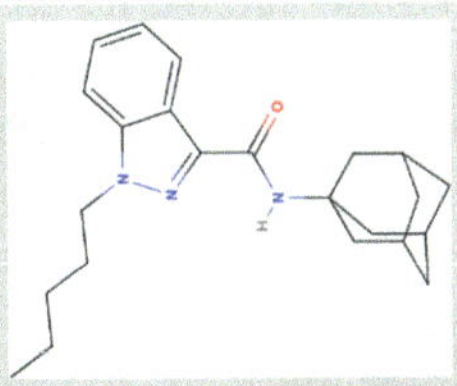	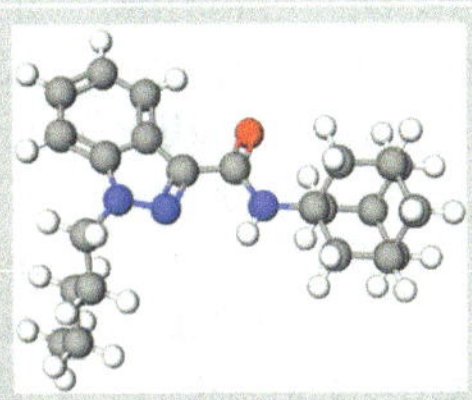	$CB_1 \approx CB_2$ agonist	Canazza et al. (2016)

(Continued)

TABLE 39.2 (Continued)

Phytocannabinoids				
Endogenous cannabinoids (eicosinoids)				
Anandamide (AEA)	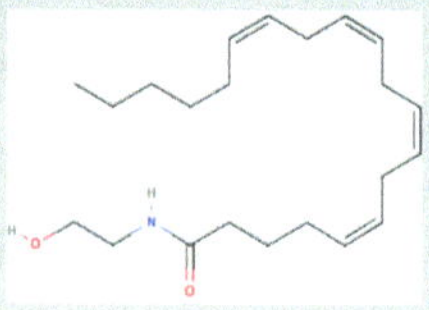	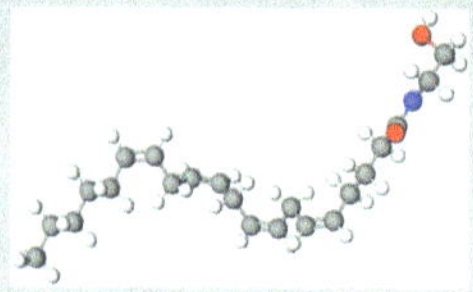	$CB_1 >> CB_2$ agonist, $TRPV_1$ agonist	Mechoulam et al. (1995), Showalter et al. (1996), Felder et al. (1995), Zygmunt et al. (1999)
2-Arachidonoyl glycerol	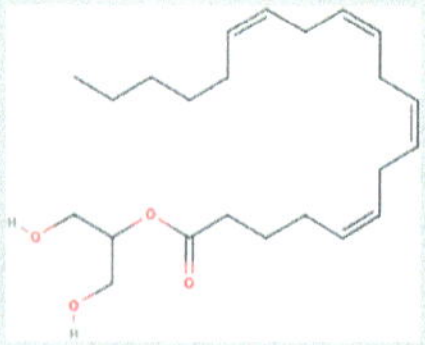	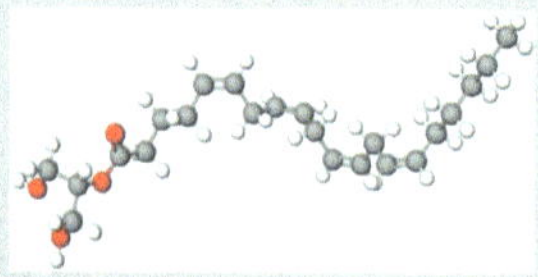	$CB_1 \approx CB_2$ agonist	Mechoulam et al. (1995), Ben-Shabat et al. (1998)
2-Arachidonoyl glycerol ether	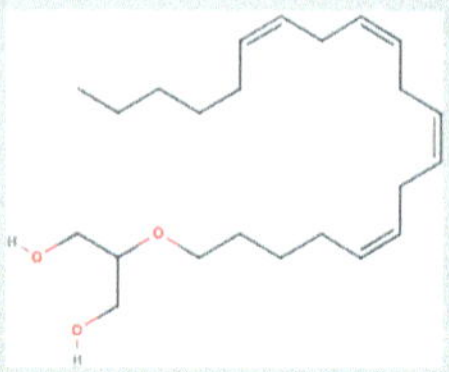	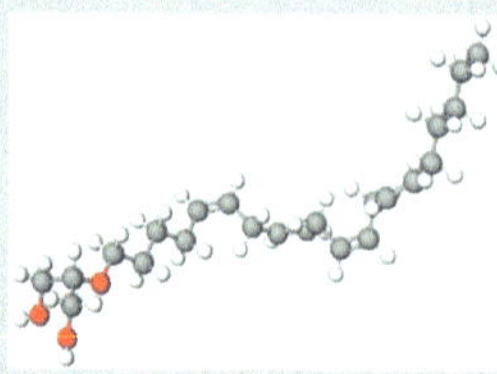	$CB_1 >> CB_2$ agonist	Hanus et al. (2001)
O-Arachidonoyl ethanolamine (virodhamine)	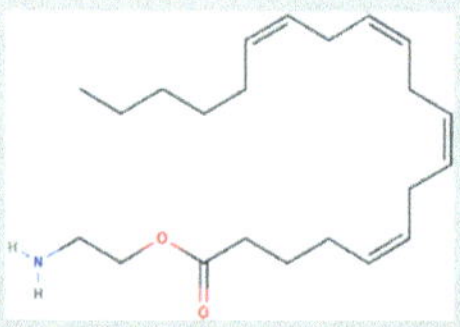	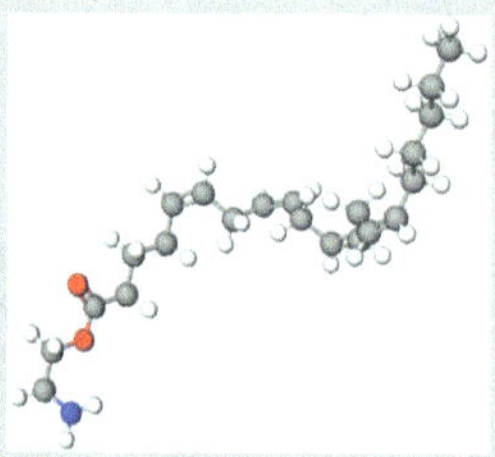	$CB_1 >> CB_2$ agonist	Porter et al. (2002)
N-Arachidonoyl dopamine	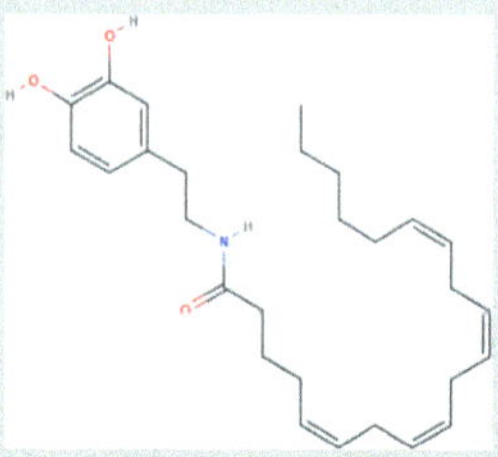	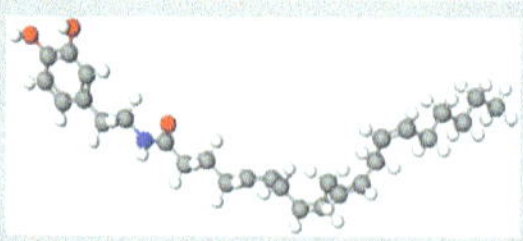	$CB_1 >> CB_2$ agonist, $TRPV_1$ agonist	Bisogno et al. (2000), Huang et al. (2002)

CB1 and *CB2*, cannabinoid 1 and 2; *TRPV1*, transient receptor potential vanilloid type 1.

CBD, and 6-OH-CBD as the main metabolites. CYPs 1A2, 2D6, 2C19, and 3A have been shown to yield other hydroxylated derivates including 1'-OH-CBD, 2'-OH-CBD, 3'-OH-CBD, 4'-OH-CBD, 6β-OH-CBD, and the carboxylic acid derivative 7-COOH-CBD (FDA, 2018a; Jiang et al., 2011). On the other hand, clinical studies reveal that CBD could be metabolized in the liver and gastrointestinal tract (GIT) to yield 6-OH-CBD, 7-OH-CBD, and 7-COOH-CBD (FDA, 2018a) with 7-OH-CBD accounting for 38% of all metabolites of CBD and 6-OH-CBD being the least abundant (less than 10%). The amount of metabolite obtained also depends on the dose of CBD ingested; hence, the plasma area under the curve (AUC) for 7-COOH-CBD could be as high as 46 times that of the parental compound, indicating a rapid metabolism to produce the carboxylic acid derivative. Based on human steady-state pharmacokinetics data, however, the main circulating human metabolites of CBD were 7-OH-CBD and 7-COOH-CBD. Finally, metabolism of CBD in humans involves the conjugative enzymes Uridine glucuronosyltransferases (UGTs) 1A7, 1A9, and 2B7 (FDA, 2017). Other preclinical animal models have shown that CBD can be absorbed into the systemic circulation when given via the transdermal route in the rat (Hammell et al.,

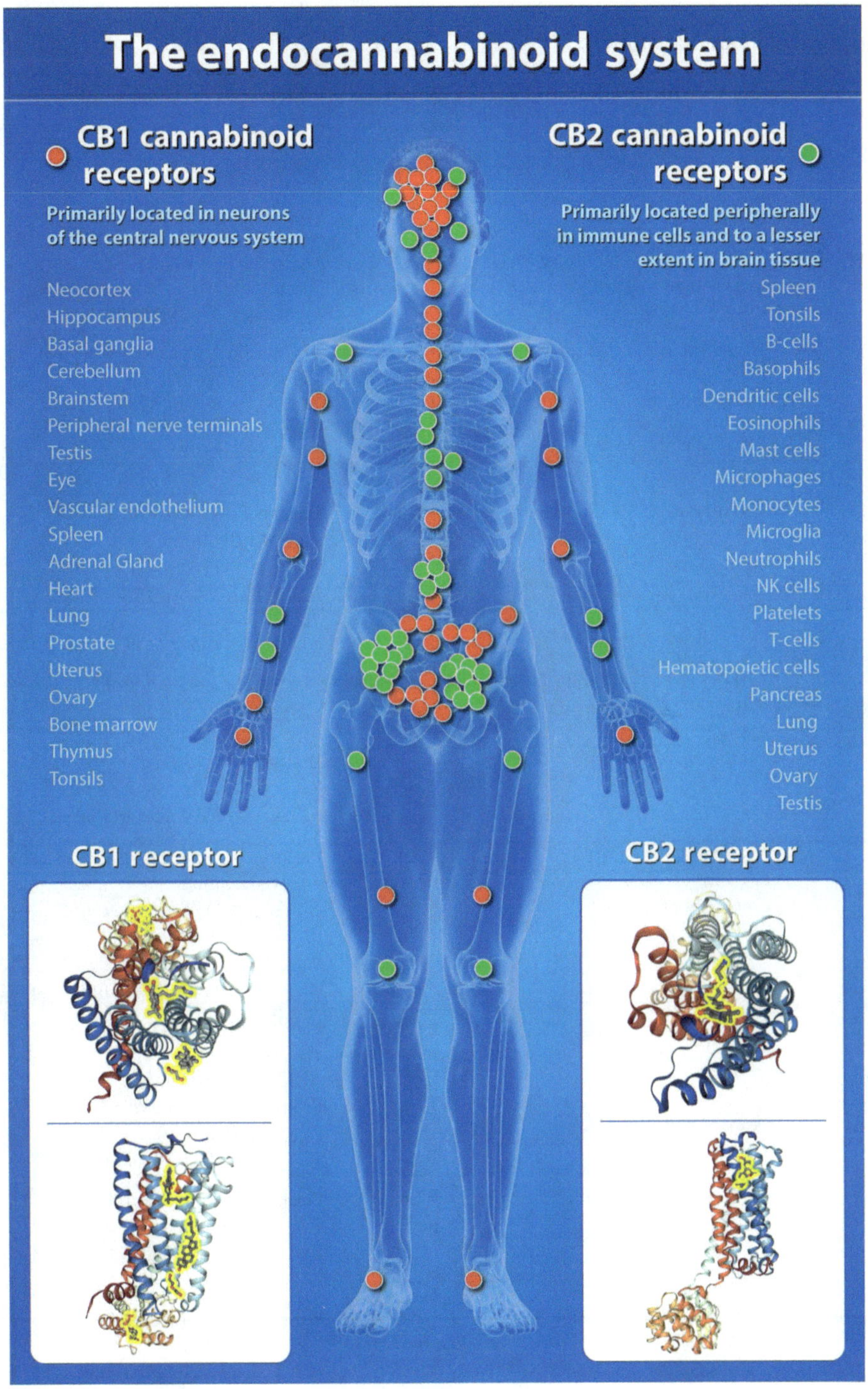

FIGURE 39.3 Cannabinoid receptor expression in human tissues. Image created by Ken Whittaker of KGK Science Inc.

2016), mouse (Giacoppo et al., 2015), and guinea pig (Paudel et al., 2010). Further, these animal studies have given insight into the distribution and elimination of CBD as orally administered CBD has been detected at high concentrations in the liver with concentrations as high as 8.2 μg/g of tissue in rats. Biliary excretion is also prominent and has been shown to account for almost 90% of the total CBD eliminated in rats (FDA, 2018a).

Toxicity of cannabidiol

Genotoxicity testing

The genotoxicity potential of CBD has been investigated in different studies. As part of a series of studies carried out in the development of the CBD-based antiepileptic drug Epidiolex by GW Pharma, an Ames reverse mutation assay was conducted on purified CBD using specific

TABLE 39.3 Pharmacokinetics of cannabidiol in animal models.

Study	Species/design	Compound/ dose	Route	Sample size	Outcome
Lodzki et al. (2003)	**Mouse**. PK of cannabidiol (CBD) from an ethosomal formulation	6 mg	Transdermal	$n=24$	$C_{max}=1.07$ μg/mL, $T_{max}=4$ h
Paudel et al. (2010)	**Guinea pig**. PK of CBD from different formulations	18 mg/mL of CBD alone	transdermal	$n=6$	C_{max}=8.6 ng/mL, $T_{max}=38.4$ h, $C_{ss}=6.3$ ng/mL, $AUC_{0\text{-}48}=276$ ng · h/mL
		18 mg/mL of CBD in transcutol	Transdermal	$n=3$	C_{max}=35.6 ng/mL, $T_{max}=31.2$ h, $C_{ss}=23.7$ ng/mL, $AUC_{0\text{-}48}=888$ ng · h/mL
		1 mg/kg CBD	iv	$n=6$	C_{max}=269 ng/mL, $AUC_{0\text{-}48}=175$ ng · h/mL
Deiana et al. (2012)	**Rat.** Evaluation of PK of CBD formulated in different solvents	120 mg/kg CBD in cremophor	Oral	$n=24$ (4/ timepoint)	Plasma $C_{max}=2$ μg/mL, $T_{max}=120$ min, $AUC_{0\text{-}\infty}=1175$ μg/mL · min Brain $C_{max}=8.6$ μg/mL, $T_{max}=240$ min, $AUC_{0\text{-}\infty}=4941$ μg/mL · min
		120 mg/kg CBD in Solutol	Oral	$n=24$ (4/ timepoint)	Plasma $C_{max}=3.2$ μg/mL, $T_{max}=360$ min, $AUC_{0\text{-}\infty}=2623$ μg/mL · min Brain $C_{max}=12.6$ μg/mL, $T_{max}=240$ min, $AUC_{0\text{-}\infty}=9574$ μg/mL · min
		120 mg/kg CBD in cremophor	ip	$n=24$ (4/ timepoint)	Plasma $C_{max}=2.6$ μg/mL, $T_{max}=120$ min, $AUC_{0\text{-}\infty}=1178$ μg/mL · min Brain $C_{max}=6.8$ μg/mL, $T_{max}=120$ min, $AUC_{0\text{-}\infty}=4532$ μg/mL · min
			ip	$n=24$ (4/ timepoint)	Plasma $C_{max}=2.4$ μg/mL, $T_{max}=30$ min, $AUC_{0\text{-}\infty}=2463$ μg/mL · min Brain $C_{max}=5.2$ μg/mL, $T_{max}=360$ min, $AUC_{0\text{-}\infty}=7031$ μg/mL · min
	Mouse. Evaluation of PK of CBD formulated in different solvents	120 mg/kg CBD	Oral	$n=24$ (4/ timepoint)	Plasma $C_{max}=2.2$ μg/mL, $T_{max}=60$ min, Brain $C_{max}=1.3$ μg/mL, $T_{max}=360$ min
		120 mg/kg CBD	ip	$n=24$ (4/ timepoint)	Plasma C_{max}=14.3 μg/mL, $T_{max}=120$ min, $AUC_{0\text{-}\infty}=3181$ μg/mL · min Brain $C_{max}=6.9$ μg/mL, $T_{max}=60$ min, $AUC_{0\text{-}\infty}=1903$ μg/mL · min
Giacoppo et al. (2015)	**Mouse.** PK of transdermal 1% CBD cream in a model of multiple sclerosis	1% CBD cream	Transdermal	$n=5$	C_{max}=8.3 ng/mL, $T_{max}=38.2$ h, $C_{ss}=6.1$ ng/mL
Hammell et al. (2016)	**Rat.** Plasma levels of CBD creams	0.62 mg CBD	Transdermal	$n=9$	$C_{max}=3.8$ ng/mL

(Continued)

TABLE 39.3 (Continued)

Study	Species/design	Compound/ dose	Route	Sample size	Outcome
		3.1 mg CBD		$n = 8$	$C_{max} = 17.5$ ng/mL
		6.2 mg CBD		$n = 8$	$C_{max} = 33.3$ ng/mL
		62.3 mg CBD		$n = 4$	$C_{max} = 1629.9$ ng/mL
FDA (2018b)	**Mouse.** TK study during a 13-week repeat-dose study. Plasma taken on day 82 of study (Epidiolex® study)	400 mg/kg CBD (Male)	Oral	$n = 18$	C_{max}=6420 ng/mL, $T_{max} = 4$ h,
		400 mg/kg CBD (female)		$n = 18$	C_{max}=8740 ng/mL, $T_{max} = 6$ h
		550 mg/kg CBD (male)		$n = 18$	C_{max}=7050 ng/mL, $T_{max} = 2$ h, $AUC_{0-\infty} = 32{,}100$ ng · h/mL
		550 mg/kg CBD (female)		$n = 18$	C_{max}=8440 ng/mL, $T_{max} = 2$ h, $AUC_{0-\infty} = 41{,}700$ ng · h/mL
		625 mg/kg CBD (male)		$n = 18$	C_{max}=7020 ng/mL, $T_{max} = 2$ h, $AUC_{0-\infty} = 41{,}900$ ng · h/mL
		700 mg/kg CBD (female)		$n = 18$	C_{max}=8110 ng/mL, $T_{max} = 2$ h, $AUC_{0-\infty} = 47{,}900$ ng · h/mL
FDA (2018b)	**Rat.** PK of CBD in rats following single-dose administration	10 mg/kg	Oral		C_{max}=295 ng/mL, $T_{max} = 1$ h
			iv	N/A	C_{max}=10,556 ng/mL, $T_{max} = 5$ min
			ip	N/A	C_{max}=549 ng/mL, $T_{max} = 15–30$ min

AUC, area under the curve; C_{max}, maximum concentration; *PK*, pharmacokinetic; *TK*, toxicokinetic; T_{max}, time to maximum concentration.

strains of *Salmonella typhimurium* (strains TA98, TA100, TA1535, TA1537, and TA102) with or without metabolic bioactivation at a maximum concentration of 5000 μg/plate. The test was negative for mutagenicity (FDA, 2018a). Further, CBD has tested negative in other genotoxicity studies including in vitro micronucleus test with Chinese hamster ovary cells, in vivo micronucleus test in rats and mice, and alkaline comet assay in male Sprague-Dawley rats (FDA, 2018a; Marx et al., 2018).

In vitro toxicology studies

There are conflicting data on the effect of CBD on cell viability, with the majority in vitro studies suggesting the potential of CBD to induce cell damage. CBD has been shown to induce oxidative stress as evidenced by reduced glutathione level accompanied by an activation of caspase 8 (Wu et al., 2008) and at concentrations ranging from 4–16 μM, CBD-induced apoptosis in mouse splenocytes. In vivo, a similar effect has been demonstrated in mouse lymphocytes following an intraperitoneal injection of 10 mg/kg CBD (Lee et al., 2008). An in ovo study indicated that 50 μM CBD induced an 80% reduction in viability of the fertilized Leghorn chicken egg embryos (Gustafsson and Jacobsson, 2019). CBD-induced apoptosis via the activation of capsase-3, capsase-8, and capsase-9 in leukemia cells and human glioma cells (Massi et al., 2006; McKallip et al., 2006), via caspase-3 activation in human myeloblastic leukemia cells (Gallily et al., 2003), via formation of reactive oxygen species (ROS) in murine thymocytes, EL-4 thymoma cells, lymphocytes, and precultured human monocytes (Lee et al., 2008; Wu et al., 2010; Wu et al., 2008), and via an increase in the expression of caveolin-1 and GM1 ganglioside in human microglial cell line (Wu et al., 2012). A follow-up study by Wu et al. (2018) suggested that CBD-induced apoptosis in human monocytes was via ROS production induced by mitochondrial permeability transition pore. CBD was recently shown to have potential to induce apoptosis by increasing the production of ROS thereby causing an inhibition of SGC-7901 cell proliferation in a human gastric cancer cell line (Zhang et al., 2019).

TABLE 39.4 Pharmacokinetics of cannabidiol in human subjects alone or in combination with other compounds.

Study	Study design	Compound/ dose	Route	Main outcomes
Ohlsson et al. (1986)	PK of cannabidiol (CBD) in male marijuana smokers ($n = 5$)	^{2}H-CBD at 20 mg single dose	iv	$T_{max} = 3$ min; $C_{max} = 686$ ng/mL, $AUC_{0\text{-}72} = 16.67$ ng · min/mL; $t_{1/2} = 24$ h
		^{2}H-CBD at 20 mg single dose	Inhalation	$C_{max} = 110$ ng/mL, $T_{max} = 3$ min, $t_{1/2} = 31$ h, $AUC_{0\text{-}72} = 4.85$ ng · min/mL
Nadulski et al. (2005)	Detection of cannabinoids method ($n = 24$)	5.4 mg of CBD	Oral	$C_{max} = 0.95$ ng/mL, $T_{max} = 59.6$ min
Guy and Flint (2004)	Tetrahydrocannabinol (THC) and CBD combinations in different delivery routes ($n = 6$)	20 mg of CBD	Sublingual	$C_{max} = 2.02$ ng/mL, $T_{max} = 130$ min, $AUC_{0\text{-}t} = 156.13$ ng · min/mL
		CBD:THC (20 mg: 20 mg)	Sublingual	$C_{max} = 2.58$ ng/mL, $T_{max} = 100$ min, $AUC_{0\text{-}t} = 209.30$ ng · min/mL
		CBD:THC (20 mg: 20 mg)	Aerosol	$C_{max} = 2.60$ ng/mL, $T_{max} = 141$ min, $AUC_{0\text{-}t} = 325.93$ ng · min/mL
		CBD:THC (20 mg: 20 mg)	Nebulized	$C_{max} = 9.49$ ng/mL, $T_{max} = 36$ min, $AUC_{0\text{-}t} = 564.35$, ng · min/mL
Karschner et al. (2011)	Comparison of PK of CBD following different doses of Sativex	Sativex CBD: THC (15 mg: 16.2 mg)	Oro-mucosal	$C_{max} = 6.7$ ng/mL, $T_{max} = 4$ h, $AUC_{0\text{-}10.5} = 18.1$ ng · h/mL
		Sativex CBD: THC (5 mg: 5.4 mg)	Oro-mucosal	$C_{max} = 1.6$ ng/mL $T_{max} = 3.7$ h, $AUC_{0\text{-}10.5} = 4.5$ ng · h/mL
Stott et al. (2013)	PK of CBD in the presence of inhibitors and inducers of CYP enzymes ($n = 12$)	Sativex @ CBD: THC (10 mg: 10.8 mg) + rifampicin	Oro-mucosal	Before Rifampicin: $C_{max} = 1.03$ ng/mL; $T_{max} = 1$ h; After rifampicin: $C_{max} = 0.5$ ng/mL; $T_{max} = 1.5$ h
		Sativex @ CBD: THC (10 mg: 10.8 mg) + Ketoconazole		Before Ketoconazole: $C_{max} = 0.66$ ng/mL; $T_{max} = 1.38$ hAfter Ketoconazole: $C_{max} = 1.25$ ng/mL; $T_{max} = 1.75$ h
Sellers et al. (2013)	PK of CBD in ultra-high dose oromucosal Sativex ($n = 258$)	Sativex @ CBD: THC (20 mg: 21.6 mg)	Oro-mucosal	$C_{max} = 1.5$ ng/mL, $T_{max} = 1.4$ h, $AUC_{0\text{-}t} = 6.1$ ng · h/mL
		Sativex @ CBD: THC (60 mg: 64.8 mg)		$C_{max} = 4.8$ ng/mL, $T_{max} = 1.5$ h, $AUC_{0\text{-}t} = 38.9$ ng · h/mL

(Continued)

TABLE 39.4 (Continued)

Study	Study design	Compound/ dose	Route	Main outcomes	
Manini et al. (2015)	PK of CBD in patients intravenously administered with fentanyl (*n* = 6/group)	400 mg CBD + 0.5 μg/ kg or 1 μg/kg of iv Fentanyl; 800 mg CBD + 0.5 μg/ kg of or 1 μg/kg of iv Fentanyl	Oral	400 mg CBD + 0.5 μg/kg Fentanyl	C_{max} = 181 ng/mL, T_{max} = 3 h, AUC = 704 μg · h/dL
				400 mg CBD + 1 μg/kg Fentanyl	C_{max} = 114.2 ng/mL, T_{max} = 3 h, AUC = 482 μg · h/dL
				800 mg CBD + 0.5 μg/kg Fentanyl	C_{max} = 221.1 ng/mL, T_{max} = 3 h, AUC = 867 μg · h/dL
				800 mg CBD + 1 μg/kg Fentanyl	C_{max} = 157.1 ng/mL, T_{max} = 4 h, AUC = 722 μg · h/dL
Haney et al. (2016)	PK (*n* = 31)	CBD @ 800 mg	Oral	C_{max} = 77.9 ng/mL, T_{max} = 3 h	
Cherniakov et al. (2017)	PK of a pro-nanoliposphere oral formulation (*n* = 9)	10 mg of CBD + 10 mg of THC	Oral	AUC = 6.9 ng · h/mL, C_{max} = 2.1 ng/mL, T_{max} = 1 h	
		Sativex 10 mg of CBD + 10.8 mg of THC	Oro-mucosal	AUC = 3.1 ng · h/mg, C_{max} = 0.5 ng/mL, T_{max} = 3 h	
Devinsky et al. (2018)	PK of CBD in pediatrics with Dravet syndrome (*n* = 32)	CBD given at 5, 10, or 20 mg/kg daily for 4 weeks	Oral	Day 1: C_{max} and T_{max} were similar at all dosesDay 22: 5 mg/kg: C_{max} = 100 ng/mL, T_{max} = 2.5 h 10 mg/kg: C_{max} = 280 ng/mL, T_{max} = 2.5 h 20 mg/kg: C_{max} = 380 ng/mL, T_{max} = 2.5 h	
Atsmon et al. (2018)	PK of CBD in an oral gelatin matrix pellet technology (PTL101) (*n* = 15)	10 mg CBD in PTL101	Oral	C_{max} = 3.22 ng/mL, T_{max} = 3 h, $AUC_{0\text{-}t}$ = 9.64 ng · h/mL	
		100 mg CBD in PTL101	Oral	C_{max} = 47.44 ng/mL, T_{max} = 3.5 h, $AUC_{0\text{-}t}$ = 149.54 ng · h/mL	
		Sativex @ CBD: THC (10 mg: 10.8 mg)	Oro-mucosal	C_{max} = 2.05 ng/mL, T_{max} = 3.5 h, $AUC_{0\text{-}t}$ = 7.3 h	
FDA (2018b) (GW)	Study of Epidiolex®: PK of CBD following single oral dose (*n* = 9)	1500 mg CBD	Oral	C_{max} = 292 ng/mL; T_{max} = 4.0 h; $AUC_{0\text{-}\infty}$ = 1618 ng/mL · min; $t_{1/2}$ = 14.4 h	
		3000 mg CBD		C_{max} = 533 ng/mL; T_{max} = 5.0 h; $AUC_{0\text{-}\infty}$ = 2809 ng/mL · min; $t_{1/2}$ = 14.4 h	

(Continued)

TABLE 39.4 (Continued)

Study	Study design	Compound/ dose	Route	Main outcomes
		4500 mg CBD		C_{max} = 722 ng/mL; T_{max} = 5.0 h; $AUC_{0-\infty}$ = 3426 ng/mL · min; $t_{1/2}$ = 16.6 h
		6000 mg CBD		C_{max} = 782 ng/mL; T_{max} = 5.0 h; $AUC_{0-\infty}$ = 3900 ng/mL · min; $t_{1/2}$ = 15.4 h
Taylor et al. (2018)	Study of the effect of food interaction with Epidiolex®: PK of CBD following single or multiple oral doses (n = 12 for multiple-dose fasting study and n = 9 for food study)	750, 1500 mg CBD single or multiple doses fasted or with high-fat food	Oral	750 mg single dose (fasted); C_{max} = 290 ng/mL 750 mg three doses (fasted); C_{max} = 330 ng/mL1500 mg single dose (fasted); C_{max} = 335 ng/mL1500 mg three doses (fasted); C_{max} = 541.2 ng/mL
				1500 mg single dose (high-fat meal); C_{max} = 1628 ng/mL

AUC, area under the curve; C_{max}, maximum concentration; $t_{1/2}$, biological half-life; T_{max}, time to maximum concentration.

It does not appear that CBD-induced apoptosis is necessarily pathogenic in nature. This is shown in a study where apoptosis only occurs in activated stellate cells which promote hepatic fibrosis, whereas quiescent stellate cells were not susceptible to CBD-induced apoptosis (Lim et al., 2011). Similarly, CBD did induce apoptosis and autophagy via ROS production in breast cancer cells (Shrivastava et al., 2011). Similar observations have been made in other cancer cell lines (Massi et al., 2004; Olivas-Aguirre et al., 2019). Other studies have suggested that CBD has vasculo-protective and antiapoptotic effects via an induction of heme oxygenase 1 (Schwartz et al., 2018). Also, a study in oligodendrocyte progenitor cells have attributed prosurvival, antiapoptotic properties to CBD whereby an increase in the antiapoptotic protein Bcl-2 was reported during inflammation (Mecha et al., 2012).

CBD was shown to inhibit the activity of drug transport protein ATP-binding cassette (ABCC1) in a human ovarian cell line (Holland et al., 2008) and ATP-binding cassette transporter (ABC) ABCG2 in mouse embryonic fibroblasts (Holland et al., 2007). Effects of CBD on other drug transport proteins have been reviewed (Ujváry and Hanuš, 2016). An inhibition of efflux drug transporters relates to an enhancement of the cell concentration of any other drug that might be coadministered with CBD; this could potentially make for increased adverse effect and toxicity of the drug (Brown and Winterstein, 2019). The effect of CBD on drug transporters also relates to its other well-reported effects on the pharmacokinetics of other drugs, including their distribution and metabolism. In vitro studies have indicated that CBD has the potential to inhibit the catalytic activities of cytochromes P450 isotypes 1A1, 1A2, 2C9, 2B6, 2D6, and 3A5/7 (Arnold et al., 2018; Ibeas Bih et al., 2015; Jiang et al., 2011; Yamaori et al., 2010, 2011, 2012). Potential for interaction between CBD and inducers and inhibitors of the CYPs enzyme family has been reviewed in Brown and Winterstein (2019).

Toxicity in animal models—evidence from preclinical studies

Acute toxicity of cannabidiol

Single-dose toxicity of CBD has been tested in a variety of animal species and several routes of administration. While some were studies specifically designed to investigate the acute toxicity of this compound, others were pharmacological studies in which CBD was applied as a single-dose treatment. For example, the oral median lethal dose (LD_{50}) of CBD was demonstrated in mice at 12.7 g/kg body weight (Loewe, 1946) and the intravenous LD_{50} was reported as 50 mg/kg (and 30 mg/kg for the levo isomer) (Uliss et al., 1975). In male and female rhesus monkeys, purified (99%) CBD had an intravenous LD_{50} of 212 mg/kg, with the leading cause of death among the test animals being tremor, convulsion, spasm, respiratory depression, and cardiac

TABLE 39.5 Acute toxicity studies of cannabidiol and derived LD_{50} values.

Study	Species/strain	Study design	Dose	Sample size	Maximum tolerable dose/LD_{50}	HED for LD_{50} or maximum tolerable dose
Loewe (1946)	Mice (unreported strain)	Single oral dose	Up to 12.7 g/kg	$n = 11$	>12.7 g/kg	>1032.5 mg/kg
Uliss et al. (1975)	Male albino CD-1 mice	Single iv dose	50 mg/kg	$n < 6$	50 mg/kg	4.07 mg/kg
Fairbairn and Pickens (1980)	LACA Tuck no.1 albino mice	Single ip and oral dose	250–1000 mg/kg ip; 500, 2000 mg/kg oral	$n = 3–10$/group	380 mg/kg ip > 2000 mg/kg oral	162.6 mg/kg oral
Steger et al. (1990)	Sprague-Dawley rats	Single oral dose, sacrificed after 1 h	0.1–10 mg/kg BW	$n = 8–12$	10 mg/kg	1.6 mg/kg
FDA (2018a)	Male Sprague-Dawley rats	Rat micronucleus test. Two doses administered	CBD @ 0, 125, 250, and 500 mg/kg	$n = 6$/group	500 mg/kg was well-tolerated	80.6 mg/kg
FDA (2018a)	Male Sprague-Dawley rats	Rat alkaline comet assay	CBD @ 0, 125, 250, and 500 mg/kg/day	$n = 6$/group	500 mg/kg was well-tolerated	80.6 mg/kg
Marx et al. (2018)	Crl:NMRI BR mice	In vivo mouse micronucleus test	CBD-rich hemp extract at 0, 500, 1000, or 2000 mg/kg oral dose	$n = 5–10$/group	500 mg/kg well-tolerated	80.6 mg/kg
Ewing et al. (2019b)	8-week-old male B6C3f1/J mice	Single oral dose, sacrificed after 24 h	246, 738, 2460 mg/kg (acute)	$n = 6$/group	CBD reported to be well-tolerated at 246 mg/kg single dose	20 mg/kg

CBD, cannabidiol; *HED*, human equivalent dose, based on conversion factors extracted from Nair and Jacob (2016); *MTD*, maximum tolerable dose.

failure (Rosenkrantz et al., 1981). While the male monkeys in the study had reduced testicular weights, the females had increases in ovarian weights. Table 39.5 summarizes the outcomes from studies reporting acute toxicity and highlights the interspecies differences that suggests that selection of the appropriate animal model is critical in assessing the safety of CBD.

Repeat-dose toxicity studies

Short-term and subchronic toxicity studies of CBD-rich hemp extract and purified CBD isolate have reported varying toxicities that appear to be influenced by both the species tested and length of exposure. The no-observed-adverse-effect level (NOAEL) established by these studies ranges from the human equivalent of 1.61 mg/kg/day to 55.5 mg/kg/day for CBD isolate and up to 161 mg/kg/day for CBD-rich hemp extract. Table 39.6 describes the NOAEL and human equivalent doses established by repeat-dose toxicity studies across several species. Despite the varying NOAELs reported, similarities in the adverse events reported in the literature have established areas of concern.

Male and female Wistar rats given CBD-rich hemp extract at oral doses up to 3000 mg/kg for 2 weeks revealed essentially normal blood chemistry and hematological findings. However, histological analysis found cytoplasmic vacuolation in hepatocytes and adrenal cortical cells as well as impairment of spermatogenesis in the high doses of 2000 and 3000 mg/day. All other major organs were normal (Marx et al., 2018). In a follow-up 90-day study in which male and female Wistar rats were orally dosed with CBD-rich hemp extract at doses up to 720 mg/kg, a dose-dependent trend of toxicity with cases of alveolar histiocytosis, kidney cysts, and cytoplasmic vacuolation in the adrenal glands of both male and female rats and abnormal hematological and clinical chemistry parameters were reported. Similarly, a 26-week study in Wistar rats demonstrated few cases of hepatocellular hypertrophy at doses of 50 and 150 mg/kg which the authors interpreted as being nonclinically significant. Liver abnormalities have also been reported in a 39-week study in male and female Beagle dogs orally dosed with CBD up to a maximum dose of 100 mg/kg/day. Moderate increases in the levels of hepatic ALT enzyme and hepatocyte hypertrophy in both sexes of the test animals were reported (FDA, 2018a). Further, a 104-week dietary carcinogenicity study was carried out in Wistar rats using 58%–67% pure CBD and daily oral doses of 5, 15, or 50 mg/kg CBD were delivered to the test animals. Aside from dose-related reduction in body weights of the animals, there were no remarkable findings from this study. Survival was not impacted by CBD administration and there was no induction of any rare tumors; there was also no test-article-related increase in the incidence of neoplasia. However, a postmarketing oral gavage study was planned by the sponsor of the study. The result from this study indicates that CBD is potentially noncarcinogenic (FDA, 2018a). Moreover, one of the earliest studies conducted in Rhesus monkeys at doses ranging between 30–300 mg/kg/day purified CBD for 90 days failed to determine a NOAEL value as there were cases of test-article-related effects on liver, kidneys, and heart in male and female monkeys, despite normal hematological and blood chemistry values (Rosenkrantz et al., 1981). Testicular sizes were reduced in male monkeys with a corresponding reduction in spermatogenesis, similar to the observations in rats described above. The reported effects on liver function and spermatogenesis across several species have raised concerns that similar toxicities may carry over to humans at physiologically relevant doses. These concerns are addressed in further detail throughout this chapter with a cautionary focus on the interpretation of results not relevant to humans.

Cannabidiol induces reproductive/developmental effects in rats and rabbits

There are conflicting data as to the effect of CBD on reproductive functions in different animal models. One of the earliest studies of the effect of CBD on reproductive function in mice which were orally dosed with CBD daily for 5 days showed no significant effect on sperm morphology (Zimmerman et al., 1979). However, repeat-dose studies in CBD have suggested that it induced potentially reproductive effects, mainly in male rats and male Rhesus monkeys (Marx et al., 2018; Rosenkrantz et al., 1981). Specific reproductive and developmental toxicity studies were conducted in rats and rabbits to assess the effect of oral exposure to CBD in these as shown in the GW Pharmaceutical study cited above (FDA, 2018a). Rat studies revealed no adverse effects in reproductive performance of male and female rats at doses up to 250 mg/kg/day of orally administered purified CBD. Similarly, the developmental toxicity of purified CBD was assessed at a dose of 250 mg/kg with no toxic effect on embryo-fetal development, leading to the determination of developmental NOAEL of 250 mg/kg/day for CBD. However, a recent study suggests that 30 mg/kg CBD inhibited sexual performance in male Swiss rats as well as a 30% reduction in the fertility rates and more than 20% decrease in the number of offspring, in comparison with a vehicle control group (Carvalho et al., 2018b). Another study by the same authors, which attempted to elucidate the mechanism of the reproductive toxicity of CBD, indicated that 30 mg/kg induced a 76% reduction in the plasma concentration of testosterone, 38% reduction in the number of spermatozoa, and sperm head abnormality was demonstrated in rats treated with CBD (Carvalho et al., 2018a). Other studies that discuss the potential reproductive toxicity of CBD were reviewed in Carvalho et al. (2020). A pre- and postnatal toxicity study carried out in Wistar rats demonstrated a potential for toxicity. After female rats were orally fed with CBD at daily doses of 75, 150, or 250 mg/kg from gestational day 6 to postnatal day 21 resulted in no maternal effects but progenies had low birth weights, delayed maturation of reproductive organs and some impairments of sexual functions at 150 and 250 mg/kg maternal dose groups. A similar outcome was demonstrated in the embryo/fetal developmental toxicity study conducted in New Zealand rabbits at levels of 50, 80, or 125 mg/kg/day CBD. Rabbits treated with 125 mg/kg/day CBD had a 10% reduction in fetal weight and developmental delays. Similar to this, an earlier study in mice indicated that maternal exposure to 50 mg/kg CBD resulted in an alteration of reproductive functions in male and female animals (Dalterio et al., 1983).

Recent evidence implicating cannabidiol-induced liver toxicity in mice

A recently published manuscript appears to demonstrate multifaceted CBD-induced liver toxicity (Ewing et al., 2019b). The authors of the Ewing et al. paper carried out their study in two phases—acute and subacute toxicity studies—to show that CBD was hepatotoxic at very high doses through demonstration of elevation in liver transaminases, other serum biomarkers of liver injury, increases in liver-to-body weight ratios, upregulations in some liver-specific genes and genes involved in oxidative response, but little evidence of histological changes. These effects were seen in the highest doses tested in the two phases of the study. Doses were selected based on the estimation of the mouse equivalent doses by allometric scaling as determined by an FDA guideline and discussed by Nair and Jacob (2016). Using the allometric scale, they selected 246 mg/kg as the starting dose; other doses

TABLE 39.6 Preclinical repeat-dose toxicity studies and derived no observed adverse effect levels (NOAELs) of CBD.

Study	Species/ strain	Study design	Oral cannabidiol dose	Sample size	Suggested NOAEL	HED for NOAEL
Rosenkrantz et al. (1981)	Mature Rhesus monkey	90-day subchronic	0, 30, 100, 300 mg/kg	$n = 8$/group	<30 mg/kg/ day	9.6 mg/kg/day
Comelli et al. (2008)	Male Wistar rats, 200–220 g	Neuropathic pain model (sham + pain induced)	0, 10 mg/kg, 0.62 mg/kg	$n = 8–10$/group	10 mg/kg/day	1.61 mg/kg/ day
FDA (2018b)	CD-1 mice	13-week subchronic toxicity	0, 400, 550, and 700 mg/ kg/day	$n = 24$/ group + $n = 36$/ group TK	Male: 400 mg/kg/ day; female: 550 mg/kg/ day	Male: 32.5 mg/kg/ day; female: 44.7 mg/kg/ day
FDA (2018b)	Wistar (HsdHan: WIST) rats	26-week chronic toxicity study in rats	0, 15, 50, or 150 mg/kg	$n = 30$/group main, 20/group C & HD recovery, 6–12/ group TK	150 mg/kg/ day	24.2 mg/kg/ day
FDA (2018b)	Beagle dogs	39-week chronic toxicity study in Beagle dogs	0, 10, 50, and 100 mg/kg	$n = 8$/group main, 4/group C & HD recovery	100 mg/kg/ day	55.5 mg/kg/ day
FDA (2018b)	Male and female Wistar [Crl: WI(Han)] rats	Reproductive and developmental toxicological evaluation in rats	0, 75, 150, or 250 mg/kg	$n = 40$/group	250 mg/kg/ day	20.3 mg/kg/ day
FDA (2018b)	Mated female Wistar (Han) rats	Embryo-fetal development in the rat	0, 75, 150, or 250 mg/kg/ day	$n = 20$/group	250 mg/kg/ day	40.3 mg/kg/ day
FDA (2018b)	Mated female NZW rabbits	Embryo-fetal development study in the rabbit	50, 80, or 125 mg/kg/day	$n = 22$/group	80 mg/kg/day	25.8 mg/kg/ day
FDA (2018b)	Female Wistar (Han) rats	Pre- and postnatal developmental toxicity in rats	75, 150, or 250 mg/kg/ day	$n = 22$/group	75 mg/kg/day	12.1 mg/kg/ day
FDA (2018b)	Neonatal/ juvenile Wistar rats	10-week toxicity study followed by 6-week follow up	0, 100, 150, 250 mg/kg/ day	$n = 100$/group + 6-60/group TK	<100 mg/kg/ day	16.1 mg/kg/ day
Marx et al. (2018)	Hsd.Han Wistar rats	14-day oral subacute study	CBD-rich hemp extract at 0, 1000, 2000, or 4000 mg/kg/day	$n = 10$/group	<1000 mg/ kg/day	<161.3 mg/ kg/day
Marx et al. (2018)	Hsd.Han Wistar rats	90-day oral subchronic study	CBD-rich hemp extract at 100, 360, 720 mg/kg/day	$n = 100$/group	Male: 100 mg/kg/ day; female: 360 mg/kg/ day	Male: 16.1 mg/kg/ day; female: 58.1 mg/kg/ day
Ewing et al. (2019b)	8-week-old male B6Cef1 mice	Hepatotoxicity	61.5, 184.5, 615 mg/kg	$n = 6$/group	184.5 mg/kg/ day	15 mg/kg/day
Ewing et al. (2019a)	9-month-old CD-1 mice	CBD–drug interaction	0, 116, or 290 mg/kg/ day + Phosphate-buffered saline (PBS) or ip APAP at 400 mg/kg	$n = 7$ or 8/group	Not relevant to humans	Indeterminate

C & HD, control and high dose; CBD, cannabidiol; HED, human equivalent dose based on conversion factors obtained from Nair and Jacob (2016); *NOAEL*, no observed adverse effect level; TK, toxicokinetics.

tested in the acute toxicity study were 738 and 2460 mg/kg. Further, a 10-day subacute toxicity study employed doses of 0, 61.5, 184.5, or 615 mg/kg CBD. The study findings are highlighted in Table 39.7.

The recent studies suggesting CBD-induced liver toxicity in mice are informative, but they are not conclusive enough to implicate liver toxicity in humans. In their analysis, the authors were led to conclude that CBD was hepatotoxic due to the observance of slight increases seen in ALT and AST in the highest CBD test arm (2460 mg/kg). They also cited the upregulation of some of the makers of oxidative stress and some key drug-metabolizing enzymes as a strong basis for their claim. The authors also cast doubts on well-documented claims for antioxidant properties of CBD by suggesting that upregulation of some markers of oxidative stress like *heme oxygenase* and *Nqo1* somehow attributed a pro-oxidant characteristic to CBD. These and other observations formed the basis to conclude that CBD posed a risk of severe hepatotoxicity to humans.

The recent liver toxicity papers should not be used to predict CBD toxicity in humans. First, the paper fails to account for the pharmacokinetic differences between humans and mice as this finding is well documented for CBD. There are several important differences between human and animal studies regarding route of

TABLE 39.7 Summary of the Ewing et al. (2019) liver toxicity studies in mice.

Study	Ewing et al. (2019b)	Ewing et al. (2019b)	Ewing et al. (2019a)
Design	*Acute toxicity* 246, 738, 2460 mg/kg (single dose)	*Subacute toxicity*61.5, 184.5, 615 mg/kg for 10 days	*CBD-acetaminophen interaction* PBS or oral CBD at 116 or 290 mg/kg daily for 3 days followed by + PBS or intraperitoneal APAP at 400 mg/kg (single dose)
Animals	8-week-old male B6C3f1/J mice[a] (inbred line)*n* = 6/group	8-week-old male B6C3f1/J mice[a] (inbred line)*n* = 6/group	9-month-old CD-1 mice (outbred line)*n* = 7 or 8/group
Body weight (BW)	No change in BW in all groups↑ liver: BW ratio at 738 mg/kg CBD	No change in BW in all groups↑ liver: BW ratio at 738 mg/kg and 2460 mg/kg CBD	↓ BW (4%) and ↑ liver: BW at 116 mg/kg CBD + 400 mg/kg APAP, but not 290 mg/kg CBD + 400 mg/kg APAP
Histological findings	No findings	Significant histological changes in the highest dose of 615 mg/kg	Liver abnormalities with various features at 116 mg/kg CBD + 400 mg/kg APAP
Glutathione assay	No change in reduced glutathione (GSH) and GSH/GSSG ratio↑ Oxidized glutathione (GSSG) at 2460 mg/kg CBD	No change in GSH, GSSG, and GSH/GSSG ratio	↓ GSH at 116 mg/kg CBD + 400 mg/kg APAP, slight ↓ GSH at all other CBD + 400 mg/kg APAP doses↑ GSH/GSSG ratio at 116 mg/kg CBD + 400 mg/kg APAP
Liver function/ toxicity	No change in ALT, AST, alkaline phosphatase (ALP) and Gamma-glutamyltransferase (GGT) up to 246 mg/kg↑ in ALT and AST at 738 mg/kg and 2460 mg/kg CBDN.S. ↑ in total bilirubin at 2460 mg/kg CBD	No change in ALT, AST, ALP, and GGT↑ in total bilirubin at 615 mg/kg CBD	↑ ALT at all APAP or APAP + CBD doses. N.S. differences between dosesNo change in ALP, GGT or bilirubinNo relationship between CBD and the formation of APAP-Cys adduct
Genetic marker of liver toxicity	No change on expression of most liver toxicity genes at 246 mg/kg↑2.9-fold and 3.2-fold reported for *cyp*3a11 and *Ls*. at 246 mg/kgGene upregulation trend at 2460 mg/kg CBD	↑ varied with treatment↑ gene expression for *cytochromes P450 1a1, 1a2, 2b10, 2d22, 2e1, 3a4,* and *3a11*↑ *Ugts 1a1* and↓ *ugt 2a3*Marginal ↑ expression of liver injury genes at 61.5 mg/kg and 184.5 mg/kgUp to 9.1-fold at 615 mg/kg CBD	Dose-dependent ↑ P450 (CYP) and UGTS gene expression

APAP, acetaminophen; *CBD*, cannabidiol; *N.S.*, not significant; *UGTS*, uridine glucuronosyltransferases.
[a]*B6C3f1/J mice are known for a high incidence of spontaneous hepatic carcinoma (average 31% in males, up to 56%) and have been found to be very sensitive to acetaminophen-induced hepatic injury.*

administration and pharmacokinetics, calling into question the most appropriate animal model for CBD toxicity and the applicability of animal toxicity studies to humans. First, CBD has been studied in humans using oral administration and inhalation. Administration in rodents typically occurs either via intraperitoneal injection or the oral route. Second, the plasma levels reached via oral administration in rodents and humans can differ (Iffland and Grotenhermen, 2017). Both observations can lead to differing active blood concentrations of CBD and hence different concentrations at target receptors in the body (Bergamaschi et al., 2011; Deiana et al., 2012).

Therefore a significant concern over determination of the most appropriate animal model for CBD research going forward persists. Toxicologists must consider the most appropriate animal model when predicting human toxicity. To assess the utility of the mouse in studying the safety and toxicity of molecules that would be used in humans, the 2017 CBD review of Iffland and Grotenhermen (Iffland and Grotenhermen, 2017) articulates a clear pharmacokinetic difference between humans and mice (Deiana et al., 2012). The plasma levels reached via oral administration in rodents and human can differ greatly (Bergamaschi et al., 2011; Consroe et al., 1991). These observations can lead to differing active blood concentrations of CBD. It is postulated that CBD targets differ between humans and animals and therefore the same blood concentration might lead to different effects.

When mice and humans are given the same CBD dose, more of the compound becomes available in the mouse. This higher bioavailability can cause larger CBD-related adverse effects in the mouse. The mouse study from Deiana et al. (2012) demonstrated that, following a single oral dose of 120 mg/kg CBD in mice, the plasma C_{max} of 2.2 μg/mL was achieved 1 hour after administration. The pharmacokinetic study from Consroe et al. (1991), where human subjects received an oral dose of 10 mg/kg daily for 6 weeks, resulted in a C_{max} of approximately 10 ng/mL. A dose of 120 mg/kg is the mouse equivalent of the human dose of 10 mg/kg after applying the relevant conversion factor (Nair and Jacob, 2016). In comparing the two studies on the basis of the equivalent doses being ingested by mice and humans, the C_{max} value for the former is 220 times that of the latter; with a much lower T_{max} in mice.

The implication of this comparison rests in the fact that there is a direct correlation between plasma C_{max} of a chemical, its bioavailability, its physiological effect, and its toxicological potential. Undoubtedly, a higher bioavailability would ultimately result in a higher toxicity. As a result of the undeniable and quantifiably large pharmacokinetic difference between humans and mice, a mouse model would not be the appropriate choice for studying toxicity/safety of a molecule with the intent to extrapolating outcomes to humans. Moreover, the study by Ewing and colleagues evaluated liver toxicity of CBD in the mouse strain B6C3F1/J. B6C3F1/J mice are known to be highly sensitive to drug-induced liver injury marked by elevation of ALT, hepatocellular necrosis, among others (Harrill et al., 2009).

Finally, the paper by Ewing et al. does not indicate any evidence of histologic or clinical pathology alterations indicative of liver toxicity. The third International European Society of Toxicologic Pathology (ESTP) expert workshop discussed the significance of hepatocellular hypertrophy findings in rodent studies, in the absence of histologic or clinical pathology (Hall et al., 2012). Applying the findings from the expert toxicology working group to these studies, these liver toxicity results would be considered as nonadverse changes secondary to liver enzyme induction, adaptive changes, and changes corresponding to little relevance to man. The Ewing et al. (2019) study fails to elaborate on the hepatic hypertrophy observed without evidence of histologic or clinical pathology.

A second journal article by this group was designed to demonstrate the hepatotoxicity of CBD-rich cannabis extract in the presence of acetaminophen (APAP) in aged female CD-1 mice (9-month-old), with the primary aim of showing a CBD–drug interaction (Ewing et al., 2019a). APAP is a routinely used analgesic medication which has been linked with cases of liver toxicity. The authors orally dosed mice ($n = 8$/group) with PBS, 116 mg/kg CBD [human equivalent dose (HED) = 9.4 mg/kg], 290 mg/kg CBD (HED = 23.6 mg/kg) daily for 4 days followed by a single intraperitoneal injection of 400 mg/kg (HED = 32.5 mg/kg) APAP or PBS. Blood samples were analyzed for key liver function biomarkers like ALT and bilirubin, and endogenous antioxidants like glutathione (GSH). Liver samples were analyzed for the expression of genes associated with oxidative and conjugative drug-metabolizing enzymes (*cyps* and *ugts*). Histological evaluation of liver tissues was carried out for structural effect of the treatment following exposure to CBD, APAP, or a combination of these. The authors reported a significant 4% reduction in body weight, a significant increase on liver-to-body-weight ratio, and hepatic effects reminiscent of sinusoidal obstruction syndrome (SOS) at the lower dose of 116 mg/kg CBD + 400 mg/kg APAP. Aside from an increase in AST, this study did not report an elevation in biomarkers of liver injury such as ALT, alkaline phosphatase (ALP), and Gamma-glutamyltransferase (GGT) in this CBD–APAP combination. They reported a dose-dependent increase in hepatic oxidative (cytochromes P450) and conjugative (uridine diphosphate glucuronosyltransferase enzymes) as well as a significant reduction in oxidized and reduced glutathione, mostly at the lower dose.

The study clearly confirmed well-reported liver toxicity of APAP and some underlying mechanisms thereof (Ito et al., 2003). It also demonstrated the well-known potential of CBD to induce hepatic-derived drug-metabolizing enzymes—cytochromes P450 isoenzyme family (CYPs) and the UGTs in humans and animal models (FDA, 2018a,b). However, the results presented in this study are not a reflection of a CBD–APAP interaction that is relevant to humans. Selecting a mouse model to simulate a potential drug–drug interaction in humans was an error repeated in this study that likely exacerbated the liver injury reported due to the pharmacokinetic difference between humans and mice discussed earlier in this chapter.

The choice of dose and route of administration of APAP (400 mg/kg given intraperitoneally) also does not reflect human use of APAP. The human equivalent dose of 400 mg/kg is 32.5 mg/kg while the maximum recommended single dose in adult humans is 1000 mg or 14.3 mg/kg (Toms et al., 2008). Mice in this study were therefore given an overdose of APAP. Additionally, the intraperitoneal route increases bioavailability of APAP and therefore increases its toxicity (Colin et al., 1986). The dose and route of administration of APAP in the study by Ewing et al. (2019) thereby resembles a model of APAP-induced acute liver injury rather than a simulation of therapeutically relevant CBD–APAP interaction. This is likely why circulating liver transaminases ALT and AST were almost 1000-fold higher than controls in the presence of APAP alone and the addition of CBD did not seem to have any significant impact on the level of ALT or other markers of liver function. The elevation of markers of liver injury following APAP administration has been widely reported across different species, particularly mice (Taguchi et al., 2015). Additionally, SOS marked by sinusoidal endothelial cell swelling in the periportal area, erythrocyte infiltration of the space of Disse, and necrosis of the parenchymal cells (hepatocytes) is one key outcome of APAP overdose (Ito et al., 2003).

In summary, the second journal article by Ewing et al. (2019) is not a good model to predict human hepatotoxicity resulting from CBD–drug interaction because: (1) the dose (400 mg/kg) of APAP and the route of administration (intraperitoneal) used in the study are not a true reflection of clinically relevant human exposure to APAP but rather a good model for studying APAP toxicity in laboratory mice; (2) the toxic effects recorded including elevation in ALT and AST, glutathione depletion, and SOS likely resulted from the intraperitoneally injected APAP itself and not from a combination of CBD and APAP; and (3) the absence of marked toxic effects following a combination of 290 mg/kg CBD + 400 mg/kg APAP is a reflection of the protective role CBD plays as already demonstrated in peer-reviewed literature.

While there remains a need to delineate the potential interactions between CBD and other xenobiotics, the study by Ewing et al. (2019) must be interpreted with caution as the design does not appropriately assess the potential toxicity associated with concomitant ingestion of CBD and APAP (Table 39.8).

Toxicity in humans—evidence from clinical studies

Safety/toxicology of cannabidiol in the general, healthy population

In the 1970s, human studies showed that oral CBD intake from 15 to 160 mg (Hollister, 1973; Karniol et al., 1974), inhalation of 0.15 mg/kg BW (Dalton et al., 1976), or intravenous injection from 5 to 30 mg (Hollister, 1973; Perez-Reyes et al., 1973) were not followed by adverse events. CBD does not interfere with several psychomotor and psychological functions in humans. CBD does not affect heart rate, blood pressure, or performance in the verbal paired-associate learning test as measured by recall score at doses up to 600 mg/day (Bergamaschi et al., 2011; Consroe et al., 1979; Karniol et al., 1974; Zuardi et al., 1982). Investigations into antipsychotic effects of CBD have not reported any side effects after CBD ingestion (Bhattacharyya et al., 2010; Hallak et al., 2011; Hallak et al., 2010). Studies have shown that CBD is safe at levels that range between 100–800 mg single dose. In a Critical Review of CBD published by the World Health Organization (WHO, 2018b), studies indicate that oral ingestion of CBD at doses up to 800 mg were devoid of adverse effects (Babalonis et al., 2017; Haney et al., 2016). This indicates that CBD can safely be taken at 800 mg per day via the oral route. Studies in which healthy volunteers were tested for the anxiolytic activity of CBD at doses of 400 mg (Crippa et al., 2004) alone or in combination with THC in order to assess the anxiolytic effect of CBD on THC-induced anxiety or other THC-specific effects (Arndt and de Wit, 2017; Fusar-Poli et al., 2009, 2010; Hollister, 1973; Karniol et al., 1974; Zuardi et al., 1982, 1993) or for the evaluation of the potential of CBD to reverse alcohol-induced motor impairment at a single dose of 100 mg (Consroe et al., 1979) reported no adverse effects. Other studies in healthy individuals including safety assessments with single daily doses of 10 mg CBD (Mincis et al., 1973) and 3 mg/kg CBD daily for 1 month (Cunha et al., 1980) also reported no serious adverse effects. In healthy volunteers, CBD has been shown to be generally safe with minimal adverse effects (Bergamaschi et al., 2011; Crippa et al., 2004). A summary of these studies can be found in Table 39.9.

TABLE 39.8 Evaluation of the Ewing et al. (2019) liver toxicity studies.

Factor	Consideration
Mouse as an appropriate animal model for studying cannabidiol (CBD) toxicity	• The most appropriate animal model selected for any toxicity study should be weighed based upon the available data set
	• There are large pharmacokinetic differences between humans and mice regarding the fate of orally administered CBD
	• Mice have greater bioavailability for CBD than humans
	• C_{max} for mice has been shown to be 220-fold greater than the C_{max} for humans at equivalent doses
Dosing and route of administration	• Dosing used in the CBD liver toxicity studies in mice was not reflective of oral exposure in humans
	• Doses administered to mice were many fold higher than human dose
	• The human equivalent dose of highest CBD dose in the Ewing et al. (2019b) paper was 200 mg/kg/day CBD which is at least 5 times higher than the highest human dose in clinical trials (50 mg/kg/day)
	• The dose of any concomitantly administered xenobiotic should also be relevant to human exposure. The dose of APAP used in the Ewing et al. (2019a) study was 400 mg/kg; human equivalent dose of 32.5 mg/kg. The recommended single dose in humans is 14.3 mg/kg
	• The route of administration must be relevant to human exposure. APAP in Ewing et al. (2019a) was administered intraperitoneally, not by the traditional oral route
Outcomes	• The outcomes from the evaluation of liver damage in the mouse study do not meet the criteria for drug-induced liver injury
Effects on metabolizing enzymes in the liver	• Some of the effects of CBD on expression of drug-metabolizing enzymes are not unique to CBD and cannot necessarily form a basis for classifying CBD as a liver toxin
Adaptive change	• Hepatic hypertrophy, observed upon examination of mouse histological specimens in the Ewing et al. (2019a,b) studies, represents nonadverse, adaptive change with little relevance to man as a result of liver enzyme induction in the absence of histologic or clinical pathology
Recommendation	• The recent 2019 Ewing et al. studies suggesting liver toxicity by CBD in mice should not be used to predict CBD toxicity in humans
	• More toxicity research is necessary to demonstrate any liver toxicity by CBD in humans
	• Studies to support selection of the most appropriate animal model for CBD toxicity studies should be undertaken

Opinion: The recent Ewing et al. (2019a,b) studies suggesting liver toxicity in mice should not be used to predict CBD toxicity and interaction with other medications in humansIn summary, the study by Ewing et al. (2019b) does not reflect the true state of CBD for the following reasons:

1. The doses used were not reflective of what humans would be exposed to
2. The outcomes from the evaluation of liver damage do not meet the criteria for drug-induced liver injury
3. Some of the effects of CBD on expression of drug-metabolizing enzymes are not unique to CBD and cannot form a basis for classifying CBD as a liver toxin.
4. Clear, documented pharmacokinetic differences between humans and mice make the mouse a poor model for studying the toxicity of CBD with a view to extrapolating the data to humans
5. Authors of the Ewing et al. (2019b) paper failed to take into account pharmacokinetic differences between humans and mice
6. Hepatic hypertrophy, observed upon examination of mouse histological specimens in the Ewing et al. (2019b) study, represents nonadverse, adaptive change with little relevance to man as a result of liver enzyme induction in the absence of histologic or clinical pathology
7. The toxic effects recorded including elevation in ALT, AST, and glutathione depletion as well as SOS were more relevant to intraperitoneally injected APAP; it does not reflect/predict outcomes obtained from a combination of clinically relevant, orally administered CBD and APAP
8. The absence of marked toxic effects following a combination of 290 mg/kg CBD + 400 mg/kg APAP reflects the protective potential of CBD rather than a toxic response

ALT; alanine aminotransferase; *APAP*, acetaminophen; *AST*, aspartate aminotransferase; *SOS*, sinusoidal obstruction syndrome.

Safety of cannabidiol in individuals with neurological and psychiatric disorders

The safety of CBD has been demonstrated across a wide range of doses and disease conditions, including seizure disorders (Devinsky et al., 2018; Devinsky et al., 2016; Thiele et al., 2018), anxiety disorders (Crippa et al., 2004), and psychosis (Leweke et al., 2012). In single-dose studies across the dose range investigated, CBD was well tolerated with no signs of toxicity. A similar trend was seen in repeat-dose studies in which a few cases of adverse effects were reported though these were minor and transient. At daily doses that ranged between 100–1200 mg, CBD has been evaluated for efficacy in psychotic disorders, pain, and Parkinson disease with different safety outcomes (Boggs et al., 2018; Consroe et al., 1991; Crippa et al., 2011; Cuñetti et al., 2018; Devinsky et al., 2016, 2018; Hallak et al., 2010; Leweke et al., 2012; McGuire et al., 2018; Thiele et al., 2018; Zuardi et al., 2009, 2010). In these studies, CBD was generally well tolerated, with most minor side effects like somnolence, headache, and some diarrhea and vomiting. One very prominent application of CBD has been in the management of epilepsy of different etiologies. With daily doses as high as 50 mg/kg or 3500 mg/day, CBD was shown to be generally tolerated with minor side effects ranging from diarrhea and fatigue to elevation of liver transaminases (Devinsky et al., 2016, 2018; Szaflarski et al., 2018; Thiele et al., 2018). Aside from the potential of high-dose CBD to induce elevated liver transaminases as shown in animal studies (FDA, 2018a), liver transaminases were mostly seen to be increased in cases where the patients were simultaneously being managed with drugs known to induce liver injury. A common medication coadministered with CBD in some of these studies was valproate, which is also indicated for the management of convulsive disorders (Devinsky et al., 2016; Thiele et al., 2018). Valproates are known to induce liver injuries (Gayam et al., 2018). Other relevant factors to consider for the onset of CBD-induced adverse reaction include the potential for interaction with other coadministered compounds. Being an inducer and inhibitor of different members of the cytochrome P450 isoenzyme family, CBD has been shown to increase the bioavailability of the benzodiazepine, clobazam. A concomitant intake of clobazam and CBD decreased the required dose of clobazam in subjects in a study of children with refractory epilepsy (Geffrey et al., 2015). Adverse events also occurred in 77% of subjects, though these were alleviated in all cases through adjustment in the dose of clobazam. Other studies have also reported excessive somnolence as a side effect of CBD treatment (Carlini and Cunha, 1981; Devinsky et al., 2016). This can be considered an on-target effect of CBD since CBD is mainly centrally acting, and has been used in the management of insomnia and other sleep disorders.

CBD is not psychoactive (Hollister, 1973); it does not produce the psychotropic effects associated with THC because it does not interact with the same set of molecular targets as THC. While THC is known to act on cannabinoid receptors 1 and 2 (CB1 and CB2), CBD mediates its physiologic action by indirect modulation of CB receptors, cell-surface and nuclear receptors, as well as ion channels (Lee et al., 2017) (see Table 39.1). Furthermore, these differences in the receptor affinity and interaction may explain reports suggesting that CBD may reverse some of the neurophysiologic and organ-system adverse effects associated with THC (Karniol et al., 1974; Zuardi et al., 1982).

Toxicity data from Epidiolex clinical trials

Safety endpoints from human clinical trials

During review of the marketing application for Epidiolex, a CBD oral solution indicated for the treatment of rare seizures associated with Lennox–Gastaut syndrome and Dravet syndrome in patients 2 years of age and older, the FDA identified safety concerns in an unhealthy population. Common adverse reactions reported in the safety data included decreased appetite, diarrhea, somnolence, lethargy, and sleep disorders. Other rare symptoms noted from the Epidiolex clinical trials included hypersensitivity reactions. GW Pharma submitted three pivotal phase III clinical trials, which evaluated a dose of 20 mg/kg/day, divided twice daily for CBD vs. placebo as adjunctive therapy. Patients were an equal mix of males and females, ranging in age from 2 to 55 years of age. One clinical study included a 10 mg/kg/day dose arm to ascertain effectiveness at a lower dose. At the time of the original New Drug Application (NDA) submission, 1756 subjects were exposed to CBD oral solution with 1391 treated for epilepsy. Approximately half (684) were exposed in the uncontrolled expanded access program for drug-resistant epilepsy. For both CBD and placebo groups, most of the adverse events reported were classified as mild to moderate. The most common adverse events that occurred in 10% or more of CBD-treated patients were somnolence, decreased appetite, diarrhea, transaminase elevations, fatigue, malaise, asthenia, rash, insomnia, sleep disorder and poor-quality sleep, and infections (FDA, 2018c). Adverse events were low for both CBD treatment and placebo groups. Thirty subjects in the CBD groups (9.3%) reported an adverse event leading to discontinuation, compared to three subjects (1.3%) in the placebo

TABLE 39.9 Human clinical studies of cannabidiol in healthy subjects.

Study	Route	Dose	Design	Duration	Population attributes	Sample size	Side effects
Hollister (1973)	Oral, intravenous	20–100 mg CBD oral, 5–30 mg CBD intravenous 20–400 mg CBN oral	Parallel	Single dose	Males; 30–50 years	$n = 15$	No significant side effect
Karniol et al. (1974)	Oral	15 mg CBD, 30 mg CBD, 60 mg CBD, 30 mg THC, 30 mg THC + 15 mg CBD, 30 mg THC + 30 mg CBD, 30 mg THC + 60 mg CBD, placebo	Parallel	Single dose	Healthy males, age 21–34 years, 50–91 kg, $n = 22$ with history of marijuana use	$n = 40$	No significant side effects on heart rate, psychological reactions, and on time production tasks
Consroe et al. (1979)	Oral	200 mg CBD, placebo, 1 g/kg ethanol, 200 mg CBD + 1 g/kg ethanol	Crossover randomized controlled trial (RCT); 1-week washout	Single dose	Healthy, 21–33 years, 52–85 kg, $n = 5$ infrequent cannabis use	$n = 10$	No significant side effects
Cunha et al. (1980)	Oral	3 mg/kg BW CBD or placebo; 200 and 300 mg/kg/day CBD or placebo	Parallel	Phase I = 30 days Phase II = 126 days	Phase I: healthyPhase II: secondary generalized epilepsy with temporal focus	Phase I: $n = 16$ phase II: $n = 15$	No significant side effects
Carlini and Cunha (1981)	Oral	Maximum single dose of 600 mg followed by 10 mg/kg/day for 20 days	Various designs	Single and repeat dosing	Acute and chronic studies: healthy 21–47 years. Sleep study: insomniacs; treatment resistance epilepsy patients	$n = 42$; $n = 15$; $n = 15$	Somnolence
Zuardi et al. (1982)	Oral	1 mg/kg BW CBD, 0.5 mg/kg BW THC, mixture 0.5 mg/kg BW THC and 1 mg/kg BW CBD, placebo, 10 mg diazepam	Crossover RCT; minimum 1-week washout	Single dose	Healthy (age 20–38 years); $n = 5$ with history of marijuana use	$n = 8$	No significant effects on heart rate and bodily symptoms
Zuardi et al. (1993)	Oral	300 mg CBD, isapirone 5 mg, diazepam 10 mg	Parallel	Single dose	Healthy, 20–30 years, students	$n = 40$	No significant effects on heart rate, blood pressure, psychomotor performance, bodily symptoms, and psychological outcomes
Crippa et al. (2004)	Oral	400 mg CBD, placebo	Crossover RCT; 1-week washout	Single dose	Healthy men, postgraduate students, 25–42 years, 67–85 kg. Lifetime cannabis use <5 times	$n = 10$	No significant effects on psychological outcomes
Borgwardt et al. (2008)	Oral	600 mg CBD, 10 mg THC, placebo	Crossover RCT, 1-month washout	Single dose	Healthy, Caucasian, right-handed men, 20–42 years	$n = 15$	No significant effects

(Continued)

TABLE 39.9 (Continued)

Study	Route	Dose	Design	Duration	Population attributes	Sample size	Side effects
Fusar-Poli et al. (2009)	Oral	600 mg CBD, 10 mg THC, placebo	Crossover RCT	Single dose	Healthy, English-native, right-handed men; cannabis use 15 × or less	$n = 15$	No significant side effects
Bhattacharyya et al. (2009)	Oral	600 mg CBD, 10 mg THC, placebo	Crossover RCT	Single dose	Healthy, right-handed males with; cannabis use 15 times or less	$n = 15$	None for CBD
Bhattacharyya et al. (2010)	Oral/iv	Phase I: 600 mg CBD oral, 10 mg THC oral, or placeboPhase II: 5 mg CBD iv or placebo preceding 1.25 mg THC iv	Crossover RCT; Phase I = 1 month washout, phase II = two week washout	Single dose	Phase I: healthy, right-handed males, age 20–42Phase II: healthy, 21–42	Phase I: $n = 15$ Phase II: $n = 6$	None for CBD
Fusar-Poli et al. (2010)	Oral	600 mg CBD, 10 mg THC, placebo	Crossover RCT	Single dose	Healthy, right-handed males, 20–42 years; cannabis use 15 × or less	$n = 15$	CBD was well-tolerated
Hallak et al. (2011)	Oral	600 mg CBD or placebo preceded by ketamine	Crossover RCT; 1-week washout	Single dose	Healthy men, 20–36 years, white or mixed race	$n = 10$	No significant side effects on heart rate and psychotic symptoms
Arndt and de Wit (2017)	Oral	300, 600, or 900 mg of CBD or placebo	Crossover RCT with 1-week washout	Single dose	Healthy, BMI = 19–30, lifetime cannabis use 100 times or less	$n = 38$	No significant effects
WHO (2018b); Haney et al. (2016); Babalonis et al. (2017).	Oral	0, 200, 400, and 800 CBD and/or inhaled THC (0.1% or 5.30–5.80%)	Crossover RCT	Single dose	Frequent marijuana smokers	$n = 31$	None for CBD

BW, body weight; *CBD*, cannabidiol; *THC*, tetrahydrocannabinol.

group. The separation between CBD and placebo was even less distinguishable for reporting of serious adverse events. The incidence of serious adverse events across the Dravet syndrome and Lennox–Gastaut syndrome CBD-treated subjects was 5%, compared to 3.1% in the placebo group. The reported serious adverse events in the CBD subjects were status epilepticus (5%), pneumonia (2.8%), convulsion (2.2%), and increased aspartate aminotransferase (AST) (2.2%). All of these serious adverse events, with the exception of increased hepatic transaminases, were considered to be typical for this patient population (Raspall-Chaure et al., 2007).

Transaminase elevation in Epidiolex studies

Transaminase elevations were observed in 14% of CBD patients versus 3% in placebo. In the clinical development program for CBD, the incidence of elevation of alanine aminotransferase (ALT) or AST greater than three times the upper limit of normal (ULN) was 2/219 (0.9%) in placebo, 2/67 (3.0%) in CBD 10 mg/kg/day, and 18/228 (18.1%) in CBD 20 mg/kg/day). Elevations in ALT were more pronounced than AST, suggesting that the liver was the source of the transaminase elevations. There also was a clear dose-dependent increase with 8% elevations overall in the 10 mg/kg CBD group and 16% in the 20 mg/kg group. None of the CBD subjects experienced liver failure as neither hyperbilirubinemia nor international normalized ratio (INR) elevation were seen in conjunction with the enzyme elevations. Although small increases in total bilirubin were seen in a few cases, the bilirubin levels generally remained within normal limits and there were no cases that met Hy's law criteria for ALT > 3 × ULN and bilirubin > 2 × ULN. Some events of transaminase elevation were serious or severe; however, there were no events of liver failure or death related to liver injury. Furthermore, risk factors for transaminase elevation included concomitant valproic acid use, elevated baseline liver function tests, and higher doses of CBD. Most events of transaminase elevation occurred within 30–90 days after initiation of CBD treatment; however, rare cases were observed up to 200 days after initiation of treatment, particularly in patients taking concomitant valproic acid. Events of transaminase elevation generally resolved with discontinuation of CBD or dose decreases in either CBD or valproic acid; however, some events resolved during continued treatment with CBD at the same dose (FDA, 2018d). In summary, 20 mg/kg/day of Epidiolex in clinical trials was associated with a potential for liver injury, as evidenced by elevated transaminase levels. The incidence of transaminase elevations after ingestion of 10 mg/kg CBD was closer to the elevated transaminase incidence found in placebo. There is a lack of data implicating CBD-induced elevations of transaminases in healthy subjects.

Deaths in Epidiolex studies

FDA noted 20 deaths at the time of their Epidiolex NDA submission. All causes of death were deemed related to sudden unexpected death in epilepsy (SUDEP), which is a fatal complication of epilepsy, and other causes typical for this patient population. SUDEP is the sudden, unexpected, nontraumatic death of a person with epilepsy, without a toxicological or anatomical cause of death, as determined through postmortem examination. Deaths in the Epidiolex trials included asphyxia, hypoxemia, respiratory failure, and complications secondary to pneumonia. The FDA reviewers were unable to attribute deaths to CBD as the numbers of deaths did not seem to differ from the numbers that would be expected in the Dravet syndrome or Lennox–Gastaut patient populations.

Absence of abuse potential by cannabidiol in preclinical and clinical studies

In humans, CBD exhibits no effects indicative of substance abuse or dependence potential. There is no evidence of abuse potential from recreational CBD use or any public, health-related problems associated with the use of hemp-derived CBD (WHO, 2018a). Abuse potential of CBD has been investigated through receptor-binding studies, and animal behavior and human abuse potential studies. Dronabinol and benzodiazepines were used as positive controls for several of the abuse potential studies. In experimental models of abuse liability, CBD appears to have little effect on conditioned place preference or intracranial self-stimulation. In an animal drug discrimination model, CBD failed to substitute for THC. In a human abuse potential study evaluating the abuse potential of single doses of CBD compared to alprazolam or THC in healthy recreational polydrug users, CBD was significantly less likely to be associated with drug liking or with a desire to take the drug again (Greenwich Biosciences Research Ltd, 2017); DEA, (2018). The active comparator groups were alprazolam (2 mg) and dronabinol (10 and 30 mg). CBD was used at doses of 750 mg (proposed therapeutic 10 mg/kg dose group) and at high therapeutic and supratherapeutic dose groups (1500 and 4500 mg). These are doses far above those listed on products in the marketplace. In other words, CBD was more comparable to placebo than it was to either active comparator groups consisting of alprazolam or dronabinol

CBD does not produce overt behaviors similar to those produced by drugs of abuse such as cocaine, phencyclidine, LSD, and benzodiazepines. It does not produce a cannabinoid agonist response in the tetrad test, which is the opposite effect produced by dronabinol. Binding studies indicated that CBD does not bind to cannabinoid receptors or any other receptor associated with drugs of abuse. It does not generalize to dronabinol or to the depressant midazolam in separate drug discrimination studies. It does not produce effects similar to cannabinoid agonists or to benzodiazepines. Finally, CBD does not produce self-administration, suggesting it does not have rewarding properties like many known drugs of abuse (Abou-Sayed et al., 2018). There are no euphoria-related adverse events in the open literature or from phase I clinical studies conducted with CBD that would be indicative of abuse potential. Drugs that have abuse potential typically produce euphoria-related adverse events in clinical studies. Phases II and III clinical studies with CBD were not assessed for euphoria-related adverse events because the epileptic patients were taking concomitant antiepileptic medications with known abuse potential that would have confounded the evaluation.

Cannabidiol–drug interactions

Liver metabolism of cannabidiol

CBD is extensively metabolized in the liver primarily by hydroxylation to 7-OH-CBD, which is metabolized further, resulting in a number of metabolites excreted in the feces and urine (Ujváry and Hanuš, 2016). A study in human liver microsomes demonstrated that CBD was metabolized by pooled human liver microsomes to 8-mono-hydroxylated metabolites. Seven recombinant human cytochrome P450 (CYP) enzymes were identified as capable of metabolizing CBD: CYP1A1, CYP1A2, CYP2C9, CYP2C19, CYP2D6, CYP3A4, and CYP3A5. The two major isoforms involved are CYP3A4 and CYP2C19 (Jiang et al., 2011). It is also a minor substrate of CYP2C19 and CYP3A4, as well as a substrate for UGT1A7, UGT1A9, and UGT2B7.

In vitro evidence

Evidence from in vitro and animal studies indicate that CBD can inhibit CYPs 1A1, 1A2, 1B1, 2A6, 2B6, 2C9, 2C19, 2D6, 3A4, and 3A7 (Stout and Cimino, 2014). Some of these pharmacokinetic drug interactions have been confirmed in studies conducted in mice whereby repeated CBD exposure resulted in the induction of CYPs 3A4 and 2B10 (Bornheim et al., 1994). While CYP1A1 can be induced by CBD (reviewed in Iffland and Grotenhermen, 2017), Bornheim et al. (1994) demonstrated the ability of CBD to increase the mRNA levels of CYPs 3A and 2C enzymes without a corresponding increase in their catalytic activities. On the other hand, CBD is a known inhibitor of hepatic drug metabolism. This effect raises the question of whether CBD can inhibit the metabolism of other drugs in vivo, affecting their metabolite concentration in the central nervous system (Jones and Pertwee, 1972; Paton and Pertwee, 1972). CBD-mediated inhibition of drug metabolism is likely a result of the covalent binding of a reactive CBD metabolite to hepatic microsomal P450 (Bornheim et al., 1994), which affects specific isozymes. CBD has been shown to inhibit CYP isozymes in vitro, but it is not clear that this occurs at concentrations achieved with doses used clinically (WHO, 2018a). CBD can inactivate human P450 3A4 and 2C19 (Jaeger et al., 1996) in addition to mouse 3a and 2c subfamilies. The metabolism of drugs by cytochrome P450s 3A, 2C, and 2b subfamilies can be affected by CBD; however, CBD extracts (full spectrum or those containing some levels of THC, other cannabinoids, and terpenes), do not seem to inhibit or induce hepatic CYP450s (Stott et al., 2007; Stott et al., 2005).

As CBD is a substrate of a number of the cytochrome P450 isoenzyme family including CYP3A4 (FDA, 2018a; Jiang et al., 2011), it could have profound implications for drugs that are biotransformed by these enzymes and may affect their bioavailability, duration of action, and inherently their safety (Pelkonen et al., 1998). Patients who are managed with drugs like warfarin, which have a narrow therapeutic window (Johnson, 2012), may experience excessive bleeding if they concomitantly ingest CBD, given the likelihood of pharmacokinetic interaction. The same principle applies to nonsteroidal antiinflammatory drugs like diclofenac, known to induce antiplatelet potential and hepatotoxic activities (Schapira et al., 1986). Since CYP3A4 metabolizes almost 60% of medications, coadministration of CBD with inducers or inhibitors of CYP3A4 will be expected to alter its pharmacokinetics. Clarithromycin, ketoconazole, itraconazole, and ritonavir are inhibitors of CPY3A4 and may increase plasma levels of CBD and hence increase toxicity. CBD biotransformation would be slowed, and it would remain bioactive for longer periods. In contrast, phenobarbital, rifampicin, phenobarbital, carbamazepine, and phenytoin induce CYP3A4, causing reduced CBD bioavailability (German Medicinal Codex, 2015) and loss of efficacy (Qian et al., 2019). In addition, since CBD is also an inhibitor of bile salt export pump (BSEP)/ABCB11 and CYP2C19, coadministration would result in the increase of plasma concentrations of cilostazol, citalopram, flibanserin, and clobazam (Jiang et al., 2013, 2011).

Interaction of CBD with the cytochrome P450 system is complex and somewhat dependent upon the presence of THC. In the presence of THC, as in CBD extracts, there does not appear to be any clinically relevant interaction (inhibition or induction) with P450 enzymes. Data from a phase I clinical trial on Sativex (1:1 combination of THC and CBD) have shown the CYP3A4 inducer, rifampicin, to reduce the C_{max} of CBD whereas the ketoconazole, which is an inhibitor of CYP2C19, did increase the peak plasma concentration of CBD by almost 100% (Stott et al., 2013). Another clinical report in children revealed that concomitant administration of CBD with clobazam induced a 60% increase in the plasma level of the parental drug and a 500% increase in the blood levels of its active metabolite—norclobazam (Geffrey et al., 2015). A 3-week coadministration of CBD and THC has been shown to result in a twofold increase in the plasma level of THC, resulting in an increase in the behavioral effect of THC in rats (Klein et al., 2011).

More clinical data are required to determine the clinical relevance of a concomitant intake of CBD and other therapeutic agents and whether its combination with low amounts of THC may mitigate any potential effect CBD might have on drug-metabolizing enzymes. However, caution is advised with a concomitant use of CBD and other pharmacologically active compounds given their reciprocal pharmacokinetic, pharmacodynamic, or toxicological effects, which are not fully understood.

Effect of foods on liver enzymes

Drug–food interactions can occur with a variety of foods, including botanical dietary ingredients, because of their actions on the P450 liver enzyme system.

Grapefruits, for example, are sold as conventional foods or as extracts in dietary supplements and can have food–drug interactions (Bailey et al., 1998). Grapefruits contain furanocoumarins and flavonoids, which can inhibit key drug-metabolizing enzymes, such as CYP3A4, a metabolizing enzyme for almost 60% of drugs. Drugs known to interact with grapefruit juice include amiodarone, dronedarone, terfenadine, diphenhydramine, astemizole, simvastatin, lovastatin, dextromethorphan, buspirone, isapride, and ivabradine. Depending on the type of food, concomitant ingestion may impact other pharmacokinetic factors of the compound in question such as its absorption, distribution, and excretion (reviewed in Koziolek et al., 2019). The review by Bushra et al. (2011) describes the effects of select food items on the pharmacokinetic parameters of different medicinal compounds, particularly their bioavailability.

In one of the trials leading to the approval of Epidiolex, intake of high-fat meals in the study participants impacted the bioavailability of CBD (FDA, 2018b). Compared with fasted subjects, CBD administered with high-fat diet resulted in almost fivefold increase in C_{max} for CBD and a two- to threefold increase in the C_{max} of its metabolites. A similar outcome was reported by Taylor et al. (2018) and Birnbaum et al. (2019). In both studies, intake of high-fat meals increased the C_{max}, and the C of CBD, without affecting its T_{max}. In congruence with the increase in plasma level of CBD and its metabolites, there was also a significant increase in the reported adverse effects such as nausea, diarrhea, abdominal discomfort, headache, and somnolence in participants who ingested a high-fat high-cholesterol diet alongside CBD. From these studies, it can be deduced that concomitant intake of CBD with food (especially those containing a high level of fat) has a potential of altering the pharmacokinetics of CBD in such a manner as to increase its bioavailability, exposure time, and ultimately contribute to its toxicity.

Conclusion—determining a safe level of cannabidiol for use in nutraceuticals

As shown above, CBD has been employed in studies across a broad range of doses and durations of exposure. For example, CBD has been used as a single administration across doses spanning 15 mg (Karniol et al., 1974), 1 mg/kg (Zuardi et al., 1982), 300 mg (Cunha et al., 1980; Hollister, 1973), 600 mg (Bergamaschi et al., 2011; Bhattacharyya et al., 2009; Borgwardt et al., 2008; Fusar-Poli et al., 2009, 2010; Hallak et al., 2010; Zuardi et al., 1993), and 1280 mg (Zuardi et al., 2006b). Further, CBD has been used in repeat-dose studies at doses of 5–50 mg/kg for 48 weeks (Devinsky et al., 2018), 10 mg/kg for 6 weeks (Consroe et al., 1991), 20 mg/kg for 10 days (Thiele et al., 2018), and 50 mg/kg for 8 weeks (Szaflarski et al., 2018). These studies have reported varying safety outcomes. Valuable information of the determination of the safe dose for CBD has been gleaned from the clinical trials for Epidiolex and those investigating indications appropriate for nutraceutical products (Bergamaschi et al., 2011; Carlini and Cunha, 1981; Linares et al., 2019; Linares et al., 2018; Nicholson et al., 2004; Zuardi et al., 1993; Zuardi et al., 2017). An estimation of NOAEL values derived from human studies can be found in Table 39.10.

In the 2018 WHO's Critical Review of CBD, the authors made reference to multiple studies indicating a safe level of use at serving levels up to 800 mg/day for oral ingestion. These serving levels were devoid of adverse effects (Babalonis et al., 2017; Haney et al., 2016). Increases in liver transaminase levels, the first sign of adverse effects for CBD, appear to occur at levels above 800 mg/day. The clinical trials carried out by GW Pharmaceuticals for the approval of Epidiolex for the treatment of people with Lennox–Gastaut and Dravet disease (2 years old and above), also reported an elevation in liver transaminases. However, this may be associated with the use of CBD, these patients were not of the healthy population and so were being managed with other medications known to have potential for hepatotoxicity. It remains to be confirmed if this is a significant concern in the healthy population. A daily dose of 1500 mg (or 21 mg/kg/day) administered orally to a healthy cohort was generally well tolerated as the only adverse effects reported were due to diarrhea, nausea, headache, and somnolence (Taylor et al., 2018). There were no reports of hepatic damage in this study, underlining the importance of correctly interpreting toxicity signals in preclinical studies. This may explain why a dose of 20 mg/kg/day CBD was approved for use clinically in the case of Epidiolex. However, there is need for further study in a wider population as well as longer duration of exposure to ascertain the safety of CBD at this dose level.

In addition to the use of CBD as a therapeutic agent, interest has grown in its application as an ingredient of nutraceuticals, including food and beverages. In the United States, an explosion of companies selling CBD products has flooded the market, with over 1500 products on the shelves. Many CBD products that are marketed as dietary supplements or conventional foods make unlawful and unsubstantiated health claims that have highlighted a need for regulation. Currently, CBD is considered a prescription drug in the United States and is not permitted to be included as an ingredient in dietary supplements or foods. The US FDA has taken a stance of enforcement discretion, only pursuing regulatory action against those companies making egregious health claims. Similar issues

TABLE 39.10 Estimated no observed adverse effect levels (NOAELs) from human clinical trials.

Study	Daily dose	Design	Population	Sample size	Side effects	Estimated NOAEL
Cunha et al. (1980)	210 mg or 300 mg/day	Parallel, phase I = 30 days or phase II = 126 days	Phase I = healthy	Phase I, *n* = 16	None	4.3 mg/kg/ day (300 mg/ day)
			Phase II + patients with generalized epilepsy	Phase II, *n* = 15		
Consroe et al. (1991)	0 or 10 mg/kg/ day	Crossover, 6-week study	Patients with Huntington disease	*n* = 15	None	700 mg/day (10 mg/kg/ day)
Zuardi et al. (1995)	1500 mg/day	Case report, 26 days	Patient with schizophrenia	*n* = 1	None significant AE	21 mg/kg/ day (1500 mg/ day)
Zuardi et al. (2006)	40–1280 mg/ day	Open label, 30 days	Males with treatment-resistant schizophrenia	*n* = 3	None significant	18 mg/kg/ day (1280 mg/ day)
Zuardi et al. (2009)	150–400 mg/ day	Open label, 4 weeks	Parkinson disease patients with psychosis	*n* = 6	None significant	5.7 mg/kg/ day (400 mg/ day)
Zuardi et al. (2010)	600–1200 mg/ day	Open label, 30 days	Patients with bipolar affective disorder	*n* = 2	None significant	17 mg/kg/ day (1200 mg/ day)
Leweke et al. (2012)	200–800 mg/ day	Parallel for 4 weeks	Patients with schizophrenia (age 18–50 years)	*n* = 42	Minor AEs	11.4 mg/kg/ day (800 mg/ day)
Chagas et al. (2014)	75 or 300 mg	Case report, 6 weeks	Patients with Parkinson disease experiencing sleep disorders	*n* = 4	None	4.3 mg/kg/ day (300 mg/ day)
Chagas et al. (2014)	75 or 300 mg	Parallel, 6-week study	Patients with Parkinson disease	*n* = 21	None	4.3 mg/kg/ day (300 mg/ day)
Devinsky et al. (2016)	2–50 mg/kg/ day	Open label, 12-week study	Patients with treatment-resistant epilepsy (age 1–30 years)	*n* = 214	Thrombocytopenia and elevated liver transaminases in patients also taking valproate	Doses responsible for AEs not stated
Cuñetti et al. (2018)	100–300 mg	Case reports, 21-day study	Kidney transplant patients with chronic pain	*n* = 7	Nausea, not related to CBD	4.3 mg/kg/ day (300 mg/ day)

(Continued)

TABLE 39.10 (Continued)

Study	Daily dose	Design	Population	Sample size	Side effects	Estimated NOAEL
McGuire et al. (2018)	1000 mg/day	Parallel, for 8 weeks	Patients with schizophrenia (age 18–65 years)	$n = 83$	Minor AEs	14 mg/kg/day (1000 mg/day)
Boggs et al. (2018)	0 or 600 mg	Parallel. 6-week study	Patients with schizophrenia	$n = 36$	None	8.6 mg/kg/day (600 mg/day)
Devinsky et al. (2018a)	0, or 5–20 mg/kg/day	Double-blind, 21 days	Children with Dravet syndrome (age 4–10 years)	$n = 34$	Pyrexia, somnolence, and decreased appetite. Increased liver transaminases in patients also taking valproate	20 mg/kg
Thiele et al. (2018)	20 mg/kg	Parallel, for 14 weeks	Patients with Lennox-Gastaut syndrome patients (2–55 years)	$n = 171$	Minor AEs. Transient elevation in liver transaminases in patients in both placebo and CBD groups (valproate coadministered in 80% of CBD cases)	≤ 20 mg/kg
Taylor et al. (2018)	0, 750, or 1500 mg twice daily	Double-blind, placebo-controlled, 7 days	Healthy adults	$n = 9$ for CBD group and $n = 6$ for placebo	Minor side effects	21 mg/kg/day (1500 mg/day)
Szaflarski et al. (2018)	Up to 50 mg/kg	Open label, for 48 weeks	Patients with treatment-resistant epilepsy	$n = 607$	None reported	50 mg/kg/day
Devinsky et al. (2018b)	2–50 mg/kg/day	Open label, 48 weeks	Patients with different epileptic conditions	$n = 55$	Fatigue, diarrhea, decreased appetite, increased convulsions	Doses responsible for AEs not stated
Pietrafusa et al. (2019)	2.5–25 mg/kg	Open label, for 6–25 months	Children with developmental and epileptic encephalopathy	$n = 29$. Male: $n = 12$; Female: $n = 17$	Mild to moderate AE including diarrhea or decreased appetite. Increased liver transaminase seen in patients also taking valproate	25 mg/kg/day
Wheless et al. (2019)	10–40 mg/kg	Open label; 6 days	Pediatrics with treatment-resistant epilepsy (age 1–17 years)	$n = 61$	Minor side effects including somnolence, anemia, diarrhea. Increase in liver transaminase seen in one patient taking 40 mg/kg/day. This resolved without intervention	40 mg/kg/day

AE, adverse effect; *CBD*, cannabidiol; *NOAEL*, no observed adverse effect level.

are being faced by governments the world over, as the spread in the popularity of CBD has been vast. Canada also currently regulates the medicinal use of CBD products as prescription drugs, although recreational use is permitted under the Cannabis Act. In June 2019, Health Canada issued a framework for the regulation of "Cannabis Health Products" not requiring the oversight of a physician for public comment. This framework remained open to the inclusion of other cannabinoids in Cannabis Health Products, but the implication was that these products would be primarily aimed at CBD and other nonintoxicating cannabinoids. Health Canada did not however define a safe level for these products, instead they solicited further safety information from industry and academic stakeholders. In January 2019, the EU commission updated the Novel Food Catalogue to include CBD and hemp flower extracts as novel foods. The Food Standards Agency of the United Kingdom subsequently set a deadline of January 2021 for all CBD companies to comply and register their products as novel foods to avoid regulatory action. At the same time, the Food Standards Agency's Committee on Toxicity of Chemicals in Food recommend that a maximum daily intake of 70 mg of CBD can be consumed by adults (Food Standards Agency, 2020). Similarly, the Australian Therapeutics Goods Administration (TGA) recently indicated the evidence supported rescheduling CBD to an over-the-counter product at a daily dose of 1 mg/kg, equivalent to 62 mg per day (TGA, 2020). In both countries, this level of intake (approximately 1 mg/kg) was deemed likely to be safe with no adverse effects based on a thorough review of the publicly available safety literature reviewed in this chapter.

Whether CBD is being included in over-the-counter medications, dietary supplements, or conventional foods, a critical consideration that regulatory toxicologists must make when recommending an acceptable daily intake is the level of exposure. Self-administration of a product without the supervision of a healthcare provider leaves the consumer susceptible to intake far beyond the intended dose of the product. Estimating the mean and upper limits (90th or 95th percentile) of consumption of the product is required. This is the challenge faced by regulators across the globe as there is no historic precedent to base these estimations on and data on current consumption levels are limited. Although doses of up to 800 mg have been recognized by the WHO as being safe, estimates of the upper limit of exposure likely exceed this, leading to the conservative recommendations of 60 or 70 mg per day issued by regulators thus far.

Although there are studies conducted on the safety and efficacy of CBD in vitro, in vivo, and in human subjects, there are still significant gaps to be filled in our understanding of both its safety and efficacy in the healthy human population. It has been established that, in human studies, CBD administration does not induce side effects across a wide range of dosages and routes of administration, including acute and chronic dose regimens, and tolerance to CBD does not develop. Human studies have shown that CBD is well tolerated and safe at levels that range between 100–800 mg/day. Although there is evidence for developmental and feto-toxicity in rats and rabbits, it is unknown whether this effect is translatable to humans. Despite the approved use of Epidiolex in a pediatric population, the toxicity of CBD in healthy children is also not established. There also remains a paucity of human efficacy data for indications of use appropriate for nutraceuticals and self-care products. As the use of CBD continues to expand, so does our knowledge of its potential benefits and toxicities through increased data collection. Our basic comprehension of its safety has thus far provided regulatory agencies with the confidence to issue cautious recommendations to the public that will be refined as the understanding of the appropriate intended uses of CBD increases.

References

Abou-Sayed, Y., Fitzgerald, D., Parker, J.W., 2018. Risk Assessment and Risk Mitigation Review. Application Number: 210365Orig1s000.

Adams, R., Hunt, M., Clark, J.H., 1940. Structure of cannabidiol, a product isolated from the marihuana extract of minnesota wild hemp. I. J. Am. Chem. Soc. 62, 196–200.

Andre, C.M., Hausman, J.-F., Guerriero, G., 2016. *Cannabis sativa*: the plant of the thousand and one molecules. Front. Plant Sci. 7. <https://doi.org/10.3389/fpls.2016.00019>.

Arndt, D.L., de Wit, H., 2017. Cannabidiol does not dampen responses to emotional stimuli in healthy adults. Cannabis Cannabinoid Res. 2, 105–113.

Arnold, W.R., Weigle, A.T., Das, A., 2018. Cross-talk of cannabinoid and endocannabinoid metabolism is mediated via human cardiac CYP2J2. J. Inorg. Biochem. 184, 88–99.

Atsmon, J., Heffetz, D., Deutsch, L., Deutsch, F., Sacks, H., 2018. Single-dose pharmacokinetics of oral cannabidiol following administration of PTL101: a new formulation based on gelatin matrix pellets technology. Clin. Pharmacol. Drug Dev. 7, 751–758. Available from: https://doi.org/10.1002/cpdd.408.

Avraham, Y., Zolotarev, O., Grigoriadis, N.C., et al., 2008. Cannabinoids and capsaicin improve liver function following thioacetamide-induced acute injury in mice. Am. J. Gastroenterol. 103, 3047–3056.

Babalonis, S., Haney, M., Malcolm, R.J., et al., 2017. Oral cannabidiol does not produce a signal for abuse liability in frequent marijuana smokers. Drug Alcohol Depend. 172, 9–13.

Bailey, D.G., Malcolm, J., Arnold, O., et al., 1998. Grapefruit juice–drug interactions. Br. J. Clin. Pharmacol. 46, 101–110.

Bar-Joseph, A., Berkovitch, Y., Adamchik, J., Biegon, A., 1994. Neuroprotective activity of HU-211, a novel NMDA antagonist, in global ischemia in gerbils. Mol. Chem. Neuropathol. 23 (2–3), 125–135. Available from: https://doi.org/10.1007/BF02815406.

Ben-Shabat, S., Fride, E., Sheskin, T., Tamiri, T., Rhee, M.H., Vogel, Z., et al., 1998. An entourage effect: inactive endogenous fatty acid glycerol esters enhance 2-arachidonoyl-glycerol cannabinoid activity. Eur. J. Pharmacol. 353, 23–31. Available from: https://doi.org/10.1016/s0014-2999(98)00392-6.

Bergamaschi, Queiroz, R.H.C., Chagas, M.H.N., et al., 2011. Cannabidiol reduces the anxiety induced by simulated public speaking in treatment-naïve social phobia patients. Neuropsychopharmacology 36, 1219–1226.

Bhattacharyya, S., Fusar-Poli, P., Borgwardt, S., et al., 2009. Modulation of mediotemporal and ventrostriatal function in humans by Delta9-tetrahydrocannabinol: a neural basis for the effects of *Cannabis sativa* on learning and psychosis. Arch. Gen. Psychiatry 66, 442–451.

Bhattacharyya, S., Morrison, P.D., Fusar-Poli, P., et al., 2010. Opposite effects of delta-9-tetrahydrocannabinol and cannabidiol on human brain function and psychopathology. Neuropsychopharmacology 35, 764–774.

Birnbaum, A.K., Karanam, A., Marino, S.E., et al., 2019. Food effect on pharmacokinetics of cannabidiol oral capsules in adult patients with refractory epilepsy. Epilepsia 60 (8), 1586–1592.

Bisogno, T., Hanuš, L., De Petrocellis, L., et al., 2001. Molecular targets for cannabidiol and its synthetic analogues: effect on vanilloid VR1 receptors and on the cellular uptake and enzymatic hydrolysis of anandamide. Br. J. Pharmacol. 134, 845–852.

Bisogno, T., Melck, D., Bobrov, M.Y., Gretskaya, N.M., Bezuglov, V.V., De Petrocellis, L., et al., 2000. N-acyl-dopamines: novel synthetic CB1 cannabinoid-receptor ligands and inhibitors of anandamide inactivation with cannabimimetic activity in vitro and in vivo. Biochem. J. 351, 817–824. Available from: https://doi.org/10.1042/bj3510817.

Boggs, D.L., Surti, T., Gupta, A., et al., 2018. The effects of cannabidiol (CBD) on cognition and symptoms in outpatients with chronic schizophrenia a randomized placebo controlled trial. Psychopharmacology (Berl.) 235, 1923–1932.

Borgwardt, S.J., Allen, P., Bhattacharyya, S., et al., 2008. Neural basis of Delta-9-tetrahydrocannabinol and cannabidiol: effects during response inhibition. Biol. Psychiatry 64, 966–973.

Bornheim, L.M., Everhart, E.T., Li, J., et al., 1994. Induction and genetic regulation of mouse hepatic cytochrome P450 by cannabidiol. Biochem. Pharmacol. 48, 161–171.

Brown, J.D., Winterstein, A.G., 2019. Potential adverse drug events and drug–drug interactions with medical and consumer cannabidiol (CBD) use. J. Clin. Med. 8. <https://doi.org/10.3390/jcm8070989>.

Buratti, S., Lavine, J.E., 2002. Drugs and the liver: advances in metabolism, toxicity, and therapeutics. Curr. Opin. Pediatr. 14, 601–607.

Bushra, R., Aslam, N., Khan, A.Y., 2011. Food-drug interactions. Oman Med. J. 26, 77–83.

Cabral, G.A., Rogers, T.J., Lichtman, A.H., 2015. Turning over a new leaf: cannabinoid and endocannabinoid modulation of immune function. J. Neuroimmune Pharmacol. 10, 193–203. <https://doi.org/10.1007/s11481-015-9615-z>.

Callén, L., Moreno, E., Barroso-Chinea, P., et al., 2012. Cannabinoid receptors CB1 and CB2 form functional heteromers in brain. J. Biol. Chem. 287, 20851–20865.

Canazza, I., Ossato, A., Vincenzi, F., Gregori, A., Rosa, F.D., Nigro, F., et al., 2016. Pharmaco-toxicological effects of the novel third-generation fluorinate synthetic cannabinoids, 5F-ADBINACA, AB-FUBINACA, and STS-135 in mice. In vitro and in vivo studies. Human Psychopharm.: Clin. Exp. 32, e2601. Available from: https://doi.org/10.1002/hup.2601.

Carlini, E.A., Cunha, J.M., 1981. Hypnotic and antiepileptic effects of cannabidiol. J. Clin. Pharmacol. 21, 417S–427S.

Carvalho, R.K., Santos, M.L., Souza, M.R., et al., 2018a. Chronic exposure to cannabidiol induces reproductive toxicity in male Swiss mice. J. Appl. Toxicol. JAT. 38, 1215–1223.

Carvalho, R.K., Souza, M.R., Santos, M.L., et al., 2018b. Chronic cannabidiol exposure promotes functional impairment in sexual behavior and fertility of male mice. Reprod. Toxicol. 81, 34–40.

Carvalho, R.K., Andersen, M.L., Mazaro-Costa, R., 2020. The effects of cannabidiol on male reproductive system: a literature review. J. Appl. Toxicol. 40, 132–150.

Chagas, M.H.N., Eckeli, A.L., Zuardi, A.W., Pena-Pereira, M.A., Sobreira-Neto, M.A., Sobreira, E.T., et al., 2014. Cannabidiol can improve complex sleep-related behaviours associated with rapid eye movement sleep behaviour disorder in Parkinson's disease patients: a case series. J. Clin. Pharm. Ther. 39, 564–566. Available from: https://doi.org/10.1111/jcpt.12179.

Cherniakov, I., Izgelov, D., Barasch, D., Davidson, E., Domb, A.J., Hoffman, A., 2017. Piperine-pro-nanolipospheres as a novel oral delivery system of cannabinoids: Pharmacokinetic evaluation in healthy volunteers in comparison to buccal spray administration. J. Control. Release 266, 1–7. Available from: https://doi.org/10.1016/j.jconrel.2017.09.011.

Colin, P., Sirois, G., Chakrabarti, S., 1986. Effects of route of administration on the dose-dependent metabolism of acetaminophen in rats: relationship with its toxicity. Arch. Int. Pharmacodyn. Ther. 281, 181–191.

Comelli, F., Giagnoni, G., Bettoni, I., Colleoni, M., Costa, B., 2008. Antihyperalgesic effect of a Cannabis sativa extract in a rat model of neuropathic pain: mechanisms involved. Phytother. Res. 22, 1017–1024. Available from: https://doi.org/10.1002/ptr.2401.

Consroe, Carlini, E.A., Zwicker, A.P., et al., 1979. Interaction of cannabidiol and alcohol in humans. Psychopharmacology (Berl.) 66, 45–50.

Consroe, P., Laguna, J., Allender, J., et al., 1991. Controlled clinical trial of cannabidiol in Huntington's disease. Pharmacol. Biochem. Behav. 40, 701–708.

Crippa, Zuardi, A.W., Garrido, G.E.J., et al., 2004. Effects of cannabidiol (CBD) on regional cerebral blood flow. Neuropsychopharmacology 29, 417–426.

Crippa, Derenusson, G.N., Ferrari, T.B., et al., 2011. Neural basis of anxiolytic effects of cannabidiol (CBD) in generalized social anxiety disorder: a preliminary report. J. Psychopharmacol. (Oxf.) 25, 121–130.

Cuñetti, L., Manzo, L., Peyraube, R., et al., 2018. Chronic pain treatment with cannabidiol in kidney transplant patients in Uruguay. Transplant. Proc. 50, 461–464.

Cunha, J.M., Carlini, E.A., Pereira, A.E., et al., 1980. Chronic administration of cannabidiol to healthy volunteers and epileptic patients. Pharmacology 21, 175–185.

Dalterio, S.L., Bartke, A., Mayfield, D., 1983. Cannabinoids stimulate and inhibit testosterone production in vitro and in vivo. Life Sci. 32, 605–612.

Dalton, W.S., Martz, R., Lemberger, L., et al., 1976. Influence of cannabidiol on delta-9-tetrahydrocannabinol effects. Clin. Pharmacol. Ther. 19, 300–309.

DEA, 2018. Schedules of Controlled Substances: Placement in Schedule V of Certain FDA-Approved Drugs Containing Cannabidiol; Corresponding Change to Permit Requirements [WWW Document]. Federal Register. https://www.federalregister.gov/documents/2018/09/28/2018-21121/schedules-of-controlled-substances-placement-in-schedule-v-of-certain-fda-approved-drugs-containing (accessed 9.28.20).

De Petrocellis, L., Ligresti, A., Moriello, A.S., Allarà, M., Bisogno, T., Petrosino, S., et al., 2011. Effects of cannabinoids and cannabinoid-enriched Cannabis extracts on TRP channels and endocannabinoid metabolic enzymes. Br. J. Pharmacol. 163 (7), 1479–1494. Available from: https://doi.org/10.1111/j.1476-5381.2010.01166.x.

Defer, N., Wan, J., Souktani, R., et al., 2009. The cannabinoid receptor type 2 promotes cardiac myocyte and fibroblast survival and protects against ischemia/reperfusion-induced cardiomyopathy. FASEB J. 23, 2120–2130.

Deiana, S., Watanabe, A., Yamasaki, Y., et al., 2012. Plasma and brain pharmacokinetic profile of cannabidiol (CBD), cannabidivarine (CBDV), Δ9-tetrahydrocannabivarin (THCV) and cannabigerol (CBG) in rats and mice following oral and intraperitoneal administration and CBD action on obsessive–compulsive behaviour. Psychopharmacology (Berl.) 219, 859–873.

Devinsky, O., Cilio, M.R., Cross, H., et al., 2014. Cannabidiol: pharmacology and potential therapeutic role in epilepsy and other neuropsychiatric disorders. Epilepsia 55, 791–802.

Devinsky, O., Marsh, E., Friedman, D., et al., 2016. Cannabidiol in patients with treatment-resistant epilepsy: an open-label interventional trial. Lancet Neurol. 15, 270–278.

Devinsky, O., Verducci, C., Thiele, E.A., et al., 2018. Open-label use of highly purified CBD (Epidiolex) in patients with CDKL5 deficiency disorder and Aicardi, Dup15q, and Doose syndromes. Epilepsy Behav. EB 86, 131–137.

Di Marzo, V., 2008. Targeting the endocannabinoid system: to enhance or reduce? Nat. Rev. Drug Discov. 7, 438–455.

Dibba, P., Li, A.A., Cholankeril, G., et al. 2018. The role of cannabinoids in the setting of cirrhosis. Medicines 5. <https://doi.org/10.3390/medicines5020052>

Dyson, A., Peacock, M., Chen, A., Courade, J.-P., Yaqoob, M., Groarke, A., et al., 2005. Antihyperalgesic properties of the cannabinoid CT-3 in chronic neuropathic and inflammatory pain states in the rat. Pain 116 (1–2), 129–137. Available from: https://doi.org/10.1016/j.pain.2005.03.037.

ElSohly, M.A., Mehmedic, Z., Foster, S., et al., 2016. Changes in cannabis potency over the last 2 decades (1995-2014): analysis of current data in the United States. Biol. Psychiatr. 79, 613–619.

Estacion, M., Gasser, A., Dib-Hajj, S.D., et al., 2010. A sodium channel mutation linked to epilepsy increases ramp and persistent current of Nav1.3 and induces hyperexcitability in hippocampal neurons. Exp. Neurol. 224, 362–368.

Ewing, L.E., McGill, M.R., Yee, E.U., et al., 2019a. Paradoxical patterns of sinusoidal obstruction syndrome-like liver injury in aged female CD-1 mice triggered by cannabidiol-rich cannabis extract and acetaminophen co-administration. Molecules Basel Switz. 24. <https://doi.org/10.3390/molecules24122256>.

Ewing, L.E., Skinner, C.M., Quick, C.M., et al., 2019b. Hepatotoxicity of a cannabidiol-rich cannabis extract in the mouse model. Molecules 24, 1694.

FDA, 2017. Clinical Pharmacology and Biopharmaceutical Reviews. https://www.accessdata.fda.gov/drugsatfda_docs/nda/2018/210365Orig1s000ClinPharmR.pdf (accessed 7.31.19).

FDA, 2018a. Cannabidiol, Epidiolex. https://www.accessdata.fda.gov/drugsatfda_docs/nda/2018/210365Orig1s000PharmR.pdf (accessed 1.28.19).

FDA, 2018b. Clinical Trial for Epidiolex. https://www.accessdata.fda.gov/drugsatfda_docs/nda/2018/210365Orig1s000OtherR.pdf (accessed 3.25.19).

FDA, 2018c. Epidiolex (Cannabidiol) Oral Solution. https://www.accessdata.fda.gov/drugsatfda_docs/label/2018/210365lbl.pdf> (Accessed 9.29.20)

FDA, 2018d. FDA Briefing Document by the Peripheral and Central Nervous System Drugs Advisory Committee Meeting on CBD. https://www.fda.gov/media/112565/download (accessed 6.24.19).

Food Standards Agency, 2020. Food standards agency sets deadline for the CBD industry and provides safety advice to consumers [WWW Document]. Food Standards Agency. <https://www.food.gov.uk/news-alerts/news/food-standards-agency-sets-deadline-for-the-cbd-industry-and-provides-safety-advice-to-consumers> (Accessed 5 January 2020).

Fusar-Poli, P., Crippa, J.A., Bhattacharyya, S., et al., 2009. Distinct effects of {delta}9-tetrahydrocannabinol and cannabidiol on neural activation during emotional processing. Arch. Gen. Psychiatry 66, 95–105.

Fairbairn, J.W., Pickens, J.T., 1980. The effect of conditions influencing endogenous prostaglandins on the activity of Δ′-tetrahydrocannabinol in mice. Br. J. Pharmacol 69, 491–493. Available from: https://doi.org/10.1111/j.1476-5381.1980.tb07039.x.

Felder, C.C., Joyce, K.E., Briley, E.M., Mansouri, J., Mackie, K., Blond, O., et al., 1995. Comparison of the pharmacology and signal transduction of the human cannabinoid CB1 and CB2 receptors. Mol. Pharmacol. 48 (3), 443–450.

Fusar-Poli, P., Allen, P., Bhattacharyya, S., et al., 2010. Modulation of effective connectivity during emotional processing by Δ9-tetrahydrocannabinol and cannabidiol. Int. J. Neuropsychopharmacol. 13, 421–432.

Gallily, R., Even-Chen, T., Katzavian, G., et al., 2003. γ-Irradiation enhances apoptosis induced by cannabidiol, a non-psychotropic cannabinoid, in cultured HL-60 myeloblastic leukemia cells. Leuk. Lymphoma 44, 1767–1773.

Gaoni, Y., Mechoulam, R., 1964. Isolation, structure, and partial synthesis of an active constituent of Hashish. J. Am. Chem. Soc. 86, 1646–1647.

Gareau, Y., Dufresne, C., Gallant, M., Rochette, C., Sawyer, N., Slipetz, D.M., et al., 1996. Structure activity relationships of tetrahydrocannabinol analogues on human cannabinoid receptors. Bioorg Med Chem Lett. 6, 189–194. Available from: https://doi.org/10.1016/0960-894X(95)00573-C.

Gayam, V., Mandal, A.K., et al., 2018. Valproic acid induced acute liver injury resulting in hepatic encephalopathy—a case report and literature review. J. Community Hosp. Intern. Med. Perspect. 8, 311–314.

Geffrey, A.L., Pollack, S.F., Bruno, P.L., et al., 2015. Drug-drug interaction between clobazam and cannabidiol in children with refractory epilepsy. Epilepsia 56, 1246–1251.

German Medicinal Codex, 2015. DAC / NRF: New in DAC / NRF: Cannabidiol [WWW Document]. <https://dacnrf.pharmazeutische-zeitung.de/index.php?id = 557> (Accessed 5 April 2020).

Giacoppo, S., Galuppo, M., Pollastro, F., et al., 2015. A new formulation of cannabidiol in cream shows therapeutic effects in a mouse model of experimental autoimmune encephalomyelitis. DARU J. Pharm. Sci. 23. <https://doi.org/10.1186/s40199-015-0131-8>.

Gustafsson, S.B., Jacobsson, S.O.P., 2019. Effects of cannabinoids on the development of chick embryos *in ovo*. Sci. Rep. 9, 1–8.

Hahn, S., 2020. Report to the U.S. House Committee on Appropriations and the U. S. Senate Committee on Appropriations, Cannabidiol (CBD). Report in Response to Further Consolidated Appropriations Act, 2020. FDA.

Guy, G.W., Flint, M.E., 2004. A single centre, placebo-controlled, four period, crossover, tolerability study assessing, pharmacodynamic effects, pharmacokinetic characteristics and cognitive profiles of a single dose of three formulations of cannabis based medicine extracts (CBMEs) (GWPD9901), plus a two period tolerability study comparing pharmacodynamic effects and pharmacokinetic characteristics of a single dose of a cannabis based medicine extract given via two administration routes (GWPD9901 EXT). Journal of Cannabis Therapeutics 3, 35–77. Available from: https://doi.org/10.1300/J175v03n03_03.

Hall, A.P., Elcombe, C.R., Foster, J.R., et al., 2012. Liver hypertrophy: a review of adaptive (adverse and non-adverse) changes—conclusions from the 3rd international ESTP expert workshop. Toxicol. Pathol. 40, 971–994.

Hallak, J.E.C., Machado-de-Sousa, J.P., Crippa, J.A.S., et al., 2010. Performance of schizophrenic patients in the Stroop Color Word Test and electrodermal responsiveness after acute administration of cannabidiol (CBD). Rev. Bras. Psiquiatr. Sao Paulo Braz 32, 56–61.

Hallak, J.E.C., Dursun, S.M., Bosi, D.C., et al., 2011. The interplay of cannabinoid and NMDA glutamate receptor systems in humans: preliminary evidence of interactive effects of cannabidiol and ketamine in healthy human subjects. Prog. Neuropsychopharmacol. Biol. Psychiatr. 35, 198–202.

Hammell, D.C., Zhang, L.P., Ma, F., et al., 2016. Transdermal cannabidiol reduces inflammation and pain-related behaviours in a rat model of arthritis. Eur. J. Pain Lond. Engl. 20, 936–948.

Haney, M., Malcolm, R.J., Babalonis, S., et al., 2016. Oral cannabidiol does not alter the subjective, reinforcing or cardiovascular effects of smoked cannabis. Neuropsychopharmacology 41, 1974–1982.

Hanuš, L., Abu-Lafi, S., Fride, E., Breuer, A., Vogel, Z., Shalev, D.E., et al., 2001. 2-Arachidonyl glyceryl ether, an endogenous agonist of the cannabinoid CB1 receptor. PNAS 98, 3662–3665. Available from: https://doi.org/10.1073/pnas.061029898.

Hanuš, L., Martin Meyer, S., Muñoz, E., et al., 2016. Phytocannabinoids: a unified critical inventory. Nat. Prod. Rep. 33, 1357–1392.

Harrill, A.H., Watkins, P.B., Su, S., et al., 2009. Mouse population-guided resequencing reveals that variants in CD44 contribute to acetaminophen-induced liver injury in humans. Genome Res. 19, 1507–1515.

Hartsel, J.A., Eades, J., Hickory, B., et al., 2016. *Cannabis sativa* and hemp. In: Gupta, R.C. (Ed.), *Nutraceuticals: Efficacy, Safety and Toxicity*. Academic Press/Elsevier, Amsterdam, pp. 735–754.

Holland, M.L., Lau, D.T.T., Allen, J.D., et al., 2007. The multidrug transporter ABCG2 (BCRP) is inhibited by plant-derived cannabinoids. Br. J. Pharmacol. 152, 815–824.

Hillard, C.J., Muthian, S., Kearn, C.S., 1999. Effects of CB(1) cannabinoid receptor activation on cerebellar granule cell nitric oxide synthase activity. FEBS Lett. 459 (2), 277–281. Available from: https://doi.org/10.1016/s0014-5793(99)01253-3.

Holland, M.L., Allen, J.D., Arnold, J.C., 2008. Interaction of plant cannabinoids with the multidrug transporter ABCC1 (MRP1). Eur. J. Pharmacol. 591, 128–131.

Hollister, L.E., 1973. Cannabidiol and cannabinol in man. Experientia 29, 825–826.

Huang, S.M., Bisogno, T., Trevisani, M., Al-Hayani, A., De Petrocellis, L., Fezza, F., et al., 2002. An endogenous capsaicin-like substance with high potency at recombinant and native vanilloid VR1 receptors. Proc. Natl. Acad. Sci. U.S. A. 99, 8400–8405. Available from: https://doi.org/10.1073/pnas.122196999.

Ibeas Bih, C., Chen, T., Nunn, A.V.W., et al., 2015. Molecular targets of cannabidiol in neurological disorders. J. Am. Soc. Exp. Neurother. 12, 699–730.

Iffland, K., Grotenhermen, F., 2017. An update on safety and side effects of cannabidiol: a review of clinical data and relevant animal studies. Cannabis Cannabinoid Res. 2, 139–154.

Ito, Y., Bethea, N.W., Abril, E.R., et al., 2003. Early hepatic microvascular injury in response to acetaminophen toxicity. Microcirculation N. Y. N 10, 391–400. 1994.

Izzo, A.A., Borrelli, F., Capasso, R., et al., 2009. Non-psychotropic plant cannabinoids: new therapeutic opportunities from an ancient herb. Trends Pharmacol. Sci. 30, 515–527.

Jaeger, W., Benet, L.Z., Bornheim, L.M., 1996. Inhibition of cyclosporine and tetrahydrocannabinol metabolism by cannabidiol in mouse and human microsomes. Xenobiotica Fate Foreign Compd. Biol. Syst. 26, 275–284.

Járai, Z., Wagner, J.A., Varga, K., Lake, K.D., Compton, D.R., Martin, B.R., et al., 1999. Cannabinoid-induced mesenteric vasodilation through an endothelial site distinct from CB1 or CB2 receptors. Proc. Natl. Acad. Sci. U.S.A. 96 (24), 14136–14141. https://doi.org/10.1073/pnas.96.24.14136.

Jeong, W., Osei-Hyiaman, D., Park, O., et al., 2008. Paracrine activation of hepatic CB1 receptors by stellate cell-derived endocannabinoids mediates alcoholic fatty liver. Cell Metab. 7, 227–235.

Jiang, R., Yamaori, S., Takeda, S., et al., 2011. Identification of cytochrome P450 enzymes responsible for metabolism of cannabidiol by human liver microsomes. Life Sci. 89, 165–170.

Jikomes, N., Zoorob, M., 2018. The cannabinoid content of legal cannabis in Washington State varies systematically across testing facilities and popular consumer products. Sci. Rep. 8, 4519. <https://doi.org/10.1038/s41598-018-22755-2>.

Johnson, J.A., 2012. Warfarin pharmacogenetics: a rising tide for its clinical value. Circulation 125, 1964–1966.

Jones, G., Pertwee, R.G., 1972. A metabolic interaction in vivo between cannabidiol and 1 -tetrahydrocannabinol. Br. J. Pharmacol. 45, 375–377.

Karniol, I.G., Shirakawa, I., Kasinski, N., et al., 1974. Cannabidiol interferes with the effects of Δ9-tetrahydrocannabinol in man. Eur. J. Pharmacol. 28, 172–177.

Karschner, E.L., Darwin, W.D., Goodwin, R.S., Wright, S., Huestis, M.A., 2011. Plasma cannabinoid pharmacokinetics following controlled oral Δ9-tetrahydrocannabinol and oromucosal cannabis extract administration. Clin. Chem. 57, 66–75. Available from: https://doi.org/10.1373/clinchem.2010.152439.

Kathmann, M., Flau, K., Redmer, A., et al., 2006. Cannabidiol is an allosteric modulator at mu- and delta-opioid receptors. Naunyn. Schmiedebergs Arch. Pharmacol. 372, 354–361.

Klein, C., Karanges, E., Spiro, A., et al., 2011. Cannabidiol potentiates Δ9-tetrahydrocannabinol (THC) behavioural effects and alters THC pharmacokinetics during acute and chronic treatment in adolescent rats. Psychopharmacology (Berl.) 218, 443–457.

Koziolek, M., Alcaro, S., Augustijns, P., et al., 2019. The mechanisms of pharmacokinetic food-drug interactions—a perspective from the UNGAP group. Eur. J. Pharm. Sci. 134, 31–59.

Kunos, G., Osei-Hyiaman, D., Bátkai, S., et al., 2009. Should peripheral CB1 cannabinoid receptors be selectively targeted for therapeutic gain? Trends Pharmacol. Sci. 30, 1–7.

Laun, A.S., Shrader, S.H., Brown, K.J., et al., 2019. GPR3, GPR6, and GPR12 as novel molecular targets: their biological functions and interaction with cannabidiol. Acta Pharmacol. Sin. 40, 300. <https://doi.org/10.1038/s41401-018-0031-9>.

Lee, C.-Y., Wey, S.-P., Liao, M.-H., et al., 2008. A comparative study on cannabidiol-induced apoptosis in murine thymocytes and EL-4 thymoma cells. Int. Immunopharmacol. 8, 732–740.

Lee, J.L.C., Bertoglio, L.J., Guimarães, F.S., et al., 2017. Cannabidiol regulation of emotion and emotional memory processing: relevance for treating anxiety-related and substance abuse disorders. Br. J. Pharmacol. 174, 3242–3256.

Leite, J.R., Carlini, E.A., Lander, N., et al., 1982. Anticonvulsant effects of the (-) and (+) isomers of cannabidiol and their dimethylheptyl homologs. Pharmacology 24, 141–146.

Leweke, F.M., Piomelli, D., Pahlisch, F., et al., 2012. Cannabidiol enhances anandamide signaling and alleviates psychotic symptoms of schizophrenia. Transl. Psychiatry 2, e94. <https://doi.org/10.1038/tp.2012.15>.

Lim, M.P., Devi, L.A., Rozenfeld, R., 2011. Cannabidiol causes activated hepatic stellate cell death through a mechanism of endoplasmic reticulum stress-induced apoptosis. Cell Death Dis. 2, e170. <https://doi.org/10.1038/cddis.2011.52>.

Linares, I.M.P., Guimaraes, F.S., Eckeli, A., et al., 2018. No acute effects of cannabidiol on the sleep-wake cycle of healthy subjects: a randomized, double-blind, placebo-controlled, crossover study. Front. Pharmacol. 9. <https://doi.org/10.3389/fphar.2018.00315>.

Linares, I.M., Zuardi, A.W., Pereira, L.C., et al., 2019. Cannabidiol presents an inverted U-shaped dose-response curve in a simulated public speaking test. Braz. J. Psychiatry 41, 9–14.

Lodzki, M., Godin, B., Rakou, L., Mechoulam, R., Gallily, R., Touitou, E., 2003. Cannabidiol-transdermal delivery and anti-inflammatory effect in a murine model. J. Control Release 93, 377–387. Available from: https://doi.org/10.1016/j.jconrel.2003.09.001.

Loewe, S., 1946. Studies on the pharmacology and acute toxicity of compounds with marihuana activity. J. Pharmacol. Exp. Ther. 88, 154–161.

Mallat, A., Teixeira-Clerc, F., Deveaux, V., et al., 2011. The endocannabinoid system as a key mediator during liver diseases: new insights and therapeutic openings. Br. J. Pharmacol. 163, 1432–1440.

Manini, A.F., Yiannoulos, G., Bergamaschi, M.M., Hernandez, S., Olmedo, R., Barnes, A.J., et al., 2015. Safety and pharmacokinetics of oral cannabidiol when administered concomitantly with intravenous fentanyl in humans. J. Addict. Med. 9, 204–210. Available from: https://doi.org/10.1097/ADM.0000000000000118.

Marx, T.K., Reddeman, R., Clewell, A.E., et al., 2018. An assessment of the genotoxicity and subchronic toxicity of a supercritical fluid extract of the aerial parts of Hemp. J. Toxicol. <https://doi.org/10.1155/2018/8143582>.

Massi, P., Vaccani, A., Ceruti, S., et al., 2004. Antitumor effects of cannabidiol, a nonpsychoactive cannabinoid, on human glioma cell lines. J. Pharmacol. Exp. Ther. 308, 838–845.

Massi, P., Vaccani, A., Bianchessi, S., et al., 2006. The non-psychoactive cannabidiol triggers caspase activation and oxidative stress in human glioma cells. Cell. Mol. Life Sci. 63, 2057–2066.

McGuire, P., Robson, P., Cubala, W.J., et al., 2018. Cannabidiol (CBD) as an adjunctive therapy in Schizophrenia: a multicenter randomized controlled trial. Am. J. Psychiatry 175, 225–231.

McKallip, R.J., Jia, W., Schlomer, J., et al., 2006. Cannabidiol-induced apoptosis in human leukemia cells: a novel role of cannabidiol in the regulation of p22phox and Nox4 expression. Mol. Pharmacol. 70, 897–908.

McPartland, J.M., Duncan, M., Di Marzo, V., Pertwee, R.G., 2015. Are cannabidiol and $\Delta(9)$ -tetrahydrocannabivarin negative modulators of the endocannabinoid system? A systematic review. Br. J. Pharmacol. 172 (3), 737–753. Available from: https://doi.org/10.1111/bph.12944.

McPartland, J.M., Glass, M., Pertwee, R.G., 2007. Meta-analysis of cannabinoid ligand binding affinity and receptor distribution: interspecies differences. Br. J. Pharmacol. 152 (5), 583–593. Available from: https://doi.org/10.1038/sj.bjp.0707399.

Mecha, M., Torrao, A.S., Mestre, L., et al., 2012. Cannabidiol protects oligodendrocyte progenitor cells from inflammation-induced apoptosis by attenuating endoplasmic reticulum stress. Cell Death Dis. 3, e331. <https://doi.org/10.1038/cddis.2012.71>.

Mechoulam, R., Shvo, Y., 1963. Hashish. I. The structure of cannabidiol. Tetrahedron 19, 2073–2078.

Mechoulam, R., Ben-Shabat, S., Hanus, L., Ligumsky, M., Kaminski, N.E., Schatz, A.R., et al., 1995. Identification of an endogenous 2-monoglyceride, present in canine gut, that binds to cannabinoid receptors. Biochem. Pharmacol. 50, 83–90. Available from: https://doi.org/10.1016/0006-2952(95)00109-d.

Mechoulam, R., Peters, M., Murillo-Rodriguez, E., et al., 2007. Cannabidiol-recent advances. Chem. Biodivers. 4, 1678–1692.

Mehmedic, Z., Chandra, S., Slade, D., et al., 2010. Potency trends of Δ9-THC and other cannabinoids in confiscated cannabis preparations from 1993 to 2008. J. Forensic Sci. 55, 1209–1217.

Mincis, M., Pfeferman, A., Guimarães, R.X., et al., 1973. Chronic administration of cannabidiol in man. Pilot study. AMB. Rev. Assoc. Med. Bras. 19, 185–190.

Montecucco, F., Lenglet, S., Braunersreuther, V., et al., 2009. CB(2) cannabinoid receptor activation is cardioprotective in a mouse model of ischemia/reperfusion. J. Mol. Cell. Cardiol. 46, 612–620.

Nadulski, T., Sporkert, F., Schnelle, M., et al., 2005. Simultaneous and sensitive analysis of THC, 11-OH-THC, THC-COOH, CBD, and CBN by GC-MS in plasma after oral application of small doses of THC and cannabis extract. J. Anal. Toxicol. 29, 782–789.

Nair, A.B., Jacob, S., 2016. A simple practice guide for dose conversion between animals and human. J. Basic Clin. Pharm. 7, 27–31.

Navarro, G., Borroto-Escuela, D., Angelats, E., Etayo, Í., Reyes-Resina, I., Pulido-Salgado, M., et al., 2018. Receptor-heteromer mediated regulation of endocannabinoid signaling in activated microglia. Role of CB_1 and CB_2 receptors and relevance for Alzheimer's disease and levodopa-induced dyskinesia. Brain Behav. Immun. 67, 139–151. Available from: https://doi.org/10.1016/j.bbi.2017.08.015.

Nicholson, A.N., Turner, C., Stone, B.M., et al., 2004. Effect of delta-9-tetrahydrocannabinol and cannabidiol on nocturnal sleep and early-morning behavior in young adults. J. Clin. Psychopharmacol. 24, 305–313.

Ohlsson, A., Lindgren, J.E., Andersson, S., et al., 1986. Single-dose kinetics of deuterium-labelled cannabidiol in man after smoking and intravenous administration. Biomed. Environ. Mass Spectrom. 13, 77–83.

Olivas-Aguirre, M., Torres-López, L., Valle-Reyes, J.S., et al., 2019. Cannabidiol directly targets mitochondria and disturbs calcium homeostasis in acute lymphoblastic leukemia. Cell Death Dis. 10, 779. <https://doi.org/10.1038/s41419-019-2024-0>.

Paton, W.D., Pertwee, R.G., 1972. Effect of cannabis and certain of its constituents on pentobarbitone sleeping time and phenazone metabolism. Br. J. Pharmacol. 44, 250–261.

Paudel, K.S., Hammell, D.C., Agu, R.U., et al., 2010. Cannabidiol bioavailability after nasal and transdermal application: effect of permeation enhancers. Drug Dev. Ind. Pharm. 36, 1088–1097.

Pelkonen, O., Mäenpää, J., Taavitsainen, P., et al., 1998. Inhibition and induction of human cytochrome P450 (CYP) enzymes. Xenobiotica Fate Foreign Compd. Biol. Syst. 28, 1203–1253.

Perez-Reyes, M., Timmons, M.C., Davis, K.H., et al., 1973. A comparison of the pharmacological activity in man of intravenously administered delta9-tetrahydrocannabinol, cannabinol, and cannabidiol. Experientia 29, 1368–1369.

Pertwee, R.G., 2004. The pharmacology and therapeutic potential of cannabidiol. In: Di Marzo, V. (Ed.), Cannabinoids. Kluwer Academic/Plenum Publishers, pp. 32–83.

Pertwee, R.G., 2008. The diverse CB1 and CB2 receptor pharmacology of three plant cannabinoids: delta9-tetrahydrocannabinol, cannabidiol and delta9-tetrahydrocannabivarin. Br. J. Pharmacol. 153, 199–215.

Pertwee, R.G., Howlett, A.C., Abood, M.E., et al., 2010. International union of basic and clinical pharmacology. LXXIX. Cannabinoid receptors and their ligands: beyond CB and CB. Pharmacol. Rev. 62, 588–631.

Pertwee, R.G., Ross, R.A., Craib, S.J., Thomas, A., 2002. (-)-Cannabidiol antagonizes cannabinoid receptor agonists and noradrenaline in the mouse vas deferens. Eur. J. Pharmacol. 456 (1–3), 99–106. https://doi.org/10.1016/s0014-2999(02)02624-9.

Pietrafusa, N., Ferretti, A., Trivisano, M., de Palma, L., Calabrese, C., Carfì Pavia, G., et al., 2019. Purified cannabidiol for treatment of refractory epilepsies in pediatric patients with developmental and epileptic encephalopathy. Pediatr. Drugs. Available from: https://doi.org/10.1007/s40272-019-00341-x.

Porter, A.C., Sauer, J.-M., Knierman, M.D., Becker, G.W., Berna, M.J., Bao, J., et al., 2002. Characterization of a novel endocannabinoid, virodhamine, with antagonist activity at the CB1 receptor. J. Pharmacol. Exp. Ther. 301, 1020–1024. Available from: https://doi.org/10.1124/jpet.301.3.1020.

Qian, Y., Gurley, B.J., Markowitz, J.S., 2019. The potential for pharmacokinetic interactions between cannabis products and conventional medications. J. Clin. Psychopharmacol. 39, 462–471.

Raspall-Chaure, M., Chin, R.F.M., Neville, B.G., et al., 2007. The epidemiology of convulsive status epilepticus in children: a critical review. Epilepsia 48, 1652–1663.

Rhee, M.H., Vogel, Z., Barg, J., Bayewitch, M., Levy, R., Hanus, L., et al., 1997. Cannabinol derivatives: binding to cannabinoid receptors and inhibition of adenylylcyclase. J. Med. Chem. 40 (20), 3228–3233. https://doi.org/10.1021/jm970126f.

Rinaldi-Carmona, M., Barth, F., Héaulme, M., Shire, D., Calandra, B., Congy, C., et al., 1994. SR141716A, a potent and selective antagonist of the brain cannabinoid receptor. FEBS Lett. 350 (2–3), 240–244. https://doi.org/10.1016/0014-5793(94)00773-x.

Ritter, J., Flower, R., Henderson, G., et al., 2011. Rang & Dale's Pharmacology, seventh ed. Elsevier.

Rosenkrantz, H., Fleischman, R.W., Grant, R.J., 1981. Toxicity of short-term administration of cannabinoids to rhesus monkeys. Toxicol. Appl. Pharmacol. 58, 118–131.

Rosenthaler, S., Pöhn, B., Kolmanz, C., Huu, C.N., Krewenka, C., Huber, A., et al., 2014. Differences in receptor binding affinity of several phytocannabinoids do not explain their effects on neural cell cultures. Neurotoxicol. Teratol. 46, 49–56. https://doi.org/10.1016/j.ntt.2014.09.003.

Ross, R.A., Gibson, T.M., Stevenson, L.A., Saha, B., Crocker, P., Razdan, R.K., et al., 1999. Structural determinants of the partial agonist-inverse agonist properties of 6'-azidohex-2'-yne-delta8-tetrahydrocannabinol at cannabinoid receptors. Br. J. Pharmacol. 128 (3), 735–743. https://doi.org/10.1038/sj.bjp.0702836.

Russo, E.B., Burnett, A., Hall, B., et al., 2005. Agonistic properties of cannabidiol at 5-HT1a receptors. Neurochem. Res. 30, 1037–1043.

Russo, E.B., Guy, G.W., Robson, P.J., 2007. Cannabis, pain, and sleep: lessons from therapeutic clinical trials of Sativex, a cannabis-based medicine. Chem. Biodivers. 4, 1729–1743.

Ryberg, E., Larsson, N., Sjögren, S., et al., 2007. The orphan receptor GPR55 is a novel cannabinoid receptor. Br. J. Pharmacol. 152, 1092–1101.

Schapira, D., Bassan, L., Nahir, A.M., et al., 1986. Diclofenac-induced hepatotoxicity. Postgrad. Med. J. 62, 63–65.

Schwartz, M., Böckmann, S., Hinz, B., 2018. Up-regulation of heme oxygenase-1 expression and inhibition of disease-associated features by cannabidiol in vascular smooth muscle cells. Oncotarget 9, 34595–34616.

Sellers, E.M., Schoedel, K., Bartlett, C., Romach, M., Russo, E.B., Stott, C.G., et al., 2013. A multiple-dose, randomized, double-blind, placebo-controlled, parallel-group QT/QTc study to evaluate the electrophysiologic effects of THC/CBD spray. Clin. Pharmacol. Drug. Dev. 2, 285–294. Available from: https://doi.org/10.1002/cpdd.36.

Showalter, V.M., Compton, D.R., Martin, B.R., Abood, M.E., 1996. Evaluation of binding in a transfected cell line expressing a peripheral cannabinoid receptor (CB2): identification of cannabinoid receptor subtype selective ligands. J. Pharmacol. Exp. Ther. 278 (3), 989–999.

Shrivastava, A., Kuzontkoski, P.M., Groopman, J.E., et al., 2011. Cannabidiol induces programmed cell death in breast cancer cells by coordinating the cross-talk between apoptosis and autophagy. Mol. Cancer Ther. 10, 1161–1172.

Sierra, S., Luquin, N., Navarro-Otano, J., 2018. The endocannabinoid system in cardiovascular function: novel insights and clinical implications. Clin. Auton. Res. Off. J. Clin. Auton. Res. Soc. 28, 35–52.

Stott, C.G., Guy, G., Wright, S., 2005. The effects of cannabis extracts tetranabinex and nabidiolex on human cytochrome P450-mediated metabolism. Presented at the International Cannabinoid Research Association, Clearwater, FL, 163.

Stott, C.G., Ayerakwa, L., Wright, S., et al. 2007. Lack of human cytochrome P450 induction by Sativex. proceedings of the Seventeenth Annual Symposium on the Cannabinoids, International Cannabinoid Research Society, Saint-Sauveur, Quebec, Canada, 211.

Steger, R.W., Murphy, L.L., Bartke, A., Smith, M.S., 1990. Effects of psychoactive and nonpsychoactive cannabinoids on the hypothalamic-pituitary axis of the adult male rat. Pharmacol. Biochem. Behav. 37, 299–302. Available from: https://doi.org/10.1016/0091-3057(90)90338-i.

Stott, C., White, L., Wright, S., et al., 2013. A Phase I, open-label, randomized, crossover study in three parallel groups to evaluate the effect of rifampicin, ketoconazole, and omeprazole on the pharmacokinetics of THC/CBD oromucosal spray in healthy volunteers. Springer Plus 2, 236. <https://doi.org/10.1186/2193-1801-2-236>.

Stout, S.M., Cimino, N.M., 2014. Exogenous cannabinoids as substrates, inhibitors, and inducers of human drug metabolizing enzymes: a systematic review. Drug Metab. Rev. 46, 86–95.

Szaflarski, J.P., Bebin, E.M., Cutter, G., et al., 2018. Cannabidiol improves frequency and severity of seizures and reduces adverse events in an open-label add-on prospective study. Epilepsy Behav. 87, 131–136.

Taguchi, K., Tokuno, M., Yamasaki, K., et al., 2015. Establishment of a model of acetaminophen-induced hepatotoxicity in different weekly-aged ICR mice. Lab. Anim. 49, 294–301.

Taylor, L., Gidal, B., Blakey, G., et al., 2018. A Phase I, randomized, double-blind, placebo-controlled, single ascending dose, multiple dose, and food effect trial of the safety, tolerability and pharmacokinetics of highly purified cannabidiol in healthy subjects. CNS Drugs 32, 1053–1067.

TGA, 2020. Review on the safety of low dose cannabidiol. Therapeutic Goods Administation. TGA. <https://www.tga.gov.au/alert/review-safety-low-dose-cannabidiol> (Accessed 5 January 2020).

Thiele, E.A., Marsh, E.D., French, J.A., et al., 2018. Cannabidiol in patients with seizures associated with Lennox-Gastaut syndrome (GWPCARE4): a randomised, double-blind, placebo-controlled phase 3 trial. Lancet 391, 1085–1096.

Thomas, A., Baillie, G.L., Phillips, A.M., et al., 2007. Cannabidiol displays unexpectedly high potency as an antagonist of CB1 and CB2 receptor agonists in vitro. Br. J. Pharmacol. 150, 613–623.

Thomas, B.F., ElSohly, M., 2015. The Analytical Chemistry of Cannabis: Quality Assessment, Assurance, and Regulation of Medicinal Marijuana and Cannabinoid Preparations. Elsevier.

Thomas, A., Stevenson, L.A., Wease, K.N., Price, M.R., Baillie, G., Ross, R.A., et al., 2005. Evidence that the plant cannabinoid Delta9-tetrahydrocannabivarin is a cannabinoid CB1 and CB2 receptor antagonist. Br. J. Pharmacol. 146 (7), 917–926. Available from: https://doi.org/10.1038/sj.bjp.0706414.

Toms, L., McQuay, H.J., Derry, S., et al., 2008. Single dose oral paracetamol (acetaminophen) for postoperative pain in adults. Cochrane Database Syst. Rev. 2008.

Udoh, M., Santiago, M., Devenish, S., McGregor, I.S., Connor, M., 2019. Cannabichromene is a cannabinoid CB_2 receptor agonist. Br. J. Pharmacol. 176 (23), 4537–4547. Available from: https://doi.org/10.1111/bph.14815.

Ujváry, I., Hanuš, L., 2016. Human metabolites of cannabidiol: a review on their formation, biological activity, and relevance in therapy. Cannabis Cannabinoid Res. 1, 90–101.

Uliss, D.B., Dalzell, H.C., Handrick, G.R., et al., 1975. Hashish. Importance of the phenolic hydroxyl group in tetrahydrocannabinols. J. Med. Chem. 18, 213–215.

WHO, 2018a. WHO Expert Committee on Drug Dependence. 2018. Critical Review. Cannabis and Cannabis Resin. https://www.who.int/medicines/access/controlled-substances/Cannabis-and-cannabis-resin.pdf (accessed 6.4.19).

WHO, 2018b. Cannabidiol (CBD) Critical Review Report. https://www.who.int/medicines/access/controlled-substances/CannabidiolCriticalReview.pdf (accessed 6.4.19).

Wu, H.-Y., Chu, R.-M., Wang, C.-C., et al., 2008. Cannabidiol-induced apoptosis in primary lymphocytes is associated with oxidative stress-dependent activation of caspase-8. Toxicol. Appl. Pharmacol. 226, 260–270.

Wheless, J.W., Dlugos, D., Miller, I., Oh, D.A., Parikh, N., Phillips, S., et al., 2019. Pharmacokinetics and tolerability of multiple doses of pharmaceutical-grade synthetic cannabidiol in pediatric patients with treatment-resistant epilepsy. CNS Drugs 33, 593–604. Available from: https://doi.org/10.1007/s40263-019-00624-4.

Wu, H.-Y., Chang, A.-C., Wang, C.-C., et al., 2010. Cannabidiol induced a contrasting pro-apoptotic effect between freshly isolated and precultured human monocytes. Toxicol. Appl. Pharmacol. 246, 141–147.

Wu, H.-Y., Goble, K., Mecha, M., et al., 2012. Cannabidiol-induced apoptosis in murine microglial cells through lipid raft. Glia 60, 1182–1190.

Wu, H.-Y., Huang, C.-H., Lin, Y.-H., et al., 2018. Cannabidiol induced apoptosis in human monocytes through mitochondrial permeability transition pore-mediated ROS production. Free Radic. Biol. Med. 124, 311–318.

Yamaori, S., Kushihara, M., Yamamoto, I., et al., 2010. Characterization of major phytocannabinoids, cannabidiol and cannabinol, as isoform-selective and potent inhibitors of human CYP1 enzymes. Biochem. Pharmacol. 79, 1691–1698.

Yamaori, S., Okamoto, Y., Yamamoto, I., et al., 2011. Cannabidiol, a major phytocannabinoid, as a potent atypical inhibitor for CYP2D6. Drug Metab. Dispos. Biol. Fate Chem. 39, 2049–2056.

Yamaori, S., Koeda, K., Kushihara, M., et al., 2012. Comparison in the in vitro inhibitory effects of major phytocannabinoids and polycyclic aromatic hydrocarbons contained in marijuana smoke on cytochrome P450 2C9 activity. Drug Metab. Pharmacokinet. 27, 294–300.

Zhang, X., Qin, Y., Pan, Z., et al., 2019. Cannabidiol induces cell cycle arrest and cell apoptosis in human gastric cancer SGC-7901 cells. Biomolecules 9, <https://doi.org/10.3390/biom9080302>.

Zimmerman, A.M., Bruce, W.R., Zimmerman, S., 1979. Effects of cannabinoids on sperm morphology. Pharmacology 18, 143–148.

Zuardi, A.W., 2008. Cannabidiol: from an inactive cannabinoid to a drug with wide spectrum of action. Rev. Bras. Psiquiatr. Sao Paulo Braz. 30, 271–280.

Zuardi, Shirakawa, I., Finkelfarb, E., et al., 1982. Action of cannabidiol on the anxiety and other effects produced by δ9-THC in normal subjects. Psychopharmacology (Berl.) 76, 245–250.

Zuardi, Cosme, R.A., Graeff, F.G., et al., 1993. Effects of ipsapirone and cannabidiol on human experimental anxiety. J. Psychopharmacol. (Oxf.) 7, 82–88.

Zuardi, Crippa, J.A.S., Hallak, J.E.C., et al., 2006a. Cannabidiol, a *Cannabis sativa* constituent, as an antipsychotic drug. Braz. J. Med. Biol. Res. Rev. Bras. Pesqui. Medicas E Biol 39, 421–429.

Zuardi, Hallak, J.E.C., Dursun, S.M., et al., 2006b. Cannabidiol monotherapy for treatment-resistant Schizophrenia. J. Psychopharmacol. Oxf. Engl. 20, 683–686.

Zuardi, Crippa, J.A.S., Hallak, J.E.C., et al., 2009. Cannabidiol for the treatment of psychosis in Parkinson's disease. J. Psychopharmacol. Oxf. Engl. 23, 979–983.

Zuardi, Crippa, J., Dursun, S., et al., 2010. Cannabidiol was ineffective for manic episode of bipolar affective disorder. J. Psychopharmacol. Oxf. Engl. 24, 135–137.

Zuardi, A.W., Hallak, J.E.C., Dursun, S.M., Morais, S.L., Sanches, R.F., Musty, R. E., et al., 2006. Cannabidiol monotherapy for treatment-resistant schizophrenia. J. Psychopharmacol. (Oxford) 20, 683–686. Available from: https://doi.org/10.1177/0269881106060967.

Zuardi, A.W., Morais, S.L., Guimarães, F.S., Mechoulam, R., 1995. Antipsychotic effect of cannabidiol. J. Clin. Psychiatry 56, 485–486.

Zuardi, A.W., Rodrigues, N.P., Silva, A.L., et al., 2017. Inverted U-shaped dose-response curve of the anxiolytic effect of cannabidiol during public speaking in real life. Front. Pharmacol. 8, 259. <https://doi.org/10.3389/fphar.2017.00259>.

Zygmunt, P.M., Petersson, J., Andersson, D.A., Chuang, H., Sørgård, M., Di Marzo, V., et al., 1999. Vanilloid receptors on sensory nerves mediate the vasodilator action of anandamide. Nature 400, 452–457. Available from: https://doi.org/10.1038/22761.

Chapter 40

St John's wort (*Hypericum perforatum* L)

Robert W. Coppock and Margitta Dziwenka

Introduction

St. John's wort (*Hypericum perforatum*) is used around the world and is one of the best-selling and best-studied phytopharmaceuticals (Mischoulon, 2009). The medical use of St. John's wort can be traced back through ancient history and, in a religious context, has been referred to as a "demon chaser" when used as an incense against spells and apparitional spirits (Gerarde, 1597; Hobbs, 1990; Falk, 1999; Ben-Eliezer and Yechiam, 2016; Belwal et al., 2019). The worldwide sales of St. John's wort are estimated at several billion dollars (Klemow et al., 2011). *H. perforatum* preparations are also among the most studied phytopharmaceuticals. St. John's wort, an aromatic perennial plant, is a member of the Hypericaceae family, identified as *H. perforatum* (L.) and is a morphological variable species. The morphological variation in *H. perforatum* is recognized as four subspecies (Dauncey et al., 2019). The observed morphological differences in *H. perforatum* subspecies are on a continuum and the subspecies define the variations that occur across geographic distribution (Table 40.1). The name St. John's wort comes from its flowering period, which begins June 25, which is celebrated as St. John's day. This plant is native to Europe, Asia, and North Africa. *H. perforatum* has been introduced into North and South America as a cultivated herb, and in some regions it has escaped from cultivation and is considered a noxious weed that causes photosensitization in ruminants (Hobbs, 1990; Belwal et al., 2019). Small oil glands in the leaves appear as perforations (translucent glands) that span from the upper to lower leaf epidermis. The translucent glands and surrounding tissues contain the phytochemicals hyperforin and adhyperforin (Kucharikova et al., 2016). The dark glands in the leaf and other parts of the plant contain hypericin and pseudohypericin (Rizzo et al., 2019). *Hyperici herba*, as described in pharmacopeias, refers to the dried aerial parts, generally in the flowering stage, of *H. perforatum* (WHO, 2003, 2010). Some authors also use *H. herba* or equivalent terminology as a reference to the wet and dried aerial tops of *H. perforatum* (Linde, 2009). The pharmacological properties and health benefits of extracts from *H. herba* (*Hypericum* extracts) are recognized and accepted as efficacious (WHO, 2003; Linde, 2009; Linde et al., 2009). The bioactive phytochemicals are concentrated in the buds, blossom, fruit, and tips of the branches

TABLE 40.1 Subspecies of *Hypericum perforatum*.

Subspecies of *Hypericum perforatum*	Geographic location
perforatum	North America. Northern and north-west Europe to central Siberia. Not continuous to China to southeast desert and steppe regions of Mongolia; going east transforms into subspecies *chinense*
chinense (N. Robson)	In China, eastern Qinghai and Gansu, east to Shandong and south to Yunnan and Guizhou. Petal laminar glands black or absent
songaricum (N. Robson)	Found in the mountains of Kazakhstan and Kyrgyzstan, and in Xinjiang in China
veronense (includes varieties *angustifolium* and *microphyllum*)	In Turkey and westward into southern Europe, the Mediterranean and Macaronesia, south to Saudi Arabia and the Sudan

Source: Adopted from Dauncey, E.A., Irving, J.T.W., Allkin, R., 2019. A review of issues of nomenclature and taxonomy of Hypericum perforatum L. and Kew's Medicinal Plant Names Services. J. Pharm. Pharmacol., 71(1), 4–14.

Nutraceuticals. DOI: https://doi.org/10.1016/B978-0-12-821038-3.00040-9

TABLE 40.2 Common preparations of *Hypericum perforatum* (L.) herba.

Herbal preparation	Solvent	*Hyperici herba* to solvent ratio	Route of administration
Dry extract	Aqueous ethanol 38% (m/m)	4–7:1	Oral as a solid
Tincture	Aqueous ethanol 55%–50% (V/V)	1:10	Oral as a liquid Cutaneous
Liquid extract	Vegetable oil	1:4–20	Oral as a liquid Cutaneous
Fresh juice		1:0.5–0.9	Oral as a liquid
Powered and macerated herb	Can use water	Form infusion	Oral; can be ingested as a solid
Oleum	Olive oil, arachis oil, other vegetable oils and oily solvents	Macerated *Hyperici herba* covered in oil, 40 days in sunlight.	Topical

Source: Adopted from EMA, 2010. Guideline on Declaration of Herbal Substances and Herbal Preparations in Herbal Medicinal Products /Traditional Herbal Medicinal Products. European Medicines Agency, Committee on Herbal Medicinal Products, Amsterdam; EMA, 2018. European Union Herbal Monograph on Hypericum perforatum L., Herba (Traditional Use). European Medicines Agency, Committee for Medicinal Products for Human Use, Amsterdam; Lyles, J.T., Kim, A., Nelson, K., et al., 2017. The chemical and antibacterial evaluation of St. John's wort oil macerates used in Kosovar traditional medicine. Front. Microbiol., 8, 1639. doi:10.3389/fmicb.2017.01639.

(Kladar et al., 2017). Root cultures of *H. perforatum* are a rich source of hyperforin (Gaid et al., 2018). The phytochemical composition of St. John's wort varies with cultivar, climate, horticultural practices, growing conditions, stage of maturity, time and method of harvest, and extraction procedures (Denke et al., 1999; Mueller et al., 2004; Zobayed et al., 2005; Schmidt and Butterweck, 2015; Bruni and Sacchetti, 2009; Linde, 2009; Heinrich et al., 2017; Kladar et al., 2017; Yao et al., 2019). As the fruit sets, there is an increase in phloroglucinols and a decrease in naphthodianthrone phytochemicals (Isacchi et al., 2007). Dried *H. herba* parts can also be ground or chopped, and are used to make tea infusions and tablets.

Extraction, chemistry, and quality control

Extraction

The extracts of *H. perforatum* are prepared from designated wet and dried plant material using different solvents (supercritical liquids such as carbon dioxide and other gases, vegetable oils, water, ethanol, methanol, and other solvents; Table 40.2). The ratio of plant material to solvent, extraction temperatures, solvent, duration of extraction procedure, and presence of light determine the chemical profile of the extract (Butterweck et al., 1997; Linde, 2009; Cossuta et al., 2012; Heinrich et al., 2017; Elena Tebrencu et al., 2020; Nicolussi et al., 2020). Tinctures are alcoholic extracts of *H. herba* with aqueous ethanol being the most common tincture. Wet and dried extracts are placed in capsules. Olive oil, arachis oil, other vegetable oils, and oily solvents can be used as extract to make *Hyperici oleum*. For example, *Hypericum* IL-160 is manufactured[1] with 80% methanol in a solvent-to-herb ratio of 3:1 to 7:1 and standardized on 0.3% (wt/wt) content of hypericin (Castor et al., 2001; Linde, 2009; Linde et al., 2009). The ratio of herb to solvent is important and is specified for products that meet manufacturing standards. Extracts can be fractionated by various methods to concentrate phytochemicals that are considered to be pharmacologically essential. Numerous products extracted from *H. perforatum* are commercially available. It is the phytochemical profile in the plant and the extraction procedure that determine the phytochemical composition of the final product manufactured from *H. perforatum* (Cossuta et al., 2012; Heinrich et al., 2017).

Chemistry

H. perforatum extracts have been partially chemically characterized and more than 150 different compounds have been identified. It is estimated that 30%–50% of the constituents need to have definitive chemical descriptions and the diverse chemistry makes this task difficult (Crupi et al., 2013; Bridi et al., 2018; Agapouda et al., 2019). The chemical nomenclature used in grouping the various compounds found in *H. perforatum* varies among authors (Yang et al., 2018). For *H. perforatum*, important chemical groups are flavonoids, naphthodianthrones, phloroglucinols, phenolics, xanthones, tannins and related compounds, and volatile compounds (Heinrich et al., 2017; Belwal et al., 2019). The largest group of compounds in *H. perforatum* is the

1. Lichtwer Pharma; DE.

TABLE 40.3 Pharmacologically active substances in *Hypericum perforatum*.

Chemical/extract group	Exemplary compound (%)	High-yielding plant part	Pharmacology/toxicology
Naphthodianthrones	Hypericin (0.09) Pseudohypericin (0.23)	Buds, flower parts, dark leaf glands	Antidepressant, antiviral, photosensitizing
Phloroglucinols	Hyperforin (2.0–4.5) Adhyperforin (0.2–1.8)	Buds, flowers, green fruit, translucent glands, root culture	Antidepressant, antimicrobial
Flavonoids	Quercetin (2.0) Hyperoside (0.7) Quercitrine (0.5) Isoquercitrin (0.3) Rutin (0.3) Miquelianin Isorhamnetin	Buds, leaves, stems	Antidepressant, antiphlogistic
Biflavonoids	Biapigenin (0.1–0.5) Amentoflavone (0.01–0.05)	Flower parts	Sedating, antiphlogistic
Procyanidins (hydrophilic)	Procyanidin Catechin Epicatechin	Buds, flowers, leaves, stems	Antiphlogistic, antioxidant
Essential oils	Terpenes Alcohols	Flowers, leaves	Aroma
Amino acids	Glutamate and γ-aminobutyric acid	Flowers, leaves	Antidepressant
Phenylpropanes	Caffeic acid Chlorogenic acid	Flowers, leaves	Antioxidant
Xanthons	Norathyriol	Roots, flowers	Antidepressant

Source: Adapted from Butterweck, V., Schmidt, M., 2007. St. John's wort: role role of active compounds for its mechanism of action and efficacy. Wien. Med. Wochenschr., 157(13–14), 356–361; Linde, K., 2009. St. John's wort—an overview. Forsch. Komplementmed., 16(3), 146–155; Greeson, J.M., Sanford, B., Monti, D.A., 2001. St. John's wort (Hypericum perforatum): a review of the current pharmacological, toxicological, and clinical literature. Psychopharmacology (Berl.), 153(4), 402–414; Bridi, H., Meirelles, G.C., von Poser, G.L., 2018. Structural diversity and biological activities of phloroglucinol derivatives from Hypericum species. Phytochemistry, 155, 203–232; Gaid, M., Biedermann, E., Fuller, J., et al., 2018. Biotechnological production of hyperforin for pharmaceutical formulation. Eur. J. Pharm. Biopharm., 126, 10–26; Nicolussi, S., Drewe, J., Butterweck, V., et al., 2020. Clinical relevance of St. John's wort drug interactions revisited. Br. J. Pharmacol., 177(6), 1212–1226.

flavonol glycosides (up to 4%) (Butterweck and Schmidt, 2007). The major flavonoids are rutin, hyperoside, isoquercitrin, quercitrin, miquelianin, and quercetin (Table 40.3) (Greeson et al., 2001). Exclusively found in the flowering parts are the biflavonoids; examples of these are amentoflavone (I3',II8-biapigenin, 3',8''-biapigenin) and 13,Ii8-biapigenin. Crude extracts of the flowers and leaves contain the naphthodianthrones hypericin (0.03%–0.3%) and pseudohypericin at 2–10 times the levels of hypericin (Table 40.3). Hyperforin (bicyclic polyprenylated acylphloroglucinol) is the most important compound in the phloroglucinol group (Beerhues, 2006). Adhyperforin and hyperforin are exclusively formed in the pistils and green fruit. Depending on the method of preparation, 2%–5% of the crude extracts can be hyperforin, refined extracts can contain 6%, and some traditional preparations have only minute amounts. Hyperforin is unstable in its pure form in the presence of light, oxygen, water, aprotic solvents, and higher temperatures (Beerhues, 2006; Friedland and Harteneck, 2015; Heinrich et al., 2017; Gaid et al., 2018; Nicolussi et al., 2020). Essential oils in *H. perforatum* ssp. *veronense* have been reported as 0.48% and 113 compounds have been identified (Zeliou et al., 2020). These are predominantly sesquiterpenes (61.3%), monoterpene 10%, and alkane 11.3%. (E)-β-caryophyllene, a cannabinoid CB2 receptor agonist, was found at 5%–10%. The authors also found α- and β-selinene at approximately 15% each. The major volatile compounds in *H. perforatum* harvested in Iran were identified by Ghiasvand et al. (2016) as β-caryophyllene (25.05%), α-pinene (15.24%), γ-cadinene (8.05%), α-selinene (7.38%), germacrene-D (6.71%), limonene (6.51%), myrcene (6.13%), δ-elemene (3.48%), α-amorphene (3.24%), valencene (2.64%), β-pinene (2.36%), and β-elemene (2.34%). It is the mixture of phytochemicals in the extracts of *H. perforatum* that are the pharmacological principal (Nicolussi et al.,

2020). It is the phytochemical profile in the plant and the extraction procedure that determines the phytochemistry of the final product.

Quality control

Regulations for manufacturing and chemical consistency generally drive the quality control and quality assurance of herbal products (Mukherjee, 2019). The European Union (EU), after a tragic incident with aristolochic acid nephropathy, instituted the Traditional Herbal Medicinal Products Directive in 2011 to protect the public and allow free trade in herbal remedies (Debelle et al., 2008; Gilbert, 2011). The European Medical Agency (EMA, 2010) regulates herbal products in the EU. In most countries outside the EU, the regulations applicable to nutraceuticals can have considerable variability. Quality control for *H. perforatum* includes accurate identification of plant material by physical appearance and genetic identification, using standardized extraction procedures, determination of purity by analytical chemistry and chemical profile of a particular species, and testing for toxic adulterants (Frommenwiler et al., 2016; Ghiasvand et al., 2016; Heinrich et al., 2017; Raclariu et al., 2017; Agapouda et al., 2019; Howard et al., 2019; Zeliou et al., 2020). Dyes can be added to St. John's wort products to mask the lack of authenticity (Frommenwiler et al., 2016). There is a genetic identification test for *H. perforatum* that likely can be adapted to test *H. herba* for purity (Raclariu et al., 2017; Zeliou et al., 2020). Assuring precise chemical standardization across multiple chemicals in plant products is difficult because their production is influenced by multiple factors inclusive of plant maturity (Kladar et al., 2017). Some manufacturers notably underappreciate quality assurance and quality control and their products do not meet the analytical standards as claimed on the packaging and also have inconsistent chemical profiles (Draves and Walker, 2003; Booker et al, 2018; Scotti et al., 2018). When St. John's wort extracts are manufactured for pharmaceutical sales, there is considerable variability in hyperforin ($<0.2\%-4.3\%$) and hypericin ($0.1\%-0.3\%$) (Butterweck, 2009; Nicolussi et al., 2020). These findings suggest that patients should not interchange products inclusive of products that have procurement restricted to pharmacies (Wurglics et al., 2001; Wurglics and Schubert-Zsilavecz, 2006). Adding to variation among standardized *H. perforatum* products is that little attention has been given to in vitro and in vivo biodissolution standards (Westerhoff et al., 2002). The European Scientific Cooperative on Phytotherapy (ESCOP) is an umbrella organization representing national herbal medicine or phytotherapy associations across Europe (ESCOP, 2020) This organization has discussions with European medicine regulators regarding standards for manufacturing preparations and efficacy criteria. WHO (2013) and Mukherjee (2019) have proposed the way forward in international harmonization of herbal remedies. There is a body of evidence that shows a need exists in North America for herbal products to comply with recognized and functioning phytopharmaceutical standards for assays and product descriptions.

Basic phytopharmacology

Background

Phytopharmaceuticals generally have complex mechanisms of action because most are a mixture of pharmacologically active phytochemicals (Houghton, 2009). Simplistic reasoning of one plant producing one phytochemical that has one mechanism of action is incorrect. Perusal of this reasoning generally leads to inaccurate conclusions regarding the desired and undesired pharmacological effects. Understanding the phytopharmacology of extracts of *H. herba* extracts and other *H. perforatum* products is hindered because there is a lack of certainty regarding the total pharmacological contribution of the individual phytochemicals in the mixture (Butterweck, 2009; Schmidt and Butterweck, 2015; Nicolussi et al., 2020). For *Herba perforatum*, the extract itself is considered to be the active drug (EMA, 2018; Nicolussi et al., 2020). The complex pharmacology of phytopharmaceutics is underappreciated because of multiple phytocompounds that are present and pharmacologically interactive with a suite of individual and collective interactions (Butterweck et al., 2003a; Butterweck, 2009; Nahrstedt and Butterweck, 2010; Di Pierro et al., 2018; Nicolussi et al., 2020). The phytochemicals in *H. herba* extracts that have attracted the most attention for their phytopharmaceutic activities are hypericin, pseudohypericin, rutin, quercetin, quercitrin, miquelianin, amentoflavone, hyperoside, hyperforin, and adhyperforin. The efficacy of *H. perforatum* extracts can vary with the input plant material, and methods of extraction, manufacturing, and product formulation. Switching products can alter the desired effect and dose requirements.

St. John's wort preparations are available as tablets, capsules, oils (*Oleum hyperici*), tinctures, and ointments. The use of *H. perforatum* products is important in complementary and alternative medicines. In Germany, extract of St. John's wort is a prescription drug and also is considered safe to prescribe by physicians for off-label use in children (Dorks et al., 2013). *Hypericum* is also used to control anxiety, as an anticonvulsive, and in pain management. *Hypericum* extracts are used topically for wound treatment and in dental practice. There are studies in laboratory animals evaluating the effectiveness of *H. perforatum* preparations in treating cancer, inflammation-linked disorders, and bacterial and viral diseases, and as an antioxidant and neuroprotective agent (Klemow et al., 2011).

Pharmacokinetics

The various preparations of *H. perforatum* are a complex mixture of phytochemicals. The phytochemicals likely have interactions. Thus the pharmacokinetics of one isolated phytochemical can be different than its behavior in a mixture of phytochemicals because of complex phytopharmacological interactions (Butterweck, 2003;Verjee et al., 2015, 2019). For example, Butterweck (2003) showed that procyanidin B2 and hyperoside improved the oral bioavailability of hypericin by increasing the plasma levels of hypericin by ca. 52% and 34%, respectively. The pharmacokinetics of hypericin[2] has been studied in male rhesus monkeys (Fox et al., 2001). Two mg hypericin/kg body weight (bwt) were administered as an intravenous bolus and the levels were determined in plasma and cerebral spinal fluid (CSF). These data fit a two-compartment open model. For the 2 mg/kg dose, the peak plasma level was 142 ± 45 μm. The α half-life ($t_{½}$, distribution phase) was 2.8 ± 0.3 h and β $t_{½}$ (elimination phase) was 26 ± 14 h. The mean clearance was 6 ± 2 mL/kg/h and the mean area under the curve was 646 ± 146 μM · h. The levels of hypericin in CSF were below the limits of detection (0.1 μM). In animal studies, the blood–brain barrier (BBB) has also been shown to essentially block hyperforin, quercetin, and pseudohypericin (Caccia and Gobbi, 2009).

A pharmacokinetic study involving male volunteers 25–30 years of age was performed using the LI-160 containing 300 mg of *H. perforatum* extract[3] (Kerb et al., 1996). Oral administration of 250, 750, and 1500 μg of hypericin and 526, 1578, and 3156 μg of pseudohypericin resulted in median peak levels (C_{max}) of 1.3, 7.2, and 16.6 μg/L plasma for hypericin, respectively, and 3.4, 12.1, and 29.7 μg pseudohypericin /L of plasma, respectively. The lowest dose was disproportionally lower for C_{max} and the area under the curve values when compared to these values for the higher doses. A sixfold increase in hypericin dose resulted in a 12-fold increase in C_{max} and area under the curve. Hypericin was measurable in plasma at 72 and 120 h for the low dose and high dose, respectively. The lag time for the appearance of hypericin in plasma as 1.9 h compared to a lag time of 0.4 h for pseudohypericin. The authors were of the opinion that the method of product formulation did not influence the time for absorption. The systemic bioavailability of hypericin and pseudohypericin from LI-160 was estimated to be 14% and 21%, respectively. The median $t_{½}$ for absorption, distribution, and elimination after administration of 750 μg hypericin were 0.6, 6.0, and 43.1 h, respectively, and 1.3, 1.4, and 24.8 h, respectively, after the administration of 1578 μg pseudohypericin. The authors also conducted a 14-day oral treatment with 250 μg hypericin and 526 μg pseudohypericin administered three times/day achieved median steady-state trough levels of 7.9 μg/L for hypericin. At 4 days, the steady-state trough was achieved for pseudohypericin at 4.8 μg/L. The steady-state C_{max} levels were 8.8 and 8.5 μg/L, respectively. At steady state, the mean distribution volumes were 19.7 L for hypericin and 39.3 L for pseudohypericin, respectively. The authors noted that high doses of the LI-160 *Hypericum* extract were tolerated well.

A study was conducted involving volunteers with a confirmed diagnosis of infection with hepatitis C virus (Jacobson et al., 2001). Patients with acquired immune deficiency syndrome (AIDS) were excluded. Twelve patients were administered 0.05 mg liquid hypericin/kg/day for 8 weeks (wk) and 12 patients were administered 0.1 mg liquid hypericin/kg/day for 8 wk. During the 8th week, pharmacokinetic parameters were determined. The mean plasma elimination $t_{½}$ of hypericin was 36.1 ± 22.6 h for the 0.1 mg/kg dose level and 33.8 ± 18.8 h for the 0.05 mg/kg dose level. The mean peak C_{max} 28.1 ± 11.8 ng/mL and the trough C_{min} was 15.4 ± 5.3 ng/mL. The pharmacokinetics of two *Hypericum* extract [STW-3, Laif [4] 600 and STW-3-VI, Laif 900] was studied in two phase I clinical trials, respectively (Schulz et al., 2005a, b). In the first trial (STW-3, Laif 600), five phytochemicals in *Hypericum* extract were studied in human volunteers for bioavailability after administration of a single dose of 612 mg dry *H. perforatum* extract or administration of the hypericin extract for 14 days. Plasma concentration/time curves were calculated for hypericin, pseudohypericin, hyperforin, quercetin, and isorhamnetin for 48 h after single dosing and for 24 h on the last study day (day 14) of the daily dosing regimen. The single oral dose showed the following: for hypericin, the plasma C_{max} was 3.14 ng/mL at 8.1 h and the elimination $t_{½}$ was 23.76 h; for pseudohypericin, the plasma C_{max} of 8.51 ng/mL occurred at 3.0 h and the elimination $t_{½}$ was 25.39 h; and for hyperforin, the plasma C_{max} was 83.5 ng/mL at 4.4 h and the elimination $t_{½}$ was 19.64 h. The flavonoids quercetin and isorhamnetin showed two peaks of maximum plasma concentration. For quercetin, the first C_{max} was 47.7 ng/mL at 1.17 h and the second C_{max} was 43.8 ng/mL at 5.47 h, and the elimination $t_{½}$ was 4.16 h. For isorhamnetin, the first C_{max} of 7.6 ng/mL occurred at 1.53 h and the second C_{max} of 9.0 ng/mL occurred at 6.42 h, and terminal $t_{½}$ was 4.45 h. A steady state for hypericin was achieved before study day 14. For pseudohypericin, multiple dosing resulted in a pharmacokinetic profile comparable to the single oral dose and the mean

2. Supplied by VIMRx Pharmaceuticals; Irvine, CA, United States.
3. Lichtwer Pharma; Berlin, DE.
4. Laif; Steigerwald Arzneimittelwerk GmbH; Darmstadt. DE.

C_{max} was 8.51 ng/mL. For hyperforin, the average C_{max} was higher at 97.4 ng/mL. In the multidose study, for quercetin, the second mean C_{max} was increased; for isorhamnetin, the concentration peaks were less distinct. In the second trial STW 3-VI, Laif 900 was administered either as a single oral dose of 900 mg or as a multiple once-daily dose over a period of 14 days (Schulz et al., 2005b). For the single-dose study, the concentration/time graphs for STW 3-VI, Laif 900 were determined at 48 h after single dosing and for 24 h at day 14 of the study. For the single dose, the parameters for hypericin C_{max} were 3.8 ng/mL at 7.9 h and the elimination $t_{½}$ was 18.71 h. For pseudohypericin, the C_{max} was 10.2 ng/mL at 2.7 h and the elimination $t_{½}$ was 17.19 h. For hyperforin, the C_{max} was 122.0 ng/mL at 4.5 h and elimination $t_{½}$ was 17.47 h. Again, quercetin and isorhamnetin had two plasma C_{max} peaks. For quercetin, the first C_{max} was 89.5 ng/mL at 1.0 h and the second of C_{max} was 79.1 ng/mL at 4.4 h with a terminal $t_{½}$ of 2.6 h. For isorhamnetin, the first C_{max} was 12.5 ng/mL and occurred at 1.4 h and the second C_{max} was 14.6 ng/mL and occurred at 4.5 h, and the elimination $t_{½}$ was 5.61 h. For the 14-day study, hypericin and hyperforin reached a steady state before day 14, and for pseudohypericin the kinetic profile was similar to the single oral dose study. For the flavonoids, quercetin had a higher second C_{max} than the second C_{max} that occurred in the single-dose study, and for isorhamnetin the first and second peaks were identifiable. These pharmacokinetic studies show that the different phytochemicals have different pharmacokinetic profiles and add to the complexity of the pharmacodynamics of *Hypericum* extract. Additionally, there are likely interactions enhancing the absorption of the differing phytochemicals (Butterweck and Schmidt, 2007).

Mechanisms of action

There is evidence that the different phytochemicals in *H. perforatum* extract have polyvalent complex pharmacological interactions (Butterweck, 2009; Houghton, 2009). It is difficult, using laboratory animal models, to reveal a specific phytochemical in *Hypericum* extracts that accounts for all of the antidepressive activity observed in laboratory models (Butterweck and Schmidt, 2007; Schmidt and Butterweck, 2015; Butterweck, 2009; Nicolussi et al., 2020). For *H. perforatum* preparations, scientific evidence does not support the concept of one chemical producing one effect. Rather the observed pharmacology supports multiple chemical interactions. A study showed that fractions containing hyperoside, isoquercitrin, miquelianin and quercitrin, hyperoside and astilbin were efficacious in the forced swim test (Butterweck et al., 2000). Studies have shown that the procyanidins, which are not found to be active antidepressants, greatly increased their antidepressant activity when added to the naphthodianthrones (Butterweck et al., 2003b). This effect is because procyanidins increase the solubility of the naphtodianthrones, which is likely their contribution to the pharmacokinetics of the *Hypericum* extract. Hyperforin has been shown to block the reuptake of multiple neurotransmitters and alter ionic conductance, including the p-type calcium channel involved in the release of neurotransmitters from the presynaptic membrane. There is disagreement in applying in vitro data to in vivo pharmacology when the plasma levels of hypericin are included in the mechanistic model. Rutin, above a threshold limit, is essential for the antidepressive action of *H. perforatum* extracts and acts in a synergistic manner with other phytochemicals (Nöldner, 2005). The naphthodianthrones by themselves are not active as antidepressants unless administered in very large doses, and by themselves, are close to being insoluble in water. Water extracts, performed at 60°C–80°C, remove approximately 40% of the naphthodianthrones from the crude drug. This simple study provides evidence there are other substances in the crude extracts that enhance the aqueous solubility of naphthodianthrones. The procyanidins, which are not found to be active antidepressants, greatly increased their antidepressant activity when added to the naphthodianthrones (Butterweck et al., 1998). This effect is because procyanidins increase the solubility of the naphtodianthrones, which is likely their contribution to the pharmacokinetics of the *Hypericum* extract. The flavonol-O-glycosides have antidepressive activity and the tests show that sugar moieties are essential for activity (e.g., the glucoside, galactoside, and glucuronide forms). These forms, when compared to the other polyphenols, have high rates of intestinal absorption (Butterweck, 2003).

Molecular neurological effects

Essentially, all neurotransmitters and a variety of neuroproteins are involved in the pathophysiology of depression (Friedland and Harteneck, 2015). In vitro studies on *H. perforatum* extract and specific phytochemicals from *H. perforatum* have shown inhibition of brain proteins, affinity for brain receptors, actions on neurotransmitters, and alterations of ion channels and inhibition of transmitter reuptake (Nöldner, 2005). Butterweck et al. (1997) showed that that changes in the dopamine neurotransmitter were likely involved in the antidepressant activity of LI-160. There is evidence that *Hypericum* extracts cause downregulation of β-adrenergic receptors and increase dopaminergic activity in the prefrontal cortex (Butterweck, 2003; Kasper et al., 2010). There is limited clinical evidence to show that *Hypericum* extracts inhibit monoamine oxidases at expected in in vivo pharmacologically active tissue levels (Nathan, 2001; Nöldner, 2005; Schmidt and Butterweck, 2015).

Leuner et al. (2013) showed that hyperforin reduces the uptake of serotonin and norepinephrine by increasing intracellular concentrations of sodium and calcium ions. They also provided histological evidence that hyperforin modulates the dendritic spine morphology in the CA1 and CA3 pyramidal neurons of the hippocampus. These actions are considered to be an effect of hyperforin on transient receptor potential cation channel 6 (TRPC6) and this effect on calcium ion influx can be blocked with lanthanum ions. The authors concluded that hyperforin acts as a brain-derived neurotrophic factor (BDNF) mimetic on hippocampal pyramidal neurons and this explains the effects on plasticity of dendritic spine morphology. Rat studies have shown that hypericin, pseudohypericin, and hyperforin have antidepressive activity, as determined by the rat forced swimming test (Butterweck and Schmidt, 2007). However, the levels used in the experimental study are considered insufficient to explain the antidepressant effect observed with pharmacokinetics of administered doses of *H. perforatum* extract. This observation suggests complex phytopharmacologic interactions of phytochemicals. Hyperforin has been shown to block the reuptake of multiple neurotransmitters and alter ionic conductance, including the p-type calcium channel involved in the release of neurotransmitters from the presynaptic membrane. Hyperforin has been shown to inhibit high-affinity choline reuptake in rat synaptosomes. Hyperforin has been shown to reduce mechanisms of inflammation in the brain and some authors are of the opinion that brain inflammation contributes to depression (Kraus et al., 2010; Schmidt and Butterweck, 2015).

Extracts high in *Hypericum* have been shown to reduce behavioral deficits induced by uncontrollable electric shock (escape deficit paradigm test model). There is disagreement in applying in vitro data to in vivo pharmacology when the plasma levels of hypericin are included in the mechanistic model. Hypericin and quercetin also have affinity for dopamine receptors, whereas miquelianin and rutin have affinity for adrenergic receptors. Hyperforin has been shown to inhibit the synaptosomal reuptake of serotonin, norepinephrine, glutamate and γ-aminobutyric acid (GABA), and dopamine (Butterweck, 2003; Kasper et al., 2010). However, it cannot be assumed that other constituents in *H. perforatum* extracts do not block or influence synaptosomal uptake. Hypericin has high binding inhibition for the corticotropin-releasing hormone receptor 1. Quercetin, miquelianin, and rutin have affinity for brain receptors at the single digit nanomolar concentrations. The activity of the purified flavonoid fraction was comparable to imipramine. Further research showed that hyperoside, isoquercitrin, and miquelianin were efficaciousness as antidepressants in the mouse forced swimming test. Amentoflavone has affinity for the κ- and δ-opioid receptor subtypes in the nanomolar range, benzodiazepine receptor, and also inhibited binding at the 5-hydroxytryptamine (5-HT; serotonin receptors) and dopamine receptors (Butterweck et al., 2002). *H. perforatum* extracts act synergistically with opioid drugs, possibly through protein kinase C (Galeotti et al., 2014a; Sanna et al., 2017). Upregulation of adrenocorticotropic hormone (ACTH) and plasma cortisol are correlated with effective therapy with synthetic antidepressant drugs. After 2 weeks of daily treatment, hyperoside, isoquercitrin, and miquelianin downregulated plasma (ACTH) and corticosterone levels in stressed laboratory animals (Butterweck et al., 2004). However, this effect was not observed after 8 weeks of administration. In the rat cerebral artery occlusion model, Lin et al. (2013) concluded that brain damage was decreased by hyperforin because TRPC6 channels were spared. Using C6 cell culture, De Marchis et al. (2006) showed that exposure of the cells to *H. perforatum* extract[5] (Ze-117) for 7 days reduced the numbers of β-adrenoceptors; a response similar to desipramine and fluoxetine. Coexposure to the cells blocked the reduction in β-adrenoceptors by Ze-117 and β-adrenoceptors. The mechanism of α-tocopherol blockage at the membrane level has not been clearly identified. *H. perforatum* extract can have a direct inhibitory effect on smooth muscle and could also modulate neurotransmission. Standardized (0.3% *Hypericum*) hydromethanolic *Hypericum* extract from flowering plants (*H. perforatum*) was studied in rats for its effects on gastric motility (Capasso et al., 2008). Oral administration of 100–600 mg of the standardized *Hypericum* extract delayed gastric emptying. Further in vitro studies showed that motility of gastric strips (fundic region) was reduced in an increasing dose-dependent manner. The mechanism of action is not known.

Phytopharmacology—depression

Background

Worldwide, depression is one of the more common debilitating diseases. Depression as defined by the WHO (2020) "a common mental disorder, characterized by sadness, loss of interest or pleasure, feelings of guilt or low self-worth, disturbed sleep or appetite, feelings of tiredness and poor concentration." Depression is also defined as a disease with a thousand faces (Gonda et al., 2019). Depressive disorders often start at a young age, and worldwide, are the leading cause of disability by affecting an estimated 260 million people (GBD, 2018). Bauer et al. (2016) estimate the lifetime per person cost of childbearing-linked depression at £34,811 to £75,728. Depression can become chronic or recurrent and lead to substantial impairments in an individual's ability to take care of his or her everyday responsibilities and is likely underdiagnosed. At its worst, depression can lead to suicide. Depression has multifactorial causation (Kohler

5. Max Zeller's Söhne AG; Romanshorn, CH.

et al., 2018; Fang, 2019). Depression in the elderly is increased by comorbidities such as stroke and other neurological diseases. These is evidence that cytokines due to neuroinflammation contribute to depression. There are changes in the balance between GABA, with GABA being lower in depressed individuals. These patients show a cortical GABA deficit (Fogaca and Duman, 2019). There is evidence that there are other molecular mechanisms that contribute to the pathophysiology of depression. Anatomically, the prefrontal cortex, limbic structures, and anterior cingulate cortex may contribute to the sense of guilt and associated emotions. Alterations in the hypothalamic areas and brain stem alter sleep, appetite, and the feeling of fatigue. The hypothalamic–pituitary axis is affected, and one effect is increased plasma levels of cortisol due to changes in adrenocorticotropic regulation. Increased oxidative stress is associated with depression. The majority of people with depression have to continue working, attend to family, perform maternal, domestic, and work duties during pregnancy, meet the demands of lactation and childcare, and perform other important activities (Avila et al., 2018). Prescribing medications is a challenge because of adverse drug reactions, risk issues with pregnancy and lactation, and concerns of patients that they may become drug-dependent and have medication-linked uncontrolled mood shifts. The second-generation antidepressants are the most commonly prescribed pharmaceutics and are reuptake inhibitors of serotonin and norepinephrine. Products made from *H. herba* are emerging as the treatment of choice for mild to moderate depression because of their fewer psychological side effects (Cui and Zheng, 2016; Ng et al., 2017).

Efficacy of *Hypericum* products for depression

H. perforatum extracts have been shown in numerous studies and reviews to be an efficacious treatment for minor to moderate depression (Woelk et al., 1994; WHO, 2003; Bilia et al., 2002; Linde, 2009; Linde et al., 2009; Ng et al., 2017; Zirak et al., 2019). Woelk et al. (1994) conducted a study with 663 medical doctors in 3250 patients on the effectiveness and acceptance of LI-160, a German standardized extract of *H. perforatum.* Of the 3250 patients, 76% were women and 24% were men and the patients ranged in age from 20 to 90 years of age. Depression diagnoses were 49% with mild depression, 46% with intermediate depression, and 3% with severe depression. The physicians considered LI-160 therapy to have normalized or improved 30% of the patients. Of all the patients, 2.4% had undesirable side effects. The side effects were gastrointestinal issues, allergic reactions, tiredness, and restlessness. Brenner et al. (2000) compared, in a randomized double-blind study, the efficacy of sertraline (a selective serotonin reuptake inhibitor, SSRI) to the efficacy of *H. perforatum* product LI-160 (product meeting the German Pharmaceutical Codex) using volunteers with mild to moderate depression. Thirty patients [19 women, 11 men, mean age of 45.5 years (18–65 years)] were enrolled in the study. For the first week, treatments were 600 mg LI 160/day or 50 mg sertraline/day. For weeks 2–6, treatments were 900 mg LI 160/day and 75 mg/day, respectively. These doses were divided with the tablets taken three times a day. Medication compliance was monitored. Both treatments were effective ($p < 0.05$) as evaluated on the Hamilton Rating Scale for Depression (HAM-D) and the Clinical Global Impression Scale (CGIS). The CGIS is an evaluation index developed by the National Institution of Metal Health for patients and physicians evaluating the efficacy of medications used in psychopharmacology. It measures social anxiety, depression, impairment, and quality of life. In a double-blind study conducted at seven different locations, *H. perforatum* extract (Ze-117) 250 mg was administered to patients twice per day for 6 weeks (Schrader, 2000). In patients with mild to moderate depression, it was compared to fluoxetine and placebo in a double-blind study. For HAM-D efficacy (50% reduction in the score), there were no significant differences ($p = 0.09$) between the patients treated with Ze-117 and those treated with fluoxetine. For the CGI item 1 (severity), Ze-117 was significantly improved ($p < 0.03$) compared with fluoxetine. The authors also reported that the adverse reactions were less in the patients receiving Ze-117. Brattstrom (2009) conducted a 1-year study on the use and safety of Ze-117 [a standardized *H. perforatum* ethanolic extract[6] containing 0.2% hypericin], as a treatment for mild to moderate depression. This was an open multicenter study enrolling 440 patients. Two-thirds of the patients were women and 33% were age 60 years or older. The daily dose was 500 mg Ze-117. The mean baseline HAM-D score was 20.58; at treatment week 26, it was 12.07; at treatment week 52, it was 11.18. The mean CGI baseline score was 3.99 and was reduced to 2.19 at treatment week 52. The author considered 6% of the patients to have adverse reactions that had a plausible link to Ze-117 treatment. These were gastrointestinal and skin complaints. There were no changes in body mass index, clinical chemistries, or electrocardiographs.

Linde et al. (2009) concluded in a metaanalysis that *H. perforatum* extracts for major depression were similarly effective to synthetic antidepressants and that *H. perforatum* extracts had fewer side effects. They also recommended that *H. perforatum* products only be taken after consultation with a physician. In a sponsored[7] study, Szegedi et al. (2005) reported on a double blinded study comparing *H. perforatum*

6. Brand names Remotiv, ReBalance and Esbericum forte, Zeller AG; Romanshorn, CH.

7. Dr. Willmar Schwabe Pharmaceuticals; Karlsruhe, DE.

extract[8] to paroxetine (a SSRI). The study was carried out in 21 German primary care psychiatric centers enrolling 251 adult patients (age 18–70 years). The patients were diagnosed with single or recurrent moderate or severe episodes of unipolar major depression without psychotic features that persisted for 2 weeks to 1 year and scored greater than 22 points on the Hamilton depression scale. The *H. perforatum* extract (WS-5570) was a hydroalcoholic extract containing 3%–6% hyperforin and 0.12%–0.28% hypericin manufactured into a coated tablet containing 300 or 600 mg of extract. Paroxetine was given in a 20 mg coated tablet. The number and identity of the tablets were blinded by number and color. The dosages were 900 mg WS-5570 divided three times a day or 20 mg paroxetine taken once a day, and both drugs were given for 6 weeks. In patients who did not respond to medications, the doses were increased to 1800 mg WS-5570 divided three times a day or 40 mg paroxetine taken once a day. Efficacy was measured by a change in the Hamilton depression scale from baseline to day 42 (primary outcome). Secondary measures were improvements in the Montgomery-Asberg depression rating scale, clinical global impressions, and Beck depression inventory (BDI). The authors concluded that WS-5570 had the same efficacy as paroxetine and was tolerated better by patients. The most common complaint for patients receiving WS-5570 was gastrointestinal disorders.

Gastpar et al. (2006) studied the efficacy and safety of a standardized extract of *H. perforatum* (STW3-VI) for the treatment of moderate depression in 388 outpatients. This was a placebo-controlled, multicenter study comparing STW3-VI to citalopram (an SSRI) and placebo. Over a 6-week period, depression in the patients was evaluated by the HAM-D scores, the von Zerssen's Adjective Mood Scale (BfS), and the CGIS. The once-daily treatments were 900 mg STW3-VI to 20 mg or citalopram or placebo. The baseline mean HAM-D scores for the patients receiving STW3-VI was 21.9 ± 1.2, for citalopram 21.8 ± 1.2, and 22.0 ± 1.2 for placebo. Over the 6 week-period, the HAM-D scores reduced to 10.3 ± 6.4, 10.3 ± 6.4, and 13.0 ± 6.9, respectively, for STW3-VI, citalopram, and placebo. There was no significant difference ($p < 0.0001$) in the HAM-D, CGI, and BfS scores between the STW3-VI and citalopram patients and both drugs were significantly different ($p < 0.05$) from placebo. The attending physicians considered that treatment responses to STW3-VI and citalopram were, respectively, 54.2% and 55.9%. The adverse reactions were 17.2%, 53.2%, and 30%, respectively, for STW3-VI, citalopram, and placebo. Using the data from the Gastpar et al. (2006) study, Singer et al. (2011) determined the duration of response for patients receiving STW3-VI, citalopram, or placebo. The median time to onset of depression relapse was 802 days, 1755 days, and 1817 days for placebo, citalopram, and STW3-VI, respectively. A sponsored study was carried out comparing WS -5570,[9] to paroxetine as a treatment for moderate depression (Seifritz et al., 2016). WR-5570 was described as a dry stabilized hydromethanolic extract (plant-to-extract ratio 3 to 7:1) containing 3%–6% hyperforin, 0.1%–0.3% hypericin, not less than 6% flavonoids, and not less than 1.5% rutin. Patients enrolled on the study had unipolar depression without psychiatric features that had persisted for a duration of 2 weeks to 1 year. The patients also had an HAM-D score greater than 22 and were diagnosed with moderate depression with the HAM-D score being less than 25. WR-5570 was administer by the patient taking one 300 mg tablet three times a day (a total of 900 mg WR-5570/day) or one 20 mg paroxetine/day concealed in one of three tablets. After 7 days of treatment, the patients medicated with WR-5570 had a significantly ($p < 0.05$) more pronounced improvement in the HAM-D scores than the patients medicated with paroxetine. There were fewer side effects in the WR-5570-medicated patients. A metaanalysis review by Rahimi et al. (2009) concluded that proprietary extracts from *H. perforatum* are as efficacious in the treatment of major depressive disorders as SSRIs and the patients receiving *H. perforatum* extracts have lower numbers of study withdrawals due to adverse events. In a systematic review, the *H. perforatum* extracts (0.3% hypericin and 1%–4% hyperforin) were more efficacious than placebo and were as efficacious as synthetic prescription medications (Apaydin et al., 2016).

A standardized *H. perforatum* extract[10] (LI-160) was studied as a treatment for major depression at 12 institutions in the United States (HDTSG, 2002). The double-blind, randomized, placebo-controlled trial compared LI-160 to sertraline[11] and placebo. Outpatients meeting the *Diagnostic and Statistical Manual of Mental Disorders, 4th Edition* criteria, a score of 20 on the HAM-D scale, a score of 60 on the Global Assessment of Functioning, and at least 18 years old, were enrolled. The CGIS was used as a secondary assessment scale. Exclusions were risks for suicide and homicide, being pregnant, planning a pregnancy, lactating, and not using medically accepted birth control. All doses of medications were divided and taken three times a day. At baseline, patients were randomized to receive total daily doses of 900 mg LI-160/day, 50 mg sertraline/day, or placebo. After 3 or 4 weeks, the physicians could increase the total daily doses to 1500 mg LI-160/day, 100 mg sertraline/day, or placebo

8. WS-5570, Dr Willmar Schwabe Pharmaceuticals; Karlsruhe, DE.

9. Neuroplant brand name, Dr. Willmar Schwabe GmbH & Co. KG; Karlsruhe, DE.

10. LI 160, Lichtwer Pharma; Berlin, DE.

11. Pfizer; New York, United States.

equivalent. After 8 weeks, patients eligible to continue on the study could be prescribed daily doses of 1800 mg LI-160/day, 150 mg sertraline/day, or placebo equivalent. Efficacy parameters were changes in the HAM-D total scores from baseline to the 8-week scores and rates of full response as determined by the HAM-D scores and CGIS. The results showed that LI-160 and sertraline were not significantly different from placebo. On the CGIS, sertraline was better than placebo ($p = 0.02$) and the CGIS was considered as a secondary measure. Both LI-160 and sertraline had more adverse reactions than placebo. Including a placebo group in a clinical study is assumed to identify therapeutic effects from the patients' psychological perception of efficacy. Double blinding of studies is done to remove the investigators' bias based on knowledge of treatment types.

Another factor in clinical studies is the association between the patient beliefs regarding the assigned treatment and clinical response (Chen et al., 2011). These authors reanalyzed data from the HDTSG (2002) study and focused on the acute phase of the study (baseline to week 8). Since the patients were asked to guess the outcome, these data could be modeled in the study. These authors found that patients guessing LI-160 or sertraline had more significant improvement ($p < 0.001$) than patients guessing placebo. There were no significant differences ($p = 0.25$) between the sertraline and LI-160 groups. The interactions between treatments and guesses were not significant. For adverse effects, there were no significant differences between guesses of patients receiving sertraline or LI-160. These are important robust observations because the HDTSG (2002) study failed to demonstrate any significant differences between the three treatment groups. There was also a high number of subjects that expressed a belief that their treatment was LI-160. In reanalysis by Chen et al. (2011), there were no significant demographics differences in patients' beliefs that they were receiving LI-160 or patients who believed they were being treated with sertraline or placebo. Analysis of plasma from weeks 3 and 8 of the HDTSG (2002) study showed that 16% of the placebo patients had measurable hyperforin and 16% of the LI-160 patients had no measurable levels of hyperforin (Vitiello et al., 2005). All of the patients taking sertraline had measurable parent drug or its metabolites. The authors stated that the reanalysis of the HDTSG (2002) data did not change the outcome of the study. Run-in-studies have a run-in phase wherein the test drug or placebo is administered. The information generated is used to exclude nonadherent patients, placebo responders, active drug nonresponders, or patients who do not tolerate the active drug. Using a run-in generally improves the statistical power of the study. Run-in designs can change the results of a study and the application of efficacy to the general population. There is considerable debate on the use of the run-in design (Fralick et al., 2018).

A multicenter double-blind trial was conducted on *H. perforatum* extract (LI-160)[12] comparing LI-120 to placebo (Shelton et al., 2001). The patients were male or female over 18 years of age and met the *Diagnostic and Statistical Manual of Mental Disorders, 4th Edition* criteria for previous 4-week occurrences of single or multiple episodes of single or recurrent major depressive disorder, without psychotic features. After the screening visit, the patient received placebo for 1 week, a baseline evaluation with placebo responders being excluded from the study. Two-hundred enrollees were randomized to receive LI-160 (300 mg tablet taken three times a day, total dose of 900 mg/day) or placebo for 4 weeks. At the end of week 4, the dose of LI-160 was increased to four tablets/day (1200 mg total dose). The primary efficacy index was rate of change on the HAM-D. Secondary indices were the BDI, Hamilton Rating Scale for Anxiety (HAM-A), the Global Assessment of Function scale, and the Clinical Global Impression Severity and Improvement scales (CGIS and CGI-I). Evaluations for all tests were done for baseline, and the HAM-D at the end of weeks 2, 4, 6, and 8. The HAM-A was repeated at the end of weeks 3 and 8. The remaining secondary evaluations were repeated at the end of week 8. Sixteen patients were removed from the study because they were initial placebo responders. The baseline scores were not significantly different ($p < 0.05$) by demographic and treatment groups. The authors reported random coefficient analyses for the HAM-D, HAM-A, CGI-S, and CGI-I showed significant effects for time but not for treatment or time-by-treatment interaction (for HAM-D scores, $p < 0.001$, $p = 0.16$, and $p = 0.58$, respectively). For LI-160, the remission of severe depression was significantly higher ($p = 0.02$) than placebo ($p = 0.02$), but occurrences were low with 14/98 (14.3%) for LI-160 and 5/102 (4.9%) for placebo. Headache, reported as a side effect for LI-160, was significantly higher than placebo.

This study was in part repeated in a complex, acute-phase, multicenter double-blind trial comparing LI-160 to placebo (Gelenberg et al., 2004). The patients were male or female over 18 years of age and met the *Diagnostic and Statistical Manual of Mental Disorders, 4th Edition* criteria for 4-week occurrences of single or multiple episodes of single or recurrent major depressive disorder, without psychotic features. The patients selected had an HAM-D score ≥ 20. After the screening visit, the patients received 1 week of placebo, a baseline evaluation, and were then randomized to receive LI-160 or placebo for 8 weeks. Response to medication was defined as HAM-D

12. Lichtwer Pharma; Berlin, DE.

<12 and CGIS score <2. Patients with a baseline HAM-D score ≤ 22 had to have a $>50\%$ score reduction. For patients considered to be responders to medication, the study was unblinded by informing the patients of their treatment. Patients taking LI-160 could continue their treatment up to 6 months with bimonthly evaluations. Patients that relapsed were referred for appropriate treatment. Patients that responded to placebo were referred for treatment pro re nata. In both treatment groups, nonresponders (>12 HAM-D and CGI scores >2), who had completed 8 weeks of double-blind study, were discontinued from treatment. At the end of week 9, they were treated with a synthetic antidepressant of the physicians' choice. Of the 200 enrollees, 158 (79%) completed the study. Forty-one patients were classified as responders and 117 as nonresponders. The nonresponders were treated with synthetic drugs with the choice being made by the attending physician. IL-160 was not included in the drugs chosen by the attending physicians. For the nonresponders, synthetic drugs significantly decreased ($p < 0.05$) the indices of depression. Seventeen patients were continued on LI-160 and five relapsed. Side effects reported by LI = 160 were diarrhea, weight loss, intermittent bruxism, and dry skin. The authors did not report side effects for the synthetic antidepressants. Haller et al. (2019) downgraded the evidence for *H. perforatum* extracts for depress from moderate evidence to very low quality of evidence. This change was made because most of the reports of higher efficacy were from German-speaking countries than the lower level of efficacy reported from studies conducted in the United States and other European countries. The authors did consider *H. perforatum* extracts as a gap-bridge between diagnosis and access to psychotherapies. They also advise medical professions to advise patients of the drug interactions that can occur when using *H. perforatum* extracts. The authors of this chapter remind readers that the large HDTSG (2002) study did not show any significant difference between standardized *H. perforatum* LI 160 extract, sertraline, and placebo.

The Canadian Network for Mood and Anxiety Treatments Clinical Guidelines for the Management of Adults with Major Depressive Disorder in 2016 reported on complementary and alternative medicine treatments (Ravindran et al., 2016). The Committee, using Level 1 Evidence (metaanalysis with narrow confidence intervals and/or two or more RCTs with adequate sample size, preferably placebo controlled and with clinical support. Clinical support refers to the application of expert opinion of the CANMAT), recommended standardized *H. perforatum* extracts as a monotherapeutic for mild to moderate depressive disorders. It can be used as second-line adjunct treatment for moderate to severe depressive disorders. The committee also cautioned that serotonin syndrome and hypomania can occur when *H. perforatum* extract is used concurrently with synthetic antidepressants and *H. perforatum* extract can alter the baseline pharmacology of other drugs. In a review of studies in depression and other neurological disorders in elderly people (≥ 60 years old), Kasper (2015) concluded that WS-5570[13] was superior to placebo and was not different from the efficacy reported for younger patients (<60 years old) and as effective as synthetic drugs prescribed for depression. In a metaanalysis of complementary therapies for clinical depression, a Cochrane review concluded that *H. perforatum* extracts have similar efficacy to synthetic drugs in treating mild to moderate depression (Linde et al., 2009). Efficacious formulations are most likely to be hydroalcoholic extracts using 50%–60% ethanol or 80% methanol with a ratio of *H. herba* to solvent of 3 to 7:1. Daily dose of the standardized *H. perforatum* extract should be 500–1200 mg/day. Linde (2009) further reported that compared with tricyclic or tetracyclic antidepressants and SSRIs, respectively, the combined response rate ratios were 1.02 (95% CI, 0.90–1.15; five trials) and 1.00 (95% CI, 0.90–1.11; 12 trials). There was a bias with the studies conducted in German-speaking countries finding standardized *H. perforatum* extracts more efficacious. The authors also stated that individuals wishing to use standardized *H. perforatum* extracts should consult a health professional. Key items for the physician to consider are intrinsic variability in standardized *H. perforatum* extracts, potentially major variations in the nonstandardized preparations, and potential interactions between *H. perforatum* extracts and other herbs and drugs.

Hypericum perforatum extracts in laboratory animal models for depression

Laboratory animal and in vitro models have been developed to increase the understanding of the pharmacodynamics of *H. perforatum* extracts in humans (Butterweck, 2003; Crupi et al., 2013). Antidepressants decrease the immobility time (the time rats limit motion to keep their head above the water level). False positives can result from this test model. Stimulants, such as caffeine and other purine alkaloids, and amphetamines also increase locomotor activity. Laboratory animals may not accurately address the multifactorial aspects of depression syndrome in humans (Krishnan and Nestler, 2011; Ayuob and Balgoon, 2018). The forced swimming test is one of a battery of tests used for simulating a state of depression in rats and other rodents (Nestler et al., 2002; Can et al., 2012; Nicolussi et al., 2020). In a confined space, with no means of self-rescue, an animal (usually a mouse or rat) is forced to swim. The animal quickly assumes an immobility posture, with their head

13. Neuroplant, Dr. Willmar Schwabe GmbH & Co KG; Karlsruhe, DE.

above the water level, which is assumed to represent a mental state of despair. Antidepressants typically decrease the immobility time in rodents. Rats were used in a forced swimming test to compare an extract[14] of *H. perforatum* (standardized to 0.3% hypericin) to the antidepressant drugs fluoxetine, imipramine, and venlafaxine (Bukhari and Dar, 2013). The antidepressants were dissolved in physiological saline and the herbal extract was dissolved in dimethyl sulfoxide. The drugs were given by intraperitoneal injection (ip). All animals were administered the drugs 1 h before the forced swimming test was conducted for a period of 5 min in water at 25°C–30°C. For *H. perforatum* extract at doses of 30–90 mg/kg bwt, the maximal immobilization time was 53% at 90 mg/kg bwt. At doses from 100 to 250 mg/kg bwt, immobilization time was decreased from control values. For the nonphytopharmaceutics, 30 mg imipramine (ip)/kg bwt caused 55% immobilization, about 50% immobilization at 70 mg fluoxetine/kg bwt, and a 34% decrease in immobilization time for a 30 mg venlafaxine/kg bwt. With the exception of fluoxetine, increasing the dose above maximal immobilization decreased immobilization. In the locomotor activity test[15], the *H. perforatum* extract decreased locomotor counts in a dose-dependent manner with the no-effect dose being 1 mg/kg bwt administered ip and the maximal dose (80% reduction) was 30 mg/kg bwt. For doses of 30 mg fluoxetine/kg bwt, 10 mg imipramine/kg bwt, and 50 mg venlafaxine/kg, all administered ip, the reduction in locomotor counts were, respectively, 89%, 73%, and 62%.

In a forced swimming test using mice, Bach-Rojecky et al. (2004) showed that 70 mg/kg bwt of a *H. perforatum* extract[16], administered intraperitoneally, increased activity by 36%. These researchers also studied the effects of *H. perforatum* in the mouse tail suspension model. In this model a mouse is suspended by its tail and the time of inactivity after the initial activity is measured. A dose of 70 mg *H. perforatum* extract/kg bwt decreased the immobility time by 24.5%. In a tail suspension model, there is synergism between *H. perforatum* extract and quercetin (Schmidt and Butterweck, 2015). The stress-induced hyperthermia model in mice was used to study in mice (male BL6/C57J) the efficacy of *H. perforatum* extract (Grundmann et al., 2006). The parameter was the difference in body temperature (ΔT) induced by stress using the open field model (10 min exposure). The *H. perforatum* extract (STW3-VI)[17] was 0.3% hypericin, 4% hyperforin, and 9.4% flavonoids. Selected constituents[18] of *H. perforatum* extract tested were isoquercitrin (quercetin 3-O-glucopyranoside), hyperoside (quercetin 3-galactoside), quercitrin (quercetin 3-rhamnoside), rutin (quercetin 3-rutinoside), amentoflavone (I3′,II8-biapigenin), 13,Ii8-biapigenin [5,5’,7,7’-tetrahydroxy-2,2’-bis(4-hydroxyphenyl)-4H,4’H-[3,8’-bichromene]-4,4’-dione], hypericin, hyperforin, and miquelianin (quercetin 3-O-glucuronide). The phytopharmaceuticals, phytochemicals, and pharmaceuticals were administered orally. At doses between 250 and 500 mg STW3-VI /kg bwt, the ΔT was significantly ($p < 0.05$) reduced. The low dose of 125 mg STW3-VI/kg bwt, and high doses of 750 and 1000 mg STW3-VI/kg bwt did not significantly change ΔT. Oral administration of 0.1 mg hypericin/kg bwt 60 min before open field testing significantly decreased ($p < 0.05$) ΔT. Oral doses of 1–10 mg hyperforin/kg bwt had no effect on ΔT. The hyperoside, isoquercitrin, and quercitrin flavonoids administered orally at doses of 0.6 mg/kg bwt and 1 mg rutin/kg bwt partially blocked ΔT. Comparing flavonoids, miquelianin administered at 1.2 mg/kg bwt was the most potent compound. Of the biflavonoids, amentoflavone decreased ΔT at a dose of 0.1 mg/kg bwt. The anxiolytic drugs diazepam (5 mg/kg bwt) and buspirone (10 mg/kg bwt) significantly ($p < 0.05$) reduced ΔT. The antidepressants imipramine and fluoxetine had no significant ($p < 0.05$) effect on ΔT. These data show that STW3-VI extract, hypericin, and miquelianin had anxiolytic pharmacodynamic activity. Stress and administration of corticosterone cause histological changes in the hippocampal region of the brain. A study was done in mice (CD1) to show the protective effects of a standardized *H. perforatum* extract[19] on the stress and corticosterone-induced histological changes in the dentate gyrus of the hippocampus (Crupi et al., 2011). The mice were administered the equivalent of 5 mg corticosterone/day in drinking water for 7 weeks. The mice were subjected to the open field, novelty-suppressed feeding and forced swim test. A group of mice were injected (ip) with the 30 mg *H. perforatum* extract/kg bwt for 3 weeks. The *H. perforatum* extract contained hypericin 0.34%, hyperforin 4.1%, flavonoids 5% (rutin, kaempferol, luteolin, myricetin, quercetin, quercitrin, and isoquercetrin), and 10% tannins. The dentate gyrus was stained with special stains and examined microscopically. Administration of the *H. perforatum* extract reversed the anxiety/depressive-like syndrome induced by corticosterone. Histologically, the corticosterone treatment reduced the proliferation of progenitor cells. This effect was reversed by administration of *H. perforatum* extract as was the corticosterone-induced increase in dendritic spine density.

Vissiennon et al. (2012) showed that the flavonols kaempferol and quercetin found in *H. perforatum* lack

14. Medics Laboratories; Karachi, PK.
15. Columbus Instruments; Columbus, OH, United States.
16. Belupo; Koprivnica, HR.
17. Steigerwald Arzneimittel GmbH; Darmstadt, DE.
18. See [https://pubchem.ncbi.nlm.nih.gov/] for IUPAC name.
19. Methanolic extract, Indena; Milan, IT.

anxiolytic activity when administered ip to mice (C57BL/6). Kaempferol and quercetin were not efficacious as anxiolytics in mice following orally administered 7 mg enrofloxacin/kg bwt/day for 4 days. These authors also identified that the *p*-hydroxyphenylacetic acid and 3,4-dihydroxyphenylacetic acid metabolites of kaempferol and quercetin, respectively, were efficacious as anxiolytics when given ip. Other studies have shown that kaempferol and quercetin are metabolized by colonic microorganisms to *p*-hydroxyphenylacetic acid and 3,4-dihydroxyphenylacetic acid, respectively (Scalbert et al., 2002). A study was carried out in rats on the effects of *H. perforatum* extract[20] on cognitive behavior and levels of neurotropic factor in the brain (Valvassori et al., 2018). Eighty rats (40 males and 40 females) were randomly assigned to orally receive 300 mg HPIT/kg bwt/day or 1 mL water/kg/day (control) for 28 days. The dried extract contained 0.32% hypericin, and before use, was reconstituted in 1 mL water/kg bwt and protected from light. Behavior tests were started 24 h after the last dose of HPIT or water control was administered. Parameters used were novel object recognition (NOR) and the step-down inhibitory avoidance task. After completing the NOR test (study day 31), 10 animals receiving HPIT and 10 water control animals were killed and the brains harvested for determining the levels of nerve growth factor (NGF), BDNF, and GDNF (glial cell-line-derived neurotrophic factor) in the hippocampus and frontal cortex, using sandwich enzyme-linked immunosorbent assay. No significant changes between the groups ($p > 0.05$) were observed in spontaneous locomotion. In the NOR test there were no differences among groups during the training trials showing that HPIT did not affect sensorimotor parameters, such as locomotion and motivation. Both males and females showed impaired retention of both short-term and long-term memory, and the acquisition of memory and induced impairment of intermediate memory in female rats. Compared with the training trial, male rats receiving HPIT showed a significantly higher preference for the novel object during long-term memory, and females showed no differences during short-term and long-term memory tests. In the step-down inhibitory avoidance task, HPIT induced impairments in both the short-term and long-term memories of both male and female rats. HPIT treatment decreased NGF in the hippocampus of female rats, but this effect was not observed in male rats. For both sexes, HPIT did not change GDNF in the hippocampus and frontal cortex. HPIT decreased BDNF in the hippocampus of male and female rats.

20. Vitalis Farmacia de Manipulacao; Criciuma, BR.

Phytopharmacology—cognation, aging, and pain

Cognation

Multiple doses of *H. perforatum* extract[21] (BNO-1385) were studied in humans (males, 23–31 years old) for its cognation and cardiac effects using a double-blind crossover design (Siepmann et al., 2002). The volunteers were considered healthy and did not consume alcoholic or caffeinated beverages during the study. Amitriptyline, an inhibitor of serotonin and noradrenaline reuptake, was used as a reference pharmaceutical. The dose of BNO-1385 was administered in capsules (255–285 mg, 900 μg hypericin) or 25 mg amitriptyline capsules or placebo; all capsules were given three times a day for 14 days. There was a 14-day washout between medication or placebo crossover. The study parameters were heart rate variability, quantitative electroencephalogram (EEG), and psychometric test. BNO-1385 did not affect heart rate variability, but amitriptyline did affect heart rate variability. Amitriptyline and BNO-1385 did not influence cognitive performance as determined by psychomotor coordination, responsiveness to distractive stimuli, choice reaction, and short-term memory. Self-rated activity was significantly decreased ($p < 0.05$) by amitriptyline. The EEG changes induced by amitriptyline and BNO-1385 were increased theta power density and amitriptyline also increased fast alpha power density. A double-blind, placebo-controlled repeated measure study was done on the acute nootropic effects of a *H. perforatum* extract [Blackmore's[22] Hyperiforte (BH)] in healthy human volunteers (7 female and 5 males) aged 18–54 years (Ellis et al., 2001). Each 300 mg tablet of BH was standardized to 0.3% hypericin and 3%–5% hyperforin. Doses of BH used were 900 mg, 1800 mg, and placebo (0 mg BH). A washout time of 7 days was taken between the single dose of BH or placebo. Parameters employed were immediate word recall, choice recognition time, delayed word recall, spatial working memory (SWM), numeric working memory, and delayed word and picture recognitions. There were no differences ($p < 0.05$) between treatments for SWM tasks, and the word recall tasks (immediate and delayed). The interactions between treatment conditions and time were observed for response speed or accuracy in identifying new items were not significant ($p < 0.05$). The accuracy in identifying the original stimuli were significantly changed ($p < 0.05$). For planned comparisons, 1800 mg BH significantly impaired the accuracy for identifying the original stimuli. In the delayed picture recognition, the interaction between

21. Bionorica; Neumarkt, DE.

22. Blackmore; Sydney, AU.

treatment condition and time was observed for accuracy in identifying the original stimuli was significant ($p < 0.05$). The dose of 1800 mg BH caused a significant impairment ($p < 0.01$) in planned comparisons. The interaction between the treatment conditions and time for accuracy in identifying the new stimuli or response time was not significant ($p > 0.50$). The authors concluded that a single dose of *H. perforatum* extract, in the commonly used dosage range, does not have an acute memory-enhancing (nootropic) effect in young, healthy humans.

In an Israeli study, the acute effects of *H. perforatum* extract on memory were shown to be dose related (Yechiam et al., 2019). The study in 82 healthy adults (18 to 40 years old with a mean age of 24.9 years, 49 males and 33 females) with attention-deficit hyperactivity disorder focused on the effects of *H. perforatum* extract[23] (Ze-117) on improved memory, sustained attention, and mood and state anxiety. This single dose placebo-controlled study was done in two sessions separated by at least 2 weeks. The dose of Ze-117 was 250 or 500 mg capsules (0.1–0.3 hypericins and <1% hyperforin) and lactose-gelatin was used a control. A battery of tests was given to assess short-term memory capacity and sustained attention. In the first two sessions, the participant received either a placebo or a 250 mg Ze-117 capsule or a 500 mg Ze-117 capsule. The volunteer receiving a dose of Ze-117 at the first session subsequently received the placebo treatment. The participant, after receiving the randomized treatment, was instructed to take a 60-min break and then return for the cognitive examinations. All cognitive testing was computer-based. There were no differences between placebo and Ze-117 in the participants who did not complete the study. For memory, there was a curvilinear response for Ze-117 on memory with the 250 mg ZE-117 improving memory and a negative effect on memory for the 500 mg Ze-117 dosage group. There was no main effect of Ze-117 for Go/No-Go omission and commission errors. These findings show that the effects of *H. perforatum* extracts on memory should be evaluated in context of the +/– dose response observed in this study. The neurocognitive effect of Ze-117 was studied during nicotine withdrawal (Camfield et al., 2013). Twenty volunteers, age 18–60 years and committed to cessation of tobacco smoking, were enrolled in the study. Randomly assigned treatments consisting of seven individuals administered 500 mg Ze-117 taken twice a day (500 mg Ze-117/day), six individuals administered transdermal nicotine[24] replacement therapy (TNRT), and seven volunteers administered 250 mg Ze-117 twice a day plus 250 mg BID plus TNRT. Administration of Ze-117 commenced 28 days before cessation of smoking. TNRT (21 mg/patch) started with cessation of smoking and continued for 6 weeks. At 6 weeks, the TNRT was reduced to a 14 mg nicotine patch for 2 weeks and then a 7 mg nicotine patch for 4 weeks. Continued cessation of smoking was confirmed by measuring carbon monoxide in exhaled air. Parameters were the SWM and steady-state visually evoked potential (SSVEP) recording. The SWM is known to be adversely affected during nicotine withdrawal. For the SWM, there were no significant differences ($p < 0.05$) in the groups for the pretest and posttest for reaction time and accuracy. A stronger SSVEP amplitude increase in posterior-parietal regions was observed for the Ze-117 and TNRT groups and there was greater frontocentral SSVEP phase advance in the Ze-117 group. The authors concluded that Ze-117 and Ze-117 + TNRT are equally effective in ameliorating the cognitive deficits associated with nicotine withdrawal. Negative cognation is considered by some authors as an early sign of depression (Warren et al., 2019). Negative cognation is when negative words can be remembered better than positive words. The effects of treatment with *H. perforatum* extract[25] (LI-160) for 7 days on emotional processing was studied in 48 subjects. The volunteers (24 females and 12 males, 19–43 years of age) were randomly assigned to receive LI-160 or placebo. The LI-160 contained 0.3% hypericin and 3%–5% hyperforin. The dose of 300 mg LI-160 was taken three times a day (total dose of 900 mg LI-160) or lactose placebo. Parameters were BDI, facial expression task, emotional categorization, emotional recall, emotional recognition, dot probe, n-back working memory, emotion-potentiated startle task, and a reward task. The participants were self-medicated with LI-160 or placebo for 6 days and on day 7 they were self-administered one 300 mg LI-160 or placebo 2 h before the test battery. The testing lasted approximately 2.5 h and was scheduled to avoid the premenstrual week in females. The authors concluded that short-term treatment with LI-160 produced the same changes in emotional responses as the changes expected with other antidepressants. In this study, there was a significant reduction in attention to unmasked fearful faces, disgusted facial expressions and increased memory for positive words. Two individuals receiving placebo and eight individuals receiving Li-160 had at least one moderate side effect. For LI-160 the most common side effect was dry mouth followed by headache. The phases of the menstrual cycle were not included in the study parameters.

The effect of *H. perforatum extracts* in rodent models on the cognitive enhancement in laboratory animals has been reviewed (Ben-Eliezer and Yechiam, 2016). The differing experiment models of evaluating the nootropic

23. Remotiv, Max Zeller Söhne AG; Romanshorn, CH.
24. Nicabate patch, GSK; Melbourne, AU.
25. Jarsin 300 mg, MCM Klosterfrau Vetriebsgesellschaft mbH; Cologne, DE.

effects of a *H. perforatum* extract were examined. In about 33% of the studies examined, administering *H. perforatum* extract improved cogitative performance in rats with experimentally induced deficiencies in cognitive function. The primary methods used to induce cognitive impairment were stress and amnestic agents. The nootropic effect was observed in tasks assessing reference memory and in tasks assessing primarily working memory. Treatment with *H. perforatum* extracts enhanced tasks in both appetitive and aversive memory. This implies that there is modulation of learning and memory processes. Tasks using appetitive and aversive stimuli were enhanced by *H. perforatum* extracts and support the phytopharmacological effect is modulation of wider learning and memory processes.

Aging neurological diseases

Alzheimer disease is a neurodegenerative disease and a common form of dementia. It is a progressive neurodegenerative disease with memory impairment, cognitive decline, altered behavior, and impaired neuromotor functions (DeTure and Dickson, 2019). In 2018, the estimated global cost of treating 47 million people affected with Alzheimer disease was US $1 trillion.[26] Alzheimer disease has characteristic pathology and can present with comorbidities. The histopathology of Alzheimer disease includes the formation of amyloid plaques. The mechanism is considered to be accumulations of Aβ40 and Aβ42 peptides of amyloid precursor protein by the β- and γ-secretases due to imbalance in the production and clearance pathways. The intracerebral levels of Aβ42 are correlated with clinical signs of dementia. The accumulation of amyloid β is linked to intracerebral inflammation. The ATP-binding cassette transporter P-glycoprotein is an important transporter in the BBB. Cao et al. (2017) investigated the prophylactic effects of *H. perforatum* extract[27] (HPC) as a treatment for aluminum chloride ($AlCl_3$)-induced Alzheimer disease in rats. Rats were dosed orally with 150 mg $AlCl_3$/kg bwt for 90 days. Starting on day 31, the rats were orally dosed with 150 or 300 mg HPC/kg bwt/day for 60 days. Compared to the $AlCl_3$-only treated rats, hippocampal levels of norepinephrine and dopamine were significantly increased ($p < 0.05$) in the $AlCl_3$-treated rats receiving 300 mg HPC/kg/day. Compared to the $AlCl_3$-only treated rats, the 300 mg HPC/day also significantly improved the Morris Water Maze and Open Field Tests. Treatment with 150 mg HPC, compared to $AlCl_3$ rats not treated with HPC, significantly reduced ($p < 0.05$) the amyloid plaques and treatment with both 150 and 300 mg HPC reduced the levels of amyloid β42 (Aβ42) peptides. Treatment with both doses of HPC reduced the mRNA levels of IL-1β, IL-6, and TNF-α (tumor necrosis factor-α). In humans, an age decrease in the activity of adenosine triphosphate-binding cassette transport protein P-glycoprotein (ABCB1) is considered to be a factor in the causation of Alzheimer disease. This protein transports β amyloid from the brain into the blood. *H. perforatum* extract (5% hyperforin) was fed to C57BL/6JAPP/ PS1$^{+/-}$ mice (mice with mutant amyloid precursor protein, APP mice) for 30 or 120 days and aged-matched C57BL/6JAPP/ PS1$^{+/-}$ mice were used as controls (Brenn et al., 2014). Brain levels of β amyloid 40 and 42 were significantly ($p < 0.05$) reduced. The expression of P-glycoprotein was significantly increased. Induction of p-glycoprotein in the cerebrovascular BBB may be a mechanism for treatment of Alzheimer disease. Hofrichter et al. (2013) studied *H. perforatum* extract[28] and hyperforin in mice expressing mutated human amyloid precursor protein and mutated presenilin 1 transgenes under control of the thy1 promoter (APP transgenic mice). This study showed that *H. perforatum* extract attenuated the amyloid β-induced histopathology and memory imprints in the APP mice. These effects were not dependent on hyperforin. The mechanism is increased excretion of amyloid β across the BBB. Radulovic et al. (2018) identified in *H. perforatum* extract a β-diketone (2,6,9-trimethyl-8-decene-3,5-dione) that is a potent acetylcholinesterase inhibitor that has potential as a phytopharmaceutical for Alzheimer disease.

Pain

H. perforatum extracts have been studied for use in pain management and both systemic (oral) and topical preparations are used (Galeotti, 2017). Neuropathic pain is from a lesion or pathophysiology of the somatosensory system. Causes include trauma, neoplasia, chronic diabetes, AIDS, and antiviral drugs. Treatment of neurogenic pain includes antidepressants, alpha 2 delta ligands (gabapentin, pregabalin), and lidocaine. It is difficult to achieve satisfactory relief in patients with neurogenic pain. There is interest in using *H. perforatum* extracts in managing patients with neurogenic pain (Galeotti and Ghelardini, 2013a,b; Galeotti et al., 2014a,b, 2017; Sanna et al., 2017). Galeotti et al. (2014a,b) studied the synergism of *H. perforatum* extract on synergism with morphine in humans and laboratory animals. The *H. perforatum* extract[29] (HPE) (standardized to 0.3% hypericin) was administered by oral gavage to mice. A dose of 5 mg/kg bwt delivered approximately 0.016 mg hypericin/kg bwt, 0.21 mg hyperforin/kg bwt, 0.0415 quercetin/kg bwt, 0.0029 mg amentoflavone/kg bwt, and 0.3175 mg

26. https://www.alz.co.uk/.
27. Xian Xiaocao Botanical Development Co. Ltd; Xian, CN.
28. Finzelberg GmbH & Co KG; Andernach, DE.
29. Indena Research Laboratories; Milan, IT.

hyperoside/kg bwt. Morphine tolerance was induced in mice. The nociceptive stimulus was a hotplate[30] test for mice and hot water emersion for human volunteers. For mice, morphine was administered subcutaneously at 5 and 7 mg/kg bwt 60 min after HPE. In mice, the threshold dose of morphine was 5 mg/kg and induced antinociceptive with maximum effect at 30 min after administration and a return to nociceptive baseline at 60 min. For HPE, the threshold dose was 10 mg/kg bwt. At this dose, antinociceptive threshold increased at 60 min, was maximum at 90 min, and had not returned to baseline at 120 min. A dose of 1 mg HPE/kg bwt potentiated morphine (5 mg/kg) antinociceptive activity. HPC potentiated morphine in mice that were tolerant to morphine and the antinociceptive response to morphine was similar to mice that had not developed tolerance to morphine. Intracerebroventricular injections of hypericin in mice did not have antinociceptive activity but did potentiate the antinociceptive effects of morphine. The human volunteers were administered oral HPE normalized to a 70 kg person (300 mg tablet) and oral morphine equivalent to the mouse dose. The antinociceptive response of morphine was increased with coadministration of HPE. In a double-blind study, a naturopathic herbal drop formulation (NHED) was evaluated in children as a treatment for acute otitis media (Sarrell et al., 2003). NHED is a multiherbal formulation using the tinctures of *Calendula officinalis* flores (28%), *H. perforatum* herba (30%), and *Verbascum thapsus* flores (25%) in olive oil and the essential oils of *Allium sativum* in 0.05% in olive oil (10%), *Lavandula officinalis* (5%), and tocopherol acetate oil. Treatments were NHEDA administered as 5 drops in the ear canal three times a day, 80 mg amoxicillin per oral/kg bwt (divided three times a day with a 500 mg maximum/dose), 5 drops amethocaine and phenazone in glycerin in the ear canal three times a day, or 5 drops topical anesthetic with amoxicillin administered three times a day. A total of 180 children were enrolled in the study. The study showed that the majority of acute otitis media in children is a self-limiting condition. Treatments without antibiotics were more efficacious than treatments with antibiotics. This study showed that, starting on study day 2, NHED was more efficacious that the amethocaine and phenazone treatment.

Burning mouth syndrome (BMS) is a disease that primarily occurs in postmenopausal women. Sardella et al. (2008) evaluated, in a double-blind placebo-controlled study, a *H. perforatum* extract (HPT, 0.31% hypericin, 3.0% hyperforin) as a treatment for BMS. Thirty-five women and four men (64.9 $\pm$ 4.7 years old) completed the study. The patients were evaluated and then given HPT or placebo. The dose of HPT was 300 mg tablets taken three times a day (900 mg HPT/day) for 12 weeks. Parameters evaluated were pain, number of oral mucosa sites with sensation of burning, and evaluation of quality of life. At the start of the study, pain scores were not statistically different ($p > 0.05$) between the treatment groups. After 4 weeks of therapy, the patients receiving HPT reported significantly fewer ($p < 0.05$) oral mucosal sites with burning sensation that patients receiving placebo. In the patients' quality of life evaluations, both groups improved and were not statistically different ($p > 0.05$). Raak et al. (2012), in a metaanalysis *of H. perforatum* preparations for pain control in dental practice, concluded that the evidence to support it use is not conclusive. A study was carried out on Dynamiclear(Global Herbal Supplies Pty Ltd; AU) (HDC), a formulated product that contains *H. perforatum* extract (0.10%), and copper sulfate pentahydrate (0.5%) for the treatment of diseases caused by herpes simplex viruses types 1 and 2 (HSV1, HSV2) (Clewell et al., 2012). This product was efficacious for relieving pain and itching.

Migraine is a primary disabling headache disorder that affects 15% of the global population and occurs more often in women than men (3:1). The pathophysiology of migraine is multifactorial and there are differing headache symptoms (Gross et al., 2019; Qubty and Patniyot, 2020). Using a mouse model of migraine, Galeotti and Ghelardini (2013a) showed that *H. perforatum* extract[31] (0.32% hypericins) was efficacious. An oral dose of 5 mg extract/kg bwt prolonged relief from pain hypersensitivity and increased the latency to the induction of hyperalgesia and reduced the duration of the painful symptomatology. The nitric oxide (NO) donors, nitroglycerin and nitroprusside, induced allodynia and hyperalgesia. The NO donors upregulated the phosphorylation of PKCγ and PKCε in the periaqueductal gray matter. This effect was blocked by an oral dose of the *H. perforatum* extract with no alteration of locomotor activity. Another study, using the same model, showed that *H. perforatum* extract also blocked the overexpression of IL-1β and inducible NO synthetase (Galeotti and Ghelardini, 2013b). The nuclear expression of these pathways was also suppressed by oral treatment with *H. perforatum* extract. Painful diabetes neuropathy is a complication of diabetes mellitus. The efficacy of *H. perforatum* extracts was determined in a rat model of painful diabetes neuropathy (Galeotti et al., 2014b). Diabetes mellitus was induced with parenteral administration of streptozotocin. Both feverfew (*Tanacetum parthenium* L.) and St. John's wort were evaluated in this study. The efficacy of *T. parthenium* was linked to the presence of parthenolide. A reference *H. perforatum* extract[32] (aerial plant parts without seeds) was compared to an ethanolic extract of *H. perforatum* seeds. Reference drugs were carbamazepine, phenobarbital, lamotrigine, and

30. KW Mechanical Workshop; Siena, IT

31. Indena Research Laboratories; Milan, IT.

32. Aboca SpA; Sansepolcro, IT.

L-acetyl-levocarnitine given at doses that did not impede locomotion. The *H. perforatum* seed extract, with a prolonged effect, reversed the mechanical-induced hyperalgesia. The extract from aerial parts without seeds was partially effective in relieving neuropathic pain. The antihyperalgesic activity of *H. perforatum* and feverfew extracts were comparable with carbamazepine, lamotrigine, and L-acetyl-levocarnitine. Both phytopharmaceuticals alter locomotion and behavior.

Phytopharmacology—women specific uses

Premenstrual syndrome and menopause

H. perforatum extract has been evaluated for the treatment of pain and mood changes in premenstrual disorder. Premenstrual disorder is a multifactorial continuum of menstrual cycle-linked psychological, physical, and behavioral symptoms that occur before menstruation. Canning et al. (2010) conducted a randomized control crossover trial on the efficacy of *H. perforatum* extract in reducing mild forms of premenstrual syndrome (PMS). Women who met the criteria of PMS were enrolled in the study which was conducted over 10 menstrual cycles. The *H. perforatum* extract[33] was LI-160 (0.18% hypericin and 3.38% hyperforin) in 450 mg tablets. The women had three screening cycles and were given placebo for cycles 4 and 5. The women then received IL-160 or placebo for cycles 6 and 7, given drug washout over cycle 8, and then crossed over to IL-160 or placebo. Parameters were the daily symptom report (DSR) comprising of 17 premenstrual symptoms with choices from 0 (not present) to 4 being severe and unable to undertake daily activities. The subsets included anxiety, irritability, depression, nervous tension, mood swing, feeling out of control, behavioral (poor coordination, insomnia, confusion, headache, crying, fatigue), pain (aches, cramps, breast tenderness), and physical (food craving, swelling). During the medication cycles, the women were instructed to take two 450 mg LI-160 tablets/day (900 mg LI-160/day) or two placebo tablets. Thirty-six women were enrolled in the study—36 completed the run-in cycles and 34 completed the study. The women were predominantly northern European origin, 18–45 years of age, and the menstrual cycles were 25–35 days. The women were required to meet with the principal investigator during the follicular and luteal phase of cycles 3, 5, 7, 8, and 10. Venous blood was collected during the visit to determine the plasma levels of follicle-stimulating hormone, luteinizing hormone, estradiol, progesterone, prolactin, testosterone, IL-1β, IL-6, IL-8, IFN-γ, and TNF-α. There were no significant effects ($p > 0.05$) of the treatments on these parameters. For comparison of DSR parameters with placebo, treatment with LI-160 significantly improved ($p < 0.05$) the behavioral (poor coordination, insomnia, confusion, headaches, crying, fatigue) and physical (food craving, swelling) symptoms. There were no significant differences ($p > 0.05$) between the treatment groups for mood and pain. The authors were of the opinion that although not statistically significant, there was a decrease in pain and this effect increased with duration of treatment. There were no adverse events attributed to LI-160 treatment. Stevinson and Ernst (2000) conducted a pilot study on *H. perforatum* extract[34] (300 mg standardized to 900 μg hypericin) as a remedy for PMS. Nineteen women met the enrollment criteria and completed the study. The parameters for efficacy were the Hospital Anxiety and Depression Scale (HADS) and the modified Social Adjustment Scale (MSAS) and 17 PMS scores. At the end of the second cycle, the women selected for the study returned for a visit with a physician, with the timing judged to be within 2 days before menstruation, and the HADS and MSAS completed. At this time the 300 mg tablets were dispensed with instructions to take one tablet a day over menstrual cycles 3 and 4. At the premenstrual parts of cycles 3 and 4, the participants visited the physician and the HADS and MSAS were completed. The overall PMS score was significantly reduced ($p < 0.01$) from a baseline of 128.42 ± 68.15 to 70.11 ± 46.10 after cycle 3. From cycle 3 to cycle 4 the scores reduced to 57.63 ± 42.74 and were not statically different ($p > 0.05$) from cycle 3. For individual scores, "feeling out of control" and "tender breasts" were significantly improved ($p < 0.05$). The HADS and MSAS scores were significantly improved ($p < 0.05$) during cycle 3 but the improvement did not increase in cycle 4. At the start of treatment, some women reported nausea, constipation, flatulence, dizziness, and heavy menstrual flow which disappeared without any changes in the medication regimen. In a review of medical plants used for cyclic mastalgia, Niazi et al. (2019) concluded that orally administered *H. perforatum* extract was efficacious. In a review of the Persian literature, Maleki-Saghooni et al. (2018) concluded that women receiving *H. perforatum* extracts, compared to placebo, had a reduction in the symptoms of PMS. Liu et al. (2014), in a metaanalysis concluded that *H. perforatum* preparations are more effective than placebo in alleviating the complaints associated with menopause.

33. Lichtwer Pharma AG; Berlin, DE.

34. St. John's Wort One-A-Day, Kira Limited, Lichtwer Pharma AG; Berlin, DE.

Birthing-associated uses

Samadi et al. (2010), in a randomized placebo control study on an *H. perforatum* ointment, extracted with grape seed oil, applied to cesarean section wounds, found a significant reduction in pain and itching. In a triple-blind trial, a proprietary *H. perforatum* extract[35] (HDC) was evaluated in primiparous women for pain control following episiotomy (Vakili et al., 2018). The chemical specifications of the extract and product formulation of the ointment were not provided. All of the patients were prescribed oral mefenamic acid. The women were randomly assigned to treatments. Pain was assessed at 4 h postpartum and the women received face-to-face instruction on perineal care, hygiene, ointment application, and how to keep a record of self-administered pain medications. The instructions were to apply 3 g of ointment twice a day on the surgical incision for 10 days. The patients were instructed to rate pain on a 1 to 10 scale at 8 and 12 h on postpartum days 5 and 10. For medical questions, the researcher had contact with the patients during the evaluation period. At 8 h postpartum, there were no differences ($p > 05$) between the HDC and vehicle groups. At 12 h, 5 days and 10 days postpartum there was a significant reduction ($p < 0.001$) in pain in the HDC group. There were no differences ($p > 0.05$) between groups across study days for the ingestion of mefenamic acid. Dried *H. perforatum* and *Achillea millefolium* extracts mixed (5% v/v) with Vaseline[36] were studied in a double-blind trial as topical treatments for episiotomy wounds (Hajhashemi et al., 2018). Patients using the *H. perforatum* extract were compared to patients using *A. millefolium* extract, patients using the vehicle (Vaseline), and patients without intervention. The enrollees were 135 nulliparous women who were considered healthy at the time of birth, were not receiving concurrent medications, and had an uneventful delivery. Women that failed to follow instructions and had had adverse effects to the treatment were removed from the study. The instructions were to wash and dry the perineal area, and on the second postpartum day, cover the entire incision surface with 1 mL of ointment for 10 days. After the test ointment was applied, they were to use a sanitary napkin as per standard medical instructions. Parameters used to evaluate the ointments were pain severity, redness, edema, ecchymosis, secretions from the incision, and wound dehiscence. These parameters were evaluated on postdelivery days 2, 7, 10, and 14. The groups were not significantly different on postpartum day 2. For pain, both phytopharmaceutics were more efficacious than the vehicle and no intervention in reducing pain on postpartum days 7, 10, and 14. On postpartum day 10, *H. perforatum* ointment treatment scores for pain were significantly lower ($p < 0.05$) than the pain scores for *A. millefolium* extract and both phytopharmaceutical extracts had significantly lower pain scores ($p < 0.05$) than vehicle and no intervention. On postpartum day 14, both phytopharmaceutical groups significantly reduced ($p < 0.05$) pain scores when compared to the vehicle and no intervention groups. On postpartum days 7 and 10, *H. perforatum* extract significantly reduced edema and was significantly different from the vehicle group and the no-intervention group. The *A. millefolium* extract group was significantly different ($p < 0.05$) from the no-intervention group. The phytopharmaceutical treatments were not significantly different ($p > 0.05$) from vehicle and no intervention in preventing wound dehiscence.

35. Dineh Company; Tehran, IR.

36. Details not provided in the reference.

Perinatal depression

Approximately 18%–19% of women suffer from perinatal depression which, if left untreated, can pose a significant risk to mother, baby, and family. Deligiannidis and Freeman (2014), in a review of herbal products for postpartum depression, pointed out that there were limited studies that investigated the safety of *H. perforatum* extracts available on fetotoxicity when pregnant women used St. John's wort. The authors discussed a study with limited observations that showed an increased rate of adverse events in infants exposed via breastfeeding. These authors recommended that additional studies are needed before St. John's wort can be recommended during the perinatal period.

Phytopharmacology—immune, antiinflammatory, and antidiabetic actions

Immune

Topical application of *H. perforatum* extract[37] has been investigated in a double-blind study as a treatment for psoriasis (Mansouri et al., 2017). The extract was formulated as (5% wt/wt) *H. perforatum* extract (HPCT), Vaseline[38] (84% wt/wt), propylene glycol (10% wt/wt), and Avicel[39] (1% wt/wt). The patients (three male and eight female) with a mean age of 41.25 ± 14.24 years were instructed to apply the HPCT or vehicle to psoriasis lesions for 4 weeks. The lesion on one side (right or left, preferably symmetrical lesions) was randomly assigned vehicle and the other lesions were assigned HPCT. Biopsies were taken before treatment and after treatments and the lesions were scored using the Psoriasis Area Severity Index (PASI) before and after treatments. The tissues were stained for TNF-α. Treatment with HPCT

37. Poursina Co.; Tehran, IR.

38. Unilever; Surry, United Kingdom.

39. Dupont; Wilmington, Delaware, United States

significantly reduced ($p < 0.05$) TNF-α concentration in epidermis, endothelial cells, and dendrite cells. The HPCT significantly reduced ($p < 0.05$) the histopathology and PASI scores of the psoriatic lesion for erythema, scaling, and thickness, and pruritis was significantly reduced. Infections with microorganisms can trigger a "cytokine storm" that causes tissue damage in multiple organs inclusive of intravascular colligation (Behrens and Koretzky, 2017; Sun et al., 2020). Huang et al. (2013), using an in vivo mouse (BALB/c) model, showed that treatment with *H. perforatum* extract increased the severity of A/PR/8/34 H1N1 influenza virus infection. The treatment with 110 mg/kg bwt of *H. perforatum* extract (6 g plant material extracted with 500 mL 95% ethanol) started the day before the viral challenge and contained to necropsy. In the BALB/C mice, transcription of the suppressor of the cytokine signaling 3 (SOCS3) was increased by H. perforatum extract. Subsequent elevation of SOCS3 may have impaired the immune defense against the influenza virus.

Using in vitro models, Huang et al. (2012) showed that pseudohypericin, amentoflavone, quercetin, and chlorogenic acid synergistically inhibit macrophage lipopolysaccharide-induced production of prostaglandin E2 (PGE2). Using small interfering RNA (siRNA) in a macrophage model, pseudohypericin, amentoflavone, quercetin, and chlorogenic acid knocked down SOCS3 and this effect was not observed with the complete *H. perforatum* extract. These four phytochemicals also decreased IL-6, TNF-α, and the four phytochemicals and the unfractionated extract lowered IL-1β. Pseudohypericin was the major contributor to the PGE2 and NO inhibition. The mechanism for knockdown of lipopolysaccharide endotoxin-induced PGE2 and NO activation was by SOCS3 activation. Cyclooxygenase-2 (COX-2) and inducible NO synthase protein expression were not altered by the treatments. However, the enzyme activity of COX-2 was decreased by the extract and the four phytochemicals and increased by SOCS3 knockdown. In addition to the immune targets, this study also demonstrates the complex interactions that occurs among phytochemicals. The immune modulation effects of *H. perforatum* extract were studied in a mouse (BALB/c) model (Abtahi Froushani et al., 2015). The extract was prepared from 6 g of plant material extracted with 500 mL ethanol (95%). The mice were given 110 mg *H. perforatum* extract (ip)/kg body weight or vehicle at the onset of immunization. Mice were given weekly injections of sheep red blood cells (SRBCs) in complete Freund's adjuvant. Forty-eight h before bleeding time the foot pad was injected with SRBCs. Treatment with *H. perforatum* extract decreased delayed-type hypersensitivity and significantly alleviated production of IL-17, IFN-γ, lymphocyte proliferation index, and nitroblue tetrazolium dye reduction test. Compared to control mice, the antibody titer to SRBC was significantly higher in the *H. perforatum* extract-treated mice. *H. perforatum* also inhibited proinflammatory genes.

Diabetes mellitus

A Danish crossover study in human volunteers focused on the interactions between metformin[40] (medication for type II diabetes) and *H. perforatum* extract[41] (HPM) (Stage et al., 2015). The crossover design was with and without 21 days of preceding treatment with HPM. To achieve steady-state kinetics, the metformin treatment regimen was: day 1, 500 mg metformin was ingested twice daily; on day 2, 500 mg was ingested in the morning and 1 g metformin in the evening; followed by 1 g twice a day (08:00 and 20:00 h) for 5 days. One HPM tablet (240–294 mg, 900 μg hypericin) was administered twice a day (08:00 and 20:00 h). On study day 7, the kinetics of metformin was determined concurrent with an oral glucose tolerance test. Serum insulin and C-peptide were measured over 120 min. After a washout of 7 days, the subjects were orally administered HPM for 21 days and metformin was administered on days 16–22. On day 22, the pharmacokinetics of metformin was determined concurrent with the oral glucose tolerance test and insulin and C-peptide determinations. Twelve individuals completed the study and the volunteer deletions were due to noncompliance with the protocol. The estimated values of the homeostasis model of insulin resistance (HOMA-IR) and of surrogate markers for insulin sensitivity (BIGTT-S_I, HOMA-IS and Stumvoll index) were not altered by administrating HPM. Treatment with HPM improved insulin secretion and did not alter the gene expression of organic cation transporter 1 (OCT1, determined in RBCs). Treatment with HPM produced a small and significant increase in renal clearance of metformin.

A study was carried out in a rat (Webster) model of diabetes mellitus on *H. perforatum* extract[42] (HPAP) preventing diabetic nephropathy (Abd El Motteleb and Abd El Aleem, 2017). Diabetes mellitus was induced with administration of nicotinamide followed by streptozotocin and rats with a blood glucose above 200 mg/dL were considered to be diabetic. There were six groups of rats, a control group with no treatments, and five groups of diabetic rats: a group with no treatment for diabetes for 8 weeks, one group receiving 10 mg gliclazide/kg bwt/day for 8 weeks, and three groups of rats receiving 50, 100, or 200 mg HPAP orally/kg bwt/day for 8 weeks, respectively. Rats treated with gliclazide or 50, 100, or 200 mg HPAP had reductions in blood glucose of 78.8%, 26.6%, 41.3%, and 62%, respectively, when compared to the

40. Actavis; Gentofte, DK.
41. Modigen, Jemo-Pharm A/S; Stege, DK.
42. Atos Pharma; Cairo, EG.

diabetic group. Treatment with gliclazide increased serum insulin by 202.7% and treatment with 100 and 200 mg HPAP/kg/ day significantly increased ($p < 0.05$) serum insulin by 116.2% and 200%, respectively. Treatment with HPAP improved renal function as determined by serum urea and creatinine and urine albumen. The malondialdehyde marker of lipid peroxidation increased by 273.5% in the untreated diabetic group and was reduced by gliclazide and HPAP treatments and reduced glutathione was increased with the antidiabetic treatments, as was superoxide dismutase. Compared to untreated diabetic mice, treatment with gliclazide, HPAP at 50, 100, and 200 mg significantly reduced NF-κB by 55.0%, 41.0%, 50.0%, and 79.1%, respectively. Treatment of diabetic mice with gliclazide and HPAP also significantly reduced the renal expression of inducible NO synthase, cyclooxygenase II, collagen IV, and fibronectin. Induction of diabetes decreased renal expression of peroxisome proliferator-activated receptor (PPARγ), treatment with gliclazide had no effect, and treatment with HPAP showed dose-dependent elevations. Comparing the untreated diabetic rats to the untreated control rats, the increases were TNF-α (600.0%), TNF- β (320.6%), IL-1β (770.8%), intracellular adhesion molecule (560.0%), and monocellular chemoattractant protein-1 (298.0%). Treatment with gliclazide and HPAP significantly decreased these parameters. Compared to untreated controls, the diabetic rats had increases in cytochrome c (627.2%) and caspase-3 (486.3%). These markers of apoptosis were significantly decreased by gliclazide and dose dependently by HPAP treatments. For histopathology, either gliclazide or HP (50, 100, and 200 mg/kg per day) caused significant reductions of fibrotic areas of the renal cortices. The authors concluded that treatment with HP in three dose levels improved both renal functional and structural changes; the kidney function markers and the histopathology reverted back to near-normal state, and the renoprotection obtained by 200 mg HPAP/kg bwt per day was superior to that afforded by gliclazide. In a study on hydroalcoholic *H. perforatum* extract (Indena; Milan, IT) (HPI) and hyperforin, it was shown to protect rat and human pancreatic β cells from cytokine challenge (Novelli et al., 2014). HPI and hyperforin downregulated mRNA expression of proinflammatory target genes, for example, iNOS, CXCL9, CXCL10, and COX2. The protective effects of HPI were present after 48 h of exposure to cytokines and, at the same time, hyperforin was partially effective. After 20 h of exposure, both HPI and hyperforin protected cells from mitochondrial alterations and loss of insulin granules, as shown by ultrastructural pathology. Novelli et al. (2019), using a different hydroalcoholic extract[43] (HPA) of *H. perforatum*, and hyperforin showed that STAT1 (signal transducer and activator of transcription 1) in pancreatic β cells is downregulated. The β cells were preincubated for 1–2 h with HPA or hyperforin and then exposed to a cytokine mixture. STAT1 was downregulated and the downregulation persisted after removal of HPA and hyperforin. The protective effect was observed if HPA or hyperforin were added after cytokine exposure (15 and 90 min). Twenty-four hours of incubation with HPA protected the β cells from cytokine-induced dysfunction and apoptosis.

43. Aboca; Sansepolcro, IT.

Phytopharmacology—antimicrobial activity

Antibacterial/antifungal

H. perforatum preparations have been shown to have antimicrobial activity (Saddiqe et al., 2010). *Hypericum* preparations are effective against Gram-positive bacteria, including meticillin-resistant (MRSA) and penicillin-resistant *Staphylococcus aureus* (PRSA) (Negut et al., 2018). Methanol extracts of *H. perforatum* were tested for antimicrobial activity using the Kirby-Bauer agar diffusion method (Cecchini et al., 2007). Antimicrobial activity was observed in the reference standard *S. aureus*. The naphthodianthrone and phloroglucinol groups have antimicrobial activity. Identification of the phytochemicals suggested that hypericin and hyperforin have the most antimicrobial activity. It is likely that there are antimicrobial interactions with the other phytochemicals in *Hypericum* extracts. Extracts from other *Hypericum* spp. also showed antimicrobial activity with Gram-negative organisms. The highly lipophilic hyperforin has antimicrobial activity (Schempp et al., 1999). Hyperforin (>90% pure) was dissolved in phosphate buffer and added to Muller-Hinton agar. The minimal inhibitory concentrations (μg/mL) observed were: *Corynebacterium diphtheriae* (1.0); *Streptococcus agalactiae* (100); and *Streptococcus pyogenes* (100). Both PRSA and MRSA were inhibited by hyperforin and *Enterococcus faecalis*, *Escherichia coli*, *Pseudomonas aeruginosa*, and *Candida albicans* were not inhibited. Light increases the sensitivity of Gram-negative bacteria and envelope viruses to hyperforin, and cream and oil extracts of *H. perforatum*. The water extract of *H. perforatum*, at 10 mg/mL had antifungal activity against *Microsporum gypsum*, *Trichophyton rubrum*, *Aspergillus flavus*, *Curvularia lunata*, and *Fusarium vasiinfectum* (Saddiqe et al., 2010). Extract of *H. perforatum*, compared to other *Hypericum* spp., has reduced antimicrobial activity against pathogenic yeasts and dimorphic fungi (Tocci et al., 2018).

Biofilms

Biofilms are microorganisms that collectively produce extracellular polymeric substances. They have a defined architecture, provide a matrix for genetic exchange, and the cells generally communicate using quorum sensing. They can form on inanimate and animate surfaces and have inherent barrier-type resistance to antimicrobial chemotherapeutics and chemical disinfection agents. *P. aeruginosa* infections can be difficult to treat and control with antibiotics because of biofilm formation. Dogan et al. (2019) studied the in vitro activity of *H. perforatum* extracts on inhibiting the quorum-sensing mechanisms of *P. aeruginosa*. The ethanol, methanol, and acetone extractions of *H. perforatum* were not effective in reducing the formation of biofilms by *P. aeruginosa*. Lyles et al. (2017) investigated the antimicrobial efficacy of different *Oleum. Hyperici* prepared by traditional methods used in the Balkans. Crafted oil macerates inhibited *S. aureus* ($MBIC_{50}$) biofilm formation in concentrations of 0.004%−0.016% (v/v), and had inhibition of quorum sensing ($QSIC_{50}$) at concentrations of 0.064%−0.512% (v/v). These samples contained detectable amounts of hyperforin, but not hypericin.

Antiviral

Envelope viruses (e.g., herpes simplex, influenza virus A, and Mo-MuLV) were shown to be sensitive to the viricidal effects of hypericin at 1.56−25 μg/mL (Tang et al., 1990). Data from an in vitro study suggest that hypericin has antiviral activity against duck hepatitis B virus (Moraleda et al., 1993). Jacobson et al. (2001) reported that a *H. perforatum* extract[44] was not effective as an antiviral drug in patients with hepatitis C. Clewell et al. (2012), in a prospective, randomized, multicentered, comparative, open-label clinical study, evaluated the efficacy of Dynamiclear[45] (HDC). The HDC formulation contains *H. perforatum* extract (0.10%) and copper sulfate pentahydrate (0.5%) for the treatment of diseases caused by HSV1 and HSV2. The disease caused by HSV1 is oral herpes, including cold sores, and genital herpes is the disease caused by HSV2. An individual generally has lifelong infections and the infection rate in a population for HSV1 can reach 90% of a population. The lesions caused by both viruses can be very painful and unsightly. The study used acyclovir[46] (a DNA polymerase inhibitor) as a reference drug. Individuals enrolled in the study were 18−55 years old and had active HSV infections (81 HSV1 and 46 HSV2). The patients visited the clinic before the study and those meeting study criteria were enrolled. The patients had complete blood counts and serum clinicopathological parameters were determined. The patients were instructed to apply the acyclovir twice daily for 7 days or the HDC preparation only once on study day 1. The patients were evaluated on study days 1−3, 8, and 14 for scabbing and crusting, burning-stinging, acute pain, erythema, induration, and vesiculation. Laboratory tests were repeated on study day 14. Eighty-one patients had orofacial lesions, 46 had genital lesions, 5 had lesions on chest, and 1 patient had lesions on a limb. On study day 1, scabbing and crusting, erythema, pain, itching, and vesiculation were not significantly different ($p > 0.05$) between groups. Scabbing and crusting scores were only significantly lower ($p = 0.005$) in the acyclovir group on study day 3 and pain was significantly lower ($p = 0.018$) in the HDC group on study day 8. Also, on study day 8, erythema and itching were significantly less ($p < 0.05$) for the HDC-treated patients compared to the acyclovir-treated patients. Vesiculation was significantly lower in the HDC- treated patients on study day 14. No patients were disenrolled from the study because of adverse effects.

44. VimRx Pharmaceuticals; Wilmington, DE, United States.
45. Global Herbal Supplies Pty Ltd; AU.
46. Zovirax, Glaxo Smithkline; IN.

Phytopharmacology—vulenaries

Hypericum has a 400-plus-year history as a phytovulnerary (Gerarde, 1597). Vulneraries made from *H. perforatum* are sold as gels, ointments, creams, sprays, and bath oils. Phytochemical characterization of many of these products is limited. There is ethnic variation in the use of *H. perforatum* preparations for topical applications and wound healing (Jaric et al., 2018). Water extracts of *H. perforatum* have historical phytopharmaceutical uses in treating infected burns and soft-tissue wounds (Whitehead et al., 2019). Gerarde (1597) described the recipe for making *Hyperici oleum.* The crushed plant material (wet leaves, flowers, and fruit) is mixed with edible vegetable oil in a clear glass container, set in the sun, and allowed to ferment for several weeks. There are modifications of the original recipe. For example, the infusion process can start in a dark area with the infusion occurring over several weeks. The infusion mixture is filtered to remove plant material, a drying agent is added to the filtrate, it is decanted and filtered, and finally placed in the sun for several weeks. In the sun, the appearance of the extract becomes rusty red, likely the production of cyclopseudohypericin, an oxidation product of pseudohypericin, and the finished *oleum. hyperici* is placed in amber bottles. This product has limited shelf-life. In some geographic regions, *O. hyperici* has common use in the treatment of bruises, burns, and other superficial injuries. There is evidence that products formulated using benchmark standards are efficacious (Wolfle et al., 2014). Traditional *Hypericum O. hyperici* is made from the flowers, leaves, and buds of *H. perforatum* (EMA, 2018; Brenner et al., 2000; Linde, 2009; Linde et al., 2009). Solvents, for example, 1,3-butylene glycol solution and

1,2-propylene glycol, are also used (Anon, 2001). Oil extracts from the flowers of *H. perforatum* were found to enhance healing of decubitus ulcers (Wolfle et al., 2014). A study was performed using 123 women volunteers (23.50 ± 4.03 years old) on the efficacy of an *H. perforatum* extract[47] (HGD) on wound healing following cesarean section (Samadi et al., 2010). The propitiatory HGD was prepared by extracting flowering *H. perforatum* with grape seed oil. There were four groups; 24 h after surgery they applied HGD ointment, control ointment, and no treatment. The ointments were applied three times a day for 16 days. Evaluation parameters were erythema, edema, ecchymosis, discharge, the approximation scale, and the Vancouver Scar Scale (VSS). On study day 40 the VSS, pain, and itching were significantly lower ($p < 0.05$) for the group treated with HGD ointment. There were no significant differences ($p > 0.05$) between the groups for patient satisfaction.

A proprietary[48] wound dressing (OPWD) was evaluated in patients with scalp wounds complicated by having exposed bone (Lauchli et al., 2014). OPWD is a proprietary phytopharmaceutical formulation consisting of a mixture of *Hypericum* oil (*H. perforatum*) and neem oil (*Azadirachta indica* A. Juss.) in a spray-type container. Parameters in the study were healing time, wound size, area of exposed bone, pain, and complications. Mean patient age was 76.87 ± 10.3 years (59–90 year), average wound size was 10.9 ± 6.84 cm^2 (0.4–22.6 cm^2) with 4.8 ± 5.9 cm^2 (0.3–20.7 cm^2) of exposed bone. The wound was managed with debridement pro re nata, wound dressings (nonwoven gauze), and the phytopharmaceutic spray applied once a day. Treatment was continued until the wound had 100% epithelization. No infections or other complications were observed. The second intention healing time with granulation tissues was observed after 4 weeks. The exposed bone was covered with granulation tissue in 11/15 (73%) of the wounds. The overall mean healing time was 56.7 days. The authors concluded that OPWD is an efficacious and economical treatment for scalp wounds. Lauchli et al. (2012), in another study on wound healing, reported satisfactory efficacy for OPWD. Ercolani et al. (2019) evaluated OPWD in the management of stoma wounds after enterotomies. The authors reported that with OPWD the dressings were easy to remove, with a rapid reduction in pain and healing in 12–14 days. OPWD was evaluated in a human pediatric study (retrospective, noncontrolled) as a treatment for partial- and full-thickness burns (Mainetti and Carnevali, 2013). The authors concluded that OPWD was efficacious for inducing rapid granulation and reepithelialization. Over the first week of treatment, children over 5 years old reported a reduction of pain from 7–8 to a pain score of 0 (1–10 scale). OPWD has been used in treatment of wound myiasis by *Wohlfahrtia magnifica* (spotted flesh fly) in domestic animals (Carnevali et al., 2019). The wounds were managed by standard veterinary care inclusive of surgical debridement. No concurrent insecticides, pharmaceuticals, or phytopharmaceuticals were used with the OPWD treatment regimen. Treatment with OPWD prevents new fly strikes, suppressed existing larval infections, and promoted wound healing.

47. Gol-Daru Company; Isfahan, IR.

48. One Primary Wound Dressing; Kerecis AG Phytoceuticals AG; Zurich, CH.

Burns

The efficacy of a proprietary product[49] (HANO) containing *H. perforatum* extract (from flowers) and neem oil (*A. indica*) was studied as a treatment for skin burns from ionizing radiation used in patients with head and neck malignancies (Franco et al., 2014). The patients ($N = 28$) had surgery and/or neck irradiation modality treatment induction and/or concurrent chemotherapy. They had a diagnosis of squamous cell histology (96.5%) and high tumor grade (46.5%). Locations of the neoplasia were 21.5% larynx, 28.5% oral cavity, and oropharynx 39.5%. The radiation dose was 60–70 Gy. The RTOG/EORTC scale was used to evaluate radiation burns (toxic reaction of the skin to ionizing radiation). The HANO treatment was started when bright erythema, moderate edema, or patchy moist desquamation were observed. The authors concluded that HANO was an efficacious treatment in the context of the study. In a follow-up study, two preparations of HANO (gel and oil) were evaluated in patients ($N = 50$) receiving ionizing radiation used for head and neck malignancies (Franco et al. (2017). All of the patients (mean = 59 years old) were concurrently receiving cisplatin. Skin lesions were scored using the RTOG/EORTC scale. HANO treatment was started when bright erythema, moderate edema, or patchy moist desquamation were observed. Patients with erythema and/or edema received the gel formulation and patients with patchy moist desquamation received the oil preparation. The authors evaluated, based on experience, the treatments as efficacious as both beneficial and prophylactic in reducing the severity of the lesions and associated pain. A study using a guinea pig model to mimic human oral mucositis induced by radiation and chemotherapy showed *Hypericum* extract to be an effective treatment (Tanideh et al., 2014). The dose of *Hypericum* extract (product unspecified) was 300 mg/kg/day.

49. Holoil; RI.MOS s.r.l.; Mirandola, IT.

Chemical burns

Ingestion of caustic substances generally causes some degree of damage to the esophagus (Hall et al., 2019). The esophageal tissues are more resilient to acid insults than caustics. The effect with caustic ingestion is liquefaction necrosis and damage to esophageal tissue layers. Healing of the esophagus can result in strictures and other complications. A rat model of caustic injury to the esophagus was used to evaluate the efficacy of an olive oil extract[50] of *H. perforatum* (OHPB) on the pathogenesis of caustic injury to an isolated segment of the esophagus (Akay et al., 2019). The treatment groups were (1) surgical isolation of the esophagus and treatment with saline, (2) caustic burn with 10% sodium hydroxide (NaOH) and no treatment, (3) caustic burn and daily treatment with 3 mg methylprednisolone/kg bwt for 21 days, (4) caustic burn and treatment with OHPB for 21 days, and (5) 3 mg methylprednisolone/kg bwt for 21 days plus OHPB. Stenosis scores for group 5 was significantly less ($p < 0.05$) that group 2. Histopathology parameters included the increase in submucosal collagen, damage and collagen deposition in the tunica muscularis, and damage to muscularis mucosa. Hydroxyproline, an indicator of collagen accumulation, was decreased in the groups receiving methylprednisolone. The OHBP treatment increased the submucosal collagen. The rats treated with methylprednisolone and OHPB had the largest luminal diameter. Lyophilized ethanolic extracts of *H. perforatum* are made into oral and topical preparations. The topical preparation was prepared by reconstitution of the lyophilized extract in olive oil which is then aged for 4 weeks (Yilmaz et al., 2019). The preparation was evaluated in a rat model of corneal burns caused by 2 molar NaOH. Both the oral and topical preparations significantly reduced ($p < 0.001$) scores for corneal neovascularization, inflammation, and fibroblastic activity. Rats receiving the oral phytomedication had significantly lower ($p < 0.05$) scores for corneal neovascularization, inflammation, fibroblastic activity, and CD31 (platelet endothelial cell adhesion molecule 1) staining rates.

Hypericum perforatum extracts and drug interactions

Phytopharmaceutics made from *H. perforatum* can interact with other medications (Table 40.4) (Russo et al., 2014). *H. perforatum* extracts upregulate P-glycoprotein [P-gp, multidrug resistance protein 1, (MDR1 gene)] functions at the cellular level in the transport of numerous structurally and pharmacologically unrelated lipophilic and amphipathic drugs, carcinogens, toxins, and other xenobiotics in the intestine, liver, kidney, brain, and other organs. It can also be upregulated and downregulated by environmental xenobiotics and other phytopharmaceutics (Zhou et al., 2004). The CYP enzymes (P450 enzymes) are multisubstrate enzymes that their present in multiple organs, especially the liver, gut, kidneys, lungs, and brain can be genetically upregulated by xenobiotics (Nair et al., 2016). Reduced nicotinamide adenine dinucleotide phosphate (NADPH) via NADPH-CYP reductase provide electrons for the reductive activation of oxygen delivered by the blood. The CYPs function as monooxygenases, through the formation of hydroxylated metabolites. Other common CYP-catalyzed reactions include O-, N-, and S-dealkylation, and epoxidation. *H. perforatum* extracts inhibit and induce CYP enzymes (Goey et al., 2013). In vitro studies showed that hyperforin, hypericin, and I3-II8-biapigenin inhibited CYP3A4-mediated metabolism of midazolam and testosterone. An in vitro study showed that I3,II8-biapigenin is a competitive inhibitor of CYP3A4, CYP2C9, and CYP1A2. Hyperforin is a noncompetitive inhibitor of CYP2D6 and competitive inhibitor of CYP2C9 and CYP3A4, and hypericin inhibits several CYP enzymes (Obach, 2000). Human hepatocytes were evaluated for the effect of hyperforin and hypericin on CYP enzymes (Komoroski et al., 2004). Incubation with hyperforin significantly increased mRNA, protein, and enzyme activity of CYP3A4 and CYP2C9 and exerted no changes on CYP1A2 or CYP2D6 parameters. Hypericin had no effect on the enzyme tested. Hypericin, hyperforin, and quercitrin have been shown by in vitro studies to induce CYP3A4 mRNA expression. Other studies have shown that *H. perforatum* extracts and its major components have both inhibition and induction depending on the in vitro incubation times. Hyperforin, a lipophilic component in *H. perforatum* extracts, activates the nuclear pregnane X receptor (steroid and xenobiotic sensing nuclear receptor) and thereby induces the expression of CYP3A4 (Moore et al., 2000; Wentworth et al., 2000). The pregnane X receptor is highly expressed in the intestine and liver, and in vitro studies have shown that *H. perforatum* extract is similar to the concentration-response observed for rifampicin. Adiwidjaja et al. (2019) developed a physiological-based pharmacokinetic model that predicted induction of CYP3A, CYP2C9, and CYP2C19 and corresponding reduction of area under the curve for synthetic metabolized by the CYPs. The model also predicted a 15.4-fold increase of enterocyte CYPs and a 1.1-fold increase in hepatic CYPs. Hyperforin was identified as an inducer of CYPs (Nahrstedt and Butterweck, 2010; Soleymani et al., 2017). A review by Chrubasik-Hausmann et al. (2019) concluded that hyperforin is the major substrate for induction of CYP3A4 and P-gp. They also recommended a daily dose of less than 1 mg hyperforin to prevent or decrease the risk of medication interactions.

Soleymani et al. (2017) state that removal of hyperforin from the *H. perforatum* extracts can decrease the risk of

50. Botalife Ltd.; Isparta, TR.

TABLE 40.4 Interactions between *Hypericum* extract and nonphytomedications.

Example drug	Interaction observed	Likely mechanism by *H. perforatum* extract
Alprazolam	Decreased area under curve	CPY3A
Atorvastatin	Increased low density lipoprotein	Upregulation of CPY3A4 and P-glycoprotein
Carbamazepine	Decreased plasma levels	CPY3A
Cyclosporine and tacrolimus	Reduction in plasma levels	Upregulation of CPY3A4 and P-glycoprotein
Digoxin	Decreased intestinal absorption	Upregulation of gut P-glycoprotein
Etinilestradiol with desogestrel or norethindrone	Oral contraception failure resulting in pregnancy, intracyclic bleeding, decreased half-life of drugs	Upregulation of CPY3A4
Dexamethasone	Lowered blood levels	Upregulation of CYP3A4
Ibuprofen	Lowered blood levels	Upregulation of P-glycoprotein; CYP2C9
Indinavir	Decreased area under curve	CYP3A4
Nifedipine	Increased metabolites and decreased parent drug	Upregulation of CPY3A4 and CYP2C19
Methadone and pethidine	Lowered blood levels	Induction of CYP2D2
Oxycodone	Lowered blood level	Induction of CYP3A4
Theophylline	Decrease plasma concentration	Up regulation of CYP enzymes
Tolbutamide	Decrease plasma levels	CYP2C9
Verapamil	Increased first-pass effect	Induction of CYP3A4
Voriconazole	Decreased area under curve	Induction of CYP3A4, CYP2C19 and CYP2C9
Warfarin	Decrease in efficacy	Upregulation of CPY3A4

Source: Adopted from Russo, E., Scicchitano, F., Whalley, B.J., et al., 2014. *Hypericum perforatum*: pharmacokinetic, mechanism of action, tolerability, and clinical drug-drug interactions. Phytother. Res., 28(5), 643–655; Adiwidjaja, J., Boddy, A.V., McLachlan, A.J., 2019. Physiologically based pharmacokinetic modelling of hyperforin to predict drug interactions with St John's wort. Clin. Pharmacokinet., 58(7), 911–926.

crucial drug interactions. A classic study in human volunteers showed that *H. perforatum* products that did not contain hyperforin failed to alter the pharmacokinetics of digoxin and products that contained hyperforin reduced the area under the curve for digoxin (Mueller et al., 2004). A study in human volunteers using a marker drug, with known mechanisms and comparing pharmacokinetics, showed that *H. perforatum* extract[51] induced CYP3A and P-gp (Dresser et al., 2003). Durr et al. (2000) showed that *H. perforatum* extract[52] increased 1.5-fold, in duodenal biopsies, the human expression of CYP3A4 and showed a 1.4-fold increase in the expression of P-gp/MDR1. These authors also observed a 2.5-fold increase in hepatic activity of CYP3A2 and a 3.8-fold increase in intestinal P-gp/MDR1 after rats were administered the same *H. perforatum* extract for 14 days. Studies in human volunteers using pharmaceutical probes to determine phenotypical expression of CYP enzymes have been reported (Markowitz et al., 2000, 2003; Goey et al., 2013). These studies show that both intestinal and hepatic CYP3A4 are upregulated by multiple brands of *H. perforatum* extract. With intestinal and hepatic activity of CYP3A4 upregulated, orally administered drugs metabolized by CYP3A4 enzymes have less parent compound reaching systemic circulation (a first-pass effect). In volunteers consuming 600 mg *H. perforatum* standardized extract[53] (1800 mg/day) for 16 days, the

51. Jarsin 300, LI-160, batch No. 99033500, Lichtwer Pharma AG; Berlin, DE.

52. LI-160, Lichtwer Pharma AG; Berlin, DE.

53. Good n' Natural.

peripheral lymphocyte expression of P-gp was increased 4.2-fold based on the changes in rhodamine efflux (Hennessy et al., 2002). The P-gp efflux of rhodamine was inhibited by ritonavir. The interactions of *H. perforatum* extract and clopidogrel were studied in human volunteers (Lau et al., 2011). Clopidogrel is metabolized to its active form by CYP enzymes. Hyporesponders to clopidogrel were administered *H. perforatum* extract (300 mg tablet/day). Treatment with *H. perforatum* extract increased the responsiveness to clopidogrel (decreased platelet aggregation) and upregulated CYP3A4. Gurley et al. (2008), using a drug probe model in human volunteers, found that oral administration of *H. perforatum* extract[54] changed the activity of CYP2D6. Hyperforin increased mRNA, protein, and activity of CYP3A4 and CYP2C9, and had no effect on CYP1A2 or CYP2D6 activities. Hypericin had no effect on CYP1A2, CYP2C9, CYP2D6, or CYP3A4. These components have similar in vitro inhibiting activity as ketoconazole (an antifungal medication). Zahner et al. (2019), in a phase I, nonrandomized, single-sequence study, investigated the drug interactions of Ze-117 (Rebalance[55] 500, containing 0.96 mg hyperforin—considered a low level) in 10 female and 10 male human volunteers (18–55 years of age). The dose of Ze-117 was a 500 mg tablet taken once a day with 240 mL of water. The pharmacokinetics of probe drugs with known CYP biotransformation pathways and/or mechanisms of absorption were used to estimate induction of CYP enzymes and P-gp. There was no upregulation of CYP1A2, CYP2B6, CYP2C9, CYP2C19, CYP3A4, or P-glycoprotein. For the probe drugs, the area under the curve and C_{max} of the used probe drugs had 90% confidence intervals of the geometric mean ratios of the drugs taken together with Ze 117 verses probe drug alone. With the dextromethorphan probe, there was a suggestion of weak inhibition of CYP2D6.

These studies have shown that *H. perforatum* extracts can alter drug metabolism and the effects vary with the chemical composition of the extracts. It is important to point out that products from *H. perforatum* can be ingredients in beverages, blended functional foods, and herbal formulations (Mouly et al., 2017). The chemistry of these preparations is dependent on extraction procedures. Upregulation of P-gp and CYPs can alter the metabolism of drugs. Unwanted pregnancies and intracyclic bleeding have occurred in women consuming *H. perforatum* extracts concurrent with hormonal contraceptives (Henderson et al., 2002; Pfrunder et al., 2003; Schwarz et al., 2003).

54. Nature's Way, Springville; UT, United States.

55. Max Zeller Soehne AG; Romanshorn, CH.

Toxicology of *Hypericum perforatum* preparations

There is a misconception in the general public that phytopharmaceuticals are inherently safe (Mischoulon, 2009). As with all drugs, toxicity can, in some circumstances, occur within the recommended dosage range. The purity of herbal remedies can be an issue, and contaminants and adulterants can cause intoxication. Some herbal products may be adulterated with prescription drugs, heavy metals, metalloids, pesticides, and other substances (Ernst, 2000; Kosalec et al., 2009).

Acute toxicity

The LD_{50} for *Olea Europaea* in rats is greater than 20 mL/kg (Anon, 2001). The mouse LD_{50} values of the polyphenol, lipophilic, and water-soluble extracts *H. perforatum* in mice were 780, 4300, and 2800 mg/kg, respectively. Mice (BALB/c) were administered *H. perforatum* extract at 0, 60, 110, or 220 mg/kg bwt for 3 weeks (Huang et al., 2013). The *H. perforatum* extract was prepared from 6 g of plant material[56] extracted with 500 mL of 95% ethanol. There were no significant differences between *H. perforatum* extract-treated and control mice for feed intake, body weights, and organ weights (liver, kidney, spleen, stomach, and intestine). No behavioral or skin abnormalities were observed. In a pharmacokinetic study, 2 mg hypericin (iv)/kg bwt, given as a bolus dose to three rhesus monkeys was well tolerated (Fox et al., 2001). A rhesus monkey that may have previously received a 2 mg hypericin (iv)/kg dose, after receiving 5 mg hypericin (iv), developed a transient rash on light-exposed skin. A mild increase in serum aspartate aminotransferase and alanine aminotransferase also occurred and these effects were resolved within 12 days.

Reproductive toxicity—humans

Depression is the most common disorder in pregnant and lactating women. There is a substantial increase in women using *H. perforatum* extract as a replacement for conventional prescription medications (Deligiannidis and Freeman, 2014; Einarson et al., 2000). There are estimates that unintentional conception accounts for 50%–75% of pregnancies. Many women use or are using self-prescribed phytopharmaceutics before they become aware that conception has occurred. Also, there is a belief that natural products are more food-like rather than drug-like and therefore are inherently safe, do not interfere with other drugs including with hormonal birth control

56. North Central Regional Plant Introduction Station; Ames, IA, United States.

medications, and are safe to use during pregnancy and breastfeeding. These observations have raised concerns about safe use of *H. perforatum* extracts by pregnant and lactating women (Dugoua et al., 2006; Deligiannidis and Freeman, 2014). Many practitioners in medicine and alternative medicine are not informed of the safety of herbal medicines in pregnancy and lactation. One reason is that few studies have been published, and there is an assumption that historical usage provides sufficient evidence that herbal products are safe for use in pregnancy and lactation. Additionally, physicians and nurses do not routinely include questions about consumption of phytopharmaceutics when taking histories of maternity patients. Moretti et al. (2009) evaluated pregnancy outcomes in women in the Motherisk Program who were using *H. perforatum* extracts for depression. The women that had previously contacted the Canadian Motherisk Program were subsequently enrolled in the study. The Motherisk Program provides pregnant and lactating women with professional advice on the safety and risk for drugs, chemical exposures, and radiation. The prospective study compared gravid women using *H. perforatum* extracts to women using conventional pharmacological therapy and women who were not using any medication. There were 54 pregnant women in the *H. perforatum* extract group and a total of 108 gravid women. There was a 5–7-year postpartum follow-up period. The average dose *of H. perforatum* extract was 615 mg/day. Dosage could not be established for *H. perforatum* teas, tinctures, or granules used by the women. Usage during pregnancy was: 76% of the women reported consuming *H. perforatum* extract in the first trimester; 5.5% reported consuming extract in the second trimester; 7.3% consumed extract during their entire pregnancy; and 9.1% reported consuming *H. perforatum* extract some time during their pregnancy. For major malformations, there were no differences between groups ($p = 0.26$), and the major malformations were 5%, 4%, and 0% in the *H. perforatum* extract group, pharmacological group, and no medications group, respectively Babies born to women in the general population have a 3%–5% risk of birth defects. There were no significant differences ($p > 0.05$) between the groups for stillbirths and low-birth-weight infants. Kolding et al. (2015), using the Danish National Birth Cohort data, investigated the pregnancy outcomes in 38 pregnancies were the mother consumed *H. perforatum* prescriptions during pregnancy. The exposed women had used an *H. perforatum* product before the 17th week of gestation and were compared to women who had not used *H. perforatum* products during their pregnancy. The comparator group ($N = 90{,}128$) also included women who had used other antidepressants during their pregnancy. Women using *H. perforatum* extract tended to be older, were multipara, and had a lower body mass index. Correcting for confounders, there were no differences for gestational age, preterm birth, head circumference, body length, birth weight, and Apgar scores. There were three occurrences of malformations in the *H. perforatum* group, giving 8.1% compared to 3.3% for the controls. There was no specific pattern to the malformations.

Human lactation

A Canadian observational cohort study was performed in women who contacted the Hospital for Sick Children teratogen/toxicant counseling service (Motherisk Program) (Lee et al., 2003). The women were concerned because they were ingesting *H. perforatum* extracts during lactation. These women continued to use *H. perforatum* extracts and were compared to morbidity-matched and control women. The study showed there were no differences for maternal adverse events, infant adverse events, infant weight gain, and estimated milk yield. Klier et al. (2002) have reported on the pharmacokinetics of the transfer of phytochemicals from *H. perforatum* extract to breast milk. A mother with postnatal depression was consuming a St. John's wort preparation (Qarsin[57] 300) three times per day. Over an 18-h period, four samples of breast milk (first and last milk) were collected and the levels of hypericin and hyperforin were determined. Low levels of hyperforin were detected in the samples. In another study, five mothers who were using 300 mg of LI-160 (Jarsin) three times/day (900 mg total dose) donated milk samples (Klier et al., 2006). Milk, before nursing and after nursing, was collected over an 18-h period. Thirty-six milk specimens were collected, five blood samples were collected from the mothers, and two blood specimens were collected from the infants. Hyperforin levels were determined using tandem mass spectrometry and the limit of detection was 0.1 ng/mL. The hyperforin was present in milk at 0.9%–2.5% of drug intake, and the milk/plasma ratio for hyperforin was 0.04/0.13. The level in infant plasma was at the limit of detection. No abnormal effects from consuming LI-160 were observed in the infants or mothers. There appears to be some consensus that *Hypericum* extracts at doses of 300 mg/kg may be safe for pregnant and nursing mothers. If achievable, it is considered prudent to avoid the use of *H. perforatum* extracts and all other phytomedications and synthetic drugs during pregnancy and lactation.

Reproductive toxicity—laboratory animals

A reproductive study was performed in Wistar rats because of concerns of the potential effects of *Hypericum* extracts on human reproductive outcomes (Gregoretti et al., 2004). *H. perforatum* extract (HPG, 0.3% hypericin) was orally administered at doses of 0, 100, and

57. Lichtwer Pharma AG; Berlin, DE.

1000 mg HPG/day starting 2 weeks before mating and continuing through gestation. No differences were observed between treatment groups for maternal weight gain during gestation, for pup weight at birth and weaning, the number of fetuses born, and sex of the fetus. Soft tissue hepatic and renal lesions were observed in the pups from dams treated with HPG. The liver lesions in the pups were described as hepatocyte hyaline degeneration, lobular fibrosis, and disorganization of hepatocyte arrays. The renal lesions were described as a reduction in glomerular size with disappearance of Bowman's space and hyaline tubular degeneration. The severity of the lesions increase with the dose of HPG administered to the dams. In the same study groups of postpartum dams with no previous exposure to HPG, the rats were orally administered 0, 100, and 1000 mg HPG/day with dosing limited to the 21 days of lactation. At the end of lactation, there were no apparent differences in the body weight of the pups. The same hepatic and renal lesions as described for in utero exposure were also observed in the pups with lactational exposure, and at both dosage levels, milk exposure caused more severe lesions than in utero exposure to HPG. This study is interesting because the renal and liver lesions were produced by both prenatal hematogenous and postnatal milk exposures. A mouse (CD-1 mice) developmental study was undertaken using a *H. perforatum* extract[58] (0.3% *Hypericum*) (Rayburn et al., 2001a). The mice in the treatment group received drug bars designed to deliver 180 mg *Hypericum*/kg bwt/day or placebo 2 weeks before mating. The mice consumed 182 ± 4 mg *Hypericum*/kg bwt/day before mating and 150 ± 3 mg *Hypericum*/kg bwt/day at gestational day 11. There were no differences between groups for food consumption and the mean gestational period was 19.0 days for both groups. The number of live pups was 11.6 per group and there were no significant differences ($p < 0.05$) between groups. The percent of males was 47% for both groups and the sex ration was not significantly different ($p < 0.05$). The male pups born to the *Hypericum*-treated females weighed significantly less ($p < 0.05$) but there were no differences in male body weights by postnatal day 3. Morphometric measurements of the pups were not different between groups at birth and postnatal day 5, and no anatomical anomalies were observed. Mothering, nursing, and behavioral development of the pups were the same in both groups. The only development delay observed was a delay in the eruption of the upper incisors in the male offspring from the *Hypericum*-treated mothers. Female offspring that were born to the *Hypericum*-treated dams had lower body weights at postnatal days 30, 37, and 45.

A study was performed to determine the effects of in utero exposure to *Hypericum* extract on the behavior of mice (Rayburn et al., 2000). Female mice were exposed as described in the developmental study by Rayburn et al. (2001a). Developmental testing parameters (locomotor activity and exploratory behavior) and juvenile and adult behavioral testing parameters (forced swim depression test, anxiety test, and evaluation plus maze test) showed no difference between the pups from control and *Hypericum*-treated dams. In the same study, mice pups exposed in utero to *H. perforatum* extract were tested for memory using an assortment of tests (Rayburn et al., 2001b). For both genders at the final trials, there were no significant differences ($p < 0.003$) for the juvenile runway with adult memory test, adult Morris maze test, adult passive avoidance tests, or adult straight water runway test followed by a dry Cincinnati maze test. The females exposed to *H. perforatum* extract in utero required significantly more ($p < 0.05$) time to learn the Morris maze task. Behavioral and physiological alterations resulting from developmental treatment were investigated in Sprague-Dawley rats exposed to diets containing 0, 180, 900, 1800, or 4500 ppm *H. perforatum* extract[59] (HPSV) beginning on gestational day 3 and ending at offspring weaning on postnatal day (PND) 21 (Cada et al., 2001). Dietary levels were estimated at 1, 5, 10, and 25 times the human dose, respectively. Postweaning behavioral assessments of male and female offspring included open field activity, acoustic startle, performance of complex and Morris water mazes, and activity in an elevated plus maze. There were no effects of the HPSV on maternal weight gain or duration of gestation; offspring body weights were similar to those of controls from PND 2 through PND 56; after which, some treated groups weighed significantly less than the controls. There were no HPSV-related behavioral alterations. Whole and regional brain weights of offspring at adulthood showed no significant adverse effects from in utero exposure to HPSV. Campos et al. (2017a) evaluated the behavior of mature male rats that had been exposed to *H. perforatum* extract[60] (CHP, 0.3% hypericin) in utero. Oral exposure of the dams, by groups, was 0 (Cont, distilled water), 36, (D1) 72 (D2), and 144 mg (D3) CHP/kg bwt for the duration of gestation. At 90 days the following tests were done on the male mice: hole-board and forced swimming test, pentobarbital-induced sleep time, rotarod, and elevated plus maze. There were no significant differences ($p < 0.05$) between groups in body weights of the F1 males. For time on the rotarod, comparing groups showed

58. Basic Organics Inc; Columbus, OH, United States.

59. Marketed by Spring Valley; Manufactured by NaturPharma; American Fork, UT, United States.

60. Aqueous Extract Imported from CN. Prepared by Mbpharma Manipulacoes Ltda; Matias Barbosa, BR.

that groups D2 and D3 were significantly lower ($p < 0.05$) that Cont and D1. For phenobarbital sleep time, F1 rats in group D3 were under the influence significantly longer ($p < 0.05$) that rats in the other groups. For the elevated maze, whole board, and forced swim tests, group D3 was significantly different ($p < 0.05$) from the other groups. The authors also tested the mature male and female F1 rats for antinociceptive and anticonvulsant activity (Campos et al., 2017b). The authors concluded that the F1 rats showed a decrease in these tests. No histopathology was done. Interpretations of in utero effects of *H. perforatum* extracts in laboratory animal studies are varied. The rodent studies on the safety of *H. perforatum* extracts during pregnancy have recently been reviewed (Avila et al., 2018). These authors concluded that additional research is needed as the studies in the public domain have large variation is experimental design, the *H. perforatum* extracts are not well defined, and the findings are not consistent for defining the effects in the F1 generation.

Endocrine and sexual dysfunctions

Sexual and endocrine dysfunction in males and females can occur concurrently with ingestion of antidepressant drugs. Failures of oral hormonal contraceptives and breakthrough bleeding have been reported with concurrent administration of *H. perforatum* extracts and limited information exists for injected hormonal contraceptives (Berry-Bibee et al., 2016). The authors also concluded that pharmacokinetic evidence is mixed but suggests that administration of *H. perforatum* extract may be associated with weak to moderate induction of the metabolism of ethinyl estradiol and progestin combined oral contraceptives. Trautmann-Sponsel and Dienel (2004) evaluated the safety of hydroalcoholic extracts[61] of *Herba hyperici* in a review of data from three double-blind, randomized multicenter trials with 594 outpatients. Three women reported premature menses by 2 and 7 days, respectively, and two woman reported dysmenorrhea. Sexual dysfunction was not reported. In vitro studies on rat and human vas deferens showed that *Hypericum* extract inhibits contractility of smooth muscle. This effect may explain the impaired ejaculation orgasm that is observed in some male patients using *Hypericum* extracts for depression (Capasso et al., 2005). There are case reports of sexual dysfunction in men taking *H. perforatum* extract and successful treatment with sildenafil (Assalian, 2000). Comorbidities were not reported. Studies have shown that sexual dysfunction in men consuming *H. perforatum* extracts is generally less than in men consuming prescribed synthetic medications (Ulbricht et al., 2010). Schule et al. (2004) in a single-blind study using healthy volunteers showed that *H. perforatum* extract (WS-5570) caused endocrine disruption. The WS-5570 was administered orally at dosages of 600, 900, and 1200 mg on 4 different days. Venous blood was withdrawn from an indwelling catheter before drug administration and then at 30-min intervals for 5 h. Growth hormone, ACTH, and prolactin and cortisol levels were determined. The area under the curve was used to quantify the hormone response. Administration of WS-5570 stimulated the release of ACTH and a spike in growth hormone. There were no changes in prolactin and cortisone. Another study of similar design showed that plasma levels of cortisol increased 30–90 min after 600 mg WS-5570 was administered orally (Schule et al., 2001). A study with mice showed that high doses of *H. perforatum* extract[62] inhibited prolactin release (Di Carlo et al., 2005). This observation is of interest as inhibition of prolactin release is a cause of agalactia in lactating mammals. A study involving human volunteers showed that *H. perforatum* extract[63] decreased plasma prolactin levels, increased growth hormone, and did not alter plasma cortisol levels (Franklin et al., 1999). A study by Donovan et al. (2005) determined the levels of testosterone, dihydrotestosterone, dehydroepiandrosterone sulfate, sex hormone-binding globulin, and the combined concentrations of androsterone sulfate and epiandrosterone sulfate in men ($n = 6$) and women ($n = 6$) volunteers (22–38 years old) before and after a 14-day regimen of LI-160 (300 mg *H. perforatum* extract[64], standardized to 0.12%–0.3% hypericin). One tablet was taken three times per day (total dose 900 mg). The change observed was a significant decline in the combined concentrations of androsterone sulfate and epiandrosterone sulfate in men and women. There was a trend for the testosterone/dihydrotestosterone ratio to decrease in both women and men. This study provided evidence that induction of CYP3A4 likely decreased the production of the androsterone sulfate and epiandrosterone sulfate metabolites (Markowitz et al., 2003). Butterweck (2003) concluded that there is evidence to show that both depression and *H. perforatum* extracts can alter the hypothalamic–pituitary axis.

Side effects and toxicity of *Hypericum perforatum* extracts

There is some evidence that *H. perforatum* extracts, compliant with applicable high-level licensing standards, are

61. WS-5570 (3%–6%) hyperforin, WS-5572 (5% hyperforin), WS®-5573 (3%–6% hyperforin), Dr. Willmar Schwabe Pharmaceuticals; Karlsruhe, DE.

62. Arkopharma; Carros, FR.

63. Jarsin 300 containing 300 mg of LI-160; Lichtwer Pharma AG; Berlin, DE.

64. Lichtwer Pharma AG; Berlin, DE.

generally safe as the side effects of *Hypericum* extracts, when consumed in moderation, are considered rare and mild (WHO, 2003, 2010). The side effects observed in large clinical trials conducted over several weeks included minor gastrointestinal irritation, allergic reactions, tiredness, and restlessness. Also suggested by the clinical studies is that *H. perforatum* extract does not affect general performance or the ability to operate an automobile. Seizures and confusion were reported in a 16-year-old female, who, as an apparent suicide attempt, consumed a large quantity of *H. perforatum* extract in tablet form (Becker et al., 2014). Neuropathy has also been reported in a woman who consumed 500 mg *Hypericum* extract/day and developed photosensitization. Autonomic arousal has been reported in a woman who consumed an undetermined amount of a St. John's wort preparation (Brown, 2000). Hepatocellular carcinoma frequently develops in patients with liver disease. A rare case of pleomorphic hepatocellular carcinoma was reported in a cirrhotic alcoholic consuming unknown amounts of *H. perforatum* extract Lampri et al., 2014). It is unknown if this rare neoplasm had links to the ingestion of St. John's wort. Bone marrow necrosis was observed in a 22-year-old male using 1 g/day of a *H. perforatum* extract (Demiroglu et al., 2005). In both of these cases the chemical composition of the *H. perforatum* extract was not examined. The Cosmetic Ingredient Review Expert Panel concluded that *H. perforatum*-derived ingredients were safe as cosmetic ingredients within the scope of the practices of use and concentration as described in their safety assessment (Anon, 2001; Becker et al., 2014).

Photosensitization

There are numerous reports in the veterinary medical literature of *H. perforatum* causing photosensitization in domestic animals and in some locations it is considered a noxious weed. Wet and/or desiccated plant tops can also cause photosensitization in humans, ruminants, horses, and other mammals. Onoue et al. (2011), using in vitro methods, showed that hypericin, pseudohypericin, and hyperforin had photoreactivity when exposed to 250 W/m^2 of sunlight. In sunlight, quercetin attenuated the generation of singlet oxygen by hyperforin and quercetin had no effect on hypericin. The photosensitizing chemical is hypericin, a phenanthroperylenequinone that has complex physical chemical behavior (Ernst, 2000; Theodossiou et al., 2009). In most bioassays hypericin is present as a monosodium salt. The maximum absorbance of hypericin salts is 548 and 591 nm in ethanol solution and it has red fluorescence at 594 and 642 nm. Hypericin associates with proteins (albumen, low-density lipoprotein, etc.) and has red fluorescence. In light, it has a high triplet quantum yield. It is an efficient singlet oxygen producer and superoxide anion generator, and these radicals can trigger apoptosis and cellular necrosis. Hypericin can also cause phototoxicity in the lens and retina, thereby damaging the eye (Ehrenshaft et al., 2013). Damage to proteins in the lens of the eye is not repairable and temporal summation results in cataracts due to damage to cx-crystallin protein in the lens. Using self-reporting information, a study showed that there is an association between cataracts and the use of *H. perforatum* extract (Booth and McGwin, 2009). Photosensitization is a dose-limiting side effect of *H. perforatum* extracts. There are reports of photosensitization occurring after oral use of phytopharmaceutics containing *H. perforatum*—both were uses of extract prepared for topical use (Ulbricht et al., 2010).

Prescribing St John's wort

Extracts of St. John's wort have been shown to be efficacious for mild to moderate depression (Woelk et al., 1994; Linde, 2009; Linde et al., 2009; Nicolussi et al., 2020). *H. perforatum* products are licensed in Germany and prescribed for treatment of depressive, anxiety, and sleep disorders. Some physicians and pharmacists recommend that all *Hypericum* products that are orally administered be sold only after consultation with physicians and pharmacists. Additionally, product labeling should include side effects from ingesting *Hypericum* products. Labeling must clearly state that untoward reactions and decreased efficacy with synthetic drugs can occur (Clauson et al., 2008; Dalwood et al., 2015). Physicians practicing conventional medicine have mixed attitudes regarding prescribing extracts of *H. perforatum* (Forsdike and Pirotta, 2019; Mischoulon and Rapaport, 2019). Influencing their decision are inconsistency in dose response among products and negative interactions with some prescription medications. Many medical school curricula do not include materia medica of plants and physicians are uncomfortable prescribing phytopharmaceuticals (Solomon et al., 2011; Solomon et al., 2013). Prescribing patterns for phytopharmaceuticals vary with geographic areas. Areas with reduced prescribing of phytopharmaceuticals include North America, Australia, and New Zealand. Prescribing phytopharmaceutical with documented efficacy can reduce patient management costs (Solomon et al., 2013; Lauchli et al., 2014). The dose of *H. perforatum* products, based on the literature, falls between 900 mg/day and 1800 mg/day that is divided into two or three daily administrations (Mischoulon, 2009). It is recommended that oral preparations of *H. perforatum* be taken with medical oversight (physician or pharmacist). Freeman et al. (2010), in the American Psychiatric Association Task Force report concluded that *H. perforatum* extracts may be a suitable treatment for mild to moderate depression. The Task Force also cautioned that orally administered St. John's wort can have negative interactions with other medications. Nicolussi et al. (2020) in a recent review concluded that drug interactions need to be part of the treatment plan. With the exception of Europe and parts of Asia, there is reluctance for *H. perforatum*-source products to be prescribed by general practitioners (Forsdike and Pirotta, 2019; Nicolussi et al., 2020).

Concluding remarks and future directions

H. perforatum phytopharmaceuticals are the most commonly used and the best studied. Pharmacopeias define its various preparations, and in some countries, standardized *H. perforatum* extracts are licensed medications for the treatment of mild to moderate depression. Outside the United States, standardized *H. perforatum* extracts are recognized as an effective treatment for mild to moderate depression. The side effects for recognized dosage regimens are considered less than the side effects of synthetic drugs used for mild to moderate depression. Orally administered *H. perforatum* extracts can alter the pharmacokinetics of synthetic drugs. It is generally accepted in European countries that *H. perforatum* should be taken under the advice of a physician or pharmacists. In the history of concurrent medications, physicians and nurses should include nutraceuticals and be knowledgeable of drug interactions that can occur. There is a need to have *H. perforatum* extracts to meet an international standard. In many countries, nutraceuticals have substantial cost savings for patients. Depression is the most common mental health disease that occurs concurrent with pregnancy and the postpartum period. Future directions should focus on the safety of standardized *H. perforatum* extracts in pregnancy and lactation because of its used in prepartum and postpartum depression.

References

Abd El Motteleb, D.M., Abd El Aleem, D.I., 2017. Renoprotective effect of *Hypericum perforatum* against diabetic nephropathy in rats: insights in the underlying mechanisms. Clin. Exp. Pharmacol. Physiol. 44 (4), 509–521.

Abtahi Froushani, S.M., Esmaili Gouvarchin Galee, H., Khamisabadi, M., et al., 2015. Immunomudulatory effects of hydroalcoholic extract of *Hypericum perforatum*. Avicenna J. Phytomed. 5 (1), 62–68.

Adiwidjaja, J., Boddy, A.V., McLachlan, A.J., 2019. Physiologically based pharmacokinetic modelling of hyperforin to predict drug interactions with St John's wort. Clin. Pharmacokinet. 58 (7), 911–926.

Agapouda, A., Booker, A., Kiss, T., et al., 2019. Quality control of *Hypericum perforatum* L. analytical challenges and recent progress. J. Pharm. Pharmacol. 71 (1), 15–37.

Akay, M.A., Akduman, M., Tataroglu, A.C., et al., 2019. Evaluation of the efficacy of *Hypericum perforatum* (St. John's wort) oil in the prevention of stricture due to esophageal corrosive burns. Esophagus 16 (4), 352–361.

Anon, 2001. Final report on the safety assessment of *Hypericum perforatum* extract and *Hypericum perforatum* oil. Int. J. Toxicol. 20 (Suppl 2), 31–39.

Apaydin, E.A., Maher, A.R., Shanman, R., et al., 2016. A systematic review of St. John's wort for major depressive disorder. Syst. Rev. 5 (1), 148. Available from: 10.1186/s13643-016-0325-2.

Assalian, P., 2000. Sildenafil for St. John wort-induced sexual dysfunction. J. Sex Marital Ther. 26 (4), 357–358.

Avila, C., Whitten, D., Evans, S., 2018. The safety of St John's wort (*Hypericum perforatum*) in pregnancy and lactation: a systematic review of rodent studies. Phytother. Res. 32 (8), 1488–1500.

Ayuob, N.N., Balgoon, M.J., 2018. Histological and molecular techniques utilized to investigate animal models of depression. An updated review. Microsc. Res. Tech. 81 (10), 1143–1153.

Bach-Rojecky, L., Kalodjera, Z., Samarzija, I., 2004. The antidepressant activity of *Hypericum perforatum* L. measured by two experimental methods on mice. Acta Pharm. 54 (2), 157–162.

Bauer, A., Knapp, M., Parsonage, M., 2016. Lifetime costs of perinatal anxiety and depression. J. Affect. Disord. 192, 83–90.

Becker, L.C., Bergfeld, W.F., Belsito, D.V., et al., 2014. Amended safety assessment of *Hypericum perforatum*-derived ingredients as used in cosmetics. Int. J. Toxicol. 33 (3 Suppl), 5S–23S.

Beerhues, L., 2006. Hyperforin. Phytochemistry 67 (20), 2201–2207.

Behrens, E.M., Koretzky, G.A., 2017. Review: cytokine storm syndrome: looking toward the precision medicine era. Arthritis Rheumatol. 69 (6), 1135–1143.

Belwal, T., Devkota, H.P., Singh, M.K., et al., 2019. St. John's wort (*Hypericum perforatum*). In: Nabavi, S.M., Silva, A.S. (Eds.), Nonvitamin and Nonmineral Nutritional Supplements. Academic Press, San Diego, CA, pp. 415–432.

Ben-Eliezer, D., Yechiam, E., 2016. *Hypericum perforatum* as a cognitive enhancer in rodents: a meta-analysis. Sci. Rep. 6, 35700. Available from: 10.1038/srep35700.

Berry-Bibee, E.N., Kim, M.J., Tepper, N.K., et al., 2016. Co-administration of St. John's wort and hormonal contraceptives: a systematic review. Contraception 94 (6), 668–677.

Bilia, A.R., Gallori, S., Vincieri, F.F., 2002. St. John's wort and depression: efficacy, safety and tolerability-an update. Life Sci. 70 (26), 3077–3096.

Booker, A., Agapouda, A., Frommenwiler, D.A., et al., 2018. St John's wort (*Hypericum perforatum*) products—an assessment of their authenticity and quality. Phytomedicine 40, 158–164.

Booth III, J.N., McGwin, G., 2009. The association between self-reported cataracts and St. John's wort. Curr. Eye Res. 34 (10), 863–866.

Brattstrom, A., 2009. Long-term effects of St. John's wort (*Hypericum perforatum*) treatment: a 1-year safety study in mild to moderate depression. Phytomedicine 16 (4), 277–283.

Brenn, A., Grube, M., Jedlitschky, G., et al., 2014. St. John's wort reduces beta-amyloid accumulation in a double transgenic Alzheimer's disease mouse model-role of P-glycoprotein. Brain Pathol. 24 (1), 18–24.

Brenner, R., Azbel, V., Madhusoodanan, S., et al., 2000. Comparison of an extract of hypericum (LI 160) and sertraline in the treatment of depression: a double-blind, randomized pilot study. Clin. Ther. 22 (4), 411–419.

Bridi, H., Meirelles, G.C., von Poser, G.L., 2018. Structural diversity and biological activities of phloroglucinol derivatives from *Hypericum* species. Phytochemistry 155, 203–232.

Brown, T.M., 2000. Acute St. John's wort toxicity. Am. J. Emerg. Med. 18 (2), 231–232.

Bruni, R., Sacchetti, G., 2009. Factors affecting polyphenol biosynthesis in wild and field grown St. John's Wort (*Hypericum perforatum* L. Hypericaceae/Guttiferae). Molecules 14 (2), 682–725.

Bukhari, I.A., Dar, A., 2013. Behavioral profile of *Hypericum perforatum* (St. John's Wort) extract. A comparison with standard antidepressants in animal models of depression. Eur. Rev. Med. Pharmacol. Sci. 17 (8), 1082–1089.

Butterweck, V., 2003. Mechanism of action of St John's wort in depression: what is known? CNS Drugs 17 (8), 539–562.

Butterweck, V., 2009. St John's wort: quality issues and active compounds. In: Cooper, R., Kronenberg, F. (Eds.), Botanical Medicine: From Bench to Bedside. Marry Ann Liebert Inc, New Rochelle, NY, pp. 69–91.

Butterweck, V., Schmidt, M., 2007. St. John's wort: role of active compounds for its mechanism of action and efficacy. Wien. Med. Wochenschr. 157 (13–14), 356–361.

Butterweck, V., Wall, A., Lieflander-Wulf, U., et al., 1997. Effects of the total extract and fractions of *Hypericum perforatum* in animal assays for antidepressant activity. Pharmacopsychiatry 30 (Suppl 2), 117–124.

Butterweck, V., Petereit, F., Winterhoff, H., et al., 1998. Solubilized hypericin and pseudohypericin from *Hypericum perforatum* exert antidepressant activity in the forced swimming test. Planta Med. 64 (4), 291–294.

Butterweck, V., Jurgenliemk, G., Nahrstedt, A., et al., 2000. Flavonoids from *Hypericum perforatum* show antidepressant activity in the forced swimming test. Planta Med. 66 (1), 3–6.

Butterweck, V., Nahrstedt, A., Evans, J., et al., 2002. In vitro receptor screening of pure constituents of St. John's wort reveals novel interactions with a number of GPCRs. Psychopharmacology (Berl.) 162 (2), 193–202.

Butterweck, V., Christoffel, V., Nahrstedt, A., et al., 2003a. Step by step removal of hyperforin and hypericin: activity profile of different *Hypericum* preparations in behavioral models. Life Sci. 73 (5), 627–639.

Butterweck, V., Lieflander-Wulf, U., Winterhoff, H., et al., 2003b. Plasma levels of hypericin in presence of procyanidin B2 and hyperoside: a pharmacokinetic study in rats. Planta Med. 69 (3), 189–192.

Butterweck, V., Hegger, M., Winterhoff, H., 2004. Flavonoids of St. John's wort reduce HPA axis function in the rat. Planta Med. 70 (10), 1008–1011.

Caccia, S., Gobbi, M., 2009. St. John's wort components and the brain: uptake, concentrations reached and the mechanisms underlying pharmacological effects. Curr. Drug Metab. 10 (9), 1055–1065.

Cada, A.M., Hansen, D.K., LaBorde, J.B., et al., 2001. Minimal effects from developmental exposure to St. John's wort (*Hypericum perforatum*) in Sprague-Dawley rats. Nutr. Neurosci. 4 (2), 135–141.

Camfield, D.A., Scholey, A.B., Pipingas, A., et al., 2013. The neurocognitive effects of *Hypericum perforatum* special extract (Ze 117) during smoking cessation. Phytother. Res. 27 (11), 1605–1613.

Campos, L.V., Vieira, V.A., Silva, L.R., et al., 2017a. Rats treated with *Hypericum perforatum* during pregnancy generate offspring with behavioral changes in adulthood. Rev. Bras. Farmacogn. 27 (3), 361–368.

Campos, L.V., Guerra, M.O., Peters, V.M., et al., 2017b. Effect of the extract of *Hypericum perforatum* on neurodevelopment of regions related to pain control and convulsion. J. Med. Plant Res. 11 (6), 107–117.

Can, A., Dao, D.T., Arad, M., et al., 2012. The mouse forced swim test. J. Vis. Exp. 59, e3638. Available from: 10.3791/3638.

Canning, S., Waterman, M., Orsi, N., et al., 2010. The efficacy of *Hypericum perforatum* (St John's wort) for the treatment of premenstrual syndrome: a randomized, double-blind, placebo-controlled trial. CNS Drugs 24 (3), 207–225.

Cao, Z., Wang, F., Xiu, C., et al., 2017. *Hypericum perforatum* extract attenuates behavioral, biochemical, and neurochemical abnormalities in aluminum chloride-induced Alzheimer's disease rats. Biomed. Pharmacother. 91, 931–937.

Capasso, R., Borrelli, F., Aviello, G., et al., 2008. Inhibitory effect of the herbal antidepressant St. John's wort (*Hypericum perforatum*) on rat gastric motility. Naunyn Schmiedebergs Arch. Pharmacol. 376 (6), 407–414. 10.1007/s00210-007-0230-2. Available from: 10.1007/s00210-007-0230-2.

Capasso, R., Borrelli, F., Montanaro, V., et al., 2005. Effects of the antidepressant St. John's wort (*Hypericum perforatum*) on rat and human vas deferens contractility. J. Urol. 173 (6), 2194–2197.

Carnevali, F., Franchini, D., Otranto, D., et al., 2019. A formulation of neem and hypericum oily extract for the treatment of the wound myiasis by *Wohlfahrtia magnifica* in domestic animals. Parasitol. Res. 118 (8), 2361–2367.

Castor, T.P., Tyler, T.A., Student, R.J., 2001. <http://www.google.com/patents/US6291241>.

Cecchini, C., Cresci, A., Coman, M.M., et al., 2007. Antimicrobial activity of seven hypericum entities from central Italy. Planta Med. 73 (6), 564–566.

Chen, J.A., Papakostas, G.I., Youn, S.J., et al., 2011. Association between patient beliefs regarding assigned treatment and clinical response: reanalysis of data from the Hypericum Depression Trial Study Group. J. Clin. Psychiatry 72 (12), 1669–1676.

Chrubasik-Hausmann, S., Vlachojannis, J., McLachlan, A.J., 2019. Understanding drug interactions with St John's wort (*Hypericum perforatum* L.): impact of hyperforin content. J. Pharm. Pharmacol. 71 (1), 129–138.

Clauson, K.A., Santamarina, M.L., Rutledge, J.C., 2008. Clinically relevant safety issues associated with St. John's wort product labels. BMC Complement. Altern. Med. 8, 42. Available from: 10.1186/1472-6882-8-42.

Clewell, A., Barnes, M., Endres, J.R., et al., 2012. Efficacy and tolerability assessment of a topical formulation containing copper sulfate and *Hypericum perforatum* on patients with herpes skin lesions: a comparative, randomized controlled trial. J. Drugs Dermatol. 11 (2), 209–215.

Cossuta, D., Vatai, T., Bathori, M., et al., 2012. Extraction of hyperforin and hypericin from St. John's wort (*Hypericum perforatum* l.) with different solvents. J. Food Process. Eng. 35 (2), 222–235.

Crupi, R., Mazzon, E., Marino, A., et al., 2011. *Hypericum perforatum* treatment: effect on behavior and neurogenesis in a chronic stress model in mice. BMC Complement. Altern. Med. 11, 7. Available from: 10.1186/1472-6882-11-7.

Crupi, R., Abusamra, Y.A., Spina, E., et al., 2013. Preclinical data supporting/refuting the use of *Hypericum perforatum* in the treatment of depression. CNS Neurol. Disord. Drug Targets 12 (4), 474–486.

Cui, Y.H., Zheng, Y., 2016. A meta-analysis on the efficacy and safety of St John's wort extract in depression therapy in comparison with selective serotonin reuptake inhibitors in adults. Neuropsychiatr. Dis. Treat. 12, 1715–1723.

Dalwood, J., Dhillon, R., Tibrewal, P., et al., 2015. St John's wort—is it safe in bipolar disorder? Aust. N. Z. J. Psychiatry 49 (12), 1226–1227.

Dauncey, E.A., Irving, J.T.W., Allkin, R., 2019. A review of issues of nomenclature and taxonomy of *Hypericum perforatum* L. and Kew's Medicinal Plant Names Services. J. Pharm. Pharmacol. 71 (1), 4–14.

De Marchis, G.M., Burgi, S., Kientsch, U., et al., 2006. Vitamin E reduces antidepressant-related beta-adrenoceptor down-regulation in cultured cells. Comparable effects on St. John's wort and tricyclic antidepressant treatment. Planta Med. 72 (15), 1436–1437.

Debelle, F.D., Vanherweghem, J.L., Nortier, J.L., 2008. Aristolochic acid nephropathy: a worldwide problem. Kidney Int. 74 (2), 158–169.

Deligiannidis, K.M., Freeman, M.P., 2014. Complementary and alternative medicine therapies for perinatal depression. Best Pract. Res. Clin. Obstet. Gynaecol. 28 (1), 85–95.

Demiroglu, Y.Z., Yeter, T.T., Boga, C., et al., 2005. Bone marrow necrosis: a rare complication of herbal treatment with *Hypericum perforatum* (St. John's wort). Acta Med. (Hradec Kralove) 48 (2), 91–94.

Denke, A., Schempp, H., Mann, E., et al., 1999. Biochemical activities of extracts from *Hypericum perforatum* L. 4th Communication: influence of different cultivation methods. Arzneim. Forsch. 49 (2), 120–125.

DeTure, M.A., Dickson, D.W., 2019. The neuropathological diagnosis of Alzheimer's disease. Mol. Neurodegener. 14 (1), 32. Available from: 10.1186/s13024-019-0333-5.

Di Carlo, G., Pacilio, M., Capasso, R., et al., 2005. Effect on prolactin secretion of *Echinacea purpurea*, *Hypericum perforatum* and *Eleutherococcus senticosus*. Phytomedicine 12 (9), 644–647.

Di Pierro, F., Risso, P., Settembre, R., 2018. Role in depression of a multi-fractionated versus a conventional *Hypericum perforatum* extract. Panminerva Med. 60 (4), 156–160. Available from: 10.23736/S0031-0808.18.03518-8.

Dogan, S., Gokalsin, B., Senkardes, I., et al., 2019. Anti-quorum sensing and anti-biofilm activities of *Hypericum perforatum* extracts against *Pseudomonas aeruginosa*. J. Ethnopharmacol. 235, 293–300.

Donovan, J.L., DeVane, C.L., Lewis, J.G., et al., 2005. Effects of St John's wort (*Hypericum perforatum* L.) extract on plasma androgen concentrations in healthy men and women: a pilot study. Phytother. Res. 19 (10), 901–906.

Dorks, M., Langner, I., Dittmann, U., et al., 2013. Antidepressant drug use and off-label prescribing in children and adolescents in Germany: results from a large population-based cohort study. Eur. Child Adolesc. Psychiatry 22 (8), 511–518.

Draves, A.H., Walker, S.E., 2003. Analysis of the hypericin and pseudohypericin content of commercially available St John's wort preparations. Can. J. Clin. Pharmacol. 10 (3), 114–118.

Dresser, G.K., Schwarz, U.I., Wilkinson, G.R., et al., 2003. Coordinate induction of both cytochrome P4503A and MDR1 by St John's wort in healthy subjects. Clin. Pharmacol. Ther. 73 (1), 41–50.

Dugoua, J.J., Mills, E., Perri, D., et al., 2006. Safety and efficacy of St. John's wort (*Hypericum*) during pregnancy and lactation. Can. J. Clin. Pharmacol. 13 (3), e268–e276.

Durr, D., Stieger, B., Kullak-Ublick, G., et al., 2000. St John's wort induces intestinal P-glycoprotein/MDR1 and intestinal and hepatic CYP3A4. Clin. Pharmacol. Ther. 68 (6), 598–604.

Ehrenshaft, M., Roberts, J.E., Mason, R.P., 2013. Hypericin-mediated photooxidative damage of alpha-crystallin in human lens epithelial cells. Free Radic. Biol. Med. 60, 347–354.

Einarson, A., Lawrimore, T., Brand, P., et al., 2000. Attitudes and practices of physicians and naturopaths toward herbal products, including use during pregnancy and lactation. Can. J. Clin. Pharmacol. 7 (1), 45–49.

Elena Tebrencu, C., Iordache, T.V., Ionescu, E., et al., 2020. Straightforward preparation of naphtodianthrone-rich ethanolic extracts from wild St. John's wort. J. Diet. Suppl, 17 (1), 88–96.

Ellis, K.A., Stough, C., Vitetta, L., et al., 2001. An investigation into the acute nootropic effects of *Hypericum perforatum* L. (St. John's Wort) in healthy human volunteers. Behav. Pharmacol. 12 (3), 173–182.

EMA, 2010. Guideline on Declaration of Herbal Substances and Herbal Preparations in Herbal Medicinal Products/Traditional Herbal Medicinal Products. European Medicines Agency (EMA), Committee on Herbal Medicinal Products, Amsterdam.

EMA, 2018. European Union Herbal Monograph on *Hypericum perforatum* L., *Herba* (Traditional Use). European Medicines Agency (EMA), Committee for Medicinal Products for Human Use, Amsterdam.

Ercolani, M.S., La Torre, F., Toma, E., 2019. Treatment of peristomal granulomatosis with a neem and red *Hypericum* oil application: case studies. J. Med. Sci. Clin. Invent. 6 (5), 4485–4489.

Ernst, E., 2000. Adverse effects of herbal drugs in dermatology. Br. J. Dermatol. 143 (5), 923–929.

ESCOP (European Scientific Cooperative on Phytotherapy). (2020). <https://escop.com/>.

Falk, H., 1999. From the photosensitizer hypericin to the photoreceptor stentorin—the chemistry of phenanthroperylene quinones. Angew. Chem. Int. Ed. Engl. 38 (21), 3116–3136.

Fang, Y. (Ed.), 2019. Depressive Disorders: Mechanisms, Measurement and Management. Springer, Singapore, <https://doi.org/10.1007/978-981-32-9271-0>.

Fogaca, M.V., Duman, R.S., 2019. Cortical GABAergic dysfunction in stress and depression: new insights for therapeutic interventions. Front. Cell Neurosci. 13, 87. Available from: 10.3389/fncel.2019.00087.

Forsdike, K., Pirotta, M., 2019. St John's wort for depression: scoping review about perceptions and use by general practitioners in clinical practice. J. Pharm. Pharmacol. 71 (1), 117–128.

Fox, E., Murphy, R.F., McCully, C.L., et al., 2001. Plasma pharmacokinetics and cerebrospinal fluid penetration of hypericin in nonhuman primates. Cancer Chemother. Pharmacol. 47 (1), 41–44.

Fralick, M., Avorn, J., Franklin, J.M., et al., 2018. Application and impact of run-in studies. J. Gen. Intern. Med. 33 (5), 759–763.

Franco, P., Potenza, I., Moretto, F., et al., 2014. *Hypericum perforatum* and neem oil for the management of acute skin toxicity in head and neck cancer patients undergoing radiation or chemo-radiation: a single-arm prospective observational study. Radiat. Oncol. 9, 297. Available from: 10.1186/s13014-014-0297-0.

Franco, P., Rampino, M., Ostellino, O., et al., 2017. Management of acute skin toxicity with *Hypericum perforatum* and neem oil during platinum-based concurrent chemo-radiation in head and neck cancer patients. Med. Oncol. 34 (2), 30. Available from: 10.1007/s12032-017-0886-5.

Franklin, M., Chi, J., McGavin, C., et al., 1999. Neuroendocrine evidence for dopaminergic actions of *Hypericum* extract (LI 160) in healthy volunteers. Biol. Psychiatry 46 (4), 581–584.

Freeman, M.P., Fava, M., Lake, J., et al., 2010. Complementary and alternative medicine in major depressive disorder: the American Psychiatric Association Task Force report. J. Clin. Psychiatry 71 (6), 669–681.

Friedland, K., Harteneck, C., 2015. Hyperforin: to be or not to be an activator of TRPC(6). Rev. Physiol. Biochem. Pharmacol. 169, 1–24.

Frommenwiler, D.A., Reich, E., Sudberg, S., et al., 2016. St. John's wort versus counterfeit St. John's wort: an HPTLC study. J. AOAC Int. 99 (5), 1204–1212.

Gaid, M., Biedermann, E., Fuller, J., et al., 2018. Biotechnological production of hyperforin for pharmaceutical formulation. Eur. J. Pharm. Biopharm. 126, 10–26.

Galeotti, N., 2017. *Hypericum perforatum* (St John's wort) beyond depression: a therapeutic perspective for pain conditions. J. Ethnopharmacol. 200, 136–146.

Galeotti, N., Ghelardini, C., 2013a. St. John's wort relieves pain in an animal model of migraine. Eur. J. Pain 17 (3), 369–381.

Galeotti, N., Ghelardini, C., 2013b. St. John's wort reversal of meningeal nociception: a natural therapeutic perspective for migraine pain. Phytomedicine 20 (10), 930–938.

Galeotti, N., Farzad, M., Bianchi, E., et al., 2014a. PKC-mediated potentiation of morphine analgesia by St. John's wort in rodents and humans. J. Pharmacol. Sci. 124 (4), 409–417.

Galeotti, N., Maidecchi, A., Mattoli, L., et al., 2014b. St. John's wort seed and feverfew flower extracts relieve painful diabetic neuropathy in a rat model of diabetes. Fitoterapia 92, 23–33.

Gastpar, M., Singer, A., Zeller, K., 2006. Comparative efficacy and safety of a once-daily dosage of hypericum extract STW3-VI and citalopram in patients with moderate depression: a double-blind, randomised, multicentre, placebo-controlled study. Pharmacopsychiatry 39 (2), 66–75.

GBD, 2018. Global, regional, and national incidence, prevalence, and years lived with disability for 354 diseases and injuries for 195 countries and territories, 1990–2017: a systematic analysis for the Global Burden of Disease Study 2017. Lancet 392 (10159), 1789–1858.

Gelenberg, A.J., Shelton, R.C., Crits-Christoph, P., et al., 2004. The effectiveness of St. John's wort in major depressive disorder: a naturalistic phase 2 follow-up in which nonresponders were provided alternate medication. J. Clin. Psychiatry 65 (8), 1114–1119.

Gerarde, J., 1597. The Herhull or General Historie of Plants. John Norton, London.

Ghiasvand, A., Shadabi, S., Hajipour, S., et al., 2016. Comparison of ultrasound-assisted headspace solid-phase microextraction and hydrodistillation for the identification of major constituents in two species of *Hypericum*. J. Chromatogr. Sci. 54 (2), 264–270.

Gilbert, N., 2011. Regulations: herbal medicine rule book. Can Western guidelines govern Eastern herbal traditions? Nature 480 (7378), S98–S99.

Goey, A.K., Mooiman, K.D., Beijnen, J.H., et al., 2013. Relevance of in vitro and clinical data for predicting CYP3A4-mediated herb-drug interactions in cancer patients. Cancer Treat. Rev. 39 (7), 773–783.

Gonda, X., Petschner, P., Eszlari, N., et al., 2019. Genetic variants in major depressive disorder: from pathophysiology to therapy. Pharmacol. Ther. 194, 22–43.

Greeson, J.M., Sanford, B., Monti, D.A., 2001. St. John's wort (*Hypericum perforatum*): a review of the current pharmacological, toxicological, and clinical literature. Psychopharmacology (Berl.) 153 (4), 402–414.

Gregoretti, B., Stebel, M., Candussio, L., et al., 2004. Toxicity of *Hypericum perforatum* (St. John's wort) administered during pregnancy and lactation in rats. Toxicol. Appl. Pharmacol. 200 (3), 201–205.

Gross, E.C., Lisicki, M., Fischer, D., et al., 2019. The metabolic face of migraine—from pathophysiology to treatment. Nat. Rev. Neurol. 15 (11), 627–643.

Grundmann, O., Kelber, O., Butterweck, V., 2006. Effects of St. John's wort extract and single constituents on stress-induced hyperthermia in mice. Planta Med. 72 (15), 1366–1371.

Gurley, B.J., Swain, A., Hubbard, M.A., et al., 2008. Clinical assessment of CYP2D6-mediated herb-drug interactions in humans: effects of milk thistle, black cohosh, goldenseal, kava kava, St. John's wort, and *Echinacea*. Mol. Nutr. Food Res. 52 (7), 755–763.

Hajhashemi, M., Ghanbari, Z., Movahedi, M., et al., 2018. The effect of *Achillea millefolium* and *Hypericum perforatum* ointments on episiotomy wound healing in primiparous women. J. Matern. Fetal Neonatal Med. 31 (1), 63–69.

Hall, A.H., Jacquemin, D., Henny, D., et al., 2019. Corrosive substances ingestion: a review. Crit. Rev. Toxicol. 49 (8), 637–669.

Haller, H., Anheyer, D., Cramer, H., et al., 2019. Complementary therapies for clinical depression: an overview of systematic reviews. BMJ Open 9 (8), e028527. Available from: 10.1136/bmjopen-2018-028527.

HDTSG (Hypericum Depression Trial Study Group), 2002. Effect of *Hypericum perforatum* (St John's wort) in major depressive disorder randomized controlled trial. J. Am. Med. Assoc. 287 (14), 1807–1814.

Heinrich, M., Daniels, R., Stintzing, F.C., et al., 2017. Comprehensive phytochemical characterization of St. John's wort (*Hypericum perforatum* L.) oil macerates obtained by different extraction protocols via analytical tools applicable in routine control. Pharmazie 72 (3), 131–138.

Henderson, L., Yue, Q.Y., Bergquist, C., et al., 2002. St John's wort (*Hypericum perforatum*): drug interactions and clinical outcomes. Br. J. Clin. Pharmacol. 54 (4), 349–356.

Hennessy, M., Kelleher, D., Spiers, J.P., et al., 2002. St John's wort increases expression of P-glycoprotein: implications for drug interactions. Br. J. Clin. Pharmacol. 53 (1), 75–82.

Hobbs, C., 1990. St. John's wort—ancient herbal protector. Pharm. Hist. 32 (4), 166–169.

Hofrichter, J., Krohn, M., Schumacher, T., et al., 2013. Reduced Alzheimer's disease pathology by St. John's wort treatment is independent of hyperforin and facilitated by ABCC1 and microglia activation in mice. Curr. Alzheimer Res. 10 (10), 1057–1069.

Houghton, P., 2009. Synergy and polyvalence: paradigms to explain the activity of herbal products. In: Mukherjee, P.K., Houghton, P.J. (Eds.), Evaluation of Herbal Medicinal Products: Perspectives on Quality, Safety and Efficacy. Pharmaceutical Press, London, pp. 85–94.

Howard, C., Hill, E., Kreuzer, M., et al., 2019. DNA authentication of St John's wort (*Hypericum perforatum* L.) commercial products targeting the ITS region. Genes (Basel) 10 (4). Available from: 10.3390/genes10040286.

Huang, N., Rizshsky, L., Hauck, C.C., et al., 2012. The inhibition of lipopolysaccharide-induced macrophage inflammation by 4 compounds in *Hypericum perforatum* extract is partially dependent on the activation of SOCS3. Phytochemistry 76, 106–116.

Huang, N., Singh, N., Yoon, K., et al., 2013. The immuno-regulatory impact of orally-administered *Hypericum perforatum* extract on Balb/C mice inoculated with H1n1 influenza A virus. PLoS One 8 (9), e76491. Available from: 10.1371/journal.pone.0076491.

Isacchi, B., Bergonzi, M.C., Carnevali, F., et al., 2007. Analysis and stability of the constituents of St. John's wort oils prepared with different methods. J. Pharm. Biomed. Anal. 45 (5), 756–761.

Jacobson, J.M., Feinman, L., Liebes, L., et al., 2001. Pharmacokinetics, safety, and antiviral effects of hypericin, a derivative of St. John's wort plant, in patients with chronic hepatitis C virus infection. Antimicrob. Agents Chemother. 45 (2), 517–524.

Jaric, S., Kostic, O., Mataruga, Z., et al., 2018. Traditional wound-healing plants used in the Balkan region (southeast Europe). J. Ethnopharmacol. 211, 311–328.

Kasper, S., 2015. Phytopharmaceutical treatment of anxiety, depression, and dementia in the elderly: evidence from randomized, controlled clinical trials. Wien. Med. Wochenschr. 165 (11–12), 217–228.

Kasper, S., Caraci, F., Forti, B., et al., 2010. Efficacy and tolerability of *Hypericum* extract for the treatment of mild to moderate depression. Eur. Neuropsychopharmacol. 20 (11), 747–765.

Kerb, R., Brockmoller, J., Staffeldt, B., et al., 1996. Single-dose and steady-state pharmacokinetics of hypericin and pseudohypericin. Antimicrob. Agents Chemother. 40 (9), 2087–2093.

Kladar, N., Mrdanovic, J., Anackov, G., et al., 2017. *Hypericum perforatum*: synthesis of active principles during flowering and fruitification-novel aspects of biological potential. Evid. Based Complement. Altern. Med. 2017, 2865610.

Klemow, K.M., Bartlow, A., Crawford, J., et al., 2011. Medical attributes of St. John's wort (*Hypericum perforatum*). In: Benzie, I.F.F., Wachtel-Galor, S. (Eds.), Herbal Medicine: Biomolecular and Clinical Aspects, second ed. CRC Press, Boca Raton, FA, pp. 211–237.

Klier, C.M., Schafer, M.R., Schmid-Siegel, B., et al., 2002. St. John's wort (*Hypericum perforatum*)—is it safe during breastfeeding? Pharmacopsychiatry 35 (1), 29–30.

Klier, C.M., Schmid-Siegel, B., Schafer, M.R., et al., 2006. St. John's wort (*Hypericum perforatum*) and breastfeeding: plasma and breast milk concentrations of hyperforin for 5 mothers and 2 infants. J. Clin. Psychiatry 67 (2), 305–309.

Kohler, C.A., Freitas, T.H., Stubbs, B., et al., 2018. Peripheral alterations in cytokine and chemokine levels after antidepressant drug treatment for major depressive disorder: systematic review and meta-analysis. Mol. Neurobiol. 55 (5), 4195–4206.

Kolding, L., Pedersen, L.H., Henriksen, T.B., et al., 2015. *Hypericum perforatum* use during pregnancy and pregnancy outcome. Reprod. Toxicol. 58, 234–237.

Komoroski, B.J., Zhang, S., Cai, H., et al., 2004. Induction and inhibition of cytochromes P450 by the St. John's wort constituent hyperforin in human hepatocyte cultures. Drug Metab. Dispos. 32 (5), 512–518.

Kosalec, I., Cvek, J., Tomic, S., 2009. Contaminants of medicinal herbs and herbal products. Arh. Hig. Rada Toksikol. 60 (4), 485–501.

Kraus, B., Wolff, H., Elstner, E.F., et al., 2010. Hyperforin is a modulator of inducible nitric oxide synthase and phagocytosis in microglia and macrophages. Naunyn Schmiedebergs Arch. Pharmacol. 381 (6), 541–553.

Krishnan, V., Nestler, E.J., 2011. Animal models of depression: molecular perspectives. Curr. Top. Behav. Neurosci. 7, 121–147.

Kucharikova, A., Kimakova, K., Janfelt, C., et al., 2016. Interspecific variation in localization of hypericins and phloroglucinols in the genus *Hypericum* as revealed by desorption electrospray ionization mass spectrometry imaging. Physiol. Plant. 157 (1), 2–12.

Lampri, E.S., Ioachim, E., Harissis, H., et al., 2014. Pleomorphic hepatocellular carcinoma following consumption of *Hypericum perforatum* in alcoholic cirrhosis. World J. Gastroenterol. 20 (8), 2113–2116.

Lau, W.C., Welch, T.D., Shields, T., et al., 2011. The effect of St John's wort on the pharmacodynamic response of clopidogrel in hyporesponsive volunteers and patients: increased platelet inhibition by enhancement of CYP3A4 metabolic activity. J. Cardiovasc. Pharmacol. 57 (1), 86–93.

Lauchli, S., Hafner, J., Wehrmann, C., et al., 2012. Post-surgical scalp wounds with exposed bone treated with a plant-derived wound therapeutic. J. Wound Care 21 (5), 228. 230, 232-223.

Lauchli, S., Vannotti, S., Hafner, J., et al., 2014. A plant-derived wound therapeutic for cost-effective treatment of post-surgical scalp wounds with exposed bone. Forsch. Komplementmed. 21 (2), 88–93.

Lee, A., Minhas, R., Matsuda, N., et al., 2003. The safety of St. John's wort (*Hypericum perforatum*) during breastfeeding. J. Clin. Psychiatr. 64 (8), 966–968.

Leuner, K., Li, W., Amaral, M.D., et al., 2013. Hyperforin modulates dendritic spine morphology in hippocampal pyramidal neurons by activating $Ca^{(2+)}$ -permeable TRPC6 channels. Hippocampus 23 (1), 40–52.

Lin, Y., Zhang, J.C., Fu, J., et al., 2013. Hyperforin attenuates brain damage induced by transient middle cerebral artery occlusion (MCAO) in rats via inhibition of TRPC6 channels degradation. J. Cereb. Blood Flow Metab. 33 (2), 253–262.

Linde, K., 2009. St. John's wort—an overview. Forsch. Komplementmed. 16 (3), 146–155.

Linde, K., Berner, M.M., Kriston, L., 2009. St John's wort for major depression. Cochrane Database Syst. Rev. 8 (4), CD000448. Available from: 10.1002/14651858.CD000448.pub3.

Liu, Y.R., Jiang, Y.L., Huang, R.Q., et al., 2014. *Hypericum perforatum* L. preparations for menopause: a meta-analysis of efficacy and safety. Climacteric 17 (4), 325–335.

Lyles, J.T., Kim, A., Nelson, K., et al., 2017. The chemical and antibacterial evaluation of St. John's wort oil macerates used in Kosovar traditional medicine. Front. Microbiol. 8, 1639. Available from: 10.3389/fmicb.2017.01639.

Mainetti, S., Carnevali, F., 2013. An experience with paediatric burn wounds treated with a plant-derived wound therapeutic. J. Wound Care 22 (12), 681–682. 684-685, 688-689.

Maleki-Saghooni, N., Karimi, F.Z., Behboodi Moghadam, Z., et al., 2018. The effectiveness and safety of Iranian herbal medicines for treatment of premenstrual syndrome: a systematic review. Avicenna J. Phytomed. 8 (2), 96–113.

Mansouri, P., Mirafzal, S., Najafizadeh, P., et al., 2017. The impact of topical Saint John's wort (*Hypericum perforatum*) treatment on tissue tumor necrosis factor-alpha levels in plaque-type psoriasis: a pilot study. J. Postgrad. Med. 63 (4), 215–220.

Markowitz, J.S., DeVane, C.L., Boulton, D.W., et al., 2000. Effect of St. John's wort (*Hypericum perforatum*) on cytochrome P-450 2D6 and 3A4 activity in healthy volunteers. Life Sci 66 (9), L133–PL139.

Markowitz, J.S., Donovan, J.L., DeVane, C.L., et al., 2003. Effect of St John's wort on drug metabolism by induction of cytochrome P450 3A4 enzyme. JAMA 290 (11), 1500–1504.

Mischoulon, D., 2009. Update and critique of natural remedies as antidepressant treatments. Obstet. Gynecol. Clin. North. Am. 36 (4), 789–807.

Mischoulon, D., Rapaport, M.H., 2019. Current role of herbal and natural preparations. Handb. Exp. Pharmacol. 250, 225–252.

Moore, L.B., Goodwin, B., Jones, S.A., et al., 2000. St. John's wort induces hepatic drug metabolism through activation of the pregnane X receptor. Proc. Natl. Acad. Sci. USA 97 (13), 7500–7502.

Moraleda, G., Wu, T.T., Jilbert, A.R., et al., 1993. Inhibition of duck hepatitis B virus replication by hypericin. Antivir. Res. 20 (3), 235–247.

Moretti, M.E., Maxson, A., Hanna, F., et al., 2009. Evaluating the safety of St. John's wort in human pregnancy. Reprod. Toxicol. 28 (1), 96–99.

Mouly, S., Lloret-Linares, C., Sellier, P.O., et al., 2017. Is the clinical relevance of drug-food and drug-herb interactions limited to grapefruit juice and Saint-John's wort? Pharmacol. Res. 118, 82–92.

Mueller, S.C., Uehleke, B., Woehling, H., et al., 2004. Effect of St John's wort dose and preparations on the pharmacokinetics of digoxin. Clin. Pharmacol. Ther. 75 (6), 546–557.

Mukherjee, P.K., 2019. Quality control and evaluation of herbal drugs. Evaluating Natural Products and Traditional Medicine. Elsevier, Cambridge, MA.

Nahrstedt, A., Butterweck, V., 2010. Lessons learned from herbal medicinal products: the example of St. John's wort (perpendicular). J. Nat. Prod. 73 (5), 1015–1021.

Nair, P.C., McKinnon, R.A., Miners, J.O., 2016. Cytochrome P450 structure-function: insights from molecular dynamics simulations. Drug Metab. Rev. 48 (3), 434–452.

Nathan, P.J., 2001. *Hypericum perforatum* (St John's wort): a non-selective reuptake inhibitor? A review of the recent advances in its pharmacology. J. Psychopharmacol. 15 (1), 47–54.

Negut, I., Grumezescu, V., Grumezescu, A.M., 2018. Treatment strategies for infected wounds. Molecules 23 (9), 2392.

Nestler, E.J., Gould, E., Manji, H., et al., 2002. Preclinical models: status of basic research in depression. Biol. Psychiatry 52 (6), 503–528.

Ng, Q.X., Venkatanarayanan, N., Ho, C.Y., 2017. Clinical use of *Hypericum perforatum* (St John's wort) in depression: a meta-analys*is*. J. Affect. Disord. 210, 211–221.

Niazi, A., Rahimi, V.B., Hatami, H., et al., 2019. Effective medicinal plants in the treatment of the cyclic mastalgia (breast pain): a review. J. Pharmacopunct. 22 (3), 131–139.

Nicolussi, S., Drewe, J., Butterweck, V., et al., 2020. Clinical relevance of St. John's wort drug interactions revisited. Br. J. Pharmacol. 177 (6), 1212–1226.

Nöldner, M., 2005. Comparative preclinical antidepressant activity of isolated constituents. In: Muller, W.E. (Ed.), St. John's Wort and Its Active Principles in Depression and Anxiety. Birkhäuser Verlag, Berlin, pp. 73–84.

Novelli, M., Beffy, P., Menegazzi, M., et al., 2014. St. John's wort extract and hyperforin protect rat and human pancreatic islets against cytokine toxicity. Acta Diabetol. 51 (1), 113–121.

Novelli, M., Beffy, P., Gregorelli, A., et al., 2019. Persistence of STAT-1 inhibition and induction of cytokine resistance in pancreatic beta cells treated with St John's wort and its component hyperforin. J. Pharm. Pharmacol. 71 (1), 93–103.

Obach, R.S., 2000. Inhibition of human cytochrome P450 enzymes by constituents of St. John's wort, an herbal preparation used in the treatment of depression. J. Pharmacol. Exp. Ther. 294 (1), 88–95.

Onoue, S., Seto, Y., Ochi, M., et al., 2011. In vitro photochemical and phototoxicological characterization of major constituents in St. John's wort (*Hypericum perforatum*) extracts. Phytochemistry 72 (14–15), 1814–1820.

Pfrunder, A., Schiesser, M., Gerber, S., et al., 2003. Interaction of St John's wort with low-dose oral contraceptive therapy: a randomized controlled trial. Br. J. Clin. Pharmacol. 56 (6), 683–690.

Qubty, W., Patniyot, I., 2020. Migraine pathophysiology. Pediatr. Neurol. 107, 1–6. Available from: 10.1016/j.pediatrneurol.2019.12.014.

Raak, C., Bussing, A., Gassmann, G., et al., 2012. A systematic review and meta-analysis on the use of *Hypericum perforatum* (St. John's Wort) for pain conditions in dental practice. Homeopathy 101 (4), 204–210.

Raclariu, A.C., Paltinean, R., Vlase, L., et al., 2017. Comparative authentication of *Hypericum perforatum* herbal products using DNA metabarcoding, TLC and HPLC-MS. Sci. Rep. 7 (1), 1291. Available from: 10.1038/s41598-017-01389-w.

Radulovic, N.S., Gencic, M.S., Stojanovic, N.M., et al., 2018. Prenylated beta-diketones, two new additions to the family of biologically active *Hypericum perforatum* L. (Hypericaceae) secondary metabolites. Food Chem. Toxicol. 118, 505–513.

Rahimi, R., Nikfar, S., Abdollahi, M., 2009. Efficacy and tolerability of *Hypericum perforatum* in major depressive disorder in comparison with selective serotonin reuptake inhibitors: a meta-analysis. Prog. Neuropsychopharmacol. Biol. Psychiatry 33 (1), 118–127.

Ravindran, A.V., Balneaves, L.G., Faulkner, G., et al., 2016. Canadian network for mood and anxiety treatments (CANMAT) 2016 clinical guidelines for the management of adults with major depressive disorder: section 5. Complementary and alternative medicine treatments. Can. J. Psychiatr. 61 (9), 576–587.

Rayburn, W.F., Christensen, H.D., Gonzalez, C.L., 2000. Effect of antenatal exposure to Saint John's wort (*Hypericum*) on neurobehavior of developing mice. Am. J. Obstet. Gynecol. 183 (5), 1225–1231.

Rayburn, W.F., Gonzalez, C.L., Christensen, H.D., et al., 2001a. Effect of prenatally administered hypericum (St John's wort) on growth and physical maturation of mouse offspring. Am. J. Obstet. Gynecol. 184 (2), 191–195.

Rayburn, W.F., Gonzalez, C.L., Christensen, H.D., et al., 2001b. Impact of hypericum (St.-John's-wort) given prenatally on cognition of mice offspring. Neurotoxicol. Teratol. 23 (6), 629–637.

Rizzo, P., Altschmied, L., Stark, P., et al., 2019. Discovery of key regulators of dark gland development and hypericin biosynthesis in St. John's wort (*Hypericum perforatum*). Plant Biotechnol. J. 17 (12), 2299–2312.

Russo, E., Scicchitano, F., Whalley, B.J., et al., 2014. *Hypericum perforatum*: pharmacokinetic, mechanism of action, tolerability, and clinical drug-drug interactions. Phytother. Res. 28 (5), 643–655.

Saddiqe, Z., Naeem, I., Maimoona, A., 2010. A review of the antibacterial activity of *Hypericum perforatum* L. J. Ethnopharmacol. 131 (3), 511–521.

Samadi, S., Khadivzadeh, T., Emami, A., et al., 2010. The effect of *Hypericum perforatum* on the wound healing and scar of cesarean. J. Altern. Complement. Med. 16 (1), 113–117.

Sanna, M.D., Ghelardini, C., Galeotti, N., 2017. St. John's wort potentiates anti-nociceptive effects of morphine in mice models of neuropathic pain. Pain Med. 18 (7), 1334–1343.

Sardella, A., Lodi, G., Demarosi, F., et al., 2008. *Hypericum perforatum* extract in burning mouth syndrome: a randomized placebo-controlled study. J. Oral Pathol. Med. 37 (7), 395–401.

Sarrell, E.M., Cohen, H.A., Kahan, E., 2003. Naturopathic treatment for ear pain in children. Pediatrics 111 (5 Pt 1), e574–e579.

Scalbert, A., Morand, C., Manach, C., et al., 2002. Absorption and metabolism of polyphenols in the gut and impact on health. Biomed. Pharmacother. 56 (6), 276–282.

Schempp, C.M., Pelz, K., Wittmer, A., et al., 1999. Antibacterial activity of hyperforin from St John's wort, against multiresistant *Staphylococcus aureus* and gram-positive bacteria. Lancet 353 (9170), 2129.

Schmidt, M., Butterweck, V., 2015. The mechanisms of action of St. John's wort: an update. Wien. Med. Wochenschr. 165 (11–12), 229–235.

Schrader, E., 2000. Equivalence of St John's wort extract (Ze 117) and fluoxetine: a randomized, controlled study in mild-moderate depression. Int. Clin. Psychopharmacol. 15 (2), 61–68.

Schule, C., Baghai, T., Ferrera, A., et al., 2001. Neuroendocrine effects of *Hypericum* extract WS 5570 in 12 healthy male volunteers. Pharmacopsychiatry 34 (Suppl 1), S127–S133.

Schule, C., Baghai, T., Sauer, N., et al., 2004. Endocrinological effects of high-dose *Hypericum perforatum* extract WS 5570 in healthy subjects. Neuropsychobiology 49 (2), 58–63.

Schulz, H.U., Schurer, M., Bassler, D., et al., 2005a. Investigation of the bioavailability of hypericin, pseudohypericin, hyperforin and the flavonoids quercetin and isorhamnetin following single and multiple oral dosing of a *Hypericum* extract containing tablet. Arzneim. Forsch. 55 (1), 15–22.

Schulz, H.U., Schurer, M., Bassler, D., et al., 2005b. Investigation of pharmacokinetic data of hypericin, pseudohypericin, hyperforin and the flavonoids quercetin and isorhamnetin revealed from single and multiple oral dose studies with a *Hypericum* extract containing tablet in healthy male volunteers. Arzneim. Forsch. 55 (10), 561–568.

Schwarz, U.I., Buschel, B., Kirch, W., 2003. Unwanted pregnancy on self-medication with St John's wort despite hormonal contraception. Br. J. Clin. Pharmacol. 55 (1), 112–113.

Scotti, F., Lobel, K., Booker, A., et al., 2018. St. John's wort (*Hypericum perforatum*) products—how variable is the primary material? Front. Plant Sci. 9, 1973.

Seifritz, E., Hatzinger, M., Holsboer-Trachsler, E., 2016. Efficacy of *Hypericum* extract WS® 5570 compared with paroxetine in patients with a moderate major depressive episode—a subgroup analysis. Int. J. Psychiatry Clin. Pract. 20 (3), 126–132.

Shelton, R.C., Keller, M.B., Gelenberg, A., et al., 2001. Effectiveness of St John's wort in major depression: a randomized controlled trial. JAMA 285 (15), 1978–1986.

Siepmann, M., Krause, S., Joraschky, P., et al., 2002. The effects of St John's wort extract on heart rate variability, cognitive function and quantitative EEG: a comparison with amitriptyline and placebo in healthy men. Br. J. Clin. Pharmacol. 54 (3), 277–282.

Singer, A., Schmidt, M., Hauke, W., et al., 2011. Duration of response after treatment of mild to moderate depression with *Hypericum* extract STW 3-VI, citalopram and placebo: a reanalysis of data from a controlled clinical trial. Phytomedicine 18 (8–9), 739–742.

Soleymani, S., Bahramsoltani, R., Rahimi, R., et al., 2017. Clinical risks of St John's Wort (*Hypericum perforatum*) co-administration. Expert Opin. Drug Metab. Toxicol. 13 (10), 1047–1062.

Solomon, D., Ford, E., Adams, J., et al., 2011. Potential of St John's wort for the treatment of depression: the economic perspective. Aust. N. Z. J. Psychiatry 45 (2), 123–130.

Solomon, D., Adams, J., Graves, N., 2013. Economic evaluation of St. John's wort (*Hypericum perforatum*) for the treatment of mild to moderate depression. J. Affect. Disord. 148 (2–3), 228–234.

Stage, T.B., Pedersen, R.S., Damkier, P., et al., 2015. Intake of St John's wort improves the glucose tolerance in healthy subjects who ingest metformin compared with metformin alone. Br. J. Clin. Pharmacol. 79 (2), 298–306.

Stevinson, C., Ernst, E., 2000. A pilot study of *Hypericum perforatum* for the treatment of premenstrual syndrome. BJOG 107 (7), 870–876.

Sun, X., Wang, T., Cai, D., et al., 2020. Cytokine storm intervention in the early stages of COVID-19 pneumonia. Cytokine Growth Factor Rev. 53, 38–42. Available from: 10.1016/j.cytogfr.2020.04.002.

Szegedi, A., Kohnen, R., Dienel, A., et al., 2005. Acute treatment of moderate to severe depression with *Hypericum* extract WS 5570 (St John's wort): randomised controlled double blind non-inferiority trial versus paroxetine. BMJ 330 (7490), 503. Available from: 10.1136/bmj.38356.655266.82.

Tang, J., Colacino, J.M., Larsen, S.H., et al., 1990. Virucidal activity of hypericin against enveloped and non-enveloped DNA and RNA viruses. Antivir. Res. 13 (6), 313–325.

Tanideh, N., Namazi, F., Andisheh Tadbir, A., et al., 2014. Comparative assessment of the therapeutic effects of the topical and systemic forms of *Hypericum perforatum* extract on induced oral mucositis in golden hamsters. Int. J. Oral Maxillofac. Surg. 43 (10), 1286–1292.

Theodossiou, T.A., Hothersall, J.S., De Witte, P.A., et al., 2009. The multifaceted photocytotoxic profile of hypericin. Mol. Pharm. 6 (6), 1775–1789.

Tocci, N., Perenzoni, D., Iamonico, D., et al., 2018. Extracts from *Hypericum hircinum* subsp. *majus* exert antifungal activity against a panel of sensitive and drug-resistant clinical strains. Front. Pharmacol. 9, 382. Available from: 10.3389/fphar.2018.00382.

Trautmann-Sponsel, R.D., Dienel, A., 2004. Safety of *Hypericum* extract in mildly to moderately depressed outpatients: a review based on data from three randomized, placebo-controlled trials. J. Affect. Disord. 82 (2), 303–307.

Ulbricht, C., Barrette, E.P., Basch, E., et al., 2010. A thorough analysis of the safety profile of St. John's wort by the natural standard research collaboration. Nat. Med. J. 2 (2), 1–11.

Vakili, F., Mirmohammadaliei, M., Montazeri, A., et al., 2018. Impact of *Hypericum perforatum* ointment on perineal pain intensity following episiotomy: a randomized placebo-controlled trial. J. Caring Sci. 7 (4), 205–211.

Valvassori, S.S., Borges, C., Bavaresco, D.V., et al., 2018. *Hypericum perforatum* chronic treatment affects cognitive parameters and brain neurotrophic factor levels. Braz. J. Psychiatr. 40 (4), 367–375.

Verjee, S., Brugger, D., Abdel-Aziz, H., et al., 2015. Permeation characteristics of hypericin across Caco-2 monolayers in the absence or presence of quercitrin—a mass balance study. Planta Med. 81 (12–13), 1111–1120.

Verjee, S., Kelber, O., Kolb, C., et al., 2019. Permeation characteristics of hypericin across Caco-2 monolayers in the presence of single flavonoids, defined flavonoid mixtures or hypericum extract matrix. J. Pharm. Pharmacol. 71 (1), 58–69.

Vissiennon, C., Nieber, K., Kelber, O., et al., 2012. Route of administration determines the anxiolytic activity of the flavonols kaempferol, quercetin and myricetin—are they prodrugs? J. Nutr. Biochem. 23 (7), 733–740.

Vitiello, B., Shader, R.I., Parker, C.B., et al., 2005. Hyperforin plasma level as a marker of treatment adherence in the National Institutes of Health Hypericum Depression Trial. J. Clin. Psychopharmacol. 25 (3), 243–249.

Warren, M.B., Cowen, P.J., Harmer, C.J., 2019. Subchronic treatment with St John's wort produces a positive shift in emotional processing in healthy volunteers. J. Psychopharmacol. 33 (2), 194–201.

Wentworth, J.M., Agostini, M., Love, J., et al., 2000. St John's wort, a herbal antidepressant, activates the steroid X receptor. J. Endocrinol. 166 (3), R11–R16.

Westerhoff, K., Kaunzinger, A., Wurglics, M., et al., 2002. Biorelevant dissolution testing of St John's wort products. J. Pharm. Pharmacol. 54 (12), 1615–1621.

Whitehead, A.J., Nelson, N.W., Brame, L.S., et al., 2019. Endemic North American plants as potentially suitable agents for wound cleaning under resource scarce conditions. Wilderness Environ. Med. 30 (4), 401–406.

WHO, 2003. WHO Monograph on Selected Medicinal Plants, Vol. 2. World Health Organization (WHO), Geneva, <https://apps.who.int/medicinedocs/en/d/Js4927e/16.html>.

WHO, 2010. Who Monographs on Medicinal Plants Commonly Used in The Newly Independent States (NIS). World Health Organization (WHO), Geneva.

WHO, 2013. WHO Traditional Medicine Strategy: 2014–2023. World Health Organization, Geneva, <https://www.who.int/medicines/publications/traditional/trm_strategy14_23/en/>.

WHO. (2020). <http://www.euro.who.int/en/health-topics/noncommunicable-diseases/pages/news/news/2012/10/depression-in-europe/depression-definition>.

Woelk, H., Burkard, G., Grunwald, J., 1994. Benefits and risks of the *Hypericum* extract LI 160: drug monitoring study with 3250 patients. J. Geriatr. Psychiatry Neurol. 7 (Suppl 1), S34–38.

Wolfle, U., Seelinger, G., Schempp, C.M., 2014. Topical application of St. John's wort (*Hypericum perforatum*). Planta Med. 80 (2–3), 109–120.

Wurglics, M., Westerhoff, K., Kaunzinger, A., et al., 2001. Batch-to-batch reproducibility of St. John's wort preparations. Pharmacopsychiatry 34 (Suppl 1), S152–S156.

Wurglics, M., Schubert-Zsilavecz, M., 2006. *Hypericum perforatum*: a 'modern' herbal antidepressant: pharmacokinetics of active ingredients. Clin. Pharmacokinet. 45 (5), 449–468.

Yang, X.W., Grossman, R.B., Xu, G., 2018. Research progress of polycyclic polyprenylated acylphloroglucinols. Chem. Rev. 118 (7), 3508–3558.

Yao, Y., Kang, T., Jin, L., et al., 2019. Temperature-dependent growth and hypericin biosynthesis in *Hypericum perforatum*. Plant Physiol. Biochem. 139, 613–619.

Yechiam, E., Ben-Eliezer, D., Ashby, N.J.S., et al., 2019. The acute effect of *Hypericum perforatum* on short-term memory in healthy adults. Psychopharmacology (Berl.) 236 (2), 613–623.

Yilmaz, U., Kaya, H., Turan, M., et al., 2019. Investigation the effect of *Hypericum perforatum* on corneal alkali burns. Cutan. Ocul. Toxicol. 38 (4), 356–359.

Zahner, C., Kruttschnitt, E., Uricher, J., et al., 2019. No clinically relevant interactions of St. John's wort extract Ze 117 low in hyperforin with cytochrome P450 enzymes and P-glycoprotein. Clin. Pharmacol. Ther. 106 (2), 432–440.

Zeliou, K., Koui, E.M., Papaioannou, C., et al., 2020. Metabolomic fingerprinting and genetic discrimination of four *Hypericum* taxa from Greece. Phytochemistry 174, 112290.

Zhou, S., Lim, L.Y., Chowbay, B., 2004. Herbal modulation of P-glycoprotein. Drug Metab. Rev. 36 (1), 57–104.

Zirak, N., Shafiee, M., Soltani, G., et al., 2019. *Hypericum perforatum* in the treatment of psychiatric and neurodegenerative disorders: current evidence and potential mechanisms of action. J. Cell Physiol. 234 (6), 8496–8508.

Zobayed, S.M., Afreen, F., Kozai, T., 2005. Temperature stress can alter the photosynthetic efficiency and secondary metabolite concentrations in St. John's wort. Plant Physiol. Biochem. 43 (10–11), 977–984.

Chapter 41

Green tea

Robert W. Coppock and Margitta Dziwenka

Introduction

History, varieties, economics, and culture

Asian mountain tribes, before written history, used green tea (GT) as a nutraceutical. Recorded history shows that the medical use of GT is more than 5000 years old (Wheeler and Wheeler, 2004; Teschke et al., 2013). Historically, GT has been part of traditional Chinese medicine and is in the Buddhism ethos. King Shen Nong chronicled GT in ancient Chinese history by promoting it to maintain and improve the health of his subjects (Kapoor et al., 2013; Nakamura, 2017). Dating back through history, the use of GT rapidly spread through Eastern countries, and trading with Western countries spread its use worldwide. GT infusions and extracts are made from the leaves and twigs of *Camellia sinensis* (varieties *sinensis* and *assamica*) and there are numerous cultivars and hybrids of these varieties (Table 41.1). The genome of *C. sinensis* is being elucidated to provide insights into its lineage divergences and biochemical pathways (Wei et al., 2018; Xia et al., 2020a,b). The leaves or twigs of various varieties and cultivars of *C. sinensis* are used to manufacture hundreds of commercial teas under the general classifications of green, white, black, oolong, and kukicha (green twig). In 2016, more than 30 countries produced approximately 5.73 billion metric tons of dried tea, valued[1] in 2018 at US$52.1 billion (FAO, 2018). The predominant teas traded were classified as black and green (Zhang et al., 2019). GT comprises 15%–20% of the tea products sold and Western countries have the highest annual increase in consumption. The manufacturing differences in leaf teas are the postharvest processes that produce a plethora of teas with differing aromas, flavors, flavor stability, and phytochemicals (Fu et al., 2020; Zhang et al., 2020a). There are geographical, regional, and ethnic preferences for the tea products that are grown and manufactured using differing methods.

Manufacturing

The leaves of *C. sinensis*, at various stages of maturity, are plucked from the plant at different seasons of the year and processed into various teas and tea extracts (Wan et al., 2009; Kapoor et al., 2013; Lee et al., 2019; Zhang et al., 2019). To produce black and oolong teas, the leaves are withered/fermented (browning reaction due to enzymatic reactions); with oolong, the leaves have a shorter withering/fermentation process. In the manufacturing of black and oolong teas, the monomeric flavan-3-ols undergo polyphenol oxidase-linked oxidative polymerization resulting in the formation of bisflavanols, theaflavins, thearubigins, and other oligomer complexes. Black and GTs from identical raw products and infusion techniques have the same levels of methylxanthines. White tea is prepared from GT buds covered with silvery hairs. After collection, both white and GTs are immediately steamed or heat-dried (firing/pan roasting or frying) to prevent enzymatic degradation and oxidation of polyphenols (Teschke et al., 2013; Bhagwat et al., 2014). GTs are classified as steamed GT or pan-fried GT. The crude teas are refined by sifting, cutting, polishing, and blending and may have a unique physical appearance, flavors, and aromas (Fu et al., 2020; Zhang et al., 2019, 2020a).

Extraction

Teapot-scale infusions and industrial-scale extractions are the two major methods used to remove phytochemicals from GT. Water infusion is commonly used to brew a pot of tea. The objective of brewing a cup of tea is to infuse water with the soluble polyphenols, purine alkaloids (methylxanthines, namely caffeine and theophylline), aromas, flavors, and other substances from dried GT. The typical brewing method for a cup of GT is 55 mL of water at 60°C–80°C and 2 g of GT in an infuser that is placed

1. https://www.statista.com/statistics/326384/global-tea-beverage-market-size/.

Nutraceuticals. DOI: https://doi.org/10.1016/B978-0-12-821038-3.00041-0

TABLE 41.1 Other names of *Camellia sinensis* (ESCO, 2009).

Parameter	Names
Recognized scientific name	*Camellia sinensis* (L.) O. Kuntze
Other scientific names	*Camellia thea* Link, *Thea sinensis* L., *Thea viridis* L, *Theaphyla lanceolata* Raf., *Theaphyla laxa* Raf., *Theaphyla viridis* Raf., *C. sinensis* (L.) Kuntze var. *sinensis* (Engl.: China-Tea, small-leaved tea plant), *C. sinensis* (L.) Kuntze var. *assamica* (J.W. Mast.), Kitam [Syn. *Camellia assamica* (Mast.) Hung T. Chang], *Camellia theifera* Griff., *Thea assamica* J.W. Masters, *C. assamica* [J.W. (Mast.) W. Wight (Engl.: Assam-Tea, Ceylon-Tea, Indian-Tea, large-leaved tea plant)]

in hot water for 1–15 min (Astill et al., 2001). Loose-leaf extraction is more efficient than the use of tea bags. The smaller GT particles increase the surface area and have higher transfer rates of phytochemicals from the plant material to the infusion solvent. To avoid a bitter taste, recommended brewing temperatures for GT are generally lower than for fermented teas. The taste of the second brew of GT is considered by some individuals to be better than the first brew. The total solids extracted from GT, using package instructions, generally are 13.6%–30.4% mass/mass. The infusion of soluble solids is high during the first minute of infusion and then gradually decreases. The leaf-to-water ratio (mass/mass), infusion time, and water temperature affect the concentration of polyphenols and purine alkaloids (caffeine and theophylline) in the brew. Lower leaf-to-water ratio and increased water temperature increase the extraction efficiency. The purine alkaloids are extracted more efficiently than the polyphenols. The grade of the GT can also affect diffusion. Brewing methods will vary with different ethnic cultures and geographic regions. Commercial extraction is used to produce various GT extracts, modified GT extracts, and dried extract products (Banerjee and Chatterjee, 2015). The objective of making a GT extract is to concentrate the phytochemicals especially the flavonoids, methylxanthines, proteins, amino acids, aromas, and flavors. Commercial extraction methods, especially those to selectively concentrate certain phytochemical profiles, are often shrouded in trade secrets and patents protecting intellectual property. The historic and current intellectual property associated with modifying GT infusions and GT extracts is enormous (Wheeler and Wheeler, 2004; Banerjee and Chatterjee, 2015). Many differing methods are used in manufacturing GT extracts, especially modified dried products (Banerjee and Chatterjee, 2015; ESCO, 2009). These practices vary in the use of extraction solvents, ultrasound, high hydrostatic pressure, temperature including microwave heating, duration of steeping, pH, and other processing procedures. Extraction procedures may utilize supercritical liquids, for example, carbon dioxide, for extraction (Banerjee and Chatterjee, 2015; ESCO, 2009). Variations in the GT itself influence the extraction, namely leaf maturity, genetics of the tea, and climate and season of the year. GT extracts are established nutraceuticals and have complex chemistry.

Chemistry of green tea infusions and extracts

There are more than 2000 different chemical substances associated with GT. Leaves harvested at differing stages of maturity do not have the same phytochemistry because the phytochemistry of tea leaves is temporally changing with bud opening and leaf growth (Zhang et al., 2019). Also, the chemical composition of GT leaves varies with cultivar, macroclimate and microclimate, soil type, horticulture practices, seasonal timing of harvest, leaf-plucking practices, steaming or fire-drying practices, drying, and final tea grades (Lee et al., 2019; Zhang et al., 2019; Engelhardt, 2020). Fresh tea leaves are rich in polyphenols, especially flavan-3-ols and their dimers and trimers. Processing of tea leaves removes varying degrees of the bitter taste. The manufacturing steps include leaf plucking, the initial fixations (steaming or pan-frying/flame-drying), and final drying methods. Differing these procedures can alter the chemistry of GT leaves. Heating of GT leaves immediately after plucking inactivates the indigenous polyphenol oxidase enzymes, maintaining the polyphenols essentially in their leaf-on-the-plant chemical form (Tables 41.2–41.4).

The polyphenol compounds in GT infusions and extracts are a group of natural aromatic compounds, which are substituted with multiple hydroxyl groups and comprise up to 30%–40% (mass/mass) of the extraction solids (Harbowy and Balentine, 1997; ESCO, 2009; Zhang et al., 2019). The polyphenols and numerous other compounds including chlorophyll are responsible for the appearance and flavor of GT. The polyphenols and other substances contribute to the nutraceutical uniqueness of GT (Table 41.4). The polyphenols are the most studied chemicals in GT for pharmaco-physiological-biochemical effects. The flavonoids are a subgroup of the polyphenols that predominate in GT. Flavonoids are secondary plant metabolites chemically containing a phenylbenzopyran structure, and they are more abundant in buds than in older leaves (Tables 41.3 and 41.4) (Harbowy and Balentine, 1997; Zhang et al., 2019; Engelhardt, 2020). The catechins are the major flavan-3-ols in tea and can be more than 30% (mass/mass) of the dissolved solids in GT infusions and extracts (Lin et al., 2003; ESCO, 2009). The catechins are nonplanar molecules and therefore exist as the *trans*-catechins and *cis*-epicatechins isomers (Rusak et al., 2008). These molecules have two optical isomers, namely (+)-catechin, (−)-catechin, (−)-epicatechin, and (+)-epicatechin. In addition, (−)-catechin esters can be formed with gallic acid. These are (−)-catechin-3-gallate, (−)-epicatechin-3-gallate, (−)-epigallocatechin-3-gallate, and (−)-gallocatechin-3-gallate. The six catechins present in GT known to possess many of the phytopharmacological properties are (−)-epigallocatechin gallate, (−)-epicatechin gallate, (−)-epigallocatechin, (+)-gallocatechin gallate, (−)-epicatechin, (+)-gallocatechin, and (+)-catechin. Various flavonols (up to 4% mass/mass) and flavones (in traces) are also present in tea leaves. The main flavonols in tea are conjugates of quercetin and kaempferol, with conjugating moiety varying from monoglycosides to diglycosides and triglycosides (Del Rio et al., 2004). These are also known to have pharmacological activities (Calderon-Montano et al., 2011; Zhang et al., 2014, 2019; Hu et al., 2018). Other related compounds found in tea are gallic and caffeic acids, and coumaric (Rusak et al., 2008). GT extract also contains

TABLE 41.2 Groups of chemicals in green tea (*Camellia sinensis*) (Harbowy and Balentine, 1997; ESCO, 2009).

Chemical groups	Percent (mass/mass)
Catechins	30
Simple polyphenols	2
Flavonols	2
Other polyphenols	6
Theanine (amino acid)	3
Amino acids	3
Peptides/proteins	6
Sugars	7
Caffeine (also called theine or teine)	3
Other methylxanthines	<1

TABLE 41.3 Polyphenol levels in dried green tea leaves (*Camellia sinensis*) (Bhagwat et al., 2014).

Group	Flavonoid	Mean (mg/g) ± std. error	Minimum–maximum
Flavan-3-ols	(−)-Epicatechin	791.46 ± 3.3	190–2000
	(−)-Epicatechin 3-gallate	1701.59 ± 16.2	500–4630
	(−)-Epigallocatechin	1695.02 ± 0.66	100–5440
	(−)-Epigallocatechin 3-gallate	8294.91 ± 91.8	1600–20320
	(+)-Catechin	33.00 ± 1.59	0.00–100
	(+)-Gallocatechin	258.11 ± 80.69	69.46–446.76
	Theaflavin	4.10 ± 0.39	2.49–6.25
	Theaflavin-3,3′-digallate	1.08 ± 0.31	0.00–2.39
	Theaflavin-3′-gallate	0.44 ± 0.13	0.00–0.99
	Theaflavin-3-gallate	1.17 ± 0.34	0.00–2.74
	Thearubigins	131.91 ± 114.23	0.00–527.63
Flavones	Apigenin	0.17 ± 0.07	0.00–0.5
	Luteolin	0.17 ± 0.07	0.00–0.5
Flavonols	Kaempferol	151.90 ± 6.08	77.61–331.00
	Myricetin	108.25 ± 1.55	52.00–159.00
	Quercetin	255.55 ± 0.46	140.00–405.00

TABLE 41.4 Polyphenol levels in green tea infusions (ESCO, 2009; Bhagwat et al., 2014).

Group	Flavonoid	Mean (mg/g) ± std error	Minimum–maximum
Flavan-3-ols	(−)-Epicatechin	8.29 ± 0.49	1.90–26.00
	(−)-Epicatechin 3-gallate	19.73 ± 2.76	4.30–139.60
	(−)-Epigallocatechin	16.71 ± 1.41	1.00–54.40
	(−)-Epigallocatechin 3-gallate	77.81 ± 6.97	2.31–203.20
	(+)-Catechin	2.55 ± 1.53	0.00–44.40
	(+)-Gallocatechin	1.54[a]	
	Theaflavin	0.50[a]	
	Theaflavin-3,3′-digallate	Vz	
	Theaflavin-3′-gallate	Vz	
	Theaflavin-3-gallate	Vz	
	Thearubigins	1.08[a]	
Flavones	Apigenin	0.17[a]	
	Luteolin	0.17[a]	0.00–0.5
Flavonols	Kaempferol	1.42 ± 0.22	0.67–3.31
	Myricetin	1.10 ± 0.11	0.52–1.60
	Quercetin	2.69 ± 0.26	1.40–4.10

Vz, values approaching zero and considered to be unreliable.
[a]*Number of samples to low to establish the std error.*

condensed and hydrolyzed tannins (Forester and Lambert, 2011; Engelhardt, 2020). The brews and extracts of GT brew contain phenolic acids and gallic acid, and its esters are also present. Some flavonoids can exist in plants as glucosides and also may be attached to sugars other than glucose. The primary purine alkaloids present in tea by descending levels of quantity are caffeine (2.9%–4.2%, 1,3,7-trimethylxanthine), theophylline, and theobromine (ESCO, 2009; Zhang et al., 2020a; Engelhardt, 2020). Infusions from fresh tea leaves contain 3.11%–4.09% (mass/mass) caffeine (Astill et al., 2001). During manufacturing, the purine alkaloids show a large increase with drying and then remain stable. The caffeine levels in green and black teas vary between cultivars. L-Theanine, a nonprotein amino acid that has phytopharmaceutical activity, is found in home-brewed GT infusions at ca. 6.56 mg/g (Boros et al., 2016; Williams et al., 2019).

Taste, aromas, and tea cream

Volatile chemicals in GT extract are important for GT aroma (Table 41.5). The major groups of chemicals, considered important for taste and aroma, are the aliphatic aldehydes, aliphatic alcohols, aromatic alcohols, aliphatic aldehydes, aromatic aldehydes, aromatic ketones, and other aromatic compounds (Lee et al., 2013, 2014; Zhang et al., 2020a). Cooling of tea extract can produce a "cloud" called "tea cream," which is a colloid of small insoluble microscopic particles that remain in suspension, that can be seen on the surface of a cup of tea. Caffeine, polyphenols, carbohydrates, lipids, proteins, mineral ions, and other compounds comprise the chemical complex forming the microscopic colloidal particles. The methods used for preparing the infusion can influence the formation of solids and tea creams in GT extract (Lin et al., 2015; Xu et al., 2012, 2015). The microscopic tea cream colloids can coalesce and form larger particles that precipitate and thereby reduce the soluble antioxidants in GT by 56% (Kim and Talcott, 2012). Most of the tea cream colloids and sediments go into solution when the infusion or extract is heated. Some mineral precipitates can remain insoluble during reheating. The oxalates in GT extract form salts with calcium, magnesium, and other mineral ions and form insoluble precipitates.

Stability of green tea extracts

The United States government and other governments have placed supplements under good manufacturing practices or equivalent regulations (Bhagwat et al., 2014;

TABLE 41.5 Groups of chemicals providing aroma to green tea.

Group of volatile compounds	Description of aroma sensation
Aliphatic alcohols	Green, sweet, balsamic, Jasmine, bergamot oil, butter-like
Aromatic alcohols	Floral, rose, lilac
Aliphatic aldehydes	Almond, herbal, green leaf, citrus
Aromatic aldehydes	Hyacinth, bitter almond
Ketones	Jasmine, mint-like, green wood-like
Aromatic compounds	Sweet clover, petroleum-like, solvent-like
Acids	Green-like

Source: Adopted from Lee, J., Chambers, D.H., Chambers, E., et al., 2013. Volatile aroma compounds in various brewed green teas. Molecules, 18(8), 10024–10041.

Engelhardt, 2020). The range of total polyphenols, catechins, and caffeine in marked basket samples of GT supplements was 3.6%–95.8%, 3.8%–70.2%, and 0.8%–11.2%, respectively. The stability of GT extract has been studied (Abd-Elsalam et al., 2014; Abourashed et al., 2014). The accelerated stability conditions in the cabinet were 40°C and 75% relative humidity. Reduction of epigallocatechin gallate levels at 1, 3, and 6 months was 24%, 28%, and 52%, respectively. For solutions of tea polyphenols, the shelf-life is improved by lowering the pH (Zeng et al., 2017).

Pharmacokinetics and biotransformation

Pharmacokinetics

GT and GT extracts contain a large number of phytochemicals. It is estimated that 8%–17% of the catechins in GT are bioavailable in the small intestine (Liu et al., 2018). The low bioavailability is attributed to low absorption by the enterocytes. There is evidence that the multidrug resistance protein transports some catechins, especially the galloylated catechins. The pharmacokinetics of some pharmacologically active chemicals in GT have been elucidated (Stalmach et al., 2009; Yashin et al., 2012; Clifford et al., 2013). For 48 h before ingesting a Choladi GT extract,[2] volunteers were on a diet low in flavonoids and other phenolic compounds (Stalmach et al., 2009). Ten volunteers were fasted overnight and given 500 mL of Choladi GT extract. Blood samples were taken at 0, 0.5, 1, 2, 4, 6, 8, and 24 h before and after ingestion of tea extract. Urine was also collected at 0–5, 5–12, and 12–24 h after ingestion of tea. The catechins levels in the Choladi GT extract consumed were: 257 μM (micromolar) (−)-epigallocatechin; 230 μM (−)-epigallocatechin-3-gallate; 58 μM (−)-epicatechin; 49 μM (−)-epicatechin-3-gallate; 36 μM (+)-gallocatechin; 9 μM (+)-catechin; and 8 μM (+)-gallocatechin-3-gallate. The total catechins were 648 μM. No catechins were detected in plasma and urine at the 0.0 h sampling. At 24 h, metabolites were identified in plasma. These were the glucuronide, sulfate, and methyl-sulfate metabolites of (epi) catechin, methyl-glucuronide, and methyl-sulfate metabolites of (epi)gallocatechin and the tea flavan-3-ols (−)-epicatechin-3-gallate and (−)-epigallocatechin-3-gallate. (Epi)gallocatechin-O-glucuronide reached its highest concentration (concentration max, C_{max}) of 126 nM (nanomolar) at 2.2 h, the C_{max} of (−)-epicatechin-3′-O-glucuronide of 29 nM at 1.7 h, (−)-epicatechin-3-gallate had a C_{max} of 55 nM at 1.9 h, and (−)-epigallocatechin-3-gallate had a C_{max} of 25 nM at 1.6 h. Only trace amounts of catechins were present in plasma at 8 h. None of the parent flavin-3-ol compounds was present in urine. Fifteen metabolites of (epi)catechin and (epi)gallocatechin were identified in urine. The two dominant forms were 4′-O-methyl-(epi)gallocatechin-O-sulfates and O-methyl-(epi)catechin-O-sulfates. Most of the compounds were eliminated in the first 12 h.

In another study on a GT beverage,[3] eight volunteers, after a 24-h fast, consumed 20 mg of dried GT/kg body weight (bwt) dissolved in 200 mL of warm water (Lee et al., 2002). Levels of catechins in the GT solids were 13.9% (−)-epigallocatechin-3-gallate, 11.0% (−)-epigallocatechin, and 3.2% (−)-epicatechin. The GT solids were decaffeinated. The major catechins detected in plasma were (−)-epigallocatechin, (−)-epicatechin,

2. Nestle Choladi; Tamil Nadu, IN.
3. Thomas J. Lipton Co.; Englewood Cliffs, NJ.

(−)-epigallocatechin-3-gallate, and (−)-epicatechin gallate, and their 4′-O-methyl-epigallocatchen, (−)-5-(3′,4′,5′-trihydroxyphenyl)-γ-valerolactone, and (−)-5-(3′,4′-dihydroxyphenyl)-γ-valerolactone metabolites. The estimated half-lives ($t_{1/2}$) of (−)-epigallocatechin gallate, (−)-epigallocatechin, and (−)-epicatechin were 3.4 ± 0.3, 1.7 ± 0.4, and 2.0 ± 0.4 h, respectively. These compounds were found in higher concentrations in urine except for (−)-epigallocatechin gallate and (−)-epigallocatechin. In plasma, the (−)-epigallocatechin gallate was in unconjugated form and (−)-epigallocatechin and (−)-epicatechin were in conjugated forms. The kinetics of GT catechins was studied in 16 volunteers (4 men and 12 women, 35−50 years of age) (Fung et al., 2013). The treatment was a single dose of 200 mL of 1.5% (w/v) of freshly brewed tea (Longjing "pre-rain" GT[4]) or 150 mL of 1.5% (w/v) infusion ingested twice per day for 7 days (crossover design with a 4-week washout). The total daily ingestion of (−)-catechin, (−)-epicatechin, (−)-epicatechin gallate, (−)-epigallocatechin, and (−)-epigallocatechin gallate was the same in the single dose study and the 7-day study. The mean levels in plasma at 1 h after dosing were 310 ± 117 nM/L free epigallocatechin gallate (no conjugated forms), 192 ± 67 nM/L epigallocatechin (30% free), and 134 ± 51 nM/L epicatechin gallate (25% conjugated). Plasma levels mirrored the concentrations in the GT brew. After 7 days of twice per day ingestion of 150 mL GT brew, the fasting plasma levels were 80 nM/L epigallocatechin gallate (baseline of 15 nM/L) and 120 nM/L epicatechin gallate (baseline of 40 nM/l); both had 90% conjugated forms and 10 nM/L total epigallocatechin.

Stalmach et al. (2010) studied the kinetics of GT brew in human volunteers with previous surgery for ileostomy construct. The GT infusion was prepared by adding boiling water to 3 g GT[5] and brewing for 3 min. The GT infusion contained 634 mmol of flavan-3-ols. The excreted ileal fluid feces contained 70% of the ingested flavan-3-ols with 33% excreted as parent compounds and 37% as 23 metabolites. The main metabolites excreted into the small intestinal tract were the O-linked sulfates and methyl-sulfates of epicatechin and epigallocatechin. Based on this study it is logical to assume that some flavan-3-ols enter the large intestine and are subject to microbial biochemical processes. Stalmach et al. (2010) also showed that 16 metabolites of epicatechin and epigallocatechin are primarily the sulfated, glucuronidated, and methylated forms. Auger et al. (2008), using human volunteers with an ileostomy, studied the pharmacokinetics of a standardized GT extract.[6] After oral administration of 200 mg polyphenon E, the average percentages of the flavan-3-ols dose recover in distal ileal materials at 0−24 h were 27 ± 2% for (−)-epicatechin and (−)-epigallocatechin, and 59 ± 3% for (−)-epicatechin gallate and (−)-epigallocatechin gallate. Polyphenon E incubation (in vitro) in gastric juice and ileal fluid showed that the flavan-3-ols are chemically stable in the stomach and small intestine. The differences suggest that (−)-epicatechin and (−)-epigallocatechin are absorbed from the small intestine more efficiently than their 3-O-galloylated analogs. The sulfates of (epi)catechin and methyl-(epi)catechin metabolites were recovered in the urine. The 3-O-galloylated flavan-3-ols or their metabolites were not detected. Urinary excretion of (epi)catechin metabolites was 47 ± 12% of the ingested (epi)catechin. Although (epi)gallocatechin metabolites were excreted in smaller quantities, 5.7 ± 1.9 mM, this nonetheless represented 26 ± 9% of (epi)gallocatechin intake. These results show that both (epi)catechins and (epi)gallocatechins are bioavailable. Studies in volunteers on decaffeinated modified GT extract (Polyphenon E, 60% epigallocatechin gallate) showed that after 4 weeks of administering a capsule powder containing 800 mg of epigallocatechin gallate/day for 4 weeks there was more than a 60% increase in systemic bioavailability (Chow et al., 2003). Administering 400 mg twice per day did not increase the bioavailability of epigallocatechin gallate and the mechanism was not elucidated. GT cetaceans have noncovalent affinity for human serum albumin (Zinellu et al., 2014). Catechins with a galloyl moiety (epicatechin gallate and epigallocatechin gallate) have a higher noncovalent affinity for human serum albumen than catechins lacking the galloyl moiety. The order of serum albumen affinity is: epicatechin < epigallocatechin < epicatechin gallate < epigallocatechin gallate. With the exception of epigallocatechin gallate, the majority of catechins found in human plasma are in the sulfated or glucuronide forms. There are conflicting reports on the effect of milk on the absorption of GT catechins.

Hollman et al. (2001) reported no effect and Egert et al. (2013) reported a reduction in the absorption of catechins. Current studies suggest that pathophysiological changes in type II diabetes influence the pharmacokinetics of absorbed phytopolyphenols by competition between glucose and polyphenols for binding to plasma proteins, diabetes-induced glycation of plasma proteins weakens the noncovalent affinities of plasma proteins and polyphenols, and the clearance of phytopolyphenols is increased in people with type II diabetes. The particle size of GT extracts affects in vitro digestion, and is likely a surface

4. Ying Kee Tea House; HK.
5. Indonesian, Tetley; Middlesex, UK
6. Polyphenon E, Mitsui Norin Co. Ltd; Minato, JP.

area phenomenon (Shu et al., 2019). Male dogs have been reported to have higher absorption of orally administered epigallocatechin gallate than female dogs and plasma levels reach steady state in week 12 of oral administration (Isbrucker et al., 2006a). Fasted dogs absorb more epigallocatechin gallate than fed dogs, and it was considered a factor in the toxicity of dehydrated GT extracts. All of the human studies on the pharmacokinetics and metabolism of GT catechins have shown considerable variation between individuals (van het Hof et al., 1998; Lee et al., 2002; Auger et al., 2008; Stalmach et al., 2009, 2010; Fung et al., 2013).

Biotransformation

Biotransformation of the catechins occurs. Catechol-O-methyltransferase (COMT) forms the O-methylation of the major catechins found in GT. There is some evidence that enterocytes methylate the catechins and this pathway forms considerably less methylated products than the hepatic methylation pathway. These and other findings show the phase II synthetic pathway is important in the metabolism of GT catechins. The uridine diphosphate-glucuronosyltransferase and sulfotransferases pathways form the glucuronide and sulfate products of catechins, respectively (Sang et al., 2011). There was rapid renal excretion of epigallocatechin and epicatechin, and negligible excretion of epigallocatechin gallate and epicatechin gallate. This study also shows that regular consumption of GT causes the levels of epigallocatechin gallate and epicatechin gallate to increase in fasting plasma.

Gut microbiome metabolism of green tea phytochemicals

The diverse microbiome in the gut has amazing biochemical activity. It can hydrolyze conjugated substances including glycosides, glucuronides, sulfates, amides, esters, and lactones. Other enzymatic activities include isomerization, demethylation, nonaromatic ring cleavage, reduction, and decarboxylation. Enzymes expressed in the microbiome include rhamnosidases, glucuronidases, glucosidases, sulfatase, and esterase. The microbiome biotransformation of catechins can be grouped by processes (Liu et al., 2018). These reactions are galloyl ester hydrolysis, opening of the C-ring, and subsequent reactions including dehydroxylation, lactonization, decarboxylation, and oxidation. The valerolactones, phenolic acids, and other metabolites of catechins are absorbed from the large intestine into the hepatic portal system. The GT catechins may modulate the gut microbiome. The phytopharmacology of the gut-formed products is most likely important for the nutraceutical effects of GT.

Pharmacology of green tea infusions and green tea extracts

There is evidence to show that the catechins in GT infusions and extracts are effective antioxidants in neutralizing free radicals (Braicu et al., 2013). The estimated consumption of the catechins in GT infusions is given in Table 41.6. The relative order of antioxidant activity is: (−)-epicatechin gallate > (−)-epigallocatechin gallate > (−)-epigallocatechin > (−)-epicatechin, > (−)-catechin (Mitscher et al., 1997). The flavonol, kaempferol, is also used to index the antioxidant beneficial effects of fruits, vegetables, herbs, teas, and other nutraceuticals (Calderon-Montano et al., 2011). The catechins likely have dual antioxidant and prooxidant activities depending on the dose, duration of exposure, and interactions with other substances in foods. In vitro studies have shown that different body cell types can have differing pharmacological responses to the catechins (Mitscher et al., 1997; Musial et al., 2020). At high concentrations, the

TABLE 41.6 Estimated intake of catechins from the consumption of traditional green tea infusion (ESCO, 2009).

Source[a]	Intake[b]	EC[c]	ECG[d]	EGC[e]	EGCG[f]	Total[g]
China	362	17.0	15.9	59.0	95.2	186.4
	1097	51.6	48.3	178.8	288.5	565.0
Japan	362	34	21.4	103.9	147.7	307.4
	1097	103.1	64.7	314.8	447.6	931.4

[a]*Country in which the tea is grown.*
[b]*Mean intake in g dry infusion/person/day and 1097 g/person/day is the 95th percentile.*
[c]*(−)-Epicatechin (EC).*
[d]*(−)-Epicatechin-3-gallate (ECG).*
[e]*(−)-Epigallocatechin (EGC).*
[f]*(−)-Epigallocatechin-3-gallate (EGCG).*
[g]*Total intake of catechins.*

catechins inhibit hepatocyte cytosolic COMT (Chen et al., 2005). Catechins with the galloyl-type D-ring [(−)-epigallocatechin-3-gallate, epicatechin-3-gallate] are the most potent inhibitors. Catechins without the D-ring have a decrease of two to three orders of magnitude in inhibition activity. The primary activity of COMT is to deactivate biologically active catechols.

Green tea and metabolic syndrome

Antioxidant activity

The health benefits of GT brew and GT extracts are primarily attributed to their antioxidant activity. Animals have indigenous biochemical pathways to remove reactive oxygen species (ROS), for example, super oxide anion (O_2^-), hydroxyl radicals (OH^-), and hydrogen peroxide (H_2O_2). When these mechanisms are overwhelmed by exposure to toxic substances or physiological stress, reparable and irreparable cellular, tissue, and organ damage can occur. The phenolic compounds in GT can reduce the free radicals and their biotransformation pathways can also produce free radicals (Zhang et al., 2014; Yan et al., 2020). There is evidence that GT brew and extracts have multicomponent in vivo mechanisms. The catechins in GT have low bioavailability and their contributions are likely at very low tissue levels. GT brew also contains prooxidants, and these can autoxidize to form H_2O_2 (Forester and Lambert, 2011; Ho et al., 2013). The antioxidant activity of GT brew and solutions of GT extract are measured in the laboratory. The ferric reducing/antioxidant power (FRAP) assay is one method that is used. The catechol moiety in GT polyphenols in the presence of ferrous ions (Fe^{2+}) or cupric ions (Cu^{2+}), also known as the Fenton conditions, produces the OH^-. These reactions can occur in vivo, are not well understood, and may be of importance in the pharmacology of GT. The oxygen radical absorption capacity assay is also used. The Folin-Ciocalteu method is also used to estimate the total phenolic content of GT. These assays, although appearing on labels, do not give information regarding the in vivo antioxidant capacity of the product. Cell culture procedures have been developed to measure protection of GT polyphenols induced by a variety of chemical agents and irradiation with ultraviolet light. Redox-sensitive transcription factors are upregulated by tea phenolics (Yan et al., 2020). The prooxidant actions of GT likely work by triggering the antioxidant response element in cells that upregulate production of cryoprotective products (Ho et al., 2013). The triggers may be the oral dose of phytocompounds contained in GT. A human study using GT brewed in hot water (tea bags containing 1.5 g "pre-rain" Loong-Cheng) for a 12-week exposure period showed that GT has a genoprotective effect (Choi et al., 2015). However, the effect may be due to phytohormesis (the effect was induction of redox-sensitive transcription). This observation may explain the efficacy of GT extracts and catechins at nanomolar plasma levels. GT extract was studied in postmenopausal volunteers for its postprandial antiglycemic and antioxidant effects (Takahashi et al., 2014). No difference between the treatment and control groups was observed for the postprandial increase in oxidative stress. GT extract did initially increase postprandial plasma thioredoxin levels, an antioxidant upregulated by epigallocatechin-3-gallate. After 4 weeks, the thioredoxin levels were the same in the treatment and control groups. GT extract did not change the postprandial increase in reactive oxygen metabolites. A study involving humans showed that consumption of 400 mL of GT brew after 24 h of fasting increased the antioxidant activity of plasma (FRAP assay) by 4% at 40 min after the brew was ingested (Benzie et al., 1999).

Green tea infusions and extracts, and diabetes mellitus

Diabetes mellitus is a disease condition of elevated blood glucose (hyperglycemia) due to a lack of insulin production by the pancreatic β islet cells or insulin resistance where the endocrine function of insulin to reduce blood glucose is compromised. Skeletal muscles, adipose tissue, and the liver are the organ/tissues most affected by insulin resistance. Abnormal insulin signaling results in the elevation of serum glucose, especially after ingestion of glycemic foods (high in starch and sugars). Causes of insulin resistance are multifactorial and risk factors are obesity, metabolic syndrome, and abnormal serum lipids. In types I and II diabetes mellitus there is a requirement for medical intervention to regulate blood glucose and thereby prevent or diminish the risk for degenerative microvascular complications inclusive of loss of appendages, the development of retinopathies, and degenerative kidney disease. The uptake, utilization, and storage of glucose by the liver, muscle, and adipose tissue is compromised in type II diabetes (Kotze-Horstmann and Sadie-Van Gijsen, 2020). The pathophysiology of type II diabetes is insulin, in varying degrees, losing its glucose-regulatory actions, resulting in diminished transmembrane transport of glucose into insulin-targeted tissues. When insulin is attached to the membrane receptor, it activates a cascade of signaling events, including the insulin phosphoinositide-3-kinase−protein kinase B/Akt (PI3K-PKB/Akt) pathway. Phosphorylation of tyrosine residues activates IRS. If IRS is phosphorylated on the serine residues, the enzymatic function of this protein is diminished, resulting in insulin resistance. Further in the cascade, IRS, with phosphorylated tyrosine residues, phosphorylates PI3K kinase, which in turn phosphorylates AKT (protein kinase B), thereby propagating the insulin signal. In adipose tissue and muscle, insulin, to cause insulin-stimulated glucose uptake, translocates the glucose transporter 4 (GLUT4) from the cytoplasm to the cell membrane. A characteristic of type II diabetes is reduced GLUT4 expression and impaired translocation. GLUT2, a major glucose transporter and glucose receptor in the liver, is important in both glucose uptake and glucose output. A record of hyperglycemia in diabetes mellitus occurs because hemoglobin is glycated and the level of glycated hemoglobin is proportional to the level of blood glucose (Rahbar, 2005). Hemoglobin A1c (HbA1c) is used as an index of the regulation of blood glucose as elevations in blood glucose are temporally "recorded" by elevated HbA1c.

Molecular targets in signaling pathways are successful therapeutic approaches in the pharmacology of type II diabetes (Sanchez et al., 2020). A type II arabinogalactan from GT increases insulin in vitro by stimulatingc AMP/protein kinase A pathway. Protein-bound polysaccharides [guanosine-5′-triphosphate (GTP), ca. 120 kDa molecular weight] are the carbohydrate components in water extracts of GT. In GTP, the polysaccharide is composed of L-arabinose, D-ribose, D-xylose, D-glucose, and D-galactose in both the α and β configurations with many branched chains. In a mouse type II diabetic model, GTP increased body weight and decreased blood glucose and low-density cholesterol with the phytopharmacological activity of upregulating PI3Kp85/p-Akt/GLUT4 and reduction of insulin resistance (Li et al., 2015). In mice fed a high-fat diet, Sampath et al. (2017) showed that GT-source epigallocatechein-3-gallate reduced weight gain and fasting blood glucose and insulin levels and the expression of Nrf2, and enhanced glutathione/glutathione disulfide ratios. The blood glycation products were also reduced. GT water is the highest in inhibiting intestinal α-glucosidase and both black and GT inhibit α-amylase. A study in C57BL/6J mice fed a high-lard diet showed that GT brew increased glucose transporter IV expression (Nishiumi et al., 2010). A study involving Wistar rats used crude catechins extracted from GT at doses of 80, 60, and 40 mg/rat, per gavage, followed 30 min later by a gavage of 4 mL of a 40% starch solution (Matsumoto et al., 1993). The study showed that catechins suppressed an increase of plasma glucose and increased plasma insulin, and the activity of α-amylase. In a study conducted in Sprague-Dawley rats, the drinking water was replaced with a GT extract reconstituted at 500 mg concentrate diluted with 100 mL of water (Wu et al., 2004). Oral glucose tolerance tests were performed at 4, 6, and 8 weeks. The rats given GT extract had increased insulin-response-stimulated uptake of glucose following the glucose challenge. The binding of glucose to adipocytes was also increased.

There are conflicting reports on the efficacy of GT brew and extract in patients with insulin resistance and frank T2DM. Hsu et al. (2011) conducted a study in Taipei that enrolled individuals (Chinese ethnicity) who had type II diabetes for more than 1 year. The 80 patients were to continue with medications for their type II diabetes management and consume an isocaloric diet. The patients in the treatment group were given a decaffeinated GT extract for 16 weeks, and 68 patients completed the study. No difference was observed between the groups for body mass index (BMI), waist circumference, fasting glucose, HbA1c level, fasting insulin, leptin, lipoproteins, adiponectin, and ghrelin. The Singapore study investigated coffee, tea, and incident type II diabetes (Odegaard et al., 2008). Study participants were from the Singapore Chinese Health Study, a population-based, prospective cohort of diet and cancer risk, and the 63,257 participants were from the Hokkiens and Cantonese dialects. There was no association between GT commutation and the risk of type II diabetes. There was an inverse relationship between black tea consumption and the risk of type II diabetes. A study involving Japanese postmenopausal women found that consumption of 350 mL of GT with breakfast reduced postpyramidal serum glucose by 3% (Takahashi et al., 2014). In another Japanese study, volunteers (54 males and 14 females) with a fasting blood glucose at or above 6.1 mmol/L and nonfasting blood glucose at or above 7.8 mmol/L were studied for the effects of GT extract on blood glucose (Fukino et al., 2008). The study design was a crossover randomized control trial without blinding or a washout period. The early intervention group consumed dried GT extract containing 544 mg polyphenols and 456 mg catechins daily for 2 months. They then entered the nonintervention group for a 2-month interval without GT extract and then consumed the GT extract for an additional 2 months. Comparing treatment groups to controls, intervention showed a significant reduction in HbA1c level and a borderline significant reduction in diastolic blood pressure. Intervention with GT did not show significant changes in body weight, BMI, body fat, systolic blood pressure, fasting serum glucose, homeostasis model assessment index, serum lipid, and hypersensitive C-reactive protein (CRP). A metaanalysis review of published randomized controlled trials (limited to English and Chinese languages) concluded that there is insufficient evidence to clearly show that GT brew and extracts are beneficial in insulin resistance and in reducing the risk of diabetes mellitus (Wang et al., 2014a).

Ninomiya et al. (2019) studied the palliative effects of GT on type II diabetes in a Japanese population (Hisayama Study). L-Theanine is a free amino acid (not incorporated into proteins) found in teas, especially GT. This amino acid is metabolized to ethylamine and glutamic acid and these metabolites in serum are biomarkers of tea consumption, especially GT. In the Hisayama Study, participants are given an annual opportunity for a glucose tolerance test. For 7 years, a total of 2253 participants were monitored for ethylamine levels. In follow-up, 282 participants developed type II diabetes. The adjusted cumulative index for type II diabetes decreased significantly (=0.04) with elevating levels of ethylamine. The adjusted multivariable hazard ratio (HR) for type II diabetes was significantly lower in the fourth quartile of serum ethylamine than in the first quartile (HR 0.69, 95% CI 0.49–0.98). The trend for decrease in diabetic risk across serum ethylamine levels was more prominent in middle-aged subjects and in subjects with prediabetes, obesity, or insulin resistance.

The effects of phytochemicals on type II diabetes may be multifactorial. Yao et al. (2019), in a Chinese dietary study, correlated the estimated intake of quercetin with type II diabetes in adults. The primary sources of quercetin are apples, oranges, and GT. The estimated mean intake of quercetin was 20.9 ± 2.32 mg/day. After adjusting for confounders, the odds ratios (95% confidence interval) for type II diabetes, across the ascending quartiles of quercetin intake, were 1.00 (reference), 0.75 (0.60–0.95), 0.76 (0.59–0.99), and 0.63 (0.51–0.94), showing an inverse relationship between dietary intake of quercetin and type II diabetes. Dietary phytopolyphenols, common in the human diet, generally are poorly absorbed, have low intrinsic activity, are highly metabolized, or are rapidly eliminated (Xiao and Hogger, 2014). The pathophysiology occurring in diabetes mellites can alter the pharmacokinetics of the polyphenols. GT extract has been shown to reduce dietary-induced developmental programming. A study in pregnant rats fed a high-fat diet showed that concurrent feeding of GT can reduce the epigenetic developmental programming effects of the high-fat diet (Li et al., 2012). The insulin resistance index was lower in the pups whose dam had received the GT extract. A review by Sanchez et al. (2020) concluded that more research is needed on the mechanisms of action of GT products in the prevention and management of type II diabetes.

Metabolic syndrome and obesity

The diagnosis of metabolic syndrome is obesity, elevated serum triglycerides, low serum levels of HDL cholesterol (high-density lipoprotein cholesterol), and elevated blood pressure. These factors increase the risk of developing insulin resistance, type II diabetes requiring medications, cardiovascular disease and nonalcoholic hepatic steatosis (fatty liver), and is a worldwide health issue. Obesity is defined as a BMI greater than 30 kg/m^2 (body mass in kg/body, height in m^2). There are improved health rewards for relatively small reductions in BMI. Obese individuals that achieve a 5%–10% reduction in BMI also have a significant reduction in their risk for developing cardiovascular disease. Compared to synthetic pharmaceuticals, the general public preferred intervention/management of obesity and its associated comorbidities are nutraceuticals and increased physical activities. There is keen interest in the use of GT and its extracts to prevent and treat obesity. Obesity is a difficult disease to manage because of multifactorial and interactive etiologies (Jurgens et al., 2012).

Complex endocrine and metabolic energy intake–expenditure interactions can stimulate hyperplasia and hypertrophy of adipose cells that result in excess lipid deposition, causing obesity (Feve, 2005; Wang et al., 2014b). Mechanisms of effective weight loss can be diminution of hunger and decreased ingestion of energy-dense foods, increased catabolism, decreased nutrient absorption, and upregulation of endocrine and metabolic factors altering the body management of lipids (Rains et al., 2011; Akter et al., 2013; Abdali et al., 2015; Sanchez et al., 2020). The GT extracts, generally promoted for weight loss or weight maintenance, are dried products as well as modified GT extracts with intake levels that exceed the intake achievable by imbibing GT infusions. Zhang et al. (2020b) studied the palliative effects of Yunkang 10 (Y-10, *C. sinensis* var. *assamica*, big leaf) GT extract and exercise on metabolic syndrome in male mice (C57BL/6J) fed a high-fat diet. Mice on the high-fat diet had higher body weight, plasma levels of glucose, insulin, total cholesterol, triglycerides, and alanine aminotransferase activity (ALT). At 10 weeks of the high-fat diet, the obese mice were treated with Y-10 and Y-10 plus forced exercise. After 8 weeks of Y-10 treatment, there was a significant decrease in plasma glucose, insulin, and total cholesterol. Y-10 did not significantly decrease body and liver weights, or plasma triglycerides. Mice receiving exercise and no Y-10 treatment had decreased body and liver weights, and decreased plasma insulin. The group of obese mice receiving exercise plus Y-10 treatment, compared to obese mice with no treatment, had decreased body and liver weights, reduction in plasma parameters; glucose decreased by 26.3%, insulin decreased by 46.8%, triglycerides decreased by 21.3%, and total cholesterol decreased by 23.7%. Hepatic mRNA expression profile of proinflammatory cytokines of IL-6, IL-1β, MCP-1, and TNF-α was significantly increased in the mice fed the high-fat diet. Groups of obese mice treated for 8 weeks with just exercise or treated with Y-10 and Y-10 plus exercise had a significant and profound downregulation of expression for the expression of cytokine genes in liver tissue. Histopathology showed that Y-10 plus exercise reduced hepatic steatosis. Both treatment with exercise and Y-10 plus exercise decreased ALT.

The mechanisms of action of GT extracts and associated catechins as antiobesity agents have been studied in cell culture, animal models, and humans. Studies in cell cultures have shown that (−)-epigallocatechin gallate and GT extract inhibit preadipocyte differentiation, decreasing adipocyte proliferation, induce adipocyte apoptosis, suppress lipogenesis, and promote lipolysis and β-oxidation of fatty acids (Huang et al., 2014; Wang et al., 2014b; Peluso and Serafini, 2017). Many of these effects have been demonstrated to be dependent on the level of exposure (dose) upregulating and downregulating cellular functions. (−)-Epigallocatechin gallate has been shown to upregulate thermogenesis at the cellular level and downregulate resistin, an inflammatory product produced by adipocytes. (−)-Epigallocatechin gallate and GT extract have demonstrated antiobesity effects in animal models. Generally accepted beneficial effects observed are decreased adipose mass, decreased total plasma lipids and cholesterol, and decreased insulin resistance. In laboratory animals fed energy-dense and high-cholesterol diets that caused obesity, there is a varying level of evidence to support that (−)-epigallocatechin gallate and GT extracts suppress intestinal absorption of lipids, increase lipid metabolism, and decrease lipogenesis.

Animal models of obesity also have provided evidence that chronic inflammation ROS of chemicals are important in the development of obesity. Some authors are of the opinion that the antiobesity effects of polyphenols in GT can be linked to scavenging of free radicals, downregulation of inflammatory cytokines, and upregulation of glutathione peroxidase (Lu et al., 2012). There is some evidence that the effects of GT on the sympathetic nervous system may be important in obesity (Rains et al., 2011). GT catechins inhibit COMT. This enzyme metabolizes the neurotransmitter norepinephrine and the synaptic actions of norepinephrine are prolonged. Caffeine inhibits phosphodiesterase and prolongs the action of cyclic adenosine monophosphate (cAMP). Both the catechins and caffeine are constituents in GT. Stimulation of the sympathetic nervous system increases voluntary activity in otherwise lethargic individuals.

Human clinical studies and green tea extracts

Studies involving obese humans and GT infusions, GT extracts, and GT catechins have variable results. The majority of studies in humans have been conducted in Asia. Yonekura et al. (2020) carried out a cross-sectional study in middle-aged women on the effects of coffee and GT consumption on BMI and cardiovascular disease. Daily consumption of coffee was inversely associated with BMI (adjusted odds ratio, 0.14; 95% confidence interval, 0.14−0.96) and body fat percentage (BF%) (0.33; 0.14−0.82). Daily consumption of GT was also inversely related to high BF% (0.36; 0.14−0.96). Daily consumption of coffee and GT was also inversely associated with high BMI (0.15; 0.05−0.50) and BF% (0.30; 0.12−0.74). A study in Oklahoma involved volunteers who had three out of five of the signs of metabolic stress syndrome (waist circumference ≥ 102 cm in men and ≥ 88 cm in women, triglycerides ≥ 150 mg/dL, high-density lipoprotein <40 mg/dL in men or <50 mg/dL in women, blood pressure ≥ 130/85 mmHg, or fasting glucose ≤ 100 mg/dL) (Basu et al., 2010). Volunteers, predominantly women, with other health issues were excluded. Subjects drank two cups of GT brew twice per day or were given capsules containing GT extract to be taken with two cups of water twice per day. Parameters for making the GT brew and extract were controlled. The daily consumption of catechins in GT brew and GT extract were 928 and 870 mg, respectively. The total consumption of epigallocatechin-3-gallate was 440 and 460 mg and of caffeine was 8.96 and 3.6 mg, respectively, for GT brew and GT extract. At 8 weeks, pairwise comparisons found that the GT beverage and GT extract caused a significant decrease in body mass and BMI. For the GT beverage, there was a decreasing trend in serum low-density lipoprotein, cholesterol, and low-density lipoprotein/high-density lipoprotein ratio. GT beverage also significantly decreased plasma lipid peroxidation [malondialdehyde (MDA) and hydroxynonenals]. Free catechins were detectable in serum from all GT groups.

Postmenopausal women are at greater risk for accumulation of abdominal fat accompanied by increased potential for metabolic syndrome and cardiovascular disease. Volunteers consisting of postmenopausal women with a BMI of 25−39.9 kg/m^2 were recruited for an intervention study (Hill et al., 2007). The subjects in the treatment group were given a modified GT extract[7] capsule (decaffeinated) to provide a daily intake of 300 mg of epigallocatechin-3-gallate/day. The women also exercised at 75% of their age indexed maximum heart rate for 45 min/day. In this study, exercise was found to be the effective treatment for reducing weight circumference, total body fat, abdominal fat, and intraabdominal adipose tissue. However, the modified GT extract was effective in reducing resting heart rate and blood glucose in volunteers with preexisting abnormal glucose tolerance. Another study involving obese women volunteers with a BMI greater than 27 kg/m^2 evaluated dried GT extract as a therapeutic agent (Hsu et al., 2008). The daily dose of dried GT extract was 400 mg given in one capsule. After 12 weeks, GT extract did not significantly reduce BMI, body mass,

7. Teavigo, Taiyo Kagaku India Pvt. Ltd.; Aurangabad, IN.

or waist circumflex when compared to the control group. There was a 0.15 kg reduction in the body mass of the GT group. Within-group comparison found a significant reduction in low-density lipoprotein. Women with a BMI of 30.0–39.9 kg/m^2 were enrolled in a Spanish study on GT catechins (Mielgo-Ayuso et al., 2014). The women were placed on a weight-reduction diet consisting of 55% carbohydrates, 30% lipids, and 15% protein to provide 600 calories less than their individually calculated energy requirements. The diet was targeted to achieve a 0.5–1.0 kg reduction in total body mass. The test group received a commercial dried GT product[8] (decaffeinated) and the test agent was administered at 300 mg epigallocatechin-3-gallate/day per woman for 12 weeks. Comparing the treatment group to controls, there was no significant difference for total body weight, fat mass, resting metabolic rate and fat oxidation rates, homeostasis assessment model for insulin resistance, total serum cholesterol, low-density lipoprotein cholesterol, serum triacylglycerol, and liver function test.

The 2012 Cochrane evaluation of GT for weight loss concluded that GT and GT extracts, when used for 12 weeks, produced a small and statistically nonsignificant reduction in weight gain, BMI, and waist-to-hip ratio. The Cochrane review did point out a country (ethnic) effect for studies reporting significant weight reductions (Jurgens et al., 2012). They also found substantial differences in the concentrations of the catechins in the products tested, and some of the products tested were decaffeinated. Phung et al. (2010), using metaanalyses to review published studies, concluded that caffeine needs to be present for GT and its extracts to have an effect on changes in the indices of obesity. Studies involving humans are inconsistent for the effects of GT extracts on the reduction of appetite (Rains et al., 2011). Studies using $^{13}CO_2$ analyses have not shown a decrease in fat absorption. GT likely has marginal benefits in the management of obesity (Abdali et al., 2015). Payab et al. (2020) has reviewed the use of herbs in obesity and metabolic syndrome. The doses of GT extract that achieved weight loss and a reduction in hip and waist circumference were from 300 to 6000 mg/day and the doses of catechins were from 150 to 1200 mg/day. For the GT extracts there was also a decrease in blood pressure. For GT catechins there was an increase in serum triglycerides and high-density lipoproteins, and a decrease in total cholesterol, low-density lipoprotein, and fasting glucose. Rothenberg et al. (2018) has provided evidence that the catechins depress carbohydrate digestion in the small intestine and the resulting increased microbiome production and absorption of volatile fatty acids (short chains) which upregulate lipid metabolism via the AMP-activated protein kinase pathways.

Green tea and cardiovascular diseases

Worldwide, approximately 31% of deaths are caused by cardiovascular disease. Coronary heart disease is due to atherosclerosis of the blood vessels linked to chronic inflammatory syndrome. The pathophysiology is complex and involves the immune system and chronic inflammation (Grassi et al., 2013; McEwen, 2015; Maranhao et al., 2018). The end result is fatty material and cholesterol being deposited inside the lumen of medium-sized and large blood vessels. The arteries lose their elasticity, and the lumen narrows, causing pathological hemodynamic changes. Events can cause blood clots to form, causing blockage and resulting in ischemia in the heart, brain, and other tissues. Arthrosclerosis is linked with high blood pressure and cerebral hemorrhage. Risk factors include elevated LDL cholesterol (low density lipoprotein cholesterol) elevated blood pressure, obesity, and reduced physical activity. The endothelia are the cells lining the blood vessels and form the capillaries. Endothelial health is important in preventing cardiovascular disease (Landberg et al., 2012; Grassi et al., 2013; Maranhao et al., 2018). The risk of cardiovascular mortality increases from prehypertensive to frank hypertension (Liu et al., 2014). At the molecular level, the endothelium secretes mediators, especially nitric oxide (NO), that regulate vascular muscle tone and reduce endothelin-1. There is evidence that green and black teas maintain NO homeostasis and reduce inflammation in the endothelial cells (Yamagata, 2020). (−)-Epigallocatechin gallate inhibits the expression of adhesion molecules, for example, vascular cell adhesion molecule-1, moderating adhesion of white blood cells, especially monocytes. It also reduces asymmetric dimethylarginine, increasing bioavailability of NO. The Nrf2/HO-1 pathway is diminished by (−)-epigallocatechin gallate, reducing oxidative stress. The endothelium secretes mediators, especially NO, that regulate vascular muscle tone and reduce endothelin-1. There is evidence that green and black teas maintain NO homeostasis and reduce inflammation in endothelial cells (Yamagata, 2020).

(−)-Epigallocatechin gallate inhibits the expression of adhesion molecules, for example, vascular cell adhesion molecule-1, moderating adhesion of white blood cells, especially monocytes. It also reduces asymmetric dimethylarginine, increasing the bioavailability of NO. The Nrf2/HO-1 pathway is diminished by (−)-epigallocatechin gallate reducing oxidative stress. Arab et al. (2013), in a review, concluded that tea consumption

8. Teavigo; Taiyo Kagaku Co., Ltd; Tokyo, JP.

lowered the risk of stroke. The Zutphen Elderly Study found a trend toward an inverse relationship between flavonoid intake and myocardial infarction and coronary heart disease mortalities (Hertog et al., 1993, 1995). The Ohsaki National Health Insurance Cohort Study in Japan in which 40,530 adults were monitored for 11 years found an inverse dose–response relationship between GT consumption and mortality from all causes of cardiovascular disease (Kuriyama et al., 2006). The inverse relationship for GT consumption and mortality was more significant in women, and in both men and women, reduction in stroke mortality had the highest relationship. Similar results were observed in Chinese and Japanese studies (Liang et al., 2009; Mineharu et al., 2011; Grassi et al., 2013). INTERHEART is an international, standardized, case-control study that was designed as an initial step to address the burden of cardiovascular disease. In this study, patients admitted to hospital with their first acute myocardial infarct were enrolled within 24 h. One or more controls (sex and age matched) without a history or signs of cardiac disease were recruited. A detailed questionnaire administered by trained personnel was used to obtain medical history, substance abuse, and consumption of tea and coffee. The data from the INTERHEART China study were used to examine the relationship with tea consumption and acute myocardial infarct (Hao et al., 2015). Compared to controls who do not drink tea, tea drinkers had a significant increased risk for myocardial infarct. A similar trend was found for GT when consumed at more than three cups per day. Women who consumed tea were at greater risk than men. Conflicting results were reported from another study designed to study GT consumption and coronary heart disease (Pang et al., 2015). Hospital enrollment was 423 male and 205 female patients with a mean age of 64.5 years. In men, there was a significant trend for the risk of coronary heart disease to decrease with increasing consumption of GT. The risk for coronary heart disease was reduced for male patients who drank GT for more than 30 years. In females, drinking more than three cups of GT per day decreased the risk of coronary heart disease. The Singapore study did not find any association between GT and mortality (Odegaard et al., 2015). Mahdavi-Roshan et al. (2020) in a metaanalysis on beverages and hypertension calculated that regularly drinking tea reduced systolic blood pressure [weighted mean difference (WMD): −4.81 mmHg, 95 % CI: −8.40 to −1.58, $P = .004$] and diastolic blood pressure (WMD:-1.98 mmHg, 95 %CI: −3.77 to −0.20, $P = 0.029$) and removing heterogeneous trials suggested that systolic and diastolic blood pressure were reduced by 3.53 and −0.99 mmHg, respectively. In a metaanalysis and systematic review of tea consumption and cardiovascular disease, Chung et al. (2020), found that an increase in tea consumption (both black and GTs) decreased the risk of a cardiovascular-linked mortality. The decrease was 4% for each 237 mL cup of tea/day and a cup of tea was estimated to contain 280 and 338 mg total flavonoids for black and GT, respectively. The magnitude of association was larger in the elderly. Studies with a tendency for bias showed increased inverse associations. In metaanalyses of random controlled trials, long-term (more than 12 weeks) consumption of tea resulted in decreased blood pressure (Liu et al., 2014). When these data were stratified, GT significantly reduced systolic blood pressure by 2.1 mm Hg (95% CI: 22.9, 21.2) and decreased diastolic blood pressure by 1.7 mm Hg (95% CI: 22.9, 20.5). Black tea reduced blood pressure to a lesser amount. There is consensus among health professionals that, in hypertensive syndrome, any reduction in blood pressure is generally considered to be beneficial.

Cancer and green tea infusions and extracts

Neoplasia (cancer), on a global scale, is the second leading cause of death. The initiation and promotion of neoplasia are multifactorial in both causation and growth promotion. Many patients with cancer use GT nutraceuticals. GT brew and extracts contain a mixture of chemicals that have been shown or have structural activity molecules that can prevent neoplasia. Of the catechins in GT brew and GT extract, (−)-epigallocatechin-3-gallate is the best studied for its antineoplastic activity (Bimonte et al., 2019). There is substantial evidence that GT catechins prevent the provoked occurrences of neoplasia in laboratory animals (Yang et al., 2011). There is some also insight into its mechanisms of action (Yang et al., 2014). GT polyphenols inhibit different types of phosphorylation, proteolytic changes, and methylations considered to be important epigenetic steps in neoplasia. The catechins in GT extracts alter receptors for tyrosine kinase in colonic and hepatic carcinoma cells. The epidermal growth factor family is often overexpressed in neoplasms. Epigallocatechin gallate has an inhibitory effect on epidermal growth factor receptor binding, and one mechanism is plasma membrane lipid organization. There is evidence to show that other growth factors that changed, including epithelial–mesenchymal transition, are associated with neoplastic invasion, metastasis, and angiogenesis (Aggarwal et al., 2020). For example, the vascular endothelial growth factor is likely inhibited by GT catechins interfering with receptor binding on endothelial cells.

In vitro studies have shown that epigallocatechin gallate can block neoplastic invasion. GT catechins cause small changes in micro-RNA, and changes in specific micro-RNAs can induce a reduction in cell division and increase apoptosis of neoplastic cells. Bimonte et al. (2019) has proposed that (−)-epigallocatechin-3-gallate be studied as a treatment for hepatocellular carcinoma because in vitro studies support its efficacy. Miyata et al. (2019) proposed that phytochemical other than (−)-epigallocatechin-3-gallate may be involved in the antineoplastic actions of GT, and phytopharmaceutical interactions among the phytochemicals may be important. GT extract itself can have both neoplastic promotion and suppression activities in endothelial cell cultures and purified catechins only induce apoptosis (Chu et al., 2018). The actions of GT catechins and other chemicals on CYP enzymes (also known as the cytochrome P450 enzymes) can be a factor in the endogenous production of xenobiotic metabolites that form cancer-causing DNA adducts. A study regarding the relationship between hepatocellular carcinoma and urinary excretion of catechins found in GT was performed in China (Butler et al., 2015). The study was a nested case-control study of hepatocellular carcinoma (211 cases and 1067 matched controls). Exposure to catechins as determined by urinary analyses did not protect individuals from hepatocellular carcinoma, and there was a trend for high levels of urinary catechins to increase the risk for hepatocellular carcinoma. GT catechins in Japanese studies have shown potential for preventing liver neoplasias associated with metabolic syndrome (Shimizu et al., 2015). The Ohsaki National Health Insurance Cohort Study in Japan did not find any reductions in cancer mortalities with the consumption of GT infusions (Kuriyama et al., 2006).

Prostate and breast cancer

Prostate cancer

Prostate cancer is one of the most common neoplasia in men. Neoplasia of the prostate gland has higher occurrences in Western than Eastern countries. Established risk factors are increasing age and family history of prostate cancer. The effect of dietary choices on risk factors for prostate neoplasia is inconclusive. There are conflicting reports on the effects of GT brew and GT extracts influencing favorable outcomes in individuals diagnosed with cancer (Boehm et al., 2009). A metaanalysis of observational studies on GT consumption and risk of prostate cancer in humans showed that tea consumption may reduce the odds of being diagnosed with prostate cancer (Guo et al., 2017). GT catechins were effective in preventing prostate cancer, with a relative risk of 0.38 ($P = .02$). These is a trend for reduced prostate cancer with each cup of GT consumed and a linear reduction in risk with more than seven cups consumed. Perletti et al. (2019) in a concise review and metaanalysis of placebo-controlled studies of patients with histopathological diagnosis of high-grade prostate intraepithelial neoplasia (most likely a precursor of prostatic carcinoma) or atypical small acinar proliferation (cannot be diagnosed as being benign or malignant) that were treated with placebo or catechins. There were 114 patients in the catechin group and 109 patients in the placebo group. For men that developed prostate cancer, 9 cases occurred in the catechin group and 24 cases occurred in the placebo group. For the catechin treatment, there was a significant reduction in the risk ratio. This study provided evidence that the intake of concentrated GT catechins provides protection from abnormal cell types developing into prostate cancer. Pulverized GT power was studied as a therapeutic in a phase II study (Zheng et al., 2011). The GT used was 1 g of pulverized GT with citric acid, sugar, and flavoring mixed with water, and this mixture was consumed six times per day. The 42 patients were diagnosed by biopsy with androgen-dependent prostate carcinoma. The results found limited antineoplastic activity of GT in the treatment of androgen-independent prostate carcinoma. A Canadian study found that drinking black tea infusions did not alter the risk for prostate cancer (Ellison, 2000). However, drinking more than 250 mL of coffee infusions caused a 40% increased risk for prostate cancer. A cohort study in Scotland involving 6016 men showed that drinking more than seven cups of tea infusion (most likely black tea) increased the risk for prostate cancer (Shafique et al., 2012). A metaanalysis by Lin et al. (2014) concluded that there is essentially no reduction in risk of prostate cancer with consumption of black or GTs. The Cochrane report found that consumption of GT may reduce the risk of prostate neoplasia (Boehm et al., 2009).

Breast cancer

Breast cancer is neoplastic proliferation of ductal or lobular cells in the mammary gland. It is the most common cancer in women and can also occur in males. Inherent risk factors for women include genetics, reproductive history, endogenous hormones, and exposure to endocrine-disrupting hormones. Breast cancer survival and treatment outcomes can be predicted from the molecular identification of neoplastic cells. Modified GT extract (Polyphenon E[9]) has been studied in 40 women with stage I–III, estrogen receptor-negative mammary neoplasia (Crew et al., 2012, 2015). The study, 6 months in duration, was a randomized design with double-blind placebo control. Twenty-six women were treated with Polyphenon E. Parameters in the study were hepatocyte growth factor, vascular endothelial growth factor, serum cholesterol,

9. Mitisui Norin Co Ltd; JP.

triglycerides, oxidative damage markers (urine 8-oxo-7,8-dihydrodeoxyguanosine and 15-F2-isoprostanes and plasma protein carbonyls), and inflammatory markers (serum high-sensitivity CRP, urine prostaglandin E2 metabolite). In the women treated with Polyphenon E, at 2 months there was a significant decrease in hepatocyte growth factor and vascular endothelial growth factor. At 6 months, this trend was not significant for the hepatocyte growth factor and vascular endothelial growth factor. The other study parameters were not different from controls.

Metaanalyses of epidemiological studies of the GT infusions reducing the risk of mammary neoplasia showed a trend for consumption of more than five cups of tea per day and may also prevent reoccurrences of stage I–II neoplasias (Seely et al., 2005). Another study showed that drinking large amounts of GT may not reduce the risk of breast cancer reoccurrence (Li et al., 2014). There is also some evidence that increased consumption of GT infusions may increase the risk for breast cancer (Ogunleye et al., 2010). Wang et al. (2020a), in a dose–response metaanalysis of Chinese data, concluded that drinking GT contributed 25% to Chinese women not developing breast cancer. In another study, Zhang et al. (2020c) reported that drinking five cups/day of black or GT reduced the risk of breast cancer in Puerto Rican and US women enrolled in the Sister Study. These 45,744 women did not have diagnosed breast cancer but had a sister diagnosed with breast cancer. The mean age of the women was 55.4 ± 8.9 years and the mean follow-up time was 8.6 years. In these 8.6 years, 2809 women developed breast cancer. Black tea consumption was 81.6% and GT consumption was 56.0%. GT consumption was significantly inversely associated with a lower rate of breast cancer. Drinking at least five cups verses no cups of GT was associated with a lower risk of breast cancer (HR = 0.82, 95% CI 0.70, 0.95 for ≥ 5 vs 0 cups/week of GT; p-trend <0.01). The association of black tea consumption with reduced breast cancer was less consistent in the model favoring GT being more beneficial. The estrogen receptor, menopausal status, and BMI were significant in the model. Triple-negative breast cancer (negative for the estrogen and progesterone receptors, and negative for the epidermal growth factor receptor 2) is frequently found in African women. Bimonte et al. (2020) have proposed that nanoparticles impregnated with (−)-epigallocatechin-3-gallate could be a therapeutic for this aggressive form of cancer.

Green tea extracts and colorectal cancer

Studies have investigated GT in preventing colorectal cancer, which is the fourth most common cause of death worldwide (Nunez-Sanchez et al., 2015). It is important to point out that colorectal neoplasia has complex causes and pathogeneses that have not been elucidated. (−)-Epigallocatechin-3-gallate, added to colorectal cancer cell cultures (HT-29), upregulated transferrin receptor and downregulated ferritin-H protein (Md Nesran et al., 2020). (−)-Epigallocatechin-3-gallate, in vitro, also has a strong affinity (−7.3 kcal/M) for the ferritin protein. These finding suggest that (−)-epigallocatechin-3-gallate interference with the ferritin protein could be important in its mechanism of action. Kwon et al. (2020) showed that (−)-epigallocatechin-3-gallate and tumor-necrosis-factor-related apoptosis-inducing ligand in colorectal cancer cell cultures (SW480 and HCT116) acted synergistically to mediate apoptosis. Studies (in vivo and in vitro) on epigallocatechin gallate have shown chemopreventive activity. Cell models show that epigallocatechin gallate and GT extracts act in blocking neoplastic proliferation by arresting the cell cycle, inducing apoptosis, damaging neoplastic DNA, and blocking the growth of blood vessels (antiangiogenesis). Molecular targets include cell cycle regulators, apoptotic proteins, transcription factors, and signaling pathways. (−)-Epigallocatechin gallate in colorectal cells inhibits the expression of cyclooxygenase-2, inhibits activation of epidermal growth factor, and inhibits the activation of insulin-like growth factor 1 receptor (Hoensch and Kirch, 2005; Shimizu et al., 2008; Liu et al., 2019).

Rodent models of colorectal cancer have shown positive improvement by reducing the number and size of aberrant crypt foci, inhibiting neoplastic cell proliferation, and inducing apoptosis (Xiao et al., 2008). Clinical trials in humans have been performed on GT extract in the treatment of colorectal neoplasia. Adenomatous polyps commonly advance to malignant colorectal neoplasms. A study in Japanese volunteers found that giving three tablets of GT extract was effective, following endoscopic removal of polyp adenomas, in preventing the reoccurrences of metachronous adenomas, reducing the number of recurring metachronous adenomas, and reducing the size of reoccurring metachronous adenomas. No serious adverse effects were observed. One 500-mg GT extract tablet contained 52.5 mg (−)-epigallocatechin gallate, 12.3 mg (−)-epicatechin, 34.6 mg (−)-epigallocatechin, 11.1 mg (−)-epicatechin gallate, and 15.7 mg caffeine (Shimizu et al., 2008). Hoensch et al. (2008) investigated, in a prospective cohort comparison, the effect of a GT flavonoid preparation on the reoccurrence of resected colorectal cancer (36 patients) or removal of polyp adenomas (20 patients). The patients received two tablets of a GT nutraceutical and each tablet contained (+)-catechin 0.3 mg, (−)-epicatechin 1.4 mg, (−)-epigallocatechin 2.1 mg, (−)-epigallocatechin gallate 10.4 mg, (−)-epicatechin gallate 4.0 mg, apigenin-7-glucoside 10.0 mg, myricetin 0.2 mg, kaempferol 0.0 mg, and quercetin 0.6 mg. Over a 3–4-year period, patients with colorectal cancer did not have reoccurrences of colorectal cancer and one patient developed a polyp adenoma. Reoccurrences in the matched control group (15 patients) had 20% reoccurrences of colorectal cancer and 27% of the patients developed polyp adenomas. Reoccurrences of cancerous polyps were 5 out of 20 in the treated group and 7/21 in the control group. There was a trend for the polyp cancers to have a higher cancerous index in the control group. In a case-controlled study, Kim et al. (2019) showed that the risk of colorectal cancer was 40% less for high tea drinkers (comparison of the high-intake tertile with the low-intake tertile). High intake of GT decreased the risk for colorectal cancer with and without lifestyle factors. Low consumption of GT increased the risk of colorectal cancer in the groups with ever-smokers and drinkers (alcoholic beverages) and consumers of a high-inflammatory diet. High BMI and physical inactivity were independent in the model. Wada et al. (2019), in the Takayama study, found there was no significant reduction in colorectal cancer in men and women from drinking GT. These was a weak trend for high GT consumption to reduce colon cancer in men.

Modified green tea extracts and genital warts

The human papillomavirus is a DNA virus that infects basal skin and mucosal cells and causes genital/perianal warts, oral-pharyngeal neoplasia, and cervical cancer in women. Sinecatechins 10% ointment[10] is a formulated GT extract from *C. sinensis* leaves (Anon, 2008; Miyoshi et al., 2020). The US Food and Drug Administration (FDA) has approved sinecatechins, formulated in an ointment base, as a botanical for topical treatment of genital and perianal warts (Abramovits and Gupta, 2010). The marketed products are not decaffeinated and have >85% catechins (Gupta and Daigle, 2015). The FDA approval was based on double-blind phase III studies of a generic sinecatechins 15% ointment. The ointment was applied three times daily for up to 16 weeks. The efficacy was based on clearance of the warts based on visual evaluations. The outcome was 53.6% clearance in the treatment group and 35.3% clearance in the control group. Similar results were observed in a European study on Polyphenon E, a product very similar to sinecatechins 15% ointment. In the clinical trials there were 5% treatment interruptions. The most common causes were application site reactions, hypersensitivity type IV, and coinfections. Miyoshi et al. (2020), in a recent review, evaluated the efficacy of the sinecatechins ointments for the treatment of anogenital warts and other related diseases caused by the human papillomavirus. For other related diseases caused by human papillomavirus, the research findings are variable. Histopathology suggests that the sinecatechins induce apoptosis and downregulate inflammation. In a prospective study, treatment with sinecatechins 10% ointment reduced the reoccurrences of genital warts after abolition surgery with CO_2 laser (Puviani et al., 2019).

Intestinal microbes and green tea

There is increasing scientific interest and understanding that the microbiota in the large intestine is important for health. Substances that are not digested and absorbed in the small intestine are the substrates for the diverse population of microbiota residing in the large intestine (Graf et al., 2015; Rothenberg et al., 2018). Diet, pharmaceutical chemotherapeutics, and nutraceuticals can have a major influence on total microbial populations and enterotypes. Diet is the most important aspect of maintaining the health of the colonic microbiome and a healthy microbiota provides a steady endogenous source of essential and health-preserving nutrients. The polyphenols in GT brew and extracts that are not absorbed in the small intestine and the majority of their metabolites excreted in bile become available to the microbes in the large intestine (Stalmach et al., 2010; Liu et al., 2018). Gut microbes hydrolyze the conjugates of polyphenols and can chemically alter the polyphenolic ring structure. Studies in human volunteers with large intestine bypass constructs have shown that a considerable amount of GT polyphenols is excreted in ileal feces (Stalmach et al., 2010).

There is considerable evidence to support that the colonic microbial population enzymatically degrades the flavan-3-ols ring structures and forms valerolactones and phenolic acids (Sang et al., 2011; Calani et al., 2012; Liu et al., 2018). The small phenolic compounds are likely absorbed from the large intestine into the portal circulation. In hepatic biotransformation, the microbial metabolites of catechins are susceptible to synthetic phase II reactions. These synthetic products are excreted in urine in substantial quantities. Finding these conjugates in the urine provided evidence that the mass of catechin microbial biotransformation products absorbed in the colon exceeds that of GT catechins absorbed by the small intestine. In vitro dynamic studies, using a proprietary human colonic model, showed that lyophilized GT brew formed 3-methoxy-4-hydroxyphenylacetic acid, 4-hydroxyphenyl acetic acid, 3,4-dihydroxyphenylacetic acid, and 3-(3-hydroxyphenyl) propionic acid (Gao et al., 2006). The 3,4-dihydroxyphenylacetic acid metabolite showed in vitro antiproliferative activity against prostate and colon cancer cells. It is reasonable to assume these

10. Veregen, Aresus Pharma; Strausberg, DE.

metabolites would be formed and absorbed after the GT phytochemicals reach the large intestine. These microbial products from fermenting the phytochemicals found in GT and GT extracts have received little attention for their potential health benefits. GT has been reported to alter the composition of human intestinal microbiota. The consumption of 1 L of GT per day showed a trend for the proliferation of *Bifidobacterium* spp. (Jin et al., 2012). There is interest in GT extract as a probiotic. The intestinal mucosa, part of the innate immune system, is the first line of defense against microbes entering the body. In the large intestine, the luminal flora crosstalk with the gut-associated lymphoid tissue. An emerging mechanism for inflammatory bowel syndrome is dysregulation between the indigenous luminal flora and the immune system. There is evidence from laboratory animal studies that epigallocatechin gallate has promise as a therapeutic in controlling inflammatory bowel syndrome and restoring the immune–enteric microbial synchrony. There is also interest in the possible linkages of the gut microbiome with autoimmune diseases.

Green tea and the immune system

The primary function of the immune system is to defend against pathogenic microorganisms, aberrant colonization by commensal microorganisms, and surveillance to defend the body against its own anomalous cells. The immune system is diffusely distributed and resides in every tissue and organ in the body, and is comprised of specialized organs and cells. The specialized cells produce numerous substances that, for immune competence, crosstalk and communicate with other cells in the body. Additionally, the receptor on the surface of the immunocytes and other cells, under their own and remote cytokine influences, can be upregulated and downregulated. The immune system has histocompatibility factors that, with other mechanisms, ensure the immune system recognizes normal cells and substances in its own body. Laboratory studies show that GT extracts, catechins, and L-theanine have modulating effects on the immune system (Saeed et al., 2017). For 6 weeks, mice (BALB/c) were orally administered the epigallocatechin gallate fraction of GT extract and selected immune system parameters were used as indicators of the effect (Kuo et al., 2014). The treatment with epigallocatechin gallate increased natural killer cell cytolysis, peritoneal cell phagocytic action, splenocyte proliferation, and secretion of interlukin-2 and interferon-γ. Other effects observed were decreased production of antigen-specific immunoglobulin E with increased proportion of CD4+ and CD25+ regulatory T lymphocytes in the spleen. These observations suggest that epigallocatechin gallate could modulate an allergic response. Studies on GT extract, epigallocatechin-3-gallate, and other catechins as potential pharmaceuticals in the management of immune-mediated diseases have been reviewed (Pae and Wu, 2013).

GT extract and its associated catechins have been shown to modulate both the innate and adaptive immune systems in autoimmune diseases. Studies in laboratory animals have shown that GT infusions and epigallocatechin-3-gallate have an effect on white blood cells, including T-cell activation, proliferation, differentiation, and production of cytokines. Autoimmune encephalomyelitis in rodents is a laboratory model of multiple sclerosis in humans. Epigallocatechin-3-gallate has been shown in a dose–response manner to attenuate the clinical and pathological indices of myelin oligodendrocyte glycoprotein-induced autoimmune encephalomyelitis in rodents.

Rheumatoid arthritis is an immune-mediated inflammatory disease wherein immunocytes and cells in synovial tissue are activated. The net results are pathological changes in joint cartilage, bone, and other tissues. Phytochemicals from GT have been shown to target inflammatory cytokines in autoimmune arthritis (Dudics et al., 2018). Epigallocatechin-3-gallate, given parenterally, has been shown to be effective in reducing adjuvant-induced arthritis in Lewis rats. There is evidence to suggest that GT and catechins alter the formation of cytokines and their receptors that are important in the pathogenesis of rheumatoid arthritis.

Chronic inflammation is associated with a number of syndromes. CRP is used as a bio-index of chronic inflammation associated with a number of diseases, including substance abuse, obesity, some types of neoplasia (cancer), and diabetes mellitus (Sproston and Ashworth, 2018). A Japanese study recruited 10,325 participants (4407 men and 5918 women) aged 49–76 years of age (Maki et al., 2010). Patients with known chronic cardiovascular disease, alcohol abuse, and chronic liver disease were excluded from the study. Self-reported consumption of GT, coffee, intake of alcoholic beverages, and smoking habits was recorded. Body mass, height, and waist and hip circumference were measured along with an index of physical activity. The CRP was measured in serum. Multivariate analyses were used to correct for variables. The CRP was not changed with increased consumption of GT. Asbaghi et al. (2019) in a systematic review and metaanalysis concluded that GT, in humans, significantly reduced serum levels of CRP and had no significant effect on MDA and total antioxidant capacity.

Drug interactions and iron absorption

Drug interactions

The use of GT extracts can interfere with the pharmacokinetics of prescription medications. Wang et al. (2020b) reported that the oral administration of ticagrelor to rats that had received multiple does of GT extract[11] significantly reduced ($p < 0.05$), for ticagrelor, the C_{max}, AUC_{0-t} (AUC = area under the curve), and $AUC_{0-\infty}$ by 64.34%, 53.18%, and 53.15%, respectively. The apparent total clearance (CL/F) of ticagrelor was increased by 2.47 times and the $t_{1/2}$ was not changed. The pharmacokinetics of ticagrelor, when given intravenously, was not changed. The authors hypothesized that the interactions are the phytochemicals acting on a transporter in the small intestine. The interactive pharmacology of GT extracts has been studied in humans. Forty-two human volunteers (10 males and 32 females, 19–73 years of age) were enrolled in a pharmacokinetic study using a decaffeinated GT extract[12] (Polyphenon E, 800 mg epigallocatechin gallate/day) to study the effects of GT extract on drug metabolism (Chow et al., 2006). The probe drugs, caffeine, dextromethorphan, buspirone, and losartan, were concurrently administered without and with the GT extract (4-week washout between crossing over). Blood and urine were assayed for the parent drug and metabolites. The phenotypical evidence for the effects of GT extract (equivalent to 8–16 cups of GT) on CYP isozyme activity included no change in the paraxanthine-to-caffeine ratio, indicating no change in the activity of CYP1A2. The changes in dextromethorphan-to-dextrorphan molar ratio were variable but not statistically different ($P = .05$), indicating no change in the activity of CYP2D6. There was no change in losartan-to-metabolite (E3174) ratio, indicating no change in the activity of the CYP2C9. There was a significant increase (20%) in the AUC for buspirone, indicating a reduction in the phenotypical activity of CYP3A4. There was considerable variation between individuals in the pharmacokinetic parameters in this study. Volunteers given 2 g GT extract[13] (3 g catechins/100 g of GT powder) concurrent with sildenafil (50 mg) had a significant increase in $AUC_{0-\infty}$ and increased C_{max} by approximately 36% and decreased its clearance (Hegazy, 2014).

Phytopharmaceuticals can change the pharmacokinetics of prescription and over-the-counter drugs (Colalto, 2010; Awortwe et al., 2019). These changes are AUC, clearance, $t_{1/2}$, C_{max}, and t_{max}. For example, GT extract[14] reduces the AUC for nadolol (β blocker) in human volunteers, most likely by competitive inhibition by (−)-epigallocatechin gallate of organic anion-transporting polypeptide (OATP)1A2 (Abe et al., 2018). Epigallocatechin-3-gallate has been demonstrated to inhibit activities of glycoprotein and organic anion-transporting polypeptide B (OATP-B) and the cytochrome P450 (CYP) mixed oxidase enzymes CYPs 1A1, 1A2, 3A4, 2C9, and 2D6 (Jodoin et al., 2002; Nishikawa et al., 2004; Chow et al., 2006; Colalto, 2010; Werba et al., 2018). Misaka et al. (2013a), using human microsomes in an in vitro model, showed the inhibitory concentrations 50 (IC_{50}) for GT extract[15] (SunphenonBG3) in μg/L were 5.9 for CYP2B6, 4.5 for CYP2C8, 48.7 for CYP2C19, 25.1 for CYP2D6, and 13.8 for CYP3A, respectively. (−)-Epigallocatechin-3-gallate has similar IC_{50} activities. The GT extract contained 9.7% (w/w) (−)-epicatechin, 1.7% (−)-epicatechin-3-gallate, 24.8% (−)-epigallocatechin, and 43.4% (−)-epigallocatechin-3-gallate and was 86.5% total catechins. Satoh et al. (2016) reported that the nongallated catechins, in an in vitro model, did not inhibit CYPs. The CYP1A2, CYP2D6, CYP2C9, and CYP3A4 enzymes are the primary phase I metabolic isozymes for the majority of currently marketed medications. Using the drug midazolam as a probe in rats, GT extract administered to rats for 1 week increased the metabolism of intravenously administered midazolam, suggesting GT extract caused upregulation of hepatic CYP3A (Nishikawa et al., 2004). In the same study, orally administered midazolam in another group of rats given GT extract for 1 week had an increase in plasma concentration, suggesting that enterocyte CYP3A activity was decreased by GT extract. Fasted rats were administered 400 mg of a GT extract (SunphenonBG3) by gavage (Misaka et al., 2013b). The drug simvastatin, a statin predominantly metabolized by CYP3A, was administered per gavage 30 min later. In the pharmacokinetic model, GT extract increased the AUC for plasma simvastatin by three- to fourfold. These findings suggest that gut enterocyte CYP3A metabolism of simvastatin was decreased. In rats, the ingestion of GT brew was shown to suppress 3-methylcholanthrene-induced hepatic aryl hydrocarbon receptor activation (Fukuda et al., 2015). The expected increase in hepatic ethoxyresorufin O-deethlyase activity did not occur. Ingestion of GT brew also showed a suppressive effect on the expression of CYP1A1. In ex vivo biochemistry, the expected aryl hydrocarbon receptor induction by 2,3,7,8-tetrachlorodibenzo-p-dioxin was suppressed. The latter suggests that those GT catechins and their metabolites were pharmacologically active, causing reduced induction of these hepatic CYP enzymes. A review of the literature on the potential interactions of GT and GT extracts with prescription and over-the-counter drugs suggested individuals should use caution when using analgesics,

11. Meilun Biotechnology; Dalian, CN.
12. Polyphenon E, Mitisui Norin Co Ltd; JP.
13. Mepaco, Anshas El-Raml; Sharqia, EG.
14. Invitrogen GmbH; Karlsruhe, DE.
15. SunphenonBG3, Taiyo Kagak; Yokkaichi, JP.

antilipemics, antiseizures, antivirals, β-adrenoceptor blockers, CYP450-metabolized drugs, potentially hepatotoxic drugs, hormonal agents, sedatives, and anticoagulants. This also raises the issue of the interactions between GT and GT extracts and occupational exposure to industrial chemicals (Posadzki et al., 2013).

Green tea and folic acid

There is evidence that GT extract reduces folic acid absorption in human. Volunteers were given 0.4 mg or 5 mg of folic acid with green or black tea extracts (Alemdaroglu et al., 2008). Both tea extracts reduced the maximal serum levels of folate compared to controls who ingested an equivalent amount of water. Female volunteers with a singleton pregnancy were studied for serum folic acid levels (Shiraishi et al., 2010). Consumption of more than 57.3 mL of GT brew per 1000 kcal in the diet was defined as high consumption of GT infusion. The multivariate regression analysis showed that high consumption of GT decreased serum folic acid levels. These findings raise concerns because folic acid is an essential dietary constituent and its deficiency in pregnancy is linked to neurotubular defects (e.g., spina bifida). A Chinese study by Liu et al. (2015) suggests that moderate to low consumption of GT infusions (cups of tea) does not lower plasma folate.

Green tea and purgatives

Polyethylene glycol (PEG) is used as an intestinal purge to remove colonic contents prior to colonoscopy. A study was carried out in China on patients scheduled for colonoscopy using a proprietary[16] PEG product (PEG-P) with and with concurrent inclusion of GT infusion[17] (Hao et al., 2020). Parameters used were medication compliance, the Aronchick scoring system for colonic evacuation, and product acceptance. Physicians carrying out the endoscopy were blinded for the medication groups. Compliance for PEG-P + GT was 93.2% and 59.6 for PEG-P group ($p < 0.05$). There were no significant differences ($p > 0.05$) between groups for Aronchick scores. Occurrences of adverse events were 55.9% and 77.2% for PEG-P + GT and PEG-P, respectively. The most common adverse events were nausea, emesis, and abdominal pain. The patients in the PEG-P + GT group were significantly ($p < 0.05$) more willing to repeat the PEG-P + GT treatment. In a review of GT extracts for weight loss, the primary adverse event, regardless of caffeine content, was abdominal discomfort and diarrhea (Wharton et al., 2020). Side effects requiring medical intervention were hypertension and hepatotoxicity.

Green tea and iron absorption

GT interferes with iron absorption. Many nutritionists recommend that, to minimize interference with iron absorption, GT or GT extract should not be taken with meals. However, there is evidence that imbibing dried GT extracts and modified GT extracts without food ingestion increases the risk of hepatotoxicity. Tea polyphenols and tannins are generally considered to inhibit iron absorption (Zijp et al., 2000; Cooper et al., 2006; Wierzejska, 2014). GT is considered to be less detrimental to nonheme iron absorption by the gut than black tea. An epidemiological study suggests that multiple factors may interact with GT in altering the absorption of iron. A cross-sectional study in France enrolling 954 men and 1639 women was corrected for other dietary factors enhancing or inhibiting iron absorption (Mennen et al., 2007). The study found that serum ferritin in the multivariate model was not related to any level of tea consumption (adult men and women before and after menopause) or the infusion times (less than 1 min to more than 5 min) for tea brewing. Fan (2016), in a case report, described iron-deficiency anemia in a 48-year-old man drinking 1.5 L of GT a day. The anemia with treatment was in remission after he stopped drinking GT; resumption of tea drinking required parenteral iron treatments to stabilize his anemia. In vitro studies have shown that (−)-epi-gallocatechin-3-gallate at 46 mg/L inhibits heme–iron absorption in Carco-2 cells, and this effect can be offset by ascorbic acid if the (−)-epigallocatechin-3-gallate is less than 4.6 ppm (Ma et al., 2011). A study in rats showed that dietary-consumed and gavage-administered (250 mg extract/mL) dried GT extract did not alter iron absorption (Beverly et al., 2012). In a rat study, dietary GT (28.6 g GT infusion powder/kg diet) was not observed to reduce iron ($^{58}Fe^{3+}$) absorption and liver stores of iron.

Toxicology

The toxicologies of GT infusions, GT extracts, and modified GT extracts have been studied in laboratory animals and observed in human trials. The chronic toxicity of Polyphenon E[18](PPE), a modified GT extract, was studied in fasted and nonfasted dogs (Kapetanovic et al., 2009). The neat (unformulated) PPE contained approximately 64% (−)-epigallocatechin gallate (CAS 989-51-5), approximately 3.4% (−)-epigallocatechin (CAS 970-74-

16. Shu Taiqing, Staidson Biopharmaceuticals Co Ltd; CN.
17. Meijiawu, Longjin; CN.
18. PPE, Mitisui Norin Co Ltd; Tokyo, JP.

1), approximately 7.0% (−)-epi-catechin gallate (CAS 1257-08-5), and approximately 0−1% (−)-epicatechin (CAS 490-46-0). Five- to 6-month-old Beagle dogs were orally administered 0, 200, 500, and 1000 (reduced to 800) mg/day of neat PPE in gelatin capsules. Dogs were also divided into two large groups, with one group receiving food 2 h before PPE administration and one group (fasted dogs) receiving food 2 h after PPE administration. Clinical signs of intoxication and mortality were observed in the fasted dogs starting at day 9 of treatment. The dose of 1000 mg/kg bwt/day was reduced to 800 mg/kg bwt/day. The study was terminated at 6.5 months and there were 16 deaths in 24 dogs. The majority of the deaths occurred in the first 13 weeks of the study. Male dogs were considered more susceptible than female dogs. In the dogs that received PPE, histopathology was observed in the liver, kidneys, gastrointestinal tract, bone marrow, and lymphoid tissues. The liver lesion was described as centrolobular necrosis and chronic-active inflammation. Infiltration of neutrophils and mononuclear cells was evident in the liver along with brown intracytoplasmic pigment in Kupffer cells. Renal lesions were tubular epithelial necrosis, tubular dilation with granular or hyaline protein casts, epithelial regeneration, acute inflammatory exudate, and transitional epithelium hyperplasia. The authors mentioned the likely contribution of the dehydration-linked decrease in renal perfusion was caused by vomiting and diarrhea. Necrosis was observed in the esophageal epithelium. Intestinal tract lesions included minimal to marked necrosis of the mucosal epithelium accompanied by hemorrhage (ulceration), acute inflammatory exudate with brown pigment, and epithelial hyperplasia that was interpreted as a regenerative response. Atrophy or necrosis was observed in the spleen and lymph nodes. Epithelial necrosis was observed in the testicles along with aspermia. Additional histopathology observed was atrophy of the prostate gland in males and ovarian atrophy in females. Dogs in the fasted group had much higher plasma levels of total (−)-epigallocatechin gallate, (−)-epigallocatechin, (−)-epicatechin gallate, and (−)-epicatechin than dogs that received food before PPE was administered. Serum folic acid levels were not reduced, and indices of oxidative stress (levels of reduced and oxidized glutathione, glutathione peroxidase, lipid peroxidation, and total antioxidant activity) were not changed. Hematological effects observed were mild to marked increase in white blood cell numbers and a decrease in red blood cell numbers with a parallel increase in reticulocyte numbers. The anemia varied from macrocytic inadequately regenerative to macrocytic nonregenerative. The changes in bone marrow reflected the changes observed in circulating red and white cell numbers. Similar lesions were observed in rats administered PPE (Johnson et al., 1999). A follow-up 13-week study in male dogs using a dose of 200 mg PPE/kg bwt/day showed that feeding the dogs before administration of PPE reduced the toxic effects. The authors also concluded that the toxicity of GT extract is greater than the predicted toxicity based on (−)-epigallocatechin gallate content. Dogs were orally administered encapsulated spray-dried formulation [80% (−)-epigallocatechin gallate] at 0, 40, 120, and 400 mg (−)-epigallocatechin gallate per day. The capsules were administrated after 15 h of fasting and food was not offered until 3−4 h after the capsules were administered. Occasional emesis was observed in dogs in the 400 mg epigallocatechin gallate group. Three dogs in the 400 mg group and two dogs in the 120 mg group, respectively, died or were euthanized because of severe health effects. In the high-dose group, dogs that died or were euthanized in extremis had necrosis of the proximal tubules, necrosis of the liver, and hemolytic anemia. In addition to anemia, and renal and liver pathology, surviving dogs in the high-dose group also had lymphoid follicular atrophy. No lesions were reported for dogs in the low-dose and control groups. A study of GT extract in rats (Wistar) with access to feed and water ad libitum did not show clinicopathological evidence of liver damage based on ALT, aspartate aminotransferase, alkaline phosphatase (AKP), lactate dehydrogenase, γ-glutamyl transferase, iron, total bilirubin, and total cholesterol parameters (Bun et al., 2006). The extracts used in the study were production-scale GT extract[19] using 80% ethanol at 30°C and laboratory-scale extraction using methylene chloride at 40°C.

In another study, hot water extracts of *C. sinensis* leaves were used to make a crystallized product[20] containing more than 90% epigallocatechin-3-gallate (EGCG-T) (Isbrucker et al., 2006a). In a 13-week rat study with dietary exposure to EGCG-T, no pathology was observed at 500 mg/kg/day. Isbrucker et al. (2006a) also studied the effects of EGCG-T in dogs. Dogs were administered daily doses of 0, 50, 150, and 500 mg/kg bwt for 13 weeks. The doses of EGCG delivered were 0, 40, 120, and 400 mg/kg bwt. The 500 mg dose caused occasional vomiting and frequent diarrhea. Dogs in the 150 mg group had less frequent occurrences. Four dogs in the 500 mg dose group and four dogs in the 150 mg group died or were euthanized. In the 500 and 150 mg groups, anorexia and anemia were observed. Liver and kidney pathology were also observed. The dogs in the 50 mg group were not adversely affected by the treatment. Isbrucker et al. (2006a) concluded that the EGCG-T NOAEL (no

19. Exolise, active ingredient is AR25, Burgundy Botanical Extracts; Reyssouze, FR.
20. Teavigo; Taiyo Kagaku Co., Ltd; Tokyo, JP.

observed adverse effect level) for rats is considered to be 500 mg/kg bwt/day, for fasted dogs is 40 mg ECCG/kg bwt/day, and for prefed dogs, the LD_{50} is five times higher at 500 EGCG/kg/day. Studies in mice have shown that a single oral dose (1500 mg/kg) of (−)-epigallocatechin gallate administered to fasted CF-1 mice causes moderate to severe liver necrosis (Lambert et al., 2010). Oxidative stress was accompanied by high blood levels of (−)-epigallocatechin gallate. Oxidative stress was measured by a fivefold increase in lipid peroxidation, 9.5-fold increase in 8-isoprostane, and increased hepatic metallothionein and c-histone2AX protein expression. These findings are in contrast to a study involving dogs in which increases in oxidative stress parameters were not observed. Subcutaneous injections of 200 mg of (−)-epigallocatechin gallate to female rats caused hepatic and renal proximal tubular necrosis (Isbrucker et al., 2006a).

Green tea extracts, modified green tea extracts, and human toxicity

The toxicity of GT extract in humans is considered greater than the toxicity indexed on (−)-epigallocatechin gallate. Many studies in humans have used relatively high doses of GT catechins to achieve high concentrations of catechins in the blood and tissues (Shimizu et al., 2015). The products containing high concentrations of catechins have been manufactured using supercritical liquid extraction, chromatography, and other methods to enrich the catechins. The side effects observed in these studies are bloating, nausea, heartburn, abdominal pain, dizziness, headache, muscle pain, hypertension, and hepatotoxicity. The side effects that most often require medical intervention are hypertension and hepatotoxicity. Oketch-Rabah et al. (2020) recently published the United States Pharmacopeia comprehensive review of the hepatotoxicity of GT extracts. The standout was a large study in postmenopausal women on the palliative effects of GT extract on the biomarkers of breast cancer (Dostal et al., 2015; Samavat et al., 2015; Yu et al., 2017). This 12-month protocol-controlled study identified hepatotoxic effects of GT extract. The dose of GT catechins was 1315 ± 115.4 mg total catechins, 843 ± 44.1 mg (−)-epigallocatechin-3-gallate, 202 ± 74.0 mg epicatechin gallate, 106 ± 118.8 mg epigallocatechin, 107 ± 23.4 mg epicatechin, and 15.8 mg caffeine. The authors (Dostal et al., 2015; Samavat et al., 2015; Yu et al., 2017) considered the dose of EGCG to be equivalent to four or five cups (equivalent to 8 fluid ounces or 240 mL) of brewed GT. The hepatic injury was determined by testing serum for ALT, aspartate aminotransferase, AKP, albumin, globulin, total proteins, albumin/globulin ratio, and indirect, direct, and total bilirubin. For the women in this study, the odds ratio for developing liver function abnormalities was 7.0 [$P = .0002$; 95% CI = 2.4−20.3] compared to placebo and these findings were significant. Stopping the ingestion of GT extract caused the clinicopathologic parameters to decrease to their normal range and restarting ingestion of GT extract caused the clinicopathologic parameters to become elevated outside their normal range. The fall−rise pattern of clinicopathologic parameters concurrent with the stop−-start of GT extract is important evidence. Since this phenomenon occurred in 5% of the women in the study, it provides evidence that genetic sensitivity or other contributing factors may be occurring. There did not appear to be a link to medications (statins, nonsteroidal antiinflammatory drugs) and alcohol consumption.

Modified GT extracts have been used as antiobesity drugs. In these trials, preexisting disease, concurrent medications, and substances of abuse could have been factors leading to toxicity of the GT extract. In 2003 a European clinical trial using a modified GT extract was suspended because of associated hepatotoxicity (ESCO, 2009). The phytotherapeutic was a proprietary extract (80% ethanol solvent) that contained 25% (−)-epigallocatechin gallate and 5%−10% caffeine in a 375-mg capsule. The recommended dose was two capsules per day. The product was marketed as an aid in weight loss to be ingested concurrent with adherence to a strict low-caloric diet. Liver toxicity appeared after 50 days of use and there was one case of hepatotoxicity per 100,000 boxes of the phytopharmaceutics sold. Discontinuation of the phytotherapeutic ingestion resulted in full recovery of all but one individual. The liver disorder was diagnosed as nonuniform hepatitis, cytolytic and cholestatic liver disorders, as well as jaundice. One person developed liver failure, and concurrent drugs and imbibing of alcohol were considered important contributing factors. Using the Naranjo scale, there was one probable causality and 11 possible causalities. Australia has reported three cases of abnormal liver function associated with GT extract in a polyherbal formulation. There are other reports of hepatotoxicity, all using concentrated or powered GT extracts. The Organization for Economic Cooperation and Development (OECD) working group concluded that the human NOAEL (no observed adverse effect level) for (−)-epigallocatechin gallate ranges from 40 to 50 mg/kg bwt/day (ESCO, 2009). The OECD also concluded that the carcinogenicity data for GT brew, GT extracts, and modified GT extracts do not raise concerns for genotoxicity and carcinogenicity.

Contaminants in green tea

There are concerns that contaminants in GT can be transferred to humans (Sanzini et al., 2011; Abd El-Aty et al., 2014; Yang et al., 2020). Pesticides are used to control weeds, mites, insects, and other pests. These pesticides may be used according to label or they may be illegally used off-label, and more than double the recommended application rates or nonapproved pesticides may be illegally used in tea production. There can be a shorter than label-approved interval between application and harvest resulting in pesticide carryover. Additionally, there can be different regulations between countries for approved pesticides and maximum allowable levels in teas with the conflict of the pesticide residue being legal in one country and illegal in a different country. Pesticide residues in GT generally partition according to their physicochemical properties (Yang et al., 2020). Pesticides used on GT can be transferred to the brewed infusion and transferred to the extract. Some pesticides, depending on their chemical and physical properties, can be reduced during the manufacturing processes. GT can be contaminated with polycyclic aromatic hydrocarbons and other environmental pollutants during growing and manufacturing processes. Persistent organic pollutants can be incorporated into tea leaves by atmospheric deposition. Radioactive isotope fallout can contaminate tea due to incidents in nuclear-powered facilities and other releases of radionuclides. Wang et al. (2014c), in a Canadian study, has shown that pesticides can be transferred to the infusion during the brewing process at levels up to 92% of the tea level. Overbrewing tea for more than 3 min is considered to increase the risk for pesticides being transferred to the tea infusion (Abd El-Aty et al., 2014).

Green tea extract and reproduction

The consumption of standard GT infusions at less than four cups per day (total caffeine less than 200 mg/day) is not associated with human infertility and undesirable birth outcomes (ESCO, 2009). A study involving rats using GT infusion observed adverse changes in the reproductive organs (Chandra et al., 2011). The infusions were prepared by adding 100 mL boiling water to 2.5 g GT and steeping for 15 min. Catechin concentrations were 1.55% epicatechin, 9% epigallocatechin gallate, 4.8% epicatechin gallate, 5% epigallocatechin, and 2.38% caffeine. Rats in each treatment group were fed 1 mL extract/kg bwt/day for 26 days and controls received water. Extracts used in the study by the groups were 1.5%, 2.5%, and 5% solids. The epididymal sperm counts in rats were significantly decreased in the 2.5% and 5% groups. Serum luteinizing hormone and testosterone were significantly decreased in a dose-dependent manner. Spermatogenesis was reduced in the 2.5% and 5% groups. A study in rats showed that GT catechins are transferred to the fetus (Chu et al., 2006). Balazi et al. (2019) in a study feeding GT extract to doe rabbits reduced conception and kindling rates and the number of weaned pups. These effects can be due to alterations of steroid hormones in the ovary. The catechin levels in the placenta were 10 times lower and fetal levels were 50–100 times lower than in maternal plasma. After gavage with GT extract, the highest level of catechins was observed in the fetal eye (Chu et al., 2007). A teratology study in rats showed that oral administration of 1000 mg (−)-epigallocatechin gallate/kg bwt/day (dosing started on day 6 of pregnancy) did not cause maternal toxicity, loss of fetuses, or teratogenic effects (Isbrucker et al., 2006b). In an OECD-compliant teratology study, rats were placed on diets containing (−)-epigallocatechin gallate. The estimated daily intake of (−)-epigallocatechin gallate was 111, 337, and 1097 mg/kg bwt/day. No skeletal malformations were observed. In the 337 mg dose group, there was a higher occurrence of fetuses with dilation of the renal pelvis (Isbrucker et al., 2006b). Isbrucker et al. (2006c) showed in the laboratory that a GT extract was not genotoxic.

Concluding remarks and future directions

The consumption of GT infusions by Asians predates recorded history. The practice is enshrined in some cultures and religions. GT is a mixture of phytochemicals that have individual and collective phytopharmacological actions. The phytochemical composition of GT extracts varies with the extraction method. GT extracts generally are not equivalent to a brewed cup of tea because of the higher concentration of phytochemicals achieved by the extraction method. There is evidence that GT is beneficial to health. Consumption of GT extracts can alter the pharmacokinetics of prescription medications and GT extracts can be toxic to the liver. Chronic inflammatory syndrome, worldwide, is an escalating chronic disease and consumption of GT has been shown to be beneficial in this disorder. There need to be more studies into the interactions of GT with exercise and the interactions of GT with colonic microbiome to reduce BMI, chronic inflammation, and insulin resistance. Studies have shown that achieving small reductions in BMI and chronic inflammation have substantial returns for improving and preserving health and quality of life.

References

Abd El-Aty, A.M., Choi, J.H., Rahman, M.M., et al., 2014. Residues and contaminants in tea and tea infusions: a review. Food Addit. Contam. Part A Chem. Anal. Control Expo. Risk Assess. 31 (11), 1794–1804.

Abdali, D., Samson, S.E., Grover, A.K., 2015. How effective are antioxidant supplements in obesity and diabetes? Med. Princ. Pract. 24 (3), 201–215.

Abd-Elsalam, H.A., Al-Ghobashy, M.A., Zaazaa, H.E., et al., 2014. Stability of catechins in green tea nutraceutical products: application of solid phase extraction-thin layer chromatography densitometry. Food Chem. 156, 94–99.

Abe, O., Ono, T., Sato, H., et al., 2018. Role of (-)-epigallocatechin gallate in the pharmacokinetic interaction between nadolol and green tea in healthy volunteers. Eur. J. Clin. Pharmacol. 74 (6), 775–783.

Abourashed, E.A., Roberson, C.L., Elsharkawy, N., 2014. Content variation of catechin markers, total phenolics and caffeine in green tea dietary supplements. J. Diet. Suppl. Available from: https://doi.org/10.3109/19390211.2014.965868.

Abramovits, W., Gupta, A.K., 2010. Veregen (sinecatechins ointment) 15%. Skinmed 8 (1), 46–49.

Aggarwal, V., Tuli, H.S., Tania, M., et al., 2020. Molecular mechanisms of action of epigallocatechin gallate in cancer: recent trends and advancement. Semin. Cancer Biol. Available from: https://doi.org/10.1016/j.semcancer.2020.05.011.

Akter, S., Nanri, A., Pham, N.M., et al., 2013. Dietary patterns and metabolic syndrome in a Japanese working population. Nutr. Metab. (Lond.) 10 (1), 30. Available from: https://doi.org/10.1186/1743-7075-10-30.

Alemdaroglu, N.C., Dietz, U., Wolffram, S., et al., 2008. Influence of green and black tea on folic acid pharmacokinetics in healthy volunteers: potential risk of diminished folic acid bioavailability. Biopharm. Drug Dispos. 29 (6), 335–348.

Anon, 2008. Veregen: a botanical for treatment of genital warts. Med. Lett. Drugs Ther. 50 (1280), 15–16.

Arab, L., Khan, F., Lam, H., 2013. Tea consumption and cardiovascular disease risk. Am. J. Clin. Nutr. 98 (6 Suppl), 1651S–1659S.

Asbaghi, O., Fouladvand, F., Gonzalez, M.J., et al., 2019. The effect of green tea on C-reactive protein and biomarkers of oxidative stress in patients with type 2 diabetes mellitus: a systematic review and meta-analysis. Complement. Ther. Med. 46, 210–216.

Astill, C., Birch, M.R., Dacombe, C., et al., 2001. Factors affecting the caffeine and polyphenol contents of black and green tea infusions. J. Agric. Food Chem. 49 (11), 5340–5347.

Auger, C., Mullen, W., Hara, Y., et al., 2008. Bioavailability of Polyphenon E flavan-3-ols in humans with an ileostomy. J. Nutr. 138 (8), 1535S–1542S.

Awortwe, C., Bruckmueller, H., Cascorbi, I., 2019. Interaction of herbal products with prescribed medications: a systematic review and meta-analysis. Pharmacol. Res. 141, 397–408.

Balazi, A., Sirotkin, A.V., Foldesiova, M., et al., 2019. Green tea can suppress rabbit ovarian functions in vitro and in vivo. Theriogenology 127, 72–79.

Banerjee, S., Chatterjee, J., 2015. Efficient extraction strategies of tea (*Camellia sinensis*) biomolecules. J. Food Sci. Technol. 52 (6), 3158–3168.

Basu, A., Sanchez, K., Leyva, M.J., et al., 2010. Green tea supplementation affects body weight, lipids, and lipid peroxidation in obese subjects with metabolic syndrome. J. Am. Coll. Nutr. 29 (1), 31–40.

Benzie, I.F., Szeto, Y.T., Strain, J.J., et al., 1999. Consumption of green tea causes rapid increase in plasma antioxidant power in humans. Nutr. Cancer 34 (1), 83–87.

Beverly, A.B., Zhu, L., Fish, T.L., et al., 2012. Green tea ingestion by rats does not affect iron absorption but does alter the composition of the saliva proteome. J. Food Sci. 77 (5), H96–H104.

Bhagwat, S., Haytowitz, D.B., Holden, J.M., 2014. USDA Database for the Flavonoid Content of Selected Foods. USDA, Beltsville, MD, Release 4.1.

Bimonte, S., Albino, V., Piccirillo, M., et al., 2019. Epigallocatechin-3-gallate in the prevention and treatment of hepatocellular carcinoma: experimental findings and translational perspectives. Drug Des. Devel. Ther. 13, 611–621.

Bimonte, S., Cascella, M., Barbieri, A., et al., 2020. Current shreds of evidence on the anticancer role of EGCG in triple negative breast cancer: an update of the current state of knowledge. Infect. Agents Cancer 15, 2. Available from: https://doi.org/10.1186/s13027-020-0270-5.

Boehm, K., Borrelli, F., Ernst, E., et al., 2009. Green tea (*Camellia sinensis*) for the prevention of cancer. Cochrane Database Syst. Rev. 2009 (3), CD005004. Available from: https://doi.org/10.1002/14651858.CD005004.pub2.

Boros, K., Jedlinszki, N., Csupor, D., 2016. Theanine and caffeine content of infusions prepared from commercial tea samples. Pharmacogn. Mag. 12 (45), 75–79.

Braicu, C., Ladomery, M.R., Chedea, V.S., et al., 2013. The relationship between the structure and biological actions of green tea catechins. Food Chem. 141 (3), 3282–3289.

Bun, S.S., Bun, H., Guedon, D., et al., 2006. Effect of green tea extracts on liver functions in Wistar rats. Food Chem. Toxicol. 44 (7), 1108–1113.

Butler, L.M., Huang, J.Y., Wang, R., et al., 2015. Urinary biomarkers of catechins and risk of hepatocellular carcinoma in the Shanghai cohort study. Am. J. Epidemiol. 181 (6), 397–405.

Calani, L., Dall'Asta, M., Derlindati, E., et al., 2012. Colonic metabolism of polyphenols from coffee, green tea, and hazelnut skins. J. Clin. Gastroenterol. 46 (Suppl), S95–S99.

Calderon-Montano, J.M., Burgos-Moron, E., Perez-Guerrero, C., et al., 2011. A review on the dietary flavonoid kaempferol. Mini Rev. Med. Chem. 11 (4), 298–344.

Chandra, A.K., Choudhury, S.R., De, N., et al., 2011. Effect of green tea (*Camellia sinensis* L.) extract on morphological and functional changes in adult male gonads of albino rats. Indian J. Exp. Biol. 49 (9), 689–697.

Chen, D., Wang, C.Y., Lambert, J.D., et al., 2005. Inhibition of human liver catechol-O-methyltransferase by tea catechins and their metabolites: structure-activity relationship and molecular-modeling studies. Biochem. Pharmacol. 69 (10), 1523–1531.

Choi, S.W., Yeung, V.T., Collins, A.R., et al., 2015. Redox-linked effects of green tea on DNA damage and repair, and influence of microsatellite polymorphism in HMOX-1: results of a human intervention trial. Mutagenesis 30 (1), 129–137.

Chow, H.H., Cai, Y., Hakim, I.A., et al., 2003. Pharmacokinetics and safety of green tea polyphenols after multiple-dose administration of epigallocatechin gallate and Polyphenon E in healthy individuals. Clin. Cancer Res. 9 (9), 3312–3319.

Chow, H.H., Hakim, I.A., Vining, D.R., et al., 2006. Effects of repeated green tea catechin administration on human cytochrome P450 activity. Cancer Epidemiol. Biomarkers Prev. 15 (12), 2473–2476.

Chu, K.O., Wang, C.C., Chu, C.Y., et al., 2006. Pharmacokinetic studies of green tea catechins in maternal plasma and fetuses in rats. J. Pharm. Sci. 95 (6), 1372–1381.

Chu, K.O., Wang, C.C., Chu, C.Y., et al., 2007. Uptake and distribution of catechins in fetal organs following in utero exposure in rats. Hum. Reprod. 22 (1), 280–287.

Chu, K.O., Chan, K.P., Chan, S.O., et al., 2018. Metabolomics of green-tea catechins on vascular-endothelial-growth-factor-stimulated human-endothelial-cell survival. J. Agric. Food Chem. 66 (48), 12866–12875.

Chung, M., Zhao, N., Wang, D., et al., 2020. Dose-response relation between tea consumption and risk of cardiovascular disease and all-cause mortality: a systematic review and meta-analysis of population-based studies. Adv. Nutr. Available from: https://doi.org/10.1093/advances/nmaa010.

Clifford, M.N., van der Hooft, J.J., Crozier, A., 2013. Human studies on the absorption, distribution, metabolism, and excretion of tea polyphenols. Am. J. Clin. Nutr. 98 (6 Suppl), 1619S–1630S.

Colalto, C., 2010. Herbal interactions on absorption of drugs: mechanisms of action and clinical risk assessment. Pharmacol. Res. 62 (3), 207–227.

Cooper, M.J., Cockell, K.A., L'Abbe, M.R., 2006. The iron status of Canadian adolescents and adults: current knowledge and practical implications. Can. J. Diet. Pract. Res. 67 (3), 130–138.

Crew, K.D., Brown, P., Greenlee, H., et al., 2012. Phase IB randomized, double-blinded, placebo-controlled, dose escalation study of Polyphenon E in women with hormone receptor-negative breast cancer. Cancer Prev. Res. (Phila.) 5 (9), 1144–1154.

Crew, K.D., Ho, K.A., Brown, P., et al., 2015. Effects of a green tea extract, Polyphenon E, on systemic biomarkers of growth factor signalling in women with hormone receptor-negative breast cancer. J. Hum. Nutr. Diet. 28 (3), 272–282.

Del Rio, D., Stewart, A.J., Mullen, W., et al., 2004. HPLC-MSn analysis of phenolic compounds and purine alkaloids in green and black tea. J. Agric. Food Chem. 52 (10), 2807–2815.

Dostal, A.M., Samavat, H., Bedell, S., et al., 2015. The safety of green tea extract supplementation in postmenopausal women at risk for breast cancer: results of the Minnesota Green Tea Trial. Food Chem. Toxicol. 83, 26–35.

Dudics, S., Langan, D., Meka, R.R., et al., 2018. Natural products for the treatment of autoimmune arthritis: their mechanisms of action, targeted delivery, and interplay with the host microbiome. Int. J. Mol. Sci. 19 (9). Available from: https://doi.org/10.3390/ijms19092508.

Egert, S., Tereszczuk, J., Wein, S., et al., 2013. Simultaneous ingestion of dietary proteins reduces the bioavailability of galloylated catechins from green tea in humans. Eur. J. Nutr. 52 (1), 281–288.

Ellison, L.F., 2000. Tea and other beverage consumption and prostate cancer risk: a Canadian retrospective cohort study. Eur. J. Cancer Prev. 9 (2), 125–130.

Engelhardt, U.H., 2020. Tea chemistry—what do and what don't we know? A micro review. Food Res. Int. 132, 109120.

ESCO, 2009. Advice on the EFSA guidance document for the safety assessment of botanicals and botanical preparations intended for use as food supplements, based on real case studies. EFSA J. 7 (9), 280.

Fan, F.S., 2016. Iron deficiency anemia due to excessive green tea drinking. Clin. Case Rep. 4 (11), 1053–1056.

FAO. Problems, Committee on Commodity. (2018). Intergovernmental group on tea. In: Twenty-Third Session. Available from: <http://www.fao.org/economic/est/est-commodities/tea/tea-meetings/tea23/en/>.

Feve, B., 2005. Adipogenesis: cellular and molecular aspects. Best Pract. Res. Clin. Endocrinol. Metab. 19 (4), 483–499.

Forester, S.C., Lambert, J.D., 2011. Antioxidant effects of green tea. Mol. Nutr. Food Res. 55 (6), 844–854.

Fu, Y.-Q., Wang, J.-Q., Chen, J.-X., et al., 2020. Effect of baking on the flavor stability of green tea beverages. Food Chem. 331, 127258.

Fukino, Y., Ikeda, A., Maruyama, K., et al., 2008. Randomized controlled trial for an effect of green tea-extract powder supplementation on glucose abnormalities. Eur. J. Clin. Nutr. 62 (8), 953–960.

Fukuda, I., Nishiumi, S., Mukai, R., et al., 2015. Catechins in tea suppress the activity of cytochrome P450 1A1 through the aryl hydrocarbon receptor activation pathway in rat livers. Int. J. Food Sci. Nutr. 66 (3), 300–307.

Fung, S.T., Ho, C.K., Choi, S.W., et al., 2013. Comparison of catechin profiles in human plasma and urine after single dosing and regular intake of green tea (*Camellia sinensis*). Br. J. Nutr. 109 (12), 2199–2207.

Gao, K., Xu, A., Krul, C., et al., 2006. Of the major phenolic acids formed during human microbial fermentation of tea, citrus, and soy flavonoid supplements, only 3,4-dihydroxyphenylacetic acid has antiproliferative activity. J. Nutr. 136 (1), 52–57.

Graf, D., Di Cagno, R., Fak, F., et al., 2015. Contribution of diet to the composition of the human gut microbiota. Microb. Ecol. Health Dis. 26, 26164.

Grassi, D., Desideri, G., Di Giosia, P., et al., 2013. Tea, flavonoids, and cardiovascular health: endothelial protection. Am. J. Clin. Nutr. 98 (6 Suppl), 1660S–1666S.

Guo, Y., Zhi, F., Chen, P., et al., 2017. Green tea and the risk of prostate cancer: a systematic review and meta-analysis. Medicine (Baltim.) 96 (13), e6426. Available from: https://doi.org/10.1097/MD.0000000000006426.

Gupta, A.K., Daigle, D., 2015. Sinecatechins 10% ointment: a green tea extract for the treatment of external genital warts. Skin Ther. Lett. 20 (1), 6–8.

Hao, G., Li, W., Teo, K., et al., 2015. Influence of tea consumption on acute myocardial infarction in China population: the INTERHEART China study. Angiology 66 (3), 265–270.

Hao, Z., Gong, L., Shen, Q., et al., 2020. Effectiveness of concomitant use of green tea and polyethylene glycol in bowel preparation for colonoscopy: a randomized controlled study. BMC Gastroenterol. 20 (1), 150.

Harbowy, M.E., Balentine, D.A., 1997. Tea chemistry. Crit. Rev. Plant Sci. 16 (5), 415–480.

Hegazy, S.K., 2014. The effect of green tea on sildenafil pharmacokinetics in Egyptian healthy volunteers. Br. J. Pharm. Res. 4 (3), 289–300.

Hertog, M.G., Feskens, E.J., Hollman, P.C., et al., 1993. Dietary antioxidant flavonoids and risk of coronary heart disease: the Zutphen Elderly Study. Lancet 342 (8878), 1007–1011.

Hertog, M.G., Kromhout, D., Aravanis, C., et al., 1995. Flavonoid intake and long-term risk of coronary heart disease and cancer in the seven countries study. Arch. Intern. Med. 155 (4), 381–386.

Hill, A.M., Coates, A.M., Buckley, J.D., et al., 2007. Can EGCG reduce abdominal fat in obese subjects? J. Am. Coll. Nutr. 26 (4), 396S–402S.

Ho, C.K., Siu-wai, C., Siu, P.M., et al., 2013. Genoprotection and genotoxicity of green tea (*Camellia sinensis*): are they two sides of the same redox coin? Redox Rep. 18 (4), 150–154.

Hoensch, H.P., Kirch, W., 2005. Potential role of flavonoids in the prevention of intestinal neoplasia: a review of their mode of action and their clinical perspectives. Int. J. Gastrointest. Cancer 35 (3), 187–195.

Hoensch, H., Groh, B., Edler, L., et al., 2008. Prospective cohort comparison of flavonoid treatment in patients with resected colorectal cancer to prevent recurrence. World J. Gastroenterol. 14 (14), 2187–2193.

Hollman, P.C., Van Het Hof, K.H., Tijburg, L.B., et al., 2001. Addition of milk does not affect the absorption of flavonols from tea in man. Free Radic. Res. 34 (3), 297–300.

Hsu, C.H., Tsai, T.H., Kao, Y.H., et al., 2008. Effect of green tea extract on obese women: a randomized, double-blind, placebo-controlled clinical trial. Clin. Nutr. 27 (3), 363–370.

Hsu, C.H., Liao, Y.L., Lin, S.C., et al., 2011. Does supplementation with green tea extract improve insulin resistance in obese type 2 diabetics? A randomized, double-blind, and placebo-controlled clinical trial. Altern. Med. Rev. 16 (2), 157–163.

Hu, J., Webster, D., Cao, J., et al., 2018. The safety of green tea and green tea extract consumption in adults—results of a systematic review. Regul. Toxicol. Pharmacol. 95, 412–433.

Huang, J., Wang, Y., Xie, Z., et al., 2014. The anti-obesity effects of green tea in human intervention and basic molecular studies. Eur. J. Clin. Nutr. 68 (10), 1075–1087.

Isbrucker, R.A., Edwards, J.A., Wolz, E., et al., 2006a. Safety studies on epigallocatechin gallate (EGCG) preparations. Part 2: dermal, acute and short-term toxicity studies. Food Chem. Toxicol. 44 (5), 636–650.

Isbrucker, R.A., Edwards, J.A., Wolz, E., et al., 2006b. Safety studies on epigallocatechin gallate (EGCG) preparations. Part 3: teratogenicity and reproductive toxicity studies in rats. Food Chem. Toxicol. 44 (5), 651–661.

Isbrucker, R.A., Bausch, J., Edwards, J.A., et al., 2006c. Safety studies on epigallocatechin gallate (EGCG) preparations. Part 1: genotoxicity. Food Chem. Toxicol. 44, 626–635.

Jin, J.S., Touyama, M., Hisada, T., et al., 2012. Effects of green tea consumption on human fecal microbiota with special reference to *Bifidobacterium* species. Microbiol. Immunol. 56 (11), 729–739.

Jodoin, J., Demeule, M., Beliveau, R., 2002. Inhibition of the multidrug resistance P-glycoprotein activity by green tea polyphenols. Biochim. Biophys. Acta Mol. Cell Res. 1542 (1–3), 149–159.

Johnson, W.D., Morrissey, R.L., Crowell, J.A., et al., 1999. Subchronic oral toxicity of green tea polyphenols in rats and dogs. Toxicologist 48, 57–58 (Abstr. 271).

Jurgens, T.M., Whelan, A.M., Killian, L., et al., 2012. Green tea for weight loss and weight maintenance in overweight or obese adults. Cochrane Database Syst. Rev. 12, CD008650.

Kapetanovic, I.M., Crowell, J.A., Krishnaraj, R., et al., 2009. Exposure and toxicity of green tea polyphenols in fasted and non-fasted dogs. Toxicology 260 (1–3), 28–36.

Kapoor, M.P., Rao, T.P., Juneja, L.R., 2013. Green tea: history, processing techniques, principles, traditions, features, and attractions. In: Juneja, L.R., Kapoor, M.P., Okubo, T., et al.,Green Tea Polyphenols: Nutraceuticals of Modern Life. CRC Press, Boca Raton, FA, pp. 1–17.

Kim, Y., Talcott, S.T., 2012. Tea creaming in nonfermented teas from *Camellia sinensis* and *Ilex vomitoria*. J. Agric. Food Chem. 60 (47), 11793–11799.

Kim, H., Lee, J., Oh, J.H., et al., 2019. Protective effect of green tea consumption on colorectal cancer varies by lifestyle factors. Nutrients 11 (11). Available from: https://doi.org/10.3390/nu11112612.

Kotze-Horstmann, L.M., Sadie-Van Gijsen, H., 2020. Modulation of glucose metabolism by leaf tea constituents: a systematic review of recent clinical and pre-clinical findings. J. Agric. Food Chem. 68 (10), 2973–3005.

Kuo, C.L., Chen, T.S., Liou, S.Y., et al., 2014. Immunomodulatory effects of EGCG fraction of green tea extract in innate and adaptive immunity via T regulatory cells in murine model. Immunopharmacol. Immunotoxicol. 36 (5), 364–370.

Kuriyama, S., Shimazu, T., Ohmori, K., et al., 2006. Green tea consumption and mortality due to cardiovascular disease, cancer, and all causes in Japan: the Ohsaki study. JAMA 296 (10), 1255–1265.

Kwon, O.S., Jung, J.H., Shin, E.A., et al., 2020. Epigallocatechin-3-gallate induces apoptosis as a TRAIL sensitizer via activation of caspase 8 and death receptor 5 in human colon cancer cells. Biomedicines 8 (4). Available from: https://doi.org/10.3390/biomedicines8040084.

Lambert, J.D., Kennett, M.J., Sang, S., et al., 2010. Hepatotoxicity of high oral dose (-)-epigallocatechin-3-gallate in mice. Food Chem. Toxicol. 48 (1), 409–416.

Landberg, R., Naidoo, N., van Dam, R.M., 2012. Diet and endothelial function: from individual components to dietary patterns. Curr. Opin. Lipidol. 23 (2), 147–155.

Lee, M.J., Maliakal, P., Chen, L., et al., 2002. Pharmacokinetics of tea catechins after ingestion of green tea and (-)-epigallocatechin-3-gallate by humans: formation of different metabolites and individual variability. Cancer Epidemiol. Biomarkers Prev. 11 (10 Pt 1), 1025–1032.

Lee, J., Chambers, D.H., Chambers, E., et al., 2013. Volatile aroma compounds in various brewed green teas. Molecules 18 (8), 10024–10041.

Lee, J., Chambers, D.H., Chambers, E., 2014. A comparison of the flavor of green teas from around the world. J. Sci. Food Agric. 94 (7), 1315–1324.

Lee, M.-K., Kim, H.-W., Lee, S.-H., et al., 2019. Characterization of catechins, theaflavins, and flavonols by leaf processing step in green and black teas (*Camellia sinensis*) using UPLCDAD-QToF/MS. Eur. Food Res. Technol. 245, 997–1010.

Li, S., Tse, I.M., Li, E.T., 2012. Maternal green tea extract supplementation to rats fed a high-fat diet ameliorates insulin resistance in adult male offspring. J. Nutr. Biochem. 23 (12), 1655–1660.

Li, M.J., Yin, Y.C., Wang, J., et al., 2014. Green tea compounds in breast cancer prevention and treatment. World J. Clin. Oncol. 5 (3), 520–528.

Li, S., Chen, H., Wang, J., et al., 2015. Involvement of the PI3K/Akt signal pathway in the hypoglycemic effects of tea polysaccharides on diabetic mice. Int. J. Biol. Macromol. 81, 967–974.

Liang, W., Lee, A.H., Binns, C.W., et al., 2009. Tea consumption and ischemic stroke risk: a case-control study in southern China. Stroke 40 (7), 2480–2485.

Lin, Y.S., Tsai, Y.J., Tsay, J.S., et al., 2003. Factors affecting the levels of tea polyphenols and caffeine in tea leaves. J. Agric. Food Chem. 51 (7), 1864–1873.

Lin, Y.W., Hu, Z.H., Wang, X., et al., 2014. Tea consumption and prostate cancer: an updated meta-analysis. World J. Surg. Oncol. 12, 38. Available from: https://doi.org/10.1186/1477-7819-12-38.

Lin, X., Chen, Z., Zhang, Y., et al., 2015. Comparative characterization of green tea and black tea cream: physicochemical and phytochemical nature. Food Chem. 173, 432–440.

Liu, G., Mi, X.N., Zheng, X.X., et al., 2014. Effects of tea intake on blood pressure: a meta-analysis of randomised controlled trials. Br. J. Nutr. 112 (7), 1043–1054.

Liu, J., Jin, L., Zhang, Y., et al., 2015. Tea consumption is not associated with reduced plasma folate concentration among Chinese pregnant women. Birth Defects Res. A Clin. Mol. Teratol. 103 (9), 747–753.

Liu, Z., Bruins, M.E., Ni, L., et al., 2018. Green and black tea phenolics: bioavailability, transformation by colonic microbiota, and modulation of colonic microbiota. J. Agric. Food Chem. 66 (32), 8469–8477.

Liu, C., Li, P., Qu, Z., et al., 2019. Advances in the antagonism of epigallocatechin-3-gallate in the treatment of digestive tract tumors. Molecules 24 (9). Available from: https://doi.org/10.3390/molecules24091726.

Lu, C., Zhu, W., Shen, C.L., et al., 2012. Green tea polyphenols reduce body weight in rats by modulating obesity-related genes. PLoS One 7 (6), e38332. Available from: https://doi.org/10.1371/journal.pone.0038332.

Ma, Q., Kim, E.Y., Lindsay, E.A., et al., 2011. Bioactive dietary polyphenols inhibit heme iron absorption in a dose-dependent manner in human intestinal Caco-2 cells. J. Food Sci. 76 (5), H143–H150.

Mahdavi-Roshan, M., Salari, A., Ghorbani, Z., et al., 2020. The effects of regular consumption of green or black tea beverage on blood pressure in those with elevated blood pressure or hypertension: a systematic review and meta-analysis. Complement. Ther. Med. 51, 102430. Available from: https://doi.org/10.1016/j.ctim.2020.102430.

Maki, T., Pham, N.M., Yoshida, D., et al., 2010. The relationship of coffee and green tea consumption with high-sensitivity C-reactive protein in Japanese men and women. Clin. Chem. Lab. Med. 48 (6), 849–854.

Maranhao, R.C., Filho, A.C., Sigal, G.A., et al., 2018. HDL and endothelium. In: da Luz, P.L., Libby, P., Chagas, A.C.P., et al.,Endothelium and Cardiovascular Diseases: Vascular Biology and Clinical Syndromes. Academic Press, San Diego, CA, pp. 297–317.

Matsumoto, M., Ishigaki, F., Ishigaki, A., et al., 1993. Reduction of blood glucose levels by tea catechin. Biosci. Biotech. Biochem. 57 (4), 525–527.

McEwen, B.J., 2015. The influence of herbal medicine on platelet function and coagulation: a narrative review. Semin. Thromb. Hemost. 41 (3), 300–314.

Md Nesran, Z.N., Shafie, N.H., Md Tohid, S.F., et al., 2020. Iron chelation properties of green tea epigallocatechin-3-gallate (EGCG) in colorectal cancer cells: analysis on Tfr/Fth regulations and molecular docking. Evid. Based Complement. Altern. Med. 2020, 7958041. Available from: https://doi.org/10.1155/2020/7958041.

Mennen, L., Hirvonen, T., Arnault, N., et al., 2007. Consumption of black, green and herbal tea and iron status in French adults. Eur. J. Clin. Nutr. 61 (10), 1174–1179.

Mielgo-Ayuso, J., Barrenechea, L., Alcorta, P., et al., 2014. Effects of dietary supplementation with epigallocatechin-3-gallate on weight loss, energy homeostasis, cardiometabolic risk factors and liver function in obese women: randomised, double-blind, placebo-controlled clinical trial. Br. J. Nutr. 111 (7), 1263–1271.

Mineharu, Y., Koizumi, A., Wada, Y., et al., 2011. Coffee, green tea, black tea and oolong tea consumption and risk of mortality from cardiovascular disease in Japanese men and women. J. Epidemiol. Commun. Health 65 (3), 230–240.

Misaka, S., Kawabe, K., Onoue, S., et al., 2013a. Effects of green tea catechins on cytochrome P450 2B6, 2C8, 2C19, 2D6 and 3A activities in human liver and intestinal microsomes. Drug Metab. Pharmacokinet. 28 (3), 244–249.

Misaka, S., Kawabe, K., Onoue, S., et al., 2013b. Green tea extract affects the cytochrome P450 3A activity and pharmacokinetics of simvastatin in rats. Drug Metab. Pharmacokinet. 28 (6), 514–518.

Mitscher, L.A., Jung, M., Shankel, D., et al., 1997. Chemoprotection: a review of the potential therapeutic antioxidant properties of green tea (*Camellia sinensis*) and certain of its constituents. Med. Res. Rev. 17 (4), 327–365.

Miyata, Y., Shida, Y., Hakariya, T., et al., 2019. Anti-cancer effects of green tea polyphenols against prostate cancer. Molecules 24 (1). Available from: https://doi.org/10.3390/molecules24010193.

Miyoshi, N., Tanabe, H., Suzuki, T., et al., 2020. Applications of a standardized green tea catechin preparation for viral warts and human papilloma virus-related and unrelated cancers. Molecules 25 (11). Available from: https://doi.org/10.3390/molecules25112588.

Musial, C., Kuban-Jankowska, A., Gorska-Ponikowska, M., 2020. Beneficial properties of green tea catechins. Int. J. Mol. Sci. 21 (5). Available from: https://doi.org/10.3390/ijms21051744.

Nakamura, Y., 2017. Characteristics of Japanese green tea. In: Hara, Y., Yang, C. S., Isemura, M., et al.,Health Benefits of Green Tea: An Evidence-Based Approach. CABI, Boston, MA, pp. 9–18.

Ninomiya, T., Kanzaki, N., Hirakawa, Y., et al., 2019. Serum ethylamine levels as an indicator of l-theanine consumption and the risk of type 2 diabetes in a general Japanese population: the Hisayama study. Diabetes Care 42 (7), 1234–1240.

Nishikawa, M., Ariyoshi, N., Kotani, A., et al., 2004. Effects of continuous ingestion of green tea or grape seed extracts on the pharmacokinetics of midazolam. Drug Metab. Pharmacokinet. 19 (4), 280–289.

Nishiumi, S., Bessyo, H., Kubo, M., et al., 2010. Green and black tea suppress hyperglycemia and insulin resistance by retaining the expression of glucose transporter 4 in muscle of high-fat diet-fed C57BL/6J mice. J. Agric. Food Chem. 58 (24), 12916–12923.

Nunez-Sanchez, M.A., Gonzalez-Sarrias, A., Romo-Vaquero, M., et al., 2015. Dietary phenolics against colorectal cancer—from promising preclinical results to poor translation into clinical trials: pitfalls and future needs. Mol. Nutr. Food Res. 59 (7), 1274–1291.

Odegaard, A.O., Pereira, M.A., Koh, W.P., et al., 2008. Coffee, tea, and incident type 2 diabetes: the Singapore Chinese Health Study. Am. J. Clin. Nutr. 88 (4), 979–985.

Odegaard, A.O., Koh, W.P., Yuan, J.M., et al., 2015. Beverage habits and mortality in Chinese adults. J. Nutr. 145 (3), 595–604.

Ogunleye, A.A., Xue, F., Michels, K.B., 2010. Green tea consumption and breast cancer risk or recurrence: a meta-analysis. Breast Cancer Res. Treat. 119 (2), 477–484.

Oketch-Rabah, H.A., Roe, A.L., Rider, C.V., et al., 2020. United States Pharmacopeia (USP) comprehensive review of the hepatotoxicity of green tea extracts. Toxicol. Rep. 7, 386–402.

Pae, M., Wu, D., 2013. Immunomodulating effects of epigallocatechin-3-gallate from green tea: mechanisms and applications. Food Funct. 4 (9), 1287–1303.

Pang, J., Zhang, Z., Zheng, T., et al., 2015. Association of green tea consumption with risk of coronary heart disease in Chinese population. Int. J. Cardiol. 179, 275–278.

Payab, M., Hasani-Ranjbar, S., Shahbal, N., et al., 2020. Effect of the herbal medicines in obesity and metabolic syndrome: a systematic review and meta-analysis of clinical trials. Phytother. Res. 34 (3), 526–545.

Peluso, I., Serafini, M., 2017. Antioxidants from black and green tea: from dietary modulation of oxidative stress to pharmacological mechanisms. Br. J. Pharmacol. 174 (11), 1195–1208.

Perletti, G., Magri, V., Vral, A., et al., 2019. Green tea catechins for chemoprevention of prostate cancer in patients with histologically-proven HG-PIN or ASAP. Concise review and meta-analysis. Arch. Ital. Urol. Androl. 91 (3). Available from: https://doi.org/10.4081/aiua.2019.3.153.

Phung, O.J., Baker, W.L., Matthews, L.J., et al., 2010. Effect of green tea catechins with or without caffeine on anthropometric measures: a systematic review and meta-analysis. Am. J. Clin. Nutr. 91 (1), 73–81.

Posadzki, P., Watson, L., Ernst, E., 2013. Herb-drug interactions: an overview of systematic reviews. Br. J. Clin. Pharmacol. 75 (3), 603–618.

Puviani, M., Galluzzo, M., Talamonti, M., et al., 2019. Efficacy of sinecatechins 10% as proactive sequential therapy of external genital warts after laser CO_2 ablative therapy: the PACT study (post-ablation immunomodulator treatment of condylomata with sinecatechins): a randomized, masked outcome assessment, multicenter trial. Int. J. STD AIDS 30 (2), 131–136.

Rahbar, S., 2005. The discovery of glycated hemoglobin: a major event in the study of nonenzymatic chemistry in biological systems. Ann. N. Y. Acad. Sci. 1043, 9–19.

Rains, T.M., Agarwal, S., Maki, K.C., 2011. Antiobesity effects of green tea catechins: a mechanistic review. J. Nutr. Biochem. 22 (1), 1–7.

Rothenberg, D.O., Zhou, C., Zhang, L., 2018. A review on the weight-loss effects of oxidized tea polyphenols. Molecules 23 (5). Available from: https://doi.org/10.3390/molecules23051176.

Rusak, G., Komes, D., Likic, S., et al., 2008. Phenolic content and antioxidative capacity of green and white tea extracts depending on extraction conditions and the solvent used. Food Chem. 110 (4), 852–858.

Saeed, M., Naveed, M., Arif, M., et al., 2017. Green tea (*Camellia sinensis*) and l-theanine: medicinal values and beneficial applications in humans-A comprehensive review. Biomed. Pharmacother. 95, 1260–1275.

Samavat, H., Dostal, A.M., Wang, R., et al., 2015. The Minnesota Green Tea Trial (MGTT), a randomized controlled trial of the efficacy of green tea extract on biomarkers of breast cancer risk: study rationale, design, methods, and participant characteristics. Cancer Causes Control 26 (10), 1405–1419.

Sampath, C., Rashid, M.R., Sang, S., et al., 2017. Green tea epigallocatechin 3-gallate alleviates hyperglycemia and reduces advanced glycation end products via nrf2 pathway in mice with high fat diet-induced obesity. Biomed. Pharmacother. 87, 73–81.

Sanchez, M., Gonzalez-Burgos, E., Iglesias, I., et al., 2020. The pharmacological activity of *Camellia sinensis* (L.) Kuntze on metabolic and endocrine disorders: a systematic review. Biomolecules 10 (4). Available from: https://doi.org/10.3390/biom10040603.

Sang, S., Lambert, J.D., Ho, C.T., Yang, C.S., 2011. The chemistry and biotransformation of tea constituents. Pharmacol. Res. 64 (2), 87–99.

Sanzini, E., Badea, M., Santos, A.D., et al., 2011. Quality control of plant food supplements. Food Funct. 2 (12), 740–746.

Satoh, T., Fujisawa, H., Nakamura, A., et al., 2016. Inhibitory effects of eight green tea catechins on cytochrome P450 1A2, 2C9, 2D6, and 3A4 activities. J. Pharm. Pharm. Sci. 19 (2), 188–197.

Seely, D., Mills, E.J., Wu, P., et al., 2005. The effects of green tea consumption on incidence of breast cancer and recurrence of breast cancer: a systematic review and meta-analysis. Integr. Cancer Ther. 4 (2), 144–155.

Shafique, K., McLoone, P., Qureshi, K., et al., 2012. Tea consumption and the risk of overall and grade specific prostate cancer: a large prospective cohort study of Scottish men. Nutr. Cancer 64 (6), 790–797.

Shimizu, M., Fukutomi, Y., Ninomiya, M., et al., 2008. Green tea extracts for the prevention of metachronous colorectal adenomas: a pilot study. Cancer Epidemiol. Biomarkers Prev. 17 (11), 3020–3025.

Shimizu, M., Shirakami, Y., Sakai, H., et al., 2015. Chemopreventive potential of green tea catechins in hepatocellular carcinoma. Int. J. Mol. Sci. 16 (3), 6124–6139.

Shiraishi, M., Haruna, M., Matsuzaki, M., et al., 2010. Association between the serum folate levels and tea consumption during pregnancy. Biosci. Trends 4 (5), 225–230.

Shu, Y., Li, J., Yang, X., et al., 2019. Effect of particle size on the bioaccessibility of polyphenols and polysaccharides in green tea powder and its antioxidant activity after simulated human digestion. J. Food Sci. Technol. 56 (3), 1127–1133.

Sproston, N.R., Ashworth, J.J., 2018. Role of C-reactive protein at sites of inflammation and infection. Front. Immunol. 9, 754. Available from: https://doi.org/10.3389/fimmu.2018.00754.

Stalmach, A., Troufflard, S., Serafini, M., et al., 2009. Absorption, metabolism and excretion of Choladi green tea flavan-3-ols by humans. Mol. Nutr. Food Res. 53 (Suppl 1), S44–53.

Stalmach, A., Mullen, W., Steiling, H., et al., 2010. Absorption, metabolism, and excretion of green tea flavan-3-ols in humans with an ileostomy. Mol. Nutr. Food Res. 54 (3), 323–334.

Takahashi, M., Miyashita, M., Suzuki, K., et al., 2014. Acute ingestion of catechin-rich green tea improves postprandial glucose status and increases serum thioredoxin concentrations in postmenopausal women. Br. J. Nutr. 112 (9), 1542–1550.

Teschke, R., Frenzel, C., Glass, X., et al., 2013. Herbal hepatotoxicity: a critical review. Br. J. Clin. Pharmacol. 75 (3), 630–636.

van het Hof, K.H., Kivits, G.A., Weststrate, J.A., et al., 1998. Bioavailability of catechins from tea: the effect of milk. Eur. J. Clin. Nutr. 52 (5), 356–359.

Wada, K., Oba, S., Tsuji, M., et al., 2019. Green tea intake and colorectal cancer risk in Japan: the Takayama study. Jpn. J. Clin. Oncol. 49 (6), 515–520.

Wan, X., Li, D., Zhang, Z., 2009. Green tea and black tea manufacturing and consumption. In: Ho, C.-T., Lin, J.-K., Shahidi, F. (Eds.), Tea and Tea Products Chemistry and Health-Promoting Properties. CRC Press, Boca Raton, FA, pp. 1–8.

Wang, X., Tian, J., Jiang, J., et al., 2014a. Effects of green tea or green tea extract on insulin sensitivity and glycaemic control in populations at risk of type 2 diabetes mellitus: a systematic review and meta-analysis of randomised controlled trials. J. Hum. Nutr. Diet. 27 (5), 501–512.

Wang, S., Moustaid-Moussa, N., Chen, L., et al., 2014b. Novel insights of dietary polyphenols and obesity. J. Nutr. Biochem. 25 (1), 1–18.

Wang, J., Cheung, W., Leung, D., 2014c. Determination of pesticide residue transfer rates (percent) from dried tea leaves to brewed tea. J. Agric. Food Chem. 62 (4), 966–983.

Wang, Y., Zhao, Y., Chong, F., et al., 2020a. A dose-response meta-analysis of green tea consumption and breast cancer risk. Int. J. Food Sci. Nutr. 1–12.

Wang, Z.T., Xue, Y., Sun, H., et al., 2020b. Effect of tea polyphenols on the oral and intravenous pharmacokinetics of ticagrelor in rats and its *in vitro* metabolism. J. Food Sci. 85 (4), 1285–1291.

Wei, C., Yang, H., Wang, S., et al., 2018. Draft genome sequence of *Camellia sinensis* var. *sinensis* provides insights into the evolution of the tea genome and tea quality. Proc. Natl. Acad. Sci. USA 115 (18), E4151–E4158.

Werba, J.P., Misaka, S., Giroli, M.G., et al., 2018. Update of green tea interactions with cardiovascular drugs and putative mechanisms. J. Food Drug Anal. 26 (2S), S72–S77.

Wharton, S., Bonder, R., Jeffery, A., et al., 2020. The safety and effectiveness of commonly-marketed natural supplements for weight loss in populations with obesity: a critical review of the literature from 2006 to 2016. Crit. Rev. Food Sci. Nutr. 60 (10), 1614–1630.

Wheeler, D.S., Wheeler, W.J., 2004. Medical history of tea. Drug Dev. Res. 61 (2), 45–65.

Wierzejska, R., 2014. Tea and health—a review of the current state of knowledge. Przegl Epidemiol. 68 (3), 501–506.

Williams, J., Sergi, D., McKune, A.J., et al., 2019. The beneficial health effects of green tea amino acid L-theanine in animal models: promises and prospects for human trials. Phytother. Res. 33 (3), 571–583.

Wu, L.Y., Juan, C.C., Ho, L.T., et al., 2004. Effect of green tea supplementation on insulin sensitivity in Sprague-Dawley rats. J. Agric. Food Chem. 52 (3), 643–648.

Xia, E.H., Tong, W., Wu, Q., et al., 2020a. Tea plant genomics: achievements, challenges and perspectives. Hortic. Res. 7, 7. Available from: https://doi.org/10.1038/s41438-019-0225-4.

Xia, E., Tong, W., Hou, Y., et al., 2020b. The reference genome of tea plant and resequencing of 81 diverse accessions provide insights into its genome evolution and adaptation. Mol. Plant . Available from: https://doi.org/10.1016/j.molp.2020.04.010.

Xiao, J., Hogger, P., 2014. Influence of diabetes on the pharmacokinetic behavior of natural polyphenols. Curr. Drug Metab. 15 (1), 23–29.

Xiao, H., Hao, X., Simi, B., et al., 2008. Green tea polyphenols inhibit colorectal aberrant crypt foci (ACF) formation and prevent oncogenic changes in dysplastic ACF in azoxymethane-treated F344 rats. Carcinogenesis 29 (1), 113–119.

Xu, Y.Q., Chen, G.S., Wang, Q.S., et al., 2012. Irreversible sediment formation in green tea infusions. J. Food Sci. 77 (3), C298–C302.

Xu, Y.Q., Hu, X.F., Tang, P., et al., 2015. The major factors influencing the formation of sediments in reconstituted green tea infusion. Food Chem. 172, 831–835.

Yamagata, K., 2020. Protective effect of epigallocatechin gallate on endothelial disorders in atherosclerosis. J. Cardiovasc. Pharmacol. 75 (4), 292–298.

Yan, Z., Zhong, Y., Duan, Y., et al., 2020. Antioxidant mechanism of tea polyphenols and its impact on health benefits. Anim. Nutr. 6 (2), 115–123.

Yang, C.S., Wang, H., Li, G.X., et al., 2011. Cancer prevention by tea: evidence from laboratory studies. Pharmacol. Res. 64 (2), 113–122.

Yang, P., He, X., Malhotra, A., 2014. Epigenetic targets of polyphenols in cancer. J. Environ. Pathol. Toxicol. Oncol. 33 (2), 159–165.

Yang, J., Luo, F., Zhou, L., et al., 2020. Residue reduction and risk evaluation of chlorfenapyr residue in tea planting, tea processing, and tea brewing. Sci. Total Environ. 738, 139613.

Yao, Z., Gu, Y., Zhang, Q., et al., 2019. Estimated daily quercetin intake and association with the prevalence of type 2 diabetes mellitus in Chinese adults. Eur. J. Nutr. 58 (2), 819–830.

Yashin, A., Nemzer, B., Yashin, Y., 2012. Bioavailability of tea components. J. Food Res. 1 (2), 281–290.

Yonekura, Y., Terauchi, M., Hirose, A., et al., 2020. Daily coffee and green tea consumption is inversely associated with body mass index, body fat percentage, and cardio-ankle vascular index in middle-aged Japanese women: a cross-sectional study. Nutrients 12 (5), 1370.

Yu, Z., Samavat, H., Dostal, A.M., et al., 2017. Effect of green tea supplements on liver enzyme elevation: results from a randomized intervention study in the United States. Cancer Prev. Res. (Phila.) 10 (10), 571–579.

Zeng, L., Ma, M., Li, C., et al., 2017. Stability of tea polyphenols solution with different pH at different temperatures. Int. J. Food Prop. 20 (1), 1–18.

Zhang, S., Shan, L., Li, Q., et al., 2014. Systematic analysis of the multiple bioactivities of green tea through a network pharmacology approach. Evid. Based Complement. Altern. Med. 2014, 512081.

Zhang, L., Ho, C.-T., Zhou, J., et al., 2019. Chemistry and biological activities of processed *Camellia sinensis* teas: a comprehensive review. Compr. Rev. Food Sci. Food Saf. 18 (5), 1474–1495.

Zhang, J., Cao, Q.-Q., Granato, D., et al., 2020a. Association between chemistry and taste of tea: a review. Trends Food Sci. Technol. 101, 139–149.

Zhang, Y., Gu, M., Wang, R., et al., 2020b. Dietary supplement of Yunkang 10 green tea and treadmill exercise ameliorate high fat diet induced metabolic syndrome of C57BL/6 J mice. Nutr. Metab. (Lond.) 17, 14. Available from: https://doi.org/10.1186/s12986-020-0433-9.

Zhang, D., Nichols, H.B., Troester, M., et al., 2020c. Tea consumption and breast cancer risk in a cohort of women with family history of breast cancer. Int. J. Cancer 147 (3), 876–886.

Zheng, J., Yang, B., Huang, T., et al., 2011. Green tea and black tea consumption and prostate cancer risk: an exploratory meta-analysis of observational studies. Nutr. Cancer 63 (5), 663–672.

Zijp, I.M., Korver, O., Tijburg, L.B., 2000. Effect of tea and other dietary factors on iron absorption. Crit. Rev. Food Sci. Nutr. 40 (5), 371–398.

Zinellu, A., Sotgia, S., Scanu, B., et al., 2014. Evaluation of non-covalent interactions between serum albumin and green tea catechins by affinity capillary electrophoresis. J. Chromatogr. A 1367, 167–171.

Chapter 42

Green coffee beans

Satish K. Garg, Amit Shukla and Soumen Choudhury

Introduction and background

Coffee is one of the most commonly consumed functional beverages throughout the world. As per the reports available, compared to the consumption of 158 million bags in 2017–18, around 165.35 million 60 kg bags of coffee were consumed worldwide during 2018–19 (https://www.statista.com/statistics/292595/global-coffee-consumption). It is the most valuable primary product globally, employing over 26 million people along the chain from cultivation mainly in developing countries to consumption in developed countries (http://www.ico.org/new_historical.asp). Coffee is prepared from the roasted or baked seeds of several species of an evergreen shrub of the genus *Coffea*. Coffee has been consumed for more than 1000 years, and today it is one of the most consumed drinks (more than 400 billion cups annually) in the world (Sobésa Café, 2008). The beverage coffee began to be consumed over the globe before 1700s and was termed as the "ancient wonder drug" due to its great potential and physiological effects (Bothiraj et al., 2020). Coffee is an important source of several nutritious and therapeutic phytoconstituents including lipids, carbohydrates, minerals, vitamins, and nitrogenous compounds along with bioactive compounds like cafestol and kahweol diterpenes, caffeine, and chlorogenic acid (CGA) (Akash et al., 2014), which all possess great therapeutic potential.

Coffee serves as one of the major cash crops in the developing world and its exports account for more than 50% of total foreign exchange earnings in some developing countries (www.en.wikipedia.org/wiki/Coffee). Therefore coffee is an economically important commodity marketed internationally as green coffee (Wintgens, 2009). The word "coffee" originated from the Turkish word "KAUVEH" and the Arabic word "QAHWAH." "KAUVEH or QAHWAH" means "infusion or beverage or wine or some type of wine" and is considered to be a derivative of the herb root "qahiya," which indicates having "no appetite," causing no appetite, or an appetite suppressant (www.en.wikipedia.org/wiki/Coffee). Some people also consider the Ethiopian Kaffa Province as the center of origin, diversification, and dissemination of coffee plant (Mekuria et al., 2004). Ethiopian mountains are home to the native forests of coffee plant, which is a type of shrub. The plant possesses petite inflorescence with bisexual white flowers called glomerules. The fruits have a smooth exocarp, pulpy mesocarp, and fibrous endocarp surrounding the seeds, decoratively termed coffee beans. The hard endocarp of the coffee fruit protects it from frugivorous animals including monkeys (Castro and Marraccini, 2006).

Coffee bean is the seed of coffee plant or shrub (*Coffea* L.; family—*Rubiaceae*) having an average height of 5–10 m (Fig. 42.1). There are more than 70 different species of this genera. The seeds found inside the red or purple fruit are called "cherries" or "berries." Coffee cherries mostly contain two stones held together, but 10%–15% of berries contain only a single seed, termed as "peaberry." It is believed that these have more flavor than the normal coffee beans. Water, sucrose, and other reducing sugars are present in mesocarp of the fruits of coffee plant. The presence of volatile as well as nonvolatile compounds is responsible for the aroma of coffee bean.

Green coffee bean (GCB) refers to the unroasted mature or immature coffee bean. Of several species of *Coffea*, *Coffea arabica*, and *Coffea caniphora* (Robusta) are the

FIGURE 42.1 Coffee plant. *(Source: http://itsnotonlyinstantcoffee.com/tag/coffee-cultivation/)*

Nutraceuticals. DOI: https://doi.org/10.1016/B978-0-12-821038-3.00042-2

FIGURE 42.2 Coffee plant bearing unripe (*green*) and ripe (*red*) berries. *(Source: http://www.canstockphoto.com/images-photos/coffee-berry.html).*

major ones, and these account for almost 75% and 25% of the total coffee produced in the world, respectively (Etienne, 2005; Belitz et al., 2009). *C. arabica* grows well at high altitudes (600–2000 m), while *C. canephora* adapts well to lower (<600 m) altitudes (Comité Français du Café, 1997). Annual rainfall of approximately 15–30 cm and temperatures of 15°C–24°C are favorable for cultivation of *Arabica* coffee, and temperatures of approximately 24°C–30°C are favorable for *Robusta coffee* (www.en.wikipedia.org/wiki/Coffee-bean). Heavy rainfall is useful during development of the fruit but it requires less rain during the ripening stage. Coffee plant usually starts producing flowers within 3–4 years of its plantation and the first usable crop is harvested during the fifth year. As the tree grows older, it branches less and less but bears more and more leaves and fruits. After emergence of flowers, *Arabica* berries ripen in 6–8 months while *Robusta* berries ripen in 9–11 months. When coffee beans are immature, they are green in color but turn brown to yellow or red when mature (Fig. 42.2); and each dry bean on an average weighing approximately 300–330 mg (www.en.wikipedia.org/wiki/Coffee-production).

Historical perspectives

Oromo people are believed to have been the first to recognize the energizing effect of coffee plant. During the ninth century, a goat herder noticed the excitatory effect of coffee on his goats after they had consumed coffee beans. Leaves of coffee plant, discovered in Ethiopia in the 11th century, considered as magical fruit, were used in concoction, and were thought to have medicinal properties. The first cultivation of coffee in the Caribbean countries (Cuba, Haiti, Jamaica, Santo Domingo, and Puerto Rico) has been reported between 1715 and 1730, and subsequently in Brazil (1727) and South America (1730). By 1788, the Dominican Republic had become the world leader by supplying half of the world's coffee; and by 1949, Haiti had become the third largest coffee exporter in the world (www.en.wikipedia.org/wiki/Coffee). Sufi Muslim monasteries around Mocha in Yemen roasted and brewed coffee seeds. From Yemen, coffee seeds were smuggled to India, and these were initially planted in Mysore. Thereafter, coffee plantation spread to Italy, rest of the Europe, Indonesia, and America (www.en.wikipedia.org/wiki/Coffee).

Coffee was not only enjoyed in homes, but also in many public coffee houses described as "qahveh khaneh," and these became the "school for wise-man talk." Later, in 1555, coffee was introduced to Istanbul during the reign of Sultan Suleiman. The position of chief coffee maker (kahvecibaşı) was added to the court functionaries' roster. In 1615, Europeans got their first organoleptic tinge of coffee with the help of Venetian merchants who carried coffee from Istanbul and later, in 1645, the first coffee-house was opened in Italy. England became acquainted with coffee for the first time in 1637 with the help of a Turk who introduced it to Oxford. The first coffee-house in Oxford opened in 1650, and was given the name "Angel." North America tasted coffee for the first time in 1668 and the first coffee-house in New York was opened in 1696 and was named as "The King's Arms."

Today, the top five coffee-producing countries in the world are Brazil, Vietnam, Indonesia, Columbia, and Ethiopia, which account for almost 65% of the total global coffee production. India ranks seventh in the world, and produces 4.1% of the total global production (www.en.wikipedia.org/wiki/Coffee). *Arabica* coffee seeds are cultivated in Latin America, Eastern Africa, Arabia and Asia while *Robusta* coffee seeds are grown in Western and Central Africa, throughout Southeast Asia, and to a limited extent in Brazil. Seeds from different countries or regions can usually be distinguished by differences in their flavor, aroma, body, and acidity. The taste characteristics depend not only on the region where they are grown but also on the genetic subspecies and processing methodologies (www.en.wikipedia.org/wiki/Coffee).

Global scenario of coffee production

As per the report of Szenthe (2019), the top 10 green coffee-producing countries of the world are listed in Table 42.1.

South America exports almost 45% of the total world consumption of coffee, but most of this coffee is grown in Brazil. The United States imports more coffee than any other nation and the total value of imported coffee exceeds US$8 billion annually.

TABLE 42.1 Major coffee-producing countries in the world.

Ranking	Country	Production (metric tons)
1	Brazil	2,595,000
2	Vietnam	1,650,000
3	Colombia	810,000
4	Indonesia	660,000
5	Ethiopia	384,000
6	Honduras	348,000
7	India	348,000
8	Uganda	288,000
9	Mexico	234,000
10	Guatemala	204,000

Processing of green coffee beans

Ripe fruits can be "hand-picked" by selectively plucking only the ripe fruits or "strip-picked," wherein all the fruits are removed from branch at once. After picking, green coffee is processed either by wet or dry process methods. In wet processing, the flesh of the cherry is separated from seeds and then seeds are soaked in water for approximately 2 days and fermented. This method is practiced in Central America and areas of Africa. Dry processing, also known as unwashed or natural coffee, is the old method of processing, wherein the entire cherry is first cleaned after harvesting and then placed or spread out in open sun on a concrete or brick floor or tables for drying for 2–3 weeks, being turned regularly for even drying. This method is inexpensive and simpler, and it is used for approximately 90% of the *Arabica* coffee produced in Brazil, and also for most of the coffee produced in Ethiopia, Haiti, and Paraguay, and also for some *Arabica* coffee grown in India and Ecuador.

Almost all *Robusta* coffee is processed by dry method (www.en.wikipedia.org/wiki/Coffee-Production). Apart from the afore-mentioned two processes, the semidry process, also termed as "wet-hulled," "semiwashed," or "Gilling Basah," is also used in Indonesia and Brazil (www.en.wikipedia.org/wiki/Coffee-Production). After processing, the husk is removed and seeds are roasted; thereafter the seeds become brown (Fig. 42.3) due to caramelization, which breaks down starches into simple sugars, and these sugars alter the color of coffee (www.en.wikipedia.org/wiki/Coffee-Production). During roasting, coffee is graded on the basis of size of the bean, its place of origin, especially altitude, process of plucking and preparation, taste, and cup quality (www.en.wikipedia.org/wiki/Coffee-Production). Aging of coffee beans also influences the taste and flavor. Some people disagree about whether overly aged coffee beans lose much of their essential oil content and thus flavor and taste, too. Most coffee experts agree that green coffee peaks in flavor and freshness within 1 year of harvesting because overly aged GCB loses much of its essential oil content (www.en.wikipedia.org/wiki/Coffee-Production).

FIGURE 42.3 Unroasted and roasted coffee beans. *(Source: http://www.123rf.com/).*

Major phytoconstituents of green coffee beans

Coffee is lauded for its aroma and flavor, and these features are imparted to it by certain constitutive alkaloids, trigonelline, proteins, amino acids, and carbohydrates. Nonvolatile and volatile compounds present in coffee beans not only contribute to its flavor when roasted but also deter insects from damaging beans (www.en.wikipedia.org/wiki/Coffee-bean). Coffee is a complex mixture of more than 1000 diverse phytoconstituents, including cellulose, soluble carbohydrates, insoluble polysaccharides, nonvolatile and volatile compounds, aliphatic acids, nitrogenous compounds, polyphenols, proteins, lipids, amino acids, vitamins, and minerals (Higdon and Frei, 2006). Bitterness in coffee beans has been positively correlated with the amount of caffeine (Farah et al., 2006a,b).

Secondary metabolites present in GCB have been reported to be endowed with certain pharmacological activities of therapeutic importance, such as central nervous system stimulant, peripheral vasoconstrictor, and diuretic (Affonso et al., 2016; Alsunni, 2015; Takahashi and Ishigami, 2017; Vasanthi et al., 2012). The antioxidant capacity of GCB is attributed to the presence of caffeine and high amount of phenolic compounds. Chemical processing, such as roasting, significantly decreases the amount of CGA, esters of the *trans*-cinnamic acids, the amount of which in green coffees is significantly reduced during the roasting process.

Mahajan and Kapoor (2018) have reported the highest amount of phenolic compounds in methanolic extract ($16.84 \pm 0.12\ \mu g$ Gallic acid equivalents (GAE)/g dry weight) compared to $4.857 \pm 0.23\ \mu g$ GAE/g dry weight in aqueous extract of the GCB. Extraction of the phenolic compounds depends on the polarity of the solvents used. Highly polar solvents extracted a greater quantity of phenolic compounds as compared to the nonpolar or less-polar solvents. Flavanoids are also present in GCB but extraction procedures do not significantly influence their quantity after extraction. It has been reported that mild roasting did not cause any difference in the flavanoids content of GCB. There is a good correlation between the phenolic compounds and flavanoids (Hecimovic et al., 2011).

Major phytochemical constituents present in GCB, which are important from nutraceutical and pharmacotherapeutic perspectives, are summarized here.

Phenolic compounds

More than 8000 phenolic compounds with benzene/aromatic rings are produced as secondary metabolites in plants and these mainly act as protecting agents for flora against abiotic stress including changes in water content, temperature, and also against invading pathogens (Monteiro and Farah, 2012). These phenolic compounds mainly have two carbon skeletons with a hydroxycinnamic and hydroxybenzoic nucleus. Among all the taxa of coffee plants, the most commonly found phenolic compounds are caffeic, vanillic, *p*-coumaric, protocatechuic, and ferulic acid (Stalikas, 2007). 5-Caffeoylquinic acid (5-CQA) is the most abundant soluble ester among the coffee species. Chemical analysis of the beans of *C. canephora* revealed the presence of feruloylquinic acids (3-, 4-, and 5-FQA) and its isomers as monoester (3-, 4-, and 5-CQA) and diester (3,4-, 3,5-, and 4,5- diCQA) CQAs. Temperature and radiation adversely affect the total phenolics content. However, the stage of fruit production does not have significant effect on phenolics content in the plant (Salgado et al., 2008).

Alkaloids

Monoterpenoid alkaloids are characteristic of the *Rubiaceae* family. Caffeine, the most important alkaloid of the coffee species, is synthesized in young leaves of *C. arabica* and immature fruits, and accumulated in mature leaves. The upper parts of leaves and stem are responsible for synthesis of active caffeine. However, it is not found in lower parts of leaves, stem, and roots (Ashihara et al., 2008). Three-dimensional high-performance liquid chromatography (HPLC) chromatogram (Fig. 42.4) and ultra-high performance liquid chromatography- mass spectrophometry (UPLC-MS) chromatogram of green coffee bean extract (GCBE) (Fig. 42.5) revealed that 3-, 4-, and 5-caffeoylquinic acids (3-, 4-, and 5-CQA), 3,4-, 3,5-, and 4,5-dicaffeoylquinic acids (3,4-, 3,5-, and 4,5-DCQA), and 3-, 4-, and 5-feruloylquinic acids (3-, 4-, and 5-FQA) are the major hydroxycinnamic acids present in GCB (Shimoda et al., 2006; Alonso-Salces et al., 2009; Gawlik-Dziki et al., 2014). The chemical structures of different CGA present in coffee are shown in Fig. 42.6. CGA is an important and valuable polyphenol and an inexpensive source of dietary phenol. It is comprised of caffeic acid, ferulic acid, coumaric acid, and 3,4-

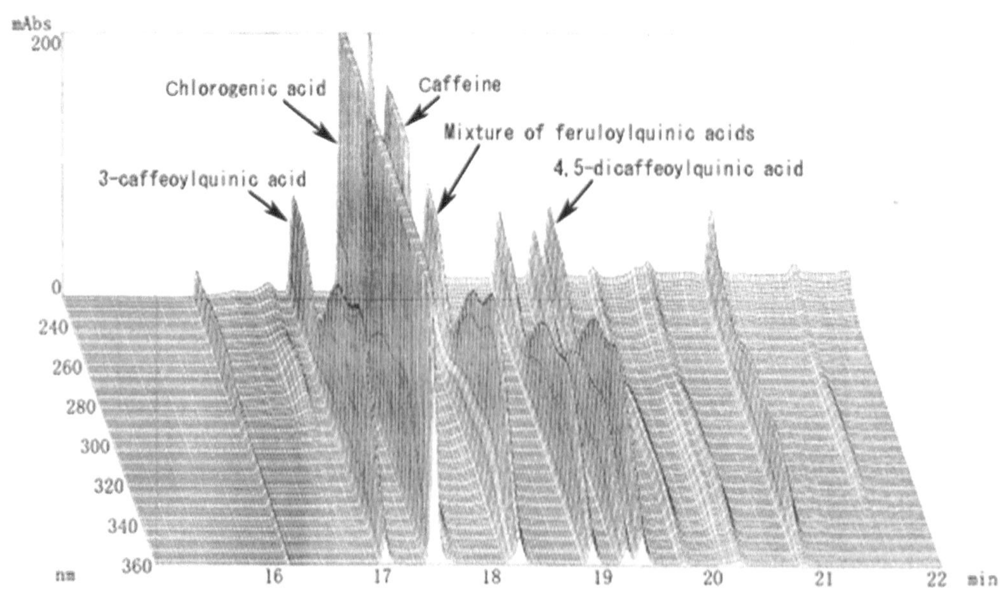

FIGURE 42.4 Three-dimensional high-performance liquid chromatography chromatogram of green coffee bean extract. *(Source: Shimoda, H., Seki, E., Aitani, M., 2006. Inhibitory effect of green coffee bean extract on fat accumulation and body weight gain in mice. BMC Complement. Altern. Med., 6, 1–9).*

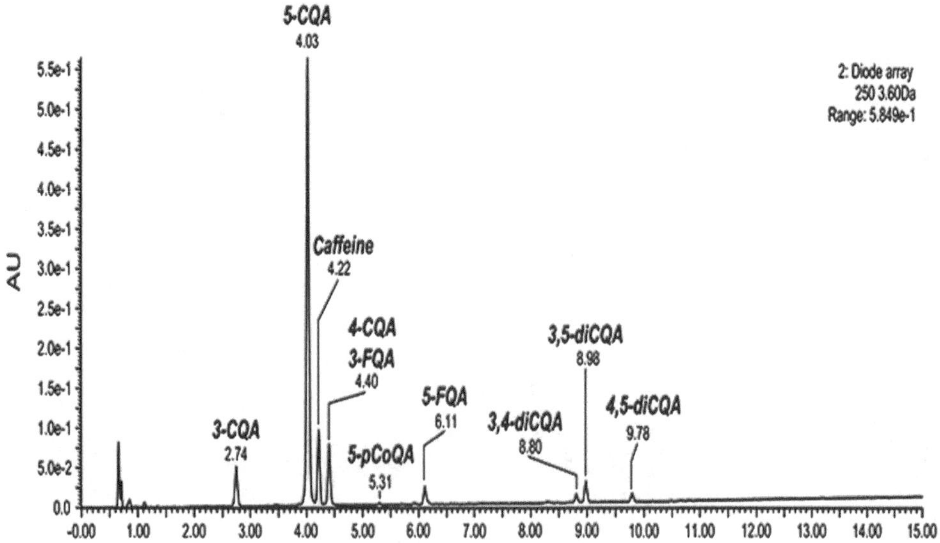

FIGURE 42.5 UPLC-MS chromatogram of green coffee samples. *3-CQA*, 3-caffeoylquinic acid; *5-CQA*, 5-caffeoylquinic acid; *4-CQA*, 4-caffeoylquinic acid; *3-FQA*, 3-feruoylquinic acid; *5-FQA*, 5-feruoylquinic acid; *5-pCoQA*, 5-p-coumaroylquinic acid; *3,4-diCQA*, 3,4-dicaffeoylquinic acid; *3,5-diCQA*, 3,5-dicaffeoylquinic acid; *4,5-diCQA*, 4,5-dicaffeoylquinic acid. *(Source: Gawlik-Dziki, U., Świeca, M., Dziki, D., et al., 2014. Lipoxygenase inhibitors and antioxidants from green coffee—mechanism of action in the light of potential bioaccessibility. Food Res. Int., 61, 48–55).*

dimethoxycinnamic acid (Belay and Gholap, 2009). It is bitter in taste at lower concentrations (50 mg/L) but is sour at higher concentrations (1 g/L). The CGA levels are 65 mg/g in dried GCB of *Robusta* and 140 mg/g in dried GC of *Arabica*, depending on the timing of harvesting. More than 70% of CGA is destroyed at roasting temperatures, leaving a residual level of almost 30 mg/g in the roasted coffee bean. In contrast to green coffee, green tea contains an average of 85 mg/g polyphenols.

Chlorogenic acid (CGA) possesses potential antioxidant activity (Kono et al., 1997; Luzia et al., 1997; Castilho et al., 2002), slows the release of glucose into the bloodstream after meals (Johnston et al., 2003; Salazar-Martinez et al., 2004; Bassoli et al., 2008; www.en.wikipedia.org/wiki/Coffee-bean), and also increases hepatic utilization of glucose (Arion et al., 1997; Hemmerle et al., 1997; Herling et al., 1998; Johnston et al., 2003; Shearer et al., 2003). In addition, it also has antispasmodic (Trute et al., 1997) and antimutagenic activities (Stich et al., 1982). Therefore it is considered to play an important role in preventing various diseases associated with oxidative stress such as cancer, cardiovascular, aging, and neurodegenerative diseases.

Caffeine (1,3,7-trimethyl-xanthine) is one of the major purine alkaloidal components of GCB and accounts for 99.78% of the total methyl-xanthines. A battery of enzymes are required for synthesis of purine alkaloids such as caffeine synthase, xanthosine 7-N-methyltransferase, 7-methylxanthine 3-N-methyltransferase, caffeine xanthine methyltransferase 1 (CaMXMT1), caffeine methylxanthine-methyltransferase 2 (CaMXMT2), caffeine dimethylxanthine methyltransferase (CaDXMT1), and theobromine 1-N-methyltransferase (Ashihara et al., 2013). The content of caffeine does not change during maturation of GCB (Clifford and Kazi, 1987). It is not destroyed by excessive roasting because it is thermostable (Mussatto et al., 2011). However, other constituents, namely proteins, sugars, CGA, trigonelline, and fat, may be preserved or destroyed and transferred into reactive substances during roasting (Trugo and Macrae, 1984; Ginz et al., 2000; Trugo, 2003; Rawel and Kulling, 2007). In addition, it also contains lower concentrations of theophylline, theobromine, paraxanthine, libertine, and methyl libertines (www.en.wikipedia.org/wiki/Coffee-bean). The level of caffeine ranges between 0.8% and 1.4% (w/w) in *Arabica* coffee and between 1.7% and 4.0% in *Robusta* coffee (Belitz et al., 2009). The concentration of aroma precursors of GCB such as trigonelline and sucrose ranges between 0.39%–1.77% and 3.8%–10.7% of dry plant material, respectively. Their highest concentration was measured in the green beans of *Coffea arabica, Coffea kapakata*, and *Coffea salvatrix* (Campa et al., 2004), while *Coffea canephora* contained more CGA and caffeine (Campa et al., 2004; Ky et al., 2001). Additionally, coffee also contains the monoamine oxidase inhibitors β-carboline and harman, which may be responsible for its psychoactive effects (Herraiz and Chaparro, 2006).

Terpenoids

The characteristic aroma of coffee is attributed to the presence of α-2-furfurylthiol, 4-vinylguaiacol, some alkyl and 3-methylbutane tyrosine derivatives, furanones, acetaldehyde, propanal, methylpropanal, and 2-acontent as analyzed by chromatographic techniques (Buffo and Freire, 2004). Cafesterol and bengalensol have also been isolated and identified by various chromatographic techniques in *Coffea benghalensis* (Ashihara and Crozier, 1999; Bilkis et al., 2003). Diterpenes are approximately 20% of the lipid fraction, and the diterpenes found in green coffee include cafestol, kahweol, 16-*O*-methylcafestol, cafestal, and kahweal. In coffee oil, diterpenes are esterified with

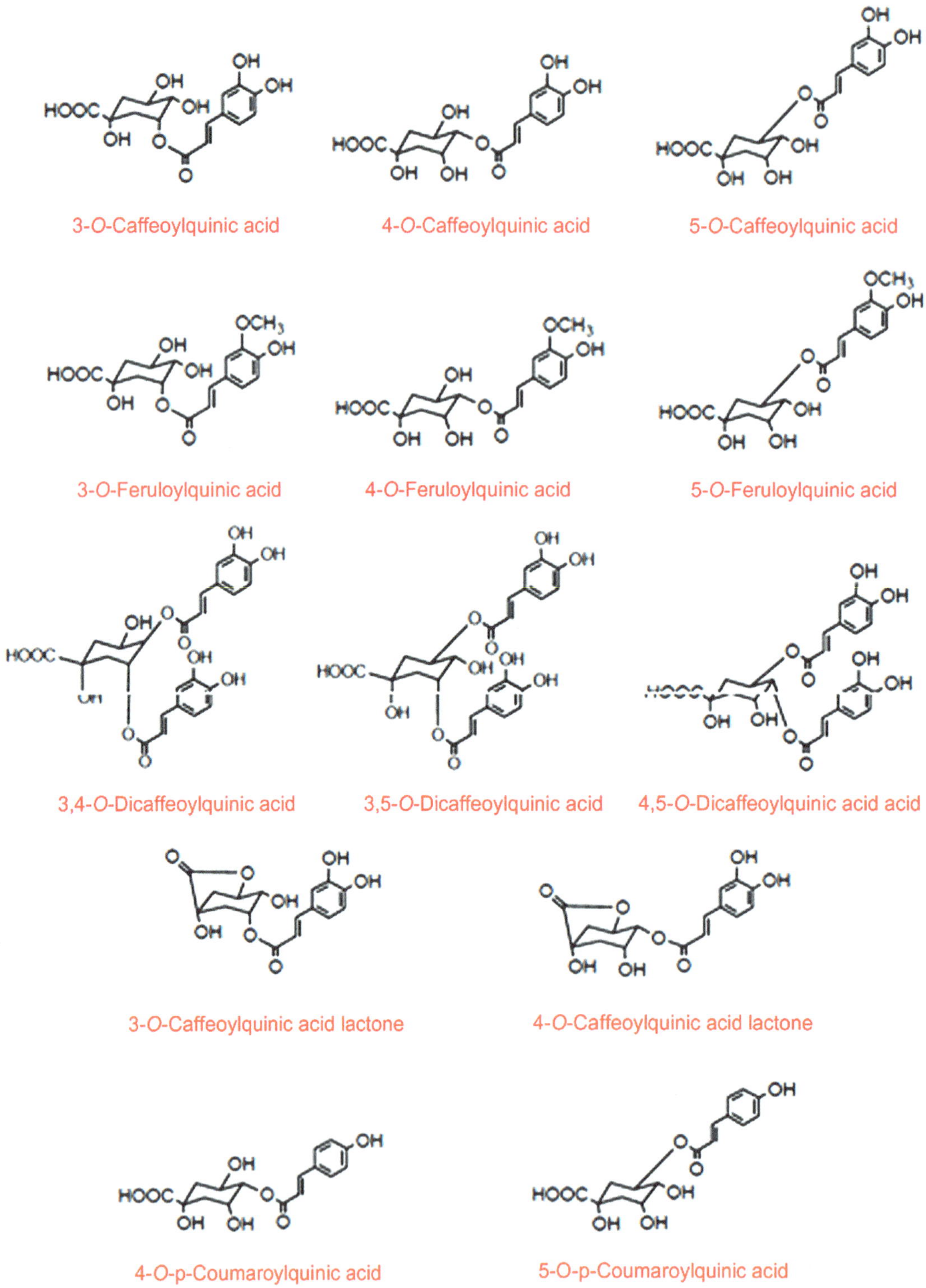

FIGURE 42.6 Structures of different chlorogenic acids present in coffee beans. *(Source: Rio, D.D., Stalmach, A., Calani, L., et al., 2010. Bioavailability of coffee chlorogenic acids and green tea flavan-3-ols. Nutrients, 2, 820–833).*

saturated long-chain fatty acids (LCFAs) (Lee and Jeong, 2007). Some of these diterpenes in in vitro experiments have revealed that they protect the liver against chemical oxidation. Cafestol and kahweol (Fig. 42.7) from GCB are isolated from ground coffee while brewing, but are mostly removed from coffee by paper filters. Their content is higher in espresso coffee, whereas filtered,

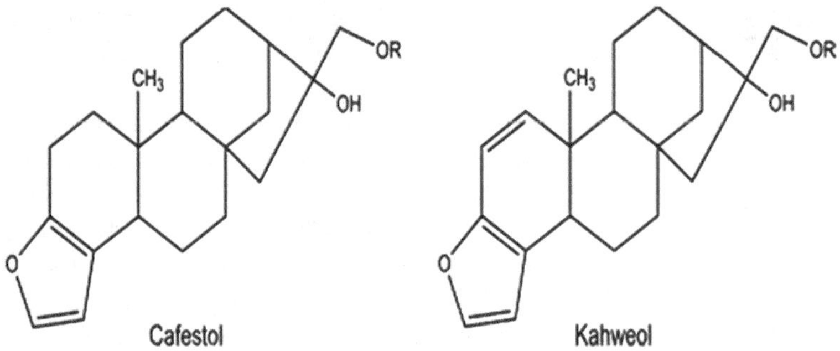

FIGURE 42.7 Chemical structures of cafestol and kahweol. *(Source: Higdon, J.V., Frei, B., 2006. Coffee and health: a review of recent human research. Crit. Rev. Food Sci. Nutr., 46, 101–123).*

percolated, and instant coffees contain low levels (Urgert et al., 1995; Gross et al., 1997).

Enzymes

Young leaves of *C. arabica* cv. *catimor* revealed maximum activity of adenosine nucelosidase at pH = 6.0 in citrate-phosphate buffer. It helps in recycling pathway of adenine and other purines. Polyphenol oxidases are active mostly during the early developmental stages of leaf and endosperm, for example, in the case of *C. arabica.*

Minerals and vitamins

Green coffee beans (GCBs) are rich in several minerals, including potassium, magnesium, calcium, sodium, iron, manganese, zinc, copper, chromium, vanadium, barium, nickel, cobalt, lead, molybdenum, and cadmium (Mussatto et al., 2011). As per the recommended daily allowance, one cup of coffee provides 1%–5% of magnesium, 1%–2% of potassium, 1–3 mg of nicotinic acid, 6%–18% niacin, and 0.1%–0.21% vitamin E (Adrian and Frangne, 1991; Higdon and Frei, 2006; http://www.nal.usda.gov/fnic/foodcomp).

Trigonelline, the second most abundant alkaloid in raw coffee beans, is *N*-methyl-nicotinate, a derivative of vitamin B6, and constitutes approximately 0.6%–1.0% of the dry matter of roasted coffee (Viani and Horman, 1974; Wu et al., 1997). At the roasting temperature (230°C), 85% of trigonelline is degraded to nicotinic acid. In GCB, it is synthesized from nicotinic acid (pyridinium-3-carboxylic acid) by methylation from methionine, a sulfur-containing amino acid (Poisson, 1979). Trigonelline is also a novel phytoestrogen (Allred et al., 2009). Coffee beans are also rich in pyridoxine phosphate, another important member of the vitamin B complex. A cup of coffee contains about 1 mg of nicotinic acid. Three cups of coffee, which is the maximum suggested quantity of coffee per day, provides 25%–50% of the daily requirement of this vitamin.

Proteins and amino acids

Proteins account for 8%–12% of dried GCBs and the majority of proteins are degraded to amino acids during maturation; this process is accelerated in the presence of organic acids such as CGA and their derivatives. Other proteins present in GCB include enzymes, such as catalase and polyphenol oxidase, which are important for maturation of GCB. Coffee beans contain a number of amino acids, namely alanine, arginine, asparagines, cysteine, glutamic acid, histidine, glycine, isoleucine, leucine, lysine, methionine, phenylalanine, proline, serine, tyrosine, threonine, and vanillin (Santos and Oliveira, 2001; Grembecka et al., 2007; Belitz et al., 2009; www.en.wikipedia.org/wiki/Coffee-bean). The free amino acids content in mature coffee is 4.0 mg amino acid/g *Robusta* coffee and up to 4.5 mg amino acid/g *Arabica* coffee. The highest concentration of alanine in *C. arabica* is 1.2 mg/g, followed by asparagine (0.66 mg/g), while the content of alanine in *C. robusta* is 0.8 mg/g and asparagine is 0.36 mg/g (Arnold et al., 1994; Murkovic and Derler, 2006). Unpleasant taste of coffee beans is due to free hydrophobic amino acids present in fresh GCB and, because of these, it is not possible to prepare a desirable beverage from fresh GCB. However, roasted coffee does not contain any free amino acids (www.en.wikipedia.org/wiki/Coffee-bean). Levels of these amino acids in fresh green coffee from Peru have been reported to be isoleucine 81 mg/kg, leucine 100 mg/kg, valine 93 mg/kg, tyrosine 81 mg/kg, and phenylalanine 133 mg/kg. GCB from Tanzania has been reported to have very high levels of γ-aminobutyric acid at 143–703 mg/kg (Teutsch, 2004).

Carbohydrates

Carbohydrates constitute around 50% of the dry weight of GCB and the major fraction of the carbohydrates present in green coffee are polysaccharides, such as arabinogalactan, galactomannan, and cellulose. Arabinogalactan alone comprises up to 17% of the dry weight. Major sugars present in green coffee are sucrose, glucose, fructose, arabinose, galactose, and mannose (Mussatto et al., 2011), and these contribute to its tasteless flavor. Mature brown to yellow coffee beans contain fewer residues of galactose and arabinose at the side chain of polysaccharides, making the GCB more resistant to physical breakdown and less soluble in water (Rodgwell et al., 2003). Free monosaccharides are present in mature brown to yellow-GCB. The free part of monosaccharides contains sucrose (gluco-fructose) up to 9000 mg/100 g in *Arabica* GCB, and up to 4500 mg/100 g in *Robusta*. In *Arabica* GCB, the content of free glucose is 30–38 mg/100 g, free fructose is 23–30 mg/100 g, free galactose is 35 mg/100 g, and mannitol is 50 mg/100 g of dried coffee beans. Mannitol is a powerful scavenger for hydroxyl radicals, which are generated during lipid peroxidation in biological membranes (www.en.wikipedia.org/wiki/Coffee-bean).

Lipids

The total content of lipids in dried green coffee is between 11.7 g and 14 g/100 g (www.en.wikipedia.org/wiki/Coffee-bean). Green coffee oil (GCO) extracted from *C. arabica* before roasting is composed of 75% triacylglycerols, 13.54% unsaponifiable matter, and 0.24% wax (1/3N). The main fatty acids present in GCO are linoleic and palmitic acids (46.3 and 30.2 g per 100 g, respectively) (Wagemaker et al., 2011). Lipids (linoleic acid, palmitic acid, oleic acid, stearic acid, arachidic acid, diterpenes, unsaturated LCFAs, esters, and amides) are present on the surface and in the interior matrix of GCB, but the major ones are linoleic acid (46% of total free lipids) and palmitic acid (30%–35% of total free lipids). Lipids of the exterior tissue form wax-like cover on the surface of coffee beans (200–300 mg lipids/100 g dried GCB), which protects the interior matrix against oxidation and insects, and such molecules have antioxidative activity due to their chemical structure. *Arabica* beans have higher lipids content (13.5–17.4 g lipids/100 g dried GCB) than *Robusta* (9.8–10.7 g lipids/100 g dried GCB). Diterpenes constitute about 20% of the lipid fraction of GCB and the important diterpenes found to be present are cafestol, kahweol, and 16-*O*-methylcafestol (www.en.wikipedia.org/wiki/Coffee-bean).

Volatile compounds

Green coffee bean (GCB) contain several volatile compounds that include short-chain fatty acids, aldehydes, and nitrogen-containing aromatic molecules, which are derivatives of pyrazine. These volatile compounds are responsible for the unpleasant odor and taste of GCB. They cause nausea and vomiting upon inhalation of the ground GCB (www.en.wikipedia.org/wiki/Coffee-bean).

Some of the volatile compounds that impart nauseating odor to GCB are acetic acid (pungent, unpleasant odor), propionic acid (odor of sour milk or butter), butanoic acid (odor of rancid butter), pentanoic acid (unpleasant fruity aroma), hexanoic acid (fatty, rancid odor), heptanoic acid (fatty odor), octanoic acid (repulsive, oily, rancid odor), nonanoic acid (mild, nut-like, fatty odor), decanoic acid (sour repulsive odor), and derivatives of such fatty acids as 3-methyl-valeric acid (sour, green/herbaceous, unpleasant odor), acetaldehyde (pungent, nauseating odor, even when highly diluted), propanal (choking effect on respiratory system, penetrating, nauseating), butanal (nauseating effect), or pentanal (very repulsive nauseating effect) (Bessiere–Thomas and Ivon, 2002). Most of the unpleasant-tasting volatile compounds present in GCB are neutralized during roasting. Unfortunately, however, other important molecules such as antioxidants and vitamins present in green coffee are also destroyed.

Phytocomposition of different parts of the coffee plant

Coffee seed

Coffee seed is the most important part of this plant from a commercial point of view. The chemical composition of the coffee seed is given in Table 42.2.

In addition to the compounds listed in above table, other active components such as quinic acid, sitosterol, dihydrositosterine, stigmasterol, coffeasterin, tannin, wax, caffeic acid, sugar, hemicellulose, cellulose, volatile acids (acetic, propanoic, butanoic, isovaleric, hexanoic, and decanoic acids), nonvolatile aliphatic acids (citric, malic, and oxalic acid), soluble carbohydrates (e.g. monosaccharides:fructose, glucose, and arabinose; oligosaccharides: sucrose, raffinose, and stachyose and polymers of galactose, mannose and glucose, etc.) are also present in coffee seeds (Buffo and Freire, 2004; Kraft and Hobbs, 2004).

Several biologically active phytoconstituents, such as polyphenol and others, for example, kaempferol, quercetin, ferulic, sinapic, nicotinic, quinolic, tannic, and pyrogallic acids, present in coffee bean extract possess therapeutic potential as hepatoprotectant, antioxidant, antiinflammatory, antihyperlipidemic, and antimicrobial agents (Narayana et al., 2001, Naidu et al., 2008). The

TABLE 42.2 Active compound(s) and their percentage in coffee seed.

Active compound(s)	Content (%)
Caffeine, theobromine, theophylline	1.25–2.50
Chlorogenic acid	4.40–7.50
Trigonelline	0.80–1.25
Choline	0.022
Fatty oil	10.0–16.00

Source: Buffo, R.A., Freire, C.C., 2004. Coffee flavor: an overview. Flavour Frag. J., 19, 99–104.

lipid fraction of the seed has waxes and other lipid-soluble ingredients which have a role in embryonic development (Wagemaker et al., 2011).

Coffee leaves

Leaves of *C. arabica* contain phenolic acids such as caffeic, *p*-coumaric, chlorogenic, ferulic, as well as quercetin, kaempferol, isoquercitrine, and rutin (Patay et al., 2016a). Leaves of *C. arabica* have a greater quantity of *cis*-isomers of CGA than the seeds. Mass spectrometry (MS) and Nuclear magnetic resonance studies also revealed the presence of mangiferin, *C*-glucosylxanthone, feruloylquinic acid esters, and isomangiferin in *C. arabica* and *C. canephora* (Campa et al., 2012). Mangiferin possesses numerous pharmacological attributes such as antiinflammatory, antihyperlipidemic, antidiabetic, and neuroprotective activities along with antioxidant and antimicrobial properties (Campa et al., 2012).

Stem

Herbaceous stem of GCB contains around 0.45 $\pm$ 0.04 mg/g of trigonelline, 2.67 $\pm$ 0.29 mg/g of 5-*O*-caffeoylquinic acid (5-CQA), the main form of CGA in GCBE, and 1.24 $\pm$ 0.35 mg/g of sucrose. However, the woody stem of this plant contains comparatively lower concentrations of all these three constituents and their respective levels have been reported to be 0.20 $\pm$ 0.08, 0.66 $\pm$ 0.23, and 0.57 $\pm$ 0.11 mg/g of the stem material (Acidri et al., 2020). Leaves have been used traditionally in many coffee-growing regions for mitigation of a number of illnesses including cardiovascular, gastrointestinal, cancer, diabetes, dermatological, and obesity (Patay et al., 2016b). In addition to the known phytochemicals present in coffee seeds, leaves also contain certain exclusive compounds such as mangiferin (1,3,6,7-tetrahydroxyxanthone-C2-β-D-glucoside) and pigments like chlorophylls and carotenoids, which possess excellent antioxidant potency (Talamond et al., 2008; Campa et al., 2012; Imran et al., 2017; Chen et al., 2018; Stahl and Sies, 2003; Lanfer-Marquez et al., 2005; Wangcharoen and Phimphilai, 2016). Although coffee leaves are not currently being used in preparation of beverages, with appropriate processing methods, beverages prepared from leaves (coffee leaf tea) could be a rich source of phytochemicals and antioxidants with therapeutic and pharmacological values for human health. Therefore in view of the presence of useful phytochemicals, their total phenolic content, and antioxidant capacity, coffee leaves can be considered of great therapeutic and pharmacological value and thus their use as a health beverage should not be ignored. Good managerial practices involved in cultivation of coffee, such as desuckering, in which new shoots are regularly pruned, can be economically utilized for "coffee leaf tea" processing; and thus commercially exploited as an extra source of income for coffee farmers (Acidri et al., 2020).

Roots

Roots also contain trigonelline, 5-CQA, and sucrose, but their content in roots is very low and their levels have been reported to be 0.14 $\pm$ 0.00, 0.71 $\pm$ 0.44, and 1.01 $\pm$ 0.30 mg/g, respectively (Acidri et al., 2020). However, there are no reports on its commercial use, possibly due to the lower phytobiomolecules content.

High-performance liquid chromatography spectra of green coffee beans

According to the reports of Acidri et al. (2020), HPLC analysis of coffee extract revealed the presence of trigonelline, 5-CQA, caffeine, and sugar at different retention times. HPLC analysis also suggested that the predescribed compounds were present in all organs except mangiferin, which was absent in roots, stems, and seeds, and caffeine, which was absent in stems and roots. Phytochemical analysis also showed significant decline of 66% following roasting procedure as compared to GCB. This reduction was attributed to decomposition of 5-CQA and sucrose by 68% and 97%, respectively (Figs. 42.8 and 42.9).

Gas Chromatography-Mass Sectrophotometry analysis of green coffee beans

As per the reports of Mehari et al. (2019); a total of 21 fatty acids were detected following GC-MS analysis of GCB samples. Among the 21 fatty acids, 12 saturated and 9 unsaturated fatty acids were identified (Mehari et al., 2019); details of which are summarized in Table 42.3.

Major pharmacokinetic attributes

There is not much information available on the pharmacokinetic profile including bioavailability of CGA and other important phytoconstituents. Following administration, CGA and some other phenolic compounds have been observed in saliva, gastrointestinal fluid, and urine of fasting subjects (Cremin et al., 2001; Farah et al., 2006a,b). Baer-Dubowska and Szaefer (1998) and Farah et al. (2008) have suggested that the major CGA compounds present in green coffee are absorbed, stored, metabolized, and recycled through excretion and reabsorption in humans. Absorption and metabolism of phytoconstituents of GCB are very complex and dynamic processes. Several variations in the bioavailability of active constituents of GCB have been reported. The bioavailability of 5-CQA has been reported to be almost

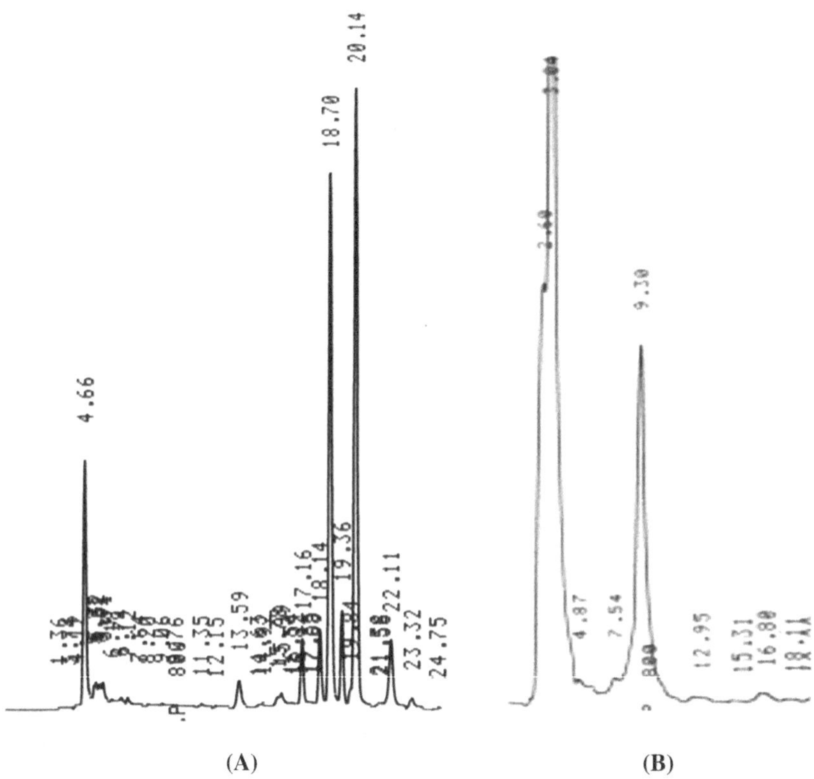

FIGURE 42.8 High-performance liquid chromatography chromatograms of an extracted coffee sample. (A) Displaying peaks for trigonelline, 5-CQA, caffeine, and mangiferin at 4.66, 18.70, 20.14, and 22.11 min, respectively, obtained using an ultraviolet detector at 270 nm; (B) displaying a sucrose peak at 9.30 min obtained using a refractive index detector. *(Source: Acidri, R., Sawai, Y., Sugimoto, Y. et al., 2020. Phytochemical profile and antioxidant capacity of coffee plant organs compared to green and roasted coffee beans. Antioxidants 9, 93. https://doi.org/10.3390/antiox9020093).*

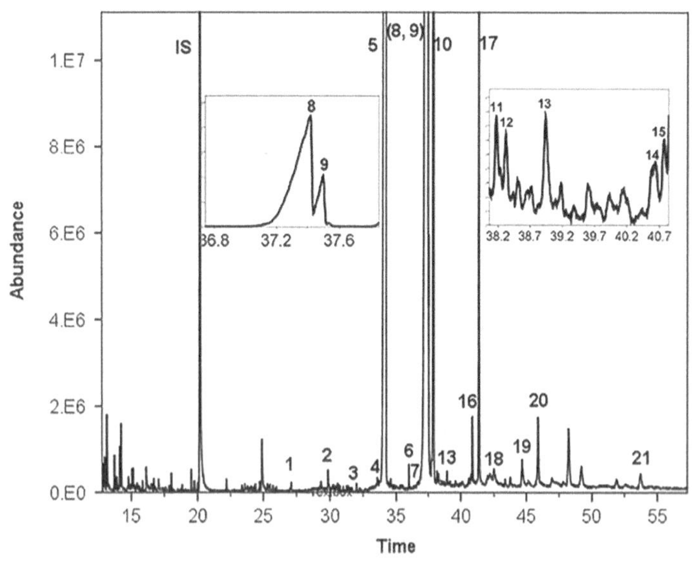

FIGURE 42.9 Gas chromatogram of a green coffee bean sample revealing 21 peaks at different retention times. *(Source: Mehari, B., Redi-Abshiro, M., Chandravanshi, B.S., Combrinck, S., McCrindle, R., Atlabachew M., 2019. GC-MS profiling of fatty acids in green coffee* (Coffea arabica *L.) beans and chemometric modeling for tracing geographical origins from Ethiopia. J. Sci. Food Agric., 99, 3811–3823).*

33 ± 27% (Bourne and Rice-Evans, 1998; Spencer et al., 1999; Dupas et al., 2006).

Recently, human colon carcinoma cell line Caco-2 has been used for evaluating the intestinal absorption of 5-caffeoylquinic acid (5-CQA), a major component of GCB. The absorption rate for 5-CQA was found to be 0.10% at physiological concentrations equivalent to gut lumen concentrations (0.1–1 mM). Transepithelial movement of the components of GCB, that is, CQA, has been reported to be concentration-dependent, but unsaturable. This might be suggestive of passive diffusion of 5-CQA across the epithelial lining of the colon (Świeca et al., 2017).

Major CGA compounds (CQA and diCQA) present in GCB are differentially absorbed and metabolized throughout the whole gastrointestinal tract (GIT) in humans (Farah et al., 2008; Rio et al., 2010). Initial plasma concentration peaks suggest early absorption of CGA from the stomach and jejunum, followed by absorption from the small intestine (Spencer et al., 1999; Konishi et al., 2006; Lafay et al.,

TABLE 42.3 Fatty acids detected in green coffee bean extract using GC-MS.

Peak no.	Chemical name	Retention time (minutes)
1	9-Oxo-decanoic acid	25.4
2	Tetradecanoic acid	30.1
3	Pentadecanoic acid	32.1
4	*cis*-7-Hexadecenoic acid	33.7
5	Palmitic acid	34.3
6	Heptadecanoic acid	36.2
7	*cis*-7,*cis*-10-Octadecadienoic acid	36.8
8	Linoleic acid	37.6
9	Oleic acid	37.8
10	Stearic acid	38.2
11	*cis*-9,*trans*-12-Octadecadienoic acid	38.4
12	*trans*-9,*trans*-12-Octadecadienoic acid	38.6
13	*trans*-7,*trans*-10-Octadecadienoic acid	39.0
14	*cis*-6,*cis*-9,*cis*-12-Octadecatrienoic acid	40.8
15	2′-Hexyl-[1,1′-bicyclopropyl]-2-octanoic acid	41.0
16	Gondoic acid	41.2
17	Arachidic acid	41.7
18	3-Pentyl oxirane undecanoic acid	42.7
19	14-Methyleicosanoic acid	43.5
20	Behenic acid	46.1
21	Lignoceric acid	54.1

Source: Mehari, B., Redi-Abshiro, M., Chandravanshi, B.S., Combrinck, S., McCrindle, R., Atlabachew M., 2019. GC-MS profiling of fatty acids in green coffee (*Coffea arabica* L.) beans and chemometric modeling for tracing geographical origins from Ethiopia. J. Sci. Food Agric., 99, 3811–3823.

2006; Monteiro et al., 2007), whereas the later plasma peaks, in some subjects, indicate absorption through the large intestine (Dupont et al., 2002), along with recycling through digestive fluids (Farah et al., 2006a,b). Rio et al. (2010) reported that 30% of the bioavailability of CGA is from the ileostomists of the small intestine and 70% pass from the small to the large intestine.

Nanoencapsulation of GCB has helped in protecting phenolics and other important phytoconstituents and enhanced the chances of improving the bioavailability of CGA. Nanoencapsulation protects the bioactive phytobiomolecules from adverse environmental circumstances, solubilization, pH variations, and other incompatibilities (Meliana et al., 2017). According to Desai et al. (2019) nanoencapsulation of GCB ensured $80 \pm 2\%$ (w/w) bioavailability of CGA following controlled release in the intestinal tract. This technological intervention is likely to curb the shortcoming of stability and poor bioavailability of natural polyphenols. Braga et al. (2016) also suggested the role of coffee oil capsule in maintaining the aroma of coffee beans as it is diminished quickly and therefore it improves the intake of coffee indirectly.

Chlorogenic acid (CGA) appears in circulatory system predominantly as glucuronide, sulfate, and methylated metabolites after reactions in the small and large intestines and in hepatic cells (Donovan et al., 2006; Crozier et al., 2009; Rio et al., 2010). Substantial amounts of parent compounds and some of the metabolites, after appearing in ileal fluid, pass to the large intestine, where they are acted on by colonic microflora. Microbial catabolites may influence the local microflora and exert prebiotic activity. In addition, the parent flavonoid molecules and catabolites present in the colon may have the potential to inhibit proliferation of cancerous cells (Rechner et al., 2001; Aura, 2008; Rio et al., 2010).

Following consumption of green coffee extract (GCE), apart from small amounts of caffeic, ferulic, isoferulic, and *p*-coumaric acids, six major CGA compounds, which account for almost 90% of the phenolic compounds, have been identified in the plasma of humans (Monteiro et al., 2007; Farah et al., 2008). In humans, approximately one-third of the ingested CGA enters the blood circulation after absorption from small intestine (Olthof et al., 2001). It is metabolized to caffeic acid, quinic acid, and ferulic acid in the liver and/or kidneys and traces of CGA that remains unmetabolized are excreted in the urine (Olthof et al., 2003). Ferulic and isoferulic acids are formed by methylation of caffeic acid (Rechner et al., 2001), whereas caffeic acid may be derived from hydrolysis of FQA (Bourne and Rice-Evans, 1998).

Urine is not the major excretion pathway of intact CGA compounds. Major phenolic compounds excreted in urine are sinapic, gallic, *p*-hydroxybenzoic, and dihydrocaffeic acids, and these are probably the metabolic products of cinnamates present in plasma, whereas vanillic and siringic acids are probably derived from the secondary metabolic pathway (Farah et al., 2008). The quantities of CGAs and their metabolites excreted in urine by healthy subjects comprise 29.2% of the intake, whereas only 8.0% of the ingested CGA in ileostomists are excreted, suggesting the importance of colon in bioavailability of dietary CGA. Further, even after absorption from small intestine, substantial quantities pass to the large intestine, where the parent compounds and their metabolites influence both the colonic health and colonic microflora. Urinary levels indicate that substantial quantities of the colonic catabolites are absorbed into the portal vein and pass through the body via circulatory system prior to excretion (Rio et al., 2010).

A study by Farah et al. (2008) on the bioavailability of CGA in human plasma and urine after consumption of green coffee supplement containing 170 mg (dry weight) in 10 healthy subjects revealed that the concentration of CGA and its metabolites in human plasma was higher than $33.1 \pm 23.1\%$, with peak levels from 0.5 to 8 h after ingestion with only $5.5 \pm 10.6\%$ of quinic and cinnamic acids and were overall highly absorbed in urine over the same period of time. CGA showed significant absorption and metabolism (Farah et al., 2008) and the bioavailability of these components and phytoconstituents is directly associated with the therapeutic efficacy of GCB against hypertension, obesity, cardiovascular diseases, and antineoplastic as well as antidiabetic activity (Kozuma et al., 2005; Ochiai et al., 2004; Blum et al., 2007).

Green coffee bean as a nutraceutical and pharmacotherapeutic agent

The modern lifestyle, including sedentary and comfortable ways of working with minimal physical activities, materialistic life, food, and sleep habits and even increasing environmental pollutants are posing greater challenges to human health. Obesity is a problem of modern society, and it is likely to worsen further and has been recognized as one of the major causes of diabetes and hypertension. Therefore it is important to think about the amount of food we eat, the time at which we eat, and also the kinds of food we eat (www.TerrytalksNutrition.com). Global health policy emphasizes promoting the consumption of plant foods, including fruits and vegetables, for prevention of several oxidative stress-induced chronic diseases, namely cancer, diabetes, and cardiovascular and neurodegenerative diseases of modern society (WHO, 2004), because plant foods are rich sources of antioxidant bioactive compounds and are natural sources of various carbohydrates, amino acids, proteins, fatty acids, and minerals. GCB and its extracts have been found to be laden with a variety of promising phytobiomolecules as enumerated in this chapter. Apart from their importance as nutraceuticals, some of these also possess promising pharmacological activities and regulate various body homeostatic mechanisms required for sound health; therefore, coffee is one of the most widely consumed beverages worldwide.

The GCB has been recognized to possess potential health benefits because it prevents various diseases, including lifestyle diseases like obesity, type II diabetes, hypertension, hepatic injury, cirrhosis, hepatocellular carcinoma, colorectal cancer, liver cancer, Parkinson disease (PD), and Alzheimer disease (AD), and also reduces suicide risk (Higdon and Frei, 2006; Barranco et al., 2007). In addition, it has several other desirable activities, including antibacterial (Almeida et al., 2006; Lou et al., 2011), antiinflammatory (De Castro et al., 2013), and immunomodulatory activities (www.makeam-ericawell.com). The antioxidant capacity of CGA is more than that of ascorbic acid (vitamin C) or mannitol, which is a selective hydroxyl-radical scavenger (Morishita and Kido, 1995). Being rich in natural antioxidants, there is vast scope for its use as a nutraceutical. Wide spectra of phytobiomolecules possessing nutritional and therapeutic potential make GCB a strong candidate for considering its fortification as a constituent of functional foods in addition to its global use in beverages.

Therapeutic uses of GCB and its various major chemical constituents against some of the most important human diseases along with their mechanistic pathways are summarized here.

Obesity and weight loss

Obesity is the excess accumulation of body fat and is also considered as a low-grade inflammatory condition (Mohammadshahi et al., 2014). As indicated by the WHO, in 2016, more than 1.9 billion adults, 18 years and older, were overweight and, of these, over 650 million were obese. A total of 38 million children under the age of 5 were overweight or obese in 2019. In the United States alone, approximately 75% of the population is either overweight or obese (www.TerryTalksNutrition.com). The worldwide prevalence of obesity has nearly tripled between 1975 and 2016. Most of the world's population lives in countries where being overweight and obesity kill more people than being underweight. Therefore obesity represents a major health challenge because it substantially increases the risk of diseases such as type II diabetes mellitus (DM), fatty liver disease, hypertension, myocardial infarction, stroke, dementia, osteoarthritis, obstructive sleep apnea, and several cancers, thereby contributing to decline in both quality of life and life expectancy.

Globally, 8% of deaths in 2017 were due to obesity, compared to just 4.5% in 1990. Therefore, scientists are looking for some effective measures to address the issues related to obesity. In that respect, GCB seems to hold promising potential as GCB extract promotes loss of visceral fat, which is the fat in tissue lining the abdominal cavity surrounding the internal organs. The effects of GCB extracts and CGA are multifaceted, and they are considered to be weight loss supplements (Shimoda et al., 2006; Cho et al., 2010) because their consumption is inversely related to weight gain (Lopez-Garcia et al., 2006).

Mechanism

Green coffee bean (GCB) exerts its antiobesity effect by modulating lipogenesis. Molecular evidence revealed that GCB downregulates SREBP-1c (Sterol regulatory element-binding protein) and related molecules, which further leads to suppression of body fat accumulation. Thus it enhances energy metabolism and reduces lipolysis, leading to less accumulation of fat culminating in an antiobesity effect (Murase et al., 2011).

Chlorogenic acid (CGA) also reduces food cravings, reduces daily calorie intake, and induces body fat loss by thermogenesis (Greenberg et al., 2006) by acting as a peroxisome proliferator-activated receptor alpha (PPARα) agonist, just like statins used in treatment of cholesterol and triglyceride disorders. GCB activates the liver to produce more bile and increase the metabolic process, and discharge glucose into the bloodstream. Further, caffeine releases fatty acids from the stored body fat and also suppresses fat absorption, whereas CGA assists the liver in processing fatty acids more efficiently and reducing hepatic triglyceride levels, thus resulting in weight loss (www.makeamericawell.com/svetol-green-coffee-bean-extract). In addition, CGA may also reduce proliferation of new fat cells through its antioxidant effects (www.mypcos.info/1/treatments/natural/chlorogenic-acid; www.en.wikipedia.org/wiki/Coffee%20bean). Further, intestinal glucose uptake is also reduced (Bidel et al., 2006; Lopez-Garcia et al., 2006), thus reducing blood and liver fats (triglycerides). CGA and caffeine work together in reducing abdominal fat by increasing lipolysis or fat cells being opened up for energy release (www.jenny-neill.com). CGA also stimulates weight loss by increasing the levels of a hormone called adiponectin.

Green coffee bean extract (GCBE) suppresses fat absorption and CGA reduces hepatic triglyceride levels. Other phenolic compounds present in GCBE, namely neoCGA and feruloylquinic acid, enhance hepatic carnitine palmitoyl transferase activity (Shimoda et al., 2006). GCBE has also been postulated to modify hormone secretion and glucose tolerance in humans by facilitating the absorption of glucose from the distal rather than the proximal part of gastrointestinal tract (Johnston et al., 2003). Antiobesity effects are mediated via alterations in plasma adipokine level, body fat distribution, downregulating fatty acids and cholesterol biosynthesis, and upregulating fatty acid oxidation and expression of PPARα in liver (Cho et al., 2010), in addition to suppressing the accumulation of hepatic triglycerides (Shimoda et al., 2006). Further studies related to the effect of green coffee on body composition changes, body mass index, adipocytokines levels, and also physical activity levels and dietary intake are essential to formulate the new drug regimen.

Oxidative stress

In oxidative stress, reactive oxygen species (ROS) have been advocated in initiation and propagation of chronic and fatal diseases such as cancer, cardiovascular diseases and diabetes. Oxidative stress plays key role in the pathogenesis of several diseases. Free radicals induce damage to macromolecules, such as proteins, lipids, and DNA, in addition to purines and pyrimidines, and lead to instability of the genome (Chobotova, 2009). Free radicals, if left unneutralized, cause prolonged oxidative stress, which ultimately results in the development of a variety of ailments including cancer, cardiovascular diseases, diabetes, and premature aging. The antioxidant activity of coffee is attributed to well-known antioxidants present in phenolic and nonphenolic bioactive compounds, namely chlorogenic, ferulic, caffeic, and *n*-coumaric acids (Nicoli et al., 1997; Henning et al., 2004). In roasted coffee, melanoidins (brown pigments) are synthesized; these are strong antioxidants (Steinhart et al., 2001; Yen et al., 2005; Murthy and Naidu, 2012; Bravo et al., 2013; Yashin et al., 2013; Bravo et al., 2013; Yashin et al., 2013). During roasting, many phenolic compounds are deteriorated; albeit CGA and caffeine survive this process and these are mainly responsible for the antioxidant potential (Gallardo et al., 2006) and geno-protective properties of coffee in human cells are due to significant reduction in ROS generation and 29%–73% of protection in DNA strand breaks (Bravo et al., 2013).

Mechanism

Green coffee bean (GCB) is an excellent source of bioaccessible and bioavailable polyphenols, namely CGA, 5-caffeoylquinic acid, and 3,5-dicaffeoylquinic acid, and all these possess multidirectional antioxidant activity (Gawlik-Dziki et al., 2014). GCB extract, 5-caffeoylquinic acid, and 3,5-dicaffeoylquinic acid significantly decrease ROS production and restore glutathione levels and activity of antioxidant enzymes in HepG2 cells. However, caffeine does not offer significant protection against oxidative stress (Baeza et al., 2014). In addition, GCB has been reported to inhibit lipooxygenase (Gawlik-Dziki et al., 2014), chelate metals, scavenge free radicals, or modulate endogenous antioxidant defenses (Alía et al., 2006; Granado-Serrano et al., 2010; Rodríguez-Ramiro et al., 2011). The antioxidant activity of cholorgenic acid is attributed to its chemical structure, which contains a phenolic hydroxyl group that counters the hydroxyl radical and superoxide anion radical (Gallus et al., 2002a; Padol and Hunt, 2004; Bohn et al., 2014).

Type II diabetes

Metabolic syndrome is an important risk factor of Diabetes Mellitus (DM) type II and cardiovascular diseases. Diabetes has a close association with obesity. According to reports from the International Diabetes Federation study, "diabesity – a naïve term was coined to refer the coexistence of diabetes mainly Type II DM and obesity" in 380 million people and counting due to unreported and undiagnosed case reports (Santos and Lima, 2016). Coffee consumption lowers the risk of DM, and there is an inverse association between coffee consumption and risk of DM, including several glycemic markers of diabetes (van Dam and Feskens, 2002; Tuomilehto et al., 2004; van Dam and Hu, 2005), and a positive association between coffee consumption and insulin sensitivity (Agardh et al., 2004; van Dam et al., 2004).

Mechanism

Green coffee bean (GCB) lowers the risk of DM and some possible mechanisms of action have been proposed, however, the precise target site and mechanism are yet to be established. Small concentrations of cafestol (10^{-10} to 10^{-6} M) showed significant increase in secretion of insulin and thereby improved glucose uptake; and thus showed preventive efficacy against type II DM (O'keefe et al., 2013). Alterations in intestinal microbiota due to chronic coffee consumption also mitigate type II DM by altering the ratio of bacteroides/Prevotella and Fermicutes/Bacteroides. This might be suggestive of the preventive efficacy of coffee consumption with a high-fat diet (Cowan et al., 2014; Choi et al., 2016). CGA appears to slow, but not prevent, the absorption of glucose from the human gut by its action as an α-glucosidase inhibitor, therefore reducing the insulin spike after meals in type II diabetic patients. CGA also acts by inhibiting glucose-6-phosphatase enzyme involved in gluconeogenesis and glycogenolysis (Herling et al., 1998) in addition to being a specific competitive inhibitor of glucose-6-phosphate translocase (Arion et al., 1997). When glycogenolysis is inhibited, the body gets its energy from fat cells and the blood sugar level decreases, along with decrease in body weight (www.makeam-ericawell.com/svetol-green-coffee-bean-extract). Coffee compounds also slow the process of intestinal glucose absorption (Johnston et al., 2003) by regulating levels of glucose-dependent insulin-tropic polypeptide and glucagon-like peptide-I (GLP-I), which augment insulin secretion after oral glucose consumption. In addition, CGA also increases the rate of glucose utilization by muscles by stimulating the AMP-activated protein kinase enzyme, specifically the A2 subunit, where uptake of glucose into muscles for energy is independent of insulin. Further, CGA also acts by stimulating glucose transporter type 4 (GLUT4) protein, which works with insulin to transport glucose into cells. The GLUT 4 protein is stimulated through phosphorylation of AKT2-type protein kinase B (www.mypcos.info/1/treatments/natural/chlorogenic-acid). CGA and magnesium in green coffee reduce insulin resistance, which plays an important role in diabetes. An inverse relationship between risk of type II diabetes and dietary intake of serum magnesium concentrations has also been reported (Kao et al., 1999; Lopez-Ridaura et al., 2004).

Cardiovascular diseases

Cardiovascular diseases are one of the most common examples of lifestyle diseases. Cardiovascular diseases include hypertension, atherosclerosis, angina pectoris, and cardiac failure. Hypertension is a primary risk factor for stroke, heart disease, and renal failure, and it is one of the most critical issues for human health (Domanski et al., 2002; Lloyd-Jones et al., 2002; Whelton et al., 2002). Endothelial dysfunction is considered an initial stage of arteriosclerosis (Ross, 1999), and a high homocysteine concentration in blood is considered a risk indicator for cardiovascular diseases (www.makeamericawell.com/svetol-green-coffee-bean-extract) because it is involved in vascular endothelial hypofunction (Rodgers and Conn, 1990). Metaanalysis data have revealed that coffee drinking significantly reduced the mortality risk in patients with myocardial infarction (Brown et al., 2016). Efficacy of coffee consumption on premorbid patients with cardiovascular diseases has not been researched well till date, albeit consumption of coffee has been attributed to curtailing the potential threat amongst Dutch people with a history of heart attack or myocardial infarction (Van Dongen et al., 2017).

Rutoside and isoquercitrin, extracted from coffee seeds, can also be used for treatment of atherosclerosis, while quercitrin exerts positive-inotropic, positive-chronotropic, and antiarrythmic effects in a guinea pig model. Quercetin and rutoside have also been suggested as ancillary therapy in treatment of phlebosclerosis and capillary fragility (Narayana et al., 2001). Rutin (rutoside or rutinoside) is a dietary flavonol and has been reported to be effective in reducing post-thrombotic syndrome, venous insufficiency, or endothelial dysfunction, but there was no very well validated evidence for its safe and effective use as of 2018 (Martinez-Zapata et al., 2016; Morling et al., 2018). However, it has low bioavailability due to poor absorption, high metabolism, and rapid excretion. It is well established that high cortisol levels contribute to hypertension by overstimulation of the hypothalamic–adrenal axis and sympathetic nervous system and such hypertensive individuals have higher levels of urinary free cortisol and catecholamine metabolites. GCB serves as an antistressor and may be responsible for a reduction in the release of stress hormone (i.e., cortisol) and thereby produces a hypotensive effect at low doses.

Mechanism

Green coffee bean extract (GCBE) reduces visceral fat and homocysteine levels in blood plasma (Ochiai et al., 2004), and inhibits lipid and low density lipoprotein degradation and clumping of blood platelets to improve blood circulation all over the body, thus preventing hardening of the arteries (www.makeamericawell.com/svetol-green-coffee-bean-extract). The pathophysiology of hypertensive crisis is associated with increased oxidative stress with elevated concentrations of superoxide anions and hydrogen peroxide and decreased antioxidant mechanisms (Prabha et al., 1990; Redon et al., 2003). Increased superoxide anions thereby reduce the bioavailability of nitric oxide (NO) to endothelial cells leading to quenching of NO-dependent vasorelaxation, which ultimately results in hypertensive state (Thomas et al., 2001). The antihypertensive action of GCBE is attributed to the CGA present in it. Ferulic acid, a metabolic component of CGA, improves the release of NO from vascular endothelium (Suzuki et al., 2002), which in turn regulates vasoconstriction and vasodilatation by producing and releasing various vasoactive factors, including NO (Swales, 1995; Vanhoutte and Boulanger, 1995). Therefore, it keeps blood vessels open through NO-induced vasodilatation, improves vasoreactivity, controls hypertension, reduces the risk of atherosclerosis, and thus helps in preventing stroke and coronary heart disease (Suzuki et al., 2002; Saito et al., 2002; Ochiai et al., 2004; Kozuma et al., 2005). Green coffee reduces blood pressure and body composition by influencing 11β-HSD1 enzyme activity along with decrease in urinary free cortisol levels, the cortisol/cortisone ratio, and blood pressure, and it is due to CGA which has an inhibitory effect on the enzyme 11-βHSD1 (Revuelta-Iniesta and Al-Dujaili, 2014). However, the active black coffee produced during the roasting process has a stimulant effect on 11-βHSD2 (Arao and Tamura, 2011). In addition, the inhibitory effect on the enzyme 11β-HSD1 is 7–10-fold higher than that of 11β-HSD2, which may explain the higher effect of GC on the urinary free cortisol/cortisone ratio. Second, cortisol reduces the production and bioavailability of vasodilators such as NO (Mitchell and Webb, 2002). Thus reduction in cortisol levels results in improvement in arterial elasticity and blood pressure.

Anticancer activity

Chlorogenic acid (CGA) neutralizes several classes of carcinogenic compounds in vitro and in vivo. Polyphenols present in coffee directly bind with the signaling molecules involved in carcinogenesis and regulate their activity. Therefore, the development of polyphenols as direct inhibitors against target proteins is regarded as a rational approach for chemoprevention (Kang et al., 2011). These lower the risk of liver, colon, breast, skin, and rectal cancers (www.makeamericawell.com). Russo et al. (2017) also reported that consumption of coffee could reduce the risk of prostatic cancer in humans. Coffee consumption has been inversely associated with the incidence of different types of cancers. Compared to noncoffee drinkers, the incidence of colon or colorectal cancer was 24%–48% lower (Giovannucci, 1998; Michels et al., 2005), and hepatocellular carcinoma was 76% lower (Inoue et al., 2005) in coffee consumers. Similarly, reduced risks of developing invasive epithelial ovarian cancer (Jordan et al., 2004), endometrial cancer (Hirose et al., 2007), and breast cancers (Le, 1985; Lubin et al., 1985; Mannisto et al., 1999) have also been documented. It also inhibits the invasion of liver cancerous cells in vitro (Hirakawa et al., 2005).

Mechanism

Precise molecular mechanisms and target(s) underlying the chemopreventive effects of coffee and its active ingredient(s) remain obscure. However, compounds present in coffee, especially diterpenes, decrease the synthesis and secretion of bile acids, which promote colon cancer

(Tavani and La Vecchia, 2004). The anticancer potential of chlorogenic acid 7 (CGA7), one of the ingredients of green coffee, in human cultured colon cell lines (HCT-116 cells) has been recently reported based on in vitro studies. CGA7 produced its anticancerous effect by upregulating proapoptotic protein Bax and downregulating antiapoptotic protein Bcl-2. It also induces apoptosis that was underpinned by DNA fragmentation, PARP-1 cleavage, and caspase-9 activation (Gouthamchandra et al., 2017). Singh et al. (2018) have also advocated the anticancerous activity of quinic acid, a phytobiomolecule of the coffee bean, against oral cancer which acts by stimulating apoptosis through downregulation of antiapoptotic genes and attenuation of Akt signaling pathway and expression of cyclin D1 in squamous cell carcinoma 4. Diterpenes promote elimination of carcinogens and improve antioxidant status by enhancing phase II biotransformation activity and glutathione synthesis (Huber et al., 2002a, b). Coffee consumption has also been proposed to increase colonic motility, decrease exposure of colonic epithelial cells to potential carcinogens (Brown et al., 1990), and decrease formation of DNA adducts by genotoxic carcinogens (Huber et al., 1997; Cavin et al., 2002).

Coffee or Chlorogenic acid (CGA) inhibited CT-26 colon cancer cell-induced lung metastasis by blocking phosphorylation of extracellular signal-regulated kinase (ERK). Coffee or caffeic acid strongly suppressed mitogen-activated protein kinase/extracellular signal-regulated protein kinase (MEK1) and T-cell-originated protein kinase (TOPK) activities and bound directly to either MEK1 or TOPK in an ATP in a noncompetitive manner. Coffee or caffeic acid, but not caffeine, inhibited ERK phosphorylation, activator protein-1(AP-1), and NF-κB transactivation, and subsequently inhibited 12-O-tetradecanoylphorbol 13-acetate, epidermal growth factor, and H-Ras-induced neoplastic transformation of JB6 P*1* cells. Therefore, these results suggest that coffee and caffeic acid target MEK1 and TOPK to suppress colon cancer metastasis and neoplastic cell transformation (Kang et al., 2011). However, no direct correlation has been observed between coffee intake and ovarian, prostate, and pancreatic cancers till date; therefore further studies are warranted in these areas.

Hepatic disorders

Epidemiological data suggest that coffee consumption helps in preventing several chronic diseases, including cirrhosis of the liver (Corrao et al., 1994, 2001; Gallus et al., 2002a) and hepatocellular carcinoma (Gallus et al., 2002b; Gelatti et al., 2005; Inoue et al., 2005), as well as gallstones (Higdon and Frei, 2006). Zelber-Sagi et al. (2016) also reported a correlation between consumption of coffee and fibrosis and revealed that three cups of coffee per day could significantly prevent hepatic fibrosis. Despite evidence of the hepato-protective effect of coffee, the exact underlying mechanism is yet to be elucidated.

Mechanism

Coffee is a complex "blend" of a large number of phytobiomolecules, and any of these chemical moieties may be responsible for its hepato-protective effects. Cafestol and kahweol not only cleanse liver by inducing the liver to purge fats, bad cholesterol, and other harmful toxic chemicals (www.makeamericawell.com), but also reduce the toxicity of a variety of carcinogens in animal models. Chemo-protective effects involve induction of phase II enzyme activity, enhancement of hepatic glutathione levels, and a decrease in liver DNA adducts due to chemical carcinogens (Huber et al., 1997, 2002a,b; Cavin et al., 2002). The protective efficacy of cafestol and kahweol against hepatocellular carcinoma has been reported to be through modulation of certain enzymes associated with the detoxification of carcinogens (Mellbye et al., 2015). In addition, diterpenes also induce conjugating enzymes (e.g., glutathione S-transferase, glucuronosyl-S-transferase), increase expression of proteins involved in cellular antioxidant defense (e.g., γ-glutamyl cysteine synthetase and heme oxygenase-1), and inhibit expression and/or activity of cytochrome P450 involved in carcinogen activation (e.g., CYP2C11, CYP3A2). In the small intestine, these inductions have been shown to be mediated by Nrf2-dependent transcriptional activation. Coffee consumption has been found to be inversely associated with serum gamma glutamyl transferase (Casiglia et al., 1993; Tanaka et al., 1998; Honjo et al., 1999; Nakanishi et al., 2000) and alanine amino transferase activities (Tanaka et al., 1998; Honjo et al., 2001; Ruhl and Everhart, 2005).

Neurodegenerative diseases including Parkinson disease and Alzheimer disease

Neurodegenerative disorders are generally associated with oxidative stress to neurons due to ROS, including hydrogen peroxide. GCB extract improves learning abilities and alertness, reduces fatigue, and increases endurance (www.makeameri-cawell.com). Caffeine exerts positive effects on cognitive and behavioral processes, especially in suboptimal conditions when arousal is low. Caffeine, along with hordenine and phenylethylamine, showed marked increase in central nervous system activity and also induced thermogenesis in individuals (Allison et al., 2018). This might be suggestive of improved neuronal activity along with reduced excitotoxic effect of glutamate and synaptic plasticity.

Mechanism

Bioactive compounds present in coffee, including caffeine and phenolic compounds (ferulic acid, caffeic acid, and CGA), have promising antioxidant properties. CGA protects neurons from hydrogen peroxide-induced stress by upregulating the antioxidant enzyme reduced form of nicotinamide adenine dinucleotide phosphate: quinine oxidoreductase 1 (NQO1). CGA also preserves dopaminergic neurons by suppressing inflammation in neurons (Cropley et al., 2012). Caffeic acid has been associated with neuroprotection against glutamate-induced excitotoxicity and also against sodium nitroprusside-induced nitrosactive stress.

Caffeine is effective against AD as it reduces beta amyloid accumulation by blocking the pathway that increases beta secretase secretion in brain. It also suppresses GSK-3 isoforms associated with *tau*-hyperphosphorylation and presenilin-1/gamma secretase activation (Arendash et al., 2009). GCBE moderates the breakdown of carbohydrate to its lipogenic substrate, glucose. Caffeine is a potent antiinflammatory, antioxidant, and mitochondrial activator. It also activates neurons and stimulates glucose utilization (Arendash and Cao, 2010). It also protects against neurodegenerative diseases such as ischemic stroke (Jeszka-Skowron et al., 2016b).

Oxidative stress is one of the mechanisms involved in the pathogenesis of PD and coffee has the ability to raise plasma antioxidants levels. Adenosine A2 receptors are important targets in basal ganglia disorders such as Parkinson's. Caffeine protects dopaminergic neurons from excitotoxic components by inactivating adenosine A2 receptors (Hu et al., 2007; Sääksjärvi et al., 2008). Metaanalysis studies have revealed that those who consume three cups of coffee a day had a lower risk of PD, but there was no change in risk with more than three cups of coffee (Li et al., 2013). Caffeine suppresses the production of proinflammatory cytokines and has neuroprotective properties through the inhibition of phosphodiesterase (PDE) by being an adenosine receptor antagonist (D'hooghe et al., 2012).

Dermatological, antiaging, and cosmetic effects

Green coffee seed extract exerts antiinflammatory, antiulcer, anticellulitic, antibacterial, and antiallergic activities due to the presence of active phytoconstituents such as kaempferol, rutoside, and quercetin (Boros et al., 2010; Narayana et al., 2001). Free radicals, the by-products of metabolism, destroy cell membranes and result in aging. CGA present in GCB destroy free radicals and thus have regenerative properties and slow the process of aging (www.makeamericawell.com). Linolenic acid, one of the active principles present in green coffee seeds, has been associated with a UV-protecting property and has a role in sunscreen and thus is useful in cosmetics industry. The sunscreen effect was between 0.0 and 4.1 sun protection factor depending on the coffee species (Wagemaker et al., 2011). The high concentration of CGA present in GCE reduces the redness associated with excessive sunlight exposure. Caffeine also contains antiaging properties. It limits photo-damage, decreases skin roughness and wrinkle formation, and reduces the appearance of crow's feet. Green coffee can benefit the skin in a number of ways. CGA increases skin hydration and relieves dryness. It also improves the barrier function of skin, which reduces the amount of water that evaporates from skin surface. By helping the skin to retain water, it addresses the problem of dry skin and results in healthier and more glowing complexion. Further, CGA also increases blood flow to skin and increased blood flow gives the face a more radiant appearance and even tone. Therefore, taking GCBE, may reduce the reliance on highlighter and foundation. Coffee extract also improves fine lines, wrinkles, and pigmentation, as it contains powerful antioxidants. Antioxidants fight free radicals that damage the skin and contribute to aging.

Improved energy levels, focus, and mood

American Psychological Association has concluded that caffeine has significant effects on psychomotor and cognitive performance, psychological well-being, blood pressure, and athletic performance. Other studies have suggested that caffeine is capable of improving attention, alertness, motivation, and reaction times.

Miscellaneous effects

Asthma, headache, and respiratory effect

Green coffee is known to provide relief from asthma by inhibition of the lipoxygenase pathway and also relieves severe headaches (www.makeamericawell.com). Coffee seeds can also be used as a respiratory stimulant due to caffeine.

Immunomodulatory effect

Caffeine modulates the immune system and acts as an antiallergic by inhibiting histamine release and improving oxygen utilization (www.makeamericawell.com). Recent studies have revealed that the immune system is modulated by phytochemicals, which prevent DNA from being damaged and also repair DNA and reduce oxidative stress-induced cellular damage (Bothiraj et al., 2020). Caffeine is a member of the methyl-xanthine family of drugs, and is one of the most widely consumed behaviorally active substances in the Western world. Caffeine modulates both innate and adaptive immune responses.

Caffeine and its major metabolite paraxanthine suppress neutrophil and monocyte chemotaxis, and suppress production of proinflammatory cytokine tumor necrosis factor-alpha from human blood. Caffeine suppresses human lymphocyte function as indicated by reduced T-cell proliferation and impaired production of Th1 [interleukin (IL)-2 and interferon (IFN)-gamma], Th2 (IL-4, IL-5), and Th3 (IL-10) cytokines. It also suppresses antibody production. Some of the immune-modulatory actions of caffeine are mediated via inhibition of cyclic adenosine monophosphate (cAMP)-PDE, and consequential increase in intracellular cAMP concentrations. Its immunomodulatory effects occur at concentrations that are relevant to normal coffee consumption. The potential of caffeine-induced immunomodulation significantly affects human health and well-being (Horrigan et al., 2006).

Antibacterial activity

Coffee fruits have been reported to have antibacterial efficacy against a wide array of microbes/bacterial population such as *Proteus vulgaris*, *Escherichia coli*, *Klebsiella pneumoniae*, *Vibrio cholerae*, *Salmonella typhimurium*, *Salmonella typhi*, *Streptococcus faecalis*, and *Staphylococcus aureus* (Shetty et al., 1994; Almeida et al., 2006; Kiran et al., 2011). Recent studies have also revealed antibacterial activity of *Coffea canephora* against *Sterptococcus mutans* (bacteria involved in tooth decay). The antibacterial activity is attributed to the presence of CGA and the Maillard reaction during the roasting process of coffee seeds (Antonio et al., 2011; Almeida et al., 2012). Additionally, CGA effectively inhibits the growth of bacterial pathogens by adversely affecting the barrier functions, significantly increasing the outer and plasma membrane permeability, and inducing slight leakage of nucleotides. Based on these findings, Lou et al. (2011) hypothesized that CGA, after binding with the outer membrane, disrupts the membrane, exhausts intracellular potential, and releases cytoplasmic macromolecules, ultimately resulting in cell death.

Other effects

Chlorogenic acid (CGA) has mild laxative, psychostimulant, mood-elevating (www.mypcos.info/1/treatments/natural/cholinergic-acid), and antiinflammatory effects (Gawlik-Dziki et al., 2014). GCE is also effective in inhibiting carrageenan-induced edema and lipopolysaccharide-induced peritonitis (De Castro et al., 2013). It has antibacterial activity and promotes the regeneration of dendrites and axons (Tohda et al., 2005). Coffee consumption has also been reported to reduce the suicide risk in humans (Klatsky et al., 1993; Kawachi et al., 1996).

Green coffee bean as a fortified food supplement

The beneficial health effects of nutraceuticals have increased the importance of certain commonly used household ingredients which have been used in traditional foods and drinks for centuries. Incorporation of one or more such components into the daily diet can prevent certain deficiency diseases and also provide additional benefits to consumers (Allen et al., 2006). This method of incorporation of some of the ingredients to produce a better quality of food is referred to as "fortification." The GCB is rich in phytobiomolecules possessing medicinal and therapeutic potential. Phenolics possess antineoplastic, antiinflammatory, and antioxidative activities, and therefore, these are commonly used in the food-fortification industry (Arts and Hollman, 2005; Wang et al., 2011). According to the report by Świeca et al. (2017), incorporation of 5% supplementation of coffee with wheat bread would increase the antioxidant property of wheat bread two- to threefold. Similarly, there seems vast scope for evolving certain GCB-fortified and value-added drinks for human consumption.

Risks to human health

Consumption of coffee has been found to have several beneficial effects on human health and reduces the incidence of several ailments compared to those who do not drink coffee. However, some of the major health risks associated with higher coffee consumption have also been documented, including coronary heart disease (Kawachi et al., 1994; Urgert and Katan, 1997), myocardial infarction or unstable angina (Panagiotakos et al., 2003), cardiac arrhythmias (Wilhelmsen et al., 2001; Frost and Vestergaard, 2005), stroke (Hakim et al., 1998), PD (Ascherio et al., 2001), osteoporosis and hip fracture (Heaney, 2002), impaired iron (Hallberg and Rossander, 1982; Fairweather-Tait, 2004) and zinc absorption (Van Dyck et al., 1996), and pregnancy-related problems including significant delay in conception (Williams et al., 1990; Hassan and Killick, 2004), spontaneous abortions (Cnattingius et al., 2000; Wen et al., 2001; Giannelli et al., 2003), and fetal growth restriction (Fortier et al., 1993; Vik et al., 2003). GCB may frequently result in sensitization and symptoms of asthma, allergic rhinitis, and allergic conjunctivitis in workers, including contact allergy (Diba and English, 2002). Other side effects of caffeine are insomnia, nervousness and restlessness, stomach upset, nausea and vomiting, and increased heart and breathing rates. Consuming large amounts of coffee might also cause headache, anxiety, agitation, ringing in the ears, and worsening of bleeding disorders and diabetes as caffeine has been reported to cause increases as well as decreases in blood sugar. A caffeine concentration of more than 1.5 g/day could impose serious secondary effects such as hypertensive episodes, muscular spasms of back muscles, numbness, epileptiform convulsions and tremors, hallucinations, tachyarrhythmia, and respiratory paralysis (Istudor, 2005). Moreover, caffeinism could be achieved on consumption of more than five cups of coffee per day. This addictive phenomenon is associated with clinical symptoms including irritability, sleep disturbances, tremors, gastrointestinal disturbances such as emesis and diarrhea, and also headache (Kraft and Hobbs, 2004).

Although beneficial effects of GCB outweigh its possible use as a nutraceutical or a component of the functional foods for humans, detailed controlled and long-term use studies are required before it is recommended as a prospective nutraceutical.

Concluding remarks and future directions

Global health policies encourage the use of plant foods, including fruits and vegetables, in prevention of several oxidative stress-induced chronic diseases (WHO, 2004). Coffee is one of the most popular drinks in the world and is known to be a rich source of antioxidants in human diet, contributing to the prevention of oxidative stress-related diseases. Therefore, it is being advocated as a weight loss supplement. Although efficacy data support its promising potential as a nutraceutical, certain adverse effects of GCB, especially if consumed in excess, also warn for its judicious use. Harvard School of Public Health, based on their 22-year study, stated that "Coffee may have potential health benefits, but more research needs to be done." The majority of the health risks are due to its caffeine content, therefore these can be avoided by drinking decaffeinated coffee.

Green coffee bean (GCB) contains the largest amount of CGA and, in view of its promising pharmacological activities, especially antioxidant activity, ability to increase hepatic glucose utilization, suppression of fat absorption, enhancement of fat metabolism in liver, inhibition of HIV-1 integrase, antispasmodic activity, inhibition of mutagenicity of carcinogens, antihypertensive effect, and several other desirable pharmacological effects, there is scope for its exploitation in drug development programs and as a food supplement. Therefore pharmacological, therapeutic, and nutraceutical properties of GCB seem to hold very promising potential for its use not only as a nutraceutical but also for development of certain herbal-based drug formulations. However, assessment of the health risks and benefits of coffee and caffeine consumption requires further experimental and controlled clinical studies, including studies on underlying mechanisms, and also dose- and time-dependent effects of coffee consumption. Genetic heterogeneity with respect to the genetic polymorphisms in phase I and II biotransformation enzymes, dietary habits, and chronic diseases would help in identifying the specific genotypes which are more susceptible to adverse or beneficial effects. Further research should focus on it's possible adverse effects and their mechanistic pathways, if any, following long-term coffee consumption in children, adolescents, and pregnant women because it is one of the most popular global beverages.

References

Acidri, R., Sawai, Y., Sugimoto, Y., et al., 2020. Phytochemical profile and antioxidant capacity of coffee plant organs compared to green and roasted coffee beans. Antioxidants 9, 93. Available from: https://doi.org/10.3390/antiox9020093.

Adrian, J., Frangne, R., 1991. Synthesis and availability of niacin in roasted coffee. Adv. Exp. Med. Biol. 289, 49–59.

Affonso, R.C., Voytena, A.P., Fanan, S., et al., 2016. Phytochemical composition, antioxidant activity, and the effect of the aqueous extract of coffee (*Coffea arabica* L.) bean residual press cake on the skin wound healing. Oxid. Med. Cell. Longev. 2016, 1923754 (2016).

Agardh, E.E., Carlsson, S., Ahlbom, A., et al., 2004. Coffee consumption, type 2 diabetes and impaired glucose tolerance in Swedish men and women. J. Intern. Med. 255, 645–652.

Akash, M.S.H., Rehman, K., Chen, S., 2014. Effects of coffee on type 2 diabetes mellitus. Nutrition 30, 755–763.

Alía, M., Mateos, R., Ramos, S., et al., 2006. Influence of quercetin and rutin on growth and antioxidant defense system system of human hepatoma cell line (HepG2). J. Biochem. Mol. Toxicol. 12, 119–128.

Allen, L., de Benoist, B., Dary, O., et al., 2006. Guidelines on Food Fortification With Micronutrients. World Health Organization, Food and Agricultural Organization of the United Nations, pp. 1–341. Available from: http://doi.org/10.1242/jeb.02490.

Allison et al., 2018. Synergistic Nutraceutical Beverage Formulations Providing Enhanced Thermogenesis, Mental Clarity, and Stamina While Minimizing Adrenaline and Dopamine Concentration Perturbations Associated With Withdrawal. United States Patent Application Publication. US 2018/0055850 A1.

Allred, K.F., Yackley, K.M., Vanamala, J., et al., 2009. Trigonelline is a novel phytoestrogen in coffee beans. J. Nutr. 139, 1833–1838.

Almeida, A.A., Farah, A., Silva, D.A.M., et al., 2006. Antibacterial activity of coffee extracts and selected coffee chemical compounds against enterobacteria. J. Agric. Food Chem. 54, 8738–8743.

Almeida, A.A.P., Naghetini, C.C., Santos, V.R., et al., 2012. Influence of natural coffee compounds, coffee extracts and increased levels of caffeine on the inhibition of *Streptococcus mutans*. Food Res. Int. 49, 459–461.

Alonso-Salces, R.M., Serra, F., Reniero, F., et al., 2009. Botanical and geographical characterization of green coffee (*Coffea arabica* and *Coffea canephora*): chemometric evaluation of phenolic and methylxanthine contents. J. Agric. Food Chem. 57, 4224–4235.

Alsunni, A.A., 2015. Energy drink consumption: beneficial and adverse health effects. Int. J. Health Sci. 9, 468–474.

Antonio, A.G., Iorio, N.L.P., Pierro, V.S.S., et al., 2011. Inhibitory properties of *Coffea canephora* extract against oral bacteria and its effect on demineralization of deciduous teeth. Arch.Oral Biol. 56, 556–564.

Arao, T., Tamura, H., 2011. Coffee induces 11β-hydroxysteroid dehydrogenase type 2 activity in human placental choriocarcinoma cells, JEG-3. J. Health Sci. 57 (5), 436–441.

Arendash, G., Mori, T., Cao, C., et al., 2009. Caffeine reverses cognitive impairment and decreases brain Ab levels in aged Alzheimer's mice. J. Alzheimers Dis. 17, 661–680.

Arendash, G.W., Cao, C., 2010. Caffeine and coffee as therapeutics against Alzhemer's disease. J. Alzheimer's Dis. 20, 117–126.

Arion, W.J., Canfield, W.K., Ramos, F.C., et al., 1997. Chlorogenic acid and hydroxynitrobenzaldehyde: new inhibitors of hepatic glucose 6-phosphatase. Arch. Biochem. Biophys. 339, 315–322.

Arnold, U., Ludwig, E., Kühn, R., et al., 1994. Analysis of free amino acids in green coffee beans. Z. für Lebensmittel-Untersuchung und -Forschung 199 (1), 22–25. Available from: 10.1007/BF01192946. PMID 8067059.

Arts, I.C.W., Hollman, P.C.H., 2005. Polyphenols and disease risk in epidemiologic studies 1 –4. Am. J. Clin. Nutr. 81, 317–325.

Ascherio, A., Zhang, S.M., Hernan, M.A., 2001. Prospective study of caffeine consumption and risk of Parkinson's disease in men and women. Ann. Neurol. 50, 56–63.

Ashihara, H., Crozier, A., 1999. Biosynthesis and catabolism of caffeine in low-caffeine-containing species of *Coffea*. J. Agric. Food Chem. 47, 3425–3431.

Ashihara, H., Sano, H., Crozier, A., 2008. Caffeine and related purine alkaloids: biosynthesis, catabolism, function and genetic engineering. Phytochemistry 69, 841–856.

Ashihara, H., Yokota, T., Crozier, A., 2013. Biosynthesis and catabolism of purine alkaloids, In: Advances in Botanical Research, 68. Elsevier Ltd.

Aura, A.M., 2008. Microbial metabolism of dietary phenolic compounds in the colon. Phytochem. Rev. 7, 407–429.

Baer-Dubowska, W., Szaefer, H., 1998. Inhibition of murine hepatic cytochrome p-450 acitivies by natural and synthetic compounds. Xenobiotica 28, 735–743.

Baeza, G., Amigo-Benavent, M., Sarriá, B., et al., 2014. Green coffee hydroxycinnamic acids but not caffeine protect human HepG2 cells against oxidative stress. Food Res. Int. 62, 1038–1046.

Barranco, Q.J.L., Allam, M.F., Serrano Del Castillo, A.R., et al., 2007. Alzheimer's disease and coffee: a quantitative review. Neurol. Res. 29, 91–95.

Bassoli, B.K., Cassolla, P., Borba-Murad, G.R., et al., 2008. Chlorogenic acid reduces plasma glucose peak in the oral glucose tolerance test: effect on hepatic glucose release and glycaemia. Cell Biochem. Funct. 26 (3), 320–328.

Belay, A., Gholap, A.V., 2009. Characterization and determination of chlorogenic acids (CGA) in coffee benas by UV-VIS spectroscopy. Afr. J. Pure Appl. Chem. 3 (11), 234–240.

Belitz, H.D., Grosch, W., Schieberle, P., 2009. Coffee, tea, cocoa. In: Belitz, H.D., Grosch, W., Schieberle, P. (Eds.), Food Chemistry, fourth ed. Springer, Leipzig, pp. 938–951.

Bessiere–Thomas, Y., Ivon, F., 2002. Coffee Flavour Chemistry. John Wiley & Sons, Chichester.

Bidel, S., Hu, G., Sundvall, J., et al., 2006. Effects of coffee consumption on glucose tolerance, serum glucose and insulin levels: a cross-sectional analysis. Horm. Metab. Res. 38 (1), 38–43.

Bilkis, B., Choudhury, M.H., Mohammad, A.R., 2003. Caffeine from the mature leaves of *Coffea bengalensis*. Biochem. Syst. Ecol. 31, 1219–1220.

Blum, J., Lemaire, B., Lafay, S., 2007. Effect of a green decaffeinated coffee extract on glycaemia. NUTRAfoods Res. 6, 13–17.

Bohn, S.K., Blomhoff, R., Paur, I., 2014. Coffee and cancer risk, epidemiological evidence, and molecular mechanisms. Mol. Nutr. Food Res. 58 (5), 915–930. Available from: 10.1002/mnfr.201300526.

Boros, B., Jakabová, S., Dörnyei, A., et al., 2010. Determination of polyphenolic compounds by liquid chromatography-mass spectrometry in Thymus species. J. Chromatogr. A 1217, 7972–7980.

Bothiraj, K.V., Murugan, Vanitha, V., 2020. Green coffee bean seed and their role in antioxidant–a review. Int. J. Res. Pharm. Sci. 11 (1), 233–240.

Bourne, L.C., Rice-Evans, C.A., 1998. Urinary detection of hydroxycinnamates and flavonoids in humans after high dietary intake of fruit. Free Radic. Res. 28, 429–438.

Braga et al., 2016. Coffee Oil Capsules. United States Patent Application Publication.US 2016/0113314 A1.

Bravo, J., Arbillaga, L., de Peña, M.P., et al., 2013. Antioxidant and genoprotective effects of spent coffee extracts in human cells. Food Chem. Toxicol. 60, 397–403.

Brown, S.R., Cann, P.A., Read, N.W., 1990. Effect of coffee on distal colon function. Gut 31, 450–453.

Brown, O.I., Allgar, V., Wong, K.Y., 2016. Coffee reduces the risk of death after acute myocardial infarction: a meta-analysis. Coron. Artery Dis. 27 (7), 566.

Buffo, R.A., Freire, C.C., 2004. Coffee flavor: an overview. Flavour Frag. J. 19, 99–104.

Campa, C., Ballester, J.F., Doulbeau, S., et al., 2004. Trigonelline and sucrose diversity in wild *Coffea* species. Food Chem. 2004 (88), 39–43.

Campa, C., Mondolot, L., Rakotondravao, A., et al., 2012. A survey of mangiferin and hydroxycinnamic acid ester accumulation in coffee (*Coffea*) leaves: biological implications and uses. Ann. Bot. 2012 (110), 595–613.

Casiglia, E., Spolaore, P., Ginocchio, G., et al., 1993. Unexpected effects of coffee consumption on liver enzymes. Eur. J. Epidemiol. 9, 293–297.

Castilho, M.D., Ames, J.M., Gordon, M., 2002. Effect of roasting on the antioxidant activity of coffee brews. J. Agric. Food Chem. 50, 3698–3703.

Castro, R.D., Marraccini, P., 2006. Cytology, biochemistry and molecular changes during coffee fruit development. Braz. J. Plant Physiol. 18 (1), 175–199.

Cavin, C., Holzhaeuser, D., Scharf, G., et al., 2002. Cafestol and kahweol, two coffee specific diterpenes with anticarcinogenic activity. Food Chem. Toxicol. 40, 1155–1163.

Chen, X.M., Ma, Z., Kitts, D.D., 2018. Effects of processing method and age of leaves on phytochemical profiles and bioactivity of coffee leaves. Food Chem. 2018 (249), 143–153.

Cho, A.S., Jeon, S.M., Kim, M.J., et al., 2010. Chlorogenic acid exhibits anti-obesity property and improves lipid metabolism in high-fat diet-induced-obese mice. Food Chem. Toxicol. 48 (3), 937–943.

Chobotova, K., 2009. Aging and cancer: converging routes to disease prevention. Integr. Cancer Ther. 8, 115–122.

Choi, B.K., Park, S.B., Lee, D.R., et al., 2016. Green coffee bean extract improves obesity by decreasing body fat in high-fat diet-induced obese mice. Asian Pac. J. Trop. Med. 9 (7), 635–643.

Clifford, M.N., Kazi, M., 1987. The influence of coffee bean maturity on the content of chlorogenic acid, caffeine and trigonelline. Food Chem. 26, 59–69.

Cnattingius, S., Signorello, L.B., Anneren, G., et al., 2000. Caffeine intake and the risk of first-trimester spontaneous abortion. N. Engl. J. Med. 343, 1839–1845.

Comité Français du Café, 1997. Café—a la découverte du café. Adexquation Publicite, Paris.

Corrao, G., Lepore, A.R., Torchio, P., et al., 1994. The effect of drinking coffee and smoking cigarettes on the risk of cirrhosis associated with alcohol consumption. A case-control study. Provincial group for the study of chronic liver disease. Eur. J. Epidemiol. 10, 657–664.

Corrao, G., Zambon, A., Bagnardi, V., et al., 2001. Coffee, caffeine, and the risk of liver cirrhosis. Ann. Epidemiol. 11, 458–465.

Cowan, T.E., Palmnäs, M.A.S., Yang, J., et al., 2014. Chronic coffee consumption in the diet-induced obese rat: impact on gut microbiota and serum metabolomics. J. Nutr. Biochem. 25 (4), 489–495. 2014.

Cremin, P., Kasim-Karakas, S., Waterhouse, A.L., 2001. LC/ES-MS detection of hydroxycinnamates in human plasma and urine. J. Agric. Food Chem. 49, 1747–1750.

Cropley, V., Croft, R., Silber, B., et al., 2012. Does coffee enriched with chlorogenic acids improve mood and cognition after acute administration in healthy elderly? A pilot study. Psychopharmacology 219 (3), 737–749.

Crozier, A., Jaganath, I.B., Clifford, M.N., 2009. Dietary phenolics: chemistry, bioavailability and effects on health. Nat. Prod. Rep. 26, 1001–1043.

D'hooghe, M.B., Haentjes, P., Nagels, G., et al., 2012. Alcohol, coffee, fish, smoking and disease progression in multiple sclerosis. Eur. J. Neurol. 19 (4), 616–624. Available from: 10.1111/j.1468-1331.2011.03596.x.

De Castro, M.E., Fonseca, R.G., Ferreira, D., et al., 2013. Anti-inflammatory effect of aqueous extract of roasted and green *Coffea Arabica* L. J. Funct. Foods 5, 466–474.

Desai, N.M., Stanley, J.G., Murthy, P.S., 2019. Green coffee nanoparticles: optimization, *invitro* bioactivity and bio-release property. J. Microencapsul. Available from: 10.1080/02652048.2019.1692946.

Diba, V.C., English, J.S., 2002. Contact allergy to green coffee bean dust in a coffee processing plant worker. Contact Dermatisis 47, 56.

Domanski, M., Mitchell, G., Pfeffer, M., et al., 2002. Pulse pressure and cardiovascular disease-related mortality: follow-up study of the multiple risk factor intervention trial (MRFIT). J. Am. Med. Assoc. 287, 2677–2683.

Donovan, J.L., Manach, C., Faulks, R.M., et al., 2006. Absorption and metabolism of dietary secondary metabolites. In: Crozier, A., Clifford, M.N., Ashihara, H. (Eds.), Plant Secondary Metabolites. Occurrence, Structure and Role in the Human Diet. Blackwell Publishing, Oxford, pp. 303–351.

Dupas, C., Baglieri, A.M., Ordonaud, C., et al., 2006. Chlorogenic acid is poorly absorbed, independently of the food matrix: a Caco-2 cells and rat chronic absorption study. Mol. Nutr. Food Res. 50, 1053–1060.

Dupont, M.S., Bennett, R.N., Mellon, F.A., et al., 2002. Polyphenols from alcoholic apple cider are absorbed, metabolized and excreted by humans. J. Nutr. 132, 172–175.

Etienne, H., 2005. Somatic embryogenesis protocol: coffee (*Coffea arabica* L. and *C. canephora* P.). In: Jain, S.M., Gupta, P.K. (Eds.), Protocol for Somatic Embryogenesis in Woody Plant. Springer, Dordrecht, pp. 167–168.

Fairweather-Tait, S.J., 2004. Iron nutrition in the UK: getting the balance right. Proc. Nutr. Soc. 63, 519–528.

Farah, A., Monteiro, M.C., Calado, V., Trugo, L.C., 2006a. Correlation between cup quality and chemical attributes of Brazilian coffee. Food Chem. 98, 373–380.

Farah, A., Monteiro, M.C., Guigon, F., et al., 2006b. Chlorogenic acids from coffee are absorbed and excreted in human digestive fluids. In: Proceedings of the Twenty-First International Conference on Coffee Science, pp. 58–64.

Farah, A., Monteiro, M., Donangelo, C.M., et al., 2008. Chlorogenic acids from green coffee extract are highly bioavailable in humans. J. Nutr. 138, 2309–2315.

Fortier, I., Marcoux, S., Beaulac-Baillargeon, L., 1993. Relation of caffeine intake during pregnancy to intrauterine growth retardation and preterm birth. Am. J. Epidemiol. 137, 931–940.

Frost, L., Vestergaard, P., 2005. Caffeine and risk of atrial fibrillation or flutter: the Danish Diet, Cancer, and Health Study. Am. J. Clin. Nutr. 81, 578–582.

Gallardo, C., Jiménez, L., García-Conesa, M.T., 2006. Hydroxycinnamic acid composition and in vitro antioxidant activity of selected grain fractions. Food Chem. 99 (3), 455–463.

Gallus, S., Tavani, A., Negri, E., Vecchia, C.L., 2002a. Does coffee protect against liver cirrhosis? Ann. Epidemiol. 12 (3), 202–205.

Gallus, S., Bertuzzi, M., Tavani, A., et al., 2002b. Does coffee protect against hepatocellular carcinoma? Br. J. Cancer 87, 956–959.

Gawlik-Dziki, U., Świeca, M., Dziki, D., et al., 2014. Lipoxygenase inhibitors and antioxidants from green coffee—mechanism of action in the light of potential bioaccessibility. Food Res. Int. 61, 48–55.

Gelatti, U., Covolo, L., Franceschini, M., et al., 2005. Coffee consumption reduces the risk of hepatocellular carcinoma independently of its aetiology: a case-control study. J. Hepatol. 42, 528–534.

Giannelli, M., Doyle, P., Roman, E., et al., 2003. The effect of caffeine consumption and nausea on the risk of miscarriage. Paediatr. Perinat. Epidemiol. 17, 316–323.

Ginz, M., Balzer, H.H., Bradbury, A.G.W., et al., 2000. Formation of aliphatic acids by carbohydrate degradation during roasting of coffee. Eur. Food Res. Technol. 211, 404–410.

Giovannucci, E., 1998. Meta-analysis of coffee consumption and risk of colorectal cancer. Am. J. Epidemiol. 147, 1043–1052.

Gouthamchandra, K., Sudeep, H.V., Venkatesh, B.J., et al., 2017. Chlorogenic acid complex (CGA7), standardized extract from green coffeebeans exerts anticancer effects against cultured human colon cancerHCT-116 cells. Food Sci. Hum. Wellness 6, 147–153.

Granado-Serrano, A.B., Martín, M.A., Haegeman, G., et al., 2010. Epicatechin induces NFκB, AP-1 and Nrf2 via PI3K/AKT and ERK signaling in HepG2 cells. Br. J. Nutr. 103, 168–179.

Greenberg, J.A., Boozer, C.N., Geliebter, A., 2006. Coffee, diabetes, and weight control. Am. J. Clin. Nutr. 84 (4), 682–693.

Grembecka, M., Malinowska, E., Szefer, P., 2007. Differentiation of market coffee and its infusions in view of their mineral composition. Sci. Total Environ. 383, 59–69.

Gross, G., Jaccaud, E., Huggett, A.C., 1997. Analysis of the content of the diterpenes cafestol and kahweol in coffee brews. Food Chem. Toxicol. 35, 547–554.

Hakim, A.A., Ross, G.W., Curb, J.D., et al., 1998. Coffee consumption in hypertensive men in older middle-age and the risk of stroke: the Honolulu Heart Program. J. Clin. Epidemiol. 51, 487–494.

Hallberg, L., Rossander, L., 1982. Effect of different drinks on the absorption of non-heme iron from composite meals. Hum. Nutr. Appl. Nutr. 36, 116–123.

Hassan, M.A., Killick, S.R., 2004. Negative lifestyle is associated with a significant reduction in fecundity. Fertil. Steril. 81, 384–392.

Heaney, R.P., 2002. Effects of caffeine on bone and the calcium economy. Food Chem. Toxicol. 40, 1263–1270.

Hecimovic, I., Belscak-Cvitanovic, A., Horzic, D., et al., 2011. Comparative study of polyphenols and caffeine in different coffee varieties affected by the degree of roasting. Food Chem. 129, 991–1000 (2011).

Hemmerle, H., Burger, H.J., Bellow, P., et al., 1997. Chlorogenic acid and synthetic chlorogenic acid derivatives: novel inhibitors of hepatic glucose-6-phosphate translocase. J. Med. Chem. 40, 137–145.

Henning, S.M., Niu, Y., Lee, N.H., et al., 2004. Bioavailability and antioxidant activity of tea flavanols after consumption of green tea, black tea or green tea extract supplement. Am. J. Clin. Nutr. 80 (5), 1558–1564.

Herling, A.W., Burger, H.J., Schwab, D., et al., 1998. Pharmacodynamic profile of a novel inhibitor of the hepatic glucose-6-phosphatase system. Am. J. Physiol. 274, G1087–G1093.

Herraiz, T., Chaparro, C., 2006. Human monoamine oxidase inhibition by coffee and β-carbolines and nonharmane and harmane isolated from coffee. Life Sci. 78 (8), 795–802.

Higdon, J.V., Frei, B., 2006. Coffee and health: a review of recent human research. Crit. Rev. Food Sci. Nutr. 46, 101–123.

Hirakawa, N., Okauchi, R., Miura, Y., et al., 2005. Anti-invasive activity of niacin and trigolline against cancer cells. Biosci. Biotechnol. Biochem. 69, 653–658.

Hirose, K., Niwa, Y., Wakai, K., et al., 2007. Coffee consumption and the risk of endometrial cancer: evidence from a case control study of female hormone-related cancers in Japan. Cancer Sci. 98, 411–415.

Honjo, S., Kono, S., Coleman, M.P., et al., 1999. Coffee drinking and serum gamma-glutamyltransferase: an extended study of Self- Defense Officials of Japan. Ann. Epidemiol. 9, 325–331.

Honjo, S., Kono, S., Coleman, M.P., et al., 2001. Coffee consumption and serum aminotransferases in middle-aged Japanese men. J. Clin. Epidemiol. 54, 823–829.

Horrigan, L.A., Kelly, J., Connor, T.J., 2006. Immunomodulatory effects of caffeine: friend or foe? Pharmacol. Ther. 111 (3), 877–892.

Hu, G., Bidel, S., Jousilahti, P., et al., 2007. Coffee and tea consumption and the risk of Parkinson's disease. Mov. Disord. 22 (15), 2242–2248. Available from: 10.1002/mds.21706.

Huber, W.W., McDaniel, L.P., Kaderlik, K.R., et al., 1997. Chemoprotection against the formation of colon DNA adducts from the food-borne carcinogen 2-amino-1-methyl-6-phenylimidazo[4,5-b]pyridine (PhIP) in the rat. Mut. Res. 376, 115–122.

Huber, W.W., Scharf, G., Rossmanith, W., et al., 2002a. The coffee components kahweol and cafestol induce gamma-glutamylcysteine synthetase, the rate limiting enzyme of chemoprotective glutathione synthesis, in several organs of the rat. Arch. Toxicol. 75, 685–694.

Huber, W.W., Prustomersky, S., Delbanco, E., et al., 2002b. Enhancement of the chemoprotective enzymes glucuronosyl transferase and glutathione transferase in specific organs of the rat by the coffee components kahweol and cafestol. Arch. Toxicol. 76, 209–217.

Imran, M., Arshad, M.S., Butt, M.S., et al., 2017. Mangiferin: a natural miracle bioactive compound against lifestyle related disorders. Lipids Health Dis. 16, 84.

Inoue, M., Yoshimi, I., Sobue, T., et al., 2005. Influence of coffee drinking on subsequent risk of hepatocellular carcinoma: a prospective study in Japan. J. Natl. Cancer Inst. 97, 293–300.

Istudor, V., 2005. Farmacognozie, Fitochimie, Fitoterapie, vol. III. Editura Medical, București, pp. 265–266.

Jeszka-Skowron, M., Stanisz, E., Paz De Pena, M., 2016b. Relationship between antioxidant capacity, chlorogenic acids and elemental composition of green coffee. LWT—Food Sci. Technol. 73, 243–250.

Johnston, K.L., Clifford, M.N., Morgan, L.M., 2003. Coffee acutely modifies gastrointestinal hormone secretion and glucose tolerance in humans: glycemic effects of chlorogenic acid and caffeine. Am. J. Clin. Nutr. 78 (4), 728–733.

Jordan, S.J., Purdie, D.M., Green, A.C., et al., 2004. Coffee, tea and caffeine and risk of epithelial ovarian cancer. Cancer Causes Control 15, 359–365.

Kang, N.J., Lee, K.W., Kim, B.H., et al., 2011. Coffee phenolic phytochemicals suppress colon cancer metastasis by targeting MEK and TOPK. Carcinogenesis 32 (6), 921–928.

Kao, W.H., Folsom, A.R., Nieto, F.J., et al., 1999. Serum and dietary magnesium and the risk for type 2 diabetes mellitus: the atherosclerosis risk in communities study. Arch. Int. Med. 159, 2151–2159.

Kawachi, I., Colditz, G.A., Stone, C.B., 1994. Does coffee drinking increase the risk of coronary heart disease? Results from a meta-analysis. Br. Heart J. 72, 269–275.

Kawachi, I., Willett, W.C., Colditz, G.A., et al., 1996. A prospective study of coffee drinking and suicide in women. Arch. Int. Med. 156, 521–525.

Kiran, B., Baruah, R., Ojha, R., et al., 2011. Antibacterial and antioxidant activity of *Coffea benghalensis* Roxb. Ex. Schult. Fruit against human bacteria. Res. J. Pharm. Biol. Chem. Sci. 2, 856–865.

Klatsky, A.L., Armstrong, M.A., Friedman, G.D., 1993. Coffee, tea, and mortality. Ann. Epidemiol. 3, 375–381.

Konishi, Y., Zhao, Z., Shimizu, M., 2006. Phenolic acids are absorbed from the rat stomach with different absorption rates. J. Agric. Food Chem. 54, 7539–7543.

Kono, Y., Kobayashi, K., Tagawa, S., et al., 1997. Antioxidant activity of polyphenols in diets. Rate constants of reactions of chlorogenic acid and caffeic acid with reactive species of oxygen and nitrogen. Biochim. Biophys. Acta 1335, 335–342.

Kozuma, K., Tsuchiya, S., Kohori, J., et al., 2005. Antihypertensive effect of green coffee bean extract on mildly hypertensive subjects. Hypertens. Res. 28, 711–718.

Kraft, K., Hobbs, C., 2004. Pocket Guide to Herbal Medicine. Georg Thieme. Verlag, Stuttgart, pp. 50–51.

Ky, C.L., Louarn, J., Dussert, S., et al., 2001. Caffeine, trigonelline, chlorogenic acids and sucrose diversity in wild *C. arabica* L. and *Coffea canephora* P. accessions. Food Chem. 2001 (75), 223–230.

Lafay, S., Gil-Izquierdo, A., Manach, C., et al., 2006. Chlorogenic acid is absorbed in its intact form in the stomach of rats. J. Nutr. 136, 1192–1197.

Lanfer-Marquez, U.M., Barros, R.M.C., Sinnecker, P., 2005. Antioxidant activity of chlorophylls and their derivatives. Food Res. Int. 38, 885–891.

Le, M.G., 1985. Coffee consumption, benign breast disease, and breast cancer. Am. J. Epidemiol. 122, 721.

Lee, K.J., Jeong, H.G., 2007. Protective effects of kahweol and cafestol against hydrogen peroxide-induced oxidative stress and DNA damage. Toxicol. Lett. 173 (2), 80–87.

Li, X., Zhuo, R., Tiong, S., et al., 2013. The Smc5/Smc6/MAGE complex confers resistance to caffeine and genotoxic stress in *Drosophila melanogaster*. PLoS One 8 (3), e59866. Available from: 10.1371/journal.pone.0059866.

Lloyd-Jones, D.M., Larson, M.G., Leip, E.P., et al., 2002. Lifetime risk for developing congestive heart failure: the Framingham Heart Study. Circulation 106, 3068–3072.

Lopez-Garcia, E., VanDam, R.M., Rajpathak, S., et al., 2006. Changes in caffeine intake and long term weight change in men and women. Am. J. Clin. Nutr. 83 (3), 674–680.

Lopez-Ridaura, R., Willett, W.C., Rimm, E.B., et al., 2004. Magnesium intake and risk of type 2 diabetes in men and women. Diabetes Care 27, 134–140.

Lou, Z., Wang, H., Zhu, S., et al., 2011. Antibacterial activity and mechanism of action of chlorogenic acid. J. Food Sci. 76 (6), M398–M403.

Lubin, F., Ron, E., Wax, Y., et al., 1985. Coffee and methylxanthines and breast cancer: a case-control study. J. Natl. Cancer Inst. 74, 569–573.

Luzia, M.R., Paixão, K.C.C., Marcilio, R., et al., 1997. Effect of the 5- caffeoylquinic acid on soybean oil oxidative stability. Int. J. Food Sci. Technol. 32, 15–19.

Mahajan, R., Kapoor, N., 2018. Phytochemical analysis and antimicrobial activity of roasted beans of *Coffea robusta*. Int. J. Pharm. Biol. Sci. 8 (1), 89–95.

Mannisto, S., Pietinen, P., Virtanen, M., et al., 1999. Diet and the risk of breast cancer in a case-control study: does the threat of disease have an influence on recall bias? J. Clin. Epidemiol. 52, 429–439.

Martinez-Zapata, M.J., Vernooij, R.W., Uriona Tuma, S.M., et al., 2016. Phlebotonics for venous insufficiency. Cochrane Database Syst. Rev. 4, CD003229. Available from: 10.1002/14651858.CD003229.pub3. PMID 27048768.

Mehari, B., Redi-Abshiro, M., Chandravanshi, B.S., et al., 2019. GC-MS profiling of fatty acids in green coffee (*Coffea arabica* L.) beans and chemometric modeling for tracing geographical origins from Ethiopia. J. Sci. Food Agric. 99, 3811–3823.

Mekuria, T., Neuhoff, D., Kopke, U., 2004. The status of coffee production and the potential for organic conversion in Ethopia. In: Proceedings of the Conference on International Agricultural Research and Development. Deutscher Tropentag, Berlin.

Meliana, Y., Harmami, S.B., Restu, W.K., 2017. Characterization of nanoencapsulated *Centella asiatica* and *Zingiber officinale* extract using combination of malto dextrin and gum arabic as matrix. IOP Conf. Ser. Mater. Sci. Eng. 172, 012065.

Mellbye, F.B., Jeppesen, P.B., Hermansen, K., Gregersen, S., 2015. Cafestol, a bioactive substance in coffee, stimulates insulin secretion and increases glucose uptake in muscle cells: studies in vitro. J. Nat. Prod. 78, 2447–2451.

Michels, K.B., Willett, W.C., Fuchs, C.S., et al., 2005. Coffee, tea, and caffeine consumption and incidence of colon and rectal cancer. J. Natl. Cancer Inst. 97, 282–292.

Mitchell, B.M., Webb, R.C., 2002. Impaired vasodilation and nitric oxide synthase activity in glucocorticoid-induced hypertension. Biol. Res. Nurs. 4 (1), 16–21.

Mohammadshahi, M., Haidari, F., Karandish, M., Ebrahimi, S., Haghighizadeh, M.H., 2014. A randomized clinical trial of nutrition education for improvement of diet quality and inflammation in Iranian obese women. J. Nutr. Metab. 605782.

Monteiro, M., Farah, A., Perrone, D., et al., 2007. Chlorogenic acid compounds from coffee are differentially absorbed and metabolized in humans. J. Nutr. 137, 2196–2201.

Monteiro, M.C., Farah, A., 2012. Chlorogenic acids in Brazilian *C. arabica* cultivars from various consecutive crops. Food Chem. 134, 611–614.

Morishita, H., Kido, R., 1995. Antioxidant derivatives and chlorogenic acid. In: Proceedings of the Sixteenth International Colloquium on Chemicum Coffee, Kyoto, April 09–14, 1995.

Morling, J.R., Yeoh, S.E., Kolbach, D.N., 2018. Rutosides for treatment of post-thrombotic syndrome. Cochrane Database Syst. Rev. (11), CD005625. Available from: 10.1002/14651858.

Murase, T., Misawa, K., Minegishi, Y., Aoki, M., et al., 2011. Coffee polyphenols suppress diet-induced body fat accumulation by downregulating SREBP-1c and related molecules in C57BL/6J mice. Am. J. Physiol. Endocrinol. Metab. 300 (1), E122–E133.

Murkovic, M., Derler, K., 2006. Analysis of amino acids and carbohydrates in green coffee. J. Biochem. Biophys. Methods 69 (1–2), 25–32.

Murthy, P., Naidu, M.M., 2012. Recovery of phenolic antioxidants and functional compounds from coffee industry by-products. Food Bioprocess Technol. 5, 897–903.

Mussatto, S.I., Machado, E.M.S., Martins, S., et al., 2011. Production, composition and application of coffee and its industrial residues. Food Bioprocess Technol. 4, 661–672.

Naidu, M.M., Sulochanamma, G., Sampathu, S.R., et al., 2008. Studies on extraction and antioxidant potential of green coffee. Food Chem. 107, 377–384.

Nakanishi, N., Nakamura, K., Nakajima, K., et al., 2000. Coffee consumption and decreased serum gamma-glutamyltransferase: a study of middle-aged Japanese Men. Eur. J. Epidemiol. 16, 419–423.

Narayana, K.R., Reddy, M.S., Chaluvadi, M.R., et al., 2001. Bioflavonoids classification, pharmacological, biochemical effects and therapeutic potential. Indian J. Pharmacol. 33, 2–16.

Nicoli, M.C., Anese, M., Manzocco, L., et al., 1997. Antioxidant properties of coffee brews in relation to the roasting degree. Lebensm. Wiss. Technol. 30, 292–297.

O'keefe, J.H., Bhatti, S.K., Patil, H.R., et al., 2013. Effects of habitual coffee consumption on cardiometabolic disease, cardiovascular health, and all-cause mortality. J. Am. Coll. Cardiol. 62 (12), 1043–1051. 2013.

Ochiai, R., Jokura, H., Suzuki, A., et al., 2004. Green coffee bean extract improves human vasoreactivity. Hypertens. Res. 27, 731–737.

Olthof, M.R., Hollman, P.C., Katan, M.B., 2001. Chlorogenic acid and caffeic acid are absorbed in humans. J. Nutr. 131, 66–71.

Olthof, M.R., Hollman, P.C.H., Buijsman, M.N.C.P., et al., 2003. Chlorogenic acid, quercetin-3-rutinoside and black tea phenol are extensively metabolized in humans. J. Nutr. 133, 1806–1814.

Padol, I.T., Hunt, R.H., 2004. Effect of Th1 cytokines on acid secretion in pharmacologically characterised mouse gastric glands. Gut 53 (8), 1075–1081.

Patay, É.B., Bencsik, T., Papp, N., 2016b. Phytochemical overview and medicinal importance of *Coffea* species from the past until now. Asian Pac. J. Trop. Med. 9, 1127–1135.

Patay, É.B., Sali, N., Koszegi, T., et al., 2016a. Antioxidant potential, tannin and polyphenol contents of the seed and pericarp of three *Coffea* species. Asian Pac. J. Trop. Med. 9 (4), 366–371.

Panagiotakos, D.B., Pitsavos, C., Chrysohoou, C., et al., 2003. The J-shaped effect of coffee consumption on the risk of developing acute coronary syndromes: the CARDIO2000 case-control study. J. Nutr. 133, 3228–3232.

Poisson, J., 1979. Aspects chimiques et biologiquesde la Composition du café vert. In: Proceedings of the Eighth International Colloquium on Chemicum Coffee, Abidjan, November 28 to December 03, 1988. ASIC, pp. 33–37.

Prabha, P.S., Das, U.N., Koratkar, R., Sagar, P.S., et al., 1990. Free radical generation, lipid peroxidation and essential fatty acids in uncontrolled essential hypertension. Prostaglandins Leukot. Essent. Fatty Acids 41 (1), 27–33.

Rawel, H.M., Kulling, S.E., 2007. Nutritional contribution of coffee, cacao and tea phenolics to human health. J. Consum. Prot. Food Saf. 2, 399–406.

Rechner, A.R., Spencer, J.P.E., Kuhnle, G., et al., 2001. Novel biomarkers of the metabolism of caffeic acid derivatives *in vivo*. Free Radic. Biol. Med. 30, 1213–1222.

Redon, J., Oliva, M.R., Tormos, C., et al., 2003. Antioxidant activities and oxidative stress byproducts in human hypertension. Hypertension 41 (5), 1096 (Dallas, TX: 1979).

Revuelta-Iniesta, R., Al-Dujaili, E.A.S., 2014. Consumption of green coffee reduces blood pressure and body composition by influencing 11β-HSD1 enzyme activity in healthy individuals: a pilot crossover study using green and black coffee. Biomed. Res. Int. 2014, <https://doi.org/10.1155/2014/482704>.

Rio, D.D., Stalmach, A., Calani, L., et al., 2010. Bioavailability of coffee chlorogenic acids and green tea flavan-3-ols. Nutrients 2, 820–833.

Rodgers, G.M., Conn, M.T., 1990. Homocysteine, an atherogenic stimulus, reduces protein C activation by arterial and venous endothelial cells. Blood 75, 895–901.

Rodgwell, R.J., Curti, D., Rogers, J., et al., 2003. Changes to the galactose/mannose ratio in galactomannan during coffee bean (*Coffea arabica*) development: implications for *in vivo* modification for galactomannan synthesis. Planta 217, 316–326.

Rodríguez-Ramiro, I., Ramos, S., Bravo, L., et al., 2011. Procyanidin B2 induces Nrf2 translocation and glutathione S-transferase P1 expression via ERKs and p38-MAPK pathways and protect human colonic cells against oxidative stress. Eur. J. Nutr. 51, 881–882.

Ross, R., 1999. Atherosclerosis—an inflammatory disease. N. Engl. J. Med. 340, 115–126.

Ruhl, C.E., Everhart, J.E., 2005. Coffee and caffeine consumption reduce the risk of elevated serum alanine aminotransferase activity in the United States. Gastroenterology 128, 24–32.

Russo, G.I., Campisi, D., Di Mauro, M., et al., 2017. Dietary consumption of phenolic acids and prostate cancer: a case-control dtudy in Sicily, Southern Italy. Molecules (Basel, Switz.) 22 (12), 2159. <https://doi.org/10.3390/molecules22122159>.

Sääksjärvi, K., Knekt, P., Rissanen, H., et al., 2008. Prospective study of the coffee consumption and risk of Parkinson's disease. Eur. J. Clin. Nutri. 62 (7), 908–915.

Saito, I., Tsuchida, T., Watanabe, T., et al., 2002. Effect of coffee bean extract in essential hypertension. Jpn. J. Med. Pharm. Sci. 47, 67–74.

Salazar-Martinez, E., Willett, W.C., Ascherio, A., et al., 2004. Coffee consumption and risk for type 2 diabetes mellitus. Ann. Int. Med. 140, 1–8.

Salgado, P.R., Favarin, J.L., Leandro, R.A., et al., 2008. Total phenol concentrations in coffee tree leaves during fruit development. Sci. Agric. 65, 354–359.

Santos, E.J., Oliveira, E., 2001. Determination of mineral nutrients and toxic elements in Brazilian soluble coffee by ICP-AES. J. Food Comp. Anal. 14, 523–531.

Santos, R.M., Lima, D.R.A., 2016. Coffee consumption, obesity and type 2 diabetes: a mini-review. Eur. J. Nutr. 55, 1345–1358.

Shearer, J., Farah, A., de Paulis, T., et al., 2003. Quinides of roasted coffee enhance insulin action in conscious rats. J. Nutr. 133, 3529–3532.

Shetty, M., Subbannayya, K., Shivananda, P.G., 1994. Antibacterial activity of tea (*Camellia sinensis*) and coffee (*C. arabica*) with special reference to *Salmonella typhimurium*. J. Commun. Dis. 26 (3), 147–150.

Shimoda, H., Seki, E., Aitani, M., 2006. Inhibitory effect of green coffee bean extract on fat accumulation and body weight gain in mice. BMC Complement. Altern. Med. 6, 1–9.

Singh, A., Chauhan, S.S., Tripathi, V., 2018. Quinic acid attenuates oral cancer cell proliferation by downregulating cyclin D1 Expression and Akt signaling. Pharmacogn. Mag. 14 (Suppl S1), 14–19.

Sobésa Café, 2008. Available from: <http://www.sobesa.com.br> (Accessed 5 March 2010).

Spencer, J.P., Chowrimootoo, G., Choudhury, R., et al., 1999. The small intestine can both absorb and glucuronidate luminal flavonoids. FEBS Lett. 458, 224–230.

Stahl, W., Sies, H., 2003. Antioxidant activity of carotenoids. Mol. Aspects Med. 24, 345–351.

Stalikas, C.D., 2007. Extraction, separation, and detection methods for phenolic acids and flavonoids. J. Sep. Sci. 2007 (30), 3268–3295.

Steinhart, H., Luger, A., Piost, J., 2001. Antioxidative effect of coffee melanoidins. In: Proceedings of the Nineteenth International Scientific Colloquium on Coffee, Trieste, Italy, March 14–18, 2001.

Stich, H.F., Rosin, M.P., Bryson, L., 1982. Inhibition of mutagenicity of a model nitrosation reaction by naturally occurring phenolics, coffee and tea. Mut. Res. 95, 119–128.

Suzuki, A., Kagawa, D., Ochiai, R., et al., 2002. Green coffee bean extract and its metabolites have a hypotensive effect in spontaneously hypertensive rats. Hypertens. Res. 25, 99–107.

Swales, J.D., 1995. Manual of Hypertension, first ed. Blackwell Science Ltd, London, pp. 32–36.

Świeca, M., Gawlik-Dziki, U., Dziki, D., Baraniak, B., 2017. Wheat bread enriched with green coffee–in vitro bioaccessibility and bioavailability of phenolics and antioxidant activity. Food Chem. 221, 1451–1457.

Szenthe, A., 2019. Top coffee producing countries. World Atlas. <worldatlas.com/articles/top-coffee-producing-countries.html>.

Takahashi, K., Ishigami, A., 2017. Anti-aging effects of coffee. Aging (Albany NY.) 9, 1863–1864 (2017).

Talamond, P., Mondolot, L., Gargadennec, A., de Kochko, A., Hamon, S., Fruchier, A., et al., 2008. First report on mangiferin (C-glucosyl-xanthone) isolated from leaves of a wild coffee plant, *Coffea pseudozanguebariae* (Rubiaceae). Acta Bot. Gall. 155, 513–519.

Tanaka, K., Tokunaga, S., Kono, S., et al., 1998. Coffee consumption and decreased serum gamma-glutamyltransferase and aminotransferase activities among male alcohol drinkers. Int. J. Epidemiol. 27, 438–443.

Tavani, A., La Vecchia, C., 2004. Coffee, decaffeinated coffee, tea and cancer of the colon and rectum: a review of epidemiological studies, 1990–2003. Cancer Causes Control 15, 743–757.

Teutsch, I.A., 2004. Einfluss der Rohkaffeeverarbeitung auf Aromastoffveränderungen in Gerösteten Kaffeebohnen Sowie im Kaffeebetränk (PDF) (Ph.D.). Department of Chemistry, Technical University Munich, Germany.

Thomas, G.D., Zhang, W., Victor, R.G., 2001. Nitric oxide deficiency as a cause of clinical hypertension: promising new drug targets for refractory hypertension. JAMA 285 (16), 2055–2057.

Tohda, C., Kuboyama, T., Komatsu, K., 2005. Search for natural products related to regeneration of the neuronal network. Neurosignals 14, 34–45.

Trugo, L., 2003. Coffee. In: Caballero, B., Trugo, L., Finglas, P. (Eds.), Encyclopedia of Food Sciences and Nutrition, second ed. Academic Press, London.

Trugo, L.C., Macrae, R., 1984. Chlorogenic acid composition of instant coffees. Analyst 109, 263–266.

Trute, A., Gross, J., Mutschler, E., et al., 1997. *In vitro* antispasmodic compounds of the dry extract obtained from *Hedera helix*. Planta Med. 63, 125–129.

Tuomilehto, J., Hu, G., Bidel, S., 2004. Coffee consumption and risk of type 2 diabetes mellitus among middle-aged Finnish men and women. J. Am. Med. Assoc. 291, 1213–1219.

Urgert, R., Katan, M.B., 1997. The cholesterol-raising factor from coffee beans. Ann. Rev. Nutr. 17, 305–324.

Urgert, R., van der Weg, G., Kosmeijer-Schuil, T.G., et al., 1995. Levels of the cholesterol-elevating diterpenes cafestol and kahweol in various coffee brews. J. Agric. Food Chem. 43, 2167–2172.

van Dam, R.M., Feskens, E.J., 2002. Coffee consumption and risk of type 2 diabetes mellitus. Lancet 360, 1477–1478.

van Dam, R.M., Hu, F.B., 2005. Coffee consumption and risk of type 2 diabetes: a systematic review. J. Am. Med. Assoc. 294, 97–104.

van Dam, R.M., Dekker, J.M., Nijpels, G., et al., 2004. Coffee consumption and incidence of impaired fasting glucose, impaired glucose tolerance, and type 2 diabetes: the Hoorn Study. Diabetologia 47, 2152–2159.

Van Dongen, L.H., Mölenberg, F.J.M., Soedamah-Muthu, S.S., et al., 2017. Coffee consumption after myocardial infarction and risk of cardiovascular mortality: a prospective analysis in the Alpha Omega Cohort. Am. J. Clin. Nutr. 106 (4), 1113–1120.

Van Dyck, K., Tas, S., Robberecht, H., et al., 1996. The influence of different food components on the *in vitro* availability of iron, zinc and calcium from a composed meal. Int. J. Food Sci. Nutr. 47, 499–506.

Vanhoutte, P.M., Boulanger, C.M., 1995. Endothelium-dependent responses in hypertension. Hypertens. Res. 18, 87–98.

Vasanthi, H.R., ShriShriMal, N., Das, D.K., 2012. Phytochemicals from plants to combat cardiovascular disease. Curr. Med. Chem. 19, 2242–2251 (2012).

Viani, R., Horman, I., 1974. Thermal behaviour of trigonelline. J. Food Sci. 39, 1216–1217.

Vik, T., Bakketeig, L.S., Trygg, K.U., et al., 2003. High caffeine consumption in the third trimester of pregnancy: gender-specific effects on fetal growth. Paediatr. Perinat. Epidemiol. 17, 324–331.

Wagemaker, T.A.L., Carvalho, C.R.L., Maia, N.B., et al., 2011. Sun protection factor, content and composition of lipid fraction of green coffee beans. Ind. Crop Prod. 33, 469–473.

Wang, S., Melnyk, J.P., Tsao, R., et al., 2011. How natural dietary antioxidants in fruits, vegetables and legumes promote vascular health. Food Res. Int. 44 (1), 14–22.

Wangcharoen, W., Phimphilai, S., 2016. Chlorophyll and total phenolic contents, antioxidant activities and consumer acceptance test of processed grass drinks. J. Food Sci. Technol. 53, 4135–4140.

Wen, W., Shu, X.O., Jacobs Jr., D.R., et al., 2001. The associations of maternal caffeine consumption and nausea with spontaneous abortion. Epidemiology 12, 38–42.

Whelton, P.K., He, J., Appel, L.J., et al., 2002. National high blood pressure education program coordinating committee: primary prevention of hypertension: clinical and public health advisory from the national high blood pressure education program. J. Am. Med. Assoc. 288, 1882–1888.

WHO, 2004. Global strategy on diet, physical activity and health. The 57th World Health Assembly. World Health Organization, Geneva. Available from: <http://www.who.int/dietphysicalactivity/strategy/eb11344/strategy>.

Wilhelmsen, L., Rosengren, A., Lappas, G., 2001. Hospitalizations for atrial fibrillation in the general male population: morbidity and risk factors. J. Int. Med. 250, 382–389.

Williams, M.A., Monson, R.R., Goldman, M.B., et al., 1990. Coffee and delayed conception. Lancet 335, 1603.

Wintgens, J.N., 2009. The coffee plant. In: Wintgens, J.N. (Ed.), Coffee: Growing, Processing, Sustainable Production—A Guidebook for Growers, Processors, Traders, and Researchers, second ed. Wiley-VCH, Weinheim, pp. 3–24.

Wu, X., Skog, K., Jagerstad, M., 1997. Trigonelline, a naturally occurring constituent of green coffee beans behind the mutagenicity activity of roasted coffee. Mut. Res. 391, 171–177.

Yashin, A., Yashin, Y., Wang, J.Y., et al., 2013. Antioxidant and antiradical activity of coffee. Antioxidants 2, 230–245.

Yen, W.J., Wang, B., Chang, L., et al., 2005. Antioxidant properties of roasted coffee residues. J. Agric. Food Chem. 53, 2658–2663.

Zelber-Sagi, S., Godos, J., Salomone, F., 2016. Lifestyle changes for the treatment of nonalcoholic fatty liver disease: a review of observational studies and intervention trials. Ther. Adv. Gastroenter. 9 (3), 392–407.

Chapter 43

Quercetin

Muhammet Ay, Adhithiya Charli, Huajun Jin, Vellareddy Anantharam, Arthi Kanthasamy and Anumantha G. Kanthasamy

Introduction

Dietary flavonoids were first discovered by Nobel Prize laureate Albert Szent Gyorgyi in 1936, and are widely recognized for their potential beneficial effects on human health. Quercetin is the major flavonoid in our daily diet (Table 43.1) (see further details at: http://www.ars.usda.gov/SP2UserFiles/Place/80400525/Data/Flav/Flav_R03-1.pdf), and its estimated daily intake is between 5 and 40 mg (Hertog et al., 1995). After absorption, quercetin is mainly metabolized in the intestine and liver. The plasma concentration of quercetin is normally in the nanomolar range, but can reach the micromolar range after consumption of quercetin-rich foods (Russo et al., 2012). Table 43.1 summarizes foods with high quercetin levels. Quercetin contains three rings and five hydroxyl groups (Fig. 43.1). It mainly occurs as glycosides, ethers, and, to a lesser extent, sulfates (Williams and Grayer, 2004), and glycosylation of quercetin increases its hydrophilicity, stability, and bioavailability (He et al., 2006). So far, quercetin has been the most extensively studied flavonoid that has been shown to exhibit antioxidant, antiviral, antibacterial, antiinflammatory, and anticarcinogenic properties (Boots et al., 2008; Duthie and Dobson, 1999; Ramos et al., 2006; Zandi et al., 2011).

Quercetin and cancer

Quercetin has been considered a potential anticancer agent because it can inhibit cell proliferation and angiogenesis and induce apoptosis and cellular senescence (Lamson and Brignall, 2000; Russo et al., 1999; Zamin et al., 2009). Quercetin can induce cell death and cell cycle arrest in cancer cells through the downregulation of oncogenes including Mcl-1, Bcl-2, Ras, MEK, and PI3K or the upregulation of tumor suppressor genes including p53 and p21 (Iwao and Tsukamoto, 1999; Lim et al., 2007; Sharma et al., 2005; Spagnuolo et al., 2011). In MDA-MB-231 human breast cancer cells, quercetin was reported to suppress cell proliferation through the upregulation of Connexin 43 (Cx43) (Conklin et al., 2007), whose expression correlates inversely with the aggressiveness of the tumor (Laird et al., 1999). Similarly, another study demonstrated that quercetin inhibited the growth of MCF-7 breast cancer cells and also downregulated the expression of survivin, which is highly expressed in most cancers (Deng et al., 2013).

Quercetin has been reported to induce apoptosis in a human prostate cell line (LNCaP) by inducing the dissociation of Bax from Bcl-xL, activating caspases, and inhibiting phosphorylation of Akt (Kim and Lee, 2007; Lee et al., 2008). Urokinase-type plasminogen activator (uPA) has been associated with cancer invasion, including prostate cancer. Quercetin significantly decreased the expressions of uPA, MMP-2, MMP-9, EGF, β-catenin, and NF-κB and thus suppressed the proliferation, migration, and invasion of prostate cancer cells (Senthilkumar et al., 2011; Vijayababu et al., 2005). The chemopreventive effects of quercetin have been studied in an in vivo model of prostate cancer (Sharmila et al., 2014). The results showed that quercetin significantly reversed the reduction in superoxide dismutase (SOD), catalase, and glutathione peroxidase (GPx) activities and the increase in Akt phosphorylation in prostate cancer-induced rats.

Activation of β-catenin/Tcf signaling plays a central role in colorectal carcinogenesis (Park et al., 2005). Interestingly, quercetin has been shown to inhibit cell proliferation through suppression of β-catenin/Tcf signaling and survivin expression in SW480 colon cancer cells (Shan et al., 2009). Quercetin has also been demonstrated to suppress cell proliferation in HT-29 colon cancer cells by activating AMPK and decreasing the expression of COX-2, which is known to be overexpressed in colon cancer (Lee et al., 2009). Moreover, quercetin inhibited colorectal carcinogenesis and reduced the size of colorectal tumors in azoxymethane-treated rats (Dihal et al., 2006).

Nutraceuticals. DOI: https://doi.org/10.1016/B978-0-12-821038-3.00043-4

TABLE 43.1 Dietary sources and concentrations of quercetin.

Food source (ref. USDA 2014)	Quercetin content (mg/100 g)	Food source (ref. USDA 2014)	Quercetin content (mg/100 g)
Capers, raw	234	Red chard leaves, raw	7
Dill weed, fresh	55	Chicory greens, raw	6
Oregano, fresh	7	Chives, raw	5
Apples with skin	4	Cilantro leaves, raw	53
Bayberries, raw	4	Corn poppy, leaves	26
Blackberries, raw	4	Dock, raw	86
Blueberries, raw	4	Drumstick leaves, raw	17
Cherries, sweet, raw	2	Fennel leaves, raw	49
Chokeberry, raw	19	Hartwort leaves	29
Cranberries, raw	15	Kale, raw	23
Crowberries, raw	5	Lettuce, red leaf, raw	8
Currants, black, raw	4	Lovage, leaves, raw	170
Elderberries, raw	27	Mizuna	9
Figs, raw	5	Mustard greens, raw	9
Goji berry, dried	14	Nalta jute, raw	24
Grapes, black	2	Okra, raw	21
Juniper berries, ripe	47	Onions, red raw	39
Lingonberries, raw	13	Peppers, yellow wax, raw	51
Pitanga, raw	6	Radicchio, raw	32
Plums, purple, raw	2	Radish leaves, raw	70
Rowanberries, raw	7	Rocket, wild, raw	66
Saskatoon berries	17	Sweet potato leaves, raw	17
Arugula, raw	8	Watercress, raw	30
Asparagus, raw	14	Chia seeds, raw	18
Broccoli, raw	3	Carob fiber	58

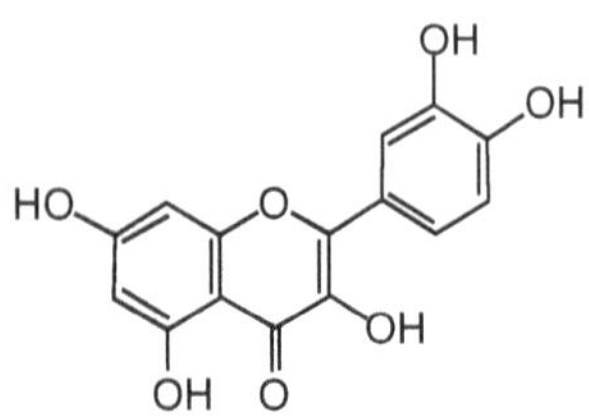

FIGURE 43.1 Quercetin.

Quercetin has been reported to induce apoptosis in A549 lung cancer cells via caspase-3 activation, poly (ADP) ribose polymerase cleavage, extracellular signal-regulated kinase activation, and c-Jun phosphorylation (Nguyen et al., 2004). The chemopreventive efficacy of quercetin was investigated using benzo(a)pyrene (B(a)P)-treated mice as an in vivo model of lung carcinogenesis, wherein pretreatment with quercetin protected against B(a) P-induced cytotoxicity by increasing the activity of antioxidants [SOD, catalase, GPx, glutathione transferase (GST)] (Kamaraj et al., 2007). More importantly, epidemiological studies reported an inverse association of quercetin and lung cancer (Cui et al., 2008; Knekt et al., 2002).

Quercetin has also been shown to reduce chromosomal aberration during 7,12-dimethylbenz(a) anthracene (DMBA)-induced mouse skin carcinogenesis (Sengupta et al., 2001). Similarly, quercetin treatment inhibited DMBA-12-o-tetradecanoylphorbol- 13-acetate (TPA)-induced tumor formation in the BK5.IGF-1 transgenic mouse model for skin carcinogenesis and suppressed activation of insulin-like growth factor-1 (IGF-1) signaling (Jung et al., 2013).

In addition to the above-mentioned studies, the anticancer effects of quercetin also have been investigated in other types of cancer, including pancreatic, ovarian, and gastric cancers (Angst et al., 2013; Maciejczyk and Surowiak, 2013; Ramachandran et al., 2013). Taken together, quercetin can act as an anticancer agent by virtue of its ability to modulate pro- and antiapoptotic proteins, cell cycle progression, angiogenesis, autophagy, inflammation, protein kinases involved in cell proliferation, and telomerase activity (Boly et al., 2011; Cosan et al., 2011; Lee et al., 2004; Psahoulia et al., 2007; Russo et al., 2012).

Quercetin and cardiovascular diseases

Quercetin is a well-known inhibitor of cellular proliferation and also an effective minimizer of DNA damage (Boyle et al., 2000; Choi et al., 2008). A natural source of quercetin from plants has been shown to increase the bioavailability of this flavonoid (Miean and Mohamed, 2001). Several studies have demonstrated a protective response by quercetin against cardiovascular disorders (Larson et al., 2012).

Quercetin exhibits a number of properties that lead to cardiovascular protection, such as being antioxidative (Chaillou and Nazareno, 2006; da Silva et al., 2000), antiplatelet, antismooth muscle cell proliferation and migration, an enhancer of cardiac cell mitochondrial function, and an inhibitor of NF-κB (Alcocer et al., 2002; Duarte et al., 1993a; Duarte et al., 1993b; Gryglewski et al., 1987; Hajra et al., 2000; Moon et al., 2003).

Many animal studies involving dietary supplementation with quercetin have been shown to be protective against cardiac problems. For example, studies conducted by Han et al. have demonstrated that a quercetin-supplemented diet attenuated the development of cardiac hypertrophy in a rat model of pressure overloads (Gao et al., 2014; Han et al., 2009). Moreover, a recent study conducted by Jimenez et al. also showed a promising effect of quercetin. Their study revealed that quercetin and its plasma metabolites effectively lowered levels of vascular superoxide by inhibiting nicotinamide adenine dinucleotide phosphate (NADPH) activity in the vascular smooth muscle cells (Jimenez et al., 2015). On the contrary, a few studies have shown that the risk and severity of cardiovascular problems were not lessened in a spontaneously hypertensive rat model receiving a quercetin-supplemented diet (Carlstrom et al., 2007).

Epidemiological evidence also indicates that quercetin is protective against cardiac diseases. One of the most famous studies was the Zutphen Elderly Study, which demonstrated that increasing flavonoid (high quercetin content) intake by dietary means decreased the risk of heart disease mortality (Hertog et al., 1993). Another study showing a promising protective effect of quercetin against cardiovascular disease risk was performed by Egert et al. (2009). This study involved overweight subjects between ages 25 and 65 years with a high cardiovascular disease risk phenotype. The subjects were randomized in a double-blind, placebo-controlled crossover trial, with the quercetin treatment group receiving about 150 mg per day for a period of 6 weeks followed by a 5-week washout period. The results of this study stated that quercetin lowered the systolic blood pressure and plasma levels of oxidized low-density lipoprotein in subjects with a high cardiovascular disease risk phenotype (Egert et al., 2009). During the past couple of decades, a number of consolidated studies have highlighted the efficiency of quercetin in preventing coronary heart disease (Formica and Regelson, 1995; Pace-Asciak et al., 1995; Saleem and Basha, 2010). Moreover, these studies also provided novel insights into understanding of the effects of dietary quercetin on body metabolism. Apart from the field of cardiovascular diseases, quercetin has also proved to be a useful supplement for athletes. Many studies have shown that intake of quercetin supplements by athletes improved their performance and generally facilitated healthy maintenance of the body (Askari et al., 2013; Nieman et al., 2010). In line with these data, several recent reports demonstrated that quercetin supplements also strengthened athletes' development of immunity (Gleeson, 2013; Walsh et al., 2011).

Quercetin and neurodegenerative diseases

Oxidative stress is recognized as an important factor in the pathogenesis of several neurodegenerative diseases such as Alzheimer disease (AD), Parkinson disease (PD), and Huntington disease (HD). The brain is highly susceptible to oxidative stress, given its high content of polyunsaturated fatty acids, high oxygen consumption, and lower antioxidant capacity (Bellissimo et al., 2001; Freitas et al., 2004; Sies, 1993). Therefore increasing the brain's antioxidant capacity may provide protection against oxidative stress-induced neurodegeneration.

Different groups of polyphenols have been tested for their disease-modifying potential for neurodegenerative diseases. For quercetin, there is now compelling evidence of its neuroprotective role in various neurodegenerative diseases (Ansari et al., 2009; Haleagrahara et al., 2013; Karuppagounder et al., 2013; Sabogal-Guaqueta et al., 2015).

Quercetin has been explored as a promising disease-modifying neuroprotective agent for AD in various experimental models. A glucoside form of quercetin, quercetin-3'-glucoside, reduced H_2O_2-induced ROS generation and also protected against Aβ-induced cell death in PC12 cells (Zhu et al., 2007). Similarly, another study investigated whether quercetin protects against oxidative stress and Aβ-induced cell death in primary neurons. It found that lower concentrations of quercetin (5 and 10 μM) reduced protein carbonyl, 3-NT, and 4-hydroxy-2-nonenal (HNE) levels, as wells as cell death in Aβ-treated primary cortical neuron cultures, whereas higher concentrations of quercetin (20 and 40 μM) showed the opposite effects (Ansari et al., 2009). Moreover, the role of ginkgo flavonols (mainly composed of quercetin, kaempferol, and isorhamnetin) in a depression-related signaling pathway mediated by BDNF/pCREB was tested in double-transgenic (Tg APP/PS1) mice (Hou et al., 2010). It was found that gingko flavonols stimulated the CREB-BDNF signaling pathway, reduced Aβ production in the hippocampus, and also reversed the spatial learning deficit in transgenic mice as evidenced by improved performance in the Morris water maze.

The potential neuroprotective effects of quercetin have also been investigated in a triple-transgenic mouse model of AD (3xTg-AD) (Sabogal-Guaqueta et al., 2015). The results revealed that quercetin treatment significantly reversed β-amyloidosis and tauopathy and reduced astrogliosis and microgliosis in 3xTg-AD mice. Furthermore, in comparison with vehicle-treated 3xTg-AD mice, quercetin-treated 3xTg-AD mice performed better in learning and memory tasks. Since acetylcholine-esterase (AchE) inhibitors that are widely used in the treatment of AD may cause notable side effects, the possible AchE-inhibitory effect of quercetin was tested and compared with conventional drugs by an in silico analysis (Islam et al., 2013). The analysis revealed that quercetin's binding strength to the active site of the enzyme was better than conventional drugs and that a 4-OH methylated form of quercetin (azaleatin) had even better binding affinity than quercetin, suggesting the natural compound quercetin has significant AchE-inhibitory effect as compared to widely used conventional drugs. By comprehensively analyzing 14 animal studies assessing the protective effect of quercetin in AD models, a recent systematic review demonstrated consistent neuroprotective effects of quercetin on AD animal models (Zhang et al., 2020). Owing to its antioxidant and antiinflammatory properties, quercetin has been gaining interest as a neuroprotective agent for PD. Isoquercitrin, a glycoside form of quercetin, has been shown to play a neuroprotective role against 6-hydroxy dopamine (6-OHDA)-induced neurotoxicity in PC12 cells (Magalingam et al., 2014). In this study, it was shown that pretreatment with isoquercitrin increased levels of ROS-scavenging enzymes (SOD, catalase, and GPx) and reduced lipid peroxidation in 6-OHDA-treated PC12 cells. Similarly, quercetin was shown to reduce protein carbonyl content and lipid hydroperoxide (LPO) levels in the striatum of 6-OHDA-treated rats (Haleagrahara et al., 2011). It was also found that quercetin treatment significantly reduced striatal dopamine depletion and total glutathione (GSH) depletion and improved neuronal survival in 6-OHDA-treated rats. Interestingly, quercetin has also been shown to enhance cognitive function in 6-OHDA-treated rats as evidenced by improved performance in the Morris water maze (Sriraksa et al., 2012). Additionally, their results demonstrated that quercetin treatment decreased hippocampal acetylcholinesterase (AChE) activity and protected against neurodegeneration in all subregions of the hippocampus. Likewise, quercetin was also reported to protect against 1-methyl-4-phenylpyridinium ion (MPP^+)-induced neurotoxicity in PC12 cells (Bournival et al., 2009). They found that pretreatment with quercetin significantly reduced MPP^+-induced cell death as evidenced by decreases in both caspase-3 activation and DNA fragmentation. Furthermore, decreased nuclear translocation of apoptosis-inducing factor (AIF) and decreased cytochrome c release into cytosol were also observed in this study. Another study explored the potential protective effect of quercetin in a 1-methyl-4- phenyl-1,2,3,6-tetrahydropyridine (MPTP) mouse model of PD (Lv et al., 2012). The results revealed that quercetin treatment reduced motor deficits and striatal dopamine depletion and also increased SOD and GPx activities in MPTP-treated mice. Furthermore, quercetin treatment reduced lipid peroxidation in MPTP-treated mice as evidenced by reduced immunoreactivity of 4-HNE in striatal neurons.

Quercetin has also been shown to protect against rotenone-induced nigral neurodegeneration (Karuppagounder et al., 2013). The results showed that quercetin treatment improved mitochondrial complex I activity and reduced striatal dopamine and nigral GSH depletion in rotenone-lesioned rats. It has also been shown that quercetin acts as an antiinflammatory agent that restricts MPP^+-induced microglial activation (Bournival et al., 2012). They found that pretreatment with quercetin reduced MPP^+-induced IL-6, IL-1β, iNOS, and TNF-α expression in N9 microglial cells. They further demonstrated that pretreatment with quercetin protected against cell death induced by microglial activation, when PC12 cells were cocultured with N9 cells treated with MPP^+. Mechanistic studies from our laboratory revealed that quercetin is able to activate the protein kinase D1 and Akt prosurvival signaling axis, enhance mitochondrial biogenesis, and augment mitochondrial bioenergetics capacity in dopaminergic neuronal cells. We also demonstrated in vivo neuroprotective effects of quercetin in the progressively degenerative MitoPark mouse model of PD (Ay et al., 2017). Another study reported that using the well-established bioavailability enhancer piperine to improve quercetin's bioavailability can increase its antioxidant and antiinflammatory properties and lead to strong neuroprotective effects against MPTP-induced neurotoxicity (Singh et al., 2017). Interestingly, quercetin nanocrystals prepared by the evaporative precipitation of nanosuspension method displayed greater bioavailability and were more effective than quercetin alone in the treatment of a Parkinson-like rat model (Ghaffari et al., 2018), suggesting nano-quercetin could be a beneficial neuroprotection applicant for treating various neurodegenerative disorders. Supporting this, solid lipid nanoparticles of quercetin have been shown to protect against pentylenetetrazole (PTZ)-induced cognitive impairment in zebrafish (Rishitha and Muthuraman, 2018).

The protective effects of quercetin have also been observed in other neurodegenerative diseases, including HD and amyotrophic lateral sclerosis (Chakraborty et al., 2014; Maher, 2019; Said Ahmed et al., 2000; Sandhir and Mehrotra, 2013). Collectively, all these studies suggest that quercetin is a promising neuroprotective agent for the treatment of neurodegenerative diseases.

Concluding remarks and future directions

This chapter discusses the potential use of quercetin in the prevention of cancer, cardiovascular diseases, and neurodegenerative diseases. As described above, there is strong evidence that quercetin can modulate several cellular signaling pathways involved in regulating the antioxidant response, cell survival, apoptosis, and inflammation. More large-scale epidemiological studies as well as well-designed clinical trials are needed to further clarify the potential benefits and to substantiate the efficacy of quercetin for the prevention and treatment of human diseases.

Acknowledgments

The writing of this chapter was supported by the National Institutes of Health RO1 grants ES026892, ES027245, NS100090, and NS088206. The W. Eugene and Linda Lloyd Endowed Chair and Eminent Scholar to A.G.K. and the Salisbury chair to A.K. are also acknowledged. The authors also acknowledge Gary Zenitsky for his assistance in the preparation of this manuscript.

Conflict of interest

A.G.K. and V.A. have an equity interest in PK Biosciences Corporation located in Ames, IA. The terms of this arrangement have been reviewed and approved by Iowa State University in accordance with its conflict-of-interest policies.

References

Alcocer, F., Whitley, D., Salazar-Gonzalez, J.F., et al., 2002. Quercetin inhibits human vascular smooth muscle cell proliferation and migration. Surgery 131 (2), 198–204.

Angst, E., Park, J.L., Moro, A., et al., 2013. The flavonoid quercetin inhibits pancreatic cancer growth in vitro and in vivo. Pancreas 42 (2), 223–229.

Ansari, M.A., Abdul, H.M., Joshi, G., et al., 2009. Protective effect of quercetin in primary neurons against Abeta(1-42): relevance to Alzheimer's disease. J. Nutr. Biochem. 20 (4), 269–275.

Askari, G., Ghiasvand, R., Paknahad, Z., et al., 2013. The effects of quercetin supplementation on body composition, exercise performance and muscle damage indices in athletes. Int. J. Prev. Med. 4 (1), 21–26.

Ay, M., Luo, J., Langley, M., Jin, H., Anantharam, V., Kanthasamy, A., et al., 2017. Molecular mechanisms underlying protective effects of quercetin against mitochondrial dysfunction and progressive dopaminergic neurodegeneration in cell culture and MitoPark transgenic mouse models of Parkinson's disease. J. Neurochem. 141 (5), 766–782.

Bellissimo, M.I., Amado, D., Abdalla, D.S., et al., 2001. Superoxide dismutase, glutathione peroxidase activities and the hydroperoxide concentration are modified in the hippocampus of epileptic rats. Epil Res. 46 (2), 121–128.

Boly, R., Gras, T., Lamkami, T., et al., 2011. Quercetin inhibits a large panel of kinases implicated in cancer cell biology. Int. J. Oncol. 38 (3), 833–842.

Boots, A.W., Wilms, L.C., Swennen, E.L., et al., 2008. In vitro and ex vivo anti-inflammatory activity of quercetin in healthy volunteers. Nutrition 24 (7-8), 703–710.

Bournival, J., Plouffe, M., Renaud, J., et al., 2012. Quercetin and sesamin protect dopaminergic cells from MPP^+-induced neuroinflammation in a microglial (N9)-neuronal (PC12) coculture system. Oxid. Med. Cell Longev. 2012, 921941.

Bournival, J., Quessy, P., Martinoli, M.G., 2009. Protective effects of resveratrol and quercetin against MPP^+-induced oxidative stress act by modulating markers of apoptotic death in dopaminergic neurons. Cell Mol. Neurobiol. 29 (8), 1169–1180.

Boyle, S.P., Dobson, V.L., Duthie, S.J., et al., 2000. Absorption and DNA protective effects of flavonoid glycosides from an onion meal. Eur. J. Nutr. 39 (5), 213–223.

Carlstrom, J., Symons, J.D., Wu, T.C., et al., 2007. A quercetin supplemented diet does not prevent cardiovascular complications in spontaneously hypertensive rats. J. Nutr. 137 (3), 628–633.

Chaillou, L.L., Nazareno, M.A., 2006. New method to determine antioxidant activity of polyphenols. J. Agr. Food Chem. 54 (22), 8397–8402.

Chakraborty, J., Singh, R., Dutta, D., et al., 2014. Quercetin improves behavioral deficiencies, restores astrocytes and microglia, and reduces serotonin metabolism in 3-nitropropionic acid-induced rat model of Huntington's disease. CNS Neurosci. Ther. 20 (1), 10–19.

Choi, E.J., Bae, S.M., Ahn, W.S., 2008. Antiproliferative effects of quercetin through cell cycle arrest and apoptosis in human breast cancer MDA-MB-453 cells. Arch. Pharmacol. Res. 31 (10), 1281–1285.

Conklin, C.M., Bechberger, J.F., MacFabe, D., et al., 2007. Genistein and quercetin increase connexin43 and suppress growth of breast cancer cells. Carcinogenesis 28 (1), 93–100.

Cosan, D.T., Soyocak, A., Basaran, A., et al., 2011. Effects of various agents on DNA fragmentation and telomerase enzyme activities in adenocarcinoma cell lines. Mol. Biol. Rep. 38 (4), 2463–2469.

Cui, Y., Morgenstern, H., Greenland, S., et al., 2008. Dietary flavonoid intake and lung cancer--a population-based case-control study. Cancer 112 (10), 2241–2248.

da Silva, E.L., Abdalla, D.S., Terao, J., 2000. Inhibitory effect of flavonoids on low-density lipoprotein peroxidation catalyzed by mammalian 15-lipoxygenase. IUBMB Life 49 (4), 289–295.

Deng, X.H., Song, H.Y., Zhou, Y.F., et al., 2013. Effects of quercetin on the proliferation of breast cancer cells and expression of survivin. Exp. Ther. Med. 6 (5), 1155–1158.

Dihal, A.A., de Boer, V.C., van der Woude, H., et al., 2006. Quercetin, but not its glycosidated conjugate rutin, inhibits azoxymethane-induced colorectal carcinogenesis in F344 rats. J. Nutr. 136 (11), 2862–2867.

Duarte, J., Perez Vizcaino, F., Utrilla, P., et al., 1993b. Vasodilatory effects of flavonoids in rat aortic smooth muscle. Structure–activity relationships. Gen. Pharmacol. 24 (4), 857–862.

Duarte, J., Perez-Vizcaino, F., Zarzuelo, A., et al., 1993a. Vasodilator effects of quercetin in isolated rat vascular smooth muscle. Eur. J. Pharmacol. 239 (1-3), 1–7.

Duthie, S.J., Dobson, V.L., 1999. Dietary flavonoids protect human colonocyte DNA from oxidative attack in vitro. Eur. J. Nutr. 38 (1), 28–34.

Egert, S., Bosy-Westphal, A., Seiberl, J., et al., 2009. Quercetin reduces systolic blood pressure and plasma oxidised low-density lipoprotein concentrations in overweight subjects with a high-cardiovascular disease risk phenotype: a double-blinded, placebo-controlled cross-over study. Br. J. Nutr. 102 (7), 1065–1074.

Formica, J.V., Regelson, W., 1995. Review of the biology of quercetin and related bioflavonoids. Food Chem. Toxicol. 33 (12), 1061–1080.

Freitas, R.M., Nascimento, V.S., Vasconcelos, S.M., et al., 2004. Catalase activity in cerebellum, hippocampus, frontal cortex and striatum after status epilepticus induced by pilocarpine in Wistar rats. Neurosci. Lett. 365 (2), 102–105.

Gao, C., Chen, X., Li, J., et al., 2014. Myocardial mitochondrial oxidative stress and dysfunction in intense exercise: regulatory effects of quercetin. Eur. J. Appl. Physiol. 114 (4), 695–705.

Ghaffari, F., Hajizadeh Moghaddam, A., Zare, M., 2018. Neuroprotective effect of quercetin nanocrystal in a 6-hydroxydopamine model of Parkinson disease: biochemical and behavioral evidence. Basic. Clin. Neurosci. 9 (5), 317–324.

Gleeson, M., 2013. Nutritional support to maintain proper immune status during intense training. Nestle Nutr. Inst. Workshop Ser. 75, 85–97.

Gryglewski, R.J., Korbut, R., Robak, J., et al., 1987. On the mechanism of antithrombotic action of flavonoids. Biochem. Pharmacol. 36 (3), 317–322.

Hajra, L., Evans, A.I., Chen, M., et al., 2000. The NF-kappa B signal transduction pathway in aortic endothelial cells is primed for activation in regions predisposed to atherosclerotic lesion formation. Proc Nat Acad. Sci. USA 97 (16), 9052–9057.

Haleagrahara, N., Siew, C.J., Mitra, N.K., et al., 2011. Neuroprotective effect of bioflavonoid quercetin in 6-hydroxydopamine-induced oxidative stress biomarkers in the rat striatum. Neurosci. Lett. 500 (2), 139–143.

Haleagrahara, N., Siew, C.J., Ponnusamy, K., 2013. Effect of quercetin and desferrioxamine on 6-hydroxydopamine (6-OHDA) induced neurotoxicity in striatum of rats. J. Toxicol. Sci. 38 (1), 25–33.

Han, J.J., Hao, J., Kim, C.H., et al., 2009. Quercetin prevents cardiac hypertrophy induced by pressure overload in rats. J. Vet. Med. Sci. 71 (6), 737–743.

He, X.Z., Wang, X., Dixon, R.A., 2006. Mutational analysis of the Medicago glycosyltransferase UGT71G1 reveals residues that control regioselectivity for (iso)flavonoid glycosylation. J. Biol. Chem. 281 (45), 34441–34447.

Hertog, M.G., Feskens, E.J., Hollman, P.C., et al., 1993. Dietary antioxidant flavonoids and risk of coronary heart disease: the Zutphen Elderly Study. Lancet 342 (8878), 1007–1011.

Hertog, M.G., Kromhout, D., Aravanis, C., et al., 1995. Flavonoid intake and long-term risk of coronary heart disease and cancer in the seven countries study. Arch. Intern. Med. 155 (4), 381–386.

Hou, Y., Aboukhatwa, M.A., Lei, D.L., et al., 2010. Anti-depressant natural flavonols modulate BDNF and beta amyloid in neurons and hippocampus of double TgAD mice. Neuropharmacology 58 (6), 911–920.

Islam, M.R., Zaman, A., Jahan, I., et al., 2013. In silico QSAR analysis of quercetin reveals its potential as therapeutic drug for Alzheimer's disease. J. Young Pharm. 5 (4), 173–179.

Iwao, K., Tsukamoto, I., 1999. Quercetin inhibited DNA synthesis and induced apoptosis associated with increase in c-fos mRNA level and the upregulation of p21WAF1CIP1 mRNA and protein expression during liver regeneration after partial hepatectomy. Biochim. et. Biophys. Acta 1427 (1), 112–120.

Jimenez, R., Lopez-Sepulveda, R., Romero, M., et al., 2015. Quercetin and its metabolites inhibit the membrane NADPH oxidase activity in vascular smooth muscle cells from normotensive and spontaneously hypertensive rats. Food Funct. 6 (2), 409–414.

Jung, M., Bu, S.Y., Tak, K.H., et al., 2013. Anticarcinogenic effect of quercetin by inhibition of insulin-like growth factor (IGF)-1 signaling in mouse skin cancer. Nutr. Res. Pract. 7 (6), 439–445.

Kamaraj, S., Vinodhkumar, R., Anandakumar, P., et al., 2007. The effects of quercetin on antioxidant status and tumor markers in the lung and serum of mice treated with benzo(a)pyrene. Biol. Pharm. Bull. 30 (12), 2268–2273.

Karuppagounder, S.S., Madathil, S.K., Pandey, M., et al., 2013. Quercetin up-regulates mitochondrial complex-I activity to protect against programmed cell death in rotenone model of Parkinson's disease in rats. Neuroscience 236, 136–148.

Kim, Y.H., Lee, Y.J., 2007. TRAIL apoptosis is enhanced by quercetin through Akt dephosphorylation. J. Cell Biochem. 100 (4), 998–1009.

Knekt, P., Kumpulainen, J., Jarvinen, R., et al., 2002. Flavonoid intake and risk of chronic diseases. Am. J. Clin. Nutr. 76 (3), 560–568.

Laird, D.W., Fistouris, P., Batist, G., et al., 1999. Deficiency of connexin43 gap junctions is an independent marker for breast tumors. Cancer Res. 59 (16), 4104–4110.

Lamson, D.W., Brignall, M.S., 2000. Antioxidants and cancer, part 3: quercetin. Altern. Med. Rev. 5 (3), 196–208.

Larson, A.J., Symons, J.D., Jalili, T., 2012. Therapeutic potential of quercetin to decrease blood pressure: review of efficacy and mechanisms. Adv Nutr. 3 (1), 39–46.

Lee, D.H., Szczepanski, M., Lee, Y.J., 2008. Role of Bax in quercetin-induced apoptosis in human prostate cancer cells. Biochem. Pharmacol. 75 (12), 2345–2355.

Lee, L.T., Huang, Y.T., Hwang, J.J., et al., 2004. Transinactivation of the epidermal growth factor receptor tyrosine kinase and focal adhesion kinase phosphorylation by dietary flavonoids: effect on invasive potential of human carcinoma cells. Biochem. Pharmacol. 67 (11), 2103–2114.

Lee, Y.K., Park, S.Y., Kim, Y.M., et al., 2009. AMP kinase/cyclooxygenase-2 pathway regulates proliferation and apoptosis of cancer cells treated with quercetin. Exp. Mol. Med. 41 (3), 201–207.

Lim, J.H., Park, J.W., Min, D.S., et al., 2007. NAG-1 up-regulation mediated by EGR-1 and p53 is critical for quercetin-induced apoptosis in HCT116 colon carcinoma cells. Apoptosis 12 (2), 411–421.

Lv, C., Hong, T., Yang, Z., Zhang, Y., et al., 2012. Effect of quercetin in the 1-Methyl-4-phenyl-1, 2, 3, 6-tetrahydropyridine-induced mouse model of Parkinson's disease. Evid. Based Complement. Altern. Med. 2012, 928643.

Maciejczyk, A., Surowiak, P., 2013. Quercetin inhibits proliferation and increases sensitivity of ovarian cancer cells to cisplatin and paclitaxel. Ginekol. Pol. 84 (7), 590–595.

Magalingam, K.B., Radhakrishnan, A., Haleagrahara, N., 2014. Protective effects of flavonol isoquercitrin, against 6-hydroxy dopamine (6-OHDA)-induced toxicity in PC12 cells. BMC Res. Notes 7, 49.

Maher, P., 2019. The potential of flavonoids for the treatment of neurodegenerative diseases. Int. J. Mol. Sci. 20 (12), 3056.

Miean, K.H., Mohamed, S., 2001. Flavonoid (myricetin, quercetin, kaempferol, luteolin, and apigenin) content of edible tropical plants. J. Agr. Food Chem. 49 (6), 3106–3112.

Moon, S.K., Cho, G.O., Jung, S.Y., et al., 2003. Quercetin exerts multiple inhibitory effects on vascular smooth muscle cells: role of ERK1/2, cell-cycle regulation, and matrix metalloproteinase-9. Biochem. Biophys. Res. Commun. 301 (4), 1069–1078.

Nguyen, T.T., Tran, E., Nguyen, T.H., et al., 2004. The role of activated MEK-ERK pathway in quercetin-induced growth inhibition and apoptosis in A549 lung cancer cells. Carcinogenesis 25 (5), 647–659.

Nieman, D.C., Williams, A.S., Shanely, R.A., et al., 2010. Quercetin's influence on exercise performance and muscle mitochondrial biogenesis. Med. Sci. Sports Exer 42 (2), 338–345.

Pace-Asciak, C.R., Hahn, S., Diamandis, E.P., et al., 1995. The red wine phenolics trans-resveratrol and quercetin block human platelet aggregation and eicosanoid synthesis: implications for protection against coronary heart disease. Clin. Chim. Acta. 235 (2), 207–219.

Park, C.H., Chang, J.Y., Hahm, E.R., et al., 2005. Quercetin, a potent inhibitor against beta-catenin/Tcf signaling in SW480 colon cancer cells. Biochem. Biophys. Res. Commun. 328 (1), 227–234.

Psahoulia, F.H., Moumtzi, S., Roberts, M.L., et al., 2007. Quercetin mediates preferential degradation of oncogenic Ras and causes autophagy in Ha-RAS-transformed human colon cells. Carcinogenesis 28 (5), 1021–1031.

Ramachandran, L., Manu, K.A., Shanmugam, M.K., et al., 2013. Isorhamnetin inhibits proliferation and invasion and induces apoptosis through the modulation of peroxisome proliferator-activated receptor gamma activation pathway in gastric cancer. J. Biol. Chem. 287 (45), 38028–38040.

Ramos, F.A., Takaishi, Y., Shirotori, M., et al., 2006. Antibacterial and antioxidant activities of quercetin oxidation products from yellow onion (Allium cepa) skin. J. Agr. Food Chem. 54 (10), 3551–3557.

Rishitha, N., Muthuraman, A., 2018. Therapeutic evaluation of solid lipid nanoparticle of quercetin in pentylenetetrazole induced cognitive impairment of zebrafish. Life Sci. 199, 80–87.

Russo, M., Palumbo, R., Tedesco, I., et al., 1999. Quercetin and anti-CD95(Fas/Apo1) enhance apoptosis in HPB-ALL cell line. FEBS Lett. 462 (3), 322–328.

Russo, M., Spagnuolo, C., Tedesco, I., et al., 2012. The flavonoid quercetin in disease prevention and therapy: facts and fancies. Biochem. Pharmacol. 83 (1), 6–15.

Sabogal-Guaqueta, A.M., Munoz-Manco, J.I., Ramirez-Pineda, J.R., et al., 2015. The flavonoid quercetin ameliorates Alzheimer's disease pathology and protects cognitive and emotional function in aged triple transgenic Alzheimer's disease model mice. Neuropharmacology 93C, 134–145.

Said Ahmed, M., Hung, W.Y., Zu, J.S., et al., 2000. Increased reactive oxygen species in familial amyotrophic lateral sclerosis with mutations in SOD1. J. Neurol. Sci. 176 (2), 88–94.

Saleem, T.S., Basha, S.D., 2010. Red wine: a drink to your heart. J. Cardiovasc. Dis. Res. 1 (4), 171–176.

Sandhir, R., Mehrotra, A., 2013. Quercetin supplementation is effective in improving mitochondrial dysfunctions induced by 3-nitropropionic acid: implications in Huntington's disease. Biochim. et. Biophys. Acta 1832 (3), 421–430.

Sengupta, A., Ghosh, S., Das, S., 2001. Modulation of DMBA induced genotoxicity in bone marrow by quercetin during skin carcinogenesis. J. Exp. Clin. Cancer Res. 20 (1), 131–134.

Senthilkumar, K., Arunkumar, R., Elumalai, P., et al., 2011. Quercetin inhibits invasion, migration and signalling molecules involved in cell survival and proliferation of prostate cancer cell line (PC-3). Cell Biochem. Funct. 29 (2), 87–95.

Shan, B.E., Wang, M.X., Li, R.Q., 2009. Quercetin inhibit human SW480 colon cancer growth in association with inhibition of cyclin D1 and survivin expression through Wnt/beta-catenin signaling pathway. Cancer Invest. 7 (6), 604–612.

Sharma, H., Sen, S., Singh, N., 2005. Molecular pathways in the chemosensitization of cisplatin by quercetin in human head and neck cancer. Cancer Biol. Ther. 4 (9), 949–955.

Sharmila, G., Bhat, F.A., Arunkumar, R., et al., 2014. Chemopreventive effect of quercetin, a natural dietary flavonoid on prostate cancer in in vivo model. Clin. Nutr. 33 (4), 718–726.

Sies, H., 1993. Strategies of antioxidant defense. Eur. J. Biochem. 215 (2), 213–219.

Singh, S., Jamwal, S., Kumar, P., 2017. Neuroprotective potential of quercetin in combination with piperine against 1-methyl-4-phenyl-1,2,3,6-tetrahydropyridine-induced neurotoxicity. Neural Regen. Res. 12 (7), 1137–1144.

Spagnuolo, C., Cerella, C., Russo, M., et al., 2011. Quercetin downregulates Mcl-1 by acting on mRNA stability and protein degradation. Brit J. Cancer 105 (2), 221–230.

Sriraksa, N., Wattanathorn, J., Muchimapura, S., et al., 2012. Cognitive-enhancing effect of quercetin in a rat model of Parkinson's disease induced by 6-hydroxydopamine. Evid. Based Complement. Altern. Med. 2012, 823206.

Vijayababu, M.R., Kanagaraj, P., Arunkumar, A., et al., 2005. Quercetin-induced growth inhibition and cell death in prostatic carcinoma cells (PC-3) are associated with increase in p21 and hypophosphorylated retinoblastoma proteins expression. J. Cancer Res. Clin. Oncol. 131 (11), 765–771.

Walsh, N.P., Gleeson, M., Pyne, D.B., et al., 2011. Position statement. Part two: maintaining immune health. Exerc. Immunol. Rev. 17, 64–103.

Williams, C.A., Grayer, R.J., 2004. Anthocyanins and other flavonoids. Nat. product. Rep. 21 (4), 539–573.

Zamin, L.L., Filippi-Chiela, E.C., Dillenburg-Pilla, P., et al., 2009. Resveratrol and quercetin cooperate to induce senescence-like growth arrest in C6 rat glioma cells. Cancer Sci. 100 (9), 1655–1662.

Zandi, K., Teoh, B.T., Sam, S.S., et al., 2011. Antiviral activity of four types of bioflavonoid against dengue virus type-2. Virol. J. 8, 560.

Zhang, X.-W., Chen, J.-Y., Ouyang, D., et al., 2020. Quercetin in animal models of Alzheimer's disease: a systematic review of preclinical studies. Int. J. Mol. Sci. 21 (2), 493.

Zhu, J.T., Choi, R.C., Chu, G.K., et al., 2007. Flavonoids possess neuroprotective effects on cultured pheochromocytoma PC12 cells: a comparison of different flavonoids in activating estrogenic effect and in preventing beta-amyloid-induced cell death. J. Agr. Food Chem. 55 (6), 2438–2445.

Chapter 44

German chamomile

Renata Kolanos and Szabina A. Stice

Introduction

Chamomile is one of the most popular traditional medicinal herbs. The name chamomile refers to several daisy-like plants in the family *Asteraceae*. The most popular species used in commercial chamomile preparations are *Matricaria chamomilla* L. (syn. *M. recutita*, *Chamomilla recutita*) and *Chamaemelum nobilis* L., known as German chamomile and Roman chamomile, respectively. While they are both members of the *Asteraceae* family, they belong to different genera and their morphology and chemical composition are markedly different (Mann and Staba, 1986). Therefore the two species possess different properties, resulting in different uses. Of the two, German chamomile is the species most commonly used for medicinal purposes. Hence, this chapter will focus on German chamomile, and references to chamomile will refer to German chamomile only.

Chamomile grows in its wild form in Southern and Eastern Europe, North Africa, and West Asia, from where it originated. It also has been introduced to South America, North America, and Australia, and belongs to a major group of medicinal plants cultivated in Europe (Singh et al., 2011). It grows 1–2 feet tall, with one or two showy daisy-like flower heads on each upper stem. Typically, fully open flower heads are harvested for use in herbal, medicinal, and cosmetic preparations and products.

Chamomile has been used in herbal medicines for thousands of years, at least since the time of Hippocrates (Zadeh et al., 2014). It is listed as an official drug in the pharmacopeias of 26 countries, including Germany, Belgium, France, and the United Kingdom; and is mentioned in the monographs of the German Commission E, European Scientific Cooperative on Phytotherapy (ESCOP), and World Health Organization (Ross, 2008; WHO, 1999).

Multitherapeutic values of chamomile have been established through traditional and scientific use and research. In traditional medicine, chamomile has been used to treat multiple diseases and ailments, including digestive disorders, conjunctivitis, dysmenorrhea, cold, cough, kidney stones, eye infection, headaches, and insomnia (Benitez et al., 2010); colic and nervousness (Gómez-Estrada et al., 2011); sinusitis, burns, wounds, ulcers, and stomachache (Jarić et al., 2007); flatulence, infections, and pharyngitis (Al-Ramahi et al., 2013); and hypertension, intestinal worms, anxiety, and hair loss (Ahmed, 2016). The European Medicine Agency's Committee on Herbal Medicinal Products concluded, based on its longstanding use, that chamomile flower-based medicines can be used for the following conditions: treating the symptoms of minor gastrointestinal (stomach and gut) complaints such as bloating and minor spasms, relieving the symptoms of common cold, treating minor ulcers (open sores) and inflammation of the mouth and throat, as add-on treatment for irritation of the skin and areas around the anus and genitals (once serious conditions have been excluded by a doctor), treating minor inflammation of the skin (e.g., sunburn) and superficial wounds and small boils (furuncles) (EMA, 2015).

Chamomile is most often used as a herbal tea, but can also be taken orally as drops, capsules, or tablets, applied topically, or inhaled. As mentioned above, the primary part of the chamomile plant used for medicinal and nutraceutical purposes is the flower head, which can be prepared and used in various ways. The dried flower heads are usually extracted with water, ethanol, or methanol, and the corresponding extracts are known as aqueous, ethanolic, and/or methanolic extracts. For extracts, the optimum ratio of ethanol to water is 1:1. The extracts are concentrated and added to gels, ointments, and creams; or dried and used in the form of tablets, capsules, or coated pills (Schilcher, 2005). Chamomile is most commonly prepared as a hot-water infusion (tea) using the dried flower heads. Chamomile tea is one of the world's most popular single-ingredient herbal teas, with about a million cups consumed every day (Srivastava et al., 2010). According to the German Commission E monograph, chamomile tea is prepared by pouring 150 mL boiling water over a heaped tablespoon (about 3 g) of flower heads and steeping in a covered cup for 5–10 min before

Nutraceuticals. DOI: https://doi.org/10.1016/B978-0-12-821038-3.00044-6

straining (Blumenthal et al., 1998). Tinctures are prepared by homogenizing the dried flower heads at room temperature in a mixture of ethanol and water at a 1:5 ratio (Schilcher, 2005).

Chamomile is also used in the form of an essential oil. Although the essential oil is found in all parts of the plant, essential oil from the flower head is most commonly used for medicinal purposes. The oil is obtained by subjecting flower heads to high pressure, temperature, and steam during the process known as steam distillation or hydrodistillation. According to many pharmacopeias, chamomile flower heads used for medicinal purposes should contain no less than 0.4% of essential oil (Schilcher, 2005).

Phytochemical constituents

Nutraceuticals based on chamomile typically contain its essential oil or extract. More than 120 secondary metabolites belonging to various classes of compounds have been identified in the essential oil and extracts of chamomile (Mann and Staba, 1986). The most important constituents contributing to the medicinal properties of chamomile are terpenoids and flavonoids (Srivastava et al., 2010).

Essential oil

Essential oil is accumulated predominantly in the flower head, comprising from 0.24%–2% of its composition (Gupta et al., 2010). The content and chemical composition of the oil in the flower head have been found to differ depending on environmental conditions, farming practices, and cultivars. The flower essential oil mainly consists of sesquiterpenes and their derivatives (75%–90%) and polyacetylenes (up to 20%) (Singh et al., 2011). Typical sesquiterpene-type compounds present in the essential oil include chamazulene, (−)-α-bisabolol, (*E*)-β-farnesene, bisabolol oxides A and B, and bisabolone oxide A (Singh et al., 2011). Polyacetylenes reported to be present in the essential oil include the isomeric spiroethers *cis*- and *trans*-en-yn-dicyloether (Szőke et al., 2004).

Chamazulene is an aromatic compound belonging to the azulene class that renders a characteristic blue color to the essential oil of chamomile flower heads. It is not found naturally in the flowers, but is formed by the breakdown of its precursor, matricin, during the hydrodistillation process used to obtain the essential oil (Povh et al., 2001). The reported content of chamazulene in the oil varies from 1.2% to 17.69% (Gosztola et al., 2010; Schilcher, 1973). (−)-α-Bisabolol, also known as levomenol, is a monocyclic sesquiterpene alcohol. It is a very lipophilic substance with a propensity to oxidize, mainly to bisabolol oxides A and B (Waleczek et al., 2003). Reported (−)-α-bisabolol content in the essential oil ranges from 1.3% to 77.21% (Franz, 1992; Schilcher, 1973). Bisabolol oxide A and bisabolol oxide B comprise from 0.4% to 52.25% and from 0.2% to 58.85%, respectively (Franz, 1992; Schilcher, 1973). Certain chamomile essential oils were found not to contain bisabolol oxide A, bisabolol oxide B, or chamazulene (Franz, 1992; Salamon et al., 2010). Salamon et al. (2010) reported farnesene content as low as 0.2% in certain cultivars of chamomile growing in Iran, while Satyal et al. (2015) found that the essential oil contained 42.2% (*E*)-β-farnesene. The spiroether content was found to range from 1.5% (*cis*-en-yn-dicycloether) to 12% (Salamon, 1994; Schilcher, 1973). Bisabolone oxide A has been reported to be present in the essential oil from 4% (Satyal et al., 2015) to 65.41% (Piri et al., 2019). Many other compounds have been found in flower essential oil including germacrene D, (*E*)-nerolidol, farnesol, spathulenol, *n*-hexanol, isoborneol, bnerol, phytol, limonene, camphene, α-terpineol, chamomillol, β-elemene, α-humulene, hexadec-11-yn-13,15-diene, β-bourbonene, and γ-cadinol (Das, 2015). An analysis of oils from various parts of the flower heads, that is, the ligulate and tubular florets, and receptacles, resulted in the identification of more than 70 compounds representing about 99% of the oil components (Tirillini et al., 2006).

Extracts

While hydrodistillation of different parts of chamomile yields the essential oil composed primarily of volatile compounds such as terpenes, the use of various solvents such as ethanol and water, glycerin and propylene glycol, toluene, pentane, sunflower oil, *n*-hexane, or supercritical carbon dioxide fluid extraction yields extracts containing nonvolatile compounds. These mainly are flavonoids, such as glucosides and/or their aglycones, coumarin and its derivatives, and other organic compounds. The major flavonoids found in chamomile flower extracts include apigenin, luteolin, quercetin, and patuletin. Mulinacci et al. (2000) performed quantitative analysis of flavonoids in methanolic extracts of receptacles, ligulate, tubular, and total flowers, and showed that the relative concentration of the flavonoids varies among the different flower parts. A high apigenin concentration was found in ligulate flowers (68.6%). Much lower amounts of apigenin were found in the tubular flowers (0.9%) and receptacles (0.8%). In contrast, quercetin accumulated mainly in the tubular flowers (12.4%) and receptacles (10%), with a lower concentration present in the ligulate flowers (2.1%). The total flower extract contained the following amounts of flavonoids and their glucosides: 16.8% apigenin, 9.9% quercetin, 6.5% patuletin, and 1.9% luteolin. The major coumarin derivatives found in chamomile flower extract are herniarin and umbelliferone. Commercial chamomile extracts (50% alcohol) were found to contain an average amount of 10–15 mg/100 mL of herniarin and 2–3 mg/100 mL of umbelliferone

(Redaelli et al., 1981). In addition, the presence of polysaccharides (inulin type fructans, a highly methyl esterified and acetylated homogalacturonan, type II arabinogalactan, fructooligosaccharides) and phenolic acids (chlorogenic, caffeic, vanillic, syringic, anisic, and coumaric acids) have been reported (Atoui et al., 2005; Chaves et al., 2019).

Pharmacological effects

Chamomile is one of the most widely used and well-documented medicinal plants in the world, with one of the longest known histories of use. Chamomile and/or its constituents have been utilized for dermatological, gastrointestinal, neurological/psychiatric, and many other wide-ranging medicinal purposes. An encyclopedic description of each of these uses is beyond the scope of this work. Instead, the aim of this chapter is to summarize the current and collective knowledge about (1) the most well-known and extensively described medicinal uses of chamomile and its constituents, and (2) those uses that have been demonstrated to have direct relevance to human health. Although many medicinal applications have been reported, few clinical studies with human subjects have been conducted. Most studies investigating the therapeutic effects of chamomile or its specific constituents have used in vitro systems or in vivo animal models.

Antiinflammatory properties

The antiinflammatory properties of chamomile are well known, and its preparations are widely used to reduce cutaneous and mucosal inflammation. Typically, topical applications in the form of creams, gels, and ointments are applied to sore and cracked skin, diaper rash, bruises, burns, and inflamed mouth and gums. Various compounds are thought to be responsible for the antiinflammatory effect of chamomile preparations, including chamazulene and (−)-α-bisabolol in essential oils; and flavonoids such as apigenin, quercetin, and luteolin in extracts.

In vitro and ex vivo antiinflammatory effects

Several studies have shown the in vitro and ex vivo anti-inflammatory capacity of chamomile extract. In an ex vivo experiment, rat colonic specimens were exposed to lipopolysaccharide (LPS) to induce inflammation, and subsequently treated with the ligulate flower extract (Menghini et al., 2016). Activities of different biomarkers of lipid peroxidation and inflammation were assessed, including myeloperoxidase, serotonin (5-HT), prostaglandin E_2 (PGE2), 8-*iso*-prostaglandin $F_{2\alpha}$, nuclear factor kappa-light-chain-enhancer of activated B cells (NF-κB), tumor necrosis factor-α (TNF-α), and interleukin-6 (IL-6). Chamomile extract was effective in reducing the oxidative stress, inflammation, and immune response biomarkers at concentrations of 100−1000 μg/mL. The efficacy was comparable to that of sulfasalazine (2 mg/mL), an antiinflammatory drug. The aqueous flower extract of chamomile was incubated with THP1 macrophages, and the production of cytokines interleukin-1β (IL-1β), IL-6, and TNF-α involved in inflammatory responses was measured (Drummond et al., 2013). The extract showed a significant reduction of IL-6 production only, and only at the higher extract concentration, providing 20 μg of apigenin. Bulgari et al. (2012) investigated the antiinflammatory effect of aqueous infusions prepared from capitula or sifted flowers of chamomile on AGS cells (epithelial cells obtained from the stomach of a person with gastric adenocarcinoma) stimulated with phorbol 12-myristate 13-acetate (PMA), and on the activity of human neutrophil elastase, for which enhanced activities have been reported following *Helicobacter pylori* infections. Both the capitula and sifted flower infusions inhibited the PMA-induced upregulation of the inflammation mediator matrix metalloproteinase-9 (MMP-9) and reduced the activity of the elastase. The decrease in the secretion of MMP-9 was attributed in part to the inhibition of NF-κB-driven transcription of the gene controlling the synthesis of this metalloproteinase. This study provided some evidence that the gastroprotective effect of chamomile shown in animal studies is mediated by MMP-9 and elastase inhibition. The authors stated that the effects occurred at concentrations that might be found in the gastric lumen of individuals consuming a cup of chamomile tea. Srivastava et al. (2009) showed that the aqueous flower extract treatment inhibited LPS-induced NO production and inducible nitric oxide synthase (iNOS) gene expression in murine macrophages, the effects mediated, at least in part, through blockage of NF-κB transcriptional activation. The extract also significantly blocked IL-1β, IL-6 and TNF-α-induced NO production. It has been speculated that the aqueous flower extract may also elicit its antiinflammatory activity by decreasing LPS-induced PGE2 release in murine macrophages due to the inhibition of cyclooxygenase-2 (COX-2) enzyme activity, a mechanism much like that attributed to nonsteroidal antiinflammatory drugs (NSAIDs).

In vivo antiinflammatory effects in animals

Tubaro et al. (1984) and Della Loggia et al. (1990) investigated the antiinflammatory activity of hydroethanolic flower extracts in a croton oil-induced mouse ear edema model. Benzydamine (an NSAID) was used as the reference drug. Both studies found that the extracts resulted in a significant reduction in edema. Based on the Della Loggia et al. (1990) study, the dried flower extract was less effective (23.7% inhibition) than the fresh flower extract (31.6% inhibition), whose activity was similar to that of benzydamine. The higher activity of the fresh flower extract was attributed to its higher concentration of matricin. The investigators also found that the antiinflammatory activity of isolated apigenin was 10 times greater than that of matricin, which, in turn, was 10 times greater than that of chamazulene. Additionally, the flower essential oil was shown to lack antiinflammatory activity (6.6% inhibition).

Antiinflammatory properties of nontopical applications of chamomile preparations also have been reported to considerably reduce carrageenan-induced edema in the rat and mouse paw. Al-Hindawi et al. (1989) showed that intraperitoneal administration of the hydroethanolic flower extract (400 mg/kg) reduced carrageenan-induced paw edema by 41.1% in rats. By comparison, the edema in rats treated with 100 mg/kg acetylsalicylic acid (an NSAID also known as aspirin) was inhibited by 32.4%. A diet containing essential oil at a concentration of 5000 ppm has been shown to considerably reduce carrageenan-induced edema in the mouse paw, as well as colonic inflammation induced by administration of trinitrobenzene sulfonic acid (Fabian et al., 2011). A significant reduction in bacterial translocation from the lumen into mesenteric lymph nodes observed in this study was hypothesized to result from synergism of the antiinflammatory and antimicrobial effects of the chamomile essential oil.

In another study, the methanolic flower extract was administered orally to mice infected with larvae of *Heligmosomoides polygyrus* (Hajaji et al., 2019). *H. polygyrus* infection was found to cause a significant increase in serum levels of proinflammatory cytokines IL-1β, IL-6, and TNF-α involved in inflammatory responses. Treatment with different doses of the methanolic flower extract produced a significant and dose-dependent decrease in the levels of these cytokines and protected against the inflammatory state associated with *H. polygyrus* infestation. Chandrashekhar et al. (2011) suggested that the methanolic extract from chamomile capitula possesses potent antiallergic activity. This study showed the extract's inhibitory activity on effects induced by the compound 48/80, one of the most potent stimulators of mast cell degranulation. In rats, oral administration of chamomile extract (1) effectively stabilized mast cells and reduced histamine release, (2) inhibited anaphylactic and pruritic activity, and (3) lowered the levels of nitric oxide (NO) in serum, peritoneal fluid, and bronchoalveolar lavage fluid.

Antiinflammatory effects in humans

A few studies on the antiinflammatory properties of chamomile in humans have been reported. A topical lukewarm compress of chamomile extract (2.5%–5%) was found to be useful in the treatment of phlebitis in cancer patients receiving peripheral intravenous antineoplastic chemotherapy (dos Reis et al., 2011). The use of topical chamomile ointment (1.5%) was shown to significantly decrease the incidence, but not the severity of, phlebitis in patients treated with amiodarone (an antiarrhythmic medication) compared to those in the control group (Sharifi-Ardani et al., 2017). Topical application of a chamomile extract solution was found to be more effective than hydrocortisone ointment 1% (an antiinflammatory drug) for relieving itching and inflammation of peristomal skin lesions in colostomy patients (Charousaei et al., 2011). Aertgeerts et al. (1985) reported that in patients with inflammatory dermatoses, a cream containing chamomile showed efficacy comparable to 0.25% hydrocortisone and better therapeutic results than 0.75% fluocortin butyl ester (an antiinflammatory corticosteroid) and 5% bufexamac (an NSAID). A mouthwash containing chamomile extract significantly lowered both dental plaque and gingival inflammation in gingivitis patients compared to the control rinse (Pourabbas et al., 2005). These effects also were reported in another clinical study investigating a mouthwash containing 1% chamomile extract (Goes et al., 2016). Valenzuela et al. (2016) concluded that the efficacy of the topically applied chamomile gel (2%) in burning mouth syndrome appeared to be questionable. In that study, the severity of oral pain decreased in patients regardless of whether they were in the treatment or placebo group.

Compounds responsible for the antiinflammatory effects

Because chamomile preparations are mixtures of many chemical compounds, numerous studies have focused on individual chamomile constituents in order to pinpoint specific compounds responsible for the antiinflammatory effect of chamomile. The investigated compounds include the terpenoids chamazulene and (−)-α-bisabolol, and the flavonoids apigenin, luteolin, and quercetin.

Jakovlev et al. (1983) reported that oral administration of chamazulene effectively reduced carrageenan-induced paw edema in rats; however, it did so less effectively than the standard drug, salicylamide. Chamazulene was also reported to show 50% inhibition of leukotriene B4 (LTB4) formation in intact rat peritoneal neutrophilic granulocytes at 10 and 15 μM, and this activity was attributed to the antiinflammatory effect of chamazulene (Safayhi et al., 1994). Miller et al. (1996) showed that chamazulene does not have an inhibitory effect on histamine release from rat mast cells; in fact, the authors concluded that chamazulene at concentrations higher than 10^{-5} M stimulated histamine release. Chamazulene was found to be nearly inactive in the treatment of croton oil-induced mouse ear edema and exhibited only a 3%

reduction of the edematous response, much lower than the 58% reduction exhibited by its proazulene matricin (Zidorn et al., 1999). Flemming et al. (2015) compared the antiinflammatory activity of matricin and its degradation product chamazulene. They reported that at 75 μM matricin, expression of the intracellular adhesion molecule-1 (ICAM-1) was reduced to 52.7% or 20.4% compared to the control in TNF-α or LPS-stimulated human microvascular endothelial cells-1 (HMEC-1), respectively. Additional functional assays confirmed the inhibitory effect of matricin on NF-κB signaling. Chamazulene was inactive in these assays. Ma et al. (2020) investigated chamazulene's potential in the prevention of osteoarthritic inflammation induced by IL-1β in rat chondrocytes and complete Freund's adjuvant (CFA)-induced osteoarthritic inflammation in rats. The study showed that chamazulene was able to protect chondrocytes in vitro by downregulating protein expressions of MMP-3, MMP-9, COX-2, iNOS, and NF-κB p65. Levels of TNF-α and IL-6 were decreased in both in vitro and in vivo experiments. Based on histopathological examination, the authors concluded that chamazulene was able to prevent the inflammatory response in CFA-induced rats.

The antiinflammatory activity of (−)-α-bisabolol was established in several studies. Experiments conducted using carrageenan-induced paw edema and adjuvant arthritis rat models have shown that (−)-α-bisabolol is more effective than (+)-α-bisabolol, (+/−)-α-bisabolol, the bisabolol oxides A and B, and bisabolon oxide (Jakovlev et al., 1979). In a separate set of experiments, (−)-α-bisabolol at doses of 100 and 200 mg/kg administered by gavage reduced mouse paw edema induced by carrageenan and dextran (Rocha et al., 2011). The same authors found that 5-hydroxytryptamine (5-HT)-induced edema was significantly reduced at 200 mg/kg (−)-α-bisabolol. However, (−)-α-bisabolol was not effective in reducing histamine-induced paw edema. At both tested doses, (−)-α-bisabolol was able to decrease both TNF-α and leukocyte migration to the peritoneal cavity in response to carrageenan (Rocha et al., 2011). These findings clarified, in part, the mechanism of the antiinflammatory activity of (−)-α-bisabolol. Kim et al. (2011) demonstrated antiinflammatory effects of (−)-α-bisabolol and characterized its mechanisms of action. The study showed that (−)-α-bisabolol inhibited LPS-induced production of NO and PGE2 in RAW264.7 murine macrophages. In addition, expression of COX-2 and iNOS genes and LPS-induced activation of activator protein-1 (AP-1) and NF-κB promoters were significantly reduced. Thus, it was concluded that (−)-α-bisabolol exerts antiinflammatory effects by downregulating expression of iNOS and COX-2 genes through inhibition of NF-κB and AP-1 signaling. Miller et al. (1996) found that (−)-α-bisabolol had an insignificant effect on protamine sulfate-induced mast cell degranulation in concentrations ranging from 10^{-9} to 10^{-5} M, and that it stimulated histamine release in concentrations higher than 10^{-5} M. The authors speculated that the high concentrations of (−)-α-bisabolol lead to cytolysis. The antiinflammatory properties of (−)-α-bisabolol were studied by de O. Leite et al. (2011). Croton oil, arachidonic acid, phenol, and capsaicin were used to induce dermatitis in the ear of mice pretreated with (−)-α-bisabolol. The compound inhibited inflammation in all cases of dermatitis, except in the group treated with capsaicin. The authors concluded that (−)-α-bisabolol properties may be useful in developing drugs to treat inflammation associated with atopic or contact dermatitis or psoriasis. (−)-α-Bisabolol orally administered in lipid-core nanocapsules was shown to have an antiinflammatory effect in LPS-induced acute lung inflammation in mice (D'Almeida et al., 2017). All tested doses (30, 50, and 100 mg/kg) were able to significantly reduce LPS-induced inflammation. The authors of the study attributed this effect, at least in part, to the inhibition of mitogen-activated protein kinase (MAPK) signaling, which is involved in the inflammatory response in lung injury and is related to the release of several proinflammatory cytokines. It was also concluded that the nanoencapsulation was a crucial factor in significant improvement of the efficacy of (−)-α-bisabolol compared to its free form. The encapsulation affected the amount of α-bisabolol in the lungs, increasing the local action against the inflammatory process. Cavalcante et al. (2020) reported that 100 mg/kg (−)-α-bisabolol was able to reduce the mortality rate of mice with cecal ligation and puncture-induced sepsis. A significant reduction (41.28%) of leukocytes in the peritoneal cavity was observed at 100 mg/kg (−)-α-bisabolol, as well as in the production of NO at 50 and 100 mg/kg. The study results suggested that (−)-α-bisabolol affected the systemic inflammatory response. A recent in vitro study showed that (−)-α-bisabolol has an antiinflammatory effect in pregnant human myometrial explants stimulated with LPS (Muñoz-Pérez et al., 2018). It decreased the production of TNF-α and IL-1β in a concentration-dependent manner without significantly increasing the production of the anti-inflammatory cytokine IL-10.

Flavonoids present in chamomile preparations have been the subject of multiple studies investigating their antiinflammatory properties. It was demonstrated that an ethyl acetate fraction of a total hydroalcoholic chamomile extract containing the flavonoids apigenin, quercetin, luteolin, rutin, and apigenin-7-glucoside was more effective than the hydroalcoholic extract itself in reducing croton oil-induced dermatitis in the mouse ear (Della Loggia et al., 1986). The same study reported that apigenin was more effective than luteolin, and it was suggested that the two flavonoids elicit their antiinflammatory effects through different mechanisms. Studies have shown that apigenin was able to reduce levels of TNF-α, IL-1β, IL-6 (Drummond et al., 2013), NO, COX-2 (Lee et al., 2007), LTB4 (Ammon et al., 1996), and neutrophil infiltration (Doseff et al., 2011). It also has been shown to inhibit molecular signaling via the extracellular signal-regulated kinase (ERK) pathway (Sawatzky et al., 2006) and suppress NF-κB (Funakoshi-Tago et al., 2011). Apigenin was found to be effective in LPS-induced airway infection in the mouse, resulting in an $\sim$80% reduction in mortality (Doseff et al., 2011).

When used topically, apigenin reduced both acute and irritant contact dermatitis and allergic contact dermatitis in mice (Man et al., 2012). Most of the relevant studies report that luteolin exerts its antiinflammatory effects partly by regulating various inflammatory mediators. The compound can inhibit Il-1β, IL-4, IL-5, IL-6, IL-8, and IL-13 in vitro (Hirano et al., 2004; Hougee et al., 2005; Kang et al., 2010; Park et al., 1999) and increase the level of antiinflammatory cytokine IL-1 (Comalada et al., 2006). Chiu-Yuan et al. (2007) reported that luteolin is capable of inhibiting in a dose-dependent manner the production of NO and PGE2 and the expression of iNOS, COX-2, and TNF-α in mouse alveolar macrophage MH-S and peripheral macrophage RAW264.7 cells. Luteolin was shown to alter signaling pathways that regulate the expression of inflammatory mediators, including the NF-κB pathway (Kim et al., 2005; Sung and Lee, 2015), the MAPK/AP-1 pathway (Choi and Lee, 2010), and the Janus tyrosine kinase pathway (JAK)-STAT (Liu et al., 2016; Lou et al., 2015).

Antiinflammatory effects of luteolin have been tested in various animal models. Li et al. (2012) reported that intraperitoneal pretreatment with luteolin inhibited LPS-induced histological changes of acute lung injury and lung tissue edema in mice. The LPS-induced inflammatory responses, including increased vascular permeability, TNF-α and IL-6 production, and iNOS and COX-2 expression, were reduced by luteolin in a concentration-dependent manner. In addition, activation of NF-κB was suppressed at the highest luteolin dose (70 μmol/kg). A significant decrease in levels of TNF-α and IL-6 and inhibition of NF-κB p65 was also observed in mice intraperitoneally injected with luteolin (100 mg/kg) prior to the induction of severe acute pancreatitis (Xiong et al., 2017). Kuo et al. (2011) found that luteolin relieved LPS-induced acute lung injury by inhibiting MAPK pathways, and concluded that this effect was due to a significant suppression of IκB degradation, which is linked to NF-κB activation. Shi et al. (2015) reported that luteolin may have the potential to prevent or treat inflammation and arthritis. In this study, gastric administration of luteolin (10 and 20 mg/kg) significantly suppressed TNF-α, IL-6, IL-1β, and IL-17, and reduced inflammatory cells in rats injected with FCA to induce arthritis compared to positive controls.

Similarly, quercetin has been investigated for its antiinflammatory activity and found to have effects on multiple proinflammatory mediators and signaling pathways involved in inflammatory responses. Quercetin has repeatedly been shown to exert antiinflammatory properties in vitro by lowering NO/iNOS levels (Chan et al., 2000; Cho et al., 2003), affecting inflammatory mediators such as TNF-α, IL-1β, IL-6, and IL-8 (Chuang et al., 2010; Min et al., 2007), and signaling pathways mediated by ERK and NF-κB (Cho et al., 2003). Quercetin administered intraperitoneally at a dose of 80 mg/kg decreased edema in the paw during the acute phase of inflammation in rats with carrageenan-induced arthritis (Guardia et al., 2010). The compound had no significant action in the chronic phase of inflammation, however. These observations were confirmed in a study by Rotelli et al. (2003), who tested quercetin in rat models of acute and chronic inflammation at a dose of 75 mg/kg. Mamani-Matsuda et al. (2006) investigated the in vivo effects of oral or intracutaneous administration of quercetin in chronic rat adjuvant-induced arthritis. Oral administration (30 mg every 2 days for 10 days) resulted in a significant decrease of clinical signs in comparison to untreated controls. Intracutaneous injections of lower doses (10 mg every 2 days for 10 days) produced similar antiarthritic effects; however, dosing at 5 mg every 2 days for 10 days did not. Injection of relatively low doses (5 mg every 2 days for 10 days) prior to the induction of arthritis by an adjuvant significantly reduced the signs of arthritis. The study showed that the antiarthritic effects of quercetin may be due, in part, to lower TNF-α, IL-1β, NO, and monocyte chemoattractant protein-1 (MCP-1) production capacity, as demonstrated in macrophages ex vivo. Another study reported that quercetin at a dose of 10 mg/kg was effective in preventing both early and late phases of the acute inflammatory response when administered simultaneously with, or 1 h after, carrageenan injection in air pouches induced subcutaneously on the backs of rats (Morikawa et al., 2003). Results showed that quercetin may be able to inhibit the production of PGE2 and TNF-α and that it regulated on activation normal T-cells expressed and secreted (RANTES), as well as expression of COX-2 mRNA. The decrease in the production of macrophage inflammatory protein-2 (MIP-2) was insignificant. Orally administered quercetin (200 mg/kg) was effective in reducing inflammatory responses in the gastric mucosa of guinea pigs infected with *H. pylori* (González-Segovia et al., 2008), suppressing eosinophil-mediated inflammation in a murine model of asthma at a dose of 10 mg/kg (Rogerio et al., 2007), and lowering the steady-state levels of inflammatory gene C-reactive protein, IL-6, and MCP-1 in mouse liver at 2 mg/g in the diet (Boesch-Saadatmandi et al., 2012). Boots et al. (2011) demonstrated that a 1-day oral supplementation with 2 g quercetin that resulted in a significant decrease of the ratios of pro- versus antiinflammatory cytokines TNF-α/IL-10 and IL-8/IL-10 in patients with sarcoidosis, suggesting that the compound could mitigate the inflammation present in sarcoidosis. In another clinical trial, pretreatment with quercetin given orally for 3 days decreased nickel-contact dermatitis by more than 50% (Weng et al., 2012). The same study showed that 1 g quercetin increased resistance to the development of skin reddening after UVB irradiation.

Antioxidant properties

Antioxidants play an important role in the prevention of cellular damage caused by oxidative stress arising from free radicals such as reactive oxygen species (ROS), the mechanism that appears to underlie numerous pathophysiological conditions such as atherosclerosis, neuronal degeneration, and tumor promotion. Essential oils, ethanolic extracts, and infusions of chamomile have been shown to possess antioxidant properties and have been used therapeutically for conditions in which oxidative stress is implicated.

Antioxidant activity of chamomile extracts

Aqueous and ethanolic extracts of chamomile have a moderate antioxidant activity when compared with other medicinal or culinary herbs. Dragland et al. (2003) tested chamomile for its antioxidant capacity using the ferric-reducing ability of a plasma assay to quantify total antioxidants. The authors concluded that the antioxidant capacity of chamomile extracts was much lower (17.7 mmol/100 g at the mean) than for other culinary herb extracts investigated in this study, including peppermint (78.5 mmol/100 g) and oregano (137.5 mmol/100 g), but higher than coriander (3.3 mmol/100 g). Lee and Shibamoto (2002) reported that chamomile extracts exhibited higher inhibitory activities (50% inhibition of hexanal oxidation) than the extracts of thyme and basil, lower than the extracts of lavender and cinnamon, and comparable to rosemary extract. Luqman et al. (2009) found that aqueous and ethanolic flower extracts of chamomile significantly protected mouse erythrocytes against the oxidative stress induced by hydrogen peroxide and *tert*-butyl hydroperoxide. Roby et al. (2013) evaluated antioxidant activities of chamomile and fennel extracts prepared using methanol, ethanol, diethyl ether, or hexane using the 2,2-diphenyl-1-picrylhydrazil (DPPH) radical-scavenging assay. The highest anti-radical power was observed for methanolic chamomile extract. Sebai et al. (2015), using the same assay, demonstrated that chamomile decoction extract presents a high radical-scavenging capacity; however, less than that of ascorbic acid. The study also showed that pretreatment with the decoction significantly reversed all ethanol-induced antioxidant enzyme (superoxide dismutase, catalase, and glutathione peroxidase) depletion in rats. Chamomile decoction extract also was effective in reversing ethanol-induced depletion of these enzymes in rat gastric mucosa (Jabri et al., 2017).

Antioxidant activity of major chamomile constituents

Data on the antioxidant activity of major constituents of chamomile essential oil, chamazulene and (−)-α-bisabolol, also are available. Rekka et al. (1996) studied the potential of chamazulene to compete with dimethyl sulfoxide (DMSO) in scavenging free hydroxyl radicals generated in lipid peroxidation induced by Fe^{2+}/ascorbate. Chamazulene was shown to inhibit lipid peroxidation in a concentration- and time-dependent manner ($IC_{50} = 18\ \mu M$ after 45 min incubation) and autoxidation of DMSO (75% inhibition at 25 mM). The study also showed that chamazulene had a low capacity to interact with DPPH free radicals. Querio et al. (2018) investigated the effects of chamazulene (25 μg/mL) on ROS levels in bovine aortic endothelial cells stressed for 3 h with either high glucose or H_2O_2. High glucose was chosen to mimic the hyperglycemic state characteristic of diabetes, and H_2O_2 represented a direct stronger insult. The results of the study showed a reduction in ROS levels produced after simultaneous treatment with high glucose or H_2O_2 and chamazulene, suggesting the antioxidant activity of the compound and its potential to prevent the rise of endothelial ROS levels and the consequent vascular damage. According to Capuzzo et al. (2014), chamazulene isolated from the essential oil of chamomile exhibited a higher total antioxidant capacity ($IC_{50} = 6.4$ μg/mL) than did ascorbic acid ($IC_{50} = 12.8$ μg/mL), α-tocopherol ($IC_{50} = 20.5$ μg/mL), and butylated hydroxytoluene (BHT) ($IC_{50} = 30.8$ μg/mL). While chamazulene was unable to react with DPPH free radicals, when assayed with 2,2′-azinobis-(3-ethylbenzothiazoline-6-sulfonic acid) (ABTS) free radical cations, the compound showed significantly higher free radical-scavenging activity ($IC_{50} = 3.7$ μg/mL) than did BHT ($IC_{50} = 6.2$ μg/mL) and α-tocopherol ($IC_{50} = 11.5$ μg/mL).

(−)-α-Bisabolol also exhibits antioxidant activity, albeit lower than that of chamazulene. The IC_{50} value determined for (−)-α-bisabolol using a DPPH assay was 449.7 μM (van Zyl et al., 2006). Agatonovic-Kustrina et al. (2015) investigated the antioxidant activity of the extracts from chamomile flower heads and leaves using high-performance thin-layer chromatography and spraying the developed plates with a solution of DPPH. Based on the comparison of band intensity relative to the applied standard, the authors concluded that the free radical-scavenging activity of chamazulene was higher than that of (−)-α-bisabolol. Braga et al. (2009) investigated the ability of (−)-α-bisabolol to interfere with the production of ROS in human polymorphonuclear cells stimulated with *Candida albicans* and *N*-formyl-methionyl-leucyl-phenylalanine (fMLP). ROS production was measured using luminol-amplified chemiluminescence (LACL). After stimulation with *C. albicans* and fMLP, (−)-α-bisabolol inhibited LACL in a concentration-dependent manner at concentrations of 7.7−31 mg/L and 3.8−31 mg/L, respectively. (−)-α-Bisabolol was shown to significantly increase levels/activities of antioxidants (superoxide dismutase, catalase, and glutathione), and decrease levels of lipid peroxidation products (thiobarbituric acid reactive substances and lipid hydroxyperoxide) in the heart mitochondria of rats with isoprotenerol-induced myocardial infarction (Meeran et al., 2019). The rats were pretreated interperitoneally with (−)-α-bisabolol (25 mg/kg bw/day) daily for a period of 8 days. Subsequently, they were cotreated interperitoneally with (−)-α-bisabolol and subcutaneously with isoprotenerol on days 9 and 10. In addition, free hydroxyl radical-scavenging activity of (−)-α-bisabolol was determined via in vitro assay. The compound showed about 79.96% scavenging activity at the concentration of 50 μM. The results of this study suggest that (−)-α-bisabolol is a potent free radical quencher and that it has the potential to protect mitochondria by increasing the mitochondrial antioxidant status and counteracting ROS-mediated oxidative stress, thus inhibiting the progression of damage after myocardial injury.

Antimicrobial properties

Antimicrobial activity of chamomile essential oil

Essential oils extracted from chamomile have exhibited some antimicrobial activity against certain species of bacteria and fungi in vitro. Kazemi (2015) tested the effect of chamomile essential oil on the growth of Gram-positive (*Staphylococcus aureus*, *Bacillus cereus*, and *Bacillus subtilis*) and Gram-negative bacteria (*Shigella shiga*, *Shigella sonnei*, *Pseudomonas aeruginosa*, and *Proteus* sp.) using the microdilution method. In general, Gram-positive bacteria were more susceptible than Gram-negative bacteria to the antimicrobial activity of the essential oil, with *B. cereus* being the most sensitive (inhibition zone 36). Based on the results obtained for components of the essential oil, the author concluded that the oil's antimicrobial activity apparently is related to its terpenoid components. The chamomile oil also was reported to have antifungal activity against *Aspergillus niger*, *Aspergillus ochraceus*, and *Fusarium culmorum* (Lis-Balchin et al., 1998). Another study reported that chamomile oil at the highest concentration tested (3000 ppm) was effective (91%−95% inhibition) against two additional *Aspergillus* strains (*A. flavus* and *A. parasiticus*) and *Fusarium moniliforme* (Soliman and Badeaa, 2002). On the other hand, in the study reported by Mekonnen et al. (2016), the flower essential oil of chamomile failed to inhibit the growth of all tested bacteria (*Salmonella parathypi*, *Salmonella typhimurium*, *Shigella* spp., *P. aeruginosa*, *S. aureus*, and *Escherichia coli*) and fungi (two *Trichophyton* spp. and two *Aspergillus* spp.) in the cavity diffusion test. The authors suggested that the lack of inhibition was possibly due to a low (−)-α-bisabolol content in the essential oil (4.1%) compared with that in another published report (Tolouee et al., 2010).

Antimicrobial activity of chamomile extracts

Ethanolic and cyclohexane extracts of the chamomile flower were tested for their antimicrobial activity using two techniques, broth dilution and diffusion in agar. The microorganisms used in the study were *P. aeruginosa*, *S. aureus*, *E. coli*, and *Salmonella enterica* subsp. *enterica* serovar *Typhimurium*. Only the ethanolic extract showed significant antibacterial activity and only against *P. aeruginosa*, presenting a minimum inhibitory concentration value of 1 g/mL and an inhibition halo 10 mm greater than the positive control. Cinco et al. (1983) studied the effect of hydroethanolic extract from chamomile flowers on bacterial strains cultured from oral and vaginal isolates. Chamomile extract completely inhibited the growth of *S. aureus*, *Streptococcus mutans*, group B *streptococci*, and *Streptococcus salivarius*, and exerted a bactericidal effect on several other bacterial strains.

Antinociceptive and analgesic effects

Several studies investigated the antinociceptive effect of chamomile preparations in animal models. Chamomile extract was reported as a potential alternative approach in the treatment of neuropathic pain caused by chemotherapeutic agents such as cisplatin or vincristine. Abad et al. (2011) investigated the effect of intraperitoneal administration of hydroalcoholic chamomile flower extract on cisplatin-induced peripheral neuropathy. Mice received cisplatin (single dose) and the extract (25 mg/kg every 24 h) 96 h before the formalin test consisting of two phases (0−5 min and 15−40 min after injection). Coadministered chamomile extract and cisplatin reduced the second phase of cisplatin-induced pain. In another study, the hydroalcoholic chamomile extract was tested against vincristine-induced peripheral neuropathy in mice (Nouri and Abad, 2012). Coadministration of the extract (25 mg/kg) and vincristine (100 μg/kg/day) for 12 days before the formalin test resulted in significantly reduced vincristine-induced pain. Both studies demonstrated that although the chamomile extract exhibits a lower analgesic effect in comparison to morphine, its antiinflammatory effect in the second phase of the formalin test is more pronounced than that of morphine.

Several clinical studies have been conducted to evaluate the efficacy of chamomile in pain management. Topical use of chamomile oil three times daily for 3 weeks in patients with knee arthritis significantly decreased the number of acetaminophen tablets used for pain, compared to patients receiving diclofenac gel or placebo (Shoara et al., 2015). All groups showed significant differences in the areas of pain, stiffness, and physical activity when compared with baseline values, but no significant differences between groups were observed. The authors suggested that this lack of difference could be due to successful pain management with more acetaminophen by patients in the diclofenac or placebo groups. One limitation of the study was that sesame oil, the vehicle used in this study, has also been reported to possess antioxidant, antiinflammatory and anesthetic activity. Therefore the authors concluded that the results of their clinical trial could not be attributed to chamomile only.

Chamomile was also studied as a potential treatment for conditions associated with premenstrual syndrome and labor. Oral administration of chamomile extract drops for 8 weeks significantly decreased pain in women with moderate cyclic mastalgia compared to the placebo group (Saghafi et al., 2018). Topical application of 0.5 g cream containing 1.3% chamomile flower extract for 10 days (every 12 h) effectively decreased the intensity of the episiotomy (incision of the perineum and the posterior vaginal wall) pain on the seventh, 10th, and 14th days after delivery (Aradmehr et al., 2017). Zafar et al. (2016) investigated the efficacy of chamomile in labor analgesia as compared to pentazocine, an opioid pain medication, using the visual analogue scale (VAS) to assess pain intensity. The difference in mean VAS scores in the pentazocine and chamomile groups was not statistically significant compared to that of the placebo group, and the authors concluded that chamomile can offer substantial pain relief during labor.

Wound-healing effects

Chamomile traditionally has been used for wound care, and as such, topically applied chamomile preparations have been investigated for their effectiveness in wound healing. Chamomile flower extract dissolved in olive oil and applied to a dorsal incision on rat skin resulted in a reduced wound area in the treatment group compared to the control group (Jarrahi et al., 2010). The rats in the treatment group also showed a faster rate of wound healing. Chamomile fluid extract ointment showed promising results in the treatment of a wound inflicted on the rat tongue (Duarte et al., 2011). It stimulated reepithelialization and the formation of collagen fibers after 10 days of treatment. No significant differences between the treated and control groups in the degree of inflammation, fibroblast count, and wound size were reported. Interestingly, a topically applied chamomile preparation was reported to be more effective than topical corticosteroids in the healing of traumatic ulcers in rat tongue (Martins et al., 2009). Glowania et al. (1987) reported that the application of a compress containing chamomile extract had beneficial effects on weeping wounds in patients who underwent tattoo dermabrasion. Specifically, it increased the wound drying area and speeded up reepithelialization.

A beneficial effect on wound healing was also reported by Maiche et al. (1991). In this study, a cream containing a standardized extract of chamomile flower heads was applied twice daily to areas above and below the scars of women who had undergone local breast cancer surgery in order to investigate its effect on acute radiation skin reaction. The extract did not prevent the skin reaction, and all patients developed grade 1 erythema; however, the number of the patients with grade 2 erythema receiving the chamomile treatment was smaller compared to the group treated with routinely used almond ointment. In addition, grade 2 erythema in the chamomile treatment groups appeared later than in the almond ointment group. A study investigating the effect of pretreatment with a chamomile extract spray on sore throat and hoarseness in patients who had undergone intubation concluded that this approach was not effective (Kyokong et al., 2002).

Sedative effects

Chamomile preparations such as tea and essential oil aromatherapy have been claimed to have calming effects and traditionally have been used to treat insomnia and induce sedation. The sedative effect of chamomile in animal models has been reported by several researchers. Della Loggia et al. (1982) observed that intraperitoneal injection of a lyophilized infusion of chamomile flowers at five dose levels (20, 40, 80, 160, and 320 mg/kg) into starved rats resulted in a significant potentiation of the hexobarbital (a sedative and hypnotic)-induced sleeping time only at the two highest doses (37% and 63%, respectively) compared to controls given 100 mg/kg of hexobarbital alone. Lower doses of the chamomile infusion exerted a depressive effect on locomotor activity. The authors noted that their chamomile infusion was more concentrated than popular chamomile infusions that typically contain about 10 mg/kg of dried product for a 70-kg man. Shinomiya et al. (2005) employed a sleep-disturbed rat model to demonstrate benzodiazepine-like hypnotic activity of an aqueous chamomile flower extract. The extract was administered orally at three different levels: 30, 100, and 300 mg/kg. A significant shortening of sleep latency was observed only at 300 mg/kg. This effect was significantly antagonized by 3 mg/kg flumazenil, a benzodiazepine receptor antagonist used as an antidote for benzodiazepine overdose. The authors hypothesized that the hypnotic effect of chamomile extract may be caused by one of the extract components, apigenin, that has been reported to act as a benzodiazepine receptor antagonist, that is, γ-aminobutyric acid (GABA) receptor antagonist (Medina et al., 1989; Viola et al., 1995).

Despite the wide use of chamomile preparations as sleep aids, the number of efficacy studies in humans is limited. Zick et al. (2011) examined the efficacy of dry ethanolic extract from chamomile flowering tops in patients with chronic primary insomnia. One capsule contained 90 mg of extract, including 3.9 mg apigenin and 1.8 mg (−)-α-bisabolol. Patients were given 270 mg chamomile extract twice daily for 28 days. Outcomes were sleep diary measures and evaluation of common daytime consequences of insomnia, including fatigue. The study showed no benefit of chamomile extract on subjective sleep efficiency and total sleep time, and no statistically significant benefit on sleep latency and night-time awakenings. While not statistically significant, the authors reported a small increase in the fatigue scale in the placebo group and a small decrease in the fatigue scale in the treatment group. They concluded that chamomile extract provided modest and mixed clinical benefit in patients with chronic insomnia. In addition, the authors noted that the dose of chamomile extract used in the study (15 g flowers/day) would be difficult to deliver by drinking tea. On the other hand, chamomile extract (400 mg twice daily for 4 weeks) has been shown to have a statistically significant effect on sleep quality improvement for elderly patients living in nursing homes (Abdullahzadeh et al., 2017). Chamomile tea was recommended as a supplementary approach to improve sleep quality in postnatal women (Chang and Chen, 2016). The study participants who drank chamomile tea for a period of 2 weeks demonstrated significantly lower scores of physical symptom-related sleep inefficiency compared to the control group, whose participants received only regular postpartum care.

Anxiolytic effects

Most testing methods for anxiolytic drugs measure affinity to the benzodiazepine site of the $GABA_A$ receptor. GABA is the main inhibitory neurotransmitter in the central nervous system and plays an important role in anxiety. Awad et al. (2007) investigated the effect of commercially available aqueous and ethanolic extracts of chamomile on enzymes responsible for GABA production, that is, glutamate decarboxylase (GAD) and GABA transaminase (GABA-T), using in vitro rat brain homogenate assays. Both aqueous and ethanolic extracts showed significant inhibition of GAD ($IC_{50} = 0.11$ and 0.22 mg/mL, respectively). GABA-T was considerably inhibited only by ethanolic extract ($IC_{50} = 0.86$ mg/mL). The authors suggested that the anxiolytic and sedative effects of chamomile may not involve GAD because chamomile extracts inhibited this enzyme. Amsterdam et al. (2009) reported on a clinical trial investigating the effect of chamomile in patients with mild to moderate generalized anxiety disorder.

During study week 1, patients were given capsules containing 220 mg of chamomile extract standardized to a content of 1.2% apigenin, to be taken daily. The dose was increased weekly by 220 mg, reaching a maximum 1100 mg/day at study week 5, which continued through the end of the study (week 8). The study results showed a significantly greater reduction over time in mean total anxiety symptom scores in the treatment group compared to the placebo group. A post hoc analysis of the same sample found that chamomile extract may also reduce comorbid depression in anxious patients (Amsterdam et al., 2012). These effects were also observed in a recent study in which anxious patients with or without comorbid depression had been treated with chamomile extract (1500 mg/day) for 8 weeks. The anxiolytic effects of the chamomile extract were similar in both diagnostic groups. In addition, a substantial reduction in depressive symptoms was reported in the group with comorbid depression (Amsterdam et al., in press). Short-term (8 weeks) and long-term (38 weeks) studies both showed that use of 1500 mg/day of chamomile extract significantly reduced moderate to severe symptoms of generalized anxiety disorder (Keefe et al., 2016; Mao et al., 2016). No statistically significant reduction in relapse rates was observed in the 38-week study (Mao et al., 2016).

Antispasmodic effects

Chamomile is known for its antispasmodic effect. Maschi et al. (2008) suggested that inhibition of cyclic adenosine monophosphate (cAMP)-phosphodiesterase (PDE) is the likely mechanism underlying the antispasmodic activity of a chamomile infusion. The study showed that the infusion, prepared from either entire dried flowers or sifted flowers, inhibited cAMP-PDE activity, with the lesser effect on cyclic guanosine monophosphate (cGMP)-PDE activity. Among the individual compounds occurring in chamomile, flavonoids ($IC_{50} = 1.3-14.9$ μM) contributed ~39% to the inhibitory effect of chamomile infusion on cAMP-PDE. The phenolic acids, (−)-α-bisabolol, and coumarins were inactive.

Among chamomile flavonoids with antispasmodic activity, apigenin is the most often mentioned.

Achterrath-Tuckermann et al. (1980) investigated the antispasmodic effect of hydrophilic and lipophilic chamomile constituents on spasms induced in isolated guinea pig ileum. The hydrophilic apigenin was 3.3 times more potent than the reference drug papaverine (an opium alkaloid antispasmodic drug). Quercetin was 72% as active as papaverine, followed by patuletin (68%) and luteolin (44%). In this study, (−)-α-bisabolol was the most potent lipophilic compound, with potency comparable to papaverine, while bisabolol oxides A and B were less effective (46%−50%). Pumnea et al. (2017) demonstrated that chamomile flower extract reduced spontaneous as well as electrical field stimulation-induced contractility in the mouse small intestine in a dose-dependent manner. In addition, an inhibitory effect on neurotransmission in the intestine was observed. The authors concluded that the chamomile extract modulates bowel smooth muscle contractility and motility, which are mechanisms underlying the symptoms of functional bowel disorders.

Hydromethanolic chamomile extract (150 and 300 mg/kg) administered orally to mice with castor oil-induced diarrhea and intestinal fluid accumulation showed effects similar to cromakalin and loperamide, which are known for their spasmolytic, antidiarrheal, and antisecretory activities (Mehmood et al., 2015). The same study showed that the extract caused a complete relaxation of low and high K^+-induced contractions in an isolated rabbit jejunum, in addition to showing a Ca^{2+} channel antagonist-like effect. The authors suggested that the chamomile extract contains spasmolytic constituents that are likely responsible for its antidiarrheal and antisecretory activities in vivo by mediating their effect primarily through K^+ channel activation along with weak Ca^{2+} channel inhibition.

Safety and toxicity of chamomile

Despite the wide uses of various chamomile preparations, they have not been convincingly linked to instances of clinically relevant liver injury (NIDDKD, 2017), and no long-term problems associated with their use have been identified (Zadeh et al., 2014).

Topical chamomile is a known sensitizing agent, and a fairly common cause of contact dermatitis, possibly due to delayed-type hypersensitivity to its sesquiterpene lactone component (Paulsen, 2002). Other sensitizers may include herniarin and bisabolol (Paulsen et al., 2010; Wilkinson and Hausen, 1995). Individuals sensitized from other members of the *Compositae* (i.e., *Asteraceae*) family may cross-react with chamomile. In such individuals, the prevalence of patch-test positivity to ether chamomile extracts has been reported to be 64% (Paulsen et al., 2001). The use of chamomile tea as a compress to treat pink eye has been reported to result in allergic conjunctivitis. It was suggested that pollens present in tea were responsible for this reaction (Subiza et al., 1990).

Allergic reactions caused by ingestion of chamomile appear to be rare; however, individuals sensitized to other *Compositae* family members may exhibit immediate-type allergies. In a study investigating in vivo cross-reactivity between chamomile and mugwort (*Artemisia vulgaris*), sensitization to mugwort was the primary risk factor for allergy symptoms in individuals who ingested a commercial chamomile infusion (de la Torre Morrin et al., 2001). The incidence and risk of mugwort−chamomile association may be underestimated (Andres et al., 2009; Reider et al., 2000). Reider et al. (2000) described both clinical and immunological cross-reactivities to mugwort and plant-derived food. In 14 patients with a history of sensitization to chamomile confirmed by skin prick test and/or radioallergosorbent test, oral intake of chamomile tea mainly resulted in bowel cramps, vomiting, and diarrhea, while inhalation mainly caused asthma. The most severe reactions were observed in two female patients after receiving an enema containing oily chamomile extract, with one developing anaphylactic shock. The authors concluded that the severe reactions cannot be explained by an extraordinarily high level of sensitization to chamomile, and that they were probably evoked by application of the chamomile preparation to the large intestinal resorption area (Reider et al., 2000). Rare cases of anaphylactic reactions to ingestion of chamomile tea have also been reported (Andres et al., 2009; Nakagawa et al., 2019; Subiza et al., 1989). Even though chamomile preparations are extensively used in medicinal products and cosmetics, the initial sensitization from chamomile is less common compared to other *Compositae* species, and the allergenic potential of chamomile is considered to be low (Paulsen et al., 2001).

It has been suggested that chamomile preparations may have inhibitory effects on major human drug-metabolizing enzymes. A 2% chamomile tea was reported to reduce the activity of CYP450 1A2 to 39% in Wistar rats when it replaced their drinking water (Maliakal and Wanwimolruk, 2001). Chamomile also showed an in vitro inhibitory effect on CYP450 1A2, 3A4, 2D6, and 2C9, with the highest inhibition against CYP1A2 (Ganzera et al., 2006). The relatively high inhibitory effect of commercial chamomile preparations on CYP3A4 was observed in an in vitro fluorometric microtiter plate assay (Budzinski et al., 2000). The clinical relevance of these results is unclear, although cases of potential interactions with some medications have been reported. A case of possible interaction of chamomile preparations with warfarin has been reported (Segal and Pilote, 2006). The authors speculated that the coumarin constituent of chamomile may have accounted for potentiation of the effects of warfarin. An interaction between chamomile and cyclosporine (a drug used to prevent rejection of organ transplants) also has been reported (Nowack and Nowak, 2005). There are theoretical reasons to suspect that chamomile might interact with other medications, such as drugs with sedative properties (Abebe, 2002).

Chamomile has been widely consumed during pregnancy as a beverage to counter morning sickness (Forster et al., 2006; Nordeng and Havnen, 2004); however, its excessive use is considered potentially harmful. A higher incidence of threatening miscarriages and preterm labors was observed among users, women taking chamomile regularly from the beginning of pregnancy, compared to nonusers (Cuzzolin et al., 2010). While chamomile taken orally was investigated and reported to be effective in inducing labor in women with postterm pregnancy (Gholami et al., 2016), its toxicity was demonstrated in cases where a chamomile-containing enema was used to induce labor, resulting in maternal anaphylaxis (Jensen-Jarolim et al., 1998).

Acute oral and repeat-dose toxicity studies in animals have been reported for chamomile preparations. The acute oral toxicity of a lyophilized chamomile flower infusion was evaluated using two groups of 12 female mice given a single oral dose of 720 and 1440 mg/kg, which were observed for 24 h postdosing. None of the animals died, and no evidence of acute toxicity was reported (Della Loggia et al., 1982). No mortality was reported in another acute study in which chamomile flower oil extract was administered orally to groups of male National Institutes of Health mice at the following doses: 10, 100, 1000, 1600, 2900, 4300, and 5600 mg/kg (Hernández-Ceruelos et al., 2002). A study investigating the activity of CYP450 enzymes in female rats found that free access to a 2% chamomile tea solution (w/v) for 4 weeks did not result in a significant effect on body weight or in signs of gross pathology of internal organs. Liver weight/body weight ratios of treated rats were not significantly different from control values (Maliakal and Wanwimolruk, 2001).

Several studies have assessed the toxicity of chamomile in humans. Wang et al. (2005) investigated the responses of 14 human volunteers to ingestion of chamomile flower tea over a 6-week period during baseline, treatment, and posttreatment periods, each lasting 2 weeks. The daily consumption of tea corresponded to approximately five cups of a commercial tea preparation. Volunteers reported no adverse effects after ingestion of chamomile tea. Chamomile at different doses, the highest being 1500 mg/day, demonstrated a favorable adverse event and tolerability profile when used to treat generalized anxiety disorder and insomnia in four clinical trials (Amsterdam et al., 2009; Keefe et al., 2016; Mao et al., 2016; Zick et al., 2011). Adverse effects were mild to moderate, and included headache, dizziness, digestive and musculoskeletal problems, diarrhea, drowsiness, fatigue, taste lingering, indigestion, nausea, and bruising. No serious adverse effects were observed in any of the four clinical trials.

Concluding remarks and future directions

Although many medicinal and nutraceutical uses of chamomile and its preparations have been reported, only a few clinical studies with human subjects have proven their effectiveness in ameliorating or treating a wide variety of symptoms, ailments, and illnesses. Most studies investigating the therapeutic effects of chamomile or its specific constituents have been conducted using in vitro systems or in vivo animal models. Therefore, due to its wide-ranging therapeutic potential, more human studies are needed to better assess the potential beneficial effects of this traditional herbal medicine and nutraceutical.

References

Abad, A.N.A., Nouri, M.H.K., Gharjanie, A., et al., 2011. Effect of *Matricaria chamomilla* hydroalcoholic extract on cisplatin-induced neuropathy in mice. Chin. J. Nat. Med. 9, 126–131.

Abdullahzadeh, M., Matourypour, P., Naji, S.A., 2017. Investigation effect of oral chamomilla on sleep quality in elderly people in Isfahan: a randomized control trial. J. Educ. Health Promot. 6, 53.

Abebe, W., 2002. Herbal medication: potential for adverse interactions with analgesic drugs. J. Clin. Pharm. Ther. 27, 391–401.

Achterrath-Tuckermann, U., Kunde, R., Flaskamp, E., et al., 1980. Pharmacological investigations with compounds of chamomile. Planta Med. 39, 38–50.

Aertgeerts, P., Albring, M., Klaschka, F., et al., 1985. Comparative testing of Kamillosan cream and steroidal (0.25% hydrocortisone, 0.75% fluocortin butyl ester) and non-steroidal (5% bufexamac) dermatologic agents in maintenance therapy of eczematous diseases. Z. Hautkr. 60, 270–277.

Agatonovic-Kustrina, S., Ortakanda, D.B., Mortona, D.W., et al., 2015. Rapid evaluation and comparison of natural products and antioxidantactivity in calendula, feverfew, and German chamomile extracts. J. Chromatogr. A 1385, 103–110.

Ahmed, H.M., 2016. Ethnopharmacobotanical study on the medicinal plants used by herbalists in Sulaymaniyah Province, Kurdistan, Iraq. J. Ethnobiol. Ethnomed. 12, 8.

Al-Hindawi, M.K., Al-Deen, I.H.S., Nab, M.H.A., et al., 1989. Anti-inflammatory activity of some Iraqi plants using intact rats. J. Ethnopharmacol. 26, 163–168.

Al-Ramahi, R., Jaradat, N., Adawi, D., 2013. Use of herbal medicines during pregnancy in a groupof Palestinian women. J. Ethnopharmacol. 150, 79–84.

Ammon, H.P.T., Sabieraj, J., Kaul, R., 1996. Mechanism of anthiplogistic activity of chamomile extracts and their contents. Dtsch. Apoth. Ztg. 136, 17–30.

Amsterdam, J.D., Li, Y., Soeller, I., et al., 2009. A randomized, double-blind, placebo-controlled trial of oral *Matricaria recutita* (chamomile) extract therapy of generalized anxiety disorder. J. Clin. Psychopharmacol. 29, 378–382.

Amsterdam, J.D., Shults, J., Soeller, I., et al., 2012. Chamomile (*Matricaria recutita*) may have antidepressant activity in anxious depressed humans—an exploratory study. Altern. Ther. Health Med. 18, 44–49.

Amsterdam, J.D., Li, Q.S., Xie, S.X., et al., in press. Putative antidepressant effect of chamomile (*Matricaria chamomilla* L.) oral extract in subjects with comorbid generalized anxiety disorder and depression. J. Altern. Complement. Med. Available from: https://doi.org/10.1089/acm.2019.0252.

Andres, C., Chen, W.C., Ollert, M., et al., 2009. Anaphylactic reaction to chamomile tea. Allergol. Int. 58, 135–136.

Aradmehr, M., Azhari, S., Ahmadi, S., et al., 2017. The effect of chamomile cream on episiotomy pain in primiparous women: a randomized clinical trial. J. Caring Sci. 6, 19–28.

Atoui, A.K., Mansouri, A., Boskou, G., et al., 2005. Tea and herbal infusions: their antioxidant activity and phenolic profile. Food Chem. 89, 27–36.

Awad, R., Levac, D., Cybulska, P., et al., 2007. Effects of traditionally used anxiolytic botanicals on enzymes of the γ-aminobutyric acid (GABA) system. Can. J. Physiol. Pharmacol. 85, 933–942.

Benitez, G., Gonzalez-Tejero, M.R., Molero-Mesa, J., 2010. Pharmaceutical ethnobotany in the western part of Granada province (southern Spain): ethnopharmacological synthesis. J. Ethnopharmacol. 129, 87–105.

Blumenthal, M., Busse, W.R., Goldberg, A., et al., 1998. Chamomile flower, German. In: B, Mark (Ed.), The Complete German Commission E monographs. Therapeutic Guide to Herbal Medicines. American Botanical Council, Austin, TX, p. 107.

Boesch-Saadatmandi, C., Wagner, A.E., Wolffram, S., et al., 2012. Effect of quercetin on inflammatory gene expression in mice liver *in vivo*—role of redox factor 1, miRNA-122 and miRNA-125b. Pharmacol. Res. 65, 523–530.

Boots, A.W., Drent, M., de Boer, V.C.J., et al., 2011. Quercetin reduces markers of oxidative stress and inflammation in sarcoidosis. Clin. Nutr. 30, 506–512.

Braga, P.C., Dal Sasso, M., Fonti, E., et al., 2009. Antioxidant activity of bisabolol: inhibitory effects on chemiluminescence of human neutrophil bursts and cell-free systems. Pharmacology 83, 110–115.

Budzinski, J.W., Foster, B.C., Vandenhoek, S., et al., 2000. An *in vitro* evaluation of human cytochrome P450 3A4 inhibition by selected commercial herbal extracts and tinctures. Phytomedicine 7, 273–282.

Bulgari, M., Sangiovanni, E., Colombo, E., et al., 2012. Inhibition of neutrophil elastase and metalloproteinase-9 of human adenocarcinoma gastric cells by chamomile (*Matricaria recutita* L.) infusion. Phytother. Res. 26, 1817–1822.

Capuzzo, A., Occipinti, A., Maffei, M.E., 2014. Antioxidant and radical scavenging activities of chamazulene. Nat. Prod. Res. 28, 2321–2323.

Cavalcante, H.A.O., Silva-Filho, S.E., Wiirzler, L.A.M., et al., 2020. Effect of (-)-α-bisabolol on the inflammatory response in systemic infection experimental model in C57BL/6 mice. Inflammation 43, 193–201.

Chan, M.M., Mattiacci, J.A., Hwang, H.S., et al., 2000. Synergy between ethanol and grape polyphenols, quercetin, and resveratrol, in the inhibition of the inducible nitric oxide synthase pathway. Biochem. Pharmacol. 60, 1539–1548.

Chandrashekhar, V.M., Halagali, K.S., Nidavani, R.B., et al., 2011. Anti-allergic activity of German chamomile (*Matricaria recutita* L.) in mast cell mediated allergy model. J. Ethnopharmacol. 127, 336–340.

Chang, S.M., Chen, C.H., 2016. Effects of an intervention with drinking chamomile tea on sleep quality and depression in sleep disturbed postnatal women: a randomized controlled trial. J. Adv. Nurs. 72, 306–315.

Charousaei, F., Dabirian, A., Mojab, F., 2011. Using chamomile solution or a 1% topical hydrocortisone ointment in the management of peristomal skin lesions in colostomy patients: results of a controlled clinical study. Ostomy Wound Manage. 57, 28–36.

Chaves, P.F.P., Iacomini, M., Cordeiro, L.M.C., 2019. Chemical characterization of fructooligosaccharides, inulin and structurally diverse polysaccharides from chamomile tea. Carbohydr. Polym. 214, 269–275.

Chiu-Yuan, C., Wen-Huang, P., Kuen-Daw, T., et al., 2007. Luteolin suppresses in-ammation-associated gene expression by blocking NF-κB and AP-1 activation pathway in mouse alveolar macrophages. Life Sci. 81, 1602–1614.

Cho, S.-Y., Park, S.-J., Kwon, M.-J., et al., 2003. Quercetin suppresses proinflammatory cytokines production through MAP kinases and NF-κB pathway in lipopolysaccharide-stimulated macrophage. Mol. Cell. Biochem. 243, 153–160.

Choi, E.M., Lee, Y.S., 2010. Luteolin suppresses IL-1beta-induced cytokines and MMPs production via p38 MAPK, JNK, NF-kappaB and AP-1 activation in human synovial sarcoma cell line, SW982. Food Chem. Toxicol. 48, 2607–2611.

Chuang, C.C., Martinez, K., Xie, G., et al., 2010. Quercetin is equally or more effective than resveratrol in attenuating tumor necrosis factor-α-mediated in-ammation and insulin resistance in primary human adipocytes. Am. J. Clin. Nutr. 92, 1511–1521.

Cinco, M., Banfi, E., Tubaro, A., et al., 1983. A microbiological survey on the activity of a hydroalcoholic extract of camomile. Int. J. Crude Drug Res. 21, 145–151.

Comalada, M., Ballester, I., Bailón, E., et al., 2006. Inhibition of pro-inflammatory markers in primary bone marrow-derived mouse macrophages by naturally occurring flavonoids: analysis of the structure–activity relationship. Biochem. Pharmacol. 72, 1010–1021.

Cuzzolin, L., Francini-Pesenti, F., Verlato, G., et al., 2010. Use of herbal products among 392 Italian pregnant women: focus on pregnancy outcome. Pharmacoepidemiol. Drug Saf. 19, 1151–1158.

D'Almeida, A.P.L., de Oliveira, M.T.P., de Souza, É.T., et al., 2017. A-Bisabolol-loaded lipid-core nanocapsules reduce lipopolysaccharide-induced pulmonary inflammation in mice. Int. J. Nanomed. 12, 4479–4491.

Das, M., 2015. Chamomile oil and extract: localization, chemical composition, extraction, and identification. In: Hardman, R. (Ed.), Chamomile: Medicinal, Biochemical, and Agricultural Aspects, first ed. CRC Press, Boca Raton, FL, pp. 155–198.

de la Torre Morrin, F., Sánchez Machín, I., García Robaina, J.C., et al., 2001. Clinical cross-reactivity between *Artemisia vulgaris* and *Matricaria chamomilla* (chamomile). J. Invest. Allergol. Clin. Immunol. 11, 118–122.

de O. Leite, G., Leite, L.H.I., de S. Sampaio, R., et al., 2011. (–)-α-Bisabolol attenuates visceral nociception and inflammation in mice. Fitoterapia 82, 208–211.

Della Loggia, R., Traversa, U., Scarcia, V., et al., 1982. Depressive effects of *Chamomilla recutita* (L.) Rauch, tubular flowers, on central nervous system in mice. Pharmacol. Res. Commun. 14, 153–162.

Della Loggia, R., Tubaro, A., Zilli, C., et al., 1986. The role of flavonoids in the antiinflammatory activity of *Chamomilla recutita*. Prog. Clin. Biol. Res. 213, 481–484.

Della Loggia, R., Carle, R., Sosa, S., et al., 1990. Evaluation of the anti-inflammatory activity of chamomile preparations. Planta Med. 56, 657–658.

dos Reis, P.E.D., de Carvalho, E.C., Bueno, P.C.P., et al., 2011. Clinical application of *Chamomilla recutita* in phlebitis: dose response curve study. Rev. Lat. Am. Enfermagem. 19, 3–10.

Doseff, A.I., Arango, D., Cardenas, H., et al., 2011. Apigenin a potent anti-inflammatory flavonoid: its efficacy in lung inflammation and target identification provides novel potential alternative therapeutic approaches. Am. J. Respir. Crit. Care Med. 183, A4508.

Dragland, S., Senoo, H., Wake, K., et al., 2003. Several culinary and medicinal herbs are important sources of dietary antioxidants. J. Nutr. 133, 1286–1290.

Drummond, E.M., Harbourne, N., Marete, E., et al., 2013. Inhibition of proinflammatory biomarkers in THP1 macrophages by polyphenols derived from chamomile, meadowsweet, and willow bark. Phytother. Res. 27, 588–594.

Duarte, C.-M.-E., Quirino, M.-R.-S., Patrocínio, M.-C., et al., 2011. Effects of *Chamomilla recutita* (L.) on oral wound healing in rats. Med. Oral Patol. Oral Cir. Bucal 16, e716–e721.

EMA, 2015. Assessment Report on *Matricaria recutita* L., flos and *Matricaria recutita* L., *aetheroleum*. European Medicines Agency, Committee on Herbal Medicinal Products. EMA/HMPC/55843/2011, <https://www.ema.europa.eu/en/medicines/herbal/matricariae-flos>. (Accessed 9 March 2020).

Fabian, D., Juhas, S., Bukovska, D., et al., 2011. Anti-inflamatory effects of chamomile essential oil in mice. Slovak J. Anim. Sci. 44, 111–116.

Flemming, M., Kraus, B., Rascle, A., et al., 2015. Revisited anti-inflammatory activity of matricine *in vitro*: comparison with chamazulene. Fitoterapia 106, 122–128.

Forster, D.A., Denning, A., Wills, G., et al., 2006. Herbal medicine use during pregnancy in a group of Australian women. BMC Pregnancy Childbirth 6, 21.

Franz, C., 1992. Genetica biochimica e coltivazione della camomilla (*Chamomilla recutita* (L.) Rausch.). Agricoltura Ricerca 14, 87–96.

Funakoshi-Tago, M., Nakamura, K., Tago, K., et al., 2011. Anti-inflammatory activity of structurally related flavonoids, apigenin, luteoli and fisetin. Int. Immunopharmacol. 11, 1150–1159.

Ganzera, M., Schneider, P., Stuppner, H., 2006. Inhibitory effects of the essential oil of chamomile (*Matricaria recutita* L.) and its major constituents on human cytochrome P450 enzymes. Life Sci. 78, 856–861.

Gholami, F., Samani, L.N., Kashanian, M., et al., 2016. Onset of labor in post-term pregnancy by chamomile. Iran. Red Crescent Med. J. 18, e19871.

Glowania, H.J., Raulin, C., Swoboda, M., 1987. Effect of chamomile on wound healing – a clinical double-blind study. Z. Hautkr. 62, 1267–1271.

Goes, P., Dutra, C.S., Lisboa, M.R., et al., 2016. Clinical efficacy of a 1% *Matricaria chamomile* L. mouthwash and 0.12% chlorhexidine for gingivitis control in patients undergoing orthodontic treatment with fixed appliances. J. Oral Sci. 58, 569–574.

Gómez-Estrada, H., Díaz-Castillo, F., Franco-Ospina, L., et al., 2011. Folk medicine in the northern coast of Colombia: an overview. J. Ethnobiol. Ethnomed. 7, 27.

González-Segovia, R., Quintanar, J.L., Salinas, E., et al., 2008. Effect of the flavonoid quercetin on inflammation and lipid peroxidation induced by *Helicobacter pylori* in gastric mucosa of guinea pig. J. Gastroenterol. 43, 441–447.

Gosztola, B., Sárosi, S., Németh, É., 2010. Variability of the essential oil content and composition of chamomile (*Matricaria recutita* L.) affected by weather conditions. Nat. Prod. Commun. 5, 465–470.

Guardia, T., Rotelli, A.E., Juarez, A.O., et al., 2010. Anti-inflammatory properties of plant flavonoids. Effects of rutin, quercetin and hesperidin on adjuvant arthritis in rat. Farmaco 56, 683–687.

Gupta, V., Mittal, P., Bansal, P., et al., 2010. Pharmacological potential of *Matricaria recutita*—a review. Int. J. Pharm. Sci. Drug Res. 2, 12–16.

Hajaji, S., Jabri, M.A., Alimi, D., et al., 2019. Chamomile methanolic extract mitigates small bowel inflammation and ROS overload related to the intestinal nematodes infection in mice. Acta Parasitol. 64, 152–161.

Hernández-Ceruelos, A., Madrigal-Bujaidar, E., de la Cruz, C., 2002. Inhibitory effect of chamomile essential oil on the sister chromatid exchanges induced by daunorubicin and methylmethanesulfonate in mouse bone marrow. Toxicol. Lett. 135, 103–110.

Hirano, T., Higa, S., Arimitsu, J., et al., 2004. Flavonoids such as luteolin, fisetin and apigenin are inhibitors of interleukin-4 and interleukin-13 production by activated human basophils. Int. Arch. Allergy Immunol. 134, 135–140.

Hougee, S., Sanders, A., Faber, J., et al., 2005. Decreased pro-inflammatory cytokine production by LPS-stimulated PBMC upon in vitro incubation with the flavonoids apigenin, luteolin or chrysin, due to selective elimination of monocytes/macrophages. Biochem. Pharmacol. 69, 241–248.

Jabri, M.-A., Aissani, N., Tounsi, H., et al., 2017. Protective effect of chamomile (*Matricaria recutita* L.) decoction extract against alcohol-induced injury in rat gastric mucosa. Pathophysiology 24, 1–8.

Jakovlev, V., Isaac, O., Thiemer, K., et al., 1979. Pharmacological investigations with compounds of chamomile. II. New investigations on the antiphlogistic effects of (−)-α-bisabolol and bisabolol oxides. Planta Med. 35.

Jakovlev, V., Isaac, O., Flaskamp, E., 1983. Pharmacological investigations with compounds of chamomile. VI. Investigations on the antiphlogistic effect of chamazulene and matricine. Planta Med. 49, 67–73.

Jarić, S., Popović, Z., Macukanović-Jocić, M., et al., 2007. An ethnobotanical study on the usage of wild medicinal herbs from Kopaonik Mountain (Central Serbia). J. Ethnobiol. Ethnomed. 111, 160–175.

Jarrahi, M., Vafaei, A.A., Taherian, A.A., et al., 2010. Evaluation of topical *Matricaria chamomilla* extract activity on linear incisional wound healing in albino rats. Nat. Prod. Res. 24, 697–702.

Jensen-Jarolim, E., Reider, N., Fritsch, R., et al., 1998. Fatal outcome of anaphylaxis to camomile-containing enema during labor: a case study. J. Allergy Clin. Immunol. 102, 1041–1042.

Kang, O., Choi, J., Lee, J., et al., 2010. Luteolin isolated from the flowers of Lonicera japonica suppresses in-ammatory mediator release by blocking NF-κB and MAPKs activation pathways in HMC-1 cells. Molecules 15, 385–398.

Kazemi, M., 2015. Chemical composition and antimicrobial activity of essential oil of *Matricaria recutita*. Int. J. Food Prop. 18, 1784–1792.

Keefe, J.R., Mao, J.J., Soeller, I., et al., 2016. Short-term open-label chamomile (*Matricaria chamomilla* L.) therapy of moderate to severe generalized anxiety disorder. Phytomedicine 23, 1699–1705.

Kim, J., Kim, D., Kang, O., et al., 2005. Inhibitory effect of luteolin on TNF-α-induced IL-8 production in human colon epithelial cells. Int. Immunopharmacol. 5, 209–217.

Kim, S., Jung, E., Kim, J.-H., et al., 2011. Inhibitory effects of (−)-a-bisabolol on LPS-induced inflammatory response in RAW264.7 macrophages. Food Chem. Toxicol. 49, 2580–2585.

Kuo, M.Y., Liao, M.F., Chen, F.L., et al., 2011. Luteolin attenuates the pulmonary inflammatory response involves abilities of antioxidation and inhibition of MAPK and NFkappaB pathways in mice with endotoxin-induced acute lung injury. Food Chem. Toxicol. 49, 2660–2666.

Kyokong, O., Charuluxananan, S., Muangmingsuk, V., et al., 2002. Efficacy of chamomile-extract spray for prevention of post-operative sore throat. J. Med. Assoc. Thai. 85, 180–185.

Lee, K.G., Shibamoto, T., 2002. Determination of antioxidant potential of volatile extracts isolated from various herbs and spices. J. Agric. Food Chem. 50, 4947–4952.

Lee, J.H., Zhou, H.Y., Cho, S.Y., et al., 2007. Anti-inflammatory mechanisms of apigenin: inhibition of cyclooxygenase-2 expression, adhesion monocytes to human umbilical vein endothelial cells, and expression of cellular adhesion molecules. Arch. Pharm. Res. 30, 1318–1327.

Li, Y.C., Yeh, C.H., Yang, M.L., et al., 2012. Luteolin suppresses inflammatory mediator expression by blocking the Akt/NFkappaB pathway in acute lung injury induced by lipopolysaccharide in mice. J. Evid. Based Complement. Alternat. Med. 2012, 383608.

Lis-Balchin, M., Deans, S.G., Eaglesham, E., 1998. Relationship between bioactivity and chemical composition of commercial essential oils. Flavour Fragr. J. 13, 98–104.

Liu, C.W., Lin, H.W., Yang, D.J., et al., 2016. Luteolin inhibits viral-induced inflammatory response in RAW264.7 cells via suppression of STAT1/3 dependent NF-kappaB and activation of HO-1. Free Radic. Biol. Med. 95, 180–189.

Lou, L., Liu, Y., Zhou, J., et al., 2015. Chlorogenic acid and luteolin synergistically inhibit the proliferation of interleukin-1β-induced fibroblast-like synoviocytes through regulating the activation of NF-κB and JAK/STAT-signaling pathways. Immunopharmacol. Immunotoxicol. 37, 499–507.

Luqman, S., Kaushik, S., Srivastava, S., et al., 2009. Protective effect of medicinal plant extracts on biomarkers of oxidative stress in erythrocytes. Pharm. Biol. 47, 483–490.

Ma, D., He, J., He, D., 2020. Chamazulene reverses osteoarthritic inflammation through regulation of matrix metalloproteinases (MMPs) and NF-kβ pathway in in-vitro and *in-vivo* models. Biosci. Biotechnol. Biochem. 84, 402–410.

Maiche, A.G., Gröhn, P., Mäki-Hokkonen, H., 1991. Effect of chamomile cream and almond ointment on acute radiation skin reaction. Acta Oncol. 30, 395–397.

Maliakal, P.P., Wanwimolruk, S., 2001. Effect of herbal teas on hepatic drug metabolizing enzymes in rats. J. Pharm. Pharmacol. 53, 1323–1329.

Mamani-Matsuda, M., Kauss, T., Al-Kharrat, A., et al., 2006. Therapeutic and preventive properties of quercetin in experimental arthritis correlate with decreased macrophage in-ammatory mediators. Biochem. Pharmacol. 72, 1304–1310.

Man, M.-Q., Hupe, M., Sun, R., et al., 2012. Topical apigenin alleviates cutaneous inflammation in murine models. J. Evid. Based Complement. Alternat. Med. 2012, 912028.

Mann, C., Staba, E.J., 1986. The chemistry, pharmacology, and commercial formulations of chamomile. In: Craker, L.E., Simon, J.E. (Eds.), Herbs, Spices, and Medicinal Plants: Recent Advances in Botany, Horticulture, and Pharmacology. Oryx Press, Phoenix, pp. 235–280.

Mao, J.J., Xie, S.X., Keefe, J.R., et al., 2016. Long-term chamomile (*Matricaria chamomilla* L.) treatment for generalized anxiety disorder: a randomized clinical trial. Phytomedicine 23, 1735–1742.

Martins, M.D., Marques, M.M., Bussadori, S.K., et al., 2009. Comparative analysis between *Chamomilla recutita* and corticosteroids on wound healing. An *in vitro* and *in vivo* study. Phytother. Res. 23, 274–278.

Maschi, O., Cero, E.D., Galli, G.V., et al., 2008. Inhibition of human cAMP-phosphodiesterase as a mechanism of the spasmolytic effect of *Matricaria recutita* L. J. Agric. Food Chem. 56, 5015–5020.

Medina, J.H., Pena, C., de Stein, M.L., 1989. Benzodiazepine-like molecules, as well as other ligands for the brain benzodiazepine receptors, are relatively common constituents of plants. Biochem. Biophys. Res. Commun. 165, 547–553.

Meeran, M.F.N., Laham, F., Azimullah, S., et al., 2019. α-Bisabolol abrogates isoproterenol-induced myocardial infarction by inhibiting mitochondrial dysfunction and intrinsic pathway of apoptosis in rats. Mol. Cell. Biochem. 453, 89–102.

Mehmood, M.H., Munir, S., Khalid, U.A., et al., 2015. Antidiarrhoeal, antisecretory and antispasmodic activities of *Matricaria chamomilla* are mediated predominantly through K^+-channels activation. BMC Complement. Altern. Med. 15, 75.

Mekonnen, A., Yitayew, B., Tesema, A., et al., 2016. In vitro antimicrobial activity of essential oil of *Thymus schimperi*, *Matricaria chamomilla*, *Eucalyptus globulus*, and *Rosmarinus officinalis*. Int. J. Microbiol. 2016, 9545693.

Menghini, L., Ferrante, C., Leporini, L., et al., 2016. An hydroalcoholic chamomile extract modulates inflammatory and immune response in HT29 cells and isolated rat colon. Phytother. Res. 30, 1513–1518.

Miller, T., Wittstock, U., Lindequist, U., et al., 1996. Effects of some components of the essential oil of chamomile, *Chamomilla recutita*, on histamine release from rat mast cells. Planta Med. 62, 60–61.

Min, Y.D., Choi, C.H., Bark, H., et al., 2007. Quercetin inhibits expression of in-ammatory cytokines through attenuation of NF-κB and p38 MAPK in HMC-1 human mast cell line. Inflamm. Res. 56, 210–215.

Morikawa, K., Nonaka, M., Narahara, M., et al., 2003. Inhibitory effect of quercetin on carrageenan-induced inflammation in rats. Life Sci. 74, 709–721.

Mulinacci, N., Romani, A., Pinelli, E., et al., 2000. Characterization of *Matricaria recutita* L. flower extracts by HPLC-MS and HPLC-DAD analysis. Chromatographia 51, 301–307.

Muñoz-Pérez, V.M., Ortiz, M.A., Ponce-Monter, H.A., et al., 2018. Anti-inflammatory and utero-relaxant effect of α-bisabolol on the pregnant human uterus. Korean J. Physiol. Pharmacol. 22, 391–398.

Nakagawa, M., Hanada, M., Amano, H., 2019. A case of anaphylactic reaction to chamomile tea in a patient with mugwort pollinosis. Allergol. Int. 68, 396–398.

NIDDKD, 2017. Chamomile. LiverTox: Clinical and Research Information on Drug-Induced Liver Injury. National Institute of Diabetes and Digestive and Kidney Diseases, Bethesda, MD, <https://www.ncbi.nlm.nih.gov/books/NBK548163/>. Last updated January 23, 2017. (Accessed March 20, 2020).

Nordeng, H., Havnen, G.C., 2004. Use of herbal drugs in pregnancy: a survey among 400 Norwegian women. Pharmacoepidemiol. Drug Saf. 13, 371–380.

Nouri, M.H.K., Abad, A.N.A., 2012. Antinociceptive effect of *Matricaria chamomilla* on vincristine-induced peripheral neuropathy in mice. Afr. J. Pharm. Pharmacol. 6, 24–29.

Nowack, R., Nowak, B., 2005. Herbal teas interfere with cyclosporin levels in renal transplant patients. Nephrol. Dial. Transplant 20, 2554–2556.

Park, K.Y., Lee, S.H., Min, B.K., et al., 1999. Inhibitory effect of luteolin 4'-*O*-glucoside from *Kummerowia striata* and other flavonoids on interleukin-5 bioactivity. Planta Med. 65, 457–459.

Paulsen, E., 2002. Contact sensitization from *Compositae*-containing herbal remedies and cosmetics. Contact Dermatitis 47, 189–198.

Paulsen, E., Andersen, K.E., Hausen, B.M., 2001. Sensitization and cross-reaction patterns in Danish *Compositae*-allergic patients. Contact Dermatitis 45, 197–204.

Paulsen, E., Otkjaer, A., Andersen, K.E., 2010. The coumarin herniarin as a sensitizer in German chamomile [*Chamomilla recutita* (L.) Rauschert, Compositae]. Contact Dermatitis 62, 338–342.

Piri, E., Sourestani, M.M., Khaleghi, M., et al., 2019. Chemo-diversity and antiradical potential of twelve *Matricaria chamomilla* L. populations from Iran: proof of ecological effects. Molecules 24, 1315.

Pourabbas, R., Delazar, A., Chitsaz, M.T., 2005. The effect of German chamomile mouthwash on dental plaque and gingival inflammation. Iran. J. Pharm. Res. 2, 105–109.

Povh, N.P., Marques, M.O.M., Meireles, A.A., 2001. Supercritical CO_2 extraction of essential oil and oleoresin from chamomile (*Chamomilla recutita* [L.] Rauschert). J. Supercrit. Fluids 21, 245–256.

Pumnea, T., Sibaev, A., Lipowicz, B., et al., 2017. The herbal extracts of myrrh, chamomile and coffee charcoal modulate intestinal neurotransmission and motility in murine small intestine. EC Gastroenterol. Dig. Syst. 4, 15–30.

Querio, G., Antoniotti, S., Foglietta, F., et al., 2018. Chamazulene attenuates ROS levels in bovine aortic endothelial cells exposed to high glucose concentrations and hydrogen peroxide. Front. Physiol. 9, 246.

Redaelli, C., Formentini, L., Santaniello, E., 1981. HPLC determination of coumarins in *Matricaria chamomilla*. Planta Med. 43, 412–413.

Reider, N., Sepp, N., Fritsch, P., et al., 2000. Anaphylaxis to camomile: clinical features and allergen cross-reactivity. Clin. Exp. Allergy 30, 1436–1443.

Rekka, E.A., Kourounakis, A.P., Kourounakis, P.N., 1996. Investigation of the effect of chamazulene on lipid peroxidation and free radical processes. Res. Commun. Mol. Pathol. Pharmacol. 92, 361–364.

Roby, M.H.H., Sarhan, M.A., Selim, K.A.-H., et al., 2013. Antioxidant and antimicrobial activities of essential oil and extracts of fennel (*Foeniculum vulgare* L.) and chamomile (*Matricaria chamomilla* L.). Ind. Crops Prod. 44, 437–445.

Rocha, N.F.M., Rios, E.R.V., Carvalho, A.M.R., et al., 2011. Anti-nociceptive and anti-inflammatory activities of (−)-α-bisabolol in rodents. Naunyn Schmiedebergs Arch. Pharmacol. 384, 525–533.

Rogerio, A.P., Kanashiro, A., Fontanari, C., et al., 2007. Anti-inflammatory activity of quercetin and isoquercitrin in experimental murine allergic asthma. Inflamm. Res. 56, 402–408.

Ross, S.M., 2008. Chamomile: a spoonful of medicine. Holist. Nurs. Pract. 22, 56–57.

Rotelli, A.E., Guardia, T., Juárez, A.O., et al., 2003. Comparative study of flavonoids in experimental models of inflammation. Pharmacol. Res. 48, 601–606.

Safayhi, H., Sabieraj, J., Sailer, E.R., et al., 1994. Chamazulene: an antioxidant type inhibitor of leukotriene B4 formation. Planta Med. 60, 410–413.

Saghafi, N., Rhkhshandeh, H., Pourmoghadam, N., et al., 2018. Effectiveness of *Matricaria chamomilla* (chamomile) extract on pain control of cyclic mastalgia: a double-blind randomised controlled trial. J. Obstet. Gynaecol. 38, 81–84.

Salamon, I., 1994. Growing conditions and the essential oil of *Chamomilla recutita* (L.) Rausch. J. Herbs Spices Med. Plants 2, 31–37.

Salamon, I., Ghanavati, M., Khazaei, H., 2010. Chamomile biodiversity and essential oil qualitative-quantitative characteristics in Egyptian production and Iranian landraces. Emir. J. Food Agric. 22, 59–64.

Satyal, P., Shrestha, S., Setzer, W.N., 2015. Composition and bioactivities of an (*E*)-β-farnesene chemotype of chamomile (*Matricaria chamomilla*) essential oil from Nepal. Nat. Prod. Commun. 10, 1453–1457.

Sawatzky, D.A., Derek, A.W., Coville-Nash, P.R., et al., 2006. The involvment of the apoptosis-modulating protein ERK 1/2, Bcl-x_L and bax in the resolution of acute inflammation *in vivo*. Am. J. Pathol. 6, 317–333.

Schilcher, H., 1973. Neuere erkenntnisse bei der qualitätsbeurteilung von kamillenblúten bzw. kamillenöl. teil 2: qualitative beurteilung des arherischen oles in flores chamomillae. aufteilung der handelskammillen in vier bzw. Fünf chemische typen. Planta Med. 23, 132–144.

Schilcher, H., 2005. Traditional use and therapeutic indications. In: Franke, R., Schilcher, H. (Eds.), Chamomile: Industrial Profies. CRC Press, Boca Raton, FL, pp. 7–38.

Sebai, H., Jabri, M.-A., Souli, A., et al., 2015. Chemical composition, antioxidant properties and hepatoprotective effects of chamomile (*Matricaria recutita* L.) decoction extract against alcohol-induced oxidative stress in rat. Gen. Physiol. Biophys. 34, 263–275.

Segal, R., Pilote, L., 2006. Warfarin interaction with *Matricaria chamomilla*. Can. Med. Assoc. J. 174, 1281–1282.

Sharifi-Ardani, M., Yekefallah, L., Asefzadeh, S., et al., 2017. Efficacy of topical chamomile on the incidence of phlebitis due to an amiodarone infusion in coronary care patients: a double-blind, randomized controlled trial. J. Integr. Med. 15, 373–378.

Shi, F., Zhou, D., Ji, Z., et al., 2015. Anti-arthritic activity of luteolin in Freund's complete adjuvant-induced arthritis in rats by suppressing P2X4 pathway. Chem. Biol. Interact. 226, 82–87.

Shinomiya, K., Inoue, T., Uysu, Y., et al., 2005. Hypnotic activities of chamomile and passiflora extracts in sleep-disturbed rats. Biol. Pharm. Bull. 28, 808–810.

Shoara, R., Hashempur, M.H., Ashraf, A., et al., 2015. Efficacy and safety of topical *Matricaria chamomilla* L. (chamomile) oil for knee osteoarthritis: a randomized controlled clinical trial. Complement. Ther. Clin. Pract. 21, 181–187.

Singh, O., Khanam, Z., Misra, N., et al., 2011. Chamomile (*Matricaria chamomilla L.*): an overview. Pharmacogn. Rev. 5, 82–95.

Soliman, K.M., Badeaa, R.I., 2002. Effect of oil extracted from some medicinal plants on different mycotoxigenic fungi. Food Chem. Toxicol. 40, 1669–1675.

Srivastava, J., Pandey, M., Gupta, S., 2009. Chamomile, a novel and selective COX-2 inhibitor with anti-inflammatory activity. Life Sci. 85, 663–669.

Srivastava, J.K., Shankar, E., Gupta, S., 2010. Chamomile: a herbal medicine of the past with bright future. Mol. Med. Rep. 3, 895–901.

Subiza, J., Subiza, J.L., Hinojosa, M., et al., 1989. Anaphylactic reaction after the ingestion of chamomile tea: a study of cross-reactivity with other composite pollens. J. Allergy Clin. Immunol. 84, 353–358.

Subiza, J., Subiza, J.L., Alonso, M., et al., 1990. Allergic conjunctivitis to chamomile tea. Ann. Allergy 65, 127–132.

Sung, J., Lee, J., 2015. Anti-inflammatory activity of butein and luteolin through suppression of NFκB activation and induction of heme oxygenase-1. J. Med. Food 18, 557–564.

Szőke, E., Máday, E., Tyihák, E., et al., 2004. New terpenoids in cultivated and wild chamomile (*in vivo* and *in vitro*). J. Chromatogr. B Anal. Technol. Biomed. Life Sci. 800, 231–238.

Tirillini, B., Pagiotti, R., Menghini, L., et al., 2006. Esseential oil composition of ligulae and tubular flowers and receptacle from wild *Chamomilla rectita* (L.) Rausch. grown in Italy. J. Essent. Oil Res. 18, 42–45.

Tolouee, M., Alinezhad, S., Saberi, R., et al., 2010. Effect of *Matricaria chamomilla* L. flower essential oil on the growth and ultrastructure of *Aspergillus niger* van Tieghem. Int. J. Food Microbiol. 139, 127–133.

Tubaro, A., Zilli, C., Redaelli, C., et al., 1984. Evaluation of antiinflamatory activity of a chamomile extract after topical application. Planta Med. 50, 359.

Valenzuela, S., Pons-Fuster, A., López-Jornet, P., 2016. Effect of a 2% topical chamomile application for treating burning mouth syndrome: a controlled clinical trial. J. Oral Pathol. Med. 45, 528–533.

van Zyl, R.L., Seatlholo, S.T., van Vuuren, S.F., 2006. The biological activities of 20 nature identical essential oil constituents. J. Essent. Oil Res. 18, 129–133.

Viola, H., Wasowski, G., de Stein, M.L., et al., 1995. Apigenin, a component of *Matricaria recutita* flowers, is a central benzodiazepine receptors-ligand with anxiolytic effects. Planta Med. 61, 213–216.

Waleczek, K.J., Cabral Marques, H.M., Hempel, B., et al., 2003. Phase solubility studies of pure (−)-α-bisabolol and camomile essential oil with β-cyclodextrin. Eur. J. Pharm. Biopharm. 55, 247–251.

Wang, Y., Tang, H., Nicholson, J.K., et al., 2005. A metabonomic strategy for the detection of the metabolic effects of chamomile (*Matricaria recutita* L.) ingestion. J. Agric. Food Chem. 53, 191–196.

Weng, Z., Zhang, B., Asadi, S., et al., 2012. Quercetin is more effective than cromolyn in blocking human mast cell cytokine release and inhibits contact dermatitis and photosensitivity in humans. PLoS One 7, e33805.

WHO, 1999. (World Health Organization) WHO monographs on selected medicinal plants, vol. 1, pp. 86-94, <https://apps.who.int/medicinedocs/pdf/s2200e/s2200e.pdf> (Accessed 19 March 2020).

Wilkinson, S.M., Hausen, B.M., 1995. Allergic contact dermatitis from plant extracts in a cosmetic. Contact Dermatitis 33, 58–59.

Xiong, J., Wang, K., Yuan, C., et al., 2017. Luteolin protects mice from severe acute pancreatitis by exerting HO-1-mediated anti-inflammatory and antioxidant effects. Int. J. Mol. Med. 39, 113–125.

Zadeh, J.B., Kor, N.M., Kor, Z.M., 2014. Chamomile (*Matricaria recutita*) as a valuable medicinal plant. Int. J. Adv. Biol. Biomed. Res. 2, 823–829.

Zafar, S., Najam, Y., Arif, Z., et al., 2016. A randomized controlled trial comparing pentazocine and *Chamomilla recutita* for labor pain relief. Homeopathy 105, 66–70.

Zick, S.M., Wright, B.D., Sen, A., et al., 2011. Preliminary examination of the efficacy and safety of a standardized chamomile extract for chronic primary insomnia: a randomized placebo controlled pilot study. BMC Complement. Altern. Med. 11, 78.

Zidorn, C., Dirsch, V.M., Rüngeler, P., et al., 1999. Anti-inflammatory activities of hypocretenolides from *Leontodon hispidus*. Planta Med. 65, 704–708.

Chapter 45

Isoflavones: toxicological aspects and efficacy

Harald L. Esch and Leane Lehmann

Introduction

Isoflavones are part of the human diet all over the world. In Asian countries, the traditional cuisine is rich in soy products that contain naturally high concentrations of isoflavones, whereas in Western countries soy is, in general, less frequently consumed.

In Eastern Asia, coronary heart disease (Zhao et al., 2017; Sacks et al., 2006) as well as breast and prostate cancer (Ferlay et al., 2020) are less frequent than in Western countries, which has been associated with the consumption of soy food. After a health claim for foods containing soy protein as protective against coronary heart disease was approved by the US Food and Drug Administration in 1999, consumption of soy gained much more attention (Xiao, 2008). In addition, isolated isoflavones are promoted now in the form of dietary supplements as relief from menopausal symptoms. However, apart from putative beneficial health effects, the consumption of high doses of isoflavones has also raised concerns about their safety. Up to now, a plethora of studies in cell-free systems, cultured mammalian cells, and experimental animals as well as epidemiological studies have investigated a multitude of biological effects (beneficial and adverse) of isolated isoflavones and also extracts and food products containing isoflavones. This chapter summarizes the occurrence and dietary intake, biokinetics, biotransformation, toxicity, carcinogenicity, and efficacy of isoflavones, focusing on key studies in humans supplemented by selected studies in experimental animals and in vitro.

Occurrence and dietary intake of isoflavones

Isoflavones are polyphenolic secondary plant metabolites that are produced primarily from members of the *Leguminosae*. Biosynthesis of isoflavones is part of the phenyl-propanoid pathway by which flavonoids are synthesized. However, isoflavone synthase, an enzyme involved in the conversion of flavone precursors to isoflavones, is expressed in legumes and only a few other species (Veitch, 2013; Barnes, 2010). Over the past 25 years, hundreds of naturally occurring isoflavones have been isolated and identified. All isoflavones share the 3-phenylchromen-4-one backbone (Table 45.1), which is always modified, mainly by *O*-substituents, glycosides, and/or prenylated derivatives (Veitch, 2013, and previous reviews of this series).

In plants consumed as part of the human diet (including dietary supplements), the highest concentrations of isoflavones have been observed in soy (*Glycine max L.*), red clover (*Trifolium pratense*), and kudzu (*Pueraria lobata*) (Mortensen et al., 2009). Soybeans and sprouts are widely consumed as fermented (miso, tempeh, natto, soy sauce, etc.) or nonfermented (tofu, soymilk, soybean sprouts, etc.) products (Barnes, 2010). In addition, isoflavone-rich extracts of soy, red clover, and kudzu root are used in dietary supplements.

The isoflavones in these sources occur as β-glucosides (i.e., conjugated to glucose), aglycones (i.e., without glucose residue), and cyclites (Table 45.1). In β-glucosides present in food products, isoflavones are bound to glucose via a β-glycosidic bond, preferentially with the hydroxyl group at position C7 (7-*O*-glucoside) (Mortensen et al., 2009). In cyclites, misleadingly called *C*-glycosides, the glucose is attached via a C—C bond to the aglycone. For example, in puerarin (daidzein-8-β-D-glucoside), the major isoflavone in kudzu root, C1 of glucose is attached to the carbon at position 8 (Table 45.1). In the plant, the glucose residue is further modified to malonyl-glucosides (Barnes, 2010; Mortensen et al., 2009). In unprocessed soybeans and red clover, 7-*O*-glucosides and 6-*O*-malonyl-7-*O*-glucosides predominate. Depending on the conditions during food processing, the malonyl residue decarboxylates, increasing the

Nutraceuticals. DOI: https://doi.org/10.1016/B978-0-12-821038-3.00045-8

TABLE 45.1 Chemical structure of the isoflavone backbone 3-phenylchromen-4-one and substitution pattern for isoflavones that occur in soy products as well as dietary supplements containing extracts of red clover and kudzu root (Lutter et al., 2014; Mortensen et al., 2009; Delmonte et al., 2006; Setchell et al., 2001)

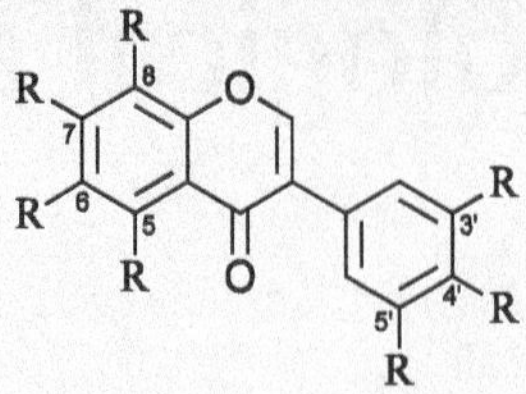

Isoflavone	R_5	R_6	R_7	R_8	$R_{3'}$	$R_{4'}$	$R_{5'}$
Biochanin A	OH	H	OH	H	H	OCH_3	H
Biochanin A-7-*O*-β-glucoside (Sissotrin)	OH	H	OGlu	H	H	OCH_3	H
Calycosin	H	H	OH	H	OH	OCH_3	H
Daidzein	H	H	OH	H	H	OH	H
Daidzein-7-*O*-β-glucoside (Daidzin)	H	H	OGlu	H	H	OH	H
Formononetin	H	H	OH	H	H	OCH_3	H
Formononetin-7-*O*-β-glucoside (Ononin)	H	H	OGlu	H	H	OCH_3	H
Genistein	OH	H	OH	H	H	OH	H
Genistein-7-*O*-β-glucoside (Genistin)	OH	H	OGlu	H	H	OH	H
Glycitein	H	OCH_3	OH	H	H	OH	H
Glycitein-7-*O*-β-glucoside (Glycitin)	H	OCH_3	OGlu	H	H	OH	H
Irilone	OH	$O–CH_2–O$		H	H	OH	H
Orobol	OH	H	OH	H	OH	OH	H
Pratensein	OH	H	OH	H	OH	OCH_3	H
Prunetin	OH	H	OCH_3	H	H	OH	H
Pseudobaptigenin	H	H	OH	H	H	$O–CH_2–O$	
Puerarin	H	H	OH	Glu	H	OH	H

percentage of acetyl-glucosides and unmodified β-glucosides (Barnes, 2010; Mortensen et al., 2009).

In general, the percentage of aglycones in nonfermented soy products (10%–40%) is lower than in fermented soy products (40%–100%; Nakamura et al., 2000). In addition, in fermented soy products, in particular soy sauce, the amount of total isoflavone equivalents can be reduced to less than 10% compared to nonfermented soy products (Nakamura et al., 2000), presumably by microbial degradation, oxidation, or derivatization. During fermentation, formation of dihydro- or 6- and 8-hydroxylated genistein and daidzein as well as their 7-tartaric acid ethers is observed (Barnes, 2010; Mortensen et al., 2009).

The main isoflavone aglycones in food products derived from soy and red clover are genistein, daidzein, glycitein, formononetin, biochanin A, and irilone, respectively (Lutter et al., 2014; Mortensen et al., 2009; Setchell et al., 2001). In kudzu root, daidzein and puerarin prevail (Mortensen et al., 2009). Minor isoflavone constituents are calycosin, pratensein, pseudobaptigenin, and prunetin (Table 45.1; Delmonte et al., 2006; Setchell et al., 2001).

Generally, there is high variability in the isoflavone content of food products, which can be attributed to the plant raw materials, which are affected by cultivar, weather conditions, soil, and stage of maturity, and also to the processing and storage conditions (Barnes, 2010). In the case of dietary supplements, the use of different plant parts for extract preparation contributes further to the observed high variability of isoflavone contents and profiles. For example, soy protein extracted from whole soybeans contains approximately equal amounts of total genistein and total daidzein equivalents (molar sum of glucosides and aglycone), whereas in extracts from soy germ, daidzein and glycitein equivalents exceed genistein equivalents by approximately fourfold (Barnes, 2010; Mortensen et al., 2009; Setchell et al., 2001; Nakamura et al., 2000).

Because the isoflavone profiles and contents of different (soy) products within the same product category are highly variable, the estimation of human exposure based on food questionnaires combined with databases of isoflavone content in different food products and dietary supplements is error-prone. Furthermore, soy food contains nutrients (polyunsaturated fats, fiber, vitamins, and minerals) and antinutrients (lectins, saponins, oxalates, enzyme inhibitors, and phytic acid) that may affect biokinetics of isoflavones (D'Adamo and Sahin, 2014; Mortensen et al., 2009; Nielsen and Williamson, 2007) and have biological effects of their own (D'Adamo and Sahin, 2014). As a consequence, chemical analysis of the intervention product with respect to the isoflavone profile and also other constituents of the food matrix is highly recommended for intervention trials (Mortensen et al., 2009), but is usually lacking.

In Asian countries, the traditional diet consists mainly of fermented soy products, leading to an estimated daily intake of 16–70 mg isoflavones (predominately aglycones). And in contrast to Western dietary habits, exposure to isoflavones starts early in life (Hüser et al., 2018). In Western countries, fermented soy products are rarely part of the traditional diet and the overall estimated daily intake of isoflavones is less 3 mg. Only certain Western subpopulations reach higher daily isoflavone intakes; for example, vegetarians and infants exclusively fed soy protein-based formula reach up to 20 mg (Zamora-Ros et al., 2012; Mortensen et al., 2009) and up to 47 mg (McCarver et al., 2011), respectively. Nevertheless, in the Western population the lifetime exposure to isoflavones and exposure to individual isoflavones generally differs in comparison to the Asian population, based on their respective traditional diet (Hüser et al., 2018). In contrast, the estimated daily uptake of isoflavones from dietary supplements available in the European Union is up to 100 mg (soy-based supplements) or up to 160 mg (red clover-based supplements), yet extremely variable; for example, 0.1–100 mg/day for soy-based supplements (Hüser et al., 2018).

The resulting mean blood levels of total isoflavone (i.e., genistein and daidzein aglycone equivalents) range from approximately 10–50 nmol/L in the general Western population compared to approximately 150–2000 nmol/L in Japanese adults. Mean blood levels of Western vegetarians and vegans are higher (35–230 nmol/L), but still lower than those of the Japanese (McCarver et al., 2011). The Western subgroup with the highest mean plasma levels (3695 nmol/L, 75th percentile approximately 7400 nmol/L) is infants fed soy protein-based formula (McCarver et al., 2011; Mortensen et al., 2009).

Biokinetics of isoflavones

The biological activity depends primarily on the bioavailability or, more precisely, the amount of free isoflavone aglycones in the blood (i.e., unconjugated and not bound to plasma proteins) unless their metabolites show a higher efficacy. The absolute bioactivity of isoflavones has not been determined yet, but many studies have described the biokinetics of individual and mixed isoflavones in different matrices, which might provide the key to understanding the biological activity of food products and dietary supplements containing isoflavones.

Isoflavone glucosides are effectively hydrolyzed in the oral cavity (Allred et al., 2001) and small intestine (Walsh et al., 2007) by mammalian glycosidases such as lactase and bacterial glycosidases of the microbiota of the small intestine (Fig. 45.1; Larkin et al., 2008; Nielsen and Williamson, 2007), whereas acidic hydrolysis in the stomach most likely does not occur (Islam et al., 2014). The total hydrolysis efficacy from ingestion to the end of the small intestine has been investigated once in humans and accounted for more than 95% (Walsh et al., 2007). Approximately 1% of the glucosides may pass the intestinal brush border unchanged (Mortensen et al., 2009). However, no isoflavone glucosides have ever been detected in peripheral human blood (Larkin et al., 2008; Nielsen and Williamson, 2007). Isoflavone aglycones are rapidly absorbed by passive diffusion (Chen et al., 2005) in the small and large intestines (Fig. 45.1; Mortensen et al., 2009; Larkin et al., 2008). Within the enterocytes and the liver, isoflavone aglycones are subjected to an extensive first-pass effect. The hydroxyl groups of the isoflavone aglycones (Fig. 45.1) render them good substrates for conjugation (i.e., glucuronidation and sulfonation), resulting in hydrophilic glucuronides and sulfates, respectively. Conjugation of isoflavones is efficiently catalyzed by human UGT1A1, 1A6, 1A8, 1A9, and 1A10 and human SULT1A1, 1A2, 1A3, 1B1, 1E1, 1C2, 2B1 (Mortensen et al., 2009), respectively. The hydroxyl group at position C7 of isoflavones including the cyclite puerarin is preferred over that at C4′. Furthermore, diconjugates and sulfoglucuronides are formed (Hosoda et al., 2011; Mortensen et al., 2009).

Thus, a proportion of the isoflavone aglycones are already conjugated within the enterocytes (Larkin et al., 2008). Because of their hydrophilic nature, glucuronides and sulfates of isoflavones require transport proteins (Grosser et al., 2015; Mortensen et al., 2009) to pass through cell membranes. The isoflavone glucuronides formed in the enterocytes either are transported back into the small intestine or are released together with the isoflavone sulfates into the portal vein. Within the liver, isoflavone aglycones and conjugates enter hepatocytes by passive diffusion and transport proteins, respectively. Aglycones are subjected to phase I and/or II metabolism. Isoflavone conjugates are exported either into the bile, thus re-entering the small intestine (Fig. 45.1), or

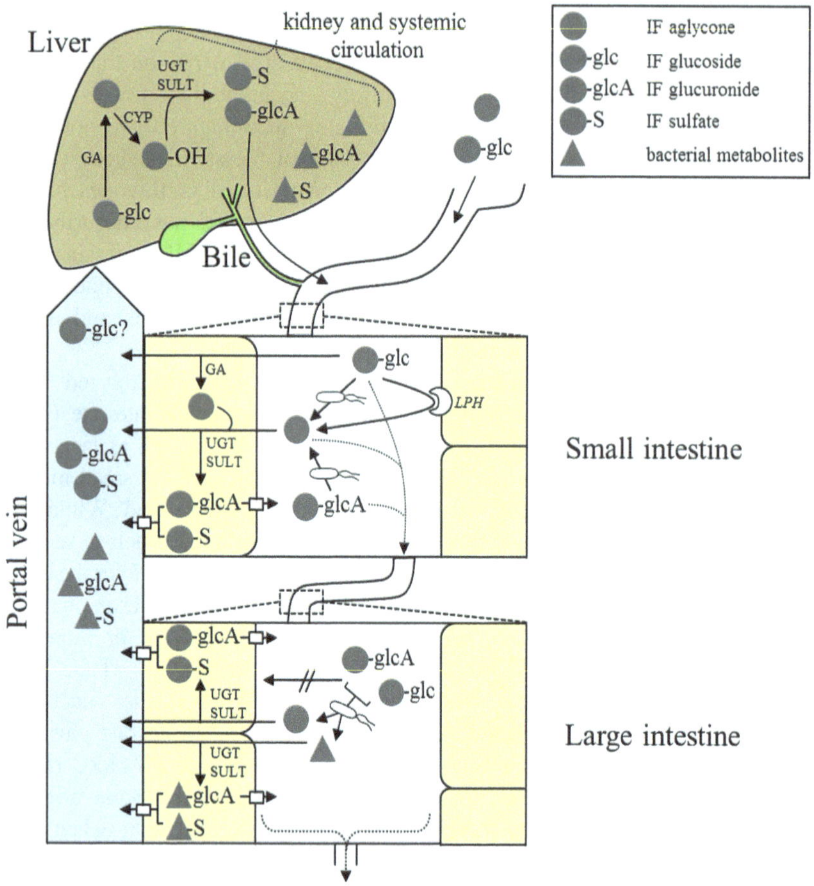

FIGURE 45.1 Absorption of dietary isoflavone (IF) aglycones and β-glucosides (not esterified) in the human gastrointestinal tract taking into consideration hydrolysis of glucosides in the small intestine by mammalian and bacterial glycosidases, fermentation in the large intestine, as well as enteric and enterohepatic circulation. For the sake of simplicity, the demethylation of methoxy-isoflavones is not depicted. Further details and references are provided in this section.

into the systemic circulation and eventually excreted via the kidney. In the small intestine, bacteria release isoflavone aglycones again from their glucuronides (Walsh et al., 2007) so that they can be absorbed again, completing enterohepatic circulation, or they migrate into the large intestine where they are metabolized by the obligate anaerobe bacteria as described below (Fig. 45.1), forming another set of potentially bioactive compounds (Mortensen et al., 2009; Larkin et al., 2008; Nielsen and Williamson, 2007), which are absorbed by passive diffusion and become subjected to phase I and II metabolism before reaching the systemic circulation (Fig. 45.1).

Consequently, isoflavones circulating in human peripheral blood are predominately conjugates (Soukup et al., 2016; Mortensen et al., 2009). For example, after consumption of kinako (baked soybean flour), genistein and daidzein sulfoglucuronides (7-*O*-glucuronide-4′-sulfate) accounted for more than 50% of the circulating species, followed by mono- and diglucuronides (Hosoda et al., 2011). More importantly, free unconjugated genistein and daidzein in peripheral blood account for less than 1%–4% (Soukup et al., 2016; Hosoda et al., 2011; Rüfer et al., 2008).

Moreover, a marked interspecies difference exists in occurrence and levels of unconjugated and conjugated genistein and daidzein in blood between humans, Wistar rats, and BALB/c mice exposed to the same isoflavone-rich soy extract (Soukup et al., 2016). Also, rodents and nonhuman primates mostly differ from humans with respect to the percentage of circulating unconjugated isoflavones (Soukup et al., 2016; Gu et al., 2006). Because formononetin, biochanin A, glycitein, and prunetin are similar or even better substrates for conjugation reactions (Chen et al., 2005), the same can be hypothesized for other isoflavones as well, despite the lack of appropriate studies. Even for the cyclite puerarin, its glucuronide is likely to be the main urinary metabolite (Hosoda et al., 2011).

Because of their lipophilicity, isoflavones are expected to bind to plasma proteins, further decreasing the availability of biologically active aglycones. Although human data are not available, one study in rats suggests that more than 80% of genistein may be bound to plasma proteins (Larkin et al., 2008; McCarver et al., 2011).

Thus, after consumption of isolated isoflavone aglycones and glucosides as well as soy products and red clover-based dietary supplements, genistein, daidzein, glycitein, biochanin A, and formononetin consumed or formed by hydrolysis of their glycosides are absorbed in the small intestine and the respective conjugates appear quickly in blood and urine, with their plasma concentrations peaking within 1–2 h after intake for the first time

(Mortensen et al., 2009; Larkin et al., 2008; Nielsen and Williamson, 2007). Thereafter, a second and higher peak in urine and blood is observed, reaching its maximum 5–9 h after isoflavone intake. Predominately enterohepatic circulation of the isoflavones and aglycone absorption in the large intestine contribute to this second peak (Mortensen et al., 2009; Nielsen and Williamson, 2007).

Accordingly, hydroxylation products of isoflavones formed by hepatic enzymes appear in the blood 2 h after consumption (Rüfer et al., 2008), whereas bacterial metabolites of the isoflavones are detected after 6–8 h (Larkin et al., 2008; Rüfer et al., 2008; Nielsen and Williamson, 2007). In contrast to *O*-glycosides, the cyclite puerarin has been detected in human blood and reaches maximum blood levels after 2 h and does not exhibit a biphasic absorption pattern (Franke et al., 2014). Release of its aglycone daidzein requires bacterial activity. Consequently, daidzein is observed in human blood not before 12 h after ingestion of puerarin (Franke et al., 2014).

The impact of esterified glucosides, in particular malonyl-glucosides, on the biokinetics of isoflavones in humans is unknown, but results obtained from rats suggest that higher plasma levels and total systemic absorption can be achieved by the consumption of malonyl-glucosides (Yonemoto-Yano et al., 2014). Also, information on the tissue levels of isoflavones in humans tissue is limited to blood, urine, female breast tissue, and male prostate fluid (Hüser et al., 2018; Mortensen et al., 2009; Larkin et al., 2008). Isoflavone levels and profiles in tissues might differ from that determined in plasma, but more studies are needed to draw a conclusive picture.

Urinary excretion is the main elimination route for isoflavones, recovering up to 40% (genistein) to 60% (daidzein, glycitein) of the ingested isoflavones (EFSA Panel on Food Additives and Nutrient Sources added to Food, 2015; Hosoda et al., 2011; Mortensen et al., 2009).

Typical fecal excretion rates account for less than 5% (genistein≥daidzein) of the ingested dose and, in contrast to urinary excretion, fecally excreted isoflavones are predominately aglycones (>90%) (Larkin et al., 2008). Moreover, both urinary and fecal excretion rates seem to not correlate with the ingested isoflavone dose but with the gut transit time (Nielsen and Williamson, 2007).

The total recovery rate (i.e., the sum of urinary and fecal excretion rates) accounts for only 10%–70% of the ingested dose of isoflavone aglycones (daidzein > genistein) but may be higher for isoflavone glucosides (Rüfer et al., 2008; Nielsen and Williamson, 2007). Interestingly, even when the usual bacterial and/or oxidative metabolites (see below, "Oxidative and Reductive/Bacterial Metabolism") are considered the total recovery still does not add up to 100% for ingestion of isolated aglycones (Rüfer et al., 2008) or soy products (Hosoda et al., 2011), indicating the absence of important metabolites in the usual panel of isoflavone metabolites quantified in human studies.

Biokinetic parameters vary considerably between studies (Nielsen and Williamson, 2007). Factors possibly contributing to this variation are: (1) composition of soy food matrix/isoflavone profile; (2) dietary habits; and (3) individual activities of enteral mammalian and bacterial enzymes due to genetic and lifestyle factors (Hüser et al., 2018; Mortensen et al., 2009; Larkin et al., 2008; Nielsen and Williamson, 2007). Depending on the grade of processing, soy-based food products are rich in malonylglucosides, glucosides, or aglycones. In human studies with isolated isoflavone aglycones and glucosides, higher maximum plasma concentrations and higher total systemic exposure were observed after ingestion of glucosides. However, plasma kinetics observed from individual isolated isoflavones may differ from that of isoflavone mixtures (Hosoda et al., 2011; Mortensen et al., 2009; Nielsen and Williamson, 2007). Moreover, aglycone-rich soy food yielded faster absorption of isoflavones than glucoside-rich soy food and liquid soy food provided higher plasma concentrations than solid soy food. In addition, the composition of the macronutrients accompanying the isoflavones in soy food is quite different and may impact the absorption pattern (Mortensen et al., 2009).

Oxidative hepatic and reductive bacterial metabolism

In the liver, isoflavone aglycones are subjected to hydroxylation catalyzed by cytochrome P450 (CYP450)-dependent monoxygenases (CYPs). Hydroxylation of isoflavone aglycones occurs predominantly in positions C6, C8, and C3′ (Fig. 45.2; Table 45.2) and hydroxylated metabolites of genistein, daidzein, formononetin, and the bacterial metabolite equol have been detected in human urine (Mortensen et al., 2009; Heinonen et al., 2004; Kulling et al., 2002). CYP1A1, 1A2, 1B1, 2E1, and 3A4 are the main isozymes contributing to isoflavone hydroxylation, with 3′-hydroxylated isoflavones being the preferred reaction products (McCarver et al., 2011). In addition, glycitein, formononetin, and biochanin A are also demethylated (Fig. 45.2; Table 45.2) by hepatic enzymes; in fact, demethylation seems to be the preferred reaction (Kulling et al., 2002). When hydroxylation or demethylation of isoflavones (e.g., daidzein or calycosin, respectively) leads to the formation of catechols (Fig. 45.2; Table 45.2), or when they already harbor catecholic hydroxyl groups (e.g., orobol), they become substrates for the catechol-*O*-methyltransferase (Kulling et al., 2002; Fig. 45.2; Table 45.2), avoiding further oxidation to quinones, which could otherwise react with cellular nucleophiles such as proteins and DNA.

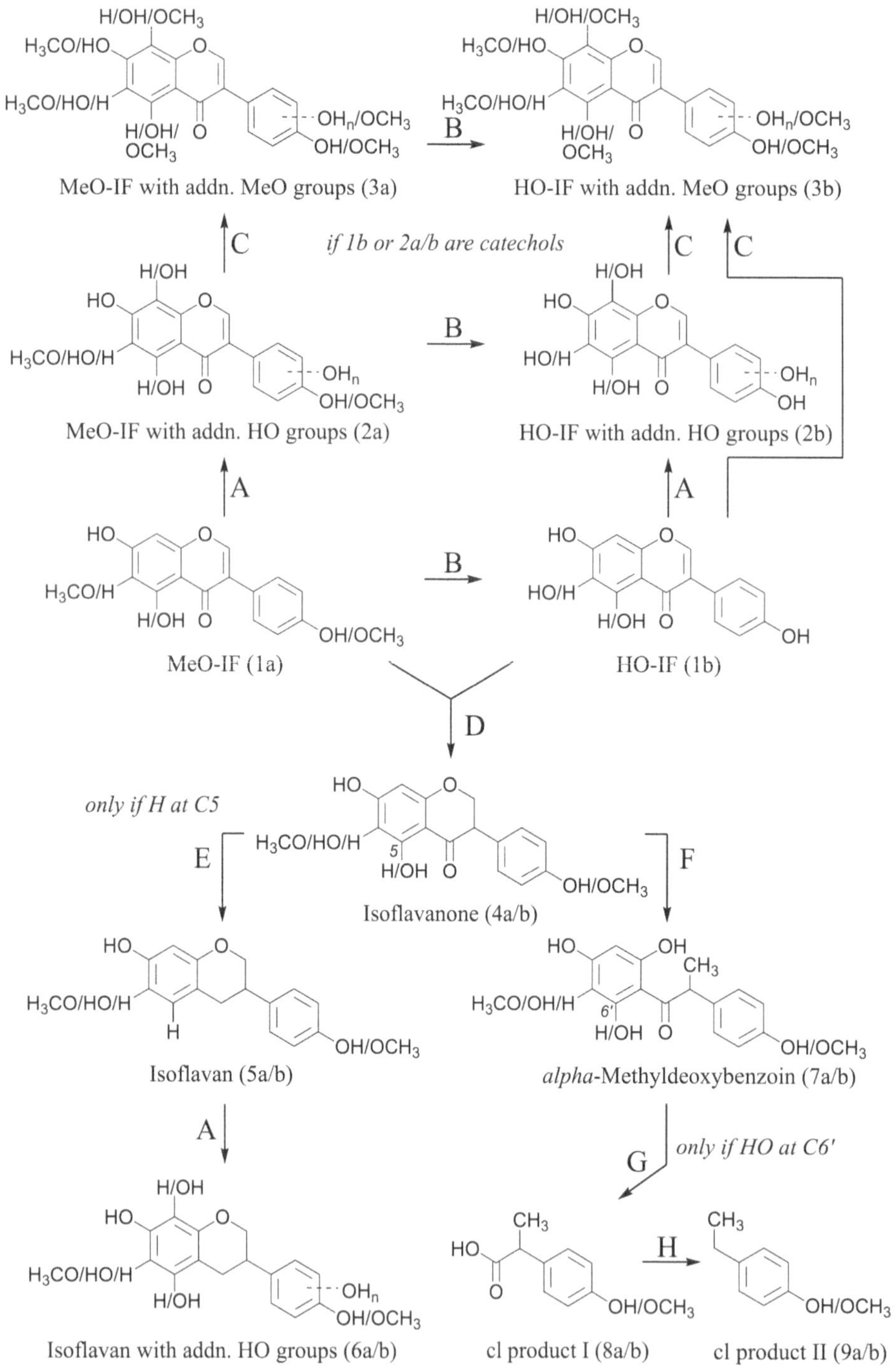

FIGURE 45.2 Oxidative and bacterial metabolites of the main dietary isoflavone aglycones (genistein, daidzein, biochanin A, formononetin). All reactions demonstrated to apply to the respective isoflavones are shown (Heinonen et al., 2004; Mortensen et al., 2009). Numbers (1–9) represent the chemical structures formed by the following metabolic reactions of aglycones with hydroxyl and methoxy groups (labeled with letter a) and isoflavones with hydroxyl groups only (labeled with letter b), respectively: hydroxylation (A); demethylation (B); methylation (C); reduction (D); deoxygenation (E); ring opening (F); ring cleavage (G); and decarboxylation (H). *addn*, additional; *cl product*, cleavage product.

The importance of the intestinal microbiota to the biokinetics of isoflavones is well established and may even impact the health effects related to soy consumption (Franke et al., 2014; Nagata, 2010; Setchell and Clerici, 2010; Mortensen et al., 2009; Larkin et al., 2008).

When unabsorbed isoflavone aglycones (Fig. 45.2; Table 45.2) reach the large intestine, the obligate anaerobic microbiota creates a whole new set of compounds by demethylation, reductions, and ring opening and ring cleavage of the heterocyclic B-ring (Mortensen et al., 2009; Heinonen et al., 2004; Fig. 45.2; Table 45.2).

However, the extent of bacterial metabolism is quite different between the isoflavones. Methoxy-isoflavones from red clover-based dietary supplements, such as formononetin and biochanin A (Table 45.1), are preferably demethylated (bacterial as well as hepatic; Fig. 45.1), resulting in high circulating blood levels of daidzein and genistein, respectively. In contrast, genistein and daidzein are fairly well metabolized, and

TABLE 45.2 Isoflavone derivatives generated by mammalian and bacterial metabolism after consumption of the main dietary isoflavone aglycones genistein (GEN), daidzein (DAI), biochanin A (BCA), and formononetin (FOR).

IF #		Isoflavone (metabolite)							
1	a	FOR	FOR	FOR	FOR	GLY	BCA	BCA	BCA
	b	DAI	DAI	DAI	DAI	6-HO-DAI	GEN	GEN	GEN
2	a		CAL	8-HO-FOR	6-HO-FOR	n.a.		3′-HO-BCA	6/8-HO-BCA
	b		3′-HO-DAI	8-HO-DAI	6-HO-DAI	?		3′-HO-GEN	*
3	a		n.a.	–	–	?		n.a.	?
	b		–	–	GLY	?		–	?
4	a	dhFOR	?	?	?	dhGLY	dhBCA	?	?
	b	dhDAI	3′-HO-dhDAI	8-HO-dhDAI	6-HO-dhDAI	6-HO-dhDAI	dhGEN	?	?
5	a	4′-OMe-EQU	?	?	?	6-MeO-EQU	n.a.	n.a.	n.a.
	b	EQU	3′-HO-EQU	8-HO-EQU	6-HO-EQU	6-HO-EQU	n.a.	n.a.	n.a.
6	a	?	?	?	?	?	n.a.	n.a.	n.a.
	b	HO-EQUs	–	?	?	?	n.a.	n.a.	n.a.
7	a	ANG	?	?	?	5′-MeO-ODMA	6′-HO-ANG	?	?
	b	ODMA	3″-HO-ODMA	3′-HO-ODMA	5′-HO-ODMA	5′-HO-ODMA	6′-HO-ODMA	?	?
8	a	n.a.	n.a.	n.a.	n.a.	n.a.	4-OMe-Ph-2-PA	?	?
	b	n.a.	n.a.	n.a.	n.a.	n.a.	4-HO-Ph-2-PA	?	?
9	a	n.a.	n.a.	n.a.	n.a.	n.a.	4-Et-A	?	?
	b	n.a.	n.a.	n.a.	n.a.	n.a.	4-Et-Ph	?	?

Isoflavone (metabolite) numbers (IF# 1–9) correspond to the chemical structures formed by the metabolic transformation of aglycones with hydroxyl and methoxy groups (labeled with letter a) and isoflavones with hydroxyl groups only (labeled with letter b), respectively, depicted in Table 45.1. Each column signifies one plausible series of reactions according to Fig. 45.2. White cells, reaction not considered in the respective series of reactions; *n.a.*, not applicable (reaction not possible); –, reaction product is not observed;?, possible reaction but formation has not been investigated yet. *MeO*, methoxy; *HO*, hydroxy; *CAL*, calycosin; *GLY*, glycitein; *EQU*, equol; *dh*, dihydro; *ODMA*, *O*-desmethylangolensin; *PA*, propionic acid; *Ph*, phenol/phenyl; *A*, anisol; *Et*, ethyl; *ANG*, angolensin; *, multiple-hydroxylated metabolites of GEN were detected. According to the multitude of possible reactions, ingestion of isolated formononetin may cause the formation of up to 23 human and bacterial metabolites, the majority of which can be additionally converted into glucuronides and/or sulfates, further increasing the spectrum of analytes putatively to be considered in human and animal studies.

their bacterial transformation products (Fig. 45.2; Table 45.2) may reach up to 25% of the total urinary isoflavone derivatives (Rüfer et al., 2008). Ring cleavage observed for genistein has been proposed for all isoflavones with a hydroxyl group at position C5 (Fig. 45.2; Table 45.2); however, irilone (containing a methylenedioxy group next to the hydroxyl group at C5; Table 45.1) is fairly resistant to microbial degradation (Braune et al., 2010). Although irilone accounts for less than 20% in red clover-based dietary supplements (Lutter et al., 2014), it turns out to be the most abundant isoflavone in human plasma together with daidzein (Maul and Kulling, 2010). This may increase the biological impact of irilone in humans even though its estrogenicity in vitro is low compared to that of genistein (Table 45.2; Lutter et al., 2014).

After consumption of soy food, dihydrodaidzein, *O*-desmethylangolensin, *S*-(−)equol, dihydrogenistein, and hydroxy-*O*-desmethylangolensin (Fig. 45.2; Table 45.2) are the most abundant bacterial metabolites in human blood and urine (Franke et al., 2014; Mortensen et al., 2009). Absorption of the products resulting from C-ring cleavage, such as 2-(4-hydroxyphenyl)-propanoic acid and ethyl phenol, has not yet been investigated in human trials (Mortensen et al., 2009).

Among the bacterial metabolites, *S*-(−)equol has received most scientific attention because it remains longer in the human body and has a higher binding affinity to estrogen receptor β than to estrogen receptor α (see "Estrogenicity") than other isoflavone aglycones (Setchell and Clerici, 2010). Not all humans are able to produce equol; in fact, the percentage of equol producers among Eastern Asians (50%–60%) is approximately twice as high as in Western countries (25%–30%). However, among Western formula-fed 3-year-old infants (Brown et al., 2014) and vegetarians (Setchell and Clerici, 2010), the percentage of equol producers might be as high as in Eastern Asians. In contrast, production of *O*-desmethylangolensin from daidzein is observed in 80%–90% of the study populations (Setchell and Clerici, 2010; Mortensen et al., 2009). Based on intervention studies with soy food, Setchell et al. (2001) formulated the equol hypothesis, proposing that "the efficiency with which adults convert daidzein to equol when consuming diets containing soy foods could enhance the clinical effectiveness of soy-based diets" (Setchell and Clerici, 2010). Also, the reduced risk of breast and prostate cancer observed in Asians and the higher percentage of equol producers has been suggested to be associated with the ability to produce equol (Maskarinec, 2013; Setchell and Clerici, 2010); however, the evidence gathered so far is insufficient to draw firm conclusions.

The reason for the high percentage of equol producers among Eastern Asians, Western children, and vegetarians (Brown et al., 2014; Degen et al., 2011; Mortensen et al., 2009) is not known. Although daidzein is required as a precursor for equol production, long-term soy food consumption seems to not make an equol producer (Setchell and Clerici, 2010).

In general, the equol producer status is considered a stable phenomenon (Franke et al., 2014; Setchell and Clerici, 2010), that is, an equol producer is always an equol producer and vice versa. However, in at least seven soy intervention and nonintervention studies, a change in individual equol producer status has been observed in both ways (Franke et al., 2014). This change observed in men, women, Westerners, Eastern Asians, adults, and children (Franke et al., 2014; Degen et al., 2011) was not related to the use of antibiotics (Franke et al., 2014). The highest percentage (up to 35%) of individuals changing their equol producer status was observed in postmenopausal women (Franke et al., 2014) and in children (Degen et al., 2011).

Only specific intestinal bacteria are capable of equol formation. Most of them belong to the genus *Bifidobacterium* (22 strains), *Eggerthella* (three strains), *Slackia* (three strains), *Escherichia* (two strains), and *Lactobacillus* (two strains) (Setchell and Clerici, 2010). The adult gut microbiota differs between the continents; for example, European and North American microbiota is significantly different from Eastern Asian. Nevertheless, the composition of the intestinal microbiota has been considered as stable from early childhood (~3 years) throughout adulthood. The cause for this regional difference is not quite clear, but dietary habits may play a significant role (Brown et al., 2014). Also, the impact of the diet on the composition of the colonic microbial community has become more obvious, both in the short term and long term (Maukonen and Saarela, 2015; Flint et al., 2012). Observations regarding the equol producer status are in line with these general considerations, such as the extraordinary percentage of equol producers (59%) among Western vegetarians (Setchell and Clerici, 2010) and the observation that 50% of formula-fed 3-year-old infants were equol producers compared to only 25% of breastfed infants (Maukonen and Saarela, 2015). Nevertheless, the part of the human diet—if any at all—that contributes to equol producer status is not well understood.

Notably, experimental rodents and nonhuman primates such as cynomolgus macaques (*Macaca fascicularis*) are 100% equol producers (Soukup et al., 2016; Cline and Wood, 2009; Gu et al., 2006).

Isoflavones as endocrine disruptors

According to the WHO definition "an endocrine disruptor is an exogenous substance or mixture that alters the function(s) of the endocrine system and consequently causes adverse effects in an intact organism, or its progeny, or (sub) populations." Alterations of the function of the endocrine system may arise from interaction with hormone receptors, changes in circulating levels of the hormone, and from the impact of chemical(s) on hormone synthesis, transport, metabolism and other factors (Solecki et al., 2017).

Estrogenicity

The first indication of the estrogenicity of isoflavones was observed in sheep and zoo animals in which isoflavone-rich red clover caused reproductive failure (*clover disease*), related to an estrogenic mode of action (Setchell and Clerici, 2010). Decades later, isoflavones and their bacterial metabolite equol were identified as a cause and their estrogenicity was further characterized by estrogen receptor binding in cell-free systems and activation of estrogen receptor in cultured human cell lines, and also in laboratory animals (rodents and nonhuman primates) and humans.

Mechanistic studies on the estrogenic activity of isoflavones

Isoflavones are structurally similar to the endogenous female sex hormone 17β-estradiol, which exerts its biological effects by binding to estrogen receptors (and activation of other transcription factors), thus altering the transcription of a multitude of estrogen-regulated genes.

The direct interaction of isoflavones and their hydroxylated metabolites with both estrogen receptors (i.e., estrogen receptors α and β, the latter existing in five different splice variants) has been investigated for many isoflavones at all possible levels: (1) receptor binding (e.g., Lehmann et al., 2005; Kuiper et al., 1998); (2) binding of isoflavone-activated estrogen receptors to estrogen responsive elements (e.g., Kostelac et al., 2003); (3) recruitment of coactivators and corepressors (e.g., Liu et al., 2003); and (4) modulation of transcription of estrogen receptor target genes in cultured mammalian cells detected at mRNA, protein, or enzyme activity level as well as phenotypic changes such as stimulation of proliferation in steroid-depleted medium (see Table 45.3 for a summary of isoflavones modulating estrogen receptor-dependent enzyme activity and stimulation of proliferation).

Compared to 17β-estradiol and synthetic estrogens used in oral contraceptives and hormone replacement therapies, the affinity of isoflavones to the estrogen receptors is low (<30%) but stretches over two orders of magnitude (Mortensen et al., 2009). Moreover, the binding affinity of isoflavones is generally higher to estrogen receptor β compared with estrogen receptor α (Mortensen et al., 2009; Kuiper et al., 1998).

In cellular assays, the estrogenic potential of isoflavones varies greatly (Table 45.3), with some of them reaching the maximum stimulation achieved by 17β-estradiol, for example, maximum activity of alkaline phosphatase in cultured human endometrial adenocarcinoma Ishikawa cells, and others not. Although different quantitative results of tests performed in different laboratories render comparison of the estrogenic activity of isoflavones difficult, among all the isoflavones relevant to human nutrition, the concentration necessary for estrogenic stimulation is lowest for genistein (Table 45.3).

Isoflavone aglycones are generally considered the biologically active species. However, recent observations in cultured mammalian cells suggest that conjugates (sulfates, glucuronides, and/or sulfoglucuronides) of genistein, daidzein, and equol may activate estrogen receptors, especially estrogen receptor β (Hüser et al., 2018). Yet, it has not been shown whether the estrogenic activity is elicited by the conjugates themselves or if intra- or extracellular hydrolysis of the conjugates and release of the respective aglycones occurs prior to binding to the estrogen receptor (Hüser et al., 2018).

Most isoflavones are able to transactivate both estrogen receptors, but some isoflavones, especially genistein and equol, exhibit a higher affinity to estrogen receptor β than to estrogen receptor α, and thus they are considered to act primarily through estrogen receptor β (Mortensen et al., 2009). However, in cultured human breast adenocarcinoma MCF-7 cells, the preference of genistein to estrogen receptor β was observed, particularly at low nanomolar concentrations, whereas at high concentrations activation of estrogen receptor α occurred (Chang et al., 2008). Because of the tissue-specific distribution of estrogen receptors α and β as well as of coactivators/corepressors, genistein in particular is considered a selective

TABLE 45.3 Estrogenic activity of isoflavone aglycones in two cultured human cell lines in relation to the endogenous estrogen, 17β-estradiol (E2).

Isoflavone	MCF-7		Ishikawa	
	RBM %	RSP %	RBM %	RSP %
Genistein	0.00070–0.10000	72–90	0.0500–10	53–100
Daidzein	0.00300–0.01000	47–108	0.0050–1.0	45–102
Biochanin A	0.01000–0.05000	57–100	0.0050	80
Equol	0.01000	85–100	0.0100–0.0500	53–100
Prunetin	0.00400	100	0.0050	45
Formononetin	0.00200	100	0.0010	90
Irilone	0.00200	56	0.0050	41
Pseudobaptigenin[a]	0.00100	94	0.0050	36
Orobol[a]	0.00100	80	0.0050	40
Pratensein[a]	0.00050	63	0.0025	23
Calycosin[a]	0.00020	91	<0.0005	n.a.
O-Desmethylangolensin	0.00100–1.00000	50–100	n.d.a.	n.d.a.
4-Ethylphenol	0.00004	40	<0.0002	n.a.
Glycitein	<0.00002–0.0002	87	<0.0002	n.a.
Dihydrodaidzein	<0.00001	n.d.a.	n.d.a.	n.d.a.

RBM, relative benchmark (100 × benchmark concentration, BC, of E2/BC of isoflavone); *RSP*, relative stimulatory potential (100 × maximum induction by IF/maximum induction by E2). Only studies containing all necessary information for comparison (e.g., benchmark concentration for E2) have been included. *n.d.a.*, no data available; *n.a.*, not applicable. Data summarized in Lutter et al. (2014) if not indicated otherwise.
[a]*Unpublished data of the authors.*

estrogen receptor modulator, exerting estrogenic (e.g., proliferative) effects only in estrogen receptor α-rich tissues (Cederroth et al., 2012; North American Menopause Society, 2011; Setchell and Clerici, 2010; Mortensen et al., 2009; Larkin et al., 2008).

For selective estrogen receptor modulators, bell-shaped dose–response curves are often observed in vitro and are difficult to interpret when cytotoxic events are present at high concentrations (Vandenberg et al., 2012). However, in experimental animals, opposing effects are observed at different doses as well (Blei et al., 2015).

Effect of isoflavones on levels of endogenous estrogens

In addition to binding and activation of the estrogen receptors, genistein has an impact on endogenous biosynthesis of male and female sex hormones in vitro (Nielsen et al., 2012). The rate-limiting step of endogenous estrogen biosynthesis is the conversion of androgens (testosterone, androstenedione) to estrogens (17β-estradiol, estrone) by aromatase and studies in cultured mammalian cells indicate that isoflavones; for example, genistein, biochanin A, and daidzein alone and in combination, decrease transcript levels and activity of aromatase (Ronis, 2016; EFSA Panel on Food Additives and Nutrient Sources added to Food, 2015; Rice et al., 2006). Conjugation of 17β-estradiol, especially at the phenolic hydroxy group, results in the formation of more hydrophilic sulfates and glucuronides, which facilitates excretion and eliminates binding to the estrogen receptors. Genistein and equol may inhibit whereas daidzein may stimulate enzymes involved in estrogen conjugation (EFSA Panel on Food Additives and Nutrient Sources added to Food, 2015). Yet, to which extent this mode of action applies to the in vivo situation is not extensively investigated.

In nonhuman primates cynomolgus macaques (*Macaca fascicularis*), whose reproductive and mammary gland physiology is much closer to humans than that of rodents, no estrogenic effect in breast and endometrium was observed at high dietary genistein and daidzein exposures (500 mg/day) in surgical postmenopausal females, whereas at lower exposure (120 mg/day) soy isoflavones reduced estrogen-induced proliferative responses of the breast and endometrium. The reduced proliferative responses were accompanied by reduced circulating estrogen levels, altered urinary excretion of estrone metabolites, and reduced levels of intramammary 17β-estradiol, yet intramammary transcript levels of genes coding for estradiol-producing and estradiol-conjugating enzymes were not affected by dietary isoflavones (Cline and Wood, 2009). In contrast, approximately 75% of the available human intervention trials report no impact of isoflavone consumption on circulating levels of estrogen in premenopausal and postmenopausal women (Fritz et al., 2013). A systematic review and meta-analysis revealed no effect of isoflavones and soy consumption on blood levels of 17β-estradiol, estrone, or their transport protein sex hormone-binding globulin in premenopausal and postmenopausal women, although in postmenopausal women a small increase in total 17β-estradiol (i.e., free and bound to sex hormone-binding globulin) with isoflavone consumption barely missed significance (Hooper et al., 2009).

Isoflavones and the thyroid

The thyroid produces and stores the hormones thyroxine (T4) and triiodothyronine (T3). T3 is the biologically active hormone and is produced enzymatically by deiodination of T4, primarily in the target cells. T4 is the transport form. Thyroid hormones control the growth and development in offspring as well as metabolic activity (e.g., for thermoregulation) in adults and children (National Research Council, 2005; Hüser et al., 2018). The production of thyroid hormones involves iodine uptake, enzymatic iodination of tyrosine, and coupling of iodinated tyrosine by thyroid peroxidase to yield T4 and T3. The function of the thyroid is controlled by the thyroid-stimulating hormone (TSH) produced by the pituitary gland.

The distribution of thyroid hormones by binding to plasma proteins is different in humans and rodents. In humans, T4 binds primarily to thyroxine-binding globulin which is not present in rats. In rats, T4 binds to albumin or transthyretin, which has a much lower (less than 1/100) binding affinity to T4 than thyroxine-binding globulin. Thus, T4 is more rapidly cleared/excreted from rats, resulting in a higher T4 production rate in rats compared to humans (National Research Council, 2005). As a consequence, studies in rats showing isoflavone-induced alterations in circulating thyroid hormones are considered less relevant for human risk assessment (Hüser et al., 2018; EFSA Panel on Food Additives and Nutrient Sources added to Food, 2015).

However, genistein and daidzein have been shown to inhibit competitively isolated thyroid peroxidase (Hüser et al., 2018) and monoiodinated genistein and daidzein has been detected in the urine of postmenopausal women after dietary isoflavone supplementation (2 × 40 mg/day) for 3 months (Sosvorova et al., 2012). Furthermore, genistein inhibits human deiodinase, catalyzing the intracellular conversion of T4 to T3 in vitro (Hüser et al., 2018). In adult female cynomolgus monkeys (*Macaca fascicularis*; nonhuman primates) fed isoflavone-containing soy protein for 34 months, increased serum T3 was associated with serum genistein and total serum isoflavones (Hüser et al., 2018; Silverstein et al., 2014). In healthy humans, however, isoflavone consumption (in food or supplements) for up to 3 years had no impact on blood levels of thyroid hormones (Hüser et al., 2018; EFSA Panel on Food Additives and Nutrient Sources added to Food, 2015; D'Adamo and Sahin, 2014). Yet, in individuals with subclinical hypothyroidism and adequate iodine intake, isoflavones (in soy protein powder) might favor the development of overt hypothyroidism (Hüser et al., 2018).

Isoflavones and reproduction and development

Isoflavones are estrogens, thus there is concern that they might have an adverse effect on endogenous estrogen (endocrine or paracrine) signaling in the female and male reproductive system. Numerous studies in experimental rodents suggest that neonatal exposure to soy extract, genistein, or genistein glucoside affects the female reproductive system by altering ovarian development, timing of vaginal opening, prolonged length of estrous cycle, subfertility, and others (Cederroth et al., 2012). A meta-analysis of human studies reported a significant positive association between the length of the menstrual cycle (by 1 day) and isoflavone consumption. This association was accompanied by a significant reduction of circulating follicle-stimulating hormone and luteinizing hormone by approximately 20% (Hooper et al., 2009). Thus, there is a potential hazard that suggests a risk for long-term effects on female fertility.

In the United States, an estimated 18%–25% of infants consume soy protein-based formula continuously or incidentally during their first year of life (McCarver et al., 2011). Even though American and European pediatric societies do not recommend the exclusive use of soy protein-based formulas, many consumers' clinicians consider it a safe alternative to breastfeeding (Cederroth et al., 2012). In infants consuming soy protein-based infant formulas, mean daily intake ranges between 20 and 47 mg/day, yielding mean blood levels of 3695 nmol/L, which is more than 70-fold higher than mean blood levels in Western adults (see "Occurrence and Dietary Intake of Isoflavones").

There are a few studies (clinical and observational) investigating infants exposed to soy protein-based formula that suggest that physiologic estrogenization of the newborn may be prolonged, because breast tissue was more prevalent and an estrogenized vaginal epithelium was observed. Furthermore, the long-term effects of soy protein exposure during infancy might lead to an increased risk of developing benign smooth muscle tumors of the uterus (Cederroth et al., 2012).

Ambiguous observations were made in male rodents exposed to isoflavones in utero and neonatally. No effects on reproduction as well as decreased testicular size and weight, spermatogenesis, smaller anogenital distance, and circulating hormone levels were observed. Exposure of adult male rodents to soy and isoflavones did not affect male fertility in the long term. However, male rodents exposed to soy isoflavones from conception to adulthood may be more susceptible to a decrease in sperm production and corresponding decrease in litter size (Cederroth et al., 2012).

There is a limited number of studies investigating the impact of isoflavones on semen quality. Unfortunately, some lack adequate exposure estimates of isoflavones and/or exhibit discrepancies in the experimental designs and/or small number of participants. The more reliable ones observed no adverse effect (Cederroth et al., 2012). Furthermore, a meta-analysis revealed that levels of free testosterone in blood are not affected by isoflavone consumption (Hamilton-Reeves et al., 2010).

In the safety assessment on the developmental toxicity of soy infant formula by the US National Toxicology Program (NTP) and the US National Institute of Environmental Health Sciences, the expert panel expressed "minimal concern for adverse developmental effects in infants fed soy infant formula. This level of concern represents a '2' on the five-level scale of concern." However, the expert panel also addressed the limited scientific insight provided by the few available human studies, especially to the long-term effects of infant isoflavone exposure (McCarver et al., 2011).

Recent observational studies support the concern of possible adverse effects in women who were exposed to soy-based infant formula or isoflavones early in life. Consumption of soy-based infant formula early was associated with estrogen-mediated vaginal tissue alterations within the first 36 weeks of life (Adgent et al., 2018) and with menstrual pain in women aged 23–35 (Upson et al., 2019), yet the exact exposure to isoflavones remains elusive. In contrast, in utero exposure to *O*-desmethylangolensin (a bacterial daidzein metabolite), indicated by maternal urinary concentrations, was associated with early menarche; that is, <11.5 years of age at onset (Marks et al., 2017).

Genotoxicity of isoflavones

Genotoxicity of most isoflavone aglycones has been investigated in vitro in a variety of standard test systems also used for preclinical hazard assessment of pharmaceuticals. In vivo, only data for genistein, daidzein, and equol are available (Table 45.4).

The outcome of genotoxicity studies in vitro is different for each isoflavone and at least one isoflavone induced topoisomerase II inhibition (daidzein and some, but not all, hydroxylated metabolites thereof), topoisomerase II poisoning (genistein, prunetin), adduct formation, induction of DNA strand breaks, bacterial and mammalian gene mutations, chromosome mutations, and genome mutations and transformation in at least one study (Table 45.4). However, there is no clear pattern. In theory, topoisomerase poisons are expected to induce DNA strand

TABLE 45.4 Genotoxicity of isoflavones in vitro and in vivo.

	Isoflavone	Test (outcome)						
		Topoisomerase II	DNA damage	Gene mutations	Chromosome/genome mutations			Transformation
		Interaction			Chromosome	Genome	Undefined	
In vitro	Biochanin A	DA(−)[1]	PL(+)[2]	BACT(−/−[a])[1]	MNT(−)[3]	MNT(+)[3]	MNT(−)[1]	SHE(+)[2]
				HPRT(−)[2,3]	CA(−)[2]	CA(+)[2]		
	Daidzein	ICE(−)[4], TDC(−)[1]	AFE(−)[1]	BACT(−/−[a])[1]	MNT(+)[1]	MNT(+)[1]	MNT(−)[1,3,5]	SHE(+)[2]
		DA(+)[1], DA(−)[4]	COMET(+)[3]	HPRT(−)[1,3]	MNT(−)[3]	MNT(−)[3]		
		TI(+)[1]	COMET(−)[1,4]	HPRT(+)[2]	CA(−)[1,2,3]	CA(−)[2]		
			PL(+)[2]					
	3′-HO-Daidzein	DA(−)[4]	N/A	N/A	MNT(+)[6]	MNT(−)[6]	MNT(+)[3]	N/A
	6-HO-Daidzein	ICE(−)[4], DA(+)[4]	COMET(+)[4]	N/A	MNT(+)[6]	MNT(+)[6]	MNT(+)[3]	N/A
	8-HO-Daidzein	ICE(−)[4], DA(+)[4]	COMET(+)[4]	N/A	N/A	N/A	N/A	N/A
	Equol	N/A	COMET(−)[1]	BACT(−/−[a])[11]	MNT(+)[6]	MNT(+)[1,6]	MNT(+)[3]	N/A
					MNT(−)[1]			
					CA(−[b])[11]			
	Formononetin	N/A	N/A	BACT(−/−[a])[1]	MNT(−)[3]	MNT(+)[3]	N/A	N/A
				HPRT(−)[3]				
	Genistein	ICE(+)[7]	AFE(+)[1]	BACT(−/+[a])[8]	MNT(+)[1,3]	MNT(−)[1,3]	MNT(+)[1]	SHE(+)[2]
		TDC (+)[1]	COMET(+)[1,3]	BACT(−[c]+[a,c,d])[1]	CA(+)[1,3,2]	CA(+)[2]	MNT(−)[5]	SHE(−)[1]
		TP(+)[1], DA(+)[1]	PL(+)[2]	BACT(−/−[a])[1]				
				TK (+/+[a]/+[c]/+[a,c])[1]				
				HPRT(+)[1,2,3]				
	Glycitein	N/A	N/A	N/A	N/A	N/A	MNT(−)[3]	N/A
	Irilone	N/A	N/A	HPRT(−)[9]	MNT(+)[9]	MNT(+)[9]	N/A	N/A
	Prunetin	TDC(+)[1]	PL(−)[2]	HPRT(−)[2]	CA(−)[2]	CA(−)[2]	N/A	SHE(−)[2]

In vivo	Daidzein	N/A	PL, m(−)[3] SCE, m(+)[3]	N/A	N/A	N/A	N/A	N/A
	Equol	N/A	N/A	N/A	N/A	N/A	MNT, r(−)[11]	N/A
	Genistein	N/A	COMET, h(− [e])[1]	BB, r(−)[10]	CA, h(− [b,e])[1]	N/A	MNT, m(−)[1]	N/A
			PL, m(−)[3]				MNT, r(−)[1]	
			SCE, m(+)[3]				MNT, h(− [e])[1]	

Summary of test result regarding topoisomerase II poisoning and catalytic inhibition, DNA damage, and various mutagenicity endpoints in bacteria and mammals. Topoisomerase II poisoning was investigated by the determination of topoisomerase covalent complexes using the in vivo complex of enzyme (ICE) assay, determination of topoisomerase II-dependent DNA cleavage (TDC) by gel electrophoresis for and cell-based assays for detecting topoisomerase II poisons (TP). Catalytic inhibition was investigated by the decatenation assay (DA), or cell-based assays for topoisomerase inhibition (TI). DNA damage was investigated by alkaline filter elution (AFE), single-cell gel electrophoresis (COMET), sister chromatid exchange (SCE) and ^{2}P-postlabeling (PL). Gene mutations were investigated in strains of *Salmonella typhimurium* (BACT), in cultured mammalian cells by the hypoxanthine guanine phosphoribosyltransferase assay (HPRT) and thymidine kinase assay (TK), or in the Big Blue transgenic rodent mutation assay in vivo. Chromosome and genome mutations were investigated by cytogenetic analysis (CA) and the micronucleus assay in vitro and in vivo in mice (m), rats (r) and humans (h). Cell transformation was investigated by the Syrian hamster embryo cell morphological transformation assay (SHE).

[a]*With S9 activation.*

[b]*Complemented with fluorescence in situ hybridization. The summary is based on reviews by NTP National Toxicology Program (2008a) and Stopper et al. (2005) and studies cited therein, as well as other studies; that is, References g–p.*

[c]*Using an isoflavone mixture containing 40%–50% genistein, 18%–25% daidzein, and 1%–4% glycitein.*

[d]*Only with* S. typhimurium *TA100.*

[e]*Study involving 20 patients with prostate cancer without current endocrine therapy exposed to 139.5 mg genistein, 74 mg daidzein, and 11 mg glycitein for 28 days and subsequently to 600 mg for 56 days. At day 0 and after several exposure times: (1) DNA damage was assessed by COMET assay in peripheral lymphocytes; (2) binucleated lymphocytes were generated in vitro under isoflavone exposure via the culture medium and the frequency of micronucleated cells was determined in 100 cytokinesis-blocked cells per patient; and (3) at days 0 and 84, cytogenesis analysis was performed in lymphocytes of three patients cultured in the presence of the isoflavone mixture in vitro for 72 h focusing on translocation of MLL gene.*

[1]*NTP National Toxicology Program, 2008a.*

[2]*Schwen et al., 2010.*

[3]*Tsutsui et al., 2003.*

[4]*Scheffler et al., 2015.*

[5]*Lehmann et al., 2005.*

[6]*Baechler et al., 2014.*

[7]*Manjanatha et al., 2005.*

[8]*NTP National Toxicology Program, 2008b. Study no A90231.*

[9]*Kalfalah et al., 2011.*

[10]*Lepri et al., 2013.*

[11]*Stopper et al., 2005.*

breaks and thus predominately chromosome mutations, whereas topoisomerase inhibitors may induce aneuploidy and, possibly, also DNA strand breaks. Yet, the topoisomerase poison prunetin induced neither DNA damage nor chromosome mutations. Furthermore, one would think testing isolated isoflavones in vitro could not cause a lot of variability between studies. However, different studies testing the same isoflavone using in vitro indicator tests and mutagenicity assays yielded different outcomes; for example, daidzein may or may not induce DNA strand breaks and gene mutations, and genistein induced transformation of Syrian hamster embryo cells in one study but not in another (Table 45.4). Comparing studies from the same laboratory, induction of aneuploidy was predominately observed after treatment with isoflavones exhibiting methoxy groups and irilone (methylenedioxy group), whereas genistein and catecholic metabolites of daidzein (but not daidzein) induced chromosome mutations (Scheffler et al., 2015; Lehmann et al., 2005; Stopper et al., 2005).

In vivo, equol did not induce micronuclei in mice and genistein did not induce micronuclei in mice, rats, or humans. The result of the latter study is based on micronuclei frequencies in binucleated lymphocytes formed ex vivo using peripheral lymphocytes from 20 prostate cancer patients before, during, and after exposure to predominately genistein for 84 days (Table 45.4). However, it is known that the sensitivity of the assay for the detection of induced micronuclei in binucleated cells is limited and it is highly unlikely that DNA damage induced in vivo is the cause of increased micronuclei frequencies in binucleated cells (Speit, 2013). Furthermore, only 100 binucleated cells were analyzed with respect to micronuclei formation. In the same study, no increase in DNA damage was observed in peripheral lymphocytes. Furthermore, no induction of gene mutations was observed in transgenic rats exposed to genistein. In contrast, one study observed induction of sister chromatid exchange, indicative for DNA damage, in mice by genistein and daidzein (Table 45.4).

Taken together, the results of the numerous in vitro studies are contradictory and do not allow for hazard assessment of the major isoflavones genistein and daidzein. The few in vivo studies available indicate no mutagenic potential of daidzein, equol, and genistein in experimental animals and in prostate cancer patients; unfortunately, the human study suffers methodical limitations.

Carcinogenicity versus chemoprevention

Carcinogenesis is a multistep process during which a cell acquires the ability of uncontrolled proliferation. Based on the initiation–promotion model, mutagenicity (tumor initiation and progression) and estrogenicity (tumor promotion) play a crucial role in the development of cancer. Individual isoflavone aglycones exhibit the potential to cause mutations (see "Mutagenicity") and/or possess estrogenic activity (see "Estrogenicity"). Thus, the activity profile of isoflavones is similar to that of the female steroid hormone 17β-estradiol, which is considered to cause breast cancer in women (Yager, 2015). Consequently, there is some concern that isoflavones might also contribute to carcinogenesis.

The hazard identification of a human carcinogen is complex and requires the evaluation of human epidemiological studies, long-term studies in animals, as well as mechanistic data, ensuring a coherent mode of action in humans and experimental animals. Because of its complexity, the hazard assessment (and risk assessment) of carcinogens is often performed by a scientific expert panel. Drawing a final conclusion is thus beyond the scope of this chapter.

Motivated by the apparent contradiction of chemoprevention versus the possible contribution to tumor formation in estrogen-sensitive tissues, this chapter discusses the carcinogenicity of genistein and the impact of isoflavones on growth of tumor cells in experimental animals, followed by modulation of chemically induced carcinogenesis of the female breast by genistein in experimental animals. This section closes with the impact of isoflavones on human breast cancer risk. The biological effects of isoflavones on hormone-sensitive cancers other than breast cancer (prostate, female reproductive system, thyroids, colon, etc.) are reviewed elsewhere (e.g., Rietjens et al., 2017; D'Adamo and Sahin, 2014; Mahmoud et al., 2014; North American Menopause Society, 2011) and are not discussed in this chapter.

Carcinogenicity of genistein in experimental animals

Long-term studies in experimental animals are available only for genistein aglycone and consist of one 2-year bioassay in rats and two 1-year studies in rats and dogs (McClain et al., 2005, 2006; NTP National Toxicology Program, 2008a). The 2-year bioassay, mimicking lifelong exposure, was performed as a multigenerational feeding study and included perinatal exposure (in utero via the placenta) and exposure via lactation (NTP National Toxicology Program, 2008a). As concluded by the NTP, "exposure to genistein for 2 years caused tumors of the mammary gland and pituitary gland in female rats. Exposure to genistein for shorter durations following birth was also possibly associated with increased rates of pituitary gland and mammary gland tumors" (NTP National Toxicology Program, 2008a). Because genistein also accelerated the onset of aberrant estrous cycles, it was suggested that genistein acts via an estrogenic mode of action (NTP National Toxicology Program, 2008a). No carcinogenic activity of genistein was observed in male rats, although male mammary gland hyperplasia was observed at shorter exposure periods (NTP National Toxicology Program, 2008a). Reversible effects attributed to the estrogenic nature of genistein were also observed after 1-year exposure of rats and dogs to higher doses of genistein than in the NTP study. Common targets in both species were the reproductive organs of males (e.g., epididymis, prostate) and females (ovaries, uterus), but reversible changes in the glandular proliferation of the male and female breast were also observed (McClain et al., 2005, 2006; NTP National Toxicology Program, 2008a).

However, with regard to carcinogenic hazard identification according to internationally recognized guidelines for the testing of chemicals, both studies might have their limitations. The exposure period in the 1-year study was too short and the highest dose applied in the 2-year study was too low. Nevertheless, these animal studies indicate that estrogen-sensitive tissues can be targeted by chronic exposure to genistein and a contribution to the carcinogenesis in estrogen-sensitive tissues cannot be excluded with certainty.

Tumor proliferation studies

Genistein, genistin, daidzein, and soy protein isolate containing different amounts of genistein have been shown to stimulate MCF-7 growth in ovariectomized athymic mice dose-dependently. The tumors regressed when mice were switched from genistein and a genistein-enriched diet to a isoflavone-free diet. Furthermore, when similar amounts of genistein aglycone equivalents (750 ppm) were accompanied by differently processed soy foods, the stimulation of MCF-7 growth correlated inversely with the complexity of the soy product matrix (i.e., genistin > mixed isoflavones > isoflavone-rich soy extract Novasoy > soy flour plus mixed isoflavones). However, the diet containing soy flour enriched with mixed isoflavones did not alter MCF-7 tumor growth in vivo, even though the serum genistein levels were similar in all diet groups (Helferich et al., 2008). Also, daidzein was less effective in this model than genistein (Helferich et al., 2008), and equol inactive (Song et al., 2020). Interestingly, plasma equol levels were highest in diets containing less processed soy. Because equol itself had no effect on the MCF-7 growth in vivo (Helferich et al., 2008), bacterial metabolism not related to equol formation might have contributed. However, this thought has not yet been further pursued. Tumor growth stimulated by genistin and mixed isoflavone accompanied activation of the estrogen receptor α. In contrast, no stimulation of estrogen receptor α was observed in daidzein-induced tumor growth.

Chemoprevention studies

Soy food, isolated genistein, and biochanin A, but not daidzein, increased the tumor latency period, decreased the promotion of tumors, and/or decreased tumor multiplicity in rodents with chemically induced mammary carcinogenesis. Chemoprotection was observed when exposure to genistein and soy products occurred prepubertally but was most effective when exposure occurred both prepubertally and during adulthood. No chemoprotection was observed after in utero exposure to soy isoflavones or when isoflavone exposure started during adulthood (Hüser et al., 2018; Mortensen et al., 2009; Whitsett and Lamartiniere, 2006). In contrast, in utero exposure to genistein was observed to stimulate chemically induced tumorigenesis (Hilakivi-Clarke et al., 1999; Mortensen et al., 2009) or alter the estrogen responsiveness of the mammary gland (Hüser et al., 2018; Blei et al., 2015).

Isoflavone exposure after tumor induction or after tumor development resulted in enhanced tumor formation, reduced angiogenesis and metastasis, or had no effect on tumor development (Moorehead, 2019; Mortensen et al., 2009).

Comparing the outcome of these chemoprevention studies, trying to deduce their relevance for humans, rodent prepubertal and pubertal development of the mammary gland should be considered. During development of the mammary gland, the number of terminal end buds increases, reaching a maximum. The subsequent decrease in the number of terminal end buds is due to differentiation of terminal end buds into alveolar buds and lobules. The time window immediately after reaching the maximum number of terminal end buds and first appearance of more developed structures is considered to be the window of highest susceptibility to carcinogenic insult (Russo and Russo, 1994). In contrast, the mature rat mammary gland is less susceptible. Accordingly, tumor incidence and multiplicity are affected by the timing of carcinogen exposure, which consists of a single dosage of carcinogen in most chemoprevention studies.

Also, the genistein-induced chemoprotection in rodents (as well as the lack thereof) has been linked to changes in the morphology of the mammary gland. In rats prepubertally treated with genistein, the number of terminal end buds in mammary gland whole mounts was significantly decreased compared to control animals of the same age, and structures resembling a more mature mammary gland were present. However, in utero exposure to genistein caused an increase in terminal end buds, which is consistent with the stimulation of chemically induced mammary carcinogenesis (Möller et al., 2016; Whitsett and Lamartiniere, 2006). These study results draw a fairly consistent picture of the mode of action of genistein in rodents challenged with a mammary carcinogen. Genistein seems to accelerate maturation of the female mammary gland, thus rendering it less susceptible to carcinogenic insult in comparison with a less mature control group exposed on the same postnatal day. However, the question remains what would have happened if the carcinogenic insult in the genistein group had been in the time window of highest susceptibility (e.g., by adjusting the timing of carcinogen exposure to the developmental stage of the animals) instead of the same postnatal day?

Nevertheless, in the case of isoflavone toxicity/chemoprevention the relevance of animal studies to humans is debatable. Apart from the experimental limitations (summarized in Hüser et al., 2018), there is always the observation that biokinetics of isoflavones in animals (higher levels of aglycones, 100% equol producer, and different conjugation pattern) differs significantly from humans.

Isoflavones and human breast cancer risk

Breast cancer is the most frequent cancer in women both in developed as well as in developing countries. However, the number of new cases per year is highest in Western Europe and Northern America, and is lowest in Eastern Asia.

Breast cancer risk increases with prolonged lifetime exposure to increased levels of the endogenous sex hormone 17β-estradiol caused, for example, by early menarche, late menopause, and obesity (Colditz and Bohlke, 2014), but also by exogenous estrogens used for hormone replacement (Collaborative Group on Hormonal Factors in Breast Cancer, 2019) and oral contraception (Grosse et al., 2009).

In premenopausal adult women, high plasma levels of estradiol stimulate cell proliferation of epithelial cells in glandular tissue of the breast during the luteal phase of the menstrual cycle. At its end, the epithelial growth partially regresses. The proliferation of epithelial cells is highest in young women and plateaus in their mid-30s (Geddes, 2007). On a molecular level, the proliferation of epithelial cells is well regulated by intramammary signaling and requires activation of estrogen receptor α (Lanigan et al., 2007). Most breast tumors arise from the epithelial cells of the glandular tissue and approximately 75% are estrogen receptor α-positive, which means that proliferation of these tumor cells can be stimulated by 17β-estradiol and other estrogens including isoflavones (Hüser et al., 2018; EFSA Panel on Food Additives and Nutrient Sources added to Food, 2015).

Most studies in humans are observational, that is, prospective cohort or nested case-control studies, or retrospective case-control studies. The strengths and weaknesses of the available study types as well as the meta-analyses of the observational studies on isoflavone consumption and breast cancer risk are summarized in Hüser et al. (2018). The most consistent result in all observational studies and meta-analyses is that no adverse effect has been associated with isoflavone consumption (Hüser et al., 2018). With regard to an inverse association between isoflavone consumption and breast cancer risk, that is, chemoprevention, results from meta-analyses are inconsistent. This inconsistency persists, even if biomarkers for isoflavone consumption (blood or urinary levels of aglycones or isoflavone metabolites including equol) are considered (Hüser et al., 2018). In the rare occasion that isoflavone-containing dietary (soy-based) supplements had been considered as a source of isoflavones, no association of isoflavone consumption and breast cancer risk has been observed (EFSA Panel on Food Additives and Nutrient Sources added to Food, 2015; Hüser et al., 2018) until recently when current use of isoflavone supplements was positively associated with an increased risk in estrogen receptor-positive breast cancer risk, but inversely with estrogen receptor-negative breast cancer risk (Touillaud et al., 2019).

The difference in breast cancer incidence between Asian (especially China and Japan) and Western countries and the observation that migration of women from low-incidence Asian countries to the United States results in an increase in breast cancer risk (Ziegler et al., 1993) gave rise to the hypothesis that environmental factors and lifestyle, in particular diet, account for the difference in breast cancer risk.

However, an inverse association between soy consumption and breast cancer risk in Asian study populations has been observed in retrospective case-control but not prospective studies and seems to apply mainly to postmenopausal women (Wei et al., 2019; Hüser et al., 2018; Zhao et al., 2019).

Isoflavone consumption in American Asians is lower than in East Asians but still higher than in Americans with no Asian ancestries (Maskarinec, 2013; Nagata, 2010). Also, second-generation Asian Americans have a lower proportion of equol producers (32% vs. 53%, not significant) and a significantly lower urinary equol excretion rate (~10%) than their ancestors (Maskarinec et al., 2007). However, whether any of these observations is ultimately connected to the factors deteriorating the breast cancer risk in migrants remains elusive. The traditional Asian diet consists of minimally processed products from whole soybeans and fermented soy products, resulting in a high intake of malonyl-glucosides and aglycones/microbial metabolites, respectively (see "Biokinetics"). Western populations consume more isoflavone-rich soy extracts and soy protein isolates (i.e., isoflavone glucosides) than Asians (Nagata, 2010; Setchell and Clerici, 2010; Mortensen et al., 2009). For the sake of completeness, it should also be mentioned that isoflavone intake might simply indicate a healthier (Asian) lifestyle and healthier nutrition (Sacks et al., 2006).

In contrast to the confusing data on isoflavone-associated breast cancer risk, two recent meta-analyses indicate that breast cancer recurrence is inversely associated with soy isoflavone consumption (Qiu and Jiang, 2019; Hüser et al., 2018). However, available data do not conclude if this association applies to Western and/or Asian women or if it is related to estrogen receptor alpha positive and/or negative breast cancer.

For obvious reasons there is no clinical intervention study on breast cancer risk in humans. However, surrogate markers for the biological activity of isoflavones in the breasts of healthy women (mammographic density and proliferation marker, namely Ki-67 staining and [^{3}H] thymidine labeling index) have been investigated. Current meta-analyses and reviews summarize that consumption of isoflavones as soy food or dietary supplements neither alters mammographic density (Hüser et al., 2018; EFSA Panel on Food Additives and Nutrient Sources added to Food, 2015; Fritz et al., 2013; Hooper et al., 2010) nor markers for cell proliferation (Hüser et al., 2018; EFSA Panel on Food Additives and Nutrient Sources added to Food, 2015).

Human data suggest that consumption of isoflavones is not positively associated with breast cancer risk. Only within the Asian population some results indicate an inverse association between isoflavone consumption and breast cancer risk. Most human studies suffer from an insufficient characterization of the isoflavone profile dependent on their source and do not take into account individual differences resulting from biotransformation, for example, equol formation (Hüser et al., 2018; Mortensen et al., 2009). Thus, a potential impact of isoflavones on breast cancer risk in the Western population cannot be answered satisfactorily. Likewise, the obstacles observed in assessing the contribution to breast cancer risk (Hüser et al., 2018; EFSA Panel on Food Additives and Nutrient Sources added to Food, 2015) extends to all other hypothesized biological effects of isoflavones.

Efficacy

The US health claim for soy protein in the prevention of coronary heart disease does not include soy isoflavones because "the evidence did not support a significant role for soy isoflavones in cholesterol-lowering effects of soy protein" (21 CFR 101.82 FR 50324; Sacks et al., 2006). Although similar health claims for soy protein have been approved in eight more countries (United Kingdom, Japan, Korea, Malaysia, Brazil, South Africa, Indonesia, and the Philippines; Xiao, 2008), an isoflavone-related health claim has not been approved up to now. In the European Union alone, a multitude of health claims concerning potentially beneficial effects of (soy) isoflavones have been evaluated by the European Food Safety Authority. Whereas some claims were rejected because they were out of the scope of Regulation (EC) 1924/2006 defining health claims (e.g., maintenance of normal skin tonicity, cardiovascular health, upper respiratory tract, treatment of prostate cancer), the majority were rejected due to lack of evidence for a cause-and-effect relationship between isoflavone consumption and the claimed health effects (protection of DNA, proteins, and lipids of oxidative damage), maintenance of blood low density lipoprotein (LDL) cholesterol levels, reduction of vasomotor symptoms associated with menopause (soy isoflavones/isoflavones), contribution to normal hair growth (soy isoflavones/soy extract containing isoflavones), maintenance of joints, and maintenance of bone mineral density (soy isoflavones/ipriflavone, a synthetic derivative of daidzein/Femarelle, a mixture of a soy derivative with unknown isoflavone content and ground flaxseed; summarized in EFSA Panel on Food Additives and Nutrient Sources added to Food, 2015).

Thus, the impact of isoflavones on cardiovascular disease and menopausal symptoms is still being heavily investigated and is therefore addressed within this section. The impact of isoflavones on breast cancer has already been addressed (see "Carcinogenesis"). For other suggested health effects, for example, bone health, neurocognitive function, glucose metabolism, and prostate and lung cancers, the reader is referred to contemporary reviews (e.g., Rietjens et al., 2017; D'Adamo and Sahin, 2014; Soni et al., 2014; North American Menopause Society, 2011).

Cardiovascular disease

In Japan, a lower incidence of cardiovascular disease compared with Western countries is observed. An acknowledged risk factor for cardiovascular disease is blood cholesterol, particularly LDL cholesterol, triglycerides, blood pressure, and atherosclerosis (D'Adamo and Sahin, 2014; Mortensen et al., 2009; Xiao, 2008; Sacks et al., 2006). A long-term reduction in serum cholesterol concentration of 10% lowers the risk of coronary heart disease by 20%–50% (Law et al., 1994).

In experimental animals, dietary exposure to soy protein instead of animal protein resulted in reduced LDL cholesterol and total cholesterol levels (Mortensen et al., 2009; Sacks et al., 2006), suggesting that soy protein can be a nutritional approach to reducing blood cholesterol levels. Epidemiological studies in Eastern Asian populations have supported this hypothesis (D'Adamo and Sahin, 2014; Mortensen et al., 2009; Xiao, 2008; Sacks et al., 2006).

In experimental animals, soy isoflavones caused a decrease in blood cholesterol and improved arterial function, and inhibited atherosclerosis, which led to the idea that only soy proteins containing high levels of isoflavone produced favorable outcomes in epidemiological and intervention studies (Sacks et al., 2006).

The impact of isoflavones on reducing blood lipids was reviewed in 2006 by the American Heart Association (Sacks et al., 2006). Among 19 studies that compared the impact of isoflavone-containing soy proteins with low-isoflavone and isoflavone-free soy proteins, only three showed significant reductions in LDL cholesterol, and the effect among all studies was null. Also, isoflavone-induced changes in high density lipoprotein cholesterol and triglycerides were not significant and isoflavone blood levels did not affect blood lipids (Sacks et al., 2006). The American Heart Association summarized that reduced blood pressure was observed sporadically; however, in the majority of the studies blood pressure remained unaffected (Sacks et al., 2006).

Two meta-analyses of randomized, controlled clinical trials involving women and men with hypercholesterolemia also found no evidence of isoflavones lowering blood cholesterol (Tokede et al., 2015; Qin et al., 2013). In a small population a slight but significant decrease in triglycerides (Qin et al., 2013) was observed but not confirmed in the later meta-analysis (Tokede et al., 2015). Interestingly, consumption of soy product (especially soymilk, soybeans, and nuts) was negatively associated with blood concentrations of LDL cholesterol, triglycerides, and total cholesterol (Tokede et al., 2015). Also, a meta-analysis of observational studies with men and women from Western and Asian countries concluded that consumption of soy but not isoflavones is inversely associated with the risk of cardiovascular disease (Yan et al., 2017).

Menopausal symptoms

Hot flashes and night sweats are menopause-related vasomotor symptoms, which occur more frequently in North American (70%–80%) than in Asian (10%–20%) women (North American Menopause Society, 2011). Prescription hormone replacement therapy relieves hot flashes, yet increases breast cancer risk (Steward and Wild, 2014). The question was whether estrogenic soy isoflavones (see "Estrogenicity") present in the soy-rich Asian diet might provide the same benefit as hormone replacement therapy, preferably without any adverse effects.

However, the outcome of clinical trials and their meta-analyses and systematic reviews investigating isoflavones (soy-based and red clover-based supplements) and menopausal vasomotor symptoms is ambiguous (Rietjens et al., 2017; D'Adamo and Sahin, 2014; Newton, 2014; Lethaby et al., 2013; Taku et al., 2012; North American Menopause Society, 2011; Sacks et al., 2006). In a systematic review of 14 randomized, controlled trials, a significant improvement in vasomotor symptoms by isoflavones (40–160 mg/day) was observed in 11 studies, reducing the daily frequencies of hot flashes by 24%–60% (North American Menopause Society, 2011). However, a strong placebo effect is generally observed in many trials on hot flashes, resulting in the reduction of daily hot flash frequencies by up to 60% (Sacks et al., 2006; Lethaby et al., 2013). Thus, a clear dose–response relationship is difficult to establish when such a high placebo effect prevails.

Nevertheless, a meta-analysis observed that soy isoflavones (median 54 mg/day) reduced the frequency and severity by 21% and 26%, respectively, and dietary supplements providing more than 19 mg genistein were more potent than supplements providing less (Taku et al., 2012). Genistein was reported to have beneficial effects on vasomotor symptoms (North American Menopause Society, 2011), and a systematic review suggests that genistein supplements providing more than 30 mg/day are effective in reducing hot flashes (Lethaby et al., 2013). This result was independently confirmed by Newton (2014), who stated that three out of five trials using high genistein extracts (30–60 mg/day) observed a significant reduction in daily vasomotor symptoms (41%–61%) when compared with placebo (7%–29%). However, the author noted that finding "studies adequate enough to merit trust in their findings" was challenging and that there was a high risk of attrition and reporting bias (Newton, 2014).

In summary, the conclusion of systematic reviews and meta-analyses for isoflavones and vasomotor symptoms remains ambiguous because even significant reductions of vasomotor symptoms lie within the placebo effect generally observed in many trials on vasomotor symptoms (up to 60%) (Lethaby et al., 2013; Sacks et al., 2006). This ambiguity is also reflected in the conclusions drawn by the American Heart Association ("it seems unlikely that soy isoflavones have enough estrogenic activity to have an important impact on vasomotor symptoms of estrogen deficiency in perimenopausal women"; Sacks et al., 2006) and by The North American Menopause Society ("Soy-based isoflavones are modestly effective in controlling hot flashes, as demonstrated to date in predominantly Caucasian women in early postmenopause who have at least four hot flashes a day"; North American Menopause Society, 2011).

Taken together, no isoflavone-related health claim has been approved worldwide up to now. Besides the possibility that isoflavones simply do not induce the claimed effect, the large variation in results between studies (particularly in clinical trials) renders the establishment of a cause-and-effect relationship particularly difficult. The heterogeneity of study results seems to be due to different isoflavone sources that were often not adequately chemically characterized, different dosages, and experimental designs, as well as large interindividual variations in the susceptibility to isoflavones.

Concluding remarks and future directions

Considering the diversity of isoflavones (see "Occurrence and Dietary Intake of Isoflavones") and their individual complex biokinetics (see "Biokinetics") and biological effects (e.g., see "Estrogenicity"), the diverse and seemingly contradictory outcomes of studies between species and even within the same species are less surprising.

Strikingly, up to now, only limited data are available on species-specific, tissue-specific, life stage-specific biokinetics, and sensitivity to isoflavones. This might be, at least in part, due to the lack of comprehensive quantitative information on isoflavone metabolite profiles in conjunction with that regarding endogenous estrogens in human tissues (and experimental animals). Likewise, correlations between plasma and tissue levels in human and experimental animals are scarce. This lack of information is, at least in part, because of too few studies applying/considering:

1. Specific detection of all isoflavone conjugates by mass spectrometry. Analysis with and without enzymatic or chemical hydrolysis of conjugates is prone to artifacts and does not account for the complexity of conjugates (monosulfates, disulfates, monoglucuronides, diglucuronides, and sulfoglucuronides);
2. The possibility of unintentional hydrolysis of conjugates during sample preparation of food samples or biological matrices;
3. Analysis of bacterial metabolites besides equol.

Because the same limitations affecting analysis of isoflavone levels (lack of specific detection, sample preparation prone to artifacts, disregard of specific conjugates) apply to analysis of endogenous estrogens as well, and because few human data are available for tissues obtained from noncancer patients, the estrogen background putatively interacting with the isoflavone(s) is not well-defined in most studies.

Thus, even after more than two decades of research on isoflavones, lack of crucial data still prevents expert panels (such as the EFSA), not to mention the authors of this chapter, from drawing a final conclusion regarding isoflavone efficacy and safety.

References

Adgent, M.A., Umbach, D.M., Zemel, B.S., et al., 2018. A longitudinal study of estrogen-responsive tissues and hormone concentrations in infants fed soy formula. J. Clin. Endocrinol. Metab. 103, 1899–1909.

Allred, C.D., Ju, Y.H., Allred, K.F., et al., 2001. Dietary genistin stimulates growth of estrogen-dependent breast cancer tumors similar to that observed with genistein. Carcinogenesis 22, 1667–1673.

Baechler, S.A., Schroeter, A., Walker, J., et al., 2014. Oxidative metabolism enhances the cytotoxic and genotoxic properties of the soy isoflavone daidzein. Mol. Nutr. Food Res. 58, 1269–1281.

Barnes, S., 2010. The biochemistry, chemistry and physiology of the isoflavones in soybeans and their food products. Lymphat. Res. Biol. 8, 89–98.

Blei, T., Soukup, S.T., Schmalbach, K., et al., 2015. Dose-dependent effects of isoflavone exposure during early lifetime on the rat mammary gland: studies on estrogen sensitivity, isoflavone metabolism, and DNA methylation. Mol. Nutr. Food Res. 59, 270–283.

Braune, A., Maul, R., Schebb, N.H., et al., 2010. The red clover isoflavone irilone is largely resistant to degradation by the human gut microbiota. Mol. Nutr. Food Res. 54, 929–938.

Brown, N.M., Galandi, S.L., Summer, S.S., et al., 2014. S-(-)Equol production is developmentally regulated and related to early diet composition. Nutr. Res. 34, 401–409.

Cederroth, C.R., Zimmermann, C., Nef, S., 2012. Soy, phytoestrogens and their impact on reproductive health. Mol. Cell. Endocrinol. 355, 192–200.

Chang, E.C., Charn, T.H., Park, S.H., et al., 2008. Estrogen Receptors alpha and beta as determinants of gene expression: influence of ligand, dose, and chromatin binding. Mol. Endocrinol. 22, 1032–1043.

Chen, J., Lin, H., Hu, M., 2005. Absorption and metabolism of genistein and its five isoflavone analogs in the human intestinal Caco-2 model. Cancer Chemother. Pharmacol. 55, 159–169.

Cline, J.M., Wood, C.E., 2009. Estrogen/isoflavone interactions in cynomolgus macaques (Macaca fascicularis). Am. J. Primatol. 71, 722–731.

Colditz, G.A., Bohlke, K., 2014. Priorities for the primary prevention of breast cancer. CA Cancer J. Clin. 64, 186–194.

Collaborative Group on Hormonal Factors in Breast Cancer, 2019. Type and timing of menopausal hormone therapy and breast cancer risk: individual participant meta-analysis of the worldwide epidemiological evidence. Lancet 394, 1159–1168.

D'Adamo, C.R., Sahin, A., 2014. Soy foods and supplementation: a review of commonly perceived health benefits and risks. Altern. Ther. Health Med. 20 (Suppl. 1), 39–51.

Degen, G.H., Blaszkewicz, M., Shi, L., et al., 2011. Urinary isoflavone phytoestrogens in German children and adolescents—a longitudinal examination in the DONALD cohort. Mol. Nutr. Food Res. 55, 359–367.

Delmonte, P., Perry, J., Rader, J.I., 2006. Determination of isoflavones in dietary supplements containing soy, Red Clover and kudzu: extraction followed by basic or acid hydrolysis. J. Chromatogr. A 1107, 59–69.

EFSA Panel on Food Additives and Nutrient Sources added to Food, 2015. Scientific opinion on the risk assessment for peri- and post-menopausal women taking food supplements containing isolated isoflavones. EFSA J. 13, 4246.

Ferlay, J., Colombet, M., Bray, F., 2020 Cancer Incidence in Five Continents, CI5plus: IARC CancerBase No. 9 [Internet]. <http://ci5.iarc.fr> (accessed 29.05.2020)

Flint, H.J., Scott, K.P., Louis, P., et al., 2012. The role of the gut microbiota in nutrition and health. Nat. Rev. Gastroenterol. Hepatol. 9, 577–589.

Franke, A.A., Lai, J.F., Halm, B.M., 2014. Absorption, distribution, metabolism, and excretion of isoflavonoids after soy intake. Arch. Biochem. Biophys. 559, 24–28.

Fritz, H., Seely, D., Flower, G., et al., 2013. Soy, red clover, and isoflavones and breast cancer: a systematic review. PLoS One 8, e81968.

Geddes, D.T., 2007. Inside the lactating breast: the latest anatomy research. J. Midwifery Womens Health 52, 556–563.

Grosse, Y., Baan, R., Straif, K., et al., 2009. A review of human carcinogens—part A: pharmaceuticals. Lancet Oncol. 10, 13–14.

Grosser, G., Doring, B., Ugele, B., et al., 2015. Transport of the soy isoflavone daidzein and its conjugative metabolites by the carriers SOAT, NTCP, OAT4, and OATP2B1. Arch. Toxicol. 89, 2253–2263.

Gu, L., House, S.E., Prior, R.L., et al., 2006. Metabolic phenotype of isoflavones differ among female rats, pigs, monkeys, and women. J. Nutr. 136, 1215–1221.

Hamilton-Reeves, J.M., Vazquez, G., Duval, S.J., et al., 2010. Clinical studies show no effects of soy protein or isoflavones on reproductive hormones in men: results of a meta-analysis. Fertil. Steril. 94, 997–1007.

Heinonen, S.M., Wahala, K., Adlercreutz, H., 2004. Identification of urinary metabolites of the red clover isoflavones formononetin and biochanin A in human subjects. J. Agric. Food Chem. 52, 6802–6809.

Helferich, W.G., Andrade, J.E., Hoagland, M.S., 2008. Phytoestrogens and breast cancer: a complex story. Inflammopharmacology 16, 219–226.

Hilakivi-Clarke, L., Cho, E., Onojafe, I., et al., 1999. Maternal exposure to genistein during pregnancy increases carcinogen-induced mammary tumorigenesis in female rat offspring. Oncol. Rep. 6, 1089–1095.

Hooper, L., Madhavan, G., Tice, J.A., et al., 2010. Effects of isoflavones on breast density in pre- and post-menopausal women: a systematic review and meta-analysis of randomized controlled trials. Hum. Reprod. Update 16, 745–760.

Hooper, L., Ryder, J.J., Kurzer, M.S., et al., 2009. Effects of soy protein and isoflavones on circulating hormone concentrations in pre- and post-menopausal women: a systematic review and meta-analysis. Hum. Reprod. Update 15, 423–440.

Hosoda, K., Furuta, T., Ishii, K., 2011. Metabolism and disposition of isoflavone conjugated metabolites in humans after ingestion of kinako. Drug. Metab. Dispos. 39, 1762–1767.

Hüser, S., Guth, S., Joost, H.G., et al., 2018. Effects of isoflavones on breast tissue and the thyroid hormone system in humans: a comprehensive safety evaluation. Arch. Toxicol. 92, 2703–2748.

Islam, M.A., Punt, A., Spenkelink, B., et al., 2014. Conversion of major soy isoflavone glucosides and aglycones in in vitro intestinal models. Mol. Nutr. Food Res. 58, 503–515.

Kalfalah, F.M., Mielke, C., Christensen, M.O., et al., 2011. Genotoxicity of dietary, environmental and therapeutic topoisomerase II poisons is uniformly correlated to prolongation of enzyme DNA residence. Mol. Nutr. Food Res. 55 (Suppl. 1), S127–S142.

Kostelac, D., Rechkemmer, G., Briviba, K., 2003. Phytoestrogens modulate binding response of estrogen receptors alpha and beta to the estrogen response element. J. Agric. Food Chem. 51, 7632–7635.

Kuiper, G.G., Lemmen, J.G., Carlsson, B., et al., 1998. Interaction of estrogenic chemicals and phytoestrogens with estrogen receptor beta. Endocrinology 139, 4252–4263.

Kulling, S.E., Lehmann, L., Metzler, M., 2002. Oxidative metabolism and genotoxic potential of major isoflavone phytoestrogens. J. Chromatogr. B Analyt. Technol. Biomed. Life Sci. 777, 211–218.

Lanigan, F., O'connor, D., Martin, F., et al., 2007. Molecular links between mammary gland development and breast cancer. Cell. Mol. Life Sci. 64, 3159–3184.

Larkin, T., Price, W.E., Astheimer, L., 2008. The key importance of soy isoflavone bioavailability to understanding health benefits. Crit. Rev. Food Sci. Nutr. 48, 538–552.

Law, M.R., Wald, N.J., Thompson, S.G., 1994. By how much and how quickly does reduction in serum cholesterol concentration lower risk of ischaemic heart disease? BMJ 308, 367–372.

Lehmann, L., Esch, H.L., Wagner, J., et al., 2005. Estrogenic and genotoxic potential of equol and two hydroxylated metabolites of Daidzein in cultured human Ishikawa cells. Toxicol. Lett. 158, 72–86.

Lepri, S.R., Luiz, R.C., Zanelatto, L.C., et al., 2013. Chemoprotective activity of the isoflavones, genistein and daidzein on mutagenicity induced by direct and indirect mutagens in cultured HTC cells. Cytotechnology 65, 213–222.

Lethaby, A., Marjoribanks, J., Kronenberg, F., et al., 2013. Phytoestrogens for menopausal vasomotor symptoms. Cochrane Database Syst. Rev. 12, CD001395.

Liu, J., Knappenberger, K.S., Kack, H., et al., 2003. A homogeneous in vitro functional assay for estrogen receptors: coactivator recruitment. Mol. Endocrinol. 17, 346–355.

Lutter, S., Schmalbach, K., Esch, H.L., Lehmann, L., 2014. The isoflavone irilone contributes to the estrogenic potential of dietary supplements containing red clover. Arch. Toxicol. 88, 309–321.

Mahmoud, A.M., Yang, W., Bosland, M.C., 2014. Soy isoflavones and prostate cancer: a review of molecular mechanisms. J. Steroid Biochem. Mol. Biol. 140, 116–132.

Manjanatha, M.G., Shelton, S.D., Rhodes, B.S., et al., 2005. 17 Beta-estradiol and not genistein modulates lacI mutant frequency and types of mutation induced in the heart of ovariectomized big blue rats treated with 7, 12-dimethylbenz[a] anthracene. Environ. Mol. Mutagen. 45, 70–79.

Marks, K.J., Hartman, T.J., Taylor, E.V., et al., 2017. Exposure to phytoestrogens in utero and age at menarche in a contemporary British cohort. Environ. Res. 155, 287–293.

Maskarinec, G., 2013. The human mammary gland as a target for isoflavones: how does the relation vary in individuals with different ethnicity? Planta Med. 79, 554–561.

Maskarinec, G., Yamakawa, R., Hebshi, S., et al., 2007. Urinary isoflavonoid excretion and soy consumption in three generations of Japanese women in Hawaii. Eur. J. Clin. Nutr. 61, 255–261.

Maukonen, J., Saarela, M., 2015. Human gut microbiota: does diet matter? Proc. Nutr. Soc. 74, 23–36.

Maul, R., Kulling, S.E., 2010. Absorption of red clover isoflavones in human subjects: results from a pilot study. Br. J. Nutr. 103, 1569–1572.

McCarver, G., Bhatia, J., Chambers, C., et al., 2011. NTP-CERHR expert panel report on the developmental toxicity of soy infant formula. Birth Defects Res. B Dev. Reprod. Toxicol. 92, 421–468.

McClain, R.M., Wolz, E., Davidovich, A., et al., 2005. Subchronic and chronic safety studies with genistein in dogs. Food Chem. Toxicol. 43, 1461–1482.

McClain, R.M., Wolz, E., Davidovich, A., et al., 2006. Acute, subchronic and chronic safety studies with genistein in rats. Food. Chem. Toxicol. 44, 56–80.

Möller, F.J., Pemp, D., Soukup, S.T., et al., 2016. Soy isoflavone exposure through all life stages accelerates 17beta-estradiol-induced mammary tumor onset and growth, yet reduces tumor burden, in ACI rats. Arch. Toxicol. 90, 1907–1916.

Moorehead, R.A., 2019. Rodent models assessing mammary tumor prevention by soy or soy isoflavones, Genes (Basel), 10.

Mortensen, A., Kulling, S.E., Schwartz, H., et al., 2009. Analytical and compositional aspects of isoflavones in food and their biological effects. Mol. Nutr. Food Res. 53 (Suppl. 2), S266–S309.

Nagata, C., 2010. Factors to consider in the association between soy isoflavone intake and breast cancer risk. J. Epidemiol. 20, 83–89.

Nakamura, Y., Tsuji, S., Tonogai, Y., 2000. Determination of the levels of isoflavonoids in soybeans and soy-derived foods and estimation of isoflavonoids in the Japanese daily intake. J. AOAC Int. 83, 635–650.

North American Menopause Society, 2011. The role of soy isoflavones in menopausal health: report of The North American Menopause Society/Wulf H. Utian Translational Science Symposium in Chicago, IL (October 2010). Menopause 18, 732–753.

National Research Council, 2005. Health Implications of Perchlorate Ingestion. The National Academies Press.

Newton, K.M., 2014. Isoflavones hold limited promise for the treatment of menopausal vasomotor symptoms. Evid. Based Med. 19, 178.

Nielsen, F.K., Hansen, C.H., Fey, J.A., et al., 2012. H295R cells as a model for steroidogenic disruption: a broader perspective using simultaneous chemical analysis of 7 key steroid hormones. Toxicol. Vitro 26, 343–350.

Nielsen, I.L., Williamson, G., 2007. Review of the factors affecting bioavailability of soy isoflavones in humans. Nutr. Cancer 57, 1–10.

NTP, 2008a. Toxicology and carcinogenesis studies of genistein (Cas No. 446-72-0) in Sprague-Dawley rats (feed study), National Toxicology Program. Natl. Toxicol. Program. Tech. Rep. Ser. 1–240.

NTP, 2008b. Genetic toxicity evaluation of genistein in Salmonella/E.coli mutagenicity test or Ames test study A90231 [Online]. National Toxicology Program. <http://tools.niehs.nih.gov/cebs3/ui/?casnumber=446-72-0> (accessed 29.05.2020).

Qin, Y., Niu, K., Zeng, Y., et al., 2013. Isoflavones for hypercholesterolemia in adults. Cochrane Database Syst. Rev. 6, CD009518.

Qiu, S., Jiang, C., 2019. Soy and isoflavones consumption and breast cancer survival and recurrence: a systematic review and meta-analysis. Eur. J. Nutr. 58, 3079–3090.

Rice, S., Mason, H.D., Whitehead, S.A., 2006. Phytoestrogens and their low dose combinations inhibit mRNA expression and activity of aromatase in human granulosa-luteal cells. J. Steroid Biochem. Mol. Biol. 101, 216–225.

Rietjens, I., Louisse, J., Beekmann, K., 2017. The potential health effects of dietary phytoestrogens. Br. J. Pharmacol. 174, 1263–1280.

Ronis, M.J., 2016. Effects of soy containing diet and isoflavones on cytochrome P450 enzyme expression and activity. Drug. Metab. Rev. 48, 331–341.

Rüfer, C.E., Bub, A., Moseneder, J., et al., 2008. Pharmacokinetics of the soybean isoflavone daidzein in its aglycone and glucoside form: a randomized, double-blind, crossover study. Am. J. Clin. Nutr. 87, 1314–1323.

Russo, J., Russo, I.H., 1994. Toward a physiological approach to breast cancer prevention. Cancer Epidemiol. Biomarkers Prev. 3, 353–364.

Sacks, F.M., Lichtenstein, A., Van Horn, L., American Heart Association Nutrition, C., et al., 2006. Soy protein, isoflavones, and cardiovascular health: an American Heart Association Science Advisory for professionals from the Nutrition Committee. Circulation 113, 1034–1044.

Scheffler, A., Albrecht, A.E., Esch, H.L., Lehmann, L., 2015. Mutagenic potential of the isoflavone irilone in cultured V79 cells. Toxicol. Lett. 234, 81–91.

Schwen, R., Jackson, R., Proudlock, R., 2010. Genotoxicity assessment of S-equol in bacterial mutation, chromosomal aberration, and rodent bone marrow micronucleus tests. Food Chem. Toxicol. 48, 3481–3485.

Setchell, K.D., Brown, N.M., Desai, P., et al., 2001. Bioavailability of pure isoflavones in healthy humans and analysis of commercial soy isoflavone supplements. J. Nutr. 131, 1362S–1375SS.

Setchell, K.D., Clerici, C., 2010. Equol: history, chemistry, and formation. J. Nutr. 140, 1355S–1362SS.

Silverstein, M.G., Kaplan, J.R., Appt, S.E., et al., 2014. Effect of soy isoflavones on thyroid hormones in intact and ovariectomized cynomolgus monkeys (*Macaca fascicularis*). Menopause 21, 1136–1142.

Solecki, R., Kortenkamp, A., Bergman, A., et al., 2017. Scientific principles for the identification of endocrine-disrupting chemicals: a consensus statement. Arch. Toxicol. 91, 1001–1006.

Song, H., Hughes, J.R., Turner, R.T., 2020. (+/−)-Equol does not interact with genistein on estrogen-dependent breast tumor growth. Food Chem. Toxicol. 136, 110979.

Soni, M., Rahardjo, T.B., Soekardi, R., et al., 2014. Phytoestrogens and cognitive function: a review. Maturitas 77, 209–220.

Sosvorova, L., Miksatkova, P., Bicikova, M., et al., 2012. The presence of monoiodinated derivates of daidzein and genistein in human urine and its effect on thyroid gland function. Food Chem. Toxicol. 50, 2774–2779.

Soukup, S.T., Helppi, J., Muller, D.R., et al., 2016. Phase II metabolism of the soy isoflavones genistein and daidzein in humans, rats and mice: a cross-species and sex comparison. Arch. Toxicol. 90, 1335–1347. Erratum in Arch. Toxicol., 90, 1349.

Speit, G., 2013. Does the recommended lymphocyte cytokinesis-block micronucleus assay for human biomonitoring actually detect DNA damage induced by occupational and environmental exposure to genotoxic chemicals? Mutagenesis 28, 375–380.

Steward, B.W., Wild, C.P., 2014. World Cancer Report 2014. International Agency for Research on Cancer, Lyon, France.

Stopper, H., Schmitt, E., Kobras, K., 2005. Genotoxicity of phytoestrogens. Mutat. Res. 574, 139–155.

Taku, K., Melby, M.K., Kronenberg, F., et al., 2012. Extracted or synthesized soybean isoflavones reduce menopausal hot flash frequency and severity: systematic review and meta-analysis of randomized controlled trials. Menopause 19, 776–790.

Tokede, O.A., Onabanjo, T.A., Yansane, A., et al., 2015. Soya products and serum lipids: a meta-analysis of randomised controlled trials. Br. J. Nutr. 114, 831–843.

Touillaud, M., Gelot, A., Mesrine, S., et al., 2019. Use of dietary supplements containing soy isoflavones and breast cancer risk among women aged >50 y: a prospective study. Am. J. Clin. Nutr. 109, 597–605.

Tsutsui, T., Tamura, Y., Yagi, E., et al., 2003. Cell-transforming activity and mutagenicity of 5 phytoestrogens in cultured mammalian cells. Int. J. Cancer 105, 312–320.

Upson, K., Adgent, M.A., Wegienka, G., et al., 2019. Soy-based infant formula feeding and menstrual pain in a cohort of women aged 23-35 years. Hum. Reprod. 34, 148–154.

Vandenberg, L.N., Colborn, T., Hayes, T.B., et al., 2012. Hormones and endocrine-disrupting chemicals: low-dose effects and nonmonotonic dose responses. Endocr. Rev. 33, 378–455.

Veitch, N.C., 2013. Isoflavonoids of the leguminosae. Nat. Prod. Rep. 30, 988–1027.

Walsh, K.R., Haak, S.J., Bohn, T., et al., 2007. Isoflavonoid glucosides are deconjugated and absorbed in the small intestine of human subjects with ileostomies. Am. J. Clin. Nutr. 85, 1050–1056.

Wei, Y., Lv, J., Guo, Y., et al., 2019. Soy intake and breast cancer risk: a prospective study of 300,000 Chinese women and a dose-response meta-analysis. Eur. J. Epidemiol. 35 (6), 567–578.

Whitsett Jr., T.G., Lamartiniere, C.A., 2006. Genistein and resveratrol: mammary cancer chemoprevention and mechanisms of action in the rat. Expert. Rev. Anticancer. Ther. 6, 1699–1706.

Xiao, C.W., 2008. Health effects of soy protein and isoflavones in humans. J. Nutr. 138, 1244S–1249SS.

Yager, J.D., 2015. Mechanisms of estrogen carcinogenesis: the role of E2/E1-quinone metabolites suggests new approaches to preventive intervention—a review. Steroids 99, 56–60.

Yan, Z., Zhang, X., Li, C., et al., 2017. Association between consumption of soy and risk of cardiovascular disease: a meta-analysis of observational studies. Eur. J. Prev. Cardiol. 24, 735–747.

Yonemoto-Yano, H., Maebuchi, M., Fukui, K., et al., 2014. Malonyl isoflavone glucosides are chiefly hydrolyzed and absorbed in the colon. J. Agric. Food Chem. 62, 2264–2270.

Zamora-Ros, R., Knaze, V., Lujan-Barroso, L., et al., 2012. Dietary intakes and food sources of phytoestrogens in the European Prospective Investigation into Cancer and Nutrition (EPIC) 24-hour dietary recall cohort. Eur. J. Clin. Nutr. 66, 932–941.

Zhao, M., Vaartjes, I., Graham, I., et al., 2017. Sex differences in risk factor management of coronary heart disease across three regions. Heart 103, 1587–1594.

Zhao, T.T., Jin, F., Li, J.G., et al., 2019. Dietary isoflavones or isoflavone-rich food intake and breast cancer risk: a meta-analysis of prospective cohort studies. Clin. Nutr. 38, 136–145.

Ziegler, R.G., Hoover, R.N., Pike, M.C., et al., 1993. Migration patterns and breast cancer risk in Asian-American women. J. Natl. Cancer Inst. 85, 1819–1827.

Chapter 46

Propolis

Tirupapuliyur Damodaran

Introduction

Propolis (PP) is a mixture of resinous substances used by honeybees as a protective structure for defending the hives in which they live. This honeybee glue is collected from several plant sources. The list of verified sources that form bee wax includes essence of flower buds, exudates of various plant parts, and salivary enzymes from the bees. The color of PP is generally dark brown, but may vary depending on the source of plant materials. Because of the high resin material content, Propolis becomes stickier at and above 20°C. However, it may become hard and brittle at lower temperatures. PP components provide reinforcements to the structure of the beehive by filling any openings created in the walls of the hive, thus making cold winter days warmer. The wax-like quality of the PP traps hive intruders and mummifies them, preventing decay-related changes in the hives. The presence of PP and wax in the beehives makes the structure more stable. Bees also have been known to use PP for lining brood cells to ensure the health of their larvae by creating a sterile environment.

PP has been in use since ancient times and has become a part of indigenous traditional medicinal practice around the globe due to its low toxicity and plurality of usages in many areas of daily life. The term "PP" comes from Greek, meaning "the entrance to the city." Because of the ubiquitous nature of different species of bees, PP (from various continents and countries with specific climates allowing the growth of a variety of plants) has a plethora of variable components. Seasonal variability between the concentrations of the various components of PP (especially components like vestitol, neovestitol, and isoliquiritigenin) has been reported. The highest contents of these ingredients and antibacterial activity are recorded during the rainy season (Isla et al., 2012).

Different species of bees seem to produce PP with differing contents. The term "geopropolis" is used to describe such variations. While comparing two ethanolic extracts of PP collected from two stingless species of bees, *Melipona quadrifasciata* and *Tetragonisca angustula*, more significant activity of geopropolis extracts against Gram-positive (*Staphylococcus aureus* MSSA and MRSA, *Enterococcus faecalis*) than Gram-negative bacteria (*Klebsiella pneumoniae*, *Escherichia coli*) was noted. From the two geopropolis which were analyzed, the *M. quadrifasciata* was more effective in its antibacterial capabilities (Bankova, 2005). Adding to the variability in the components are the different extraction methodologies employed for processing raw PP. As extraction of the raw PP is required to dissolve and release the active ingredients, several solvents have been used, including ethanol, methanol, water, hexane, acetone, dichloromethane, and chloroform. The extraction procedure may yield an approximate amount of 70% concentration of PP. Depending on the solvent used, variations in biological activity have been noted. For example, higher concentrations of phenolic compounds in ethanolic extracts of brown, green, and red PP than in extracts obtained by supercritical extraction have been reported. At the same time, the levels of flavonoids were higher in ethanolic extracts of green and red PP and lower in brown PP, compared to supercritical extraction (Devequi-Nunes et al., 2018).

Based on the consistencies of certain components, PP has been shown to have the following types of molecules as core components: (1) resin and balsam (50%); (2) wax (30%); (3) essential and aromatic oils (10%); (4) pollen (5%); and (5) other substances such as organic debris (5%) (Anjum et al., 2019). Irrespective of the variations in the chemical composition, various types of PP from different geographical locations have similar functionalities. One unifying quality of PP from different origins lies in its ability to function in all or most of the following biomedical aspects: (1) antiinflammatory; (2) antiproliferative; (3) antioxidant; (4) antiseptic; (5) immunomodulatory; (6) antimicrobial; and (7) wound-healing properties. PP also contains persistent lipophilic acaricides, a natural pesticide that deters mite infestations (WHO, 2008).

Nutraceuticals. DOI: https://doi.org/10.1016/B978-0-12-821038-3.00046-X

In order to maximize the utilization of various beneficial aspects of PP in the biomedical, nutritional, skin care, and other industries, a thorough discussion on the classification of PP subgroups, identification of its components, and a delineation of the molecular cell biological pathways is warranted. This chapter mostly focuses on the biomedical and molecular basis of the biological outcomes following PP treatment.

Chemical compounds of propolis

Due to the extensive variability that is influenced by geographical, seasonal, and other aspects of the PP synthesis process, the chemical composition and molecular components that arise from it are quite diverse in nature. Overall, the spectrum of the chemical composition of PP is closely related to the plant sources used and species of bees involved in its synthesis. It is estimated that more than 300 chemical components make up the molecular profile of PP (Table 46.1).

Table 46.2 provides details of the salient chemical compounds that make up PP, which can be grouped into (1) flavonoids such as chrysin, pinocembrin, apigenin, galangin, kaempferol, quercetin, tectochrysin, pinostrobin, and others; (2) aromatic acids such as ferulic, cinnamic, caffeic, benzoic, salicylic, and *p*-cumaric acids; (3) terpene-related compounds such as artepillin C, terpenes, camphor, terpineol, geraniol, and nerol (Przybyłek and Karpiński, 2019). Polyphenols and terpenoids are also considered to be the most active components responsible for the fragrance and nutritional/biomedical aspects of PP (Zabaiou et al., 2017).

Geographical origin and composition of propolis

The geographical distribution of various plant kingdom members around the world and the plant components used by the bees determine the content of the PP in a particular region. Season-related changes in various countries also contribute to the variability of the components that make up the PP varieties. North America, Europe, nontropical parts of Asia, and New Zealand have a tree variety called *Populus nigra* (the poplar tree). Middle Eastern countries like Egypt are also known to have the poplar tree. The components of PP include phenols, esters of caffeic acid, and long-chain fatty alcohols including tetradecanol, hexadecanol, and dodecanol (Bankova et al., 2000). Russian PP comes from birch flavonols and flavones emanating from other plant sources (*Betula verrucosa*).

Brazilian PP comes from *Baccharis dracunculifolia* leaf resin. Its components include diterpenes, lignans, prenylated derivatives of *p*-coumaric acid, acetophenone, and flavonoids. Brazilian PP also contains artepillin C in large amounts as compared to caffeic acid phenethyl ester (CAPE) (Chan et al., 2013; Ferreira et al., 2017). PP from tropical zones has other components such as sesquiterpenoids, including germacren-D, ledol, and spatulenol (Bankova et al., 2000). *Clusia rosea* is the main plant source for Cuban PP, which contains polyisoprenylated benzophenone (Bankova, 2005; Bankova et al., 2000). The complexity of the activities of various components of PP in modulating various functional aspects of platelet aggregation provides the prototypical scenario of variability not just in the ratio and distribution of PP components but also in their molecular behavior within any organ system or parts of the organ system (Table 46.3).

Biomedical applications of propolis

Because of the numerous and wide variety of useful components accompanied by its low toxicity, PP is used for a huge number of applications within many aspects of human life. Every major organ system has been found to have some beneficial use of PP. PP has a wide spectrum of antimicrobial effects (antibacterial, antifungal, antiviral, etc.). It has an excellent capability to function as an antiinflammatory agent and its antioxidant function is well known (Sforcin, 2016; Omar et al., 2017). Complex organ systems also benefit from PP treatment as follows: cardiovascular (antihypertensive), hepatic system (antihepatotoxic), nervous system (antineuroinflammatory and antineural death), gastrointestinal tract (GIT) (antimicrobial/microbiome homeostasis), reproductive system (optimization of gender-specific reproductive function and protection from toxicity), integumentary/dermatological (wound healing), urinary system (optimization of renal function), and bones (antioxidant and protection against degenerative changes). PP also functions as an anticarcinogenic agent as it can increase cytotoxicity and apoptosis, while decreasing excessive proliferation of cancer cells (Toreti et al., 2013). PP can also effectively minimize the toxicity damage to various organ systems which is a side effect of chemotherapy (Akyol et al., 2016). Enhanced functioning of the immune system as a direct outcome of PP treatment has been demonstrated. The emergence of PP as an adjuvant for vaccine delivery is a recent concept. Complex disorders such as diabetes have been shown to respond well to PP treatment, thus minimizing target organ damage. Although the benefits of PP treatment are numerous, a small percentage of humans may be allergic to some of its components.

Antioxidant activity of propolis

PP has been shown to have an antioxidant function due to its components galangin and pinocembrin (El-Guendouz

TABLE 46.1 Gross composition of propolis.

Compounds	References
Flavonoids, flavanones, flavones, and flavonols	Walker and Crane (1987)
Islapinin, ermanin, pectolinarigenin, sakuranetin, isosakuranetin, quercetin 3,30-dimethyl ether, 3-acetyl pinobanksin, betuletol, isorhamnetin, kaempferide, rhamnazin, rhamnetin, alnusin, alpinetin, alnusitol, pinostrobin, pinocembrin, chrysin, tectochrysin, acacetin, rhamnocitrin, quercetin, galangin, apigenin, pinobanksin, kempferol, rutin, catechin, luteolin, naringenin	Lotfy (2006)
Benzoic acid and derivatives	Walker and Crane (1987)
Benzoic acid, salicylic acid, gentisic acid, gallic acid, phenylmethyl ester of benzoic acid, phenylmethyl ester of salicylic acid, *trans*-coniferyl benzoate, *trans-p*-coumaryl benzoate, protocatechuic acid	
Benzaldehyde derivative:	
Vanillin, caproic aldehydes, isovanillin *p*-hydroxybenzaldehyde, protocatechualdehyde	Abdulkhani et al. (2017), Akbay et al. (2017), Marcucci (1995), Walker and Crane (1987)
Cinnamyl alcohol, cinnamic acid, and its derivatives	Walker and Crane (1987)
Cinnamyl alcohol, hydrocaffeic acid, isoferulic acid, cinnamic acid methyl ester, cinnamic acid ethyl ester, cinnamylidene acetic acid, cinnamic acid, caffeic acid, ferulic acid	
Aliphatic hydrocarbons	
Eicosine, 1-octadecene, tricosane, pentacosane, eicosane, heneicosane	
Sugar	
d-Ribofuranose, d-fructose, d-glucitol, d-gulose, talose, sucrose, d-glucose	
Vitamins	
B1, B2(complex), B6, C, E	Kuropatnicki et al. (2013)
Nicotinic acid, pantothenic acid, chalcones, and dihydrochalcones	
Alpinetin chalcone, naringinen chalcone, pinobanksin chalcones, pinobanksin-3-acetate chalcone, pinostrobin chalcone, pinocembrin chalcones, sakuranetin chalk one, 20, 60, a-trihydroxy-40-methoxy chalcone, 20, 6, dihy droxy-40-methoxydihydro chalcone, 20, 40, 6- trihydroxydihydro chalcone	Marcucci (1995)
Amino acids	
Alanine, b-alanine, a-amino butyric acid, d-amino butyric acid, arginine, asparagine, aspartic acid, cystine, cystein, glutamic acid, glycine, histidine, hydroxyproline, isoleucine, leucine, lysine, methionine, ornithine, phenylalanine, proline, pyroglutamic acid, sarcosine, serine, threonine, tryptophane, tyrosine, valine	
Esters	El Hady and Hegazi (2002)
Palmitate, cinnamyl-*trans*-4-coumarate, ethyl palmitate, stearic acid methyl ester, phthalate ester, benzyl benzoate, benzyl-trans-4- coumarate, 3-methyl-3-butenyl isoferulate, 3-methyl-2-butenyl isoferulate, 3-methyl-3- butenyl caffeate, 2-methyl-2-butenyl caffeate, 3-methyl-2-butenyl caffeate, benzyl caffeate, phenylethyl caffeate, cinnamyl caffeate, tetradecyl caffeate, tetradecenyl caffeate, tetradecenyl caffeate (isomer)b, tetradecanyl caffeateb, hexadecyl caffeate	
Other acids and derivatives	Walker and Crane (1987)
Phenylmethyl ester of 14-methylpentadecanoic acid, ethyl ester of palmitic acid, myristic acid, sorbic acid, butyl-2-methylpropyle ester of phthalic acid, stearic acid, methyl ester of alnustic acid	

(Continued)

TABLE 46.1 (Continued)

Compounds	References
Alcohol, ketones, phenols, and heteroaromatic compounds	
Benzyl alcohol, hexadecanol acetate, coumarine, pterostilbene, xanthorrhoeol, scopoletol	
Terpene, sesquiterpene, alcohol, and derivatives	
Geraniol, neroledol, b-bisabolol, guaiol, farnisol, dihydroeudesmol, aacetoxybetulenol	
Sesquiterpene and triterpene hydrocarbons	
b-Patchoulene, b-bisabolene, squalene, b-bourbonene, copaene, calarene, calamenene, caryophyllene, patchoulane, selenene, aromadendrene	
Sterols and steroid hydrocarbons	
Cholestrilene, cholinasterol, stigmasterol, b-dihydrofucosterol, lanosterol, cholesterol	
Minerals	
Sr, Ba, Cd, Sn, Pb, Ti, Ag, Co, Mo, Al, Si, V, Ni, Mn, Cr Na, Mg, Cu, Ca, Zn, Fe, K	Lotfy (2006), Pasupuleti et al. (2017)
Enzymes	Walker and Crane (1987)
Glucose-6-phosphatase, acid phosphatase, adenosine triphosphatase, succinic dehydrogenase	
Ketones	Marcucci (1995)
Acetophenone, *p*-acetophenolacetophenone, dihydroxy-acetope9i9inone, methylacetophenone, hept-5-en-2-one, 6-methylketone	
Archid acid, behenic acid, cerotic acid, lauric acid, linoleic acid, lignoceric acid, montanic acid	
Aliphatic acids and aliphatic esters	
Acetic acid, angelic acid, butyric acid, crotonic acid, fumaric acid, isobutyric acid, methylbutyric acid, isobutyl acetate, isopentyl acetate, isopentinyl acetate	
Alcohol	
Benzene methanol, cinnamyl alcohol, glycerol, a-glycerophosphate, phenethyl alcohol, isobutenol, hydroquinone, prenyl alcohol	
Aliphatic acids	El Hady and Hegazi (2002)
Lactic acid, hydroxyacetic acid, malic acid, 5-hydroxy-n-valeric acid, 2,3-dihydroxypropanoic acid, pentonic acid- 2-deoxy-3,5-dihydroxy-c-lactoneb, pentonic acid- 2-deoxy-3,5- dihydroxy-c-lactone (isomer)b, succinic acid, 2,3,4,5-tetrahydroxypentanoic acid-1,4-lactone, 2,3,4,5-tetrahydroxypentanoic acid-1,4-lactone(isomer)b, nonanoic acid, palmitic acid, oleic acid, decanoic acid, dodecanoic acid, tetradecanoic acid, heptadecanoic acid, octadecenoic acid, tetracosanoic acid, eicosanoic acid, hexacosanoic acid, 2-hydroxy hexacosanoic acid b	
Fatty acids (C7-C18 acids) and other compounds	
Phosphoric acid, 1,4-dihydroxy benzene, 4-hydroxy-benzaldehyde, 4-hydroxy acetophenone, 1,2,4-trihydroxy butane, 1,2,3-trihydroxy butanal, 1,2,3-trihydroxy butanal (isomer)b, myristicin, 2,4-bis(dimethyl benzyl)-6-t-butyl phenol, 1,8-dihydroxy-3-methyl anthraquinone, myristicin (isomer)	

Murtaza, G., Karim, S., Akram, M.R., et al., 2014. Caffeic acid phenethyl ester and therapeutic potentials. BioMed. Res. Int. 2014, 19. Article ID 145342.

TABLE 46.2 Chemical structures of common components of propolis of all geographical origin.

Flavonoids
1. Chrysin
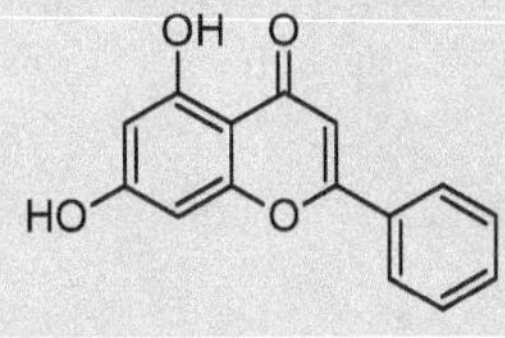
2. Pinocembrin
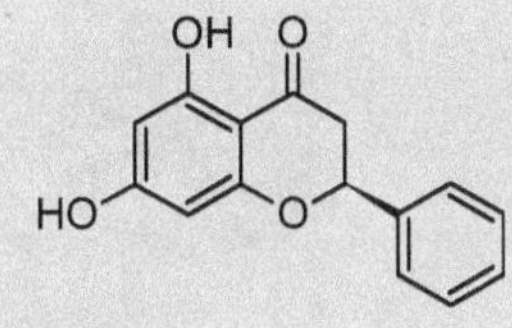
3. Apigenin
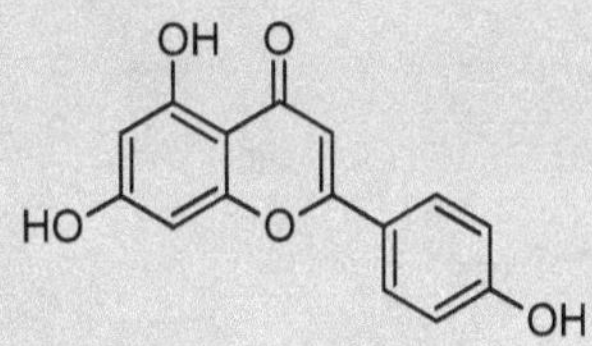
4. Galangin
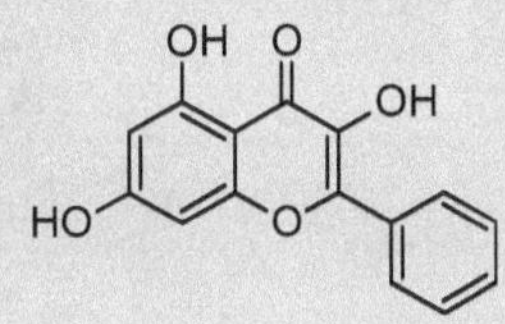
5. Kaempferol
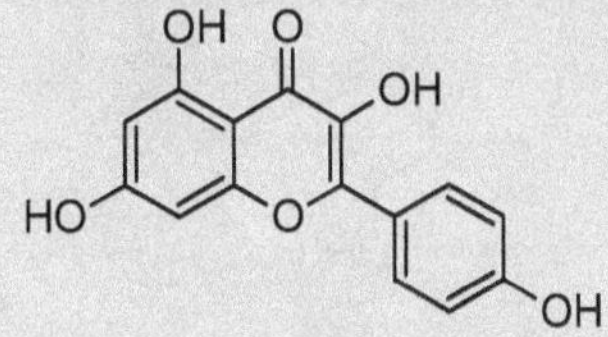
6. Quercetin
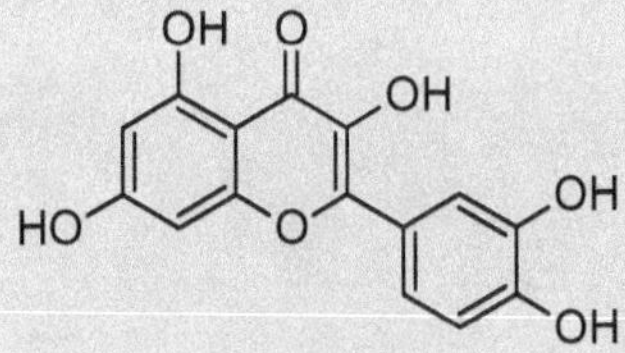

(Continued)

TABLE 46.2 (Continued)

7. Tectochrysin

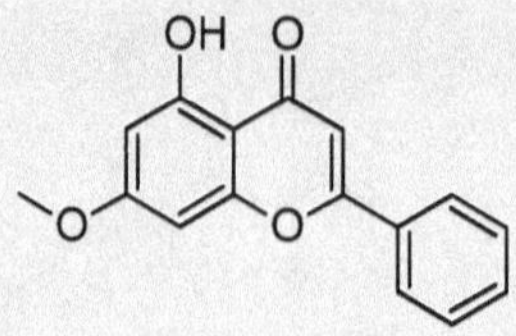

8. Pinostrobin

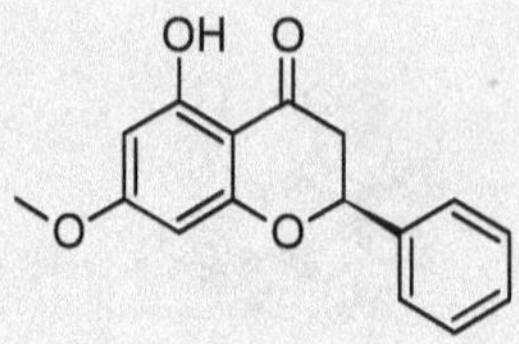

Aromatic acids

1. Ferulic acid

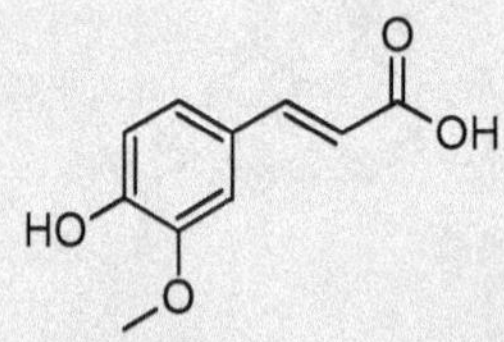

2. Cinnamic acid

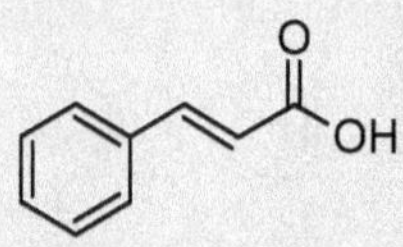

3. Caffeic acid

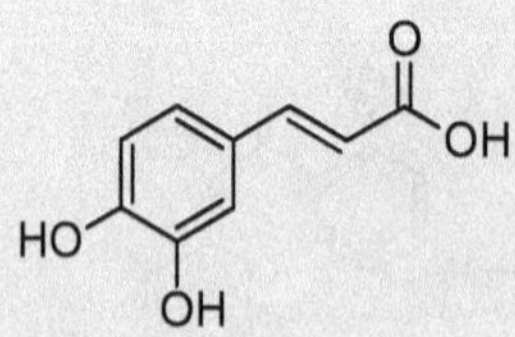

4. Benozoic acid

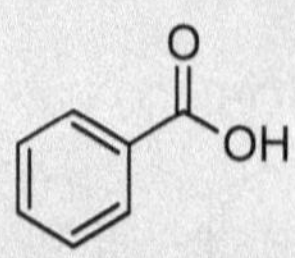

5. Salicylic acid

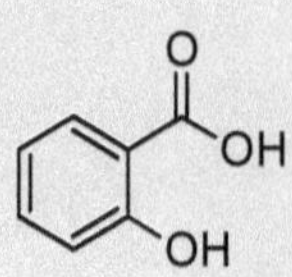

(Continued)

TABLE 46.2 (Continued)

6. *p*-Cumaric acid

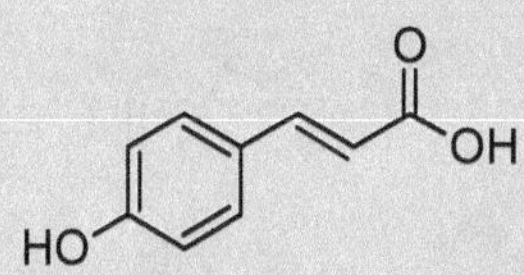

Other compounds

1. Artepillin C

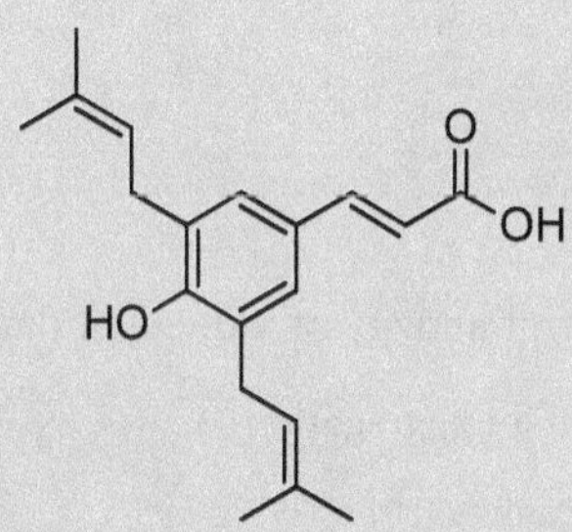

2. Terpenes

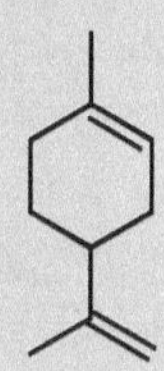

Limonene (monoterpene)

3. Camphor

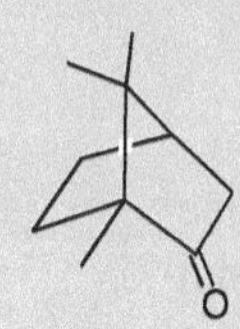

(this is one of the terpenoids which are derivatives of terpenes)

4. Terpineol

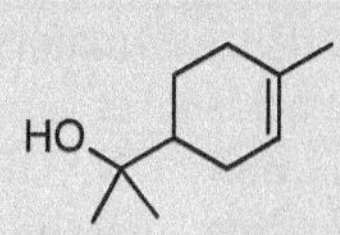

(this is also one of the monoterpenes)

4. Geraniol

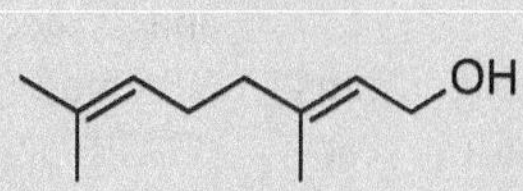

(this is also one of the monoterpenes)

(*Continued*)

TABLE 46.2 (Continued)

6. Nerol

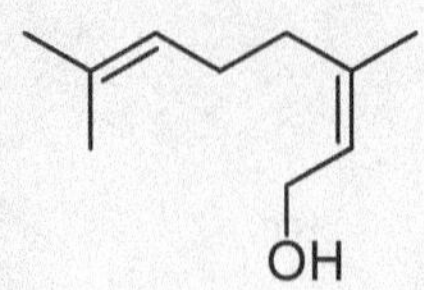

(this is also one of the monoterpenes)

TABLE 46.3 Complex mechanisms of antiplatelet activity of individual flavonoids (Bojić et al., 2019).

Mechanism of action	Influence on platelet aggregation
Change of bilayer function	↓ Quercetin, naringenin, hesperetin
Oxidative stress	↓ Epicatechin ↓ hesperetin ↔ hesperetin ↑hesperetin (via xanthine oxidoreductase) ↑ baicalein, morin, myricetin, quercetin, rutin
Intracellular Ca^{2+} concentration	↓ Hesperetin ↓ 3,6-dihydroxyflavone↑ isorhamnetin, isosakuranetin
Inhibition of phospholipase C	↓ Quercetin, apigenin, catechin, tamarixetin
Inhibition of cAMP phosphodiesterase	↓ Hesperetin ↓ luteolin, quercetin, apigenin
Inhibition of cyclooxygenase	↓ Hesperetin, quercetin ↓ luteolin, kaempferol, hesperetin, naringenin
Inhibition of thromboxane A2 synthase	↓ Epicatechin

et al., 2017; Machado et al., 2017). The following molecular mechanisms have been identified as operating simultaneously or based on the need of the tissues in their local milieu of the organ systems involved:

1. Antioxidants have the ability to refuse free radicals, thus preventing destruction of vitamin C, lipids, and other compounds by oxidization (Anjum et al., 2019).
2. Components of PP such as vanillin and phenolic acids have the capability of gaining entry through the epidermis as well as dermis and protect them from free radicals produced due to radiation or before maturation of dermal cells (Król et al., 2013). This also probably holds true for any chemical exposure-related scenarios.
3. Phenolic compounds of PP may donate hydrogen ions to free radicals to protect cells from oxidation reactions via stress-related physiological changes and poisoning. PP has the ability to remove free radicals, thereby preventing oxidation of lipids, nucleic acids, and proteins (Chandna et al., 2014).

Antiinflammatory activity of propolis

The flavonoid component of PP provides its antiinflammatory capability (Martinotti and Ranzato, 2015; Anjum et al., 2019). This component modulates the activities of NADPH oxidase ornithine decarboxylase, myeloperoxidase, hyaluronidase, and tyrosine-protein kinase (Lotfy, 2006). The production of chemical mediators such as leukotrienes and prostaglandins is induced by white blood cells in response to PP treatment (Machado et al., 2017) and is accompanied by inhibition of the aforementioned enzymes (Lotfy, 2006; Anjum et al., 2019). Components (CAPE and galangin) of poplar PP demonstrated antiinflammatory property by inhibiting carrageenan-induced pleurisy and edema, and have the ability to reduce arthritis-related inflammations in rats (Wagh, 2013). Brazilian and Chinese PP were shown to reduce the pathogenesis of collagen-induced arthritis in mice (Sforcin, 2016). PP downregulates inflammation accumulating in cells due to exposure to poisonous materials or other harmful conditions (Król et al., 2013).

Antibacterial activity of propolis

In an extensive study, more than 600 bacterial strains were analyzed and the outcome indicated that PP inhibited the Gram-positive bacteria to a greater extent than the Gram-negative bacteria (Przybyłek and Karpiński, 2019).

As a part of the study, the antimicrobial activity of PP from different regions of the world was compared. The potency of PP from the Middle East was greater in killing both Gram-positive (*S. aureus*) and Gram-negative (*E. coli*) strains. The potency of PP from Germany, Ireland, and Korea was low (Przybyłek and Karpiński, 2019).

The antibacterial activity of PP may be connected to the following two broad approaches:

1. Direct action on the microorganism, which is achieved by increasing the membrane permeability, disruption of membrane potential, interference in adenosine triphosphate (ATP) production, and decreasing bacterial mobility, and
2. Stimulation of the immune system resulting in activation of the natural defenses of the organism.

The mode of action of PP is due to the interaction between phenolic and other compounds such as pinocembrin, galangin, and pinobanksin (Wagh, 2013). Pinocembrin shows antibacterial activity towards *Streptococcus* spp., while artepillin C, *p*-coumaric acid, and 3-phenyl-4-di hydrocinnamylocinnamic acid inhibited *Helicobacter pylori* and apigenin greatly by restraining bacterial glycosyltransferase (Martinotti and Ranzato, 2015). Similarly, antibacterial activity is also mediated by aromatic compounds (caffeic acid) and flavonoids (Parolia et al., 2010). Moreover, PP acts as a bactericidal agent to stop division of the bacterial cell, destroy the cell wall and bacterial cytoplasm (Parolia et al., 2010), and causes cessation of protein synthesis (Machado et al., 2017). Many studies have documented the antimicrobial activity of cinnamic acid against *Aeromonas* spp., *Vibrio* spp., *E. coli*, *Listeria monocytogenes*, *Mycobacterium tuberculosis*, *Bacillus* spp., *Staphylococcus* spp. *Streptococcus pyogenes*, *Micrococcus flavus*, *Pseudomonas aeruginosa*, *Salmonella enterica* serotype *Typhimurium*, *Enterobacter cloacae*, and *Yersinia ruckeri* (Machado et al., 2017).

Cinnamic acid and its derivatives inhibit bacteria by damaging the cell membrane, and inhibiting ATPases, cell division, and biofilm formation. Moreover, they have shown antiquorum-sensing activity (Machado et al., 2017). There is limited research on the action of PP on anaerobic bacteria. A higher degree of efficacy against *Clostridium*, *Bacteroides*, *Porphyromonas*, *Prevotella*, *Fusobacterium*, *Actinomyces*, and *Propionibacterium* species has been noted (Machado et al., 2017). An active PP component, CAPE, inhibits *H. pylori* peptide deformylase activity (Cui et al., 2013). PP also has a significant effect against bacteria such as *Enterococcus* spp., *E. coli* and *S. aureus* (AL-Waili et al., 2012; Kasiotis et al., 2017).

Antifungal activity of propolis

Various components of PP have shown antifungal activity (Franchin et al., 2016). These components were shown to diminish conidial growth in *Aspergillus flavus*, *Aspergillus fumigates*, and aflatoxigenic fungi. Components of PP such as, 3-acetylpinobanksin, pinobanksin-3-acetate, pinocembrin, *p*-coumaric acid, and caffeic acid have been identified as key molecules that exhibit antifungal activity (Anjum et al., 2019). PP components have been shown to be effective against several *Candida* spp., such as *Candida guilliermondii*, *Candida krusei*, *Candida albicans*, *Candida pelliculosa*, *Candida parapsilosis*, *Pichia ohmeri*, *Candida famata*, and *Candida glabrata* (Wagh, 2013). Pinocembrin showed activity against *Penicillium italicum* by diminishing mycelial growth.

Respiratory and energy homeostasis of the fungal pathogen was impaired due to pinocembrin treatment, leading to rupturing of the cell membrane and metabolic disorder (Sforcin, 2016). PP also showed a fungicidal effect against yeast (Sforcin, 2016). Another component of PP (caffeic acid) showed antimycotic activity toward *Helminthosporium carbonum* (Özcan, 1999). PP has been effective against *Mycobacteria*, *Candida*, *Trichophyton*, and *Fusarium* as well as other fungi that cause dermatological disorders (Fokt et al., 2010). *Fusarium oxysporum*, *Fusarium sporotrichioides*, *Fusarium subglutinans*, and *Fusarium proliferatum* were shown to be very efficiently inhibited by PP components (Ristivojević et al., 2020). Biological control of yeast contamination of industrial foods by PP treatment has been shown to be an effective and safer alternative for food preservation than chemical methods (Alsayed et al., 2020).

Antiprotozoal activity of propolis

Several components of PP, such as caffeic acid, chrysin, moronic acid, protocatechuic acid, *p*-coumaric acid, apigenin, terpenoids, esters, and phenols have been shown to be effective against protozoal and other microbes (Sforcin, 2016). The list of protozoans that are destroyed by PP include the following: *Leishmania donovani*, *Trypanosoma cruzi*, *Giardia lamblia*, *Trichomonas vaginalis*, *Toxoplasma gondii*, and *Giardia duodenalis* (Wagh, 2013; Anjum et al., 2019). Other phenolic compounds from Brazilian PP, such as 2,2-dimethyl-6- carboxyethenyl-2H-1-benzo pyran, 3,5-diprenyl-4-hydroxycinnamic acid, 3-(2,2-dimethyl-8-p renylbenzopyran-6-yl) propanoic acid, and 3,5-diprenyl-4hydroxycinnamicacid, have been shown to be effective antimicrobial compounds against *T. cruzi* (Anjum et al., 2019). Thus, giardiasis, Chagas disease, leishmaniasis, trichomoniasis, and toxoplasmosis can be treated effectively by PP.

Antiviral activity of propolis

Several flavonoid components of PP, which include kaempferol, acacetin, quercetin, galangin, and chrysine, were effective in treating cells infected with viruses, as they were shown to induce cytotoxicity (Marcucci, 1995). Other compounds of PP such as 3-methyl-but-2-enyl caffeate have been shown to inhibit viral titer by diminishing DNA synthesis of herpes simplex virus (type 1) (Bankova et al., 2014). Virus entry into cells has been shown to be inhibited by PP (Amoros et al., 1992), while it was shown to be very effective against replication by destroying virus RNA (Sforcin, 2016). Antiviral activities against HSV-2 (genital herpes virus) have been shown (Kuropatnicki et al., 2013). Influenza virus A1 has been inhibited by isopentyl ferulate, a PP component (Lotfy, 2006). Other viruses identified as being inhibited by components of PP include Rift Valley fever, Newcastle disease, herpes bursal disease, and influenza (El Hady and Hegazi, 2002).

Neuroprotective mechanism conferred by propolis in neurodegenerative diseases

Aging of the global population has been a subject of interest and concern as age-related ailments and the need for specialized healthcare approaches are on the rise. It is well established that many organ systems tend to lose functional units (such as neurons in the nervous system, nephrons in the kidney, alveoli in the lungs, muscles fibers in the muscular systems, and osteons in bone), thereby increasing their susceptibilities for developing disorders and diseases. Several neurodegenerative disorders such as Parkinson disease (PD), Alzheimer disease, amyotrophic lateral sclerosis, Huntington disease, and multiple sclerosis have been known to have the following commonalities: (1) progressive neuronal loss, (2) impaired nervous system, (3) chronic neuronal inflammation, (4) oxidative stress, and (5) mitochondrial dysfunctions (Wang et al., 2018). Even though therapeutic interventions using various approaches to combat some or many of the aspects of disease progression have been underway, there is always a need for novel approaches that may enhance the possibility of a better outcome with fewer side effects. Nutrition-based approaches are always considered to be safer and less expensive. Many of the indigenous approaches may have an impact on multiple molecular pathways that improve the various pathological conditions affecting the nervous system.

Research around the globe on dietary phytochemicals, such as PP, mulberry, and green tea has identified at least three therapeutic attributes: gastrointestinal function improvement, immunity enhancement, and neuroprotective outcomes (Wang et al., 2018). Various components of the central nervous system, peripheral nervous system, and vagal reflex pathway acting via the enteric nervous system are involved in recruiting gastrointestinal signaling to the brain, besides initiating the inflammatory infiltration pathway through the blood–brain barrier (BBB) and gateway reflex reaction (Wang et al., 2018). PP treatment has been shown to enhance the efficiency of various injury responses and neuroprotection pathways, operating via glial cells (astroglial and microglial) of the nervous system (Wang et al., 2018). Exposure to neurotoxic chemicals during accidents, terrorist attacks, and chemical warfare has been a matter of concern for all age groups. Other scenarios such as hypoxia and ischemia may have a drastic effect on the nervous system. A limited amount of research addressing these issues and assessing the usage of PP as a treatment option for protection against neurotoxicity as well as a treatment for such an exposure has been reported.

Oxygen-derived free radicals have been implicated in the pathogenesis of cerebral injury after ischemia reperfusion (Lotfy, 2006). PP components such as CAPE and α-tocopherol have been shown to suppress ischemia–reperfusion-induced cerebral lipid peroxidation and injury in a rat model (Lotfy, 2006). CAPE was equally protective for the spinal cord from ischemia–reperfusion injury (Lotfy, 2006). In an experimental autoimmune encephalomyelitis (EAE) model, CAPE significantly inhibited reactive oxygen species (ROS) production induced by EAE and ameliorated clinical symptoms in rats by inhibiting ROS production at the transcriptional level through the suppression of nuclear factor kappa B (NF-κB) activation, and by directly inhibiting the catalytic activity of inducible nitric oxide synthase (NOS) (Lotfy, 2006)

Gastrointestinal health by propolis and its connection with the brain's innate immunity via the inflammatory reflex

Increased expression of tight junction-related genes such as occludin and zonula occludens-1, activating AMP-activated protein kinase and extracellular regulated protein kinase signaling, has been shown to protect the intestinal barrier by PP treatment in Caco-2 cells and in rats (Wang et al., 2018). Furthermore, intestinal health has been shown to be protected by CAPE by its ability to inhibit NF-κB signaling and TNF-induced and IFN-induced protein-10 (IP-10) expression in intestinal epithelial cells. This antiinflammatory effect is achieved through PP treatment, independent of p38 mitogen-activated protein kinase (MAPK) and HO-1 signaling pathways (Khan et al., 2018; Swamy et al., 2014).

Systemic immunity regulation by propolis and its connection with brain innate immunity across the blood–brain barrier

PP components (primarily flavonoids, phenolic compounds, and CAPE) have been shown to have a wide spectrum of pharmacological activities such as antiinflammation and antioxidation in systemic immunity. Cellular immunity [via peritoneal macrophages and cytotoxicity of natural killer (NK) cells] as an initial step of the immune response is mediated by increased phagocytosis and expression of key inflammatory response molecules such as IL-1β, IL-6, TLR2, and TLR4 following PP treatment (Gao et al., 2014). Downregulation of TLR2 and TLR4 signaling along with diminished levels of IL-4, IL-10, and transforming growth factor-β (TGF-β), as well as attenuation of proinflammatory mediators (IL-12, GM-CSF, IFN-γ, and IL-1β), accompanied with moderate upregulation of TNF-α and IL-6 by Brazilian red PP, has been effective as an antiinflammatory pathway protecting the BBB (Bueno-Silva et al., 2015). Proinflammatory cytokines (e.g., TNF-α, IL-1α, IL-1β, IL-6, IL-4, and ICAM-1) have been shown to be downregulated by CAPE. This has been accompanied by upregulation of antiinflammatory cytokines such as IL-10 (Cho et al., 2014).

Brain innate immunity regulation by phytochemicals and its neuroprotective implication

The neuroprotective effects of PP have been confirmed in numerous studies. In experimental hypoxia conditions, PP treatment inhibited the proinflammatory cytokines IL-1β, TNF-α, and IL-6 in microglia, while diminishing the generation of ROS from the mitochondria. PP treatment further downregulated the activation of NF-κB (Wang et al., 2018). Furthermore, CAPE has been shown to inhibit NOS and Cox-2 expression and NO production, while increasing the expression of HO-1 and erythropoietin in microglia (Tsai et al., 2015). Similarly, PP treatment conferred neuroprotection in kainic acid-induced excitotoxicity (Wang et al., 2018).

CAPE has been shown to protect dopaminergic neurons against ROS-mediated mitochondrial permeability transition, leading to cytochrome C- and caspase-3-mediated neuronal death, thereby effectively preventing neurodegenerative conditions like PD (Wang et al., 2018). Chrysin also has been shown to inhibit proinflammatory cytokines (Ha et al., 2010). Furthermore, inhibition of acetylcholinesterase activity by intragastric administration of PP has been observed in the hippocampus of scopolamine-treated mice, indicating that PP may protect brain function through a vagal reflex (Chen et al., 2008).

Propolis and cardiovascular function

Myocardiopathy has multiple etiologies, of which oxidative stress has been a verified factor, occurring through free radical formation. In a rat model, induced myocardiopathy by doxorubicin and the modifying effect of PP treatment were investigated. Elevated levels of serum creatine phosphokinase, aspartate aminotransferase, blood and tissue glutathione (GSH), and thiobarbituric acid reactive substances in the heart after doxorubicin administration have been shown to be restored closer to normal levels following PP treatment. Histopathological examination also confirmed the protective role of PP against induced myocardiopathic conditions. The protection offered by PP has been comparable to known myocardial protective agents such as rutin (Chopra et al., 1995). PP also showed an antihypertensive effect in rats (Kubota et al., 2004).

Similarly, administration of PP to diabetic rats showed several beneficial changes in the levels of various molecules across the spectrum of cardiovascular health-related scenarios. The following molecules were shown to have reached closer to normal levels: blood glucose (FBG), fructosamine, malondialdehyde (MDA), nitric oxide, NOS, total cholesterol, triglyceride (TG), low-density lipoprotein cholesterol, and very low-density lipoprotein cholesterol in the serum of fasting rats. This is accompanied by increased serum levels of beneficial molecules such as high-density lipoprotein cholesterol and superoxide dismutase (Lotfy, 2006). This study also confirmed that modulation of glucose and blood lipid levels by PP helped to maintain the correct amount of these molecules at any given time either during the absorptive or post-absorptive period after a meal. Decreased lipid peroxidation and free radical scavenging have been identified to be key factors involved in restoring homeostatic normalcy. Thus, PP treatment may have a direct beneficial impact on cardiovascular disorders that may happen due to complex diseases such as high blood pressure, diabetic conditions, and potentially other idiopathic metabolic disorders.

PP can influence other cardiovascular components in humans, as it can interfere in platelet aggregation in nanomolar concentrations and consequently impair interpretation of functional tests of primary hemostasis. Common nonspecific mechanisms could be related to the stabilization of platelet membranes based on a strong negative correlation of platelet aggregation (MINaAC) and lipophilicity (ClogP) (Bojić et al., 2019).

Gastrointestinal health and propolis

PP has been shown to have direct and indirect effects on the GIT. Direct action includes its ability to inhibit the growth of pathogens of microbial origin such as bacteria and trophozoites. It may also modify various symptom-specific changes. Generally, most infectious processes involve abdominal pain, diarrhea, bloating, and nausea. Because of its antimicrobial, antiinflammatory, and other beneficial features, PP treatment may bring about the delicate homeostasis of the GIT. PP has been shown to inhibit the growth and adherence of trophozoites, as it promotes the detachment of these parasitic organisms. Several protozoal diseases, including giardiasis, have been shown to be treated more effectively by PP than conventional approaches (Anjum et al., 2019). Other studies showed that PP has antihistaminergic, antiinflammatory, antiacid, and anti-*H. pylori* activities that can be used to treat gastric ulceration (Pasupuleti et al., 2017; Anjum et al., 2019).

The investigation of optimal gut health and gastrointestinal bacterial characterization identified the intestinal microbiota connected with good health. These microbial populations release microbial bioactive molecules, such as short-chain fatty acids and vitamins (Braakhuis, 2019). PP components have recently been defined as probiotics by an international consortium as they are selectively metabolized by gut microbiota (Braakhuis, 2019). PP-derived phenolics provide protection to the GIT by inhibiting the growth of pathogenic bacteria such as *Clostridium* spp., *S. aureus*, and *Bacteroides* spp., besides suppressing the adhesion of pathogens to the intestinal wall and enhanced systemic immune function including NK cell activity and cytokine secretion (Braakhuis, 2019). Intestinal barrier function of the GIT has been shown to be strengthened with PP treatment via activated adenosine monophosphate-activated protein kinase, extracellular signal-regulated kinase 1 and 2 (ERK), p38, ZO-1, and protein kinase B. Thus, PP may serve as an important treatment approach for inflammatory and ulcerative colitis (Braakhuis, 2019).

Effect of propolis on oral and dental health

PP has been postulated to act through different mechanisms for periodontal health. Some of the suggested mechanisms include the following: (1) inhibition of cyclooxygenase, PGE2, and proinflammatory cytokines (e.g., IL-6, IL-8); (2) inhibition of cell migration and macrophage activation (Bueno-Silva et al., 2017); (3) reduction of iNOS (inducible NOS) activity (Franchin et al., 2016); and (4) reduction of leukocyte adhesion (Bersan et al., 2014). CAPE is believed to attenuate osteoclastogenesis and bone resorption via the suppression of RANKL-induced NF-κB (Ang et al., 2009). Toker et al. (2008) demonstrated that systemic administration of PP prevented periodontitis-related bone loss in a ligature-induced model.

Oral health

Poor oral health can lead to many other illnesses, including cardiovascular dysfunction. Bacterial and fungal overgrowth as a by-product of health conditions and as a therapeutic response to harsh antibiotic treatment protocols are not uncommon. PP treatment has been shown to restrict bacterial-plaque development and the inhibition of periodontitis-causing pathogens because of its antibacterial properties. PP solutions exert selectively lower cytotoxic action on human gum fibroblasts compared to chlorhexidine. In addition, mouthwash containing PP has shown effectiveness in the healing of surgical wounds. This encourages the use of PP in solutions used as mouthwash (Jain et al., 2014). PP solution can also be used to disinfect toothbrushes. A 3% ethanolic extract of PP toothpaste gel showed greater potency against gingivitis caused by dental plague in a group of patients (Pasupuleti et al., 2017). PP extracts have also helped cure halitosis (bad breath), a condition where an individual experiences unpleasant breath, predominantly due to poor oral hygiene. PP toothpaste or mouthwash is used for its ability to reduce the growth of bacterial plaque and pathogenic microflora that causes gingivitis and periodontitis. Thus, PP also plays a role as a therapeutic agent (Jain et al., 2014). Due to the presence of flavonoids, such as galangin, chrysin, pinobanksin, quercetin, naringenin, galangine, and aromatic acids in PP, they are more effective against mouth abnormalities (especially dental diseases) (De Castro, 2001; Sforcin, 2016). It plays a significant role in the repair of dental pulp (Chandna et al., 2014; Kuropatnicki et al., 2013). The combination of ethanolic extract of PP with mouthwash and toothpastes enhanced the prevention of microbial infection as well as the treatment of gum inflammation (De Castro, 2001; Pasupuleti et al., 2017).

Dermatological care and propolis

Wide usage of PP components to treat skin conditions in traditional medical practice around the globe led to scientific investigations which eventually provided a good amount of experimental data confirming its wound-healing properties. Today, many dermatological products, such as creams and ointments, contain PP components. Various components of PP have been shown to have a direct impact on the skin via mechanisms of antiallergy, antiinflammation, and antimicrobial properties, in addition to tissue repair capacities such as promoting collagen synthesis. A case-controlled clinical study on patients affected by *Acne vulgaris* demonstrated the high efficacy of treatment with PP. While the collagen content of the tissues has been shown to be increased by PP, wound closure has been hastened in human diabetic foot ulcer (Pasupuleti et al., 2017).

Fibronectin (FN) is a multifunctional glycoprotein and an important structural molecule providing stability to various organs and tissues (Stoffels et al., 2013). The FN matrix and its accumulation are essential for cell migration, cell proliferation, cell differentiation, cell adhesion, apoptosis, angiogenesis, collagen biosynthesis, reepithelialization, clot formation, and platelet activity (Stoffels et al., 2013). FNs are also important in repair mechanisms as several repairing components such as collagen types I and III, teasing, laminin, and fibrillin are known to collaborate to complete the process of wound healing. PP has been shown to modify the metabolism of FN by developing a fibrous network of extracellular matrices and by inhibiting FN disintegration (Pasupuleti et al., 2017)

Hepatoprotective activity of propolis

PP acts as a hepatoprotective agent. It increases the GSH level while stopping levels of lipid peroxidation and oxidized GSH (Wagh, 2013). PP showed a hepatoprotective effect toward liver damage in rats due to allyl alcohol CCL4 and paracetamol (Wagh, 2013). The protective effects of PP on hepatotoxicity induced by acetaminophen (AA, paracetamol) and the mechanisms of this hepatoprotective effect were investigated. In mice, pretreatment with PP decreased mortality and the incidence and severity of hepatic necrosis induced by AA. Differential alteration of hepatic enzyme activities of cytochrome P450 monooxygenases (P450s), UDP-glucanosyltransferase, phenolsulfotransferase, and GSH S-transferase was identified as being responsible for the protective action of PP (Lotfy, 2006). PP treatment caused a marked increase in the myofibril, protein fraction, and muscle total protein, along with differential alteration of cholesterol, transaminase activity, total protein, gamma globulins, and free amino acids. Thus, the increases in gamma globulins and proteins were suggested to have a stimulatory effect on the body's immune response (Lotfy, 2006). Aqueous PP extract was found to protect the liver in rats against carbon tetrachloride (CCl4) and it afforded its protection by a decrease in the leakage of the cytosolic enzyme lactate dehydrogenase, decreased generation of lipid peroxide, and maintenance of cellular reduced GSH content (Lotfy, 2006). PP extract was found to have a protective effect against alcohol liver injury by preventing elevations of total cytochrome P-450 enzymes, NADPH-dependent cytochrome C reductase, aniline hydroxylation, 7-ethoxyresorufin hydroxylation, 7-penthoxyresorufin hydroxylation, and lipid peroxidation (Lin et al., 1999).

Propolis and wound healing

Tissue repair is a very complex process involving a multitude of cellular (fibroblasts), immune system cells (T and B cells), white blood cells (such as neutrophil), collagen, and several chemical mediators to recruit various cell types and molecules for sequential changes that ensure fast and effective reconstruction of injured tissue. Several components of PP have antiinflammatory, antimicrobial, immunomodulatory, and antioxidative capabilities. These PP components (bioflavonoids, arginine, vitamin C, provitamin A, B complex, as well as some minerals) also accelerate enzyme reactions, cellular metabolism, blood circulation, synthetic processes for formation of collagen fibers, and tissue repair-related cellular and tissue architectural components (Sforcin, 2016). Animal studies showed the ability of PP to promote keratinocyte proliferation, stimulate glycosaminoglycan deposition in the wound, and to modify the chondroitin/dermatan sulfate structure. Promotion of wound debridement, stimulation of collagen synthesis, and reduction of wound inflammation have been observed after PP treatment in patients presenting with second-degree burns (Sforcin, 2016). In a case-control study to evaluate the efficacy of PP treatment for nonhealing venous leg ulcers, healing was noted to be much faster as compared to when untreated (Sforcin, 2016).

Propolis in gynecological care

The female reproductive system changes as a function of aging, connected with post-menopausal conditions that predispose aging females to developing vaginal infections. There are two major categories: (1) bacterial vaginosis and (2) vulvovaginal candidiasis. Depletion of natural inhabitants such as *Lactobacillus* spp. in the vagina is a key event that allows the overgrowth of harmful bacteria. Overgrowth of certain *Candida* spp., such as *C. albicans*, may be a side effect of prolonged antibiotic treatment or in diabetic conditions. These infections may be accompanied by an overgrowth of vaginal pathogens such as yeast-like fungi, leading to an elevated vaginal pH (Pasupuleti et al., 2017).

Treatment with a 5% aqueous PP solution has been shown to improve vagina health (Imhof et al., 2005).

Antibiotic, antimycotic, antiinflammatory, antioxidant, and anesthetic actions of PP thus provide early symptomatic relief of vaginosis (Pasupuleti et al., 2017). Recurrent vulvovaginal candidiasis (RVVC) may be treated more effectively with PP in patients who are unable to take antibiotics due to a concurrent pharmacological treatment or other contraindications (Pasupuleti et al., 2017). PP extract solution (PES) also has low toxicity in human cells and can be an alternative treatment for chronic vaginitis (Pasupuleti et al., 2017). PES may also function as an antibiofilm material for RVVC to counteract the growth of *C. albicans* and resistance to antifungal drugs (Capoci et al., 2015). The ameliorative effect of PP against methoxychlor-induced ovarian toxicity in the rat confirms the protective actions of PP against chemical-induced damage in the female reproductive tract (El-Sharkawy et al., 2014).

Propolis in human male fertility

A beneficial impact of PP treatment on human sperm motility, mitochondrial respiratory activity, and membrane potential has been demonstrated (Cedikova et al., 2014). This study provided data that confirmed the increased mitochondrial respiratory complexes II and IV without affecting mitochondrial membrane potential in sperm motility in native ejaculates. PP protects human spermatozoa from DNA damage caused by benzo[a]pyrene and exogenous ROS (Russo et al., 2002). PP has been shown to have a protective effect against oxidative stress, thus preserving sperm count and epididymal morphology (Capucho et al., 2012). However, ingestion of Brazilian green PP for several days displayed different cellular events according to concentration, demonstrating a dose-dependent response mechanism. A concern about the disruption of male reproductive homeostasis, in spite of other beneficial effects of PP on the male reproductive system has been raised, warning about the dosage and period of exposure to ensure beneficial outcome and avoid deleterious effects (Severi-Aguiar et al., 2017).

Propolis and cancer

The main mechanisms of action of PP and its compounds regarding their anticarcinogenic action, are related to apoptosis, cell cycle arrest, and interference of metabolic pathways (Watanabe et al., 2011; Anjum et al., 2019). Various phenolic and other components of PP are involved in the inhibition of matrix metalloproteinases, antiangiogenesis effects, and also the inhibition of disease transfer from one body system to another (Sforcin, 2016). PP components can inhibit DNA synthesis in tumor cells and enhance innate and adaptive immune responses by regulating the function of B, T, and NK cells, respectively (Salomão et al., 2011; Wagh, 2013). Uncontrolled cell division is one of the hallmarks of cancer formation. PP components such as galangin, cardanol, nemorosone, and chrysin have been shown to prevent the rapid division of tumor cells (Sforcin, 2016). Damaged cells that refuse to die is a key feature connected to abnormal apoptosis and eventually cancer. Apoptosis, when it functions as a natural phenomenon, is expected to execute the death of damaged cells. The presence of tumor suppressor proteins in CAPE causes the C6 glioma cells to undergo rapid apoptosis, thus avoiding the formation of cancer clones (Sforcin, 2016; Watanabe et al., 2011). Components of PP have been shown to induce apoptotic cell death in vitro in several transformed (cancer) human cell lines, including those derived from renal, lung, liver, prostate, bladder, and lymphoid cancers (Zabaiou et al., 2017). Thus, PP seems to have a universal effect on cancers of all tissues in the human body (Sforcin, 2016).

PP components seem to have a specific effect on various cancers and cancer cell lines. CAPE and artepillin C were shown to inhibit colon carcinogenesis (Sforcin, 2016). The absence of ACF (aberrant crypt foci) in healthy and untreated animals and their induction by colon carcinogens, but not by noncarcinogens, has suggested that aberrant crypts are early preneoplastic lesions in the colon (Sforcin, 2016). PP components inhibit ACF development and have been proven to prevent colon cancer in rodents by suppressing the development of preneoplastic lesions, with no effect on the initiation step of carcinogenesis (Sforcin, 2016).

Cytotoxic effects of an isolated compound (PRF-1) within PP on human hepatocellular carcinoma have been noted. A protective effect of PP against renal adenocarcinoma in CD-1 and ddY mice that were treated with ferric nitrilotriacetate has been reported (Sforcin, 2016). It has been proposed that the inhibitory effect of PP on human prostate cancer proliferation was achieved by regulation of cyclin D1, B1, and cyclin-dependent kinase, as well as by p21 expression (Li et al., 2007). Cuban PP exerted a significant antiproliferative activity on the MCF-7 human breast cancer cell line (estrogen receptor positive, ER +) rather than MDA-MB 231 (ER−) (Popolo et al., 2009). Seven different PP extracts from Turkey were investigated on the human breast cell line MCF-7, and it was concluded that PP may exert antitumor effects by increasing apoptosis through the caspase pathway (Seda et al., 2010). Both artepillin and green PP extract selectively blocked PAK1 signaling, without affecting another kinase known as AKT, in NF tumors (Messerli et al., 2009). Propolin G was isolated from Taiwanese PP, which exerted a broad spectrum of biological activities and could efficiently induce apoptosis in brain cancer cell lines (glioma and glioblastoma). Apoptosis might have occurred through a mitochondrial- and caspase-dependent

TABLE 46.4 Applications of propolis (caffeic acid phenethyl ester components) against chemotherapy-induced toxicities (Murtaza et al., 2014).

Toxicity-inducing drugs	Toxicity/subject
Doxorubicin toxicity	Nephrotoxicity/rats Cardiotoxicity/rats Neuronal oxidant injury/rats Medulloblastoma cell toxicity/human
Cisplatin toxicity	Nephrotoxicity/rats Bone marrow cell toxicity/rats Hepatotoxicity/rats Ototoxicity Cerebellar oxidative stress/rats Testicular toxicity/rats Hepatorenal oxidative injury/rats
Bleomycin toxicity	Lung fibrosis/rats
Tamoxifen toxicity	Hepatotoxicity/rats

pathway (Huang et al., 2007). Chrysin reduced melanoma cell proliferation in both human and murine melanoma cells due to intracellular accumulation of protoporphyrin IX. Chrysin also induced cell death in human and murine melanoma cells through caspase-dependent mechanisms, involving downregulation of ERK 1/2 and activation of p38 MAP kinases (Pichichero et al., 2011). Mucronulatol is one of the most cytotoxic substances present in Caribbean PP. It was shown to decrease cell division in bladder carcinoma.

CAPE-mediated radiosensitivity and proliferation of medulloblastoma inhibited the growth of Daoy (human medulloblastoma) cells. This was proposed to be an outcome of the depletion of GSH, the increase in ROS activity, and the inhibition of NF-κB activity (Lee et al., 2008). The effects of Chinese PP and its constituents (chrysin, galangin, pinocembrin, caffeic acid, and CAPE) were evaluated against tunicamycin-induced neuronal cell death in SH-SY5Y cells (Watanabe et al., 2011).

PP exposure could reduce the expression of isoform *RASSF1C* or disrupt the RASSF1A/RASSF1C ratio in cancer cells. This finding is relevant because some studies have indicated that RASSF1C has oncogenic properties and could promote cell survival and proliferation. Components of PP extract reverted the DNA methylation of an important tumor suppressor gene. PP exposure could reduce the expression of isoform *RASSF1C* or disrupt the RASSF1A/RASSF1C ratio in cancer cells, thereby reducing the oncogenic conversion of this loci. Thus, epi-drugs, such as PP, which target DNA methyltransferases, are becoming a promising alternative to improve cancer therapy. The combined use of DNMTi at low doses might revert resistance to cytotoxic agents in part or even remove the acquired epigenetic alterations associated with resistance to therapy (Assumpção et al., 2020).

Propolis as a protective agent against therapeutic toxicities

The protective effect of CAPE on chemotherapy- and radiotherapy-induced toxicity has been shown (Akyol et al., 2016). CAPE's protective effects on oxidative stress-induced toxicities of doxorubicin (nephrotoxicity), cisplatin (nephrotoxicity, ototoxicity, and hepatotoxicity), and bleomycin (lung fibrosis) have been reported (Akyol et al., 2016). Table 46.4 depicts the utility of CAPE against toxicity by doxorubicin, cisplatin, bleomycin, and tamoxifen in various organs/systems (Murtaza et al., 2014).

Propolis in diabetes

Diabetes mellitus is one of the major complex metabolic diseases impacting about half of the world's population. It is characterized by high levels of blood sugar. A variety of pathogenic processes are involved which led to the classification of the disorder into several subgroups. Irrespective of its etiology, diabetes is widely believed to have increased levels of oxidative stress during onset, development, and progression, as well as during its related complications. Studies in animal and cellular models have also indicated that PP modulates oxidative stress, the accumulation of advanced glycation end products (AGEs), and adipose tissue inflammation, all of which contribute to insulin resistance or defects in insulin secretion. CAPE may mitigate diabetic complications such as nephropathy, retinopathy, foot ulcers, and nonalcoholic fatty liver disease (Pittalà et al., 2018; Kitamura, 2019).

Propolis as an adjuvant

Successful immunization of many vaccines depends on efficient adjuvants for increasing their immunogenicity and ensuring long-term propagation and protection. Many of the adjuvants have adverse effects due to their toxicity and reactogenicity. Different extracts and formulations of PP have been used as adjuvants for bacterial, viral, and parasitic vaccines. PP has been found to be a safe immunostimulator with low toxicity. Several varieties of PP's preparations have been proven to be potent adjuvants. An encapsulated PP flavonoids mixture is the most promising formulation of all that have been developed (Ashry and Ahmad, 2012).

Safety concerns for usage of propolis as a nutritional and/or health care product

Allergies

PP may cause allergic reactions, especially in those who are allergic to bees, bee products such as honey, conifers, poplars, Peru balsam, and salicylates. Allergic contact dermatitis from PP is due to the presence of allergens, such as 3-methyl-2-butenyl caffeate and phenylethyl caffeate, which are constituents of LB-1, the first allergen identified in PP. Phenylethyl caffeate leads to strong reactions in PP-sensitive patients, while benzyl salicylate and benzyl cinnamate result in very weak-to-moderate reactions.

Pregnancy and breastfeeding

The scarcity of research data on the efficacy and safety of PP as a supplement or medication during pregnancy and lactation makes precautionary prevention mandatory until further data are available.

Asthma

Certain components in PP may cause asthma and some cases may worsen already-existing asthmatic conditions.

Bleeding conditions

Some components of PP may slow down the process of clotting, which may increase the risk of excessive bleeding in patients with blood-clotting disorders. Discontinuing PP consumption for 2 weeks before any type of surgery is strongly recommended.

Contradictions with other medications

Taking PP along with medications such as aspirin, clopidogrel (Plavix), dalteparin (Fragmin), enoxaparin (Lovenox), heparin, ticlopidine (Ticlid), and warfarin (Coumadin) may increase the chances of bruising and bleeding.

Concluding remarks and future directions

PP is multifaceted in its composition, activities, and benefits to human health. The use of PP in daily life is becoming ubiquitous and can be listed as follows: (1) for treating skin conditions such as dermatitis, acne, burns, and cuts; (2) for treating cold sores, genital herpes, and viral disorders; (3) for oral health and dental plaques; (4) for sore throat or inflamed tonsils; (5) for promoting cardiovascular health; (6) for preventing lung dysfunctions such as asthma; (7) for treating various ulcers; (8) for strengthening the immune system to fight against all kinds of microbes such as bacteria, fungi, protozoa, and viruses; (9) for treating arthritis, joint pain, and urinary infections; (10) for prevention of colds and cancer cell formation; (11) for maintenance of neural health in the advancement of aging; and (12) for cardiovascular function. Human health and longevity can be greatly enhanced by this naturally occurring, low-cost, and low-toxicity nutrient. There is an urgent need to form an international depository of PP and details of its molecular components from various countries around the world. The key for more successful utilization of the various useful components of PP lies in the fact that more research is required to explore the beneficial effects of combinations of PP for human health, animal health, and other areas. Joint ventures among academicians, industries, and government agencies around the globe can make it possible to research, share, and deliver the benefits of this ancient and important nutraceutical.

References

Abdulkhani, A., Hosseinzadeh, J., Ashori, A., Esmaeeli, H., 2017. Evaluation of the antibacterial activity of cellulose nanofibers/polylactic acid composites coated with ethanolic extract of propolis. Polym. Compos. 38, 13–19.

Akbay, E., Özenirler, Ç., Çelemli, Ö.G., Durukan, A.B., Onur, M.A., Sorkun, K., 2017. Effects of propolis on warfarin efficacy. Kardiochirurgia I Torakochirurgia Polska = Polish. J. Cardio-Thorac. Surg 14, 43.

Akyol, S., Gulec, M.A., Erdemli, H.K., et al., 2016. Can propolis and caffeic acid phenethyl ester be promising agents against cyclophosphamide toxicity? J. Intercult. Ethnopharmacol. 5 (1), 105–107.

Alsayed, M.F.S., Hashem, A., Al-Hazzani, A.A., et al., 2020. Biological control of yeast contamination of industrial foods by propolis. Saudi J. Biol. Sci. 27 (3), 935–946.

AL-Waili, N., Al-Ghamdi, A., Ansari, M.J., et al., 2012. Synergistic effects of honey and propolis toward drug multi-resistant *Staphylococcus aureus*, *Escherichia coli* and *Candida albicans* isolates in single and polymicrobial cultures. Int. J. Med. Sci. 9, 793–800.

Amoros, M., Simõs, C., Girre, L., et al., 1992. Synergistic effect of flavones and flavonols against *Herpes simplex* virus type 1 in cell culture. Comparison with the antiviral activity of propolis. J. Nat. Prod. 55, 1732–1740.

Ang, E.S., Pavlos, N.J., Chai, L.Y., et al., 2009. Caffeic acid phenethyl ester, an active component of honeybee propolis attenuates osteoclastogenesis and bone resorption via the suppression of RANKL-induced NF-kappaB and NFAT activity. J. Cell Physiol. 221 (3), 642–649.

Anjum, S.I., Ullah, A., Khan, K.A., et al., 2019. Composition and functional properties of propolis (bee glue): a review. Saudi J. Biol. Sci. 26 (7), 1695–1703.

Assumpção, J.H.M., Takeda, A.A.S., Sforcin, J.M., et al., 2020. Effects of propolis and phenolic acids on triple-negative breast cancer cell lines: potential involvement of epigenetic mechanisms. Molecules 25 (6), 12.

Bankova, V., 2005. Recent trends and important developments in propolis research. Evid. Based Complement. Alternat. Med. 2, 29–32.

Bankova, V., De Castro, S., Marcucci, M., 2000. Propolis: recent advances in chemistry and plant origin. Apidologie 31, 3–15.

Bankova, V., Galabov, A.S., Antonova, D., et al., 2014. Chemical composition of propolis extract ACF® and activity against herpes simplex virus. Phytomedicine 21 (11), 1432–1438.

Bersan, S.M., Galvão, L.C., Goes, V.F., et al., 2014. Action of essential oils from Brazilian native and exotic medicinal species on oral biofilms. BMC Complement. Altern. Med. 14, 451.

Bojić, M., Maleš, Ž., Antolić, A., et al., 2019. Antithrombotic activity of flavonoids and polyphenols rich plant species. Acta Pharm. 69 (4), 483–495.

Braakhuis, A., 2019. Evidence on the health benefits of supplemental propolis. Nutrients 11, 2705. Available from: 10.3390/nu11112705. 2019.

Bueno-Silva, B., Kawamoto, D., Ando-Suguimoto, E.S., et al., 2015. Brazilian red propolis attenuates inflammatory signaling cascade in LPS-activated macrophages. PLoS One 10 (12), article e0144954.

Bueno-Silva, B., Marsola, A., Ikegaki, M., Alencar, S.M., Rosalen, P.L., 2017. The effect of seasons on Brazilian red propolis and its botanical source: chemical composition and antibacterial activity. Nat. Prod. Res. 31, 1318–1324.

Capoci, I.R., Bonfim-Mendonça Pde, S., Arita, G.S., et al., 2015. Propolis is an efficient fungicide and inhibitor of biofilm production by vaginal *Candida albicans*. Evid. Based Complement. Alternat. Med. 2015, 9.

Capucho, C., Sette, R., De Souza, P.F., et al., 2012. Green Brazilian propolis effects on sperm count and epididymis morphology and oxidative stress. Food Chem. Toxicol. 50, 3956–3962.

Cedikova, M., Miklikova, M., Stachova, L., et al., 2014. Effects of the Czech propolis on sperm mitochondrial function. Evid. Based Complement. Alternat. Med. 2014, 248768.

Chan, G.C.-F., Cheung, K.-W., Sze, D.M.-Y., 2013. The immunomodulatory and anticancer properties of propolis. Clin. Rev. Allergy Immunol. 44, 262–273.

Chandna, P., Adlakha, V.K., Das, S., et al., 2014. Complementary and alternative medicine (CAM): a review of propolis in dentistry. Technology 4, 6.

Chen, J., Long, Y., Han, M., et al., 2008. Water-soluble derivative of propolis mitigates scopolamine-induced learning and memory impairment in mice. Pharmacol. Biochem. Behav. 90 (3), 441–446.

Cho, M.S., Park, W.S., Jung, W.K., et al., 2014. Caffeic acid phenethylester promotes anti-inflammatory effects by inhibiting MAPK and NF-κB signaling in activated HMC-1 human mast cells. Pharm. Biol. 52 (7), 926–932.

Chopra, S., Pillai, K.K., Husain, S.Z., et al., 1995. Propolis protects against doxorubicin-induced myocardiopathy in rats. Exp. Mol. Pathol. 62, 190–198.

Cui, K., Lu, W., Zhu, L., et al., 2013. Caffeic acid phenethyl ester (CAPE), an active component of propolis, inhibits *Helicobacter pylori* peptide deformylase activity. Biochem. Biophys. Res. Commun. 435, 289–294.

De Castro, S., 2001. Propolis: biological and pharmacological activities. Therapeutic uses of this bee-product. Ann. Rev. Biomed. Sci. 3, 49–83.

Devequi-Nunes, D., Machado, B.A.S., Barreto, G.A., et al., 2018. Chemical characterization and biological activity of six different extracts of propolis through conventional methods and supercritical extraction. PLoS One 13.

El Ashry, S.H., Ahmad, T.A., 2012. The use of propolis as vaccine's adjuvant. Vaccine 31 (1), 31–39. 17.

El-Guendouz, S., Al-Waili, N., Azza, S., et al., 2017. Antioxidant and diuretic activity of co-administration of *Capparis spinosa* honey and propolis in comparison to furosemide. Asian Pac. J. Trop. Med.

El Hady, F.K.A., Hegazi, A.G., 2002. Egyptian propolis: 2. Chemical composition, antiviral and antimicrobial activities of East Nile Delta propolis. Z. für Naturforschung C. 57, 386–394.

El-Sharkawy, E.E., Kames, A.O., Sayed, S.M., et al., 2014. The ameliorative effect of propolis against methoxychlor induced ovarian toxicity in rat. Exp. Toxicol. Pathol. 66, 415–421.

Ferreira, J.M., Fernandes-Silva, C.C., Salatino, A., et al., 2017. New propolis type from north-east Brazil: chemical composition, antioxidant activity and botanical origin. J. Sci. Food Agric. 97 (11), 3552–3558.

Fokt, H., Pereira, A., Ferreira, A., et al., 2010. How do bees prevent hive infections? The antimicrobial properties of propolis. Curr. Res. Technol. Educ. Top. Appl. Microbiol. Microbial. Biotechnol. 1, 481–493.

Franchin, M., Cólon, D.F., Castanheira, F.V., et al., 2016. Vestitol isolated from Brazilian red propolis inhibits neutrophils migration in the inflammatory process: elucidation of the mechanism of action. J. Nat. Prod. 79, 954–960.

Gao, W., Wu, J., Wei, J., et al., 2014. Brazilian green propolis improves immune function in aged mice. J. Clin. Biochem. Nutr. 55 (1), 7–10.

Ha, S.K., Moon, E., Kim, S.Y., 2010. Chrysin suppresses LPS stimulated proinflammatory responses by blocking NF-κB and JNK activations in microglia cells. Neurosci. Lett. 485 (3), 143–147.

Huang, W.J., Huang, C.H., Wu, C.L., et al., 2007. Propolin G, a prenylflavanone, isolated from Taiwanese propolis, induces caspase-dependent apoptosis in brain cancer cells. J. Agric. Food Chem. 5, 7366–7376.

Imhof, M., Lipovac, M., Kurz, C., 2005. Propolis solution for the treatment of chronic vaginitis. Int. J. Gynecol. Obstetr 89 (2), 127–132.

Isla, M.I., Dantur, Y., Salas, A., et al., 2012. Effect of seasonality on chemical composition and antibacterial and anticandida activities of Argentine propolis. Design of a topical formulation. Nat. Prod. Commun. 10, 1315–1318.

Jain, S., Rai, R., Sharma, V., et al., 2014. Propolis in oral health: a natural remedy. World J. Pharm. Sci. 2 (1), 90–94.

Kasiotis, K.M., Anastasiadou, P., Papadopoulos, A., et al., 2017. Revisiting Greek propolis: chromatographic Analysis and antioxidant activity study. PLoS One 12, e0170077.

Khan, M.N., Lane, M.E., McCarron, P.A., et al., 2018. Caffeic acid phenethyl ester is protective in experimental ulcerative colitis via reduction in levels of pro-inflammatory mediators and enhancement of epithelial barrier function. Inflammopharmacology 26 (2), 561–2569.

Kitamura, H., 2019. Effects of propolis extract and propolis-derived compounds on obesity and diabetes: knowledge from cellular and animal models. Molecules 24 (23), E4394. pii.

Król, W., Bankova, V., Sforcin, J.M., et al., 2013. Propolis: properties, application, and its potential. Evid. Based Complement. Alternat. Med.

Kubota, Y., Umegaki, K., Kobayashi, K., et al., 2004. Anti-hypertensive effects of Brazilian propolis in spontaneously hypertensive rats. Clin. Exp. Pharmacol. Physiol. 31 (S2), S29–S30.

Kuropatnicki, A.K., Szliszka, E., Krol, W., 2013. Historical aspects of propolis research in modern times. Evid. Based Complement. Alternat. Med.

Lee, Y.Y., Kao, C.L., Tsai, P.H., et al., 2008. Caffeic acid phenethyl ester preferentially enhanced radiosensitizing and increased oxidative stress in medulloblastoma cell line. Childs Nerv. Syst. 24, 987–994.

Li, H., Kapur, A., Yang, J.X., et al., 2007. Antiproliferation of human prostate cancer cells by ethanolic extracts of Brazilian propolis and its botanical origin. Int. J. Oncol. 31, 601–606.

Lin, G., Nnane, I.P., Cheng, T.Y., 1999. The effects of pretreatment with glycyrrhizin and glycyrrhetinic acid on the retrorsine-induced hepatotoxicity in rats. Toxicon 37 (9), 1259–1270.

Lotfy, M., 2006. Biological activity of bee propolis in health and disease. Asian Pac. J. Cancer Prev. 7, 22–31.

Machado, B., Pulcino, T.N., Silva, A.L., et al., 2017. Propolis as an alternative in prevention and control of dental cavity. Immunity 19, 24.

Marcucci, M.C., 1995. Propolis: chemical composition, biological properties and therapeutic activity. Apidologie 26, 83–99.

Martinotti, S., Ranzato, E., 2015. Propolis: a new frontier for wound healing? Burn. Trauma. 3, 9.

Messerli, S.M., Ahn, M.R., Kunimasa, K., et al., 2009. Artepillin C (ARC) in Brazilian green propolis selectively blocks oncogenic PAK1 signaling and suppresses the growth of NF tumors in mice. Phytother. Res. 23, 423–427.

Murtaza, G., Karim, S., Akram, M.R., et al., 2014. Caffeic acid phenethyl ester and therapeutic potentials. BioMed. Res. Int. 2014, 1–9. Article ID 145342.

Omar, R., Igoli, J.O., Zhang, T., et al., 2017. The chemical characterization of Nigerian propolis samples and their activity against *Trypanosoma brucei*. Sci. Rep. 7, 923.

Özcan, M., 1999. Antifungal properties of propolis. Grasas Aceites 50, 395–398.

Park, Y.K.

Parolia, A., Thomas, M.S., Kundabala, M., Mohan, M., 2010. Propolis and its potential uses in oral health. Int. J. Med. Med. Sci. 2, 210–215.

Pasupuleti, V.R., Sammugam, L., Ramesh, N., et al., 2017. Propolis, and royal jelly: a comprehensive review of their biological actions and health benefits. Oxid. Med. Cell. Longev. 2017, 1–21. Article ID 1259510.

Pichichero, E., Cicconi, R., Mattei, M., et al., 2011. Chrysin-induced apoptosis is mediated through p38 and Bax activation in B16-F1 and A375 melanoma cells. Int. J. Oncol. 38, 473–483.

Pittalà, V., Salerno, L., Romeo, G., et al., 2018. Therapeutic potential of caffeic acid phenethyl ester (CAPE) in diabetes. Curr. Med. Chem. 25 (37), 4827–4836.

Popolo, A., Piccinelli, L.A., Morello, S., et al., 2009. Antiproliferative activity of brown Cuban propolis extract on human breast cancer cells. Nat. Prod. Commun. 4, 1711–1716.

Przybyłek, I., Karpiński, T.M., 2019. Antibacterial properties of propolis. Molecules 24, 2047. Available from: 10.3390/molecules24112047.

Russo, A., Longo, R., Vanella, A., 2002. Antioxidant activity of propolis: role of caffeic acid phenethyl ester and galangin. Fitoterapia. 73, S21–S29.

Ristivojević, P., Stević, T., Starović, M., et al., 2020. Phenolic composition and biological activities of geographically different type of propolis and black cottonwood resins against oral streptococci, vaginal microbiota and phytopathogenic *Fusarium* species. J. Appl. Microbiol. Available from: 10.1111/jam.14633.

Salomão, K., de Souza, E.M., Henriques-Pons, A., et al., 2011. Brazilian green propolis: effects in vitro and in vivo on *Trypanosoma cruzi*. Evid. Based Complement. Alternat. Med.

Seda, V.H., Sorkun, K., Ismet Deliloğlu Gurhan, S., et al., 2010. Propolis from Turkey induces apoptosis through activating caspases in human breast carcinoma cell lines. Acta Histochem. 112, 546–556.

Severi-Aguiar, G.D., Pinto, S.J., Capucho, C., Oliveira, C.A., et al., 2017. Chronic intake of green propolis negatively affecting the rat testis. Pharmacogn. Res. 9 (1), 27–33.

Sforcin, J.M., 2016. Biological properties and therapeutic applications of propolis. Phytother. Res. 30, 894–905.

Stoffels, J.M., Zhao, C., Baron, W., 2013. Fibronectin in tissue regeneration: timely disassembly of the scaffold is necessary to complete the build. Cell. Mol. Life Sci. 70 (22), 4243–4253.

Swamy, M., Suhaili, D., Sirajudeen, K.N., et al., 2014. Propolis ameliorates tumor nerosis factor-α, nitric oxide levels, caspase-3 and nitric oxide synthase activities in kainic acid mediated excitotoxicity in rat brain. Afr. J. Tradit. Complement. Altern. Med. 11 (5), 48–53.

Toker, H., Ozan, F., Ozer, H., et al., 2008. A morphometric and histopathologic evaluation of the effects of propolis on alveolar bone loss in experimental periodontitis in rats. J. Periodontol. 79, 1089–1094.

Toreti, V.C., Sato, H.H., Pastore, G.M., et al., 2013. Recent progress of propolis for its biological and chemical compositions and its botanical origin. Evid. Based Complement. Alternat. Med.

Tsai, C.F., Kuo, Y.H., Yeh, W.L., et al., 2015. Regulatory effects of caffeic acid phenethyl ester on neuroinflammation in microglial cells. Int. J. Mol. Sci. 16 (12), 5572–5589.

Walker, P., Crane, E., 1987. Constituents of propolis. Apidologie. 18, 327–334.

Wagh, V.D., 2013. Propolis: a wonder bees product and its pharmacological potentials. Adv. Pharmacol. Sci. 2013, 308249.

Wang, J., Song, Y., Chen, Z., et al., 2018. Connection between systemic inflammation and neuroinflammation underlies Neuroprotective mechanism of several phytochemicals in neurodegenerative diseases. Oxid. Med. Cell. Longev. 2018, 1–16. Article ID 1972714.

Watanabe, M.A., Amarante, M.K., Conti, B.J., et al., 2011. Cytotoxic constituents of propolis inducing anticancer effects: a review. J. Pharm. Pharmacol. 63 (11), 1378–1386. 2011.

WHO. Joint FAO/WHO Expert Committee on Food Additives. Meeting (2008: Geneva, Switzerland). Evaluation of certain veterinary drug residues in food: 70th report of the Joint FAO/WHO Expert Committee on Food Additives (WHO technical report series; no. 954).

Zabaiou, N., Fouache, A., Trousson, A., et al., 2017. Biological properties of propolis extracts: Something new from an ancient product. Chem. Phys. Lipids 207 (Pt B), 214–222. 2017 Oct.

Chapter 47

Bee products as nutraceuticals to nutraceuticals for bees

Robert W. Coppock

Introduction

Bee products that are used as nutraceuticals include propolis, royal jelly (RJ), pollen, wax, and venom. Honey, using a broad definition, refers to a product made by *Apis* spp. from the nectar of flowers of various plant species and the excretions of plant-sucking insects. Honey is classified as a functional food because of its phytochemical profile and is part of the Mediterranean diet (Battino et al., 2019). The nutraceutical, medical, and cosmetic use of bee products has been recorded since 4500 BC. In traditional and modern medicine, bee products have historical and modern uses as remedies for human and animal maladies. Honey, used as a prophylactic, is considered to be protective from metabolic and infectious diseases (Cianciosi et al., 2018). There is evidence that beeswax has historical use in dentistry (Bernardini et al., 2012). Material removed from a Neolithic human tooth has the infrared spectrum of beeswax. Beeswax has historical use as an ingredient in embalming substances and as a propping agent. There is evidence of the use of beeswax in ancient medicines and the preparation of the dead for burial (Tchapla et al., 2004; Gamberini et al., 2011; Stacey, 2011; Fratini et al., 2016).

Honey

Honey is considered a nutraceutical and is made-up of components that bees collect from plants and other sources. Regulations give honey a specific definition. For example, honey, as defined by the European Codex Alimentarius (2001), "is the natural sweet substance, produced by honeybees from the nectar of plants or from secretions of living parts of plants, or excretions of plant-sucking insects on the living parts of plants, which the bees collect, transform by combining with specific substances of their own, deposit, dehydrate, store and leave in honeycombs to ripen and mature." Some governments further restrict the term "honey" to the product being produced by the honeybee *Apis mellifera*. Honey can be classified by plant species making up blossom and honeydew and/or by geographic locations (Santos-Buelga and González-Paramás, 2017). Honey retains the odors from the blossoms and honeydew excreted by sap-sucking insect, for example, aphids (da Silva et al., 2016). Honey also has plant-specific phytochemicals.

Chemistry of honey

The chemistry of honey is complex, and this complex chemistry is essential for the health benefits that come from honey consumption. The specific chemistry of honey varies among botanical ecosystem, season of the year, and geographical location because bees are opportunistic harvesters (Gheldof et al., 2002; Islam et al., 2014; Salonen et al., 2017; Tanleque-Alberto et al., 2019). Honey, when considered by bees to be a storable product, is capped with wax and allowed to ripen (Ayton et al., 2019). Ripened honey is a stable product and is resistant to microbial degradation. Chemically, honey is a water solution containing 80%–83% sugars, primarily the monosaccharides glucose and fructose, and about 30 other disaccharides and oligosaccharides. The sucrose content of honey is 5%–15% depending on the floral source (Sodré et al., 2011). Honeys from canola (*Brassica napus*) and dandelion (*Taraxacum officinale*), have a higher fraction of glucose and crystallization generally occurs after capping (da Silva et al., 2016). Enzymes added by bees to nectar and honeydew hydrolyze polysaccharides to monosaccharides. Honey contains organic acids including gluconic, lactic, formic, butyric, tartaric, pyruvic, acetic, citric, oxalic, succinic, malic, maleic, α-ketoglutaric, glucose-6-phosphate, pyroglutamic, and glycolic acids with gluconic being the most common (Siddiqui et al., 2017). The pH of honey is generally 3.5–5.5. Organic

Nutraceuticals. DOI: https://doi.org/10.1016/B978-0-12-821038-3.00047-1

acids can be formed enzymatically from sugars by bee-source enzymes. The mean protein content of honey is 0.2%–1.6% and originates from plants and the cephalic gland of honeybees (da Silva et al., 2016). The proteins, including enzymes, are important in the honey-making process and breakdown of pollen proteins. Pollen in honey is a botanical biomarker. Vitamins, at low levels, identified in honey include the B vitamins [thiamine (B1), riboflavin (B_2), nicotinic acid (B_3), pantothenic acid (B_5), pyridoxine (B_6), biotin (B_7 or H) and folate (B_9)] and vitamin C (da Silva et al., 2016). The acidic pH of honey preserves the B vitamins. Most of the vitamins are found in the pollen content of honey. Honey contains phenolic compounds and the flavonoids are considered the most import group in honey (Tables 47.1 and 47.2) (da Silva et al., 2016; Afrin et al., 2019). Volatile compounds give honey its aroma and flavor. Bees, using their own enzymes, chemically convert phytochemicals into substances that add gustatory pleasantries to honey. The source of the honey can drastically change its desired medical properties (Hixon et al., 2019).

Beeswax

Background and chemistry

Beeswax has historical uses. Using high-temperature gas chromatography and mass spectrometry patterns in extracts of ancient pottery, it has been determined that beeswax was harvested and used by Neolithic peoples (Roffet-Salque et al., 2015). Beeswax is a complex chemical mixture secreted by the abdominal wax glands of honeybees (Fratini et al., 2016; Bogdanvov, 2016). Bees, between 12–18 days of age (postemergence), secrete beeswax as a liquid which hardens on contact with air. Beeswax is used to make the foundation and honeycomb hexagonal cells used for raising the brood and storage of honey and pollen (Bogdanvov, 2016). The wax cells are used to cradle larvae, and are storage cells for pollen and honey, and the entire wax foundation supports the mass of working bees. In beekeeping, the foundations generally are provided upon which bees build wax hexagonal cells. Wax is strong, for example, approximately 100 g of wax is used to construct all the cells in one Langstroth deep frame. The chemistry of wax varies between genera and species of bees. It is estimated that beeswax from *A. mellifera* consists of over 300 different compounds (Aguilar et al., 2007). The major groups of compounds found in beeswax are alkanes, free fatty acids, monoesters, diesters, and hydroxy-monoesters. Fatty alcohols and hydroxy-diesters are minor constituents. Beeswax is purified by different methods; these include hot water extraction, steam extraction, and centrifugation. Beeswax obtained by pressure filtration is referred to as yellow beeswax and bleached beeswax is identified as white beeswax (Aguilar et al., 2007).

Nutraceutical, pharmacological, and food-processing uses of beeswax

There are pharmacopeia standards for beeswax. Beeswax has uses as a vehicle for drugs, slow-release medium for drugs, and as an emulsion and foundation for cosmetics. Beeswax is added to skin care products and has anticomedogenic properties. Beeswax is approved as a glazing agent for baked goods, candies, as an adjuvant for flavoring agents, and is used in sealants to prolong the shelf-life of fruits (Formiga et al., 2019). It is added to water-based drinks as a clouding agent and carrier for flavors. Various extracts of beeswax are used as nutraceuticals and ingredients in cosmetics. The medical use of beeswax dates back to the second century BC (Mosleh and Badr, 2016). Beeswax has been and currently is used for managing abscesses, treating burns, and various types of soft-tissue wounds, topical applications for various maladies, in dental remedies, and in the formulation of remedies and cosmetics (Bernardini et al., 2012; Fratini et al., 2016). It is used as an emulsifying agent, an agent to increase absorption, in wound management, and as a poultice. In alternative medicine, it is used for its benefits on blood, energy systems including body balance, and as an antiaging product.

Propolis

Propolis is a honeybee product produced by mixing plant gums, resins, and oils with beeswax and bodily secretions (Santos et al., 2020). Bees use propolis to seal frames in position, plug cracks and other openings in the hive, and for its hospitable and aseptic contribution to the hive environment (Castaldo and Capasso, 2002). The color, and physical and chemical properties of propolis vary with the phytogeographical location of the bees and the season of the year (Khurshid et al., 2017; Galeotti et al., 2018; Blicharska and Seidel, 2019; Santos et al., 2020). Over 500 chemicals have been identified in propolis, demonstrating its complex chemistry (Viuda-Martos et al., 2008; Blicharska and Seidel, 2019). For example, bees in temperate regions have limited choices. They collect secretions from trees that include poplar (*Populus* spp.), birch (*Betula* spp.) willow (Salix spp.), oak (*Quercus* spp.), ash (*Fraxinus* spp.), spruce (*Picea* spp.), fir (*Abies* spp.), alder (*Alnus* spp.), and pine (*Pinus* spp.). In tropical regions, bees collect secretions from a large number of plants and produce propolis with more complex chemistry and more diverse pharmacology. Propolis has low solubility in water. Purification of the flavonoids and phenolic acids in propolis is done by solvent extractions and the chemistry

TABLE 47.1 Common flavonoids in honey (da Silva et al., 2016; Afrin et al., 2019; PubChem, 2020).

Flavonols/IUPAC name	Flavanols/IUPAC name	Flavones/IUPAC name	Flavanones/IUPAC name
Quercetin 2-(3,4-dihydroxyphenyl)-3,5,7-trihydroxychromen-4-one	Catechin (2R,3S)-2-(3,4-dihydroxyphenyl)-3,4-dihydro-2H-chromene-3,5,7-triol	Chrysin 5,7-dihydroxy-2-phenylchromen-4-one	Pinocembrin (2S)-5,7-dihydroxy-2-phenyl-2,3-dihydrochromen-4-one
Myricetin 3,5,7-trihydroxy-2-(3,4,5-trihydroxyphenyl) chromen-4-one	Epicatechin (2R,3R)-2-(3,4-dihydroxyphenyl)-3,4-dihydro-2H-chromene-3,5,7-triol	Apigenin 5,7-dihydroxy-2-(4-hydroxyphenyl) chromen-4-one	Pinobanksin (2R,3R)-3,5,7-trihydroxy-2-phenyl-2,3-dihydrochromen-4-one
Kaempferol 3,5,7-trihydroxy-2-(4-hydroxyphenyl) chromen-4-one	Gallocatechin (2R,3S)-2-(3,4,5-trihydroxyphenyl)-3,4-dihydro-2H-chromene-3,5,7-triol	Luteolin 2-(3,4-dihydroxyphenyl)-5,7-dihydroxychromen-4-one	Pinostrobin (2S)-5-hydroxy-7-methoxy-2-phenyl-2,3-dihydrochromen-4-one
Galangin 3,5,7-trihydroxy-2-(3,4,5-trihydroxyphenyl) chromen-4-one	Epigallocatechin gallate [(2R,3R)-5,7-dihydroxy-2-(3,4,5-trihydroxyphenyl)-3,4-dihydro-2H-chromen-3-yl] 3,4,5-trihydroxybenzoate	Diosmetin 5,7-dihydroxy-2-(3-hydroxy-4-methoxyphenyl) chromen-4-one	Hesperetin (2S)-5,7-dihydroxy-2-(3-hydroxy-4-methoxyphenyl)-2,3-dihydrochromen-4-one
Myricetin-3′-methylether 5,7-dihydroxy-3-methoxy-2-(3,4,5-trihydroxyphenyl) chromen-4-one	Rutin 2-(3,4-dihydroxyphenyl)-5,7-dihydroxy-3-[(2S,3R,4S,5S,6R)-3,4,5-trihydroxy-6-[[(2R,3R,4R,5R,6S)-3,4,5-trihydroxy-6-methyloxan-2-yl]oxymethyl]oxan-2-yl] oxychromen-4-one	Tricetin 5,7-dihydroxy-2-(3,4,5-trihydroxyphenyl) chromen-4-one	Naringenin (2S)-5,7-dihydroxy-2-(4-hydroxyphenyl)-2,3-dihydrochromen-4-one
Dimethylquercetin 2-(3,4-dihydroxyphenyl)-3,5,7-trihydroxy-6,8-dimethylchromen-4-one	Isorhamnetin-*O*-pentosyl-hexoside	Baicalein 5,6,7-trihydroxy-2-phenylchromen-4-one	Naringin (2S)-7-[(2S,3R,4S,5S,6R)-4,5-dihydroxy-6-(hydroxymethyl)-3-[(2S,3R,4R,5R,6S)-3,4,5-trihydroxy-6-methyloxan-2-yl] oxyoxan-2-yl]oxy-5-hydroxy-2-(4-hydroxyphenyl)-2,3-dihydrochromen-4-one
Isorhamnetin 3,5,7-trihydroxy-2-(4-hydroxy-3-methoxyphenyl) chromen-4-one	Isorhamnetin-*O*-hexosyl-*O*-rutinoside		
Fisetin 2-(3,4-dihydroxyphenyl)-3,7-dihydroxychromen-4-one			

IUPAC, International Union of Pure and Applied Chemistry.

of the extracts varies with the extraction method (Wang, 2013; Galeotti et al., 2017, 2018). Solvents used include ethanol, methanol, water, hexane, acetone, dichloromethane, chloroform, and supercritical fluids. Propolis extract contains phenolic acid and esters, flavonoids, aromatic compounds, and caffeic acid phenethyl ester. This mixture of chemicals has a wide range of biological activities, including antioxidant, antibacterial, antiviral, antifungal, antiinflammatory, and antitumor properties. Many of the antimicrobial active substances in propolis are soluble in water and alcohol extracts (Galeotti et al., 2018). Important biologically active compounds in propolis are phenolic acids (caffeic, ferulic,

TABLE 47.2 Phenolics in honey (Afrin et al., 2019; PubChem, 2020).

Hydroxybenzoic acid/IUPAC name	Hydroxycinnamic acid/IUPAC name
Vanillic acid 4-hydroxy-3-methoxybenzoic acid	Caffeic acid 3-(3,4-dihydroxyphenyl)acrylic acid
Syringic acid 4-hydroxy-3,5-dimethoxybenzoic acid	*p*-Coumaric acid (E)-3-(4-hydroxyphenyl)prop-2-enoic acid
Gallic acid 3,4,5-trihydroxybenzoic acid	Ferulic acid (E)-3-(4-hydroxy-3-methoxyphenyl)prop-2-enoic acid
Salicylic acid 2-hydroxybenzoic acid	Sinapic acid (E)-3-(4-hydroxy-3,5-dimethoxyphenyl)prop-2-enoic acid
Ellagic acid 6,7,13,14-tetrahydroxy-2,9-dioxatetracyclo[6.6.2.04,16.011,15] hexadeca-1(15),4,6,8(16),11,13-hexaene-3,10-dione	p-Hydroxycinnamic acid (E)-3-(4-hydroxyphenyl)prop-2-enoic acid
p-Hydroxybenzoic acid 4-hydroxybenzoic acid	Chlorogenic acid (1S,3R,4R,5R)-3-[(E)-3-(3,4-dihydroxyphenyl)prop-2-enoyl]oxy-1,4,5-trihydroxycyclohexane-1-carboxylic acid
Homogentisic acid 2-(2,5-dihydroxyphenyl)acetic acid	Rosmarinic acid (2R)-3-(3,4-dihydroxyphenyl)-2-[(E)-3-(3,4-dihydroxyphenyl) prop-2-enoyl]oxypropanoic acid
Gentisic acid 2,5-dihydroxybenzoic acid	*Trans*-cinnamic acid (E)-3-phenylprop-2-enoic acid

IUPAC, International Union of Pure and Applied Chemistry.

chlorogenic, 4-hydroxycinnamic acid), benzoic acid, cinnamic acid, and flavonoids. Important flavonoids are chrysin, luteolin, apigenin, galangin, kaempferol, quercetin, pinostrobin, pinocembrin, and terpene compounds and alcohols (cetyl, myricyl, mannitol, and inositol) (Naz et al., 2019; Kurek-Gorecka et al., 2020). Propolis is an important dietary source of chrysin.

Nutraceutical, medical, and food preparation uses of propolis

Propolis products are added to foods and beverages. Ethanol extracts of propolis made into an emulsion have been shown to be effective as a preservative for orange juice (Yang et al., 2017). The medical uses of propolis are being investigated. These include antimicrobial (antimicrobial-resistant pathogens, fungi, and viruses) treatment for cancer, and its antioxidant activities. Propolis in various forms is used as a treatment for upper respiratory tract infections, numerous dermatological conditions, wounds and burns, ulcers, genital infections, gingivitis, and stomatitis. Propolis has a low order of oral toxicity with oral dosing of mice with 4 g/kg/day for 14 days did not result in clinical or pathological evidence of intoxication (Burdock, 1998). A study using drinking water administration of approximately 4.0 g propolis/day for 90 days did not produce evidence of intoxication. Cuban propolis was shown in laboratory animals to provide hepatoprotective effects from carbon tetrachloride and galactosamine intoxication (Gonzalez et al., 1995; Rodriguez et al., 1997). Its historical use as a remedy is recorded in ancient medical codices and recipes for embalming formulations.

Royal jelly

Honeybees are unique among the bee species because they secret RJ and feed it to the brood (larvae) (Wright et al., 2018). RJ is secreted in the mandibular and hypopharyngeal glands located in the head. It has epigenetic activity in genetically female bees. The phenotype of the female honeybee larva (genetically 2N), laid in a standard size brood cell, is determined by worker bees limiting the presentation of RJ. For female eggs that develop into worker bees, the larvae are fed RJ for 3 days posthatching and then fed a combination of honey and pollen. For a female larva in a large queen cell, RJ is presented by the worker bees until pupation. This high level of nutrition is essential to trigger the epigenetic effects leading to the development of a queen. Queen bees are mouth fed by the workers, and this is assumed to be primarily RJ (Martin et al., 2019). Chemically RJ is approximately 50% protein (dry weight), 7.5%–25% carbohydrate, and 10% fat predominantly existing as free monosaturated fatty acids containing 8–12 carbons (Xue et al., 2017; Wright et al., 2018; Martin et al., 2019). The predominant protein in RJ is MRJP1-5. Decenoic acid in various forms is the predominant fatty acid and these are considered important contributors to the antibacterial activity of RJ. Unique components in RJ are 10-hydroxy-2-decenoic acid, 3,10-dihydroxydecanoic acid, and sebacic acid (Moutsatsou et al., 2010). There is evidence that the monosaturated fatty acids in RJ contribute to the longevity of the queen. The RJ proteins differ between *Apis* spp., with the Chinese *Apis cerana cerana* producing proteins that are less polymorphic compared to European bees (*Apis mellifera ligustica*) and Africanized bees (*Apis mellifera scutellata*).

Pollen

Pollen is the main source of protein, fats, vitamins, and other noncarbohydrate nutrients for bees. The source and nutrient value of pollen collection by bees varies with the season, geographic region, and microclimate (Lau et al., 2019). Bees collect pollen from plant anthers and mix it with salivary secretions and nectar. The pollen is then rolled into small balls that are placed on their hind legs and transported to the hive. Pollen is stored in wax cells and undergoes fermentation to bee bread that is considered to improve its nutrient value. Pollen stores are generally found close to the brood cells in Langstroth frames. Bee pollen is recognized as a human food and medicine by many governments (Denisow and Denisow-Pietrzyk, 2016). Nutraceutical uses of pollen are antioxidant, antineoplastic, antiinflammatory, antimicrobial, and beneficial in preventing and curing chronic inflammatory syndrome (Margaoan et al., 2019). Pollen is also used in treating allergies and other immune-mediated diseases. It is added to foodstuffs as an antioxidant to enhance shelf and freezer storage times (Kostic et al., 2020).

Antimicrobial use of bee products

Viral infections

The use of honey and propolis as a treatment for genital herpes virus has been reviewed (Munstedt, 2019). A cross-over study on genital herpes in men and women compared honey treatment to acyclovir treatment (Al-Waili, 2004). The evaluators considered honey a more effective treatment than acyclovir. Acyclovir inhibits viral DNA polymerase. Duration of herpes attacks were significantly decreased by honey. For the occurrence of crusts and pain duration the authors considered honey to be the best treatment. The antimicrobial activity of honey varies with the honey type (Hixon et al., 2019). Kanuka honey (from *Kunzea ericoides*) in a commercial[1] preparation (in a 10% glycerol cream) was studied as a treatment for oral herpes (HSV-1) infections (Semprini et al., 2019). Patients with viral infections on the oral labia (cold sore) were selected for the study. The interventions to the HSV-1 infections were 5% acyclovir[2] cream and a commercial 90% kanuka honey:10% glycerol cream formulation. There were no significant differences ($p = 0.56$) between treatment groups. Munstedt (2019) in a literature review concluded that propolis may be superior to honey as topical treatment for HSV-1 and HSV-2 and both honey and propolis are superior to acyclovir. There are studies that show propolis ointment is superior to a placebo and acyclovir in healing and pain reduction for herpes simplex types 1 and 2, and varicella zoster virus (chickenpox and shingles), and the optimum concentration of propolis extract is 3% (Munstedt, 2019).

Propolis antimicrobial activity

Red and green Brazilian propolis (from different plant sources) are used as antimicrobials for their bacteriostatic and bactericidal actions. Propolis acts at two levels to have antimicrobial activity. These are direct antibacterial activity and stimulation of the immune response to microbial invasion (Przybylek and Karpinski, 2019). Antimicrobial mechanisms include alteration of cell membrane permeability, disruption of membrane potential, disruption of energy production, and inhibition of motility. A number of chemicals have been identified in propolis to have antimicrobial activity (Blicharska and Seidel, 2019). A study using ethanolic extracts of propolis from differing phytogeographical regions showed that propolis from the tropical rainforest regions had the highest antibacterial activity (Seidel et al., 2008; Santos et al., 2020). The antimicrobial activity of propolis varies with the extraction solvent and method (Przybylek and Karpinski, 2019). The efficacy of propolis was studied in a rabbit model for *Staphylococcus aureus* septic arthritis (Oner et al., 2011). An ethanolic extract of propolis (Kayseri region of Turkey) was dissolved in propylene glycol. The in vitro minimum inhibitory concentration for *S. aureus* was 0.12 mg extracted propolis/mL. All the rabbits in the study were inoculated in the right knee (intraarticular) with *S. aureus* (2×10^5 cells/mL). There were four treatments groups and surgical drainage of the knees was done on study day 7. Rabbits ($n = 7$) in group I only had surgical drainage of the knee on day 7. Treatments with cefazolin sodium, propolis, and cefazolin sodium plus propolis extract started after surgery on study day 7. Rabbits ($n = 7$) in group II received daily intramuscular injections of 75 mg cefazolin sodium/kg body weight

1. Honevo, Honeylab Ltd; Tauranga, NZ.
2. Viraban, AFT Pharmaceuticals; Auckland, NZ.

(bwt) for 7 days. Rabbits ($n = 8$) in group III received intraarticular injections (right knee) of 0.5 mg of propolis extract on study days 7, 14, and 21. Group IV rabbits ($n = 8$) received daily intramuscular injections of 75 mg cefazolin sodium/kg bwt for 7 days plus intraarticular injections (right knee) of 0.5 mg of propolis extract on study days 7, 14, and 21. For the histopathology parameters of chondrocyte loss, loss of matrix, and pannus ingrowth, the rabbits in group IV had the best score ($p > 0.05$).

Helicobacter pylori

Helicobacter pylori is a bacterium that colonizes the gastric mucosa and muscularis and is a common cause of peptic ulcers. Worldwide, it is the most common bacterial disease and the infection rate can be 40%–90% of the population, with the highest rate of infection being in developing countries. Infection with *H. pylori* is an important cause of gastric cancer and this organism is a class I carcinogen. *H. pylori* can live inside yeast cells (Siavoshi et al., 2018). A study in Iran on osmotolerant yeasts found that 83.3% of the osmotolerant yeasts isolated from honeybees and honey tested positive for *H. pylori*. The authors concluded that *H. pylori*-positive fungi originated in soil and were transported by bees to floral nectars, fruits, honeydews, the beehive, and honey. The *H. pylori* inside the yeast cell and yeast could have a symbiotic relationship and be a likely reservoir of infection to humans.

Green Brazilian propolis extract (ethanolic extract) was studied in 18 patients with confirmed *H. pylori* infections (Coelho et al., 2007). Twenty drops of the extract were administered orally three times a day. Eighty percent of the patients were positive for *H. pylori* infection at 40 days posttreatment. This commonly used a dose of green propolis extract that was not an ineffective treatment. Antibiotic resistance of *H. pylori* is an important cause of treatment failure. A study evaluated the in vitro antimicrobial activity of propolis and ginger (*Zingiber officinale*) (Roscoe) ethanolic extract combined with clarithromycin against *H. pylori* (Nostro et al., 2006). The study showed that the combinations of propolis extract + clarithromycin and *Z. officinale* extract + clarithromycin exhibited improved inhibition of *H. pylori* with synergistic or additive activity. The efficacy of the combinations was independent of the *H. pylori* sensitivity to clarithromycin. Caffeic acid phenethyl ether is considered an important antibacterial compound in propolis (Cui et al., 2013). It has been shown to be a potent inhibitor of peptide deformylase and its activity is essential for bacterial growth, including *H. pylori* (Cui et al., 2013). A study was carried out in Bulgaria to determine the influence of diet on immunological evidence of *H. pylori* infection (Yordanov et al., 2017). Blood was collected from 294 patients (males and females, age 18–69 year) that were asymptomatic for *H. pylori*. Diet and lifestyle were determined by the subjects completing a questionnaire on the use of tobacco and the consumption of alcoholic beverages excluding red wine, yogurt, green tea, coffee, raw garlic and onion, honey, fresh fruits and vegetables, red wine, and olive oil. The sera were tested for antibodies to *H. pylori* using commercial kits for enzyme-linked immunosorbent assay,[3] including anti-*H. pylori* ELISA (IgG) and anti-*H. pylori* CagA ELISA (IgG). CagA is a protein injected into gastric cells by *H. pylori* using the bacteria type IV secretion system. Statistical analysis showed that frequent (>5 days per week) honey consumption was associated with both reduced *H. pylori* seroprevalence [OR 0.68 with 95% confidence interval (CI) 0.473–0.967] and reduced CagA IgG seroprevalence (OR 0.65 with 95% CI 0.486–0859). Also, yogurt consumption (>5 days per week) was associated with lower *H. pylori* virulence (CagA IgG, OR 0.56 with 95% CI 0.341–0.921). Use of tobacco, consumption of all alcoholic beverages and other foodstuffs had no influence on *H. pylori* infections. Propolis has been shown to have activity against the nascent polypeptide chains on *H. pylori* (Ayala et al., 2014).

The efficacy of *Lactobacillus rhamnosus* as a treatment for *H. pylori* was studied in C57BL/6 mice that had established infections of *H. pylori* (Asgari et al., 2018). The *L. rhamnosus* was isolated from honey obtained in the mountains of northern Iran. Infected mice were divided into treatment groups: group I, with no treatment; group II, treated with a bismuth preparation + omeprazole; group III, treated with omeprazole + clarithromycin; and group IV, omeprazole + 10^9 colony-forming units of *L. rhamnosus*. Treatments were given orally for 2 weeks. Treatments with omeprazole + *L. rhamnosus* and omeprazole + clarithromycin gave negative stool samples for *H. pylori* and similar reductions in gastric inflammation as determined by histopathology.

Vulvovaginitis

Vulvovaginitis is an infection of the vulva and vagina and, over their lifetimes, 75% of mature women have at least one infection. Infections can reoccur and 5% of mature women have multiple repeated infections. Infections with *Candida* spp. are the most common etiology, with *Candida albicans* being the most common isolate. Antimicrobial resistance and side effects from antimicrobial therapy are incentives for alternative

3. ELISA, Euroimmun; Medizinische Laboratordiagnostika AG; Lübeck, GE.

therapies. Honey was evaluated as a remedy for vulvovaginitis in 100 women (Banaeian et al., 2017). Iranian honey was mixed with a neutral cream (70:30 ratio) and 1% clotrimazole was used as a control. In a double-blind study, both preparations were administered intravaginally at 5 g/ treatment for 7 days, with treatment occurring at night. The mean age of the patients entered on the study was 34.3 ± 8.6 year and there were no differences between groups for age, interval from last infection, severity of infection, and use of oral contraceptives. On study day 8, reductions in inflammation and vaginal discharge were better in the clotrimazole group ($n = 36$) than the honey group ($n = 44$, $p < 0.05$). There were no significant differences ($p < 0.05$) between groups in severity of irritation and patient satisfaction and no side effects were reported. Propolis wax is the waxy material that is insoluble in the extraction solvents. A study was done on the efficacy of propolis wax from stingless bees (*Tetragonula* spp.) as a treatment for vaginal candidiasis (Farida et al., 2020). The propolis treatment was prepared by mixing propolis wax with oleum cacao and other ingredients and molded in an ovule mold to make a suppository. Forty women were treated for 7 days with propolis wax preparation ($n = 18$) or nystatin ($n = 17$) with self-treatment occurring before sleep. Nystatin is an antifungal, including yeasts, chemotherapeutic. Using clinical signs and microbiology parameters, the authors concluded that there were no differences between treatment groups for remediation parameters.

Antimicrobial activity of beeswax

The antimicrobial activity of crude beeswax is recognized. Beeswax is a historical folk remedy for dermatitis caused by dermatophytes. Extracts of beeswax (70% ethanol at 37°C for 48 h) were shown to have in vitro antimicrobial activity against *C. albicans*, *S. aureus*, *Bacillus subtilis*, *Escherichia coli*, and some *Aspergillus* spp. (Ghanem, 2011). Of these, crude beeswax was shown to be effective against *C. albicans*. This study also showed that extracts of propolis + extracts of beeswax had possible synergistic antimicrobial interactions for activity against *C. albicans* and *S. aureus*. In another study, beeswax was extracted at 80°C for 1 h with 80 mL of four different solvents [99.9% and 70% (v/v) aqueous methanol and 96% and 70% (v/v) aqueous ethanol] (Kacaniova et al., 2012). The 70% aqueous alcohols used for extraction were pH adjusted to 1.5 for methanol and 2.0 for ethanol. The microbes most sensitive to beeswax extracts were *E. coli*, *Aspergillus niger*, and *Candida tropicalis*. A mixture of beeswax, honey, and olive oil was shown in vitro to inhibit the growth of *S. aureus* and *C. albicans* (Al-Waili, 2005a).

Bee products in dentistry and mouth care

Avulsed teeth

Propolis extracts are used in dentistry as a medium to preserve the viability of avulsed teeth, in endodontic procedures, as a treatment for oral sores, and in the management of gingivitis. The success of tooth replacement is dependent on the viability of periodontal ligament cellular components located on the roots. The most common natural substances used for storage and transport of avulsed teeth are milk, propolis, and coconut water (Adnan et al., 2018). Human teeth extracted as part of orthodontic procedures were used to evaluate propolis extract (95% ethanol) as a medium for storage and transport of avulsed teeth (Prueksakorn et al., 2016). For maintaining a percentage of viability ligament cells, a solution containing 2.5 mg propolis extract/mL was not significantly different ($p = 1$) from Hank's balanced salt solution, 78.84% ± 3.34% and 80.14% ± 2.44%, respectively. The 2.5 mg propolis extract/mL was significantly better ($p < 0.001$) than milk (71.27% ± 2.79%) in maintaining the viability of ligament cells.

Endodontics

Propolis is used in endodontics as an intracanal medication for its antimicrobial activity (Ahangari et al., 2018). Propolis has less toxicity that calcium hydroxide on dental pulp and the periodontal ligament. Propolis was compared to calcium hydroxide for its efficacy for postoperative amelioration of endodontic pain (Shabbir et al., 2020). Patients enrolled on the study were diagnosed with necrotic teeth (single root). The propolis was 200 mg of 95% propolis powder[4] in 3 mL of saline. The intracanal treatments were standard care calcium hydroxide[5] and China-sourced propolis. Overall, these were no significance differences ($p < 0.025$) between groups for patients (78%) with no or mild postoperative pain between the endodontic appointments. For the patients with flareups, the occurrence was slightly higher in the propolis group. The authors attributed pain with postsurgical infection. A study in guinea pigs showed propolis to be superior to calcium hydroxide for dental pulp capping (Ahangari et al., 2012). An ethanolic extract of propolis collected in Azerbaijan was used in the study. The propolis-treated teeth had superior dentin regeneration ($p < 0.05$) compared to the calcium hydroxide-treated teeth and it is

4. Henan Fumei Biotechnology Co., Ltd.; Changge, CN.
5. Calcipulpe, Septodont; Saint-Maur-des-Fossés, FR.

considered a candidate capping agent in pulp therapy. Ethanolic extracts of propolis mixed with glycerin were shown in dogs to be equivalent ($p < 0.05$) to triple antibiotic paste (mixture of ciprofloxacin, metronidazole, and minocycline or doxycycline) (El-Tayeb et al., 2019). Parameters used were root length, dentin thickness, hard tissue formation, tissue formation inside the pulp canal, and apical closure.

Mucositis

Propolis was studied as treatment for oral mucositis associated with chemotherapy (Tomazevic and Jazbec, 2013). A double-blind study was done on the use of propolis to treat oral mucositis that developed as a complication of chemotherapy. The subjects were 1–19 years of age and were receiving antineoplastic chemotherapy. Patient with preexisting mucositis before the start of chemotherapy or who were receiving other treatments for mucositis were excluded from the study. The test substance was the ethanolic (76%) extract from crude propolis that had a chemical profile constant with poplar tree sources. The propolis was in a sticky form to maximize adherence to the oral mucosa. The propolis was applied with a microtip applicator. In this study, there were no statistical differences between the propolis and placebo treatments. Honey and a compounded formulation of honey, olive oil–propolis extract, and beeswax (HOPBW) or just raw honey were studied as a treatment for oral mucositis (Abdulrhman et al., 2012). The mucositis was caused by antineoplastic chemotherapy with methotrexate. The patients were randomized into three groups that received topical treatment with honey, HOPBW, and control (received 7.5% benzocaine jell) applied to the oral mucosa three times a day for 10 days. The mean age of the 90 patients on this study was 6.9 years. For grade 2 mucositis, honey significantly ($p < 0.05$) reduced the healing time. Honey and HOPBW significantly reduced ($p < 0.01$) healing time for grade 3 mucositis when compared to control, and honey and HOPBW were not statistically significantly different from each other. In all grades of mucositis, honey significantly reduced the healing time. Recovery time in grade 2 mucositis was significantly reduced in the honey group as compared with either HOPBW or controls ($p < 0.05$). In grade 3 mucositis, recovery time did not differ significantly between honey and HOPBW ($p = 0.61$) but compared with controls, healing was faster with either honey or HOPBW ($p < 0.01$). Generally, in both grades of mucositis, honey produced faster healing than either HOPBW or controls ($p < 0.05$).

Periodontitis

Periodontitis is a disease that weakens and destroys periodontal and boney tissues that support the teeth. Its causes are multifactorial and the most common aspects are immune response to biofilms on the teeth. A review by Freires et al. (2018) compared laboratory studies in rats and concluded that the gel of *Myracrodruon urundeuva* (5%) plus *Lippia sidoides* (0.5%) (topical application) was 45%–65% effective in preventing bone loss, propolis 100–200 mg/kg (administered orally) was 40%–44% effective in preventing bone loss, and the extract of *Ginkgo biloba* 28–56 mg/kg (administered orally) prevented bone destruction by 50%. Doxycycline antibiotic was 74.3% effective in preventing bone destruction.

Mouth care

A propolis formulation was evaluated as a treatment for denture stomatitis (Santos et al., 2008). Patients with completed dentures and longstanding denture-linked stomatitis were treated with miconazole gel[6] or the propolis gel formulation. No statistical differences were observed between treatment and control groups. Propolis is used in oral hygiene to control the growth of plaque-causing organisms (Sabanovic et al., 2019). Oral hygiene is the most common dental use of propolis (Carvalho et al., 2019).

Skin care

Honey, RJ, and propolis have multiple uses for skin care and remedies (Burlando and Cornara, 2013; Kurek-Gorecka et al., 2020). Traditional uses of honey in commercial and homemade cosmetics include for skin softening and for unesthetic conditions. Honey ingredient in cosmetics is identified as "mel" or "mel extract." A common concentration of honey is 10% and mel can be 70% when dispersed in oils and polymers. Honey in various concoctions is used for freckle removal. In folk medicine, honey is mixed with flours and used as a poultice for furuncles and abscesses and other skin infections. In some regions, honey is mixed with horsetail (*Equisetum* spp.) and used as an emollient. Honey is also mixed with local herbs, fats, and vegetable oils to make remedies for skin wounds. Beeswax is blended with herbal oils and other natural products and is successfully used for a plethora of skin conditions.

Al-Waili (2005b) studied an ointment preparation consisting of 50% honey, 29% olive oil, and 21% beeswax (w/v) on diaper rash in infants. The ointment was applied

6. Janssen-Cilag; Beerse, BE.

four times a day for 7 days. Fifty percent efficacy was achieved in infants that had *C. albicans* isolated from the diaper rash. Topical application of honey, olive oil, and beeswax mixture (1:1:1, v/v) has been studied as a treatment for hemorrhoids (10 patients) and anal fissures (five patients) (Al-Waili et al., 2006). The mixture was applied topically every 12 h for 4 weeks. The treatment was considered to be efficacious. The efficacy of the same formulation (honey, olive oil, and beeswax) was studied in patients with pityriasis versicolor, tinea corporis, tinea cruris, and tinea faciei (Al-Waili, 2004). The formulation was applied topically every 8 h for 3–4 weeks. Treatment was discontinued in patients that were assessed as having no improvement after 3 weeks of treatment. A favorable clinical response was obtained in patients with pityriasis versicolor (86%), patients with tinea cruris showed 78% improvement, and 75% of patients with tinea corporis showed improvement. Mycological cure (no dermatophyte observed on microscopic examination of skin scraping after treatment with KOH) was obtained in 75%, 71%, and 62% of patients with pityriasis versicolor, tinea cruris, and tinea corporis, respectively. No adverse effects were observed.

Bee products use in wound healing and dermatology

Nonhealing wounds have a high cost to health care and are painful for the patient. Bee products, especially honey, have historical use in wound healing (Shah, 2011). An important cause of delayed wound healing is infections, especially infections caused by organisms resistant to antimicrobial chemotherapeutics. The first pathophysiological response to a wound is both biochemical and physiological actions that impeded and eventually stop blood loss. The next response is innate and acquired immune responses (Rodrigues et al., 2019). Wound healing follows, and includes cell proliferation, scaffolding, angiogenesis, collagen formation, and reepithelization. In wound healing, honey has antimicrobial activity, high osmolarity, acid pH, and can stimulate angiogenesis (Takzaree et al., 2016; Hixon et al., 2019). Honey has high osmolarity and reduces free water activity in the wound. The hygroscopic activity of honey is in part dependent on a functional capillary bed under the wound to provide an outflow of lymph that cleanses the wound and bathes the proliferating cells in extracellular fluid. Honey contains glucose oxidase (notatin) which oxidizes glucose to gluconic acid and hydrogen peroxide. The acidic effect of gluconic acid and high redox potential of hydrogen peroxide have antibacterial activity. Honey also encourages oxygenation, and stimulates macrophages and fibroblast proliferation concurrent with collagen production. Honey upregulates proinflammatory cytokines and other biochemical reactions that contribute to wound healing. Honey-based wound dressings are commercially available (Hixon et al., 2019)

Wang et al. (2019), in a review of the literature on the use of honey for diabetic foot ulcers, showed that honey decreases the time for bacterial clearance, debriding, and healing time. The efficacy of Brazilian propolis extracts to promote wound healing was studied in diabetic mice evaluating iatrogenic lesions (Picolotto et al., 2019). The substances studied were obtained through a sequential series of solvent extractions. The red Brazilian propolis was first extracted with 70% ethanolic solvent, which was then extracted with hexane and the hexane extract was, to achieve the final product, extracted with either ethyl acetate or butanol. The extracts were solubilized in bacterial cellulose membranes (*Komagataei bacterrhaeticus*, K-membrane). The wounds were topically treated with physiological saline, K-membrane solubilized in saline, K-membrane solubilized in ethanol, K-membrane solubilized in 1% propolis from the butanol extraction, or with K-membrane solubilized in 1% propolis from the ethyl acetate extraction. The mice treated with both propolis extracts had better healing as demonstrated by histopathology. Also, TNF-α was significantly higher in wounds stimulated with both red propolis extracts. Formononetin was the predominant phytocompound in the propolis ethyl acetate extract. A study in rats (Wistar) on the efficacy of propolis and honey showed that honey plus propolis promotes wound healing more than honey or propolis treatments alone (Takzaree et al., 2016). Filtered honey (85.5 total dissolved solids) from the thymus plant was used in the study and the propolis was extracted with ethanol. Using surgical procedures, a dorsal cervical wound was incised and the efficacy of honey, honey plus propolis extract, phenytoin, and control wounds were evaluated for healing. Healing was evaluated by macroscopic and microscopic parameters. The microscopic parameters (fibroblast, macrophage and neutrophil numbers, and the formation of collagen) were significantly higher for the honey + propolis extract than the control group. Ethanolic extracts of propolis from poplar trees has been shown to improve fracture healing in rats (Guney et al., 2011). Surgical fractures of the femur were fixed using internal fixation. Rats in the treatment group received (gavage) 200 mg of the propolis extract in propylene glycol for 3 or 6 weeks. The parameters consisting of mineral density, radiological, and histological evaluation scores in rats receiving propolis were significantly improved ($p < 0.05$) and plasma myeloperoxidase levels were significantly lower among the rats that received oral propolis. In a Danish case report by Sylvester-Hvid et al.

(2019), manuka honey dressing (Activon[7]) was used as a wound dressing following surgery to remove penile epithelial hyperplasia (pseudocarcinomatous) without malignancy. The manuka honey dressing with antibiotics was initiated to achieve granulation tissue. After a failed split-thickness skin graft, sterile manuka honey (only treatment) was used in a chimney-type gauze bandage. Forty-two days later the penile lesion was healed. For wound healing, methylglyoxal (2-oxopropanal) is considered an important ingredient in manuka honey. Honey is used in tissue-engineered scaffolds. Electrospinning drawing of a polymer into a nanofiber mesh with macropores maintains a surface for drug delivery including honey (Hixon et al., 2019).

Burns

Burns are a common injury and are caused by excessive heat, radiation, cold, and chemicals. These agents cause necrosis of the skin layers and can involve the subcutaneous tissues. This discussion is limited to thermal burns. The degree of a burn is defined by the layers of skin damaged. In first-degree burns, the epidermis is the primary damaged layer of skin; in second-degree burns, both the epidermis and the dermis are the primary layers affected; in third-degree burns, the skin plus subcutaneous tissues are damaged. Beeswax is used in the management of burns (Gumus and Ozlu, 2017; Bayir et al., 2019). The efficacy of a mixture of beeswax, olive oil, and *Alkanna tinctoria* (L.) was evaluated in human patients with second-degree burns (Gumus and Ozlu, 2017). The mixture was prepared by blending 1 L of medical olive oil, 30 g of beeswax, and 50 g of *A. tinctoria* with the mixture being sterilized in amber bottles. Patients entered on the study had second-degree burns that were not infected. Control patients received topical applications of nitrofurazone and rifamycin in a sterile compress. The study patients had a sterile gauze saturated with the phytomixture and placed on the burn wound. Patients in the control group had an average age of 5.52 ± 0.64 years, and 6.68 ± 1.09 years in the experimental group. The most common cause of the burns was exposure to boiling liquids. There were no significant differences ($p < 0.08$) between the treatment and control groups for the pretreatment evaluations. Wound epithelialization time was significantly ($p < 0.05$) less (3.0 ± 0.85 days) in the patients receiving the natural product dressing versus (6.90 ± 1.77 days) for the control patients. Pain scores for the natural product group were significantly less ($p < 0.05$) than the control group and the mean hospital (8.22 ± 3.05) days was significantly less ($p < 0.05$) than the 14.42 ± 7.79 hospital days for the control group. A study was carried out to compare Medilixir[8] to a standard moisturizing cream dispensed at the Professor Stuart Pegg Adult Burns Centre[9] for controlling postburn purities (Lewis et al., 2012). Medilixir is a blend of coconut oil (*Cocos nucifera*), palm oil (*Elaeis guineensis*), castor oil (*Ricinus communis*), olive fruit oil (*Olea europaea*), beeswax (*Cera alba*), hemp seed oil (*Eupatorium cannabinum*), wheat germ oil (*Triticum vulgare*), and canola seed oil (*B. napus*). The Medilixir was significantly better ($p < 0.05$) than the standard moisturizing cream in reducing purities and medication usage.

Beeswax (5 g), olive oil (10 mL), and butter (10 g) (BOB) were prepared by mixing while heating in a water bath (bain-marie) and the mixture used to impregnate bandages (Bayir et al., 2019). The impregnated bandages were sterilized by gamma radiation (25 kGy). The bandages were applied to rats that previously had induction of second-degree burns under anesthesia. The treatment groups were control (no burns), group 2 (no burn treatment), group 3 (treated with silver sulfadiazine), and group 4 (treated with BOB). The burn lesions were evaluated for healing on days 3, 7, and 14 by macroscopic observation, histopathology, and histochemistry. The BOB treatment increased transforming growth factor-β1 (TGF-β1) expressions compared to groups 2 and 3. Compared to group 2, the vascular endothelial growth factor-α (VEGF-α) was increased in groups 3 and 4, and groups 3 and 4 were not significantly different from each other. In the authors' opinion, the histopathology showed that treatment with the BOB-impregnated bandage improved burn healing. The conclusion was that BOB-impregnated bandages are a treatment for second-degree burns. Ointments consisting of beeswax, olive oil, and *A. tinctoria* (L. Tausch) extract (OBA) have been used as a remedy for burns (Gumus and Ozlu, 2017). A clinical study was carried out on OBA ointment as a dressing for burns (Gumus and Ozlu, 2017). The majority of the patients entered on the study were children (mean of 5.52 ± 0.64 year for control group and 6.68 ± 1.09 year for the OBA ointment) and the majority of burns (93.9% of controls and 83.9% of the treatment groups) were caused by scalding of the extremities with boiling liquids. For preparation of the OBA ointment, 100 mL of medical olive oil was heated to 200°C–210°C, 30 g of beeswax were dissolved, and 50 g of *A. tinctoria* were added to the mixture (OBA). The OBA mixture was heated and mixed for 5 min and then filtered using a fine mesh filter and placed in 100 mL amber bottles. The OBA ointment was

7. Advancis Medical; UK.
8. Medilixir Pty Ltd.; Brisbane, AU.
9. Royal Brisbane and Women's Hospital; Brisbane, AU.

steam sterilized (1.5 atmospheric and 121°C). Immediately prior to dressing the burn, the bandages were saturated with the OBA ointment. The average time for epithelialization in the OBA ointment-treated patients was 3.00 ± 0.85 days and was significantly shorter ($p < 0.05$) than control group patients (6.90 ± 1.77 days). Additionally, the mean pain scores experienced by the patients during dressing changes were significantly lower ($p < 0.05$) for the patients receiving the OBA dressings than control patients. The duration of hospitalization for the patients receiving the OBA dressings was significantly shorter ($p < 0.05$) than for control patients; 8.22 ± 3.05 and 14.42 ± 7.79, respectively. In vitro studies show that by-products of beeswax are protective to cells in preventing oxidative damage (Giampieri et al., 2018).

Wijesinghe et al. (2009), in a metaanalysis, found that most studies using honey for burn care had a Jadad score of 1. In most studies, a sterile gauze was covered with unprocessed honey and silver sulfadiazine-impregnated gauze were compared for healing time. The authors concluded that fixed effects odds ratio for healing at 15 days was 6.1 (95% CI 3.7–9.9) and the random effects pooled odds ratio was 6.7 (95% CI 2.8–15.8) with both showing honey to be superior treatment. There is evidence that burn wounds have fewer adverse effects when a honey dressing is used (Jull et al., 2015).

Bee products and neoplasia (cancer)

A study of Jordanian oncology patients showed that honey was the most common nutraceutical consumed by cancer patients (Akhu-Zaheya and Alkhasawneh, 2012). The phenolics and flavonoids are considered important in the anticancer activity of honey and other bee products (Table 47.1) (Afrin et al., 2019; Naz et al., 2019). Honey has been shown to have in vitro antiproliferative effects on differing types of cancer cells grown in tissue culture (Battino et al., 2019). The mechanisms of honey as an antineoplastic agent have been reviewed (Afrin et al., 2019; Battino et al., 2019; Waheed et al., 2019). In vitro studies suggest that honey arrests cell cycle at the G_o/G_1 phase and chrysin (5,7-dihydroxyflavone) is a flavone of interest. Changes in mitochondrial membrane permeability leading to mitochondrial collapse and cell death have been proposed as the in vitro mechanism for antineoplastic activity for colon cancer cells (HT-29 and HCT-15). Honey has been shown in vitro to induce apoptosis of neoplastic cells and by immunomodulatory mechanisms (elevating cytokines TNF-α and IL-1β) and increased phagocytic activity of macrophages (Attia et al., 2008; Battino et al., 2019). There is some evidence that honey can both promote and inhibit metastasis (Attia et al., 2008). Honey treatment of surgical wounds significantly decreased ($p < 0.05$) the rate of Ehrlich ascites tumor transplantation in BALB/c mice (Hamzaoglu et al., 2000). Honey and RJ were studied as a recourse in cancer-related fatigue (Mofid et al., 2016). Cancer-related fatigue is estimated to occur in 60%–96% of cancer patients receiving treatment and to persist for 5–10 years in cancer survivors. The pathophysiology of cancer-related fatigue is considered multifactorial. A double-blind study was done on 4 weeks of orally administered 5 mL of honey (control) and 5 mL honey + RJ (treatment) to ameliorate cancer-related fatigue. Fatigue was measured using a visual analog scale and fatigue severity scale. Patients receiving honey + RJ had significantly improved ($p < 0.001$) fatigue indices scores. A study in mice showed that Tualang honey (*Apis dorsata*) reduced the carcinogenicity of 1-methyl-1-nitrosourea using the Thompson model (Thompson and Adlakha, 1991; Ahmed and Othman, 2017). An in vitro study by Yaacob and Ismail (2014) using cancer cells MCF-7 and MDA-MB-231, and MCF-10A breast epithelial cells showed that 4-hydroxytamoxifen and Tualang honey have positive interactions for inhibition of cancer cells. Tualang honey was protective of the adverse effects of 4-hydroxytamoxifen on the MCF-10A cells.

Bee products and ophthalmology

Consumption of honey may improve eye health. There are reports of honey being used to treat different ophthalmological maladies (Majtanova et al., 2016). These conditions include dry eye syndrome, bullous keratopathy, corneal opacities, hepatic keratitis, conjunctivitis, and chemical and thermal burns of the cornea. The causation of dry eye syndrome is multifactorial and the syndrome is tear film instability that results in pathophysiology of the eye, especially the cornea. A study was done in patients using a multifloral honey diluted (20%) with distilled water (Jankauskiene et al., 2007). Thirty-six patients (21 females, 15 males, 29–62 years old) were enrolled. Nineteen patients were treated with the honey formulation and 17 were treated with control eye drops. The patients were examined for eye pathology. The authors concluded that honey eyedrop treatment could be used as an alternate therapy for dry eye syndrome. Honey eyedrops have been studied as a treatment for vernal keratoconjunctivitis (Salehi et al., 2014). Sixty patients with uncomplicated keratoconjunctivitis were enrolled on the study. Thirty patients in the treatment group received fluorometholone (1%) and sodium cromolyn along with honey drops (60%) and the control group consisting of 30 patients received fluorometholone and sodium cromolyn along with artificial tear. The treatments were applied four times a day. In the treatment group, eye pressure was significantly increased ($p < 0.01$, within the normal range) and, compared to the control group, redness and limbal papillae were significantly reduced ($p < 0.05$). After the trial, one patient in the treatment group and seven in the placebo group had limbal papillae. Infections within the eye (endophthalmitis) are a complication of eye surgery inclusive of cataract surgery. Prophylactic use of antibiotics before interocular surgery may contribute to antibiotic resistance. Honey made from honeydew (*Abies* spp. source) was studied in 101 patients as a replacement for ofloxacin prophylaxis (Cernak et al., 2012). The honey was from an apiary located in Bardejov, Slovakia, was sterilized by gamma irradiation, and diluted in distilled water to 25% (w/v) and filtered (0.22 mm PES filter) to remove microorganisms and debris. Treatment groups were honey drops (49 patients) and controls ($n = 52$) receiving ofloxacin[10] 0.3% ophthalmic. The interior conjunctival fornices were aggressively swabbed, and the swabs cultured for bacteria. The eye drops were administered five times a day for 7 days before surgery and postsurgery days 1–5 days concurrent with fluorometholone.[11] Before treatment, 18 and 25 isolates were detected in the ofloxacin and honey groups, respectively. After 7 days of treatment, four isolates of coagulase-negative *Staphylococcus* spp. were detected in swabs from each group. A cyclodextrin-complexed manuka honey microemulsion (MHM) cream was evaluated[12] for safe use in humans (Craig et al., 2020). Volunteers applied the MHM cream to the right or left eye, with the untreated eye serving as a control. The MHM cream was applied periocularly once each night for 14 days. Two volunteers reported transient ocular stinging. During the treatment period, no significant changes in the treated eyes ($p > 0.05$) were observed for visual acuity, eyelid irritation symptoms, ocular surface characteristics, tear film parameters, and inflammatory marker expression. These measurements were not significantly different ($p < 0.05$) between treated and control eyes.

Reproductive and gynecological uses

Reproduction

RJ has been shown to improve the viability of semen from domestic animals used in artificial insemination (Abdelnour et al., 2020). Oral administration of RJ to buck rabbits improved the conception rate of does receiving their semen by artificial insemination. Pollen added to rabbit diets improved reproductive parameters (Abdelnour et al., 2019). Honey was evaluated as an additive (10%) to commercial cryoprotection medium for human semen. The postthawing viability parameters were improved (Fakhrildin and Alsaadi, 2014). Banday et al. (2017) concluded that honey could not be used as a replacement for antibiotics in semen extender (tris-buffer) used to preserve ram semen. RJ has been shown to improve the cryopreservation of honeybee semen (Alcay et al., 2019). Vitrification of embryos is a process of rapid cooling of embryos with cryopreservative that prevents the formation of damaging intercellular ice crystals. Honey was evaluated as a nonpenetrating cryopreserve for embryos (Sarmadi et al., 2019). At 1 Osm/L, honey was considered to be the optimum level for mouse embryos and was comparable to conventional medium. Haron and Mohamed (2016) studied the ameliorative effects of administering honey to pregnant rats to reduce the adverse effects of stress on the fetus. Treatment groups were control, honey, stress only, and honey + stress ($n = 10$). The bred females in the honey groups orally received 1.2 g Tualang honey/kg bwt from mating to whelping, and the control and stress only group received an equivalent volume of water per gavage. The stress and honey + stress groups received stress by restrain (three times per day) from day 11 of pregnancy until whelping. At 10 weeks of age the male offspring (F_1) were mated with receptive females. There were no significant differences ($p < 0.05$) between groups for sexual performance of the F_1 male rats. There were no significant differences between groups for the relative weights of the seminal vesicles. The relative weights of the testicles and epididymides in the F_1 male rats from mothers in the honey + stress group were significantly higher ($p < 0.05$) than these organ weights in F_1 males from mothers in the control, stress, and honey groups.

Menstrual distress and labor pain

Honey has been studied to relieve pain associated with menstruation (dysmenorrhea) and parturition (Amiri Farahani et al., 2017; Taavoni et al., 2019). Primary dysmenorrhea is considered to be multifactorial in causation. A crossover design was used to compare the efficacy of honey and mefenamic acid on dysmenorrhea in single women with regular menstruation cycles (21–35-day interval) and having cycles predominantly with types 2 to 3 dysmenorrhea with onset of pain occurring at menstruation bleeding (Amiri Farahani et al., 2017). The mean age of the patients was 22.01 ± 1.78 years. Honey was produced in Iran, had a moisture content of 14%, and the dosage was 1.2 mg honey/kg bwt given from the 15th day of the cycle to the onset of menstruation for 2 consecutive months. Mefenamic acid[13] was a 250 mg capsule with instructions to take two capsules at the onset of menstruation and then one capsule every 6 h for 2 days of the first

10. Uniflox, Unimed Pharma; Bratislava, SK.
11. Efflumidex, Allergan Pharmaceuticals; Westport, IR.
12. Clinical Trial No. ACTRN12616000540415, Manuka Health; Auckland, NZ.
13. Amin Medicine Company; Isfahan Province, Iran.

and second menstrual cycles. The study showed, between groups, there were no significant differences ($p > 0.05$) between honey and mefenamic acid for the most severe level of pain in the first and second cycles of the first treatment period, and the first and second cycles of the second treatment period. The authors consider honey a replacement for mefenamic acid for the treatment of dysmenorrhea. Premenstrual syndrome (PMS) is characterized by somatic and psychological symptoms occurring during the luteal phase of a woman's menstrual cycle. To meet the criteria, onset must occur during the last week of the luteal phase, lessen during menstruation, and be absent in the follicular phase. The symptoms must be severe enough to interfere with a patient's usual activities or relationships (Appleton, 2018). RJ was studied as a treatment for premenstrual syndrome in women (Taavoni et al., 2014). The subjects were medical science students living in university dormitories. The premenstrual profiles for each woman were scored for approximately 70 days. The students, single women 18–35 years of age, were assigned to intervention (RJ) or a control group (medical paraffin). One-gram RJ or 1 g of paraffin were in capsules and were taken orally as assigned starting on the first day of menstruation and continued daily throughout two consecutive menstrual cycles. For the RJ group, the PMS scores decreased from 23.17 ± 17.43 to 11.42 ± 14.58 with a mean change of 11.75 (95% confidence interval 8.31–15.19). In the paraffin control group, PMS score changed from 21.48 ± 16.39 to 20.27 ± 15.76 with a mean change of 1.20 (95% confidence 1.69–4.10). The PMS scores between the RJ and paraffin groups were significantly different ($p < 0.006$). Although the mechanism of action is not known, Mishima et al. (2005) have shown that RJ has estrogenic activities through interaction with estrogen receptors followed by gene expression. Another possible molecular mechanism for the estrogenic activities of RJs is unique fatty acids that mediate estrogen signaling, at least in part, by modulating the recruitment of estrogen receptor-α and ER-β at the co-activator binding site (Moutsatsou et al., 2010). Taavoni et al. (2019) conducted a study on the effects of honey in labor pain in primigravida women. In an Iranian study, 80 women considered to be in the early stage of labor were divided into two groups. The treatment group received 2.5 teaspoons of Iranian honey in 150 mL of water and the control group only received water. The honey water was offered at the beginning of the study period and continued at 30-minute intervals. Compared to the control group, honey treatment significantly reduced ($p < 0.05$) labor pain during the 30 to 120 min interval. At 150 min there was no significant differences between the treatment groups.

Postmenopausal uses

Shimizu et al. (2018) showed in a rat menopausal model (ovariectomized) that treatment with RJ did not prevent bone loss but did improve bone strength. After ovariectomy, the rats were given 10 mL RJ orally/rat for 12 weeks. Bone mass, bone histomorphometry, and bone mechanics were analyzed at the end of the 12-week period. For group comparisons, femur bone mineral density in the RJ group was significantly lower ($p < 0.05$) than the control group (surgery without ovariectomy) showing by this effect that RJ was not an effective treatment in preventing bone mineral loss. However, femur stiffness (evaluated by a three-point bending test) was significantly higher ($p < 0.05$) in the ovariectomized rats that were given RJ. This showed that RJ prevent weakening of the bone despite mineral loss. Hidaka et al. (2006) studied ovariectomized rats for the effects of RJ, protease-treated RJ (pRJ), and 17β-estradiol on bone mineral density. The control group underwent abdominal surgery without ovariectomy. Compared to controls, the ovariectomized rats, over 7 weeks, had a 24% decrease in tibia mineral bone density. The subcutaneously administered dose of 17β-estradiol was 10 μg/kg bwt given once a day, 5 days/week for 7 weeks. The rats were fed MF[14] powdered pellets. Four groups received the RJ or pRJ by mixing RJ or pRJ into the diet at 0.5% (w/w, 0.5 g/100 feed) and two ovariectomized groups received RJ or pRJ at 2% (w/w). The bone mineral density of ovariectomized rats that did not receive any treatment decreased by 24%. Ovariectomized rats treated with 17β-estradiol recovered 100% of their bone mineral density. The mineral bone density of the ovariectomized rats that received the 2.0% (w/w) RJ diet, 0.5% (w/w) pRJ diet, and 2.0% (w/w) pRJ diet were 85%, 87%, and 93% of the values, respectively, for the rats that received 17β-estradiol. Unlike 17β-estradiol, the RJ and pRJ did not decrease feed consumption or block the decrease in uterine weights. pRJ has a higher content of free amino acids and peptides than RJ, and pRJ increases the intestinal absorption of calcium. In vitro cultures of tibial bone showed that both RJ and pRJ stimulated bone formation and did not inhibit parathyroid-induced bone loss (Hidaka et al., 2006). In humans, atrophy of the tissues lining the vulva, vagina, urethra, and urinary bladder occur during menopause due to a decline in estrogen and this is known as the genitourinary syndrome of menopause (Davis et al., 2015). RJ was studied in Iranian women as a remedy for the genitourinary syndrome of menopause (Seyyedi et al., 2016). Postmenopausal women (53.57 ± 4.89 years of age) were assigned to treatments group consisting of lubricant

14. Oriental Yeast Co. Ltd; Tokyo, JP.

($n = 27$), vaginal conjugated estrogen ($n = 22$), and RJ ($n = 24$). The intravaginal lubricant was a commercial lubricating jelly[15] (glycerol and hydroxyethylcellulose) and the intravaginally administered estrogen was 0.625 mg estrone sulfate sodium[16] in a vaginal cream, and RJ was formulated (15%) in a vaginal cream. Treatments were blinded for patients and staff doing the evaluations. Patients were instructed to self-administer (intravaginally) one tube/night for weeks 1–2; for weeks 3–4, they were to self-administer one tube/night for 4 nights/week; and for weeks 5–12, they were to self-administer one tube/night for 2 nights/week. The treatment period was 3 months. For all women, average onset of menopause was 48.33 ± 4.97 years of age and the average parity was 5.21 ± 2.66. There were no significant differences ($p < 0.05$) between groups for onset of menopause, parity, age, and the evaluations for the genitourinary syndrome of menopause. Treatment with RJ and estrone sulfate sodium was significantly ($p < 0.05$) more effective for treatment of genitourinary syndrome of menopause than the lubricating jelly. Estrone sulfate sodium was the most effective treatment for improvement of vaginal atrophy and both RJ and vaginal lubricant groups were not significantly different ($p = 0.89$) from each other.

Honey (Tualang) studied in a rat model considered to mimic the postmenopausal women (Al-Rahbi et al., 2014). The Malaysian Tualang honey is a multifloral honey produced by *A. dorsata*. This study consisted of six treatments groups with six female rats/group. The rats (Sprague-Dawley) underwent sham ovariohysterectomy (surgery with no **organ** removal) or ovariohysterectomy. Rats in group 1 were sham surgery controls, group II rats were sham surgery plus stress, group III rats underwent ovariohysterectomy and no stress, group IV rats had ovariohysterectomy plus stress, rats in group V had ovariohysterectomy surgery, and were treated with 17-β-estradiol (E2) plus stress, and rats in group V underwent ovariohysterectomy plus treatment with Tualang honey. Ten days was allowed for wound healing and then subject to 15 days of stress (alternating crowding and isolation). The doses of E2 and Tualang honey were 20 μg/kg bwt and 200 mg/kg bwt, respectively, started 3 days before stress, and continued during stressing interval. The rats were also given a novel recognition test. The rats were sacrificed and blood was collected for serum corticosterone and areas of the hippocampus were collected for examination of the pyramidal neurons. Both E2 and Tualang honey-treated groups had improved short-term memory comparable to the sham-operated control group and no improvement for long-term memory. Treatment with E2 or Tualang honey significantly decreased ($p < 0.05$) serum corticosterone as comparable to stressed sham surgery controls. The authors concluded that rats treated with E2 or Tualang honey had a higher number of Nissl-positive cells in hippocampal CA2, CA3, and DG regions compared to untreated stressed OVX rats.

Protective effects of honey

Honey has been shown to protect rats from the toxic effects of tobacco smoke (cigarettes) (Mohamed et al., 2012). Four groups of male rats were exposed to no honey and tobacco smoke (control), honey without tobacco smoke, honey plus tobacco smoke. The oral dose of honey was 1.2 g/kg bw/day and whole-body exposure to tobacco smoke was done three times a day for 18 weeks by exposing the rats to the smoke from 10 cigarettes for 8 min. After 18 weeks of exposure the rats were sacrificed and the epididymis, prostate, seminal vesicles, and coagulating gland were evaluated by histopathology. Exposure to cigarette smoke significantly ($p < 0.05$) increased the relative weight of the epididymis and prostate, and the number of clear cells and height of the epithelium in the cauda epididymis. Additionally, there was severe interstitial edema and decreased epithelial height in the prostate gland. Concurrent exposure to honey significantly reversed the histopathology caused by cigarette smoke.

Bee venom

Honeybee venom, also known as apitoxin, is produced by the poison glands in the abdomen of the honeybee. Sensitive individuals can have life-threatening reactions to honeybee stings. More than 50 biologically active compounds, including the enzymes phospholipases and phosphoesterases, the polypeptide melittin, and numerous other substances are found in the venom (Cornara et al., 2017; Pascoal et al., 2019). When stinging, a honeybee injects between 50 and 140 μg of venom. The majority of individuals have a local dermal reaction to a honeybee sting. Melittin is approximately 50% w/w of dry honeybee venom (Lee and Bae, 2016). This polypeptide has 26 amino acids and is amphiphilic. Honeybee venom is used in acupuncture (Cherniack and Govorushko, 2018). Acupuncture applications include lumbar disk disease, osteoarthritis, rheumatoid arthritis, and neuropathies. Small parenteral doses of melittin have antiinflammatory actions. Purified honeybee venom[17] (PHBV) reconstructed in 0.5% lidocaine was studied in patients with knee pain from osteoarthritis (Conrad et al., 2019). The

15. Aburaihan Pharmaceutical Company; Tehran, IR.
16. Premarin (estrone sulfate sodium), Aboreihan Bironi; Tehran, IR.
17. Jubilant HollisterStier LLC; Spokane, WA.

patients were given biotreatment injected at five acupuncture sites based on traditional Chinese medicine. The number of injection sites increased from 2 to 15 according to a weekly schedule. Then the maximum dose was given in 15 acupuncture sites over 84 days. At the maximum dose, 0.1 μg PHBV/site was administered. Control patients received histamine. Comparing the two groups, there was statistical improvement in physical function and pain reduction in the group treated with PHBV. In laboratory animals and tissue cultures, melittin has been shown to be a candidate drug for acne vulgaris, amyotrophic lateral sclerosis, atherosclerosis, arthritis, and chronic liver inflammation (Lee and Bae, 2016). Melittin inhibits the transcription of inflammatory genes through various mechanisms. Melittin also lyses red blood cells. There is some evidence that professional beekeepers have a lower risk for arthritis.

Toxicology of bee products

Adulteration of honey

Honey can be adulterated with the addition of sugar sirups. The most commonly used are corn, beet, rice, and inulin sirups (Siddiqui et al., 2017; Se et al., 2019). Of these, rice sirup is the adulterant that is the most difficult to identify. The amino acid proline, at lower than expected levels, is a key marker for adulteration of honey with sirups. Using ion chromatography for Na^{1+}, Ca^{2+}, Mg^{2+}, NH_4^{1+}, Cl^{1+}, Br^{1-}, SO_4^{2-}, NO_3^{1-}, and PO_4^{3-}, Fermo et al. (2013) showed that these ions are indicators of environmental pollution, botanical origin, and authenticity. The levels of environmental contaminants can vary among bee products. In the landscape context, French honey is less contaminated with lead than pollen from the same nectar sources (Lambert et al., 2012).

5-Hydroxymethylfurfural

Honey can contain 5-hydroxymethylfurfural (HMF), which is a cyclic aldehyde with both aldehyde and alcohol functional groups. HMF is produced by the degradation of sugars through the nonenzymatic Maillard reaction. HMF is present in all heat-processed sugar-containing foods and beverages (Al-Jahdali and Carbonero, 2019). The levels of HMF in foodstuffs are set by the regulations and guidelines. The levels of simple sugars, organic acids, and minerals, and heating and storage time can enhance the production of HMF. In honey, the formation of HMF is linked to pH, free acid content, mineral content, and lactones, and all of these are linked to the floral sources of the honey (da Silva et al., 2016). The presence of iron (II), magnesium, manganese, and zinc in honey increase HMF formation. HMF production in honey increases with processing temperature and duration of heating (Fallico et al., 2004). In stored honey, water activity (a_w) and storage time and temperature also influence the formation of HMF. More HMF is produced in honey with increased a_w, longer storage times, and warm storage conditions. HMF is also produced by oligo- and polysaccharides that hydrolyze into hexoses. In honey, HMF is selectively formed from fructose and glucose, with fructose having the higher enolization rate (Kuster, 1994). Yang, et al. (2019a) recently showed that formation of HMF in rapeseed honey was dependent on high a_w, Ca^{2+}, and Mg^{2+} and low pH. The free acids in honey can catalyze the dehydration of fructose and glucose to form HMF instead of the Maillard reaction.

Biological activity of 5-hydroxymethylfurfural

HMF has been shown to be genotoxic, an indirect mutagen, causes DNA-damage, and is a carcinogen (Shapla et al., 2018). It also inhibits DNA polymerase γ and is nephrotoxic. HMF has also been shown in vitro to be beneficial in blocking adverse effects of hypoxia (Li et al., 2011). Paradoxically, HMF has antioxidant properties, blocks immune-mediated allergic reactions, downregulates xanthine oxidase, and aids in preventing sickle hemoglobin (Shapla et al., 2018). HMF has been shown to be toxic to bee larva (*Apis mellifera carnica*) and has an LC_{50} of 4280 ppm (7 day posthatching larva) and 2424 ppm (for 22-day larva) (Krainer et al., 2016). There appears to be a residual effect as bees exposed to HMF in the larval stage have a lower LD_{50} to HMF at postemergence day 22 than bees that were not exposed as larva. High-fructose corn sirup containing 250 ppm HMF decreased bee survival (Zirbes et al., 2013). Crystallization of the sirup increased bee mortality. HMF also decreases the resistance of bees to diseases and decreases the margin of safety of some medications. This is likely due to inhibition of CYP enzymes.

Toxic substances in honey

Poisoning by honey is recorded in ancient medical history. Toxic substances in honey can be plant toxins secreted into the nectar, toxic substances contaminating materials brought to the hive by bees, and substances formed by the redox potential of honey (Islam et al., 2014). Plant toxins in honey can be toxic to humans and have no apparent ill effects in bees. Factors affecting toxins being incorporated into honey are season of the year, availability of honeydew, and foraging habits of bees.

Mad honey syndrome

Mad honey poisoning is recorded in ancient history. This poisoning honey-linked syndrome is caused by the presence of grayanotoxins (GTXs) in honey. It occurs primarily in Turkey and is also reported in North America, Korea, and Nepal. GTXs are present in nectar and pollen from *Rhododendron* spp. and are incorporated into honey. GTXs do not appear to adversely affect bees but cause a cholinergic toxidrome in humans (Gunduz et al., 2008). This toxidrome is characterized by dizziness, diplopia, altered metal status, abdominal pain, bradycardia, hypotension, and respiratory depression, and the toxic effects of GTXs can be life-threatening. Some patients can have signs of acute coronary syndrome (Chen et al., 2013). GTXs are diterpenes and at least 60 different toxins have been identified. Of these GTXs, I and III are the most common. These is evidence that GTXs act on the voltage-dependent sodium channel to cause a shift toward hyperpolarization. GTXs I and III can be detected in blood, urine, and honey using liquid chromatography with mass spectrometry or time-of-flight mass spectrometry instrumentation (Aygun et al., 2018). In affected patients, mean blood levels (ng/mL) were 4.82 ng/mL GTX-I and 6.56 ng/mL GTX-III. Mean urine levels were 0.036 mg/mL GTX-I and 0.391 mg/mL for GTX-III. The GTXs in mad honeys consumed by patients were 8.73 mg/g GTX-I and 27.60 mg/g GTX-III. Mad honey is also used as a natural remedy and intoxications can occur from this use (Demircan et al., 2009).

Tutin and 14-(R)-hydroxy-gelsenicine

Poisoning from honey containing tutin, chemically similar to picrotoxinin, has been reported (Beasley et al., 2018). Tutin is a phytotoxin in passion vine (*Coriaria arborea*). Tutin is concentrated in honeydew secretions by the passion vine hopper (*Scolypopa australis*). This sweet honeydew is collected by honeybees and further concentrated in ripened honey. Tutin inhibits of $\alpha 1$ and $\alpha 2$ glycine receptors in a concentration-dependent manner causing vomiting, headache, and grand mal seizures (Fuentealba et al., 2011; Beasley et al., 2018). Regional poisoning linked to tutin in honey has occurred in New Zealand (Beasley et al., 2018). The physician-observed clinical signs were headache, emesis, and grand mal seizures. The neurotoxin 14-(R)-hydroxy-gelsenicine (HGE) present in honey is made from plant products bees obtain from *Gelsmium* spp. (Yang et al., 2019b). Honey containing HGE has caused human fatalities in China. The LD_{50} of HGE in male and female mice is 0.125 and 0.295 mg/kg, respectively. The mechanism is antagonism of γ-aminobutyric acid, the lethal effects of HGE are partially blocked by flumazenil, a γ-aminobutyric antagonist.

Pyrrolizidine alkaloids

Pyrrolizidine alkaloids (PAs) are a class of over 600 different compounds that include hepatotoxins, carcinogens, and genotoxins (Fu et al., 2004). The EFSA (2017) has expressed a concern that children could be at risk from PAs in honey. They are found in flowering plants of the families Boraginaceae, Fabaceae (tribe Crotalarieae), Asteraceae (tribe Senecioneae), and some Lamiaceae and Poaceae (Fu et al., 2004; Carpinelli de Jesus et al., 2019; Picron et al., 2020). Pyrrolizidine alkaloids have been identified in honey from regional markets in Australia at levels up to 2.0 μg PAs/g of honey (Carpinelli de Jesus et al., 2019). Picron et al. (2020) identified pyrrolizidine alkaloids in honey and pollen with pollen having the highest levels. Mitterrandian honey contained a maximum of 182 ng PAs/g of honey and Belgium honey contained a maximum of 60 ng PAs/g of honey. The highest level of PAs in pollen collected by bees was 1672 ng/g of pollen. The primary concern of PAs in honey is human exposure to carcinogens and the use of honey in foods that appeal to children. In humans, the most commonly diagnosed sources in PA-linked hepatotoxicity are herbal teas.

Allergic reactions

There are reports of individuals developing contact dermatitis and severe systemic allergies to bee products (Bauer et al., 1996; Jagdis and Sussman, 2012; Gunawardana, 2017). The risk is multifactorial depending on previous exposure, sensitivity acquired from environmental and occupational exposures, subsequent exposure to honeybee products, geographic location, season of the year, and antigens shared across plant species, especially the family Compositae (Jagdis and Sussman, 2012). Severe systemic reactions can occur to honeybee products in individuals with no history of allergies. Substances secreted by the honeybees themselves may be involved in allergies to honey (Bauer et al., 1996). Acute allergic reactions have occurred after oral ingestion of honeybee pollen (Choi et al., 2015). Topical products containing propolis can cause contact dermatitis and also can trigger acute systemic immune-mediated reactions (Hsu et al., 2004; Gunawardana, 2017). The highest risk in honeybee venom therapy is a systemic anaphylactic reaction (Rueff et al., 2017). Of the stinging bees (Apidae and Vespidae), honeybees inject the most venom (50–140 μg). Fifty percent of individuals with fatal bee sting reactions had no history of a previous systemic reaction. Persons at increased risk to bee venom are those taking prescribed medications, especially acetylcholinesterase inhibitors and β-blockers. Also at increased risk are geriatric patients, individuals with concomitant autoimmune diseases, and individuals with neoplasias in remission (Pitsios et al., 2019). In Korea, bee venom accounted for 55% of the anaphylaxis cases presented to the emergency room (Cho et al., 2019).

Toxicology of beeswax

Undulated beeswax, in an acute oral toxicity study, was administered to rats at a dose of 5 g beeswax/kg bwt (Anon, 1984). Four of the 10 rats died on the second day after dosing and depression and ataxia were observed in the surviving animals. No details were provided on the dosing procedures and no pathological information was provided. The European Commission Panel on Food Additives, Flavorings, Processing Aids and Materials in Contact with Food considers beeswax to be of low acute and chronic toxicity (Alexander et al., 2012). Beeswax has a low level of digestibility in the small intestine of mammals (Aguilar et al., 2007). There is some evidence that it may undergo some degradation by the microbiome in the large intestine. Beeswax can contain environmental chemicals and drugs used in beekeeping. Beeswax has not been shown to be mutagenic and has a low order of ocular and dermal irritation. There are reports of contact dermatitis of the breast, face, and hands linked to beeswax (Camarasa, 1975; Garcia et al., 1995; Nyman et al., 2019). These reports include a female art worker molding beeswax, allergic reaction to a breast nipple protectant, and cheilitis and facial eczema caused by cosmetics containing beeswax. Beeswax and refined beeswax have been used to control hemorrhage in sternal incisions and as a bone sealant. Sudmann et al. (2006) studied postsurgical histopathology in patients that had beeswax and its refined products used as a bone sealant. There was histopathological evidence that beeswax–base bone wax was not absorbed 10 years after sternal surgery and granulomas were observed in sternal specimens collected from incision site of 18 cadavers. Beeswax can contain environmental contaminants (Barganska et al., 2016).

Nutraceuticals and probiotics for bees

The health of honeybees is of increasing concern because of the dependence on bees for pollination of crops essential for human welfare and a supply of honeybee-source products. Good honeybee nutrition is an important aspect of controlling bee diseases and improving the foraging and overall productivity of bees. Microalgae have substantial promise as a health protein supplement for honeybees, and honeybees may collect microalgae when foraging (Ricigliano, 2020). A devastating pathogen of honeybees is *Paenibacillus larvae*, the cause of American foulbrood (Daisley et al., 2020; Tsourkas, 2020). *P. larvae* is a Gram-positive spore-forming bacterium that only infects bees and this bacterium is becoming antibiotic resistant. The spores of *P. larvae* are refractory to environmental conditions and removal requires harsh disinfection procedures (Ohashi et al., 2020). The spores are also highly infective, with as few as 10 spores triggering an infection in honeybee larva. Daisley et al. (2020) evaluated, under field and laboratory conditions, a probiotic administered to bees in a protein supplement. The probiotic patties contained *Lactobacillus plantarum* (Lp39), *L. rhamnosus* (GR-1), and *Lactobacillus kunkeei* (BR-1). In treated hives that had spontaneous *P. larvae* infection, the probiotic reduced the pathogen load in the larvae. A study using artificially reared bee larvae showed that treatment with Lp39, GR-1, and BR-1 could reduce pathogen load, upregulate expression of key immune genes and improve survival during *P. larvae* infection. Alberoni et al. (2018) studied the effect of exposing honeybees to a microbial cocktail containing *Bifidobacterium asteroides* DSM 20431, *Bifidobacterium coryneforme* C155, *Bifidobacterium indicum* C449, *L. kunkeei* Dan39 (KP114141), *L. plantarum* Dan91 (KP114145), and *Lactobacillus johnsonii* Dan92 (KP114146). Exposure to the cocktail significantly increased hive brood population by 46.2%, store pollen by 53.4%, and stored honey by 59.21%. Extracts of *Hypericum* spp. have been shown in vitro to be effective against *P. larvae* (Hernandez-Lopez et al., 2014). Six compounds isolated from the extracts showed antibacterial activity (0.168–220 mm inhibition) against vegetative *P. larvae*. These are hyperforin, uliginosin A, uliginosin B, 7-epiclusianone, albaspidin AA, and drummondin E.

Nosema ceranae is a microsporidium that is an obligate parasite of invertebrates, including honeybees. It infects the midgut epithelium and impairs digestion by downregulating genes related to nutrient absorption and intestinal health. Fumagillin has been the drug of choice for treating nosemosis but currently has limited availability. Borges et al. (2020), as a remedy for nosemosis, investigated 10 nutraceuticals, namely oregano oil, thymol, carvacrol, *trans*-cinnamaldehyde, tetrahydrocurcumin, sulforaphane, naringenin, embelin, allyl sulfide, and hydroxytyrosol. Two immunostimulatory compounds, chitosan and poly I:C, were also investigated. Sulforaphane was the most effective in reducing *N. ceranae* spore counts and was also toxic to bees at the effective dose range. For example, at 100% spore reduction the mortality was 100%. Naringenin reduced the *N. ceranae* spore counts by 49% ± 7.09% with 15% bee mortality and the treated bees had the highest longevity. For carvacrol, the maximum reduction in spore counts was 57% at 2.1 mg carvacrol/mL and the maximum bee mortality was 23%. Roussel et al. (2015) investigated sulfated polysaccharides as a nutraceutical to control nosemosis in honeybees. They found iota carrageenan and a sulfated polysaccharide isolated from a *Porphyridium* spp. decreased the nosemosis and were not toxic to bees.

Concluding remarks and future directions

Bees, in addition to being pollinators, produce products that have increasing interest as nutraceuticals and medical remedies. New technologies in apiculture are making products other than honey available at reduced costs. There are improved extraction procedures to concentrate substances from bee products that have specific biological activities. Honey made from specific sources of nectar and honeydew are being sought after for specific actions as nutraceuticals and medical remedies. The increase in consumption of honey, in addition to being a sweeting agent, is to increase wellness primarily because of its phytochemical content. There is a growth of cottage industries making beeswax-based skincare products. The biomedical industries are using honey and beeswax as ingredients in wound-healing products. Many of these products also contain plant extracts. It is most likely there will be a rapid growth in the use of bee products as lower cost replacement than synthetic pharmaceuticals. There is increasing concern that honeybee colony collapse syndrome could impact the supply of human foods. Honeybee health is driving environmental concerns, and the development and use of nutraceuticals, probiotics, and phytochemicals in maintaining colony health.

References

Abdelnour, S.A., Abd El-Hack, M.E., Alagawany, M., et al., 2019. Beneficial impacts of bee pollen in animal production, reproduction and health. J. Anim. Physiol. Anim Nutr. 103 (2), 477–484.

Abdelnour, S.A., Abd El-Hack, M.E., Alagawany, M., et al., 2020. Useful impacts of royal jelly on reproductive sides, fertility rate and sperm traits of animals. J. Anim. Physiol. Anim. Nutr. Available from: 10.1111/jpn.13303 (in this issue).

Abdulrhman, M., Elbarbary, N.S., Ahmed Amin, D., et al., 2012. Honey and a mixture of honey, beeswax, and olive oil-propolis extract in treatment of chemotherapy-induced oral mucositis: a randomized controlled pilot study. Pediatr. Hematol. Oncol. 29 (3), 285–292.

Adnan, S., Lone, M.M., Khan, F.R., et al., 2018. Which is the most recommended medium for the storage and transport of avulsed teeth? A systematic review. Dent. Traumatol. 34 (2), 59–70.

Afrin, S., Haneefa, S.M., Fernandez-Cabezudo, M.J., et al., 2019. Therapeutic and preventive properties of honey and its bioactive compounds in cancer: an evidence-based review. Nutr. Res. Rev. 1–27. Available from: 10.1017/S0954422419000192.

Aguilar, F., Autrup, H., Barlow, S., et al., 2007. Beeswax (E 901) as a glazing agent and as carrier for flavours. Scientific opinion of the panel on food additives, flavourings, processing aids and materials in contact with food (AFC). EFSA J. 615, 1–28.

Ahangari, Z., Naseri, M., Jalili, M., et al., 2012. Effect of propolis on dentin regeneration and the potential role of dental pulp stem cell in Guinea pigs. Cell J. 13 (4), 223–228.

Ahangari, Z., Naseri, M., Vatandoost, F., 2018. Propolis: chemical composition and its applications in endodontics. Iran. Endod. J. 13 (3), 285–292.

Ahmed, S., Othman, N.H., 2017. The anti-cancer effects of Tualang honey in modulating breast carcinogenesis: an experimental animal study. BMC Complement. Altern. Med. 17 (1), 208.

Akhu-Zaheya, L.M., Alkhasawneh, E.M., 2012. Complementary alternative medicine use among a sample of Muslim Jordanian oncology patients. Complement. Ther. Clin. Pract. 18 (2), 121–126.

Alberoni, D., Baffoni, L., Gaggia, F., et al., 2018. Impact of beneficial bacteria supplementation on the gut microbiota, colony development and productivity of *Apis mellifera* L. Benef. Microbes 9 (2), 269–278.

Alcay, S., Cakmak, S., Cakmak, I., et al., 2019. Successful cryopreservation of honey bee drone spermatozoa with royal jelly supplemented extenders. Cryobiology 87, 28–31.

Alexander, J., Benford, D., Boobis, A.R., et al., 2012. Scientific opinion on the evaluation of the substances currently on the list in the annex to Commission Directive 96/3/EC as acceptable previous cargoes for edible fats and oils – Part II of III. EFSA J. 10 (5), 2703.

Al-Jahdali, N., Carbonero, F., 2019. Impact of Maillard reaction products on nutrition and health: current knowledge and need to understand their fate in the human digestive system. Crit. Rev. Food Sci. Nutr. 59 (3), 474–487.

Al-Rahbi, B., Zakaria, R., Othman, Z., et al., 2014. Tualang honey supplement improves memory performance and hippocampal morphology in stressed ovariectomized rats. Acta Histochem. 116 (1), 79–88.

Al-Waili, N.S., 2004. An alternative treatment for pityriasis versicolor, tinea cruris, tinea corporis and tinea faciei with topical application of honey, olive oil and beeswax mixture: an open pilot study. Complement. Ther. Med. 12 (1), 45–47.

Al-Waili, N.S., 2005a. Mixture of honey, beeswax and olive oil inhibits growth of *Staphylococcus aureus* and *Candida albicans*. Arch. Med. Res. 36 (1), 10–13.

Al-Waili, N.S., 2005b. Clinical and mycological benefits of topical application of honey, olive oil and beeswax in diaper dermatitis. Clin. Microbiol. Infect. 11 (2), 160–163.

Al-Waili, N.S., Saloom, K.S., Al-Waili, T.N., et al., 2006. The safety and efficacy of a mixture of honey, olive oil, and beeswax for the management of hemorrhoids and anal fissure: a pilot study. Sci. World J. 6, 1998–2005.

Amiri Farahani, E.L., Hasanpoor-Azghdy, S.B., Kasraei, H., et al., 2017. Comparison of the effect of honey and mefenamic acid on the severity of pain in women with primary dysmenorrhea. Arch. Gynecol. Obstet. 296 (2), 277–283.

Anon, 1984. Final report on the safety assessment of candelilla wax, carnauba wax, Japan wax, and beeswax. J. Am. Coll. Toxicol. 3 (3), 1–41.

Appleton, S.M., 2018. Premenstrual syndrome: evidence-based evaluation and treatment. Clin. Obstet. Gynecol. 61 (1), 52–61.

Asgari, B., Kermanian, F., Derakhshan, N., et al., 2018. Honey-derived *Lactobacillus rhamnosus* alleviates *Helicobacter pylori*-induced gastrointestinal infection and gastric inflammation in C57bl/6 mice: an immuno-histologic study. Arq. Gastroenterol. 55 (3), 279–282.

Attia, W.Y., Gabry, M.S., El-Shaikh, K.A., et al., 2008. The anti-tumor effect of bee honey in Ehrlich ascite tumor model of mice is coincided with stimulation of the immune cells. Egypt. J. Immunol. 15 (2), 169–183.

Ayala, G., Escobedo-Hinojosa, W.I., de la Cruz-Herrera, C.F., et al., 2014. Exploring alternative treatments for *Helicobacter pylori* infection. World J. Gastroenterol. 20 (6), 1450–1469.

Aygun, A., Sahin, A., Karaca, Y., et al., 2018. Grayanotoxin levels in blood, urine and honey and their association with clinical status in patients with mad honey intoxication. Turk. J. Emerg. Med. 18 (1), 29–33.

Ayton, J., Prenzler, P., Raman, H., et al., 2019. Review of Chemistry Associated With Honey. Wagga Wagga, NSW, AU: AgriFutures. <www.agrifutures.com.au>.

Banaeian, S., Sereshti, M., Rafieian, M., et al., 2017. Comparison of vaginal ointment of honey and clotrimazole for treatment of vulvovaginal candidiasis: a random clinical trial. J. Mycol. Med. 27 (4), 494–500.

Banday, M.N., Lone, F.A., Rasool, F., et al., 2017. Does natural honey act as an alternative to antibiotics in the semen extender for cryopreservation of crossbred ram semen? Iran. J. Vet. Res. 18 (4), 258–263.

Barganska, Z., Slebioda, M., Namiesnik, J., 2016. Honey bees and their products: bioindicators of environmental contamination. Crit. Rev. Environ. Sci. Technol. 46 (3), 235–248.

Battino, M., Forbes-Hernandez, T.Y., Gasparrini, M., et al., 2019. Relevance of functional foods in the Mediterranean diet: the role of olive oil, berries and honey in the prevention of cancer and cardiovascular diseases. Crit. Rev. Food Sci. Nutr. 59 (6), 893–920.

Bauer, L., Kohlich, A., Hirschwehr, R., et al., 1996. Food allergy to honey: pollen or bee products? Characterization of allergenic proteins in honey by means of immunoblotting. J. Allergy Clin. Immunol. 97 (1 Pt 1), 65–73.

Bayir, Y., Un, H., Ugan, R.A., et al., 2019. The effects of beeswax, olive oil and butter impregnated bandage on burn wound healing. Burns 45 (6), 1410–1417.

Beasley, M., Hood, D., Anderson, P., et al., 2018. Poisoning due to tutin in honey—a report of an outbreak in New Zealand. N. Z. Med. J. 131 (1473), 59–71.

Bernardini, F., Tuniz, C., Coppa, A., et al., 2012. Beeswax as dental filling on a Neolithic human tooth. PLoS One 7 (9), e44904.

Blicharska, N., Seidel, V., 2019. Chemical diversity and biological activity of African propolis. Prog. Chem. Org. Nat. Prod. 109, 415–450.

Bogdanvov, S., 2016. Beeswax: production, properties, composition, control. The Wax Book. Muehlethurnen, Switzerland. <www.bee-hexagon.net/download/download.php?key = 6z8CvmqHkeyBr1QQ>.

Borges, D., Guzman-Novoa, E., Goodwin, P., 2020. Control of the microsporidian parasite *Nosema ceranae* in honey bees (*Apis mellifera*) using nutraceutical and immuno-stimulatory compounds. PLoS One . Available from: 10.1371/journal.pone.0227484.

Burdock, G.A., 1998. Review of the biological properties and toxicity of bee propolis (propolis). Food Chem. Toxicol. 36 (4), 347–363.

Burlando, B., Cornara, L., 2013. Honey in dermatology and skin care: a review. J. Cosmet. Dermatol. 12 (4), 306–313.

Camarasa, G., 1975. Occupational dermatitis from beeswax. Contact Dermat. 1 (2), 124.

Carpinelli de Jesus, M., Hungerford, N.L., Carter, S.J., et al., 2019. Pyrrolizidine alkaloids of blue heliotrope (*Heliotropium amplexicaule*) and their presence in Australian honey. J. Agric. Food Chem. 67 (28), 7995–8006.

Carvalho, C., Fernandes, W.H.C., Mouttinho, T.B.F., et al., 2019. Evidence-based studies and perspectives of the use of Brazilian green and red propolis in dentistry. Eur. J. Dent. 13 (3), 459–465.

Castaldo, S., Capasso, F., 2002. Propolis, an old remedy used in modern medicine. Fitoterapia 73 (Suppl 1), S1–6. Available from: 10.1016/s0367-326x(02)00185-5.

Cernak, M., Majtanova, N., Cernak, A., et al., 2012. Honey prophylaxis reduces the risk of endophthalmitis during perioperative period of eye surgery. Phytother. Res. 26 (4), 613–616.

Chen, S.P., Lam, Y.H., Ng, V.C., et al., 2013. Mad honey poisoning mimicking acute myocardial infarction. Hong Kong Med. J. 19 (4), 354–356.

Cherniack, E.P., Govorushko, S., 2018. To bee or not to bee: the potential efficacy and safety of bee venom acupuncture in humans. Toxicon 154, 74–78.

Cho, H., Kim, D., Choo, Y., et al., 2019. Common causes of emergency department visits for anaphylaxis in Korean community hospitals: a cross-sectional study. Medicine (Baltim.) 98 (4), e14114. Available from: 10.1097/MD.0000000000014114.

Choi, J.H., Jang, Y.S., Oh, J.W., et al., 2015. Bee pollen-induced anaphylaxis: a case report and literature review. Allergy Asthma Immunol. Res. 7 (5), 513–517.

Cianciosi, D., Forbes-Hernandez, T.Y., Afrin, S., et al., 2018. Phenolic compounds in honey and their associated health benefits: a review. Molecules 23 (9). Available from: 10.3390/molecules23092322.

Codex Alimentarius, 2001. Revised codex standard for honey. Codex Stan 12:1982 7. <https://www.google.com/search?client = firefox-b-d&q = Alimentarius + C + %282001% 29 + Revised + codex + standard + for + hone>.

Coelho, L.G., Bastos, E.M., Resende, C.C., et al., 2007. Brazilian green propolis on *Helicobacter pylori* infection. A pilot clinical study. Helicobacter 12 (5), 572–574.

Conrad, V.J., Hazan, L.L., Latorre, A.J., et al., 2019. Efficacy and safety of honey bee venom (*Apis mellifera*) dermal injections to treat osteoarthritis knee pain and physical disability: a randomized controlled trial. J. Altern. Complement. Med. 25 (8), 845–855.

Cornara, L., Biagi, M., Xiao, J., et al., 2017. Therapeutic properties of bioactive compounds from different honeybee products. Front. Pharmacol. 8, 412.

Craig, J.P., Cruzat, A., Cheung, I.M.Y., et al., 2020. Randomized masked trial of the clinical efficacy of MGO manuka honey microemulsion eye cream for the treatment of blepharitis. Ocul. Surf. 18 (1), 170–177.

Cui, K., Lu, W., Zhu, L., et al., 2013. Caffeic acid phenethyl ester (CAPE), an active component of propolis, inhibits *Helicobacter pylori* peptide deformylase activity. Biochem. Biophys. Res. Commun. 435 (2), 289–294.

da Silva, P.M., Gauche, C., Gonzaga, L.V., et al., 2016. Honey: chemical composition, stability and authenticity. Food Chem. 196, 309–323.

Daisley, B.A., Pitek, A.P., Chmiel, J.A., et al., 2020. Novel probiotic approach to counter *Paenibacillus larvae* infection in honey bees. ISME J. 14 (2), 476–491.

Davis, S.R., Lambrinoudaki, I., Lumsden, M., et al., 2015. Menopause. Nat. Rev. Dis. Prim. 1, 15004. Available from: 10.1038/nrdp.2015.4.

Demircan, A., Keles, A., Bildik, F., et al., 2009. Mad honey sex: therapeutic misadventures from an ancient biological weapon. Ann. Emerg. Med. 54 (6), 824–829.

Denisow, B., Denisow-Pietrzyk, M., 2016. Biological and therapeutic properties of bee pollen: a review. J. Sci. Food Agric. 96 (13), 4303–4309.

EFSA Panel on Contaminants in the Food Chain, 2017. Risk for human related to the presence of pyrrolizidine alkaloids in honey, tea, herbal infusions and food supplements. EFSA J. 15 (7), 4908. <www.efsa.europa.eu/en/efsajournal/pub/4908>.

El-Tayeb, M.M., Abu-Seida, A.M., El Ashry, S.H., et al., 2019. Evaluation of antibacterial activity of propolis on regenerative potential of necrotic immature permanent teeth in dogs. BMC Oral Health 19 (1), 174. Available from: 10.1186/s12903-019-0835-0.

Fakhrildin, M.B., Alsaadi, R.A., 2014. Honey supplementation to semen-freezing medium improves human sperm parameters post-thawing. J. Family Reprod. Health 8 (1), 27–31.

Fallico, B., Zappala, M., Arena, E., et al., 2004. Effects of conditioning on HMF content in unifloral honeys. Food Chem. 85 (2), 305–313.

Farida, S., Sahlan, M., Rohmatin, E., et al., 2020. The beneficial effect of Indonesian propolis wax from *Tetragonula* sp. as a therapy in limited vaginal candidiasis patients. Saudi J. Biol. Sci. 27 (1), 142–146.

Fermo, P., Beretta, G., Maffei Facino, R., et al., 2013. Ionic profile of honey as a potential indicator of botanical origin and global environmental pollution. Environ. Pollut. 178, 173–181.

Formiga, A.S., Pinsetta Jr, J.S., Pereira, E.M., et al., 2019. Use of edible coatings based on hydroxypropyl methylcellulose and beeswax in the conservation of red guava 'Pedro Sato'. Food Chem. 290, 144–151.

Fratini, F., Cilia, G., Turchi, B., et al., 2016. Beeswax: a minireview of its antimicrobial activity and its application in medicine. Asian Pac. J. Trop. Med. 9 (9), 839–843.

Freires, I.A., Santaella, G.M., de Cassia Orlandi Sardi, J., et al., 2018. The alveolar bone protective effects of natural products: a systematic review. Arch. Oral Biol. 87, 196–203.

Fu, P.P., Xia, Q., Lin, G., et al., 2004. Pyrrolizidine alkaloids—genotoxicity, metabolism enzymes, metabolic activation, and mechanisms. Drug Metab. Rev. 36 (1), 1–55.

Fuentealba, J., Munoz, B., Yevenes, G., et al., 2011. Potentiation and inhibition of glycine receptors by tutin. Neuropharmacology 60 (2-3), 453–459.

Galeotti, F., Crimaldi, L., Maccari, F., et al., 2017. Selective treatment to reduce contamination of propolis by polycyclic aromatic hydrocarbons (PAHs) still preserving its active polyphenol component and antioxidant activity. Nat. Prod. Res. 31 (17), 1971–1980.

Galeotti, F., Maccari, F., Fachini, A., et al., 2018. Chemical composition and antioxidant activity of propolis prepared in different forms and in different solvents useful for finished products. Foods 7 (3). Available from: 10.3390/foods7030041.

Gamberini, M.C., Baraldi, C., Freguglia, G., et al., 2011. Spectral analysis of pharmaceutical formulations prepared according to ancient recipes in comparison with old museum remains. Anal. Bioanal. Chem. 401 (6), 1839–1846.

Garcia, M., del Pozo, M.D., Diez, J., et al., 1995. Allergic contact dermatitis from a beeswax nipple-protective. Contact Dermat. 33 (6), 440–441.

Ghanem, N.B., 2011. The antimicrobial activity of some honey bee products and some Saudi folkloric plant extracts. JKAU Sci. 23 (2), 47–62.

Gheldof, N., Wang, X.H., Engeseth, N.J., 2002. Identification and quantification of antioxidant components of honeys from various floral sources. J. Agric. Food Chem. 50 (21), 5870–5877.

Giampieri, F., Gasparrini, M., Forbes-Hernandez, T.Y., et al., 2018. Beeswax by-products efficiently counteract the oxidative damage induced by an oxidant agent in human dermal fibroblasts. Int. J. Mol. Sci. 19 (9), 2842.

Gonzalez, R., Corcho, I., Remirez, D., et al., 1995. Hepatoprotective effects of propolis extract on carbon tetrachloride-induced liver injury in rats. Phyto Res. 9, 114–117.

Gumus, K., Ozlu, Z.K., 2017. The effect of a beeswax, olive oil and *Alkanna tinctoria* (L.) tausch mixture on burn injuries: an experimental study with a control group. Complement. Ther. Med. 34, 66–73.

Gunawardana, N.C., 2017. Risk of anaphylaxis in complementary and alternative medicine. Curr. Opin. Allergy Clin. Immunol. 17 (5), 332–337.

Gunduz, A., Turedi, S., Russell, R.M., et al., 2008. Clinical review of grayanotoxin/mad honey poisoning past and present. Clin. Toxicol. (Phila.) 46 (5), 437–442.

Guney, A., Karaman, I., Oner, M., et al., 2011. Effects of propolis on fracture healing: an experimental study. Phytother. Res. 25 (11), 1648–1652.

Hamzaoglu, I., Saribeyoglu, K., Durak, H., et al., 2000. Protective covering of surgical wounds with honey impedes tumor implantation. Arch. Surg. 135 (12), 1414–1417.

Haron, M.N., Mohamed, M., 2016. Effect of honey on the reproductive system of male rat offspring exposed to prenatal restraint stress. Andrologia 48 (5), 525–531.

Hernandez-Lopez, J., Crockett, S., Kunert, O., et al., 2014. In vitro growth inhibition by *Hypericum* extracts and isolated pure compounds of *Paenibacillus* larvae, a lethal disease affecting honeybees worldwide. Chem. Biodivers. 11 (5), 695–708.

Hidaka, S., Okamoto, Y., Uchiyama, S., et al., 2006. Royal jelly prevents osteoporosis in rats: beneficial effects in ovariectomy model and in bone tissue culture model. Evid. Based Complement. Altern. Med. 3 (3), 339–348.

Hixon, K.R., Klein, R.C., Eberlin, C.T., et al., 2019. A critical review and perspective of honey in tissue engineering and clinical wound healing. Adv. Wound Care (N. Rochelle) 8 (8), 403–415.

Hsu, C.Y., Chiang, W.C., Weng, T.I., et al., 2004. Laryngeal edema and anaphalactic shock after topical propolis use for acute pharyngitis. Am. J. Emerg. Med. 22 (5), 432–433.

Islam, M.N., Khalil, M.I., Islam, M.A., et al., 2014. Toxic compounds in honey. J. Appl. Toxicol. 34 (7), 733–742.

Jagdis, A., Sussman, G., 2012. Anaphylaxis from bee pollen *supplement*. CMAJ 184 (10), 1167–1169.

Jankauskiene, J., Jarushaitiene, D., Cheksteryte, V., et al., 2007. Using 20% honey solution eye drops in patients with dry eye syndrome. J. Apic. Res. 36 (4), 232–235.

Jull, A.B., Cullum, N., Dumville, J.C., et al., 2015. Honey as a topical treatment for wounds. Cochrane Database Syst. Rev. (3), CD005083. Available from: 10.1002/14651858.CD005083.pub4.

Kacaniova, M., Vukovic, N., Chlebo, R., et al., 2012. The antimicrobial activity of honey, bee pollen loads and beeswax from Slovakia. Arch. Biol. Sci. Bel. 64 (3), 927–934.

Khurshid, Z., Naseem, M., Zafar, M.S., et al., 2017. Propolis: a natural biomaterial for dental and oral healthcare. J. Dent. Res. Dent. Clin. Dent. Prospect. 11 (4), 265–274.

Kostic, A.Z., Milincic, D.D., Barac, M.B., et al., 2020. The application of pollen as a functional food and feed ingredient—the present and perspectives. Biomolecules 10 (1). Available from: 10.3390/biom10010084.

Krainer, S., Brodschneider, R., Vollmann, J., et al., 2016. Effect of hydroxymethylfurfural (HMF) on mortality of artificially reared honey bee larvae (*Apis mellifera carnica*). Ecotoxicology 25 (2), 320–328.

Kurek-Gorecka, A., Gorecki, M., Rzepecka-Stojko, A., et al., 2020. Bee products in dermatology and skin care. Molecules 25 (3). Available from: 10.3390/molecules25030556.

Kuster, B.F.M., 1994. 5-Hydroxymethylfurfural (HMF). A review focusing on its manufacture. Starch/Stärke 42 (8), 314–321.

Lambert, O., Piroux, M., Puyo, S., et al., 2012. Bees, honey and pollen as sentinels for lead environmental contamination. Environ. Pollut. 170, 254–259.

Lau, P., Bryant, V., Ellis, J.D., et al., 2019. Seasonal variation of pollen collected by honey bees (*Apis mellifera*) in developed areas across four regions in the United States. PLoS One 14 (6), e0217294. Available from: 10.1371/journal.pone.0217294.

Lee, G., Bae, H., 2016. Anti-Inflammatory applications of melittin, a major component of bee venom: detailed mechanism of action and adverse effects. Molecules 21 (5). Available from: 10.3390/molecules21050616.

Lewis, P.A., Wright, K., Webster, A., et al., 2012. A randomized controlled pilot study comparing aqueous cream with a beeswax and herbal oil cream in the provision of relief from postburn pruritus. J. Burn Care Res. 33 (4), e195–e200. Available from: 10.1097/BCR.0b013e31825042e2.

Li, M.M., Wu, L.Y., Zhao, T., et al., 2011. The protective role of 5-HMF against hypoxic injury. Cell Stress Chaperones 16 (3), 267–273.

Majtanova, N., Cernak, M., Majtan, J., 2016. Honey: a natural remedy for eye diseases. Forsch. Komplementmed. 23 (6), 364–369.

Margaoan, R., Strant, M., Varadi, A., et al., 2019. Bee collected pollen and bee bread: bioactive constituents and health benefits. Antioxidants (Basel) 8 (12). Available from: 10.3390/antiox8120568.

Martin, N., Hulbert, A.J., Brenner, G.C., et al., 2019. Honeybee caste lipidomics in relation to life-history stage and the long life of the queen. J. Exp. Biol. 222 (Pt 24). Available from: 10.1242/jeb.207043.

Mishima, S., Suzuki, K.M., Isohama, Y., et al., 2005. Royal jelly has estrogenic effects in vitro and in vivo. J. Ethnopharmacol. 101 (1-3), 215–220.

Mofid, B., Rezaeizadeh, H., Termos, A., et al., 2016. Effect of processed honey and royal jelly on cancer-related fatigue: a double-blind randomized clinical trial. Electron. Physician 8 (6), 2475–2482.

Mohamed, M., Sulaiman, S.A., Jaafar, H., 2012. Histological changes in male accessory reproductive organs in rats exposed to cigarette smoke and the protective effect of honey supplementation. Afr. J. Tradit. Complement. Altern. Med. 9 (3), 329–335.

Mosleh, G., Badr, P., 2016. Qairooti (cerate or cera beeswax salve) in Medieval Persia. Pharm. Hist. (Lond.) 46 (4), 72–74.

Moutsatsou, P., Papoutsi, Z., Kassi, E., et al., 2010. Fatty acids derived from royal jelly are modulators of estrogen receptor functions. PLoS One 5 (12), e15594. Available from: 10.1371/journal.pone.0015594.

Munstedt, K., 2019. Bee products and the treatment of blister-like lesions around the mouth, skin and genitalia caused by herpes viruses—a systematic review. Complement. Ther. Med. 43, 81–84.

Naz, S., Imran, M., Rauf, A., et al., 2019. Chrysin: pharmacological and therapeutic properties. Life Sci. 235, 116797. Available from: 10.1016/j.lfs.2019.116797.

Nostro, A., Cellini, L., Di Bartolomeo, S., et al., 2006. Effects of combining extracts (from propolis or *Zingiber officinale*) with clarithromycin on *Helicobacter pylori*. Phytother. Res. 20 (3), 187–190.

Nyman, G.S.A., Tang, M., Inerot, A., et al., 2019. Contact allergy to beeswax and propolis among patients with cheilitis or facial dermatitis. Contact Dermat. 81 (2), 110–116.

Ohashi, I., Kato, K., Okamoto, M., et al., 2020. Microbicidal effects of slightly acidic hypochlorous acid water and weakly acidified chlorous acid water on foulbrood pathogens. J. Vet. Med. Sci. 82 (3), 261–271.

Oner, M., Kafadar, I., Guney, A., et al., 2011. Effect of intraarticular propolis in an experimental septic arthritis model. J. Pediatr. Orthop. B 20 (1), 8–13.

Pascoal, A., Estevinho, M.M., Choupina, A.B., et al., 2019. An overview of the bioactive compounds, therapeutic properties and toxic effects of apitoxin. Food Chem. Toxicol. 134, 110864. Available from: 10.1016/j.fct.2019.110864.

Picolotto, A., Pergher, D., Pereira, G.P., et al., 2019. Bacterial cellulose membrane associated with red propolis as phytomodulator: improved healing effects in experimental models of diabetes mellitus. Biomed. Pharmacother. 112, 108640. Available from: 10.1016/j.biopha.2019.108640.

Picron, J.F., Herman, M., Van Hoeck, E., et al., 2020. Monitoring of pyrrolizidine alkaloids in beehive products and derivatives on the Belgian market. Environ. Sci. Pollut. Res. Int. 27 (6), 5693–5708.

Pitsios, C., Tsoumani, M., Bilo, M.B., et al., 2019. Contraindications to immunotherapy: a global approach. Clin. Transl. Allergy 9, 45. Available from: 10.1186/s13601-019-0285-4.

Prueksakorn, A., Puasiri, S., Ruangsri, S., et al., 2016. The preservative effect of Thai propolis extract on the viability of human periodontal ligament cells. Dent. Traumatol. 32 (6), 495–501.

Przybylek, I., Karpinski, T.M., 2019. Antibacterial properties of propolis. Molecules 24 (11). Available from: 10.3390/molecules24112047.

PubChem 2020 <https://pubchem.ncbi.nlm.nih.gov/>.

Ricigliano, V.A., 2020. Microalgae as a promising and sustainable nutrition source for managed honey bees. Arch. Insect Biochem. Physiol. e21658. Available from: 10.1002/arch.21658.

Rodrigues, M., Kosaric, N., Bonham, C.A., et al., 2019. Wound healing: a cellular perspective. Physiol. Rev. 99 (1), 665–706.

Rodriguez, S., Ancheta, O., Ramos, M.E., et al., 1997. Effects of Cuban red propolis on galactosamine-induced hepatitis in rats. Pharmacol. Res. 35 (1), 1–4.

Roffet-Salque, M., Regert, M., Evershed, R.P., et al., 2015. Widespread exploitation of the honeybee by early Neolithic farmers. Nature 527 (7577), 226–230.

Roussel, M., Villay, A., Delbac, F., et al., 2015. Antimicrosporidian activity of sulphated polysaccharides from algae and their potential to control honeybee nosemosis. Carbohydr. Polym. 133, 213–220.

Rueff, F., Kroth, J., Przybilla, B., 2017. Risk factors in Hymenoptera venom allergy. Allergol. Sel. 1 (1), 53–58.

Sabanovic, M., Saltovic, S., Mujkic, A.M., et al., 2019. Impact of propolis on the oral health. Balk. J. Dent. Med. 23 (1), 1–9.

Salehi, A., Jabarzare, S., Neurmohamadi, M., et al., 2014. A double blind clinical trial on the efficacy of honey drop in vernal keratoconjunctivitis. Evid. Based Complement. Altern. Med. 2014, 287540. Available from: 10.1155/2014/287540.

Salonen, A., Virjamo, V., Tammela, P., et al., 2017. Screening bioactivity and bioactive constituents of Nordic unifloral honeys. Food Chem. 237, 214–224.

Santos, V.R., Gomes, R.T., de Mesquita, R.A., et al., 2008. Efficacy of Brazilian propolis gel for the management of denture stomatitis: a pilot study. Phytother. Res. 22 (11), 1544–1547.

Santos, L.M., Fonseca, M.S., Sokolonski, A.R., et al., 2020. Propolis: types, composition, biological activities, and veterinary product patent prospecting. J. Sci. Food Agric. 100 (4), 1369–1382.

Santos-Buelga, C., González-Paramás, A.M., 2017. Chemical composition of honey. In: Alvarez-Suarez, J.M. (Ed.), Bee Products—Chemical and Biological Properties. Springer, Cham, pp. 43–82.

Sarmadi, F., Kazemi, P., Tirgar, P., et al., 2019. Using natural honey as an antioxidant and thermodynamically efficient cryoprotectant in embryo vitrification. Cryobiology 91, 30–39.

Se, K.W., Wahab, R.W., Yaacob, S.N.S., et al., 2019. Detection techniques for adulterants in honey: challenges and recent trends. J. Food Compost. Anal. 80, 16–32.

Seidel, V., Peyfoon, E., Watson, D.G., et al., 2008. Comparative study of the antibacterial activity of propolis from different geographical and climatic zones. Phytother. Res. 22 (9), 1256–1263.

Semprini, A., Singer, J., Braithwaite, I., et al., 2019. Kanuka honey versus aciclovir for the topical treatment of herpes simplex labialis: a randomised controlled trial. BMJ Open 9 (5), e026201. Available from: 10.1136/bmjopen-2018-026201.

Seyyedi, F., Rafiean-Kopaei, M., Miraj, S., 2016. Comparison of the effects of vaginal royal jelly and vaginal estrogen on quality of life, sexual and urinary function in postmenopausal women. J. Clin. Diagn. Res. 10 (5), . Available from: 10.7860/JCDR/2016/17844.7715QC01-05.

Shabbir, J., Qazi, F., Farooqui, W., et al., 2020. Effect of Chinese propolis as an intracanal medicament on post-operative endodontic pain: a double-blind randomized controlled trial. Int. J. Environ. Res. Public Health 17 (2). Available from: 10.3390/ijerph17020445.

Shah, J.B., 2011. The history of wound care. J. Am. Col. Certif. Wound Spec. 3 (3), 65–66.

Shapla, U.M., Solayman, M., Alam, N., et al., 2018. 5-Hydroxymethylfurfural (HMF) levels in honey and other food products: effects on bees and human health. Chem. Cent. J. 12 (1), 35.

Shimizu, S., Matsushita, H., Minami, A., et al., 2018. Royal jelly does not prevent bone loss but improves bone strength in ovariectomized rats. Climacteric 21 (6), 601–606.

Siavoshi, F., Sahraee, M., Ebrahimi, H., 2018. Natural fruits, flowers, honey, and honeybees harbor *Helicobacter pylori*-positive yeasts. Helicobacter 23 (2), e12471. Available from: 10.1111/hel.12471.

Siddiqui, A.J., Musharraf, S.G., Choudhary, M.I., et al., 2017. Application of analytical methods in authentication and adulteration of honey. Food Chem. 217, 687–698.

Sodré, G.S., Marchini, L.C., Moreti, A.C.C., et al., 2011. Physico-chemical characteristics of honey produced by *Apis mellifera* in the Picos region, state of Piauí, Brazil. R. Bras. Zootec. 40 (8), 1837–1843.

Stacey, R.J., 2011. The composition of some Roman medicines: evidence for Pliny's punic wax? Anal. Bioanal. Chem. 401 (6), 1749–1759.

Sudmann, B., Bang, G., Sudmann, E., 2006. Histologically verified bone wax (beeswax) granuloma after median sternotomy in 17 of 18 autopsy cases. Pathology 38 (2), 138–141.

Sylvester-Hvid, A., Avnstorp, M.B., Fjeld, T., et al., 2019. Benign tumours leading to total penile denudation treated with manuka honey dressings: a case report and review of literature. Int. J. Surg. Case Rep. 61, 191–194.

Taavoni, S., Barkhordari, F., Goushegir, A., et al., 2014. Effect of royal jelly on premenstrual syndrome among Iranian medical sciences students: a randomized, triple-blind, placebo-controlled study. Complement. Ther. Med. 22 (4), 601–606.

Taavoni, S., Fathi, L., Nazem-Ekbatani, N., et al., 2019. The effect of oral intake of honey syrup on the pain intensity of active phase of parturition of nulliparous women: a randomized clinical trial. Casp. J. Intern. Med. 10 (1), 98–101.

Takzaree, N., Hadjiakhondi, A., Hassanzadeh, G., et al., 2016. Synergistic effect of honey and propolis on cutaneous wound healing in rats. Acta Med. Iran. 54 (4), 233–239.

Tanleque-Alberto, F., Juan-Borras, M., Escriche, I., 2019. Quality parameters, pollen and volatile profiles of honey from north and central Mozambique. Food Chem. 277, 543–553.

Tchapla, A., Mejanelle, P., Bleton, J., et al., 2004. Characterization of embalming materials of a mummy of the Ptolemaic era. Comparison with balms from mummies of different eras. J. Sep. Sci. 27 (3), 217–234.

Thompson, H.J., Adlakha, H., 1991. Dose-responsive induction of mammary gland carcinomas by the intraperitoneal injection of 1-methyl-1-nitrosourea. Cancer Res. 51 (13), 3411–3415.

Tomazevic, T., Jazbec, J., 2013. A double blind randomised placebo controlled study of propolis (bee glue) effectiveness in the treatment of severe oral mucositis in chemotherapy treated children. Complement. Ther. Med. 21 (4), 306–312.

Tsourkas, P.K., 2020. *Paenibacillus larvae* bacteriophages: obscure past, promising future. Microb. Genom. 6 (2). Available from: 10.1099/mgen.0.000329.

Viuda-Martos, M., Ruiz-Navajas, Y., Fernandez-Lopez, J., et al., 2008. Functional properties of honey, propolis, and royal jelly. J. Food Sci. 73 (9), R117–R124.

Waheed, M., Hussain, M.B., Javed, A., et al., 2019. Honey and cancer: a mechanistic review. Clin. Nutr. 38 (6), 2499–2503.

Wang, A., 2013. Propolis: a wonder bees product and its pharmacological potentials. Adv. Pharmacol. Sci. 2013 (308249), <https://doi.org/10.1155/2013/308249>.

Wang, C., Guo, M., Zhang, N., et al., 2019. Effectiveness of honey dressing in the treatment of diabetic foot ulcers: a systematic review and meta-analysis. Complement. Ther. Clin. Pract. 34, 123–131.

Wijesinghe, M., Weatherall, M., Perrin, K., et al., 2009. Honey in the treatment of burns: a systematic review and meta-analysis of its efficacy. N. Z. Med. J. 122 (1295), 47–60.

Wright, G.A., Nicolson, S.W., Shafir, S., 2018. Nutritional physiology and ecology of honeybees. Annu. Rev. Entomol. 63, 327–344.

Xue, X., Wu, L., Wang, K., 2017. Chemical composition of royal jelly. In: Alvarez-Suarez, J.M. (Ed.), Bee Products—Chemical and Biological Properties. Springer, Cham, pp. 181–190.

Yaacob, N.S., Ismail, N.F., 2014. Comparison of cytotoxicity and genotoxicity of 4-hydroxytamoxifen in combination with Tualang honey in MCF-7 and MCF-10A cells. BMC Complement. Altern. Med. 14, 106. Available from: 10.1186/1472-6882-14-106.

Yang, W., Wu, Z., Huang, Z.Y., et al., 2017. Preservation of orange juice using propolis. J. Food Sci. Technol. 54 (11), 3375–3383.

Yang, W., Zhang, C., Li, C., et al., 2019a. Pathway of 5-hydroxymethyl-2-furaldehyde formation in honey. J. Food Sci. Technol. 56 (5), 2417–2425.

Yang, S., Liu, Y., Sun, F., et al., 2019b. Gelsedine-type alkaloids: discovery of natural neurotoxins presented in toxic honey. J. Hazard. Mater. 381, 120999.

Yordanov, D., Boyanova, L., Markovska, R., et al., 2017. Influence of dietary factors on *Helicobacter pylori* and CagA seroprevalence in Bulgaria. Gastroenterol. Res. Pract. 2017, 9212143. Available from: 10.1155/2017/9212143.

Zirbes, L., Nguyen, B.K., de Graaf, D.C., et al., 2013. Hydroxymethylfurfural: a possible emergent cause of honey bee mortality? J. Agric. Food Chem. 61 (49), 11865–11870.

Chapter 48

Ginkgo biloba

Margitta Dziwenka and Robert W. Coppock

Introduction

The *Ginkgo biloba* tree is a very hardy tree and is believed to be the world's oldest living tree, dating back over 250 million years ago and is described as a living fossil (Omidkhoda et al., 2019; Dugoua et al., 2006). It is native to China, Japan, and Korea but is now cultivated in Asia, Europe, North America, New Zealand, and Argentina. The leaf extract has been employed for treating diseases for centuries and has been well studied in recent years for its bioactive constituents, such as flavonoid compounds, diterpene lactone compounds, and polysaccharides (Zeng et al., 2013; Omidkhoda et al., 2019). The medicinal use of *Ginkgo* can be traced back to 2800 BC in Chinese medicine where it was used for brain disorders, circulatory disorders, respiratory diseases such as asthma, urinary tract disease, and as an antiparasitic (Dugoua et al., 2006). *G. biloba* is one of the most commonly used herbal medicines in both the United States and Europe today (Heinonen and Gaus, 2015; Hermann and von Richter, 2012). Since the early 1990s, EGb761 has been considered to be one of the most popular supplements used for memory enhancement in the United States (Smith and Luo, 2004). Diamond and Bailey (2013) list it as one of the most frequently prescribed herbal preparations in Germany and as an over-the-counter herbal preparation in the United States. Li et al. (2019) report that, in 2012, the total worldwide sales of *Ginkgo* products was USD 1.26 billion, the majority of which was sold in China. Chertkow (2020) reported that in 2020 that sales of *G. biloba* in North America was more than US$1 billion. It is difficult to assess the current market in the United States as herbal medication preparations are sold in a variety of ways including in health food stores and online, which makes accurate collection of information difficult. *G. biloba* leaf extract is available in tablet, oral liquid, and injectable forms, and it is very important that consumers are aware that *G. biloba* is offered in many different preparations and by many different vendors, sometimes without any standardization of the active ingredients. In the United States, herbal products are regulated as foodstuff, food supplements, or dietary supplements. In the European Union, they are offered as "rational" medicines and licensed to a defined medical indication "traditional use," dietary supplements, as food, or as homeopathic products (Ude et al., 2013). The European Union and other governing bodies are leaders in recognizing standards for herbal medicines and publishing monographs on their use (HMPC, 2015; Zöllner and Schwarz, 2013). The Chinese government has taken regulatory action in the production, labeling, and sale of *Ginkgo* extracts to assure safe products (Blumenthal, 2015). In the United States, *Ginkgo* is marketed as a dietary supplement and is therefore regulated by the US Food and Drug Administration.

Manufacturing standards and medical indications

Common commercial formulations of *G. biloba* include capsule or tablet forms. The Chinese have developed an intravenous form of a modified *G. biloba* extract for treating ischemic stroke (Fan et al., 2020). Standardization of extraction methods and phytochemical levels for leaf extracts can vary significantly among manufacturers, especially when national standards do not exist. It is important that manufacturers adhere to a standard basis on the phytochemical composition. The United States Pharmacopeia (USP) 32 and the European Pharmacopoeia 6.1 published monographs for *G. biloba* extract and formulations provide standards for herbal preparations. The USP 32 has monographs for *Ginkgo*, powdered *Ginkgo* extract, *Ginkgo* capsules, and tablets, and the European Pharmacopoeia 6.1 has a monograph for *G. biloba* extract. Both pharmacopeias list the standardized dry extracts which are produced with acetone 60%, drug/extract ratio of 35:37:1, such as EGb 761 and Li 1370, both of which have a defined composition. Variations in the levels of active ingredients are likely due to the multistep

Nutraceuticals. DOI: https://doi.org/10.1016/B978-0-12-821038-3.00048-3

extraction and concentration process used in manufacturing (Heinonen and Gaus, 2015).

G. biloba is used to treat cognitive impairment in various types of dementia, peripheral arterial diseases, ischemic stroke, tinnitus, macular degeneration, and autism, and to moderate the undesirable effects of some cancer therapies (Heinonen and Gaus, 2015). According to recent pharmacological studies, *G. biloba* extracts have been used to treat diseases such as neurodegenerative diseases, cerebral disorders, vascular problems, age-related memory loss, and oxidative stress (Omidkhoda et al., 2019). EGb 761 has been shown to have antistress effects, which supports its use to prevent or impede the development of psychiatric disorders in which stress is a pathogenic factor (Montes et al., 2015). In Gurley (2012), the therapeutic uses were listed as being for insufficient blood flow, memory deficits, cognitive disorders, Alzheimer disease, depression, vertigo, tinnitus, and intermittent claudication. Dugoua et al. (2006) state additional uses which include memory enhancement, asthma, varicose veins, mountain sickness, idiopathic cyclic edema, sexual dysfunction secondary to the use of selective serotonin reuptake inhibitors, and connective tissue disorders. They also stated that the World Health Organization (WHO) recommended *G. biloba* for Raynaud disease. The WHO has accepted the standardized extracts such as EGb 761 and Li 1370 as antidementia drugs based on the pharmacological studies done in vitro and in vivo and the numerous clinical studies supporting the efficacy of EGb 761 in the central nervous system (CNS) when taking a 240 mg daily dose. In Chan et al.(2007), current uses in medical applications include improving brain function, strengthening the cerebrovascular and cardiovascular systems by the inhibition of platelet aggregation, and the increase of blood flow and oxygen supply, neutralizing free radicals, stabilizing cellular energy production, and suppressing hemorrhoids, inflammation, migraines, allergies, and asthma. Opinions on the efficacy of *G. biloba* differ as Birks and Grimley Evans (2009) concluded that *G. biloba* extracts are no different from placebo for efficacy and side effects. It is the opinion of the authors of this chapter that the acceptance of *G. biloba* as an efficacious treatment varies between countries and has highest acceptance in Asian countries.

Chemistry/mechanisms

There are many phytochemical constituents in *G. biloba* leaves. The primary active components which are considered to be responsible for the pharmacological properties of *Ginkgo* include flavonoids, including kaempferol, quercetin, myricetin, apigenin, isorhamnetin, luteolin, and tamarixetin, terpene trilactones—ginkgolide A (GA), ginkgolide B (GB), ginkgolide C (GC), ginkgolide J (GJ), ginkgolide M (GM), ginkgolide K (GK), ginkgolide L (GL), ginkgolide P (GP), ginkgolide Q (GQ), and bilobalide (BB) and proanthocyanidins (Omidkhoda et al., 2019). Terpene lactones are unique to *G. biloba* and include the ginkgolides and BB, a sesquiterpene lactone, while the flavonoids found in *G. biloba* occur primarily as glycoside derivatives (Gurley et al., 2012). Kaempferol is a metabolite of quercetin and isorhamnetin. The majority of *G. biloba* extracts available are enriched water–acetone or water–ethanol extracts from the leaves. Also, there is supercritical liquid extract of *G biloba*. The extracts are often standardized by the flavonoid or content of terpene trilactones. Chan et al. (2007) stated that the commercial extracts contain 22%–27% flavone glycosides, 5%–7% terpene lactones; of which 2.8%–3.4% are the ginkgolides A, B, and C and 2.6%–3.2% was BB; and less than 5 ppm ginkgolic acids. They also noted that the alkylphenol and alkylbenzoic acid derivatives are removed from the extract as they have allergenic, immunotoxic, and other undesirable properties. The standardized extract most frequently described in the literature is EGb 761 which contains 5.4%–6.6% terpene trilactones and 22%–27% flavonoids, as stated earlier. Studies on the metabolism and absorption of orally consumed *G. biloba* extract indicate that the flavonoid constituents are absorbed in the form of metabolites following their transformation by the intestinal flora (DeFeudis and Drieu, 2000). These authors reported that the bioavailability of ginkgolides A and B are almost complete, whereas that of C is very low and the bioavailability of BB was about 72% following oral administration in humans. They also report that ginkgolides A and B and BB were bioavailable after a single oral dose in rats. The terpene trilactones, as a compound class, are rare in the plant kingdom and have only been identified so far in *G. biloba* (Ude et al., 2013). The ginkgolides all have the same molecular geometrical skeleton but differ in the number and location of the hydroxyl functional groups (Chan et al., 2007). Ginkgolides are C20 diterpenes and consist of six five-membered rings; a spiro[4,4]nonane carbocylic ring, three lactones, and a tetrahydrofuran, and BB is a sesquiterpene and are the only known natural products with a *tert*-butyl group (Ude et al., 2013).

The quality of the extracts depends not only on the specific constituents but also on the biopharmaceutical properties of those constituents. These biopharmaceutical properties are characterized by the dissolution rate and bioavailability of these components released into the digestive system (Unger, 2013). In solid dosage forms, at least 75% of the pharmacologically active compounds in *G. biloba*, the flavonoids and terpene trilactones, should be released in the dissolution medium within 30 minutes as this is an important prerequisite for absorption in the gastrointestinal tract (Unger, 2013). Kressmann et al. (2002) tested various *G. biloba* extract products from the

US market for their dissolution rates. The authors analyzed 27 preparations sold as capsules or tablets for the flavonol glycoside, terpene lactone, and ginkgolic acid contents as well as determining the dissolution rates as per official pharmacopeia methods. The contents of the majority of the preparations were very similar for the flavonol glycosides and terpene lactones but some did differ from the concentration ranges recommended by the German Commission E, 1994. The majority of the preparations dissolved to >75% within 30 minutes, several of the extracts did not match the quality criteria of the USP and Commission E monographs.

Piazza et al. (2019) investigated if ethanol and acetone extracts, with comparable standardization, could be considered comparable in terms of biological activity. They focused on the radical-scavenging and antiinflammatory activities. The two extracts evaluated were an aqueous ethanol extract of *G. biloba*, G4E and an aqueous acetone extract, EPG246 or G24. For the preparation of G4E, the *Ginkgo* leaves were submitted to solid/liquid extraction using aqueous ethanol. The extracted fractions were then concentrated under reduced pressure to remove the extraction solvent, purified by filtration on resin column and liquid/liquid extractions. To prepare the G24, the leaves were submitted to solid/liquid extraction using aqueous acetone as extraction solvent; the extracted fractions were concentrated under reduced pressure and purified by liquid/liquid extractions. In both, the organic solvent was then replaced with water and the aqueous phase was concentrated and dried to obtain the final powdered extract. The flavonoid and terpene profiles were evaluated in both preparations using HPLC. Human endothelial cells (HUVECs, CRL-1730, ATCC, Virginia, United States) were used and the cytotoxicity of the extracts was evaluated using the 3,4,5-dimethylthiazol-2-yl-5-diphenyltetrazolium bromide (MTT) assay. The HUVECs were also used for the measurement of soluble intercellular adhesion molecule-1 (ICAM-1), VACM-1, and E-selectin adhesion molecules, cell surface adhesion molecules, gene expression following exposure to the extracts, the effect of the extracts on reactive oxygen species (ROS) production, on NF-κB (p65) nuclear translocation, IκBα phosphorylation, and measurement of cyclic AMP. Both extracts showed high inhibition of TNF-α-induced VCAM-1 release, which was partly due to the NF-κB pathway impairment. No statistical difference was observed between the extracts, and the authors concluded that ethanol and acetone extracts showed comparable biological activities in human endothelial cells and that this provides new insights into the use of ethanol extracts when the allowable amount of acetone residues is restricted in food supplements. The authors note that the concentrations of extracts used in these studies may be considered high and not easily attainable in vivo, there is evidence that lactones as well as flavonoids or their metabolites could show biological activity in endothelial cells in vivo.

Pharmacokinetics

The advancement in analytical methods in the last few years has allowed for more thorough studies of the pharmacokinetic behavior of the important phytochemical constituents in many herbal preparations including the terpene trilactones and flavonoids in *G. biloba* extract in both animals and humans. This has led to the conclusion that a target tissue of EGb 761 is the CNS and recent studies on CNS bioavailability and tissue levels of these compounds in rats allowed a pharmacokinetic characterization at the target tissue (Ude et al., 2013). There are many studies in the literature on the in vitro and in vivo effects and mechanisms of *G. biloba* extract and its constituents which allows for extensive knowledge of its absorption, excretion, and metabolism. It is important to remember that botanical extracts can vary widely in composition based on the method of manufacture or preparation. As noted by Catlin et al. (2018), this is a challenge in determining how the diverse products in the marketplace relate to the chemical and toxicological information reported from well-characterized reference samples. The growing region, climate, and time of harvest significantly influence ($P < .05$) the content of active compounds in *G. biloba* and its extracts (Rimkiene et al., 2017).

Cao et al. (2019) developed a model in rats to describe the metabolome of gut mucosal-associated metabolites. Rats were orally administered 600 mg *G. biloba* extract (prepared as a Chinese herbal medicine)/kg body weight. Three hours after dosing plasma, liver, brain, and intestinal mucosa were collected. The metabolites were identified in intestinal mucosa and the plasma, brain, and liver were then assayed for target ginkgolides. Phase I metabolites were identified in the intestinal mucosa, plasma, and liver and only phase II metabolites were detected in the brain. The flavonoids are rapidly hydrolyzed by the intestinal microflora to aglycones, metabolized, and then absorbed following oral ingestion (Heinonen and Gaus, 2015). The primary metabolic route is thought to be glucuronidation in the intestines but methylation, sulfation, and oxidative degradation may also occur. This type of microbial metabolism has not been identified for the terpene trilactones. Following oral ingestion, ginkgolide A and B as well as BB are absorbed as the parent compounds with a half-life of between 2 and 4 hours. Kaempferol was found in both the free form (40% of total plasma levels) as well as its glucuronide conjugated form (60% of total plasma levels) (Heinonen and Gaus, 2015). Huang et al. (2014) showed in rats that ingestion of food (versus fasting) and gender are important in the pharmacokinetics of the *Ginkgo* terpene lactones. For fed rats, the plasma $t_{1/2}$ values for BB, ginkgolide A (GA), ginkgolide B (GB), and ginkgolide C (GC) were significantly increased ($P < .05$). Feeding also significantly increased ($P < .05$) the area under the concentration−time curve (AUC_{0-t}) and $AUC_{0-\infty}$ for BB, GA, and GB and significantly decreased ($P < .05$) the maximum concentration (C_{max}) of BB, GA, GB, and GC. For sex comparisons, the female rats $t_{1/2}$ and AUC_{0-t} values for BB, GA, GB, and GC were significantly higher ($P < .05$). For BB, GA, GB, and GC, there were no significant differences between sexes for the T_{max} (time to maximum concentrations) values. Qian et al. (2019) observed that dosing rats with ginkgotoxin (4-*O*-methylpyridoxine, found in seeds) concurrent with ginkgolides delayed the T_{max} of ginkgotoxin. The hydroxylation of certain ginkgolides was reduced and the rate of glucuronidation for some ginkgolides was increased. Oral administration of *G. biloba* extract mixed with sesame seed extract (*Sesamum indicum* L.) and turmeric oil (*Curcuma longa* L.) increased serum levels of terpene lactones in a dose-dependent manner. The brain levels of terpene lactones were not increased. Coadministration of sesame seed extract, turmeric oil, and *G. biloba* extract increased the brain level of ginkgolide A.

The pharmacokinetics of radiolabeled EGb 761 showed the radioactivity fit a two-compartment model with an apparent first-order phase and a half-life ($t_{½}$) of approximately 4−5 hours (Ude et al., 2013). Absorption of the EGb 761 was at least 60% and the specific blood activity peaked after 1−5 hours with the highest values for specific radioactivity measured in the stomach and small intestine at 3 hours, which may indicate that this was the site of absorption. Neuronal tissue, eyes, and glandular tissue had high levels of the labeled substance and after oral dosing, and after 72 hours, the exhaled $^{14}CO_2$ accounted for approximately 38% of the administered dose. Excretion in the urine and feces at 72 hours was 22% and 29%, respectively. Preclinical studies in rats showed that terpene trilactones reached t_{max} very quickly with ginkgolide B reaching t_{max} a little later than A or BB (Ude et al., 2013). The $t_{½}$ of the terpene trilactones ranges from 1 to 11 hours and in the majority of humans the $t_{½}$ values are longer than those in animals with the $t_{½}$ of ginkgolide B being longer than that of A and BB. It is suggested that the short $t_{½}$ is an indicator of a lack of accumulation of the terpene trilactones in humans and animals, which was confirmed by Biber (2003). Oral dosing of *G. biloba* extract for 8 days did not change the values for t_{max}, and concentration max (C_{max}) and the area under the curve value when days 1 and 8 were compared. Given that the majority of effects of *G. biloba* extract are in the central CNS, it is important to know that it crosses the blood−brain barrier (BBB). Significant levels of the terpene trilactones and flavonoids in *G. biloba* cross the BBB and enter the CNS in rats following oral administration (Ude et al., 2013). Ude et al. (2013) characterized ginkgolides A and B as well as BB in the CNS of rats following oral dosing of 600 mg EGb 761/kg. It showed that ginkgolides A and B and BB could be detected at concentrations sufficient to exert a pharmacological effect in the brain tissue. The authors used a fast and sensitive LC-MS method, which allowed them to show that these substances reached the CNS rapidly with a t_{max} range of 1−2 hours. No terpene trilactones could be detected after 24 hours and native ginkgolide C was not found in the brain. Further investigation in this study using a combination of microdialysis and LC-MS methods showed that BB was present in defined regions of the mouse brain. This study showed that BB is present extracellularly in an unbound state and does not bind in the membrane.

Flavonoids undergo extensive metabolism and neither flavonoid glucosides nor their corresponding aglycones are detectable in plasma (Ude et al., 2013). After absorption through the intestinal wall, the flavonoid aglycones are excreted as glucuronides, sulfate conjugates, and in the hydro-, dihydro-, or methylated forms in the urine. Ude et al. (2013) concluded that due to the metabolism of the flavonoids described in the literature, the genuine flavonoid glycosides and their corresponding aglycones are not responsible for the pharmacological effects, but rather it is their metabolites which are pharmacologically active. In a study where healthy human volunteers were given three different formulations—drops, capsules, and tablets—of *G. biloba* extract to compare the oral bioavailability of the flavonoid glycosides, a one-compartment model with zero-order absorption without a lag time was described (Wojcicki et al., 1995). A t_{max} of 2 hours was noted for each flavonoid and the zero value was reached at 24 hours after a single dose. It was suggested that there was moderately extensive penetration into the tissues based on the volume of distribution and the elimination half-life was of a short, 3-hour duration. The total body clearance of approximately 39−53 L/h demonstrates rapid elimination of the flavonoids from the blood. The study concludes that the bioavailabilities of the three different forms were bioequivalent based on the comparable pharmacokinetic characteristics after oral intake. In a study where the C_{max} following a single oral dose of EGb 761 at 600 mg/kg or repeated daily doses at 100 and 600 mg/kg for 8 days were compared, the C_{max} of quercetin, kaempferol, and isorhamnetin/tamarixetin were compared (Ude et al., 2013). Following the single oral dose, the C_{max} of quercetin, kaempferol, and isorhamnetin/tamarixetin were 176, 341, and 183 ng/mL, respectively, and in the repeated dose animals there was a 4.5-fold, 11.5-fold, and 10-fold increase in the C_{max}, respectively, as compared to the single-dose C_{max}. In the single-dose study, the C_{max} for kaempferol was achieved after 8 hours, remained constant for several hours, and then slowly decreased after 24 hours. The plasma levels of the quercetin metabolites increased rapidly, achieving the C_{max} within 2 hours after dosing. The C_{max} of quercetin decreased slowly over the first 72 hours, with 78% of the C_{max} still detected after this 72-hour period. The plasma levels of isorhamnetin/tamarixetin increased over the first 8 hours, peaked, and then decreased continuously with 9.5% of the C_{max} still detectible in the plasma after 72 hours. Although the literature shows that flavonoids cross the BBB, in this study only kaempferol and isorhamnetin/tamarixetin were detected in brain samples following a single oral dose of 600 mg/kg of EGb 761with maximum concentrations of 291 and 161 ng/g protein, respectively. In the repeated-dose animals, there was an approximately twofold increase in the two compounds. The C_{max} of isorhamnetin/tamarixetin was lower than that for kaempferol, and it could be detected longer in the brain tissue. Liu et al. (2019), using an in vitro system of SH-SY5Y cells, showed that cytosolic ginkgolides GA and GB were increased in a dependent manner with increasing concentrations in the incubation fluid. The levels were the highest in cytosolic organelles. Incubating with verapamil increased the entry rate and cytosolic and organelle concentrations of GA and GB. The cells were oxygen-deprived and reoxygenated. The incubation with ginkgolides decreased apoptosis and loss of mitochondrial membrane potential. The pharmacokinetics of *Ginkgo* diterpene lactone meglumine injection (GDLMI) was determined in beagle dogs (Wang et al., 2017). GA, GB, and GK had a $t_{1/2}$ of approximately 2.7, 3.4, and 1.2 hours, respectively. The GA, GB, and GK in *Ginkgo* diterpene lactone meglumine exist in the hydrolyzed form. In a rat intravenous infusion model using GDLMI, Wang et al. (2016) showed that GA, GB, and GK lactones and hydrolyzed forms existed in plasma is a dose-dependent manner. The ginkgolides existed in plasma primarily as the hydrolyzed form, with <20% of the total amount existing as the prototype lactone. For tissue levels, the lowest level was in brain, and the highest levels were identified in intestine, kidney, and liver. These findings may suggest that important pharmacodynamics may occur outside the CNS or ginkgolides that penetrate the brain have a very potent effect.

Pharmacology

The pharmacological modes of action of *G. biloba* leaf extract are polyvalent and linked to the phytopharmacological interactions of terpene trilactones ginkgolides A, B, and C as well as BB and the flavonoid glycosides, quercetin, kaempferol, and isorhamnetin (Heinonen and Gaus, 2015; Houghton, 2009). Chan et al. (2007) noted that the therapeutic benefit depends on the combination of the constituents of *G. biloba* leaf extract and that the total extract is more active than any one of the single, isolated components, which suggests that the various components act synergistically and may be why it is effective in treating more than one condition. Extracts of *G. biloba* have antioxidant effects, radical scavenging, inhibition of platelet-activating factor (PAF), alterations in membrane fluidity/signal transduction, and inhibition of corticoid synthesis (Heinonen and Gaus, 2015). The variation in clinical efficacy reported in the literature is likely due to the variation in phytochemical content of the extracts tested as well as the various biopharmaceutical characteristics of the extracts (Gurley, 2012).

G. biloba is reported to increase cerebral and peripheral blood circulation, reduce vascular permeability, cause vascular contraction, improve venous tone, inhibit phosphodiesterase type 4, relax vascular smooth muscle via a nitric oxide pathway and improve blood flow to the corpus cavernosum of the penis, reduce platelet aggregation, decrease systolic and diastolic blood pressure, increase fasting plasma insulin and C-peptide, decrease cortisol secretion, and decrease secretion of corticotropic-releasing hormone (Dugoua et al., 2006). Diamond and Bailey (2013) state that the clinical effects attributed to *G. biloba* are via the antioxidant effects of the flavones and the antiplatelet-activating factor activity of the terpene lactones and that it has a modulatory effect on the cerebrovasculature tone, receptor/transmitter activity, glucose metabolism, and electroencephalographic activity. Some epidemiological data show that high intake of flavonoids has a preventative effect on cardiovascular disease and the flavonoid glycosides may also be effective in the treatment of chronic venous insufficiency (Ude et al., 2013). These protective effects are thought to be due to the antioxidant effects and free radical scavenger effects of the flavonoids. In EGb 761 the flavonoid fraction has higher radical scavenger potency than the terpene trilactones, and in cellular systems, the flavone glycosides scavenge superoxide anions, hydroxyl radicals, and peroxyl radicals and prevent lipid peroxidation (Ude et al., 2013).

Inflammation

Oxidative stress occurs when the cellular antioxidant defenses cannot keep the levels of ROS below the toxic threshold and is the key factor in a number of acute and chronic conditions such as neurodegenerative diseases. Ginkgolides have very specific and potent antagonist activity against platelet aggregation factor, which may help increase peripheral blood flow. The inhibition of cyclooxygenase and lipoxygenase activity by *G. biloba* extract may be the mechanism of the antiinflammatory effect (Chan et al., 2007). Some flavonoids promote the loss of innate immune cell functions, a reduction of inflammatory cytokine production, particularly TNF-α, IL-1β, prostaglandin E_2, and a decrease in NF-κB signaling (Ude et al., 2013). Lichota et al. (2019) discuss the risk factors, pathogenesis of chronic venous disease, markers of the epithelial and leukocyte activation, state of hypoxia, inflammation, ROS generation, and oxidative stress, as well as substances of plant origin which are used in the treatment of venous insufficiency. In this discussion, they state that *G. biloba* extract inhibits ICAM-1 and cellular vascular adhesion molecule 1 (VCAM-1) expression in vitro, which are molecules that promote the recruitment of leukocytes, platelets, and erythrocytes and their adhesion to the venous wall. In addition, it is also important in the peripheral circulation and microcirculation. Ginkgolide B is a PAF antagonist involved in the platelet aggregation process as well as in inflammatory reactions, and also has a wide range of direct and indirect antioxidant activities while both GA and GB inhibit free radical formation (Lichota et al., 2019). Ginkgolide terpenes also contribute to the control of inflammation, they also are powerful antioxidants which scavenge ROS.

In vitro experiments have shown that treatment with EGb 761 reduced the membrane lipid peroxidation in isolated rat retina, blocked production of free radicals in a reaction mixture, protected cultured neurons from oxidative stress caused by hydrogen peroxide, iron sulfate, and apolipoprotein E, and protected neural tissue against cerebral ischemia (Ahlemeyer and Krieglstein, 2003). *G. biloba* extract has both free radical-scavenging and antilipoperoxidative effects, which may protect both nitric oxide (NO) and prostaglandin I_2 from being degraded (Chan et al., 2007). *G. biloba* leaf extract, which was applied to the skin of rats, induced superoxide dismutase (SOD) and catalase enzyme activities locally in the epidermis and also systemically in the liver, heart, and kidney (Chan et al., 2007). It has also been shown to inhibit or reduce the functional and morphological impairments in the retina following lipoperoxide release and also protect pancreatic cells against the effects of alloxan, likely by free radical-scavenging effects and the stimulation of glucose utilization (Chan et al., 2007). The longevity and antiaging effects attributed to *G. biloba* leaf extract may be linked with the free radical-scavenging activity. In a study where male Fischer 344 rats consumed 50 mg/kg per day of EGb 761 in the drinking water for their entire life, they had significantly higher survival rates than the controls (Chan et al., 2007). In another study with old rats where they consumed EGb 761 in the drinking water (100 mg/kg per day) for 3 months, the morphological changes and the indices of oxidative damage in brain and liver mitochondria were partially prevented (Chan et al., 2007).

Stress can cause an increase in plasma levels of epinephrine, norepinephrine, and corticosterone and *G. biloba* can significantly reduce the concentrations of these as well as adrenocorticotrophic hormone (ACTH) in rats (Chan et al., 2007). It also acts on the CNS to decrease the levels of corticotropin-releasing hormone and arginine vasopressin in the hypothalamus and ACTH in the pituitary (Chan et al., 2007). The leaf extract also has been shown to inhibit monoamine oxidase and elevate brain concentrations of serotonin, norepinephrine, and other biogenic amines in rats (Chan et al., 2007).

Cardiovascular effects

Cardiovascular disease is a significant healthcare concern around the world. The use of herbal medicines is an increasingly popular alternative to the generally more expensive synthetic medicines (Shaito et al., 2020). Herbal medicines have historical use in human and veterinary medicine. *G. biloba* extracts are used to prevent chronic inflammation with associated endothelial changes and intravascular clots. The mechanism and pathophysiology of varicose vein formation is complex but includes hypotensive blood vessels, hypoxia, and other mechanisms associated with inflammation (Crawford et al., 2017; Lichota et al., 2019). Physical activity is required to propel the blood from the veins, especially the lower limbs, and prevent venous dilation and dysfunction of the venous valves. Inflammation of the venous endothelium leads to thickening and stiffening of the veins and essentially destruction of the physiological function of the venous valves. Pooling of blood can add to the inflammatory response and pressure on the large abdominal veins during pregnancy can contribute to varicose veins. In addition to a discussion on risk factors, pathogenesis of chronic venous disease, markers of the epithelial and leukocyte activation, state of hypoxia, inflammation, ROS generation, and oxidative stress, Lichota et al. (2019) discuss substances of plant origin which are used in the treatment of venous insufficiency. As noted in an earlier section, they discuss *G. biloba* extract inhibiting ICAM-1 and VCAM-1 expression in vitro as these are molecules which promote the recruitment of leukocytes, platelets, and erythrocytes and their adhesion to the venous wall. In addition, it is also important in peripheral circulation and in microcirculation. Ginkgolide B is a PAF antagonist involved in the platelet aggregation process as well as in inflammatory reactions, and also has a wide range of direct and indirect antioxidant activities, while both GA and GB inhibit free radical formation (Lichota et al., 2019). Of all the ginkgolides, GB is reported to be the strongest antagonist of the PAF, but it also has other properties including antiischemic, antioxidant, and anticonvulsant properties. Key to the prevention and treatment of varicose veins is vein "milking" from muscular movement and prevention of inflammation and its consequences leading to dilation of the veins. *G biloba* extracts provide the antioxidant and mild anticlotting activity in managing the pathophysiology of this malady.

Essential hypertension has an idiopathic cause. Hypertension increases the physical work the heart must do to pump blood. *G. biloba* extract is often used to treat essential hypertension as a complementary therapy in China and some European countries, and Xiong et al. (2014) conducted a systemic review to critically assess the clinical evidence of *G. biloba* extract for essential hypertension. The authors reviewed a total of nine randomized clinical trials which included 1012 hypertensive patients and concluded that most of these had a high risk of bias with flawed study design and poor methodological quality. The conclusions of the review were that no confirmative conclusions could be made on the efficacy and safety of the extract for essential hypertension and that more rigorous trials are required. In a more recent systematic review and meta-analysis, the aim of the review was to assess the preclinical evidence and possible mechanisms of GB for myocardial ischemia–reperfusion injury (Zhu et al., 2019). The authors ultimately reviewed 19 studies with a total of 437 animals and concluded that GB exerted potential cardioprotective function primarily through antioxidant, antiinflammation, antiapoptosis, and improvement of energy metabolism. Protecting the heart and reduction of endocardial and myocardial inflammation is important in preventing degenerative changes in the heart.

Neuroprotective effects and oxygen deprivation

The neuroprotective effects attributed to *G. biloba* have been shown in a number of in vitro and in vivo studies with a number of different models. Neuron death occurs by two processes, apoptosis or necrosis, and apoptosis has been shown to be the primary process in the majority of neurodegenerative diseases. Oxygen deprivation followed by reoxygenation is also damaging to neurons. In vitro studies with EGb761, protection of cultured neurons from cell death induced by hypoxia, hydrogen peroxide, glutamate, verapamil, amyloid β, NO, 1-methyl-4-phenyl-1,2,3,6-tetrahydropyridine, and cyanide at concentrations of 10–100 μg/mL were demonstrated (Ahlemeyer and Krieglstein, 2003). EGB 761 showed a reduction of neuronal damage at dosages of 10–100 mg/kg when given per os or by intraperitoneal injection in a transient middle cerebral artery occlusion model in rats and gerbils, a focal cerebral ischemia model in rats and mice, in heat stress, subchronic cold stress, and hypoxia conditions, with amphetamine-induced behavioral sensitization as well as in a transgenic mouse model of amyotrophic lateral sclerosis (Ahlemeyer and Krieglstein, 2003). It has been reported that *G. biloba* can improve cerebral blood flow, increase glucose uptake and utilization, reduce corticosteroid production and improve ATP production and mitochondrial metabolism as well as affect intra- and extracellular ionic gradients. It has been shown that EGb 761 increased local cerebral blood flow by 50–100% in almost all of the brain regions when evaluated by ^{14}[C] iodoantipyrine autoradiography (Krieglstein et al., 1986). Di Renzo (2000) discussed studies which suggest that treatment with EGB 761 can prevent the occurrence of processes which lead to neuronal degeneration in gerbil models of cerebral ischemia. In one study, the unilateral ligation of the carotid artery in the gerbil was used which showed that oral or intraperitoneal administration of the EGb 761 induced the normalization of mitochondrial respiration, a decrease in cerebral edema, correction of the associated ionic changes, and almost total restoration of function as determined by a neurological index. In a second gerbil model, bilateral forebrain ischemia, ischemia was induced following daily administration of EGb 761 for 14 days. There was a significant increase in the area of preserved neurons. Following global ischemia, the cerebral blood flow initially increases, known as postischemic hyperperfusion, and then decreases for several hours to 40%–60% of the preischemic levels (known as postischemic hypoperfusion) (Ahlemeyer and Krieglstein, 2003). Kriegelstein et al. (1995) showed that dosing rats with EGb 761 reversed the hypoperfusion 15 and 45 minutes after a 10-minute global cerebral ischemia to nearly control levels, but that the infarct volume was not reduced.

Feng et al. (2019) conducted a literature review into the neuroprotective mechanisms of ginkgolides (GA and GB) and BB in cerebral injury following ischemic stroke. A variety of mechanisms are discussed in the review including a variety of physiological changes which occur following a cerebral ischemic event. These changes include increased permeability and dysfunction of the BBB, cerebral edema, brain metabolic disturbances, inflammation, oxidative stress, and ginkgolides decreased apoptosis and loss of mitochondrial membrane potential. The authors concluded that the neuroprotective effects of ginkgolides and BB have been validated in several preclinical studies. They also concluded that advancements in animal models and imaging techniques allow for a better appreciation of the significance of these bioactive constituents of *G. biloba* in the cellular and signaling events of cerebral ischemia (Feng et al., 2019). The pharmacokinetic and tissue distribution study by Wang et al. (2016) suggests that the ginkgolides may have phytopharmacological actions outside the brain that are protective of the CNS.

EGb 761 has been used to reduce cognitive dysfunction in a model of vascular dementia in gerbils in which the studies show a significant recovery of spatial memory on regular postischemic treatment (Mohanta et al., 2014). When examined at a cellular level, the extract has been shown to reduce plasma SOD activity and to protect the hippocampal CA1 neuron. Ihl et al. (2011) conducted a study to test the efficacy and safety of a once-daily formulation of EGb 761 in the treatment of patients with dementia accompanied with neuropsychiatric features. The patients ($n = 410$) were randomly allocated to a double-blinded treatment with 240 mg of EGb 761 or placebo once daily for 24 weeks. The patients were all outpatients with mild to moderate dementia (Alzheimer disease, vascular dementia, or a mixed form) with a score of between 9 and 23 on the Short Cognitive Performance Test (Syndrom Kurz Test, SKT), at least 5 on the Neuropsychiatric Inventory (NPI), and 3 or higher in at least one item of the NPI. The primary outcomes of the study were changes from baseline in the SKT total score and the NPI total score, secondary outcomes were also used. The results of the trial showed that the EGb 761-treated patients ($n = 202$) improved significantly on the SKT and on the NPI total score when compared to the placebo group ($P < .001$ for both drug–placebo comparisons). EGb 761 treatment was also significantly superior with respect to the secondary outcome measures, as compared to the placebo. Adverse event rates were comparable for both groups. In a meta-analysis with expert opinion, Kandiah et al. (2019) concluded that, compared to placebo, EGb 761 treatment significantly improved quality of life, activities in daily living, and cognitive functions, and decreased neuropsychiatric symptoms. These effects had the same strength of evidence as acetylcholinesterase inhibitor and *N*-methyl-D-aspartate antagonists. Compared to other medications, EGb 761, for Alzheimer disease and vascular-linked dementia, is the only medication with Level B evidence for improved cognitive behavior and activities of daily living by the World Federation of Societies of Biological Psychiatry guidelines. Kaschel (2011) conducted a study in which 188 healthy patients received either EGb 761 at 240 mg/day or a placebo once daily for 6 weeks. The outcome measures for this study were the change in memory performance in a demanding standardized free recall paradigm and a less demanding standardized recognition test. After 6 weeks of treatment, the patients in the EGb 761 group improved significantly in the quality of the recall but not in the recognition test.

In vivo studies in aging mice have shown that EGb 761 can increase brain membrane fluidity and that both the flavonoid and terpenoid constituents may be responsible for this; flavonoids by scavenging free radicals and other ROS and possibly acting as Fe^{2+}-chelators and the terpenoids by decreasing the generation of free radicals (DeFeudis and Drieu, 2000). A review of studies on the antiapoptotic effects of EGb 761 shows some evidence that the effects exist but they require confirmation (DeFeudis and Drieu, 2000). Confirming these effects could be important for determining the effects of *G. biloba* on Alzheimer disease as the effects are likely linked to opposing oxidative damage and oxidative stress is thought to be a major process of Alzheimer disease pathology. *G. biloba* is reported to reduce oxidative stress and improve mitochondrial respiration, and therefore may be useful in preventing or slowing down the progression of Alzheimer disease (Eckert, 2012). In a study with aged mice (18 months of age) which were treated with EGb 761 for 14 days, the cortical brain cells were removed and evaluated for the presence of ROS using fluorescent dyes (Eckert, 2012). In the EGb 761-treated mice, the levels of ROS and O_2^- were reduced in brain cells as compared to the placebo controls, which led the authors to conclude that EGB 761 provides an antioxidant effect. Postmortem analyses show an increased lipid peroxidation, protein oxidation, and oxidation of DNA and RNA in the CNS of Alzheimer patients (Markesbery and Lovell, 2007).

Lejri et al. (2019) prepared a review article on mitochondria- and oxidative stress-targeting substances in cognitive decline-related disorders. They reviewed standardized *G. biloba* extract, resveratrol, and phytoestrogens in this review and noted that all showed promising effects on mitochondria. Mitochondria are affected during aging as well as in pathologies such as age-related neurodegenerative diseases including when ROS are produced in excess. Several studies have shown that *G. biloba* extract has been shown to attenuate mitochondrial dysfunction via several mechanisms of action including an antioxidant effect and free radical-scavenging properties (Lejri et al., 2019). In in vitro and preclinical studies, *G. biloba* extracts have been shown to be effective in reducing the deposition of Aβ as well as its toxicity and Aβ plaque deposition being one of the hallmarks of Alzheimer disease. *G. biloba* extracts have been shown to have beneficial effects on neuroplasticity pathways. As stated in Muller et al. (2012), impaired mitochondrial function represents an important aspect of brain aging and age-related glucose consumption, reduced energy production, impaired synaptic function, and cell death. Synaptic function has high ATP demand due to the ability to adapt dynamically to variations in functional demand or neuroplasticity. The high ATP demand makes synaptic function very sensitive to mitochondrial dysfunction. In numerous preclinical studies, EGb761 has been shown to increase nearly all aspects of impaired neuroplasticity including long-term potentiation, spine density, neuritogenesis, and neurogenesis (Muller et al., 2012).

In the review conducted by Lejri et al. (2019), the authors note that *G. biloba* extract has been investigated in clinical trials in a range of age-associated cognitive conditions. Lejri et al. (2019) reviewed the clinical evidence in patients with subjective memory impairment and mild cogmitive impairment, and dementia including Alzheimer disease and vascular dementia. They also discussed the use of *G. biloba* extract in the prevention of age-associated cognitive conditions and concluded that it has been proven more effective in patients with impairment at baseline than preventing the onset. Pohl and Kong Thoo Lin (2018) also conducted a review into the potential use of plant natural products and plant extracts for the prevention and treatment of neurodegenerative diseases. The review gave an overview of the importance of oxidative stress and its importance in neurodegenerative diseases. The authors state that although pure natural products or plant extracts with antioxidant activity show good results in both in vitro and preclinical studies, the outcomes in clinical studies are inconclusive and show limited success. They suggest one possible reason for this is that in clinical trials, mostly single compounds are studied, while those studies conducted prior to clinical studies are conducted on plant extracts containing a variety of secondary metabolites which may have additive or synergistic effects. The authors suggest that closer investigation of the results from a single individual, or a small group of individuals which showed significant improvement in the trial and determine why they may have responded to the treatment while others did not. This type of investigation could lead to a better understanding of the potential use of antioxidants in specific groups of patients and to a better understanding of neurological disorders. Pohl and Kong Thoo Lin (2018) conclude that further research into the pathology and mechanisms of neurodegenerative diseases may assist in understanding the lack of success of plant extracts and natural products in human patients. Sarris (2018) conducted an update to a previous review regarding herbal medicines in the treatment of psychiatric disorders and concluded that there was only preliminary evidence for the use of *G. biloba* as an adjunctive treatment in schizophrenia.

The literature reports that *G. biloba* can protect against neuronal damage which may be by direct action on the neurons or by an indirect effect via modulation of blood flow and antioxidant action (Chan et al., 2007). The uptake of radiolabeled norepinephrine, dopamine, and 5-hydroxytryptamine into the synaptosome-enriched areas of the rat brain is inhibited by high concentrations of EGb 761 (DeFeudis and Drieu, 2000). The Ki values for these effects were reported to be 0.12 mg/mL for norepinephrine in the occipital cortex, 0.21 mg/mL for dopamine in the corpus striatum, and 0.44 mg/mL for 5-hydroxytryptamine in the frontal cortex. While these effects may be seen to show antidepressant activity for *G. biloba*, the concentrations required to demonstrate the effects make the in vivo relevance questionable. DeFeudis and Drieu (2000) also reported the results of in vitro experiments using mouse cerebral cortex which showed that EGb 761 modified 5-hydroxytryptamine uptake in a biphasic fashion but was not a specific effect as it also occurred with choline and dopamine uptake. However, the effect on 5-hydroxytryptamine could be relevant in vivo as it was also demonstrated in ex vivo studies in which animals had been given EGb 761. It has been shown that EGb 761 influences the neuromodulation of adrenergic neurotransmission and this could explain the positive effect seen in mice and humans on cognition. Fehske et al. (2009) showed that the flavonoid fraction of EGb 761 led to an inhibition of norepinephrine uptake following 14 days of daily oral treatment with 100 mg/kg, which was assumed to be linked to the positive effect on cognition and mood in NMRI mice. When treated for 14 days with oral EGb 761 at 100 mg/kg, the dopamine concentration in the extracellular fluid of the prefrontal cortex of awake rats was increased in a dose-dependent manner (Yoshitake et al., 2010). This effect was shown to be caused primarily by the flavonol glycosides and the ginkgolide fractions while BB treatment did not have any effect.

Flavonoids are seen to be potent antioxidants and radical scavengers in preclinical studies, which led to a decrease in ROS levels in tissues and they chelate prooxidant transitional metal ions and modify the expression of antioxidants (Ude et al., 2013). EGb 761 has also been shown to protect the mitochondria from age-related damage and to improve mitochondrial function and energy metabolism, which has been attributed primarily to the ginkgolides and BB (Ude et al., 2013). EGb 761, at a concentration of 10 μg/mL which contained approximately 300 ng/mL BB and 300 ng/ml ginkgolides, was shown to protect PC12 cells against H_2O_2 in vitro and this same concentration also protected the complexes of the mitochondrial respiratory chain in PC12 cells against thenoyltrifluoroacetone, NaN_3 at 10 mmol/L, and oligomycin at 10 μmol/L (Ude et al., 2013). Two possible molecular mechanisms for the mitochondrial protection by EGb 761 were discussed in Ude et al. (2013); the first was that the terpene trilactones in EGb 761 inhibit the effect of PAF at the PAF receptor with ginkgolides B being the strongest PAF antagonist of all the terpene trilactones in *G. biloba*. By comparison, ginkgolide B is 25 times stronger than ginkgolide C. Given that PAF is a potent phospholipid mediator of leukocyte functions, platelet aggregation, and proinflammatory signaling and that the terpene trilactones act as antagonists of PAF, they improve the blood circulation. It has been shown that excessive PAF leads to neuronal damage, and inhibition of PAF plays a critical role in neuronal survival as well as in the prevention of ischemic brain injury due to the PAF antagonist activity of the ginkgolide. The second mechanism of mitochondrial protection discussed in Ude et al. (2013) was that the terpene trilactones also act as selective use-dependent blockers of glycine-activated chloride channels in the rat hippocampal neurons. Ginkgolides A and J are less effective for the saturating blocking action of ginkgolides B and C, which is evidenced by the lower inhibitory concentration fifty (IC_{50}) values for ginkgolides A and J as compared to B and C. Ginkgolide B inhibits glycine currents in a noncompetitive manner and also has increased sensitivity for the inhibitory glycine receptors if they contain a β-subunit.

Marx et al. (2020) conducted a systematic review of the effect of nutraceutical interventions for clinical and biological outcomes in multiple sclerosis including *G. biloba*. For this review the studies needed to meet selected eligibility criteria including that it need to have utilized a randomized, parallel or crossover trial study design, have investigated a nutraceutical as a stand-alone intervention, recruited only participants with multiple sclerosis, and measured any biological or clinical outcome that was related to multiple sclerosis pathology or symptoms. Five studies utilizing *G. biloba* were reviewed and improvement was only reported with measurements of fatigue and/or depression but not in clinical improvement in the Expanded Disability Status Scale or inflammatory markers. The authors concluded that most of the reviewed trials had a relatively small sample size and were short term. They state that the existing literature provides preliminary support for the use of several nutraceutical interventions in multiple sclerosis, but sufficiently powered long-term trials are required to expand the currently limited literature.

Ginkgo diterpene lactone meglumine

The ginkgolides are natural antagonists of the platelet activation factor (PAF). GDLMI has been approved as a treatment for ischemic stroke in China (Fan et al., 2020). The active ingredient is an ethanolic extract of *G. biloba* leaves in meglumine (an amino sugar), formulated for intravenous injection. The primary ginkgolides are GB, GB, and Gk. The meglumine increases the solubility of the ginkgolides by hydrolyzing the lactone rings. The mechanisms of action are modulating glycolysis and other biochemical pathways, and mitochondrial functions. Using a rat model, Li et al. (2020) showed that GDMLI reduced infarct size and brain edema and improved neurodeficiency scores. Platelet aggregation, astrocyte activation, and proinflammatory cytokines were suppressed, as was TLR4 mRNA expression and transfer of NF-κB from cytoplasm to nucleus.

Metabolic syndrome and related conditions

Eisvand et al. (2020) conducted a review on the effects of *G. biloba* on metabolic syndrome which has also been evaluated for its effects on weight loss as well as its antidiabetic, antihypertensive, and antilipidemic properties. The authors state that in vitro and in vivo studies showed the effects of *G. biloba* and *G. biloba* extract on the different parts of metabolic syndrome such as cardiovascular disease, hypertension, obesity, dyslipidemia, and type II diabetes. The review found that *G. biloba* lowers the levels of low-density lipoprotein and triglycerides and increases the levels of high-density lipoprotein. It also showed that *G. biloba* extract decreases blood pressure through the increased expression of endothelial NO synthase and reduced blood glucose by increasing glucose use and insulin sensitivity. The authors concluded that the literature reviewed indicated that *G. biloba* may improve metabolic syndrome but that more studies, in particular clinical trials, are needed to evaluate its safety and efficacy.

Other areas of interest

Many protective effects against induced toxicity by drugs, chemicals, and heavy metals have been attributed to *G. biloba* leaf extract. It has been attributed with a protective effect against doxorubicin-induced cardiotoxicity in mice, bleomycin-induced lung fibrosis, amelioration of gentamicin-induced nephrotoxicity in rats, cisplatin-induced nephrotoxicity in mice, acetaminophen-induced toxicity in mice, mercury (II)-induced oxidative tissue damage in rats, vancomycin-induced nephrotoxicity. and endotoxin-induced oxidative renal tissue damage in rats (Chan et al., 2007). EGb 761 has also been shown to have a neuroprotective effect in cortical cultured neurons from staurosporine-induced neuronal death as well as caspase activity which has been linked with the antioxidant effects (Chan et al., 2007). EGb 761 has also been shown to have a protective effect on hippocampal cells exposed to amyloid-β peptides and inhibited the cytotoxicity and ROS accumulation when rat hippocampal cells were exposed to sodium nitroprusside (Chan et al., 2007). Ginkgolide B treatment of rat adrenocortical cells reduced the mRNA, protein, and ligand-binding levels of the adrenal peripheral-type benzodiazepine receptor, which is a mitochondrial cholesterol-binding protein—this leads to decreased corticosteroid synthesis (Ude et al., 2013). It was shown that EGb 761, especially ginkgolide B, decreased stress-induced increases in serum corticosterone but did not affect the basal physiological levels. BB has a protective effect on mitochondria and shows a dose-dependent increase in respiratory control ratio in vitro (Ude et al., 2013). This effect is shown to be via protection of complex I and III activities which allow mitochondria to maintain respiratory activity in ischemic conditions because some oxygen is still present which delays the onset of ischemia-induced damage. In certain stroke models, BB is able to reduce triethyl tin-induced cerebral edema and damage from cerebral ischemia as well as decrease cortical infarct volume in some models of stroke (Ude et al., 2013). The hepatoprotective effects of *G. biloba* have been documented against carbon tetrachloride (CCl_4)-induced hepatic damage in Sprague-Dawley rats with the elevated levels of serum alanine aminotransferase, aspartate aminotransferase, alkaline phosphatase, albumin, total protein, hyaluronic acid, laminin, triglyceride, and total cholesterol significantly restored to almost normal values in a dose-dependent manner (Mohanta et al., 2014). The CCl_4-diminished activities of the antioxidant enzymes were reversed and the CCl_4-elevated levels of malondialdehyde were also reduced. Omidkhoda et al. (2019) conducted a comprehensive literature review on studies which investigated the protective effects of *G. biloba* L. The review was conducted using some keywords in combination—*G. biloba* or ginkgolide, and toxicity or toxin or toxic or venom or poison or chemical or radiation or metal or pesticide. The authors reviewed natural toxins, chemical toxicities, and radiation. They concluded that taking all the information together, the effectiveness of *G. biloba* L. extract was established against a wide variety of intoxications. They also noted that fermentation of the *Ginkgo* leaves increased the bioactivity in two of the studies which were reviewed. The authors concluded that few clinical studies have been conducted on the protective effects and that this lack of data makes the potential effectiveness useless.

In a meta-analysis to evaluate the treatment effects on tinnitus and dizziness in patients with dementia, randomized, placebo-controlled trials of EGb 761 were selected based on the following criteria: diagnosis of dementia according to generally accepted criteria, treatment period of at least 20 weeks, outcome measures covering at least two of the three conventional domains of assessment, presence and severity of dizziness and tinnitus were assessed, and assessment was done before and after randomized treatment (Spiegel et al., 2018). Five trials were found that met all of the selection criteria and it was found that, overall, EGb 761 at daily doses of 240 mg, was superior to the placebos in patients with dementia.

Safety

The European Food Safety Authority has published a toolkit to assist risk assessors and food manufacturers in ensuring the products they put on the market are safe. This toolkit includes a Compendium of Botanicals which has been reported to contain substances which may be of health concern when used in food or food supplements. *Ginkgo* is listed in this compendium but all of the listed studies showing adverse effects come from one reference—Leistner and Drewke (2010). Leistner and Drewke (2010) report on acute toxicity from the *G. biloba* seed in humans and list adverse effects, including seizures and unconsciousness, however there is no associated dose reported which limits the value of this report in that regard. They discuss the fact that phytopharmaceutical preparations and botanical dietary supplements contain an array of natural products and that the German and European Pharmacopoeias limit ginkgolic acids to 5 ppm of the dried extract because of their allergenic potential. Leistner and Drewke (2010) also discuss that *G. biloba* tea may contain very high amounts of ginkgolic acids and would not fall within the recommended daily dosage of a standardized preparation.

Studies conducted in animals are currently the primary method of assessing the safety of herbal extracts (Heinonen and Gaus, 2015). The predictive value of rodent and nonrodent experiments for assessing human toxicity has been placed at 43% and 63%, respectively (Heinonen and Gaus, 2015). An example for these differences is the variations in liver xenobiotic metabolism activity between animals and humans, including the differences in type and activity of the cytochromes and conjugation enzyme (Heinonen and Gaus, 2015). The mechanisms by which substances are metabolized or transported in the gut, liver, or kidney can all be inhibited or induced by other drugs or natural products which can then increase or decrease the bioavailability of simultaneously ingested pharmaceutical agents. The National Toxicology Program (NTP) of the US Department of Health and Human Services conducted testing of *G. biloba* extract (CASRN: 90045-36-6) (NTP, 2020). Short-term toxicity studies completed include a 5-day gavage study (data in review) and a 13-week gavage study whose results are reported in Hoenerhoff et al. (2013) and Rider et al. (2014). Both of these references also report on a long-term carcinogenicity study. Hoenerhoff et al. (2013) report that exposure of male and female B6C3F1 mice to doses of up to 2000 mg of *G. biloba* extract/kg bw for 90 days resulted in an extract-related increase in hepatocellular hypertrophy and focal hepatocellular necrosis. The hepatocellular hypertrophy was characterized by centrilobular and midzonal hepatocyte enlargement. The hepatocytes showed increased amounts of homogeneous to finely granular eosinophilic to amphophilic cytoplasm and enlarge nuclei, while the focal necrosis showed loss of the hepatocyte architecture, hypereosinophilia and nuclear pyknosis, karyorrhexis, and karyolysis. Hoenerhoff et al. (2013) also reported on the 2-year bioassay that was conducted where exposure to *G. biloba* extract (200, 600, and 2000 mg/kg bw) caused a dose-dependent increase in the severity and incidence of centrilobular hepatocellular hypertrophy in both sexes. Dose-related increases in hepatocellular necrosis were seen in all extract-treated males, while in the mid- and high-dose females, increases in inflammatory cell infiltrates were reported. The authors concluded that oral exposure to *G. biloba* extract may result in hepatocelluar hypertrophy, which may increase in severity over time with continued exposure. Rider et al. (2014) report on the same set of studies as Hoenerhoff et al. (2013) but also discuss the studies conducted in rats and focus on identifying the target organs following exposure to *G. biloba* extract. They report that the target organs were the liver, thyroid, and nose and these were consistent across exposure period, sex, and species.

Waidyanatha et al. (2019) report that questions arose about the *G. biloba* extract used in the NTP studies; specifically, whether it was representative of other *Ginkgo* extract products and how the rodent doses were related to human doses. To address this, the authors generated systemic exposure data for terpene trilactones in male rats from a 2-year bioassay. In this study, the rat strain, route of administration, and *Ginkgo* extract used were all the same as those used in the NTP 2-year bioassay. The two highest doses used in the NTP study, 100 and 300 mg/kg, were used in the assay as well as a lower dose, 30 mg/kg, which was included to generate data closer to the human recommended daily dose. Animals were dosed via a single oral gavage and blood collected at set timepoints to collect plasma for quantitation of terpene trilactones and flavonols. The authors concluded that systemic exposure to *Ginkgo* specific terpene trilactones, following a 100 mg *G. biloba* extract/kg per day dose in rats was similar to a single clinically relevant dose of 240 mg in humans, which demonstrates the relevance of NTP rodent toxicity data to humans. They also caution that the mechanism of toxicity and observed effects in animal studies and the relevance to humans requires further consideration by the risk assessment community.

Unger (2013) concluded that the intake of a standardized extract such as EGb 761 together with other synthetic drugs appears to be safe as long as the daily dose does not exceed 240 mg. However, the same could not be said for poorly standardized extracts. The variations in the composition and bioavailability of different formulations can lead to variations in the pharmacological activity, efficacy, metabolism, and adverse effects (Heinonen and Gaus, 2015). The same variation in phytochemical content and biopharmaceutical characteristics which result in the variation in efficacy reported in clinical studies are likely responsible for the different herb–drug interactions reported (Gurley, 2012). Gurley et al. (2012) concluded that dosage forms that have standardized *G. biloba* extracts and are administered at 240 mg/day or less do not have clinically relevant herb–drug interactions but, if dosed at higher levels, a potential for interactions exists.

The potential for interactions between botanical extracts/preparations with pharmaceuticals is an important consideration. Pharmacokinetic and pharmacodynamic components may both be involved in the interactions with metabolic induction, or inhibition is commonly an underlying mechanism. Chen et al. (2011) discussed the mechanisms and clinical implications of these types of interactions and noted that drugs that have a high potential to interact with herbal medicines usually have a narrow therapeutic index and that many of them are substrates of cytochrome P450s and/or P-glycoprotein. They report that case reports have shown that *Ginkgo* may potentiate bleeding when taken with warfarin or aspirin, can increase blood pressure when taken with thiazide diuretics, and can lead to a coma when combined with trazodone. Herb–drug interactions may either interfere with the systemic clearance of the drug or compete with the drug target and the clinical outcomes can range from mild to lethal (Chen et al., 2011). Often the absorption, distribution, metabolism, and excretion details of herbal medicines are not as well defined as for pharmaceutical agents, however with the increased interest in herbal remedies, studies to evaluate this information have started to be reported (Choi et al., 2011). These authors also state that ginkgolides A, B, C, J, and BB may weakly or negligibly inhibit CYP1A2, CYP3A, and CYP2C9, while kaempferol, quercetin, apigenin, myricetin, and tamarixetin inhibited CYP1A2 and CYP3A. Several studies report that *G. biloba* extract induces several cytochrome P-450 (CYP) isoenzymes in mice and rats, however when 400 mg/day was given orally to healthy human volunteers for 13 days, there were no changes in hepatic microsomal enzyme activity as demonstrated by measuring plasma antipyrine concentrations (Heinonen and Gaus, 2015). The conclusion was that the xenobiotic biotransformation differs between laboratory rodents and man and that hepatic enzyme activity was not induced in humans even at higher than therapeutic levels. Ude et al. (2013) stated that using published data it was not possible to assess whether the interactions of *G. biloba* compounds and CYP enzymes have any clinical relevance for humans and that further investigations in humans are needed. Zhou et al. (2014), using human hepatocytes in an in vitro model, showed that 0.1–10 μL hydrolyzed ginkgolides (hydrolyzed using meglumine) did not inhibit CYP1A2, CYP2B6, CYP2C8, CYP2C9, CYP2C19, CYP2D6, CYP2E1, or CYP3A4m. At 10 μL, the exposure is expressed as the equivalent sum of GA + GB + GK. At 10 μL, the relative activity and expression of messenger RNA for CYP3A4 was increased by 4.59 ± 3.67 and 17.2 ± 9.16 over control values, respectively.

Zadoyan et al. (2012) assessed the human in vivo metabolic drug interaction profile of EGb 761 as related to the activities of major CYP enzymes in a single-center, open-label, randomized, threefold crossover, cocktail phenotyping design study. Subjects were given either placebo twice daily, EGb 761 120 mg twice daily, or EGb 761 240 mg in the morning and placebo in the evening. On day 8, dosing was completed in the same manner but was performed together with an orally administered phenotyping cocktail: 150 mg caffeine (CYP1A2), 125 mg tolbutamide (CYP2C9), 20 mg omeprazole (CYP2C19), 30 mg dextromethorphan (CYP2D6), and 2 mg of midazolam (CYP3A). They concluded that EGb 761 had no relevant effect on the in vivo activity of the major CYP enzymes in humans and has no relevant potential to cause metabolic drug–drug interactions. *Ginkgo biloba* extracts can alter the expected pharmacology of antileptics, antiepileptics, acetylsalicylic acid, diuretics, ibuprofen, risperidone, rofecoxib, trazodone, and warfarin (Shaito et al., 2020).

Gurley et al. (2012) noted that depending upon which experimental model is used to evaluate the herb–drug interaction potential of *G. biloba*, different interpretations of the interaction capabilities are reported. The authors report that in many in vitro experimental studies, *Ginkgo* extracts, individual terpene lactones, and flavonoid glycosides are shown to inhibit various xenobiotic metabolizing enzymes and transporters but usually at high doses. This is in contrast to studies that show that *Ginkgo* extracts and individual terpene lactones can induce the expression of several CYP, UGT, and ABC isoforms in rat and human hepatocytes and human mammary epithelial cells. The evidence from in vitro and in vivo studies shows that *G. biloba* extracts and their constituents are inducers of xenobiotic metabolizing enzymes and transporter activity, but clinical evidence is not compelling (Gurley et al., 2012). Hermann and von Richter (2012) conducted a systematic review of the available clinical evidence of herbal drugs as perpetrators of pharmacokinetic drug interactions. They evaluated a number of herbal remedies including *Ginkgo* and stated that the total number of clinical herb–drug interaction trials with pharmacokinetic objectives has increased. They noted that, in comparison to a review from 2005, there is a more than 3.5-fold increase in studies over the period from 2005 to 2012. Hermann and von Richter (2012) also stressed that it is important to note that it is not just the number of studies that have increased but the overall quality in study design and analysis points reflects a trend toward conceptual improvements. They also commented that there is a lack of data evaluating the possibility that herbal drugs can be the objects of drug interactions as well but that gathering these data would require identification of the major active constituents in herbal products as well as the development of sensitive bioanalytical assays to quantify alterations of the exposure characteristics of herbal drugs when administered with inducer/inhibitor drugs.

Mandery et al. (2012) conducted a study to investigate the impact of the *G. biloba* flavonoids apigenin, kaempferol, and quercetin on the organic anion-transporting polypeptide 1B1 (OATP1B1), OATP1B3, and organic cation transporter 1. These are drug uptake transporters that mediate the uptake of some drugs and are located on the basolateral membranes of hepatocytes. They found that the *G. biloba* flavonoids showed competitive inhibition of the OATP1B1- and OATP1B3-mediated [^{3}H]sulfobromophthalein and the OATP1B1-mediated [^{3}H] sulfobromophthalein uptake with quercetin being the most potent. In a study to investigate the effect of *G. biloba* coadministration on the pharmacokinetics of tamoxifen, anastrozole, and letrozole, coadministration did not significantly affect the pharmacokinetics of the chemotherapy agents (Vardy et al., 2013). In this prospective open-label crossover study, 60 women with early-stage breast cancer who were taking either tamoxifen, anastrozole, or letrozole received EGb 761 for 3 weeks (120 mg twice daily). The trough concentrations of drugs were measured using LC-MS/MS before and after the *G. biloba* treatment and no significant differences were found. The authors concluded that coadministration of *G. biloba* did not significantly affect the pharmacokinetics of tamoxifen, anastrozole, or letrozole. Toxicities were graded according to the National Cancer Institute Common Terminology Criteria for Adverse Events and showed that *G. biloba* extract was well tolerated with no difference in toxicity reported.

Woron and Siwek (2018) analyzed 147 cases of adverse events with evidence indicating that they resulted from the inclusion of an herbal preparation into an applied pharmacotherapy, primarily with psychotropic drugs. The reactions were evaluated in four groups of drugs used in psychiatry: antidepressants, antipsychotics, anxiolytics, and Z-drugs. Herbal preparations in this paper included plant medicines and nutritional supplements containing plant extracts. The authors reported that in all 147 cases, a probable ($n = 15$) or proven causal relationship ($n = 132$) was found between the addition of medication and/or nutritional supplements to the regular therapy and the occurrence of clinical complications. In addition, the authors found that 86 cases were associated with the use of nutritional supplements, while 61 cases resulted from the administration of plant medicines. In several cases, 132 of the 147 cases, when the plant preparations were discontinued, the severity of the reported adverse reactions decreased or there was a complete reversal of the symptoms. While *G. biloba* was only one of the herbal preparations evaluated, the authors concluded that it was associated with a significant risk of pharmacokinetic and pharmacodynamic interactions with psychotropic drugs.

Posadzki et al. (2013) conducted an overview and critical evaluation of the evidence from systematic reviews of herb–drug interactions. They concluded that due to the limited quality and scarcity of any primary data, any conclusions of these systemic reviews should be critically evaluated. From a review of the literature it is apparent that this is an area that would benefit from additional research. Posadzki et al. (2012) had conducted a previous overview of systematic reviews of herb–drug interactions and noted that moderately severe interactions were noted for *G. biloba* (Table 48.1). These included interactions with anesthetics, analgesics, anticoagulants, and antiplatelet agents. There is a possible interaction reported between nicardipine and *Ginkgo* in an animal model, which could result in the attenuation of the hypotensive action of nicardipine through a mechanism which involves the induction of CYP3A2 and other liver-metabolizing enzymes (Diamond and Bailey, 2013). The dosage levels of *Ginkgo* were higher in the animal model than the standard human dosage. These clinical interactions are not well defined, but caution should be employed when using long-term high-dose *Ginkgo* and calcium channel antagonists as they are metabolized by the same enzyme systems. The American Society of Anesthesiologists recommends that all herbal medications be stopped 2–3 weeks before elective surgery and *G. biloba* has been specifically identified.

In a recent review (Heinonen and Gaus, 2015), it was stated that the majority of the animal, clinical, and epidemiological studies have been conducted on the standardized *G. biloba* extract, EGb 761 (Heinonen and Gaus, 2015). In a study where patients were treated with the *Ginkgo* formulation, EGb 761, at oral dosage levels of 240 mg/kg per day, only a low number (7/205) stopped treatment due to adverse effects. Heinonen and Gaus (2015) evaluated 88 publications on 75 different clinical studies for *G. biloba* which was primarily the standardized *Ginkgo* leaf extract EGb761. They found that *G. biloba* is used to treat cognitive impairment in various types of dementia, peripheral arterial diseases, ischemic stroke, tinnitus, macular degeneration, and autism, and to moderate the undesirable effects of some cancer therapies. In the 75 studies this review paper evaluated, a total of 7115 patients were treated with *G. biloba* under controlled observations. There were no specific and serious undesired reactions noted and there were as many adverse events reported in the placebo groups as in the treated groups.

Gastrointestinal upset, headaches, dizziness, and bleeding are listed as adverse effects of *G. biloba* in humans (Heinonen and Gaus, 2015). Other side effects linked to *G. biloba* include allergic skin reactions and occasional anaphylaxis-like reactions following intravenous administration (De Smet, 2002). In a reporter gene mutation assay using *gpt* delta mice and a combined liver comet assay and bone marrow micronucleus assay using C3H-derived constitutive androstane receptor knockout (CARKO) and wild-type mice, there was no evidence of genotoxicity (Maeda et al., 2014). The *gpt* mice were exposed to a *Ginkgo biloba* extract at levels up to 2000 mg/kg bw/day for 13 weeks for the reporter gene mutation assays and the CARKO and wild-type mice were exposed to doses up to 2000 mg/kg by daily gavage for 3 days. Lin et al. (2014) studied the in vitro genotoxicity of *G. biloba* leaf extract using the mouse lymphoma assay and Comet assay to determine the underlying mechanisms of the genotoxicity. There was a dose-dependent increase in the mutant frequency and DNA double-strand breaks noted in the study and western blot analysis confirmed that *G. biloba* leaf extract, quercetin, and kaempferol activated the DNA damage signaling pathway with increased expression of γ-H2AX and phosphorylated Chk2 and Chk1. *G. biloba* extract also produced ROS and decreased glutathione levels in L5178Y cells. The authors also concluded that the loss of heterozygosity analysis of the mutants indicated the treatment

TABLE 48.1 Key data from the evaluated systemic review.

Drugs which interact	Type of interaction	Clinical outcomes	Mechanism of action	Overall judgment	Comments by authors
Acetaminophen	Inhibition	Bleeding, apraxia, death, hematoma, hyphemia, permanent neurological deficit	Inhibition of platelet aggregation	Some concerns	Use cautiously in patients taking anesthetics, analgesics, and anticoagulant and antiplatelet agents
Lisinopril	NA				
Anesthetics	Inhibition				
Aspirin	Inhibition				
Warfarin	NA				
Ibuprofen	Inhibition				

NA, not available.
Source: Adapted from Posadzki, P., Watson, L. and Ernst, E., 2012. Herb-drug interactions: an overview of systematic reviews, Br. J. Clin. Pharmacol. 75(3), pp. 603–618.

with *G. biloba* leaf extract, quercetin, and kaempferol resulted in extensive chromosomal damage. There were no neoplastic or preneoplastic effects reported in mice and rats following oral doses of 200 mg/kg per day and 100 mg/kg per day, respectively, in a carcinogenicity study (Heinonen and Gaus, 2015). This dose translated to 600 mg/m^2 in both the mice and rats, which is four times higher than the highest daily recommended dose for humans. The NTP carcinogenicity studies showed that the target organs of toxicity were the liver, thymus, and nose when doses of up to 1000 mg/kg and 2000 mg/kg in corn oil were tested in the rat and mouse. Dunnick and Nyska (2013) also reviewed studies conducted by the NTP which provided information on the potential for toxicity following long-term use of various herbal remedies and stated that there was no clear evidence of carcinogenicity in rats. In mice, however, there was clear evidence for carcinogenic activity in male and female mice, with a significant increase in incidences of hepatocellular adenoma, hepatocellular carcinoma, or hepatoblastoma in all dosed groups.

The potential in vitro and in vivo toxicity of the bioflavonoids in *G. biloba*, amentoflavone, bilobetin, ginkgetin, isoginkgetin, and sciadopitysin, was investigated by Li et al. (2019). These authors evaluated the cytotoxicity using HK-2 and L-02 cells and an MTT assay. For the in vivo study, BALB/c mice were dosed with either the vehicle control or one of the bioflavonoids by oral gavage once daily for 7 days at a dose of 20 mg/kg per day. On the 8th day, animals were euthanized, and blood samples were taken for serum collection and biochemical analysis. The heart, liver, spleen, lung, and kidneys were collected and fixed for histopathological examinations. The kidney was evaluated with a TUNEL assay and immunohistochemistry analysis. All five biflavoinoids reduced cell viability in the human renal tubular epithelial cells and human normal hepatocytes in a dose-dependent manner. The authors stated that these results may indicate that the biflavoinoids may have potential liver and kidney toxicity. Biochemical analysis indicated that the activity of alkaline phosphatase was significantly increased following all bioflavonoid exposure. Widespread hydropic degeneration of hepatocytes was noted in histopathological examination following ginkgetin- or bilobetin-treated animals. In addition, all five biflavonoids induced acute kidney injury and histopathology confirmed that the injury was to the tubule, glomeruli, and interstitium. The authors concluded that these five bioflavonoids from *G. biloba* have the potential to induce hepatic and renal toxicity. Yao et al. (2018) conducted a study to evaluate the metabolic characteristics and metabolic pathways of ginkgolic acids and to evaluate the hepatotoxicity in vitro. The MTT assays utilized indicated that the cytotoxicity in HepG2 cells occurred in a time- and dose-dependent manner and had less cytotoxicity in primary rat hepatocytes than in the HepG2 cells. The toxicity was also enhanced through CYP1A- and CYP3A-mediated metabolism.

G. biloba was not found to be genotoxic in vivo in multiple studies reported by Heinonen and Gaus (2015). When the findings from the NTP studies and other published studies were evaluated, the underlying mechanism of the thyroid and liver tumors reported was due to increased activity of hepatic drug-metabolizing enzymes, which leads to accelerated cell proliferation and neoplasm (Heinonen and Gaus, 2015). No evidence of carcinogenicity was observed following low doses in rodents and no indication has been reported in humans.

Diamond and Bailey (2013) reported that following an analysis of random controlled trials with adults with dementia, peripheral artery disease, or diabetes mellitus, the results showed that there was a positive effect on blood perfusion following use of *G. biloba* and there was no higher risk of bleeding demonstrated by the lack of a significant effect on adenosine 5′-diphosphate-induced platelet aggregation, fibrinogen concentration, activated partial thromboplastin time, and prothrombin time. There is a case report in the literature in which a healthy 73-year-old man presented to medical personnel with a complaint of recent episodes of spontaneous bleeding and had never been diagnosed with cognitive dysfunction (Bent et al., 2005). The patient was not on any prescription medication but took vitamin supplements and had used *G. biloba* at 756 mg/day for the previous 6 months. The product used was standardized to 27% ginkgo flavone glycosides and 10% terpene lactones. The patient had discontinued the *G. biloba* but continued with the other vitamin supplements and noted that the bleeding episodes stopped. Following discussions with his physician, the patient restarted the *G. biloba* but had monitoring laboratory tests completed including complete blood count, prothrombin time/international normalized ratio, partial thromboplastin time, and bleeding time which were all within the normal range prior to restarting the *G. biloba*. During the next 10 weeks he noticed ecchymosis but no other spontaneous bleeding. At 10 weeks after beginning the *G. biloba* again, the complete blood count and coagulation studies were normal, but the bleeding time was elevated from 5 minutes when not consuming *Gingko* to greater than 15 minutes. Bent et al. (2005) reviewed a total of 15 reports of spontaneous bleeding in patients taking *G. biloba* which were published between 1996 and 2003, the majority of which involved serious medical conditions. In eight of the 12 reports which identified duration of *Ginkgo* consumption, the patients had been taking the *Ginkgo* for more than 6 months before any report of a bleeding event, in two cases of intracranial bleeding the *Ginkgo* use was less than 2 months and in the two cases of spontaneous hyphema the use was less than 2 weeks prior to the onset of symptoms. In the majority of the cases reviewed the patients were at high risk for bleeding for reasons other than ginkgo consumption, with age being the most common. In only 6 of the 15 cases was it reported that the *Ginkgo* was stopped and none of these had recurrent bleeding, however the follow-up time was between 3 months and 4 years. In three of the cases the bleeding times were reported, and the times were shorter when the patients were not on *Ginkgo* extract and two cases had abnormally high bleeding times when taking *Ginkgo*. Bent et al. (2005) concluded that there is a possible causal association between *Ginkgo* and bleeding events and that until additional well-designed studies are completed, patients should be warned about the possible increased risk of bleeding. Koch (2005) stated that under the experimental conditions which were employed in the study, the concentrations at which the PAF-mediated aggregation of human platelets was half-maximally inhibited were generally more than 100 times higher than the peak plasma values measured after the oral consumption of EGb 761 at the recommended doses, which were between 120 and 240 mg. It was concluded that there should be serious doubts that the PAF antagonistic effect of ginkgolides are responsible for the hemorrhage documented in patients taking EGb 761.

There is disagreement between scientists, clinicians, herbalists, and consumers regarding the safety of *G. biloba* in women of reproductive age as well as during pregnancy and lactation. The German Institute for Quality and Efficiency in Health Care published a meta-analysis of clinical studies investigating the effects of EGb 761. They concluded that there was a positive effect, particularly in relation to activity in daily life and cognitive efficiency, and showed that it was well tolerated with few side effects. In a study where male Swiss albino mice were treated with different doses of an aqueous suspension of *G. biloba* (General Nutrition Corporation, USA) at 25, 50, and 100 mg/kg per day for 90 days via oral gavage, the effects on reproductive, cytological, and biochemical toxicity were examined (Al-Yahya et al., 2006). The treated animals showed significant changes in the weight of caudate epididymis, prostate, chromosomal aberrations, rate of pregnancy, and preimplantation loss, and the percent motility, sperm count, and morphology of spermatozoa were not affected. The females in the study were not exposed to *G. biloba*. The evaluation of biochemical parameters showed a depletion of nucleic acids, NP-SH, and increase of malondialdehyde. The authors state that while the exact mechanism is not known, the activation of gamma amino butyric acid, glycine, and glutamate following treatment with *G. biloba* may have generated free radicals and depleted cellular glutathione by calcium influx and membrane depolarization. The observed toxicity was attributed to the toxic constituents of the extract including ginkgolic acids, bioflavones, cardanols, cardols, BBs, and quercetin. They go on to caution that *G. biloba* should be used cautiously for impotence and/or erectile dysfunction therapy. A study in mice showed that high doses of *G. biloba* standardized extract (General Nutrition Centre) caused teratogenic lesions in mice (Zehra et al., 2010). Bred mice (day 0.5 after breeding) were administered 0, 78, or 100 mg *G. biloba* extract/kg body weight until euthanized on day 18 of pregnancy. There were no differences between groups for litter size. The body mass of the fetus was significantly decreased in the 78 and 100 mg groups. The crown–rump length was significantly decreased in the 100 mg group and facial deformities were also observed in this group. The authors considered 78 mg *G. biloba* extract/kg body weight to be consistent with the recommended human dosage. Dugoua et al. (2006) conducted a literature review of the safety of *G. biloba* extract in pregnancy and lactation. The use of *Ginkgo* extracts is not advised before delivery because of the increased risk of postpartum hemorrhage due to antiplatelet activity. The authors of this chapter are of the opinion that there are insufficient data to support the use of *Gingko* extracts during pregnancy. In addition, *Ginkgo* extracts should not be used during lactation because there is insufficient information to establish safety.

Fraunfelder (2004) reviewed cases of ocular side effects from herbal medicines and nutritional supplements. In this report, cases were identified from spontaneous reports submitted to the WHO, the FDA, and the National Registry of Drug-Induced Ocular Side Effects as well as from a review of literature using MEDLINE and PharmaNewsFeed. From the 27 spontaneous reports to the National Registry regarding *G. biloba*, there were two cases of spontaneous hyphema and seven reports of retinal hemorrhages. Some of the cases of spontaneous hyphema were linked with retrobulbar injection but some did not have any link with a procedure or trauma.

Ginkgotoxins are found predominately in the seeds of *G. biloba* and not in the leaves which are used to produce *G. biloba* extract. The ginkgotoxins are thought to be responsible for the epileptiform seizures, unconsciousness, and paralysis of the legs sometimes reported (Ude et al., 2013).

Concluding remarks and future directions

There is a large volume of data in the literature regarding *Ginkgo* with a large number of clinical trials being reported and there is a significant volume of safety data available. It is important to note that there are also significant amounts of information showing potential adverse effects with long-term consumption of *Ginkgo* nutraceuticals. A data gap exists in that the dosages used in animal and in vitro studies are often significantly higher than the dosages used in humans and the pharmacokinetics of the phytochemicals in plasma and tissues. It is well known from pharmaceutical clinical trials that the predictive value of animal data to relevance in humans is often poor, particularly if only rodent data are used. Reporting of new, well-designed clinical trials with well-characterized *G. biloba* extracts will provide additional information to support the use of this herbal preparation in humans. In addition, the available data on herb–drug interactions, not limited to *Ginkgo*, is an area that requires additional research. Future clinical trials and safety studies need to employ recognized design standards and protocols for efficacy and safety. A reduction in the variation between products in phytochemical levels needs to be addressed. The way forward most likely is multinational acceptance of manufacturing standards and enforcement. Safe and efficacious natural products are generally more affordable to the world population and the raw materials for their manufacturing are readily available.

References

Ahlemeyer, B., Krieglstein, J., 2003. Neuroprotective effects of *Ginkgo biloba* extract. Cell Mol. Life Sci. 60 (9), 1779–1792.

Al-Yahya, A.A., Al-Majed, A.A., Al-Bekairi, A.M., et al., 2006. Studies on the reproductive, cytological and biochemical toxicity of *Ginkgo Biloba* in Swiss albino mice. J. Ethnopharmacol. 107 (2), 222–228.

Bent, S., Goldberg, H., Padula, A., et al., 2005. Spontaneous bleeding associated with *Ginkgo biloba*: a case report and systematic review of the literature: a case report and systematic review of the literature. J. Gen. Intern. Med. 20 (7), 657–661.

Biber, A., 2003. Pharmacokinetics of *Ginkgo biloba* extracts. Pharmacopsychiatry 36 (Suppl. 1), S32–S37.

Birks, J. and Grimley Evans, J., 2009. *Ginkgo biloba* for cognitive impairment and dementia, *Cochrane Database Syst. Rev.* (1).

Blumenthal, M., 2015. Chinese governemtn takes action against companies producing substandard and adulterated Ginkgo extracts. HerbalGram 108, 24–26.

Cao, G., Wang, N., He, D., et al., 2019. Intestinal mucosal metabolites-guided detection of trace-level *Ginkgo biloba* extract metabolome. J. Chromatogr. A 1608, 460417.

Catlin, N.R., Collins, B.J., Auerbach, S.S., et al., 2018. How similar is similar enough? A sufficient similarity case study with *Ginkgo biloba* extract. Food Chem. Toxicol. 118, 328–339.

Chan, P.C., Xia, Q., Fu, P.P., 2007. *Ginkgo biloba* leave extract: biological, medicinal, and toxicological effects. J. Environ. Sci. Health C Environ. Carcinog. Ecotoxicol. Rev. 25 (3), 211–244.

Chen, X.W., Serag, E.S., Sneed, K.B., et al., 2011. Clinical herbal interactions with conventional drugs- from molecules to maladies. Curr. Med. Chem. 18, 4836–4850.

Chertkow, H., 2020. Emerging pharmacological therapies for mild cognitive impairment. In: Tuokko, H.A., Hultsch, D.F. (Eds.), Mild Cognitive Impairment: Inienational Perspectives. Taylor and Francis, New York, pp. 217–243.

Choi, Y.H., Chin, Y.W., Kim, Y.G., 2011. Herb-drug interactions: focus on metabolic enzymes and transporters. Arch. Pharm. Res. 34 (11), 1843–1863.

Crawford, J.M., Lal, B.K., Duran, W.N., et al., 2017. Pathophysiology of venous ulceration. J. Vasc. Surg. Venous Lymphat. Disord. 5 (4), 596–605.

De Smet, P.A., 2002. Herbal remedies. N. Engl. J. Med. 347 (25), 2046–2056.

DeFeudis, F.V., Drieu, K., 2000. *Ginkgo biloba* extract (EGb 761) and CNS functions: basic studies and clinical applications. Curr. Drug Targets 1 (1), 25–58.

Di Renzo, G., 2000. Ginkgo biloba and the central nervous system. Fitoterapia 71 (Suppl. 1), S43–S47.

Diamond, B.J., Bailey, M.R., 2013. *Ginkgo biloba*: indications, mechanisms, and safety. Psychiatr. Clin. North Am. 36 (1), 73–83.

Dugoua, J.J., Mills, E., Perri, D., et al., 2006. Safety and efficacy of Ginkgo (*Ginkgo biloba*) during pregnancy and lactation. Can. J. Clin. Pharmacol. 13 (3), e277–e284.

Dunnick, J.K., Nyska, A., 2013. The toxicity and pathology of selected dietary herbal medicines. Toxicol. Pathol. 41 (2), 374–386.

Eckert, A., 2012. Mitochondrial effects of *Ginkgo biloba* extract. Int. Psychogeriatr. 24 (Suppl. 1), S18–S20.

Eisvand, F., Razavi, B.M., Hosseinzadeh, H., 2020. The effects of *Ginkgo biloba* on metabolic syndrome: a review. Phytother. Res.

Fan, Q., Zhou, J., Wang, Y., et al., 2020. Chip-based serum proteomics approach to reveal the potential protein markers in the sub-acute stroke patients receiving the treatment of Ginkgo diterpene lactone meglumine injection. J. Ethnopharmacol. 260, 112964.

Fehske, C.J., Leuner, K., Muller, W.E., 2009. *Ginkgo biloba* extract (EGb761) influences monoaminergic neurotransmission via inhibition of NE uptake, but not MAO activity after chronic treatment. Pharmacol. Res. 60 (1), 68–73.

Feng, Z., Sun, Q., Chen, W., et al., 2019. The neuroprotective mechanisms of ginkgolides and bilobalide in cerebral ischemic injury: a literature review. Mol. Med. 25 (1), 57.

Fraunfelder, F.W., 2004. Ocular side effects from herbal medicines and nutritional supplements. Am. J. Ophthalmol. 138 (4), 639–647.

Gurley, B.J., 2012. Pharmacokinetic herb-drug interactions (part 1): origins, mechanisms, and the impact of botanical dietary supplements. Planta Med. 78 (13), 1478–1489.

Gurley, B.J., Fifer, E.K., Gardner, Z., 2012. Pharmacokinetic herb-drug interactions (part 2): drug interactions involving popular botanical dietary supplements and their clinical relevance. Planta Med. 78 (13), 1490–1514.

Heinonen, T., Gaus, W., 2015. Cross matching observations on toxicological and clinical data for the assessment of tolerability and safety of *Ginkgo biloba* leaf extract. Toxicology 327, 95–115.

Hermann, R., von Richter, O., 2012. Clinical evidence of herbal drugs as perpetrators of pharmacokinetic drug interactions. Planta Med. 78 (13), 1458–1477.

HMPC, 2015. European Union Herbal Monograph on Ginkgo biloba L. Folium. European Medicines Agency, LondonAccessed: June 5, 2020 Available at:. Available from: https://www.ema.europa.eu/en/documents/herbal-summary/ginkgo-leaf-summary-public_en.pdf.

Hoenerhoff, M.J., Pandiri, A.R., Snyder, S.A., et al., 2013. Hepatocellular carcinomas in B6C3F1 mice treated with *Ginkgo biloba* extract for two years differ from spontaneous liver tumors in cancer gene mutations and genomic pathways. Toxicol. Pathol. 41 (6), 826–841.

Houghton, P., 2009. Synergy and polyvalence: paradigms to explain the activity of herbal products. In: Mukherjee, P.K., Houghton, P.J. (Eds.), Evaluation of Herbal Medicinal Products: Perspectives on Quality, Safety and Efficacy. Pharmaceutical Press, London.

Huang, P., Zhang, L., Chai, C., et al., 2014. Effects of food and gender on the pharmacokinetics of ginkgolides A, B, C and bilobalide in rats after oral dosing with ginkgo terpene lactones extract. J. Pharm. Biomed. Anal. 100, 138–144.

Ihl, R., Bachinskaya, N., Korczyn, A.D., et al., 2011. Efficacy and safety of a once-daily formulation of *Ginkgo biloba* extract EGb 761 in dementia with neuropsychiatric features: a randomized controlled trial. Int. J. Geriatr. Psychiatry 26 (11), 1186–1194.

Kandiah, N., Ong, P.A., Yuda, T., et al., 2019. Treatment of dementia and mild cognitive impairment with or without cerebrovascular disease: expert consensus on the use of *Ginkgo biloba* extract, EGb 761(R). CNS Neurosci. Ther. 25 (2), 288–298.

Kaschel, R., 2011. Specific memory effects of *Ginkgo biloba* extract EGb 761 in middle-aged healthy volunteers. Phytomedicine 18 (14), 1202–1207.

Koch, E., 2005. Inhibition of platelet activating factor (PAF)-induced aggregation of human thrombocytes by ginkgolides: considerations on possible bleeding complications after oral intake of *Ginkgo biloba* extracts. Phytomedicine 12 (1-2), 10–16.

Kressmann, S., Muller, W.E., Blume, H.H., 2002. Pharmaceutical quality of different *Ginko biloba* brands. J. Pharm. Pharmacol. 54, 661–669.

Krieglestein, J., Ausmeier, F., El-Abhar, H., et al., 1995. Neuroprotective effects of *Ginkgo biloba* constituents. Eur. J. Pharm. Sci. 3, 39–48.

Krieglstein, J., Beck, T., Seibert, A., 1986. Influence of an extract of *Ginkgo biloba* on cerebral blood flow and metabolism. Life Sci. 39, 2327–2334.

Leistner, E., Drewke, C., 2010. *Ginkgo biloba* and Ginkgotoxin. J. Nat. Prod. 73, 86–92.

Lejri, I., Agapouda, A., Grimm, A., et al., 2019. Mitochondria- and oxidative stress-targeting substances in cognitive decline-related disorders: from molecular mechanisms to clinical evidence. Oxid. Med. Cell. Longev. 2019, 9695412.

Li, X., Huang, L., Liu, G., et al., 2020. Ginkgo diterpene lactones inhibit cerebral ischemia/reperfusion induced inflammatory response in astrocytes via TLR4/NF-kappaB pathway in rats. J. Ethnopharmacol. 249, 112365.

Li, Y.-Y., Lu, X.-Y., Sun, J.-L., et al., 2019. Potential hepatic and renal toxicity induced by the biflavonoids from *Ginkgo biloba*. Chin. J. Nat. Med. 17 (9), 672–681.

Lichota, A., Gwozdzinski, L., Gwozdzinski, K., 2019. Therapeutic potential of natural compounds in inflammation and chronic venous insufficiency. Eur. J. Med. Chem. 176, 68–91.

Lin, H., Guo, X., Zhang, S., et al., 2014. Mechanistic evaluation of *Ginkgo biloba* leaf extract-induced genotoxicity in L5178Y cells. Toxicol. Sci. 139 (2), 338–349.

Liu, J.Y., Geng, T., Duan, K., et al., 2019. Cellular pharmacokinetics and pharmacodynamics mechanisms of ginkgo diterpene lactone and its modulation of P-glycoprotein expression in human SH-SY5Y cells. Biomed. Chromatogr. 33 (12), e4692.

Maeda, J., Kijima, A., Inoue, K., et al., 2014. *In Vivo* genotoxicity of *Ginkgo biloba* extract in gpt delta mice and constitutive androstane receptor knockout mice. Toxicol. Sci. 140 (2), 298–305.

Mandery, K., Balk, B., Bujok, K., et al., 2012. Inhibition of hepatic uptake transporters by flavonoids. Eur. J. Pharm. Sci. 46 (1-2), 79–85.

Markesbery, W.R., Lovell, M.A., 2007. Damage to lipids, proteins, DNA, and RNA in mild cognitive impairment. Arch. Neurol. 64, 954–956.

Marx, W., Hockey, M., McGuinness, A.J., et al., 2020. The effect of emerging nutraceutical interventions for clinical and biological outcomes in multiple sclerosis: a systematic review. Mult. Scler. Relat. Disord. 37, 101486.

Mohanta, T.K., Tamboli, Y., Zubaidha, P.K., 2014. Phytochemical and medicinal importance of *Ginkgo biloba* L. Nat. Prod. Res. 28 (10), 746–752.

Montes, P., Ruiz-Sanchez, E., Rojas, C., et al., 2015. *Ginkgo biloba* extract 761: a review of basic studies and potential clinical use in psychiatric disorders. CNS Neurol. Disord. Drug Targets .

Muller, W.E., Heiser, J., Leuner, K., 2012. Effects of the standardized *Ginkgo biloba* extract EGb 761(R) on neuroplasticity. Int. Psychogeriatr. 24 (Suppl. 1), S21–S24.

NTP, 2020. *Testing Status of Ginkgo biloba extract M990050*. Available at: https://ntp.niehs.nih.gov/whatwestudy/testpgm/status/ts-m990050.html?utm_source = direct&utm_medium = prod&utm_campaign = ntpgolinks&utm_term = ts-m990050 (accessed 01.06.20.).

Omidkhoda, S.F., Razavi, B.M., Hosseinzadeh, H., 2019. Protective effects of *Ginkgo biloba* L. against natural toxins, chemical toxicities, and radiation: a comprehensive review. Phytother. Res. 33 (11), 2821–2840.

Piazza, S., Pacchetti, B., Fumagalli, M., et al., 2019. Comparison of two *Ginkgo biloba* L. extracts on oxidative stress and inflammation markers in human endothelial cells. Mediators Inflamm. 2019, 6173893.

Pohl, F., Kong Thoo Lin, P., 2018. The potential use of plant natural products and plant extracts with antioxidant properties for the prevention/treatment of neurodegenerative diseases: *in vitro*, *in vivo* and clinical trials. Molecules 23 (12).

Posadzki, P., Watson, L., Ernst, E., 2012. Herb-drug interactions: an overview of systematic reviews. Br. J. Clin. Pharmacol. 75 (3), 603–618.

Posadzki, P., Watson, L.K., Ernst, E., 2013. Adverse effects of herbal medicines: an overview of systematic reviews. Clin. Med. 13 (1), 7–12.

Qian, Y., Su, S., Wei, M., et al., 2019. Interactions of pharmacokinetic profiles of Ginkgotoxin and Ginkgolic acids in rat plasma after oral administration. J. Pharm. Biomed. Anal. 163, 88–94.

Rider, C.V., Nyska, A., Cora, M.C., et al., 2014. Toxicity and carcinogenicity studies of Ginkgo biloba extract in rat and mouse: liver, thyroid, and nose are targets. Toxicol. Pathol. 42 (5), 830–843.

Rimkiene, L., Kubiliene, A., Zevzikovas, A., et al., 2017. Variation in flavonoid composition and radical-scavenging activity in *Ginkgo biloba* L. due to the growth location and time of harvest. J. Food Qual. 2017, 1–8.

Sarris, J., 2018. Herbal medicines in the treatment of psychiatric disorders: 10-year updated review. Phytother. Res. 32 (7).

Shaito, A., Thuan, D.T.B., Phu, H.T., et al., 2020. Herbal medicine for cardiovascular diseases: efficacy, mechanisms, and safety. Front. Pharmacol. 11, 422.

Smith, J.V., Luo, Y., 2004. Studies on molecular mechanisms of *Ginkgo biloba* extract. Appl. Microbiol. Biotechnol. 64 (4), 465–472.

Spiegel, R., Kalla, R., Mantokoudis, G., et al., 2018. *Ginkgo biloba* extract EGb 761® alleviates neurosensory symptoms in patients with dementia: a meta-analysis of treatment effects on tinnitus and dizziness in randomized, placebo-controlled trials. Clin. Interv. Aging 13, 1121–1127.

Ude, C., Schubert-Zsilavecz, M., Wurglics, M., 2013. *Ginkgo biloba* extracts: a review of the pharmacokinetics of the active ingredients. Clin. Pharmacokinet. 52 (9), 727–749.

Unger, M., 2013. Pharmacokinetic drug interactions involving *Ginkgo biloba*. Drug Metab. Rev. 45 (3), 353–385.

Vardy, J., Dhillon, H.M., Clarke, S.J., et al., 2013. Investigation of herb-drug interactions with ginkgo biloba in women receiving hormonal treatment for early breast cancer. Springerplus 2 (1), 126.

Waidyanatha, S., Mutlu, E., Gibbs, S., et al., 2019. Systemic exposure to *Ginkgo biloba* extract in male F344/NCrl rats: relevance to humans. Food Chem. Toxicol. 131, 110586.

Wang, S., Ouyang, B., Aa, J., et al., 2016. Pharmacokinetics and tissue distribution of ginkgolide A, ginkgolide B, and ginkgolide K after intravenous infusion of ginkgo diterpene lactones in a rat model. J. Pharm. Biomed. Anal. 126, 109–116.

Wang, S.-Y., A, J.-Y., Fei, F., et al., 2017. Pharmacokinetics of the prototype and hydrolyzed carboxylic forms of ginkgolides A, B, and K administered as a ginkgo diterpene lactones meglumine injection in beagle dogs. Chin. J. Nat. Med. 15 (10), 775–784.

Wojcicki, J., Gawronska-Szklarz, B., Bieganowski, W., et al., 1995. Comparative pharmacokinetics and bioavailability of flavonoid glycosides of *Ginkgo biloba* after a single or administration of three formulations to healthy volunteers. Mater. Med. Pol. 27, 141–146.

Woron, J., Siwek, M., 2018. Unwanted effects of psychotropic drug interactions with medicinal products and diet supplements containing plant extracts. Psychiatr. Pol. 52 (6), 983–996.

Xiong, X.J., Liu, W., Yang, X.C., et al., 2014. *Ginkgo biloba* extract for essential hypertension: a systemic review. Phytomedicine 21 (10), 1131–1136.

Yao, Q.-Q., Li, L., Xu, M.-C., et al., 2018. The metabolism and hepatotoxicity of ginkgolic acid (17: 1) in vitro. Chin. J. Nat. Med 16 (11), 829–837.

Yoshitake, T., Yoshitake, S., Kehr, J., 2010. The *Ginkgo biloba* extract EGb 761 (R) and its main constituent flavonoids and ginkgolides increase extracellular dopamine levels in the rat prefrontal cortex. Br. J. Pharmacol. 159 (3), 659–668.

Zadoyan, G., Rokitta, D., Klement, S., et al., 2012. Effect of *Ginkgo biloba* special extract EGb 761(R) on human cytochrome P450 activity: a cocktail interaction study in healthy volunteers. Eur. J. Clin. Pharmacol. 68 (5), 553–560.

Zehra, U., Tahir, M., Lone, K.P., 2010. Ginkgo biloba induced malformations in mice. J. Coll. Physicians Surg. Pak. 20, 117–121.

Zeng, Z., Zhu, J., Chen, L., et al., 2013. Biosynthesis pathways of ginkgolides. Pharmacogn. Rev. 7 (13), 47–52.

Zhou, X.W., Ma, Z., Geng, T., et al., 2014. Evaluation of in vitro inhibition and induction of cytochrome P450 activities by hydrolyzed ginkgolides. J. Ethnopharmacol. 158 (Pt A), 132–139.

Zhu, P.C., Tong, Q., Zhuang, Z., et al., 2019. Ginkgolide B for myocardial ischemia/reperfusion injury: a preclinical systematic review and meta-analysis. Front. Physiol. 10, 1292.

Zöllner, T., Schwarz, M., 2013. Herbal reference standards: applications, definitions and regulatory requirements. Rev. Bras. Farmacogn. 23 (1), 1–21.

Chapter 49

Chinese ginseng

Mildred S. Yang

Characteristics of ginseng

Chinese ginseng refers to the root of a perennial plant, *Panax ginseng*, which grows in the mountains of northern China, eastern Siberia, Japan, and Korea (Gillis, 1997; Yun, 2001; Park et al., 2012). It is a slow-growing plant with one leaf that grows in the first year and one leaflet appearing in each subsequent year until the sixth year. The root of the plant has two to five ramified rootlets that also increase in size as the plant matures (Fig. 49.1). The thickness of the rhizome has been considered an important indicator of the quality and value of the ginseng (World Health Organization, 1999).

Eleven species of the genus *Panax* have been identified (Porcher et al., 2006) (Table 49.1). Asian ginseng includes *Panax notoginseng*, *Panax japonicus*, and *P. ginseng*, while North American ginseng belongs to the genus *Panax quinquefolium* (Natural Medicines Comprehensive Database, 2014; Porcher et al., 2006; McGuffin, 2000). *P. notoginseng* is commonly found in China, while *P. ginseng* is grown in areas around the borders between China and Korea (Porcher et al., 2006; Park et al., 2012; Shim et al., 2005). Current literature often regards *P. ginseng* grown in China to be Chinese ginseng, or renshen, Ginseng cultivated in China, Korea, Japan, Russia, and Germany is also known as *P. ginseng* C A Meyer, after the German–Russian botanist Carl Anton Meyer who described it in 1843 (Yun, 2001). The genus *Panax* contains the active ingredient ginsenoside. Other genera such as Siberian ginseng (*Eleutherococcus senticosus*), are not considered as ginseng as they do not contain this active ingredient (Yun, 2001). In the following discussion, ginseng will refer mainly to the species *P. ginseng* unless otherwise indicated.

The Ginseng plant

FIGURE 49.1 Illustration of ginseng from the *Compendium of Materia Medica* (Ben Cao Gang Mu).

The chemical contents of ginseng

Ginseng contains many valuable substances. Besides proteins and carbohydrates, ginseng also contains volatile oils, ginsenosides (ginseng saponins), amino acids, vitamins, and fatty acids (Jia and Zhao, 2009). Among these, ginsenosides have the highest medicinal value (Liu and Xiao, 1992; Yun, 2001; Gui and Ryu, 2014).

Ginsenosides are saponins which are derivatives of triterpene dammarane. Each ginsenoside is characterized by the position of its sugar moiety on the dammarane and triterpene rings and can be divided into three categories, oleanolic, panaxadiol, and panaxatriol, which are distributed throughout the different parts of the plant (Liu and Xiao, 1992). The most abundant saponins are protopanaxadiol (PPD, e.g., Ra, Rb, Rc, Rd, Rg3, Rh2) and protopanaxatriol (PPT, e.g., Re, Rf, Rg1, Rg2, Rh1) (Choi, 2008; World Health Organization, 2009). Ginsenoside Ro and polyacetylene ginsenoside Ro are the only oleanoic acid-type saponins identified (Zhang et al., 2002). Fig. 49.2 shows the chemical structure of ginsenosides. Ginsenosides can be found in the roots, rootlets, leaves, and flower buds (Fuzzati et al., 1999; Li et al., 2012; Kang and Kim, 2016; Shi et al., 2007; Kim et al., 2014). The amount and distribution of ginsenosides differs in the different species, with age, as well as in the different parts of the plant (Xiao et al., 2015; Xiu et al., 2019; Lee et al., 2019a,b). These profiles have been used as markers for

Nutraceuticals. DOI: https://doi.org/10.1016/B978-0-12-821038-3.00049-5

TABLE 49.1 Classification of species in genus *Panax*.

Genus	Sections	Series	Species	Common names
Panax	*Panax*	*Panax*	*P. bipinnatifidus* Seem	Zhu zishen, feather-leaf bamboo ginseng
			P. ginseng C.A. Mey	Chinese ginseng, Korean ginseng, red ginseng, renshen
			P. japonicus C.A. Mey	Japanese ginseng
			P. quinquefolius L.	American ginseng
			P. vietnamensis Ha et Grushv	Vietnamese ginseng, bamboo ginseng
			P. wangianus S.C. Sun	Narrow-leaved pseudoginseng
			P. zingiberensis C.Y. Wu et Feng	Ginger ginseng, ginger-like pseudo-ginseng
		Notoginseng	*P. notoginseng* (Burk.) F.H. Chen	Tien-qi, Sanchi ginseng, three-seven root
	Pseudoginseng		*P. pseudoginseng* Wall	Himalayan ginseng
			P. stipuleanatus H.T. Tsai et Feng	Pingpien ginseng
Trifolius			*P. trifolius* L.	Dwarf ginseng, groundnut

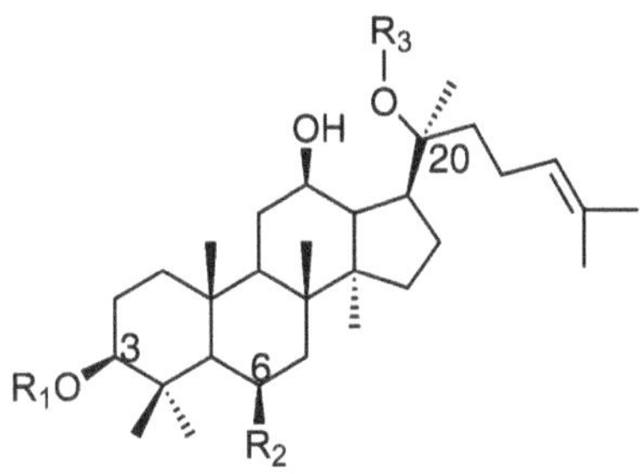

Ginsenoside	R_1	R_2	R_3
Protopanaxatriol			
Rg_1	H	-O-*glc*	*glc*
Rg_2	H	-O-*glc*(2-1)*rha*	H
Re	H	-O-*glc*(2-1)*rha*	*glc*
Rf	H	-O-*glc*(2-1)*glc*	H
Rh_1	H	-O-*glc*	H
F1	H	-OH	*glc*
Protopanaxadiol			
Rb_1	*glc*(2-1)*glc*	H	*glc*(6-1)*glc*
Rb_2	*glc*(2-1)*glc*	H	*glc*(6-1)*ara*(p)
Rg_3	*glc*(2-1)*glc*	H	H
Rh_2	*glc*	H	H
Rc	*glc*(2-1)*glc*	H	*glc*(6-1)*ara*(f)
Rd	*glc*(2-1)*glc*	H	*glc*
Ra	*glc*(2-1)*glc*	H	*glc*(6-1)*glc*
F2	*glc*	H	glc

FIGURE 49.2 Chemical structure of different ginsenosides.

identification and quality control purposes (Choi, 2008; Kim et al., 2014; Liu et al., 2009). For example, the content of PPT in the rootlets is high in *P. quinquefolium* while *P. ginseng* contains a higher level of PPD by a ratio of 2:1 (Choi, 2008; Kim et al., 2014). Ginsenoside Rf is only found in *P ginseng* but not in *P. quinquefolius* (Fuzzati, 2004). *P. quinquefolius* is found to have a higher level of the ginsenosides Rb1, Re, and Rd than *P. ginseng*. *P. ginseng* contains 38 different ginsenosides, while *P. quinquefolius* only has 19 (Choi, 2008). Furthermore, *P. ginseng* also contains more nonsaponin active substances such as compound K, acid polysaccharides, and polyethylene compounds which are considered more beneficial to health (Attele et al., 1999).

Age, root dry weight, soil fertility, light, and location of growth are factors that can affect ginsenoside contents (Li et al., 2012; Kim et al., 2014; Xiao et al., 2015; Lee et al., 2019a,b; Xiu et al., 2019). Fresh ginseng harvested at less than 4 years contains little ginsenoside (Xiao et al., 2015) and thus, is of least pharmacological interest. They can be used as cooking ingredients (Shi et al., 2007). The content of ginsenosides in roots and rootlets increases with increasing age from 1 to 5 years. The roots of *P ginseng*, harvested after 5 years or more contain around 1%–2% ginsenosides (Hong et al., 2012). The content and compositions of ginsenosides in the roots also vary with processing methods (Hong et al., 2011; Gui and Ryu, 2014; Kim et al., 2014; In et al., 2017; Xu et al., 2018). Ginseng for medicinal use is made into white and red ginseng, each with different properties (Awang and

TABLE 49.2 Ginsenosides in *Panax ginseng*.

Types of aglycone	Ginsenosides	Presence in ginseng	
		White	Red
Protopanaxadiol	Ra1, Ra2, Ra3, Rb1, Rg3, Rh2	+	++
	Rb2, Rb3, Rc, Rd, Rw1, F2, Ro	++	+
Protopanaxatriol	Rg4, Rh1	+	++
	Re, Rf, Rg1, 20(S)Rg2, Noto-R1	++	+
Malonyl ginsenosides		++	+
Other	Quinquenoside R1 Rs1, Rs2, Rk, 20(R)-Rg2, Rg5 Rg6 Rs, Rs4,	–	++
	F4, F5	+	++

–, Not detected; +, detected with relatively low concentration; ++, detected with relatively high concentration.
Source: Adopted from He, M., Huang, X., Liu, S., et al., 2018. The difference between white and Red ginseng: variations in ginsenosides and immunomodulation. Planta Med., 84, 845–854.

Li, 2008; Gui and Ryu, 2014; He et al., 2018). White ginseng is produced by peeling followed by sun-drying a 4–6-year-old ginseng root. Red ginseng, harvested at 6 years or older, requires repetitive steaming of the fresh root at 98°C–100°C for 2–3 hours, and then dried until the moisture content is reduced to below 15% (Awang and Li, 2008). Table 49.2 shows the different types of ginsenoside present in white and red ginseng.

Ginsenosides are generally acid-labile; therefore acidic hydrolysis usually occurs with other side reactions like cyclization of side chains and deglycosylation, followed by chemical transformations during steaming (Yun, 2003). The deglycosylated derivatives are reported to be more bioavailable in humans and result in red ginseng having better pharmacological functions (Hong et al., 2011; Hasegawa, 2004). Another compound, malonyl ginsenoside, including malonyl ginsenosides Rb1, Rb2, Rc, and Rd, is thermally unstable and these can be completely transformed into the corresponding ginsenosides Rb1, Rb2, Rc, and Rd after steaming (Fuzzati, 2004). Therefore malonyl ginsenosides are more abundant in white ginseng than red ginseng. Treatment at different temperatures may alter the ginsenoside profile (In et al., 2017; Xu et al., 2018).

Health benefits of ginseng

The ancient Chinese pharmacopeia, *Shen Nong Ben Cao Jing* (300 BC–200 AD), described ginseng as a root with some resemblance to the human form (Fig. 49.1) and thus it is able to nourish all body organs, stabilize the soul, calm the nerves and spirit, remove bad Qi (basic element to generate energy and power), improve eyesight, maintain a cheerful mind, and enhance cognition and intelligence. Chronic use of Chinese ginseng could help reduce body weight and prolong life span. According to traditional Chinese medicinal knowledge, American ginseng (*P. quinquefolius*) can produce a "cool" or "yin" tonic used to treat those suffering from "hot" symptoms such as stress, insomnia, palpitations, and headache, whereas Asian ginseng (*P. ginseng*) produces a tonic that is "hot" or "yang" in nature to invigorate a weakened body (Awang, 1998; Chen et al., 2008; Park et al., 2014). Ginseng is considered a panacea acting by balancing the basic elements of the body for optimal health (Awang, 1998; Attele et al., 1999; Park et al., 2005a).

As early as the 1970s, ginseng was used for treating diabetes mellitus (Yokozawa et al., 1975). Presently, ginseng and many ginsenosides have been found to modulate blood glucose levels in animal models (Ng and Yeung, 1985; Choi, 2008; Kim et al., 2011; Lai et al., 2006; Lee et al., 2007; Park et al., 2003, 2005a). Ginseng could also protect against cardiovascular diseases through regulating blood pressure (Caron et al., 2002; Kim, 2012), inhibiting platelet aggregation (Kitts and Hu, 2000), preventing the formation of atherosclerotic plaques, preventing vascular injuries (Zhou et al., 2004), modulating blood lipoprotein levels (Park et al., 2005b), preventing cancer (Nga et al., 2012) and various nervous system diseases (Ong et al., 2015; Kim et al., 2013a; Jakaria et al., 2018), which includes antidepressive (Ge et al., 2017), antistress (Kim et al., 2013b), antiepileptic (Lian et al., 2005), and antipsychotic (Chatterjee et al., 2012) activities, and improves cognitive and memory performance (Lee et al., 2009b; Geng et al., 2010; Jakaria et al., 2018). Ginseng could also improve immune function to prevent viral infection, inflammation, AIDS, allergic reactions and autoimmune diseases (Hofseth and Wargovich, 2007; Lu et al., 2009). As a functional food, ginseng could also help improve body strength in exercise and sports (Maughan et al., 2007; Murphy and Lee, 2002; Lee and Son, 2011), sexual performance (Jang et al., 2008; Lee and Son, 2011), and relieve menopausal symptoms (Yue et al., 2007; Kim et al., 2013a,b).

As ginseng, and other traditional Chinese herbs, were introduced into Western countries for managing long-term debilitating diseases, they were marketed under the category of "Chinese medicine." To comply with regulations for controlling drug use, research began, hoping to identify the active substances responsible for their therapeutic actions. Ginsenosides were considered the prime candidates. Over the years, research into ginsenosides was carried out to identify the different ginsenosides, examining their biochemical actions, and testing their therapeutic efficacies through animal experiments and clinical trials.

Biochemical actions of ginseng

Antioxidation is one of the major biochemical actions of ginsenosides (Gillis, 1997; Kitts and Hu, 2000). Experimentally, Zhang et al. (1996) found that a concentration of 0.05% ginseng extract can inhibit the conversion of 2,2-diphenyl-1-picrylhydrazyl (DPPH) to its oxidized form. The same level of ginseng extract was also able to interrupt the propagation of a free radical chain reaction. Kim et al. (2002) reported that aqueous extract of red ginseng could quench hydroxyl radical-induced DPPH oxidation. Hu and Kitts (2001) demonstrated that a 30 mg/mL ginseng extract can suppress the oxidation of human low-density lipoprotein when in the presence of Cu^{2+}, as well as inhibit the cleavage of DNA via scavenging hydroxyl radicals generated through the Fenton reaction. Ginsenosides Rb1, Rb2, Rb3, Rc, Re, Rg1, Rg2, and Rh1 all possess significant antioxidation property (Zhong and Jiang, 1997). Both ginseng extracts and the ginsenoside Rb1 alone were able to prevent breakage of the supercoiled DNA from peroxyl radical attack in a dose-dependent manner (Hu and Kitts, 2001). Ginsenosides Rb1, Rb2, Rb3, and Rc, but not Rd, at 30 mg/mL can protect against superoxide radical-induced damage of cultured cardiac myocytes (Kitts and Hu, 2000), and reactive oxygen species-induced damage to pulmonary endothelium (Gillis, 1997). Besides scavenging free radicals, ginsenosides Rb1 and Rg1 were able to alter hepatic glutathione peroxidase and superoxide dismutase activities and decrease malondialdehyde levels (Kitts and Hu, 2000). Ng and Yang (2008) showed that Re and Rg3 could alter the cellular redox state of cultured C6 glioma cells in opposite directions by changing the activities of glutathione peroxidase and γ-glutamylcystienyl synthase resulting in a change in the intracellular reduced/oxidized glutathion ratio, which serves to reflect the oxidative status of the cell. Despite the increasing evidence that supports the antioxidant function of ginsenosides, evidence that ginseng could act as antioxidant in vivo was not clear. The antioxidizing effects of *P. ginseng* on human studies were limited. Kim and Lee (2001) were unable to find a significant effect of inhibiting malondialdehyde-induced lipid peroxidation after administering 1.8 g red ginseng to smokers for 4 weeks, while Kim et al. (2011) reported a significant reduction of lipid peroxidation following administration of 1–2 g ginseng to healthy participants for the same period. It must be noted that each ginsenoside might have unique and different potency and function (Shergis et al., 2013a,b), and interactions among ginsenosides would confound the studies, making it difficult to draw any meaningful conclusions (Zhou et al., 2004).

Besides antioxidant properties, ginsenosides can also modulate signaling pathways that regulate cell proliferation (i.e., the cyclin-dependent protein kinases and cyclins, c-myc, epidermal growth factor, vascular endothelial growth factor, tumor suppressors p53 and p21, oncogenes mouse double minute 2 homolog), cell death (Bcl-2, Bcl-xL, X-linked inhibitor of apoptosis protein, caspases, and death receptors), inflammatory response molecules (NF-κB and cyclooxygenase-2), and many protein kinases (c-Jun N-terminal kinase, Akt, and AMP-activated protein kinase) (Nga et al., 2012). Due to their steroid-like structure, ginsenosides could also act through binding to hormonal receptors (Yue et al., 2007), thus achieving the action of homeostasis. In addition, Lee et al., (2019) proposed that ginsenoside Rg3 was able to protect against UV-induced aging of human dermal fibroblasts by restoring mitochondria transmembrane potential and improving ATP production. Other actions including the ability to reduce platelet adhesion, influencing ion channels, altering autonomic neurotransmitter release, and improving lipid profiles also have been reported (Kim, 2012; Lee and Kim, 2014). Ginsenosides can also activate cellular release of nitrous oxide, a potent smooth muscle relaxant, to improve physiological functions ranging from angiogenesis (Park et al., 2005b; Yue et al., 2007) to sexual performance (Murphy and Lee, 2002; Leung and Wong, 2013).

In vivo and clinical studies

While biochemical studies were conducted to reveal the underlying mechanisms of action of ginsenosides, extensive animal studies and clinical trials were also conducted in order to confirm the therapeutic efficacies. In 2014, Van Hateren (2013) published an article that documented the clinical studies of ginseng in the treatment of diabetes mellitus. Through summarizing the results of studies retrieved from Medline, PubMed, and other Korean, Japanese, and Chinese databases, the conclusion failed to show ginseng's ability to control fasting blood glucose and HbA1c in patients with type II diabetes mellitus. A similar review also failed to provide conclusive support for ginseng on the treatment of cancer, cardiovascular diseases, and neurological functions (Yang and Wu, 2016).

One major factor that literature reports failed to confirm regarding the action of ginseng in clinical studies was the lack of consistencies of the ginseng extracts used in the different experiments. Ginseng is a plant material harvested and processed crudely for use. The chemical constituents may differ between different batches grown at different locations and harvested at different times. Heterogeneity and lack of standardization of materials render the evaluation of data on human studies difficult to compare. G115 is a concentrated ethanol extract (40%v/v) of *P. ginseng* C.A. Meyer to a content of 4% major ginsenosides (Scaglione et al., 2005). The European Medical Agency for Science, Medicines, and Health reviewed clinical trials performed using G115 ginseng preparation in 2013 and found that these studies were also not conclusive because of heterogeneity of the experimental designs. A more recent review conducted by Bilia and Bergonzi (2019) further evaluated the literature results and revealed that in some double-blind, placebo-controlled studies, ginseng's effectiveness on blood glucose and lipid profile regulation, managing chronic obstructive pulmonary diseases, improving energy and physical performance, immunological and cognitive functions began to emerge.

Bioavailability of ginsenosides

Another difficulty in studying the therapeutic efficacies of ginseng is to understand its bioavailability after oral administration. Most ginsenosides are bulky hydrophilic molecules with low absorption rates (Han and Fang, 2006; Han et al., 2006). While ginsenosides may cross the intestinal membrane via the sodium-dependent glucose transporter (Xiong et al., 2009), the amount that can be detected in blood is too small to demonstrate significant therapeutic efficacies (Han and Fang, 2006; Han et al., 2006). With improving analytical techniques, it was found that in rats fed with ginseng extract, blood ginsenoside levels were quite different from those found in the ginseng extracts administered. One example was the appearance of the deglycosylated metabolite, compound K [20-O-β-D-glucopyranosyl-20-(S)protopanaxadiol], detected after rats were fed the ginsenoside Rb1. Rb1 is one of the most abundant ginsenosides in ginseng extracts (Ji et al., 2001) and is known to possess antiinflammatory, antiobesity, and neuroprotective actions (Park et al., 2005a). Analysis of ginsenosides in the blood of rats showed that while the amount of Rb1 appeared in small amounts in blood, there was a proportional increase in compound K with an increasing amount of Rb1 administered (Akao et al., 1998a; Choi et al., 2011), and the efficacies of Rb1 can be seen in nonsterile rats but not sterile rats with few intestinal microorganisms (Akao et al., 1998b). Compound K can be detected when Rb1 was exposed to intestinal microbiota. The conversion could be related to the glycosidase enzymes (Tawab et al., 2003; Hasegawa et al., 1996, 2000; Trinh et al., 2007; Kim, 2012; Noh et al., 2009; Shin et al., 2017; Kim, 2018). Another example is the metabolism of ginsenoside Rg3 as it travels along the intestinal tracts. It is believed that Rg3 could undergo hydrolysis in the stomach acid followed by the action of intestinal microbial flora (Kanaoka, 1994; Bae et al., 2002) and converted to the 20(S)-Rg3 which is known to be the more bioactive metabolite.

Besides Rb1, other ginsenosides including Rb2, Rc, and Rd could also be metabolized to compound K (Kim, 2018). Compound K has been found to possess antitumor (Oh et al., 2004), hepatoprotective (Lee et al., 2005) and antiaging (Shin et al., 2014) actions. Thus, it might be the ultimate molecule responsible for the therapeutic efficacies of ginseng. Currently, an investigation is in progress to improve the quantity and stability of compound K in ginseng products (Shin et al., 2017). As more chemicals are identified in ginseng, other novel metabolites may be discovered.

Ginseng and antiinflammation

While it remains difficult to verify the therapeutic efficacies of ginseng, recent new developments on the understanding of the cause of diseases may shed light on the mechanism of action of ginseng. Inflammation is a complex biological response fundamental to how the body deals with injury and infection. An acute inflammatory response enables the human body to eliminate the initial cause of diseases and to effectively repair the problem, while chronic inflammation can lead to gradual tissue damage and organ injuries (Hofseth and Wargovich, 2007; Arulselvan et al., 2016). It is believed that most chronic diseases are in some ways related to the occurrence of persistent low-grade inflammation leading to tissue damage and eventually, onset of systemic diseases such as diabetes mellitus (Tsalamandris et al., 2019), artherosclerosis and cardiovascular diseases (Hansson, 2005; Kaptoge et al., 2010, 2013; Katsiari et al., 2019), cancer (Hofseth and Wargovich, 2007; Munn, 2017), and neurological diseases including neuropathic pain, Alzheimer disease, Parkinson disease, multiple sclerosis, motor neuron disease, ischemia and traumatic brain injury, depression, and autism spectrum disorder (Skaper et al., 2018). Oxidative stress causes an increase in free radical production in cells, and slowly leads to injury to vital cellular components such as DNA, proteins, and membrane lipids, resulting in inflammation (Arulselvan et al., 2016). Many natural compounds commonly used for health maintenance act by balancing the body cellular oxidation state and prevent excessive free radical production and chronic inflammation (Arulselvan et al., 2016). It was also suggested that many of the conventional drugs used for treatment of diabetes mellitus (Pollack et al., 2016) or cardiovascular diseases (Ridker, 2013) may show antiinflammatory property.

Ginseng extracts, as well as many ginsenosides, have been found to possess antioxidative (Lee et al., 2017) and antiinflammatory (Riaza et al., 2019) actions. Studies of in vitro and in animal models have demonstrated that a polysaccharide extract from ginseng berry exhibits an immunostimulatory effect through enhancing natural killer cell activity which, in turn, inhibits tumor cell metastasis (Lee et al., 2019b). Song et al. (2019) found that ginseng could also suppress the immune response upon heat stress through decreasing the production of inflammatory cytokines in muscle and spleen. Hong et al. (2016) reported that ginseng could reduce fatigue in patients suffering from nonalcoholic fatty liver disease through its action in reducing proinflammatory cytokines and reverse the inflammatory reaction. Ginsenoside Rg3, Rb1, and compound K were found to slow down inflammatory action (Jia and Zhao, 2009). Joh et al. (2011) showed that compound K was able to inhibit IRAK-1 activation, which is a key step in inflammation. By using sophisticated purification technologies to identify individual constituents and define their bioactivity using genomics and bioactivity assays, Lee et al. (2009a) treated human immune cells with different extracts of ginseng and found that ginsenosides could selectively inhibit expression of the inflammatory gene CXCL-10. Kim (2018) concluded that the hydrophobic metabolites such as compound K also exhibit potential pharmacological effects such as antitumor, antiinflammatory, antidiabetic, antiallergic, and neuroprotective functions.

Toxicity of ginseng

Toxic effects

Ginseng is a relatively safe substance with few adverse effects (IARC Working Group on the Evaluation of Carcinogenic Risks to Humans, 2002) at a recommended daily dose of between 3 and 9 g of powdered ginseng root (Bucci, 2000), or 200–400 mg ginseng extract (Dunnick and Nyska, 2013). Toxic effects may appear with an increase in administered doses. A report by Siegel (1979) showed that 14 of 133 participants suffered from adverse symptoms following exposure to 15 g/day ginseng over a 2-year period. Nausea and vomiting were reported by a healthy participant 3 weeks after taking 200 mg ginseng in a study on cardiac function (Caron et al., 2002). Hypoglycemia, tachycardia, headache, blurry vision, insomnia, and irritability were reported in type II diabetic patients taking red ginseng (Kim et al., 2011). Symptoms ranging from mild discomfort to gastrointestinal problems such as nausea, vomiting, gastric upset, constipation, and diarrhea have been reported in human studies (Lee and Son, 2011). In addition, hypoglycemic symptoms have been reported in participants with type II diabetes following administration of 6 g of red ginseng for 12 weeks (Lee and Son, 2011; Vuksan et al., 2008). Other adverse events such as hyperglycemia (Sievenpiper et al., 2004) and an increase in heart rate and blood pressure (Posadzki et al., 2013a) were also reported. Nevertheless, the withdraw rate from patients enrolled in a clinical trial due to adverse events was less than 10% (Table 49.3), comparable to that from the placebo group. Other rare reports of patients experiencing severe symptoms like inflammation of the arteries in the brain (cerebral arteritis), severe skin reactions (Stevens–Johnson syndrome), inflammation of the liver (cholestatic hepatitis), and anaphylactic allergic reactions were also documented by the National Capital Poison Center (2019).

Drug–herb interaction

Ginseng is commonly used in combination with other herbs to obtain a soup or decoction in traditional Chinese medicinal practice. The different combinations can be used to manage the different types of diseases (Park et al., 2012). The synergistic actions among the different herbal ingredients is not well understood. As increasing numbers of patients are using some form of ginseng product as supplementary treatment for diseases, the interaction of ginseng with conventional drugs is of paramount importance as they may alter the bioavailability of therapeutic pharmacological agents. Induction of drug-metabolizing enzymes such as the liver cytochrome P450s and drug transporters such as the P-glycoproteins, may affect the disposition and bioavailability of the therapeutic agents (Meijermana et al., 2006; Lin and Yamazaki, 2003; Kim et al., 2018). Kim et al. (2018) showed that ginseng and ginsenosides could induce proteins and mRNA of different cytochrome P450 isoenzymes and transport proteins, but the induction may depend on the dose of ginseng relative to the drug administered.

The wide range of known health benefits of ginseng may also pose some concern for coadministration of ginseng with conventional foods and drugs. Posadzki et al. (2013b) showed that ginseng can interact synergistically with drugs that are used for treatment of depression and diabetic mellitus, some hormones, and those for lowering

TABLE 49.3 Reports of adverse effects.

References	Cognition	Exercise	Sex-related functions	Glucose metabolism	CVD	Cancer	Others
Jang et al. (2008)			5 (7)[a] studies, 337M, 12 AE[b]				
Lee et al. (2009b)	2 (2) study, 88 M and F, 16 AE						
Kim et al. (2011)				1 (4) studies, 19 M and F, 1 AE			
Lee and Son (2011)	2 (8) studies, 158 M and F, 7 AE	3 (8) studies, 88 M and F, 6 AE	4 (7) studies, 190 M, 6 AE	2 (7) studies, 58 M and F, 1 AE	4 (5) studies, 590 M and F, 1 AE. 1 (2) study for cerebrovascular, 10 M, 0 AE.		1(2) study for pulmonary, 100 M & F, 0 AE
Choi et al. (2013)	1 (1) study, 50 M and F, 0 AE		4 (6) studies, 119 M, 14 AE	0 (2) study, 34 M and F	0 (1) study, unknown subject number. 0 (1) study for metabolic syndrome, 23 M and F	1 (4) study, 23 M and F; 0 AE	1 (1) study for weight management, 45 F, 0 AE. 1 (1) study for dry mouth, 90 M and F; 7 AE. 0 (1) study for glaucoma, 18 M and F. 0 (1) study for gastric problem, 67 M and F
Kim et al. (2013a,b)			3 (4) studies, 442 F, 131 AE				
Shergis et al. (2013a)	6 (17) studies, 232 M and F, 12 AE. 1 (4) study for quality of life, 240 M and F, 6 AE	8 (10) studies, 137 M and F, 10 AE	4 (4) studies, 180 M, 7 AE. 2 (3) studies, 82 F, 9 AE.	4 (9) studies, 99 M and F, 4 AE	7 (8) studies, 160 M and F, 8 AE	2 (2) studies, 347 M and F, 19 AE	3 (5) studies for respiratory, 198 M and F, 9 AE. 0 (5) study for immunomodulation, 69 M and F. 1 (2) study, 55 M and F, 3 AE

AE, adverse effect; *CVD*, cardiovascular diseases; *F*, female; *M*, male.

[a] *n (N) number with adverse effects reported in (number of studies).*
[b] *AE not specified from ginseng and control group.*

lipids. It can antagonize the action of anticoagulants, antiplatelet agents, and diuretics. Ginsenoside Rh2 was able to sensitize breast cancer cells during therapeutic treatment with paclitaxel due to ginseng's ability to induce the exporter P-glycoprotein (Szakacs et al., 2006). However, the poor oral absorption of ginsenoside Rh2 may limit its use as a chemoprevention agent (Yang et al., 2011). Ginseng can act synergistically and antagonistically with antiarrhythmic agents by interfering with calcium channel blockers. These interactions, however, may only raise minor concerns rather than a serious threat according to the limited data available (Qi et al., 2011). Also, because of the stimulating effects of ginseng, caution must also be taken upon coadministration with food containing caffeine or alcohol (National Capitol Poison Center, 2019). Table 49.4 shows some interactions between ginseng and agents used in the treatments of some diseases.

Quality control

Due to the high demand, wild ginseng has become rare (Changbai Mountain Academy of Sciences, 2009). Nowadays, ginseng is mainly cultivated in northeastern China, Korea, and Japan (Park et al., 2005a,b; Tang and Eisenbrand, 1992; World Health Organization, 1999). To ensure that cultivated ginseng can meet the global market demands, pesticides and herbicides are used to minimize possible losses from insect infestations and weeds. A list of pesticide residues was detected in ginseng-containing dietary supplements (Durgnat et al., 2005; Food and Drug Administration, 2008) in a pesticide residue monitoring program in 2003 and 2004 in the United States. Twenty-six of 49 ginseng samples purchased over the Internet or at a retail establishment contained violative residues. A number of pesticides were detected in a Korean domestic product made from 4–6-year-old ginseng grown in four different agricultural areas of Jeonnam province of South Korea, but only a few, including Tolclofos-m, were declared (Park et al., 2007). The US FDA also found that 22 of 31 imported ginseng samples originating from Canada, Hong Kong, China, and South Korea in the form of dried root, ground or powdered root, liquid or powdered extracts or concentrates, or finished products shipped in bulk for repacking in the United States, all contained violative pesticide residues. Pentachloroaniline, quintozene, and pentachlorobenzene were at the top of the list in both domestic and imported samples (Table 49.5). Their levels were comparable in both the domestic and imported products (Food and Drug Administration, 2008). Other contaminants also include trace metals (Durgnat et al., 2005) and aflatoxin (D'Ovidio et al., 2006).

Adulteration is another important issue in the ginseng market. In October 2014, the US FDA issued a warning against a ginseng product "Ginseng KianpiPil" (Food and Drug Administration, 2014). The product claimed to contain ginseng root oil as the active ingredient to relieve fatigue, improve the natural resistance of the body against diseases, and develop an ideal physique (U.S. National Library of Medicine, 2014), all consistent with the beneficial effect of ginseng on the central nervous system, immune function, and weight management. Analysis from an FDA laboratory, however, also detected dexamethasone (a synthetic corticosteroid) and cyproheptadine (an antihistamine drug) in the product that were not declared on the label (Food and Drug Administration, 2014).

Ginseng as a nutraceutical

In Asia and throughout the world, *P. ginseng* has been considered one of the most precious herbs. It has a wide scope of medicinal actions and is relatively safe. It can be obtained over-the-counter without prescription. The Chinese Food and Drug Administration classifies ginseng as "food that can be used as medicine." While on-going efforts are still underway to identify some active compounds that will enable this herb to satisfy the Western perspective of nutraceuticals, it can be used safely by patients who believe that the herb can benefit their health

TABLE 49.4 Interactions between *Panax ginseng* and synthetic drugs.

Class of medication	Synergism	Antagonism
Antiarrhythmias	*	*
Anticoagulants and antiplatelet agents		*
Antidepressant	*	
Stimulants (caffeine)	*	
Antidiabetics	*	
Diuretics		*
Hormonal agents	*	
Lipid-lowering agents	*	
Immunomodulation	*	*

Source: Adopted from Posadzki, P., Watson, L., and Ernst, E., 2013b. Herb-drug interactions: an overview of systematic reviews. Br. J. Clin. Pharmacol., 75, 603–618.

TABLE 49.5 Violative residues in domestic and imported ginseng in United States.

Violative residues	Domestic (26[a])	Imported (22[a])
	No. of samples detected to contain the violative residue	
Qunitozene	20	16
Pentachloroaniline	16	15
Total Benzene Hexachloride (BHCs alpha, beta, delta lindane)	13	12
Pentachlorobenzene	12	15
Pentachlorophenyl methyl sulfide	11	10
Tecnazene	10	7
HCB (hexachlorobenzene)	12	9
2,3,5,6-Tetrachloroaniline	8	7
1,2,3,5-Tetrachlorobenzene	2	1
Others:		
Difenoconazole		
Chinaymidone		
Dichlorodiphenyldichloro ethylene (DDE)		
Fluvalinate		

[a]*Total number of samples tested.*
Source: Adapted from Food and Drug Administration, 2008. Ginseng dietary supplements special survey (CFSAN assignments and field regulatory monitoring for FY 2003 and FY 2004). Available from: <http://www.fda.gov/Food/FoodborneIllnessContaminants/Pesticides/ucm116752.htm#attach1>. (Accessed 11 February 2015).

and improve their quality of life (Chen et al., 2008; Choi et al., 2013; Coleman et al., 2003). According to Market Watch, as published in August 2019, the global Ginseng Supplements market was valued at 2660 million USD in 2017 and is expected to reach 4140 million USD by the end of 2023, growing at an annual compound rate of 7.7% (Market watch press release, 2019). Around 100 ginseng products are advertised for sales in the form of tablets, powder, drinks, and whole root through Amazon on-line purchasing. Ginseng is now approved for use for its potential antioxidative and antiinflammatory actions, possible benefit on improving brain functions, erectile dysfunction, boosting the immune system, and possessing potential benefits against cancer, fighting tiredness, increasing energy levels, as well as possibly lowering blood sugar. The safe dose of ginseng consumed can be up to 9 g/day for 12 weeks for alleviating Alzheimer disease, 1–3 g/day in two to three divided doses for up to 12 weeks for health maintenance in women, and to improve sexual function in men. For concentrated products (e.g., G115 or microionized preparations), smaller doses (200 mg/day) should be used. The product is safe for adults and the elderly, but caution must be taken when used by pregnant women and children (Seely et al., 2008).

Concluding remarks and future directions

While ginseng has been used for thousands of years in Eastern cultures, scientific research has only began to demonstrate its medical efficacies since its introduction to the Western world. Ginseng contains many chemical substances, among which, ginsenosides are the major active ingredients. The diversity in chemical structure and biological activities has led to some challenges. The design of standard ginseng products, such as G115, is now considered a reference for commercial use (European Medical Agencies, 2013). G115 is produced from *P. ginseng* concentrated to one containing 4% ginsenosides. However, it is still not clear whether the ginsenoside content is consistent among different batches harvested at different times and from different places. In order to meet the market requirement, Kim et al. (2014) had attempted to use a hydroponic culture technique to produce pesticide-free and quality product containing predictable ginsenoside composition under a controlled environment.

Another new development of ginseng research is based on the advancements in analytical technologies from which the minor component, compound K, was found to be the molecule responsible for ginseng's medicinal actions. For that, applications of bioengineering to improve the plant's ability to produce ginseng with elevated amount of compound K (Noh et al., 2009), or by microorganisms through fermentation (Trinh et al., 2007; Piao et al., 2011; Shin et al., 2017), are also underway. Finally, with the general approval of ginseng as a dietary supplement, firm regulations must also be instituted within the pharmaceutical industry to prevent unscrupulous manufacturers from adding, substituting, or adulterating unwanted drugs or chemicals to achieve the required function of ginseng.

References

Akao, T., Kanaoka, M., Kobashi, K., 1998a. Appearance of compound K, a major metabolite of ginsenoside Rb1 by intestinal bacteria, in rat plasma after oral administration—measurement of compound K by enzyme immunoassay. Biol. Pharm. Bull. 21 (1998), 245–249.

Akao, T., Kida, H., Kanaoka, M., et al., 1998b. Intestinal bacterial hydrolysis is required for the appearance of compound K in rat plasma after oral administration of ginsenoside Rb1 from *Panax ginseng*. J. Pharm. Pharmacol. 50 (1998), 1155–1160.

Arulselvan, P., Fard, M.T., Tan, W.S., et al., 2016. Role of antioxidants and natural products in inflammation. Oxi. Med. Cell. Longev. 2016, Article ID 5276130, 15 pages <https://doi.org/10.1155/2016/5276130>.

Attele, A.S., Wu, J.A., Yuan, C.S., 1999. Ginseng pharmacology: multiple constituents and multiple actions. Biochem. Pharmacol. 58 (11), 1685–1693.

Awang, D.V.C., 1998. The anti-stress potential of North American ginseng. J. Herbs Spices Med. Plants 6, 87–91.

Awang, D.V.C., Li, M.Z.C., 2008. The pharmacologically active constituents of white and red ginseng root. HerbalGram 80, 38–49.

Bae, E.A., Han, M.J., Choo, M.K., Park, S.Y., Kim, D.H., 2002. Metabolism of 20 (S)-and 20 (R)-ginsenoside Rg3 by human intestinal bacteria and its relation to in vitro biological activities. Biol. Pharm. Bull. 25, 58–63.

Bilia, A., Bergonzi, M., 2019. The G115 standardized ginseng extract: an example for safety, efficacy, and quality of an herbal medicine. J. Ginseng Res. 44, 179–193. <https://doi.org/10.1016/j.jgr.2019.06.003>.

Bucci, L.R., 2000. Selected herbals and human exercise performance. Am. J. Clin. Nutr. 72 (2 Suppl), 624s–636s.

Caron, M.F., Hotsko, A.L., Robertson, S., et al., 2002. Electrocardiographic and hemodynamic effects of *Panax ginseng*. Ann. Pharmacother. 36, 758–763.

Changbai Mountain Academy of Sciences, 2009. Impacts and Countermeasures of Global Climate Changes on Alpine Areas. UNESCO, Jilin province.

Chatterjee, M., Singh, S., Kumari, R.G., et al., 2012. Evaluation of the antipsychotic potential of *Panax quinquefolium* in ketamine induced experimental psychosis model in mice. Neurochem. Res. 37, 759–770.

Chen, C.F., Chiu, W.F., Zhang, J.T., 2008. Comparison of the pharmacological effects of *Panax ginseng* and *Panax quinquefolium*. Acta Pharmacol. Sin. 29, 1103–1108.

Choi, K.T., 2008. Botanical characteristics, pharmacological effects and medicinal components of Korean *Panax ginseng* C A Meyer. Acta Pharmacol. Sin. 29, 1109–1118.

Choi, J.R., Hong, S.W., Kim, Y., et al., 2011. Metabolic activities of ginseng and its constituents, ginsenoside rb1 and rg1, by human intestinal microflora. J. Ginseng Res. 35, 301–307.

Choi, J., Kim, T.H., Choi, T.Y., et al., 2013. Ginseng for health care: a systematic review of randomized controlled trials in Korean literature. PLoS One 8, e59978.

Coleman, C.I., Hebert, J.H., Reddy, P., 2003. The effects of *Panax ginseng* on quality of life. J. Clin. Pharm. Ther. 28, 5–15.

D'Ovidio, K., Trucksess, M., Weaver, C., et al., 2006. Aflatoxins in ginseng roots. Food Addit. Contam. 23 (2), 174–180.

Dunnick, J.K., Nyska, A., 2013. The toxicity and pathology of selected dietary herbal medicines. Toxicol. Pathol. 41, 374–386.

Durgnat, J.M., Heuser, J., Andrey, D., et al., 2005. Quality and safety assessment of ginseng extracts by determination of the contents of pesticides and metals. Food Addit. Contam. 22 (12), 1224–1230.

European Medical Agencies, 2013. Assessment Report on Panax ginseng C.A. Meyer, Radix. Available from: EMA/HMPC/321232/2012.

Food and Drug Administration, 2008. Ginseng dietary supplements special survey (CFSAN assignments and field regulatory monitoring for FY 2003 and FY 2004). Available from: <http://www.fda.gov/Food/FoodborneIllnessContaminants/Pesticides/ucm116752.htm#attach1>. (Accessed 11 February 2015).

Food and Drug Administration, 2014. Ginseng Kianpi Pil: Public Notification - Undeclared Drug Ingredients. Available from: <http://www.fda.gov/safety/medwatch/safetyinformation/safetyalertsforhumanmedicalproducts/ucm421303.htm>. (Accessed 2 November 2015).

Fuzzati, N., 2004. Analysis methods of ginsenosides. J. Chromatogr. B 812 (1–2), 119–133.

Fuzzati, N., Gabetta, B., Jayakar, K., et al., 1999. Liquid chromatography-electrospray mass spectrometric identification of ginsenosides in *Panas ginseng* roots. J. Chromatogr. A 854, 69–79.

Ge, W., Li, H., Zhao, Y., et al., 2017. Study on antidepressant activity of sesquiterpenoids from ginseng root. J. Funct. Foods 33, 261–267.

Geng, J., Dong, J., Ni, H., et al., 2010. Ginseng for cognition. Cochrane Database Syst. Rev. (12), Cd007769.

Gillis, C.N., 1997. Panax ginseng pharmacology: a nitric oxide link? Biochem. Pharmacol. 54 (1), 1–8.

Gui, Y., Ryu, G.-H., 2014. Effects of extrusion cooking on physicochemical properties of white and red ginseng (powder). J. Ginseng Res. 38, 146–153.

Han, M., Fang, X.L., 2006. Difference in oral absorption of ginsenoside Rg1 between in vitro and in vivo models. Acta Pharmacol. Sin. 27, 499–505.

Han, M., Sha, X., Wu, Y., et al., 2006. Oral absorption of ginsenoside Rb1 using in vitro and in vivo models. Planta Med. 72, 398–404.

Hansson, G.K., 2005. Inflammation, atherosclerosis, and coronary artery disease. N. Engl. J. Med. 352, 1685–1695.

Hasegawa, H., 2004. Proof of the mysterious efficacy of ginseng: basic and clinical trials: metabolic activation of ginsenoside: deglycosylation by intestinal bacteria and esterification with fatty acid. J. Pharmacol. Sci. 95, 153–157.

Hasegawa, H., Sung, J.H., Matsumiya, S., et al., 1996. Main ginseng metabolites formed by intestinal bacteria. Planta Med. 62, 453–455.

Hasegawa, H., Lee, K.S., Nagaoka, T., et al., 2000. Pharmacokinetics of ginsenoside deglycosylated by intestinal bacteria and its transformation to biologically active fatty acid esters. Biol. Pharm. Bull. 23 (2000), 298–304.

He, M., Huang, X., Liu, S., et al., 2018. The difference between white and Red ginseng: variations in ginsenosides and immunomodulation. Planta Med. 84, 845–854.

Hofseth, L.J., Wargovich, M.J., 2007. Inflammation, cancer, and targets of ginseng. J. Nutr. 137 (1), 183S–185S. January 2007.

Hong, S.Y., Oh, J.H., Lee, I., 2011. Simultaneous enrichment of deglycosylated ginsenosides and monacolin K in red ginseng by fermentation with *Monascus pilosus*. Biosci. Biotechnol. Biochem. 75, 1490–1495.

Hong, H.D., Cho, C.W., Kim, Y.C., et al., 2012. Morphological characteristics of Korean dried ginseng products. J. Ginseng Res. 36, 314–321.

Hong, M., Lee, Y.H., Kim, S., et al., 2016. Anti-inflammatory and antifatigue effect of Korean Red Ginseng in patients with nonalcoholic fatty liver disease. J. Ginseng Res. 40, 203–210.

Hu, C., Kitts, D.D., 2001. Free radical scavenging capacity as related to antioxidant activity and ginsenoside composition of Asian and North American ginseng extracts. J. Am. Oil Chemists' Soc. 78, 249–255.

IARC Working Group on the Evaluation of Carcinogenic Risks to Humans, 2002. Some traditional herbal medicines, some mycotoxins, naphthalene and styrene. IARC Monogr. Eval. Carcinog. Risks Hum. 82, 1–556.

In, G., Ahn, N., Bae, B., et al., 2017. In situ analysis of chemical components induced by steaming between fresh ginseng, steamed ginseng, and red ginseng. J. Ginseng Res. 41, 361–369.

Jakaria, M., Haque, M.E., Kim, J., et al., 2018. Active ginseng components in cognitive impairment: therapeutic potential and prospects for delivery and clinical study. Oncotarget 9, 33601–33620.

Jang, D.J., Lee, M.S., Shin, B.C., et al., 2008. Red ginseng for treating erectile dysfunction: a systematic review. Br. J. Clin. Pharmacol. 66, 444–450.

Ji, Q.C., Harkey, M.R., Henderson, G.L., et al., 2001. Quantitative determination of ginsenosides by high-perfromance liquid chromatography-tandem mass spectrometry. Phytochem. Anal. 12, 320–326.

Jia, L., Zhao, Y., 2009. Current evaluation of the milleninum phytomedicine – ginseng (I): etymology, pharmacognesy, phytochemistry, market and regulation. Curr. Med. Chem. 16, 2475–2484.

Joh, E.H., Lee, I.A., Jung, I.H., et al., 2011. Ginsenoside Rb1 and its metabolite compound K inhibit IRAK-1 activation—the key step of inflammation. Biochem. Pharmacol. 82, 278–286.

Kanaoka, M., 1994. Metabolism of ginseng saponins, ginsenosides, by human intestinal bacteria. J. Trad. Med. 11, 241–245.

Kang, O., Kim, J., 2016. Comparison of ginsenoside contents in different parts of Korean Ginseng (*P.ginseng* C.A. Meyer). Prev. Nutr. Food Sci. 21, 389–392.

Kaptoge, S., Di Angelantonio, E., Lowe, G., et al., 2010. Emerging risk factors collaboration: C-reactive protein concentration and risk of coronary heart disease, stroke, and mortality: an individual participant meta-analysis. Lancet 375, 132–140.

Kaptoge, S., Seshasai, S.R.K., Gao, P., et al., 2013. Inflammatory cytokines and risk of coronary heart disease: mew prospective study and updated meta-analysis. Eur. Heart J. 35, 578–589.

Katsiari, C.G., Bogdanos, D.P., Sakkas, L.T., 2019. Inflammation and cardiovascular disease. World J. Transl. Med. 8, 1–8.

Kim, J.H., 2012. Cardiovascular diseases and *Panax ginseng*: a review on molecular mechanisms and medical applications. J. Ginseng Res. 36, 16–26.

Kim, D., 2018. Gut microbiota-mediated pharmacokinetics of ginseng saponins. J. Ginseng Res. 42, 255–263.

Kim, H.S., Lee, B.M., 2001. Protective effects of antioxidant supplementation on plasma lipid peroxidation in smokers. J. Toxicol. Environ. Health A 63, 583–598.

Kim, Y.K., Guo, Q., Packer, L., 2002. Free radical scavenging activity of red ginseng aqueous extracts. Toxicology 172, 149–156.

Kim, S., Shin, B.C., Lee, M.S., et al., 2011. Red ginseng for type 2 diabetes mellitus: a systematic review of randomized controlled trials. Chin. J. Integr. Med. 17, 937–944.

Kim, H.J., Kim, P., Shin, C.Y., 2013a. A comprehensive review of the therapeutic and pharmacological effects of ginseng and ginsenosides in central nervous system. J. Ginseng Res. 37, 8–29.

Kim, E.H., Kim, I.H., Ha, J.A., et al., 2013b. Antistress effect of red ginseng in brain cells is mediated by TACE repression via PADI4. J. Ginseng Res. 37, 315–323.

Kim, Y., Jion, J., Jang, M., et al., 2014. Ginsenoside profiles and related gene expression during foliation in *Panax ginseng* Meyer. J. Ginseng Res. 38, 66–72.

Kim, S., Choi, S., Kim, M., et al., 2018. Effect of Korean Red Ginseng extracts on drug-drug interactions. J. Ginseng Res. 42, 370–378.

Kitts, D., Hu, C., 2000. Efficacy and safety of ginseng. Public Health Nutr. 3, 473–485.

Lai, D.M., Tu, Y.K., Liu, I.M., et al., 2006. Mediation of beta-endorphin by ginsenoside Rh2 to lower plasma glucose in streptozotocin-induced diabetic rats. Planta Med. 72, 9–13.

Lee, N.H., Son, C.G., 2011. Systematic review of randomized controlled trials evaluating the efficacy and safety of ginseng. J. Acupunct. Meridian Stud. 4, 85–97.

Lee, C.H., Kim, J.H., 2014. A review on the medicinal potentials of ginseng and ginsenosides on cardiovascular diseases. J. Ginseng Res. 38, 161–166.

Lee, H.U., Bae, E.A., Han, M.U., et al., 2005. Hepatoprotective effect of ginsenoside Rb1 and compound K on tert-butyl hydroperoxide-induced liver injury. Liv. Int. 25, 1069–1073.

Lee, W.K., Kao, S.T., Liu, I.M., et al., 2007. Ginsenoside Rh2 is one of the active principles of *Panax ginseng* root to improve insulin sensitivity in fructose-rich chow-fed rats. Horm. Metab. Res. 39, 347–354.

Lee, D.C.W., Yang, C.L.H., Chik, S.C.C., et al., 2009a. Bioactivity-guided identification and cell signaling technology to delineate the immunomodulatory effects of *Panax ginseng* on human promonocytic U937 cells. J. Transl. Med. 7, 34–43. 2009.

Lee, M.S., Yang, E.J., Kim, J.I., et al., 2009b. Ginseng for cognitive function in Alzheimer's disease: a systematic review. J. Alzheimers Dis. 18, 339–344.

Lee, Y., Yoon, H., Park, H., et al., 2017. Implications of red *Panax ginseng* in oxidative stress associated chronic diseases. J. Ginseng Res. 41, 113–119.

Lee, H., Jeong, J., Alves, A.C., et al., 2019a. Metabolomic understanding of intrinsic physiology in *Panax ginseng* during whole growing seasons. J. Ginseng Res. 43, 654–665.

Lee, D.Y., Park, C.W., Lee, S.J., et al., 2019b. Immunostimulating and antimestastatic effects of polysaccharides purified from ginseng berry. Am. J. Clin. Med. 47, 823–839.

Lee, H., Hong, Y., Tran, Q., Cho, H., Kim, M., Kim, C., et al., 2019. A new role for the ginsenoside RG3 in antiaging via mitochondria function in ultraviolet-irradiated human dermal fibroblasts. J Ginseng Res. 43, 431–441.

Leung, K.W., Wong, A.S.T., 2013. Ginseng and male reproductive function. Spermatogenesis 3 (3), e26391. Available from: https://doi.org/10.4161/spmg.26391.

Li, X., Yan, Y., Jin, X., et al., 2012. Ginsenoside content in the leaves and roots of *Panax ginseng* at different ages. Life Sci. J. 9, 679–683.

Lian, X.Y., Zhang, Z.Z., Stringer, J.L., 2005. Anticonvulsant activity of ginseng on seizures induced by chemical convulsants. Epilepsia 46, 15–22.

Lin, J.H., Yamazaki, M., 2003. Role of P-glycoprotein in pharmacokinetics: clinical implications. Clin. Pharmacokinet. 42 (1), 59–98. 2003.

Liu, C.X., Xiao, P.G., 1992. Recent advances on ginseng research in China. J. Ethnopharmacol. 36, 27–38.

Liu, H., Yang, J., Du, F., et al., 2009. Absorption and disposition of ginsenosides after oral administration of *Panax notoginseng* extract to rats. Drug Metab. Dispos. 37, 2290–2298.

Lu, J.M., Yao, Q.Z., Chen, C.Y., 2009. Ginseng compounds: an update on their molecular mechanisms and medical applications. Curr. Vasc. Pharmacol. 7, 293–302.

Maughan, R.J., Depiesse, F., Geyer, H., 2007. The use of dietary supplements by athletes. J. Sports Sci. 25 (Suppl 1), S103–S113.

Market watch press release: Ginseng Supplements Market, 2019. Worldwide industry share, size, gross margin, trend, future demand, analysis by top leading player and forecast till 2024. Published: Aug 19, 1:44 a.m. ET <https://www.marketwatch.com/press-release/ginseng-supplements-market-2019-worldwide-industry-share-size-gross-margin-trend-future-demand-analysis-by-top-leading-player-and-forecast-till-2024-2019-08-19>.

McGuffin, M., 2000. The American Herbal Products Association's Herbs of Commerce. American Herbal Products Association.

Meijermana, I., Beijnen, J.H., Schellens, J.H.M., 2006. Herb-drug interactions in oncology: focus on mechanisms of induction. Oncologist 11, 742–752.

Munn, L., 2017. Cancer and inflammation. Wiley Interdiscip. Rev. Syst. Biol. Med. 9 (2). Available from: https://doi.org/10.1002/wsbm.1370.

Murphy, L.L., Lee, T.J., 2002. Ginseng, sex behavior, and nitric oxide. Ann. N. Y. Acad. Sci. 962, 372–377.

National Capital Poison Center, 2019. Side effects of ginseng supplements. <https://www.poison.org/articles/side-effects-of-ginseng-supplements-191>.

Natural Medicines Comprehensive Database, 2014. Ginseng, Panax. Available from: <http://www.nlm.nih.gov/medlineplus/druginfo/natural/1000.html>. (Accessed 25 January, 2015).

Ng, T.B., Yeung, H.W., 1985. Hypoglycemic constituents of *Panax ginseng*. Gen. Pharmacol. 16, 549–552.

Ng, W.Y., Yang, M.S., 2008. Effects of ginsenosides Re and Rg3 on intracellular redox state and cell proliferation in C6 glioma cells. Chin. Med. 3, 8–16.

Nga, S.A., Qin, J.J., Wang, W., et al., 2012. Ginsenosides as anticancer agents: in vitro and in vivo activities, structure–activity relationships, and molecular mechanisms of action. J. Front. Pharmacol. 3, 25.

Noh, K., Son, J., Kim, H., Oh, D., 2009. Ginsenosdie compound K production from ginseng root extract by a thermostable β-glycosidase from *Sulfolobus solfataricus*. Biosci. Biotechnol. Biochem. 73, 316–323.

Oh, S.H., Yin, H.Q., Lee, B.H., 2004. Role of the Fas/Fas ligand death receptor pathway in ginseng sponin metabolite-induced apoptosis in HeptG2 cells. Arch. Pharm. Res. 27, 402–406.

Ong, W.Y., Farooqui, T., Koh, H.L., et al., 2015. Protective effects of ginseng on neurological disorders. Front. Aging Neurosci. 7, 129.

Park, K.S., Ko, S.K., Chung, S.H., 2003. Comparisons of antidiabetic effect between Ginseng Radix Alba, Ginseng Radix Rubra and Panax Quinquefoli Radix in MLD STZ-induced diabetic rats. J. Ginseng Res. 27, 56–61.

Park, J., Rhee, D., Lee, Y., 2005a. Biological activities and chemistry of saponins from *Panax ginseng* C. A. Meyer. Phytochem. Rev. 4, 159–175.

Park, E.K., Shin, Y.W., Lee, H.U., et al., 2005b. Inhibitory effect of ginsenoside Rb1 and compound K on NO and prostaglandin E2 biosyntheses of RAW264.7 cells induced by lipopolysaccharide. Biol. Pharm. Bull. 28, 652–656.

Park, Y.S., Abd El-Aty, A.M., Choi, J.H., et al., 2007. Pesticide multiresidue analysis in *Panax ginseng* (C. A. Meyer) by solid-phase extraction and gas chromatography with electron capture and nitrogen-phosphorus detection. Biomed. Chromatogr. 21, 29–39.

Park, H.J., Kim, D.H., Park, S.J., et al., 2012. Ginseng in traditional herbal prescriptions. J. Ginseng Res. 36, 225–241.

Park, E.Y., Kim, M.H., Kim, E.H., et al., 2014. Efficacy comparison of Korean ginseng and American ginseng on body temperature and metabolic parameters. Am. J. Chin. Med. 42 (1), 173–187.

Piao, J.Y., Kim, Y.J., Quan, L.H., et al., 2011. Bioconversion of ginsenoside Rb1 to compound K using *Leuconostoc lactis* DC201. Korean J. Plant. Res. 24, 712–718.

Pollack, R.M., Donath, M.Y., LeRoth, D., et al., 2016. Anti-inflammatory agents in the treatment of diabetes and its vascular complications. Diabetes Care 39, S244–S252.

Porcher M.H. et al., 2006. Sorting panax names. Multilingual multiscript plant name database – a work in progress. Available from: <http://www.plantnames.unimelb.edu.au/Sorting/Panax.html> (1995-2020, 29/04/2006). (Accessed 25 January, 2015).

Posadzki, P., Watson, L.K., Ernst, E., 2013a. Adverse effects of herbal medicines: an overview of systematic reviews. Clin. Med. 13, 7–12.

Posadzki, P., Watson, L., Ernst, E., 2013b. Herb-drug interactions: an overview of systematic reviews. Br. J. Clin. Pharmacol. 75, 603–618.

Qi, L.W., Wang, C.Z., Du, G.J., et al., 2011. Metabolism of ginseng and its interactions with drugs. Curr. Drug Metab. 12, 818–822.

Riaza, M., Rahman, N.U., Zia-Ul-Haq, M., 2019. Ginseng: a dietary supplement as immune-modulator in various diseases. Trends Food Sci. Technol. 83, 12–30.

Ridker, P.M., 2013. Targeting inflammatory pathways for the treatment of cardiovascular disease. Eur. Heart J. 35, 540–543.

Scaglione, F., Pannacci, M., Petrini, O., 2005. The standardised G115® *Panax ginseng* C.A. Meyer extract – a review of its properties and usage. Evid-Based-Integrative-Med. 2, 195. <https://doi.org/10.2165/01197065-200502040-00003>.

Seely, D., Dugoua, J.J., Perri, D., et al., 2008. Safety and efficacy of panax ginseng during pregnancy and lactation. Can. J. Clin. Pharmacol. 15, e87–e94.

Shergis, J.L., Zhang, A.L., Zhou, W., et al., 2013a. Panax ginseng in randomised controlled trials: a systematic review. Phytother. Res. 27, 949–965.

Shergis, J.L., Zhang, A.L., Zhou, W., et al., 2013b. Quality and risk of bias in *Panax ginseng* randomized controlled trials: a review. Am. J. Chin. Med. 41, 231–252.

Shi, W., Wang, Y., Li, J., et al., 2007. Investigation of ginsenosides in different parts and ages of *Panax ginseng*. Food Chem. 102 (3), 664–668.

Shim, Y.H., Park, C.D., Kim, D.H., et al., 2005. Identification of Panax species in the herbal medicine preparations using gradient PCR method. Biol. Pharm. Bull. 28, 671–676.

Shin, D.J., Kim, J.E., Lim, T.G., et al., 2014. 20-O-β-D-glycopyranosyl-20(S)-protopanaxadiol suppresses UV-induced MMP-1 expression through AMPK-mediated mTOR inhibition as a downstream of the PKA-LKB1 pathway. J. Cell Biochem. 115, 1702–1711.

Shin, K., Choi, H., Seo, M., Oh, D., 2017. Improved conversion of ginsenoside Rb1 to compound K by semi-rational design of *Sulfolobus solfataricus* β-glycosidase. AMB. Express 7, 186–197.

Siegel, R.K., 1979. Ginseng abuse syndrome. Problems with the panacea. JAMA 241, 1614–1615.

Sievenpiper, J.L., Arnason, J.T., Leiter, L.A., et al., 2004. Decreasing, null and increasing effects of eight popular types of ginseng on acute postprandial glycemic indices in healthy humans: the role of ginsenosides. J. Am. Coll. Nutr. 23, 248–258.

Skaper, S.D., Facci, L., Zusso, M., et al., 2018. An inflammation-centric view of neurological disease: beyond the neuron. Front. Cell. Neurosci. 12, 1–26.

Song, J.H., Kim, K.J., Choi, S.Y., et al., 2019. Korean ginseng extract ameliorates abnormal immune response through the regulation of inflammatory constituents in Sprague Dawley rat subjected to environmental heat stress. J. Ginseng Res. 43, 252–260.

Szakacs, G., Paterson, J.K., Ludwig, J.A., et al., 2006. Targeting multidrug resistance in cancer. Nat. Rev. Drug Discov. 5, 219–234.

Tang, W., Eisenbrand, G., 1992. Panax ginseng C.A. Mey. Chinese Drugs of Plant Origin. Springer, Berlin Heidelberg, pp. 711–737.

Tawab, M.A., Bahr, U., Karas, M., et al., 2003. Degradation of ginsenosides in humans after oral administration. Drug Metab. Disp. 31, 1065–1071.

Trinh, H.T., Han, S.J., Kim, S.W., et al., 2007. Bifidus fermentation increases hypolipidemic and hypoglycemic effects of red ginseng. J. Microbiol. Biotechnol. 17, 1127–1133.

Tsalamandris, S., Antonopoulos, A.S., Oikonomou, E., et al., 2019. The role of inflammation in diabetes: current concepts and future perspectives. Eur. Cardiol. Rev. 14, 50–59.

U.S. National Library of Medicine, 2014. GINSENG KIANPI PIL- angelica sinensis whole, morinda officinalis whole and panax ginseng root oil capsule. Available from: <http://dailymed.nlm.nih.gov/dailymed/fda/fdaDrugXsl.cfm?setid = ee656e9a-abd3-488b-949a-fca2d1a95bf0&type = display>. (Accessed 2 October 2015).

Van Hateren, H., 2013. Clinical use of ginseng in treatment of diabetes. J. Ginseng Res. 37, 8–29.

Vuksan, V., Sung, M.K., Sievenpiper, J.L., et al., 2008. Korean red ginseng (*Panax ginseng*) improves glucose and insulin regulation in well-controlled, type 2 diabetes: results of a randomized, double-blind, placebo-controlled study of efficacy and safety. Nutr. Metab. Cardiovasc. Dis. 18, 46–56.

World Health Organization, 1999. WHO Monographs on Selected Medicinal Plants, Vol. 1. WHO press, Malta.

World Health Organization, 2009. WHO Monographs on Selected Medicinal Plants, Vol. 4. WHO Press, Geneva.

Xiao, D., Ye, H., Yang, X., et al., 2015. Accumulation characteristics and correlation analysis of five ginsenosides with different cultivation ages from different regions. J. Ginseng Res. 39, 338–344.

Xiong, J., Sun, M., Guo, J., et al., 2009. Active absorption of ginsenoside Rg1 in vitro and in vivo: the role of sodium-dependent glucose co-transporter 1. J. Pharm. Pharmacol. 61, 381–386.

Xiu, Y., Li, X., Sun, X., et al., 2019. Simultaneous determination ad difference evaluation of 14 ginsenosides in *Panax ginseng* roots cultivated in different areas and ages by high-performance liquid chromatography coupled with triple quadrupole mass spectormeter in the multiple reaction – monitoring mode combined with multivariate statistical analysis. J. Ginseng Res. 43, 508–516.

Xu, X., Gao, Y., Xy, S.Y., et al., 2018. Remarkable impact of steam temperature on ginsenosides transformation from fresh ginseng to red ginseng. J. Ginseng Res. 42, 277–287.

Yang, M.S., Wu, M.Y., 2016. Chinese ginseng. In: Gupta, R.C. (Ed.), Nutraceuticals: Efficacy, Safety and Toxicity. Academic Press/Elsevier, Amsterdam, pp. 693–705.

Yang, Z., Gao, S., Wang, J., et al., 2011. Enhancement of oral bioavailability of 20 (S)-ginsenoside Rh2 through improved understanding of its absorption and efflux mechanisms. Drug Metab. Dispos. 39, 1866–1872.

Yokozawa, T., Seno, H., Oura, H., 1975. Effect of ginseng extract on lipid and sugar metabolism. Chem. Pharmacol. 23, 3095–3100.

Yue, P.Y., Mak, N.K., Cheng, Y.K., et al., 2007. Pharmacogenomics and the Yin/Yang actions of ginseng: anti-tumor, angiomodulating and steroid-like activities of ginsenosides. Chin. Med. 2, 6.

Yun, T.K., 2001. Brief introduction of *Panax ginseng* C.A. Meyer. J. Korean Med. Sci. 16, S3–S5.

Yun, T.K., 2003. Experimental and epidemiological evidence on non-organ specific cancer preventive effect of Korean ginseng and identification of active compounds. Mutat. Res. 523-524, 63–74.

Zhang, D., Yasuda, T., Yu, Y., et al., 1996. Ginseng extract scavenges hydroxyl radical and protects unsaturated fatty acids from decomposition caused by iron-mediated lipid peroxidation. Free Radic. Biol. Med. 20, 145–150.

Zhang, H.J., Lu, Z.Z., Tan, G.T., et al., 2002. Polyacetyleneginsenoside-Ro, a novel triterpene saponin from *Panax ginseng*. Tetrahedron Lett. 43, 973–977.

Zhong, G., Jiang, Y., 1997. Calcium channel blockage and anti-free-radical actions of ginsenosides. Chin. Med. J. (Engl.) 110, 28–29.

Zhou, W., Chai, H., Lin, P.H., et al., 2004. Molecular mechanisms and clinical applications of ginseng root for cardiovascular disease. Med. Sci. Monit. 10, Ra187–Ra192.

Chapter 50

Ashwagandha: multiple health benefits

Vijay K. Bharti, Jitendra K. Malik and Ramesh C. Gupta

Introduction

Medicinal plants are rich in secondary metabolites and essential oils of therapeutic importance. The important advantages claimed for therapeutic uses of medicinal plants in various ailments include their economic feasibility, effectiveness, ease of availability, and safety (Atal and Kapoor, 1989; Siddiqui, 1993). Ashwagandha (*Withania somnifera*) is a perennial plant belonging to the family Solanaceae. Ashwagandha, a popular Ayurvedic herb, is commonly known as "Indian winter cherry." The root smells like a horse ("ashwa"), which is why it is named Ashwagandha (on consuming it gives the power of a horse). The species name *somnifera* means "sleep-inducing" in Latin, indicating its sedating properties. Some herbalists refer to Ashwagandha as Indian ginseng, because it is used in Ayurvedic medicine in a similar way to Chinese ginseng (*Panax ginseng*) in traditional Chinese medicine (TCM). It has been used as an antibacterial, antioxidant, adaptogen, aphrodisiac, liver tonic, and anti-inflammatory agent (Mehrotra et al., 2011). *W. somnifera* has been used in different medicinal systems for centuries. Writings indicate that the therapeutic use of plants is as far back as 4000–5000 BCE and the Chinese used the first natural herbal preparations as medicines. In India, however, the earliest references to the use of plants as medicine appear in Rigveda, which is said to have been written between 3500 and 1600 BC. Later, the properties and therapeutic uses of medicinal plants were studied in detail and recorded empirically by the ancient physicians in Ayurveda, which is a basic foundation of ancient medical science in India (Sirkar, 1989). This chapter describes various aspects of Ashwagandha, including its multiple health effects.

Botany of *Withania somnifera*

W. somnifera, commonly known as Ashwagandha, is an important medicinal plant that has been used in Ayurvedic and indigenous medicine for more than 3000 years. Ashwagandha (*W. somnifera*) belongs to the genus *Withania* and family Solanaceae. Two species, *Withania coagulans* Dunal and *W. somnifera* Dunal, are found in India. *W. coagulans* is a rigid gray undershrub that grows to 60–120 cm. *W. somnifera* is an erect, evergreen, tomentose shrub 30–75 cm in height. Its roots are stout, fleshy, cylindrical, 1–2 cm in diameter, and whitish brown in color. The leaves are simple, ovate, glabrous, and opposite. The flowers are bisexual, inconspicuous, greenish or dull yellow in color, borne on axillary umbellate cymes, and comprise five sepals, petals, and stamens each; the two-celled ovary has a single style and a bilobed stigma. The petals are united and tubular. The stamens are attached to the corolla tube and bear erect anthers that form a close column or cone around the style. Pollen production is poor. The fruit is a small, globose, orange-red berry when mature and is enclosed in a persistent calyx. The seeds are small, flat, yellow, reniform in shape, and very light in weight. The cultivated plants have sizeable differences from the wild forms, not only in their morphological characteristics but also in their therapeutic action, although the alkaloids present are the same in both (Kaul, 1957; Atal and Schwarting, 1961; Schonbeck-Temesy, 1972; The Ayurvedic Pharmacopoeia of India, 1990a,b,c; Hepper et al., 1991; Mozaffarian Mozaffarian, 2003; Durg et al., 2015).

The 23 known *Withania* species are widely distributed in the drier parts of tropical and subtropical zones, ranging from the Canary Islands, the Mediterranean region, northern Africa, India, Sri Lanka, Afghanistan, Baluchistan, and Sindh (Schonbeck-Temesy, 1972; Hepper et al., 1991; Warrier et al., 1996; Hunziker Hunziker, 2001). It is found at high altitudes, ascending to 5500 feet in the Himalayas (Fig. 50.1). It has various names, including Kaknaj-e-Hindi (Arabic), winter cherry (English), Asgandh or Punir (Hindi), Kaknaj-e-Hindi or Asgand Nagaori (Persian, Urdu), and Ashwagandha (Sanskrit).

Nutraceuticals. DOI: https://doi.org/10.1016/B978-0-12-821038-3.00050-1

FIGURE 50.1 Global distribution of *Withania somnifera*.

Important bioactive molecules present in *Withania somnifera*

The pharmacological effects of the roots of *W. somnifera* are attributed to the presence of withanolides, a group of steroidal lactones (Budhiraja et al., 1987). This plant is commonly used in formulations for its wide range of health benefits. In Ayurveda, *Withania* is widely claimed to be a potent aphrodisiac, sedative, and rejuvenative, and to have life-prolonging properties. It is also used as a general energy-enhancing tonic known as Medharasayana, which means "it promotes learning and a good memory" (Nadkarni, 1976; Williamson, 2002).

Laboratory analysis has revealed more than 35 chemical constituents contained in the roots of *W. somnifera* (Rastogi and Mehrotra, 1998). The biologically active chemical constituents are alkaloids (isopellertierine, anferine), steroidal lactones (withanolides, withaferins), saponins containing an additional acyl group (sitoindoside VII and VIII), and withanoloides with a glucose at carbon 27 (sitonidoside XI and X). *W. somnifera* is also rich in iron. The roots of *W. somnifera* consist primarily of withanolides, which are believed to account for its extraordinary medicinal properties. Withanolides are steroidal and bear a resemblance, both in their action and appearance, to the active constituents ginsenosides present in Asian ginseng (*Panax ginseng*). Ashwagandha's withanolides have been researched in a variety of animal studies examining their effect on numerous conditions, including immune function and even cancer (Grandhi et al., 1994; Durg et al., 2015). Chemical analysis of Ashwagandha shows its main constituents to be alkaloids and steroidal lactones. Among the various alkaloids, withanine is the main constituent.

The other alkaloids are somniferine, somnine, somniferinine, withananine, pseudo-withanine, tropine, pseudotropine, 3-a-gloyloxytropine, choline, cuscohygrine, isopelletierine, anaferine, and anahydrine. The withanolides within the leaves have a C28 steroidal nucleus with C9 side chain, and a six-membered lactone ring. Twelve alkaloids, 35 withanolides, and several sitoindosides from *W. somnifera* have been isolated and studied. A sitoindoside is a withanolide containing a glucose molecule at carbon 27. Much of Ashwagandha's pharmacological activity has been attributed to two main withanolides, withaferin A and withanolide D (Bone, 1996; Elsakka et al., 1990; Durg et al., 2015). Further chemical analysis has shown the presence of the following:

Alkaloids: anaferine, anahygrine, cuscohygrine, scopoletin, withanine, withaninine, somniferine, tropeltigloate, somniferinine, somninine, nicotine, visamine, withasomine, and pseudotropine.

Salts: cuscohygrine, anahygrine, tropine, pseudotropine, and anaferine.

Steroidal lactones: Withaferin-A, withanone, WS-1, withanolide E, withanolide F, withanolide G, withanolide H, withanolide I, withanolide J, withanolide K, withanolide L, and withanolide M.

Nitrogen-containing compounds: withanol, somnisol, and somnitol.

Steroids: cholesterol, β-sitosterol, stigmasterol, diosgenin, stigmastadien, sitoinosides VII, sitoinosides VIII, sitoinosides IX, and sitoinosides X.

Flavonoids: kaempferol and quercetin (Nadkarni, 1954).

The plant contains carbohydrate-D-glucose, maltose, rhamnose, sucrose, and starch. It also contains proteins, amino acids, and the alkaloid shankhpushpine ($C_{17}H_{25}NO_2$), with a melting point of 162–164°C. The most notable constituents are tropine alkaloids. Only convolamine has been identified, but other alkaloids (convoline, convolidine, convolvine, confoline, convosine, etc.) found in other species from this family are most likely present as well (Prasad et al., 1974; Lounasmaa and Brossi, 1988; Singh and Bhandari, 2000). The fresh plant contains volatile oils, fatty acids, fatty alcohols, and hydrocarbons, such as myristic acid (30.9%), palmitic acid (66.8%), linoleic acid (2.3%), and straight-chain hydrocarbon hextriacontane (Deshpande and Srivastava, 1975). A study was performed for a chemical examination of the whole plant (*C. pluricaulis*) and reported the presence of scopoletin, β-sitosterol, and ceryl alcohol. The chloroform fraction of this plant extract contains 20-oxodotriacontanol, tetratriacontanoic acid, and 29-oxodotriacontanol. The flavonoid kaempferol and steroids phytosterol and β-sitosterol were also found in significant amounts (Singh and Bhandari, 2000). An estimation of scopoletin (Nahata and Dixit, 2008) content was determined by spectrofluorometry (Zafar et al., 2005) and high-performance thin-layer chromatography (HPTLC) (Kapadia et al., 2006).

Ashwagandha roots contain alkaloids, starch, reducing sugar, hentriacontane, glycosides, dulcital, withaniol acid, and a neutral compound. Wide variation (0.13%–0.31%) is observed in alkaloid content. Majumdar (1955) isolated eight amorphous bases (withanine, somniferine, somniferinine, somnine, withananine, withananinine, pseudowithanine, and withasomnine). Other alkaloids reported are nicotine, tropine, pseudotropine, 3,a-tigloyloxytropine, choline, cuscudohygrine, anaferine, anahygrine, and others. Free amino acids in the roots include aspartic acid, glycine, tyrosine, alanine, proline, tryptophan, glutamic acid, and cystine. The leaves contain 12 withanolides, alkaloids, glycosides, glucose, and free amino acids. The berries contain a milk-coagulating enzyme, two esterases, free amino acids, fatty oil, essential oils, and alkaloids. Methods for the analysis of alkaloid in Ashwagandha roots have also been reported (Majumdar, 1955; Maheshwari, 1989; Mishra, 1989; Table 50.1).

Major medicinal value in animals and humans

Withaferin A and withanolide D are the two main withanolides that contribute to most of the biological actions of *Withania somnifera* (Matsuda et al., 2001; Sharma et al., 2011). The active ingredients of *W. somnifera* are alkaloids (isopelletierine, anaferine, cuscohygrine, anahygrine, etc.), steroidal lactones (withanolides and withaferins), and saponins (Mishra et al., 2000). Sitoindosides and acylsterylglucosides in Ashwagandha are antistress agents. Active principles of Ashwagandha, for example, sitoindosides VII–X and withaferin A, have been shown to have significant antistress activity against acute models of experimental stress (Bhattacharya et al., 1987). The aerial parts of *W. somnifera* yield 5-dehydroxy withanolide-R and withasomniferin-A (Rahman et al., 1991).

The biological activities of *W. somnifera* are anxiolytic, antidepressant, antifungal (Bhattacharya et al., 2000b), antimicrobial (Girish et al., 2006), antimalarial (Dikasso et al., 2006), apoptotic (Senthil et al., 2007), chondroprotective (Sumantran et al., 2007), cardioprotective (Hamza et al., 2008), immunomodulatory (Davis and Kuttan, 2000), and neuroprotective (Sankar et al., 2007; Durg et al., 2015), promote inhibition of COX-2 enzyme (Jayaprakasam et al., 2003), and promote learning and memory in Alzheimer disease (AD) (Bhattacharya et al., 1995). Numerous studies have indicated that Ashwagandha possesses antioxidant, antitumor, antistress, antiinflammatory, immunomodulatory, hematopoietic, antiaging, anxiolytic, antidepressive, and rejuvenating properties, and that it also influences various neurotransmitter receptors in the central nervous system (Pattipati et al., 2003; Durg et al., 2015).

Animals

Use of medicinal plants for the treatment of various diseases has been a part of human culture since ancient times. Medicinal properties of plants were mostly discerned through trial and error, but they were also influenced by the belief systems of the people involved and often became entangled with religious and mythical practices (Mathias et al., 1996). Medicinal plant use evolved into an art and a science practiced according to the experience, traditions, and disease theory of the healer. Treatment of animal diseases developed in parallel with the treatment of human diseases. This knowledge was passed on verbally, by example, and sometimes through writing (Mathias et al., 1996). Ethnoveterinary medicine (EVM) is important in animal health care in developing countries (Cunningham and Zondi, 1991). It has become a recognized field of research that includes traditional veterinary theory, medicines, surgical methods, diagnostic procedures, and animal husbandry practices (Mathias et al., 1996). Veterinary aspects of ethnobotany are included in the field of EVM.

TABLE 50.1 Sources of bioactive molecules from different parts of *Withania somnifera* and their pharmacological importance in animals

Sl. no.	Type of bioactive molecules	Bioactive molecules	Plants parts	Medicinal value of the bioactive molecules	Benefited/experimented animals	References
1.	Steroidal lactones	Withanolides	Leaves (Atta-ur-Rahman et al., 1991; Choudhary et al., 1996; Bandyopadhyay et al., 2007)	Antioxidant activity	Rat	Bhattacharya et al. (1997a,b)
				Anxiolytic and antidepressant actions	Rat	Bhattacharya et al. (2000b)
				Treat AD and associated problems	Isolated rabbit jejunum	Choudhary et al. (2005)
			Leaves (Atta-ur-Rahman et al., 1991; Choudhary et al., 1996; Bandyopadhyay et al., 2007)	Antimicrobial activities	Balb/C mice	Owais et al. (2005)
		Withaferin A	Leaves (Jayaprakasam et al., 2003)	Stimulating tumor cell apoptosis	MCF-7 breast cancer cells	Zhang et al. (2011, 2012)
			Leaves (Jayaprakasam et al., 2003)	Causes G2 and M phase cell cycle arrest	Human breast cancer cell lines, MDA-MB-231 and MCF-7	Stan et al. (2008)
			Leaves (Jayaprakasam et al., 2003)	Chemoprevention	Golden Syrian hamsters	Manoharan et al. (2009), Panjamurthy et al. (2008, 2009)
			Leaves (Jayaprakasam et al., 2003)	Chemosensitization	Nu/nu mice	Fong et al. (2012)
			Leaves (Jayaprakasam et al., 2003)	Radiosensitization	Swiss albino mice	Ganasoundari et al. (1997)
			Leaves (Jayaprakasam et al., 2003)	Chemotherapeutic	Nu/nu mice, Balb/c mice, SCID mice	Samadi et al. (2010), Thaiparambil et al. (2011), Lahat et al. (2010)
			Leaves (Jayaprakasam et al., 2003)	Antiangiogenesis	Human umbilical vein endothelial cells	Mohan et al. (2004)
			Leaves (Jayaprakasam et al., 2003)	Antiinflammation	Swiss albino mice	Sabina et al. (2008)
			Leaves (Jayaprakasam et al., 2003)	NF-κB-inhibiting activity cell line	Spinal cord tissue	Swarup et al. (2011)
			Leaves (Jayaprakasam et al., 2003)	NF-κB-inhibiting activity cell line	Lung epithelial cell line (A549)	Oh and Kwon (2009)
			Leaves (Jayaprakasam et al., 2003)	NF-κB-inhibiting activity cell line	Macrophage cell line (RAW 264.7)	Oh et al. (2008b)
			Leaves (Jayaprakasam et al., 2003)	Antiproliferative activity	Leukemia cell line	Oh et al. (2008a)
			Leaves (Jayaprakasam et al., 2003)	Antiproliferative activity	Renal carcinoma cell line	Swarup et al. (2011)
			Leaves (Jayaprakasam et al., 2003)	Antiproliferative activity	Ovarian carcinoma cell line	Zhang et al. (2012)
			Leaves (Jayaprakasam et al., 2003)	Antiproliferative activity	Breast carcinoma cell line	Zhang et al. (2011), Hahm et al. (2011)

			Leaves (Jayaprakasam et al., 2003)	Antiangiogenic activity	Primary endothelial cells, HUVEC and HCEC	Yokota et al. (2006), Mohan et al. (2004), Bargagna-Mohan et al. (2006)
			Leaves (Jayaprakasam et al., 2003)	Activation of caspase-3, increased translocation of cytochrome C from mitochondria to cytosol	Human leukemia U937 cells	Oh et al. (2008a,b)
			Leaves (Jayaprakasam et al., 2003)	Tumor-inhibitory potential in both in vitro and in vivo studies	Balb/c nude mice	Widodo et al. (2008)
			Leaves (Jayaprakasam et al., 2003)	Cell cycle arrest	U2OS cells	Widodo et al. (2008)
		Withaferin A	Root (Sinha and Ostrand-Rosenberg, 2013)	WA reduces macrophage production of proinflammatory cytokines, WA reduces MDSC accumulation in tumor-bearing mice, WA minimizes MDSC production of the protumor cytokine IL-10	BALB/c mice	Sinha and Ostrand-Rosenberg (2013)
2.	Alkaloides	Nicotine, somniferine, somniferinine, withanine, withananine, pseudowithanine, tropine, pseudotropine, 3α-tigloyloxytropane, choline, cuscohygrine, *dl*-isopelletierine	Roots (Gupta and Rana, 2007)	Prolonged hypotensive, bradycardic, and respiratory-stimulant action	Dogs	Malhotra et al. (1981)
3.	Glycosides	Sitoindosides VII and VIII	Roots (Bhattacharya et al., 1987)	Antistress and antistress adaptogenic activity	Wistar rats	Jayaprakasam et al. (2003), Dhuley (2000), Singh et al. (2001)
4.	Glycowithanolide	Sitoindosides VII–X	Roots (Bhattacharya et al., 1995)	Significantly reversed ibotenic acid-induced cognitive defects in AD model	Rats	Bhattacharya et al. (1995)
		Glycowithanolides and sitoindosides IX and X	Whole plant	Significant antistress activity	Albino mice and rats	Ghosal et al. (1989)
		Sitoindosides VII–X, and withaferin-A	Roots (Schliebs et al., 1997)	Attenuate cerebral functional deficits, including amnesia, in geriatric patients	Rat	Schliebs et al. (1997)
5.	Withanolides	Withanolide A, withanoside IV, and withanoside VI	Leaves (Chaurasiya et al., 2008)	Anticarsinogenic effect	Human BJ-5ta fibroblasts, MCF-7 cells	Grin et al. (2012)
		Ashwagandhanolide	Roots (Subaraju et al., 2006; Mirjalili et al., 2009)	Abrogation of TNF-induced NF-κB activation	Human tumor cells	Mulabagal et al. (2009)
		Withanoside I	Root (Jeyanthi and Subramanian, 2009)	Ameliorate neuronal dysfunction in AD	Rat	Kuboyama et al. (2006)

(*Continued*)

TABLE 50.1 (Continued)

Sl. no.	Type of bioactive molecules	Bioactive molecules	Plants parts	Medicinal value of the bioactive molecules	Benefited/experimented animals	References
		Withanoside	Root (Jeyanthi and Subramanian, 2009)	Improved memory deficits in β amyloid-injected mice (to induce dendritic and axonal atrophy) and prevented loss of axons, dendrites, and synapses in the cerebral cortex and hippocampus	Mice	Kuboyama et al. (2005, 2006)
		Withanolide sulfoxide	Roots (Mulabagal et al., 2009)	Inhibited COX-2 enzyme and suppressed human tumor cell proliferation	Human tumor cells	Mulabagal et al. (2009)
		L-asparaginase	Fruits (Oza et al., 2010)	Inhibitory effect against lymphoblastic leukemia	Human leukemia cells	Oza et al. (2010)
		Withaferin A and withanolide D	Root (Leyon and Kuttan, 2004)	Significant antitumor and radio-sensitizing withanolides	Mice	Leyon and Kuttan (2004)
6.	Withanolides	Withanolide A	Root (Bani et al., 2006)	Increases the expression levels of T-helper 1 (Th1) cytokines, as well as CD4 and CD8 counts, enhances natural killer (NK) cell activity	Mice	Bani et al. (2006)
		Withanoside IV	Root (Kuboyama et al., 2006)	Attenuated the axonal, dendritic, and synaptic losses and memory deficits induced by amyloid peptide Aβ(25−35)	Mice	Kuboyama et al. (2006)
		Withanolide A, withanoside IV, and withanoside VI	Whole plants (Zhao et al., 2002)	Induce neurite outgrowth in human neuroblastome SHSY5Y	Human	Zhao et al. (2002)
		5-dehydroxy withanolide-R, withasomniferin-A, 1-oxo-5*b*, 6*b*-epoxy-witha-2-ene-27-ethoxy-olide, 2,3-dihydrowithaferin A, 24,25-dihydro-27-desoxy withaferin A, 27-α-*b*-D-glucopyranosyl physagulin D, physagulin D, withanoside I−VII,27-O-*b*-D-glucopyranosylviscosalactoneB, 4,16-dihydroxy-5*b*,6*b*-epoxyphysagulinD, viscosa lactone B, and diacetyl withaferin A	Root (Chaurasiya et al., 2008; Jeyanthi and Subramanian, 2009)	Nephroprotective effect	Rats	Jeyanthi and Subramanian (2009)

HUVEC = Human umbilical vein endothelial cell, HCEC = Human corneal epithelial cells, SCID = Severe combined immunodeficiency, AD = Alzheimer disease, WA = Withaferin A.

Humans

Withania roots are astringent, bitter, acrid, somniferous, thermogenic, stimulant, aphrodisiac, diuretic, and tonic. The leaves contain antibiotic, antitumorus, antihepatotoxic, and antiinflammatory properties. The seed contains milk-coagulating, hypnotic, and diuretic properties. *W. somnifera* has been used as an antibacterial, antioxidant, adaptogen, aphrodisiac, liver tonic, and antiinflammatory agent (Mehrotra et al., 2011; Durg et al., 2015). It is a reputed health food and herbal tonic that is used for cardiovascular diseases in ethnomedicine. It is available for human use either as a single herb or an ingredient of polyherbal or herbomineral formulations. The human doses of *Ashwagandha* are generally in the range of 4–6 g/day and are expected to be safe and nontoxic. Stress, as a major cardiovascular risk factor, leads to activation of sympathoadrenal and hypothalamic–pituitary–adrenal (HPA) axis and causes oxidative stress. Ashwagandha possesses a potent antistressor effect and is reported to alleviate stress-induced changes and provide cardioprotection in ischemic rats similar to the properties ascribed to adaptogens such as *Panax ginseng*. It also increases heart weight and glycogen in the myocardium and liver, indicating intensification of the anabolic process, and enhances the duration of contractility as well as coagulation time (Dhuley, 1998, 2000).

Experimental studies supporting the medicinal value of *Withania somnifera*

The leaves and roots of this plant are used as an abortifacient, aphrodisiac, diuretic, nerve tonic, narcotic, sedative, astringent, growth promoter, and anthelmintic. It has antiarthritic, antibacterial, antistress, antitumor, and anticancer activities. It is an antidote for scorpion stings. It is used for toning the uterus, consumption, dropsy, leucoderma, impotence, rheumatism, debility from old age, ulcer, sexual and genital weakness, assumption, rheumatic swelling, loss of memory, loss of muscular energy, spermatorrhea, syphilis, sterility in women, blood discharge, leucorrhea, anemia with emaciation, multiple sclerosis, neoplasia, cancer, and fatigue. The fruits and seeds are diuretics and are used in the coagulation of milk (Nadkarni, 1954).

Ashwagandha (*W. somnifera*) has been recommended for the treatment of various ailments, which include polyarthritis (*Waja-ul-Mafasil*), rheumatoid arthritis (*Hudar*), lumbago (*Wajaul- Qutn*), painful swellings (*Tawwarum-e-Alami*), spermatorrhea (*Jaryan-e-Mani*), asthma (*Zeeq-un-Nafas*), leucoderma (*Bars*), general debility (*Zof-e-Aam*), sexual debility (*Zof-e-Bah*), (Ali, 1997; Anon, 2007a,b; Khare, 2007; Kirtikar and Basu, 1980; Nadkarni, 1982), amnesia (*Nisyan*) (Ali, 1997), anxiety neurosis (*Qalaq-e-Usabi*), (Ali, 1997; Khare, 2007), scabies (*Jarb*), ulcers (*Qurooh*), marasmus (*Saghal*), and leucorrhea (*Sailan-ur-Rahem*) (Anon, 2007a,b; Tables 50.2–50.4).

Potential medicinal application in ameliorating high-altitude stress

Administration of the active principles of *W. somnifera* at equimolar concentrations of sitoindosides VII–X and withaferin A was found to increase superoxide dismutase (SOD), catalase (CAT), and glutathione peroxidase (GPx) activity in rat brain frontal cortex and striatum. The antioxidant effect of active glycowithanolides of *W. somnifera* (WSG) may explain, at least in part, the reported antiinflammatory, immunomodulatory, antistress, antiaging, and cognition-facilitating effects produced in experimental animals and in clinical situations (Bhattacharya et al., 1997a,b).

Another study reported by Bhattacharya et al. (2000a, b, 2001) investigated the antioxidant activity of WSG in chronic foot shock stress-induced changes in rat brain frontal cortex and striatum. The stress procedure, administered once daily for 21 days, induced an increase in SOD and lipid peroxidation (LPO) activities, with a concomitant decrease in CAT and GPx activities in both brain regions. WSG administered orally 1 h prior to the stress procedure for 21 days (10, 20, and 50 mg/kg) induced a dose-related reversal of the stress effects. Thus, they concluded that WSG tended to normalize the augmented SOD and LPO activities and enhance the activities of CAT and GPx. Their results indicate that at least part of the chronic stress-induced pathology may be due to oxidative stress, which is mitigated by WSG, lending support to the clinical use of the plant as an antistress adaptogen (Bhattacharya et al., 2000b).

The brain is relatively more susceptible to free radical damage than other tissues because it is rich in lipids and iron, both of which are known to be important in generating reactive oxygen species (ROS). The brain also uses nearly 20% of the total oxygen supply. Free radical damage to the brain may contribute to neuronal loss in cerebral ischemia and may be involved in normal aging and neurodegenerative diseases (epilepsy, schizophrenia, PD, AD, and others). Because traditional Ayurvedic use of *W. somnifera* has included many diseases associated with free radical oxidative damage, it has been considered likely that the effects may be due to a certain degree of antioxidant activity. Decreased activity of these enzymes leads to accumulation of toxic oxidative free radicals, resulting in degenerative effects. An increase in these enzymes would represent increased antioxidant activity and a protective effect on neuronal tissue. Active WSG (10 or 20 mg/kg, intraperitoneal)

TABLE 50.2 In vivo study of different bioactive molecules of *Withania somnifera* with their special medicinal value

Sl. no.	Extract name	Dosages	Route of dosages	Duration	Experimental animals	Pharmacological effects	References
1.	Glycowithanolides	20 and 50 mg/kg	Oral	5 days	Wistar rats	Antidepressant effect, induced an anxiolytic effect, antianxiety effects	Bhattacharya et al. (2000b)
2.	Glycowithanolides	10, 20, and 50 mg/kg	Oral	10 days	Wistar rats	Antioxidative effect	Bhattacharya et al. (2000a)
3.	*W. somnifera* root extract	25 and 50 mg/kg	Oral	21 days	Wistar rats	Significant antistress adaptogenic activity	Bhattacharya and Muruganandam (2003)
4.	*W. somnifera* root extract (withaferin A, sitoindosides VII–X, 5-dehydroxy withanolide-R, withasomniferin-A, 1-oxo-5b, 6b-epoxy-witha-2-ene-27-ethoxy-olide, 2,3- dihydrowithaferin A, 24,25-dihydro-27-desoxy withaferin A, 27-O-b-D-glucopyranosyl physagulin D, physagulin D, withanoside I–VII,27-O-b-D-glucopyranosylviscosalactoneB, 4,16-dihydroxy-5b,6b-epoxyphysagulinD, viscosa lactone B, and diacetyl withaferin A)	250, 500, and 750 mg/kg	Oral	14 days	Rats	Nephroprotective effect	Jeyanthi and Subramanian (2009)
5.	Withaferin A	2–4 mg/kg	Intraperitoneal (IP)	10 days	Mice	Prolonged survival of S-180 ascites	Lahat et al. (2010)
6.	Withaferin A	10–60 mg/kg		24 h	Swiss albino mice	Inhibited growth of mouse Ehrlich ascites carcinoma cells and increased tumor-free survival	Byun et al. (2001)
7.	*W. somnifera* root extract	100 mg/kg	Oral	28 days	Mice	Improves catecholamines and physiological abnormalities seen in PD	RajaSankar et al. (2009)
8.	*W. somnifera* root extract	100 and 200 mg/kg	Intraperitoneal (IP)	14 days	Rat	Improves Huntington HD	Kumar and Kumar (2009)
9.	*W. somnifera* leaf extract	100 mg/kg	Oral administration	7 days	Balb/C mice	Antimicrobial activities	Owais et al. (2005)
10.	Withaferin A	1, 2, 4, 8 mg/kg body weight	Oral administration	7 days	Balb/C mice	WA reduces macrophage production of proinflammatory cytokines, withaferin A reduces MDSC accumulation in tumor-bearing mice, WA minimizes MDSC production of the protumor cytokine IL-10	Sinha and Ostrand-Rosenberg (2013)

11.	Sitoindosides VII and VIII	25 and 50 mg/kg	Oral administration	21 days	Wistar rats	Antistress and antistress adaptogenic activity	Bhattacharya and Muruganandam (2003)
12.	Withaferin A	4 or 8 mg/kg	Intraperitoneal (IP)	24 days	PC-3 (prostate)	↓Tumor growth	Yang et al. (2007)
13.	Withaferin A	8 mg/kg	Intraperitoneal (IP)	21 days	DRO81-1 (medullary thyroid)	↓Tumor growth	Samadi et al. (2010)
14.	Withaferin A	2 and 4 mg/kg	Intraperitoneal (IP)	30 days	4T1 (mouse breast)	↓Tumor growth	Thaiparambil et al. (2011)
15.	Withaferin A	5 mg/kg	Intraperitoneal (IP)	17 days	AB12 (mesothelioma)	↓Tumor growth	Yang et al. (2012)
16.	Withaferin A	3 and 6 mg/kg	Intraperitoneal (IP)	28 days	Panc-1 (pancreatic)	↓Tumor growth	Yu et al. (2010)
17.	Withaferin A	20 mg/kg	Intraperitoneal (IP)	105 days	DMBA (oral)	↓Oral cancer (100%)	Manoharan et al. (2009)
18.	Withaferin A	100 μg/mouse	Intraperitoneal (IP)	196 days	MMTV-neu (breast)	↓Macroscopic and ↓microscopic tumor burden	Hahm et al. (2014)
19.	Glycowithanolides	1 mL/kg	Intraperitoneal (IP)	21 days	Rat	Increase the antioxidant activity	Bhattacharya et al. (1997a,b)
20.	Leaf powder	1.1 μg/mL	Oral	20 days	Mice	↓Tumor growth	Widodo et al. (2008)
21.	Leaf extract	0.3 mL of 24 μg/mL leaf extract	Subcutaneous	20 days	BALB/c nude mice	↓Tumor growth	Widodo et al. (2007)
22.	Withaferin A	10–60 mg/kg	Intraperitoneal (IP)	120 days	Mice	↓Tumor growth	Devi et al. (1992)
23.	Sitoindosides IX and X	50–200 mg/kg	Oral and intraperitoneal	4 days	Mice	↑Immunomodulatory and CNS effects	Ghosal et al. (1989)
24.	Root extract of *W. somnifera*	20 mg/kg	Intragastric	45 days	Mice	Antioxidative effect	Hussain et al. (2013)

PD = Parkinson disease, HD = Hutchington disease, CNS = Central nervous system.

TABLE 50.3 In vitro study of different bioactive molecules of *Withania somnifera* with their special medicinal value

Sl. no.	Extract name	Dosages	Duration	Experimental cell lines	Pharmacological effects	References
1.	27-Acetoxy-4b, 6a-dihydroxy-5b-chloro-1-oxowitha-2, 24-dienolide. 5b,6b,14a,15a-diepoxy-4b,27-dihydroxy-1-oxowitha-2,24-dienolide, and withaferin A	40 mg/mL, 9.4 mg/mL	4 days	Human lung cancer cell line (NCI-H460)	Growth inhibition and cytotoxic activity against human lung cancer cell line	Choudhary et al. (2010)
2.	L-asparaginase	0.01, 0.05, 0.1, 0.2, 0.5, 1, 2, 3, and 5 IU	3 days	Human leukemia cells	Inhibitory effect against lymphoblastic leukemia	Oza et al. (2010)
3.	Withanolides	0.003–1.0 μg/mL	3 h	Isolated rabbit jejunum	Treat Alzheimer disease (AD) and associated problems	Choudhary et al. (2005)
4.	Withanolides	200 μg/mL	3 h	Isolated human neutrophils	Treat Alzheimer disease (AD) and associated problems	Choudhary et al. (2005)
5.	Withaferin A	0, 0.156, 0.313, 0.625, 1.25, 2.5, 5 μM	72 h	MCF-7 breast cancer cells	Stimulating tumor cell apoptosis	Zhang et al. (2011, 2012)
6.	Withaferin A	2 μM	24 h	Human breast cancer cell lines MDA-MB-231 and MCF-7	Causes G2 and M phase cell cycle arrest	Stan et al. (2008)
7.	Withaferin A	0.25, 0.5, 1.0, 1.5, 2.0 μM	24 h	Human leukemia U937 cells	Activation of caspase-3, increase translocation of cytochrome C from mitochondria to cytosol	Oh et al. (2008a)
8.	Withaferin A	0.5 μM	24 h	Human STS cell lines	Anticancerous effect	Lahat et al. (2010)
9.	Withaferin A	2 mg/kg	24 h	HT-1080, SKLMS-1 (soft tissue sarcoma)	↓Tumor growth	Lahat et al. (2010)
10.	Withaferin A	8 mg/kg	48 h	CaSki (cervical)	↓Tumor growth	Munagala et al. (2011)
11.	Withaferin A	8 or 12 mg/kg	72 h	Human uveal melanoma cell lines	Induce apoptosis	Samadi et al. (2012)
12.	Withaferin A	2 μM	24 h	Human umbilical vein endothelial cells	Antiangiogenic activity	Mohan et al. (2004)
13.	Withaferin A	1.5 μM	48 h	Osteogenic sarcoma (U2OS) and fibrosarcoma (HT1080) cells	Anticancerous effect	Widodo et al. (2008)

(Continued)

TABLE 50.3 (Continued)

Sl. no.	Extract name	Dosages	Duration	Experimental cell lines	Pharmacological effects	References
14.	Leaf extract, withaferin A, Withanone, Withanolide A	(0.8–5.0 μg/mL), (0.1–0.5 μM), (5–10 μg/mL), (5–10 μg/mL)	48–72 h	Glioma cell lines C6 (rat) and YKG1 (human)	Effective glioma therapy	Shah et al. (2009)
15.	Withaferin A	4.0 and 5.0 μM	3 h	Human lung cancer cells, A549	Alters intermediate filament organization, cell shape and behavior	Grin et al. (2012)
16.	Withaferin A	1.0–4.0 μM	3 h	MCF-7 cells		
17.	Withaferin A	2 μM	24 h	BJ-5ta cells		
18.	Withaferin A	2.5 and 5.0 μM	16 h	Human breast cancer cells (MCF-7 and SUM159 cells)	Inducts the apoptotic process	Hahm et al. (2014)
19.	Withaferin A	5 and 10 μM	48 h	Leukocyte-depleted erythrocytes	Suicidal effectiveness on erythrocyte	Jilani et al. (2013)
20.	Water extract of *W. somnifera* leaves	0.1%–2.5% of the crude extract	72 h	Rat C6 glioma cell line	Useful for complimentary therapy for glioma	Kataria et al. (2011)
				Human glioma cell lines (YKG1, U118MG and A172)		
21.	Ashwagandha leaf extracts	0.3 μg/mL	24 h	Human neuroblastoma (IMR32) and rat glioblastoma (C6) cells	Protect brain-derived cells against oxidative stress and induce differentiation	Shah et al. (2015)

was given once daily for 21 days to groups of six rats. Dose-related increases in all enzymes were observed; the increases were comparable to those seen with deprenyl (a known antioxidant) administration (2 g/kg per day, intraperitoneal). This implies that *W. somnifera* does have an antioxidant effect in the brain that may be responsible for its diverse pharmacological properties. Further studies on other parts of the brain (e.g., cerebellum, medulla, and hypothalamus) may provide information with respect to the effects of *W. somnifera* on cognitive behavior and other functions of the brain in both healthy and diseased individuals.

In another study, an aqueous suspension of *W. somnifera* root extract was evaluated for its effect on stress-induced lipid peroxidation (LPO) in mice and rabbits (Dhuley, 1998). LPO blood levels were increased by intravenous administration of 0.2 mg/kg of lipopolysaccharides (LPS) from *Klebsiella pneumoniae* and 100 mg/kg of peptidoglycans (PGN) from *Staphylococcus aureus*. Simultaneous oral administration of *W. somnifera* extract (100 mg/kg) prevented an increase in LPO. The authors indicated that the almost innocuous doses of LPS and PGN used in this study were comparable to a mild bacteremia that may follow tooth extraction or *Streptococcal angina*.

Antistressor effects of Ashwagandha were investigated in rats using a cold water swimming stress test. The treated animals showed better stress tolerance (Archana and Namasivayam, 1999). A withanolide-free aqueous fraction isolated from the roots of *W. somnifera* exhibited antistress activity in a dose-dependent manner in mice (Khare, 2007). Ashwagandha has been evaluated for its adaptogenic activity. Administration of Ashwagandha with other drugs in experimental animals exposed to a variety of biological, physical, and chemical stressors was

TABLE 50.4 Different types of curable disease after application of different bioactive molecules of *Withania somnifera*

Sl. no.	Curable disease	Bioactive molecules	References
1.	Oxidative stress	Withanolides	Bhattacharya et al. (1997a,b)
2.	Alzheimer disease	Withanolides	Choudhary et al. (2005)
		Sitoindosides VII–X	Bhattacharya et al. (1995)
		Withanoside I	Kuboyama et al. (2006)
		Withanoside	Kuboyama et al. (2005, 2006)
		Withanolide sulfoxide	Mulabagal et al. (2009)
3.	Neurodegenerative disease	Withanoside IV	Kuboyama et al. (2006)
		Withanolide A, withanoside IV, and withanoside VI	Zhao et al. (2002)
4.	Microbial disease	Withanolides	Owais et al. (2005)
5.	Cancer	Withaferin A	Zhang et al. (2011, 2012), Stan et al. (2008), Fong et al. (2012), Panjamurthy et al. (2009), Samadi et al. (2010), Thaiparambil et al. (2011), Lahat et al. (2010)
		Withanolide A, withanoside IV, and withanoside VI	Grin et al. (2012)
6.	Bacterial disease	Withaferin A and withanolide D	Leyon and Kuttan (2004)
		Withaferin A and 3-b-hydroxy-2,3-dihydrowithanolide F	Budhiraja et al. (1987)
7.	Stress relief	Sitoindosides VII and VIII	Bhattacharya and Muruganandam (2003), Dhuley (2000), Singh et al. (2001)
		Glycowithanolides and sitoindosides IX and X	Ghosal et al. (1989)
8.	Amnesia	Sitoindosides VII–X, and withaferin-A	Schliebs et al. (1997)
9.	Lymphoblastic leukemia	L-asparaginase	Oza et al. (2010)
10.	Poor immunity	Withanolide A	Bani et al. (2006)
11.	Nephronic disease	5-Dehydroxy withanolide-R, withasomniferin-A, 1-oxo-5*b*, 6*b*-epoxy-witha-2-ene-27-ethoxy-olide, 2,3- dihydrowithaferin A, 24,25-dihydro-27-desoxy withaferin A, 27-α-*b*-D-glucopyranosyl physagulin D, physagulin D, withanoside I–VII,27-O-*b*-D-glucopyranosylviscosalactoneB, 4,16-dihydroxy-5*b*,6*b*-epoxyphysagulinD, viscosa lactone B, and diacetyl withaferin A	Jeyanthi and Subramanian (2009)

found to offer protection against these stressors (Bhattacharya, 1992; Rege et al., 1999).

Available drug formulation using *Withania somnifera*

Ashwagandha is advocated as a protective drug against atherosclerosis, hypertension, and coronary heart diseases (Mehra et al., 2009). It reduces the sensitivity of the heart to adrenergic stimulation and thereby protects the heart against sympathetic outbursts. Moharana (2008) reported that the roots and leaves of *Ashwagandha* are traditionally used in the form of powder, decoction, or oil. These have been used in folk medicine against general disability, hypertension, inflammation, and wounds. Thirunavukkarasu et al. (2006) found Ashwagandha to have energy-boosting properties and

TABLE 50.5 List of some available drugs formulated with *Withania somnifera*

Sl. no.	Product name	Manufacturer	Utility of the product
1.	Stresswin	Baidynath Ayurved Bhawan	→ Combating exertion, reduction in anxiety, strain, and stress, improvement of stamina, relief from disturbed sleep, mental alertness
2.	Stresscom	Dabur India Ltd.	→ Relieves anxiety neurosis, physical and mental stress, and relieves general debility and depression
3.	Brento	Zandu Pharmaceutical Works Ltd.	→ Nerve tonic
4.	Ashvagandha	Morpheme Remedies	→ Combating stress
5.	Dabur Ashwagandha Churna	Dabur	→ Combating stress
6.	Ashwagandha	Ayurceutics	→ Stress reliever
7.	Himalaya Massage oil	The Himalaya Drug Co.	→ Stress relief and relief from insomnia
8.	Ashwagandharista	Baidynath Ayurved Bhawan	→ Nerve tonic, memory and cognition improvement, better power of concentration, relieves mental tension, natural sleep induction, and recovery from nervous and general debility
9.	Arshadi pills	Dehlvi Remedies	→ Stress, depression, cardiac tonic
10.	Ashwagandha extract	Nanjing Zelang Medical Technology Co., Ltd.	→ It is called "Indian ginseng" and contains alkaloids, steroidal lactones, withanaloids, and iron; it has antiallergy, antihistamine, antibacterial, local anesthetic, antipyretic, and pain-relieving functions
			→ Alkaloids could be used in sedative and pain relief, lowering blood pressure. The function of anti-inflammation of withanolides can inhibit cancer cell growth, treat chronic inflammation, such as lupus and rheumatoid arthritis, reduce vaginal discharge, improve sexual function, etc.
11.	Ashwagandha/*W. somnifera* extract withanolides	Wuxi Gorunjie Natural-Pharma Co., Ltd.	→ Ashwaghanda extract can treat different diseases like inflammation, arthritis, stress and anxiety, and even mental disorders→ Improves the white blood cell in the body in a way that increases the phagocytosis or the process of removing dead cells→ Enhances the results of radiation and chemotherapy→ Regulates the nerve signals and makes sure that it is balanced→ A great source of energy; it develops stamina by managing the process of metabolism→ Best source for antioxidants and hormonal forerunners→ Maintains necessary bodily functions→ Protects the immunity cells that are subjected to chemicals that may prevent the cells to work properly
12.	Nutramax-AE (100%)	Hunan Nutramax Inc.	→ Antiallergy→ Antihistamine→ Antipyretic and pain-relieving→ Local anesthetic→ Antibacterial
13.	Nutramax *W. somnifera* extract (10:1)	Hunan Nutramax Inc.	→ Winter cherry medicinally works as an adaptogen, antistress agent, aphrodisiac, and in all cases of general debility→ Ashwangandha extract is traditionally used in case of spermatorrhoes, loss of strength, seminal debility, and as a growth promoter→ The root extracts exhibits antistress, hypotensive, antispasmodic, bradicardic, and respiratory stimulant activities→ The herb promotes sound sleep, provides protection against environmental free radicals, nourishment of the cells, and works as rejuvenate
14.	Natural 80 mesh American ginseng root extract	Qingdao Fraken International Trading Co., Ltd.	→ Withanolides possess remarkable antibacterial, antitumor, antiarthritic, anti-inflammatory, and immunosuppressive properties
15.	100% Natural *W. somnifera* extract Alkaloids, Withanolides Ashwagandha extract	Xi'an Saina Biological Technology Co., Ltd	→ Antiallergy, antihistamine, antipyretic, pain-relieving, local anesthetic, antibacterial

recommended its use as a dietary supplement for cardioprotection. The effect of Ashwagandha root was evaluated for lipid peroxidation in stress. The herb was found to have very good antioxidant activity, which may partly explain the antistress, congestion-facilitating, antiinflammatory, and antiaging effects of this herb (Moharana, 2008; Table 50.5).

Concluding remarks and future directions

In Ayurveda, the use of medicinal plants for better health management has been performed up to the present time, and its utility has increased across the globe. Medicinal plants contain several types of medicinally and pharmacologically important bioactive molecules that maintain the health and vitality of individuals and also cure diseases. Several studies of *W. somnifera* have indicated its use as an aphrodisiac, liver tonic, antiinflammatory agent, and astringent, and to treat bronchitis, asthma, ulcers, emaciation, insomnia, and senile dementia. Clinical trials and animal research support the use of Ashwagandha for anxiety, immunomodulation, hematopoiesis, cognitive and neurological disorders, inflammation, tumors, and PD. This plant is also has an influence on the endocrine, nervous, and cardiopulmonary systems. The most well-known bioactive molecules of *W. somnifera* are withaferins and sitoindosides, which reduce the oxidative damage to the cells and help in the prevention of several diseases. Studies on *W. somnifera* have indicated that supplementing with this plant can reduce lipid peroxidation, possibly by scavenging free radicals, and can help to build up a potent antioxidant defense system. This chapter describes that *W. somnifera* contains several bioactive molecules that are important for good health and for maintaining life at high altitudes, because it is responsible for a reduction of oxidative stress. It can also be used as a multipurpose medicinal agent. More research is needed to determine the potential dosage range for achieving its multipurpose drug properties.

Acknowledgments

The authors thank Ms. Robin B. Doss for her technical assistance, and Mr. Arup Giri for his help in the preparation of this chapter.

References

Ali, S.S., 1997. Unani Advia-e-Mufradah, eighth ed. National Council for Promotion of Urdu Language (NCPUL), New Delhi, p. 33.

Anon, 2007a. Standardisation of Single Drugs of Unani Medicine. Part III, first ed. Central Council for Research in Unani Medicine (CCRUM), New Delhi, pp. 9–14.

Anon, 2007b. The Unani Pharmacopoeia of India. Part I, vol. I. Department of AYUSH, Ministry of Health and Family Welfare, Govt. of India, New Delhi, pp. 7–8.

Archana, R., Namasivayam, A., 1999. Anti-stressor effect of *Withania somnifera*. J. Ethnopharmacol. 64 (1), 91–93.

Atal, C.K., Schwarting, A.E., 1961. Aswagandha, an ancient Indian drug. Econ. Bot. 15 (3), 256–263.

Atal, C.K., Kapoor, B.M., 1989. Cultivation and utilization of medicinal plants (Eds. PID CSIR).

Atta-ur-Rahman, J.A.S., Choudary, M.I., Asif, I., 1991. Two withanolides from *Withania somnifera*. Phytochemistry 30, 3824–3825.

Bandyopadhyay, M., Jha, S., Tepfer, D., 2007. Changes in morphological phenotypes and withanolide composition of Ri-transformed roots of *Withania somnifera*. Plant. Cell Rep. 26 (5), 599–609.

Bani, S., Gautam, M., Sheikh, F.A., 2006. Selective Th1 up-regulating activity of *Withania somnifera* aqueous extract in an experimental system using flow cytometry. J. Ethnopharmacol. 107 (1), 107–115.

Bargagna-Mohan, P., Ravindranath, P.P., Mohan, R., 2006. Small molecule anti-angiogenic probes of the ubiquitin proteasome pathway: potential application to choroidal neovascularization. Invest. Ophthalmol. Vis. Sci. 47 (9), 4138–4145.

Bhattacharya, S.K., 1992. Evaluation of adaptogenic activity of some Indian medicinal plants *Withania somnifera* and *Ocimum sanctum* with special reference to stress-induced gastric ulcer in albino rats. Proc. Intl. Semin. Trad. Med. Calcutta 7–9.

Bhattacharya, A., Ramanathan, M., Ghosal, S., 2000a. Effect of *Withania somnifera* glycowithanolides on iron-induced hepatotoxicity in rats. Phytother. Res. 14 (7), 568–570.

Bhattacharya, S.K., Bhattacharya, A., Sairam, K., 2000b. Anxiolytic–antidepressant activity of *Withania somnifera* glycowithanolides: an experimental study. Phytomedicine 7 (6), 463–469.

Bhattacharya, A., Ghosal, S., Bhattacharya, S.K., 2001. Anti-oxidant effect of *Withania somnifera* glycowithanolides in chronic footshock stress-induced perturbations of oxidative free radical scavenging enzymes and lipid peroxidation in rat frontal cortex and striatum. J. Ethnopharmacol. 74 (1), 1–6.

Bhattacharya, S.K., Muruganandam, A.V., 2003. Adaptogenic activity of *Withania somnifera*: an experimental study using a rat model of chronic stress. Pharmacol. Biochem. Behav. 75 (3), 547–555.

Bhattacharya, S.K., Goel, R.K., Kaur, R., 1987. Anti-stress activity of sitoindosides VII and VIII, new acylsterylglucosides from *Withania somnifera*. Phytother. Res. 1, 32–37.

Bhattacharya, S.K., Kumar, A., Ghosal, S., 1995. Effects of glycowithanolides from *Withania somnifera* on an animal model of Alzheimer's disease and perturbed central cholinergic markers of cognition in rats. Phytother. Res. 9, 110–113.

Bhattacharya, S.K., Satyan, K.S., Chakrabarti, A., 1997a. Effect of transina (TR), an Ayurvedic herbal formulation, on pancreatic islet superoxide dismutase (SOD) activity in hyperglycaemic rats. Indian. J. Exp. Biol. 35 (3), 297–299.

Bhattacharya, S.K., Satyan, K.S., Ghosal, S., 1997b. Antioxidant activity of glycowithanolides from *Withania somnifera*. Indian. J. Exp. Biol. 35 (3), 236–239.

Bone, K., 1996. Clinical applications of Ayurvedic and Chinese herbs: Monograph for the Western Herbal Practitioner. Phytotherapy Press, Australia.

Budhiraja, R.D., Sudhir, S., Garg, K.N., 1987. Review of biological activity of withanolides. J. Sci. Ind. Res. 46, 488–491.

Byun, Y., Chen, F., Chang, R., 2001. Caspase cleavage of vimentin disrupts intermediate filaments and promotes apoptosis. Cell Death Differ. 8 (5), 443–450.

Chaurasiya, N.D., Uniyal, G.C., Lal, P., 2008. Analysis of withanolides in root and leaf of *Withania somnifera* by HPLC with photodiode array and evaporative light scattering detection. Phytochem. Anal. 19 (2), 148–154.

Choudhary, I., Abbas, S.M., Arshad, J.S., 1996. *Withania somnifera*: a source of exotic withanolides. Heterocycles 42 (2), 555–563.

Choudhary, M.I., Nawaz, S.A., Lodhi, M.A., 2005. Withanolides, a new class of natural cholinesterase inhibitors with calcium antagonistic properties. Biochem. Biophys. Res. Commun. 334 (1), 276–287.

Choudhary, M.I., Hussain, S., Yousuf, S., 2010. Chlorinated and diepoxy withanolides from *Withania somnifera* and their cytotoxic effects against human lung cancer cell line. Phytochemistry 71 (17), 2205–2209.

Cunningham, A.B., Zondi, A.S., 1991. Cattle Owners and Traditional Medicines Used for Livestock. Investigational Report No. 69. Institute of Natural Resources, University of Natal, Pietermaritzburg.

Davis, L., Kuttan, G., 2000. Immunomodulatory activity of *Withania somnifera*. J. Ethnopharmacol. 71, 193–200.

Deshpande, S.M., Srivastava, D.N., 1975. Gas chromatographic identification of fatty acids, fatty alcohols, and hydrocarbons of *Convolvulus pluricaulis* (Chois). J. Am. Oil Chem. Soc. 52 (8), 318–319.

Devi, P.U., Sharada, A.C., Solomon, F.E., 1992. *In vivo* growth inhibitory effect of *Withania somnifera* (Ashwagandha) on a transplantable mouse tumor, Sarcoma 180. Indian J. Exp. Biol. 30 (3), 169–172.

Dhuley, J.N., 1998. Effect of *Ashwagandha* on lipid peroxidation in stress-induced animals. J. Ethnopharmacol. 60, 173–178.

Dhuley, J.N., 2000. Adaptogenic and cardioprotective action of Ashwagandha in rats and frogs. J. Ethnopharmacol. 70 (1), 57–63.

Dikasso, D., Makonnen, E., Debella, A., 2006. Antimalarial activity of *Withania somnifera* L. Dunal extracts in mice. Ethiop. Med. J. 44, 279–285.

Durg, S., Dhadde, S.B., Vandal, R., Shivakumar, B.S., 2015. *Withania somnifera* (Ashwagandha) in neurobehavioral disorders induced by brain oxidative stress in rodents: a systematic review and meta-analysis. J. Pharm. Pharmacol. 67, 879–899.

Elsakka, M., Grigorescu, E., Stanescu, U., 1990. New data referring to chemistry of *Withania somnifera*. Rev. Med. Chir. Soc. Med. Nat. lasi 94, 385–387.

Fong, M.Y., Jin, S., Rane, M., 2012. Withaferin A synergizes the therapeutic effect of doxorubicin through ROS-mediated autophagy in ovarian cancer. PLoS One 7 (7), 42265.

Ganasoundari, A., Zare, S.M., Devi, P.U., 1997. Modification of bone marrow radiosensitivity by medicinal plant extracts. Br. J. Radiol. 70 (834), 599–602.

Ghosal, S., Lal, J., Srivastava, R., 1989. Immunomodulatory and CNS effects of sitoindosides IX and X, two new glycowithanolides from *Withania somnifera*. Phytother. Res. 3 (5), 201–206.

Girish, K.S., Machiah, K.D., Ushanandini, S., 2006. Antimicrobial properties of a non-toxic glycoprotein (WSG) from *Withania somnifera* (Ashwagandha). J. Basic. Microbiol. 46, 365–374.

Grandhi, A., Mujumdar, A.M., Patwardhan, B., 1994. A comparative pharmacological investigation of Ashwagandha and ginseng. J. Ethnopharmacol. 44, 131–135.

Grin, B., Mahammad, S., Wedig, T., 2012. Withaferin-A alters intermediate filament organization, cell shape and behavior. PLoS One 7 (6), 39065.

Gupta, G., Rana, A.C., 2007. *Withania somnifera* (Ashwagandha): a review. Pharmacogn. Rev. 1 (1), 129.

Hahm, E.R., Lee, J., Huang, Y., 2011. Withaferin a suppresses estrogen receptor-α expression in human breast cancer cells. Mol. Carcinog. 50 (8), 614–624.

Hahm, E.R., Lee, J., Singh, S.V., 2014. Role of mitogen-activated protein kinases and Mcl-1 in apoptosis induction by withaferin A in human breast cancer cells. Mol. Carcinog. 53, 907–916.

Hamza, A., Amin, A., Daoud, S., 2008. The protective effect of a purified extract of *Withania somnifera* against doxorubicin-induced cardiac toxicity in rats. Cell Biol. Toxicol. 24 (1), 63–73.

Hepper, F.N., Hawkes, J.G., Lester, R.N., Nee, M., Estrada, E., 1991. Solanaceae III: Taxonomy, Chemistry, Evolution. Royal Botanic Gardens, Kew.

Hunziker Hunziker, A.T., 2001. Genera Solanacearum: The Genera of the Solanaceae Illustrated, Arranged According to a New System. Gantner Verlag, Ruggell, Liechtenstein.

Hussain, S.M., Reyahd, M.R., Noah, A.M., 2013. Effect of alcoholic extract of *Withania Somnifera* on antioxidant enzymes in mice with induced mammary carcinoma. Al-Mustansiriyah J. Sci. 24 (4).

Jayaprakasam, B., Zhang, Y., Seeram, N.P., 2003. Growth inhibition of human tumor cell lines by withanolides from *Withania somnifera* leaves. Life Sci. 74 (1), 125–132.

Jeyanthi, T., Subramanian, P., 2009. Nephroprotective effect of *Withania somnifera*: a dose-dependent study. Ren. Fail. 31 (9), 814–821.

Jilani, K., Lupescu, A., Zbidah, M., 2013. Withaferin A-stimulated Ca^{2+} entry, ceramide formation and suicidal death of erythrocytes. Toxicol. Vitro 27, 52–58.

Kapadia, N.S., Acharya, N.S., Acharya, S.A., 2006. Use of HPTLC to establish a distinct chemical profile for Shankhpushpi and for quantification of scopoletin in *Convolvulus pluricaulis* and in commercial formulations of Shankhpushpi. J. Planar Chromatogr. Mod. TLC 19 (109), 195–199.

Kataria, H., Shah, N., Kaul, S.C., 2011. Water extract of Ashwagandha leaves limits proliferation and migration, and induces differentiation in Glioma cells. Evid. Based Complement. Alternat. Med. 1–12.

Kaul, K.N., 1957. On the origin, distribution and cultivation of Aswagandha, the so called *Withania somnifera* Dunal, of Indian Literature on Materia Medica. In: Pharmaceutical and Drug Committee Symposium on the Utilisation of Indian medicinal plants, Lucknow. pp. 7–8.

Khare, C.P., 2007. Indian Medicinal Plants – An Illustrated Dictionary. First Indian Reprint. Springer (India) Pvt. Ltd, New Delhi, pp. 717–718.

Kirtikar, K.R., Basu, B.D., 1980. Indian Medicinal Plants, vol. III, second ed. Lalit Mohan Basu, Allahabad, pp. 1774–1777.

Kuboyama, T., Tohda, C., Komatsu, K., 2005. Neuritic regeneration and synaptic reconstruction induced by withanolide A. Br. J. Pharmacol. 144 (7), 961–971.

Kuboyama, T., Tohda, C., Komatsu, K., 2006. Withanoside IV and its active metabolite, sominone, attenuate Aβ (25–35)-induced neurodegeneration. Eur. J. Neurosci. 23 (6), 1417–1426.

Kumar, P., Kumar, A., 2009. Possible neuroprotective effect of *Withania somnifera* root extract against 3-nitropropionic acid-induced behavioral, biochemical, and mitochondrial dysfunction in an animal model of Huntington's disease. J. Med. Food. 12 (3), 591–600.

Lahat, G., Zhu, Q.S., Huang, K.L., 2010. Vimentin is a novel anti-cancer therapeutic target; insights from *in vitro* and *in vivo* mice xenograft studies. PLoS One. 5 (4), 10105.

Leyon, P.V., Kuttan, G., 2004. Effect of *Withania somnifera* on B16F-10 melanoma induced metastasis in mice. Phytother. Res. 18 (2), 118–122.

Lounasmaa, M., Brossi, A., 1988. Tropane Alkaloids. In: Brossi, A. (Ed.), The Alkaloids, Chemistry and Pharmacology. Academic Press, New York.

Maheshwari, S.K., 1989. The other method for estimation of various constituents of Withania roots. In: Proc. All India Workshop on M&AP, pp. 439–441.

Majumdar, D.N., 1955. *Withania somnifera* Dunal. Indian. J. Pharmacol. 17 (8), 158.

Malhotra, C.L., Das, P.K., Dhalla, N.S., 1981. Studies on *Withania* Ashwagandha, Kaul. III. The effect of total alkaloids on the cardiovascular system and respiration. Indian. J. Med. Res. 49, 448–460.

Manoharan, S., Panjamurthy, K., Menon, V.P., 2009. Protective effect of withaferin-A on tumour formation in 7, 12-dimethylbenz[a]anthracene induced oral carcinogenesis in hamsters. Indian. J. Exp. Biol. 47 (1), 16–23.

Mathias, E., McCorkle, C.M., Schillhorn, V.V.T.W., 1996. Introduction: ethnoveterinary research and development 1. In: McCorkle, C.M., Mathias, E., Schillhorn van Veen, T.W. (Eds.), Ethnoveterinary Research and Development. p. 16.

Matsuda, H., Murakami, A.K., Yoshikawa, M., 2001. Sturctures of withanosides I, II, III, IV, V, VI and VII new withanolide glycosides from the roots of Indian *Withania somnifera* Dunal and inhibitory activity for tachphylaxis to clonidine in isolated guinea pig ileum. Bioorg. Med. Chem. 9, 1499–1507.

Mehra, R., Prasad, M., Lavekar, G.S., 2009. An approach of *Ashwagandha* + *Guggulu* in atheromatous CHD associated with obesity. J. Res. Ayu. 30 (2), 121–125.

Mehrotra, V., Mehrotra, S., Kirar, V., 2011. Antioxidant and antimicrobial activities of aqueous extract of *Withania somnifera* against methicillin-resistant *Staphylococcus aureus*. J. Microbiol. Biotechnol. Res. 1 (1), 40–45.

Mirjalili, M.H., Moyano, E., Bonfill, M., 2009. Steroidal lactones from *Withania somnifera*, an ancient plant for novel medicine. Molecules 14 (7), 2373–2393.

Mishra, L.C., Singh, B.B., Dagenais, S., 2000. Scientific basis for the therapeutic use of *Withania somnifera* (Ashwagandha): a review. Altern. Med. Rev. 5, 334–346.

Mishra, S.N. 1989. Analytical methods for analysis of total alkaloids in root of *Withania* spp. In: Proc. All India Workshop on M&AP, pp. 492–495.

Mohan, R., Hammers, H., Bargagna-Mohan, P., 2004. Withaferin A is a potent inhibitor of angiogenesis. Angiogenesis 7 (2), 115–122.

Moharana, D., 2008. *Shatavari, Jastimadhu* and *Aswagandha* the Ayurvedic Therapy. Orissa Rev. 72–77.

Mozaffarian Mozaffarian, V., 2003. Trees Shrubs Iran. Farhange Moaser, Tehran, Iran, pp. 874–877.

Mulabagal, V., Subbaraju, G.V., Rao, C.V., 2009. Withanolide sulfoxide from Aswagandha roots inhibits nuclear transcription factor-kappa-B, cyclooxygenase and tumor cell proliferation. Phytother. Res. 23 (7), 987–992.

Munagala, R., Kausar, H., Munjal, C., 2011. Withaferin A induces p53-dependent apoptosis by repression of HPV oncogenes and upregulation of tumor suppressor proteins in human cervical cancer cells. Carcinogenesis 32 (11), 1697–1705.

Nadkarni, A.K., 1954. Dr. K.M. Nadkarni's Indian Materia Medica two vols., third ed. Popular Prakashan Ltd, Bombay.

Nadkarni, K.M., 1976. Indian Materia Medica. Popular Prakshan Limited, Bombay.

Nadkarni, K.M., 1982. Indian Materia Medica vol. I, third ed. Popular Prakashan Ltd, Bombay, pp. 1292–1294.

Nahata, A., Dixit, V.K., 2008. Spectrofluorimetric estimation of scopoletin in *Evolvulus alsinoides* Linn. and *Convulvulus pluricaulis* Choisy. Indian. J. Pharm. Sci. 70, 834–837.

Oh, J.H., Kwon, T.K., 2009. Withaferin A inhibits tumor necrosis factor-α-induced expression of cell adhesion molecules by inactivation of Akt and NF-κB in human pulmonary epithelial cells. Int. Immunopharmacol. 9 (5), 614–619.

Oh, J.H., Lee, T.J., Kim, S.H., 2008a. Induction of apoptosis by withaferin A in human leukemia U937 cells through down-regulation of Akt phosphorylation. Apoptosis. 13 (12), 1494–1504.

Oh, J.H., Lee, T.J., Park, J.W., 2008b. Withaferin A inhibits iNOS expression and nitric oxide production by Akt inactivation and down-regulating LPS-induced activity of NF-κB in RAW 264.7 cells. Eur. J. Pharmacol. 599 (1), 11–17.

Owais, M., Sharad, K.S., Shehbaz, A., 2005. Antibacterial efficacy of *Withania somnifera* (Ashwagandha) an indigenous medicinal plant against experimental murine salmonellosis. Phytomedicine 12 (3), 229–235.

Oza, V.P., Parmar, P.P., Kumar, S., 2010. Anticancer properties of highly purified L-asparaginase from *Withania somnifera* L. against acute lymphoblastic leukemia. Appl. Biochem. Biotechnol. 160 (6), 1833–1840.

Panjamurthy, K., Manoharan, S., Menon, V.P., 2008. Protective role of withaferin-A on 7, 12-dimethylbenz (a) anthracene-induced genotoxicity in bone marrow of Syrian golden hamsters. J. Biochem. Mol. Toxicol. 22 (4), 251–258.

Panjamurthy, K., Manoharan, S., Nirmal, M.R., 2009. Protective role of withaferin-A on immunoexpression of p53 and bcl-2 in 7, 12-dimethylbenz (a) anthracene-induced experimental oral carcinogenesis. Invest. N. Drugs. 27 (5), 447–452.

Pattipati, S., Amanpreet, S., Shrinivas, K., 2003. Effect of *Withania somnifera* root extract on haloperidol induced orofacial dyskinesia: possible mechanism of action. J. Med. Food. 6 (2), 107–114.

Prasad, G.C., Gupta, R.C., Srivastava, D.N., 1974. Effect of shankhpushpi on experimental stress. J. Res. Ind. Med. 9 (2), 19–27.

Rahman, A., Abbas, S., Shahwar, D., 1991. New withanolides from Withania spp. J. Nat. Prod. 56, 1000–1006.

RajaSankar, S., Manivasagam, T., Sankar, V., 2009. *Withania somnifera* root extract improves catecholamines and physiological abnormalities seen in a Parkinson's disease model mouse. J. Ethnopharmacol. 125 (3), 369–373.

R.P. Rastogi, B.N. Mehrotra, Central Drug Research Institute, New Delhi 1998.

Rege, N.N., Thatte, U.M., Dahanukar, S.A., 1999. Adaptogenic activity of six rasayana herbs used in Ayurvedic medicine. Phytother. Res. 13 (4), 275–291.

Sabina, E.P., Chandal, S., Rasool, M.K., 2008. Inhibition of monosodium urate crystal-induced inflammation by withaferin A. J. Pharm. Pharm. Sci. 11 (4), 46–55.

Samadi, A.K., Mukerji, R., Shah, A., 2010. A novel RET inhibitor with potent efficacy against medullary thyroid cancer *in vivo*. Surgery. 148 (6), 1228–1236.

Samadi, A.K., Cohen, S.M., Mukerji, R., 2012. Natural withanolide withaferin A induces apoptosis in uveal melanoma cells by suppression of Akt and c-MET activation. Tumour Biol. 33 (4), 1179–1189.

Sankar, S.R., Manivasagam, T., Krishnamurti, A., 2007. The neuroprotective effect of *Withania somnifera* root extract in MPTP-intoxicated mice: an analysis of behavioral and biochemical variables. Cell. Mol. Biol. Lett. 12 (4), 473–481.

Schliebs, R., Liebmann, A., Bhattacharya, S.K., 1997. Systemic administration of defined extracts from *Withania somnifera* (Indian ginseng) and Shilajit differentially affects cholinergic but not glutamatergic and GABAergic markers in rat brain. Neurochem. Int. 30 (2), 181–190.

E. Schonbeck-Temesy, In: Akademische Druck-u. Verlagsanstalt, Graz, Austria 1972.

Senthil, V., Ramadevi, S., Venkatakrishnan, V., 2007. Withanolide induces apoptosis in HL-60 leukemia cells via mitochondria mediated cytochrome c release and caspase activation. Chem. Biol. Interact. 167, 19–30.

Shah, N., Kataria, H., Kaul, S.C., 2009. Effect of the alcoholic extract of Ashwagandha leaves and its components on proliferation, migration, and differentiation of glioblastoma cells: combinational approach for enhanced differentiation. Cancer Sci. 100 (9), 1740–1747.

Shah, N., Singh, R., Sarangi, U., 2015. Combinations of Ashwagandha leaf extracts protect brain-derived cells against oxidative stress and induce differentiation. PLoS One. 10 (3), 1–14.

Sharma, V., Sharma, S., Pracheta, P.R., 2011. *Withania somnifera*: a rejuvenating Ayurvedic medicinal herb for the treatment of various human ailments. Int. J. Pharm. Tech. Res. 3 (1), 187–192.

Siddiqui, H.H., 1993. Safety of herbal drugs – an overview. Drugs N. & Views 1 (2), 7–10.

Singh, G.K., Bhandari, A., 2000. Textbook of Pharmacognosy, first ed. CBS Publishers, New Delhi.

Singh, B., Saxena, A.K., Chandan, B.K., 2001. Adaptogenic activity of a novel, withanolide-free aqueous fraction from the roots of *Withania somnifera* Dun. Phytother. Res. 15 (4), 311–318.

Sinha, P., Ostrand-Rosenberg, S., 2013. Myeloid-derived suppressor cell function is reduced by withaferin A, a potent and abundant component of *Withania somnifera* root extract. Cancer Immunol. Immunother. 62 (11), 1663–1673.

Sirkar, N.N., 1989. Pharmacological basis of Ayurvedic therapeutics. In: Atal, C. K., Kapoor, B.M. (Eds.), Cultivation and Utilization of Medicinal Plants. Published by PID CSIR.

Stan, S.D., Hahm, E.R., Warin, R., 2008. Withaferin A causes FOXO3a- and Bim-dependent apoptosis and inhibits growth of human breast cancer cells *in vivo*. Cancer Res. 68 (18), 7661–7669.

Sumantran, V.N., Kulkarni, A., Boddul, S., 2007. Chondroprotective potential of root extracts of *Withania somnifera* in osteoarthritis. J. Biosci. 32, 299–307.

Swarup, V., Phaneuf, D., Dupre, N., 2011. Deregulation of TDP-43 in amyotrophic lateral sclerosis triggers nuclear factor κB–mediated pathogenic pathways. J. Exp. Med. 208 (12), 2429–2447.

Thaiparambil, J.T., Bender, L., Ganesh, T., 2011. Withaferin A inhibits breast cancer invasion and metastasis at sub-cytotoxic doses by inducing vimentin disassembly and serine 56 phosphorylation. Int. J. Cancer. 129 (11), 2744–2755.

The Ayurvedic Pharmacopoeia of India, 1990a. Part- I, vol-I, first ed. Delhi: Government of India, Controller of Publication, pp. 45–48.

The Ayurvedic Pharmacopoeia of India, 1990b. Part- I, vol-I, first ed. Delhi: Government of India, Controller of Publication, pp. 113–116.

The Ayurvedic Pharmacopoeia of India, 1990c. Part- I, vol-I, first ed. Delhi: Government of India, Controller of Publication, pp. 127–128.

Thirunavukkarasu, M., Penumathsa, S., Juhasz, B., 2006. Enhanced cardiovascular function and energy level by a novel chromium (III)-supplement. Biofactors 27, 53–67.

Warrier, P.K., Nambiar, V.P.K., Ramankutty, C., 1996. Indian Medicinal Plants: A Compendium of 500 Species, 5. Orient Longman, Hyderabad, India, p. 409.

Widodo, N., Kaur, K., Shrestha, B.G., 2007. Selective killing of cancer cells by leaf extract of Ashwagandha: identification of a tumor-inhibitory factor and the first molecular insights to its effect. Clin. Cancer Res. 13 (7), 2298–2306.

Widodo, N., Takagi, Y., Shrestha, B.G., 2008. Selective killing of cancer cells by leaf extract of Ashwagandha: components, activity and pathway analyses. Cancer Lett. 262 (1), 37–47.

Williamson, E.M., 2002. Major Herbs of Ayurveda. Churchill Livingstone, London, UK.

Yang, H., Shi, G., Dou, Q.P., 2007. The tumor proteasome is a primary target for the natural anticancer compound withaferin A isolated from "Indian winter cherry". Mol. Pharmacol. 71 (2), 426–437.

Yang, H., Wang, Y., Cheryan, V.T., 2012. Withaferin A inhibits the proteasome activity in mesothelioma *in vitro* and *in vivo*. PLoS One 7 (8), 412–414.

Yokota, Y., Bargagna-Mohan, P., Ravindranath, P.P., 2006. Development of withaferin A analogs as probes of angiogenesis. Bioorg. Med. Chem. Lett. 16 (10), 2603–2607.

Yu, Y., Hamza, A., Zhang, T., 2010. Withaferin A targets heat shock protein 90 in pancreatic cancer cells. Biochem. Pharmacol. 79 (4), 542–551.

Zafar, R., Ahmad, S., Mujeeb, M., 2005. Estimation of scopoletin in leaf and leaf callus of *Convolvulus microphyllus* Sieb. Indian. J. Pharm. Sci. 67, 600–603.

Zhang, X., Mukerji, R., Samadi, A.K., 2011. Down-regulation of estrogen receptor-alpha and rearranged during transfection tyrosine kinase is associated with withaferin A-induced apoptosis in MCF-7 breast cancer cells. BMC Complement. Altern. Med. 11 (1), 84.

Zhang, X., Samadi, A.K., Roby, K.F., 2012. Inhibition of cell growth and induction of apoptosis in ovarian carcinoma cell lines CaOV3 and SKOV3 by natural withanolide withaferin A. Gynecol. Oncol. 124 (3), 606–612.

Zhao, J., Nakamura, N., Hattori, M., 2002. Withanolide derivatives from the roots of *Withania somnifera* and their neurite outgrowth activities. Chem. Pharm. Bull. 50 (6), 760–765.

Chapter 51

Astaxanthin: health benefits and toxicity

Tetsuo Satoh and Ramesh C. Gupta

Introduction

Oxidative stress is caused by the excessive formation of reactive oxygen species (ROS), and the imbalance between antioxidants and ROS can result in damage to cells, tissues, and human/animal bodily functions (Davies, 1995; Dore et al., 2005). Carotenoid group members are produced from microalgae *Haematococcus pluvialis* (*Asta carox*) (Querin et al., 2003), as well as a variety of bacteria and fungi (Yuan et al., 2011). Among the various carotenoids, astaxanthin (ATX) is currently the most potent known natural antioxidant against oxidative stress (Hussein et al., 2006). However, unlike other carotenoids such as zeaxanthin, ATX does not function as a prooxidant. Cooper et al. (1975), who first identified the chemical structure of ATX by synthesis, found that ATX is a metabolite of zeaxanthin and/or canthaxanthin, and that it contains both hydroxyl and ketone functional groups. ATX (3,3′-dihydroxy-β,β-carotene-4,4′-dione) has a chemical formula of $C_{40}H_{52}O_4$ and molecular weight of 596.84 [CAS, 472-61-7]. The structural formula of ATX is shown in Fig. 51.1.

The blood-red coloration of ATX is due to the extended chain of conjugated double bonds at the center of its chemical structure. This chain of conjugated double bonds is also responsible for ATX's potent antioxidant effects, which have been verified in numerous studies and that makes it useful in the treatment of cardiovascular, inflammatory, immune, and neurodegenerative diseases, and cancer (Fassett et al., 2008; Fig. 51.2).

To date, ATX has been used in the treatment of Alzheimer disease (AD), Parkinson disease (PD), stroke, hypercholesterolemia, and ocular diseases, including age-related macular degeneration (AMD) (Yang et al., 2013; Higuera-Ciapara et al., 2006). In addition, it has been determined that ATX exerts chemopreventive effects by concurrently inhibiting the phosphorylation of transcription factors and signaling kinases, and by inducing intrinsic apoptosis (Kavitha et al., 2013). Furthermore, while acute ultraviolet B (UVB) exposure causes photokeratitis and induces apoptosis in mouse corneal cells, ATX has been found to reduce oxidative stress in UVB-irradiated corneas (Lennikov et al., 2012).

In this chapter, current research developments related to ATX are described, focusing mainly on its human health benefits such as gastrointestinal, neurodegenerative, diabetes, and immune response. The prodrug benefits of ATX, its pharmacokinetics, the singlet oxygen scavenging effect, and toxicity are also highlighted.

Pharmacokinetics (absorption, distribution, metabolism, and excretion)

Absorption and distribution

In its free form, ATX is water-insoluble and possesses low bioavailability. To improve its solubility, ATX prodrug esters have been developed. When administered orally, these prodrugs are readily absorbed into tissues and facilitate higher bioavailability (Mercke Odeberg et al., 2003; Rufer et al., 2008; Okada et al., 2009; Miyazawa et al., 2011). The hydrophobicity of ATX is a key factor affecting its bioaccessibility levels in naturally produced mixed micelles, as well as their postprandial blood response. In contrast, the physicochemical properties of ATX do not significantly affect its uptake efficiency into intestinal cells or their accumulation in adipose tissue (Sy et al., 2012).

A number of pharmacokinetic studies with orally administered ATX in humans have been reported. In one such study (Ohi et al., 2011), healthy male volunteers received a single 40-mg ATX dose in one of three lipid-based formulations or from a commercially available food supplement. From the results, it was found that the elimination half-life of ATX was 15.9 h ($n = 532$) and that the three lipid-based formulations enhanced bioavailability from 1.7 to 3.7 times that of the reference formulation (Mercke Odeberg et al., 2003). In another study, Rao et al. (2010) reported that peak ATX levels in plasma, liver, and eyes after administration of ATX were observed at 2, 4, and 6 h, respectively. From these findings, it was

Nutraceuticals. DOI: https://doi.org/10.1016/B978-0-12-821038-3.00051-3

concluded that ATX from *H. pluvialis* showed better bioavailability than carotene and lutein, and that the microalgae could be considered a good source of high bioavailability and nutraceutical value of ATX.

In their in vivo study, Takizawa et al. (2011) showed the inhibitory effects of ATX on intestinal ischemia–reperfusion (I/R) injuries. They also noted its inhibitory effects on mucosal lesions and tight junction openings, as well as a decrease in the protein expression level of P-glycoprotein. Previously, Coral-Hinostroza et al. (2004) studied the appearance, pharmacokinetics, and distribution of ATX all-E-, 9Z-, and 13Z-geometrical, along with (3R, 3VR)-, (3R, 3VS)-, and (3S, 3VS)-optical isomers in plasma fractions obtained from volunteers. The pharmacokinetic and distribution data obtained via direct resolution of the geometrical and optical isomers showed that a selective process increased the relative proportion of ATX Z-isomers in comparison to the all-E ATX isomers before uptake in blood, and that ATX esters are hydrolyzed selectively during absorption.

As part of efforts to determine the organ distribution of ATX in rats after oral application, 2-week experiments were conducted by Petri and Lundebye (2007) and Rao et al. (2010). The results of both studies showed that low ATX concentrations, distributed over a wide range, were detected in the viscera. However, they also reported that levels did not increase from day 7 to day 14. These results indicate that there was rapid elimination, or catabolism, and no significant long-term accumulation. Furthermore, liver concentrations were unexpectedly low and the highest concentrations were found in the spleen, kidneys, eyes, and adrenals.

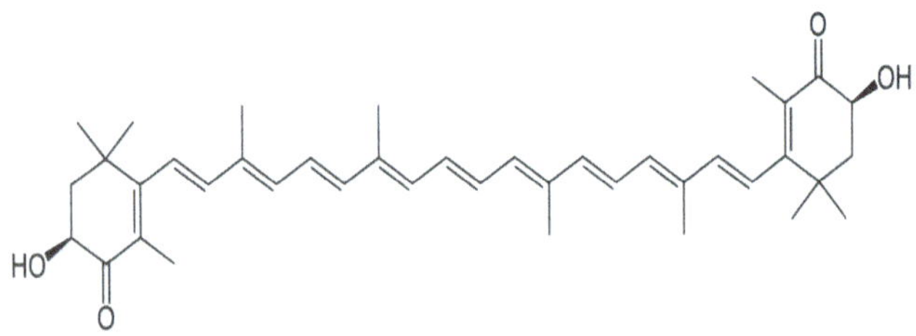

FIGURE 51.1 Structural formula of astaxanthin.

Choi et al. (2011a) reported that, due to the saturable hepatic metabolism of ATX, its pharmacokinetic parameters were dose-dependent after intravenous administration but dose-independent after oral administration, and that the gastrointestinal absorption of ATX followed the flip-flop model. Additionally, the hepatic gastrointestinal first-pass extraction ratios of ATX were approximately 0.490 and 0.901, respectively.

Metabolism

Choi et al. (2011a) determined that ATX was metabolized primarily by hepatic cytochrome P450 (CYP) 1A1 and 1A2 in rats, whereas another ATX metabolism study (Wolz et al., 1999) involving primary rat hepatocyte cultures found one major metabolite, (*rac*)-3-hydroxy-4-oxo-β-ionone, which was successively conjugated primarily into glucuronides. Deconjugation of the polar conjugates with glusulase identified two metabolites, (*rac*)-3-hydroxy-4-oxo-β-ionone and its reduced form (*rac*)-3-hydroxy-4-oxo-7,8-dihydro-β-ionone, thus indicating that the former was reduced in the conjugated form. Because there was no evidence of CYP involvement in ATX-induced

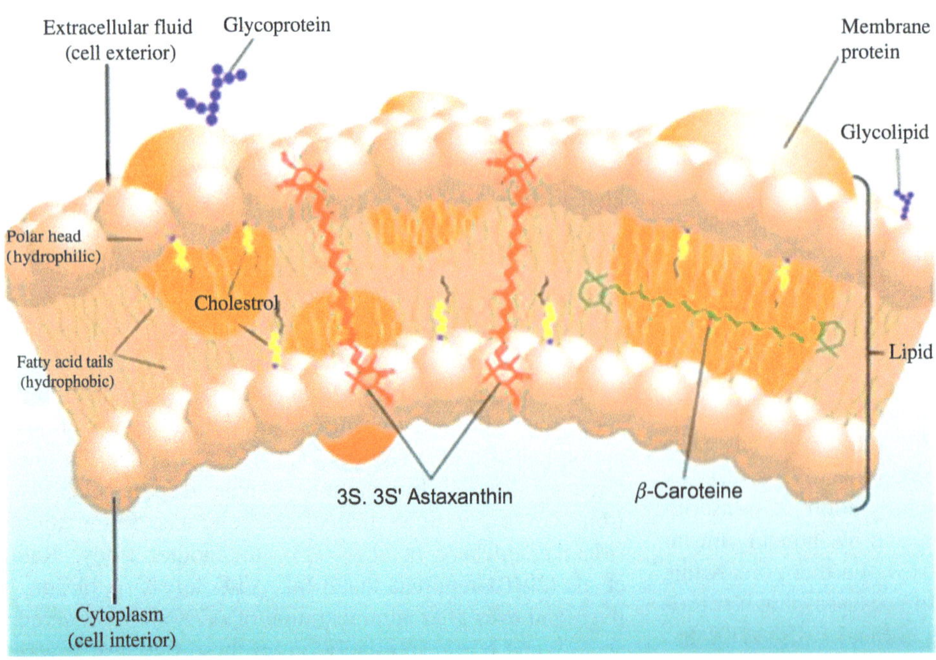

FIGURE 51.2 Astaxanthin (ATX) chain. ATX belongs to the carotenoid group and shares a part of the same biosynthetic pathway as carotene; it also possesses a polyene chain that quenches free radicals. ATX is part of the xanthophyll subgroup because of the additional keto- and hydroxyl-groups that are attached to the isoprene rings; this gives ATX the ability to orient into the optimal hydrophilic and hydrophobic position (as shown) and provides a lipid bilayer that provides superior protection against peroxidation. *(Reproduced with permission from AstaREAL website)*

induction, it is clear that ATX was metabolized into two metabolites by non-P450 enzymes.

In their study of the metabolism of ATX as well as its capacity to induce CYP genes, Kistler et al. (2002) identified four metabolites in human plasma after oral administration of 100 mg ATX to volunteers. They also found that, in human cultured hepatocytes, ATX was a potent inducer of CYP3A4 and CYP2B6, but not other CYPs, including the CYP1A and CYP2C families. Because no ATX metabolism autoinduction protocol was found, neither CYP3A4 nor CYP2B6 was thought to contribute to the formation of metabolites. In summary, different CYP isozymes are solely responsible for the metabolism and induction of ATX in humans.

The effect of ATX on the metabolic activation of a typical promutagen, benzo[a]pyrene (B[a]P), by CYP1A was investigated in rats by Ohno et al. (2011). Their results showed that ATX significantly increased the expression of CYP1A1 mRNA, protein, and their activities. In contrast, the activities of phase II enzymes, including glucuronosyltransferase and GSH *S*-transferase, were not modulated by ATX, and B[a]P was found to be more enhanced in ATX-treated rats when compared with controls in the Ames assay.

Prodrugs

To improve the water solubility of free ATX, Hawaii Biotech, Inc. (HBI) successfully synthesized a novel synthetic ATX carotenoid derivative called disodium disuccinate derivative (DDA; Cardax). This derivative exhibits higher water solubility than free ATX, and thus it is more suitable for parenteral injection in aqueous formulations. Intravenous administration of Cardax for 4 days in a Sprague–Dawley rat infarct model was found to significantly improve cardiac events (Gross and Lockwood, 2004). These results suggest that parenteral Cardax™ may be effective when administered as a cardioprotective pretreatment agent.

Single administration of another novel ATX derivative, Heptax (a disodium disuccinate diester of ATX), to C57BL/6 mice resulted in a significant manifestation of free ATX in plasma ($C_{max} = 0.2$ mg/L; 381 nM) and its accumulation in solid organs (e.g., liver $C_{max} = 0.9$ mg/L; 1735 nM) at levels not previously reported after single ATX doses in rodents (Showalter et al., 2004). At each point in the area under the curve, free ATX levels in liver were greater than the corresponding concentrations in plasma, thus suggesting concentrative liver uptake.

Oral administration of yet another novel ATX prodrug, CDX-085, rapidly resulted in high ATX levels in plasma, which were effectively distributed in lipoproteins and associated with reduced total cholesterol and triglycerides (Ryu et al., 2012). These results indicate that ATX has inhibitory effects on macrophage activation, such as scavenger receptor upregulation and proinflammatory cytokines secretion. Furthermore, in comparison to free ATX, pharmacokinetic data support the in vivo superiority of orally administered CDX-085 for systematic ATX delivery.

Jackson et al. (2004) synthesized a novel derivative of ATX by esterification to the amino acid L-lysine and observed its subsequent conversion to tetrahydrochloride salt. Administration of this ATX prodrug represented a significant improvement over the parent compound. Furthermore, the novel derivative proved to be an extremely potent aqueous phase scavenger that provides nearly complete suppression of the superoxide anion signal (Rao et al., 2013).

Toxicity health benefits

Oxidative stress

Disturbances in the normal redox state of cells can cause toxic effects through the production of free radicals that damage the cell, including proteins, lipids, and DNA. Some ROS disrupt cellular messengers in redox signaling. Singlet oxygen ($^1O_2$) is a nonradical ROS with one of the strongest activities.

Oxidative stress leading to mitochondrial dysfunction is thought to be a trigger for various diseases, including cancer (Halliwell, 2007), AD, PD (Valko et al., 2007; Pohanka, 2013), atherosclerosis, heart failure (Fassett and Coombes, 2011), and myocardial infarction (Dean et al., 2011). At the nanomolar level, ATX is known to improve maintenance of high mitochondrial membrane potential by restraining mitochondria in a reduced state (Wolf et al., 2010; Fakhri et al., 2018).

ATX is specifically known for its ability to quench singlet oxygen (Nishida et al., 2007) and to provide significant protection against oxidative stress, disease, and age-related degeneration (Halliwell, 1992; Ames et al., 1993; Davies, 1995; Fakhri et al., 2018). ATX and vitamin E (α-tocopherol) work synergistically to protect against lipid peroxidation (Raneva et al., 2003). More specifically, Choi et al. (2011a,b) found that ATX supplementation for 3 weeks suppressed lipid peroxidation and stimulated the activity of the antioxidant defense system. Accordingly, because the free radicals induced by cigarette smoking have been strongly linked to increased oxidative stress in vivo, it is felt that ATX supplementation might prevent oxidative damage in smokers by suppressing lipid peroxidation and stimulating their antioxidant system activities (Kim et al., 2011).

ATX is available commercially in bulk as a "racemic" mixture of the stereoisomers 3S, 30S, meso(3R,30S), and 3R,30R mixed in a 1:2:1 ratio. Most studies involving ATX stereoisomers have been limited to the all-*trans* isomer because all-*trans* natural ATX is readily isomerized to *cis–trans*, especially the 9-*cis* and 13-*cis* isomers. Furthermore, Liu and Osawa (2007) first reported that *cis* ATX isomer ATX, especially 9-*cis* ATX, has much higher antioxidant potency than that of the all-*trans* isomer. Capelli et al. (2013) reported that natural ATX from *H. pluvialis* microalgae was more than 50 times stronger than synthetic ATX in singlet oxygen quenching, and approximately 20 times stronger in free radical elimination in vitro.

Brain

ATX provides noticeable neuroprotection against ischemia–reperfusion-induced brain damage in rats by inhibiting the oxidative stress reduction activities of glutamate release and by antiapoptosis, by which infarct volumes are significantly diminished and the neurological deficit is improved (Lu et al., 2010; Shen et al., 2009; Fakhri et al., 2018). The induction of various factors involved in oxidative stress processes, such as inducible nitric oxide synthase (iNOS), was suppressed by ATX treatment in rats (Lee et al., 2010). Furthermore, it was found that ATX pretreatment significantly inhibited apoptosis, mitochondrial abnormalities, and intracellular ROS generation in docosahexaenoic acid, hydroperoxide, or 6-hydroxydopamine-treated cells (Liu et al., 2009).

Administration of ATX-rich *H. pluvialis* extract to 10 subjects with age-related forgetfulness for 12 weeks resulted in no abnormality and showed effectiveness for improving cognitive and psychomotor functions (Satoh et al., 2009a; Katagiri et al., 2012). As for the efficacy of ATX on AD, ATX was found to decrease red blood cell (RBC) amyloid β-peptide (Aβ) concentrations, and the RBC Aβ levels were positively correlated with phospholipid hydroperoxide levels in RBCs (Kiko et al., 2012). These findings suggest that therapeutic application of ATX as an Aβ-lowering agent in RBCs could be considered as a possible antidementia agent. As a neurodegenerative disorder, PD is characterized by a selective loss of dopaminergic neurons in the substantia nigra pars compacta. Ikeda et al. (2008) reported that ATX significantly suppressed 6-hydroxydopamine-induced apoptosis in human neuroblastoma SH-SY5Y cells, presumably through attenuating intracellular ROS generation and mitochondrial dysfunctions, and ATX was found to have antithrombotic and antihypertensive effects in stroke-prone spontaneously hypertensive rats (Sasaki et al., 2011).

Cancer

In numerous epidemiological laboratory studies, carotenoids such as ATX have been implicated in protective actions against many forms of cancer. More specifically, ATX feeding has been found capable of suppressing the expression of inflammatory cytokines, including nuclear factor (NF)-κB, tumor necrosis factor (TNF)-α, and interleukin (IL)-1β, as well as inhibiting proliferation and inducing apoptosis of colonic adenocarcinomas (Yasui et al., 2011). ATX-rich *H. pluvialis* extract was also found to inhibit the growth of HCT-116 colon cancer cells by arresting cell cycle progression and promoting apoptosis (Palozza et al., 2009).

Furthermore, Hix et al. (2005) showed the ability of tetrasodium diphosphate astaxanthin (pAST), an ATX prodrug, to: (1) upregulate connexin 43 (Cx43) protein expression; (2) increase the formation of Cx43 immunoreactive plaques; (3) upregulate gap junctional intercellular communication; and (4) totally inhibit methylcholanthrene-induced neoplastic transformation at 10^{-6} M.

Rao et al. (2013) investigated the anticancer potency of ATX monoester (AXME) and diesters (AXDE) against the UV-7,12-dimethylbenz[a]anthracene (DMBA)-induced skin cancer model in rats and found that at 200 μg/kg body weight, AXDE and AXME reduced UV-DMBA-induced tumor incidences up to 96% and 88%, respectively. They also speculated that the anticancer potency of ATX esters could be due to increased bioavailability.

Diabetes and cardiovascular diseases

Metabolic syndrome is a disease marked by hypertriglyceridemia and low levels of high-density lipoprotein (HDL) cholesterol. The dysregulation of adipocytokines, secreted from adipose tissue, can affect insulin sensitivity involved in glucose and lipid metabolism. In a clinical study, 16 volunteers with metabolic syndrome showed significant decreases in HbAlc and TNF-α levels, along with a significant increase in adiponectin after treatment with ATX (Uchiyama and Okada, 2009). These results suggest that ATX has beneficial effects in patients with borderline diabetes mellitus or persons at risk for metabolic syndrome. For the ex vivo study, 24 volunteers receiving ATX for 14 days were found to have inhibited LDL oxidation, which might indicate a potential contribution by ATX to the prevention of atherosclerosis (Iwamoto et al., 2000).

Otton et al. (2010) reported that the hypolipidemic and antioxidant effects of ATX in plasma of ATX-fed/diabetic rats were apparently reflected in the circulating lymphocytes, because lower activities of catalase restored the ratio between glutathione peroxidase and glutathione reductase activities.

In the first-ever randomized, placebo-controlled human study, the results suggested that ATX consumption ameliorates triglyceride and HDL cholesterol in correlation with increased adiponectin (Yoshida et al., 2010).

Experimental studies in several species using an ischemia–reperfusion myocardial model have demonstrated that ATX protects the myocardium when administered either orally or intravenously prior to the induction of the ischemic event (Fassett and Coombes, 2011). Furthermore, it has been shown that ATX alleviates endothelial dysfunction and protects against cardiovascular disease. Sasaki et al. (2011) suggested that the antithrombotic and antihypertensive effects of ATX may be related to an increase in bioavailable nitric oxide (NO), possibly mediated by decreased NO inactivation by ROS. ATX treatment was able to partially reduce ROS and reactive nitrogen species (RNS) production by reducing the availability of nicotinamide adenine dinucleotide phosphate and by facilitating recovery of the phagocytic capacity of neutrophils (Guerra and Otton, 2011).

In their study, Kim et al. (2009) reported that ATX protects against oxidative stress in proximal tubular epithelial cells by preventing the progression of diabetic nephropathy, primarily through the ROS-scavenging effect in the mitochondria of mesangial cells. Thus, it is expected that ATX will be very useful for the prevention of diabetic nephropathy (Naito et al., 2004).

Kishimoto et al. (2010) indicated that ATX has inhibitory effects on macrophage activation in areas such as scavenger receptor upregulation, matrix metalloprotein activation, and proinflammatory cytokine secretion. It was also found that ATX supplementation decreased oxidative damage, restored defective antioxidant systems, and upregulated transforming growth factor-1 expression in obese mice displaying hyperglycemia, hyperinsulinemia, and insulin resistance (Bhuvaneswari et al., 2010). In these mice, insulin sensitivity was promoted and liver injuries were reduced by decreasing CYP2E1 levels. In addition, myeloperoxidase and nitrosative/oxidative stress was prevented.

ATX enhanced the autophosphorylation of insulin receptor-β (IR-β), IRS-1-associated PI3-kinase step, phospho-Akt/Akt ratios, and insulin-stimulated glucose transporter type four translocation and glucose uptake in skeletal muscle (Arunkumar et al., 2012; Ishiki et al., 2013). Bhuvaneswari and Anuradha (2012) demonstrated that ATX promotes the IRS-PI3K-Akt insulin-signaling pathway by decreasing serine phosphorylation of IRS proteins and improves glucose metabolism by modulating metabolic enzymes. In summary, ATX is a very effective antioxidant for ameliorating insulin resistance by protecting cells from the oxidative stress generated via various stimuli, including TNF-α. Additionally, Leite et al. (2010a,b) reported antioxidant alteration by ATX in the dental pulp and salivary glands of alloxan-induced diabetic rats.

Eyes

AMD is a progressive eye condition that is the number one cause of severe vision loss and legal blindness in adults older than 60. It is currently believed that ATX may be the most effective carotenoid for preventing various types of eye damage, including AMD. In an open-label randomized study with 145 patients (Piermarocchi et al., 2012), ATX supplementation was found to improve visual acuity and visual function in persons suffering from AMD.

In 2008, it was found that ATX treatment led to significant suppression of the choroidal neovascularization development associated with AMD (Izumi-Nagai et al., 2008). In addition, selective dysfunction in the central retina in nonadvanced AMD eyes was also improved by supplementation with carotenoids and antioxidants (Parisi et al., 2008).

As an antiinflammatory agent, 100 mg/kg ATX was found to be as effective as that of 10 mg/kg prednisolone. Furthermore, ATX decreased production of NO, PGE_2, and TNF-α in RAW264.7 cells in vitro (Ohgami et al., 2003). ATX was also found to reduce ocular inflammation with endotoxin-induced uveitis by downregulating proinflammatory factors and by inhibiting the NF-κB-dependent signaling pathway (Suzuki et al., 2006).

Eye fatigue, reduced eye accommodation, is a common eye disease ailment. ATX was found to be effective at improving accommodation during the recovery process by rapidly relieving related fatigue (Takahashi and Kajita, 2005; Nagaki et al., 2006; Iwasaki and Tahara, 2006). Additionally, uncorrected far visual acuity and asthenopia were significantly improved and the accommodation time was significantly shortened following ATX administration (Nakamura et al., 2004; Hiratori et al., 2005).

Xanthophylls are more potent than α-tocopherol for protecting human lens epithelial cells against UVB insult and ATX reduced oxidative stress in UVB-irradiated corneas (Lennikov et al., 2012).

ATX supplementation clearly suppressed total hydroperoxide production in human aqueous humor (Hashimoto et al., 2013) and inhibited the neurotoxicity induced by hydrogen peroxide or serum deprivation in cultured retinal ganglion cells in vitro and in vivo (Nakajima et al., 2008). The DNA damage in human neuroblastoma cells was also reduced by ATX supplementation (Santocono et al., 2007).

Gastrointestinal tract

Pretreatment with ATX esters in rats significantly increased antioxidant enzyme levels, including catalase, superoxide dismutase, and glutathione peroxidase in stomach homogenate and free radical-scavenging activity in vitro (Kamath et al., 2008). These results revealed that *H. pluvialis*-derived ATX might protect against gastric mucosal injury by stimulating the antioxidative mechanism.

ATX treatment significantly decreased the extent of HCl/ethanol-induced and acidified aspirin-induced gastric ulcers in mice and rats (Murata et al., 2012; Kim et al., 2005b). Electron spin resonance measurement showed that ATX had radical-scavenging activities against 1,1-diphenyl-2-picrylhydrazyl and superoxide anion radicals. These results suggest that ATX has antioxidant properties and exerts a protective effect against ulcer formation in murine models.

When compared with a proton pump-blocking antiulcer drug, omeprazole, ATX esters provide similarly beneficial effects on gastric ulcers by inhibiting H^+-ATPase and K^+-ATPase activity. Additionally, ATX induces a potential gastroprotective effect by regulating the gastric mucosal injury and gastric acid secretion by the gastric cells (Kamath et al., 2008; Kim et al., 2005a,b).

ATX significantly reduced reflux symptoms in dyspepsia patients (Kupcinskas et al., 2008). This response was more pronounced in *H. pylori*-infected patients. There was a significant upregulation of lymphocyte CD4 and a downregulation of CD8 in *H. pylori* patients treated with ATX (Andersen et al., 2007).

ATX has also been reported to have beneficial effects on liver conditions. For example, ATX treatment in rats for 14 consecutive days significantly decreased total histopathological scoring of cellular damage in hepatic ischemia–reperfusion injury (Curek et al., 2010). Specifically, electron microscopy of postischemic tissue demonstrated that parenchymal cell damage, swelling of mitochondria, and disarrangement of rough endoplasmic reticulum could be partially reduced by ATX supplements.

ATX has considerable potential for controlling the oxidative stress associated with nonalcoholic steatohepatitis in obese mice fed an unbalanced high-fat/high-fructose diet (HFFD) (Bhuvaneswari et al., 2010). Additionally, ATX supplementation for 60 days prevented liver injury by decreasing hepatic cytochrome P450 2E1 (CYP2E1), myeloperoxidase, and nitrosative/oxidative stress, and by improving the antioxidant status in mice. The increased transforming growth factor-β expression induced by HFFD was also abolished by ATX. Furthermore, ATX protected the liver from the toxic effects of CCl_4 and could be used as a hepatoprotectant against reactive oxygen radicals in mice (Kim et al., 2009).

Immune system

When consumed as a dietary supplement, ATX has been noted to reduce DNA damage level biomarkers and increase IFN-γ and IL-6 immune responses in young healthy females (Park et al., 2010). Both high glucose and a mixture of free fatty acids are potent inducers of ROS/RNS production in neutrophils, as observed by higher levels of superoxide anion, hydrogen peroxide, and NO production (Guerra and Otton, 2011). Furthermore, ATX treatment improves the human neutrophil function by stimulating increased phagocytic and fungicide capacity, as well as by reducing superoxide anion and hydrogen peroxide production without affecting the neutrophils' capacity to kill *Candida albicans* (Macedo et al., 2010).

Fucoxanthin, ATX, zeaxanthin, and β-carotene significantly inhibited the antigen-induced release of β-hexosaminidase in rat basophilic leukemia 2H3 cells and mouse bone marrow-derived mast cells (Sakai et al., 2009). These carotenoids also inhibited antigen-induced aggregation of the high-affinity IgE receptor (Fc epsilon RI), which is the most upstream of the degranulating mast cell signals.

In their clinical study, Satoh et al. (2009b) found that 27 male and female patients with mild to moderate atopic dermatitis displayed improved pruritus, state anxiety, immune function (Th1/Th2 ratio), and oxidative stress after 4 weeks of ATX intake without encountering any significant side effects. Since the immunoregulatory activity of ATX has been gradually revealed at the mechanistic level, it is expected that the severity of atopic dermatitis can be mitigated by long-term ATX intake.

Toxicity

ATX is used as a dietary supplement worldwide, and no adverse effects have been reported. As a component of combined nutraceuticals, ATX is safe, well tolerated, and not known for inducing side effects in patients (Spiller and Dewell, 2003; Marazzi et al., 2011). An acute toxicity study of the ATX-rich biomass obtained from *H. pluvialis* showed that the oral median lethal dose (LD_{50}) and the no observed adverse effect level (NOAEL) of the biomass were greater than 12,000 mg/kg and 465 mg/kg per day in rats, respectively (John et al., 2008). Using ATX produced from *H. pluvialis*, the LD_{50} and NOAEL of ATX were found to be greater than 2000 mg/kg and 50 mg/kg per day, respectively (Takahashi et al., 2005). In addition, no serious toxicity in mutagenicity (Ames test) was observed (Takahashi et al., 2005).

According to numerous human studies utilizing ATX from a variety of sources and in a wide range of dosages, there are no side effect concerns. ATX 6 mg per day from an *H. pluvialis* algal extract can be safely consumed by healthy adults (Spiller and Dewell, 2003; Satoh et al., 2009a), and ATX is generally recognized as safe in the United States.

Concluding remarks and future directions

Natural ATX has been marketed as a human dietary supplement since the late 1990s, when the US Food and Drug Administration first approved it for sale. Much of the ATX used and investigated to date has been obtained from *H. pluvialis* (Capelli et al., 2013). Comparative in vitro and in vivo studies have demonstrated that ATX is currently the most potent known natural antioxidant, and that it specifically quenches singlet oxygen. To improve the relatively low water solubility of free ATX, several ATX prodrugs with higher solubility have been synthesized and are now commercially available.

Due to its potent antioxidant activity, ATX is currently used in the treatment of cardiovascular, gastrointestinal, and neurodegenerative diseases, and cancer. Although several clinical studies reporting on the beneficial effects of ATX have been published, additional well-designed human clinical trials are still required to more thoroughly clarify the therapeutic potential of ATX against various diseases.

Acknowledgments

The author thanks Dr. Hidekuni Takahagi and Dr. Jiro Takahashi for their technical assistance in the preparation of this chapter.

References

Ames, B., Shigenara, M.K., Hagen, T.M., 1993. Oxidants, antioxidants, and the degenerative diseases of aging. Proc. Natl. Acad. Sci. USA 90, 7915–7922.

Andersen, L.P., Holck, S., Kupcinskas, L., 2007. Gastric inflammatory markers and interleukins in patients with functional dyspepsia treated with astaxanthin. FEMS Immunol. Med. Microbiol. 50, 244–248.

Arunkumar, E., Bhuvaneswari, S., Anuradha, C.V., 2012. An intervention study in obese mice with astaxanthin, a marine carotenoid—effects on insulin signaling and pro-inflammatory cytokines. Food Funct. 3, 120–126.

Bhuvaneswari, S., Anuradha, C.V., 2012. Astaxanthin prevents loss of insulin signaling and improves glucose metabolism in liver of insulin resistant mice. Can. J. Physiol. Pharmacol. 90, 1544–1552.

Bhuvaneswari, S., Arunkumari, E., Viswanathan, P., 2010. Astaxanthin restricts weight gain, promotes insulin sensitivity and curtails fatty liver disease in mice fed an obesity-promoting diet. Proc. Biochem. 45, 1406–1414.

Capelli, B., Bagehi, D., Cysewski, G.R., 2013. Synthetic astaxanthin is significantly inferior to alga-based astaxanthin as an antioxidant and may not be suitable as a human nutraceutical supplement. Nutrafoods 12, 145–152.

Choi, H.D., Kang, H.E., Yang, S.H., 2011a. Pharmacokinetics and first pass metabolism of astaxanthin in rats. Br. J. Nutr. 105, 220–227.

Choi, H.D., Kim, J.H., Chang, M.J., 2011b. Effects of astaxanthin on oxidative stress in overweight and obese adults. Phytother. Res. 25, 1813–1818.

Cooper, R.D.G., Davis, J.B., Leftwick, A.P., 1975. Carotenoids and related compounds. XXXII. Synthesis of astaxanthin, hoenicoxanthin, hydroxyechinenone, and the corresponding diosphenols. J. Chem. Soc. Perkin. Trans. 1 2195–2204.

Coral-Hinostroza, G.N., Ytrestoyl, T., Ruyter, B., 2004. Plasmâ appearance of unesterified astaxanthin geometrical E/Z and optical R/S isomers in men given single doses of a mixture of optical 3 and 3′R/S isomers of astaxanthin fatty acyl diesters. Comp. Biochem. Physiol. C Toxicol. Pharmacol. 139, 99–110.

Curek, G.D., Cort, A., Yucel, G., 2010. Effect of astaxanthin on hepatocellular injury following ischemia/reperfusion. Toxicology. 267, 147–153.

Davies, K.J., 1995. Oxidative stress: the paradox of aerobic life. Biochem. Soc. Symp. 61, 1–31.

Dean, O.M., van den Buuse, M., Berk, M., 2011. *N*-acetyl cysteine restores brain glutathione loss in combined 2-cyclohexene-1-one and d-amphetamine-treated rats: relevance to schizophrenia and bipolar disorder. Neurosci. Lett. 499, 149–153.

Dore, J.E., Bagchi, D., Preuss, H.G., 2005. Astaxanthin and cancer chemoprevention. In: Bagchi, D., Preuss, H.G. (Eds.), Phytopharmaceuticals in Cancer Chemoprevention. CRC Press, Boca Raton, FL2005555–570.

Fakhri, S., Abbaszadeh, F., Dargahi, L., Jorjani, M., 2018. Astaxanthin: a mechanistic review on its biological activities and health. Pharmacol. Res. 136, 1–20.

Fassett, R.G., Coombes, J.S., 2011. Astaxanthin: a potential therapeutic agent in cardiovascular diseae. Mar. Drugs. 9, 447–465.

Fassett, R.Q., Healy, H., Driver, R., 2008. Astazxanthin vs placebo on arterial stiffness, oxidative stress and inflammation on renal transplant patients (Xanthin): a randomized controlled trial. BMC Nepgrol 18, 9–17.

Gross, G.J., Lockwood, S.F., 2004. Cardioprotection and myocardial salvage by a disodium disuccinate astaxanthin derivative (Cardax™). Life Sci. 75, 215–224.

Guerra, B.A., Otton, R., 2011. Impact of the carotenoid Astaxanthin on phagocytic capacity and ROS/RNS production of human neutrophils treated with free fatty acids and high glucose. Int. Immunopharmacol. 11, 2220–2226.

Halliwell, B., 1992. Reactive oxygen species and the central nervous system. J. Neurochem. 59, 1609–1623.

Halliwell, B., 2007. Oxidative stress and cancer: have we moved forward? Biochem. J. 401, 1–11.

Hashimoto, H., Arai, K., Hayashi, S., 2013. Effects of astaxanthin on antioxidation in human aqueous humor. J. Clin. Biochem. Nutr. 53, 1–7.

Higuera-Ciapara, I., Felix-Valenzuela, L., Goyoolea, F.M., 2006. Astaxanthin: a review of its chemistry and applications. Crit. Rev. Food Sci. Nutr. 40, 185–196.

Hiratori, K., Ogami, K., Nitta, T., 2005. Effect of astaxanthin on accommodation and asthenopia-efficacy-identification study in healthy volunteers. J. Clin. Ther. Med. 21, 637–650.

Hix, L.M., Frey, D.A., McLaws, M.D., 2005. Inhibition of chemically-induced neoplastic transformation by a novel tetrasodium diphosphate astaxanthin derivative. Carcinogenesis. 26, 1634–1641.

Hussein, G., Goto, H., Oda, S., 2006. Antihypertensive potential and mechanism of action of astaxanthin. III. Antioxidant and histopathological effects in spontaneously hypertensive rats. Biol. Pharm. Bull. 29, 684–688.

Ikeda, Y., Tsuji, S., Satoh, A., 2008. Protective effects of astaxanthin on 6-hydroxydopamine-induced apoptosis in human neuroblastoma SH-SY5Y cells. J. Neurochem. 107, 1730–1740.

Ishiki, M., Nishida, Y., Ishibashi, H., 2013. Impact of divergent effects of astaxanthin on insulin signaling in L6 cells. Endocrinology. 154, 2600–2612.

Iwamoto, T., Hosoda, K., Hirano, R., 2000. Inhibiition of low-density lipoprotein oxidation by astaxanthin. J. Atheroscler. Thromb. 7, 216–222.

Iwasaki, T., Tahara, A., 2006. Effects of astaxanthin on eyestrain by accommodative dysfunction. J. Eye. 23, 829–834.

Izumi-Nagai, K., Nagai, N., Ohgami, K., 2008. Inhibition of choroidal neovascularization with an anti-inflammatory carotenoid astaxanthin. Invest. Ophthalmol. Vis. Sci. 49, 1679–1685.

Jackson, H.L., Cardoune, A.J., Zweier, J.L., 2004. Characterization, and direct aqueous superoxide anion acavenging of a highly water-dispersible astaxanthin-amino acid conjugate. Bioorg. Med. Chem. Lett. 14, 3985–3991.

John, S., Stewart, J.S., Lingnell, Å., 2008. Safety assessment of astaxanthin-rich microalgae biomass: acute and subchronic toxicity studies in rats. Food Chem. Toxicol. 46, 3030–3036.

Kamath, B.S., Srikanta, B.M., Dharmesh, S.M., 2008. Ulcer preventive and antioxidative properties of astaxanthin from *Haematococcus pluvialis*. Eur. J. Pharmacol. 590, 387–395.

Katagiri, M., Satoh, A., Tsuji, S., 2012. Effects of astaxanthin-rich *Haematococcus pluvialis* extract on cognitive function:a randomised, double-blind, placebo-controlled study. J. Clin. Biochem. Nutr. 51, 102–107.

Kavitha, K., Kowshik, J., Kishore, T.K., 2013. Astaxanthin inhibits NF-κB and Wnt/β-catenin signaling pathways via inactivation of Erk/MA PK and PI3K/Akt to induce intrinsic apoptosis in a hamster model of oral cancer. Biochem. Biophys. Acta. 1830, 4433–4444.

Kiko, T., Nakagawa, K., Satoh, A., Tsuduki, T., Furukawa, K., Arai, H., 2012. Amyloid β levels in human red blood cells. PLoS One. 7 (e-49620), 1–6.

Kim, J.H., Kim, Y.S., Song, G.G., 2005a. Protective effect of astaxanthin on naproxen-induced gastric antral ulceration in rats. Eur. J. Pharmacol. 514, 53–59.

Kim, J.H., Choi, S.K., Choi, S.Y., 2005b. Suppressive effect of astaxanthin isolated from the *Xanthophyllomyces dendrorhous* mutant on ethanol-induced gastric mucosal injury in rats. Biosci. Biotechnol. Biochem. 69, 1300–1305.

Kim, S.H., Jean, D.I., Lim, Y.P., Lee, C., An, G., 2009. Weight gain limitation and liver protection by long-term feeding of astaxanthin in murines. J. Korean Soc. Appl. Biol. Chem. 52, 180–185.

Kim, J.H., Chang, M.J., Choi, H.D., 2011. Protective effect of *Haematococcus* astaxanthin on oxidative stress in healthy smokers. J. Med. Food. 14, 1–7.

Kishimoto, Y., Tami, M., Uto-Kondo, H., 2010. Astaxanthin suppresses scavenger receptor expression and matrix metalloproteinase activity in macrophages. Eur. J. Nutr. 49, 119–126.

Kistler, A., Liechti, H., Pichard, L., 2002. Metabolism and CYP-inducer properties of astax anthin in man and primary human hepatocytes. Arch. Toxicol. 75, 665–675.

Kupcinskas, L., Lafolie, P., Lignell, A., 2008. Efficacy of the natural antioxidant astaxanthin in the treatment of functional dyspepsia in patients with or without *Helicobacter pylori* infection: a prospective, randomized, double blind, and placebo-controlled study. Phytomedicine. 15, 391–399.

Lee, D.H., Lee, Y.J., Kwon, K.H., 2010. Neuroprotective effects of astaxanthin in oxygen-glucose deprivation in SH-SY5Y cells and global cerebral ischemia in rat. J. Clin. Biochem. Nutr. 447, 121–129.

Leite, M.F., de Lima, A., Massuyama, M.M., 2010a. *In vivo* astaxanthin treatment partially prevents antioxidant alterations in dental pulp from alloxan induced diabetic rats. Int. Endod. J. 43, 959–967.

Leite, M.F., de Lima, A.M., Massuyama, M., 2010b. Astaxanthin restores the enzymatic antioxidant profile in salivary gland of alloxan-induced diabetic rats. Arch. Oral Biol. 55, 479–485.

Lennikov, A., Nobuyoshi, K., Fukase, R., 2012. Amelioration of ultraviolet-induced photokeratins in mice treated with astaxanthin eye drop. Mol. Vis. 18, 455–464.

Liu, X., Osawa, T., 2007. *Cis* astaxanthin and especially 9-*cis* astaxanthin exhibits a higher antioxidant activity *in vitro* compared to the all-trans isomer. Biochem. Biophys. Res. Commun. 357, 187–193.

Liu, X., Shibata, T., Hisaka, S., 2009. Astaxanthin inhibits reactive oxygen species-mediated cellular toxicity in dopaminergic SH-SY5Y cells via mitochondria-targeted protective mechanism. Brain Res. 1254, 18–27.

Lu, Y.P., Liu, S.Y., Sun, H., 2010. Neuroprotective effect of astaxanthin on H_2O_2-induced neurotoxicity *in vitro* and on focal cerebral ischemia *in vivo*. Brain Res. 1360, 40–48.

Macedo, R.C., Bolin, A.P., Marin, D.P., 2010. Astaxanthin addition improves human neutrophils function: *in vitro* study. Eur. J. Nutr. 49, 447–457.

Marazzi, G., Cacciotti, L., Pelliccia, F., 2011. Long-term effects of nutraceuticals (berberine, red yeast rice, policosanol) in elderly hypercholesterolemic patients. Adv. Ther. 28, 1105–1113.

Mercke Odeberg, J., Lignell, A., Pattersson, A., 2003. Oral bioavailability of the antioxidant astaxanthin in humans is enhanced by incorporation of lipid based formulations. Eur. J. Pharm. Sci. 19, 299–304.

Miyazawa, T., Nakagawa, K., Kimura, F., 2011. Plasma cartenoid concentrations before and after supplementation with astaxanthin in middle-aged and senior subjects. Biosci. Biotechnol. Biochem. 75, 1856–1858.

Murata, K., Oyagi, A., Takahira, D., 2012. Protective effects of astaxanthin from *Paracoccus carotinifaciens* on murine gastric ulcer models. Phytother. Res. 26, 1126–1132.

Nagaki, Y., Mihara, M., Tsukahara, H., 2006. The supplementation effect of astaxanthin on accommodation and asthenopia. J. Clin. Ther. Med. 22, 41–54.

Naito, Y., Ushiyama, K., Aoi, W., 2004. Prevention of diabetic nephropathyl by treatment with astaxanthin in diabetic *db/db* mice. Biofactors. 20, 49–59.

Nakajima, Y., Inokuchi, Y., Shimazawa, M., 2008. Astaxanthin, a dietary carotenoid, protects retinal cells against oxidative stress *in-vitro* and in mice *in-vivo*. Pharm. Pharmacol. 60, 1365–1374.

Nakamura, A., Isobe, R., Otaka, Y., 2004. Changes in visual function following peroral astaxanthin. Jpn. J. Clin. Ophthalmol. 58, 1051–1054.

Nishida, Y., Yamashita, E., Miki, W., 2007. Quenching activities of common hydrophilic and lipophylic antioxidants against singlet oxygen using chemiluminescence detection system. Carotenoid Sci. 11, 16–20.

Ohgami, K., Shiratori, K., Kotake, S., 2003. Effects of astaxanthin on lipopolysaccharide-induced inflammation *in vitro* and *in vivo*. Invest. Ophthalmol. Vis. Sci. 44, 2694–2701.

Ohi, Y., Kitamura, A., Tsukahara, H., Takahashi, J., Itakura, H., 2011. Pharmacokinetic of astaxanthin after oral administration of a soft gel capsule preparation (in Japanese). Rinshoiyaku 27, 297–303.

Ohno, M., Darwish, W.S., Ikenaka, Y., 2011. Astaxanthin can alter CYP1A-dependent activities via two different mechanisms: induction of protein expression and inhibition of NADPH P450 reductase dependent electron transfer. Food Chem. Toxicol. 49, 1285–1291.

Okada, Y., Ishukura, M., Maoka, T., 2009. Bioavailability of astaxanthin in *Haematococcus* algal extract: the effects of timing of diet and smoking habits. Biosci. Biotechnol. Biochem. 73, 1928–1932.

Otton, B., Marin, D.P., Bolin, A.P., 2010. Astaxanthin ameliorates the redox imbalance in lymphocytes of experimental diabetic rats. Chem. Biol. Interact. 186, 306–315.

Palozza, P., Torelli, C., Boninsegna, A., 2009. Growth-inhibitory effects of the astaxanthin-rich alga *Haematococcus pluvialis* in human colon cancer cells. Cancer Lett. 283, 108–117.

Parisi, V., Piermarocchi, S., Tedeschi, M., Gallinaro, G., Varano, M., Saviano, S., CARMIS Study Group, 2008. Carotenoids and antioxidants in age-related maculopathy italian study: multifocal electroretinogram modifications after 1 year. Ophthalmology. 115, 324–333.

Park, J.S., Chyun, J.H., Kim, Y.K., 2010. Astaxanthin decreased oxidative stress and inflammation and enhanced immune response in humans. Nutr. Metab 5, 7–18.

Petri, D., Lundebye, A.K., 2007. Tissue distribution of astaxanthin in rats following exposure to graded levels in the feed. Comp. Biochem. Physiol. C. 145, 202–209.

Piermarocchi, S., Saviano, S., Parisi, V., 2012. Carotenoids in age-related maculopathy italian study (CARMIS): two-year results of a randomized study. Eur. J. Ophthalmol. 22, 216–225.

Pohanka, M., 2013. Alzheimer's disease and oxidative stress: a review. Curr. Med. Chem. 21, 356–364.

Querin, M., Huntley, M.E., Olaizola, M., 2003. *Haematococcus* astaxanthin: applications for human health and nutrition. Trends Biotechnol. 21, 210–216.

Raneva, V.G., Shimasaki, H., Ueta, N., 2003. Interaction between α-tocopherol, tocotrienols and astaxanthin in liposome, subjected to lipid peroxidation. J. Oleo Sci. 7, 347–352.

Rao, A.R., Reddy, R.L., Baskaran, V., 2010. Characterization of microalgal cartenoids by mass spectrometry and their bioavailability and antioxidant properties elucidated in rat model. J. Agric. Food Chem. 58, 8553–8559.

Rao, A.R., Sindhuja, H.N., Dharmesh, S.M., 2013. Effective inhibition of skin cancer, tyrosinase, and antioxidative properties by astaxanthin and astaxanthin esters from the green alga *Haematococcus pluvialis*. J. Agric. Food Chem. 61, 3842–3851.

Rufer, C.E., Moesender, J., Jarlis, B., 2008. Bioavailability of astaxanthin stereoisomers from wild (*Oncorhynchus* spp.) and aquacultured (*salmo salar)* samon and healthy men. A randomized, double-blind study. Br. J. Nutr. 99, 1048–1054.

Ryu, S.K., King, T.J., Fujioka, K., 2012. Effect of an oral astaxanthin prodrug (CDX-085) on lipoprotein levels and progression of atherosclerosis in $LDLR^{-/-}$ and $ApoE^{-/-}$ mice. Atherosclerosis. 222, 99–105.

Sakai, S., Sugawara, T., Matsubara, K., 2009. Inhibitory effect of carotenoidson the degranulation of mast cells via suppression of antigen-induced aggregation of high affinity IgE receptors. J. Biol. Chem. 284, 28172–28179.

Santocono, M., Zurria, M., Berrettini, M., 2007. Lutein, zeaxanthin and astaxanthin protect against DNA damage in SK-N-SH human neuroblastoma cells induced by reactive nitrogen species. Photochem. Photobiol. B 88, 1–10.

Sasaki, Y., Kobara, N., Higashino, S., 2011. Astaxanthin inhibits thrombosis in erebral vessels of stroke-prone spontaneously hypertensive rats. Nutr. Res. 31, 784–789.

Satoh, A., Tsuji, S., Okada, Y., 2009a. Preliminary clinical evaluation of toxicity and efficacy of a new astaxanthin-rich *Haematococcus pluvialis* extract. J. Clin. Biochem. Nutr. 44, 280–284.

Satoh, A., Kawamura, T., Horibe, T., 2009b. Effect of the intake of astaxanthin-containing *Haematococcus pluvialis* extract on the severity, immunofunction and physiological function in patients with atopic dermatitis. J. Environ. Cutan. Allergol. 3, 429–438.

Shen, H., Kuo, C.C., Chou, J., 2009. Astaxanthin reduces ischemic brain injury in adult rats. FASEB J. 23, 1958–1968.

Showalter, L.A., Weinman, S.A., Østerlie, M., 2004. Plasma appearance and tissue accumulation of non-esterified, free astaxanthin in C57BL/6 mice after oral dosing of a disodium disuccinate diester of astaxanthin (Heptax™). Comp. Biochem. Physiol. C. Toxicol. Pharmacol. 137, 227–236.

Spiller, G.A., Dewell, A., 2003. Safety of an astaxanthin-rich *Haematococcus pluvialis* algal extract: a randomized clinical trial. J. Med. Food. 6, 51–56.

Suzuki, Y., Ohgam, K., Shiratori, K., 2006. Suppressive effects of astaxanthin against rat endotoxin-induced uveitis by inhibiting the NF-kappa B signaling pathway. Exp. Eye Res. 82, 275–281.

Sy, C., Gleize, B., Dangles, O., 2012. Effects of physicochemical properties of carotenoids on their bioaccessibility, intestinal cell uptake, and blood and tissue concentrations. Mol. Nutr. Food Res. 56, 1385–1397.

Takahashi, N., Kajita, M., 2005. Effects of astaxanthin on accommodative recovery. J. Clin. Therap. Med. 21, 431–436.

Takahashi, J., Tsukahara, H., Minato, S., 2005. Toxicological studies of astaxanthin from *Haematococcus pluvialis*-Ames test, oral single dose and 90-days subchronic toxicity studies in rats. J. Clin. Therap. Med. 20, 867–881.

Takizawa, Y., Kitazato, T., Kishimoto, H., 2011. Effects of antioxidants on drug absorption in *in vivo* intestinal ischemia/reperfusion. Eur. J. Drug Metab. Pharmacokinet. 35, 89–95.

Uchiyama, A., Okada, Y., 2009. Clinical efficacy of astaxanthin containing *Haematococcus pluvialis* extract for the volunteers at risk of metabolic syndrome. J. Clin. Biochem. Nutr. 44, 280–284.

Valko, M., Leibfritz, D., Moncol, J., 2007. Free radicals and antioxidants in normal physiological functions and human disease. Int. J. Biochem. Cell Biol. 39, 44–84.

Wolf, A.M., Asoh, S., Hiranuma, H., 2010. Astaxanthin protects mitochondrial redox state and functional integrity against oxidative stress. J. Nutr. Biochem. 21, 381–389.

Wolz, E., Liechti, H., Notter, B., 1999. Characterization of metabolites of astaxanthin in primary cultures of rat hepatocytes. Drug Metab. Dispos. 27, 456–462.

Yang, Y., Kim, B., Lee, J.Y., 2013. Astaxanthin structure, metabolism and health benefits. J. Hum. Nutr. Food Sci. 1, 1003–1013.

Yasui, Y., Hosokawa, M., Mikami, N., 2011. Dietary astaxanthin inhibits colitis and colitis-associated colon carcinogenesis in mice via modulation of the inflammatory cytokines. Chem. Biol. Interact. 193, 79–87.

Yoshida, H., Yanai, H., Ito, K., 2010. Administration of natural astaxanthin increases serum HDL-cholesterol and adiponectin in subjects with mild hyperlipidemia. Atherosclerosis. 209, 520–523.

Yuan, J.P., Peng, J., Yin, K., 2011. Potential health-promoting effects of astaxanthin: a high-value carotenoid mostly from microalgae. Mol. Nutr. Food Res. 55, 150–165.

Chapter 52

Thymoquinone

Amul Jain, Leena Dhruw, Priyank Sinha, Anchal Pradhan, Rahul Sharma and Bhanushree Gupta

Introduction

Medicinal plants are a vital source of flavors and pigments, and contain potent pharmacological properties. *Nigella sativa* (NS) is a promising nutraceutical plant and is used to remedy various ailments. This plant has been widely grown in western Asia, eastern Europe, the Middle East and Mediterranean countries (Sharma et al., 2019). The plant has been known for its fragrance which is attributed to its essential oils. These essential oils can be obtained from the seeds (Alhaj et al., 2008). Various encouraging properties of the essential oils of NS are attributed to its major phytochemical, thymoquinone (TQ) (Fig. 52.1). The presence of TQ has made NS a fascinating plant and is responsible for its various biological activities which include antioxidant, antiinflammatory, antidiabetic, drug detoxification, anticancer, etc. (Nagi et al., 2010). TQ is chemically known as 2-methyl-5-isopropyl-1,4-benzoquinone, and comes under the class of monoterpene.

TQ is affiliated to a terpenoids family consisting of a quinone framework. It was first isolated by El-Dakhakhny (1963), using thin-layer chromatography. The shelf-life of TQ is affected by several factors such as pH, solvent, and light. It is unstable at higher pH, and is sensitive to light exposure. Degradation of TQ was observed in aqueous media, thus there is a need to increase its stability in an aqueous medium which could be achieved through nanoencapsulation or derivatization.

Thymoquinone derivatives and its biological activity

In order to obtain enhanced efficacy and lipophilicity of TQ for cellular acceptance, a variety of derivatives have been synthesized. Derivatization can improve various pharmacological properties of the parent compounds such as increase biological efficacy, reduce undesired side effects and lower toxicity, better bioavailability, and addition of completely new features that were absent in the parent compound. Coupling of two or more natural product fragments is one of the standard approaches during derivatization, that is, hybridization of bioactive natural products. To achieve better pharmacological activities, the development of new analogs is one of the most fundamental and novel approaches.

Fröhlich et al. (2017) reported seven novel analogs of TQ in combination with artemisinin (Art) (an anticancer drug) and artesunic acid. All seven hybrids were tested for anticancer activity against colon, breast, prostate, liver, and leukemia cell lines. All tested analogs show high specificity for colon cancer cells, among them analog **1** (Fig. 52.2) was found to be a 20-fold more effective analog than Art and TQ itself against the colorectal cancer cell lines with an IC_{50} of 2.4 μM in HCT116 cells and 2.8 μM in HT29 cells and nontoxic for nonmalignant colon epithelial cells with IC_{50} >100 μM. They continued their work using the same methodology and reported another five new analogs of TQ-Art hybrids and further examined there in vitro biological activity against two leukemia cell lines CEM/ADR5000 (and CCRF-CEM), malaria parasites (*Plasmodium falciparum* 3D7), and HCMV (human cytomegalovirus) (Fröhlich et al., 2018). Among all five analogs, only **2a** and **2b** (Fig. 52.3) were recognized as the most active compounds toward all tested cell lines. It showed an EC_{50} value of 0.2 μM against the doxorubicin-sensitive as well as the multidrug-resistant leukemia cells, which is greater than the effect of doxorubicin. Both analogs exhibited outstanding antimalarial activity with improved EC_{50} values of 5.9 and 3.7 nM, respectively, than those for artesunic acid

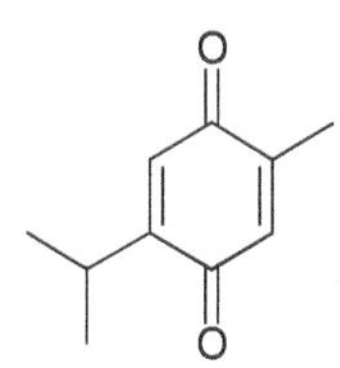

FIGURE 52.1 Structure of thymoquinone (2-methyl-5-isopropyl-1, 4-benzoquinone).

Nutraceuticals. DOI: https://doi.org/10.1016/B978-0-12-821038-3.00052-5

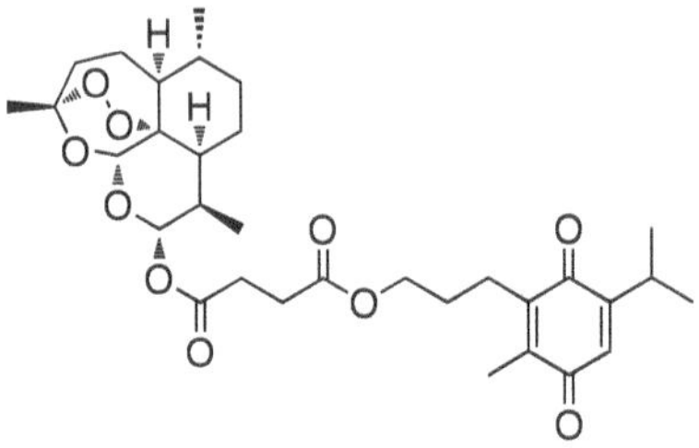

FIGURE 52.2 Analog 1.

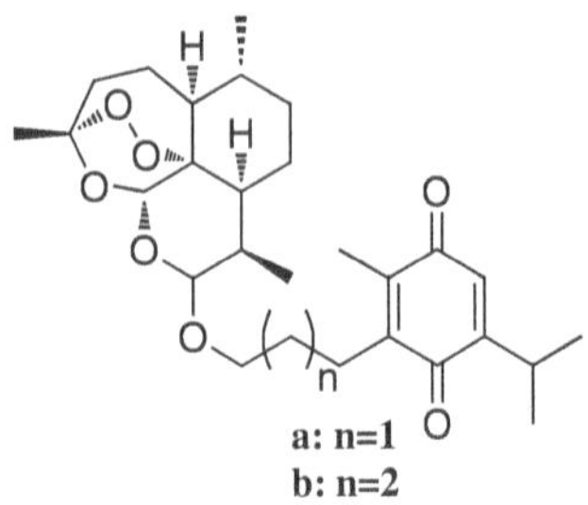

FIGURE 52.3 Analog 2.

(EC_{50} = 8.2 nM), chloroquine (EC_{50} = 9.8 nM), and dihydro-artemisinin (EC_{50} = 2.4 nM). It exhibited great antiviral activity against HCMV than the other reference drug.

Johnson-Ajinwo et al. (2018) synthesized 21 analogs of TQ and studied their antiproliferative activities against human malarial parasite and ovarian cancer cell lines. Among those, six were nitrogen substituted, five were reduced hydroquinone, and another 10 analogs were substituted with different alkyl chain and halogen groups. After analyzing the structure–activity relationships (SARs) of all TQ derivatives, they found that the IC_{50} value of 11 analogs was less than 10 μM in the A2780 cell line, which also showed strong cytotoxicity in CIS-A2780 (cisplatin-resistant cell line) and OVCAR-8 cancer cell lines. In comparison with the commonly used ovarian cancer drug carboplatin, most of the analogs were found to be more potent. Analogs with halogen atoms demonstrated similar potential to TQ. On the other hand, activity of an alkylamino or amino group was found to be decreased while their cytotoxic activity was doubled against the cancer cell line. Sybr Green (fluorescent dye) i-based fluorescence assay was used to study the antiplasmodial activity of TQ analogs against *P. falciparum*, that is, human malarial parasite. Of 21 analogs, 10 compounds showed more effective antiplasmodial activity than TQ itself. Surprisingly, nitrogen-substituted TQ compound **3b** (Fig. 52.4) exhibited twice as much antiplasmodial activity. The IC_{50} values of TQ analogs **3a** and **3b** (Fig. 52.4) were less than 10 μM against *P. falciparum* cell line and showed twofold more inhibitory activities than TQ.

Biotransformation of bioactive compounds is also one of the aspects to obtain a TQ derivative. Mohammad et al. (2018) reported the microbial transformation of TQ to its analogs by *Aspergillus niger*, a plant pathogenic fungus in a suspended cell-culture. Among three synthesized analogs, analog **4a** (Fig. 52.5) was reported for the first time and analog **4c** (Fig. 52.5) was found to be a powerful antioxidant agent concerning ascorbic acid and TQ. This study showed that hydroxylation and lactonization of monoterpene can be easily obtained through biotransformation by *Aspergillus niger*.

Anticancer potentials of thymoquinone in combination with various drugs

Various chemotherapeutic agents are used to reduce the effect of a variety of cancers (breast, hepatic, etc.), some of these include cisplatin (*cis*-diamminedichloroplatinum II) (CIS), resveratrol (3,5,4'-trihydroxy-trans-stilbene) (RES), doxorubicin (DOX), gemcitabine (GCB), and others (Fig. 52.6). These chemotherapeutic agents are characterized by high cytotoxicity. CIS has an inhibitory action on DNA replication, RNA formation, apoptosis, and activating cell cycle arrest by binding to guanine bases of DNA strands (Huang et al., 2017). TQ has been tested against various commercial drugs to diminish these cytotoxic effects. It was found that TQ was able to alter the activities of these chemotherapeutic drugs.

Shahid et al. (2017) investigated the effect of TQ with CIS on intestinal brush-border membrane (BBM). Treatment with CIS has demonstrated extensive intestinal damage, which includes marked degeneration, a fusion of villi, villus hypertrophy, and apoptosis. Interestingly, no such adverse effect was observed upon administration of TQ. Lysosomal marker enzyme and BBM marker enzyme assay were performed through the method reported by Goldstein and Oliver Lampen (1975). Salim et al. (2017) reported a method to determine the activity of acid phosphatase and lysosomal enzyme. A significant decline was observed in the activity of BBM marker enzyme. However, the activity of the lysosomal enzyme and acid phosphatase was increased by 59.38%. Moreover, no significant effect was observed in BBM marker enzyme and lysosomal enzyme by TQ administration only. The coadministration of CIS and TQ prevented the CIS-triggered decrease in activities. Alobaedi et al (2017) determined the effect of TQ in combination with resveratrol against epithelial breast cancer cell in mice, namely EMT6/P. T47D and MCF-7 are two human epithelial cancer cell lines which were also used in this study. Vero cell line, which is the kidney cells obtained from the African green monkey, was used as a normal control. Antiproliferative studies were performed through antiproliferative assay.

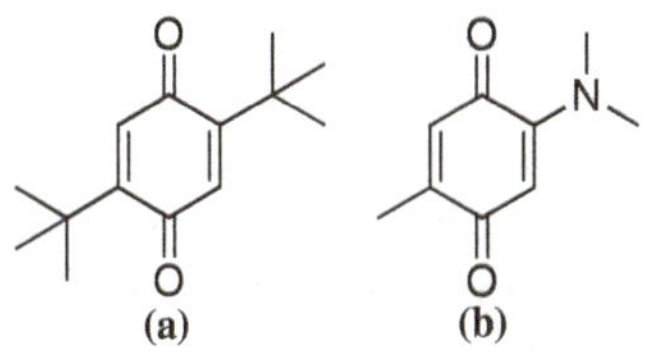

FIGURE 52.4 Analog 3.

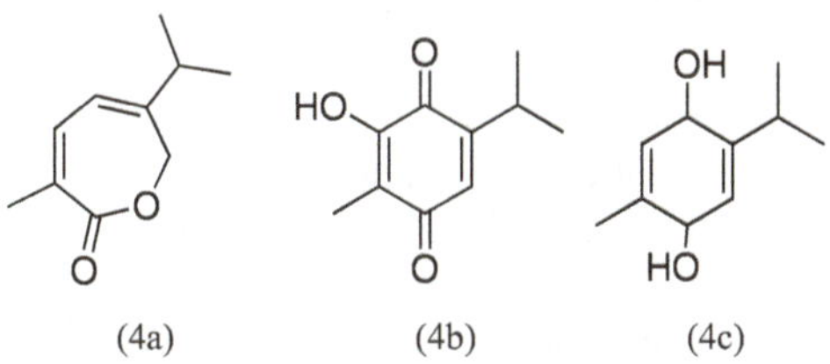

FIGURE 52.5 Analog 4.

The findings revealed that upon treatment with TQ, Vero cell lines had the highest activity, while MCF-7 cells had the lowest activity. When these cells were treated with RES, EMT6/P cells showed the highest activity, while T47D cells showed the lowest. When a combination of RES and TQ (50 μM and 140 μM, respectively) was applied against T47D cells, the highest inhibition rate of 53% was observed. Based on the above results, EMT6/P was chosen for inducing tumor in mice. The highest antitumor activity was recorded for the combination of TQ and RES, which resulted in an increment of 0.37% ± 0.05% in the tumor size. Among the treated mice, 40% were cured with TQ administration, while 60% were cured with the combination of TQ and RES. Thus, the combination has high ability to prevent the growth of breast cancer cells both in vitro and in vivo. The studies

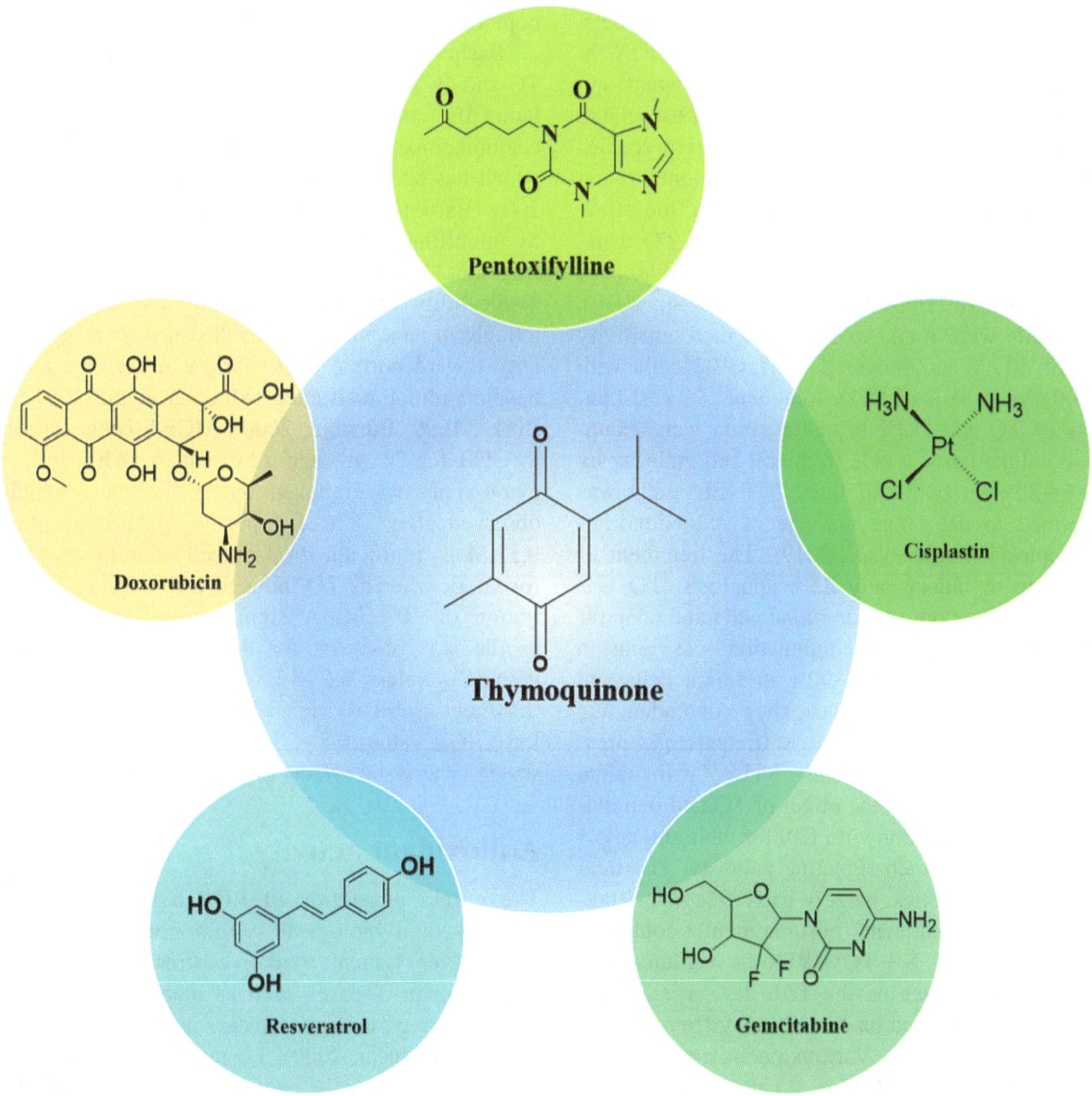

FIGURE 52.6 Structure of various commercially available drugs used.

revealed that the mice treated with TQ exhibited approximately a 47% delay in tumor growth compared with the untreated mice. TQ was able to initiate DNA fragmentation and apoptosis in the mice (Woo et al., 2013). The combination has the highest ability to induce apoptosis. Doses of RES, such as 3000 mg/kg, were considered toxic and nephrotoxic (Crowell et al., 2004). The study used 50 mg/kg of RES to preserve the anticancer potential with less toxic effects (Harper et al., 2007). Thus, TQ and RES work synergistically to inhibit the cancerous cell growth in breast cancer cell lines.

Another class of cancer attributed to human T-cell lymphotropic virus HTLV-1 is the Adult T-cell leukemia. Fatfat et al. (2019) demonstrated an enhancement in the anticancer activity of DOX towards adult T-cell leukemia when used in combination with TQ. A time- and dose-dependent caspase-triggered apoptosis was induced by TQ in both HTLV-1 negative CD + T-cell and HTLV-1-positive cells (Dergarabetian et al., 2013). TQ has been found to improve the anticancer capacity of DOX in TOPO breast cancer cell/MCF-7 multidrug-resistant and Hl-60 myeloblastic T-cell cancer cells through associated mitochondrial apoptosis and reactive oxygen species (ROS)-mediated mechanism (Effenberger-Neidnicht and Schobert, 2011). The group further investigated the effect of DOX individually and in combination with TQ using MTT assay and against HTLV-1-negative (Jurkat cell) and HTLV-1-positive (Hut-102 cell) (Fatfat et al., 2019). The Jurkat cells were found to have a higher sensitivity with IC_{50} of 50 nM as compared to Hut-102 cells with IC_{50} of 100 nM at 48 h of DOX treatment. Several concentrations of TQ and DOX in combination were examined. The combination was able to inhibit cell viability by up to 70%–80% in both cell lines. TUNEL assay was also performed to determine the rate of mitochondrial-induced apoptosis (Fatfat et al., 2019). The treatment of DOX was able to induce only 25% apoptosis. TQ was able to induce 55% apoptosis for Jurkat cells and 65% for HuT-102 cells, whereas the combination was able to enhance this effect up to 70%–90% in Jurkat cells and HuT-102 cells, respectively. Hence, they concluded that twofold lower doses of DOX were sufficient to express the same inhibitory activity against ATL. Zarringhalam and Toghraie (2019) studied the effect of TQ and pentoxifylline (PTX) in combination with CIS for their anticancer activities through the Notch signaling pathway. The treated mice in two subgroups, namely early treatment (pretreatment) and late treatment (posttreatment), contained CIS, CIS + TQ, and CIS + TQ + PTX as variants. The early treatment was given on the 12th day, whereas the last treatment was provided on the 19th day from the day of tumor induction in mice. Various parameters, such as Hes 1, Jagged 1, Notch 1, TNF-α, β-catenin, IL-6, IL-2, IFN-γ, VEGF, CD4, CD8, and apoptosis, were determined. In contrast to CIS treatment, CIS + TQ showed a significant decrease in tumor weight and tumor volume in early-treated mice, but they were not able to exhibit the same effect in the late-treated mice. Similarly, a combination of CIS + TQ + PTX was able to inhibit tumor growth (weight and volume) more significantly than CIS + TQ combination. The combination (CIS + TQ + PTX) was highly active for pretreated mice but showed much less response in the post-treated mice. CIS + TQ was able to suppress the gene expression of HES 1 by 46% and Notch 1 by 49%, whereas the post-treated mice exhibited gene suppression by 22% for both genes. The CIS + TQ + PTX combination suppresses the gene to 80% in posttreatment mice, while in the early treatment the inhibition was found to be 36% for HES 1 and 32% for Notch 1. Through hypoxia-inducible factor, the gene Notch 1 can induce cancer (Yuan et al., 2015). This gene is also able to suppress p53 pathway and proapoptotic gene (Xiao et al., 2019).

Bashmail et al. (2018) studied the synergistic effect of TQ and GCB as an antibreast cancer agent through modulating the autophagy and apoptosis activities. GCB is a chemotherapeutic agent with a nucleoside framework which has been approved for treating breast cancer since 2004 (Barton-Burke, 1999). It exhibits anticancer activity by inhibiting DNA synthesis via incorporation into DNA fragment upon triphosphate activation. Despite wide applicability of GCB in treating cancerous cells, it is thought to have many downsides such as exaggerated toxicity toward normal cells, narrow selectivity, and it may develop tumor resistance (Moorsel et al., 1997; Barton-Burke, 1999; Burstein, 2000). MCF-7 cells were exposed to GCB for 24, 48, and 72 h. Up to 48 h a low level of cytotoxicity was induced but gradient cytotoxicity was observed after 72 h of exposure. At a concentration of 0.3 μM a significant drop in cell viability was observed for GCB, whereas TQ induced cell viability at a concentration of 100 μM. A significantly improved cytotoxic profile was observed for the combination (GCB + TQ) with IC_{50} values 9.5–15.5-fold higher than for a monotreatment. Similarly, for T47D cells, the combination had lower IC_{50} values 6.2–25-fold greater. These values show strong synergistic effects of GCB and TQ.

Antioxidant activity

The excess generation of ROS beyond the antioxidant capacity of a biological system gives rise to oxidative stress. Free radical oxidative stress is responsible for many human diseases such as diabetes mellitus, atherosclerosis, hypertension, cancer, inflammation, and AIDS (Burits and Bucar, 2000). TQ regulates such reactions that are responsible for the excess generation of ROS by scavenging them (Dergarabetian et al., 2013). TQ undergoes some oxidation and reduction by reacting with amino acids, resulting in the production of semiquinone and thymohydroquinone on reduction of 1 and 2 e^- from TQ, respectively (Cremer et al., 1987; Gupta et al., 2016). TQ effectively inhibited the lipid peroxidation producing superoxide radicals and behaves as a potent antioxidant (Banerjee et al., 2010). It reinforces the enzymatic activity of antioxidant enzymes such as superoxide dismutase, catalase, glutathione, and glutathione transferase. TQ reacts with antioxidant enzymes and produces two antioxidants—dihydrothymoquinone and glutathionyl-dihydrothymoquinone (Khalife and Lupidi, 2007). TQ acts as both an antioxidant and prooxidant due to the cellular microenvironment. It was reported that suppression of Fe-dependent microsomal lipid peroxidation is modulated by TQ in a dose-dependent manner (Badary et al., 2003). The above-mentioned process was monitored with TQ and its analog *tert*-butylhydroquinone (TBHQ). However, TQ was recognized as a more potent superoxide anion inhibitor. However, it does not scavenge hydroxyl and 2, 2′-diphenyl-p-picrylhydrazyl radicals very effectively (Badary et al., 2003). TQ belongs to the terpenoids, which are volatile and quite sensitive to high temperature. The presence of high temperature causes a drop in the concentration of TQ. To overcome these problems, there is a need for encapsulation of TQ and preparation of TQ analogs (Mohammad et al., 2018). Some derivatives of TQ have been reported to enhance its antioxidant activity.

Derivatization of TQ was carried out through some chemical and bacterial interactions which enhanced its antioxidant activity (Fig. 52.7). Iqbal et al. (2019) reported the effect of sodium azide on the antioxidant property of TQ in a concentration-dependent manner. They found an increase in antioxidant activity with a decreasing content of TQ, following the treatment of a high concentration of sodium azide. This could be due to some conformational changes that occur in TQ structure or to the formation of different antioxidants because of salt stress produced by sodium azide. Reaction with sodium azide results in the rearrangement reaction, followed by production of different analogs (Fig. 52.8). These analogs are reported have greater antioxidant property (Iqbal et al., 2019).

Derivatization of TQ (**1**) by *Aspergillus niger* has been reported (Mohammad et al., 2018). Three derivatives are formed: 5-isopropyl-2-methylbenzene-1,4-diol (**4c**), 3-hydroxy-5-isopropyl-2-methylcyclohexa-2,5-diene-1,4-dione (**4a**), and 5-isopropyl-2methyloxepin-1-one (**4b**) (Fig. 52.5). Analog **4a** was reported for the first time and analog **4c** showed antioxidant activity. Antioxidant activity of analog **4c** was compared to ascorbic acid and found to be a potent antioxidant (Mohammad et al., 2018). TQ was reported to rectify the gastrointestinal and renal toxicity induced by a high dose of diclofenac (DCLF). Diclofenac is lipotropic, a nonsteroidal antiinflammatory drug (Aycan et al., 2018). Normal healing doses of DCLF are effective, safe, and universally used, but a high dose of DCLF for a long time causes nephrotoxicity. DCLF treatment causes excess formation of ROS/RNS (reactive nitrogen species) and decreases the total antioxidant capacity (TAC) capacity in the stomach, duodenal tissue, and kidney. Administration of TQ increased renal and gastrointestinal TAC in DCLF- treated rats. These indicate the ability of TQ to reduce renal and gastrointestinal toxicity caused by high-dose DCLF treatment (Aycan et al., 2018).

Antidiabetic activity

TQ was reported for its antihyperglycemic property on diabetic rats. The effect of nanoformulated TQ was studied in streptozotocin-induced type II diabetic rats and its activity was compared with pure TQ and metformin (a standard antidiabetic drug). A decrease in the secretion of insulin causes diabetes, which further results in elevated lipid profile parameters, that is, dyslipidemia, triglycerides, cholesterol, and nonplanar lipid proteins. These are essential for the normal functioning of all animal cells. Upon induction of type II diabetes, there is an increase in glucose secretion, and through protein wasting body weight is reduced (Rani et al., 2018). Surprisingly, TQ increased the bodyweight of diabetic rats through activation of a key enzyme in carbohydrate metabolism suggesting its antihyperglycemic potential (Pari and Sankaranarayanan, 2009). TQ normalizes the increased glucose level by boosting glucose utilization and suppressing hepatic glucose production mediated through insulin. When TQ was administrated orally it had lower cellular uptake and was eliminated rapidly. Thus, nanoformulation/nanocapsulation of TQ is essentially required to improve its bioavailability, absorption, metabolism, distribution, excretion, and the minimization of possible side effects at the cellular level (Grama et al., 2013; Xue et al., 2013). TQ and TQ nanocapsules (TQnc) have significantly helped to regain body weight loss due to diabetes type II. TQnc have better activity compared to only TQ. Thus, TQnc can be used to replace metformin as this drug is characterized by side effects such as nausea, anorexia, diarrhea, and abdominal discomfort. Surprisingly, the antidiabetic effect of TQnc was comparable to the effect of metformin. Therefore TQnc can be used as a potent antihyperglycemic drug (Rani et al., 2018).

Antiinflammatory action

TQ was shown to have notable antiinflammatory activity against pathogens and targets various sites (Fig. 52.9). It is involved in the inhibition of transcription factor NF-κB

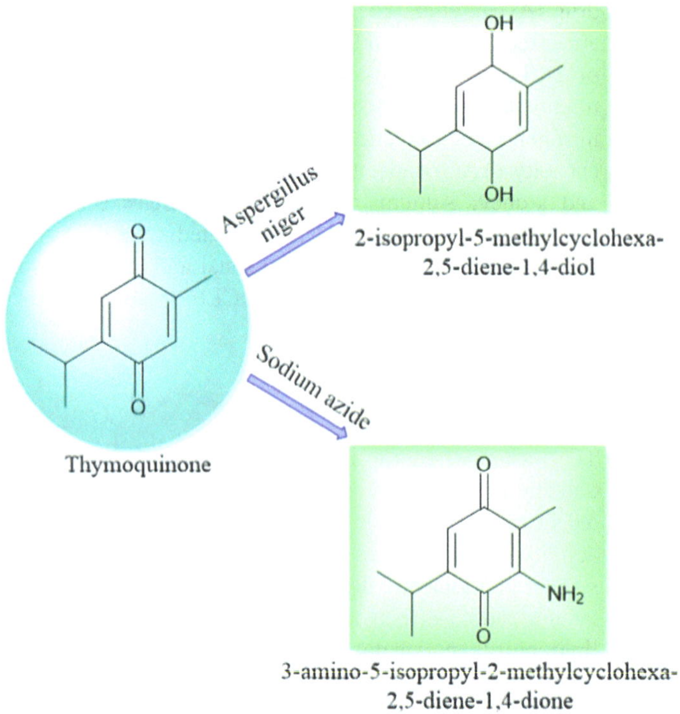

FIGURE 52.7 Derivatives of thymoquinone having anti-oxidant property.

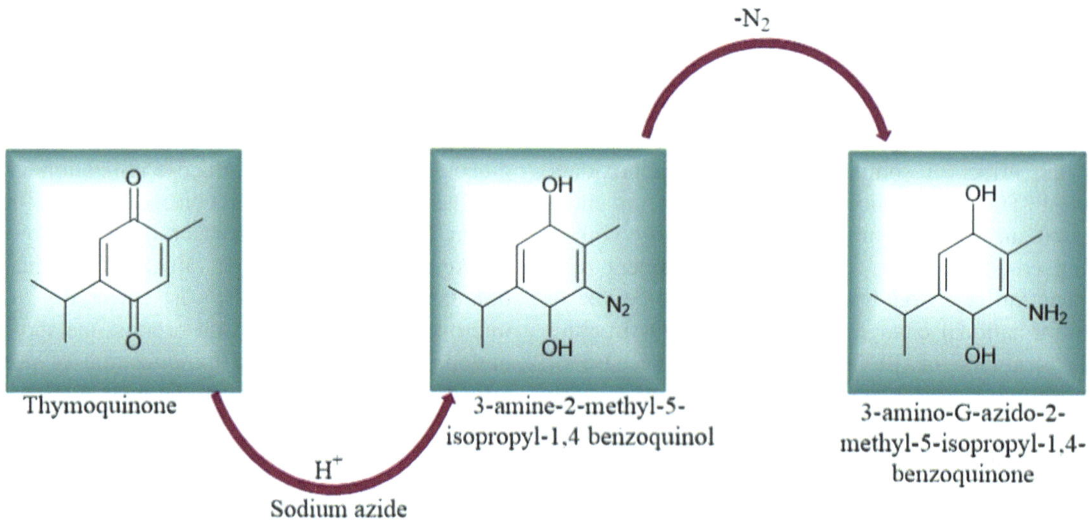

FIGURE 52.8 Structural changes in thymoquinone reacting with sodium azide.

that binds to DNA and activates genes of transcription, which encourages neuroinflammation. Lepiarz et al. (2017) have reported the importance of nuclear erythroid 2 related factor 2 (Nrf2)/antioxidant response element (ARE) signaling activation by TQ for the inhibition of neuroinflammation. Nrf2 is a regulatory protein that protects the cell against oxidative stress. To initiate the transcription of numerous cytoprotective genes, Nrf2 binds to ARE DNA sites. Sandberg et al. (2014) studied the antiinflammatory activity by activating Nrf2 which involves NF-κB.

Induction of neuroinflammation by lipopolysaccharide (LPS) was inhibited by TQ through interference with NF-κB signaling. TQ also stimulated (Nrf2)/(ARE) signaling by boosting transcriptional activity, DNA binding, and nuclear localization of Nrf2, as well as increasing protein levels of NQO1 and HO-1. Loss of antiinflammatory activity by TQ was observed due to suppression of Nrf2

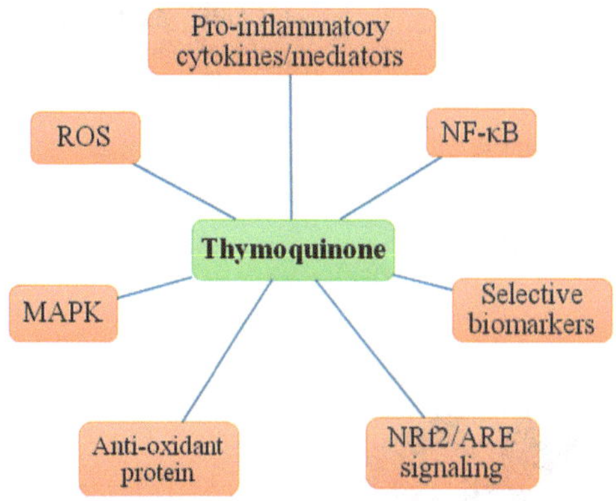

FIGURE 52.9 Different targets of thymoquinone in inflammatory diseases.

activity through small interfering RNA (siRNA) or with the use of trigonelline. This study revealed that neuroinflammation due to transcription factor NF-κB was inhibited by TQ along with activation of Nrf2/ARE signaling (Velagapudi et al., 2017).

The excessive inflammatory response due to microbial infection causes a severe systemic inflammatory response syndrome known as sepsis (Alkharfy et al., 2018). Alkharfy et al. (2018) examined the effect of TQ on Gram-negative bacteria induced in mice. Increased levels of interleukin (IL)-1, IL-2, IL-6, and IL-10, including tumor necrosis factor (TNF)-α, cytokines are observed after the injection of *E. coli*. After *E. coli* inoculation, increased levels of selective biomarkers for sepsis such as endothelial cell-specific molecule 1(ESM-1), vascular endothelial growth factor (VEGF), and C-reactive protein were noted. A relationship between higher circulating levels of VEGF and severe septic shock in humans was reported. A higher concentration of VEGF in plasma is directly correlated with the severity of disease (Pickkers et al., 2005). TQ decreases the concentration of circulating protein and is effective in reducing mortality in mice due to sepsis (Alkharfy et al., 2018).

Prolonged activation of microglial cells releases ROS and proinflammatory cytokines disproportionately. Cellular oxidative stress is caused by an elevated level of reactive oxygen. BV-2 microglial cells were triggered by LPS and interferon gamma (INF-γ) and treated with TQ. The level of antioxidant neuroprotective protein such as billiverdin reductase A, mitochondrial ion protease, 3-mercaptopyruvate sulfotransferase, and glutaredoxin-3 is increased as compared to activated cells which are untreated with TQ. The effects of TQ in BV-2 microglial cells were performed using an orbitrap/Q executive LCMS/MS proteomics. As a result, TQ reduces neuroinflammation and neurodegeneration that suppresses the inflammatory response caused by microglia as well as promoting the formation of genes and proteins (Cobourne-duval et al., 2018). Alzheimer disease (AD) is a serious neurodegenerative disease that results in cognitive decline, memory loss, and neural death (Elibol et al., 2019). Elibol et al. (2019) have established a rat model of AD by intrahippocampal infusion which contained agreggated $A\beta_{(1-42)}$. After amyloid beta-infused neurotoxicity, treatment was administered by TQ and it was observed that the inhibition of cultured cortical and hippocampal neurons of rats were able to enhance the viability of cells by decreasing neurotoxicity. This was achieved by inhibition of mitochondrial membrane potential depolarization and generation of ROS. Decreased levels of inflammatory markers NO metabolites, TNF-α, and IL-6 helped to understand how TQ improved memory in rat hippocampus (Elibol et al., 2019).

Cell cycle and apoptosis

Apoptosis is programmed cell death. Interestingly, this process is nonfunctional in cancerous cells (Yano et al., 2018). Some morphological alternations, such as DNA fragmentation, chromatin condensation, membrane blebbing, and cellular shrinkage are characteristic features of apoptosis (Wyllie, 1997; Gali-Muhtasib et al., 2004). However, this process can be induced through instigators (such as TQ). During apoptosis mitochondrial proteins like cytochrome c (cyt c) are released into cytosol when triggered with instigators such as stress and drugs (TQ) This phenomenon occurs with the rupture of the mitochondrial membrane (Samarghandian et al., 2019). Bcl-2 is a class of protein, which regulates the release of cyt c. Bcl-2 includes antiapoptotic (e.g., Bcl-2) and proapoptotic (e.g., Bax) proteins. The permeability of mitochondria is increased by Bax, which also facilitates the release of cyt c. The transportation of cyt c is inhibited by Bcl-2 through stabilization of the mitochondrial membrane. Release of an apoptogenic factor known as caspase is regulated by Bcl-2, which decides the fate of cells as caspases are the chief contributors to the caspase-apoptotic cascade involving successive proteolytic cleavage (Fig. 52.10).

Using the MTT test, the antiviability effect of TQ, apoptotic properties and toxicity were investigated in A549 lung cancer cells in a dose- and time-dependent manner. Growth inhibitory effects and cytotoxic effects of TQ were confirmed with the decreased formation of formazan crystals after TQ exposure. Exposure to TQ for a day at 50 and 100 μM concentrations contributed to a survival rate of >60% in A549 cancer cells, which was decreased by 10% when the concentration was kept at 25 μM. After an incubation period of 72 h, the highest IC_{50} of TQ was found to be 47 ± 0.09 μM for neoplastic cells. Upon administration of TQ, through annexin V/PI test, the quantitative assay for the apoptosis was

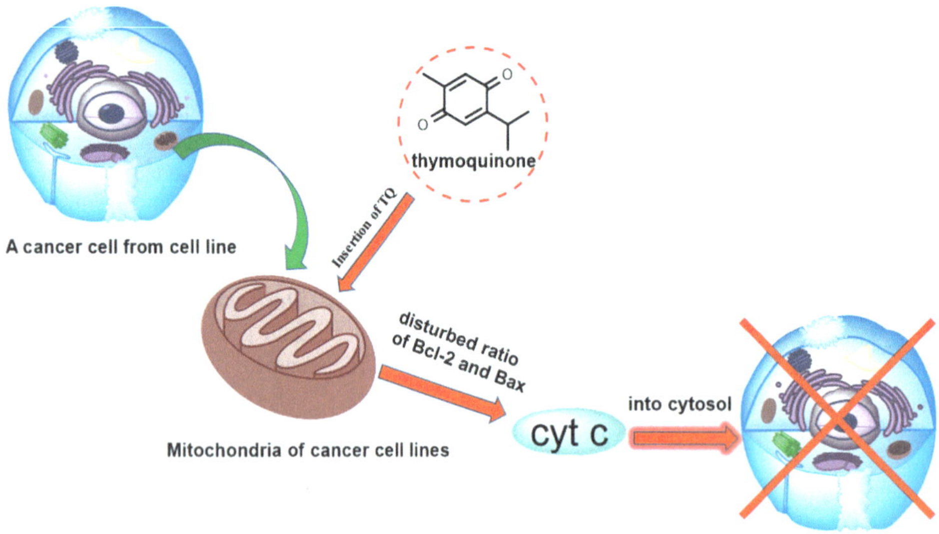

FIGURE 52.10 The apoptosis pathway induced by thymoquinone.

performed in the A549 cells which indicated a hike in the ratio of both early- and late-stage apoptotic cells from 14.9% to 34.5% and from 20.5% to 39.2%, respectively. At concentrations of 25 and 50 μM of TQ the rate of apoptosis was most significant. The rate of apoptosis was greater at higher concentrations, hence signifying the dose-dependent pattern (Samarghandian et al., 2019).

Park et al. (2019) studied the apoptotic effect of TQ on human epidermoid carcinoma A431 cells. They found that the TQ shows a great cytotoxic effect on A431 cells. Cells were stained with Annexin V and PI to differentiate the cells undergoing apoptosis, and analysis was carried out through cytometry. TQ reduces the capability of the cell with the exposure time and different concentrations (1, 15, or 30 μM) of TQ. Intercellular ROS accumulation causes cell death. The group also analyzed the ROS generation activity of TQ in A431 cells (Park et al., 2019). This property was examined by FACS analysis and immunofluorescence analysis with DCF-DA staining. The ROS scavenger NAC was used to investigate ROS involvement in TQ-induced apoptosis.

In A431 cells, they also analyzed the downregulation of antiapoptotic Bcl-2 due to TQ by examining the regulation of the STAT3 enzyme which is responsible for upregulation of prosurvival genes. Upon treatment with TQ, the phosphorylation of enzyme STAT3 was decreased while the level of STAT3 was not affected. The decreased phosphorylation of STAT3 resulted in a lowered DNA-binding activity which reduced the expression of gene products such as cyclin D1 and surviving. TQ-induced inactivation of STAT3 was elucidated for which the effect of TQ upon phosphorylation of upstream kinase, Src (phosphorylate STAT3), was examined (Olayioye et al., 1999; Park et al., 2019). TQ treatment showed lowered phosphorylation of Src kinase in a concentration-dependent manner when examined for A431 cells (Park et al., 2019).

Encapsulation of thymoquinone

Encapsulation of drugs with different drug carriers is one of the approaches to enhance the efficiency of drug delivery. Encapsulation has emerged as one of the best alternatives in the preservation of such oil components. In encapsulation, one compound is entrapped into another wall material compound that can be of different particle sizes (Burgain et al., 2011). There are different encapsulation systems, such as fluidized-bed coating, emulsion, spray-drying, liposome entrapment, and coacervation. These techniques have shown proficiency in entrapping the active compound to increase stability during storage (Pham-Hoang et al., 2013).

Bishop and co-workers demonstrated the use of yeast cells as a popular encapsulating system due to the presence of various organelles, antioxidant mechanism, and multilayer cell wall (Bishop et al., 1998). Yeast cells have proved to be a better coating material as they behave like a protecting capsule through which cellular exchanges and osmotic pressure are controlled (Pham-Hoang et al., 2013). Fat-soluble materials can be encapsulated to a good extent using yeast cells. This encapsulation is supported by the presence of a cell wall which permits molecular diffusion and provides mechanical support. Various studies have been performed to check the encapsulation of various bioactive compounds using yeast cells such as anthocyanins (Nguyen et al., 2018), curcumin (Young and Nitin, 2019), and menhaden fish oil (Czerniak et al., 2015).

TQ could be encapsulated successfully using the yeast cell of *S. cerevisiae* and increase stability against unsuitable conditions and could be preserved within the yeast capsules as a microcarrier against heat and humidity. A decrement in the biofunctionality of TQ can be prevented through encapsulation. Plasmolysis has emerged as a vital pretreatment during encapsulation because it enhances the stability of encapsulated material against adverse conditions and increases the oil absorption capacity. Thermal and radiation (light)-mediated degradation of TQ could be minimized through the microcapsules of yeast cells (Karaman, 2020). Fakhria et al. (2019) reported encapsulation of TQ with chitosan vesicle as an effective oral delivery system for the treatment of hyperlipidemia and studied its histopathological effect on liver. The TQ–chitosan vesicle formulation was developed by the combination of nonionic surfactants, cholesterol, phospholipid, and chitosan using probe sonication and a solvent evaporation method. The size of all the developed TQ-loaded chitosan vesicles (TQCHVs) varies from 372.8 to 432.55 nm. They showed great encapsulation efficiency within the range of 67.56%–82.23% (Fakhria et al., 2019). It enhanced the drug release profile of TQ twofold. The in vitro drug release scheme of TQCHVs behaves in two ways: TQ is initially release from the surface of vesicles and later prolonged release from the inner core of the vesicle.

The initial fast release may be due to the release of TQ from the surface of vesicles, whereas at a later stage, TQ may be constantly released from the inner core, which is responsible for the prolonged release. TQCHVs has 4.4 times higher mucin-binding capacity and is 1.9 times more permeable to the intestinal tract as compared to TQ. TQCHVs showed improved antihyperlipidemic activity. No type of inflammation or irritancy was observed in its histopathology study. Due to the blood–brain-barrier, the delivery of many toxic materials as well as various novel drugs to the different brain regions are restricted or filtered out. Fahmy et al. (2019) worked on the encapsulation of TQ with mesoporous silica nanoparticles (MSNs) and examined the distribution of TQ MSNs to different brain parts compared to TQ alone. The physical examination of TQ MSNs showed an almost spherical shape with a size of about 90 nm, encapsulation efficiency of about 95.10%, and zeta potential value of about −15.90 mV (Fahmy et al., 2019). It showed biphasic drug release profile initially from the surface of TQ MSNs for the first 5 h and later from the inner vacancies of MSNs. The chromatographic quantification of TQ and the study of various oxidative stress biomarkers showed that the prepared TQ MSNs hold a great drug-delivery profile with enhanced efficiency of about 71% in the hypothalamus, 11% in the thalamus, 20% in the midbrain, and 13% in the cortex (Fahmy et al., 2019). Drug distribution to the hippocampus, medulla, and striatum showed no significant change, while targeting to the cerebellum was reduced by 20% (Fahmy et al., 2019).

Concluding remarks and future directions

The extraordinary properties and broad applicability of phytochemicals as a potent source of drugs has long been a hot topic of research around the globe. TQ is well known for its biological and therapeutic activities. It has multiple applications in treating various diseases alone or in combination with different drugs. Several in vitro and in vivo studies are well documented for accounting for the expansive pharmaceutical properties which include anticancer, antioxidant, antidiabetic, antiinflammatory effects on the cell cycle, encapsulation, and many more.

Various structural modifications have also been carried out on TQ to enhance its stability, efficiency, bioavailability, and other pharmacological properties. It is reported to be an effective agent in diminishing the cytotoxic effects of various commercially available drugs. The nanoencapsulation of TQ has boosted its stability. Therefore it can be summarized that TQ is an ideal candidate with extraordinary ability to counteract various diseases.

Acknowledgment

The authors are grateful to Pt. Ravishankar Shukla University, Raipur, India, for their continued support.

References

Alhaj, N.A., Shamsudin, M.N., Zamri, H.F., et al., 2008. Extraction of essential oil from Nigella sativa using supercritical carbon dioxide: study of antibacterial activity. Am. J. Pharmacol. Toxicol. 3 (4), 225–228. Available from: 10.3844/ajptsp.2008.225.228.

Alkharfy, K.M., Ahmad, A., Jan, B.L., et al., 2018. Biomedicine and pharmacotherapy thymoquinone reduces mortality and suppresses early acute inflammatory markers of sepsis in a mouse model. Biomed. Pharmacother. 98, 801–805. Available from: 10.1016/j.biopha.2018.01.028.

Alobaedi, O.H., Talib, W.H., Basheti, I.A., 2017. Antitumor effect of thymoquinone combined with resveratrol on mice transplanted with breast cancer. Asian Pac. J. Trop. Med. 10 (4), 400–408. Available from: 10.1016/j.apjtm.2017.03.026. Elsevier B.V.

Aycan, İ.Ö., Elpek, Ö., Akkaya, B., et al., 2018. Diclofenac induced gastrointestinal and renal toxicity is alleviated by thymoquinone treatment. Food Chem. Toxicol. 118, 795–804. Available from: 10.1016/j.fct.2018.06.038.

Badary, O.A., Taha, R.A., Gamalel-Din, A.M., et al., 2003. Thymoquinone is a potent superoxide anion scavenger. Drug. Chem. Toxicol. 26 (2), 87–98. Available from: 10.1081/DCT-120020404.

Banerjee, S., Padhye, S., Azmi, A., et al., 2010. Review on molecular and therapeutic potential of thymoquinone in cancer. Nutr. Cancer. 62 (7), 938–946. Available from: 10.1080/01635581.2010.509832.

Barton-Burke, M., 1999. Gemcitabine: a pharmacologic and clinical overview'. Cancer Nurs. 22 (2), 176–183. Available from: 10.1097/00002820-199904000-00011.

Bashmail, H.A., Alamoudi, A.A., et al., 2018. Thymoquinone synergizes gemcitabine anti-breast cancer activity via modulating its apoptotic and autophagic activities'. Sci. Rep. 8 (1), 1–11. Available from: 10.1038/s41598-018-30046-z.

Bishop, J.R.P., Nelson, G., Lamb, J., 1998. Microencapsulation in yeast cells. J. Microencapsul. 761–773. Available from: 10.3109/02652049809008259.

Burgain, J., Gaiani, C., Linder, M., et al., 2011. Encapsulation of probiotic living cells: from laboratory scale to industrial applications. J. Food Eng. 467–483. Available from: 10.1016/j.jfoodeng.2010.12.031.

Burits, M., Bucar, F., 2000. Antioxidant activity of *Nigella sativa* essential oil. Phytother. Res. 14 (5), 323–328.

Burstein, H.J., 2000. Side effects of chemotherapy. J. Clin. Oncol. 18 (3), 693. Available from: 10.1200/JCO.2000.18.3.693.

Cobourne-duval, M.K., et al., 2018. Thymoquinone increases the expression of neuroprotective proteins while decreasing the expression of pro-in fl ammatory cytokines and the gene expression NF κ B pathway signaling targets in LPS / IFN γ -activated BV-2 microglia cells. J. Neuroimmunol. 320, 87–97. Available from: 10.1016/j.jneuroim.2018.04.018.

Cremer, D., Hausen, B.M., Schmalle, H.W., 1987. Toward a rationalization of the sensitizing potency of substituted p-benzoquinones: reaction of nucleophiles with p-benzoquinones. J. Med. Chem. 30 (9), 1678–1681.

Crowell, J.A., Korytko, P.J., Morrissey, R.L., et al., 2004. Resveratrol-associated renal toxicity. Toxicol. Sci. 82 (2), 614–619. Available from: 10.1093/toxsci/kfh263.

Czerniak, A., Kubiak, Biatas, W., et al., 2015. Improvement of oxidative stability of menhaden fish oil by microencapsulation within biocapsules formed of yeast cells. J. Food Eng. 167, 2–11. Available from: 10.1016/j.jfoodeng.2015.01.002.

Dergarabetian, E.M., Ghattass, K.I., El-Sitt, S.B., et al., 2013. Thymoquinone induces apoptosis in malignant T-cells via generation of ROS. Front. Biosci. 5 (E(2)), 706–719. Available from: 10.2741/e651.

Effenberger-Neidnicht, K., Schobert, R., 2011. Combinatorial effects of thymoquinone on the anti-cancer activity of doxorubicin. Cancer Chemother. Pharmacol. 67 (4), 867–874. Available from: 10.1007/s00280-010-1386-x.

El-Dakhakhny, M., 1963. Studies on the chemical constituents of Egyptian *Nigella sativa* L. seeds. II. The essential oil. Planta Medica. 11 (04), 465–470. Available from: 10.1055/s-0028-1100266.

Elibol, B., et al., 2019. Thymoquinone (TQ) demonstrates its neuroprotective effect via an anti-inflammatory action on the $A\beta_{(1-42)}$-infused rat model of Alzheimer's disease. Psych. Clin. Psychpharmacol. 29 (4). Available from: 10.1080/24750573.2019.1673945.

Park, J.E., Kim, D.H., Ha, E., et al., 2019. Thymoquinone induces apoptosis of human epidermoid carcinoma A431 cells through ROS-mediated suppression of STAT3. Chemico-Biol Interact. 312, 108799. Available from: 10.1016/j.cbi.2019.108799.

Fahmy, H.M., Fathy, M.M., Abd-elbadia, R.A., et al., 2019. Targeting of thymoquinone-loaded mesoporous silica nanoparticles to different brain areas: in vivo study. Life Sci. 222, 94–102. Available from: 10.1016/j.lfs.2019.02.058.

Fakhria, A., et al., 2019. Formulation of thymoquinone loaded chitosan nano vesicles: in vitro evaluation and in-vivo anti-hyperlipidemic assessment. J. Drug. Deliv. Sci. Technol. 50, 339–346. Available from: 10.1016/j.jddst.2019.01.033.

Fatfat, M., Fakhoury, I., Habli, Z., et al., 2019. Thymoquinone enhances the anticancer activity of doxorubicin against adult T-cell leukemia in vitro and in vivo through ROS-dependent mechanisms. Life Sci. 232, 1–9. Available from: 10.1016/j.lfs.2019.116628.

Fröhlich, T., Ndreshkjana, B., Muenzner, J.K., et al., 2017. Synthesis of novel hybrids of thymoquinone and artemisinin with high activity and selectivity against colon cancer. Med. Chem. Chem 12 (3), 226–234. Available from: 10.1002/cmdc.201600594.

Fröhlich, T., Reiter, C., et al., 2018. Synthesis of thymoquinone-artemisinin hybrids: new potent antileukemia, antiviral, and antimalarial agents. ACS Med. Chem. Lett. 9 (6), 534–539. Available from: 10.1021/acsmedchemlett.7b00412.

Gali-Muhtasib, H.U., Abou, K., Wassim, G., et al., 2004. Molecular pathway for thymoquinone-induced cell-cycle arrest and apoptosis in neoplastic keratinocytes. Anticancer. Drugs 15 (4), 389–399. Available from: 10.1097/00001813-200404000-00012.

Goldstein, A., Oliver Lampen, J., 1975. [76] β-D-fructofuranoside fructohydrolase from yeast. Methods Enzymol, 42(C) 504–511. Available from: 10.1016/0076-6879(75)42159-0.

Grama, C.N., Suryanarayana, P., et al., 2013. Efficacy of biodegradable curcumin nanoparticles in delaying cataract in diabetic rat model. PLoS ONE. 8 (10), e78217. Available from: 10.1371/journal.pone.0078217.

Gupta, B., Ghosh, K.K., Gupta, R.C., 2016. Thymoquinone. In: Gupta, R.C. (Ed.), Nutraceuticals: Efficacy, Safety and Toxicity. Elsevier Inc, pp. 541–550.

Harper, C.E., Patel, B.B., Wang, J., et al., 2007. Resveratrol suppresses prostate cancer progression in transgenic mice. Carcinogenesis 28 (9), 1946–1953. Available from: 10.1093/carcin/bgm144.

Huang, C.-Y., Ju, D.-T., Chang, C.-F., et al., 2017. A review on the effects of current chemotherapy drugs and natural agents in treating non–small cell lung cancer. BioMedicine. 7 (4). Available from: 10.1051/BMDCN/2017070423.

Iqbal, M.S., Ansari, I., Jafri, A., et al., 2019. Stress response due to sodium azide treatment inside Nigella sativa L. plant and its effect on antioxidative property. Biocatal. Agri Biotechnol. 19, 101171. Available from: 10.1016/j.bcab.2019.101171.

Johnson-Ajinwo, O.R., Ullah, I., Mbye, H., et al., 2018. The synthesis and evaluation of thymoquinone analogues as anti-ovarian cancer and antimalarial agents. Bioorg Med. Chem. Lett. 28 (7), 219–1222. Available from: 10.1016/j.bmcl.2018.02.051.

Karaman, K., 2020. Characterization of *Saccharomyces cerevisiae* based microcarriers for encapsulation of black cumin seed oil: stability of thymoquinone and bioactive properties'. Food Chem. 313, 126129. Available from: 10.1016/j.foodchem.2019.126129.

Khalife, K.H., Lupidi, G., 2007. Nonenzymatic reduction of thymoquinone in physiological conditions. Free. Rad. Res. 41 (2), 153–161. Available from: 10.1080/10715760600978815.

Lepiarz, I., Fiebich, B.L., Olajide, O.A., 2017. Inhibition of neuroinflammation by thymoquinone requires activation of Nrf2/ARE signalling. Int. Immunopharmacol. 48, 17–29. Available from: 10.1016/j.intimp.2017.04.018.

Mohammad, M.Y., Shakya, A., Al-Bakain, R., et al., 2018. New monoterpenoid by biotransformation of thymoquinone using *Aspergillus niger*. Bioorg Chem. 80, 212–215. Available from: 10.1016/j.bioorg.2018.06.019.

Moorsel, C.J.A., Peters, G.J., Pinedo, H.M., 1997. Gemcitabine: future prospects of single-agent and combination studies. Oncologist 2 (3), 127–134. Available from: 10.1634/theoncologist.2-3-127.

Nagi, M.N., Almakki, H.A., Sayed-Ahmed, M.M., et al., 2010. Thymoquinone supplementation reverses acetaminophen-induced oxidative stress, nitric oxide production and energy decline in mice liver. Food Chem. Toxicol. 48 (8–9), 2361–2365. Available from: 10.1016/j.fct.2010.05.072.

Nguyen, T.T., Phan-Thi, H., Pham-Hoang, B.N., et al., 2018. Encapsulation of *Hibiscus sabdariffa* L. anthocyanins as natural colours in yeast. Food Res. Int. 107, 275–280. Available from: 10.1016/j.foodres.2018.02.044.

Olayioye, M.A., Beuvink, I., Horsch, K., et al., 1999. ErbB receptor-induced activation of stat transcription factors is mediated by Src tyrosine kinases. J. Biol. Chem. 274 (24), 17209–17218. Available from: 10.1074/jbc.274.24.17209.

Pari, L., Sankaranarayanan, C., 2009. Beneficial effects of thymoquinone on hepatic key enzymes in streptozotocin-nicotinamide induced diabetic rats. Life Sci. 85 (23–26), 830–834. Available from: 10.1016/j.lfs.2009.10.021.

Pham-Hoang, B.N., Romero-Guido, C., Phan-Thi, H., et al., 2013. Encapsulation in a natural, preformed, multi-component and complex capsule: yeast cells. Appl. Microbiol. Biotechnol. 6635–6645. Available from: 10.1007/s00253-013-5044-1.

Pickkers, P., Sprong, T., van Eijk, L.T., et al., 2005. Vascular endothelial growth factor is increased during the first 48 hours of human septic shock and correlates with vascular permeability. Shock. 24 (6), 508–512. Available from: 10.1097/01.shk.0000190827.36406.6e.

Rani, R., Dhingra, D., Dahiya, S., et al., 2018. Improvement of antihyperglycemic activity of nano-thymoquinone in rat model of type-2 diabetes. Chem Biol Interact. 295, 119–132. Available from: 10.1016/j.cbi.2018.02.006.

Salim, S., Farooqui, Z., Asghar, M., et al., 2017. Effect of fasting on enzymes of carbohydrate metabolism, brush border membrane and on transport functions in superficial and juxta-medullary cortex of rat kidney. J. Nutr. Food Sci. 07 (02). Available from: 10.4172/2155-9600.1000588.

Samarghandian, S., Azimi-Nezhad, M., Farkhondeh, T., 2019. Thymoquinone-induced antitumor and apoptosis in human lung adenocarcinoma cells. J. Cell Physiol. 234 (7), 10421–10431. Available from: 10.1002/jcp.27710.

Sandberg, M., Patil, J., et al., 2014. NRF2-regulation in brain health and disease: implication of cerebral inflammation. Neuropharmacology. 298–306. Available from: 10.1016/j.neuropharm.2013.11.004.

Shahid, F., Farooqui, Z., Abidi, S., et al., 2017. Oral administration of thymoquinone mitigates the effect of cisplatin on brush border membrane enzymes, energy metabolism and antioxidant system in rat intestine. Biomed.

Pharmacother. 94, 1111–1120. Available from: 10.1016/j.biopha.2017.08.044.

Sharma, R., Sahu, P., Jain, A., et al., 2019. Nigella sativa. In: Gupta, R.C., Srivastava, A., Lall, R. (Eds.), Nutraceuticals in Veterinary Medicine. Springer Nature, Cham, Switzerland, pp. 91–101.

Velagapudi, R., Kumar, A., Bhatia, H.S., et al., 2017. Inhibition of neuroinflammation by thymoquinone requires activation of Nrf2/ARE signalling. Intl Immunopharmacol. 48, 17–29. Available from: 10.1016/j.intimp.2017.04.018.

Woo, C.C., Hsu, A., et al., 2013. Thymoquinone inhibits tumor growth and induces apoptosis in a breast cancer xenograft mouse model: the role of p38 MAPK and ROS. PLoS ONE 8 (10). Available from: 10.1371/journal.pone.0075356.

Wyllie, A.H., 1997. Apoptosis: an overview. Brit Med. Bull. 53 (3), 451–465. Available from: 10.1093/oxfordjournals.bmb.a011623.

Xiao, Y.S., Zeng, D., Liang, Y.-K., et al., 2019. Major vault protein is a direct target of Notch1 signaling and contributes to chemoresistance in triple-negative breast cancer cells. Cancer Lett. 440–441, 156–167. Available from: 10.1016/j.canlet.2018.09.031.

Xue, M., Yang, M.-X., et al., 2013. Characterization, pharmacokinetics, and hypoglycemic effect of berberine loaded solid lipid nanoparticles. Int. J. Nanomed. 8, 4677–4687. Available from: 10.2147/IJN.S51262.

Yano, S., Wu, S., Sakao, K., et al., 2018. Wasabi 6-(methylsulfinyl)hexyl isothiocyanate induces apoptosis in human colorectal cancer cells through p53-independent mitochondrial dysfunction pathway. BioFactors. 44 (4), 361–368. Available from: 10.1002/biof.1431.

Young, S., Nitin, N., 2019. Thermal and oxidative stability of curcumin encapsulated in yeast microcarriers. Food Chem. 275, 1–7. Available from: 10.1016/j.foodchem.2018.08.121.

Yuan, X., Wu, H., Xu, H., et al., 2015. Notch signaling: an emerging therapeutic target for cancer treatment. Cancer Lett. 20–27. Available from: 10.1016/j.canlet.2015.07.048.

Zarringhalam, M., Toghraie, D., 2019. The effects of suspending copper nanoparticles into argon base fluid inside a microchannel under boiling flow condition by using of molecular dynamic simulation. J. Mol. Liq. 293, 111474. Available from: 10.1016/j.molliq.2019.111474.

Chapter 53

Glucosinolates

Karyn L. Bischoff

Introduction

The *Brassica* (*Brassicaceae*) family includes cruciferous vegetables like cabbage, broccoli, cauliflower, and mustards. It also includes the oil seed crops rapeseed and canola (Burel et al., 2000). *Brassica* have been cultivated since before the time of the Roman Empire (Šamec et al., 2018).

Many *Brassica* spp. are frost tolerant, easy, and inexpensive to cultivate, and different varieties can be harvested early spring through late autumn, thus they are available when other plants are dormant (Burrows and Tyrl, 2013; Šamec et al., 2018). There are at least 341 genera and 3977 species within the family *Brassicaceae* (Šamec et al., 2018).

Plants in the *Brassicaceae* family contain several bioactive chemicals that can be beneficial or harmful, including glucosinolates, other sulfur-containing compounds, and erucic acid. These compounds evolved as generalized defensive chemicals against microorganisms, insects, and herbivores (Abbaoui et al., 2018; Cavaiuolo and Ferrante, 2014; Hanschen et al., 2014). Some insects have adapted specifically to feed on glucosinolate-containing plants (Prieto et al., 2019). As natural pesticides, crushed *Brassicae* plants have been used to decrease plant pathogen load in soils (Hanschen et al., 2014). *Brassica* spp. have been used in traditional medicines for gastritis and ulcer, diabetes mellitus, rheumatism, liver disease, eye disease, anemia, and obesity (Šamec et al., 2018). This chapter describes some of the documented beneficial health effects and toxic effects of glucosinolates.

Classification of glucosinolates

About 200 glucosinolates have been identified (Šamec et al., 2018; Hanschen et al., 2014). Species in the *Brassica* family typically produce between 10 and 40 different glucosinolates, with approximately four at significant concentrations (Capuano et al., 2017; Halkier and Gershenzon, 2006; Šamec et al., 2018). The glucosinolate composition of a plant depends on genetic factors, stage of growth, season, and local environmental conditions (Abbaoui et al., 2018; Šamec et al., 2018).

Glucosinolates consist of a β-D-glucopyranose residue linked to a hydroxylamine sulfate ester by a sulfur bridge, plus an R-group. The R-group is derived from one of eight amino acids and can be aliphatic (alanine, leucine, isoleucine, methionine, or valine), aromatic (phenylalanine or tyrosine), or indole (tryptophan) (Collett et al., 2014). Aliphatic glucosinolates are the most common, making up about 50% of the total known glucosinolates, and can be classified based on straight or branched chains, with or without hydroxyl groups, in addition to the oxidation state of sulfur atoms in side chains (Hanschen et al., 2014). The indole glucosinolates are associated with bitter flavors (Halkier and Gershenzon, 2006).

Sources of glucosinolates

Glucosinolate concentrations in plants vary based on genetic and environmental factors. Glucosinolates can make up about 1% of the dry matter of some cruciferous plants (Abbaoui et al., 2018). Glucosinolates are present in all species in the *Brassica* family and can be low to moderate in foliage, ranging from 1000 ppm (0.1%) in some plants to up to 3000 ppm (0.3%) in Brussels sprouts (*Brassica oleracea*). Concentrations in roots and seeds can be higher, up to 30,000 ppm (3%) in horseradish root (*Armoracia rusticana*) and 60,000 ppm (6%) in mustard seed (*Brassica nigra*) (Burrows and Tyrl, 2013). However, plants differ, and glucosinolate concentrations in *Brassica* plants are usually highest in the seeds and inflorescence and lowest in the roots (Hanschen et al., 2014; Oliviero et al., 2018). Environmental conditions affect plant glucosinolate concentrations; factors that can increase concentrations include water limitation and decreased temperature with increased UVB radiation (Hanschen et al., 2014).

Nutraceuticals. DOI: https://doi.org/10.1016/B978-0-12-821038-3.00053-7

Stability and metabolism of glucosinolates

There is evidence that short-term storage can increase glucosinolate concentrations, but long-term storage leads to decreased glucosinolate concentrations by up to 99% in 10 days (Abbaoui et al., 2018; Cavaiuolo and Ferrante, 2014). Different methods of food processing impact isothiocyanate formation. Cold storage and freezing decrease glucosinolate concentrations over time (Oliviero et al., 2018). Freezing followed by thawing decreases glucosinolates in cruciferous vegetables (Abbaoui et al., 2018). Glucosinolate concentrations are reported to be conserved calcium chloride treatment, commonly used to increase the shelf-life of various vegetables with the benefits of delaying aging or ripening and improving the nutritional value (Niu et al., 2015; Sun et al., 2015). Fermentation also reduces glucosinolate concentrations in vegetables (Hanschen et al., 2014; Oliviero et al., 2018). Ensiling decreases glucosinolate concentration, with a 10-fold decrease after 30 days reported for *Brassica napus* (Burrows and Tyrl, 2013).

Glucosinolates have low bioavailability, thus only a small fraction of the parent glucosinolate compounds are detectable in blood or urine after ingestion ; Abbaoui et al., 2018; Capuano et al., 2017; Oliviero et al., 2018). Based on urine studies, approximately 5% of the ingested glucosinolate dose is absorbed as parent compound. Low pH denatures myrosinase and produces degradation of glucosinolates, decreasing concentrations in food by up to 40% at pH = 1.5 (Oliviero et al., 2018). Cooking can inactivate myrosinase and epithiospecifier protein and degrade glucosinolates (Capuano et al., 2017; Oliviero et al., 2018). Loss of myrosinase reduces the benefit of dietary glucosinolates, but enzymes of the intestinal microbiome offer some compensatory effects (Oliviero et al., 2018). Cooking can decrease glucosinolate concentrations by up to 60% (Abbaoui et al., 2018). Cell lysis during cooking can lead to leaching of glucosinolates, especially when boiling, leading to glucosinolate losses of up to 75% (Oliviero et al., 2018).

Glucosinolates are glucoside compounds. They are stored in intracellular vacuoles within S-cells in plants, which are located between the endodermis and phloem (Grob and Matile, 1979; Cavaiuolo and Ferrante, 2014; Oliviero et al., 2018). Myrosin cells, located in the plant's vascular and epithelial cells, contain idioblasts filled with myrosinase, which has β-thioglucoside glucohydrolase activity (Burel et al., 2000; Cavaiuolo and Ferrante, 2014; Collett et al., 2014; Halkier and Gershenzon, 2006; Oliviero et al., 2018). The S-cells and myrosin cells are physically separated within the intact plant (Oliviero et al., 2018). Damage to the plant releases both glucosinolates and myrosinase, which, at a neutral pH, leads to β-D-glucose hydrolysis (Cavaiuolo and Ferrante, 2014; Ahlin et al., 1994). Myrosinase, with rate-limiting cofactor L-ascorbate, catalyzes hydrolysis of glucosinolates to unstable thiohydroximate- O-sulfonate aglycones (Hanschen et al., 2014; Oliviero et al., 2018; Prieto et al., 2019). The resulting aglycones go on to form isothiocyanates at neutral or high pH or release sulfur to form nitriles at pH <3.0 or in the presence of divalent iron (Oliviero et al., 2018; Hanschen et al., 2014; Prieto et al., 2019). When the glucosinolate side chain has a terminal double bond, sulfur released during hydrolysis is captured, resulting in the formation of epithionitriles (Prieto et al., 2019). Some plants produce nitrile specifier proteins or epithiospecifier proteins (ESPs), which interact with divalent iron ions and favor nitrile formation, even at higher pH (Hanschen et al., 2014; Oliviero et al., 2018; Narbad and Rossiter, 2018). The ESPs can also cause epithionitrile formation (Hanschen et al., 2014; Oliviero et al., 2018). Some plants produce thiocyanate-forming proteins (Hanschen et al., 2014).

Glucosinolate metabolism to aglycones and other byproducts by phytoenzymes can occur during mastication or in the rumen of cattle and other ruminants (Burrows and Tyrl, 2013; Capuano et al., 2017). Colonic microbes in monogastric animals also produce β-thioglucosidase, which also acts on glucosinolates to form active metabolites (Capuano et al., 2017; Hanschen et al., 2014). Metabolism of glucosinolates in the intestine is thus dependent on the microbiota (Capuano et al., 2017). Frequent intake of *Brassica* vegetables may increase the population of intestinal bacteria capable of metabolizing glucosinolates (Prieto et al., 2019). *Bifidobacterium* spp. capable of metabolizing glucosinolates have been identified in the human colon (Prieto et al., 2019). Metabolism of glucosinolates to isothiocyanates by intestinal microflora is reported, but metabolism to nitriles and epithionitriles is suspected but as of yet undocumented (Prieto et al., 2019).

Isothiocyanate and possibly nitrile and epithionitrile metabolites of glucosinolates are readily absorbed by gastric and intestinal epithelium by passive diffusion (Abbaoui et al., 2018; Hanschen et al., 2014; Prieto et al., 2019). Isothiocyanates can reach peak blood concentrations between half an hour and 6 hours postingestion (Abbaoui et al., 2018; Hanschen et al., 2014). Isothiocyanates and epithionitriles absorbed in the cytoplasm of epithelial cells are rapidly conjugated to glutathione within enterocytes and these conjugates are transported to the extracellular space by multidrug resistance-associated protein (MRP)1, MRP2 and by P-glycoprotein, either into blood or back to the intestinal lumen (Hanschen et al., 2014; Oliviero et al., 2018). Glutathione dissociates in the blood and some isothiocyanates become bound to serum albumin, while unbound circulating isothiocyanates are distributed to the tissues (Oliviero et al., 2018).

Absorbed isothiocyanates have tissue half-lives of around 2 hours (Abbaoui et al., 2018). Isothiocyanates are potent inducers of phase II enzymes (Prieto et al., 2019). Isothiocyanates and epithionitriles proceed to the liver by the portal circulation where the thiol groups become conjugated to glutathione and *N*-acetylcysteine for urinary elimination (Abbaoui et al., 2018; Capuano et al., 2017; Hanschen et al., 2014; Oliviero et al., 2018). Aliphatic nitriles are metabolized to aldehydes and cyanide ions, which are excreted in the urine as thiocyanate ions (Hanschen et al., 2014). Metabolites and conjugates have been detected in the plasma, kidneys, liver, urine, and other tissues (Capuano et al., 2017). Between 12% and 88% of ingested glucosinolates, depending on meal preparation, are excreted as metabolites in the urine (Hanschen et al., 2014; Prieto et al., 2019). Urine concentrations of isothiocyanates are up to 50-fold greater than plasma concentrations (Abbaoui et al., 2018). Glucosinolates and their metabolites are minimally excreted in milk and eggs, where they produce off-flavors (Bell, 1984; Burrows and Tyrl, 2013). Goitrogenic glucosinolates in particular are excreted in milk (Hanschen et al., 2014).

Major glucosinolates

The five glucosinolates most common in the human diet are glucobrassicin, sinigrin, glucoraphasatin, glucoraphanin, and glucoiberin (Melrose, 2019). Of these, glucobrassicin, sinigrin, and glucoraphanin are relatively well characterized in the literature. These are predominantly metabolized to reactive isothiocyanates, which impart the mustard-like or garlic-like odors associated with mustard and horseradish (Burrows and Tyrl, 2013). Another compound of interest is progoitrin.

Glucobrassicin

Glucobrassicin is a 3-indolylmethyl glucosinolate found in most *Brassicae* plants. Glucobrassicin is hydrolyzed to an unstable isothiocyanate. Metabolites of glucobrassicin include indole-3 carbinol, indol-3-acitonitrile, and 3,3'-diindolylmethane (Katz et al., 2018; Šamec et al., 2018; Wagner et al., 2013). These metabolites act on cell-cycle regulation in cancer cells, with actions on nuclear factor-κB (NF-κB) signaling, caspase activation, estrogen metabolism, estrogen receptors (ERs), endoplasmic reticulum stress, and breast cancer gene expression (Šamec et al., 2018; Wagner et al., 2013; Katz et al., 2018). Indole-3-carbinol induced apoptosis in breast cancer cells but not normal breast epithelium in vitro (Katz et al., 2018). Other anticancer effects of indole-3-carbamol relate to inhibiting neovascularization, tissue invasion, and metastasis (Katz et al., 2018). Diidolylmethane and indole-3-carbinol induce phase I enzymes, and thus can affect metabolism of carcinogens (Katz et al., 2018).

Sinigrin

Sinigrin is allyl-glucosinolate. The first glucosinolate discovered, it was originally found in mustard seeds about 200 years ago and is also present in cabbage, Brussels sprouts, and broccoli (Hanschen et al., 2014; Mazumder et al., 2016). Sinigrin is hydrolyzed to allyl isothiocyanate by plant myrosinase and a number of Gram-positive intestinal flora, including *Lactobacillus agilis* and *Streptomyces*, *Bacillus*, and *Staphylococcus* spp. (Narbad and Rossiter, 2018).

Allyl isothiocyanate is antimicrobial and prevents food spoilage (Mazumder et al., 2016). Antimicrobial effects may be through sulfhydryl binding to microbial thioredoxin, leading to inhibition of DNA replication (Mazumder et al., 2016). Allyl isothiocyanate has anticancer activity through downregulation of Bcl-2 genes, upregulation of tumor protein P53, and causing cell cycle arrest (Abbaoui et al., 2018; Mazumder et al., 2016). Antiinflammatory effects of allyl isothiocyanate were determined to be through inhibition of tumor necrosis factor-α (TNF-α) and interleukin-6 (IL-6), decreased phosphorylation of c-Jun N terminal kinases (JNKs), and decreased expression of transcription factor P65 and other proteins (Mazumder et al., 2016). Allyl isothiocyanate has broad fungicidal activity in foods, preventing spoilage (Melrose, 2019).

Allyl isothiocyanate has been studied for prevention of atherosclerosis through decreased expression of vascular cell adhesion molecule-1 (VCAM-1) and intracellular adhesion molecule-1 (ICAM-1) in experimental animals, along with modulation of circulating low-density lipoprotein (LDL), oxidized LDL, high-density lipoprotein (HDL) triglycerides, and cholesterol (Mazumder et al., 2016). Allyl isothiocyanate is hypothesized to inhibit microRNA expression and activates a transcription factor, nuclear factor erythroid 2-related factor 2 (Nrf2) (Wagner et al., 2013).

Sinigrin is also metabolized, at low pH, to allyl nitrile. Allyl nitrile is an important modifier of phase II enzymes, leading to increased activity of glutathione S-transferase, quinone reductase, glutathione, thioredoxin reductase, glutathione peroxidase, and superoxide dismutase, but reduced activity of catalase and glutathione reductase in various tissues, possibly contributing to the anticancer and antiinflammatory effects of sinigrin (Tanii, 2017).

Glucoraphanin

Glucoraphanin, or 4-methylsulfinylbutyl glucosinolate, is a major component of broccoli, kale, Brussels sprouts, cabbage, and other vegetables. Glucoraphanin is hydrolyzed to sulforaphane, or 4-methyl-sulfinylbutyl isothiocyanate, other isothiocyanates, and inactive nitrile compounds (Vanduchova et al., 2019). *Lactobacillus* spp. have been implicated in this process (Narbad and Rossiter, 2018). Sulforaphane has antimicrobial properties effective against *Helicobacter pylori* (Šamec et al., 2018; Melrose, 2019; Prieto et al., 2019).

Sulforaphane induces glutathione synthesis via the glutamate–cysteine ligase catalytic subunit, and induces phase II metabolic enzyme glutathione S-transferase in mammals, thus it enhances the metabolism and excretion of other chemicals (Melrose, 2019; Vanduchova et al., 2019). This increase in phase II enzyme activity from sulforaphane is implicated in decreased risk of urinary bladder cancers documented in people who regularly consume cruciferous vegetables (Abbaoui et al., 2018). Additionally, sulforaphane is protective against reactive oxygen species through induction of protonated nicotinamide adenine dinucleotide phosphate (NADPH) quinone reductase and through interaction with thiols on Kelch-like ECH-associated protein 1 (KEAP1), part of the KEAP1-Nrf2-antioxidant response signaling pathway (Abbaoui et al., 2018; Melrose, 2019). Experimentally, the antioxidant effects of sulforaphane via the NrF2 pathway prevent obesity in laboratory animals on a high-fat diet, decrease oxidative stress and insulin resistance, and reduce hepatic steatosis (Xu et al., 2018).

Sulforaphane has tumor prevention properties: it blocks the cell cycle and promotes apoptosis (Halkier and Gershenzon, 2006; Prieto et al., 2019). Cell cycle arrest is through disruption of microtubule assembly (Melrose, 2019). The mechanisms by which sulforaphane promotes apoptosis are through inhibiting histone deacetylase, inducing caspase-mediated apoptosis, downregulation of B-cell lymphoma 2 regulator protein (Bcl-2), and increased Bcl-2-associated X protein (BAX) (Abbaoui et al., 2018; Melrose, 2019).

Sulforaphane has been studied for antiarthritic effects. Sulforaphane protects cartilage by inducing NAD(P)H: quinone oxidoreductase 1 activity and inhibiting matrix metalloproteinase (MMP)-1, MMP-3, MMP-13 (Vanduchova et al., 2019). There are antiinflammatory effects of sulforaphane through inhibition of IL-6, IL-27, TNF-α, and interferon-γ, and suppression of cyclooxygen-2 (COX-2) mRNA can also be protective against arthritis (Prieto et al., 2019; Vanduchova et al., 2019).

Progoitrin

Progoitrin is hydrolyzed to goitrin, or l-5-vinyl-2-thioöxazolidine, which is named for its antithyroid effects. The colonic flora can contribute to the degradation of progoitrin, and some *Lactobacillus* spp. and a strain of enterobacterium actively degrade progoitrin in mammals (Narbad and Rossiter, 2018). Goitrin blocks tyrosine iodination and inhibits T4 formation (Burrows and Tyrl, 2013). Long-term ingestion is associated with goiter formation. Progoitrin is found in rapeseed, kale, and a variety of vegetables and seeds.

Beneficial effects of glucosinolates

Metabolites of glucosinolates have various beneficial biological activity and chemopreventive actions (Das et al., 2000; Hayes et al., 2008; Sarikamis, 2009). Cruciferous vegetables are reported to be protective against numerous types of cancer, including urinary bladder, gastric, colorectal, prostate, thyroid, ovarian, breast, endometrial, renal, pulmonary, pancreatic, hepatic, and lymphoid (Capuano et al., 2017). Various mechanisms have been ascribed to the anticancer activities of isothiocyanates; including preventing DNA adduct formation, proapoptotic activity, decreased histone deacetylation, downregulation of inflammation, suppression of angiogenesis, and decreased metastasis (Razis et al., 2018). There is some evidence in cancer prevention from consumption of normal dietary quantities of *Brassica* vegetables; some of the effects appear to require higher concentrations of thiocyanates than possible through dietary glucosinolates (Hanschen et al., 2014; Razis et al., 2018).

Glucosinolate metabolites can cause cell cycle arrest and regulate apoptosis in cancer cells (Capuano et al., 2017; Hanschen et al., 2014). Isothiocyanates affect regulation of angiogenesis and tumor metastasis, possibly through hypoxia inducible factors (Capuano et al., 2017; Hanschen et al., 2014). Some isothiocyanates inhibit antiapoptotic telomerase enzymes in cancer cells (Hanschen et al., 2014). Other metabolites appear to modulate certain cancers through suppression of expression of androgen and ERs (Capuano et al., 2017; Hanschen et al., 2014). Some glucosinolates are reported to regulate ERα and ERβ signaling and increase P450-mediated estrogen breakdown (Hanschen et al., 2014). Glucosinolate metabolites are also epigenetic modulators (Capuano et al., 2017; Hanschen et al., 2014). Isothiocyanates can inhibit histone deacetylase transferases, DNA methylation, and microRNA expression (Capuano et al., 2017; Hanschen et al., 2014; Wagner et al., 2013).

Isothiocyanates are important regulators of phase I and II metabolic enzymes, thus they affect metabolism of xenobiotics (Razis et al., 2018). Because many carcinogenic chemicals are, in and of themselves, relatively inert, phase I metabolic enzymes are important in carcinogenesis through production of electrophilic intermediates from parent compounds (Razis et al., 2018). Because isothiocyanates inhibit the activity of some of the cytochrome P450 enzymes involved in phase I metabolism, they can decrease the production of these reactive metabolites (Abbaoui et al., 2018; Capuano et al., 2017; Prieto et al., 2019). Aside from phenethyl isothiocyanate, which has

P450 effects at low concentrations, relatively high isothiocyanate concentrations are required to suppress the activity of phase I metabolic enzyme, making this mechanism unlikely to be the primary cause of the anticarcinogenic effects (Razis et al., 2018).

Isothiocyanates can activate phase II enzymes, which are responsible for conjugating reactive metabolites of phase I enzymes to less reactive molecules that enhance excretion (Capuano et al., 2017; Hanschen et al., 2014; Prieto et al., 2019). Thiocyanates promote synthesis of Nrf2, which regulates expression of antioxidants and phase II microsomal enzymes (Wagner et al., 2013; Cavaiuolo and Ferrante, 2014). Some glucosinolate metabolites activate aryl hydrocarbon receptors, important in phase I and II enzyme transcription regulation, which mediates cell response to environmental pollutants including polycyclic aromatic hydrocarbons such as dioxin (Capuano et al., 2017).

Some isothiocyanates inhibit activity of NF-κB, which is involved in stress response and inflammation (Capuano et al., 2017; Hanschen et al., 2014; Wagner et al., 2013; Prieto et al., 2019). NF-κB regulates the expression of COX-2, which produces inflammatory prostaglandins (Prieto et al., 2019). Indole glucosinolates can inhibit the secretion of TNF-α (Prieto et al., 2019).

Glucosinolate metabolites, erucic acids, and oleic acids, are reported to have direct antimicrobial effects (Cavaiuolo and Ferrante, 2014). Dietary glucosinolates alter the microbiota of the colon, where glucosinolate metabolism takes place (Capuano et al., 2017). Isothiocyanates inhibit *H. pylori* bacteria and have been demonstrated to kill *Escherichia coli*, *Pseudomonas aeruginosa*, *Listeria monocytogenes*, and *Staphylococcus aureus* (Capuano et al., 2017).

Cruciferous vegetable consumption is associated with decreased risk of neurodegenerative diseases, most likely through oxidant scavenging related to activation of glyoxalase 1 to promote formation of S-lactoglutathione (Capuano et al., 2017).

Toxic effects of glucosinolates

Exclusive or excessive feeding of vegetables and/or seeds from the *Brassica* family has been associated with toxic effects in livestock. Some effects are nutritional. Kale contains a high concentration of sulfur and can inhibit uptake of copper and selenium, causing micronutrient deficiencies (Taljaard, 1993). Strong flavors of some plants can produce a tainted flavor in eggs and milk (Bell, 1984; Burrows and Tyrl, 2013). While glucosinolates stimulate appetite at low concentrations and contribute to the flavor of foods such as horseradish and mustard, plants containing high concentrations of glucosinolates have been associated with reduced feed intake in livestock (Burrows and Tyrl, 2013). Allyl isothiocyanates can irritate the gastrointestinal tract (Abbaoui et al., 2018; Mason and Lucas, 1983; Taljaard, 1993; Woyengo et al., 2011). Clinical signs occur soon after ingestion and include colic in horses and abdominal pain in ruminants, though they are less sensitive. Pigs are more severely affected by dietary glucosinolates compared to rabbit, poultry, and fish (Tripathy and Mishra, 2007). Glucosinolates interfere with liver function by altering phase I and II enzyme expression (Williamson et al., 1996). Other toxic effects of glucosinolates include goiter, reduced circulating thyroid hormone, reduced growth, decreased reproductive performance, and even mortality (VanEtten et al., 1983; Tripathy and Mishra, 2007). The decreased growth and egg production reported in poultry could be associated with decreased feed intake (Woyengo et al., 2011; Martland et al., 1984; Wight et al., 1987). Lesions in ruminants include sloughing of ruminal epithelium and hyperemia of gastrointestinal mucosa, which can be severe enough to cause death. Treatment is symptomatic and includes a clean diet and pain control (Burrows and Tyrl, 2013).

Thyroid enlargement is associated with prolonged ingestion of *Brassica* containing goitrin (Burrows and Tyrl, 2013). Thiocyanates competitively inhibit the thyroid hormone axis (Hanschen et al., 2014). Goitrin prevents tyrosine iodination, thus causing a noniodine responsive inhibition of T4 production (Burrows and Tyrl, 2013). Thiocyanate causes iodine depletion in the thyroid by blocking uptake of iodide; therefore it is an iodine-responsive cause of T4 inhibition (Taljaard, 1993). Antithyroid effects of glucosinolates can result in subclinical signs such as decreased reproductive performance and growth or, in more severe cases, clinically evident goiter (Taljaard, 1993). Thyroid enlargement and fetal deaths were described in laboratory rodents. Histologic lesions are characterized by hypertrophy of follicular epithelium and reduced colloid (Burrows and Tyrl, 2013). Thyroid hypertrophy is reported in poultry (Martland et al., 1984; Wight et al., 1987). Decreased thyroid function is reported in fish (Burel et al., 2000). Vegetables with high goitrogenic potential in the human diet, like Brussel's sprouts, are normally cooked, which helps mitigate these effects.

Anemia is also a common sequela to overfeeding of *Brassica* to livestock. Monogastric animals are susceptible to anemia due to Heinz body hemolysis (Cox-Ganser et al., 1994). Ruminants are less susceptible due to degradation of glucosinolates by adapted rumen microflora (Burrows and Tyrl, 2013). However, ruminal metabolism of sulfoxides to dimethyl sulfide is also associated with Heinz body hemolysis (Taljaard, 1993). Anemia is also reported in poultry (Woyengo et al., 2011).

Nitriles are associated with several hepatic effects, including bile duct hyperplasia, megalocytosis, zonal hepatocyte necrosis, and fibrosis, and sometimes renal megalocytosis (Collett et al., 2014). Hepatocellular damage and bile stasis are evident on serum chemistry. Liver hemorrhage is reported in laying hens fed rapeseed meal, even low-glucosinolate varieties (Wight et al., 1987). Hemorrhage could be secondary to lysis of vascular reticulin, causing pressure necrosis of hepatic parenchyma (Martland et al., 1984).

The effects of glucosinolates on phase I and II metabolic enzymes may affect drug metabolism (Vanduchova et al., 2019). Sulforaphane inhibits cytochrome P450 (CYP)3A4, CPY3A5, and CYP2D6, which are important in drug metabolism, causing changes in metabolism of furosemide, verapamil, and ketoprofen (Vanduchova et al., 2019).

Glucosinolate metabolites can be genotoxic, and possibly carcinogenic (Abbaoui et al., 2018; Capuano et al., 2017; Hanschen et al., 2014). Isothiocyanates are electrophilic, thus they can form adducts with DNA, and contribute to the formation of reactive oxygen species (Capuano et al., 2017). Epithionitriles are epoxide-like, and thus also have the potential to form DNA adducts (Capuano et al., 2017).

Other compounds in *Brassica*

Erucic acid

Rapeseed, mustards (*B. napus*, *Brassica campestris*, *Brassica juncea*, and others), and some other related plants contain high concentrations of erucic acid. Erucic acid has been associated with cardiac injury in cattle and experimental rodents (Burrows and Tyrl, 2013). Canola oil contains less than 5% erucic acid (Burrows and Tyrl, 2013). Canola meal, which is high in protein, has a variable erucic acid content depending on the method of extraction (Woyengo et al., 2011).

Sulfur-containing compounds

High dietary sulfur can limit the absorption of copper and other trace minerals, predisposing to deficiency, and has been associated with polioencephalomalacia in ruminants (Burrows and Tyrl, 2013; Taljaard, 1993).

Concluding remarks and future directions

Glucosinolates are biologically active compounds found in the *Brassicaceae* family of plants, including common vegetable plants such as broccoli, cabbage, cauliflower, rapeseed, mustard, and horseradish. Strong flavors correlate with high glucosinolate concentrations. Recent studies have found beneficial effects of glucosinolates, including regulatory functions in inflammation, stress response, xenobiotic metabolism, and antioxidant production, as well as direct antimicrobial properties. Future studies will no doubt find even more beneficial biological effects. However, adverse effects have been reported in a variety of livestock species fed rations high in *Brassicaceae*, including off-flavors in animal products, reduced feed consumption and growth, gastrointestinal irritation, goiter, anemia, and hepatic lesions. High sulfur can be associated with trace mineral deficiencies and polioencephalomalacia. Therefore while *Brassicaceae* are an important part of the diet, overconsumption is not recommended.

References

Abbaoui, B., Lucas, C.R., Riedl, K.M., et al., 2018. Cruciferous vegetables, isothiocyanates, and bladder cancer prevention. Mol. Nutr. Food. Res. 62 (18), e1800079. Available from: 10.1002/mnfr.201800079.

Ahlin, K.A., Emmanuelson, M., Wiktorsson, H., 1994. Rapeseed products from double cultivars as feed for dairy cows: effects of long-term feeding on thyroid function, fertility and animal health. Acta Vet. Scand. 35 (1), 37–53.

Bell, J.M., 1984. Nutrients and toxicants in rapeseed meal: a review. J. Anim. Sci. 58 (4), 996–1010.

Burel, C., Boujard, T., Escaffre, A.-M., 2000. Dietary low-glucosinolate rapeseed meal affects thyroid status and nutrient utilization in rainbow trout (*Oncorhynchus mykiss*). Br. J. Nutr. 83 (6), 653–664.

Burrows, G.E., Tyrl, R.J., 2013. Toxic Plants of North America, second ed. Wiley-Blackwell, Ames, IA.

Capuano, E., Dekker, M., Verkerk, R., et al., 2017. Food as pharma? The case of glucosinolates. Curr. Pharm. Des. 23 (19), 2697–2721. Available from: 10.2174/1381612823666170120160832.

Cavaiuolo, M., Ferrante, A., 2014. Nitrates and glucosinolates as strong determinants of the nutritional quality in rocket leafy salads. Nutrients 6 (4), 1519–1538.

Collett, M.G., Stegelmeier, B.L., Tapper, B., 2014. Could nitrile derivatives of turnip (*Brassica rapa*) glucosinolates be hepato- or cholangiotoxic in cattle? J. Agric. Food Chem. 62 (30), 7370–7375.

Cox-Ganser, J.M., Jung, G.A., Pushkin, R.T., 1994. Evaluation of *Brassicas* in grazing systems for sheep: II. Blood composition and nutrient status. J. Anim. Sci. 72 (7), 1832–1841.

Das, S., Tyagi, A.K., Kaur, H., 2000. Cancer modulation by glucosinolates: a review. Curr. Sci. 79, 1665–1671.

Grob, K., Matile, P., 1979. Vacuolar location of glucosinolate in horseradish root cells. Plant Sci. Lett. 14, 327–335.

Halkier, B.A., Gershenzon, J., 2006. Biology and biochemistry of glucosinolates. Ann. Rev. Plant. Biol. 57, 303–333.

Hanschen, F.S., Lamy, E., Schreiner, M., Rohn, S., 2014. Reactivity and stability of glucosinolates and their breakdown products in food. Agnew. Chem. Int. Ed. 53, 11430–11450.

Hayes, J.D., Kelleher, M.O., Eggleston, I.M., 2008. The cancer chemopreventive actions of phytochemicals derived from glucosinolates. Eur. J. Nutr. 47, 73–88.

Katz, E., Nisani, S., Chamovitz, D.A., 2018. Indole-3-carbinol: a plant hormone combatting cancer. F1000 Faculty Rev. 689. Available from: 10.12688/f1000research.14127.1.

Martland, M.F., Butler, E.J., Fenwick, G.R., 1984. Rapeseed induced liver hemorrhage, reticulolysis and biochemical changes in laying hens: the effects of feeding high and low glucosinolate meals. Res. Vet. Sci. 36 (3), 298–309.

Mason, R.W., Lucas, P., 1983. Acute poisoning in cattle after eating old non-viable seed of chou moellier (*Brassica oleracea* convar. acephala). Aust. Vet. J. 60 (9), 272–273.

Mazumder, A., Dwivedi, A., du Plessis, J., 2016. Sinigrin and its therapeutic benefits. Molecules 21, 416.

Melrose, J., 2019. The glucosinolates: a sulphur glucoside family of mustard anti-tumor and antimicrobial phytochemicals of potential therapeutic application. Biomedicines 7 (62), 1–28. Available from: 10.3390/biomedicines7030062.

Narbad, A., Rossiter, J.T., 2018. Gut glucosinolate metabolism and isothiocyanate production. Mol. Nutr. Food Res. 62, 1700991.

Niu, Y.X., Rogiewicz, A., Wan, C.Y., 2015. Effect of microwave treatment on the efficacy of expeller pressing of *Brassica napus* rapeseed and *Brassica juncea* mustard seeds. J. Agric. Food Chem. 63, 3078–3084.

Oliviero, T., Verkerk, R., Dekker, M., 2018. Isothiocyanates from *Brassica* vegetables—effects of processing, cooking, mastication, and digestion. Mol. Nutr. Food Res. 62, 1701069.

Prieto, M.A., López, C.J., Simal-Gandara, J., 2019. Glucosinolates: molecular structure, breakdown, genetic bioavailability properties and healthy and adverse effects. Adv. Food Nutrit. Res. 90, 305–350.

Razis, A.F.A., Konsue, N., Ioannides, C., 2018. Isothiocyanates and xenobiotic detoxification. Mol. Nutr. Food Res. 62, 1700916.

Šamec, D., Urlić, B., Salopek-Sondi, B., 2018. Kale (*Brassica olecaracea* var *acephala*) as a superfood: review of the scientific evidence behind the statement. Crit. Rev. Food Sci. Nutr. 59, 2411–2422.

Sarikamis, G., 2009. Glucosinolates in crucifers and their potential effects against cancer: review. Can. J. Plant Sci. 89, 953–959.

Sun, J., Kou, L., Geng, P., 2015. Metabolic assessment reveals an elevated level of glucosinolate content in $CaCl_2$ treated Broccoli microgreens. J. Agric. Food Chem. 63.

Taljaard, T.L., 1993. Cabbage poisoning in ruminants. J. S. Afr. Vet. Assoc. 64 (2), 96–100.

Tanii, H., 2017. Allyl nitrile: toxicity and health effects. J. Occup. Health 59, 104–111.

Tripathy, M.K., Mishra, A.S., 2007. Glucosinolates in animal nutrition. Anim. Feed. Sci. Technol. 132, 1–27.

Vanduchova, A., Azenbacher, P., Anzenbacherova, E., 2019. Isothiocyanate from broccoli, sulforaphane, and its properties. J. Med. Food 22 (2), 121–126.

VanEtten, C.H., Tookey, H.L., Rechcigl, M., 1983. Glucosinolates handbook of naturally occurring food toxicants. In: Rechcigl, M. (Ed.), Handbook of Naturally Occurring Food Toxicants. CRC Press, Boca raton, FL, p. 198315.

Wagner, A.E., Terschluessen, A.M., Rimbach, G., 2013. Health promoting effects of brassica-derived phytochemicals: from chemopreventive and anti-inflammatory activities to epigenetic regulation. Oxid. Med. Cell. Longev. 2013, 1–12.

Wight, P.A., Scougall, R.K., Shannon, D.W., 1987. Role of glucosinolates in the causation of liver haemorrhages in laying hens fed water-extracted or heat-treated rapeseed cakes. Res. Vet. Sci. 43 (3), 313–319.

Williamson, G., Wang, H., Griffiths, S., 1996. Glucosinolates as bioactive components of *Brassica* vegetables: induction of cytochrome P450 1A1 in Hep G2 cells as assessed using transient transfection. Biochem. Soc. Trans. 24, 383S.

Woyengo, T.A., Kiarie, E., Nyachoti, C.M., 2011. Growth performance, organ weights, and blood parameters of broilers fed diets containing expeller-extracted canola meal. Poult. Sci. 90 (11), 2520–2527.

Xu, L., Nagata, N., Ota, T., 2018. Glucoraphanin: a broccoli sprout extract that ameliorates obesity-induced inflammation and insulin resistance. Adipocyte 7 (3), 218–225.

Chapter 54

Organosulfur compounds as nutraceuticals

Nikolay V. Goncharov, Daria A. Belinskaia, Anton I. Ukolov, Richard O. Jenkins and Pavel V. Avdonin

Introduction and background

Sulfides, disulfides, and other organosulfur compounds (OSCs) are widely present in our bodies and the natural environment. Three sulfur-containing amino acids (cysteine, cystine, and methionine) are important components of proteins, peptides, enzymes, vitamins, and hormones. They can interact with or neutralize reactive oxygen species (ROS) and determine the thiol/disulfide redox states in signaling and sensing (Jones, 2010), thus maintaining our bodies health or contributing to the development of diseases (Goncharov et al., 2015a). Disulfide bridges that are formed during the posttranslational modifications of proteins are highly important in keeping their tertiary structure and maintaining their functional activity (Patil et al., 2015). The principal reaction by which disulfide bonds are formed and rearranged in a protein is thiol–disulfide exchange, also known as disulfide shuffling, which does not change the number of disulfide bonds and is generally much faster than oxidation/reduction reactions, which change the number of disulfide bonds within a protein (Gilbert, 1995). The in vivo oxidation and reduction of protein disulfide bonds by thiol–disulfide exchange is facilitated by a protein called thioredoxin (Trx), and this property could be exploited to elaborate a new type of therapeutics (Pirazzini et al., 2014). Ultimately, disulfide bond formation is possible due to oxidative potential existing in specific cellular compartments, and it was wittily noted by Bardwell that "the long-running race to find the source of oxidizing potential for disulfide bond formation is over, and the winner is one of the first contestants to enter: oxygen" (Bardwell, 2002). Nevertheless, oxygen is not always available in plentiful amounts, for example, in ischemic states. In such cases some artificial or natural electrophiles (electron acceptors) could be helpful, and this property has been exploited for elaborating a new type of adjuvant therapeutics with the help of sulfur-containing redox compounds (Valachová et al., 2015; Goncharov et al., 2015c). Both biogenic and abiogenic disulfides can serve as electrophiles and redox mediators of plasma membrane oxidoreductase activity (Prokofieva and Goncharov, 2014). There are many plants that contain disulfides and other OSCs, which play a significant role in the smell and taste of food. Some OSCs can be toxic at high doses, though at low doses they can provide beneficial and even therapeutic effects by increasing tissue activities of enzymes that protect against electrophiles (Munday, 2012). In considering nutraceuticals, there are two principal groups of vegetables that contain OSCs with special properties. Garlic, onion, shallot, leek, and chives are well-known representatives of the *Allium* genus (family *Amaryllidaceae*), that contain *S*-alk(en)yl-L-cysteine sulfoxides. Cabbage, cauliflower, Brussels sprouts, kale, etc. are representatives of the *Brassica* genus, and rucola (rocket salad) of the *Eruca* genus of the mustard or cruciferous family (*Brassicaceae*), which contain *S*-methyl cysteine-L-sulfoxide (Munday, 2012). These vegetables are popular in cuisines worldwide and are valued for their potential medicinal properties. Epidemiological studies indicate a positive association of their consumption with decreased risk of cancer and other diseases (Nicastro et al., 2015). This chapter mostly deals with OSCs that are found in the vegetables of these two families and could be used as nutraceuticals: original sulfoxides, their derivatives, mechanisms of action, and some therapeutic effects.

Chemistry and occurrence of organosulfur compounds

Onion (*Allium cepa*) is found in various regions of Europe, America, Asia, and Africa, and is the most widely cultivated species of the genus *Allium*, though other species are also referred to as onions, such as the

Nutraceuticals. DOI: https://doi.org/10.1016/B978-0-12-821038-3.00054-9

Japanese bunching onion (*A. fistulosum*), the Egyptian onion (*A. proliferum*), and the Canada onion (*Allium canadense*). It is one of the classic examples of the *Allium* species used not only for culinary preparations but also for medicinal purposes, and there are two chemical groups present that have a marked effect on human health. These are flavonoids and sulfur-containing compounds (Suleria et al., 2015). Yellow onions contain up to 1187 mg of flavonols per kilogram of fresh weight (FW), whereas red onions contain up to 1917 mg/FW of flavonols, which are the predominant pigments of onions. At least 25 different flavonols have been characterized, and quercetin glycosyl derivatives are the most important ones in all onion cultivars (Slimestad et al., 2007). Among the onion-derived substances, dipropyl sulfide, dipropyl disulfide, dipropenyl sulfide, and dipropenyl disulfide are the best known organosulfur substances (Munday and Munday, 2001), though cycloalliin, *S*-methyl-L-cysteine, *S*-propyl-L-cysteine sulfoxide, dimethyl trisulfide, and *S*-methyl-L-cysteine sulfoxide are also thought to be responsible for the beneficial properties of onion extract (Yoshinari et al., 2012) (Table 54.1).

Garlic (*Allium sativum*) is a well-known vegetable plant in many countries. The active ingredients of garlic include sulfur-containing compounds such as the nonproteinogenic amino acid alliin (*S*-allyl-L-cysteine sulfoxide) and other *S*-alkyl-L-cysteine sulfoxides, which are precursors of allicin and related compounds produced enzymatically through alliinase (Table 54.1). Allicin (allyl 2-propenethiosulfinate or diallyl thiosulfinate) is the principal bioactive compound present in the aqueous extract of garlic or raw garlic homogenate. Allicin, as well as many other OSCs, is also referred to as a reactive sulfur species and undergoes a redox reaction with thiol groups in glutathione and proteins that is thought to be essential for its biological activity (Borlinghaus et al., 2014). Ajoene (4,5,9-trithiadodeca-1,6,11-triene-9-oxide) is one of the major natural compounds derived from garlic through the conversion of alliin into allicin, by an alliinase-induced cleavage. Ajoene has the advantage of greater chemical stability than allicin (Viswanathan et al., 2014). Also, other important compounds present in garlic homogenate include 1-propenyl allyl thiosulfonate, allyl methyl thiosulfonate, and γ-L-glutamyl-*S*-alkyl-L-cysteine. Interestingly, the adenosine concentration increases several-fold as the homogenate is incubated at room temperature for several hours (Bayan et al., 2014). Sodium 2-propenylthiosulfate is another water-soluble organosulfane sulfur compound isolated from garlic. Steam distillation of garlic bulbs yields garlic oil, with allyl (poly) sulfides being the most abundant compounds present, accounting for nearly 94% of the total amount with the following proportions: 4.7%–8% diallyl sulfide (DAS), 21.9%–40% diallyl disulfide (DADS), and 39%–41.5% diallyl trisulfide (DATS), depending on the extraction conditions (Wang et al., 2012). In addition, sliced raw garlic stored in 15%–20% ethanol for more than 1.5 years is known as aged garlic extract (AGE), which does not contain allicin though it has a significant biological effect in the treatment of various diseases due to increased concentration of certain new OSCs (*S*-allylcysteine, *S*-allylmercaptocysteine, allixin), as well as selenium (Bayan et al., 2014). In optimizing procedures of isolation of active compounds from garlic, it was found that allicin transformation under the effect of microwaves is faster than transformations performed under the influence of ultrasound or by conventional methods. Pharmacologically active compounds of (E)-ajoene, (Z)-ajoene, 3-vinyl-4H-1,2-dithiin, 2-vinyl-4H-1,3-dithiin, and DADS were isolated from the mixture of transformation products of allicin under the influence of microwaves in methanol at 55°C (Ilić et al., 2012).

Glucosinolates (*S*-glucopyranosyl thiohydroximates) are naturally occurring *S*-linked glucosides found mainly in cruciferous vegetables (family *Brassicaceae* or *Cruciferae*). They are enzymatically hydrolyzed by myrosinase to produce sulfate ions, D-glucose, and characteristic degradation products such as isothiocyanates (sulforaphane, erucin, etc.), thiocyanate, epithionitrile, nitrile, and goitrin. Isothiocyanates are believed to be responsible for the physiological activity of these plants. Sulforaphane was identified in broccoli sprouts, which have the highest concentration of sulforaphane in the cruciferous vegetables (Zhang et al., 1992). It was also found in other vegetables or cultivars of *Brassica oleracea* species (Brussels sprouts, white cabbage, kohlrabi, cauliflower, collard greens, savoy, Chinese kale), *Brassica rapa* species (Chinese cabbage, turnip, broccoli raab), mustard (*Sinapis hirta*, *Brassica juncea*, *Brassica nigra*), radish (*Raphanus sativus*), arugula (rucola, *Eruca sativa*), and watercress (*Nasturtium officinale*). Erucin [1-isothiocyanato-4-(methylthio) butane], which is metabolically and structurally related to sulforaphane, is present in large quantities in arugula, kohlrabi, and Chinese cabbage (Ishida et al., 2014; Azarenko et al., 2014). Berteroin (5-methylthiopentyl isothiocyanate) is a sulforaphane analog present in various cruciferous vegetables, including Chinese cabbage, arugula, and mustard (Jung et al., 2014). Sinigrin, which is the precursor of allyl isothiocyanate, is stored in plentiful amounts in mustard seed powder and can be readily hydrolyzed by the accompanying endogenous myrosinase (Bhattacharya et al., 2010). Glucosinolates in cruciferous vegetables can be divided into several subgroups based on the chemical structure of the side chains: indole, aliphatic, alcoholic, alkylthioalkyl, aromatic. Their occurrence differs significantly in the cruciferous vegetables (Table 54.2).

TABLE 54.1 Organosulfur compounds of vegetables from *Allium*, *Eruca*, and *Brassica* genera.

Chemical classes	Structure	Source of OSC
Cycloalliin	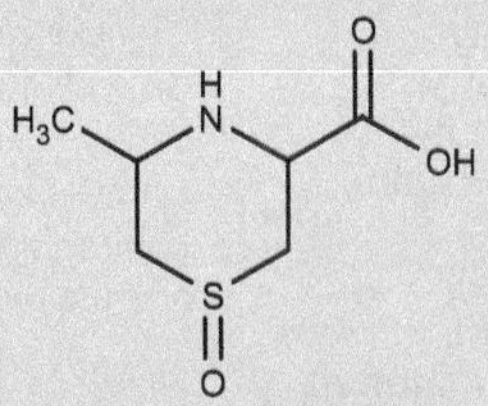	Onion, chives, shallots, leek
S-alkyl-L-cysteinesulfoxides	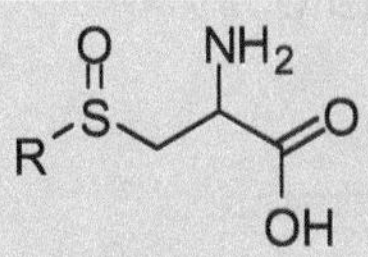 R = methyl, propyl, allyl (alliin), propenyl-1 (isoalliin)	Onion, chives, shallots, leek, garlic, ramsons
Mono-, di-, and trisulfides	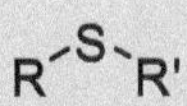 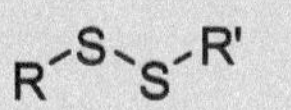 1. R = R' = methyl, propyl, allyl, propenyl-1, 2. R = Allyl, R' = methyl 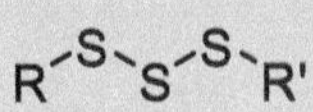	
Thiosulfonates	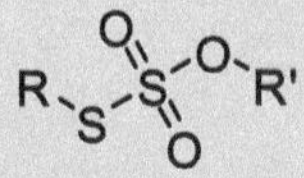 1. R = propenyl-1, R' = allyl 2. R = methyl, R' = allyl 3. R = allyl, R' = H	
S-alkylcysteines	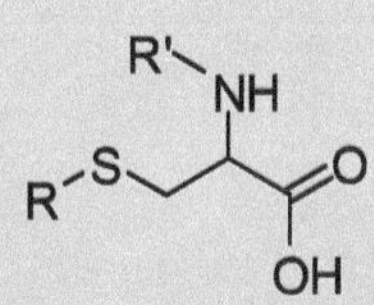 1. R = alkyl, R' = glutamyl 2. R = allyl, R' = H	Garlic, ramsons
Cysteine alkyl disulfides	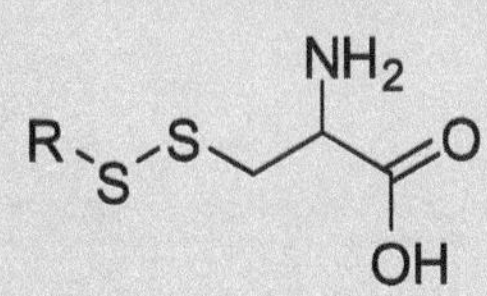 R = allyl	
3-Vinyl-4H-1,2-dithiin	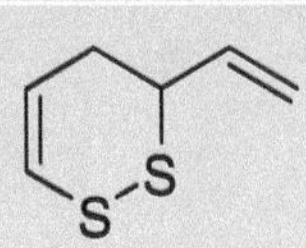	
2-Vinyl-4H-1,3-dithiin	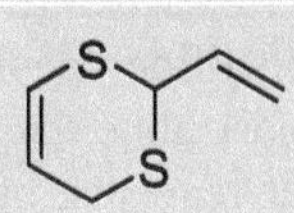	
E- and *Z*-4,5,9-trithiadodeca-l,6,11-triene 9-oxides (ajoenes)	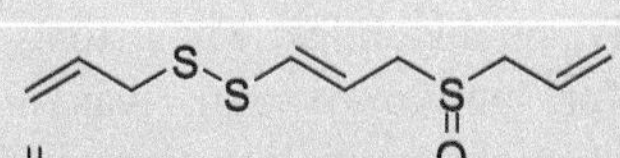 *E*-ajoene 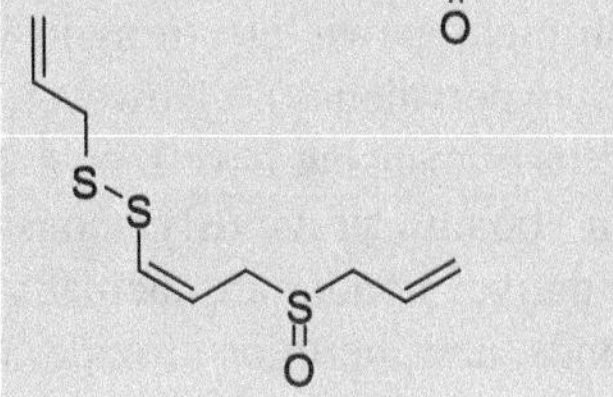 *Z*-ajoene	

(*Continued*)

TABLE 54.1 (Continued)

Chemical classes	Structure	Source of OSC
Glucosinolates	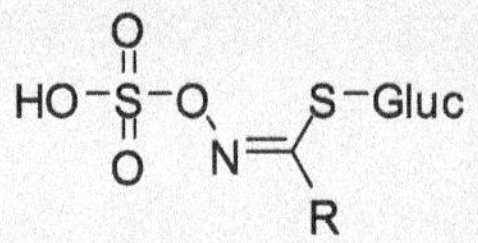 R = allyl (sinigrin), 4-methylthiobutyl (glucoerucin), 4-methylthiobut-3-enyl, (dehydroerucin), 4-methylsulfinylbutyl (glucoraphanin), indolyl (glucobrassicin), 2-phenylethyl (gluconasturtiin), 3-methylthiopropyl (glucoiberverin), 3-methylsulfinylpropyl (glucoiberin), 4-methylsulfinyl-3-butenyl (glucoraphenin), 3-butenyl (gluconapin), 3-hydroxy-3-butenyl (progoitrin), 5-methylthiopentyl (glucoberteroin), 5-methylsulfinylpentyl (glucoalyssin), 4-pentenyl (glucobrassicanapin), 2-hydroxy-4-pentenyl (gluconapoleiferin), 4-hydroxy-3-indolylmethyl (4-hydroxyglucobrassicin), 4-methoxy-3-indolylmethyl (4-methoxyglucobrassicin), N-methoxy-3-indolylmethyl (neoglucobrassicin)	Chinese cabbage (bok choy), Brussels sprouts, cabbage, cauliflower, kale, collards, Chinese broccoli, broccoli raab, kohlrabi, mustard, turnip, radish, arugula, watercress, rapeseed oil
Isothiocyanates	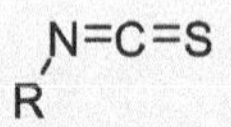 R = allyl, butyl, 2-phenylethyl, phenylhexyl, benzyl, 3-methylsulfinylpropyl (iberin), 8-methylsulfinyloctyl, 5-methylsulfinylpentyl (berteroin), 4-methylsulfinylbutyl (sulforaphane), 4-methylthiobutyl (erucin), 1-phenylethyl	
Goitrin	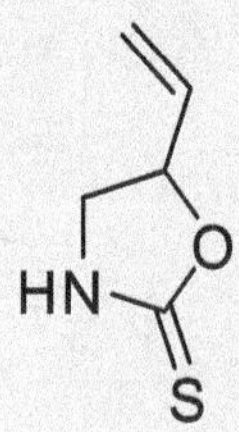	
Epithionitrile	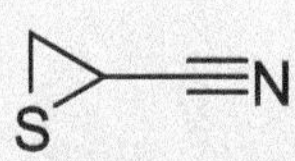	

OSC, organosulfur compound.

Biochemistry and molecular interactions of organosulfur compounds

Formally, a disulfide bond is a covalent bond derived by the coupling of two thiol groups. The connection is a persulfide, analogous to peroxide, but this terminology is obscure and is used only in some rare cases. The disulfide bond is about 2.05 Å in length, with a bond dissociation energy of about 60 kcal/mol (251 kJ/mol), and standard redox potential of about −250 mV (Cremlyn, 1996). Disulfides show a distinct preference for dihedral angles approaching 90 degrees. When the angle approaches 0 or 180 degrees, then the disulfide becomes a better oxidant. As divalent sulfur is polarizable, the S−S bond is susceptible to scission by polar reagents, both electrophiles and, especially, nucleophiles. For a long time, it was thought that sulfur, unlike oxygen, was able to accommodate more than eight Lewis electrons in its valence shell (the so-called hypervalency). However, recent studies of electron-density mapping have shown that many chalcogen–chalcogen bonds previously considered to be four-electron dative bonds are actually two-electron polar dative bonds, and the thiosulfoxide bond turned out to be much weaker than the previously assumed double bond.

TABLE 54.2 Percent occurrence of organosulfur glucosinolates in some cruciferous vegetables (Navarro et al., 2011).

OSC	Cruciferous vegetable			
	Broccoli	Green cabbage	Brussels sprouts	Watercress
Indole glucosinolates	19	31	19	—
Aliphatic glucosinolates	4	39	42	—
Alcoholic glucosinolates	37	1	22	—
Alkylthioalkyl glucosinolates	40	29	16	15
Aromatic glucosinolates	—	—	1	85

OSC, organosulfur compound.

Therefore, sulfane sulfur, which is sulfur in the thiosulfoxide form (represented as S^0) is relatively reactive, and this greatly contributes to the regulatory functions in biological systems (Toohey and Cooper, 2014). There is a family of carrier proteins that includes rhodanese (thiosulfate−cyanide sulfur transferase), mercaptopyruvate sulfur transferase, cystathionine γ-lyase, and serum albumin. The first two are specialized sulfane sulfur carriers, and the sulfur atom is carried as a persulfide on a cysteine residue in a specific domain (rhodanese homology domain, RHOD) (Cipollone et al., 2007). There are also proteins not containing the RHOD domain, to which sulfane sulfur can bind nonspecifically: antibodies, cystathionine γ-lyase, Cu-Zn superoxide dismutase 1 (SOD1), and some others (Liu and May, 2012; Nielsen et al., 2011). The binding and enzymatic properties of albumin were recently revised (Goncharov et al., 2015b, 2017a); its Cys34 plays a significant role in the redox status of blood plasma.

Protein oxidative or redox modifications can generally be classified into two categories: irreversible oxidation and reversible oxidation. While irreversible oxidation usually leads to protein aggregation and degradation, reversible oxidation that usually occurs on protein cysteine residues can often serve as an "on and off" switch that regulates protein function and redox signaling pathways upon stress challenges. In the context of ischemic tolerance, including preconditioning and postconditioning, increasing evidence has indicated that reversible cysteine redox modifications such as *S*-sulfenation, *S*-nitrosylation, *S*-glutathionylation, and disulfide bond formation can serve as a cellular defense mechanism against tissue ischemic injury (Yan, 2014). It could be especially important for protein phosphatases (PPs), one of the most abundant proteins encoded by the mammalian genome (Almo et al., 2007). There are two broad families, the protein tyrosine PPs (PTPs) comprising about 112 human proteins, and the serine/threonine PPs (Ser/Thr PPs) encompassing about 31 proteins. A signature cysteine, responsible for redox-modulation of PTPs, is deprotonated at physiological pH, making it very vulnerable to oxidation by ROS and resulting in loss of its capacity for catalytic dephosphorylation. The Cys-S^- is easily oxidized to the sulfenic acid form (−cys-SOH) and then, under continued oxidant exposure, to more stable −cys-SO_2 (sulfinic) and −cys-SO_3 (sulfonic) acids, which are not easily reversible. Some PTPs, such as PTP1B, have also been shown to form a sulfonamide bond, following oxidation, between the cysteine and a nearby nitrogen on serine. Other oxidation states, such as disulfide bonds and glutathiolation, are common occurrences with other specific PTPs (Salmeen, Barford, 2005).

Ajoene was shown to inhibit cholesterol biosynthesis by directly affecting 3-hydroxy-3-methyl-glutaryl coenzyme A (HMG-CoA) reductase and late enzymatic steps of the mevalonate (MVA) pathway (Ferri et al., 2003). MVA is the precursor of cholesterol and a number of nonsterol isoprenoids, such as farnesyl and geranylgeranyl groups. Covalent attachment of these MVA-derived isoprenoid groups (prenylation) is a required function of several proteins that regulate cell proliferation. Similar to the HMG-CoA reductase inhibitor simvastatin, ajoene inhibited cholesterol biosynthesis. However, in contrast to simvastatin, the antiproliferative effect of ajoene was not prevented by the addition of MVA, farnesol, and geranylgeraniol. In vitro assays for protein farnesyltransferase and protein geranylgeranyltransferase type I (PGGTase-I) confirmed that ajoene inhibits protein prenylation, via a covalent modification of the cysteine SH group of a peptide substrate for protein PGGTase-I (Ferri et al., 2003).

Bioactive peptides and proteins often possess disulfide bonds, which are important for their structure, function, and stability. The main pathways that catalyze disulfide bond formation in peptides and proteins in prokaryotes and eukaryotes are remarkably similar and share several mechanistic features (Patil et al., 2015). The endoplasmic reticulum (ER) provides an environment that is highly optimized for oxidative protein folding. Disulfide formation is driven by a protein relay involving Ero1, a conserved FAD-dependent enzyme, and protein disulfide isomerase (PDI); Ero1 is oxidized by molecular oxygen and in turn acts as a specific oxidant of PDI, which then directly oxidizes disulfide bonds in folding proteins (Tu and Weissman, 2004). While providing a robust driving force for disulfide formation, the use of molecular oxygen as the terminal electron acceptor can lead to oxidative stress through the production of ROS and oxidized glutathione. Healthy cells have a negative cytosolic redox potential, which means the cytosol is in a reduced/reducing state with a redox potential from −220 to −320 mV (Nagy, 2013). The cellular redox potential is controlled largely by the ratio of reduced to oxidised glutathione (GSH/GSSG) redox couple (glutathione pool), along with the $NAD(P)H/NAD(P)^+$ couples and thioredoxin (Borlinghaus et al., 2014). It seems that thiosulfoxides are able to oxidize thiols in cells, for example, glutathione and cysteine residues in proteins, and the oxidized glutathione pool leads to a higher cellular redox potential. The redox-triggered structural changes can lead to loss or gain of function of proteins. Moreover, a shift in the equilibrium from GSH toward a higher proportion of GSSG may promote the glutathiolation of free cysteines on protein surfaces, thus shielding these cysteine residues during oxidative stress conditions from over-oxidation (Borlinghaus et al., 2014).

Also, there is an alternative mechanism leading to disulfide formation. A protein thiol may be oxidized via Trx, first becoming oxidized. Trx has a lower p*K*a than most other proteins and so is more likely to be preferentially oxidized by H_2O_2 to form an intramolecular disulfide. The Trx disulfide is then attacked by a thiol of a second protein with a higher p*K*a, which is then reduced by a second thiol. Thus, Trx essentially picks up and passes on the oxidation state to the less reactive target protein thiol (Rudyk and Eaton, 2014).

OSCs are often regarded as antioxidants (Colín-González et al., 2012). It is thought that OSCs have no reducing or radical-quenching groups, and their so-called antioxidant properties are probably indirect and mediated through the sulfane sulfur generated from these compounds (Toohey and Cooper, 2014). There are several mechanisms of these indirect effects. First, elevated activity of glutathione was described without changing its concentration, and it was found that glutathione persulfide (G-S-SH) is responsible for this effect, being a much more powerful reductant (Ida et al., 2014); Second, covalent modification of SOD1 with formation of a novel heptasulfane bridge connecting the two Cys111 residues in the SOD1 homodimer (You et al., 2010). On the other hand, this modification may endow SOD1 with increased toxicity, because polysulfanes and polysulfides can catalyze the generation of ROS. However, if the elevated toxicity is within cancer cells, it would undoubtedly be a positive event for the organism. For example, it was found that DATS of garlic was effective in stimulating H_2O_2 production in human hepatoblastoma cells (Iciek et al., 2012). Among various OSCs, DATS showed the highest biological activity in HepG2 cells. It was proposed that sulfane sulfur containing DATS can be reduced in cancer cells to hydroperthiol that leads to H_2O_2 generation, thereby influencing cell proliferation. Another possibility is that the increased production of H_2O_2 in the cancer cells treated with garlic-derived sulfur compounds can be a result of increased activity of SOD1, since H_2O_2 is the product of SOD1.

Another antioxidant mechanism of OSCs is generation of the deoxyadenosyl radical (ado•) by the enzyme MoaA. MoaA, a radical *S*-adenosylmethionine enzyme, catalyzes the first step in molybdopterin biosynthesis (Mehta et al., 2013). This reaction involves a complex rearrangement in which C8 of guanosine triphosphate is inserted between C2' and C3' of the ribose. The ado• radical is formed when one electron is transferred from the iron of an iron sulfur cluster to *S*-adenosylmethionine, resulting in the release of methionine. In mammals, this radical participates in the biosynthesis of molybdenum cofactor (MoCo) from guanosine triphosphate (Hänzelmann and Schindelin, 2004). MoCo, in its turn, participates in disposing of sulfite ions, the end-product of sulfur metabolism, as well as purine catabolites (xanthine) (Mendel, 2013).

The next mechanism of the antioxidative effect of OSCs is stimulation of expression of ROS-neutralizing factors such as GSH, catalase, and SOD (Bayan et al., 2014). It was found that raw garlic homogenate increased myocardial catalase and GPx activity, GSH and H_2S levels, and reduced cardiac hypertrophy and fructose-induced myocardial oxidative stress through a PI3K/AKT/Nrf2-Keap1-dependent pathway (Padiya et al., 2014). *S*-allylcysteine (SAC), another product of garlic that is produced from γ-glutamyl-*S*-allylcysteine by a γ-glutamyltransferase during the process of garlic "aging," is the most abundant compound of AGE that exhibits various antioxidant properties: scavenging of free radicals and prooxidant species; inhibition of prooxidant and induction of antioxidant enzymes; chelating properties on Fe^{2+} and Cu^{2+} ions; inhibition of NF-κB translocation into the nucleus; and activation of Nrf2 factor (Colín-González et al., 2012). The effect of AGE, SAC, and isothiocyanates of *Brassicaceae* vegetables on the Nrf2/Keap1 complex leading to activation of Nrf2 factor—a master regulator of the cellular redox state—is regarded as the most important event in the antioxidative properties of AGE. Under normal (unstressed) conditions, this complex is dissociated and Nrf2 can either suffer proteosomal degradation or respond to stimuli typical of basal cell metabolism, being phosphorylated and translocated to the nucleus, then forming heterodimers with Maf and acting on the antioxidant response element (ARE). Under oxidative conditions, or in the presence of inducers (AGE or SAC), several cysteine residues of Keap1 are oxidized, releasing Nrf2 and allowing its transactivation and further translocation to the nucleus, where it will induce phase II gene transcription: NAD(P)H:quinone oxidoreductase 1 (NQO1), glutathione-*S*-transferase (GST), heme oxygenase-1 (HO-1), and glutamate cysteine ligase (Colín-González et al., 2012; Ishida et al., 2014). While Nrf2 is mostly described as an antiapoptotic factor regulating the expression of antiapoptotic proteins of the Bcl-2 family such as Bcl-2 and Bcl-xL (Niture, Jaiswal, 2013), it was shown that Nrf2 is involved in allicin-induced apoptosis (Bat-Chen et al., 2010), thus revealing the "Janus face" of Nrf2 (Loboda et al., 2008). Also, the ERK1/2 map kinases were shown to be influenced by allicin in immune cells; these kinases are also important for apoptosis induction by allicin (Cha et al., 2012). Berteroin (5-methylthiopentyl isothiocyanate, a sulforaphane analog present in cruciferous vegetables) exhibits potent antiinflammatory properties through several reciprocal mechanisms: (1) inhibition of LPS-induced phosphorylation of p38 MAPK, ERK1/2, and AKT; (2) inhibition of LPS-induced degradation of inhibitor of κBα (IκBα) and nuclear factor-κB p65 translocation to the nucleus and DNA binding activity; and (3) downregulation of iNOS and COX-2 expression (Jung et al., 2014). A similar mechanism of antiinflammatory properties could also be attributable to erucin, another isothiocyanate (Cho et al., 2013).

As the thiosulfoxide bond is weak, the sulfur is easily ejected as elemental sulfur, transferred to another sulfur atom, or reduced to H_2S by thiols. H_2S, in turn, generates polysulfides, which are much more active signaling molecules that stimulate TRP channels (Kimura et al., 2013). Importantly, it seems that H_2S can replicate the effects of OSCs. Administration or upregulation of H_2S by pharmacological agents leads to an increase in the Bcl-2/Bax ratio and preservation of mitochondrial function. Simultaneously, H_2S may also enhance antioxidant enzymes SOD and GPx (Szabó and Papapetropoulos, 2011; Pan et al., 2012). Inhibition of oxidative stress by H_2S is coupled with inhibition of mitochondrial complex IV activity and increased activities of CuZn-SOD (SOD1) and Mn-SOD (SOD2). H_2S upregulated the expression of SOD2, but activated SOD1 directly by interacting with this enzyme (Sun et al., 2012). H_2S activates the Akt signaling pathways and simultaneously enhances vascular endothelial growth factor (VEGF) production and VEGFR2 phosphorylation, leading to downstream Akt activation. Activation of the Akt pathway leads to Nrf-2 nuclear translocation, resulting in increased cell protection. H_2S may cause an inhibition of inflammatory cytokine production through NF-κB-dependent pathways, leading to tissue protection. H_2S also exerts angiogenic effects through activation of ATP-sensitive potassium channels (K_{ATP} channels), which in turn facilitate activation of the MAPK (ERK1/2 and p38) pathways, leading to new blood vessel formation (Szabó and Papapetropoulos, 2011). H_2S can activate K_{ATP} channels via direct interaction with the extracellular N terminal of the rvSUR1 subunit. H_2S promotes VEGF synthesis and angiogenesis by inducing matrix metalloproteinase MMP-2, while it suppresses MMP-9 and tissue inhibitor of metalloproteinase TIMP-3 levels, thus inhibiting antiangiogenic factors. VEGF may activate CSE activity and H_2S release that contributes to VEGF-stimulated angiogenesis-related properties of endothelial cells. An increase in H_2S inhibits proliferation of smooth muscle cells (SMCs) via upregulation of ERK1/2 and $p21^{Cip/WAF-1}$ phosphorylation and downregulation of cyclin D1. H_2S might control the balance of pro- and antiinflammatory cytokines in circulatory shock, and this balance plays a major role in inflammatory reactions and tissue injury (Pan et al., 2012).

In mammals, desulfhydration of cysteine is the major source of H_2S and is catalyzed by the trans-sulfuration pathway enzymes cystathionine β-synthase (CBS), cystathionine γ-lyase (CSE) and 3-mercaptopyruvate sulfurtransferase (3-MST). Cystathionine can be converted by CSE to form H_2S. CBS can form cystathionine from serine and homocysteine, and additionally can form H_2S from cysteine. Cysteine, along with α-ketoglutarate (α-KG), is converted into 3-mercaptopyruvate (3MP) by cystine aminotransferase. 3MP can then be broken down by 3MST to form H_2S (Predmore et al., 2012a,b). H_2S physiologically modifies cysteines in a large number of proteins by *S*-sulfhydration. About 10%–25% of many liver proteins, including actin, tubulin, and glyceraldehyde-3-phosphate dehydrogenase (GAPDH), are sulfhydrated under physiological conditions. Sulfhydration augments GAPDH activity and enhances actin polymerization (Mustafa et al., 2009). Persulfidation (a synonym for sulfhydration) of proteins requires attack of H_2S to an oxidized side chain, for example, a sulfenic acid or disulfide. Once formed, persulfides need to be protected from small-molecule thiols (e.g., glutathione) or thiol groups on the same or different proteins. Conversely, the same chemical reactions can be exploited to liberate H_2S from the "stored sulfur" pool (Kabil and Banerjee, 2010). Garlic-derived organic disulfides as well as pure DADS permeate the red blood cells (RBCs) membrane and modify deoxyhemoglobin at beta-93C or beta-112C. Deoxygenated human RBCs converted garlic-derived polysulfides into hydrogen sulfide, which in turn produced vasorelaxation in aortic ring preparations. The vasoactivity was proposed to occur via glucose- and thiol-dependent acellular reactions (Bonaventura et al., 2010). The RBCs of garlic-fed mice demonstrated shorter half-lives than the control, though they had normal blood chemistry and RBC indices. Their spleens manifested increased HO-1, higher levels of iron and bilirubin, and presumably higher CO, a pleiotropic gasotransmitter (Akgül et al., 2010). Heat shock genes and those critical for erythropoiesis were elevated in spleens but not in bone marrow. However, the garlic-fed mice had lower plasma erythropoietin than the controls. Thus, garlic consumption not only causes increased energy demand from the faster RBC turnover but also increases the production of CO, which in turn stimulates splenic erythropoiesis by an erythropoietin-independent mechanism (Akgül et al., 2010).A similar effect of sulforaphane on RBCs was observed in a model of eryptosis, a suicidal cell death characterized by cell shrinkage and phosphatidylserine translocation to the erythrocyte surface (Alzoubi et al., 2014). Sulforaphane (100 μM) stimulated eryptosis, that was accompanied by stimulation of Ca^{2+}entry and ceramide formation. As sulforaphan can stimulate apoptosis of tumor cells, the cytophysiological experiments with RBCs—and probably other cells—involved in tumorigenesis can serve not only for alternative and easy screening of nutraceuticals, but also help to discover and/or explain the multitarget strategy for treatment of complex diseases and forecast side effects. Development of a new methodology for multitargeted cytophysiological screening is an urgent issue for experimental biology and medicine.

Organosulfur compounds in the treatment of diseases

To continue with what has been said above, and before considering pathological conditions in which OSCs could be helpful as nutraceutical compounds, it should be mentioned that all the vegetables considered here mostly as sources of OSCs also contain many other nutraceuticals with their own specific properties and molecular targets. For example, onion peel extract (OPE) contains quercetin, one of the major flavonoids, which is believed to insure the antiplatelet effect of OPEs due to blocking cyclooxygenase-1 and thromboxane A2 synthase activities, and elevated formation of cyclic adenosine monophosphate (cAMP) (Ro et al., 2015); quercetin derivatives of onion have also been shown to possess radical-scavenging potential and hyaluronidase-inhibiting activity (González-Peña et al., 2013). Furthermore, when taken together, both sulfur compounds and flavonoids are mainly responsible for the activities of liver hexokinase, glucose 6-phosphatase, and HMG coenzyme-A reductase, and finally for hypoglycemic and other activities of *A. cepa* (Akash et al., 2014). Epidemiological studies indicate some associations of *Allium* vegetable consumption with decreased risk of cancer, particularly cancers of the gastrointestinal tract (Nicastro et al., 2015). Enhanced toxicity of onion extracts in multidrug-resistant erythroleukemic cell lines is also explained by the combined effects of quercetin and propyl disulfide (Votto et al., 2010). On the other hand, upregulation of paraoxonase 1 by onion extract was shown to be equal to that of quercetin (Jaiswal and Rizvi, 2014). Nonetheless, due to their distribution to all tissues analyzed and their multiple molecular targets suggesting the potential for systemic benefits, the sulfur-containing vegetables can prevent or help to treat very different pathologies: they are thought to reduce the risk for cardiovascular diseases, have antitumor and antimicrobial effects, show benefit on high blood glucose concentration, and to reduce chemically induced hepatotoxicity; they have been employed to treat infections, wounds, diarrhea, rheumatism, the common cold, and as an insect repellent (Bricker et al., 2014; Bayan et al., 2014). Some researchers consider OSCs as ideal nutraceutical agents as they can serve not only as direct antioxidant-trapping electrons, but also have nonantioxidant effects such as antiplatelet, fibrinolytic, antiinflammatory, immunomodulatory, and antiaging actions (Augusti et al., 2012).

Cardiovascular diseases

Garlic consumption has significant effects on lowering blood pressure, prevention of atherosclerosis, reduction of serum cholesterol and triglyceride, inhibition of platelet aggregation, and increasing fibrinolytic activity (Bayan et al., 2014). New drugs containing garlic OSCs, Allicor and Karinat, were developed that possess direct antiatherosclerotic activity. Two-year treatment with Allicor (garlic powder) has a direct antiatherosclerotic effect on carotid atherosclerosis in asymptomatic men. The phytoestrogen-rich drug Karinat (garlic powder, extract of grape seeds, green tea leaves, hop cones, β-carotene, α-tocopherol, and ascorbic acid) prevents the development of carotid atherosclerosis in postmenopausal women (Orekhov, 2013). In a study of the effect of ajoene—a component of garlic powder—on rat aortic SMC proliferation as related to protein prenylation, it was shown that ajoene (1−50 μM) interfered with the progression of the G1 phase of the cell cycle, and inhibited rat SMC proliferation (Ferri et al., 2003).

Sulfur deficiency might play a part in the etiology of hypertension, which could be alleviated with supplementation of organosulfur compounds derived from garlic.

Garlic supplements have shown promise in the treatment of uncontrolled hypertension, lowering blood pressure. Garlic-derived polysulfides stimulate the production of H_2S and enhance the regulation of endothelial NO, which induces SMC relaxation, vasodilation, and blood pressure reduction (Ried and Fakler, 2014). In a model of myocardial ischemia–reperfusion injury in vivo, DATS was shown to be a cardioprotective agent which was a donor of H_2S, it activated eNOS and increased NO metabolites, reduced mitochondrial respiration in a concentration-dependent manner, and significantly improved mitochondrial coupling after reperfusion (Predmore et al., 2012a, b). In this experiment, DATS did not appear to significantly induce the Nrf2 pathway.

The AGE inhibited the binding of ADP-activated platelets to immobilized fibrinogen. This suggested that AGE inhibited platelet aggregation via inhibition of the GPIIb/IIIa receptor and an increase in cAMP (Allison et al., 2012). It has been stated by many researchers that garlic was highly tolerable and had minimal side effects in the trials with cardiovascular patients (Bayan et al., 2014).

Cancer

Some OSCs of *Allium* and/or cruciferous vegetables exert chemopreventive effects on chemical carcinogenesis, but they may also demonstrate promotion potential, depending on the organ examined (Fukushima et al., 1997). The OSCs have been reported to inhibit the development of a wide range of tumors: hepatic (Shaarawy et al., 2009), lung (Hu and Singh, 1997), prostate (Chandra-Kuntal and Singh, 2010), breast (Azarenko et al., 2014), stomach and small intestine (Kaneko et al., 2012), colon (Lai et al., 2012), esophagus (Gao et al., 1999), bladder (Munday et al., 2008), pancreas (Ma et al., 2014), skin (Wang et al., 2012), and leukemia (Hassan, 2004).

The ability of OSCs to hinder cancer cell proliferation and viability tightly correlates with the length of the sulfur chain (Cerella et al., 2011). The exact mechanisms of the cancer-preventive effects of OSCs are not clear, though there is much evidence indicating that they modulate the activity of several metabolizing enzymes that activate (phase I enzymes CYP2E1, CYP1A1, and CYP1A2) or detoxify [phase II enzymes NQO1, glutathione *S*-transferases, UDP-glucuronosyltransferases, etc.] carcinogens, through inhibition of histone deacetylase activity and inhibition of the formation of DNA adducts in several target tissues (Zhang et al., 2013; Navarro et al., 2011). Also, current data support a mechanism of mitotic arrest of cancer cells due to the alteration of the microtubule network, possibly as a consequence of the high reactivity of sulfur atoms against the thiol groups of different cellular macromolecules controlling crucial regulatory functions (Cerella et al., 2011). It is worth mentioning that DADS synergizes the effect of eicosapentaenoic acid, a breast cancer suppressor, and antagonizes the effect of linoleic acid, a breast cancer enhancer; moreover, garlic extract reduces the side effects caused by anticancer agents (Tsubura et al., 2011). The proportion of allyl sulfides in garlic oil consists of approximately 4.7%–8% DAS, 21.9%–40% DADS, and 39%–41.5% DATS, which depends on the extraction conditions, and DATS is more potent than mono- and disulfides against skin cancer: it inhibits cell growth of human melanoma A375 cells and basal cell carcinoma cells by increasing the levels of intracellular ROS and DNA damage and by inducing G_2/M arrest, ER stress, and mitochondria-mediated apoptosis, including the caspase-dependent and -independent pathways (Wang et al., 2012). It was also found that DADS and DATS are important in the anticancer action of garlic, while dipropenyl sulfide could be involved in the beneficial action of onions (Munday and Munday, 2001).

It is now recognized that the cystine/glutamate transporter (antiporter) is intimately associated with ferroptosis, an iron-dependent nonapoptotic form of cell death observed in cancer cells, and also with cancer stem cells; the former with transporter blockage but the latter with its stabilization. Excess iron in the presence of oxygen appears the most common known mutagen (Toyokuni, 2014). Ironically, the persistent activation of antioxidant systems via genetic alterations in Nrf2 and Keap1 also contributes to carcinogenesis. H_2O_2 is now recognized as a signaling molecule whose main regulator is the peroxiredoxin/sulfiredoxin systems, which are at least partially under the control of the Nrf2/Keap1 system. Some cancers hijack the Nrf2/Keap1 system with mutation in these genes, which persistently activate antioxidant systems in the cancer cells. Overexpression of CD44v(8-11) stabilizes the cystine/glutamate antiporter to increase cysteine and GSH in cancer stem cells. Conversely, erastin blocks the cystine/glutamate antiporter, lowers cysteine, and finally induces ferroptosis, which can be blocked with deferoxamine (Toyokuni, 2014). It was recently found that 10AT-Her2 cells, a new breast cancer cell line which is highly enriched in cells with stem/progenitor cell-like character, are highly sensitive to the antiproliferative apoptotic effects of I3C, a natural anticancer agent from cruciferous vegetables of the *Brassica* genus such as broccoli and cabbage (Tin et al., 2014). It would be interesting to study a possible impact of OSC from *Brassica* and *Allium* genera on this kind of cancer cell death, via either ROS generation or downregulation of CD44.

Neurodegenerative disorders

A wide variety of acute and chronic neurodegenerative diseases, including ischemic/traumatic brain injury,

Alzheimer disease (AD), and Parkinson disease, share common characteristics such as oxidative stress, misfolded proteins, excitotoxicity, inflammation, and neuronal loss. As no drugs are available to prevent the progression of these neurological disorders, intervention strategies using phytochemicals have been proposed as an alternative form of treatment. Modulation of endogenous cellular defense mechanisms represents an innovative approach to therapeutic intervention in diseases causing chronic tissue damage, such as in neurodegeneration (Calabrese et al., 2010). Among phytochemicals, isothiocyanate sulforaphane has demonstrated neuroprotective effects in several in vitro and in vivo studies. In particular, evidence suggests that sulforaphane beneficial effects could be mainly ascribed to its peculiar ability to activate the Nrf2/ARE pathway (Tarozzi et al., 2013). Sulforaphane has also been shown to stimulate the ubiquitin proteasome system activity in vitro; moreover, it activates protein degradation machineries (autophagic activities) in both the brain and peripheral tissues and may be a therapeutic reagent for Huntington disease, preventing accumulation of mutant huntingtin protein (Liu et al., 2014). SAC prevents the progression of AD, and there are multiple mechanisms underlying this capacity. As an antioxidant, SAC scavenges oxidant species (direct antioxidant) and restores GPx, GR, and SOD levels (indirect antioxidant), resulting in diminished lipid peroxidation, DNA fragmentation, protein oxidation, and ER stress. The decrease in ER stress attenuates Ca^{2+} release and the subsequent activation of calpain and the caspase-12-dependent pathway, which altogether decrease the cell death. As an antiamyloidogenic agent, SAC decreases Aβ formation and/or increases Aβ clearance, lowers amyloid precursor protein (APP) mRNA expression, BACE (β-site APP cleavage enzyme 1) expression and activity, and restores protein kinase C activity under AD-like condition, which benefits APP cleavage and decreases the available APP for Aβ; in addition, SAC can bind to Aβ-inhibiting, Aβ fibrillation, and destabilizing preformed Aβ-peptide fibrils. As an antiinflammatory agent, SAC decreases IL-1β and TNF-α levels and IL-1β-positive plaque-associated microglia. Antitangle effects of SAC consist of a reduction of tau2 reactivity and its phosphorylation; this reduction in tau appears to involve GSK-3β protein (Colín-González et al., 2012).

Diabetes

Garlic and onion sulfoxide amino acids were effective in the reduction of blood glucose in streptozotocin- as well as alloxan-induced diabetes mellitus in rats and mice (Sheela et al., 1995; Lee et al., 2013). Garlic extracts have been reported to be effective in reducing insulin resistance, and the beneficial effect is attributed to the presence of volatile sulfur compounds, such as alliin, allicin, DADS, DATS, DAS, SAC, ajoene, and allyl mercaptan (Padiya and Banerjee, 2013). SAC is considered an antiglycative agent, reducing both the activity and mRNA expression of aldose reductase, which subsequently reduces the production of sorbitol and prevents advanced glycation end-product formation, such as carboxymethyllysine and pentosidine, decreasing glycative stress (Colín-González et al., 2012). Raw garlic homogenate is effective in reducing cardiac hypertrophy and fructose-induced myocardial oxidative stress through a PI3K/AKT/Nrf2-Keap1-dependent pathway (Padiya et al., 2014). Modulation of NADPH oxidase subunit expression is another way of preventing ROS formation and restoring erectile function by SAC in diabetic rats; it was suggested that the poor efficacy of conventional insulin treatment for diabetic erectile dysfunction may be associated with an elevated level of ROS in penile tissue (Yang et al., 2013).

Supplementation of type II diabetics with high sulforaphane content broccoli sprouts resulted in increased total antioxidant capacity of plasma and decreased oxidative stress index, lipid peroxidation, serum triglycerides, oxidized low-density lipoprotein (LDL)/LDL-cholesterol ratio, serum insulin, insulin resistance, and serum high-sensitive C-reactive protein (Bahadoran et al., 2013). Sulforaphane could prevent nephropathy, diabetes-induced fibrosis, and vascular complications. The effects of sulforaphane are attributed to activation of the Nrf2-dependent antioxidant response-signaling pathway, induction of phase II enzymes and peroxisome proliferator-activated receptors, attenuation of oxidative stress, and inactivation of NF-κB (a key modulator of inflammatory pathways). The potential efficacy of sulforaphane and probably other bioactive components of young broccoli sprouts makes it as an excellent choice for supplementary treatment in type II diabetes (Bahadoran et al., 2013).

Antibacterial, antiviral, antifungal, and antiparasitic activities

Garlic has been proven to be effective against a plethora of Gram-positive, Gram-negative, and acid-fast bacteria, including *Salmonella*, *Escherichia coli*, *Pseudomonas*, *Proteus*, *Staphylococcus aureus*, *Klebsiella*, *Micrococcus*, *Bacillus subtulis*, *Clostridium*, *Mycobacterium*, and *Helicobacter* (Bayan et al., 2014). Lactic acid bacteria are the least sensitive microorganisms to the inhibitory effects of garlic. In mixed-culture studies of *Lactobacillus acidophilus* and *E. coli*, garlic prevented the establishment of the latter species (Rees et al., 1993). *A. sativum* offers hope for developing alternative drugs to treating *Mycobacterium tuberculosis* and particularly of multidrug-resistant tuberculosis (Dini et al., 2011; Viswanathan et al., 2014). Many fungi are sensitive to garlic, including *Candida*, *Torulopsis*, *Trichophyton*,

Cryptococcus, *Aspergillus*, *Scedosporium*, *Tinea*, *Trichosporon*, and *Rhodotorula*. Garlic extract was also effective against a host of protozoa including *Opalina ranarum*, *Balantidium entozoon*, *Entamoeba histolytica*, *Trypanosoma*, *Leishmania*, *Leptomonas*, and *Crithidia*. As for viral infections, garlic extract showed activity against influenza A and B, cytomegalovirus, rhinovirus, HIV, herpes simplex virus 1, herpes simplex virus 2, viral pneumonia, and rotavirus (Bayan et al., 2014).

As compared with other OSCs, allicin has drawn much more attention, since it has potent antimicrobial activity against a wide range of microorganisms, including meticillin-resistant *S. aureus*. Allicin readily diffuses across artificial and natural phospholipid membranes (Miron et al., 2000). In terms of antibiotic potency against Gram-positive and Gram-negative bacteria, authentic allicin had roughly 1%–2% of the potency of streptomycin (vs *S. aureus*), 8% of that of vancomycin (vs *S. aureus*), and only 0.2% of that of colistin (vs *E. coli*) (Fujisawa et al.,2009). The oxygen in the structure [− S(=O) − S −] of allicin functions to liberate the *S*-allyl moiety, which might be an offensive tool against bacteria. The garlic extract had more potent antistaphylococcal activity than an equal amount of allicin. These findings indicate either synergistic effects of allicin with other components in the extract, or the additional action of other antimicrobial compounds. A greater chemical stability of ajoene or other components as compared with allicin may also have the advantage of complex extracts (Viswanathan et al., 2014)

It was found long ago by Cavallito and Bailey and reaffirmed by many recent data that the active principles of *A. sativum*, *Erythronium americanum*, *Asarum reflexum*, *Arctium minus*, *Ranunculus acris*, *Ranunculus bulbosus*, and of *Brassica* species—as well as the nonplant-derived antibiotics penicillin, citrinin, gliotoxin, clavacin, and pyocyanines—involves reaction with cysteine. The antibacterial activity of allicin and other OSCs was completely abolished by cysteine, glutathione, and coenzyme A, but not by non-SH compounds (Cavallito and Bailey, 1944; Borlinghaus et al., 2014; Fujisawa et al., 2009).

Concluding remarks and future directions

Accumulating data indicate that sulfane sulfur has important functions in cells. The broad diversity of effects suggests that its functions are general and not specific to any tissue or process. Moreover, it should not be called a "signaling agent" since there is no evidence that it acts in a controlled rise and fall pattern (as with neurotransmitters or hormones) (Toohey and Cooper, 2014). Dean Jones suggested and consistently substantiated the idea of "orthogonal control," which means that there are "redox-sensing" mechanisms as more global biologic redox control mechanisms, in contrast to "redox signaling" mechanisms that involve conveyance of discrete activating or inactivating signals (Jones, 2010). Both redox sensing and redox signaling use sulfur switches, especially Cys residues in proteins, which are sensitive to reversible oxidation, nitrosylation, glutathionylation, acylation, sulfhydration, or metal binding. Attempts to regulate redox state have focused on electrophiles, which activate potent cellular defense systems against oxidative stress. Nutraceuticals could serve not only as electrophiles, but as a kind of "pro-electrophilic drugs" that become electrophilic in response to oxidation and then activate the Keap1/Nrf2/ARE transcription pathway to synthesize endogenous antioxidant "phase II" enzymes (Satoh et al., 2014). Another important prerequisite of the new concept in redox biology/medicine is the use of compounds that has been termed "pathologically activated therapeutics." that is, relatively small molecules are chemically converted to their active form by the very oxidative stress that they are intended to then combat (Satoh et al., 2014). This "electrophilic counterattack" could be organized by administration of OSC-containing nutraceuticals, which exert their effects not only through their original properties quickly reacting with or generating ROS, but rather through mobilization and/or coordination of necessary "weapons" in definite areas of the "battlefield." According to Jones, there are eight such areas, or cellular compartments, including the extracellular one (Jones, 2010). In the extracellular space, the cysteine/cystine redox couple is quantitatively the most significant. In normal human plasma, the concentrations of cysteine (8–10 μM) and cystine (40–50 μM) are considerably higher than those of GSH (2.8 μM) and GSSG (0.14 μM). These correspond to redox potentials of −80 mV for the cysteine/cystine couple and −140 mV for the GSH/GSSG couple (Banerjee, 2012). A linear age-dependent increase in the plasma cysteine/cystine redox potential (at a rate of 0.14 mV/year) is observed for individuals between 19 and 85 years of age, whereas the GSH/GSSG redox potential is stable for the first 45 years and thereafter becomes more oxidized at a linear rate of 0.7 mV/year. It would be very interesting to conduct research to find out if these patterns are different in countries with different cuisines, or within a country to find a difference between different social strata. It is also important to reveal a possible link between a cuisine peculiarity and genetic polymorphism of enzymes participating in regulation of the redox potential.

Since different cells are functionally and structurally different, intracellular thiol-disulfide redox systems should be regulated differently by the same external or internal OSCs. Trying to mobilize the "true weapon" in the "true areas," one should understand "alignment of forces" to avoid "friendly fire." For example, garlic can elevate intracellular H_2O_2 and free iron and induce anemia into erythrocytes characterized by increased MDA, carbonyl protein, and antioxidant enzyme activities as catalase, GPx, and SOD; on the other hand, grape seed and skin extract treatment counteract almost all deleterious effects of garlic (Hamlaoui et al., 2012). Complex preparations containing, besides garlic powder, antioxidant extracts and vitamins were developed that possess direct antiatherosclerotic activity minimizing possible adverse effects of garlic (Orekhov, 2013). Diet quality indices that consider overall diet quality rather than single nutrients have been drawing increasing attention (Wang et al., 2013). In spite of major advances in understanding oxidative stress and nutrition that have occurred with broad characterization of dietary oxidants and antioxidants, randomized controlled trials in humans with free-radical-scavenging antioxidants have provided limited or inconsistent evidence for health benefits. This indicates that holistic models are needed to understand the interplay of nutrition and oxidative stress (Dennis et al., 2019). With this framework, integrated redox network models will support optimization of diet to protect against oxidative stress and disease.

On the other hand, long-term and large trials are needed to evaluate the differences in mortality, adverse events, and morbidity of cancer and cardiovascular diseases after garlic therapy (Bayan et al., 2014). It is particularly important from the point of view of seeking new biomarkers of various pathological states: the first results have shown that thiol-containing albumin is among the few selected biomarkers that provide insight into the risk of death from all causes (Fischer et al., 2014; Hui et al., 2014). Albumin is a major plasma protein and contributes greatly to the plasma protein-SH pool. Homocysteinylation of albumin at cysteine-34 may make it increasingly susceptible to oxidative degradation (Prakash et al., 2008). A new field of research named "Cys34 adductomics" has emerged recently, focused on modifications to Cys34 in human serum albumin, which is responsible for scavenging small reactive electrophiles that might initiate cancers (Grigoryan et al., 2019). Thus, the search for new properties of albumin under the influence of OSC is also important, both with in silico and kinetic studies (Singh and Singh, 2010; Goncharov et al., 2015b, 2017a,b).

There are also problems of stability and/or bioavailability of OSCs. For example, allicin is known to have major stability issues due to rapid degradation even at low temperatures, whereas garlic oil, being lipophilic, shows poor bioavailability after oral administration. Buccal tablets as well as capsules containing garlic oil self-nanoemulsifying systems provide promising strategies to overcome the difficulties associated with formulation of allicin and garlic oil (Phadatare et al., 2014). Future trials on the effect of OSC-containing vegetables should include information on the dosage of active ingredients of standardized preparations for better comparison of trials. It would also be interesting to explore the effects of different forms of garlic and other extracts on standard drug therapy, especially when used as adjuvant therapy (Bayan et al., 2014).

Indeed, studies of OSCs—and vegetables containing them—is an exciting field of research, and we believe that new discoveries concerning redox biology and human health are not far off.

Acknowledgments

This work was supported by the State Program AAAA-A18-118012290142-9, and Russian Foundation for Basic Research Grant #18-015-00304.

References

Akash, M.S., Rehman, K., Chen, S., 2014. Spice plant *Allium cepa*: dietary supplement for treatment of type 2 diabetes mellitus. Nutrition 30 (10), 1128–1137.

Akgül, B., Lin, K.W., Ou Yang, H.M., et al., 2010. Garlic accelerates red blood cell turnover and splenic erythropoietic gene expression in mice: evidence for erythropoietin-independent erythropoiesis. PLoS One 5 (12), e15358.

Allison, G.L., Lowe, G.M., Rahman, K., 2012. Aged garlic extract inhibits platelet activation by increasing intracellular cAMP and reducing the interaction of GPIIb/IIIa receptor with fibrinogen. Life Sci. 91, 1275–1280.

Almo, S.C., Bonanno, J.B., Sauder, J.M., et al., 2007. Structural genomics of protein phosphatases. J. Struct. Funct. Genomics 8, 121–140.

Alzoubi, K., Calabrò, S., Faggio, C., et al., 2014. Stimulation of suicidal erythrocyte death by sulforaphane. Basic Clin. Pharmacol. Toxicol. 116, 229–235.

Augusti, K.T., Jose, R., Sajitha, G.R., et al., 2012. A rethinking on the benefits and drawbacks of common antioxidants and a proposal to look for the antioxidants in allium products as ideal agents: a review. Indian J. Clin. Biochem. 27 (1), 6–20.

Azarenko, O., Jordan, M.A., Wilson, L., 2014. Erucin, the major isothiocyanate in arugula (*Eruca sativa*), inhibits proliferation of MCF7 tumor cells by suppressing microtubule dynamics. PLoS One 9 (6), e100599.

Bahadoran, Z., Mirmiran, P., Azizi, F., 2013. Potential efficacy of broccoli sprouts as a unique supplement for management of type 2 diabetes and its complications. J. Med. Food 16 (5), 375–382.

Banerjee, R., 2012. Redox outside the box: linking extracellular redox remodeling with intracellular redox metabolism. J. Biol. Chem. 287 (7), 4397–4402.

Bardwell, J.C., 2002. Disulfide bond formation, a race between FAD and oxygen. Dev. Cell 3 (6), 758–760.

Bat-Chen, W., Golan, T., Peri, I., et al., 2010. Allicin purified from fresh garlic cloves induces apoptosis in colon cancer cells via Nrf2. Nutr. Cancer 62, 947–957.

Bayan, L., Koulivand, P.H., Gorji, A., 2014. Garlic: a review of potential therapeutic effects. Avicenna J. Phytomed. 4 (1), 1–14.

Bhattacharya, A., Li, Y., Wade, K.L., et al., 2010. Allyl isothiocyanate-rich mustard seed powder inhibits bladder cancer growth and muscle invasion. Carcinogenesis 31 (12), 2105–2110.

Bonaventura, J., Rodriguez, E.N., Beyley, V., et al., 2010. Allylation of intraerythrocytic hemoglobin by raw garlic extracts. J. Med. Food 13 (4), 943–949.

Borlinghaus, J., Albrecht, F., Gruhlke, M.C., et al., 2014. Allicin: chemistry and biological properties. Molecules 19 (8), 12591–12618.

Bricker, G.V., Riedl, K.M., Ralston, R.A., et al., 2014. Isothiocyanate metabolism, distribution, and interconversion in mice following consumption of thermally processed broccoli sprouts or purified sulforaphane. Mol. Nutr. Food Res. 58 (10), 1991–2000.

Calabrese, V., Cornelius, C., Dinkova-Kostova, A.T., et al., 2010. Cellular stress responses, the hormesis paradigm, and vitagenes: novel targets for therapeutic intervention in neurodegenerative disorders. Antioxid. Redox Signal. 13 (11), 1763–1811.

Cavallito, C.J., Bailey, J.H., 1944. Preliminary note on the inactivation of antibiotics. Science 100, 390.

Cerella, C., Dicato, M., Jacob, C., et al., 2011. Chemical properties and mechanisms determining the anti-cancer action of garlic-derived organic sulfur compounds. Anticancer Agents Med. Chem. 11 (3), 267–271.

Cha, J.H., Choi, Y.J., Cha, S.H., et al., 2012. Allicin inhibits cell growth and induces apoptosis in U87MG human glioblastoma cells through an ERK-dependent pathway. Oncol. Rep. 28, 41–48.

Chandra-Kuntal, K., Singh, S.V., 2010. Diallyl trisulfide inhibits activation of signal transducer and activator of transcription 3 in prostate cancer cells in culture and in vivo. Cancer Prev. Res. 3 (11), 1473–1483.

Cho, H.J., Lee, K.W., Park, J.H., 2013. Erucin exerts anti-inflammatory properties in murine macrophages and mouse skin: possible mediation through the inhibition of NFκB signaling. Int. J. Mol. Sci. 14 (10), 20564–20577.

Cipollone, R., Acsenzi, P., Visca, P., 2007. Common themes and variations in the rhodanese superfamily. IUBMB Life 59, 51–59.

Colín-González, A.L., Santana, R.A., Silva-Islas, C.A., et al., 2012. The antioxidant mechanisms underlying the aged garlic extract- and S-allylcysteine-induced protection. Oxid. Med. Cell. Longev. 2012: Article ID 907162.

Cremlyn, R.J., 1996. An Introduction to Organosulfur Chemistry. John Wiley and Sons, Chichester, p. 250.

Dennis, K.K., Go, Y.M., Jones, D.P., 2019. Redox systems biology of nutrition and oxidative stress. J Nutr 149 (4), 553–565.

Dini, C., Fabbri, A., Geraci, A., 2011. The potential role of garlic (*Allium sativum*) against the multi-drug resistant tuberculosis pandemic: a review. Ann. Ist. Super. Sanita 47 (4), 465–473.

Ferri, N., Yokoyama, K., Sadilek, M., et al., 2003. Ajoene, a garlic compound, inhibits protein prenylation and arterial smooth muscle cell proliferation. Br. J. Pharmacol. 138 (5), 811–818.

Fischer, K., Kettunen, J., Würtz, P., et al., 2014. Biomarker profiling by nuclear magnetic resonance spectroscopy for the prediction of all-cause mortality: an observational study of 17,345 persons. PLoS Med. 11 (2), e1001606.

Fujisawa, H., Watanabe, K., Suma, K., et al., 2009. Antibacterial potential of garlic-derived allicin and its cancellation by sulfhydryl compounds. Biosci. Biotechnol. Biochem. 73 (9), 1948–1955.

Fukushima, S., Takada, N., Hori, T., et al., 1997. Cancer prevention by organosulfur compounds from garlic and onion. J. Cell. Biochem. Suppl. 27, 100–105.

Gao, C.M., Takezaki, T., Ding, J.H., et al., 1999. Protective effect of allium vegetables against both esophageal and stomach cancer: a simultaneous case-referent study of a high-epidemic area in Jiangsu Province, China. Jpn. J. Cancer Res. 90 (6), 614–621.

Gilbert, H.F., 1995. Thiol/disulfide exchange equilibria and disulfide bond stability. Methods Enzymol. 251, 8–28.

Goncharov, N.V., Avdonin, P.V., Nadeev, A.D., et al., 2015a. Reactive oxygen species in pathogenesis of atherosclerosis. Curr. Pharm. Des. 21 (9), 1134–1146.

Goncharov, N.V., Belinskaia, D.A., Razygraev, A.V., et al., 2015b. On the enzymatic activity of albumin. Bioorg. Khim. 41 (2), 131–144 [Article in Russian].

Goncharov, N., Savelieva, E., Zinchenko, V., et al., 2015c. Fluoroacetate. In: Gupta, R.C. (Ed.), Handbook of Toxicology of Chemical Warfare Agents, 2nd edition Academic Press, pp. 193–214.

Goncharov, N.V., Belinskaia, D.A., Shmurak, V.I., et al., 2017a. Serum albumin binding and esterase activity: mechanistic interactions with organophosphates. Molecules 22 (7), pii: E1201. doi: 10.3390/molecules22071201.

Goncharov, N.V., Terpilovskii, M.A., Shmurak, V.I., Belinskaya, D.A., Avdonin, P.V., 2017b. Comparative analysis of esterase and paraoxonase activities of different serum albumin species. J. Evol. Biochem. Physiol. 53 (4), 271–281. Available from: https://doi.org/10.1134/S0022093017040032.

González-Peña, D., Colina-Coca, C., Char, C.D., et al., 2013. Hyaluronidase inhibiting activity and radical scavenging potential of flavonols in processed onion. J. Agric. Food Chem. 61 (20), 4862–4872.

Grigoryan, H., Schiffman, C., Gunter, M.J., et al., 2019. Cys34 adductomics links colorectal cancer with the gut microbiota and redox biology. Cancer Res. 79 (23), 6024–6031. Available from: https://doi.org/10.1158/0008-5472.CAN-19-1529.

Hamlaoui, S., Mokni, M., Limam, N., et al., 2012. Protective effect of grape seed and skin extract on garlic-induced erythrocyte oxidative stress. J. Physiol. Pharmacol. 63 (4), 381–388.

Hänzelmann, P., Schindelin, H., 2004. Crystal structure of the S-adenosylmethionine-dependent enzyme MoaA and its implications for molybdenum cofactor deficiency in humans. Proc. Natl. Acad. Sci. USA 101, 12870–12875.

Hassan, H.T., 2004. Ajoene (natural garlic compound): a new anti-leukaemia agent for AML therapy. Leuk. Res. 28 (7), 667–671.

Hu, X., Singh, S.V., 1997. Glutathione S-transferases of female A/J mouse lung and their induction by anticarcinogenic organosulfides from garlic. Arch. Biochem. Biophys. 340 (2), 279–286.

Hui, L., Qigui, L., Sashuang, R., et al., 2014. Nonspecific changes in clinical laboratory indicators in unselected terminally ill patients and a model to predict survival time based on a prospective observational study. J. Transl. Med. 12, 78.

Iciek, M., Kwiecień, I., Chwatko, G., et al., 2012. The effects of garlic-derived sulfur compounds on cell proliferation, caspase 3 activity, thiol levels and anaerobic sulfur metabolism in human hepatoblastoma HepG2 cells. Cell Biochem. Funct. 30, 198–204.

Ida, T., Sawa, T., Ihara, H., et al., 2014. Reactive cysteine persulfides and S-polythiolation regulate oxidative stress and redox signaling. Proc. Natl. Acad. Sci. USA 111, 7606–7611.

Ilić, D., Nikolić, V., Stanković, M., et al., 2012. Transformation of synthetic allicin: the influence of ultrasound, microwaves, different solvents and temperatures, and the products isolation. Sci. World J. 2012, Article ID 561823.

Ishida, M., Hara, M., Fukino, N., et al., 2014. Glucosinolate metabolism, functionality and breeding for the improvement of Brassicaceae vegetables. Breed Sci. 64 (1), 48–59.

Jaiswal, N., Rizvi, S.I., 2014. Onion extract (*Allium cepa* L.), quercetin and catechin up-regulate paraoxonase 1 activity with concomitant protection against low-density lipoprotein oxidation in male Wistar rats subjected to oxidative stress. J. Sci. Food Agric. 94 (13), 2752–2757.

Jones, D.P., 2010. Redox sensing: orthogonal control in cell cycle and apoptosis signalling. J. Intern. Med. 268 (5), 432–448.

Jung, Y.J., Jung, J.I., Cho, H.J., et al., 2014. Berteroin present in cruciferous vegetables exerts potent anti-inflammatory properties in murine macrophages and mouse skin. Int. J. Mol. Sci. 15 (11), 20686–20705.

Kabil, O., Banerjee, R., 2010. Redox biochemistry of hydrogen sulfide. J. Biol. Chem. 285 (29), 21903–21907.

Kaneko, T., Shimpo, K., Chihara, T., et al., 2012. Inhibition of ENNG-induced pyloric stomach and small intestinal carcinogenesis in mice by high temperature- and pressure-treated garlic. Asian Pac. J. Cancer Prev. 13 (5), 1983–1988.

Kimura, Y., Mikami, Y., Osumi, K., et al., 2013. Polysulfides are possible H2S-derived signaling molecules in rat brain. FASEB J. 27 (6), 2451–2457.

Lai, K.C., Kuo, C.L., Ho, H.C., et al., 2012. Diallyl sulfide, diallyl disulfide and diallyl trisulfide affect drug resistant gene expression in colo 205 human colon cancer cells in vitro and in vivo. Phytomedicine 19 (7), 625–630.

Lee, C.W., Lee, H.S., Cha, Y.J., et al., 2013. In vivo investigation of anti-diabetic properties of ripe onion juice in normal and streptozotocin-induced diabetic rats. Prev. Nutr. Food Sci. 18 (3), 169–174.

Liu, H., May, K., 2012. Disulfide bond structures of IgG molecules: structural variations, chemical modifications and possible impacts to stability and biological function. MAbs 4 (1), 17–23.

Liu, Y., Hettinger, C.L., Zhang, D., et al., 2014. Sulforaphane enhances proteasomal and autophagic activities in mice and is a potential therapeutic reagent for Huntington's disease. J. Neurochem. 129 (3), 539–547.

Loboda, A., Was, H., Jozkowicz, A., et al., 2008. Janus face of Nrf2-HO-1 axis in cancer—friend in chemoprevention, foe in anticancer therapy. Lung Cancer 60, 1–3.

Ma, H.B., Huang, S., Yin, X.R., et al., 2014. Apoptotic pathway induced by diallyl trisulfide in pancreatic cancer cells. World J. Gastroenterol. 20 (1), 193–203.

Mehta, A.P., Hanes, J.W., Abdelwahed, S.H., et al., 2013. Catalysis of a new ribose carbon-insertion reaction by the molybdenum cofactor biosynthetic enzyme MoaA. Biochemistry 52 (7), 1134–1136.

Mendel, R.R., 2013. The molybdenum cofactor. J. Biol. Chem. 288, 13165–13172.

Miron, T., Rabinkov, A., Mirelman, D., et al., 2000. The mode of action of allicin: its ready permeability through phospholipid membranes may contribute to its biological activity. Biochim. Biophys. Acta Biomembr. 1463, 20–30.

Munday, R., 2012. Harmful and beneficial effects of organic monosulfides, disulfides, and polysulfides in animals and humans. Chem. Res. Toxicol. 25 (1), 47–60.

Munday, R., Munday, C.M., 2001. Relative activities of organosulfur compounds derived from onions and garlic in increasing tissue activities of quinone reductase and glutathione transferase in rat tissues. Nutr. Cancer 40 (2), 205–210.

Munday, R., Mhawech-Fauceglia, P., Munday, C.M., et al., 2008. Inhibition of urinary bladder carcinogenesis by broccoli sprouts. Cancer Res. 68 (5), 1593–1600.

Mustafa, A.K., Gadalla, M.M., Sen, N., et al., 2009. H2S signals through protein S-sulfhydration. Sci. Signal. 2 (96), ra72.

Nagy, P., 2013. Kinetics and mechanisms of thiol-disulfide exchange covering direct substitution and thiol oxidation-mediated pathways. Antioxid. Redox Signal. 18, 1023–1041.

Navarro, S.L., Li, F., Lampe, J.W., 2011. Mechanisms of action of isothiocyanates in cancer chemoprevention: an update. Food Funct. 2 (10), 579–587.

Nicastro, H.L., Ross, S.A., Milner, J.A., 2015. Garlic and onions: their cancer prevention properties. Cancer Prev. Res. 8 (3), 181–189.

Nielsen, R.W., Tachibana, C., Hansen, N.E., et al., 2011. Trisulfides in proteins. Antioxid. Redox Signal. 15 (1), 67–75.

Niture, S.K., Jaiswal, A.K., 2013. Nrf2-induced antiapoptotic Bcl-xL protein enhances cell survival and drug resistance. Free Radic. Biol. Med. 57, 119–131.

Orekhov, A.N., 2013. Direct anti-atherosclerotic therapy; development of natural anti-atherosclerotic drugs preventing cellular cholesterol retention. Curr. Pharm. Des. 19 (33), 5909–5928.

Padiya, R., Banerjee, S.K., 2013. Garlic as an anti-diabetic agent: recent progress and patent reviews. Recent Pat. Food Nutr. Agric. 5, 105–127.

Padiya, R., Chowdhury, D., Borkar, R., et al., 2014. Garlic attenuates cardiac oxidative stress via activation of PI3K/AKT/Nrf2-Keap1 pathway in fructose-fed diabetic rat. PLoS One 9 (5), e94228.

Pan, L.L., Liu, X.H., Gong, Q.H., et al., 2012. Role of cystathionine γ-lyase/hydrogen sulfide pathway in cardiovascular disease: a novel therapeutic strategy? Antioxid. Redox Signal. 17 (1), 106–118.

Patil, N.A., Tailhades, J., Hughes, R.A., et al., 2015. Cellular disulfide bond formation in bioactive peptides and proteins. Int. J. Mol. Sci. 16 (1), 1791–1805.

Phadatare, A.G., Viswanathan, V., Mukne, A., 2014. Novel strategies for optimized delivery of select components of *Allium sativum*. Pharmacognosy Res. 6 (4), 334–340.

Pirazzini, M., Azarnia Tehran, D., Zanetti, G., et al., 2014. Thioredoxin and its reductase are present on synaptic vesicles, and their inhibition prevents the paralysis induced by botulinum neurotoxins. Cell Rep. 8 (6), 1870–1878.

Prakash, M., Shetty, J.K., Rao, L., et al., 2008. Serum paraoxonase activity and protein thiols in chronic renal failure patients. Indian J. Nephrol. 18 (1), 13–16.

Predmore, B.L., Kondo, K., Bhushan, S., et al., 2012a. The polysulfide diallyl trisulfide protects the ischemic myocardium by preservation of endogenous hydrogen sulfide and increasing nitric oxide bioavailability. Am. J. Physiol. Heart Circ. Physiol. 302 (11), H2410–H2418.

Predmore, B.L., Lefer, D.J., Gojon, G., 2012b. Hydrogen sulfide in biochemistry and medicine. Antioxid. Redox Signal. 17 (1), 119–140.

Prokofieva, D.S., Goncharov, N.G., 2014. Effects of biogenic and abiogenic disulphides upon endothelial cells in culture: comparison of three methods of viability assessment. Tsitologiya 56 (6), 410–418.

Rees, L.P., Minney, S.F., Plummer, N.T., et al., 1993. A quantitative assessment of the antimicrobial activity of garlic (*Allium sativum*). World J. Microbiol. Biotechnol. 9 (3), 303–307.

Ried, K., Fakler, P., 2014. Potential of garlic (*Allium sativum*) in lowering high blood pressure: mechanisms of action and clinical relevance. Integr. Blood Press. Control 7, 71–82.

Ro, J.Y., Ryu, J.H., Park, H.J., et al., 2015. Onion (*Allium cepa* L.) peel extract has anti-platelet effects in rat platelets. Springerplus 4, 17.

Rudyk, O., Eaton, P., 2014. Biochemical methods for monitoring protein thiol redox states in biological systems. Redox Biol. 2, 803–813.

Salmeen, A., Barford, D., 2005. Functions and mechanisms of redox regulation of cysteine-based phosphatases. Antioxid. Redox Signal. 7, 560–577.

Satoh, T., McKercher, S.R., Lipton, S.A., 2014. Reprint of: Nrf2/ARE-mediated antioxidant actions of pro-electrophilic drugs. Free Radic. Biol. Med. 66, 45–57.

Shaarawy, S.M., Tohamy, A.A., Elgendy, S.M., et al., 2009. Protective effects of garlic and silymarin on NDEA-induced rats hepatotoxicity. Int. J. Biol. Sci. 5 (6), 549–557.

Sheela, C.G., Kumud, K., Augusti, K.T., 1995. Anti-diabetic effect of onion and garlic sulfoxide amino acids in rats. Planta Medica 61, 356–357.

Singh, Y.P., Singh, R.A., 2010. *In silico* studies of organosulfur-functional active compounds in garlic. Biofactors 36 (4), 297–311.

Slimestad, R., Fossen, T., Vågen, I.M., 2007. Onions: a source of unique dietary flavonoids. J. Agric. Food Chem. 55 (25), 10067–10080.

Suleria, H.A., Butt, M.S., Anjum, F.M., et al., 2015. Onion: nature protection against physiological threats. Crit. Rev. Food Sci. Nutr. 55 (1), 50–66.

Sun, W.H., Liu, F., Chen, Y., et al., 2012. Hydrogen sulfide decreases the levels of ROS by inhibiting mitochondrial complex IV and increasing SOD activities in cardiomyocytes under ischemia/reperfusion. Biochem. Biophys. Res. Commun. 421 (2), 164–169.

Szabó, C., Papapetropoulos, A., 2011. Hydrogen sulphide and angiogenesis: mechanisms and applications. Br. J. Pharmacol. 164 (3), 853–865.

Tarozzi, A., Angeloni, C., Malaguti, M., et al., 2013. Sulforaphane as a potential protective phytochemical against neurodegenerative diseases. Oxid. Med. Cell. Longev. 2013, Article ID 415078.

Tin, A.S., Park, A.H., Sundar, S.N., et al., 2014. Essential role of the cancer stem/progenitor cell marker nucleostemin for indole-3-carbinol anti-proliferative responsiveness in human breast cancer cells. BMC Biol. 12, 72.

Toohey, J.I., Cooper, A.J., 2014. Thiosulfoxide (sulfane) sulfur: new chemistry and new regulatory roles in biology. Molecules 19 (8), 12789–12813.

Toyokuni, S., 2014. Iron and thiols as two major players in carcinogenesis: friends or foes? Front. Pharmacol. 5, 200.

Tsubura, A., Lai, Y.C., Kuwata, M., et al., 2011. Anticancer effects of garlic and garlic-derived compounds for breast cancer control. Anticancer Agents Med. Chem. 11 (3), 249–253.

Tu, B.P., Weissman, J.S., 2004. Oxidative protein folding in eukaryotes: mechanisms and consequences. J. Cell Biol. 164 (3), 341–346.

Valachová, K., Baňasová, M., Topoľská, D., et al., 2015. Influence of tiopronin, captopril and levamisole therapeutics on the oxidative degradation of hyaluronan. Carbohydr. Polym. 134, 516–523. Available from: https://doi.org/10.1016/j.carbpol.2015.07.029.

Viswanathan, V., Phadatare, A.G., Mukne, A., 2014. Antimycobacterial and antibacterial activity of *Allium sativum* bulbs. Indian J. Pharm. Sci. 76 (3), 256–261.

Votto, A.P., Domingues, B.S., de Souza, M.M., et al., 2010. Toxicity mechanisms of onion (*Allium cepa*) extracts and compounds in multidrug resistant erythroleukemic cell line. Biol. Res. 43 (4), 429–437.

Wang, H.C., Pao, J., Lin, S.Y., et al., 2012. Molecular mechanisms of garlic-derived allyl sulfides in the inhibition of skin cancer progression. Ann. NY Acad. Sci. 1271, 44–52.

Wang, Y., Chun, O.K., Song, W.O., 2013. Plasma and dietary antioxidant status as cardiovascular disease risk factors: a review of human studies. Nutrients 5 (8), 2969–3004.

Yan, L.J., 2014. Protein redox modification as a cellular defense mechanism against tissue ischemic injury. Oxid. Med. Cell. Longev. 2014, Article ID 343154.

Yang, J., Wang, T., Yang, J., et al., 2013. S-allyl cysteine restores erectile function through inhibition of reactive oxygen species generation in diabetic rats. Andrology 1 (3), 487–494.

Yoshinari, O., Shiojima, Y., Igarashi, K., 2012. Anti-obesity effects of onion extract in Zucker diabetic fatty rats. Nutrients 4 (10), 1518–1526.

You, Z., Cao, X., Taylor, A.B., et al., 2010. Characterization of a covalent polysulfane bridge in copper-zinc superoxide dismutase. Biochemistry 49 (6), 1191–1198.

Zhang, Y., Talalay, P., Cho, C.G., et al., 1992. A major inducer of anticarcinogenic protective enzymes from broccoli: isolation and elucidation of structure. Proc. Natl. Acad. Sci. USA 89 (6), 2399–2403.

Zhang, C.L., Zeng, T., Zhao, X.L., et al., 2013. Garlic oil attenuated nitrosodiethylamine-induced hepatocarcinogenesis by modulating the metabolic activation and detoxification enzymes. Int. J. Biol. Sci. 9 (3), 237–245.

Chapter 55

Arginine and citrulline as nutraceuticals: efficacy and safety in diseases

Shilpa N. Kaore and Navinchandra M. Kaore

Introduction

L-citrulline is a naturally occurring nonessential amino acid (AA) present in mammals and also in each living organism, but conditionally essential in stress and diseases; while Arg is semiessential. Citrulline (Cit) is a by-product of the formation of nitric oxide (NO) from arginine (Fig. 55.1). Cit may be recycled to arginine (Arg), thereby maintaining adequate Arg levels required for the production of NO (Norris et al., 1995; Marx et al., 2000). Cit supplementation administered orally increases plasma Arg concentration and augments NO-dependent signaling in a dose-dependent manner (Schwedhelm et al., 2008). Unlike L-arginine, Cit bypasses hepatic metabolism and is not a substrate of arginase (Romero et al., 2006), but rather it is diverted to three metabolic pathways. Thus Cit has a very specific metabolism: it is not subjected to presystemic elimination, but rather to systemic metabolism. Arg is a conditionally essential AA in children, but is nonessential in adults. Arg is subjected to extensive presystemic as well as systemic metabolism by bacteria in the gut and arginases in the gut and liver, and it is converted to NO by NO synthase (NOS).

Citrulline is produced by the body naturally and is also found naturally in certain foods like watermelons, cucumbers, pumpkins, muskmelons, bitter melons, squashes, and gourds. In fact, the name "citrulline" was derived from the Latin word for watermelon *Citrullus vulgaris* (Wada, 1930) because it was first isolated from it (Collins et al., 2007), and watermelon is unusually rich in citrulline (Rimando and Perkins-Veazie, 2005). It is a key intermediate in the urea cycle, the pathway by which mammals excrete ammonia (Takeda et al., 2011). In plants, citrulline functions as a potent hydroxyl radical scavenger, affording protection from oxidative injuries and maintenance of water status, which are also evident in adverse conditions such as in the Kalahari desert (Kawasaki et al., 2000; Yokota et al., 2002; Rimando and Perkins-Veazie, 2005). Arg and Cit are essential in health and can be used as nutraceuticals or as dietary supplements in conditions in which they are deficient or reduced as a consequence of dietary deficiency or diseased states. Cit and Arg are used as nutraceutical agents for a variety of clinical conditions, and there is a scientific basis for their use.

According to the Dietary Supplement Health and Education Act, dietary supplements are products intended to supplement the diet and bear or contain one or more of the following dietary ingredients: a vitamin, a mineral, a herb or other botanical, an AA, a dietary substance for use by humans to supplement the diet by increasing the total daily intake, or a concentrate, metabolite, constituent, extract, or combinations of these ingredients (Zeisel, 1999). It may be taken in the form of a capsule, tablet, or liquid. It is not represented for use as a conventional food or as the sole item of a meal or diet. It is labeled as a "dietary supplement" (Chauhan et al., 2013). Under the Dietary Supplement Health and Education Act of 1994, the manufacturer of a dietary supplement is responsible for ensuring that the dietary supplement is safe before it is marketed.

Nutraceuticals. DOI: https://doi.org/10.1016/B978-0-12-821038-3.00055-0

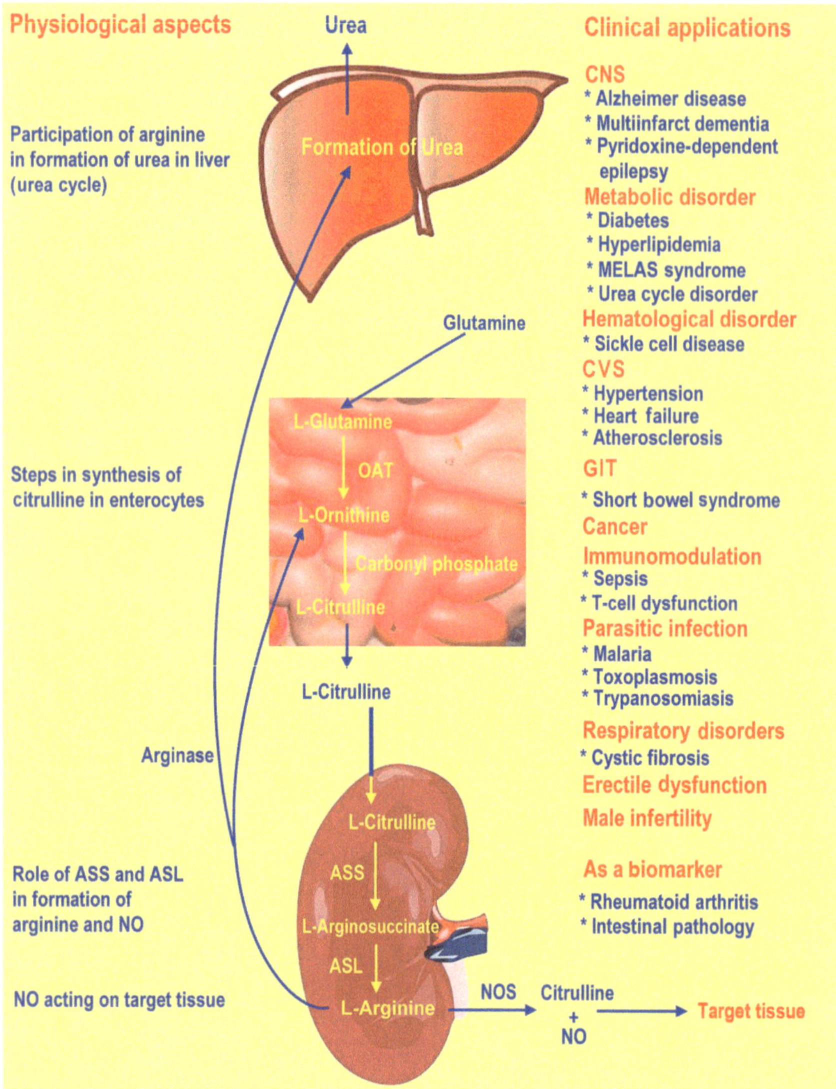

FIGURE 55.1 Schematic representation of synthesis of citrulline and arginine with potential clinical applications and role as biomarker.

Biochemical aspects

L-citrulline has the molecular formula $C_6H_{13}N_3O_3$ (see its chemical structure in Fig. 55.2) and a molar mass of 175.19 g/mol. Cit (2-amino-5-ureidovaleric acid) is an α-AA. Citrulline exists in two forms: (1) free citrulline, which is a product of the NOS enzyme family, and (2) citrulline, which results from the posttranslational modification of certain proteins at arginine residues, which is catalyzed by peptidyl arginine deiminases (PADs) (Herrera-Esparza et al., 2013). Its exploration has established citrulline as a biomarker in a variety of disorders, and many more may be established in the future (Kaore et al., 2013).

Arginine has the molecular formula $C_6H_{14}N_4O_2$, (see its chemical structure in Fig. 55.3) and a molar mass of 174.2 g/mol. Arg (2-amino-5-guanidinovaleric acid) is the precursor of NO, an endogenous messenger molecule involved in a variety of endothelium-mediated physiological effects in the vascular system (Boger and Bode-Boger, 2001).

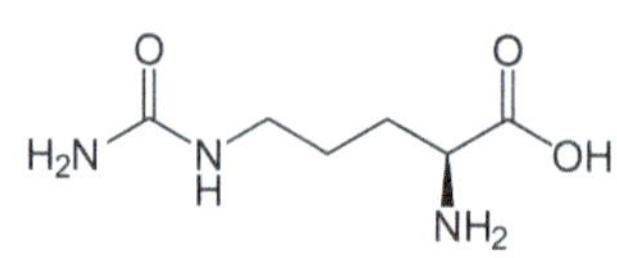

FIGURE 55.2 Chemical structure of citrulline.

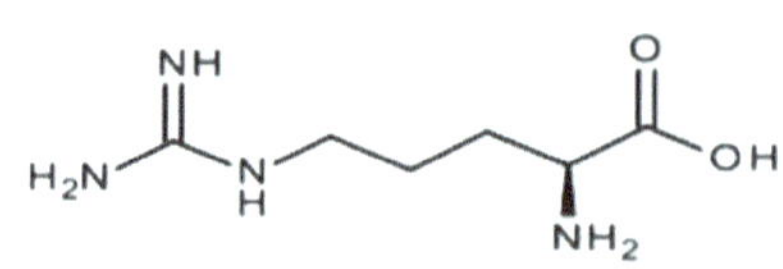

FIGURE 55.3 Chemical structure of arginine.

Synthesis

Citrulline, made from ornithine and carbamoyl phosphate, is a component of the urea cycle in the liver (Endo et al., 2004). It is synthesized from Arg and glutamine in enterocytes. Cit enters the kidneys, vascular endothelium, macrophages, and other tissues and can be readily converted to Arg and NO (Romero et al., 2006). Once Cit is converted to Arg, it is made available for utilization by the peripheral tissues (Fig. 55.1). All three isoforms of nitric oxide synthase (NOS), neuronal NOS, inducible NOS (iNOS), and endothelial (eNOS), convert Arg to NO and Cit (Schwedhelm et al., 2008). In addition, asymmetric dimethyl arginine (ADMA) has very low circulating levels in humans and is found to be an endogenous inhibitor of all three isoforms of NOS. The Arg-to-ADMA ratio is one of the determinants of NO production. In turn, NO activates soluble guanylylcyclase (sGC) in smooth muscles, leading to an increase in intracellular cyclic guanosine monophosphate pathway (cGMP), causing vasodilation. This process is essential for endothelial function, and disturbed NO production in the human endothelium is attributed to endothelial dysfunction. Under ischemic conditions, NO is synthesized from nitrite; in acidic conditions, it is synthesized via a nonenzymatic pathway mainly in tissues (Kaore et al., 2013).

Cit is formed exclusively in enterocytes, with the liver having only a minimal effect on Cit production, but a decrease in the functional mass of enterocytes decreases plasma Cit levels (Oliverius et al., 2010). Approximately 80% of Cit is converted in the kidney to Arg (Luiking et al., 2009; Oliverius et al., 2010), finally converting/recycling Arg to Cit and NO and serving as a potent Arg precursor (Waugh et al., 2001; Urschel et al., 2006; Flam et al., 2007; Moinard et al., 2007; Luiking et al., 2009; Fig. 55.1). This is carried out via argininosuccinate synthase (ASS) converting L-citrulline to L-argininosuccinate, which is subsequently converted to Arg by argininosuccinate lyase (ASL). The importance of the argininosuccinate pathway has been demonstrated in the pig for the Cit–Arg conversion in the cerebral vascular endothelium, which is a constant neuronal source of NO for inducing cerebral vasodilatation (Yu et al., 1997; Lee and Yu, 2002). Therefore Cit is essential for Arg and NO production; the latter acts as an endogenous vasodilator (Flam et al., 2007) and both Cit and Arg act as NO donors.

Salient aspects of nitric oxide

NO is a free gaseous chemical messenger thought to be involved in a number of physiological processes in the mammalian central nervous system, including memory, respiration, skeletal muscle activity, and cardiovascular homeostasis (Smith et al., 1996; Milatovic et al., 2005). In the brain, NO has a half-life of less than a few seconds and the distance of NO diffusion is 300 μm, enough to reach neighboring neurons (Gupta, 2004). Estimates of the half-life of NO in the blood range from 0.05 to 1.8 ms. Once it enters the circulation, it is not only destroyed quickly but also transported to and interchanged with other compartments (Rassaf et al., 2002).

There is growing evidence for the role of NO in the regulation of insulin sensitivity and oxidation of energy substrates, and also for its key role in the immune response, neurological functions, and hemodynamics in animals and humans. NO released from the endothelium is an important regulator of vascular tone, and it is also an inhibitor of platelet and leukocyte aggregation and adhesion (Kaore et al., 2013). It undoubtedly plays a crucial role in the regulation of vascular disease. Moreover, Cit supplementation is more efficacious and reliable for increasing NO availability as compared to Arg; therefore Cit is a more promising and successful nutraceutical or dietary supplementation. Supplementation with Cit is a novel therapeutic approach in conditions of Arg and NO deficiency (Luiking et al., 2010).

However, excessive NO [as reactive nitrogen species (RNS)] is implicated in seizures and neurotoxicity because it causes impairment of mitochondrial functions, especially inhibiting the mitochondrial respiratory chain (enzymes like cytochrome c oxidase) as well as causing DNA damage due to NO-induced increased reactive oxygen species (ROS). Thus the accompanying increase in Cit levels (also NO/NO synthase and RNS) can be used as markers in certain toxicities (Gupta, 2004). This chapter describes various aspects of Arg and Cit, their pharmacological perspectives, and details about their exploration for therapeutic benefits, safety, and efficacy.

Citrulline: more efficacious than arginine—a justification

Cit supplements are superior for therapeutic purposes as compared to Arg (Kuhn et al., 2002), because data support elevated plasma Arg levels on Cit supplementation and augmentation of NO signaling in a dose-dependent manner (Schwedhelm et al., 2008). Arg undergoes extensive presystemic and systemic metabolism by bacteria in the gut and arginases in the gut and liver (Schwedhelm et al., 2008). Although Cit has a very specific metabolism, not subjected to presystemic elimination but rather to systemic metabolism, it is not metabolized in liver or intestines (Romero et al., 2006). It is diverted to three metabolic pathways: (1) Arg biosynthesis, (2) the NO cycle, which can involve local recycling of Cit, and (3) participation in the urea cycle for detoxification of ammonia in the liver (Curis et al., 2005). This makes Cit an effective substitute to restore NO production in situations of limited Arg availability. Acute oral Cit administration efficiently increases plasma Arg levels better than long-term Arg supplementation, increasing plasma Arg levels to approximately 227% with 3.8 g Cit (within 4 h) compared to only a 90% increase with an equivalent dose of Arg. Cit (oral) supplementation studies in healthy volunteers raised blood Arg levels more effectively than Arg supplementation (Kaore et al., 2013), therefore supporting its promising role as a nutraceutical. Further, acute Arg supplementation does not increase plasma concentrations of NO in healthy individuals with normal plasma concentrations of ADMA (Alvares et al., 2012).

The beneficial effects of Arg do not endure with chronic therapy. In diseased states such as heart failure, atherosclerosis, diabetic vascular disease, and ischemia–reperfusion injury, Arg supplementation enhances arginase expression and activity, thereby reducing the effectiveness of Arg therapy. In contrast, Cit inhibits arginase activity. For these obvious reasons, Cit supplementation seems to be a better alternative for supplying Arg than Arg supplementation itself (Romero et al., 2006). Interestingly, dietary Arg supplementation is also used to increase pork quality in terms of an increase in intramuscular fat content in finishing pigs (Ma et al., 2015).

Cit causes a greater increase in Arg, and this has a major role in driving NO production. This concept is supported by the observation that the three enzymes responsible for recycling citrulline to NO (ASS, ASL, and NOS) function as a complex that can result in compartmentalizing NO synthesis and channeling Cit efficiently to NO synthesis (El-Hattab et al., 2012).

Therapeutic utility of nutraceuticals

Citrulline supplementation holds promise as a therapeutic adjunct in clinical conditions associated with Arg/NO deficiency. Thus supplementation is likely to benefit clinical conditions such as hypertension, heart failure, atherosclerosis, diabetes, vascular disease, and ischemia–reperfusion injury. Future studies will evaluate its therapeutic potential in clinical conditions such as sickle cell anemia, hypertension, hyperlipidemia, erectile dysfunction (ED), male infertility, sepsis, arginase-associated T-cell dysfunction, short bowel syndrome (SBS), cancer, hyperoxic lung damage, urea cycle disorders, mitochondrial encephalomyopathy, lactic acidosis, and stroke-like episodes (MELAS) syndrome, Alzheimer disease (AD), and intrauterine growth restriction (IUGR). Its role will also be investigated in the treatment of multiinfarct dementia (MID), intestinal pathologies, and parasitic diseases (such as severe falciparum malaria). These aspects, as well as the basis for therapeutic benefits, are discussed in the following sections. Supplements targeted toward certain major health problems in humans should go through rigorous clinical trials before being turned into established strategies accepted worldwide.

Another area of interest is the exploration of Cit as a predictor of diseases by detection of the antibody to citrullinated peptides, serving as a biomarker of various diseases. This is due to protein citrullination (i.e., post-translational modification of Arg, Arg to Cit conversion, deaminated Arg) of certain body proteins that produce novel epitopes and give rise to autoantibodies, the detection of which provides a novel way to diagnose a wide variety of diseases and serves as a biomarker for monitoring certain clinical conditions.

Plasma Cit concentration has been extensively researched and supportive results have established its concentration as a reliable quantitative biomarker of enterocyte absorptive capacity (Papadia et al., 2014). This capacity is used by clinicians to assess functional enterocyte mass in various chronic and acute small bowel pathologies, such as short bowel syndrome (Oliverius et al., 2010). Plasma Cit values can aid in diagnosis and measurement of the severity of intestinal failure, and possibly monitor bowel function. Thus a decreased plasma Cit concentration reflects a decrease of functional mass of enterocytes provided that renal function is normal (Moinard and Cynober, 2007).

Citrullination occurs with specific proteins, like filaggrin, producing novel epitopes that produce autoantibodies. Therefore anticitrullinated protein antibodies are now recognized as biomarkers for rheumatoid arthritis, and assays that use cyclic citrullinated peptide (CCP) as an artificial antigen are showing high specificity and sensitivity (Kaore et al., 2013). Currently, the combination of anti-CCP antibodies with RF IgM and RF IgA is recommended either in parallel or for stepwise determination for rheumatoid arthritis (Conrad et al., 2010) and erosive arthritis, suggesting that protein citrullination plays a role in the etiology of RA and that PAD inhibitors might play a therapeutic role (Duskin and Eisenberg, 2010). Its inclusion as a criterion in the 2010 American Rheumatoid Arthritis Classification emphasizes its importance and significance (Centers for Disease Control, 2014). Thus these antibodies could be used as predictors of disease.

Pharmacokinetics of citrulline and arginine

Pharmacokinetics of oral citrulline

Citrulline supplementation shows a significant increase in plasma Arg and augmentation of NO-dependent signaling mechanisms in a dose-dependent manner. Oral Cit exhibits a dose-dependent and more efficient increase in AUC and C_{max} of plasma Arg concentration as compared to Arg supplementation ($P < 0.01$). The highest dose of Cit (3 g twice a day) increased the C_{min} of plasma Arg and improved the Arg-to-ADMA ratio significantly from baseline [$P < 0.01$, 95% confidence interval], in addition to increasing urinary nitrate and cGMP significantly. Improvement of flow-mediated vasodilation (FMD) over baseline was not demonstrated, but a pooled analysis of all FMD data revealed a correlation between the increase of the arginine-to-ADMA ratio and improvement of FMD (Schwedhelm et al., 2008).

Because Cit is formed mostly in enterocytes (80%) and is converted to Arg in the kidneys (Oliverius et al., 2010), chronic renal insufficiency results in decreased uptake of Cit and release of Arg (by 60%–70%). Research depicts a linear relationship between Cit extraction and Arg release in subjects with normal renal function (Tizianello et al., 1980); hence, decreased Cit levels are clinically relevant because they reflect a decrease in the functional mass of enterocytes (Bailly-Botuha et al., 2009).

A sustained release (SR) Cit formulation was compared with immediate release (IR) Cit in experimental studies that concluded that SR Cit led to a sustained increase in citrullinemia and argininemia compared to IR Cit, and on day 6 argininemia was significantly ($P < 0.01$) higher with SR compared to IR Cit (Berthe et al., 2011). Also, argininemia was significantly higher in the SR Cit group than in controls throughout the study. SR citrulline maintained high argininemia and citrullinemia, at least over 12 h throughout the study. Thus there is a strong rationale to use the SR formulation for the treatment of atherosclerosis and the benefit can be extended to other clinical conditions as well.

Pharmacokinetics of arginine

Studies performed in three species (sheep, pigs, and rats) indicate that all of them rapidly catabolize supplemental Arg. Elevated circulating Arg concentrations generally returned to baseline levels within 4–5 h of administration, with rates varying with the age and physiological status of the animal. Further, Arg clearance was found to be greater in pregnant than in nonpregnant animals, greater in young than in adult animals, greater in lean than in obese animals, and greater in type I diabetic than in nondiabetic animals (Wu et al., 2007a).

Normal plasma Arg concentrations are approximately 80–120 μM; intracellular concentrations are even greater (up to 1 mM). The michaelis constant (km) for arginine as a substrate for NOS is in the region of 1–10 μM; thus there would appear to be a surplus of substrate. However, reports indicate that administration of exogenous Arg may enhance the generation of NO (Gad, 2010). Further, a study found that acute Arg supplementation does not increase plasma concentrations of NO in healthy individuals with normal plasma concentrations of ADMA (Alvares et al., 2012).

Safety considerations

Citrulline

Citrulline is generally recognized as safe for oral use (Romero et al., 2006). In contrast to Arg, Cit at high doses (i.e., >10 g in one bolus) may induce gastrointestinal side effects, but Cit is well tolerated. This could be due to rapid saturation of intestinal absorption of Arg causing osmotic diarrhea at high loads. This difference between these AA suggests that the intestinal absorption of Cit is not a limiting step in Cit bioavailability, even at high Cit loads (Romero et al., 2006; Collins et al., 2007; Moinard et al., 2007).

Short-term administration of Cit is safe and well tolerated. No adverse effects were observed in healthy subjects given different loading doses of 2, 5, 10, or 15 g (Collins et al., 2007). There is a dearth of reliable scientific data regarding its use during pregnancy and lactation.

Cit ingestion decreases QT intervals in healthy subjects, which means it shortens the time required for completion of myocardial depolarization and repolarization; therefore Cit may be associated with cardiovascular risks (Kameda et al., 2011; Meldrum et al., 2011). Further studies may clarify its actual incidence in practice.

Arginine

Among studies available on oral administration of Arg in humans, only a few reported any adverse effects following acute or chronic treatment. Data cover dose ranges from 3 g/day to >100 g/day, with the most common adverse effects being nausea, vomiting, and diarrhea. These signs appeared infrequently at 15–30 g/day; 30 g/day was well tolerated, and at >9 g/day athletes were more susceptible to gastrointestinal tract (GIT) side effects than diabetics, which may relate to effects of the disease on gastrointestinal (GI) motility. The average human subject should be able to tolerate long-term parenteral and enteral supplemental doses of 6 and 15 g/day Arg, respectively, in addition to a basal amount of Arg (4–6 g/day) from their regular diets (Grimble, 2007; Wu et al., 2007a). At a dose of >30 g/d, Arg may have a laxative effect, and this can be lessened by administering in divided doses. Arg may induce bronchoconstriction when nebulized and inhaled (Chambers and Ayres, 2001). Long-term studies are needed to further evaluate its safety.

Regarding the safety of chronic dietary Arg supplementation in animals, it was found that the Arg concentrations in serum increased until 3 h after administration of chronic intravenous or dietary supplementation. However, in all animals studied, serum Arg concentrations did not differ at 5 h after intravenous or oral administration between Arg-supplemented and nonsupplemented groups.

Further, Arg administered daily via the intravenous route to adult pregnant ewes (81 mg/kg body weight per day) for 87 days did not result in any undesirable effects. Also, neonatal pigs, growing finishing pigs, pregnant gilts, and adult rats tolerated large amounts of chronic supplemental Arg administered via enteral diets without the appearance of any adverse effects. In these experiments, neither oral nor intravenous administration of Arg produced any sickness or death in any animal (Wu et al., 2007a).

Previous reviews document that side effects are rare, mostly mild, and dose-dependent with Arg administration. Long-term studies with chronic oral Arg administration or intermittent infusion therapy imply that it can improve clinical symptoms of cardiovascular disease in humans. A case report documents an accidental fatal overdose with Arg while the patient was being treated for growth hormone deficiency that resulted in cardiopulmonary arrest, and, after 36 h of resuscitation, she developed fatal myelinolysis (Boger and Bode-Boger, 2001; Gerard and Luisiri, 1997).

No changes in liver function, blood glucose, or plasma electrolytes have been reported. In the absence of appropriately designed safety studies, caution should be taken if Arg is used in infants, pregnant or lactating individuals, and those with viral infections and seriously compromised renal or hepatic function (Tankersley, 1964; Grimble, 2007; Gad, 2010). Care must also be taken in renal/hepatic failure and in patients with hypotension and myocardial infarction (MI). Use with caution with antihypertensive drugs, nitrates, and phosphodiesterase-5 inhibitors. The literature contains one study on patients with MI with ST elevation who were started on oral Arg (3 g three times daily) within 24 h of MI for 30 days. This was well tolerated; there were no serious adverse effects during the study and no significant beneficial clinical outcome (Bednarz et al., 2005). Further safety evaluation needs to be geared toward these nutraceutical agents.

Pharmacodynamics of citrulline and arginine

The following section discusses the clinical utility and therapeutic uses of Cit and Arg in certain disease conditions. The main pharmacological basis of their actions in clinical conditions can be basically attributed to the generation of NO, leading to its beneficial actions in vivo.

Citrulline in sickle cell anemia

In sickle cell disease, vaso-occlusion leads to organ damage, decreased quality of life, and decreased life expectancy (Schnog et al., 2004). Cit (0.1 g/kg, orally twice daily) given in sickle cell disease as an Arg precursor is valuable in palliative therapy, for which it has cleared phase II trials; it decreases complications and increases overall well-being of these patients. Arg is required as a substrate in the Arg–NO pathway for endogenous nitrovasodilation and vasoprotection (inhibition of adhesion and activation of leukocytes). The basis for use is to provide benefit of vasoprotection by Arg, partly mediated by NO-induced inhibition of endothelial damage and inhibition of adhesion and activation of leukocytes, decreasing the vaso-occlusive crisis. When treatment is continued with Cit, it causes symptomatic improvement and also maintains nearly normal total leukocyte and neutrophil counts (Waugh et al., 2001). Thus Arg is a conditionally essential AA in children and adolescents with sickle cell disease wherein the need can be fulfilled by Cit supplementation. Large-scale clinical trials are needed in this area before it can be widely used.

Citrulline and cardiovascular effects

NO is synthesized by the endothelial cells using Arg as a substrate in a process catalyzed by the NOS. The contribution of NO to the regulation of vascular tone in humans has been demonstrated by showing that inhibition of its synthesis results in significant vasoconstriction and contributes to impaired endothelial regulation of vascular tone in patients with systemic arterial hypertension (Cardillo and Panza, 1998). Thus increasing Arg availability either by Arg/Cit therapy or by arginase inhibition may provide therapeutic benefits in patients with hypertension. NO activates sGC in smooth muscles, leading to an increase in intracellular cGMP, causing vasodilation. This process is essential for endothelial function, and disturbed NO production in the human endothelium is attributed to endothelial dysfunction.

Citrulline and arginine in hypertension

It has been observed that hypertensive patients had a lower Arg–Cit ratio than normotensive patients, and this is generally associated with NO deficiency, oxidative stress, increased ROS, and a disturbed constrictor–dilator balance in the kidneys. A recent meta-analysis of clinical trials showed that oral Arg supplementation could lower BP by 5.39/2.66 mmHg, which was comparable with the effects of diet changes and exercise implementation. Further, the antihypertensive properties of Cit are more questionable, and in range of 4.1/2.08 to 7.54/3.77 mmHg, since both Arg and Cit are effective at increasing plasma Arg (David et al., 2019).

Oral L-citrulline supplementation (6 g/day) given for 4 weeks attenuated the brachial systolic blood pressure (BP), aortic SBP, and aortic pulse pressure responses to cold pressor test (CPT) in young normotensive men. Increased wave reflection time contributes to the reduction in aortic systolic BP response to CPT (Figueroa et al., 2010).

Watermelon supplementation also produced a reduction in arterial stiffness and aortic systolic BP in obese postmenopausal women with hypertension (Figueroa et al., 2013).

In spontaneously hypertensive rats (SHR), there are defects in the renal Cit−Arg pathway or Arg reabsorption, potentially causing renal NO deficiency and hence hypertension (Tain et al., 2014). Research shows that if NO availability is increased perinatally, it reduces BP in SHR, and Cit supplementation causes a significant increase in renal NO (Kaore et al., 2013). Melatonin was found to provide restoration of the NO pathway by a reduction of plasma ADMA, preservation of renal Arg availability, and attenuation of oxidative stress in young SHR (Tain et al., 2010).

The effects of maternal caloric restriction in rat offspring were evaluated and found to have fewer nephrons, renal dysfunction, and development of hypertension that was reversed with maternal Cit supplementation with long-term antihypertensive effects (Tain et al., 2010). An experimental study showed that programmed hypertension developed in the male offspring of pregnant rats exposed to NO deficiency was prevented by maternal Cit supplementation due to expression levels of genes related to BP control and oxidative stress (Tain et al., 2014). The ADMA−NO pathway serves as a therapeutic target and more studies are needed to explore mechanistic insights.

Intravenous infusion of Arg (15 mg/kg body weight/min for 35 min) improved pulmonary vascular resistance index and cardiac output in infants with pulmonary hypertension (PH). Another metaanalysis included 11 randomized, double-blind, placebo-controlled trials involving 387 participants who received oral Arg ranging from 4 to 24 g/day, which significantly lowered systolic BP and diastolic BP in comparison to placebo (Dong et al., 2011).

Beneficial effects of Cit supplementation include improvement in left ventricular ejection fraction as well as right ventricular ejection fraction both at rest and with stress with Cit 3 g/day in two doses of 1.5 g, in addition to conventional therapy using the dose suggested in the literature (Orozco-Gutiérrez et al., 2010; Balderas-Munoz et al., 2012). There is a decrease in systolic pulmonary artery pressure due to beneficial effects on endothelial function in heart failure patients with preserved ejection fraction (Kaore et al., 2013).

Citrulline is a safer way of delivering Arg for endothelial and immune cells and preventing excessive uncontrolled NO production (Cynober et al., 2010) that exerts deleterious effects of its own. Thus Cit can be an important co-adjuvant in the treatment of stable and unstable systolic heart failure patients who also have hypertension. The ADMA−NO pathway serves as a therapeutic target and more studies are needed to explore mechanistic insights. Further clinical trials are needed to evaluate Cit/Arg versus conventional antihypertensive treatment and to support the notion of their therapeutic utility in hypertension and/or heart failure.

Arginine in pregnancy-induced hypertension

Preeclampsia is a complicating health problem in approximately 2%−8% of pregnancies. Systemic reviews regarding Arg show a reduction in preeclampsia when compared with placebo and a reduction in risk of preterm birth (Wu et al., 2007a; Dorniak-Wall et al., 2014). Few studies are suggestive of a beneficial response of Arg, with one study involving women with pregnancy-induced hypertension given an intravenous infusion of Arg (20 g over a 2-h period daily for 5 consecutive days) showing reduced BP. Thus it might be possible that Arg exerts its nutritional and physiological functions partly through changes in cellular expression of related key genes (Wu et al., 2007a). This needs further exploration to evaluate whether Arg supplementation has the potential for a preventative role in pregnancy-related hypertensive disease and whether it can improve maternal and infant health outcomes in large-scale clinical trials.

Citrulline and arginine in hyperlipidemia

A study analyzing the role of Cit and Arg in hypercholesterolemic rats found reversal of increased serum levels of aspartate aminotransferase, alanine aminotransferase, and urea with a concomitant increase of high-density lipoprotein cholesterol and NO. The data indicate Cit, as well as Arg, has good efficacy as a hypolipidemic agent in rats (El-Kirsh et al., 2011).

Studies show watermelon juice (rich in Cit, an effective Arg precursor) to ameliorate metabolic syndrome in Zucker diabetic fatty rats in terms of reduced fat accretion, and lowered serum concentrations of glucose and free fatty acids, thus reducing cardiovascular risk factors, improved glycemic control, and amelioration of vascular dysfunction in obese animals with type II diabetes mellitus (T2DM) (Collins et al., 2007; Gad, 2010). A recent study showed improvement in coronary risk factors as endothelial function, aging, and atherosclerosis in ZFDR with Cit/Arg alone and Cit plus Arg (Tsuboi et al., 2018); showing Cit, Arg scavenging action on ROS, Cit better than Arg in significantly increasing NO levels by inhibition of ROS and arginase 2 protein expression, while Cit and Cit plus Arg supplementation reversed high glucose-induced endothelial senescence in vitro. Further studies may enlighten their use as therapeutic agents in hyperlipidemia or other indications in the future.

Citrulline in respiratory conditions

Citrulline supplementation prevents hyperoxia-induced lung injury as well as PH in newborn rats, suggesting an alternative to inhaled NO, in addition to a benefit in bronchopulmonary dysplasia (BPD). BPD is characterized by arrested alveolar development and is complicated by PH.

A recent case report suggested single-dose oral supplementation of Cit continued for 70 days, which enabled a 6-month old baby to be weaned in the first 3 weeks from mechanical ventilation, with a significant decrease in brain natriuretic peptide in the first month and then becoming stable. This suggests a beneficial alternative in the prophylaxis or therapy of chronic PH in newborns (Lauterbach et al., 2018).

A previous study supported disrupted NO–cyclic guanosine monophosphate pathway (NO-cGMP) signaling in neonatal pups, and thus Arg supplementation or arginase inhibition may be a therapeutic tool to prevent or treat neonatal lung injury (Ali et al., 2012). NO may promote alveolar growth, so Cit supplementation might represent a novel therapeutic alternative to inhaled NO for prevention of BPD. Cit supplementation had been shown earlier to preserve alveolar and vascular growth in O_2-induced lung injury and decreased pulmonary arterial medial wall thickness and right ventricular hypertrophy in pups. Plasma Arg and Cit concentrations were significantly increased by Cit treatment, with increased ASL protein expression in hyperoxia converting Cit to Arg (Arg in turn converted to Cit + NO) (Vadivel et al., 2010; Sopi et al., 2012).

In patients with cystic fibrosis, Arg bioavailability is reduced due to increased arginase levels. Repeated Arg inhalation was found to be safe and well tolerated, with evidence of an increase in FEV_1 that was statistically insignificant and without evidence of changes in airway inflammation (Grasemann et al., 2013).

It is noteworthy to mention here severe acute respiratory syndrome coronavirus (SARS CoV-2), the novel coronavirus causing the coronavirus disease 2019 (COVID-19) pandemic (Guo et al., 2020) and a global threat in 2019 and 2020 with 1,279,722 cases confirmed and 72,614 deaths as per the situation report of WHO on World Health Day of April 7, 2020 and contributing with increased counts of mortality across the globe (WHO, 2020). A study (Akerström et al., 2005) pointed out that the organic NO donor, *S*-nitroso-*N*-acetyl penicillamine, has significant inhibition of the replication cycle of SARS CoV in a concentration-dependent manner, inhibiting viral protein and RNA synthesis. It also demonstrated that NO generated by inducible NOS, an enzyme that produces NO, inhibits the SARS CoV replication cycle.

Additional in vitro evidence (Keyaerts et al., 2004) suggests that NO and NO donors may have an antiviral effect against SARS-CoV and inhalation of NO gas could have an antiviral effect against SARS-CoV based on which they have encouraged the inclusion of NO donors in the treatment of SARS. This is probably an area of research where exploration is required to reveal the roles of citrulline and arginine, since it is well established that either supplementation increases NO signaling.

Studies can prove or refute the role of Cit as a novel therapeutic alternative to inhaled NO for prevention of BPD and Arg inhalation for cystic fibrosis. Further exploration is indicated for the therapeutic indications of concern.

Citrulline and arginine: use in MELAS syndrome

Mitochondrial diseases arise as a result of dysfunction in the respiratory chain. MELAS syndrome is a common mitochondrial disorder in which there is insufficient adenosine triphosphate (ATP) generation to meet the energy demands of various organs. In addition, there is NO deficiency (due to NO sequestration by cytochrome c oxidase, NO shunting into ROS formation, and decreased availability of the NO precursors Arg and Cit), leading to impaired blood perfusion that contributes significantly to several complications. Hence, both nutraceuticals may be used in this condition as an NO precursor. Cit raises NO production to a greater extent than Arg and significantly decreases lactic acidosis, whereas data for Arg are insignificant (El-Hattab et al., 2014).

The first pilot study done in patients supported that impaired NO production occurs since childhood in MELAS syndrome and therefore the initiation of the NO donor treatment may be beneficial earlier in life. Citrulline supplementation shows a significant increase in NO production, arginine flux, and arginine concentration, and it may have a better therapeutic effect than Arg (El-Hattab et al., 2016). Further evaluation of their therapeutic effectiveness should be determined by further controlled clinical trials.

Citrulline in diabetes

Citrulline benefits the underlying endothelial dysfunction (Hecker et al., 1990) in diabetes, in addition to producing a reduction in glucose levels. In autoimmune conditions, activation of the Arg–Cit cycle in macrophages is observed, thus increasing Arg and NO, leading to NO-induced apoptosis of the vulnerable beta cells, leading to destruction of pancreatic beta cells in insulin-dependent diabetes mellitus (Kaore et al., 2013). Moreover, lower levels of NO protect cells from apoptosis, whereas excessive NO is toxic. NO also regulates insulin sensitivity and oxidation of energy substrates, along with playing a key role in the immune response. Further, increased arginase activity is also suggested for the underlying endothelial dysfunction in diabetes (arginase competes with NO synthase for their common substrate Arg), thus decreasing Arg availability to NOS and, hence, NO availability.

Watermelon juice exerts beneficial effects by increasing Arg availability, and improving glycemic control and vascular dysfunction in T2DM through intake of Cit (Wu et al., 2007b). Therapy with Cit (50 mg/kg/day, orally) and arginase inhibitors has been found to benefit the endothelial dysfunction and improve impaired vasodilation of coronary arteries in diabetes (Romero et al., 2008). Studies support the beneficial actions of Cit in diabetes and metabolic syndrome, which acts by improving glycemic control and ameliorating underlying endothelial dysfunction.

Chronic Arg supplementation for 3 months (9 g three times daily) in patients with visceral obesity improved insulin sensitivity with no impact on tumor necrosis factor alpha (TNF-α; one of the adipose tissue products) concentration, a role that is certain in the pathogenesis of insulin resistance (Bogdanski et al., 2012). Arginase inhibitors may prove useful because there is increased arginase activity in diabetes, thus increasing Arg availability to NOS. Studies emphasize that restoring eNOS activity is key to preventing or ameliorating cardiovascular complications of diabetes, and newer strategies for increasing NO bioavailability may prove fruitful in the future (Hoang et al., 2013). This area remains unexplored in large-scale clinical trials to evaluate the therapeutic strategies.

Citrulline for prevention of kidney failure

Citrulline was evaluated for protection against glycerol-induced acute renal failure in rats, and the findings were suggestive of improvement in intrarenal oxygenation and kidney function. In addition, it prevented renal oxidative stress as well as severe functional and morphological renal deterioration; thus it may prevent rhabdomyolysis-induced acute renal failure (Liu et al., 2013). One study revealed reversal of glycerol-induced deterioration of renal functions (degeneration, necrosis, sloughed brush borders, tubular degeneration, and renal failure) by Arg supplementation (Shaalan and Khalil, 2009). This explains that overall upregulation of iNOS due to pathological conditions is accompanied by decreased NO availability from the downregulation of eNOS. Another study showed that Cit induced protective effects in kidneys of STZ-rats, with a marked reduction in urinary albumin excretion, tubulo-interstitial fibrosis, and kidney hypertrophy observed in untreated diabetic mice; and antiinflammatory action in human proximal tubular cells with increased IL-10 and reduced IL-1β and IL-12 (Romero et al., 2013). Therefore Cit supplementation is likely to be of benefit in this condition, but needs greater exploration.

Citrulline in erectile dysfunction and male infertility

Cit administered 1.5 g/day for a month in men with ED (erection hardness score of 3) proved to be safe, psychologically well accepted, and reported as very satisfying by patients, although less effective than phosphodiesterase type-5 enzyme inhibitors, with improvement in erection hardness score from 3 to 4 (Cormio et al., 2011). Studies are required to prove its effectiveness in ED.

A double-blind, randomized, placebo-controlled, crossover study reported significant improvement in sperm quality in terms of motility, vitality, morphology, enhanced sperm volume, and concentration versus placebo with a polyherbal combination (Prelox) containing a French maritime pine bark extract, Arg, Cit, and roburins in male fertility administered for a period of 1 month (Stanislavov and Rohdewald, 2014). The fertility index increased to normal values during treatment with elevation of eNOS activity in sperm, without any significant adverse effects. A recent review suggests limited studies with small sample size explored Arg and its combinations with B-complex vitamins, ginseng, and other adjunctive agents rather than Cit; with positive evidence with Arg though (Koolwal et al., 2019). Further exploration with a large sample size in well-designed trials for efficacy and safety may open new horizons in ED and male infertility.

Citrulline as an immunomodulator

Citrulline production is significantly reduced in sepsis due to diminished de novo conversion of Arg and NO, thus reducing Cit and Arg availability in septic patients (and transgenic mice), in part due to ADMA release. Higher ADMA levels are correlated with a higher mortality rate in sepsis.

In sepsis, there is a complex interplay between proinflammatory and antiinflammatory cytokines, and endothelial NO may be a key factor in regulating this response. Cit supplementation seems to be a valuable means of supplementing Arg, favorably increasing NO availability and decreasing proinflammatory mediator response (IL-6 and resistin) without impairing the secretion of antiinflammatory mediators (IL-10 and adiponectin), hence acting as an immunomodulator (Kaore et al., 2013). Thus Cit may be a safe means of immunomodulation that preserves the antiinflammatory mediator response.

A recent study found that Arg and Cit supplementation (200 mg/kg/day intraperitoneally) exerted different programming effects on regulatory T-Cells (Treg) function of infantile rats. Arg therapy increased IL-10 and also enhance TGF-β (tumor growth factor) signaling by decreasing SMAD7 (an acronym from the fusion of *Caenorhabditis elegans* Sma genes and the *Drosophila* Mad) expression and enhanced SIRT-1 (Sirtuins) abundance, while Cit enhanced TGF-β signaling by decreasing SMAD7 expression and enhanced FOXP3/SIRT-1 (Foxhead proteins) molecule, that is critical to B-cell development. Apart from the different effects observed, it seems that Cit supplementation had a more potent effect than Arg in the modulation of the immune response for newborn rats (Lee et al., 2018).

Arginine and citrulline: role in immunonutrition

This novel approach may be useful in some critical conditions, such as surgery on cancer patients, for providing immunoenhancement. Malnutrition in head and neck cancer is found to be associated with immunosuppression; therefore patients are more susceptible to postoperative infections and complications. Arg becomes a conditionally essential AA, as stores could become depleted in times of stress (including sepsis). In addition, it plays an important role in T-cell and B-cell immunity and in NO production, reduces the production of inflammatory mediators such as IL-1β, IL-6, and TNF-α at the site of injury in septic models, and accelerates tissue growth after trauma or infection (Rodera et al., 2012). Thus dietary Arg supplementation has positive effects on immune function and reparative collagen synthesis. Arg, used as a part of immunonutrition in combination with nucleotides and omega-3-polyunsaturated fatty acids in postsurgical adult patients, reduced postoperative infections and length of hospital stay (Drover et al., 2011).

Randomized clinical trials show that immunonutrition with formulations containing medium-dose Arg and high-dose Arg (20 g/day) significantly lessened overall complications, improved local wound issues, and also caused a reduction in the length of hospital stay in postoperative cancer patients (De Luis et al., 2009; Rodera et al., 2012). Another study supported the use of immunonutrition (containing Arg, zinc, and antioxidants within a high-calorie, high-protein formula) to improve healing in pressure ulcers in malnourished patients, resulting in greater reduction in ulcer area at 8 weeks (Cereda et al., 2015).

Because Arg deficiency is presumed to be attributable to reduced Cit availability, either Arg or Cit supplementation seems logical. Arg is of specific interest as an immunonutrient due to the assumed deficiency thereof, especially in surgical patients, so it may be of consideration in the preoperative period (Kotze, 2011).

In spite of the many studies that have been conducted in this regard, its role as immunonutrition remains controversial, but has been proven to some extent to reduce infectious complications and the risk of multiorgan failure. However, Arg supplementation does raise concerns for increased mortality, increased risk of hemodynamic instability, and increased oxidative stress (Luiking and Deutz, 2007; Bertolini et al.,2007). Further robust studies could help in making a decision regarding proving its routine use as being beneficial or controversial.

Citrulline for acceleration of wound healing

Citrulline supplementation increases Arg (also produced through protein breakdown), and NO synthesis has a role in cellular regeneration and wound healing; it also increases secretion of growth hormone, insulin, and glucagon, thereby modulating blood flow, angiogenesis, epithelialization, and tissue granulation, collagen deposition (crucial to proliferative phase) (Norris et al., 1995; Stechmiller et al., 2005; Luiking et al., 2009; Oh, et al., 2017). NO-induced vasodilatation acts as a host-protective agent by killing pathogens in an in vitro study (Norris et al., 1995). This notion is supported by clinical studies in graft patients, suggesting Arg metabolism involvement in tissue repair, including angiogenesis, epithelialization, and collagen formation, and increased levels of Cit and other metabolites found in wound fluid versus plasma (Debats et al., 2009). Hence, immunonutrition with Arg may help in improving local wound healing and decreasing complications.

Citrulline in sepsis

Sepsis is a clinical condition that includes high morbidity and mortality rates and may be considered to be an Arg deficiency state (Luiking et al., 2004). Arg availability is increased in sepsis mainly by protein breakdown, but Arg catabolism is also increased due to its enhanced utilization for NO synthesis and higher arginase activity. Thus Arg availability is reduced in sepsis due to increased Arg catabolism and a slower Arg/Cit production rate (Luiking et al., 2009), and several studies show Cit supplementation to be a more productive arginine precursor than arginine (Lee et al., 2018). Because data for Cit in intensive care units are not available, large clinical trials are warranted before any reliable recommendations can be made for Cit-enriched diets. Its utility as a marker in sepsis or critically ill patients is doubtful because of associated vascular abnormalities, acute renal failure and/or multiple organ failure, or comorbid conditions (Cynober, 2013). Further studies are warranted to either prove or disprove the use of Cit therapy as an Arg precursor in this condition.

Citrulline in T-cell dysfunction

It is interesting to note that T lymphocytes depend on Arg for their proliferation, function, zeta-chain peptide formation, T-cell receptor complex expression, and development of memory; hence, Arg deficiency may result in T-cell abnormalities. Further, decreased T-cell proliferation and loss of zeta chain are observed in cancer and after trauma (Kaore et al., 2013). Thus Cit has the potential to serve as a substitute for arginase-associated T-cell dysfunction. The underlying molecular basis appears to be the capability of T cells to increase Cit membrane transport and upregulate ASS expression, and thus convert Cit to Arg in in vitro studies, thereby escaping the ill effects of Arg depletion (Bansal et al., 2004). Cit supplementation can preserve T-cell proliferation and prevent the loss of the CD3 zeta chain under conditions of low Arg.

A recent study showed that T cells can replenish intracellular Arg allowing antimycobacterial T cells to overcome arginase-mediated suppression and mediate inflammatory reaction. Further, it was also concluded that there was a contribution for Arg synthesis by T cells during mycobacterial infection, and Cit was implicated as a potential immunonutrient to modulate host immunity (Lange et al., 2017).

Therefore new drug targets for improving T-cell-associated dysfunction may be explored in the future, and further research may open novel mechanistic insights in this regard.

Citrulline in cancer

Cancer chemotherapy is a very challenging area, and resistance to anticancer drugs multiplies the concerns. Decreased ASS expression is correlated with decreased sensitivity of tumor cells, and Cit is found to increase sensitivity due to a concomitant increase in ASS expression. Combination therapy of Cit supplementation with human recombinant arginase I cobalt coupled with polyethylene glycol 5000 [i.e., HuArgI (Co)-PEG5000] is selectively cytotoxic to a fraction of human cancer cell lines in tissue culture, including some melanomas, mesotheliomas, acute myeloid leukemias, hepatocellular carcinomas, pancreas adenocarcinomas, prostate adenocarcinomas, lung adenocarcinomas, osteosarcomas, and small cell lung carcinomas. Unfortunately, subsets of normal human tissues are also found to be sensitive to human recombinant arginases with Cit supplementation, including umbilical endothelial cells, bronchial epithelium, neurons, and renal epithelial cells. Therefore these in vitro findings suggest combination therapy to be a potential therapeutic agent for some ASS-deficient tumors (Agrawal et al., 2012). In a recent study, colorectal cell migration and invasion were inhibited by arginase and Cit, which significantly restored intracellular arginine levels, reversing the effect of HuArgI(Co)-PEG5000. Further, it was concluded that colorectal cancer is partially auxotrophic to Arg and that arginine depletion may be a potential selective inhibitory approach for motility and invasion in colon cancer cells (Al-Koussa et al., 2019).

Another study further evaluated the maximum tolerated dose (MTD), pharmacodynamics, and dose-limiting toxicities (DLTs) of twice-weekly intraperitoneal HuArg I (Co)-PEG5000 in Balb/c mice along with supplemental Cit. For, HuArgI (Co)-PEG5000, MTD was 5 mg/kg twice weekly, and DLTs included weight loss and marrow necrosis, but no other organ damage or changes in blood cell counts or chemistries were observed with reduced Arg levels. Interestingly, other findings showed that Cit given once daily orally/subcutaneously reduced and delayed toxicities, whereas Cit given twice daily subcutaneously completely prevented animal toxicities. On the basis of these results, it may be hypothesized that HuArg I (Co)-PEG5000, particularly with supplemental Cit, may be an attractive therapeutic agent for ASS-deficient tumors (Mauldin et al., 2012).

Depletion of Arg generally is well tolerated and might become a targeted therapy for arginine auxotrophic cancers; but the modulatory effect of Arg on cancer cells is very complicated and remains controversial.

Citrulline in dementia of Alzheimer disease and in multiinfarct dementia

Citrulline levels are significantly elevated in the cerebrospinal fluid in MID when compared to a healthy control group (Mochizuki et al., 1996). Further, there is also elevated NOS activity in microvessels of the brain in AD, raising vascular NO production, exerting neurotoxicity, and increasing susceptibility to neuronal injury and cell death in AD (Kaore et al., 2013). Some studies contradict this notion and suggest decreased NO production plays a role in neurodegenerative disorders.

Also, ADMA that is hydrolyzed by the dimethylarginine-dimethyl aminohydrolase enzyme into Cit and dimethylamine is elevated specifically in neurons displaying cytoskeletal abnormalities and oxidative stress in the later stages of AD. Experimental studies indicate that pharmacological interventions using NO donors and/or NO suppressors could delay or at least minimize brain lesion development and further progression of brain pathology and dementia (Kaore et al., 2013).

Arg fortification with a lysine-restricted diet is one of the emerging strategies for a rare autosomal recessive disorder, pyridoxine-dependent epilepsy, which is resistant to conventional antiepileptic drugs but responsive to pharmacological dosages of pyridoxine (van Karnebeek and Jaggumantri, 2015). Arg levels were inversely associated in patients of neurodegenerative pathology, hence supporting that an Arg decrease is associated with dementia progression and AD (Fleszar et al., 2019). Observations from a study suggest that ketogenic and/or glucogenic AA such as aspartate, glutamate, valine, phenylalanine, and tyrosine might progressively decrease, while urea cycle AA such as Cit and ASA might progressively increase, along the continuum from cognitively normal individuals over subjective memory complaint and mild cognitive impairment to AD patients; these conditions are identified as the earliest stages of disease and are associated with an increased risk for AD. Hence these results suggest that several AA and their ratios might help to identify patients before phenotype conversion to AD with a good diagnostic accuracy.

Citrulline: can it improve aerobic function/energy production?

Findings suggest high-intensity exercises increase ammonia levels in skeletal muscles, which in turn activate phosphofructokinase and prevent the oxidation of pyruvate to acetyl CoA, thus leading to exhaustion and exercise-induced fatigue. Further, the urea cycle is responsible for detoxification of ammonia (Takeda et al., 2011), whereas skeletal muscle buffers ammonia via transamination reactions. Experimental studies prove that Arg, Cit, and ornithine supplementation suppress the accumulation of ammonia after exercise and prolong the time taken for exhaustion. Therefore there is a possibility of improvement in the length of treadmill time with Cit (Kaore et al., 2013). Citrulline malate may promote beneficial effects on the elimination of ammonia in the course of recovery from exhaustive muscular exercise (Sureda and Pons, 2012).

Cit is a more suitable substrate than Arg to restore NO production in microcirculation during endotoxemia (Wijnands et al., 2012). It enhances protein expression of the myofibrillar constituents (Faure et al., 2013), and protein synthesis occurs through the mammalian target of rapamycin signaling pathway (Bahri et al., 2013).

In a clinical trial, fatigue sensation was significantly reduced by Cit ingestion due to an increase in the rate of oxidative ATP (by 34%) production during exercise and an increase in the recovery rate of phosphocreatine after exercise, suggesting a larger contribution of oxidative ATP synthesis to energy production. The improvement in aerobic energy production in muscles and exercise tolerance significantly enhances aerobic performance and prolongation of onset of muscular fatigue (Sureda and Pons, 2012; Kaore et al., 2013). Cit also limits muscle fatigue or skeletal muscle dysfunction induced by bacterial endotoxins (Kaore et al., 2013). Studies show ergogenic effects on Cit supplementation with improvement of muscular contraction efficiency (Giannesini et al., 2011; Sureda and Pons, 2012) and with improvement of athletic performance in mice (Takeda et al., 2011). This may be due to facilitation of ammonia detoxification via the urea cycle and inhibition of additional glycolysis by Cit supplementation that is suggestive of improvement in athletic performance (Takeda et al., 2011). Watermelon juice given to athletes helped to reduce muscle soreness after 24 h, and Cit bioavailability was increased in vitro when it was contained in a matrix of watermelon juice and not heated (Tarazona-Díaz et al., 2013).

Another clinical trial supports improvement in athletic performance in high-intensity anaerobic exercises after Cit ingestion (8 g single dose) with short rest times and relief from postexercise muscle soreness, indicating Cit supplementation for those undergoing intensive preparation [significantly decreased muscle soreness (40%) at 24 and 48 h] (Perez-Guisado and Jakeman, 2010) and also significantly decreases the time oral L-citrulline supplementation is taken to complete a cycle ergometer exercise trial (Suzuki et al., 2016). However, another study found conflicting results where CM supplementation (single 6 g dose preworkout) does not improve the muscle recovery process following a high-intensity resistance exercise session in untrained young adult men (Silva et al., 2017). Citrulline may be used as an agent to increase exercise capacity for various reasons, but Arg does not improve exercise performance.

Citrulline: an antioxidant?

Cit acts as an antioxidant in plants and is also supported as an antioxidant by studies on healthy cyclists, sepsis, and sickle cell anemia (Waugh et al., 2001; Sureda et al., 2009), an effect due to increased NO availability and dependent on the nature of oxidative stress (Allerton et al., 2018). This notion is supported in a study involving healthy cyclists where prior Cit supplementation (6 g) reduced oxidative damage by priming polymorphonuclear neutrophils for oxidative burst without evidence of oxidative damage and a progressive decrease in ROS production and markers of oxidative damage (Sureda et al., 2009), which is likely due to increased Arg and NO availability, thus preventing oxidative damage. Therefore it is postulated that Cit, probably a potent hydroxyl radical scavenger, tends to protect DNA and metabolic enzymes against oxidative stress.

Citrulline for urea cycle disorders

Citrulline may be used to treat the inborn errors of metabolism of the urea cycle, responsible for removal of ammonia, that are fortunately quite rare. The first step in the urea cycle is the reaction of ammonia and bicarbonate using ATP, which is then catalyzed by the enzyme carbamoyl phosphate synthetase to produce carbamoyl phosphate. The second step is the reaction of carbamoyl phosphate and L-ornithine to produce L-Cit, catalyzed by the enzyme ornithine transcarbamylase.

It is noted that deficiencies of either of these two enzymes lead to low serum levels of L-Cit, which can be treated with oral Cit supplementation. Cit supplementation is recommended for patients diagnosed with a deficiency of CPS; daily recommended intake is 0.17 g/kg/day or 3.8 g/m^2/day, and for acute decompensation the recommended dose is 100–200 mg/kg/day (maximum, 6 g/day). Arg supplementation is needed for patients diagnosed with deficiency of ASS. The recommended daily intake of arginine (free base) is 0.4–0.7 g/kg/day or 8.8–15.4 g/m^2/day; for acute decompensation, Arg is given intravenously in a dose of 250 mg/kg as a bolus in 90–120 min, and then a maintenance dose of 250 mg/kg/day is used. Arg and/or Cit administration aims at maximizing ammonia excretion through the urea cycle. Cit administration would appear preferable over L-arginine because it should allow for the incorporation of one nitrogen atom into arginine and urea. However, Cit is more expensive than Arg, but with similar efficacy (National Institutes of Health, 2015; Haberle et al., 2012). In long-term management, dietary treatment is also a part of treatment that must be individualized (Kaore et al., 2013).

Citrulline in parasitic infections

Malaria is an entirely preventable and treatable mosquito-borne illness. In 2018, total estimated cases were 228 million cases worldwide. Children accounted for 67% (272,000) of all malaria deaths with 93% of malaria cases and 94% if deaths in the African region (World Health Organization, 2019).

An insight into the pathophysiology of cerebral malaria suggests endothelial dysfunction and its restoration may improve disease outcome (Sobolewski et al., 2005; Kim et al., 2011). Cit is being evaluated and explored as an adjuvant therapy for malaria. Replenishment of Arg and NO should diminish parasitized red blood cell adherence to endothelium, reducing the sequelae of these interactions, such as severe or cerebral malaria. Carbon monoxide (which shares certain functional properties with NO) protects against cerebral malaria in mice (Weinberg et al., 2008). However, extensive clinical trials are required because findings suggest that unfortunate consequences of elevated NO levels may include impaired neuronal signaling, oxidant damage, and RBC damage that may cause anemia (Sobolewski et al., 2005). A recent study found increasing dietary Cit/Arg partially protected mice from cerebral malaria, suggesting that ameliorating hypoargininemia with citrulline plus superoxide scavenging may decrease mortality in cerebral malaria cases. However, extrapolation is tricky since there are differences between humans and mice and between *Plasmodium* species infecting humans and those infecting mice.

Cit supplementation induced NO production in macrophages which promoted intracellular death of *Trypanosoma cruzi* as observed in in vitro studies. This suggests that Arg/Cit supplementation significantly enhances the trypanocidal activity of macrophages and expression of the highly active Cit–NO cycle in activated macrophages (Norris et al., 1995). This aspect has been further explored in vivo in BALB/c mice, where Arg supplementation decreased parasitemia and heart parasite burden, thus improving clinical score and survival by increased NO production in heart tissue and plasma. Also, Arg improved electrocardiograms in mice, suggesting that it modulates infection and heart function and is thus a potential biomarker of pathology and might be useful in treating *T. cruzi* infection, either alone or in combination with other antiparasitic drugs (Carbajosa et al., 2018).

It is of interest to note that physiological levels of Arg were able to induce sufficient intrinsic NO production to inhibit intracellular multiplication of a toxoplasmatic tachyzoite but unable to protect the cell from infection. Additionally, increasing the substrate levels for NO biosynthesis may enhance in vitro the toxoplasmastatic or toxoplasmacidal activity of activated macrophages (Zheng and Lin, 1998), which may enlighten researchers regarding the role of Cit supplementation as an alternative for parasitic infection. Another in vitro study (Fox et al., 2004) has shown that Arg deprivation effectively blocked the replication of intracellular *Toxoplasma gondii*. Therefore it may be hypothesized that the absence of gene activities, essential for the biosynthesis of arginine from carbamoyl phosphate, confers a selective advantage by increasing bradyzoite switching during the host response to *T. gondii* infection. This is very similar to a defense mechanism on the part of a parasite that allows host control for bradyzoite induction by trading virulence for increased transmission. This aspect stresses the need for further studies because the elimination of the cyst form can help to overcome the carrier stage at large and curb the transmission of infection. Further investigation is needed for assessing the therapeutic utility of Cit in these parasitic diseases.

Citrulline and arginine in intrauterine growth restriction amelioration

IUGR is a major health problem worldwide that lacks an effective therapeutic solution. Physiological levels of NO and polyamines, products of Arg catabolism, play important roles in implantation, embryogenesis, and uterine quiescence throughout gestation. Further, it can be stated that NO is formed from its precursor Arg by a family of NOS in the presence of O_2 and the cofactor tetrahydrobiopterin with the production of Cit. NO is found to have a role in the transfer of nutrients from mother to fetus as well as fetal growth via regulation of placental angiogenesis and uteroplacental blood flow. Further, administration of Arg ameliorated IUGR in women, whereas dietary Arg supplementation enhanced fetal survival and growth in animal studies. These findings provide support for the clinical use of Arg and Cit as therapeutic means to prevent or ameliorate fetal growth restriction in mammals (Lassala et al., 2009). Findings from a recent study too support the notion, where Cit supplementation produced a 5- and 2-fold increase in fetal plasma Cit and Arg respectively, whereas Arg supplementation only increased fetal Arg supplementation in IUGR model. This confirms that Arg/Cit supplementation enhances fetal growth. (Bourdon et al., 2020). Another study (Lassala et al., 2010) also supports the utility of parenteral Arg administration to underfed ewes in preventing fetal growth restriction and enhancing birth weight of lambs when compared with saline-infused underfed ewes, thus supporting for its clinical use to ameliorate IUGR in humans. Further, administration of Arg to underfed ewes increased ($P < 0.01$) concentrations of Arg (69%) along with some other AA, and free fatty acids (FFA) (43%) in maternal serum, decreased maternal circulating levels of ammonia (18%) and triglycerides (32%), and enhanced birth weights of lambs by 21% compared with saline-infused underfed ewes. This also lay a new framework for studying cellular and molecular mechanisms responsible for the beneficial effects of Arg in regulating conceptus growth and development.

Because oral Arg treatment is hampered by presystemic elimination (e.g., hepatic and intestinal arginase), and because Cit can be converted to Arg, oral Cit supplementation is a potential means to raise plasma Arg concentration and augment NO. All these findings support the role of NO as a major vasodilator of the placenta and that it is involved in the regulation of fetoplacental vascular reactivity, placental bed vascular resistance, trophoblast invasion and apoptosis, platelet adhesion, and aggregation in the intervillous space (Huang et al., 2012). Data seem favorable for use of Cit/Arg in the prevention of IUGR, but extensive studies for assessment of its safety and efficacy need to be performed.

Citrulline in intestinal pathology

Citrulline is easily taken up by the gut with the help of a broad set of transporters and then by enterocytes. Pharmacokinetic studies in animals also confirm that Cit supplementation is found to be more efficacious than Arg supplementation in providing Arg (Kaore et al., 2013). Studies suggest that Cit has a very good potential in total parenteral nutrition (TPN) after massive intestinal resection (SBS), causing an increase in the Arg pool and acting to restore nitrogen balance. Long-term Arg supplementation intraduodenally in rats failed to show beneficial results in intestinal ischemia and reperfusion injury (Lee et al., 2012).

Citrulline pretreatment improves the integrity of the gut barrier and helps to preserve ileum mucosa in experimental studies, thus reducing bacterial translocation; similar effects are seen with Arg as well. Cit, rather than Arg, is the limiting AA in situations of intestinal failure, which is proven in a rat model of SBS, because Cit supplementation improved protein synthesis and Arg availability (Kaore et al., 2013). This understanding of the molecular mechanisms may help in developing new strategies that could be extended to malnourished patients with compromised intestinal functions.

Citrulline: effects of amino acids on hair strength

Studies report considerable amounts of Arg deposition on, or in, hair fibers from coloring agents, whereas decreased amounts of Arg and Cit were found in damaged hairs in users of "relaxers"; decreased Cit has been associated with inflammation. One study indicated that when coloring agents were partially replaced with Arg, it decreased the oxidative change in tensile strength of the hair by preventing the undesirable attack by hydrogen peroxide on hair proteins and hair surface lipids. Therefore prospective studies need to be undertaken to understand whether or how "relaxers" induce inflammation and whether Arg or Cit can substantially reduce hair damage and fragility (Khumalo et al., 2010).

Citrulline: effect on protein synthesis

Cit is found to increase protein synthesis during refeeding in rodents with SBS, aging, and malnutrition, and improves nitrogen balance in healthy humans (Thibault et al., 2011). Cit is a conditionally essential AA in stress or when intestinal function is compromised. This notion is proven in SBS in rats, where Cit is able to restore nitrogen balance, increase mucosal protein content in the ileum, increase Arg levels, and increase muscle protein content and synthesis (190%) in elderly malnourished rats. Moreover, it also has potential as a supplement for TPN in SBS patients because it additionally prevents muscle atrophy, which is not seen with Arg supplementation (Osowska et al., 2008; Kaore et al., 2013). Cit plays a crucial role in maintaining protein homeostasis and improving muscle mass related to malnutrition, and further understanding of molecular mechanisms may help in developing new strategies for malnourished patients with compromised intestinal functions.

Concluding remarks and future directions

Citrulline (Cit) is a ubiquitous AA in mammals that is strongly related to Arg and has a scope beyond that normally associated with an AA. It has generated tremendous interest and curiosity in the scientific world due to its plethora of effects, and Arg may follow in due course. Cit has the potential to be appraised and has been proved to be an important signaling molecule apart from being used as a pharmaconutrient. Both AA have a role in the urea cycle and protein homeostasis, and further exploration is opening new horizons for Cit as well as Arg as nutraceuticals in various diseases. Practically, Cit supplementation is considered to be more efficacious than its precursor, Arg. Cit metabolism has two parts, free Cit and citrullinated proteins. The three key enzymes involved in metabolism of free Cit are NO synthase

BOX 55.1 Salient features of citrulline, major pharmacological actions, and its clinical applications.

Citrulline

Natural sources

- Watermelon (*Citrullis vulgaris*), cucumbers, pumpkins, muskmelons, bitter melons, squashes, gourds

Salient features

- A nonessential amino acid
- Donor of arginine and nitric oxide
- Conditionally essential amino acid in intestinal pathology and the critically ill

Major pharmacological actions

- Antioxidant
- Vasodilatation
- Epithelialization, angiogenesis, collagen deposition
- Decreases leukocyte migration
- Restores nitrogen balance and increases muscle protein content as well as muscle protein synthesis
- Improves endothelial dysfunction
- Increases arginosuccinate synthase expression in cancer cells
- Preserves antiinflammatory mediator response (in sepsis)

Clinical applications

- Alzheimer disease
- Multiinfarct dementia
- Urea cycle disorder
- Diabetes
- Sickle cell disease
- Hypertension
- Heart failure
- Atherosclerosis
- Intrauterine growth restriction
- Short bowel syndrome
- Sepsis
- T-cell dysfunction
- Cancer
- Erectile dysfunction
- Male infertility
- Cystic fibrosis
- Mitochondrial encephalomyopathy, lactic acidosis, and stroke-like episodes syndrome

(NOS), ornithine carbamoyl-transferase (OCT), and ASS. In summary, regarding the metabolic pathways in tissues: (1) liver Cit is locally synthesized from OCT and metabolized by ASS to urea; (2) in most tissues, Cit is recycled into Arg and NO via ASS, increasing NO availability; and (3) Cit is synthesized in the gut from glutamine and released into the circulation, with 80% being converted back into Arg in the kidneys by ASS and made available for utilization by peripheral tissues. Circulating Cit is in fact a masked form of Arg that degrades Cit in the liver (Cit is readily taken up by the liver and degraded). Cit is actually a better method of getting supplemental Arg into the blood than Arg itself, proving that Cit is a novel therapeutic agent in conditions of Arg and NO deficiencies. Cit supplementation may offer an advantageous nutritional therapeutic objective that may prove beneficial in various diseases due to its multiple effects. Cardiovascular risk may be associated with Cit supplementation, which needs rigorous evaluation (see Box 55.1).

Cit supplementation is exploited in athletics training due to its capability of reduction of oxidative stress and improvement in exercise performance. Dietary supplements are dispensed in a variety of forms, including tablets, capsules, liquids, powders, and bars to enhance exercise and athletic performance (National Institutes of Health, 2020), with some listed here. Even with proper nutritional preparation, observations are that the results of taking any dietary supplement(s) for exercise and athletic performance vary by level of training; the nature, intensity, and duration of the activity; and the environmental conditions (National Institutes of Health, 2020) (see Box 55.2).

Supplementation of Cit/Arg needs further robust studies before it could be used therapeutically in certain areas in the future, such as ED, male infertility, hypertension, sickle cell anemia, SBS, hyperlipidemia, cancer chemotherapy, IUGR, cystic fibrosis, and PH, and to increase protein content in malnourished as well as elderly patients, to increase endurance for athletics training, for acceleration of wound healing, AD, MID, immunonutrition, as an antioxidant; and various fields are being explored where it might find a place in management in the future (Box 55.3).

Cit alters the systemic response of mediators and cytokines in sepsis and acts as an immunomodulator. It has also been tried as an adjunctive therapy in parasitic infections such as toxoplasma, trypanosomiasis, and malaria. Further studies are needed to explore the pleiotropic effects of Cit supplementation in these areas.

Another area of interest regarding Cit is the detection of an antibody to citrullinated peptides that serve as predictors for diseases, thus serving as biomarkers in various diseases. Until now, Cit has been used on a large scale for diseases such as rheumatoid arthritis, intestinal pathology,

BOX 55.2 Preparations of arginine and citrulline.

1. Arginine preparations: capsule (500 mg), SR tablets (350 mg)
2. Citrulline preparations: capsule (0.5 g, 1 g), powder (100 g, 300 g)
3. Combinations: capsule, arginine 500 mg + citrulline 250 mg
   a. Tablet, arginine 500 mg + citrulline 250 mg
   b. Tablet, arginine 750 mg + citrulline 250 mg
4. Fortification with vitamins (powder): arginine 5 g + citrulline 1 g (fortified with vitamins C, D, K, folic acid, B12)
5. Infusion: contains arginine 5 g + citrulline 1 g

BOX 55.3 Therapeutic applications of citrulline.

Citrulline supplementation is exploited for benefits in a variety of conditions:

Cardiovascular conditions
Hypertension
Heart failure
Hyperlipidemia
Hematological disorders
Sickle cell anemia
Metabolic disorders
Diabetes
Urea cycle disorders
Neurodegenerative conditions
Alzheimer disease
Multiinfarct dementia
Pyridoxine-dependent epilepsy
Parasitic infections
Malaria
Trypanosoma cruzi
Toxoplasma gondii
Others
Mitochondrial encephalomyopathy, lactic acidosis, and stroke-like episodes syndrome
Wound healing
Immunonutrition
Athletic training
Erectile dysfunction
Male infertility
Sepsis (as an immunomodulator)
Arginase-associated T-cell dysfunction
Cancer chemotherapy (in certain tumors)

and SBS as biomarkers. In the future, it is likely to emerge as a biomarker for other conditions such as parenteral nutrition in HIV patients, congenital anomalies, acute rejections in transplantations, IUGR, critically ill patients, and acute kidney failure (Kaore and Kaore, 2014). Its role in toxicities is realized as a biomarker of nitrosative stress (NO/NOS) and RNS in conditions such as anticholinesterase poisoning and kainic acid-induced neurotoxicity, because oxidative stress increases RNS and ROS (when the rate of ATP demand exceeds production), causing lipid peroxidation and DNA damage.

Greater exploration in these clinical conditions is indicated in all aspects; therefore large-scale clinical trials are needed for wider acceptance of these nutraceuticals (Cit and Arg) as an established strategy.

Abbreviations

ACPA anticitrullinated protein antibodies
AD Alzheimer disease
ADMA asymmetric dimethyl arginine
ALT alanine aminotransferase
ARDS acute respiratory distress syndrome
Arg arginine
ASL arginosuccinate lyase
ASS arginosuccinate synthase
AST aspartate aminotransferase
ATP adenosine triphosphate
AUC area under the curve
BP blood pressure
BPD bronchopulmonary dysplasia
BNP brain natriuretic peptide
CD3 cluster of differentiation
cGMP cyclic guanosine monophosphate
Cit citrulline
CNS central nervous system
CPS carbamoyl phosphate synthetase
CPT cold pressor test
DNA deoxyribonucleic acid
DSHEA Dietary Supplement, Health, and Education Act
eNOS endothelial nitric oxide synthase
FOXP Forkhead box P
HDL high-density lipoprotein cholesterol

HuArgI	(Co)-PEG5000 human recombinant arginase
IgA	immunoglobulin A
IgM	immunoglobulin M
IL-10	interleukin 10
IL-6	interleukin 6
iNOS	inducible nitric oxide synthase
IR	immediate release
IUGR	intrauterine growth restriction
L-NAME	L-arginine derivative, NG-nitro-L-arginine methyl ester
MCI	mild cognitive impairment
MELAS	mitochondrial encephalomyopathy, lactic acidosis, and stroke-like episodes syndrome
MID	multiinfarct dementia
mTOR	mammalian target of rapamycin
nNOS	neuronal nitric oxide synthase
NO	nitric oxide
NOS	nitric oxide synthase
O_2	oxygen
OAT	ornithine-δ-aminotransferase
OTC	ornithine transcarbamylase
PADs	peptidyl arginine deaminases
PDE	pyridoxine-dependent epilepsy
PH	pulmonary hypertension
PP	pulse pressure
PU	pressure ulcers
RF	rheumatoid factor
RNS	reactive nitrogen species
ROS	reactive oxygen species
RVH	right ventricular hypertrophy
sGC	soluble guanylylcyclase
SBS	short bowel syndrome
SHR	spontaneously hypertensive rats
SIRT	Sirtuin 3
SMAD	an acronym from the fusion of *Caenorhabditis elegans* Sma genes and the *Drosophila* Mad
SR	sustained release
TGF-β	tumor growth factor beta
TNF-α	tumor necrosis factor alpha
TPN	total parenteral nutrition
Treg	regulatory T-Cells

References

Agrawal, V., Woo, J.H., Mauldin, J.P., 2012. Cytotoxicity of human recombinant arginase I (Co)-PEG5000 in the presence of supplemental L-citrulline is dependent on decreased arginino-succinate synthetase expression in human cells. Anticancer Drugs 23, 51–64.

Akerström, S., Mousavi-Jazi, M., Klingström, J., et al., 2005. Nitric oxide inhibits the replication cycle of severe acute respiratory syndrome coronavirus. J. Virology 79 (3), 1966–1969.

Ali, N.K., Jafri, A., Sopi, R.B., 2012. Role of arginase in impairing relaxation of lung parenchyma of hyperoxia-exposed neonatal rats. Neonatology 101, 106–115.

Al-Koussa, H., Al-Haddad, M., Abi-Habib, R., et al., 2019. Human recombinant arginase I [HuArgI (Co)-PEG5000]-induced arginine depletion inhibits colorectal cancer cell migration and invasion. Int. J. Mol. Sci. 20 (23), 6018. <https://doi.org/10.3390/ijms20236018>.

Allerton, T.D., Proctor, D.N., Stephens, J.M., et al., 2018. l-Citrulline supplementation: Impact on cardiometabolic health. Nutrients 10 (7), 921.

Alvares, T.S., Conte-Junior, C.A., Silva, J.T., 2012. Acute L-Arginine supplementation does not increase nitric oxide production in healthy subjects. Nutr. Metab. (Lond.) 9, 54.

Bahri, S., Zerrouk N., Aussel, C., et al., 2013. Citrulline: from metabolism to therapeutic use. Nutrition 29, 479–484.

Bailly-Botuha, C., Colomb, V., Thioulouse, E., 2009. Plasmacitrulline concentration reflects enterocyte mass in children with short bowel syndrome. Pediatr. Res. 65, 559–563.

Balderas-Munoz, K., Castillo-Martínez, L., Orea-Tejeda, A., et al., 2012. Improvement of ventricular function in systolic heart failure patients with oral L-citrulline supplementation. Cardiol. J. 19, 612–617.

Bansal, V., Rodriguez, P., Wu, G., 2004. Citrulline can pre-serve proliferation and prevent the loss of CD3 zeta chain under conditions of low arginine. J. Parenter. Enteral. Nutr. 28, 423–430.

Bednarz, B., Jaxa-Chamiec, T., Maciejewski, P., 2005. Efficacy and safety of oral l-arginine in acute myocardial infarction. Results of the multicenter, randomized, double-blind, placebo-controlled ARAMI pilot trial. Kardiol. Pol. 62, 421–427.

Berthe, M.C., Darquy, S., Breuillard, C., 2011. High plasma citrulline and arginine levels ensured by sustained-release citrulline supplementation in rats. Nutrition 27, 1168–1171.

Bertolini, G., Luciani, D., Biolo, G., 2007. Immunonutrition in septic patients: a philosophical view of the current situation. Clin. Nutr. 26, 25–29.

Bogdanski, P., Suliburska, J., Grabanska, K., 2012. Effect of 3-month L-arginine supplementation on insulin resistance and tumor necrosis factor activity in patients with visceral obesity. Eur. Rev. Med. Pharmacol. Sci. 16, 816–823.

Boger, R.H., Bode-Boger, S.M., 2001. The clinical pharmacology of L-arginine. Annu. Rev. Pharmacol. Toxicol. 41, 79–99.

Bourdon, A., Hannigsberg, J., Misbert, E., Tran, T.N., Amarger, V., Ferchaud-Roucher, V., et al., 2020. Maternal supplementation with citrulline or arginine during gestation impacts fetal amino acid availability in a model of intrauterine growth restriction (IUGR). Clin Nutr. S0261-5614(20)30154-0.

Carbajosa, S., Rodríguez-Angulo, H.O., Gea, S., et al., 2018. L-arginine supplementation reduces mortality and improves disease outcome in mice infected with *Trypanosoma cruzi*. PLoS Negl. Trop. Dis. 12 (1), e0006179.

Cardillo, C., Panza, J.A., 1998. Impaired endothelial regulation of vascular tone in patients with systemic arterial hypertension. Vasc. Med. 3, 139–144.

Centers for Disease Control, 2014. Centers for disease control (Internet), Rheumatoid Arthritis, Centers for Disease Control and Prevention website. Updated 2014 November. Available from: <http://www.cdc.gov/arthritis/basics/rheumatoid.htm> (Accessed 2 March 2015).

Cereda, E., Klersy, C., Serioli, M., 2015. A nutritional formula enriched with arginine, zinc, and antioxidants for the healing of pressure ulcers: a randomized trial. Ann. Intern. Med. 162, 167–174.

Chambers, D.C., Ayres, J.G., 2001. Effect of nebulised L- and D-arginine on exhaled nitric oxide in steroid naive asthma. Thorax 56, 602–606.

Chauhan, B., Kumar, G., Kalam, N., 2013. Current concepts and prospects of herbal nutraceuticals: a review. J. Adv. Pharm. Technol. Res. 4, 4–8.

Collins, J.K., Wu, G., Perkins-Veazie, P., Spears, K., 2007. Watermelon consumption increases plasma arginine concentrations in adults. Nutrition 23, 261–266.

Conrad, K., Roggenbuck, D., Reinhold, D., 2010. Profiling of rheumatoid arthritis associated autoantibodies. Autoimmun. Rev. 9, 431–435.

Cormio, L., De Siati, M., Lorusso, F., 2011. Oral L-Citrulline supplementation improves erection hardness in men with mild erectile dysfunction. Urology 77, 119–122.

Curis, E., Nicolis, I., Moinard, C., 2005. Almost all about citrulline in mammals. Amino Acids 29, 177–205.

Cynober, L., 2013. Citrulline: just a biomarker or a conditionally essential amino acid and a pharmaconutrient in critically ill patients? Crit. Care 17, 122.

Cynober, L., Moinard, C., De Bandt, J.P., 2010. The 2009ESPEN Sir David Cuthbertson. Citrulline: a new major signaling molecule or just another player in the pharmaconutrition game? Clin. Nutr. 29, 545–551.

David, K., et al., 2019. The effects of oral l-arginine and l-citrulline supplementation on blood pressure. Nutrients 11, 1679.

De Luis, D.A., Izaola, O., Cuellar, L., 2009. High dose of arginine enhanced enteral nutrition in postsurgical head and neck cancer patients. A randomized clinical trial. Eur. Rev. Med. Pharmacol. Sci. 13, 279–283.

Debats, I.B., Wolfs, T.G., Gotoh, T., 2009. Role of arginine in superficial wound healing in man. Nitric Oxide 21, 175–183.

Dong, J.Y., Qin, L.Q., Zhang, Z., 2011. Effect of oral L-arginine supplementation on blood pressure: a meta-analysis of randomized, double-blind, placebo-controlled trials. Am. Heart J. 162, 959–965.

Dorniak-Wall, T., Grivell, R.M., Dekker, G.A., 2014. The role of L-arginine in the prevention and treatment of pre-eclampsia: a systematic review of randomised trials. J. Hum. Hypertens. 28, 230–235.

Drover, J.W., Dhaliwal, R., Weitzel, L., 2011. Perioperative use of arginine-supplemented diets: a systematic review of the evidence. J. Am. Coll. Surg. 212, 385–399.

Duskin, A., Eisenberg, R.A., 2010. The role of antibodies in inflammatory arthritis. Immunol. Rev. 233, 112–125.

El-Hattab, A.W., Emrick, L.T., Craigen, W.J., 2012. Citrulline and arginine utility in treating nitric oxide deficiency in mitochondrial disorders. Mol. Genet. Metab. 107, 247–252.

El-Hattab, A.W., Emrick, L.T., Chanprasert, S., 2014. Mitochondria: role of citrulline and arginine supplementation in MELAS syndrome. Int. J. Biochem. Cell Biol. 48, 85–91.

El-Hattab, A.W., Emrick, L.T., Hsu, J.W., et al., 2016. Impaired nitric oxide production in children with MELAS syndrome and the effect of arginine and citrulline supplementation. Mol. Genet. Metab. 117 (4), 407–412.

El-Kirsh, A.A., Abd El-Wahab, H.M., Abd-Ellah Sayed, H.F., 2011. The effect of L-arginine or L-citrulline supplementation on biochemical parameters and the vascular aortic wall in high-fat and high-cholesterol-fed rats. Cell Biochem. Funct. 29, 414–428.

Endo, F., Matsuura, T., Yanagita, K., 2004. Clinical manifestations of inborn errors of the urea cycle and related metabolic disorders during childhood. J. Nutr. 134, 1605S–1609S.

Faure, C., Morio, B., Chafey, P., 2013. Citrulline enhancesmyofibrillar constituents expression of skeletal muscle and induces a switch in muscle energy metabolism in malnourished aged rats. Proteomics 13, 2191–2201.

Figueroa, A., Trivino, J.A., Sanchez-Gonzalez, M.A., 2010. Oral L-citrulline supplementation attenuates blood pressure response to cold pressor test in young men. Am. J. Hypertens. 23, 12–16.

Figueroa, A., Wong, A., Hooshmand, S., 2013. Effects of watermelon supplementation on arterial stiffness and wave reflection amplitude in postmenopausal women. Menopause 20, 573–577.

Flam, B.R., Eichler, D.C., Solomonson, L.P., 2007. Endothelialnitric oxide production is tightly coupled to the Citrulline-NO cycle. Nitric Oxide 17, 115–121.

Fleszar, M.G., Wiśniewski, J., Zboch, M., et al., 2019. Targeted metabolomic analysis of nitric oxide/L-arginine pathway metabolites in dementia: association with pathology, severity, and structural brain changes. Sci. Rep. 9 (1), 13764.

Fox, B.A., Gigley, J.P., Bzik, D.J., 2004. *Toxoplasma gondii* lacks the enzymes required for de novo arginine biosynthesis and arginine starvation triggers cyst formation. Int. J. Parasitol. 34, 323–331.

Gad, M.Z., 2010. Anti-aging effects of l-arginine. J. Adv. Res. 1, 169–177.

Gerard, J.M., Luisiri, A., 1997. A fatal overdose of arginine hydrochloride. J. Toxicol. Clin. Toxicol. 35, 621–625.

Giannesini, B., Le Fur, Y., Cozzone, P.J., 2011. Citrulline malate supplementation increases muscle efficiency in rat skeletal muscle. Eur. J. Pharmacol. 667, 100–104.

Grasemann, H., Tullis, E., Ratjen, F., 2013. A randomized controlled trial of inhaled L-arginine in patients with cystic fibrosis. J. Cyst. Fibros. 12, 468–474.

Grimble, K.G., 2007. Adverse gastrointestinal effects of arginine and related amino acids. J. Nutr. 137, 693S–1701S.

Guo, R.Y., Cao, Q.-D., Hong, Z.-S., et al., 2020. The origin, transmission and clinical therapies on coronavirus disease 2019 (COVID-19) outbreak – an update on the status. Mil. Med. Res. 7 (11), 1–10.

Gupta, R.C., 2004. Brain regional heterogeneity and toxicological mechanisms of organophosphates and carbamates. Toxicol. Mech. Methods 14, 103–143.

Haberle, J., Boddaert, N., Burlina, A., 2012. Suggested guidelines for the diagnosis and management of urea cycle disorders. Orphanet. J. Rare Dis. 7, 32–62.

Hecker, M., Mitchell, J.A., Harris, H.J., 1990. Endothelial cells metabolize NG-monomethyl-L-arginine to L-citrulline and subsequently to L-arginine. Biochem. Biophys. Res. Commun. 167, 1037–1043.

Herrera-Esparza, R., Rodríguez-Rodríguez, M., Pérez-Pérez, M.E., et al., 2013. Posttranslational protein modification in the salivary glands of Sjögren's syndrome patients. Autoimmune Dis. 2013, 548064.

Hoang, H.H., Padgham, S.V., Meininger, C.J., 2013. L-Arginine, tetrahydrobiopterin, nitric oxide and diabetes. Curr. Opin. Clin. Nutr. Metab. Care 16, 76–82.

Huang, L.T., Hsieh, C.S., Chang, K.A., 2012. Roles of nitric oxide and asymmetric dimethylarginine in pregnancy and fetal programming. Int. J. Mol. Sci. 13, 14606–14622.

Kameda, N., Okigawa, T., Kimura, T., 2011. The effect of L-citrulline ingestion on ECG QT interval and autonomic nervous system activity. J. Physiol. Anthropol. 30, 41–45.

Kaore, S.N., Amane, H.S., Kaore, N.M., 2013. Citrulline: pharmacological perspectives and its role as an emerging biomarker in future. Fundam. Clin. Pharmacol. 27, 35–50.

Kaore, S.N., Kaore, N.M., 2014. Citrulline: pharmacological perspectives and role as a biomarker in diseases and toxicities. In: Gupta, R.C. (Ed.), Biomarkers in Toxicology. Academic Press/Elsevier, Amsterdam, pp. 883–905.

Kawasaki, S., Miyake, C., Kohchi, T., 2000. Responses of wild watermelon to drought stress: accumulation of an ArgE homologueand citrulline in leaves during water deficits. Plant Cell Physiol. 41, 864–873.

Keyaerts, E., Vijgena, L., Chenb, L., et al., 2004. Inhibition of SARS-coronavirus infection in vitro by S-nitroso-N-acetylpenicillamine, a nitric oxide donor compound. Int. J. Infect. Dis. 8, 223–226.

Khumalo, N.P., Stone, J., Gumedze, F., 2010. 'Relaxers' damage hair: evidence from amino acid analysis. J. Am. Acad. Dermatol. 62, 402–408.

Kim, H., Higgins, S., Liles, W.C., 2011. Endothelial activation and dysregulation in malaria: a potential target for novel therapeutics. Curr. Opin. Hematol. 18, 177–185.

Koolwal, A., Manohar, J.S., Sathyanarayana Rao, T.S., Koolwal, D.G., 2019. L-arginine and erectile dysfunction. J. Psychosexual Health 1 (1), 37–43.

Kotze, V., 2011. Perioeperative nutrition: what do we know? S. Afr. J. Clin. Nutr. 24, S19–S22.

Kuhn, K.P., Harris, P.A., Cunningham, G.R., 2002. Oral citrulline effectively elevates plasma arginine levels for 24 hours in normal volunteers. Circulation 106, II1–766S.

Lange, S.M., McKell, M.C., Schmidt, S.M., et al., 2017. l-Citrulline metabolism in mice augments CD4+ T cell proliferation and cytokine production in vitro, and accumulation in the Mycobacteria-infected lung. Front. Immunol. 8, 1561.

Lassala, A., Bazer, F.W., Cudd, T.A., 2009. Intravenous administration of L-citrulline to pregnant ewes is more effective than L-arginine for increasing arginine availability in the fetus. J. Nutr. 139, 660–665.

Lassala, A., Bažer, F.W., Cudd, A.T., 2010. Parenteral administration of l-Arginine prevents fetal growth restriction in undernourished ewes. J. Nutr. 140, 1242–1248.

Lauterbach, R., Pawlik, D., Lauterbach, J.P. 2018. L-citrulline supplementation in the treatment of pulmonary hypertension associated with bronchopulmonary dysplasia in preterm infant: A case report. SAGE Open Med Case Report. Available from: https://doi.org/10.1177/2050313X18778730.

Lee, T.J., Yu, J.G., 2002. L-Citrulline recycle for synthesis of NO in cerebral perivascular nerves and endothelial cells. Ann. N.Y. Acad. Sci. 962, 73–80.

Lee, C.H., Hsiao, C.C., Hung, C.Y., 2012. Long-term enteral arginine supplementation in rats with intestinal ischemia and reperfusion. J. Surg. Res. 175, 67–75.

Lee, Y.C., Su, Y.T., Liu, T.Y., et al., 2018. L-Arginine and L-citrulline supplementation have different programming effect on regulatory T-cells function of infantile rats. Front. Immunol. 9, 2911.

Liu, Y., Fu, X., Gou, L., 2013. L-citrulline protects against glycerol-induced acute renal failure in rats. Ren. Fail. 35, 367–373.

Luiking, Y.C., Deutz, N.E., 2007. Exogenous arginine in sepsis. Crit. Care Med. 35, S557–S563.

Luiking, Y.C., Poeze, M., Dejong, C.H., 2004. Sepsis: an arginine deficiency state? Crit. Care Med. 32, 2135–2145.

Luiking, Y.C., Poeze, M., Ramsay, G., 2009. Reduced citrulline production in sepsis is related to diminished de novo arginine and nitric oxide production. Am. J. Clin. Nutr. 89, 142–152.

Luiking, Y.C., Engelen, M.P., Deutz, N.E., 2010. Regulation of nitric oxide production in health and disease. Curr. Opin. Clin. Nutr. Metab. Care 13, 97–104.

Ma, X., Zheng, C., Hu, Y., 2015. Dietary L-arginine supplementation affects the skeletal longissimus muscle proteome in finishing pigs. PLoS One 10, e0117294.

Marx, S., Vedernikov, Y., Saade, G.R., Garfield, R.E., 2000. Citrulline does not relax isolated rat and rabbit vessels. Br. J. Pharmacol. 130, 713–716.

Mauldin, P.J., Zeinali, I., Kleypas, K., 2012. Recombinant human arginase toxicity in mice is reduced by citrulline supplementation. Transl. Oncol. 5, 26–31.

Meldrum, D.R., Gambone, J.C., Morris, M.A., 2011. The link between erectile and cardiovascular health: the canary in the coal mine. Am. J. Cardiol. 108, 599–606.

Milatovic, D., Gupta, R.C., Dekundy, A., 2005. Carbofuran-induced oxidative stress in slow and fast skeletal muscles: prevention by memantine and atropine. Toxicology 208, 13–24.

Mochizuki, Y., Oishi, M., Hara, M., 1996. Amino acid concentration in dementia of the Alzheimer type and multi-infarct dementia. Ann. Clin. Lab. Sci. 26, 275278.

Moinard, C., Cynober, L., 2007. Citrulline: a new player in the control of nitrogen homeostasis. J. Nutr. 137, 1621S–1625S.

Moinard, C., Nicolis, I., Neveux, N., 2007. Dose-ranging effects of citrulline administration on plasma amino acids and hormonal patterns in healthy subjects: the citru dose pharmacokinetic study. Br. J. Nutr. 99, 855–862.

National Institutes of Health, 2015. Urea cycle disorder consortium on urea cycle disorders treatment guidelines, Rare Diseases Clinical Research Network. Available from: <http://www.rarediseasesnetwork.org/ucdc/physicians/guidelines-main.htm> (Accessed 9 October 2015).

National Institutes of Health, 2020. U.S. Department of Health and Human Services. Fact sheet for health professionals: dietary supplements for exercise and athletic performance. Updated: October 17, 2019. Available from: <https://ods.od.nih.gov/factsheets/ExerciseAndAthleticPerformance-HealthProfessional/>. (Accessed 13 April 2020).

Norris, K.A., Schrimpf, J.E., Flynn, J.L., 1995. Enhancement of macrophage microbicidal activity: supplemental arginine and citrulline augment nitric oxide production inmurine peritoneal macrophages and promote intracellular killing of *Trypanosoma cruzi*. Infect. Immun. 63, 2793–2796.

Oh, H.S., Oh, S.K., Lee, J.S., et al., 2017. Effects of l-arginine on growth hormone and insulin-like growth factor 1. Food Sci. Biotechnol. 26 (6), 1749–1754.

Oliverius, M., Kudla, M., Baláz, P., et al., 2010. Plasma citrulline concentration. À reliable noninvasive marker of functional enterocyte mass. Cas. Lek. Cesk. 149, 160–162.

Orozco-Gutiérrez, J.J., Castillo-Martínez, L., Orea-Tejeda, A., et al., 2010. Effect of L-arginine or L-citrulline oral supplementation on blood pressure and right ventricular function in heart failure patients with preserved ejection fraction. Cardiol. J. 17, 612–618.

Osowska, S., Neveux, N., Nakib, S., 2008. Impairment of arginine metabolism in rats after massive intestinal resection: effect of parenteral nutrition supplemented with citrulline compared with arginine. Clin. Sci. (Lond.) 115, 159–166.

Papadia, C., Sabatino, A.D., Corazza, G.R., 2014. Diagnosing small bowel malabsorption: a review. Intern. Emerg. Med. 9, 3–8.

Perez-Guisado, J., Jakeman, P.M., 2010. Citrulline malate enhances athletic anaerobic performance and relieves muscle soreness. J. Strength Cond. Res. 24, 1215–1222.

Rassaf, T., Preik, M., Kleinbongard, P., 2002. Evidence for in vivo transport of bioactive nitric oxide in human plasma. J. Clin. Invest. 109, 1241–1248.

Rimando, A.M., Perkins-Veazie, P.M., 2005. Determination of citrulline in watermelon rind. J. Chromatogr. A. 1078, 196–200.

Rodera, P.C., De Luis, D.A., Candela, C.G., 2012. Immuno enhanced enternal nutrition formulas in head and neck cancer surgery; a systematic review. Nur. Hosp. 27, 681–690.

Romero, M., Platt, D., Caldwell, R., 2006. Therapeutic use of citrulline in cardiovascular disease. Cardiovasc. Drug Rev. 24, 275–290.

Romero, M.J., Platt, D.H., Tawfik, H.E., 2008. Diabetes induced coronary vascular dysfunction involves increased arginase activity. Circ. Res. 102, 95–102.

Romero, M.J., Yao, L., Sridhar, S., et al., 2013. L-Citrulline protects from kidney damage in Type 1 diabetic mice. Front. Immunol. 4, 480.

Schnog, J.J., Jager, E.H., van der Dijs, F.P., et al., 2004. Evidence for a metabolic shift of arginine metabolism in sickle cell disease. Ann. Hematol. 83, 371–375.

Schwedhelm, E., Maas, R., Freese, R., 2008. Pharmacokinetic andpharmacodynamic propertiesof oral citrulline and arginine: impact on nitric oxide metabolism. Br. J. Clin. Pharmacol. 65, 51–59.

Shaalan, A.A.M., Khalil, M.S., 2009. Histological and immunohistochemical study of the postulated protective effect of L-arginine on glycerol induced acute renal failure in adult male albino rats. Egypt. J. Histol. 32, 333–345.

Silva, D.K., Jacinto, J.L., de Andrade, W.B., et al., 2017. Citrulline malate does not improve muscle recovery after resistance exercise in untrained young adult men. Nutrients 9 (10), 1132.

Smith, S.E., Man, C.M., Yip, P.K., 1996. Anticonvulsant effects of 7-nitroindazole in rodents with reflex epilepsy mayresult from L-arginine accumulation or a reduction in nitric oxide or L-citrulline formation. Br. J. Pharmacol. 119, 165–173.

Sobolewski, P., Gramaglia, I., Frangos, J., Intaglietta, M., van der Heyde, H.C., 2005. Nitric oxide bioavailability in malaria. Trends Parasitol. 21, 415–422.

Sopi, R.B., Zaidi, S.I., Mladenov, M., 2012. L-Citrulline supplementation reverses the impaired airway relaxation in neonatal rats exposed to hyperoxia. Respir. Res. 13, 68.

Stanislavov, R., Rohdewald, P., 2014. Sperm quality in men is improved by supplementation with a combination of L-arginine, L-citrullin, roburins and Pycnogenol®. Minerva Urol. Nefrol. 66, 217–223.

Stechmiller, J.K., Childress, B., Cowan, L., 2005. Arginine supplementation and wound healing. Nutr. Clin. Pract. 20, 52–61.

Sureda, A., Pons, A., 2012. Arginine and citrulline supplementation in sports and exercise: ergogenic nutrients? Med. Sports Sci. 59, 18–28.

Sureda, A., Cordova, A., Ferrer, M.D., 2009. Effects of L-citrulline oral supplementation on polymorphonuclear neutrophils oxidative burst and nitric oxide production after exercise. Free Radic. Res. 43, 828–835.

Suzuki, T., Morita, M., Kobayashi, Y., et al., 2016. Oral L-citrulline supplementation enhances cycling time trial performance in healthy trained men: double-blind randomized placebo-controlled 2-way crossover study. J. Int. Soc. Sports Nutr. 13, 6.

Tain, Y.L., Hsieh, C.S., Lin, I.C., 2010. Effects of maternal L-citrulline supplementation on renal function and blood pressure in offspring exposed to maternal caloric restriction: the impact of nitric oxide pathway. Nitric Oxide 23 (1), 34–41.

Tain, Y.L., Lee, C.T., Huang, L.T., 2014. Long-term effects of maternal citrulline supplementation on renal transcriptome prevention of nitric oxide depletion-related programmed hypertension: the impact of gene-nutrient interactions. Int. J. Mol. Sci. 15 (12), 23255–23268.

Takeda, K., Machida, M., Kohara, A., Omi, N., Takemasa, T., 2011. Effects of citrulline supplementation on fatigue and exercise performance in mice. J. Nutr. Sci. Vitaminol. 57 (3), 246–250.

Tankersley, R.W., 1964. Amino acid requirements of herpes simplex virus in human cells. J. Bacteriol. 87, 609–613.

Tarazona-Díaz, M.P., Alacid, F., Carrasco, M., 2013. Watermelon juice: potential functional drink for sore muscle relief in athletes. J. Agric. Food Chem. 61 (31), 7522–7528.

Thibault, R., Flet, L., Vavasseur, F., 2011. Oral citrulline does not affect whole body protein metabolism in healthy human volunteers: results of a prospective, randomized, double-blind, cross-over study. Clin. Nutr. 30 (6), 807–811.

Tizianello, A., De Ferrari, G., Garibotto, G., Gurreri, G., Robaudo, C., 1980. Renal metabolism of amino acids and ammonia in subjects with normal renal function and in patients with chronic renal insufficiency. J. Clin. Invest. 65 (5), 1162–1173.

Tsuboi, T., Maeda, M., Hayashi, T., 2018. Administration of L-arginine plus L-citrulline or L-citrulline alone successfully retarded endothelial senescence. PLoS One 13 (2), e0192252. eCollection 2018.

Urschel, K.L., Shoveller, A.K., Uwiera, R.R., 2006. Citrulline is an effective arginine precursor in enterally fed neonatal piglets. J. Nutr. 136, 1806–1813.

Vadivel, A., Aschner, J.L., Rey-Parra, J.G., et al., 2010. L-citrulline attenuates arrested alveolar growth and pulmonary hypertension in oxygen-induced lung injury in newborn rats. Pediatr. Res. 68, 519–525.

van Karnebeek, C.D., Jaggumantri, S., 2015. Current treatment and management of pyridoxine-dependent epilepsy. Curr. Treat. Options Neurol. 17, 335.

Wada, M, Über Citrullin, 1930. Über Citrullin, eine neue Aminosäure im presssaft der Wassermelone, *Citrullus vulgaris* schrad. Biochem. Zeit. 224, 420–429.

Waugh, W.H., Daeschner, C.W., Files, B.A., 2001. Oral citrulline as arginine precursor may bebeneficial in sickle cell disease: early phase two results. J. Natl. Med. Assoc. 93, 363–371.

Weinberg, J.B., Lopansri, B.K., Mwaikambo, E., 2008. Arginine, nitric oxide, carbon monoxide, and endothelial function in severe malaria. Curr. Opin. Infect. Dis. 21, 468–475.

WHO, 2020. Situation report 78 on COVID-19. Available from: <https://www.who.int/docs/default-source/coronaviruse/situation-reports/20200407-sitrep-78-covid-19.pdf?sfvrsn = bc43e1b_2> (Accessed 9 April 2020).

Wijnands, K.A., Vink, H., Briedé, J.J., 2012. Citrulline a more suitable substrate than arginine to restore NO production and the microcirculation during endotoxemia. PLoS One 7, e37439.

World Health Organization, 2019. Factsheets on malaria: key facts. World Health organization, Geneva, Switzerland. Available from: <https://www.who.int/en/news-room/fact-sheets/detail/malaria> (Accessed 13 April 2020).

Wu, G., Bazer, F.W., Cudd, T.A., Jobgen, W.S., 2007a. Pharmacokinetics and safety of arginine supplementation in animals. J. Nutr. 137, 1673S–1680S.

Wu, G., Collins, J.K., Perkins-Veazie, P., 2007b. Dietary supplementation with watermelon pomace juice enhances arginine availability and ameliorates the metabolic syndrome in Zucker diabetic fatty rats. J. Nutr. 137, 2680–2685.

Yokota, A., Kawasaki, S., Iwano, M., 2002. Citrulline andDRIP-1 protein (ArgE homologue) in drought tolerance of wild watermelon. Ann. Bot. 89, 825–832.

Yu, J.G., O'Brien, W.E., Lee, T.J., 1997. Morphologic evidence for L-citrulline conversion to L-arginine via the argininosuccinate pathway in porcine cerebral perivascular nerves. J. Cereb. Blood Flow Metab. 17, 84–93.

Zeisel, S.H., 1999. Regulation of nutraceuticals. Science 285, 1853–1855.

Zheng, C., Lin, J., 1998. The role of L-arginine and L-citrulline inactivated macrophage against Toxoplasma gondii infection in vitro. Zhongguo Ji Sheng Chong Xue Yu Ji Sheng Chong Bing. Za Zhi 16, 326–330.

Chapter 56

Neem Extract

Anu Rahal, Dinesh Kumar and Jitendra K. Malik

Introduction and background

Neem is an evergreen tree found in seasonally dry, tropical woodlands such as in northeast India, parts of Asia, the northern part of Nigeria, Australia, Africa, and Central and South America. Till date, two species of neem have been reported: *Azadirachta indica* A. Juss, a native of the Indian subcontinent and *Azadirachta excels* Kack, indigenous to Indonesia and the Philippines. It is a fast-growing tree, reaching up to 30 m, with leafy spreading branches, and it adapts well to all climatic, geographical, and topographic conditions, including saline, alkaline, and other wastelands. It remains green throughout the year and bears white flowers that smell like honey at an age of 2–3 years and fruits 1–2 years later. The ripe fruit is oval, approximately 2 cm long, with a light-colored seed approximately 1.5 cm long. It grows in boulevard, patterned, and agroforestry plantations or as shade trees along roadsides; it has held high esteem because of its medicinal and insecticidal properties in the Ayurvedic and Unani systems of medicine.

Neem is a renewable resource of various useful domestic and agricultural products, including cattle feed, medicines, soap industry, nitrification inhibition, slow nutrient-release manure, fuel, energy, pest control, and others. All parts of the tree, including the leaf, flower, fruit, seed, kernel, bark, wood, and twigs, are biologically active, and the maximum activity is associated with the seed kernel. As per National Academy of Science (1992), India produces 442,300 tonnes of seeds and 88,400 tonnes of neem oil annually. Several international publications, such as *Neem: A Tree for Solving Global Problems* published by the (US) National Academy of Sciences in 1992, *Neem: Research and Development* in 1993, and *The Neem Tree* in 1995, projected the prospects for this wonder tree. A large number of neem-based pest control and healthcare products are already available in the international market.

Several of these are now registered in the United States and other countries for diverse applications in health and agriculture. Because of its features of diverse activity and relative safety to nontarget organisms, it is now recognized as the single most important source of biopesticides and allied medicinal products.

Phytoconstituents

Neem has a rich repository of more than 300 primary and secondary metabolites (Table 56.1). The primary compounds include fat or carbohydrate derivatives and protein molecules, whereas the secondary metabolites include bioactive molecules such as flavonoids, steroid, saponins, and alkaloids (Fig. 56.1).

The bioactive phytoconstituents of neem are broadly classified into two major groups: isoprenoids and nonisoprenoids. The tetranortriterpenoids form the main active ingredient and are commonly referred to as C-seco meliacins or limonoids. The bioactivity-related research in this plant has mainly focused on azadirachtin because of its abundance and unique mode of action. The azadirachtinoids are a mixture of 12 closely related isomeric meliacins (azadirachtin A–G), which constitute 0.3%–0.6% of seed kernel. Of these, azadirachtin E is relatively more effective. Other compounds that have biological activity include a number of salannin, nimbin, nimbinin, nimbidin, 6-desacetylnimbin, phenolic compounds, carotenoids, steroids, ketones, and others. The main trunk bark is a rich source of many tannins, with important ones including tricyclic diterpenoids, NB-II peptidoglycan, gallic acid, gallocatechin, epigallocatechin, catechin, epicatechin, margolone, margolonone, and isomargolonone.

In addition, the pollen of neem contains several amino acids including glutamic acid, phenylalanine, tyrosine, arginine, methionine, histidine, isoleucine and aminocaprylic acid. Its gum, on hydrolysis, yields L-fucose, L-arabinose, D-galactose, and D-glucuronic acid. The older trunks exude a sap rich in amino acids (alanine, arginine, aminobutyric acid, asparagines, aspartic acid, proline, glycine, norvaline, etc.), free sugars (glucose, mannose, fructose, and xylose), and organic acids (malonic, citric,

Nutraceuticals. DOI: https://doi.org/10.1016/B978-0-12-821038-3.00056-2

TABLE 56.1 Some pharmacologically active molecules present in neem (Parida et al., 2002; Badam et al., 1999).

Active principle	Part of the plant	Pharmacological activity
Azadirachtin	Seed	Inhibited Dengue virus type-2 replication group B Coxsackieviruses potato virus X
Nimbin	Seed oil	Spermicidal, antiviral, antifungal infections
Nimbidin	Seed oil	Antiinflammatory, antiarthritic, antipyretic, spermicidal, antigastric ulcer, antibacterial hypoglycemic, diuretic
Nimbolide	Seed oil	Antibacterial, antimalarial
Sodium nimbate		Antiinflammatory
Mahmoodin	Seed oil	Antibacterial
Margolone, margolonone and isomargolonone	Bark	Antibacterial
Cyclic trisulfide and cyclic tetrasulfide	Leaf	Antifungal
Gedunin	Neem oil	Antimalarial
Gallic acid, epicatechin and catechin	Bark	Immunomodulatory, antiinflammatory
Meliacin and meliacinol	Seed	Insecticidal
Meliatetraolenone	Fruit coat	Insecticidal
Sesquiterpene	Fruit coat	Insecticidal
Polysaccharides GIa, GIb	Bark	Antitumor
Polysaccharides GIIa, GIIIa	Bark	Antiinflammatory
NB-II peptidoglycan	Bark	Immunomodulatory

succinic, and fumaric), and can treat general weakness and skin diseases.

Extraction procedures

Neem has a rich content of bitter active compounds that needs to be removed to make its consumption as a nutraceutical feasible. Different extraction procedures can be adopted depending on the use of ethanol, methanol, methyl ethyl ketone, methyl *tert*-butyl ether, or water, or their azeotropic mixtures as solvents. Different liquid chromatographic modifications such as column chromatography, preparative thin-layer chromatography, vacuum liquid chromatography, and reverse-phase medium- or high-pressure preparative liquid chromatography have been described in the literature for isolating major azadirachtins and other bioactive meliacins on an industrial scale. The choices of solvent and method depend on the compound to be concentrated.

Technologies for the preparation of azadirachtin concentrates of varying strengths have been developed over time for both household and industrial uses. The indigenous methods include cold maceration, steam distillation, aqueous extraction, and organic solvent extraction. Maceration is a cold extraction process in which defatted kernels are soaked in methanol with intermittent stirring for 3 days, followed by filtration and evaporation of the solvent under vacuum to obtain the dried crude extract. In batch stirring extraction, a magnetic pellet is dropped in the flask and the mixture is stirred for 8 h on a magnetic stirrer plate. Finally, the solvent is evaporated in vacuo to obtain crude extract.

The aqueous extracts are as effective as other extracts and can be prepared with minimal facilities. Powdered seed, kernels, or leaves are soaked overnight and strained, and the cold extract can easily be prepared in the household. Alternatively, a cloth bag containing the powder can be placed in a bucket and steeped overnight. The second

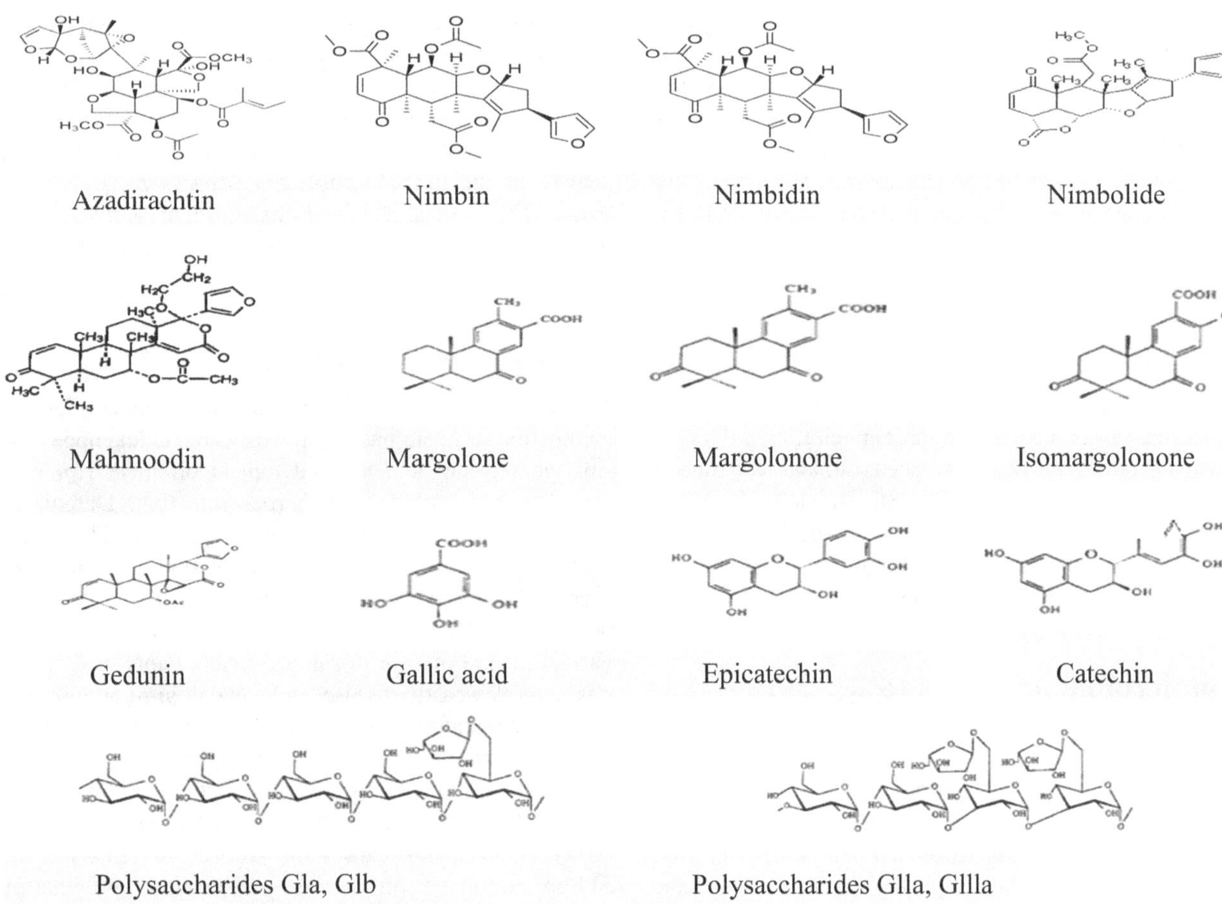

FIGURE 56.1 Structure of bioactive principles in neem.

method gives a more concentrated extract, which is generally rich in water-soluble amino acids, bitters, and carbohydrates. Azadirachtin and other limonoids have better solubility in alcohol as compared to water. Therefore alcoholic extracts are generally 50 times as concentrated as aqueous extracts and may result in 0.2%–6% active ingredients. The procedure for the alcoholic extraction can be similar to aqueous extraction. Organic solvents like dimethyl sulfoxide and ethers are used to procure highly concentrated extracts. Even better concentration of neem oil may be obtained using carbon dioxide supercritical flow at high temperature and high pressure.

The indigenous method of neem oil extraction involves placing smashed seed kernels in a vessel of water and collecting the oil that floats on top of the water, however this results in compromised quality and quantity yield. Industrial extraction is done by steam distillation with petroleum ether (boiling point 60°C–80°C). The oil is removed from filtrate by using regular distillation at 70°C, and the final traces of solvent are removed by placing it in a rotary vacuum evaporator for 20 h at 60°C–70°C. Oil extraction can also be done using *n*-hexane and ethanol as solvent. The oil quality is affected by the temperature, hydrolysis, oxidation, lipase enzyme, and presence of undesirable compounds in the oil.

Pharmacological activities and therapeutic uses

Different active principles of neem exert diverse remedial effects on several aspects of life. The leaves are a bitter tonic that reduces inflammation and clears contaminants, while promoting health and improving all bodily functions (Rahal et al., 2019). A variety of preparations made from different parts of neem exhibit antimicrobial (antibacterial, antiviral, and antifungal), antimalarial, antioxidant, antimutagenic, anticarcinogenic, antiinflammatory, contraceptive, and antiulcer activities. The leaves are useful for potentiating the immune system of the body, reducing the incidence of malarial fever, and treating neuromuscular pain, foot fungi and termites. Acetone extract of the leaves has a significant central nervous sytem depressant activity (Shah et al., 2009).

Dried leaves are effective in slowing down fungi growth on smoked dried fish *Scoliodon nigrodigitatus*

(Dodia et al., 2020). Neem bark and roots are excellent bioinsecticides for the control of ectoparasites including fleas and ticks, as well as for treating skin infections such as acne, scabies, psoriasis, and eczema (Abdel-Ghaffar et al., 2012). Other conditions indicated for its use include tetanus, otitis, abscess in the ear, orchitis, swelling of the kidney, stoppage of urination, metritis, mastitis, rinderpest, and rheumatism. The hormone-mimicking action of neem extracts causes interference with the parasitic life cycle, thereby inhibiting their nutrition intake as well as the hatching of eggs (Kusamran et al., 1998; Mahmoudi et al., 2002; Kumar and Navaratnam, 2013; Luong et al., 2012). Neem leaves are also effective in treating snake poisons and insect bites due to their anticlotting phytoconstituents. Products derived from seed, oil, and leaf such as eczema cream, antiseptic cream, hair shampoo, hair oils, toothpaste, neem leaves tea, vegetarian capsules, powders, soaps, insect repellents (spray and lotion), and candles have unique qualities for daily use.

Antimicrobial activity

Antibacterial activity

In folklore medicine, hot water extract of neem flower and leaf is used internally as an antihysteric remedy and externally to treat wounds. Hot water extract of dried fruit is used orally for piles and topically for skin disease and ulcers. Fine powder of fruit, leaf, and root mixed with dried ginger and "Triphala" is taken orally with tepid water to treat common fever, whereas its anthraquinone fraction is taken orally for leprosy. Leaf juice is a household remedy for gonorrhea and leucorrhea.

Neem stick has a rich content of alkaloid margosine, resins, gum, chloride, fluoride, silica, sulfur, tannins, oils, saponins, flavonoids, sterols, and calcium. Even at 5% concentration, the neem stick extract shows antimicrobial activity against oral streptococcal pathogens and has resulted in microscopically observable bacterial aggregation, indicating that neem stick extract can reduce the ability of some streptococci to colonize tooth surfaces (Wolinsky et al., 1996).

Neem oil preparations are effective against a wide spectrum of bacteria, including *Bacillus cereus*, *Bacillus pumilus*, *Staphylococcus aureus*, *Mycobacterium tuberculosis*, *Escherichia coli*, *Pasteurella vulgaris*, *Salmonella typhi*, *Klebshiella pneumonae*, *Salmonella dysenterae*, *Enterococcus faecalis*, *Streptococcus mutans*, *Streptococcus salivarius*, *Streptococcus mitis*, *Streptococcus sanguis*, and even streptomycin-resistant strains. Some researchers have raised doubts regarding the antibacterial activity of neem oil against *P. vulgaris* (Parekh and Chanda, 2007). Neem leaf extract inhibits the formation of biofilm in *Pseudomonas aeruginosa* (Harjai et al., 2013; Hore et al., 1999), and reduces the bacterial morbidity caused by *Streptococcus* spp., *Aeromonas hydrophila*, *Enterobacter* spp., *E. coli*, *Pseudomonas* spp., *Proteus* spp., *Vibrio* spp., and *Yersinia enterocolitica* in marine ornamental fish (Durrani et al., 2008). The phosphate buffer, ether, and alcoholic extracts of the leaves inhibit the activity of the microorganism *Micrococcus pyrogenes* var. *aureus*. The essential oil inhibits the growth of *M. tuberculosis*, *S. typhi*, *S. paratyphi*, *Vibrio cholera* Pacini, and *K. pneumonia* (Schroter) Trevisan.

Antiviral activity

Neem leaves can eliminate virus by contact. Topical application of neem paste removes warts, chickenpox, and smallpox. Neem leaf tea and topical ointment (for cold sores) provide relief from herpes infection. Due to its antiviral activity, boiled neem leaves are also added to the baths of individuals suffering from skin ailments like eczema, cold sores, or inflammatory conditions. An aqueous extract of tender leaves has antiviral potency against vaccinia (viral disease in cattle), variola (smallpox), fowlpox, and Newcastle diseases, whereas in vitro activity has been documented against dengue virus, Chikungunya virus, poliovirus, measles viruses, fowlpox viruses, and Newcastle disease viruses (Badam et al., 1999; Parida et al., 2002; Girish and Bhat, 2008; Faccin-Galhardi et al., 2012). Azadirachtin causes the absence of virus-specific 511 bp amplicon in reverse-transcriptase polymerase chain reaction (Parida et al., 2002). The methanolic extract of neem leaves has better virucidal efficacy in coxsackie B-4 viruses, along with its influence during the early events of its replication (Biswas et al., 2002). The polysaccharides have also shown antiviral activity against poliovirus in vitro (Galhardi et al., 2012).

As an antiviral agent, neem extracts significantly inhibit poliovirus, herpes infectious virus, and dengue virus at the early stage of viral genome replication (Badam et al., 1999; Parida et al., 2002; Rai and Sethi, 1972; Rao et al., 1969; Reddy and Sethi, 1974; Upadhyay et al., 1990), perhaps by virus inactivation and also by interfering during an early event of its replication cycle. Neem bark extract (NBE) significantly blocked herpes simplex virus type-1 (HSV-1) access into cells at concentrations of 50−100 μg/mL (Tiwari et al., 2010). Inhibition of HSV-1 glycoprotein-mediated cell-to-cell fusion and polykaryocyte formation also indicate an additional role of NBE at the viral fusion step, allowing a prospective possibility for the development of NBE as a novel antiherpetic microbicide.

Antifungal activity

Neem seed oil is fungitoxic and reduces fungal lung infections as well as oral thrush. In addition, several in vitro investigations have revealed the inhibitory effect of neem extracts on the production of mycotoxins, such as patulin produced by *Penicillium expansum* (Mossini et al., 2004). Neem extract has fungistatic and fungicidal effects against plant pathogenic fungi like *Alternaria solani*, *Curvularia lunata*, *Fusarium oxysporum*, *Helminthosporium* spp., and *Schlerotium rolfsii*, and against dermatophytes such as *Epidermophyton floccosum*, *Trichophyton rubrum*, *Trichophyton mentagrophytes*, *Trichophyton violaceum*, and *Microsporum nanum* (Natarajan, 2002). A range of fungal infections including *Candida albicans*, *Aspergillus fumigatus*, *Aspergillus flavus*, *Aspergillus niger*, *Microsporum gypseum*, *Microsporum canis*, *T. mentagrophytes*, *T. rubrum*, *Cladosporium* spp., *F. oxysporum*, *Penicillium notatum*, and *Penicillium citrinum* can also be treated using neem oil (Natarajan, 2002; Asif, 2012; Al-Samarrai et al., 2012). Neem oil has distinctive antiplaque activity (Elavarasu et al., 2012), and chewing neem twigs is effective in controlling dental tartar and dental caries.

Immunostimulant activity

Neem promotes specific (humoral and cell-mediated) as well as nonspecific (cytotoxic and phagocytic activity of macrophages) immune responses (Dholi et al., 2011; Mohammed et al., 2007). Adding 4% infusion of neem leaves to fresh drinking water (50 mL/L) of poultry is a potential natural growth promoter and an immune stimulant resulting in better body weight gain, feed conversion ratio, lower mortality, and maintaining a superior antibody titer against infectious bursal disease (IBD).

Tumor necrosis factor (TNF) inhibitors are widely used in the management of inflammatory and autoimmune diseases. Azadirachtin acts with the TNF-binding domain of the receptor and hinders TNF-induced biological responses (Thoh et al., 2011). Nimbidin, another phytoconstituent found in the seed, is quite effective in the treatment of skin conditions such as eczema, furunculosis, arsenical dermatitis, scabies, and seborrheic dermatitis. The epoxyazadiradione content also has antiinflammatory activity against the macrophage migration inhibitory factor (Alam et al., 2012).

Retinoic acid is an essential biomolecule in mucosal immune responses. It is produced by gut-associated dendritic cells and is required for generating gut-tropic lymphocytes and IgA antibody-secreting cells. The retinoic acid produced by dendritic cells with surface phenotypic biomarker CD103 and alveolar macrophages along with transforming growth factor-β (TGF-β) functions to support translation of naive T cells into regulatory T cells with Foxp3 biomarker, and thereby contributes to maintenance of mucosal tolerance. It binds to its nuclear receptors that are expressed in lymphoid cells and acts as a transcription factor to regulate cell homing and differentiation. Azadirachtin also inhibits retinoic acid receptor-mediated biological responses (Alam et al., 2012).

Antidiabetic activity

The dried neem flower is recommended orally for diabetes. In diabetic rats, neem extract (250 mg/kg) demonstrated reduced glucose (18%), triglycerides (32%), cholesterol (15%), lipids (15%), urea (13%), and creatinine (23%) (Chattopadhyay et al., 2004; Patil et al., 2013). In normal and streptozotocin-induced diabetic animals, aqueous leaf extracts displayed an antihyperglycemic effect, but little effect occurred on peripheral utilization of glucose and hepatic glycogen metabolism. The blood sugar-lowering effect takes place through removal of the inhibitory effect of serotonin on insulin release (mediated by glucose) (Chattopadhyay, 1999; Clausen et al., 1985). The prophylactic treatment with leaf or seed oil partially prevented the alloxan-induced increase in blood sugar (Kroes et al., 1993). In addition, nimbidiol, present in root and bark, can inhibit intestinal glucosidases and thus can be beneficial in controlling diabetes (Mukherjee and Sengupta, 2013).

Antioxidant and antitumor activities

Neem exerts its antitumor effects through modulation of several cellular processes. The anticancerous bioactive compounds from neem oil include several steroids and triterpenoids like nimbin, nimbinin, nimbidinin, nimbinene, nimbandiol, salannin, 3-deacetylsalannin, salannol, nimbidic acid, meliantriol, epoxyazadiradione, azadirone, azadiradione, gedunin, 7-deacetyl-gedunin, salannin-lactone, nimbin-lactone, meldenin, vepinin, 6-o-acetylnimbandiol, 1,3-diacetyl-vilasinin, and 6-deacetylnimbinene (Patel et al., 2016). Azadirachtin and a range of limonoids stimulate cancer cell-death via multiple mechanisms including the intrinsic pathway of apoptosis (Priyadarsini et al., 2010). Nimbolide induces extrinsic as well as intrinsic pathways of apoptosis in human breast cancer cells (Elumalai et al., 2012). In fact, neem inhibits cell proliferation, causes free radical scavenging; induces cell death together with apoptosis and autophagy, causes immune surveillance, antiinflammatory and antiangiogenic effects; reduces cellular oxidative stress; and prevents invasion and metastasis (Patel et al., 2016). Probably, neem employs the pleiotropic effects of various phytoconstituents to produce its broad chemopreventive and anticancer activities.

NF-κB has a direct role in the desensitization of tumor toward several chemotherapeutics and radiation therapy (Elumalai et al., 2012). Nimbolide inhibits the NF-κB pathway through inhibition of IKK activation, leading to suppression of NFKBIA phosphorylation, degradation and subsequent negative regulation of the expression of many tumorigenic proteins (Gupta et al., 2010). Phosphorylation of NF-κB leading to its activation is one of the downstream targets of the PI3K/Akt pathway, leading to decreased ABCB1/MDR1 gene expression (Ma et al., 2014).

Nimbolide-mediated collateral sensitivity through PTEN and its downstream components results in significant downregulation of the ABCB1/MDR1 gene and protein expression. The strong collateral sensitivity toward P-glycoprotein (ABCB1/MDR1)-expressing cells suggests involvement of essential cellular metabolic-regulating elements (HIF1α, FoxO1, MYC, and reactive oxygen species (ROS)) in manipulation of nimolide-induced tumor sensitivity.

Aqueous leaf extracts of neem boost blood antioxidant status and offer protection against paracetamol-induced hepatic damage in rats. Neem and its derivatives have been shown to exert their antioxidant properties by decreasing TNF-α, increasing interferon (IFN)-γ, and modulating antioxidant enzymes such as glutathione *S*-transferase (GST) and certain hepatic cytochrome P450 (CYP450)-dependent monooxygenases (Seddiek et al., 2013). The flowers contain compounds capable of inducing monofunctional phase II enzymes and repressing some monooxygenases, especially those involved in the metabolic activation of chemical carcinogens. Feeding diets containing 12.5% neem flowers for 2 weeks strongly enhanced GST activity by almost 2.7-fold and caused a marked reduction in the levels of phase I reactions (Lai et al., 1990).

The aqueous leaf extract is also used prophylactically against cancer development at extrahepatic sites. The bark and leaves contain polysaccharides and limonoids that are beneficial in alleviating the risk of cancer development. The leaf extract of neem enhances hepatic glutathione-dependent enzymes in gastric carcinogenesis and offers protection against in vivo oxidative stress and clastogenic effects induced by carcinogens. Similar chemopreventive potential of neem leaf extract has been demonstrated in the murine carcinogenesis model system and against dimethyl benz(a)-anthracene (DMBA)-induced hamster buccal pouch carcinogenesis in model systems (Subapriya et al., 2004) as well as in hepatocarcinogenesis. It also causes immune-mediated prophylactic growth inhibition of murine Ehrlich carcinoma and B16 melanoma (Baral and Chattopadhyay, 2004). Azadirone, a limonoid, has good in vivo antitumor activity, whereas prenylated flavones isolated from flowers are potent antimutagenic compounds against heterocyclic compounds. Prenylated flavones also inhibit DMBA-induced mammary gland carcinogenesis in female Sprague–Dawley rats as well as aflatoxin B_1-induced hepatocarcinogenesis in Wistar rats (Tepsuwan et al., 2002).

Ethanolic extract of neem leaves differentially suppresses human breast and cervical cancerous cells in a dose- and time-dependent manner through apoptosis. It significantly alters the expression of bax, cyclin D1, and CYP450 monooxygenases (CYP1A1 and CYP1A2) in a time-dependent manner in these cells (Sharma et al., 2014). It induces apoptosis via both the intrinsic and extrinsic pathways and produces cell cycle arrest via p53-dependent p21 accumulation and downregulation of the cell-cycle regulatory proteins cyclin B, cyclin D1, p53, and proliferating cell nuclear antigen (Arakaki et al., 2006; Ghosh et al., 2006; Kumar et al., 2010; Priyadarsini et al., 2010). Extracts rich in nimbolide and azadirachtin produce radiosensitization and lessen tumor progression in varied solid tumor models. Pretreatment with neem extract can result in a dose-dependent inhibition of radiation-induced stemness-related molecules. Even 24 h after irradiation, no recovery of neem leaf extract-inhibited transcription has been recorded with complete inhibition of 12, 66, and 64 genes at levels of 1, 10, and 100 μg dose, respectively. Both dose-independent (20 genes at 3 h and 12 genes at 24 h) and time-independent (6, 50, and 61 genes at 1 μg, 10 μg, and 100 μg, respectively) inhibition of genes was observed. Interestingly, six genes, namely *BMP4*, *CDH1*, *EGF3*, *FOXA2*, *GSK3β*, and *HDAC2*, showed both time- and dose-independent inhibition. Coadministration with cisplatin results in a further synergistic response. Thus, neem may be a prospective prophylactic and therapeutic agent to combat cancers.

Neem leaf glycoprotein treatment in murine sarcoma causes alterations in the cytokine profile in the tumor microenvironment, such as from interleukin (IL)-10 and TGF-β. IL-6-rich type 2 characters switched to the type 1 microenvironment with dominance of IFN-γ secretion (Barik et al., 2013). It increases the $CD8^+$ T-cell population with a higher expression of cytotoxicity-related molecules, perforin, and granzyme B, along with a low expression of the FasR gene in cells, symbolizing prevention from angiogenic and immunosuppressive mediator-induced cell death. The glycoprotein of neem leaf prevents in vivo and in vitro growth of murine carcinoma and melanoma. It activates T cells and NK cells, inhibits suppressor Tregs, promotes type 1 cytokines, and maintains the type 1 antitumor chemokine milieu, thus instigating antigen-specific tumor killing. The efficacy of neem glycoprotein to restrict sarcoma growth has also been established in therapeutic settings, where glycoprotein-activated recognition and killing of tumor cells via cytotoxic T lymphocytes plays a pivotal role (Barik et al., 2013).

Antifertility and spermicidal activities

Hot aqueous extract of neem bark is an oral energizer and emmenagogue. A single intrauterine application of neem oil induces a preimplantation block in fertility, which is reversible after 5–6 months (Upadhyay et al., 1990), suggestive of its possible use as a long-term female contraceptive. The postcoital contraceptive effect of neem seed oil has also been demonstrated in rats (Prakash et al., 1988). Moreover, neem seed oil lacks estrogenic, antiestrogenic, or progesteronal activity, so lesser side effects are expected. Similarly, purified neem seed extract, administered orally, has been found to be successful in terminating pregnancy in rodents and primates, perhaps through activation of cell-mediated immune reactions. A polyherbal vaginal cream containing extracts of dried neem seeds has shown contraceptive efficacy that has been attributed to a mixture of six components containing mono- and diunsaturated free fatty acids and their methyl esters (Garg et al., 1998).

Neem also has antiandrogenic effects. Various neem extracts produce a statistically significant decrease in serum testosterone and phosphatase activities accompanied by biochemical and histological regression in seminal vesicles and ventral prostrate. This is probably through a reduction in FSH levels which disrupts spermatogenesis (Akpantah et al., 2011). Using the Sander–Cramer test, the minimum effective spermicidal concentrations of aqueous extracts of old and tender neem leaves were 2.91 ± 0.669 and 2.75 ± 0.754 mg/million sperm, respectively (Khillare and Shrivastav, 2003). The male antifertility activity of neem seed oil and leaves extract is well proven in rats and mice (Aladakatti and Ahmed, 1999; Aladakatti et al., 2001), as well as in monkeys (Bardhan et al., 1991). Several antispermatogenic and antiandrogenic histoarchitectural changes might have a role to play in the development of male infertility. These changes include development of intracellular spaces and vacuolization in Sertoli cells, diminished Leydig cells and cytoplasmic inclusions of testis, hypoandrogenic effects as characterized by enlarged nuclei of principal cells, missing microvilli and mitochondrial cristae, reduced number of coated micropinocytotic vessels of the apical cytoplasm, Golgi complex disruption, abounding lysosomal bodies in the cytoplasm, decreased chromatin content of nuclei, vacuolized cytoplasm, and bulging nuclear membrane in cauda epididymis. Physiological maturation of sperm might also be affected due to diminished levels of fructose.

Antiparasitic activity

Hot water extract of the entire neem plant is used as an anthelmintic and insecticide. The antiplasmodial property is perhaps the earliest and most popular pharmacological activity. Neem bark juice is taken orally for filariasis. Some neem extracts are, at best, effective in malaria prophylaxis. Chewing neem leaves has been a folklore remedy for malaria. Gedunin, a lesser limonoid found in neem, is very effective in vitro but poor in vivo for treating malaria (Omar et al., 2003). It is effective in treating malaria through teas and infusions of the leaves. Another compound isolated from neem leaf, Irodin A, is also toxic to causative strains of malaria (Anyaehie, 2009).

Oral dosing with methanolic leaf extracts (500 and 125 mg/kg body weight) causes a statistically significant decline in parasitic load in *Plasmodium berghei*-infested mice with no benefit on the period of protection. The alcoholic extracts of the leaves and seeds are effective against chloroquine-resistant as well as chloroquine-sensitive strains of the malaria parasite, whereas purified fractions inhibit growth and development of sexual and asexual stages of drug-resistant strains (Biswas et al., 2002). A single intravenous dose of azadirachtin has a blocking effect on the development of *Typanosoma cruzi* and induces permanent resistance of the vector against reinfection with *T. cruzi* (De Azambuja and Garcia, 1992), and thus may help in preventing the spread of infection (Dhayanithi et al., 2010).

In an experimental study of antinematodal activity of neem leaf supplementation in sheep of the Morada Nova breed, the results varied with the season and composition (protein content) of diet offered. Results were significantly more positive in the rainy season in comparison to the dry season (Chagas et al., 2008). Neem kernel powder decreases the nematodal fecal egg count and causes an increase in body weight gain. In rabbits, the aqueous leaf extract of neem has shown acaricidal efficiency on a par with ivermectin (Seddiek et al., 2013).

Antidrug resistance activity

Novel agents that selectively kill ABCB1/MDR1-expressing cells are today being looked upon as an alternative strategy to overcome MDR through resensitization (Pluchino et al., 2012). Nimbolide as a potent collateral sensitivity promoting agent has been suggested by the reduction in ABCB1/MDR1 gene expression detected in microarray data. MYC and FoxO genes are well recognized in cancer metabolic adaptation and reprogramming, and the regulation of this axis is connected with the Akt/mTOR signaling pathway (Stine et al., 2015). Nimbolide potentially targets HIFα (Kowshik et al., 2017). Nimbolide reduces ROS, downregulates MYC oncogene, and upregulates FoxO1 transcription factor expressions, suggesting a novel mechanism for collateral sensitivity. Therefore, it can be supposed that nimbolide induces collateral sensitivity in P-glycoprotein-overexpressing cells by targeting tumor suppressor gene PTEN and modulating tumor cellular metabolic elements (HIF1α, FoxO1, c-MYC, and ROS). Changes in TP53 or epidermal growth factor receptor (EGFR) genes do not confer resistance toward nimbolide. The EGFR gene is also reported to be correlated with cancer development and poor survival in addition to resistance to cytotoxic agents (Baselga, 2002).

Neem extract as a nutraceutical

Besides a wide range of activities of the neem leaf, including antibacterial, anticarcinogenic, antifungal, antihyperglycemic, antiinflammatory, antimalarial, antimutagenic, antioxidant, antiulcer, antiviral, and immunomodulatory, chewing fresh leaves acts as a sedative and relaxant. Aqueous extract of neem leaves shows noteworthy antiulcer activity and reduction in severity of gastric injury, and prevents mast cell degranulation and mucus depletion. The bark extract can effectively revert the increased lipid peroxidation, carbonyl content, and reduced endogenous levels of GSH that usually occur during ulceration. In addition, nimbidin suppresses basal as well as stimulated gastric acid release along with blockade of histamine H_2 receptors. Neem leaves are relatively rich in nutritious elements as compared to any other similar vegetation that has been subjected to chemical analysis. They are a rich source of carbohydrates (22.9%), protein (7.1%), vitamin C, minerals, calcium, phosphorus, carotene, etc., in addition, they also contain glutamic acid, aspartic acid, tyrosine, alanine, glutamine, proline, and cystine amino acids, and several fatty acids (dodecanoic, tetradecanoic, eicosanic, etc.). A cup of neem furnishes about 28.4% of the daily iron requirement.

Neem shows antiaging properties. It protects the skin from harmful UV rays, pollution, and other xenobiotics. The vitamins and fatty acids content in neem promote the elasticity of the skin and reduce wrinkles. It is customary to drink a small quantity of neem juice on Guddipadwa in Maharashtra. It is also used for making herbal caffeine-free tea. Neem leaf chutney was a regular feature of Mahatma Gandhi's diet. A nutraceutical tea now being manufactured would indisputably have been Gandhi's favorite drink. Neem can also be blended with green tea or black tea for a milder taste. The neem leaf extract has also been reported to be advantageous in the treatment of carbon tetrachloride-induced liver damage (Mujumdar et al., 1998).

In the Ayurvedic system of medicine, neem is always mixed with other herbs to augment its efficacy and to make its taste more palatable. Herbs good for the pitta dosha such as licorice, honey, sugar, lemon juice, and/or spices like cardamom may be used to amplify the efficacy or to reduce the side effects. The bitterness can also be counterbalanced with herbs and spices like cinnamon, orange peel, licorice root, and fennel seed. Neem has been indicated in contemporary Ayurveda for diabetes mellitus, perhaps by enhancing insulin receptor sensitivity. Oral administration of leaf extract has been found to significantly reduce insulin requirements for noninsulin-dependent diabetes patients. An active principle, Nimbidiol, which is present in the root and bark, can inhibit intestinal glucosidases, and thus can be helpful in controlling diabetes (Mukherjee and Sengupta, 2013). Fresh mature leaves, along with the seeds of *Psoralea corylifolia* and *Cicer arietinum*, are effective in leucoderma. Tender leaves, along with black pepper, are effective in intestinal helminthic infections.

A mature neem tree can produce 300–400 kg of leaves per year, which may be used for feeding cattle during famine. Deoiled seeds may be used as fertilizer or as unconventional animal feed. Neem seed cake is a good source of protein (30%–40% crude protein) for livestock; there are approximately 0.9 million tons available in India (Singh, 1993). Its protein is relatively balanced in its amino acid content and mineral profile (Gowda and Sastry, 2000), and it has tremendous potential to compensate for protein supplement shortages in the livestock industry. The neem seed is readily available in northeast India and perhaps in other parts of Asia, the northern part of Nigeria, Australia, Africa, and Central and South America.

Although immunostimulants are known to trigger body immunity, their regular consumption may cause intolerance because of overstimulation of the immune system, disrupting the normal metabolic processes in the body. Supplementation of specific immunostimulants in feed can improve animal health and thereby reduce management costs. Among immunostimulants, neem leaf can be used as a feed additive to replace antibiotics and/or disinfectants to develop a new concept of organic biotechnology for sustainable aquaculture. The therapeutic potential of the neem leaf as a dietary supplement improves the nonspecific immunity of fish. Dietary doses of neem leaf significantly influenced the immune parameters, hematological parameters, and blood biochemical indices and increased the survival rate in *Lates calcarifer* fingerlings against *Vibrio harveyi* infection. It increased the phagocytic activity, superoxide anion production, serum lysozyme, serum bactericidal activity, and serum antiprotease activity of treated fish (Talpur and Ikhwanuddin, 2013).

Feeding neem kernel cakes without treatment has been shown to have a negative impact on nutrient digestibility and performance of ruminants. Poor palatability, reduced weight gain, and lower nutrient digestibility were observed in cross-bred calves fed concentrated mixtures containing neem seed cake (Bedi et al., 1975). Lowered rumen protozoan count and reduced enzyme activity have been reported in calves fed neem kernel cake (Mondal et al., 1995), suggesting a reduction in rumen digestibility. Poor growth has also been recorded in adult cattle when 27% of the protein requirement in the diet was provided through neem seed cake (Pradhan, 1975).

Overnight soaking of neem kernel cake in 0.8% sodium hydroxide and water followed by washing with water two or three times leads to significantly higher nitrogen balance (Nath et al., 1983). In addition, cattle productivity remained unaffected qualitatively as well as quantitatively. There was no significant variation in milk yield, milk fat content, sensory evaluation of milk, dry matter intake, or digestibility when treated neem kernel cake was incorporated at a level of 40% in the concentrated diet of dairy cows for 300 days (Nath et al., 1989).

The palatability of neem seed cake can be improved with appropriate treatment to remove the bioactive principles or by feeding along with barley, molasses, and peanut meal, but reduced consumption (from 79% to 39%) of the concentrated mixture may result when neem seed cake levels are increased from 59% to 90% (Bhandari and Joshi, 1974). Yearling sheep can completely utilize concentrate mixture with 75 parts neem seed cake and 25 parts maize, but their feed consumption reduces to one-third if neem seed cake is fed alone (Gupta and Bhaid, 1980). Buffalo calves continued to consume neem seed cake after its proportion was increased from 5 to 15 parts when fed along with 7 parts of molasses and 20 parts of peanut meal; the overall feed consumption reduced to half following the withdrawal of molasses (Baxi and Kektar, 1976).

Inclusion of neem seed cake is economical in ruminant farming. Urea-ammoniated neem seed kernel meal (UANSKM) has been proven to be a satisfactory, economical, and wholesome substitute for complete replacement of deoiled groundnut cake (DGNC) in rations of growing goats. Approximately 20% lower costs per weight gain have been reported when DGNC was replaced with urea-ammoniated neem kernel meal in goats (Baxi and Kektar, 1976).

A growth trial was undertaken to study the feasibility of total replacement of DGNC by UANSKM on the basis of nutritive value, biochemical parameters, and feeding economics (Anandan et al., 1996). Dietary variation did not significantly influence the efficiency of dry matter (8.7 vs 9.1 g DM), crude protein (1.2 vs 1.2 g CP), and metabolic energy (22.9 vs 22.8 kcal ME) utilization per unit gain. The total body weight gain (kg), average daily gain (g), and feed conversion efficiency (feed/gain) in the experimental group were 5.6 ± 0.59, 31.0 ± 3.25, and 9.1 ± 0.57, respectively, as compared to the corresponding values of 6.0 ± 0.56, 32.8 ± 3.20, and 8.7 ± 0.55, respectively, in the DGNC group. The hematobiochemical profile (hemoglobin, glucose, and urea nitrogen) and the activity of various enzymes (transaminases and alkaline phosphatase) did not differ significantly between the treatments, but the feed cost per kilogram gain of the UANSKM diet was cheaper than that of the DGNC diet by 14.2%. The feeding cost per unit weight gain and average daily gain in males was marginally superior to that in females. Glucose concentrations were higher ($P < .01$) in males, whereas other blood constituents did not differ between the sexes.

With a ban on four commonly used feed antibiotic growth promoters—monensin, salinomycin, avilamycin, and flavophospholipol—by the European Commission, neem leaf meal (NLM) may be a useful ingredient in the diet of broilers as a medicinal and nutritional element (Bonsu et al., 2012), but the inclusion level varies with age and physiological status. Neem and *Ocimum sanctum* (tulsi) leaves extract activate the cell-mediated immune response, therefore creating an enhanced response to any future challenges due to disease organisms. The feeding of neem and tulsi leaves to immunosuppressed poultry birds increases their humoral and cell-mediated immune responses (Sadekar et al., 1998). A low dose of neem leaf extract has an inhibitory action on a wide spectrum of microorganisms. A 4% infusion of neem leaves at a concentration of 50 mL per liter of fresh drinking water could be effectively used as a potential natural growth promoter and an immune stimulant, contributing to better body weight gain, feed conversion ratio, gross return, lower mortality, and higher antibody titer against IBD (Durrani et al., 2008).

The proximate composition of NLM indicated that it is a good source of protein, with crude protein values ranging from 18.10% to 20.68%. The maximum tolerance level of NLM as a protein source described in the literature is 10% in starter broilers and 15% in laying birds and rabbits, although it contains bioactive compounds such as azadirachtin, nimbin, salanin, limonoids, and tannin that may affect nutrient utilization (Aruwayo and Maigandi, 2013). The bioactive toxic principles may be reduced by sun drying. The only concern in this regard is its acceptability, because of its pungent smell and bitterness due to the bioactive principles. Despite high crude protein content, its incorporation in animal diets was discouraged due to the presence of bitter and toxic triterpenoids, mainly nimbin, nimbidin, azadirachtin, and salanin (Paul et al., 1996). Furthermore, feeding such ingredients may impart an unpleasant taste or smell to meat (Clausen et al., 1985). Removal of these bitter phytoconstituents through solvent extraction, water washing, alkali soaking, and urea ammoniation has gained appreciable success in improving its acceptability (Gowda and Sastry, 2000). In addition, the palatability and utilization of such unconventional feedstuffs can be further improved by incorporating them in complete diets and processing into pellets (Reddy and Reddy, 1999).

Efficacy

Neem leaves are considered safe to be taken internally on a regular basis. Neem is one of the most powerful blood purifiers and detoxifiers in the Ayurvedic system of medicine. Neem leaves relieve the symptoms associated with viral infections, including common cold, herpes, influenza, and chickenpox. Eating neem leaves reduces fever associated with viral infection. Neem leaves also contain phospholipase A_2 inhibitor, which can be used as a snake venom inhibitor.

Fusarium is a filamentous fungi widely distributed in soil. Several species of *Fusarium* produce mycotoxins such as trichothecenes and zearalenone in cereal crops that are used as animal and poultry feed and can affect human and animal health if they enter the food chain. Zearalenone triggers reproductive disorders, including hyperestrogenic syndromes and tumor production, whereas trichothecenes are sesquiterpenoid epoxides that act as potent inhibitors of eukaryotic protein synthesis. Neem oil extract decreases zearalenone production at 0.1%–0.5% concentrations, but the highest inhibition (59.05%) of mycotoxin synthesis occurred at 0.1% (Geraldo et al., 2011).

Safety and toxicity considerations

Safety

Neem has attracted worldwide distinction due to its vast range of medicinal properties without showing any adverse effects (Kale et al., 2003). Based on its extensive traditional use, neem seems to be quite safe. More than 75% of Ayurvedic remedies contain neem, usually in the form of leaf (or extract) and rarely in the form of oil. The majority of scientific studies have been performed with neem leaf or neem leaf extracts. Neem leaves taken internally on a regular or daily basis are considered safe unless any physiological stress is indicated. In the millennial history of indigenous use, there have never been any reports of side effects from neem leaves. However, a comprehensive safety evaluation of the different formulations of neem has never been performed.

Formal safety testing has only involved neem oil as an insecticide. Being a natural antioxidant, neem can protect both the liver and kidneys at low dosages. However, the neem phytoconstituents, taken in high quantities or for extended periods, can overwork the liver and kidneys and cause damage. Common signs of problem with the kidneys and liver include jaundice or yellowing of the skin and reduced urine production. It may also cause infertility and miscarriage. Hypoglycemia along with vomiting, diarrhea, provocation of immune response, and interference with other types of medications may also be observed. Serious encephalopathy may develop in infants and small children within hours after taking neem oil. Serious side effects include metabolic acidosis, vomiting, diarrhea, drowsiness, blood disorders, loss of consciousness, brain disorders and seizures, coma, and death (Lai et al., 1990).

Neem leaf extract at 100 mg/kg orally for 20 days decreases serum triiodothyronine (T_3) and increases serum thyroxine (T_4) concentrations, but no such changes have been reported at a lower dose (40 mg/kg for 20 days), indicating that the high concentrations of neem extract can be inhibitory to thyroid function, particularly in the conversion of T_4 to T_3, the major source of T_3 generation (Panda and Kar, 2000). Other side effects at a higher dose include an increase in hepatic lipid peroxidation and a decrease in glucose-6-phosphatase activity, along with an enhancement in the activities of two defensive enzymes, superoxide dismutase (SOD) and catalase. Thus, it appears that the higher concentration of neem extract may not be safe with respect to thyroid function and the generation of oxidative stress coupled to biomembrane damage.

The volatile, odorous fraction of neem oil obtained by steam distillation is spermicidal. A minimum concentration of 0.25 and 25 mg/mL inhibits spermatozoal motility in rat and human semen, respectively. The effect on spermatozoal motility is dose-dependent and remains unaffected by the presence of vaginal or cervical mucus. Intravaginal application of oil in rabbits also does not produce any signs of irritation to the vaginal mucosa. An oral dose of 15 mg/kg neem extract delays reproduction in male rats up to 60 days (normal reproduction cycle is 20–23 days). Approximately 50% mortality has been observed in rats at an oral dose of 25 mg/kg extract, whereas in the remaining 50%, reproduction was delayed up to 3 months. Histological studies of testes indicate aberrations in spermatogenesis and sperm production in some of the seminiferous tubules. However, these changes are reversible and reproduction can be observed after 60–90 days of the treatment (Mahmoudi et al., 2002).

In vivo studies have demonstrated that neem oil is 100% effective in preventing pregnancy when used as a vaginal lubricant prior to coitus (Jacobson and Schummtterer, 1995). In an experimental study on rabbits, application of neem with another herbal preparation before coitus prevented pregnancy (Raghuvanshi et al., 2011), whereas, in rats, the disruption of the estrous cycle leading to a partial block in ovulation has been reported after treatment with alcoholic extract of neem flower (Gbotolorun et al., 2008).

Joshi et al. (2005) assessed the safety of Praneem polyherbal tablets (containing purified extracts of neem) as an intravaginal microbicide. No serious adverse effects were observed after 14 consecutive days of treatment. Praneem also exhibited contraceptive activity. Of 20 volunteers, nine reported 17 episodes of genital irritation. Transient itching was reported by eight; two participants reported burning during urination; and one reported lower abdominal pain, genital burning, and intermenstrual spotting. Currently, Praneem polyherbal tablets are considered safe for once-daily intravaginal use for 14 consecutive days in sexually active HIV-uninfected women and in low-risk women for up to 6 months (Joshi et al., 2005, 2008).

Toxicity

Neem oil

The nonaqueous extracts of neem are perhaps the most toxic neem-based products, with an estimated safe dose of 0.002 and 12.5 μg/kg body weight per day. The unprocessed materials like seed oil and the aqueous extracts are less toxic. For all preparations, reversible reproductive toxicity was evident in both male and female mammals upon subacute or chronic exposure. Acute tubular necrosis is a major finding, along with intravascular hemolysis, hepatic damage, and hypersensitivity reactions (Chakraborty et al., 1998). In female rats, neem oil disturbs the estrus cycle, resulting in higher abortive rates and reduced fertility. A reduction in sperm count and motility, weight of seminal vesicle and ventral prostate (Kasturi et al., 1997), and an increase in percentage of malformed sperms (Parveen et al., 1993) have also been reported. The decline in the testosterone level is thought to be responsible for the delay in maturation of the sperm (Aladakatti et al., 2001). A dose-dependent reduction in the height of epithelium in caput and cauda epididymis has also been reported and the testosterone level was also reduced (Kasturi et al., 1995).

Neem leaf

Acute toxic effects have been reported for indigenous neem leaf medications used to treat fever and abdominal upset or for inducing infertility (Kadiri et al., 1999). Unprocessed neem leaves may cause acute toxicity in sheep and produce nervous symptoms, including head movements, circular walking, hyperthermia, hepatic failure, and tympanitis, finally terminating in death (Ali and Salih, 1982). The leaf sap has an antianxiety effect at low doses, but no such activity has been reported at higher doses (Jaiswal et al., 1994). Subacute and chronic leaf toxicity are mainly observed in reproductive function, although it is reversible and gradually subsides as the treatment is terminated. Biochemical alterations such as reduced protein content, reduced acid phosphatase activity, and an increase in alkaline phosphatase and lactate dehydrogenase were also observed (Kasturi et al., 1997). The no adverse effect level (NAEL) for seminal vesicle protein content is 6.6 mg/rat or 26.4 mg/kg body weight (Kasturi et al., 1997). Using the standard safety factor of 10 for interspecies and intraspecies extrapolation, the acceptable daily intake (ADI) of unprocessed leaves for humans may be calculated as $26.4 \times 0.1 \times 0.1 = 0.264$ mg/kg body weight/day.

Compared to synthetic pesticides, neem extracts appear to be less toxic to fish (Wan et al., 1996). Tilapia fish are more sensitive than catfish. It affects the crustacean zooplankton population at all the concentrations tested. At higher concentrations of neem leaf extract, fish showed signs of respiratory distress, erratic swimming, and nervous symptoms. At all concentrations, fish showed signs of damaged gills and kidney tissue (Winkaler et al., 2007). Hepatocellular damage followed by complete inhibition of aspartate aminotransferase and alanine aminotransferase enzyme synthesis has also been demonstrated at higher concentrations.

Although neem has been found to be adequately safe for use as an insecticide, animal studies suggest that long-term oral use of neem oil might produce toxic effects. Acute toxicity of neem oil is evident at doses of 14 and 24 mL/kg body weight for rats and rabbits, respectively. The oil is nontoxic to mice at lower doses, but at high doses, hyperexcitability to sound and touch, labored respiration, and convulsive jerks were induced, and some animals died (Tandan et al., 1995). Neem oil increases tail flick reaction time and reduces induced writhing (Khosla et al., 2000). In normal and hyperglycemic rats, administration of oil caused lowering of blood glucose (Dixit et al., 1986). The most relevant adverse effect reported in neem oil-exposed animals seems to be the antifertility effect in female rats observed on subacute exposure to 2.0–4.6 mL/kg body weight (Dhaliwal et al., 1998). Neem oil shows higher abortive potential during the first few days of pregnancy in rats as compared to during the later stages. The lowest observed adverse effect level for the antifertility effect in female rats is 2.0 mL/kg body weight. Taking a standard safety factor of 10, a NAEL would be a value of 0.2 mL/kg body weight. Extrapolating this to humans using the standard safety factors for intraspecies and interspecies extrapolations results in an ADI of $0.2 \times 0.1 \times 0.1$ mL/kg body weight, amounting to 0.002 mL/kg body weight. This value of 0.002 mL/kg body weight is 100 times less than the dose of 0.2 mL/kg body weight reported to be lethal in children. Moreover, whole neem extract may produce genotypic damage with prolonged use at higher doses. Therefore the use of neem is generally not advised in clinically unhealthy or immunocompromised animals.

Aqueous extracts

Fermented decoction of neem bark inhibits the human complement system and polymorphonuclear cells (Kroes et al., 1993). Leaf extract causes hypoglycemia, increased clotting time and serum cholesterol, reduced liver fat, and a dose-dependent decrease in liver proteins (Mossa, 1985; El-Hawary and Kholief, 1990). The level of T_3 was reduced but thyroxine was increased, along with liver peroxidation and SOD and catalase activities (Panda and Kar, 2000). At a subchronic exposure level, rats showed inappetance, weakness, loss of condition, depression, and a decrease in body weight and papillary reflex. Congestion in the liver, kidney, lung, and brain has also been observed (Hore et al., 1999).

Neem flower

Neem flowers are monooxygenase inhibitors and phase II enzyme inducers. Feeding of flowers causes an increase in hepatic GST activity and a reduction in the activities of CYP450, aniline hydroxylase, and aminopyrine-*N*-demethylase. This may lead to altered pharmacokinetics of chemotherapeutic agents, leading to their therapeutic failure or adverse effects.

Concluding remarks and future directions

Neem appears to be a tree tailor-made for the struggle against the emerging health issues of humankind. Its bioactivity, mammalian safety, and ecofriendly qualities are quite encouraging for researchers in the field of nutraceuticals. Its multiple activities are highly beneficial for the poor economies of underdeveloped and developing countries. Combining the strengths of the knowledge base of traditional systems such as Ayurveda with the dramatic power of combinatorial sciences and high-throughput screening will be a great step forward in the generation of structure–activity libraries of nutraceutical biomolecules. Ayurvedic knowledge and experiential databases can provide new functional leads to reduce time, money, and toxicity. Extensive work is ongoing worldwide to reveal the biochemical nature of the biomolecules, particularly, the limonoids. Still, great efforts are needed to fill the critical gaps and to realize new but unrecognized biomolecules hidden in neem so that this wonder tree may be exploited to its full potential for use in humans and animals. The demand for finished products like neem tea is increasing in the modern world. More target-specific research is required for studying its pharmacotherapeutic profile and toxicity. Proper standardization and clinical evaluation of the final products need to be accomplished. Neem as a nutraceutical will continue to be of interest to pharmaceutical companies, researchers, and, ultimately, the global community for the promotion of holistic health.

References

Abdel-Ghaffar, F., Al-Quraishy, S., Al-Rasheid, K.A.S., 2012. Efficacy of a single treatment of head lice with a neem seed extract: an *in vivo* and *in vitro* study on nits and motile stages. Parasitol. Res. 110, 277–280.

Akpantah, A.O., Ekong, M.B., Obeten, K.E., et al., 2011. Hormonal and histomorphologic effects of *Azadirachtaindica* leaf extract on the pars anterior of wistar rats. Int. J. Morphol. 29, 441–445.

Aladakatti, R.H., Ahmed, R.N., 1999. Effect of *Azadirachtaindica* leaves on rat spermatozoa. Indian J. Exp. Biol. 37, 1251–1254.

Aladakatti, R.H., Ahmed, R.N., Ahmed, M., 2001. Sperm parameters changes induced by *Azadirachtaindica* in albino rats. J. Basic Clin. Physiol. Pharmacol. 12, 69–76.

Alam, A., Haldar, S., Thulasiram, H.V., 2012. Novel anti-inflammatory activity of epoxyazadiradione against macrophage migration inhibitory factor: inhibition of tautomerase and proinflammatory activities of macrophage migration inhibitory factor. J. Biol. Chem. 287, 24844–24861.

Ali, B.H., Salih, A.M.M., 1982. Suspected *Azadirachtaindica* toxicity in a sheep. Vet. Rec. 111, 494.

Al-Samarrai, G., Singh, H., Syarhabil, M., 2012. Evaluating eco-friendly botanicals (natural plant extracts) as alternatives to synthetic fungicides. Ann. Agric. Environ. Med. 19, 673–676.

Anandan, S., Sastry, V.R.B., Musalia, L.M., 1996. Growth rate and nutrient efficiency of growing goats fed urea ammoniated neem (*Azadirachtaindica*) seed kernel meal as protein supplement. Small Rumin. Res. 22, 205–212.

Anyaehie, U.B., 2009. Medicinal properties of fractionated acetone/water neem (*Azadirachtaindica*) leaf extract from Nigeria: a review. Niger. J. Physiol. Sci. 24, 157–159.

Arakaki, J., Suzui, M., Morioka, T., 2006. Antioxidative and modifying effects of a tropical plant *Azadirachtaindica* (neem) on azoxymethane-induced preneoplastic lesions in the rat colon. Asian Pac. J. Cancer Prev. 7, 467–471.

Aruwayo, A., Maigandi, S.A., 2013. Neem (*Azadirachtaindica*) seed cake/kernel as protein source in ruminants feed. Am. J. Exp. Agric. 3 (3), 482–494.

Asif, M., 2012. Antimicrobial potential of *Azadirachtaindica* against pathogenic bacteria and fungi. J. Pharmacog. Phytochem. 1, 78–83.

Badam, L., Joshi, S.P., Bedekar, S.S., 1999. *In vitro* antiviral activity of neem (*Azadirachtaindica*, A. Juss) leaf extract against group B Coxsackieviruses. J. Commun. Dis. 31, 79–90.

Baral, R.N., Chattopadhyay, U., 2004. Neem (*Azadirachtaindica*) leaf mediated immune activation causes prophylactic growth inhibition of murine Ehrlich carcinoma and B16 melanoma. Int. Immunopharmacol. 4, 355–366.

Bardhan, J., Riar, S.S., Sawhney, R.C., 1991. Neem oil—a fertility controlling agent in rhesus monkey. Ind. J. Physiol. Pharmacol. 35 (4), 278–280.

Barik, S., Banerjee, S., Mallick, A., 2013. Normalization of tumor microenvironment by neem leaf glycoprotein potentiates effector T cell functions and therapeutically intervenes in the growth of mouse sarcoma. PLoS One. 8 (6), e66501.

Baselga, J., 2002. Why the epidermal growth factor receptor? The rationale for cancer therapy. Oncologist. 7, 2–8.

Baxi, M.R., and Kektar, C.M., 1976. Deoiled neem cake trial. In: C.M. Kektar, Utilization of Neem and Its By Product. Khadi and Village Industries Commission, Bombay, India.

Bedi, S.P.S., Vijjan, V.K., Ranjhan, S.K., 1975. Utilization of neem (*Azadirachta indica*) seed cake and its influence on nutrients digestibility in buffaloes. Indian J. Dairy Sci. 28, 104–107.

Bhandari, A.S., Joshi, M.S., 1974. The effect of feeding deoiled neem cake on health of sheep. Indian Vet. J. 51, 659–660.

Biswas, K., Chattopadhyay, I., Banerjee, R.K., 2002. Biological activities and medicinal properties of neem (*Azadirachta indica*). Curr. Sci. 82, 1336–1345.

Bonsu, F.R.K., Kagya-Agyemang, J.K., Kwenin, W.K.J., 2012. Medicinal response of broiler chickens to diets containing neem (*Azadirachta indica*) leaf meal, hematology and meat sensory analysis. World Appl. Sci. J. 19, 800–805.

Chagas, A.C.S., Vieira, L.S., Freitas, A.R., 2008. Anthelmintic efficacy of neem (*Azadirachta indica* A. Juss) and the homeopathic product Fator Vermes® in Morada Nova sheep. Vet. Parasitol. 151, 68–73.

Chakraborty, P., Gupta, B.S., Chakraborty, C., 1998. Airborne allergenic pollen grains on a farm in West Bengal, India. Grana. 37, 53–57.

Chattopadhyay, R.R., 1999. A comparative evaluation of some blood sugar lowering agents of plant origin. J. Ethnopharmacol. 67, 367–372.

Chattopadhyay, I., Nandi, B., Chatterjee, R., et al., 2004. Mechanism of antiulcer effect of neem (*Azadirachta indica*) leaf extract: effect on H + -K + -ATPase, oxidative damage and apoptosis. Inflammopharmacology. 12, 153–176.

Clausen, S., Larsen, L.M., Ploger, A., 1985. Advances in the Production and Utilization of Cruciferous Crops. Junk Publications, Netherlands.

De Azambuja, P., Garcia, E.S., 1992. Effects of azadirachtin on *Rhodnius prolixus*: immunity and trypanosoma interaction. Mem. Inst. Oswaldo Cruz. 87 (Suppl. 5), 69–72.

Dhaliwal, P.K., Roop, J.K., Guraya, S.S., 1998. Antifertility activity of neem-seed oil in cyclic female rats. In: Indian Ecological Society and Centre for Research in Rural and Industrial Development, Chandigarh, India, pp. 340–346.

Dhayanithi, N.B., Ajith Kumar, T.T., Kathiresan, K., 2010. Effect of neem extract against the bacteria isolated from marine fish. J. Environ. Biol. 31, 409–412.

Dholi, S.K., Raparla, R., Mankala, S.K., et al., 2011. *In vivo* antidiabetic evaluation of neem leaf extract in alloxan induced rats. J. Appl. Pharm. Sci. 1, 100–105.

Dixit, V.P., Sinha, R., Tank, R., 1986. Effect of neem seed oil on the blood glucose concentration of normal and alloxan diabetic rats. J. Ethnopharmacol. 17, 95–98.

Dodia, A.R., Poojaben, T., Durga, F., 2020. Dried shark in vacuum packaging with Neem leaves (*Azadirachta indica*) for preservation. Int. J. Trend Sci. Res. Dev. 4, 208.

Durrani, F.R., Chand, N., Jan, M., 2008. Immunomodulatory and growth promoting effects of neem leaves infusion in broiler chicks. Sarhad J. Agric. 24, 655–659.

Elavarasu, S., Abinaya, P., Elanchezhiyan, S., 2012. Evaluation of anti-plaque microbial activity of *Azadirachta indica* (neem oil) *in vitro*: a pilot study. J. Pharm. Bioallied Sci. 4, S394–S396.

El-Hawary, Z.M., Kholief, T.S., 1990. Biochemical studies on hypoglycemic agents. I. Effect of *Azadirachta indica* leaf extract. Arch. Pharmacol. Res 13, 108–112.

Elumalai, P., Gunadharini, D., Senthilkumar, K., et al., 2012. Induction of apoptosis in human breast cancer cells by nimbolide through extrinsic and intrinsic pathway. Toxicol. Lett. 215, 131–142.

Faccin-Galhardi, L.C., Yamamoto, K.A., Ray, S., 2012. The *in vitro* antiviral property of *Azadirachta indica* polysaccharides for poliovirus. J. Ethnopharmacol. 142, 86–90.

Galhardi, L.C., Yamamoto, K.A., Ray, S., 2012. The *in vitro* antiviral property of *Azadirachta indica* polysaccharides for poliovirus. J. Ethnopharmacol. 142 (1), 86–90.

Garg, S., Talwar, G.P., Upadhyay, S.N., 1998. Immunocontraceptive activity guided fractionation and characterization of active constituents of neem (*Azadirachta indica*) seed extract. J. Ethnopharmacol. 60 (3), 235–246.

Gbotolorun, S.C., Osinubi, A.A., Noronha, C.C., 2008. Antifertility potential of neem flower extract on adult female Sprague-Dawley rats. Afr. Health Sci. 8, 168–173.

Geraldo, M.R.F., Da Costa, C.L., Arrotéia, C.C., 2011. The neem [*Azadirachta indica* a. juss (*meliaceae*)] oil reduction in the *in vitro* production of zearalenone by *Fusarium graminearum*. Braz. J. Microbiol. 42, 707–710.

Ghosh, D., Bose, A., Haque, E., 2006. Pretreatment with neem (*Azadirachta indica*) leaf preparation in Swiss mice diminishes leukopenia and enhances the antitumor activity of cyclophosphamide. Phytother. Res. 20, 814–818.

Girish, K., Bhat, S.S., 2008. Neem – a green treasure. J. Biol. 4, 102–111.

Gowda, S.K., Sastry, V.R.B., 2000. Neem (*Azadirachta indica*) seed cake in animal feeding – scope and limitations – review. Asian-Australas. J. Anim. Sci. 13, 720–728.

Gupta, R.S. and Bhaid, M.V., 1980. A study on the effect of feeding different levels of deoiled neem fruitcake in the concentrate mixture on water consumption in lambs. A review of Agro-Animal Science and Health, p. 121.

Gupta, S.C., Prasad, S., Reuter, S., et al., 2010. Modification of cysteine 179 of Ikappa B alpha kinase by nimbolide leads to down-regulation of NF-kappaB regulated cell survival and proliferative proteins and sensitization of tumor cells to chemotherapeutic agents. J. Biol. Chem. 285, 35406–35417.

Harjai, K., Bala, A., Gupta, R.K., 2013. Leaf extract of *Azadirachta indica* (neem): a potential antibiofilm agent for *Pseudomonas aeruginosa*. Pathog. Dis. 69, 62–65.

Hore, S.K., Maiti, S.K., Gupta, N., 1999. Effect of subacute exposure to neem (*Azadirachta indica*) leaf extract in rats. Indian Vet. J. 76, 1011–1012.

Jacobson, M., Schummtterer, H., 1995. Antifertility effects and population control the neem tree. In: Schummtterer, H. (Ed.), The Neem Tree. VCH Publication, Hoshenheim, Germany, pp. 526–530.

Jaiswal, A.K., Bhattacharya, S.K., Acharya, S.B., 1994. Anxiolytic activity of *Azadirachta indica* leaf extract in rats. Indian J. Exp. Biol. 32, 489–491.

Joshi, S.N., Katti, U., Godbole, S., 2005. Phase I safety study of Praneem polyherbal vaginal tablet use among HIV-uninfected women in Pune, India. Trans. R. Soc. Trop. Med. Hyg. 99, 769–774.

Joshi, S.N., Dutta, S., Kumar, B.K., 2008. Expanded safety study of Praneem polyherbal vaginal tablet among HIV-uninfected women in Pune, India: a phase II clinical trial report, Sex. Transm. Infect., 84. pp. 343–347.

Kadiri, S., Arije, A., Salako, B.L., 1999. Traditional herbal preparations and acute renal failure in South West Nigeria. Trop. Doct. 29, 244–246.

Kale, B.P., Kothekar, M.A., Tayade, H.P., 2003. Effect of aqueous extract of *Azadirachta indica* leaves on hepatotoxicity induced by antitubercular drugs in rats. Indian J. Pharmacol. 35, 177–180.

Kasturi, M., Manivannan, B., Ahamed, R.N., 1995. Changes in epididymal structure and function of albino rat treated with *Azadirachta indica* leaves. Indian J. Exp. Biol. 33, 725–729.

Kasturi, M., Ahamed, R.N., Pathan, K.M., 1997. Effects of *Azadirachta indica* leaves on the seminal vesicles and ventral prostate in albino rats. Indian J. Physiol. Pharmacol. 41, 234–260.

Khillare, B., Shrivastav, T.G., 2003. Spermicidal activity of *Azadirachta indica* (neem) leaf extract. Contraception. 68 (3), 225–259.

Khosla, P., Bhanwra, S., Singh, J., 2000. A study of hypoglycemic effects of *Azadirachta indica* (neem) in normal and alloxan diabetic rabbits. Indian J. Physiol. Pharmacol. 44, 69–74.

Kowshik, J., Mishra, R., Sophia, J., et al., 2017. Nimbolide upregulates RECK by targeting miR-21 and HIF-1 alpha in cell lines and in a hamster oral carcinogenesis model. Sci. Rep. 7, 2045.

Kroes, B.H., Van den Berg, A.J.J., Labadie, R.P., 1993. Impact of the preparation process on immunomodulatory activities of the ayurvedic drug Nimbaarishta. Phytother. Res. 7, 35–40.

Kumar, V.S., Navaratnam, V., 2013. Neem (*Azadirachta indica*): prehistory to contemporary medicinal uses to humankind. Asian Pac. J. Trop. Biomed. 3, 505–514.

Kumar, G.H., Vidya Priyadarsini, R., Vinothini, G., 2010. The neem limonoids azadirachtin and nimbolide inhibit cell proliferation and induce apoptosis in an animal model of oral oncogenesis. Invest. New Drugs 28, 392–401.

Kusamran, W.R., Ratanavila, A., Tepsuwan, A., 1998. Effects of neem flowers, Thai and Chinese bitter gourd fruits and sweet basil leaves on hepatic monooxygenases and glutathione S-transferase activities, and *in vitro* metabolic activation of chemical carcinogens in rats. Food Chem. Toxicol. 36 (6), 475–484.

Lai, S.M., Lim, K.W., Cheng, H.K., 1990. Margosa oil poisoning as a cause of toxic encephalopathy. Singapore Med. J. 31, 463–465.

Luong, K., Dunkel, F.V., Coulibaly, K., 2012. Potential use of neem leaf slurry as a sustainable dry season management strategy to control the malaria vector *Anopheles gambiae* (Diptera: Culicidae) in West African villages. J. Med. Entomol. 49, 1361–1369.

Ma, H.Y., Cheng, L., Hao, K.J., et al., 2014. Reversal Effect of ST6GAL 1 on multidrug resistance in human leukemia by regulating the PI3K/Akt pathway and the expression of P-gp and MRP1. PLoS One. 9, e85113.

Mahmoudi, M.M., Morowati, M., Ghazi Khansari, M., 2002. Sterility effects of neem (*Azadirachta indica*) extract on male rat. J. Reprod. Infertil. 3 (2), 4–13.

Mohammed, A., Tanko, Y., Okasha, M.A., et al., 2007. Effects of aqueous leaves extract of *Ocimum gratissimum* on blood glucose levels of streptozotocin induced diabetic Wistar rats. Afr. J. Biotechnol. 6, 20872090.

Mondal, D.P., Garg, A.K., Mondal, B.C. et al., 1995. Effect of feeding neem kernel meal on ruminal protozoan assay. Abst. No. 137, 7th Animal Nutrition Research Workers Conf. Bombay, India.

Mossa, J.S., 1985. A study on the crude antidiabetic drugs used in Arabian folk medicine. Int. J. Crude Drug Res. 23, 137–145.

Mossini, S.A., De Oliveira, K.P., Kemmelmeier, C., 2004. Inhibition of patulin production by *Penicillium expansum* cultured with neem (*Azadirachta indica*) leaf extracts. J. Basic Microbiol. 44 (2), 106–113.

Mujumdar, A.M., Upadhye, A.S., Pradhan, A.M., 1998. Effect of *Azadirachta indica* leaf extract on CCl_4 induced hepatic damage in albino rats. Indian J. Pharm. Sci. 60, 363–367.

Mukherjee, A., Sengupta, S., 2013. Characterization of nimbidiol as a potent intestinal disaccharidase and glucoamylase inhibitor present in *Azadirachta indica* (neem) useful for the treatment of diabetes. J. Enzyme Inhib. Med. Chem. 28, 900–910.

National Research Council. 1992. National Academy of Sciences: Constitution and Bylaws, April 28, 1992. Washington, DC: The National Academies Press. https://doi.org/10.17226/21331.

Natarajan, V., 2002. *Azadirachta indica* in the treatment of dermatophytosis. J. Ecobiol. 14, 201–204.

Nath, K., Rajgopal, S., Garg, A.K., 1983. Water washed neem (*Azadirachta indica*) seed kernel cake as a cattle feed. J. Agric. Sci. 101, 323–326.

Nath, K., Agrawal, D.K., Hassan, Q.Z., 1989. Water washed neem (*Azadirachta indica*) seed kernel cake in feeding of milk cows. Anim. Prod. 48, 497–502.

Omar, S., Zhang, J., MacKinnon, S., 2003. Traditionally-used antimalarials from the Meliaceae. Curr. Top. Med. Chem. 3 (2), 133–139.

Panda, S., Kar, A., 2000. How safe is neem extract with respect to thyroid function in male mice? Pharmacol. Res. 41, 419–422.

Parekh, J., Chanda, S.V., 2007. *In vitro* antimicrobial activity and phytochemical analysis of some Indian medicinal plants. Turk. J. Biol. 31, 53–58.

Parida, M.M., Upadhyay, C., Pandya, G., 2002. Inhibitory potential of neem (*Azadirachta indica* Juss) leaves on Dengue virus type-2 replication. J. Ethnopharmacol. 79, 273–278.

Parveen, D.S., Manivannan, B., Pathan, K.M., 1993. Antispermatic activity of *Azadirachta indica* leaves in albino rats. Curr. Sci. 64, 688–689.

Patel, S.M., Nagulapalli Venkata, K.C., Bhattacharyya, P., et al., 2016. Potential of neem (*Azadirachta indica* L.) for prevention and treatment of oncologic diseases. Semin. Cancer Biol. 40-41, 100–115.

Patil, P., Patil, S., Mane, A., et al., 2013. Antidiabetic activity of alcoholic extract of neem (*Azadirachta Indica*) root bark. Natl. J. Physiol. Pharm. Pharmacol. 3, 142–146.

Paul, B.N., Haque, N., Garg, A.K., 1996. Effect of feeding neem seed kernel cake (*Azadirachta indica*) on rumen enzyme of buffalo. Indian Vet. J. 73, 720–728.

Pluchino, K.M., Hall, M.D., Goldsborough, A.S., Callaghan, R., Gottesman, M.M., 2012. Collateral sensitivity as a strategy against cancer multidrug resistance. Drug Resist. Updat 15, 98–105.

Pradhan, K., 1975. Utilization of neem and its by-products. In: Ketkat, C.M. (Ed.), Final Report, Khadi and Village Industries Commission, Bombay, India.

Prakash, A.O., Tewari, R.K., Mathur, R., 1988. Non-hormonal postcoital contraceptive action of neem oil in rats. J. Ethnopharmacol. 23 (1), 53–59.

Priyadarsini, R.V., Murugan, R.S., Sripriya, P., et al., 2010. The neem limonoids azadirachtin and nimbolide induce cell cycle arrest and mitochondria-mediated apoptosis in human cervical cancer (HeLa) cells. Free Radic. Res. 44 (6), 624–634.

Raghuvanshi, P., Bagga, R., Malhotra, D., 2011. Spermicidal and contraceptive properties of Praneem polyherbal pessary. Indian J. Med. Res. 113, 135–141.

Rahal, A., Kumar, D., and Malik, J.K., 2019. Neem extract. In: Gupta, R.C., Srivastava, Ajay, Lall, Rajiv (Eds.), *Nutraceuticals in Veterinary Medicine*, pp. 37–50.

Rai, A., Sethi, M.S., 1972. Screening of some plants for their activity against vaccinia and fowl pox viruses. Indian J. Anim. Sci. 42, 1066–1070.

Rao, A.R., Kumar, S.S.V., Paramsivam, T.R., 1969. Study of antiviral activity of leaves of margosa tree on vaccinia and variola viruses – a preliminary report. Indian J. Med. Res. 57, 495–502.

Reddy, G.V.N., Reddy, M.R., 1999. Effect of feeding extruded complete diet containing maize cobs in Ongole bull calves. Indian J. Anim. Nutr. 16, 210–214.

Reddy, A.B., Sethi, M.S., 1974. Antiviral effects of some indigenious plant extracts on vaccinia and fowlpox viruses on chick embryo fibroblasts. Indian J. Exp. Biol. 12, 572–579.

Sadekar, R.D., Kolte, A.Y., Barmase, B.S., 1998. Immunopotentiating effects of *Azadirachta indica* (neem) dry leaves powder in broilers, naturally infected with IBD virus. Indian J. Exp. Biol. 36, 1151–1153.

Seddiek, S.A., Khater, H.F., El-Shorbagy, M.M., 2013. The acaricidal efficacy of aqueous neem extract and ivermectin against *Sarcoptesscabiei* var. *Cuniculi* in experimentally infested rabbits. Parasitol. Res. 112, 2319–2330.

Shah, A.S., Gunjal, M.A., Juvekar, A.R., 2009. Immunomostimulatory activity of aqueous extract of *Azadirachta indica* flowers on specific and nonspecific immune response. J. Nat. Remedies 9, 35–42.

Sharma, C., Vas, A.J., Goala, P., et al., 2014. Ethanolic neem (*Azadirachta indica*) leaf extract prevents growth of MCF-7 and HeLa cells and potentiates the therapeutic index of cisplatin. J. Oncol. 321754. Available from: https://doi.org/10.1155/2014/321754.

Singh, K., 1993. Livestock production and health. In: Randhawa, N.S., Parmar, B.S. (Eds.), Neem Research and Development. Society of Pesticide Science, India, pp. 187–198.

Stine, Z.E., Walton, Z.E., Altman, B.J., Hsieh, A.L., Dang, C.V., 2015. MYC, metabolism, and cancer. Cancer Discov 5 (10), 1024–1039. Available from: https://doi.org/10.1158/2159-8290.CD-15-0507.

Subapriya, R., Kumaraguruparan, R., Abraham, S.K., 2004. Protective effects of ethanolic neem leaf extract on N-methyl-N 1562 0-nitro-N-nitrosoguanidine-induced genotoxicity and oxidative stress in mice. Drug Chem. Toxicol. 27 (1), 15–26.

Talpur, A.D., Ikhwanuddin, M., 2013. *Azadirachta indica* (neem) leaf dietary effects on the immunity response and disease resistance of Asian seabass, *Latescalcarifer* challenged with *Vibrio harveyi*. Fish Shellfish Immunol. 34, 254–264.

Tandan, S.K., Gupta, S., Chandra, S., 1995. Safety evaluation of *Azadirachta indica* seed oil, a herbal wound dressing agent. Fitoterapia. 66, 69–72.

Tepsuwan, A., Kupradinun, P., Kusamran, W.R., 2002. Chemopreventive potential of neem flowers on carcinogen-induced rat mammary and liver carcinogenesis. Asian Pac. J. Cancer Prev. 3 (3), 231–238.

Thoh, M., Babajan, B., Raghavendra, P.B., 2011. Azadirachtin interacts with retinoic acid receptors and inhibits retinoic acid-mediated biological responses. J. Biol. Chem. 286 (6), 4690–4702.

Tiwari, V., Darmani, N.A., Yue, B.Y., 2010. *In vitro* antiviral activity of neem (*Azardirachta indica* L.) bark extract against herpes simplex virus type-1 infection. Phytother. Res. 24, 1132–1140.

Upadhyay, S.N., Kaushic, C., Talwar, G.P., 1990. Antifertility effects of neem (*Azadirachta indica*) oil by single intrauterine administration: a novel method for contraception. Proc. Biol. Sci. 242 (1305), 175–179.

Wan, M.T., Watts, R.G., Isman, M.B., 1996. Evaluation of the acute toxicity to juvenile Pacific Northwest salmon of azadirachtin, neem extract, and neem based products. Bull. Environ. Contam. Toxicol. 56, 432–439.

Winkaler, E.U., Santos, T.R.M., Machado-Neto, J.G., 2007. Acute lethal and sublethal effects of neem leaf extract on the neotropical freshwater fish *Prochilodus lineatus*. Comp. Biochem. Physiol. C 145, 236–244.

Wolinsky, L.E., Mania, S., Nachnani, S., 1996. The inhibiting effect of aqueous *Azadirachta indica* (neem) extract upon bacterial properties influencing *in vitro* plaque formation. J. Dent. Res. 75 (2), 816–822.

Chapter 57

Spirulina

Dan Wan, Qinghua Wu and Kamil Kuča

Introduction

Spirulina, also called *Arthrospira*, is a multicellular and filamentous cyanobacteria (blue-green algae). It has gained considerable popularity in the health food industry and is considered a food supplement to human, livestock, poultry, and aquaculture diets. *Spirulina* grows in water, can be harvested and processed easily, and has very high macro- and micronutrient contents, such as proteins, amino acids, unsaturated fatty acids, minerals, vitamins, etc. It has been used as a dietary supplement for centuries by people living close to the alkaline lakes where it is naturally found—for instance those living adjacent to Lake Chad in the Kanem region and Lake Texcoco in Mexico (Sotiroudis and Sotiroudis, 2013; Chei et al., 2020).

Currently, this traditional food is used as a major source of protein in Africa. Due to its potential applications, it is now widely cultivated and commercially produced in large outdoor or greenhouse ponds under controlled conditions throughout the world. In Asia, it is used as a complementary dietary ingredient in feed for fish, shrimp, and poultry, and is also increasingly used as a protein and vitamin supplement in aquafeeds. For example, in China *Arthrospira* spp. is used as a partial substitute for imported forage to promote the growth and immunity of shrimp (Habib et al., 2008). Recently, great attention and extensive studies have been devoted to evaluating its therapeutic benefits in an array of disease conditions including hypercholesterolemia, hyperglycerolemia, cardiovascular diseases, inflammatory diseases, cancer, and viral infections (Deng and Chow, 2010).

The primary objective of this chapter is therefore to assess and evaluate the existing knowledge of *Arthrospira*'s activities, side effects, and safety for both human consumption and animal feeds.

The consumption of *Arthrospira*

Arthrospira is a photosynthesizing cyanophyte that grows vigorously in strong sunshine under high temperatures and highly alkaline conditions. In the 16th century, Spanish invaders discovered that the Aztecs living in the Valley of Mexico in the capital Tenochtitlan were collecting a blue-colored *Arthrospira* from the lagoons and making a blue-green cake from it (Sasson, 1997). The Kanembu population living along the shores of Lake Chad employed it for the same purpose. The semidried algae is cut into small squares and taken to the villages where the drying is completed on mats in the sun (Abdulqader et al., 2000). When dry, women take these algae cakes "dihé" to sell in the local market. It has been observed that the local trading value of the dihé annually harvested from Lake Kossorom in Chad (about 40 tons) amounted to more than US $100,000, which represents an important contribution to the economy of the area (Abdulqader et al., 2000). However, in the remnant today, only Lake Texcoco has living algae *Arthrospira*.

In 1940, the French phycologist Dangeard (1940) noted that the Kanembu people near Lake Chad consumed an algae-cake named "dihé," and that these same algae populated a number of lakes in the Rift Valley of East Africa. "Dihe" was also the main food for the flamingos living around those lakes. During 1964–65, the botanist Jean Léonard, reported edible cakes of "dihé" being sold in native markets of Fort-Lamy (now N'Djamena) in Chad. When Léonard rediscovered *Arthrospira* in Africa, the first systematic and detailed study of the growth requirements and physiology of *Arthrospira* was performed by Zarrouk, which was the basis for establishing the first large-scale production plant of *Spirulina*. In 1967, *Arthrospira* was established as a "wonderful future food source." As no microorganisms have fulfilled their promise of cheap protein, *Arthrospira* has continued to give rise to research and increasing production, reflecting its perceived nutritional assets.

General characteristics of *Arthrospira*

Morphology

Arthrospira is multicellular and filamentous blue-green microalgae with symbiotic bacteria that fix nitrogen from

Nutraceuticals. DOI: https://doi.org/10.1016/B978-0-12-821038-3.00057-4

the air. They are photosynthetic and therefore autotrophic, and reproduced by binary fission. *Arthrospira* trichomes (filaments), which contain cylindrical cells, are aligned together in spirals or straight lines. The helical shape of the filaments is characteristic of the genus and is maintained only in a liquid environment or culture medium. The presence of gas-filled vacuoles in the cells, together with the helical shape of the filaments, result in floating mats. These filaments have variable lengths (usually 50–500 μm) and a diameter close to 3–12 μm. However, the cell dimensions, degree of coiling, and length of filaments vary with species. The cell organization of *Arthrospira* is typical of a prokaryote Gram-negative bacterium with a lack of membrane-bound organelles. The cell wall constitutes a weak envelope that is composed of a number of layers, mostly of a peptidoglycan and lipopolysaccharide nature. The *Arthrospira* cells have a number of inclusions, such as thylacoid membranes with phycobilisomes, carboxysomes, ribosomes, DNA fibrils, and gas vacuoles, as well as polyglycan, polyphosphate, and cyanophycin granules. The body surface of *Arthrospira* is smooth and has no covering, therefore it is readily digestible by simple enzymatic systems (Habib et al., 2008).

Taxonomy

In 1827, P.J. Turpin isolated *Spirulina* from a freshwater sample. In 1844, Wittrock and Nordstedt reported a helical, septal, and green-blue microalgae near the city of Montevideo, which was then named *Spirulina platensis*. It was not until 1852, however, that the first taxonomic report written by Stizenberger appeared. He gave this new genus the name *Arthrospira* based on its septa, helical form, and multicellular structure. In 1892, Gomont discovered the aseptate form of the *Spirulina* genus, and the septal form of the *Arthrospira* genus. However, because of the common helical morphology, the members of the *Spirulina* genus and the *Arthrospira* genus were reunified under the designation *Spirulina* in 1932. In 1989, these microorganisms were separately classified into two genera, *Spirulina* and *Arthrospira*, and this classification is currently accepted (Tomaselli et al., 1996). In the various *Arthrospira* genera, the two species *Arthrospira platensis* and *Arthrospira maxima* are the most important (Fig. 57.1) (Small, 2011).

For historical purposes, most of the worldwide investigation into these microalgae has been carried out under the name of "*Spirulina*." Therefore the name of "*Spirulina*" is the most common designation among scientists, manufacturers, and consumers. Most recently, the reports on *Spirulina* in peer-reviewed journals have been corrected to *Arthrospira*. Considering both the taxonomy of this microalgae and the historic designation, we refer to both collectively in this chapter. Specifically, *Arthrospira* is used when referring to the taxonomy, and *Spirulina* is used when referring to the composition of this microalgae.

Natural habitat, source, and growth

Besides Lake Texcoco, the largest *Arthrospira* lakes are in Central Africa around Lakes Chad and Niger, and in East Africa along the Great Rift Valley. Under normal water conditions, *Arthrospira* may be one of many algal species present. It thrives in alkaline lakes where it is difficult or impossible for other microorganisms to survive (Kebede and Ahlgren, 1996). Lakes Bodou and Rombou in Chad have a stable monoculture of *Arthrospira* dating back centuries. Kenya's Lakes Nakuru and Elementeita and Ethiopia's Lakes Aranguadi and Kilotes, also contain *Arthrospira* as a major microorganism. Alkaline water (> 30 g/L) with high pH (8.5–11.0) favors good production of *Arthrospira*, especially where there is a high level of solar radiation at tropical altitudes. The higher the pH and conductivity of the water, the greater the predominance of *Arthrospira* spp. *A. platensis* was isolated from waters containing salt from 85 to 270 g/L, and optimum growth occurred between 20 and 70 g/L. A relatively high cytoplasmic pH (4.2–8.5) may account for the ability of this microorganism to utilize ammonia as a source of nitrogen at high-alkaline pH values (Sasson, 1997).

Like most cyanobacteria, *Arthrospira* is an obligate photoautotroph. It reduces carbon dioxide in the light and assimilates mainly nitrates. The main assimilation product of *Arthrospira* photosynthesis is glycogen. *Arthrospira* shows optimum growth between 35°C and 37°C under laboratory conditions and an increase in temperature up to 39°C for a few hours does not harm blue-green algae or its photosynthetic ability. However, this algae does not tolerate low temperatures (15°C), and has relatively high resistance to ultraviolet rays.

Biochemical compositions

Generally, *Spirulina* consists of 55%–70% protein, 15%–25% polysaccharides, 5%–6% total lipids, 6%–13% nucleic acids, and 2.2%–4.8% minerals (Hosseini et al., 2013). The branched polysaccharide (25%) in *A. platensis* is structurally similar to glycogen. The polyunsaturated fatty acids (PUFAs) constitute 1.5%–2.0% of the total lipid content of this algae. *Arthrospira* spp. is rich in linolenic acid (LA, 30%–35% of total PUFAs), and also provides stearidonic acid, eicosapentaenoic acid, docosahexaenoic acid, and arachidonic acid. γ-Linolenic acid (GLA) is an essential fatty acid rarely available in ingredients or the diet. Moreover, *Arthrospira* spp. is low in cholesterol (32.5 mg/100 g). For instance, 10 g of

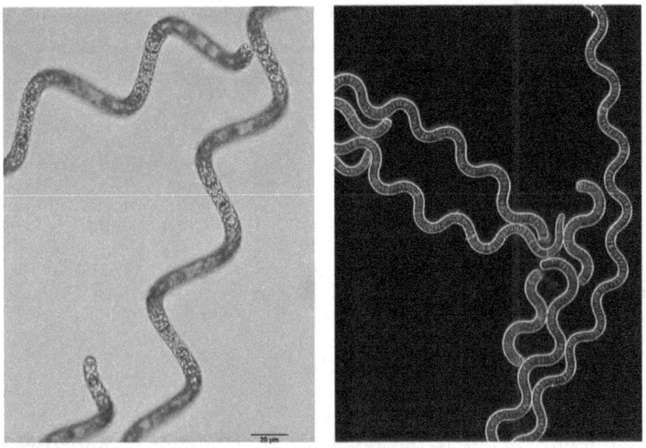

FIGURE 57.1 *Spirulina (Arthrospira platensis* and *Arthrospira maxima). Reproduced from Small, E., 2011. Spirulina—food for the universe. Biodiversity, 12, 255–265.*

Spirulina powder provides only 1.3 mg of cholesterol, whereas the equivalent quantity of protein from an egg would provide 300 mg of cholesterol. In addition, *Arthrospira* spp. is readily digested due to the absence of cellulose in its cell walls. After 18 hours, more than 85% of its protein is digested and assimilated (Sasson, 1997). This high nutritional content and low cholesterol are very important for its function as a food component.

Arthrospira spp. contain a variety of vitamins (B_1, B_2, B_3, B_6, B_9 and B_{12}), minerals (Ca, Cr, Cu, Fe, Mn, Mg, P, Se, Na, and Zn), photosynthetic pigments, and insulin-like proteins. Therefore the biomass of this rich source of elements is employed as a food and feed additive in many industries, for example, agriculture, food, and pharmaceuticals.

Despite its nutritional function, some of the species-typical chemicals (i.e., c-phycocyanin) in *Spirulina* powder have been shown to have pharmacological activity and a clinically curative effect in research (Reddy et al., 2000; Li et al., 2006; Soheili and Khosravi-Darani., 2011; Ou et al., 2012; Chen et al., 2014). Others include zeaxanthin (Yu et al., 2012), GLA (Tanticharoen et al., 1994; Parages et al., 2012), sulfated polysaccharides (Hayashi et al., 1996b; Ghosh et al., 2009), certain sulfolipids (Barron et al., 2008), and insulin-like protein (Anwer et al., 2012). Among all the natural compositions of *Arthrospira*, the most interesting biochemical compound is c-phycocyanin. It provides anticancer effects by selectively inhibiting cyclooxygenase-2 (COX-2) due to its conformation and large structure (Fig. 57.2). The structure of c-phycocyanin facilitates proper binding to the active site of COX-2 (Reddy et al., 2000).

However, the chemical characteristics of species belonging to the same microalgal category differ according to the specific source, culture condition, harvest time, and extraction method, even if their appearance is similar. Vonshak et al. (1996) reported that salt-adapted cells had a modified biochemical composition with reduced protein and chlorophyll content, and increased carbohydrate content. *A. platensis* is one of the most promising sources of GLA which can be enhanced by cultivation under light–dark cycles in the laboratory or outdoors (Tanticharoen et al., 1994). Therefore the quality of commercial *Spirulina* powder varies dependent on the different sources.

Healthy aspects of *Spirulina*

Although historically *Spirulina* was used as a food component, it has been thoroughly investigated using in vitro and in vivo experiments for its role in human health management. A large number of publications covering the health aspects of *Spirulina* in cell and tissue culture, in animal testing, as well as in human clinical trials have appeared since the 1990s. These articles describe experimental approaches involving whole-cell preparations, various cell extracts, and purified biomolecules (c-phycocyanin, GLA, sulfated polysaccharides, etc.), aimed at elucidating the potential health benefits of the consumption of this microalgae. Thus far, the results have been exciting. Potential health effects include immunomodulation, antioxidant, anticancer, antiviral, and antibacterial activities, as well as positive effects against malnutrition, hyperlipidemia, diabetes, obesity, inflammatory allergic reactions, heavy metal/chemical-induced toxicity, radiation damage, and anemias (Lee et al., 1998; Deng and Chow, 2010; Karkos et al., 2011; Soheili and Khosravi-Darani., 2011).

Proinflammatory and antioxidant effects

Both in vivo and in vitro studies indicate that *Spirulina* supplementation is an antioxidant but mostly proinflammatory agent. It enhances the antioxidant activity and promotes the production of antibodies and cytokines in both healthy animals and disease models. *Spirulina* can prevent skeletal muscle damage and stimulate antibodies and cytokines to induce an immunomodulatory response (Wu et al., 2016).

It has been shown that *Spirulina* (*A. platensis*) is a dietary supplement valued for its immune-enhancing properties probably via a high-molecular-weight polysaccharide fraction. It activates nuclear factor kappa-B (NF-κB) and increases expression of proinflammatory cytokines, including IL-8, IL-1β, TNF-α, MCP-1, MIP-1α, MIP-1β, and IP-10, as well as the expression of COX-2 in human monocytic THP-1 cells (Grzanna et al., 2006). In murine macrophage cell line J774A.1 cells, the immunomodulating activities of *Spirulina* are attributed to the c-phycocyanin component. It has the capability to induce

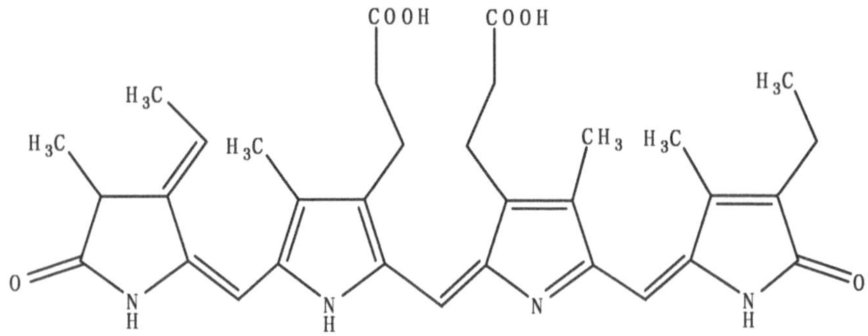

FIGURE 57.2 Chemical structure of c-phycocyanin.

secretion of inflammatory cytokines, including TNF-α, IL-1β, and IL-6, as well as proIL-1β and COX-2 protein expression. Furthermore, it rapidly stimulates phosphorylation of inflammatory-related signaling molecules, including extracellular regulated protein kinases (ERK), c-Jun N-terminal kinase (JNK), p38, and IκB (Chen et al., 2014). Moreover, the c-phycocyanin, which is extracted from *A. platensis*, could reduce IL-4 production and significantly increase interferon-γ (IFN-γ) production when the peripheral blood mononuclear cells (PBMC) from allergic patients are incubated with and without heat-killed *Lactobacillus rhamnosuseither* and *L. rhamnosus* (El Sheikh et al., 2014).

Spirulina has a synergistic effect with prophylaxis and induces the production of cytokines TNF-α and IFN-γ in Balb/C mice (Soltani et al., 2012). A 10-week feeding trial was conducted to examine the effects of dietary *Spirulina* and quercetin on growth, innate immunity, disease resistance, and dietary antioxidant capacity in the juvenile olive flounder. The results revealed a significant increase in lysozyme activity and respiratory burst activity with a dietary supplementation of 3.4% or 6.8% *Spirulina* in olive flounder (Kim et al., 2013). Moreover, *Spirulina* decreases liver weight, cytochrome P450 7A1 expression, and cholesterolemia, improving the glutathione redox index (superoxide dismutase, catalase, glutathione peroxidase, and glutathione reductase activities and endothelial nitric oxide synthase (eNOS), inducible nitric oxide synthase (iNOS), and TNF-α levels) in Zucker Fa/Fa rats (Vázquez-Velasco et al., 2014).

A number of investigations have been performed on disease models to evaluate the immune-enhancing properties of *Spirulina*. Evidence shows that *Spirulina* also has the ability to moderate antibody levels in response to tetanus toxoid. Supplementation of *Spirulina* could enhance the primary immune response in terms of antibody production via significantly enhanced IgG levels when vaccinated with tetanus toxoid (Chu et al., 2013). In tumor-bearing (breast cancer) mice, *A. platensis* decreased the secretion of IL-4 and IL-10 infected with *Candida albicans*, whereas the levels of IL-17, TNF-α, and IFN-γ increased to 93.4, 316.2, and 137.1 pg/mL, respectively. These findings clearly suggest that *A. platensis* has a remarkable immunomodulatory effect and this property provides a potential application in cancer treatment (Shokri et al., 2014). Moreover, *Spirulina* exhibits a significant modulatory effect on acetic acid-induced colitis in rats in macroscopic damage, bloody diarrhea scores, and colon weight/length, which may be due to a significant increase in antioxidant enzyme activity and significant inhibition of lipid peroxidation and inflammation markers (TNF-α and interleukins) (Abdel-Daim et al., 2015). However, *Spirulina* administration to nonalcoholic steatohepatitis (NASH) model rats resulted in an alleviative inflammatory response through antioxidative and antiinflammatory mechanisms, breaking the crosstalk between oxidative stress and inflammation. It effectively inhibited NASH progression by decreasing the production of reactive oxygen species (ROS) from liver mitochondria and leukocytes, activating the NF-κB signaling pathway, and modifying the lymphocyte surface antigen ratio (Pak et al., 2012).

An enhancement of natural killer (NK) cell activity following administration of Immulina (a commercial extract of *A. platensis*) was observed in a clinical study. It presented that Braun-type lipoproteins of the *Spirulina* commercial extract were responsible for the major portion of the in vitro monocyte activation (Nielsen et al., 2010). Indoleamine 2,3-dioxygenase enzyme activity, which is a sign of immune function, was induced by a *Spirulina* supplement (3 g per day, 12 weeks) in 40 volunteers of both sexes (Selmi et al., 2011). *Spirulina* consumption also effectively reduced antiinflammatory and pain-related infectious diseases by decreasing the production of TNF-α, IL-1β, and IL-6 (Abu-Taweel et al., 2019). *Spirulina* shows protective effects against ulcerative colitis in rats by reducing the inflammation effect (Morsy et al., 2019). A very recent study shows that *Spirulina* provided protection against hemato-biochemical alterations and nephrotoxicity, due to its antioxidant effect (Gargouri et al., 2020). The immunomodulation and antioxidant properties of *Spirulina* in preclinical studies are summarized in Table 57.1.

Hypolipidemic, hypoglycemic, and hypotensive effects

Hyperlipidemia involves abnormally elevated levels of lipids or lipoproteins in the blood. High levels of cholesterol, especially those high in low-density lipoproteins (LDLs) and very low-density lipoproteins (VLDLs), are critical reasons for these two diseases. Increased LDL, VLDL, and triglycerides (TG) levels, and decreased high-density lipoproteins (HDL) are the major independent risk factors for cardiovascular events. Although differences in study design, sample size, and patient conditions resulted in minor inconsistencies in response to *Spirulina* supplementation, the cumulative data from those studies clearly demonstrate the hypolipidemic and hypertensive activities of *Spirulina* in humans (Table 57.2).

The hypolipidemic effect of *Spirulina* or its extracts has been demonstrated in various animal models including mice, rats, hamsters, and rabbits. In a mouse model, supplementation of 16% *Spirulina* in a high-fat, high-cholesterol

TABLE 57.1 The immunomodulation and antioxidative properties of *Spirulina* in preclinical studies.

Subject	Specific effects	Component	References
Human (~50 year old)	Increase indoleamine 2,3-dioxygenase enzyme activity and white blood cell count	*Spirulina*	Selmi et al. (2011)
Human	Enhance the natural killer cell activity	Immulina (commercial extracts of *Arthrospira platensis*)	Nielsen et al. (2010)
Murine macrophage cell line J774A.1	Induce secretion of TNF-α, IL-1β, and IL-6 Increase proIL-1β and COX-2 protein expression Phosphorylation of ERK, JNK, p3,8 and IκB Decrease ROS	C-phycocyanin	Chen et al. (2014)
Mice	Induce secretion of TNF-α and IFN-γ	*Spirulina* supplement	Soltani et al. (2012)
Olive flounder	Increase lysozyme activity and respiratory burst activity	*Spirulina* supplement	Kim et al. (2013)
THP-1 cells	Induce secretion of IL-8 and MIP-1β		Grzanna et al. (2006)
Zucker Fa/Fa rats	Decrease liver weight, cytochrome P450 7A1 expression and cholesterolemia Improve superoxide dismutase, catalase, glutathione peroxidase, and GR activities and eNOS, iNOS, and TNF-α levels	*Spirulina* supplement	Vázquez-Velasco et al. (2014)
Nonalcoholic steatohepatitis model rats	Decrease in production of reactive oxygen species from liver mitochondria and from leukocytes Activate nuclear factor-kappa B Change in the lymphocyte surface antigen ratio ($CD4^+$/$CD8^+$)	*Spirulina*	Pak et al. (2012)
Mice	Enhance primary immune response in terms of IgG antibody production	*Spirulina* (50 and 150 mg/kg bw)	Chu et al. (2013)
Breast cancer mice	Decrease the secretion of IL-4 (45.1 pg/mL) and IL-10 (208.4 pg/mL) in tumor-bearing mice infected with *Candida albicans* Increase the secretion of IL-17, TNF-α, and IFN-γ to 93.4, 316.2, and 137.1 pg/mL, respectively.	*A. platensis*	Shokri et al. (2014)
Peripheral blood mononuclear cells	Markedly reduce the IL4 production and significantly increase the IFNγ production	*Spirulina*	El Sheikh et al. (2014)

GR, glutathione reductase.

TABLE 57.2 The hypolipidemic, hypoglycemic, and hypotensive effects of *Spirulina* in studies.

Subject	Specific effects	Bioactive component	References
Mouse	Reduce total serum cholesterol, LDL, VLDL cholesterol, and phospholipids Increase serum HDL cholesterol Reduce high hepatic lipids induced by a high-fat and high-cholesterol diet	16% *Spirulina* in a high-fat and high-cholesterol diet	Kato et al. (1984)
Wistar rats	Increase lipoprotein lipase and hepatic triglyceride lipase activity	*Spirulina* (5, 10, and 15%)	Iwata et al. (1990)
CD-1 mice	Pretreatment of *Spirulina* before fatty liver induction resulted in decrease of hepatic total lipids, triacylglycerols, and serum triacylglycerols, accompanied by an increase in serum HDL cholesterol	*Spirulina*	Blé-Castillo et al. (2002)
Diabetic mice	Prevent fatty liver production Decrease serum and hepatic triacylglycerols Fully or partially normalize HDL, LDL, and VLDL cholesterol levels	5% *Spirulina*	Rodríguez-Hernández et al. (2001)
Hamsters	Exhibit lower total cholesterol, LDL, and VLDL cholesterol, whereas HDL cholesterol was not affected Reduce aortic fatty streak area	*Spirulina* or phycocyanin	Riss et al. (2007)
Rabbits	Serum total cholesterol was decreased by 49% while HDL cholesterol was increased by 25%	*Spirulina*	Colla et al. (2008)
Healthy volunteers (60 men)	Total serum cholesterol and LDL were reduced significantly Triglyceride levels decreased slightly, whereas HDL-cholesterol showed no significant changes The reduction of serum cholesterol was even greater in those men with the highest cholesterol levels	*Spirulina*	Nakaya et al. (1988), Park and Kim (2003), Torres-Duran et al. (2007), Park et al. (2008)
52 adult Cretan outpatients, 3 months (32 men, 20 women)	The levels of triglycerides, low-density lipoprotein cholesterol, total cholesterol, nonhigh-density lipoprotein cholesterol levels, and the ratio of total cholesterol to high-density lipoprotein cholesterol were significantly decreased by 16.3%, 10.1%, 8.9%, 10.8%, and 11.5%, respectively	*Arthrospira maxima*	Mazokopakis et al. (2014)
30 patients with ischemic heart disease	Total plasma cholesterol, LDL, VLDL, and triglycerides were significantly reduced by 22.4% (2 g group)/33.5% (4 g group), 31%/45%, 22%/23%, and 22%/23%, respectively. HDL was significantly increased by 11.5%/12.8%. In addition, a significant reduction in body weight was achieved in both treatment groups	*Spirulina*	Ramamoorthy and Premakumari (1996)
Patients with type II diabetes mellitus	Decrease in total serum cholesterol, triglycerides, free fatty acid levels, LDL and VLDL, blood sugar, and glycated serum protein levels	*Spirulina*	Mani et al. (2000), Parikh et al. (2001), Lee et al. (2008), Kaur et al. (2008)
Patients with nephritic syndrome	Total serum cholesterol, LDL cholesterol, and triglycerides were all significantly decreased by 46, 33, and 45 mg/dL, respectively. The ratios	*Spirulina*	Samuels et al. (2002)

(Continued)

TABLE 57.2 (Continued)

Subject	Specific effects	Bioactive component	References
	of LDL/HDL and total cholesterol/HDL were also decreased significantly		
Elderly women with hypercholesterolemia	Serum levels of total cholesterol, LDL cholesterol, and oxidized LDL were significantly reduced. In addition, apolipoprotein B, IL-6, and IL-6 production by peripheral blood lymphocyte were also decreased	*Spirulina*	Kim and Kim (2005)
41 runners, 15 days (21 men and 20 women)	The levels of triglycerides and the postprandial area under the curve of TAG concentrations were decreased	*Arthrospiramaxima*	Torres-Durán et al. (2012)
3 patients	The levels of triacylglycerols, total cholesterol, high-density lipoprotein cholesterol, alanine aminotransferase, and low-density lipoprotein cholesterol levels were modified	*A. maxima*	Ferreira-Hermosillo et al. (2010)
Insulin-resistant rat	Decrease in total bone mineral density by Rosiglitazone was improved Decrease in fasting serum glucose, HDL, LDL, and triglycerides levels	*Spirulina*	Gupta et al. (2010)
Volunteers, rats	Decrease in levels of hyperlipidemia	*Spirulina*	El-Sheekh et al. (2014)
Alloxan-injured mice	Inhibited p53 pathway that led to the protection of pancreatic islets from alloxan injury Promoted liver glycogen synthesis Reduced blood glucose level	Phycocyanin	Ou et al. (2012)
Diabetic rats	Increase in body weight Decrease in the levels of blood glucose, glycosylated hemoglobin, serum glutamic oxaloacetic transaminase, serum glutamic pyruvic transaminase, alkaline phosphatase, total bilirubin, serum creatinine, serum uric acid, and blood urea nitrogen	Crude, aqueous extract, ethanolic extract, insulin-like protein of *Spirulina*	Anwer et al. (2012, 2013)

HDL, high-density lipoproteins; *LDL*, low-density lipoproteins; *TAG*, triacylglycerol.

diet resulted in a significant reduction in blood and hepatic lipids (Kato et al., 1984; Deng and Chow, 2010). In another study (Blé-Castillo et al., 2002), mice were fed with *Spirulina* for 2 weeks and fatty liver induction decreased hepatic total lipids by 40%, TG by 50%, and serum TG by 45%, accompanied by a 45% increase in serum HDL cholesterol. The hypolipidemic activity of *Spirulina* was also confirmed in a diabetic mouse model which was induced by administration of alloxan (Rodríguez-Hernández et al., 2001). Mice receiving a diet containing 5% *Spirulina* for 1 week after the administration of alloxan for 4 weeks totally prevented fatty liver production, decreased serum and hepatic TG, and fully or partially normalized HDL, LDL, and VLDL cholesterol levels. In another trial in alloxan-injured mice, oral administration of c-phycocyanin provided notable hypolipidemic, hypoglycemic, and antioxidant effects (Ou et al., 2012; Deng and Chow, 2010). *Spirulina* decreases fasting blood glucose, glycosylated serum protein, cholesterol (TC), and TG levels, maintains total antioxidative capability, averts malondialdehyde (MDA) formation, increases the levels of hepatic glycogen level, and maintains glucokinase expression in the liver. Very recently, *Spirulina* was shown to enhance the hypolipidemic effect in men with excess body weight and dyslipidemia (Hernandez-Lepe et al., 2019).

Hyperlipidemia induced in Wistar rats by a high-fructose diet was significantly improved by increasing percentages of *Spirulina* in the diet (Iwata et al., 1990). Moreover, *Spirulina* feeding resulted in a significant increase in the lipoprotein lipase and hepatic triglyceride lipase activity, which suggests that *Spirulina* could be used to improve the hyperlipidemia induced by a high-fructose diet. In another report, treatment of diabetic rats with crude, aqueous extract, ethanolic extract, and insulin-like protein of *Spirulina* successfully ameliorated diabetic complications by increasing body weight and significantly decreasing the levels of blood glucose, glycosylated hemoglobin, serum glutamic oxaloacetic transaminase, serum glutamic pyruvic transaminase, alkaline phosphatase, total bilirubin, serum creatinine, serum uric acid, and blood urea nitrogen (Anwer et al., 2013). The hypolipidemic effects of *Spirulina* have also been observed in insulin-resistant rats (Gupta et al., 2010). Fasting serum glucose, HDL, LDL, and TG levels clearly improved following *Spirulina* administration. It is suspected that the chromium and gamma-linoleic acid in *Spirulina* helped to decrease the fasting serum glucose, HDL, LDL, and TG levels in insulin-resistant rats. The hypolipidemic effects of *Spirulina* observed in mice and rats were verified in hamsters (Riss et al., 2007) and rabbits (Colla et al., 2008).

A group of hamsters fed an atherogenic diet supplemented with *Spirulina* or c-phycocyanin exhibited lower total cholesterol, LDL, and VLDL cholesterol (Riss et al., 2007). In a study with hypercholesterolemic rabbits, a diet of *Spirulina* (0.5 g daily) for 30 and 60 days decreased serum total cholesterol by 49%, while HDL cholesterol was increased by 25% (Colla et al., 2008). *Spirulina* decreased the levels of hyperlipidemia in these animals probably via a marked reduction of aminotransferase through correcting the lipid profile and increasing high-density lipoprotein (El-Sheekh et al., 2014).

Clinical studies

The first human study was carried out in 1988 with 30 healthy male volunteers with mild hyperlipidemia or hypertension (Nakaya et al., 1988). The results of this study indicated that intake of *Spirulina* for 4 and 8 weeks significantly decreased total serum cholesterol and the decrease was more marked in mild hypercholesterolemic than in normocholesterolemic subjects. However, discontinuation of *Spirulina* supplement for 4 weeks resulted in a return to the previous cholesterol level. In a recent clinical trial with 36 healthy volunteers between ages 18 and 65 (Torres-Duran et al., 2007; Deng and Chow, 2010), ingestion of *Spirulina* at a dose of 4.5 g/day for 6 weeks decreased total plasma cholesterol and TG by 10% and 28%, respectively. Meanwhile, the HDL cholesterol was increased by 15%, whereas LDL cholesterol was significantly decreased. In addition, both systolic and diastolic blood pressures were significantly reduced in both men and women.

The hypolipidemic effect of *Spirulina* was also demonstrated in ischemic heart disease patients with hypercholesterolemia (Ramamoorthy and Premakumari, 1996). After 3 months of *Spirulina* supplementation, total plasma cholesterol of the patients was significantly decreased by 22.4% and 33.5% in groups receiving 2 and 4 g *Spirulina*, respectively, whereas no significant change was detected in the control group. Meanwhile, the LDL and VLDL cholesterol levels were significantly reduced, whereas HDL was significantly increased by 11.5% and 12.8%.

In addition, hypolipidemic effects of *Spirulina* in type II diabetic patients were observed. In a before-and-after study with 15 type II diabetes patients (Mani et al., 2000), supplementation of *Spirulina* at a dose of 2 g/day for 2 months resulted in a significant decrease in total serum cholesterol, TG, and free fatty acid levels, as well as LDL and VLDL cholesterol levels in blood. Moreover, blood sugar and glycated serum protein levels were also significantly decreased. In a second randomized and controlled study (Parikh et al., 2001), subjects in the study group received *Spirulina* at a dose of 2 g/day for 2 months. Besides reduced LDL, VLDL, and TG levels, the level of apolipoprotein B showed a significant drop with a concurrent significant increase in the level of apolipoprotein A1. These findings from earlier studies were confirmed in two recent human clinical trials with type II diabetic patients (Lee et al., 2008; Kaur et al., 2008). Both trials were randomized, controlled studies with a relatively large sample size. One study enrolled 37 patients and showed significantly reduced total serum cholesterol, LDL fraction, TG levels, and blood pressure after a dose of 8 g/day *Spirulina* for 12 weeks (Lee et al., 2008). The second trial included 60 male patients aged from 40 to 60 years, who received 1 or 2 g *Spirulina* daily for 2 months (Kaur et al., 2008). A significant decrease was observed in total serum cholesterol, TG, LDL and VLDL cholesterol, and fasting and postprandial blood glucose.

The hypolipidemic benefit of *Spirulina* was also reported in patients with nephritic syndrome and hyperlipidemia (Samuels et al., 2002). Supplementation of *Spirulina* at a dose of 1 g/day for 2 months resulted in a reduction in total serum cholesterol, LDL fraction, and TG by 46, 33, and 45 mg/dL, respectively. The ratios of LDL/HDL and total cholesterol/HDL were also decreased significantly. Moreover, three human clinical studies have been carried out to investigate the therapeutic effects of *Spirulina* in an elderly population (Park and Kim, 2003; Kim and Kim, 2005; Park et al., 2008). In one study with 12 subjects (six male and six female) between the ages 60 and 75 (Park and Kim, 2003), subjects received a supplement of *Spirulina* at a dose of 7.5 g/day for 24 weeks. Plasma concentrations of TG, total cholesterol, and LDL fraction were decreased after 4 weeks of the supplementation. The second before-and-after trial included 26 women over 60 years of age with hypercholesterolemia (Kim and Kim, 2005).

Intake of *Spirulina* at a dose of 7.5 mg/day for 8 weeks resulted in a significant reduction in serum levels of total cholesterol, LDL cholesterol, and oxidized LDL, as well as apolipoprotein B levels. Seventy-eight subjects between the ages of 60 and 87 (Park et al., 2008) consumed *Spirulina* at a dose of 8 g/day for 16 weeks. Total plasma cholesterol and LDL fraction were only significantly reduced in female subjects. Therefore, these data largely support the notion that *Spirulina* supplementation is beneficial for the elderly. In Cretan Greek dyslipidemic patients (Mazokopakis et al., 2014), a 3-month intervention period of *Spirulina* supplementation decreased the TG, LDL, total cholesterol, and nonhigh-density lipoprotein cholesterol levels.

Effects against heavy metal/chemical-induced toxicity

Apart from its importance as a food additive for supplementary dietary proteins and enhanced innate immunity, there is much potential for medical and therapeutic applications. *Spirulina* was shown to protect animals from sublethal levels of some chemicals. It could reduce the toxicity from heavy metals or chemicals, for example, Cd, aflatoxins, CCl_4, etc. Studies on the antitoxic properties of *Spirulina* are summarized in Table 57.3.

Spirulina significantly protected against cisplatin-induced nephrotoxicity through its antioxidant properties. The raised plasma urea, creatinine, urinary b-NAG, plasma and kidney tissue MDA, and histomorphological changes in cisplatin-induced rats were significantly attenuated by *Spirulina* (Mohan et al., 2006). Both the ethanol extract of *Spirulina laxissima* and *Spirulina* powder have protective effects against carbon tetrachloride (CCl_4)-induced hepatotoxicities in rats (Gad et al., 2011). In another study with rats (Torres-Durán et al., 1999), fatty liver was induced by intraperitoneal injection of CCl_4, resulting in an increase in

TABLE 57.3 The antitoxic properties of *Spirulina*.

Subject	Inducing agents	Specific effects	Component	Reference
Rat	Acetic acid	Increase in antioxidant enzymes activity Inhibited lipid peroxidation and inflammation markers	*Spirulina* (500 mg/kg bw)	Abdel-Daim et al. (2015)
Rat	Aflatoxins	Decrease in lipid peroxidation Increase in glutathione content Upregulated PHGPx gene expression Inhibited DNA damage	*Spirulina* (50 mg/kg bw)	Hassan et al. (2012)
Rat	Cisplatin	Reduced plasma urea, creatinine, urinary b-NAG Decrease in plasma and kidney tissue malondialdehyde Prevented histomorphological changes	*Spirulina* (1000 mg/kg)	Mohan et al. (2006)
Rat	CCl4	Decrease in liver total cholesterol and triacylglycerols levels	*Spirulina*	Torres-Durán et al. (1999)
Rat	CCl4	Increase in the concentration of antioxidant enzymes Prevented liver damage	*A. laxissima* (ethanol extract); *Spirulina*	Gad et al. (2011)
Rat	Cyclophosphamide	Increase in the malondialdehyde, superoxide dismutase and catalase levels in tissues Inhibited the histomorphologic alteration in urinary bladder Inhibited cell apoptosis in kidney	*Spirulina* (1 g/kg bw/ day, 7 days)	Sinanoglu et al. (2012)
Rat	6-OHDA	Increase in locomotor activity Decrease in nitric oxide, ROS, and lipoperoxidation	*Arthrospira maxima* (0.7 g/kg/day, 40 days)	Tobón-Velasco et al. (2013)
Female mice	Cd	Decrease in the frequency of fetuses with exencephaly, micrognathia, and skeletal abnormalities Decrease in lipid peroxidation	*Spirulina* (62.5, 125, 250, or 500 g/kg bw/ day)	Paniagua-Castro et al. (2011)
Pregnant rats	Sodium fluoride	Facilitated antioxidant formation, improved behavior, and protected Purkinje cells in offspring	*Spirulina* (250 and 500 mg/kg)	Banji et al. (2013)

PHGPx, phospholipid hydroperoside glutathione peroxidast.

liver total cholesterol and triacylglycerols. However, such increases were significantly reduced by feeding oil extracts of *Spirulina* or the defatted fraction of *Spirulina*. In addition, a CCl_4-induced increase in total cholesterol was completely prevented by feeding a diet containing whole *Spirulina*. It is also evidenced that the cyclophosphamide-induced nephrotoxicity and hemorrhagic cystitis in rats can be proved by pretreatment with *Spirulina* and its antioxidant properties (Sinanoglu et al., 2012).

There is evidence to support that an impaired energy metabolism and excessive generation of ROS contribute to brain injury in neurodegenerative disorders such as Parkinson disease, whereas diets enriched in foods with *A maxima* could modulate its progression. The protective effect of *A. maxima* against 6-OHDA-induced toxicity in rat striatum was via restored locomotor activity, which is associated with the decreased levels of nitric oxide, ROS, and lipoperoxidation in the striatum, although mitochondrial functions and dopamine levels remained preserved (Tobón-Velasco et al., 2013). Moreover, it prevents cadmium teratogenicity in imprinting control region mice by decreasing lipid peroxidation, which was dramatically increased by administration of the metal (Paniagua-Castro et al., 2011). When pretreating pregnant rats with *Spirulina* before inducing thyroid, neurodevelopment, and oxidative alterations by sodium fluoride, it augmented the displacement of fluoride, facilitated antioxidant formation, improved behavior, and protected Purkinje cells in offspring (Banji et al., 2013).

Anticancer

Strong evidence has shown that *A. platenesis* is imbued with antitumor and anticancer properties. Evidence for the multidimensional applications of the c-phycocyanin content of *A. platensis* is summarized in Table 57.3. The *A. platensis* exerts stimulatory effects in inhibiting tumor cell proliferation (Basha et al., 2008; Ismail et al., 2009; Ouhtit et al., 2014). It can promote the secretion of TNF-α and IFN-γ, induce tumor cell apoptosis mostly via the p53/p21 pathway, and modulate the Bcl2/BAX ratio in human hepatoma cell lines (HepG2), human chronic myeloid leukemia cell line (K562), macrophages, and HeLa cells (Subhashini et al., 2004; Ismail et al., 2009; Parages et al., 2012; Shokri et al., 2014). Hot-water extract of *A. platensis* facilitated enhanced antitumor activity of NK cells and augmented IFN-γ production in rats (Akao et al., 2009). High production of TNF-α in macrophages was recorded in the presence of acidic polysaccharides from *A. platensis* (Parages et al., 2012). In a 7,12-dimethylbenz[α]anthracene (DMBA)-induced rat breast cancer model, *A. platensis* cleared DMBA-induced rat mammary tumors and reduced tumor incidence from 87% to 13% (Grawish et al., 2010).

Recently, complex polysaccharides from *Spirulina* have brought about suppression of glioma cell growth by downregulating angiogenesis via partial regulation of IL-17 production (Kawanishi et al., 2013). Li et al. (2006) have shown that with increased c-phycocyanin concentration, expression of CD59 proteins in HeLa cells was promoted while Fas protein-induced apoptosis. A recent study showed that Fe_3O_4/Ag nanocomposite biosynthesized from *S. platensis* extract was effective in the inhibition of breast cancer (Salehzadeh et al., 2019). The anticancer properties of *Spirulina* are summarized in Table 57.4.

Antiviral and antibacterial activities

The antiviral and antibacterial activities of *A. platensis* are summarized in Table 57.4. This antiviral activity, in large part, is attributable to the richness of *A. platensis* in vital proteins, fatty acids, minerals, and other important constituents. It was previously found that aqueous extracts from *A. platensis* inhibited HSV-1 (Hayashi et al., 1993). Subsequently, Ayehunie et al. (1998) determined that an aqueous extract of *A. platensis*, at concentrations that were nontoxic to human cells, inhibited syncytium formation and HIV-1 replication in human T-cell lines, PBMC, and Langerhans cells. Previously, calcium spirulan (Ca-SP), a novel sulfated polysaccharide that was isolated from hot water extract of *A. platensis*, showed antiviral activities against different enveloped viruses such as herpes simplex virus type-1, human cytomegalovirus, measles virus, mumps virus, influenza A virus, and HIV-1 (Hayashi et al., 1996a,b; Hernández-Corona et al., 2002). The initial analyses indicated that Ca-SP was composed of rhamnose, ribose, mannose, fructose, galactose, xylose, glucose, glucuronic acid, galacturonic acid, sulfate, and calcium. This highly sought after antiviral activity has been suggested to be due to the effect that the chelation of calcium ions to sulfate groups has on molecular conformation.

However, *Spirulina* is not without antibacterial activity. Microalgal cultures of *A. platensis* have displayed significant antibacterial activity against *Staphylococcus aureus*, *Streptococcus pyogenes*, *Escherichia coli*, *Aspergillus niger*, *Salmonella typhimurium*, *S. aureus*, *Bacillus cereus*, and *C. albicans* (Ozdemir et al., 2004; Mendiola et al., 2007; Kumar et al., 2011; Sarada et al., 2011; Najdenski et al., 2013). The antiviral and antibacterial components of *Spirulina* and their properties are summarized in Table 57.5.

Effects against malnutrition

Arthrospira is among the richest sources of protein. In a study that attempted to use *Spirulina* as a protein

TABLE 57.4 The anticancer properties of *Spirulina*.

Subject	Specific effects	Bioactive component	References
Breast cancer mice	Decrease in the secretion of IL-4 and IL-10 Increase in the secretion of IL-17, TNF-α, and IFN-γ	*Arthrospira platensis*	Shokri et al. (2014)
Human hepatoma cell lines (HepG2)	Inhibited the proliferation of cells	C-Phycocyanin	Basha et al. (2008)
	Reduced the PCNA and p53 expression induced by dibutyl nitrosamine precursors Inhibited cell proliferation Increase in p21 and decrease in Rb expression Increase in Bax and decrease in Bcl-2 expression and induced apoptosis	*A. platensis*	Ismail et al. (2009)
Human chronic myeloid leukemia cell line (K562)	Induced apoptosis by release of cytochrome c into the cytosol and poly (ADP) ribose polymerase (PARP) cleavage, and downregulation of antiapoptotic Bcl-2 but without any changes in proapoptotic Bax	C-Phycocyanin	Subhashini et al. (2004)
Mammary tumors of DMBA-induced rat	Regression of cancer progression and cell proliferation Clear DMBA-induced rat mammary tumors Reduced the incidence of breast tumors from 87% to 13% Reduced expression of both Ki-67 and estrogen alpha Increased p53 expression and its downstream target gene, Cdkn1a [alias p21 or p21(Waf1/Cip1)] Increased Bax but decreased Bcl-2 expression Induced apoptosis	*A. platensis*	Grawish et al. (2010), Ouhtit et al. (2014)
Syngeneic tumor-implant mice (melanoma)	Augmented IFN-γ production Enhanced tumoricidal natural killer activation through the MyD88 pathway	*A. platensis*	Akao et al. (2009)
Syngeneic tumor-implant mice (glioma)	Suppressed glioma growth by downregulating angiogenesis via a Toll-like receptor 4 signal	*Spirulina* complex polysaccharides	Kawanishi et al. (2013)
Macrophages	Increase in TNF-α production	Acidic polysaccharide fraction from *A. platensis*	Parages et al. (2012)
HeLa cells	Promoted the expression of CD59 protein and Fas protein	C-phycocyanin	Li et al. (2006)

DMBA, dimethylbenz[α]anthracene; *PCNA*, proliferating cell nuclear antigen.

supplement, it replaced up to 40% of the protein content in tilapia diets (Olvera-Novoa et al., 1998). In a study conducted by Kapoor and Mehta (1998), dietary supplementation of *A. platensis* was found to increase the iron storage of rats, greater than that achieved from the combination of casein and wheat gluten diets, during the first half of pregnancy and lactation. Most recently, a study has shown that *Spirulina* could be a protein source in the nutritional recovery of rats subjected to protein malnutrition. It demonstrated that the nutritional recovery of animals was enabled by different concentrations (8.8%–26.4%) of *Spirulina* (Moreira et al., 2013).

Effects against anemias

Dietary supplementation of *Spirulina* has helped in alleviating the incidence of anemia experienced during pregnancy, lactation, aging, and toxicosis. As reported by Simsek et al. (2009), the consumption of *Spirulina* modulated the decrease of red blood cells, white blood cells, and α-naphthyl acetate esterase-positive T lymphocytes, as well as the values of hemoglobin, packed cell volume, and mean corpusular hemoglobin concerntration by lead and chrome. Over the 12-week study period, there was a steady increase in average values of mean corpuscular hemoglobin in subjects of both sexes. In addition, mean

TABLE 57.5 The antiviral and antibacterial components of *Spirulina* and their properties.

Specific effects	Bioactive component	References
Inhibit the replication of HIV-1 in human peripheral blood mononuclear cells (EC_{50} 0.3–1.2 μg/mL) and Langerhans cells (EC_{50} 0.8–3.1 mg/mL)	Tannins and polysaccharide fraction and *Arthrospira platensis*	Ayehunie et al. (1998)
Antiviral activity in the polysaccharide fraction and also in a fraction depleted of polysaccharides and tannins		
Inhibit the replication of HSV-1 in HeLa cells (EC_{50} 0.08–50 mg/mL)	Water-soluble extract of *A. platensis*	Hayashi et al. (1993)
Inhibit the replication of HSV-1, human cytomegalovirus, measles virus, mumps virus, influenza A virus, and HIV-1	Sulfated polysaccharide (calcium spirulan)	Hayashi et al. (1996a,b)
Inhibit the replication of HSV-2 (ED50: 0.069 mg/mL), pseudorabies virus (ED50: 0.103 mg/mL), human cytomegalovirus (ED50: 0.142 mg/mL), and HSV-1 (ED50: 0.333 mg/mL)	Hot-water extract of a commercial preparation of *Arthrospira maxima*	Hernández-Corona et al. (2002)
Inhibit *Staphylococcus aureus, Escherichia coli, Candida albicans, Aspergillus niger*	Supercritical fluid extraction of *A. platensis* at 220 bar and 26.7°C with 10% ethanol	Mendiola et al. (2007)
Inhibit *S. aureus* and *Salmonella typhimurium*	Methanol extract from *A. platensis* (mostly fatty acid compounds)	Ozdemir et al. (2004), Kumar et al. (2011)
Inhibit the growth of *Streptococcus pyogenes, S. aureus, Bacillus cereus, E. coli,* and *C. albicans* (MIC ranging from 1.56 to 12.5 mg/mL)	Phycobiliproteins from *Arthrospirafusiformis*	Najdenski et al. (2013)
Inhibit the growth of drug-resistant bacteria *E. coli, Klebsiella pneumoniae, Pseudomonas aeruginosa,* and *S. aureus*	Phycocyanin	Sarada et al. (2011)

corpuscular volume and mean corpuscular hemoglobin concentration also increased in male participants. Older women appeared to benefit more rapidly from *Spirulina* supplements. Similarly, the majority of subjects manifested increased indoleamine 2,3-dioxygenase activity and white blood cell count at 6 and 12 weeks of *Spirulina* supplementation (Selmi et al., 2011).

Others

Spirulina has proven to be effective in treating memory dysfunction in senescence-accelerated-resistant (SAMR1) mice by reducing amyloid β-protein (Aβ) deposition and lipid peroxidation in the hippocampus, striatum, and cortex and by inducing catalase and glutathione peroxidase activity. Oral submucous fibrosis is a habit-associated insidious precancerous condition of the oral cavity commonly found in Asian countries. A recent study has shown that *Spirulina* displayed promising effects in the treatment of oral submucous fibrosis and that it was superior in reducing the characteristic burning sensation (Mulk et al., 2013).

After rats were administered with *Spirulina* orally once daily for 5 consecutive weeks, there were significant downregulations in hepatic expression levels and inhibition in enzymatic activities of CYP1A2 and CYP2E1 compared to controls. In addition, a significant decrease was observed in CYP2C6-associated enzyme activity with no remarkable changes in messenger RNA (mRNA)/protein levels. The spirulan (SP) application resulted in significant increases in mRNA/protein levels of both CYP2B1 and CYP3A1 without a significant change in enzyme activities (Savranoglu and Tumer, 2013).

Two bioactive peptides P1 (LDAVNR; 686 Da) and P2 (MMLDF; 655 Da) purified from gastric enzymatic hydrolysate of *A. maxima* have shown protective effects against early atherosclerotic responses induced by histamine in EA.hy926 endothelial cells. Interestingly, both P1 and P2 have exhibited inhibitory activities on the production and expression of IL-6 and MCP-1 and have inhibited the production of adhesion molecules including P-selectin and E-selectin, and thus reducing in vitro cell adhesion of monocytes onto endothelial cells. Notably, the inhibitory activities of P1 and P2 are probably via the downregulation of Egr-1 expression in the mitogen-activated protein kinase activation pathway (Vo and Kim, 2013). However, supplementation of *Spirulina* in arsenical palmer keratosis patients' diet failed to return the vitamin E and linoleic acid levels back to normal.

Safety profile

Spirulina has been consumed by humans as a food for a long time and it is currently widely used as a food supplement, especially in Asia. Over the history of *Spirulina* consumption, it has been generally considered to be nontoxic to humans. However, the safety evaluation data of *Spirulina* are mostly from animals in preclinical studies. Few clinical studies have been carried out to systemically establish the safety profile of *Spirulina* for humans.

To evaluate the relative safety of feeding *Spirulina* long-term, mice were fed with a diet containing increasing percentages (0%, 10%, 20%, and 30% w/w) of *Spirulina* for 13 weeks (Salazar et al., 1998). Ingestion of *Spirulina* had no effects on behavior, food/water intake, growth, or survival. Hematologic and clinical chemistry analyses revealed no abnormalities. In addition, no gross or microscopic changes were detected upon histological evaluation. In an in vivo cytogenetic assay, the dietary supplementation of *Spirulina* to mice at doses of 10.5, 21, and 42 mg/kg bw did not cause clastogenic effects or any adverse effects (Ghoshal et al., 2001). In a recent study with short- and long-term *Spirulina* feeding in rodents (Hutadilok-Towatana et al., 2008), feeding male Swiss mice with a high dose of *Spirulina* (30 g and 10 g/kg body weight of fresh and dried *A. platensis*, respectively) for 7 days resulted in no signs of toxicity. In a subchronic toxicity study, Sprague-Dawley rats were administered with *Spirulina* at various doses for 12 weeks. Consumption of *Spirulina* also showed no effects on behavior, food/water intake, growth, health status, and measurements of clinical chemistry (Hutadilok-Towatana et al., 2008).

To determine the effects of *Spirulina* feeding on the growth and development of the embryo and fetus, studies with pregnant rats have been conducted (Chamorro et al., 1989; Chamorro and Salazar, 1990). In one study (Chamorro et al., 1989), no teratogenicity was detected in rats following the consumption of *Spirulina* even at the highest dose and longest duration. Meanwhile, *Spirulina* in the diet did not change maternal nor fetal weight. Consistent results were obtained with a similar study in pregnant rats (Chamorro and Salazar, 1990). Supplementation of *Spirulina* in the diet at doses much higher than any anticipated human consumption did not cause any signs of embryotoxic effects. A study with pregnant rats to assess general reproductive performance showed that *Spirulina* feeding did not change body weight, showed no signs of toxicity, and was not associated with adverse effects including for fertility, gestation, and abnormal pups (Salazar et al., 1996). However, weight gain and litter size were found to be associated with *Spirulina*/wheat gluten diet in rats. Taken together, it is concluded that *Spirulina* diet has no detectable adverse effects on reproductive performance, or embryo and fetal development and growth.

Despite a favorable safety profile in rodents, there were reports that raised concerns about the safety of *Spirulina* consumption. For example, consumption of *Spirulina* from a water source polluted with mercury and other heavy metals would result in increased deposits of these heavy metals causing toxic effects (Johnson and Shubert, 1986). Certain cyanobacterial species produce cyanotoxin (anatoxin-α) and contamination of those species within *Spirulina* products may be deleterious for consumers (Rellán et al., 2009). A case of rhabdomyolysis was recently reported as a result of *Spirulina* intake (Mazokopakis et al., 2008). Finally, an association of *Spirulina* consumption and the development of a mixed immunoblistering disorder with characteristic features of bullous pemphigoid and pemphigus foliaceus was reported in an 82-year-old healthy woman (Kraigher et al., 2008).

In summary, *Spirulina* is generally considered safe for human consumption as supported by its long history of use as a food source and its favorable safety profile in animal studies. However, rare cases of side effects have been reported and should be taken into consideration. In addition, further clinical studies are required to systemically establish the safety profile of *Spirulina* in humans. Finally, quality control in the growth and processing of *Spirulina* to avoid contamination is mandatory to guarantee the safety of *Spirulina* products.

Concluding remarks and future directions

Spirulina is used widely throughout the world as a dietary supplement. Today, it is considered a potential solution for combating malnutrition, oxidative stress, cancer, viruses, bacteria, hyperlipidemia, diabetes, obesity, inflammatory allergic reactions, heavy metal/chemical-induced toxicity, radiation damage, and anemia in clinical and preclinical research. It seems to offer significant health advantages to the consumer.

Recently, great attention and extensive studies have been devoted to evaluating the therapeutic benefits of *Spirulina* on various disease conditions including hypercholesterolemia, hyperglycerolemia, cardiovascular diseases, inflammatory diseases, cancer, and viral infections. Data from preclinical studies with various animal models consistently demonstrate the therapeutic benefits of *Spirulina*. Although there are differences in study design, sample size, and patient conditions, the findings from human clinical trials are largely consistent with the effects of *Spirulina* observed in the preclinical studies. However, the effects of *Spirulina* in relation to these benefits require larger clinical trials.

Spirulina is generally considered safe for human consumption supported by its long history of use and its favorable safety profile in animal studies. However, rare cases of side effects have been reported. Additional preclinical and clinical studies should also be taken into consideration. Finally, quality control in the growth and processing of *Spirulina* to avoid contamination is mandatory to guarantee the safety of *Spirulina* products.

In addition, our understanding of the underlying mechanisms for *Spirulina*'s activities is still limited. Future studies to identify the active ingredients in *Spirulina* and uncover the mechanistic insights into its therapeutic effects will provide the basis for developing new drugs for preventing or treating diseases, such as cardiovascular diseases and hypercholesterolemia.

References

Abdel-Daim, M.M., Farouk, S.M., Madkour, F.F., et al., 2015. Anti-inflammatory and immunomodulatory effects of *Spirulina platensis* in comparison to *Dunaliella salina* in acetic acid-induced rat experimental colitis. Immunopharmacol. Immunotoxicol. 37 (2), 126–139.

Abdulqader, G., Barsanti, L., Tredici, M.R., 2000. Harvest of *Arthrospira platensis* from Lake Kossorom (Chad) and its household usage among the Kanembu. J. Appl. Phycol. 12, 493–498.

Abu-Taweel, G.M., Mohsen, G.A.M., Antonisamy, P., et al., 2019. Spirulina consumption effectively reduces anti-inflammatory and pain related infectious diseases. J. Infect. Public Health 12, 777–782.

Akao, Y., Ebihara, T., Masuda, H., et al., 2009. Enhancement of antitumor natural killer cell activation by orally administered Spirulina extract in mice. Cancer Sci. 100, 1494–1501.

Anwer, R., Khursheed, S., Fatma, T., 2012. Detection of immunoactive insulin in Spirulina. J. Appl. Phycol. 24, 583–591.

Anwer, R., Alam, A., Khursheed, S., et al., 2013. Spirulina: possible pharmacological evaluation for insulin-like protein. J. Appl. Phycol. 25 (3), 883–889.

Ayehunie, S., Belay, A., Baba, T.W., et al., 1998. Inhibition of HIV-1 replication by an aqueous extract of *Spirulina platensis* (*Arthrospira platensis*). J. Acquir. Immune Defic. Syndr. Hum. Retrovirol. 18, 7–12.

Banji, D., Banji, O.J., Pratusha, N.G., et al., 2013. Investigation on the role of *Spirulina platensis* in ameliorating behavioural changes, thyroid dysfunction and oxidative stress in offspring of pregnant rats exposed tofluoride. Food Chem. 140, 321–331.

Barron, B.L., Torres-Valencia, J.M., Chamorro-Cevallos, G., et al., 2008. In: Gershwin, M.E., Belay, A. (Eds.), Human Nutrition and Health. CRC Press, Taylor and Francis Group, Boca Raton, FL, p. 227.

Basha, O.M., Hafez, R.A., El-Ayouty, Y.M., et al., 2008. C-Phycocyanin inhibits cell proliferation and may induce apoptosis in human HepG2 cells. Egypt. J. Immunol. 15 (2), 161–167.

Blé-Castillo, J.L., Rodríguez-Hernández, A., Miranda-Zamora, R., et al., 2002. Arthrospira maxima prevents the acute fatty liver induced by the administration of simvastatin, ethanol and a hypercholesterolemic diet to mice. Life Sci. 70, 2665–2673.

Chamorro, G., Salazar, M., 1990. Teratogenic study of Spirulina in mice. Arch. Latinoam. Nutr. 40, 86–94.

Chamorro, G., Salazar, M., Salazar, S., 1989. Teratogenic study of Spirulinain rats. Arch. Latinoam. Nutr. 39, 641–649.

Chei, S., Oh, H., Song, J., et al., 2020. *Spirulina maxima* extract prevents activation of the NLRP3 inflammasome by inhibiting ERK signaling. Sci. Rep. 10, 2075.

Chen, H.W., Yang, T.S., Chen, M.J., et al., 2014. Purification and immunomodulating activity of C-phycocyanin from *Spirulina platensis* cultured using power plant flue gas. Process Biochem. 49, 1337–1344.

Chu, W.L., Quynh le, V., Radhakrishnan, A.K., 2013. Effect of Spirulina (Arthrospira) supplementation on the immune response to tetanustoxoid vaccination in a mouse model. J. Diet. Suppl. 10 (3), 229–240.

Colla, L.M., Muccillo-Baisch, A.L., Costa, J.A.V., 2008. *Spirulina platensis* effects on the levels of total cholesterol, HDL and triacylglycerols in rabbits fed with a hypercholesterolemic diet. Braz. Arch. Biol. Technol. 51, 405–411.

Dangeard, P., 1940. Sur une algue bleue alimentaire pour l'homme: *Arthrospira platensis* (Nordstedt) Gomont. Actes Soc. Linn. Boreaux Extr. Procés-verbaux 91, 39–41.

Deng, R., Chow, T.J., 2010. Hypolipidemic, antioxidant and antiinflammatory activities of microalgae spirulina. Cardiovasc. Ther. 28, 33–45.

El-Sheekh, M.M., Hamad, S.M., Gomaa, M., 2014. Protective effects of Spirulina on the liver function and hyperlipidemia of rats and human. Braz. Arch. Biol. Technol. 57, 77–86.

El Sheikh, S.M., Shalaby, M.A.M., Hafez, R.A., et al., 2014. The immunomodulatory effects of probiotic bacteria on peripheral blood mononuclear cells (PBMCS) of allergic patients. Am. J. Immunol. 10 (3), 116–130.

Ferreira-Hermosillo, A., Torres-Duran, P.V., Juarez-Oropeza, M.A., 2010. Hepatoprotective effects of *Spirulina maxima* in patients with non-alcoholic fatty liver disease: a case series. J. Med. Case Rep. 4, 103.

Gad, A.S., Khadrawy, Y.A., El-Nekeety, A.A., et al., 2011. Antioxidant activity and hepatoprotective effects of whey protein and Spirulina in rats. Nutrition 27 (5), 582–589.

Gargouri, M., Akrouti, A., Magne, C., et al., 2020. Protective effects of spirulina against hemato-biochemical alterations, nephrotoxicity, and DNA damage upon lead exposition. Hum. Exp. Toxicol. 39, 855–869. Available from: 10.1177/0960327120903490.

Ghosh, T., Chattopadhyay, K., Marschall, M., et al., 2009. Focus on antivirally active sulfated polysaccharides: from structure-activity analysisto clinical evaluation. Glycobiology 19 (1), 2–15.

Ghoshal, S., Mukhopadhyay, M.J., Mukherjee, A., 2001. Clastogenic effects of dietary *supplement-Spirulina* alga, and some medicinal plant products from *Boswellia serrata, Withania smnnifera* on mice. Indian J. Exp. Biol. 39, 1068–1070.

Grawish, M.E., Zaher, A.R., Gaafar, A.I., et al., 2010. Long-term effect of *Spirulina platensis* extract on DMBA-induced hamster buccal pouch carcinogenesis (immunohistochemical study). Med. Oncol. 27 (1), 20–28.

Grzanna, R., Polotsky, A., Phan, P.V., et al., 2006. Immolina, a high-molecular-weight polysaccharide fraction of Spirulina, enhances chemokine expression in human monocytic THP-1 cells. J. Altern. Complement. Med. 12 (5), 429–435.

Gupta, S., Hrishikeshvan, H.J., Sehajpal, P.K., 2010. Spirulina protects against rosiglitazone induced osteoporosis in insulin resistance rats. Diabetes Res. Clin. Pract. 87 (1), 38–43.

Habib, M.A.B., Parvin, M., Huntington, T.C. et al., 2008. A Review on Culture, Production and Use of Spirulina as Food for Humans and Feeds for Domestic Animals and Fish. FAO Fisheries and Aquaculture Circular No. 1034.

Hassan, A.M., Abdel-Aziem, S.H., Abdel-Wahhab, M.A., 2012. Modulation of DNA damage and alteration of gene expression during aflatoxicosis via dietary supplementation of Spirulina (Arthrospira) and whey protein concentrate. Ecotoxicol. Environ. Saf. 79, 294–300.

Hayashi, K., Hayashi, T., Morita, N., et al., 1993. An extract from *Spirulina platensis* is a selective inhibitor of herpes simplex virus type 1 penetration into HeLa cells. Phytother. Res. 7, 76–80.

Hayashi, T., Hayashi, K., Maeda, M., et al., 1996a. Calcium spirulan, an inhibitor of enveloped virus replication, from a blue-green alga *Spirulina platensis*. J. Nat. Prod. 59, 83–87.

Hayashi, K., Hayashi, T., Kojima, I., 1996b. A natural sulfated polysaccharide, calcium spirulan, isolated from *Spirulina platensis*: In vitro and ex vivo evaluation of anti-herpes simplex virus and anti-human immunodeciency virus activities. AIDS Res. Hum. Retroviruses 12, 1463–1471.

Hernández-Corona, A.1, Nieves, I., Meckes, M., et al., 2002. Antiviral activity of *Spirulina maxima* against herpes simplex virus type 2. Antivir. Res. 56 (3), 279–285.

Hernandez-Lepe, M.A., Wall-Medrano, A., Lopez-Diaz, J.A., et al., 2019. Hypolipidemic effect of arthrospira (*Spirulina*) *maxima* supplementation and a systematic physical exercise program in overweight and obese men: a double-blind, randomized, and crossover controlled trial. Mar. Drugs 17, 270.

Hosseini, S.M., Khosravi-Darani, K., Mozafari, M.R., 2013. Nutritional and medical applications of spirulina microalgae. Mini Rev. Med. Chem. 13 (8), 1231–1237.

Hutadilok-Towatana, N., Reanmongkol, W., Satitit, S., et al., 2008. A subchronic toxicity study of *Spirulina platensis*. Food Sci. Technol. Res. 14, 351.

Ismail, M.F., Ali, D.A., Fernando, A., et al., 2009. Chemoprevention of rat liver toxicity and carcinogenesis by Spirulina. Int. J. Biol. Sci. 5 (4), 377–387.

Iwata, K., Inayama, T., Kato, T., 1990. Effects of *Spirulina platensis* on plasma lipoprotein lipase activity in fructose-induced hyperlipidemic rats. J. Nutr. Sci. Vitaminol. 36, 165–171.

Johnson, P.E., Shubert, L.E., 1986. Accumulation of mercury and other elements by Spirulina (*cyanophyceae*). Nutr. Rep. Int. 34, 1063–1070.

Kapoor, R., Mehta, U., 1998. Supplementary effect of spirulina on hematological status of rats during pregnancy and lactation. Plant Food Hum. Nutr. 52, 315–324.

Karkos, P.D., Leong, S.C., Karkos, C.D., et al., 2011. Spirulina in clinical practice: evidence-based human applications. Evid. Based Complement. Alternat. Med. Available from: 10.1093/ecam/nen058.

Kato, T., Takemoto, K., Katayama, H., et al., 1984. Effects of Spirulina (*Spirulina platensis*) on dietary hypercholesterolemia in rats. J. Jpn. Soc. Nutr. Food Sci. 37, 323–332.

Kaur, K., Sachdeva, R., Grover, K., 2008. Effect of supplementation of Spirulina on blood glucose and lipid profile of the non-insulin dependent diabetic male subjects. J. Dairy Foods Home Sci. 27, 3–4.

Kawanishi, Y., Tominaga, A., Okuyama, H., et al., 2013. Regulatory effects of Spirulina complex polysaccharides on growth of murine RSV-M glioma cells through Toll-like receptor-4. Microbiol. Immunol. 57, 63–73.

Kebede, E., Ahlgren, G., 1996. Optimum growth conditions and light utilization efficiency of *Spirulina platensis* (*Arthrospira fusiformis*) (Cyanophyta) from Lake Chitu. Ethiopia Hydrobiol. 332, 99–109.

Kim, M.H., Kim, W.Y., 2005. The change of lipid metabolism and immune function caused by antioxidant material in the hypercholesterolemin elderly women in Korea. Korean J. Nutr. 38, 67–75.

Kim, D.H., Park, M.H., Choi, Y.J., et al., 2013. Molecular study of dietary heptadecane for the anti-inflammatory modulation of NF-kB in the aged kidney. PLoS One 8 (3), e59316.

Kraigher, O., Wohl, Y., Gat, A., et al., 2008. A mixed immunoblistering disorder exhibiting features of bullous pemphigoid and pemphigus foliaceus associated with Spirulina algae intake. Int. J. Dermatol. 47, 61–63.

Kumar, V., Bhatnagar, A.K., Srivastava, J.N., 2011. Antibacterial activity of crude extract of *Spirulina platensis* and its structural elucidation of bioactive compound. J. Med. Plants Res. 5 (32), 7043–7048.

Lee, J.B., Hayashi, T., Hayashi, K., et al., 1998. Further purification and structural analysis of calcium spirulan from *Spirulina platensis*. J. Nat. Prod. 61, 1101–1104.

Lee, E.H., Park, J.E., Choi, Y.J., et al., 2008. A randomized study to establish the effects of Spirulinain type 2 diabetes mellitus patients. Nutr. Res. Pract. 2, 295–300.

Li, B., Gaok, M.H., Zhang, X.C., et al., 2006. Molecular immune mechanism of C-phycocyanin from *Spirulina platensis* induces apoptosis in HeLa cells in vitro. Biotechnol. Appl. Biochem. 43, 155–164.

Mani, U.V., Desai, S., Iyer, U., 2000. Studies on the long-term effect of Spirulina supplementation on serum lipid profile and glycated proteins in NIDDM patients. J. Nutraceut. Funct. Med. Foods 2, 25–32.

Mazokopakis, E.E., Karefilakis, C.M., Tsartsalis, A.N., et al., 2008. Acute rhabdomyolysis caused by Spirulina (*Arthrospira platensis*). Phytomedicine 15, 525–527.

Mazokopakis, E.E., Starakis, I.K., Papadomanolaki, M.G., et al., 2014. The hypolipidaemic effects of Spirulina (*Arthrospira platensis*) supplementation in a Cretan population: a prospective study. J. Sci. Food Agric. 94 (3), 432–437.

Mendiola, J.A., Jaime, L., Santoyo, S., et al., 2007. Screening of functional compounds in supercritical fluid extracts from *Spirulina platensis*. Food Chem. 102, 1357–1367.

Mohan, I.K., Khan, M., Shobha, J.C., et al., 2006. Protection against cisplatin-induced nephrotoxicity by Spirulina inrats. Cancer Chemother. Pharmacol. 58 (6), 802–808.

Moreira, L.M., Behling, B.S., Rodrigues, R.S., et al., 2013. Spirulina as a protein source in the nutritional recovery of Wistar rats. Braz. Arch. Biol. Technol. 56, 447–456.

Morsy, M.A., Gupta, S., Nair, A.B., et al., 2019. Protective effect of spirulina platensis extract against dextran-sulfate-sodium-induced ulcerative colitis in rats. Nutrients 11, 2309.

Mulk, B.S., Deshpande, P., Velpula, N., et al., 2013. Spirulina and pentoxyfilline-novel approach for treatment of oral submucousfibrosis. J. Clin. Diagn. Res. 7 (12), 3048–3050.

Najdenski, H.M., Gigova, L.G., Iliev, I.I., et al., 2013. Antibacterial and antifungal activities of selected microalgae and cyanobacteria. Int. J. Food Sci. Tech. 48, 1533–1540.

Nakaya, N., Homa, Y., Goto, Y., 1988. Cholesterol lowering effect of Spirulina. Nutr. Rep. Int. 37, 1329–1337.

Nielsen, C.H., Balachandran, P., Christensen, O., et al., 2010. Enhancement of natural killer cell activity in healthy subjects by Immulina®, a Spirulina extract enriched for Braun-type lipoproteins. Planta Med. 76 (16), 1802–1808.

Olvera-Novoa, M.A., Dominguez-Cen, L.J., Olivera-Castillo, L., et al., 1998. Effect of the use of the microalga *Spirulina maxima* as fish meal replacement in diets for tilapia. Aquac. Res. 29, 709–715.

Ou, Y., Lin, L., Pan, Q., et al., 2012. Preventive effect of phycocyanin from *Spirulina platensis* on alloxan-injured mice. Environ. Toxicol. Pharmacol. 34 (3), 721–726.

Ouhtit, A., Ismail, M.F., Othman, A., et al., 2014. Chemoprevention of rat mammary carcinogenesis by spirulina. Am. J. Pathol. 184 (1), 296–303.

Ozdemir, G., Ulku Karabay, N., Dalay, M.C., et al., 2004. Antibacterial activity of volatile component and various extracts of *Spirulina platensis*. Phytother. Res. 18, 754–757.

Pak, W., Takayama, F., Mine, M., et al., 2012. Anti-oxidative and anti-inflammatory effects of spirulina on rat model of non-alcoholic steatohepatitis. J. Clin. Biochem. Nutr. 51, 227–234.

Paniagua-Castro, N., Escalona-Cardoso, G., Hernández-Navarro, D., et al., 2011. Spirulina (Arthrospira) protects against cadmium-induced teratogenic damage inmice. J. Med. Food 14 (4), 398–404.

Parages, M.L., Rico, R.M., Abdala-Diaz, R.T., et al., 2012. Acidic polysaccharides of Arthrospira (Spirulina) platensis induce the synthesis of TNF-α in RAW macrophages. J. Appl. Phycol. 24, 1537–1546.

Parikh, P., Mani, U., Iyer, U., 2001. Role of Spirulinain the control of glycemia and lipidemia in type 2 diabetes mellitus. J. Med. Food 4, 193–199.

Park, J.Y., Kim, W.Y., 2003. The effect of Spirulina on lipid metabolism, antioxidant capacity and immune function in Korean elderly. Korean J. Nutr. 36, 287–297.

Park, H.J., Lee, Y.J., Ryu, H.K., et al., 2008. A randomized double-blind, placebo controlled study to establish the effects of Spirulina in elderly Koreans. Ann. Nutr. Metab. 52, 322–328.

Ramamoorthy, A., Premakumari, S., 1996. Effect of supplementation of Spirulina on hypercholesterolemic patients. J. Food Sci. Technol. 33, 124–128.

Reddy, C.M., Bhat, V.B., Kiranmai, G., et al., 2000. Selective inhibition of Cyclooxygenase-2 by C-phycocyanin, a biliprotein from *Spirulina platensis*. Biochem. Biophys. Res. Commun. 277, 599–603.

Rellán, S., Osswald, J., Saker, M., et al., 2009. First detection of anatoxin-a in human and animal dietary supplements containing cyanobacteria. Food Chem. Toxicol. 47, 2189–2195.

Riss, J., Décordé, K., Sutra, T., et al., 2007. Phycobiliprotein C-phycocyanin from *Spirulina platensisis* powerfully responsible for reducing oxidative stress and NADPH oxidase expression induced by an atherogenic diet in hamsters. J. Agric. Food Chem. 55, 7962–7967.

Rodríguez-Hernández, A., Blé-Castillo, J.L., Juárez-Oropeza, M.A., et al., 2001. *Spirulina maxima* prevents fatty liver formation in CD-1 male and female mice with experimental diabetes. Life Sci. 69, 1029–1037.

Salazar, M., Chamorro, G.A., Salazar, S., et al., 1996. Effect of *Spirulina maxima* consumption on reproduction and peri- and postnatal development in rats. Food Chem. Toxicol. 34, 353–359.

Salazar, M., Martínez, E., Madrigal, E., et al., 1998. Subchronic toxicity study in mice fed *Spirulina maxima*. J. Ethnopharmacol. 62, 235–241.

Samuels, R., Mani, U.V., Iyer, U.M., et al., 2002. Hypocholesterolemic effect of Spirulina in patients with hyperlipidemic nephrotic syndrome. J. Med. Food 5, 91–96.

Sarada, D.V.L., Kumar, C.S., Rengasamy, R., 2011. Purified C-phycocyanin from *Spirulina platensis* (Nordstedt) Geitler: a novel and potent agent against drug resistant bacteria. World J. Microb. Biot. 27, 779–783.

Salehzadeh, A., Naeemi, A.S., Khaknezhad, L., et al., 2019. Fe_3O_4/Ag nanocomposite biosynthesised using *Spirulina platensis* extract and its enhanced anticancer efficiency. IET Nanobiotechnology 13, 766–770.

Sasson, A., 1997. Micro Biotechnologies: Recent Developments and Prospects for Developing Countries. BIOTEC Publication. 1/2542, 11–31.

Savranoglu, S., Tumer, T.B., 2013. Inhibitory effects of *Spirulina platensis* on carcinogen-activating cytochrome P450 isozymes and potential for drug interactions. Int. J. Toxicol. 32 (5), 376–384.

Selmi, C., Leung, P.S., Fischer, L., et al., 2011. The effects of Spirulina on anemia and immune function in senior citizens. Cell Mol. Immunol. 8 (3), 248–254.

Shokri, H., Khosravi, A., Taghavi, M., 2014. Efficacy of *Spirulina platensis* on immune functions in cancer mice with systemic candidiasis. J. Mycol. Res. 1, 7–13.

Simsek, N., Karadeniz, A., Kalkan, Y., et al., 2009. *Spirulina platensis* feeding inhibited the anemia- and leucopenia-induced lead and cadmium in rats. J. Hazard. Mater. 164, 1304–1309.

Sinanoglu, O., Yener, A.N., Ekici, S., et al., 2012. The protective effects of spirulina in cyclophosphamide induced nephrotoxicity andurotoxicity in rats. Urology 80 (6), 1392.e1-6.

Small, E., 2011. Spirulina–food for the universe. Biodiversity 12, 255–265.

Soheili, M., Khosravi-Darani, K., 2011. The potential health benefits of algae and micro algae in medicine: a review on *Spirulina platensis*. Curr. Nutr. Food Sci. 7, 279–285.

Soltani, M., Khosravi, A.R., Asadi, F., et al., 2012. Evaluation of protective efficacy of *Spirulina platensis* in Balb/C mice with candidiasis. J. Mycol. Med. 22, 329–334.

Sotiroudis, T.G., Sotiroudis, G.T., 2013. Health aspects of Spirulina (Arthrospira) microalga food supplement. J. Serb. Chem. Soc. 78 (3), 395–405.

Subhashini, J., Mahipal, S.V., Reddy, M.C., et al., 2004. Molecular mechanisms in C-Phycocyanin induced apoptosis in human chronicmyeloid leukemia cell line-K562. Biochem. Pharmacol. 68 (3), 453–462.

Tanticharoen, M., Reungjitchachawali, M., Boonag, B., et al., 1994. Optimization of gamma-linolenic acid (GLA) production in *Spirulina platensis*. J. Appl. Phycol. 6, 295–300.

Tobón-Velasco, J.C., Palafox-Sánchez, V., Mendieta, L., et al., 2013. Antioxidant effect of *Spirulina (Arthrospira) maxima* in a neurotoxic model caused by6-OHDA in the rat striatum. J. Neural Transm. 120 (8), 1179–1189.

Tomaselli, L., Palandri, M., Tredici, M., 1996. On the correct use of Spirulina designation. Algological Studies/Archiv für Hydrobiologie 83, 539–548.

Torres-Duran, P.V., Ferreira-Hermosillo, A., Juarez-Oropeza, M.A., 2007. Antihyperlipemic and antihypertensive effects of *Spirulina maxima* in an open sample of mexican population: a preliminary report. Lipids Health Dis. 6 (33), 1–8.

Torres-Durán, P.V., Miranda-Zamora, R., Paredes-Carbajal, M.C., et al., 1999. Studies on the preventive effect of *Spirulina maxima* on fatty liver development induced by carbon tetrachloride, in the rat. J. Ethnopharmacol. 64, 141–147.

Torres-Durán, P.V., Ferreira-Hermosillo, A., Ramos-Jiménez, A., et al., 2012. Effect of *Spirulina maxima* on postprandial lipemia in young runners: a preliminaryreport. J. Med. Food 15 (8), 753–757.

Vázquez-Velasco, M., González-Torres, L., López-Gasco, P., et al., 2014. Liver oxidation and inflammation in Fa/Fa rats fed glucomannan/spirulina-surimi. Food Chem. 159, 215–221.

Vo, T.S., Kim, S.K., 2013. Down-regulation of histamine-induced endothelial cell activation as potential anti-atherosclerotic activity of peptides from *Spirulina maxima*. Eur. J. Pharm. Sci. 50 (2), 198–207.

Vonshak, A., Chanawongse, L., Bunnag, B., et al., 1996. Light acclimation and photoinhibition in three *Spirulina platensis* (Cyanobacteria) isolates. J. Appl. Phycol. 8, 35–40.

Wu, Q., Liu, L., Miron, A., et al., 2016. The antioxidant, immunomodulatory, and anti-inflammatory activities of Spirulina: an overview. Arch. Toxicol. 90, 1817–1840.

Yu, B., Wang, J., Suter, P.M., Russell, R.M., et al., 2012. Spirulina is an effective dietary source of zeaxanthin to humans. Br. J. Nutr. 108 (4), 611–619.

Chapter 58

Garcinia cambogia

Rajinder Raina, Pawan K. Verma, Ila Taku, JItendra K. Malik and Ramesh C. Gupta

Introduction and background

Garcinia belongs to family *Guttifereae* (*Clusiaceae*) and the genus comprises around 200 species throughout the world, with most of them found in Asia and Africa. More than 36 species of this genus are reported to grow in India, including *Garcinia cambogia* L., which is a tropical dicotyledonous tree and is also native to other countries of Southeast Asia. It is a small to medium-sized tree which grows to a height of 20 m with a rounded crown and horizontal or drooping branches. The leaves of the tree are dark green, shining, ovate in shape, ranging between 5.5–9 cm in length and 2.5–4 cm in breadth. The tree bears small green fruits that are ovoid in shape with a diameter of 2.5–4 cm which when ripened are yellow or red in color with 6–8 grooves. The fruit resembles a small yellow or reddish pumpkin, or it may have a unique purple color. The fruit seeds are surrounded by a succulent aril part. The tree flowers during the hot season, and fruits ripen during the rainy season.

G. cambogia fruit is dried by a conventional sun drying method after cutting it into thin slices and dehydrating in the sun. However, the quality of the fruit is difficult to control as drying takes place in an open environment. The application of a new super-heated steam-drying technique for the fruit under controlled conditions (150°C for 46.6 min) has been recommended as it is a faster drying process with better quality of dried *G. cambogia* fruit as compared to conventional sun dried fruits (Lim et al., 2020). The dried fruit rind of *G. cambogia* has been used for centuries in Asian countries for culinary purposes as a condiment and flavoring agent without any adverse effects. It possesses a distinct sour taste which it imparts to food and is used as an alternative to tamarind or lemon in curries and other ethnic food preparations to make them more filling and satisfying (Sergio, 1988; Jena et al., 2002). Fruit rind is also eaten raw, sometimes at meal times as an appetizer in parts of India. The dried fruit rind of *G. cambogia* combined with salt and other organic acids can lower the pH and thus provide a bacteriostatic effect for curing fish. This method of fish preservation is known as Colombo curing. *G. cambogia*, *Garcinia indica*, and *Garcinia atroviridis*, being rich in hydroxycitric acid (HCA), are at times used interchangeably in food preparation and fish preservation.

Apart from its use as a preservative and condiment in cuisines, *Garcinia* extracts have been used in the treatment of ailments in the traditional Ayurvedic system of medicine (Khare, 2007). A decoction of fruit rind is given as a purgative in the treatment of intestinal parasites, bilious digestive conditions, dysentery, rheumatism, and tumors. The extracts are also employed as a cardio-tonic to treat angina. In veterinary medicine, the extract is also used as a rinse for diseases of the mouth (Jena et al., 2002). The fruit rind is also used in rickets, enlargement of the spleen, and to aid healing of bone fractures (Khare, 2007).

G. cambogia, being rich in HCA, is a popular component of dietary supplements for the promotion of weight loss and maintenance of body weight, as has been demonstrated in various studies on experimental animals (Sullivan et al., 1974a,b; Leonhardt et al., 2001; Ohia et al., 2002; Shara et al., 2003; Asghar et al., 2007). The antiobesity effect of *Garcinia*/HCA has been elucidated with overwhelming evidence over the last few decades. *Garcinia*/HCA has also been found to be efficacious in ameliorating obesity-related complications such as inflammation, oxidative stress, and insulin resistance (Asghar et al., 2007; Yonei et al., 2008). Several studies have supported the positive effect of HCA administration alone or in combination with other ingredients on body weight loss, reduced food intake due to its anorexigenic effect, increased fat oxidation, and/or energy expenditure (Ishihara et al., 2000; Leonhardt et al., 2004; Roy et al., 2004). However, many randomized clinical trials in humans have not reported the same outcome (Heymsfield et al., 1998; Kriketos et al., 1999; Kovacs et al., 2001). None of these studies have shown long-term beneficial effects of *Garcinia* on weight loss. Therefore there is no conclusive evidence to support the potential effectiveness

Nutraceuticals. DOI: https://doi.org/10.1016/B978-0-12-821038-3.00058-6

and long-term benefits of *G. cambogia* on weight loss (Marquez et al., 2012). Several explanations were advanced for such a lack of positive effects, which include low dose of HCA, short duration of study, poor patient compliance, and low bioavailability of HCA.

Other beneficial health effects of the extract in experimental studies are antiulcerogenic, antioxidative, antidiabetic, antiinflammatory, anticancer, erythropoietic, and other miscellaneous effects. These effects are also contributed to by polyisoprenylated benzophenones, namely garcinol and guttiferones, in addition to the HCA present in *G. cambogia*. Previously, HCA was considered to be hepatoprotective but with the rise in cases of hepatic injuries after its supplementation questions have been raised about such an effect.

Phytochemical ingredients

G. cambogia is endowed with diverse phytochemical ingredients such as organic acids, xanthones, polyisoprenylated benzophenones, and isoxanthohumols, some of which are depicted in Fig. 58.1. Various organic acids including HCA, citric, tartaric, mallic, and succinic acid are isolated from this plant. HCA is the principal acid not only present in fruit rind of *G. cambogia* but also in the rinds of *G. indica* and *Garcinia atrovidis*. HCA is the primary medicinal component present at up to 30% by weight in the pericarp of *G. cambogia* fruits. Chemically HCA is 1,2-di-hydroxy propane 1,2,3-tricarboxylic acid, and is very similar to citric acid found in oranges and other citrus fruits (Lewis and Neelakantan, 1965; Jena et al., 2002).

The HCA molecule exists in four different isomers, which are (−)-HCA, (+)-HCA, (−)-allo-HCA, and (+)-allo-HCA. The most potent inhibitor of ATP citrate lyase (ACLY) is the (−)-HCA isomer also known as 2S,3S−HCA (Stallings et al., 1979). Commercially available *G. cambogia* extracts are prepared from the fruit rinds and contain 20%−60% (−)-HCA (Jena et al., 2002). HCA of high purity can be chemically synthesized from citric acid. It can exist as a free acid or lactone form. (−)-HCA as free acid, though biologically active, is chemically unstable and gets converted into a less biologically active but highly stable lactone form. The lactone form is not therapeutically preferred because of its low bioavailability. Therefore (−)-HCA is combined with various counter ions to form single, double, or triple stable salts to prevent cyclization of HCA into the lactone form which increases its solubility and bioavailability (Downs et al., 2005). Sodium salt of HCA is physiologically more effective than the lactone form but is highly hygroscopic and unfavorable for use in dietary supplements in dry form. Double or triple salts of calcium, magnesium, and potassium are now used instead of single salt of HCA because of their higher solubility and stability. HCA−SX or super CitriMax, which is a Ca^{+2}/K^{+} salt of (−)-HCA, is completely soluble in water which confers higher bioavailability (Downs et al., 2005). The fruit rind of *G. cambogia* is also rich in polyisoprenylated benzophenones (Garcinol/comboginol), tetracyclic xanthones (oxyguttiferones), and isoxanthohumol (guttiferones I, J, K, M, and N) (Masullo et al., 2008). Garcinol is also found in the latex and bark of the plant (Iinuma et al., 1998; Masullo et al., 2008). The presence of phenolic, hydroxyl groups and β-diketone moiety in the garcinol molecule imparts strong antioxidant activity to it (Padhye et al., 2009). Garbogiol, a xanthone, is found in the roots, whereas another xanthone, Rheediaxanthone A, is present in the bark of the plant (Iinuma et al., 1998). An analysis of *G. cambogia* fruit indicated that it contained various amino acids and minerals. The amino acids detected in the fruits include arginine, asparagine, glutamine, threonine, glycine, proline, γ-aminobutyric acid, leucine, isoleucine, ornithine, and lysine. The elemental analysis of *G. cambogia* has shown the presence of macro, micro, and trace essential elements in appreciable concentrations, which are able to meet human nutritional requirements. Furthermore, the trace toxic elements were found to be present within safe limits (Jamila et al., 2019).

Pharmacological effects

G. cambogia extract is a popular antiobesity herbal supplement for weight loss and weight management (Ohia et al., 2002; Preuss et al., 2005). Other beneficial health effects of *Garcinia* in experimental studies are antiulcerogenic (Mahendran et al., 2002a,b), antioxidative (Asghar et al., 2007), antidiabetic (Wielinga et al., 2005), antiinflammatory (dos-Reis et al., 2009), anticancerous (Varalakshmi et al., 2011), erythropoietic (Oluyemi et al., 2007), diuretic (Mathew et al., 2011), and as a mineral supplement (Gurse et al., 2012).

Antiobesity effect

The antiobesity effect of *G. cambogia* is largely attributed to its HCA content, the primary acid in the fruit rind of the plant (Shara et al., 2003). HCA is also beneficial in ameliorating obesity-related complications such as inflammation, oxidative stress, and insulin resistance (Asghar et al., 2007). Various studies with rodent models have conclusively proved the ability of HCA to promote weight loss and maintain body weight with no significant adverse effects (Ohia et al., 2002; Shara et al., 2003, 2004; Deshmukh et al., 2008a,b). The effectiveness of HCA is dependent on dose, form of HCA used, duration of treatment, strain of rat, and the general experimental design. The antiobesity effects of *G. cambogia*/HCA are as a

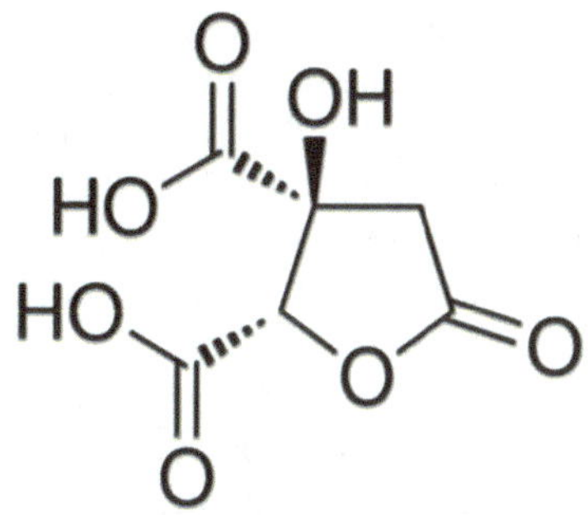

Hydroxyycitric acidd (HCA)

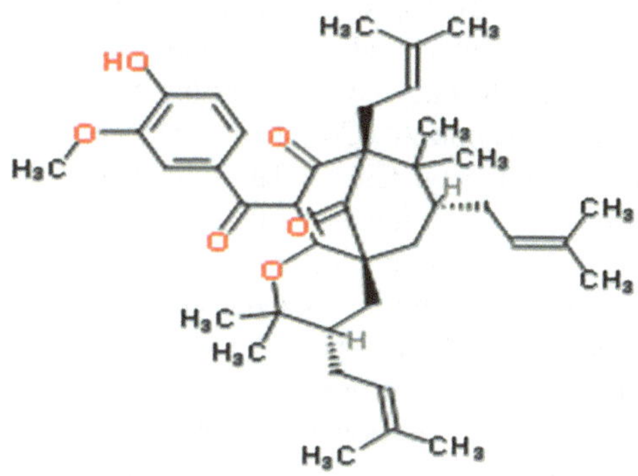

Garcinol

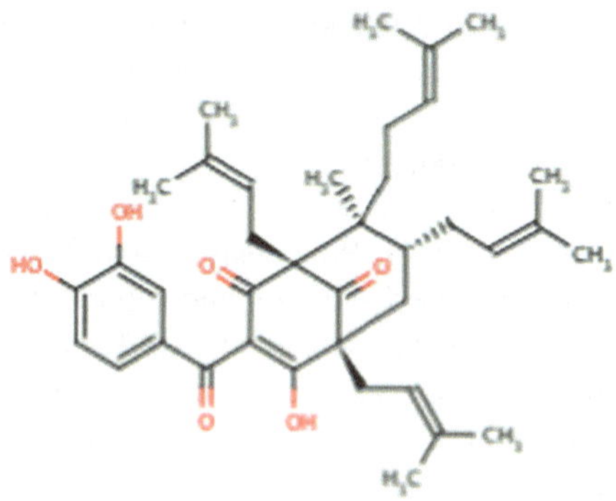

Guttiferrone A

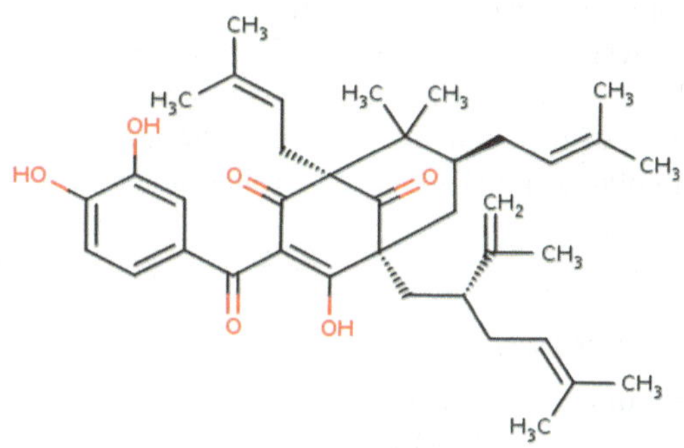

Guttiferone E

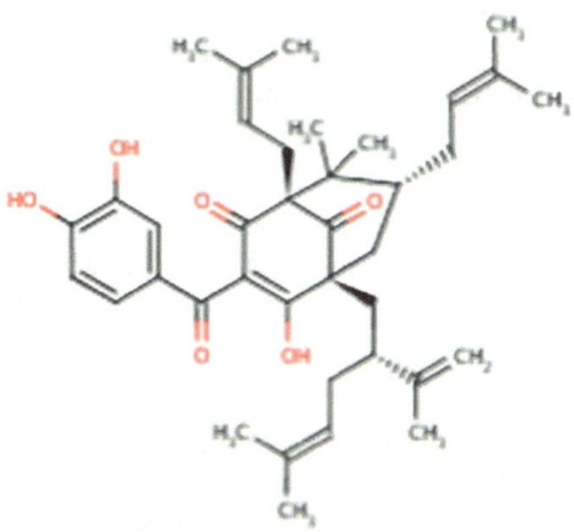

Guttiferrone F

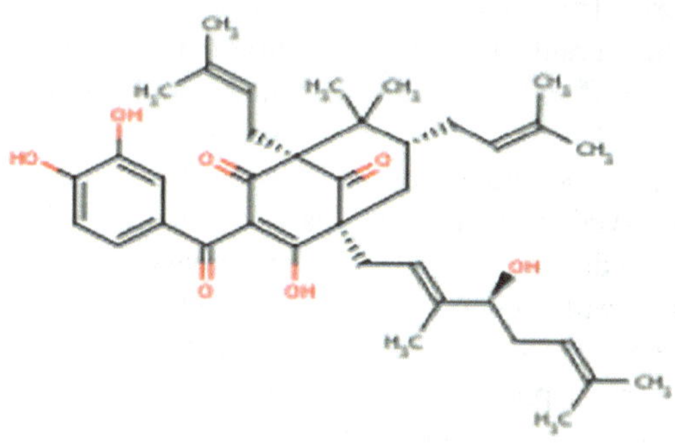

Guttiferone M

FIGURE 58.1 Chemical structures of active phytochemical ingredients isolated from *Garcinia cambogia*.

result of the combined actions of several mechanisms including suppressing de novo fatty acid biosynthesis and appetite (Lowenstein, 1971; Sullivan et al., 1974a,b; Leonhardt and Langhans, 1977), increasing energy expenditure (Leonhardt et al., 2004), and subsequently reducing body fat accumulation and weight gain in experimental animals (Kim et al., 2008). However, a meta-analysis of nine randomized controlled clinical trials of *G. cambogia* extract for weight loss in humans by Onakpoya et al. (2011) concluded that it leads to only a small short-term weight loss (average dose of GC extract used was 1–3 g).

Suppressing de novo fatty acid biosynthesis

The primary mechanism of *G. cambogia* is believed to involve the inhibition of fatty acid synthesis (Kovacs and

Westerterp-Plantenga, 2006). HCA acts as a competitive inhibitor of the extra-mitochondrial enzyme ACLY (Sullivan et al., 1977a). The enzyme catalyzes the conversion of citrate and coenzyme A (Co-A) to oxaloacetate and acetyl coenzyme A (acetyl Co-A). The inhibition of this reaction limits the availability of acetyl Co-A required for fatty acid and cholesterol biosynthesis, thereby suggesting that HCA inhibits lipogenesis (Kovacs and Westerterp-Plantenga, 2006). Several experimental studies in rats have confirmed the administration of HCA inhibiting lipogenesis in several tissues such as liver, adipose tissue, and small intestine, where carbohydrate is converted to fatty acid (Lowenstein, 1971; Sullivan et al., 1974a; Sullivan et al., 1977b). A large number of random clinical trials with *G. cambogia* in obese human subjects have demonstrated its hypotriglyceridemic effect (Preuss et al., 2004; Hayamizu et al., 2003b; Vasques et al., 2014). Recently, it was reported that extracts of pear pomace in conjunction with *G. cambogia* prevent adipogenesis and boosted lipolysis in 3T3-L1 cells (Sharma et al., 2018).

Suppression of appetite

A growing body of evidence demonstrates the efficacy of *G. cambogia*-derived natural (−)-HCA in weight management by curbing appetite besides inhibiting body fat biosynthesis. The consumed carbon source from carbohydrate feeding is diverted for the synthesis of glycogen in liver after HCA inhibits fatty acid and cholesterol biosynthesis. This alteration in metabolism triggers a signal for the brain to release serotonin or 5-hydoxytryptamine (5-HT), a neurotransmitter implicated in the regulation of eating behavior and appetite control. Therefore the rise in serotonin results in reduced appetite. It has been observed in microarray analysis that Ca2 + / K + salt of (-)-hydroxycitric acid (HCA-SX) have a distinct effect on appetite suppression due to selective upregulation of genes encoding serotonin receptors (Roy et al., 2004). The suppression of appetite concomitant with a rise in serotonin has been reported in all the weight control studies of HCA on both animal and human subjects (Ohia et al., 2002; Preuss et al., 2005; Downs et al., 2005; Asghar et al., 2007). HCA also inhibits serotonin uptake in isolated rat brain slices in a manner similar to selective serotonin reuptake inhibitors, thereby suggesting serotogenic involvement in the effect of HCA on food intake suppression (Ohia et al., 2001). HCA caused a significant reduction in appetite, weight, and plasma leptin level with a concomitant rise in serum serotonin level and favorable lipid profile in a human clinical trial (Preuss et al., 2004). This suggests the apparent ability of HCA to reduce body fat and body weight gain due to its combined effect on metabolic and serotonergic pathways.

HCA may bring about the anorectic effect by a second possible mechanism, namely reducing acetyl CoA, subsequently decreasing malonyl CoA levels, and thereby reducing negative feedback on carnitine acyltransferase (CPT-1). Long-chain acyl Co-As, as a substrate for CPT-1, are also known to be a mediator(s) of appetite (Rederman et al., 1999; Lopaschuk et al., 2010). Reduced output of neuropeptide Y (NPY) from the brain has also been implicated in appetite suppression. Supplementation of HCA has been reported to significantly reduce the level of NPY in the hypothalamic tissues (Downs et al., 2005). The satiety-inducing property of *G. cambogia*-containing nutraceutical supplements in human volunteers has also been reported (Mayer et al., 2014). Exposure of rat brain slices to HCA resulted in a 20% decrease, but a combination of serotonin reuptake inhibitors (fluoxetine and clomipramine) caused a 30% decrease in serotonin reuptake (Ohia et al., 2001). The increased availability of serotonin may suppress appetite, which could be another mechanism for weight loss with *G. cambogia* use in humans (Mayer et al., 2014). Previously, contradictory results in humans on increased satiation and suppression of appetite were obtained by Kovacs et al., 2001, who observed no such appetite-suppressant effect after 2-week supplementation of HCA alone or combined with medium-chain triglycerides.

Increasing energy expenditure

HCA-SX either alone or in combination with chromium and *Gymnema sylvestre* extract (GSE) has been demonstrated to effectively promote breakdown of body fats (Preuss et al., 2004, 2005). A significant increase in levels of serum free fatty acid and glycogen content in gastrocnemius muscle of mice after acute administration of 10 mg HCA/mice was observed (Ishihara et al., 2000). However, chronic administration of 10 mg HCA/mice/day for 3 weeks augmented the endurance exercise performance in mice during running exercise by promoting lipid oxidation and attenuating glycogen consumption.

Short-term administration of HCA has also been observed to enhance fat oxidation significantly with lowering of carbohydrate oxidation during exercise in athletes and untrained women (Lim et al., 2002, 2003). Enhancement of lipid metabolism by HCA in human test subjects can also be inferred from many-fold increased urinary concentration of metabolites like malondialdehyde (MDA), acetaldehyde, formaldehyde, and acetone. Increased urinary excretion of these metabolites is thought to be a consequence of enhanced β-oxidation of fat in body tissues (Preuss et al., 2004). Similar enhanced urinary excretion of fat metabolites has also been observed in obese human subjects as well as Sprague-Dawley rats receiving HCA-SX alone or in combination with niacin-bound chromium and *GSE*, thereby suggesting that HCA-SX effectively promotes breakdown of fats (Preuss et al., 2005).

The administration of *G. cambogia* in conjunction with glucomannan in obese human patients has been reported to result in reduced fat mass, visceral fat, blood glucose and lipid profile, and increased basal metabolic rate (Maia-Landim et al., 2018).

Downregulation of a spectrum of obesity-associated genes

The effects of HCA-SX on body fat and abdominal fat gene expression in rats have been determined using a gene chip high-density microarray. HCA supplementation selectively influenced 1% of 9960 genes in fat tissue without affecting the genes responsible for transcribing mitochondrial and nuclear proteins (Roy et al., 2004). The genes encoding for abdominal fat leptin are downregulated by HCA salts in rats, while the expression of plasma leptin remains unaltered. Studies on cultured mature adipocytes revealed significant upregulation of 366 and downregulation of 348 fat- and obesity-related genes (Roy et al., 2007). Additionally HCA-SX was also found to be capable of activating hypoxia inducible factor, a transcription factor involved in energy metabolism (Roy et al., 2007).

Kim et al. (2008) studied the effect of *G. cambogia* on the expression of multiple genes associated with adipogenesis and other visceral obesity-related biomarkers in a high-fat diet (HFD)-induced obese mouse model. *G. cambogia* was found to ameliorate HFD-induced obesity by modulating multiple genes associated with adipogenesis such as adipocytes protein aP2 (aP2), sterol regulatory element-binding factor 1c (SREBP1c), peroxisome proliferator-activated receptor 2 (PPAR-γ2), and CCAT/enhancer-binding protein α (C/EBPα) in the visceral fat tissue of mice. Therefore the extract was effective in modulating multiple genes associated with visceral adipogenesis such as PPAR-γ2, SREBP1c, and C/EBPα, and aP2 in visceral fat tissue of mice. Further, in microarray analysis HCA-SX have a distinct effect on appetite suppression due to selective upregulation of genes encoding serotonin receptors (Roy et al., 2004).

The clinical efficacy of *G. cambogia* in promoting weight loss and lowering cholesterol level thereby producing an antiobesity effect has been extensively studied in humans. Various researchers have reviewed and summarized the results of efficacy of *G. cambogia*/HCA of different human clinical trials (Soni et al., 2004; Downs et al., 2005; Onakpoya et al., 2011; Chuah et al., 2013). *G. cambogia*/HCA had a statistically significant effect on weight loss and weight management in the majority of clinical trials. However, the results of some randomized placebo-controlled trials have not reported the same outcomes (Marquez et al., 2012). Several explanations can be given for the lack of a positive effect including low dose of HCA, short duration of study, poor patient compliance, and low bioavailability of the form of HCA (single salt) used (Heymsfield et al., 1998; Kriketos et al., 1999). Little evidence supports the long-term benefits of *G. cambogia* extract as studies have shown that its effects on body weight and fat loss do not persist beyond 12 weeks of intervention (Onakpoya et al., 2011; Marquez et al., 2012).

Antiinflammatory effect

G. cambogia extract has been experimentally assessed for its antiinflammatory activity and such experimental studies are listed in Table 58.1. Clouatre and Preuss (2008) observed that HCA K^+ Mg^{2+} administration in rats decreases markers of inflammation like c-reactive protein (CRP) and tumor necrosis factor-alpha (TNF-α) besides reducing the induced paw edema. Similarly, administration of HCA-SX in developing obese Zucker rats also lowered the CRP and IL-6 markers of inflammation (Asghar et al., 2007). The antiinflammatory activity of a *G. cambogia* extract has also been assessed in experimentally induced colitic rats. Treatment with extract significantly improved the macroscopic damage and caused a substantial reduction in increased myeloperoxidase activity, cyclooxygenase (COX-2), and inducible nitric oxide synthase (iNOS) expression. The extract also reduced the levels of prostaglandin E_2 (PGE_2) and colonic levels of interleukin 1β (IL-1β) (dos-Reis et al., 2009). The extract of *G. cambogia* has also been found to reduce the markers of inflammation in brain, intestine, kidney, and serum (Amin et al., 2011b).

Similarly, the antiinflammatory activity of extract was also assessed in in vitro system (human RBC membrane stabilization method) and in vivo (carrageenan-induced rat paw edema model) studies. It was concluded that the ethanolic extract of *Garcinia cambogia* possesses significant in vivo antiinflammatory activity, and moderate in vitro antiinflammatory activity (Prasanth et al., 2013). Garcinol and its derivatives isolated form *G. cambogia* produce antiinflammatory effects by inhibiting the activities of 5-lipoxygenase and prostaglandin PGE2 synthase (Koeberle et al., 2009). These two enzymes are often over-expressed in many tumors and inflammatory processes. Further studies have also shown that garcinol exerts antiinflammatory activity in LPS-stimulated murine macrophages, through inhibition of NF-κB and/or JAK/STAT-1 activation and also lowered the LPS-induced intracellular reactive oxygen species (Liao et al., 2005). The release of arachidonic acid metabolites from human colon adenocarcinomas cell lines and rat normal

TABLE 58.1 Antiinflammatory and antioxidant activities of *Garcinia cambogia* or its active constituents in experimental models.

Active constituent(s)	Model	Mechanism of action	Reference
Antiinflammatory			
Garcinol, oxyguttiferons, guttiferones K and M from fruits of *Garcinia cambogia*	Cultured cell lines	Modulate cytokine signaling by inhibiting STAT-1 nuclear transfer and DNA binding	Masullo et al. (2008)
Garcinol	Human colon adenocarcinoma cells	Potent inhibitor of 5-lipooxygenase and PGE2 synthetase—crucial enzymes of inflammation	Koeberle et al. (2009)
HCA-SX	Developing obese rats	Lowered C-reactive protein and interleukin-6 (markers of inflammation in plasma)	Asghar et al. (2007)
HCA K^{+} Mg^{2+} salt	Rats	Decrease various markers of inflammation like c-reactive protein levels and tumor necrosis factor-α, also decreased paw edema	Clouatre and Preuss (2008)
Garcinol	TPA induced ear inflammation in mice	Topically applied garcinol inhibited induced expression of proinflammatory cytokine IL-6 protein, thereby modulating ear inflammation	Huang et al. (2008)
Garcinol	UVB-induced ear inflammation	Oral administration modulates induced ear inflammation by inhibiting proinflammatory IL-1 and IL-6	Huang et al. (2008)
G. cambogia extract	TNBS-induced colitic rats	Improved macroscopic damage and reduced increase in MPO, COX-2, inducible nitric oxide synthase (iNOS) expression, also reduces PGE_2 and IL-1β colonic levels	dos-Reiss et al. (2009)
Antioxidant			
Garcinol	Rat cortical astrocyte culture	Lowers nitric oxide production and expression of lipopolysaccharide-induced inflammatory mediators (iNOS and Cox-2)	Liao et al. (2005)
Garcinol and guttiferone isolated from *G. cambogia*	Human platelets and plasma exposed to 100 μM peroxynitrite	Protects lipid peroxidation revealed by distinct reduction in the formation of carbonyl groups in plasma and platelet proteins with a decrease in thiobarbituric acid-reactive species	Kolodziejczyk et al. (2009)
Garcinol	In vitro antioxidant activity	Free radical-scavenging activity against the hypoxanthine/xanthine oxidase system, superoxide anion, peroxide, nitric oxide, and hydroxyl radicals	Rawri et al. (2013)
HCA-SX	Obese Zucker rats	Attenuates oxidative stress biomarkers including malondialdehyde (MDA) and protein carbonyl (DNPH) and protein tyrosine nitration [Tyr-NO(2)] in liver and kidney tissues	Asghar et al. (2007)
HCA	Obese Zucker rats fed high-fat diet (HFD)	Protected brain from oxidative dysfunction and lowered brain catalase and MDA levels	Amin et al. (2011a)
G. cambogia extracts	Obese Zucker rats	Ameliorated the damaging effects of feeding HFD or HSD rats by decreasing renal oxidative markers like MDA level, nitric oxide, and catalase activity	Amin et al. (2011b)

COX-2, cyclooxygenase-2; *HSD*, high sucrose diet; *MPO*, myeloperoxidase; *TNBS*, 2,4,6-trinitrobengene sulphonic acid; *TPA*-tetradecanoylphorbol 13-acetate.

immortalized intestinal cells was inhibited by 40%–50% by garcinol at a concentration of 1 μM concentration. This was thought to be due to the ability of garcinol to modulate arachidonic metabolism by blocking the phosphorylation of cPLA2 and decrease iNOS protein levels by inhibiting STAT-1 activation (Hong et al., 2007). Masullo et al. (2008) reported that fruits of *G. cambogia*, along with two other polyisoprenylated benzophenones, guttiferone K and M, were able to modulate cytokine signaling in different cultured cell lines mainly by inhibiting STAT-1 nuclear transfer and DNA binding. Oral administration or topical application of garcinol inhibited induced ear inflammation in a dose-dependent manner in CD-1 mice. Topical application of garcinol to the ears of CD-1 mice inhibited induced expression of proinflammatory cytokine IL-6 protein. Oral administration of garcinol inhibited UVB-induced ear inflammation and proinflammatory IL1 and IL6 (Huang et al., 2008).

Anticancer effects

Various studies on the anticancer properties of *G. cambogia* extract and its active constituents are depicted in Table 58.2. ACLY is a key enzyme of de novo fatty acid synthesis responsible for generating cytosolic acetyl-CoA and oxaloacetate. Enhanced glucose and lipid metabolism is one of the most common properties of malignant cells. ACLY over-expression or activation have been reported in urinary bladder, breast, liver, stomach, colon, and prostate tumors (Szutowicz et al., 1979; Halliday et al., 1988; Yahagi et al., 2005).

It has also been reported that chemical inhibitors or ACLY RNAi suppress tumor cell growth (Hatzivassiliou et al., 2005; Bauer et al., 2005). ACLY is also responsible for the phosphorylation of other kinases at different sites such as nucleoside diphosphate kinase (Wagner and Vu, 1995) and cyclic adenosine monophosphate-dependent protein kinase (Pierce et al., 1981), which contributes to protein stabilization and tumor cell growth. Selective inhibition of ACLY by *Garcinia* extract/HCA may help in the suppression of tumor cell growth (Migita et al., 2008). Garcinol and guttiferones present in the extracts of *G. cambogia* have been evaluated for antiproliferative effects of the on Balb/c 3T3 mouse fibroblasts and human peripheral lymphocytes. The results showed that *Garcinia* extracts inhibited lymphocytes and 3T3 fibroblast cell survival because of pronounced cytotoxic effects (Varalakshmi et al., 2011). Similarly, guttiferone K has a potent antitumor effect on colon cancer caspase-3-mediated apoptosis and antiproliferative activity against human ovarian cancer cell line A2780 (Cao et al., 2007; Kan et al., 2013). Studies show that garcinol represses p300 acetylation of p53 gene and promotes apoptosis through caspase-3 activation besides predominantly downregulation of global gene expression in HeLa cells (Balasubramanyam et al., 2004). Garcinol also exhibits an anticarcinogenic activity due to its capability to modulate key regulatory cell-signaling pathways. Garcinol is an effective inhibitor of several key regulatory pathways (e.g., NF-κB, STAT3, etc.) in cancer cells, thereby being able to control malignant growth of solid tumors in vivo. Despite its high potential as an antineoplastic modulator of several cancer types such as head and neck cancer,

TABLE 58.2 Anticancer activity of *Garcinia cambogia* or its active constituents in experimental models.

Active constituent	Mechanism of cytotoxicity	References
Garcinol	Promotes apoptosis through caspase-3 activation and downregulates global gene expression in HeLa cells	Balasubramanyam et al. (2004)
Garcinol	Blocks NF-κB signaling with reduced activation of antiapoptotic Bcl-2 in breast cancer	Ahmad et al. (2010)
Garcinol	Suppression of TGF-activated kinase 1 (TAK1) and inhibitor of IκB kinase (IKK) head and neck carcinoma	Li et al. (2013)
Xanthones	Inhibits preneoplastic lesions in mammary and colon cancers. Also induces apoptosis in mouth, breast, gastric, and lung cancer cells in vitro	Mazzio and Soliman (2009)
Garcinol	Inhibits release of arachidonic acid metabolites in human colon adenocarcinoma	Koeberle et al. (2009)
Guttiferone K	Potent antitumor effect on colon cancer caspase-3-mediated apoptosis; antiproliferative activity against human ovarian cancer cell line A2780	Cao et al. (2007), Kan et al. (2013)
Garcinol and guttiferones	Inhibited lymphocyte and Balb/C3T3 mouse fibroblast cell survival due to pronounced cytotoxic effects	Varalakshmi et al. (2011)

breast cancer, hepatocellular carcinoma, prostate cancer, and colon cancer, it is still in the preclinical stage due to the lack of systematic and conclusive evaluation of pharmacological parameters (Behera et al., 2016). The additional important targets of garcinol are 5-lipoxygenase (5-LOX) and microsomal prostaglandin E2 synthase-1 (Koeberle et al., 2009), as suppression of 5-LOX and PGE2 was observed in human neutrophils and interleukin-stimulated human lung cancer cells. Inhibition of 5-LOX activity seems to be a particularly efficient strategy to combat metastases, and COX-2 catalyzed PGE2 formation is often related to enhanced metastasis by upregulation of matrix metalloproteinases such as MMP-2 (Flavin, 2007). 5-LOX inhibition by topical garcinol treatment suppressed 7,12-dimethylbenz[a]anthracene (DMBA)-induced hamster cheek pouch carcinogenesis by inhibition of leukotriene B4 (LTB4) biosynthesis (Xin et al., 2012). The fruit also contains xanthones which also inhibit preneoplastic lesions in mammary and colon cancer. The xanthones may also induce apoptosis in mouth, leukemia, breast, gastric, and lung cancer cells in vitro (Mazzio and Soliman, 2009). Several case series and a case report describe the treatment of patients with advanced metastatic cancer with the combination regimen of lipoic acid and oral HCA in addition to standard chemotherapy. However, well-designed randomized clinical trials on this aspect are required before the regimen can be recommended for cancer treatment (Schwartz et al., 2014; Guais et al., 2012).

Antioxidant activity

Various studies on the antioxidant activity of *G. cambogia* and its active constituents are listed in Table 58.1. Different phytochemical ingredients present in the fruit extract of *Garcinia* species have high in vitro antioxidant potential due to their potent scavenging activity of superoxides, peroxides, nitric oxide (NO), and hydroxyl radicals in a dose-dependent manner (Rawri et al., 2013). Bioactive components of *G. cambogia* including benzophenones are reported to reduce oxidative stress levels based on in vitro experiments in human plasma, suggesting the protective potential of *G. cambogia* against diseases associated with oxidative stress. Polyisoprenylated benzophenones such as garcinol and guttiferone K were found to protect human platelets and plasma against lipid and protein oxidation indicating good antioxidant activity under in vitro models and thus may have some promising in vivo effects as well (Kolodziejczyk et al., 2009). Similarly, guttiferone (32-hydroxy-*ent*guttiferone M) isolated from the plant showed significantly high antioxidant activity in both 2,2-diphenyl-1-picrylhydrazyl hydrate and 2,2′-Azino-bis(3-ethylbenzothiazoline-6-sulfonic acid) diammonium salt free radical-scavenging assays (Acuna et al., 2010). Oxidative stress markers including MDA and protein tyrosine nitration (tyr-NO2) were attenuated after supplementation of HCA-SX to developing obese Zucker rats (Asghar et al., 2007). Similar protective effects of HCA in relation to brain oxidative dysfunction as a result of feeding rats a HFD have also been reported (Amin et al., 2011b).

The protective effects of *G. cambogia* against renal oxidative stress biomarkers and renal function induced by a HFD and high-sucrose diet were also demonstrated. *G. cambogia* improved levels of blood lipid profile and kidney function and also reversed the oxidative markers by decreasing MDA, NO, and catalase activity (Amin et al., 2011b). Garcinol has also been reported to provide a neuroprotective effect by lowering NO production and expression of lipopolysaccharide-induced inflammatory mediators, iNOS and COX-2, in rat cortical astrocyte culture (Liao et al., 2005).

Antidiabetic effect

The administration of *G. cambogia* extract was observed to improve glucose metabolism and lower serum insulin and leptin levels in mice (Hayamizu et al., 2003a). HCA treatment was also observed to delay and reduce intestinal glucose absorption in rats (Wielinga et al., 2005). Studies have shown the antidiabetic effect of HCA-SX due to its attenuation of insulin resistance in developing obese Zucker rats (Asghar et al., 2007). Glucose tolerance and insulin challenges indicated that administration of HCA K^{+} Mg^{+2} enhanced the insulin sensitivity in rats (Clouatre and Preuss, 2008). Daily supplementation of HCA in humans at the dose rate of 500 mg/day for 7 days resulted in an enhanced glycogen synthesis rate and fatty acid translocase/CD36mRNA expression in exercised skeletal muscle. This supplementation was also found to improve postmeal insulin sensitivity (Cheng et al., 2012). Administration of a polyherbal formulation containing *G. cambogia* as one of the constituent was reported to reduced blood glucose, and lipids. The effect of the formulation in diabetes was very similar to glibenclamide (Patil et al., 2010).

Hepatoprotective effect

Studies in experimental animal models and humans indicate that HCA-SX per se is neither hepatotoxic nor hepatoprotective. *G. cambogia* in a rat model has been tested against toxic challenge to the liver by both ethanol and dexamethasone. Mahendran and Devi (2001a) demonstrated that *G. cambogia* supplementation prevented dexamethasone-induced undesirable alterations in the lipid profile and also protected normal liver phospholipid levels. In another study, cotreatment of rats with ethanol and *G. cambogia* significantly inhibited the rise in lipid levels and ethanol-induced peroxidative damage of hepatocytes as evidenced from improved antioxidant status [superoxide dismutase, catalase, and glutathione (GSH) levels] and maintaining levels of serum aspartate and alanine aminotransferase and alkaline phosphatase to near normalcy (Mahendran and Devi, 2001b). Studies have also shown that administration of *Garcinia* fruit extract protects the hepatic damage in rats induced by antitubercular drugs (isoniazid, rifampicin, and pyrazinamide). The hepato-protective effect of the extract may thus be attributed to its underlying antioxidant activity, that is, inhibit MDA formation and prevention of GSH depletion, which prevents the process of initiation and progress of hepatocellular injury (Vandana et al., 2013). However, the hepatoprotective potential of *G. cambogia* in humans is under a cloud in view of reports of hepatotoxicity in different human clinical case reports and studies (Crescioli et al., 2018; Sharma et al., 2018).

Antiulcerative effect

G. cambogia has shown promise to treat chemical/drug- and bacteria (*Helicobacter pylori*)-induced gastric ulcer. Oral pretreatment of rats with *G. cambogia* extract (1 g/kg BW) at intervals of days 7 and 15 prior to ulcer induction with hydrochloric acid and ethanol, significantly decreased the volume and acidity of gastric juice, reversed increased lipid peroxidation, and altered the activities of antioxidant enzymes, proteins, and glycoproteins in ulcerative mucosa. The results suggest that *G. cambogia* decreased acidity and increased mucosal defense (Mahendran et al., 2002a). In another similar supportive study, oral pretreatment with *G. cambogia* fruit extract (1 g/kg BW/day) protected gastric mucosa against the damage induced by indometacin (20 mg/kg BW) for up to 15 days. The antiulcerogenic effect of extract was due to ability of the extract to decrease the acidity and increase the mucosal defense in gastric areas (Mahendran et al., 2002b). Similarly, garcinol has been reported to treat indometacin-induced acute ulceration and such an effect can be explained by the radical-scavenging and antiinflammatory properties of garcinol (Yamaguchi et al., 2000). Garcinol has also been reported to kill *H. pylori*, the bacteria responsible for duodenal ulcers in both a time- and dose-dependent manner (Chatterjee et al., 2003, 2005). *G. cambogia* extract has also been found to protect rats against induced colitis due to its anti-inflammatory activity (dos-Reis et al., 2009).

Erythropoietic effect

Ethanolic extract of *G. cambogia* seeds in rats has also been reported to have an erythropoietic effect besides the antiobesity effects (Oluyemi et al., 2007). *G. cambogia* fruit has been analyzed and found to be rich in iron, zinc, cobalt, copper, phosphorus, and other trace elements (Jamila et al., 2019). Iron and other trace elements (zinc, cobalt, and copper) are required for the erythropoiesis. Furthermore, the levels of bioflavonoids in the plant may increase the level of peripheral testosterone, which also stimulates erythropoiesis. Additionally, due to the antioxidant activity of *G. cambogia* there is a decreased rate of oxidant-induced hemolysis, which enhances the life span of RBCs. The extract can be useful in phosphorus and copper deficiencies as the extract is rich in these minerals (Gurse et al., 2012).

Antimicrobial activity

Various extracts of fruit rind have shown significant antimicrobial activity against *Escherichia coli*, *Klebsiella pneumoniae*, *Pseudomonas aeruginosa*, *Bacillus subtilis*, and *Staphylococcus aureus*. The ethyl acetate extract was found to be the most active, followed by ethanol and hydroalcoholic extracts, whereas hexane extracts were found to be inactive against all the test pathogens (Shivakumar et al., 2013). Ethanolic and water extracts of *G. cambogia* leaves exhibited inhibitory activity against HIV-1 protease and HIV-1 integrase—major HIV enzymes. The IC_{50} values for the water extract were determined to be 67 and 70 μg/mL against protease and integrase, respectively, whereas the ethanolic extract showed comparatively weak activity with an IC_{50} value of 100 μg/mL against both enzymes (Narayan et al., 2011).

Miscellaneous effects

The diuretic activity of leaves of *G. cambogia* has also been reported. Ethanolic and aqueous leaves of *G. cambogia* in rats demonstrated a dose-dependent increase in output of urine along with excretion of electrolytes, such as sodium, potassium, and chloride compared to control animals (Mathew et al., 2011). The aqueous extract of *G. cambogia* fruit had a significant anticholinesterase activity by inhibiting cholinesterase activity by 30% and 60% at concentrations of 500 and 1000 μg/mL, respectively, in vitro. However, the inhibitory activity of this enzyme by the extract was less than neostigmine (Subhashini et al., 2011).

G. cambogia extract was also found to have potency to inhibit CYP2B6, an isozyme of CYP450. However, HCA alone could not significantly inhibit CYP2B6 (Yu et al., 2017).

Safety

G. cambogia extract has been considered to be safe for human consumption and has a long history of traditional use as a preservative and flavoring agent of curries.

G. cambogia extract has been available as an antiobesity herbal supplement around the world for decades and has been claimed to be safe. No deaths or other serious adverse effects were reported prior to 2005 with its regular consumption. Some mild side effects such as headache, nausea, dizziness, some laxative effects, cold, and skin rashes, though rare, may occur when the amount taken is more than the recommended dose. Allergic reactions are also rare (Heymsfield et al., 1998; Onakpoya et al., 2011). The supplement has no reported teratogenic effect but still should not be used in pregnant or breastfeeding women. The *G. cambogia* extract supplement should not be used in patients taking medication for diabetes mellitus as it may cause a fall in blood sugar levels (Roche et al., 2014). In view of the anticholinergic activity of *G. cambogia* (Subhashini et al., 2011) the supplement should also not be used in patients with Alzheimer and similar dementia diseases as such conditions may worsen. *G. cambogia*, being rich in iron, may have an additive adverse effect in patients taking medications for anemia (Oluyemi et al., 2007). The extract has some interactions with green tea, alcohol, anabolic steroids, acetaminophen, meperidine, tramadol, montelukast (leukotrienes receptor antagonist), serotonin reuptake inhibitors (fluoxetine), tricyclic, and typical antidepressants (amitriptyline, imipramine, etc.). Therefore coconsumption of the extract with these agents should be avoided to circumvent adverse drug interactions. Coconsumption of green tea, acetaminophen, or alcohol were the probable causes of hepatotoxicity associated with Hydroxycut, an over-the-counter weight management supplement in the United States in which *G. cambogia* was one of the ingredients (Dara et al., 2008). Use of a nutritional supplement containing *G. cambogia* in patients concomitantly with montelukast (leukotrienes receptor antagonist) and escitalopram (selective serotonin reuptake inhibitor) resulted in toxicity (Actis et al., 2007; Annette et al., 2014). HCA, due to its serotogenic and muscarinic-enhancing properties via inhibition of serotonin reuptake and acetylcholinesterase, has been reported to induce mania and associated psychotic disorders in various clinical case reports (Narasimha et al., 2013; Lopez et al., 2014; Hendrickson et al., 2016; Nguyen et al., 2017).

Toxicity

Acute oral, acute dermal, primary dermal irritation, and primary eye irritation and chronic toxicity studies have been reported in animals using different doses of HCA-SX. Toxicity studies in animals demonstrate that HCA-SX has a low acute oral toxicity with the LD_{50} being greater than 5000 mg/kg in fasted male and female albino rats (Ohia et al., 2002). This dose is equivalent to 350 g or 233 times the maximum recommended dose of 1.5 g/day of HCA in humans. Long-term safety of HCA-SX has been assessed in rats treated with various doses of the salt for 90 days. Shara et al. (2003) did not observe any significant alterations in organ histopathology, hepatic and testicular lipid peroxidation, or DNA fragmentation. Further, no remarkable dose- or time-dependent toxicity after chronic administration of HCA was observed on hepatic and testicular lipid peroxidation and DNA fragmentation, hematology, clinical chemistry, and histopathology in male and female rats (Shara et al., 2004).

Safety of HCA-SX has also been assessed in two-generation reproductive and developmental toxicity in male and female rats. Deshmukh et al. (2008a) conducted a two-generation reproduction toxicity study to evaluate the effects of HCA-SX on the reproductive systems of male and female rats, postnatal maturation, and reproductive capacity of their offspring, and possible cumulative effects through multiple generations. Sprague-Dawley rats were provided with diets containing 0, 1000, 3000, or 10,000 ppm of HCA-SX for a period of 10 weeks prior to mating, during mating, and for females through gestation and lactation across two generations. No treatment-related adverse effects on reproductive performance in terms of fertility and mating in both sexes, gestation, parturition, litter size, lactation, sexual maturity, and development of offspring in females were observed during HCA-SX exposure from F_0 and F_1 generations. The developmental toxicity conducted in continuation of the two-generation reproductive toxicity studies using HCA-SX was not found to be teratogenic to rats at a dietary exposure of 1000, 3000, or 10,000 ppm, equivalent to dose levels of 103, 352, and 1240 mg/kg/BW/day, respectively. At these levels HCA-SX was not found to induce any maternal toxicity or external abnormalities, soft tissue, or skeletal abnormality in fetuses (Deshmukh et al., 2008b).

Rats orally administered with crude ethanolic seed extract of *G. cambogia* at the rate of 100 and 200 mg/kg BW once a day, 6 days a week for 6 weeks revealed an increase in the interstitial spaces, degeneration of the Leydig cells, and distortion in the arrangement of the cells of spermatogenic series in histological examination of testes at the end of the experiment. However, the sperm counts were significantly higher in the experimental groups when compared with the control (Kayode et al., 2007). The histological preparations of the testis and epididymis revealed toxic effects which were more pronounced in the group that received 200 mg/kg BW extract, thereby indicating a dose-dependent toxic effect. An increase in sperm density may be due to the presence of bioflavonoid and xanthone in the plant extract, as these are potent antioxidants capable of increasing the production of testosterone, the key hormone involved in the production and maturation of spermatozoa in the seminiferous tubules of the testis (Ganong, 2003). The seeds of another plant of *Garcinia* genus, *Garcinia kola* have been reported to increase the peripheral testosterone level (Akpantah et al., 2003), which may cause increases in the sperm counts as observed in rabbits after feeding 35 mg/kg of *G. kola* (Akinloye et al., 1999). However, *G. cambogia* extract administration to overweight human subjects for 12 weeks neither significantly altered serum testosterone, estrone, or estradiol levels nor induced any adverse effects with respect to hematology and serum clinical pathology parameters (Hayamizu et al., 2008).

Dietary feeding of high doses of *G. cambogia* (containing 102 and 154 mmol of HCA/kg of diet) to 6-week-old male Zucker obese rats for more than 90 days suppressed body fat accumulation, increased concentration of liver glycogen, but was highly toxic to testis of the rats. The weight of testes was reduced by half and histopathologically the testes of developing rats revealed marked testicular atrophy and impairment of spermatogenesis. The highest dose of *G. cambogia* (154 mmol HCA/kg of diet) also caused severe diarrhea in all the rats. Lower doses (10 and 51 mmol/HCA/kg diet) did not induce any significant testicular toxicity in male developing Zucker obese rats (Saito et al., 2005). However, various aspects of the study were questioned by Burdock et al. (2005) who noted unacceptable levels of uncertainty about the HCA salts tested, animal model used, and various other aspects of the study design.

In another study on obesity-prone mice fed a HFD along with *G. cambogia* (1%) for 16 weeks the body weight was controlled, visceral fat accumulation, adipocyte size, and serum glucose lowered, yet purportedly hepatic fibrosis, inflammation, lipid peroxidation and increased mRNA levels of genes related to oxidative stress (Superoxide dismutase and glutathione peroxidase) and proinflammatory cytokines like TNF-α and monocyte chemo-attractant protein-1 were exhibited (Kim et al., 2013). These findings were at variance with a large number of published studies conducted on the safety and efficacy of HCA in animals and humans.

Several experimental studies have indicated that HCA actually exerts a hepatoprotective effect. Earlier human trials also do not document any adverse effects on the liver. Hayamizu et al. (2008) in a trial on humans providing 1000 mg HCA per day (i.e., 1666 mg of 60% HCA salt) for 3 months observed no significant changes in hepatic function parameters. Stohs et al. (2010) quoted the summarized results of Soni et al. (2004) and Downs et al. (2005) involving 750 subjects of 14 double-blind and placebo-controlled studies and one single arm of an open trial of HCA and observed no significant adverse effects reported in any of these trials. None of these studies reported any clinically significant alterations in hematology, blood chemistry, including liver enzymes, or any other pathological changes and from these studies it was concluded that dietary intake of HCA up to 2800 mg/kg in humans could be regarded as the no observed adverse effect level (NOAEL).

Consumption of a dietary supplement containing *G. cambogia* for weight reduction by humans was reported to induce hepatotoxicity in various case reports and case studies (Actis et al., 2007; Elinav et al., 2007; Schoepfer et al., 2007). Hepatotoxicity/acute liver injury was also reported from the United States following consumption of Hydroxycut, a popular weight-reducing dietary supplement containing HCA in addition to other ingredients (Stevens et al., 2005; Jones and Andrews, 2007; Dara et al., 2008). Three female patients reported acute liver injury after taking *G. cambogia* extract for a period varying from 3 months to 1 year and these patients showed marked clinical improvement after cessation of exposure (Chen et al., 2010). Similarly, Fong et al. (2010) reported induction of severe liver injury in eight patients in the United States after being exposed for varying periods to a weight-reducing dietary supplement (Hydroxycut). Hepatic injury was so severe that three patient had to undergo liver transplantation, whereas others improved with conventional treatment during their hospitalization. These reported cases of hepatic injury were severe enough to necessitate liver transplant and a single fatality prompted the US-FDA in 2009 to withdraw the Hydroxycut dietary supplement from the market and warn consumers about its potential to cause fatal hepatotoxicity.

The combined human and animal data suggest that HCA per se was not responsible with respect to hepatotoxicity in the case studies associated with the use of Hydroxycut products. Neither HCA nor other ingredients present in the Hydroxycut formulation induce any hepatotoxicity in animal toxicity studies. The hepatic safety of HCA and other ingredients present in Hydroxycut was evaluated comprehensively by Stohs et al. (2009) and no evidence of toxicity with HCA was found. However, the hepatotoxicity was attributed to the presence of (−)epigallocatechin-3-gallate, a polyphenol of green tea extract which is a constituent of Hydroxycut and this polyphenol at high doses induces hepatotoxicity in animal models (Lambert et al., 2010). Consumption of green tea has been associated with liver damage in isolated cases in humans which may be an idiosyncratic reaction (Mazzanti et al., 2009). Further, toxicity of Hydroxycut in different human patients may also have resulted from coconsumption of substances like alcohol or drugs (acetaminophen, anabolic steroids) or a combination of these (Dara et al., 2008). Various other clinical toxic reports elucidating clinical evidence of hepatitis and acute liver failure with submassive necrosis in human patients taking dietary supplements containing *G. cambogia* have been reported (Melendez-Rosado et al., 2015; Kothadia et al.,

2016; Lunsford et al., 2016; Vitalone et al., 2011; Crescioli et al., 2018). A case of acute liver failure in a female patient progressing to fulminant hepatic failure requiring liver transplantation has been reported after exposure to *G. cambogia* extract (Corey et al., 2016). In all cases of fulminant hepatic failure, biopsy showed significant necrosis and collapse of the liver parenchymal architecture.

All these clinical reports and case studies suggest a potential casual association between consumption of *G. cambogia*-containing products and the development of acute liver injury. However, the length of *G. cambogia* treatment was heterogeneous in different case reports. Hepatotoxicity in different patients may have also resulted from individual factors such as comorbidities, concomitant treatment with other drugs, and coconsumption of several other ingredients present in dietary supplements which makes it difficult to identify the substance especially responsible for the liver damage. However, Lunsford et al. (2016) described the first acute liver failure following the administration of purified extract of *G. cambogia*, with HCA being the main active principle of this purified extract and it was, therefore, proved to be directly responsible for such liver damage.

The exact mechanism by which HCA causes liver injury remains unclear. However, some information unveiling the mechanism of hepatic damage can be inferred from an experimental study on obesity-prone mice fed a HFD supplemented with *G. cambogia* (1%) for 16 weeks (Kim et al., 2013). In these mice the supplement controlled the weight, significantly lowering visceral fat accumulation and adipocyte size by inhibiting fatty acid synthesis and β-oxidation. However, the supplemented mice exhibited elevated plasma levels of alanine and aspartate aminotransferase, indicating impaired liver function. The liver of such mice had increased hepatic collagen accumulation, lipid peroxidation, and mRNA levels of genes related to oxidative stress (superoxide dismutase and glutathione peroxidase) and proinflammatory cytokines such as TNF-α and monocyte chemo-attractant protein-1. Oxidative stress is known to play an important role in the progression of steatohepatitis and hepatocellular injury (Kim et al., 2013). Another probable mechanism could be a genetic predisposition in certain patients such as those with CYP polymorphism which promotes toxic accumulation of metabolites leading to hepatotoxicity (Brewer and Chen, 2017; Nicolatti et al., 2017; Krupa et al., 2017). Concomitant intake of herbal supplements like *G. cambogia* extract and prescribed regular medications is very common in patients suffering from underlying chronic diseases and therefore unpredictable herb–drug interactions may resulting in herb-induced liver injury cannot be ruled out (Lee and Fermo, 2006).

Other clinical adverse effects in humans after exposure to *G. cambogia* have been reported. A single case report of acute necrotizing eosinophilic myocardities in a female patient exposed to *G. cambogia* extract for 3 weeks was reported and the patient was successfully treated with a high dose of corticosteroids (Allen et al., 2014). In another case report, a female without any clinical history developed stress-induced cardiopathy after ingesting *G. cambogia* extract for about a week (Joseph and Levine, 2014). A female patient with a history of hypertension developed hypoglycemia following exposure to *G. cambogia*-containing supplement for 2 days (Roche et al., 2014). Continuous administration of *G. cambogia* for 1 month resulted in development of drug-induced immune thrombocytopenia (Sikka et al., 2016). Serotonin toxicity has also been observed in a female patient exposed to *G. cambogia* supplement more than 2 months (Lopez et al., 2014). Diabetic ketoacidosis and cardiopathy developed in a female patient after ingesting *G. cambogia* extract daily for 1 month (Bystrak et al., 2017). Three separate case studies reported rhabdomyolysis after exposure to a *G. cambogia*-containing supplement (Mansi and Huang, 2004; Dehoney and Wellein 2009; Hines et al., 2015). A number of recent reports have described hypo(mania) and/or psychosis following *G. cambogia* consumption (Narasimha et al., 2013; Wong et al., 2016; Nguyen et al., 2017; Cotovio and Oliveira-Maia, 2017).

Concluding remarks and future directions

Obesity has reached epidemic proportions globally, and millions of peoples are dying each year as a result of this malady and associated health problems. Commonly prescribed synthetic antiobesity drugs come with a variety of adverse effects which gives impetus to the discovery of new drugs from herbal sources purportedly with minimal side effects. Many herbal formulations are promoted for weight loss and these are generally sold as self = medications that are often used by people under the misconception that being natural in origin guarantees their safety. Therefore their use is widespread and increasing throughout the world.

G. cambogia has become a popular herbal supplement for its weight-loss properties. The primary active ingredient in *G. cambogia* extract is HCA, which exhibits antiobesity activity by a multitude of effects such as suppressing appetite, reducing food intake, preventing lipogenesis, and increasing fat oxidation. Large numbers of well-designed appropriately controlled studies in animals and humans have indicated that *G. cambogia*-containing HCA is safe and efficacious. Collective results from human clinical studies demonstrate the safety of HCA or its salts, except minor gastrointestinal adverse events (Onakpoya et al., 2011). These studies provide scientific evidence that intake of HCA-SX alone did not produce any adverse effects and a dietary dose of up to 2800 mg/day was regarded as the NOAEL in humans. However, a metaanalysis of many random clinical trials in humans suggests that *Garcinia* extract/HCA induces only small but statistically significant reduction in body weight and fat loss on a short-term basis (Onakpoya et al., 2011). Such effects were of short duration and there is not much evidence to support the long-term benefits of *G. cambogia* extracts (Marquez et al., 2012). Therefore future trials of *G. cambogia* should be more rigorous and better reported.

Other beneficial health effects, like antioxidant, antiinflammatory, antidiabetic, antiulcerogenic, hepatoprotective, and anticancer are still in an experimental stage and need to be scientifically evaluated in humans. Evidence from various case studies and case reports in humans suggests a potential causal association between *G. cambogia* exposure and development of herb-induced liver injury, mania, psychosis, and other toxic manifestations.

In view of the lack of manufacturing regulations on *G. cambogia* products as in other herbal supplements, quality assurance and standardization of the product needs to be undertaken to help to quantify the uniform quality taken by patients. Therefore continuous monitoring of *G. cambogia*-containing dietary supplements should be encouraged to characterize their risk profile which will strengthen the regulations and pharmacovigilance of this natural antiobesity supplement.

References

Actis, G.C., Bugianesi, E., Ottobrelli, A., et al., 2007. Fatal liver failure following food supplements during chronic treatment with montelukast. Dig. Liver Dis. 39 (10), 953–955.

Acuna, U.M., Figueroa, M., Kavalier, A., et al., 2010. Benzophenones and biflavonoids from from *Rheedia edulis*. J. Nat. Prod. 73 (11), 1775–1779.

Ahmad, A., Wang, Z., Ali, R., Maitah, M.Y., et al., 2010. Apoptosis-inducing effect of garcinol is mediated by NF-kB signalling in breast cancer. J. Cell Biochem. 109, 1134–1141.

Akinloye, A.K., Igharna, O.O., Olaniyi, M.O., et al., 1999. Preliminary investigations on the effects of *Garcinia kola*, (bitter kola) on the rabbit's testis and epididymis. Trop. Vet. 18, 49–54.

Akpantah, A.O., Oremosu, A.A., Ajala, M.O., et al., 2003. The effect of crude extract of *Garcinia Kola* seed on the histology and hormonal milieu of male Sprague-Dawley rats' reproductive organs. Niger. J. Health Biomed. Sci. 2 (1), 40–46.

Allen, S.F., Godley, R.W., Evron, J.M., et al., 2014. Acute necrotizing eosinophilic myocarditis in a patient taking *Garcinia cambogia* extract successfully treated with high-dose corticosteroids. Can. J. Cardiol. 30, 1732.

Amin, K.A., Kamel, H.H., Abd, Eltawab, M.A., 2011a. The relation of high fat diet, metabolic disturbance and brain oxidative dysfunction; modulation by hydroxyl citric acid. Lipids Health Dis 10, 74.

Amin, K.A., Kamel, H.H., Abd Eltawab, M.A., 2011b. Protective effect of *Garcinia* against renal oxidative stress and biomarkers induced by high fat and sucrose diet. Lipids Health Dis. 10, 6.

Annette, M.L., Kornegay, J., Hendrickson, R.G., 2014. Serotonin toxicity associated with *Garcinia cambogia* over the counter supplement. J. Med. Toxicol. 10, 399–401.

Asghar, M., Monjok, E., Kouamou, G., et al., 2007. Super CitriMax (HCA-SX) attenuates increases in oxidative stress, inflammation, insulin resistance, and body weight in developing obese Zucker rats. Mol. Cell Biochem. 304, 93–99.

Balasubramanyyam, K., Altaf, M., Varier, R.A., et al., 2004. Polyisoprenylated benzophenones, garcinol, a natural histone acetyl transferase inhibitor, represses chromatin transcription and alters global gene expression. J. Biol. Chem. 32, 33716–33726.

Bauer, D.E., Hatzivassiliou, G., Zhao, F., et al., 2005. ATP citrate lyase is an important component of cell growth and transformation. Oncogene 24, 6314–6322.

Behera, A.K., Swamy, M.M., Natesh, N., et al., 2016. Garcinol and its role in chronic diseases. In: Gupta, S., Prasad, S., Aggarwal, B. (Eds.), Anti-Inflammatory Nutraceuticals and Chronic Diseases. Advances in Experimental Medicine and Biology, vol 928. Springer, Cham, pp. 435–452.

Brewer, C.T., Chen, T., 2017. Hepatotoxicity of herbal supplements mediated by modulation of cytochrome P450. Int. J. Mol. Sci. 18, 2353. <https://doi.org/10.3390/ijms18112353>.

Burdock, G., Soni, M., Bagchi, M., Bagchi, D., 2005. *Garcinia cambogia* toxicity is misleading. Food Chem. Toxicol. 43, 1683–1684.

Bystrak, T., Cervera-Hernandez, M.E., Reddy, N., King, Z., Bratberg, J., 2017. *Garcinia Cambogia*, diabetic ketoacidosis, and pancreatitis. R. I. Med. J. 100, 48–50.

Cao, S., Brodie, P.J., Miller, J.S., et al., 2007. Guttiferones K and L, Antiproliferative compounds of *Rheedia calcicola* from the Madagascar rain forest. J. Nat. Prod. 70 (4), 686–688.

Chatterjee, A., Yaasmin, T., Bagchi, D., et al., 2003. The bactericidal effects of *Lactobacillus acidophilus*, garcinol and Protykin® compared to clarithromycin, on *Helicobacter pylori*. Mol. Cell Biochem. 243, 29–35.

Chatterjee, A., Bagchi, D., Yasmin, T., et al., 2005. Antimicrobial effects of antioxidants with and without clarithromycin on *Helicobacter pylori*. Mol. Cell Biochem. 270, 125–130.

Chen, G.C., Ramanathan, V.S., Law, D., Funchain, P., Chen, G.C., French, S., et al., 2010. Acute liver injury induced by weight-loss herbal supplements. World J. Hepatol. 2, 410–415.

Cheng, I.S., Huang, S.W., Lu, H.C., et al., 2012. Oral hydroxycitrate supplementation enhances glycogen synthesis in exercised human skeletal muscle. Br. J. Nutr. 107, 1048–1055.

Chuah, Li Oon, Ho, W.Y., et al., 2013. Updates on Antiobesity Effect of Garcinia Origin (-)-HCA. Evidence Based Complimentary and Alternative Medicine. Article ID. 751658.

Clouatre, L.D., Preuss, H.G., 2008. Potassium magnesium hydroxycytrate at physiologic levels influences various metabolic parameters and inflammation in rats. Curr. Top. Nutraceut. Res. 6, 201–210.

Corey, R., Werner, K.T., Singer, A., et al., 2016. Acute liver failure associated with *Garcinia cambogia* use. Ann. Hepatol. 15 (1), 123–126.

Cotovio, G., Oliveira-Maia, A.J., 2017. Hypomania induced by a *Garcinia cambogia* supplement. Aust. N. Z. J. Psychiatry 51, 641–642.

Crescioli, G., Lombardi, N., Bettiol, A., et al., 2018. Acute liver injury following *Garcinia cambogia* weight-loss supplementation: case series and literature review. Intern. Emerg. Med. 13 (6), 857–872.

Dara, L., Hewett, J., Lim, J.K., 2008. Hydroxycut hepatotoxicity: a case series and review of liver toxicity from herbal weight loss supplements. World J. Gastroenterol. 14 (45), 6999–7004.

Deshmukh, N.S., Bagchi, M., Yasmin, T., et al., 2008a. Safety of a novel calcium/potassium salt of hydroxycitric acid (HCA-SX): I. Two-generation reproduction toxicity study. Toxicol. Mechan. Methods 18 (5), 433–442.

Deshmukh, N.S., Bagchi, M., Yasmin, T., et al., 2008b. Safety of a novel calcium/potassium salt of hydroxycitric acid (HCA-SX): II. Developmental toxicity study in rats. Toxicol. Mechan Methods 18 (5), 442–451.

dos-Reis, S.B., De Oliveira, C.C., Acedo, S.C., et al., 2009. Attenuation of colitis injury in rats using *Garcinia cambogia* extract. Phytother. Res. 23 (3), 324–329.

Dehoney, S., Wellein, M., 2009. Rhabdomyolysis associated with the nutritional supplement Hydroxycut. Am. J. Health Syst. Pharm. 66, 142–148.

Downs, B.W., Bagchi, M., Subbaraju, G.V., et al., 2005. Bioefficiency of a novel calcium-potassium salt of (−)-hydroxycitric acid. Mutat. Res. 579, 149–162.

Elinav, E., Pinsker, G., Safadi, R., Pappo, O., Bromberg, M., Anis, E., et al., 2007. Association between consumption of Herbalife nutritional supplements and acute hepatotoxicity. J. Hepatol. 47, 514–520.

Flavin, D.F., 2007. A lipoxygenase inhibitor in breast cancer brain metastases. J. Neurooncol. 82, 91–93.

Fong, T.L., Klontz, K.C., Canas-Coto, A., et al., 2010. Hepatotoxicity due to hydroxycut: a case series. Am. J. Gastroenterol. 105 (7), 1561–1566.

Ganong, W.F., 2003. Twenty first ed. Review of Medical Physiology., 23. Lange/ McGraw-Hill Companies, New York, pp. 429–430.

Guais, A., Baronzio, G., Sanders, E., et al., 2012. Adding a combination of hydroxycitrate and lipoic acid (METABLOC™) to chemotherapy improves effectiveness against tumor development: experimental results and case report. Invest. New Drugs 30 (1), 200–211.

Gurse, F.E., Ates, A., Bilal, T., et al., 2012. Effect of dietary *G. cambogia* extract on serum essential minerals (calcium phosphorus, magnesium) and trace elements (iron, copper, zinc) in rats fed with high lipid diet. Biol. Trace El Res. 148 (3), 378–382.

Halliday, K.R., Fenoglio-Preiser, C., Sillerud, L.O., 1988. Differentiation of human tumors from non-malignant tissue by natural-abundance 13C NMR spectroscopy. Magn. Reson. Med. 7, 384–411.

Hatzivassiliou, G., Zhao, F., Bauer, D.E., et al., 2005. ATP citrate lyase inhibition can suppress tumor cell growth. Cancer Cell 8, 311–321.

Hayamizu, K., Hirakawa, H., Oikawa, D., et al., 2003a. Effect of *Garcinia cambogia* extract on serum leptin and insulin in mice. Fitoterapia 74, 267–273.

Hayamizu, K., Ishii, Y., Kaneko, I., et al., 2003b. Effects of *Garcinia cambogia* (hydroxycitric acid) on visceral fat accumulation: a double-blind, randomized, placebo-controlled trial. Curr. Ther. Res. 64 (8), 551–567.

Hayamizu, K., Tomi, H., Kaneko, I., et al., 2008. Effect of *Garcinia cambogia* extract on serum sex hormones in over weight objects. Fitoterapia 79, 267–273.

Hendrickson, B.P., Shaikh, N., Occhiogrosso, M., Penzner, J.B., 2016. Mania induced by Garcinia cambogia: a case series. Prim care companion CNS disorder. https://doi.org/10.4088/PCC.15101890.

Heymsfield, S.B., Allison, D.B., Vasselli, J.R., et al., 1998. *Garcinia cambogia* (hydroxycitric acid) as a potential antiobesity agent: a randomized controlled trial. JAMA 280, 1596–1600.

Hong, J., Kwon, S.J., Sang, S., et al., 2007. Effects of garcinol and its derivatives on intestinal cell growth: inhibitory effects and autoxidation dependent growth-stimulatory effects. Free Radic. Biol. Med. 42, 1211–1221.

Huang, M.T., Liu, Y., Badmaev, V., et al., 2008. Anti-inflammatory and anticancer activities of garcinol. Dietary Supplements. American Chemical Society, Washington DC, pp. 293–303.

Hines, E.Q.D.T.E., Melville, L.D., Su, M.K., 2015. Severe rhabdomyolysis associated with *Garcinia cambogia*. Clin. Toxicol. 53, 746–747.

Iinuma, M.T., Ito, R., Miyake, H., et al., 1998. A xanthone from *Garcinia cambogia*. Phytochemistry 47 (6), 1169–1170.

Ishihara, K.S., Oyaizu, K., Onuki, K., et al., 2000. Chronic (–)-hydroxycitrate administration spares carbohydrate utilization and promotes lipid oxidation during exercise in mice. J. Nutr. 130 (12), 2990–2995.

Jamila, N., Khan, N., Hwang, I.M., et al., 2019. Determination of macro, micro, trace essential, and toxic elements in *Garcinia cambogia* fruit and its anti-obesity commercial products. J. Sci. Food Agric. 99 (5), 2455–2462.

Joseph, D.J.M., Levine, M., 2014. Stress induced cardiomyopathy after ingestion of *Garcinia cambogia*. Clin. Toxicol. 52, 739–740.

Jena, B.S., Jayaprakasha, G.K., Singh, R.P., et al., 2002. Chemistry and biochemistry of hydroxycitric acid from *Garcinia*. J. Agric. Food Chem. 50 (1), 10–22.

Jones, F.J., Andrews, A.H., 2007. Acute liver injury associated with the herbal supplement hydroxycut in a soldier deployed to Iraq. Am. J. Gastroenterol. 102, 2357–2358.

Kan, W.L., Yin, C., Xu, H.X., et al., 2013. Antitumor effects of novel compound, guttiferone K, on colon cancer by p21Waf1/Cip1-mediated G(0)/G(1) cell cycle arrest and apoptosis. Int. J. Cancer 132 (3), 707–716.

Kayode, O.A., Jimoh, Olusegun, R., et al., 2007. Effects of crude ethanolic extract of *Garcinia cambogia* on the reproductive system of male wistar rats (*Rattus novergicus*). Afr. J Biotechnol. 6 (10), 1236–1238.

Khare, C.P., 2007. Indian Medicinal Plants: An Illustrated Dictionary. Springer, Berlin.

Kim, K.Y., Lee, H.N., Kim, Y.J., et al., 2008. *Garcinia cambogia* extract ameliorates visceral adiposity in C57BL/6J mice fed on a high-fat diet. Biosci. Biotechnol. Biochem. 72, 1772–1780.

Kim, Y.J., Choi, M.S., Park, Y.B., et al., 2013. *Garcinia cambogia* attenuates diet induced adiposity but exacerbates hepatic collagen accumulation and inflammation. World J. Gastroenterol. 19, 4689–4701.

Kothadia, J.P.R.R., Olivera-Martinez, M., 2016. Hepatoxicity associated with *Garcinia cambogia* used as a weight loss supplement. Am. J. Gastroenterol 111, S881.

Koeberle, A., Northoff, H., Werz, O., 2009. Identification of 5-lipoxygenase and microsomal prostaglandin E_2 synthase 1 as functional targets of the anti-inflammatory and anti-carcinogenic garcinol. Biochem. Pharmacol. 77 (9), 1513–1521.

Kolodziejczyk, J., Masullo, M., Olas, B., et al., 2009. Effects of garcinol and guttiferone K isolated from *Garcinia cambogia* on oxidative/nitrative modifications in blood platelets and plasma. Platelets 20 (7), 487–492.

Kovacs, E.M.R., Westerp-Plantenga, M.S., De Vries, M., et al., 2001. Effects of 2-week ingestion of (-)-hyrdoxycitrate and (-)-hyrdoxycitrate combined with medium-chain triglycerides on satiety and food intake. Physiol. Behav. 74, 543–549.

Kovacs, E.M.R., Westerterp-Plantenga, M.S., 2006. Effects of (−)-hydroxycitrate on net fat synthesis as *de novo* lipogenesis. Physiol. Behav. 88, 371–381.

Kriketos, A.D., Thompson, H.R., Greene, H., et al., 1999. (–)-Hydroxycitric acid does not affect energy expenditure and substrate oxidation in adult males in a post-absorptive state. Int. J. Obes. Relat. Metab. Disord. 23 (8), 867–873.

Krupa, R., Czarny, P., Wigner, P., et al., 2017. The relationship between single-nucleotide polymorphisms, the expression of DNA damage response genes, and hepatocellular carcinoma in a polish population. DNA Cell Biol. 36, 693–708.

Lambert, J.D., Kennett, M.J., Sang, S., et al., 2010. Hepatotoxicity of high oral dose (-)-epigallocatechin -3-gallate in mice. Food Chem. Toxicol. 48, 409–416.

Lee, N.J., Fermo, J.D., 2006. Warfarin and royal jelly interaction. Pharmacotherapy 26, 583–586.

Leonhardt, M., Langhans, W., 1977. Hydroxycitrate has long-term effects on feeding behavior, body weight regain and metabolism after body weight loss in male rats. J. Nutr. 132 (7), 1977–1982.

Leonhardt, M., Hrupka, B., Langhans, W., 2001. Effect of hydroxycitrate on food intake and body weight regain after a period of restrictive feeding in male rats. Physiol. Behav. 74, 191–196.

Leonhardt, M., Balkan, B., Langhans, W., 2004. Effect of hydroxycitrate on respiratory quotient, energy expenditure, and glucose tolerance in male rats after a period of restrictive feeding. Nutrition 20, 911–915.

Lewis, Y.S., Neelakantan, S., 1965. -)-Hydroxycitric acid—the principal acid in the fruits of *Garcinia cambogia* desr. Phytochemistry 4, 619–625.

Li, F., Shanmugam, M.K., Chen, L., et al., 2013. Garcinol, a polyisoprenylated benzophenone modulates multiple proinflammatory signaling cascades leading to the suppression of growth and survival of head and neck carcinoma. Cancer Prev. Res. 6, 843–854.

Liao, C., Sang, S., Ho, C., et al., 2005. Garcinol modulates tyrosine phosphorylation of FAK and subsequently induces apoptosis through down-regulation of Src, ERK, and Akt survival signaling in human colon cancer cells. J. Cell Biochem. 96 (1), 155–169.

Lim, K., Ryu, S., Ohishi, Y., et al., 2002. Short term (−)-hydroxycitrate ingestion increases fat oxidation during exercise in athletes. J. Nutr. Sci. Vitaminol. 48, 128–133.

Lim, K., Ryu, S., Nho, H.S., et al., 2003. (−)-Hydroxycitric acid ingestion increases fat utilization during exercise in untrained women. J. Nutr. Sci. Vitaminol. 49, 163–167.

Lim, G.W., Jafarzadeh, S., Norazatul Hanim, M.R., 2020. Kinetic study, optimization and comparison of sun drying and superheated steam drying of asam gelugor (*Garcinia cambogia*). Food Res. 4 (2), 396–406.

Lobb, A., 2009. Hepatoxicity associated with weight-loss supplements: a case for better post-marketing surveillance. World J. Gastroenterol. 15 (14), 1786–1787. 19360927.

Lopaschuk, G.D., Ussher, J.R., Jaswal, J.S., 2010. Targeting intermediary metabolism in the hypothalamus as a mechanism to regulate appetite. Pharmacol. Rev. 62 (2), 237–264.

Lopez, A.M., Kornegay, J., Hendrickson, R.G., 2014. Serotonin toxicity associated with *Garcinia cambogia* over-the-counter supplement. J. Med. Toxicol. 10 (4), 399–401.

Lowenstein, J.M., 1971. Effect of (-)-hydroxycitrate on fatty acid synthesis by rat liver *in vivo*. J. Biol. Chem. 246, 629–632.

Lunsford, K.E., Bodzin, A.S., Reino, D.C., et al., 2016. Dangerous dietary supplements: *Garcinia cambogia*-associated hepatic failure requiring transplantation. World J. Gastroenterol. 22 (45), 10071–10076.

Mahendran, P., Devi, C.S., 2001a. Effect of *Garcinia cambogia* extract on lipids and lipoprotein composition in dexamethasone administered rats. Indian J. Physiol. Pharmacol. 45, 345–350.

Mahendran, P., Devi, C.S., 2001b. The modulating effect of *Garcinia cambogia* extract on ethanol-induced peroxidative damage in rats. Indian J. Pharmacol. 33, 87–91.

Mahendran, P., Sabitha, K.E., Devi, C.S., 2002a. Prevention of HCl ethanol induced gastric mucosal injury in rats by *Garcinia cambogia* extract and its possible mechanism of action. Indian J. Exp. Biol. 40, 58–62.

Mahendran, P., Vanisree, A.J., Devi, C.S., 2002b. Antiulcer activity of *Garcinia cambogia* extract against indomethacin induced gastric ulcer in rats. Phytother. Res. 16 (1), 80–83.

Maia-Landim, A., Ramirez, J.M., Lancho, C., et al., 2018. Long-term effects of *Garcinia cambogia*/Glucomannan on weight loss in people with obesity, PLIN4, FTO and Trp64Arg polymorphisms. BMC Complement. Altern. Med. 18, 26.

Mansi, I.A., Huang, J., 2004. Rhabdomyolysis in response to weightloss herbal medicine. Am. J. Med. Sci. 327, 356–357.

Marquez, F., Babio, N., Bullo, M., et al., 2012. Evaluation of the safety and efficacy of hydroxycitric acid or *Garcinia cambogia* extracts in humans. Crit. Rev. Food Sci. Nutr. 52, 585–594.

Masullo, M., Bassarello, C., Suzuki, H., et al., 2008. Polyisoprenylated benzophenones and an unusual polyisoprenylated tetracyclic xanthone from the fruits of *Garcinia cambogia*. J. Agric. Food Chem. 56, 5205–5210.

Mathew, G.E., Mathew, B., Shaneeb, M.M., et al., 2011. Diuretic activity of leaves of *Garcinia cambogia* in rats. Indian J. Pharm. Sci. 73 (2), 228–230.

Mayer, M.A., Finlayson, G., Fischman, D., et al., 2014. Evaluation of the satiating properties of a nutraceutical product containing *Garcinia cambogia* and *Ascophyllum nodosum* extracts in healthy volunteers. Food Funct. 5, 773–779.

Mazzanti, G., Menniti-Ippolito, F.-M., et al., 2009. Hepatotoxicity from green tea a review of the literature and two unpublished cases. Euro J. Clin. Pharmacol. 65, 331–341.

Mazzio, E.A., Soliman, K.F., 2009. *In-vitro* screening for the tumoricidal properties of international medicinal herbs. Phytother. Res. 23 (3), 385–398.

Melendez-Rosado, J., Snipelisky, D., Matcha, G., et al., 2015. Acute hepatitis induced by pure *Garcinia cambogia*. J. Clin. Gastroenterol. 49 (5), 449–450.

Migita, T., Narita, T., Kimie, et al., 2008. ATP citrate lyase: activation and therapeutic implications in non–small cell lung cancer. Cancer Res. 68, 8547–8554.

Narasimha, A., Shetty, P.H., Nanjundaswamy, M.H., Viswanath, B., Bada Math, S., 2013. Hydroxycut-dietory supplement for weight loss: can they induce mania? Aust. N. Z. J. Psychiatry 47, 1205–1206.

Narayan, L.C., Rai, V.R., Tewtrakul, S., 2011. A screening strategy for selection of anti-HIV-1 integrase and anti-HIV-1 protease inhibitors from extracts of Indian medicinal plants. Int. J. Phytomed. 2011 (3), 312–318.

Nguyen, D.C., Timmer, T.K., Davison, B.C., McGrane, I.R., 2017. Possible Garcinia cambogia-induced mania with psychosis: a case report. J. Pharm. Pract. 1, 897190017734728.

Nicoletti, P., Aithal, G.P., Bjornsson, E.S., et al., 2017. Association of liver injury from specific drugs, or groups of drugs, with polymorphisms in HLA and other genes in a genome-wide association study. Gastroenterology 152, 1078–1089.

Ohia, S.E., Awe, S.O., LeDay, A.M., et al., 2001. Effect of hydroxycitric acid on serotonin release from isolated rat brain cortex. Res. Commun. Mol. Pathol. Pharmacol. 109 (3-4), 210–216.

Ohia, S.E., Opere, C.A., LeDay, A.M., et al., 2002. Safety and mechanism of appetite suppression by a novel hydroxycitric acid extract (HCA-SX). Mol. Cell. Biochem. 238, 89–103.

Oluyemi, K.A., Omotuyi, I.O., Jimoh, O.R., et al., 2007. Erythropoietic and anti-obesity effects of *G. cambogia* (bitter kola) in Wistar rats. Biotechnol. Appl. Biochem. 46 (1), 69–72.

Onakpoya, I., Hung, S.K., Perry, R., et al., 2011. The use of garcinia extract (hydroxycitric acid) as a weight loss supplement: a systemic review and meta analysis of randomized clinical trials. J. Obesity 2011, Article ID509038: pages 9.

Padhye, S., Ahmad, A., Oswal, N., et al., 2009. Emerging role of Garcinol, the antioxidant chalcone from *Garcinia indica* Choisy and its synthetic analogs. J. Hematol. Oncol. 2, 38.

Patil, M.N., Kagathara, V.G., Harle, U.N., et al., 2010. Effect of polyherbal formulation in obesity associated diabetes. Int. J. Pharm. Sci. 2, 180–186.

Pierce, M.W., Palmer, J.L., Keutmann, H.T., et al., 1981. ATPcitrate lyase. Structure of a tryptic peptide containing the phosphorylation site directed by glucagon and the cAMP-dependent protein kinase. J. Biol. Chem. 256, 8867–8870.

Prasanth, N.V., Shebina, P.R., Tina, T., et al., 2013. Evaluation of *in vitro* and *in vivo* anti- inflammatory activity of *Garcinia cambogia*. Int. J. Pharm. Pharm. Sci. 5 (2), 263–264.

Preuss, H.G., Bagchi, D., Bagchi, M., et al., 2004. Effects of a natural extract of (−)-hydroxycitric acid (HCA-SX) and a combination of HCA-SX plus niacin-bound chromium and *Gymnema sylvestre* extract on weight loss. Diab. Obes. Metab. 6, 171–180.

Preuss, H.G., Garis, R.I., Bramble, J.D., et al., 2005. Efficacy of a novel calcium/potassium salt of (−)-hydroxycitric acid in weight control. Int. J. Clin. Pharmacol. Res. 25, 133–144.

Rawri, R.K., Bharati, K., Jayaveera, K.N., et al., 2013. *In-vitro* antioxidant properties of *Garcinia indica* Linn. alcoholic fruit extract. J. Pharm. Sci. Innov. 2 (3), 9–12.

Roche, B., Chenoweth, J.A., Radke, J.B., Poppenga, R.H., Sutter, M.E., 2014. Hypoglycemia associated with Garcinia cambogia ingestion. J. Clin. Toxicol. 52, 536.

Rederman, N.B., Saha, A.K., Vavvas, D., et al., 1999. Malonyl-CoA, fuel sensing, and insulin resistance. Am. J. Physiol. 276, E1–E18.

Roy, S., Rink, C., Khanna, S., et al., 2004. Body weight and abdominal fat gene expression profile in response to a novel hydroxycitric acid based dietary supplement. Gene Expr. 11, 252–263.

Roy, S., Shah, H., Rink, C., et al., 2007. Transcriptome of primary adipocytes from obese women in response to a novel hydroxycitric acid-based dietary supplement. DNA Cell Biol. 26, 627–639.

Saito, M., Ueno, M., Ogino, S., et al., 2005. High dose of *Garcinia cambogia* is effective in suppressing fat accumulation in developing male Zucker obese rats, but highly toxic to the testis. Food Chem. Toxicol. 43, 411–419.

Schoepfer, A.M., Engel, A., Fattinger, K., Marbet, U.A., Criblez, D., Reichen, J., et al., 2007. Herbal does not mean innocuous: ten cases of severe hepatotoxicity associated with dietary supplements from Herbalife products. J. Hepatol. 47, 521–526.

Schwartz, L., Buhler, L., Icard, P., et al., 2014. Metabolic treatment of cancer: intermediate results of a prospective case series. Anticancer Res. 34 (2), 973–980. 2014.

Sergio, W., 1988. A natural food, the Malabar Tamarind, may be effective in the treatment of obesity. Med. Hypothesis 29, 39–40.

Shara, M., Ohia, S.E., Yasmin, T., et al., 2003. Dose- and time dependent effects of a novel (−)-hydroxycitric acid extract on body weight, hepatic and testicular lipid peroxidation, DNA fragmentation and histopathological data over a period of 90 days. Mol. Cell. Biochem. 254, 339–346.

Shara, M., Ohia, S.E., Schmidt, R.E., et al., 2004. Physicochemical properties of a novel hydroxy citric acid extract and its effect on body weight, selected organ weights, hepatic lipid peroxidation and DNA fragmentation, hematology and clinical chemistry, and histopathological changes over a period of 90 days. Mol. Cell. Biochem. 260, 171–186.

Sharma, K., Kang, S., Gong, D., et al., 2018. Combination of *Garcinia cambogia* extract and pear pomace extract additively suppresses adipogenesis and enhances lipolysis in 3T3-L1 cells. Pharmacogn. Mag. 14, 220–226.

Shivakumar, S., Sandhiya, S., Subhasree, N., et al., 2013. *In vitro* assessment of antibacterial and antioxidant activities of fruit rind extracts of *Garcinia cambogia* L. Int. J. Pharm. Pharm. Sci. 5, 254–257.

Sikka, G., Gupta, A., Bachan, M., Khan, Z., 2016. "Uber-trim" causes immune thrombocytopenic purpura: a life threatening case report. Chest 150, 392A.

Soni, M.G., Burdock, G.A., Preuss, H.G., et al., 2004. Safety assessment of (−)-hydroxycitric acid and Super CitriMax, a novel calcium/potassium salt. Food Chem. Toxicol. 42, 1513–1529.

Stallings, W.C., Blount, J.F., Srere, P.A., et al., 1979. Structural studies of hydroxycitrates and their relevance to certain enzymatic mechanisms. Arch. Biochem. Biophys. 193 (2), 431–448.

Stevens, T., Qadri, A., Zein, N.N., 2005. Two patients with acute liver injury associated with use of the herbal weight-loss supplement hydroxycut. Ann. Intern. Med. 142 (6), 477–478.

Stohs, S.J., Preuss, H.G., Ohia, S.E., et al., 2009. No evidence demonstrating hepatotoxicity associated with hydroxycitric acid. World J. Gastroenterol. 15 (32), 4087–4089.

Stohs, S.J., Preuss, H.G., Ohia, S.E., et al., 2010. Safety and efficacy of hydroxycitric acid derived from *Garcinia cambogia* - a literature review. HerbalGram 85, 58–63.

Subhashini, N., Nagarajan, G., Kaviman, S., 2011. *In vitro* antioxidant and anticholinesterase activities of *Garcinia cambogia*. Int J. Pharm. Pharm. Sci. 3, 129–132.

Sullivan, C., Triscari, J., 1977a. Metabolic regulation as a control for lipid disorders. I. Influence of (-)-hydroxycitrate on experimentally induced obesity in the rodent. Am. J. Clin. Nutr. 30, 767–776.

Sullivan, A.C., Singh, M., Srere, P.A., et al., 1977b. Reactivity and inhibitor potential of hydroxycitrate isomers with citrate synthase, citrate lyase, and ATP citrate lyase. J. Biol. Chem. 252, 7583–7590.

Sullivan, A.C., Triscari, J., Neal, M.O., 1974a. The influence of (−)-hydroxycitrate on in vivo rates of hepatic glycogenesis: lipogenesis and cholesterol-genesis. Fed. Proc. 33, 656.

Sullivan, A.C., Triscari, J., Hamilton, J.G., et al., 1974b. Effect of (-)-hydroxycitrate upon the accumulation of lipid in the rat. II. Appetite. Lipids 9, 129–134.

Szutowicz, A., Kwiatkowski, J., Angielski, S., 1979. Lipogenetic and glycolytic enzyme activities in carcinoma and nonmalignant diseases of the human breast. Br. J. Cancer 39, 681–687.

Vandana, S., Panda, V.S., Ashar, H.S., et al., 2013. Antioxidant and hepatoprotective effects of *Garcinia indica* fruit rind in anti-tubercular drug-induced liver injury in rats. Botanics Targets Ther. 23, 29–37.

Varalakshmi, K.N., Sangeetha, C.G., Samee, U.S., et al., 2011. *In vitro* safety assessment of the effect of five medicinal plants on human peripheral lymphocytes. Trop. J. Pharm. Res. 10 (1), 33–40.

Vasques, C.A., Schneider, R., Klin-Junior, L.C., et al., 2014. Hypolipidemic effect of *Garcinia cambogia* in obese women. Phytother. Res. 28 (6), 887–891.

Vitalone, A., Menniti-Ippolito, F., Moro, P.A., Firenzuoli, F., Raschetti, R., Mazzanti, G., 2011. Suspected adverse reactions associated with herbal products used for weight loss: a case series reported to the Italian national institute of health. Eur. J. Clin. Pharmacol. 67, 215–224.

Wagner, P.D., Vu, N.D., 1995. Phosphorylation of ATP-citrate lyase by nucleoside diphosphate kinase. J. Biol. Chem. 270, 21758–21764.

Wielinga, P.Y., Wachters-Hagedoorn, R.E., Bouter, B., et al., 2005. Hydroxycitric acid delays intestinal glucose absorption in rats. Am. J. Physiol. 288 (6), G1144–G1149.

Wong, M.K., Darvishzadeh, A., Maler, N.A., Bota, R.G., 2016. Five supplements and multiple psychotic symptoms: a case report. Prim Care Companion CNS Disord. https://doi.org/10.4088/PCC.15br01856.

Xin, C., Zhang, X., Lu, Y., et al., 2012. Chemoprevention of 7,12- dimethylbenz [a]anthracene (DMBA)-induced hamster cheek pouch carcinogenesis by a 5-lipoxygenase inhibitor, garcinol. Nutr. Cancer 64, 1211–1218.

Yahagi, N., Shimano, H., Hasegawa, K., et al., 2005. Co-ordinate activation of lipogenic enzymes in hepatocellular carcinoma. Eur. J. Cancer 41, 1316–1322.

Yamaguchi, F., Saito, M., Ariga, T., et al., 2000. Free radical scavenging activity and antiulcer activity of garcinol from fruit rind. J. Agric. Food Chem. 48, 2320–2325.

Yonei, Y.Y., Takahashi, S., Hibino, M., et al., 2008. Effects on the human body of a dietary supplement containing L-carnitine and *Garcinia cambogia* extract: a study using double-blind tests. J. Clin. Biochem. Nutr. 42 (2), 89–103.

Yu, J.S., Choi, M.S., Park, J.S., et al., 2017. Inhibitory effects of *Garcinia cambogia* extract on CYP2B6 enzyme activity. Planta Med. 83, 895–900.

Chapter 59

Vitamin E TPGS and its applications in nutraceuticals

Andreas M. Papas

Introduction

Vitamin E TPGS (d-α-tocopheryl polyethylene glycol 1000 succinate) is a water-soluble derivative of the naturally occurring d-α-tocopherol. Its polar hydrophilic head (polyethylene glycol 1000) and lipophilic tail (phytyl chain of d-α-tocopherol) provide amphiphilic properties. As an amphiphile, it is an emulsifier of lipids and helps solubilize poorly soluble compounds. In addition, it has been shown to form micelle-like particles and affect the absorption mechanism, resulting in increased absorption and bioavailability (Wu and Hopkins, 1999).

Vitamin E TPGS was initially used to overcome severe malabsorption and correct vitamin E deficiency in cholestatic children (Sokol et al., 1987). The observation that vitamin E TPGS increased the absorption of vitamin D in cholestatic children (Argao et al., 1992) provided impetus for further research on its role in increasing the solubility and absorption of lipophilic and poorly soluble compounds. Significant research and clinical evidence, including major multilocation human clinical trials, have evaluated its safety (Krasavage and Terhaar, 1977; NCI National Cancer Institute, 1994a,b; Zondlo Fiume, 2002; EFSA, 2007) and efficacy, leading to commercial applications. Its first application in pharmaceuticals was the formulation of cyclosporin (Sokol et al., 1991) and was followed by other drugs approved by the Food and Drug Administration (FDA), the European Medicines Agency, and other international regulatory agencies.

Vitamin E TPGS also has been used in the formulation of disease-specific nutritional products for conditions associated with malabsorption (Sokol et al., 1987; Traber et al., 1994; Socha et al., 1997; Papas et al., 2007, 2008; Sagel et al., 2011, 2018; Perin et al., 2018). There have been increasing interest and commercial applications of vitamin E TPGS in dietary supplements, food and beverage, personal care and cosmetics, and animal nutrition and health products. This chapter describes the properties, safety, and efficacy of vitamin E TPGS with a focus on its applications in nutraceuticals and the emerging field of cannabinoids.

Properties of vitamin E TPGS

Chemical properties

Vitamin E TPGS is the polyethylene glycol 1000 ester of d-α-tocopheryl succinate. It is produced by the esterification of crystalline d-α-tocopheryl succinate with polyethylene glycol 1000 (Wu and Hopkins, 1999; USP United States Pharmacopeia and National Formulary, 2016). Its chemical structure and properties are summarized in Table 59.1.

Physical properties

Vitamin E TPGS is a water-soluble waxy solid with a low melting point—the physical properties are summarized in Table 59.2.

Nutritional properties

Vitamin E TPGS is a water-soluble form of the naturally occurring d-α-tocopherol form of vitamin E (Table 59.3). Its enhanced absorption and bioavailability and role in malabsorption are discussed below.

Stability

The chemical structure of vitamin E TPGS, as ester of d-α-tocopherol, reduces oxidation due to air, light, and oxidizing agents. Its low melting temperature and high

Nutraceuticals. DOI: https://doi.org/10.1016/B978-0-12-821038-3.00059-8

TABLE 59.1 Chemical properties of vitamin E TPGS.

Chemical structure

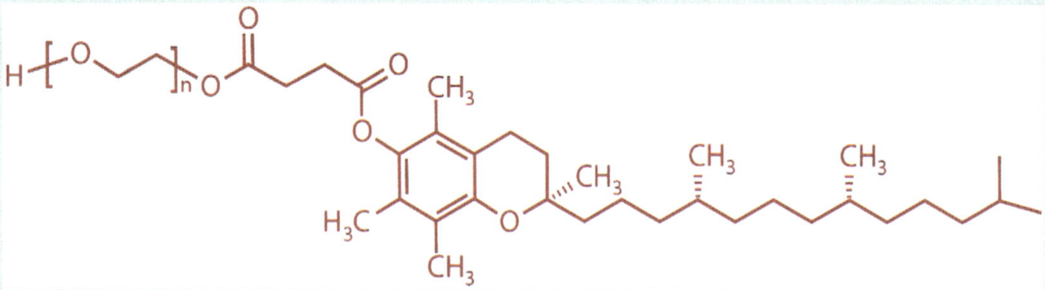

Chemical name	d-α-Tocopheryl polyethylene glycol 1000 succinate (TPGS)
Chemical abstract index name	α-[4-[[(2r)-3,4-Dihydro-2,5,7,8-tetramethyl-2-[(4r,8r)- 4,8,12-trimethyltridecyl]-2h-1-benzopyran-6-yl]oxy]-1,4- dioxobutyl]-ω-hydroxy-poly(oxy-1,2-ethanediyl)
CAS number	9002-96-4 (Chemical Abstracts Service registration number)
Empirical formula	$C_{33}O_5H_{54}(CH_2CH_2O)n$
Synonyms/acronyms	Vitamin E TPGS or TPGS Tocophersolan (International Nomenclature of Cosmetic Ingredients—INCI and United States Adopted Names—USAN)
Molecular weight	1513 (approx.)
d-α-Tocopherol	25% minimum weight basis; range 25–30%
Acid value	0.027 meq 0.10 N NaOH/g max or 1.5 mg KOH/g max
Reactivity	Reacts with alkali, very low reactivity with air

TABLE 59.2 Physical properties of vitamin E TPGS.

Physical form	Waxy solid
Color	White to light tan (as solid)
Gardner color	Less than 10 (generally less than 5)
Specific gravity	1.06 at 50°C to 1.03 at 90°C
Viscosity = f(T°C)	~390 cP at 50°C
Dynamic viscosity	~1.5 MPa*s at 22°C
Solubility in water	~20 weight% at 20°C. Forms gel in the range >20–90 weight%
Specific rotation (α)	Not less than +24°
Melting point	38°C approximate (range 37°C–41°C)
Heat of melting	99.8 J/g
Heat capacity	1.7 J/g.K
Taste	Neutral

thermal degradation temperature and testing indicate good stability during storage at room temperature. TPGS it is stable during processing or formulation including sterilization, melt granulation, or hot-melt extrusion for solid dosage formulations such as tableting and encapsulation (Feiyan and Tatavarti, 2010; Pandey et al., 2013).

TABLE 59.3 Nutrient composition of vitamin E TPGS.

Proximate, per 100 g		Vitamins, per 100 g	
Water, g	0.33	d-α-Tocopherol, g (minimum)	25 (range 25–30)
Energy, kcal	51		
Protein, g	0	Vitamin E, IU (minimum)	37,250 (range 37,250–44,700)
Total lipid (fat), g	1		
Saturated fat, g	0	Other tocopherols and tocotrienols, g	Less than 0.5
Cholesterol, g	0	Other vitamins, g	0
Carbohydrates, g	12.5	**Minerals, per 100 g**	
Sugars, g	0		
Dietary fiber, g	0	Sodium, calcium iron copper magnesium, potassium zinc, g	0
Ash, g	0		

TABLE 59.4 Stability properties of vitamin E TPGS.

Sterilization	Stable to exposure of approx. 125°C for 1 h
Thermal degradation temperature	199°C. No exotherm up to 300°C
Oxidative thermal degradation	219°C
Flash point	278°C
Stability in heat/cool cycles	Stable 20 cycles (between 0°C to 85°C)
Stability in aqueous solutions	Stable at pH 4.5–7.5
International Council for Harmonization stability study	48 months at 25°C (6 months at 40°C)
Stability in commercial packaging	4 years when stored at ambient temperature in the original unopened container

The stability properties of vitamin E TPGS are summarized in Table 59.4.

Safety and regulatory status

The safety of vitamin E TPGS has been evaluated extensively, including studies by the National Cancer Institute and others supported by the National Institutes of Health (NIH), the Cystic Fibrosis Foundation, and research organizations (Krasavage and Terhaar, 1977; NCI National Cancer Institute, 1994a,b; Zondlo Fiume, 2002). These studies have evaluated the acute and chronic toxicity in two species, reproduction in rats, and developmental toxicity in rodents and rabbits. Major clinical studies in humans have confirmed its safe use in foods and beverages, dietary supplements, personal care, medical foods, and drug formulations. An extensive review is available in the Opinion of the Scientific Panel European Food Safety Authority (EFSA, 2007), which led to its approval for use in foods for special medical purposes. Table 59.5 summarizes the safety characteristics.

The diverse functions and applications of vitamin E TPGS are reflected in its regulatory status as summarized in Table 59.6, which ranges from a nutrient form of vitamin E and inactive excipient to API (active pharmaceutical ingredient). It has been used in commercial products for over 20 years including pharmaceuticals, dietary supplements for babies and young children, and other products evaluated for safety in human clinical studies.

TABLE 59.5 Summary of toxicology and safety characteristics of vitamin E TPGS.

Skin irritation	No effect Zondlo Fiume (2002)
Eye irritation	Slight effect Zondlo Fiume (2002)
Skin sensitization	None (guinea pig) Zondlo Fiume (2002)
Oral LD_{50}	Higher than 7000 mg/kg in rat (highest dose tested), Krasavage and Terhaar (1977)
Skin LD_{50}	Higher than 2000 mg/kg in rat (highest dose tested)
Chronic exposure	No toxicity in rats with dietary exposures of up to 16% for 10 weeks. Two studies by the National Cancer Institute showed no toxicity in rats or dogs following a 1-year oral gavage dosing as high as 1000 mg/kg NCI National Cancer Institute (1994a,b)
Reproduction and developmental toxicity	No adverse effects in rats at dietary levels as high as 2.0% or gavage exposure up to 1000 mg/kg. No developmental toxicity in rabbits with gavage exposures of up to 1000 mg/kg Krasavage and Terhaar (1977). The no observed adverse effect level for maternal effects was 300 mg/kg
Genotoxicity	Results of three studies reviewed by the European Food Safety Authority EFSA (2007) showed no mutagenic nor clastogenic effects
Dosing tested in humans	Multilocation clinical studies Sokol et al. (1987) evaluated doses of 20–25 IU/kg in cholestatic children (equivalent to 800–1000 IU/day for a 40 kg child or 2.1–2.7 g vitamin E TPGS/day). Other studies evaluated lower levels but significantly higher than the recommended daily allowance for vitamin E Socha et al. (1997), Papas et al. (2007), (2008); Sagel et al. (2011), (2018)
	The EFSA estimated potential TPGS intakes varying from 5 mg/kg in teenagers to 13 mg TPGS/kg in 1-month old infants EFSA (2007)

TABLE 59.6 Summary of regulatory status of vitamin E TPGS.

United States	Vitamin E TPGS has a self-affirmed GRAS (generally recognized as safe) status when used as an oral dietary vitamin E supplement. The United States Pharmacopeia (USP) published a monograph for vitamin E TPGS and is included in its current edition USP monographs (2016). The Food and Drug Administration FDA/Center for Drug Evaluation and Research (2020) includes vitamin E TPGS in the list of inactive ingredients as tocophersolan (its unique ingredient identifier) in the USP/FDA Substance Registration System
Canada	In Canada, vitamin E TPGS is listed as tocophersolan by Health Canada in its *Natural Health Products Ingredients Database* with two roles Health Canada (2020): 1. Medicinal role: Classified as a Natural Health Product under Schedule 1, item 3 (a vitamin) of the Natural Health Products Regulations 2. Nonmedicinal role: Classified as an antioxidant
European Community	Vitamin E TPGS is approved as an API (active pharmaceutical ingredient). The European Medicines Agency granted on July 24, 2009, marketing authorization for vitamin E TPGS for the treatment of vitamin E deficiency due to digestive malabsorption in pediatric patients suffering from congenital chronic cholestasis or hereditary chronic cholestasis EMA (2020). The European Food Safety Authority on April 17, 2007, approved its use in foods for special medical purposes EFSA (2007).
Japan	Vitamin E TPGS was approved in 2005 as a pharmaceutical excipient for oral drugs
Other countries	A number of products including pharmaceuticals, medical foods, and dietary supplements are approved on a product and application basis

Functionality of vitamin E TPGS

Vitamin E TPGS has amphiphile properties with a polar hydrophilic head (polyethylene glycol 1000) and a lipophilic tail (phytyl chain of d-α-tocopherol). This amphiphilic characteristic leads to its self-association in water when the concentration exceeds a threshold known as the critical micelle concentration (CMC). For surfactant molecules, the CMC value is a key parameter that characterizes their surface activity. Vitamin E TPGS combines a low CMC with good hydrophile/lipophile balance and bulky lipophilic and hydrophilic portions surface areas. These characteristics, summarized in Table 59.7, make vitamin E TPGS a good emulsifier for wide range of lipophilic and water–oil immiscible systems (Wu and Hopkins, 1999; USP United States Pharmacopeia and National Formulary, 2016; Yang et al., 2018).

Digestion and absorption of lipids

A brief review will help understand the role of vitamin E TPGS. Water accounts for over 50% and up to 60% of the human body. The absorption, transport, and secretion systems are water based. The body has developed special mechanisms for the digestion, absorption, transport, and secretion of lipophilic and poorly soluble compounds. The key mechanism for absorption is the formation of micelles (Fig. 59.1), microsphere-shaped particles composed of a hydrophilic exterior encompassing a lipophilic or poorly water-soluble interior (Reboul, 2017; Papas, 1998, 1999). The reported sizes of micelles range from less than one to several microns. The formation of micelles requires two major components.

Bile is produced in the liver and secreted into the duodenum, the upper part of the small intestine. Its main components, the bile salts and phospholipids, help emulsify the lipids in the diet and facilitate enzymatic hydrolysis. Bile also provides the key components for the outer layer of micelles. Bile plays a key role in removing byproducts of lipid digestion from the liver into the gut (Macierzanka et al., 2019).

Pancreatic juice is also secreted into the duodenum. This secretion includes pancreatic enzymes that have a critical role in the digestion and absorption of the major dietary components including proteins, carbohydrates, and lipids. Lipases and esterases in the pancreatic juice are key enzymes in the digestion and absorption of lipids (Keller and Layer, 2005).

Micelles are the key vehicle for absorption of lipids by entering the enterocytes of the intestinal wall. Further transport of lipids is facilitated by lipoproteins which include chylomicrons, very low-density lipoproteins (VLDLs), low-density lipoproteins (LDLs), and high-density lipoproteins (HDLs). Like the micelles, lipoproteins combine a hydrophilic exterior composed of phospholipids and cholesterol, and a hydrophobic interior, thus enabling the transport of lipids. In the intestinal wall, the micelle contents are formulated into chylomicrons and secreted into the lymph. Lipoprotein lipases catabolize chylomicrons rapidly forming chylomicron remnants. During this process, apolipoprotein E binds to chylomicron remnants. The liver, through specific apolipoprotein E receptors, retains and clears most of the chylomicron remnants. Contents of the remnants are secreted into VLDL and circulated through the plasma. VLDL is hydrolyzed by lipoprotein lipase to LDL, which carries the lipids and appears to exchange them readily with HDL (Fig. 59.2).

Factors affecting the absorption of lipids

Optimal absorption of lipophilic compounds, including nutrients and nutraceuticals, depends on optimum function of the biliary system of the liver, the pancreas, and the gastrointestinal tract (GIT). Disease conditions, which affect these organs, cause malabsorption. Examples of disease conditions associated with severe malabsorption include the following.

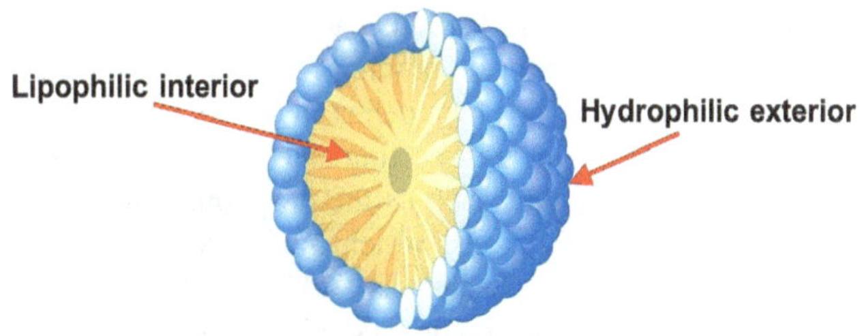

Formation of micelles is essential in the emulsification and absorption of lipids.

FIGURE 59.1 Structure of micelles.

TABLE 59.7 Amphiphilic properties of vitamin E TPGS.

HLB	13 (approximate hydrophile/lipophile balance)
CMC	0.02 weight% (wt%) at 37°C (critical micellar concentration). Vitamin E TPGS forms low-viscosity solutions with water up to approximately 20 wt.%. At higher concentrations, the structure of the vitamin E TPGS/water liquid crystalline phase evolves gradually from isotropic globular micellar to isotropic cylindrical micellar, hexagonal, and finally to the lamellar phase
Particle size	Reported particle size diameter ranged from nano and up to several microns for liquid emulsions and solid formulations with vitamin E TPGS which forms various liquid crystalline forms with water

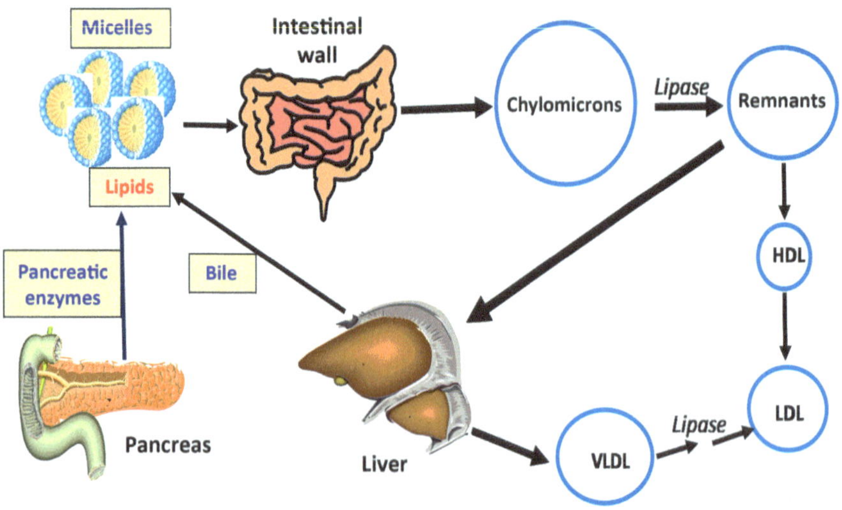

FIGURE 59.2 Absorption of lipids in humans.

Cholestasis which may be caused by inherited genetic defects (familial cholestatic syndromes), diseases of the liver (idiopathic neonatal hepatitis, chronic hepatitis, and metabolic liver diseases), and injury. Cholestasis stops or reduces the bile flow from the liver to the gut and prevents the formation of micelles. It therefore prevents the absorption of fat-soluble nutrients and important nutraceuticals. Cholestatic children, and even adults, develop very serious and often fatal degenerative neurological diseases due to vitamin E deficiency. Other liver diseases such as hepatitis, alcoholic hepatitis, and damage or inflammation of the liver from the use of powerful medicines or chemotherapy, can affect the production of bile (Keller and Layer, 2005; Sokol et al., 1987; Papas, 1998, 1999).

Cystic fibrosis, the most common fatal genetic disease in white people, is caused by a defect in the CFTR protein which is responsible for transporting chloride from epithelial cells that line organs such as the lungs and pancreas. This causes thick secretions, which block the ducts and eventually destroy the pancreas in 85% of patients. The combined effect of pancreatic and biliary insufficiency impairs the formation of micelles and absorption. These patients develop severe malabsorption which includes β-carotene (Feranchak et al., 1999).

Inflammation of the gut caused by diseases, medications, and other conditions affects absorption. Inflammatory bowel disease, which includes Crohn disease and ulcerative colitis, is associated with serious inflammation and diarrhea and interferes with absorption (Fabisiak et al., 2017). Other disease conditions such as HIV (The Italian Paediatric Intestinal/HIV Study Group, 1993; van der Heide, 2016) and medications (Eusebi et al., 2017), especially those associated with gut inflammation and diarrhea, impair absorption (Wehrmann et al., 1993).

Bariatric surgery and other surgical procedures of the gut including gastric banding, sleeve gastrectomy, biliopancreatic diversion, duodenal switch, and Roux-en-Y gastric bypass can cause malabsorption including lipid-soluble nutrients and natural compounds (Stroh et al., 2017; Stein et al., 2014). Also, short bowel syndrome, the removal of part of the small intestine, primarily in order to treat Crohn disease, affects absorption (Traber et al., 1994; Fabisiak et al., 2017).

Absorption, even in healthy people, can be seriously affected by dietary and physiological factors. Major examples are the effects of low/fat-free diets and aging.

Low- and fat-free diets illustrate the role of dietary fat in signaling the liver to produce bile and release it from the gall bladder for secretion into the duodenum. These diets are associated with reduced bile secretion, impaired emulsification of lipids and formation of micelles, and poor absorption. The fat replacement product Olestra (Thomson et al., 1998), and the drug Olistat (Filippatos et al., 2008), which inhibits lipid digestion and absorption, are associated with malabsorption of lipid-soluble nutrients including vitamins. The FDA mandated the label statement that "Olestra inhibits the absorption of some vitamins and other nutrients. Vitamins A, D, E and K have been added."

Aging affects taste, esophageal sphincter motility, gastric emptying, and neurons of the myenteric plexus related to gut transit. It is also associated with structural and functional mucosal defects, diminished abilities to generate protective immunity, and increased incidence of inflammation and oxidative stress (Soenen et al., 2016). These effects contribute to diminished digestion and absorption capabilities leading to nutritional deficiencies.

Medications often affect absorption directly by injuring the intestinal wall and digestion process or indirectly by causing diarrhea, constipation, change in the gut microflora, and the function of the liver and pancreas. These effects in some cases are severe and lead to deficiencies. Examples include proton pump inhibitors (Eusebi et al., 2017), antibiotics, and others (Wehrmann et al., 1993). Lipase inhibitors reduce absorption of lipids including fat-soluble vitamins, essential fatty acids, and nutraceuticals (van der Heide, 2016).

Mechanism of action

Key to the mechanism is the amphiphilic properties of vitamin E TPGS (Table 59.7), which make it an emulsifier. Vitamin E TPGS increases solubility of lipids and poorly soluble compounds and forms micelle-like particles which, as discussed above, are key vehicles for absorption (Yang et al., 2018). Its effect on bioavailability is further mediated by its modulation of the P-glycoprotein (P-gp) efflux pump (Collnot et al., 2007). This mechanism is functionally expressed in the intestinal epithelium and constitutes, along with intestinal metabolism, an important part of the biochemical barrier function of the intestinal mucosa. This efflux carrier limits the drug and other compound accumulations, leading to a significant percentage of these compounds being effluxed back into the lumen. This efflux reduces the bioavailability of nutraceuticals by:

1. Decreasing the net amount absorbed and available to the transport system;
2. Increasing the extent of intestinal metabolism. Because absorbed compounds including nutraceuticals can be effluxed numerous times, the extent of intestinal metabolism can be extensive.

Vitamin E TPGS has been shown to be a stronger inhibitor of the P-gp efflux mechanism than the surfactants Pluronic (poloxamers, a class of synthetic block copolymers) Tweens (polysorbates), and Spans (sorbitan esters).

The combined properties of vitamin E TPGS on solubility, formation of micelles, and inhibition of P-gp support its role on absorption and bioavailability in malabsorption but also in normal absorption for compounds of very poor solubility and extensive metabolism. Vitamin E TPGS is used as the single emulsifier or as a major component of self-emulsifying drug-delivery systems (SEDDSs; Gupta et al., 2013; Kanaujia et al., 2014; Knaub et al., 2019). SEDDSs are mixtures of oils and surfactants, and optionally contain hydrophilic solvents. Upon contact with an aqueous phase, such as gastric or intestinal fluids, they spontaneously emulsify under the conditions of gentle agitation in the GIT.

Efficacy of vitamin E TPGS

The efficacy of vitamin E TPGS has been evaluated in animal and major human clinical studies which may be grouped into the following categories.

Source of vitamin E for enhanced absorption

The team of Dr. Ron Sokol at the University of Colorado demonstrated that vitamin E TPGS overcame corrected vitamin E deficiency in cholestatic children with severe malabsorption. The studies included an eight-location clinical trial with 60 cholestatic children unresponsive to very large doses of other oral vitamin E forms (70–212 IU/kg/day equivalent to 124–400-fold the recommended daily allowance) (Sokol et al., 1987). These children developed severe vitamin E deficiency which caused differing degrees of neurological degeneration. An oral vitamin E TPGS dose of 25 IU/kg/day normalized the vitamin E status of all children, improved neurological function in 25, stabilized it in 27, and it became worse in two children. Separate studies have shown that the concentration of vitamin E in plasma, erythrocytes, and adipose tissues was higher with vitamin E TPGS even when tocopheryl acetate, an oil form of vitamin E, was emulsified with medium-chain triglycerides and polysorbate-80. Dosing with vitamin E TPGS has become the standard of nutritional care for cholestatic patients with the products Aqua-E (Papas et al., 2007) and Vedrop.

Increased absorption and bioavailability of lipophilic and poorly soluble compounds

Researchers at the University of Cincinnati reported that children with severe chronic cholestasis exhibited enhanced absorption of vitamin D_3 coadministered with vitamin E TPGS (Argao et al., 1992). Following this observation, Dr. Sokol clinically demonstrated that vitamin E TPGS enhanced absorption of the poorly soluble immunosuppressant drug cyclosporin (Sokol et al., 1991). These reports provided the impetus for significant clinical research, product development, and commercialization of pharmaceuticals, dietary supplements, topical, personal care, animal nutrition, and health products. These products include:

1. The *drugs* cyclosporine (immunosuppressant), ibuprophen (various trade names including Advil, NeoProfen, Caldolor -antiinflammatory), Aptivus (tipranavir, HIV protease inhibitor), Agenerase (amprenavir, HIV protease inhibitor), Viekira (dasabuvir, ombitasvir, paritaprevir and ritonavir, treatment of hepatitis C), and Vedrop (Tocophersolan, for vitamin E deficiency).
2. The *dietary supplements* Aqua-E, Aqua-D, Aqua-K, and DEKAs. Essential formulations of water-soluble forms of vitamins A, D, E, and K and a combination of the four fat-soluble vitamins, for correcting deficiency in malabsorbing patients. A family of

multivitamin disease-specific products has been introduced and become part of the standard of nutritional care for cystic fibrosis and cholestatic patients. These include the DEKAs and AquaDEKs Multivitamins and Mineral supplements (Papas et al., 2007, 2008; Sagel et al., 2011, 2018). Similar supplements have also been developed for bariatric patients including DEKAs Bariatric and Advanced Multi EA (Perin et al., 2018). These supplements contain β-carotene and CoQ_{10} which are of interest in the applications of vitamin E TPGS in nutraceuticals. The development of these products was supported by extensive testing in humans including multilocation human clinical trials supported by the NIH and the Cystic Fibrosis Foundation. These studies focused on the absorption and bioavailability of nutrients and natural compounds including nutraceuticals in malabsorbing patients.

In cystic fibrosis patients, blood β-carotene levels are very low, often at the limit of detection (Feranchak et al., 1999). In addition, the levels of important lipophilic nutrients such as vitamins A, D, and K and natural antioxidants such as CoQ_{10}, lutein, and lycopene are significantly below the normal range. Several clinical studies with cystic fibrosis patients (Papas et al., 2007, 2008; Sagel et al., 2011) culminated in a 16-week multicenter, randomized, double-blind, controlled trial with 73 patients (Sagel et al., 2011, 2018). This study evaluated an antioxidant-enriched multivitamin with a control multivitamin (Fig. 59.3). Systemic β-carotene, CoQ_{10}, and γ-tocopherol concentrations were significantly increased in the antioxidant-treated group. Antioxidant treatment was also associated with lower oxidative stress and risk of first pulmonary exacerbation, indicating health benefits.

Animal health and nutrition

Elephants and rhinoceros in captivity develop severe vitamin E deficiency which is not corrected even with very high levels of supplementation with the oil forms of vitamin E (Dierenfeld et al., 1988). In contrast, vitamin E TPGS supplied orally increased blood levels rapidly (Papas et al., 1990, 1991). The data, summarized in Fig. 59.4, show that vitamin E TPGS produces a rapid and significant increase in blood levels while very large doses of the oil form (d-α-tocopherol) and oil forms formulated as dispersible solid forms (d-α-tocopheryl acetate) produced a marginal increase even at ninefold or higher doses. Research suggested that the diet of these species in their natural habitat includes natural emulsifiers which reduce the significance and, with evolution, the strength of their biliary system (Sharma and Hiller, 2018; Hofmann et al., 2010). This is reflected in the composition of their bile and its emulsifying capacity. In captivity the diets of these species do not support optimum emulsification and absorption of lipids. Preliminary research suggested that vitamin E TPGS may have a role as a highly absorbable form of vitamin E in horses and, based on their bile composition, in other animals including

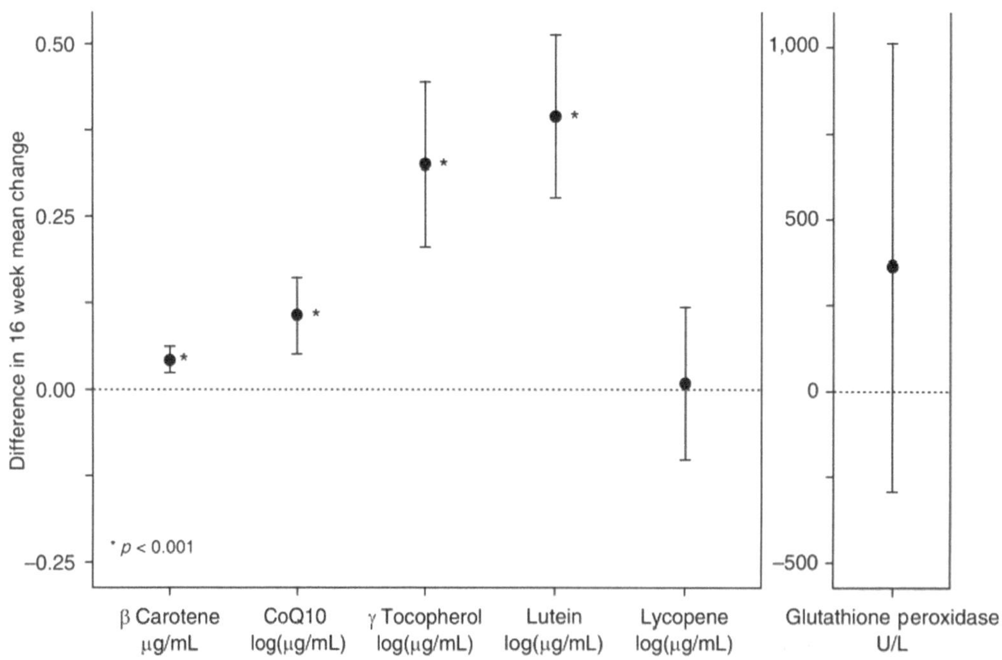

FIGURE 59.3 Multilocation clinical study with cystic fibrosis patients with vitamin E TPGS-formulated antioxidants. Difference in 16-week change in mean circulating antioxidant levels between the antioxidant and control groups with 95% confidence intervals.

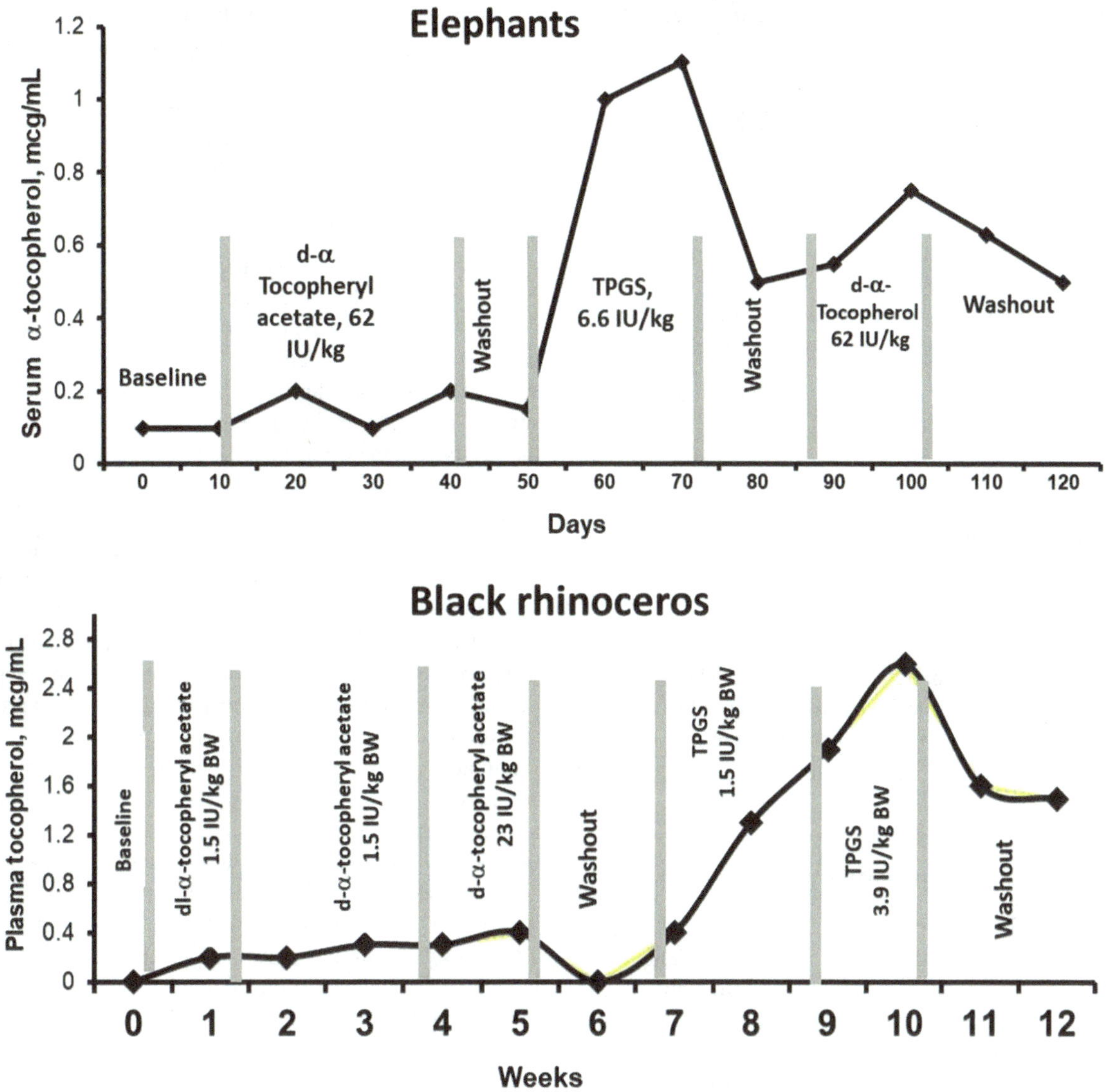

FIGURE 59.4 Effect of supplementation with vitamin E TPGS, fat-soluble, and water-dispersible forms of vitamin E on the circulating serum levels of α-tocopherol of elephants and black rhinoceros.

manatees and tapirs. Commercial products include the Mazuri vitamin E supplement.

A new strategy for animal health combines the properties of iron transport protein ovotransferrin and vitamin E TPGS (Srivastava et al., 2019). Ovotransferrin helps reduce gut pathogens and supports gastrointestinal health, which has a major role in the efficient digestion and absorption process. The formulation of ovotransferrin with vitamin E TPGS (ITPGS) aims to combine the benefit of a healthy gut with the functional properties of vitamin E TPGS in enhancing the absorption and bioavailability of therapeutic drugs, phytomedicines, nanomedicines, and nutrients. A line of commercial products formulated with ITPGS has been introduced under the Clinics Choice label for pet and companion animals and includes oral, topical, and sprays.

Dermal uptake

The dermal uptake of topically applied compounds is generally much lower than the oral route due in large part to the barrier function of the stratum corneum and the absence of the digestion and absorption mechanisms of the gut. Approaches for enhancing dermal permeability and uptake include use of surfactants and cosolvents to disrupt lipid structures and/or interact with the intracellular proteins of the stratum corneum and increase permeability of solubilized drug formulations (Aggarwal et al., 2012; Ghosh and Michniak-Kohn, 2012). Vitamin E TPGS is used alone or in combination with other components and processing technologies to produce micro- and nanoemulsions which aim to increase dermal penetration and uptake of active agents for both topical and systemic applications (Boakye et al., 2016; Telò et al., 2017; Benigni et al., 2018).

Applications of vitamin E TPGS in nutraceuticals

The main applications include:

- Source of water-soluble vitamin E;
- Emulsifier, solubilizer, and formulation excipient;
- Enhancer of absorption and bioavailability;
- Component in delivery/controlled-release systems.

These applications include the following product categories:

- Pharmaceuticals;
- Dietary supplements;
- Food and beverage;
- Personal care and cosmetics;
- Animal nutrition and health;
- Cannabinoids.

The functional properties of vitamin E TPGS and its neutral taste have facilitated its use in a wide range of product formats. Below is a discussion of the major applications of vitamin E TPGS with a focus on major nutraceuticals.

Tocopherols and tocotrienols

Vitamin E TPGS is derived from d-α-tocopherol, one of the eight members of the vitamin E family. The other vitamin E members are d-β, d-γ and d-δ-tocopherol and α-, β-, γ-, and δ-tocotrienols. The tocopherols consist of a head (chroman ring) and a tail (phytyl). The chroman ring carries the active antioxidant group. Each tocotrienol has an identical chroman ring as the corresponding tocopherol (Fig. 59.5). Tocotrienols differ from tocopherols on their tail; the tocotrienols have three unsaturated sites while tocopherols have none. In many countries, including the United States, the typical diet supplies higher levels of non-α-tocopherols, especially γ-tocopherol (Papas, 1998, 1999).

Although tocopherols and tocotrienols are absorbed with the same mechanisms as lipids, α-tocopherol is by far the predominant form in the blood due to highly specific transfer protein in the liver which preferentially transfers α-tocopherol in the plasma lipoproteins.

In the past vitamin E became synonymous with α-tocopherol and, as a result, this is the only form assigned vitamin E activity and the predominant form used in food fortification and dietary supplements. The mixed tocopherols have extensive use as food antioxidants. α-Tocopherol is available commercially in the naturally occurring d-α and the synthetic dl-α forms. The synthetic dl-α-tocopherol is a mixture of eight stereoisomers resulting from the three asymmetric carbons on the phytyl chain of the molecule. The naturally occurring d-α- is officially assigned a higher potency (1.49 IU/mg) than the synthetic dl-α-tocopherol (1.10 IU/mg). Vitamin E TPGS is derived from the naturally occurring form.

Significant ongoing research shows that tocopherols and tocotrienols, and probably their bioactive metabolites, exhibit individually and in combination responses that are different that those of α-tocopherol. These responses include inflammation, cell signaling and gene expression, and tumor cell-specific toxicity and apoptosis. The antiinflammatory properties of γ-tocopherol-rich concentrate were evaluated in animal models and human studies for allergic rhinitis and asthma. The results showed a reduction of sputum eosinophils and mucins, as well as an acute airway response to inhaled liposaccharide challenge, suggesting a therapeutic effect (Burbank et al., 2017). Tocopherols and tocotrienols are evaluated for their effect in other conditions associated with inflammation including some forms of cancer (Jiang, 2014).

Tocotrienols, and particularly γ-tocotrienol, are actively evaluated as countermeasures for acute radiation exposure. The Armed Forces Radiobiology Research Institute and their collaborators reported that γ-tocotrienol exhibits major protection from radiation exposure. Additional research showed that both γ- and δ-tocotrienols reduce injury to bone marrow and the hematopoietic system and stimulate hematopoiesis. The implications of these findings are evaluated for radiation

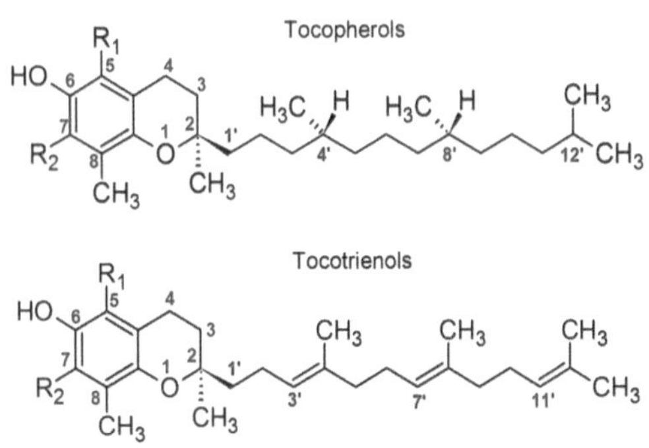

Congener	R_1	R_2
α	CH_3	CH_3
β	CH_3	H
γ	H	CH_3
δ	H	H

FIGURE 59.5 Chemical structure of naturally occurring tocopherols and tocotrienols.

exposure to nuclear accident or attack but also in radiation therapy and wound healing (Singh and Hauer-Jensen, 2016; Aykin-Burns et al., 2019).

Tocopherols and tocotrienols combine their roles as a source of the essential nutrient vitamin E with the characteristics and applications of nutraceuticals. Vitamin E TPGS has been used to solubilize and increase their absorption and bioavailability. The multivitamin mineral disease-specific dietary supplements DEKAs and AquaDEKs for cystic fibrosis and DEKAs Bariatric and Advanced Multi EA for bariatric patients contain mixed tocopherols and are formulated with vitamin E TPGS. Aqua-E is available as a concentrate containing only vitamin E TPGS or as a water formulation of tocopherols plus tocotrienols. The latter form of Aqua-E was compared with an oil-based softgel format in a clinical study with cystic fibrosis patients who had documented malabsorption. Eight patients were enrolled in the study and randomized to Aqua-E or softgels. The absorption of γ-tocopherol and total tocopherols in Aqua-E was greater than that of oil-based softgels (Fig. 59.6; Papas et al., 2007).

Carotenoids

Carotenoids are a group of over 600 naturally occurring plant pigments that provide the yellow, orange, and red colors seen in many fruits and vegetables. β-Carotene, the most well-known carotenoid, can be converted within the intestinal mucosal cell to two identical molecules of the retinal form of vitamin A. Most carotenoids are devoid of vitamin A activity. Carotenoids can be classified into two major groups based on their structure (Khachik et al., 1997; Fig. 59.7):

1. Xanthophylls are oxygenated carotenoids containing carboxyl and/or hydroxyl group(s) as end-group substituents. Lutein is the major xanthophyll found in human serum, but zeaxanthin and cryptoxanthin are also present.
2. Carotenes or hydrocarbon carotenoids are carotenoids not containing oxygen molecules or carotenes. The major provitamin A carotenoids, β-carotene and α-carotene, as well as lycopene, are found in this group.

To date, 34 dietary carotenoids have been found in human serum. The hydrocarbon structure of the carotenoids prevents them from being water soluble and, like other nonpolar lipids, their intestinal absorption depends on the biliary function and formation of micelles. After uptake of the micelles in enterocytes, carotenoids are incorporated into chylomicrons and released into the lymphatic system. Carotenoids possessing vitamin A activity may either be absorbed intact or cleaved to form vitamin A prior to secretion into the lymph. Carotenoids that are taken up by the mucosal cells but not incorporated into chylomicrons are sloughed off when the enterocytes turn over and spill their contents into the lumen of the GIT (Papas, 1998).

Natural carotenoids are available from their food matrix or as oil-soluble extracts of varying concentrations. Their absorption varies significantly between carotenoids, method of extraction, dietary fat, and other factors. In general, their absorption is low as evidenced by the potential for significant increase by processing and cooking. The part of carotenoids released from their food matrix and incorporated in micelles, which represents the fraction of carotenoids that is likely to be absorbed, is low. It was reported that the amount of lycopene released from the food matrix was very limited (from 0.1% in raw tomatoes to 1.5% in tomato puree). For β-carotene it was also low (from about 4% in carrot puree to 14% in carrot juice), while for lutein it was the highest (from 37% in raw spinach

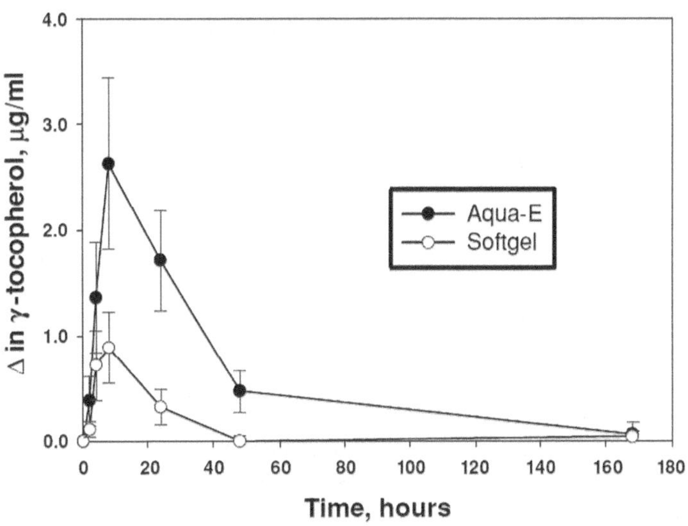

FIGURE 59.6 Changes in γ-tocopherol plasma levels above baseline over time in cystic fibrosis patients receiving a single dose of oil-based (softgel) and vitamin E TPGS-formulated tocols (Aqua-E).

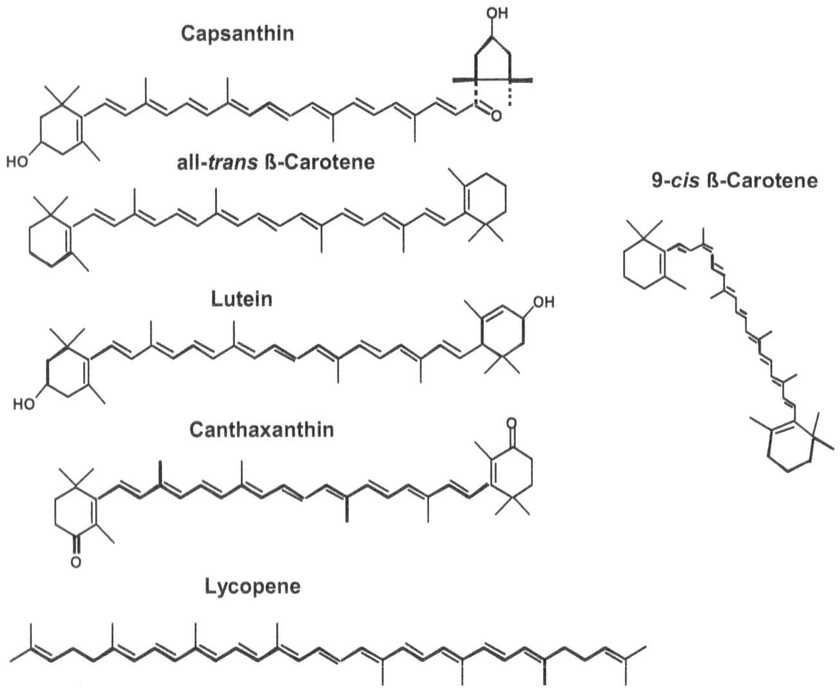

FIGURE 59.7 Chemical structure of the major carotenoids.

leaves to 48% in boiled spinach). These values correlate with in vivo data and show that, although disruption of the food matrix by thermal treatment or processing can increase carotenoid absorption, their overall absorption and bioavailability are low (Reboul, 2019).

β-Carotene, lutein, lycopene, astaxanthin, and zeaxanthin have major commercial use in dietary supplements, food and beverage, personal care and cosmetics, and animal nutrition and health products. In addition to their nutritional and health benefits, carotenoids are used as food colorants. Thus solubilization of carotenoids, which is key to increasing absorption, also facilitates applications as colorants and others.

Clinical studies in malabsorbing cystic fibrosis patients showed the potential and significance of the carotenoid formulation. As discussed earlier, blood levels of β-carotene in cystic fibrosis patients are very low, in some cases at the limit of detection. Several studies showed that very large doses, over 15-fold the dietary intake, were needed to normalize blood levels and reduce oxidative stress (de Vries et al., 2018; Renner et al., 2001).

An initial 8-week clinical study with 10 cystic fibrosis patients showed that a formulation of β-carotene, tocopherols, and CoQ_{10} with TPGS, in a water-based liquid formulation, normalized β-carotene blood levels from levels at or below the limit of detection (Fig. 59.8; Papas et al., 2008). It also increased the γ-tocopherol and CoQ_{10} blood levels from baseline. These

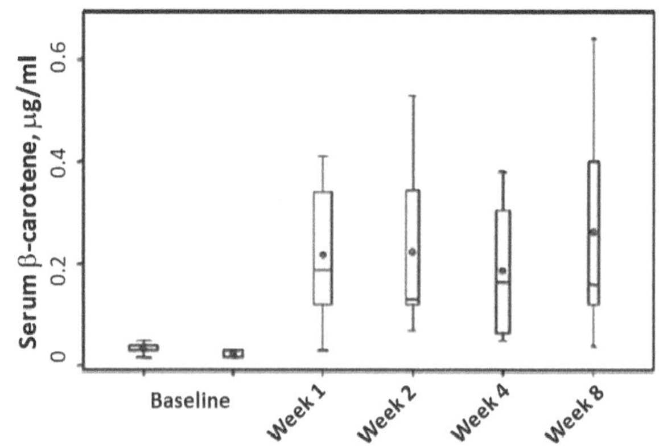

FIGURE 59.8 Change in serum β-carotene lower baseline levels of malabsorbing patients dosed with β-carotene formulated with vitamin E TPGS. Visits 1 and 2 established the baseline followed by 8 weeks of dosing.

improvements in antioxidant plasma levels were associated with reductions in airway inflammation. Another open-label 12-week study with 14 cystic fibrosis patients evaluated a vitamin E TPGS multivitamin formulation in the form of softgels (Sagel et al., 2011). Daily supplementation significantly increased blood levels of β-carotene, γ-tocopherol, and CoQ_{10}. These results were further confirmed in the multicenter, randomized, double-blind, controlled trial discussed above (Fig. 59.3). Antioxidant treatment was also associated with lower oxidative stress and risk of first pulmonary exacerbation (Sagel et al., 2018).

Coenzyme Q_{10}

Although not an essential nutrient, because it is synthesized in the body, Coenzyme Q_{10} (CoQ_{10}) has become an important ingredient in dietary supplements, food, and personal care products. CoQ_{10} plays a vital role in the production of energy from the terminal oxidation of products of carbohydrate, fat, and protein metabolism in the mitochondria. As a component of the electron transport system and donor of electrons, the reduced form of CoQ_{10} as ubiquinol functions as an antioxidant and plays a role in regenerating vitamins C and E. These properties make CoQ_{10} a key component for muscular function including the heart (Ernster and Dallner, 1995; Papas, 1998). Statins, the hypercholesterolemia treatment drugs, inhibit 3-hydroxy-3-methylglutaryl coenzyme-A reductase, the key enzyme in the endogenous production of CoQ_{10}, causing lower blood and tissue levels. Studies have suggested that myopathy, associated with statin use, was caused by the lower endogenous CoQ_{10} production which, in some studies, was corrected with supplementation (Littlefield et al., 2014; Marcoff and Thompson, 2007). CoQ_{10} is available commercially as ubiquinone and its reduced form ubiquinol (Fig. 59.9).

The efficiency of absorption of CoQ_{10} is limited because of its poor solubility and large molecular weight (Miles, 2007). In conditions associated with malabsorption, such as cystic fibrosis, cholestasis, bariatric surgery, and inflammatory bowel disease, blood levels are significantly below the normal range (Mancini et al., 2008; Laguna et al., 2008). Vitamin E TPGS increased CoQ_{10} absorption in three clinical trials with malabsorbing fibrosis patients (Papas et al., 2008; Sagel et al., 2011, 2018). The results of the first pilot of the three trials, summarized in Fig. 59.10, show that dosing increased blood serum levels. These results were confirmed in the other trials including the double-blind placebo control (Fig. 59.3). The multivitamin mineral disease-specific DEKAs and AquaDEKs supplements for cystic fibrosis and DEKAs Bariatric and Advanced Multi EA for bariatric patients contain CoQ_{10} formulated with vitamin E TPGS.

Commercial formulations of CoQ_{10} for increased absorption and bioavailability include Solu Q_{10}, All-Q, Ultra Q-Gel, CoQsol, QSorb, and VESIsorb (Schulz et al., 2006; Ullmann et al., 2005; Liu and Artmann, 2009). Some formulations use mechanical processing to produce smaller particle size, diluents such as oils and glycerin, solubilizers such polysorbate-80, and emulsifiers such as lecithins.

Curcumin

Curcumin is a polyphenolic compound derived from the rhizome of *Curcuma longa* Linn of the *Zingiberaceae* family. Curcumin is the principal curcuminoid in turmeric and has generally been considered its most active constituent. It is composed of two tautomers the bis-keto form in acid, solid conditions, and the enolote form in alkali conditions. Other curcuminoids in turmeric include demethoxycurcumin and bisdemethoxycurcumin (Fig. 59.11). In addition to its use as a spice and pigment, turmeric has been used for medicinal purposes for centuries. Recently, curcumin has received increased attention for its antiinflammatory and other properties and is evaluated for its effects on the prevention and treatment of diseases including cardiovascular, diabetes, and some forms of cancer (Kotha and Luthria, 2019; Hewlings and Kalman, 2017; Salehi et al., 2019; Anand et al., 2007; Guorgui et al., 2018).

Commercial curcumin is usually a mixture of curcumin, demethoxycurcumin, and bisdemethoxycurcumin. Curcuminoid extracts are available as dietary supplements often claimed to be standardized to contain 95% curcuminoids. Some curcumin preparations also contain piperine which has been suggested to enhance curcumin

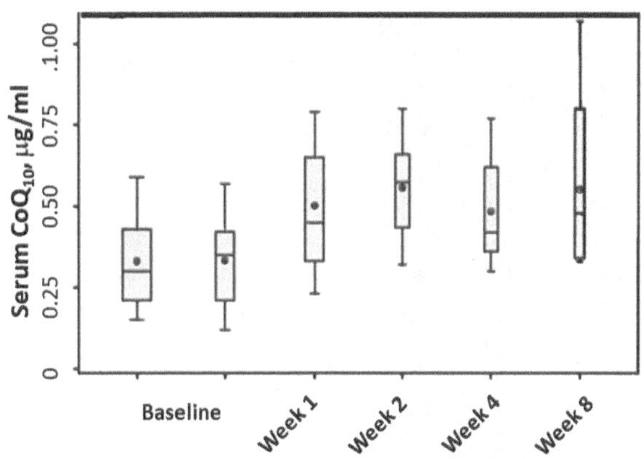

FIGURE 59.10 Change in serum Coenzyme Q_{10} baseline levels of malabsorbing patients dosed with Coenzyme Q_{10} in an antioxidant formulation with vitamin E TPGS.

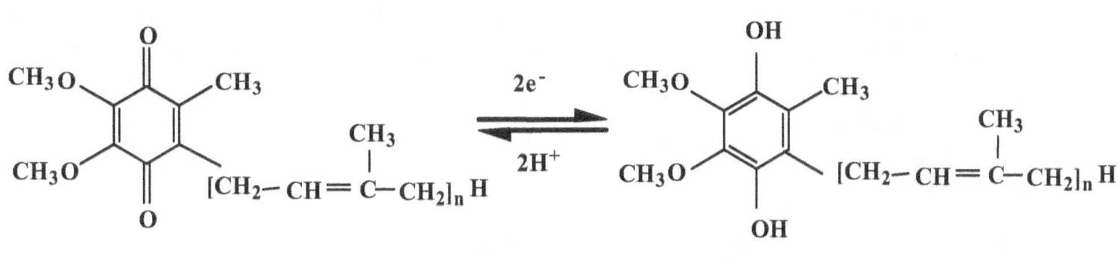

FIGURE 59.9 Oxidized (ubiquinone, left) and reduced (ubiquinol, right) Coenzyme Q_{10}.

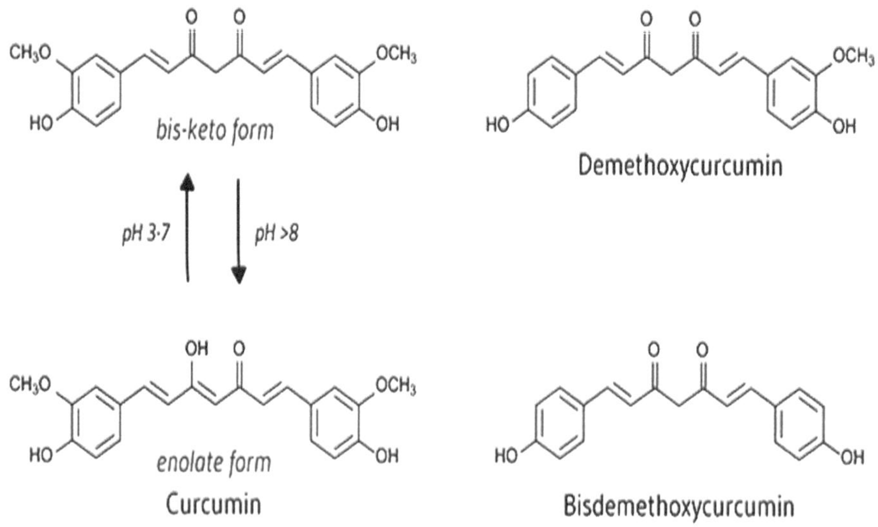

FIGURE 59.11 Chemical structure of curcuminoids in turmeric.

bioavailability (Shoba et al., 1998) by preventing its metabolism (Suresh and Srinivasan, 2006). In the United States, turmeric is generally recognized as safe by the FDA as a food additive.

Curcumin is classified in group 4 of BCS (Biopharmaceutical Class System), which indicates poor solubility and permeability. It is practically insoluble in water, with a maximum solubility as low as 11 ng/mL in pH 5.0 aqueous buffer. Low permeability of curcumin in Caco-2 cells, enterocyte-based metabolism by CYP450 3A4, and involvement of P-gp in curcumin transport have also been reported. In the intestine and liver, curcumin is readily conjugated to form curcumin glucuronide and curcumin sulfate or, alternately, reduced to tetrahydrocurcumin, hexahydrocurcumin, and octahydrocurcumin (Kotha and Luthria, 2019; Rotimi et al., 1998; Vareed et al., 2008; Jamwal, 2018; Anand et al., 2007).

The combined poor solubility and permeability and high first-pass metabolism of curcumin cause its poor absorption and bioavailability. Clinical trials in animals and humans confirmed very low absorption and bioavailability of orally administered curcumin and that mostly metabolites of curcumin, instead of curcumin itself, are detected in plasma or serum and tissues (Vareed et al., 2008; Jamwal, 2018; Salehi et al., 2019).

Curcumin provides a good example of a nutraceutical whose potential is severely restricted due to its poor water solubility, chemical instability first-pass metabolism, and the P-gp-mediated efflux mechanism. The very low bioavailability of curcumin is indicated by the presence of only 10 ng/mL in plasma after oral administration of 2 g of pure curcumin (Shoba et al., 1998). To address this issue, researchers have developed and evaluated pharmaceutical and other formulations such as liposomes, solid dispersions, complexes, emulsions, micelles, nanogels, and microspheres. In addition to common solubilizers/emulsifiers, these formulations included polylactic-co-glycolic acid (PLG), which has been approved by the FDA as a drug-delivery system, and TPGS. Several commercial curcumin formulations such as Vesisorb, NovaSol, CurcuWin, and LongVida have claimed higher bioavailability relative to reference unformulated curcumin (den Haan et al., 2018; Knaub et al., 2019).

Vitamin E TPGS has received attention in the formulation of curcumin especially in its clinical evaluation for medical applications. Its effects on solubility and formation of micelles for enhanced absorption, combined with its effect on P-gp and first-pass metabolism are particularly important for curcumin. Vitamin E TPGS has been used as the single formulation agent but, more often as the surfactant of delivery systems which include other components such as PLGA (poly (d,l-lactic-co-glycolic acid), other biodegradable polymers, phospholipids, cyclodextrins, and others (Chen et al., 2019; den Haan et al., 2018; Knaub et al., 2019). Such formulations are being actively investigated in a wide range of health and disease conditions including cardiovascular, neurodegenerative, cancer, diabetes, gastrointestinal, and periodontic (Kotha and Luthria, 2019; Hewlings and Kalman, 2017; Salehi et al., 2019; Anand et al., 2007; Guorgui et al., 2018).

Cannabinoids

Cannabis, a species of the *Cannabaceae* family, has a history of over 2600 years of medical and recreational use. Although removed from the British (in 1932) and United States (in 1942) pharmacopoeias and subjected to stringent regulatory and legal restrictions, its use has continued. The isolation of major cannabinoids, as well as the discovery of endocannabinoids, and their cognate receptors in humans, facilitated the evaluation of the therapeutic potential of cannabis in humans. These advances support gradual legalization and a rapid increase in the use of cannabis for therapeutic and other purposes (Pertwee, 2006; Zgair et al., 2016).

Cannabis may be classified as fiber-type or drug-type. The fiber-type cannabis, referred to as hemp, is rich in cannabidiol (CBD) or related compounds and almost devoid of the psychoactive delta-9-tetrahydrocannabinol (THC). Hemp is also used for textile or food purposes. The drug-type, often referred to as marijuana, is rich in THC and is used for medicinal or recreational purposes. In addition to the two main subspecies of *Cannabis indica* and *Cannabis sativa*, crossbreeding has produced subspecies with varying compositions and properties.

Cannabis contains over 500 distinct compounds, which include cannabinoids (over 100 have been identified), terpenoids, flavonoids, and omega fatty acids. Terpenoids are responsible for the aroma of cannabis and other flowering plants. It is estimated that at least 80 cannabis components have biological activity. The best known and most studied are CBD and the psychoactive THC; the prevalent terpene is myrcene (Fig. 59.12).

The regulatory and legal status of cannabis use has been changing at a fast pace internationally. In early 2020 in the Unites States a total or 33 states and the District of Columbia legalized use of cannabis, of which 22 were for medical use only. Internationally the number of countries which allow the use of medical cannabis under controlled conditions is increasing and includes Canada, Israel, the Netherlands, New Zealand, and Switzerland. The FDA approved one cannabis-derived and three cannabis-related drug products. Epidiolex, a purified form of CBD, has been approved for the treatment of seizures associated with Lennox-Gastaut or Dravet syndrome. The drugs Marinol and Syndros have been approved for the treatment of anorexia associated with weight loss in AIDS patients. Marinol and Syndros include the active ingredient dronabinol, a synthetic delta-9-THC which is a psychoactive component of cannabis. Cesamet contains the active ingredient nabilone, a synthetic THC analog (VanDolah et al., 2019; Zgair et al., 2016).

Cannabis is used in a wide range of forms, from the flower to pure isolates. The raw materials used to formulate end-user products are primarily oils, distillates or extracts, concentrates, and isolates. Depending on the processing method and concentration, the physical forms range from oils of varying viscosity, to solids, including in the form of crystals—especially the isolates. The oils, extracts, and concentrates include a large number of the lipophilic components, while the isolates are primarily purified forms of CBD, THC, and terpenes.

The absorption and bioavailability studies of cannabinoids, focused on CBD and THC, have shown low absorption in the range of 5%–19% (Millar et al., 2018). This was attributed to its low solubility in water and significant first-pass metabolism. Dietary fats and pharmaceutical lipid-based excipients are commonly used in the formulation of cannabinoids. These include sesame oil and medium-chain triglycerides. Although such formulations have been shown to increase absorption, probably in part by stimulating the biliary system, their effect is modest. It was reported that CBD and THC, diluted in sesame oil and subjected to in vitro lipolysis, were solubilized by approximately one-third in the micellar layer, the fraction readily available for absorption. The remaining two-thirds of both compounds were distributed in the undigested lipid fraction and the sediment layer and are not available for absorption (Zgair et al., 2016).

The first-pass metabolism of CBD and THC is substantial and reduces the amount that enters the systemic circulation. Several cytochrome P450 (CYP) isoforms, primarily CYP3A4 and CYP2C19, have been identified as capable of metabolizing CBD and THC (Stout and Cimino, 2014). The catabolism of CBD, THC, and other cannabinoids is probably significantly enhanced by the effect of these cannabinoids on the ATP-dependent drug efflux transporter P-gp. THC and CBD are among the cannabinoids stimulating P-gp ATPase activity, which causes higher effluxing into the gut and opportunity for catabolism. Furthermore, CBD appears to significantly inhibit P-gp-mediated transport, suggesting that CBD could reduce bioavailability (Zhu et al., 2006).

The combined functions of vitamin E TPGS to increase solubility leading to formation of micelle-like particles and its inhibition of P-gp have attracted strong interest in the formulation of cannabinoids. Initial work indicates that the strong clinical evidence from other lipophilic compounds is applicable to CBD and THC. A study comparing a SEDDS formulation with CBD showed higher absorption and bioavailability in humans than

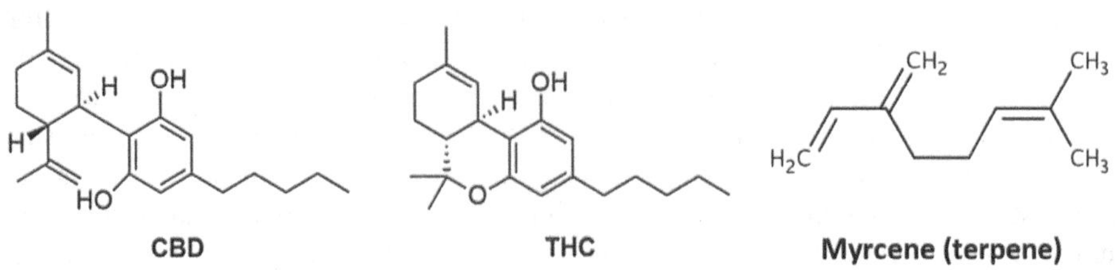

FIGURE 59.12 Chemical structure of the major cannabinoids cannabidiol and 9-tetrahydrocannabinol and the terpene myrcene.

formulation with medium-chain triglycerides. The difference in maximum blood concentration (C_{max}) was 4.4-fold higher and the area under the curve after 24 hours was 1.7-fold higher. The time to C_{max} was 2 hours earlier (1 vs 3 hours). These results showed faster absorption, which contributes to reduced first-pass metabolism and increased bioavailability (Knaub et al., 2019).

Cannabinoids formulated with vitamin E TPGS are available in a large number of product formats including bottled water, beverages, and sports drinks. While vitamin E TPGS has a neutral taste, its function as a solubilizer enhances components of cannabinoids with a bitter taste. For applications where taste is important, taste-masking agents are included in the formulation.

Vaping remains a popular administration method for cannabis users, often in the form of e-cigarettes. The United States Centers for Disease Control (CDC) investigated cases of lung illnesses and death linked to e-cigarette and vaping products including cannabis (Cherian et al., 2020). CDC has identified dl-α-tocopheryl acetate (vitamin E acetate) in vaping products associated with lung injury as a potential contributor (Blount et al., 2020). Vitamin E acetate is a viscous oil and has a long and strong record of safety for oral and topical use. Its effect on the lungs, when used in vaping, is very likely similar to the established damage caused by oil vapors on the surfactant properties of the lung tissue (Hadda and Khilnani, 2010). Unlike vitamin E acetate, vitamin E TPGS is a water-soluble waxy solid surfactant. Its molecular weight is over three times higher and has different vapor-forming properties. Vitamin E TPGS has not been evaluated for use in vaping products.

Other nutraceuticals

The nutraceuticals and cannabinoids discussed above provide examples of compounds which have or are being evaluated for formulation with vitamin E TPGS. Others include the essential nutrients vitamins A, D, E, and K, essential fatty acids, and a large number of phenolic and other nutraceuticals such as resveratrol, daidzein, daidzin, genistein, and quercetin (Jadhav et al., 2016; Wang et al., 2019; Zou and Gu, 2013; Katekar et al., 2020; Mittal et al., 2019; Patra et al., 2018); and also, extracts such as grape, rosemary, Cruciferae, and herb extracts, colorants, taste marking agents, and others.

Product formats and routes of administration

Vitamin E TPGS is used in a wide range of product formats and delivery routes. A brief overview of its use, the special requirements, challenges, and associated research in developing products which combine stability, safety, and targeted delivery of the active for enhanced absorption and bioavailability is given here.

Solid products

Capsules and tablets dominate this category due to ease of use for adults and children above 8 years of age, stability, and the absence of taste concerns (except for chewables).

Gelatin *capsules*, both as soft and hard gels, are major product formats. Their popularity provided the impetus for the development of capsules from vegetarian ingredients and from acid-resistant polymers or coated to withstand the low pH of the stomach and delivery to the small intestine. New technology allows the filling of capsules with liquids, powders, and mixtures of ingredients.

Vitamin E TPGS, as a waxy material with a low melting point, requires special handling and mixing with the material formulated for enhanced absorption and bioavailability (Feiyan and Tatavarti, 2010; Pandey et al., 2013). Nutraceuticals such as carotenoids, CoQ_{10}, curcumin, and the cannabinoids are mixed well in melted TPGS prior to encapsulation. In addition, diluents and carriers such as medium-chain triglycerides, beeswax, and others may be used to facilitate the flowing of the mixture for encapsulation. For powder filling, the mixture containing vitamin E TPGS is formulated in free-flowing solid forms as discussed below.

Tablets including chewable formulations containing vitamin E TPGS present challenges resulting from its waxy nature and low melting point. Critical formulation variables are the levels of vitamin E TPGS, active materials to be formulated, binder (such as hydroxypropyl cellulose), tensile strength, heat stability, disintegration time, and dissolution (Feiyan and Tatavarti, 2010). These challenges, which vary significantly depending on the properties of the formulated active and the expected site and rate of dissolution, are being addressed with a number of technologies which include beadlets, tablet-in-capsule, and lipid multiparticulate technologies. For chewable tablets the texture and taste are addressed with the selection of inactive ingredients and taste agents.

Powder formulations with vitamin E TPGS are widely used for pharmaceutical formulations, capsules, and tablets, but also for direct addition to foods and beverages. Significant ongoing research aims to develop free-flowing powder formulations of nutraceuticals. A key functionality requirement is dissolution leading to the formation of micelle-like particles which support the efficacy of vitamin E TPGS in enhancing absorption and bioavailability. These powders incorporate established and new innovative carriers and formulation agents and processing technologies. The carriers and formulating agents include naturally occurring compounds such as lecithins, chitosan, starch, dextrans, alginates, cellulose, monoglycerides, and FDA-approved biodegradable synthetic polymers such as poly(lactic-co-glycolic acid) (PLGA) (Chen et al., 2019), poly(glycolic acid) (PGA), and poly(lactic acid) (PLA). Sonication, homogenization, high shear mixing, and spray drying are among the processing technologies used. Powders are custom formulated for a wide range of applications, such as drug formulation and delivery, to direct addition to beverages.

Other solid product formats include gummies, lozenges, nutrition bars, and foods. The formulation challenges and solutions for the solids above and other formats below are applicable to these products.

Liquid products

Liquid products include, drops, sprays, sirups, sport drinks, beverages and bottled water, foods, and pharmaceuticals. The USP TPGS monograph details the water solubility test producing a 20% vitamin E TPGS concentrate, which can be used as stock solution for liquid product formulation (USP United States Pharmacopeia and National Formulary, 2016). Vtamin E TPGS is also melted and mixed with lipophilic and poorly soluble materials and then slowly added while mixing in heated water at 60°C–90°C (Wu and Hopkins, 1999). The ratio of vitamin E TPGS to the lipophile/poorly soluble compound needed to achieve solubilization in water ranges from 1:1 to 10:1 or higher depending on the solubility properties of the lipophile, presence of other components (such as in wide-spectrum extracts), use of other emulsifiers, and the desired properties of the final formulation. For example, in liquid dietary supplements, sports drinks, and some beverages, turbidity and milky emulsion are acceptable and even preferable, while in bottled water and other formulations they are unacceptable. Fig. 59.13 shows the emulsification of an insoluble in water/viscous oil and further dilution to produce water emulsion of varying concentrations.

Liquid dietary supplements are particularly suited for children, the elderly, and people who have swallowing difficulties. In addition to its main functions, Vitamin E TPGS serves to solubilize and attenuate the effects of taste-masking agents and colorants. Products include:

1. Vitamin E supplements containing only vitamin E TPGS or formulation of vitamin E TPGS with mixed tocopherols and tocotrienols;
2. Fat-soluble and multivitamin/mineral supplements, especially disease-specific ones, such as those for cystic fibrosis, cholestasis, and bariatric patients, as discussed above;
3. Formulations which include nutraceuticals as part of multivitamin/mineral supplements or as single-component formulations with vitamin E TPGS. These include the nutraceuticals discussed above and others such as sterols and estrogens.

Sport drinks and beverages. Vitamin E TPGS is used primarily as a water-soluble source of vitamin E and for the solubilization of nutrients, nutraceuticals, cannabinoids, and coloring and taste agents.

Bottled water. Vitamin E TPGS is used in fortified bottled water primarily as a water-soluble source of vitamin E and for the solubilization of added nutrients, nutraceuticals, and cannabinoids. In order to fully prevent turbidity and haziness, full solubilization and very low particle size are required. A higher ratio of vitamin E TPGS to the fortification agents and special processing such as sonication and high shear mixing may be needed.

Liquid pharmaceutical and personal care formulations, sirups, and sprays. Use of vitamin E TPGS is primarily in medical applications for the solubilization and enhanced absorption of actives but also for solubilization of color, taste, and fragrance agents. Such formulations, especially of pharmaceuticals, often include other agents such as thickeners, coemulsifiers, and processing technologies (Kaur et al., 2015).

A	B	C	D	E	F

A: Mixed tocopherol concentrate (viscous oil)	**D: Blend C in water, 10% weight basis**
B: Mixed tocopherol concentrate in water	**E: Blend C in water, 2.0% weight basis**
C: TPGS/Mixed tocopherol concentrate ratio 7:3	**F: Blend C in water, 0.2% weight basis**

FIGURE 59.13 Use of vitamin E TPGS to emulsify tocopherol concentrate (viscous oil) and produce water emulsion of varying concentrations.

Creams and gels

These forms are more common in topical pharmaceutical, personal care, and cosmetic products. In some products, the objective is limited to solubilizing components such as colorants and aromatic agents. In the majority, especially pharmaceuticals and personal care formulations, vitamin E TPGS is used alone or in combination with other components and processing technologies to produce micro- and nanoemulsions, which aim to increase dermal penetration and uptake of active agents for both topical and systemic application (Aggarwal et al., 2012; Boakye et al., 2016; Telò et al., 2017; Benigni et al., 2018).

Parenteral and other administration routes and drug delivery

TPGS is currently approved for oral, topical, and other methods of administration, which exclude parenteral applications. Vitamin E TPGS has been used in approved animal and human clinical studies, with parenteral administration of pharmaceuticals and other formulations. These studies included intravenous infusion of curcumin (Katekar et al., 2020). Significant ongoing research and clinical testing are evaluating intramuscular and intraperitoneal administration.

Although the safety evaluation of vitamin E TPGS and clinical studies included parenteral administration, its use in commercial products may require additional testing of the product such as for endotoxins and pyrogens. Because parenteral administration is generally intended for medical applications with health claims, regulatory approval of the formulation may also be required.

The reactivity of the hydroxy group of polyethylene glycol in vitamin E TPGS facilitated its conjugation with therapeutic agents to form vitamin E TPGS-based prodrugs, in which it serves as a drug carrier and P-gp inhibitor to reduce drug resistance, especially in cancer treatments. It has also been used to produce biodegradable and biocompatible polymers for drug delivery. Conjugates with FDA-approved poly (lactide) and polylactic-co-glycolic acid have received significant attention in drug delivery and for medical applications of nutraceuticals. Curcumin formulated with such delivery systems is evaluated in the treatment of cancer (Chen et al., 2019).

Vitamin E TPGS has been used in the formulation of ophthalmic products and in the development of products for nasal, vaginal, and otic administration. These include pharmaceuticals, extracts, and nutraceuticals.

References

Aggarwal, N., Goindi, S., Mehta, S.D., 2012. Preparation and evaluation of dermal delivery system of griseofulvin containing vitamin E-TPGS as penetration enhancer. AAPS Pharm. Sci. Tech. 13 (1), 67–74.

Anand, P., Kunnumakkara, A.B., Newman, R.A., et al., 2007. Bioavailability of curcumin: problems and promises. Mol. Pharm. 4 (6), 807–818.

Argao, E.A., Heubi, J.E., Hollis, B.W., et al., 1992. d-Alpha-tocopheryl polyethylene glycol-1000 succinate enhances the absorption of vitamin D in chronic cholestatic liver disease of infancy and childhood. Pediatr. Res. 31 (2), 146–150.

Aykin-Burns, N., Pathak, R., Boerma, M., et al., 2019. Utilization of vitamin E analogs to protect normal tissues while enhancing antitumor effects. Semin. Radiat. Oncol. 29 (1), 55–61.

Benigni, M., Pescina, S., Grimaudo, M.A., et al., 2018. Development of microemulsions of suitable viscosity for cyclosporine skin delivery. Int. J. Pharm. 545 (1-2), 197–205.

Blount, B.C., Karwowski, M.P., Shields, P.G., et al., 2020. Vitamin E acetate in bronchoalveolar-Lavage fluid associated with EVALI. N. Engl. J. Med. 382 (8), 697–705.

Boakye, C.H.A., Patel, K., Doddapaneni, R., et al., 2016. Ultra-flexible nanocarriers for enhanced topical delivery of a highly lipophilic antioxidative molecule for skin cancer chemoprevention. Colloids Surf. B Biointerfaces 143, 156–167.

Burbank, A.J., Duran, C.G., Almond, M., et al., 2017. A short course of gamma-tocopherol mitigates LPS-induced inflammatory responses in humans ex vivo. J. Allergy Clin. Immunol. 140 (4), 1179–1181.

Chen, X.P., Li, Y., Zhang, Y., et al., 2019. Formulation, Characterization and evaluation of curcumin- loaded PLGA- TPGS nanoparticles for liver cancer treatment. Drug Des. Devel. Ther. 13, 3569–3578.

Cherian, S.V., Kumar, A., Estrada-Y-Martin, R.M., 2020. E-cigarette or vaping-product associated lung injury: a review. Am. J. Med. 133, 657–663.

Collnot, E.M., Baldes, C., Wempe, M.F., et al., 2007. Mechanism of inhibition of P-glycoprotein mediated efflux by vitamin E TPGS: influence on ATPase activity and membrane fluidity. Mol. Pharm. 4 (3), 465–474.

de Vries, J.J., Chang, A.B., Bonifant, C.M., et al., 2018. Vitamin A and beta (β)-carotene supplementation for cystic fibrosis. Cochrane Database Syst. Rev. 8 (8), CD006751.

den Haan, J., Morrema, T.H.J., Rozemuller, A.J., et al., 2018. Different curcumin forms selectively bind fibrillar amyloid beta in post mortem Alzheimer's disease brains: implications for in-vivo diagnostics. Acta Neuropathol. Commun. 6 (1), 75.

Dierenfeld, E.S., du Toit, R., Miller, R.E., 1988. Vitamin E in captive and wild black rhinoceros (*Diceros bicornis*). J. Wildl. Dis. 24 (3), 547–550.

EFSA, 2007. Opinion of the Scientific Panel on Food Additives, Flavourings, Processing Aids and Materials in Contact with Food on a request from the Commission related to D-alpha-tocopheryl polyethylene glycol 1000 succinate (TPGS) in use for food for particular nutritional purposes. EFSA J. 490, 1–20.

EMA (European Medicines Agency), 2020. <http://webprod.hc-sc.gc.ca/nhpid-bdipsn/ingredReq.do?id = 7267&lang = eng>.

Ernster, L., Dallner, G., 1995. Biochemical, physiological and medical aspects of ubiquinone function. Biochim. Biophys. Acta 1271 (1), 195–204.

Eusebi, L.H., Rabitti, S., Artesiani, M.L., et al., 2017. Proton pump inhibitors: risks of long-term use. J. Gastroenterol. Hepatol. 32 (7), 1295–1302.

Fabisiak, N., Fabisiak, A., Watala, C., et al., 2017. Fat-soluble vitamin deficiencies and inflammatory bowel disease: systematic review and meta-analysis. J. Clin. Gastroenterol. 51 (10), 878–889.

FDA/Center for Drug Evaluation and Research, 2020. <https://www.accessdata.fda.gov/scripts/cder/iig/index.cfm?event = BasicSearch.page>.

Feiyan, J., Tatavarti, A., 2010. Tabletability assessment of conventional formulations containing Vitamin E tocopheryl polyethylene glycol succinate. Int. J. Pharm. 389, 58–65.

Feranchak, A.P., Sontag, M.K., Wagener, J.S., et al., 1999. Prospective, long-term study of fat-soluble vitamin status in children with cystic fibrosis identified by newborn screen. J. Pediatr. 135 (5), 601–610.

Filippatos, T.D., Derdemezis, C.S., Gazi, I.F., et al., 2008. Orlistat-associated adverse effects and drug interactions: a critical review. Drug Saf. 31 (1), 53–65.

Ghosh, I., Michniak-Kohn, B., 2012. A comparative study of vitamin E TPGS/HPMC supersaturated system and other solubilizer/polymer combinations to enhance the permeability of a poorly soluble drug through the skin. Drug Dev. Ind. Pharm. 38 (11), 1408–1416.

Guorgui, J., Wang, R., Mattheolabakis, G., et al., 2018. Curcumin formulated in solid lipid nanoparticles has enhanced efficacy in Hodgkin's lymphoma in mice. Arch. Biochem. Biophys. 648, 12–19.

Gupta, S., Kesarla, R., Omri, A., 2013. Formulation strategies to improve the bioavailability of poorly absorbed drugs with special emphasis on self-emulsifying systems. ISRN Pharm. 2013, 848043.

Hadda, V., Khilnani, G.C., 2010. Lipoid pneumonia: an overview. Expert Rev. Respir. Med. 4 (6), 799–807.

Health Canada, 2020. <http://webprod.hc-sc.gc.ca/nhpid-bdipsn/ingredReq.do?id = 7267&lang = eng>.

Hewlings, S.J., Kalman, D.S., 2017. Curcumin: a review of its' effects on human health. Foods. 6 (10), 92. 3.

Hofmann, A.F., Hagey, L.R., Krasowski, M.D., 2010. Bile salts of vertebrates: structural variation and possible evolutionary significance. J. Lipid Res. 51 (2), 226–246.

Jadhav, P., Bothiraja, C., Pawar, A., 2016. Resveratrol-piperine loaded mixed micelles: formulation, characterization, bioavailability, safety and in vitro anticancer activity. RSC Adv. 6, 112795–112805.

Jamwal, R., 2018. Bioavailable curcumin formulations: a review of pharmacokinetic studies in healthy volunteers. J. Integr. Med. 16 (6), 367–374. Nov.

Jiang, Q., 2014. Natural forms of vitamin E: metabolism, antioxidant, and anti-inflammatory activities and their role in disease prevention and therapy. Free Radic. Biol. Med. 72, 76–90.

Kanaujia, P., Ng, W.K., Tan, R.B., 2014. Solid self-emulsifying drug delivery system (S-SEDDS) for improved dissolution rate of fenofibrate. J. Microencapsul. 31 (3), 293–298.

Katekar, R., Thombre, G., Riyazuddin, M., et al., 2020. Pharmacokinetics and brain targeting of trans-resveratrol loaded mixed micelles in rats following intravenous administration. Pharm. Dev. Technol. 25 (3), 300–307.

Kaur, L., Jain, S.K., Singh, K., 2015. Vitamin E TPGS based nanogel for the skin targeting of high molecular weight anti-fungal drug: development and in vitro and in vivo assessment. RSC Adv. 5, 53671–53686.

Keller, J., Layer, P., 2005. Human pancreatic exocrine response to nutrients in health and disease. Gut 54 (Suppl 6), vi1–vi28.

Khachik, F., Spangler, C.J., Smith Jr, J.C., et al., 1997. Identification, quantification, and relative concentrations of carotenoids and their metabolites in human milk and serum. Anal. Chem. 69 (10), 1873–1881.

Knaub, K., Sartorius, T., Dharsono, T., et al., 2019. A novel self-emulsifying drug delivery system (SEDDS) based on VESIsorb® formulation technology improving the oral bioavailability of cannabidiol in healthy subjects. Molecules 24 (16), 2967.

Kotha, R.R., Luthria, D.L., 2019. Curcumin: biological, pharmaceutical, nutraceutical, and analytical aspects. Molecules 24 (16), 2930.

Krasavage, W.J., Terhaar, C.J., 1977. D-α-tocopheryl poly(ethylene glycol) 1000 succinate. Acute toxicity, subchronic feeding, reproduction and teratology studies in the rat. J. Agric. Food Chem. 25, 273–278.

Laguna, T.A., Sontag, M.K., Osberg, I., et al., 2008. Decreased total serum coenzyme-Q10 concentrations: a longitudinal study in children with cystic fibrosis. J. Pediatr. 153 (3), 402–407.

Littlefield, N., Beckstrand, R.L., Luthy, K.E., 2014. Statins' effect on plasma levels of coenzyme Q10 and improvement in myopathy with supplementation. J. Am. Assoc. Nurse Pract. 26 (2), 85–90.

Liu, Z.X., Artmann, C., 2009. Relative bioavailability comparison of different coenzyme Q10 formulations with a novel delivery system. Altern. Ther. Health Med. 15 (2), 42–46.

Macierzanka, A., Torcello-Gómez, A., Jungnickel, C., et al., 2019. Bile salts in digestion and transport of lipids. Adv. Colloid Interface Sci. 274, 102045.

Mancini, A., Leone, E., Festa, R., et al., 2008. Evaluation of antioxidant systems (coenzyme Q10 and total antioxidant capacity) in morbid obesity before and after biliopancreatic diversion. Metabolism 57 (10), 1384–1389.

Marcoff, L., Thompson, P.D., 2007. The role of coenzyme Q10 in statin-associated myopathy: a systematic review. J. Am. Coll. Cardiol. 49 (23), 2231–2237.

Miles, M.V., 2007. The uptake and distribution of coenzyme Q10. Mitochondrion 7 (Suppl), S72–S77.

Millar, S.A., Stone, N.L., Yates, A.S., et al., 2018. A systematic review on the pharmacokinetics of cannabidiol in humans. Front. Pharmacol. 9, 1365.

Mittal, P., Vrdhan, H., Ajmal, G., et al., 2019. Formulation and characterization of genistein-loaded nanostructured lipid carriers: pharmacokinetic, biodistribution and in vitro cytotoxicity studies. Curr. Drug Deliv. 16 (3), 215–225.

NCI National Cancer Institute, 1994a. One-Year Chronic Study in Rats. Conducted at International Research and Development Corporation, November1, 1994, Mattawan, MI, IRDC Report No. 560-040, National Cancer Institute, Bethesda, MD.

NCI National Cancer Institute, 1994b. One-Year Chronic Study in Dogs. Conducted at International Research and Development Corporation, November1, 1994, Mattawan, MI, IRDC Report No. 560-041, National Cancer Institute, Bethesda, MD.

Pandey, P., Sinko, P.D., Bindra, D.S., et al., 2013. Processing challenges with solid dosage formulations containing Vitamin E TPGS. Pharm. Dev. Tech. 18 (1), 296–304.

Papas, A.M., 1998. Major dietary and other antioxidants. In: Papas, A.M. (Ed.), Antioxidant Status, Diet, Nutrition, and Health. CRC Press, pp. 133–248.

Papas, A., 1999. The Vitamin E Factor. HarperCollins Publishers, New York.

Papas A.M., Cambre R.C., Citino S.B., et al., 1990. Species differences in the utilization of various forms of vitamin E. In: Proceedings of the Annual Meeting on American Association of Zoo Veterinarians, pp. 186–190.

Papas, A.M., Cambre, R.C., Citino, S.B., et al., 1991. Efficacy of absorption of various vitamin E forms by captive elephants and black rhinoceroses. J. Zoo. Wildl. Med. 22 (3), 309–317.

Papas, K., Kalbfleisch, J., Mohon, R., 2007. Bioavailability of a novel, water-soluble vitamin E formulation in malabsorbing patients. Dig. Dis. Sci. 52 (2), 347–352.

Papas, K.A., Sontag, M.K., Pardee, C., et al., 2008. A pilot study on the safety and efficacy of a novel antioxidant rich formulation in patients with cystic fibrosis. J. Cyst. Fibros. 7 (1), 60–67.

Patra, A., Satpathy, S., Shenoy, A.K., et al., 2018. Formulation and evaluation of mixed polymeric micelles of quercetin for treatment of breast, ovarian, and multidrug resistant cancers. Int. J. Nanomed. 13, 2869–2881.

Perin, J., Prokopowicz, G., Furtado, M., Papas, K., et al., 2018. A randomized trial of a novel chewable multivitamin and mineral supplement following Roux-en-Y Gastric Bypass. Obes. Surg. 28 (8), 2406–2420.

Pertwee, R.G., 2006. Cannabinoid pharmacology: the first 66 years. Br. J. Pharmacol. 147 (Suppl 1), S163–S171.

Reboul, E., 2017. Vitamin E bioavailability: Mechanisms of intestinal absorption in the spotlight. Antioxidants (Basel) 6 (4), 95.

Reboul, E., 2019. Mechanisms of carotenoid intestinal absorption: where do we stand? Nutrients 11 (4), 838.

Renner, S., Rath, R., Rust, P., et al., 2001. Effects of beta-carotene supplementation for six months on clinical and laboratory parameters in patients with cystic fibrosis. Thorax 56 (1), 48–52.

Rotimi, N., Tongiani, R., Cervelli, F., et al., 1998. Effects of curcumin on P-glycoprotein in primary cultures of rat hepatocytes. Life Sci. 62 (25), 2349–2358.

Sagel, S.D., Sontag, M.K., Anthony, M.M., Emmett, P., Papas, K.A., 2011. Effect of an antioxidant-rich multivitamin supplement in cystic fibrosis. J. Cyst. Fibros 10 (1), 31–36.

Sagel, S.D., Khan, U., Jain, R., et al., 2018. Effects of an antioxidant-enriched multivitamin in cystic fibrosis. A randomized, controlled, multicenter clinical trial. Am. J. Respir. Crit. Care Med. 198 (5), 639–647.

Salehi, B., Stojanović-Radić, Z., Matejić, J., et al., 2019. The therapeutic potential of curcumin: a review of clinical trials. Eur. J. Med. Chem. 163, 527–545.

Schulz, C., Obermüller-Jevic, U.C., Hasselwander, O., et al., 2006. Comparison of the relative bioavailability of different coenzyme Q10 formulations with a novel solubilizate (Solu Q10). Int. J. Food Sci. Nutr. 57 (7-8), 546–555.

Sharma, V., Hiller, M., 2018. Loss of enzymes in the bile acid synthesis pathway explains differences in bile composition among mammals. Genome Biol. Evol. 10 (12), 3211–3217.

Shoba, G., Joy, D., Joseph, T., et al., 1998. Influence of piperine on the pharmacokinetics of curcumin in animals and human volunteers. Planta Med. 64 (4), 353–356.

Singh, V.K., Hauer-Jensen, M., 2016. γ-tocotrienol as a promising countermeasure for acute radiation syndrome: current status. Int. J. Mol. Sci. 17 (5), 663.

Socha, P., Koletzko, B., Pawlowska, J., et al., 1997. Treatment of cholestatic children with water-soluble vitamin E (α-tocopherol polyethylene glycol succinate): effects on serum vitamin E, lipid peroxides, and polyunsaturated fatty acids. J. Pediatr. Gastroenterol. Nutr. 24, 189–193.

Soenen, S., Rayner, C.K., Jones, K.L., et al., 2016. The ageing gastrointestinal tract. Curr. Opin. Clin. Nutr. Metab. Care 19 (1), 12–18.

Sokol, R.J., Butler-Simon, N.A., Bettis, D., et al., 1987. Tocopheryl polyethylene glycol 1000 succinate therapy for vitamin E deficiency during chronic childhood cholestasis: neurologic outcome. J. Pediatr. 111 (6 Pt 1), 830–836.

Sokol, R.J., Johnson, K.E., Karrer, F.M., et al., 1991. Improvement of cyclosporin absorption in children after liver transplantation by means of water-soluble vitamin E. Lancet 338 (8761), 212–214.

Srivastava, A., Lall, R., Talukder, J., DuBourdieu, D., Gupta, R.C., 2019. Iron transport tocopheryl polyethylene glycol succinate in animal health and diseases. Molecules 24 (23), 4289.

Stein, J., Stier, C., Raab, H., et al., 2014. Review article: the nutritional and pharmacological consequences of obesity surgery. Aliment. Pharmacol. Ther. 40 (6), 582–609.

Stout, S.M., Cimino, N.M., 2014. Exogenous cannabinoids as substrates, inhibitors, and inducers of human drug metabolizing enzymes: a systematic review. Drug Metab. Rev. 46 (1), 86–95.

Stroh, C., Manger, T., Benedix, F., 2017. Metabolic surgery and nutritional deficiencies. Minerva Chir. 72 (5), 432–441.

Suresh, D., Srinivasan, K., 2006. Influence of curcumin, capsaicin, and piperine on the rat liver drug-metabolizing enzyme system in vivo and in vitro. Can. J. Physiol. Pharmacol. 84 (12), 1259–1265.

Telò, I., Favero, E.D., Cantù, L., et al., 2017. Gel-like TPGS-based microemulsions for imiquimod dermal delivery: role of mesostructure on the uptake and distribution into the skin. Mol. Pharm. 14 (10), 3281–3289.

The Italian Paediatric Intestinal/HIV Study Group, 1993. Intestinal malabsorption of HIV-infected children: relationship to diarrhoea, failure to thrive, enteric microorganisms and immune impairment. AIDS 7 (11), 1435–1440.

Thomson, A.B., Hunt, R.H., Zorich, N.L., 1998. Review article: olestra and its gastrointestinal safety. Aliment. Pharmacol. Ther. 12 (12), 1185–1200.

Traber, M.G., Schiano, T.D., Steephen, A.C., et al., 1994. Efficacy of water-soluble vitamin E in the treatment of vitamin E malabsorption in short-bowel syndrome. Am. J. Clin. Nutr. 59, 1270–1274.

Ullmann, U., Metzner, J., Schulz, C., et al., 2005. A new Coenzyme Q10 tablet-grade formulation (all-Q) is bioequivalent to Q-Gel and both have better bioavailability properties than Q-SorB. J. Med. Food 8 (3), 397–399.

USP (United States Pharmacopeia and National Formulary), 2016. 41-NF 36 Monographs: Vitamin E Polyethylene Glycol Succinate. Rockville, MD.

van der Heide, F., 2016. Acquired causes of intestinal malabsorption. Best Pract. Res. Clin. Gastroenterol. 30 (2), 213–224.

VanDolah, H.J., Bauer, B.A., Mauck, K.F., 2019. Clinicians' guide to cannabidiol and hemp oils. Mayo Clin. Proc. 94 (9), 1840–1851.

Vareed, S.K., Kakarala, M., Ruffin, M.T., et al., 2008. Pharmacokinetics of curcumin conjugate metabolites in healthy human subjects. Cancer Epidemiol. Biomarkers Prev. 17 (6), 1411–1417.

Wang, H., Xiao, Y., Wang, H., et al., 2019. Development of daidzein nanosuspensions: preparation, characterization, in vitro evaluation, and pharmacokinetic analysis. Int. J. Pharm. 566, 67–76.

Wehrmann, T., Lembcke, B., Caspary, W.F., 1993. Drug-induced malabsorption. Drug-Induced Injury to the Digestive System. Springer, Milano.

Wu, S.H., Hopkins, W.K., 1999. Characteristics of D-α-tocopheryl PEG 1000 succinate for applications as an absorption enhancer in drug delivery systems. Pharmacol. Technol. 23, 52–68.

Yang, C., Wu, T., Qi, Y., et al., 2018. Recent advances in the application of vitamin E TPGS for drug delivery. Theranostics 8 (2), 464–485.

Zgair, A., Wong, J.C., Lee, J.B., et al., 2016. Dietary fats and pharmaceutical lipid excipients increase systemic exposure to orally administered cannabis and cannabis-based medicines. Am. J. Transl. Res. 8 (8), 3448–3459.

Zhu, H.J., Wang, J.S., Markowitz, J.S., et al., 2006. Characterization of P-glycoprotein inhibition by major cannabinoids from marijuana. J. Pharmacol. Exp. Ther. 317 (2), 850–857.

Zondlo Fiume, M., 2002. Final report on the safety assessment of tocopherol, tocopheryl acetate, tocopheryl linoleate, tocopheryl linoleate/oleate, tocopheryl nicotinate, tocopheryl succinate, dioleyl tocopheryl methylsilanol, potassium ascorbyl tocopheryl phosphate, and tocophersolan. Int. J. Toxicol. 21 (Suppl 3), 51–116.

Zou, T., Gu, L., 2013. TPGS emulsified zein nanoparticles enhanced oral bioavailability of daidzein: in vitro characteristics and in vivo performance. Mol. Pharm. 10 (5), 2062–2070.

Chapter 60

Role of transferrin: an iron-binding protein in health and diseases

Jamil Talukder

Introduction

Transferrin (Tf) is the major iron-transport protein of blood plasma, with a molecular weight ~80 kDa. It is predominantly synthesized in the liver but also other tissues, such as the reticuloendothelial system (Huebers and Finch, 1987). Tf has N- and C-lobes and each can bind one molecule of iron (Sun et al., 1999). Tf has a central role in iron metabolism, returning iron derived from the catabolism of hemoglobin and other proteins to hematopoietic tissue (Elsayed et al., 2016). Tf is different from other proteins as it returns to the circulation after unloading its iron at the cell membrane (Arosio et al., 2017). The concentration of Tf is altered due to different pathophysiological states of the body, for example, the plasma Tf concentration is increased in pregnancy (Singhal et al., 1993; Kurhade et al., 1994). The plasma Tf concentration is elevated with the use of oral contraceptives, indicating that hormonal changes alter plasma Tf levels (Kurhade et al., 1994). In addition, plasma Tf levels may increase if there is a deficiency of iron and it is augmented via hepatic synthesis (de Jong et al., 1990; Nakajima et al., 2014). A more severe and prolonged iron deficiency would lead to even higher plasma Tf levels (de Jong et al., 1990). In addition, iron overload may lead to a high level of iron saturation with normal Tf, such as in hemochromatosis (de Jong et al., 1990; van Eijk and de Jong,1992). In contrast, some disease conditions reduce plasma Tf concentration such as malnutrition, liver disease, inflammatory diseases, and rheumatoid arthritis (Abdel-Noor et al., 2019). A negative acute-phase response is observed due to the severity and chronicity of rheumatoid arthritis (Abdel-Noor et al., 2019).

Furthermore, plasma Tf levels and saturation with iron help in the differential diagnosis of anemia. For example, iron-deficiency anemia is usually associated with an increased serum Tf level with a low saturation, whereas anemia due to failure to incorporate iron into the red cells has a normal or low Tf level with a high saturation (Abdel-Noor et al., 2019). Tf levels can be measured by different methods such as colorimetry, turbidimetry, nephelometry, and immunochemistry (Elsayed et al., 2016). Altogether, measurement of plasma Tf level and iron saturation is more important for monitoring iron therapy and differential diagnosis of anemia than the measurement of serum ferritin (Abdel-Noor et al., 2019).

Structure of transferrin

Tf is a polypeptide chain consisting of two carbohydrate chains and almost 700 amino acids (Hall et al., 2002; Gomme et al., 2005; Wally and Buchanan, 2007). Tf has two homologous globular lobes, the N- and C-terminals comprised of α-helices and β-sheets, with an iron-binding site in between (Testa, 1985). The site itself is a six-iron coordinate site occupied by a carbonate anion and four residues (Wally and Buchanan, 2007; Gomme et al., 2005). Each lobe is further divided into two clefts or domains. This specific structure lends Tf's ability to undergo large conformational changes when iron needs to be taken up or released (Testa, 1985; Hall et al., 2002; Wally and Buchanan, 2007). This was made possible by domains that rotate around a screw axis. Through X-ray crystallography, scientists have uncovered the mechanism for iron release (Hall et al., 2002). This lies in how two of the basic residues from two of the domains create a special hydrogen bond under neutral pH. However, this bond will break and thus release iron in the acidic pH of the endosome at its delivery site (Grant and Donaldson, 2009; Steere et al., 2012). Each Tf molecule can carry two iron molecules in the bloodstream, and we discuss in more detail the importance of sheltering iron until it is needed. The structure of human transferrin protein is shown in Fig. 60.1.

Nutraceuticals. DOI: https://doi.org/10.1016/B978-0-12-821038-3.00060-4

Biochemistry of transferrin

Tf consists of a single polypeptide chain of 679 amino acids, 0–2 N-linked oligosaccharide units (N-glycans), and two iron-binding sites (Wally and Buchanan, 2007; Gomme et al., 2005). The N-glycans show a complex structure and different Tf glycoforms are named based on the total number of terminals, negatively charged sialic acid residues. Tetrasialotransferrin, which contains two biantennary N-glycans each terminated by a sialic acid, normally accounts for ~75%–80% of total Tf in human serum (Stibler, 1991; Wally and Buchanan, 2007; Gomme et al., 2005). Other common glycoforms are pentasialo- (~15%), trisialo- (~5%), hexasialo- (~2%), and disialotransferrin (~2%). Asialo-, monosialo-, heptasialo-, and octa-sialotransferrin are usually found in trace amounts or are not detectable at all (Stibler, 1991; Weykamp et al., 2013). There are also several genetic B, C, and D homozygous and heterozygous variants of Tf, of which Tf C is the most common phenotype. Kasvosve et al. (2002) studied vitamin C status in a Zimbabwean population according to Tf phenotype because vitamin C metabolism is influenced by iron-driven oxidative stress. They observed that Tf polymorphism may affect vitamin C status in the black population.

Tf protein synthesis is controlled by the Tf gene. Lu et al. (2018) investigated the influence of Tf gene polymorphism on carbohydrate-deficient Tf and concluded that a *Tf*-Dchi variant influences the carbohydrate-deficient Tf. Zakin (1992) described that a number of studies have been completed to characterize the transcriptional elements implicated in the regulation of Tf gene expression in different organs. Their results support the hypothesis that the Tf gene makes use of different combinations of nuclear proteins in different subsets of cells to achieve tissue-specific expression, which interestingly occurs in tissues arising from different embryological origin. Uzan et al. (1984) performed molecular cloning and sequence analysis of cDNA for human Tf. They identified a cDNA clone for human Tf from a human liver cDNA library by prescreening with different ss-cDNA probes against length-fractionated liver mRNAs, positive hybridization selection, and nucleotide sequence analysis. They observed an insert of 1 kb, encoding human Tf from amino acid 403 through the COOH terminus, with a 3′ noncoding region of 166 nucleotides. Furthermore, that insert was hybridized with a single major mRNA species of about 2.4 kb and several genomic DNA restriction fragments. Hybridization of the Southern blots with different parts of the Tf insert and at different stringencies suggest that the various bands observed correspond to splice sites inside one gene rather than to hybridization of several related genes. Finally, they concluded that a single or low number of Tf gene copies seem to exist in the human genome. Hershberger et al. (1991) cloned the human Tf gene that encodes Tf and determined that the deduced protein contains 678 residues and 19 disulfide bonds, with a molecular mass of ~80 kDa. Yang et al. (1984) mapped the Tf gene to chromosome 3q21-q25 by in situ hybridization and analysis of somatic cell hybrids. Huerre et al. (1984) confirmed these findings. The location of human Tf gene on chromosome 3 is shown in Fig. 60.2.

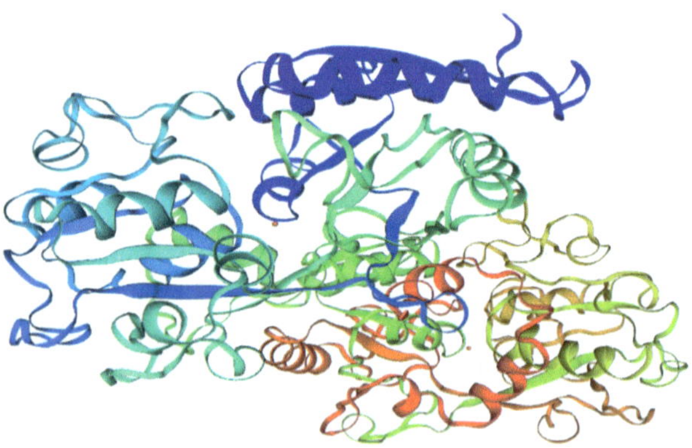

FIGURE 60.1 Structure of human transferrin protein.

Transferrin receptors and mechanism of iron transport

Despite the presence of divalent metal transporters, there are Tf receptors (TfRs). The TfR gene encodes the TfR, which is important for cellular iron uptake. There are two types of TfRs, TfR1 (also known as CD71) and TfR2, and both are transmembrane proteins with 66% homology

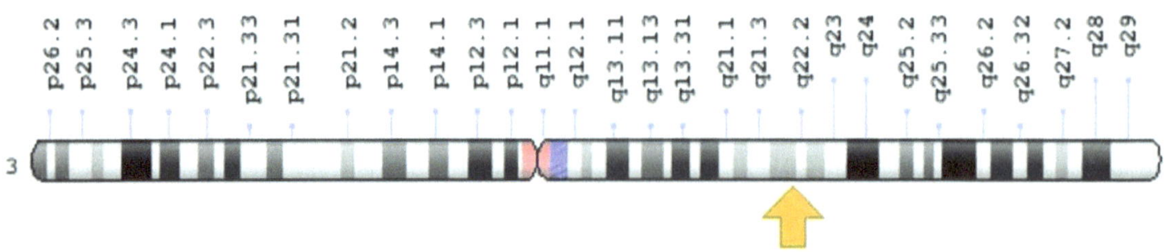

FIGURE 60.2 The human Tf gene is located on chromosome 3 at 3q22.1 position, which is the long (q) arm with a base pair of 133,661,998 to 133,796,641 (annotation release 109.20200522, GRCh38.p13, NCBI).

(Gammella et al., 2017). TfR2 expression is largely restricted to hepatocytes (Daniels et al., 2006). TfRs are highly expressed on immature erythroid cells, placental tissue, and rapidly dividing cells, both normal and malignant (Ponka and Lok, 1993). The expression of TfRs is negatively regulated posttranscriptionally by intracellular iron through iron-responsive elements (IREs) in the 3′ untranslated region of TfR-mRNA of proliferating nonerythroid cells (Ponka and Lok, 1993). The specific cytoplasmic proteins recognize IREs in the absence of iron in the labile pool, bind to the IREs of TfR-mRNA, and prevent degradation. In contrast, the expansion of the labile iron pool leads to a rapid degradation of Tf-mRNA, which is not protected since IRPs are not bound to it (Chan et al., 1994; Ponka and Lok, 1993). However, Ponka and Lok (1993) described that some cells and tissues with specific requirements for iron probably evolved mechanisms that can override the IRE/IRP-dependent control of Tf receptor expression. Erythroid cells, which are the most avid consumers of iron in an organism, use a transcriptional mechanism to maintain very high TfR levels. Transcriptional regulation is also involved in the TfR expression during T- and B-lymphocyte activation. Macrophages are another example of a cell type that shows "unorthodox" responses in terms of the IRE/IRP paradigm since elevated iron levels in these cells increase rather than decrease TfR mRNA and protein levels (Chan et al., 1994; Ponka and Lok, 1993). Further, Ponka and Lok (1993) described that erythroid cells contain the highest mass of the total organismal TfRs which are released from reticulocytes during their maturation to erythrocytes. Hence, plasma contains small amounts of TfRs which represent a soluble fragment of the extracellular receptor domain. Measurements of serum TfR concentrations are clinically useful since their levels correlate with the total mass of immature erythroid cells (Ponka and Lok, 1993). The location of human TfR gene on chromosome 7 is shown in Fig. 60.3.

Circulating apo-transferrin binds two Fe^{3+} ions to form holo-transferrin, which binds to TfR. The TfR–holo-transferrin complex is internalized by receptor-mediated endocytosis (Jabara et al., 2016). TfR1 has a high affinity for iron and is ubiquitously expressed on the cell membrane. TfR1s are abundant in the GI tract, hepatocytes, erythrocyte precursors, and rapidly dividing cells. TfRs are also found in the gray matter, mainly on the luminal side of the brain capillary endothelial cells, with very little on neurons or glial cells in the central nervous system (Roelcke et al., 1996). TfRs are expressed on the cell membrane and Tf is absorbed through a receptor-mediated endocytosis mechanism (Luck and, Mason, 2012). Tf binds to a specific receptor (TfR) to deliver iron to various tissues via the blood and cerebrospinal fluid. Iron-loaded Tf is called holo-transferrin, which binds with high affinity to its receptor site to form the receptor–Tf complex (Tf–TfR) that triggers the formation of clathrin-coated pit and finally an endocytic vesicle, which is internalized into the cytoplasm as a coated vesicle (Ponka and Lok, 1993; Gkouvatsos et al., 2012). Once the endosome is internalized, vacuolar-ATPase (V-ATPase) is recruited to the endosomal membrane. V-ATPase pumps H^+ ions into the endosome and makes the internal environment acidic (pH 5.5). The Tf–TfR complex is exposed to a low pH within the endosome, which allows release of iron from Tf, while Tf remains bound to its receptor as a Tf–TfR complex. Tf is then turned into apo-Tf. Apo-Tf remains bound to its receptor because it has a high affinity for its receptors at a reduced pH (Ponka et al., 1998; Grant and Donaldson, 2009). Therefore the Tf–TfR complex is also known as an iron-scavenging system. Iron can then be utilized for storage or physiological processes, such as incorporation into hemoglobin. Some Tfs are degraded within the cytoplasm after dissociation from the receptors. Those Tf–TfR that are not degraded, finally arrive at the basolateral part of the cell membrane and the exocytosis mechanism for apo-Tf–TfRs takes place. Tf dissociates from its receptor at neutral pH and is released intact into the extracellular space and circulation as iron-free Tf. This is known as the recycling pathway of Tf (Ciechanover et al., 1983; Ponka et al., 1998; Arosio et al., 2017). All TfRs eventually follow the degradation pathway for receptor turnover, for example, erythroid precursor cells in the bone marrow. The cellular uptake of transferrin-iron, its intracellular utilization, and cycle of Tf and TfR are presented in Figs. 60.4 and 60.5.

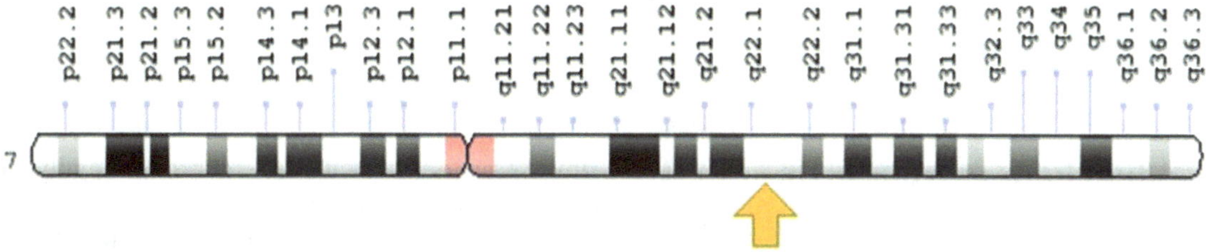

FIGURE 60.3 The human TfR gene is located on chromosome 7 at position 7q22.1, which is the long (q) arm with a base pair of 100,620,416 to 100,641,552 (updated annotation release 109.20200522, GRCh38.p13, NCBI).

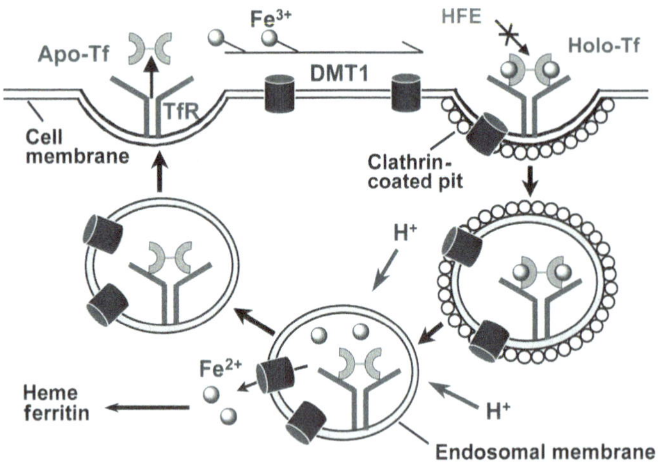

FIGURE 60.4 The cycle of transferrin and TfR1-mediated cellular iron uptake. Differic transferrin (holo-transferrin) binds to TfR1 on the cell surface. The resulting complex is internalized and acidified through the action of a proton pump. Iron is subsequently released from transferrin and transported out of endosomes via the divalent metal transporter DMT1. Hemochromatosis protein (HFE) appears to inhibit transferrin receptor 1 endocytosis, probably through binding to TfR1. Apo-Tf and TfR1 both returned to the cell surface, where they dissociate at neutral pH, in another round cycle of iron uptake. Intracellular iron is either incorporated into heme or stored in ferritin. *DMT1*, divalent metal transporter 1; *TfR1*, transferrin receptor 1. *(Adapted from Qian et al 2002).*

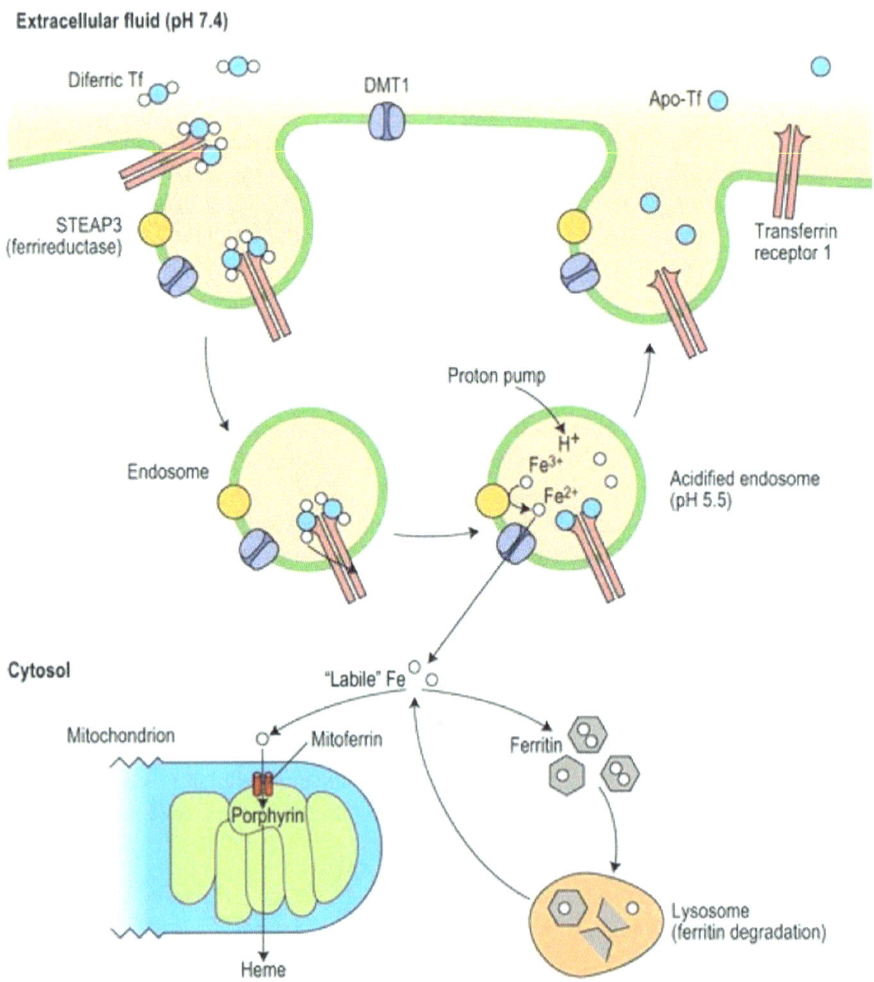

FIGURE 60.5 Cellular uptake of transferrin iron and its intracellular utilization. The iron released into the cytoplasm plays a role in the translational regulation of transferrin receptor, ferritin, and erythroid ALA-synthase. Heme synthesis dominates in erythroid cells, whereas ferritin metabolism is crucial in iron-storage cells (macrophages and hepatocytes). Ferritin is constantly catabolized and the iron is either released back to the cytosol or remains in lysosomes as insoluble hemosiderin. Source: *https://basicmedicalkey.com/iron-deficiency-anemia-anemia-of-chronic-disorders-and-iron-overload.*

Biological functions of transferrin

Iron is one of the most important elements for sustaining life (Ponka and Lok, 1993) because of its ability to bind oxygen and be utilized by several proteins as a cofactor for major biological processes (Gkouvatsos et al., 2012). However, its capacity to generate highly reactive hydroxyl free radicals via the Fenton reaction can also render it toxic to cells. Humans use iron for many cellular processes including carrying oxygen in hemoglobin for cellular respiration without allowing radical oxygen to roam freely. Tf is the most valuable source of iron for red blood cells, with the highest turnover. Plasma Tf plays an important role in iron metabolism in the body. It limits the levels of free iron in the blood. Free iron is hazardous in that it carries the risk of triggering free radical reactions, which set off lipid oxidation and the destruction of thousands of molecules (Brieland et al., 1992). The chelation of free iron turns into soluble and nontoxic iron as it is being delivered to the tissues. Thus Tf renders iron soluble, prevents iron-triggered free radical damage, and transports iron. Consequently, Tf has evolved in such a way that it can recycle and store iron for future use in the body. This is an important task accomplished by Tf due to its extremely high-affinity iron (Fe^{3+})-binding sites (Maguire et al., 1982). Lending to this affinity is an anion cofactor (carbonate), that in its absence will make iron and Tf binding negligible. The remaining four coordination sites are those from the Tf molecule including an aspartic carboxylate oxygen, two tyrosine phenolate oxygens, and a histidine nitrogen. At any given time, about one-third of Tfs-binding sites are filled. Upon radioactively labeling Tf, it was found that about 80% of its iron was delivered to the bone marrow and then integrated into newly formed red blood cells. Other sites of delivery included the liver and spleen, which are major storage sites. It is said that of the 3 g of iron found in adult human males, only about 0.1% of it ends up circulating in the plasma (Lane et al., 2015).

Determination of plasma TfR concentration is also important for a differential diagnosis of anemia and iron stores in the body. Proteolytic cleavage produces a truncated form of the extracellular domain of the TfR which is present in the plasma in direct proportion to the total number of receptors expressed in all body tissues. More TfRs appear on cell surfaces when functional iron depletion occurs. Concomitantly, it leads in the increase of extracellular domain concentration in the plasma produced by proteolytic cleavage. The magnitude of the increase is proportional to the size of the functional iron deficit. Saturation of Tf with iron and free erythrocyte protoporphyrin levels are both abnormal in the anemia of chronic diseases even if iron stores are adequate because the iron supply is reduced by impaired release. The plasma TfR concentration may help to distinguish these two common causes of anemia, since plasma TfR levels are normal in the anemia of chronic disease (Bermejo and García-López, 2009; Abdel-Noor et al., 2019).

Tf, particularly Beta-2-Tf, is a protein found only in cerebrospinal fluid (CSF) and perilymph. It is also an important protein constituent of human CSF, with some 30% being present in an unsialylated form (Nandapalan et al., 1996). Beta-2-Tf has been used extensively by otolaryngologists in the diagnosis of CSF rhinorrhea and skull-base cerebrospinal fluid fistulas (Meurman et al., 1979). The glycan primary structure of the so-called β_2-Tf or τ-globulin has been recently discovered. The major structure is an *N*-acetyllactosaminic type agalacto diantennary oligosaccharide with bisecting *N*-acetylglucosamine and proximal fucose. Analysis of a second Tf preparation containing both asialo- and sialotransferrin revealed another major glycan species derived from the sialylated transferrin variant which is galactosylated, lacks fucose, and lacks bisecting *N*-acetylglucosamine. The first result supports the hypothesis of a de novo "brain-type" glycosylation of intrathecally synthesized cerebrospinal fluid proteins and the second one is in favor of an enzymatic desialylation of serotransferrin (Irie et al., 1989). The pathways of cellular iron (Fe) uptake are depicted in Figs. 60.6 and 60.7.

The TfRs have also been identified in multiple cell layers in the eye including the retinal pigment epithelium (RPE), the inner segments of photoreceptors, the outer plexiform layer, the inner nuclear layer, and the ganglion cell layer (Yefimova et al., 2000; Picard et al., 2008). Thus Tf can pass across the blood–retina barrier (BRB). In addition, Tf is also produced by the RPE and neural retina. All of this implies that Tf also play an important role within the eye (Picard et al., 2008).

Briefly, the important functions of Tf are: (1) free Fe^{3+} is insoluble at a neutral pH, when iron binds to Tf it becomes soluble; (2) deliver and transfer iron to all the various biological tissues between sites of absorption, utilization, and storage; (3) iron and Tf are required for DNA synthesis and cell division; (4) prevent the formation of reactive oxygen species; (5) chelate free toxic iron and act as a protective scavenger; (6) deliver WBC macrophages to all tissues; (7) transferrin is a part of the innate immune system, the binding of Tf to iron impedes bacterial survival; (8) transferrin acts as a marker for inflammation—the level of Tf decreases during inflammation; (9) upstream regulator of hepcidin, a liver-derived peptide hormone that controls systemic iron traffic; and (10) it may inhibit inflammatory cytokines. The supply, storage, and role of iron in cellular and mitochondrial oxidative stress are revealed in Figs. 60.8–60.10.

Although their extracellular structures are quite similar, the two forms of the receptor have important functional differences. The binding affinity of TfR1 for diferric transferrin is about 25–30-fold greater than that of TfR2. Expression of TfR1 is regulated by intracellular iron levels (see below), but that of TfR2 is not. TfR1 is expressed by all iron-requiring cells, whereas TfR2 is most highly expressed on hepatocytes and developing erythroid cells. TfR1 is required for life in mammals; TfR2 cannot compensate for the absence of TfR1. TfR2 seems to have other functions in maintaining iron homeostasis. The delivery of Tf-bound iron begins with the binding of two molecules of monoferric or diferric Tf to a TfR on the cell surface. The iron–Tf-T fR complex then moves

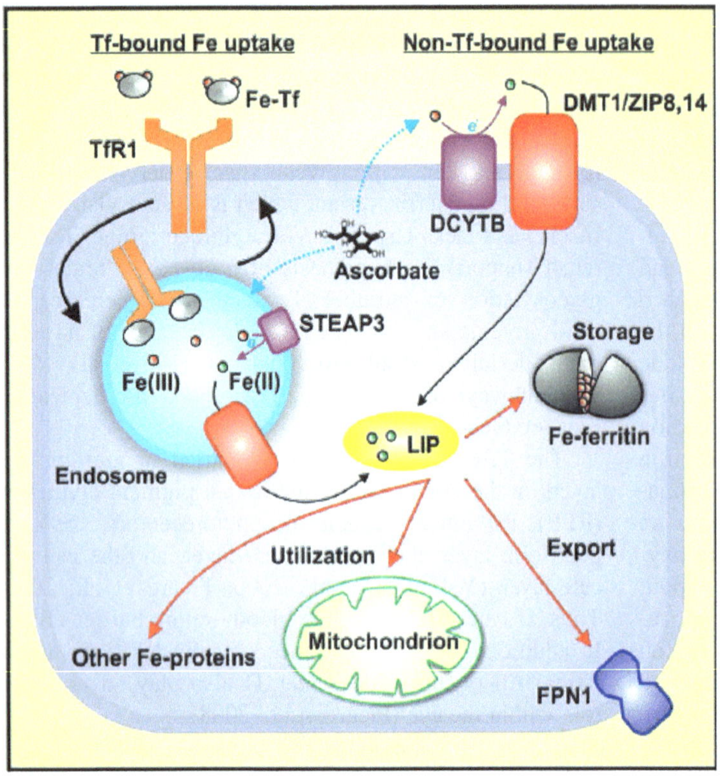

FIGURE 60.6 There are two major cellular Fe uptake pathways, namely transferrin-bound Fe uptake and NTBI uptake. Under physiological conditions, most Fe is bound to transferrin which binds to the transferrin receptor 1 on the cell surface that is then involved in receptor-mediated endocytosis with the Fe being released from transferrin by a decrease in endosomal pH and reduction by an endosomal reductase (e.g., six transmembrane epithelial antigen of the prostate 3) or, potentially, by ascorbate. The Fe (II) is then transported across the endosomal membrane by divalent metal transporter 1, where it then becomes part of the poorly characterized labile iron pool in the cytosol. Iron in the labile iron pool acts as an intermediate and can be utilized for storage in the iron storage protein, ferritin, or used for synthesis of heme and iron–sulfur clusters in the mitochondrion or cytosol. Iron can also be exported from the cell by ferroprotein 1. In conditions of iron overload, NTBI exists in the blood, and may be taken up by processes that include cell surface reduction by ferrireductases such as duodenal cytochrome b, or by reflux reductants, such as ascorbate. Intracellular ascorbate can also supply electrons for duodenal cytochrome b-dependent ferrireduction. Such enzymes reduce ferric NTBI to its ferrous state, which can be imported from the plasma membrane by transporters such as divalent metal transporter 1. *Fe*, iron; *NTBI*, non-Tf-bound Fe. *With permission of Dr. Richardson.*

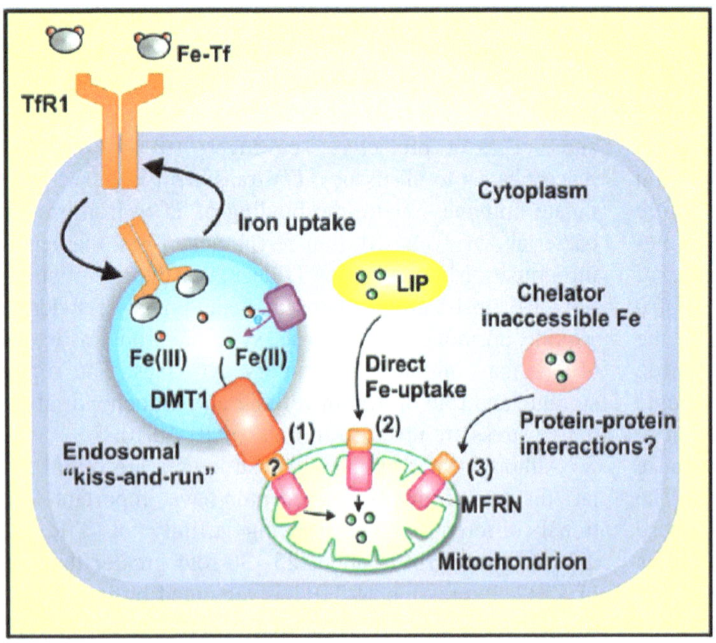

FIGURE 60.7 Mitochondrial iron import pathways. There are three potential iron uptake pathways that provide the metal for transport across the outer mitochondrial membrane via an unknown transporter (designated "?"), and into the mitochondrion, all of which may involve mitoferrin isoforms in the inner mitochondrial membrane. *With permission of Dr. Richardson.*

to the interior of the cell within an endosome, where it releases its iron. The dissociated iron is then taken up by the divalent metal transporter-1 (DMT1) and transported across the endosomal membrane for utilization or storage in the cell. Freed of its iron, the Tf (now apo-Tf) binds avidly to the TfR and is carried back to the cell membrane and released into the plasma.

Aberrations of transferrin

Abnormal protein glycosylation is observed in many common disorders like cancer, inflammation, Alzheimer, disease and diabetes. Transferrin has two glycosylation sites. Each site has an asparagine-linked biantennary complex (Van Scherpenzeel et al., 2016). Over 100 human genetic

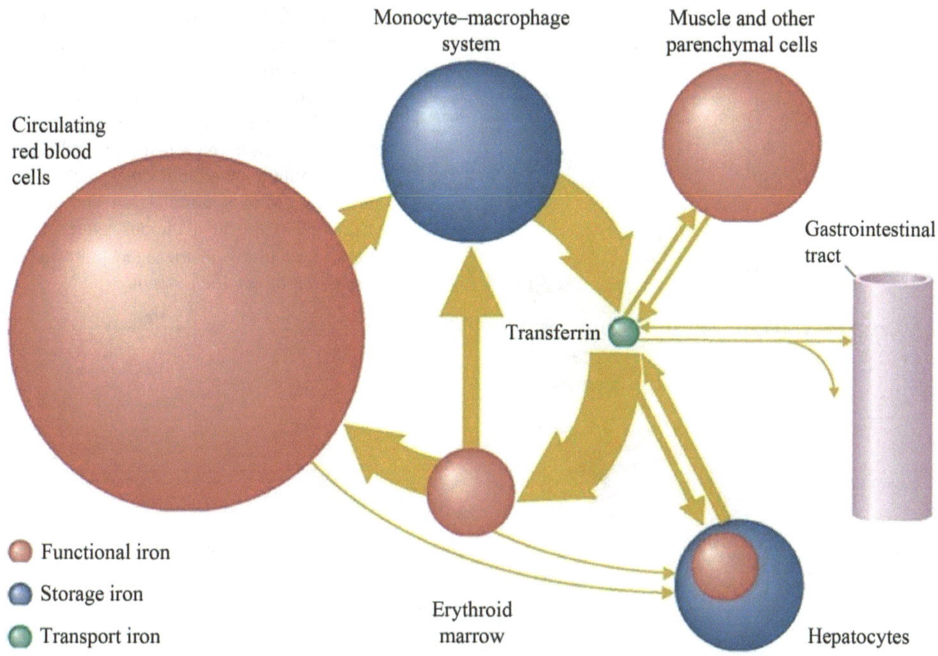

FIGURE 60.8 Body iron supply and storage. This figure shows a schematic representation of the routes of iron movement in the adult. The area of each circle is proportional to the amount of iron contained in the compartment, and the width of each arrow is proportional to the daily flow of iron from one compartment to another. The major portion of iron is found in the erythron as hemoglobin iron (28 mg/kg in women; 32 mg/kg in men) dedicated to oxygen transport and delivery. Small amounts of erythron iron (<1 mg/kg) are also present in heme and nonheme enzymes in developing red cells. The remainder of the functional iron is found as myoglobin iron (4 mg/kg in women; 5 mg/kg in men) in muscle and as iron-containing and iron-dependent enzymes (1–2 mg/kg) throughout the cells of the body. Small amounts of iron are deposited within ferritin in erythroid cells, but most storage iron (5–6 mg/kg in women; 10–12 mg/kg in men) is held in reserve by hepatocytes and macrophages in the liver, bone marrow, spleen, and muscle. The small fraction of transport iron (about 0.2 mg/kg) in the plasma and extracellular fluid is bound to the protein transferrin, which carries iron to meet tissue needs throughout the body. Source: *http://what-when-how.com/acp-medicine/red-blood-cell-function-and-disorders-of-iron-metabolism-part-1/.*

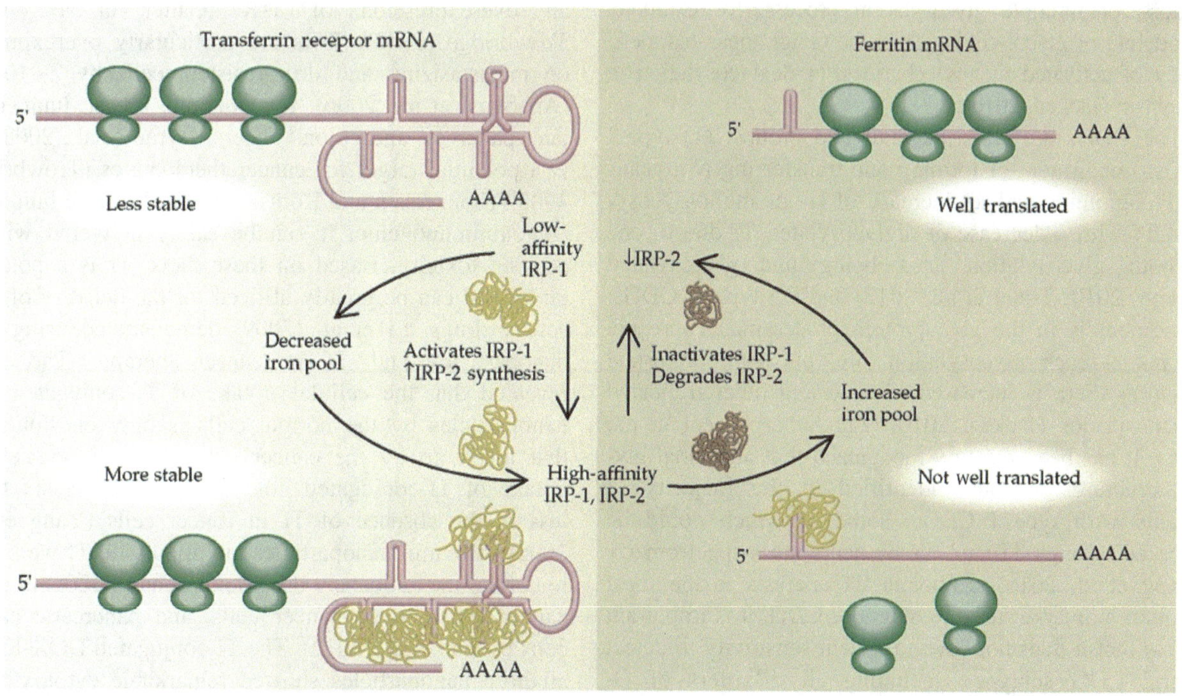

FIGURE 60.9 Regulation of transferrin receptor and ferritin expression by the iron regulatory proteins IRP-1 and IRP-2. Source: *http://what-when-how.com/acp-medicine/red-blood-cell-function-and-disorders-of-iron-metabolism-part-1/.*

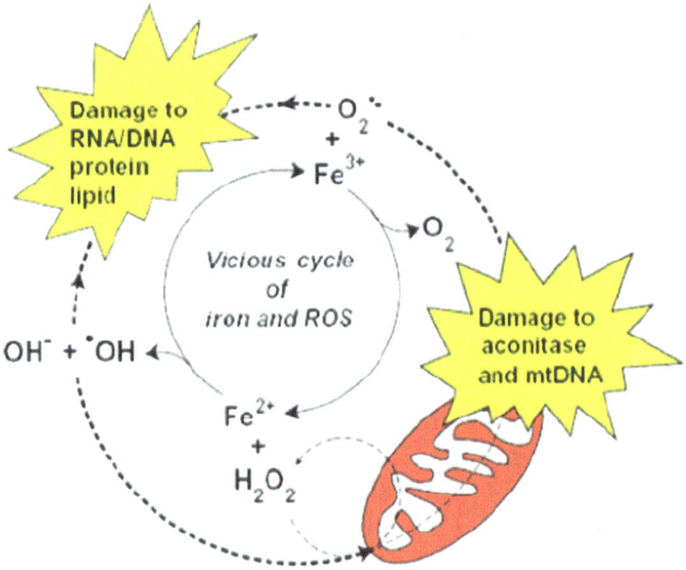

FIGURE 60.10 Postulated role of iron in cellular and mitochondrial oxidative stress. Accumulation of redox-active labile iron, possibly due to altered mitochondrial iron transport mechanisms, can shift cellular redox status toward oxidative stress. This state increases the production of cellular and mitochondrial reactive oxygen species and accelerates oxidative damage to cellular macromolecules including RNA/DNA, proteins, and lipids. Mitochondrial DNA and aconitase can be markedly affected, leading to a vicious cycle of Fe-associated oxidative stress, particularly in the aged brain and skeletal muscle. *With permission of Dr. Richardson.*

disorders have been associated with abnormal glycosylation (Hennet, 2012; Freeze et al., 2014), of which at least 50 types are linked to deficient *N*-glycosylation of serum transferrin, historically referred to as congenital disorders of glycosylation (CDG) (Scott et al., 2014). The CDGs are a diverse and rapidly expanding category of disorders with a key defect in *N*-glycosylation. In other words, diseases affect O-glycosylation, Glycosylphosphatidylinositol (GPI) anchors, or multiple glycosylation process, for example, remodeling of glycol-conjugate in the Golgi apparatus, deficiency of activated monosaccharides, or deglycosylation in cytoplasm (Jaeken, 2010).

The CDGs are classified into two groups: (1) type I CDGs: impairment of forming and transferring *N*-glycans to glycoproteins and an increase of α- or monoglycosylated Tf with a decrease of diglycosylated Tf due to one or both glycosylation sites being underglycosylated (Jaeken, 2010, Timal et al., 2012) and (2) type II CDGs: impairment is in the glycoprotein processing. There are glycans at each glycosylation site, but with abnormal structures there is increased tri-sialo and other truncated glycan species (Jaeken, 2010; Van Scherpenzeel et al., 2016). It has been recently recognized that abnormal glycan structures are also identified in the majority of patients with type I CDGs, some of which could be observed through Tf analysis by acute-mass spectrometry (Zhang et al, 2016). Although Tf analysis is the most common biomarker used to diagnose CDG, it is important to note that it does not have sufficient sensitivity to detect several CDG subtypes, including all disorders of O-glycosylation (Trinidad et al., 2013). Different subtypes of congenital disorders of glycosylation are summarized in Table 60.1.

Application of transferrin in drug delivery

Cancer cells have a high rate of proliferation and a greatly increased need for iron to carry out cellular functions including faster cell growth and division, which results in an overexpression of TfR (Gatter et al., 1983; Trowbridge, 1988). TfRs are particularly overexpressed on metastasizing and drug-resistant cells by 2–10-fold (Anabousi et al., 2006), for example, breast, lung, prostate, pancreas, and colon cancers (Byrne et al., 2008). Tf is a potential target for cancer therapeutics (Trowbridge, 1988; Thorstensen and Romslo, 1993). Because human Tf is nonimmunogenic, it can be safely delivered without causing toxicity. Based on these facts, Tf is a potential target and can be greatly utilized for the delivery of anticancer drugs. Li et al. (2009) demonstrated using gold nanoparticles and Tf in cancer therapy. The study revealed that the cellular uptake of Tf-conjugated gold nanoparticles by the normal cells is only one-fourth of that taken up by the cancer cells. In addition, cellular uptake of Tf-conjugated gold nanoparticles is six times less in the absence of Tf in cancer cells (Yang et al., 2005). Albumin nanoparticles modified with Tf were used to deliver the anticancer drug doxorubicin (DOX) to colon cancer cells, breast cancer cells, and pancreatic cancer cells by Bae et al. (2012). The Tf-conjugated DOX-loaded albumin nanoparticles showed remarkable cytotoxic and apoptotic effects. Additionally, the nanoparticles were

TABLE 60.1 Examples of biomarkers in type I, type II, and mixed types I and II congenital disorders of glycosylation subtypes.

Disorders	Plasma or serum CDT	Plasma or serum N-glycan	Plasma or serum O-glycan	Other testing
ALG12-CDG	Abnormal	Abnormal	Normal	
ALG1-CDG	Abnormal	Abnormal	Normal	
ALG3-CDG	Abnormal	Abnormal	Normal	
ALG8-CDG	Abnormal/ normal	Normal	Normal	
ALG9-CDG	Abnormal	Abnormal	Normal	
ATP6V0A2-CDG	Abnormal/ normal	Abnormal	Abnormal	
DDOST-CDG and other OST complex subunit	Abnormal/ normal	Normal	Normal	Reduced ICAM-1 cell surface expression in fibroblast by FACS
Disorders of COG	Abnormal/ normal	Abnormal	Abnormal	
Disorders of O-mannosylation	Normal	Normal	Normal	Glycosylated dystroglycan staining by immunohistochemistry
Disorders of O-xylosylation	Normal	Normal	Normal	
DPAGT1-CDG	Abnormal	Normal	Normal	
GNE-CDG	Normal	Normal	Abnormal	
MAN1B1-CDG	Abnormal	Abnormal	Normal	
MOGS-CDG	Normal/ abnormal	Abnormal	Normal	Abnormal urine free glycans
MPI-CDG	Abnormal	Abnormal	Normal	Enzyme in fibroblast
NGLY1-CDG	Normal	Normal	Normal	ddVenus assay in fibroblast
PGM1-CDG	Abnormal	Abnormal	Abnormal	
PGM3-CDG	Normal	Abnormal/normal	Abnormal	
PMM2-CDG	Abnormal	Abnormal	Normal	Enzyme in fibroblast

CDT, carbohydrate-deficient Tf; *ALG12*, Asparagine-Linked Glycosylation 12; *ALG1*, Asparagine-Linked Glycosylation 1; *ALG3*, Asparagine-Linked Glycosylation 3; *ALG8*, Asparagine-Linked Glycosylation 8; *ALG9*, Asparagine-Linked Glycosylation 9; *ATP6V0A2*, autosomal recessive cutis laxa type 2 / vacuolar ATPase; *DDOST-CDG*, Dolichyl-diphosphooligosaccharide congenital disorders of glycosylation; *OST*, Oligosaccharyl transferase; *ICAM*, Intercellular adhesion molecule; *FACS*, Fanconi anemia complementation syndrome; *COG*, Conserved Oligomeric Golgi Complex disorder congenital disorder of glycosylation; *DPAGT1-CDG*, Dolichyl-phosphate N-acetylglucosamine phosphotransferase congenital disorder of glycosylation; *GNE-CDG*, UDP-N-acetylglucosamine 2-epimerase/N-acetylmannosamine kinase congenital disorder of glycosylation; *MAN1B1-CDG*, mannosidase alpha class 1B congenital disorder of glycosylation; *MOGS-CDG*, Mannosyl-oligosaccharide glucosidase deficiency congenital disorder of glycosylation; *MPI-CDG*, Mannosephosphate isomerase deficiency congenital disorder of glycosylation; *NGLY1-CDG*, N-glycanase 1 deficiency congenital disorder of glycosylation; *PGM1-CDG*, Phosphoglucomutase deficiency type 1 congenital disorder of glycosylation; *PGM3-CDG*, Phosphoglucomutase deficiency type 3 congenital disorder of glycosylation; *PMM2-CDG*, Phosphomannomutase 2 deficiency congenital disorder of glycosylation .

effectively localized to solid tumors in a colon carcinoma-xenografted nude mouse over 32 h. Tf divides into three subgroups: serum Tf, lactotransferrin (lactoferrin), and melanotransferrin (Wally and Buchanan, 2007). The concentration of lactoferrin (Lf), an iron-binding glycoprotein member of the Tf family, was measured in the CSF and receptors were discovered in the choroid plexus of the brain (Talukder et al., 2003). Findings showed that Lf crosses the BBB. One water-soluble antiviral drug that has poor BBB permeability is azidothymidine (AZT). Therefore Tf ligand has been employed for brain targeting with the delivery of AZT. Mishra et al. (2006) prepared Tf-anchored PEGylated albumin nanoparticles to encapsulate AZT, which provided a significant enhancement of brain localization of AZT when compared with unanchored nanoparticles from an in vivo experiment. Tf conjugation to the diphtheria toxin was shown to significantly enhance its efficacy (O'Keefe and Draper, 1985). A nearly eightfold increase in the circulatory half-life of the chemotherapeutic drug neocarzinostatin was also demonstrated following its conjugation to Tf (Kohgo et al., 1990). These early studies demonstrated the potential for TfR-targeted delivery of chemotherapeutic drugs.

The nanoparticle formulation containing *N*-glutaryl phosphatidylethanolamine (NGPE) liposomes encapsulating oxaliplatin and conjugated to the human Tf ligand, with potential antineoplastic activity. Upon infusion of oxaliplatin-encapsulated Tf-conjugated NGPE liposomes, the Tf moiety targets and binds to the Tf receptor, which is overexpressed in a variety of human cancer cells. Upon binding and internalization, oxaliplatin is released and its active derivatives, alkylate macromolecules, form both inter- and intrastrand platinum–DNA cross-links, resulting in inhibition of DNA replication and transcription. By extending the circulation time and specifically targeting TfRs, this formulation may improve the efficacy and safety of oxaliplatin therapy, compared to administration of oxaliplatin alone. DOX-loaded, lipid-coated PLGA NPs modified with Tf were tested against lung cancer cell line A549 and found to be more efficiently endocytosed than the nontargeting DOX PLGA NPs (Guo et al., 2013). Tf-modified DOX-PLGA NPs showed greater cytotoxicity as compared to nontargeted NPs, while in vivo studies confirmed better tumor-inhibiting effects with the Tfr-targeted DOX-PLGA NPs than with the nontargeted NPs in mouse-bearing A549 tumor (Guo et al., 2013).

Recently, Shao et al. (2015) reported the fabrication of Tf-conjugated nanostructured lipid carriers (NLCs) for the codelivery of paclitaxel (PTX), a taxane obtained from Pacific yew (*Taxus brevifolia*) which is a microtubule stabilizer, and enhanced green fluorescence protein plasmid. The in vitro and in vivo transfection efficiency of this drug delivery system were evaluated on human non-small cell lung carcinoma cell line NCl-H460 cells and mouse bearing NCl-H460 cells, respectively. The results of this study indicated that Tf functionalization could augment the active targeting of Tf-PTX-Plasmid-NLC in the NCl-H460 cell line. Liang et al. (2014) have reported that human ferritin (HFn) (a small iron-storage protein) nanocages can carry an ample amount of DOX and can be targeted to Tf without any need for surface modification, since HFn inherently binds with Tf. The study indicated that DOX-loaded HFn specifically bound and was internalized by the Tf-overexpressing tumor cells, and resulted in a release of DOX into lysosomes. In vivo, the intratumoral drug concentration was found to be 10-fold higher than free Dox and it significantly inhibited the growth of the tumor after a single-dose injection (Liang et al., 2014). The side effects of HFn-DOX toward healthy cells were less than free DOX. This preparation holds promise for success in clinical trials.

TfRs are expressed on the cells of the mucosal layer of the GI tract (Andrews, 2000). It has been demonstrated that TfR are expressed in enterocytes from the duodenum (Pietrangelo et al., 1992; Andrews, 2000). Targeting the TfR has been shown to be effective in delivering many different therapeutic agents and causing cytotoxic effects in cancer cells. Thus, oral delivery of drugs is considered using Tf–drug conjugates (Lim and Shen, 2005; Xia et al., 2000). Du et al. (2013) confirmed that TfRs are responsible for receptor-mediated endocytosis for the absorption of TfR-targeted drug-delivery systems. This suggests the potential of Tf as a targeting ligand for devising drug–Tf conjugates. More than a few Tf-conjugated drugs including adriamycin, cisplatin, and diphtheria toxin have been tested and have been shown to have improved outcomes in clinical trials (Faulk et al., 1990; Head et al., 1997; Rainov and Söling, 2005). Hu-Lieskovan et al. (2005) used Tf–polymer–drug conjugation to treat several types of malignant cells in mouse models. In addition, Sahoo and Labhasetwar (2005) demonstrated that Tf-conjugated particles loaded with paclitaxel have enhanced antiproliferative activity compared to free drug and nontargeted particles due to increased cellular uptake and reduced exocytosis. In vitro studies revealed that these particles have proven effective against breast and prostate cancer cells in mouse models (Sahoo and Labhasetwar, 2005; Sahoo et al., 2004). Furthermore, increased uptake of lipid particles conjugated with Tf was observed in leukemia cells in vitro (Khajavinia et al., 2012). Pulkkinen et al. (2008) demonstrated that Tf-conjugated particles can cross the BBB and target rat brain tumor glioma cells in in vitro models. Tf-conjugated drugs have been shown to prevent cardiotoxicity and drug resistance. It is conceivable that the binding of Tf to tumor cells is higher than with nonmalignant cells because TfR on tumor cells can be recycled more quickly (Choudhury et al., 2018). The role of Tf and TfR in iron transport across the BBB are shown in Fig. 60.11.

Tf can also be easily obtained from either human or recombinant sources. However, there are some disadvantages to using Tf as a targeting ligand. Delivery of Tf-conjugated drugs in high doses may result in damage to other cells that express low levels of TfR (Laske et al., 1997). Transferrin ligands may target liver cells and highly proliferative nonmalignant cells (Daniels et al., 2006). Further, free Tf in the blood may competitively bind to malignant cell receptors, preventing uptake and limiting efficacy (Xu et al., 1999; Alexander-Bryant et al., 2013). Transferrin-targeted

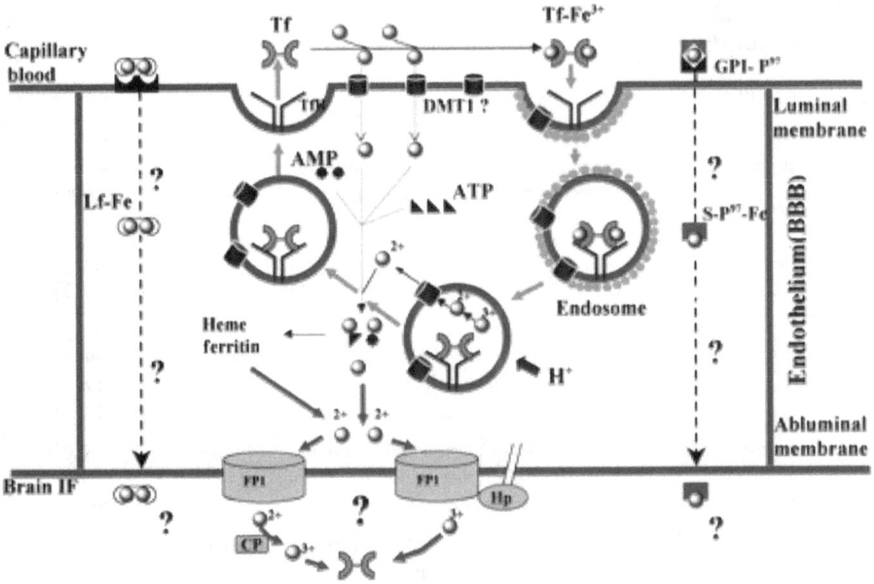

FIGURE 60.11 The role of Tf and TfR in iron transport across the BBB. The Tf/TfR pathway may be the major route of iron transport across the luminal membrane of the BBB. The Tf-Fe (Tf-bound iron) uptake by endothelial cells is not different in nature from the uptake into other cell types. DMT1 might play a role in translocation of iron from endosome to cytosol. Then iron (Fe^{2+}) crosses the abluminal membrane probably via FP1/HP and/or HP-independent export systems. LfR/Lf and GPI-P97/S-P97 pathways might also involve iron transport across the BBB. (GPI-P97, GPI-anchored P97; Lf/LfR, lactoferrin/lactoferrin receptor; S-P97, secreted P97.) *BBB*, blood−brain barrier; *DMT1*, divalent metal transporter 1; *FP1*, ferroprotein 1; *HP*, hephaestion. *(Adapted from Qian et al 2002)*

nanocarriers for drug delivery to brain tumors are summarized in Table 60.2.

Clinical implications of transferrin

One of the most prevalent nutritional deficits is iron. There is a correlation between the amount of iron and Tf concentration in the blood. A high serum Tf level indicates low iron, which means they are inversely proportional. When there is less iron bound to Tf, this will lead to a high circulation of apo-Tf in the body, indicating a possible iron-deficiency anemia. In order to maintain iron homeostasis in the body, the liver increases the production of Tf to enable the Tf to bind to iron and transport it to the cells. Hence upregulation of TfR occurs in iron-deficiency anemia (Bermejo and García-López, 2009). Low iron-bound Tf indicates low iron levels in the body, which affects hemoglobin and erythropoiesis. The significance of Tf is that it can detect iron deficiency and be used to monitor erythropoiesis. Tf concentration in the blood is about 35 μM and ~30% is saturated with iron with a half-life of 7−8 days.

Anemia is defined as low levels of total hemoglobin or hematocrit or low numbers of red blood cells or its content. The most common type is iron-deficiency anemia. Iron is essential for hemoglobin synthesis and low levels of available iron will impede hematopoiesis. Symptoms of anemia may include pale coloration of mucous linings, fatigue, irritability, and shortness of breath.

There are multiple causes of anemia, one of which may lead to determination of the serum Tf level and Tf saturation, as well as total iron-binding capacity (TIBC). Tf concentration and TIBC are currently used to assess iron status. This test provides an insight into the underlying problem at the level of Tf. This test will also reveal how many of the possible Tf binding sites end up "saturated," or filled. In healthy individuals, Tf levels range between 170−370 mg/dL and the percent saturated should be between 20%−30%. It may show lower than 10%. The Tf-iron saturation percentage is usually low in patients with iron deficiency, and treatment options may include iron supplements or even blood transfusions. There are different causes of low serum Tf levels, such as hepatitis, nephritis, chronic infection, and malignancy (cancer). A genetic mutation resulting in the absence of Tf, known as atransferrinemia, which leads to hemosiderosis in the heart and liver, may also lead to heart and liver failure. Atransferrinemia can be treated by plasma transfusion. A low level of plasma Tf may indicate iron overload, which means that the binding site of Tf is highly saturated with iron. Iron overload may also lead to hemochromatosis.

Conditions other than the above, which may be associated with Tf and TfR, include diminishing tumor cells, while TfR are used to attract antibodies and high Tf saturation which increases the risk of cardiovascular mortality in patients with high Tf saturation (>55%) and low density lipoprotein levels (Wells et al., 2004). The TIBC can be calculated by measuring the Tf concentration. Each mole of Tf (MW 79.5 kDa) is capable of binding to two moles of iron (MW 55.8 kDa). This means that if a Tf concentration of 1 g/L is equivalent to 12.57 μmol/L, it would be able to carry 25.1 μmol/L of iron. Hence, TIBC can be estimated by multiplying the calculated Tf concentration by a converting factor (usually approximate to 25.1 for TIBC in μmol/L or 1.4 for TIBC in μg/L). Therefore TIBC = transferrin × conversion factor, and Tf saturation = serum iron transferrin × conversion factor × 100.

TABLE 60.2 Transferrin-targeted nanocarriers for drug delivery to brain tumors.

Drug	Tf-nanocarriers	Preparation technique	Key outcomes	Reference
5-Fluorouracil	Tf-liposomes	Thin-film hydration	10- and 17-fold increase in the brain uptake of 5-fluorouracil by the nonmodified and Tf-coupled liposomes	Soni et al. (2008)
Artesunate	Tf-liposomes	Ammonium sulfate transmembrane gradient method	Within 32 days of treatment, the growth of U87 cells was well inhibited by Tf-ART-LPs without significant toxicity	Tong et al. (2015)
Cis	Tf-liposomes	Thin-film hydration	Higher transport across the BBB and stronger cytotoxicity to C6 cells	Iinuma et al. (2002)
DOX	Tf-PEG-graphene oxide NPs	Carbodiimide coupling	Higher rates of glioma cell death and prolonged survival of glioma-bearing rats compared with nontargeted therapy	Dong et al. (2016)
DOX	Tf-C-dots	Carbodiimide coupling	Greater uptake of the C-dots-Trans-Dox conjugate compared to Dox alone	Li et al. (2016)
DOX	Tf-CPP-PEG liposomes	Thin-film hydration	Highest targeting efficacy for brain microvascular endothelial and C6 cell uptake but avoided capture by normal cells	Liu et al. (2017)
DOX	Tf/tamoxifen PAMAM dendrimers	Chemical coupling	The tumor volume was significantly inhibited and deceased to 38% after 47 days	Li et al. (2012)
DOX/PTX	Tf-magnetic silica PLGA NPs	Double emulsion S/O/W solvent evaporation	DOX-PTX-NPs-Tf treatment resulted in the strongest inhibition of tumor growth	Cui et al. (2013)
DTX	Tf-PLA-TPGS NPs	Nanoprecipitation	IC_{50} of Tf-PLA-TPGS NPs was 23.4%, 16.9%, and 229% more efficient than PLGA NPs, PLA-TPGS NPs, and taxotere, respectively	Agrawal et al. (2017)
pDNA	Tf-PEI-ChA nanoplexes	Polyelectrolyte complexation	The tumors of mice treated with Tf-nanoplexes were five times smaller in size than the tumors of untreated animals	Dube et al. (2017)
PTX	Tf-polyphosphoester micelles	Dimethylsuberimidate crosslinking	The mean survival time of U-87 MG glioma mice treated with TPM (39.5 days) was longer than with Taxol (33.6 days)	Zhang et al. (2012)
RSV	Tf-PEG-PLA NPs	Ring opening polymerization	Significantly decreased tumor volume and accumulated in brain tumor, prolonged survival of C6 glioma-bearing rats	Guo et al. (2013)
RBT	RBT@MRN-SS-Tf/Apt	Tf and aptamer AS1411Apt grafted on surface of ruthenium nanoparticles	Significantly increased apoptosis of tumor cells of gliomas, and prolonged the survival period	Zhu et al. (2018)
TMZ	Tf-PEG-PLGA NPs	Emulsification/solvent evaporation	Enhanced drug accumulation in the brain and higher antitumor efficacy	Jain et al. (2011)
Vincristine/TET	Tf-liposomes	Film dispersion method	Significantly prolonged circulation time, efficiently accumulated in brain tumor, thus leading to a robust anticancer efficacy	Song et al. (2017)
ZOL	Tf-liposome-CaP NPs	Thin-film hydration	The sequential therapy of TMZ and Tf-NPs led to superior therapeutic activity compared to their single administration	Salzano et al. (2016)

BBB, blood– brain barrier; *DOX*, doxorubicin; *Tf*, transferrin; *DTX*, Docetaxel; *PAMAM*, Poly(amido)amine dendrimers; *pDNA*, Plasmid DNA; *PTX*, Paclitaxel; *PLGA* NPs, Poly D, L-lactic-co-glycolic acid-nanoparticles; *RBT*, $Ru(bpy)_2(tip)]^2$; *RSV*, Respiratory syncytial virus; *TET*, Ten-eleven translocation; *Tf-PLA-TPGS*, Transferrin conjugated-poly-lactic acid-D-α-tocopheryl polyethylene glycol; *TfCPP-PEG'* Transferrin conjugated Cell Penetrating Peptides Polyethylene glycol; *TMZ*, Temozolomide; *TPM*, Tetramethylpyrazine; *ZOL*: Zoledronate.

Concluding remarks and future directions

Tf is an iron-binding protein. Each molecule of Tf carries two molecules of iron and is absorbed by receptor-mediated endocytosis. Iron is indispensable for life and essential for such processes as oxygen transport, electron transfer, and DNA synthesis. Thus Tf plays an important role in iron homeostasis in the body. In addition, Tf has different functions including DNA synthesis, cellular growth, and proliferation. TfRs are expressed ubiquitously in the body, which has intrigued scientists to target TfR for drug-delivery strategies in cancer patients. Many clinical trials are in progress using different methods and technology including conjugation of nanoparticles with Tf. Recently, recombinant human Tf was produced from plants as a nonglycosylated, polypeptide chain containing 679 amino acids and having a molecular mass of 76 kDa. It is available on the market as Optiferrin. It has been claimed that this recombinant human transferrin is completely blood-free, and its performance is equivalent to human holo-transferrin. It has also been stated that Optiferrin is suitable for a broad range of cell types that include stem, primary, and bioproduction cells. The recombinant human Tf needs more studies and can also be used as a supplement. Moreover, bovine lactoferrin and ovotransferrin are largely available in the market as supplements. Taken together, Tf is an important molecule, and some of the remarkable progress that has been made toward its application in diverse areas warrants further investigations for health and controlling diseases in human and animals.

References

Abdel-Noor, R.A., Abu-Zaid, M.H., Elshweikh, S.A., et al., 2019. The importance of log ferritin and transferrin /log ferritin in differentiating iron deficiency anemia from anemia of chronic disease in rheumatoid arthritis patients. Egypt. J. Immunol. 26 (1), 141–150.

Agrawal, P., Singh, R.P., Sonali, et al., 2017. TPGS-chitosan cross-linked targeted nanoparticles for effective brain cancer therapy. Mater. Sci. Eng. C. Mater. Biol. Appl. 74, 167–176.

Alexander-Bryant, A.A., Vanden Berg-Foels, W.S., Wen, X., 2013. Bioengineering strategies for designing targeted cancer therapies. Adv. Cancer Res. 118, 1–59.

Anabousi, S., Bakowsky, U., Schneider, M., et al., 2006. *In vitro* assessment of transferrin-conjugated liposomes as drug delivery systems for inhalation therapy of lung cancer. Eur. J. Pharm. Sci. 29 (5), 367–374.

Andrews, N.C., 2000. Intestinal iron absorption: current concepts circa 2000. Dig. Liver Dis. 32 (1), 56–61.

Arosio, P., Elia, L., Poli, M., 2017. Ferritin, cellular iron storage and regulation. IUBMB Life 69, 414–422.

Bae, S., Ma, K., Kim, T.H., et al., 2012. Doxorubicin-loaded human serum albumin nanoparticles surface-modified with TNF-related apoptosis-inducing ligand and transferrin for targeting multiple tumor types. Biomaterials 33 (5), 1536–1546.

Bermejo, F., García-López, S., 2009. A guide to diagnosis of iron deficiency and iron deficiency anemia in digestive diseases. World J. Gastroenterol. 15 (37), 4638–4643.

Brieland, J.K., Clarke, S.J., Karmiol, S., et al., 1992. Transferrin: a potential source of iron for oxygen free radical-mediated endothelial cell injury. Arch. Biochem. Biophys. 294 (1), 265–270.

Byrne, J.D., Betancourt, T., Brannon-Peppas, L., 2008. Active targeting schemes for nanoparticle systems in cancer therapeutics. Adv. Drug Deliv. Rev. 60 (15), 1615–1626.

Chan, R.Y., Seiser, C., Schulman, H.M., et al., 1994. Regulation of transferrin receptor mRNA expression. Distinct regulatory features in erythroid cells. Eur. J. Biochem. 220 (3), 683–692.

Choudhury, H., Pandey, M., Chin, P.X., et al., 2018. Transferrin receptors-targeting nanocarriers for efficient targeted delivery and transcytosis of drugs into the brain tumors: a review of recent advancements and emerging trends. Drug Deliv. Transl. Res. 8 (5), 1545–1563.

Ciechanover, A., Schwartz, A.L., Lodish, H.F., 1983. Kinetics of internalization and recycling of transferrin and the transferrin receptor in a human hepatoma cell line. Effect of lysosomotropic agents. J. Biol. Chem. 32, 267–275.

Cui, Y., Xu, Q., Chow, P.K., et al., 2013. Transferrin-conjugated magnetic silica PLGA nanoparticles loaded with doxorubicin and paclitaxel for brain glioma treatment. Biomaterials 34 (33), 8511–8520.

Daniels, T.R., Delgado, T., Rodriguez, J.A., et al., 2006. The transferrin receptor part I: Biology and targeting with cytotoxic antibodies for the treatment of cancer. Clin. Immunol. 121, 144–158.

de Jong, G., van Dijk, J.P., van Eijk, H.G., 1990. The biology of transferrin. Clin. Chim. Acta 190 (1-2), 1–46.

Dong, H., Jin, M., Liu, Z., et al., 2016. *In vitro* and *in vivo* brain-targeting chemo-photothermal therapy using graphene oxide conjugated with transferrin for Gliomas. Lasers Med. Sci. 31 (6), 1123–1131.

Du, W., Fan, Y., Zheng, N., He, B., Yuan, L., Zhang, H., et al., 2013. Transferrin receptor specific nanocarriers conjugated with functional 7peptide for oral drug delivery. Biomaterials 34 (3), 794–806.

Dube, B., Pandey, A., Joshi, G., et al., 2017. Hydrophobically modified polyethylenimine-based ternary complexes for targeting brain tumor: stability, in vitro and in vivo studies. Artif. Cell Nanomed. Biotechnol. 45 (8), 1685–1698.

Elsayed, M.E., Sharif, M.U., Stack, A.G., 2016. Transferrin saturation: a body iron biomarker. Adv. Clin. Chem. 75, 71–97.

Faulk, W.P., Taylor, C.G., Yeh, C.J., McIntyre, J.A., 1990. Preliminary clinical study of transferrin-adriamycin conjugate for drug delivery to acute leukemia patients. Mol. Biother 2 (1), 57–60.

Freeze, H.H., Chong, J.X., Bamshad, M.J., et al., 2014. Solving glycosylation disorders: fundamental approaches reveal complicated pathways. Am. J. Hum. Genet. 94 (2), 161–175.

Gammella, E., Buratti, P., Cairo, G., et al., 2017. The transferrin receptor: the cellular iron gate. Metallomics 9 (10), 1367–1375.

Gatter, K.C., Brown, G., Trowbridge, I.S., et al., 1983. Transferrin receptors in human tissues: their distribution and possible clinical relevance. J. Clin. Pathol. 36 (5), 539–545.

Gkouvatsos, K., Papanikolaou, G., Pantopoulos, K., 2012. Regulation of iron transport and the role of transferrin. Biochim. Biophys. Acta 1820 (3), 188–202.

Gomme, P.T., McCann, K.B., Bertolini, J., 2005. Transferrin: structure, function and potential therapeutic actions. Drug Discov. Today 10 (4), 267–273.

Grant, B.D., Donaldson, J.G., 2009. Pathways and mechanisms of endocytic recycling. Nat. Rev. Mol. Cell Biol. 10 (9), 597–608.

Guo, W., Li, A., Jia, Z., et al., 2013. Transferrin modified PEG-PLA-resveratrol conjugates: *in vitro* and *in vivo* studies for glioma. Eur. J. Pharmacol. 718 (1-3), 41–47.

Hall, D.R., Hadden, J.M., Leonard, G.A., et al., 2002. The crystal and molecular structures of diferric porcine and rabbit serum transferrins at resolutions of 2.15 and 2.60 A, respectively. Acta Crystallogr. D. Biol. Crystallogr 58 (Pt 1), 70–80.

Head, J.F., Wang, F., Elliott, R.L., 1997. Antineoplastic drugs that interfere with iron metabolism in cancer cells. Adv. Enzyme Regul. 37, 147–169.

Hennet, T., 2012. Diseases of glycosylation beyond classical congenital disorders of glycosylation. Biochim. Biophys. Acta (BBA)-Gen. Subj. 1820 (9), 1306–1317.

Hershberger, C.L., Larson, J.L., Arnold, B., et al., 1991. A cloned gene for human transferrin. Ann. N. Y. Acad. Sci. 646, 140–154.

Huebers, H.A., Finch, C.A., 1987. The physiology of transferrin and transferrin receptors. Physiol. Rev. 67, 520.

Huerre, C., Uzan, G., Grzeschik, K.H., et al., 1984. The structural gene for transferrin (TF) maps to 3q21-3qter. Ann. Genet. 27, 5–10.

Hu-Lieskovan, S., Heidel, J.D., Bartlett, D.W., Davis, M.E., Triche, T.J., 2005. Sequence-specific knockdown of EWS-FLI1 by targeted, nonviral delivery of small interfering RNA inhibits tumor growth in a murine model of metastatic Ewing's sarcoma. Cancer Res. 65 (19), 8984–8992.

Iinuma, H., Maruyama, K., Okinaga, K., et al., 2002. Intracellular targeting therapy of cisplatin-encapsulated transferrin-polyethylene glycol liposome on peritoneal dissemination of gastric cancer. Int. J. Cancer 99 (1), 130–137.

Irie, S., Minguell, J.J., Tavassoli, M., 1989. Comparison of desialylation of rat transferrin by cellular and non-cellular methods. Biochem. J. 259 (2), 427–431.

Jabara, H.H., Boyden, S.E., Chou, J., et al., 2016. A missense mutation in TFRC, encoding transferrin receptor 1, causes combined immunodeficiency. Nat. Genet. 48, 74–78.

Jaeken, J., 2010. Congenital disorders of glycosylation. Ann. N. Y. Acad. Sci. 1214 (1), 190–198.

Jain, A., Chasoo, G., Singh, S.K., et al., 2011. Transferrin-appended PEGylated nanoparticles for temozolomide delivery to brain: *in vitro* characterization. J. Microencapsul. 28 (1), 21–28.

Kasvosve, I., Delanghe, J.R., Gomo, Z.A.R., Gangaidzo, I.T., Khumalo, H., Langlois, M.R., et al., 2002. Effect of transferrin polymorphism on the metabolism of vitamin C in Zimbabwean adults. Am. J. Clin. Nutr. 75 (2), 321–325. Available from: https://doi.org/10.1093/ajcn/75.2.321.

Khajavinia, A., Varshosaz, J., Dehkordi, A.J., 2012. Targeting etoposide to acute myelogenous leukaemia cells using nanostructured lipid carriers coated with transferrin. Nanotechnology 23 (40), 405101.

Kohgo, Y., Kondo, H., Kato, J., et al., 1990. Kinetics of internalization and cytotoxicity of transferrin-neocarzinostatin conjugate in human leukemia cell line, K562. Jpn. J. Cancer Res. 81 (1), 91–99.

Kurhade, G.A., Khanorkar, S.V., Puranik, B.M., et al., 1994. Serum level of iron and transferrin in pregnancy and postpartum period. Indian J. Physiol. Pharmacol. 38 (1), 34–38.

Lane, D.J., Merlot, A.M., Huang, M.L., et al., 2015. Cellular iron uptake, trafficking and metabolism: key molecules and mechanisms and their roles in disease. Biochim. Biophys. Acta 1853 (5), 1130–1144.

Laske, D.W., Youle, R.J., Oldfield, E.H., 1997. Tumor regression with regional distribution of the targeted toxin TF-CRM107 in patients with malignant brain tumors. Nat. Med. 3 (12), 1362–1368.

Liang, M., Fan, K., Zhou, M., et al., 2014. H-ferritin-nanocaged doxorubicin nanoparticles specifically target and kill tumors with a single-dose injection. Proc. Natl. Acad. Sci. U.S.A 111 (41), 14900–14905.

Li, J.L., Wang, L., Liu, X.Y., et al., 2009. *In vitro* cancer cell imaging and therapy using transferrin-conjugated gold nanoparticles. Cancer Lett. 274 (2), 319–326.

Li, Y., He, H., Jia, X., et al., 2012. A dual-targeting nanocarrier based on poly(amidoamine) dendrimers conjugated with transferrin and tamoxifen for treating brain gliomas. Biomaterials 33 (15), 3899–3908.

Li, S., Amat, D., Peng, Z., et al., 2016. Transferrin conjugated nontoxic carbon dots for doxorubicin delivery to target pediatric brain tumor cells. Nanoscale 8 (37), 16662–16669.

Lim, C.J., Shen, W.C., 2005. Comparison of monomeric and oligomeric transferrin as potential carrier in oral delivery of protein drugs. J. Control. Rel. 106 (3), 273–286.

Liu, C., Liu, X.N., Wang, G.L., et al., 2017. A dual-mediated liposomal drug delivery system targeting the brain: rational construction, integrity evaluation across the blood-brain barrier, and the transporting mechanism to glioma cells. Int. J. Nanomed. 12, 2407–2425.

Lu, Y., Li, J., He, Y., et al., 2018. The influence of transferrin gene polymorphism on measuring carbohydrate deficient transferrin by capillary electrophoresis. Sichuan Da Xue Xue Bao Yi Xue Ban. 49 (1), 98–101.

Luck, A.N., Mason, A.B., 2012. Transferrin-mediated cellular iron delivery. Curr. Top. Membr. 69, 3–35.

Maguire, J.J., Kellogg 3rd, E.W., Packer, L., 1982. Protection against free radical formation by protein bound iron. Toxicol. Lett. 14 (1-2), 27–34.

Meurman, O.H., Irjala, K., Suonpaa, J., et al., 1979. A new method for identification of cerebrospinal fluid leakage. Acta Otolaryngol. 87 (3-4), 366–369.

Mishra, V., Mahor, S., Rawat, A., et al., 2006. Targeted brain delivery of AZT via transferrin anchored pegylated albumin nanoparticles. J. Drug Target. 14 (1), 45–53.

Nakajima, M., Ohno, K., Goto-Koshino, Y., et al., 2014. Plasma transferrin concentration as a nutritional marker in malnourished dogs with nutritional treatment. J. Vet. Med. Sci. 76 (4), 539–543.

Nandapalan, V., Watson, I.D., Swift, A.C., 1996. Beta-2-transferrin and cerebrospinal fluid rhinorrhea. Clin. Otolaryngol. 21 (3), 259–264.

O'Keefe, D.O., Draper, R.K., 1985. Characterization of a transferrin-diphtheria toxin conjugate. J. Biol. Chem. 260 (2), 932–937. 1985.

Picard, E., Fontaine, I., Jonet, L., et al., 2008. The protective role of transferrin in Müller glial cells after iron-induced toxicity. Mol. Vis. 14, 928–941.

Pietrangelo, A., Rocchi, E., Casalgrandi, G., et al., 1992. Regulation of transferrin, transferrin receptor, and ferritin genes in human duodenum. Gastroenterology 102 (3), 802–809.

Ponka, P., Lok, C.N., 1993. The transferrin receptor: role in health and disease. Int. J. Biochem. Cell Biol. 31 (10), 1111–1137.

Ponka, P., Beaumont, C., Richardson, D.R., 1998. Function and regulation of transferrin and ferritin. Semin. Hematol. 35 (1), 35–54.

Pulkkinen, M., Pikkarainen, J., Wirth, T., Tarvainen, T., Haapa-aho, V., Korhonen, H., et al., 2008. Three-step tumor targeting of paclitaxel using biotinylated PLA-PEG nanoparticles and avidin-biotin technology: Formulation development and in vitro anticancer activity. Eur. J. pharmaceutics biopharmaceutics: Off. J. Arbeitsgemeinschaft fur Pharmazeutische Verfahrenstechnik e.V 70 (1), 66–74.

Qian, Z.M., Li, H., Sun, H., Ho, K., 2002. Targeted drug delivery via the transferrin receptor-mediated endocytosis pathway. Pharmacol. Rev. 54 (4), 561–587.

Rainov, N.G., Söling, A., 2005. Technology evaluation: TransMID, KS Biomedix/Nycomed/Sosei/PharmaEngine. Curr. Opin. Mol. Ther. 7 (5), 483–492.

Roelcke, U., Leenders, K.L., von Ammon, K., et al., 1996. Brain tumor iron uptake measured with positron emission tomography and 52Fe-citrate. J. Neurooncol. 29 (2), 157–165.

Sahoo, S.K., Labhasetwar, V., 2005. Enhanced antiproliferative activity of transferrin-conjugated paclitaxel-loaded nanoparticles is mediated via sustained intracellular drug retention. Mol. Pharm. 2 (5), 373–383.

Sahoo, S.K., Ma, W., Labhasetwar, V., 2004. Efficacy of transferrin-conjugated paclitaxel-loaded nanoparticles in a murine model of prostate cancer. Int. J. Cancer 112 (2), 335–340.

Salzano, G., Zappavigna, S., Luce, A., et al., 2016. Transferrin-targeted nanoparticles containing zoledronic acid as a potential tool to inhibit glioblastoma growth. J. Biomed. Nanotechnol. 12 (4), 811–830.

Scott, K., Gadomski, T., Kozicz, T., 2014. Congenital disorders of glycosylation: new defects and still counting. J. Inherit. Metab. Dis. 37 (4), 609–617.

Shao, Z., Shao, J., Tan, B., et al., 2015. Targeted lung cancer therapy: preparation and optimization of transferrin-decorated nanostructured lipid carriers as novel nanomedicine for co-delivery of anticancer drugs and DNA. Int. J. Nanomed. 10, 1223–1233.

Singhal, U., Mohsin, S., Saini, K., et al., 1993. Serum level of iron and transferrin in normal and anemic pregnant women. Indian J. Physiol. Pharmacol. 37 (1), 71–74.

Song, X.L., Liu, S., Jiang, Y., et al., 2017. Targeting vincristine plus tetrandrine liposomes modified with DSPE-PEG_{2000}-transferrin in treatment of brain glioma. Eur. J. Pharm. Sci. 96, 129–140.

Soni, V., Kohli, D.V., Jain, S.K., 2008. Transferrin-conjugated liposomal system for improved delivery of 5-fluorouracil to brain. J. Drug Target. 16 (1), 73–78.

Steere, A.N., Byrne, S.L., Chasteen, N.D., et al., 2012. Kinetics of iron release from transferrin bound to the transferrin receptor at endosomal pH. Biochim. Biophys. Acta 1820 (3), 326–333.

Stibler, H., 1991. Carbohydrate-deficient transferrin in serum: a new marker of potentially harmful alcohol consumption reviewed. Clin. Chem. 37, 2029–2037.

Sun, H., Li, H., Mason, A.B., Woodworth, R.C., et al., 1999. N-lobe versus C-lobe complexation of bismuth by human transferrin. Biochem. J. 337 (Pt1), 105–111 (Pt 1).

Talukder, M.J.R., Takeuchi, T., Harada, E., 2003. Receptor-mediated transport of lactoferrin into the cerebrospinal fluid via plasma in young calves. J. Vet. Med. Sci. 65 (9), 957–964.

Testa, U., 1985. Transferrin receptors: structure and function. Curr. Top. Hematol. 5, 127–161.

Thorstensen, K., Romslo, I., 1993. The transferrin receptor: its diagnostic value and its potential as therapeutic target. Scand. J. Clin. Lab. Invest. Suppl. 215, 113–120.

Timal, S., Hoischen, A., Lehle, L., et al., 2012. Gene identification in the congenital disorders of glycosylation type I by whole-exome sequencing. Hum. Mol. Genet. 21 (19), 4151–4161.

Tong, H., Wang, Y., Lu, X., et al., 2015. On the preparation of transferrin modified artesunate nanoliposomes and their glioma-targeting treatment in-vitro and in-vivo. Int. J. Clin. Exp. Med. 8 (12), 22045–22052.

Trinidad, J.C., Schoepfer, R., Burlingame, A.L., et al., 2013. *N*-and *O*-glycosylation in the murine synaptosome. Mol. Cell. Proteomics 12 (12), 3474–3488.

Trowbridge, I.S., 1988. Transferrin receptor as a potential therapeutic target. Prog. Allergy 45, 121–146.

Uzan, G., Frain, M., Park, I., et al., 1984. Molecular cloning and sequence analysis of cDNA for human transferrin. Biochem. Biophys. Res. Commun. 119, 273–281.

van Eijk, H.G., de Jong, G., 1992. The physiology of iron, transferrin, and ferritin. Biol. Trace Elem. Res. 35 (1), 13–24.

Van Scherpenzeel, M., Willems, E., Lefeber, D.J., 2016. Clinical diagnostics and therapy monitoring in the congenital disorders of glycosylation. Glycoconj. J. 33, 345–358Published online January 7. Available from: http://doi.org/10.1007/s10719-015-9639-X.

Wally, J., Buchanan, S.K., 2007. A structural comparison of human serum transferrin and human lactoferrin. Biometals 20 (3-4), 249–262.

Wells, B.J., Mainous, A.G., King, D.E., et al., 2004. The combined effect of transferrin saturation and low density lipoprotein on mortality. Fam. Med. 36 (5), 324–329.

Weykamp, C., Wielders, J.P., Helander, A., et al., 2013. Toward standardization of carbohydrate-deficient transferrin (CDT) measurements: III. Performance of native serum and serum spiked with disialotransferrin proves that harmonization of CDT assays is possible. Clin. Chem. Lab. Med. 51, 991–996.

Xia, C.Q., Wang, J., Shen, W.C., 2000. Hypoglycemic effect of insulin-transferrin conjugate in streptozotocin-induced diabetic rats. J. Pharmacol. Exp. Ther. 295 (2), 594–600.

Xu, L., Pirollo, K.F., Tang, W.H., et al., 1999. Transferrin-liposome-mediated systemic p53 gene therapy in combination with radiation results in regression of human head and neck cancer xenografts. Hum. Gene Ther. 10 (18), 2941–2952.

Yang, F., Lum, J.B., McGill, J.R., et al., 1984. Human transferrin: cDNA characterization and chromosomal localization. Proc. Natl. Acad. Sci. USA 81, 2752–2756.

Yang, P.H., Sun, X., Chiu, J.F., et al., 2005. Transferrin-mediated gold nanoparticle cellular uptake. Bioconjug. Chem. 16 (3), 494–496.

Yefimova, M.G., Jeanny, J.C., Guillonneau, X., et al., 2000. Iron, ferritin, transferrin, and transferrin receptor in the adult rat retina. Invest. Ophthalmol. Vis. Sci. 41 (8), 2343–2351.

Zakin, M.M., 1992. Regulation of transferrin gene expression. FASEB J. 6 (14), 3253–3258.

Zhang, P., Hu, L., Yin, Q., et al., 2012. Transferrin-conjugated polyphosphoester hybrid micelle loading paclitaxel for brain-targeting delivery: synthesis, preparation and in vivo evaluation. J. Control. Rel. 159 (3), 429–434.

Zhang, W., James, P.M., Ng, B.G., et al., 2016. A novel *N*-tetrasaccharide in patients with congenital disorders of glycosylation, including asparagine-linked glycosylation protein 1, phosphomannomutase 2, and phosphomannose isomerase deficiencies. Clin. Chem. 62 (1), 208–217.

Zhu, X., Zhou, H., Liu, Y., et al., 2018. Transferrin/aptamer conjugated mesoporous ruthenium nanosystem for redox-controlled and targeted chemo-photodynamic therapy of glioma. Acta Biomater. 82, 143–157.

Chapter 61

Glutamine supplementation: hope, hype, or stay tuned?

Dan DuBourdieu

Introduction

Glutamine is an amino acid that has received a fair amount of attention from the scientific community for its various biological roles in the body as well as from the general public for its roles when marketed as a nutraceutical or health supplement. The biological roles of glutamine in humans and animals are widely known to the scientific community but how effective supplementing additional glutamine is for any particular health application can vary. Glutamine supplements can be considered simply a part of standard nutrition, as drugs, or as nutraceuticals (Kelly and Wischmeyer, 2003). Regardless of how one might want to classify glutamine, it has been widely used in supplements for a variety of reasons. This chapter will attempt to summarize some of the wide range of biological functions attributed to glutamine and the efficacy of supplementation of additional glutamine for various health conditions.

Biological functions of glutamine

Glutamine is an amino acid that is used by the body as one of the building blocks for the biosynthesis of proteins. It is normally thought of as a nonessential amino acid because the body usually can create sufficient amounts of it. A normal diet provides enough glutamine for most people as it is not essential under normal physiological conditions, as the body will synthesize it. Animal food sources including eggs and meat protein are very rich in glutamine. However, various plant sources such as nuts, rice, cabbage, beans, corn, and other vegetables also contain glutamine. Industrially produced glutamine found in supplements is made by fermentation techniques typically using mutants of *Brevibacterium flavum*. During stress, the demand for glutamine can increase in the body, and glutamine then needs to be obtained from the diet. With severe infections, major burns, physical trauma, and radiation-induced damage, the physiological level of glutamine can become inadequate. It is during these times of stress or disease that glutamine is thought to become conditionally essential.

Glutamine is generally made from glutamate and ammonia using an enzyme called glutamine synthetase. This is commonly done in the muscles but also in the lungs, brain, and liver. The structure of glutamine is a linear molecule and it is also one of the uncharged polar amino acids, as it carries a polar amide group at the end of the side chain. The polar aspect makes glutamine water soluble (Fig. 61.1).

Like other amino acids, glutamine is used in protein synthesis and without it, a protein may not function properly due to improper folding. When a protein does not fold correctly, death or disease conditions can occur. Likewise, when a protein folds but cannot function because of changes in amino acid structure, body functions can change, thus affecting the overall health of the person.

Glutamine transport

Glutamine is water soluble. One gram will dissolve in 20.8 mL of water at 30°C. A family of N and A glutamine transporters has been identified that helps glutamine cross cell membranes (Chaudhry et al., 2002). The SNAT3 transport system is important for glutamine uptake in cells of the liver and kidney and in astrocytes (McGivan and Bungard, 2007). Sodium–glutamine transport systems have also been identified (Talukder et al., 2008a) that help bring sodium across cell membranes. Increases of glutamine in the blood have been shown in both animal models and human studies that occur from about 30–120 min after oral/enteral free glutamine supplementation (Rogero et al., 2004) or alanine–glutamine supplementation (Rogero et al., 2004; Klassen et al., 2000). Other studies have shown about 46% of an 8-g oral dose

Nutraceuticals. DOI: https://doi.org/10.1016/B978-0-12-821038-3.00061-6

of glutamine reaches the bloodstream, while the rest appears to be metabolized in the gut, liver, and kidneys (Bowtell et al., 1999). How much of the glutamine supplement is metabolized by these organs may ultimately play a role in how effective glutamine is for any particular intended purpose during supplementation.

Glutamine has many functions in the body (Table 61.1) such as protein synthesis, donation of nitrogen and carbon, and many other roles. For example, glutamine supplementation can also increase antioxidant potential in the body and may also attenuate renal oxidative damage (Tsai et al., 2012). Glutamine is also a precursor to adenosine triphosphate (ATP) which, in turn, will be used by cells for energy to make biological reactions occur. Many fast-growing cells like intestinal cells and cells in the immune system will utilize glutamine as their energy source and for other purposes (Newsholme et al., 1985; Alpers, 2000). Mammalian cells will generate ATP via glycolysis or mitochondrial respiration. Living systems typically use the sugar glucose to make ATP but when glucose is not available, cells can use glutamine. Carbon groups from the side chain of glutamine can be used to form ATP to ensure cellular energy. Glutamine-driven oxidative phosphorylation is a major means of ATP production in cells, even in hypoxic cancer cells (Fan et al., 2013). Glutamine has the highest concentration of free amino acids in the blood and can cross the blood–brain barrier where its role in energy formation is important with other related aspects still being determined. Indeed, glutamine blood levels appear to play a role in patients with recent onset of schizophrenia where an increased glutamine/glutamate ratio occurs, while patients with chronic schizophrenia show decreased glutamine/glutamate ratios (Madeira et al., 2018).

Glutamine, in certain anabolic processes, can donate nitrogen to create new molecules. One of these processes is the biosynthesis of purines found in deoxyribonucleic acid (DNA). Glutamine donates the nitrogen for certain nitrogen atoms located in purine nucleobases, adenine, and guanine, as well as donating the amino group of guanine. The relevance of glutamine in DNA biosynthesis is the fact that the enzyme glutamine 5-phosphoribosyl-1-pyrophosphate amidotransferase is the rate-limiting step in making new purines (Cory and Cory, 2006).

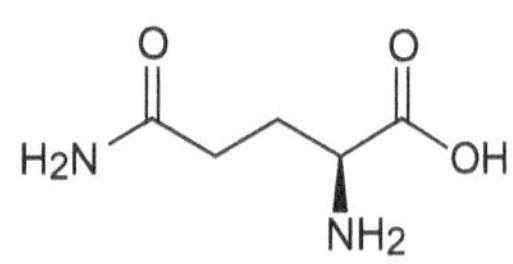

FIGURE 61.1 Structure of glutamine.

Cell proliferation and cancer

Cancer is one of most complicated health conditions known to science and is full of uncertainties and ambiguities. The use of glutamine supplementation for cancer patients falls into the ambiguity category of science with hope but uncertainties. Glutamine given by intravenous methods can significantly reduce vomiting, nausea, and diarrhea in gastric and colorectal cancer patients (Li et al., 2009). Enteral nutrition with a mixture of glutamine, arginine, and omega -3 fatty acids can improve short-term survival in stage IV gastric cancer patients (Klek et al., 2017). Glutamine supplementation appears to be effective in helping breast cancer patients survive when receiving radiation treatment (Rubio et al., 2013). At the cellular level in cancer patients, however, what glutamine is doing is not always clear.

Cell proliferation is dependent on glutamine. This fact has been known for at least 60 years. It was noticed that the glutamine supplementation in tissue-culturing medium at millimolar concentrations could greatly enhance cell growth and proliferation (Eagle et al., 1956). As such, glutamine is one of the most required nutrients used in mammalian cell culture media. Just like normal cells, rapidly proliferating cells such as cancer cells also use glutamine as cellular building blocks and energy. They also need the ability to maintain the redox balance, which glutamine provides. A number of studies have shown that glutamine is involved in multiple aspects of cancer metabolism as it is a major nutrient (Hensley et al., 2013). Glutamine is used by cancer cells to increase proliferation as well as to help survival under metabolic stress conditions (Choi and Park, 2018). Transformed cells will consume glutamine faster than

TABLE 61.1 Some functions of glutamine.

Protein synthesis	Energy production	Nucleotide synthesis	Indirect antioxidant potential
Tissue protection	Immune functions	Cell proliferation	Glycogenesis
Ammonia buffering	Acid–base balance	Nitrogen transport	Precursor to excitatory neurotransmitters glutamate and aspartate; and the inhibitory gamma-aminobutyric acid (GABA)

nontransformed cells. However, it was shown that decreasing the glutamine concentration resulted in a decreased growth rate of the transformed cells (Rubin, 2019). Starving cancer cells of glutamine and targeting glutamine metabolism is a new approach being used in cancer research (Jiang et al., 2019). This will perhaps bear fruit someday as a practical cancer treatment and resolve some of the ambiguities regarding glutamine for cancer treatments for the whole person versus just the cancer cell.

Ammonia regulation

The pH of cerebral spinal fluid (CSF) and brain is vital for homeostatic functions for the body to maintain. Ammonia is a neurotoxic substance that plays a role in determining pH. Increases of ammonia in the brain have been associated with Alzheimer disease (Adlimoghaddam et al., 2016). Ammonia content in the central nervous system (CNS) is regulated through the glutamine–glutamate pathway (Kelley and Kazemi, 1974). CNS ammonia production therefore is an important buffering mechanism for controlling both CSF and brain pH. There is a large increase in glutamine content in the brain when hyperammonia occurs, but only small changes in α-ketoglutarate and glutamate since glutamate is converted to glutamine. As such, glutamate uses the ammonia to convert into glutamine and therefore buffers the brain against the increased toxic ammonia (Cooper and Jeitner, 2016). Glutamine synthetase is the enzyme that catalyzes ammonia condensation with glutamate to glutamine. It has been hypothesized that glutamine supplementation may be a new strategy to treat hyperammonia (Hakvoort et al., 2017). This would be an interesting future development if it could be shown to benefit Alzheimer patients (Table 61.2).

Oral rehydration solutions

Glutamine can play a role in oral rehydration solutions (ORSs). Diarrhea is a leading cause of sickness and death all over the world with a billion cases and up to 2.2 million deaths every year among children less than 5 years of age (Black et al., 2003). Oral rehydration therapy that uses ORSs can effectively treat mild-to-moderate dehydration that results from diarrhea. The principle of how ORS works is fairly simple as the ability to transport sodium-coupled glucose across the small intestinal epithelial membrane remains largely intact in secretory diarrhea and continues to stimulate absorption of water and electrolytes (Field, 2003). ORS formulations typically contain glucose, salts, and sometimes amino acids or other ingredients. The transport of water and the electrolytes Na^+, K^+, Cl^-, and HCO_{3-} is a key function of the intestinal tract. Water molecules are moved in and out of the lumen by passive osmotic processes. However, electrolytes are transported by an active process that glucose and certain amino acids like glutamine can help with. Transcellular pathways that are involved in electrolyte transport consist of pumps, carriers, and ion channels. Functional polarity of the intestinal epithelial cells occurs by having the location of the pumps, carriers, and channels only in certain locations of the cell, thereby driving this transcellular movement of electrolytes through the cell. Na^+ is the primary ion that drives water absorption in the intestine. In the small intestine, the cotransport of Na^+ with food-derived products and the electroneutral Na^+, Cl^- absorptive mechanisms are responsible for most of the water and electrolyte absorption. The large intestine absorbs Na^+ through both electrogenic mechanisms involving apical Na^+ channels and an electroneutral Na^+, and Cl^- absorptive mechanisms. With some of the mechanisms of Na^+ absorption in intestines being elucidated, a new pharmacological therapy concept for the treatment of diarrhea

TABLE 61.2 Roles of glutamine in body organs.

Brain	Kidney	Intestine	Liver	Muscle	Other tissues
Ammonia buffering	Energy substrate	Fluid and electrolyte absorption	Energy substrate	Energy substrate	Energy substrate
	Ammonia excretion		Glycogen and glutathione synthesis	Glycogen and glutathione synthesis	Antiinflammatory, cytoprotective, and immunomodulatory roles
Glutamine			Ammonia metabolism	Avoids ammonia metabolism	Indirect antioxidant by increasing glutathione synthesis
				Attenuation of muscle damage	

Source: Adapted from Coqueiro, A.Y., Rogero, M.M., Tirapegui, J., 2019. Glutamine as an anti-fatigue amino acid in sports nutrition. Nutrients 11 (4), 863.

with an ORS emerged by stimulating neutral NaCl absorptive processes with glutamine. Neutral NaCl absorptive mechanisms remain intact despite intestinal villus atrophy in at least *Cryptosporidia* (Argenzio et al., 1994) and *Rotavirus* (Rhoads, et al., 1991) infection in the neonatal piglet model. It would appear that sodium–glutamine cotransporters uptake mechanisms are located in the crypt cells at the base of the intestinal villi, such as in rabbits (Talukder et al., 2008a,b) and are not destroyed during infections. Other mammals also have the capacity for neutral NaCl absorptive processes. The presence of sodium–glutamine cotransport mechanisms in the gut allows for sodium and water uptake even in severe infections that destroy intestinal villi. Because glutamine is able to stimulate electroneutral NaCl absorption (Rhoads et al., 1990) and function as a Na^+ cotransport substrate in an electrogenic manner, glutamine generated interest as a potentially useful component in ORS. Experimental trials showed that glutamine is able to promote the absorption of Na^+ and water even better than glucose alone (Silva et al., 1998; Abely et al., 2000). In an animal model using a perfused rabbit ileum that was exposed to cholera toxin, a glutamine ORS reduced about 90% of the Na^+ and water loss that was caused by the cholera toxin (Silva et al., 1998; Abely et al., 2000). Other studies (Lima et al., 2002) in a rat model of secretory diarrhea induced by cholera also showed that glutamine worked well to have a positive effect upon Na^+ absorption. However, it appeared that an alanine–glutamine-based ORS was even better for increasing Na^+ absorption (Lima et al., 2002). The approach of using the dipeptide of alanine–glutamine may prove to be an even better approach in ORS technology.

The World Health Organization (WHO) established ORSs as its principal tool to fight diarrhea (WHO, 2002). The WHO formulation, without glutamine, has been quite successful in reducing child mortality rates from 5 million/year to 1.3 million/year. This many deaths is still too many and so trials have been carried out to see if improvements might occur if glutamine is placed into the formulation. In adult cholera patients it appeared that glutamine in ORS did help reduce stool volume by 25% and total fecal loss by 30% (Punjabi et al., 1991). Dehydrated infants with noncholera diarrhea were also found to be well tolerated and worked as well as the standard WHO-ORS (Ribeiro et al., 1994). Unfortunately, glutamine by itself is not always an improvement as other studies in children with acute noncholera diarrhea using a glutamine-containing glucose-free formulation did not work any better than the standard WHO-ORS formulation (GutiErrez et al., 2007). The success of an ORS will depend on various factors such as low cost and ease of use (Bardhan, 2007). How glutamine meets these criteria for routine use in humans as well as in nonhuman use of an ORS is still being determined.

Calves are also given ORS when diarrhea occurs for the identical reasons that humans are given an ORS. The use of glutamine in ORS for calves has been examined in a number of clinical trials. Although not all trials have been positive for including glutamine in ORS (Naylor et al., 1997), the results are generally favorable for wanting to include glutamine in ORS when given to calves. Therapeutic treatments with ORS containing glutamine were concluded to be efficacious in terms of improved serum electrolytes and serum protein fraction (Jaiswal et al., 2019) in calves that had acute diarrhea. Other studies showed that adding glutamine to a high-caloric ORS was beneficial for improved clinical recovery in preweaned calves with diarrhea (Turner et al., 2016; Brooks et al., 1997), with improvements in plasma volume and packed cell volume occurring. However, if glutamine was added to a non-nutritious ORS, the results were not the same. Glutamine in nutritious ORS appeared to repair intestinal villi as the loss of villus length was worse in calves treated with normal ORS that did not contain glutamine (Brooks et al., 1998). As such, the use of glutamine in an ORS makes sense from a technical standpoint for improving intestinal villi height and helping to rehydrate calves.

Glutamine and the critically ill patients controversy

In critically ill patients, glutamine depletion can be common. Levels lower than 420 μmol/L are considered to be an independent predictive factor of mortality (Costa et al., 2017), and supplementing with exogenous glutamine remains controversial. A number of small trials initially showed positive results of glutamine supplementation in hospital settings. A meta-analysis of 22 studies (McRae, 2017) indicated that glutamine supplementation for critically ill or surgical patients by oral or enteral routes reduces length of hospital stay and hospital-acquired infections. The American Society for Parenteral and Enteral Nutrition (ASPEN) guidelines have suggested that glutamine from 0.2 to 0.5 g/kg per day could be given (Vanek et al., 2011) to particular patients. However, later ASPEN guidelines based on larger trials suggest that IV and enteral glutamine should not be used (Mundi et al., 2016), and in particular, should not be used for ventilated critical care patients (van Zanten, 2015). The use of glutamine in pediatric patients is also unclear (Mok and Hankard, 2011) due to a lack of evidence. This would include premature infants, Crohn disease in children with diarrhea, short bowel syndrome, malnutrition, cystic fibrosis, and many other types of conditions in children. More research is obviously required in specific situations to determine whether certain types of critical care conditions actually benefit from glutamine supplementation. The use of glutamine for burn patients is recommended. Clinical data indicate that glutamine improves the outcome in burn-injured patients by showing reduced mortality and reduced length of hospital stay, without apparent clinical risk (Wischmeyer, 2019a). These clinical trials demonstrated significantly less Gram-negative bacteremia occurring with burn injuries. This is presumably because of glutamine maintaining the gut barrier and/or immune function. For the intensive care unit, glutamine is most likely still relevant to positive outcomes and certainly deserves further investigation (Wischmeyer, 2019b).

The gut and glutamine

The health of people starts with the gut as it is the site in the body where diseases typically start. Certain gut diseases have been shown to improve with the use of glutamine (Kim and Kim, 2017). Among these diseases are inflammatory bowel disease (IBD). IBD is a general term used to describe digestive disorders involving chronic inflammation. Types of IBD include ulcerative colitis and Crohn disease. Intestinal walls become inflamed, swollen, and develop ulcers. All of this causes discomfort and serious digestive problems. The related inflammatory bowel syndrome (IBS) of the large intestine is known to be a functional disorder as it has no identifiable cause and is not an actual disease. However, it also gives symptoms of bloating, gas, diarrhea, cramping, and abdominal pain. Another digestive issue of the intestines is leaky gut syndrome. In leaky gut syndrome, bacteria and toxins can enter the bloodstream in gaps in the intestinal walls. The epithelial lining of the intestine gets compromised and causes increased permeability for bacteria and toxins. Increased intestinal epithelial permeability is believed to be part of the pathogenesis of various gut diseases (Berkes et al., 2003; Rao, 2009). A likely intervention strategy involves the use of glutamine as it reduces intestinal permeability from various stress factors and can help maintain the intestinal barrier (Yoshida et al., 1998; Foitzik et al., 1997; Panigrahi et al., 1997; Rao and Samak, 2012).

Clinical studies have shown that patients with IBS with diarrhea dramatically and safely improved their condition when oral dietary supplements were given (Zhou et al., 2019). Because the human gut has little capacity to synthesize glutamine, the gut relies on the supply of glutamine by other tissues and diet under physiologic and pathophysiologic conditions. As such, it would appear that glutamine supplementation does help with barrier function and mucosal integrity of the gut by serving as a protective essential nutrient for gut mucosal epithelial cell growth. It is thought that glutamine activates epidermal growth factor receptors, which in turn leads to stabilization of tight junctions and actomyosin between cells by mechanisms involving various kinase enzymes such as protein kinase C and mitogen-activated protein kinase (Rao and Samak, 2012).

Gut microbiota

Besides the known functions of glutamine supplementation and how it acts in the body for energy production or cell proliferation, etc., new functions are being elucidated. One of these functions is a role for changing the types of gut microbiota normally found in humans and animals (Catinean et al., 2018). Gut microbiota play an important role in the overall health of people. Some types of microorganisms are associated with certain health issues and the ability to change gut microbiota can have beneficial health results. For example, the ratio of *Firmicutes/Bacteriodes* can be an indicator of obesity and type II diabetes. It has been shown (De Souza et al., 2015) that oral glutamine supplementation can reduce the amount of *Firmicutes*. While the exact mechanisms are unclear, glutamine may act by reducing proinflammatory cytokines in the gut which can affect microbes such as *Firmicutes* and alter the ratio of *Firmicutes/Bacteriodes*, as seen in weight loss programs. Similar reductions of *Firmicutes* in the gut were also seen in a mouse study (Ren et al., 2014) when glutamine supplementation was given. In addition, different parts of the gut had different microbe levels being altered by the glutamine supplementation. These data suggest that dietary glutamine supplementation can beneficially alter the intestinal bacterial community as well as activate innate immunity in the small intestine, presumably through signal pathways for NF-κB, MAPK, and PI3K-Akt.

Immunity and glutamine

It is believed that during severe stress situations, where the body starts to catabolize or hypercatabolize molecules, glutamine can be essential to metabolic functions. However, glutamine availability might be compromised due to impairments of homeostasis and metabolism of intertissue amino acids. For immune-suppressed patients, glutamine supplementation can be part of their nutritional regimen. Glutamine is required for cell proliferation in the immune system for such cells as lymphocytes (Yaqoob and Calder, 1997). This information has been used to help AIDS patients in treating their symptoms. HIV-infected patients with AIDS typically experience weight and muscle loss. However, some studies have shown that glutamine supplementation along with antioxidants such as vitamin C, E, and β-carotene along with other nutrients like selenium and *N*-acetyl cysteine, can increase weight and lead to better intestinal absorption (Shabert et al., 1999). Glutamine supplementation can also increase glutathione levels to near normal and allow glutathione to protect against viral, fungal, and bacterial infections in AIDS patients (Burini et al., 2015). However, AIDS is a complicated disease so while the $CD4^+$ T cells require glutamine to proliferate, glutamine is also required for HIV-1 replication (Hegedus et al., 2017) and the infection leads to changes of $CD4^+$ T-cell glutamine metabolism. How glutamine helps the patient and also potentially helps the virus remains to be determined.

Alanine–glutamine peptide

L-alanyl–L-glutamine is a dipeptide molecule, consisting of L-glutamine and L-alanine. Interestingly, use of the alanine–glutamine combination appears to be superior for the uptake of glutamine when compared to using just free glutamine in a supplement. Some 50%–80% of free-form glutamine appears to be used by enterocytes (D'Souza and Powell-Tuck, 2004; Déchelotte et al., 1991), thus making glutamine less available to other tissues or organs. Studies in rodents (Petry et al., 2014) suggest that glutamine concentrations in muscles increase when using an alanine–glutamine supplement more than with just glutamine supplementation. This may be due to the absorption of glutamine as a dipeptide of alanine–glutamine by the liver and intestines that allows for more glutamine to reach the muscles.

Dipeptide forms of glutamine appear to be more slowly metabolized by enterocytes and therefore more glutamine can reach the plasma and elsewhere. This slower metabolism is presumably due to Pep-1 found exclusively in the luminal side cells of the intestine that transports dipeptides and tripeptides (Adibi, 2003), although other mechanisms may be involved (Petry et al., 2014). In addition, the alanine–glutamine dipeptide appears to be more thermally stable in acid media and more soluble than free-form glutamine (Fürst et al., 2004). The use of alanine–glutamine in total parenteral nutrition for critically ill patients may actually be a better approach than glutamine alone. It has been shown (Grau et al., 2011) that patients in intensive care had lower rates of infections and better control of glycemia when supplemented with alanine–glutamine. Studies in liver transplant patients with complications of diabetes or NODM seemed to have reduced hospital stays, reduced insulin requirements, and infections when supplemented with alanine–glutamine (Korraa et al., 2016). Patients with AIDS have also seen benefits when using glutamine–alanine supplementation. The dipeptide form helped treat diarrhea while increasing antiretroviral drug levels in AIDS patients in tropical areas (Oluma et al., 2004). The glutamine dipeptide also seems to help women with HIV/AIDS in cognitive functions after resistance training (De Souza et al., 2018).

While the dipeptide form of glutamine–alanine has efficacy in hospital settings, its efficacy may not be uniform for all uses. The dipeptide form does not necessarily improve actual performance in sports (Coqueiro et al., 2018) despite its popularity in athletes (Raizel et al., 2016; Leite et al., 2016; Coqueiro et al., 2017). While there are no magic bullets, the use of alanine–glutamine dipeptide supplements may prove to be more efficacious for certain kinds of health applications than free glutamine supplements as more glutamine can potentially reach sites in the body where it might be needed. More research may ultimately show whether this hope is accurate.

Sports, athletes, and glutamine

In situations where glutamine levels in the blood are low, it is difficult to say whether oral supplementation will be useful or not. For athletes, this is particularly true. Beyond the medical reasons for taking glutamine supplements by burn patients or other patients, the marketing of glutamine to athletes to combat fatigue, develop strong muscles, and boost immunity has received considerable consideration by athletes. Unfortunately, while this rationale is based on novel ideas, actual sports performance results have yet to be proven (Eichner, 2013). The marketing strategy takes its rationale from the idea that glutamine is found to be the most abundant amino acid in blood and muscle, and that skeletal muscle is where glutamine is mostly synthesized. Muscle growth needs proteins but exercise can metabolize protein and therefore deplete nitrogen in the muscle and change the nitrogen balance (Mori, 2014). It is further believed that a positive nitrogen balance can promote muscle growth (Burd et al., 2009; Chesley et al., 1992). Because glutamine has ample nitrogen, it is understandable why athletes would want to supplement with glutamine.

Since glutamine in muscle may buffer acids and help build glycogen and protein, and be an energy source for immune cells, supplementing with glutamine should help fight fatigue in athletes and help improve physical performance. It was initially noted that glutamine levels in blood decreased after strenuous exercise such as marathon running (Kingsbury et al., 1998) and this decrease would be a "window" when impaired immunity occurred (Gleeson, 2008; Hiscock and Pedersen, 2002). This basic rationale established the marketing strategy of glutamine to athletes.

Supplementation of glutamine to athletes has been taking place since at least the 1980s. The Internet has greatly expanded the marketing and sales of glutamine to body builders and athletes. Marketing claims to athletes regarding glutamine include that it is one of the "best post-workout recovery supplements" and you can "protect your immune system" or "enhance muscularity" or that it is "great for muscle repair and recovery" or can "increase energy levels and endurance", etc. One of the jobs of marketing is to take elements from the science or background of an item and spin it into soundbites that help convince someone to purchase the item. To that end, the success of the marketing of glutamine has resulted in it being a very common supplement food in health food stores. According to a 2012 survey regarding fitness and bodybuilding Internet sites (Mason and Lavallee, 2012), glutamine was among the six most common supplements being touted to athletes. Therefore the question to address is whether glutamine supplementation for athletes really has any effect on sports performance? While the answer from commercial sellers of glutamine supplements is that of course it "works" to improve performance, reduce fatigue, and you should buy it, the answer from the actual research literature does not appear to support this claim. This general conclusion has been known for more than 10 years. The suggested reasons for taking glutamine supplements such as supporting the immune system, increasing glycogen synthesis, or having anticatabolic effects have not been supported in well-controlled scientific studies in the healthy, well-nourished humans (Gleeson, 2008) that athletes tend to be.

While glutamine levels in the blood do drop after exhaustive exercise (Kingsbury et al., 1998), it is not low enough to affect lymphocyte proliferation. Other research has suggested this same basic conclusion. While ingesting up to 28 g/day of glutamine for 2 weeks in healthy adults appears safe (Gleeson, 2008), a meta-analysis of 47 studies in athletes indicated that glutamine supplementation had no effect on aerobic performance, the immune system, or body composition (Ahmadia et al., 2019). Another meta-analysis of 55 studies in athletes concluded that physical performance in athletes is only of limited use when glutamine is supplemented (Coqueiro et al., 2019) but there were some improvements. For example, when soccer players were placed on a treadmill for intermittent exercise and given glutamine peptide or carbohydrate supplement, they saw an increase of 21% in time and 22% in distance compared to a control of a carbohydrate supplement (Favano et al., 2008). Other intermittent running also saw improvements for maximal and minimal power in sprinters that were supplemented with glutamine and carbohydrate (Khorshidi-Hosseini and Nakhostin-Roohi, 2013). However, other studies (Krieger et al., 2004) did not show any improvements in interval training with chronic glutamine supplementation. It should be noted that athletes in these studies were mostly 18–35 years old and in presumably good body condition to begin with and most likely had adequate nutrition. The measurement of the physical sports performance was subjective as it asked each athlete how they felt after workouts or if they performed well in the particular sports activity. It is interesting to note that these studies did suggest that combining glutamine with carbohydrates is a better approach than using glutamine alone, thus indicating that a synergy between glutamine and carbohydrates in supplements might be a possibility for future studies. Perhaps the bioavailability of glutamine may account for its perceived lack of performance efficacy in athletes. However, feeding glutamine intervention studies showed that the plasma glutamine concentration can be kept constant during and after strenuous exercise, but supplementation did not prevent post-exercise changes in various aspects of immune function (Gleeson, 2008). As such, to date the use of glutamine alone as a supplement for actual sports performance still appears to be of limited use. Perhaps in the future different formats of glutamine supplementation may change this current general opinion.

Does glutamine relieve any kind of measurable symptoms of fatigue in athletes? The quick answer is yes. Certain aspects of fatigue can be reduced. While these may not help much at the moment of actual sports performance, perhaps it can be of use in recovery, to a degree. There are a number of aspects that cause fatigue in exercise. Some of these aspects include reducing pH in muscle cells, a reduction of energy sources such as glycogen, accumulations of ammonia in tissues and blood, oxidative issues, and changes in neurotransmitter levels that in turn cause lethargy (Parry-Billings et al., 1990; Katz et al., 1986; Sewell et al., 1994; Smriga et al., 2006).

When glutamine was initially marketed to athletes, it was in response to reports that fewer respiratory infections or illnesses occurred in athletes after taking glutamine supplements (Castell et al.,1996). Other reports (Bowtell et al., 1999) indicated that oral glutamine alone promoted muscle glycogen storage at a similar rate as oral glucose. As stated earlier, some 46% of an 8-g oral dose of glutamine went into the blood while the rest was metabolized in the liver, kidneys, and gut. Bowtell et al. (1999) concluded that glutamine appeared to promote muscle glycogen resynthesis after exercise. Since glycogen is a storage form of glucose that can be metabolized to glucose and subsequently to energy when needed by the body, athletes found this to be quite appealing as a method to potentially increase energy via glutamine supplementation.

The prevention of ammonia accumulation in the body would also be an important method for preventing fatigue during sports activity since ammonia can affect the activity of various enzymes, cell permeability to ions, and NAD^+/NADH ratios (Bassini-Cameron et al., 2008; Meneguello et al., 2003). Glutamine may help to minimize these processes by its effects on energy metabolism, as ammonia production occurring during exercise is due to amino acid oxidation and deamination of adenosine monophosphate to release ammonia. Glutamine synthase activation occurs as a detoxification mechanism in response to increased ammonia levels (Guezennec et al., 1998) and increases of glutamine occur in the brain during exhaustive exercise (Blomstrand et al., 2005). Because glutamine is the main transporter of nitrogen and ammonia contains nitrogen, it may prevent ammonia accumulation in muscles and help favor metabolism of ammonia in the liver and subsequent excretion from the body. Glutamine is of fundamental importance to interorgan nitrogen exchange via ammonia (Cruzat et al., 2018). Endurance running athletes that have been supplemented with glutamine had reduced levels of ammonia in the blood (Carvalho-Peixoto et al., 2007). Muscle soreness can also be reduced by using glutamine supplementation after eccentric exercise (Legault et al., 2015) with this being more effective in men than in women. Therefore it is possible that glutamine may help athletes in the recovery of certain aspects of fatigue after exercise.

Concluding remarks and future directions

Glutamine supplementation has a history of success in many health applications, including for burn patients, IBD patients, AIDs patients, and others. However, not all health applications have seen successful with glutamine supplementation and details as to which conditions are best suited for it are still being determined. Likewise, while glutamine supplementation may reduce certain markers of fatigue, it seems to only have limited effects on actual performance for athletes. The use of dipeptides of alanine–glutamine may hold promise as an alternative form of glutamine supplementation. Time will tell whether there is more hope in the future for the use of glutamine in whatever form to help with various health conditions. Stay tuned.

References

Abely, M., Dallet, P., Boisset, M., et al., 2000. Effect of cholera toxin on glutamine metabolism and transport in rabbit ileum. Am. J. Physiol. Gastrointest. Liver Physiol 278, G789–G796.

Adibi, S.A., 2003. Regulation of expression of the intestinal oligopeptide transporter (pept-1) in health and disease. Am. J. Physiol. Gastrointest. Liver Physiol. 285, G779–G78.

Adlimoghaddam, A., Sabbir, M.G., Albensi, B.C., 2016. Ammonia as a potential neurotoxic factor in Alzheimer's disease. Front. Mol. Neurosci. 9, 57Article. Available from: doi.org/10.3389/fnmol.2016.00057.

Ahmadia, A.R., Rayyanic, E., Bahreinib, M., et al., 2019. The effect of glutamine supplementation on athletic performance, body composition, and immune function: a systematic review and a meta-analysis of clinical trials. Clin. Nutr. 38 (3), 1076–1091.

Alpers, D.H., 2000. Is glutamine a unique fuel for small intestinal cells? Curr. Opin. Gastroenterol. 16 (2), 155.

Argenzio, R.A., Rhosads, M.J., Armstrong, M., et al., 1994. Glutamine stimulates prostaglandin-sensitive Na^+ H^- exchange in experimental porcine cryptosporidiosis. Gastroenterology 106, 1418–1428.

Bardhan, P.K., 2007. Improving the ORS: does glutamine have a role? J. Health Popul. Nutr. 25 (3), 263–266.

Bassini-Cameron, A., Monteiro, A., Gomes, A., et al., 2008. Glutamine protects against increases in blood ammonia in football players in an exercise intensity-dependent way. Br. J. Sports Med. 42, 260–266.

Berkes, J., Viswanathan, V.K., Savkovic, S.D., et al., 2003. Intestinal epithelial responses to enteric pathogens: effects on the tight junction barrier, ion transport, and inflammation. Gut. 52 (3), 439–451.

Black, R.E., Morriss, S.S., Bryce, J., 2003. Where and why are 10 million children dying every year. Lancet. 361, 2226–2234.

Blomstrand, E., Møller, K., Secher, N., et al., 2005. Effect of carbohydrate ingestion on brain exchange of amino acids during sustained exercise in human subjects. Acta Physiol. Scand. 185, 203–209.

Bowtell, J.L., Gelly, K., Jackman, M.L., et al., 1999. Effect of oral glutamine on whole body carbohydrate storage during recovery from exhaustive exercise. J. Appl. Physiol. 86 (6), 1770–1777.

Brooks, H.W., White, D.G., Wagstaff, A.J., et al., 1997. Evaluation of a glutamine containing oral rehydration solution for the treatment of calf diarrhea using an *Escherichia coli* model. Vet. J. 153, 163–170.

Brooks, H.W., Hall, G.A., Wagstaff, A.J., et al., 1998. Detrimental effects on villus form during conventional oral rehydration therapy for diarrhea in calves; alleviation by a nutrient oral rehydration solution containing glutamine. Vet. J. 155, 263–274.

Burd, N.A., Tang, J.E., Moore, D.R., Phillips, S.M., 2009. Exercise training and protein metabolism: influences of contraction, protein intake, and sex-based differences. J. Appl. Physiol. 106, 1692–1701.

Burini, R.C., Moreto, F., Yu, Y.M., 2015. HIV-Positive patients respond to dietary supplementation with cysteine or glutamine. In: Watson, R.R. (Ed.), Health of HIV Infected People: Food, Nutrition and Lifestyle Without Antiretroviral Drugs. Academic Press, pp. 245–269. , Chapter 15.

Carvalho-Peixoto, J., Alves, R.C., Cameron, L.C., 2007. Glutamine and carbohydrate supplements reduce ammonemia increase during endurance field exercise. Appl. Physiol. Nutr. Metab. 32 (6), 1186–1190.

Castell, L.M., Newsholme, E.A., Poortmans, J.R., 1996. Does glutamine have a role in reducing infections in athletes? Eur. J. Appl. Physiol. Occup. Physiol 73 (5), 488–490.

Catinean, A., Neag, M.A., Muntean, D.M., et al., 2018. An overview on the interplay between nutraceuticals and gut microbiota. PeerJ. 6, e4465.

Chaudhry, F.A., Reimer, R.J., Edwards, R.H., 2002. The glutamine commute: take the N line and transfer to the A. J. Cell Biol. 157 (3), 349–355.

Chesley, A., MacDougall, J.D., Tarnopolsky, M.A., et al., 1992. Changes in human muscle protein synthesis after resistance exercise. J. Appl. Physiol. 73, 1383–1388.

Choi, Y.K., Park, K.G., 2018. Targeting glutamine metabolism for cancer treatment. Biomol. Ther. 26 (1), 19–28.

Cooper, A.J.L., Jeitner, T.M., 2016. Central role of glutamate metabolism in the maintenance of nitrogen homeostasis in normal and hyperammonemic brain. Biomolecules 6 (2), 16.

Coqueiro, A.Y., Raizel, R., Hypólito, T.M., et al., 2017. Effects of supplementation with L-glutamine and L-alanine in the body composition of rats submitted to resistance exercise. Rev. Bras. Cienc. Esporte 39, 417–423.

Coqueiro, A.Y., Raizel, R., Bonvini, A., et al., 2018. Effects of glutamine and alanine supplementation on central fatigue markers in rats submitted to resistance training. Nutrients 10 (2), 119.

Coqueiro, A.Y., Rogero, M.M., Tirapegui, J., 2019. Glutamine as an anti-fatigue amino acid in sports nutrition. Nutrients. 11 (4), 863.

Cory, J.G., Cory, A.H., 2006. Review critical roles of glutamine as nitrogen donors in purine and pyrimidine nucleotide synthesis: asparaginase treatment in childhood acute lymphoblastic leukemia. Vivo 20, 587–590.

Costa, B.P., Martins, P., Verissimo, C., et al., 2017. Glutaminemia prognostic significance in critical surgical patients – an analysis of plasma aminogram profile. Nutr. Hosp. 34 (4), 799–807.

Cruzat, V., Rogero, M.M., Keane, K.N., et al., 2018. Glutamine: metabolism and immune function, supplementation and clinical translation. Nutrients 10 (11), 1564.

De Souza, A.Z.Z., Zambom, A.Z., Abboud, K.Y., et al., 2015. Oral supplementation with l-glutamine alters gut microbiota of obese and overweight adults: a pilot study. Nutrition 31, 884–889.

De Souza, D.C., da Silva, J.C., de Oliveira Matos, F., et al., 2018. The effect of a short period of supplementation with glutamine dipeptide on the cognitive responses after a resistance training session of women with HIV/AIDS: a randomized double-blind placebo-controlled crossover study. BioMed. Res. Int. 2525670. Article ID.

Déchelotte, P., Darmaun, D., Rongier, M., et al., 1991. Absorption and metabolic effects of enterally administered glutamine in humans. Am. J. Physiol. 260 (5 Pt 1), G677–G682.

D'Souza, R., Powell-Tuck, J., 2004. Glutamine supplements in the critically ill. J. R. Soc. Med. 97, 425–427.

Eagle, H., Oyama, V.I., Levy, M., et al., 1956. The growth response of mammalian cells in tissue culture to L-glutamine and L-glutamic acid. J. Biol. Chem. 218, 607–616.

Eichner, E.R., 2013. Glutamine supplementation: overstating its welcome. Curr. Sports Med. Rep. 12 (4), 211–212.

Fan, J., Kamphorst, J.J., Mathew, R., et al., 2013. Glutamine-driven oxidative phosphorylation is a major ATP source in transformed mammalian cells in both normoxia and hypoxia. Mol. Syst. Biol. 9, 712.

Favano, A., Santos-Silva, P., Nakano, E., et al., 2008. Peptide glutamine supplementation for tolerance of intermittent exercise in soccer players. Clinics 63, 27–32.

Field, M., 2003. Intestinal ion transport and the pathophysiology of diarrhea. J. Clin. Invest. 111, 931–943.

Foitzik, T., Stufler, M., Hotz, H.G., et al., 1997. Glutamine stabilizes intestinal permeability and reduces pancreatic infection in acute experimental pancreatitis. J. Gastrointest. Surg. 1 (1), 40–46. discussion 6–7.

Fürst, P., Alteheld, B., Stehle, P., 2004. Why should a single nutrient – glutamine – improve outcome? The remarkable story of glutamine dipeptides. Clin. Nutr. Suppl. 1, 3–15.

GutiErrez, C., Villa, S., Mota, F.R., et al., 2007. Does an L-glutamine-containing, glucose-free, oral rehydration solution reduce stool output and time to rehydrate in children with acute diarrhea? A double-blind randomized clinical trial. J. Health Popul. Nutr. 25, 278–284.

Gleeson, M., 2008. Dosing and efficacy of glutamine supplementation in human exercise and sport training. J. Nutr. 138 (10), 2045S–2049S.

Grau, T., Bonet, A., Miñambres, E., et al., 2011. Metabolism, nutrition working the effect of L-alanyl-L-glutamine dipeptide supplemented total parenteral nutrition on infectious morbidity and insulin sensitivity in critically ill patients. Crit. Care Med. 39 (6), 1263–1268.

Guezennec, C., Abdelmalki, A., Serrurier, B., 1998. Effects of prolonged exercise on brain ammonia and amino acids. Int. J. Sports Med. 19 (5), 323–327.

Hakvoort, T., He, Y., Kulik, W., et al., 2017. Pivotal role of glutamine synthetase in ammonia detoxification. Hepatology 65 (1), 2071. 281.

Hegedus, A., Kavanagh Williamson, M., Khan, M.B., et al., 2017. Evidence for altered glutamine metabolism in human immunodeficiency virus Type 1 infected primary human CD4 + T cells. AIDS Res. Hum. Retroviruses. 33 (12), 1236–1247.

Hensley, C.T., Wasti, A.T., DeBerardinis, R.J., 2013. Glutamine and cancer: cell biology, physiology, and clinical opportunities. J. Clin. Invest. 123 (9), 3678–3684.

Hiscock, N., Pedersen, B.K., 2002. Exercise-induced immunodepression-plasma glutamine is not the link. J. Appl. Physiol. 93, 813–822.

Jaiswal, M., Shukla, P.C., Mishra, A., et al., 2019. Effect of glutamine and zinc containing rehydration electrolyte solution on serum protein and electrolytes in cattle calves having acute diarrhea. Int. J. Curr. Microbiol. App. Sci. 8 (2), 2165–2175.

Jiang, J., Srivastava, S., Zhang, J., 2019. Starve cancer cells of glutamine: break the spell or make a hungry monster? Cancers 11, 804. Available from: 10.3390/cancers11060804.

Katz, A., Broberg, S., Sahlin, K., et al., 1986. Muscle ammonia and amino acid metabolism during dynamic exercise in man. Clin. Physiol. 6, 365–379.

Kelley, M.A., Kazemi, H., 1974. Role of ammonia as a buffer in the central nervous system. Resp. Physiol. 22 (3), 345. 35.

Kelly, D., Wischmeyer, P.E., 2003. Role of L-glutamine in critical illness: new insights. Curr. Opin. Clin. Nutr. Metab. Care 6 (2), 217–222.

Khorshidi-Hosseini, M., Nakhostin-Roohi, B., 2013. Effect of glutamine and maltodextrin acute supplementation on anaerobic power. Asian J. Sports Med. 4, 131–136.

Kim, M.-H., Kim, H., 2017. The roles of glutamine in the intestine and its implication in intestinal diseases. Int. J. Mol. Sci. 18 (5), 1051.

Kingsbury, K., Kay, L., Hjelm, M., 1998. Contrasting plasma free amino acid patterns in elite athletes: association with fatigue and infection. Br. J. Sports Med. 32, 25–33.

Klassen, P., Mazariegos, M., Solomons, N.W., et al., 2000. The pharmacokinetic responses of humans to 20 g of alanyl-glutamine dipeptide differ with the dosing protocol but not with gastric acidity or in patients with acute dengue fever. J. Nutr. 130, 177–182.

Klek, S., Scislo, L., Walewska, E., et al., 2017. Enriched enteral nutrition may improve short-term survival in stage IV gastric cancer patients: a randomized, controlled trial. Nutrition 36, 46–53.

Korraa, A.A., Labib, H.A., Salah, D., 2016. Studying the effect of parenterally administered L-alanyl L-glutamine dipeptide in diabetes and new onset diabetes in liver transplantation. Egypt. J. Anaesth. 32 (3).

Krieger, J., Crowe, M., Blank, S., 2004. Chronic glutamine supplementation increases nasal but not salivary IgA during 9 days of interval training. J. Appl. Physiol. 97, 585–591.

Legault, Z., Bagnall, N., Kimmerly, D.S., 2015. The influence of oral L-glutamine supplementation on muscle strength recovery and soreness following unilateral knee extension eccentric exercise. Int. J. Sport. Nutr. Exerc. Metab. 25 (5), 417–426.

Leite, J.S.M., Raizel, R., Hypólito, T.M., et al., 2016. L-Glutamine and L-alanine supplementation increase glutamine-glutathione axis and muscle HSP-27 in rats trained using a progressive high-intensity resistance exercise. Appl. Physiol. Nutr. Metab. 41, 842–849.

Li, Y., Ping, X., Yu, B., et al., 2009. Clinical trial: prophylactic intravenous alanyl-glutamine reduces the severity of gastrointestinal toxicity induced by chemotherapy—a randomized crossover study. Aliment. Pharmacol. Ther. 30 (5), 452–458.

Lima, A.A., Carvalho, G.H., Figueiredo, A.A., et al., 2002. Effects of an alanyl-glutamine-based oral rehydration and nutrition therapy solution on electrolyte and water absorption in a rat model of secretory diarrhea induced by cholera toxin. Nutrition 18, 458–462.

Madeira, C., Alheira, F.V., Calcia, M.A., et al., 2018. Blood levels of glutamate and glutamine in recent onset and chronic Schizophrenia. Front. Psychiatr. 9, 713. Available from: 10.3389/fpsyt.2018.00713.

Mason, B.C., Lavallee, M.E., 2012. Emerging supplements in sports. Sports Health 4, 142–146.

McGivan, J.D., Bungard, C., 2007. The transport of glutamine into mammalian cells. Front. Biosci. 12, 874–882.

McRae, M.P., 2017. Therapeutic benefits of glutamine: an umbrella review of meta-analyses. Biomed. Rep. 6, 576–584.

Meneguello, M., Mendonça, J., Lancha Jr, A., et al., 2003. Effect of arginine, ornithine and citrulline supplementation upon performance and metabolism of trained rats. Cell Biochem. Funct. 21, 85–91.

Mok, E., Hankard, R., 2011. Glutamine supplementation in sick children: is it beneficial? J. Nutr. Metab. 41. Article ID 617597.

Mori, H., 2014. Effect of timing of protein and carbohydrate intake after resistance exercise on nitrogen balance in trained and untrained young men. J. Physiol. Anthropol. 33 (1), 24. Available from: 10.1186/1880-6805-33-24.

Mundi, M.S., Shah, M., Hurt, R.T., 2016. When is it appropriate to use glutamine in critical illness? Nutr. Clin. Pract. 31 (4), 445–450.

Naylor, J.M., Leibel, T., Middleton, T.M., 1997. Effect of glutamine or glycine containing oral electrolyte solutions on mucosal morphology, clinical and biochemical findings, in calves with viral induced diarrhea. Can. J. Vet. Res. 61 (1), 43–48.

Newsholme, E.A., Crabtree, B., Ardawi, M.S., 1985. Glutamine metabolism in lymphocytes: its biochemical, physiological and clinical importance. Q. J. Exp. Physiol. 70 (4), 473–489.

Oluma, Y., Bushen, J.A., Davenport, et al., 2004. Diarrhea and reduced levels of antiretroviral drugs: improvement with glutamine or alanyl-glutamine in a randomized controlled trial in northeast Brazil. Clin. Infect. Dis. 38 (12), 1764–1770.

Panigrahi, P., Gewolb, I.H., Bamford, P., et al., 1997. Role of glutamine in bacterial transcytosis and epithelial cell injury. J. Parenter. Enter. Nutr. 21 (2), 75–80.

Parry-Billings, M., Blomstrand, E., McAndrew, N., et al., 1990. A communicational link between skeletal muscle, brain, and cells of the immune system. Int. J. Sports Med. 11, S122–S128.

Petry, E.R., Cruzat, V.F., Heck, T.G., et al., 2014. Alanyl-glutamine and glutamine plus alanine supplements improve skeletal redox status in trained rats: involvement of heat shock protein pathways. Life Sci. 94 (2), 130–136.

Punjabi, N.H., Kumala, S., Rasidi, C., et al., 1991. Glutamine supplemented ORS is superior to standard citrate glucose ORS for the maintenance therapy of adult cholera patients in Jakarta (abstract) DAm. J. Trop. Med. Hyg. 45 (Suppl), 114–115.

Raizel, R., Leite, J.S.M., Hypólito, T.M., et al., 2016. Determination of the anti-inflammatory and cytoprotective effects of L-glutamine and l-alanine, or dipeptide, supplementation in rats submitted to resistance exercise. Br. J. Nutr. 116, 470–479.

Rao, R., 2009. Endotoxemia and gut barrier dysfunction in alcoholic liver disease. Hepatology. 50 (2), 638–644.

Rao, R.K., Samak, G., 2012. Role of glutamine in protection of intestinal epithelial tight junctions. J. Epithel. Biol. Pharmacol. 5 (Suppl 1-M7), 47–54.

Ren, W., Duan, J., Yin, J., et al., 2014. Dietary l-glutamine supplementation modulates microbial community and activates innate immunity in the mouse intestine. Amino Acids. 46, 2403–2413.

Rhoads, J.M., Keku, E.O., Bennett, L.E., et al., 1990. Development of L-glutamine-stimulated electroneutral sodium absorption in piglet jejunum. Am. J. Physiol. 259, G99–G107.

Rhoads, M.J., Keko, E.O., Quin, J., et al., 1991. L-glutamine stimulate jejunal sodium and chloride absorption in pig rotavirus enteritis. Gastroenterology 100, 683–691.

Ribeiro Jr, H., Ribeiro, T., Mattos, A., et al., 1994. Treatment of acute diarrhea with oral rehydration solutions containing glutamine. J. Am. Coll. Nutr. 13, 251–255.

Rogero, M.M., Tirapegui, J., Pedrosa, R.G., et al., 2004. Plasma and tissue glutamine response to acute and chronic supplementation with L-glutamine and L-alanyl-L-glutamine in rats. Nutr. Res. 24, 261–270.

Rubin, H., 2019. Deprivation of glutamine in cell culture reveals its potential for treating cancer. PNAS 116, 6964–6968.

Rubio, I., Suva, L.J., Todorova, V., et al., 2013. Oral glutamine reduces radiation morbidity in breast conservation surgery. J. Parenter. Enter. Nutr. 37 (5), 623–630.

Sewell, D., Gleeson, M., Blannin, A., 1994. Hyperammonemia in relation to high-intensity exercise duration in man. Eur. J. Appl. Physiol. 69, 350–354.

Shabert, J.K., Winslow, C., Lacey, J.M., et al., 1999. Glutamine-antioxidant supplementation increases body cell mass in AIDS patients with weight loss: a randomized, double-blind controlled trial. Nutrition 15 (11-12), 860–864.

Silva, A.C., Santos-Neto, M.S., Soares, A.M., et al., 1998. Efficacy of a glutamine-based oral rehydration solution on the electrolyte and water absorption in a rabbit model of secretory diarrhea induced by cholera toxin. J. Pediatr. Gastroenterol. Nutr. 26, 513–519.

Smriga, M., Kameishi, M., Torii, K., 2006. Exercise-dependent preference for a mixture of branched-chain amino acids and homeostatic control of brain serotonin in exercising rats. J. Nutr. 136, 548–552.

Talukder, J.R., Kekuda, R., Arthur, P.S.A., et al., 2008a. Identification and characterization of rabbit small intestinal villus cell brush border membrane Na-glutamine cotransporter. Am. J. Physiol. Gastrointest. Liver Physiol 295, G7–G15.

Talukder, J.R., Kekuda, R., Saha, P., et al., 2008b. Functional characterization, localization, and molecular identification of rabbit intestinal N-amino acid transporter. Am. J. Physiol. Gastrointest. Liver Physiol 294, G1301–G1310.

Tsai, P.H., Liu, J.J., Yeh, C.L., et al., 2012. Effects of glutamine supplementation on oxidative stress-related gene expression and antioxidant properties in rats with streptozotocin-induced type 2 diabetes. Br. J. Nutr. 107 (8), 1112–1118.

Turner, A.E., Rees, G., Barrett, D.C., et al., 2016. Does inclusion of glutamine in oral rehydration solutions improve recovery from mild to moderate diarrhea in preweaned calves? Vet. Rec. 179 (11), 283–284.

Vanek, V.W., Matarese, L.E., Robinson, M., Novel Nutrient Task Force, Parenteral Glutamine Workgroup, American Society for Parenteral and Enteral Nutrition (ASPEN) Board of Directors, et al., 2011. A.S.P.E.N. position paper: parenteral nutrition glutamine supplementation. Nutr. Clin. Pract. 26, 479–494.

van Zanten, A.R., 2015. Glutamine and antioxidants: status of their use in critical illness. Curr. Opin. Clin. Nutr. Metab. Care 18 (2), 179–186.

WHO, 2002. Drug. Inf. 16 (2), 121–122.

Wischmeyer, P.E., 2019a. Glutamine in burn Injury. Nutr. Clin. Pract. 34 (5), 681–687.

Wischmeyer, P.E., 2019b. The glutamine debate in surgery and critical care. Curr. Opin. Crit. Care 25 (4), 322–328.

Yaqoob, P., Calder, P., 1997. Glutamine requirement of proliferating T lymphocytes. Nutrition. 13 (7-8), 646–651.

Yoshida, S., Matsui, M., Shirouzu, Y., et al., 1998. Effects of glutamine supplements and radiochemotherapy on systemic immune and gut barrier function in patients with advanced esophageal cancer. Ann. Surg. 227 (4), 485–491.

Zhou, Q., Verne, M.L., Fields, J.Z., et al., 2019. Randomized placebo-controlled trial of dietary glutamine supplements for postinfectious irritable bowel syndrome. Gut 68 (6), 996–1002.

Chapter 62

What to do with placenta after birth: is it a waste?

Jamil Talukder

Introduction

The placenta is a temporary organ that develops in the uterus of mammals during pregnancy. The maternal side of the placenta is usually dark red, while the fetal side is shiny and almost translucent in color. The placenta attaches to the wall of the uterus and the fetal umbilical cord forms from it. The placenta of all eutherian mammals provides common functional features, but there are striking differences among species in the gross and microscopic structure of the placenta. The placenta consists of three layers. The innermost layer surrounding the fetus is called the amnion, the middle layer is called the allantois, and the outermost layer is the chorion. It comes into contact with the endometrium. Based on the fetal membrane, placentas are classified into two groups: (1) choriovitelline (yolk sac placenta) and (2) chorioallantoic. In most mammals, the yolk sac placenta becomes vestigial after the first trimester, except in rodents and rabbits. The mammalian placenta can be classified into four types based on the number of layers between maternal and fetal blood (Pelkonen et al., 2006): (1) epitheliochorial (swine, horse), (2) endotheliochorial (dog, cat), (3) hemochorial (human, primates, rat, rabbit, guinea pig), and (4) syndesmochorial (sheep, ruminants). The placentas reveal striking differences in their shape and the area of contact between fetal and maternal tissue. They are classified as follows (Fig. 62.1): (1) diffuse: almost the entire surface of the allantochorion is involved in the formation of the placenta, for example, horses and pigs; (2) cotyledonary (multicotyledonary): multiple, discrete areas of attachment called cotyledons are formed by interaction of patches of allantochorion with the innermost lining of the uterus (endometrium). The fetal portions of this type of placenta are called cotyledons, the maternal contact sites (caruncles), and the cotyledon–caruncle complex is called a placentome. This type of placenta is found in ruminants like cattle, goats, sheep, and deer. (3) Zonary: the placenta takes the form of a complete or incomplete band of tissue surrounding the fetus. This type of placenta is found in dogs, cats, ferrets, seals, bears, and elephants. (4) Discoid: a single placenta is formed and is discoid in shape. This type of placenta is found in humans, primates, and rodents.

The human placenta is dark maroon in color without any specific odor and has a discoid shape. There are approximately 15–25 cotyledon separations of the decidua basalis of the placenta, separated by placental septa. The placenta forms the interface between the microcirculatory systems of the mother and the fetus during gestation. The placenta protects the growing fetus, transports gases (oxygen and carbon dioxide), nutrients (water and lipid-soluble vitamins), minerals, electrolytes (sodium and potassium), and removes metabolic waste material from the fetus. In addition, many drugs and toxicants are transferred across the placenta by simple diffusion (Waddell and Marlowe, 1981; Welsch, 1982; Gupta, 2009; Gupta and Gupta, 2017). In contrast, simple carbohydrates and certain trace minerals (e.g., iron) are transported to the fetus by facilitated diffusion. Furthermore, amino acids, water-soluble vitamins (B complex, C), and other ions (e.g., calcium and phosphorus) are transported to the fetus by receptor-mediated and active transport mechanisms, respectively. Proteins are most readily transferred via pinocytosis or diffusion to the fetal circulation in those species with a scant placental barrier. The placenta produces some hormones, thus serving as an endocrine organ, and de novo synthesis of progesterone from cholesterol ensues throughout pregnancy. The placenta also synthesizes chorionic gonadotropins, estrogen (estriol), relaxin, lactogens, hypothalamic-like peptides, and opiates. The placenta produces estradiol just before the birth process.

After delivery, the umbilical corid and placenta are no longer needed by the fetus or the mother. Delivery of the

Nutraceuticals. DOI: https://doi.org/10.1016/B978-0-12-821038-3.00062-8

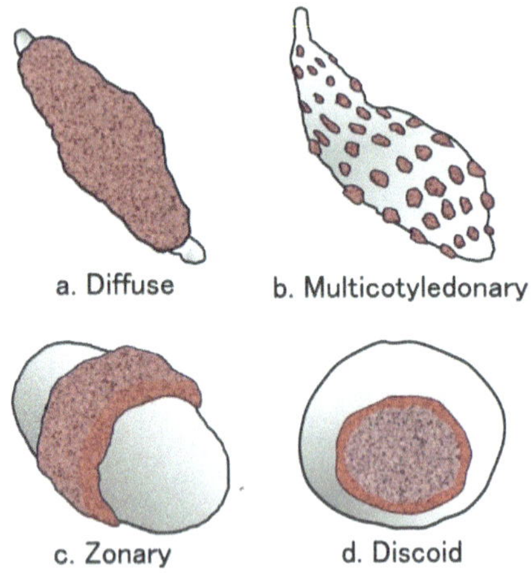

FIGURE 62.1 Types of placenta.

placenta is usually a quick process, often within 5 minutes of the birth process. Human umbilical cord and cord blood are discarded as medical waste in the hospital. Placenta, cord blood, and other birth tissues from animals are incinerated as medical waste. It should be noted, however, that the blood in the umbilical cord contains hematopoietic stem cells, and the immature blood-forming cells which are usually found in the bone marrow. These important cells are transformed into adult red or white blood cells and platelets. The umbilical cord and/or its blood offer a potential source of hematopoietic stem cells for transplant, known as cord blood transplant. Stem cells may aid in the treatment of people with certain life-threatening diseases, including some types of cancers. This is why some individuals may preserve their cord blood or placenta for future use. More than 25,000 transplants using umbilical cord blood have taken place worldwide. Cord blood can be preserved in a Cord Blood Bank for years for future use. In this regard, someone can reach out to the nonprofit National Marrow Donor Program, which has a nationwide bone marrow and cord blood registry, for a list of hospitals that work with cord banks. The banked tissue is strictly regulated for purity and safety and is FDA approved for wound healing.

Contents of placenta and associated structures

In general, the placenta is a complex structure formed by a heterogeneous population of cells (Huppertz et al., 2006; Griffiths and Campbell, 2015; Renaud, 2017). It contains many undifferentiated mesenchymal stem cells (MSCs) which are multipotent in nature (Fukuchi et al., 2004; In't Anker et al., 2004; Macias et al., 2010; Hass et al., 2011). The MSCs can be differentiated into specialized cells with specific functions, and investigators have successfully repaired bones, cartilage, and fat tissues using MSCs. Many phase I clinical trials are still ongoing and MSCs are being used for different conditions, including liver fibrosis, Parkinson's disease, rheumatoid arthritis, sports injuries (cartilage), gastrointestinal disease, respiratory, neurological disease, and injury, transplant complications, wounds, burns, and ulcers. In addition, some clinical trials with MSCs are in phase II trials, such as type I diabetes, diabetes mellitus, autoimmune and inflammatory disease, cancers, heart and vascular disease, and ocular surface disease.

The placenta is a rich source of protein or amino acids, vitamins, minerals, and hormones (Lewis et al., 2017). Eating the placenta is very common in the animal kingdom, known as "placentophagy", and most likely a way of eradicating the aroma of their newborn to protect them against predators. In addition, many people around the globe partake of the benefits of the placenta either raw or in an altered form such as dried, cooked, or enzymatic and chemical extractions. For humans, the most common way to consume the placenta is to have it made into capsules. In this regard, the placenta is steamed, dried, and then ground into a fine powder. The resulting capsules are taken during the postpartum period. Many celebrities in the United States have taken the capsules because they believe in the numerous benefits purportedly offered to the mother. It is believed that consumption of processed placenta prevents postpartum depression, reduces pain and postpartum bleeding, increases breast milk production, and improves mother–infant bonding. Furthermore, it is also purported to balance hormones, replenish depleted iron levels, assist the uterus in returning to its prepregnancy state (involution), reduce postnatal bleeding, and increase energy levels. However, to establish the scientific basis of consumption or utilization of placenta, more investigations are warranted.

Very recently, porcine placentas were collected aseptically from different farms in Wisconsin, United States, and processed under the good manufacturing practice (GMP) guidelines. Analysis of the processed placenta (analyzed at Vanderbilt University Medical Center, Nashville, Tennessee, USA; Analytical Services Core, supported by NIH grants DK059637 and DK020593) provided levels of various biomolecules. These are presented in Table 62.1.

Data from proximate analysis of processed placenta using GMP are presented in Table 62.2.

Applications of placental tissue as a nutraceutical

Beyond the physiological functions, the biochemical composition of the placenta and its associated structures (umbilical cord/blood) hold great promise for science and technology to benefit humans in health and disease conditions.

TABLE 62.1 Biomolecules in the porcine placenta

Ingredient	Name	Amount (ng/g)
Hormones	Growth hormone	846.0
	Cortisol	84.0
Amino acid/ protein	Name	Amount (mg/ g)
	Total protein	200
	Soluble protein	120
	Taurine	3.9
	Phosphoethanolamine	BDL
	Urea	2.6
	Aspartic acid	21.5
	Threonine	16.7
	Serine	15.7
	Asparagine	BDL
	Glutamic acid	36.7
	Glutamine	BDL
	Glycine	38.1
	Alanine	75.3
	Citrulline	7.4
	α-Amino-n-butyric acid	7.0
	Valine	35.8
	Cystine	0.5
	Methionine	6.9
	Cystathionine-1 and -2	0.3
	Isoleucine	14.8
	Leucine	41.2
	Tyrosine	2.2
	β-Alanine	3.0
	Phenylalanine	14.9
	α-Aminoisobutyric acid	BDL
	Homocysteine	0.07
	γ-Aminobutyric acid	7.0
	Ethanolamine	5.5
	Hydroxylysine-1 and -2	2.7
	Ornithine	11.9
	Lysine	18.0

(Continued)

TABLE 62.1 (Continued)

Ingredient	Name	Amount (ng/g)
	1-Methylhistidine	0.24
	Histidine	19.0
	Tryptophan	19.0
	3-Methylhistidine	0.06
	Anserine	0.19
	Carnosine	0.3
	Arginine	0.2
	Hydroxyproline	1.2
	Proline	23.0

*BDL, below detectable limits.

TABLE 62.2 Analytes in processed porcine placenta

Analyte	Amount
Moisture (%)	4.54
Ash (%)	4.75
Calcium (%)	0.74
Copper (ppm)	10.32
Fat (%)	4.86
Fiber (%)	0.09
Iron (ppm)	422.8
Magnesium (ppm)	898.8
Manganese (ppm)	8.66
Phosphorous (%)	0.8922
Potassium (%)	0.85
Protein (%)	72.9
Sodium (%)	2.17
Sulfur (ppm)	7057
Zinc (ppm)	57.59

Severe rheumatoid arthritis (RA) patients were treated with human placenta-eluted gamma globulins (PEGG) composed of purified IgC eluted from trophoblastic cells. A control group of RA patients received retroplacental gamma globulins without clinical effects. Seven of 11 treatments with PEGG showed good (four patients) or very good (three patients) clinical results. All patients had immunologic modifications demonstrated by a significant increase in active E-rosette-forming cells and mitogen responsiveness (Sany et al., 1982).

The placenta makes different hormones and growth factors (Gude et al., 2004; Lewis et al., 2017). Menzies et al. (2011) characterized pituitary GH, GH-R, IGF-2, PRL, and LHb in a macropodid marsupial, the tammar wallaby (*Macropus eugenii*). They expressed these genes in the tammar placenta during the last third of gestation when most fetal growth occurs and active organogenesis is initiated. In addition, the mRNA of key growth genes GH, GH-R, IGF-2, and PRL were expressed during late pregnancy. These authors found significant upregulation of GH, GH-R, and IGF-2 after the start of the rapid growth phase of organogenesis, which suggests that the placental growth hormones regulate the rapid phase of fetal growth. Our own GMP-processed porcine placenta showed the presence of GH (846 ng/g).

Lee et al (2016a,b) investigated the mechanisms underlying porcine placenta hydrolysates (PPH)-mediated inhibition of hydrogen peroxide (H_2O_2)-induced endoplasmic reticulum (ER) stress and reactive oxygen species (ROS) production. They found that PPHs also regulated osteoblast differentiation via the upregulation of alkaline phosphatase (ALP) expression in MC3T3-E1 osteoblastic cells. Treatment with PPHs also enhanced the transcription of osteocalcin, Runx2, and osterix. These effects were all associated with the antioxidant actions of PPHs. Moreover, PPHs reversed the decrease in superoxide dismutase (SOD) activity, decreased ROS release, and inhibited NADPH oxidase activity in H_2O_2-treated MC3T3-E1 osteoblastic cells. It was concluded that PPHs protect cells against H_2O_2-induced cell damage when ER stress is involved. In addition, PPHs enhance osteoblast differentiation. This enhancement likely explains the regulatory effect of PPHs on bone metabolism disturbances, that is, PPHs control ER stress and the related ROS production in osteoblasts.

Lee et al. (2016a,b) investigated the effect of PPH on calcium regulation in MC3T3-E1 osteoblastic cells. They observed that PPHs protect MC3T3-E1 osteoblastic cells against thapsigargin (Tg)-induced ER stress. Moreover, PPHs regulate caspase-12 and -3 activities, thereby protecting against cell death, and also regulating Tg-induced Ca^{2+} release. The Ca^{2+} chelator BAPT/AM also regulated caspase-12 and -3 activities and prevented Ca^{2+} stress-induced cell death. In the presence of PPHs or BAPT/AM [1,2-bis(o-aminophenoxy) ethane-*N,N,N′,N′*-tetraacetic acid, calcium-specific aminopolycarboxylic acid], Ca^{2+}-related ROS were also regulated, as demonstrated by alterations in NADPH oxidase and SOD activity. They concluded that PPHs regulate bone metabolism disturbances by controlling Ca^{2+} concentrations, ER stress, and ROS.

Antifatigue effect

Kim et al. (2016) evaluated the antifatigue effects of fermented porcine placenta (FPP) and its main constituents, lysine (Lys) and leucine (Leu), with treadmill stress test and forced swimming test (FST) in mice. The mice received FPP, Lys, and Leu for 21 days. They observed that FPP, Lys, and Leu significantly decreased fatigue-related parameters, including lactate, lactate dehydrogenase, glucose, creatine kinase, urea nitrogen, cortisol, and proinflammatory cytokines after treadmill exercise. In contrast, SOD activity and glycogen levels were significantly increased by FPP, Lys, and Leu. In the FST, FPP, Lys, and Leu significantly decreased immobility times after treadmill exercise. In addition, FPP, Lys, and Leu upregulated brain-derived neurotrophic factor expression in brain. Furthermore, FPP, Lys, and Leu significantly decreased production of tumor necrosis factor-α (TNF-α), interleukin (IL)-6, IL-1b, and IL-4 through blockade of caspase-1/nuclear factor-κB (NF-κB) pathway in stimulated splenocytes. In addition, FPP, Lys, and Leu significantly promoted proliferation of splenocytes. These findings suggest the potential of FPP as a novel functional food for the regulation of fatigue.

Jayabalan et al. (2017) observed the crosstalk between adipose tissue and placenta in obese and gestational diabetes mellitus pregnancies via exosomes. Their findings warrant further investigations.

The antiaging properties of human placental hydrolysate (HPE) and dieckol (DE) from *Ecklonia cava* against free radical scavenging, muscle hypertrophy-related follistatin mRNA expression, amelioration of cognition-related genes and proteins, inhibition of collagenase-regulating genes, and elastinase activity were examined by Jang et al. (2015). They used human fibroblast (CCD986sk), mouse myoblast (C2C12), and neuroblastoma (N2a) cell models, by using various assays such as 2,2-diphenyl-1-picrylhydrazyl hydrate (DPPH) scavenging, hydroxyl radical-mediated oxidation, quantitative real-time polymerase chain reaction, enzyme activity, and immunocytochemistry observation. They observed that HPE combined with DE (HPE:DE) strongly scavenged DPPH radicals and protected proteins against degradation by hydroxyl radical attack. HPE:DE effectively inhibited matrix metalloproteinase-1 expression, protein kinase C alpha expression, and elastinase activity. Furthermore, HPE:DE improved the expression of cognition-related genes (choline acetyltransferase and vesicular acetylcholine transporter). These events may proactively retard the aging process and the abrupt physiological changes induced by the mitochondrial dysfunction of aging. They concluded that the accumulation of oxidative damage is the main driving force and the combined treatment of HPE:DE would be a useful antiaging therapy.

Iron is a vital element for several metabolic pathways and physiological processes in the human body. It is one of the most important elements for sustaining life (Ponka and Lok, 1999) because of its ability to bind oxygen and its utilization by several proteins as a cofactor for major biological processes (Gkouvatsos et al., 2012). However, its capacity to generate highly reactive hydroxyl free radicals via the Fenton reaction can also render it toxic to cells. Humans use iron for many cellular processes including carrying oxygen into hemoglobin for cellular respiration without letting radical oxygen roam freely. However, an overload in the body leads to steatohepatitis. Yamauchi et al. (2019) reported that human placenta extract (HPE) ameliorated the pathology of nonalcoholic steatohepatitis (NASH) in a mouse model of NASH with hepatic iron deposition by combining a methionine- and choline-deficient (MCD) diet with an iron-overload diet. In this study, HPE treatment led to a marked reduction of hepatic iron deposition with a corresponding increase in biliary iron excretion. In addition, macrophage accumulation was greatly reduced by HPE treatment, as was the serum oxidation-reduction potential, an index of oxidative stress. These data indicated that by suppressing inflammation, oxidative stress, and iron deposition, and enhancing iron excretion, HPE effectively ameliorates iron overload-induced liver injury. HPE administration may therefore be an effective strategy for treating NASH.

Liver fibrosis results from sustained injury which can be inflicted by various factors such as viruses, drugs, alcohol, metabolic diseases, and autoimmune attacks (Eom et al., 2015). Prolonged exposure to these harmful factors causes hepatocyte apoptosis, inflammatory cell recruitment, endothelial cell impairment, and activation of hepatic stellate cells, the major cells involved in liver fibrosis. Liver fibrosis is a kind of scar tissue that is formed in response to liver damage (Domitrović and Jakovac, 2011; Snowdon and Fallowfield, 2011). Jung et al. (2011) investigated the therapeutic effects of HPE in a carbon tetrachloride (CCl_4)-injured rat liver model in vivo and in damaged rat hepatic cells exposed to CCl_4 in vitro. It was observed that serum levels of glutamic-oxaloacetic transaminase (GOT) or aspartate aminotransferase (AST), and glutamate-pyruvate transaminase (GPT) or alanine transaminase (ALT). were significantly reduced and uptake/excretion of indocyanine green in serum was significantly induced 3 weeks after intravenous HPE administration. Furthermore, expression of type I collagen (Col I) and α-smooth muscle actin (α-SMA) was decreased, whereas matrix metalloproteinase-9 (MMP-9) was increased, resulting in an improvement of score for fibrotic grade in the HPE group. Albumin, proliferation activities, and molecules associated with liver regeneration (e.g., interleukin-6, gp130, ATP-binding cassette transporters, and cyclin A) were also more increased in the HPE administration group than the non-HPE group. Moreover, HPE administration suppressed activated T-cell proliferation via increasing antiinflammatory cytokines and decreasing proinflammatory cytokines. These results suggested that HPE could be effective in the treatment of liver disease through reduction of fibrosis, induction of liver regeneration, and regulation of inflammatory responses. These findings are important for understanding the roles of HPE and provide evidence for the therapeutic effects of HPE in hepatic diseases which could lead to potential clinical applications.

MSCs have great potential for cell therapy in regenerative medicine, particularly for liver disease. It is not yet known whether MSCs can trigger autophagy to enhance regeneration and/or provide a therapeutic effect as cellular survival promoters. Jung et al. (2013) investigated autophagy activation in carbon tetrachloride (CCl_4)-injured rat liver following transplantation with chorionic plate-derived MSCs (CP-MSCs) isolated from human placenta.

Researchers noticed that caspase 3/7 activities were reduced, and that the expression levels of hypoxia-inducible factor-1α (HIF-1α) and factors for autophagy, survival, and regeneration were significantly increased by CP-MSC transplantation. In addition, they observed a decrease in necrotic cells and increased autophagic signals in CCl_4-treated primary rat hepatocytes during in vitro coculture with CP-MSCs. Furthermore, the upregulation of HIF-1α promoted the regeneration of damaged hepatic cells through an autophagic mechanism marked by increased levels of light-chain 3 II (LC 3II). These results suggested that the administration of CP-MSCs promotes repair by systemically concomitant mechanisms involving HIF-1α and autophagy. These findings provided further understanding of the mechanisms involved in these processes and will help develop new cell-based therapeutic strategies for regenerative medicine in liver disease.

It has been claimed that "the placental membrane" has over 50 different soluble growth factors, cytokines, and chemokines that are involved in the growth of hair follicles and the repair of aging or struggling blood vessels that supply the hair follicles. The growth factors stimulate dermal stem cells to proliferate and migrate to the area of injection. One important growth factor is the vascular-derived growth factor (VDGF). This is the factor responsible for the growth of new blood vessels. Increasing the blood supply to aging hair follicles brings oxygen and vital nutrition to them. Not only does this help prevent further hair loss, it also stimulates new and enriched hair growth. Similar procedures have been performed using a patient's own blood (platelet-rich plasma) but the results are not nearly as impressive as using the growth factor-rich placental extracts (https://southflderm.com/cosmetic-services/placenta-for-hair-growth/).

"Placenta Cream" was created to protect the skin from sun damage (specifically UV), signs of aging, and dry skin. One of its main ingredients is from cattle or sheep placenta. Many brands of placenta cream are made in Australia and New Zealand, where sheep placentas are collected and used (http://www.carevitaminsonline.com.au/sheep-placenta-cream-review-the-benefits-you-need-to-know).

Placenta extract is a "material" extracted from the placenta of certain mammals, including humans. Some professionals believe that placenta extract has several wellness benefits. Some studies have revealed that this extract could possibly relieve menopause signs. Some cosmetics manufacturers have included this extract in their hair and skin treatment products. Placenta extract can be provided in different forms, such as capsules to be taken orally. It can additionally be made into a solution that is either infused intravenously or dropped onto the tongue. Certain clinics in Japan offer intravenous drips of placenta extract. Some scientists believe that these drips could assist in a reduction of aging effects and provide greater energy. Some researchers also believe that this extract may cause several unfavorable responses.

Heo et al. (2018) investigated the antiinflammatory and antioxidative activities of porcine placenta extracts (PPE) against contact dermatitis in vivo. 2,4-Dinitrochlorobenzene (DNCB)-induced skin lesions displayed clear contact dermatitis-like symptoms, and topical application of PPE dramatically alleviated both local and systemic inflammatory responses. Inflammatory epidermal thickening was completely abrogated and allergen-specific serum IgE levels significantly reduced in the presence of PPE. Moreover, antioxidative activities of PPE were observed both in vitro and in vivo, which may lead to attenuation of inflammatory responses. Prolonged treatment with PPE strongly inhibited production of DNCB-induced ROS and subsequently prevented oxidative degradation of hyaluronic acid (HA), which triggers innate inflammatory responses. This study provided valuable insights into the mechanisms underlying the antiinflammatory effects of PPE and a functional basis for the clinical application of PPE in inflammatory diseases.

Kim et al. (2010) examined the preventive and therapeutic effects of placenta extract in contact hypersensitivity (CHS), a mouse model of allergic contact dermatitis. They found that administration of placental extract prior to the sensitization of allergic antigen (Ag) significantly inhibited the severity of CHS induced by Ag challenge. This effect was associated with reduced numbers of $CD4^+$ T cells in peripheral blood, decreased tissue-infiltrating lymphocytes, and preferential production of Th2-type cytokines in Ag-challenged sites. In addition, CHS caused by repetitive challenges of allergic Ag was also prevented and treated by administration of placental extract. Finally, administration of cyclo-trans-4-L-hydroxyprolyl-L-serine, a dipeptide derived from placental extract, also alleviated CHS, suggesting its potential role in CHS. Taken together, they demonstrated experimental evidence supporting immunoregulatory effects of placental extract in allergic skin diseases and elucidated its potential mechanisms.

Furthermore, more benefits of placenta extracts have been claimed such as "Placenta serum," a type of skin treatment product normally made from sheep placenta which is applied topically as a skin restorative. Serum made from placenta could commonly include other advantageous active ingredients, such as collagen, vitamin E, or green herbal tea extract. Producers generally claim that placenta serum can help slow and lessen the aging process, that it can enhance immunity, assist with basic health, and lead to a very high level of energy. Numerous cosmetic manufacturers also include placental extract in some of their products, such as placenta shampoo and conditioner. Because it consists of numerous supplements and proteins, it is believed that using placenta extract for hair care could make hair stronger and shinier. Placenta cream (http://www.carevitaminsonline.com.au/sheep-placenta-cream-review-the-benefits-you-need-to-know) is one of several placenta extracts being utilized in cosmetics. Producers claim that this sort of cream can help reduce wrinkles and make skin smoother and more supple. Studies warn, however, that several of these creams could induce major negative responses in some people, particularly women. Placenta extract can be extracted from different mammals, though human placenta extract is believed to be the most beneficial. Human placentas can be acquired from hospitals, or a new mother can select to keep her placenta after giving birth. Placenta cream is another example of placenta extract being used in cosmetics (https://www.wisegeek.com/what-is-placentaserum.html).

In addition, many products are available from a Japanese company (Japan Bio Products Co., Ltd), including "Grace Hai" which contains both porcine placental extract (PPE) and other essential nutrients designed to maintain healthy skin.

JBP also lists the essential ingredients of placenta including amino acids/proteins, carbohydrates, vitamins, nucleic acids, biologically active peptides, lipids and fatty acids, mucopolysaccharides, minerals, and enzymes. They also present the benefits of placenta extracts on autonomic nervous system control, endocrine hormone control, liver function enhancement and detoxification, improvement of basal metabolism, activation of the immune system, improvement of physical activity, anemia improvement, recovery of tiredness, blood circulation and hematogenesis effect, antioxidant action, and wound healing. JBP has a major product (LAENNEC P.O.) made with 100% human placental extract. It is designed for those who feel a physical change with age and would like to keep their youthfulness through at-home self-care.

Furthermore, the JPB internet link (http://jbpglobal.placenta.co.jp/product/laennec) mentions that Laennec is an ethical drug manufactured by JBP's unique technologies for effective extraction of a variety of growth factors, cytokines, and other physiologically active substances from the human placenta. For example, hepatocyte growth factor (HGF) promotes the proliferation of hepatic parenchymal cells for recovery of damaged liver. Their product safety claims to be ensured by the most rigid safety measures among existing scientific standards.

Concluding remarks and future directions

To explore the unknown medical potential of the placenta, samples (either placenta and/or its associated structures, umbilical cord, blood, etc.) collected from any sources should be free of disease and contamination. The highest possible level of safety is required in each step of handling, from collection to manufacture of the finished placental products. Over the last few years, many advancements have been made in exploring new components and/or uses of these organs to improve quality of life. Further investigations including safety, clinical trials, and validations are required.

References

Domitrović, R., Jakovac, H., 2011. Effects of standardized bilberry fruit extract (Mirtoselect®) on resolution of CCl4-induced liver fibrosis in mice. Food Chem Toxicol 49 (4), 848–854.

Eom, Y.W., Shim, K.Y., Baik, S.K., 2015. Mesenchymal stem cell therapy for liver fibrosis. Korean J Intern Med 30 (5), 580–589.

Fukuchi, Y., Nakajima, H., Sugiyama, D., et al., 2004. Human placenta-derived cells have mesenchymal stem/progenitor cell potential. Stem Cells. 22, 649–658.

Gkouvatsos, K., Papanikolaou, G., Pantopoulos, K., 2012. Regulation of iron transport and the role of transferrin. Biochim Biophys Acta 1820 (3), 188–202.

Griffiths, S.K., Campbell, J.P., 2015. Placental structure, function and drug transfer. Crit Care Pain. 15 (2), 84–89.

Gude, N.M., Roberts, C.T., Kalionis, B., et al., 2004. Growth and function of the normal human placenta. Thromb Res. 114, 397–407.

Gupta, R.C., 2009. Toxicology of the placenta. In: Ballanyne, B., Marrs, T.C., Syversen, T. (Eds.), General and Applied Toxicology, third ed. John Wiley & Sons, Chichester, UK, pp. 2003–2039.

Gupta, R.K., Gupta, R.C., 2017. Placental toxicology. In: Gupta, R.C. (Ed.), Reproductive and Developmental Toxicology, second ed. Academic Press/ Elsevier, Amsterdam, pp. 1301–1325.

Hass, R., Kasper, C., Böhm, S., et al., 2011. Different populations and sources of human mesenchymal stem cells (MSC): a comparison of adult and neonatal tissue-derived MSC. Cell Comm Signal 9, 12.

Heo, J.H., Heo, Y., Lee, H.J., et al., 2018. Topical anti-inflammatory and anti-oxidative effects of porcine placenta extracts on 2,4-dinitrochlorobenzene-induced contact dermatitis. BMC Complement Alternat Med. 18 (1), 331. Available from: 10.1186/s12906-018-2396-1.

Huppertz, B., Burton, G., Cross, J.C., et al., 2006. Placental morphology: from molecule to mother—a dedication to Peter Kufman. A review. Placenta 27 (Suppl. A), S3–S8.

In't Anker, Pieternella, et al., 2004. Isolation of mesenchymal stem cells of fetal or maternal origin from human placenta. Stem cells (Dayton, Ohio), 22 (4), 1338–1345. In this issue. Available from: https://doi.org/10.1634/stemcells.2004-0058.

In 't Anker, Scherjon, S. A., Kleijburg-van der Keur, C., de Groot-Swings, Claas, F. H., Fibbe, W. E., et al., 2004. Isolation of mesenchymal stem cells of fetal or maternal origin from human placenta. Stem cells (Dayton. Ohio) 22 (7), 1338–1345.

Jang, S.K., Lee, D.I., Kim, S.T., et al., 2015. The anti-aging properties of a human placental hydrolysate combined with dieckol isolated from *Ecklonia cava*. BMC Complement Alternat Med. 15, 345. Available from: https://doi.org/10.1186/s12906-015-0876-0.

Jayabalan, N., Nair, S., Nuzhat, Z., et al., 2017. Cross talk between adipose tissue and placenta in obese and gestational diabetes mellitus pregnancies via exosomes. Front Endocrinol. 8, 239. Available from: https://doi.org/10.3389/fendo.2017.00239.

Jung, J., Lee, H.J., Lee, J.M., et al., 2011. Placenta extract promote liver regeneration in CCl4-injured liver rat model. Int Immunopharmacol. 11 (8), 976–984. Available from: https://doi.org/10.1016/j.intimp.2011.02.012.

Jung, J., Choi, J.H., Lee, Y., et al., 2013. Human placenta-derived mesenchymal stem cells promote hepatic regeneration in CCl4 -injured rat liver model via increased autophagic mechanism. Stem Cells. 31 (8), 1584–1596. Available from: https://doi.org/10.1002/stem.1396.

Kim, Y.S., Park, J.J., Sakoda, Y., et al., 2010. Preventive and therapeutic potential of placental extract in contact hypersensitivity. Int Immunopharmacol. 10 (10), 1177–1184. Available from: https://doi.org/10.1016/j.intimp.2010.06.024.

Kim, H.Y., Han, N.R., Kim, N.R., et al., 2016. Effect of fermented porcine placenta on physical fatigue in mice. Exp Biol Med (Maywood) 241 (17), 1985–1996.

Lee, H.Y., Chae, H.J., Park, S.Y., et al., 2016a. Porcine placenta hydrolysates enhance osteoblast differentiation through their antioxidant activity and effects on ER stress. BMC Complement Alternat Med 16 (1), 291. Available from: https://doi.org/10.1186/s12906-016-1274-y.

Lee, H.Y., Kim, H.R., Park, S.Y., et al., 2016b. Porcine placenta hydrolysates regulate calcium disturbance in MC3T3-E1 osteoblastic cells. BMC Complement Altern Med 16, . Available from: https://doi.org/10.1186/s12906-016-1202-1Article No. 237.

Lewis, R.M., Cleal, J.K., Godfrey, K.M., 2017. The placental role in developmental programming. In: Gupta, R.C. (Ed.), Reproductive and Developmental Toxicology, second ed. Academic Press/Elsevier, Amsterdam, pp. 1245–1258.

Macias, M.I., Grande, J., Moreno, A., et al., 2010. Isolation and characterization of true mesenchymal stem cells derived from human term decidua capable of multilineage differentiation into all 3 embryonic layers. Am J Obstet Gynecol 203 (5), e9-495.e23.

Menzies, B.R., Pask, A.J., Renfree, M.B., 2011. Placental expression of pituitary hormones is an ancestral feature of therian mammals. Evodevo. 2, 16. Available from: https://doi.org/10.1186/2041-9139-2-16.

Pelkonen, O., Vähäkangas, K., Gupta, R.C., 2006. Placental toxicity of organophosphate and carbamate pesticides. In: Gupta, R.C. (Ed.), Toxicology of Organophosphate and Carbamate Compounds. Academic Press/Elsevier, Amsterdam, pp. 463–479.

Ponka, P., Lok, C.N., 1999. The transferrin receptor: role in health and disease. Int J Biochem Cell Biol 31 (10), 1111–1137.

Renaud, S.J., 2017. Strategies for investigating hemochorial placentation. In: Gupta, R.C. (Ed.), Reproductive and Developmental Toxicology, second ed. Academic Press/Elsevier, Amsterdam, pp. 1259–1273.

Sany, J., Clot, J., Bonneau, M., Andary, M., 1982. Immunomodulating effect of human placenta-eluted gamma globulins in rheumatoid arthritis. Arthritis Rheum. 25 (1), 17–24.

Snowdon, V.K., Fallowfield, J.A., 2011. Models and mechanisms of fibrosis resolution. Alcoholism: Clin Exp Res 35 (5), 794–799.

Waddell, W.J., Marlowe, C., 1981. Transfer of drugs across the placenta. Pharmacol Ther. 14, 374.

Welsch, F., 1982. Placental transfer and fetal uptake of drugs. J Vet Pharmacol Ther 5, 91–104.

Yamauchi, A., Kamiyoshi, A., Sakurai, T., et al., 2019. Development of a mouse iron overload-induced liver injury model and evaluation of the beneficial effects of placenta extract on iron metabolism. Heliyon 5 (5), e01637. Available from: https://doi.org/10.1016/j.heliyon.2019.e01637.

Chapter 63

Chitin, chitosan, and their derivatives

Ramesh C. Gupta, Robin B. Doss, Rajiv Lall, Ajay Srivastava and Anita Sinha

Introduction

Chitin was extracted for the first time in 1811 from mushroom by a French chemist and pharmacist Henri Braconnot, then Professor of Natural History in France (Labrude and Becq, 2003). The word "chitin" is derived from the Greek word "XITών," which means "tunic" or "envelope." Chitin is a homopolymer of β1→4-linked *N*-acetyl *D*-glucosamine residue, and chitosan is a linear polysaccharide composed of randomly distributed β-(1→4)-linked *D*-glucosamine (deacetylated unit) and *N*-acetyl-*D*-glucosamine (acetylated unit). Chitosan is processed (deacetylated) from chitin, which is usually obtained from the rough exoskeleton of shellfish, including lobster, crab, crayfish, and shrimp. Chitin can also be obtained from yeasts. Chitin and chitosan have broad applications in bioadhesives (Ryu et al., 2015), improved drug-delivery systems (Baldrick, 2010; Iglesias et al., 2019), development of nanomaterials (Agnihotri et al., 2004; Shukla et al., 2013), enteric coatings (Aiedeh et al., 2005), medical devices (Elsabee and Abdou, 2013; Shukla et al., 2013; Ryu et al., 2015), and the food industry (Shahidi et al., 1999; Gómez-Rivas et al., 2004; Raafat and Sahl, 2009).

For more than three decades, chitosan has been used in water purification plants for its unique ability to bind to toxic substances. In recent years, new forms of chemically modified chitosan have been developed in order to improve the properties of chitosan for various biological activities, and these substances have gained increasing attention. Representative members of these novel polymers include ammonium chitosan, carboxymethyl chitosan, and derivatives (Dai et al., 2011). Chitosan is a bioactive cationic polysaccharide with antimicrobial, antioxidative, and antiinflammatory properties. It is commonly used for diabetes, obesity, hypercholesterolemia, hypertension, dry eye disease (DED), wound healing, and cancer (Goosen, 1996). Future investigations will discover newer derivatives with novel applications in the biomedical, pharmaceutical, and engineering fields. This chapter describes the physicochemical properties, pharmacokinetics, applications of chitin, chitosan and chitosan derivatives, and their safety and toxicity.

Sources and physicochemical properties of chitin and chitosan

The main industrial source of (α)-chitin is the shell (exoskeleton) of crabs or shrimps, while that of (β)-chitin is the squid pen. γ-Chitin is a combination of α- and β-chitins. Chitin can also be extracted from cultivated commercially available edible mushrooms (such as *Agaricus bisporus*), and the radulas of mollusks. Chemically, chitin is a homopolymer of β1→4-linked *N*-acetyl *D*-glucosamine residue. The degree of deacetylation (DDA) of chitosan from different sources such as the shells of fish, shrimp, and crab are reported to be 75%, 78%, and 70%, respectively (Kumari et al., 2017). The physicochemical properties of chitin, chitosan, and chitosan derivatives are well documented (Dutta et al., 2004;

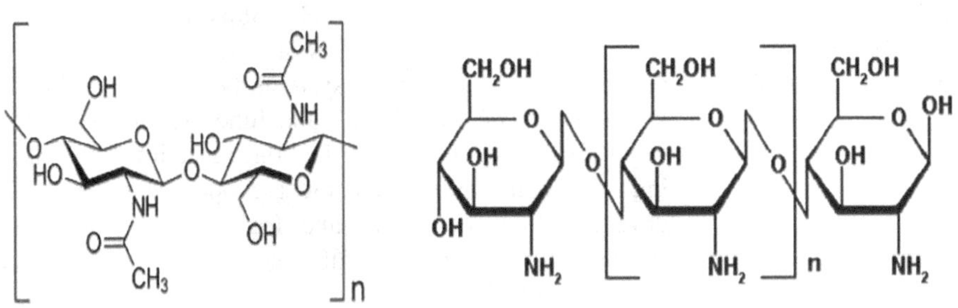

FIGURE 63.1 Structural formulas of chitin (left) and chitosan (right).

Nutraceuticals. DOI: https://doi.org/10.1016/B978-0-12-821038-3.00063-X

Rampino, 2011; Castro and Paulín, 2012; Bellich et al., 2016; Kumari et al., 2017; Divya and Jisha, 2018). Chitosan is a linear polysaccharide composed of randomly distributed β-(1→4)-linked *D*-glucosamine (deacetylated unit) and *N*-acetyl-*D*-glucosamine (acetylated unit). It is a deacetylated derivative of chitin (class α, β, and γ). Chitosan containing 100% glucosamine has a 100% DDA, whereas chitosans with 50%–85% DDA are partly acetylated (Hoemann and Fong, 2017). The structural formulas of chitin and chitosan are shown in Fig. 63.1.

The solubility of chitosan(s), that is, the entire family of partially deacetylated chitins with average DDA $\leq 50\%$, depends on the molar mass, and number and distribution of acetylation sites. The molecular weight (MW) of commercially available chitosan is approximately 50,000–190,000 daltons based on viscosity. Chitosan is a polymer with a number of basic amino groups and hence it possesses an overall cationic charge, especially at acidic pH. This is due to the presence of primary amines on the molecule that bind protons according to the following equation:

$$\text{Chit - NH}_2 + \text{H}_3\text{O}^+ \text{«} \text{ Chit -NH}_2{}^+ + \text{H}_2\text{O}$$

The amino group in chitosan has a pK_a value of ~6.5, which leads to significant protonation in neutral solutions, increasing with higher acidity. This makes chitosan water-soluble and bioadhesive. In many studies, the physicochemical properties of chitosan have been modulated to: (1) improve water solubility and blending ability with anionic biopolymers, enhancing and manipulating the rheological, hydration, and compatibility properties of the new materials, and (2) increase hydrophobic or emulsifying properties (reviewed in Bellich et al., 2016). Chitosan derivatives were also modulated to enhance antimicrobial activity (Muzzarelli et al., 1990; Ravi Kumar, 2000, Qin et al., 2006; Jeon et al., 2014).

Chitosan is only soluble in an aqueous acidic solution below pH 6.5 in which the primary amino groups are protonated. Therefore, the extended applications of chitosan are limited by its insoluble nature in a biological solution (pH 7.4). Carboxymethyl chitosan (CM-chitosan) is one of the water-soluble chitosan (WSC) derivatives and a versatile functional biomaterial that has many attractive physical and biological properties, such as moisture retention, gel-forming capability, and antibacterial activity (Dong et al., 2010). The MW and viscosity development in aqueous solution play a significant role in the biomedical and pharmacological/pharmaceutical applications of chitosan. Chemically modified chitosan and its derivatives have varying solubility, biological activity, and degradation rate.

Pharmacokinetics

In a rat model, Dong et al. (2010) investigated the tissue distribution, pharmacokinetics, and excretion of the fluorescein isothiocyanate (FITC)-labeled CM-chitosan (a water-soluble derivative of chitin with an MW ~ 300 kDa). Following intraperitoneal administration of FITC-CM-chitosan (50 mg/kg), it was quickly and efficiently absorbed by the abdominal cavity. However, FITC was absorbed more rapidly than FITC-CM-chitosan. FITC-CM-chitosan was widely and rapidly distributed into the blood and many tissues (including the liver, spleen, and kidney). The liver appears to play a central role in the biodegradation of CM-chitosan. Both FITC and FITC-CM-chitosan displayed similar elimination trends. Findings revealed that 85% of the administered dose was excreted over an 11-day period. Onishi and Machida (1999) investigated the body distribution and urinary excretion of FITC-chitosan at 1, 14, and 24 h after IP injection to mice. The findings revealed that FITC-chitosan moved quickly to the kidney and urine, and was scarcely distributed in the liver, spleen, abdomen, and plasma. Most of the FITC-chitosan was excreted into urine after 14 h, and the MW was low, suggesting chitosan to be highly biodegradable.

Chitosan is a natural polysaccharide that is cleared by enzymatic hydrolysis (Hirano et al., 1989; Sashiwa et al., 1990). The degradation of protonated chitosan is mediated largely by lysozymes in the body to glucosamine and the conjugate base of the acid, such as lactate and succinate (Muzzarelli et al., 1988). It is dependent on the DDA (Nishimura et al., 1985) and most likely MW and concentration as well (Zaharoff et al., 2007).

Pharmacokinetic studies on chitosan and its derivatives are limited due to wide variations in their source, purity, DDA, and intended applications.

Applications of chitosan and chitosan derivatives

Chitosan and its derivatives have a wide range of biological, biomedical, and pharmaceutical applications, such as drug-delivery vehicles, tissue engineering, wound healing, and enzyme immobilization for biosensing (Illum, 1998; Dash et al., 2011; Castro and Paulín, 2012; Cheung et al., 2015; Bellich et al., 2016; Hamed et al., 2016; Vunain et al., 2017; Liaqat and Eltem, 2018). Chitosan has become especially popular in the field of nanobiotechnology (Rampino et al., 2013; Shukla et al., 2013; Rajalakshmi et al., 2014; Divya and Jisha, 2018). It is employed in drug-delivery systems because of its biodegradability, biocompatibility, and functional moieties (N-H, O-H) for chemical modification, in addition to tailoring of specific physicochemical properties essential in drug targeting (Fernandez and Ingber, 2014; Bose and Wong, 2018). Chitosan is also commonly used in food science and technology.

Chitosan, a polycationic polymer is comprised of different functional groups that can be modified with a range of ligands (Vunain et al., 2017). The quaternized chitosan derivative is known to exhibit improved water solubility and stronger antibacterial activity over a broad pH range relative to chitosan. Quaternized chitosan is prepared by introducing a quaternary ammonium group on a dissociative hydroxyl group or amino group of the chitosan. Some of the applications of chitosan and its derivatives are described below.

Antioxidative activity

Oxidative stress adversely affects human and animal health and performance. In addition, it spoils food quality and causes discoloration. Food/feed additives with antioxidant abilities are known to alleviate oxidative stress. In a number of studies, chitosan and its derivatives (of different MW, DDA, and the nanofibers) have been reported to exert antioxidant activity against various types of oxidative stressors (Xie et al., 2001; Shon et al., 2002; Xu et al., 2010; Ngo and Kim, 2014; Kerch, 2015; Laokuldilok et al., 2017; Anraku et al., 2018). Chitosan is reported to scavenge free radicals including hydroxyl radicals, and chelate metal ions including ferrous ions (Yen et al., 2008; Lin et al., 2009), thereby exerting antioxidative activity (Rajalakshmi et al., 2013; Divya and Jisha, 2018). With regard to antioxidant properties, the effectiveness of chitosans C60, C90, and C120 correlated with their *N*-deacetylation times, and chelating abilities on ferrous ions were 82.9%–96.5% (Yen et al., 2008).

Recently, Lan et al. (2020) investigated the antioxidative capacity of dietary supplement chitosan oligosaccharides (COS) in Sprague-Dawley rats challenged with hydrogen peroxide (H_2O_2). The results revealed that COS provided better radical-scavenging capacity of 1,1-diphenyl-2-picrylhydrazyl (DPPH), superoxide anion (O_2^-), H_2O_2, and ferric ion-reducing antioxidant power than butylated hydroxy anisole. COS increased the activity of superoxide dismutase (SOD), catalase, glutathione peroxidase, glutathione, and total antioxidant capacity, as well as decreasing the malondialdehyde level in serum, liver, spleen, and kidney. The study concluded that COS can protect Sprague-Dawley rats from H_2O_2 challenge by reducing lipid peroxidation and restoring antioxidant capacity.

In an in vitro study, Anraku et al. (2018) reported that low-MW chitosan with high DDA has more potent antioxidant properties than those of high-MW chitosan. On the other hand, high- MW chitosan has higher adsorption properties than those of low-MW chitosan. The study suggested that the ingestion of chitosan and its nanofiber derivative, with moderate MW and DDA, results in a significant reduction in oxidative stress in several chronic oxidative stress-related diseases such as metabolic syndrome and renal failure. It needs to be pointed out that the MW and deacetylation ratio of COS also exert some synergistic effects on the biological capacities. In general, low MW provides stronger scavenging activity of DPPH, H_2O_2, and O_2^- (Anraku et al., 2018).

In a number of studies, chitosan and its derivatives have also been used in the prevention and treatment of oxidative stress in age-related diseases (Xie et al., 2001; Chen et al., 2003; Jeon et al., 2003; Feng et al., 2007). In addition, because of their antioxidant properties, chitosan and its derivatives are commonly used to protect food from spoilage and discoloration (Wang, 1992; Shahidi et al., 1999; Gómez-Rivas et al., 2004).

Antiinflammatory activity

Inflammation is an underlying complex mechanism in many chronic diseases and disorders, including diabetes, Alzheimer's, Parkinson's, cardiac, respiratory, renal, hepatic, and cancer. Oxidative stress and excessive production of reactive oxygen species (ROS) are associated with inflammation, leading to the synthesis and release of proinflammatory cytokines (Lan et al., 2020). Currently, there are many natural products that have antiinflammatory activities. Among their many biological properties, chitosan and its derivatives exert an antiinflammatory effect, and the pathways involved are moderately understood. In a recent study, Lan et al. (2020) reported that supplementation of rats with COS showed that the levels of TNF-α, IL-1β, and IL-6 were comparatively lower than in H_2O_2-challenged rats, and the level of IL-10 was higher, which may be due to the antiinflammatory activity of COS. There are many studies in which chitosan and its derivatives have been shown to exhibit an antiinflammatory effect (Yang et al., 2010; Chung et al., 2012; Zhang et al., 2014; Azuma et al., 2015).

Immunostimulatory/adjuvant activity

Adjuvants are compounds that stimulate the immune system and increase the response to a vaccine, without having any specific antigenic effect of their own (Li et al., 2013). Immunostimulatory adjuvants affect the immune system, thereby increasing the immune response to antigens (Schijns, 2003). The development of safe and novel adjuvants is necessary to maximize the efficacy of new and/or available vaccines (Zaharoff et al., 2007). In a number of studies, chitosan and its derivatives have been found to be safe and effective adjuvants due to their unique properties of biodegradability, biocompatibility, immune-enhancing activity, and high viscosity for vaccination via subcutaneous, intranasal, or other parenteral routes (Illum et al., 2001; Seferian and Martinez, 2001;

van der Lubben et al., 2001; Stoker et al., 2008; Arca et al., 2009; Li et al., 2013; Smith et al., 2013; Choi et al., 2016). Due to their mucoadhesive properties, chitin and chitosan are widely applied for mucosal routes (oral, nasal, and ocular) of administration. Nishimura et al. (1985) proved the immunological activity of chitin derivatives and compared their effect with other adjuvants (Nishimura et al., 1984, 1986). Seferian and Martinez (2001) found that chitosan containing two adjuvant formulations (a zinc-chitosan particle formulation and an emulsion formulation containing chitosan) stimulated immune activity by activating both B and T lymphocytes in mice and guinea pigs for antigen-specific delayed-type hypersensitivity. Zaharoff et al. (2007) reported that chitosan enhanced both humoral and cell-mediated immune responses following subcutaneous vaccination. Chitosan significantly elevated antigen-specific antibody titers (IgG_1, 5.9-fold; and IgG_{2a}, 8.0-fold) implying a mixed T_H1/T_H2 response, and enhanced antigen-specific splenic $CD4^+$ proliferation >6-fold. Mice vaccinated with β-galactosidase formulated with chitosan had an increase of 116% in ear thickness, indicating a robust cell-mediated immune response. Chitosan exhibited two characteristics of the enhanced immune response. First, subcutaneous injections of chitosan solution led to a 67% cellular expansion in lymph nodes and stimulated natural killer (NK) cells and macrophages. Second, chitosan's ability to form an antigen depot, thereby generating an adaptive immune response (reviewed in Zaharoff et al., 2007).

The mechanisms of vaccine enhancement by chitosan through mucosal administration are believed to be due to both retention of vaccine in the nasal passages via mucoadhesion and opening of endothelial cell junctions for paracellular transport of vaccine (Illum et al., 2001; Li et al., 2013). Clinical studies suggest that chitosan is a promising adjuvant for intranasal vaccination (Read et al., 2005). Archimedes Development Ltd. has developed a novel delivery technology based on chitosan (ChiSys), which has been shown to enhance the intranasal delivery of conventional drugs as well as proteins, peptides, and vaccine antigens.

Tokura et al. (1999) reported that macrophages were activated to various extents by chitin derivatives, and 30% deacetylated chitin and chitin sulfate stimulated the production of circulating antibodies. Accumulation of carboxymethyl chitin takes place in granulocytes and macrophages. Chitin has complex and size-dependent effects on innate and adaptive immune responses, which include the ability to recruit and activate innate immune cells and induce cytokine and chemokine production (Lee et al., 2008). Hoemann and Fong (2017) reported that MW and DDA strongly influence innate immune responses, however, the physical structure of a chitosan implant can also dramatically alter the immunological response. In an in vitro study, macrophages that phagocytose chitosan were shown to secrete increased levels of chemokines, anabolic cytokines, antiinflammatory factors including IL-10 and IL-1 receptor antagonist (IL-1ra), and to undergo inflammasome activation accompanied by IL-β release (Hoemann and Fong, 2017). However, no study has supported that chitin, chitosan, or chitosan derivatives can enhance or attenuate the release of inflammatory mediators (such as TNF-α and NO). In all these studies, chitosan displayed positive adjuvant properties with a potential for clinical benefit.

Antimicrobial activity

Chitosan and its derivatives are known to exert bacteriostatic and bactericidal activities against Gram-negative and Gram-positive bacteria (Choi et al., 2001; Goy et al., 2009; Raafat and Sahl, 2009; Kong et al., 2010; Tang et al., 2010; Dai et al., 2011). Chitosan is widely used as an antimicrobial agent, alone or in combination with other natural polymers, due to its high biodegradability, and antimicrobial and nontoxic properties (Kong et al., 2010). Chitosan and its derivatives are also reported to exert antiviral (Makita-Chingombe et al., 2016; Tachaboonkiat, 2017) and antifungal (Kendra and Hadwiser, 1984; Tikhonov et al., 2006; Seyfarth et al., 2008) activities.

Raafat and Sahl (2009) suggested that the molecular mechanisms of action of chitosan in inhibiting or killing bacteria is definitely a complex process, which involves a series of interrelated events. Chitosan and its derivatives are reported to kill bacteria by disrupting the outer and inner membranes (Helander et al., 2001; Rabea et al., 2003; Zivanovic et al., 2004; Liu et al., 2004; Je and Kim, 2006; Tachaboonkiat, 2017). The electrostatic interaction between the polycationic nature of chitosan and the negatively charged microbial surface emerges as a fundamental feature of the killing potential. Of course, different organisms expose different molecular patterns and respond to chitosan differently. Chitosan permeates the bacterial cell wall and interacts with nucleic acid, thus interfering with protein synthesis. Alternatively, chitosan may deposit on the cell surface and could lead to a blockage of nutrients, reducing microbial growth. In addition, chitosan can mediate antimicrobial activity by chelating with various metal ions, such as Ni^{2+}, Zn^{2+}, Co^{2+}, Fe^{2+}, Mg^{2+}, and Cu^{2+} (Rabea et al., 2003).

Kong et al. (2010) reported that various factors can influence the bactericidal efficacy of chitosan. These factors are classified into four categories: (1) microbial factors, related to microorganism species and cell age; (2) intrinsic factors of chitosan, including MW, positive charge density, concentration, hydrophilic/hydrophobic characteristics, and chelating capacity; (3) physical state, namely water-soluble and solid state of chitosan; and (4) environmental factors, involving ionic strength in medium, pH, temperature, and reaction time. MW and DDA appear to greatly influence chitosan's antimicrobial activity (Tokura et al., 1997; No et al., 2002; Zivanovic et al., 2004; Tachaboonkiat, 2017). Tachaboonkiat (2017) stressed that the critical MW for bacterial growth inhibition by chitosan has been shown to be above 10 kDa against Gram-positive and Gram-negative bacteria. With increasing MW of chitosan, antibacterial activity was increased, especially against *Escherichia coli*. However, increasing MW of chitosan over 100 kDa has been shown to reduce antibacterial activity against *E. coli*, while this trend was not observed against Gram-positive bacteria. Interestingly, Tokura et al. (1997) demonstrated that 9.3 kDa chitosan inhibits the growth of *E. coli* while 2.2 kDa chitosan promotes its growth. Chitosan with higher DDA (higher in positive charges) is expected to have a stronger inhibitory effect against both Gram-positive and Gram-negative bacteria (Takahashi et al., 2008; Tachaboonkiat, 2017). Takahashi et al. (2008) demonstrated that the growth of Gram-positive *Staphylococcus aureus* was more strongly inhibited by chitosan than the Gram-negative *E. coli*. However, Goy et al. (2009) remained uncertain whether Gram-positive bacteria were more or less sensitive than Gram-negative bacteria to chitosan. Antimicrobial activity of chitin and chitosan can be further enhanced by derivatization (quaternization) with quaternary ammonium groups.

Chitosan has emerged as a material for biodental applications due to its unique properties such as antimicrobial activity, bioactivity, and biocompatibility (Husain et al., 2017; Gupta et al., 2019a). It is directly applied to gums to treat inflammation and to prevent dental caries. Chitosan is known to exert antimicrobial activity against *Porphyromonas gingivalis*, especially with a high-MW formulation. Chitosan has been proven to be an effective antibacterial and antifungal agent (Uraz et al., 2012; Costa et al., 2014). Inkinci et al. (2002) reported that when chitosan was given in combination with chlorhexidine gluconate, it showed a greater activity compared to chlorhexidine alone, and the dose of chlorhexidine needed was lower. Costa et al. (2014) demonstrated that mouthwash containing chitosan was capable of interfering with all microorganisms' adherence and biofilms with superior activity to other commercially available mouthwashes. Chitosan, by having antimicrobial, antifungal, and adhesive properties, appears to be ideal for local delivery systems for the prevention or therapy of periodontal diseases.

Due to selective antimicrobial action and against spoilage yeasts, chitosan has broad applications in food preservation (Shahidi et al., 1999; Gómez-Rivas et al., 2004). Wang (1992) reported inhibition and inactivation of many foodborne pathogens (such as *Salmonella typhimurium*, *Listeria monocytogenes*, *Yersinia enterocolitica*, *S. aureus*, and *E. coli*) by chitosan.

Antiobesity activity

The prevalence of obesity is increasing globally (World Health Organization, 2015). Several natural products have been used in the management of obesity and weight loss. Chitosan is a great source of dietary insoluble fiber that is commonly used to reduce dietary fat absorption to control body weight and lipids. Chitosan fiber can bind to fat up to six times its weight. The mechanism of interaction between chitosan and fat is not well understood. It has also not been clinically proven that chitosan is an effective complement to help lose weight or to stabilize body weight (Gades and Stern, 2005: Kaats et al., 2006).

Some experimental studies suggest that chitosan exerts an antiobesity effect (Sumiyoshi and Kimura, 2006; Huang et al., 2015). Jull et al. (2008) reported that there is some evidence that chitosan is more effective than placebo in the short-term treatment of overweight individuals and obesity. Trivedi et al. (2016) demonstrated that administration of chitosan (KiOnutrime-CsG capsules, 500 mg, five capsules/day in three divided doses) resulted in a significant mean weight loss of about 3 kg without diet restriction over a period of 90 days. Many trials to date have been of poor quality and the results have been variable. Results from high-quality trials indicate that the effect of chitosan on body weight is minimal and unlikely to be of clinical significance. In a recent metaanalysis, based on clinical trials, Moraru et al. (2018) concluded that the usage of chitosan as a dietary supplement can lead to a short-term effect on body weight loss and to the improvement of serum lipid profile and cardiovascular factors. Therefore based on experimental and clinical studies, chitosan remains a controversial supplement for weight loss (Pittler and Ernst, 2004; Mhurchu et al., 2005; Macchi, 1996; Ríos-Hoyo and Gutiérrez-Salmeán, 2016; Trivedi et al., 2016; Bahijri et al., 2017).

Antihypertensive activity

Hypertension is one of the most common cardiovascular diseases worldwide, affecting 15%–20% of the adult population. According to the World Health Organization, more than 17.5 million people die from cardiovascular diseases per year, and this number is projected to rise to > 23.6 million by 2030 (Smith et al., 2012; Huang et al., 2018). The number of people with hypertension is expected to grow to more than 1.5 billion by 2025. It is the most common chronic disease because it is a significant risk factor for the development of arteriosclerosis, stroke, myocardial infarction, and end-stage renal disease (Kannel, 1996; Williams et al., 2004). The mechanism of hypertension has been explained by the physiological and biochemical reactions of the peripheral renin–angiotensin system, and its treatment is focused on the inhibition of angiotensin-converting enzyme (ACE) activities (Park et al., 2009).

In age-related hypertension, chloride ions play a significant role in severity, and therefore decreasing the dietary intake of sodium chloride (NaCl) is recommended (Tanaka et al., 1997; Williams et al., 2004). Many natural products (such as *Apocynum venetum*, soybean, quercetin, chitin and chitin derivatives, fermented oyster sauce, and others) have a potential antihypertensive effect (Kim et al., 2000; Yang et al., 2004; Je et al., 2005, 2006; Serban et al., 2016; Zoltani, 2019). Park et al. (2009) suggested that chitosan is an ideal candidate as an ACE inhibitor because it is milder and safer when compared to mainstay pharmacologic therapies (Hong et al., 1998; Koide, 1998; Je et al., 2006). These authors evaluated the effect of NaCl plus 3% chitosan on the systolic blood pressure (SBP) in spontaneously hypertensive rats (SHR) and compared with NaCl (49.36%) plus KCl (49.36%) and chitosan or NaCl treatment alone. In SHR, administration of NaCl plus chitosan (44 mM Na/day) for 2 months significantly decreased the SBP greater than NaCl plus KCl and NaCl alone. NaCl plus chitosan resulted in decreased urinary Na^+ and blood urea nitrogen, although the values were statistically insignificant. The results demonstrated that diets having supplementary NaCl plus chitosan had antihypertensive effects in an experimental model. In a recent study, Allaert (2017) also demonstrated that 3% NaCl plus chitosan can prevent hypertension by controlling blood pressure elevation in subjects presenting with prehypertensive status, which is the target population of the food supplement.

Je et al. (2006) developed three ACE-inhibiting aminoethyl-chitin (AEC) derivatives (AEC with 10% DDA; AEC with 50% DDA; and AEC with 90% DDA) as antihypertensive agents. The IC_{50} values of these three chitin derivatives were 0.064, 0.038, and 0.103 μM, respectively. The results of Dixon plots revealed that AEC50 was a competitive ACE inhibitor, and the inhibition constant (Ki) value was 0.021 μM. In addition, AEC50 produced decreased SBP in SHR rats.

Du et al. (2016) investigated the regulatory effects of WSC on vascular remodeling in hypertension, and further to confirm the roles of nuclear factor c1 of activated T cells (NFATc1) during the effect. WSC significantly reduced cell viability in primary cultured rat abdominal aortic smooth muscle cells when compared with phosphate buffered saline and AngII-treated cells. Compared with control rats, the abnormally high blood pressure and membrane thickness/lumen diameter ratio of the abdominal aorta were noted in SHR model rats, which were strongly reduced after WSC administration. Hypertension resulted in an increased expression of both NFATc1 protein and mRNA in the abdominal aorta of Wistar-Kyoto rats compared to control groups. WSC-inhibited vascular remodeling may be related to the modulation of NFATc1 expression, thereby exerting an antihypertensive effect.

Huang et al. (2018) evaluated the effects of chitosan on SBP and diastolic blood pressure (DBP) through a pooled analysis of available randomized controlled trials. Findings of this metaanalysis indicated that chitosan consumption significantly decreases DBP at higher dosages ($>$2.4 g/day) and in short-term ($<$12 weeks) interventions. In a previous systemic review by Jull et al. (2008), evidence suggested that chitosan significantly lowered the levels of both SBP and DBP. Taken together, it can be suggested that findings related to chitosan or its derivatives for antihypertensive effects should be interpreted with great caution.

Wound-healing activity

Wound healing is a complex process that occurs in five overlapping stages (hemostasis, inflammation, migration, proliferation, and maturation). The main goals of wound care and management are prevention of infection, maintenance of a moist environment, protection of the wound, and achievement of rapid and complete healing with minimum scar formation (Dai et al., 2011). The degree of healing is mainly regulated by cytokines and growth factors (Periayah et al., 2015).

Currently, there are many biomaterials and nutraceuticals that can promote wound healing due to their hemostatic properties and/or due to activation of growth factors and aggregation of platelets. Chitosan, as a cationic natural polymer, has been widely used as a topical dressing in wound and burn management owing to its hemostatic, stimulation of healing, antimicrobial, biocompatible, biodegradable, and nontoxic properties (Brandenberg et al., 1984; Fukasawa et al., 1992; Klokkevold et al., 1991, 1999; Okamoto et al., 2003; Dai et al., 2011; Mohiti-Asli and Loboa, 2016). In fact, chitosan has been described as a wound-healing accelerator. Platelet adhesion and activation are crucial early events in the processes after traumatic bleeding, leading to thrombus formation (Lord et al., 2011). When chitosan-derived biomaterials come in contact with blood they stimulate platelets to adhere, activate, and aggregate at the sites of vascular injury, thereby enhancing wound-healing capacity (Periayah et al., 2015). Chou et al. (2003) tested chitosan (MW = 50,000) for its enhanced platelet activity in rabbit platelet suspensions and explored the underlying mechanism(s). The results revealed that after initial (5 min) and long-term (30 min) contact of platelets with chitosan, platelet adhesion to chitosan-coated microtiter plates was dose-dependently increased compared to that of solvent control. Also, chitosan dose-dependently increased platelet aggregation. Chitosan significantly enhanced the expression of platelet glycoprotein IIb/IIIa complex assayed by a flow cytometer. The study concluded that chitosan is an effective inducer for platelet adhesion and aggregation. The mechanism of action of chitosan involves the increasing of $[Ca^{2+}]_i$ mobilization and enhancing the expression of GPIIb/IIIa complex on platelet membrane surfaces.

In another study, Lord et al. (2011) reported that platelet adhesion to chitosan was found to be enhanced in the presence of absorbed plasma and extracellular matrix proteins and was found to be primarily mediated by $\alpha_{IIb}\beta_3$ integrins, while $\alpha_2\beta_1$ integrins were found to be involved in platelet adhesion to collagen and perlecan. Further, platelets were found to be activated by chitosan, as evidenced by an increase in the expression of $\alpha_{IIb}\beta_3$ integrins and P-selectin, while the extent of activation was modulated by the presence of proteins including perlecan and fibrinogen. Taken together, the findings of this investigation suggested that chitosan promotes wound healing by platelet adhesion and activation.

Periayah et al. (2015) investigated the changes in platelet morphology and the expression level of transforming growth factor-β1 (TGF-β1) and platelet-derived growth factor-AB (PDGF-AB) in the adherence of two different types of chitosans in von Willebrand disease (vWD): *N,O*-carboxymethyl chitosan (NO-CMC) and oligo-chitosan (O-C). The findings demonstrated that the greater expression level of O-C assists in mediating the cytokine complex networks of TGF-β1 and PDGF-AB and induces platelet activities toward wound healing in vWD. PDGF is known to stimulate the migration and proliferation of macrophages and fibroblasts. Furthermore, PDGF activates the synthesis of TGF-β in macrophages, which also activates the synthesis of collagen in fibroblasts (Kojima et al., 2004). In an in vitro study, Periayah et al. (2013) determined the capacity of different grades of chitosan derivatives for platelet adhesion and aggregation. The results showed that O-C 52 and O-C 53 have potential to induce the release of platelets engaged in the process of hemostasis, thus promoting clotting.

Recently, Yan et al. (2018) introduced succinyl groups, carboxymethyl groups, and quaternary ammonium groups onto the surface of chitosan to elevate hemostatic activity. The results demonstrated that these surface-modified chitosan nonwoven dressings could be used as a promising hemostatic intervention. Chitosan has also been shown to increase the tensile strength of wounds (Degim et al., 2002). It needs to be pointed out that the wound-healing effects of chitosan could be affected by such factors as MW and DDA, as well as the state of the chitosan (Azad et al., 2004; Kojima et al., 2004; Minagawa et al., 2007; Yang et al., 2007; Alsarra, 2009). Burns treated with chitosan of a high MW and high DDA showed greater granulation tissue formation and epithelialization in wounds, and the fastest wound closure.

Chitosan can be used to prevent or treat wound and burn infections not only because of its intrinsic antimicrobial properties, but also by virtue of its ability to deliver extrinsic antimicrobial agents to wounds and burns (Dai et al., 2011). Furthermore, a variety of growth factors (such as epidermal growth factor, PDGF, fibroblast growth factor, TGF-β1, IGF-1, human growth hormone, and granulocyte-macrophage colony-stimulating factor) are involved in cell division, migration, differentiation, protein expression, and enzyme production. Chitosan can be used as a delivery vehicle for these factors to enhance wound healing (Shen et al., 2006; Dai et al., 2011).

Currently, the chitosan-based HemCon dressing (approved by the US FDA) is available as a hemostatic agent used in combat environments for the external temporary control of severely bleeding wounds, and it has demonstrated efficacy in both humans and animals (Wedmore et al., 2006; Pusateri et al., 2003). There are many other chitosan-based bandages and formulations (CELOX, QuikClot, Omiderm, etc.) for hemostasis and wound healing on the market (Ueno et al., 2001; Kozen et al., 2008; Millner et al., 2009; Dai et al., 2011). In some studies, chitosan is indicated in combination with other nutraceuticals, such as hyaluronic acid, thyme oil, taurine, or green tea polyphenol for wound and ulcer therapy (Degim et al., 2002; Rossi et al., 2007; Altiok et al., 2010; Qin et al., 2010). Blending chitosan with other polymers and cross-linking with glutaraldehyde vapor has also been found to be successful in the treatment of wounds.

Chitosan in Crohn's disease

Crohn's disease is a chronic inflammatory disease that can affect a part or the entire gastrointestinal tract. The common symptoms are abdominal pain, diarrhea, nausea, vomiting, inflammation, ulcers, lack of appetite, and weight loss. Crohn's disease of the intestine can cause severe obstruction of the small intestine. The causes of Crohn's disease may include immune system disorder, previous infection, genetic predisposition, and environmental factors. Dietary fat is also thought to exacerbate intestinal inflammation. Diagnosis of Crohn's disease rests with GI endoscopy, and the US FDA-approved treatment includes the use of steroids, immunosuppressants, and antiinflammatory drugs.

The signs and symptoms of Crohn's disease can also be managed by alternative medicines. Currently, there are several complementary medicines, such as *Glycyrrhiza glabra* (licorice), *Asparagus racemosus*, *Boswellia serrata*, *Psyllium* seed/husk, and bromelain. Chitosan can prevent Crohn's disease by reducing the levels of homocysteine in the body. In a clinical study, Tsujikawa et al. (2003) indicated that oral administration of a chitosan and ascorbic acid mixture in patients with Crohn's disease increased fecal fat excretion and was well tolerated without affecting disease activity. Chitosan improves the growth of commensal organisms and thereby functions of the digestive system.

Chitosan in dry eye disease

DED is a major ophthalmological condition affecting a large number of humans and animals. It is a multifactorial disease of the ocular surface characterized by a loss of homeostasis of the tear film, accompanied by symptoms of ocular surface inflammation and damage, and neurosensory abnormalities (Nelson et al., 2017, Lorenz et al., 2018). A malfunction of any of the lacrimal functional unit components (cornea, conjunctiva, accessory lacrimal and meibomian glands, the main lacrimal glands, and the sensory and motor nerves) can lead to DED (Stern et al., 2004; Messina and Dua, 2019). Predisposing factors for DED include age, gender, medications, exposure to cold, dry or polluted air, autoimmune diseases, and extensive computer use (Malet et al., 2014; Lorenz et al., 2018). DED is divided into two subtypes: aqueous tear-deficient dry eye and evaporative dry eye.

The diagnosis of DED is based on clinical history and symptoms supported by a number of tests. Artificial tears are the mainstay for treatment of DED. One drop twice daily in each eye (approximately 12 h apart) of XIIDRA (lifitegrast 5% solution, Novartis Pharmaceutical Corp.) is indicated for the treatment of the signs and symptoms of DED. XIIDRA works as a lymphocyte function-associated antigen-1 (LFA-1) antagonist. However, it may produce side effects, such as blurred vision, conjunctival hyperemia, eye irritation, headache, increased lacrimation, eye discharge, eye pruritis, and sinusitis. Hyaluronic acid (0.1%–0.13% solution) is also commonly used in many formulations of artificial tears for humans and animals with DED (Williams and Mann, 2013; Fallacara et al., 2017; Lee et al., 2017; Gupta et al., 2019b).

In recent studies, novel treatments consisting of chitosan or its derivative(s) have been implicated in DED or related ophthalmological conditions (Schmidt et al., 2017; Lorenz et al., 2018; Schuerer et al., 2018; Messina and Dua, 2019). Lacrimera (CHROMA-PHARMA GmbH, Leobendorf, Austria), which is based on chitosan biopolymeric backbone, modified by the introduction of *N*-acetylcysteine (NAC) via nucleophilic substitution, is commonly used in the management of DED. Messina and Dua (2019) reported that a single-dose instillation (one drop in the morning) of chitosan-*N*-acetylcysteine (C-NAC) for 5 consecutive days improved signs and symptoms in patients affected with DED from a variety of causes. In a long-term study, once-daily use of Lacrimera (C-NAC) was found to be a highly effective, tolerable, safe, and convenient treatment option for mild-to-moderate DED (Lorenz et al., 2018). In this clinical trial, DED symptoms were alleviated by 50% after 6 of weeks of treatment and 70% after 6 months of use. C-NAC works through a unique mechanism of action, which involves the formation of an artificial glycocalyx-like structure (protection shield) on the ocular surface (Lorenz et al., 2018). In a more recent study, Yu et al. (2019) developed a novel eye drop called water-soluble glycol chitosan cerium oxide nanoparticle (GCCNP). It scavenged ROS and exerted an antioxidative effect in mice primary corneal and conjunctival cells in vitro. GCCNP also improved DED in a murine model by stabilizing the tear film, scavenging ROS, upregulating SOD, and promoting and maintaining corneal and conjunctival cell growth and integrity. In many other in vitro and in vivo experimental studies, chitosan and its derivatives have been found effective and safe in DED and related ophthalmological conditions (Cheung et al., 2015; Sambursky, 2016; Fischak et al., 2017).

Anticancer activity

Cancer is reported to be one of the leading causes of death worldwide and is associated with mortality and morbidity (Seigel et al., 2016). Chitosan and chitosan derivatives have been used in drug delivery and cancer therapy for decades. Chitosan and its derivatives (trimethyl chitosan, carboxymethyl chitosan, *N*-succinyl chitosan, polyethylene glycated chitosan, hyaluronic acid-coupled chitosan, thiolated chitosan, folic acid-conjugated chitosan, and others) have been transformed into nanocomposites/nanoparticles (NPs) as a drug carrier for cancer therapy (Bose and Wong, 2018). Recently, Guo et al. (2018) developed a multifunctional nanocarrier through electrostatic interaction between fluorescent gold nanocluster-conjugated chitosan and the nucleolin targeting AS1411 aptamer. In an in vivo study, intravenous administration of nanodrug carrier systems into BALB/c mice led to an accumulation of methotrexate at the tumor site and significantly inhibited tumor growth without any toxicity (Guo et al., 2018). These authors suggested that this multifunctional nanocarrier can be used as an effective drug-delivery system for anticancer drugs that exhibits great potential in clinical applications.

Chitosan exhibits anticancer activity via immunomodulatory, antiangiogenic, and antiinflammatory actions. Zhou et al. (2011) designed glycated chitosan for immunostimulation in combination with phototherapies in cancer treatment. In vitro results showed that glycated chitosan could enter into macrophages to stimulate NO generation and TNF-α secretion. In vivo results also showed immunological effects of glycated chitosan, particularly in inducing tumor-specific immune responses. Future clinical applications require a thorough understanding of the properties of glycated chitosan.

Synthetic polymers, such as polylactic-*co*-glycolic acid (PGLA) NPs, have been extensively employed to target many anticancer drugs through attachment of specific ligands to the surface of the particles (Chen et al., 2006). Badran et al. (2017) prepared 5-fluorouracil (5-FU) loaded chitosan (C)-coated PLGA NPs (C-5-FU PGLA NPs) and polycaprolactone (PCL) (C-5-FU PCL NPs) and employed them as carriers for cancer treatment. The in vitro studies showed an initial rapid 5-FU release followed by a sustained-release profile. The in vitro cytotoxicity of C-5-FU PLGA NPs showed significant inhibition of colon cancer cells (HT-29) compared to the other NPs and drug solutions. The findings of this study suggested that C-5-FU PLGA NPs can be considered as a promising carrier for cancer therapy.

Hasegawa et al. (2001) observed induction of apoptosis (evidenced by elevated DNA fragmentation and caspase-3-like activity) in chitosan-treated bladder tumor cells. The findings highlighted that the biological effect of chitosan is characteristic of the high-MW form and not the low-MW form. Since chitosan is a highly positive charged molecule, it may adhere to the cell membrane and evoke some biological change in cell membrane (Schipper et al., 1996: Carreno-Gomez and Duncan, 1997). Chitosan efficiently attracts to the negatively charged membrane, which enhances the retention and permeation of NPs (Sunderland et al., 2006). Chitosan induces apoptosis through a signal transduction pathway (s) other than membrane lipid peroxidation. The second messenger of the apoptosis pathway activated by chitosan is yet to be clarified. This study suggested that chitosan has potential in cancer therapy.

Murata et al. (1991) determined the effect of antimetastatic sulfated chitin derivatives (SCM-chitin III) on angiogenesis induced by B16-BL-6 cells in syngeneic mice. SCM-chitin III markedly decreased the number of vessels toward tumor mass without affecting the tumor cell growth when coinjected with tumor cells. Findings also revealed that the invasion of endothelial cells through reconstituted basement membrane (Matrigel) toward tumor-conditioned media was significantly inhibited by SCM-chitin III in a Transwell chamber assay. The study suggested that inhibition of lung tumor metastasis by SCM-chitin III may in part be due to the inhibition of tumor-associated angiogenesis.

Chitosan has excellent physicochemical properties that appear to be favorable for nucleic acid delivery by overcoming the systemic barriers of gene delivery. It readily forms complexes, microspheres, or NPs upon electrostatic interaction with nucleic acids (Mao et al., 2010). These authors emphasized that the charge-to-charge ratio between chitosan and DNA is a critical factor for successful electrostatic binding of DNA or siRNA to chitosan. Furthermore, the nitrogen-to-phosphate (N/P) ratio, that is, the ratio between the positively charged nitrogen of free amines in chitosan to negatively charged phosphates in nucleic acids potentially influences the efficiencies of the chitosan polymer to condense and protect the DNA or siRNA. Because of these promising characteristics, chitosan has been increasingly studied as a gene, DNA, siRNA delivery system in cancer therapy, and the transfection has been heavily influenced by MW, pH, and DDA (Sato et al., 2001; Huang et al., 2005; Lavertu et al., 2006; Nimesh et al., 2010; Agirre et al., 2014; Li et al., 2015; Patil et al., 2016; Babu and Ramesh, 2017). Studies suggest that chitosan used for gene transfection should have an intermediate degree of stability and appropriate MW, which appears to be better achieved with low-MW chitosan, that provides a balance between DNA protection ability and intracellular release (Köping-Höggård et al., 2004; Mao et al., 2010). The DDA in chitosan is important in determining its efficacy for stable complex formation with DNA or siRNA (Babu and Ramesh, 2017).

Most of the chitosan used in gene-delivery applications has a high DDA, since it corresponds to more free amines and increased positive charge for efficient DNA binding. Köping-Höggård et al. (2001) reported that $>65\%$ DDA in chitosan is needed for efficient complex formation with DNA. Furthermore, Huang et al. (2005) reported that the particle size is decreased with increasing DDA percentage of chitosan when complexed with DNA. The low acidic pH (range 5–6) and presence of serum are important factors in determining the stability of a cationic gene-delivery system (Nimesh et al., 2010). Taking all findings into consideration, biodegradability, biocompatibility, and NP formulation, chitosan or its derivatives is a promising gene/chemo drug-delivery system for in vitro and in vivo applications in cancer therapy.

Highton et al. (2016) assigned an additional application of chitosan as a potential immune-adjuvant for cancer vaccines, due to its unique properties, such as cationic nature, biocompatibility, safety, and its ability to be used as an antigen carrier (Tokura et al., 1999). Zaharoff et al. (2010) reported the use of chitosan as an adjuvant for IL-12 therapy in colorectal (MC33a) and pancreatic (Panc02) solid tumors in mice. Following an intratumoral injection, chitosan prolonged the retention of IL-12 in the injection site resulting in tumor regression in more than 80% of mice. Zaharoff et al. (2007, 2010) also demonstrated the efficacy of chitosan/IL-12 adjuvant therapy in superficial bladder cancer treatment. The chitosan/IL-2 adjuvant system enhances IgG2a and IgG2b (immunoglobulin G2a and b) antibody responses to ovalbumin. Wen et al. (2011) reported that chitosan NPs enhanced the Th1 and

Th2 immune responses induced by ovalbumin in mice. Chitosan NPs improved not only Th1 (IL-2 and INF-γ) and Th2 (IL-10) cytokine levels but also increased the killing activity of NK cells. Thus, chitosan appears to be a safe and promising immune-adjuvant for a cancer vaccine, by promoting both humoral and cellular immune responses.

Toxicity and safety of chitosan

Chitosan has been used safely in humans for topical, intranasal, eye drop, and oral applications (Pittler and Ernst, 2004; Mhurchu et al., 2005; Read et al., 2005; Wedmore et al., 2006; Lorenz et al., 2018; Messina and Dua, 2019). By having a low toxicity (LD_{50} >16 g/kg) and nonimmunogenic activity, chitosan is safe for people, pets, wildlife, and the environment when used according to label directions (Arai et al., 1968; Illum, 1998; Baldrick, 2010; Kean and Thanou, 2010; Yong et al., 2015). Some nonpeer-reviewed literature has indicated that, in some people, chitosan may cause stomach upset, constipation, nausea, headaches, or swelling of wrists and heels. Chitosan may also interfere with absorption of fat-soluble vitamins (A, D, E, and K) and magnesium. Chitosan from a crustacean source may also cause allergic reactions, increased pulse rate, and anaphylactic shock. Patients who are on diabetic medicines or blood thinners (such as warfarin) should consult their physician before taking these medications.

Concluding remarks and future directions

The uses of chitin, chitosan, and their derivatives are worldwide in chemistry, biotechnology, human and veterinary medicine, dentistry, food processing, and agriculture and environmental protection. Among the many applications, drug delivery, hemostasis and wound healing, antimicrobial, and DED are noteworthy. In immunotherapy, the roles of chitosan and its derivatives as adjuvants and immunostimulants are well recognized in vaccination. These compounds exert antibacterial, antiviral, and antifungal activities. Chitosans have unique properties of absorption, solution stability, biodegradability, biocompatibility, nonimmunogenic, and nontoxicity. NP-based formulations of chitosan and its derivatives are more effective and popular than microparticle-based compounds. Future investigations will envisage new derivatives and novel applications in the fields of nutraceuticals, pharmaceuticals, biomedicals, and food engineering.

References

Agirre, M., Zarate, J., Ojeda, E., et al., 2014. Low molecular weight chitosan (LMWC)-based polyplexes for pDNA delivery: from bench to bedside. Polymers 6, 1727–1755.

Agnihotri, S.A., Mallikarjuna, N.N., Aminabhavi, T.M., 2004. Recent advances on chitosan-based macro- and nanoparticles in drug delivery. J. Control. Rel. 100 (1), 5–28.

Aiedeh, K.M., Taha, M.O., Al-Khatib, H., 2005. Evaluation of chitosan succinate and chitosan phthalate as enteric coating polymers for diclofenac sodium tablets. J. Drug Deliv. Sci. Technol. 15 (3), 207–211.

Allaert, F.A., 2017. Effect of a NaCl + chitosan 3% vs. NaCl on high blood pressure parameters of healthy volunteers with hypertension. Minerva Cardioangiol. 65, 563–576.

Alsarra, I.A., 2009. Chitosan topical gel formulation in the management of burn wounds. Int. J. Biol. Macromol. 45 (1), 16–21.

Altiok, D., Altiok, E., Tihminlioglu, F., 2010. Physical, antibacterial and antioxidant properties of chitosan films incorporated with thyme oil for potential wound healing applications. J. Mater. Sci. Mater. Med. 21 (7), 2227–2236.

Anraku, M., Gebicki, J.M., Iohara, D., et al., 2018. Antioxidant activities of chitosans and its derivatives in *in vitro* and *in vivo* studies. Carbohydr. Polym. 199, 141–149.

Arai, K., Kinumaki, T., Fujita, T., 1968. Toxicity of chitosan. Bull. Tokai Reg. Fish. Lab. 43, 89–94.

Arca, H.C., Günbeyaz, M., Senel, S., 2009. Chitosan-based systems for the delivery of vaccine antigens. Exp. Rev. Vaccines 8, 937–953.

Azad, A.K., Sermsintham, N., Chandrkrachang, S., et al., 2004. Chitosan membrane as a wound-healing dressing: characterization and clinical application. J. Biomed. Mater. Res. B Appl. Biomater. 69 (2), 216–222.

Azuma, K., Osaki, T., Minami, S., et al., 2015. Anticancer and anti-inflammatory properties of chitin and chitosan oligosaccharides. J. Funct. Biomater. 6, 33–49.

Babu, A., Ramesh, R., 2017. Multifaceted applications of chitosan in cancer drug delivery and therapy. Mar. Drugs 15, 96. Available from: https://doi.org/10.3390/md15040096.

Badran, M.M., Mady, M.M., Ghannam, M.M., et al., 2017. Preparation and characterization of polymeric nanoparticles surface modified with chitosan for target treatment of colorectal cancer. Int. J. Biol. Macromol. 95, 643–649.

Bahijri, S.M., Alsheikh, L., Ajabnoor, G., et al., 2017. Effect of supplementation with chitosan on weight, cardiometabolic, and other risk indices in Wistar rats fed normal and high-fat/high-cholesterol diets *ad libitum*. Nutr. Metab. Insights 10, 1–8.

Baldrick, P., 2010. The safety of chitosan as a pharmaceutical excipient. Regul. Toxicol. Pharmacol. 56 (3), 290–299.

Bellich, B., D'Agostino, I., Semeraro, S., et al., 2016. The good, the bad and the ugly of chitosan. Mar. Drugs 14, 99. Available from: https://doi.org/10.3390/md14050099.

Bose, A., Wong, T.W., 2018. Oral colon cancer targeting by chitosan nanocomposites. Appl. Nanocomp. Mater. Drug Deliv. 2018, 409–429. Available from: https://doi.org/10.1016/B978-0-12-813741-3.00018-2.

Brandenberg, G., Leibrock, L.G., Shuman, R., et al., 1984. Chitosan: a new topical hemostatic agent for diffuse capillary bleeding in brain tissue. Neurosurgery 15, 9–13.

Carreno-Gomez, B., Duncan, R., 1997. Evaluation of the biological properties of soluble chitosan and chitosan microspheres. Int. J. Pharm. 148, 231–240.

Castro, S.P.M., Paulín, E.G.L., 2012. Is chitosan a new panacea? Areas of applications. In: Karunaratne, D.N. (Ed.), In *Complex World of Polysaccharides*. InTech Publ, Rijeka, pp. 3–46.

Chen, A.S., Taguchi, T., Sakai, K., et al., 2003. Antioxidant activities of chitobiose and chitotriose. Biol. Pharm. Bull. 26, 1326–1330.

Chen, W., Shen, H., Li, X., et al., 2006. Synthesis of immune-magnetic nanoparticle and their application in the separation and purification of CD34 hematopoietic stem cells. Appl. Surf. Sci. 253, 1762–1769.

Cheung, R., Ng, T., Wong, J., et al., 2015. Chitosan: an update on potential biomedical and pharmaceutical applications. Mar. Drugs 13, 5156–5186.

Choi, B.-K., Kim, K.-Y., Yoo, Y.-J., et al., 2001. *In vitro* antimicrobial activity of a chitooligosaccharides mixture against *Actinobacillus actenomycetemcomitans* and *Streptococcus mutans*. Int. J. Antimicrob. Agents 18, 553–557.

Choi, B., Jo, D.-H., Mostafa Anower, A.K.M., et al., 2016. Chitosan as an immunomodulating adjuvant on T-cells and antigen-presenting cells in Herpes Simplex virus type 1 infection. Mediat. Inflamm. Available from: https://doi.org/10.1155/2016/4374375. Article ID 4374375.

Chou, T.C., Fu, E., Wu, C.J., et al., 2003. Chitosan enhances platelet adhesion and aggregation. Biochem. Biophys. Res. Commun. 302, 480–483.

Chung, M.J., Park, J.K., Park, Y.I., 2012. Anti-inflammatory effect o low-molecular weight chitosan oligosaccharides in IgE-antigen complex-stimulated RBL-2H3 cells and asthma model mice. Int. Immunopharmacol. 12, 453–459.

Costa, E.M., Silva, S., Madureira, A.R., et al., 2014. A comparative study into the impact of a chitosan mouthwash upon microorgansims' biofilm formation *in vitro*. Carbohydr. Polym. 101, 1081–1086.

Dai, T., Tanaka, M., Huang, Y.Y., et al., 2011. Chitosan preparations for wounds and burns: antimicrobial and wound-healing effects. Expert. Rev. Anti Infect. Ther. 9, 857–879.

Dash, M., Chiellini, F., Ottenbrite, R.M., et al., 2011. Chitosan-A versatile semi-synthetic polymer in biomedical applications. Prog. Polym. Sci. 36, 981–1014.

Degim, Z., Celebi, N., Sayan, H., et al., 2002. An investigation on skin wound healing in mice with a taurine-chitosan gel formulation. Amino Acids 22 (2), 187–198.

Divya, K., Jisha, M.S., 2018. Chitosan nanoparticles preparation and applications. Environ. Chem. Lett. 16, 101–112.

Dong, W., Han, B., Feng, Y., et al., 2010. Pharmacokinetics and biodegradation mechanisms of a versatile carboxymethyl derivative of chitosan in rats: *In vivo* and *in vitro* evaluation. Biomacromolecules 11, 1527–1533.

Du, M., Kou, L., Li, S., 2016. Water-soluble chitosan regulates vascular remodeling in hypertension via NFATc1. Eur. Heart J. 18 (Suppl. F), 38. Available from: https://doi.org/10.1039/eurheart/suw037.

Dutta, P.K., Dutta, J., Tripathi, V.S., 2004. Chitin and chitosan: chemistry, properties and applications. J. Sci. Ind. Res. 63, 20–31.

Elsabee, M.Z., Abdou, E.S., 2013. Chitosan based edible films and coatings: a review. Mater. Sci. Eng. C 33 (4), 1819–1841.

Fallacara, A., Vertuani, S., Panozzo, G., et al., 2017. Novel artificial tears containing cross-linked hyaluronic acid: an *in vitro* re-epithelialization study. Molecules 22, 2104. Available from: https://doi.org/10.3390/molecules22122104.

Feng, T., Du, Y., Li, J., et al., 2007. Antioxidant activity of half *N*-acetylated water-soluble chitosan in vitro. Eur. Food Res. Technol. 225, 133–138.

Fernandez, J., Ingber, D., 2014. Manufacturing of large-scale functional objects using biodegradable chitosan bioplastic. Macromol. Mater. Eng. 299 (8), 932–938.

Fischak, C., Klaus, R., Werkmeister, R.M., et al., 2017. Effect of topically administered chitosan-*N*-acetylcysteine on corneal wound healing in a rabbit model. J. Ophthalmol. 2017, 5192924.

Fukasawa, M., Abe, H., Masaoka, T., et al., 1992. The hemostatic effect of deacetylated chitin membrane on peritoneal injury in rabbit model. Surg. Today 22, 333–338.

Gades, M.D., Stern, J.S., 2005. Chitosan supplementation and fat absorption in men and women. J. Am. Diet. Assoc. 105 (1), 72–77.

Gómez-Rivas, L., Escudero-Abarca, B.I., Guadalupe, M., et al., 2004. Selective antimicrobial action of chitosan against spoilage yeasts in mixed culture fermentations. J. Ind. Biotechnol. 31, 16–22.

Goosen, M.F.A., 1996. Applications of Chitin and Chitosan. CRC Press, Boca Raton, FL, pp. 132–139.

Goy, R.C., de Britto, D., Assis, O.B.G., 2009. A review of the antimicrobial activity of chitosan. Polimeros Cienc. e Tecnol. 19 (3), 241–247.

Guo, X., Zhuang, Q., Ji, T., et al., 2018. Multi-functionalized chitosan nanoparticles for enhanced chemotherapy in lung cancer. Carbohydr. Polym. 195, 311–320.

Gupta, R.C., Gupta, D.M., Lall, R., Srivastava, A., Sinha, A., 2019a. Nutraceuticals in periodontal health and diseases. In: Gupta, R.C., Srivastava, A., Lall, R. (Eds.), Nutraceuticals in Veterinary Medicine. Springer Nature, Cham, pp. 447–466.

Gupta, R.C., Lall, R., Srivastava, A., Sinha, A., 2019b. Hyaluronic acid: molecular mechanisms and therapeutic trajectory. Front. Vet. Sci. 6, 192. Available from: https://doi.org/10.3389/fvets.2019.00192.

Hamed, I., Özogul, F., Regenstein, J.M., 2016. Industrial applications of crustacean by-products (chitin, chitosan, and chitooligosaccharides): a review. Trends Food Sci. Technol. 48, 40–50.

Hasegawa, M., Yagi, K., Iwakawa, S., et al., 2001. Chitosan induces apoptosis via caspase-3 activation in bladder tumor cells. Jpn. J. Cancer Res. 92, 459–466.

Helander, I.M., Nurmiaho-Lassila, E.L., Ahvenainen, R., et al., 2001. Chitosan disrupts the barrier properties of the outer membrane of Gram-negative bacteria. Int. J. Food Microbiol. 71, 235–244.

Highton, A.J., Girardin, A., Bell, G.M., et al., 2016. Chitosan gel vaccine protects against tumor growth in an intercaecal mouse model of cancer by modulating systemic immune responses. BMC Immunol. 17, 39.

Hirano, S., Tsuchida, H., Nagao, N., 1989. *N*-acetylation in chitosan and the rate of its enzymatic hydrolysis. Biomaterials 10 (8), 574–576.

Hoemann, C.D., Fong, D., 2017. Immunological responses to chitosan for biomedical applications. Chitosan Based Biomater. 1, 45–79.

Hong, S.P., Kim, M.H., Oh, S.W., et al., 1998. ACE inhibitory and antihypertensive effect of chitosan oligosaccharides in SHR. Korean. J. Food Sci. Technol. 30 (1), 1476–1479.

Huang, M., Fong, C.W., Khor, E., et al., 2005. Transfection efficiency of chitosan vectors: effect of polymer molecular weight and degree of deacetylation. J. Control. Rel. 106, 391–406.

Huang, L., Chen, J., Cao, P., et al., 2015. Anti-obese effect of glucosamine and chitosan oligosaccharide in high-fat diet-induced obese rats. Mar. Drugs 13, 2732–2756.

Huang, H., Zou, Y., Chi, H., 2018. Quantitative assessment of the effects of chitosan intervention on blood pressure control. Drug Des. Dev. Ther. 12, 67–75.

Husain, S., Al-Samadani, K.H., Najeeb, S., et al., 2017. Chitosan biomaterials for current and potential dental applications. Materials 10, 602. Available from: https://doi.org/10.3390/ma10060602.

Iglesias, N., Galbis, E., Días-Blanco, M., et al., 2019. Nanostructured chitosan-based biomaterials for sustained and colon-specific resveratrol release. Int. J. Mol. Sci. 20, 398. Available from: https://doi.org/10.3390/ijms2002398.

Illum, L., 1998. Chitosan and its use as a pharmaceutical excipient. Pharm. Res. 15, 1326–1331.

Illum, L., Jabaal-Gill, I., Hinchcliffe, M., et al., 2001. Chitosan as a novel nasal delivery system for vaccines. Adv. Drug Deliv. Rev. 51, 81–96.

Inkinci, G., Şenel, S., Akincibay, H., et al., 2002. Effect of chitosan on a periodontal pathogen *Porphyromonas gingivalis*. Int. J. Pharm. 235 (1-2), 121–127.

Je, J.Y., Kim, S.K., 2006. Chitosan derivatives killed bacteria by disrupting the outer and inner membrane. J. Agric. Food Chem. 54, 6629–6633.

Je, J.Y., Park, J.Y., Jung, W.K., et al., 2005. Isolation of angiotensin I converting enzyme (ACE) inhibitor from fermented oyster sauce, *Crassostrea gigas*. Food Chem. 90, 809–814.

Je, J.Y., Park, P.J., Kim, B., et al., 2006. Antihypertensive activity of chitin derivatives. Biopolymers 83, 250–254.

Jeon, T.I., Hwang, S.G., Park, N.G., et al., 2003. Antioxidative effect of chitosan on chronic carbon tetrachloride induced hepatic injury in rats. Toxicology 187, 67–73.

Jeon, S.J., Oh, M., Yeo, W.-S., et al., 2014. Underlying mechanism of antimicrobial activity of chitosan microparticles and implications for the treatment of infectious diseases. PLoS One 9 (3), 92723.

Jull, A.B., Mhurchu, C.N., Bennett, D.A., et al., 2008. Chitosan for overweight or obesity. Cochrane Database Syst. Rev. 3, CD003892.

Kaats, G.R., Michalek, J.E., Preuss, H.G., 2006. Evaluating efficacy of a chitosan product using a double-blinded, placebo-controlled protocol. J. Am. Coll. Nutr. 25 (5), 389–394.

Kannel, W.B., 1996. Blood pressure as a cardiovascular risk factor: prevention and treatment. JAMA 275, 1571–1576.

Kean, T., Thanou, M., 2010. Biodegradation, biodistribution and toxicity of chitosan. Adv. Drug Deliv. Rev. 62, 3–11.

Kendra, D.F., Hadwiser, L.A., 1984. Characterization of the smallest chitosan oligomer that is maximally antifungal to *Fusarium solani* and elicits pisatin formation in *Pisum sativum*. Exp. Mycol. 8, 276–281.

Kerch, G., 2015. The potential of chitosan and its derivatives in prevention and treatment of age-related diseases. Mar. Drugs 13, 2158–2182.

Kim, D., Yokozawa, T., Hattori, M., et al., 2000. Effect of aqueous extracts of *Apocynum venetum* leaves on spontaneously hypertensive, renal hypertensive and NaCl-fed-hypertensive rats. J. Ethnopharmacol. 72, 53–59.

Klokkevold, P.R., Lew, D.S., Ellis, D.G., et al., 1991. Effect of chitosan on lingual hemostasis in rabbits. J. Oral Maxillofac. Surg. 49, 858–863.

Klokkevold, P.R., Fukayama, H., Sung, E.C., et al., 1999. The effect of chitosan (poly-*N*-acetyl glucosamine) on lingual hemostasis in heparinized rabbits. J. Oral Maxillofac. Surg. 57, 49–52.

Koide, S.S., 1998. Chitin-chitosan: properties, benefits and risks. Nutr. Res. 18, 1091–1101.

Kojima, K., Okamoto, Y., Kojima, K., et al., 2004. Effects of chitin and chitosan on collagen synthesis in wound healing. J. Vet. Med. Sci. 66 (12), 1595–1598.

Kong, M., Chen, X.G., Xing, K., et al., 2010. Antimicrobial properties of chitosan and mode of action: a state of the art review. Int. J. Food Microbiol. 144, 51–63.

Köping-Höggård, M., Tubulekas, I., Guan, H., et al., 2001. Chitosan as a nonviral gene delivery system. Structure-property relationships and characteristics compared with polyethylenimine *in vitro* and after lung administration *in vivo*. Gene Ther. 8, 1108–1121.

Köping-Höggård, M., Varum, K.M., Issa, M., et al., 2004. Improved chitosan-mediated gene delivery based on easily dissociated chitosan polyplexes of highly defined chitosan oligomers. Gene Ther. 11, 1441–1452.

Kozen, B.G., Kircher, S.J., Henao, J., et al., 2008. An alternative hemostatic dressing: comparison of CELOX, HemCon, and QuikClot. Acad. Emerg. Med. 15 (1), 74–81.

Kumari, S., Annamareddy, S.H.K., Abanti, S., et al., 2017. Physicochemical properties and characterization of chitosan synthesized from fish scales, crab and shrimp shells. Int. J. Biol. Macromol. B 104, 1697–1705.

Labrude, P., Becq, C., 2003. Pharmacist and chemist Henri Braconnot. Rev. Hist. Pharm. (Paris.) 51 (337), 61–78.

Lan, R., Chang, Q., An, L., et al., 2020. Dietary supplementation with chitosan oligosaccharides alleviates oxidative stress in rats challenged with hydrogen peroxide. Animals 10, 55. Available from: https://doi.org/10.3390/ani10010055.

Laokuldilok, T., Potivas, T., Kanha, N., et al., 2017. Physicochemical, antioxidant, and antimicrobial properties of chitooligosaccharides produced using three different enzyme treatments. Food Biosci. 18, 28–33.

Lavertu, M., Méthot, S., Tran-Khanh, N., et al., 2006. High efficiency gene transfer using chitosan/DNA nanoparticles with specific combinations of molecular weight and degree of deacetylation. Biomaterials 27, 4815–4824.

Lee, C.G., Da Silva, C.A., Lee, J.Y., et al., 2008. Chitin regulation of immune responses: an old molecule with new roles. Curr. Opin. Immunol. 20 (6), 684–689.

Lee, D., Lu, Q., Sommerfeld, S.D., et al., 2017. Targeted delivery of hyaluronic acid to the ocular surface by a polymer peptide conjugate system for dry eye disease. Acta Biomater. 55, 163–171.

Li, X., Min, M., Du, N., et al., 2013. Chitin, chitosan, and glycated chitosan regulate immune responses: the novel adjuvants for cancer vaccine. Clin. Dev. Immunol 2013. Available from: https://doi.org/10.1155/2013/387023. Article ID 387023.

Li, G.F., Wang, J.C., Feng, X.M., et al., 2015. Preparation and testing of quaternized chitosan nanoparticles as gene delivery vehicles. Appl. Biochem. Biotechnol. 175, 3244–3257.

Liaqat, F., Eltem, R., 2018. Chitooligosaccharides and their biological activities: a comprehensive review. Carbohydr. Polym. 184, 243–259.

Lin, S.B., Chen, S.H., Peng, K.C., 2009. Preparation of antibacterial chito-oligosaccharide by altering the degree of deacetylation of β-chitosan in a *Trichoderma harzianum* chitinase-hydrolyzing process. J. Sci. Food Agric. 89 (2), 238–244.

Liu, H., Du, Y.M., Wang, X.H., et al., 2004. Chitosan kills bacteria through cell membrane damage. Int. J. Food Microbiol. 95, 147–155.

Lord, M.S., Cheng, B., McCarthy, S.J., et al., 2011. The modulation of platelet adhesion and activation by chitosan through plasma and extracellular matrix proteins. Biomaterials 32 (28), 6655–6662.

Lorenz, K., Garhofer, G., Hoeller, S., et al., 2018. Long-term management of dry eye by once-daily use of chitosan-*N*-acetylcysteine (Lacrimera®) eye drops. J. Clin. Ophthalmol. 2 (1), 47–54.

Macchi, G., 1996. A new approach to the treatment of obesity: chitosan's effects on body weight reduction and plasma cholesterol's levels. Acta Toxicol. Ther. 17, 301–320.

Makita-Chingombe, F., Hilliard, L.K., Sara, L.D., et al., 2016. Ply(lactic-*co*-glycolic) acid-chitosan dual loaded nanoparticles for antiretroviral nanoformulations. J. Drug Deliv. 2016, E381017.

Malet, F., Le Goff, M., Colin, J., et al., 2014. Dry eye disease in French elderly subjects: the Alienor study. Acta Ophthalmol. 92, e429-6.

Mao, S., Sun, W., Kissel, T., 2010. Chitosan-based formulations for delivery of DNA and siRNA. Adv. Drug Deliv. Rev. 62, 12–27.

Messina, M., Dua, H.S., 2019. Early results on the chitosan-*N*-acetylcysteine (Lacrimera®) in the management of dry eye disease of varied etiology. Int. Ophthalmol. 39, 693–696.

Mhurchu, C.N., Dunshea-Mooij, C., Bennett, D., et al., 2005. Effect of chitosan on weight loss in overweight and obese individuals: a systematic review of randomized controlled trials. Obes. Rev. 6 (1), 35–42.

Millner, R.W., Lockhart, A.S., Bird, H., et al., 2009. A new hemostatic agent: initial life-saving experience with Celox (chitosan) in cardiothoracic surgery. Ann. Thorac. Surg. 8792, 113–e14.

Minagawa, T., Okamura, Y., Shigemasa, Y., et al., 2007. Effect of molecular weight and deacetylation degree of chitin/chitosan on wound healing. Carbohydr. Polym. 67 (4), 640–644.

Mohiti-Asli, M., Loboa, E.G., 2016. Nanofibrous smart bandages for wound care. Wound Healing Biomater. 483–499.

Moraru, C., Mincea, M.M., Frandes, M., et al., 2018. A meta-analysis on randomized controlled clinical trials evaluating the effect of a dietary supplement chitosan on weight loss, lipid parameters and blood pressure. Medicina 54, 109. Available from: https://doi.org/10.3390/medicina54060109.

Murata, J., Saiki, I., Makabe, T., et al., 1991. Inhibition of tumor-induced angiogenesis by sulfated chitin derivatives. Cancer Res. 51, 22–26.

Muzzarelli, R., Baldassarre, V., Conti, F., et al., 1988. Biological activity of chitosan: ultrastructural study. Biomaterials 9 (3), 247–252.

Muzzarelli, R., Tarsi, R., Filippini, O., et al., 1990. Antimicrobial properties of *N*-carboxybutyl chitosan. Antimicrob. Agents Chemother. 34, 2019–20123.

Nelson, J.D., Craig, J.P., Akpek, E.T., et al., 2017. TFOS DEWS II introduction. Ocul. Surf. 15, 269–275.

Ngo, D.H., Kim, S.-K., 2014. Antioxidant effect of chitin, chitosan and their derivatives. In: Part, B., Kim, S.-K. (Eds.), Marine Carbohydrates: Fundamentals and Applications. Elsevier Inc, Oxford, pp. 15–31.

Nimesh, S., Thibault, M.M., Lavertu, M., et al., 2010. Enhanced gene delivery mediated by low molecular weight chitosan/DNA complexes: effect of pH and serum. Mol. Biotechnol. 46, 182–196.

Nishimura, K., Nishimura, S., Nishi, N., 1984. Immunological activity of chitin and its derivatives. Vaccine 2 (1), 93–99.

Nishimura, K., Nishimura, S.I., Nishi, N., 1985. Adjuvant activity of chitin derivatives in mice and guinea-pigs. Vaccine 3 (5), 379–384.

Nishimura, K., Ishihara, C., Ukei, S., 1986. Stimulation of cytokine production in mice using deacetylated chitin. Vaccine 4 (3), 151–156.

No, H.K., Park, N.Y., Lee, S.H., et al., 2002. Antibacterial activity of chitosans and chitosan oligomers with different molecular weights. Int. J. Food Microbiol. 74, 65–72.

Okamoto, Y., Yano, R., Miyatake, K., et al., 2003. Effects of chitin and chitosan on blood coagulation. Carbohydr. Polym. 53, 337–342.

Onishi, H., Machida, Y., 1999. Biodegradation and distribution of water-soluble chitosan in mice. Biomaterials 20 (2), 175–182.

Park, S.-H., Dutta, N.K., Baek, M.-W., et al., 2009. NaCl plus chitosan as a dietary salt to prevent the development of hypertension in spontaneously hypertensive rats. J. Vet. Sci. 10 (2), 141–146.

Patil, S., Bhatt, P., Lalani, R., et al., 2016. Low molecular weight chitosan-protamine conjugate for siRNA delivery with enhanced stability and transfection efficiency. RSC Adv. 6, 110951–110963.

Periayah, M.H., Halim, A.S., Hussein, A.R., et al., 2013. *In vitro* capacity of different grades of chitosan derivatives to induce platelet adhesion and aggregation. Int. J. Biol. Macromol. 52, 244–249.

Periayah, M.H., Halim, A.S., Mat Saad, A.Z., et al., 2015. Chitosan scaffold enhances growth factor release in wound healing in von Willebrand disease. Int. J. Clin. Exp. Med. 8 (9), 15611–15620.

Pittler, M.H., Ernst, E., 2004. Dietary supplements for body-weight reduction: a systematic review. Am. J. Clin. Nutr. 79 (4), 529–536.

Pusateri, A.E., McCarthy, S.J., Gregory, K.W., et al., 2003. Effect of a chitosan-based hemostatic dressing on blood loss and survival in a model of severe venous hemorrhage and hepatic injury in swine. J. Trauma 54, 177–182.

Qin, C., Li, H., Xiao, Q., et al., 2006. Water-solubility of chitosan and its antimicrobial activity. Carbohydr. Polym. 63, 367–374.

Qin, Y., Wang, H.W., Karuppanapandian, T., et al., 2010. Chitosan green tea polyphenol complex as a released control compound for wound healing. Clin. J. Traumatol. 13 (2), 91–95.

Raafat, D., Sahl, H.-G., 2009. Chitosan and its antimicrobial potential-A critical survey. Microb. Biotechnol. 2, 186–201.

Rabea, E.I., Badawy, M.E.T., Stevens, C.V., et al., 2003. Chitosan as antimicrobial agent: applications and mode of action. Biomacromolecules 4, 1457–1465.

Rajalakshmi, A., Krithiga, N., Jayachitra, A., 2013. Antioxidant activity of the chitosan extracted from shrimp exoskeleton. Middle East J. Sci. Res. 16 (10), 1446–1451.

Rajalakshmi, R., Indira Muzib, Y., Aruna, U., et al., 2014. Chitosan nanoparticles—an emerging trend in nanotechnology. Int. J. Drug Deliv. 6, 204–229.

Rampino A., 2011. Polysaccharide-Based Nanoparticles for Drug Delivery. (Ph.D. thesis). University of Trieste, Trieste, Italy.

Rampino, A., Borgogna, M., Blasi, P., et al., 2013. Chitosan nanoparticles: preparation, size evaluation and stability. Int. J. Pharm. 455, 219–228.

Ravi Kumar, M.N., 2000. A review of chitin and chitosan applications. React. Funct. Polym. 46, 1–27.

Read, R.C., Naylor, S.C., Potter, C.W., et al., 2005. Effective nasal influenza vaccine delivery using chitosan. Vaccine 23 (35), 4367–4374.

Ríos-Hoyo, A., Gutiérrez-Salmeán, G., 2016. New dietary supplement for obesity: what we currently know. Curr. Obes. Rep. 59 (2), 262–270.

Rossi, S., Marciello, M., Sandri, G., et al., 2007. Wound dressings based on chitosans and hyaluronic acid for the release of chlorhexidine diacetate in skin ulcer therapy. Pharm. Dev. Technol. 12 (4), 415–422.

Ryu, J.H., Hong, S., Lee, H., 2015. Bio-inspired adhesive catechol-conjugated chitosan for biomedical applications: a mini review. Acta Biomater. 27, 101–115.

Sambursky, R., 2016. Presence or absence of ocular surface inflammation directs clinical and therapeutic management of dry eye. Clin. Ophthalmol. 2016 (10), 2337–2343.

Sashiwa, H., Saimoto, H., Shigemasa, Y., et al., 1990. Lysozyme susceptibility of partially deacetylated chitin. Int. J. Biol. Macromol. 12 (5), 295–296.

Sato, T., Ishii, T., Okahata, Y., 2001. *In vitro* gene delivery mediated by chitosan. Effect of pH, serum, and molecular mass of chitosan on the transfection efficiency. Biomaterials 22, 2075–2080.

Schijns, V.E.J.C., 2003. Mechanisms of vaccine adjuvant activity: initiation and regulation of immune responses by vaccine adjuvants. Vaccine 21 (9-10), 829–831.

Schipper, N.G.M., Varum, K.M., Artursson, P., 1996. Chitosan as absorption enhancers for poorly absorbable drugs. 1: influence of molecular weight and degree of acetylation on drug transport across human intestinal epithelial (Caco-2) cells. Pharmacol. Res. 13, 1686–1692.

Schmidt, D., Werkmeister, R., Kaya, S., et al., 2017. A controlled, randomized double-blind study to evaluate the safety and efficacy of chitosan-*N*-acetylcysteine for the treatment of dry eye syndrome. J. Ocul. Pharmacol. Ther. 33, <https://doi.org/10.1089/jop.2016.0123>.

Schuerer, N., Stein, E., Inic-Kanada, A., et al., 2018. Effects of chitosan and chitosan *N*-acetylcysteine solutions on conjunctival epithelial cells. J. EuCornea 1 (1), 12–18.

Seferian, P.G., Martinez, M.L., 2001. Immune stimulating activity of two new chitosan containing adjuvant formulations. Vaccine 19 (6), 661–668.

Seigel, R.L., Miller, K.D., Jemal, A., 2016. Cancer statistics. CA Cancer J. Clin. 66, 7–30.

Serban, M.C., Sahebkar, A., Zanchetti, A., et al., 2016. Lipid and blood pressure meta-analysis collaboration (LBPMC) group. Effects of quercetin on blood pressure: a systematic review and meta-analysis of randomized controlled trials. J. Am. Heart Assoc. 5 (7), e002713.

Seyfarth, F., Schliemann, S., Elsner, P., et al., 2008. Antifungal effect of high- and low- molecular weight chitosan hydrochloride, carboxymethyl chitosan, chitosan oligosaccharide and *N*-acetyl-*D*-glucosamine against *Candida albicans*, *Candida krusei* and *Candida glabrata*. Int. J. Pharm. 353 (1-2), 139–148.

Shahidi, F., Vidana, J., Jeon, Y., 1999. Food applications of chitin and chitosans. Trends Food Sci. Technol. 10, 37–51.

Shen, E.C., Chou, T.C., Gau, C.H., et al., 2006. Releasing growth factors from activated human platelets after chitosan stimulation: a possible biomaterial for platelet-rich plasma preparation. Clin. Oral Implants Res. 17, 572–578.

Shon, Y.H., Park, I.K., Moon, I.S., et al., 2002. Effect of chitosan oligosaccharide on 2,3,7,8-tetrachlorodibenzo-*p*-doxin- induced oxidative stress in mice. Biol. Pharm. Bull. 25, 1161–1164.

Shukla, S.K., Mishra, A.K., Arotiba, O.A., et al., 2013. Chitosan-based nanomaterials: a state-of-the-art review. Int. J. Biol. Macromol. 59, 46–58.

Smith Jr, S.C., Collins, A., Ferrari, R., et al., 2012. Our time: a call to save preventable death from cardiovascular disease (heart disease and stroke). Circulation 126 (23), 2769–2775.

Smith, A., Perelman, M., Hinchcliffe, M., 2013. Chitosan: a promising safe and immune-enhancing adjuvant for intranasal vaccines. Hum. Vaccine Immunother. 10 (3), 797–807.

Stern, M.E., Gao, J., Siemasko, K.F., et al., 2004. The role of the lacrimal functional unit in the pathophysiology of dry eye. Exp. Eye Res. 78, 409–416.

Stoker, D.G., Reber, K.R., Waltzman, L.S., et al., 2008. Analgesic efficacy and safety of morphine-chitosan nasal solution in patients with moderate to severe pain following orthopedic surgery. Pain Med. 9, 3–12.

Sumiyoshi, M., Kimura, Y., 2006. Low molecular weight chitosan inhibits obesity induced by feeding a high-fat diet long-term in mice. J. Pharm. Pharmacol. 58, 201–207.

Sunderland, C.J., Steiert, M., Talmadge, J.E., et al., 2006. Targeted nanoparticles for detecting and treating cancer. Drug Dev. Res. 67, 70–93.

Tachaboonkiat, W., 2017. Antimicrobial applications of chitosan. Chitosan Based Biomater. 2, 245–274.

Takahashi, T., Imai, M., Suzuki, I., et al., 2008. Growth inhibitory effect on bacteria of chitosan membranes regulated with deacetylation degree. Biochem. Eng. J. 40 (3), 485–491.

Tanaka, M., Schmidlin, O., Yi, S.L., et al., 1997. Genetically determined chloride-sensitive hypertension and stroke. Proc. Natl. Acad. Sci. (USA) 94, 14748–14752.

Tang, H., Zhang, P., Kieft, T.L., et al., 2010. Antibacterial action of a novel functionalized chitosan-arginine against Gram-negative bacteria. Acta Biomater. 6 (7), 2562–2571.

Tikhonov, V.E., Stepnova, E.A., Babak, V.G., et al., 2006. Bactericidal and antifungal activities of a low molecular weight chitosan and its N-/2(3)-(dodec-2-enyl)succinoyl/-derivatives. Carbohydr. Polym. 64, 66–72.

Tokura, S., Ueno, K., Miyazaki, S., et al., 1997. Molecular weight dependent antimicrobial activity by chitosan. Macromol. Symp. 120, 1–9.

Tokura, S., Tamura, H., Azuma, I., 1999. Immunological aspects of chitin and chitin derivatives administered to animals. EXS 87, 279–292.

Trivedi, V.R., Satia, M.C., Deschamps, A., et al., 2016. Single-blind, placebo-controlled randomized clinical study of chitosan for body weight reduction. Nutr. J. 15, 3.

Tsujikawa, T., Kanauchi, O., Andoh, A., et al., 2003. Supplement of a chitosan and ascorbic acid mixture for Crohn's disease: a pilot study. Nutrition 19 (2), 137–139.

Ueno, H., Mori, T., Fuginaga, T., 2001. Topical formulations and wound healing applications of chitosan. Adv. Drug Deliv. Rev. 52 (2), 105–115.

Uraz, A., Boynueğri, D., Özcan, G., et al., 2012. Two percent chitosan mouthwash: a microbiological and clinical comparative study. J. Dent. Sci. 7 (4), 342–349.

van der Lubben, I.M., Verhoef, J.C., Borchard, G., et al., 2001. Chitosan for mucosal vaccination. Adv. Drug Deliv. Rev. 52 (2), 139–144.

Vunain, E., Mishra, A.K., Mamba, B.B., 2017. Fundamentals of chitosan for biomedical applications. Chitosan Based Mater. 1, 3–30.

Wang, G., 1992. Inhibition and inactivation of five species of food-borne pathogens by chitosan. J. Food Prot. 55, 916–919.

Wedmore, I., McManus, J.G., Pusateri, A.E., et al., 2006. A special report on the chitosan based hemostatic dressing: experience in current combat operations. J. Trauma 60 (3), 655–658.

Wen, Z.S., Xu, Y.L., Zou, X.T., et al., 2011. Chitosan nanoparticles act as an adjuvant to promote both Th1 and Th2 immune responses induced by ovalbumin in mice. Mar. Drugs 9, 1038–1055.

Williams, D.L., Mann, B.K., 2013. A crosslinked HA-based hydrogel ameliorates dry eye symptoms in dogs. Int. J. Biomater. 2013, 460435. Available from: https://doi.org/10.1155/2013/460437.

Williams, B., Poulter, N.R., Brown, M.J., et al., 2004. Guidelines for management of hypertension: report of the fourth working party of British Hypertension Society, 2004-BHS IV. J. Hum. Hypertens. 18, 139–185.

World Health Organization, 2015. Office of Health Communication and Public Relations: Obesity and Overweight. World Health Organization, Geneva.

Xie, W., Xu, P., Liu, Q., 2001. Antioxidant activity of water-soluble chitosan derivatives. Bioorg. Med. Chem. Lett. 11, 1699–1701.

Xu, Q., Ma, P., Yu, W., et al., 2010. Chitooligosaccharides protect human embryonic hepatocytes against oxidative stress induced y hydrogen peroxide. Mar. Biotechnol. 12, 292–298.

Yan, D., Hu, S., Zhou, Z., et al., 2018. Different chemical groups modification on the surface of chitosan nonwoven dressing and the hemostatic properties. Int. J. Biol. Macromol. 107, 463–469.

Yang, H.Y., Yang, S.C., Chen, J.R., et al., 2004. Soybean protein hydrolysate prevents the development of hypertension in spontaneously hypertensive rats. Br. J. Nutr. 92, 507–512.

Yang, J., Tian, F., Wang, Z., et al., 2007. Effect of chitosan molecular weight and deacetylation degree on hemostasis. J. Biomed. Mater. Res. B Appl. Biomater. 84, 131–137.

Yang, E.J., Kim, J.G., Kim, J.Y., et al., 2010. Anti-inflammatory effect of chitosan oligosaccharides in RAW 264.7 cells. Cent. Eur. J. Biol. 5, 95–102.

Yen, M.T., Yang, J.H., Mau, J.L., 2008. Antioxidant properties of chitosan from crab shells. Carbohydr. Polym. 74 (4), 840–844.

Yong, S.K., Shrivastava, M., Srivastava, P., et al., 2015. Environmental applications of chitosan and its derivatives. Rev. Environ. Contam. Toxicol. 233, 1–43.

Yu, F., Zheng, M., Zhang, A.Y., et al., 2019. A cerium oxide loaded glycol chitosan nano-system for the treatment of dry eye disease. J. Control. Rel. 315, 40–54.

Zaharoff, D.A., Rogers, C.J., Hance, K.W., et al., 2007. Chitosan solution enhances both humoral and cell-mediated immune responses to subcutaneous vaccination. Vaccine 25, 2085–2094.

Zaharoff, D.A., Hance, K.W., Rogers, C.J., et al., 2010. Intratumoral immunotherapy of established solid tumors with chitosan/IL-12. J. Immunother. 33, 697–705.

Zhang, P., Liu, W., Peng, Y., et al., 2014. Toll like receptor 4 (TLR4) mediates the stimulating activities of chitosan oligosaccharide on macrophages. Int. Immunopharmacol. 23, 254–261.

Zhou, F., Song, S., Chen, W.R., et al., 2011. Immunostimulatory properties of glycated chitosan. J. X-Ray Sci. Technol. 19 (2), 285–292.

Zivanovic, S., Basurto, C.C., Chi, S., et al., 2004. Molecular weight of chitosan influences antimicrobial activity in oil-in-water emulsions. J. Food Prot. 67, 952–959.

Zoltani, C.K., 2019. Nutraceuticals in cardiovascular diseases. In: Gupta, R.C., Srivastava, A., Lall, R. (Eds.), Nutraceuticals in Veterinary Medicine. Springer Nature, Cham, pp. 427–435.

Section IV

Prebiotics and probiotics

Chapter 64

Prebiotics: safety and toxicity considerations

Arturo Anadón, Irma Ares, María-Rosa Martínez-Larrañaga and María-Aránzazu Martínez

Introduction

Prebiotics are of interest because they can reduce the risk for certain conditions and promote better health. In recent years, consumer interest in the relationship between diet and health has increased the demand for correct information on prebiotics and has become a growing segment in the global market. This growth has been enhanced by technological innovations, development of new products, and an increasing number of health-conscious consumers interested in products that improve quality of life. Prebiotics have a long history of safe use and have been known for their health benefits in humans, including an increase in the bioavailability of minerals, modulation of the immune system, prevention of gastrointestinal (GI) infections, modification of inflammatory conditions, regulation of metabolic disorders, and reduction of the risk for cancer (Roberfroid et al., 2010). Health benefits of the prebiotics include improved lactose tolerance, antitumor properties, neutralization of toxins, and stimulation of the intestinal immune system, and reduction of constipation, blood lipids, and blood cholesterol levels (Das et al., 2012).

Prebiotics are a special form of dietary fiber, which are not affected by heat, cold, acid, or time. They provide a wide range of health benefits, and nourish the good bacteria that everyone already has in their gut. A prebiotic is a nondigestible food ingredient that beneficially affects the host by selectively stimulating growth and/or activity of one or a limited number of bacteria species already residing in the colon (Nagpal and Kaur, 2011). When prebiotics are used to increase bifidobacteria or lactobacilli toward being the numerically predominant genus in the colon, improved colonization resistance results.

Prebiotic compounds are able to modulate both the luminal and mucosal microbial composition and activities as well as beneficially regulate host–microbe interactions. Moreover, they change the gut microbiota composition (especially the number of bifidobacteria), thus contributing to modulation of metabolic processes associated with obesity and type II diabetes (Roberfroid et al., 2010). Prebiotics induce changes not only in the intestinal microbiota and the mucosal surface of the colon, but also the transepithelial transport of the short-chain fatty acids (SCFAs), stimulating shifts of fluid to and from the lumen; furthermore, the transport of cationic minerals is stimulated by the lowered pH of the lumen. Intraluminal colonic propionate induces the non-neuronal release of acetylcholine synthesized by the epithelial crypt cells to the serosal surface, especially in the distal colon. This is associated with modifications of the electrical parameters of the mucosa and chloride excretion (Yajima et al., 2011).

Gut microbiota is now perceived as a key player in health and well-being, with a composition in which potentially health-promoting dominant microorganisms (particularly the saccharolytic genera/species, e.g., bifidobacteria) are elevated and/or are more active than the potentially harmful ones (especially the proteolytic/putrefactive genera/species); therefore the composition and activity of the intestinal microbiota can influence health and disease through its involvement in nutrition, host physiological functions, and pathogenesis of certain disease conditions (Ringel and Carroll, 2009; Gibson et al., 1997).

The development of normal intestinal microflora is a gradual process. It is accepted that resident microbes contribute fundamentally to infant and childhood health and immunity. The effects of prebiotics on local and systemic immunity have been attributed to the stimulation of lymphocytes by SCFA. Healthy babies are sterile prior to birth. During delivery the human gastrointestinal tract (GIT) is sterile, but in the first days of life different types of bacteria rapidly colonize the gut, mainly environmental and maternal (i.e., vaginal, fecal, and oral flora). Coliforms and streptococci (Tannock et al., 1990). Full-term infants delivered vaginally are colonized very early by anaerobic bacteria, predominantly *Bacteroides* (Long and Swenson, 1977). The process of bacterial colonization

Nutraceuticals. DOI: https://doi.org/10.1016/B978-0-12-821038-3.00064-1

begins in utero with exposure of fetal mucosal surfaces to amniotic fluid microflora, and progresses rapidly, such that within days of birth, the bacterial cells contained within the lumen of the intestinal fat outnumber the infant's cells (Indrio and Neu, 2011). In contrast, infants delivered by cesarean section exhibit delayed colonization by anaerobes and by enterobacteria and Gram-negative organisms (Rotimi and Duerden, 1981).

Prebiotics act like growth factors in particular commensal bacteria (e.g., *Lactobacillus* spp.), which can both improve gut barrier function through protection of the epithelial tight junction during external stress and activate immune responses in the gut-associated lymphoid tissue (GALT) (Seth et al., 2008). Most human lymphoid tissue is located in and around the GIT, and the bacterial colonization of its mucosa is a key immunological stimulus for the development of the immune system. The significance of colonization for the development of a protective immune system is illustrated by experimental studies in which sterile mice lack sufficient development of lymphoid tissue, immune effector cells, and molecules (Smith et al., 2007).

Colonization of the sterile gut by commensal bacteria is also essential for tolerance induction to oral allergens. Commensals stimulate the GALT with their cell wall components, such as lipopolysaccharide from Gram-negative bacteria and peptidoglycan from Gram-positive and Gram-negative bacteria. Human breast milk contains a pattern recognition receptor, CD14, to which commensals bind. As a result of binding, immune cells differentiation into the Th1 type, production of regulatory cytokines (i.e., IL-10 and TGF-β) is upregulated, and Th2 type immunity as a promoter of allergy is downregulated (Labeta et al., 2000; Kelly et al., 2005). Human milk contains relatively high levels of oligosaccharides (3–6 g/L) and breastfeeding was associated with a reduced prevalence of allergy in numerous epidemiological studies. Epidemiological studies from the 1990s showed an association of higher counts of probiotic bacteria and less pathogenic bacteria in the stool of less allergic populations. In a prospective study of infants, it was shown that early gut colonization with nonprobiotic *Clostridiae* is associated with the development of allergy later in life (Björkstén et al., 2001).

The "gold standard" of early baby nutrition is human milk. Breast milk contains both digestive and indigestive carbohydrates and promotes colonization of the gut with nonpathogenic commensal bacteria (probiotics). *Bifidobacterium* as a probiotic representative also affects the microbiota composition and is more prevalent in the stool of breastfed infants, whereas *Clostridiae*, as representatives of more pathogenic bacteria are more abundant among formula-fed infants. Human maternal milk may also protect the mucosal surface of the GIT from its colonization by enteropathogens and contains a high amount of nondigestible oligosaccharides with more than 1000 different oligosaccharide structures. It has been shown that human milk, as well as specific dietary fibers like chicory (*Cichorium intybus*)-derived inulin and lactose-derived short-chain galactooligosaccharides, selectively supports the growth of host microorganisms conferring a health benefit, particularly *Lactobacillus* and *Bifidobacterium* strains (Boehm et al., 2005). Traces of polysaccharides appear in the urine of breastfed infants, indicating that they cross the intestinal epithelial layer. It is highlighted that the nucleotides in human milk or those added to infant formula have bifidogenic effects and may be considered in some to act as prebiotics.

Breastfed infants are primarily colonized with bifidobacteria, with subpopulations of *Lactobacillus* and *Streptococcus*. In contrast, formula-fed infants are primarily colonized with *Bacteroides* and bifidobacteria, but in fewer numbers compared to breastfed infants, with subpopulations of clostridia, *Staphylococcus*, and *Escherichia coli*. Their relevance to health and disease is questionable, particularly as the intestinal microbiota remains mutable during the first 6–12 months of life (Pietzak, 2004).

In human milk, oligosaccharides account for approximately 10% of the solid milk components and stimulate the growth of probiotic bacteria (prebiotic). Human milk contains a complex mixture of more than 130 different oligosaccharides, comprising a total concentration of 15–23 g/L in colostrum and 8–12 g/L in transitional and mature milk (Kunz et al., 2000). Quantitatively, oligosaccharides are the third largest solute in human milk after lactose and fat. Mature milk contains over 15 g/L of human milk oligosaccharides compared with only approximately 9 g/L of protein. The oligosaccharide content of human milk varies with the duration of lactation, diurnally, and with the genetic make-up of the mother (Coppa et al., 1993). Many of the 130 different human milk oligosaccharides may be important in infant brain development and resistance to infection. Their presence may be one of the reasons that breastfed infants appear to have higher intelligence quotients and greater resistance to infection than formula-fed infants. Moreover, it is known that healthy children receiving human milk oligosaccharides showed a lower rate of bronchitis compared with control. It must be highlighted that sialylated human milk oligosaccharides can enhance liver, muscle, and brain metabolism by increasing utilization of nutrients in infants and prevent neonatal infections by reducing the adhesion of pathogens to the intestinal epithelium (Wang et al., 2019). Human milk oligosaccharides may be a similar low-osmolar source of energy for the infant. However, the glycosidic linkages between the monosaccharide residues of human milk oligosaccharides are unlikely to be hydrolyzed by lactase and other known brush-border disaccharidases. These enzymes have no action on linkages involving fucose, sialic acid, or *N*-acetylglucosamine (Rings et al., 1994). Furthermore, the brush-border neutral lactase cannot hydrolyze the lactose moiety of trisaccharides or higher oligosaccharides.

Whereas the probiotic approach adds one or two species to a spectrum of hundreds of species in the gut flora, the prebiotic approach aims at fertilization of the intestinal ecosystem. Functional imitations of the naturally occurring prebiotics, galactooligosaccharides (GOS) and fructooligosaccharides (FOS), stimulate intestinal growth of bifidobacteria as a marker of probiotics in a dose-dependent manner and transform the microbiota profile of formula-fed infants toward that of breastfed infants (Kol et al., 2005).

Average life expectancy continues to increase in Western societies. Aging predisposes humans to a natural degeneration in GI function, epithelial barrier integrity, GI microbiota composition, and immune system function (adaptative and innate), elevating the risk of infections. Aging has relatively little effect on the GIT function, but involves multiple small changes that may increase the incidence of common symptoms (i.e., constipation, diverticulosis, malnutrition). The human GI microbiota goes through transitional stages throughout life, with the most profound changes detected before 3 years of age or after 100 years of age (Yatsunenko et al., 2012). In elderly subjects, it has been commonly documented that their microbiota may be different from that of younger adults. Commonly encountered effects on the GI microbiota due to aging have been brought forward, although the elderly subject groups studied varies by age, health status, and lifestyle.

Prebiotics may influence the immune system; this is likely to happen primarily through modulation of the microbiota activity and/or composition. Also, immune function is reduced in the elderly as compared to younger adults and therefore is an appropriate target for prebiotics. Various prebiotics have been observed to improve bowel function in the elderly. The fermentation of prebiotics by colonic bacteria increases the bacteria biomass leading to increased fecal output. During the fermentation of prebiotics like GOS, active by-products such as SCFA, butyrate, propionate, and acetate (90%–95% of all microbiota metabolites produced in the colon) are physiologically generated and efficiently absorbed by epithelial cells and in circulation. SCFA can target G-protein couple receptors (Yajima et al., 2011; Cantu-Jungles et al., 2019).

Prebiotic concept/prebiotic effects

Prebiotics are naturally found in many foods and can also be isolated from plants or synthesized from lactose or sucrose by enzymatic methods. They allow the selective growth of certain indigenous gut bacteria. A prebiotic was defined by Gibson and Roberfroid (1995) as "a nondigestive food ingredient that beneficially affects the host by selectively stimulating the growth and/or activity of one or a limited number of bacteria in the colon, and thus improves host health." These authors revised this concept and proposed a new prebiotic definition to be "a selectively fermented ingredient that allows specific changes; both in the composition and/or activity in the gastrointestinal microbiota that confers benefits upon host well-being and health" (Gibson, 2004; Roberfroid, 2007). The latest definition results in an equalization of "*prebiotic*" and "*bifidogenic*" and includes the "*prebiotic index*" (i.e., gives the absolute increase of the fecal bifidobacteria concentrate on per gram of daily consumed prebiotics). Because the prebiotic effects, rather *bifidogenic effects* depend on the type and concentration of the prebiotic and on the bifidobacteria concentration in the intestine of the host, there is not a simple dose–effect relationship. FOS, which are nondigestible in the small intestine, could be utilized by only a few bacteria species, notably bifidobacteria, and including such oligosaccharides in the food may increase the count of bifidobacteria in intestinal content. It should be stated that it is difficult to test the selective stimulation of individual bacterial strains between the more than 400 cultivable and noncultivable bacterial strains in the human gut. For that reason, prebiotic effect has been defined as "the selective stimulation of growth and/or activity of one or a limited number of microbial genera/species in the gut microbiota that confer health benefits to the host" (Roberfroid et al., 2010).

Although the concept of prebiotics has been developed over time, it is proposed that a candidate prebiotic must fulfill the following criteria that are to be proven by in vitro and in vivo studies to be considered effective: (1) nondigestibility (i.e., resistance to low pH gastric acid, enzymatic digestion, and intestinal absorption in the upper part of the GIT); (2) good fermentation by the large intestinal microbiota which can be investigated by measuring breath hydrogen or fecal recovery of the administered carbohydrate after a single prebiotic meal; and (3) selective stimulation of growth and activity of intestinal bacteria (i.e., measuring bacterial counts in fecal samples, or intestinal content, before and during exposure to the test material in batch or multichamber fermentation systems) with associated health-promoting effects. Prebiotics are short-chain polysaccharides that have unique chemical structures that are not digested by humans; in particular fructose-based oligosaccharides that exist naturally in food or are added to food (Macfarlane et al., 2006; De Vrese and Schrezenmeir, 2008).

Prebiotics undergo fermentation by beneficial microflora in the large intestine, and these are sources of energy for the microflora. The carbohydrates (i.e., inulin and oligofructose, GOS/TOS (trans-galacto-oligosaccharides), lactulose), which are nondigestible, can be fermented by the intestinal flora.

Because other investigators produced definitions that preferentially emphasized one or more different functional characteristics of prebiotics, and to further clarify and unify this concept, a meeting convened in 2007 by the Food and Agriculture Organization (FAO) in Rome agreed on a definition that focused on what are considered fundamental aspects for clinical, research, and regulatory purposes: "Prebiotics are nonviable food components that confer a health benefit on the host associated with modulation of the microbiota" (FAO, 2007). The definition arose from observations that dietary fibers in particular bring about a specific modulation of the gut microbiota, especially increased numbers of bifidobacteria and/or lactobacilli cell counts or a decrease in potential harmful bacteria. The selectivity for bifidobacteria may be promoted by the ingestion of substances such as FOS, inulin, TOS, and soybean oligosaccharides (SOS) (Mitsuoka et al., 1987). In regular terms, prebiotics are "*food components*" for live microorganisms that are considered beneficial for health and well-being. It is scientifically accepted that prebiotics are valuable dietary additions for modulating the growth and activity of specific bacterial species in the colon that are considered health-supporting.

Source of dietary oligosaccharides

Oligosaccharides are defined as molecules containing a small number (3–10) of monosaccharide residues connected by glycosidic linkages. Oligosaccharides, consisting of a mixture of hexose oligomers with a variable extent of polymerization, are food products with interesting nutritional properties. They may be naturally present in food, mostly in fruits, vegetables, or grains, or produced by biosynthesis from natural sugars or polysaccharides and added to food products because of their nutritional properties or organoleptic characteristics (Delzenne, 2003).

Most prebiotics are short-chain carbohydrates with a degree of polymerization of two or more that are not susceptible to digestion by pancreatic and brush-border enzymes (Steed and Macfarlane, 2009). Marine algae have been described as rich sources of dietary fiber (De Borba Gurpilhares et al., 2019).

Prebiotic oligosaccharides may be manufactured by extraction from plant materials, microbial/enzymatic synthesis, and by enzymatic hydrolysis of polysaccharides. In recent years, oligosaccharides have been increasingly used by the food industry (beverages, sweets) for modifying viscosity, emulsification capacity, gel formation, freezing point, and food coloring. They show nutrition- and health-relevant properties like moderate sweetness (i.e., 30%–60% of the sucrose value), low carcinogenicity, low caloric values, and a low glycemic index. Prebiotics must be chemically stable to withstand food-processing treatments such as heat, low pH, and Maillard reaction conditions.

Potential prebiotic oligosaccharides can be classified according to their chemical constituents and degree of polymerization and these include mannooligosaccharides, pectic-oligosaccharides, SOS, isomaltooligosaccharides (IMOs), xylooligosaccharides (XOS), oligosaccharide-sialic acid (OS), and lactulose. However, the majority of studies on prebiotics have focused on inulin, FOS, and GOS/TOS (Macfarlane et al., 2006; Anadón et al., 2016; Wang et al., 2019; Anadón et al., 2019).

Fructans

Fructans are important storage carbohydrates in many plant families, with each having its own fructan profile. They are also produced by algae, fungi, and bacteria. Fructans are a group of naturally occurring oligosaccharides and FOS found in milligram quantities in onions, bananas, asparagus, leeks, wheat, artichokes, garlic, alliums, and other whole foods (Chow, 2002). They are also extracted from chicory or manufactured from sucrose for use in the food industry. Despite their similarities, the fructans remain distinct from each other in origin, structure, and fermentation characteristics (Douglas and Sanders, 2008). Fructan molecules consist of monosaccharide units, usually one glucosyl unit and a varying number of fructosyl units, and are composed of one or as many as 70 units of fructose linked or not linked to a terminal sucrose molecule such as inulin and oligofructosyl-saccharose (oligofructose, FOS derived from inulin) and GOS. Fructose-based oligosaccharide is a fermentable dietary fiber (Anadón et al., 2019).

The most important commercially available types of fructans are inulin (2–60 units of fructose) and oligofructosyl-saccharose (Nordic Committee of Ministers, 2000). The absence of digestibility of these oligosaccharides has been demonstrated by in vitro and animal studies, as well as human studies.

Prebiotics can be found in wheat, bananas, and onions. The main candidates for prebiotic (recognized and emergent) status reported in the scientific literature are the following: inulin, FOS, GOS, OS, XOS, IMO, SOS, galactan raffinose, stachyose, arabinoxylans, lactulose, polydextrose (synthetic fiber), lactosucrose, lactilol, chitin, and polysaccharides from fungi (Anadón et al., 2016; Anadón et al., 2019). Raffinose oligosaccharides are found in lentils, peas, beans, and chickpeas.

- *Lactulose*: Lactulose is nonabsorbable disaccharides composed of galactose and fructose, whereas XOS are sugar oligomers formed by units of xylose (Gibson et al., 2004). Lactulose (4-0-β-D-galactopyranosyl-D-fructofuranose) is a galacto-fructose isomerization product derived from lactose with multiple effects on gut flora. It is believed to be a fermentable substrate for bacteria, leading to improved mineral absorption. Lactulose is regarded as an intestinal prebiotic that is not digestible in the small bowel. Lactulose effectively evades degradation in the stomach and small intestine and reaches the colon intact where it is fermented by anaerobic bacteria (*Bacteroides* spp., $10^{11}-10^{12}$ cfu/g feces). Fermentation of lactulose by these anaerobic bacteria results in lactic, acetic, and butyric acid and gas, particularly hydrogen (H_2). These weak acids are excellent substrates for the colonocytes. A total daily dose of 10–20 g is small compared to 500–1000 g feces/day, and the impact on the fecal flora is limited (Bongaerts et al., 2005). The massive H_2 volumes cause rapid intestinal hurry and thus massive amounts of colonic bacteria (not only urease-positive bacteria, but also deaminating bacteria) are removed. Also, intestinal uptake of toxic bacterial metabolites, for example, ammonia, is reduced, thereby lactulose is used in hepatic encephalopathy. Lactulose is usually found in milk and milk products that undergo thermal treatments and is used in the baby food and pharmaceutical industries. Traditionally, it has been used as a laxative in clinical treatment of constipation. The main effect of lactulose is a rapid stool transit due to the bacterial gas-mediated activation of intestinal peristalsis. It also applies to other low-molecular-weight sugar-like compounds such as lactiol (Bongaerts et al., 2005).

- *FOS and fructose-polysaccharide (inulin)*: FOS and inulin are polymers of D-fructose joined by β-2-1 bonds with an α-1–2-linked D-glucose at the terminal end of the molecule. Molecules with a degree of polymerization greater than 20 are referred to as inulin. Inulin is present in nature in a range of plant sources utilized as foods (fruits and vegetables), including chicory, onion, leek, garlic, Jerusalem artichoke, tomato, wheat, and banana. Inulin is β-(21)-fructan and a mixture of FOS and polysaccharides. It may be obtained by synthesis from saccharose, or through controlled and partial hydrolysis from chicory (*Cichorium intybus*) roots. FOS are prepared commercially from chicory in a hydrolysis reaction using inulinase, and may also be derived in an enzymatic synthetic reaction via transfer of fructosyl units from sucrose molecules (Tuohy et al., 2005). To date, FOS are the best prebiotic studied. The application of FOS as a prebiotic ingredient in dairy products and other foods such as breakfast cereal and soft drinks is increasing.
- *GOS*: One of the earliest sources of GOS was human milk, which contains approximately 7% carbohydrates, 90% of which is lactose, and a variety of oligosaccharides based on lactose (ESPGAN, 1977). GOS comprise the third largest component of breast milk (5–10 g/L) after lactose and lipids; GOS are unique in that they contribute greatly to the protective microbiome that persists in the intestine of breastfed infants. The human microbiome is associated with immunological function and metabolism. Each individual milk oligosaccharide is based on a variable combination of glucose, galactose, sialic acid (also known as *N*-acetylneuraminic acid), fucose, and/or *N*-acetylglucosamine, with many and varied linkages between them, thus accounting for the enormous number of different oligosaccharides in human milk (at least 130 and rising). Almost all of these oligosaccharides have a lactose moiety at their reducing end, whereas fucose and sialic acid (when present) occur at the nonreducing end (Kobata et al., 1978). High levels of GOS are found in colostrum where these substances constitute up to 24% of the total colostrum carbohydrates. Galactose-containing oligosaccharides of the form Glu α 1–4[β Gal 1–6]$_n$, where $n = 2–5$, are termed GOS and are present in both human and cow milk. GOS is a galactose oligomer and some glucose/lactose/galactose units. GOS may also be produced synthetically from lactose syrup using β-galactosidase. GOS is not broken down in the stomach or small intestine and, thus, reaches the colon largely intact (Tuohy et al., 2001a, b). β-Galactosidases are essential enzymes for bifidobacteria to be able to grow on milk or milk-based substrates, such as lactose and lactose-derived products containing β-galactosidic linkages. The diversity of *Bifidobacterium* strains has been studied with respect to their β-galactosidase activity in more detail. Galactans can also be commercially prepared through the enzymatic conversion of lactose with β-D-galacosidase, resulting in lactose connected to a chain of galactose monomers ranging between two and eight monomers in length. These polymers vary in chain length and linkages, making them indigestible by human enzymes and thus giving them their prebiotic effect.
- *Isomaltooligosaccharides (IMO)*: IMO is an α(1–4)-glucose and branched α(1–6)-glucose (Kohmoto et al., 1988). IMO (starches such as wheat, barley, corn, oats, tapioca, rice, potato) are comprised of glucose monomers linked by α-1–6 glucosidic linkages and are produced commercially from the enzymatic action of α-amylase, pullulanase, and α-glucosidase on corn starch. Commercially available products contain a mixture of isomaltose, isomaltotriose, and panose. Maltose and maltooligosaccharides are breakdown products from starch.

- *SOS*: The main sugars in SOS extracted from soy whey are raffinose [a trisaccharide, a-D-Gal-(1−6)-a-DGlu-(1−2)-b-D-Fru] and tetrasaccharide stachyose [a-DGal-(1−6)-a-D-Gal-(1−6)-a-D-Glu-(1−2)-b-D-Fru]. Both of these have been shown to reach the colon and are readily fermented by bifidobacteria.
- *XOS*: XOS are sugar oligomers formed by units of xylose and made up of xylose molecules linked by β-1−4 bonds and mainly consist of xylobiose, xylothiose, and xylo-tetraose. XOS can be effectively fermented by bifidobacteria found naturally in bamboo, shoots, fruits, vegetables, milk, and honey, and can be used for various purposes, among which are applications in the food and pharmaceutical industries. Moreover, XOS are moderately sweet, stable over a wide range of pH and temperatures, and have organoleptic characteristics suitable for incorporation into foods.
- *OS*: GOSs, one important component in OS, can benefit the maturation of the immune system in infants by increasing *Bifidobacterium* directly. Recently, the effect of OS on maternal/newborn gut microbiome, glucose metabolism, and systematic immunity in pregnancy were studied. The specific composition of OS compounds is GOS (6 g/100 g) and sialic acid (3 g/100 g). This compound, used at a dose of 60 g per day, two times a day (once in the morning and evening) induced modifications of maternal/neonatal health through the gut microbiome beginning in the first trimester (Wang et al., 2019). The follow-up period covers the entire gestation and is extended to 42 days postpartum to evaluate the long-term effects through broad modification of gut microbiota. Prebiotics might improve immune functions and correct disturbances in lipid and glucose metabolism.

Resistant starch (nonfructans)

Nonfructan prebiotics or resistant starches are also under investigation for their fermentation characteristics (some specific physical properties), prebiotic effect, and health benefits. Resistant starch is the sum of starch itself and the products of its decomposition that are neither being digested nor absorbed in the small bowel. This includes the portion of starch that can resist digestion by human pancreatic amylase in the small intestine, and thus reach the colon. The general behavior of resistant starch is physiologically similar to that of soluble and fermentable fiber, that is, guar gum. Resistant starch has been the subject of numerous studies that document a prebiotic effect, both as a single ingredient and in combination with FOS (Anadón et al., 2019).

Resistant starch has a long history of safe consumption and is a natural component of many common foods, including grains, cereals, vegetables (e.g., raw potatoes), legumes, seeds, some nuts, and unripe fruits (e.g., bananas), as well as cooked and cooled starchy products (retrograde starch). Appreciable amounts of resistant starch exist in many commercial food products due to the effect of processing (Douglas and Sanders, 2008). Resistant starch is also specifically manufactured for its use in the food industry. Doses are approximately 20 g/day, but lower dose ranges of 2.5−5 g/day have demonstrated a prebiotic effect. The difference in dosing is due to the varying fermentation profiles of prebiotic ingredients. A minimum intake of resistant starch of 5−6 g/day appears to be needed for beneficial reductions in insulin response. As a food ingredient, resistant starch has a lower calorific value compared with fully digestive starch, which can be a substitute for digestive carbohydrates, thus lowering the energy content of the final formulation.

The bread and cereal categories are filled with products that include meaningful amounts of resistant starch or inulin for fiber content and sometimes energy reduction. As previously indicated, it has been reported that 20 g/day of resistant starch is a minimum healthy dose (Cassidy et al., 1994). Bouhnik et al. (2004) found that short-chain FOS, SOS, GOS, and type III resistant starch measurably raised fecal counts of *Bifidobacterium* species at reasonable dose ranges of 2.5−5 g/day within 7 days of administration.

The resistant dextrins are defined as short-chain glucose polymers without sweetness that have strong resistance to the hydrolytic action of human digestive enzymes. There are three major groups of dextrins: maltodextrins produced by hydrolysis of dispersed starch by action of liquefying enzymes such as amylase; degradation products produced by acid hydrolysis of dispersed starch, cyclodextrin, and pirodextrins. The prebiotics usually used and the candidate ones are listed in Table 64.2.

Effects of prebiotics

Prebiotic oligosaccharides have recently been recognized as components of dietary fiber because of their interesting physiological effects, which are similar to those of well-known "*soluble fibers*" (Flamm et al., 2001; Anadón et al., 2019). The prebiotic carbohydrates are not digested by human enzymes but are fermented by the flora of the large intestine. Thus, they increase biomass, feces weight, feces frequency, and have a positive effect on constipation and the health of the mucosa of the large intestine (Cherbut, 2002; Nyman, 2002). Dietary fibers mostly include nonstarch polysaccharides such as celluloses (general components of plant cell walls), hemicelluloses, gums and pectins (cell wall components of many fruits), lignin, resistant dextrins, and resistant starches. Foods rich in soluble fiber include fruits, oats, barley, and beans.

Chemically dietary fiber means carbohydrate polymers with a degree of polymerization not lower than 3, which are neither digested nor absorbed in the small intestine. Based on their water solubility, dietary fibers may be divided into two forms: (1) insoluble dietary fiber (i.e., celluloses, some hemicelluloses, lignins) which is fermented to a limited extent in the colon; and (2) soluble dietary fiber (i.e., β-glucans, pectins, gums, mucilages, and hemicelluloses) that are fermented in the colon (Das et al., 2012). β-Glucans are found in oats and barley cereals. Dietary fiber increases the fecal bulk by two mechanisms: incompletely fermented types of fiber bind water throughout the GIT, whereas readily fermented types of fiber contribute by increasing the microbial mass (Cummings, 1993). In adults, it is recommended that dietary fiber intake be in the range of 20–35 g/day. Inulin and FOS seem to have dose-related laxative effects. Studies suggest that soluble fibers help to regulate the digestive system by increasing micronutrient absorption, stabilizing blood glucose, lowering serum lipids, and possibly preventing cardiovascular disease. Fermentation of oligosaccharides in the ceco-colon could contribute to preventing colon cancer (De Vrese and Schrezenmeir, 2008; Delzenne, 2003).

A summary of mechanisms on how prebiotics influence human health is presented in Table 64.1 and Fig. 64.1.

There are several prebiotic functions that are summarized in Fig. 64.2.

Gastrointestinal tract

One important characteristic of oligosaccharides, once ingested, is their relative resistance to digestion by hydrolytic enzymes secreted into, or active in, the intestine (e.g., α-glucosidase, maltase, and isomaltase). This is dependent on the extent of polymerization. IMO have good resistance with an extent of polymerization >3. The links between the hexose moieties of β-1–2 fructans are completely resistant, whereas β-2-6-linked FOS may be partially hydrolyzed at low pH. Oligosaccharides, which mostly escape digestion in the upper GIT, are important sources of energy for bacteria in the ceco-colon that express enzymes such as β-fructosidase, β-galactosidase, xylanase, or any other hydrolases (Bernalier et al., 1999). The ingestion of some oligosaccharides may lead to the proliferation of certain types of bacteria that are generally considered to be beneficial (e.g., bifidobacteria, lactobacilli, nonpathogenic *E. coli*, and decreasing *Bacteroidaceae*) to the detriment of harmful bacteria. The (re)equilibration of the colonic biotope, first defined as the "*prebiotic effect*" by Gibson and Roberfroid (1995), has been demonstrated for dietary fructans in many studies in animals and in humans. Promising results have also been obtained with other oligosaccharides (Delzenne and Williams, 2002; Anadón et al., 2019). Recent data obtained with FOS demonstrate that the dose and duration of oligosaccharide intake, the location of fermentation (proximal or distal colon), as well as the initial composition of fecal flora, are important factors influencing the extent of the prebiotic effect, that is, the increase in bifidobacteria (Rao, 2001; Tuohy et al., 2001a,b).

The SCFA have important effects in the intestinal tract. It is largely accepted that butyrate plays an essential role in maintaining the metabolism, proliferation, and different epithelial cell types (Blottière et al., 1999), and it has also proven to exert an antiinflammatory action. The butyrate produced during colonic dietary fiber fermentation supports gut barrier function through the stimulation of mucus production and tight junction assemblies, minimizing antigen translocation and inflammation.

Interestingly, short-chain FOS known to produce substantial amounts of butyrate in the colon, increase mucosal crypt height and epithelial cell density in neonatal pigs. "Protective" sulfomucins in the ceco-colonic mucosa

TABLE 64.1 Mechanisms of prebiotic action.

- Increase in the expression or change in the composition of short-chain fatty acids (SCFAs) (lactic, acetic, propionic, and butyric acids) in colonocytes during fermentation of prebiotic carbohydrates
- Increased fecal weight and a mild reduction in luminal colon pH
- A more acidic pH and modulation of the intestinal flora, especially growth stimulation of carbohydrate-fermenting bacteria, decreased the concentration of putrefactive, toxic, mutagenic, or genotoxic substances and bacterial metabolites, as well as of secondary bile acids and cancer-promoting enzymes
- Bifidobacteria and lactobacilli (increased by oligosaccharides) exhibit low β-glucuronidase and nitroreductase activity
- Decreased nitrogenous end products and reductive enzymes
- Production of butyric acid reinforces the regeneration of intestinal epithelium (i.e., through its proapoptotic potency)
- Increased expression of the binding proteins or active carriers associated with mineral absorption
- Enhanced immunity and modulation of mucin production

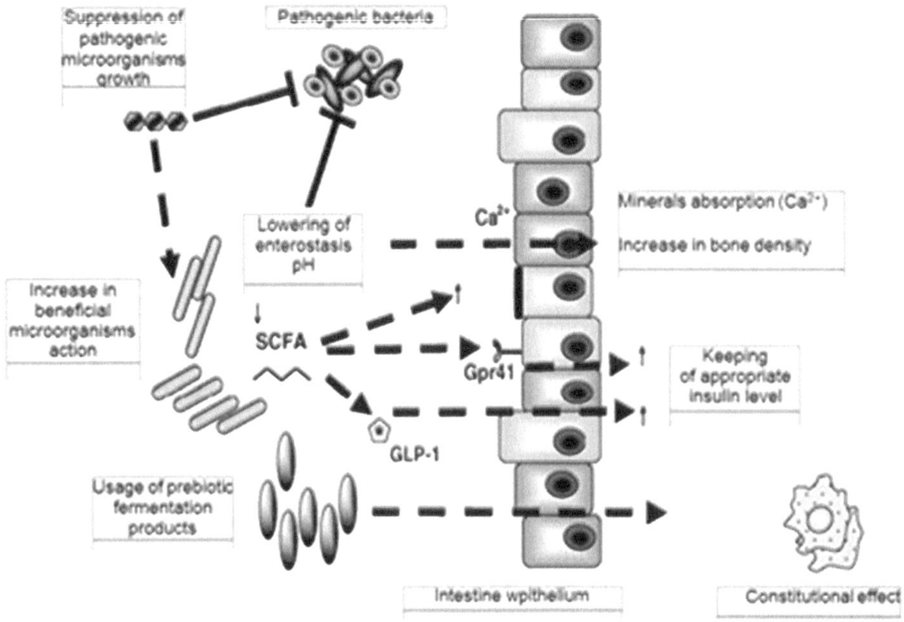

FIGURE 64.1 Mechanism of prebiotic action.

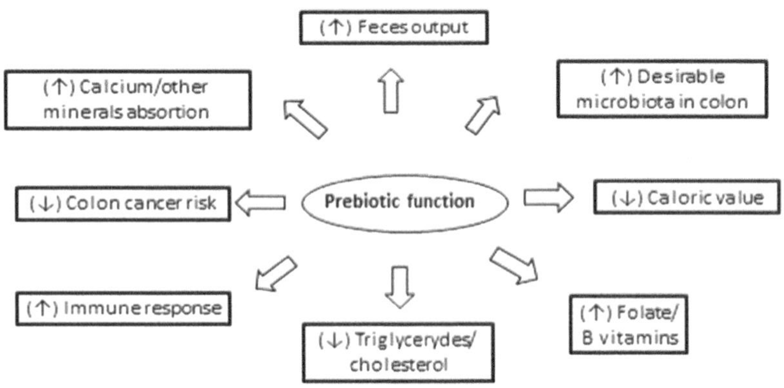

FIGURE 64.2 Prebiotic function.

are higher in rats given human fecal flora and fed a diet enriched in FOS (Cherbut, 2002; Cherbut et al., 2003). All these effects could have contributed to the amelioration of symptoms associated with colitis in several experimental models (Butel et al., 2002; Cherbut, 2002). Butyrate also supports gut barrier function through the stimulation of tight junction assemblies and mucus production.

Intestinal immunity is also enhanced by dietary oligosaccharides (Fig. 64.1; Slizewska et al., 2012). Synthetic fructans at a dose of 50–80 g/kg diet increase Peyer's patches in mice, and at a dose of 100 g/kg promote cecal and colonic macrophages in rats (Schley and Field, 2002). The mechanism of such an effect is unknown but could be attributed to the prebiotic effect of oligosaccharides; bifidobacteria are able to stimulate production of cytokine (TNF-α, IL-6) and reactive molecules (NO, H_2O_2) by macrophages in vitro. Inulin consumption increases the phagocytic capacity of macrophages and the production of secretory immunoglobulin A (IgA-s), which plays an important role in the defense of the GIT. In the elderly, prebiotics GOS can promote antiinflammatory factors (IL-10) or reduce proinflammatory cytokines (IL-6, IL-1β, TNF-α) (Vulevic et al., 2015).

The accessibility of bacteria to Peyer's patches remains an argument against a direct effect of intact lactic acid bacteria (LAB) as stimulators of immunity via GALT. Bacterial cell by-products (cell wall peptidoglycans, cytoplasmic antigens) are more credible candidates (Schley and Field, 2002). Whatever the mechanism, it is quite interesting that an overall improvement in immunity (fewer physician visits, fewer fever episodes, higher immunological response) has also been reported in human subjects (infants and adults) ingesting a diet enriched in short-chain FOS (Saavedra and Tschernia, 2002).

Results suggest that fermentation of oligosaccharides in the ceco-colon could contribute to the prevention of colon cancer in several ways: (1) enhanced immunity and modulation of mucin levels. Increased mucin production improves intestinal barrier function because epithelial cells of human intestine express and/or secrete mucin glycoprotein; (2) the production of butyrate, which could exert a role in the regulation of altered cell proliferation through its proapoptotic potency; and (3) the "*prebiotic effect*" of bifidobacteria and lactobacilli, whose growth is promoted by several oligosaccharides, exhibit low β-glucuronidase and nitroreductase activity. This phenomenon could contribute to lower activation of carcinogenic molecules in the colon (Delzenne, 2003; Slizewska et al., 2012) (Fig. 64.1).

Prebiotics and resistance to gastrointestinal infections

The gut microflora and the mucosa may act as barriers against invasion by potential pathogens. Bifidobacteria and lactobacilli can inhibit pathogens like *E. coli*, *Campylobacter*, and *Salmonella* spp. The lactic microflora of the human GIT is thought to play a significant role in improved colonization resistance (Gibson et al., 2005). These authors stated different mechanisms of: (1) metabolic end products, such as acids, excreted by these microorganisms may lower the gut pH, in a microniche, to levels below those at which pathogens are able to effectively compete; (2) competitive effects from occupation of normal colonization sites; (3) direct antagonism through natural antimicrobial excretion (LAB produce inhibitory peptides); (4) competition for nutrients and blocking of pathogen adhesion sites in the gut; and (5) enhancement of the immune system. Moreover, many lactobacilli and bifidobacteria species are able to excrete natural antibiotics, which can have a broad spectrum of activity (Gibson et al., 2005).

A potential correlation exists with reduced pathogen resistance, decreased numbers of beneficial commensal bacteria genus of bifidobacteria in the elderly, and the production of natural resistance factors. In essence, the natural gut flora may have been compromised through reduced bifidobacteria numbers and may have a diminished ability to deal with pathogens. If prebiotics are used to increase bifidobacteria or lactobacilli so that they are the numerically predominant genus in the colon, then improved colonization resistance will result.

Several studies have been conducted using human subjects, although the dose, substrate, duration, and number of volunteers vary. A general observation was a greater bifidogenic effect of substrates in subjects with a low initial bifidobacteria count (107/g feces) than in those with a higher initial number (109.5/g feces) (Hidaka et al., 1986). A negative correlation between bifidobacteria and *Clostridium perfringens* was also observed, suggesting that the former may inhibit growth of the latter in the intestine, supporting earlier studies (Wang and Gibson, 1993; Gibson and Wang, 1994).

Although inulin and oligofructose have no direct immunogenic effects, by influencing the intestinal flora, they can indirectly modulate various parameters of the immune system (i.e., NK-cell activity, secretion of IL-10 and interferon, and lymphocyte proliferation).

Mineral homeostasis and nitrogen-containing substrates

The distal part of the intestine plays an important and often underestimated role in mineral absorption, and is also important in the excretion of nitrogen. Fermentation of the oligosaccharides (FOS, GOS) reaching the ceco-colon contributes to increased cation solubility (i.e., through the decrease in pH). This effect may facilitate the dissociation of bivalent cation–phytate complexes, and is thus the "*antinutrient effect*" of dietary components (Greger, 1999; López et al., 2000). Dietary intake of particular food products with a prebiotic effect has been shown, particularly in adolescents but also tentatively in postmenopausal women, to increase Ca^{2+} absorption as well as bone Ca^{2+} accretion and bone mineral density. The effect on other minerals (Mg^{2+}, Fe^{2+}, and Zn^{2+}) is unclear.

Several factors influence the potential of oligosaccharides to promote ion absorption: (1) the effect is only associated with certain cations (mainly Ca^{2+} and Mg^{2+}, and, to a much lower and less well established extent, Zn^{2+}, Cu^{2+}, and Mn^{2+}); (2) the structure and extent of polymerization, together with the pH-modulating effect of oligosaccharides; (3) the duration of oligosaccharide treatment (e.g., in rats supplemented with arabinoxylans there is an increase in the soluble cecal Ca^{2+} pool up to day 20 of treatment); and (4) the presence of other nutrients (Ca^{2+} and Mg^{2+}) absorption is increased by FOS in the presence of resistant starch; FOS decrease the "*antinutrient effect*" of phytate (López et al., 2000; Younes et al., 2001).

The improvement in Ca^{2+} availability associated with some oligosaccharides (i.e., inulin-type oligosaccharides) is now considered an interesting "*functional effect*" of these nutrients and has been reported to stimulate calcium absorption in both experimental animals and humans (Scholz-Ahrens et al., 2001). Lowering the pH in the gut improves absorption of calcium, iron, and magnesium in the large intestine, probably due to an increased mineral solubility.

Oligosaccharide feeding may also influence polyamine metabolism. The fermentation of FOS in rats is associated with an increase in putrescine and a decrease in spermidine in cecal contents, a phenomenon that probably reflects modulation of the bacterial ecosystem (Delzenne and Williams, 2002). However, the increase in total polyamine concentration in the ceco-colonic tissue itself reflects an increase in ornithine decarboxylase activity (related to cell proliferation) rather than absorption of bacterial-derived polyamines. Dietary FOSs effectively enhance both fecal and renal nitrogen excretion in rats (Roberfroid and Delzenne, 1998; Fig. 64.1). It has been proposed that their osmotic effect in the small intestine allows the transfer of urea into the distal ileum and the large intestine, where a highly ureolytic microflora proliferates. When fermentable oligosaccharide intake is high, the amount of NH_4 required to sustain maximal bacterial growth may become insufficient, and blood urea is then required as a ready source of nitrogen for protein synthesis by cecal bacteria (Scheppach et al., 2001). The relationship between fermentable FOSs and hyperammonemia, which appears to be a fermentation-dependent process, has been demonstrated for lactulose (usually found in milk and milk products which undergo thermal treatment and are used in the baby food and pharmaceutical industries), which is used as a pharmacological agent in the treatment of hepatic encephalopathy. The treatment relies on suppressing the production of toxic substances in the intestine and is most commonly done with the laxative lactulose or nonabsorbable antibiotics (Delzenne, 2003; Marteau and Boutron-Ruault, 2002).

Systemic effects

There are preliminary indications of a hypotriglyceridemic effect of inulin-type fructans in humans with mild hypertriglyceridemia (Jackson et al., 1999), which might point to a possible interference with lipid absorption. The high proportion of propionate produced in the cecum (which reaches the liver through the portal vein) is, at least in animals, a key factor in the reduction of hepatic triacylglycerol synthesis observed when oligosaccharides are fed to normal and obese Zucker rats. SCFA are known to modulate hepatocyte metabolism. Acetate may be considered to be both a lipogenic and cholesterogenic substrate, whereas propionate acts as an inhibitor of hepatic lipid synthesis (Demigné et al., 1999). Propionate decreases lipogenic gene expression in cultured rat hepatocytes and competitively inhibits monocarboxylate transporters specific to acetate transfer to hepatocytes, therefore, to a large extent, acetate is not incorporated into liver cholesterol and fatty acids (Delzenne and Williams, 2002). The relevance of such a mechanism in humans remains controversial and will require further studies with appropriate techniques (e.g., the use of a stable isotope substrate).

Reported effects of oligosaccharides on circulating blood lipids (lipid metabolism) in human subjects are variable (Mitsuoka et al., 1987; Kohmoto et al., 1988). Both positive and negative outcomes have been obtained in a small number of well-designed human studies that investigated the effect of dietary supplementation with FOS (doses from 8 to 20 g/day) exhibiting prebiotic properties (Delzenne and Williams, 2002).

Three of 11 studies reported significant reductions in serum triacylglycerols, whereas five studies reported modest reductions in total cholesterol and low density lipoproteins (LDL) cholesterol. Studies have been conducted in both normohyperlipidemic and moderately hyperlipidemic subjects. There appear to be no obvious differences in the type of oligosaccharide, dosages, or duration of treatment between studies reporting negative and positive outcomes. Subjects with serum cholesterol higher than 2500 mg/L tended to have the greatest reduction in cholesterol after supplementation with inulin-type oligosaccharides. Research regarding the putative effect of FOS supplementation on hepatic steatosis and lipogenesis in obese subjects and in nonalcoholic steatohepatic patients is in progress because effects on plasma triacylglycerol concentrations remain to be elucidated for inulin or oligofructose (Agerholm-Larsen et al., 2000).

In another study, Roberfroid (2000) reviewed the possible effect of inulin-type fructans on the modulation of triacylglycerol metabolism. Two effects were discussed in this study. The first effect was the modification of glucose or insulin concentrations, because dietary modulation of lipogenesis is often linked to such physiologic changes. The second effect was the production of SCFA in the large bowel, which results in an increase in the portal concentration of both acetate and propionate.

There is less evidence of the beneficial effects of prebiotics on other symptoms and diseases associated with the metabolic syndrome (overweight/obesity, disorders of lipid metabolism, atherosclerosis, hypertension, insulin resistance/diabetes). Prebiotics targeting the gut microbiota and modulating metabolic and immune syndrome have been very effective in some populations, particularly in the nonpregnant population with metabolic syndrome.

Some effects of oligosaccharides on glucose metabolism and hormone metabolism have been described. Feeding FOS at 100 g/kg diet to male Wistar rats for 30 days reduces postprandial glycemia without affecting the glycemic response during a glucose tolerance test after overnight fasting (Kok et al., 1998; Daubioul et al., 2000). Feeding streptozotocin-treated (diabetic) rats a diet containing 100 g oligofructose/kg for 3 weeks decreases postprandial glycemia and partially restores insulin secretion. This effect seems to be linked to a higher secretion of glucagon-like peptide 1 in the portal vein as a result of an increase in the colonic production of this peptide

(Kok et al., 1998; Cani et al., 2007). The fact that nondigestible or fermentable carbohydrates may influence intestinal hormone production is interesting, and it could be suggested that these hormones may be a link between the outcome of fermentation in the lower part of the gut and systemic consequences of "colonic food," such as oligosaccharide intake.

In vitro testing is not sufficient for prebiotic qualification or claims of efficacy because this method cannot approach the dynamic nature of colonic metabolism. There are also method limitations that involve the metabolism of the resident microflora as well as that of the host. These factors contribute to wide variations in measurable colony-forming unit counts, SCFA and enzyme levels, and other measurements of outcome (Blaut, 2002). Many factors can confound results, including the chemical composition of the proposed prebiotic, its fermentation profile, study design, the baseline distribution of a subject's colonic microbiota, the methodologies used in observing an effect in a particular subject group, and the statistical constructs used to interpret the data (Scholz-Ahrens et al., 2001).

Recent evidence supports the involvement of the GIT in Parkinson disease pathogenesis, including alterations in microbiota and intestinal permeability. The communication between the GIT and the central nervous system is bidirectional and it is hypothesized that Parkinson disease arises in the gut and spreads to the brain via the vagus nerve, which helps to propagate the α-synuclein that targets the brain. One report indicates that oxidative stress, mitochondrial failure, neuroinflammation, protein aggregates, failure of clearance mechanisms, excitotoxicity, gene mutation, and prion-like phenomena are the prime factors involved in the pathogenesis of Parkinson disease and are interconnected to initiate pathological insult (Sharma et al., 2019).

With the exception of being the preferred energy source for colonic epithelial cells, butyrate is involved in antiinflammatory, enteroendocrine, and epigenetic mechanisms that influence colonic and systemic health, including brain function. A few studies using oral administration of sodium butyrate indicate beneficial effects in Parkinson disease animal models (Cantu-Jungles et al., 2019). Insoluble substrates such as chitin–glucan complexes, β-1,3/1,6-D-glucan, whole-grain barley, wheat bran, acetylated galactoglucomannan, highly acetylated arabinoglucuronoxylan, wheat bran, and coarse wheat bran are capable of promoting *butyrogenic colonic bacteria* such as, respectively: *Clostridium* cluster XIVa, including *Roseburia* spp., *Eubacterium rectale*, *Roseburia faecis*, and *Roseburia intestinalis*, members of the *Lachnospiraceae* family, *Coprococcus eutactus*, *Roseburia*, and other *Lachnospiraceae* family members (Cantu-Jungles et al., 2019). Proinflammatory dysbiosis of patients with Parkinson disease suggests that this dysbiosis could provoke inflammation-induced misfolding of the protein α-synuclein and development of Parkinson disease pathology (Keshavarzian et al., 2015). Therefore prebiotic fiber attributes (for example, solubility degree, and chemical and physical structures) may be important in permitting butyrogenic bacteria to compete against Gram-negative carbohydrate-utilizing bacteria for a more targeted prebiotic approach (Cantu-Jungles et al., 2019).

Effects of prebiotics on infants

In infants, it has been demonstrated that the prebiotic effect includes a significant change of gut microbiota composition, particularly an increase of fecal concentration of bifidobacteria accompanied by improvement in stool quality (i.e., pH, SCFA, frequency, and consistency), reduction in the risk of gastroenteritis and infections (Mitsuoka, 1990), improvement in general well-being, and a reduction in the occurrence of allergic symptoms such as atopic eczema (Roberfroid et al., 2010).

Currently, there are only limited studies evaluating prebiotic substances in dietetic products for infants. Although administration of prebiotic oligosaccharides has the potential to increase the total number of bifidobacteria in feces and may also soften stools, there is no published evidence of clinical benefits to adding prebiotic oligosaccharides to dietetic products for infants.

Effects of prebiotics on adults and the elderly

Although the difference in fecal microbiota composition between the elderly and younger adults is not always reported to be the same, reduction in bifidobacteria is a recurring topic. Many probiotics have been selected for their so-called bifidogenic activity and because they may increase fecal bifidobacteria levels (Tuohy et al., 2001a, b). Changing the microbiota by increasing levels of bifidobacteria is not necessarily a health benefit and thus should always be correlated with other health benefits. The effect of a prebiotic on potential intestinal pathogens has not been sufficiently investigated in the elderly. It can be speculated that colonic pathogens in particular may be affected by prebiotics. For prebiotics, a typical health benefit for older individuals would be improved bowel function. Although this has been a well-documented benefit in adults, it has not been reported for the elderly.

Use of prebiotics

Although prebiotics and probiotics probably share common mechanisms of action (especially modulation of endogenous flora), they differ in their composition and metabolism. The fate of prebiotics in the GIT is better known than that of probiotics. It is generally accepted that prebiotics have a selective effect on the microbiota that results in improved health of the host. Therefore, a substance is considered prebiotic when it has the following benefits: (1) resistance to digestion, (2) fermentation by the large intestinal microbiota, and (3) a selective effect on the microbiota. Prebiotics, like other low-digestible carbohydrates, exert an osmotic effect in the GIT as long as they are not fermented. When they are fermented by endogenous flora (i.e., at the place where they exhibit their prebiotic effect), they also increase intestinal gas production (Roberfoid and Slavin, 2000). The prebiotic or rather bifidogenic effects depend on the type and concentration of the prebiotic and bifidobacteria concentration in the intestine of the host; no simple dose–effect relationship exists. Only carbohydrates like inulin [β(2-1-)-fructans], oligofructose [β(2-1-)-fructans], GOS/TOS (galactose oligomers and some glucose/lactose/galactose units), or lactulose, synthetic, nondigestible sugar, which are nondigestible but can be fermented by the intestinal flora, fulfill the criteria (De Vrese and Schrezenmeir, 2008). Inulin-type fructans, are the best documented oligosaccharides for their effect on intestinal bifidobacteria and are considered important prebiotic substrates.

Commonly used prebiotics and prebiotic candidates are shown in Table 64.2.

With the exception of inulin (a mixture of FOS and polysaccharides), known prebiotics are mixtures of undigestible oligosaccharides (i.e., chains consisting of 3–10 carbohydrate monomers). Oligosaccharides are carbohydrates of low molecular weight with polymerization values (2 and 9) that exhibit properties typical of dietary fibers and are found in several vegetables as fructans (i.e., asparagus, onions, garlic, and leeks), stachyose in soybeans, and oligosaccharides in human breast milk and bovine milk. Oligosaccharides are readily water-soluble and exhibit some sweetness that decreases with increasing chain length. Water-binding and gelling properties, and therefore use as a fat substitute, increase with the number of hexose molecules and reticulation (Delzenne, 2003).

The glycosidic bonds of oligosaccharides are resistant to hydrolysis by intestinal digestive enzymes and therefore are poorly degraded in the upper regions of the GIT. Therefore intact oligosaccharides that reach the colon serve as fermentable substrate. The colonic microbes ferment nondigestible oligosaccharides to produce SCFA (i.e., acetic, propionic, and butyric acid), lactic acid, and gases (i.e., carbon dioxide, methane, hydrogen). Finally, it is known that the ingestion of prebiotics can elevate indigenous *Bifidobacterium* and *Lactobacillus* levels in the colon. Due to fermentation in the large intestine, ingestion of higher quantities of prebiotics may lead to flatulence, abdominal disorders, and diarrhea. High levels of oligosaccharides (i.e., >10 g/day) may produce intestinal discomfort and flatulence.

Several oligosaccharide have been assessed for their prebiotic potential. The dose and duration for nutrition purposes are as follows: inulin, 8–40 g/day, 15–64 days;

TABLE 64.2 Therapeutic use of prebiotics.

	Prebiotics to be used	Possible mechanism	References
Alleviation of constipation	Lactulose, fructooligosaccharides (FOS), galactooligosaccharides (GOS)	Osmotic effect and modulation of indigenous microflora	Gibson et al. (2004) and Conway (2001)
Treatment of hepatic encephalopathy	Lactulose, Lactilol	Bacterial incorporation of nitrogen and acidification of the colonic environment which in turn reduces the breakdown of nitrogen-containing compounds to ammonia and other potential cerebral toxins	Delzenne (2003) and Marteau and Boutron-Ruault (2002)
Inflammatory bowel diseases (IBD)	Inuline, fructooligosaccharides (FOS), galactooligosaccharides (GOS)	Regulating immune responses to commensal and pathogenic bacteria	Cherbut et al. (2003), Schultz et al. (2004), Furrie et al. (2005), and Kelly et al. (2005)
Prevention of cholesterol gallstones	Oligosaccharides: fructooligosaccharides (FOS), isomaltooligosaccharides (IMO), galactooligosaccharides (GOS), palatinose condensate, raffinose, and soybean oligosaccharides (SOS)	Stimulating the growth of bifidobacteria in vitro and in vivo	Mitsuoka et al. (1987) and Kohmoto et al. (1988)
Prevention of infections of intestinal origin	Oligosaccharides	Contributing to a greater resistance to infection. Most *Bifidobacterium* species have a scavenging function	Mitsuoka (1990)

FOS, 4–12.5 g/day, 8–12 days; GOS, 7.5–15 g/day, 7–21 days; SOS, 10 g/day, 21 days; and lactulose, 3–20 g/day, 14–28 days (Conway, 2001). When fed to patients with chronic constipation at doses greater than 20 g/day, lactulose increased fecal output. At lower doses, lactulose acts as a prebiotic within the colonic microflora, increasing numbers of bifidobacteria. Overall, dosage levels for the most health benefits will range from 3 g/day for short-chain FOS to 8 g/day for mixed short- and long-chain inulin, although more may be safely consumed according to individual tolerance (Marteau and Flourie, 2001; Douglas and Sanders, 2008). Rao (1999) reviewed the dose in relation to the extent of the elevation of bifidobacteria and indicated that, even with a dose of saccharide that ranged from 8 to 40 g/day, there was no correlation with the resultant elevation of bifidobacteria.

Prebiotics can also be used as a supplement and special food. Supplements may provide an easy way to boost prebiotic fiber consumption, giving consumers a clear, convenient, and fool-proof way of obtaining a particular type of prebiotic and dose level. Prebiotic supplements can be sprinkled directly on food; stirred into beverages, or taken as capsules, tablets, or chewables. Because the most commonly available prebiotics are water-soluble and completely clear in water, they are easily incorporated into most foods and are undetectable. Special foods such as sports drinks, weight-loss powders, ready-to-drink protein meal replacers, and nutrition bars provide a popular way for people to obtain prebiotic fiber. These food items often contain FOS, some form of inulin, or resistant starch for their fiber content and prebiotic advantages, although there may be no prebiotic-associated label claims (Douglas and Sanders, 2008)

New prebiotic compounds are emerging because of interesting physiological effects (e.g., low energy value, low carcinogenicity, prebiotic effect, improvement of mineral absorption). The addition of some oligosaccharides to foodstuffs that normally contain low or negligible amounts induces an increase of bifidobacteria, especially in feces.

Use of prebiotic for medical purposes

Prebiotics have been used for medical purposes. Frequently, they are used in enteral nutrition products. They are used in adult and pediatric patients presenting with a wide range of medical conditions, including diabetes, cancer, renal failure, pressure ulcers, metabolic stress, trauma, and immunosuppression (Ross Products Division, 2005; Anadón et al., 2016). Prebiotic enrichment of these liquid products is used as a means to provide SCFA to colonocytes via fermentation, to normalize and maintain bowel function and colon integrity, and to build colonization resistance in the hospital setting. These characteristics make prebiotics appropriate for use in patients with antibiotic-associated diarrhea, various irritable bowel conditions (no effects on symptoms of this condition were reported with oligofructose and FOS), including colitis, and for general bowel maintenance while receiving a formulated diet for medical nutrition therapy (Seidner et al., 2005). When used in appropriate amounts, the effect of prebiotic fiber may also lead to an alteration in nitrogen excretion that is advantageous to renal patients (Younes et al., 1995, 2001; Table 64.2).

Intestinal bacteria are useful in the promotion of human health, but certain components of microflora in genetically susceptible individuals contribute to various pathological disorders, including inflammatory bowel disease (IBD). Changes in the gut microbiota composition are classically considered as one of the many factors involved in the pathogenesis of either IBD or irritable bowel syndrome. For these reasons, the use of prebiotics in IBD such as Crohn disease, ulcerative colitis, and pouchitis are very important because they restore the balance of the GI microflora, reducing and preventing intestinal inflammation (Cherbut et al., 2003; Schultz et al., 2004; Furrie et al., 2005; Kelly et al., 2005). These diseases are characterized by persistent mucosal inflammation at different levels of the GIT (Guarner, 2007). In the GIT, the inflammatory capacity of commensal bacteria varies because some bacteria are proinflammatory, whereas others attenuate inflammatory responses.

Wolf et al. (2005) provide an overview of the medical uses of FOS at levels found in enteral products. The use of these products to provide total nutrition will deliver efficacious amounts of prebiotic fiber to hospitalized patients, generally in the range of 10–15 g/day. For patients not receiving formulated diets, simply start with 1 g/day for the first week, increasing by 1 g/week until a 3 g dose is attained. The maximum dose that is generally recognized as safe for all persons older than 1 year of age is 20 g, although much higher doses have also been suggested as safe (Douglas and Sanders, 2008). Finally, it must be highlighted that the use of lactulose, resistant starch, and probiotics may replace antibiotics for hepatic encephalopathy. Lactulose has become an alternative to the use of antibiotics in this pathology (Bongaerts et al., 2005).

Prevention of allergic diseases

Allergic diseases (i.e., atopic dermatitis, respiratory allergies, and food allergies) are now classified as the fourth most common global disease by the World Health Organization. The development of allergic diseases has been linked to the dysfunction of three interdependent biological systems (i.e., microbiota, epithelial barrier, and immune system), operating as an interdependent functional unit to maintain both host protection and immune tolerance. In this context, future strategies based on prebiotic supplementation to prevent allergies may be promising. The prebiotic approach seems to be a powerful tool for the prevention of atopic dermatitis when used as food supplements (Gruber et al., 2010). Whether there is a role for prebiotics in the therapy of established allergic disease remains to be elucidated. Preclinical models have proven the efficacy of prebiotic supplementation to overcome skin, food, and respiratory allergic symptoms. However, infant supplementation with prebiotics has shown limited effects. Maternal prebiotic supplementation during pregnancy to prevent allergy in children seems to be the best window of opportunity to modulate biological systems (Brosseau et al., 2019). Further research options include the therapeutic use of bacterial metabolites or fermentation products, polyunsaturated fatty acids, or possibly in combination with probiotics, prebiotics, or hypoallergenic formulas.

Guidelines of prebiotic safety

According to the FAO Technical Meeting on Prebiotics (FAO, 2007), how to evaluate a product and substantiate that it is a prebiotic is indicated in Fig. 64.3.

Taking into account the flowchart in Fig. 64.3, the steps to be followed are:

- *Product specification/characteristics of the prebiotic.* The component to which the claim of being prebiotic is attributed must be characterized for any given product. This includes source and origin, purity, chemical composition and structure, vehicle, concentration, and amount which is to be delivered to the host.
- *Functionality.* At a minimum, there needs to be evidence of a correlation between measurable physiological outcomes and modulation of microbiota at a specific site (primarily the GIT, but potentially other sites such as the vagina and skin). A specific function at a specific site with the physiological effect and its associated timeframe needs to be correlated. Within a study, the target variable should change in a statistically significant way and change should be biologically meaningful for the target group, consistent with the claim to be supported. Substantiation of a claim should be based on studies with the final product type tested in the target host. A suitably sized randomized control trial (compared to placebo or a standard control substance) is required, preferably with a second independent study. Examples of physiological outcomes due to administration of prebiotics could be: (1) satiety (measured in carbohydrates, fats, total energy intake), (2) endocrine mechanisms regulating food intake and energy usage in the body, (3) effects on absorption of nutrients (e.g., calcium, magnesium, trace elements, protein), (4) reduced incidence or duration of infection, (5) blood lipid and classic endocrine parameters, (6) bowel movement and regularity, (7) markers for cancer risk, and (8) and changes in innate and acquired immunity that are evidence of a health benefit.
- *Qualifications.* The qualifications for a prebiotic can be: (1) *component* (chemical substance or a food-grade component), (2) *health benefit* (measurable and not due to absorption of the component, or due to the

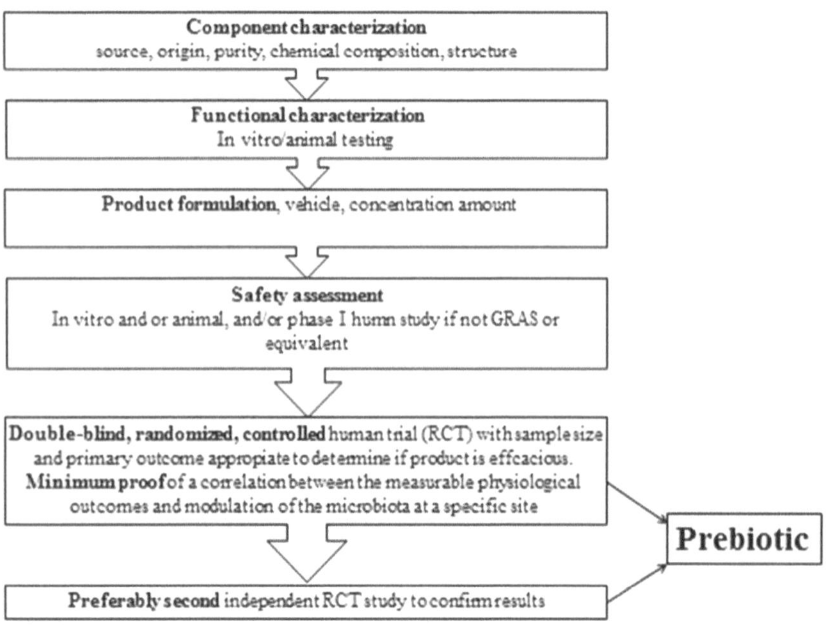

FIGURE 64.3 Guidelines for the evaluation and substantiation of prebiotics.

component acting alone and overriding any adverse effects), or (3) *modulation* (changes in the composition or activities of the microbiota in the target host). A prebiotic can be a fiber, but a fiber is not necessarily a prebiotic. It has been stated that bifidogenic effects are not sufficient without demonstrated physiological health benefits. It is also recognized that determining events that take place within compartments of the intestine are often difficult. Specific site sampling or more sophisticated methods can reliably link microbiota modulation with health benefits. Generally, fecal analysis is suitable.

- *Safety*. It is recommended that the following issues be covered in any safety assessment of a prebiotic final product formulation (Fig. 64.3):
    - If, according to the legislation of the country, the history of safe use of the product in the target host is known, such as generally recognized as safe or its equivalents [i.e., the qualified perception of safety in Europe], then it is suggested that further toxicological tests on animals and human studies may not be necessary.
    - Safe, allowable norms for consumption levels with minimal symptoms and adverse effects should be established.
    - The product must not contain contaminants and impurities. The contaminants should be identified and measured, and the impurities should be well characterized and submitted for toxicity evaluation if needed.
    - The prebiotic cannot change the microbiota in such a way as to cause a long-lasting detrimental effect on the host.

For functional ingredients, animal models can be used to ascertain the target organs and effects that are produced as a result of toxicity. The extent of testing necessary for a functional ingredient is increased in response to the lack of understanding of potential for toxicity as a result of inadequately characterized products. The following criteria must be met to derive a safe level of exposure without additional toxicology testing (Kruger and Mann, 2003): (1) active component(s) and related substances are well characterized and there is adequate understanding of the lack of potential for toxicity at the human dose levels recommended based on existing data from the literature; (2) the impurities are well characterized and there is an adequate understanding of the lack of potential for toxicity based on existing data from the literature; and (3) the manufacturing process is standardized and reproducible.

When either the active component(s) or impurities are not fully characterized or there are not enough data available to evaluate potential for toxicity, the following preclinical toxicological information is needed to assess the functional ingredient: (1) toxicity studies in vitro and in vivo, including mutagenicity studies, reproduction, and teratogenicity studies, (2) pharmacokinetics and special pharmacology studies, and (3) long-term feeding studies following a tiered approach on a case-by-case basis. An element that must be considered in the design of animal studies for functional ingredients is the margin of safety between the no-observed-adverse-effect level (NOAEL) determined in the animal studies and the anticipated human level of intake (Anadón et al., 2014, 2016).

Safety and toxicity considerations

There are certainly safety concerns for the consumer with regards to the selection and dosage of nondigestive substances, mainly carbohydrates, and their ability to be tolerated and the selection of nonpathogenic bacteria strains. However, there is consensus on the prebiotic metabolic substrates (e.g., digestibility, composition, dosage, specificity of metabolization) and on the selection of bacterial strains (e.g., counts, survival of GI passage, growth conditions, nonpathogenicity, nontoxinogenicity, stability, identity) (Przyrembel, 2001; Anadón et al., 2014, 2016). A prebiotic should be an inert and harmless compound that should not react with the body and the prebiotics must be definitely safe.

The safety of inulin-type fructans and galactose and their sources as food or food ingredients is not debated based on their long-term use. However, the Nordic Working Group on Food Toxicology and Risk Evaluation (Nordic Committee of Ministers, 2000) performed a safety evaluation of fructans. The report concluded from the 1–3-week human studies available that adverse effects like flatulence, abdominal pain, bloating, cramps, or diarrhea are unlikely to occur with consumption of 20 g/day or less of FOS by a person weighing 60 kg. This is in agreement with a previous evaluation by the Scientific Committee on Food (SCF) of the European Commission (SCF Scientific Committee for Food, 2003), which defined the no-effect-level (NEL) for laxative effect in humans to be 0.3–0.4 g/kg body wt. Prebiotics also have disadvantages because they may exacerbate adverse effects of simple sugar absorption during active diarrhea. The overall conclusion of this safety evaluation of oligosaccharides was that FOS had no significant effects other than GI symptoms at doses 5–40 times higher than the NEL for laxative effects in humans. Although not sufficiently studied, it is assumed that inulin is better tolerated than FOS (Andersson et al., 2001).

Doses of 10–20 g inulin or oligofructose (hydrolysis product of inulin), regardless of whether ingested in a liquid or a solid meal, are considered to be without side effects. Health-relevant effects of inulin and oligofructose were demonstrated only with daily intakes higher than 8 g/day. In a trial with 80 health probands, the ingested quantity after which at least one of the tested symptoms (headache, belching, flatulence, bowel contractions, or liquid stools) had been observed, was between 31 and 41 g oligofructose, corresponding to 0.04–0.06 g/kg body wt (De Vrese and Schrezenmeir, 2008). The consumption of 80 g/day of oligofructose in one study caused diarrhea in four of 12 test subjects (Clausen et al., 1998).

There has been no study on fructans other than oligofructosyl saccharose as an ingredient of infant or toddler formula. Young infants have an immature gut barrier and are considered at increased risk for allergic symptoms. Considering a report on anaphylactic reactions to inulin and oligofructosyl saccharose in a single 39-year-old man after ingestion of vegetables (salsify, artichoke) and processed food containing inulin, the potential for sensitization should be monitored in future evaluations (Gay-Crosier et al., 2000). The SCF Committee (SCF Scientific Committee for Food, 2003) concludes that based on the data available at this time, fructans other than oligofructosyl saccharose should not be included in infant formulae and toddler formula.

Regarding toxicity of GOS, a 90-day oral (gavage) study was conducted in male and female Sprague–Dawley rats to investigate the safety of GOS syrup at 2500 or 5000 mg/kg body wt/day. Measurements included clinical observations, body weights, food consumption, hematology, clotting parameters, blood chemistries, urinalysis, ophthalmologic examinations, gross necropsies, organ weights, and histological examinations. There were no effects of feeding GOS syrup at either concentration on any parameter except food consumption. Statistically significant decreases (7%–13%) in food consumption were seen in both sexes in GOS syrup-treated animals at 5000 mg/kg body wt/day and animals treated with the FOS control when compared to the reverse osmosis deionized water controls. Thus, a NOAEL for GOS syrup of 5000 mg/kg body wt/day was identified (Anthony et al., 2006).

The novel emerging prebiotics still need further consideration for their safety in humans. It has been noticed that alterations in the intestinal microflora could result in adverse effects, depending on what bacterial populations are stimulated. Moreover, the shift of metabolism toward enhanced butyrate formation may have some disadvantageous consequences, such as supporting the growth of undesirable clostridia (Wang, 2009).

Because of fermentation in the large intestine, ingestion of higher quantities of prebiotics may lead to flatulence, abdominal disorders, and diarrhea. In general, 10–20 g oligofructose or inulin, ingested in a liquid or solid meal, is considered to have no adverse effects.

A recent study was carried out to assess the tolerance and safety of a formula containing a mixture of oligosaccharides in early infancy. Formula-fed infants were randomly fed with a regular formula containing a mixture of neutral oligosaccharides and pectin-derived acidic oligosaccharides (prebiotic formula group). The authors demonstrated the tolerability and long-term safety of such a prebiotic formula in a large cohort of health infants (Piemontese et al., 2011)

Data on oligosaccharide mixtures in infant formula do not demonstrate adverse effects, but further evaluation is recommended. Combinations and dosages of those studied so far need to be fully evaluated with respect to both safety and efficacy before their use in commercial infant food products. Well-designed and carefully conducted randomized controlled trials with relevant inclusion/exclusion criteria, adequate sample sizes, and validated clinical outcome measures are needed for preterm and term infants. Future trials should define optimal quantity and types of oligosaccharides with prebiotic function, optimal dosages and duration of intake, short- and long-term benefits, and safety.

The SCF of the European Commission reviewed the use of nondigestible carbohydrates (i.e., oligofructosyl-saccharose and oligogalactosyl-lactose) in infant and toddler formulas (SCF Scientific Committee for Food, 2003). The report stated that the inclusion of up to 0.8 g/100 mL of a combination of 90% oligogalactosyl-lactose and 10% high-molecular-weight oligofructosyl-saccharose in infant and toddler formulas caused no major concerns. This committee noted that the available trials comprise only a limited number of infants. Some compared formula, whereas others compared components of formula, not just short-chain carbohydrate content. Therefore, further information should be collected regarding the combination of short-chain carbohydrates (or any others) and suitability and safety in infant and toddler formulas. Some trials concluded that, based on available data, fructans other than oligofructosyl-saccharose should not be included in infant and toddler formulas. Moreover, the committee noted that particular attention should be given to the effects on growth and body composition and nutrient bioavailability in infants, particularly with respect to protein and amino acid utilization, and to the effects on water balance, urine output, and urine osmolarity in infants and neonates (SCF Scientific Committee for Food, 2003).

The French Agency for Food Safety (AFFSA, actually termed ANSES) Working Group on Infant Food and Modification of the Intestinal Flora concluded that prebiotics used in infant formula should be fully characterized (i.e., origin, identification, and concentration). The safety of prebiotics should be assessed according to the criteria used for authorized ingredients. Conditions of storage (temperature, humidity) should be mentioned on the product label. Although allergic reactions to prebiotics have not been described in infants and young children, AFSSA recommended follow-up of infants at risk for allergy that were fed prebiotic-containing formula in view of one reported anaphylactic reaction to inulin and oligofructose in an adult (Gay-Crosier et al., 2000). AFSSA also raised concerns regarding the use of prebiotics in infants with congenital or acquired immunodeficiency and in premature newborns. In summary, knowledge on the clinical effects of prebiotic oligosaccharide administration in adults is very limited. Many claims for the potential health benefits of prebiotics remain unsubstantiated. There is no scientific basis for generalizing data obtained in adults to infants and children without further consideration of age-related physiological differences. Prebiotics can be generally regarded as safe, although it was reported that in adults fructans may cause dose-dependent GI side effects (ESPGHAN Committee on Nutrition, 2004). Furthermore, the occurrence of repeated anaphylactic reaction to inulin and oligofructose was reported in one adult (Gay-Crosier et al., 2000). A study in rats showed that dietary FOS and lactulose improve the colonization resistance to *Salmonella enteritidis*, but in contrast to most expectations, concomitantly impair the intestinal resistance of rats to translocation (i.e., the passage of viable bacteria from the GIT through epithelial mucosa) (Van Leeuwen et al., 1994) of this invasive pathogen. No such effects were observed with resistant starch, wheat fiber, or cellulose supplementation in the diet (Gay-Crosier et al., 2000).

Inulin is a fructan and storage carbohydrate found in more than 36,000 kinds of plants, including chicory, artichokes, and salsify. Inulin has been used as a food ingredient for many years, but only recently has the use of inulin and oligofructose, a hydrolysate of inulin, been considerably increased because of their many postulated health benefits and documented properties, such as dietary fiber effect, and their ability to increase levels of certain enteric bacteria, including bifidobacteria. These fructans are now being added to an increasing number of industrially processed foods, such as candies, beverages, yogurts, ice cream, chocolate, butter, and breakfast cereals.

It has been described that a 39-year-old butcher had four episodes of anaphylaxis (generalized wheal-and-flare reaction, laryngeal edema, nasal itching, cough, and breathing difficulties) a few minutes after the ingestion of salsify, artichoke leaves, margarine (Brunch, Migros, Geneva) containing inulin extracted from chicory (Raftiline HP, Orafti, Tienen, Belgium), and a candy (Actilife Toffee orange–carrot, Migros) containing inulin (Raftiline HP) or oligofructose (Raftilose P95). He also had local wheal-and-flare reactions after touching artichokes. All four episodes occurred within a 2-year period.

Skin-prick testing, in which the lancet is first stuck in the material suspected of causing immediate hypersensitivity and then into the patient's skin, disclosed a very strong reaction (diameter of wheal, 12 $\times$ 20 mm) to inulin (Raftiline HP) and strong reactions to salsify, artichoke, a margarine containing inulin (Brunch), candy (Actilife Toffee), and oligofructose (Raftilose P95). The reactions to intradermally injected undiluted inulin for tests of renal function (Inutest 25%) and oligofructose (Raftilose P95) were strongly positive. The results of skin-prick tests with inulin extract (Raftiline HP) in 10 control subjects and intradermal tests with Inutest 25% and oligofructose (Raftilose P95) in three controls were negative. In a double-blind, placebo-controlled food challenge, generalized urticaria, rhinoconjunctivitis, and a 20% decrease in peak expiratory flow rate occurred within 10 min after the ingestion of 10 g of inulin (Raftiline HP) in mint syrup containing rice flour, but not after the ingestion of mint syrup containing rice flour alone. The patient received 8 mg of acrivastine, and the symptoms regressed. An open oral challenge with 40 g of oligofructose (Raftilose P95) was negative in the patient.

Inulin and oligofructose are classified as food ingredients (not as additives) and are considered to be safe to eat. The labeling laws of most countries require the declaration of inulin and oligofructose in processed foods. There are no other reports of anaphylaxis from inulin, including inulin used for renal-clearance studies. However, because of its expanding use in processed foods, allergic reactions to this dietary ingredient may be more frequent than currently recognized. The identification of an association with allergic reactions to natural foods with high inulin content, or to inulin with a high degree of polymerization may provide a lead in such cases.

Oligosaccharides that selectively promote some interesting bacterial species (e.g., lactobacilli, bifidobacteria), and thus equilibrate intestinal microflora, are now termed prebiotics. The pattern of SCFA production in the cecocolon and the prebiotic effect, if demonstrated, are dynamic processes that vary with the type of oligosaccharide (e.g., extent of polymerization, nature of hexose moieties), the duration of treatment, the initial composition of flora, or the diet into which they are incorporated.

Concluding remarks and future directions

Prebiotics are cost-effective and efficient tools to promote the growth and/or activity of certain bacteria in the indigenous flora of the human GIT to beneficially affect host health and well-being. Prebiotics are useful in infants, children, and the elderly. The prebiotic effect has tentatively been investigated for potential health benefits. Prebiotics have potential nutritional advantages leading to preventive and occasionally curative effects against certain diseases and may directly or indirectly induce similar immunomodulation effects. Due to methodological difficulties and insufficient knowledge of the composition of a "*healthy intestinal microbiota*" and the complex interactions between its members, it is hard to deduce concrete preventive or curative health effects from changes in bacterial cell counts (e.g., "*bifidogenic effect*," regarded as positive). As a consequence, proof of health-relevant effects in controlled human intervention studies should be analyzed. Oligosaccharides have been recognized as a component of dietary fiber due to their physiological effects, which are similar to those of well-known "*soluble fibers*." There is a high level of evidence for positive effects of some prebiotic carbohydrates (*dietary fibers*) because they are not digested by human enzymes, but rather are fermented by flora of the large intestine; therefore they increase biomass, feces weight, and frequency, help alleviate constipation, and have a positive effect on the health of mucosa of the large intestine. FOS, inulins, IMO, lactilol, lactosucrose, lactulose, oligofructose, pyrodextrins (i.e., British gums, white dextrins, yellow dextrins), soligosaccharides, TOS, and XOS are oligosaccharides classified as prebiotics that are added to processed foods and supplements. Fermentation of prebiotics produces lactic, acetic, and butyric acids as well as hydrogen (H_2). Inulin and FOS seem to have dose-related laxative effects, indicating potential use of these prebiotics in the treatment of constipation. Promising sources of prebiotics are starch products, particularly the carbohydrate-like compound-resistant starch and products of the partial degradation of starch.

All beneficial effects of prebiotics are more pronounced in youth, which is not surprising given the role of the gut microbiota for the development of normal immune function and the fact that it is established early in life. The individual gut microbiota are surprisingly stable.

In the elderly, due to the age-related decrease in bowel function, reduced immune function and nutrient metabolism are noted. Although the difference in fecal microbiota composition between elderly and younger adults is not always reported to be the same, reduction in bifidobacteria is often a recurring topic.

Many prebiotics have been selected for their so-called "*bifidogenic activity*" and may increase fecal *Bifidobacterium* levels, although this depends on the starting level of the bifidobacteria. Inulin and oligofructose could modulate hepatic lipid metabolism (so-called "*Western-style diet*"), which is rich in fat and calories, and poor in dietary fiber. Supplementation with soluble fibers (e.g., wheat, dextrin) may be useful in individuals with dietary fiber intake that is lower than recommended. Beneficial effects on calcium absorption and bone mineralization were also demonstrated. The use of lactulose, resistant starch, and probiotics could replace antibiotics for hepatic encephalopathy. Lactulose has become an alternative to the use of antibiotics in this pathology.

Future strategies based on prebiotic supplementation to prevent allergic diseases (i.e., atopic dermatitis, respiratory allergies and food allergies) may be promising.

Acknowledgment

This work was supported by Project Ref. RTA2015-00010-C03-03 from Ministerio de Economía, Industria y Competitividad, Spain.

Conflict of interest

None of the authors of this chapter has a financial or personal relationship with other people or organizations that could inappropriately influence or bias the content of the chapter.

References

Agerholm-Larsen, L., Bell, M.L., Grunwald, G.K., et al., 2000. The effect of a probiotioc milk product on plasma cholesterol: a meta-analysis of short-term intervention studies. Eur. J. Clin. Nutr. 54, 856–860.

Anadón, A., Castellano, V., Martínez-Larrañaga, M.R., 2014. Regulation and guidelines of probiotics and prebiotics. In: Ötles, S. (Ed.), Probiotics and Prebiotics in Food, Nutrition and Health. CRC Press, Boca Raton, pp. 91–113. (ISBN: 978-1-4665-8623-9).

Anadón, A., Martínez-Larrañaga, M.R., Ares, I., Martínez, M.A., 2016. Prebiotics and probiotics: an assessment of their safety and health benefits. In: Watson, R.R., Preedy, V.R. (Eds.), Bioactive Foods in Promoting Health: Probiotics and Prebiotics. Elsevier Inc., Academic Press, pp. 3–23. , Chapter 2 (ISBN: 978-0-12-802189-7).

Anadón, A., Martínez-Larrañaga, M.R., Ares, I., Martínez, M.A., 2019. Prebiotics and probiotics in feed and animal health. In: Gupta, R.C., Srivastava, A., Lall, R. (Eds.), Nutraceuticals in Veterinary Medicine. Springer Nature Switzerland AG 2019, pp. 261–285. Available from: https://doi.org/10.1007/978-3-030-04624-8-19.

Andersson, H., Asp, N.-G., Bruce, A., et al., 2001. Health effects of probiotics and prebiotics. A literature review on human studies. Scand. J. Nutr. 45, 58–75.

Anthony, J.C., Merriman, T.N., Heimbach, J.T., 2006. 90-Day oral (gavage) study in rats with galactooligosaccharides syrup. Food Chem. Toxicol. 44, 819–826.

Bernalier, A., Dore, J., Durand, M., 1999. Biochemistry of fermentation. In: Gibson, G.R., Roberfroid, M.B. (Eds.), Colonic Microbiota, Nutrition and Health. Kluwer Academic Publishers, Dordrecht, The Netherlands, pp. 37–53.

Björkstén, B., Sepp, E., Julge, K., et al., 2001. Allergy development and the intestinal microflora during the first year of life. J. Allergy Clin. Immunol. 108, 516–520.

Blaut, M., 2002. Relationship of prebiotics and food to intestinal microflora. Eur. J. Nutr. 41 (Suppl 1), S11–S16.

Blottière, H.M., Champ, M., Hoebler, C., et al., 1999. Les acides gras à chaîne courte: de la production aux effets physiologiques gastro-intestinaux (production and digestive effects of short-chain fatty acids: from production towards gastrointestinal physiological effects). Sci. Aliment. 19, 269–290.

Boehm, G., Stahl, B., Jelinek, J., et al., 2005. Prebiotic carbohydrates in human milk and formulas. Acta Paediatr. Suppl. 94, 18–21.

Bongaerts, G., Severijnen, R., Timmermand, H., 2005. Effect of antibiotics, prebiotics and probiotics in treatment for hepatic encephalopathy. Med. Hypotheses 64, 64–68.

Bouhnik, Y., Raskine, L., Simoneau, G., et al., 2004. The capacity of non-digestible carbohydrates to stimulate fecal bifidobacteria in health humans: a double-blind, randomized, placebo-controlled, parallel-group, dose-response relation study. Am. J. Clin. Nutr. 80, 1658–1664.

Brosseau, C., Selle, A., Palmer, D.J., et al., 2019. Prebiotics: mechanisms and preventive effects in allergy. Nutrients 11, 1841. Available from: https://doi.org/10.3390/nu11081841.

Butel, M.-J., Waligora-Dupriet, A.-J., Szylit, O., 2002. Oligofructose and experimental model of neonatal necrotising enterocolitis. Br. J. Nutr. 87 (Suppl. 2), S213–S219.

Cani, P.D., Neyrinck, A.M., Fava, F., et al., 2007. Selective increases of bifidobacteria in gut microflora improve high-fat-diet-induced diabetes in mice through a mechanism associated with endotoxaemia. Diabetologia 50, 2374–2383.

Cantu-Jungles, T.M., Rasmussen, H.E., Hamaker, B.R., 2019. Potencial of prebiotics butyrogenic fibers in Parkinson's disease. Front. Neurol. 10, 663. Available from: https://doi.org/10.3389/fneur.2019.00663.

Cassidy, A., Bingham, S.A., Cummings, J.H., 1994. Starch intake and colorectal cancer risk. An international comparison. Br. J. Cancer 69, 119–125.

Cherbut, C., 2002. Inulin and oligofructose in the dietary fibre concept. Br. J. Nutr. 87, 159–162.

Cherbut, C., Michel, C., Lecannu, G., 2003. The prebiotic characteristics of fructooligosaccharides are necessary for reduction of TNBS-induced colitis in rats. J. Nutr. 133, 21–27.

Chow, J., 2002. Probiotics and prebiotics. A brief overview. J. Ren. Nutr. 12, 76–86.

Clausen, M.R., Jorgensen, J., Mortensen, P.B., 1998. Comparison of diarrhea induced by ingestion of fructooligosaccharide Idolax and disaccharide lactulose: role of osmolarity versus fermentation of malabsorbed carbohydrate. Dig. Dis. Sci. 43, 2696–2707.

Conway, P.L., 2001. Prebiotics and human health: the state-of-the-art and future perspectives. Scand. J. Nutr. 45, 13–21.

Coppa, G.V., Gabrielli, O., Pierani, P., et al., 1993. Changes in carbohydrate composition of human milk over 4 months of lactation. Pediatrics 91, 637–641.

Cumming, J.H., 1993. The effects of dietary fiber on fecal weight and composition. In: Spiller, G.A. (Ed.), Handbook of Dietary Fibre in Human Nutrition, second ed. CRC, Boca Raton, FL, pp. 263–349.

Das, L., Bhaumik, E., Raychaudhuri, U., et al., 2012. Role of nutraceuticals in human health. J. Food Sci. Technol. 49 (2), 173–183.

Daubioul, C.A., Taper, H.S., De Wispelaere, L.D., et al., 2000. Dietary oligofructose lessens hepatic steatosis, but does not prevent hypertriglyceridemia in obese Zucker rats. J. Nutr. 130, 1314–1319.

De Borba Gurpilhares, D., Paes Cinelli, L., Kato Simas, N., et al., 2019. Marine prebiotics: polysaccharides and oligosaccharides obtained by using. Food Chem. 280, 175–186.

De Vrese, M., Schrezenmeir, J., 2008. Probiotics, prebiotics, and synbiotics. Adv. Biochem. Eng. Biotechnol. 111, 1–66.

Delzenne, N.M., 2003. Oligosaccharides: state of the art. Proc. Nutr. Soc. 62, 177–182.

Delzenne, N.M., Williams, C.M., 2002. Prebiotics and lipid metabolism. Curr. Opin. Lipidol. 13, 61–67.

Demigné, C., Rémésy, C., Morand, C., 1999. Short chain fatty acids. In: Gibson, G.R., Roberfroid, M.B. (Eds.), Colonic Microbiota, Nutrition and Health. Kluwer Academic Publishers, Dordrecht, The Netherlands, pp. 55–69.

Douglas, L.C., Sanders, M.E., 2008. Probiotics and prebiotics in dietetics practice. J. Am. Diet. Assoc. 108 (3), 510–521.

ESPGHAN Committee on Nutrition, 2004. Prebiotic oligosaccharides in dietetic products for infants: a commentary by the ESPGHAN committee on nutrition. J. Pediatr. Gastroenterol. Nutr. 39, 465–473.

FAO, 2007. Technical Meeting Report. FAO Technical Meeting on Prebiotics, Food Quality and Standards Service (AGNS).

Flamm, G., Glinsmann, W., Kristchevsky, D., et al., 2001. Inulin and oligofructose as dietary fiber: a review of the evidence. Crit. Rev. Food Sci. Nutr. 41, 353–362.

Furrie, E., Macfarlane, S., Kennedy, A., et al., 2005. Synbiotic therapy (*Bifidobacterium longum*/Synergy 1) initiates resolution of inflammation in patients with active ulcerative colitis: a randomised controlled pilot trial. Gut 54, 242–249.

Gay-Crosier, F., Schreiber, G., Hauser, C., 2000. Anaphylaxis from inulin in vegetables and processed food. N. Engl. J. Med. 18, 1372–1373.

Gibson, G.R., 2004. Fibre and effects on probiotics (the prebiotic concept). Clin. Nutr. (Suppl. 1), 25–31.

Gibson, G.R., Roberfroid, M.B., 1995. Dietary modulation of the human colonic microflora: introducing the concept of prebiotics. J. Nutr. 125, 1401–1412.

Gibson, G.R., Wang, X., 1994. Bifidogenic properties of different types of fructo-oligosaccharides. Food Microbiol. 11, 491–498.

Gibson, G.R., Saavedra, J.M., Macfarlane, S., et al., 1997. Gastrointestinal microbial disease. In: Fuller, R. (Ed.), Probiotics 2: Applications and Practical Aspects Chapman and Hall. Andover, pp. 10–39.

Gibson, G.R., Probert, H.M., Van Loo, J., et al., 2004. Dietary modulation of the human colonic microbiota: updating the concept of prebiotics. Nutr. Res. Rev. 17, 259–275.

Gibson, G.R., McCartney, A.L., Rastall, R.A., 2005. Prebiotics and resistance to gastrointestinal infections. Br. J. Nutr. 93 (Suppl. 1), S31–S34.

Greger, J.L., 1999. Nondigestible carbohydrates and mineral bioavailability. J. Nutr. 129 (7), 1434S–1435S.

Gruber, C., van Stuijvenberd, M., Mosca, F., et al., 2010. Reduced occurrence of early atopic dermatitis because of immunoactive prebiotics among low-atopy-risk infants. J. Allergy Clin. Immunol. 126 (4), 791–797.

Guarner, F., 2007. Prebiotics in inflammatory bowel disease. Br. J. Nutr. 98 (Suppl.1), S85–S89.

Hidaka, H., Eida, T., Takizawa, T., et al., 1986. Effects of fructooligosaccharides on intestinal flora and human health. Bifidobacteria Microflora 5, 37–50.

Indrio, F., Neu, J., 2011. The intestinal microbiome of infants and the use of probiotics. Curr. Opin. Pediatr. 23, 145–150.

Jackson, K.G., Taylor, G.R.J., Clohessy, A.M., et al., 1999. The effect of the daily intake of inulin on fasting lipid, insulin and glucose concentrations in middle-aged men and women. Br. J. Nutr. 82, 23–30.

Kelly, D., Conway, S., Aminov, R., 2005. Commensal gut bacteria: mechanisms of immune modulation. Trends Immunol. 26, 326–333.

Keshavarzian, A., Green, S.J., Engen, P.A., et al., 2015. Colonic bacterial composition in Parkinson's disease. Mov. Disord. 30, 1351–1360.

Kobata, A., Yamshita, K., Tachibana, Y., 1978. Oligosaccharides from human milk. Methods Enzymol. 50, 216–220.

Kohmoto, T., Fukui, F., Takaku, H., et al., 1988. Effect of isomalto-oligosaccharides on human fecal flora. Bifidobacteria Microflora 1, 61–68.

Kok, N.N., Taper, H.S., Delzenne, N.M., 1998. Oligofructose modulates lipid metabolism alterations induced by a fat-rich diet in rats. J. Appl. Toxicol. 18, 47–53.

Kol, J., Scholtens, P., Kafka, C., et al., 2005. Colon microflora in infants fed formula with galacto- and fructo-oligosaccharides: more like breast-fed infants. J. Pediatr. Gastroenterol. Nutr. 40, 36–42.

Kruger, C.L., Mann, S.W., 2003. Safety evaluation of functional ingredients. Food Chem. Toxicol. 41, 793–805.

Kunz, C., Rudloff, S., Baier, W., et al., 2000. Oligosaccharides in human milk: structural, functional, and metabolic aspects. Annu. Rev. Nutr. 20, 699–722.

Labeta, M.O., Vidal, K., Nores, J.E., et al., 2000. Innate recognition of bacteria in human milk is mediated by a milk-derived highly expressed pattern recognition receptor, soluble CD14. J. Exp. Med. 191, 1807–1812.

Long, S.S., Swenson, R.M., 1977. Development of anaerobic fecal flora in healthy newborn infants. J. Pediatr. 91, 298–301.

López, H.W., Ouvry, A., Bervas, E., et al., 2000. Strains of lactic acid bacteria isolated from sour doughs degrade phytic acid and improve calcium and magnesium solubility from whole wheat flour. J. Agric. Food Chem. 48 (6), 2281–2285.

Macfarlane, S., Macfarlane, G.T., Cummings, J.H., 2006. Review article: prebiotics in the gastrointestinal tract. Aliment. Pharmacol. Ther. 24, 701–714.

Marteau, P., Boutron-Ruault, M.C., 2002. Nutritional advantages of probiotics and prebiotics. Br. J. Nutr. 87 (Suppl. 2), S153–S157.

Marteau, P., Flourie, B., 2001. Tolerance to low-digestible carbohydrates: symptomatology and methods. Br. J. Nutr. 85 (Suppl. 1), S17–S21.

Mitsuoka, T., 1990. Bifidobacteria and their role in human health. J. Ind. Microbiol. 6, 263–268.

Mitsuoka, T., Hidaka, H., Eida, T., 1987. Effect of fructo-oligosaccharides on intestinal microflora. Nahrung 31, 5–6.

Nagpal, R., Kaur, A., 2011. Synbiotic effects of various prebiotics on in vitro activities of probiotic lactobacilli. Ecol. Food Nutr. 50 (1), 63–68.

Nordic Committee of Ministers, 2000. Nordic Committee of Senior Officials for Food Issues. Nordic Working Group on Food Toxicology and Risk Evaluation. Safety Evaluation of Fructans. Copenhagen. TemaNord, pp. 523.

Nyman, M., 2002. Fermentation and bulking capacity of indigestible carbohydrates: the case of inulin and oligofructose. Br. J. Nutr. 87, 163–168.

Piemontese, P., Gianni, M.L., Braegger, C.P., et al., 2011. Tolerance and safety evaluation in a large cohort of healthy infants fed an innovative prebiotic formula: a randomized controlled trial. PLoS One 6, e28010.

Pietzak, M., 2004. Bacterial colonization of the neonatal gut. J. Pediatr. Gastroenterol. Nutr. 38, 389–391.

Przyrembel, H., 2001. Consideration of possible legislation within existing regulatory frameworks. Am. J. Clin. Nutr. 73 (Suppl), 471S–475S.

Rao, A.V., 1999. Dose-response effects of inulin and oligofructose and intestinal bifidogenesis effects. J. Nutr. 129 (7S), 1442S−1445S.

Rao, V.A., 2001. The prebiotic properties of oligofructuse at low intake level. Nutr. Res. 6, 843−848.

Ringel, Y., Carroll, I.M., 2009. Alterations in the intestinal microbiota and functional bowel symptoms. Gastrointest. Endosc. Clin. N. Am. 19 (1), 141−150.

Rings, E.H.H.M., van Beers, E.H., Krasinski, S.D., et al., 1994. Lactase: origin, gene expression, localization, and function. Nutr. Res. 14, 775−797.

Roberfroid, M.B., 2000. Prebiotics and probiotics: are they functional foods? Am. J. Clin. Nutr. 71, 1682−1687.

Roberfroid, M., 2007. Prebiotics: the concept revisited. J. Nutr. 137, 830−837.

Roberfroid, M.B., Delzenne, N.M., 1998. Dietary fructans. Ann. Rev. Nutr. 18, 117−143.

Roberfoid, M., Slavin, J., 2000. Non-digestible oligosaccharides. Crit. Rev. Food Sci. Nutr. 40, 461−480.

Roberfroid, M., Gibson, G.R., Hoyles, L., et al., 2010. Prebiotic effects: metabolic and health benefits. Br. J. Nutr. 104 (Suppl. 2), 1−63.

Ross Products Division, 2005. Ross Medical Nutrition Pocket Guide. Abbot Laboratories, Columbus, OH.

Rotimi, V.O., Duerden, B., 1981. The development of the bacterial flora in normal neonates. J. Med. Microbiol. 14, 51−62.

Saavedra, J.M., Tschernia, A., 2002. Human studies with probiotics and prebiotics: clinical implications. Br. J. Nutr. 87 (Suppl. 2), S241−S246.

SCF (Scientific Committee for Food), 2003. Report of the Scientific Committee on Food on the revision of essential requirements of infant formula and follow-up formula (adopted on 4 April 2003). SCF/CS/NUT/IF/65 Final 18 May 2003.

Scheppach, W., Luehrs, H., Menzel, T., 2001. Beneficial health effects of low digestible carbohydrate consumption. Br. J. Nutr. 85, S23−S30.

Schley, P.D., Field, C.J., 2002. The immune-enhancing effects of dietary fibres and prebiotics. Br. J. Nutr. 87 (Suppl. 2), S221−S230.

Scholz-Ahrens, K.E., Schaafama, G., van den Heuvel, E.G., 2001. Effects of prebiotics on mineral metabolism. Am. J. Clin. Nutr. 73 (Suppl.), 459S−464S.

Schultz, M., Munro, K., Tannock, G.W., et al., 2004. Effects of feeding a probiotic preparation (SIM) containing inulin on the severity of colitis and on the composition of the intestinal microflora in HLA-B27 transgenic rats. Clin. Diagn. Lab. Immunol. 11, 581−587.

Seidner, D.L., Lashner, B.A., Brzezinski, A., et al., 2005. An oral supplement enriched with fish oil, soluble fiber, and antioxidants for corticosteroid sparing in ulcerative colitis: a randomized, controlled trial. Clin. Gastroenterol. Hepatol. 3, 358−369.

Seth, A., Yan, F., Polk, D.B., et al., 2008. Probiotics ameliorate the hydrogen peroxide-induced epithelial barrier disruption by a PKC- and MAP kinase-dependent mechanism. Am. J. Physiol. Gastrointest. Liver Physiol 294, G1060−G1069.

Sharma, S., Awasthi, A., Singh, S., 2019. Altered gut microbiota and intestinal permeability in Parkinson's disease: Pathological highlight to management. Neurosci. Lett. 712, 134516.

Slizewska, K., Kapusniak, J., Barczynska, R., et al., 2012. Resistant dextrins as prebiotic. Licensee InTech (http//creativecommons.org/licenses/by/3.0).

Smith, K., McCoy, K.D., Macpherson, A.J., 2007. Use of anexic animals in studying the adaptation of mammals to their commensal intestinal microbiota. Semin. Immunol. 19, 59−69.

Steed, H., Macfarlane, S., 2009. Mechanisms of prebiotic impact on health. In: Charalampopoulos, D., Rastall, R.A. (Eds.), Prebiotics and Probiotics Science and Technology. Springer, New York, USA, pp. 135−161.

Tannock, G.W., Fuller, R., Smith, S.L., et al., 1990. Plasmid profiling of members of the family *Enterobacteriaceae*, *Lactobacilli*, and *Bifidobacteria* to study the transmission of bacteria from mother to infant. J. Clin. Microbiol. 28, 1225−1228.

Tuohy, K.M., Finlay, R.K., Wynne, A.G., et al., 2001a. A human volunteer study on the prebiotic effects of HP-inulin − Faecal bacteria enumerated using fluorescent in situ hybridisation. Anaerobe 7 (3), 113−118.

Tuohy, K.M., Kolida, S., Lustenberger, A.M., et al., 2001b. The prebiotic effects of biscuits containing partially hydrolysed guar gum and fructo-oligosaccharides: a human volunteer study. Br. J. Nutr. 86, 341−348.

Tuohy, K.M., Rouzaud, G.C.M., Bruck, W.M., Gibson, G.R., 2005. Modulation of the human gut microflora towards improved health using prebiotics-assessment of efficacy. Curr. Pharm. Des. 11 (1), 75−90.

Van Leeuwen, P.A., Boermeester, M.A., Houdijk, A.P., et al., 1994. Clinical significance of translocation. Gut 35 (Suppl.), S28−S34.

Vulevic, J., Juric, A., Walton, G.E., et al., 2015. Influence of galactooligosaccharide mixture (B-GOS) on gut microbiota, immune parameters and metabonomics in elderly persons. Br. J. Nutr. 114, 586−595.

Wang, Y., 2009. Prebiotics: present and future in food science and technology. Food Res. Int. 42, 8−12.

Wang, X., Gibson, G.R., 1993. Effects of the in vitro fermentation of oligofructose and inulin by bacteria growing in the human large intestine. J. Appl. Bacteriol. 75, 373−380.

Wang, S., Peng, R., Qin, S., et al., 2019. Effects of oligosaccharide-sialic acid (OS) compound on maternal-newborn gut microbiome, glucose metabolism and systematic immunity in pregnancy: protocol for a randomised controlled study. BMJ 9 (9), e026583.

Wolf, B.W., Chow, J.M., Garleb, K.A., 2005. Medical foods and fructooligosaccharides. In: Dimitriu, S. (Ed.), Polysaccharides: Structural Diversity and Functional Versatility, 2nd ed. Marcel Dekker, New York, pp. 853−866.

Yajima, T., Inoue, R., Yajima, M., et al., 2011. The G-protein in colesterol-rich membrane microdomains mediates mucosal sensing of short-chain fatty acids and secretory response in rat colon. Acta Physiol. 203, 381−389.

Yatsunenko, T., Rey, F.E., Manary, M.J., et al., 2012. Human gut microbiome viewed across age and geography. Nature 486 (7402), 222−227.

Younes, H., Garleb, K., Behr, S., et al., 1995. Fermentable fibers or oligosaccharides reduce urinary nitrogen excretion by increasing urea disposal in the rat caecum. J. Nutr. 3, 139−148.

Younes, H., Alphonse, J.C., Hadj-Adbelkader, M., et al., 2001. Fermentable carbohydrates and digestive nitrogen excretion. J. Res. Nutr. 3, 139−148.

Chapter 65

Probiotics: safety and toxicity considerations

Arturo Anadón, Irma Ares, María-Rosa Martínez-Larrañaga and María-Aránzazu Martínez

Introduction

The term "probiotic" is derived from the Greek and means "*for life*"; however, it has had several different meaning over the years. The history of probiotics began with the consumption of fermented foods by the ancient Greeks and Romans. Probably the first foods that contained living microorganisms were the fermented milks that are recorded in the Old Testament (Genesis 18:8), which reported that "Abraham owed his longevity to the consumption of sour milk." The consumption of fermented milks in many different forms has continued until the present day. The beneficial effects of yogurt were discovered scientifically in the early 20th century. Ilya Ilyich Metchnikoff (1908, 1910), a Nobel Prize winner, working at the Pasteur Institute in Paris, France, was the first to propose the beneficial effects of fermented milk on human health and regarded the microflora of the lower gut as having an adverse effect on the health of the human adult. He put forward the theory that the effect of detrimental microbes in the intestinal tract could be alleviated by the ingestion of beneficial microbes. Metchnikoff also postulated that the longevity of the Bulgar mountain people was attributable to the consumption of the yogurt produced by fermentation with *Lactobacillus bulgaricus* (previously called the *Bulgarian bacillus*). It should be emphasized that Metchnikoff was not concerned with sour milk but rather what we now call yogurt; subsequently, when pure cultures became available he advocated the use of milk fermented with a single strain of *Lactobacillus*. Therefore probiotics and the phenomenon of "probiosis" were discovered by Metchnikoff. "Probiosis" can be defined as the positive effect of consumption of fermented dairy products with cultures of lactic acid bacteria (LAB) on the equilibrium of intestinal microflora. It is presently known that viable LAB implantation in the digestive tract acts directly on the composition of the microbial population. Metchnikoff's work can be regarded as the birth of probiotics, that is, microbes ingested with the aim of promoting good health. The habit was given added support by the publication of the book *The Bacillus of Long Life*, where the author, Loudon Douglas, reiterated the connection between fermented milk and longevity (Douglas, 1911). The term probiotic, as is used today, was first used in 1974 by Parker, who defined probiotics "as substances and organisms that contribute to intestinal microbial balance." During this period there grew a realization that the gut microflora was involved in protection of the host against disease. These studies reinforced the view that not all bacteria were having an adverse effect on the host and that there was a population of bacteria in the gut necessary for the continuing positive health and well-being of the host by improving its intestinal microbial balance (Fuller, 1992).

This definition was broadened by Havenaar et al. (1992), who included in the definition the following "a viable mono- or mixed culture of microorganisms that, applied to animal or man, beneficially affects the host by improving the properties of the indigenous microflora." This implies that the term *probiotic* is restricted to products that (1) contain live microorganisms (such as freeze-dried cells or in a fermented product); and (2) improve the health status of humans or animals and exert their effects in the mouth or gastrointestinal tract (GIT) (aerosol) or urogenital tract (by local application).

Probiotics are defined in human nutrition "as live microorganisms that can provide benefits to human health when administered in adequate amounts that confer a beneficial health effect on the host" (FAO/WHO, 2002). There is an increasing amount of evidence indicating health benefits by their consumption (Sullivan and Nord, 2002). Temporary colonizing of the gut with an appropriate probiotic strain not only promotes the state of *eubiosis* (favorable balance of the gut flora) but may have a favorable immunomodulatory effect (Hamilton-Miller, 2003).

Nutraceuticals. DOI: https://doi.org/10.1016/B978-0-12-821038-3.00065-3

The probiotic products can contain one or several species of probiotic bacteria; however, there is still a significant way to go before the association of specific probiotic health benefits with specific strains of bacteria.

Probiotics are commonly defined as "viable microorganisms (yeast and LAB) that exhibit a beneficial effect on the health of the host when they are ingested, although its health benefits are strain-specific and not species-specific or genus-specific." Many physiological responses and health-promoting effects in the consumer attributed to probiotic microorganisms are related, among others, to the GIT, and show the ability to survive through the upper GIT and the capability of surviving and growing in the intestine (acid and bile resistant) in a strain-specific manner safe for human consumption. They produce antimicrobial substances like bacteriocins (nisin and pediocin), can adhere to human intestinal cell lines, and colonize the intestine. Probiotics produce short-chain fatty acids (SCFAs) that lower the pH, thus favoring the growth of harmless microorganisms, including low-molecular-weight carboxylic acids with six to eight carbon atoms (i.e., acetate, propionate, butyrate, are typically found in a proportion of 1:1:3 in the GIT). With this acidic pH, probiotics can easily penetrate the bacterial cell. The large intestine has a higher density of microbes because content passes through the intestine more slowly, allowing microbes greater time to multiply; compared with the small intestine, the large intestine characteristically has a more neutral-to-alkaline pH, providing a more well-disposed environment for many microbes (Savage, 1977).

Probiotics are living microorganisms, generally bacteria but also yeast, that interact with the gut microflora and host when ingested in a sufficient amount. They have a positive effect on the health of an individual beyond the commonly known nutritional ones. These bacteria can help to maintain internal microbial balance and defend against harmful bacteria; three mechanisms of promoting human health has been described: (1) providing end-products of anaerobic fermentation of carbohydrates such as organic acids that can be absorbed by the host, these end-products being able to influence human mood, energy level, and even cognitive abilities; (2) successfully competing with pathogens; and (3) stimulating host immune responses by producing specific polysaccharides (Saier and Mansour, 2005). Probiotics for consumption should not cause disease in humans, should be completely nonpathogenic, nontoxic, resistant to gastric acid, adhere to gut epithelial tissues, and should not be able to evolve into pathogenic variants.

Probiotics are readily available to consumers and are commonly found as food probiotics (e.g., yogurts, cheeses, milk-based beverages, fermented fish, meats, and vegetables, among others) and as food supplements (e.g., tablets, capsules, pills, powders, liquid concentrates in vials, and soft gels among others) (Anadón et al., 2010, 2016, 2019). Although probiotics can technically be any type of beneficial microorganism, certain bacteria are more commonly found on the food market. Currently, the most widely used food-grade probiotics in the majority of countries include LAB from the genera *Lactobacillus* and *Bifidobacterium*, with some strains of *Enterococcus* and *Saccharomyces* species being among the exceptions. LAB are Gram-positive, anaerobic, aero-tolerant, nonspore-forming rods and cocci, that are indigenous inhabitants of the human GIT, vagina, and human skin.

Lactobacillus acidophilus and *Bifidobacterium* represent two distinct phyla of bacteria, the Firmicutes and Actinobacteria; even though they are genetically quiet distinct, they have many phenotypic similarities (e.g., producers of lactic acid that can easily be added to traditionally fermented foods, such as cheeses, yogurts, and other dairy products). For this reason, bifidobacteria are often included in the LAB, but because the LAB are a taxonomic group of genetically similar bacteria, the bifidobacteria cannot be included from a taxonomic point of view. Both are inhabitants of GI flora of humans and animals, with the lactobacilli believed to be important for small intestine functionality and bifidobacteria for colon functionality (Lee and O'Sullivan, 2010; Anadón et al., 2010, 2016, 2019).

Probiotics

Probiotics and their sources

The probiotic bacteria generally belong to the *Lactobacillus* and *Bifidobacterium* genera. However, other bacteria and some yeast also have probiotic properties. Common bacteria include (Mercenier et al., 2003; Seth and Maulik, 2011):

- Lactic acid bacteria (LAB): Genus *Lactobacilli* spp.; Species: *L. acidophilus*, *Lactobacillus amylovorus*, *Lactobacillus brevis*, *Lactobacillus bulgaricus*, *Lactobacillus casei*, *Lactobacillus cellobiosus*, *Lactobacillus crispatus*; *Lactobacillus curvatus*; *Lactobacillus delbrueckii* spp., *Lactobacillus bulgaris*, *Lactobacillus fermentum*, *Lactobacillus gallinarum*, *Lactobacillus helveticus*, *Lactobacillus johnsonii*, *Lactobacillus lactis*, *Lactobacillus paracasei*, *Lactobacillus plantarum*, *Lactobacillus reuteri*, *Lactobacillus rhamnosus;* Genus: *Streptococcus* spp.; Species: *Streptococcus salivaris* spp. *thermophiles*; Genus: *Lactococcus* ssp., Species: *Lactococcus lactis cremoris*; Genus: *Leuconostoc*, Species: *Leuconostoc mesenteroides*; Genus: *Pediococcus* spp., Species: *Pediococcus pentosaceus*, *Pediococcus acidilactici*.

- Bifidobacteria: Genus: *Bifidobacterium* spp., Species: *Bifidobacterium adolescentis, Bifidobacterium animalis, Bifidobacterium bifidum, Bifidobacterium breve, Bifidobacterium essensis, Bifidobacterium infantis, Bifidobacterium laterosporum, Bifidobacterium thermophilus, Bifidobacterium longum.*
- Propionibacteria: Genus: *Propionibacterium* spp.; Species: *Propionibacterium acidipropionici, Propionibacterium freudenreichii, Propionibacterium jensenii, Propionibacterium thoenii.*
- Enterobacteria: Genus: *Enterococcus* spp., Species: *Enterococcus fecalis, Enterococcus faecium.*
- Sporulated bacteria: Genus: *Bacillus* spp.; Species: *Bacillus alcalophilus, Bacillus cereus, Bacillus clausii, Bacillus coagulans, Bacillus subtilis.*
- Other bacteria: Genus: *Escherichia coli*; Species: *E. coli*; Genus: *Sporolactobacillus* spp; Species: *Sporolactobacillus inulinus.*
- Yeasts: Genus: *Saccharomyces* spp.; Species: *Saccharomyces cerevisiae (boulardii)*; that isolated from litchi fruit in Indonesia, have also been accepted and used as probiotics.

To be considered as probiotics, the different strains should be normal inhabitants of a healthy intestinal tract, survive the upper digestive tract, be capable of surviving and growing in the intestine (acid and bile resistant), be safe for human consumption, produce antimicrobial substances (i.e., bacteriocins), and have the ability of adherence to human intestinal cell lines and colonization.

Tomasik and Tomasik (2003) established the criteria for microorganisms to be included in the probiotic group as follows: (1) surviving on passing thought the GIT at low pH and on contact with bile; (2) adhesion to intestinal epithelial cells; (3) stabilization of the intestinal microflora; (4) nonpathogenic; (5) survival in foodstuffs and possibility for production of pharmacopoeia lyophilized preparations; (6) fast multiplication, with either permanent or temporary colonization of the GIT; and (7) generic specificity of probiotics. Numerous studies suggest that probiotics have beneficial effects on gut microbial disorders, although this can be difficult to demonstrate with clinical trials.

Mechanism of action of probiotics

The mechanistic approach to probiotics first established that many GIT dysfunctions are based on disturbances or imbalances of intestinal microflora. The production of inhibitory substances by probiotics such as organic acids, hydrogen peroxide, and bacteriocins is of great interest. However, the mechanisms whereby lactobacilli function as antiinfective defenses are still not fully understood (Anadón et al., 2016). Hydrogen peroxide is produced by many species of *Lactobacillus* because antimicrobial agents are oxygen-dependent. Because *Lactobacillus* does not produce catalase, the hydrogen peroxide produced cannot be degraded and acts as an oxidant by forming free radicals. Bacteriocins are proteinaceous antimicrobial substances that are sometimes associated with lipids and carbohydrates; they have demonstrated inhibitory actions on both Gram-positive and Gram-negative bacteria.

The actions of microorganisms are useful to assist the GIT by breaking down sugars and carbohydrates to promote good digestion, boost the immune system, maintain proper intestinal pH, and successfully compete with pathogens. Among the expectations for probiotics, many strains have been shown to modulate the intestinal microflora and to prevent the duration and clinical symptoms of rotavirus-induced diarrhea; probiotic bacteria also reinforce the intestinal wall by crowding out pathogenic microorganisms, therefore helping to prevent their attachment to the human gut, where they have been shown to be safe. The consumption of probiotics has been shown to influence various aspects of the innate nonspecific immune system such as promotion of mucin production, inhibition of pathogenic bacteria, decrease in gut permeability, macrophage activation and phagocytic capacity, and natural killer (NK) cell activity. Regarding the adaptive immune system, the effects observed are an increase in the production of antibodies (IgA, IgM, and IgG), and also an influence in the arrangement of both branches of the immune system by the production of cytokines and other regulatory elements.

Probiotics usually modify the resident microflora. It is known that probiotic bacteria alter the physical environment so that the pathogenic bacteria cannot survive. Probiotic bacteria act in two ways. First, they compete with pathogenic bacteria for food and energy sources. Second, they produce an inhibitory substance that impedes the growth of the pathogen by consuming the nutrients that pathogens need. Probiotics and pathogenic bacteria are in competition; probiotics inhibit the pathogens by adhering to the intestinal epithelial surfaces and blocking the adhesion sites (Anadón et al., 2016).

The second generation of probiotics is genetically modified microorganisms that provide the host with some necessary components (e.g., production of immunomodulators, such as interleukines, or *Helicobacter pylori* and rotavirus antigens) (Mercenier et al., 2003). Probiotics have several putative mechanisms such as modulation of immune or sensory–motor function, enhancement of mucosal barrier function, and antipathogen effects. The epithelial barrier consists of a thick mucus layer containing immunoglobulins, mainly IgA and antimicrobial components; their dynamic functional role is to regulate permeability between cells. Intestinal mucosal barrier

TABLE 65.1 Mechanism of action of probiotics in the gastrointestinal tract.

1. Module intestinal immune function	• Reduce proinflammatory cytokines • Promote tolerogenic cytokine profiles and regulatory pathways • Increase secretory IgA
2. Promote epithelial cell homeostasis	• Enhance barrier function • Promote cytoprotective responses, improve cell survival, and increase mucin production
3. Neuromodulatory effects	• Induce mu-opioid and cannabinoid receptors on epithelial cells • Reduce visceral hypersensitivity and stress response
4. Block effects of pathogenic bacteria	• Reduce pathogen binding • Decrease luminal pH • Produce antibacterial bacteriocins
5. Nutritional benefits	• Assist in the breakdown of undigestible foods to produce usable nutrients

function is formed by a common mucosal immune system that provides communication between the different mucosal surfaces of the body. In the future, greater understanding of probiotic-specific mechanisms could allow for precise selection of a particular probiotic strain to target a patient's specific pathogen's defects and clinical problems (Ciorba, 2012) (see Table 65.1).

Gastrointestinal microbiota

The gut is the largest internal organ and has the most extensive exposed surface within the body. It is a selective barrier with physiological and physical, chemical, immunological, and microbiological aspects where digested food and water are absorbed but disease-causing agents should be kept under control. The gut barrier is highly regulated, and responsive to internal and external stimuli, such as hormones and nutrients. The gut is also the largest immune organ of the body and the location of approximately 70% of the cells of the immune system, being the body's first line of active defense against harmless and pathogenic antigens.

A healthy gut is therefore not only a disease-free gut, but also an effective digestive organ that can give good protection against disease and subsist with changes and stresses. The GI microbiota are crucial in the host immune system, its physiological development, health, nutrition, and productivity. A healthy gut with a balanced microbial composition is critical for optimal digestion and nutrient uptake.

The presence of disease-causing microorganisms (pathogens) in the gut is not sufficient to cause disease. Trillions of microorganisms inhabit the intestinal tract (collectively called "the gut microflora"), forming a complex ecosystem that can influence the immune system both inside and outside the gut and the animal. When this gut microflora is disturbed (dysbiosis), there is a microbial imbalance (Anadón et al., 2019).

The human GIT performs many different functions, facilitating digestion and nutrient provision and shaping the immune system. Numerous functions are ascribed to the microbiota in the human gut. The functions of microbiota include *nutrition* [fermentation of nondigestible substrates (i.e., carbohydrates) to generate SCFA], absorption of ions, production of amino acids and vitamins, *protection* (the barrier effect that prevents invasion by nonnative microbes), and *trophic effects* on the intestinal epithelium and immune system (development and homeostasis of local and systemic immunity) (Guarner, 2007). The gut immune system is influenced by many factors (i.e., dietary components, indigenous/commensal bacteria), such as the relationship between diet, microbiota, and gut immune system. There is part of the intestinal mucosal barrier function that is formed by a common mucosal immune system that provides communication between the different mucosal surfaces of the body.

Over the last few years, the intestinal microflora/microbiota has emerged as a new target for the management of functional digestive alterations. It is a large bacterial community that colonizes the gut, with metabolic activity that affects the physiology and pathology of the host's mucosal immune system. In general, an adult's GI microbiota contains species belonging to four major bacterial phyla: Firmicutes, Bacteroidetes, Actinobacteria, and Proteobacteria.

Collectively, bacteria from these four phyla constitute more than 90% of the adult GI microbiota and reside mainly in the large intestine. It hosts approximately 100 trillion bacteria (more than 500 bacterial species), with the greatest density and diversity of bacteria being found in the distal small bowel and colon, is associated with our GIT, and the rich gut microbial community referred to as the *microbiota* has codeveloped in a symbiotic relationship with the human intestinal mucosa in such a way that the indigenous microbiota is essential for gut homeostasis. For this gut homeostasis, there is a delicate balance between these bacteria, the small intestinal and colonic epithelia, and the mucosa-associated lymphoid tissue. Several B vitamins, vitamin K, folate, and SCFA are produced by these bacteria. Up to 10% of an individual's daily energy needs can be derived from the by-products of bacterial fermentation. The total mucosal surface area of the adult human GIT is up to 300 m^2, making it the largest body area interacting with the environment. It is colonized with more than 10^{14} microorganisms, weighing more than 1 kg, and corresponds to more than 10 times the total number of cells in the body (Turnbaugh et al., 2007). Our gut microbiota can be delineated as a microbial organ placed within a host organ. It is composed of different cell lineages with the capacity to communicate with one another and the host. The gut "microbiome" (i.e., the natural microbial population of the host) contains more than 100 times the number of genes in our genome (genetic material) and confers us with functional features that we have not evolved ourselves. The average human body contains 100 trillion microbes, primarily residing in the gut (approximately 1–2 kg—the weight of the human brain) (Turnbaugh et al., 2007). Dysfunctional microbiomes are associated with chronic human diseases such as obesity, diabetes, and asthma.

It is important to note that the GIT of a newborn baby is sterile. Soon after birth, it is colonized by numerous types of microorganisms. Colonization is complete after approximately 1 week, but the numbers and species of bacteria fluctuate markedly during the first months of life (Rautava et al., 2012). The composition of the intestinal microflora remains relatively constant until later life.

The *microbiota* is considered a superorganism and is an integral part of the GIT. Intestinal bacteria are useful in promoting human health, but certain components of microflora, in genetically susceptible individuals, contribute to various pathological disorders, including Crohn disease and ulcerative colitis, the two main types of inflammatory bowel disease (IBD). These diseases are characterized by persistent mucosal inflammation at different levels of the GIT. Clinical and experimental observations indicate an imbalance in protective and harmful microflora components in these disorders. Probiotic bacteria are being studied for the treatment of irritable bowel syndrome.

Bacterial colonization

To prevent pathogen colonization either by binding directly or by "competitive exclusion" is important for probiotics. Bacterial colonization early in life has been shown to affect cytokine production by T-helper cell subsets (Th cells), also known as $CD4^+$ cells, implicating that "dysbiosis" at an early age may increase the risk of developing food allergy. Likewise, infants that have developed eczema by the age of 12 months show a lower diversity in the gut microbiota during the early postnatal period (Ismail et al., 2012). Thus, it seems that a lower abundance of bifidobacteria, enterococci, and *Bacteroides* and a higher abundance of *Clostridium* spp. and *Staphylococcus* are associated with a loss of tolerance and an exaggerated allergic response to food-derived antigens. The tolerance to GIT conditions may be measured by test of the sensitivity to bile salts and to gastric acidity based on bacterial plate systems assay.

Microbial colonization of GIT is linked with lifestyle and/or geographic factors. There are now numerous studies demonstrating differences in the composition of the gut microbiota between allergic and nonallergic individuals, between infants living in countries with a high or low prevalence of allergy, and between healthy and allergic infants (Zmora et al., 2018).

Probiotics operate by "competitive exclusion," meaning that when adequate populations of probiotic bacteria are present, they reduce the ability of pathogenic bacteria to get out of control and overwhelm the host. The available probiotics can be classified into (1) "colonizing" species, such as *Lactobacillus* and *Enterococcus* spp., and (2) free flowing "noncolonizing" species, such as *Bacillus* spp. (spores) and *S. cerevisiae.* Competitive exclusion describes the treatment of day-old chicks with an undefined microbiota derived from adult animals resulting in colonization resistance against pathogenic microorganisms (Huyghebaert et al., 2011; Anadón et al., 2019).

Inflammatory immune responses

The mucosal tissue of the intestines contains the largest part of the immune system present in the human body and is constantly exposed to many antigens, which are derived specially from food and microorganisms (i.e., resident commensals, pathogens), among others. Nevertheless, the majority of oral foreign antigens do not result in inflammatory responses in healthy individuals. This phenomenon is known as oral tolerance. The gut microbiota is essential for the development of the immune system. Certain commensal bacteria appear to preferentially drive T-regulatory lymphocyte development, whereas other bacteria promote Th17 development in the gut (Kamada and Nunez, 2013). Gut microbiota "dysbiosis" (i.e., disruption of the intestinal ecosystem equilibrium) is usually associated with inflammatory disorders. Local or systemic pathological inflammation may occur when oral tolerance is lost. This is seen in food allergies, which are characterized by an inflammatory immune response toward generally harmless food-derived antigens. Intestinal epithelial cells provide a physical and chemical barrier between the intestinal lumen and the lamina propria. As an antigen ends up in the lumen of the intestine, it is recognized by dendritic cells (i.e., sense antigen in tissues before migrating to draining lymph nodes) present in Peyer's patches. It is becoming clear that certain species of gut commensal microbiota are required for the regulation of immune responses and that perturbations in the microbiota could result in a lack of immune regulation, the outgrowth of more pathogenic microbes, and the promotion of inflammation. The microorganisms to which the newborn child is exposed during the first years of life will mainly determine the microbial composition of the microbiota in the adult human gut.

The pattern recognition receptors system [toll-like receptors (TLRs)] contains proteins, spanning, and noncatalytic entities that recognize structurally conserved molecules derived from pathogens and therefore recognize antigens derived from the microbiota as well as invading pathogens. TLRs are now regarded as the key molecules that alert the immune system to the presence of microbial infections. TLRs recognize antigens of microbial origin, which are referred to as pathogen-associated molecular patterns. TLRs are expressed by a variety of cells, including intestinal epithelial cells. TLR2 can dimerize with TLR1 or TLR6 to recognize bacterial cell wall lipoproteins. TLRs recognize microbial products such as lipopolysaccharide (LPS), peptidoglycan (PGN), lipoprotein, and DNA. These receptors are expressed on different immune cells like lymphocytes, macrophages, and dendritic cells, and are found close to epithelial cells. It is also recognized that interactions with the normal microbial flora of the GIT are the principal environmental signal for postnatal maturation of T-cell function (particularly the Th1 component). Although TLRs are found principally on cells of the innate immune system (including granulocytes, monocytes, and NK cells), they are also present on cells involving in programming and regulating *adaptative* immune responses (APC and regulatory T cells). It has been proposed that early microbial activation of both APC and regulatory T cells may promote Th1 maturation and play an important role in reducing the risk of Th2-mediated allergic responses (Wills-Karp et al., 2001). LPS produced by Gram-negative bacteria are recognized by TLR4 in conjunction with CD14 (a type of T cell) and MD2, whereas unmethylated CpG motifs of bacterial DNA are recognized by TLR9 (Lee et al., 2006).

Probiotics affect the immune system in different ways, such as producing cytokines, stimulating macrophages, and increasing secretory IgA concentrations (Dugas et al., 1999). The innate immune system is the first line of defense against pathogenic microorganisms. Epithelial cells also produce different types of chemokines and cytokines, which are important for the activation of innate immune cells (Kayisli et al., 2002).

Under homeostatic conditions, intestinal epithelial cells show low expression of TLR2 and TLR4, and are therefore unresponsive to TLR stimuli. However, under inflammatory conditions, epithelial TLR expression is increased, which contributes to both inflammation as well as immune tolerance. Increased epithelial TLR2 and TLR4 expression is associated with IBD (Ewaschuck et al., 2008).

In addition to regulating intestinal immunity, TLR activation on intestinal epithelial cells is also known to modulate the expression of tight junction proteins (i.e., claudins and occludins). In many inflammatory disorders, including food allergy, epithelial tight junctions are impaired and increased bacterial translocation occurs (Iemoli et al., 2012). This increased bacterial translocation (invasion) into the lamina propria may sustain the inflammatory response.

Disturbances in the commensal bacterial composition in the gut, reflected by increased colonization with *E. coli* or *Clostridium difficile*, is associated with an increased risk in the development of allergic disease and IBD in humans (Loh and Blaut, 2012). The fecal microbiota of allergic infants shows a higher prevalence of *Clostridium* spp. and *Staphylococcus aureus*.

It is important to note that changes in the microbiota are present in individuals with food allergy and that the response of immune cells toward the microbiota has also been described to be different. The so-called *beneficial bacteria* are not necessarily associated with antiinflammatory responses in allergic patients. For example, although an increased prevalence of bifidobacteria is rendered as beneficial, specific *Bifidobacterium* strains isolated from the feces of allergic infants were shown to induce increased production of the proinflammatory cytokines IL-1β, IL-6, and TNF-α. This is supported by the observation that the allergic infants showed an increased IL-6 and TNF-α response toward TLR2, TLR4, and TLR5 stimuli (Prescott et al., 2008).

The consumption of probiotics has been shown to influence various aspects of the innate nonspecific immune system, such as promotion of mucin production, inhibition of pathogens, decrease in gut permeability, macrophage activation, phagocytic capacity, and NK cell activity.

The benefits of regulating the gut microbiota to induce intestinal homeostasis largely extend to other systems in the host. Microbiota is the largest source of microbial stimulation in the gut and is necessary for development of the intestinal immune system (i.e., deregulation of the microbiota needs specific microbial stimulation for gut health). Certain species of bacteria can have large effects on the gut immune system, and the balance of these influences is important to the maintenance of homeostasis and the development of novel disease treatment and management strategies. In this context, probiotics are a powerful strategy for manipulating the microbial composition and immune responses of the host. Homeostasis of the gut microbiota maintains various functions that are vital to the maintenance of human health.

The microbiota is essential to prevent the attachment, growth, and penetration of pathogenic microorganisms on the gut surface. The intestinal microbiota plays an important role in pathogen resistance, both by direct interaction with pathogenic bacteria and by influencing the immune system (Kamada et al., 2013). Resident commensal microorganisms (normal microflora, indigenous microbiota) are known to promote gut health through the induction of mucus production and enterocyte turnover (Kamada et al., 2013; Anadón et al., 2019) and also play an important role in host immunity, nutrient absorption, and metabolism. The commensal bacteria that reside in the gut are diverse, and in certain cases individual species appear to have distinct and opposing roles in the gut immune response.

Neural function and the stress response

The brain–gut–microbiota axis comprises an extensive communication network between the brain, the gut, and the residing microbiota. Development of the nervous system, the microbiota, and the elements of the microbiota–gut–brain axis occurs in parallel during interacting time periods, and prenatal, early postnatal, and adolescence phases are particularly important for brain development (Wiley et al, 2017). When the composition, diversity, or assembly of the gut microbiota is damaged, it may produce a negative impact on host health and lead to disorders including potentially neuropsychiatric illnesses, such as anxiety and depression. A complete microbiota from conventionally colonized animals, animal models of disease, or human patients can be transferred via fecal transplantation to determine if disease characteristics are transferrable via the microbiota (Luczynski et al., 2016). It is known that gut microbiota secretes neurotransmitters and neuromodulators, such as γ-aminobutyric acid, serotonin, catecholamines, and acetylcholine, which impact host psychiatric health (O'Mahony, et al., 2015). In addition to having significant potential for treating human disorders, the role of the gut microbiota in stress has implications for animal science. Remarkably, maternal separation studies have also shown that neonatal stress results in long-term alterations in both the diversity and composition of gut microbiota, which may play a role in the long-term alterations in stress-related behavior in these animals (O'Mahony et al., 2015).

A predominant pathway in the microbiome–gut–brain axis is the vagus nerve. Certain *Lactobacillus* species communicate with the brain via this route and do not produce their central nervous system (CNS) effects if an animal has been vagotomized (Bravo et al., 2011). It has also been recognized that the hypothalamic–pituitary–adrenal axis and cytokines from the immune system play a role in microbiota–gut–brain axis communication (Jones et al., 2006).

The translational potential of *B. longum* 1714 was recently demonstrated in healthy volunteers, in whom it modulated stress, electrophysiology, and neurocognition (Allen et al., 2016). In human trials where it was used as a probiotic capsule consisting of *L. acidophilus*, *L. casei*, and *B. bifidum* demonstrated positive effects in patients with major depressive disorder. Various microbiota-directed inventions, including probiotics such as *L. rhamnosus* and *B. infantis*, reduce depressive-like behaviors (Bravo et al., 2011). The production of SCFA is important for brain health and behavior, and bacteria capable of producing them include *Bacteroides*, *Bifidobacterium*, *Propionibacterium*, *Eubacterium*, *Lactobacillus*, *Clostridium*, *Roseburia*, and *Prevotella* (Macfarlane and Macfarlane, 2012).

Probiotics used in food or as medicines

Most probiotics are marketed as foodstuffs or medicines. *Lactobacillus*, *Leuconostoc*, and *Pediococcus* species have been used extensively in food processing throughout human history, and ingestion of foods containing live bacteria, dead bacteria, and metabolites of these microorganisms has taken place for a long time (Mäyrä-Mäkien and Bigret, 1993). Currently, the most widely used probiotics include lactobacilli, bifidobacteria, and some nonpathogenic strains, mostly of human origin, that confer a health benefits on the host and enable prevention of or improvement in some diseases when administered in adequate amounts. An important fact is that probiotics must retain their viability during the storage, manufacturing process, and transit through the stomach and small intestine. Prior to being categorized as probiotics, organisms need to follow a process of testing, including strain testing, identification by genotype and phenotype, functionalized characterization and safety assessment testing, and double-blind, placebo-controlled human trials to verify their health benefits and so the guidelines for the evaluation of probiotics in food have been proposed (FAO/WHO, 2002)

Most probiotic foods contain lactobacilli and/or bifidobacteria. Enterococci are infrequently used. Microorganisms used as probiotics in animals are mainly bacterial strains of members of the heterogeneous group of LAB; lactobacilli (*L. acidophilus, L. casei, L. plantarum, L. reuteri, L. rhamnosus, L. salivarus*), bifidobacteria (*B. breve, B. longum, B. lactis*), *Bacillus* (*B. subtilis*), and *Enterocococcus* (*E. faecium*), among others (Anadón et al., 2016). The yeast *Saccharomyces boulardi* is also used as a human probiotic, although it is delivered in capsules or powders rather than by food. It has been noticed that *Bacillus* and *Lactobacillus* differ in many characteristics and that *Bacillus* and the yeasts are not usual components of the gut microflora. It has been shown that most *Lactobacillus* species (*L. acidophilus, L. rhamnosus*, and *L. reuteri*) have no pathogenicity or acute oral toxicity in animals (Perdigón and Alvarez, 1992; Anadón et al., 2014, 2016).

The use of probiotics in medical practice is important. The effects and applications of probiotics in various infectious diseases have been summarized by Goldin and Gorbach (2008), who categorized into four groups the clinical applications for probiotics by level of evidence of efficacy. The first group includes the application to cases of acute/antibiotic-associated gastroenteritis, for which benefits of probiotics are well proven. The second group includes allergic reactions, specifically atopic dermatitis I, for which there is substantial evidence of efficacy. The third group includes applications that have shown promise (e.g., childhood respiratory infections, dental caries, IBD, combating nasal pathogens, and the prevention of relapsing *C. difficile*-induced gastroenteritis). The fourth group covers potential future applications for rheumatoid arthritis, inflammatory bowel syndrome, cancer, alcohol-induced liver disease, diabetes, and graft-versus-host disease. Recent studies have investigated the use of prebiotics, probiotics, and fecal transplantation in treating IBD, diabetes, and obesity. More research is needed in respect of the indications for and methodologies of fecal transplantation and some is currently under way.

Clostridium difficile-induced gastroenteritis

Lactobacillus probiotics as single species or combination probiotic products have been tested for the prevention of *C. difficile*-induced gastroenteritis, a common nosocomial and community-based medical condition that has had increased incidence, morbidity, and mortality in past years. Treatment with antibiotics results in disturbance of the GI flora, which is associated with diarrhea and abdominal discomfort in a variable fraction of patients, depending on the age group and antibiotic used; approximately 15%–30% of patients experience recurrence of symptoms after discontinuation of antibiotics. A treatment period with antibiotics only temporarily changes the composition of the microbiota, causing environmental changes. In most cases the cause of the diarrhea is unknown, but a varying proportion of the cases are caused by *C. difficile*. The *C. difficile* toxins may cause anything from mild diarrhea, which can be cured simply by terminating the antibiotic treatment, to the life-threatening disease pseudomembranous colitis (Vanderhoof et al., 1999). Probiotics may also maintain or restore the gut microecology during or after antibiotic treatment; bifidobacteria or lactobacilli will decrease the duration and/or severity of acute diarrheal disease in infants and children (Oberhelman et al., 1999). Probiotics are beneficial for people with primary and secondary shortage of endogenic lactase (Marteau et al., 1993).

Brain disorders

Development of the nervous system, the microbiota, and the elements of the microbiota–gut–brain axis occurs in parallel during interacting time windows, and prenatal, early postnatal, and adolescence, phases that are particularly important for brain development (Wiley et al., 2017). Moreover, clinical and preclinical evidence implicates the microbiome as a possible key susceptibility factor for neurological disorders (Alzheimer disease, autism spectrum disorder, multiple sclerosis, Parkinson disease, stroke) (Cryan et al., 2020). Studies in animals have shown the crucial role of the microbiota in critical aspects of neurodevelopment, neuroinflammation, and behavior. Research was focused on elucidate the bidirectional communication pathways between gut bacteria and the CNS: the "microbiota–gut–brain axis." In this regard, a recent systematic review evaluated the effects of probiotics, prebiotics, and synbiotics on cognitive function, behavioral and psychological symptoms of dementia, quality of life, functional performance in activities of daily living, and compliance with the intervention and safety in patients with dementia (Li et al., 2020). This review concluded that these compounds may intervene in dementia through the microbiota–gut–brain axis, although their effectiveness and safety deserve further investigation (Li et al., 2020) Given the close symbiotic relationship existing between the gut microbiota and the host, it is not surprising to observe a variance from the normal microbiota composition "dysbiosis" in an abundance of disease states ranging from chronic GI diseases to neurodevelopmental disorders. Indeed, microbes within the human gut may play a significant role in the regulation of various elements of "gut–brain axis," via their influence on inflammatory cytokines and production of antimicrobial peptides that affect the epigenome through their involvement in generating SCFA, vitamin synthesis, and nutrient absorption. In addition, they may participate in gut production of many common neurotransmitters. In addition, long-chain fatty acids are involved in many important functions, including neurogenesis, neurotransmission, and neurological development, as well as having structural and signaling roles (Alam et al., 2017). It should be noted that recent study findings indicate that SCFA also regulates the production of certain neurotransmitters (including serotonin), which influence cognition, mood, and behavior (O'Mahony, et al., 2015).

Probiotics for humans, particularly the genera *Lactobacillus* and *Bifidobacterium*, have shown in clinical trials that probiotics favorably prevent or improve the symptoms of various disorders, including IBD and irritable bowel syndrome (Alam et al., 2017). Furthermore, they are believed to be crucial for mental health across the life span. Probiotics like lactobacilli, *Bifidobacterium breve*, and *Lactobacillus plantarum*, have been shown to originate conjugated linoleic acid that exerts antiangiogenic actions, and protect cortical neurons from excitotoxicity in the brain. Therefore, these bacteria have the potential to be used as probiotics for brain health. If the composition, diversity, or assembly of the gut microbiota are impaired, this impairment can have a negative impact on host health and lead to disorders such as obesity, diabetes, inflammatory diseases, and even potentially neuropsychiatric illnesses, including anxiety and depression. Therefore much research effort in recent years has focused on understanding the potential of targeting the intestinal microbiota to prevent and treat such disorders (Wiley et al., 2017). It is predominantly composed of anaerobic bacteria, but viruses, protozoa, archaea, and fungi are also present.

Synbiotics (eubiotics)

The term *probiotic* should be used for foods that contain an adequate dose of live microbes with a scientifically documented ability to confer a health benefit to the host. Probiotic-containing foods can be categorized as functional foods and are often associated with prebiotics, which are nondigestible carbohydrates that act as food for probiotics. Supplementation with a single probiotic strain is unlikely to have a major influence on the very diverse intestinal microbiota and the complex interaction between the gut bacteria and the host. When probiotics and prebiotics (dietary substrate) are simultaneously present in a product they are described as either *symbiotic* or *eubiotic*. Because the term alludes to synergism, the word should be reserved for products in which the prebiotic compound selectively favors the probiotic compound (Schrezenmeir and de Vrese, 2001) and facilitates its inoculation into the colon. In this strict sense, a product containing oligofructose and *Bifidobacterium* probiotic strain or lactilol in combination with *Lactobacillus* would fulfill the definition. A combination of probiotics and prebiotics has synergistic effects because, in addition to promoting growth of existing strains of beneficial bacteria in the colon, synbiotics also acts to improve the survival, implantation, and growth of newly added probiotic strains. The potential synergies between probiotics and prebiotics have been considered efficient due to the improvement of the survival and implantation of live microorganisms in the GIT by stimulating growth or by metabolically activating the health-promoting bacteria. It would appear unlikely that supplementation with a single probiotic strain would be sufficient to have a major influence on the very diverse intestinal microbiota and the complex interaction between the gut bacteria and the host. This has led to an interest in food components that could have a more global effect on gut microbiota, namely prebiotics (nondigestible, fermentable oligosaccharides that promote the growth of allegedly "beneficial" bacteria, particularly of *Bifidobacterium* and *Lactobacillus* species), to a similar extent to commercial xylo-oligosaccharides (XOS). Altering the intake of foods containing these products could conceivably directly influence the composition and activity of intestinal microbiota. There are many other examples of synbiotics reported in the scientific literature with data from in vitro observations [e.g., *L. casei* strain Shirota + oligomate 55^{TM}; *L. acidophilus* ATCC 4962 + mannitol, fructo-oligosaccharides (FOS), and inulin; *Lactobacillus sakei* JCM + FOS and trehalose; *L. plantarum* and *L. acidophilus* + XOS and FOS] or from in vivo observations [i.e., *B. longum* + oligofructose; *B. breve* strain Yakult + galacto-oligosaccharides (GOS); Lafti B94 + resistant starch; *Lactobacillus gasseri* + inulin and unspecified oligosaccharides].

Combinations of probiotic bacteria and oligosaccharides have been tried in the treatment of eczema in infants. Treatment of atopic dermatitis in infants with a symbiotic mixture of *B. breve* and GOS and FOS has been tested in clinical trials without confirming its effectiveness.

There are certainly safety concerns for the consumer with regards to the selection and dosage of nondigestive substances, mainly carbohydrates, and their ability to be tolerated and the selection of nonpathogenic bacteria strains, but there is consensus on the prebiotic metabolic substrates (e.g., digestibility, composition, dosage, specificity of metabolization) and on the selection of bacterial strains (e.g., counts, survival of gastrointestinal passage, growth conditions, nonpathogenicity, nontoxinogenicity, stability, identity) (Przyrembel, 2001; Anadón et al., 2014). Recently, it has been published that probiotics, prebiotics, and synbiotics may intervene on cognitive function, behavioral and psychological symptoms of dementia, quality of life, functional performance in activities of daily living, and compliance with the intervention and safety in patients with dementia through the microbiota–gut–brain axis (Li et al., 2020).

Toxicity

Few reports have appeared in the literature about the side effects of probiotics. In some cases, pneumonia and endocarditis were reported under the influence of *Lactobacillus* bacteria as well as a nonsymptomatic fungal infection caused by *Saccharomyces boulardii* (Zunic et al., 1991).

Although most of the species and genera are apparently safe, certain microorganisms may be problematic, particularly the enterococci (*E. faecium* and *Enterococcus faecalis*), which have emerged as opportunistic pathogens in hospital environments, causing nosocomial infections such as endocarditis, bacteremia, and intraabdominal, urinary tract, and CNS infections. They may also harbor transmissible antibiotic-resistance determinants (i.e., vancomycin-resistant *Enterococcus* strains) and bacilli, especially those belonging to the *B. cereus* group, which are known to produce enterotoxins and an emetic toxin (Anadón et al., 2006, 2014).

Potential for production of harmful metabolites/toxins

Most food-poisoning incidents are attributed to *Bacillus* species and are associated with *B. cereus*; this bacterium is known to cause a variety of non-GIT diseases as well as two different types of food poisoning, which are characterized by either diarrhea or emesis (Granum and Lund, 1997; Kotiranta et al., 2000). The diarrheal type is attributed to heat-labile enterotoxins that cause cytotoxicity, fluid accumulation in the ligated ileal loop of experimental animals, and dermonecrosis; it is lethal in mice (Notermans and Batt, 1998). Two protein complexes from *B. cereus* strains hemolysin BL and nonhemolytic enterotoxin and an enterotoxin protein, enterotoxin T (BceT), with these diarrheagenic properties have been previously characterized (Kotiranta et al., 2000; Schoeni and Wong, 1999). The emetic toxin (cereulide) is a cyclic dodecadepsipeptide, which is resistant to heat, proteases, and both low and high pH (Agata et al., 1995).

The relevance of other *Bacillus* species as food-poisoning organisms is being increasingly recognized, with recent epidemiological evidence linking *Bacillus licheniformis*, *B. subtilis*, *Bacillus pumilis*, and *Bacillus thuringiensis* with incidents of foodborne illness (Drobniewski, 1993). However, *B. licheniformis*, which does not belong to the *B. cereus* group, has occasionally been implicated in food poisoning and immunological cross-reactions with toxins of *B. cereus*.

Some *Bacillus* species such as *B. cereus*, *B. subtilis*, and *B. licheniformis* are used in the animal feed industry, and their addition to diets of animals (i.e., pigs, poultry, and calves) improves the zootechnical parameters and health of livestock. Due to certain specific conditions, some *Bacillus* species have been shown to be able to produce toxins and may enter the human food chain; the Scientific Committee on Animal Nutrition (SCAN) (EU, 2000; EFSA, 2014) has issued an opinion concerning the safety of the use of *Bacillus* species in animal nutrition. Two species of interest in human health problem among the *Bacillus* species are the *B. cereus* group (including *Bacillus anthracis*, *B. cereus*, *Bacillus mycoides*, *B. thuringiensis*, *Bacillus pseudomycoides*, and *Bacillus weihenstephanensis*) and the *B. subtilis* group (including *Bacillus amyloliquefaciens*, *B. licheniformis*, *Bacillus pumilus*, and *B. subtilis* itself).

There are bacilli among animal microorganisms and organisms commonly used as feed additives; in human microorganisms, lactobacilli and bifidobacteria dominate. Scientific data suggest that toxin production occurs among *Bacillus* species other than *B. cereus* and, more specifically, among strains of *B. licheniformis*. Suspicion of toxin production in strains of these species implicated *B. licheniformis* occasionally and *B. subtilis* more rarely as a cause of food poisoning, and it is considered necessary to check the toxigenicity of the bacilli microorganisms and emetic toxins (Pedersen et al., 2002).

In the case of microorganisms belonging to the *B. cereus* group, methods of detection involving commercial test kits, laboratory bioassays (rabbit ileal loop test, vascular permeability reaction in rabbits, cell cytotoxicity assays), and PCR-based methods are required to identify toxins and virulence factors. These studies are also required for bacilli other than those in the *B. cereus* group. Cross-reactions with *B. cereus* enterotoxins have been shown by commercial immunological kits for some strains of *B. licheniformis*. Various cell lines have been used with different responses to *Bacillus* toxins. The cytotoxicity potential of *B. licheniformis* DSM 9552 was determined in Chinese hamster ovary cells (CHO K1) by measuring total cellular metabolic activity using 3-(4,5-dimethylthiatzol-2-yl)-2,5-diphenyl tetrazolium bromide (MTT). Initially, the MTT assay was validated against toxigenic strains of *B. cereus* to define the exact criteria for a toxigenic versus a nontoxigenic response. No toxic effect on CHO K1 cells was detected in the culture supernatant of *B. licheniformis* DSM 9552. Although no information was provided on the presence in *B. licheniformis* DSM 9552 of genes coding for *Bacillus* toxins, the MTT test is a well-established short-term cytotoxicity assay suitable for the detection of *Bacillus* toxins (Pedersen et al., 2002; Toh et al., 2004). Previous research has shown that a variety of different *Bacillus* species isolated from the dairy environment may have the potential to produce diarrheal enterotoxins (Beattie and Williams, 1999).

The SCAN opinion (EU, 2000) examined the extent to which toxin production may be an unrecognized problem among some species of *Bacillus* and its implications for the continuing commercial use of this type of probiotic additive. The knowledge of the genetic and biochemical basis for toxin production, the methods for the detection of *Bacillus* toxins, and the recommendations for how best to ensure the absence of toxins (or a capacity for toxin production) were reviewed. The SCAN recommends (EU, 2000) that, for all future applications involving the addition of living organisms to animal feed, the use of strains from *B. cereus* taxonomic group should be strongly discouraged. The required studies must start with the taxonomy of the strain.

Presence of transmissible antibiotic resistance markers

The potential impact of the use of multidrug-resistant probiotics entering the food chain and disseminating resistance to human or animal pathogens is always considered. Bacteria may bear transferable resistances. In LAB, *Enterococcus* is perhaps the genus that is potentially most concerning. This genus has the ability to transfer genetic information, including antibiotic resistance, and the association with disease is clear. Although some traditional dairy products contain species of this genus, some communities have consumed them for generations without apparent negative health effects. Some enterococcal strains have shown resistance to vancomycin, a glycopeptide antibiotic with the capacity to transfer this kind of resistance to other bacteria species. Vancomycin is often used for treating patients with infections caused by Gram-positive microorganisms, particularly nosocomial infections due to meticillin-resistant staphylococci. Enterococci are normal inhabitants of the intestines of animals and humans, but they are also in the environment and food; they are frequent opportunistic pathogens known for their resistance to antibiotics. There is evidence that the widespread use of avoparcin, a glycopeptide antibiotic growth promoter, in pigs and poultry, which is banned in the European Union (EU), has contributed to a pool of enterococci that are resistant to vancomycin, a critically important antibiotic (CIA) for the treatment of various infections in humans (Anadón and Martínez-Larrañaga, 1999; Anadón, 2006). The SCAN Committee of EC adopted an opinion regarding the criteria for assessing the safety of microorganisms resistant to antibiotics (EU, 2001) that was later revised (EU, 2003). The European Food Safety Authority (EFSA) through the FEEDAP Panel updated the opinion and published new recommendations (EFSA, 2005a, 2012a,b). The aim of these opinions was to provide guidance for developing studies to show the potential of each bacterial strain (and not yeast) to bear resistance and to transfer resistance. The basis of such evaluation starts with the determination of in vitro antibacterial activity [minimum inhibitory concentration (MIC)] for a relevant range of antibiotics of human or veterinary importance, a list of antibiotics for which the MIC should be tested, and also MIC breakpoints for several bacteria and relevant antimicrobials.

Although the detection of the MIC values above the breakpoint levels identified for one or more antimicrobials requires further investigations to make the distinction between acquired and intrinsic resistance, the microbial breakpoints categorizing bacteria are fixed (Anadón et al., 2006, EFSA, 2012a,b) and periodically revised. It is stated that when a bacterial strain is known to be resistant to a specific antibiotic and other species are normally susceptible to this antibiotic, then the applicant should evaluate the reason for such resistance. If acquired resistance may be transferred or if known exogenous resistance genes are present, then the probiotic strain is not considered suitable for use as a feed additive. For all of these reasons, all bacterial products intended for use as feed additives must be examined to establish the susceptibility of the component strain(s) to a relevant range of clinically relevant antibiotics of human and animals.

There are two types of bacterial resistance to antibiotics: intrinsic and acquired. When resistance to an antimicrobial is inherent to a bacterial species it is generally referred to as intrinsic or natural resistance (that would be best designated as insensitivity), and it is typical of all strains of a given genus or species and based on the physiological or structural peculiarities of the strain (cell wall characteristics, lack of antibiotic target function, among others). It defines the limits in terms of the spectrum of activity of a given antibiotic. It is chromosomal and not transferable to other bacteria. In contrast, when a strain of a typically susceptible genus or species is resistant to a given antimicrobial drug, it is considered to be acquired resistance. Acquired resistance can be due to two distinct mechanisms: (1) mutation in indigenous ("*house keeping*") genes, in which case resistance is not transferable; and (2) acquisition of an exogenous resistance determinant from another bacterium by horizontal (lateral) gene transfer, in this case being retransferable. An explicative example is that of vancomycin resistance, a feature of some lactobacilli species, which is due to structural peculiarities of the PGN (Danielsen and Wind, 2003). However, in vancomycin resistance, the enterococcal strain (e.g., *E. faecium*) is acquired and plasmid-mediated (Courvalin, 2006). Clearly, the impact in terms of safety differs between the two cases.

The nature of any antibiotic resistance determinant present in a candidate microorganism to use as a feed additive needs to be determined. Intrinsic resistance may arise because the candidate microorganism may lack the target for the antibiotic. This will raise less concern in a food safety evaluation than a microorganism that has acquired mobile DNA encoding of an enzyme that modifies or destroys an antibiotic. It is also important to recognize that antibiotics differ in their human or veterinary importance. In some situations, the resistance gene may be transmitted from one bacteria to another through mobile genetic elements. Such spread can occur between bacteria of the same or different species and genera (O'Brien, 2002). Three major mechanisms of antimicrobial resistance have been described: (1) decreasing the intracellular accumulation of the antibiotic by either altering the outer membrane permeability or diminishing inner membrane transport or active efflux; (2) inducing an enzymatic detoxification of the drug; and (3) modifying the target of antibiotic action (Guardabassi and Courvalin, 2006). The antibiotic defense mechanisms of intrinsic resistances are, in most cases, related to the presence of low-affinity targets, absence of targets, decreased uptake, and accumulation or efflux of drugs.

The recognition that transmissibility of DNA has implications for risk assessment should be highlighted. It has been highlighted that transmission of DNA is a two-way process. Strains that harbor the genes for antibiotic resistance or virulence determinants may act as a reservoir for DNA encoding those traits. Strains that are initially devoid of such characteristics may, however, acquire and express DNA that encodes undesirable characteristics. In conclusion, there are close parallels between the issue of virulence determinants and the consideration of antibiotic resistance. Molecular tools are not the only methods by which virulence determinants may be characterized and, as with antibiotic resistance, they may only be applied to establish whether virulence determinants that have been characterized previously are present. As yet uncharacterized virulence determinants are not amenable to detection using molecular tools (EFSA, 2005b).

Assessment of antibacterial activity and microbial breakpoints

It is essential that tests for determining in vitro antibacterial activity are made in a consistent manner using internationally recognized and standardized methods. Susceptibility testing to antibiotics is most commonly performed by both dilution techniques (microbroth or macrobroth) and agar techniques, including the E-test method and the classical disk diffusion method. As a basic requirement, the MIC of the antimicrobial expressed as mg/L or μg/mL should be determined for each of the following substances: ampicillin, vancomycin, gentamicin, kanamycin, streptomycin, erythromycin, clindamycin, tetracycline, and chloramphenicol (Table 65.2).

These antimicrobials were chosen to maximize the identification of resistance genotypes to the most commonly used antimicrobials by assessing the resistance phenotypes. The detection of the MIC above the breakpoint levels, as identified, for one or more antimicrobials requires further investigations to make the distinction between acquired and intrinsic resistance (Fig. 65.1).

For the purpose of distinguishing strains harboring acquired antimicrobial resistance from susceptible strains, the microbiological breakpoints are set by studying the distribution of MICs of the chosen antimicrobials in bacterial populations belonging to a single taxonomic unit (species or genus). The part of the population that clearly deviates from the normal susceptible populations is categorized as resistant. The data used for the definition of microbiological breakpoints were derived from the body of research published and from national and European monitoring programs. The use of microbial breakpoints as reference data in the quality control and the genetic tools to investigate for intrinsic or acquired resistance in the research and development for new and improved cultures is of much interest.

The actual possibility of transfer of resistance to human or animal pathogenic bacteria, which could result from the use of microbial products based on drug-resistant strains, is related to the genetic basis of resistance. Although it is reasonable to assume that gene transfer from viable microorganisms will occur to other microorganisms in an open environment, such as the GIT, intrinsic resistance is presumed to have minimal potential for horizontal spread, whereas acquired resistance mediated by added genes is considered to have high potential for lateral spread.

For the assessment of susceptibility of bacterial strains, serial twofold dilution procedures in agar or broth should be used and include relevant quality control strains. The tests should be performed according to standards such as those of the Clinical and Laboratory Standards Institute (formerly the National Committee for Clinical Laboratory Standards). After incubation, the MIC is defined as the lowest concentration of the antimicrobial that inhibits bacterial growth. The data used for the definition of microbiological cut-off values (Table 65.2) were derived from the published body of research, the European Committee on Antimicrobial Susceptibility Testing, and from national and European monitoring programs (EFSA, 2012a,b).

The existing body of scientific information related to that specific or related species must be considered when the procedure for MIC determination (dilution method, growth media, and incubation conditions) is chosen, keeping in mind the possible interference of media and growth conditions. Media interference has been reported, for example, for LAB.

TABLE 65.2 Microbiological cut-off values (mg/L).

	Ampicillin	Vancomycin	Gentamicin	Kanamycin	Streptomycin	Erythromycin	Clindamycin	Tetracycline	Chloramphenicol	
Lactobacillus obligate homofermentative[a]	1	2	16	16	16	1	1	4	4	
Lactobacillus acidophilus group	1	2	16	64	16	1	1	4	4	
Lactobacillus obligate heterofermentative[b]	2	n.r.	16	32	64	1	1	8	4	
Lactobacillus reuteri	2	n.r.	8	64	64	1	1	16	4	
Lactobacillus facultative heterofermentative[c]	4	n.r.	16	64	64	1	1	8	4	
Lactobacillus plantarum/pentosus	2	n.r.	16	64	n.r.	1	2	32	8	
Lactobacillus rhamnosus	4	n.r.	16	64	32	1	1	8	4	
Lactobacillus casei/ paracasei	4	n.r.	32	64	64	1	1	4	4	
Bifidobacterium	2	2	64	n.r.	128	1	1	8	4	
Pediococcus	4	n.r.	16	64	64	1	1	8	4	
Leuconostoc	2	n.r.	16	16	64	1	1	8	4	
Lactococcus lactis	2	4	32	64	32	1	1	4	8	
Streptococcus thermophilus	2	4	32	64	64	2	2	4	4	
Bacillus spp.	n.r.	4	4	8	8	4	4	8	8	
Propionibacterium	2	4	64	64	64	0.5	0.25	2	2	
Other Gram-positive	1	2	4	16	8	0.5	0.25	2	2	
	Ampicillin	Vancomycin	Gentamicin	Kanamycin	Streptomycin	Erythromycin	Clindamycin	Tylosine	Tetracycline	Chloramphenicol
Enterococcus faecium	2	4	32	1024	128	4	4	4	4	16
	Ampicillin	Gentamicin	Kanamycin	Streptomycin	Tetracycline	Chloramphenicol	Nalidixic Acid	Siufonamide	Trimethroprim	Apramycin
Escherichia coli	8	2	8	16	8	16	16	256	2	8

n.r., not required.
[a]*Including* Lactobacillus delbrueckii, Lactobacillus helveticus.
[b]*Including* Lactobacillus fermentum.
[c]*Including the homofermentative species* Lactobacillus salivarius.

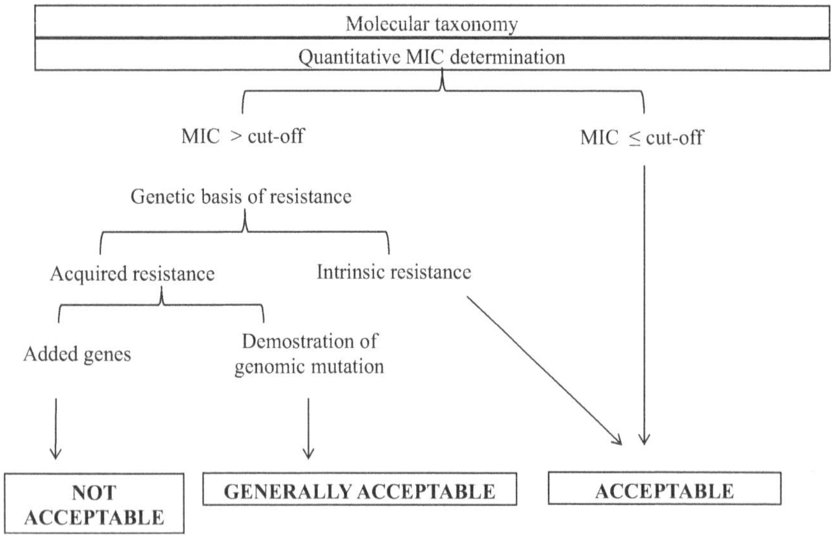

FIGURE 65.1 Proposed scheme for the antimicrobial resistance of a bacterial strain used as a feed additive as the base of quantitative MIC determination. *MIC*, minimum inhibitory concentration.

The intake of feed with millions to billions of LAB per milliliter is not suspected to be a major concern; however, from a theoretical point of view, horizontal transfer of antibiotic resistance genes can take place in the feed during production or in the animal intestinal system. LAB are used in a large variety of feed applications. Due to their reported health benefits, lactobacilli and bifidobacteria are increasingly used as probiotics. LAB (opportunist pathogens) and bifidobacteria have a long history of safe use and do not usually create any safety concerns. However, extensive use of antibiotics for treatment and prophylaxis of microbial infections in humans, animals, and even in plants has led to an increase in antibiotic resistance in bacteria. Therefore monitoring of antibiotic susceptibility profiles of nonpathogenic bacteria linked to the food chain has also become relevant. In contrast to clinically important bacteria species, there are no established standard methods and breakpoints available for susceptibility testing of nonpathogenic food-associated bacteria. Therefore various culture media and test procedures have been used in susceptibility testing of these bacteria.

Safety aspects of probiotics

Assessment of the safety of probiotics is not an easy task (Anadón et al., 2014, 2019). Although most of the effects of probiotics are beneficial, several negative effects should be considered before introducing them in food products or therapeutic applications (Suez et al., 2019). In most cases, the safety of novel strains has been deduced mainly from the common occurrence of the species either in foods or as normal commensals in the human gut. The selection of new probiotic organisms is targeted on new strains and even genera that are more beneficial or specific. The selection criteria of new probiotic strains is determined by many factors such as resistance to pancreatic enzymes, acid, and bile, preferably of human origin (although *S. boulardi* is not of human origin), documented health effects, known safety, and good technological properties, especially the potential probiotics (Ouwehand et al., 2002).

When novel microbes and GMO are introduced, their safety and the risk-to-benefit ratio must be carefully studied and assessed. Also, new probiotics should be of genera and strains commonly found in the healthy human intestinal microflora. The microbes could be classified as nonpathogenic (*Lactobacillus*, *Lactococcus*, *Bifidobacterium*, *Saccharomyces*), pathogens (*B. cereus*), and opportunistic pathogens (*Enterococcus* and also other general LAB) (Donohue, 2006). The pathogenic microbes can potentially cause an infection even in a healthy host. LAB and bifidobacteria are the most common bacteria that attach to the human intestinal mucosa and are commonly regarded as having Generally Recognized as Safe (GRAS) status; although lactobacilli and bifidobacteria are regarded as GRAS, for novel products it is necessary to perform premarketing determination of the safety of the strain and for the strains to be correctly identified.

Moreover, certain strains of probiotic bacteria have been tested to ensure they are free of risk factors like transferable antibiotic resistances, cancer-promoting and/or putrefactive enzymes and metabolites, hemolysis, activation of thrombocyte aggregation, or mucus degradation in the mucus layer of the GIT. Moreover, the toxicity profile may be strain-dependent, and among lactobacilli, one of the most important genera of LAB, few strains have been involved in septicemia. *Lactobacillus* infections are extremely rare and few infections have been associated with food-borne lactobacilli. Despite the absence of pathogenic potential, LAB were found in <0.1% (enterococci 1%) of clinical samples from severe infections (endocarditis, meningitis, or bacteremia)

(Gasser, 1994). These bacteria probably originated from the indigenous microflora, whereby in many cases the translocation (invasion) was facilitated by underlying disease, lesions, or inflammation in the oral cavity and the GIT, or by an impaired immune system. Every viable microbe that is able to grow under the conditions encountered in a host can cause an infection under certain circumstances, especially in immunocompromised humans. The factors that must be addressed in the evaluation of safety of probiotics include the following: (1) pathogenicity; (2) infectivity and virulence factors comprising toxicity; (3) metabolic activity; and (4) the intrinsic properties of the microbes. The absence of pathogenicity and infectivity is required for probiotic safety. The frequent isolation of LAB from clinical infections has raised debate over the safety of these bacteria and whether they are actually infective. Another requirement of probiotics is that the probiotic bacteria should not produce harmful substances by metabolic activity. Platelet-aggregating activity, mucus degradation activity, and antibiotic resistance also should be tested (Ishibashi and Yamazaki, 2001). It remains unknown whether strains with platelet aggregation properties enhance the risk of infection to a relevant extent, and whether they should be considered undesirable as a probiotic.

Members of the genera *Lactococcus* and *Lactobacillus* are most commonly given GRAS status, whereas members of the genera *Streptococcus* and *Enterococcus* and some other genera of LAB contain some opportunistic pathogens (Salminen and von Wright, 1998).

In terms of efficacy and adverse effects, the survival of the probiotics in the GIT, their translocation and colonization properties, and the fate of their active components need to be known (Marteau et al., 1993). Regarding the context of potential adverse effects of probiotics, there are four types of side effects or adverse reactions: (1) systemic infections; (2) risk of deleterious metabolic activities; (3) risk of adjuvant side effects and immunomodulation; and (4) risk of gene transfer (Salminen and von Wright, 1998). Additional characteristics that probiotics must possess are genetic stability and the capability to not develop antibiotic residence.

According to Salminen and von Wright (1998), three approaches can be used to assess the safety of a probiotic strain: (1) studies on the intrinsic properties of the strain; (2) studies on the pharmacokinetics of the strain survival, activity in the intestine, dose–response relationship, and fecal and mucosal recovery; and (2) studies searching for interactions between the strain and the host. The probiotic pharmacokinetics can be measured in vivo using fecal collection of intestinal intubation, and colonic biopsy techniques (Marteau, 2001). It is very important to know the survival of the probiotics within the GIT, their translocation and colonization properties, and the fate of their active components to predict the positive effects and/or side effects. The survival of ingested probiotics at different levels of the GIT differs between strains (Marteau et al., 1993). Some strains are killed in the stomach, whereas others (e.g., bifidobacteria or *L. acidophilus*) can pass through the entire gut at very high concentrations (Marteau et al., 1993). Thus, the probiotics must retain their viability during transit through the stomach and small intestine as well as during the storage and manufacturing process of the functional food. In conclusion, the safety and stability would be important criteria for probiotic selection.

In vitro studies

Estimation of the in vitro infective properties of the probiotic microorganisms is performed using cell lines (e.g., cultured intestinal cells to study the invasion capacity of a specific strain) and human intestinal mucus degradation. The minimal infective dose is very difficult to realize because of the large number of microbial and host factors involved and the high potential for individual differences. In addition, the assessment of infectivity can be tested in animal models (e.g., immunocompromised animals or lethally irradiated animals). In vitro models can also help to predict the fate of ingested strains, testing the sensitivity of the probiotic to acid or bile, and dynamic multicompartmental models simulating the dynamics of the transit and secretions in the GIT (Marteau, 2001).

Genotoxicity studies including mutagenicity

At least two different genotoxicity tests, a bacterial reverse mutation assay and an in vivo assay for clastogenicity in mammalian cells (e.g., a metaphase cytogenetic assay), should be performed; if these initial tests give an indication of mutagenicity, then two additional in vivo studies should be performed using two somatic tissue sites to demonstrate that in vitro mutagenicity is not expressed in vivo (Anadón et al., 2006, 2014, 2019).

The existence of genes that may be associated with virulence functions or antibiotic resistance has always been a concern, particularly if they were associated with a mobile element, such as a plasmid, transposon, or insertion element. The advent of genome sequences has now provided a tool to address this concern such that cultures with the potential to transfer disagreeable genes can be recognized and not used in foods.

Animal studies

Animal models are generally of limited value in microbiological risk assessment concerning probiotic bacteria (Binns, 2013). There is a high variability in response between animal species, which makes extrapolation of results obtained in animals to humans uncertain. However, the *maximum dose levels* (tolerance study) in feed that can be given to different animal species that have not caused any adverse effects are of interest. Acute toxicity studies were conducted for several strains of probiotics using the same procedures as acute toxicity studies for chemicals. Translocation of probiotics can be studied in mice and increased using a lethal irradiation model simulating an immunocompromised state (Dong et al., 1987). The assessment of the acute and subacute effects of ingestion of large amounts of the probiotic microorganisms should be done using acute oral and 90-day rodent-feeding toxicity studies performed under EU or OCDE guidelines, which are required (especially for the new probiotics). The optimal mode of administration is by incorporation into the feed, but if this is impractical, then administration in drinking water or by oral *gavage* may be used (Anadón et al., 2006, 2014, 2019).

Standard toxicology and safety evaluations for chemical and biological compounds are of limited value in assessing the safety of probiotic bacteria. Therefore appropriate safety testing for novel probiotic strains proposed for human consumption is needed.

Noninvasive tests in animal models and humans

Probiotics can have multiple effects on a host (e.g., either directly or indirectly on the pathogenesis and progress of disease), and better ways to determine both their safety and toxicity. There are noninvasive ways to apply dynamic function testing in animal models and humans to provide reference points to which other measurements can be related (e.g., altered circulating cytokines, altered gene expression). As such, this phenotypic scaffold, alone and combined with newer molecular parameters, will improve our understanding of the interaction of luminal factors within the alimentary tract and the impact that these have on physiologically challenged mucosa and in disease, both at the gastrointestinal level and in remote organs (Butler, 2008).

Studies in humans

A number of short-term clinical trials on healthy volunteers attest to the safety of current probiotics. In most clinical studies, the probiotic did not induce more adverse effects than the placebo and the tolerance studies were generally acceptable. In some studies, the presence or absence of GIT disorders has been especially studied, which seems rational because the first and probably only contact between bioproducts and the host occurs in the GIT (Salminen and von Wright, 1998). In a few studies, biological parameters were analyzed because it was thought that the probiotic might have an influence on them.

Epidemiological and postmarketing surveillance

The long history of the use of probiotic microorganisms has proven their safety. Because the risk is very low, the best approach for assessing them is probably to analyze them in epidemiology studies and prospectively using post-marketing pharmaco-toxico vigilance systems in populations ingesting large amounts of newly introduced probiotic microorganisms for infections in comparison to the use of traditional strains (Salminen and von Wright, 1998). Undoubtedly, safety assessment for probiotics should always be associated with those studies. Epidemiological and clinical studies are useful for proper monitoring of consumer safety.

The value of such studies depends on their statistical power, that is, on the number of cases studied and the attitude of the healthcare system to comply with the reported tasks (Adams and Marteau, 1995; Aguirre and Collins, 1993; Dhodapkar and Henry, 1996).

The most important concern about probiotic use is the risk of bacteremia, fungemia, and sepsis (Theodorakopoulou et al., 2013). Consumption of probiotics should be clarified in every new case of severe infection caused by lactobacilli, enterococci, or yeast, and the strains involved should be compared with the probiotic ones. Lactobacilli have been reported in adults to cause cases of sepsis, meningitis, and infections localized in different organs (Mackay et al., 1999). Thus, the use of probiotics in immunocompromised patients could generate serious health risks, although this has not been scientifically proven. The use of *S. boulardii* and *L. rhamnosus* has not been banned, despite the few cases of infections; many studies have proven their general safety and even efficacy in humans. A risk–benefit assessment could indicate that the absolute risk is very low and the risk-to-benefit ratio appears to be favorable (Marteau, 2001).

To minimize the risks of side effects, more research should be done with inactivated or nonviable preparations. Probiotic research is moving forward due to basic science and clinical trials evaluating the safety and efficacy of probiotics for various medical conditions. However, clinical trials of probiotics still have limitations and further studies about the amount and interval time are still required because the benefits of probiotic action vary.

Probiotic efficacy evidence

Probiotics have been studied to varying degrees in vitro and in animal feeding studies, but less so in human studies involving food. Available data indicate that no harmful effects have been observed in controlled clinical studies with *L. acidophilus* and bifidobacteria. The degradation of intestinal mucus has been used as the first marker of toxicity and specific commercial probiotic strains were shown to be inactive in mucosal degradation (Donohue et al., 1998).

In vitro evidence

In vitro tests can be used as the first step of screening for probiotic safety and efficacy. Useful in vitro efforts include genomic analysis, DNA-based and phenotypic strain identification, and measurement of viability. These approaches are usually applied with the following aims: (1) quantifying bacteria in a sample/product; (2) identifying and characterizing the strain(s) being studied; (3) recognizing the characterization of strain-specific or species-specific differences among a range of probiotic bacteria; (4) insuring product quality and consistency; (5) screening for survivability in the upper GIT; (6) conducting mechanistic studies with cellular models; and (7) identification of potential safety risks (Sanders et al., 2005).

Animal models

Numerous animal models have been developed for the study of the physiological effects of a number of bioactive components and diets. However, there are several important differences such as the anatomy, physiology, and metabolism, between animal species and humans, therefore the results obtained from animal experiments cannot be used as proof of efficacy but only as indications, particularly when doses used in animal studies are not reflective of realistic doses to be used in humans (Sanders et al., 2005).

Human case studies

Observations from a single case study are, at best, only suggestive of a more general effect. Most often, they only reflect peculiar effects in a specific condition and are not representative of the general population. The sample population size is always important for proving efficacy in the general population. Although human case studies can raise public awareness, they should always be confirmed by well-designed, randomized, double-blind, controlled trials. Caution should also be exercised in evaluating case studies as they relate to safety. Individual reports of rare adverse incidents can be difficult to interpret without context for evaluating the relative risk (Rautio et al., 1999).

Human trials

Well-designed, randomized, double-blind, controlled trials are the core of efficacy for probiotics (Kalliomäki et al., 2001; Oli et al., 1998). Open-label studies might provide useful information. Epidemiology was considered to be valuable, but the large degree of experimental "*noise*" in these studies makes it difficult to detect small effects. Observational studies can also be valuable but do not provide conclusive evidence.

Long-term intervention trials are also important, especially to observe the improvement of wellness. Most studies are short-term (<12 week) studies. If, for example, a risk factor can be reduced with probiotic administration, long-term trials are necessary to investigate whether the effect will persist with time. Postmarket surveillance is important in order to monitor the long-term beneficial (or adverse) effects (Sanders et al., 2005).

FAO/WHO guidelines to assess the safety of probiotics

The Joint Food and Agriculture Organization of the United Nations and World Health Organization Expert Consultation on *Evaluation of Health and Nutrition Properties of Probiotics* met in Córdoba (Argentina) in October 2001 to establish a set of guidelines (Joint FAO/WHO, 2002) and to set up a systematic approach for the evaluation of probiotics in food, which led to the substantiation of health claims (Fig. 65.2). As a consequence of this meeting, a Joint FAO/WHO Working Group on Drafting Guidelines for the Evaluation of Probiotics in Food was created and met for the first time in London, Ontario, Canada, on April 30 and May 1, 2002, to generate guidelines and recommend criteria and methodology for the evaluation of probiotics, and to identify and define what data need to be available to accurately substantiate health claims (Anadón et al., 2014, 2016, 2019; Sniffen et al., 2018). The aims of the Working Group were to identify and outline the minimum requirements needed for probiotic status. In this document no reference was made either to the term "bio-therapeutic agents" (i.e., microorganisms with therapeutic effects in humans) or "nonfood beneficial microorganisms". GMOs were also excluded.

The proposed guidelines recommended for the evaluation of probiotics for food involve the following steps:

1. *Identification of the genus/species/strain.* Nomenclature of the bacteria must conform to the current, scientifically recognized names (Anadón et al., 2014).

The taxonomy of LAB and other bacteria has changed significantly with the approach of genetic methods of classification. Strains previously thought to be dissimilar have merged, whereas other strains have been added or reassigned to different genera. Therefore the persistent use of incorrect or nonexistent species names on product

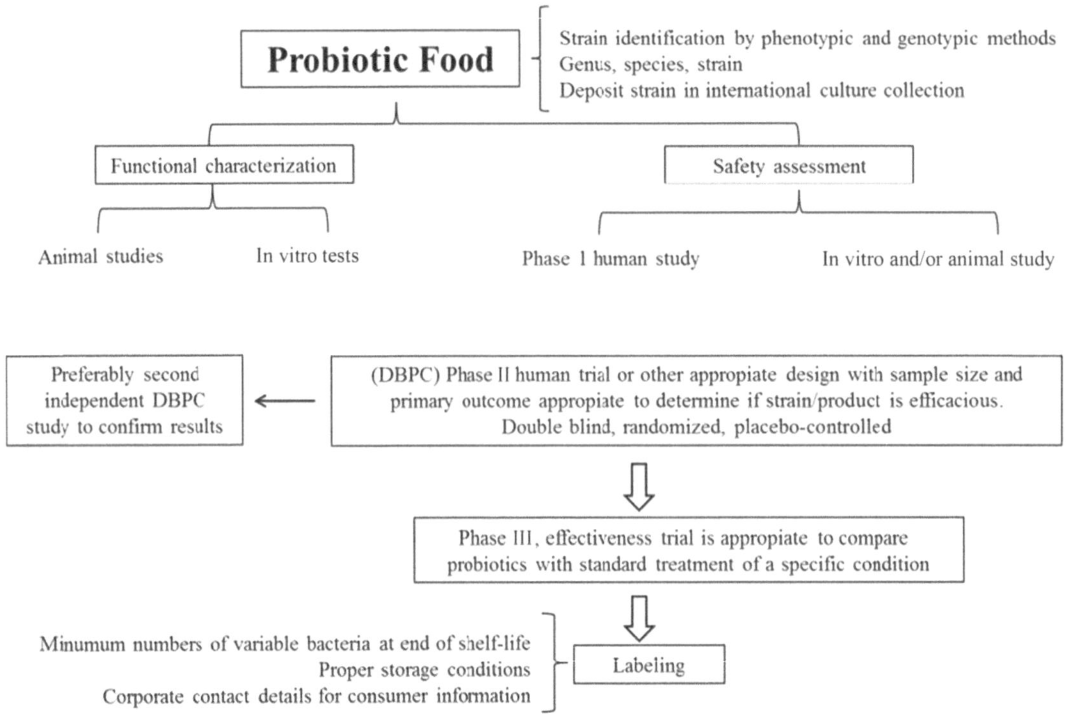

FIGURE 65.2 Systematic approach for the evaluation of probiotics in food. *DBPC*, double-blind, randomized placebo-controlled.

labels despite taxonomic reassigning is a significant issue for the safety and credibility of probiotics.

The speciation of the bacteria must be established using a combination of phenotypic and genotypic tests [DNA–DNA hybridization or 16S rRNA gene sequence analysis for species identification, DNA macrorestriction followed by pulsed field gel electrophoresis, which is the gold standard, and randomly amplified polymorphic DNA analysis, that is less reproducible among others for strain identification]. The presence of extrachromosomal genetic elements (i.e., plasmids) can contribute to strain typing and characterization. It is recommended that all strains be deposited in an internationally recognized culture collection.

Therefore, when incorporating live and active bacteria in foods, it is of paramount importance to ensure the bacteria are safe and unlikely to contribute to the evolution of more virulent pathogenic bacteria due to horizontal gene transfer. Traditional approaches to evaluate safety relied heavily on extended use in human foods without documented deleterious effects.

To conduct a thorough genome safety analysis of probiotic bacteria, a functional assessment of the genome sequence should be generated. A common functional assessment method is provided by the Clusters of Orthologous Groups database at the NCBI. This can annotate the genome and predict functions due to motif similarities with known proteins. A genome analysis can readily uncover known virulence or potential antibiotic resistance genes that may be present. One example is that the analysis of the genome annotations for the probiotic lactobacilli and bifidobacteria deposited at GenBank did not show any obvious virulence genes that present a safety concern. However, there are usually several genes predicted to be involved in antibiotic resistance in the majority of strains. These include ABC-type multidrug transporters and genes predicted to encode resistance to antibiotics (e.g., tetracycline, bacitracin, lincomycin, meticillin, or glycopeptides).

2. *In vitro tests to screen potential probiotics* (these tests are essentials to gain knowledge of strains and species, and the mechanisms of probiotic effects). At this point, it is essential to highlight that probiotics for human use will require substantiation of clinical efficacy with human trials, and appropriate target-specific in vitro tests correlating with in vivo results are recommended. The main available in vitro tests used in developing probiotics are the following: (1) resistance to gastric acidity; (2) bile acid resistance (colonization); (3) adherence to mucus and/or human epithelial cells and cell lines; (4) antimicrobial activity against potentially pathogenic bacteria; (5) ability to reduce pathogen adhesion to surfaces; (6) bile salt hydrolase activity (as a probiotic marker); and (7) resistance to spermicides (applicable to probiotics for vaginal use).
3. *Safety considerations: requirements for proof that a probiotic strain is safe and without contamination in its delivery form.* Historically, lactobacilli and bifidobacteria

associated with food have been considered to be safe; however, the probiotics may theoretically be responsible for four types of side effects (Marteau, 2001): (1) systemic infections, (2) deleterious metabolic activities, (3) excessive immune stimulation in susceptible individuals, and (4) gene transfer. Although most of the species and genera are apparently safe, certain microorganisms may be problematic, particularly the enterococci (*E. faecium and E. faecalis*). These have emerged as opportunistic pathogens in hospital environments and have caused nosocomial infections such as endocarditis, bacteremia, and intraabdominal, urinary tract, and CNS infections, and may also harbor transmissible antibiotic resistance determinants (i.e., vancomycin-resistant *Enterococcus* strains) and bacilli, especially those belonging to the *B. cereus* group that are known to produce enterotoxins and an emetic toxin (Anadón et al., 2006). LAB have been isolated in immunosuppressed patients with subacute endocarditis, with increasing frequency in recent years, and thus this safety aspect should be monitored (Marteau, 2001). Other specific properties in the safety assessment are the biogenic amine formation.

Some groups of bacteria are considered as GRAS, but the FAO/WHO working group recommends that strains of probiotics should be characterized at a minimum with the following safety assessment tests: (1) determination of antimicrobial resistance patterns; (2) assessment of certain metabolic activities (i.e., undesirable) (e.g., D-lactate production and bile salt deconjugation); (3) assessment of side effects during human studies; (4) post-market epidemiological surveillance of adverse incidents in consumers; and (5) if the strain undergoing evaluation belongs to a species known to be either a mammalian toxin producer or to have hemolytic potential, then it must be tested for toxin production or hemolytic activity.

The lack of infectivity in immunocompromised animal models can be of importance.

4. In vivo studies using animals and humans. The FAO/WHO working group (FAO/WHO, 2002) encourages use of animal models to provide substantiation of in vitro effects and for determining the probiotic mechanism prior to human trials. The principal outcome of efficacy studies on probiotics should have proven benefits in human trials, such as statistically and biologically significant improvement in condition, symptoms, signs, well-being, or quality of life, reduced risk of disease or longer time to next occurrence, or faster recovery from illness.

   Probiotics have been tested in a variety of clinical conditions to prove the efficacy and safety of products. Standard methods for clinical evaluations are comprised of phase I (focused on safety), phase II (efficacy), phase III (effectiveness), and phase IV (surveillance) (Anadón et al., 2014).
5. *Health claims and labeling*. The consultation recommends that specific health claims should be permitted on the label and promotional material. The information needed to be described on the label is as follow: (1) genus, species, and strain designation (strain designation should not mislead consumers about the functionality of the strain); (2) minimum viable numbers of each probiotic strain at the end of the shelf-life; (3) the suggested serving size must deliver the effective dose of probiotics related to the health claim; (4) health claim(s); (5) proper storage conditions; and (6) corporate contact details for consumer information.

Qualified presumption of safety

There are two comparative assessment procedures used in the United States and Europe, although the Qualified presumption of safety (QPS) assessment has a different aim. One assessment is used for food supplements in the United States (FDA GRAS system) and another for the safety of bacterial dietary supplements in Europe (EFSA QPS system). In terms of safety for probiotic products, when the product has a history of safe use in the target host, such as GRAS or its equivalents (i.e., the QPS) in Europe, then it is suggested that further animal and human toxicological studies may not be necessary. Overall, QPS is a preassessment intended to be considered and complemented by the safety assessment of a specific notification. QPS always assesses the highest taxonomic unit possible, which is usually the species. It never assesses notified strains because this is within the responsibility of the Scientific Panel of the EFSA, which is mandated for the safety evaluation (EFSA, 2013).

The EFSA has proposed a scheme based on the concept of QPS, defined as *an assumption based on reasonable evidence* and qualified to allow certain restrictions to apply. The scheme aims to have consistent generic safety assessment of microorganisms through the food chain without compromising safety standards. Individual evaluations would be limited to aspects particular to the organism, such as acquired antibiotic resistance determinants in LAB. QPS status would not apply to a microorganism that commonly causes pathogenicity (e.g., species or subspecies with pathogenic potential). Therefore assessment on a case-by-case basis (individual evaluations) would be always required. A microorganism would not necessarily be considered a potential pathogen where there are infrequent reports of clinical isolates from severely ill people (EFSA, 2005b; Anadón et al., 2010).

QPS has recently entered EU law with the publication of a new Commission Implementing Regulation (EU) No 562/2012 (EU, 2012) amending Commission Regulation

(EU) No 234/2011 (EU, 2011) with regard to specific data required for risk assessment of food enzymes.

The EFSA has proposed a scheme based on the concept of QPS, defined in Europe as *a belief or assumption based on reasonable evidence* and qualified to allow certain restrictions to apply. The QPS approach regarding microorganisms in food and feed is a system similar to the GRAS definition used in the United States but modified to take into account the different regulatory practices in Europe. This was considered necessary because issues of importance to Europe would not necessarily influence a GRAS listing. Consequently, it was proposed that any *generic listing* of a microorganism should be qualified, allowing the general safety of the organism/group of organisms to be concluded provided that certain specific criteria were met. It is intended to provide a mechanism to recognize and give weight to prior knowledge (whether gained through formal investigation or by experience).

The QPS schema is a qualified approval system that aims to have consistent generic safety assessment of microorganisms throughout the food chain without compromising the safety standards. Individual evaluations would be limited to aspects particular to the organism (i.e., presence of virulence factors, toxic metabolites, and antibiotic-resistant determinants in LAB).

In the case of novel use of a microorganism that also has a traditional use in food or feed production, the Novel Food Regulation [Regulation (EC) No. 258/97] (EC, 1997) covers the safety assessment of these products.

To establish the QPS status for a given microorganism, several requirements should be fulfilled which are rested on four pillars: taxonomy, familiarity, pathogenicity, and end use. The main requirements are relating to taxonomy and familiarity of the candidate organisms for QPS assessment and included taxonomic entity (i.e., the taxon or taxonomic unit such a genus, species, subspecies, or other grouping such as homofermentative lactobacilli should be considered).

The term *body of knowledge* (or familiarity) should be used following the recommendation of EFSA Scientific Colloquium (EFSA, 2005b). This term is required for the QPS safety assessment of probiotics and who should decide what defines the boundaries to that *body of knowledge*. Several elements will comprise the *body of knowledge* (Fig. 65.3). In addition to the peer-reviewed scientific literature, these include an understanding of the history of use of a microorganism, its industrial applications, its ecology, any clinical reports concerning the microorganisms, and information from the scientific literature (Anadón et al., 2010, 2016).

Pathogenicity involves whether the grouping considered for QPS contains known pathogens and, if so, whether sufficient data are known about their virulence determinants or toxigenic potential to exclude pathogenic

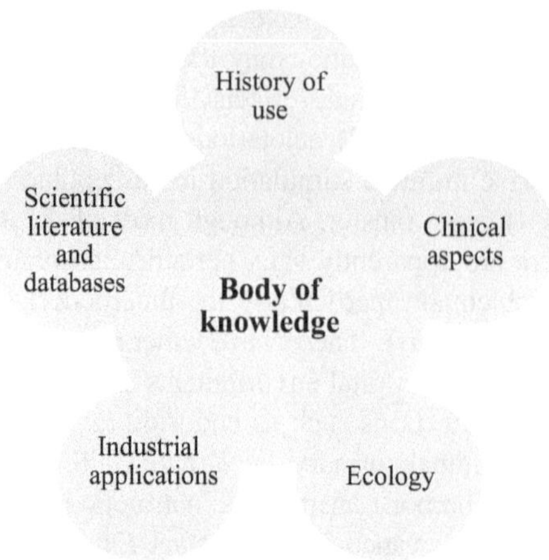

FIGURE 65.3 Body of knowledge.

strains (e.g., the *Bacillus subtilis* group to exclude the occasional strain with a toxigenic potential or fungal species to exclude mycotoxin producers or to distinguish between the virulent and avirulent forms of enterococci) and their end use, whether viable organisms enter the food chain, or whether they are used to produce other products. As a principle, it is desirable to conclude regarding the safety of the taxonomic unit for all purposes. This is likely to be the case with many fungal species for which safety might be established for a particular use (e.g., for plant protection purposes) but insufficient is known about metabolic activities to allow extrapolation to other uses (Table 65.3).

Antimicrobial resistance was introduced as a possible safety concern for assessment of the inclusion of bacterial species in the QPS list. *E. faecium* is not recommended for the QPS list despite the recent scientific knowledge allowing a differentiation between pathogenic and nonpathogenic strains.

The QPS guidelines regarding GRAS are applied to: (1) microorganisms; (2) a positive list that will be available; (3) the inclusion of the microorganism based on the history of use and adverse effects; and (4) defined taxonomic unit (e.g., genus, species, or strain).

European regulations of probiotics

The current EU legislation covers substances with a physiological effect, such as probiotic bacteria and prebiotic compounds. The new food probiotics are regulated by the Novel Foods Regulation [Regulation (EC) No. 258/97] (EC, 1997) and the Nutrition and Health Claims regulation [Regulation (EC) No. 1924/2006] (EC, 2006). Microbial strains will have to pass one or both of these regulations. Any claims proposed for probiotic bacteria to be used in food must be based on, and substantiated by, the generally accepted scientific data (EFSA, 2011). The

TABLE 65.3 Substantiation of claims by generally acceptable scientific evidence [Article 13 of Regulation (EC) No. 1924/2006].

Substantiation of claims by generally acceptable scientific evidence
• The claimed effect of the food is beneficial for human health
• A cause-and-effect relationship is established between consumption of the food and the claimed effect in humans such as the strength, consistency, specificity, dose–response, and biological plausibility of the relationship
• The quantity of the food and pattern of consumption required to obtain the claimed effect could reasonably be achieved as part of the balanced diet
• The specific study group(s) in which the evidence is presented is representative of the target population for which the claim is intended

EU regulations will prohibit any claims referring to the prevention, treatment, or cure of a human disease by a food, which is in contrast to that proposed by other countries such as Canada and the United States (Sanders et al., 2005). In the United States, probiotic products may be marketed and regulated to a generally health population as food or dietary supplement or biological product (i.e., drugs) depending on the intended use of a probiotic.

Probiotic substantiation domains of efficacy are needed to support claims of health benefits. Probiotic products that claim specific nutritional, functional, or therapeutic characteristics make the borderlines between food, nutrition and health claims, and dietary supplement or human medicines ambiguous, thereby producing questions for regulators (EFSA, 2011).

Regulation of probiotics

Depending on the intended use of a probiotic, the legislative background requirements differ because the probiotic can be considered a drug, novel food ingredient, nutritional and health claim, or a dietary supplement, and this sometimes depends on the relevant countries and agencies. The regulations of these products share may similarities with those for drugs.

Drugs

Human drugs are regulated by the EU under Directive 2001/83/EC (EC, 2001), amended by Directive 2004/27/EC (EC, 2004a,b). A medicinal product may only be placed on the market in the European Economic Area (EEA) when a marketing authorization has been issued by the competent authority of a member state (or EEA country) for its own territory (national authorization) or when an authorization has been granted in accordance with Regulation (EC) No. 726/2004 (EC, 2004a,b)

Novel food ingredients

To date, probiotic foods are not governed under specific EU regulatory frameworks; however, Regulation (EC) No. 258/97 (EC, 1997) may cover other more novel types of probiotic species that need to be discussed and assessed in the light of the Novel Food Guidelines (Jonas et al., 1996) and states safety rules for authorization of novel food/ingredients. Commission Recommendation 97/618/EC (July 29, 1997) concerns the scientific aspects and presentation of information necessary to support applications to place novel foods and novel food ingredients on the market, and the preparation of initial assessment reports under Regulation (EC) No. 258/97 (EC, 1997) provides the instructions necessary to sustain an application to fulfill the safety assessment.

The Novel Food Regulation defines novel foods as foods and food ingredients that were not used for human consumption to a significant degree within the Community before May 15, 1997. Human consumption to a significant degree within the Community, in this context, has been interpreted as being demonstrated by a food having been generally available within the community (i.e., food/food ingredient that does not have a significant history of human consumption within the EU prior to May 15, 1997).

Nutrition and health claims

In the EU, the claims are governed by Regulation (EC) No. 1924/2006 (EC, 2006). Article 6 of this regulation indicates that health claims shall be based on and substantiated by generally accepted scientific evidence, and Article 13(1) indicates that health claims should be evaluated by the EFSA after a scientific assessment of the highest possible standard (see Table 65.4). Many claims were evaluated on a case-by-case basis (some were clustered into one opinion) regarding uniform scientific criteria for consistency. The opinions released by EFSA reflect the varying quality of the information submitted (characterization, dose, health relationship, scientific evidence) and the evaluation results were: (1) approximately one-third were favorable with sufficient scientific evidence; (2) half of the nonfavorable evaluations had insufficient information regarding the food/constituent (e.g.,

TABLE 65.4 Representation of the organization of the application

PART 1. ADMINISTRATIVE AND TECHNICAL DATA
1.1 Table of contents**1.2** Application form**1.3** General information**1.4** Health claims particulars**1.5** Summary of the application**1.6** References
PART 2. FOOD/CONSTITUENTS CHARACTERISTICS
2.1 Food constituent**2.2** Food or category of food**2.3** References
PART 3. OVERALL SUMMARY OF SCIENTIFIC DATA
3.1 Tabulated summary of all pertinent studies identified**3.2** Tabulated summary of data from pertinent human studies**3.3** Written summary of data from pertinent human studies**3.4** Written summary of data from pertinent non-human studies**3.5** Overall conclusions
PART 4. BODY OF PERTINENT SCIENTIFIC DATA IDENTIFIED
4.1 Identification of pertinent scientific data**4.2** Pertinent data identified
PART 5. ANNEXES TO THE APPLICATION
5.1 Glossary/abbreviations**5.2** Copies of article/reprint of references, review articles**5.3** Full study reports of unpublished studies or unpublished reviews**5.4** Others

probiotic claims); and (3) probiotic claims regarding characterized strains had a nonfavorable outcome. The difficulties encountered with probiotic claims can be summarized as follows: identification and characterization of strains (phenotypic and genetic tests); impact of carrier or vehicle food matrix; stability (processes and storage); and clinical intervention studies often performed in diseased subjects/populations. Regarding the claim application challenges, the main ones are: component should be well-defined; basis for effects should be sufficiently characterized; the clinical studies should be performed in the target population; a clear rationale for extrapolation from specific populations should be presented; the cause and effects must be demonstrated; and the proposed health claim hypothesis and demonstration should be clearly presented through the whole application.

In accordance with the requirements of Regulation (EC) No. 1924/2006 (EC, 2006), the layout of the submission dossier is divide into five parts and the corresponding content of the application (Anadón et al., 2014). The five parts are listed in Table 65.4.

Concluding remarks and future directions

Probiotics are generally accepted as being living organisms that, when administered in adequate amounts, confer a health benefit to the host. Knowledge of probiotic effects that are strain-specific and characterize the relationship between the dose, its duration, and effect, on a strain-by-strain basis is needed. The organisms must arrive in effective levels at the site of action to determine the dose–response relationship. Bacterial species that have traditionally been regarded as safe are used in probiotics. Very few cases of adverse events have been reported in humans consuming probiotics. However, reports of frequent isolation of bacteria used in probiotics from infection sources in recent years have raised much debate over the safety of probiotics. Novel probiotics based on new species and more specific strains of bacteria products are constantly being sought, although their safety cannot always be assumed. Antimicrobial production is an important feature for probiotic organisms functioning in the gut. Although production of lactate as well as acetate by bifidobacteria is important in antimicrobials, bacteriocins can also provide a competitive side for bacteria. The simultaneous administration of probiotics and prebiotics (named synbiotics) may synergistically improve their health-promoting effects in organisms. The use of probiotics in medical practice is rapidly increasing. They have been used in the prevention and treatment of various diseases, but are not limited to intestinal infectious diseases. Various mechanisms, such as enhanced immune response, reduction of mutagenic compounds, reduction of intestinal inflammation, and production of SCFA, have varying levels of supporting evidence. The in vitro testing and animal and human studies used for efficacy and toxicity of probiotics sometimes contain limitations that must be considered in the assessment. Although most of the species and genera are apparently safe, certain microorganisms may be problematic, particularly the enterococci, which may harbor transmissible antibiotic resistance determinants and bacilli, especially those belonging to the *Bacillus cereus* group that are known to produce enterotoxins and an emetic toxin. Probiotics are regulated in the EU with respect to their safety and efficacy. At present, probiotic human foods are not governed under specific EU regulatory frameworks, although Novel Food Regulation (EC) No. 258/9 could be relevant in some specific applications. The qualified perception of safety (QPS) concept would allow strains with established safety to enter the market without extensive toxicity and efficacy testing requirements. In the case of human probiotics, the safety requirements could be more rigorous, considering that humans with variable health status would probably use these products as self-medication, and thus a case-by-case safety approach should be used. For QPS status, the EFSA plays a primary role in the scientific opinions on probiotics and their respective regulations.

Acknowledgment

This work was supported by Project Ref. RTA2015-00010-C03-03 from Ministerio de Economía, Industria y Competitividad, Spain.

Conflict of interest

None of the authors of this chapter has a financial or personal relationship with other people or organizations that could inappropriately influence or bias the content.

References

Adams, M.R., Marteau, P., 1995. On the safety of lactic acid bacteria from food. Int. J. Food Microbiol. 27, 263–264.

Agata, N., Ohta, M., Mori, M., Isobe, M., 1995. A novel dodecadepsipeptide, cereulide, is an emetic toxin of *Bacillus cereus*. FEMS Microbiol. Lett. 129, 17–20.

Aguirre, M., Collins, M.D., 1993. Lactic acid bacteria and human clinical infection. J. Appl. Bacteriol. 75, 95–107.

Alam, R., Abdolmaleky, H.M., Zhou, J.-R., 2017. Microbiome, inflammation, epigenetic alterations, and mental diseases. Am. J. Med. Genet. 1–10.

Allen, A.P., Hutch, W., Borre, Y.E., et al., 2016. *Bifidobacterium longum* 1714 as a translational psychobiotic: Modulation of stress, electrophysiology and neurocognition in healthy volunteers. Transl. Psychiatry 6, e939.

Anadón, A., 2006. The EU ban of antibiotics as feed additives. Alternatives and consumer safety. J. Vet. Pharmacol. Ther. 29 (Suppl 1), 41–44.

Anadón, A., Castellano, V., Martínez-Larrañaga, M.R., 2014. Regulation and guidelines of probiotics and prebiotics. In: Ötles, S. (Ed.), Probiotics and Prebiotics in Food. Nutrition and Health CRC Press, LLC Taylor & Francis Group, Boca Raton FL, USA, pp. 91–113. (ISBN: 978-1-4665-8623-9).

Anadón, A., Martínez-Larrañaga, M.R., 1999. Residues of antimicrobial drugs and feed additives in animal products: regulatory aspects. Livest. Prod. Sci. 59 (2-3), 183–198.

Anadón, A., Martinez-Larrañaga, M.R., Martinez, M.A., 2006. Probiotics for animal nutrition in the European Union. Regulation and safety assessment. Reg. Toxicol. Pharmacol. 45, 91–95.

Anadón, A., Martínez-Larrañaga, M.R., Caballero, V., et al., 2010. Assessment of prebiotics and probiotics: an overview. In: Watson, R., Preedy, V.R. (Eds.), Bioactive Foods in Promoting Health: Probiotics and Prebiotics. Academic Press, Oxford, UK, pp. 19–41. (Chapter 2) (ISBN: 978-0-12-374938-3).

Anadón, A., Martínez-Larrañaga, M.R., Ares, I., et al., 2016. Prebiotics and probiotics: an assessment of their safety and health benefits. In: Watson, R.R., Preedy, V.R. (Eds.), Bioactive Foods in Promoting Health: Probiotics and Prebiotics. Elsevier Inc., Academic Press, Oxford, UK, pp. 3–23. (ISBN: 978-0-12-802189-7).

Anadón, A., Martínez-Larrañaga, M.R., Ares, I., Martínez, M.A., 2019. Prebiotics and probiotics in feed and animal health. In: Gupta, R.C., Srivastave, A., Lall, R. (Eds.), Nutraceuticals in Veterinary Medicine, 2019. Springer Nature, Switzerland AG, pp. 261–285. Available from: https://doi.org/10.1007/978-3-030-04624-8_19.

Beattie, S.H., Williams, A.G., 1999. Detection of toxigenic strains of *Bacillus cereus* and other *Bacillus* spp. with an improved cytotoxicity assays. Lett. Appl. Microbiol. 28, 221–225.

Binns, R.N., 2013. Probiotics, Prebiotics and the Gut Microbiota. ILSI Europe, Brussels, Belgium.

Bravo, J.A., Forsythe, P., Chew, M.V., et al., 2011. Ingestion of *Lactobacillus* strain regulates emotional behavior and central GABA receptor expression in a mouse via the vagus nerve. Proc. Natl. Acad. Sci. USA 108, 16050–16055.

Butler, R.N., 2008. Non-invasive tests in animal models and humans: a new paradigm for assessing efficacy of biologics including prebiotics and probiotics. Prebiotics and probiotics delivering therapeutics as dietary components. Curr. Pharm. Des. 14, 1341–1350.

Ciorba, M.A., 2012. A gastroenterologist's guide to probiotics. Clin. Gastroenterol. Hepatol. 10, 960–968.

Courvalin, P., 2006. Vancomycin resistance in gram-positive cocci. Clin. Infect. Dis. 42 (Suppl. 1), S25–S34.

Cryan, J.F., O'Riordan, K.J., Sandhu, K., et al., 2020. The gut microbiome in neurological disorders. Lancet Neurol. 19 (2), 179–194.

Danielsen, M., Wind, A., 2003. Susceptibility of *Lactobacillu* spp. to antimicrobial agents. Int. J. Food Microbiol. 82, 1–11.

Dhodapkar, K.M., Henry, N.K., 1996. Leuconostoc bacteriemia in an infant with short-gut syndrome: case report and literature review. Mayo Clin. Proc. 71, 1171–1174.

Dong, M.Y., Chang, T.W., Gorbach, S.L., 1987. Effects of feeding Lactobacilluis GG on lethal irradiation in mice. Diag. Microbiol. Infect. Dis. 7, 1–7.

Donohue, D.C., 2006. Safety of probiotics. Asia Pac. J. Clin. Nutr. 15 (4), 563–569.

Donohue, D.C., Salminen, S., Marteau, P., 1998. Safety of probiotic bacteria. In: Salminen, S., von Wright, A. (Eds.), Lactic Acid Bacteria, second ed. Marcel Dekker Inc, New York, pp. 369–383.

Douglas, L., 1911. The Bacillis of Long Life. G.P. Putnam's Sons, New York, US.

Drobniewski, F.A., 1993. Bacillus cereus and related species. Clin. Microbiol. Rev. 65, 4470–4474.

Dugas, B., Mercenier, A., Lenoir-Wijnkoop, C., et al., 1999. Immunotoxicity and probiotics. Trends Immunol. Today 20, 387–390.

EC, 1997. Regulation (EC) No. 258/97 of the European Parliament and of the Council of 27 January 1997 concerning novel foods and novel food ingredients (OJ No. L 43, 14.2.1997).

EC, 2001. Directive 2001/83/EC (EC, 2001) of the European Parliament and of the Council of 6 November 2001 on the Community code relating to medicinal products for human use lays down harmonized rules for the authorization, supervision and pharmacovigilance of medicinal products for human use within the Union (OJ No. L 311, 28.11.2004).

EC, 2004a. Directive 2004/27/EC of the European Parliament and of the Council of 31 March 2004 amending Directive 2001/83/EC on the Community code relating to medicinal products for human use (OJ L 136, 30.4.2004).

EC, 2004b. Regulation (EC) No. 726/2004 (EC, 2004) of the European Parliament and of the Council of 31 March 2004 laying down Community procedures for the authorization and supervision of medicinal products for human and veterinary use and establishing a European Medicines Agency (OJ No. L 136, 30.4.2004).

EC, 2006. Regulation (EC) No. 1924/2006 (EC, 2006) of The European Parliament, and The Council of 20 December 2006 on nutrition and health claims made on foods (OJ No. L 404, 30.12.2006), (Corrigendum OJ No. L 12, 18.1.2007).

EFSA, 2005a. Opinion of the Scientific Panel on additives and products or substances used in animal feed on the updating of the criteria used in the assessment of bacteria for resistance to antibiotics of human or veterinary importance. EFSA J. 223, 1–12.

EFSA, 2005b. EFSA Scientific Colloquium Summary Report. OPS Qualified Presumption of Safety of Microorganisms in Food and Feed, 13–14 December 2004, Brussels, Belgium. ISBN 92-9199-012-4 http://www.efsa.eu.int/science/colloquium_series/no2_qps/948_en.html/.

EFSA, 2011. Scientific and technical guidance for the preparation and presentation of an application for authorization of a health claim (revision 1). EFSA J. 9 (5), 2170.

EFSA, 2012a. Guidance on the safety assessment of *Enterococcus faecium* in animal nutrition. EFSA J. 10 (5), 2682.

EFSA, 2012b. Guidance on the assessment of bacterial susceptibility to antimicrobials of human and veterinary importance. EFSA J. 10 (6), 2740.

EFSA, 2013. Scientific opinion on the maintenance of the list of QPS biological agents intentionally added to food and feed (2013 update). EFSA J. 11 (11), 3449.

EFSA, 2014. Guidance on the assessment of the toxigenic potential of *Bacillus* species used in animal nutrition. EFSA J. 12 (5), 3665.

EU, 2000. Opinion of the Scientific Committee on Animal Nutrition on the safety of use of *Bacillus* species in animal nutrition. 17 February. http://europa.eu.int/comm/food/fs/sc/scan/outcome_en.html.

EU, 2001. Opinion of the Scientific Committee on Animal Nutrition on the criteria for assessing the safety of micro-organisms resistant to antibiotics of human clinical and veterinary. (adopted on 3 July 2001, revised on 24 January 2003)

EU, 2003. Opinion of the Scientific Committee on Animal Nutrition on the criteria for assessing the safety of micro-organisms resistant to antibiotics of human clinical and veterinary importance. (adopted on 3 July 2001, revised on 24 january 2003)

EU, 2011. Commission Regulation (EU) No 234/2011 of 10 March 2011 implementing Regulation (EC) No 1331/2008 of the European Parliament and of the Council establishing a common authorisation procedure for food additives, food enzymes and food flavourings (OJ No. L 64, 11.3.2011)

EU, 2012. Commission Implementing Regulation (EU) No 562/2012 of 27 June 2012 amending Commission Regulation (EU) No 234/2011 with regard to specific data required for risk assessment of food enzymes (OJ No. L 168, 28.6.2012).

Ewaschuk, J.B., Diaz, H., Meddings, L., et al., 2008. Secreted bioactive factors from Bifidobacterium infantis enhance epithelial cell barrier function. Am. J. Physiol. Gastrointest. Liver. Physiol 295, G1025–G1034.

FAO/WHO (2002). Joint FAO/WHO Group report on drafting guidelines for the evaluation of probiotics in food, London Ontario, Canada, April 30 and May 1, 2002, pp 1–11.

Fuller, R., 1992. History and development of probiotics. In: Fuller, R. (Ed.), Probiotics. The Scientific Basis. Chapman and Hall, London, pp. 1–8.

Gasser, F., 1994. Safety of lactic acid bacteria and their occurrence in human clinical infections. Bull. Inst. Pasteur 92, 45–67.

Goldin, B.R., Gorbach, S.L., 2008. Clinical indications for probiotics: an overview. Clin. Infect. Dis. 46 (Suppl. 2), S96–S100.

Granum, P.E., Lund, T., 1997. *Bacillus cereus* and its food poisoning toxins. FEMS Microbiol. Lett. 157, 223–228.

Guardabassi, L., Courvalin, P., 2006. Mode of action and mechanisms of resistance. In: Aarestrup, F., Wegener, H.K. (Eds.), Antimicrobial Resistance in Bacteria of Animal Origin. American Society for Microbiology, Washington DC, pp. 1–18.

Guarner, F., 2007. Prebiotics in inflammatory bowel diseases. Br. J. Nutr. 98 (Suppl. 1), S85–S89.

Hamilton-Miller, J.M.T., 2003. The role of probiotics in the treatment and prevention of *Helicobacter pylori* infection. Int. J. Antimicrob. Agents 22, 360–366.

Havenaar, R., Ten Brink, B., Huis in't Veld, J.H.J., 1992. Selection of strains for probiotic use. In: Fuller, R. (Ed.), Probiotics. The Scientific Basis. Chapman and Hall, London, pp. 209–224.

Huyghebaert, G., Ducatelle, R., Van Immerseel, F., 2011. An update on alternatives to antimicrobial growth promoters for broilers. Vet. J. 187, 182–188.

Iemoli, E., Trabattoni, D., Parisotto, S., et al., 2012. Probiotics reduce gut microbial translocation and improve adult atopic dermatitis. J. Clin. Gastroenterol. 46 (Suppl), S33–S40.

Ishibashi, N., Yamazaki, S., 2001. Probiotics and safety. Am. J. Clin. Nutr. 73, 465S–470S.

Ismail, I.H., Oppedisano, F., Joseph, S.J., et al., 2012. Prenatal administration of *Lactobacillus rhamnosus* has no effect on the diversity of the early infant gut microbiota. Pediatr. Allergy Immunol. 23, 255–258.

Jonas, D.A., Abtignac, E., Antoine, J.M., et al., 1996. The safety assessment of novel foods. Guidelines prepared by ILSI Europe novel foods task force. Food Chem. Toxicol. 34, 931–940.

Jones, M.P., Dilley, J.B., Drossman, D., et al., 2006. Brain-gut connections in functional GI disorders: anatomic and physiologic relationships. Neurogastroenterol. Motil. 18, 91–103.

Kalliomäki, M., Salminen, S., Arvilommi, H., et al., 2001. Probiotics in primary prevention of atopic disease: a randomised placebo-controlled trial. Lancet 357 (9262), 1076–1079.

Kamada, N., Nunez, G., 2013. Role of the gut microbiota in the development and function of lymphoid cells. J. Immunol. 190, 1389–1395.

Kamada, N., Chen, G.Y., Inohara, N., et al., 2013. Control of pathogens and pathobionts by the gut microbiota. Nat. Immunol. 14, 685–690.

Kayisli, U.A., Mahutte, N.G., Arici, A., 2002. Uterine chemokines in reproductive physiology and pathology. Am. J. Reprod. Immunol. 147, 213–221.

Kotiranta, A., Lounatmaa, K., Haapasalo, M., 2000. Epidemiology and pathogenesis of *Bacillus cereus* infections. Microbes Infect. 2, 189–198.

Lee, J.-H., O'Sullivan, D.J., 2010. Genomic insights into bifidobacteria. Microbiol. Mol. Bio. Rev. 74, 538–616.

Lee, J., Mo, J.H., Katakura, K., et al., 2006. Maintenance of colonic homeostasis by distinctive apical TLR9 signalling in intestinal epithelial cells. Nat. Cell Biol. 8 (12), 1327–1336.

Li, W., Guo, J., Shen, Y., et al., 2020. Probiotics, prebiotics and synbiotics for the treatment of dementia: Protocol for a systematic review. Medicine 99 (5), e18608.

Loh, G., Blaut, M., 2012. Role of commensal gut bacteria in inflammatory bowel dis eases. Gut Microbes 3, 544–555.

Luczynski, P., Whelan, S.O., O'Sullivan, C., et al., 2016. Adult microbiota-deficient mice have distinct dendritic morphological changes: differential effects in the amygdala and hippocampus. Eur. J. Neurosci. 44, 2654–2666.

Macfarlane, G.T., Macfarlane, S., 2012. Bacteria, colonic fermentation, and gastrointestinal health. J. AOAC Int. 95, 50–60. Available from: 10.5740/jaoacint. SGE_Macfarlane.

Mackay, A.D., Taylor, M.B., Kibbler, C.C., et al., 1999. *Lactobacillus* endocarditis caused by a probiotic organism. Clin. Microbiol. Infect 5, 290–292.

Marteau, P., 2001. Safety aspects of probiotic products. Scand. J. Nutr. 45, 22–24.

Marteau, P., Pochart, P., Bouhnik, Y., et al., 1993. Fate and effects of some transiting microorganisms in the human gastrointestinal tract products. World Rev. Nutr. Diet. 74, 1–21.

Mäyrä-MäKinen, A., Bigret, M., 1993. Industrial use and production of lactic acid bacteria. In: Salminen, S., von Wright, A. (Eds.), Lactic Acid Bacteria. Marcel Dekker, Inc., New York, NY, pp. 65–96.

Mercenier, A., Pavan, S., Pot, B., 2003. Probiotics as biotherapeutic agents: present knowledge and future prospects. Curr. Pharm. Des. 9, 175–191.

Notermans, S., Batt, C.A., 1998. A risk assessment approach for food-borne Bacillus cereus and its toxin. J. Appl. Microbiol. 51S–61S (Suppl. 84).

O'Brien, T.F., 2002. Emergence, spread, and environmental effect of antimicrobial resistance: how use of an antimicrobial anywhere can increase resistance to any antimicrobial anywhere else. Clin. Infect. Dis. 34 (Suppl. 3), S78–S84.

O'Mahony, S.M., Clarke, G., Borre, Y.E., et al., 2015. Serotonin, tryptophan metabolism and the brain-gut-microbiome axis. Behav. Brain Res. 2015 (277), 32–48.

Oberhelman, R.A., Gilman, R.H., Sheen, P., et al., 1999. A placebo-controlled trial of *Lactobacillus GG* to prevent diarrhea in undernourished Peruvian children. J. Paediatrics 134, 15–20.

Oli, M.W., Petschow, B.W., Buddington, R.K., 1998. Evaluation of fructooligosaccharide supplementation of oral electrolyte solutions for treatment of diarrhea. Recovery of the intestinal bacteria. Dig. Dis. Sci. 43, 138–147.

Ouwehand, A.C., Salminen, S., Isolauri, E., 2002. Probiotics: an overview of beneficial effects. Antonie van. Leeuwenhoek 82, 279–289.

Pedersen, P.B., Bjornvad, M.E., Rasmussen, M.D., et al., 2002. Cytotoxic potential of industrial strains of *Bacillus* sp. Regul. Toxicol. Pharmacol. 36 (2), 155–161.

Perdigón, G., Alvarez, S., 1992. Probiotics and the immune state. In: Füller, R. (Ed.), Probiotics: The Scientific Basis. Chapman and Hall, London, pp. 145–179.

Prescott, S.L., Noakes, P., Chow, B.W., et al., 2008. Pre-symptomatic differences in Toll-like receptor function in infants who have allergy. J. Allergy Clin. Immunol. 122, 91–99.

Przyrembel, H., 2001. Consideration of possible legislation within existing regulatory frameworks. Am. J. Clin. Nutr. 73 (Suppl), 471S–475S.

Rautava, S., Luoto, R., Salminen, S., et al., 2012. Microbial contact during pregnancy, intestinal colonization and human disease. Nat. Rev. Gastroenterol. Hepatol. 9, 565–576.

Rautio, M., Jousimies-Somer, H., Kauma, H., et al., 1999. Liver abscess due to a *Lactobacillus rhamnosus* strain indistinguishable from *L. rhamnosus* strain GG. Clin. Infect. Dis. 28 (5), 1159–1160.

Saier, M.H., Mansour, N.M., 2005. Probiotics and prebiotics in human. J. Mol. Microbiol. Biotechnol. 10, 22–25.

Salminen, S., von Wright, A., 1998. Current probiotic-safety assured? Microb. Ecol. Health Dis. 10, 68–77.

Sanders, M.E., Tompkins, T., Heimbach, J., et al., 2005. Weight of evidence needed to substantiate a health effect for probiotics and prebiotics. Regulatory considerations in Canada, EU, and US. Eur. J. Nutr. 44, 303–310.

Savage, D.C., 1977. Microbial ecology of the gastrointestinal tract. Annu. Rev. Microbiol. 31, 107–133.

Schoeni, J.L., Wong, A.C.L., 1999. Heterogenity observed in the components of hemolyisn BL, an enterotoxin produced by *Bacillus cereus*. Int. J. Food Microbiol. 53, 159–167.

Schrezenmeir, J., de Vrese, M., 2001. Probiotics, prebiotics, and synbiotics-approaching a definition. Am. J. Clin. Nutr. 73 (Suppl 2), 361S–364S.

Seth, S.D., Maulik, M., 2011. Probiotics a Pharmacologist's Perspective. In: Balakrish Nair, G., Yoshifumi, T. (Eds.), Probiotic Foods in Health and

Disease. Science Publishers, Enfield, New Hampshire, USA, pp. 41–47. , CRC Press, Taylor & Francis Group, Boca Raton, New York, Oxon.

Sniffen, J.C., McFarland, L.V., Evans, C.T., et al., 2018. Choosing an appropriate probiotic product for your patient: an evidence-based practical guide. PLoS One 13, e0209205.

Suez, J., Zmora, N., Segal, E., et al., 2019. The pros, cons, and many unknowns of probiotics. Nat. Med. 716 (25), 716–729.

Sullivan, A., Nord, C.E., 2002. Probiotics in human infections. J. Antimicrob. Chemother. 50, 625–627.

Theodorakopoulou, M., Perros, E., Giamarellos-Bourboulis, E.J., 2013. Controversies in the management of the critically ill: the role of probiotics. Int. J. Antimicrob. Agents 42 (Suppl), S41–S44.

Toh, M., Moffitt, M.C., Henrichsen, L., et al., 2004. Cereulide, the emetic toxin of bacillus cereus, putatively a product of nonribosomal peptide synthesis. J. Appl. Microbiol. 97 (5), 992–1000.

Tomasik, P.J., Tomasik, P., 2003. Probiotics and prebiotics. Cereal Chem. 80, 113–117.

Turnbaugh, P.J., Ley, R.E., Hamady, M., et al., 2007. The human microbiome project. Nature 449, 804–810.

Vanderhoof, J.A., Whitney, D.B., Antonson, D.L., et al., 1999. *Lactobacillus* GG in the prevention of antibiotic-associated diarrhea in children. J. Pediat 135, 564–568.

Wiley, N.C., Dinan, T.G., Ross, R.P., et al., 2017. The microbiota-gut-brain axis as a key regulator of neural function and the stress response: Implications for human and animal health. J. Anim. Sci. 95, 3225–3246.

Wills-Karp, M., Santeliz, J., Karp, C.L., 2001. The germless theory of allergic disease: revisiting the hygiene hypothesis. Nat. Rev. Immunol. 1, 69–75.

Zmora, N., Zilberman Schapira, G., Suez, J., 2018. Personalized gut mucosal colonization resistance to empiric probiotics is associated with unique host and microbiome features. Cell 174, 1388–1405.e21.

Zunic, P., Lacotte, J., Pegoix, M., et al., 1991. Fongemie a *Saccharomyces boulardi*. Therapie 45, 497–501.

Chapter 66

Synbiotics: safety and toxicity considerations

Jitendra K. Malik, Rajinder Raina, Abul H. Ahmad, Starling Kalpana, Atul Prakash and Ramesh C. Gupta

Introduction and background

Synbiotics are generally considered as products that contain both probiotics and prebiotics, and thereby exert both prebiotic and probiotic effects. Thus a synbiotic product can enhance the isolated beneficial effects of probiotics and prebiotics. The definition of probiotics is "live microorganisms that, when administered in adequate amounts, confer a health benefit on the host" (Hill et al., 2014). Probiotics are nonpathogenic microorganisms which favor the growth of beneficial intestinal bacteria over that of harmful ones. Prebiotics are nondigestible food ingredients that beneficially affect the host by selectively stimulating the growth and/or activity of a selected group of microorganisms living in the colon. An important rationale for using a synbiotic is that a true probiotic, without its prebiotic food, does not survive well in the digestive system. Without the necessary food source for the probiotic, it will have a greater intolerance for oxygen, low pH, and temperature. By harnessing the advantages of both prebiotics and probiotics into synergy, the number of good bacteria increases several fold in the digestive system (Sekhon and Jairath, 2010; Malik et al., 2016, 2019). The health effects of synbiotics may be associated with the individual combination of a probiotic and a prebiotic. Taking into account a large number of possible combinations, the application of synbiotics for the modulation of gut microbiota in humans and animals appears promising (Kearney and Gibbons, 2018; Markowiak and Śliżewska, 2017, 2018).

Although the most commonly used probiotics include species of *Lactobacillus* and *Bifidobacterium*, the yeast, *Saccharomyces cerevisiae*, and some *Escherichia coli* and *Bacillus* species are also used as probiotics. Lactic acid bacteria (LAB), including species of *Lactobacillus*, serve as agents of food fermentation and impart potential health benefits. With more advanced and focused research efforts, new genera and strains of probiotic microorganisms are constantly emerging (Malik et al., 2019). Commonly known prebiotics include fructo-oligosaccharides (FOS), galacto-oligosaccharides (GOS), galacto-oligosaccharides/transgalactosylated oligosaccharides (GOS/TOS), inulin, isomalto-oligosaccharides (IMO), lactulose, pyrodextrins, and soy oligosaccharides. The emergent prebiotics are genti-oligosaccharides, gluco-oligosaccharides, IMO, lactosucrose, levans, pectic oligosaccharides, resistant starch, sugar alcohols, and xylo-oligosaccharides (Anadón et al., 2010; Malik et al., 2019). The next-generation probiotics (NGPs) are defined as live microorganisms identified on the basis of comparative microbiota analyses that, when administered in adequate amounts, confer a health benefit on the host. Examples of current NGP include *Faecalibacterium* spp., *Akkermansia* spp., *Bacteroides fragilis* strain ZY-312, *Bacteroides xylanisolvens* DSM 23964, *Clostridium butyricum* MIYAIRI 588, and *Faecalibacterium prausnitzii* (Maftei, 2019; Martin and Langella, 2019). Fermentation of FOS in the colon increases the numbers of bifidobacteria and the absorption of calcium and increases fecal weight. In addition, it leads to a decrease in gastrointestinal transit time. The increase in colonic bifidobacteria produces compounds to inhibit potential pathogens by lowering blood ammonia levels and producing vitamins and digestive enzymes.

Because drugs are not capable of reducing the morbidity and mortality associated with acute and/or chronic diseases, there is increasing awareness for a search scenario and exploiting the potential of alternative agents. Among the various alternative agents, probiotics, prebiotics, and synbiotics should have direct and/or indirect effects on pathogenesis and disease progression. The significant increase of short-chain fatty acids (SCFAs), ketones, carbon disulfide, and methyl acetate following the intake of synbiotics has been suggested as their potential health-promoting effects (Vitali et al., 2010). In recent years, new probiotics, prebiotics, and synbiotic products have

Nutraceuticals. DOI: https://doi.org/10.1016/B978-0-12-821038-3.00066-5

been developed. These are being proposed as novel therapeutic options and their supplementation may be valuable in the prevention and management of nutrition and health.

It has only been during the past few years that studies on synbiotics have started to emerge, with the main focus being applications against diseases (Kolida and Gibson, 2011; Malik et al., 2016, 2019). Synbiotics lead to a modulation of the gut metabolic activities with maintenance of the gut biostructure, and they act by delivering specific health benefits. Compared to probiotics and prebiotics, synbiotics are the least investigated substances in regard to their health effects, safety, and toxicity. This chapter describes the current knowledge regarding the safety, toxicity, and applications of synbiotics in humans and animals.

General considerations

The emerging concept of synbiotics is to modulate the targeted gut microflora/ecology. Synbiotics are a combination of both probiotics and prebiotics which can fortify the lactate-producing microorganisms of the human or animal gut (Sekhon and Jairath, 2010; Malik et al., 2016). A prebiotic is a dietary substrate that must not be hydrolyzed or absorbed in the stomach or small intestine, must be selective for beneficial commensal bacteria in the large intestine, and fermentation of the substrate should induce beneficial luminal/systemic effects within the host. The effects of prebiotics on the upper gastrointestinal tract (GIT) include restriction of gastric emptying, enhancement of orocecal transit time, reduction of glucose absorption and low glycemic index, hyperplasia of the small intestinal epithelium, and stimulation of secretion of intestinal hormonal peptides. The effects of prebiotics on the lower GIT include acting as food for colonic microbiota and as substrates for colonic fermentation, production of fermentation end products, stimulation of saccharolytic fermentation, acidification of the colonic content, hyperplasia of the colonic epithelium, stimulation of secretion of colonic hormonal peptides, bulking effect on stool production, regularization of stool production, and hastening of ceco-anal transit (Scantlebury-Manning and Gibson, 2004; Gaggia et al., 2010; Malik et al., 2016).

The expected health-promoting characteristics and safety criteria of probiotics have been discussed by Gaggia et al. (2010). These include nontoxic and nonpathogenic accurate taxonomic identification, normal inhabitants of the targeted species, survival, colonization, and being metabolically active in the targeted site. This implies resistance to gastric juice and bile, persistence in the GIT, adhesion to epithelium or mucus, and competition with resident microbiota. Other criteria include the production of antimicrobial substances, antagonism of pathogenic bacteria, modulation of immune responses, an ability to exert at least one scientifically supported health-promoting property, genetic stability, amenability of the strain and stability of the desired characteristics during processing, storage, and delivery, viability in high populations, and desirable organoleptic and technological properties when are included in industrial processes (Gaggia et al., 2010; Malik et al., 2016).

It has been suggested that the choice of prebiotics and probiotics must be on the basis of scientific evidence. In vitro studies have been performed to examine the growth properties of different probiotic microorganisms with the addition of different prebiotics for the purpose of obtaining appropriate decisions about combining probiotics and prebiotics in synbiotic formulations. However, such in vitro bacterial growth studies with a single bacterial strain and extremely high concentrations of prebiotics cannot be directly correlated with the in vivo situation, where a plethora of different bacterial (and other microbial) species compete for nutrients and ecological niches. Other factors, such as acid stability and the capability to adhere to the intestinal mucus layer, may also affect the survival and growth of externally administered probiotic strains in the GIT (Kearney and Gibbons, 2018; Salavati Schmitz and Allenspach, 2017; Markowiak and Śliżewska, 2017, 2018; Malik et al., 2019). Examples of common synbiotics are: (1) lactobacilli plus lactitol; (2) lactobacilli plus inulin; (3) lactobacilli plus FOS; (4) *Lactobacillus rhamnosus* GG plus inulin; (5) bifidobacteria plus FOS; (6) bifidobacteria plus GOS; (7) bifidobacteria and lactobacilli plus FOS; and (8) bifidobacteria and lactobacilli plus inulin.

The precise mechanisms of action of probiotics which are varied, heterogeneous, and strain-specific, have not been fully elucidated until now and further research is warranted. These mechanisms were recently reviewed by Plaza-Diaz et al. (2019). The modes of action involve: (1) colonization and normalization of perturbed intestinal microbial communities, competitive exclusion of pathogenic microorganisms and production of bacteriocin; (2) modulation of fecal enzymatic activities associated with the metabolization of biliary salts and inactivation of carcinogens and other xenobiotics, production of SCFAs and branched-chain fatty acids, which have a wide range of effects not only in the intestine but also in peripheral tissues via interactions with SCFA receptors, modulating mainly tissue insulin sensitivity; (3) cell adhesion and mucin production; and (4) modulation of the immune system, which results mainly in the differentiation of T-regulatory cells and upregulation of antiinflammatory cytokines such as interleukin (IL)-10, transforming growth factor, and interaction with the brain–gut axis by regulation of endocrine and neurologic functions.

Synbiotics impact host health by manipulating the composition of gut microbiota and by reducing the levels of pathogenic microorganisms. In synbiotics, probiotics use the prebiotics as a food source, which enables them to survive for an extended period within the intestine. The improvement in the viability of probiotics facilitates delivering the projected health benefits. These events decrease intestinal inflammation, alleviate allergy and diarrhea, produce antimicrobials, enhance and regulate immune function, elicit antitumorigenic or anticarcinogenic activity, bind to potential food carcinogens and toxins, and demote bacterial enzymes that hydrolyze precarcinogenic compounds, such as β-glucuronidase (Kolida and Gibson, 2011; Malik et al., 2016).

Synbiotics in humans

Studies on the effects of synbiotics on human health remain limited. Although synbiotics appear to be promising for the modulation of gut microbiota composition, the effects remain to be verified by human trials. Table 66.1 summarizes studies on some synbiotics in humans. Scavuzzi et al. (2014) reviewed the impact of synbiotics on the components of metabolic syndrome in humans and found inconsistent results. Only two of the five studies that considered the impact of synbiotic supplementation on body composition or energy intake documented improvements of the selected markers. Six of seven studies supported improvement of at least one blood lipid. Two of three studies documented improvements in glucose homeostasis after the consumption of dietary synbiotics. Most studies documented improvements in at least one inflammatory marker after the consumption of synbiotics.

Although the available data suggest that infant formula supplemented with synbiotics is not associated with adverse effects, only a limited number of synbiotic preparations administered in infant formula have been studied in the context of formal randomized controlled trials. However, the paucity of data calls for caution in overinterpretation of these results. The Committee on Nutrition of the European Society for Pediatric Gastroenterology, Hepatology, and Nutrition (ESPGHAN) systematically reviewed published studies related to the safety and health effects of the administration of formula supplemented with probiotics, prebiotics, or synbiotics compared with unsupplemented formula (Braegger et al., 2011). The scientific data suggested that the administration of currently evaluated formula supplemented with probiotics, prebiotics, or synbiotics to healthy infants raises no safety concerns with regard to growth and adverse effects. However, the safety and clinical effects of one product should not be extrapolated to other products. The data were found to be inadequate to recommend the routine use of probiotic-, prebiotic-, or synbiotic-supplemented formula. Accordingly, it was recommended that well-designed and carefully conducted randomized controlled trials with relevant inclusion/exclusion criteria and adequate sample sizes are needed. Furthermore, these studies should use validated clinical outcome measures to assess the effects of probiotic, prebiotic, or synbiotic supplementation of formula and define their optimal doses and intake durations, as well as provide more information about the long-term safety of probiotics, prebiotics, and synbiotics. The ESPGHAN Committee considered that recommendations for using probiotic-, prebiotic-, and synbiotic-supplemented formula are determined by their nutritional adequacy, potential short-term and long-term benefits, and safety related to the continued administration of such formula. In view of this, the Committee did not recommend the routine use of formula supplemented with synbiotics in infants and further suggested that in the future, the efficacy and safety of each synbiotic-containing infant formula should be established (Braegger et al., 2011).

The safety aspects in clinical trials conducted between 2008 and 2013 with probiotics and synbiotics in infants younger than 2 years of age have been systematically evaluated by van den Nieuwboer et al. (2014). The safety assessment was performed along the Common Terminology Clinical Adverse Events (CTCAE) version 4.0 scale. Analysis of 57 clinical trials with probiotics and synbiotics in combination with eight follow-up studies indicated that probiotic administration to infants between 0 and 24 months was safe with regard to the evaluated strains in infants with a particular health status or susceptibility. Almost all studies concluded that none of the adverse effects were related to the probiotics and synbiotics used, there were no major safety concerns, and all products were generally well tolerated. In another study, van den Nieuwboer et al. (2015) systematically evaluated the safety of synbiotics and probiotics in immune-compromised adults aged 18 years or older. Safety was analyzed using the CTCAE version 4.0 classification. Analysis of 57 clinical studies available between 2008 and 2013 indicated that probiotic and/or synbiotic administration in immune-compromised adults (HIV-infected, critically ill, surgical patients, or those with an organ or autoimmune disease) is safe with regard to the currently evaluated probiotic strains, dosages, and duration. There were no major safety concerns in the study because none of the serious adverse events (AEs) was related, or suspected to be related, to the probiotic or synbiotic product and the study products were well tolerated. Overall, AEs occurred less frequently in immunocompromised subjects receiving probiotics and/or synbiotics compared to the control group. In both studies, a deficiency in the precise reporting and classification of adverse events was evident. The inconsistent, imprecise, and potentially incomplete reporting as well as the variation in probiotic strains,

TABLE 66.1 Studies on some synbiotics in humans.

Synbiotic	Study design	Subject	Effect	Safety status	Reference
Bifidobacterium longum plus inulin and oligofructose	Randomized, double-blind placebo-controlled, crossover study for 4 weeks	43 older persons	Improvement in the composition and metabolic activities of colonic bacterial communities and immune parameters	No effect on bowel habit or any clinical parameters	Macfarlane et al. (2013)
Lactobacillus sporogenes plus inulin	Randomized placebo-controlled clinical trial for 9 weeks	52 pregnant women, primigravida, 18–35 years old, in their third trimester	Beneficial effects on insulin actions	No serious adverse reactions	Taghizadeh and Asemi (2014)
Protexin[a]	Double-blind, randomized, placebo-controlled trial for 4 weeks	66 adult men with functional constipation	Increased stool frequency and improved consistency	No adverse effects	Fateh et al. (2011)
Lactobacillus casei, Lactobacillus rhamnosus, Lactobacillus plantarum, Bifidobacterium lactis plus fiber, polydextrose, fructo-oligosaccharides (FOS), and galacto-oligosaccharides (GOS)	Randomized double-blind placebo-controlled study for 4 weeks	146 children aged 4–18 years	Beneficial in the treatment of functional constipation	No side effects such as vomiting and diarrhea	Baştürk et al. (2017)
Bifidobacterium animalis spp. *lactis* (*B. lactis*) GCL2505 plus inulin	Randomized, double-blind, placebo-controlled, crossover trial consisting of six periods of 2 weeks each	60 healthy subjects (17 males and 43 females) aged 20–64 years with a tendency for constipation (3–5 days a week)	Increased the numbers of total bifidobacteria and *B. lactis* in feces and higher numbers of total and endogenous bifidobacteria	No adverse events	Anzawa et al. (2019)
L. rhamnosus, Lactobacillus acidophilus, L. casei, Lactobacillus bulgaricus, L. plantarum, Lactobacillus salivarius, Bifidobacterium bifidum, B. longum plus fructo-oligosaccharides, *Ulmus davidiana, Geum urbanum* powder, and inulin	Single-center, randomized, double-blind, and placebo-controlled clinical trial for 8 weeks	28 irritable bowel syndrome patients aged ≥19 years	Improved bowel symptoms and fatigue	No adverse reactions	Lee et al. (2019)
Synbiotic 2000[b]	Prospective randomized monocentric double-blind trial	80 patients following pylorus-preserving pancreatoduodenectomy (PPPD);	Reduced bacterial infection rates and antibiotic therapy following PPPD	Well tolerated and all side effects disappeared under temporary reduction of the amount of enteral nutrition containing the synbiotic	Rayes et al. (2007)

L. casei strain Shirota; *Bifidobacterium breve* strain Yakult; and GOS	Randomized controlled trial	81 patients undergoing high-risk hepatobiliary resection	Enhanced immune responses, attenuated systemic postoperative inflammatory responses, and improved intestinal microbial environment	No problems related to synbiotic treatment	Sugawara et al. (2006)
B. longum plus psyllium	Randomized controlled trial for 4 weeks	40 ulcerative colitis (UC) patients	Decrease in CRP, synbiotic therapy may have a synergistic effect in the treatment of UC	No adverse events	Fujimori et al. (2009)
B. longum plus FOS and lifestyle modification	Randomized, double-blind, placebo-controlled study for 24 weeks	66 patients with nonalcoholic steatohepatitis (NASH)	Reduced TNF-α, CRP, serum AST levels, HOMA-IR, serum endotoxin, steatosis, and NASH activity index	No clinically relevant changes in hematology, clinical chemistry, and renal function; well tolerated	Malaguarnera et al. (2012)
L. sporogenes plus inulin with isomalt, sorbitol, and stevia as sweetener	Randomized double-blinded cross-over controlled clinical trial for 6 weeks	62 patients with type II diabetes, aged 35–70 years	Increased serum concentrations of calcium and iron and decreased total bilirubin		Asemi et al. (2017)
L. casei strain Shirota, *B. breve* strain Yakult plus GOS	Prospective randomized study	32 critically ill patients, who had undergone emergency surgery	Enhanced host immune function, improved the intestinal milieu, and decreased the incidence of infectious complications	No adverse events or side effects of any severity during the study period	Kataoka et al. (2013)
L. acidophilus NCFM, *L. rhamnosus* HN001, *Lactobacillus paracasei* LPC-37, *B. lactis* HN019 plus FOS	Randomized, double-blinded, placebo-controlled clinical trial for 5 days prior to the surgical procedure and for 14 days after surgery	91 patients with colorectal cancer undergoing elective surgery	Reduction in postoperative infection rates		Flesch et al. (2017)
Protexin[a] and lifestyle modification	Randomized, double-blind, placebo-controlled clinical trial conducted as a pilot study for 28 weeks	52 patients with nonalcoholic fatty liver disease (NAFLD)	Synbiotic and lifestyle modification are superior to lifestyle modification alone for the treatment of NAFLD, at least partially through attenuation of inflammatory markers	No serious adverse events	Eslamparast et al. (2014a)
Protexin[a] and healthy lifestyle	Pilot, randomized, double-blind, placebo-controlled, clinical trial for 28 weeks	50 patients, 18 years or older, with body mass index ≤ 25 with NAFLD	Reduced hepatic steatosis and fibrosis, and serum levels of fasting blood sugar, triacylglycerol and most of the inflammatory mediators	No serious adverse events	Mofidi et al. (2017)
Protexin[a]	Randomized, double-blind, placebo-controlled pilot study	38 subjects with metabolic syndrome	Significant improvement in the levels of fasting blood sugar and insulin resistance	No serious adverse events	Eslamparast et al. (2014b)

(Continued)

TABLE 66.1 (Continued)

Synbiotic	Study design	Subject	Effect	Safety status	Reference
Protexin[a]	Randomized, double-blind, placebo-controlled study for 8 weeks	40 infants and children aged 3 months to 6 years with atopic dermatitis	Clinical improvement in the severity of atopic dermatitis	No significant side effects	Farid et al. (2011)
Protexin[a]	Randomized triple-masked controlled trial	70 apparently healthy children and adolescents, aged 6–18 years	Positive influence on inflammation factors, which were dependent on its effect on weight reduction in overweight and obese children	No adverse effects or symptoms	Kelishadi et al. (2014)
B. longum BL999 plus a mixture containing GOS and FOS	Prospective, randomized, reference-controlled, double-blinded trial for 7 months	138 infants	Fewer incidences of constipation and trend toward fewer respiratory tract infections	Safe and well tolerated	Puccio et al. (2007)
B. longum BL999, *L. rhamnosus* LPR plus GOS/short-chain FOS (SCFOS) or *B. longum* BL999, *L. paracasei* ST11 plus GOS/SCFOS	Controlled, double-blind, randomized trial	284 healthy full-term infants	Similar rate in weight gain compared with those fed a control formula	Well tolerated	Chouraqui et al. (2008)
L. plantarum plus fructo-oligosaccharides	Randomized, double-masked, controlled trial for 7 days	31 healthy inborn newborns >35 weeks of gestational age	Synbiotic preparation colonized quickly, increase in bacterial diversity and Gram-positive organisms and decrease in Gram-negative bacterial load	Synbiotic was well tolerated	Panigrahi et al. (2008)
L. paracasei spp. *paracasei* and *B. animalis* spp. *lactis* plus GOS	Randomized controlled trial	126 newborns for first 3 months of life; 80 infants continued the study until 6 months old	Higher stool frequency and stool consistency score during the first 3 months	Safe, well tolerated and no adverse effects on growth and infant behavior	Vlieger et al. (2009)
L. casei, *L. rhamnosus*, *S. thermophilus*, *B. breve*, *L. acidophilus*, *Bifidobacterium infantis* plus FOS	Double-blind randomized placebo controlled study	102 children aged 4–12 years with functional constipation	Positive effects on symptoms of childhood constipation	No side effects	Khodadad and Sabbaghian (2010)
L. acidophilus, *L. Casei*, *B. bifidum* plus inulin	Randomized, double-blinded, placebo-controlled clinical trial for 12 weeks	60 women aged 18–40 years with polycystic ovary syndrome	Increased serum sex hormone-binding globulin, plasma nitric oxide, and decreased modified Ferriman Gallwey scores and serum hs-CRP	No side effects	Nasri et al. (2018)

AST, aspartate transaminase; *CRP*, C-reactive protein; *HOMA-IR*, homeostasis model of assessment-insulin resistance; *hs-CRP*, high sensitivity C-reactive protein; *TNF-α*, tumor necrosis factor-α.
[a]*Protexin:* Lactobacillus casei, Lactobacillus rhamnosus, Streptococcus thermophilus, Bifidobacterium breve, Lactobacillus acidophilus, Bifidobacterium longum/infantis, *and* Lactobacillus bulgaricus *plus fructo-oligosaccharides.*
[b]*Synbiotic 2000:* Pediacoccus pentosaceus, Leuconostoc mesenteroides, Lactobacillus paracasei *subspecies* paracasei *F19, and* Lactobacillus plantarum *2362 plus betaglucan, inulin, pectin, and resistant starch.*

dosages, administration regimens, study populations, and reported outcomes, greatly limited the generalization of conclusions. Obligatory and standardized reporting on adverse events (CTCAE) has been suggested in studies with probiotics and synbiotics.

Synbiotics in animals

Although the beneficial effects of synbiotics have not been extensively investigated in animals and poultry, the results on in vivo trials are promising and recent developments and applications of synbiotics have focused on evaluation of the beneficial effects in animal health and production (Gadde et al., 2017). The impact of some synbiotics investigated on health and performance of broiler chickens and domestic animals is presented in Table 66.2. The synergy mechanism of probiotics and prebiotics has not been fully elucidated. The combination of probiotics and prebiotics could improve survival and persistence of the health-promoting organisms in the gut of birds because of the availability of its specific substrate for fermentation (Gadde et al., 2017; Markowiak and Śliżewska, 2018).

Dietary supplementation with *Bacillus subtilis* or FOS in comparison to the antibiotic aureomycin had selective effects on increasing cecal concentrations of beneficial bacteria such as *Lactobacillus* and decreasing concentrations of harmful bacteria such as *E. coli* and *Salmonella*, whereas treatment with aureomycin had nonselective effects on cecal microflora that inhibited all bacteria. *B. subtilis* plus FOS produced greater improvement in the cecal micro-ecosystem and had better effects on reducing diarrhea rates and promoting growth in broilers than their administration alone (Li et al., 2008). Daily body weight gain and feed efficiency were significantly improved by the synbiotic treatment and therefore it has been recommended that synbiotics can be used as a nonantibiotic growth promoter feed additive for poultry (Ashayerizadeh et al., 2011).

The positive influence of dietary supplementation of synbiotics on the performance and ileal microbiota has also been confirmed. *S. cerevisiae* and mannan-oligosaccharides (MOS) combination supplementation substantially increased weight gain as well as the population of LAB and yeast in the cecum content of broiler chicks. The population of *E. coli* was significantly decreased in the cecum. Dietary supplementation of *S. cerevisiae* without MOS also produced these effects, but the effects were more pronounced following combined administration (Koc et al., 2010). Tayeri et al. (2018) compared the effects of antibiotics, probiotics, prebiotics, and synbiotics on the performance and carcass characteristics of broilers. The birds fed synbiotics had greater relative gizzard and spleen weights, lighter kidneys, and also had thinner walls of the caudal gut segments. Supplementation with synbiotics also reduced the feed conversion ratio compared with the control and antibiotic flavomycin treatment. It was suggested that there were significant performance and health benefits with using probiotics, prebiotics, and synbiotics for broilers, rather than antibiotics.

Recently, the effect of intraamniotic administration of synbiotic comprised of *Enterococcus faecium* and inulin, and its continued dietary supplementation, was investigated for performance, intestinal epithelium integrity, and cecal microflora of broiler chickens (Calik et al., 2017). Intraamniotic synbiotic administration to embryonated eggs on day 17 of incubation had no effect on the hatchability or hatching weight of the birds. However, it resulted in a positive effect on villus height, goblet cells, and proliferating cell nuclear antigen-positive cell counts. Intraamniotic synbiotic injection followed by dietary supplementation with the synbiotic was shown to significantly increase *Lactobacillus* colonization and decrease coliform population in the cecum of broiler chickens. In addition, cecal butyric acid concentration was increased proportionally to the cecal *Lactobacillus* count.

On the contrary, some of the trials conducted with dietary inclusion of synbiotics did not show that broiler performance was affected. Mookiah et al. (2014) reported that synbiotics did not show a twofold synergistic effect, as compared to those of the probiotic consisting of 11 *Lactobacillus* strains or IMO alone, on performance, cecal bacterial populations, or concentrations of cecal volatile fatty acids (VFAs) and non-VFAs of broiler chickens. Dietary supplementation with a combination of a commercial probiotic (PrimaLac) and a prebiotic (TechnoMos) had no significant effect on daily feed intake, daily body weight gain, feed conversion ratio, carcass traits, intestinal morphology, or bacterial populations in the ileum of broiler chickens (Salehimanesh et al., 2016). Such a difference may be due to variations in the strain of probiotics and/or type of prebiotics chosen, methods of preparation, the dietary supplementation level, and duration of use, diet composition, bird age, and hygiene conditions. In several cases, the environmental and stress statuses of the birds are not reported or considered, because the experimental settings are often too far removed from farm conditions (Gaggia et al., 2010; Wang et al., 2018).

Recent studies have shown that in ovo administration of selected synbiotics is a promising approach for enhancement of the chicken immune system because it combines merits of probiotics and prebiotics, and by early administration into the embryo it supports development of the immune system. In ovo administration of *Lactococcus lactis* IBB SL1 plus raffinose family oligosaccharides (RFO), *L. lactis* IBB SC1 plus RFO, or *Lactobacillus acidophilus*, and *Streptococcus faecium* plus lactose into the developing chicken embryo provided immune system stimulation of the growing chickens, but its efficiency depended on

TABLE 66.2 Effects of some synbiotics on health and performance of poultry and domestic animals.

Synbiotic	Main outcome	Reference
Eleven *Lactobacillus* strains plus isomalto-oligosaccharides	Increased the cecal populations of lactobacilli and bifidobacteria and decreased the cecal *Escherichia coli*, improved the performance and increased the cecal volatile fatty acids	Mookiah et al. (2014)
Bacillus subtilis, Bacillus licheniformis plus mannan-oligosaccharide	Improved growth performance, epithelial barrier function, enhanced *Bacillus* sp., *Lactobacillus* sp., and *Clostridium* sp. colonization, and decreased *E. coli* and *Salmonella* spp. in the intestinal gut of broiler chickens raised under high stocking density condition	Kridtayopas et al. (2019)
BiominIMBO[a]	Improved the humoral immune responses following vaccination against Newcastle disease, infectious bronchitis, and infectious bursal disease	Talebi et al. (2015)
Lactobacillus acidophilus, Lactobacillus casei, Streptococcus faecium, B. subtilis plus yeast-derived carbohydrates	Showed a balanced T-helper (Th)-1/Th-2 response locally and a more Th-2-dependent response systemically	Yitbarek et al. (2015)
BiominIMBO[a]	Improved body weight gain, feed conversion, intestinal morphology, and intestinal microbial ecology and the highest antibody response to Newcastle disease vaccine	Al-Sultan et al. (2016)
L. acidophilus, L. casei, S. faecium, B. subtilis, Saccharomyces cerevisiae plus yeast-derived carbohydrates	Enhanced humoral immunity by increasing IgG concentration in serum, and modulated the adaptive antibody-mediated immune response against infectious bronchitis virus	Alizadeh et al. (2017)
PoultryStar meUS[b]	Improved production parameters, increased the amounts of bile anti-CP IgA and decreased cecal *Clostridium perfringens* load during an necrotic enteritis infection in broilers	Shanmugasundaram et al. (2020)
Lactobacillus rhamnosus HN001, *Pediococcus acidilactici* MA18/5M plus *Agave tequilana* fructans	Stimulated bursa and IgA production, increased the size of its follicles and promoted the ability to resist infections caused by *Salmonella typhimurium* in broilers	Villagrán-de la Mora et al. (2020)
B. longum PCB133 plus xylo-oligosaccharides	Decreased the cecal *Campylobacter jejuni* and *Campylobacter* sp. upon lifelong administration of the synbiotic in broiler chickens infected with *Campylobacter jejuni* strain M1	Baffoni et al. (2017)
Lactobacillus reuteri, Enterococcus faecium, B. animalis, P. acidilactici plus fructo-oligosaccharides	Improved body weight, enhanced performance, and protected against *Salmonella enterica* Enteritidis infection	Luoma et al. (2017)
PoultryStar meUS[b]	Decreased *Salmonella enterica* ser. Enteritidis load in the cecal contents and broiler carcass	Shanmugasundaram et al. (2019)
L. casei, L. acidophilus, Bifidobacterium thermophilum, Enterococcus faecium plus mannan-oligosaccharide	Increased specific growth rate, growth efficiency, energy efficiency ratio, protein efficiency ratio, and decreased serum cholesterol concentration	Ashayerizadeh et al. (2011)
B. subtilis, B. licheniformis, Clostridium butyricum plus yeast cell wall and xylo-oligosaccharides	Increased average daily gain, breast yield and decreased feed/gain ratio, abdominal fat and lowered malondialdehyde content in the thigh muscle, resulting in the production of meat with a favorable quality and oxidative stability	Cheng et al. (2017)
L. acidophilus, L. casei, Bifidobacterium bifidum, S. faecium, Aspergillus oryzae plus isomalto-oligosaccharides	Decreased serum levels of total cholesterol at 36 weeks of age, low-density lipoprotein cholesterol, alanine aminotransferase and alkaline phosphatase, and heterophil percentage, heterophil to lymphocyte ratio and increased lymphocyte percentage at 36 and 52 weeks of age	Tang et al. (2017)

(Continued)

TABLE 66.2 (Continued)

Synbiotic	Main outcome	Reference
B. subtilis, Saccharomyces cerevisiae, S. faecium plus mannan-oligosaccharide and *β*-glucan	Improved performance, carcass yields, and serum lipid and antioxidant profiles, and lowered the polyunsaturated fatty acids (FA) n-6:n-3 ratio and increased hypocholesterolemic to hypercholesterolemic FA ratios in breast muscles of broiler chickens	Salah et al. (2019)
L. acidophilus, L. casei, B. bifidum, S. faecium, A. oryzae plus isomalto-oligosaccharides	Decreased the egg yolk cholesterol, and total saturated fatty acids and increased total unsaturated fatty acids, total omega 6 and polyunsaturated fatty acids including linoleic and alpha-linolenic acid concentrations in eggs	Tang et al. (2015)
BiominIMBO[a]	Enhanced the consumption of metabolizable energy and crude protein and higher production index, live weight, and feed efficiency	Aziz Mousavi et al. (2015)
B. subtilis plus xylo-oligosaccharides and mannan-oligosaccharide	Increased daily weight gain, feed efficiency, villus height and villus:crypt ratio in duodenum, jejunum and ileum, intestinal mucosa secretory IgA content and antioxidant capabilities	Min et al. (2016)
S. cerevisiae, E. faecium, B. subtilis, B. licheniformis plus *β*-glucans, mannan-oligosaccharide and fructo-oligosaccharides	Increased body weight, decreased mortality-adjusted feed conversion ratio and necrotic enteritis-associated mortality with no decrease in the severity of intestinal lesion scores in broiler chickens challenged with *Clostridium perfringens*	Krueger et al. (2017)
B. subtilis plus mannan-oligosaccharide	Decreased serum triglyceride and total cholesterol levels and heterophils and heterophils-to-lymphocytes ratio and increased lymphocytes in broiler chickens challenged with *C. perfringens*	Al-Baadani et al. (2018)
Lactobacillus plantarum 15-1 plus fructo-oligosaccharides	Improved the intestinal morphology, enhanced the immune response, and increased the SCFA concentration in the cecum and decreased the mortality in broilers challenged with *E. coli* O78, without affecting growth performance	Ding et al. (2019)
PoultryStar meUS[b]	Improved behavioral patterns and production profiles of broiler chickens reared under heat stress	Mohammed et al. (2018)
Biomin-PoultryStar[b]	Improved feed conversion values, decreased cecal *E. coli, Salmonella*, and *Shigella* counts, reduced excreta ammonia production, decreased drip loss and cook loss percentage of breast and leg muscles, increased dressing, breast, and leg percentages, and decreased abdominal fat percentage in broiler chickens under hot climatic conditions	Abdel-Wareth et al. (2019)
PoultryStar me[b]	Increased body weight, bone mineral density, bone mineral content, and bone area and improved walking ability in broilers exposed to a daily cyclic heat stress episodes	Yan et al. (2019)
PoultryStar meUS[b]	Decreased liver and hypothalamus HSP70 levels, increased villus height in the duodenum, and villus height and villus height:crypt depth ratios in the ileum of broiler chickens reared under cyclic heat stress	Jiang et al. (2020)
Pigs		
B. subtilis, L. casei, Pichia anomala plus oligosaccharides	Increased apparent nutrient digestibility, decreased diarrhea rates, increased fecal counts of LAB and fecal lipase activity	Fan et al. (2015)

(Continued)

TABLE 66.2 (Continued)

Synbiotic	Main outcome	Reference
E. faecium, Lactobacillus salivarius, L. reuteri, B. thermophilum plus inulin	Beneficial effect on the gut microbiota, decreased relative abundance of *Escherichia* in ileum, cecum and colon, and increased bifidobacterial numbers in the ileum	Sattler et al. (2015)
L. plantarum—Biocenol LP96 (CCM 7512), *L. fermentum*—Biocenol LF99 (CCM 7514) plus flaxseed	Decreased lactate dehydrogenase leakage in the blood serum and tissue extracts and likely improvement in the immune status and the integrity of jejunum mucosa during infection	Andrejčáková et al. (2016)
E. faecium NCIMB 11181 plus lactulose	Decreased proteobacteria abundances, increased the average population of *Lactobacillaceae* and large decreases in the proportions of *Enterobacteriaceae* in feces	Chae et al. (2016)
L. plantarum—Biocenol LP96 (CCM 7512), *L. fermentum*—Biocenol LF99 (CCM 7514) plus flaxseed	Positive effects on the blood serum levels of total lipids, and the ratio of n-3 polyunsaturated fatty acids (PUFAs)/n-6 PUFAs, and gut health and adaptation process after weaning	Sopková et al. (2017)
Calves		
E. faecium plus lactulose	Improved growth performance, decreased ileal villus height, depth of the crypts in the cecum and the surface area of lymph follicles from Peyer's patches	Fleige et al. (2007)
Lambs		
L. casei plus *A. tequilana* inulin	Increased weight gain and fecal lactobacilli, decreased fecal total coliforms, diarrhea incidence, and serum cholesterol concentration	Ayala-Monter et al. (2019)
Dogs		
E. faecium NCIMB 10415 4b1707 plus fructo-oligosaccharide and acacia	Decreased the incidence of diarrhea	Rose et al. (2017)

[a] *BiominIMBO* (Enterococcus faecium *plus prebiotic derived from chicory and immune modulating substances derived from sea algae).*
[b] *Biomin-PoultryStar/PoultryStar me/PoultryStar meUS* (Lactobacillus reuteri, Enterococcus faecium, Bifidobacterium animalis, Pediococcus acidilactici *plus fructo-oligosaccharide).*

chicken genotype. *L. lactis* probiotics survived in the chicken gut throughout their lifespan (Sławińska et al., 2014a). In addition, the effects of in ovo administration of *L. lactis* spp. *lactis* IBB SL1 plus RFOs, *L. lactis* spp. *cremoris* IBB SC1 plus RFOs, *L. acidophilus*, and *S. faecium* plus lactose have been investigated on immune-related gene expression in adult chickens. The intestinal tissues had downregulation of expression for cytokines IL-4, IL-6, IL-12p40, IL-18, interferon (IFN)-β, IFN-γ, and one chemokine (IL-8) in the cecal tonsils of synbiotic-treated chickens. In the spleen, which is a peripheral part of the immune system, the expression of IL-4 and IL-6 was upregulated, suggesting that in ovo administration of synbiotics activates the immune system in adult chickens (Sławinska et al., 2014b).

In ovo administration of synbiotics may also be an effective method to increase body weight, improve the SCFA cecal profile, and increase the villus length:crypt depth ratio in the jejunal mucosa (Miśta et al., 2017). These effects were more marked after administration of *L. lactis* spp. *lactis* IBB SL1 plus inulin as compared to another synbiotic *L. lactis* spp. *cremoris* IBB SC1 plus transgalacto-oligosaccharides (TGOS). Furthermore, in ovo administration of these two synbiotics at day 12 of incubation has been shown to modulate central and peripheral lymphatic organ development in broilers. The synbiotic *L. lactis* spp. *cremoris* IBB SC1 plus TGOS decreased the cortex/medulla ratio in the thymus and slowed the development of the cortex in bursal follicles on day 21 posthatching, with consequent impacts on the primary lymphatic organs. Synbiotic treatment also stimulated formation of germinal centers in the spleens of 21- and 35-day-old chicks, indicating enhanced B-cell proliferation in secondary lymphatic organs. It produced an age-dependent increase in the spleen/bursa of Fabricius ratio (Madej et al., 2015). In yet another study, the synbiotics *L. lactis* spp. *lactis* IBB SL1 plus inulin and *L. lactis* spp. *cremoris* IBB SC1 plus TGOS administered in ovo were found to have a stimulatory effect on the gut-associated lymphoid tissue development after hatching (Madej and Bednarczyk, 2016).

In an interesting study, Kolodziejski et al. (2018) reported that incretins are involved in the action of synbiotics or that they may even be their target in poultry. In ovo administration of two different synbiotics namely *Lactobacillus salivarius* IBB3154 plus GOS and *Lactobacillus plantarum* IBB3036 plus RFO caused a lowering of the level of incretins glucagon-like peptide 1 (GLP-1) and of gastric-inhibitory polypeptide or glucose-dependent insulinotropic peptide (GIP) in the blood/serum of broiler chickens. Both synbiotics downregulated GLP-1 and GIP mRNA expression in the duodenum and GLP-1 receptors in the pancreas. In addition, *L. plantarum* IBB3036 plus RFO increased trypsin and lipase activities in the duodenum content, simultaneously decreasing amylase activity, which is likely to promote digestion of proteins and lipids. The in ovo injection of these two synbiotics was also found to significantly influence the gut structure, which should contribute to an improvement in nutrient absorption by the gut (Sobolewska et al., 2017).

Studies on the beneficial effects of synbiotics in other domestic animal species, including companion animals, have received only minimal attention. Krause et al. (2010) investigated the efficacy of synbiotics consisting of a 50:50 mixture of probiotic *E. coli* strains UM-2 and UM-7 and prebiotic raw potato starch in young pigs challenged with enterotoxigenic *E. coli* (ETEC) K88. The synbiotics produced a beneficial effect on piglet growth performance and resulted in a reduction of diarrhea and increased microbial diversity in the gut. In another study, the potential of *L. plantarum*, lactulose, and their combination to control post-weaning colibacillosis in pigs was evaluated using an ETEC K88 oral challenge (Guerra-Ordaz et al., 2014). Lactulose improved the average daily gain and increased lactobacilli and the percentage of butyric acid in the colon. It also produced an increase in the ileum villous height and a reduction of the major acute-phase protein in serum. The probiotic increased the numbers of *L. plantarum* bacteria in the ileum and colon and the total lactobacilli in the colon, and showed a trend to reduce diarrhea. The concentrations of ammonia in ileal and colonic digesta were decreased, whereas the villous height and number of ileal goblet cells increased. The combined effects of *L. plantarum* and lactulose have potential to control postweaning colibacillosis in pigs. Tanner et al. (2015) evaluated the effect of *Bifidobacterium thermophilum* RBL67 (RBL67) alone and combined with FOS on the gut microbiota of Göttingen minipigs. During treatment and at the time of euthanasia, RBL67 was consistently detected in feces, cecum, and colon after feeding RBL67 and synbiotic diets. At the time of euthanasia, significantly higher *Bifidobacterium* numbers in the cecum and colon of synbiotic-fed minipigs were measured compared with RBL67, suggesting that RBL67 and FOS combination represents a valuable strategy to increase probiotic bacteria levels and survival in the GIT for feed and food applications.

Although synbiotics have the potential to introduce beneficial bacterial populations as well as to encourage proliferation of autochthonous-specific strains in the intestinal tract, there are limited studies using synbiotics for the prevention and treatment of various conditions in small animals. The prebiotic compound within a synbiotic product should support the growth of the probiotic microorganism and enhance its proliferation within the GIT. Obviously, prebiotic additives have to be selected depending on the probiotic strain. Although combinations of prebiotics with dietary probiotics have not been adequately explored in the domestic canine and feline species, such information might have implications for the design and development of synbiotic formulations for companion animals. In an in vitro study, Salavati Schmitz and Allenspach (2017) examined the growth properties of *E. faecium* NCIMB 10415 E1707, *E. faecium* NCIMB 30183, *Bifidobacterium longum* NCIMB 30182, and *Bifidobacterium infantis* NCIMB 30181, with the addition of FOS, MOS, and FOS plus gum Arabic. The most commonly used probiotic strain for small animals, *E. faecium* NCIMB 10415 E1707, was not affected by any of the prebiotics used. The growth of bifidobacteria was enhanced with commonly used prebiotic oligosaccharides, whereas the addition of gum Arabic had a stronger effect on growth acceleration than FOS alone, indicating that bifidobacteria might be a promising probiotic candidate in small animals.

There are limited studies using synbiotics as a supplement to prevent or treat diarrhea in companion animals. Gagné et al. (2013) examined the effects of a synbiotic containing *E. faecium* SF68, *Bacillus coagulans*, and *L. acidophilus* plus FOS and MOS and other ingredients on fecal quality, SCFA concentrations, and the microbiome, and examined its potential to decrease episodes of diarrhea in training healthy sled dogs. The use of a synbiotic in dogs resulted in a significant increase in *Lactobacillaceae*, a positive correlation between *Lactobacillaceae* and overall butyrate concentration, and an improvement in fecal score, which was associated with a decrease in the prevalence of diarrhea. On the contrary, in a previous study no changes in the major bacterial phyla were identified in healthy dogs and cats following administration of a synbiotic containing a mixture of seven probiotic strains and a blend of FOS and arabinogalactans daily for 21 days. No adverse gastrointestinal effects were recorded and no significant changes in gastrointestinal function or immune markers were observed during the study period. However, the synbiotic led to an increased abundance of probiotic bacteria in the feces of healthy dogs and cats (Garcia-Mazcorro et al., 2011). In client-owned adult cats with naturally occurring chronic diarrhea, a synbiotic product featuring multispecies probiotics (*E. faecium* NCIMB 30183, *Streptococcus salivarus* spp. *thermophilus* NCIMB 30189, *Bifidobacterium bifidum* NCIMB 30179, *L. acidophilus* NCIMB 30184, *Lactobacillus casei* spp. *rhamnosus* NCIMB 30188, *L. plantarum* NCIMB 30187, *Lactobacillus delbrueckii* spp. *bulgaricus* NCIMB 30186 combined with two prebiotics) improved stool character, and the mean fecal score decreased from 6.0 to 4.4, representing a significantly firmer stool character. After a 21-day course of synbiotic supplementation, 72% of owners perceived an improvement in diarrhea of their cats. This synbiotic product was found to be safe, well tolerated, and had no adverse effects (Hart et al., 2012).

Recently, a randomized, double-blinded, placebo-controlled, two-way, two-period, crossover trial was conducted to assess the prevention of clindamycin-induced gastrointestinal signs using a synbiotic (a commercial mixture of probiotics and prebiotics) in healthy research cats (Stokes et al., 2017). Administration of the synbiotic 1 h after clindamycin administration decreased hyporexia and vomiting in cats. In addition, the clinical benefits of synbiotic administration persisted for at least 6 weeks after discontinuation, decreasing the severity of antibiotic-associated gastrointestinal signs in cats that subsequently received clindamycin with placebo. However, the antibiotic-associated diarrhea was not decreased following administration of synbiotic.

In comparison to probiotics and prebiotics, studies of synbiotics on cattle health and performance have received only minimal attention. The efficacy and suitability of the use of synbiotics have been generally assessed on the basis of weight gain, feed intake and utilization, and health status (Markowiak and Śliżewska, 2018; Malik et al., 2019). Preruminant calves fed milk containing a synbiotic consisting of a multistrain probiotic (seven bacteria strains and two yeast strains) and a prebiotic (*S. cerevisiae* cell wall polysaccharides) resulted in an increase in average daily weight gain and a decrease in the number of pathogenic *E. coli* bacteria in the feces (Roodposhti and Dabiri, 2012). Supplementation of feed with a synbiotic product consisting of *L. casei* spp. *casei* and dextran improved milk production as a result of a positive change in the bovine intestinal microbiota and greater resistance to infectious diseases. The cows evaluated also showed a reduction in the number of somatic cells and a decrease in the incidence of mastitis (Yasuda et al., 2007).

Safety assessment

Because synbiotics are a combination of probiotics and prebiotics, it would be prudent to briefly mention their safety criteria. In general, LAB strains carry a very low infection risk. Numerous related products have been traditionally used over generations and have been established as safe (Markowiak and Śliżewska, 2017, 2018; Malik et al., 2016, 2019). In vitro and in vivo studies of LAB indicate that they are safe for human and livestock consumption (Lee et al., 2009; Urbanska and Szajewska, 2014; Yang et al., 2015). Nevertheless, plasmids in some strains of LAB have been demonstrated to encode for antibiotic-resistant genes (Hou et al., 2014; Heavens et al., 2011; Rosander et al., 2008). For example, a commercially available probiotic strain, *Lactobacillus reuteri* ATCC 55730, was demonstrated to carry potentially transferable resistance traits for the antibiotics tetracycline and lincomycin. However, this strain has been replaced by *L. reuteri* DSM 17938, in which the two resistance plasmids have been removed without loss of any probiotic characteristics (Rosander et al., 2008).

Although probiotics are designated as generally recognized as safe (GRAS), there are several concerns about the potential risk associated with the use of probiotics, particularly in premature neonates or immune-deficient individuals. The bacteria may translocate through the gastrointestinal barrier and cause invasive infections, leading to bacteremia or sepsis. Another concern is the likelihood that the metabolic activity of the microbial products might be toxic to the host. Moreover, probiotics could even have an undesirable immunologic effect (Hibberd and Davidson, 2008). It is thus imperative that profound safety aspects and caution should be taken into account before considering a probiotic therapy in preterm infants, neonates, and immune-compromised and immune-deficient individuals. The effects of probiotics cannot be generalized because their properties are strain-specific. Therefore confirmation studies are needed. Due to an inadequate understanding of the properties of probiotics and their interaction in the gut, safety predictions remain inaccurate. At the present time, it is difficult to conclude that probiotics are safe in infants because each individual strain has to be evaluated for safety at high dosages and with chronic use (van den Nieuwboer et al., 2014). A detailed assessment by the NIH on safety aspects has been published (United States Department of Health and Human Services, 2011). There is a need for systematic reporting of adverse events and better documentation of interventions (Malik et al., 2016).

Regulatory considerations

Substantial progress in legislation for the safety evaluation of probiotics has been made in the United States, Canada, and Europe (European Food Safety Authority-EFSA, 2005a; HC, 2006; FAO/WHO, 2002, von Wright, 2005). In the United States, the Food and Drug Administration regulates specific utilization of microorganisms for human consumption and indicates that these should possess GRAS status. In Europe, the European Food Safety Authority (EFSA) has initiated the concept of qualified presumption of safety (QPS), which has a similar purpose as the GRAS approach. The QPS concept provides a generic assessment system for use within the EFSA that, in principle, can be applied to all requests received for the safety assessments of microorganisms deliberately introduced into the food chain (European Food Safety Authority-EFSA, 2005b; Wassenaar and Klein, 2008; Gaggia et al., 2010). A safety assessment of selected groups of microorganisms from a defined taxonomic group including genera or groups of related species can be made on the basis of four pillars of information: identity, body of knowledge, possible pathogenicity, and end use. If the taxonomic group and characterization to strain level do not raise safety concerns, or if any safety concerns can be defined and excluded, then the organism may be granted QPS status. Therefore for any strain of microorganism that can be unambiguously demonstrated to be from a qualified QPS group such as *Lactobacillus* or *Bifidobacterium*, further assessment of safety is restricted to tests for resistance to antibiotics. If a microorganism is not considered suitable for QPS, then a full safety assessment is likely to be required.

Safety assessment of probiotics

Introduction of novel microorganisms and GMO requires well-designed studies and careful assessment of their safety and risk-to-benefit ratio. Moreover, new probiotic microorganisms should be of genera and strains commonly found in healthy human intestinal microflora. Although *Lactobacillus*, *Lactococcus*, *Bifidobacterium*, and *Saccharomyces* are nonpathogenic, *Bacillus cereus* is a pathogen and *Enterococcus* and other general LAB are opportunistic pathogens. LAB and bifidobacteria attach to the human intestinal mucosa and are generally regarded as having GRAS status.

Various factors considered in the safety assessment of probiotics include pathogenicity, infectivity, and virulence factors comprising toxicity, metabolic activity, and the intrinsic properties of the microbes. The probiotic bacteria should be tested for pathogenicity, infectivity, ability to produce harmful substances by metabolic activity, platelet-aggregating activity, mucus degradation activity, and antibiotic resistance (Ishibashi and Yamazaki, 2001; Peivasteh-Roudsari et al., 2019). In addition to a broader screening of antibiotic resistance in probiotic supplements, the use of computational simulations, live imaging, and functional genomics to gather information on the evolutionary behavior, adaptations, and dynamics of probiotics should be studied in conditions that best represent the human gut, including in the presence of antibiotics (Zheng et al., 2017).

The key issues concerning the safety aspects of probiotics were recently reviewed by Peivasteh-Roudsari et al. (2019). In terms of adverse effects of probiotics, survival of the probiotics in the GIT, their translocation and colonization properties, and the fate of their active components need to be known (Marteau et al., 1993). Four types of side effects or adverse reactions, namely systemic infections, risk of deleterious metabolic activities, risk of adjuvant side effects and immunomodulation, and risk of gene transfer have been reported (Salminen and von Wright, 1998).

Three approaches can be used to evaluate the safety of a probiotic strain, including studies on: (1) intrinsic properties of the strain; (2) pharmacokinetics of the strain incorporating survival, activity in the intestine, dose–response relationship, fecal and mucosal recovery; and (3) searching for interactions between the strain and the host. It is very important to identify the survival of the probiotics within the GIT, their translocation and colonization properties, and the fate of their active components to predict the positive effects and/or the side effects. The survival of ingested probiotics at different levels of the GIT differs between strains. Important criteria in the selection of probiotic microbes are safety and stability (Marteau et al., 1993; Salminen et al., 1998; Anadón et al., 2010; Peivasteh-Roudsari et al., 2019).

In vitro and in vivo studies conducted to evaluate safety of the probiotic microorganisms have been reviewed by Anadón et al. (2010). These include assessment of infective properties of the probiotic microorganisms in vitro using cell lines and human intestinal mucus degradation and in vivo using animal models, mainly immunocompromised or lethally irradiated animals. Animal studies are performed to determine the maximum dose levels of probiotic bacteria given in feed to different animal species without causing any side or adverse effects. An acute oral study and a 90-day rodent feeding toxicity study of probiotic microorganisms was performed as per guidelines of the EU or OCDE, and this is particularly needed for new probiotics (ILSI Europe, 1993; Anadón et al., 2006). Noninvasive tests in animal models and humans are used to provide reference points to which other measurements can be related, such as altered circulating cytokines and altered gene expression (Butler, 2008). Studies in humans involving short-term clinical trials with healthy volunteers for evaluation of the safety of probiotics (Salminen and von Wright, 1998) have been performed. Epidemiological and postmarketing surveillance is used to assess the risk and to analyze it retrospectively in epidemiology studies and prospectively by using the postmarketing pharmaco-toxicovigilance systems in populations ingesting large amounts of newly introduced probiotic microorganisms for infections and comparing their use to the use of traditional strains (Salminen and von Wright, 1998). Additional studies could be required, particularly for some new probiotics, and these include studies on genotoxicity including mutagenicity, toxin production, and virulence factors, antibiotic resistance profile, and transferability of resistance (Anadón et al., 2010).

Safety assessment of prebiotics

The safety of prebiotics has been discussed by Anadón et al. (2010). For assessing the safety of the final product formulation of a prebiotic, the issues to be considered include: (1) further toxicological studies in animals and humans may not be required if the product has a history of safe use and is granted GRAS or QPS status; (2) safe consumption levels with minimal symptoms and side effects should be determined; (3) contaminants and impurities must not be present in the product; however, if they are present, then the contaminants should be identified and measured, and the impurities should be well characterized and subjected to toxicity evaluation if required; and (4) on the basis of recent knowledge, the prebiotic should not change the microbiota in such a way as to have long-term harmful effects on the host.

Animal models can be used to determine the target organs and toxic effects of functional ingredients. The extent of testing needed for a functional ingredient is increased if there is lack of knowledge of potential for toxicity due to insufficiently characterized products. The following criteria must be met by a functional ingredient to derive its safe level of exposure without further toxicology testing (Kruger and Mann, 2003): (1) active component(s) and related substances are well demonstrated and there is sufficient knowledge of the lack of potential for toxicity at the human dose levels recommended on the basis of existing data from the literature; (2) impurities are well investigated and there is sufficient knowledge of the lack of potential for toxicity on the basis of existing data from the literature; and, (3) the process of manufacturing is standardized and reproducible.

If the active component(s) or impurities are not fully characterized, or if there is not adequate data available to assess the potential for toxicity, then various preclinical toxicological information is required to evaluate the functional ingredient. These are in vitro and in vivo toxicity studies, including mutagenicity studies, reproduction and teratogenicity studies, pharmacokinetics and special pharmacology studies, and long-term feeding studies, following a tiered approach on a case-by-case basis. The margin of safety between the no observed adverse effect level determined in the animal studies and the anticipated human level of intake must be considered while planning the animal studies for functional ingredients.

It is assumed that prebiotics lack life-threatening or severe side effects. The enzymes present in the intestines cannot break down oligosaccharides and polysaccharides. They are transported to the colon to be fermented by the gut microbiota. Therefore the side effects of prebiotics are mostly the consequence of their osmotic functions. Osmotic diarrhea, bloating, cramping, and flatulence could be experienced in prebiotic recipients (Davani-Davari et al., 2019).

Concluding remarks and future directions

Synbiotics and other functional foods are now considered important tools in maintaining human and animal health, for prevention of disease, and/or as alternatives for reducing the risk associated with diseases. Studies conducted on the relationship of the microbiota of the human and animal gut and immunology have highlighted the role of synbiotics in the health of mammals. There is evidence to suggest that synbiotics influence the microbial ecology of the intestines of humans and animals and play a role in alleviating various ailments. The majority of studies dealing with synbiotics are conducted in humans, but increasing research has been focused on animals. Improvement in the number of beneficial bacteria and the reduction of the potential pathogen load have been demonstrated following administration of synbiotics in farm animals. Despite the fact that the understanding of the effects of synbiotics has increased, important information relating to their impact on the host is so far incomplete. Potential combinations of the most appropriate probiotics and prebiotics may reduce the risks associated with intestinal diseases and eliminate specific microbial disorders.

At present, the focus is on assessing new synbiotic products aimed at improving human and animal health. The health claims of synbiotics need to be substantiated and firmly established by properly designed studies in humans and define their optimal doses, intake durations, and long-term safety, as well as their precise mechanisms of action. Obligatory and standardized reporting on adverse events has been suggested in studies with synbiotics (van den Nieuwboer et al., 2015). Well-designed and carefully conducted randomized, placebo-controlled clinical trials of synbiotics are needed (Maftei, 2019). Future emphasis will be on the development of newer beneficial, more effective and safe synbiotics by employing the most appropriate combinations of probiotics and prebiotics for use in humans and animals.

References

Abdel-Wareth, A.A.A., Hammad, S., Khalaphallah, R., et al., 2019. Synbiotic as eco-friendly feed additive in diets of chickens under hot climatic conditions. Poult. Sci. 98, 4575–4583.

Al-Baadani, H.H., Abudabos, A.M., Al-Mufarrej, S.I., et al., 2018. Dietary supplementation of *Bacillus subtilis*, *Saccharomyces cerevisiae* and their symbiotic effect on serum biochemical parameters in broilers challenged with *Clostridium perfringens*. J. Appl. Anim. Res. 46, 1064–1072.

Alizadeh, M., Munyaka, P., Yitbarek, A., et al., 2017. Maternal antibody decay and antibody- mediated immune responses in chicken pullets fed prebiotics and synbiotics. Poult. Sci. 96, 58–64.

Al-Sultan, S.I., Abdel-Raheem, S.M., El-Ghareeb, W.R., et al., 2016. Comparative effects of using prebiotic, probiotic, synbiotic and acidifier on growth performance, intestinal microbiology and histomorphology of broiler chicks. Jpn. J. Vet. Res. 64 (Suppl 2), S187–S195.

Anadón, A., Martinez-Larranaga, M.R., Martinez, M.A., 2006. Probiotics for animal nutrition in the European Union. Regulation and safety assessment. Regul. Toxicol. Pharmacol. 45, 91–95.

Anadón, A., Martínez-Larranaga, M.R., Caballero, V., et al., 2010. Assessment of prebiotics and probiotics: an overview. In: Watson, R.R., Preedy, V.R. (Eds.), Bioactive Foods in Promoting Health. Probiotics and Prebiotics Academic Press, Oxford, pp. 19–41.

Andrejčáková, Z., Sopková, D., Vlčková, R., et al., 2016. Synbiotics suppress the release of lactate dehydrogenase, promote non-specific immunity and integrity of jejunum mucosa in piglets. Anim. Sci. J. 87, 1157–1166.

Anzawa, D., Mawatari, T., Tanaka, Y., et al., 2019. Effects of synbiotics containing *Bifidobacterium animalis* subsp. *lactis* GCL2505 and inulin on intestinal bifidobacteria: a randomized, placebo-controlled, crossover study. Food Sci. Nutr. 7, 1828–1837.

Asemi, Z., Aarabi, M.H., Hajijafari, M., et al., 2017. Effects of synbiotic food consumption on serum minerals, liver enzymes, and blood pressure in patients with type 2 diabetes: a double-blind randomized cross-over controlled clinical trial. Int. J. Prev. Med. 8, 43.

Ashayerizadeh, A., Dabiri, N., Mirzadeh, Kh, et al., 2011. Effects of dietary inclusion of several biological feed additives on growth response of broiler chickens. J. Cell Anim. Biol. 5, 61–65.

Ayala-Monter, M.A., Hernández-Sánchez, D., González-Muñoz, S., et al., 2019. Growth performance and health of nursing lambs supplemented with inulin and *Lactobacillus casei*. Asian-Australas. J. Anim. Sci. 32, 1137–1144.

Aziz Mousavi, S.M.A., Seidavi, A., Dadashbeiki, M., et al., 2015. Effect of a synbiotic (Biomin®IMBO) on growth performance traits of broiler chickens. Eur. Poult. Sci. 79. Available from: 10.1399/eps.2015.78.

Baffoni, L., Gaggia, F., Garofolo, G., et al., 2017. Evidence of *Campylobacter jejuni* reduction in broilers with early synbiotic administration. Int. J. Food Microbiol. 251, 41–47.

Baştürk, A., Artan, R., Atalay, A., et al., 2017. Investigation of the efficacy of synbiotics in the treatment of functional constipation in children: a randomized double-blind placebo- controlled study. Turk. J. Gastroenterol. 28, 388–393.

Braegger, C., Chmielewska, A., Decsi, T., et al., 2011. Supplementation of infant formula with probiotics and/or prebiotics: a systematic review and comment by the ESPGHAN committee on nutrition. J. Pediatr. Gastroenterol. Nutr. 52, 238–250.

Butler, R.N., 2008. Non-invasive tests in animal models and humans: a new paradigm for assessing efficacy of biologics including prebiotics and probiotics. Curr. Pharm. Des. 14, 1341–1350.

Calik, A., Ceylan, A., Ekim, B., et al., 2017. The effect of intra-amniotic and posthatch dietary synbiotic administration on the performance, intestinal histomorphology, cecal microbial population, and short-chain fatty acid composition of broiler chickens. Poult. Sci. 96, 169–183.

Chae, J.P., Pajarillo, E.A., Oh, J.K., et al., 2016. Revealing the combined effects of lactulose and probiotic enterococci on the swine fecal microbiota using 454 pyrosequencing. Microb. Biotechnol. 9, 486–495.

Cheng, Y., Chen, Y., Li, X., et al., 2017. Effects of synbiotic supplementation on growth performance, carcass characteristics, meat quality and muscular antioxidant capacity and mineral contents in broilers. J. Sci. Food Agric. 97, 3699–3705.

Chouraqui, J.P., Grathwohl, D., Labaune, J.M., et al., 2008. Assessment of the safety, tolerance, and protective effect against diarrhea of infant formulas containing mixtures of probiotics or probiotics and prebiotics in a randomized controlled trial. Am. J. Clin. Nutr. 87, 1365–1373.

Davani-Davari, D., Negahdaripour, M., Karimzadeh, I., et al., 2019. Prebiotics: definition, types, sources, mechanisms, and clinical applications. Foods 8, 92.

Ding, S., Wang, Y., Yan, W., et al., 2019. Effects of *Lactobacillus plantarum* 15-1 and fructooligosaccharides on the response of broilers to pathogenic *Escherichia coli* O78 challenge. PLoS One 14, e0212079. <https://doi.org/10.1371/journal.pone.0212079>.

Eslamparast, T., Poustchi, H., Zamani, F., et al., 2014a. Synbiotic supplementation in nonalcoholic fatty liver disease: a randomized, double-blind, placebo-controlled pilot study. Am. J. Clin. Nutr. 99, 535–542.

Eslamparast, T., Zamani, F., Hekmatdoost, A., et al., 2014b. Effects of synbiotic supplementation on insulin resistance in subjects with the metabolic syndrome: a randomized, double-blind, placebo-controlled pilot study. Br. J. Nutr. 112, 438–445.

European Food Safety Authority-EFSA, 2005a. Opinion of the scientific committee on a request from EFSA related to a generic approach to the safety assessment by EFSA of microorganisms used in food/feed and the production of food/feed additives. EFSA J. 226, 1–12.

European Food Safety Authority-EFSA, 2005b. Opinion of the scientific panel on additives and products or substances used in animal feed on the updating of the criteria used in the assessment of bacteria for resistance to antibiotics of human or veterinary importance. EFSA J. 223, 1–12.

Fan, G., Chang, J., Yin, Q., et al., 2015. Effects of probiotics, oligosaccharides, and berberine combinations on growth performance of pigs. Turk. J. Vet. Anim. Sci. 39, 637–642.

FAO/WHO, 2002. Joint FAO/WHO (Food and Agriculture Organization/World Health Organization) Working Group Report on Drafting Guidelines for the Evaluation of Probiotics in Food. London, Ontario, Canada. Guidelines for the Evaluation of Probiotics in Food. Joint Working Group Report on Drafting. London, Ontario, pp. 1–11.

Farid, R., Ahanchian, H., Jabbari, F., et al., 2011. Effect of a new synbiotic mixture on atopic dermatitis in children: a randomized-controlled trial. Iran. J. Pediatr. 21, 225–230.

Fateh, R., Iravani, S., Frootan, M., et al., 2011. Synbiotic preparation in men suffering from functional constipation: a randomized controlled trial. Swiss Med. Wkly. 141, w13239.

Fleige, S., Preißinger, W., Meyer, H.H.D., et al., 2007. Effect of lactulose on growth performance and intestinal morphology of pre-ruminant calves using a milk replacer containing *Enterococcus faecium*. Animal 1, 367–373.

Flesch, A.T., Tonial, S.T., Contu, P.C., et al., 2017. Perioperative synbiotics administration decreases postoperative infections in patients with colorectal cancer: a randomized, double-blind clinical trial. Rev. Col. Bras. Cir. 44, 567–573.

Fujimori, S., Gudis, K., Mitsui, K., et al., 2009. A randomized controlled trial on the efficacy of synbiotic versus probiotic or prebiotic treatment to improve the quality of life in patients with ulcerative colitis. Nutrition 25, 520–525.

Gadde, U., Kim, W.H., Oh, S.T., et al., 2017. Alternatives to antibiotics for maximizing growth performance and feed efficiency in poultry: a review. Anim. Health Res. Rev. 18, 26–45.

Gaggia, F., Mattarelli, P., Biavati, B., 2010. Probiotics and prebiotics in animal feeding for safe food production. Int. J. Food Microbiol. 141 (Suppl. 1), S15–S28.

Gagné, J.W., Wakshlag, J.J., Simpson, K.W., et al., 2013. Effects of a synbiotic on fecal quality, short-chain fatty acid concentrations, and the microbiome of healthy sled dogs. BMC Vet. Res. 9, 246. <https://doi.org/10.1186/1746-6148-9-246>.

Garcia-Mazcorro, J.F., Lanerie, D.J., Dowd, S.E., 2011. Effect of a multi-species synbiotic formulation on fecal bacterial microbiota of healthy cats and dogs as evaluated by pyrosequencing. FEMS Microbiol. Ecol. 78, 542–554.

Guerra-Ordaz, A.A., González-Ortiz, G., La Ragione, R.M., et al., 2014. Lactulose and *Lactobacillus plantarum*, a potential complementary synbiotic to control postweaning colibacillosis in piglets. Appl. Environ. Microbiol. 80, 4879–4886.

Hart, M.L., Suchodolski, J.S., Steiner, J.M., et al., 2012. Open-label trial of a multi-strain synbiotic in cats with chronic diarrhea. J. Feline Med. Surg. 14, 240–245.

HC, 2006. Health Canada. Evidence for safety and efficacy of finished natural health products. [online]. Ottawa (ON): Natural Health Products Directorate, Health Canada. Available from: <http://www.hc-sc.gc.ca/dhpmps/prodnatur/legislation/docs/efe-paie_e.html>.

Heavens, D., Tailford, L.E., Crossman, L., et al., 2011. Genome sequence of the vertebrate gut symbiont *Lactobacillus reuteri* ATCC 53608. J. Bacteriol. 193, 4015–4016.

Hibberd, P.L., Davidson, L.E., 2008. Safety of probiotics. AgroFOOD Ind. Hi-Tech 19, 30–33.

Hill, C., Guarner, F., Reid, G., et al., 2014. Expert consensus document. The international scientific association for probiotics and prebiotics consensus statement on the scope and appropriate use of the term probiotic. Nat. Rev. Gastroenterol. Hepatol. 11, 506–514.

Hou, C., Wang, Q., Zeng, X., et al., 2014. Complete genome sequence of *Lactobacillus reuteri* I5007, a probiotic strain isolated from healthy piglet. J. Biotechnol. 179, 63–64.

ILSI Europe, 1993. A scientific basis for regulations on pathogenic microorganisms in foods. Summary of a Workshop Held in May 1993 and Organized by the Scientific Committee on Microbiology. ILSI Press.

Ishibashi, N., Yamazaki, S., 2001. Probiotics and safety. Am. J. Clin. Nutr. 73 (Suppl. 2), 465S–470S.

Jiang, S., Mohammed, A.A., Jacobs, J.A., et al., 2020. Effect of synbiotics on thyroid hormones, intestinal histomorphology, and heat shock protein 70 expression in broiler chickens reared under cyclic heat stress. Poult. Sci. 99, 142–150.

Kataoka, Y., Kan'o, T., Asahara, T., et al., 2013. Synbiotics reduce infectious complications by improving the intestinal milieu and enhancing the immune function in critically ill emergency surgical patients. Kitasato Med. J. 43, 17–25.

Kearney, S.M., Gibbons, S.M., 2018. Designing synbiotics for improved human health. Microb. Biotechnol. 11, 141–144.

Kelishadi, R., Farajian, S., Safavi, M., et al., 2014. A randomized triple-masked controlled trial on the effects of synbiotics on inflammation markers in overweight children. J. Pediatr. (Rio J) 90, 161–168.

Khodadad, A., Sabbaghian, M., 2010. Role of synbiotics in the treatment of childhood constipation: a double-blind randomized placebo controlled trial. Iran. J. Pediatr. 20, 387–392.

Koc, F., Samli, H., Okur, A.M., et al., 2010. Effects of *Saccharomyces cerevisiae* and/or mannanoligosaccharide on performance, blood parameters and intestinal microbiota of broiler chicks. Bulg. J. Agric. Sci. 16, 643–650.

Kolida, S., Gibson, G.R., 2011. Synbiotics in health and disease. Annu. Rev. Food Sci. Technol. 2, 373–393.

Kolodziejski, P.A., Sassek, M., Chalupka, D., et al., 2018. GLP1 and GIP are involved in the action of synbiotics in broiler chickens. J. Anim. Sci. Biotechnol. 9, 13.

Krause, D.O., Bhandari, S.K., House, J.D., et al., 2010. Response of nursery pigs to a synbiotic preparation of starch and an anti-*Escherichia coli* K88 probiotic. Appl. Environ. Microbiol. 76, 8192–8200.

Kridtayopas, C., Rakangtong, C., Bunchasak, C., et al., 2019. Effect of prebiotic and synbiotic supplementation in diet on growth performance, small intestinal morphology, stress, and bacterial population under high stocking density condition of broiler chickens. Poult. Sci. 98, 4595–4605.

Krueger, L.A., Spangler, D.A., Vandermyde, D.R., et al., 2017. Avi-Lution® supplemented at 1.0 or 2.0 g/kg in feed improves the growth performance of broiler chickens during challenge with bacitracin-resistant *Clostridium perfringens*. Poult. Sci. 96, 2595–2600.

Kruger, C.L., Mann, S.W., 2003. Safety evaluation of functional ingredients. Food Chem. Toxicol. 41, 793–805.

Lee, D.Y., Seo, Y.S., Rayamajhi, N., et al., 2009. Isolation, characterization, and evaluation of wild isolates of *Lactobacillus reuteri* from pig feces. J. Microbiol. 47, 663–672.

Lee, S.-H., Cho, D.-Y., Lee, S.-H., et al., 2019. A randomized clinical trial of synbiotics in irritable bowel syndrome: dose- dependent effects on gastrointestinal symptoms and fatigue. Korean J. Fam. Med. 40, 2–8.

Li, X., Liu, L.Q., Xu, C.L., 2008. Effects of supplementation of fructooligosaccharide and/or *Bacillus subtilis* to diets on performance and intestinal microflora in broilers. Arch. Tierz., Dummerstorf 51, 64–70.

Luoma, A., Markazi, A., Shanmugasundaram, R., et al., 2017. Effect of synbiotic supplementation on layer production and cecal *Salmonella* load during a *Salmonella* challenge. Poult. Sci. 96, 4208–4216.

Macfarlane, S., Cleary, S., Bahrami, B., et al., 2013. Synbiotic consumption changes the metabolism and composition of the gut microbiota in older people and modifies inflammatory processes: a randomized, double-blind, placebo-controlled crossover study. Aliment. Pharmacol. Ther. 38, 804–816.

Madej, J.P., Bednarczyk, M., 2016. Effect of *in ovo*-delivered prebiotics and synbiotics on the morphology and specific immune cell composition in the gut-associated lymphoid tissue. Poult. Sci. 95, 19–29.

Madej, J.P., Stefaniak, T., Bednarczyk, M., 2015. Effect of *in ovo*-delivered prebiotics and synbiotics on lymphoid-organs' morphology in chickens. Poult. Sci. 94, 1209–1219.

Maftei, M.N., 2019. Probiotic, prebiotic and synbiotic products in human health. In: Solis- Oviedo, R.L. (Ed.), Frontiers and New Trends in the Science of Fermented Food and Beverages. IntechOpen, London.

Malaguarnera, M., Vacante, M., Antic, T., et al., 2012. *Bifidobacterium longum* with fructo- oligosaccharides in patients with non alcoholic steatohepatitis. Dig. Dis. Sci. 57, 545–553.

Malik, J.K., Ahmad, A.H., Kalpana, S., et al., 2016. Synbiotics: safety and toxicity considerations. In: Gupta, R.C. (Ed.), Nutraceuticals: Efficacy, Safety and Toxicity. Academic Press/Elsevier, Amsterdam, pp. 811–822.

Malik, J.K., Prakash, A., Srivastava, A.K., et al., 2019. Synbiotics in animal health and production. In: Gupta, R.C., Srivastava, A., Lall, R. (Eds.), Nutraceuticals in Veterinary Medicine. Springer Nature, Cham, pp. 287–301.

Markowiak, P., Śliżewska, K., 2017. Effects of probiotics, prebiotics, and synbiotics on human health. Nutrients 9, 1021.

Markowiak, P., Śliżewska, K., 2018. The role of probiotics, prebiotics and synbiotics in animal nutrition. Gut Pathog. 10, 21. Available from: 10.1186/s13099-018-0250-0.

Marteau, P., Pochart, P., Bouhnik, Y., et al., 1993. Fate and effects of some transiting microorganisms in the human gastrointestinal tract. World Rev. Nutr. Diet. 74, 1–21.

Martín, R., Langella, P., 2019. Emerging health concepts in the probiotics field: streamlining the definitions. Front. Microbiol. 10, 1047.

Min, Y.N., Yang, H.L., Xu, Y.X., et al., 2016. Effects of dietary supplementation of synbiotics on growth performance, intestinal morphology, sIgA content and antioxidant capacities of broilers. J. Anim. Physiol. Anim. Nutr. (Berl.) 100, 1073–1080.

Miśta, D., Króliczewska, B., Pecka-Kiełb, E., et al., 2017. Effect of *in ovo* injected prebiotics and synbiotics on the caecal fermentation and intestinal morphology of broiler chickens. Anim. Prod. Sci. 57, 1884–1892.

Mofidi, F., Poustchi, H., Yari, Z., et al., 2017. Synbiotic supplementation in lean patients with non-alcoholic fatty liver disease: a pilot, randomised, double-blind, placebo-controlled, clinical trial. Br. J. Nutr. 117, 662–668.

Mohammed, A.A., Jacobs, J.A., Murugesan, G.R., et al., 2018. Effect of dietary synbiotic supplement on behavioral patterns and growth performance of broiler chickens reared under heat stress. Poult. Sci. 97, 1101–1108.

Mookiah, S., Sieo, C.C., Ramasamy, K., et al., 2014. Effects of dietary prebiotics, probiotic and synbiotics on performance, caecal bacterial populations and caecal fermentation concentrations of broiler chickens. J. Sci. Food Agric. 94, 341–348.

Nasri, K., Jamilian, M., Rahmani, E., et al., 2018. The effects of synbiotic supplementation on hormonal status, biomarkers of inflammation and oxidative stress in subjects with polycystic ovary syndrome: a randomized, double-blind, placebo-controlled trial. BMC Endocr. Disord. 18, 21.

Panigrahi, P., Parida, S., Pradhan, L., et al., 2008. Long-term colonization of a *Lactobacillus plantarum* synbiotic preparation in the neonatal gut. J. Pediatr. Gastroenterol. Nutr. 47, 45–53.

Peivasteh-Roudsari, L., Pirhadi, M., Karami, H., et al., 2019. Probiotics and food safety: an evidence-based review. J. Food Saf. Hyg. 5, 1–9.

Plaza-Diaz, J., Ruiz-Ojeda, F.J., Gil-Campos, M., et al., 2019. Mechanisms of action of probiotics. Adv. Nutr. 10 (Suppl 1), S49–S66.

Puccio, G., Cajozzo, C., Meli, F., et al., 2007. Clinical evaluation of a new starter formula for infants containing live *Bifidobacterium longum* BL999 and prebiotics. Nutrition 23, 1–8.

Rayes, N., Seehofer, D., Theruvath, T., et al., 2007. Effect of enteral nutrition and synbiotics on bacterial infection rates after pylorus-preserving pancreatoduodenectomy: a randomized, double-blind trial. Ann. Surg. 246, 36–41.

Roodposhti, P.M., Dabiri, N., 2012. Effects of probiotic and prebiotic on average daily gain, fecal shedding of *Escherichia coli*, and immune system status in newborn female calves. Asian-Aust. J. Anim. Sci. 25, 1255–1261.

Rosander, A., Connolly, E., Roos, S., 2008. Removal of antibiotic resistance gene-carrying plasmids from *Lactobacillus reuteri* ATCC 55730 and characterization of the resulting daughter strain, *L. reuteri* DSM 17938. Appl. Environ. Microbiol. 74, 6032–6040.

Rose, L., Rose, J., Gosling, S., et al., 2017. Efficacy of a probiotic-prebiotic supplement on incidence of diarrhea in a dog shelter: a randomized, double-blind, placebo-controlled trial. J. Vet. Intern. Med. 31, 377–382.

Salah, A.S., Ahmed-Farid, O.A., El-Tarabany, M.S., 2019. Carcass yields, muscle amino acid and fatty acid profiles, and antioxidant indices of broilers supplemented with synbiotic and/or organic acids. J. Anim. Physiol. Anim. Nutr. (Berl.) 103, 41–52.

Salavati Schmitz, S., Allenspach, K., 2017. Effects of different oligosaccharides on growth of selected probiotic bacterial strains. J. Microb. Biochem. Technol. 9, 054–058.

Salehimanesh, A., Mohammadi, M., Roostaei-Ali Mehr, M., et al., 2016. Effect of dietary probiotic, prebiotic and synbiotic supplementation on performance, immune responses, intestinal morphology and bacterial populations in broilers. J. Anim. Physiol. Anim. Nutr. (Berl.) 100, 694–700.

Salminen, S., von Wright, A., 1998. Current probiotics-safety assured? Microb. Ecol. Health D. 10, 68–77.

Salminen, S., von Wright, A., Morelli, L., et al., 1998. Demonstration of safety of probiotics – a review. Int. J. Food Microbiol. 44, 93–106.

Sattler, V.A., Bayer, K., Schatzmayr, G., et al., 2015. Impact of a probiotic, inulin, or their combination on the piglets' microbiota at different intestinal locations. Benef. Microbes 6, 473–483.

Scantlebury-Manning, T., Gibson, G.R., 2004. Microbial-gut interactions in health and disease. Prebiotics. Best Pract. Res. Clin. Gastroenterol. 18, 287–298.

Scavuzzi, B.M., Henrique, F.C., Miglioranza, L.H.S., et al., 2014. Impact of prebiotics, probiotics and synbiotics on components of the metabolic syndrome. Ann. Nutr. Disord. Ther. 1, 1009.

Sekhon, B.S., Jairath, S., 2010. Prebiotics, probiotics and synbiotics: an overview. J. Pharm. Educ. Res. 1, 13–36.

Shanmugasundaram, R., Mortada, M., Cosby, D.E., et al., 2019. Synbiotic supplementation to decrease *Salmonella* colonization in the intestine and carcass contamination in broiler birds. PLoS One 14, e0223577. <https://doi.org/10.1371/journal.pone.0223577>.

Shanmugasundaram, R., Markazi, A., Mortada, M., et al., 2020. Research note: effect of synbiotic supplementation on caecal *Clostridium perfringens* load in broiler chickens with different necrotic enteritis challenge models. Poult. Sci. 99, 2452–2458.

Sławińska, A., Siwek, M., Zylińska, J., et al., 2014a. Influence of synbiotics delivered *in ovo* on immune organs development and structure. Folia Biol. (Kraków) 62, 277–285.

Sławinska, A., Siwek, M.Z., Bednarczyk, M.F., 2014b. Effects of synbiotics injected *in ovo* on regulation of immune-related gene expression in adult chickens. Am. J. Vet. Res. 75, 997–1003.

Sobolewska, A., Bogucka, J., Dankowiakowska, A., et al., 2017. The impact of synbiotic administration through in ovo technology on the microstructure of a broiler chicken small intestine tissue on the 1st and 42nd day of rearing. J. Anim. Sci. Biotechnol. 8, 61.

Sopková, D., Hertelyová, Z., Andrejčáková, Z., et al., 2017. The application of probiotics and flaxseed promotes metabolism of n-3 polyunsaturated fatty acids in pigs. J. Appl. Anim. Res. 45, 93–98.

Stokes, J.E., Price, J.M., Whittemore, J.C., 2017. Randomized, controlled, crossover trial of prevention of clindamycin-induced gastrointestinal signs using a synbiotic in healthy research cats. J. Vet. Intern. Med. 31, 1406–1413.

Sugawara, G., Nagino, M., Nishio, H., et al., 2006. Perioperative synbiotic treatment to prevent postoperative infectious complications in biliary cancer surgery: a randomized controlled trial. Ann. Surg. 244, 706–714.

Taghizadeh, M., Asemi, Z., 2014. Effects of synbiotic food consumption on glycemic status and serum hs-CRP in pregnant women: a randomized controlled clinical trial. Hormones (Athens) 13, 398–406.

Talebi, A., Amani, A., Pourmahmod, M., et al., 2015. Synbiotic enhances immune responses against infectious bronchitis, infectious bursal disease, Newcastle disease and avian influenza in broiler chickens. Vet. Res. Forum 6, 191–197.

Tang, S.G.H., Sieo, C.C., Kalavathy, R., et al., 2015. Chemical compositions of egg yolks and egg quality of laying hens fed prebiotic, probiotic, and synbiotic diets. J. Food Sci. 80, C1686–C1695.

Tang, S.G.H., Sieo, C.C., Ramasamy, K., et al., 2017. Performance, biochemical and haematological responses, and relative organ weights of laying hens fed diets supplemented with prebiotic, probiotic and synbiotic. BMC Vet. Res 13, 248.

Tanner, S.A., Lacroix, C., Del'Homme, C., et al., 2015. Effect of *Bifidobacterium thermophilum* RBL67 and fructo-oligosaccharides on the gut microbiota in Göttingen minipigs. Br. J. Nutr. 114, 746–755.

Tayeri, V., Seidavi, A., Asadpour, L., et al., 2018. A comparison of the effects of antibiotics, probiotics, synbiotics and prebiotics on the performance and carcass characteristics of broilers. Vet. Res. Commun. 42, 195–207.

United States Department of Health and Human Services, 2011. Evidence Report/Technology Assessment Number 200, Publication No. 11-E007, AHRQ.

Urbanska, M., Szajewska, H., 2014. The efficacy of *Lactobacillus reuteri* DSM 17938 in infants and children: a review of the current evidence. Eur. J. Pediatr. 173, 1327–1337.

van den Nieuwboer, M., Claassen, E., Morelli, L., et al., 2014. Probiotic and synbiotic safety in infants under two years of age. Benef. Microbes 5, 45–60.

van den Nieuwboer, M., Brummer, R.J., Guarner, F., et al., 2015. The administration of probiotics and synbiotics in immune compromised adults: is it safe? Benef. Microbes 6, 3–17.

Villagrán-de la Mora, Z., Vázquez-Paulino, O., Avalos, H., et al., 2020. Effect of a synbiotic mix on lymphoid organs of broilers infected with *Salmonella typhimurium* and *Clostridium perfringens*. Animals (Basel) 10, 886.

Vitali, B., Ndagijimana, M., Cruciani, F., et al., 2010. Impact of a symbiotic food on the gut microbial ecology and metabolic profiles. BMC Microbiol. 10, 4. <https://doi.org/10.1186/1471-2180-10-4>.

Vlieger, A.M., Robroch, A., Van Buuren, S., et al., 2009. Tolerance and safety of *Lactobacillus paracasei* ssp. *paracasei* in combination with *Bifidobacterium animalis* ssp. *lactis* in a prebiotic-containing infant formula: a randomized controlled trial. Br. J. Nutr. 102, 869–875.

von Wright, A., 2005. Regulating the safety of probiotics—the European approach. Curr. Pharm. Des. 11, 17–23.

Wang, Y., Dong, Z., Song, D., et al., 2018. Effects of microencapsulated probiotics and prebiotics on growth performance, antioxidative abilities, immune functions, and cecal microflora in broiler chickens. Food Agric. Immunol. 29, 859–869. Available from: 10.1080/09540105.2018.1463972.

Wassenaar, T.M., Klein, G., 2008. Safety aspects and implications of regulation of probiotic bacteria in food and food supplements. J. Food Prot. 71, 1734–1741.

Yang, F., Hou, C., Zeng, X., et al., 2015. The use of lactic acid bacteria as a probiotic in swine diets. Pathogens 4, 34–45.

Yasuda, K., Hashikawa, S., Sakamoto, H., et al., 2007. A new synbiotic consisting of *Lactobacillus casei* subsp. *casei* and dextran improves milk production in Holstein dairy cows. J. Vet. Med. Sci. 69, 205–208.

Yan, F.F., Mohammed, A.A., Murugesan, G.R., et al., 2019. Effects of a dietary synbiotic inclusion on bone health in broilers subjected to cyclic heat stress episodes. Poult. Sci. 98, 1083–1089.

Yitbarek, A., Echeverry, H., Munyaka, P., et al., 2015. Innate immune response of pullets fed diets supplemented with prebiotics and synbiotics. Poult. Sci. 94, 1802–1811.

Zheng, M., Zhang, R., Tian, X., et al., 2017. Assessing the risk of probiotic dietary supplements in the context of antibiotic resistance. Front. Microbiol. 8, 908. Available from: 10.3389/fmicb.2017.00908.

Section V

Nutraceuticals in veterinary medicine

Chapter 67

Nutraceuticals in animal health and diseases

Ramesh C. Gupta, Robin B. Doss, Rajiv Lall, Ajay Srivastava and Anita Sinha

Introduction

Nutraceuticals have been used for thousands of years for the prevention and treatment of human and animal diseases. By the turn of the 21st century, the use of nutraceuticals in animals had become as popular as it was in humans. The worldwide market size for nutraceuticals in 2017 was 383 billion US dollars, and it is expected to reach more than $561 billion by 2023. In the United States, the market value of nutraceuticals was worth approximately $72 billion in 2017, and is projected to reach more than $133 billion by 2025. Nutraceuticals are used in animals against diseases of the central nervous system (CNS), cardiovascular, respiratory, musculoskeletal, hepatic, renal, immune, and other systems. Some diseases are more prevalent in certain species than others, such as osteoarthritis in horses and dogs, and type II diabetes in dogs and cats, and therefore nutraceuticals for these diseases are discussed accordingly. This chapter briefly describes some commonly used nutraceuticals in animals for the prevention and treatment of diseases.

Nutraceuticals in cognitive dysfunction

Like chronic neurodegenerative Alzheimer disease (AD) in humans, canines and felines suffer from impairment of cognition and memory as they age. The prevalence of canine cognition dysfunction (CCD)/cognitive dysfunction syndrome (CDS) is high. According to a study from the University of California Davis, 28% of dogs aged 11–12 years and 68% of dogs aged 15–16 years showed one or more signs of cognitive impairment (Neilsen et al., 2001). The prevalence of CDS in cats was found to be 36% in a population of 11-to-21-year-old cats (Moffat and Landsberg, 2003).

The underlying mechanisms in AD or CCD are complex. Some of the important mechanisms include (1) a decline in acetylcholine (ACh), (2) decreases in acetylcholinesterase (AChE) and choline acetyltransferase (ChAT) activities, (3) increases in *N*-methyl-D-aspartate (NMDA) and NMDA receptors, (4) oxidative/nitrosative stress, (5) neuroinflammation, (6) increase in butyrylcholinesterase (BuChE) activity, and (7) microvascular changes (reviewed in Gupta et al., 2019a). Currently, a number of therapeutic drugs, such as selegiline, gabapentin, buspirone, and memantine, are available for the treatment of CCD/CDS. Dogs and cats with cognition and memory impairment/loss are also treated with bioactive compounds from plant or animal origins. These bioactive compounds possess anticholinesterase (anti-AChE), antioxidative, anti-inflammatory, and immunomodulatory activities, as well as anti-Aβ aggregation and deposition in the brain (Gupta et al., 2019a). Some of these bioactive compounds have been shown to restore mitochondrial function and normalize the blood–brain barrier (BBB). Some phytoconstituents are also known to inhibit monoamine oxidase B (MAO-B) activity. Although neurodegenerative changes occur in many regions of the brain, in the context of memory, the cerebral cortex and hippocampus are of particular importance. Accumulation of $A\beta_{1-42}$ peptide and hyperphosphorylation of tau, leading to the formation of senile plaques and neurofibrillary tangles in the cortex and hippocampus, are the two pathological hallmarks of AD in humans and CDS in canines and felines.

Diagnosis of CCD/CDS is often based on clinical signs with the acronym DISHA (*D*isorientation, *I*nteraction changes, *S*leep/awake cycle, *H*ouse soiling, *A*ctivity level changes). Dogs with CCD commonly display clinical signs such as aimless wandering, staring blankly into space, avoiding being petted, and difficulty with finding dropped food. Biomarkers, such as plasma concentrations of $A\beta_{1-40}$ and $A\beta_{1-42}$ and the $A\beta_{1-42}$/$A\beta_{1-40}$ ratio, aid in the diagnosis of CCD/CDS.

A large number of molecules and extracts from plants and other biological origins have potential for managing

Nutraceuticals. DOI: https://doi.org/10.1016/B978-0-12-821038-3.00067-7

and treating CDS, and some of these are summarized in Table 67.1.

Many other substances (such as calcium-buffering proteins, ascorbic acid, lycopene, vitamin E, omega fatty acids, phosphatidylserine, s-adenosylmethionine, medium-chain triglycerides, etc.) have also been shown to improve cognition and memory (reviewed in Gupta et al., 2019a)

Nutraceuticals in cardiovascular diseases

Structural and functional abnormalities of the heart and blood vessels often lead to cardiovascular diseases (CVDs), such as coronary heart disease (CHD), heart attack (myocardial infarction), atherosclerosis, thrombosis, dyslipidemias, hypertension, hypotension, bradycardia, tachycardia, stroke, and heart failure. Nutraceuticals have potential for use in CVDs (Alissa and Fern, 2012; Sosnowska et al., 2017; Alves et al., 2019; Aquila et al., 2019; Colitti et al., 2019; Nikhra, 2019; Zoltani, 2019; Li et al., 2020; Shaito et al., 2020). Interestingly, nutraceuticals also have applications in statin-intolerant patients (Banach et al., 2018). Many medicinal drugs have been extracted from plants, such as aspirin from the *Salix alba* L. tree, digoxin from *Digitalis purpurea*, ephedrine from *Ephedra sinica*, lovastatin from *Monascus purpureus* L., taxol from *Taxus brevifolia*, reserpine from *Rauvolfia serpentine*, and many others (Cragg and Newman, 2013; Shaito et al., 2020). Table 67.2 lists plants/nutraceuticals that can be indicated in the prevention and treatment of CVDs.

Nutraceuticals can improve cardiovascular health by multiple mechanisms: (1) polyphenols and other antioxidants improve arterial function and thereby prevent arterial diseases, (2) flavonoids, proanthocyanidins, xanthones, and secoiridoids inhibit angiotensin-converting enzyme (ACE), thereby regulating arterial blood pressure and strengthening tiny capillaries that carry oxygen to all cells, and (3) polyunsaturated fatty acids (PUFAs), vitamins (A, D, and C), and minerals (zinc) may prevent and ameliorate CVDs (Hodgson et al., 2014; Zoltani, 2019). For further details on efficacy, mechanisms, applications, safety, and toxicity of nutraceuticals in the CV system, readers are referred to recent publications (Brown, 2017a; Jain et al., 2018; Poli and Visioli, 2019; Zoltani, 2019; Shaito et al., 2020).

Nutraceuticals in diabetes and metabolic syndrome

Currently, there are more than 93 million dogs and 96 million cats in the United States. One in every 400–500 dogs and one in every 100–200 cats suffer from diabetes. Dogs of certain breeds (dachshunds, schnauzers, poodles, Cairn and Australian terriers, springer spaniels, keeshonds, samoyeds, and golden retrievers) are at higher risk of developing this disease. Complementary and alternative medicines (CAMs) are used worldwide for the management of diabetes in humans (Amaeze et al., 2018; Grossman et al., 2018; Mekuria et al., 2018; Peltzer and Pengpid, 2019; WHO, 2019; Alqathama et al., 2020) and animals (Gupta et al., 2019b). In general, adult dogs and cats develop type II diabetes mellitus (T2DM), and the morbidity and mortality rates are high. Proportionately, more female dogs and male cats develop diabetes than male dogs and female cats. T2DM is characterized by pancreatic β-cell dysfunction. This causes hyperglycemia and may lead to several secondary complications, such as nephropathy, neuropathy, retinopathy, cardiopathy, pancreatitis, and wounds. Recently, Gupta et al (2019b) described the pathophysiology of T2DM in dogs, cats, humans, and experimental animals.

Diabetes in dogs and cats is usually managed by dietary choices, moderate exercise, and medication (including insulin injections). Compared to costly medicines and their side effects, as well as invasive insulin therapy, the use of nutraceuticals appears to provide an alternative option. Some nutraceuticals and plant extracts with antidiabetic potential are listed in Table 67.3.

In general, nutraceuticals may provide antidiabetic effects via multiple mechanisms: (1) increased insulin secretion and insulin sensitivity, (2) reduced gluconeogenesis, (3) activated glycogenolysis, (4) increased hepatic glycolysis, (5) increased hepatic glycogen synthesis, (6) adrenomimeticism, (7) pancreatic β-cell potassium channel blocker activity, (8) activated cyclic adenosine monophosphate (cAMP), and (9) modulation of glucose absorption from the intestine (Hanhineva et al., 2010; Patel et al., 2012; Gupta et al., 2017).

Periodontal health and diseases

Periodontal diseases are commonly encountered in dogs and cats due to a lack of oral health hygiene, drug resistance, and rising veterinary care costs. According to the American Veterinary Dental College, 80% of dogs and 70% of cats suffer from periodontal diseases, such as sore, swollen, bleeding gums, infected teeth, caries, plaque buildup, and tartar. Periodontitis can be defined as an inflammatory process, initiated by the formation of plaque biofilm, which leads to loss of periodontal attachment to the root surface and adjacent alveolar bone and may ultimately result in tooth loss (Muhammad and Lawal, 2010; Gupta et al., 2019c). In both gingivitis and periodontitis, inflammation is the result of microorganisms present in dental plaque (Newman et al., 2006; Gupta et al., 2019c). In periodontitis, oxidative stress due to excessive generation of reactive oxygen species (ROS) and reduced antioxidant capacity induces inflammation (Soni et al., 2012;

TABLE 67.1 Extracts of plants or other biologicals, their ingredients, and their effects on cognitive dysfunction.

Biological source	Active ingredient(s)	Major effects
Anchovy peptides	Anchovy protein hydrolysate (APH), and peptide (PAYCS and CVGSY)	Antioxidative, anti-AChE, antiapoptotic, alleviates memory loss
Astragalus mongholicus Bunge	Astragaloside IV	Antioxidative, anti-inflammatory, antiapoptotic, promotes hippocampal neurogenesis and cognition
Polygonum hydropiper L.	Flowers—caryophyllene; leaves—decahydronaphthalene	Antioxidative, anti-AChE and anti-BuChE
Bacopa monnieri (Brahmi)	Bacopasides	Antioxidative, antilipid peroxidation, anti-AChE, neuroprotective, elevation of neurotransmitters (ACh, serotonin, GABA, and glutamate) in the hippocampus
Berberis vulgaris, Berberis aristata, B. aquifolium, Hydrastis canadensis, Coptis chinensis, Argemone mexicana	Berberine	Antiapoptotic, neuroprotective, antioxidative, antiinflammatory, analgesic, immunomodulatory, normalizes AChE, ChAT, SOD, and MAO-B activities, and anti-AD and anti-PD
Chinese club moss (*Huperzia serrata*)	Huperzine A	Anti-AChE, NMDA receptor antagonist, neuroprotective, anti-AD in humans and anti-CDS in dogs
Piper species: *P. austrosinense, P. puberulum, P. flaviflorum, P. betle, P. hispidinervum*	Asaricin	Anti-AChE, neuroprotective, and anti-AD
Angelica archangelica	Seeds—imperatorin, xanthotoxin	Anti-AChE
Curcuma longa (turmeric)	Curcuminoids (curcumin, demethoxycurcumin, and bisdemothoxycurcumin)	Antiinflammatory, antioxidative, cytoprotective, reduces amyloids and plaques, improves cognition and memory
Ginseng: American ginseng, Chinese ginseng, Japanese ginseng, Korean ginseng, Vietnamese ginseng	Ginsenosides (Rb1, Rc, Rg1, Rg2, and Rg3)	Cognition and memory stimulant, reduces β-amyloid peptide in animal brain
Ginkgo biloba	Terpene trilactones (ginkgolides A, B, and C, and bilobalide), ginkgolic acid, and flavonoid glycosides (quercetin, kaempferol, and isorhamnetin)	Antiapoptotic, antioxidative, anti-AChE, improves mitochondrial function and ATP production, improves cerebral blood flow, improves cognition and memory, antidementia
Glycyrrhiza glabra, Glycyrrhiza uralensis, Cyamopsis psoralioides, Psoralea corylifolia	Multiple ingredients	Inhibit MAO-B activity, enhance catecholamine neuroactivity, manage symptoms of dementia and sleep–wake disorders
Melatonin	Melatonin	Maintains dendritic complexity and improves neurogenesis, improves sleep–wake pattern in dementic dogs
Monsonia angustifolia	Justicidin A, 5-methoxyjusticidin A, chinensinaphthol, retrochinensinaphthol methyl ester, and suchilactone	Reduces formation of Aβ, neuroprotective, anti-AD in humans, and anti-CDS in canines and felines
Quercetin	Quercetin	Antioxidative, free radical scavenging, antiinflammatory, improves mitochondrial function, improves learning and memory, and prevents AD

(Continued)

TABLE 67.1 (Continued)

Biological source	Active ingredient(s)	Major effects
Resveratrol	*Cis-/trans*-resveratrol	Increases mitochondrial biogenesis, antiamyloidogenic, influences SIRT1 and MAPK signaling pathways, neuroprotective, prevents AD
Rosmarinus officinalis, Salvia miltiorrhiza	Rosmarinic acid	Antioxidative, antiinflammatory, antiproliferative, cytoprotective, neuroprotective, antidepressant, and anti-AD
Royal jelly proteins	Proteins, peptides, free amino acids, sugars, lipids, and vitamins	Antioxidative, antiinflammatory, immunomodulatory, improvement of cognition and memory
Saffron/red gold (*Crocus sativus* L.)	Crocin, crocetin, dimethylcrocetin, safranal	Antioxidative, antiinflammatory, anti-AChE, antiamyloidogenic, prevents Aβ fibril and plaque formation and aggregation, antinociceptive, improves cognition, anti-AD in humans and anti-CDS in pets
Soybean isoflavone and peptides	Soy isoflavone genistein and peptides	Anti-inflammatory, immuno-stimulatory, antiapoptotic, upregulates neurotrophic factors (NGF, BDNF, and NT-3), neuroprotective, promotes neurogenesis, improves learning, memory, and cognition

ACh, acetylcholine; *anti-AChE*, anticholinesterase; *AD*, Alzheimer disease; *BDNF*, brain-derived neurotrophic factor; *BuChE*, butyrylcholinesterase; *CDS*, cognitive dysfunction syndrome; *MAO*, monoamine oxidase, *MAPK*, 5-adenosine monophosphate-activated protein kinase; *NMDA*, *N*-methyl-D-aspartate; *NGF*, nerve growth factor; *NT-3*, neurotrophin-3; SIRT1, sirtuins 1; *SOD*, superoxide dismutase.

TABLE 67.2 Plant extracts/nutraceuticals that can be indicated in cardiovascular health and diseases.

Plant/nutraceutical	Active ingredient(s)	Mechanism of action	Major effects
Acacia nilotica	Terpenoids, tannins, saponins, niloticane, lupenone, polyphenols, and flavonoids	Antiplatelet aggregatory, antioxidative, antiinflammatory, antispasmodic, hypolipidemic, and anticholesterolemic	Antihypertensive, improves lipid profile
Allium sativum (Garlic)	Allicin, ajoene, S-allylcysteine, S-allyl mercapto-L-cysteine	Antioxidative, antiinflammatory, and reduces total cholesterol and low-density lipoprotein (LDL)	Antihypertensive, antiatherosclerosis, antithrombotic, improves lipid profile
Argemone mexicana, Berberis spp., *Coptis* spp., *Hydrastis canadensis*	Berberine	Increases action potential, positive inotropic activity, mild systemic vasodilation, and cardiac output, blockage of K^+ channels and stimulation of Na^+–Ca^+ exchanger, inhibits adenylyl cyclase activity, and tyrosinehydroxylase activity	Prevents ventricular fibrillation, tachycardia, and sudden coronary death after myocardial ischemia, antiarrhythmic, antihypertensive, and improves lipid profile
Astragalus membranaceous	Astragaloside IV	Antioxidative, positive inotropic effect	Protects against ischemia-associated cardiovascular diseases (CVDs)–
Carotenoids	Lycopene	Antioxidative, and decreases LDL	Reduces systolic blood pressure
Coenzyme Q10	Coenzyme Q10	Increases mitochondrial function and ATP production, and improves endothelial cells function	Prevents congestive heart failure
Colostrum	Proline-rich peptide	Immunostimulant	Regenerates cardiac tissue, cardioprotective

(*Continued*)

TABLE 67.2 (Continued)

Plant/nutraceutical	Active ingredient(s)	Mechanism of action	Major effects
Crocus sativus L. (saffron)	Crocin, crocetin, safranal, sterols (campesterol, stigmasterol, and β-sitosterol)	Increases nitric oxide (NO) by endothelial Nitric Oxide Synthase (eNOS) activation, calcium channel blocker, antioxidative, and anti-inflammatory	Vasodilation, antihypertensive, and antiatherosclerosis
Curcuma longa (turmeric)	Curcumin (enol- and keto- form), demethoxycurcumin, bisdemethoxycurcumin	Antioxidative, antiinflammatory, and antiplatelet aggregatory	Antihypertensive, antiatherosclerosis, and prevents coronary heart disease
Digitalis purpurea	Digoxin	Increases vagal tone, inotropy, and cardiac output	Strengthens heart beat and treats atrial fibrillation, supraventricular tachycardias, and heart failure
Ganoderma lucidum (Reishi or Lingzhi)	Ganoderic acid A, ganoderenic acid, ganoderiol, ganodermadiol, β-glucan, coumarin, lucidadiol	Immunomodulatory, antioxidative, antiproliferative, antiapoptotic, improves lipid profile, hypocholesterolemic, and angiotensin-converting enzyme (ACE) inhibitory	Improves endothelial and cardiac functions, cardioprotective, and treatment of CVDs, antihypertensive
Ginkgo biloba	Terpenoids (ginkgolides A, B, C, J, and M, and bilobalide), ginkgolic acids, flavonoids (quercetin, kaempferol and isorhamnetin), terpene trilactone	Antioxidative, antilipoperoxidative, antiinflammatory, inhibits platelet activating factor (PAF) at PAF receptor, protects mitochondria, inhibits cyclooxygenase and lipoxygenase activities, vasodilation, and immunostimulating	Improves blood circulation, antihypertensive, antithrombotic, and cardioprotective
Ginseng (*Panax ginseng, P. notoginseng, P. japonicas, P. quinquefolium, P. vietnamensis, P. trifolius* L.)	Ginsenosides (protopanaxadiol and protopanaxatriol)	ACE inhibitory, antioxidative, suppresses LDL oxidation, reduces platelet aggregation, reduces LDL and increases high density lipoprotein (HDL), increases NO and cyclic guanosine monophosphate (cGMP), and blocks calcium channels causing vasorelaxation, ameliorates mitochondrial dysfunction	Antiatherosclerosis, improves lipid profile, and antihypertensive
Grape extract	Polyphenols	Lowers LDL cholesterol and total cholesterol, and elevates HDL cholesterol	Antihypertensive
Green tea	Catechins (epicatechin, epigallocatechin, epigallocatechin gallate, and epigallocatechin-3-gallate)	Inhibits NF-κB, activates eNOS, suppresses platelet adhesion, antioxidative, antiinflammatory, and lowers LDL cholesterol and total cholesterol	Antihypertensive, antiatherosclerosis, and cardioprotective
Gynostemma pentaphyllum	Gypenosides, flavonoids	Antioxidative, antiinflammatory, antihyperlipidemic, antiplatelet aggregatory	Antihypertensive, antiatherosclerotic, and cardioprotective
Haematococcus pluvialis	Astaxanthin	Antioxidative, suppresses lipid peroxidation, and improves mitochondrial membrane potential	Antihypertensive, antithrombotic, antiatherosclerosis, and improves lipid profile

(*Continued*)

TABLE 67.2 (Continued)

Plant/nutraceutical	Active ingredient(s)	Mechanism of action	Major effects
Hawthorn (*Crataegus oxyacantha*)	Anthocyanidins and proanthocyanidins	Antioxidative, improves integrity of arteries and veins, antihypercholesteremia, and enhances circulation	Antihypertensive, controls arrhythmias, and antiatherosclerosis
Hibiscus sabdariffa (Roselle)	Anthocyanins (hibiscin and chrysanthenin), gossypicyanin, flavonoids (hibiscitrin, sabdaritrin, gossypitrin, quercetin, luteolin, and β-sitosterol)	ACE inhibitory, antioxidative, decreases LDL-C, total cholesterol, and triglycerides, and increases HDL-C	Antihypertensive, antiatherosclerosis, anticoronary heart disease, lowers heart rate
Papaver somniferum	Papaverine	Antispasmodic and coronary vasodilation	Antihypertensive
Plant sterols/stanols	Sterols/stanols	Lowers LDL-C and total cholesterol	Prevention of CVDs, including myocardial infarction
Pomegranate	Polyphenols	Antioxidative and lowers lipids	Antihypertensive, antiatherosclerosis, and improves cardiovascular health
Coenzyme Q10	Coenzyme Q10	Lowers total cholesterol, LDL-C, and triglycerides	Improves lipid profile and cardiovascular health
Rauvolfia serpentina	Reserpine	Irreversibly blocks the H^+-coupled vesicular monoamine transporters (VMAT1 and VMAT2), and reduces uptake of monoamine neurotransmitters (norepinephrine, dopamine, serotonin and histamine)	Antihypertensive and improves lipid profile
Resveratrol	Resveratrol	Activates AMPK, SIRT1 and Nrf2, decreases cytokines, upregulates eNOS, antioxidative, antiinflammatory, immunomodulatory, antiplatelet aggregatory, antithrombotic, and improvement of endothelial functions	Cardioprotective against atherosclerosis, hypertension, stroke, myocardial infarction, and heart failure
Rosemarinus officinalis (rosemary)	Rosmarinic acid, caffeic acid, carsonic acid, and carnosol	Antioxidative, and antiinflammatory	Cardioprotective
Salvia miltiorrhiza (red sage)	Tanshionones, danshen, and salvianolic acid B	Scavenges free radicals, and antioxidative	Protects cardiovascular system
Seaweed	Omega-3 and omega-6 fatty acids	Antioxidative, antiinflammatory, lowers total cholesterol, LDL, and total triglycerides	Antihypertensive, dyslipidemia, and cardioprotective
Soy	Isoflavone genistein, and PUFAs	Inhibits LDL oxidation and lowers plasma lipids	Improves vascular functions and cardiac health, antiatherosclerosis, improves lipid profile, and normalizes blood pressure
Spirulina (*Arthrospira platensis* and *Arthrospira maxima*)	γ-Linolenic acid, eicosapentaenoic acid, docosahexaenoic acid, arachidonic acid, c-phycocyanin, zeaxanthin, and insulin-like protein	Antioxidative, antiinflammatory, inhibition of lipid peroxidation, improves endothelial function, lowers LDL-C and total cholesterol, and immunostimulatory	Antiatherosclerosis, antihypertensive, and improves lipid profile

TABLE 67.3 Plant extracts/nutraceuticals with anti-diabetic potential.

Plants/nutraceutical	Active ingredient(s)	Major effects
Argemone maxicana, Berberis spp., *Coptis* spp., *Hydrastis canadensis*	Berberine	Hypoglycemic and antidiabetic
Barleria prionitis Linn.	Barlerinoside	Antihyperglycemic and antidiabetic
Bitter melon (*Momordica charantia*)	Momordin I, momordin II, alkaloids, saponins, and triterpenes	Antihyperglycemic, hypoglycemic, and antidiabetic
Bovine casein hydrolysate	Leucine, and 11 casein medium-sized peptides	Insulinotropic and antihyperglycemic
Butea monosperma	Butein, butin, butrin, isobutrin, and palastrin	Hypoglycemic and antidiabetic
Catharanthus roseus	Catharanthine, ursolic acid, daucosterol, β-sitosterol, lochnericine, quercetin, and kaempferol	Antidiabetic
Chromium trivalent	Chromium trivalent	Antihyperglycemic and antidiabetic
Cinnamon (*Cinnamomum zeylanicum*)	(−)-Cinnamaldehyde	Antidiabetic and increases insulin secretion and insulin sensitivity
Citrus cinensis	Flavonoid diosmin	Antihyperglycemic, antidiabetic, insulinotropic, antioxidative, and antiinflammatory
Costus pictus D and *Costus igneus* Nak (insulin plants)	Diosgenin, lupeol, quercetin, ergastanol, and stigmasterol	Antioxidative and anti-diabetic
Fenugreek (*Trigonella foenum-graecum* L.)	Diosgenin, galactomannan,4-hydroxyisoleucine, trigonelline, gentianine, and carpaine	Antihyperglycemic, increased insulin secretion, decreased insulin resistance, decreased gluconeogenesis, antihyperlipidemic, antioxidative, and antiinflammatory
Fish oil	Eicosapentaenoic acid and docosahexaenoic acid	Antihyperglycemic and improves insulin sensitivity
Hippophae rhamnoides L.	Proanthocyanidins, isorhamnetin, and quercetin	Hypoglycemic
Mangifera indica	Antioxidants	Antioxidative, antiinflammatory, immunomodulatory, and antidiabetic
Moringa oleifera Lam.	Quercetin-3-glycoside, rutin, and kaempferol	Antidiabetic, antihyperglycemic, α-amylase inhibiting, and α-glucosidase inhibiting
Panax ginseng, Panax notoginseng, Panax japonica, Panax vietnamensis, Panax quinquefolius	Ginsenosides (Rb2, Re, and Rh2), (20R)-protopanaxadiol, protopanaxatriol, pyroglutamic acid, and peptidoglycan panaxan B	Antidiabetic, hypoglycemic, and antihyperglycemic
Polygonum multiflorum Thunb	Tetrahydroxy stilbene glucoside	Balances gut microbiota, alleviates insulin resistance, and exerts antihyperglycemic and antidiabetic effects
Rehmannia glutinosa	Catalpol	Antihyperglycemic, antidiabetic, antiinsulin resistance, antigluconeogenesis, elevates glycogen synthesis, antioxidative, and antiapoptotic
Resveratrol	Resveratrol	Antihyperglycemic, antidiabetic, and antioxidative
Royal jelly	Royal jelly proteins	Antihyperglycemic
Shilajit	Dibenzo-α-pyrones, chromoproteins, and fulvic acid	Antioxidative and antihyperglycemic
Sophora tonkinensis	Flavonoids	Antidiabetic

(Continued)

TABLE 67.3 (Continued)

Plants/nutraceutical	Active ingredient(s)	Major effects
Stevia (*Stevia rebaudiana* Bertoni)	Stevioside, steviolbioside, rebaudiosides (A, B, C, D, and E), and dulcoside A	Antidiabetic, insulinomimetic, antihyperglycemic, antioxidative, immunomodulatory, hepatoprotective, renoprotective, and antihypertensive
Terminalia arjuna	Phytosterol, flavonoids, lactones, tannins, glycosides, and phenolics	Antidiabetic
Turmeric (*Curcuma longa*)	Curcumin (enol- and keto- forms), demethoxycurcumin, and bisdemethoxycurcumin	Antioxidative, antidiabetic, antiinflammatory, and antihyperlipidemic
Virgin olive oil	Tyrosol, hydroxytyrosol, and secoiridoids	Antidiabetic, antioxidative, antiinflammatory, antiinsulin resistance, and antimitochondrial dysfunction
Zingiber officinale	Gingerol, gingerenone A, shogaols, paradol, 1-dehydro-10-gingerdione	Antidiabetic, antidiabetic nephropathy, antioxidative, and antiinflammatory

TABLE 67.4 Plant extracts, their ingredients, and major effects in periodontal diseases

Plants	Active ingredient(s)	Major biological/pharmacological effects
Abrus precatorius Linn.	Abrusogenin, stigmasterol, β-sitosterol, monoglycerides, and triglycerides	Antibacterial, antifungal, dentifrice, and strengthening of gums and teeth
Acasia arabica Wild.	Polyphenols, catechin, gallic acid, chlorogenic acid, and robidandiol	Antioxidative, antimicrobial, antiviral, antifungal, antigingivitis, and antiplaque
Acacia nilotica (babool/thorny acacia)	Androstene steroid, D-pinitol, kaempferol, gallic acid, (+/−)catechin, (−)-epigallocatechin, and rutin	Antioxidative, antibacterial, antiinflammatory, antifungal, antigingivitis, and antiplaque
Allium sativum L.	Allicin (diallyl disulfide sulfoxide), alliin (S-allyl cysteine sulfoxide), and ajoene	Antitoothache
Aloe vera L.	Salicylic acid, enzymes, vitamins, minerals, and fatty acids	Dentifrice, antigingivitis, and antiplaque
Areca catechu Linn.	Arecoline, guvacoline, procyanidins, catechin, and epicatechin	Dentifrice
Azadirachta indica (Neem)	Gallotanins	Antibacterial and antiplaque
Calendula officinalis L. (pot marigold)	Auroxanthin, flavoxanthin, carotenoids, lycopene, oligoglycosides, glucoside, lutein, quercetin, saponin, rutin, sesquiterpene, and triterpine	Antioxidative, antiinflammatory, immunostimulatory, antimicrobial, antifungal, antigingivitis, antiplaque, and wound healing
Cannabis sativa L., *Cinnamomum* spp., *Origanum vulgare* L., *Piper nigrum* L.	β-Caryophyllene	Antiinflammatory, antibacterial, and antiplaque
Ceylon cinnamon powder	Cinnamaldehyde, cinnamyl acetate, and cinnamyl alcohol	Antiinflammatory, and antimicrobial
Chamomile (*Matricaria chamomilla* L.)	Chamazulene, α-bisabolol, apigenin, luteolin, matricin, quercetin, and umbelliferone	Antioxidative, antiinflammatory, antiseptic, fungistatic, antisore gum, and antiperiodontitis
Chitosan	Chitosan	Antibacterial, antifungal, antigingivitis, and antiplaque
Clove (*Syzygium aromaticum*)	Eugenin, eugenol, acetyl eugenol, eugenitin, bicornin, β-caryophyllene, crategolic acid, caempesterol, vanillin, and sesquiterpene	Antioxidative, antiinflammatory, immunostimulatory, antibacterial, antiviral, antigingivitis, and antitoothache

(Continued)

TABLE 67.4 (Continued)

Plants	Active ingredient(s)	Major biological/pharmacological effects
Commiphora myrrha (Myrrh)	Sesquiterpenes (furanoeudesma-1,3-diene, and cuzarene)	Antiinflammatory, astringent, antispasmodic, antibacterial, antipyorrhea, analgesic, and antigum diseases and mouth ulcers
Echinacea purpurea (purple coneflowers)	Chicoric acid, alkylamides, caffeic acid, glycoproteins, and polysaccharides	Antiinflammatory, immunostimulatory, antimicrobial, analgesic, antigingivitis, and antiperiodontitis
Eucalyptus extract/ eucalyptol oil	Borneol, 1,8-cineol, globulol, *p*-cymene, limonene, α-pinene, cryptone, and α-terpineol	Antibacterial, antigingivitis, antigum bleeding, and antiplaque
Green tea	Epicatechin, epicatechin gallate, epigallocatechin, (−)-epigallocatechin-3-gallate	Antioxidative, antiinflammatory, antibacterial, preventing bone resorption, and anticarcinogenic
Hyaluronic acid	Hyaluronic acid	Bacteriostatic and periodontal regeneration
Mentha piperita L. (Peppermint)	Menthol, menthone, menthyl acetate, 1,8-cineole, limonene, β-pinene, β-caryophyllene, and flavonoids	Antioxidative, antiinflammatory, analgesic, antigingivitis, antiulcer, cytoprotective, and chemopreventive
Pau D'Arco	Quinods, quercetin, benzenoids, flavonoids, α-lapachone, β-lapachone, lapachol, lapachenol, deoxylapachol, tabebuin, anisic acid, and anthraquinones	Antiinflammatory, antimicrobial, anticancer, antiviral, antifungal, and antiabscess
Propolis (*Apis mellifera* L.)	Chalcones, flavonoids, phenolic acids, terpenes, and aromatic acids	Antioxidative, antimicrobial, antitartar, antiplaque, cavity cleaning, and wound healing
Royal jelly (Rojelexin, Vets Plus, Inc.)	10-Hydroxy-δ2-decenoic acid, apismin, royalactin, royalisin, jelleines, and apolipophorin III	Antibacterial, antifungal, and antiplaque
Salvia officinalis L. (Sage)	A-thujone, β-thujone, camphor, α-pineno, borneol, chlorogenic acid, cornsole, cornsolic acid, and tannic acid	Antibacterial, antifungal, anticandidal, anticaries, antigingivitis, antiplaque, and astringent
Tea tree oil (*Melaleuca alternifolia*)	Terpine-4-ol, γ-terpinene, 1,8-cineole, α-cymene, δ-cadinene, and globulol	Antibacterial
Thyme (*Thymus vulgaris*)	Thymol	Antioxidative, antimicrobial, antifungal, antigingivitis, anticavity, and antiplaque

Lima et al., 2017). Periodontitis can be classified into four stages:

Stage 1. Gingivitis only, without attachment loss; but the height and architecture of the alveolar margin are normal.

Stage 2. Early periodontitis with <25% of attachment loss or, at most there is a stage 1 function involvement in multirooted teeth. There are early radiographic signs of periodontitis.

Stage 3. Moderate periodontitis, with 25−50% of attachment loss, and there is *stage 2* involvement in multirooted teeth.

Stage 4. Advanced periodontitis, with >50% of attachment loss, and there is *stage 3* function involvement in multirooted teeth.

A large number of plant products have been claimed to have potential for use in periodontal diseases (Kumar et al., 2015). Many of these products exert antioxidative, antiinflammatory, antiseptic, and antimicrobial activities. Ganesan (2008) reported that plant leaves are maximally used in oral care (25.44%), followed by root (20.17%), seed/nut/fruits (18.42%), bark (14.03%), stem (12.28%), whole plant (9.65%), and gum/latex (8.77%). Plant-based products are used in various forms such as mouthwash, paste, or powder to relieve toothache, gingivitis, ulcers, gum bleeding, etc. Some nutraceuticals and plant extracts, which can be used for various periodontal diseases in dogs and cats are summarized in Table 67.4.

Biological or phytoconstituents, such as alkaloids, flavonoids, essential oils, and tannins exert antioxidative, antiinflammatory, immunomodulatory, antimicrobial, antiseptic, antiviral, and antifungal activities (Gupta et al.,

2019c). These properties help improve oral hygiene and prevent gum diseases, tooth decay, and periodontitis.

Nutraceuticals for hepatic diseases

The liver plays a vital role in nutrient homeostasis, filtration of particulates, protein synthesis, bioactivation and detoxication, formation of bile, and biliary secretion and excretion (Jaeschke, 2008). There are a number of risk factors (age, gender, dose, malnutrition, genetics, concomitant drug exposure, etc.) for acute or chronic liver injury. Biomarkers such as alanine aminotransferase, aspartate aminotransferase, alkaline phosphatase, gamma-glutamyltransferase, total bilirubin, and microRNAs play a significant role in determining the degree of liver injury (Kaur, 2019). The liver has an extensive capacity for regeneration and repair following acute or chronic insult. The extent of regeneration depends on the balance between a variety of hepatocellular factors including cell cycle regulation, metabolism, angiogenesis, cell adhesion, and extracellular matrix components (Duarte et al., 2015). Nutraceuticals can prevent or treat hepatic diseases/toxicosis via multiple mechanisms, such as antioxidative, antiinflammatory, immunomodulatory, antiproliferative, antifibrotic, and antineoplastic (Gwaltney-Brant, 2019). It needs to be mentioned that the liver is also one of the most sensitive organs in the body for the toxic effects of many nutraceuticals, such as *Aloe vera*, arrowroot, black cohosh, cascara, chapparal, comfrey, gota kolu, green tea extract, groundsel, kava, rattlebox, senna, usnic acid, valerian, and *Garcinia cambogia* (Teschke et al., 2014; Brown, 2017b; Gupta et al., 2018; Yousaf et al., 2019). Nutraceuticals that may prevent or treat liver diseases are summarized in Table 67.5.

It can be suggested that by having antioxidative, anti-inflammatory, antifibrotic, and many other properties, nutraceuticals and phytochemicals may inhibit or minimize acute injury induced by endogenous and exogenous hepatotoxic agents or they may prevent or slow the progression of chronic liver disorders, such as hepatic lipidosis, chronic hepatitis, liver storage diseases, and hepatic carcinogenesis (Gwaltney-Brant, 2019).

Nutraceuticals for wound healing

Wounds are usually due to acute or chronic physical injuries (Eming et al., 2002; Ali-Seyed and Vijayarahavan, 2019). Acute wounds are generally from surgery, bites, burns, minor cuts, abrasions, or gunshots. Chronic wounds are often caused by local infection, hypoxia, trauma, foreign bodies, diabetes mellitus, malnutrition, immunodeficiency, or medications (Menke et al., 2008). Wound healing is a complicated process initiated in response to an injury that restores the function and integrity of the damaged tissue (Wietecha and Dipietro, 2013). Wound healing takes place in four phases: (1) hemostasis, (2) inflammation, (3) proliferation, and (4) remodeling (Burford et al., 2007; Ali-Seyed and Vijayarahavan, 2019). Each phase is characterized by the infiltration of specific cell types, in addition to cytokines and growth factors, into the wound site. Epithelialization, antioxidant defense, granulation, nonvascularization, and wound contraction are also involved in wound healing. For details of wound healing pathophysiology, readers are referred to a recent publication by Ali-Seyed and Vijayarahavan (2019).

Nutraceutical formulations, especially from aromatic plants, have been used around the world for treating wounds and burns for a long time. The essential oils from various plants are also very effective in treating small to medium-sized wounds, skin abrasions, and skin infections. Some common wound-healing plants are listed in Table 67.6.

Various plant extracts and nutraceuticals have been reported to heal wounds via multiple mechanisms, such as upregulation of vascular endothelial growth factor (VEGF) and transforming growth factor $\beta1$ (TGF-$\beta1$), activation of NF-κB and IL-8, expression of iNOS and $\alpha1$ type 1 collagen, and antioxidant activity (Mirmalek et al., 2016; Shen et al., 2017; Ali-Seyed and Vijayarahavan, 2019). Interestingly, the aqueous extract of Korean red ginseng stimulates angiogenesis through activation of the phosphatidylinositol 3-kinase (PI3K)/Akt pathways (Kim et al., 2007). Finally, since infection is often an impediment to wound healing, nutraceuticals with antimicrobial potential appear to be a critical component of the healing process.

Mastitis

Mastitis can be defined as an inflammatory condition of the mammary glands or the udder in some species. There is widespread use of antibiotics as prophylactics in dry cows and therapeutics in mastitis (Coppock, 2019). Due to the resistance of microbial pathogens to antimicrobial drugs, treatment of mastitis has become a major public health issue. The pattern of antibiotic resistance has been found to be bacterial species-dependent. For example, *Staphylococcus gallinarum* is resistant to quinupristin/dalfopristin combination at a prevalence of 98%. *S. cohnii* and *S. arlettae* are resistant to erythromycin with a prevalence of 63 and 100%, respectively. The resistance of bacteria from milk to vancomycin, fluoroquinolones, linezolid, and daptomycin is prevalent at $<1\%$. Currently, nutraceuticals or alternative medicines are preferred over antibiotics for the prevention and treatment of mastitis as they contain multiple phytoconstituents, and thereby overcome the issue of drug resistance (Kumar and Bharti,

TABLE 67.5 Nutraceuticals that can be used to prevent or treat liver diseases.

Nutraceuticals	Major mechanism(s)	Major effects
S-Adenosylmethionine	Functions as a methyl group donor in transmethylation pathways, as a precursor to sulfur compounds in transsulfuration pathways, major source of hepatic glutathione (GSH), contributor to systemic redox status, and improves mitochondrial and hepatocellular functions	Antioxidative, antiinflammatory, antineuroinflammatory, anticholestatic hepatopathies, and improves liver functions
N-Acetylcysteine	Serves as a source of sulfhydryl (SH) groups, enhances glutathione-*S*-transferase activity, and stimulates GSH synthesis	Antioxidative and antihepatotoxicity
Vitamin C	Reduces reactive oxygen species (ROS), regenerates vitamin E and GSH, and induces collagen formation	Antioxidative
Vitamin E	Mitigates propagation of membrane peroxidation damage, inhibits cyclooxygenase and 5-lipoxygenase activities, inhibits platelet aggregation and adhesion, suppresses inflammatory cells, improves liver enzyme values and hepatic GSH status	Antioxidative and antiinflammatory
Alpha-lipoic acid	Metabolites of alpha-lipoic acid serve as cofactors of enzyme system, primarily within the mitochondria	Antioxidative and indicated in the management of liver diseases
Ubiquinol (Coenzyme Q_{10})	Functions in oxidation-reduction reactions in the mitochondrial respiratory chain and protects against lipid peroxidation	Antioxidative
Lecithin (phosphatidylcholines, fatty acids, and carbohydrates)	Inhibits fibrogenesis and oxidative stress, and restores hepatic and mitochondrial GSH	Antioxidative and hepatoprotective
Milk thistle (*Silybum marianum*)	Silymarin mitigates oxidative damage to hepatocellular and mitochondrial membranes, exerts antifibrotic activity, and inhibits reactive collagen formation	Antioxidative, enhances hepatocellular regeneration, and hepatoprotective
Ursodeoxycholic acid	Cytoprotection of hepatocytes and biliary epithelium, and stimulation of bile secretion	Antioxidative, immunomodulatory, indicated in hepatobiliary disorders, including necroinflammatory, and cholestasis
Anthocyanins	Decrease ROS generation, lipid peroxidation, and hepatocellular lipid accumulation, and induce antioxidant enzymes	Antioxidative, hepatoprotective, hepatotropic, and antifibrotic
Isoflavones	Decrease lipid peroxidation	Antioxidative, antiinflammatory, antifibrotic, antilipogenic, and antihepatocarcinogenesis
Quercetin	Reduces NO generation, inhibits secretion of TGF-β1 and collagen, and fibrogenesis	Antioxidative, hepatoprotective, and antifibrotic
Ashwagandha (*Withania somnifera*)	Reduces ROS and lipid content	Antioxidative and antiinflammatory
Astaxanthin (*Haematococcus pluvialis*)	Scavenges free radicals, inhibits hepatocellular lipid accumulation, enhances immune functions, and improves glucose metabolism	Antioxidative, antiischemia–reperfusion injury, and antifibrosis
Caffeine	Inhibits action of adenosine on hepatic stellate cells by adenosine A2A receptors and by blocking the expression of TGF-β, and improves mitochondrial bioenergetic function	Antifibrotic, hepatoprotective, decreases cirrhosis and carcinoma development
Curcumin (*Curcuma longa*)	Decreases ROS generation	Antioxidative and antiinflammatory

(*Continued*)

TABLE 67.5 (Continued)

Nutraceuticals	Major mechanism(s)	Major effects
Fenugreek (*Trigonella foenum-graecum*)	Downregulates lipogenic enzymes, and reduces hepatocellular lipid accumulation	Antioxidative, antiinflammatory, and hepatoprotective
Garlic (*Allium sativum*)	Reduces LDL, inducible Nitric Oxide (iNOS)	Antioxidative, antiinflammatory, and hepatoprotective
Ginseng (*Panax ginseng, Panax quinquefolius* L.)	Inhibits TNF-α, exerts antiapoptotic activity	Antiinflammatory and antifibrotic
Ginkgo biloba	Ameliorates dyslipidemia by decreasing enzymes in hepatocellular lipid biosynthesis	Hepatoprotective against hepatocarcinogenesis, antihepatotoxicity

TABLE 67.6 **Plant extracts/nutraceuticals that can be indicated for wound healing.**

Plants/ nutraceuticals	Active ingredient(s)	Effects
Acalypha indica	Polyphenols, saponins, sterols, and triterpenoids	Wound healing
Aloe vera	β-Sitosterol and glycoprotein fraction	Angiogenesis
Anredera diffusa	Oleanolic acid	Wound healing
Butea monosperma	Flavonoids, glycosides, saponins, steroids, tannins, and triterpenoids	Wound healing
Camellia sinensis (green tea)	Epigallocatechin-3-gallate	Antiinflammatory
Centaurea sadleriana	Flavonoids	Wound healing
Centella asiatica	Asiaticoside, asiatic acid, and madecassic acid	Wound healing
Ceylon cinnamon	(E)-Cinnamaldehyde	Wound healing
Chromolaena odorata	Alkaloids, phenolic acids (protocatechuic, *p*-hydroxybenzoic, *p*-coumaric, ferulic, and vanillic acids), anthraquinones, flavonoids, phytosterols, saponins, tannins, terpenoids, and chromomoric acid C-1	Antioxidative, antiinflammatory, increased collagen synthesis, burn and wound healing, and antipyretic
Cimicifuga spp.	Fukinolic acid and cimicifugic acid	Wound healing
Copaifera langsdorffii	Oleoresin	Wound healing
Ficus racemose L.	β-Sitosterol and lupeol acetate	Antifungal, antibacterial, enhanced cell migration, and wound healing
Gentiana lutea	Gentiopicroside, sweriside, and swertiamarin	Wound healing
Heliotropium indicum	Heliotrine, indicine-*N*-oxide, saponins, and tannins	Wound healing
Hippophae rhamnoides	Quercetin, isorhamnetin, and kaempferol	Wound healing
Ixora coccinea	Alkaloids, fixed oils, flavonoids, glycosides, steroids, tannins, and terpenoids	Wound healing
Jungle geranium	Flavonoids (formononetin, rutin, and β-sitosterol), lupeol, saponins, and tannins	Wound healing

(Continued)

TABLE 67.6 (Continued)

Plants/ nutraceuticals	Active ingredient(s)	Effects
Morinda pubescens (Indian mulberry)	Alkaloids, carboxylic acids, phenols, steroids, and triterpenoids	Wound healing
Opuntia ficus-indica	Polysaccharides	Wound healing
Onosma argentanum	Shikonin derivatives	Wound healing
Peacock chaste tree	Alkaloids, flavonoids, glucosides, resin, terpenes, and volatile oils	Wound healing
Peperomia galioides	(+)-bisabolol	Wound healing
Phyllanthus	Phyllanthin and hypophyllanthin	Wound healing
Plumbago zeylanica	Alkaloids and terpenoids	Wound healing
Royal jelly	Royal jelly proteins	Wound healing
Scrophularia nodosa	Acetylated iridoid glycosides	Wound healing
Tamarindus indica	Xyloglucan	Wound healing
Terminalia arjuna	Tannins	Wound healing
Turmeric (*Curcuma longa*)	Curcuminoids	Antioxidative, antiinflammatory, antimicrobial, and wound healing
Vitex altissima, Vitex trifolia	Alkaloids, flavonoids, glucosides, resin, terpenes, and volatile oil	Antioxidative, antiinflammatory, and wound healing

2013; Pinedo et al., 2013: Mason et al., 2017; Mustaq et al., 2018). Intermingled treatments with nutraceuticals and pharmaceuticals are reported to be more efficacious than either treatment alone (Coppock, 2019). Some plant extracts/nutraceuticals that have potential or can be used for the treatment of mastitis are listed in Table 67.7.

Nutraceuticals from plants and other biological sources have been found to exert antimicrobial activity against mastitis pathogens (Ananda Baskaran et al., 2009; Boorn et al., 2010; Alekish et al., 2017; Abdalhamed et al., 2018; Amber et al., 2018; Coppock, 2019). In an in vitro study, Kher et al. (2018) reported that the *Terminalia chebula* Retz. fruit extract (500 μg/mL) was as effective as amoxicillin. The crude extracts, alkaloids, saponins, and flavonoids from many plants (*Allium sativum* L., *Bunium persicum* Boiss., *Oryza sativa* L., *Triticum aestivum* L., and others) have been found to exert antibacterial activity against multidrug-resistant strains of *Escherichia coli*, *Klebsiella pneumoniae*, and *Staphylococcus aureus* (Amber et al., 2018). Alkaloids exerted greater antibacterial activity than flavonoids and saponins.

There are many other plants (oregano oil, eugenol from oil of clove, *trans*-cinnamaldehyde from cinnamon bark, *Eucalyptus globulus*, *Glycyrrhiza glabra*, *Curcuma longa*, *Cedrus deodara*, *Paederia foetida*, *Panax ginseng*, *Tinospora cordifolia* Wild., and others) and nutraceuticals (honey, sulfur, vitamin E, etc.) which have antimicrobial properties against mastitis pathogens (Boorn et al., 2010; Mukherjee et al., 2010; Bhatt et al., 2014; Almasaudi et al., 2017; Mason et al., 2017; Silvestrini et al., 2017; Abdalhamed et al., 2018). In addition to antibacterial activity, many of these plant extracts/nutraceuticals have antioxidative, antiinflammatory, and immunomodulatory properties.

Concluding remarks and future directions

Currently, the use of nutraceuticals for the prevention and treatment of diseases in animals is as common as it is in humans. Veterinarians and animal owners prefer using

TABLE 67.7 Plant extracts/phytoconstituents that have potential in prevention and treatment of mastitis in bovine and ovine species.

Plant/phytoconstituent	Route of administration
Achillea millefolium L.	Applied on udder
Allium cepa L.	Oral
Allium sativum L.	Oral
Avena sativa L.	Poultice
Bunium persicum Boiss.	Oral
Ficus carica L.	Oral
Foeniculum vulgare Miller.	Oral
Galium aparine L.	Oral
Gloriosa superba L.	Oral
Maus sylvetris L.	Oral
Origanum vulgare L.	Intramammary infusion
Oryza sativa L.	Oral
Panax ginseng	Intramammary infusion
Salix alba	Oral
Salvia officinalis	Intramammary infusion
Sambucus nigra L.	Topical
Teucrium scorodonia L.	Intramammary infusion
Vitex negundo Linn.	Oral

nutraceuticals over synthetic drugs for multiple reasons, including: (1) easy access, (2) low cost, (3) safety, and (4) lack of drug resistance. A few downsides of veterinary nutraceuticals include: (1) lack of quality control, (2) adulteration, (3) herb–drug interactions, and (4) regulatory guidelines. This chapter describes in brief the nutraceuticals for health and diseases of CNS, CVS, hepatic system, periodontal, diabetes, wound healing, and mastitis. For details of veterinary nutraceuticals, readers are referred to a recent publication by Gupta et al (2019d). Future investigations needed to be focused on the mechanistic rationale and clinical trials to make judicious use of nutraceuticals in animal health and diseases.

References

Abdalhamed, A.M., Zeedan, G.S.G., Zeina, H., 2018. Isolation and identification of bacteria causing mastitis in small ruminants and their susceptibility to antibiotics, honey, essential oils, and plant extracts. Vet. World. 11 (3), 355–362.

Alekish, M.O., Ismail, Z.B., Awawdeh, M.S., et al., 2017. Effects of intramammary infusion of sage (*Salvia officinalis*) essential oil on milk somatic cell count, milk composition parameters and selected hematology and serum biochemical parameters in Awassi sheep with subclinical mastitis. Vet. World. 10 (8), 895–900.

Ali-Seyed, M., Vijayarahavan, K., 2019. Nutraceuticals for wound healing: a special focus on *Chromolaena ordorata* as guardian of health with broad spectrum of biological activities. In: Gupta, R.C., Srivastava, A., Lall, R. (Eds.), Nutraceuticals in Veterinary Medicine. Springer Nature, Cham, Switzerland, pp. 541–562.

Alissa, E.M., Fern, G.A., 2012. Functional foods and nutraceuticals in the primary prevention of cardiovascular diseases. J. Nutr. Metab. 2012, 569486.

Almasaudi, S.B., Al-Nahari, A.A.M., Abd El-Ghani, E.S.M., et al., 2017. Antimicrobial effect of different types of honey on *Staphylococcus aureus*. Saudi J. Biol. Sci. 24 (6), 1255–1261.

Alqathama, A., Alluhiabi, G., Baghdadi, H., et al., 2020. Herbal medicine from the perspective of type II diabetic patients and physicians: what is the relationship? BMC Complement. Med. Ther. 20, 65. Available from: https://doi.org/10.1186/s12906-020-2854-4.

Alves, Q.L., Camargo, S.B., Silva, D.F., 2019. Role of nutraceuticals in the prevention and treatment of hypertension and cardiovascular diseases. J. Hypertens. Manag. 5, 037. Available from: https://doi.org/10.23937/2474-3690/1510037.

Amaeze, O.U., Aderemi-Williams, R.I., Ayo-Vaughan, M.A., et al., 2018. Herbal medicine use among type 2 diabetes mellitus patients in Nigeria: understanding the magnitude and predictors of use. Int. J. Clin. Pharmacol. 40 (3), 580–588.

Amber, R., Adan, M., Tariq, A., et al., 2018. Antibacterial activity of selected medicinal plants of northwest Pakistan traditionally used against mastitis in livestock. Saudi J. Biol. Sci. 25 (1), 154–161.

Ananda Baskaran, S., Kazmer, G.W., Hinkley, L., 2009. Antibacterial effect of plant-derived antimicrobials on major bacterial mastitis pathogens *in vitro*. J. Dairy. Sci. 92 (4), 1423–1429.

Aquila, G., Marracino, L., Martino, V., et al., 2019. The use of nutraceuticals to counteract atherosclerosis: the role of the Notch pathway. Oxid. Med. Cell Long. Article ID5470470.

Banach, M., Patti, A.M., Giglio, R.V., et al., 2018. The role of nutraceuticals in statin intolerant patients. J. Am. Coll. Cardiol. 72 (1), 96–118.

Bhatt, V.D., Shah, T.M., Nauriyal, D.S., et al., 2014. Evaluation of a topical herbal drug for its *in vivo* immunomodulatory effect on cytokines production and antibacterial activity in bovine subclinical mastitis. Ayu 35 (2), 198–205.

Boorn, K.L., Khor, Y.Y., Sweetman, E., et al., 2010. Antimicrobial activity of honey from the stingless bee *Trigona carbonaria* determined by agar diffusion, agar dilution, broth microdilution and time-kill methodology. J. Appl. Microbiol. 108 (5), 1534–1543.

Brown, A.C., 2017a. Heart toxicity related to herbs and dietary supplements: online table of case reports. Part 4 of 5. J. Diet. Suppl. 2017. Available from: https://doi.org/10.1080/19390211.2017.1356418.

Brown, A.C., 2017b. Liver toxicity related to herbs and dietary supplements: online table of case reports. Part 2 of 5. Food Chem. Toxicol. 107, 472–501.

Burford, G., Bodeker, G., Ryan, T.J., 2007. Skin and wound care: traditional, complementary and alternative medicine in public health dermatology. In: Burford, B.G. (Ed.), Traditional, Complementary and Alternative Medicine. Imperial College Press, London.

Colitti, M., Stefanon, B., Gabai, G., et al., 2019. Oxidative stress and nutraceuticals in the modulation of the immune function: current knowledge in animals of veterinary interest. Antioxidants. 8, 28. Available from: https://doi.org/10.3390/antiox8010028.

Coppock, R.W., 2019. Nutraceuticals in mastitis. In: Gupta, R.C., Srivastava, A., Lall, R. (Eds.), Nutraceuticals in Veterinary Medicine. Springer Nature, Cham, Switzerland, pp. 569–585.

Cragg, G.M., Newman, D.J., 2013. Natural products: a continuing source of novel drug leads. Biochim. Biophys. Acta 1830 (6), 3670–3695.

Duarte, S., Baber, J., Fujii, T., et al., 2015. Matrix metalloproteinases in liver injury, repair and fibrosis. Matrix Biol. 44-46, 147–156.

Eming, S.A., Smoler, H., Krieg, T., 2002. Treatment of chronic wounds: state of the art and future concepts. Cell Tissues Organs 172 (2), 105–117.

Ganesan, S., 2008. Traditional oral care medicinal plants survey of Tamil Nadu. Nat. Prod. Rad. 7 (2), 166–172.

Grossman, L.D., Roscoe, R., Shack, A.R., 2018. Complementary and alternative medicine for diabetes. Can. J. Diabetes 42, S154–S161.

Gupta, R.C., Chang, D., Nammi, S., et al., 2017. Interactions between antidiabetic drugs and herbs: an overview of mechanisms of action and clinical implications. Diabetol. Metab. Syndr. 9, 59.

Gupta, R.C., Doss, R.B., Srivastava, A., Lall, A., Sinha, A., 2019a. Nutraceuticals for cognitive dysfunction. In: Gupta, R.C., Srivastava, A., Lall, R. (Eds.), Nutraceuticals in Veterinary Medicine. Springer Nature, Cham, Switzerland, pp. 393–415.

Gupta, R.C., Gupta, D.M., Lall, R., Srivastava, A., Sinha, A., 2019c. Nutraceuticals in periodontal health and diseases in dogs and cats. In: Gupta, R.C., Srivastava, A., Lall, R. (Eds.), Nutraceuticals in Veterinary Medicine. Springer Nature, Cham, Switzerland, pp. 447–466.

Gupta, R.C., Lall, A., Sinha, A., Srivastava, A., 2019b. Nutraceuticals for diabetes in dogs and cats. In: Gupta, R.C., Srivastava, A., Lall, R. (Eds.), Nutraceuticals in Veterinary Medicine. Springer Nature, Cham, Switzerland, pp. 523–539.

Gupta, R.C., Srivastava, A., Lall, R., 2019d. Nutraceuticals in Veterinary MedicineIn: Gupta, R.C., Srivastava, A., Lall, R. (Eds.), Springer Nature, Cham, Switzerland, 877 pages.

Gupta, R.C., Srivastava, A., Lall, R., 2018. Toxicity potential of nutraceuticals. In: Nicolotti, O. (Ed.), Computational Toxicology: Methods and Protocols, Methods in Molecular Biology. Springer Nature, New York, pp. 367–394.

Gwaltney-Brant, S.M., 2019. Nutraceuticals in hepatic and pancreatic diseases. In: Gupta, R.C., Srivastava, A., Lall, R. (Eds.), Nutraceuticals in Veterinary Medicine. Springer Nature, Cham, Switzerland, pp. 437–446.

Hanhineva, K., Torronen, R., Bondia-Pons, et al., 2010. Impact of dietary polyphenols on carbohydrate metabolism. Int. J. Mol. Sci. 11, 1365–1402.

Hodgson, J.M., Croft, K.D., Woodman, R.J., et al., 2014. Effects of vitamin E, vitamin C and polyphenols on the rate of blood pressure variation: results of two randomized controlled trials. Br. J. Nutr. 112, 1551–1561.

Jaeschke, H., 2008. Toxic responses of the liver. In: Klaasen, C.D. (Ed.), Casarett and Doul's Toxicology, The Basic Science of Poisons. McGraw-Hill, New York, pp. 557–582.

Jain, S., Buttar, H.S., Chintameneni, M., et al., 2018. Prevention of cardiovascular diseases with anti-inflammatory and antioxidant nutraceuticals and herbal products: an overview of pre-clinical and clinical studies. Recent. Pat. Inflamm. Allergy Drug. Discov. 12 (2), 145–157.

Kaur, G., 2019. Hepatic toxicity biomarkers. In: Gupta, R.C. (Ed.), Biomarkers in Toxicology, second ed Academic Press, Elsevier, Amsterdam, pp. 251–266.

Kher, M.N., Sheth, N.R., Bhatt, V.D., 2018. *In vitro* antibacterial evaluation of *Terminalia chebula* as an alternative of antibiotics against bovine subclinical mastitis. Anim. Biotechnol. 1-8. Available from: http://doi.org/10.1080/10495398.2018.1451752.

Kim, Y.M., Namkoong, S., Yun, Y.G., et al., 2007. Water extract of Korean red ginseng stimulates angiogenesis by activating the PI3K/Akt dependent ERK1/2 and eNOS pathways in human umbilical vein endothelial cells. Biol. Pharm. Bull. 30, 1674–1679.

Kumar, A., Lather, A., Kumar, V., et al., 2015. Pharmacological potential of plant used in dental care: a review. J. Herb. Drug. 5 (4), 179–186.

Kumar, R., Bharti, K.A., 2013. New claims in folk veterinary medicines from Uttar Pradesh, India. J. Ethnopharmacol. 146 (2), 581–593.

Li, H., Sureda, A., Devkota, H.P., et al., 2020. Curcumin, the golden spice in treating cardiovascular diseases. Biotechnol. Adv. 38, 107343. Available from: https://doi.org/10.1016/j.biotechadv,2019.01.010.

Lima, M.D.R., Lopes, A.P., Martins, C., et al., 2017. The effect of *Calendula officinalis* on oxidative stress and bone loss in experimental periodontitis. Front. Physiol. 8, 440.

Mason, S.E., Mullen, K.A.E., Anderson, K.L., et al., 2017. Pharmacokinetic analysis of thymol, carvacrol and diallyl disulfide after intramammary and topical applications in healthy organic dairy cattle. Food Addit. Contam. Part. A. 34 (5), 740–749.

Mekuria, A.B., Belachew, S.A., Tegegn, H.G., et al., 2018. Prevalence and correlates of herbal medicine use among type 2 diabetic patients in teaching hospital in Ethiopia: a cross-sectional study. BMC Complement. Altern. Med. 18 (1), 85.

Menke, M.N., Menke, N.B., Boardman, C.H., et al., 2008. Biologic therapeutics and molecular profiling to optimize wound healing. Gynecol. Oncol. 111 (2), S87–S91.

Mirmalek, S., Parsa, T., Parsa, Y., et al., 2016. The wound healing effect of *Iris forentina* on full thickness excisional skin wounds: a histomorphometrical study. Bangladesh J. Pharmacol. 11, 86–90.

Moffat, K.S., Landsberg, G.M., 2003. An investigation of the prevalence of clinical signs of cognitive dysfunction syndrome (CDS) in cats. JAAHA 39 (5), 512.

Muhammad, S., Lawal, M.T., 2010. Oral hygiene and the use of plants. Sci. Res. Essays 5 (14), 1788–1795.

Mukherjee, R., De, U.K., Ram, G.C., 2010. Evaluation of mammary gland immunity and therapeutic potential of *Tinospora cordifolia* against bovine subclinical mastitis. Trop. Anim. Health Prod. 42 (4), 645–651.

Mustaq, S., Shah, A.M., Shah, A., et al., 2018. Bovine mastitis: an appraisal of its alternative herbal cure. Microb. Pathog. 114, 357–361.

Neilsen, J.C., Hart, B.L., Cliff, K.D., et al., 2001. Prevalence of behavioral changes associated with age-related cognitive impairment in dogs. J. Am. Vet. Med. Assoc. 218, 1787–1791.

Newman, M.G., Takai, H.H., Carranza, F.A., 2006. Carranza's Clinical Periodontology, tenth ed. WB Saunders Co, Philadelphia, PA.

Nikhra, V., 2019. Nutraceuticals for improving cardiovascular health and prognosis in cardiovascular disease. EC Nutr. 14 (5), 434–448.

Patel, D.K., Prasad, S.K., Kumar, R., et al., 2012. An overview on antidiabetic medicinal plants having insulin mimetic property. Asian Pacif. J. Trop. Biomed. 2 (4), 320–330.

Peltzer, K., Pengpid, S., 2019. The use of herbal medicines among chronic disease patients in Thailand: a cross-sectional survey. J. Multidiscip. Health 12, 573.

Pinedo, P., Karreman, H., Bothe, H., et al., 2013. Efficacy of a botanical preparation for the intramammary treatment of clinical mastitis on an organic dairy farm. Can. Vet. J. 54 (5), 479–484.

Poli, A., Visioli, F., 2019. Pharmacology of nutraceuticals with lipid lowering properties. High. Blood Press. Cardiovasc. Prev. 26 (2), 113–118.

Shaito, A., Thuan, D.T.B., Phu, H.T., et al., 2020. Herbal medicine for cardiovascular diseases: efficacy, mechanisms, and safety. Front. Pharmacol. 11. Available from: https://doi.org/10.3389/fphar.2020.00422. Article422.

Shen, H.M., Chen, C., Jiang, J.Y., et al., 2017. The *N*-butyl alcohol extract from *Hibiscus rosasinensis* L. flowers enhances healing potential on rat excisional wounds. J. Ethnopharmacol. 198, 291–301.

Silvestrini, P., Beccaria, C., Pereyra, E.A.L., et al., 2017. Intramammary inoculation of *Panax ginseng* plays an immunoprotective role in *Staphylococcus aureus* infection in a murine model. Res. Vet. Sci. 115, 211–220.

Soni, S., Aggarwal, P.K., Sharma, N., et al., 2012. Coenzyme Q10 and periodontal health: a review. IntJ Oral. Maxillofac. Pathol. 3 (2), 21–26.

Sosnowska, B., Penson, P., Banach, M., 2017. The role of nutraceuticals in the prevention of cardiovascular disease. Cardiovasc. Diagn. Ther. 7 (Suppl 1), S21–S31.

Teschke, R., Wolff, A., Frenzel, C., et al., 2014. Review article: herbal hepatotoxicity—an update on traditional Chinese medicine preparations. Aliment. Pharmacol. Ther. 40, 32–50.

WHO. (2019) World Health Organization global report on traditional and complementary medicine. Geneva, Switzerland.

Wietecha, M.S., Dipietro, L.A., 2013. Therapeutic approaches to the regulation of wound angiogenesis. Adv. Wound Care 2, 81–86.

Yousaf, M.N., Chaudhary, F.S., Hodanazary, S.M., et al., 2019. Hepatotoxicity associated with *Garcinia cambogia*: a case report. World J. Hepatol. 11, 735–742.

Zoltani, K.C., 2019. Nutraceuticals in cardiovascular diseases. In: Gupta, R.C., Srivastava, A., Lall, R. (Eds.), Nutraceuticals in Veterinary Medicine. Springer Nature, Cham, Switzerland, pp. 427–435.

Section VI

Toxicity and toxic interactions

Chapter 68

Toxic contamination of nutraceuticals and food ingredients

Fernando Gil, Antonio F. Hernández and M. Concepción Martín-Domingo

Introduction and background

Medicinal plants have been used therapeutically throughout the world as tools of traditional medicine, from Ayurveda to Chinese traditional medicine. In Western countries, despite the development and production of synthetic medicines, the popularity of over-the-counter health foods, nutraceuticals, and medicinal products from plants or other natural sources has recently increased because of the belief that they could be more effective than conventional therapies for preventing or treating diseases. This attitude indirectly indicates the lack of confidence of the general public in conventional medical treatments.

Several reasons may account for the worldwide growing use of nutraceuticals and dietary supplements, among them being the lower rate of adverse effects as compared to conventional drugs and the higher costs of many traditional pharmaceutical formulations (Meena et al., 2010; Rao et al., 2011; Martínez-Domínguez et al., 2014).

Consumers increasingly rely on dietary supplements to maintain mental acuity and to overcome age-related problems, such as menopause, benign prostate hypertrophy, and elevated blood pressure and cholesterol levels, and also to relieve stress (Raman et al., 2004). In addition, a variety of nutritional products are used by track-and-field athletes, including sports foods, performance supplements, and therapeutic nutritional aids (Peeling et al, 2019).

The term nutraceutical was coined by DeFelice to define any substance that may be considered a food or part of a food providing medical or health benefits. Nutraceuticals may be useful for the prevention and treatment of disease or as an alternative to nonconventional medicine in primary health care. They include herbal medicines (Ayurvedic, Chinese, Tibetan, African, Amazonian, Herbalism), naturopathy, vitamin and mineral therapy, and homeopathy. However, nutraceuticals is a misleading term that encompasses a wide array of products, ranging from herbal products/botanicals to dietary supplements, functional food, and enriched/fortified food, among others. The lack of a clear regulatory definition or specific legislation for nutraceutical products represents an impediment for having sufficient quality controls to ensure their quality and safety.

Nutraceuticals and food ingredients (NFI) consist of dietary supplements (e.g., vitamins, minerals, co-enzyme Q, carnitine), herbal products (e.g., flavonoids, ginseng, *Ginkgo biloba*, Saint John's wort, saw palmetto), and bioengineered and processed foods (e.g., probiotics, omega-3 fatty acids, fish oil, transgenic plants) purported to benefit health (DeFelice, 1995; Pirotta et al., 2000) and to enhance and improve athletes' health and performance.

Lockwood (2007) defined nutraceutical as a term used to describe a medicinal or nutritional component that includes a food, plant, or naturally occurring material that may have been purified or concentrated and that is used for the improvement of health by preventing or treating a disease. Dolan et al. (2003) used the term "dietary supplement" to describe a product that contains one or more of the following ingredients: vitamins; minerals; herbs or other botanicals; amino acids; dietary substances used as diet supplements to increase the total daily intake; or concentrates, metabolites, constituents, extracts, or combinations of these ingredients. The Dietary Supplement Health and Education Act (DSHEA) of 1994 considered nutraceuticals as a dietary supplement that may contain a herb or other botanical, or a concentrate, metabolite, constituent, extract, or combination of any ingredient from the other categories (Frankos et al., 2010). However, toxic contamination of NFI during any stage of production can lead to changes in their quality and safety. The health hazard of these products largely depends on the presence of unusually high concentrations of chemical ingredients that may result in toxicity or even fatality if they are consumed (Chan, 2003).

Furthermore, the consumption of NFI contaminated with environmental pollutants by vulnerable subgroups of

Nutraceuticals. DOI: https://doi.org/10.1016/B978-0-12-821038-3.00068-9

the population, such as pregnant women, children, and elderly adults may become a serious problem if the tolerable level of exposure is exceeded. For example, children and toddlers have greater susceptibility to poisoning by heavy metals and metalloids in comparison with adults because of their lower body weight, resulting in higher dosage of these substances.

Risk assessment paradigms might underestimate the effects of these chemicals on children and the elderly (Gomez et al., 2007). For instance, these subpopulations may be more susceptible to adverse effects associated with ingestion of low doses of heavy metals found in *Hypericum perforatum*, commonly known as St. John's wort, widely used for the treatment of mild to moderate forms of depression. Herbal remedies are also frequently used by pregnant women regardless of the scarce information about their safety during pregnancy and breastfeeding. In addition to the risk of miscarriage from several herbs (e.g., nettle, passionflower, and aloe), a number of epidemiologic studies have suggested that certain herbs [e.g., those rich in unsaturated pyrrolizidine alkaloids (PA)] are associated with embryotoxic or fetotoxic effects (de Smet, 2002; Gurib-Fakim, 2006; Rodriguez-Fragoso et al., 2008).

Because potential interactions between pharmaceutical formulations and herbal products (or other NFI) cannot always be predicted, unexpected effects can be observed as a result of a change in the magnitude of the effect of conventional drugs (Shaw et al., 1997). These types of interactions can be of either a pharmacokinetic or pharmacodynamic nature (Rodriguez-Fragoso et al., 2008). Pharmacokinetic interactions occurs as a result of induction or inhibition of drug-metabolizing enzyme systems [particularly the cytochrome P450 (CYP450) isoenzymes] or membrane transporters [e.g., organic anion transporter proteins, and the P-glycoprotein (P-gp) ATP-binding cassette transporter]. This type of interaction has great importance in clinical practice and may warrant contraindications by healthcare providers. A number of herb--drug interactions have been well characterized in clinical studies of human volunteers; these involve CYP3A4, CYP2D6, CYP1A2, CYP2E1, CYP2C19, CYP2C9, OAT1A2, and P-glycoprotein (Ronis et al., 2018). In addition, important genetic polymorphisms of CYP450s and membrane transporters have been reported to modify drug disposition in different populations. For instance, the interaction between peppermint oil and some CYP450 isoforms may modify levels of drugs metabolized by these isoforms (Maniacal and Wanwimolruk, 2001; Unger and Frank, 2004).

Good manufacturing practices and other legal requirements must be met to avoid adverse reactions or product quality deficiencies as a consequence of the rapid growth of health-related foods and supplements (Genuis et al., 2012).

In the United States, the DSHEA established regulations and limits label claims on dietary supplements. In Europe, food supplements are covered by Directive 2002/46/EC and herbal medicinal products by Directive 2004/24/EC. While there is no formal legislation regulating nutraceutical products (Gulati and Berry Ottaway, 2006; Martínez-Domínguez et al., 2014), the European Union (EU) adopted Regulation (EU) No 1924/2006 on the use of nutritional and health claims made on food. However, this regulation does not properly cover botanicals products, which fall under the general EU food legislation Regulation 178/2002. Although conditions for importation of pharmaceutical products are rigorous, the guidelines for importation of certain NFI, including herbal medicines and raw herbs, are quite general and only related to microbiological properties and registration of items themselves. In turn, there are no specific guidelines for active constituents or toxic contaminants (Cooper et al., 2007).

Little information is available on the toxic contamination of nutraceuticals and dietary supplements except for herbal drugs. Hence, this chapter is based largely on medicinal plants, assuming that they comprise the majority of NFI.

Herbs can be collected indiscriminately from noncultivated and nonenvironmentally friendly areas by untrained people and placed into the market without any control. This means that consumers might be exposed to herbal products potentially contaminated with pesticides, heavy metals and metalloids, mycotoxins, or radioactivity, or be adulterated with drugs (Fig. 68.1). The presence of banned pesticides or excessive amounts of regulated pesticides and heavy metals depends on the source of herbal materials and whether or not they are grown in a contaminated area. Moreover, chemical toxins may come from unfavorable or wrong storage conditions or chemical

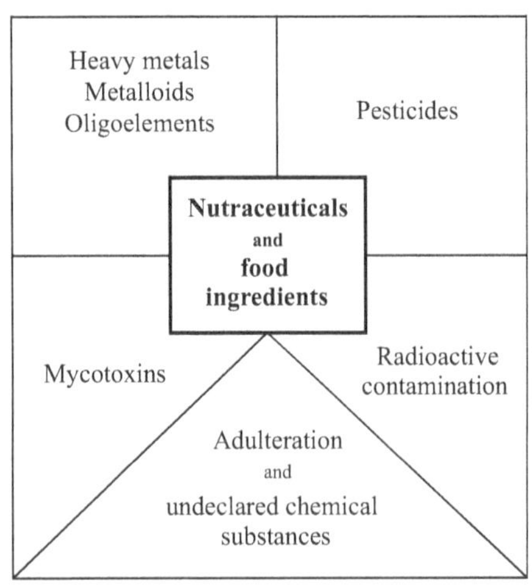

FIGURE 68.1 Main toxic contaminants in nutraceuticals and food ingredients.

treatment during their storage. In turn, the presence of drugs could be related to unprofessional practices of manufacturers. Ideally, the consumption of NFI products should be strictly controlled and is pertinent to better knowledge of the levels of different contaminants (specifically heavy metals and pesticides) in raw materials (Chan, 2003; Meena et al., 2010; Harris et al., 2011).

Phytotherapy has a very long tradition and has been popular for centuries. Medicinal plants have a long history of therapeutic use throughout the world and still constitute an important part of traditional medicine. In the last quarter of the 20th century, the therapeutic use of herbal products increased in developed countries because of their popularity among consumers and the widespread opinion that "natural products" implies "harmless products" and the lack of toxic effects. Their easy accessibility and relatively low cost also contributed to increased use; however, their popularity and global expansion have raised concerns about the safety and quality of herbal products for public health (Arpadjan et al., 2008; Kosalec et al., 2009).

Consumers are increasingly careful in choosing components for their diet and intend to incorporate high nutrient levels into their standard diet, preferably from natural sources such as plants (Bhat et al., 2010). This increased scientific interest and consumer demand have promoted the development of herbal products such as NFI. Medicinal plants behave as authentic medicines because the chemical substances of which they are formed can have a biological activity in humans. For example, because phenolic acid and flavonoids are natural antioxidants and free radical scavengers, there is growing interest in their pharmacological applications (Choudhury et al., 2006).

In many countries traditional herbal preparations can be sold on the market as food supplements, which do not require prior safety evaluation. The safety and benefit of plant products are directly related to the quality of the raw materials from which they are derived (Salgueiro et al., 2010).

Herbal medicines are widely used in the United States, with approximately one-quarter of adults reporting their use for the treatment of a medical illness. Herbs are considered dietary supplements in the United States and therefore are subjected to a very limited form of regulation and oversight. Although herbs are often believed to be safe, many side effects have been reported, including direct toxic effects, allergic reactions, toxicity from contaminants, and interactions with drugs and other herbs (Bent and Ko, 2004). Other adverse effects include possible mutagenicity and mistaken plant identities (Ernst, 1998). For instance, liver problems have been reported following the use of Chinese herbal medicine for skin disorders, there have been allergic reactions to royal jelly and propolis, and heavy metal poisoning was found in the Indian subcontinent from Ayurvedic remedies contaminated with these compounds (Shaw et al., 1997).

Overall, adverse effects of herbal drugs can be divided into two main groups: intrinsic and extrinsic. The former includes predictable toxicity, overdose, pharmacological interactions, idiosyncratic reactions (e.g., allergy and anaphylaxis), and delayed effects such as carcinogenicity and teratogenicity. The extrinsic undesirable effects largely involve the quality of the herbal medicinal products, which can be challenged by substitution, adulteration, contamination, misidentification, lack of standardization, and inappropriate labeling (Koh and Woo, 2000; Mazzanti et al., 2008; Rao et al., 2011).

Although raw materials are typically and supposedly free of substances that are currently added to patent medicines, they could contain relatively low levels of toxic compounds, particularly heavy metals and pesticides. The majority ($>95\%$) of the samples analyzed by Harris et al. (2011) contained concentrations of heavy metals or pesticide residues that were considered of negligible concern. Nevertheless, the elevated number of samples showing detectable levels of contaminants makes it advisable to monitor the concentration of heavy metals (especially cadmium and chromium) and pesticide residues (especially organophosphorus (OP) and organochlorine compounds) in raw materials.

However, if medicinal plants are used for infusion preparations, then the extractable component of toxic compounds would only be available for consumers. In this case, the concentration of contaminants (e.g., heavy metals) dissolved in the infusions should usually be below safety levels for human consumption (Martín-Domingo et al., 2017).

Chemical contamination in nutraceuticals and food ingredients

The quality and safety of nutraceuticals, food ingredients and herbal medicines accounts not only for their toxic components, but also for potential contaminants, such as heavy metals, pesticides, mycotoxins, polycyclic aromatic hydrocarbons (PAH), and other environmental contaminants.

Heavy metals, metalloids, and oligoelements

The sources of environmental pollution from heavy metals are quite diverse and mainly consist of industrial and traffic emissions, agricultural effluents, batteries containing cadmium, organic mercury fungicides, and lead arsenate as an insecticide. The concentration of these contaminants in ecosystems has increased in past decades as a result of anthropogenic activities. Because of the intrinsic toxicity of heavy metals, their presence in the environment may pose a potential threat to terrestrial and aquatic biota. In fact, metal pollution adversely affects the density and diversity of biotic communities, including people (Gomez et al., 2007).

The contents of essential and trace elements in medicinal plants vary depending on various factors (Haider et al., 2004). Among these factors are geoclimatic conditions, geochemical characteristics of the soil, anthropogenic activities (e.g., chemical industries in the vicinity), plant species (some can selectively accumulate toxic elements), and the part of the plant used for preparing the herbal medicine. The level to which metals accumulate in plants is also influenced by the physicochemical properties of the soil where plants grow (characteristics of soil or sediments, pH level, exposure period, dispersion range, and presence or absence of other elements). These properties determine the nature of the association of trace elements with soil components and are key elements for the bioavailability of metal elements (Sarma et al., 2011). While plants readily assimilate trace elements through the roots, these compounds can also be absorbed through the leaves. Rainfall, atmospheric dusts, plant protection agents, and fertilizers are additional sources of trace elements for plants (Łozak et al., 2002).

Although some trace metals present in foods play a physiological role as essential compounds and cofactors, contamination or adulteration of NFI with heavy metals, metalloids, and mineral nutrients (Table 68.1) is a matter of concern. For essential trace metals there is a safe range of intake that corresponds to the mid-dose range below or above which adverse effects may occur, thus featuring a U-shaped dose–response relationship. Conversely, heavy metal concentrations could reach levels potentially leading to hazardous effects on human health because of their cumulative properties.

Cadmium (Cd), mercury (Hg), lead (Pb), and arsenic (As) are nonessential toxic elements of special concern because of their toxicity even at low concentrations (Hsu et al., 2006; Rao et al., 2011). Lead exposure has been associated with renal tumors, reduced cognitive development, and increased blood pressure and cardiovascular toxicity. Cd may induce kidney dysfunction, osteomalacia, and reproductive system disturbances. Mercury may cause neurological disorders and kidney damage (Goyer and Clarkson, 2001; Garcia-Rico et al., 2007). Aluminum (Al) can be accumulated in certain tissues and has been associated with serious health problems, such as Alzheimer disease, dysfunction of the blood–brain barrier, and inhibition of hydroxyapatite formation, which may lead to decreased

TABLE 68.1 Toxic contaminants reported in nutraceuticals and food ingredients.

Trace metals	Heavy metals	Al, Cd, Cr, Hg, Ni, Pb
	Metalloids	As
	Oligoelements	Cu, Zn, Mo, Se
Pesticides	Organophosphorus pesticides	Chlorpyrifos, methyl-chlorpyrifos, coumaphos, diazinon, dichlorvos, dimethoate, ethion, fenchlorphos, malathion, parathion, methyl-parathion propenophos
	Organochlorine pesticides	Hexachlorocyclohexane, lindane, dichlorodiphenyl trichloroethane, benzene hexachloride, pentachloronitrobenzene, tecnazene
	Pyrethroid pesticides	Cypermethrin, esfenvalerate, fenvalerate, permethrin
	Nitrogen-containing pesticides	Atrazine
Mycotoxins	Aflatoxin, ochratoxin A, fumonisins, citrinin	
Pyrrolizidine alkaloids	Lasiocarpine, retrorsine, riddelliine, monocrotaline, heliotrine, senecionine, integerrimine, usaramine, seneciphylline	
Poycyclic aromatic hydrocarbons	Benzo(a)pyrene, benz(a)anthracene, benzo(b)fluoranthene, chrysene	
Radioactive contamination	^{137}Cs, ^{210}Pb, ^{238}U, ^{234}U, ^{226}Ra	
Adulteration/undeclared chemical substances	Caffeine, ephedrine, norpseudoephedrine, synephrine, phosphodiesterase type 5 inhibitors	

skeletal mineralization (osteopenia) (Rubio et al., 2012). In addition, Pb and Hg may reach and cross the placental barrier and interfere with placental transport systems, thus increasing prenatal exposure to these toxic compounds (Gupta, 2009, 2011). Despite these well-known adverse effects, the interpretation of total metal concentration (e.g., all ionic forms) and toxicity should be based on analytical results for the more toxic form.

The International Agency for Research on Cancer has classified certain metal elements and metalloids as carcinogenic (As, Be, Cd, Cr, and, Ni) or potentially carcinogenic to humans (Sb, Co, and V) because of their potential to induce DNA damage, although their chemical speciation plays a key role. Accordingly, it is important to monitor the presence of such metal elements in medicinal plants used as NFI to prevent excessive human exposure (Sarma et al., 2011).

Medicinal plants are good sources of mineral elements (Özcan et al., 2008) and they might be used as food supplements or for food fortification of new NFI with health-promoting properties. Thus measuring mineral elements and heavy metal concentrations in NFI is important not only from a nutritional point of view but also for the assessment of their quality and safety (Bhat et al., 2010).

In contrast to the extensive research performed on nutrient elements to define their role in the human diet, studies reporting the mineral content of NFI (e.g., spices and herbs) are limited. Malik et al. (2008) reported that infusions of some nontraditional plant species used for the preparation of stimulant beverages could be a valuable source of nutrient elements for the human diet. In particular, because zinc (Zn), copper (Cu), and molybdenum (Mo) are vital for many physiological functions, their intake is considered good for human health. Zinc acts as a catalyst, coactive, or structural unit for some enzymes and is a cofactor of metalloenzymes. Zinc deficiency can impair normal growth and development, reproduction, and immune function. Copper acts as a cofactor in cuproenzymes, which are essential enzymes for normal functioning of the body. Molybdenum is a component of the sulfite oxidase enzyme, a molybdoenzyme that catalyzes the last step in the degradation pathway of sulfur amino acids (Kosalec et al., 2009).

However, these essential metals are toxic above certain thresholds. High supplementation of oligoelements such as Cu or Zn has been related to adverse effects; for example, high Cu levels induce liver damage and high Zn concentrations reduce immune function and high-density lipoprotein levels, and also produce adverse interactions with Cu. Thus eating contaminated or adulterated foods that are not considered usual dietary sources of essential compounds may result in an inadvertent overdose, especially because consumers are often recommended to use food supplements as sources of essential elements. Exposure to these compounds from various sources also could lead to additive effects (Tongesayi et al., 2013).

Although herbal medicines may be as efficient as synthetic drugs for certain health disorders, the presence of unacceptably high levels of heavy metals makes them less safe (Blicharska et al., 2010). For this reason, herbal medicines may be responsible for the growing number of cases of adverse health consequences linked to their use (WHO, 2003). Heavy metals have been widely found in raw herbal products (Wong et al., 1993; Rai et al., 2001), and some plant species are known to be heavy metal hyperaccumulators (Pollard et al., 2002). This property has made some plants an effective source for bioaccumulation of heavy metals from contaminated soils (phytoremediation). Thus the consumption of these plants as food or therapeutic agents in traditional medicine may prove to be hazardous (Barthwal et al., 2008).

Patent herbal medicines (which consist of a mix of different substances in either pill or extract form) may contain intentionally added heavy metals (Saper et al., 2008). This is the case of Ayurvedic traditional herbal preparations, which may contain high levels of metals (Pb, Hg) or metalloids (As) as a result of the intentional incorporation of certain metallic preparations such as lead monoxide, red mercury sulfide, mercurous chloride, red mercury oxide, arsenic sulfide, disulfide or trisulfide, and arsenic trioxide (Martena et al., 2010).

Heavy metal contamination has been well documented and numerous studies have identified traditional medicines containing high levels of toxic heavy metals (Ernst and Thompson Coon, 2001; Dwivedi and Dey, 2002; Caldas and Machado, 2004; Saper et al., 2004; Kauffman et al., 2007; Rai et al., 2007; Meena et al., 2010; Rubio et al., 2012), sometimes exceeding permissible levels for some analyzed species (Sarma et al., 2011).

Toxic amounts of heavy metals in herbal preparations and case reports of heavy metal poisoning from traditional herbal products have been reported in the literature (Ernst, 2002; Tait et al., 2002; van Vonderen et al., 2000; Fuh et al., 2003). The potential severe consequences from side effects of certain herbal products have also been noted (Gurib-Fakim, 2006; Bush et al., 2007).

Trace metal monitoring in herbal products represents an important field of research that continuously provides relevant information for risk assessment and for setting permissible levels (Rubio et al., 2012). One continuing problem in protecting consumers of plant-based medicines is the lack of standardization of permissible levels of trace metals in herbal medicines by regulatory governmental entities (Sarma et al., 2011). The World Health Organization (WHO) has regulated maximum permissible limits of toxic metals like As, Hg, Cd, and Pb in herbal medicines, which amount to 10, 1.0, 0.3, and 10 ppm, respectively (WHO, 1998). However, the WHO has not set permissible limits for essential metals because many of them are considered micronutrients. Rai et al. (2007) analyzed herbal formulations for Pb and Cd, and results were within the aforementioned permissible limits. Markert (1994) reported the normal Cr, Mn, and Zn compositions of plants, which were 1.5, 200, and 50 ppm, respectively.

The United States has standardized recommended daily dietary allowances for essential dietary trace elements, but not for toxic metals (Sarma et al., 2011). In Europe, the European Commission discussed the need for setting maximum levels of Pb, Cd, and Hg in food supplements to amend the Commission Regulation (EC) No. 1881/2006, (2006), because monitoring studies found high levels of these metals in certain food supplements. The following limits were then set: 3.0 mg/kg for Pb; 1.0 mg/kg for Cd (except for seaweed products, where the limit was set at 3.0 mg/kg); and 0.10 mg/kg for Hg (Commission Regulation EC No. 629/2008, 2008; Gasser et al., 2009).

In 2009, the WHO and the Food and Agriculture Organization (FAO) jointly proposed acceptable levels for toxic substances that can be ingested on a weekly basis—the Provisional Tolerable Weekly Intake (PTWI). The following PTWI levels were set: Cd (7 μg/kg bw; 420 μg/person); Pb (25 μg/kg bw; 1500 μg/person); inorganic As (15 μg/kg bw; 900 μg/person); and Hg (5 μg/kg bw; 300 μg/person) (JECFA, 1988, 1999, 2005).

Few reports have addressed metal intakes from dietary supplements. A limited study on the safety of the botanical dietary supplements (*Echinacea*, garlic, *G. biloba*, ginseng, grape seed extract, kava kava, *Saw palmetto*, and St. John's wort) evaluated the presence of undesired heavy metals and did not find unacceptable concentrations that might pose a health risk to consumers, including pregnant women and children (Raman et al. (2004). Garcia-Rico et al. (2007) investigated the presence of Cu, Zn, Cd, Pb, and Hg in dietary supplements and the estimated daily intakes were below those recommended by WHO (1989), thus suggesting that little intake of metals is associated with the consumption of dietary supplements. However, daily intake of certain metals, like Pb, might increase due to the number of consumed dietary supplements, so the intake of combined supplements plus fortified foods may be a cause for health concern. Martín-Domingo et al. (2017) studied levels of metallic (Hg, Pb, Cr, and Cd), metalloids (As), and mineral elements (Fe, Mn, Cu, and Zn) in 12 common herbal teas (mentha, thyme, common sage, chamomile, lime flower, valerian, field horsetail, senna, lemon verbena, rooibos, green tea, and lime). Overall, levels were found to be in compliance with the maximum permissible limits set by WHO except for Cd in a few samples of thyme and chamomile. Regarding mineral elements, the highest Mn levels were found in red and green tea. However, considering the percentage of solubilization in herbal infusions and the bioavailability of each compound, the intake of commercial herbal teas did not represent a human health concern.

Cooper et al. (2007) found that certain traditional Chinese medicines (TCMs) were severely contaminated with As, Pb, and Hg. This unacceptable finding can place consumers at risk for severe or even fatal heavy metal/metalloid poisoning from the consumption of these natural therapies. However, the calculated dose of heavy metals/metalloids ingestion from these products is expected to be lower if their bioavailability is less than 100%. Thus the risk assessment of nonprescription medicines or supplements needs to consider data on bioavailability to ensure that future guidelines are achievable and efficient in preventing undue intake of dangerous ingredients.

The importance of good-quality control for medicinal herbs to protect consumers from heavy metal contamination should be emphasized (Başgel and Erdemoğlu, 2006). The estimation of trace metal residues in nutraceuticals and food formulations should be mandatory for pharmaceuticals and food ingredients to assure consumers of their quality.

Pesticides

Pesticides are among the most widely used chemicals in the world, and are also one of the most dangerous contaminants to humans. The application of pesticides in modern crops has become indispensable due to the growing demand for quantity and quality in products, particularly to increase crop productivity and minimize any possible losses due to uncontrollable pests.

Exposure to pesticide residues can result in many different adverse health consequences, from acute problems such as skin rashes and asthma attacks to chronic problems including cancer, neurological, reproductive, and respiratory disorders (Zuin and Vilegas, 2000; Calvert et al., 2004; Leung et al., 2005; Parrón et al., 2011, 2014).

Consumers of medicinal plants expect these herbs to be produced under good agriculture practices, with no residues of environmental pollutants, including pesticides. However, medicinal plants can be contaminated by pesticides in the process of cultivation, harvesting, and storage. As these compounds are often used to improve the yield and quality of agricultural products, including Chinese medicinal plants, the presence of pesticide residues in these materials becomes a pitfall for safety (Qing et al., 2009). Therefore it becomes absolutely essential to ensure the quality of medicinal plants and to detect the potential presence of pesticides to assess the safety of consumers and avoid the chronic toxicity related to long-term use (Table 68.1). However, analyses of crude herbal materials have often shown the presence of pesticide residues, which by themselves do not necessarily represent a health concern, unless the tolerances or maximum residue levels (MRLs) proposed by the international regulatory agencies or pharmacopeias are exceeded. When MRLs are exceeded, a more refined risk assessment should be conducted by comparing the intake of pesticide residues (on a body weight basis) with the acceptable daily intake (ADI), a reference dose that serves to protect the health of consumers.

High levels of pesticide residues in medicinal plants can impact the marketing of these products, especially those prepared for exports, because regulations in most importing countries are very strict. Nevertheless, in practice, the large-scale cultivation of medicinal and food plants is not possible without using pesticides (Ahmed et al., 2001; Sarkhail et al., 2012).

Based on their activity, pesticides are grouped as insecticides, fungicides, nematocides, herbicides, among others. According to their chemical structure, pesticides are classified as OP pesticides (e.g., chlorpyrifos and chlorpyrifos-methyl, coumaphos, dichlorvos, ethion, fenchlorphos, malathion, parathion), carbamates (*N*-substituted of carbamic acid, such as methomyl, oxamyl, carbaryl, carbofuran), organochlorine (OC) pesticides [e.g., hexachlorocyclohexane (HCH) isomers α-, β- and γ-(lindane), dichlorodiphenyl trichloroethane (DDT)], triazine herbicides (e.g., atrazine), and insecticides of plant origin (e.g., pyrethrins and rotenoids) (Kosalec et al., 2009).

Although the use of pesticides contributes to productivity and higher agricultural yield, there is growing concern about their adverse effects on human health. Because of their lipophilicity, OP insecticides are readily absorbed through ingestion, dermal absorption, or inhalation. Although their primary toxicity results from acetylcholinesterase inhibition in neural tissue, OPs, carbamate, and OC pesticides can also induce cellular oxidative stress by affecting mitochondrial function and may disrupt the neuronal and hormonal status of the body (Karami-Mohajeri and Abdollahi, 2011; Sarkhail et al., 2012; Gupta and Milatovic, 2012; Milatovic et al., 2013; Gupta et al., 2015).

Unfortunately, incidents concerning relatively high levels of pesticide residues have repeatedly been found in herbs exported from different countries (Leung et al., 2005). Commonly encountered pesticides in herb products include OP, OC, carbamates, pyrethroids, and paraquat. Residues of malathion, dimethoate, chlorpyrifos, and propenophos have been reported in medicinal herbs from Egypt, with malathion being the OP insecticide most often detected (Ahmed et al., 2001). Malathion, followed by profenofos, was also the OP most frequently found in leafy vegetables and some aromatic medicinal plants collected from different areas of Egypt (Dogheim et al., 2004). Residues of malathion and diazinon have also been reported in herbal drugs from Iran (Sarkhail et al., 2012). The relatively high occurrence of malathion residues may be accounted for by the fact that it is usually the recommended pesticide to be used on medicinal plants worldwide (Ahmed et al., 2001).

Although OC pesticides possess comparatively long residual actions, most are now banned because of their low biodegradability and persistence in the environment, where they can be detected for long periods of time and pass up the food chain. These pesticides also show long-term toxicity, particularly cancer and endocrine disruption potential (Mnif et al., 2011).

Xue et al. (2008) analyzed different TCMs for OC pesticides, and those most commonly found were benzene hexachloride (α-BHC), pentachloronitrobenzene (PCNB), hexachlorobenzene, and tecnazene. Approximately three-quarters (75.8%) of the herbal medicines studied by these authors contained at least one OC insecticide, and more than 50% of samples contained two of them. Nevertheless, their concentration was below the MRL set in the regulatory pharmacopeia. Quintozene, endosulfan, and BHC were the pesticide residues most commonly detected in ginseng products (Sohn et al., 2004). Despite DDT series and γ-BHC being pesticides frequently encountered in Chinese herbal medicinal products, their concentrations were, in general, below the allowable limits set by different countries (Wong et al., 2007).

Harris et al. (2011) analyzed pesticide content in commonly prescribed individual raw Chinese herbal medicines and reported the following proportion of samples with an elevated level of background pesticide exposure: 26% for chlorpyrifos; 1.4% for methyl-parathion; and 0.3% for esfenvalerate, fenvalerate, fipronil, lindane, and quintozene. Overall, pesticides most often found in Chinese herbal products include BHC, DDT, and PCNB, with pyrethroids and aminoformins rarely being found. Among the potentially polluting pyrethroids, attention has been focused on permethrin and cypermethrin (Qing et al., 2009). In contrast, there are also studies in which residues of OC, OP, and pyrethroid pesticides have not been detected in any herbal medicinal plants (Rao et al., 2011).

Residues of HCH isomers and metabolites have been detected in some herbal Ayurvedic formulations (Rai et al., 2007). Although some DDT metabolites were also found, HCH concentrations were comparatively higher than those of DDT. γ-HCH was the HCH isomer found in the greatest concentrations in all samples, as it is one of the more persistent isomers. Rai et al. (2008) also reported the presence of total HCH and its isomers in "*Dashmoola*," a popular Ayurvedic herbal formulation with immunomodulatory and febrifugal properties.

α-HCH and γ-HCH, the main constituents of commercial HCH formulations, were detected in 97.5% samples, with γ-HCH being more prominent in comparison to the rest of the isomers. Only 5% of the samples contained DDT (or metabolites) residues. These findings indicate that residual HCH build-up is more prevalent in plant samples than DDT.

With regard to nutraceuticals other than herbal medicines, residues of HCB, HCH isomers, and DDT as well as polychlorinated biphenyls have been found in cod liver oil used as a dietary supplement (Storelli et al., 2004). HCB contributed very little to the overall contaminant burden of dietary supplement oils, whereas HCH isomers were generally below instrument detection limits (Bengtson Nash et al., 2014). In light of these findings, and considering that these and other persistent organic contaminants are accumulated in the lipid stores of organisms from marine ecosystems, there is a need for strict and continuous monitoring of their content in fish and krill oil products to reduce as much as possible the risks for human health (Bengtson Nash et al., 2014). Studies monitoring multipesticide residues in different types of Chinese and one Korean medicinal herbs revealed that 3.1% of samples contained residues of five pesticides, namely, methoxychlor, DDT, γ-HCH, endosulfan, and procymidone (Oh, 2009).

Mycotoxins

Mycotoxins are secondary metabolites produced by toxigenic fungi, such as *Aspergillus*, *Alternaria*, *Penicillium*, and *Fusarium*, which are capable of causing serious toxicity and even death in humans. They are low-molecular-weight toxic metabolites produced by molds that can contaminate medicinal herbs and their preparations and products, particularly those cultivated in tropic and subtropic areas. From a food safety perspective the mycotoxins of greatest concern include aflatoxins, ochratoxin A, patulin, and toxins produced by *Fusarium* molds (fumonisins, trichothecenes, zearalenone, and zearalenol, among others). Nonetheless, aflatoxins and ochratoxin A are the most important and most frequently detected mycotoxins in raw medicinal herbs (Tripathy et al., 2015). Mycotoxin production depends on several factors, including genetic features, substrate, humidity, CO_2:O_2 ratio, and the presence of fungicides or other competitive microbial species (Kosalec et al., 2009; Martins et al., 2001).

Although the presence of toxicogenic molds in a herbal plant does not imply the presence of mycotoxins in the product, it represents a potential risk of contamination with mycotoxins. Microbiological and mycotoxicological quality assessment of medicinal herbs should include mycotoxin contamination (Table 68.1), especially of herbs grown in hot and humid climates, and in the herbal parts showing higher risk of contamination (Kabelitz and Sievers, 2004). Nevertheless, many bioactive components from medicinal herbs can regulate the fungal growth and mycotoxin production and also afford protection against mycotoxin absorption, metabolism, distribution, excretion, and toxicity (Do et al., 2015).

Raman et al. (2004) evaluated several botanical supplements and found the presence of molds in a great number of samples. The most frequently isolated molds in medicinal plants were *Penicillium* sp., *Aspergillus niger*, and *Fusarium* sp. (Abou-Arab et al., 1999). Packed samples of medicinal plants have a higher probability of being infected with molds than nonpacked samples, because of the increased humidity inside the pack and unsuitable storage methods.

Mycotoxin occurrence in herbal medicines has been extensively investigated. Although 15.5% of medicinal Chinese herbs have shown aflatoxin contamination, the high levels found in some of them are enough to cause serious health problems (Liu et al., 2012). Bugno et al. (2006) studied medicinal herbs collected from a Brazilian market and found molds of the *Aspergillus* and *Penicillium* genera to be the most common contaminants of raw medicinal herbs. Santos et al. (2009) found that all the analyzed medicinal herb samples collected in Spain were contaminated with several mycotoxins, and nearly 87% showed the combination of four or more mycotoxins. Conversely, Romagnoli et al. (2007) studied aflatoxin B_1 contamination in different kinds of spices, aromatic herbs, and medicinal plants from Italy, and none of the plants analyzed showed detectable levels of aflatoxins.

The European legislation has set maximum levels of certain mycotoxins (aflatoxin B_1 and the sum of B_1, B_2, G_1, and G_2) for a variety of foodstuffs and spices. The European Pharmacopeia has also provided a limit of 2 μg/kg for aflatoxin B_1 and 4 μg/kg for the sum of aflatoxins B_1, B_2, G_1, and G_2 for some medicinal herbs. For ochratoxin A (OTA), a limit of 20 μg/kg has been adopted in liquorice root (*Glycyrrhiza glabra*) [Commission Regulation EC No 472/2002, 2002; EDQM (Directorate for the Quality of Medicines and Health Care of the Council of Europe), 2007a,b].

Aflatoxin B_1 is the most common and toxic metabolite produced by *Aspergillus flavus*, and its toxic effects include immunosuppressive, mutagenic, teratogenic, and hepatocarcinogenic activity (Prado et al., 2012). OTA is a mycotoxin with nephrotoxic, hepatotoxic, embryotoxic, teratogenic, neurotoxic, immunotoxic, genotoxic, and carcinogenic properties. It is produced by some fungal species, notably *Penicillium verrucosum* and *Aspergillus ochraceus*, and to a lesser extent *Aspergillus carbonarius* and *A. niger*. All these mold species can contaminate medicinal herbs with OTA, particularly in temperate and colder zones (Pitt, 2000; Ostry et al., 2013).

Fumonisins are mycotoxins mainly produced by *Fusarium verticillioides* (*F. moniliforme*) and *Fusarium proliferatum*, with the most abundant being fumonisin B_1 (FB_1), which is considered as possibly carcinogenic (class 2B). Only a few studies have investigated FB_1 contamination of medicinal herbs and herbal products. Omurtag and Yazicioğlu (2004) measured levels of FB_1 contamination in herbal teas and medicinal plants regularly consumed in Turkey. FB_1 was detected in two samples (0.160 and 1.487 μg/g). Sewram et al. (2006) investigated the presence of FB_1 in 19 dietary and 30 medicinal wild plants used by residents of eastern South Africa. Eight plants, four dietary and four medicinal, were positive for FB_1 at levels ranging from 34 to 524 μg/kg and from 8 to 1553 μg/kg, respectively.

Citrinin is another mycotoxin produced by *Monascus* sp. (*M. purpureus*, *M. ruber*) and *Penicillium* sp. (*P. citrinum*, *P. expansum*, *P. radicicola*, and *P. verrucosum*). Although this mycotoxin has a powerful nephrotoxic effect, it is devoid of mutagenic and carcinogenic potential. Citrinin has been found in foodstuffs of vegetable origin (e.g., cereals, pomaceous fruits, black olive, roasted nuts, spices), food supplements based on rice fermented with the red microfungi *M. purpureus*, and in foodstuffs of animal origin (e.g., cheese) (Ostry et al., 2013). *Monascus*-fermented rice has recently become a popular dietary supplement to reduce serum cholesterol level because many of its bioactive constituents, particularly monacolins, are inhibitors of 3-hydroxy-3-methylglutaryl-coenzyme A reductase. However, controversy about its safety has been raised because citrinin is produced along with the *Monascus* secondary metabolites by certain strains or under certain cultivation conditions (Lin et al., 2008).

Other environmental contaminants

PA are heterocyclic organic compounds found in more than 6000 plants (around 3% of world flora). These alkaloids can thus contaminate foodstuff, beverages (milk, teas, and herbal infusions), and herbal teas and medicines, including St. John's wort. Nonetheless, the contamination is attributed to raw materials that are harvested along with weeds containing PAs. Some PAs have been shown to have genotoxic and carcinogenic effects and also possess subchronic toxic effects, particularly veno-occlusive liver disease. Due to reduced glutathione content in sinusoidal cells, the detoxification capacity by glutathione conjugation is lower than that of hepatocytes, resulting in higher intercellular content of free radicals. At present, and according to the Herbal Medicinal Products Committee at the European Medicines Agency, exposure to PA from herbal medicinal products should not exceed a daily intake of 1 μg (Habs et al., 2018).

PAH are a large group of organic compounds consisting of two or more fused aromatic carbon rings that may originate from combustion or pyrolysis of organic matter (e.g. coal, oil, petrol, and wood) during various industrial processes and human activities. These compounds are delivered to the atmosphere from where they can be transported over a long distance to undergo deposition in sediments, soils, and plants, where they can be accumulated. Hence, PAH are considered ubiquitous environmental pollutants that raise health concerns because many of them have carcinogenic and/or mutagenic properties.

Potential sources of occurrence of PAHs in herbal medicines include smoking, roasting, drying processes, and environmental influences. PAHs can also accumulate in herbal plants during the thermal treatment (drying or smoking) in the manufacturing process. Although the contribution of herbal medicines to total PAH intake is considered to be very low, potential sources of contamination should be identified and their potential impact on the quality of the herbal product should be examined (Steinhoff, 2019). While PAH limits do not exist for medicinal products, Regulation (EU) No. 2015/1933 describes maximum levels for PAHs in food supplements, dried herbs, and dried spices.

Radioactive contamination of nutraceuticals and food ingredients

Mushrooms are a good source of digestible proteins and fiber, are low in fat and calories, and provide a valuable vitamin and mineral intake. For this reason, there is a growing interest in the use of dried mushroom extracts as NFI for health promotion. Some mushroom species have been shown to lower cholesterol levels, have antioxidant activity, and may modulate mononuclear cell activation and the phenotypic expression of cytokines, thus enhancing immune system function. Despite these beneficial effects, there are concerns because of the high heavy metal concentrations that may be found in wild edible fungi. These compounds are a known source of chronic poisoning and can also contribute markedly to contamination with radioactive chemicals such as radioactive cesium (^{137}Cs) (Table 68.1) (Borchers et al., 2004). Higher radiocesium concentrations are found in fungi than in other foodstuffs. In addition, accumulation of radiocesium was reported in wild mushrooms from Eastern Europe after the Chernobyl nuclear accident in 1986. However, in countries not affected by this nuclear accident the dose of radiocesium associated with fungi consumption does not constitute a risk as it is below the recommended limit (de Román et al., 2006).

Phosphate feed supplements may contribute to the radiation exposure of the population because phosphates usually contain appreciable quantities of uranium (U) and its daughters (^{238}U, ^{234}U, ^{226}Ra, ^{210}Po, and ^{210}Pb). The exposure of consumers depends on the degree of equilibrium of the decay chain in the feed and through the metabolic process. Poultry products, particularly chicken meat, organs, and eggs, have been shown to be contaminated through poultry feed supplements containing phosphorus (Izak-Biran et al., 1989).

Adulteration and undeclared chemical substances in nutraceuticals and food ingredients

Terms such as "natural," "herbal," and "dietary supplement" are sometimes used as a means to mislead consumers, health professionals, and health authorities, and to obscure the potential adulteration of these products with synthetic active compounds (Table 68.1).

Dietary supplements have widespread use in sports, and most athletes competing at the highest levels use some form of dietary supplementation. Many multiingredient performance supplements (e.g., caffeine, creatine, nitrate/beetroot juice, β-alanine, bicarbonate, and phosphate) may represent a risk for athletes because of the risk of ingesting banned substances sometimes present as contaminants. While in most cases this contamination comes from poor manufacturing practices or inadequate storage conditions, there is some evidence of deliberate adulteration (Maughan, 2005). Inadvertent contamination or deliberate addition of stimulants, estrogenic compounds, diuretics, or anabolic agents may occur with the aim of increasing the efficacy of dietary supplements in sports medicine (Deldicque and Francaux, 2016).

A doping case associated with the use of an ephedra-labeled dietary supplement has been reported (Ros et al., 1999), which illustrates that the potential adulteration of herbal food supplements with undeclared agents has to be considered. The predominance of norpseudoephedrine over ephedrine indicates deliberate spiking with the former compound. While manufacturers usually meet their labeling claims for ephedra-free products (Tam et al., 2006), special attention should be given to the presence of drugs like caffeine, synephrine, and botanical sources of caffeine because these ingredients have replaced ephedra in some dietary supplements.

A number of synthetic drugs used for weight loss, hypoglycemic, antihypertensive, and antiinflammatory agents, or anabolic steroids (e.g., testosterone, nandrolone) are added to dietary supplements or herbal remedies to enhance their biological effects (Vaclavik et al., 2014). Many products labeled "herbal" or "all natural" (herbal/natural) are marketed as over-the-counter dietary supplements for the treatment of erectile dysfunction because they claim to enhance sexual performance. However, adulteration with undeclared phosphodiesterase type 5 inhibitors (e.g., sildenafil and its structurally modified analogs), which have proven efficacy in the treatment of erectile dysfunction, appears to be widespread among the Chinese over-the-counter medicines labeled for the treatment of sexual dysfunction in men (Campbell et al., 2013).

Risk assessment of toxic contaminants in nutraceuticals and food ingredients

The regulatory framework for the quality of herbal medicines consists of pharmacopoeia requirements, related to limits of potential contaminants and analytical methodologies for their determination. The maximum amounts of toxic metals and nonmetals in medicinal plant materials can be given based on the provisional tolerable intake values. The limits for toxic metals in herbal medicines and products vary throughout the world. The use of herbal medicinal products is not generally expected to contribute significantly to the exposure of the population to heavy metal contaminants. However, it should be understood that the heavy metal content of herbal medicines adds to the burden originating from food, therefore it is recommended that heavy metal contamination is minimized (WHO, 2007).

Limits for pesticide residues should be established following the recommendations of the Joint FAO/WHO Meeting on Pesticide Residues (JMPR), which includes an ADI and the analytical methodology for the assessment of pesticide residues. However, there are no standard procedures for assignment of MRL to medicinal plants. The methodology used for food commodities could be applied when a botanically identical medicinal plant is used as food. If so, then the established MRL for a specific pesticide in the latter could be regarded as the relevant MRL for the specific raw medicinal plant material. The toxicological evaluation of pesticide residues in herbal materials should be based on the likely intake of the material by consumers. In the absence of a full risk assessment, it is recommended that the intake of residues from herbal materials should account for no more than 1% of total intake from all sources (WHO, 2007).

MRL have been set for food and animal feed by different world organizations, including WHO, the FAO, the US Environmental Protection Agency (EPA), and the EU. In the EU, pesticide residues are regulated in Chapter 2.8.13 of the European Pharmacopoeia and Regulation (EC) 396/2005 (2005). The latter defines MRL as the upper legal concentration limit for a pesticide residue in or on food or feed, based on good agricultural practices and the lowest consumer exposure. Because this regulation only concerns raw materials, there is a need to define MRL for NFI to assure the safety of this type of product (Martínez-Domínguez et al., 2014).

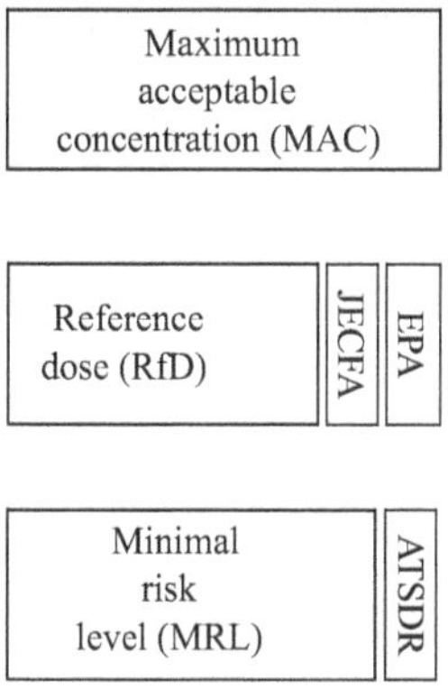

FIGURE 68.2 Proposed values for risk assessment of toxic contamination in nutraceuticals and food ingredients.

Harris et al. (2011) interpreted the toxicological significance of levels of contaminants found in NFI (Fig. 68.2) following three different comparisons. First, levels of contaminants were expressed as percent of the reference dose or population-adjusted dose from the Joint FAO/WHO Expert Committee on Food Additives (JECFA) and EPA IRIS (Integrated Risk Information System). Second, concentration of contaminants was also compared with the minimal risk levels provided by the Agency for Toxic Substances and Disease Registry, which are based on toxicological studies in animals and on reports of human occupational exposure. Minimal risk level is an estimate of the daily human exposure to a hazardous substance (based on an average body mass of 70 kg) that is likely to be without appreciable risk of adverse noncancer health effects over a specified duration of exposure. Third, levels of contaminants were also compared with the maximum acceptable concentration limits proposed for dietary supplements from the NSF/ANSI Standard 173 and the European Pharmacopoeia for dried herbal infusions (Harris et al., 2011). Based on the assumptions of the mode of consumption of raw Chinese herbal medicines, the 99% and 95% of samples tested by Harris et al. (2011) for heavy metals and pesticide residues, respectively, were likely to be of negligible concern.

Concluding remarks and future directions

The general public as well as healthcare professionals should be better informed regarding the basic concept of NFI and their usefulness, and should also be warned about the potential adverse effects associated with their use due to the potential for toxic contamination. Because of the increased use of a variety of alternative remedies, physicians should also be aware of these potential sources of toxic contamination when diagnosing conditions of uncertain etiology that have similarities to toxic syndromes.

Given that many circles foster the use of herbal medicines as an alternative to synthetic pharmaceutical compounds for the prevention and treatment of particular diseases, there is a need for strict guidelines and more focus by the international regulatory legislation (Ahmed et al., 2001). Thus an appropriate regulatory framework should be developed for medicinal herbs and herbal products to prevent and screen for contamination and to ensure that safety and quality standards are met (Kosalec et al., 2009). Implementation of a regulatory system will require evidence of safety of ingredients, quality assurance (including authentication of herbal ingredients), and encouraging appropriate research on efficacy (Shaw et al., 1997).

Phytomedicinal safety assessments necessarily involve the development of adequate analytical procedures for the analysis of toxic compounds in medicinal plants. These procedures must be implemented by the pharmaceutical industry and official regulatory laboratories as screening protocols to identify heavy metals, pesticides, and mycotoxins in herbal drugs and other NFI (Zuin et al., 2003). In this regard, consumers should be advised to minimize long-term consumption of herbal infusions until data on the levels and varieties of different toxic contaminants are made widely available. Care must be taken to pick medicinal plants from areas free of any contamination source (Manteiga et al., 1997).

If the methods for the correct and appropriate use of medicinal plants are correctly applied, then they can contribute to protecting and improving consumer health and well-being. The correct use of such methods should follow the criteria of safety, efficacy, and quality, which characterize modern medical practice and are the basis for consumer protection.

The efficacy and safety of herbal remedies or food supplements have been assessed by many studies and adverse effects have been reported. These data need to be brought together, along with data collected from adverse reaction monitoring systems, to enable patients and physicians to make the best risk–benefit assessment before using any herbal or traditional medicine. Definitive and greater control over the safety and quality of NFI could be achieved through good manufacturing practice, regulatory control, and research efforts, as well as by reporting of adverse events (Koh and Woo, 2000). Regulators should also take action to minimize these risks.

Finally, additional research is needed to define the pharmacology, stability, and bioavailability of dietary supplements, including medicinal herbs, to improve the safety and efficacy of these products (Bent and Ko, 2004).

References

Abou-Arab, A.A.K., Kawther, M.S., Tantawy, M.E.E.I., 1999. Quantity estimation of some contaminants in commonly used medicinal plants in the Egyptian market. Food Chem. 67, 357–363.

Ahmed, M.T., Loutfy, N., Yousef, Y., 2001. Contamination of medicinal herbs with organophosphorus insecticides. Bull. Environ. Contam. Toxicol. 66, 421–426.

Arpadjan, S., Celik, G., Taşkesen, S., 2008. Arsenic, cadmium and lead in medicinal herbs and their fractionation. Food Chem. Toxicol. 46 (8), 2871–2875.

Barthwal, J., Nair, S., Kakkar, P., 2008. Heavy metal accumulation in medicinal plants collected from environmentally different sites. Biomed. Environ. Sci. 21, 319–324.

Başgel, S., Erdemoğlu, S.B., 2006. Determination of mineral and trace elements in some medicinal herbs and their infusions consumed in Turkey. Sci. Total Environ. 359 (1–3), 82–89.

Bengtson Nash, S.M., Schlabach, M., Nichols, P.D., 2014. A nutritional-toxicological assessment of antarctic krill oil versus fish oil dietary supplements. Nutrients 6 (9), 3382–3402.

Bent, S., Ko, R., 2004. Commonly used herbal medicines in the United States: a review. Am. J. Med. 116 (7), 478–485.

Bhat, R., Kiran, K., Arun, A.B., 2010. Determination of mineral composition and heavy metal content of some nutraceutically valued plant products. Food Anal. Methods 3, 181–187.

Blicharska, E., Komsta, L., Kocjan, R., 2010. A preliminary study on the effect of mineralization parameters on determination of metals in Viscum album species. Cent. Eur. J. Chem. 8, 264–268.

Borchers, A.T., Keen, C.L., Gershwin, M.E., 2004. Mushrooms, tumors, and immunity: an update. Exp. Biol. Med. 229 (5), 393–406.

Bugno, A., Almodovar, A.A.B., Pereira, T.C., 2006. Occurrence of toxigenic fungi in herbal drugs. Braz. J. Microbiol. 37, 47–51.

Bush, T.M., Rayburn, K.S., Holloway, S.W., 2007. Adverse interactions between herbal and dietary substances and prescription medications: a clinical survey. Altern. Ther. Health Med. 13, 30–35.

Caldas, E.D., Machado, L.L., 2004. Cadmium, mercury and lead in medicinal herbs in Brazil. Food Chem. Toxicol. 42, 599–603.

Calvert, G.M., Plate, D.K., Das, R., 2004. Acute occupational pesticide-related illness in the US, 1998–1999: surveillance findings from the SENSOR-pesticides program. Am. J. Ind. Med. 45 (1), 14–23.

Campbell, N., Clark, J.P., Stecher, V.J., 2013. Adulteration of purported herbal and natural sexual performance enhancement dietary supplements with synthetic phosphodiesterase type 5 inhibitors. J. Sex. Med. 10 (7), 1842–1849.

Chan, K., 2003. Some aspects of toxic contaminants in herbal medicines. Chemosphere 52 (9), 1361–1371.

Choudhury, R.P., Kumar, A., Garg, A.N., 2006. Analysis of Indian mint (*Mentha spicata*) for essential, trace and toxic elements and its antioxidant behaviour. J. Pharm. Biomed. Anal. 41 (3), 825–832.

Commission Regulation (EC) No 472/2002, 2002. Commission Regulation (EC) No 472/2002 of 12 March 2002 amending Regulation (EC) No 466/2001 setting maximum levels for certain contaminants in foodstuffs. Off. J. Eur. Commun. 45 (L075), 18–20.

Commission Regulation (EC) No. 1881/2006, 2006. Commission Regulation (EC) No. 1881/2006 of 19 December 2006 setting maximum levels for certain contaminants in foodstuffs as regards heavy metals. Off. J. Eur. Union. 49 (L364), 5–24.

Commission Regulation (EC) No. 629/2008, 2008. Commission Regulation (EC) No. 629/2008, amending Regulation (EC) No. 1881/2006 of 19 December 2006 setting maximum levels for certain contaminants in foodstuffs. Off. J. Eur. Union. 51 (L173), 6–9.

Cooper, K., Noller, B., Connell, D., 2007. Public health risks from heavy metals and metalloids present in traditional Chinese medicines. J. Toxicol. Environ. Health A. 70 (19), 1694–1699.

de Román, M., Boa, E., Woodward, S., 2006. Wild-gathered fungi for health and rural livelihoods. Proc. Nutr. Soc. 65, 190–197.

de Smet, P.A., 2002. Herbal remedies. N. Engl. J. Med. 347, 2046–2056.

DeFelice, S.L., 1995. The nutraceutical revolution: its impact on food industry R&D. Trends Food Sci. Technol. 61, 59–61.

Deldicque, L., Francaux, M., 2016. Potential harmful effects of dietary supplements in sports medicine. Curr. Opin. Clin. Nutr. Metab. Care 219, 439–445.

Do, K.H., An, T.J., Oh, S.K., Moon, Y., 2015. Nation-based occurrence and endogenous biological reduction of mycotoxins in medicinal herbs and spices. Toxins (Basel) 7, 4111–4130.

Dogheim, S.M., Ashraf, El. M.M., Alla, S.A., 2004. Pesticides and heavy metals levels in Egyptian leafy vegetables and some aromatic medicinal plants. Food Addit. Contam. 21 (4), 323–330.

Dolan, S.P., Nortrup, D.A., Bolger, P.M., 2003. Analysis of dietary supplements for arsenic, cadmium, mercury, and lead using inductively coupled plasma mass spectrometry. J. Agric. Food Chem. 51, 1307–1312.

Dwivedi, S.K., Dey, S., 2002. Medicinal herbs: a potential source of toxic metal exposure for man and animals in India. Arch. Environ. Health 57 (3), 229–231.

EDQM (Directorate for the Quality of Medicines and Health Care of the Council of Europe), 2007a. Determination of aflatoxin B1 in herbal drugs, European Pharmacopoeia, sixth ed. EDQM, Strasbourg, 2007.

EDQM (Directorate for the Quality of Medicines and Health Care of the Council of Europe), 2007b. Determination of ochratoxin A in herbal drugs, European Pharmacopoeia, sixth ed. EDQM, Strasbourg, 2007.

Ernst, E., 1998. Harmless herbs? A review of the recent literature. Am. J. Med. 104, 171–178.

Ernst, E., 2002. Toxic heavy metals and undeclared drugs in Asian herbal medicines. Trends Pharmacol. Sci. 23 (3), 136–139.

Ernst, E., Thompson Coon, J., 2001. Heavy metals in traditional Chinese medicines: a systematic review. Clin. Pharmacol. Ther. 70, 497–504.

Frankos, V.H., Street, D.A., O'Neill, R.K., 2010. FDA regulation of dietary supplements and requirements regarding adverse event reporting. Clin. Pharmacol. Ther. 87, 239–244.

Fuh, C., Lin, H., Tsai, H., 2003. Determination of lead, cadmium, chromium and arsenic in 13 herbs of tocolysis formulation using atomic absorption spectrometry. J. Food Drug Anal. 11, 39–45.

Garcia-Rico, L., Leyva-Perez, J., Jara-marini, M.E., 2007. Content and daily intake of copper, zinc, lead, cadmium and mercury from dietary supplements in Mexico. Food Chem. Toxicol. 45, 1599–1605.

Gasser, U., Klier, B., Kuhn, A.V., 2009. Current findings on the heavy metal content in herbal drugs. Pharmeur. Sci. Notes 1, 37–49.

Genuis, S.J., Schwalfenberg, G., Siy, A.K., 2012. Toxic element contamination of natural health products and pharmaceutical preparations. PLoS One 7 (11), e49676.

Gomez, M.R., Cerruti, S., Sombra, L.L., 2007. Determination of heavy metals for the quality control in Argentinean herbal medicines by ETAAS and ICP-OES. Food Chem. Toxicol. 45, 1060–1064.

Goyer, R.A., Clarkson, T.W., 2001. Toxic effects of metals. In: Amdur, M.O., Doull, J., Klaassen, C.D. (Eds.), Toxicology the Basic Science of Poisons. McGraw-Hill Press, pp. 623–680.

Gulati, O.P., Berry Ottaway, P., 2006. Legislation relating to nutraceuticals in the European Union with a particular focus on botanical-sourced products. Toxicology 221 (1), 75–87.

Gupta, R.C., 2009. Toxicology of the placenta. In: Ballantyne, B., Marrs, T.C., Syversen, T. (Eds.), General and Applied Toxicology. John Wiley and Sons, Chichester, pp. 2003–2039.

Gupta, R.C., 2011. Placental toxicity. In: Gupta, R.C. (Ed.), Reproductive and Developmental Toxicology. Academic Press/Elsevier, Amsterdam, pp. 1067–1085.

Gupta, R.C., D. Milatovic, 2012. Toxicology of organophosphates and carbamates. In: T.C. Marrs. Mammalian Toxicology of Insecticid, 136.

Gupta, R.C., Milatovic, D., Fan, A.M., Fan, A.M., Khan, E.M., Alexeeff, G.V., 2015. Cholinesterase–inhibiting pesticides: toxicity, mechanisms, and implications for risk assessment. In: Fan, A.M., Khan, E.M., Alexeeff, G.V. (Eds.), Toxicology and Risk Assessment, Principles and Applications. Pan Stanford Publishing, Singapore, pp. 383–452.

Gurib-Fakim, A., 2006. Medicinal plants: traditions of yesterday and drugs of tomorrow. Mol. Aspects Med. 27, 1–93.

Habs, M., Binder, K., Krauss, S., et al., 2018. A balanced risk-benefit analysis to determine human risks associated with pyrrolizidine alkaloids (PA)—the case of herbal medicinal products containing St. John's Wort Extracts (SJW). Nutrients 10, 204.

Haider, S., Naithani, V., Barthwal, J., 2004. Heavy metal content in some therapeutically important medicinal plants. Bull. Environ. Contam. Toxicol. 72 (1), 119–127.

Harris, E.S., Cao, S., Littlefield, B.A., 2011. Heavy metal and pesticide content in commonly prescribed individual raw Chinese herbal medicines. Sci. Total Environ. 409 (20), 4297–4305.

Hsu, M.J., Selvaraj, K., Agoramoorthy, G., 2006. Taiwan's industrial heavy metal pollution threatens terrestrial biota. Environ. Poll. 143, 327–334.

Izak-Biran, T., Schlesinger, T., Weingarten, R., 1989. Concentrations of U and Po in animal feed supplements, in poultry meat and in eggs. Health Phys. 56 (3), 315–319.

JECFA, 1988. Joint FAO/WHO expert committee on food additives. In: Thirty-Third Meeting. Summary and Conclusions. World Health Organization, Geneva.

JECFA, 1999. Joint FAO/WHO expert committee on food additives. In: Fifty-Third Meeting. Summary and Conclusions. World Health Organization, Geneva.

JECFA, 2005. Joint FAO/WHO expert committee on food additives. In: Sixty-Fourth Meeting. Summary and Conclusions. World Health Organization, Geneva.

Kabelitz, L., Sievers, H., 2004. Contaminants of medicinal and food herbs with a view to EU regulations. Innov. Food Technol. 25–27.

Karami-Mohajeri, S., Abdollahi, M., 2011. Toxic influence of organophosphate, carbamate, and organochlorine pesticides on cellular metabolism of lipids, proteins, and carbohydrates: a systematic review. Hum. Exp. Toxicol. 30, 1119–1140.

Kauffman, J.F., Westenberger, B.J., Robertson, J.D., 2007. Lead in pharmaceutical products and dietary supplements. Regul. Toxicol. Pharmacol. 48 (2), 128–134.

Koh, H.L., Woo, S.O., 2000. Chinese proprietary medicine in Singapore: regulatory control of toxic heavy metals and undeclared drugs. Drug Saf. 23 (5), 351–361.

Kosalec, I., Cvek, K.J., Tomic, S., 2009. Contaminants of medicinal herbs and herbal products. Arch. Ind. Hyg. Toxicol. 60, 485–501.

Leung, K.S.Y., Chan, K., Chan, C.L., 2005. Systematic evaluation of organochlorine pesticide residues in Chinese materia medica. Phytother. Res. 19, 514–518.

Lin, Y.L., Wang, T.H., Lee, M.H., 2008. Biologically active components and nutraceuticals in the Monascus-fermented rice: a review. Appl. Microbiol. Biotechnol. 77 (5), 965–973.

Liu, L., Jin, H., Sun, L., Ma, S., Lin, R., 2012. Determination of aflatoxins in medicinal herbs by high-performance liquid chromatography-tandem mass spectrometry. Phytochem. Anal. 23, 469–476.

Lockwood, B., 2007. Nutraceuticals: A Guide for Healthcare Professionals. Pharmaceutical Press, Great Britain.

Łozak, A., Sołtyk, K., Ostapczuk, P., 2002. Determination of selected trace elements in herbs and their infusions. Sci. Total Environ. 289, 33–40.

Malik, J., Szakova, J., Drabek, O., 2008. Determination of certain micro and macroelements in plant stimulants and their infusions. Food Chem. 111, 520–525.

Maniacal, P.P., Wanwimolruk, S., 2001. Effect of herbal teas on hepatic drug metabolizing enzymes in rats. J. Pharm. Pharmacol. 53, 1323–1329.

Manteiga, R., Park, D.L., Ali, S.S., 1997. Risks associated with consumption of herbal teas. Rev. Environ. Contam. Toxicol. 50, 1–30.

Markert, B., 1994. Plants as biomonitors. Potential advantages and problems. In: Adriano, D.C., Chen, Z.S., Yang, S.S. (Eds.), Biogeochemistry of Trace Elements. Science & Technology Letters, Northwood, pp. 601–613.

Martena, M.J., Van Der Wielen, J.C.A., Rietjens, I.M.C.M., 2010. Monitoring of mercury, arsenic, and lead in traditional Asian herbal preparations on the Dutch market and estimation of associated risks. Food Addit. Contam. Part. A Chem. Anal. Control Expo. Risk Assess. 27, 190–205.

Martín-Domingo, M.C., Pla, A., Hernández, A.F., Olmedo, P., Navas-Acien, A., Lozano-Paniagua, D., et al., 2017. Determination of metalloid, metallic and mineral elements in herbal teas. Risk assessment for the consumers. J. Food Comp. Anal. 60, 81–89.

Martínez-Domínguez, G., Plaza-Bolaños, P., Romero-González, R., 2014. Analytical approaches for the determination of pesticide residues in nutraceutical products and related matrices by chromatographic techniques coupled to mass spectrometry. Talanta 118, 277–291.

Martins, H.M., Martins, M.L., Dias, M.I., 2001. Evaluation of microbiological quality of medicinal plants used in natural infusions. Int. J. Food Microbiol. 68, 149–153.

Maughan, R.J., 2005. Contamination of dietary supplements and positive drug tests in sport. J. Sports Sci. 23 (9), 883–889.

Mazzanti, G., Battinelli, L., Daniele, C., 2008. Purity control of some Chinese crude herbal drugs marketed in Italy. Food Chem. Toxicol. 46, 3043–3047.

Meena, A.K., Bansal, P., Kumar, S., 2010. Estimation of heavy metals in commonly used medicinal plants: a market basket survey. Environ. Monit. Assess. 170 (1–4), 657–660.

Milatovic, D., Zaja-Milatovic, S., Gupta, R.C., 2013. Biomarkers of oxidative/nitrosative stress and neurotoxicity. In: Gupta, R.C. (Ed.), Biomarkers in Toxicology. Academic Press/Elsevier, Amsterdam, pp. 863–882.

Mnif, W., Hassine, A.I., Bouaziz, A., 2011. Effect of endocrine disruptor pesticides: a review. Int. J. Environ. Res. Public Health 8, 2265–2303.

Oh, C.H., 2009. Monitoring of residual pesticides in herbal drug materials of Korea and China. Bull. Environ. Contam. Toxicol. 82, 639–643.

Omurtag, G.Z., Yazicioğlu, D., 2004. Determination of fumonisins B1 and B2 in herbal tea and medicinal plants in Turkey by high-performance liquid chromatography. J. Food Prot. 67, 1782–1786.

Ostry, V., Malir, F., Ruprich, J., 2013. Producers and important dietary sources of ochratoxin A and citrinin. Toxins 5 (9), 1574–1586.

Özcan, M.M., Unver, A., Uc, T., 2008. Mineral content of some herbs and herbal teas by infusion and decoction. Food Chem. 106, 1120–1127.

Parrón, T., Requena, M., Hernández, A.F., 2011. Association between environmental exposure to pesticides and neurodegenerative diseases. Toxicol. Appl. Pharmacol. 256, 379–385.

Parrón, T., Requena, M., Hernández, A.F., 2014. Environmental exposure to pesticides and cancer risk in multiple human organ systems. Toxicol. Lett. 230, 157–165.

Peeling, P., Castell, L.M., Derave, W., de Hon, O., Burke, L.M., 2019. Sports foods and dietary supplements for optimal function and performance enhancement in track-and-field athletes. Int. J. Sport Nutr. Exerc. Metab. 29, 198–209.

Pirotta, M.V., Cohen, M.M., Kotsirilos, V., 2000. Complementary therapies: have they become accepted in general practice? Med. J. Aust. 172 (3), 105–109.

Pitt, J.I., 2000. Toxigenic fungi and mycotoxins. Br. Med. Bull. 56, 184–192.

Pollard, A.J., Powell, K.D., Harper, F.A., 2002. The genetic basis of metal hyperaccumulation in plants, Crit. Rev. Plant Sci., 21. pp. 539–566.

Prado, G., Altoé, A.F., Gomes, T.C., 2012. Occurrence of aflatoxin B1 in natural products. Braz. J. Microbiol. 43, 1428–1436.

Qing, G., Xia, L., Li, T., 2009. Simultaneous determination of 26 pesticide residues in 5 Chinese medicinal materials using solid-phase extraction and GC-ECD method. Chin. J. Nat. Med. 7, 210–216.

Rai, V., Kakkar, P., Khatoon, S., 2001. Heavy metal accumulation in some herbal drugs. Pharm. Biol. 39, 384–387.

Rai, V., Kakkar, P., Misra, C., 2007. Metals and organochlorine pesticide residues in some herbal Ayurvedic formulations. Bull. Environ. Contam. Toxicol. 79 (3), 269–272.

Rai, V., Kakkar, P., Singh, J., 2008. Toxic metals and organochlorine pesticides residue in single herbal drugs used in important ayurvedic formulation – 'Dashmoola'. Environ. Monit. Assess. 143 (1–3), 273–277.

Raman, P., Patino, L.C., Nair, M.G., 2004. Evaluation of metal and microbial contamination in botanical supplements. J. Agric. Food Chem. 52, 7822–7827.

Rao, M.M., Kumarmeena, A., Galib, 2011. Detection of toxic heavy metals and pesticide residue in herbal plants which are commonly used in the herbal formulations. Environ. Monit. Assess. 181 (1–4), 267–271.

Regulation EC 396/2005, 2005. Of the European Parliament and of the Council of 23 February 2005 on maximum residue levels of pesticides in or on food and feed of plant and animal origin and amending Council Directive 91/414/EEC. <https://eur-lex.europa.eu/LexUriServ/LexUriServ.do?uri = OJ:L:2005:070:0001:0016:en:PDF> (Accessed 09 January 2020).

Rodriguez-Fragoso, L., Reyes-Esparza, J., Burchiel, S.W., 2008. Risks and benefits of commonly used herbal medicines in Mexico. Toxicol. Appl. Pharmacol. 227 (1), 125–135.

Romagnoli, B., Menna, V., Gruppioni, N., 2007. Aflatoxins in spices, aromatic herbs, herb-teas and medicinal plants marketed in Italy. Food Control 18, 697–701.

Ronis, M.J.J., Pedersen, K.B., Watt, J., 2018. Adverse effects of nutraceuticals and dietary supplements. Annu. Rev. Pharmacol. Toxicol. 58, 583–601.

Ros, J.J., Pelders, M.G., De Smet, P.A., 1999. A case of positive doping associated with a botanical food supplement. Pharm. World Sci. 21 (1), 44–46.

Rubio, C., Lucas, J.R., Gutiérrez, A.J., 2012. Evaluation of metal concentrations in mentha herbal teas (*Mentha piperita, Mentha pulegium* and *Mentha* species) by inductively coupled plasma spectrometry. J. Pharm. Biomed. Anal. 71, 11–17.

Salgueiro, L., Martins, A.P., Correia, H., 2010. Raw materials: the importance of quality and safety. A review. Flavour Fragr. J. 25, 253–271.

Santos, L., Marin, S., Sanchis, V., 2009. Screening of mycotoxin multicontamination in medicinal and aromatic herbs sampled in Spain. J. Sci. Food Agric. 89, 1802–1807.

Saper, R.B., Kales, S.N., Paquin, J., et al., 2004. Heavy metal content of ayurvedic herbal medicine products. J. Am. Med. Assoc. 292, 2868–2873.

Saper, R.B., Phillips, R.S., Sehgal, A., 2008. Lead, mercury, and arsenic in US- and Indian-manufactured Ayurvedic medicines sold via the Internet. J. Am. Med. Assoc. 300, 915–923.

Sarkhail, P., Yunesian, M., Ahmadkhaniha, R., 2012. Levels of organophosphorus pesticides in medicinal plants commonly consumed in Iran. Daru. J. Pharm. Sci. 20 (1), 9.

Sarma, H., Deka, S., Deka, H., 2011. Accumulation of heavy metals in selected medicinal plants. Rev. Environ. Contam. Toxicol. 214, 63–86.

Sewram, V., Shephard, G.S., van der Merwe, L., 2006. Mycotoxin contamination of dietary and medicinal wild plants in the Eastern Cape Province of South Africa. J. Agric. Food Chem. 54, 5688–5693.

Shaw, D., Leon, C., Kolev, S., 1997. Traditional remedies and food supplements: a 5-year toxicological study (1991–1995), Drug Saf., 17. pp. 342–356.

Sohn, S.H., Kim, S.K., Kang, H.G., 2004. Two-phase partition chromatography using soybean oil eliminates pesticide residues in aqueous ginseng extract. J. Chromatogr. A 1042, 163–168.

Steinhoff, B., 2019. Challenges in the quality of herbal medicinal products with a specific focus on contaminants. Phytochem. Anal. 1–7. Available from: https://doi.org/10.1002/pca.2879.

Storelli, M.M., Storelli, A., Marcotrigiano, G.O., 2004. Polychlorinated biphenyls, hexachlorobenzene, hexachlorocyclohexane isomers, and pesticide organochlorine residues in cod-liver oil dietary supplements. J. Food Prot. 67 (8), 1787–1791.

Tait, P.A., Vora, A., James, S., 2002. Severe congenital lead poisoning in a preterm infant due to a herbal remedy. Med. J. Aust. 177, 193–195.

Tam, J.W., Dennehy, C.E., Ko, R., 2006. Analysis of ephedra-free labeled dietary supplements sold in the San Francisco Bay area in 2003. J. Herb. Pharmacother. 6 (2), 1–19.

Tongesayi, T., Fedick, P., Lechner, L., 2013. Daily bioaccessible levels of selected essential but toxic heavy metals from the consumption of non-dietary food sources. Food Chem. Toxicol. 62, 142–147.

Tripathy, V., Basak, B.B., Varghese, A., Saha, T.S., 2015. Residues and contaminants in medicinal herbs—a review. Phytochem. Lett. 15, 67–78.

Unger, M., Frank, A., 2004. Simultaneous determination of the inhibitory potency of herbal extracts on the activity of six major cytochrome P450 enzymes using liquid chromatography/mass spectrometry and automated online extraction. Rapid Commun. Mass Spectrom. 18, 2273–2281.

Vaclavik, L., Krynitsky, A.J., Rader, J.I., 2014. Mass spectrometric analysis of pharmaceutical adulterants in products labeled as botanical dietary supplements or herbal remedies: a review. Anal. Bioanal. Chem. 406, 6767–6790.

van Vonderen, M.G.A., Klinkenberg-Knot, E.C., Craanen, M.E., 2000. Severe gastrointestinal symptoms due to lead poisoning from Indian traditional medicine. Am. J. Gastroenterol. 95, 1591–1592.

WHO, 1989. Evaluation of Certain Food Additives and Contaminants; 33rd Report of the Joint FAO/WHO Expert Committee on Food Additives; Technical Report Series 776. World Health Organization, Geneva, Switzerland. Available from: <https://apps.who.int/iris/bitstream/handle/10665/39252/WHO_TRS_776.pdf> (Accessed 09 January 2020).

WHO (World Health Organization), 1998. WHO, Quality Control Methods for Medicinal Plant Materials. World Health Organization, Geneva, 1998.

WHO, 2003. WHO Guidelines on Good Agricultural and Collection Practices for Medicinal Plants. World Health Organization, Geneva, 2003.

WHO, 2007. WHO Guidelines for Assessing Quality of Herbal Medicines With Reference to Contaminants and Residues. World Health Organization, Geneva, 2007.

Wong, M., Tan, P., Wee, Y., 1993. Heavy metals in some Chinese herbal plants. Biol. Trace Elem. Res. 36, 135–142.

Wong, T.C., Lee, F.S., Hu, G.L., 2007. A survey of heavy metal and organochlorine pesticide contaminations on commercial Lingzhi products. J. Food Drug Anal. 15, 472–479.

Xue, J., Hao, L., Peng, F., 2008. Residues of 18 organochlorine pesticides in 30 traditional Chinese medicines. Chemosphere 71, 1051–1055.

Zuin, V.G., Vilegas, J.H.Y., 2000. Pesticide residues in medicinal plants and phytomedicines. Phytother. Res. 14, 73–88.

Zuin, V.G., Yariwake, J.H., Lanças, F.M., 2003. Analysis of pesticide residues in Brazilian medicinal plants: matrix solid phase dispersion versus conventional (European Pharmacopoeia) methods. J. Braz. Chem. Soc. 14, 304–309.

Chapter 69

Nutraceuticals and adverse outcome pathways

Antonio F. Hernández and Fernando Gil

Introduction

Nutraceuticals provide essential elements (e.g., vitamins and essential minerals, polyunsaturated and monounsaturated fatty acids, and herbal products) that should be present in a healthy diet. Their administration contributes to a better quality of life and to the prevention and treatment of numerous disorders, such as cardiovascular diseases (CVDs), diabetes, osteoporosis, and cancer. Many nutraceuticals are phytochemicals (i.e., compounds present in fruits and vegetables), which are not essential nutrients but are commonly consumed as herbal remedies or dietary supplements. The antioxidant and antiinflammatory activities of phytochemicals constitute the biological basis for their growing use by consumers.

Although diet-derived compounds, including nutraceuticals, are generally regarded as safe, there is a lack of scientific data regarding their toxicity profile. Moreover, because these products are not subjected to extensive toxicity testing, potential harmful effects for consumers' health cannot be entirely ruled out. In particular, there is growing evidence that chronic administration of polyphenols at pharmacological or higher doses found in dietary supplements can exert detrimental effects through prooxidative and/or other mechanisms. A better understanding of the biological plausibility of the mechanisms underlying the beneficial and toxic effects of nutraceuticals would contribute to better protecting consumers from potential long-term adverse effects.

Adverse outcome pathways (AOPs) comprise an efficient and effective framework to describe the linkage between mechanistic data at varying levels of biological organization and apical toxicity endpoints used for risk assessment (LaLone et al., 2017). The development of an AOP requires a clear presentation of the scientific evidence for each event occurring from the first chemico–biological interaction to the final adverse outcome. A putative AOP for polyphenols toward the development of hepatotoxicity is proposed; however, its practical applicability is hampered by the scarce randomized clinical trials available, which makes it difficult to assess the consistency of clinical data with results from experimental and epidemiological studies.

Concept and regulatory aspects of nutraceuticals

The term "nutraceutical," coined by DeFelice in 1989 from the words "nutrition" and "pharmaceutical" refers to a wide array of bioactive components found in food or food products that provide health benefits beyond basic nutrition. Although the term nutraceutical is often used interchangeably with other terms, such as functional food and dietary supplement, there are distinctions. Nutraceuticals are generally marketed in a form that resembles a medicinal product. These products are gaining interest by consumers, producers, and healthcare professionals as useful remedies for the long-term prevention and treatment of major chronic diseases, including cancer, neurodegenerative, and inflammation-related disorders (Kalra, 2003; Davies et al., 2005; Schmitt and Ferro, 2013).

Nutraceuticals are proving to be increasingly popular among consumers as they are claimed to be devoid of adverse effects and offer a safe and easy healthcare option. However, the lack of clear-cut differences between nutraceuticals, dietary supplements, and functional foods across countries has made their regulation minimal and vague. In contrast to pharmaceuticals, little or no research regarding safety and quality control is required by law for nutraceuticals, which are exempt from extensive preclinical efficacy and toxicity testing by regulatory agencies. Hence, the potential of harmful effects of these products for consumer health cannot be ruled out. This potential can only be characterized by using standard

Nutraceuticals. DOI: https://doi.org/10.1016/B978-0-12-821038-3.00069-0

toxicity testing methods for assessing acute high-dose and chronic low-dose toxicity tests, as well as other specific toxicity assays (Ifeoma et al., 2013).

In the United States, the Food and Drug Administration does not strictly regulate herbs and supplements, so there is no guarantee of the purity or safety of these products. According to the Dietary Supplement, Health, and Education Act of 1994, they are monitored as "dietary supplements." In the European Union (EU), there is no regulatory framework for nutraceuticals or functional foods, and claims regarding their beneficial effects can only be "health claims" and not "medicinal claims." Any voluntary nutrition or health claim must comply with the requirements of European Regulation (EC) No 1924/2006 on the use of nutrition and health claims for food. This piece of legislation ensures that any claim made on a food label in the EU is clear, accurate, and scientifically substantiated to protect consumers from inaccurate or misleading information and to enable them to make informed and meaningful choices.

Toxicity pathways and adverse outcome pathways

Advances in biological and experimental technologies (including "omics" tools) have allowed different approaches to understanding disease and toxicity. One of these approaches deals with the consequences of normal biology disturbed by chemicals, which enables a hypothesis-based prediction of the resultant adverse outcomes. Various terms have been used for this general approach, such as mode of action (MoA), toxicity pathway, and adverse outcome pathway (AOP), which support the plausibility and causal understanding of how chemical exposures lead to toxicity/adverse outcomes (Perkins et al., 2019). While an AOP is chemical agnostic and does not account for toxicokinetic considerations in determining differential toxicity, an MoA encompasses how a chemical reaches the site of toxic effect (toxicokinetic).

The current World Health Organization definition for MoA is a biologically plausible sequence of key events leading to an observed effect supported by robust experimental observations and mechanistic data (Boobis et al., 2009). MoA does not necessarily imply adversity because it can refer to the beneficial health effects of nutraceuticals or the therapeutic effects of drugs. Evidence for the MoA hypothesis is based on Hill's criteria for establishing causality (Meek et al., 2014).

MoA does not imply full understanding of the mechanism of action and both concepts must be distinguished from each other. Mechanism of action is a more detailed description of the pathway that includes individual biochemical and physiological key events leading to a toxic or adverse effect (Boobis et al., 2006; Meek et al., 2014). For risk assessment, it is critical to accurately distinguish between adverse and adaptive responses to chemical perturbation. Cells have evolved a number of stress response pathways to provide an adaptive capacity to maintain or return to homeostasis with no loss of function at the cellular, tissue, or organ level. Conversely, adversity features a perturbation sufficient to result in an irreversible change resulting in loss of normal function (Krewski et al., 2010). An adverse effect therefore occurs only when homeostasis or adaptive responses are exceeded, so the cell or organism will not survive with such impaired function(s). Hence, the concept of "toxicity pathway" refers to a normal biological pathway that becomes perturbed beyond the point of homeostatic correction, leading to a predictive, hypothesis-driven assessment paradigm for toxicity (NRC, 2007).

An AOP has a broader scope than MoA because it starts with the chemical exposure and goes up to the individual level and even the population level. The approach is based on the concept that toxicity results from the chemical first reaching and then interacting with an initial target or targets in the organism. Once pathways are perturbed in specific ways, adverse effects can ensue.

The AOP methodology provides a framework to collect and evaluate relevant chemical, biological, and toxicological information in such a way that is useful for risk assessment (OECD, 2017). It should be noted that AOPs are pragmatic simplifications of complex biology, focused on critical and measurable key events at different biological organization levels (cellular, tissue, organ) leading to adverse outcomes (Bal-Price and Meek, 2017). An AOP has been defined as "a sequence of events from the exposure of an individual or population to a chemical substance through a final adverse (toxic) effect at the individual level (from a human health perspective) or population level (from an environmental perspective)" (Ankley et al., 2010; OECD, 2017; Meek et al., 2014). AOPs are characterized by a molecular initiating event

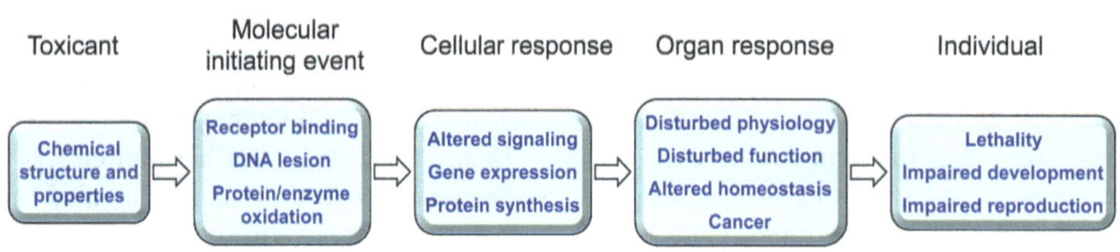

FIGURE 69.1 Conceptual framework of adverse outcome pathway (AOP).

(MIE) in which a chemical interacts with a biological target and continues through a sequential series of biological activities termed intermediate key events until producing an in vivo adverse effect relevant to human health or ecological equilibrium (Fig. 69.1). All these key events are necessary elements of the MoA and should be empirically observable or constitute biologically based markers for such an event (Boobis, 2005; OECD, 2017). Usually, the final adverse outcome is associated with an in vivo OECD test guideline, although in some cases it may be at a biological level of organization downstream of an apical endpoint described in a test guideline (OECD, 2017).

A particular MIE may lead to several final adverse effects and, conversely, several MIEs may lead to the same final adverse effect. However, each AOP will have only one MIE and one final adverse effect, but may involve an unlimited number of intermediate steps (Vinken, 2013). In practice, AOP conform networks as separate AOP can converge on the same adverse outcome or can share key events, ultimately leading to a common adverse outcome. Hence, AOP networks better represent the biological and toxicological complexity within organisms (OECD, 2018). Before the identification of intermediate events, an understanding of the normal physiological pathways of the AOP is essential. It should be noted that key events at different levels of biological organization provide a greater weight-of-evidence than multiple events at the same level of organization (OECD, 2017).

In response to the increasing use of the AOP tool, the OECD has published the second version of the Revised Guidance Document on Developing and Assessing AOPs (Table 69.1; OECD, 2017). In developing an AOP, available information on the macromolecular, cellular/tissue, organ, and organism to the population/ecosystem level needs to be identified and collected. This information is then presented in a structured way that aids in identifying gaps of knowledge. Once the MIE and the final adverse effect are identified, the task is filling in the intermediate events between them (OECD, 2017). These essential steps are identified and retrieved from an in-depth survey of relevant scientific literature or from experimental studies. Any type of information can be incorporated into an AOP, including structural data, "omics-based" data, and in vitro, in vivo, or in silico data. However, in vivo data are preferred over in vitro data, and endpoints of interest are also preferred over surrogate endpoints (Vinken, 2013). The AOPs identified must not contradict any steps of normal biological processes because they need to be biologically plausible. An AOP may be considered either plausible or probable, depending on the extent of the available scientific evidence supporting the AOP and the extent to which the key events have been experimentally tested and found to be consistent with data for other key events (OECD, 2017). A partial AOP (i.e., one where not all key events are known) may be useful in hazard identification or in priority-setting for further testing and development. A qualitative AOP is one where the key events have been identified but methods for assessing these events have not been identified. AOPs must be considered as open and flexible structures that should be continuously refined by using old and new data (Vinken, 2013).

Type, sources, and health benefits of nutraceuticals

Nutraceuticals have been classified in many ways based on their food sources, mechanism of action, chemical nature, and specific health benefits. They can be macronutrients (e.g., omega-3 fatty acids), micronutrients (e.g., vitamins, minerals), or ingredients with little or no nutritive value (e.g., phytochemicals) (Ameye and Chee, 2006). According to their food sources, nutraceuticals can be categorized as dietary fiber, polyunsaturated fatty acids (PUFAs), probiotics, prebiotics, antioxidant vitamins, polyphenols, and spices (Das et al., 2012; Chauhan et al., 2013).

TABLE 69.1 Steps for the development of an adverse outcome pathway.

Identification of the main blocks of information
1. Definition of the molecular initiating event [at the site(s) of action]
2. Identification of the adverse outcome
3. Identification of key events (KEs) leading to the adverse outcome
4. Identification of the key event relationships (KERs)
5. AOP assessment: • Biological plausibility of KERs • Essentiality of the KEs • Empirical support of KERs

Dietary fiber

Dietary fiber is found only in plant foods, particularly fruits, vegetables, nuts, and grains. Because fiber is not hydrolyzed by digestive enzymes, it is further digested by the intestinal microflora. The intake of high-fiber food improves serum lipoprotein levels, lowers blood pressure, improves blood glucose control for diabetes, contributes to weight loss, and improves bowel habits. Certain soluble fibers appear to enhance the immune function in humans. However, dietary fiber has some potential negative effects, including reduced absorption of vitamins, minerals, proteins, and calories (Das et al., 2012).

Polyunsaturated fatty acids

PUFAs are essential to the diet and cannot be synthesized de novo. The main dietary PUFAs are omega-3 (also known as n-3; e.g., linolenic acid and eicosapentaenoic acid) and omega-6 fatty acids (n-6; e.g., linoleic acid and arachidonic acid). Omega-3 fatty acids are found in soybean and canola oils, flaxseeds, walnuts, and fish oils, whereas n-6 are found in safflower, corn, soybean, and sunflower oils as well as in meat (Ameye and Chee, 2006). Fish rich in omega-3 fatty acids include anchovies, mackerel, salmon, sardines, sea bass, swordfish, and trout. These fatty acids act by inhibiting the conversion of arachidonic acid to the proinflammatory eicosanoids prostaglandin E_2 (PGE_2), leukotriene B_4 (LTB_4), and thromboxane A_2 (TXA_2). Thus, omega-3 fatty acids exert major effects on CVDs and protect against the development of arteriosclerosis because of their antiarrhythmic, hypolipidemic, and antithrombotic activities (Das et al., 2012).

Probiotics

Probiotics are food supplements based on live microbials that improve the intestinal microbial balance. They include the following categories of bacteria: lactobacilli, bifidobacteria, and Gram-positive cocci. Specific probiotics are generally used for lactose intolerance, acute diarrhea, and antibiotic-associated gastrointestinal side effects. The administration of probiotics decreases the risk of systemic conditions, such as allergy, asthma, cancer, and several infections of the ear and urinary tract (Das et al., 2012).

Prebiotics

Prebiotics are dietary ingredients, particularly short-chain polysaccharides, that are not digested by humans. They are found in vegetables like chicory roots, banana, tomato, and alliums that beneficially affect the host by selectively altering the composition and activity of the gut microbiota. Prebiotics change the composition and activity of the gastrointestinal microbiota through suppressing the proliferation of pathogens and promoting the growth of beneficial microorganisms (e.g., *Lactobacillus* and *Bifidobacterium*) in the host intestine (Gao et al., 2020). The major health benefits of prebiotics include improved lactose tolerance, antitumor properties, neutralization of toxins, stimulation of the intestinal immune system, reduction of constipation, and reduction of blood cholesterol levels (Das et al., 2012).

Interactions between prebiotics, probiotics, and phytochemicals with intestinal microbiota have consequences on human health; for instance, phytochemicals inhibit pathogenic bacteria and stimulate the growth of beneficial bacteria, thus exerting prebiotic-like effects. Conversely, gut microbiota can transform and influence the bioavailability and effects of polyphenols (Cencic and Chingwaru, 2010).

Antioxidant vitamins

Antioxidant vitamins include vitamin C, vitamin E, and carotenoids. These vitamins are abundant in many fruits and vegetables and exert their protective action by scavenging free radicals. They act both singly and synergistically for the prevention of oxidative reactions involved in the development of degenerative diseases like cancer, CVDs, and some neurological diseases. L-ascorbic acid constitutes the majority of vitamin C in food and is found in rose hips, blackcurrants, and citrus fruits; it functions as a chain-breaking antioxidant, scavenging reactive oxygen species (ROS) directly (Ameye and Chee, 2006; Davì et al., 2010). Natural vitamin E is an essential fat-soluble vitamin that comprises different forms of tocopherol and tocotrienol and is mainly found in edible plant oils. Vitamin E functions primarily as an antioxidant; by reacting directly with ROS, it can protect membranes from lipid peroxidation and protect low-density cholesterol (LDL) from oxidation. In the human body, the synthetic α-tocopherol esters are rapidly cleaved by cellular esterases, making natural vitamin E available (Ameye and Chee, 2006; Davì et al., 2010; Das et al., 2012). Carotenoids like β-carotene, lycopene, lutein, and zeaxanthin are well-known for excellent ROS-scavenging properties and are effective agents in reducing the risk of age-related macular degeneration, a major cause of impaired vision and blindness in the elderly, and cataracts (Fiedor and Burda, 2014).

Naturally occurring polyphenols

Polyphenols are secondary metabolites of plants found largely in fruits, vegetables, cereals, and plant-derived beverages such as tea and wine. In contrast to classic nutrients, these compounds are not required for vital body functions in humans such as growth, reproduction, wound repair, and development, so they are considered nonnutrients (Martin and Appel, 2010). Plant polyphenols have important antioxidant activities and can be classified into different groups according to the number of phenol rings and the structure that links these rings (Ranzato et al., 2014). The most important polyphenols are phenolic acids and flavonoids (flavonols, flavones, catechins, flavonones, anthocyanidins, and isoflavonoids) (Das et al., 2012).

Red wine contains resveratrol and green tea is a rich source of epigallocatechin 3-gallate (EGCG). Soybean is the most significant source of dietary isoflavones and contains a relatively high concentration of genistein and daidzein, which are considered phytoestrogens. Because of their antioxidant, antiinflammatory, antimicrobial, anticarcinogenic, antiatherogenic, and cardioprotective activities, polyphenols may have a role in the prevention of aging-associated chronic diseases, such as cancer, CVDs, and neurodegenerative disorders. No beneficial effects have been observed for antioxidant supplementation in well-nourished people with a balanced diet. However, a high intake of polyphenols, even from dietary sources, can result in toxic effects (Das et al., 2012).

Spices

Most of the spice components of the diet are terpenes and essential oils. Although they are present in food in minute quantities, they exert an important influence on human health because of their antioxidative, chemopreventive, antimutagenic, antiinflammatory, and immunomodulatory effects (Singh et al., 2006). They also exhibit a wide range of beneficial effects on human health by acting on the gastrointestinal, cardiovascular, respiratory, neural, reproductive, and other systems. When used as medicine, spices may exhibit toxicity or result in drug interactions (Das et al., 2012).

Mechanisms underlying nutraceuticals' beneficial effects

Inflammation and oxidative stress as common mechanisms for various diseases

Nutraceuticals and dietary bioactive substances exert their activities at the systemic level by interacting with multiple targets in different organs. Some molecular mechanisms may overlap between different disease conditions, as occurs with the nuclear factor-kappa B (NF-κB) signaling pathway, which is the key mediator of inflammation and hence it is involved in atherosclerosis and cancer (Omar et al., 2010). Better knowledge of the mechanisms of action of nutraceuticals may be useful for the prevention and treatment of a number of diseases, such as obesity, diabetes, atherosclerosis, hypertension, cancer, and neurodegenerative disorders (Ranzato et al., 2014).

Inflammation is a cellular response to the damage induced by several stressors, including chemical exposures. It can lead to a healthy restorative process or contribute to further tissue injury and disease when the response becomes chronic. Acute inflammation is initiated by the release of proinflammatory mediators (e.g., cytokines, chemokines, leukotrienes, and adhesion molecules), which are responsible for vasodilatation, increased capillary permeability, and leukocyte migration at the site of tissue damage. The reaction orchestrated in this acute inflammatory response may be sufficient to remove the damaged cells and resolve the inflammation. In contrast, prolonged exposure to stimuli can lead to persistent inflammation from long-term elevation of inflammatory mediators and tissue damage. Chronic inflammation ultimately contributes to numerous health effects, including multiple cardiovascular conditions, cancer, osteoarthritis, diabetes, and neurodegenerative diseases (Kunnumakkara et al., 2018; Schwarze et al., 2013). Molecular pathways linked to inflammation involve a number of molecular mediators including transcription factors [e.g., NF-κB and signal transducer and activator of transcription 3 (STAT3)], inflammatory enzymes [e.g., cyclooxygenase-2 (COX-2), matrix metalloproteinase-9 (MMP-9)], and inflammatory cytokines [e.g., tumor necrosis factor alpha (TNF-α)], interleukins (e.g., IL-1, IL-6, IL-8), and chemokines (Kunnumakkara et al., 2018).

The increase in oxidative stress from elevated intracellular levels of ROS can activate the NF-κB transcription factor, which is responsible for encoding numerous genes expressing proinflammatory cytokines, adhesion molecules, chemokines, and some inflammatory-related enzymes (Fig. 69.2). Because scientific evidence has demonstrated that oxidative stress and NF-κB activation are closely related, nutraceuticals with antioxidant activity may act as important modulators of systemic inflammation (Omar et al., 2010).

Molecular mechanisms for diabetic vascular complications

Systemic inflammation and oxidative stress play an important role in the pathogenesis of diabetes mellitus and metabolic syndrome, as well as in their vascular complications (Davì et al., 2010). The molecular mechanisms underlying diabetes complications arise from vascular inflammation and endothelial cell injury attributed to long-term hyperglycemia-induced ROS. Vascular inflammation is characterized by: (1) inflammation-related

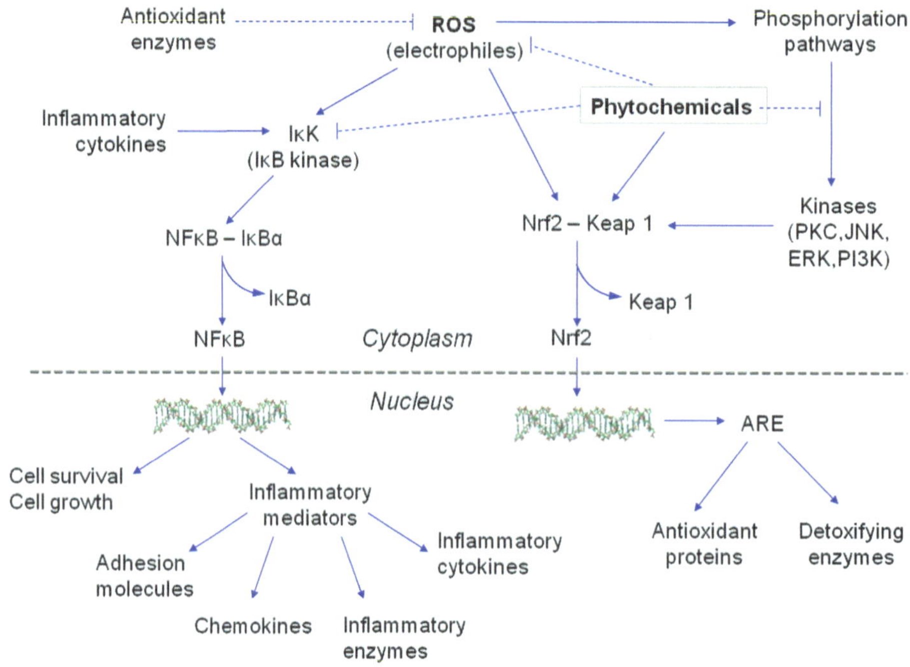

FIGURE 69.2 Influence of reactive oxygen species (ROS) and phytochemicals on NF-κB and Nrf2 activation. NF-κB is a heterodimeric transcription factor involved in the regulation of inflammatory responses of many cell types. It is usually retained in latent form in the cytoplasm coupled with the inhibitory protein IκBα. Upon stimulation with oxidative stress and proinflammatory cytokines, the IκB kinase (IKK) phosphorylates the IκBα protein, which then dissociates from NF-κB. This results in translocation of the activated NF-κB to the nucleus, where it binds to the target gene region and stimulates the expression of genes encoding proinflammatory cytokines (e.g., TNF-α, IL-1β, and IL-6), adhesion molecules (VAM-1, ICAM-1) and chemokines (MCP-1), inflammation-related enzymes (iNOS, phospholipase A2, cyclooxygenase and lipoxygenase), and cell survival and growth factors (cyclin D1). Nrf2 is a transcription factor that regulates the expression of many antioxidant proteins and detoxification enzymes. Nrf2 is held in an inactive state bound to Keapl in the cytoplasm. Some phytochemicals and pro-oxidants can cause oxidation of the cysteine residues of Keapl, thus releasing Nrf2. In addition, phosphorylation of Nrf2 by kinases such as phosphatidylinositol 3-kinase (PI3K), protein kinase C (PKC), c-Jun N-terminal kinase (JNK), and extracellular signal-regulated kinase (ERK) may also dissociate Nrf2 from KEAP1. Subsequently, Nrf2 translocates to the nucleus, where it binds to the antioxidant-responsive element (ARE) to stimulate gene expression and transcription of several antioxidant and detoxifying enzymes. Activation of Nrf2 also suppresses NF-κB activity, thereby inhibiting inflammation, indicating potential cross-talk between Nrf2 and NF-κB pathways.

enzyme production [e.g., inducible nitric oxide synthase (iNOS), phospholipase A2 (PLA2), cyclooxygenase (COX), and lipoxygenase (LOX)]; (2) proinflammatory cytokine production (e.g., TNF-α, IL-1β, and IL-6); and (3) leukocyte recruitment initiated by adhesion molecules [e.g., vascular cell adhesion molecule, intercellular adhesion molecule (ICAM-1), E-selectin], and chemokines (e.g., MCP-1) (Omar et al., 2010).

The production of cytokines by adipose tissue decreases insulin sensitivity and contributes to hyperglycemia. Sustained hyperglycemia ultimately leads to vascular complications through multiple interrelated pathways and mechanisms, including oxidative stress, protein kinase C (PKC), polyol/aldose reductase, and advanced glycation end product–receptor pathways. Nuclear transcription factors such as activated protein-1 (AP-1), mitogen-activated protein kinases (MAPK), and NF-κB signaling pathways also play a significant role in promoting diabetic vascular complications (Omar et al., 2010). Pathological changes in the diabetic vasculature can impair tissue perfusion and affect organs heavily dependent on their microvascular (retina and the kidney) and macrovascular supply (heart and brain).

Phytochemicals such as chrysin may be beneficial for diabetes and its complications because they interact with various mechanisms involved in diabetic vascular complications (Farkhondeh et al., 2019). Hypoglycemic herbs may increase insulin secretion, enhance glucose uptake by adipose and muscle tissues, inhibit glucose absorption from intestine, and decrease glucose production by hepatocytes (Pandey and Vijayakumar, 2011).

In addition to antiinflammatory properties, most phytochemicals also possess significant antioxidative activities, mainly by scavenging oxidants, preventing lipid peroxidation, and modulating antioxidant enzyme levels [e.g., glutathione peroxidase, glutathione reductase, glutathione-S-transferase (GST), superoxide dismutase (SOD), and catalase] (Omar et al., 2010). One mechanism for the inhibitory effects of antioxidants on vascular inflammation involves the interference of ROS-mediated NF-κB activation (Fig. 69.2). NF-κB is a major heterodimeric transcription factor involved in the regulation of inflammatory responses of many cell types. It is usually retained in a latent form in the cytoplasm coupled with the inhibitory protein IκBα. On stimulation with high glucose level, oxidative stress, and proinflammatory cytokines, the IκB kinase (IκK) phosphorylates the IκBα protein, which dissociates from NF-κB. Then, NF-κB translocates to the nucleus to stimulate the transcription of genes encoding proinflammatory cytokines, inflammation-related enzymes, and adhesion molecules and chemokines involved in leukocyte recruitment. All these factors are important inflammatory markers of many chronic inflammatory conditions. Peroxisome proliferator-activated receptors also can be activated, which may decrease the inflammatory response of endothelial cells by negatively regulating the NF-κB signaling pathway (Davì et al., 2010; Omar et al., 2010).

In general, dietary plant polyphenols can modulate transcription factors and subsequent gene expression, and regulate different signaling pathways in the muscle, liver, pancreatic β-cells, hypothalamus, and adipose tissue, thereby contributing to glucose homeostasis (Guasch-Ferré et al., 2017). Polyphenol-rich products exert inhibitory effects on NF-κB signaling pathways, modulate carbohydrate and lipid metabolism, ameliorate hyperglycemia, dyslipidemia, and insulin resistance, improve adipose tissue metabolism, and prevent the development of long-term diabetes complications (Postal et al., 2014).

Control over the physiological mechanism involved in the intestinal absorption and renal reabsorption of glucose contributes to achieving glucose homeostasis. Two specific transporters are required for glucose intestinal absorption, the Na^+/glucose cotransporter (SGLT1) at the enterocyte apical membrane and the facilitative glucose transporter (GLUT2) at the basolateral membrane, in proximity to blood capillaries. Polyphenols, such as phlorizin, myricitrin, quercetin, catechin, naringenin, caffeic acid, and rutin, may compete with glucose for binding to the glucose transporters (SGLT1 and/or GLUT2), thus decreasing the glucose absorption in the intestine and the glucose reabsorption in the renal proximal tubule. Polyphenols, such as caffeic acid, play an additional protective role in diabetes by attenuating hepatic glucose output and increasing glucose uptake in adipocytes after stimulating insulin secretion (Postal et al., 2014).

Regarding vitamins, vitamin C can protect against the development of the long-term complications associated with diabetes by inhibiting protein glycosylation (Pandey and Vijayakumar, 2011). Vitamin D may modulate the pathogenesis of type I diabetes by reducing the inflammatory reaction in the pancreatic islets and the autoimmune insulinitis. In turn, in type II diabetes, vitamin D decreases insulin resistance and increases insulin secretion (Davì et al., 2010). A deficiency in vitamin E is associated with increased peroxides and aldehydes in many tissues and might increase the incidence of diabetes (Pandey and Vijayakumar, 2011). Biotin, a member of the B vitamin complex, is important for patients with diabetes because it stimulates liver glucokinase activity, increases insulin production, and enhances glucose uptake in muscle cells (Pandey and Vijayakumar, 2011).

Regarding minerals and oligoelements, chromium (Cr^{+3}) is an essential trace element required for normal glucose and lipid metabolism. Chromium may be deficient in persons with diabetes and its deficiency has been correlated with an elevation in cardiometabolic risk, which results from increased inflammation (Moradi et al., 2019). Chromium appears to enhance insulin's actions by increasing the number of insulin receptors, the binding of insulin to the insulin receptor, and the activation of the insulin receptor (Pandey and Vijayakumar, 2011). Thus, chromium supplements may increase insulin sensitivity and improve glucose tolerance in patients with type II diabetes (Davì et al., 2010). Magnesium (Mg^{+2}) is an essential cofactor for numerous enzymes that may decrease the risk of diabetes by improving insulin sensitivity (Davì et al., 2010; Pandey and Vijayakumar, 2011).

Molecular mechanisms for cancer induction

Nutraceuticals exert pleiotropic effects on a number of signaling pathways involved in the regulation of cell growth, differentiation, migration, and survival (Fig. 69.3). Other effects are mediated by interacting with microRNA (miRNA) expression, as miRNAs are involved in invasion, metastasis, angiogenesis, and drug resistance (McCubrey et al., 2017). Phytochemicals can affect long noncoding RNA expression either directly or indirectly through the involvement of miRNAs, protein kinases, enzymes, and transcription factors (Mishra et al., 2019)

During cancer development, NF-κB is activated by different stimuli such as environmental carcinogens, tumor promoters, and inflammatory agents. These stimuli phosphorylate and activate IκK complexes that are responsible for the phosphorylation of the IκBα protein, which dissociates from NF-κB. Although IκBα undergoes further ubiquitination, the activated NF-κB form enters into the nucleus to promote the expression of target genes involved in inflammation and adhesion molecules, and cell survival, differentiation, and proliferation (Fig. 69.2). Activated target genes include cyclin D1 (which regulates cell-cycle progression), apoptosis suppressor proteins (Bcl-2 and Bcl-xL), matrix metalloproteases, and vascular endothelial growth factor (VEGF), which are required for metastasis and angiogenesis (Angel Nivya et al., 2012).

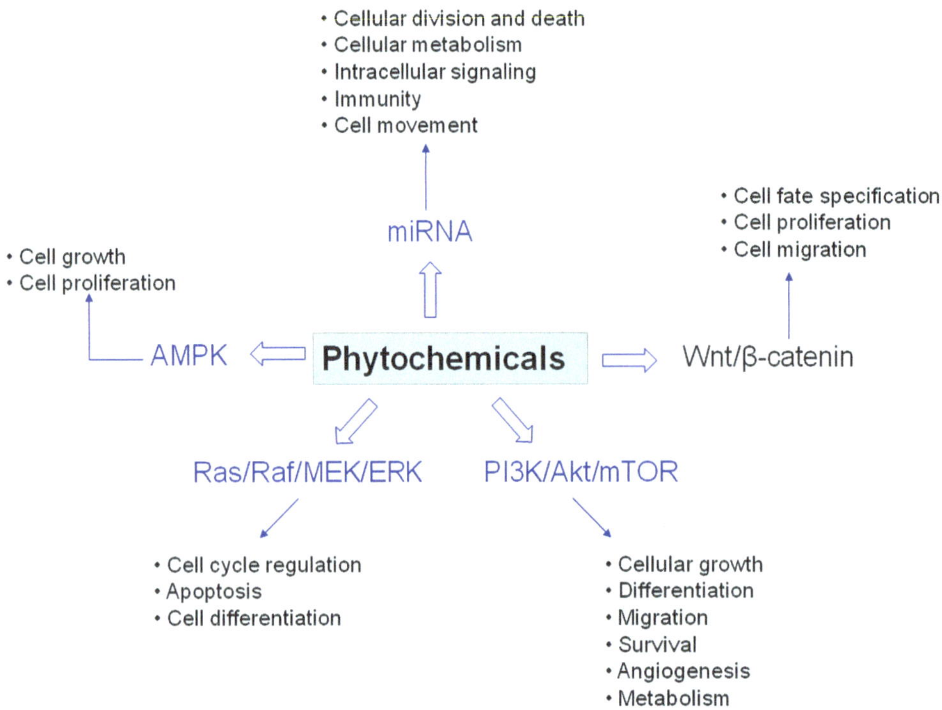

FIGURE 69.3 Signaling pathways modulated by phytochemicals of nutraceutical importance [AMP-activated protein kinase (AMPK), WNT/beta-catenin, phosphatidylinositol 3-kinase/protein kinase B/mammalian target of rapamycin (PI3K/Akt/mTOR), Ras/Raf/Mitogen-activated protein kinase/extracellular signal-regulated kinase (Ras/Raf/MEK/ERK), also known as MAPK/ERK pathway].

Daily consumption of numerous dietary nutraceuticals has shown cancer chemopreventive effects in both preclinical animal models and human epidemiological studies (Lee et al., 2013). These effects are exerted at different cellular levels by blocking the initiation of carcinogenic process (e.g., the mutagenic activity of carcinogens) and reducing the rate of cancer cell growth due to inhibition of cell proliferation and induction of apoptosis. Nutraceuticals also contribute to reducing the toxicity associated with chemotherapy and radiation therapy and provide better life conditions by reducing cancer cachexia (Young, 2008; Ranzato et al., 2014). Phytochemicals, in particular, prevent the initial steps of carcinogenesis by scavenging highly reactive species and inducing cellular defense detoxifying/antioxidant enzymes via the nuclear factor erythroid 2-related factor 2 (Nrf2) antioxidative signaling pathway (Angel Nivya et al., 2012). Nrf2 is a regulator of cellular resistance to oxidants that controls the basal and induced expression of an array of antioxidant response element (ARE)-dependent genes that regulate the physiological and pathophysiological outcomes of oxidant exposure (Fig. 69.2; Ma, 2013).

Phytochemicals also kill initiated/transformed cancer cells by promoting apoptosis as a result of downregulation of antiapoptotic molecules (e.g., Bcl-2 or Bcl-xL) and upregulation of proapoptotic molecules (e.g., such as Bax or Bak). In addition, phytochemicals regulate antiinflammatory and proliferating signaling pathways by inhibiting the activation of the antiapoptotic NF-κB and by activating the proapoptotic c-jun n-terminal kinase (JNK), a member of the MAPK family involved in cellular proliferation, differentiation, and development (Angel Nivya et al., 2012).

Inflammation plays a role in the development of several types of cancers by inadequate upregulation of enzymes involved in the inflammatory response, such as COX-2 and NOS. Dietary phytochemicals from medicinal plants have antiinflammatory and antioxidative properties and protect against lipid peroxidation, which are largely responsible for their chemopreventive activities. For example, curcumin, EGCG, and resveratrol can inhibit the expression of COX-2 and NOS by blocking NF-κB activation (Angel Nivya et al., 2012). Nutraceuticals may also improve the immunological condition of cancer patients by inducing cytokine production (e.g., TNF-α, interleukins, and interferons) and activating macrophages, T lymphocytes, and natural killer cells in patients with late-stage cancer (Angel Nivya et al., 2012; Ranzato et al., 2014).

Phytochemicals also play a crucial chemopreventive role by preventing the formation of new blood vessels (angiogenesis) necessary for cancer development. This is the case of certain polyphenols such as EGCG, resveratrol, curcumin, and genistein. For angiogenesis, endothelial cells need to be stimulated by various growth factors such as VEGF, fibroblast growth factor, or platelet-derived growth factor (Angel Nivya et al., 2012). Another molecular mechanism underlying the anticarcinogenic and antiinflammatory effects of phytochemicals is the activation of sirtuin proteins (e.g., SIR2, silent information regulatory) (Nakata et al., 2012). These proteins possess either mono-ADP-ribosyltransferase or deacylase activity and may influence cellular processes like transcription, apoptosis, inflammation, aging, and stress resistance (Preyat and Leo, 2013). Phytochemicals may also have anticancer effects through the reversion of epigenetic alterations associated with oncogene activation and inactivation of tumor suppressor genes. The best studied epigenetic alterations include inhibition of histone deacetylases, short noncoding miRNAs, and modification of the methylation of cytosine bases, particularly cytosines preceding guanines (CpG dinucleotides) of cancer-related genes (Carlos-Reyes et al., 2019; Lee et al., 2013).

In addition to phytochemicals, antioxidant vitamins (A, C, and E) and trace elements (Se) may play a role in cancer prevention and treatment. Supplementation of cancer patients with these vitamins has proven to be helpful because they induce cancer cell apoptosis through cell-cycle arrest and activation of the apoptosis factors (Ranzato et al., 2014). High levels of selenium also play a potential role in anticancer activity by inducing apoptosis, inhibiting cellular proliferation, and protecting cells from peroxide damage, because selenium is a key component of the antioxidant enzyme glutathione peroxidase (Angel Nivya et al., 2012).

Molecular mechanisms for cardiovascular diseases

Nutraceuticals exert their cardioprotective effects mainly through lipid-lowering effects, antioxidant actions, and/or by decreasing homocysteine levels. Vegetable and fruit fibers (with pectin), garlic, oily seeds (walnut, almonds, etc.), and fish oils have lipid-lowering effects in humans through inhibition of fat absorption and suppression of hepatic cholesterol synthesis. A significant cardiovascular benefit of phytochemicals (polyphenols in wine, grapes, and teas), vitamins (ascorbate, tocopherol), and minerals (selenium, magnesium) present in foods may result from scavenging free radicals produced during atherogenesis (Alissa and Ferns, 2012; Zoltani, 2019). Homocysteine increases the risk of both cardiovascular and cerebrovascular disorders by enhancing arteriolar constriction and decreasing endothelial vasodilation. A higher intake of folate, antioxidant vitamins, whole grains, and phytochemicals may ameliorate the deleterious vascular effects of homocysteine in the heart (Alissa and Ferns, 2012). Although the most efficient nutraceuticals impact the LDL cholesterol metabolism in liver, fibers and phytosterols are less efficient and interfere with the intestinal absorption of cholesterol (Cicero et al., 2015).

Monacolin, a fungal secondary metabolite very similar to natural statins, decreases plasma cholesterol levels by competitively inhibiting 3-hydroxy-3-methyl-glutaryl coenzyme A reductase (the hepatic enzyme that limits cholesterol synthesis) and by inducing the expression of LDL receptors in the liver. Monacolins also reduce high-sensitivity C-reactive proteins and improve flow-mediated dilation of the blood vessels (Cicero et al., 2015). Through these mechanisms monacolins reduce the risk of atherosclerosis.

Polyphenols show antiatherosclerotic effects in the early stages of atherosclerosis development as a result of their antioxidant and antiinflammatory effects. These compounds improve antioxidant status, decrease LDL levels and oxidation, improve endothelial function, increase nitric oxide (NO) release (potent vasodilator), modulate the activity of eicosanoid-generating enzymes in inflammatory cells, modulate lipid metabolism (i.e., hypolipidemic effect), and protect against atherothrombotic episodes such as myocardial ischemia and platelet aggregation. Oxidized LDL is a key contributor to atherogenesis through multiple mechanisms, including the ROS-mediated NF-κB signaling pathway (Alissa and Ferns, 2012). Fruits and vegetables may decrease LDL cholesterol levels, very likely by the binding of bile acids and inhibition of cholesterol synthesis. Plant sterols are compounds structurally similar and functionally analogous to cholesterol, so they reduce the intestinal absorption of cholesterol (Alissa and Ferns, 2012).

Regarding omega-3 fatty acids, potential mechanisms for their cardiovascular protective effects include antiinflammatory, antithrombotic (reduced platelet aggregability), and antiarrhythmic (reducing the risk of potentially fatal cardiac arrhythmias) effects, lowering of heart rate and blood pressure, hypotriglyceridemic effects, and improved endothelial function. The favorable effects on the lipid profile appear to be due to decreased hepatic triglyceride secretion together with enhanced clearance of triglycerides from plasma (Alissa and Ferns, 2012; Zoltani, 2019).

Vitamin C reduces tissue ROS and oxidized LDL, and contributes to preserving endothelial function and inhibiting vascular smooth muscle proliferation. Carotenoids may enhance the immune function and reduce the risk of developing degenerative diseases such as CVD and cancer. These effects are mediated by reducing oxidative stress and blood markers of inflammation. Vitamin E (tocopherol) is a free radical scavenger and an inhibitor of COX, thus playing an antiinflammatory effect that contributes to disease prevention (Bast and Haenen, 2002; Alissa and Ferns, 2012). Vitamin D improves endothelial function by reducing vascular inflammation by suppressing NF-κB signaling and production of proinflammatory cytokines (Kim et al., 2020).

Molecular mechanisms for neurodegenerative diseases

Currently, the mechanisms leading to neuronal damage and further neurodegeneration include, but are not limited to, the combination of any of the following: damage to neuronal mitochondria, intracellular calcium overload, increased generation of ROS, sustained inflammatory condition. These four mechanisms trigger changes in the expression and function of hundreds to thousands of genes (Dadhania et al., 2016).

Neurons are particularly vulnerable to oxidative stress because of their relatively low expression of antioxidant proteins and their limited regenerative capacity. Oxidative stress can induce neuronal damage by itself and by activating excitotoxic and neuroinflammatory processes. Excitotoxicity results from the hyperactivation of glutamatergic receptors, particularly *N*-methyl-D-aspartate, followed by intracellular calcium overload that triggers ROS formation and the release of proapoptotic factors (Milani et al., 2013). The activation of neuroinflammatory processes is mediated by glial cells, astrocytes, and particularly microglia. After a primary neuronal insult, microglia remains persistently activated and contributes to the release of free radicals, which exacerbate the neurodegenerative process and accelerate its progression (Milani et al., 2013).

Dietary phytochemicals, particularly flavonoids, exert beneficial effects on the central nervous system by limiting neurodegeneration and preventing or reversing age-dependent losses in cognitive performance. The mechanisms underlying these effects include: (1) the reduction of oxidative stress in the brain by scavenging of highly reactive molecules; (2) improvement of the antioxidant reserve; (3) chelation of transition metals; (4) antiinflammatory effects resulting in suppression of neuroinflammation; and (5) improvement of cognitive function (Dadhania et al., 2016; Spencer, 2012). At the molecular level, phytochemicals also exert neuroprotective actions by modulating prosurvival, proapoptotic, and stress response signaling pathways. These include NF-E2-related factor 2 antioxidant response element (Nrf2/ARE), MAPK, PKC, Janus kinase-signal transducer and activator of transcription (JAK-STAT), mitogen-activated protein kinase/extracellular signal-regulated kinase/cAMP response element-binding protein, phosphatidylinositol 3-kinase (PI3K/Akt), and insulin-signaling pathway. The interaction with these signaling cascades ultimately leads to inhibition of apoptosis and neuroinflammation and to promotion of synaptic plasticity (Dadhania et al., 2016). Also, phytochemicals improve blood flow to the brain, causing angiogenesis, neurogenesis, and changes in neuronal morphology (Spencer, 2012).

Organosulfur compounds found in garlic and cruciferous vegetables (e.g., *allicin* and *L-sulforaphane*) play a role in neuroprotection directly by their antioxidant effect against reactive species and indirectly through the activation of endogenous antioxidant systems. One of these systems is the Nrf2/ARE transcriptional pathway (Fig. 69.2), which shows neuroprotective effects against neuronal injury or neurodegeneration. Nrf2 is an unstable protein associated with its negative regulator Kelch-like ECH-associated protein 1 (Keap1), which acts as a molecular sensor of cellular redox homeostasis disturbance. Nrf2 is activated via oxidative modification of Keap1 cysteine residues and/or Keap1 ubiquitination and proteasomal degradation, or by phosphorylation mediated by the activation of several signaling pathways (e.g., MAPK, PKC, and PI3K). Once activated, Nrf2 translocates to the nucleus to induce some antioxidant genes, including γ-glutamylcysteine ligase (GCL), the rate-limiting enzyme in the synthesis of glutathione (GSH), GST, superoxide dismutase 2, NAD(P)H:quinone oxidoreductase 1 (NQO1), and heme oxygenase 1 (Kelsey et al., 2010).

Phytochemicals such as sulforaphane, polyphenols, EGCG, and curcumin can protect neurons in neurodegenerative diseases by increasing cellular defenses against oxidative stress through activation of the Nrf2/ARE pathway (Milani et al., 2013).

Nutraceuticals also confer neuroprotection by alternative mechanisms, such as sirtuin (SIRT) proteins, which are part of the histone deacetylase family. Induction of SIRT1 appears to be a principal mechanism underlying the neuroprotective effects of resveratrol. The prosurvival effects of SIRT1 activation involve the suppression of Bax-dependent apoptosis and repression of multiple proapoptotic transcription factors (Kelsey et al., 2010).

Molecular mechanisms for osteoarthritis

Osteoarthritis (OA) is the result of an imbalance in the physiologic repair process featured by increased chondrocyte metabolism, bone remodeling, osteophyte production, and synovial response (Singh et al., 2006). In osteoarthritis pathophysiology, chondrocytes undergo abnormal proliferation, senescence, cell death, and increased expression of genes encoding for inflammatory cytokines, matrix proteins, and proteolytic enzymes that disturb the articular cartilage homeostasis (Leong et al., 2013). During the development of OA, chondrocytes decrease the synthesis of extracellular matrix components, and increase that of proteases [e.g., metalloproteases (MMPs) and A Disintegrin and Metalloproteinase with Thrombospondin motifs (ADAMTs)], nitric oxide, and other proinflammatory mediators (Oliviero et al., 2018). The increased inflammation observed in this disorder is attributed to the proinflammatory cytokines IL-1β and TNF-α, which adversely affect cartilages by suppressing anabolic activity, stimulating catabolic breakdown of the articular cartilage, and increasing the production of inflammatory mediators (PGE_2 and COX-2) and ROS. These mechanisms, together, are responsible for the loss of the homeostatic balance of chondrocytes and they are partially mediated by members of the MAPK and NF-κB transcription factors (Leong et al., 2013).

Because the mechanisms involved in cartilage degradation are multifactorial, nutraceuticals and natural compounds found in fruits, teas, spices, wine, and vegetables may provide health benefits by targeting different pathways (Ameye and Chee, 2006). The effect of some phytochemicals is associated with their high content of soluble polyphenols that exhibit antioxidant, antiinflammatory, and anticatabolic effects. The antioxidant properties of many nutraceuticals decrease lipid peroxidation and enhance activities of the antioxidant enzymes catalase, SOD, glutathione peroxidase, and glutathione reductase. As a result, they decrease the high levels of oxygen free radicals produced in chondrocytes stimulated by proinflammatory cytokines. Overall, phytochemicals rich in polyphenols decrease the induction of proinflammatory mediators and enzymes, particularly by downregulating the NF-κB pathway in synoviocytes and chondrocytes, thus reducing the cartilage damage and proteoglycan loss (Leong et al., 2013).

The preventive role of polyphenols on articular damage may occur by inhibiting signal transduction pathway activation, which involves kinases and transcriptional factors, and by decreasing the production of autoantibodies and autoantigens which are involved in the development of some rheumatic diseases (Oliviero et al., 2018).

Glucosamine (derived from calf trachea) and chondroitin (derived from shellfish) are nutraceuticals commonly used for treating osteoarthritis. Both compounds stimulate articular glycosaminoglycan production and exert antiinflammatory effects by inhibiting the NF-κB pathway, resulting in a reduction in the COX-2 synthesis (Leong et al., 2013). Glucosamine has an additional antiinflammatory effect by stabilizing cell membranes and by increasing proteoglycan production (Singh et al., 2006).

Fish oils are omega-3 fatty acids that act by inhibiting the conversion of arachidonic acid to the proinflammatory eicosanoids PGE_2, LTB_4, and TXA_2, and thus have beneficial effects by modulating both pain and inflammatory cytokines (Singh et al., 2006).

Omega-3 and omega-6 fatty acids are metabolized by COXs and LOXs into distinct eicosanoids with antiinflammatory and proinflammatory activities, respectively. The antiinflammatory effect of omega-3 on cartilage metabolism could also be mediated by their oxygenated derivatives called resolvins, which are potent antagonists of inflammation through their binding to G-protein-coupled receptors. Omega-3 also has anticatabolic activity by decreasing aggrecanase-mediated proteoglycan catabolism in articular cartilage (Ameye and Chee, 2006).

Vitamin C may reduce the risk of cartilage loss and osteoarthritis progression due to its antioxidant effect. In chondrocytes, ascorbic acid serves as a cofactor for prolyl and lysyl hydroxylases, which play a key role in collagen synthesis. Ascorbic acid also modulates lysosomal enzymes, decreasing the activities of arylsulfatase A and arysulfatase B and increasing the activity of acid phosphatase in chondrocytes (Ameye and Chee, 2006; Leong et al., 2013). Suboptimal levels of vitamin D may lead to osteoarthritis by impairing bone metabolism, and the expression of vitamin D receptors is upregulated in osteoarthritis chondrocytes (Ameye and Chee, 2006).

Toxicity of nutraceuticals

Physiological versus pharmacological concentrations of nutraceuticals

Although nutraceuticals have become more popular there have been increasing concerns about the safety and potential toxicity of many of these compounds due to misuse or abuse. Plant-derived compounds are generally regarded to be safe for consumption because they are produced naturally and have a long history of use. However, because of the complex chemical composition of some natural products, there is a lack of information regarding their toxicity profile. Hence, it should not be presumed that these compounds are biologically safe and devoid of any health risk. It is becoming apparent that commonly used dietary compounds have beneficial effects at physiologic (nutritional) doses, but they can exert deleterious effects at pharmacological doses (Martin and Appel, 2010).

Two important factors need to be considered in evaluating the safety of any phytochemical are the nature and significance of the adverse effect and the exposure level at which the effect is observed. Adverse reactions can be either acute or chronic. Acute toxicity is usually dose-related and is related to the pharmacological properties of phytochemicals. Chronic adverse effects can go unrecognized for months or years (e.g., aristolochic acid nephropathy). Toxicity testing can reveal some of the risks that may be associated with the use of phytochemicals, especially in sensitive populations (Ifeoma et al., 2013). The new methodologies developed for toxicological assessment, such as predictive toxicology and omics, can help to reduce the uncertainty in decision-making with respect to natural products (Omar et al., 2010).

Although antioxidants, particularly polyphenols, have been linked to numerous health benefits and have been used to prevent oxidative stress-induced disorders, they may lead to detrimental effects when taken in larger doses found in dietary supplements and fortified foods, very likely as a result of pro-oxidative mechanisms (Martin and Appel, 2010). Specifically, reactive molecules can interact with the following molecular targets (Milani et al., 2013):

- Proteins, particularly the backbone and side chains of proteins, causing the formation of carbonyl groups and methionine sulfoxide, and often determining protein misfolding and aggregation.
- Nucleic acids, leading to DNA single-strand and double-strand breaks, DNA–protein cross-links, modification of purine and pyrimidine bases, and oxidative modification in both protein-coding RNAs and non-coding RNAs.
- Membrane lipids, causing lipid peroxidation as a result of their interaction with PUFAs, yielding highly reactive products such as malondialdehyde, 4-hydroxy-2-trans-nonenal, and acrolein, among others.

Oxidative stress causes a cascade of damaging processes eventually resulting in cell death, and it has been implicated in several pathophysiological disorders underlying many common diseases such as CVDs, diabetes, cancer, neurodegenerative disorders, chronic obstructive pulmonary disease, and the physiological aging process (Angel Nivya et al., 2012).

Polyphenol toxicity

During the scavenging of highly reactive species, antioxidants donate an electron or a hydrogen atom to the radical involved, which converts into a relatively stable nonradical. However, once an antioxidant scavenges a reactive species, it is converted into an oxidation product that takes over part of the reactivity of the radical and might cause damage to vital cellular components. Hence, the intake of relatively high doses of antioxidants as natural remedies may be harmful for human health, largely due to their pro-oxidative properties, whereas at low doses an antioxidant effect usually occurs (Boots et al., 2008; Jacobs et al., 2010). The controversial results observed in the literature may be due to this biphasic (hormetic) effect.

There are some concerns regarding deleterious effects of polyphenol consumption at levels far above natural occurrence as a result of polyphenol fortification and supplementation. The intake of polyphenolic compounds alone at such high doses may not produce the same benefits observed in epidemiological studies, but may also be associated with adverse effects, including increased mortality or stroke, pharmacological interactions with medications, and heterogeneous effects in subpopulations (Cory et al., 2018).

Some polyphenol-rich compounds have been assessed for different in vitro toxicity profiles and elicited cytotoxic effects on mitochondrial membrane potential, plasma membrane integrity, and nuclear morphology abnormalities. This observation was most pronounced when analyzing the cumulative effect of polyphenol-rich compound (Boncler et al., 2017).

Chemical structures and molecular size are important determinants of effect, and these pro-oxidative activities are attributable to the catechol or pyrogallol rings (Galati and O'Brien, 2004; Murakami, 2014). These pharmacophores are present in various antioxidants, like quercetin or EGCG, and while exerting their antioxidant activity they are converted into oxidized products, such as toxic semiquinone or phenoxyl radicals. Because these radicals are very reactive toward thiols, they are instantaneously conjugated with GSH. However, when GSH is consumed, the highly reactive radicals are no longer neutralized, and react and inactivate enzymes containing a critical thiol group, and co-oxidize unsaturated lipids, leading to increased membrane permeability (Jacobs et al., 2010). Phenoxyl radicals have also been involved in the initiation stage of atherosclerosis because myeloperoxidase/H_2O_2 readily catalyze the oxidation of tyrosine, a plasma phenol, to a tyrosyl radical, which then co-oxidizes LDL (Galati and O'Brien, 2004).

Hazards related to polyphenols include carcinogenic or genotoxic effects, likely because at high concentrations these compounds may become pro-oxidants and can enhance tumor promotion or increase proliferative signaling via ROS production and interruption of second messenger signaling (Martin and Appel, 2010). Several flavonoids can inhibit thyroid peroxidase and interfere with thyroid hormone biosynthesis. Because isoflavones exhibit estrogen-like activity, the consumption of high doses has been associated with reduced fertility in animals and with antiluteinizing hormone effects among premenopausal women. Polyphenols may also have antinutritional effects as a result of inhibition of nonheme iron absorption, particularly when polyphenol-rich foods (e.g., coffee, tea, and wine) are regularly consumed with meals lacking vitamin C, which is an enhancer of nonheme iron absorption (Martin and Appel, 2010; Mennen et al., 2005). High polyphenol intake could also increase the risk of CVD through alterations in homocysteine processing (Martin and Appel, 2010). Finally, polyphenols may affect the bioavailability and pharmacokinetics of several drugs because of inhibition of the drug-metabolizing enzymes CYP450, GST, and UDP-glucuronosyltransferase, which may lead to increased serum concentrations of some drugs, ultimately causing toxicity (Burgos-Morón et al., 2010; Martin and Appel, 2010). This occurs with St. John's wort (*Hypericum perforatum*), ginkgo (*Ginkgo biloba*), ginseng (*Panax ginseng*), kava (*Piper methysticum*), and garlic (*Allium sativum*) (Izzo and Ernst, 2009).

The ingestion of high doses of green tea polyphenols and other related supplements increases the risk of hepatotoxicity because of their pro-oxidative properties. In contrast, there is scant evidence that oral consumption of green tea, but not supplements, even in large amounts has side effects (Murakami, 2014). Although the precise mechanisms underlying that toxicity have not been fully elucidated, increased levels of liver transaminases and bilirubin and marked necrosis and inflammation in liver biopsies indicate serious liver dysfunction. Some individuals may be particularly susceptible because of genetic variations, or polymorphisms, in critical biotransformation pathways for tea polyphenols, such as catechol-*O*-methyltransferase (Martin and Appel, 2010).

In conclusion, the use of pharmacological doses of polyphenols in prevention, treatment, and dietary supplement situations can potentially cause adverse effects through pro-oxidative effects (Martin and Appel, 2010).

Vitamin toxicity

Fat-soluble vitamins have the ability to accumulate in the body, and thus they show higher potential for toxicity than water-soluble vitamins. High doses of antioxidant vitamins may exert adverse vascular effects, and the supplementation of vitamins C, E, and/or β-carotene has been associated with an increased risk of all-cause mortality (Davì et al., 2010; Alissa and Ferns, 2012).

Because vitamin A is fat-soluble and can be stored in the liver, the consumption of large amounts of this vitamin over a period of time can result in liver, bone, and skin toxicity. Chronic vitamin A toxicity causes changes in skin, hair, and nails, abnormal liver function test results, osteoporosis, and birth defects. Although carotene is converted to vitamin A in the body, excessive ingestion of carotene causes carotenosis (yellow skin), but not vitamin A toxicity. However, when taken as a supplement (not when consumed in fruits and vegetables) over long periods of time, β-carotene has been associated with increased cancer risk (The Merck Manual – Professional Edition, 2019). It is plausible that the β-carotene-induced decrease in retinoic acid level in the lung might reduce the inhibitory effect of retinoids on activator protein 1 (AP-1) via the nuclear retinoic acid receptors. Consequently, signal transduction pathways that control cell proliferation can be affected, potentially leading to tumor formation, particularly lung tumors (Bast and Haenen, 2002).

Long-term supplementation with vitamin E appears to increase the incidence of heart failure and mortality, especially in patients with chronic diseases (Miller et al., 2005). The major adverse effect of high levels of vitamin E is the exacerbation of impaired blood coagulation due to vitamin K deficiency caused by malabsorption or anticoagulant therapy (Bast and Haenen, 2002). Paradoxically, a pro-oxidant effect of tocopherol has been described similar to that occurring with vitamin C. The antioxidant activity of vitamin E may lead to the generation of quinones that exert cytotoxic effects by oxidizing cellular components with thiol groups. Moreover, the reducing power of tocopherol is responsible for the reduction of transition state metals, which facilitates the formation of free radicals. This pro-oxidant effect may be involved in the increase in fatal myocardial infarctions observed in clinical studies after vitamin E supplements (Bast and Haenen, 2002).

Vitamin D toxicity results from the ingestion of either a single megadose or excessive amounts (pharmacologic doses) over prolonged periods of time, with marked hypercalcemia being the main toxic effect. Vitamin D toxicity may be the result of elevated serum levels of calcifediol (25-OH-D_3) instead of calcitriol [1,25-$(OH)_2$-D_3, the hormonal form of vitamin D], because large amounts of calcifediol replace calcitriol and stimulate bone calcium metabolism (Alshahrani and Aljohani, 2013).

Although vitamin C has antioxidant activity, it may become a pro-oxidant under conditions of high oxidative stress (Alissa and Ferns, 2012). Excessive doses of supplemental vitamin C produce toxic effects on the digestive and urinary systems. Excessive daily amounts of vitamin C produce diarrhea from the osmotic effects of the unabsorbed vitamin in the intestinal lumen. Because oxalate is an end product of ascorbate, hyperoxaluria and potential kidney stone formation may ensue after ingestion of high doses of vitamin C, particularly in subjects with high amounts of urinary calcium. Vitamin C may also precipitate hemolysis in people with glucose-6-phosphate dehydrogenase deficiency or paroxysmal nocturnal hemoglobinuria. Because vitamin C reduces iron, and reduced iron is more easily absorbed through the gut, vitamin C supplementation may lead to iron overload in patients with thalassemia or hemochromatosis (FAO, 2001).

Fish oil toxicity

Fish oil supplements have been associated with decreased plasma insulin and increased serum glucose and cholesterol levels (Singh et al., 2006). A recent randomized, placebo-controlled trial found that supplementation with omega-3 fatty acids did not result in a lower incidence of major cardiovascular events or cancer (Manson et al., 2019). High doses of omega-3 fatty acids may increase the risk of bleeding, reduce blood pressure, and alter liver function tests. However, the aforementioned primary prevention trial did not find evidence of gastrointestinal symptoms, major bleeding episodes, or other serious adverse events. Fish-derived omega-3 supplements should be avoided in individuals allergic or sensitive to fish or fish oil enriched in omega-3 fatty acids (Mayo Clinic, 2017).

Can nutraceuticals be subjected to an adverse outcome pathway approach?

AOPs are a practical solution to the integration and dissemination of toxicological knowledge, and are a tool to

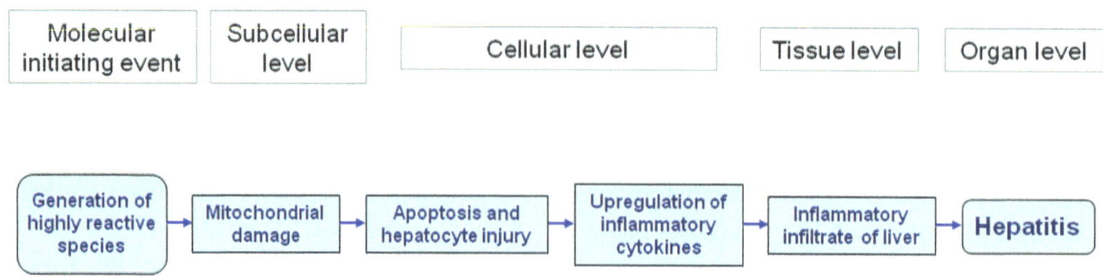

FIGURE 69.4 Putative adverse outcome pathway (AOP) for oxidative stress-based phytochemical toxicity.

apply systems biology thinking into toxicological and biomedical research. AOPs allow for expanding the use of mechanistic toxicological data for risk assessment and regulatory applications. In this framework, it is important to carefully select appropriate tools to quantitatively and objectively assess key events such as mitochondrial damage, apoptosis induction, and cell toxicity.

Development of an AOP for nutraceuticals needs to include the toxicity pathways underlying potential disorders and provide a tool to organize and integrate all levels of toxicity data. AOPs can be useful in identifying data gaps and uncertainties in the pathway linking nutraceuticals intake to the molecular and cellular response and then to potential adverse outcomes. This process needs a clear presentation of the experimental support and the strength of the evidence for each of the events in the AOP and the associated linkages as well as a discussion of the confidence in the AOP. As previously mentioned, some nutraceuticals exhibit a biphasic effect (antioxidant and pro-oxidant), depending on the doses used. Only the intake of relatively high doses of some nutraceuticals may be harmful for human health, likely because of consequences of their pro-oxidative properties. However, the available evidence supporting the adverse effects as well as the underlying molecular mechanisms is limited, thus representing a limitation to building AOPs.

In spite of that, AOP can be developed on the basis of the known effects of a wide range of nutraceuticals on biological pathways and to evaluate the evidence for specific biomarkers of toxicity (at a different level of biological organization) of nutraceuticals-induced toxicity pathways linked to final outcomes. This approach may be particularly useful in identifying potential molecular initiating events and cellular responses that may be biomarkers of certain toxicity pathways (e.g., oxidative stress, inflammation) mechanistically linked to the final adverse outcomes. For instance, the activation of the NF-κB signaling pathway by many polyphenols can be assessed in cellular assays by measuring the NF-κB/DNA-binding activity, the level of NF-κB subunit expression in the nucleus, the level of NF-κB phosphorylation, the level of NF-κB-mediated gene expression, NF-κB luciferase activity, and the presence of IκBα degradation and IκB kinase (IKK) activity, among others (Omar et al., 2010).

With a sufficient literature base for a particular substance, the evaluation should reach exposure-specific hazard identification conclusions for the association between exposure-induced toxicity pathways and final adverse outcomes. For a particular nutraceutical, the evaluation should indicate whether the compound may be of concern for inducing oxidative stress, inflammation, or any other toxicity pathway that contributes to final adverse outcomes based on the current human nutraceuticals intake.

Fig. 69.4 shows a putative AOP for polyphenols toward the development of hepatitis as an adverse liver outcome. The formation of ROS can trigger various cellular effects (mitochondrial damage, apoptosis induction), which induce inflammatory key events (upregulation of inflammatory cytokines) and further inflammatory infiltrate of liver that, if sustained over time, eventually results in hepatitis and respective systemic consequences of impaired liver function. However, the case study proposed shows limitations for practical applicability in a regulatory framework. The limited data available from randomized clinical trials assessing antioxidant-induced liver toxicity do not lend consistency to results obtained from experimental and epidemiological studies. Thus, the lack of a solid basis precludes evaluating the contribution of the identified toxicity pathways to potential polyphenols-induced liver toxicity as well as a wider array of health effects.

Concluding remarks and future directions

The intake of diet-derived nutraceuticals is generally regarded as safe and may contribute to the prevention and treatment of numerous common disorders because of their antioxidant and antiinflammatory activities. Conversely, chronic administration at higher doses of certain formulated products of phytochemicals can exert detrimental effects through pro-oxidative and/or other mechanisms. Because AOPs are conceptual frameworks that display the knowledge base for a toxicological response concerning the linkage between MIE and an adverse outcome, future evaluations should consider two complementary approaches. First, a systematic review allows evaluation of the evidence of how nutraceuticals contribute to oxidative stress or inflammation, potentially leading to several toxic health effects. Second, the development of an AOP for the influence of nutraceuticals on toxicity pathways leading to overt disorders will aid in the identification of molecular biomarkers, uncertainties, and data gaps. The establishment of qualitative and quantitative links between molecular initiating events, key events, and adverse outcomes will foster their application for regulatory decisions (Tollefsen et al., 2014).

One limitation to the practical applicability of AOPs on polyphenols-induced adverse effects is that much of the research has been conducted in vitro, and the use of animal models to investigate their biological activities has the drawback of interspecies differences in the metabolism and bioavailability of the active compounds, which may give contradicting experimental results. In addition, clinical trials intended to evaluate the physiological effects in humans are scarce and results are controversial (Angel Nivya et al., 2012). Careful consideration of the specific effect of nutraceutical intervention on the disease mechanism is critical. Therefore, the pharmacodynamic/pharmacokinetic actions of the nutraceutical intervention must be evaluated, along with its source(s), side effect(s), and potential drug interaction(s). Finally, available clinical outcomes data must be reviewed regarding efficacy and safety to support an evidence-based approach (Singh et al., 2006).

References

Alissa, E.M., Ferns, G.A., 2012. Functional foods and nutraceuticals in the primary prevention of cardiovascular diseases. J. Nutr. Metab. 2012 (2012), 569486.

Alshahrani, F., Aljohani, N., 2013. Vitamin D: deficiency, sufficiency and toxicity. Nutrients 5 (2013), 3605–3616.

Ameye, L.G., Chee, W.S., 2006. Osteoarthritis and nutrition. From nutraceuticals to functional foods: a systematic review of the scientific evidence. Arthritis Res. Ther. 8 (2006), R127.

Angel Nivya, M., Raja, K., Kumaravel, M., 2012. Role of nutraceuticals in cancer. Int. J. Pharm. Pharm. Sci. 4 (Suppl. 4), 415–420 (2012).

Ankley, G.T., Bennett, R.S., Erickson, R.J., 2010. Adverse outcome pathways: a conceptual framework to support ecotoxicology research and risk assessment. Environ. Toxicol. Chem. 29 (2010), 730–741.

Bal-Price, A., Meek, M.E.B., 2017. Adverse outcome pathways: application to enhance mechanistic understanding of neurotoxicity. Pharmacol. Ther. 179 (2017), 84–95.

Bast, A., Haenen, G.R., 2002. The toxicity of antioxidants and their metabolites. Environ. Toxicol. Pharmacol. 11 (2002), 251–258.

Boncler, M., Golanski, J., Lukasiak, M., et al., 2017. A new approach for the assessment of the toxicity of polyphenol-rich compounds with the use of high content screening analysis. PLoS One 12 (2017), e0180022.

Boobis, A.R., 2005. A tiered approach to assessing the toxicology of pesticides. Toxicol. Lett. 158 (2005), S30–S31.

Boobis, A.R., Cohen, S.M., Dellarco, V., 2006. IPCS framework for analysing the relevance of a cancer mode of action for humans. Toxicol. Lett. 164 (2006), S254–S255.

Boobis, A.R., Daston, G.P., Preston, R.J., 2009. Application of key events analysis to chemical carcinogens and noncarcinogens. Crit. Rev. Food Sci. Nutr. 49 (2009), 690–707.

Boots, A.W., Haenen, G.R., Bast, A., 2008. Health effects of quercetin: from antioxidant to nutraceutical. Eur. J. Pharmacol. 585 (2008), 325–337.

Burgos-Morón, E., Calderón-Montaño, J.M., Salvador, J. 2010. The dark side of curcumin. Int. J. Cancer 126 (7), 1771–1775.

Carlos-Reyes, Á., López-González, J.S., Meneses-Flores, M., et al., 2019. Dietary compounds as epigenetic modulating agents in cancer. Front. Genet. 10 (2019), 79.

Cencic, A., Chingwaru, W., 2010. The role of functional foods, nutraceuticals, and food supplements in intestinal health. Nutrients 2 (2010), 611–625.

Chauhan, B., Kumar, G., Kalam, N., 2013. Current concepts and prospects of herbal nutraceutical: a review. J. Adv. Pharm. Technol. Res. 4 (2013), 4–8.

Cicero, A.F.G., Parini, A., Rosticci, M., 2015. Nutraceuticals and cholesterol-lowering action. Int. J. Cardiol. Metab. Endocr. 6, 1–4.

Cory, H., Passarelli, S., Szeto, J., Tamez, M., Mattei, J., 2018. The role of polyphenols in human health and food systems: a mini-review. Front. Nutr. 5 (2018), 87.

Dadhania, V.P., Trivedi, P.P., Vikram, A., Tripathi, D.N., 2016. Nutraceuticals against neurodegeneration: a mechanistic insight. Curr. Neuropharmacol. 14 (2016), 627–640.

Das, L., Bhaumik, E., Raychaudhuri, U., 2012. Role of nutraceuticals in human health. J. Food Sci. Technol. 49 (2012), 173–183.

Davì, G., Santilli, F., Patrono, C., 2010. Nutraceuticals in diabetes and metabolic syndrome. Cardiovasc. Ther. 28 (2010), 216–226.

Davies, E., Greenacre, D., Lockwood, G.B., 2005. Adverse Effects and Toxicity of Nutraceuticals. CRC Press, Boca Raton, FL, pp. 165–195, 2005.

FAO, 2001. Human vitamin and mineral requirements. Report of a Joint FAO/WHO Expert Consultation. Chapter 6 (Vitamin C). Food and Agriculture Organization of the United Nations, Food and Nutrition Division FAO, Rome. <http://www.fao.org/3/y2809e/y2809e06.pdf> (available 27-03-20).

Farkhondeh, T., Samarghandian, S., Roshanravan, B., 2019. Impact of chrysin on the molecular mechanisms underlying diabetic complications. J. Cell. Physiol. 234 (2019), 17144–17158.

Fiedor, J., Burda, K., 2014. Potential role of carotenoids as antioxidants in human health and disease. Nutrients 6 (2014), 466–488.

Galati, G., O'Brien, P.J., 2004. Potential toxicity of flavonoids and other dietary phenolics: significance for their chemopreventive and anticancer properties. Free Radic. Biol. Med. 37 (2004), 287–303.

Gao, J., Azad, M.A.K., Han, H., Wan, D., et al., 2020. Impact of prebiotics on enteric diseases and oxidative stress. Curr. Pharm. Des. Available from: https://doi.org/10.2174/1381612826666200211121916 (2020).

Guasch-Ferré, M., Merino, J., Sun, Q., et al., 2017. Dietary polyphenols, Mediterranean diet, prediabetes, and type 2 diabetes: a narrative review of the evidence. Oxid. Med. Cell. Longev. 2017 (2017), 6723931.

Ifeoma, O., Oluwakanyinsola, S., Gowder, S., 2013. Screening of herbal medicines for potential toxicities. In: Gowder, S. (Ed.), New Insights into Toxicity and Drug Testing. InTech. <https://www.intechopen.com/books/new-insights-into-toxicity-and-drug-testing/screening-of-herbal-medicines-for-potential-toxicities> 2013. Doi: 10.5772/54493, (available 27.03.20).

Izzo, A.A., Ernst, E., 2009. Interactions between herbal medicines and prescribed drugs: an updated systematic review. Drugs 69 (2009), 1777–1798.

Jacobs, H., Moalin, M., Bast, A., 2010. An essential difference between the flavonoids monoHER and quercetin in their interplay with the endogenous antioxidant network. PLoS One 5 (11), e13880 (2010).

Kalra, E.K., 2003. Nutraceutical – definition and introduction. AAPS Pharm. Sci. 5 (3), E25 (2003).

Kelsey, N.A., Wilkins, H.M., Linseman, D.A., 2010. Nutraceutical antioxidants as novel neuroprotective agents. Molecules 15 (2010), 7792–7814.

Kim, D.H., Meza, C.A., Clarke, H., et al., 2020. Vitamin D and endothelial function. Nutrients 12 (2020), 2.

Krewski, D., Acosta, D., Andersen, M., 2010. Toxicity testing in the 21st century: a vision and a strategy. J. Toxicol. Environ. Health B Crit. Rev. 13 (2010), 51–138.

Kunnumakkara, A.B., Sailo, B.L., Banik, K., et al., 2018. Chronic diseases, inflammation, and spices: how are they linked? J. Transl. Med. 16 (2018), 14.

LaLone, C.A., Ankley, G.T., Belanger, S.E., et al., 2017. Advancing the adverse outcome pathway framework—an international horizon scanning approach. Environ. Toxicol. Chem. 36 (2017), 1411–1421.

Lee, J.H., Khor, T.O., Shu, L., 2013. Dietary phytochemicals and cancer prevention: Nrf2 signaling, epigenetics, and cell death mechanisms in blocking cancer initiation and progression. Pharmacol. Ther. 137 (2013), 153–171.

Leong, D.J., Choudhury, M., Hirsh, D.M., 2013. Nutraceuticals: potential for chondroprotection and molecular targeting of osteoarthritis. Int. J. Mol. Sci. 14 (2013), 23063–23085.

Ma, Q., 2013. Role of Nrf2 in oxidative stress and toxicity. Annu. Rev. Pharmacol. Toxicol. 53 (2013), 401–426.

Manson, J.E., Cook, N.R., Lee, I.M., VITAL Research Group, et al., 2019. Marine n-3 fatty acids and prevention of cardiovascular disease and cancer. N. Engl. J. Med. 380 (2019), 23–32.

Martin, K.R., Appel, C.L., 2010. Polyphenols as dietary supplements: a double-edged sword. Nutr. Diet. Suppl. 2010 (2010), 1–12.

Mayo Clinic, 2017. Drugs and supplements. Fish oil. <https://www.mayoclinic.org/drugs-supplements-fish-oil/art-20364810> (available 27.03.20.).

McCubrey, J.A., Lertpiriyapong, K., Steelman, L.S., et al., 2017. Effects of resveratrol, curcumin, berberine and other nutraceuticals on aging, cancer development, cancer stem cells and microRNAs. Aging 9 (2017), 1477–1536.

Meek, M.E., Boobis, A., Cote, I., 2014. New developments in the evolution and application of the WHO/IPCS framework on mode of action/species concordance analysis. J. Appl. Toxicol. 34 (2014), 1–18.

Mennen, L.I., Walker, R., Bennetau-Pelissero, C., 2005. Risks and safety of polyphenol consumption. Am. J. Clin. Nutr. 81 (Suppl. 1), 326S–329S (2005).

Milani, P., Ambrosi, G., Gammoh, O., 2013. SOD1 and DJ-1 converge at Nrf2 pathway: a clue for antioxidant therapeutic potential in neurodegeneration. Oxid. Med. Cell. Longev. 2013 (2013), 836760.

Miller, E.R., Pastor-Barriuso, R., Dalal, D., 2005. Meta-analysis: high-dosage vitamin E supplementation may increase all-cause mortality. Ann. Intern. Med. 142 (2005), 37–46.

Mishra, S., Verma, S.S., Rai, V., et al., 2019. Long non-coding RNAs are emerging targets of phytochemicals for cancer and other chronic diseases. Cell. Mol. Life Sci. 76 (2019), 1947–1966.

Moradi, F., Maleki, V., Saleh-Ghadimi, S., et al., 2019. Potential roles of chromium on inflammatory biomarkers in diabetes: a systematic. Clin. Exp. Pharmacol. Physiol. 46 (2019), 975–983.

Murakami, A., 2014. Dose-dependent functionality and toxicity of green tea polyphenols in experimental rodents. Arch. Biochem. Biophys. 557 (2014), 3–10.

Nakata, R., Takahashi, S., Inoue, H., 2012. Recent advances in the study on resveratrol. Biol. Pharm. Bull. 35 (2012), 273–279.

National Research Council (NRC), 2007. Toxicity Testing in the 21st Century: A Vision and a Strategy. National Academy Press, Washington, DC2007.

OECD (Organization for Economic Co-operation and Development), 2017. Revised Guidance Document on Developing and Assessing Adverse Outcome Pathways Series on Testing & Assessment No. 184, Environment Directorate. <http://www.oecd.org/officialdocuments/publicdisplaydocumentpdf/?cote = env/jm/mono(2013)6&doclanguage = en> (available 03.03.20.).

OECD (Organization for Economic Co-operation and Development), 2018. Considerations for Assessing the Risks of Combined Exposure to Multiple Chemicals, Series on Testing and Assessment No. 296, Environment, Health and Safety Division, Environment Directorate. <http://www.oecd.org/chemicalsafety/risk-assessment/considerations-for-assessing-the-risks-of-combined-exposure-to-multiple-chemicals.pdf> (available 03.03.20.).

Oliviero, F., Scanu, A., Zamudio-Cuevas, Y., et al., 2018. Anti-inflammatory effects of polyphenols in arthritis. J. Sci. Food Agric. 98 (2018), 1653–1659.

Omar, E.A., Kam, A., Alqahtani, A., 2010. Herbal medicines and nutraceuticals for diabetic vascular complications: mechanisms of action and bioactive phytochemicals. Curr. Pharm. Des. 16 (2010), 3776–3807.

Pandey, M., Vijayakumar, B., 2011. Nutraceutical supplementation for diabetes: a review. Int. J. Pharm. Pharm. Sci. 3 (2011), 33–40.

Perkins, E.J., Ashauer, R., Burgoon, L., et al., 2019. Building and applying quantitative adverse outcome pathway models for chemical hazard and risk assessment. Environ. Toxicol. Chem. 38 (2019), 1850–1865.

Postal, B.G., Guesser, S.M., Kappel, V.D., 2014. Mechanism of action of nutraceuticals on intestine to ameliorate glucose homeostasis: follow-up studies by an *in situ* approach. J. Cell Sci. Ther. 5 (2014), 162.

Preyat, N., Leo, O., 2013. Sirtuin deacylases: a molecular link between metabolism and immunity. J. Leukoc. Biol. 93 (2013), 669–680.

Ranzato, E., Martinotti, S., Calabrese, C.M., 2014. Role of nutraceuticals in cancer therapy. J. Food Res. 3 (2014), 18–25.

Schmitt, J., Ferro, A., 2013. Nutraceuticals: is there good science behind the hype? Br. J. Clin. Pharmacol. 75 (2013), 585–587.

Schwarze, P.E., Totlandsdal, A.I., Låg, M., 2013. Inflammation-related effects of diesel engine exhaust particles: studies on lung cells *in vitro*. Biomed. Res. Int. 2013 (2013), 685142.

Singh, M., Ansani, N.T., Starz, T.W., 2006. Nutraceuticals and arthritis: weighing the evidence. Female Patient 31 (2006), 14–18.

Spencer, J.P.E., 2012. Flavonoids and brain health: multiple effects underpinned by common mechanisms. Ann. Nutr. Metab. 61 (2012), 325.

The Merck Manual – Professional Edition, 2019. <https://www.merckmanuals.com/professional/nutritional-disorders/vitamin-deficiency,-dependency,-and-toxicity/vitamin-a-deficiency> (available 27.03.20.).

Tollefsen, K.E., Scholz, S., Cronin, M.T., 2014. Applying adverse outcome pathways (AOPs) to support integrated approaches to testing and assessment (IATA). Regul. Toxicol. Pharmacol. 70 (2014), 629–640.

Vinken, M., 2013. The adverse outcome pathway concept: a pragmatic tool in toxicology. Toxicology 312 (2013), 158–165.

Young, J.S., 2008. NK-kB and Nrf2 as potential chemopreventive targets in some anti-inflamatory and antioxidative phytonutrients with anti-inflammatory and anti oxidative activities. Asia Pac. J. Clin. Nutr. 17 (2008), 269–272.

Zoltani, C.K., 2019. Nutraceuticals in cardiovascular diseases. In: Gupta, R., Srivastava, A., Lall, R. (Eds.), Nutraceuticals in Veterinary Medicine. Springer, Cham (Switzerland), pp. 427–435. 2019.

Chapter 70

Interactions between nutraceuticals/nutrients and nutrients and therapeutic drugs

Arturo Anadón, Irma Ares, María-Rosa Martínez-Larrañaga and María-Aránzazu Martínez

Introduction

Interactions between foods/dietary supplements and drugs present complex and challenging problems. They may arise from either alteration of the absorption, distribution, biotransformation, or excretion of one drug by a food/dietary supplement or from a combination of their actions or effects. These interactions can have a marked influence on the success of drug treatment and on the adverse or side effect profiles of many drugs. However, these interactions are not always detrimental to therapy. In some cases, they can be used to improve drug absorption or to minimize adverse effects. Most food/dietary supplements and drug interactions occur through three mechanisms: (1) reduced rate or extent of absorption; (2) increased rate or extent of absorption; or (3) chemical/pharmacologic effects.

Macronutrients are nutrients that humans consume in the largest quantities and that provide bulk energy. They are needed for a wide range of body functions and processes. These three macronutrients are fat, protein, and carbohydrate. Micronutrients are essential vitamins and minerals required by the body in miniscule amounts. Drug–nutrient interactions are defined as alterations to the pharmacokinetics or pharmacodynamics of a drug or nutritional element or a compromise in nutritional status as a result of the addition of a drug. Failure to identify and properly manage drug–nutrient interactions can lead to serious consequences. For instance, food–nutrient interactions can result in reduced absorption of some oral antibiotics and may lead to suboptimal antibiotic concentrations at the site of infection. This predisposes the patient to treatment failure.

Food–drug interactions can result in two main clinical effects: (1) decreased bioavailability of a drug, which predisposes to treatment failure or (2) increased bioavailability, which increases the risk of adverse events and may even precipitate toxicities. Patient populations who have increased risks of suffering from adverse events associated with drug–nutrient interactions are elderly patients, patients with cancer and/acquired immunodeficiency syndrome, those receiving enteral nutrition, and transplant recipients. Elderly patients are particularly at risk because more than 30% of all prescription drugs are taken by this population (Chan, 2006).

The continuing aging process depends on the improvement of several factors, such as hygiene, nutritional status, sanitary conditions, and the prevention of some chronic diseases. Among the most common health problems in elderly people are cancer, spine or back diseases, arthritis/rheumatism, osteoporosis, heart diseases (i.e., congestive heart failure, coronary heart disease), chronic obstructive pulmonary disease, hypertension, dementia, diabetes mellitus, and depression. There are food–drug interactions and careful drug selection is needed (Leibovitch et al., 2004; Schmidt and Dalhoff, 2002). Accordingly, the most common drugs taken by this age group are nonsteroidal antiinflammatory drugs, antihypertensives, and antidepressants (Akamine et al., 2007). Multiple underlying chronic diseases require long-term nutritional and pharmacotherapeutic interventions. The practice of polypharmacy (i.e., using multiple drugs to manage different disease states) increases patients' risks of food–drug interactions (Chan, 2006).

Diuretics vary in their interactions with food and specific nutrients. Some diuretics cause loss of potassium, calcium, and magnesium. Triamterene (known as a "potassium-sparing" diuretic), blocks the kidneys' excretion of potassium, which can cause hyperkalemia. Excess potassium may result in irregular heartbeat and heart palpitations. When taking triamterene, one should avoid eating large amounts of potassium-rich foods (i.e., bananas,

Nutraceuticals. DOI: https://doi.org/10.1016/B978-0-12-821038-3.00070-7

oranges, and green leafy vegetables), or salt substitutes that contain potassium (NCL National Consumers League, 2015).

Angiotensin-converting enzyme (ACE) inhibitors, such as captopril, enalapril, quinapril, and moexipril, relax blood vessels by preventing angiotensin II. Food can decrease the absorption of captopril and moexipril (these are taken 1 h before or 2 h after meals). ACE inhibitors may increase the amount of potassium in the body and excess potassium can be harmful (NCL National Consumers League, 2015).

Elderly patients are more likely to experience adverse events because several physiologic functions are reduced with age, such as the sense of taste and smell, chewing and swallowing, and gut motility. Pharmacokinetic peculiarities associated with aging are absorption (changes in gastric pH, decreased gastrointestinal and splanchnic blood flow), distribution [decreased lean body mass, water, serum albumin concentration, and binding serum proteins; increased total body fat, increased blood–brain barrier (BBB) permeability], and elimination (reduced renal function). Therefore drug bioavailability, volume of distribution, clearance, and half-life of drugs are modified in the elderly. Water-soluble drugs become more concentrated and fat-soluble drugs may have longer half-lives because of a slower release of the drug from fatty tissues. Activity of cytochrome P450 (CYP 450) oxidase systems is also modified in older patients. Because many of these age-related factors are difficult to predict, the ideal prescription of a drug is complex (Akamine et al., 2007). There are, however, a minority of drugs for which concomitant diet is important because the bioavailability of some drugs may be enhanced by food. For example, an acid environment is necessary for the absorption of ketoconazole. The absorption of griseofulvin, which is lipid-soluble, shows increased absorption and higher plasma concentrations when taken with a high-fat meal. The use of antifungal (fluconazole, griseofulvin, ketoconazole, itraconazole) medications should be avoided with dairy products (milk, cheeses, yogurt, ice cream) or antacids (NCL National Consumers League, 2015). Fenofibrate, mebendazole, isotretinoin, tamsulosin, carbamazepine, and labetalol are examples of drugs that will be better absorbed when taken with food. Improved absorption of a drug may or may not have a significant effect on the drug's efficacy (Bland, 1998).

Drug–herb interactions are becoming a source of serious concern in relation to adverse effects because those who consume herbal supplements often take prescribed drugs concomitantly and health professionals are often unaware of possible interactions. This issue has generated significant concern within the pharmaceutical industry and among regulatory authorities in recent years. The medical community has also been concerned about the risk to patients using inaccurate dosages and poor-quality products, as well as drug–nutrient and nutrient–nutrient interactions. One may assume that these interactions occur primarily with oral medications and only with those that share the CYP3A4 metabolism pathway. Consequences include increased oral bioavailability, higher serum drug concentrations, and associated adverse effects.

Drug metabolism

The ability to halt drug action is an important aspect of drug therapy in humans. The amount of drug that eventually reaches the target tissue in an active form depends not only on its physicochemical characteristics but also on the extent of metabolism occurring mainly in the gastrointestinal tract (GIT), liver, and lung. Metabolism and excretion by these organs before a drug can reach the systemic arterial circulation is referred to as *presystemic drug elimination*. Extensive presystemic (first-pass) metabolism or *presystemic clearance* is of considerable clinical, pharmacological, and toxicological importance and can limit the usefulness of certain drugs. Thus, many drugs are normally given by injection because, when given orally, most are destroyed by presystemic metabolism. There are marked interindividual differences in the extent of presystemic metabolism, and metabolites generated in the liver can in some instances be more toxic than the parent compound.

The drug-metabolizing enzyme systems are primarily located in the liver and small intestine, but they are also present to a lesser extent in other organs (e.g., lungs, adrenals). The activity of the hepatic enzymes is in general higher than that of drug-metabolizing enzymes located elsewhere in the body.

Access of drugs to hepatic drug-metabolizing enzymes will be determined not only by their physicochemical characteristics, which include their lipid solubility at physiological pH and extent of protein binding, but also by the pattern of blood flow through the liver. Metabolism of drugs by the liver, as well as clearance, depends on the relative affinity of plasma and hepatic proteins to bind the drug, the concentration gradient between blood and liver, and the activity of drug-metabolizing enzymes. Drugs are metabolized by phase I (catalyzed by CYP 450 enzymes) and phase II [catalyzed by conjugation enzymes such as glutathione *S*-transferase and uridine 5'-diphospho-glucuronyltransferase] enzymes.

When drugs enter the gut they are exposed to a number of systems with the potential to chemically modify or chelate drug molecules not found within the tissues of the host. These include extremes of pH, unabsorbed nutrients and xenobiotics, digestive enzymes, and intestinal microflora. Interactions between drugs and other nutrients in the gut lumen can result in the formation of a chemical complex that is poorly absorbed. Chemical interactions of this type have been described between tetracycline antibiotics and divalent or trivalent metal ions such as calcium, magnesium, and iron. Reduction of iron absorption from food occurs in the presence of antiacid drugs. The Maillard reaction is an example of the chemical interaction between dextrose and amino acids in parenteral nutrition. The intestine has the ability to metabolize drugs and other foreign chemicals by numerous pathways involving both phase I (oxidation, reduction, and hydrolysis) and phase II (conjugation) reactions between xenobiotic and endogenous substrates (glycine, glucuronic acid, acetic acid, etc.). Although the activities of these intestinal enzymes are generally lower than in the liver, because it is the total volume of the organ, the gut may still make a significant contribution to the overall metabolism of certain compounds.

Approximately half of all drugs used in humans are handled by CYP3A4 (Bressler and Bahl, 2003). Due to the lack of absolute specificity of the cytochrome P (CYP) enzymes, the metabolism of a drug may change with an increase in drug concentration or an inhibition of the CYP enzyme, which usually metabolizes the drug. Inhibition of drug metabolism by specific drugs, chemicals, or herbal medications causes increased levels of the parent drug, prolonged drug activity, and increased potential for drug toxicity. Competition for the active site on a CYP enzyme by two or more drugs can result in decreased inactivation of one of the drugs and an increase of its activity and a prolongation of its effects.

CYP3A4 is closely associated with P-glycoprotein (P-gp). P-gp, like CYP, is active in the liver and small intestines. P-gp was discovered during a course of investigations on drug resistance in the treatment of cancer (Cordon-Cardo et al., 1989). With an efflux pump which is functionally expressed in the apical surface of epithelial cells (enterocytes) that actively secretes absorbed drugs back into the gut lumen ("*efflux*" transporter), an important part of the biochemical barrier function of the intestinal mucosa. P-gp is also functionally expressed in the enterocytes that border the epithelium of the intestinal tract, where it plays a role in intestinal metabolism and is an important part of the biochemical barrier function of the intestinal mucosa. The "*efflux carrier*" may be responsible for limiting the bioavailability of several drugs after oral intake by pumping them out into the intestinal lumen. Not all 3A4-metabolized drugs are P-gp substrates. The induction of P-gp activity is frequently the result of exposure to the same drugs that induce CYP3A4. This provides a mechanism to limit cytoplasmic concentrations of drug metabolites formed by the action of CYP3A4. Moreover, the inhibition of CYP3A4 by drugs or other xenobiotics is often accompanied by inhibition of P-gp (Soldner et al., 1999), which can slow drug absorption. Absorption is also a factor in drug handling. The organic anion-transporting polypeptide (OATP) system is an intestinal system that plays a role in absorption of specific drugs and may be involved in drug interactions with grapefruit juice (Dresser et al., 2002).

CYP is a large multigene family of heme-containing enzymes located in the endoplasmic reticulum (ER) of cells throughout the body. It is especially concentrated in the liver and intestinal wall, where it is involved in oxidative biotransformation of various endogenous and exogenous substances. CYP3A isoforms constitute 70% of CYP enzymes in enterocytes. P-gp is a 170–180 kDa transmembrane protein that acts as a multidrug resistance factor in tumor cells, thereby reducing the intracellular accumulation of drugs. This phosphorylated glycoprotein belongs to a member of the *ABC* (ATP-binding cassette) transporter superfamily, which is another membrane transporter located in the apical brush border of enterocytes and other important barriers such as the kidney, liver, and BBB. The transporter protein P-gp is a major transmembranous efflux transporter of the intestinal mucosa, which shares its tissue distribution and substrate specificity with many CYP enzymes (Wacher et al., 1995). The *ABC* transporters hydrolyze ATP to drive the flux of their substrates against a concentration gradient. CYP enzymes and P-gp are regulated via the pregnane X-receptor (PXR), which controls the transcription of their genes [P-gp serves as a key factor in conferring the multidrug resistance (MDR) phenotype to cancer cells]. Pytochemicals can modify the first-pass effect of drugs through changes to the PXR activity or by direct competitive or allosteric interference at the active substrate-binding regions of CYP enzymes and transport protein (Zhou et al., 2004). P-gp can affect oral bioavailability, biliary or renal clearance, and brain uptake of a drug.

The substrate range for these proteins is diverse and includes drugs, nutrients, amino acids, sugars, and peptides. Once taken up by the enterocytes, a lipophilic drug may be metabolized by CYP3A4 or be pumped back into the lumen by P-gp. Hence, the oral delivery of many drugs is limited by the actions of CYP3A4 or P-gp. Metabolism by CYP3A4 will also occur in the liver before the drug finally enters the systemic circulation.

Induction of drug metabolism

Induction or inhibition of enzymes in the gut by nutrients may lead to a significant change in oral bioavailability of drugs or vice versa. The enzyme induction effect is the ability of numerous foods or drugs to stimulate the production of drug-metabolizing enzymes in the liver, which may result in increased metabolism of the administered drug as well as other related or even unrelated drugs. The activity of drug-metabolizing enzymes in liver microsomes, as well as the structure and amount of ER and even the size of the liver, are influenced to a great extent by the administration of drugs and hormones, and by age, sex, temperature, nutritional status, and psychological and pathological states of the subject. Enzyme induction involves an adaptive increase in the number of molecules of a specific enzyme in response to an enzyme-inducing agent. Inducers of drug oxidation have several features in common, such as lipophilicity, the ability to bind to CYP450 enzymes, and relatively long biological half-lives. Many drugs share these properties without inducing enzyme synthesis.

Drugs metabolized after entering the GIT are exposed to a number of systems with the potential to chemically modify or chelate the drug molecules that are not found within the tissues of the host. These include extremes of pH, unabsorbed nutrients and xenobiotics, digestive enzymes, and intestinal microflora. The intestinal microorganisms represent a potent, diverse, and adaptable metabolizing force that has a large potential for the metabolism of drugs and environmental nutrients. The gut flora consists of a complex, dynamic mixture of aerobic organisms that may show large variations (enterobacteria, lactobacilli, enterococci, bacteroides, clostridia, and bifidobacteria). This gut flora may be altered by changes in the diet, by diseases, and by the administration of foreign compounds. Gut microorganisms, therefore, represent an adaptable changing source of metabolic activity that may show large interindividual and intraindividual variations in the numbers and types of organisms present. In addition, intestinal transit time and defecation frequency may affect the duration of exposure of a drug to the gut flora in vivo and produce pronounced temporal variations in levels of metabolites. The bacteria of intestinal flora have been shown to be capable of performing a large number of different metabolic reactions, most of which are degradative in some way (i.e., hydrolysis, reduction or removal of functional groups). However, the use of various tissue preparations has provided recognition of a wide variety of drug metabolism reactions in the gut such as C-oxidation, hydroxylation, dealkylation, N-oxidation, S-oxidation, desulfuration reduction and hydrolysis, and conjugation reactions. The enzyme activity in the small intestine is less than that detected in the liver, but the ratio of hepatic to small intestinal activity is highly variable (Anadón, 1983).

Enzyme inducers may modify *toxicological effects*, both short term and long term, produced by foreign compounds. Certain types of drug metabolites, so-called chemically reactive metabolites, may produce a range of toxic drug effects by reacting covalently with essential cellular components. However, many enzyme-inducing agents can enhance their own metabolism as well as that of other drugs. When prolonged treatment with a drug results in a decrease in its action, a phenomenon known as *tolerance* (enzyme induction) may be suspected, but it is not always the cause. The induction of microsomal enzymes by the concomitant use of another food/drug may lead to increased pharmacological effects or even toxicity caused by the active metabolites (Anadón, 1983).

Inhibition of drug metabolism

Variously termed Kcat inhibitors, suicide enzyme inactivators, enzyme-activated irreversible inhibitors, and suicide enzyme inhibitors represent a relatively new approach to specific irreversible inactivation of enzymes. Simply stated, this approach requires that the inhibitor contains a latent reactive grouping and be accepted as a substrate by the target enzyme, following which the normal catalytic activity of the enzyme results in its own irreversible inactivation or "*suicide*." A number of food--drug reactions are based on the inhibition of metabolism of certain drugs by food. The result of such interactions is an increase in the duration and intensity of pharmacological activity. Two points of view are seen regarding the clinical value of enzyme inhibitors: (1) potential food/drugs, and (2) causing side effects or food–drug interactions (Anadón, 1983).

Types of drug–nutrient interactions

Several types of drug–nutrient interactions are categorized based on their nature and mechanisms. They may arise either from alterations of the absorption, distribution, biotransformation, or excretion of one drug by another nutrient or from a combination of their actions or effects.

Drug–nutrient interactions in vitro

These are ex vivo bioinactivations that refer to interactions between the drug and the nutritional element or formulation through biochemical or physical reactions, such as hydrolysis, oxidation, neutralization, precipitation, or complexation. They usually occur in the delivery device. These interactions are most common with drugs and nutrients administered intravenously. There are several recommendations to minimize these interactions. For instance, drugs should not be directly mixed with enteral or parenteral formulas. Tubes should be flushed with water before and after drug administration. Preformulated oral solutions or suspensions should be selected instead of crushing tablets when administering drugs through enteral feeding tubes. For drugs with a narrow therapeutic range, monitoring of drug levels should be taken into consideration.

Drug–nutrient interactions in the gastrointestinal tract

These interactions affect absorption. They cause either an increase or decrease in oral bioavailability. The precipitant agents may modify the function of enzymes or transport mechanisms that are responsible for biotransformation. Complexation, binding, and/or other deactivating processes occur in the GIT and reduce absorption.

The effect of diet on drug absorption is also an important factor. An example of this type of interaction is the influence of oral absorption of a drug by concurrent meal intake. The rate of absorption, the magnitude of absorption, or both can be changed. Meal intake stimulates gastric and intestinal secretions, which usually improve the dissolution of drugs and facilitate absorption. Meals with a higher fat content stimulate the release of bile salts, which they increase the intestinal uptake of highly lipophilic drugs or of substances that require bile salts for optimal absorption.

In addition, high fat content of the food stimulates the release of cholecystokinin, which slows gastrointestinal motility and increases contact time between the drug and the intestine and possibly also increases absorption.

However, the potential physicochemical interactions, the potential binding of drug and food contents, the dose of the drug administered, and the composition of the meals make drug absorption in the presence of food unpredictable in specific cases. For that reason, the bioavailability of drugs should be tested with and without concurrent meal intake. It has been shown that certain drugs should be taken with food to maximize absorption. Among these drugs are the antibiotics cefuroxime and erythromycin ethylsuccinate, the HMGCoA-reductase inhibitor lovastatin, and lithium (used for psychiatric diseases). Lovastatin should be taken with the evening meal to enhance absorption (NCL National Consumers League, 2015). However, several drugs should not be taken with food to allow optimal absorption. These include the antibiotics ampicillin, ciprofloxacin, doxycycline, or tetracycline, captopril (used for elevated blood pressure), and the HIV protease-inhibitor indinavir (Chan, 2006).

It should be mentioned, however, that delayed absorption of a drug by food does not necessarily lead to reduced absorption and that pharmacokinetic changes do not necessarily have clinically relevant effects (Schmidt and Dalhoff, 2002). Conversely, there is considerable evidence that specific drug therapy may cause malabsorption of many of the essential components of the diet. One example is oral contraceptives. These drugs have initiated serious anemia in apparently normal women due to a significant suppression of their intestinal absorption of dietary polyglutamic folate. A similar complication can result from chronic anticonvulsant therapy with phenytoin (Hoffbrand and Necheles, 1968). Malabsorption of fat and vitamin B12 may be caused by *para*-aminosalicylic acid (Levine, 1968). There are other examples of drugs altering the normal rates of vitamin uptake and metabolism. One of the simplest of these is the deficiency of the fat-soluble vitamins A and D caused by chronic consumption of liquid paraffin as a laxative. Liquid paraffin is poorly absorbed in the gut. By local action it takes up fat-soluble vitamins, thus preventing their absorption.

A more serious example of this type of reaction is related to the interference of vitamin K with anticoagulant therapy (Booth et al., 1997). Oral dosage with some antibiotics (e.g., chloramphenicol, tetracyclines, or neomycin) reduces the intestinal population of microorganisms and thereby reduces the microbial synthesis of vitamin K. Vitamin K synthesis by intestinal bacteria can be inhibited by this alteration of the GI flora, which may result in an increased anticoagulant effect in patients using oral anticoagulants. Thus, the hypoprothrombinemia of patients on chronic anticoagulant treatment might be dangerously potentiated by concomitant dosage with intestinal flora-destroying antibiotics. It is also of importance that the actual intestinal absorption of vitamin K can be reduced by cholestyramine, *para*-aminosalicylic acid, and other drugs that induce malabsorption states (Koch-Weser and Sellers, 1971).

Changes in the intestinal flora by antibiotic therapy may markedly alter the state of digitalization. This phenomenon occurs in a minority of patients who have the ability to substantially convert digoxin to cardio-inactive metabolites in the gut. The digoxin is inactivated by GI bacteria, and the coadministration of erythromycin or tetracycline appears to reduce this process.

Components of food may antagonize the desired effect of the drug, as in the case of warfarin. Foods that are high in vitamin K, or that enhance vitamin K production by intestinal microorganisms, can reduce the effectiveness of warfarin by diminishing the body's supply of vitamin K, which is needed to activate clotting factors. Changing to a diet with increased consumption of leafy and/or dark green vegetables, such as broccoli, spinach, kale, turnip greens, cauliflower, and Brussels sprouts, could lessen the degree of anticoagulation made possible by warfarin by supplying additional vitamin K. High doses of vitamin E (400 IU or more) may prolong clotting time and increase the risk of bleeding (NCL National Consumers League, 2015).

Drugs and certain foods may affect enzyme transport systems and thereby alter the intestinal absorption of specific drugs. For example, allopurinol and iron preparations should not be administered simultaneously because allopurinol blocks the enzyme that prevents iron absorption. Overabsorption and iron overload in patients may occur, resulting in hemosiderosis.

Drug interactions during transport

These interactions affect systemic or physiologic disposition and occur after the drug or nutritional element has been absorbed from the GIT and has entered the systemic circulation. Changes to the cellular or tissue distribution, systemic transport, or penetration to specific organs or tissues can occur.

Drug–nutrients interactions associated with metabolism

The metabolism of orally ingested drugs before reaching the systemic circulation (first-pass metabolism) has been shown to have clinically relevant influences on the potency and efficacy of drugs. Both the intestine and liver account for presystemic metabolism in humans. One of the major groups of drug-metabolizing enzymes is the cytochrome CYP4503A4 enzymes. Grapefruit juice is a classic example of a selective intestinal CYP3A4 inhibitor. It causes an interaction with certain drugs by deactivating and destroying intestinal CYP3A4 enzymes. Grapefruit juice has pharmacokinetic effects on drugs that are substrates for metabolism by the CYP3A4 intestinal drug-metabolizing enzyme. This drug class constitutes approximately half of all prescription drugs. Grapefruit juice contains chemicals that inhibit intestinal CYP3A4 (the inhibition does not affect the liver enzymes). This inhibition of the intestinal drug-metabolizing enzymes CYP3A4 and P-gp, which decreases first-pass metabolism, causes an increase in maximal drug concentration in plasma (C_{max}) and in drug bioavailability [area under the curve (AUC)]. The increased plasma drug concentrations constitute a potential for adverse reactions. The finding that grapefruit juice can markedly augment oral drug bioavailability was originally based on an unexpected observation from an interaction study between the dihydropyridine calcium channel antagonist felodipine and ethanol, in which grapefruit juice was used to mask the taste of the ethanol (Bailey et al., 1998). Grapefruit juice has inhibitory actions on intestinal OATP, decreasing the absorption of some drugs. The effects of grapefruit juice persist for at least 24 h. Some compounds in grapefruit juice responsible for inhibitory actions on CYP3A4 have been identified. Grapefruit juice administration did not inhibit CYP3A4 substrate drugs given intravenously, whereas it inhibited the metabolism of these drugs given orally (Durcharme et al., 1995).

The overall exposure of some drugs can be increased by more than fivefold when taken with grapefruit juice. It has been shown that furanocoumarin derivatives in the juice are primarily responsible for enzymatic inhibition (Paine et al., 2006). The onset of the interaction is immediate and the magnitude of the enzymatic inhibition increases with repeated grapefruit juice consumption. The effect continues for several days after the discontinuation of grapefruit juice consumption. The increased oral absorption of certain drugs can lead to increased risk for adverse reactions of the specific drugs. For instance, drowsiness and prolonged sedation may result from taking the tranquilizers diazepam or midazolam in combination with grapefruit juice. Hypotension can result in people being treated with calcium-channel blockers (nifedipine, felodipine, verapamil). Patients taking the antibiotic erythromycin and consuming grapefruit juice have an increased risk for cardiovascular symptoms, including cardiac dysrhythmias. Other drugs with increased absorption when taken with grapefruit juice are atorvastatin, cyclosporine, and sertralin (Chan, 2006).

Drugs may also be associated with altered metabolic function or macronutrient status. Some atypical antipsychotics (clozapine, olanzapine, risperidone) are associated with glucose intolerance. Particularly when risk factors such as an underlying diabetic condition or an increase in weight are present or certain other medications are concomitantly used, more frequent monitoring of blood glucose should be considered. Alternatively, in these situations antipsychotics for which no association with glucose intolerance has been demonstrated (haloperidol, chlorpromazine) should be chosen when possible (Hedenmalm et al., 2002).

Drug interactions associated with renal excretion

Interactions that influence renal excretion of drugs or nutrients are clinically significant only when the drug, nutrients, or its active metabolite is appreciably eliminated by the kidney. Urinary pH can influence the activity of a drug or nutrient by altering the rate of renal clearance. When a drug is in its un-ionized form, it more readily diffuses from glomerular filtrate back into the blood. Thus, for a basic drug, a larger proportion of drug is in the un-ionized form in basic (or less acidic) urine than in normally acidic (pH 5.0–5.5) urine. Acid drugs are excreted faster when the urinary pH is alkaline. Even the ingestion of nonsystemic acids (e.g., magnesium hydroxide, aluminum and magnesium hydroxide, and calcium carbonate–glycine suspensions) may increase the urinary pH by 0.5–0.9 units, and these antacids in normal doses have the potential to influence the elimination kinetics of amphetamine, and other drugs. However, basic drugs (e.g., antihistamines, meperidine, and imipramine) are excreted faster when the urinary pH is acid. This acidification may be brought about by the administration of ammonium chloride or glutamic acid hydrochloride. In general, if the drug remains ionized in the urine, then the half-life is decreased and vice versa. These interactions refer to the elimination or clearance of drugs or nutrients which may involve the antagonism, impairment, or modulation of renal and/or enterohepatic elimination (Chan, 2006). These types of interactions may involve the antagonism, impairment, or modulation of renal and/or enterohepatic elimination. It should be noted that it is best to take prescription antihistamines (to relieve or prevent the symptoms of colds, hay fever, and allergies) on an empty stomach to increase their effectiveness.

Drugs that interfere with food intake

Medications can lead to altered food choices. Many drugs are reported to directly affect the sense of taste and smell, and some drugs themselves have an unpleasant taste that might interfere with food intake (Brownie, 2006).

It is well known that chemotherapeutic drugs (and to a lesser extent other drugs) induce nausea and vomiting, and therefore are associated with an increased risk of malnutrition. However, drugs that decrease acute and delayed emesis (e.g., aprepitant) may have adverse effects such as fatigue, dysphagia, taste disturbance, constipation, diarrhea, or anorexia, and could therefore by themselves adversely affect nutritional status (Santos and Boullata, 2005).

Limiting drug prescriptions to essential medications for the shortest period possible and periodic reevaluations of the treatment chosen are essential to minimize adverse drug–nutrient interactions (Schmidt and Dalhoff, 2002).

Nutrient–drug interactions

Fortified foods are also recommended for those with chewing difficulty. In an intermediate stage of obstruction (e.g., insufficient swallowing), an enteral formula is necessary in the diet to provide proper nutritional requirements. Foods are most often fortified with multivalent cationic minerals, such as calcium, iron, magnesium, and aluminum, and vitamins, like C, D, E, and B-complex. Although it is clear that certain drugs should not be taken with antacids, multivitamins, and mineral supplements, many do not consider the implications of taking their daily medications with food. The current Food and Drug Administration (FDA, 2002) standardized meal used in product labeling of drug–food interactions is a high-fat, high-caloric diet that provides only a small amount of dietary vitamins and minerals (FDA, 2002). High-fat meals may increase the amount of theophylline in the body, whereas high-carbohydrate meals may decrease it. It is important to check which form the patient is taking because food can have different effects depending on the dose form (e.g., regular release, sustained release, or sprinkles). Food increases the absorption of theophylline, which can result in side effects of nausea, vomiting, headache, and irritability. As a result, many drugs are labeled "*may be taken with or without food*" while they may also be labeled "*do not take with antacids*." The biochemical mechanisms that cause drug–antacid interactions are the same mechanisms that cause drug interactions with fortified foods. Chelation and adsorption interactions, which cause decreased drug absorption, will certainly occur between fortified foods and drugs. However, due to the fact that some fortified foods contain a quantity of polyvalent ions that approaches or exceeds that contained in antacid formulations, changes in gastric pH, changes in urinary pH, and otherwise unspecified decreases in absorption are also possible. The implications of these interactions can range from clinically insignificant to severe.

There are several mechanisms by which minerals interact with drugs. When described in the literature, they are often listed as drug–antacid interactions. However, as emphasized here, it is important to consider the nutrient content of foods, especially those that have been fortified with minerals so as to contain sufficient quantities to be equivalent to antacids with interaction capability. Chelation, adsorption, changes in gastric pH, changes in urinary pH with resultant changes in renal clearance (CL_R), and uncharacterized decreases in absorption are the five causes of moderate to major drug–mineral interactions described in the literature. Foods containing significant quantities of multiple minerals could cause moderate or major interactions, although each mineral individually may cause only a minor interaction (Wallace and Amsden, 2002).

Chelation is the formation of a complex involving a metal ion and two or more polar groupings of a single molecule. The overwhelming majority of the drugs affected by chelation with multivalent ions are antibiotics, particularly the quinolones, tetracyclines, and some oral cephalosporins. Many studies have characterized the decrease in bioavailability of ciprofloxacin due to chelation with multivalent ions, in particular aluminum-containing antacids and dialysate (Anadón and Martínez-Larrañaga, 1992; Wallace and Amsden, 2002). Most of these involved the administration of ciprofloxacin with concomitant antacids (containing calcium, magnesium, and/or aluminum), iron supplements, and multivitamin/mineral tablets (Albert and Rees, 1956). Quinolones (e.g., ciprofloxacin, levofloxacin, ofloxacin, and trovafloxacin), macrolides (azithromycin, clarithromycin, erythromycin), and sulfonamides (sulfamethoxazole + trimethoprim) should be taken on an empty stomach 1 h before or 2 h after meals. If the stomach becomes upset, these medications must be taken with food. However, avoid calcium-containing products like milk, yogurt, vitamins or minerals containing iron, and antacids because they significantly decrease drug concentration (NCL National Consumers League, 2015).

Drug–food and drug–antacid interactions with all the tetracyclines (minocycline, doxycycline, tetracycline, oxytetracycline) have been well described in the scientific literature. The complexity of tetracycline antibiotics may occur when these drugs are administered along with dairy products or aluminum, calcium, and magnesium antacid products and vitamin supplements containing iron because they can interfere with the medication's effectiveness. It has been described as decreased bioavailability of demethylchlortetracycline when administered with aluminum hydroxide gel and with whole milk. Ferrous sulfate has been shown to impair the absorption of tetracycline, oxytetracycline, methacycline, and doxycycline. Aluminum- and magnesium-containing antacid products can interfere with the absorption of digoxin and digitoxin. Tetracyclines should therefore be taken on an empty stomach 1 h before or 2 h after meals. It is also important to avoid taking tetracycline with dairy products, antacids, and vitamins containing iron because these can interfere with the medication's effectiveness (NCL National Consumers League, 2015).

Iron absorption is also decreased. It is established that the absorption of doxycycline or minocycline is not markedly influenced by simultaneous administration of antacids, and ferrous sulfate has been shown to decrease their absorption. If gastric irritation occurs when taking these tetracycline products, it is recommended that they be taken with food and milk (NCL National Consumers League, 2015).

Adsorption is the phenomenon in which a solid substance attaches other substances to its surface without covalent binding. Multivalent cationic minerals can adsorb some drugs, causing less drug to be bioavailable from the GIT. Two classic examples of this interaction are with chloroquine and propranolol.

Herb–drug interactions

Since efficacy and safety of conventional prescription drugs could be changed by coadministration with herbal medicines, there is an effective way to identify the potential herb–drug interactions. Concurrent use of herbs may mimic, magnify, or oppose the effect of drugs. Many medicinal herbs and pharmaceutical drugs are therapeutic at one dose and toxic at another. Interactions between herbs and drugs may increase or decrease the pharmacological or toxicological effects of either component. Although herbal remedies are perceived as being natural and therefore safe, many have adverse effects that sometimes produce life-threatening consequences. Herb–drug interactions, like drug–drug interactions, occur when the uptake, efflux, or metabolism of a drug is impaired or enhanced by a second drug that interacts with the same drug metabolism enzyme (inhibition or induction of cytochromes CYP enzymes) or transporter pathway such as P-gp. P-gp, a transmembrane ATP-binding cassette transporter, can affect drug absorption in the intestinal tract, distribution to the brain, and elimination by the liver and kidney (Ho and Kim, 2005). There are numerous uptake transporters located on the liver plasma membrane including the organic anion-transporting polypeptides (OATP1B1, 1B3, and 2B1), an organic cation transporter1 (OCT1), the organic anion transporters (OAT2 and OAT7), and the sodium taurocholate cotransporting polypeptide (NTCP). As a result, inhibition or induction of P-gp may cause potential herbal–drug interactions.

Oxidation metabolism represents a major route of elimination for many drugs, and because many herbals compounds/drugs compete for the same enzyme, inhibition of CYPs is one of the main reasons for drug interactions. Herbs can inhibit CYPs via three mechanisms: competitive, noncompetitive, and irreversible inhibition.

The clinical relevance of drug inhibition will depend on a number of considerations; one of the most important considerations is the therapeutic index of the drug. The lower the therapeutic index, the greater the risk that increased plasma drug concentrations will induce toxicity. For example, patients receiving anticoagulants (e.g., warfarin), antidepressants, or cardiovascular drugs (e.g., digoxin) are at a much greater risk due to the narrow therapeutic index of these drugs than patients receiving other kinds of drugs. Many dietary supplements and vitamins can interact with anticoagulants and can reduce the benefit or increase the risk of warfarin. When taking warfarin, these foods must be avoided: garlic, ginger, glucosamine, ginseng, and ginkgo, as they can increase the chance of bleeding.

Aconite (*Aconitum napellus* L.)

Aconite is a crude extract of dried leaves and roots from various species of *Aconitum* plants (or monkshood) that contain aconitine and other diterpenoid ester alkaloids (aconitine, mesaconitine, jesaconitine, hypaconitine). Aconite was used as a medicinal drug as well as a homicidal agent and arrow poison in Asia. The use of aconite may account for some deaths in history attributed to stroke, even though substantial physical pain accompanies fatal aconite poisoning. Traditional Chinese practitioners use aconite for pain relief caused by trigeminal and intercostal neuralgia, rheumatism, migraine, and general debilitation. Aconite initially stimulates and then paralyzes nerves that communicate pain, feeling, and temperature, producing anesthesia mediated by numerous different alkaloids blocking the sodium current (voltage-gated sodium channels). The aconitine alkaloid binds with high-affinity to site 2 of the α-subunit of the Na^+ channel protein. This binding to the open Na^+ channels shifts the voltage dependence of activation to more hyperpolarized potentials and reduces channel inactivation from the open state. Thus, aconitine initially stimulates and then activates the voltage-gated sodium channels in the heart and nervous system, resulting in interference with the propagation of the action potential. Aconite is also used as a mild diaphoretic and to slow the pulse rate by its effect on brainstem centers. Atrial or ventricular fibrillation, however, may result from the direct effect of aconite on the myocardium. Side effects occur even after contact with leaves or sap from *Aconitum* plants, and can range from bradycardia and hypotension to fatal ventricular arrhythmia induced by triggered activity (Lowe et al., 2005).

Black cohosh (*Actaea racemose* L.)

Black cohosh contains triterpene glycosides and has been used in remedies for the relief of symptoms of menopause, premenstrual tension, and other gynecological problems. The mechanism of action is unclear. It may bind to estrogen and serotonin receptors. After estrogen replacement therapy, it was shown to increase the risk of thromboembolic and cardiovascular events and breast cancer. Commercially available dietary supplements made from black cohosh inhibit CYP3A4 and potentially increase the risk of adverse effects from some drugs. Hepatotoxicity has been reported (Chow et al., 2008), and black cohosh should not be used during pregnancy or lactation.

Danshen (*Salvia miltiorrhiza* Bunge)

Danshen is used in traditional Chinese medicine (TCM) for treatment of coronary artery disease and menstrual abnormalities. Danshen reduces elimination of warfarin and inhibits cyclic adenosine monophosphate phosphodiesterase, which results in additive antiplatelet effects and increased risk of bleeding. Concomitant use with warfarin increases prothrombin time (Izzat et al., 1998). Danshen may also interfere with digoxin assay. In the absence of signs or symptoms of digoxin toxicity, the possibility of a false elevation of digoxin concentration should be explored.

Ginkgo tree (*Ginkgo biloba* L.)

The seeds and fruits of *Ginkgo biloba* are used as part of TCM, particularly for the treatment of asthma, indigestion, cough, and chilblains (i.e., pedal edema related to cold exposure). *G. biloba* is one of the world's oldest living tree species, dating back to the Permian period. During food shortages, ginkgo seeds were an important source of food and its ingestion has produced numerous case reports of ginkgo seed poisoning (Ginnan food poisoning). In recent decades, medicinal uses of the extract of *G. biloba* include Alzheimer disease (AD), memory loss, cerebral ischemia, cardiovascular disease, premenstrual syndrome, impotence, stress, depression, dementia, and altitude sickness.

The active ingredient of ginkgo extracts has not been identified. However, the extracts are standardized to total flavonoid glycosides and terpenoid content, and the most clinically important compounds are these flavonoids or polyphenolic compounds (kaempferol, quercetin, isorhamnetin) and terpenoids (ginkgolides A/B, bilobalide) (Kleijnen and Knipschild, 1992). Consumption of gingko seeds containing 4'-*O*-methylpyridoxine inhibits the formation of gamma amino butyric acid and increases the formation of glutamate, resulting in the occurrence of seizures.

In vitro animal and clinical studies on the effect of ginkgo extracts on platelet aggregation, coagulation, and various CYP450 isoenzymes are conflicting. Ginkgo increased bleeding risk or had significant effects on hematologic parameters (Asher et al., 2017). The concurrent use of ginkgo with antiplatelet, anticoagulant (e.g., warfarin), or antithrombotic agents increases the risk of bleeding. Hyphema, subphrenic hematoma, and intracranial hemorrhage have been reported (Gertz and Kiefer, 2004). In clinical trials, ginkgo has also been shown to reduce the effectiveness of nicardipine by interacting with the CYP450 system. The administration of a 12-day course of 140 mg ginkgo extract twice daily to healthy Chinese subjects induced CYP2C19 activity. This interaction could reduce the effects of drugs (e.g., omeprazole) metabolized by this enzyme (Yin et al., 2004). However, the clinical significance of the interactions of ginkgo extract with CYP450 isoenzymes remains undetermined.

Garlic (*Allium sativum* L.)

Garlic has been mentioned in medicinal texts since the *Ebers papyrus* (c. 1550 BC). Garlic has been widely used as a flavoring agent, in traditional medicine, and in functional food to improve physical or mental well-being. Garlic extracts are commonly used by HIV patients because these extracts possess antiseptic, bacteriostatic, antiviral, immune-enhancing, hypotensive, and antihelmintic properties. Traditionally, garlic has been used to treat respiratory catarrh, recurrent colds, bronchial asthma, influenza, and chronic bronchitis. Currently, garlic and garlic preparations are being investigated for their antihypertensive, antiatherogenic, antithrombotic, antimicrobial, fibrinolytic, cancer-preventive, and lipid-lowering effects (Reuter, 1995) and they are being used for preventing cardiovascular disease.

The active component ajoene in garlic inhibits collagen-induced platelet aggregation (Apitz-Castro et al., 1983), and garlic is used for its antiplatelet and fibrinolytic effects in patients with cardiovascular disease. However, the risk of bleeding in people using anticoagulant or antiplatelet agents increases, so its concomitant use should be avoided (Rose et al., 1990). The use of large amounts of garlic can cause platelet disorders and/or hemorrhage. Garlic supplements should be discontinued about 10 days before elective surgical procedures, especially by patients using aspirin or warfarin (German et al., 1995). The use of dried garlic powder causes some modest short-term reduction (8–12 weeks) in total cholesterol concentrations, but these effects are not sustained over 6 months.

The presence of garlic supplements in human studies decreased concentrations of drugs that are transported by P-gp, and blood concentrations of saquinavir decreased by approximately 50% among the study participants. Medications that are transported by P-gp (e.g., colchicine, digoxin, doxorubicin, quinidine, rosuvastatin, tacrolimus, verapamil) should not be combined with garlic supplements (Asher et al., 2017). Garlic contains a large number of biologically active constituents. These constituents can be divided simply into two groups: sulfur-containing and nonsulfur-containing compounds. Most of the medicinal effects of garlic are related to its sulfur compounds and the alliin-splitting enzyme alliinase, which converts alliin into allicin (responsible for the characteristic garlic odor). The major flavoring constituents of garlic are sulfur compounds (diallyl disulfide, allyl sulfide, and diallyl trisulfide). Garlic is available in different pharmaceutical preparations, such as dry powder products, oil macerates, volatile garlic oil (obtained by water vapor distillation), and juices of fresh garlic. Researchers have found that garlic supplements can cause a potentially harmful side effect when combined with a type of medication used to treat HIV/AIDS. In healthy volunteers, oral administration of a garlic preparation for 3 weeks decreased the AUC and C_{max} of HIV-1 protease inhibitors saquinavir (Piscitelli et al., 2002) and, to a lesser extent, ritonavir. Although the exact mechanism of these effects was not clear, the similarity in the reduction of the AUC, the mean maximum concentration, and the 8-h concentration of saquinavir suggested that garlic affected the bioavailability of saquinavir and its systemic clearance. Besides the observation that garlic components can affect CYP2C-, CYP2D-, and CYP3A-mediated metabolism, only slight effects on P-gp could be observed (Deferme and Augustijns, 2003).

Ginseng (*Panax* species)

Ginseng refers to the roots of the *Panax* species. The most commonly examined species are *Panax ginseng* (Asian ginseng), *Panax quinquefolius* (American ginseng), and *Panax japonicus* (Japanese ginseng). The origin of the ginseng root and its manner of extraction can produce wide variations in ginseng products. The terms "red" and "white" refer to different methods of ginseng preparation, not different species. It is administered as a whole dried root, extract, tea, or capsule. Ginseng is advertised as an immune system stimulant that increases vigor, sexual potency, well-being, and longevity, and for its use as an antihyperglycemic agent. Ginseng has both hypertensive and hypotensive effects, with the latter caused by the enhanced synthesis of nitric oxide (NO) (Sung et al., 2000). In Chinese medicine, ginseng is used for myocardial infarction, congestive heart failure, and angina pectoris. However, current evidence does not support its use for cardiovascular conditions. Ginseng abuse syndrome causes hypertension, behavioral changes, and diarrhea (Siegel, 1979). When administered with warfarin, ginseng produced decreased prothrombin time and demonstrated a 0.2-point drop in the recommended international normalized ratio (INR) range (Asher et al., 2017).

Panax ginseng (Asian ginseng) has been shown to induce CYP3A4, which could decrease the effectiveness of many drugs, including calcium channel blockers, many chemotherapy and HIV agents, certain antihypertensive and statin medications, and some antidepressants (Lee et al, 2008). In patients having difficulty maintaining adequate anticoagulation with warfarin, clinicians may suggest avoiding Asian ginseng products.

Ginseng may also produce effects similar to those of estrogen because its active components, ginsenosides, have a chemical structure similar to that of testosterone, estrogen, and glucocorticoids. The active compounds are heterogeneous triterpene saponin glycosides, collectively termed *ginsenosides*. Typical doses are 100–400 mg ginseng extract. Ginseng should not be used by women who are pregnant or receiving hormone replacement therapy. Neonatal death has been related to maternal use (Awang, 1991). Increased levels of digoxin are associated with Siberian ginseng, which interferes with the digoxin assay (Dasgupta, 2003).

Goldenseal (*Hydrastis canadensis* L.)

The root of goldenseal is the main portion of the plant used for herbal medicines, primarily as a tea. This is a popular herbal component of dietary supplements used for treatment of GI disturbances, anorexia, urinary symptoms, inflammation, and infections of the skin, mouth, respiratory tract, and eye. It has been shown to inhibit CYP2D6 and CYP3A4 enzymes, which are responsible for metabolism of more than one-half of currently used pharmaceutical agents (Gurley et al., 2018). Clinicians should strongly advise against the use of goldenseal in combination with most over-the-counter and prescription medications (Asher et al., 2017).

Green tea extract (*Camellia sinensis*)

Green tea is rich in polyphenols that have health benefits, such as reducing inflammation and helping to reduce tumor formation in the skin, lung, liver, pancreas, and GIT (Deferme and Augustijns, 2003). The polyphenols from green tea could have chemopreventive, antiatherogenic, anticarcinogenic, and antioxidant properties, and they can reduce the formation of free radicals in the body, protecting cells and molecules from damage. These free radicals play a role in aging and many types of diseases. Green tea extract has been investigated for potential drug interactions in human studies with conflicting results. Available data on green tea–cardiovascular drug interactions in humans is so far limited to warfarin, simvastatin, and nadolol, and the mean magnitude of these interactions seems to be mild to modest

(Werba et al., 2008). These authors believe that a low-to-moderate consumption of traditional green tea infusion could lead to severe problems of interaction with cardiovascular drugs, probably with the exception of a subset of susceptible individuals, which warrants a certain clinical suspicion of green tea intake in cases of reduced drug response or intolerance. Yet in patients convinced that "more is better," who consume "high-strength catechin" beverages, more concentrated infusions or large doses of green tea, health risks of green tea–cardiovascular drug interactions may be more substantial. However, in reported experimental conditions, an Italian study showed that green tea did not significantly increase simvastatin levels. However, it is noteworthy that, in one-fourth of the volunteers, green tea consumption increased simvastatin lactone AUC 0–6 h by about twofold (Werba et al., 2008). Green tea components may affect critical targets involved in the pharmacokinetics of simvastatin through CYP3A4 and/or the intestinal multidrug P-gp efflux pump. Statins are associated with oxidative stress as a plausible mechanism for statin-induced toxicity, as well as its metabolic interactions; statin-induced metabolism involves various CYP450 enzymes (Liu et al., 2019). In vitro and in vivo studies suggested green tea extract (*Camellia sinensis*) and/or (−)-epigallocatechin-3-gallate may significantly increase the bioavailability of simvastatin by inhibiting the activity of CYP1A1, CYP1A2, CYP2B6, CYP2C8, CYP2C9, CYP2D6, and CYP3A4 (Albassam and Markowitz, 2017). Two additional studies have demonstrated that green tea extract may inhibit the drug transporters OATP1A1 and OATP1A2, which are involved in the transport of many medications, including statins, fluoroquinolones, some β-blockers (imatinib), and antiretrovirals (Knop et al., 2015; Misaka et al., 2014). Recently, green tea was shown to inhibit the absorption of nadolol (β-blocker antihypertensive) by inhibiting the intestinal uptake transporter OATP1A2 (Misaka et al., 2014). This green tea–nadolol interaction was believed to be unexpected of the observed loss in hypertension control. In a Japanese study, after a 14-day pretreatment with a commercial green tea beverage (intake 700 mL with epigallocatechin gallate: 322 mg/day) or water (control phase), a single oral tablet of 30 mg of nadolol was administered for each of both kinetic studies with 700 mL of green tea or water (350 mL twice at 0 and 0.5 h). Green tea markedly reduced the plasma concentrations of nadolol. In particular, the C_{max} and the AUC 0–48 h of nadolol were decreased by 85.3% and 85.0%, respectively (Misaka et al., 2014). In conclusion, green tea extract should be avoided in combination with drugs that are transported by P-gp, OATP1A1, or OATP1A2.

Gynura (*Gynura procumbens*)

Widely used in Chinese folk medicine, gynura purportedly improves the microcirculation and relieves pain; however, it has been associated with hepatic toxicity (Chitturi and Farrell, 2008). Resultant conditions include hepatic veno-occlusive disease, which is characterized by painful hepatomegaly, fluid avidity, weight gain, and jaundice (Dai et al., 2007).

Hawthorn (*Crataegus* species)

Chinese hawthorn (*Crataegus pinnatifida* Bunge, *Crataegus cuneata* Siebold & Zucc.) contains tartaric acid. Most data demonstrated that the active ingredients in hawthorn extracts are flavonoid compounds. Hawthorn extract is commonly used by herbalists for the treatment of angina, congestive heart failure, bradyarrhythmia, and cerebral insufficiency. Hawthorn has positive inotropic, negative chronotropic, and coronary artery vasodilatory effects similar to the phosphodiesterase inhibitors (amrinone, milrinone), and is thought to increase myocardial perfusion and reduce afterload. As an adjunct treatment for congestive heart failure, hawthorn has been reported to have beneficial effects on symptom control and physiologic outcomes (Tauchert, 2002), but the efficacy and safety of its supposed inotropic activity and effect on morbidity and mortality have not been systematically assessed. Hawthorn enhances the activity of digitalis (Mashour et al., 1998), and its concomitant use should be carefully monitored for potential toxic effects. Hawthorn also inhibits the biosynthesis of thromboxane A_2, and it could potentially increase the risk of bleeding in patients taking antiplatelet or anticoagulant agents. Without additional data on safety and efficacy, clinicians should discourage unsupervised use of hawthorn in patients with congestive heart failure who are taking heart failure medications.

The ingestion of hawthorn extract causes mild induction of P-gp activity, but this induction does not necessarily result in decreased absorption of P-gp dependent substrates, such as digoxin.

Kava Kava (*Piper methysticum*)

Kava kava is used in conditions of nervous anxiety, stress, and restlessness. The rhizome of this plant contains a concentration of >3.5% cavalactones and chalcones. Patients taking medications metabolized by CYP2C9 or CYP2C19 should be closely monitored for clinical adverse effects and laboratory abnormalities (e.g., glucose level, A1C level, INR) or be instructed to not use kava-containing supplements. Caution should be exercised in patients using substances acting on the CNS (benzodiazepines, alcohol, barbiturates, psychopharmacological agents), because of the increased risk of drowsiness and motor reflex depression (Asher et al., 2017). Patients taking kava should do so for at least 5 days prior to any surgery requiring general anesthesia. Kava is not recommended for use during pregnancy and lactation.

Licorice (*Glycyrrhiza glabra*)

Licorice, an extract of the root of *Glycyrrhiza glabra*, is used as a sweetening and flavoring agent. It is also used as a herbal remedy for gastritis and upper respiratory tract infections (as an expectorant). Modern cough syrups often include licorice extract. Its use can result in pseudoaldosteronism with concomitant hypokalemia, hypertension, and edema that may reduce the effectiveness of antihypertensive drugs (Mansoor, 2001). The active constituent of licorice is glycyrrhizic acid. A metabolite, glycyrrhetinic acid, inhibits renal 11β-hydroxysteroid dehydrogenase and causes a state of mineralocorticoid excess by impeding the inactivation of cortisol (Walker and Edwards, 1994). Case reports link licorice to hypertension, hypertensive encephalopathy, pulmonary edema, edema, hypokalemia, arrhythmias, congestive heart failure, muscle weakness, and acute renal failure (Olukoga and Donaldson, 2000). Licorice-induced hypokalemia can lead to increased risk for ventricular arrhythmia, particularly *torsades de pointes* (i.e., a ventricular tachycardia that is characterized by fluctuation of the QRS complexes around the electrocardiographic baseline and is typically caused by a long QT interval) (Bryer-Ash et al., 1987). It can also potentiate the effects of spironolactone and digoxin and can cause hyperglycemia, rendering antidiabetic agents less effective. Its ability to inhibit thrombin and platelet aggregation enhances the risk of bleeding with antiplatelet and anticoagulant agents. Susceptibility to licorice varies greatly. Subjects with underlying hypertension and women may be more sensitive. Adverse effects may take weeks to reverse because of suppression of the renin–angiotensin–aldosterone axis and because glycyrrhetinic acid has a large volume of distribution (Walker and Edwards, 1994).

Motherwort (*Leonurus cardiaca*)

Motherwort has a long history of use in both European and Asian traditional medicine because of its purported sedative and antispasmodic properties. Traditionally, it has been used for "cardiac debility," tachycardia, anxiety, insomnia, and amenorrhea. It is also used as a hypotensive and a diuretic. When administered intravenously, motherwort reduces platelet aggregation and fibrinogen levels (Zou et al., 1989). It potentiates antithrombotic and antiplatelet effects and increases the risk of bleeding when taken with benzodiazepines, Motherwort can have a synergistic sedative effect and may result in coma.

Purple coneflower (*Echinacea purpura* L.)

The three main medicinal species of *Echinacea* are *E. purpurea*, *E. angustifolia* (narrow-leaved *Echinacea*), and *E. pallida* (pale-flowered *Echinacea*), with the cultivated species, *E. purpurea* being the most common constituent in herbal preparations containing *Echinacea*. The chemical composition of extracts from medicinal *Echinacea* species is complex, consisting of alkamides (straight-chain fatty acids with olefinic and/or acetylenic bonds), polyacetylenes, phenolic caffeic acid conjugates (phenylpropanoids), polysaccharides, and glycoproteins. This herb was considered an immunostimulant that promoted wound healing and increased resistance to colds. Additionally, it was a treatment for hemorrhoids, wound infections, syphilis, gangrene, malaria, typhoid, flu, and skin infections. Currently, it is a common herbal treatment for upper respiratory tract infections based on potential immunomodulatory properties, such as macrophage activation and enhanced neutrophil phagocytosis. The results of a controlled double-blind study indicated that *Echinacea* had no clinically significant effects on rhinovirus infection (Turner et al., 2005). The administration of *Echinacea* (400 mg *E. purpurea* root four times daily for 8 days) reduced the mean oral clearance of caffeine (CYP1A2) by approximately 25% (Gorski et al., 2004). In this respect, caution should be taken when combining *Echinaceae* with medicines (e.g., antipsychotic, antidepressant) that are metabolized by CYPA2 and CYP3A4 enzymes. Persistent use may result in or potentiate the hepatotoxic effects of other medications (e.g., statins, fibrates, niacins, or amiodarone). Side effects include nausea, dizziness, dyspnea, rash, and dermatitis. Flavonoids from *Echinacea* may inhibit or induce CYP450 enzymes, depending on their structure and assay conditions.

Rosemary (*Rosmarinus officinalis* L.)

Rosemary (*Rosmarinus officinalis*) is an evergreen plant of the *Lamiaceae* family, and its fresh or dried leaves are collected while flowering (Anadón et al., 2008). Rosemary has been used as an aromatic herb and medicine. In folk medicine, rosemary extract is a treatment for urinary ailments, chronic weakness, nervous disorder, hair loss, and peripheral vascular diseases. In addition, rosemary is a traditional astringent, carminative, tonic, rubefacient, antispasmodic, antiinflammatory, expectorant, emmenagogue, digestive aid, diaphoretic, choleretic, and diuretic. Increasing interest in rosemary plants is due to their antioxidant and health-enhancing properties. Rosemary leaf contains phenolic acids, phenolic diterpenoid bitter substances, triterpenoid acid, flavonoids, volatile oils, and tannins (Anadón et al., 2008). Rosemary extracts were not associated with negative toxicological effects at a single dose of 2000 mg/kg of body weight in both sexes of Wistar rats (Anadón et al., 2008). Most of the adverse reactions reported from the use of rosemary are allergic reactions. Rosemary is not recommended during pregnancy. Herb–drug interactions at the level of P-gp can be found in the inhibiting effect of a methanolic rosemary extract on the P-gp-mediated efflux of three anticancer drugs, doxorubicin (anthracycline antibiotic), paclitaxel (or taxol), and vinblastine. In the presence of rosemary extract, an increased intracellular accumulation of these P-gp substrates, and a decreased efflux of doxorubicin, could be observed (Plouzek et al 1999).

St. John's wort (*Hypericum perforatum*)

St. John's wort (*Hypericum perforatum*) is a well-known herb used worldwide for the treatment of mild to moderate forms of depression in humans (Butterweck and Schmidt, 2007; Shelton, 2009), anxiety, sleep disorders, the common cold, herpes, and human immunodeficiency virus. It is used as a topical analgesic, and even as an enema for ulcerative colitis. Although its mechanisms of action have not been fully elucidated, inhibition of the uptake of serotonin and norepinephrine reportedly contributes to the antidepressant effect of St. John's wort (Neary and Bu, 1999). Several case reports have identified a clinically relevant interaction between St. John's wort and prescription drugs. St. John's wort and pharmacokinetics drug interactions are as follows: digoxin (18% decrease of digoxin plasma concentrations after a single oral dose of 0.5 mg), indinavir (reduced indinavir plasma concentrations), theophylline (the dosage should be changed), and cyclosporine (reduction in cyclosporine plasma concentration and acute transplant rejection) (Ha et al., 1995; Koudriakowa et al., 1988).

A potential mechanism for these interactions includes induction of the hepatic cytochrome CYP450 system, particularly CYP3A4, which is the most abundant hepatic and intestinal phase I enzyme and is involved in oxidative metabolism of more than 50% of all prescription medications. This is mediated by activating the PXR, a transcription factor that is activated by structurally diverse foreign chemicals. Both CYP3A4 and P-gp have been found to be highly expressed in enterocytes and hepatocytes, which are important in drug absorption and metabolism (Durr et al., 2000). Many P-gp inhibitors studied in vitro and in vivo are also known or suspected to be the substrates and/or inhibitors of CYP3A4 (Wandel et al., 1999). Although there is a considerable overlap in the activity of the inhibitors of P-gp on CYP3A4, a varying degree of selectivity was present in the investigated herbal constituents. Clinical studies have shown St. John's wort as a potent inducer of CYP3A4 showing reductions of efficacy in concomitant treatment with cyclosporine, tacrolimus (immunosuppressive), warfarin, protease inhibitors (indinavir), irinotecan (antineoplastic), theophylline, digoxin, venlafaxine (antidepressant), and oral contraceptives, decreasing their levels in blood (Asher et al., 2017).

This herb was able to induce P-gp transporter activity in the GIT after chronic treatment. To assess the clinical interaction potential of the herbal inhibitors with P-gp, the *International Transporter Consortium* recommended a cutoff value of $[I]_1/IC_{50}$ (or Ki) ≥ 0.1 or $[I]_2/IC_{50}$ (or Ki) ≥ 10, where $[I]_1$ represents the mean steady-state total (free and bound) C_{max} following the administration of the highest proposed clinical dose and $[I]_2$ = dose of inhibitor (in mol)/250 mL.

It has been shown that hyperforin, a phloroglucinol derivative found in St John's wort, is a potent inducer of CYP3A4 in primary human hepatocytes. Theophylline is metabolized by CYP1A2 and CYP3A, cyclosporine is metabolized by CYP3A, and indinavir is metabolized by CYP3A. The digoxin interactions were due to induction of P-gp, although a potential role for CYP3A in digoxin metabolism has been reported. Plasma AUC and C_{max} of dioxin decreased after 10 days of treatment with *Hypericum* extract. St John's wort was found to increase duodenal P-gp expression by 1.4-fold in healthy volunteers after multiple administrations. Reactions resulting from induction of metabolism by St. John's wort can potentially result in therapeutic failure if two drugs are given concomitantly. Conversely, toxicity could occur if the dosage is not adjusted appropriately following discontinuation of St. John's wort. Coadministration of St. John's wort and drugs metabolized by CYP3A4 should be avoided because it may result in reduced bioavailability and effectiveness with subsequent recurrence of arrhythmia, hypertension, or other undesirable effects. It is strongly recommended to avoid the concurrent use of this herb with over-the-counter and prescription drugs.

Saw palmetto (*Serenoa repens*)

The extracts of saw palmetto have been used for the treatment of abdominal disorders and dysentery, whereas the fruit of the plant is a food and nutrient. The crude extracts have been used for centuries to improve breast size, sperm production, and sexual vigor. Currently, saw palmetto is one of the top five most popular herbal products to treat the symptoms related to benign prostatic hypertrophy. It is also a diuretic and urinary antiseptic. The primary use of saw palmetto was the short-term treatment of urinary symptoms associated with benign prostatic hypertrophy (BPH). The drug is a complex mixture primarily of free fatty acids (90%) and related esters (7%), along with small amounts of phytosterols (campesterol, β-sitosterol, stigmasterol), aliphatic alcohols, and various other compounds (arabinose, flavonoids, galactose, glucose, and uronic acid). The unsaturated fatty acids include capric, caproic, caprylic, lauric, linolenic, myristic, isomyristic, oleic, palmitic, and stearic. The phytosterols include β-sitosterol, campesterol, stigmasterol, cycloartenol, lupeol, and lupenone. The major free fatty acids are oleic acid, lauric acid, myristic acid, and palmitic acid. The exact biological mechanism of action of saw palmetto is not clear. Despite claims that saw palmetto helps relieve BPH symptoms, recent clinical trials did not demonstrate any beneficial effects on such symptoms or on postvoid residual bladder volume (Bent et al., 2006). Additional prospective studies are needed to establish the role of herbal extracts in alleviating BPH symptoms. When saw palmetto is used for BPH, the prothrombin time (PT) and activated partial thromboplastin time (aPTT) were normal, but the bleeding time was prolonged to 21 min (normal: 2–10 min). After cessation of saw palmetto, the bleeding time returned to normal values within 5 days. Saw palmetto inhibits cyclooxygenase and increases bleeding with warfarin (Bressler, 2005). In addition, its unsupervised use can result in cholestatic hepatitis, acute pancreatitis, and intraoperative floppy iris syndrome during cataract removal because of loss of iris tone (Yeu and Grostern, 2007). Ophthalmologists should be aware of this important association so that they can take the necessary steps to prevent surgical complications.

Tetrandrine (*Stephania tetrandra*)

Tetrandrine is a vasoactive alkaloid used in Chinese medicine to treat hypertension and angina. Its vasodilative effect is due to inhibition of the L-type calcium channels and possible competition with other calcium channel blockers. Tetrandrine lowers plasma glucose and causes hepatotoxicity and renal toxicity (Seeff, 2007).

Yohimbe bark and yohimbine (*Pausinystalia yohimbe*)

Yohimbine is isolated from yohimbe, the bark of the tree *Pausinystalia yohimbe*. Yohimbe bark is a traditional treatment for a variety of disorders, including sexual disorders (i.e., erectile dysfunction). These uses include the treatment of dementia, diabetic complications (e.g., neuropathy), exhaustion, fevers, insomnia, leprosy, low blood pressure, obesity, and syncope. Analysis of yohimbe bark indicates that the average total indole alkaloid content is approximately 3%–6%, with approximately 10%–15% of the alkaloids being yohimbine. In addition to yohimbine and its isomers (α-yohimbine, β-yohimbine, alloyohimbine), these alkaloids include ajmaline, dihydroyohimbine, corynantheine, dihydrocorynantheine, and corynanthine (rauhimbin). Currently, yohimbe bark extract is used to enhance athletic performance and weight loss. The main medical indication for yohimbine hydrochloride is the treatment of male impotence. Many of its effects are attributed to its α_2-adrenergic receptor antagonist activity, which increases central sympathetic outflow and raises blood pressure, heart rate, and norepinephrine levels. Yohimbine increases the release of norepinephrine, resulting in inadequate blood pressure control in people also using antihypertensive and diuretic agents (Tam et al., 2001). It also interacts with tricyclic antidepressants so that pressor effects occur at lower doses and may potentiate the α-adrenergic blocking properties of phenothiazines. The recommended dose is 5.4 mg three times daily. Pressor effects are associated with doses of 15–20 mg (De Smet and Smeets, 1994). Use of yohimbine is contraindicated in patients with hypertension, angina, and renal impairment.

Herb–micronutrient interactions

The contribution of botanical dietary supplements to alter micronutrient (i.e., vitamins, minerals, trace metals) disposition is currently a research area of much interest. To date, it has been described that herb–drug interactions can also affect micronutrient ADME (absorption, distribution, metabolism, and excretion) and, as a consequence, micronutrient efficacy. Specific botanicals such as ginseng, green tea, and milk thistle can produce herb–micronutrient interactions. In addition, it is known that independent of being a botanical source of micronutrient metals, botanicals can also modulate metal absorption from the GIT (Gurley et al., 2018). These natural modulators of metal absorption are often polyphenolic phytochemicals present in various botanical dietary supplement formulations. Polyphenolic substances contain multiple hydroxyl and carbonyl groups within their structure, and are commonly found in botanical dietary supplement single entities (e.g., phenolic acids, flavonoids, catechins) and/or polymers (e.g., tannins). Inhibitory effects of polyphenolics on iron absorption stem from adjacent hydroxyl and/or carbonyl groups that bind ferric iron to form chelates and these iron chelates are not readily absorbed across the GI mucosa. Phytic acid, green tea catechins, and milk thistle silymarins are among the more widely recognized phytochemical chelators of essential dietary metals commonly found in botanical dietary supplements. Phytic acid has an avidity to bind metals (calcium, iron, zinc) (Gurley et al, 2018).

Green tea (*Camellia sinensis*)

Green tea contains a catechin, a flavan-3-ol, called epigallocatechin-3-gallate (EGCG). Catechin is a natural phenol and antioxidant that helps prevent cell damage and provides other benefits. Some green tea supplements are decaffeinated, while others have varying caffeine content. The caffeine content of tea can fortify catechin polyphenol absorption, thereby it is possible that green tea supplements with high caffeine content have a greater impact on iron (iron-chelating properties) and/or folate absorption. One of the most popular marketing categories of botanical dietary supplements is that of weight loss, and green tea and other natural caffeine sources (e.g., guarana, maté, kola nut) are common ingredients in these formulations (Gurley et al, 2018).

Catechins are also recognized for their ability to complex iron and prevent its absorption. In a clinical trial involving healthy women, green tea-extract supplementation significantly decreased nonheme iron absorption, as measured by whole-body retention and isotope activity of extrinsically radiolabeled iron (Samman et al., 2001). In a clinical study in which 35 obese participants with the metabolic syndrome were randomly assigned to receive one of the following for 8 weeks: green tea (four cups per day), control (four cups water per day), or green tea extract (two capsules and four cups water per day). Green tea extract significantly reduced plasma iron vs. baseline (128 to 92 μg/dL, $P < 0.02$). These results support the hypothesis that green tea may provide antioxidant protection in metabolic syndrome (Basu et al., 2013). From these few clinical investigations, the potential impact of green tea on iron absorption is concerning, as green tea extract or purified green tea catechins are common components of multiingredient botanical dietary supplements marketed for weight loss (Gurley et al., 2015).

Additionally, it was reported that daily administration of 900 mg of green or black tea catechin polyphenolics to healthy volunteers (male and female) in a fasted state significantly reduced the bioavailability of coadministered folic acid. The influence of food on the green tea–folate interaction may be especially important in pregnant women. A recent survey of pregnant women in Japan noted that a significant association existed between high tea catechin consumption and low serum folate concentrations, despite regular use of folic acid supplements (Alemdaroglu et al., 2008).

Milk thistle (*Silybum marianum*)

Silymarin is a source of flavonolignans and flavonoids. Silibinin, also known as silybin, is the major active component of silymarin. Preparations are usually standardized to a concentration of 70%–80% of three flavonolignans (silybin, silychristin, and silydianin), collectively known as silymarin, that have antioxidant activity, and are commonly used for toxic liver damage, for supportive treatment in chronic inflammatory liver disease and hepatic cirrhosis. In 2001, silibinin, a principal component of silymarin, was identified as a natural iron chelator (Borsari et al., 2001).

Another silibinin trial conducted in 37 patients with chronic hepatitis C and Batts–Ludwig fibrosis stage II, III, or IV were given doses of 360, 720, and 1080 mg/day of standardized silybin and soy phosphatidylcholine complex for 12 weeks. This produced a significant decrease in serum ferritin from baseline to the end of treatment (mean 244 vs. 215 μg/L; median 178 vs. 148 μg/L; 78% of subjects had a decrease in serum ferritin level). Treatment with silybin and soy phosphatidylcholine complex is associated with reduced body iron stores, particularly in patients with advanced stages of fibrosis (Bares et al., 2008).

Food and nutrient–drug interactions

Diet–drug interactions

A dietary component may influence the absorption of drugs from the intestine. The presence of food in the stomach may also exert a nonspecific effect in reducing or slowing the absorption of some drugs, thus lowering peak plasma concentration. With some drugs, the presence of increased amounts of stomach acid results in the destruction of acid-labile drugs, such as penicillin G, ampicillin, and dicloxacillin. In other cases, the components of food, such as calcium or iron, may form complexes with the drug that are less easily absorbed. Examples include tetracycline, sodium fluoride, and ciprofloxacin.

Grapefruit juice

The interaction between grapefruit juice and a variety of drugs has been widely reported (Bressler, 2006) (Table 70.1). It appears that one or more flavonoids found in grapefruit juice inhibit CYP enzymes. This results in reduced metabolism of drugs that are cleared by the same system. Drug bioavailability can be markedly augmented by as much as 200%. Patients should avoid drinking grapefruit juice for 2 h before and 4 h after taking drugs in this category. If the drug is in an extended-release dosage form, patients should wait 6 h before drinking grapefruit juice.

Many drugs interact with grapefruit (juice, segments, extract, and certain related citrus fruits, e.g., Seville oranges, pomelos, and some exotic orange varieties) because it can increase the oral absorption of several drugs. Several components in grapefruit called furanocoumarins (two of the most common are bergamottin and 6'7'-dihydroxybergamottin) irreversibly inhibit CYP 3A4 in the small intestine, which results in a significant reduction in drug presystemic metabolism of affected drugs taken up to 72 h after grapefruit consumption. Intestinal CYP3A4 activity can remain inhibited during this time as the body produces more enzymes. Greater amounts of CYP3A4-metabolized drugs can then enter the systemic circulation. The resulting increase in drug levels can lead to an increase in therapeutic effect, adverse effects, and/or toxicity. An additional mechanism for this interaction is the inhibition of P-gp, a transporter that carries the drug from the enterocyte back to the lumen, resulting in a further increase in the fraction of drug absorbed (Lown et al., 1997a). After uptake by the enterocyte, many

TABLE 70.1 Effect on drug metabolism and possible interactions of grapefruit juice on specifically affected drugs.

Drug class	Drug	Metabolism	Effect	Possible adverse effects	Increased oral bioavailability	Management	Mechanism
Antiarrhythmics	Amiodarone	CYP 3A4CYP 2C8	Increased C_{max} 84%Increased AUC 50%	Arrhythmias	Yes	Avoid GJ	CYP 3A4 inhibition
	Quinidine	CYP 3A4CYP 2D6P-glycoprotein	Delayed absorptionIncreased C_{max} from 1.6 to 3.3 hDecreased AUCReduced quinidine ECG effects	None	No	None	Inhibition of hepatic CYP 3A4 and quinidine absorption by OATP
	Digoxin	P-glycoproteinOATP	No effect on C_{max} or AUC				CYP 3A4 inhibition
Antibiotics	Clarithromycin			None	No	None	CYP 3A4 inhibition
	Erythromycin	CYP 3A4P-glycoprotein	Increased C_{max} 52%Increased AUC 49%				CYP 3A4 inhibition
Antihistamines	Terfenadine	Not metabolized	Decreased C_{max}Decreased AUC	Arrhythmias, prolonged Q–T interval	Yes	Avoid GJ	Inhibition of OATP decreasing absorption
Anxiolytics	Buspirone	CYP 3A4	Increased C_{max} 4-foldIncreased AUC 9-foldDelayed time to C_{max} 188%	Decreased psychomotor performance, increased sedation	Yes	Avoid GJ	CYP 3A4 inhibitionSlowed absorption by inhibition of OATP
	Diazepam	CYP 3A4		Decreased psychomotor performance, increased sedation	Yes	Avoid GJ	CYP 3A4 inhibition
	Midazolam	CYP 3A4	Increased C_{max} 56%Increased AUC 52%	Decreased psychomotor performance, increased sedation	Yes	Avoid GJ	CYP 3A4 inhibition
	Triazolam	CYP 3A4P-glycoprotein	Increased C_{max} 1.3-foldIncreased AUC 1.5-fold	Decreased psychomotor performance, increased sedation	Yes	Avoid GJ	CYP 3A4 inhibition
Calcium-channel blockers	Amlodipine	CYP 3A4	Increased C_{max} 15%Increased AUC 14%	Tachycardia, hypotension	Yes	Avoid GJ	CYP 3A4 inhibition
	Felodipine	CYP 3A4	Increased C_{max} 127% to 310%Increased AUC 123% to 330%	Tachycardia, hypotension	Yes	Avoid GJ	CYP 3A4 inhibition
	Nifedipine	CYP 3A4CYP 2D6	Increased C_{max} 15%Increased AUC 40%	Tachycardia, hypotension	Yes	Avoid GJ	CYP 3A4 inhibition

	Nimodipine			Tachycardia, hypotension	Yes	Avoid GJ	CYP 3A4 inhibition
	Diltiazem			None	No	None	CYP 3A4 inhibition
	Verapamil	CYP 3A4 lessCYP 1A2 lessCYP 2E1CYP 2C8	Increased AUC 41%S 36%, R 28%C_{max} S isomer 57%, R isomer 40%Increase of PR interval to 350 ms in 2 of 24 subjects	None	No	None	CYP 3A4 inhibition
Corticosteroids	Ethinyl estradiol	CYP 3A4P-glycoprotein	Increased C_{max} 37%Increased AUC 28%	Unknown	Yes	Monitor for side effects	CYP 3A4 inhibition
	Progesterone	CYP 3A4		Unknown	Possible	Monitor for side effects	CYP 3A4 inhibition
	Prednisone	CYP 3A4		None	No	None	CYP 3A4 inhibition
HMG–CoA reductase inhibitors	Atorvastatin	CYP 3A4		Myopathy, headache, rhabdomyolysis	Yes	Avoid GJ	CYP 3A4 inhibition
	Cerivastatin	CYP 3A4		Myopathy, headache, rhabdomyolysis	Possible	Monitor for side effects	CYP 3A4 inhibition
	Lovastatin	P-glycoprotein	Increased C_{max} 12-foldIncreased AUC 15-fold	Myopathy, headache, rhabdomyolysis	Yes	Avoid GJ	CYP 3A4 inhibition
	Pravastatin	Small % CYP 3A4Most eliminated unchanged	No change C_{max}, AUC	Myopathy, headache, rhabdomyolysis	Yes	Avoid GJ	CYP 3A4 inhibition
	Simvastatin	CYP 3A4P-glycoprotein	Increased C_{max} 9-foldIncreased AUC 16-fold	Myopathy, headache, rhabdomyolysis	Yes	Avoid GJ	CYP 3A4 inhibition
HIV protease inhibitors	Saquinavir	CYP 3A4P-glycoprotein	Increased C_{max} 50%Increased AUC 100%	Unknown	Yes[a]	Monitor for side effects	CYP 3A4 inhibition
Immunosuppressants	Cyclosporine	CYP 3A4P-glycoprotein	Increased AUC 8% to 71%Increased C_{max} 0% to 78% Delayed time to C_{max} 93%	Renal/hepatic dysfunction, increased immunosuppression	Yes	Avoid GJ	CYP 3A4 inhibition
	Tacrolimus	CYP 3A4		Renal/hepatic dysfunction, increased immunosuppression	Yes	Avoid GJ	CYP 3A4 inhibition
Neuropsychiatrics	Carbamazepine	CYP 3A4P-glycoprotein	Increased C_{max} 40%Increased AUC 40.8%	Drowsiness, ataxia, nausea	Yes	Avoid GJ	CYP 3A4 inhibition
	Clomipramine	CYP 3A4		Drowsiness, respiratory depression	Yes	Monitor for side effects	CYP 3A4 inhibition
	Phenytoin	CYP 3A4		None	No	None	
	Sertraline	CYP 3A4CYP 2D6CYP 2C19	Increased plasma trough levels 47%				CYP 3A4 inhibition
	Fluoxamine	CYP 3A4CYP 2D6CYP 2C19	Increased C_{max} 1.3-foldIncreased AUC 1.6-fold				CYP 3A4 inhibition

(Continued)

TABLE 70.1 (Continued).

Drug class	Drug	Metabolism	Effect	Possible adverse effects	Increased oral bioavailability	Management	Mechanism
Other	Carvedilol	CYP 3A4		Bradycardia, hypotension	Possible	Monitor for side effects	CYP 3A4 inhibition
	Methadone	CYP 3A4	Increased C_{max} 17%Increased AUC 17%	Respiratory depression, hypotension	Possible	Monitor for side effects	CYP 3A4 inhibition
	Dextromethorphan	CYP 3A4	Increased AUC 5.4-fold				CYP 3A4 inhibition
	Sildenafil	CYP 3A4	Increased AUC 23%Primary metabolite N-demethylated compound increased 24%No change C_{max} of either compound	Headache, flushing, dyspepsia	Possible	Monitor for side effects	CYP 3A4 inhibition
	Theophylline	CYP 3A4		None	No	None	CYP 3A4 inhibition
	Warfarin	S isomer CYP 2C9R isomer CYP 1A2CYP 3A4, small contributor to metabolism	No effect on Prothrombin time or international normalized ratio	None	No	None	No GJ effect on warfarin metabolism
	Cilostazol	CYP 3A4 somoCYP 2C19	Increased C_{max} 50%No AUC effect				CYP 3A4 inhibition
	Losartan	CYP 3A4CYP 2C9	Increased AUC 41%Losartan metaboliteDecreased hepatic metabolismLess conversion of losartan to its active metaboliteLess blood pressure fall				CYP 3A4 inhibition

AUC, area under the curve; C_{max}, maximal drug concentration in plasma; *CYP*, cytochrome P; *ECG*, epicatechin-3-gallate; *GJ*, grapefruit juice and the whole fruit; *OATP*, organic anion-transporting polypeptide (a system in the intestines involved in the absorption of certain drugs).
[a]*Clinical significance unknown.*

lipophilic drugs are either metabolized by CYP3A4 or pumped back into the lumen by the P-gp transporter. In consequence, CYP3A4 and P-gp may act in tandem as a barrier to oral delivery of many drugs (e.g., decreased bioavailability of ciclosporin A in the presence of hyperforin-rich St John's wort preparations). Interactions are more pronounced for drugs that normally undergo a large amount of presystemic metabolism (low oral bioavailability). Since the effect of grapefruit juice can last 24 h, repeated juice consumption can result in a cumulative increase in drug concentrations. Clinically relevant interactions seem likely for most dihydropyridine calcium channel antagonists, terfenadine, saquinavir, ciclosporin, midazolam, triazolam, and verapamil, and may also occur with lovastatin, cisapride, and astemizole (Kronbach et al., 1988; Saito et al., 2005). Elevated plasma concentrations of these compounds are particularly worrisome because grapefruit juice and medications are commonly consumed together at breakfast. The increased drug concentrations might be associated with an increased frequency of dose-dependent adverse effects. The importance of these interactions appears to be influenced by individual patient susceptibility, the type and amount of grapefruit juice, and dosing. Although the clinical consequence of grapefruit juice interaction with most of the listed drugs has not been evaluated, increased plasma concentrations of many of these drugs could result in adverse outcomes. Examples include excessive lowering of blood pressure with calcium channel antagonists, rhabdomyolysis and the potential for renal impairment with the HMG-CoA reductase inhibitors or statins (i.e., atorvastatin, cerivastatin, fluvastatin, lovastatin, pravastatin, simvastatin) (statins specifically inhibit the enzyme HMG-CoA in the liver, thereby inhibiting the rate-limiting step in cholesterol biosynthesis and so reducing plasma cholesterol levels), and adverse pulmonary effects caused by amiodarone (Dreier and Endres, 2004; Saito et al., 2005). With regards to grapefruit juice, large amounts of this juice can enhance the levels of statins (atorvastatin, lovastatin, or simvastatin) and increase the chance of side effects. The fruit has proven to be a good source of vitamin C and the B complexes, as well as calcium, potassium, and magnesium. The three major types of grapefruit that exist today are white, pink/red, and ruby/rio red varieties. Traditionally, grapefruit juice has been found to contain antioxidant, antinitrosaminic, antiseptic, aperitif, cardiotonic, detoxicant, hypocholesterolemic, sedative, and stomachic activities.

As little as 250 mL of grapefruit juice can change the metabolism of some drugs. This drug–food interaction occurs because of a common pathway involving a specific isoform of CYP450, CYP3A4, present in both the liver and the intestinal wall. Studies suggest that grapefruit juice exerts its effect primarily at the level of the intestine, inhibiting metabolism of those drugs that are substrates for intestinal CYP3A4 and P-gp. It is an inhibition of the first-pass metabolism of the CYP3A4 substrates leading to an increase in plasma C_{max} and the amount of drug, that is, the AUC. Grapefruit juice administration did not inhibit CYP3A4 substrate drugs given intravenously, whereas it inhibited the metabolism of these drugs given orally. This action is, in essence, similar to that caused by CYP-inhibiting drugs like itraconazole, ketoconazole, and erythromycin.

After ingestion, a substrate contained in the grapefruit binds to the intestinal isoenzymes, impairing first-pass metabolism directly and causing a sustained decrease in CYP3A4 protein expression. Within 4 h of ingestion, a reduction in effective CYP3A4 concentration occurs, with effects lasting up to 24 h (Lundahl et al., 1995). Individuals express CYP3A4 in different proportions, and those with the highest intestinal concentration being most susceptible to grapefruit juice–drug interactions (Lown et al., 1997b)

However, in recent years, reports of grapefruit juice–drug interactions have been published, raising questions about potential adverse effects derived from these interactions. The elderly patient population has multiple diseases and uses multiple drugs (Bressler and Bahl, 2003). Moreover, like the general population, elderly patients may consume grapefruit juice as a popular breakfast food.

Monoamine oxidase inhibitors

Perhaps the most feared food–drug interaction is between monoamine oxidase (MAO) inhibitors and the amino acid tyramine, which is found in a variety of aged, fermented, overly ripe, or pickled foods and beverages and, to a lesser extent, chocolate and yeast-containing foods. Examination of various foods and drinks has shown that many are rich in tyramine content. For example, certain cheeses, Chianti wine, some beers, yeast products, and schmaltz pickled herring, and citrus fruits and caffeine-containing foods and drinks (i.e., coffee, cola drinks, teas, chocolate, some high-energy drinks, and other soft drinks), may cause bizarre effects in patients on MAO inhibitor therapy. MAO inhibitors are used to treat depression, emotional and anxiety disorders, due to their dopamine, tyramine, or serotonin (5-hydroxytryptamine) content (Walker et al., 1966). Additionally, antianxiety medications (lorazepam, diazepam, alprazolam) used to treat depression, and emotional and anxiety disorders can interact with caffeine, which may cause excitability, nervousness, and hyperactivity and lessen the antianxiety effects of the drugs (NCL National Consumers League, 2015). There was an effect of individuals consuming cheese or other foodstuffs high in tyramine content while using MAO inhibitors. The result was that without the detoxifying activity of MAO, the ingested tyramine entered the bloodstream and resulted in a potentially fatal hypertensive crisis (Table 70.2). Because more than

TABLE 70.2 Interactions of monoamine oxidase (MAO) inhibitors with tyramine in food.

Tyramine in food			
Normal subject =>	MAO production normal (substrates: dopamine, tyramine, serotonin) =>	Tyramine detoxified before reaching systemic circulation	
MAO inhibitors-treated patient =>	MAO production inhibited =>	Toxic tyramine reaches systemic circulation =>	Hypertensive crisis (headache, heart failure, intracerebral hemorrhage)

Warning: patients being treated with MAO inhibitors and their families should be given a list of foods to avoid diet–drug interactions due to the use of any medication by patients without permission

10 mg of tyramine seems to be required to produce significant hypertension, the most dangerous foods are aged cheeses and yeast products used as food supplements. Tyramine is indirectly sympathomimetic and is a pressor amine or substance capable of liberating stored catecholamines. When its metabolism is suppressed, as it is by MAO inhibitors (e.g., phenelsine, tranylcypromine), it can cause a significant release of norepinephrine, resulting in markedly increased blood pressure, cardiac arrhythmias, hyperthermia, and cerebral hemorrhage. The occurrence of these hypertensive side effects (forceful heartbeat, severe headache, and hypertension) depends on the duration and intensity of action of the MAO inhibitor antidepressant, the variation in tyramine content in the diet, and the marked individual variation in responsiveness of the patient. In conclusion, foods high in tyramine that should be avoided include: (1) American processed cheddar, blue, brie, mozzarella, and Parmesan cheese, yogurt, and sour cream; (2) beef or chicken liver, cured meats such as sausage and salami, game meat, caviar, and dried fish; (3) avocados, bananas, yeast extracts, raisins, sauerkraut, soy sauce, and miso soup; and (4) broad (fava) beans, ginseng, caffeine-containing products (colas, chocolate, coffee, and tea) (NCL National Consumers League, 2015).

Herbs during pregnancy

When consuming herbal therapies, it should be ensured that women are healthy prior to and during pregnancy. Herbs are extracts of plants or plant roots, and they contain numerous compounds. Different forms of herbal preparations (i.e., teas or infusions, capsules, dried extracts, and tinctures) will have different compounds as well as differing concentrations. The way that the herb is prepared is very important in relation to the effect and safety of pregnant women and fetuses. The effects and safety of herbs will depend on the trimester. One of the most important concepts is that herbs should be used with caution in the first trimester. The rapid cellular development in organogenesis can be altered by any compound, and some herbs may increase uterine tone, increasing the risk of pregnancy loss.

There are numerous herbs that are thought to be contraindicated during pregnancy. These herbs can be classified into six subgroups based on their potential effects on pregnant women (Blumenthal et al., 1998; Low Dog and Micozzi, 2005; Weed, 1986; see Table 70.3):

1. Herbs traditionally used to stimulate menstruation. These herbs may stimulate the smooth muscle of the uterus and cause a pregnancy loss.
2. Herbs controversially used during menstruation and herbs traditionally used to stimulate menstruation (angelica, celandine, goldenseal, shepherd's purse, barberry, dong quai, motherwort, southernwood, black cohosh, ephedra, mugwort, tansy, blue cohosh, feverfew, rue, yarrow, nettle root, baldo, andrographis).
3. Alkaloid-containing herbs. These herbs are a diverse group of chemical plant constituents that have a wide range of pharmacological impacts on the body. Some alkaloids have been shown to be hepatotoxic and potentially carcinogenic.
4. Essential oils. Essential oils are frequently used by patients in many situations. Some essential oils are potentially very dangerous when ingested during pregnancy. All essential oils should be appropriately diluted when used, and none should be taken internally.
5. Anthraquinone laxatives. These are very potent compounds that can stimulate bowel peristalsis. In pregnancy, overstimulation of the bowel or bladder has the potential to irritate/stimulate the uterus in some women, and may cause premature labor.
6. Herbs thought to have an effect on the hormonal system. Some herbs may have an effect on the hormonal system, and they have potential estrogen-like properties, which can cause concern regarding the possible effects on the fetus.

Concluding remarks and future directions

Human medicines can treat and cure many health problems. However, they must be appropriately taken to

TABLE 70.3 Herbs contraindicated during pregnancy.

1. Herbs used traditionally to stimulate menstruation:
- Ginger (*Zingiber officinale*), cranberry (*Vaccinium oxycoccos*), evening primrose oil (*Oenothera biennis*), aloe vera gel (*Aloe vera*), Echinacea (Echinacea), St. John's wort (*Hypericum perforatum*), valerian root (*Valeriana officinalis*)
2. Alkaloid-containing herbs:
- Autumn crocus (*Colchicum autumnale*), broom (*Cytisus scoparius*), comfrey (*Symphytum officinale*), mandrake (podophyllin) (*Podophyllum peltatum*), barberry (*Berberis vulgaris*), coffee (*Coffea canephora/arabica*), goldenseal (*Hydrastis canadensis*), tansy (*Tanacetum vulgare*), blood root (*Sanguinaria canadensis*), colt's foot (*Tussilago farfara*)
3. Essential oils:
- Arbor vitae (*Arbor vitae*), juniper (*Juniperus*), pennyroyal (*Mentha pulegium*), nutmeg (*Myristica fragran*), catnip (*Nepata cataria*), rosemary (*Rosmarinus officinalis*), boldo (*Peumus boldus*)
4. Anthraquinone laxatives:
- Alder buckthorn (*Rhamnus frangula*), cascara (*Rhamnus frangula/purshiana*), purging buckthorn (*Rhamnus cathartica*), senna (*Senna alexandrina*)
5. Herbs thought to have an effect on the hormonal system:
- Ginseng (*Panax ginseng*), licorice (*Glycyrrhiza glabra*), chasteberry (*Vitex agnus-castus*), saw palmetto (*Serenoa repens*), passion flower (*Passiflora* spp.), isoflavones (members of the Fabaceae), red clover (*Trifolium pratense*), flaxseed (*Linum usitatissimum*), hops (*Humulus lupulus*)

ensure that they are safe and effective. Certain foods can interact with medicines. Food–drug interactions are defined as alterations of the pharmacokinetics or pharmacodynamics of a drug or nutritional element or a compromise in nutritional status as a result of the addition of a drug. The most relevant nutrients to drug–food interactions that have been described in this chapter are the multivalent minerals such as calcium. More attention has been given to the characterization of food/dietary supplement–drug interactions, particularly with any dietary supplements such as vitamins, minerals, and herbs. The various stages in which food can interact with a coadministered drug occur before and after gastrointestinal absorption, and during distribution, metabolism, and elimination. Drug–acid and other drug–mineral interactions should be extrapolated to interactions with fortified foods as well as other foods that may not be listed as fortified but contain combinations of several multivalent ions. Additional drug–food interaction studies are needed to better characterize the extent of these interactions with fortified and nonfortified foods, as well as combinations of them. Herbal extracts or single, isolated compounds can be either inhibitors or inducers of various CYP enzymes or transporters. Dual inhibition of P-gp and CYP3A4/5 activities could occur due to the overlapping substrate specificities between herbal medicines and drugs. Flavonoids are polyphenolic compounds, which are abundant in fruits (especially in *Citrus*), vegetables, nuts, flowers, wine, and tea, and indicate that more research is required to identify their interactions with the intestinal absorption (and metabolism) of P-gp substrates. Interactions between herbal products and prescribed drugs are a major safety concern, especially with respect to drugs with a narrow therapeutic index. Such an interaction may cause adverse effects, which may be life-threatening. Most food–drug interactions occur through three mechanisms: (1) reduced rate or extent of absorption; (2) increased rate or extent of absorption; or (3) chemical/pharmacologic effects. The presence of multiple diseases, polypharmacy, malnutrition status, and impaired metabolism, common in elderly individuals, increases the risks of adverse events related to food–drug interactions. Special attention is focused on critically relevant herb–micronutrient interactions as well as herbal therapy during pregnancy.

Acknowledgments

This work was supported by Project Ref. RTA2015-00010-C03-03 from Ministerio de Economía, Industria y Competitividad, Spain.

Conflict of interest

None of the authors of this chapter has a financial or personal relationship with other people or organizations that could inappropriately influence or bias the content of the chapter.

References

Akamine, D., Filho, M.K., Peres, C.M., 2007. Drug-nutrient interactions in elderly people. Curr. Opin. Clin. Nutr. Metab. Care 10, 204–310.

Albassam, A.A., Markowitz, J.S., 2017. An appraisal of drug-drug interactions with green tea (*Camellia sinensis*). Planta Med. 83, 496–508.

Albert, A., Rees, C.W., 1956. Avidity of the tetracyclines for the cations of metals. Nature 177, 433.

Alemdaroglu, N.C., Dietz, U., Wolffram, S., et al., 2008. Influence of green and black tea on folic acid pharmacokinetics in healthy volunteers: potential risk of diminished folic acid bioavailability. Biopharm. Drug. Dispos. 29 (6), 335–348.

Anadón, A., 1983. Enzyme induction and inhibition. In: Ruckebusch, Y., Toutain, P.L., Koritz, G.D. (Eds.), Veterinary Pharmacology and Toxicology. MTP Press Limited, Lancaster, UK, pp. 523–537. (ISBN: O-85200-709-4).

Anadón, A., Martínez-Larrañaga, M.R., 1992. Pharmacology and toxicology of quinolones. In: Sunshine, I. (Ed.), Recent Developments in Therapeutic Drug Monitoring and Clinical Toxicology. Marcel Dekker, Inc, New York, pp. 193–198. (ISBN: 0-8247-8586-X).

Anadón, A., Martinez-Larrañaga, M.R., Martínez, M.A., et al., 2008. Acute oral safety study of rosemary extracts in rats. J. Food Prot. 71 (4), 790–795.

Apitz-Castro, R., Cabrera, S., Cruz, M.R., et al., 1983. Effects of garlic extract and of three pure components isolated from it on human platelet aggregation, arachidonate metabolism, release reaction and platelet ultrastructure. Thromb. Res. 32, 155–169.

Asher, G.N., Corbett, A.H., Hawke, R.L., 2017. Common herbal dietary supplement-drug interactions. Am. Fam. Physicians 96 (2), 101–107.

Awang, D.V., 1991. Maternal use of ginseng and neonatal androgenization. J. Am. Med. Assoc. 266, 363.

Bailey, D.G., Malcolm, J., Arnold, O., et al., 1998. Grapefruit juice-drug interactions. Br. J. Clin. Pharmacol. 46 (2), 101–110.

Bares, J.M., Berger, J., Nelson, J.E., et al., 2008. Silybin treatment is associated with reduction in serum ferritin in patients with chronic hepatitis C. J. Clin. Gastroenterol. 42 (8), 937–944.

Basu, A., Betts, N.M., Mulugeta, A., et al., 2013. Green tea supplementation increases glutathione and plasma antioxidant capacity in adults with the metabolic syndrome. Nutr. Res. 33 (3), 180–187.

Bent, S., Kane, C., Shinohara, K., et al., 2006. Saw palmetto for benign prostatic hyperplasia. N. Engl. J. Med. 354, 557–566.

Bland, S.E., 1998. Drug-food interactions. J. Pharm. Soc. Wis. 28. 28–35.

Blumenthal, M., Gruenwald, J., Hall, C., et al., 1998. EDS, the complete German Commission E Monographs: Therapeutic Guide to Herbal Medicine. Integrative Medicine Communications, Boston, MA.

Booth, S.L., Charnley, J.M., Sadowski, J.A., et al., 1997. Dietary vitamin K1 and stability of oral anticoagulation: proposal of a diet with constant vitamin K1 content. Thromb. Haemost. 77, 503–509.

Borsari, M., Gabbi, C., Ghelfi, F., et al., 2001. Silybin, a new iron-chelating agent. J. Inorg. Biochem. 85 (2-3), 123–129.

Bressler, R., 2005. Herb-drug interactions: interactions between saw palmetto and prescription medications. Geriatrics 60, 32–34.

Bressler, R., 2006. Grapefruit juice and prescription drug interactions. Exploring mechanisms of this interaction and potential toxicity for certain drugs. Geriatrics 61 (11), 12–18.

Bressler, R., Bahl, J.J., 2003. Principles of drug therapy for the elderly patient. Mayo Clin. Proc. 78 (12), 1564–1577.

Brownie, S., 2006. Why are elderly individuals at risk of nutritional deficiency? Int. J. Nurs. Pract. 12, 110–118.

Bryer-Ash, M., Zehnder, J., Angelchik, P., et al., 1987. Torsades de pointes precipitated by a Chinese herbal remedy. Am. J. Cardiol. 60, 1186–1187.

Butterweck, V., Schmidt, M., 2007. St John's wort: role of active compounds for its mechanism of action and efficacy. Wien. Med. Wochenschr. 157, 356–361.

Chan, L.N., 2006. Drug-nutrient interactions. In: Shils, M.E., Shike, M., Ross, A. C., Caballero, B., Cousins, R.J. (Eds.), Modern Nutrition in Health and Disease. Lippincott Williams & Wilkins, Baltimore, MD, pp. 1540–1553.

Chitturi, S., Farrell, G.C., 2008. Hepatotoxic slimming aids and other herbal hepatotoxins. J. Gastroenterol. Hepatol. 23, 366–373.

Chow, E.C., Teo, M., Ring, J.A., et al., 2008. Liver failure associated with the use of black cohosh for menopausal symptoms. Med. J. Aust. 188, 420–422.

Cordon-Cardo, C., O'Brien, J.P., Casals, D., et al., 1989. Multidrug resistance gene (P-glycoprotein) is expressed by endothelial cells at blood-brain barrier sites. Proc. Natl. Acad. Sci. USA 86 (2), 695–698.

Dai, N., Yu, Y.C., Ren, T.H., et al., 2007. Gynura root induces hepatic veno-occlusive disease: a case report and review of the literature. World J. Gastroenterol. 13, 1628–1631.

Dasgupta, A., 2003. Review of abnormal laboratory test results and toxic effects due to use of herbal medicines. Am. J. Clin. Pathol. 120, 127–137.

Deferme, S., Augustijns, P., 2003. The effect of food components on the absorption of P-gp substrates: a review. J. Pharm. Pharmacol. 55, 153–162.

De Smet, P.A.G.M., Smeets, O.S.N.M., 1994. Potential risks of health food products containing yohimbe extracts (letter). Br. Med. J. 309, 958.

Dreier, J.P., Endres, M., 2004. Statin-associated rhabdomyolysis triggered by grapefruit consumption. Neurology 62 (4), 670.

Dresser, G.K., Bailey, D.G., Leake, B.F., et al., 2002. Fruit juices inhibit organic anion transporting polypeptide-mediated drug uptake to decrease the oral bioavailability of fexofenadine. Clin. Pharmacol. Ther. 71 (1), 11–20.

Durcharme, M.P., Warbasse, L.H., Edwards, D.C., 1995. Disposition of intravenous and oral cyclosporine after administration with grapefruit juice. Clin. Pharmacol. Ther. 57 (5), 485–491.

Durr, D., Stieger, B., Kulak-ublick, G.A., et al., 2000. St John's wort induces intestinal P-glycoprotein/MDR1 and intestinal and hepatic CYP3A4. Clin. Pharmacol. Ther. 68, 598–604.

FDA, 2002. Food-Effect Bioavailability and Bioequivalence Studies. Draft Guide for Industry, Bethesda, MD.

German, K., Kumar, U., Blackford, H.N., 1995. Garlic and the risk of TURP bleeding. Br. J. Urol. 76, 518.

Gertz, H.J., Kiefer, M., 2004. Review about *Ginkgo biloba* special extract EGb 761 (Ginkgo). Curr. Pharm. Des. 10 (3), 261–264.

Gorski, J.C., Huang, S.-M., Pinto, A., et al., 2004. The effect of Echinacea (*Echinacea purpurea* root) on cytochrome P450 activity *in vivo*. Clin. Pharmacol. Ther. 75, 89–100.

Gurley, B.J., Steelman, S.C., Thomas, S.L., 2015. Multi-ingredient, caffeine containing dietary supplements: history, safety, and efficacy. Clin. Ther. 37 (2), 275–301.

Gurley, B.J., Tonsing-Carter, A., Thomas, S.L., et al., 2018. Clinically relevant herb-micronutrient interactions: when botanicals, minerals, and vitamins collide. Adv. Nutr. 9, 524S–532S.

Ha, H.R., Chen, J., Freiburghaus, A.U., et al., 1995. Metabolism of theophylline by cDNA-expressed human cytochromes P450. Br. J. Clin. Pharmacol. 39, 321–326.

Hedenmalm, K., Hägg, S., Stahl, M., et al., 2002. Glucose intolerance with atypical antipsychotics. Drug. Saf. 25, 1107–1116.

Ho, R.H., Kim, R.B., 2005. Transporters and drug therapy: implications for drug disposition and disease. Clin. Pharmacol. Ther. 78, 260–277.

Hoffbrand, A.V., Necheles, T.F., 1968. Mechanism of folate deficiency in patients receiving phenytoin. Lancet 292, 528–530.

Izzat, M.B., Yim, A.P., El-Zufari, M.H., 1998. A taste of Chinese medicine!. Ann. Thorac. Surg. 66, 941–942.

Kleijnen, J., Knipschild, P., 1992. Ginkgo biloba. Lancet 340, 1136–1139.

Knop, J., Misaka, S., Singer, K., et al., 2015. Inhibitory effects of green tea and (−)-epigallocatechin gallate on transport by OATP1B1, OATP1B3, OCT1, OCT2, MATE1, MATE2-K and P-Glycoprotein. PLoS One 10 (10), e0139370.

Koch-Weser, J., Sellers, E.M., 1971. Drug interactions with coumarin anticoagulant. N. Engl. J. Med. 285 (487-498), 547–558.

Koudriakowa, T., Iatsimirskaia, E., Utkin, I., et al., 1988. Metabolism of the human immunodeficiency virus protease inhibitors indinavir and ritonavir by human intestinal microsomes and expressed cytochrome P4503A4/3A5: mechanism-based inactivation of cytochrome P4503A by ritonavir. Drug. Metab. Dispos. 26, 552–561.

Kronbach, T., Fischer, V., Meyer, U.A., 1988. Cyclosporine metabolism in human liver: identification of a cytochrome P-450III gene family as the mayor cyclosporine-metabolizing enzyme explains interactions of cyclosporine with other drugs. Clin. Pharmacol. Ther. 43, 630–635.

Lee, L.S., Wise, S.D., Chan, C., et al., 2008. Possible differential induction of phase 2 enzyme and antioxidant pathways by American ginseng, *Panax quinquefolius*. J. Clin. Pharmacol. 48 (5), 599–609.

Leibovitch, E.R., Deamer, R.L., Sanderson, L.A., 2004. Food-drug interaction: careful drug selection and patients counseling can reduce the risk in older patients. Geriatrics 59, 19–22.

Levine, R.A., 1968. Steatorrhea induced by para-aminosalycilic acid. Ann. Inter. Med. 68, 1265–1270.

Low Dog, T., Micozzi, M.S., 2005. Women's Health in Complementary and Integrative Medicine: A clinical Guide. Elsevier, Oxford.

Lowe, L., Matteucci, M.J., Schneir, A.B., 2005. Herbal aconite tea and refractory ventricular tachycardia. N. Engl. J. Med. 353, 1532.

Lown, K.S., Mayo, R.R., Leichtman, A.B., et al., 1997a. Role of intestinal P-glycoprotein (*mdr*1) in interpatient variation in the oral bioavailability of cyclosporine. Clin. Pharmacol. Ther. 62, 248–260.

Lown, K.S., Bailey, D.G., Fontana, R.J., et al., 1997b. Grapefruit juice increases felodipine oral availability in human by decreasing intestinal CYP3A protein expression. J. Clin. Invest. 99 (10), 2545–2553.

Liu, A., Wu, Q., Guo, J., et al., 2019. Statins: adverse reactions, oxidative stress and metabolic interactions. Pharmacol. Ther. 195, 54–84.

Lundahl, J., Regardh, C.G., Edgar, B., et al., 1995. Relationship between time of intake of grapefruit juice and its effect on pharmacokinetics and pharmacodynamics of felodipine in healthy subjects. Eur. J. Clin. Pharmacol. 49, 61–67.

Mansoor, G.A., 2001. Herbs and alternative therapies in the hypertension clinic. Am. J. Hypertens. 14, 971–975.

Mashour, N.H., Lin, G.I., Frishman, W.H., 1998. Herbal medicine for the treatment of cardiovascular disease: clinical considerations. Arch. Intern. Med. 158, 2225–2234.

Misaka, S., Yatabe, J., Müller, F., et al., 2014. Green tea ingestion greatly reduces plasma concentrations of nadolol in healthy subjects. Clin. Pharmacol. Ther. 95 (4), 432–438.

NCL (National Consumers League), 2015. Food & Drug Interactions, Washington, DC.

Neary, J.T., Bu, Y., 1999. Hypericum LI 160 inhibits uptake of serotonin and norepinephrine in astrocytes. Brain Res. 816, 358–363.

Olukoga, A., Donaldson, D., 2000. Liquorice and its health implications. J. R. Soc. Health 120, 83–89.

Paine, M.F., Widmer, W.W., Hart, H.L., et al., 2006. A furanocoumarin-free grapefruit juice establishes furanocoumarins as the mediators of the grapefruit juice-felodipine interaction. Am. J. Clin. Nutr. 83, 1097–1105.

Piscitelli, S.C., Burstein, A.H., Welden, N., et al., 2002. The effect of garlic supplements on the pharmacokinetics of saquinavir. Clin. Infect. Dis. 34 (2), 234–238.

Plouzek, C.A., Ciolino, H.P., Clarke, R., et al., 1999. Inhibition of P-glycoprotein activity and reversal of multidrug resistance *in vitro* by rosemary extract. Eur. J. Cancer 35, 1541–1545.

Reuter, H.D., 1995. *Allium sativum* and *Allium ursinum*: part 2—pharmacology and medicinal application. Phytomedicine 2 (1), 73–91.

Rose, K.D., Croissant, P.D., Parliament, C.F., et al., 1990. Spontaneous spinal epidural hematoma with associated platelet dysfunction from excessive garlic ingestion: a case report. Neurosurgery 26, 880–882.

Saito, M., Hirata-Koizumi, M., Matsumoto, M., et al., 2005. Undesirable effects of citrus juice on the pharmacokinetics of drugs: focus on recent studies. Drug. Saf. 28 (8), 677–694.

Samman, S., Sandstrom, B., Toft, M.B., et al., 2001. Green tea or rosemary extract added to foods reduces nonheme-iron absorption. Am. J. Clin. Nutr. 73 (3), 607–612.

Santos, C.A., Boullata, J.I., 2005. An approach to evaluating drug-nutrient interactions. Pharmacotherapy 25, 1789–1800.

Schmidt, L.E., Dalhoff, K., 2002. Food-drug interactions. Drugs 62, 1481–1502.

Seeff, L.B., 2007. Herbal hepatotoxicity. Clin. Liver Dis. 11, 577–596.

Shelton, R.C., 2009. St John's wort (*Hypericum perforatum*) in major depression. J. Clin. Psychiatr. 70 (Suppl. 5), 23–27.

Siegel, R.K., 1979. Ginseng abuse syndrome: problems with the panacea. J. Am. Med. Assoc. 241, 1614–1615.

Soldner, A., Christians, U., Susanto, M., et al., 1999. Grapefruit juice activates P-glycoprotein-mediated drug transport. Pharm. Res. 16 (4), 478–485.

Sung, J., Han, K.H., Zo, J.H., et al., 2000. Effects of red ginseng upon vascular endothelial function in patients with essential hypertension. Am. J. Chin. Med. 28, 205–216.

Tam, S.W., Worcel, M., Wyllie, M., 2001. Yohimbine: a clinical review. Pharmacol. Ther. 91, 215–243.

Tauchert, M., 2002. Efficacy and safety of crataegus extract WS 1442 in comparison with placebo in patients with chronic stable New York Heart Association class-III heart failure. Am. Heart J. 143, 910–915.

Turner, R.B., Bauer, R., Woelkart, K., et al., 2005. An evaluation of *Echinacea angustifolia* in experimental rhinovirus infections. N. Engl. J. Med. 353, 341–348.

Wacher, V.J., Wu, C.Y., Benet, L.Z., 1995. Overlapping substrate specificities and tissue distribution of cytochrome P450 3A and P-glycoprotein: Implications for drug delivery and activity in cancer chemotherapy. Mol. Carcinog. 13, 129–134.

Walker, B.R., Edwards, C.R.W., 1994. Licorice-induced hypertension and syndromes of apparent mineralocorticoid excess. Endocrinol. Metab. Clin. North. Am. 23, 359–377.

Walker, S.E., Shulman, K.I., Tailor, S.A.N., et al., 1966. Tyramine content of previously restricted foods in monoamine oxidase inhibitor diets. J. Clin. Psychopharmacol. 16, 383–388.

Wallace, A.W., Amsden, G.W., 2002. Is it really OK to take this with food? Old Interactions with a new twist. J. Clin. Pharmacol. 42, 437–443.

Wandel, C., Kim, R.B., Kajiji, S., et al., 1999. P-glycoprotein and cytochrome P-450 3A inhibition: dissociation of inhibitory potencies. Cancer Res. 59, 3944–3948.

Weed, S., 1986. Wise Woman Herbal For the Childbearing Year. Ash Tree Publishing, Woodstock, NY.

Werba, J.P., Giroli, M., Cavalca, V., et al., 2008. The effect of green tea on simvastatin tolerability. Ann. Intern. Med. 149 (4), 286–287.

Yeu, E., Grostern, R., 2007. Saw palmetto and intraoperative floppy-iris syndrome. J. Cataract. Refract. Surg. 33, 927–928.

Yin, O.Q., Tomlinson, B., Waye, M.M., et al., 2004. Pharmacogenetics and herb-drug interactions: experience with *Ginkgo biloba* and omeprazole. Pharmacogen 14, 841–850.

Zhou, S., Lim, L.Y., Chowbay, B., 2004. Herbal modulation of P-glycoprotein. Drug. Metab. Rev. 36, 57–104.

Zou, Q.Z., Bi, R.G., Li, J.M., et al., 1989. Effect of motherwort on blood hyperviscosity. Am. J. Chin. Med.

Chapter 71

Genotoxicity evaluation of nutraceuticals

Nan Mei, Xilin Li, Si Chen, Lei Guo and Xiaoqing Guo

Introduction

Dietary supplements refer to products consumed by humans that contain ingredients intended to add further nutrients to the diet. These supplemental dietary ingredients include, but are not limited to, minerals, vitamins, amino acids, herbs or botanicals, and enzymes; either alone or in combination. When a dietary supplement has a medical or health benefit, including the prevention and treatment of disease, it can be classified as a nutraceutical (Aronson, 2017). However, since nutraceuticals are not regulated, the manufacturers of nutraceuticals need not prove their efficacy nor evaluate any potential risks (Kalra, 2003). In 1994 the United States Congress issued the Dietary Supplement Health and Education Act. This Act defines both dietary supplement products and dietary ingredients as a special category of foods, rather than drugs which are regulated by the Food and Drug Administration (FDA). According to the 2012 National Health Interview Survey, about 18% of Americans used botanical dietary supplements (Clarke et al., 2015). Of these dietary supplements, herbal products (classified as nonvitamin, nonmineral, natural products) are the most common complementary and alternative medicines used among adults and children in the United States. In 2012 US consumers spent an estimated $12.8 billion out-of-pocket for natural product supplements (Nahin et al., 2016). Currently, there are an estimated 85,000 dietary supplement products on the US market (Dwyer and Coates, 2018).

Herbal supplements (as a type of dietary supplement containing one or more herbs) and their plants have a long history in many cultures for use in health improvement and the treatment of various illnesses and ailments (Petrovska, 2012). Recent reports supporting their beneficial effects include improving the immune system, decreasing oxidative damage, enhancing memory, and as antidepressants (Dwyer and Coates, 2018; Mei et al., 2017; WHO, 2013). However, there are also emerging reports of adverse events following herbal supplement use, such as hepatotoxicity, hypertension, seizures, infarctions, coagulation disorders, and death (IARC, 2016; Jalili et al., 2013; Navarro and Lucena, 2014; Palmer et al., 2003). As a result, herbal product-associated toxicity has drawn increasing concern from US regulatory agencies, including the FDA, the National Cancer Institute (NCI), and the National Toxicology Program (NTP). As early as the 1990s a number of herbal dietary supplements, such as *Aloe vera* gel and ginseng were nominated by the NCI for in-depth toxicity and carcinogenicity studies under the NTP, due to their widespread use and the limited availability of toxicity and carcinogenicity data. Since that time, the NTP has performed a series of 2-year rodent studies on numerous popular herbal medicines, including *Aloe vera*, *Ginkgo biloba*, and goldenseal, with evident carcinogenicity in rodents treated with these products (NTP, 2016).

The herbal product-related carcinogenicity and other health problems following their use prompted additional toxicity research, including genotoxicity, to better understand what risks to humans may be associated with these products. Genotoxicity refers to the potential for an agent to cause DNA lesions that can result in cell death, or mutations which can lead to cancer, and is an important category in hazard identification studies. Genotoxicity is usually evaluated using a battery of standard assays agreed upon by several international organizations and regulatory agencies, such as the Organization for Economic Co-operation and Development, the International Council for Harmonization (ICH), and the US FDA (FDA, 2007; ICH, 2012). The most commonly used genotoxicity test batteries contain a number of in vitro and in vivo assays, including the Ames bacterial mutagenicity test, the mouse lymphoma gene mutation assay (MLA), the micronucleus assay, the Comet assay, sister chromatid exchange assay, and the chromosome aberration test (Chen et al., 2010; Mei et al., 2014). These genotoxicity endpoints assess DNA damage via different mechanisms, including DNA damage, intragenic mutations, chromosomal rearrangements or deletions, and loss or gain of whole chromosomes (aneuploidy) or chromosomal segments (clastogenic damage).

Nutraceuticals. DOI: https://doi.org/10.1016/B978-0-12-821038-3.00071-9

This chapter summarizes the genotoxicity of three commonly used herbal dietary supplements, *Aloe vera*, *G. biloba*, and goldenseal, focusing on standard genotoxicity assays. Also introduced are the herbal products' alleged active compounds and a discussion of their perceived beneficial effects and genotoxicity findings, for a comprehensive understanding of these popular herbal products. Problems and perspectives concerning the safe application of these herbal products are addressed.

Aloe vera

Of the more than 400 aloe species belonging to the *Liliaceae* family, *Aloe vera* is the most popular species used in herbal supplements (Guo and Mei, 2016). Aloe is a stemless, or very short-stemmed, xerophyte with elongated leaves that store large amounts of water. The green fleshy leaves have three evident layers. The outer layer is a thick cuticle or rind where carbohydrate, fat, and protein synthesis occur. Beneath the cuticle lies the thin, mucilage leaf pulp that produces latex. The inner leaf pulp contains a large volume of aloe gel that makes up the majority of the plant mass (Boudreau and Beland, 2006). The whole leaf, gel, and latex are the primary aloe products used in herbal supplements and they have distinct chemical components and biological effects. *Aloe vera* is one of the best-selling botanical supplements in the United States with ~\$22 million products sold in mainstream multioutlet stores and ~\$14 million products sold in nature and health food stores in 2018 (Smith et al., 2019).

Chemical components of *Aloe vera*

Aloe vera whole-leaf extract (WLE), which includes the gel and latex, contains more than 200 chemicals, including amino acids, vitamins, minerals, lignin, and phytosterols (Rodriguez Rodriguez et al., 2010). The latex contains approximately 80 chemical constituents and is yellow-brownish in appearance. Most of the latex compounds are phenolic in nature, mainly anthrones and anthraquinones (Park et al., 1998; Rebecca et al., 2003). Barbaloin, also known as aloin A, is identified as the major constituent in the latex (Boudreau and Beland, 2006; Reynolds, 1985), followed by isobarbaloin (aloin B), aloesin (aloeresin B), and aloeresin A (Saccu et al., 2001). Aloe gel is a clear mucilaginous jelly-like substance with 99%–99.5% water content (Hamman, 2008). The remaining solids in aloe gel are comprised of dietary fibers (nonstarch polysaccharides and lignin), sugars, proteins, enzymes, lipids, and mineral elements (Femenia et al., 1999; Grindlay and Reynolds, 1986). The main active component in the mucilaginous gel is acemannan, an acetylated glucomannan (Femenia et al., 1999). On the other hand, apigenin glycoside derivatives are identified as the main component in *Aloe vera* flower extract (Anibarro-Ortega et al., 2019). However, chemical constituents of aloe plants are significantly influenced by environmental conditions, cultivation methods, harvesting and processing, and storage conditions (Boudreau and Beland, 2006; Rodriguez Rodriguez et al., 2010).

Clinical, therapeutic, and adverse effects of *Aloe vera*

Aloe vera has been used historically for a variety of medical purposes in Arab, Chinese, Egyptian, Greek, Indian, and Roman cultures (Atherton, 1998). Pharmacologically, *Aloe vera* is an immunity booster and it shows antibiotic properties as well as antitumor effects (Kumar et al., 2019). Traditionally, it has been used for the treatment of a variety of diseases such as arthritis, asthma, *Candida*, chronic fatigue syndrome, digestive and bowel disorders, lupus erythematodes, skin problems, and external/internal ulcers (Vogler and Ernst, 1999). Several of these empirical therapeutic claims have been confirmed through randomized controlled clinical trials, including accelerating wound and recurrent aphthous stomatitis healing (Fulton, 1990; Mansour et al., 2014), relieving psoriasis and the first episode of genital herpes (Syed et al., 1997; Syed et al., 1996), preventing traumatic ulcers in patients with fixed orthodontic appliances (Leiva-Cala et al., 2019), and improving the symptom index for acute radiation proctitis (Sahebnasagh et al., 2017). In addition, the ingestion of aloe supplements has been correlated with decreases in blood lipid levels, blood pressure, and glucose levels (Choudhary et al., 2014; Yongchaiyudha et al., 1996).

On the other hand, both severe and minor clinical adverse effects of the application of *Aloe vera* have also been observed (Guo and Mei, 2017). The most severe side effect is hepatitis. Since 2005, a total of 10 patients from various countries, that is, Turkey, Korea, the United States, and Argentina, have been reported as being hospitalized for acute hepatitis after taking oral aloe products; their conditions were promptly improved after discontinuing the use of aloe. The most commonly reported side effect following dermal or oral exposure to aloe is hypersensitivity, that is, urticaria, erythema, and edema. A case of severe renal dysfunction and nephritis was reported in a 52-year-old Greek man due to a hypersensitivity reaction to aloe juice (Cholongitas et al., 2005). Aloe-associated hypokalemia was also observed in an Italian man after oral consumption of an aloe drink for 2 weeks (Baretta et al., 2009).

Genotoxicity of *Aloe vera* and its components

In vitro genotoxicity

In the absence of S9, *Aloe vera* WLE was found to be a weak mutagen in the *Salmonella typhimurium* TA98 strain as measured by the Ames test; and in human peripheral blood lymphocytes WLE did not increase sister chromatid exchange frequencies, but significantly increased micronuclei production and structural chromosome aberration frequencies (Kayraldiz et al., 2010). Human pulp fibroblasts (FP6) were exposed to lyophilized extract (obtained by scraping the parenchyma of *Aloe vera* leaves) for 24 h. The lyophilized *Aloe* extract induced low cytotoxicity, but significantly increased the production of micronuclei in cells treated with 4% extract suspended in distilled water (Carvalho et al., 2018). The authors investigated the genotoxicity of WLE and decolorized whole leaf extract (WLD) using the MLA (Guo et al., 2014). WLD is WLE processed through activated carbon filtration (1%, w/w) to remove the majority of anthraquinones, the components endowing *Aloe vera* with its laxative properties. A 24-h treatment without S9 of both extracts produced dose-dependent cytotoxicity and mutagenicity in mouse lymphoma cells, with WLE being positive for mutagenicity at lower concentrations compared to WLD (Fig. 71.1). In addition, WLD induced about threefold higher levels of intracellular reactive oxygen species (ROS) than WLE. Molecular analysis showed that the primary type of genetic damage for both extracts was large chromosome mutations (deletions and/or mitotic recombination). These results indicate that removing anthraquinones from aloe preparations may not eradicate *Aloe vera*-induced genotoxicity and that activated carbon-filtration results in increased pro-oxidant activities of the WLE (Guo et al., 2014). However, *Aloe vera* decolorized extract beverage at up to 21 × concentrations did not show positive responses in the Ames test and DNA repair assays (Sehgal et al., 2013a).

The genotoxicity of aloe gel and the main chemical components in latex, such as aloe-emodin, aloin, emodin, and danthron, have also been studied. Aloe gel is generally considered less toxic compared to latex. A high-purity *Aloe vera* gel prepared from inner leaf fillet was not mutagenic in the Ames test and did not increase chromosome aberration frequencies in Chinese hamster lung cells at concentrations up to 10 mg/mL (Williams et al., 2010). A commercial *Aloe* gel juice did not induce a significant increase in mutations in *S. typhimurium* TA100 as compared to the control (Sehgal et al., 2013b). Positive responses were reported from only one study where *Aloe vera* pulp extract produced genotoxicity in *Escherichia coli* repair-deficient mutants and dose-dependent single-strand breaks in pUC 9.1 plasmid DNA (Paes-Leme et al., 2005). In contrast, studies into aloe latex genotoxicity mainly focused on a number of anthraquinones, which were mutagenic in most in vitro assay systems. Danthron and aloe-emodin induced significant increases in revertant colonies in several *Salmonella* strains such as TA1537, TA1538, and TA98 (Heidemann et al., 1996; Westendorf et al., 1990). At micromolar concentrations three anthraquinones, emodin, danthron, and aloe-emodin induced dose-dependent increases in micronuclei formation and moderately increased mutant frequencies in mouse lymphoma L5178Y cells. Danthron and aloe-emodin at a concentration of 50 μM also induced DNA strand breaks in the Comet assay. Inhibition of the catalytic activity of topoisomerase II (Topo II) proved responsible for the genotoxic and mutagenic effects of these anthraquinones (Mueller and Stopper, 1999; Muller et al., 1996). Aloe-emodin induced DNA damage in human lung carcinoma cells through production of ROS (Lee et al., 2006), increased chromosomal aberration frequencies in Chinese hamster ovary (CHO) cells (Heidemann et al., 1996), and induced micronuclei in TK6 human lymphoblastoid cells (Nesslany et al., 2009). Additionally, Danthron was shown to induce DNA damage and caspase cascade-mediated apoptosis through mitochondrial permeability

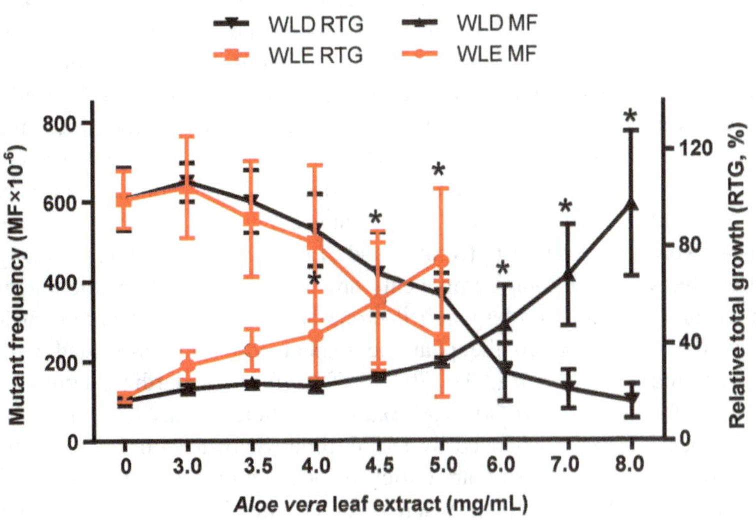

FIGURE 71.1 The mutagenicity and cytotoxicity in mouse lymphoma cells treated with *Aloe vera* for 24 h. Data points represented as the means ± standard deviation are from Guo et al. (2014). Red and black lines represent mutant frequency (MF) and relative total growth (RTG) in cultures treated with WLE and WLD, respectively. *indicates a positive mutagenic response by the induced MF exceeding the global evaluation factor of 126×10^{-6} when compared with the control.

transition pores and Bax-triggered pathways in SNU-1 human gastric cancer cells (Chiang et al., 2011).

In vivo genotoxicity

Aloe vera WLE-induced structural chromosome aberration frequencies in bone marrow cells of Sprague-Dawley rats dosed at 750, 1000, and 1250 mg/kg body weight at 12 and 24 h following exposure (Kayraldiz et al., 2010). In contrast, a high-purity *Aloe vera* inner leaf fillet preparation did not increase chromosome aberration frequencies and micronuclei induction in the bone marrow of rats administered 5000 mg/kg body weight/day for 90 days (Williams et al., 2010). An in vivo mouse Comet assay was performed on isolated kidney and colon cells in order to demonstrate the possible organo-specific genotoxicity of aloe-emodin (Garnica-Garza, 2009; Nesslany et al., 2009). The primary DNA damage in the colon was observed between 3 and 6 h after two oral administrations at 500, 1000, and 2000 mg/kg body weight, suggesting an in vivo genotoxic mechanism of action. These results suggest that aloe-emodin should be considered as an in vivo genotoxin and the risk for human exposure should be taken into account.

Carcinogenicity of *Aloe vera* and its components

A 2-year oral administration of 0%–1.5% (w/w) *Aloe vera* WLE in drinking water in F344/N rats demonstrated clear evidence of its carcinogenic activity (NTP, 2013b). Dose-related increases in the incidences of adenomas and/or carcinomas were observed in the proximal, cecum, ascending, and transverse colon of male and female rats in the 1% and 1.5% groups (Boudreau et al., 2013). A mechanistic study demonstrated similar morphologic and molecular pathway changes, such as mitogen-activated protein kinases, WNT, and TGF-β signaling, for WLE-induced large intestinal tumors in F344 rats and human colorectal cancers (Pandiri et al., 2011). To determine whether aloin, the major component of *Aloe vera* plant leaf, is a significant contributor to WLE-induced pathophysiological effects, F344/N male rats were given 6.95–446 mg/kg aloin (55.4% aloin A and 43.3% aloin B) in drinking water for 13 weeks (Boudreau et al., 2017). Similar effects, that is, increased incidences and severities of mucosal and goblet cell hyperplasia, were observed in rats administered aloin or WLE. These effects might result from the modulation of the gut microbiota composition. There are no carcinogenicity data available for *Aloe* gel. However, when male ACI/N rats were fed with 1% naturally occurring 1-hydroxyanthraquinone for 480 days, 25 of 29 rats developed adenomas or adenocarcinomas in the cecum or upper portion of the colon. In addition, liver neoplasms and benign stomach tumors were found in 12 and 5 rats, respectively (Mori et al., 1990). When 65 female F344/N rats were dosed with 0–2500 ppm emodin for 105 weeks, three Zymbal's gland carcinomas were diagnosed in rats exposed to 2500 ppm (NTP, 2001). In addition, aloe-emodin demonstrated photocarcinogenicity in C3H/HeN mice treated with aloe-emodin in 25% ethanol topically three times per week for 2 weeks and exposed to 15 kJ/m^2 UVB (280–320 nm) radiation (Strickland et al., 2000).

Ginkgo biloba

G. biloba (ginkgo) is a unique species with no other surviving relatives in the division of Ginkgophyta in the Ginkgoaceae family. Due to the discovery of plant fossils that possess ginkgo-like leaves and seed-bearing structures from 180 million years ago, ginkgo is referred to as a "living fossil" (Tredici, 2000). Ginkgo trees can live for over 1000 years and the mechanisms of how they maintain this remarkable longevity are a topic of considerable interest (Wang et al., 2020). Ginkgo has fan-shaped, leathery, bilobed leaves and is a diecious gymnosperm plant. As such, it has separate male and female trees and the seeds are not covered by an ovary wall (Chan et al., 2007; NTP, 2013a). This plant originated in China (Chinese name "*Yin-xing*") and was recorded as a traditional medicine in a Chinese *Materia Medica* about 2000 years ago (Chassagne et al., 2019; Mei et al., 2017). Currently, ginkgo is produced in various forms and can be purchased worldwide. Its leaf extract is regulated as either a dietary supplement (in the United States) or a prescription drug (in Germany and France) (NTP, 2013a).

Chemical components of *Ginkgo biloba*

G. biloba seeds and leaves contain numerous chemical constituents (Ding et al., 2008). Important constituents include terpene trilactones (i.e., ginkgolide A, B, C, J, and bilobalide), flavonol glycosides (i.e., kaempferol, quercetin, and isorhamnetin), biflavones, proanthocyanidins, alkylphenols (i.e., ginkgolic acid), simple phenolic acids, 6-hydroxykynurenic acid, 4-O-methylpyridoxine, and polyprenols (van Beek, 2002; van Beek and Montoro, 2009). The percentages of individual constituents may vary depending on growing conditions, strains, and time of harvest (Smith and Luo, 2004). Although many *G. biloba* components are biologically active, the flavonoid and terpene trilactone are considered primary contributors to the pharmacological activities associated with *G. biloba* (NTP, 2013a). Therefore standardized commercial *G. biloba* leaf extracts (denoted as EGb 761), originally produced by Dr. Willmar Schwabe Pharmaceuticals in Germany, are enriched for flavonoid and terpene trilactone content, whereas other chemicals (i.e., biflavones and ginkgolic acids) are removed (van Beek and Montoro, 2009). The standardized extract contains 22%–27% flavone glycosides (primarily quercetin, kaempferol, and isorhamnetin), 5%–7% terpene lactones (2.8%–3.4% ginkgolides A, B, and C; and 2.6%–3.2% bilobalide), less than 5 mg/kg (5 ppm) ginkgolic acids, and other constituents (Chan et al., 2007).

Therapeutic effects of *Ginkgo biloba*

G. biloba is one of the most commonly used herbal supplements in the United States with more than $20 million total sales annually (Smith et al., 2019). It has been reported to have cardioprotective, neuroprotective, hepatoprotective, antioxidative, antiinflammatory, and anticancer effects. Therefore *G. biloba* is used to treat a variety of ailments and disorders, including memory loss, brain function impairment, poor circulation, and cardiovascular, bladder, ear, nose, throat, and lung problems (Chan et al., 2007; Mei et al., 2017). More than 90% of published research articles on *G. biloba* are related to investigating its beneficial properties (Fig. 71.2).

G. biloba leaf extract has been reported to exhibit neuroprotective and/or therapeutic effects on Alzheimer disease, dementia, learning and memory deficit, traumatic brain injury, stroke, anxiety, macular degeneration, tinnitus, and stress (Diamond and Bailey, 2013; Elovic and Zafonte, 2001; Feng et al., 2019; Ihl, 2013; Ward et al., 2002). EGb 761 was also reported to promote the proliferation of endogenous neural stem cells in rats with vascular dementia (Wang et al., 2013). For cardio-cerebral vascular health, *G. biloba* extract can improve blood flow and decrease platelet aggregation and whole blood viscosity (Chan et al., 2007). Standardized *G. biloba* extract significantly increased the pain-free walking distance in patients suffering from intermittent claudication and peripheral occlusive arterial disease (Peters et al., 1998; Schweizer and Hautmann, 1999). *G. biloba* supplements may also be beneficial in the treatment or prevention of myocardial infarction and hypercoagulation (Mahady, 2002). Due to its antioxidant and positive effects on vascular systems, some evidence has shown *G. biloba* could be an adjuvant treatment for normal tension glaucoma, an optic neuropathy that leads to permanent vision loss (Adeghate et al., 2019; Cybulska-Heinrich et al., 2012). In addition, *G. biloba* extract used for other medical applications has demonstrated stabilization of cellular energy production as well as suppression of hemorrhoids,

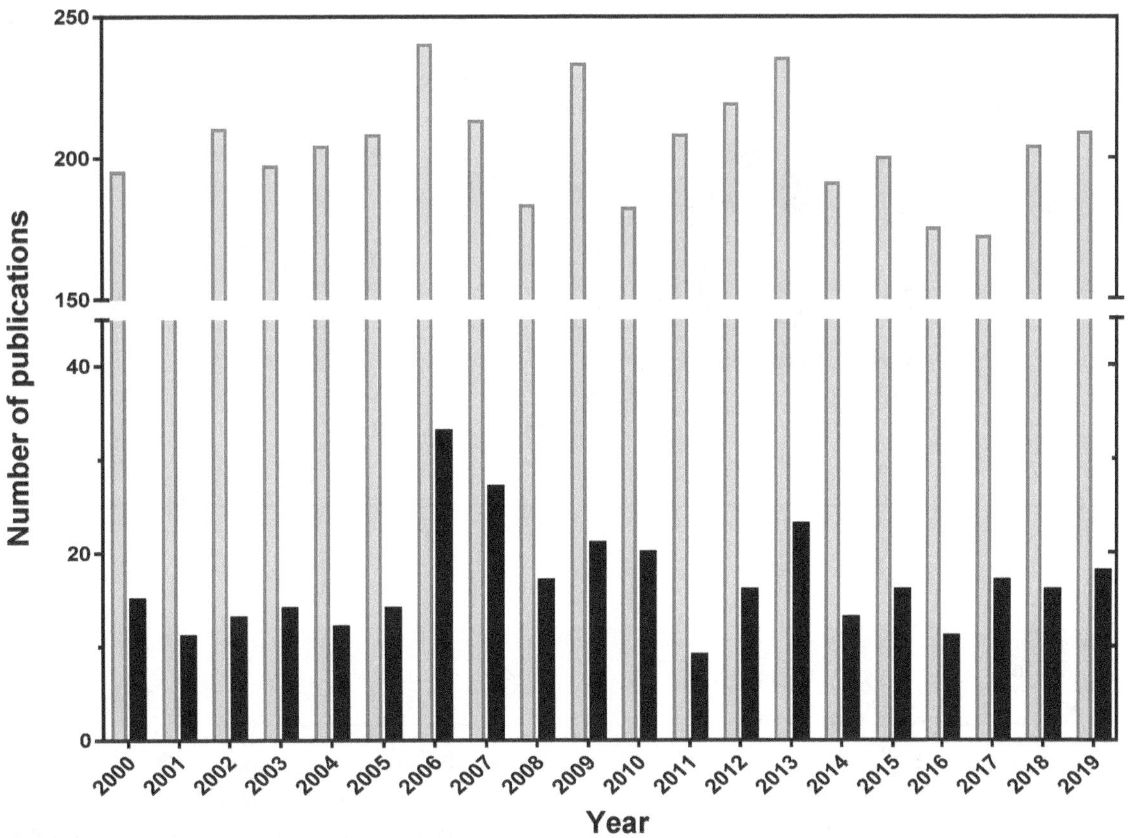

FIGURE 71.2 Pharmacological and toxicological research on *Ginkgo biloba* based on PubMed online search. Gray bars represent numbers of all research on ginkgo while black bars indicate the research related to toxicity (searches conducted on February 6, 2020).

inflammation, migraine, allergies, and asthma (Chan et al., 2007). Although the beneficial effects of *G. biloba* were reported in several small clinical trials, the claimed health effects of *G. biloba* leaf extracts have not been confirmed in large-scale clinical trials with relatively large numbers of patients, longer duration, and better study designs in France, the United Kingdom, and the United States (Mei et al., 2017).

Genotoxicity of *Ginkgo biloba* and its components

The toxic properties of *G. biloba* have also been a point of concern reported in the literature (Fig. 71.2). Both genotoxic and carcinogenic effects of *G. biloba* extract were reported by the NTP (NTP, 2013a). The 2-year bioassay, which is considered the "gold standard" for studying chemical carcinogenesis, showed clear evidence of carcinogenicity of the extract in B6C3F1/N mice based on increased incidence of hepatocellular carcinoma and hepatoblastoma. In the bacterial Ames test, *G. biloba* extract at concentrations of 1–10 mg/plate was mutagenic in *S. typhimurium* strains TA98 and TA100, and in *E. coli* strain WP2 *uvrA*/pKM101, both in the absence or presence of exogenous S9 metabolic activation. These findings indicated the induction of frame shift and base substitution mutations (NTP, 2013a). The same batch of extract used in the NTP study also demonstrated genotoxicity in the MLA and the Comet assay. Following a 4-h treatment, *G. biloba* extract at concentrations of 0.2–1.2 mg/mL induced dose-dependent cytotoxicity and mutagenicity in the MLA, as well as a dose-dependent increase in the percentage of tail DNA intensity in the Comet assay, with positive responses observed starting at 1.0 mg/mL dose for both assays. In addition, the extract produced oxidative stress and induced extensive chromosomal damage as demonstrated by the loss of heterozygosity analysis (Lin et al., 2014). In human hepatic HepG2 cells, *G. biloba* extract caused concentration-dependent DNA damage as measured by the Comet assay (4-h treatment) (Fig. 71.3) and the protein expression of γ-H2A.X (2–24 h treatments), a hallmark of DNA double-strand breaks (Zhang et al., 2015). Phosphorylation of two DNA damage-responsive cell cycle checkpoint-related proteins, CHK1 and CHK2, were also significantly increased upon *G. biloba* exposure in HepG2 cells. However, it has also been reported that a *G. biloba* extract with 24.3%

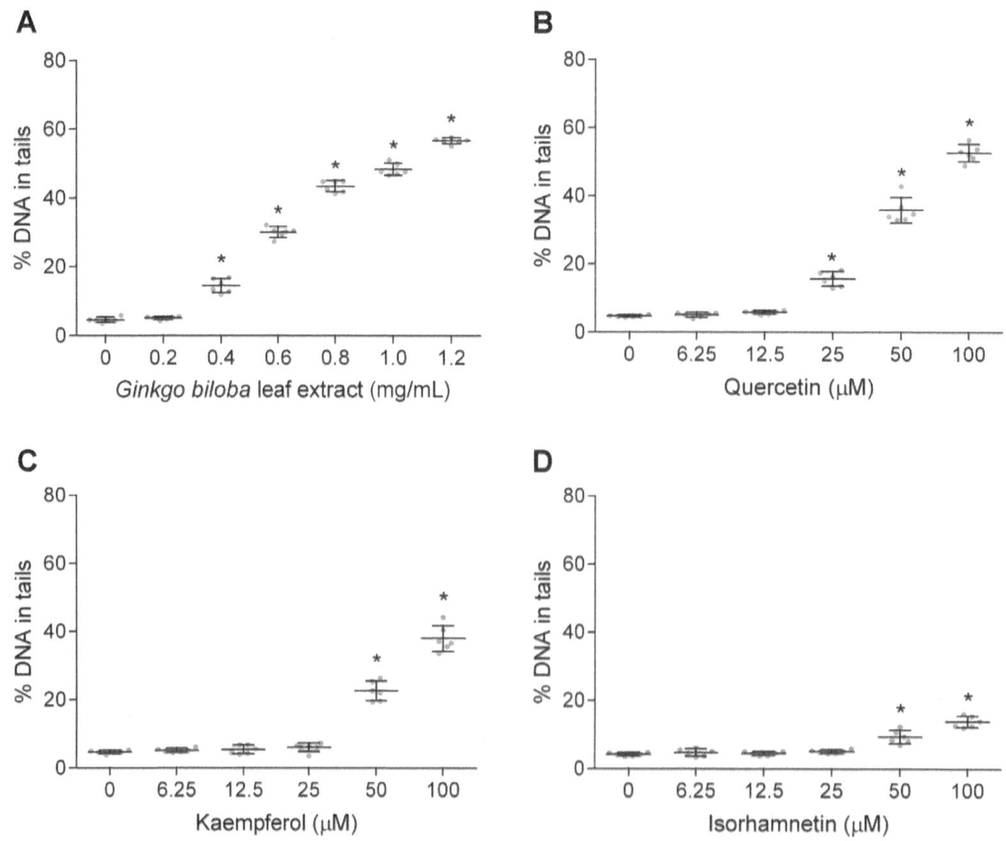

FIGURE 71.3 DNA damage induced by *Ginkgo biloba* leaf extract and its three constituents in HepG2 cells. HepG2 cells were exposed to various concentrations of *Ginkgo biloba* leaf extract (A), quercetin (B), kaempferol (C), or isorhamnetin (D) for 4 h. The data points represented as the means ± S.D. are from Zhang et al. (2015). $*p < 0.05$ *vs.* control.

flavonoids did not induce DNA damage as measured by the Comet assay in HepG2 and THLE-2 cells after 24-h treatment at concentrations up to 0.6 mg/mL (Grollino et al., 2017). In addition, a study in male and female B6C3F1/N mice showed an overall negative result in the in vivo micronucleus test using peripheral blood cells to evaluate *G. biloba* extract genotoxicity at doses of 125–2000 mg/kg body weight per day. *G. biloba* extract induced equivocal or negative responses for the micronucleus induction in female or male B6C3F1/N mice, respectively, after 3 months of treatment (NTP, 2013a). Similarly, genotoxicity was not observed in the in vivo liver Comet assay and the bone marrow micronucleus assay from *gpt* delta mice and constitutive androstane receptor knockout mice following treatment of *G. biloba* extract by gavage up to 2000 mg/kg body weight per day for 3 months (Maeda et al., 2014).

Numerous *G. biloba* extract components have been investigated individually for genotoxicity. Quercetin, a flavonol, is the most extensively studied component found in ginkgo extract and was reported to be genotoxic in vitro. Quercetin exhibited mutagenicity in various strains of *S. typhimurium* that detect base-pair substitution and frameshift mutagens (Bjeldanes and Chang, 1977; Rueff et al., 1986). Several laboratories, including that of the authors, demonstrated that quercetin induced concentration-dependent increases in the frequency of mouse lymphoma L5178Y mutant colonies (Lin et al., 2014; Meltz and MacGregor, 1981; van der Hoeven et al., 1984). Quercetin also induced chromosome aberrations in CHO cells and Chinese hamster V79 cells (Carver et al., 1983; Gaspar et al., 1994; NTP, 1992), caused sister chromatid exchanges in CHO cells (NTP, 1992), and induced micronucleus formation in peripheral human lymphocytes and V79 cells both with and without S9 metabolic activation (Caria et al., 1995). Exposure to quercetin also induced DNA strand breaks in L5178Y cells, Caco-2 (colon), HepG2 (liver), HeLa (epithelial) cells, human peripheral blood lymphocytes, and human sperm (Anderson et al., 1997; Duthie et al., 1997; Lin et al., 2014; Meltz and MacGregor, 1981). Recently, we reported that quercetin acts as a topoisomerase II inhibitor that causes time- and concentration-dependent increases in DNA damage in HepG2 cells (Fig. 71.3) (Zhang et al., 2015). In contrast to the clear genotoxic effects observed in vitro, short-term in vivo genotoxicity tests (the Comet, micronucleus, and mutation assay) produced overall negative results in mice and rats exposed orally to quercetin (Caria et al., 1995; Harwood et al., 2007; Utesch et al., 2008). Multiple long-term in vivo studies also did not show carcinogenicity for quercetin, except that a 2-year bioassay conducted by NTP found an increase in renal lesions in male F344/N rats in the high-dose group (NTP, 1992). Later, the induction of renal lesions was attributed to a chronic nephropathy instead of a genotoxic mode of action.

In addition to quercetin, other flavanols found in *G. biloba* extract also produced genotoxicity, mostly based on data from in vitro studies (Fig. 71.3). Kaempferol was positive for mutagenicity in the Ames test using *S. typhimurium* strains TA98 and TA100 (Nagao et al., 1981; Resende et al., 2012), induced dose-related increases in mutant frequencies in the MLA, and caused DNA strand breaks in HepG2 and L5178Y cells as determined by the Comet assay (Lin et al., 2014; Meltz and MacGregor, 1981; Zhang et al., 2015). Exposures to kaempferol also increased micronucleus and chromosome aberration frequencies in V79 cells (Silva et al., 1997) and induced significant chromosomal breakage and mitotic delay in CHO cells (Carver et al., 1983). The addition of S9 increased kaempferol genotoxicity following its biotransformation to quercetin, primarily by cytochrome P450 1A1 (Silva et al., 1997). Although it has a lower potency, isorhamnetin has been shown to induce DNA double-strand breaks in HepG2 cells (Zhang et al., 2015). Rutin, another flavonol found in ginkgo, also induced DNA damage in human peripheral blood lymphocytes in vitro (Anderson et al., 1997) and in bone marrow cells of Swiss-Webster mice, as measured by the in vivo Comet assay (da Silva et al., 2002).

Goldenseal

Goldenseal (*Hydrastis canadensis*), a perennial plant in the family Ranunculaceae, is native to Canada and the eastern United States (Zeiger and Tice, 1997). The goldenseal plant consists of one long-stalked basal leaf, a single stem, and two rounded, lobed, and double-toothed leaves on a forked branch (Zeiger and Tice, 1997). As a herbal product, goldenseal is widely available on the international market and demand has been increasing dramatically. The worldwide sales of goldenseal root in pharmaceutical outlets approached $25 million US dollars (IARC, 2016). Combination herbal supplements containing goldenseal and echinacea have emerged as among the top-selling herbal supplements sold in the US natural and health food stores in 2018 with more than $6 million in sales (Smith et al., 2019).

Chemical components of goldenseal

Goldenseal contains a complex mixture of components, including amino acids, carbohydrates, and isoquinoline alkaloids (Weber et al., 2003). Isoquinoline alkaloids found in goldenseal include hydrastine, berberine, berberastine, hydrastinine, tetrahydroberberastine, canadine, and canalidine (IARC, 2016). However, the alkaloid content of goldenseal products sold varies (Edwards and Draper, 2003). The major alkaloid constituents of goldenseal products are (−)-hydrastine (1.5%–4%, w/w), berberine (0.5%–6%), and canadine (0.5%–1%) (Brown and Roman, 2008; IARC, 2016). Generally, the alkaloid content in goldenseal preparations is used as a biomarker for standardization and quality control (Weber et al., 2003).

Therapeutic effects of goldenseal

Goldenseal is used both internally and externally for treating skin, mouth, and eye infections and for treatment of gastrointestinal and urinary diseases (Hamon, 1990; NTP, 2010). In recent years, pharmacological studies of goldenseal have focused on the alkaloid constituents, particularly on berberine because it contains antiamoebic, antifungal, antibacterial, antiparasitic, and antituberculous properties (Bae et al., 1998; Cicero and Baggioni, 2016; Habtemariam, 2016; Soffar et al., 2001). In addition, reports assert that berberine has the potential to lower cholesterol. One study by Kong et al (2004) reported that administration of berberine in both hypercholesterolemic patients and hypercholesterolemic hamsters reduced serum levels of cholesterol and triglycerides. The cholesterol-lowering effect of berberine likely occurs through upregulating expression of the liver low-density lipoprotein receptor (Kong et al., 2004). Berberine also proved effective in decreasing blood levels of glucose, glycated hemoglobin (HbA1c), triglycerides, total cholesterol, and low-density lipoprotein-cholesterol from patients with type II diabetes (Zhang et al., 2008). Recently, the antitumor effects of berberine were investigated and studies reported that berberine inhibited proliferation and induced apoptosis of cancer cells. For example, berberine induced cell cycle arrest and apoptosis in cervical cancer cells by regulating Activator Protein-1 (AP-1) activity (Mahata et al., 2011). Berberine also inhibited the proliferation and metastasis of hepatocellular carcinoma by suppressing Id-1 (Inhibitor of differentiation/DNA binding) expression (Tsang et al., 2014). Moreover, in combination with cisplatin, berberine increased cisplatin-induced cell death, apoptosis, and DNA damage in human breast cancer MCF-7 cells (Zhao et al., 2016).

Genotoxicity of goldenseal and its components

A standard NTP 2-year toxicology and carcinogenesis study in rodents showed that oral administration of goldenseal root powder resulted in increased incidence of hepatocellular adenoma and/or carcinoma in F344/N rats and male B6C3F1 mice (NTP, 2010). However, there was not a significant increase in micronucleated erythrocyte frequencies when evaluated by in vivo micronucleus assay using peripheral blood samples collected from male and female B6C3F1 mice fed goldenseal root powder (3125–50,000 ppm) for 3 months. In addition, the same root powder at concentrations ranging from 1 to 10 mg/plate was not mutagenic in *S. typhimurium* strains TA98 and TA100 or in *E. coli* strain WP2 *uvrA*/pKM101, both in the absence and presence of S9 (NTP, 2010). Recently, exposure to several commercial goldenseal extract products induced DNA damage in HepG2 liver cells, as measured by the induction of γ-H2A.X. It is worth noting that the levels of γ-H2A.X were closely associated with the berberine content of the goldenseal extracts (Fig. 71.4A, correlation coefficient = 0.976). These data suggest that berberine is the primary contributor to goldenseal-induced DNA damage (Chen et al., 2013).

The mutagenicity of berberine was tested in *S. typhimurium* tester strains TA98 and TA100 (Nozaka et al., 1990). In the absence of S9, berberine produced slight mutagenic activity in TA98, but did not produce mutagenicity in strain TA100, with or without S9. An NTP study also tested berberine chloride for mutagenicity using the

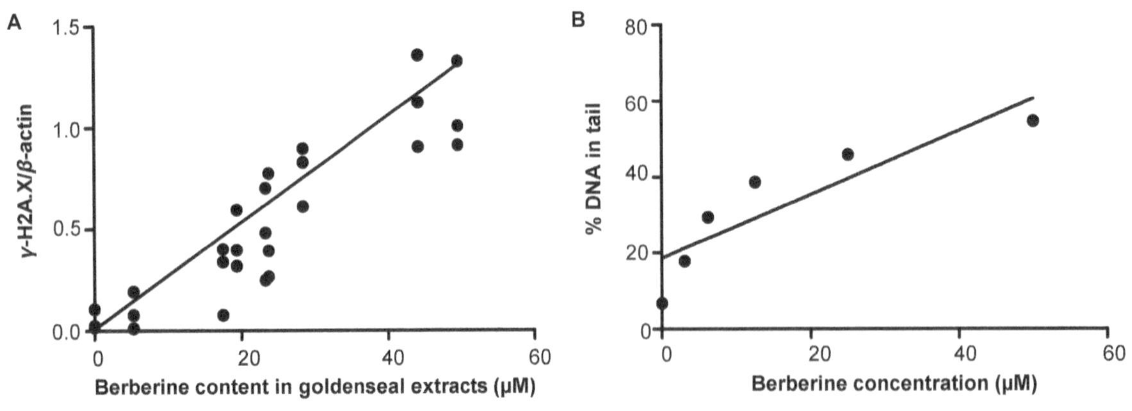

FIGURE 71.4 Commercial goldenseal extracts and berberine induce DNA damage in HepG2 cells. (A) HepG2 cells were treated with 20 μL of eight commercially available goldenseal extracts for 6 h in 10 mL culture medium and then analyzed for γ-H2A.X by Western blotting. Data points from three independent experiments are from Chen et al. (2013) and plot represents quantitative expression levels of γ-H2A.X that were normalized by β-actin. (B) HepG2 cells were exposed to increased concentrations of berberine for 24 h and the percentage of tail DNA intensity was determined using the Comet assay. Data points represented as the means are from Chen et al. (2013).

standard screening assays in *S. typhimurium* strains TA97, TA98, TA100, and TA1535, with no significant mutagenicity detected at concentrations of 0.33–1000 μg/plate, with or without S9 addition (NTP, 2010). Assessment of micronucleus frequencies using bone marrow samples from male mice administered 41–658 mg/kg berberine by three intraperitoneal injections at 24-h intervals, did produce a detectable increase in micronucleated polychromatic erythrocytes (NTP, 2010).

Berberine genotoxicity in *E. coli* strain PQ37 was examined and no obvious genotoxic activity in the SOS chromotest was found in the presence or absence of S9 (Pasqual et al., 1993). Berberine also did not induce significant genotoxic effects in haploid yeast cells tested under nongrowth conditions. However, treatment of dividing yeast cells with berberine caused significant increases in frameshift mutations and mitotic recombination, in both DNA-repair-proficient and -deficient *Saccharomyces cerevisiae* strains (Pasqual et al., 1993). Interestingly, among the different DNA-repair-deficient mutants studied, a mutant showing DNA strand-break repair deficiency was the most sensitive strain to the cytotoxic effect of berberine. The authors suggested that the observed genotoxic effects only occurred in dividing yeast cells, possibly due to berberine–DNA intercalation and subsequent disruption of topoisomerase enzyme activities (NTP, 2010; Pasqual et al., 1993).

The DNA damage potential resulting from berberine exposure was also studied in mammalian cells. UVA-irradiated berberine induced DNA single-strand breaks following treatment of NIH-3T3 cells, Ehrlich ascites carcinoma cells, and HaCaT keratinocytes (transformed human epidermal cells) (Inbaraj et al., 2001; Jantova et al., 2006). The UVA irradiation of berberine generated a superoxide anion radical and singlet oxygen capable of damaging cellular components, including DNA. In contrast, DNA damage was not detected in HaCaT cells incubated with other goldenseal constituents including hydrastine or canadine (Inbaraj et al., 2001; Jantova et al., 2006), indicating the potency of photo-irradiated goldenseal alkaloids on DNA damage is different for different goldenseal constituents.

The DNA damaging effect of five goldenseal alkaloid and their toxic potencies were compared in HepG2 cells (Chen et al., 2013), and berberine was the most potent inducer of DNA damage measured by the Comet assay and the induction of γ-H2A.X. Berberine induced concentration-dependent DNA damage in treated cells ($R^2 = 0.885$) (Fig. 71.4B). In addition, berberine inhibited the activity of topoisomerase in a cell-free system; and in Topo II gene-silenced cells, berberine reduced induction of γ-H2A.X, indicating a direct association between Topo II inhibition and DNA damage. These results suggest that the Topo II inhibition may be the underlying mechanism contributing to berberine- and goldenseal-associated genotoxicity and tumorigenicity (Chen et al., 2013).

Concluding remarks and future directions

The use of herbal dietary supplements, including *Aloe vera*, *G. biloba*, and goldenseal, is increasing both in the United States and worldwide. Despite some discrepancies in the current literature, evidence of their efficacy in promoting health has been demonstrated in a number of experimental studies and clinical trials. However, there is also published evidence for toxicity, genotoxicity, and carcinogenicity of some *Aloe vera*, *G. biloba*, and goldenseal herbal products. These findings suggest that herbal dietary supplements, which are commonly thought to be safe, require additional toxicological evaluation. Specifically, all three *Aloe vera*, *G. biloba*, and goldenseal herbal dietary supplements reviewed in this chapter demonstrated carcinogenic activity in the NTP 2-year rodent studies (NTP, 2010, 2013; NTP, 2013b). Based on available data, the International Agency for Research on Cancer (IARC) classified these three herbal supplements (i.e., WLE of *Aloe vera*, *G. biloba* extract, and golden root powder) as possible human carcinogens (Group 2B) in 2016 (IARC, 2016). However, the mechanisms of their carcinogenicity are not well understood. Further studies on the potential mechanisms of genotoxicity and carcinogenicity are warranted. There are several remaining important issues associated with the genotoxicity and safe application of herbal products that need addressing.

First, plant identification is a particular challenge and the accessibility of high-quality and contaminant-free raw herbal plants is of primary importance for understanding the risk–benefit profiles of herbal dietary supplements. In 2003 the World Health Organization developed guidelines on good agricultural and collection practices for medicinal plants. These guidelines provide general technical guidance on the sustainable cultivation, collection, harvest, post-harvest processing, handling, and storage of medicinal plants for quality assurance (Fu et al., 2009). In 2010 the US FDA published good manufacturing practice guidelines for industry in manufacturing, packaging, labeling, or holding operations for dietary supplements (FDA, 2010). These guidelines are strongly recommended to ensure the production of high-quality herbs without contamination of microorganism, toxic heavy metals, and pesticide residue (Fu et al., 2009).

Second, standardization of the herbal product's chemical components is pivotal for evaluating the beneficial effects, genotoxicity, or other adverse properties. Most herbal dietary supplements are a mixture of dozens or hundreds of chemical components. Sufficient characterization of the active components in herbal extracts is critical to recognizing which ingredients are responsible for the therapeutic and/or toxic effects. However, standardization generally involves only a few key ingredients, while variations in other ingredients may also influence both the efficacy and toxicity of the products (Ernst, 2005).

Third, genotoxicity studies on whole herbal extracts are insufficient, with most studies focused on individual ingredients contained in the extracts. While key active ingredients may largely contribute to the parent herb's biological effects, chemical interactions and the content of each ingredient may be contributing factors as well. For example, berberine was demonstrated to be the most potent inducer of DNA damage among five goldenseal constituents tested in HepG2 cells, while several commercial goldenseal powders with low berberine content ($<17.5\ \mu M$) failed to induce DNA damage as measured by the γ-H2A.X assay (Chen et al., 2013).

Fourth, fast-growing technologies provide new platforms for a comprehensive evaluation of herbal product-induced genotoxic effects and the underlying mechanisms. The integration of high-throughput genotoxicity approaches (i.e., flow cytometry-based micronucleus assay and CometChip assay) and quantitative benchmark dose analysis provides an effective tool to rank herbal products based on biological effects (Guo and Mei, 2018; Seo et al., 2019). Pathway-based assays provide further insight into the mode or mechanism of action for induced genotoxicity (Krewski et al., 2010). For example, black cohosh extract, a widely used dietary supplement widely used to alleviate symptoms of gynecological ailments, has been demonstrated to induce micronuclei via an aneugenic mode of action (Smith-Roe et al., 2018). Furthermore, next-generation sequencing and novel in vitro human cell models, along with complex multiorgan 3D spheroids in a microfluidic-based system, show promise for improving the effectiveness of risk assessment (Guo et al., 2018, 2020).

Finally, when evaluating the adverse effects of herbal dietary supplements, considering the role of herb–drug interactions and the potential long-term impacts is crucial (Asher et al., 2017). A clinical study of herb–drug interactions concluded that nine of 38 herbs, including *G. biloba*, were considered to have a major interaction with warfarin, sometimes leading to severe outcomes (Ge et al., 2014). In addition, there is a shortage of adequately sized controlled clinical trials for adequate size investigating the long-term impacts of herbal product use (Ernst, 2002).

In summary, current published research demonstrates both beneficial and adverse effects of herb nutraceuticals sold on the market. Generally, research on herbal products, especially toxicological research, is much less active compared to that of conventional drugs. The lack of research into herbal dietary supplements is in part due to the legal status of these products, as herbal products do not require safety testing prior to reaching the market. Emerging evidence for the genotoxicity and/or carcinogenicity of some herbal dietary supplements warrants further investigation regarding their safety as well as their efficacy for a safe application.

Acknowledgments

The authors thank Drs. Dayton Petibone and Qiangen Wu for their critical review of this manuscript. The information in this chapter is not a formal dissemination of information by the US FDA and does not represent agency position or policy.

References

Adeghate, J., Rahmatnejad, K., Waisbourd, M., et al., 2019. Intraocular pressure–independent management of normal tension glaucoma. Surv. Ophthalmol. 64, 101–110.

Anderson, D., Basaran, N., Dobrzynska, M.M., et al., 1997. Modulating effects of flavonoids on food mutagens in human blood and sperm samples in the comet assay. Teratog. Carcinog. Mutagen. 17, 45–58.

Anibarro-Ortega, M., Pinela, J., Barros, L., et al., 2019. Compositional features and bioactive properties of *Aloe vera* leaf (fillet, mucilage, and rind) and flower. Antioxid. (Basel) 8, 444.

Aronson, J.K., 2017. Defining 'nutraceuticals': neither nutritious nor pharmaceutical. Br. J. Clin. Pharmacol. 83, 8–19.

Asher, G.N., Corbett, A.H., Hawke, R.L., 2017. Common herbal dietary supplement-drug interactions. Am. Fam. Phys. 96, 101–107.

Atherton, P., 1998. *Aloe vera*: magic or medicine? Nurs. Stand. 12, 49–52. 54.

Bae, E.A., Han, M.J., Kim, N.J., et al., 1998. Anti-*Helicobacter pylori* activity of herbal medicines. Biol. Pharm. Bull. 21, 990–992.

Baretta, Z., Ghiotto, C., Marino, D., et al., 2009. Aloe-induced hypokalemia in a patient with breast cancer during chemotherapy. Ann. Oncol. 20, 1445–1446.

Bjeldanes, L.F., Chang, G.W., 1977. Mutagenic activity of quercetin and related compounds. Science 197, 577–578.

Boudreau, M.D., Beland, F.A., 2006. An evaluation of the biological and toxicological properties of *Aloe barbadensis* (miller), *Aloe vera*. J. Environ. Sci. Health C. Environ. Carcinog. Ecotoxicol. Rev. 24, 103–154.

Boudreau, M.D., Mellick, P.W., Olson, G.R., et al., 2013. Clear evidence of carcinogenic activity by a whole-leaf extract of *Aloe barbadensis* miller (*Aloe vera*) in F344/N rats. Toxicol. Sci. 131, 26–39.

Boudreau, M.D., Olson, G.R., Tryndyak, V.P., et al., 2017. From the cover: aloin, a component of the *Aloe vera* plant leaf, induces pathological changes and modulates the composition of microbiota in the large intestines of F344/N male rats. Toxicol. Sci. 158, 302–318.

Brown, P.N., Roman, M.C., 2008. Determination of hydrastine and berberine in goldenseal raw materials, extracts, and dietary supplements by high-performance liquid chromatography with UV: collaborative study. J. AOAC Int. 91, 694–701.

Caria, H., Chaveca, T., Laires, A., et al., 1995. Genotoxicity of quercetin in the micronucleus assay in mouse bone marrow erythrocytes, human lymphocytes, V79 cell line and identification of kinetochore-containing (CREST staining) micronuclei in human lymphocytes. Mutat. Res. 343, 85–94.

Carvalho, N.C., Guedes, S.A.G., Albuquerque-Junior, R.L.C., et al., 2018. Analysis of *Aloe vera* cytotoxicity and genotoxicity associated with endodontic medication and laser photobiomodulation. J. Photochem. Photobiol. B, Biol. 178, 348–354.

Carver, J.H., Carrano, A.V., MacGregor, J.T., 1983. Genetic effects of the flavonols quercetin, kaempferol, and galangin on Chinese hamster ovary cells in vitro. Mutat. Res. 113, 45–60.

Chan, P.C., Xia, Q., Fu, P.P., 2007. *Ginkgo biloba* leave extract: biological, medicinal, and toxicological effects. J. Environ. Sci. Health C. Environ. Carcinog. Ecotoxicol. Rev. 25, 211–244.

Chassagne, F., Huang, X., Lyles, J.T., et al., 2019. Validation of a 16th century traditional Chinese Medicine use of *Ginkgo biloba* as a topical antimicrobial. Front. Microbiol. 10, 775.

Chen, T., Mei, N., Fu, P.P., 2010. Genotoxicity of pyrrolizidine alkaloids. J. Appl. Toxicol. 30, 183–196.

Chen, S., Wan, L., Couch, L., et al., 2013. Mechanism study of goldenseal-associated DNA damage. Toxicol. Lett. 221, 64–72.

Chiang, J.H., Yang, J.S., Ma, C.Y., et al., 2011. Danthron, an anthraquinone derivative, induces DNA damage and caspase cascades-mediated apoptosis in SNU-1 human gastric cancer cells through mitochondrial permeability transition pores and Bax-triggered pathways. Chem. Res. Toxicol. 24, 20–29.

Cholongitas, E., Katsoudas, S., Dourakis, S., 2005. Henoch-Schonlein purpura associated with *Aloe vera* administration. Eur. J. Intern. Med. 16, 59–60.

Choudhary, M., Kochhar, A., Sangha, J., 2014. Hypoglycemic and hypolipidemic effect of *Aloe vera* L. in non-insulin dependent diabetics. J. Food Sci. Technol. 51, 90–96.

Cicero, A.F., Baggioni, A., 2016. Berberine and its role in chronic disease. Adv. Exp. Med. Biol. 928, 27–45.

Clarke, T.C., Black, L.I., Stussman, B.J., et al., 2015. Trends in the use of complementary health approaches among adults: United States, 2002-2012. Natl. Health Stat. Rep. 1–16.

Cybulska-Heinrich, A., Mozaffarieh, M., Flammer, J., 2012. *Ginkgo biloba*: an adjuvant therapy for progressive normal and high tension glaucoma. Mol. Vis. 18, 390.

da Silva, J., Herrmann, S.M., Heuser, V., et al., 2002. Evaluation of the genotoxic effect of rutin and quercetin by comet assay and micronucleus test. Food Chem. Toxicol. 40, 941–947.

Diamond, B.J., Bailey, M.R., 2013. *Ginkgo biloba*: indications, mechanisms, and safety. Psychiatr. Clin. North Am. 36, 73–83.

Ding, S., Dudley, E., Plummer, S., et al., 2008. Fingerprint profile of *Ginkgo biloba* nutritional supplements by LC/ESI-MS/MS. Phytochemistry 69, 1555–1564.

Duthie, S.J., Johnson, W., Dobson, V.L., 1997. The effect of dietary flavonoids on DNA damage (strand breaks and oxidised pyrimdines) and growth in human cells. Mutat. Res. 390, 141–151.

Dwyer, J.T., Coates, P.M., 2018. Why Americans need information on dietary supplements. J. Nutr. 148, 1401S–1405S.

Edwards, D.J., Draper, E.J., 2003. Variations in alkaloid content of herbal products containing goldenseal. J. Am. Pharm. Assoc. (2003) 43, 419–423.

Elovic, E.P., Zafonte, R.D., 2001. *Ginkgo biloba*: applications in traumatic brain injury. J. Head. Trauma. Rehabilitation 16, 603–607.

Ernst, E., 2002. The risk-benefit profile of commonly used herbal therapies: Ginkgo, St. John's Wort, Ginseng, Echinacea, Saw Palmetto, and Kava. Ann. Intern. Med. 136, 42–53.

Ernst, E., 2005. The efficacy of herbal medicine—an overview. Fundam. Clin. Pharmacol. 19, 405–409.

FDA, 2007. Toxicological principles for the safety assessment of food ingredients. <https://www.fda.gov/regulatory-information/search-fda-guidance-documents/redbook-2000-i-introduction> (Accessed 6 March 2020).

FDA, 2010. Guidance for industry: current good manufacturing practice in manufacturing, packaging, labeling, or holding operations for dietary supplements; small entity compliance guide. <http://www.fda.gov/Food/GuidanceRegulation/GuidanceDocumentsRegulatoryInformation/DietarySupplements/ucm238182.htm> (Accessed 8 October 2020).

Femenia, A., Sánchez, E.S., Simal, S., et al., 1999. Compositional features of polysaccharides from *Aloe vera* (*Aloe barbadensis* Miller) plant tissues. Carbohydr. Polym. 39, 109–117.

Feng, Z., Sun, Q., Chen, W., et al., 2019. The neuroprotective mechanisms of ginkgolides and bilobalide in cerebral ischemic injury: a literature review. Mol. Med. 25, 57.

Fu, P.P., Chiang, H.M., Xia, Q., et al., 2009. Quality assurance and safety of herbal dietary supplements. J. Environ. Sci. Health C. Environ. Carcinog. Ecotoxicol. Rev. 27, 91–119.

Fulton Jr., J.E., 1990. The stimulation of postdermabrasion wound healing with stabilized *Aloe vera* gel-polyethylene oxide dressing. J. Dermatol. Surg. Oncol. 16, 460–467.

Garnica-Garza, H.M., 2009. Monte Carlo-derived TLD cross-calibration factors for treatment verification and measurement of skin dose in accelerated partial breast irradiation. Phys. Med. Biol. 54, 1621–1631.

Gaspar, J., Rodrigues, A., Laires, A., et al., 1994. On the mechanisms of genotoxicity and metabolism of quercetin. Mutagenesis 9, 445–449.

Ge, B., Zhang, Z., Zuo, Z., 2014. Updates on the clinical evidenced herb-warfarin interactions. Evid. Based Complement. Altern. Med. 2014, 957362.

Grindlay, D., Reynolds, T., 1986. The *Aloe vera* phenomenon: a review of the properties and modern uses of the leaf parenchyma gel. J. Ethnopharmacol. 16, 117–151.

Grollino, M.G., Raschella, G., Cordelli, E., et al., 2017. Cytotoxicity, genotoxicity and gene expression changes elicited by exposure of human hepatic cells to *Ginkgo biloba* leaf extract. Food Chem. Toxicol. 109, 486–496.

Guo, X., Mei, N., 2016. *Aloe vera*: a review of toxicity and adverse clinical effects. J. Envirn. Sci. Health C Environ. Carcinog. Ecotoxicol. Rev. 34, 77–96.

Guo, X., Mei, N., 2017. Safety and toxicity concerns of *Aloe vera*. In: Taylor, J.C. (Ed.), Advances in Chemistry Research. Nova Science Publishers, New York, pp. 227–247.

Guo, X., Mei, N., 2018. Benchmark dose modeling of in vitro genotoxicity data: a reanalysis. Toxicol. Res. 34, 303–310.

Guo, X., Zhang, S., Dial, S.L., et al., 2014. In vitro investigation of the mutagenic potential of *Aloe vera* extracts. Toxicol. Res. 3, 487–496.

Guo, X., Pan, B., Seo, J.E., et al., 2018. Whole genome sequencing analysis of small and large colony mutants from the mouse lymphoma assay. Arch. Toxicol. 92, 3585–3595.

Guo, X., Seo, J.E., Li, X., et al., 2020. Genetic toxicity assessment using liver cell models: past, present, and future. J. Toxicol. Environ. Health. B, Crit. Rev. 23, 27–50.

Habtemariam, S., 2016. Berberine and inflammatory bowel disease: a concise review. Pharmacol. Res. 113, 592–599.

Hamman, J.H., 2008. Composition and applications of *Aloe vera* leaf gel. Molecules 13, 1599–1616.

Hamon, N.W., 1990. Goldenseal. Can. Pharm. J. 123, 508–510.

Harwood, M., Danielewska-Nikiel, B., Borzelleca, J., et al., 2007. A critical review of the data related to the safety of quercetin and lack of evidence of in vivo toxicity, including lack of genotoxic/carcinogenic properties. Food Chem. Toxicol. 45, 2179–2205.

Heidemann, A., Volkner, W., Mengs, U., 1996. Genotoxicity of aloeemodin in vitro and in vivo. Mutat. Res. 367, 123–133.

IARC, 2016. Some drugs and herbal products. IARC Working Group on the Evaluation of Carcinogenic Risks to Humans 108, <http://monographs.iarc.fr/ENG/Monographs/vol108/index.php> (Accessed 8 October 2020).

ICH, 2012. Guidance for industry S2(R1) genotoxicity testing and data interpretation for pharmaceuticals intended for human use. <https://www.fda.gov/media/71980/download> (Accessed 8 October 2020).

Ihl, R., 2013. Effects of *Ginkgo biloba* extract EGb 761 (R) in dementia with neuropsychiatric features: review of recently completed randomised, controlled trials. Int. J. Psychiatry Clin. Pract. 17 (Suppl 1), 8–14.

Inbaraj, J.J., Kukielczak, B.M., Bilski, P., et al., 2001. Photochemistry and photocytotoxicity of alkaloids from Goldenseal (*Hydrastis canadensis* L.) 1. Berberine. Chem. Res. Toxicol. 14, 1529–1534.

Jalili, J., Askeroglu, U., Alleyne, B., et al., 2013. Herbal products that may contribute to hypertension. Plast. Reconstr. Surg. 131, 168–173.

Jantova, S., Letasiova, S., Brezova, V., et al., 2006. Photochemical and phototoxic activity of berberine on murine fibroblast NIH-3T3 and Ehrlich ascites carcinoma cells. J. Photochem. Photobiol. B Biol. 85, 163–176.

Kalra, E.K., 2003. Nutraceutical—definition and introduction. AAPS PharmSci 5, E25.

Kayraldiz, A., Yavuz Kocaman, A., Rencuzogullari, E., et al., 2010. The genotoxic and antigenotoxic effects of *Aloe vera* leaf extract in vivo and in vitro. Turk. J. Biol. 34, 235–246.

Kong, W., Wei, J., Abidi, P., et al., 2004. Berberine is a novel cholesterol-lowering drug working through a unique mechanism distinct from statins. Nat. Med. 10, 1344–1351.

Krewski, D., Acosta Jr., D., Andersen, M., et al., 2010. Toxicity testing in the 21st century: a vision and a strategy. J. Toxicol. Environ. Health. B Crit. Rev. 13, 51–138.

Kumar, R., Singh, A.K., Gupta, A., et al., 2019. Therapeutic potential of *Aloe vera*-a miracle gift of nature. Phytomedicine 60, 152996.

Lee, H.Z., Lin, C.J., Yang, W.H., et al., 2006. Aloe-emodin induced DNA damage through generation of reactive oxygen species in human lung carcinoma cells. Cancer Lett. 239, 55–63.

Leiva-Cala, C., Lorenzo-Pouso, A.I., Centenera-Centenera, B., et al., 2019. Clinical efficacy of an *Aloe Vera* gel versus a 0.12% chlorhexidine gel in preventing traumatic ulcers in patients with fixed orthodontic appliances: a double-blind randomized clinical trial. Odontology 108, 470–478.

Lin, H., Guo, X., Zhang, S., et al., 2014. Mechanistic evaluation of *Ginkgo biloba* leaf extract-induced genotoxicity in L5178Y cells. Toxicol. Sci. 139, 338–349.

Maeda, J., Kijima, A., Inoue, K., et al., 2014. In vivo genotoxicity of *Ginkgo biloba* extract in gpt delta mice and constitutive androstane receptor knockout mice. Toxicol. Sci. 140, 298–306.

Mahady, G.B., 2002. *Ginkgo biloba* for the prevention and treatment of cardiovascular disease: a review of the literature. J. Cardiovasc. Nurs. 16, 21–32.

Mahata, S., Bharti, A.C., Shukla, S., et al., 2011. Berberine modulates AP-1 activity to suppress HPV transcription and downstream signaling to induce growth arrest and apoptosis in cervical cancer cells. Mol. Cancer 10, 39.

Mansour, G., Ouda, S., Shaker, A., et al., 2014. Clinical efficacy of new *Aloe vera*- and myrrh-based oral mucoadhesive gels in the management of minor recurrent aphthous stomatitis: a randomized, double-blind, vehicle-controlled study. J. Oral Pathol. Med. 43, 405–409.

Mei, N., Guo, X., Moore, M.M., 2014. Methods for using the mouse lymphoma assay to screen for chemical mutagenicity and photo-mutagenicity. In: Caldwell, G.W., Yan, Z. (Eds.), Optimization in Drug Discovery: In Vitro Methods. Humana Press, New York, pp. 561–592.

Mei, N., Guo, X., Ren, Z., et al., 2017. Review of *Ginkgo biloba*-induced toxicity, from experimental studies to human case reports. J. Environ. Sci. Health C Environ. Carcinog. Ecotoxicol. Rev. 35, 1–28.

Meltz, M.L., MacGregor, J.T., 1981. Activity of the plant flavanol quercetin in the mouse lymphoma L5178Y TK +/- mutation, DNA single-strand break, and Balb/c 3T3 chemical transformation assays. Mutat. Res. 88, 317–324.

Mori, H., Yoshimi, N., Iwata, H., et al., 1990. Carcinogenicity of naturally occurring 1-hydroxyanthraquinone in rats: induction of large bowel, liver and stomach neoplasms. Carcinogenesis 11, 799–802.

Mueller, S.O., Stopper, H., 1999. Characterization of the genotoxicity of anthraquinones in mammalian cells. Biochim. Biophys. Acta 1428, 406–414.

Muller, S.O., Eckert, I., Lutz, W.K., et al., 1996. Genotoxicity of the laxative drug components emodin, aloe-emodin and danthron in mammalian cells: topoisomerase II mediated? Mutat. Res. 371, 165–173.

Nagao, M., Morita, N., Yahagi, T., et al., 1981. Mutagenicities of 61 flavonoids and 11 related compounds. Environ. Mutagen. 3, 401–419.

Nahin, R.L., Barnes, P.M., Stussman, B.J., 2016. Expenditures on complementary health approaches: United States, 2012. Natl. Health Stat. Rep. 95, 1–11.

Navarro, V.J., Lucena, M.I., 2014. Hepatotoxicity induced by herbal and dietary supplements. Semin. Liver Dis. 34, 172–193.

Nesslany, F., Simar-Meintieres, S., Ficheux, H., et al., 2009. Aloe-emodin-induced DNA fragmentation in the mouse in vivo comet assay. Mutat. Res. 678, 13–19.

Nozaka, T., Watanabe, F., Tadaki, S., et al., 1990. Mutagenicity of isoquinoline alkaloids, especially of the aporphine type. Mutat. Res. 240, 267–279.

NTP, 1992. Toxicology and carcinogenesis studies of quercetin (CAS No. 117-39-5) in F344 rats (feed studies). Natl. Toxicol. Program Tech. Rep. Ser. 409, 1–171.

NTP, 2001. NTP toxicology and carcinogenesis studies of EMODIN (CAS NO. 518-82-1) feed studies in F344/N rats and B6C3F1 mice. Natl Toxicol. Program Tech. Rep. Ser. 493, 1–278.

NTP, 2010. Toxicology and carcinogenesis studies of goldenseal root powder (*Hydrastis canadensis*) in F344/N rats and B6C3F1 mice (feed studies). Natl. Toxicol. Program Tech. Rep. Ser. 562, 1–188.

NTP, 2016. NTP botanical dietary supplements program. <https://www.niehs.nih.gov/health/materials/botanical_dietary_supplements_program_ntp_508.pdf> (Accessed 8 October 2020).

NTP, 2013a. Toxicology and carcinogenesis studies of *Ginkgo biloba* extract (CAS No. 90045-36-6) in F344/N rats and B6C3F1/N mice (gavage studies). Natl. Toxicol. Program Tech. Rep. Ser. 578, 1–183.

NTP, 2013b. Toxicology and carcinogenesis studies of a nondecolorized whole leaf extract of Aloe vera in F344/N rats and B6C3F1 mice. Natl. Toxicol. Program. Tech. Rep. Ser. 577, 1–266.

Paes-Leme, A.A., Motta, E.S., De Mattos, J.C., et al., 2005. Assessment of *Aloe vera* (L.) genotoxic potential on *Escherichia coli* and plasmid DNA. J. Ethnopharmacol. 102, 197–201.

Palmer, M.E., Haller, C., McKinney, P.E., et al., 2003. Adverse events associated with dietary supplements: an observational study. Lancet 361, 101–106.

Pandiri, A.R., Sills, R.C., Hoenerhoff, M.J., et al., 2011. *Aloe vera* non-decolorized whole leaf extract-induced large intestinal tumors in F344 rats share similar molecular pathways with human sporadic colorectal tumors. Toxicol. Pathol. 39, 1065–1074.

Park, M.K., Park, J.H., Kim, N.Y., et al., 1998. Analysis of 13 phenolic compounds in aloe species by high performance liquid chromatography. Phytochem. Anal. 9, 186–191.

Pasqual, M.S., Lauer, C.P., Moyna, P., et al., 1993. Genotoxicity of the isoquinoline alkaloid berberine in prokaryotic and eukaryotic organisms. Mutat. Res. 286, 243–252.

Peters, H., Kieser, M., Holscher, U., 1998. Demonstration of the efficacy of ginkgo biloba special extract EGb 761 on intermittent claudication--a placebo-controlled, double-blind multicenter trial. VASA. Z. fur Gefasskrankheiten 27, 106–110.

Petrovska, B.B., 2012. Historical review of medicinal plants' usage. Pharmacogn. Rev. 6, 1–5.

Rebecca, W., Kayser, O., Hagels, H., et al., 2003. The phytochemical profile and identification of main phenolic compounds from the leaf exudate of *Aloe secundiflora* by high-performance liquid chromatography-mass spectroscopy. Phytochem. Anal. 14, 83–86.

Resende, F.A., Vilegas, W., Dos Santos, L.C., et al., 2012. Mutagenicity of flavonoids assayed by bacterial reverse mutation (Ames) test. Molecules 17, 5255–5268.

Reynolds, T., 1985. Observations on the phytochemistry of the Aloe leaf-exudate compounds. Bot. J. Linn. Soc. 90, 179–199.

Rodriguez Rodriguez, E., Darias Martin, J., Diaz Romero, C., 2010. *Aloe vera* as a functional ingredient in foods. Crit. Rev. Food Sci. Nutr. 50, 305–326.

Rueff, J., Laires, A., Borba, H., et al., 1986. Genetic toxicology of flavonoids: the role of metabolic conditions in the induction of reverse mutation, SOS functions and sister-chromatid exchanges. Mutagenesis 1, 179–183.

Saccu, D., Bogoni, P., Procida, G., 2001. Aloe exudate: characterization by reversed phase HPLC and headspace GC-MS. J. Agric. Food Chem. 49, 4526–4530.

Sahebnasagh, A., Ghasemi, A., Akbari, J., et al., 2017. Successful treatment of acute radiation proctitis with *Aloe vera*: a preliminary randomized controlled clinical trial. J. Altern. Complement. Med. 23, 858–865.

Schweizer, J., Hautmann, C., 1999. Comparison of two dosages of *Ginkgo biloba* extract EGb 761 in patients with peripheral arterial occlusive disease Fontaine's stage IIb. A randomised, double-blind, multicentric clinical trial. Arzneimittel-Forschung 49, 900–904.

Sehgal, I., Winters, W.D., Scott, M., et al., 2013a. Toxicologic assessment of a commercial decolorized whole leaf *Aloe vera* juice, lily of the desert filtered whole leaf juice with aloesorb. J. Toxicol. 2013, 802453.

Sehgal, I., Winters, W.D., Scott, M., et al., 2013b. An in vitro and in vivo toxicologic evaluation of a stabilized *Aloe vera* gel supplement drink in mice. Food Chem. Toxicol. 55, 363–370.

Seo, J.E., Tryndyak, V., Wu, Q., et al., 2019. Quantitative comparison of in vitro genotoxicity between metabolically competent HepaRG cells and HepG2 cells using the high-throughput high-content CometChip assay. Arch. Toxicol. 93, 1433–1448.

Silva, I.D., Rodrigues, A.S., Gaspar, J., et al., 1997. Involvement of rat cytochrome 1A1 in the biotransformation of kaempferol to quercetin: relevance to the genotoxicity of kaempferol. Mutagenesis 12, 383–390.

Smith, J.V., Luo, Y., 2004. Studies on molecular mechanisms of *Ginkgo biloba* extract. Appl. Microbiol. Biotechnol. 64, 465–472.

Smith, T., Gillespie, M., Eckl, V., et al., 2019. Herbal supplement sales in US increased 9.4% in 2018. HerbalGram 123, 62–73.

Smith-Roe, S.L., Swartz, C.D., Shepard, K.G., et al., 2018. Black cohosh extracts and powders induce micronuclei, a biomarker of genetic damage, in human cells. Environ. Mol. Mutagen. 59, 416–426.

Soffar, S.A., Metwali, D.M., Abdel-Aziz, S.S., et al., 2001. Evaluation of the effect of a plant alkaloid (berberine derived from *Berberis aristata*) on *Trichomonas vaginalis* in vitro. J. Egypt. Soc. Parasitol. 31, 893–904.

Strickland, F.M., Muller, H.K., Stephens, L.C., et al., 2000. Induction of primary cutaneous melanomas in C3H mice by combined treatment with ultraviolet radiation, ethanol and aloe emodin. Photochem. Photobiol. 72, 407–414.

Syed, T.A., Ahmad, S.A., Holt, A.H., et al., 1996. Management of psoriasis with *Aloe vera* extract in a hydrophilic cream: a placebo-controlled, double-blind study. Trop. Med. Int. Health TM. & IH 1, 505–509.

Syed, T.A., Afzal, M., Ahmad, S.A., et al., 1997. Management of genital herpes in men with 0.5% *Aloe vera* extract in a hydrophilic cream: a placebo-controlled double-blind study. J. Dermatol. Treat. 8, 99–102.

Tredici, P.D., 2000. The evolution, ecology, and cultivation of *Ginkgo biloba*. Ginkgo Biloba. CRC Press, pp. 7–23.

Tsang, C.M., Cheung, K.C., Cheung, Y.C., et al., 2014. Berberine suppresses Id-1 expression and inhibits the growth and development of lung metastases in hepatocellular carcinoma. Biochim. Biophys. Acta 1852, 541–551.

Utesch, D., Feige, K., Dasenbrock, J., et al., 2008. Evaluation of the potential in vivo genotoxicity of quercetin. Mutat. Res. 654, 38–44.

van Beek, T.A., 2002. Chemical analysis of *Ginkgo biloba* leaves and extracts. J. Chromatogr. A 967, 21–55.

van Beek, T.A., Montoro, P., 2009. Chemical analysis and quality control of *Ginkgo biloba* leaves, extracts, and phytopharmaceuticals. J. Chromatogr. A 1216, 2002–2032.

van der Hoeven, J.C., Bruggeman, I.M., Debets, F.M., 1984. Genotoxicity of quercetin in cultured mammalian cells. Mutat. Res. 136, 9–21.

Vogler, B.K., Ernst, E., 1999. *Aloe vera*: a systematic review of its clinical effectiveness. Br. J. Gen. Pract. 49, 823–828.

Wang, J., Chen, W., Wang, Y., 2013. A *Ginkgo biloba* extract promotes proliferation of endogenous neural stem cells in vascular dementia rats. Neural Regen. Res. 8, 1655–1662.

Wang, L., Cui, J., Jin, B., et al., 2020. Multifeature analyses of vascular cambial cells reveal longevity mechanisms in old *Ginkgo biloba* trees. Proc. Natl. Acad. Sci. U S A 117, 2201–2210.

Ward, C.P., Redd, K., Williams, B.M., et al., 2002. *Ginkgo biloba* extract: cognitive enhancer or antistress buffer. Pharmacol. Biochem. Behav. 72, 913–922.

Weber, H.A., Zart, M.K., Hodges, A.E., et al., 2003. Chemical comparison of goldenseal (*Hydrastis canadensis* L.) root powder from three commercial suppliers. J. Agric. Food Chem. 51, 7352–7358.

Westendorf, J., Marquardt, H., Poginsky, B., et al., 1990. Genotoxicity of naturally occurring hydroxyanthraquinones. Mut. Res. 240, 1–12.

WHO, 2013. WHO traditional medicine strategy 2014-2023. <http://apps.who.int/iris/bitstream/10665/92455/1/9789241506090_eng.pdf?ua = 1> (Accessed 8 October 2020).

Williams, L.D., Burdock, G.A., Shin, E., et al., 2010. Safety studies conducted on a proprietary high-purity *Aloe vera* inner leaf fillet preparation, qmatrix. Regul. Toxicol. Pharmacol. 57, 90–98.

Yongchaiyudha, S., Rungpitarangsi, V., Bunyapraphatsara, N., et al., 1996. Antidiabetic activity of *Aloe vera* L. juice. I. Clinical trial in new cases of diabetes mellitus. Phytomedicine 3, 241–243.

Zeiger, E., Tice, R., 1997. Goldenseal (*Hydrastis canadensis* L.) and two of its constituent alkaloids. Integrated Laboratory Systems, <http://ntp.niehs.nih.gov/ntp/htdocs/chem_background/exsumpdf/goldenseal_508.pdf> (Accessed 8 October 2020).

Zhang, Y., Li, X., Zou, D., et al., 2008. Treatment of type 2 diabetes and dyslipidemia with the natural plant alkaloid berberine. J. Clin. Endocrinol. Metab. 93, 2559–2565.

Zhang, Z., Chen, S., Mei, H., et al., 2015. *Ginkgo biloba* leaf extract induces DNA damage by inhibiting topoisomerase II activity in human hepatic cells. Sci. Rep. 5, 14633.

Zhao, Y., Jing, Z., Li, Y., et al., 2016. Berberine in combination with cisplatin suppresses breast cancer cell growth through induction of DNA breaks and caspase-3-dependent apoptosis. Oncol. Rep. 36, 567–572.

Section VII

Regulatory aspects

Chapter 72

The regulatory framework for nutraceuticals: North America

Andrew Charrette, Najla Guthrie and Corey J. Hilmas

Introduction

The nutraceutical industry stands at a pivotal crossroads. The regulatory framework of nutraceuticals has been targeted for upgrades by federal agencies. Legislators are demonstrating greater interest in the commodity and ingredients in particular. On February 11, 2019, the US Food and Drug Administration's (FDA's) Commissioner Scott Gottlieb announced a "new plan for policy advancements with the goal of implementing one of the most significant modernizations of dietary supplement regulation and oversight in more than 25 years." US lawmakers have increased their engagement and dialog with the US FDA over the regulation of cannabidiol (CBD), one of the most popular ingredients found in nutraceuticals in the 21st century. In Canada, the Natural and Nonprescription Health Products Directorate (NNHPD) continues to progress plans in 2020 of further integrating the regulation of natural health products (NHP) with other "self-care" products, such as nonprescription drugs and cosmetics. Whether nutraceutical regulations tighten, or restrictions loosen for some ingredients, the regulation of nutraceuticals in North America is headed for change.

"Nutraceutical" is a generic but popular term that has not yet found its way into the laws of most countries. A "nutraceutical" is a coined term that refers to the use of food-based products, typically in pharmaceutical forms (i.e., capsules, tablets, powders, tinctures, etc.), to maintain health or to treat health-related conditions. A nutraceutical can be a functional food, dietary supplement, natural health product, or novel food, depending on the country of jurisdiction. "Functional foods" are nothing more than conventional foods (e.g., cereals, baked goods, juice drinks, frozen dinners, etc.) that have ingredients added to them for which health-related claims are made. Oat fiber, phytosterols, ginseng, lactobacilli, and soy by-products are examples of functional food ingredients that have been placed into conventional foods. The term "functional food" has no formal definition or understanding in the United States or Canada so we will discuss nutraceuticals in North America mostly in the context of dietary supplements, NHP, and novel foods. This chapter will discuss the regulatory framework for nutraceuticals in the United States first, followed by Canada. It provides a comprehensive roadmap for industry, regulatory professionals, healthcare professionals, and consumers to follow and better understand how this commodity is regulated in North America.

Regulation of nutraceuticals in the United States

When the Dietary Supplement Health and Education Act (DSHEA) of 1994 created a new category within the Federal, Food, Drug, and Cosmetic Act, it redefined the food–drug paradigm that stood for well over half a century. The passage of DSHEA reshaped heath regulation, struck a balance between FDA priorities and consumer access, and launched a category of food (dietary supplements) that would expand from \$4 billion to \$43 billion in a relatively short period of time. The success of DSHEA in catapulting the United States as a global leader in providing finished nutraceutical products domestically and abroad continues to guide decision-making over health and economic policy.

Under US law, there are no recognized "nutraceuticals" or "functional foods," or traditional medicines for that matter. There are "dietary supplements" which were defined by Congress for the first time in DSHEA over 25 years ago.

The Federal Food, Drug, and Cosmetic Act (FDCA) is the law of the land that governs foods, drugs, medical devices, and cosmetics in the United States. There are many state laws that also govern these products, but most states have adopted Title 21 of the Code of Federal

Nutraceuticals. DOI: https://doi.org/10.1016/B978-0-12-821038-3.00072-0

Regulations and therefore follow the FDCA lead. The FDCA was amended by DSHEA to add the category of dietary supplements to the existing food–drug paradigm. No state has passed a dietary supplement law, so that category is exclusively recognized only at the federal level.

Dietary ingredients for use in dietary supplements (nutraceuticals)

DSHEA defined the term dietary supplement for the first time as well as the types of ingredients that could be lawfully considered as dietary ingredients (see Fig. 72.1).

In order to illustrate this point, a bottled water could not be a dietary supplement if the only ingredient it contained was water. It would have to contain at least one dietary ingredient in order to be considered a dietary supplement for the US market, and water is not a dietary ingredient. Section 201(ff)(1)(E) has met with the most debate. Industry has viewed (E) as more of an intended "catch-all" by Congress to foster innovative ingredients, while FDA considers a narrower interpretation of this statute. It is not surprising to note that the vast majority of nonbotanical new dietary ingredient (NDI) notifications are submitted under this subcategory because it has the most ambiguity of all dietary ingredient subcategories.

The six elements for determining a dietary supplement

I. Intended for humans

There are six essential elements to consider when determining whether you have a dietary supplement. The first

DIETARY INGREDIENTS DEFINED

BY DSHEA

Within the meaning of 201(ff)(1), "the term 'dietary supplement' — (1) means a product (other than tobacco) intended to supplement the diet that bears or contains one or more of the following dietary ingredients:

(A) A vitamin;
(B) a mineral;
(C) an herb or other botanical;
(D) an amino acid;
(E) a dietary substance for use by man to supplement the diet by increasing the total dietary intake; or
(F) a concentrate, metabolite, constituent, extract, or combination of any ingredient described in clause (A), (B), (C), (D), or (E)."

FIGURE 72.1 Dietary Ingredients defined by DSHEA.

element is that it must be intended for human use. DSHEA does not on its face apply to pet foods or animal food. Products intended as "pet supplements" are a unique category all their own, are not considered "dietary supplements" as defined by the standard of identity in 21 CFR 101.3, and therefore not regulated by the Office of Dietary Supplement Programs within FDA's Center for Food Safety and Applied Nutrition. FDA regulates pet supplement products as "drugs of low regulatory priority" if the product does not pose a health or safety risk to the animal, is not marketed with disease claims, and the claims are not generally false and misleading. "Low regulatory priority" means that the Agency has not been active with enforcement actions in that regulatory space. In 1996, FDA's Center for Veterinary Medicine announced that DSHEA does not apply to finished animal products and therefore substances permitted in human dietary supplements may not be legally sold in animal supplements unless they provide nutritive value to the target diet of the intended species. While many dietary ingredients for use in dietary supplements are acceptable in humans, due to their extensive, documented historical use that can be used to establish safe levels for human consumption, the same does not hold for animals. Animal species may react very differently to substances like botanicals, and even small doses could cause adverse events after consumption. Each animal species has a unique set of nutritive requirements with significant differences in how they absorb and metabolize nutrients to lesser or greater extents. Therefore pets can be susceptible to toxic metabolites of ingredients that humans seem to handle more readily. For example, dogs are sensitive to chocolate or even Δ-9-tetrahydrocannabinol (Δ-9-THC). While most pet owners are careful not to feed chocolate to their dogs, they may be unaware of adverse events related to cannabinoids in cannabis- or hemp-derived pet supplement products (Meola et al., 2012). Very minute amounts of the cannabinoid Δ-9-THC can be found in CBD-containing pet supplement products.

II. Must contain a dietary ingredient (to supplement the diet)

The second element for a dietary supplement is that it must contain an ingredient that arguably could be used to supplement the diet (see Fig. 72.1 for allowable dietary ingredients). To further illustrate the bottled water example above, water does not qualify as a dietary ingredient under 201(ff). However, if a dietary ingredient like a vitamin or one from one of the other categories was added to the water supplement, then it could qualify as a dietary supplement. In fact, the FDA has opined on its belief that "bottled water" products are exclusive to conventional foods. Bottled water is defined in 21 CFR 165.110 as water that contains no added ingredients, except for optional ingredients (e.g., ones used for a technical effect) antimicrobial agents (e.g., ozone) or fluoride. If ingredients (minerals) or flavors are added, the resulting product could be considered a multicomponent liquid food product, which could be regulated as either a conventional food or dietary supplement. If a manufacturer called its product a "water beverage," it should bear a nutrition facts label as a conventional food. If the water contained a mineral or vitamin added, the manufacturer could use "Dietary Supplement" as the statement of identity for the product as minerals or vitamins are dietary ingredients for use in dietary supplements under the Federal Food, Drug, and Cosmetic Act (FFDCA). Dietary supplement products should bear a supplement facts label, which signals to the FDA and consumers that the use of the product is intended as a dietary supplement.

III. Form

Can nutraceutical foods in capsule or tablet form be conventional foods, dietary supplements, or either?

The third element of form appears to be clear and prescriptive, but ambiguities exist. A dietary supplement product should be in tablet, capsule, powder, softgel, gelcap, liquid, or some other form that is "not represented as conventional food … [or] for use as a sole item of a meal or of the diet." Tablets, capsules, powders, softgels, gelcaps, and liquids are certainly pharmaceutical forms, indicative of nutraceuticals. This element, however, means that products such as skin creams, nasal jelly, sublingual absorption products, and intravenous products can never be dietary supplements in the United States. At first glance, it may appear that these typical pharmaceutical forms delineated above are reserved for dietary supplements alone, and dietary supplements can only exist in these forms, but the statute says that it can be in some other form as long as it is not represented as a conventional food while it is intended for use as a softgel dietary supplement, for example. In other words, dietary supplements can be in any form, even looking like bars, cakes, and sodas, as long as they are not represented as conventional foods on the label. Representations about a product through labeling are a critical modifier for form.

These pharmaceutical forms are also apparently not restricted only to dietary supplement foods but can be used for conventional foods. The Agency has commented on whether a conventional food product can also be one of these other forms (e.g., tablet, capsule, powder, softgel, gelcap, liquid, etc.). In the FDA's food labeling statement of identity final rule of 1997, "dietary supplements may be similar to conventional foods in composition and form. Whether a product is a dietary supplement or a conventional food, however, will depend on how it is represented" (62 FR 49837, 1997). The FDA goes on to state that "[p]roducts that are not represented as dietary supplements will be subject to regulation as conventional foods." In other words, if a gelcap food product does not have a statement of identity as a "dietary supplement," it could be regulated by the FDA as a conventional food. The FDA usually considers all elements of labeling in making the determination that a product will be held to conventional food labeling and part 110 food current good manufacturing practices (cGMPs) or dietary supplement labeling and part 111 dietary supplement cGMPs. To better clarify, the FDA provides an example in their 1997 final rule to illustrate this point that it is up to the manufacturer to determine whether their product is sold as a conventional food or as a dietary supplement:

> *[T]he agency does advise that the decision whether a product is sold as a dietary supplement is made by the manufacturer … For example, the manufacturer of a product that is in the form of a tablet or capsule that has nutritive value or a powdered herbal product with no nutritive value may choose to market the product as a conventional food that bears nutrition labeling in accordance with § 101.9. In that situation, the nutrition labeling on the package of tablets with nutritive value would use the title 'Nutrition Facts,' while the herbal product with no nutritive value would be exempt from nutrition labeling under § 101.9(j)(4). Should the manufacturer choose to do this, however, the label or labeling could not represent the food as a 'dietary supplement,' and the product could not rely on any of the special provisions for dietary supplements that were added by the DSHEA.*

This last part means that if the manufacturer chose to label their tablet or capsule food product with "Nutrition Facts" as a conventional food, this food product is intended by the manufacturer to be sold as a conventional food and therefore would not be eligible for the exception for dietary ingredients from the definition of a "food additive," and the product could not bear statements under the authority of section 403(4)(6) of the Act. In other words, conventional foods, if intended to be sold by the manufacturer, could not include structure–function claims on labels or in labeling as these are only for dietary supplement products.

Another place where the FDA suggests that conventional foods or dietary supplement can be liquid, powders, tablets, capsules or other forms is in Title 21 of the Codified Federal Regulations. In accordance with 21 CFR 101.17(d) Protein products, FDA states that:

> *[t]he label and labeling of any food product in liquid, powdered, tablet, capsule, or similar forms that derives more than 50 percent of its total caloric value from either whole protein, protein hydrolysates, amino acid mixtures, or a combination of these, and that is represented for use in reducing weight shall bear the following warning:*

The only problem here with this warning is that conventional foods are not permitted to make weight reduction claims, which would fall under Section 403(r)(6) of the Federal FDCA. Weight reduction claims are structure–function claims, which are reserved for dietary supplements only (65 FR 1000, 2000). In the final analysis, the form of the food product (e.g., tablet, capsule, caplet, gelcap, powder, etc.) does not determine whether it is a dietary supplement or a conventional food. The product form as a unit dose (pill, capsule) does not dictate whether it is one (dietary supplement) and not the other (conventional food). The manufacturer must decide the intended use through the label and labeling of the product as either food categories can contain these forms.

Simulate versus represented as

Is the converse true? If conventional foods can be sold in forms that are typically associated with drugs and dietary supplements (e.g., capsules, tablets, powders, gelcaps, etc.), can dietary supplements take the form of conventional foods like cakes, brownies, candies, gummies, or cereal? Prior to DSHEA, the statute stated that food for humans, if not intended for ingestion in tablet, capsule, or liquid form, could neither *simulate* nor be *represented* as conventional food. This says if it looks, walks, quacks like a conventional food, it can only be a conventional food. DSHEA made one small change by amending Section 411(c)(1)(B) [21 U.S.C. 350(c)(1)(B)(d) Form of Ingestion] of the Federal FDCA by striking "does not simulate and" in clause (ii). This small change is not so insignificant. Prior to DSHEA, dietary supplement foods could not look like brownies, cakes, candies, or gummies or take the form of any conventional food product. Today, dietary supplements can look like these conventional food forms because they are permitted to "simulate" these forms. They just cannot be "represented as" these forms in statements and representations made about the product on labels and in labeling. In other words, a company could manufacture a dietary supplement brownie, they just couldn't call themselves a dietary supplement brownie because use of the term "brownie" or "cake" on a label or in labeling are conventional food terms, according to the FDA (FDA, 2011).

In the FDA's warning letter to the manufacturer of "Lazy Larry" products (formerly "Lazy Cakes"), the Agency considered the product to be a conventional food and not a dietary supplement, despite being labeled with a "dietary supplement" statement of identity in accordance with 21 CFR 101.3 and containing nutrition information within a "Supplement Facts" label. The product contained the ingredient melatonin, which was permitted as a dietary ingredient for use in dietary supplements. The FDA held that the product was a conventional food and therefore applied the unapproved food additive statute under section 409 of the Federal FDCA [21 U.S.C. § 348] and within the meaning of 21 CFR Part 170 to melatonin. FDA held that "Lazy Larry" products were conventional foods not because they looked like a brownie but that they used the word "brownie" in the statement of identity on the package label and referred to the product using the word "cakes." This pitfall shows that when a dietary supplement makes representations as a conventional food through statements on the label or in labeling, use of those terms can cause the product to be evaluated by the Agency as a conventional food. In turn, it can cause a permissible dietary ingredient for use in dietary supplements to be evaluated as an unapproved food additive. Furthermore, the Agency has considerable case law behind it when the unapproved food additive charge is applied.

IV. Oral ingestion in the alimentary canal

The fourth element for a dietary supplement is that it must be taken into the alimentary tract by mouth and swallowed. Topical creams, transdermal patches, nasal sprays, and some sublingual products would never qualify as a dietary supplement because they would not pass this requirement. Supplements must be ingestible, regardless of form, meaning taken through the mouth and swallowed for entry into the digestive tract. Only orally administered products could qualify as dietary supplements. Nasogastric tube administration of nutrition, gastric or duodenal (e.g., feeding tube) feeding, intravenous, intramuscular, subcutaneous, percutaneous, rectal, vaginal, ocular, otic, nasal applications, nasal/lung aerosol for inhalational administration, cutaneous/topical, and transdermal would not qualify as ingestion; therefore dietary supplement-labeled products with directions for use involving these routes of administration could cause the food product to be an unapproved new drug. Among the specialized oral routes outside of general food ingestion via the oral cavity, there is also buccal (relating to the cheek), sublingual (below the tongue), glossal (tongue strip), and oral spray delivery. These specialized routes could qualify as ingestion into the alimentary canal for dietary supplements because they include oral delivery; however, the directions of use for such a dietary supplement product would need to be accompanied by language and terminology that the material is swallowed for oral ingestion. Even gums can be a dietary supplement as gums are nothing more than chewable resins containing dietary ingredients that are administered to the oral cavity and swallowed for consumption. In those cases, the gum resin, while not ingested, is simply a delivery form for the oral ingestion of the actual dietary ingredients that are swallowed during the course of product mastication. Gums are also defined as food by US statute.

There is 1995 case law (United States vs TEN CARTONS), which established that dietary supplements had to be "administered through the mouth and esophagus into the alimentary canal for digestion and absorption" (E.D.N.Y., 1995). This litigation involved Ener-B nasal gel, which directed consumers to swab the substance, which contained vitamins, into the nasal cavity. The courts found that nasal administration, rather than oral administration, of vitamins and minerals caused the product to be a drug and not a food (dietary supplement). Therefore route of administration is a major determinant of whether a product can qualify as a dietary supplement, a category of food, or whether the product should be regulated by the Agency as a drug.

In United States versus TEN CARTONS, Hon. Arthur Spatt provided the court's opinion on the definition of a "substance." In section 201(ff)(1)(E) of the FDCA, a dietary ingredient can include a dietary substance for use by man to supplement the diet by increasing the total dietary intake. Hon. Spatt wrote that the "circuit has recognized that the FDA's determination regarding classification of a substance as a 'food' or a 'drug' under the Act is generally to be accorded such deference." The term "dietary" means of or relating to diets and diet typically refers to the usual or habitual consumption of food or drink of man or the kinds and amounts of food available to or eaten by an individual, group, or population. In nutrition, the diet often implies the use of specific intake of nutrition for health. Taken together, a dietary substance in the United States can be a habitually consumed food or drug consumed by an individual, group, or population, as long as the "drug" mentioned here is not a US drug. However, if someone was consuming a traditionally prepared medicine that is not a US drug, then it could be considered a dietary substance. In other words, drugs or traditional medicines habitually consumed for health purposes in countries outside the United States could be considered a dietary substance for use by man in the United States as long as they were not already considered a US drug. If such a substance was considered a drug in the United States, it would not be considered a food or dietary ingredient for use in dietary supplements unless it was a dietary ingredient prior to it being a drug. This is the race-to-market provision.

V. Race-to-market provision

The fifth element is called the race-to-market provision. If a product was not a new drug or biologic prior to being marketed as a dietary supplement or food in this country, then it can qualify as a dietary supplement. Stated another way, if a drug company can demonstrate that its product was first approved, or subject to an investigational or new drug application for which "substantial clinical investigations" were instituted and made public, the product cannot be sold as a dietary supplement. Conversely, if the product was on the market first as a dietary supplement or food, or was grandfathered as an old dietary ingredient by the implementation date of DSHEA (October 15, 1994), then it is permitted to remain as a dietary supplement or dietary ingredient for use in a dietary supplement, even if a new drug with the same active ingredient enters later. An example of a compound with dual status as both a dietary ingredient and a drug is galantamine (galanthamine).

The reason Congress inserted the race-to-market provision is the same reason Congress is having to move carefully at creating a pathway for CBD as a dietary ingredient today. Congress wishes to retain incentives for companies to engage in expensive safety and efficacy research to bring important drug products to market. If a dietary supplement company were permitted to sell the same ingredient being sold as a drug, which underwent expensive clinical investigations before FDA approval, there would be no incentive to drug development. When a pharmaceutical company obtains FDA approval and statutory exclusivity after years of testing, and if the dietary supplement or food product was not on the market first, the exclusivity must be incentivizing. The false belief here is that there are no cost burdens to dietary supplement manufacturers. On the contrary, NDI safety dossiers for novel dietary ingredients entering the market can cost anywhere between \$500,000 to \$1 million. Studies of efficacy to substantiate structure–function claims made on dietary supplements, a requirement for both the FDA and the Federal Trade Commission (FTC), are also costly. Currently, there is no incentivization to dietary supplement companies like there are for drug manufacturers. Agency enforcement will be a critical component to maintaining consumer confidence in dietary supplements, regarding both safety and efficacy, in the foreseeable future. The FDA will need to develop new strategies to encourage NDI notifications, the only premarket gate for dietary supplements, through enforcement against companies relying on the intellectual property of other's acknowledged NDIs displayed in the public docket. Similarly, FTC enforcement of all companies, even those earning less than \$100 million annually and start-ups, will help to level the playing field.

VI. Labeling statements and claims as intended use

Statement of identity

A sixth element is whether the product contains representations, through the label or in labeling (see Fig. 72.2), indicative of intended use as a dietary supplement, unapproved new drug, or conventional food. Whether a product is a dietary supplement, or a conventional food largely depends on how it is represented.

Principle Factors of Marketer's Intent

Claims Made in Labeling

Labeling includes the label or package, as well as promotional pieces, brochures, flyers, package inserts, accompanying "shelf-talkers," and internet claims (if products are offered for sale on the website). The term "labeling" extends to all materials that physically accompany the product in an integrated distribution scheme in interstate commerce. The key point here is that when the material physically goes with the product during its interstate journey and/or is with the product at the point of retail sale, it constitutes labeling. FDA has primary jurisdiction over claims made in labeling.

Claims Made in Advertising

Advertising, although not explicitly defined by law, refers to promotional material that does not physically accompany the product, such as print advertisements in magazines, radio programs, claims made about products at conference presentations, or internet claims if the manufacturer or marketer controls the content and products are not offered for sale on the website. FTC has primary jurisdiction over advertising claims for foods, dietary supplements, and nonprescription drugs, but FDA will often use advertising with labeling *en toto* to determine the overall intended use of the manufacturer or marketer.

FIGURE 72.2 Principle Factors of Marketer's Intent.

To be a dietary supplement, a product must bear the statement of identity term "dietary supplement"[1] on the principal display panel as part of its common or usual name. Another aspect of food labeling to show intended use is choice of the heading "Nutrition Facts" or "Supplement Facts" for the nutrition label. The former is reserved for conventional foods, while the latter denotes a dietary supplement product. The statement of identity on the principal display panel of the finished food product is a major determinant of intended use. If the product states "Dietary Supplement" in accordance with 21 CFR 101.3, this is clear evidence that the product is intended to be used as a dietary supplement. If a dietary supplement product declares the less clear statement of identity "Food Supplement" or contains a "Nutrition Facts" label, this leaves the intended use of the product in limbo and in the hands of federal and state regulators to decide whether it is a dietary supplement or conventional food. Again, this decision has dire consequences if the finished product contains an acceptable dietary ingredient that would otherwise be an unapproved food additive when evaluated as a conventional food (e.g., melatonin).

Unlike products regulated in most other countries, FDA-regulated products are evaluated and categorized based on their intended use. While there are some exceptions, any company considering the introduction of a dietary supplement should decide its marketing plans first when determining whether the product will meet DSHEA requirements. If the product fails the claims review process, it can be seized as an unapproved new drug, an unapproved food additive (Agency belief that the product is intended as a conventional food and not a dietary supplement), or a misbranded conventional food, supplement, cosmetic or other product.

The wording of a claim on a product label will always be taken in a greater context from all other claims made in labeling and advertising about the product. For example, if a claim states "to help regulate or maintain blood sugar," this claim may be a gray area for the FDA. However, other claims provide further clarification as to how the product is intended for use by the marketer. If the product also claims in labeling or advertising "to modulate hemoglobin A1C (HgA1C) levels," the claim statements in totality cross the line as disease claims. The product is intended to lower blood glucose and lower HgA1C levels, a clinical marker of disease used by physicians to diagnose and monitor disease progression in both prediabetics and diabetics. The FDA and FTC never evaluate the claim text by itself as to how it can be interpreted by the reasonable consumer. Both federal agencies will evaluate the explicit as well as implicit meanings of the claim made in the context of other claims made about the product.

The FDA and FTC have separate jurisdictions, but they also collaborate through a Memorandum of Understanding (MOU 225-71-8003) to prevent injury and deception of the consumer (see Fig. 72.2). The FDA's jurisdiction is predominantly to evaluate food labels but will also evaluate labeling and advertising to decide whether claims made by the marketer are permissible structure–function claims, other acceptable statements of nutritional support, unapproved disease claims. The FTC conducts investigations to evaluate evidence provided by the marketer as to whether the claim is unfair, deceptive, or fraudulent. The FTC evaluates evidence to the substantiation standard of "competent and reliable scientific evidence." The FTC has subpoena authority to compel marketers to provide their evidence of meeting the substantiation standard. The FDA typically sends warning letters or courtesy letters, if a marketer crosses the line of making disease claims as their first step toward giving prior notice. While warning letters are deemed to be administrative actions, they are not final Agency actions of enforcement; however, the FDA has seized supplement products just on the basis of their disease claims without

1. This term may be modified to include the name or general description of the types of ingredients in the dietary supplement product.

providing a warning letter. The FTC can take enforcement action over a claim whether it is a structure–function claim or a disease claim. This Commission conducts investigations, sues companies and people violating the law, and educates consumers and businesses about their rights and responsibilities. The two agencies work together, especially in cases where a marketer makes an unsubstantiated disease claim. The FTC may take enforcement action more readily in these cases because it may result in greater harm as an unfair, deceptive, or fraudulent claim to consumers.

Statements of nutritional support, health claims, and nutrient content claims

Statements of nutritional support: structure–function claims.

DSHEA authorized statements of nutritional support to be made for dietary supplements (see Fig. 72.3). DSHEA added section 403(r)(6) to the FDCA Act, which states

SPOTLIGHT:
STATEMENTS OF NUTRITIONAL SUPPORT

Statements of Nutritional Support

DSHEA established regulated boundaries for structure/function claims and two related types of dietary supplement labeling claims, (claims of general well-being and claims related to nutrient deficiency disease.

Structure/function claims describe the role of a nutrient or dietary ingredient intended to affect the normal structure or function of the human body, for example, "calcium builds strong bones." They may characterize the means by which a nutrient or dietary ingredient acts to maintain such structure or function, for example, "maintains bowel regularity."

General well-being claims describe an overall support for a specific organ or whole-body health from consumption of the nutrient or dietary ingredient, for example, "supports heart health."

Nutrient deficiency disease claims describe or offer a benefit for specific disease that are related to lack of a nutrient in the diet (vitamin C for scurvy, niacin for pellagra, iron for iron-deficiency anemia). Marketers must be careful not to generalize the nutrient deficiency claim to include other areas which fall outside the boundaries of this claim to other forms of anemia. There are many more disease-related causes of anemia, which would not be permissible under the Act.

FIGURE 72.3 Statements of Nutritional Support.

that statements may be made for dietary supplements (nutraceuticals) if:

> *[t]he statement claims a benefit related to a classical nutrient deficiency disease and discloses the prevalence of such disease in the United States, describes the role of a nutrient or dietary ingredient intended to affect the structure or function in humans, characterizes the documented mechanism by which a nutrient or dietary ingredient acts to maintain such structure or function, or describes general well-being from consumption of a nutrient or dietary ingredient.*

These were referred to as "statements of nutritional support" in the proposed and final rules (60 FR 67176, 1995; 62 FR 49859, 1997). Today, they are known more by the term structure–function claims. The manufacturer must also have substantiation that the statement is truthful and not misleading [section 403(r)(6)(B)], and the statement must display the FDA disclaimer "This statement has not been evaluated by the Food and Drug Administration. This product is not intended to diagnose, treat, cure, or prevent any disease." All statements of nutritional support, while not preapproved by the FDA, must have substantiation that the claim is truthful and not misleading. Marketers must also submit a 30-day notification to the FDA's e-portal with the text of the claim no later than 30 days after marketing the dietary supplement with the claim. If a dietary supplement label includes such a claim, it must state in a "disclaimer" that the FDA has not evaluated the claim (see below).

Dietary supplements and conventional foods contain fundamental differences regarding structure-function claims. While both categories of food can make these claims, there are key differences. Conventional foods can make structure–function claims as long as they focus on effects derived from taste, aroma, or nutritive value. Dietary supplements may make structure–function claims that focus on nonnutritive or nutritive effects. The FDA reaffirmed this point about structure–function claims for conventional foods in the September 23, 1997 (62 FR 49859), final rule entitled "Food Labeling: Nutrient Content Claims, Health Claims, and Statements of Nutritional Support for Dietary Supplement." As explained in that rule (62 FR 49859 at 49860, 49861, and 49864), the fact that structure–function claims for conventional foods are limited to effects derived from nutritional value, while structure–function claims for dietary supplements are not, is a result of differences in the language of the exemption for foods in section 201(g)(1)(C) of the Act, as interpreted by the courts ("Nutrilab, Inc. v. Schweiker, 713 F.2d 335," 1983). Furthermore, the FDA does not require conventional food manufacturers to notify them about their structure–function claims (related to nutritive effects), and disclaimers are not required for claims on conventional foods. Dietary supplements, on the other hand, must submit 30-day structure–function notifications to the FDA regarding the exact text of the claim made in labeling and advertising. Dietary supplements must also place the disclaimer (see above) on every page that a claim is made in labeling and advertising as well as the physical label.

On January 6, 2000, more than 5 years after the passage of DSHEA, the FDA issued their structure–function claim final rule (65 FR 1000, 2000), and the rules governing these claims have been codified in Title 21 of the Code of Federal Regulations Section 101.93. The final rule title, which mentions only "structure" and "function," ironically does not tell marketers how to craft an acceptable structure–function claim. It merely defined the term "disease" for the first time. "Disease" is defined in 21 CFR 101.93(g) as "damage to an organ, part, structure or system of the body such that it does not function properly" or "state of health leading to such dysfunction." The final rule also describes the universe of claims that fall into the category of disease claims using 10 criteria (see Fig. 72.4) (CFR, 2019).

In other words, claims made about body structure or function that do not mention a disease or imply an effect of a disease are permissible structure–function claims. While many believe disease claims and drug claims are synonymous, they are not exactly the same because of the types of claims permitted as structure–function claims. For example, a drug purporting to treat male baldness would be a drug claim. A dietary supplement product designed to mitigate baldness, as long as it is substantiated, could be an acceptable structure–function claim. This conundrum arises because baldness is not a disease. Claims for occasional or mild gastrointestinal symptoms (acid indigestion), sleeplessness, weight control, and laxation are also classified as drug claims under certain over-the-counter (OTC) monographs as well as permissible structure–function claims. This blurring of the distinction between drugs/supplements allows flexibility for marketers but also makes regulatory classification of some products murkier, for example, sleep aids and weight control products.

Finally, structure–function claims have additional requirements. They must be truthful and not misleading. They must not be unfair or deceptive in advertising to consumers. Marketers are required to have efficacy data to support the claim, in accordance with the FTC and FDA substantiation standard of "competent and reliable scientific evidence." Claims must be accompanied by a two-part disclaimer (see Fig. 72.5), within the meaning of 21 CFR 101.93. Marketers are also subject to the 30-day notification provision whereupon they must submit the complete text statement of the claim made through the electronic portal system at the FDA for receiving 30-day notifications. All claims must be submitted to the FDA

SPOTLIGHT:
FDA's TEN CRITERIA FOR DISEASE CLAIMS

Ten Criteria

A statement is a disease claim if it:

I. Has an effect on a specific disease or class of diseases (includes explicit and implied diseases);
II. an effect on the characteristic signs of symptoms of a specific disease or class of diseases, using scientific or lay terminology;
III. an effect on an abnormal condition associated with a natural state or process (claimed effect on menstruation is a structure/function claim, claimed effect on menorrhagia is a disease claim);
IV. an effect on a disease through:
    a. a product name;
    b. statements about a product formulation or ingredient FDA regulates as a drug;
    c. a citation of a publication or reference, if the citation refers to a disease use and if it implies treatment or prevention of a disease
    d. use of the term disease or diseased, except general statement about disease prevention, which are OK
V. belongs to a class of drugs;
VI. is a substitute for a product that is a therapy for a disease;
VII. augments a therapy or drug action;
VIII. has a role in the body's response to a disease or disease vector;
IX. treat, prevents, or mitigates adverse events;
X. otherwise suggests an effect on a disease or diseases.

FIGURE 72.4 FDA's Ten Criteria for Disease Claims.

SPOTLIGHT:
REQUIREMENTS FOR STRUCTURE-FUNCTION CLAIMS

I. Adequate Substantiation
    a. Truthful and not misleading
    b. Substantiation Standard of FTC and FDA
    c. "Competent and Reliable Scientific Evidence
II. Every structure function claim must be accompanied by the following 2-part disclaimer statement of 21 CFR 101.93.
    a. "This statement has (or .... These statements have) not been evaluated by the Food and Drug Administration.
    b. "This product is not intended to diagnose, cure, mitigate, treat or prevent any disease."
III. 30-Day Notification Provision
    a. Submission to FDA's e-portal with text of the claim
    b. Company must certify that they have adequate substantiation to make the claim

FIGURE 72.5 Requirements for Structure-Function Claims.

within 30 days of going to market with that statement made on labels, labeling, or advertising.

Health claims

Structure–function claims are only one subset of claims that can be made for products. See Fig. 72.6 for a listing of other claims. Health claims are a special subset of disease claims that have been reviewed and authorized by the FDA to be made on food products or dietary supplements with the following caveat. They must involve risk reduction. They are used to indicate that a food or food component may reduce the risk of a disease or health-related condition. Such claims require scientific evidence support to characterize the relationship between a substance (specific food component or food) and a disease (e.g., diabetes) or health-related condition (e.g., high blood pressure). The Nutrition Labeling and Education Act (NLEA) directed the FDA to issue regulations providing for use of health claims, which undergo review by the FDA through a formal citizen petition process. The petition allows for a notice and comment period from stakeholders prior to the Agency amending the codified federal regulations to allow for the health claim. In order to receive authorization to make a US health claim, there must be significant scientific agreement (SSA) among qualified experts that the claim is supported by the totality of publicly available scientific evidence for a substance/disease relationship. The SSA standard is intended to be a strong standard that provides for a high level of confidence in the validity of the substance/disease relationship. There are 12 published health claims on food components for the reduction in risk of diseases and health-related conditions, including osteoporosis, cancer, coronary heart disease, dental caries (cavities), neural tube defects, and hypertension. SSA typically requires evidentiary support from two randomized controlled clinical trials.

SPOTLIGHT:

TYPES OF CLAIMS MADE ON US PRODUCTS

I. Structure function claims
   a. Effect on structure of the body (e.g., mechanism of action claim to bind unique proteins in the body)
   b. Effect on a function of the body
II. Classic nutrient deficiency disease states
III. General well-being claims
IV. Nutrient content claims
V. Health claims
VI. Qualified health claims
VII. Claims for safety
VIII. Comparative claims
IX. Scientific level of support claims (e.g., clinically proven/tested, scientific experts agree, recommended by scientists, studied for years, subject of university research)
X. Made in the USA Claims
XI. All natural/100% natural claims

FIGURE 72.6 Types of Claims Made on US Products.

Qualified health claims

One can also petition the Agency to make a qualified health claim (QHC). QHCs are supported by scientific evidence but do not meet the more rigorous SSA standard of a health claim. QHCs are sought when manufacturers or marketers petition the Agency to consider the use of enforcement discretion over a QHC. As such, QHCs are not approved, and the process, unlike health claims, does not involve rulemaking; however, the QHC process does involve the FDA's petition process. The FDA will write a Letter of Enforcement Discretion using very prescriptive language reflecting the overall level of supporting scientific evidence for the claim. Rather than amending the CFR, QHC Letters of Enforcement Discretion are posted to the FDA's website.

Quality manufacturing standards

The FDA released the dietary supplement cGMP final rule on June 25, 2007. The rule provided for rigorous testing, reporting, and record-keeping requirements in relation to quality. The cGMP final rule for US dietary supplements was patterned after pharmaceutical cGMPs and not food cGMPs. It requires dietary supplement manufacturers and own label distributors to test for identity, purity, strength, and composition of their dietary supplements. The details of the cGMP final rule are outside the scope of this primer but can be reviewed elsewhere in the Federal Register (FDA, 2007).

Safety

Nutraceuticals in the United States have premarket regulations for safety, despite claims to the contrary. In fact, DSHEA created greater safety protection, not less, for the public with few safety sentinel events occurring when one compares the number of years this law has been in effect. The issue has been a lack of enforcement on dietary ingredients that are considered NDI requiring notification. The example of aegeline highlights the need for more NDI enforcement. In the case of aegeline, the statute's NDI requirement was already in place to protect consumers against the introduction of a NDI without first demonstrating reasonable expectation of safety to FDA.

Aegeline was introduced into the market without filing an NDI notification with the FDA. It was found to induce liver disease post-market (Klontz et al., 2015). In fact, an acute toxicity study indicated significant mortality observed after administration of 10 × and 3 × mouse-equivalent doses (MED) of aegeline product. Increases in liver/body weight ratios as well as elevations in alanine aminotransferase and aspartate aminotransferase were observed in female $B6C3F_1$ mice after 2 × and 1.5 × MED. Similar findings were also observed in a subchronic 90-day feeding study (Miousse et al., 2017).

Old dietary ingredients

Enforcement of companies failing to file NDIs with the FDA through use of the FDA's NDI adulteration provision in the FFDCA is critical to safeguarding the industry, food supply, and consumers. The FDA does not have a good sense of what dietary ingredients in the marketplace are old dietary ingredients (ODIs), also called grandfathered, pre-DSHEA dietary ingredients. These ODIs are exempt from the NDI premarket notification requirement. The task of regulators to monitor all products and ingredients entering the marketplace and the determination as to which ingredient is an NDI or ODI is a significant challenge. Currently, the statute requires companies to perform due diligence on the safety of the ingredients. Companies should have independent and verifiable evidence in the form of physical product labels, including a listing of their ingredients, sold in interstate commerce prior to October 15, 1994, to assert that they are ODIs and therefore exempt from notification. The frustration and challenge to federal regulators lies in the burden placed on the Agency to demonstrate that products are unsafe.

If the FDA determines that a dietary supplement or nutraceutical poses a safety issue, the government may seize the product or seek an injunction against the company to prohibit any further manufacturing, using any of the following provisions:

1. General food safety standard (e.g., if it bears or contains any poison or deleterious substance which may render it injurious to health);

2. The FDA believes that the nutraceutical "presents a significant or unreasonable risk of illness or injury";

3. The product contains a "new dietary ingredient" or one marketed after October 15, 1994, and the FDA has not been provided adequate information to "reasonable assure" that the ingredient does not present a significant or unreasonable risk of illness or injury;

4. The product poses an "imminent hazard to public health or safety" such that the Secretary of the Department of Health and Human Services imposes an administrative ban;

5. The nutraceutical contains a food additive (filler, coating, binder, or other added ingredient/excipient)—not a dietary ingredient—that is unapproved for such use, not generally recognized by qualified experts as safe for its intended use (GRAS), or is otherwise adulterated (e.g., use of pesticides with zero tolerance limits, or use of a stevia form that is not one of the Joint FAO/WHO Expert Committee on Food Additives approved varieties accepted by the US FDA);

6. Failure to meet current good manufacturing practices (GMP) regulations for dietary supplements. This is a technical adulteration standard, meaning that while the food did not contain an adulterant per se, there was a technical adulteration or deficiency, in accordance with cGMP regulations, in how it was found to be manufactured.

Failure to demonstrate substantiation on product claims made for safety may also come into play as well as product liability/punitive damage laws.

New dietary ingredients

The NDI notification process is the lone premarket safety gate and safeguard for nutraceuticals. It can be thought of as a prescreen for new ingredients that have no history of use in the US market. The statute requires the filing of an identity and safety dossier (NDI notification) if a dietary ingredient contained in the nutraceutical dietary supplement is "new," meaning it was not marketed and sold in the United States before October 15, 1994. This filing must be made 75 days before marketing a nutraceutical supplement containing the NDI. The filing date is the date at which the FDA receives the notification, not the date at which the notifier submitted the notification. The notifier must submit to the FDA all of the safety data to support the notifier's (contract manufacturer, distributor, ingredient supplier) determination that the ingredient "will reasonably be expected to be safe based on history of use or other evidence of safety." Approximately half of FDA inadequate letters to notifiers are based upon not having adequate evidence to establish safety of the serving level proposed in the NDI. More specifically, 25% are due to identity issues alone, 25% due to safety issues alone, and 50% are the result of a combination of failures to establish identity and safety within the notification.

NDI notifications are as much about identity as they are safety. If history of use is used to support the safety of the NDI in the supplement, notifiers must conduct chemistry studies to establish the identity of the historically used materials and document serving levels provided in historical use. Otherwise the application of historical use or safety study found in a paper or recipe may not quantitatively or qualitatively be related to the substance in the notification. In order to adequately use historical use, the chemical composition of historically consumed ingredients or components must be compared to the composition and other identifying characteristics of the NDI. Identity is a critical component of NDI notifications and failures to adequately establish identity are the cause of half of the inadequate FDA response letters to notifiers.

The FDCA provides that a dietary supplement containing a NDI is considered adulterated under section 402(f) (21 U.S.C. 342(f)) unless it meets one of the two requirements.

1. The dietary supplement contains only dietary ingredients which have been present in the food supply as an article used for food in a form in which the food has not been chemically altered; or
2. There is a history of use or other evidence of safety establishing that the dietary ingredient when used under the conditions recommended or suggested in the labeling of the dietary supplement will reasonably be expected to be safe and, at least 75 days before being introduced or delivered for introduction into interstate commerce, the manufacturer or distributor of the dietary ingredient or dietary supplement provides the FDA with information, including any citation to published articles, which is the basis on which the manufacturer or distributor has concluded that a dietary supplement containing such dietary ingredient will reasonably be expected to be safe [FDCA 402(f), 2006; FDCA 413, 2006].

If the dietary ingredient has not been present in the food supply as an article used for food in the same chemical form that a manufacturer or distributor plans to use in the dietary supplement (e.g., an alcohol extract of botanical A is chemically different from a supercritical CO_2 fluid extract of botanical A), they are required to demonstrate reasonable expectation of safety under the conditions of use recommended or suggested in the labeling.

The FDA published administrative requirements on September 23, 1997, to be followed by notifiers submitting at least 75 days before introducing the NDI into interstate commerce (62 FR 49886, 1997). The requirements detailed in the codified regulations 21 CFR 190.6 are detailed in Fig. 72.7.

SPOTLIGHT: REQUIREMENTS FOR NDI NOTIFICATION

- Name and complete address
- Name of NDI, including Latin binomial name (including author) for herbs or other botanicals
- Description of the dietary supplement(s) that contains the NDI, including serving level of the NDI in the product, conditions of use recommended or suggested in labeling, ordinary conditions of use
- History of use OR other evidence of safety, including full copy reprints or photocopies of any references to published materials, and full English translations of any foreign language references and publications.
- The signature of the person designated who is responsible for the content of the notification if there are questions.

FIGURE 72.7 Requirements for NDI Notification.

Postmarket surveillance—serious adverse event reports

US dietary supplements are the only commodity of food for which there is mandatory reporting of serious adverse events (SAEs) to the FDA. The Dietary Supplement and Nonprescription Drug Consumer Protection Act, signed into law on December 22, 2006, requires the reporting of "serious" adverse events for both OTC drugs and dietary supplements. An adverse event is "any health-related event associated with the use of a dietary supplement that is adverse" in accordance with Section 761(a)(1) of the FFDCA. An SAE is "an adverse event that results in death, a life-threatening experience, in-patient hospitalization, a persistent or significant disability or incapacity, or a congenital anomaly or birth defect; or requires, based on reasonable medical judgment, a medical or surgical intervention to prevent an outcome" in accordance with Section 761(a)(2). Dietary supplement manufacturers or distributors who release products into interstate commerce must keep adverse event reports and maintain those records for a minimum of 6 years and submit all SAE reports to the FDA.

Competent medical authorities and healthcare practitioners play a vital role in this process. Companies must designate a company name and either a full US domestic address or domestic phone number on the label as the "responsible person," who receives reports of SAE with a dietary supplement, in accordance with Section 403(y) of the FFDCA. SAE reports must be completed and submitted using the FDA's MedWatch form within 15 days of receiving such report or information. Determination of whether or not a report is serious should be accomplished by a competent medical authority or healthcare practitioner. These individuals are not only responsible for obtaining more information from the patient to determine if the event is serious but also to report any SAEs to the FDA within 15 days. The law also requires companies to obtain additional follow-up information for up to a year from the initial SAE report filed to the FDA. Any new medical information obtained upon follow-up must be submitted to the FDA.

Regulation of nutraceutical foods for pets

In the United States, an animal food (feed) ingredient is defined as a component part, constituent, or any combination/mixture added to and comprising animal food. Animal food ingredients might include grains, milling by-products, added vitamins, minerals, fats/oils, and other nutritional and energy sources. Animal foods provide a practical outlet for plant and animal by-products not suitable for human consumption. Under the FDCA, any substance added to or expected to become a component of animal food, either directly or indirectly, must be used in accordance with a food additive regulation unless it is generally recognized as safe for that use (GRAS).

The term "food" as defined in Section 201(f) means (1) articles used for food or drink for man or *other animals*, (2) chewing gum, and (3) articles used for components of any such article, which describes food additives. The Food, Drug, and Cosmetic Act further requires that animal feeds be pure and wholesome, contain no harmful or deleterious substances, and be truthful on labels. The substances permitted and sold in animal feeds, including animal and pet supplements, include (1) common food ingredients, such as water, salt, and pepper; (2) ingredients listed as GRAS in 21 CFR parts 582 or 584; (3) ingredients listed as approved food additives in 21 CFR part 573 and 579 (Center for Veterinary Medicine, 2020); or (4) approved color additives for food in 21 CFR parts 73, 74, or 81. The FDA has also agreed to exercise enforcement discretion to permit defined ingredients in the Official Publication of the Association of American Feed Control Officials (AAFCO book)[2] to be included in animal food, so long as they are used for taste, aroma, or nutritive value (Nutrilab, 1983) and there are no safety concerns about their use or ingredient composition (Office of the Commissioner, 2019).

Enforcement of pet food products is handled at the state level, which unfortunately relies on the patchwork of state laws. Some states are more permissive when it comes to ingredients not included on the FDA lists in 21 CFR.

Regulation of nutraceuticals in Canada

In the major nutraceutical markets of the United States and Europe, many ingredients are regulated as an extension of foods with limited health claims permitted. Their primary classification is based on the notion that these products are intended to supplement the diet of the consumer by providing additional nutritional value or another benefit associated with the diet. Nutraceutical products making health claims outside of those permitted for foods are generally considered drugs. The framework by which Health Canada regulates nutraceuticals takes a different approach by establishing a standalone category in which most nutraceutical products fall: NHPs. Food, Drugs, and NHPs are subject to the *Food and Drug Act (FDA)* and *Regulations (FDR)* (Legislative Services Branch, Government of Canada, 2019a,b). Like food regulation in the United States and Europe, limited health claims are permitted for foods in Canada. NHPs on the other hand, are classified as an extension of drugs and are required to make at least one health claim on the label. This is because rather than serving a purpose associated with food, such as providing caloric content or addressing hunger, NHPs are intended to convey a specific health benefit to the user. NHPs range from low-risk health products meant for self-care to higher risk products that require the intervention of a healthcare practitioner.

NHPs are defined as substances included in Schedule I of the *Natural Health Product Regulations (NHPR)*; vitamins and minerals, herbal remedies, homeopathic medicines, traditional medicines, probiotics, and products like amino acids and essential fatty acids (see Table 72.1). Schedule II of the regulations details substances which may not be included in NHPs such as controlled substances, radiopharmaceuticals, cannabis, etc. (Legislative Services Branch, Government of Canada, 2018a).

The oversight of NHPs by the NNHPD is achieved through controls over the quality, safety, and efficacy. The *NHPR* provide a means for these controls primarily through licensing of sites producing or importing NHPs, GMPs requirements, premarket review and licensing of finished products, and mandatory adverse event reporting. Interestingly, the *NHPR* do not differentiate between raw materials and finished products in the definition of an NHP. The *NHPR* therefore are not intended to regulate raw materials but finished NHPs only. A material becomes an NHP at the point that it is manufactured, represented, or sold as an NHP. It is at this point that the product and site licensing regulations, including GMPs, apply. The role of the NNHPD is to develop and enforce policies which facilitate industry compliance to these regulations, licensing activities for sites and finished products, review applications for clinical trials to be conducted on NHPs in Canada, monitor adverse events, and provide enforcement actions.

In 1997, Health Canada established an Advisory Panel on NHP in response to feedback from the public concerning the regulation and accessibility of herbal remedies. In 1999, the Office of NHP was introduced which would later become the Natural Health Products Directorate (NHPD). The *NHPR* were published in June 2003 and came into force

2. The Official Publication of the Association of American Feed Control Officials (AAFCO) contains the most complete list of animal food ingredients with their definitions. While many of the ingredients in the AAFCO Official Publication are not approved food additives and may not meet criteria under GRAS (21 CFR 570.30), the FDA currently accepts the listing of ingredients in AAFCO's Official Publication for their marketing in interstate commerce, provided there are no apparent safety concerns about the use or composition of the ingredient.

TABLE 72.1 As per the Natural Health Product Regulations.

	Schedule I—included Natural Health Product substances		Schedule II—excluded Natural Health Product substances
1	A plant or a plant material, an alga, a bacterium, a fungus, or a nonhuman animal material	1	A substance set out in Schedule C to the Act
2	An extract or isolate of a substance described in item 1, the primary molecular structure of which is identical to that which it had prior to its extraction or isolation	2	A substance set out in Schedule D to the Act, except for the following: a) A drug that is prepared from any of the following microorganisms, namely, an alga, a bacterium, or a fungus; and b) Any substance set out on Schedule D when it is prepared in accordance with the practices of homeopathic pharmacy
3	Any of the following vitamins: biotin, folate, niacin, pantothenic acid, riboflavin, thiamine, vitamin A, vitamin B6, vitamin B12, vitamin C, vitamin D, vitamin E, vitamin K1, vitamin K2	3	(Repealed, SOR/2018-132, s.3)
4	An amino acid	4	A substance set out in Schedules I to V of the *Controlled Drugs and Substances Act*
5	An essential fatty acid	5	A substance that is administered by puncturing the skin
6	A synthetic duplicate of a substance described in any of items 2–5	6	An antibiotic prepared from an alga, a bacterium, or a fungus or a synthetic duplicate of that antibiotic
7	A mineral	7	*Cannabis* as defined in subsection 2(1) of the *Cannabis Act*, except for a derivative or a product made from a derivative that is exempt from the application of the *Cannabis Act* under the *Industrial Hemp Regulations* and that does not contain an isolated or concentrated phytocannabinoid or a synthetic duplicate of that phytocannabinoid
8	A probiotic	8	Anything referred to in Schedule 2 to the *Cannabis Act* that contains more than 10 μg/g THC, an isolated or concentrated phytocannabinoid or synthetic duplicate of that phytocannabinoid

in January 1, 2004. Implementation of the requirements set out in the regulations were gradually enforced over a transition period of 2 years for site licensing and 6 years for finished product licensing (Health Canada, 2002).

In 2014, as part of an ongoing effort to consolidate the regulation of self-care products in Canada, the mandate of the NHPD was expanded to include the oversight of nonprescription drugs and disinfectants. This resulted in a change in the name of the office to the Nonprescription and Natural Health Product Directorate (NNHPD). In 2020, the regulatory framework for self-care products continues to evolve as new measures are introduced to harmonize the oversight of NHPs, nonprescription drugs, and cosmetics (Health Canada, 2019a).

Natural health product licensing

A consequence of the classification of NHPs as drug products is the requirement for premarket review of the formulation, quality specifications, and labeling. This process is intended to ensure the consumer is only presented with products which are of a high quality, safe, and efficacious, without false or misleading labeling as provisioned by subsection 9(1) of the *FDA*. As per Section 8 of the *NHPR*, every product that is marketed for sale in Canada must first obtain a Natural Product Number (NPN) which is assigned when a product license is granted. All products issued a product license and NPN are listed in the publicly available Licensed Natural Health Product Database (LNHPD). LNHPD entries include the NPN, brand name, license holder, dosage form and route of administration, dose, use/purpose, medicinal ingredients including quantity and potency, and nonmedicinal ingredients.

An integral tool that the NNHPD utilizes is the Natural Health Product Ingredient Database (NHPID) (Health Canada, 2004). The NHPID is a publicly accessible and searchable database of ingredients that are permitted for use in NHPs sold in Canada. The database includes both "medicinal" or active ingredients and "nonmedicinal" or excipient ingredients (see Fig. 72.8). For an

SPOTLIGHT:
NHP INGREDIENTS DATABASE (NHPID)

Medicinal Ingredient: Meets the definition of an NHP
Non-Medicinal Ingredient: Any ingredient added to an NHP that confers an appropriate form or consistency to the medicinal ingredient

To add a medicinal ingredient, the requestor must demonstrate the ingredient meets the definition of an NHP. To add a non-medicinal ingredient, the requestor must demonstrate the ingredient to be safe at the proposed levels of use, do not possess pharmacological activity at the levels of use, do not interact with medicinal ingredient or interfere with assays or tests for the medicinal ingredient

An ingredient may be classified as both medicinal and non-medicinal in the database depending on its use in products. For example, ascorbic acid (vitamin C) may be included in a multivitamin formulation as a medicinal ingredient or may be included in small quantity in an herbal extract product as an antioxidant preservative

FIGURE 72.8 Natural Health Product Ingredient Database (NHPID).

NHP to be licensed, all of the medicinal and nonmedicinal ingredients must be listed in the NHPID according to their use in the product.

Accompanying the NHPID is the compendium of NHP monographs (Health Canada, 2003). Compendial monographs have been developed for ingredients that the NNHPD has determined that enough evidence exists to support the safe use of the ingredient for specific uses without full review of products that fall within the bounds set out by the monograph. Any restrictions specific to the monograph will be described in detail including those on dose and recommended use. The compendium of monographs is divided into ingredient monographs, which detail precleared information for a single ingredient (i.e., curcumin), and product monographs, which detail precleared information for an entire class of finished product (i.e., probiotics). A product that falls within the bounds established by an ingredient or product monograph may be considered to rely solely upon precleared information and therefore be eligible for expedited premarket review. Thereby, the monograph system benefits both industry and Health Canada by providing a regulatory fast-track for approval of products while reducing use of agency resources.

There are three classes of product license application (PLA) that can be filed with the NNHPD (see Table 72.2). Classes I and II rely completed on the precleared information detailed in compendial monographs. Class III applications require that sufficient evidence be provided to substantiate the safety of all medicinal ingredients at the dose and dose duration indicated, as well as the efficacy of the product as claimed in the recommended use/purpose. For all classes, details of the quality testing specifications must be provided, indicating what testing is completed at which stage of manufacture and attesting to the specifications meeting the standard of the *NHPR*. The NNHPD review of a PLA consists of three phases: administrative verification, regulatory screening, and safety and efficacy assessment. In 2019, the NNHPD issued new guidance entitled "Natural Health Product Management of Applications Policy," which describes in detail the review process (Health Canada, 2018a). This document was developed in response to years of "shotgun" submissions that frequently omitted required elements of an application and tied up NNHPD resources. Although the licensing system employed by the *NHPR* is often lauded for its premarket review of claims and safety data, PLAs rely on attestations to meeting quality standards rather than requiring submission of quality data and details of testing methodologies. This gap in the regulations is meant to be addressed through agency enforcement via site GMP audits; however, lack of resources has yet to see this manifest appropriately.

Modern health claims

The classification of NHPs as drugs allows for health claims to be included on the label following approval during the premarket review phase. Health claims can include diagnostic, treatment, cure, risk reduction, and prevention claims. General health claims are also permitted so long as the health and safety of the consumer are not at risk. These include claims regarding the modification of organic functions, structure–function claims on nutrients, quality of life claims such as general health maintenance, support and promotion claims, as well as "source of" claims. The submitter of a PLA must support the safety and efficacy of the health claim through a narrative presenting scientific data. At least one claim is required to be included on the label of a licensed NHP. The health claims made in the Recommended Use/Purpose section of an NHP label are assessed by the NNHPD according to a risk-based system which considers the ingredient's physical and chemical form, the seriousness of the health claim (see Table 72.3) and conditions of use and the impact that a lack of efficacy would have on the health and safety of the consumer (Health Canada, 2012a).

High-risk claims are for products that treat, prevent, or cure serious health conditions; those conditions that require supervision by a healthcare practitioner and if left untreated are debilitating or life threatening. Medium-risk claims have slightly less stringent requirements for evidence than high-risk claims. These claims indicate that

TABLE 72.2 Classification of product license applications.

Class	Explanation	Review period
I	Finished product that complies with a single compendial monograph exactly	60 days[a]
II	Finished product that is supported by a combination of two or more compendial monographs. Some minor deviations in wording may be permitted. Potential interaction between medicinal ingredients does not have to be addressed by data substantiating safety/efficacy. Natural and Nonprescription Health Products Directorate reserves the right to reclassify as class III and request data supporting the safety/efficacy of ingredient interaction	90 days
III	Single or multiple medicinal ingredient product that has some element which does not comply with a compendial monograph or raises additional concern. Examples: • medicinal ingredient without monograph • ingredient has monograph but its intended use differs in terms of dose and/or claims being made	210 days

[a]*Section 6 of the* NHPR *protects the timeline for Class I submissions.*

TABLE 72.3 Classes and Examples of NHP Health Claims.

Health condition class	Example claim
Serious disease or condition	Helps prevent rheumatoid arthritis For the treatment of high blood pressure For the treatment of depressive disorders
Major disease or condition	Helps to lower blood/plasma cholesterol levels Helps in the prevention of nausea and vomiting associated with conventional cancer management Helps to regulate blood glucose levels Improves joint function in osteoarthritis of the knee Helps treat erectile dysfunction
Minor disease or condition	Helps relieve minor pain associated with menstruation Used as a mild sedative (for jet lag) Soothes sore throat Used as a decongestant to relieve nasal congestion due to hay fever
General health claims	Maintains healthy gums Source of fiber Provides antioxidants A source of essential amino acids for the maintenance of good health Supports digestion

the NHP treats, cures, or prevents major health conditions; those conditions that do not resolve naturally in a timely fashion and require prompt treatment to mitigate undesirable or worsening effects. Low-risk claims are those made on diseases or abnormal states that naturally resolve on their own in a timely manner and for which a lack of efficacy does not pose a major risk to the consumer. This category also includes symptoms of serious or major health conditions as well as general health claims. Most claims made on licensed NHPs are low-risk claims. The evidentiary standard for low-risk claims is the least stringent.

General health claims are intrinsically low risk, and therefore high-risk or controversial ingredients or evidence are not considered for these claims. These claims describe the effect of the NHP on a condition in very broad or limited terms. They can include health maintenance, support, and promotion language, and often include a qualifier phrase like "helps to"; which implies a mild therapeutic effect. Function claims related to the modification of the organic functions of the body are also considered general health claims and often include qualifying language like "Supports...". General health claims can be supported by patterns of evidence rather than a confirmatory clinical trial or study. Claims relating to the NHP providing a "source of" a nutrient or constituent are also considered general health claims that must be supported by some evidence of quantification of the constituent during the quality testing and/or in preclinical or clinical research. If the constituent references in a claim has an associated recommended dietary allowance or adequate intake, the NHP must provide at least 5% of that value in order for a "source of" claim to be made.

For health claims of any risk category, the PLA should present not only the evidence required by the standard but any additional data available in order to provide a summary of the totality of the evidence that exists to support the claim. Based on the evidence available, qualifying statements may be required to be included in the claim language to ensure that the consumer is not misled about the effectiveness of the product.

Health claims, including general health claims, are considered modern health claims by the NNHPD. As the definition of an NHP includes provisions for both traditional products and homeopathic products, there exist two alternative pathways to the modern health claim route for licensing an NHP: traditional use claims and homeopathic medicines.

Traditional use claims

A significant proportion of ingredients used in NHPs have also been used in traditional modes of medicine for many years. As the *NHPR* were being developed, it was thought prudent to ensure that this traditional medicinal use, unique to NHPs, was preserved in the legislation; thereby allowing people to continue using the products as they have for decades or centuries. The definition of traditional medicine used by the NNHPD is "medicine based on the sum total of knowledge, skills and practices based on the theories, beliefs and experiences indigenous to different cultures, used in the maintenance of health, as well as in the prevention, diagnosis, improvement or treatment of physical and mental illness" (Health Canada, 2012b).

The licensing process for an NHP that is pursuing a traditional use claim is the same as that described above for modern health claims, with the exception of the standard used to assess the health claim itself. Traditional use claims must be founded on the principles and practices of the respective traditional mode of medicine and the supporting evidence must demonstrate that the ingredient has been used in a single mode of the traditional medicine. Traditional use claims are permitted on all NHPs, even those also making modern health claims, if they meet the requirements for traditional use claims. Traditional use claims however, cannot be made on serious health conditions nor on symptoms or actions that are broad in scope. Likewise, they cannot be made on conditions that are not possible to diagnose within the mode of traditional medicine. An example of this is that a traditional use claim cannot be made on a product which "increases hemoglobin levels."

For an NHP to be considered acceptable for a traditional use claim, it must be formulated such that the preparation is representative of how the ingredient has been used in a traditional mode of medicine (i.e., Ayurvedic medicine, traditional Chinese medicine, herbal medicine, etc.). For instance, if a herb is described as having only been used in a decoction in an ancient medicinal text, then the finished product that is pursuing a product license should also be in a form that is representative of a decoction, that is, water extract or tea. The dose must also be representative of what is traditionally used. The evidence supporting the claim must specifically support the dose and method of preparation, including the parts of the plant used, the extraction method used, and the mode of administration. Although this information may not be exactly the same, it must be comparable. This is also true of the information used to support the safety of the ingredient. History of use is a critical consideration and must have occurred at least two generations previously and been continuously used since. Traditional warnings regarding the use of the ingredient are also important. It should be noted that if any modern evidence exists which raises safety concerns, this evidence will be considered and may result in the ingredient not being appropriate for traditional use claims.

One item of note is that traditional health claims on NHPs do not apply to products that are compounded by traditional medicine practitioners. Compounded health products do not fall under the scope of the *NHPR* and therefore are not required to be licensed, nor are they required to be produced at a licensed facility. The healthcare practitioner instead assumes the responsibility for the safety and efficacy of the product.

Homeopathic medicines

Under the *NHPR*, homeopathic medicines must be licensed following the same product licensing process as NHPs making modern health and traditional use claims. Rather than being issued an NPN, homeopathic products are issued a Homeopathic Medicine Number (DIN-HM). The definition of a homeopathic medicine has two components. (1) A product that is manufactured from, or contains as medicinal ingredients, only substances referenced in a homeopathic monograph from an accepted homeopathic pharmacopeia. (2) A product prepared in accordance with the methods outlined in an acceptable homeopathic pharmacopeia (Health Canada, 2006a). Homeopathic medicines may include medicinal ingredients that are otherwise not permitted for use in NHPs, such as tobacum, some prescription drugs and biologics. Injectable solutions are not permitted for use as an NHP, nor as a homeopathic medicine. The NNHPD lists the minimum required potency of homeopathic ingredients on the NHPID and therefore if a minimum required potency is listed, the product must comply with this OTC limit. Ingredients considered as controlled substances or radiopharmaceuticals are not permitted for use in homeopathic medicine.

Claims are not permitted on single ingredient homeopathic medicines, but they must be labeled "homeopathic medicine" or "homeopathic remedy." These products are intended for use only by a homeopathic practitioner. Products containing multiple homeopathic ingredients are permitted to make claims so long as they have a specific use or purpose and fall between the minimum required potency and 30 CH or equivalent. These products must be intended for self-care and therefore the intended use must be appropriate. Evidence must be submitted which supports the safety of the product at the listed potency.

Homeopathic medicines must be represented as such on the label with a "Homeopathic Medicine" statement. Dosing for homeopathic medicines is standard for all products, regardless of the recommended purpose, and is based on the dosage form and subpopulation treated. A duration of use statement is required. Risk information is also required to be denoted, similar to labeling of NHPs. Homeopathic medicines fall under the *NHPR* but are treated with their own specific set of requirements.

Natural health product labeling

Part 5 Sections 86 to 98 of the *NHPR* detail the required elements for labeling and packaging of any licensed NHP. All NHPs must be labeled according to these requirements unless the NHP is being sold to a manufacturer or distributer. This allows for the sale of bulk NHP ingredients without being marked with an NPN as is required for finished products.

Mandatory labeling elements must be displayed prominently and be readily discernible (Health Canada, 2018a,b,c). The conditions of use must include at least one health claim to justify the classification of the NHP as a health product. As mentioned above, this appears in a section titled "recommended use or purpose." The brand name, manufacturer name, and/or instructions for use of the NHP can also be considered to make a claim on the product if reference is made within the name to effects outside of the approved indication. This would be considered a false or misleading statement, contravening subsection 9(1) of the *FDA* (see Table 72.4).

Section 94 of the *NHPR* details the reduced requirements for labeling products whose packaging is too small to accommodate all the required labeling elements in a font size that is reasonable for a consumer to read. If the product does not have an outer label (i.e., box), the inner label must include a statement directing the consumer to a leaflet that is required to accompany the product. Except for lozenges, all NHPs are required to be packaged in security packaging and the label must draw attention to the security feature should it not be self-evident. Security packaging refers to features that indicate to the consumer whether the package has been opened previously. Child-resistant packaging is not mandatory for all NHPs but may be required if the ingredients of the NHP are harmful to children. In these cases, warnings will also be required on the label.

The labeling requirements for NHPs are scheduled to be updated in 2020 as an NHP "Product Facts" table will be introduced as a mandatory labeling element (Health Canada, 2019a). This is phase I of an initiative by the NNHPD to streamline the self-care products (NHPs, OTCs, cosmetics). The facts table is similar to the Drug Facts Table included as part of the plain language labeling requirements for OTC drug products. Phase I of the self-care products initiative also will introduce requirements displaying risk information prominently and in plain language. Phases II and III concern streamlining the review process for OTC drugs and consolidating the evidentiary standards for all self-care products. As these more substantial changes are consulted upon and new regulations introduced, new labeling requirements are likely to be implemented for NHPs to ensure that the consumer is well-informed and not misled.

Natural health product site licensing

In order to sell a licensed NHP in Canada, the site manufacturing the finished NHP in Canada or the site importing the finished NHP into Canada must be licensed by the NNHPD (see Fig. 72.9). Part 2 of the *NHPR* describes the regulations surrounding a site license. Section 27 states that site licenses are required for any business or individual who manufactures, packages, labels, or imports an NHP for sale in Canada. It further states that anyone conducting these activities must do so according to Part 3 of the *NHPR* which details the standard for GMPs. The holder of a site license must only conduct activities permitted under their license and must renew the license prior to expiration or amend or notify the NNHPD to remain compliant. Those businesses or people importing NHPs into Canada for the purpose of sale must be licensed, whether they are having product shipped directly for retail or they are the retailer themselves. Importers of all types must ensure that they are GMP compliant. It is also the responsibility of the importer to ensure that the products are manufactured to the GMP standard under part 3 of the *NHPR*. Foreign manufacturing sites are reviewed by the NNHPD and are listed on importer site licenses.

Section 28 of the *NHPR* outlines the requirements for applying for a site license. The proposed activities of the site (manufacturing, packaging, labeling, etc.), and a report from the designated quality assurance person (QAP) demonstrating the compliance of the site facility and processes with regulations for each activity performed at the site are required (Health Canada, 2015a). This includes proof of GMP compliance. Site license applications can utilize precleared

TABLE 72.4 NHP Labeling Requirements.

Label area*	Labeling element	Language requirement
Principal display panel	Brand name	English or French
	Product number (natural product number or DIN-HM)	English or French
	Dosage form	Bilingual required
	"Sterile" if applicable	English or French
	Net amount in the immediate container	English or French
Any panel	Name and address of product license holder	English or French
	Common name of each medicinal ingredient	Bilingual required
	Proper name of each medicinal ingredient	Bilingual required
	Quantity of each medicinal ingredient per dosage unit including potency, extract ratio, and quantity crude equivalent	English or French
	Description of source material of medicinal ingredient	Bilingual required
	Recommended use or purpose	Bilingual required
	Route of administration (if not self-evident)	Bilingual required
	Recommended dose	Bilingual required
	Recommended duration of use	Bilingual required
	Risk information	Bilingual required
	Nonmedicinal ingredients	Bilingual required
	Storage conditions	Bilingual required
	Lot number	English or French
	Expiry date	English or French
	Security feature if not self-evident	English or French
	Quantity of mercury (if applicable)	English or French
Other	If no outer package label, a statement referring consumer to a leaflet	English or French
	Cautionary statements and child-resistant packaging (if applicable)	English or French

*Section 94 of *NHPR* describes provisions for extremely small packaging requirements.

information known as qualified authority certificates that are acceptable to the NNHPD as proof of facility and process compliance to GMP standards. If precleared information is not available, a quality assurance form must be completed, and full review conducted. Additional standards and third-party audit reports may be used in conjunction with a quality assurance form. These proofs of compliance, including pre-cleared information, have limited periods of validity and need to be reassessed every 1–3 years, depending on the type of evidence used.

Quality standards

Good manufacturing practice standards are outlined in Part 3 of the *NHPR* (sections 43–62). GMPs in any industry are designed to ensure that products are manufactured under conditions that effectively control quality and risk through established standards for production, packaging, labeling, storing, and importing. NHP GMPs are based on the same principle as drug GMPs with provisions made to account for issues such as batch-to-batch variation in botanical materials. Responsibility for meeting requirements of NHP GMPs fall throughout the entire supply chain, from harvesting of raw materials to storing the finished products prior to distribution at retail (Health Canada, 2015b). The holder of an NHP site license must have a QAP who must have the relevant training and experience to conduct the required quality activities and holds the responsibility for ensuring that GMP requirements are met for a licensed site. The myriad types of NHPs prevent the NNHPD from prescribing specific quality requirements, rather guidelines and practices inform NHP manufacturers on how to themselves establish appropriate production standard

SPOTLIGHT:
WHO NEEDS AN NHP SITE LICENSE?

License Required:

- Manufacturing NHPs
- Importing NHPs
- Packaging NHPs
- Labeling NHPs

License Not Required:

- NHP Distributors (No Manufacturing or Importing)
- NHP Compounding Pharmacists and Traditional Medicine Practitioners
- Producers or Importers of raw materials (not yet in form fit for sale as NHP)
- Manufacturers of NHPs solely for export
- Testing labs

FIGURE 72.9 Activities Requiring and Not Requiring an NHP Site License.

operating procedures (SOPs), quality control programs, and quality testing specifications (see Table 72.5).

Quality guidance issued by the NNHPD is based on the GMPs but provides additional specificity for testing methodologies and reporting limits based on the type of medicinal ingredient(s) included in the NHP (Health Canada, 2015c). Guidance is provided for characterization, identification and quantification, purity and impurity testing, as well as performance and stability testing. NHPs must be produced with quality testing performed both at the ingredient stage (i.e., inputs) and at the finished product stage (i.e., outputs). Section 44(2) of the *NHPR* states that product specifications must include tests for identity and quantity of the medicinal ingredient as well as purity and potency (if applicable). Quality testing should be conducted according to a compendial method whenever possible; however, noncompendial methods are acceptable with proper scientific rationale. The NNHPD has an approved list of pharmacopeias that are accepted in their entirety.

By providing several options for the conduct of quality testing, the NNHPD provides the product license holder with alternative approaches to ensure that regulatory requirements are met. Product and site license holders take on significant responsibility under this system of regulation. It is their responsibility to ensure that they are compliant with all the standards they attested to in their license applications. The safety of the NHPs that they manufacture and sell is their responsibility. For this type of regulation to provide adequate oversight, compliance auditing and enforcement action by Health Canada is required.

Compliance and enforcement

The Health Product and Food Branch Inspectorate is responsible for all compliance and enforcement activities dealing with any NHPs including but not limited to issues relating to health claims, GMP compliance, and counterfeiting. The NHP Compliance and Enforcement Policy (POL-0044) details the approach that the inspectorate takes to noncompliance in the NHP industry (Health Canada, 2010a). A risk-based approach is taken to triage noncompliant activities and set priorities for mitigation activities. Common risk mitigation activities carried out by Health Canada include requests for stopping sales or recalls and public advisory notices about specific products. These activities generally follow confirmation of products that are counterfeit, contaminated, or adulterated. From 2010 to 2015, Health Canada recalled 265 products that were found to be noncompliant; 128 of which posed a risk that could cause death. Many of these enforcement actions were derived from internal company whistleblowers leading Health Canada to investigate. Many of the products tested by Health Canada during this time were found to be adulterated with pharmaceuticals (Health Canada, 2016a,b). Health Canada investigations into contaminated products have often led to license suspensions and public warnings to avoid use of specific products. In some instances, provincial courts have also levied fines against companies for illegally selling NHPs after Heath Canada action or without the proper licensing (Government of Canada, 2017, 2019).

The sale of unlicensed NHPS is an issue that online sales have amplified in recent years. Health Canada has taken the approach of working with major online retailers, placing the onus on them to ensure that only compliant products are marketed online in Canada (Health Canada, 2015d). Health Canada may also recommend that shipments of these products may be refused at the border. Policy POL-0060 details the import and export policy for health products (Health Canada, 2010b). Import enforcement actions may also be taken for GMP violations identified at foreign licensed NHP sites. Noncompliance with Part 5 of the *NHPR* concerning labeling and packaging will also draw enforcement action deemed proportional to the risk the product poses. False or misleading statements on the label of NHPs, or statements not approved in the product license, are considered in contravention of Part 5 and/or Section 9(1) of the *FDA*.

Adverse event reporting

The *NHPR* Section 24 requires that any licensee who sells an NHP must report any serious adverse reactions to the NHP to the Health Minister within 15 days after they become aware of the reaction. This is both for serious

TABLE 72.5 NHPR sections and Application to NHP Activities.

Section of regulations	Manufacturing	Packaging	Labeling	Importing	Distributing
43. Prohibition	✔	✔	✔	✔	✔
44. Specifications	✔	✔*	X	✔	✔*
45. Premises	✔	✔	✔	✔	✔
46. Equipment	✔	✔	✔	✔*	X
47. Personnel	✔	✔	✔	✔	✔
48. Sanitation program	✔	✔	✔	✔	✔
49. SOPs	✔	✔	✔	✔*	✔*
50. Recall	✔	✔	✔	✔	✔
51. Quality assurance	✔	✔	✔	✔	✔
52. Stability	✔	X	X	✔*	X
53–57. Records	✔*	✔*	✔*	✔*	✔*
58. Records maintenance	✔	✔	✔	✔	✔
59. Sterile natural health products	✔	✔	X	X	X
60. Ophthalmic use	✔	✔	X	X	X
61. Lot or batch samples	✔	X	X	✔	✔
62. Recall reporting	✔	X	X	✔	✔

✔, applicable; *, as applicable; X, not applicable. Adapted from NNHPD "Good Manufacturing Practices Guidance Document" (Health Canada, 2015).

adverse reactions occurring in Canada and serious unexpected adverse reactions occurring inside or outside of Canada. A serious adverse reaction is defined as "noxious and unintended response to a natural health product that occurs at any dose and that requires in-patient hospitalization or a prolongation of existing hospitalization, that causes congenital malformation, that results in persistent or significant disability or incapacity, that is life threatening or that results in death." A serious unexpected adverse reaction is a serious adverse reaction that is "not identified in nature, severity or frequency in the risk information set out on the label of the natural health product." The adverse reaction program implements the recommendations of the International Council for Harmonisation of Technical Requirements for Pharmaceuticals for Human Use regarding standards for expedited reporting, individual case safety reports, and pharmacovigilance, and includes the Report of the Council for International Organizations of Medical Sciences, and thereby closely resembles health product reporting in other jurisdictions across the globe (Health Canada, 2018b).

Serious adverse reaction case reports are submitted to Health Canada through the Canada Vigilance Program operated by the Marketed Health Products Directorate. This program oversees adverse event reporting for NHPs, drugs, and biologics. Reports are generally submitted electronically and should indicate the country in which the adverse event occurred and the details of the circumstances surrounding the event in the narrative. Overdose, occupational exposure, and pregnancy exposure are just some of the situations of note and for which special considerations for follow up must be taken.

In addition to the mandatory reporting required for serious adverse reactions, a licensee who sells an NHP in Canada must prepare and maintain an annual summary report of all adverse reactions and serious adverse reactions that have occurred in Canada. Health Canada may request that this annual summary be provided to support investigations into an NHP or ingredient and may also request interim summary reports. Adverse events occurring in clinical trials of the NHP should also be reported and included in annual or interim summary reports.

Natural health products and food classification

One of the distinguishing features of Canadian regulation of nutraceuticals is the classification of NHPs as drugs. However, there are many nutraceutical products that fall into a gray area of classification when examined in this framework. Nutraceuticals in a food format, such as a drink or a bar, represent a confluence of food and NHP and thereby require overarching guidance which dictates which regulations should be applied. See Fig. 72.10 for the four criteria to determine the appropriate Canadian jurisdiction. The NNHPD has issued guidance for industry which details the principles the agency applies when determining whether a product is considered an NHP or food (Health Canada, 2017). Previously, many products in food format had been approved as NHPs. In 2010, the NNHPD reconsidered the classification of these types of products and began to transition them to food products to avoid confusion on the part of the consumer. The most notable example of this has been energy drinks, which were transitioned from NHPs to foods over an 8-month period in 2012. The prevailing principle guiding classification is that "products that look like foods and are consumed as foods are regulated as foods." This means that labeling requirements and regulations for foods should also be applied to these products.

It is recommended that a party seeking to market in Canada that is unsure of its classification submit a classification request to the NNHPD. The submission of a classification request must include the formulation of the product, a product label, and any marketing materials that will establish how the product is being positioned in the market. The NNHPD and Food Directorate coordinate the review of the product and provide a determination which will dictate the regulations that must be satisfied.

SPOTLIGHT:

FOOD OR NHP: FOUR CONSIDERATIONS

Decisions are made by a joint committee of the NNHPD and Food Directorate. These criteria are not considered equally. Certain classification decisions rely more heavily on certain criteria.

Composition:
- Are ingredients included solely to provide nutrition, satisfy hunger/thirst or provide taste, texture or flavor?
- Does an ingredient only provide a medicinal purpose and does not serve a purpose as a food?

Representation:
- Do label claims, instructions and marketing imply highlight food qualities?
- Is the product sold with groceries or medicines?

Format:
- Is the product intended for *ad libitum* consumption?

Perception:
- Does the product have a history of use as a food or medicine?

FIGURE 72.10 Consideration for the Classification of Products as NHPs or Food.

Supplemented and fortified foods

The previous section alluded to nutraceutical products which possess some characteristics of NHPs but may be regulated as foods due to characteristics of the form or presentation. These foods require special consideration when regulated as foods as they do not conform with the *FDA* and *FDR*. Temporary market authorization (TMA) is the primary mechanism for allowance of such products on the market despite contravention of the regulations. For a noncompliant food to be permitted for sale in Canada, an amendment to the *Food and Drug Regulations* is required. TMA is a provision made in Sections B.01.054 and B.01.055 of the *Food and Drug Regulations* that allows a noncompliant product access to the market for the purpose of generating data in order to support a regulatory amendment. TMA must be applied for and reviewed by the Food Directorate in a process that is akin to a PLA for NHPs but often more scrutinous. TMA is only granted when the food is determined to not pose a risk to the public and when the required data to support a regulatory amendment can only be gleaned through real-world marketing of the product (Health Canada, 2013). Once the application has been reviewed and all requirements are met, a letter of agreement between the Food Directorate and the applicant is executed. This letter describes attestations made by the applicant regarding labeling and advertising and also the prerogative of the Food Directorate to request data gathered on the product and to request that sales cease.

The example of caffeinated energy drinks acts as a prototypical example of a product granted TMA. Caffeinated energy drinks are not compliant with the *Food and Drug Regulations* due to the addition of caffeine, vitamins, minerals, and amino acids to the formulations. This drove companies to initially market caffeinated energy drinks as NHPs, and TMA has been utilized as the tool that facilitates their current market position as foods. As there are no pressing safety concerns with caffeinated energy drinks but there is a need for a better understanding of consumption patterns and the effectiveness of labeling requirements, TMA guidance specific for caffeinated energy drinks was issued (Health Canada, 2014). This guidance defines specific limits for caffeine and added nutrients as well as other restrictions on ingredients, physical attributes, and labeling. Similar category-specific guidance has been developed for supplemented foods in general.

Health Canada defines supplemented foods as a subset of food defined in Section 2 of the *FDA* that is a "prepackaged product that is manufactured, sold or represented as a food, which contains added vitamins, minerals, amino acids, herbal or bioactive ingredients. These ingredients may perform a physiological role beyond the provision of nutritive requirements" (Health Canada, 2016a,b). If a food naturally contains the above-listed nutrients and is compliant with the FDR, it is not considered a supplemented food. If a food that is not listed in Part D Division 3 of the *FDR* contains added vitamins, minerals, or amino acids, or if a food that is listed is supplemented beyond the levels permitted by the regulations, TMA is required. The food must be prepackaged, as there are labeling requirements set in the TMA granted by Health Canada. Herbal and bioactive ingredients may be typically used in NHPs but have not traditionally been used in foods or serve a food purpose. These ingredients may be considered novel foods and require premarket authorization before they are permitted to be sold as foods.

Novel foods

Division 28 of *FDR* describes the regulations surrounding novel foods. Novel foods are products and ingredients that have not historically been used safely as foods, foods that have themselves changed due to major changes to a process used to manufacture the food, and foods that have been modified by biotechnology. Nutraceutical products that are classified as foods are often considered novel foods as they contain ingredients that are commonly used in NHPs or are the result of novel technology. Novel foods require premarket approval by the Food Directorate before they are permitted to be sold in Canada. This submission must establish the ingredient or product as safe for consumption. Health Canada has issued guidance for establishing the safety of a plant-derived and microorganism-derived novel food based on the principles established by international agencies such as the OECD, FAO, WHO, and Codex Alimentarius Commission (Health Canada, 2006b). This guidance details the information that should be included in the premarket notification and encourages presubmission consultation with the Food Directorate.

Novel food notifications should be focused on the changes or novel characteristics of the food and present scientific data and rationale that support the safe use of the ingredient. The submission is required to include the intended use of the ingredient or product as well as the labeling information for the ingredient or product. If the label contains certain claims, these are required to be reviewed as part of a health and nutrition claim review submission.

When the review is complete, a summary document is presented to the Food Rulings Committee which makes the final decision on the permissibility of sale. If the product is found acceptable for sale, a letter is drafted, and the submitter is notified there is no objection to the sale of the ingredient or product as food under the conditions listed in the notification. A decision document is published on all novel food submissions which detail the product and relevant data submitted and ensure transparency in the decisions made by the Food Rulings Committee. The submitter of the novel food notification has the opportunity to review and comment on the document prior to its being published.

Cannabis and natural health products

Over the past several years, the popularity of CBD, a nonintoxicating cannabinoid derived from hemp and marijuana has exploded onto the scene in the nutraceutical industry. In October 2018, Canada federally legalized cannabis use for recreational purposes through the introduction of the *Cannabis Act* (Legislative Services Branch, Government of Canada, 2018b).

Despite the introduction of the *Cannabis Act*, phytocannabinoids remain on the prescription drug list. Therefore any product containing a cannabinoid and making a health claim is considered a prescription drug and must undergo premarket review. As such, CBD is considered a prescription drug if a health claim is made, whether it is derived from marijuana (cannabis containing delta-9-THC $>0.3\%$) or hemp (cannabis containing THC $<0.3\%$).

It is important to distinguish the difference between hemp-seed oil and CBD oil. Hemp-seed oil contains only trace amounts of naturally occurring CBD, whereas CBD oil contains concentrated amounts of CBD resuspended in oil. NHPs permit the use of hemp-derived ingredients if they do not meet the definition of cannabis under the *Cannabis Act*. Therefore the ingredients must be derived from nonviable seeds, hemp-seed derivatives that are compliant with the *Industrial Hemp Regulations*, or mature stalks that do not include any leaves, flowers, seeds, or branches.

A controversial point of the *Cannabis Act* is that CBD is subject to the same regulatory controls as THC; the primary component responsible for cannabis intoxication. Many have expressed that due to the lack of intoxicating properties, lack of abuse potential, and favorable safety profile, CBD should not be subject to the same regulations as THC. Furthermore, based on the potential health benefits of CBD and its availability on the legal recreational and medicinal markets, Canadians should be granted access to legal purpose-designed self-care health products containing CBD (CHFA, 2019; CHTA, 2019). With an illicit CBD nutraceutical market thriving, Health Canada has recognized the benefits of introducing a legal framework permitting a market for OTC health products containing cannabinoids. In June 2019, Health Canada issued a "Consultation on Potential Market for Cannabis Health Products that would not Require Practitioner Oversight" for public comment (Health Canada, 2019b). This document is the first stage in the implementation of what has become known as Cannabis 3.0 products. These are OTC self-care products that contain cannabis-derived ingredients and make health claims on minor ailments, like NHPs or nonprescription drugs. According to the proposed framework, these products may also include NHP or nonprescription drug ingredients. All products would require premarket review and approval, like what currently exists for NHPs or OTC drugs. It is expected that draft regulations for cannabis health products will be released sometime in 2020. From there, regulations could come into force as early as 2021. Although the details of the regulations remain unclear, introducing a cannabis health products market would place Canada at the global forefront of the cannabis containing nutraceuticals market.

Concluding remarks and future directions

It is an undeniable fact that nutraceuticals are regulated in North America. A frequently repeated fallacy states that nutraceuticals are not regulated in the United States. Likewise, many companies believe that there are burdensome regulatory hurdles that forestall the nutraceutical market in Canada. While they are not regulated as drugs as some would like, nutraceuticals and their ingredients are regulated under greater scrutiny than conventional foods in both the United States and Canada. The tenets of DSHEA and the NHPR arose out of the same needs and are both rooted in public access to safe and high-quality products. While the regulatory frameworks for nutraceuticals in the United States and Canada have been established for many years, the statutes concerning supplements, NHPs, and foods in both countries have continually been amended through the enactment of new laws to cast a wide net in order to safeguard the North American food supply, one of the safest in the world.

A comparison of US and Canadian nutraceutical regulations is provided in Table 72.6. Examination of this comparison demonstrates that although the approach differs, similar controls are in place in both countries. All of these requirements place economic burdens on ingredient suppliers, importers, contract manufacturers, own label distributors, and retailers to operate in compliance with federal laws. While the US nutraceutical industry has exploded because of the regulatory framework designed to balance consumer access with safety, critics have highlighted the need for more regulatory tools and greater Congressional oversight. Across the border, the Canadian nutraceutical market has seen a similar growth trajectory (albeit, much less in magnitude) under regulation that inherently restricts the market in the name of safety, but which critics have declared as excessive considering the safety profile of NHPs. Despite the great economic success of DSHEA and the NHPR, the regulatory frameworks operate under constant change and scrutiny. It will be interesting to see whether the legislative bodies in the United States and Canada make complete overhauls of those frameworks over the next decade, given their

TABLE 72.6 Comparison of Nutraceutical Regulation in the United States and Canada.

Regulations for nutraceuticals	United States	Canada
Facility registration	✔ (Biennial renewal;Biohazard Act of 2002, FSMA)	✔ NHP Site Licensing (license period based on years held; Part 2, NHPR)
Suspension of facility registration	✔ (FSMA)	✔ (Section 39, NHPR)
Mandatory stop sale/recall authority	✔ (FSMA)	✔ (Section 17, NHPR)
GMPs for ingredient suppliers	✔ (FSMA)	X Finished Product GMPs only (Part 3, NHPR). Ingredient GMP required as part of Finished Product GMP
Foreign supplier verification program	✔ (FSMA)	✔ NHP Site Licensing (Foreign site qualification; Part 2, NHPR)
Program to qualify importers	✔ (FSMA)	✔ NHP Site License required for Importers and Foreign Manufacturing Site
Food defense plan for ingredient suppliers and manufacturers (mitigating strategies)	✔ (FSMA)	X Food tampering provisions (SFCA)
HACCP and preventive controls	✔ (FSMA)	✔ (SFCA)
Nutraceutical labeling standards	✔ (Labeling standards for Conventional Foods and Dietary Supplements; NLEA and DSHEA)	✔ (Labeling requirements for NHPs; Part 5, NHPR)
General food cGMPs	✔ (21 CFR part 110)	✔ (SFCA equivalent)
Nutraceutical cGMPs	✔ (21 CFR part 111 for dietary supplements)	✔ (Part 3, NHPR)
Federal certificates for export	✔ (review of label and ingredients; FDA Certificate of Free Sale Export)	✔ (Section A.01.045 FDR)

(Continued)

TABLE 72.6 (Continued)

Regulations for nutraceuticals	United States	Canada
Mandatory reporting of serious adverse events for nutraceuticals	✔ (Dietary Supplements are the only commodity of food in US with mandatory SAE Reporting)	✔ (Section 24 NHPR)
Premarket chemistry review of ingredients	✔ For Novel Ingredients only (NDI premarket notification to FDA)	✔ Product License Application Review (Part 1 NHPR)
Premarket safety review of ingredients	✔ For Novel Ingredients only (NDI premarket notification to FDA)	✔ Product License Application Review (Part 1 NHPR)
Review of claims on nutraceutical products	✔ (30-day notice to notify FDA of any structure-function claims made on dietary supplement products prior to market)	✔ Product License Application Review (Part 1 NHPR)
Active domestic GMP finished product inspection program	✔ (over 600 dietary supplement contract manufacturers and own label distributors inspected annually)	✔ (NHP site license holders for importing and manufacturing NHPs)
Active GMP inspection program for ingredient suppliers	✔ (over 600 inspections annually)	X (Ingredient suppliers not regulated under NHPR)
Enforcement program over GMP violations	✔ (WLs, final Agency action: seizure, injunction, state embargo)	✔ (WLs, Product and Site License suspension and/or cancellation, stop sale order)
Enforcement of misbranded labels and unlawful claims	✔ (WLs, other final Agency action)	✔ (WLs, Product License suspension and/or cancellation, stop sale order)
Active GMP inspection program of international finished product manufacturers and distributors	✔ (over 150 international facilities inspected annually)	X Foreign Site Reference Number renewals required every 1–3 years, GMP evidence assessed
Federal oversight of clinical trials involving nutraceuticals	X Research Ethics approval only required	✔ Notice of Approval required prior to commencing study. Research Ethics approval required

FSMA, Food Safety Modernization Act; *cGMPs*, current good manufacturing practices; *FDR*, Food and Drug Regulations; *GMPs*, good manufacturing practices; *HACCP*, hazard analysis and critical control points; *SFCA*, Safe Foods for Canadians Act; *NDI*, new dietary ingredient; *NHP*, natural health products; *NHPR*, Natural Health Product Regulations; *WL*, warning letter.

successful growth from a cottage industry into an economic force in both economies.

References

21 CFR 101.93, 2019. Food labeling: Certain types of statements for dietary supplements. Available from: <https://www.accessdata.fda.gov/scripts/cdrh/cfdocs/cfcfr/CFRsearch.cfm?fr=101.93> (Accessed 6 October 2020).

60 FR 67176, 1995. Food labeling; nutrient content claims, health claims, and statements of nutritional support for dietary supplements. Proposed rule. Fed. Regist. 60, 67176–67184.

62 FR 49837, 1997. Food labeling; statement of identity, nutrition labeling and ingredient labeling of dietary supplements; compliance policy guide, revocation. Final rule. Fed. Regist. 62, 49757-49837.

62 FR 49859, 1997. Food labeling; requirements for nutrient content claims, health claims, and statements of nutritional support for dietary supplements. Final rule. Fed. Regist. 62, 49859–49868.

62 FR 49886, 1997. Premarket Notification for a New Dietary Ingredient. Final Rule. Fed. Regist. 62, 49886–49892.

65 FR 1000, 2000. Regulations on statements made for dietary supplements concerning the effect of the product on the structure or function of the body. Final rule. Fed. Regist. 65, 1000–1050.

Center for Veterinary Medicine, 2020. Ingredients & additives. FDA. Available from: <http://www.fda.gov/animal-veterinary/animal-food-feeds/ingredients-additives> (Accessed 19 March 2020).

CHFA, 2019. CHFA's position on cannabis. Available from: <https://chfa.ca/en/Regulatory/cannabis-regulations/chfa-position-on-cannabis> (Accessed 19 February 2020).

CHTA and CHFA, 2019. A call for smart regulation of CBD. Available from: <https://chfa.ca/Portals/30/RegAffairs/white-paper/2019-05-01%20CBD%20White%20Paper%20(Final).pdf?ver=2019-07-16-155719-820> (Accessed 19 February 2020).

E.D.N.Y. 1995. United States v. Ten Cartons, Ener-B Nasal Gel, 888 F. Supp. 381 (E.D.N.Y. 1995). Justia Law. Available from: <https://law.justia.com/cases/federal/district-courts/FSupp/888/381/1644175/> (Accessed 19 March 2020).

FDA 2007. Current good manufacturing practice in manufacturing, packaging, labeling, or holding operations for dietary supplements. Fed. Regist. Available from: <https://www.federalregister.gov/documents/2007/06/25/07-3039/current-good-manufacturing-practice-in-manufacturing-packaging-labeling-or-holding-operations-for> (Accessed 14 April 2020).

FDA 2011. FDA warning letters: HBB, LLC dba Baked World 7/28/11. FDA Warn. Lett. Available from: <http://fda-warning-letters.blogspot.com/2011/07/department-of-health-and-human-services_28.html> (Accessed 19 March 2020).

FDCA 402(f), 2006. United States Code, 2006 Edition, Supplement 5, Title 21—Food and Drugs. Section 342.

FDCA 413, 2006. United States Code, 2006 Edition, Supplement 4, Title 21—Food and Drugs.

Government of Canada, 2017. Unauthorized natural health products including "Optimum Performance Nutrition," "SmartBrain Formulations" and "Cutting Edge Naturals" sold by Robert Lamberton Consulting may pose health risks. Available from: <https://www.healthycanadians.gc.ca/recall-alert-rappel-avis/hc-sc/2017/65400a-eng.php> (Accessed 14 April 2020).

Government of Canada, H.C., 2019. Health Canada reminds Canadians of the dangers of unauthorized health products; it will not hesitate to take enforcement action if needed. Available from: <https://healthycanadians.gc.ca/recall-alert-rappel-avis/hc-sc/2019/71800a-eng.php> (Accessed 14 April 2020).

Health Canada, 2002. Natural Health Products in Canada—a history. AEM. Available from: <https://www.canada.ca/en/health-canada/services/drugs-health-products/natural-non-prescription/regulation/history.html> (Accessed 19 February 2020).

Health Canada, 2003. Compendium of monographs. AEM. Available from: <https://www.canada.ca/en/health-canada/services/drugs-health-products/natural-non-prescription/applications-submissions/product-licensing/compendium-monographs.html> (Accessed 20 February 2020).

Health Canada, 2004. Natural health products ingredients database. Available from: <http://webprod.hc-sc.gc.ca/nhpid-bdipsn/search-rechercheReq.do?lang = eng> (Accessed 19 February 2020).

Health Canada, 2006a. Evidence for homeopathic medicines. AEM. Available from: <https://www.canada.ca/en/health-canada/services/drugs-health-products/natural-non-prescription/legislation-guidelines/guidance-documents/evidence-homeopathic-medicines.html> (Accessed 19 February 2020).

Health Canada, 2006b. Guidelines for the safety assessment of novel foods. Available from: <https://www.canada.ca/en/health-canada/services/food-nutrition/legislation-guidelines/guidance-documents/guidelines-safety-assessment-novel-foods-derived-plants-microorganisms/guidelines-safety-assessment-novel-foods-2006.html> (Accessed 19 February 2020).

Health Canada, 2010a. Health Products and Food Branch Inspectorate—Natural Health Products Compliance and Enforcement Policy (POL-0044). AEM. Available from: <https://www.canada.ca/en/health-canada/services/drugs-health-products/compliance-enforcement/information-health-product/natural-health-products/health-products-food-branch-inspectorate-policy-0044.html> (Accessed 19 February 2020).

Health Canada, 2010b. Import and Export Policy for Health Products under the Food and Drugs Act and its Regulations (POL-0060). AEM. Available from: <https://www.canada.ca/en/health-canada/services/drugs-health-products/compliance-enforcement/importation-exportation/policy-health-products-under-food-drugs-act-regulations-0060.html> (Accessed 19 February 2020).

Health Canada, 2012a. Pathway for licensing Natural Health Products making modern health claims. AEM. Available from: <https://www.canada.ca/en/health-canada/services/drugs-health-products/natural-non-prescription/legislation-guidelines/guidance-documents/pathway-licensing-making-modern-health-claims.html> (Accessed 19 February 2020).

Health Canada, 2012b. Pathway for licensing Natural Health Products used as traditional medicines. AEM. Available from: <https://www.canada.ca/en/health-canada/services/drugs-health-products/natural-non-prescription/legislation-guidelines/guidance-documents/pathway-licensing-traditional-medicines.html> (Accessed 19 February 2020).

Health Canada, 2013. General guidance document for temporary marketing authorization for foods. Available from: <https://www.canada.ca/en/health-canada/services/food-nutrition/legislation-guidelines/guidance-documents/general-guidance-document-temporary-marketing-authorization-foods-2012.html> (Accessed 19 February 2020).

Health Canada, 2014. Category specific guidance for temporary marketing authorization - caffeinated energy drinks. Available from: <https://www.canada.ca/en/health-canada/services/food-nutrition/legislation-guidelines/guidance-documents/category-specific-guidance-temporary-marketing-authorization-caffeinated-energy-drinks.html> (Accessed 19 February 2020).

Health Canada, 2015a. Site licensing guidance document. Available from: <https://www.canada.ca/en/health-canada/services/drugs-health-products/natural-non-prescription/legislation-guidelines/guidance-documents/site-licensing-guidance-document.html> (Accessed 19 February 2020).

Health Canada, 2015b. Good manufacturing practices guidance document. Available from: <https://www.canada.ca/en/health-canada/services/drugs-health-products/natural-non-prescription/legislation-guidelines/guidance-documents/good-manufacturing-practices.html> (Accessed 19 February 2020).

Health Canada, 2015c. Quality of Natural Health Products Guide. Available from: <https://www.canada.ca/en/health-canada/services/drugs-health-products/natural-non-prescription/legislation-guidelines/guidance-documents/quality-guide.html> (Accessed 19 February 2020).

Health Canada, 2015d. Unauthorized health product sold on amazon.ca. Available from: <https://www.healthycanadians.gc.ca/recall-alert-rappel-avis/hc-sc/2015/55926a-eng.php> (Accessed 19 February 2020).

Health Canada, 2016a. Evaluation of the Natural Health Products Program 2010-2011 to 2014-2015. Available from: <https://www.canada.ca/en/health-canada/corporate/transparency/corporate-management-reporting/evaluation/2010-2011-2014-2015-natural-health-products-program.html#fnaxi> (Accessed 14 April 2020).

Health Canada, 2016b. Category specific guidance for temporary marketing authorization: supplemented food. Available from: <https://www.canada.ca/en/health-canada/services/food-nutrition/legislation-guidelines/guidance-documents/category-specific-guidance-temporary-marketing-authorization-supplemented-food.html> (Accessed 19 February 2020).

Health Canada, 2017. Guidance document: classification of products at the food-natural health product interface: products in food formats. Available from: <https://www.canada.ca/en/health-canada/services/drugs-health-products/natural-non-prescription/legislation-guidelines/guidance-documents/classification-products-at-food-natural-health-product-interface.html> (Accessed 19 February 2020).

Health Canada, 2018a. Natural Health Products Management of Applications Policy. AEM. Available from: <https://www.canada.ca/en/health-canada/services/drugs-health-products/natural-health-products/legislation-guidelines/guidance-documents/management-product-licence-applications-attestations.html> (Accessed 19 February 2020).

Health Canada, 2018b. Reporting adverse reactions to marketed health products—guidance document for industry. AEM. Available from: <https://www.canada.ca/en/health-canada/services/drugs-health-products/reports-publications/medeffect-canada/reporting-adverse-reactions-marketed-health-products-guidance-industry.html> (Accessed 19 February 2020).

Health Canada, 2018c. Good label and package practices guide for non-prescription drugs and Natural Health Products. Available from: <https://www.canada.ca/en/health-canada/services/drugs-health-products/reports-publications/medeffect-canada/good-label-package-practices-guide-non-prescription-drugs-natural-health-products-guidance.html#a254> (Accessed 26 November 2018).

Health Canada, 2019a. Next steps on the self-care framework. Available from: <https://www.canada.ca/en/health-canada/services/self-care-framework.html> (Accessed 19 February 2020).

Health Canada, 2019b. Document: consultation on potential market for cannabis health products that would not require practitioner oversight. Available from: <https://www.canada.ca/en/health-canada/programs/consultation-potential-market-cannabis/document.html> (Accessed 19 February 2020).

Klontz, K.C., DeBeck, H.J., LeBlanc, P., Mogen, K.M., Wolpert, B.J., Sabo, J.L., et al., 2015. The role of adverse event reporting in the FDA response to a multistate outbreak of liver disease associated with a dietary supplement. Public Health Rep. 130, 526–532. Wash. DC 1974. Available from: <https://doi.org/10.1177/003335491513000515>.

Legislative Services Branch, Government of Canada, 2018a. Consolidated Federal Laws of Canada, Natural Health Products Regulations. Available from: <https://laws-lois.justice.gc.ca/eng/regulations/sor-2003-196/> (Accessed 19 February 2020).

Legislative Services Branch, Government of Canada, 2018b. Consolidated Federal Laws of Canada, Cannabis Act. Available from: <https://laws-lois.justice.gc.ca/eng/acts/c-24.5/> (Accessed 19 February 2020).

Legislative Services Branch, Government of Canada, 2019a. Consolidated Federal Laws of Canada, Food and Drugs Act. Available from: <https://laws-lois.justice.gc.ca/eng/acts/f-27/> (Accessed 19 February 2020).

Legislative Services Branch, Government of Canada, 2019b. Consolidated Federal Laws of Canada, Food and Drug Regulations. Available from: <https://laws.justice.gc.ca/eng/regulations/c.r.c.,_c._870/index.html> (Accessed 19 February 2020).

Meola, S.D., Tearney, C.C., Haas, S.A., Hackett, T.B., Mazzaferro, E.M., 2012. Evaluation of trends in marijuana toxicosis in dogs living in a state with legalized medical marijuana: 125 dogs (2005–2010). J. Vet. Emerg. Crit. Care 22, 690–696. Available from: <https://doi.org/10.1111/j.1476-4431.2012.00818.x>.

Miousse, I.R., Skinner, C.M., Lin, H., Ewing, L.E., Kosanke, S.D., Williams, D.K., et al., 2017. Safety assessment of the dietary supplement OxyELITE™ Pro (New Formula) in inbred and outbred mouse strains. Food Chem. Toxicol. Int. J. Publ. Br. Ind. Biol. Res. Assoc. 109, 194–209. Available from: <https://doi.org/10.1016/j.fct.2017.08.025>.

Nutrilab, Inc. v. Schweiker (1983), 713 F.2d 335, USCA 7th Circuit.

Office of the Commissioner, 2019. MOU 225-07-7001. FDA. <http://www.fda.gov/about-fda/domestic-mous/mou-225-07-7001> (Accessed 19 March 2020).

Chapter 73

Evaluation and regulation of food supplements: European perspective

Arturo Anadón, Irma Ares, María-Rosa Martínez-Larrañaga and María-Aránzazu Martínez

Introduction

The regulation of food/dietary supplements is one of the most complex topics in food law. For many years food supplements were simply food items intended to balance the diet or provide for special dietary needs. Healthy nutrition requires appropriate dietary intake of energy in the form of macronutrients (i.e., protein, fat, carbohydrates, and total calories) as well as the adequate intake (AI) of essential nutrients, which comprise vitamins, trace elements, minerals, essential fatty acids, and essential amino acids. A person's requirements for different nutrients are related to his or her energy requirements, age, height, and weight. They will differ according to different physiological conditions, such as whether the person is growing or not and whether a woman is pregnant or lactating. Lifestyle factors such as level of physical activity, stress, smoking, and alcohol consumption also may affect nutrient requirements.

Nutrients are essential for human well-being and health and/or to reduce the risk of a disease within a certain range of intake. There is a long history of safe consumption of nutrients at the levels found in balanced human diets. Additionally, some nutrients have been widely chronically consumed (e.g., from food/dietary supplements) at levels significantly above those obtained from endogenous nutrients in foods without reported adverse effects. On the whole, nutrients possess some characteristics that distinguish themselves from other food chemicals for the purpose of risk assessment.

There is a growing market for food supplement products with perceived and real health benefits. This development combined with the consumers' general and mistaken perception that *natural equals safe* or *natural equals healthy* results in a tendency for increased use of food supplemental micronutrients and also botanical products as bioactive ingredients in food supplements and herbal teas. Nevertheless, a long history of use does not guarantee safety (e.g., botanical preparations may contain individual ingredients known to be genotoxic and carcinogenic). However, there is a global concern over the rising rate of *noncommunicable diseases* (NCDs), also known as chronic diseases, which are not transmitted from person to person. NCDs can progress slowly and persist in the body for decades. An unhealthy, unbalanced diet rich in unsaturated fats, sugars, and salts may contribute to a wide range of NCDs. The main types of NCDs include cardiovascular disease, cancers, respiratory diseases, and diabetes, and together they are responsible for three out of five deaths worldwide (Department of Health, 1998). These diseases are linked by common lifestyle-related risk factors, including unhealthy dietary choices and physical inactivity; in that regard, the rising trend in the rates of overweight individuals and obesity is of concern. Clearly, effective and coordinated action is needed, and much of the disease burden is preventable by interventions such as the provision of clear, simple, and logical information to the public regarding nutrition.

Accompanying the new food products with healthy benefits are new terms. New terms such as *nutraceutical*, created from the words "nutrition" and "pharmaceutical," to describe foods that contain a supplement or enhancement to provide a specific health benefit (e.g., a combination of vitamins, amino acids, and plant preparations in a capsule). Another term is *functional food*, which was introduced in Japan in the mid-1980s and used as a synonym for *nutraceuticals*; these are designer food, pharmafood, medifood, vitafood, and the more traditional food/dietary supplements and fortified foods. To conclude, nutraceutical is a term applied to many products, from functional foods and food supplements, to dietary foods able to provide health benefits. Functional foods, food supplements, and nutraceuticals are at the interface between nutrition and medicines. The variety of terms, more or less related to the Japanese Foods for Specified

Nutraceuticals. DOI: https://doi.org/10.1016/B978-0-12-821038-3.00073-2

Health Use (FOSHU), has appeared worldwide. In 1991, the functional food concept was integrated into the FOSHU Japanese system and the term "health food" was recognized and replaced the term "functional food." However, *functional food* more often refers to conventional food products that are enhanced to provide specific health benefits (as opposed to pills or capsules, liquids, and powers but part of a normal food pattern, e.g., calcium-enriched orange juice or higher calcium yogurt). Functional foods can be classified into different groups according to their effects: vitamins, minerals, antioxidants, polyphenols, and polyunsaturated fatty acids (PUFAs) [omega-3-(n-3) fatty acids and omega-6-(n-6) fatty acids], among others. The major omega-3 fatty acids are α-linolenic acid, eicosapentaenoic acid, and docosahexaenoic acid, and the mega-6-PUFAs mainly consist of linoleic acid, γ-linolenic acid, and arachidonic acid (Das et al., 2012). All are foods or food products marketed with the message of a benefit to health. For example, nutraceuticals include most of the therapeutics such as antiarthritic, cold and cough, sleeping disorders, digestion and prevention of certain cancers, osteoporosis, blood pressure, cholesterol control, pain killers, depression, and diabetes (Das et al., 2012).

In Europe, Japan, and the United States, functional products as functional foods, food supplements, and nutraceuticals satisfy these requirements and are offered for reducing the risk and preventing cardiovascular diseases. These products are difficult to distinguish between nutrition and medicines. The European regulation does not consider *functional food* or *nutraceuticals* as specific food categories and they must meet general food laws. A number of legislative texts that may be of importance to such foods have already been elaborated, including the General Food Regulation, the Food Supplements Directive (FSD), the Novel Food Regulation, and the Nutrition and Health Claims, regarding vitamins, minerals, and other substances.

The European Union's regulatory framework on food supplements and other similar nutritional products

The European Union (EU), in an effort to harmonize the internal market for food supplements and similar nutritional products, has adopted some key pieces of legislation. It has taken steps to harmonize the regulation of nutritional supplements, health foods, and herbal medicines across the EU. Although some diverging regulations among EU Member States arise, causing challenges for companies in the sector, the harmonization of EU rules overall provides an opportunity for exporters entering the single market of 27 EU Member States. These products are regulated in EU Member States by differing national rules that may restrain their free movement, create unequal conditions of competition, and therefore have a direct impact on the functioning of the EU internal market. It should be noted that the EU is a free market and anything that infringes the free movement of goods and services within the EU will attract the attention of the European Commission on the grounds that it could indicate a contravention of the Treaty of Rome (EC, 2002a). The European food safety legislation has as its goal *a high level of protection for human life and health* whereby *mutatis mutandis*, the potential benefits of micronutrients by definition cannot be neglected.

Each European country varies in the extent and process of regulatory control and the market for food supplements is extremely varied from one Member State to another regarding the substances used. Some EU Member States use a considerable amount of vitamin and mineral supplements and have introduced regulation, whereas others use them to a much lesser extent. The market share of products containing vitamins and minerals and that of products containing other substances are almost identical. The use of these substances is therefore subject to the rules enforced in national legislation without prejudice to any other Community provisions that are applied to them. The majority of EU Member States have drawn up positive or negative lists of substances other than vitamins and minerals that can be used in food supplements. However, the European Commission does not rule out the possibility of performing a supplementary analysis examining the conditions for the addition of these substances to foodstuffs in general.

One of the most important EU legislations is the FSD, Directive 2002/46/EC (EC, 2002b), adopted on June 10, 2002, which establishes harmonized rules for the labeling of food supplements and introduces specific rules on vitamins and mineral salts in food supplements. The FSD concerns food supplements, defined as concentrated sources of nutrients (vitamins and mineral salts) or other substances with a nutritional or physiological effect, alone or in combination, whose purpose is to supplement the normal diet. They are marketed in *dose form* (i.e., as pills, tablets, capsules, sachets, and liquids in measured doses, among others). Food supplements may be used to correct nutritional deficiencies or to maintain an AI of certain nutrients. As a consequence, the FSD was implemented to safeguard human health because in some cases excessive intake of vitamins and minerals may be harmful or cause unwanted side effects; therefore maximum levels are necessary to ensure their safe use in food supplements.

The FSD, in large part, is based on the risk assessment paradigm: the toxicological risks of excess intake of food-derived micronutrients are elucidated.

An important element in the FSD is that food supplements are considered as foodstuffs. Therefore food supplements are defined in the Directive 2002/46/EC (EC, 2002b) as "foodstuffs, the purpose of which is to supplement the normal diet and which are concentrated sources of nutrients or other substances with a nutritional or physiological function, alone or in combination, marketed in dose form, namely forms such as capsules, pastilles, tablets, pills and other similar forms, sachets of powder, ampules of liquids, drop dispensing bottles, and other similar forms of liquids and powders designed to be taken in measured small unit quantities." Food supplements are marketable finished products that are precisely presented to the public for supplementation of the diet and may or may not exceed the intake of micronutrients present in the consumed diet. Micronutrients (vitamins and minerals) are required by the body throughout the lifecycle in extremely small amounts. Food supplements cannot be presented as medicines or as a substitute for medicines, and the Food Supplement Directive does not apply to medicinal products as defined by Directive 2001/83/EC (EC, 2001a) (latest amendment in 2011). It is important to be explicit that the European Directive applies only to products sold in the EU; it does not apply to retailers in the United States. Therefore retailers operating in the United States will be able to continue selling products. Although Directive only applies to EU Member States, manufacturers, distributors, and retailers that have exported dietary supplements containing ingredients that are not on the Directive 2002/46/EC (EC, 2002b) list are no longer able to sell those products in the EU effective August 1, 2005.

The European Commission has established harmonized rules to help ensure that food supplements are safe and properly labeled. In the EU, food supplements are regulated as foods and legislation focuses on vitamins and minerals used as ingredients of food supplements. Rules for nutrients other than vitamins and minerals should be developed in the future, and for now they are governed as previously described by Member State legislation.

Vitamins and (essential) minerals are micronutrients which are essential components of the human diet and the human body, and these are only required in relatively small amounts. Vitamins and mineral deficiency affect the lives of perhaps one-third of the world's people. Like other chemical substances, micronutrients may have adverse (toxicological) effects if consumed in excessive amounts; hypervitaminosis is usually associated with vitamins A and D, and cases of acute vitamin A intoxication have been reported in polar regions after eating plentiful amounts of polar bear liver. However, when evaluating the adverse effects of micronutrients, it is necessary to take into account that in contrast to nonessential chemical substances, there is a (lower) level of intake below which risk of deficiency conditions (diseases) or suboptimal functioning arises. Micronutrient deficiency is a lack or shortage of a micronutrient that is essential in small amounts for the proper growth and metabolism of a human or other living organism. Some micronutrients are isolated or extracted from natural materials; others are produced by way of fermentation or chemical synthesis. All these food content chemicals have their own specific pharmaco-toxicological profile, both individually and interactively (synergism and antagonism) (Hanekamp and Bast, 2007).

Directive 2002/46/EC (EC, 2002b) requires that EU-wide maximum and minimum levels are set for each vitamin and mineral added to supplements. Because excessive intake of vitamins and minerals may result in adverse effects (i.e., toxicological risks of excess intake), the Directive provides for the setting of maximum amounts of vitamins and minerals added to food supplements. This task has been delegated to the Commission and is currently ongoing. However, a number of scientific bodies and stakeholders have considered the issue of setting maximum levels of vitamins and minerals in foods and have elaborated several models; examples of some of these models have been developed by the French Agency of Food Safety (AFSSA, actually termed ANSES, French Agency for Food, Environmental and Occupational Health & Safety), the Danish Institute of Food and Veterinary Research, and the Federal Institute for Risk Assessment (BfR) of Germany. The application of divergent maximum permitted levels of vitamins and minerals in EU Members States could cause serious problems to the free circulation of products on the European market.

Labeling

The products covered by Directive 2002/46/EC (EC, 2002b) are sold under the name "food supplements" without prejudice to the provisions of Directive 2000/13/EC (EC, 2000) relating to the labeling, presentation, and advertising of foodstuffs. The labeling for food supplements must contain, according with article 6 of Directive 2002/46/EC (EC, 2002b), the following requirements: (1) the names of the categories of the nutrients or substances that characterize the product or an indication of the nature of those nutrients or substances; (2) the portion of the product recommended for daily consumption; (3) a warning not to exceed the stated recommended daily dose; (4) a declaration to the effect that the supplement is not a substitute for a varied diet (the reference "this is not a medicinal product," when the presentation of the products is similar to that of a medicinal products); and (5) a statement to the effect that food supplements should not be used as a substitute for a varied diet.

The labeling of food supplements according to Directive 2002/46/EC (EC, 2002b) must not contain: (1) any statement attributing to the product properties of preventing, treating, or curing a human disease; and (2) any mention stating or implying that a balanced and varied diet cannot provide appropriate quantities of nutrients in general.

Vitamin and mineral levels in food supplements

Directive 2002/46/EC on food supplements (EC, 2002b) calls for the establishment of minimum dosage amounts for a product to be considered a nutritional supplement as well as upper safe limits to protect the consumer. However, the European Commission and the European Food Safety Authority (EFSA) have not made progress establishing these limits, and this competence remains with the Member States. Directive 2002/46/EC on food supplements (EC, 2002b) provides for the setting of maximum and minimum amounts of vitamins and minerals in these products via the Standing Committee procedure. However, the criteria for setting the minimum and maximum levels for such nutrients added to food should be on a scientific basis and the responsibility under this Directive lies with the EFSA, which is also tasked with determining the safe maximum and minimum levels of vitamins and minerals.

Article 5 of Directive 2002/46/EC (EC, 2002b) reads: "(1) Maximum amounts of vitamins and minerals present in food supplements per daily portion of consumption as recommended by the manufacturer shall be set, taking the following into account: (a) upper safe levels of vitamins and minerals established by scientific risk assessment based on generally accepted scientific data, taking into account, as appropriate, the varying degrees of sensitivity of different consumer groups; (b) intake of vitamins and minerals from other dietary sources; (2) When the maximum levels referred to in paragraph 1 are set, due account should also be taken of reference intakes (RI) of vitamins and minerals for the population; (3) To ensure that significant amounts of vitamins and minerals are present in food supplements, minimum amounts per daily portion of consumption as recommended by the manufacturer shall be set, as appropriate; and (4) The maximum and minimum amounts of vitamins and minerals referred to in paragraphs 1, 2 and 3 shall be adopted in accordance with the procedure referred to in Article 13(2)."

Similar provisions for the establishment of maximum levels for vitamins and minerals in foodstuffs are present in the proposal for a Regulation regarding the addition of vitamins and minerals and of certain other substances to foods. Article 6 of this proposal for a Regulation reads: "(1 When a vitamin or a mineral is added to foods, the total amount of the vitamin or mineral present, for whatever purpose, in the food as sold shall not exceed maximum amounts that shall be set in accordance with the procedure referred to in Article 14(2). For concentrated and dehydrated products, the maximum amounts set shall be those present in the foods when prepared for consumption according to the manufacturer's instructions; (2) Any conditions restricting or prohibiting the addition of a specific vitamin or mineral to a food or a category of foods shall be adopted in accordance with the procedure referred to in Article 14(2); (3) The maximum amounts referred to in paragraph 1 and the conditions referred to in paragraph 2 shall be set taking the following into account: (a) upper safe levels of vitamins and minerals established by scientific risk assessment based on generally acceptable scientific data, taking into account, as appropriate, the varying degrees of sensitivity of different groups of consumers; (b) intakes of vitamins and minerals from other dietary sources; and (4) when the maximum amounts referred to in paragraph 1 and the conditions referred to in paragraph 2 are set for vitamins and minerals whose RI for the population are close to the upper safe levels, the following shall also be taken into account, as necessary: (a) the contribution of individual products to the overall diet of the population in general or of subgroups of the population; (b) the nutrient profile of the product established as provided for by Regulation (EC) No. 1924/2006 on nutrition and health claims" (EC, 2006a).

Directive 2002/46/EC (EC, 2002b) calls for the establishment of harmonized minimum and maximum dosage amounts; however, this has yet to be done and remains a task for the EU Member States. Also, substances other than vitamins and minerals are not directly covered by the Directive and rules regulating these substances are still governed by individual EU Member States. Commission Regulation (EU) 2017/1203 (EU, 2017) amending Directive 2002/46/EC (EC, 2002b) (last amendment) regards to "silicic acid":organic silicon (monomethylsilanetriol) and calcium sulfate:calcium phosphoryl oligosaccharides added to foods and used in the manufacture of food supplements. [Commission Regulation (EU) 2017/1203] (EU, 2017) provides two lists with the vitamins and minerals (Annex I) and vitamin and mineral substances (Annex II) permitted in the manufacture of food supplements.

Food supplements: vitamins and minerals

Directive 2002/46/EC (EC, 2002b) establishes a definition for food supplements and a positive list of allowable vitamins and minerals, and sets labeling requirements, presentation, and advertising. The Directive set up a positive list of allowable vitamins and minerals and their associated forms to be included in food supplements which may be sold in Europe, and these are listed in Annexes I and II of the Directive (Tables 73.1 and 73.2). Annex I of Directive 2002/46/EC (EC, 2002b) contains vitamins and minerals that may be used in the manufacture of food supplements, and Annex II of this Directive comprises a list of permitted vitamin or mineral preparations (i.e., the sources of vitamins and minerals) that may be added for specific nutritional purposes in food supplements. Annex II has been amended by Commission Directive 2006/37/EC (EC, 2006b), Commission Regulation (EC) 1170/2009 (EC, 2009a), Commission Regulation (EU) No. 1161/2011 (EU, 2011a), and Commission Regulation (EU) No. 119/2014 (EU, 2014). The list as amended by Regulation (EC) 1170/2009 (EC, 2009a) includes vitamins with allowable sources and minerals with permissible sources (Tables 73.3 and 73.4). From August 1, 2005, manufacturers, distributors, and retailers of food supplements in the EU are not allowed to sell food supplements containing vitamins and minerals and forms other than those listed in Annexes I and II of the Directive. Annex I of Regulation (EC) No. 1925/2006 (EC, 2006c) is a list of vitamins and minerals which may be added to foods. Annex II is a list of the sources of vitamins and minerals which may be added to foods. Annexes I and II have been amended by three Commission Regulations to include additional substances (Tables 73.3 and 73.4). Annex III is a list of substances other than vitamins or minerals whose use in foods is prohibited, restricted, or under Community scrutiny. Annex III has been amended by Commission Regulation (EU) No 2015/403, prohibiting the use of *Ephedra* herb and its preparations from *Ephedra* species and subjecting *Yohimbe* bark and its preparations originating from Yohimbe [*Pausinystalia yohimbe* (K. Schum) Pierre ex Beille] to Community scrutiny. Certain EU Member States may provide for variation from this list. However, these variations were abolished on January, 2010. Therefore the manufacturers or suppliers whose products had not been evaluated were not on the positive list until the end of 2009.

Additional ingredients may be included in Annexes I and II of Directive 2002/46/EC (EC, 2002b) on the request of an interested party. To do this, a party must

TABLE 73.1 Vitamins that may be used in the manufacture of food supplements.

Vitamin A (μg RE)[a]	Retinol; retinyl acetate; retinyl palmitate; β-carotene
Vitamin D (μg)[a]	Cholecalciferol; ergocalciferol
Vitamin E (mg α-TE)[a,b]	D-α-tocopherol; DL-α-tocopherol; D-α-tocopheryl acetate; DL-α-tocopheryl acetate; D-α-tocopheryl acid succinate; Mixed tocopherols*; Tocotrienol tocopherol***α-tocopherol <20 %, β-tocopherol < 10%, γ-tocopherol 50%–70% and δ-tocopherol 10%–30%**Typical levels of individual tocopherols and tocotrienols:115 mg/g α-tocopherol (101 mg/g minimum)5 mg/g β-tocopherol (<1 mg/g minimum)45 mg/g γ-tocopherol (25 mg/g minimum)12 mg/g δ-tocopherol (3 mg/g minimum)67 mg/g α-tocotrienol (30 mg/g minimum) < 1 mg/g β-tocotrienol (<1 mg/g minimum)82 mg/g γ-tocotrienol (45 mg/g minimum)5 mg/g δ-tocotrienol (<1 mg/g minimum)
Vitamin K (μg)[a,b]	(a)Phylloquinone (phytomenadione); menaquinone (occurring principally as menaquinone-7 and, to a minor extent, menaquinone-6)
Vitamin B1[a,b]	Thiamin hydrochloride; thiamin mononitrate; thiamine monophosphate chloride; thiamine pyrophosphate chloride
Vitamin B2[a]	Riboflavin; riboflavin 5′-phosphate, sodium
Niacin (mg NE)[a,b]	Nicotinic acid; nicotinamide; inositol hexanicotinate (inositol hexaniacinate)
Pantothenic acid (mg)[a,b]	D-pantothenate, calcium; D-pantothenate, sodium; dexpanthenol; pantethine
Vitamin B6 (mg)[a,b]	Pyridoxine hydrochloride; pyridoxine 5′-phosphate; pyridoxal 5′-phosphate
Folate (μg)[a,b,c]	(a)Pteroylmonoglutamic acid; calcium-L-methylfolate
Vitamin B12 (μg)[a,b]	Cyanocobalamin; hydroxocobalamin; 5′-deoxyadenosylcobalamin; methylcobalamin
Biotin (μg)[a]	D-biotin
Vitamin C (mg)[a,b]	L-ascorbic acid; sodium-L-ascorbate; calcium-L-ascorbate[b] (may contain up to 2% of threonate); potassium-L-ascorbate; L-ascorbyl 6-palmitate; magnesium L-ascorbate; zinc L-ascorbate

[a]*Directive 2002/46/EC.*
[b]*Commission Regulation (EC) No 1170/ 2009.*
[c]*Commission directive 2006/37/EC.*

TABLE 73.2 Minerals that may be used in the manufacture of food supplements.

Calcium (mg)[a,b]	Calcium acetate; calcium L-ascorbate; calcium bisglycinate; calcium carbonate; calcium chloride; calcium salts of citric acid; calcium gluconate; calcium glycerophosphate; calcium lactate; calcium pyruvate; calcium salts of orthophosphoric acid; calcium succinate; calcium hydroxide; calcium L-lysinate; calcium malate; calcium oxide; calcium L-pidolate, calcium L-threonate; calcium sulfate
Magnesium (mg)[a,b]	Magnesium acetate; magnesium L-ascorbate; magnesium bisglycinate; magnesium carbonate; magnesium chloride; magnesium salts of citric acid; magnesium gluconate; magnesium glycerophosphate; magnesium salts of orthophosphoric acid; magnesium lactate; magnesium L-lysinate; magnesium hydroxide; magnesium malate; magnesium oxide; magnesium L-pidolate; magnesium potassium citrate; magnesium pyruvate; magnesium succinate; magnesium sulfate; magnesium taurate; magnesium acetyl taurate
Iron (mg)[a,b]	Ferrous carbonate; ferrous citrate; ferric ammonium citrate; ferrous gluconate; ferrous fumarate; ferric sodium diphosphate; ferrous lactate; ferrous sulfate; ferric diphosphate (ferric pyrophosphate); ferric saccharate; elemental iron (carbonyl + electrolytic + hydrogen reduced); ferrous bisglycinate; ferrous L-pidolate; ferrous phosphate; iron (II) taurate; ferrous bisglycinate
Copper (μg)[a,b]	Cupric carbonate; cupric citrate; cupric gluconate; cupric sulfate; copper L-aspartate; copper bisglycinate; copper lysine complex; copper (II) oxide
Iodine (μg)[a]	Sodium iodide; sodium iodate; potassium iodide; potassium iodate
Zinc (mg)[a,b]	Zinc acetate; zinc L-ascorbate; zinc L-aspartate; zinc bisglycinate; zinc chloride; zinc citrate; zinc gluconate; zinc lactate; zinc L-lysinate; zinc malate; zinc mono-L-methionine sulfate; zinc oxide; zinc carbonate; zinc L-pidolate; zinc picolinate; zinc sulfate
Manganese (mg)[a,b]	Manganese ascorbate; manganese L-aspartate; manganese bisglycinate; manganese carbonate; manganese chloride; manganese citrate; manganese gluconate; manganese glycerophosphate; manganese pidolate; manganese sulfate
Sodium (mg)[a]	Sodium bicarbonate; sodium carbonate; sodium chloride; sodium citrate; sodium gluconate; sodium lactate; sodium hydroxide; sodium salts of orthophosphoric acid
Potassium (mg)[a,b]	Potassium bicarbonate; potassium carbonate; potassium chloride; potassium citrate; potassium gluconate; potassium glycerophosphate; potassium lactate; potassium hydroxide; potassium L-pidolate; potassium malate; potassium salts of orthophosphoric acid
Selenium (mg)[a,b]	L-selenomethionine; selenium-enriched yeast*; selenious acid sodium selenate; sodium hydrogen selenite; sodium selenite
	*Selenium-enriched yeasts produced by culture in the presence of sodium selenite as selenium source and containing, in the dried form as marketed, not more than 2.5 mg Se/g. The predominant organic selenium species present in the yeast is elenomethionine (between 60% and 85% of the total extracted selenium in the product). The content of other organic selenium compounds including selenocysteine shall not exceed 10% of total extracted selenium. Levels of inorganic selenium normally shall not exceed 1% of total extracted selenium
Chromium (μg)[a,b,c]	Chromium (III) chloride; chromium-enriched yeast**; chromium (III) lactate trihydrate; chromium nitrate; chromium picolinate; chromium(III) lactate trihydrate; chromium (III) sulfate
	**Chromium-enriched yeast produced by culture of *Saccharomyces cerevisiae* in the presence of chromium (III) chloride as a source of chromium and containing, in the dried form as marketed, 230–300 mg of chromium/kg. The content of chromium (VI) shall not exceed 0.2% of total chromium
Molybdenum (μg)[a,b]	Ammonium molybdate [molybdenum (VI)]; potassium molybdate [molybdenum (VI)]; sodium molybdate [molybdenum (VI)]
Fluoride (mg)[a,b]	Calcium fluoride; potassium fluoride; sodium fluoride; sodium monofluorophosphate
Chloride (mg)[a,b]	–
Phosphorus (mg)[a]	–
Boron (mg)[b]	Boric acid; sodium borate
Silicon (mg)[b]	Choline-stabilized orthosilicic acid; silicon dioxide; silicic acid (in the form of gel)

[a]*Directive 2002/46/EC.*
[b]*Commission Regulation (EC) No 1170/ 2009.*
[c]*Commission Regulation (EU) No 119/2014.*

TABLE 73.3 Vitamin formulations and mineral substances that may be added to foods; vitamin and mineral substances which may be used in the manufacture of food supplements.

Vitamin formulations	
Vitamin A	Retinol, retinyl acetate, retinyl palmitate, β-carotene
Vitamin D	Cholecalciferol, ergocalciferol
Vitamin E	D-α-tocopherol, DL-α-tocopherol, D-α-tocopheryl acetate, DL-α-tocopheryl acetate, D-α-tocopheryl acid succinate
Vitamin K	Phylloquinone (phytomenadione) menaquinone[a]
Vitamin B1	Thiamin hydrochloride, thiamin mononitrate
Vitamin B2	Riboflavin, riboflavin 5′-phosphate, sodium
Niacin	Nicotinic acid, nicotinamide
Pantothenic acid	D-pantothenate, calcium, D-pantothenate, sodium, dexpanthenol
Vitamin B6	Pyridoxine hydrochloride, pyridoxine 5′-phosphate, pyridoxine dipalmitate
Folic acid	Pteroylmonoglutamic acid, calcium-L-methylfolate
Vitamin B12	Cyanocobalamin, hydroxocobalamin
Biotin	D-biotin
Vitamin C	L-ascorbic acid, sodium-L-ascorbate, calcium-L-ascorbate, potassium-L-ascorbate, L-ascorbyl 6-palmitate

[a]*Menaquinone occurring principally as menaquinone-7 and, to a minor extent, menaquinone-6.*

provide a scientific dossier concerning the safety and bioavailability of the individual ingredients. Bioavailability of the nutrient or other ingredient from its source following oral consumption (from human data, from in vitro or animal studies, and from information on analogous substances) should be take into account. Furthermore, the ingredients concerned must have been positively evaluated by the EFSA. The Commission is responsible for establishing the purity criteria for substances contained in food supplements and the maximum and minimum quantities authorized, with the assistance of the Standing Committee on the Food Chain and Animal Health (SCFCAH).

Table 73.2 lists the minerals that may be used in the manufacture of food supplements following the Community regulations.

Vitamin formulations and mineral substances that may be added to foods following Commission Regulation (EC) No. 1170/2009 (EC, 2009b) are set down in Table 73.3.

Mineral formulations and mineral substances that may be added to foods following Commission Regulation (EC) No. 1170/2009 (EC, 2009a) are set down in Table 73.4.

Special considerations for essential nutrients are the derivation of maximum levels of vitamins and minerals added to foods based on risk assessment (Domke et al., 2005): (1) when evaluating the adverse effects of micronutrients, it has to be taken into account that, in contrast to nonessential substances, there is a level of intake below which the risk of deficiency or suboptimal function increases; (2) there is a long history of safe consumption of nutrients at the levels found in balanced human diets; (3) for some nutrients there may be chronic consumption at levels significantly above those obtained from endogenous nutrients in foods without reported adverse effects; (4) many nutrients are subject to homeostatic regulation of body content through adaptation of absorptive, excretory, or metabolic processes; (5) added nutrients may differ from endogenous nutrients in foods in a number of ways (e.g., chemical form, timing of intake and amount consumed in a bolus dose, effect of the food matrix, bioavailability, and interaction of the nutrient with other constituents of the diet); and (6) nutrients can be classified according to the risk of adverse effects through excessive intake and the degree of risk of IA (Table 73.5).

The following steps could be applied in establishing maximum levels of vitamins and minerals for use in food supplements and fortified food (Domke et al., 2005): (1) *hazard identification (nutrient)* [nutrient description, substance characterization; metabolism, function; recommended daily amount (RDA)]; (2) *intake assessment* (food intake data, biomarkers; groups susceptible to suboptimal intake/deficiency or excessive intake/toxicity); (3) *hazard characterization, dose−response assessment* (NOAEL/LOAEL), uncertainty factors, and upper intake level (UL) by SCF/EFSA; (4) *risk characterization* (conclusions from steps 1−3, information gaps/lack of data,

TABLE 73.4 Mineral substances which may be used in the manufacture of food supplements; mineral formulations and mineral substances that may be added to foods.

Mineral substances	
Calcium	Calcium carbonate, calcium chloride, calcium citrate malate, calcium salts of citric acid, calcium gluconate, calcium glycerophosphate, calcium lactate, calcium salts of orthophosphoric acid, calcium hydroxide, calcium malate, calcium oxide, calcium sulfate, calcium phosphoryl oligosaccharides
Magnesium	Magnesium acetate, magnesium carbonate, magnesium chloride, magnesium salts of citric acid, magnesium gluconate, magnesium glycerophosphate, magnesium salts of orthophosphoric acid, magnesium lactate, magnesium hydroxide, magnesium oxide, magnesium sulfate, magnesium potassium citrate
Iron	Ferrous bisglycinate, ferrous carbonate, ferrous citrate, ferric ammonium citrate, ferrous gluconate, ferrous fumarate, ferric sodium diphosphate, ferrous lactate, ferrous sulfate, ferrous ammonium phosphate ferric sodium EDTA, ferric diphosphate (ferric pyrophosphate), ferric saccharate, elemental iron (carbonyl + electrolytic + hydrogen reduced) cupric carbonate
Copper	Cupric citrate cupric gluconate cupric sulfate copper lysine complex
Iodine	Sodium iodide, sodium iodate, potassium iodide, potassium iodate
Zinc	Zinc acetate, zinc bisglycinate, zinc chloride, zinc citrate, zinc gluconate, zinc lactate, zinc oxide, zinc carbonate, zinc sulfate
Manganese	Manganese carbonate, manganese chloride, manganese citrate, manganese gluconate, manganese glycerophosphate, manganese sulfate
Sodium	Sodium bicarbonate, sodium carbonate, sodium citrate, sodium gluconate, sodium lactate, sodium hydroxide, sodium salts of orthophosphoric acid
Potassium	Potassium fluoride, potassium bicarbonate, potassium carbonate, potassium chloride, potassium citrate, potassium gluconate, potassium glycerophosphate, potassium lactate, potassium hydroxide, potassium salts of orthophosphoric acid
Selenium	Selenium-enriched yeast,* sodium selenite, sodium hydrogen selenite, selenite sodium fluoride
	*Selenium-enriched yeasts produced by culture in the presence of sodium selenite as selenium source and containing, in the dried form as marketed, not more than 2.5 mg Se/g. The predominant organic selenium species present in the yeast is selenomethionine (between 60% and 85% of the total extracted selenium in the product). The content of other organic selenium compounds including selenocysteine shall not exceed 10% of total extracted selenium. Levels of inorganic selenium normally shall not exceed 1% of total extracted selenium
Chromium (III)	Chromium (III) chloride and its hexahydrate, chromium (III) sulfate and its hexahydrate Chromium (III) chloride and its hexahydrate chromium (III) sulfate and its hexahydrate chromium picolinate chromium (III) lactate trihydrate
Molybdenum (VI)	Ammonium molybdate [molybdenum (VI)]; sodium molybdate [molybdenum (VI)]
Boron	Boric acid, sodium borate

and derivation of maximum levels of micronutrients for addition to supplements and normal foods).

Under article 4(6) of Directive 2002/46/EC (EC, 2002b), dossiers have been submitted to the European Commission. According to this article EU Member States may allow in their territory the use of vitamins and minerals not listed in Annex I of Directive 2002/46/EC (EC, 2002b) or in forms not listed in Annex II of Directive 2002/46/EC, provided that: (1) the substance in question is used in one or more food supplements marketed in the Community on the date of enforcement of this Directive, and (2) the EFSA has not given an unfavorable opinion in regards to the use of that substance, or its use in that form, in the manufacture of food supplements on the basis of a report document supporting the use of the substance in question to be submitted to the Commission by the EU Member State no later than July 12, 2005. The trade of products containing vitamins and minerals not listed in Annex II has been prohibited from August 1, 2005.

Food supplements: substances other than vitamins and minerals

Because Community law does not include specific provisions on the use of substances other than vitamins or

TABLE 73.5 Risk of excessive and inadequate intake.

Risk of excessive intake		Risk of inadequate intake	
Categories	**Criteria**	**Categories**	**Criteria**
High	Nutrients for which the margin between the RDA (or actual intake) and UL is very narrow (≤5) (e.g., vitamin A, vitamin D, cooper, iron, zinc)	*1*	Risk of clinically manifest deficiency or a depletion of body stores in specific age groups, with specific physiological conditions with specific eating habits/in specific regions (e.g., folate, iodine, vitamin D)
Moderate	Nutrients for which the UL is 5–100 times above the RDA (or actual intake) (e.g., vitamins E, K, C, and B6, magnesium, molybdenum)	2	Uncertainty about the risk of a clinically manifest deficiency or a depletion of body stores because of the lack of or the questionable validity of biomarkers, inadequate food tables, or a lack of epidemiological studies (e.g., vitamin K, biotin, fluoride, zinc, selenium)
Low	Nutrients for which a UL cannot be set as no adverse effects have been observed till now with an event at intake 100 times the RDA (e.g., vitamins B1, B2, and B12, chromium)	3	No indication of inadequate nutrient intake, or nutrient intake in the range of recommended intake (e.g., vitamins B1 and B2, copper)
		4	Indication of nutrient intake above recommended intake (e.g., vitamins B6 and B12, sodium, chloride, phosphorous)

RDA, recommended daily amounts; *UL*, upper level of intake.

minerals in food supplements, the free movement of such products is governed by Articles 28 to 30 thereof (article 11.2) of the EC Treaty (EC, 2002a) and can thus be subject to national restrictions or bans within the limits laid down by Article 30.

The General Food Law Regulation or, as it is officially called, Regulation (EC) No. 178/2002 of the European Parliament and of the Council of January 28, 2002, laid out the general principles and requirements of food law, establishing the EFSA, laid out procedures for matters of food safety (EC, 2002c), and contains the foundation of European food law. Its aim was to provide the basis for the assurance of a high level of protection of human health and consumer interest in relation to food, taking into account, in particular, diversity in the supply of food, including traditional products, while ensuring the effective functioning of the internal market. Currently there is a position of the European Parliament with a view to the adoption of regulation on the transparency and sustainability of the EU risk assessment in the food chain and amending Regulations (EC) No. 178/2002 (EP, 2019).

Before the General Food Law Regulation came into force, the risk assessment was performed by the Scientific Committee on Food (SCF) within the European Commission.

Food supplements containing substances other than vitamins or minerals are foodstuffs within the meaning of Article 2 of Regulation (EC) No. 178/2002 (EC, 2002c), which laid out the procedures in the area of food safety, with the regulatory committees consisting of representatives of the EU Member States that have a key role in decision-making regarding food safety issues reorganized within a single new structure, the SCFCAH. The new SCFCAH regulatory committee will assist the European Commission in the risk management process and in the development of food safety measures. Its mandate covers the entire food supply chain, ranging from animal health issues on the farm to the product that arrives on the consumer's table, thus significantly enhancing its ability to target risks to health wherever they arise in the production of food. The SCFCAH regulatory committee will consist of representatives of the EU Member States and be chaired by a European Commission representative. With the creation of the EFSA, risk assessment (*science-based*) was separate to risk management (*policy-based*). When the relevant EFSA scientific panel delivers its opinion on the application file, the European Commission will then present the opinion to the SCFCAH regulatory committee. If the SCFCAH regulatory committee approves the food supplement or a novel food ingredient, then it can be placed on the EU market.

Because the Regulation applies to all foodstuffs, its general principles also cover foods with added functional properties (e.g., *functional foods*, *nutraceuticals*, *dietetic foods*, and *food supplements*). Article 2 of the said

Regulation also states that *foodstuff* (or *food*) means any substance or products, whether processed, partially processed, or unprocessed, intended to be, or reasonably expected to be, ingested by humans. This Article 2 excludes from the definition of foodstuffs a series of product categories, including medicinal products within the meaning of Directive 2001/83/EC (EC, 2001a). Nutraceutical is a term derived from nutrition and pharmaceutical; nutraceuticals are substances that can be used in the prevention of nutrition-related diseases.

With regards to food supplements, and particularly those containing substances other than vitamins and minerals, a certain number of borderline cases have given rise, or could give rise to, situations in which a given product is authorized for marketing as a food in some EU Member States while the same product is classified as a medicinal product in other EU Member States.

The European Commission has prepared a report on the use of substances other than vitamins and minerals in food supplements and concluded that it was not feasible or necessary to lay out specific EU rules, given the already existing EU legislation that regulates their use and Member State legislation that should be mutually recognized throughout the EU.

Directorate-General Health and Consumers (DG SANCO) [become Directorate-General for Health and Food Safety (DG SANTÉ) in 2015] has commissioned a study on the use of substances with nutritional or physiological effects other than vitamins and minerals in food supplements. Taking into account this study and other available information, the Commission, in accordance with the requirement set out in Article 4(8) of Directive 2002/46/EC (EC, 2002b) on food supplements, has prepared a report for the Council and the European Parliament on the use of substances other than vitamins and minerals in food supplements. The report is accompanied by two Commission staff working documents: (1) characteristics and perspectives of the market for food supplements containing substances other than vitamins and minerals and (2) available scientific information on the use of substances other than vitamins and minerals in food supplements.

The number of substances other than vitamins and minerals used in food supplements on the EU market is estimated to be over 400 with a nutritional or physiological effect. Some of the most commercially significant, including fish oils, probiotics, and herbal ingredients, are listed in Table 73.6.

The scope of Directive 2002/46/EC (EC, 2002b) encompasses all food supplements; it does not cover the use of natural ingredients in food supplements. Accordingly, a natural food source of vitamins or minerals (e.g., cod liver oil) may be included in nutritional supplements as an ingredient even though the substance *cod liver oil* is not included in any of the Annexes to this Directive.

The discussion of the EU Directive for Food Supplements regarding foodstuffs including substances other than vitamins and minerals cannot be seen in total isolation from the other EU Directives for general application in the area of food safety legislation.

Nutrition and health claims

Although the legislation in the United States, Japan, and Europe are somewhat similar in nature, the approval and use procedures and the regulatory framework are different. In the European Union, a very important piece of legislation for the food supplements sector is Regulation (EC) No. 1924/2006 on nutrition and health claims (EC,

TABLE 73.6 Substances other than vitamins and minerals.

Amino acids	L-arginine, other essential amino acids, nonessential amino acids
Enzymes	Lactase, papaine
Pre- and probiotics[a]	Inulin, *Lactobacillus acidophilus*, *Bifidobacterium* species, yeast species
Essential fatty acids	γ-Linoleic acid, EPA/DHA, evening primrose oil [*Oenothera biennis* (L.)], borage oil (*Borago officinalis*), flax seed oil [*Linum usitatissimum* (L.)]
Botanicals[b] and botanical extracts	Aloe [*Aloe vera* (L.)], ginkgo (*Ginkgo biloba*), ginseng (*Panax ginseng*), garlic [*Allium sativum* (L.)], green tea extract (*Camellia sinensis*), garcinia extract (*Garcinia cambogia*), guarana extract (*Paullinia cupana*)
Miscellaneous bioactive substances[c]	Lycopene, lutein, coenzyme Q10, taurine, carnitine, inositol, glucosamine, chitosan, spiruline, soy isoflavone

[a]*Only a very limited number of microbial strains that are currently used in food products or as food supplements were considered as probiotics; probiotic preparations used can consist of one single strain or are mixed cultures of two (Lactobacillus acidophilus + Lactobacillus rhamnosus) or even more (several Lactobacillus species).*
[b]*Includes herbs, plants, fungi, algae, and extracts thereof.*
[c]*Any substance with a nutritional or physiological effect.*

2006a), which sets EU-wide conditions for the use of nutrition and health claims. The Regulation covers all possible claims relating to health. Article 6 of this Regulation indicates that health claims shall be based on and substantiated by generally accepted scientific evidence, and Article 13(1) indicates that health claims should be evaluated by the EFSA after a scientific assessment of the highest possible standard. The updated final list of 4637 claims was the result of a consolidation process carried out by the Commission, after examining over 44,000 claims supplied by the Member States. As we can see, in order to obtain a health claim there are difficulties because of the scientific data requirements.

The claims were evaluated on a case-by-case basis (some were clustered into one opinion) with respect to uniform scientific criteria for consistency. The opinions released by EFSA reflect the varying quality of the information submitted (characterization, dose, health relationship, scientific evidence) and the evaluation results were as follows: (1) about one-third was favorable with sufficient scientific evidence; (2) half of the nonfavorable evaluations had insufficient information on the food/constituent; and (3) probiotic claims on characterized strains with a nonfavorable outcome.

Regulation (EC) No. 1924/2006 (EC, 2006a) covers nutrition claims [claims that state, suggest, or imply that a food has particular nutrition properties due to the energy (caloric value) it provides, it provides at a reduced or increased rate, or does not provide, and/or the nutrients or other substances it contains, it contains in reduced or increased proportions, or does not contain]. A nutritional claim suggests that a food has beneficial nutritional properties, such as *low fat*, *no added sugar*, and *high in fiber*. Health claims are defined as claims that state, suggest, or imply that a relationship exists between a food category, a food, or one of its constituents and health, and as reduction of disease risk claims (claims that state, suggest, or imply that the consumption of a food category, a food, or one of its constituents significantly reduces a risk factor in the development of a human disease) (Coppens et al., 2006). A health claim is a statement that suggests a relationship between food and health, such as that a food can *help lower cholesterol*, *help reinforce the body's natural defenses*, or *enhance learning ability*. Regulation (EC) No. 1924/2006 (EC, 2006a) applies to any food or drink product produced for human consumption that is sold on the EU market. Nutrition and health claims will only be allowed on food labels if they are included on one of the EU positive lists. Food products carrying claims must comply with the provisions of nutritional labeling Directive 90/496/EC and its amended version that come into effect in 2011 (EEC, 1990) [Regulation (EU) No. 1169/2011] (EU, 2011b)

The EFSA has issued a scientific and technical guidance for the preparation and presentation of the application for authorization of a health claim (EFSA, 2011) under Regulation (EC) No. 1924/2006 (EC, 2006a), and requested by the European Commission. This guidance applies to health claims related to the consumption of a food category, a food, or its constituents (including a nutrient or other substance, or a combination of nutrients/other substances), hereafter referred to as a food/constituent (Anadón et al., 2014).

The purpose of this guidance is to assist applicants in preparing and presenting their applications for authorization of health claims that fall under Article 14 of Regulation (EC) No. 1924/2006 (EC, 2006a) (i.e., reduction of disease risk claims and claims referring to children's development and health). This guidance will be updated at a later stage to cover applications for authorization of the health claims that fall under Article 18 of the Regulation (EC) No. 1924/2006 (EC, 2006a). In other words, applications for inclusion of health claims in the Community list of permitted claims provided for in Article 13(3) based on newly developed scientific evidence and/or include a request for the protection of proprietary data. As specified in Regulation (EC) No. 1924/2006 (EC, 2006a), health claims should be substantiated by taking into account the totality of the available scientific data and by weighing the evidence and subject to the specific conditions of use. Scientific substantiation is the main aspect to be taken into account to authorize health claims. Claims should be scientifically substantiated by taking into account the totality of the available scientific data and by weighing the evidence, and should demonstrate the extent to which: (1) the claimed effect of the food/constituent is relevant for human health; (2) a cause-and-effect relationship is established between the consumption of the food/constituent and the claimed effect in humans (such as the strength, consistency, specificity, dose–response, and biological plausibility of the relationship); (3) the quantity of the food/constituent and pattern of consumption required to obtain the claimed effect could reasonably be achieved as part of a balanced diet; and (4) the specific study group(s) in which the evidence was obtained is representative of the target population for which the claim is intended.

The Scientific Panel on Dietetic Products, Nutrition and Allergies (NDA) of EFSA prepared a draft opinion that was published for public consultation. After considering all comments received, the Panel adopted its opinion on July 6, 2007. In 2011, the NDA Panel was requested by EFSA to revise the opinion with regard to the forms to be used for the submission of an application for authorization of health claims pursuant to Articles 13(5) and 14 of Regulation (EC) No. 1924/2006, and for the modification of an existing authorization in accordance with Article 19 of the same Regulation. The revision of the guidance, adopted by the NDA Panel on May 13, 2011, was of a purely

TABLE 73.7 Authorization of certain health claims referring to children's development and health.

Nutrient, substance, food, or food category	Claim	Conditions of use of the claim
α-Linolenic acid (ALA) and linoleic acid (LA), essential fatty acids[a,b]	Essential fatty acids are needed for normal growth and development of children	Information to the consumer that the beneficial effect is obtained with a daily intake of 2 g of ALA and a daily intake of 10 g of LA
Calcium and vitamin D[a]	Calcium and vitamin D are needed for normal growth and development of bone in children	The claim can be used only for food which is at least a source of calcium and vitamin D as referred to in the claim *source of (name of vitamin/s) and/or (name of mineral/s)* as listed in the Annex to Regulation (EC) No. 1924/2006
Calcium[a]	Calcium is needed for normal growth and development of bone in children	The claim can be used only for food which is at least a source of calcium as referred to in the claim *source of (name of vitamin/s) and/or (name of mineral/s)* as listed in the Annex to Regulation (EC) No. 1924/2006
Vitamin D[a]	Vitamin D is needed for normal growth and development of bone in children	The claim can be used only for food which is at least a source of vitamin D as referred to in the claim *source of (name of vitamin/s) and/or (name of mineral/s)* as listed in the Annex to Regulation (EC) No. 1924/2006
Protein[a]	Protein is needed for normal growth and development of bone in children	The claim can be used only for food which is at least a source of protein as referred to in the claim *source of protein* as listed in the Annex to Regulation (EC) No. 1924/2006
Phosphorus[c]	Phosphorus is needed for the normal growth and development of bone in children	The claim can be used only for food which is at least a source of phosphorus as referred to in the claim *source of (name of vitamin/s) and/or (name of mineral/s)* as listed in the Annex to Regulation (EC) No. 1924/2006

[a]*Regulation (EC) No 983/2009.*
[b]*Regulation (EC) No 376/2010.*
[c]*Regulation (EC) No 957/2010.*

administrative nature and concerned Parts 1–4, as well as the Appendices, of the guidance to simplify the presentation of an application (EFSA, 2011; Anadón et al., 2014).

In accordance with the requirements of Regulation (EC) No. 1924/2006, the guidance imposes the layout of the submission dossier based on five parts and the correspondent content of the application.

Part 1. Administrative and technical data: This contains the specific requirements for the administrative and technical data, such as the application form, information related to the applicant, and the nature of the application (including the national and international regulatory status of the health claim), health claim particulars, and the summary of the application.

Part 2. Food/constituent characteristics: This contains information specific to the food/constituent and its characteristics (such as the composition, physical and chemical characteristics, manufacturing process, stability, and bioavailability data).

Part 3. Overall summary of scientific data: This contains summaries (tabulated summaries of all pertinent studies identified and written summaries of data from pertinent human and nonhuman studies) and overall conclusions, which follow the scope and outline of the body of scientific data identified under Part 4. All pertinent studies (human studies, animal studies, in vitro studies, other) identified should be included (published and unpublished); individual studies included in any review publication should be counted separately.

Part 4. Body of pertinent scientific data identified: This contains all identified pertinent scientific data (published and unpublished, data in favor and not in favor) that form the basis for substantiation of the health claim.

Part 5. Annexes to the application: This comprises the glossary or abbreviation of terms quoted throughout the different sections, copies/reprints of pertinent publications identified, full study reports of unpublished pertinent data, and scientific opinions of national/international regulatory bodies.

Authorization is necessary for certain health claims made on foods and referring to children's development and health following the Community Regulations (Tables 73.7 and 73.8).

TABLE 73.8 Authorization of certain health claims referring to the reduction of disease risk and to children's development and health.

Nutrient, substance, food, or food category	Claim	Conditions of use of the claim
Iodine[a]	Iodine contributes to the normal growth of children	The claim can be used only for food which is at least a source of iodine as referred to in the claim *source of (name of vitamin/s) and/or (name of mineral/s)* as listed in the Annex to Regulation (EC) No. 1924/2006
Iron[a]	Iron contributes to normal cognitive development of children	The claim can be used only for food which is at least a source of iron as referred to in the claim *source of (name of vitamin/s) and/or (name of mineral/s)* as listed in the Annex to Regulation (EC) No. 1924/2006
Docosahexaenoic acid (DHA)[b]	DHA intake contributes to the normal visual development of infants up to 12 months of age	Information shall be given to the consumer that the beneficial effect is obtained with a daily intake of 100 mg of DHA. When the claim is used on follow-on formula, the food shall contain at least 0.3% of the total fatty acids as DHA
DHA[b]	DHA maternal intake contributes to the normal development of the eye of the fetus and breastfed infants	Information shall be given to pregnant and lactating women that the beneficial effect is obtained with a daily intake of 200 mg of DHA in addition to the recommended daily intake for omega-3 fatty acids for adults [i.e., 250 mg DHA and eicosapentaenoic acid (EPA)]. The claim can be used only for food which provides a daily intake of at least 200 mg DHA
DHA[b]	DHA maternal intake contributes to the normal brain development of the fetus and breastfed infants	Information shall be given to pregnant and lactating women that the beneficial effect is obtained with a daily intake of 200 mg of DHA in addition to the recommended daily intake for omega-3 fatty acids for adults (i.e., 250 mg DHA and EPA). The claim can be used only for food which provides a daily intake of at least 200 mg DHA

[a]*Regulation (EC) No. 1024/2009.*
[b]*Regulation (EC) No 440/2011.*

Health claims regarding botanicals have been considered after 2011. Disease risk-reduction claims and claims referring to the health and development of children require authorization on a case-by-case basis following the submission of a scientific dossier to EFSA. Health claims based on new scientific data will have to be submitted to EFSA for evaluation, but a simplified authorization procedure has been established. In conclusion, health claims express a relationship between a food component or a food supplement ingredient and a disease or health-related condition.

Novel food ingredients

Food supplements also need to comply with the EU rules on novel foods (i.e., nutraceuticals or functional foods). The Novel Foods Regulation [Regulation (EC) No 258/97] (EC, 1997a) provides rules for novel foods that were not used before May 15, 1997. For food and supplements containing novel ingredients to be allowed on the EU market, an application for authorization must be submitted to the competent authorities. The EU is currently revising the rules on novel foods that will cover foods produced by new techniques/technologies such as cloning, nanotechnology, and foods with a *history of safe use* in other countries such as noni juice. The proposal for a Regulation on novel foods (EC, 2013) was adopted in 2013, although the European Commission has adopted three draft laws on animal cloning and on novel food that will provide legal certainty in this field. Two proposals will ban the use of the cloning technique in the EU for farm animals and the imports of these animal clones; no such clone will be imported as long as animal welfare concerns persist. The marketing of food from animal clones will also be prohibited and no cloning for farming purposes will be performed in the EU.

The EU Novel Foods Regulation defines novel foods as foods and food ingredients that were not used for human consumption to a *significant degree* within the Community before May 15, 1997. If a food or food ingredient falls within one or more of the four mentioned food categories, then the decisive factor on whether it will be regarded as novel is whether it has already been sold under food law on the EU market before 1997.

Human consumption to a significant degree within the Community in this context has been interpreted as being demonstrated by a food having been generally available within the Community (i.e., food/food ingredient that does not have a significant history of human consumption within the EU prior to May 15, 1997). Foods that were consumed before this date are considered to be safe if no dangerous health effects were reported in the past. For new food products that enter the marker after this chosen date, safety is not automatically assumed.

A health claim is any statement used on labels, in marketing or in advertising that health benefits can result from consuming a given food or from one of its components such as vitamins and minerals, fiber, and "probiotic" bacteria. There are different types of health claims. For instance, statements that a food can help reinforce the body's natural defenses or enhance learning ability are called "general function" claims. An example includes products with phytosterols that were first marketed in 2001 and were therefore considered as novel. New applications of phytosterols need to go through the authorization or notification procedure. Examples also include claims on the reduction of disease risk and other substances that may improve or modify the normal functions of the body, for example, "Plant sterol have shown to reduce cholesterol levels, a risk factor in the development of coronary heart disease" or "Calcium may help improve bone density."

To date the novel food ingredients are not governed under specific EU regulatory frameworks, although Regulation (EC) No. 258/97 (EC, 1997b) states safety rules for authorization of novel food/ingredients. The principal objective of Regulation (EC) No. 258/97 (EC, 1997b) was to guarantee that new foods and food ingredients: (1) protect the functioning of the internal market within the Community and (2) protect public health and the basic criteria for the authorization of novel foods in that they must not: (a) present a danger for the consumer; (b) mislead the consumer; or (c) differ from foods or food ingredients that they are intended to replace to an extent that their normal consumption would be nutritionally disadvantageous for the consumer. Commission Recommendation 97/618/EC of July 29, 1997 (EC, 1997a) concerning the scientific aspects and presentation of information necessary to support applications for the placing on the market of novel foods and novel food ingredients and the preparation of initial assessment reports under Regulation (EC) No. 258/97 (EC, 1997b) recommends instruction necessary to sustain an application to fulfill the safety assessment.

There are four novel food/food ingredient categories: (1) presenting a new or intentionally modified primary molecular structure; (2) consisting of or isolated from microorganisms, fungi, or algae; (3) consisting of or isolated from plants or isolated from animals, except for foods and food ingredients obtained by traditional propagating or breeding practices and having a history of safe food use; and (4) those that have been applied in a production process not currently used, where that process gives rise to *significant changes* in the composition or structure that affect their nutritional value, metabolism, or level of undesirability.

Regulation (EU) 2015/2283 (EU, 2015) on novel food, repealing Regulation (EC) No 258/97 (EC, 1997b), demands a scientific assessment (prior to being released on the market) of all novel foods and new food ingredients without a safe and significant consumption history before May 15, 1997. This Regulation improves a number of important issues such as nanomaterials definition and labeling, a centralized and quicker authorization procedure for novel foods, and specific measures for traditional foods from developing countries. Functional food with novel ingredients or those which are made by novel processes fall under Regulation (EU) 2015/2283 (EU, 2015). This Regulation, which was enforced beginning January 1, 2018, is aimed at facilitating the authorization of novel food. EFSA is responsible for the scientific risk assessment, ensuring an individual assessment by the national competent authorities to a generic assessment. The new regulation also governs the EFSA to deliver scientific opinions within 9 months of receiving all valid applications. The newly introduced data protection for approved novel food authorization will also help companies successfully compete in the food market (Brodmann et al., 2017).

To ensure a high level of protection of human health and consumers' interests, food consisting of engineered nanomaterials should also be considered a novel food under this Regulation. In line with this, the European Commission has been seeking feedback on how to create a more streamlined authorization procedure (including the decision) that takes into account, for example, the particular needs of traditional exotic food from developing countries and that is adjusted to applications that cover several food uses (EC, 2013).

Initially, foods and food ingredients containing or consisting of genetically modified organisms within the meaning of Directive 90/220/EEC (EEC, 1990) and foods and food ingredients produced from, but not containing, genetically modified organisms (GMO) were also part of the scope of Regulation (EC) No. 258/1997 (EC, 1997b), but with Regulation (EC) No. 1829/2003 (EC, 2003a) there is a specific authorization procedure for GMO food and feed products, and Regulation (EC) No. 1830/2003 (EC, 2003b) concerns the traceability and labeling of genetically modified organisms and the traceability of food and feed products produced from genetically modified organisms. The reader is referred to the paper of Roda et al. (2009) reporting a review on scientific risk assessment on genetically modified (GM) foods.

The Novel Foods Regulation provides two different procedures for authorization. First, *simplified procedure* is based on a notification system and can only be followed if the applicant can provide evidence that the novel food/food ingredient is substantially equivalent to an existing food ingredient regarding composition, nutritional value, metabolism, intended use, level of undesirable substances, and, second, if the food/food ingredient consists of or is isolated from microorganisms, fungi, or algae or consists of or is isolated from plants or animals, except for food ingredients obtained by traditional propagating or breeding practices and having a *history of safe food use*. This is widely used around the world and constitutes the cornerstone of the safety evaluation of novel food. Once generally agreed to have a *history of safe use*, the traditional food or crop may then be used as a comparator to assist in the safety evaluation of a novel GM food or to determine its regulatory status. This comparative approach to bring out the safety evaluation of a novel food is sometimes referred to as the *concept of substantial equivalence*. It is seen as a starting point from which to structure a program to demonstrate any potential differences from the comparator (or material of reference) that, if detected, can be evaluated in terms of their safety implications A *history of safe use* may also be applicable to the novel food itself, if that food is already traditionally consumed in countries outside the country where it is to be marketed. In the case of plant sources from which a novel food is obtained, the plants must be traditional foods or food sources with a *history of safe use*. In these events the plant source becomes the comparator. Even if the plant source cannot be considered as a food in a classical sense, it may still have a *history of safe use* in a different context. This is the case for many of the herbal products used in food supplements. Their use has been traditional and any information on their safety for use in the context of traditional consumption is a key element for the purpose of a food safety assessment (Constable et al., 2007).

If the organism (plant, animal, and microorganism) used as the source of a novel food has a *history of safe consumption* under conditions of traditional use, then any toxicological concerns about new uses are reduced. However, such organisms may still be a source of potentially harmful components (i.e., toxins, allergens, and antinutritional factors) that are not of concern in traditional use because of processing methods or consumption patterns. These compounds would have to be addressed in the safety assessment of any novel food derived from the source. If the organism has not been used in the human food chain, or to an extent to provide adequately documented assurance of safety, then the potential for toxicological concern increases and must be considered on a case-by-case basis (Howlett et al., 2003). A distinction can be made between *traditional* use (for which limited scientific data are available) and *established* use (supported by scientific publications). Ultimately, for all novel foods, a judgment regarding safety should be made. The safety assessment program considers: analytical, compositional, and nutritional data; previous history of human exposure (not only as food); expected applications and predicted exposure; necessity, appropriateness, and outcome of animal and human studies; and results of postmarket monitoring, if conducted. The assessment of the history of human exposure will take into account the scientific rigor of the available data and their applicability to the new use and/or target groups (Constable et al., 2007).

When placing a *substantially equivalent* food/food ingredient on the EU market, a dossier providing evidence of its *substantial equivalence* needs to be submitted to the European Commission. The Novel Foods Regulation foresees two different routes to proving substantial equivalence, namely on the basis of generally recognized scientific evidence or on the basis of an opinion of a competent food assessment body in the Member States. If the evidence of *substantial equivalence* is accepted by the Commission, then the novel food/food ingredient is approved. In case the evidence is regarded as insufficient by the Commission or if objections/doubts are raised by certain Member States, then the Commission will ask a competent Member State food assessment body of its own choice to provide an opinion on the product. A request for marketing approval together with a dossier needs to be submitted to one of the 27 EU Member State authorities of the applicant's own choice for initial assessment (Roda et al., 2009).

The novel food or comparator should be fully characterized, including a precise biological identification (e.g., taxonomy, phenotype, and genotype) using adequate methodologies. The origin, geographical distribution, and genetic diversity of the food source should be reported. Compositional data for complex foods should include: moisture, protein fat, and ash; amino acid and fatty acid profiles; vitamin, mineral, and trace mineral composition; nutrients; toxicants, antinutrients, allergens, mycotoxins, and heavy metals; microorganisms; and phytoestrogens/androgens. For purified ingredients, the focus should be given to chemical identity and potential impurities arising from manufacture. Special attention should be given to compounds that may have implications for the health of any groups of the general population (e.g., infants,

TABLE 73.9 Substances that may be added for specific nutritional purposes in foods for particular nutritional uses (condition of use in dietetic foods[a]).

Vitamins (category 1)	
Vitamin A	Retinol, retinyl acetate, retinyl palmitate, β-carotene
Vitamin D	Cholecalciferol
Vitamin E	D-α-tocopherol, DL-α-tocopherol, D-α-tocopheryl acetate, DL-α-tocopheryl acetate, D-α-tocopheryl acid succinate[b], D-α-tocopheryl polyethylene glycol-1000 succinate
Vitamin K	Phylloquinone (phytomenadione), menaquinone (occurring principally as menaquinone-7 and, to a minor extent, menaquinone-6)
Vitamin B1	Thiamin hydrochloride, thiamin mononitrate
Vitamin B2	Riboflavin, riboflavin 5′-phosphate, sodium
Niacin	Nicotinic acid, nicotinamide
Pantothenic acid	D-pantothenate, calcium, D-pantothenate, sodium, dexpanthenol
Vitamin B6	Pyridoxine hydrochloride, pyridoxine 5′-phosphate, pyridoxine dipalmitate
Folate	Pteroylmonoglutamic acid, calcium-L-methylfolate
Vitamin B12	Cyanocobalamin, hydroxocobalamin
Biotin	D-biotin
Vitamin C	L-ascorbic acid, sodium-L-ascorbate, calcium-L-ascorbate, potassium-L-ascorbate, L-ascorbyl 6-palmitate

[a]*Dietetic foods means foods for particular nutritional uses including foods for special medical purposes but excluding infant formulae, follow-on formulae, processed cereal-based foods, and baby foods intended for infants and young children.*
[b]*Condition of use in foods for special medical purposes means dietary foods for special medical purposes as defined in Commission Directive 1999/21/EC of 25 March 1999 on dietary foods for special medical purposes.*

children, elderly, and pregnant women, among others) (Constable et al., 2007).

Foods for particular nutritional purposes and dietetic foods

A number of nutritional substances such as vitamins, minerals, amino acids, and others may be added to foods for particular nutritional uses to ensure that the particular nutritional requirements of the persons for whom those foods are intended are fulfilled and/or to satisfy legal requirements specified in specific directives adopted pursuant to Article 4(1) of Directive 2009/39/EC (EC, 2009b). Such specific directives may cover: (1) essential requirements regarding the nature or composition of the products; (2) provisions regarding the quality of raw materials; (3) hygiene requirements; (4) permitted changes within the meaning of complying with any mandatory provisions applicable to foodstuffs for normal consumption; and (5) a list of additives.

The range of foods for particular nutritional uses is very wide and diversified, and the technological processes used for their manufacture are varied. The choice of substances is primarily based on their safety and subsequently on their availability for use by humans, and on their organoleptic and technological properties.

These foods for special dietary uses are those foods that are specially processed or formulated to satisfy particular dietary requirements that exist because of a particular physical or physiological condition or specific diseases and disorders and that are presented as such. These include foods for infants and young children. The composition of these foodstuffs must differ significantly from the composition of ordinary foods of comparable nature, if such ordinary foods exist. A number of nutritional substances such as vitamins, minerals, amino acids, and others may be added to foods for particular nutritional uses to ensure that the particular nutritional requirements for whom those foods are intended. The range of foods for particular nutritional uses is very wide and diversified and the technological processes used for their manufacture are varied. For this reason, the widest possible choice of substances that can be safely used in the manufacture of foods for particular nutritional uses should be available for the six categories of nutritional substances to be listed in this Regulation (EC) No. 953/2009 (EC, 2009c). The choice of substances should be based primarily on their

TABLE 73.10 Substances that may be added for specific nutritional purposes in foods (condition of use in dietetic foods[a]).

Minerals (category 2)	
Calcium	Calcium carbonate, calcium chloride, calcium salts of citric acid, calcium gluconate, calcium glycerophosphate, calcium lactate, calcium salts of orthophosphoric acid, calcium hydroxide, calcium oxide, calcium sulfate, calcium bisglycinate, calcium citrate malate, calcium malate, calcium L-pidolate
Magnesium	Magnesium acetate, magnesium carbonate, magnesium chloride, magnesium salts of citric acid, magnesium gluconate, magnesium glycerophosphate, magnesium salts of orthophosphoric acid, magnesium lactate, magnesium hydroxide, magnesium oxide, magnesium sulfate, magnesium bisglycinate, magnesium L-pidolate, magnesium potassium citrate[b], magnesium L-aspartate
Iron	Ferrous carbonate, ferrous citrate, ferric ammonium citrate, ferrous gluconate, ferrous fumarate, ferric sodium diphosphate, ferrous lactate, ferrous sulfate, ferric diphosphate (ferric pyrophosphate), ferric saccharate, elemental iron (carbonyl + electrolytic + hydrogen, reduced), ferrous bisglycinate, ferrous L-pidolate
Copper	Cupric carbonate, cupric citrate, cupric gluconate, cupric sulfate, copper lysine complex
Iodine	Potassium iodide, potassium iodate, sodium iodide, sodium iodate
Zinc	Zinc acetate, zinc chloride, zinc citrate, zinc gluconate, zinc lactate, zinc oxide, zinc carbonate, zinc sulfate, zinc bisglycinate
Manganese	Manganese carbonate, manganese chloride, manganese citrate, manganese gluconate, manganese glycerophosphate, manganese sulfate
Sodium	Sodium bicarbonate, sodium carbonate, sodium chloride, sodium citrate, sodium gluconate, sodium lactate, sodium hydroxide, sodium salts of orthophosphoric acid
Potassium	Potassium bicarbonate, potassium carbonate, potassium chloride, potassium citrate, potassium gluconate, potassium glycerophosphate, potassium lactate, potassium hydroxide, potassium salts of orthophosphoric acid, magnesium potassium citrate
Selenium	Sodium selenite, sodium hydrogen selenite, sodium selenite, selenium-enriched yeast**Selenium-enriched yeasts produced by culture in the presence of sodium selenite as selenium source and containing, in the dried form as marketed, not more than 2.5 mg Se/g. The predominant organic selenium species present in the yeast is selenomethionine (between 60 and 85% of total extracted selenium in the product). The content of other organic selenium compounds including selenocysteine shall not exceed 10% of total extracted selenium. Levels of inorganic selenium normally shall not exceed 1% of total extracted selenium
Chromium (III)	Chromium (III) chloride and its hexahydrate, chromium (III) sulfate and its hexahydrate
Molybdenum (VI)	Ammonium molybdate, sodium molybdate
Fluorine	Potassium fluoride, sodium fluoride
Boron	Sodium borate, boric acid

[a]Dietetic foods means foods for particular nutritional uses including foods for special medical purposes but excluding infant formulae, follow-on formulae, processed cereal-based foods, and baby foods intended for infants and young children.
[b]Condition of use in foods for special medical purposes means dietary foods for special medical purposes as defined in Commission Directive 1999/21/EC of 25 March 1999 on dietary foods for special medical purposes.

safety and subsequently on their availability for use by humans and on their organoleptic and technological properties. Pending the adoption of purity criteria for the rest of the substances at the Community level, and to ensure a high level of protection for public health, generally acceptable purity criteria recommended by international organizations or agencies including but not limited to the Joint FAO/WHO Expert Committee on Food Additives (JECFA) and EUP (European Pharmacopoeia) should apply.

This category of products has a firmly established legal framework called *Foods for Particular Nutritional Purposes*, also called PARNUTS or *dietetic food*. It offers a good example of a legislation that balances scientific risk assessment, consumer protection, manufacturer responsibility, and market innovation in a rather proportionate way. The PARNUTS legislation is applicable to some functional foods or nutraceuticals if such foods are intended for particular nutrition uses (Coppens et al., 2006).

The so-called PARNUTS framework Council Directive 89/398/EEC (EEC, 1989) concerns foodstuffs for particular nutritional uses. Foodstuffs for particular nutritional uses are foodstuffs that, because of their special composition or manufacturing process, are clearly distinguishable from foodstuffs for normal consumption, are suitable for their claimed nutritional purposes, and are marketed in such a way as to indicate such suitability. The nature or composition of the products referred to in Article 1 must be such that the products are appropriate for the particular nutritional use: (1) foodstuffs for particular nutritional uses are foodstuffs that, because of their special composition or manufacturing process, are clearly distinguishable from foodstuffs for normal consumption, are suitable for their claimed nutritional purposes, and are marketed in such a way as to indicate such suitability; (2) a particular nutritional use must fulfill the particular nutritional requirements: (a) of certain categories of persons whose digestive processes or metabolism are disturbed; or (b) of certain categories of persons who are in a special physiological condition and who are therefore able to obtain special benefit from controlled consumption of certain substances in foodstuffs; or (c) of infants or young children in good health.

The PARNUTS creates the framework for a number of vertical directives with specific provisions for a number of PARNUTS categories.

There are a series of directives related to foodstuffs intended to satisfy particular nutritional requirements of specific groups of the populations. Directive 2009/39/EC (EC, 2009b) sets out a framework of rules for the composition, marketing, and labeling requirements of these *dietetic foods*. Commission Regulation (EC) No. 953/2009 (EC, 2009c) lists the substances in six categories (vitamins; minerals; amino acids; carnitine and taurine; nucleotides; choline and inositol) that may be added for specific nutritional purposes in foodstuffs for particular nutritional uses.

Substances that may be added for specific nutritional purposes in foods for particular nutritional uses are included in Commission Regulation (EC) No. 953/2009 (Tables 73.9–73.11).

A frame of different directives on dietary foods has been published within the EU and is described as follows:

1. Processed cereal-based foods and baby foods for infants and young children (Commission Directive 2006/125/EC) (EC, 2006d). This Directive does not apply to milks intended for young children. This Directive covers foodstuffs for particular nutritional use fulfilling the particular requirements of infants and young children in good health in the Community and it is intended for use by infants while they are being weaned and by young children as a supplement to their diet and/or for their progressive adaptation to ordinary food. They comprise: (1) processed cereal-based foods that are divided into the following four categories: (a) simple cereals that are or have to be reconstituted with milk or other appropriate nutritious liquids; (b) cereals with an added high-protein food that are or have to be reconstituted with water or other protein-free liquid; (c) pastas that are to be used after cooking in boiling water or other appropriate liquids; (d) rusks and biscuits that are to be used either directly

TABLE 73.11 Substances that may be added for specific nutritional purposes in foods (condition of use in dietetic foods[a]).

*Amino acids** (category 3)
L-alanine, L-arginine, L-cysteine, Cystine, L-histidine, L-glutamic acid, L-glutamine, L-isoleucine, L-leucine, L-lysine, L-lysine acetate, L-methionine, L-ornithine, L-phenylalanine, L-threonine, L-tryptophan, L-tyrosine, L-valine[b]L-aspartic acid, L-citrulline, Glycine, L-proline, L-serine, L-arginine-L-aspartate, L-lysine-L-aspartate, L-lysine-L-glutamate, N-acetyl-L-cysteine[c]N-acetyl-L-methionine*For amino acids, as far as applicable, sodium, potassium calcium, and magnesium salts as well as their hydrochlorides also may be used
Carnitine and taurine (category 4)
L-carnitine, L-carnitine hydrochloride, taurine, L-carnitine-L-tartrate
Nucleotides (category 5)
Adenosine 5′-phosphoric acid (AMP), sodium salts of AMP, cytidine 5′-monophosphoric acid (CMP), sodium salts of CMP, guanosine 5′-phosphoric acid (GMP), sodium salts of GMP, inosine 5′-phosphoric acid (IMP), sodium salts of IMP, uridine 5′-phosphoric acid (UMP), sodium salts of UMP
Choline and inositol (category 6)
Choline, choline chloride, choline bitartrate, choline citrate, inositol

[a]*Dietetic foods means foods for particular nutritional uses including foods for special medical purposes but excluding infant formulae, follow-on formulae, processed cereal-based foods, and baby foods intended for infants and young children.*
[b]*Condition of use in foods for special medical purposes means dietary foods for special medical purposes as defined in Commission Directive 1999/21/EC of 25 March, 1999 on dietary foods for special medical purposes.*
[c]*Foods for special medical purposes in products intended for persons over 1 year of age.*

or, after pulverization, with the addition of water, milk, or other suitable liquids; and (2) *baby foods* other than processed cereal-based foods.

Commission Directive 2006/125/EC (EC, 2006d) contains several annexes dedicated mainly to: essential composition of processed cereal-based foods for infants and young children; essential composition of baby foods for infants and young children; amino acid composition of casein; nutritional substances; reference values for nutrition labeling for foods intended for infants and young children; specific maximum residue levels (MRLs) of pesticides or metabolites of pesticides in processed cereal-based foods and baby foods and pesticides that shall not be used in agricultural production intended for the production of processed cereal-based foods and baby foods.

2. Foods intended for use in energy-restricted diets for weight reduction (Directive 96/8/EC) (EC, 1996). This Directive provides compositional and labeling requirements for foods presented for use in energy-restricted diets that are total daily diet replacement or individual meal replacement for weight control. According to Article 1(2), foods for use in energy-restricted diets for weight reduction are specially formulated foods that, when used as instructed by the manufacturer, replace the whole or part of the total daily diet. They are divided in two categories: (1) products presented as a replacement for the whole of the daily diet; and (2) products presented as a replacement for one or more meals of the daily diet. This Directive contains two annexes: Annex I, "essential composition of foods for energy-restricted diets" (energy, protein, fat, dietary fiber, vitamins, and minerals), and Annex II, "amino acid requirement pattern" (cystine plus methionine, histidine, isoleucine, leucine, lysine, phenylalanine plus tyrosine, threonine, tryptophan, valine).
3. Dietary foods for special medical purposes (Commission Directive 1999/21/EC) (EC, 1999). According to 1(3), dietary foods for special medical purposes are classified in the following three categories: (1) nutritionally complete foods with a nutrient formulation that, used in accordance with the manufacturer's instructions, may constitute the sole source of nourishment for the persons for whom they are intended; (2) nutritionally complete foods with a nutrient-adapted formulation specific for a disease, disorder, or medical condition that, used in accordance with the manufacturer's instructions, may constitute the sole source of nourishment for the persons for whom they are intended; and (3) nutritionally incomplete foods with a standard formulation or a nutrient-adapted formulation specific for a disease, disorder, or medical condition that are not suitable to be used as the sole source of nourishment. The foods referred to in points (2) and (3) may also be used as a partial replacement or as a supplement to the patient's diet. This Directive contains one annex on essential composition of foods for special medical purposes and two tables (Tables 73.1 and 73.2). This Directive has been amended by Commission Directive 2006/141/EC (EC, 2006e), and this, in turn, has been amended by Commission Directive 2013/46/EU (EU, 2013).
4. Composition and labeling of foodstuffs suitable for people intolerant to gluten [Commission Regulation (EC) No. 41/2009] (EC, 2009d). This Directive provides new EU harmonized rules for the composition and labeling of foodstuffs suitable for people who are gluten intolerant. Gluten is a protein fraction from wheat, rye, barley, oats, or their crossbred varieties and derivatives thereof, to which some persons are intolerant and that is insoluble in water and 0.5 M sodium chloride solution. The food industry has developed a range of products presented as *gluten-free* or similar equivalent terms. Differences between national provisions concerning the conditions for the use of such product descriptions may impede the free movement of the concerned products and may fail to ensure the same high level of protection for consumers. This Regulation, applicable since January 1, 2012, sets conditions for the use of the terms *very low gluten* and *gluten-free*. It is known that people with celiac disease comprise a specific group of the population suffering from a permanent gluten intolerance. Article 3 of Regulation (EC) No. 41/2009 states composition and labeling of foodstuffs for people intolerant to gluten: (1) foodstuffs for people intolerant to gluten, consisting of or containing one or more ingredients made from wheat (i.e., all *Triticum* species such as durum wheat, spelt, and kamut), rye, barley, oats, or their crossbred varieties that have been especially processed to reduce gluten, shall not contain a level of gluten exceeding 100 mg/kg in the food as sold to the final consumer; (2) the labeling, advertising, and presentation of the products referred to in paragraph 1 shall bear the term "very low gluten" and may bear the term "gluten-free" if the gluten content does not exceed 20 mg/kg in the food as sold to the final consumer; (3) oats contained in foodstuffs for those intolerant to gluten must have been specially produced, prepared, and/or processed in a way to avoid contamination by wheat, rye, barley, or their crossbred varieties and the gluten content of such oats must not exceed 20 mg/kg; (4) foodstuffs for those intolerant to gluten, consisting of or containing one or more ingredients that substitute wheat, rye, barley, oats, or their crossbred varieties shall not contain a level of gluten exceeding 20 mg/kg in the food as sold to the final consumer (labeling, presentation, and advertising of

those products shall bear the term *gluten-free*); and (5) where foodstuffs for people intolerant to gluten contain both ingredients that substitute wheat, rye, barley, oats, or their crossbred varieties and ingredients made from wheat, rye, barley, oats, or their crossbred varieties that have been specially processed to reduce gluten.

Specific directives on foods and beverages for sports people or on foods intended for diabetics are still subject to EU Member State legislation.

Food supplements from botanicals

It is a common situation to find products containing ingredients that can be regarded either as medicines or as foods or cosmetics, depending on presentation and claims. There are a number of herbal ingredients that have accepted usage in a range of different categories besides medicines, including food, cosmetics, or general consumer products.

Food supplements from botanicals and derived preparations made from plants, algae, fungi, or lichens have become widely available on the EU market. Such products are typically labeled as natural foods and a variety of claims are made regarding possible health benefits. They can be bought over the counter in pharmacies, supermarkets, specialty shops, and via the Internet. There is currently no EU legislation specifically for botanicals other than the general EU food legislation Regulation 178/2002 (EC, 2002c). EFSA has adopted the scientific report on the compendium of botanicals reported to contain naturally occurring substances of possible concern for human health when used in food and food supplements (EFSA, 2012). This compendium is intended to help with the safety assessment of botanicals and botanical preparations intended for use as food supplements by facilitating hazard identification. The compendium aims at flagging plants or parts of plants or compounds of possible concern for human health naturally present in the botanicals listed in the compendium and that therefore require specific attention while assessing the safety of products containing such botanical(s). For some of the compounds flagged in the compendium, health-based guidance values (e.g., acceptable daily intake) have been established, but not mentioned in this compendium. This compendium lists botanicals in alphabetical order, including: (1) botanical name; (2) family; (3) parts of plants of possible concern; (4) chemical of concern; (5) remarks on toxic/adverse effect(s) not known to be related to the identified chemical(s) of concern; and (6) specific references.

The European Medicines Agency (EMA) only works with the herbal medicinal products and not in the areas of food or cosmetics.

Traditional herbal medicines

All medicinal products, including herbal medicinal products, need a marketing authorization to be placed on the EU market. Herbal medicines are very popular, but there have been growing concerns that there are insufficient controls on their sale. It has been reported in a study of the sale of common herbal remedies such as *Echinacea*, ginseng, garlic, and kava kava (Barceloux, 2008; Table 73.12) that there were more than 400 Internet sites, of which 149 claimed that their products would treat, prevent, or even cure particular diseases (Morris and Avorn, 2003).

The traditional herbal medicinal products, a so-called simplified registration procedure, is provided by Herbal Directive 2004/27/EC (EC, 2004) amending Directive 2001/83/EC (EC, 2001a) on the Community code relating to medicinal products for human use. The EMA provides secretariat support to the Committee on Herbal Medicinal Products (HMPC), which was established by Regulation (EC) No. 76/2004 and as result of Directive 2004/27/EC (EC, 2004). The HMPC of EMA is the committee responsible for preparing the Agency's opinions on herbal medicines. The two main tasks of the HMPC are: (1) establishing EU herbal monographs that cover the therapeutic uses and safe conditions of use for herbal substances and preparations and they may cover their well-established use and/or their traditional use (*European Union monographs*); and (2) drafting the EU list of herbal substances, preparations, and combinations thereof for use in traditional herbal medicinal products (*European Union list*).

The aim of this EU Directive is to improve patient public safety and ensure that herbal medicinal products that are available within the EU are made with agreed ingredients and are clearly labeled without exaggerated claims being made for the product. This process involved amendments to the EU Directive 2001/83/EC on Medicinal Products for Human Use (EC, 2001a), amended by 2004/27/EC (EC, 2004).

Directive 2004/27/EC (EC, 2004) was adopted to facilitate the placing on the EU market of traditional herbal medicinal products through a simplified procedure. Since May 1, 2011, applicants must submit the corresponding application to the competent authorities in the EU Member States where they want to market. Medicinal products are defined by Directive 2004/27/EC (EC, 2004), amending Directive 2001/83/EC on the Community code relating to medicinal products for human use (EC, 2001a), as: "Any substance or combination of substances presented as having properties for treating or preventing disease in human beings; or any substance or combination of substances which may be used in or administered to human beings either with a

TABLE 73.12 Common medicinal herbs.

Medicinal herbs	Botanical description	Medicinal uses	Toxicity/adverse reactions
Garlic (*Allium sativum* L.)	Perennial herb, has a bulb consisting of a cluster of smaller bulbs	Treatment of parasites, skin disease, wounds, gastrointestinal tract (GIT) illnesses, flatulence, respiratory infection, fever, aging	Breath and body odor. Flatulence, nausea, and abdominal discomfort when the stomach is empty
Ginseng (*Panax ginseng*)	There are several species. Herbal plants with similar appearance and uses	Treatment for a variety of clinical conditions: cardiovascular collapse, edema, fatigue, sexual dysfunction, immune deficiencies, cardiac disease, dyspepsia, neurosis, anemia, leukopenia	Headache, anxiety, GIT distress
Echinacea (*Echinacea angustifolia*)	There are several species. Perennial herbs with either simple and branched stems	Inmunostimulants, promote wound healing and increase resistance to colds	Allergic reactions in atopic individual
Kava kava (*Piper methysticum*)	Perennial shrub (prefers well-drained, rich soil in the sun when mature)	Treatment of asthma, rheumatism, urological problems, menopausal symptoms, gonorrhea, vaginitis, nocturnal incontinence, weight loss, insomnia, anxiety	Large doses produced cerebellar dysfunction with reduction in cognitive abilities, hepatotoxicity (idiopathic reactions)

view to restoring, correcting, or modifying physiological functions by exerting a pharmacological, immunological, or metabolic action, or to making a medical diagnosis." There are directives related to traditional herbal medicines such as the food labeling Directive 2003/89/EC (EC, 2003c) and the conduct of clinical trials for medicinal products for human use (Directive 2001/20/EC) (EC, 2001b).

For registration as a traditional herbal medicinal product, the applicant must provide sufficient evidence of the medicinal use of the product throughout a period of at least 30 years, including at least 15 years in the EU. Others conditions or criteria for traditional herbal medicinal products are as follows: (1) their indication makes them appropriate for a use without prescription; (2) their strength and posology are specified and traditionally derived; and (3) their route of administration is oral, external, and/or inhalatory.

The EMA does not have a role in the registration of traditional herbal medicinal products. This means that applications for registration need to be submitted in each EU Member State where the product is to be marketed. These applications are handled by the competent authority in each EU Member State. The European Commission shall establish an EU list of herbal preparations or herbal substances so that EU Member States shall recognize registration of traditional herbal medicinal products granted by another EU Member State whenever it is based on the EU list.

The EMA has produced a guideline on Good Agricultural and Collection Practice for starting materials of herbal origin in order to ensure appropriate and consistent quality of medicinal plant/herbal substances and an inventory of herbal substances for assessment. It is noted that being of natural origin, herbal substances generally have higher contents of microorganisms when compared to chemically defined drug substances. In addition, spores and toxic mycotoxins generated by the microorganisms may also be present. For these reasons, a number of critical points need to be considered and taken into account such as the source of the herbal substance, knowledge about the microorganisms, the manufacturing processes and any decontamination procedure used, microbiological purity of excipients, the protective capacity of the packaging material chosen, the dosage form, administration route, posology, and patient population groups.

Herbal products may be classified and placed on the market as food provided that they do not fulfill the definition of medicinal products and that they comply with the applicable food law. In particular, herbal products marketed in the form of food supplements should comply with Directive 2002/46/EC on food supplements (EC, 2002b) and Regulation (EC) No. 1924/2006 on nutrition and health claims made on foods (EC, 2006a).

After May 1, 2011, all unlicensed herbal medicinal products must be marketed as medicines or they must be *withdrawn* from the market. They could be launched, if correctly labeled, as food supplements carrying no claims unless those claims have been approved according to the Nutrition and Health Claims Regulation.

Fortified foods

Although food supplements and foods to which vitamins and minerals are added are covered by different measures, the considerations for setting maximum levels for vitamins and minerals in the different products are inevitably interrelated. In particular, the distribution of these nutrients in the two broad categories of food products, food supplements and fortified foods, has to be considered together if we are to have a clear picture of the overall food offering. Vitamins and minerals may be ingested from a variety of foods that are eaten as part of the daily diet. They can be naturally present in foods or added to them both for technological or nutritional purposes. Innovation in the food sector tends to, increasingly, take into account the relationship between diet and health and may lead to an increase in the percentage of foodstuffs fortified with vitamins and minerals on the market.

The estimates of intakes are essentially derived through dietary surveys data and relevant food composition data. The main types of dietary surveys are individual and household surveys. Individual dietary surveys include duplicate meal surveys, weighed intakes, estimated intakes, diet histories, 24-h recall, and food frequency questionnaires. Individual surveys might be conducted on specific population groups by age or other characteristics (i.e., pregnant women, low-income groups, ethnic origin). The data derived from dietary surveys of a population may be presented as mean intakes (i.e., possibly by age and sex). In some cases there may also be information on the extremes of intakes, both the lower and higher levels, and on intakes of specific population groups (i.e., infants, adolescents, elderly). In addition, some surveys may differentiate intakes from food supplements while others may give only an estimate of intake from food sources, either including or excluding food supplements (SANCO, 2006).

Products that contribute to the intake of vitamins and minerals beyond the levels that would be provided by the natural content of foods have increased considerably on the market. These may be food supplements or foods to which vitamins and minerals have been added (*fortified food*). Food fortification is the process of adding micronutrients to foodstuffs, during or after processing, to bring micronutrient levels over and above amounts in the original food product. The vitamins and minerals added to fortified foods are called fortificants. Fortified blended foods as a mixture of cereals and other ingredients (such as soya beans preferable dehulled, pulses, oil seeds, dried skimmed milk, and possibly sugar) that have been milled, blended, precooked by extrusion or roasting, and fortified with a premix of adequate amount and with a wide range of micronutrients. Fortified blended foods include corn soya blend and wheat soy blend.

European Regulation (EC) No. 1925/2006 (EC, 2006c) concerns food products fortified with vitamins, minerals, and certain other substances. According to Article 3 of this Regulation, only vitamins and/or minerals listed in Annex I in the forms listed in Annex II may be added to foods, subject to the rules specified by this Regulation. Vitamins and minerals in a form that is bioavailable to the human body may be added to foods, whether or not they are usually contained therein, to take into account, in particular: (1) a deficiency of one or more vitamins and/or minerals in the population or specific population groups that can be demonstrated by clinical or subclinical evidence of deficiency or indicated by estimated low levels of intake of nutrients; (2) the potential to improve the nutritional status of the population or specific population groups and/or correct possible deficiencies in dietary intakes of vitamins or minerals due to changes in dietary habits; or (3) evolving generally acceptable scientific knowledge on the role of vitamins and minerals in nutrition and consequent effects on health. Article 5 indicates that the maximum amounts shall be set taking into account: (a) upper safe levels of vitamins and minerals established by scientific risk assessment based on generally acceptable scientific data, taking into account, as appropriate, the varying degrees of sensitivity of different groups of consumers; and (b) intakes of vitamins and minerals from other dietary sources.

Regulation (EC) No. 1925/2006 (EC, 2006c) harmonizes the different rules enforced in EU Member States by establishing a list of vitamins and minerals that are authorized to be added to foods. This Regulation aims to regulate the addition of vitamins and minerals to foods and the use of certain other substances or ingredients containing substances other than vitamins or minerals that are added to foods or used in the manufacture of foods under conditions that result in the ingestion of amounts greatly exceeding those reasonably expected to be ingested under normal conditions of consumption of a balanced and varied diet and/or would otherwise represent a potential risk to consumers. The Regulation sets minimum amounts based on the notion of *significant amount* as defined in the Annex to Council Directive 90/496/EEC on nutrition labeling (EEC, 1990). It also deals with *other substances* and provides a list of substances *permitted*, *restricted*, or *under scrutiny*. In the absence of specific Community rules regarding prohibition or restriction of the use of substances or ingredients containing substances other than vitamins or minerals under this Regulation or under other specific Community provisions, relevant national rules may apply without prejudice to the provisions of the

Treaty. Vitamins and minerals may be added to foods voluntarily by food manufacturers or must be added as nutritional substances as provided for by specific Community legislation. They may also be added for technological purposes as additives, colorings, flavorings, or other such uses including authorized enological practices and processes provided for by relevant Community legislation. This Regulation should apply without prejudice to the specific Community rules concerning the addition of vitamins and minerals to or their use in specific products or groups of products or their addition for purposes other than those covered by this Regulation. Vitamins and minerals are added to foods by manufacturers for a number of purposes, including to restore their content when this has been reduced during manufacturing, storage, or handling procedures or to provide a similar nutritional value as foods for which they are intended to be alternatives (Table 73.13).

Mutual recognition

Mutual recognition remains an important tool for ensuring the free movement of products, including foodstuffs, on the Community market. *Mutual recognition* is not free from the risk that technical obstacles to the free movement of the products concerned can be maintained or created. However, these risks should be put into perspective because the Court of Justicial Supervision, has set precise limits within which the UE Member States may validly exempt themselves from mutual recognition by availing themselves of Article 30 EC, including in the area of foodstuffs. Accordingly, the Court frequently reiterates that it is for the Member States, in the absence of harmonization and to the extent that uncertainties continue to exist in the current state of scientific research, to decide on their intended level of protection of human health and life and on whether to require prior authorization for the marketing of foodstuffs, always taking into account the requirements of the free movement of goods within the Community. However, in exercising their discretion relating to the protection of public health, the UE Member States must comply with the principle of proportionality. The means that what they choose must therefore be confined to what is actually necessary to ensure the safeguarding of public health; they must be proportional to the objective thus pursued, which could not have been attained by measures that are less restrictive of intra-Community trade. In other words, the EU Member States are entitled to invoke the need to protect the interests referred to in Article 30 EC, including health protection, only when the conditions provided by the Court and referred to here have been met, and to the extent that there is no harmonized Community legislation capable of protecting the same interests (Coppens et al., 2006).

In this context, it should be stressed that, since May 13, 2009, refusals of mutual recognition have been subject to the conditions specified in Regulation (EC) No. 764/2008, which established a procedure for the exchange of information on national measures derogating from the principle of the free movement of goods within the Community (EC, 2008). This Regulation aims to improve the functioning of the mutual recognition of goods. It establishes procedures aimed at framing how authorities in the EU Member States monitor compliance with national technical rules on goods not covered by the harmonized Community rules. EU Member States which refuse access for these goods to their market are obliged to make contact with the enterprise and to produce detailed objective reasons for refusal. The objective is to strengthen the internal market by improving the free movement of goods within the Community.

The Member States may not prohibit or restrict trade (*safeguard clause*) in food supplements that comply with Directive 2002/46/EC (EC, 2002b), except when they find that the products pose a public health risk. When such risk occurs, an EU Member State may temporarily suspend or restrict application of the provisions of the Directive. It shall immediately inform the other EU Member States and the European Commission thereof and give reasons for its decision. The European Commission shall examine the grounds adduced by the EU Member State for temporarily suspending or restricting the trade in food supplements and shall consult the SCFAH before delivering its opinion and taking appropriate measures.

The European Food Safety Authority: role and activities

The EFSA was sets out in January 2002 by Regulation (EC) No. 178/2002 (EC, 2002c) as an independent source of scientific advice that produces opinions that then are used by the European Commission to adopt legislation. The EFSA has an important role to play in supporting policymakers at European and national levels in developing policies and setting diet-related public health targets. Although food safety was the primary force for the establishment of EFSA, Founding Regulation (EC) No. 178/2002 (EC, 2002c) gives us an important mandate in nutrition. Furthermore, the increasing burden of nutrition-related diseases has had a significant impact on the evolution of EFSA work program. The objective is: (1) to know what Europeans are eating and (2) to provide policymakers with the scientific basis for recommendations that will help consumers to make healthy dietary choices. That is why EFSA work in this area currently has three main pillars: (1) the collection of food consumption data

TABLE 73.13 Vitamin formulations and mineral substances that may be added to foods.

Vitamin formulations	
Vitamin A	Retinol; retinyl acetate; retinyl palmitate (d) β-carotene
Vitamin D	Cholecalciferol; ergocalciferol
Vitamin E	D-α-tocopherol; DL- α-tocopherol; D- α-tocopheryl acetate; DL- α-tocopheryl acetate; D- α-tocopheryl acid succinate
Vitamin K	Phylloquinone (phytomenadione)
Vitamin B1	Thiamin hydrochloride; thiamin mononitrate
Vitamin B2	Riboflavin; riboflavin 5′-phosphate, sodium
Niacin	Nicotinic acid; nicotinamide
Pantothenic acid (mg)[a,b]	D-pantothenate, calcium; D-pantothenate, sodium; dexpanthenol
Vitamin B6	Pyridoxine hydrochloride; pyridoxine 5′-phosphate; pyridoxine dipalmitate
Folic acid	Pteroylmonoglutamic acid
Vitamin B12	Cyanocobalamin; hydroxocobalamin
Biotin (μg)[a]	D-biotin
Vitamin C	L-ascorbic acid; sodium-L-ascorbate; calcium-L-ascorbate; potassium-L-ascorbate; L-ascorbyl 6-palmitate
Mineral substances	
Calcium	Calcium carbonate; calcium chloride; calcium salts of citric acid; calcium gluconate; calcium glycerophosphate; calcium lactate; calcium salts of orthophosphoric acid; calcium hydroxide; calcium oxide; calcium sulfate
Magnesium	Magnesium acetate; magnesium carbonate; magnesium chloride; magnesium salts of citric acid; magnesium gluconate; magnesium glycerophosphate; magnesium salts of orthophosphoric acid; magnesium lactate; magnesium hydroxide; magnesium oxide; magnesium sulfate
Iron	Ferrous carbonate; ferrous citrate; ferric ammonium citrate; ferrous gluconate; ferrous fumarate; ferric sodium diphosphate; ferrous lactate; ferrous sulfate; ferric diphosphate (ferric pyrophosphate); ferric saccharate; elemental iron (carbonyl + electrolytic + hydrogen reduced)
Copper	Cupric carbonate; cupric citrate; cupric gluconate; cupric sulfate; copper lysine complex
Iodine	Sodium iodide; sodium iodate; potassium iodide; potassium iodate
Zinc	Zinc acetate; zinc chloride; zinc citrate; zinc gluconate; zinc lactate; zinc oxide; zinc carbonate; zinc sulfate
Manganese	Manganese carbonate; manganese chloride; manganese citrate; manganese gluconate; manganese glycerophosphate; manganese sulfate
Sodium	Sodium bicarbonate; sodium carbonate; sodium citrate; sodium gluconate; sodium lactate; sodium hydroxide; sodium salts of orthophosphoric acid; sodium selenite; sodium hydrogen selenite; sodium selenite; sodium fluoride
Potassium	Potassium fluoride; potassium bicarbonate; potassium carbonate; potassium chloride; potassium citrate; potassium gluconate; potassium glycerophosphate; potassium lactate; potassium hydroxide; potassium salts of orthophosphoric acid
Chromium	Chromium (III) chloride and its hexahydrate; chromium (III) sulfate and its hexahydrate
Molybdenum	Ammonium molybdate [molybdenum (VI)]; sodium molybdate [molybdenum (VI)]

[a]*Directive 2002/46/EC.*
[b]*Commission Regulation (EC) No 1170/2009.*

and the harmonization of collection methodologies across EU Member States; (2) the establishment of dietary reference values (DRVs) and advising risk managers on how to translate them into food-based dietary guidelines; and (3) the scientific evaluation of health claims made on foods.

Most relevant for companies in the food supplements sector, the EFSA has been asked by the European Commission to evaluate proposals for the addition of vitamins and minerals to the FSD and to evaluate nutrition and health claims. The EFSA has also worked with the European Commission on assessing how to establish maximum limits for vitamins and minerals in food supplements and fortified foods and provided opinions on substances other than vitamins and minerals. The EFSA works in close collaboration with national authorities of the European countries and in open consultation with its stakeholders.

Vitamin and mineral substances may be considered for inclusion in the lists following the EFSA evaluation of an appropriate scientific dossier concerning the safety and bioavailability of the individual substance. Companies wishing to market a substance not included in the permitted list need to submit an application to the European Commission.

The bioavailability of a nutrient relates to its absorption and may be defined as its accessibility to normal metabolic and physiological processes. Bioavailability determines a nutrient's beneficial effects at physiological levels of intake and the nature and severity of adverse effects at excessive intakes. Because of the considerable variation in nutrient bioavailability in humans, bioavailability data for specific nutrients must be considered when deriving upper level of intake (ULs). In particular, the chemical form of a nutrient may have a large influence on bioavailability and should be specified in deriving the UL. Other modulating factors include: (1) nutritional status of the individual; (2) nutrient dose; (3) interaction with other dietary components; and (4) the food matrix (e.g., consumption with or without food).

In vivo studies will also be needed when it is appropriate to determine: (1) the availability of vitamins and minerals in the novel food in comparison with the food it would replace and (2) any interaction the novel food might have with other items of the diet that would reduce the whole diet's nutritional value. If the novel food is expected to play an important role in the diet, then it may be necessary to verify that the results of laboratory animal studies can be extrapolated to humans by measuring the availability of nutrients to human subjects.

Human studies should not be engaged until there has been a full appraisal of the safety of the novel food using all available data (e.g., history of safe use, data on chemical and microbiological impurities, composition, and toxicology). After the launch of a novel food on the market, postmarketing surveillance studies may also be helpful in providing confirmation of anticipated usage patterns and exposure levels.

Dietary reference values of vitamins and minerals

Many countries around the world have produced quantitative dietary recommendations under a variety of names such as recommended daily amounts, recommended daily allowances, recommended daily intakes, recommended dietary intakes, and recommended nutrient intakes, among others. In the SCF report [SCF (Scientific Committee for Food), 1993] of the European Commission the generic term RDA is established. In this SCF report it is highlighted that a major problem in formulating the RDA was that nutrient requirements differ between individuals; current RDAs for micronutrients are based on the prevention of diseases of deficiency. RDA is the average daily dietary intake level that is sufficient to meet the nutrient requirements of nearly all (approximately 98%) healthy individuals.

The term RDA was replaced by DRV. The European Commission asked the EFSA to review and update the values for nutrient and energy intakes that were established in 1993 by the SCF [SCF (Scientific Committee for Food) 1993]. Article 5.2 of Directive 2002/46/EC (EC, 2002c) on food supplements stipulates that: "When the maximum levels referred to in paragraph 1 are set, due account should also be taken of RI of vitamins and minerals for the population." DRVs represent the highest level of daily intake of a nutrient that is likely to pose no risk of adverse health effects.

The population reference intakes (PRIs), also called RDAs, are based on the principle that most, if not all, individuals of a population or a specific population group should obtain adequate nutrient intake to satisfy their requirements. Such recommended values are generally based on the principle of the average requirement (AR) plus two standard deviations for the nutrient in the population group.

The task was allocated to the EFSA Scientific Panel on Nutrition, Dietetic Products, and Allergies (NDA Panel) (EFSA, 2010), which, in its work, is taking into consideration the latest scientific evidence and recent recommendations issued at national and international levels.

The DRVs are defined as a complete set of reference values and nutrient recommendations. The NDA Panel proposes to derive the following DRVs: PRIs (i.e., the level of nutrient intake that is adequate for virtually all people in a population group); *AR* (i.e., the level of nutrient intake that is adequate for half of the people in a population group, given a normal distribution of requirement); and *lower threshold intake* (LTI; i.e., the level of intake below which, on the basis of current knowledge, almost all individuals will be unable to maintain *metabolic integrity* according to the criterion chosen for each nutrient). In addition, the NDA Panel also proposes to derive the following DRVs: *AI* (i.e., the value estimated when a PRI cannot be established because an AR cannot be determined). An *AI* is the average observed daily level of intake by a population group (or groups) of apparently healthy people that is assumed to be adequate and the *RI ranges for macronutrients* (RI; i.e., the reference intake ranges for macronutrients, expressed as percent of the energy intake). These apply to ranges of intakes that are adequate for maintaining health and associated with a low risk of selected chronic diseases.

Some of the DRVs (AR, PRI, and LTI) are related to nutrient requirements that are defined by specific criteria of nutrient adequacy. Because DRVs are developed for different life stages and sex groups, the NDA Panel proposes to define the age ranges used for each nutrient on a case-by-case basis depending on the available data. Interpolation or extrapolation between population groups will be used in instances where no data are available for defined age and sex groups. Scaling methods using isometric (linear with body weight) or allometric (body weight to the power of a chosen exponent) or interpolation based on other nonpredefined parameters are being used. Which method is the most appropriate will be decided on a case-by-case basis. Reference heights and weights are useful when more specificity about body size and nutrient requirements are needed than that provided by life stage categories. For nutrients such as vitamins, minerals, and protein, the PRI can be a practical starting point (EFSA, 2010). They are used to define the amount of a nutrient that must be consumed on a regular basis to maintain health and therefore to provide a scientific basis for establishing nutrient recommendations and dietary guidelines. DRVs are calculated based on a defined set of scientific criteria including, for example, the amount of a nutrient required to prevent or reverse clinical deficiencies and maintain the functional competence of the body. To ensure a consistent approach to the calculations, the NDA Panel provided general principles for establishing DRVs in an opinion published in 2010. Since then, the EFSA has published opinions on DRVs for energy, macronutrients (carbohydrates, dietary fiber, fats, and protein), and water, and for micronutrients [fluoride, molybdenum, vitamin C, manganese, biotin, pantothenic acid, iodine, niacin, zinc, chromium, selenium, folate, vitamin A, calcium, vitamin E (as α-tocopherol), phosphorus, vitamin B12 (cobalamin)], and all EFSA work in this area is subject to extensive consultation with EU Member States, the scientific community, and other stakeholders.

The EFSA's advice on nutrient intakes provides an important evidence base to underpin nutritional policies, the setting of diet-related public health targets, and the development of consumer information and educational programs on healthy diets (Table 73.14).

In addition, the Commission has asked EFSA to help public authorities in EU Member States to translate nutrient-based recommendations into food-based dietary guidelines and practical science-based policy recommendations that are easily understood and culturally appropriate. In this regard, EFSA opinion on establishing food-based dietary guidelines recommends a step-wise, multidisciplinary approach that is integrated with other food-related public policies (EFSA, 2010).

There is a wide range of vitamin preparations and mineral substances used in the manufacture of food supplements that are currently marketed in Member States and that have not undergone a scientific safety evaluation. To allow the necessary time for this safety evaluation, EU Member States could provide derogations until December 31, 2009, for vitamins and minerals and their forms not included in the Directive, in view of the future addition to the Annexes of the Directive, as long as the following conditions are fulfilled: (1) the substance in question was used in food supplements marketed in the Community

TABLE 73.14 Activities of the European Food Safety Authority.

In theory	In practice
1. European Food Safety Authority (EFSA) identifies and reviews scientific studies, including reports of national and international authorities, for possible health effects of a specific nutrient	1. Health effects of dietary fiber are identified by reviewing scientific studies
2. EFSA evaluates evidence for relationships between the intake of a nutrient and defined health outcomes	2. The evidence for the relationship between dietary fiber intake and bowel function or diet-related diseases is evaluated
3. Where nutrient–health relationships are established, EFSA provides scientific advice which may be used by policy makers	3. A daily intake of 25 g of dietary fiber is set as a dietary reference value because it is adequate for normal bowel function in adults, while consumption above 25 g dietary fiber per day may reduce risk of coronary heart disease and type II diabetes and may improve weight maintenance

prior to July 12, 2002; and (2) the EFSA has not given an unfavorable opinion regarding the use of the substance, or its use in that form, in the manufacture of food supplements on the basis of a dossier supporting use of the substance that has to be submitted to the Commission by the Member State by July 12, 2005. Petitioners should contact the competent authorities in the individual Member States if they wish to obtain such derogations.

Tolerable upper intake level

The *tolerable* UL is the maximum level of total chronic daily intake of a nutrient (from all sources) judged to be unlikely to pose a risk of adverse health effects to humans. *Tolerable intake* in this context connotes what is physiologically tolerable and is a scientific judgment as determined by assessment of risk (i.e., the probability of an adverse effect occurring at some specified level of exposure). ULs may be derived for various life stage groups in the population (SCF Scientific Committee for Food, 2000). The UL is not a recommended level of intake but is an estimate of the highest level of intake that carries no appreciable risk of adverse health effects. To establish whether an exposed population is at risk requires a risk assessment to determine what is the fraction (if any) of the population whose intake exceeds the UL and the magnitude and duration of the excessive intake. The steps in the development of the ULs are the following (SCF Scientific Committee for Food, 2000):

1. *Hazard identification*. This step outlines the adverse health effects that have been demonstrated to be caused by the nutrient. Human studies provide the most relevant data for hazard identification. Other experimental studies (in vivo and in vitro) may also be used. Six key issues that can be addressed in the data evaluation of human and animal studies are:
   a. *Evidence of adverse effects on humans*: All human, animal, and in vitro published evidence addressing the likelihood of a nutrient eliciting an adverse effect in humans is examined. Decisions on which observed effects are adverse are based on scientific judgments.
   b. *Causality*: Criteria for judging the causal significance of an exposure–effect association indicated by epidemiological studies have been adopted in various reports (e.g., NRC National Research Council, 1982, 1989; Department of Health, 1998). These include demonstration of a temporal relationship, consistency, strength of association (narrow confidence intervals for risk estimates), a dose–response relationship (a biological gradient), specificity, biological plausibility, and coherence.
   c. *Relevance of experimental data*: All animal data should be considered, taking into account: interspecies differences; route of exposure (ingestion exposure is more relevant than other routes); and duration of exposure and relevance of exposure to dietary intakes by human populations (e.g., chronic daily vs short-term bolus exposure).
   d. *Mechanisms of adverse effects*: Knowledge of the molecular or cellular events underlying the adverse effect.
   e. *Quality and completeness of the database.*
   f. *Identification of distinct and highly sensitive subpopulations*: These may or may not be included in the derivation of the UL, subject to judgment applied on a case-by-case basis.
2. *Hazard characterization.* This step includes a dose–response assessment that addresses the relationship between nutrient intake (dose) and adverse effect (in terms of intake and severity) and involves a number of key components (FNB, 1997):
   a. *Data selection*: Selecting the 1925 critical data set includes the following considerations: (i) human data are preferable to animal data, and human studies should be considered in relation to hazards identified in animal studies; (ii) in the absence of appropriate human data, information from an animal species whose biological responses are most like those of humans is most valuable; (iii) if it is not possible to identify such a species or to select such data, data from the most sensitive animal species, strain, or gender combination are given the greatest emphasis; (iv) the route of exposure that most resembles the route of expected human intake is preferable; (v) the critical data set defines the dose–response relationship between intake and the extent of the adverse effect known to be the most relevant to humans, and data on bioavailability need to be considered, and (vi) the critical data set should document the route of exposure and magnitude and duration of intake, and the intake that does not produce adverse effects as well as the intake that produces adverse effects.
   b. *Identification of no observed adverse effect (NOAEL) [or lowest observed adverse effect level (LOAEL)] and critical endpoint*: The NOAEL is the highest intake of a nutrient at which no adverse effects have been observed. If there are not adequate data demonstrating a NOAEL, then a LOAEL (the lowest intake at which an adverse effect has been demonstrated) can be used. Where different adverse effects (or *endpoints*) occur for a nutrient, the NOAELs (or LOAELs) for these endpoints will differ. The critical endpoint is the adverse effect exhibiting the lowest NOAEL (e.g.,

the most sensitive indicator of a nutrient's adverse effects). The derivation of a UL based on the most sensitive *endpoint* will ensure protection against all other adverse effects.

c. *Uncertainty assessment*: There are usually several scientific uncertainties associated with extrapolating from the observed data to the general population and several judgments must be made in deriving uncertainty factors to account for the individual uncertainties. The individual uncertainty factors may be combined into a single composite uncertainty factor for each nutrient and applying this (composite) uncertainty factor to a NOAEL (or LOAEL) will result in a value for the derived UL that is less than the experimentally derived NOAEL, unless the uncertainty factor is 1.0. The larger the uncertainty, the larger the uncertainty factors and the lower the UL, which represents a lower estimate of the threshold above which the risk of adverse effects may increase. In the application of uncertainty factors there should be cognizance of nutritional needs (e.g., the derived UL should not be lower than the recommended intake).

Because imprecision of the data, lack of data, and adequacy of the data on variability are major limitations of risk assessment, uncertainty factors are used. Generally, there are values for uncertainty factors of 10 for interhuman variations, 10 for animal to human (interspecies) extrapolations, and less than 10 for LOAEL to NOAEL extrapolations (usually 3). Because data are generally available in human populations, and because studies on human populations may cover part of the variability inherent in the population, the data on adverse effects of nutrients may not be associated with the same uncertainties as with nonessential chemical substances, resulting in uncertainty factors for nutrients typically less than 10. The uncertainty factors are lower with higher quality data and when the adverse effects are extremely mild and reversible. The availability of toxicokinetic data in humans may permit a lower uncertainty factor. In general, when determining an uncertainty factor, the following potential sources of uncertainty are considered: (1) *interindividual variation and sensitivity:* a small uncertainty factor is used if it is judged that little population variability is expected for the adverse effect, and a larger uncertainty factor (close to 10) may be used if variability is expected to be great (NRC, 1989); (2) *experimental animal to human:* an uncertainty factor is generally applied to the NOAEL to account for the uncertainty in extrapolating from animal data to humans; however, a larger uncertainty factor may be used if it is believed that the animal responses will underpredict average human responses (NRC, 1989); (3) *LOAEL to NOAEL*: if a NOAEL is not available, then an uncertainty factor may be applied to account for the uncertainty in deriving a UL from the LOAEL; the size of the uncertainty factor involves a judgment based on the severity and incidence of the observed effect at the LOAEL and the steepness (slope) of the dose–response curve; and (4) *subchronic NOAEL to predict chronic NOAEL:* when data are lacking on chronic exposures, scientific judgment is necessary to determine whether chronic exposure is likely to lead to adverse effects at lower intakes than those producing effects after subchronic exposures.

d. *Derivation of ULs*: The UL is a health-based guidance value and necessitates that the homeostatic mechanisms specific to essential nutrient substances be taken into account. UL can be derived for nutrients using the principles of risk assessment similar to those that have been developed for biological and chemical agents. The relationship between intake and risk for nutrient substances (i.e., deficiency and adverse health effects as the intake dose moves from low to high, i.e., the deficiency–excess model) should be highlighted. For most essential nutrients, homeostatic mechanisms that maintain the amount of nutrient substance in the body within a physiological range are associated with both low and high levels of intake. If intakes increase or decrease, then it is assumed that homeostatic responses of some type occur and that the responses may vary by age, sex, or life stage. However, homeostatic adaptations have a limited capacity and can be overwhelmed by excessive intake. At the extremes, as the capacity of a homeostatic mechanism is exceeded, the incidence or impact of specific adverse health effects is likely to increase. The distinctions between essentiality and a demonstrated favorable health impact require further elucidation and clarification as data evolve (FAO/WHO, 2006).

A pivotal point in the assessment process is the selection of the critical adverse health effect. This is the effect on which the UL is based or, more specifically, the effect on which a set of ULs for various age, sex, and life stage subpopulations is based. The critical adverse health effect is usually the effect that occurs at the lowest level of excessive intake within the (sub)population of interest or at the lowest experimental dose if only laboratory animal data are available (FAO/WHO, 2006). The UL is derived by dividing the NOAEL (or LOAEL) by the (composite) uncertainty factor. Uncertainty (safety) factors are used to convert a

TABLE 73.15 Reference body weights of population groups in Europe [SCF (Scientific Committee for Food) 1993].

Age (years)	Mean weight (kg)	
	Male	Female
1–3	13.0	12.5
4–6	20.0	19.0
7–10	28.5	29.0
11–14	44.5	45.0
15–17	61.5	53.5
18–29	74.6	62.1
30–59	74.6	62.1
60–74	73.5	66.1
≥75	73.5	66.1

surrogate for the threshold, such as the NOAEL in animals, into a "safe" intake for humans. ULs are derived for different life stage groups using relevant data. In the absence of data for a particular life stage group, extrapolations are made from the UL for other groups on the basis of known differences in body size, physiology, metabolism, absorption, and excretion of a nutrient. When data are not available for children and adolescents, extrapolations are made on the basis of body weight using the reference weights in Table 73.15.

3. *Characterization of risk*. This may include a description of the scientific uncertainties associated with the UL estimates to indicate the degree of scientific confidence that can be placed in these estimates. It may also include an estimate of intake for population groups, where data are available as well as an indication of the margin between recommended or actual intakes and the UL, and an indication of circumstances, if any, in which risk is likely to arise. It should indicate whether subpopulations with distinct and exceptional sensitivities to the adverse effects of the nutrient have been excluded and whether more research is needed. For nutrients for which there are no, or insufficient, data on which to base the establishment of a UL, an indication should be given on the highest level of intake when there is reasonable confidence in data on the absence of adverse effects.

Concluding remarks and future directions

Food is essential to the maintenance, development, functioning, and reproduction of human life. Regulation for food/dietary supplements and other related foods is one of the more complex subjects in food law. This chapter describes legislation that is relevant to the marketing of functional foods in the EU, the ways that this legislation was developed, as well as some practical consequences for manufacturers, marketers, and consumers. A number of EU food legislative texts have been elaborated on, including the General Food Regulation, the FSD, the Novel Food Regulation, the Novel Food Ingredients, the Foods for Particular Nutritional Purposes, the Foodstuffs Intended to Satisfy Nutritional Requirement of Specific Groups of the Populations, The Botanicals, and The Fortified Foods, among others. In this chapter the new regulatory frameworks and the implications for the food supplement are evaluated. The EFSA through Scientific Committees and their respective panels is responsible for producing opinions that are used by the European Commission to adopt legislation related to food and nutrition. The FSD in large part is based on the risk assessment paradigm: the toxicological risks of excess intake of food-derived micronutrients must be demonstrated. The increasing burden of nutrition-related diseases had a significant impact on the evolution of the EFSA work program. Nutrients possess some features that distinguish them from other chemicals with respect to risk assessment. With the creation of the EFSA, risk assessment was separated from risk management. When the relevant EFSA scientific panel delivers its opinion on the application file, the European Commission will then present the opinion to the SCFCAH regulatory committee. If the SCFCAH regulatory committee approves the food supplement or a novel food ingredient, then it can be placed on the EU market. The key elements for the purpose of a food safety assessment in food supplements and other related foods are described taking into account the requirements for each regulation and are science based.

Conflict of interest statement

None of the authors of this chapter has a financial or personal relationship with other people or organizations that could inappropriately influence or bias the content of the chapter.

Acknowledgments

This work was supported by Project Ref. RTA2015-00010-C03-03 from Ministerio de Economía, Industria y Competitividad, Spain.

References

Anadón, A., Castellano, V., Martínez-Larrañaga, M.R., 2014. Regulation and guidelines of probiotics and prebiotics. In: Ötles, S. (Ed.), Probiotics and Prebiotics in Food, Nutrition and Health. CRC Press, LLC Taylor & Francis Group, Boca Raton FL, pp. 91–113. (ISBN: 978-1-4665-8623-9).

Barceloux, D.G., 2008. Medical Toxicology of Natural Substances. John Wiley & Sons, Inc., Hoboken, N.J.

Brodmann, T., Endo, A., Gueimonde, M., et al., 2017. Safety of novel microbes for human consumption: practical examples of assessment in the European Union. Front. Microbiol. 8, 1721.

Constable, A., Jonas, D., Cockburn, A., et al., 2007. History of safe use as applied to the safety assessment of novel foods and foods derived from genetically modified organisms. Food Chem. Toxicol. 45, 2513–2525.

Coppens, P., Fernandes da Silva, M., Pettman, S., 2006. European regulations on nutraceuticals, dietary supplements and functional foods: a framework based on safety. Toxicology 221, 59–74.

Das, L., Bhaumik, E., Raychaudhuri, U., et al., 2012. Role of nutraceuticals in human health. J. Food Sci. Technol. 49 (2), 173–183.

Department of Health, 1998. Nutritional aspects of the development of cancer. Report of the working group on diet and cancer of the committee on medical aspects of food and nutrition policy. Rep. Health Soc. Subj. (Lond.) 48, 1–274. HMSO, London.

Domke, A., Großklaus, R., Niemann, B., et al., 2005. Use of Vitamins in Foods. Toxicological and Nutritional-Physiological Aspects. Part I. Federal Institute for Risk Assessment (BfR), Press and Public Relations Office, Thielallee 88-92, D-14195 Berlin.

EC, 1996. Commission Directive 96/8/EC of 26 February 1996 on Foods Intended for Use in Energy-Restricted Diets for Weight Reduction (OJ L 55, 6.3.1996).

EC, 1997a. Commission Recommendation of 29 July 1997 Concerning the Scientific Aspects and the Presentation of Information Necessary to Support Applications for the Placing on the Market of Novel Foods and Novel Food Ingredients and the Preparation of Initial Assessment Reports Under Regulation (EC) No. 258/97 of the European Parliament and of the Council (97/618/EC) (OJ L 253, 16.9.1997).

EC, 1997b. Regulation (EC) No. 258/97 of the European Parliament and of the Council of 27 January 1991 concerning novel foods and novel food ingredients (OJ L 43, 14.2.1997).

EC, 1999. Commission Directive 1999/21/EC of 25 March 1999 on Dietary Foods for Special Medical Purposes (OJ L 91, 7.4.1999).

EC, 2000. Directive 2000/13/EC of the European Parliament and of the Council of 20 March 2000 on the Approximation of the Laws of the Member States Relating to the Labelling, Presentation and Advertising of Foodstuffs (OJ L 109, 6.5.2000).

EC, 2001a. Directive 2001/83/EC of the European Parliament and of the Council of 6 November 2001 on the Community Code Relating to Medicinal Products for Human Use (OJ L 311, 28.11.2001).

EC, 2001b. Directive 2001/20/EC of the European Parliament and of the Council of 4 April 2001 on the Approximation of the Laws, Regulations and Administrative Provisions of the Member States Relating to the Implementation of Good Clinical Practice in the Conduct of Clinical Trials on Medicinal Products for Human Use (OJ L 121, 1.5.2001).

EC, 2002a. Consolidated Version of the Treaty Establishing the European Community. (OJ C325/33, 24.12.2002.)

EC, 2002b. Directive 2002/46/EC of the European Parliament and of the Council of 10 June 2002 on the Approximation of the Laws of the Member States Relating to Food Supplements (OJ L 183, 12.7.2002).

EC, 2002c. Regulation (EC) No. 178/2002 of the European Parliament and of the Council of 28 January 2002 Laying Down the General Principles and Requirements of Food Law, Establishing the European Food Safety Authority and Laying Down Procedures in Matters of Food Safety (OJ L 21, 1.2.2002).

EC, 2003a. Regulation (EC) No. 1829/2003 of the European Parliament and of the Council of 22 September 2003 on Genetically Modified Food and Feed (OJ L 268, 18.10.2003).

EC, 2003b. Regulation (EC) No. 1830/2003 of the European Parliament and the Council of 22 September 2003 Concerning the Traceability and Labelling of Genetically Modified Organisms and the Traceability of Food and Feed Products Produced From Genetically Modified Organisms and Amending Directive 2001/18/EC (OJ L 264, 18.10.2003).

EC, 2003c. Directive 2003/89/EC of the European Parliament and of the Council of 10 November 2003 Amending Directive 2000/13/EC as Regards Indication of the Ingredients Present in Foodstuffs (OJ L 308, 25.11.2003).

EC, 2004. Directive 2004/27/EC of the European Parliament and of the Council of 31 March 2004 Amending Directive 2001/83/EC on the Community Code Relating to Medicinal Products for Human Use (OJ L 136, 30.4.2004).

EC, 2006a. Regulation (EC) No. 1924/2006 of the European Parliament and of the Council of 20 December 2006 on Nutrition and Health Claims Made on Foods (L 404, 30.12. 2006).

EC, 2006b. Commission Directive 2006/37/EC of 30 March 2006 Amending Annex II to Directive 2002/46/EC of the European Parliament and of the Council as Regards the Inclusion of Certain Substances (OJ L 94, 1.4.2006).

EC, 2006c. Regulation (EC) No. 1925/2006 of the European Parliament and of the Council of 20 December 2006 on the Addition of Vitamins and Minerals and of Certain Other Substances to Foods (OJ L 404, 30.12.2006).

EC, 2006d. Commission Directive 2006/125/EC of 5 December 2006 on Processed Cereal-Based Foods and Baby Foods for Infants and Young Children (OJ L 339, 6.12.2006).

EC, 2006e. Commission Directive 2006/141/EC of 22 December 2006 on Infant Formulae and Follow-on Formulae and Amending Directive 1999/21/EC (L 401, 30.12.206).

EC, 2008. Regulation (EC) No. 764/2008 of the European Parliament and of the Council of 9 July 2008 Laying Down Procedures Relating to the Application of Certain National Technical Rules to Products Lawfully Marketed in Another Member State and Repealing Decision No. 3052/95/EC (L 218, 13.8.2008)

EC, 2009a. Commission Regulation (EC) 1170/2009 of 30 November 2009 Amending Directive 2002/46/EC of the European Parliament and of Council and Regulation (EC) No. 1925/2006 of the European Parliament and of the Council as Regards the Lists of Vitamin and Minerals and Their Forms That can be Added to Foods, Including Food Supplements (L 314, 1.12.2009).

EC, 2009b. Directive 2009/39/EC of the European Parliament and of the Council of 6 May 2009 on Foodstuffs Intended for Particular Nutritional Uses (L 124, 20.5.2009).

EC, 2009c. Commission Regulation (EC) No. 953/2009 of 13 October 2009 on Substances that may be Added for Specific Nutrition (L 269, 14.10.2009).

EC, 2009d. Commission Regulation (EC) No. 41/2009 of 20 January 2009 Concerning the Composition and Labelling of Foodstuffs Suitable for People Intolerant to Gluten (L 16, 21.1.2009).

EC, 2013. Proposal for a Regulation of the European Parliament and of the Council on Novel Foods. Brussels, 18.12.2013, COM(2013) 894 Final, 2013/0435 (COD).

EEC, 1989. Council Directive 89/398/EEC of 3 May 1989 on the Approximation of the Laws of the Member States Relating to Foodstuffs Intended for Particular Nutritional Uses (OJ L 186, 30.6.1989).

EEC, 1990. Council Directive of 23 April 1990 on the Deliberate Release Into the Environment of Genetically Modified Organisms (90/220 / EEC) (OJ L 117, 8.5.1990).

EFSA, 2010. Scientific opinion on principles for deriving and applying dietary reference values. EFSA panel on dietetic products, nutrition, and allergies (NDA). EFSA J. 8 (3), 1458.

EFSA, 2011. Scientific and technical guidance for the preparation and presentation of an application for authorization of a health claim (revision 1). EFSA J. 9 (5), 2170.

EFSA, 2012. Compendium of botanicals reported to contain naturally occuring substances of possible concern for human health when used in food and food supplements. EFSA J. 10 (5), 2663.

EP, 2019. Position of the European Parliament Adopted at First Reading on 17 April 2019 With a View to the Adoption of Regulation (EU) 2019/... of the European Parliament and of the Council on the Transparency and Sustainability of the EU Risk Assessment in the Food Chain and Amending Regulations (EC) No 178/2002, (EC) No 1829/2003, (EC) No 1831/2003, (EC) No 2065/2003, (EC) No 1935/2004, (EC) No 1331/2008, (EC) No 1107/2009, (EU) 2015/2283 and Directive 2001/18/EC (Text With EEA Relevance). EP-PE_TC1-COD (2018)0088. 17.4.2019.

EU, 2011a. Commission Regulation (EU) No. 1161/2011 of 14 November 2011 amending Directive 2002/46/EC of the European Parliament and of the Council, Regulation (EC) No. 1925/2006 of the European Parliament and of the Council and Commission Regulation (EC) No 953/2009 as Regards the Lists of Mineral Substances That Can be Added to Foods (L 296, 15.11. 2011).

EU, 2011b. Regulation (EU) No. 1169/2011 of the European Parliament and of the Council of 25 October 2011 on the Provision of Food Information to Consumers, Amending Regulations (EC) No 1924/2006 and (EC) No. 1925/2006 of the European Parliament and of the Council, and Repealing Commission Directive 87/250/EEC, Council Directive 90/496/EEC, Commission Directive 1999/10/EC, Directive 2000/13/EC of the European

Parliament and of the Council, Commission Directives 2002/67/EC and 2008/5/EC and Commission Regulation (EC) No. 608/2004 (L 304, 20.11.2011).

EU, 2013. Commission Directive 2013/46/EU of 28 August 2013 Amending Directive 2006/141/EC With Regard to Protein Requirements for Infant Formulae and Follow-on Formulae (L 230, 29.8.2013).

EU, 2014. Commission Regulation (EU) No. 119/2014 of 7 February 2014 Amending Directive 2002/46/EC of the European Parliament and of the Council and Regulation (EC) No. 1925/2006 of the European Parliament and of the Council as Regards Chromium Enriched Yeast Used for the Manufacture of Food Supplements and Chromium (III) Lactate Tri-Hydrate Added to Foods (OJ L 39, 8.2.14).

EU, 2015. Regulation (EU) 2015/2283 of the European Parliament and of the Council of 25 November 2015 on Novel Foods, Amending Regulation (EU) No. 1169/2011 of the European Parliament and of the Council and Repealing Regulation (EC) No 258/97 of the European Parliament and of the Council and Commission Regulation (EC) No 1852/2001 (L 327/1, 11.12.2015).

EU, 2017. Commission Regulation (EU) 2017/1203 of 5 July 2017 Amending Directive 2002/46/EC of the European Parliament and of the Council and Regulation (EC) No. 1925/2006 of the European Parliament and of the Council as Regards Organic Silicon (Monomethylsilanetriol) and Calcium Phosphoryl Oligosaccharides (POs-Ca®) Added to Foods and Used in the Manufacture of Food Supplements is the Last Amendment (L 173/1, 6.7.2017).

FAO/WHO, 2006. A Model for Establishing Upper Levels of Intake for Nutrients and Related Substances: Report of a Joint FAO/WHO Technical Workshop on Nutrient Risk Assessment, 2–6 May 2005. Geneva, World Health Organization <http://www.who.int/ipcs/highlights/full_report.pdf>.

FNB (Food and Nutrition Board, Institute of Medicine, National Academy of Sciences), 1997. Dietary Reference Intakes: Calcium, Phosphorus, Magnesium, Vitamin D, and Fluoride. National Academy Press, Washington DC.

Hanekamp, J.C., Bast, A., 2007. Food supplements and European regulation within a precautionary context: a critique and implications for nutritional, toxicological and regulatory consistency. Crit. Rev. Food Sci. Nutr. 47, 267–285.

Howlett, J., Edwards, D.G., Cockburn, A., et al., 2003. The safety assessment of novel foods and concepts to determine their safety in use. J. Food Sci. Nutr. 54, S1–S32.

Morris, C.A., Avorn, J., 2003. Internet marketing of herbal products. J. Am. Med. Assoc. 290, 1505–1509.

NRC (National Research Council), 1982. Diet, Nutrition and Cancer. Report of the Committee on Diet, Nutrition and Cancer, Assembly of Life Sciences. National Academy Press, Washington DC.

NRC (National Research Council), 1989. Diet and Health: Implications for Reducing Chronic Disease Risk. Report of the Committee on Diet and Health, Food and Nutrition Board, Commission on Life Sciences. National Academy Press, Washington DC.

Roda, L., Anadón, A., Recuerda, M.A., et al., 2009. Risk analysis and GM foods: scientific risk assessment. Eur. Food Feed Law Rev. 4, 235–250.

SANCO, 2006. Discussion Paper on the Setting of Maximum and Minimum Amounts for Vitamins and Minerals in Foodstuffs. Health & Consumer Protection, Directorate-General, European Commission.

SCF (Scientific Committee for Food), 1993. Nutrient and Energy Intakes for the European Community. Reports of the Scientific Committee for Food, Thirty First Series. European Commission, Luxembourg.

SCF (Scientific Committee for Food), 2000. Guidelines of the Scientific Committee on Food for the Development of Tolerable Upper Intake Levels for Vitamins and Minerals (adopted on 19 October 2000). SCF/CS/NUT/UPPLEV/11 Final 28 November 2000.

Chapter 74

Regulatory guidelines for nutraceuticals in India: an overview

Kavita Gulati, Suresh Kumar Thokchom, Jagdish Chandra Joshi and Arunabha Ray

Introduction

Humankind has made medicines from natural sources and has used these for various purposes since ancient times. Dr. Stephen L. De Felice, Chairman of the Foundation of Innovation Medicine, New Jersey, United States, coined the term "nutraceutical," which is a combination of two words "nutrient" (a nourishing food component) and "pharmaceutical" (a medical drug) and defined the term as "food or part of food that provides medical or health benefits, including the prevention and/or treatment of a disease" (DeFelice, 1992; Gupta et al., 2010). Several international agencies have defined this term in their own way but generally speaking it as a product which is isolated or purified from foods and then prepared in pharmaceutical doses form for its beneficial role in various chronic diseases and for physiological advantage (Zeisel, 1999; Thomsen et al., 1999). They include vitamins, minerals, amino acids, fatty acids, prebiotics, probiotics, herbs, botanicals, and dietary substances that humans use as supplements, for example, β-carotene (in eye disorder) and ginseng (for rejuvenation and common cold) (L'Abbe, 2008). Other popular nutraceuticals commonly prescribed by physicians include glucosamine (for arthritis), lutein (for macular degeneration) and echinacea (as immunomodulator). Nutraceutical categories include:

Dietary supplements (*including botanical*): A dietary supplement is a product that contains nutrients derived from food products and is usually concentrated in liquid, capsule, powder, or pill form. Dietary supplements are regulated by the Food and Drug Administration (FDA) as foods but their regulation differs from drugs and other foods (Rao, 2017). Examples include vitamins, minerals, coenzyme Q, carnitine, ginseng, *Gingko biloba*, St John's wort, ashwagandha, shilajit, etc.

Functional foods: As per the definition of the Institute of Medicine's Food and Nutrition Board, United States, functional food has been defined as "any food and food ingredients that may provide health benefit beyond the traditional nutrition that it contains. The sources of functional foods can be from plants or animals (Rao, 2017). Examples include oats, barley, wheat bran, psyllium, and lignins are used for heart disease and colon cancer; prebiotics, oligofructose in Jerusalem artichokes, shallots, onion powder, etc. for control of intestinal flora; omega-3 fatty acids in prevention of heart disease, canola oil for cholesterol reduction, and stanols (benecol) for reduction of cholesterol adsorption (Chaturvedi et al., 2011).

Medicinal foods: Medicinal food are products to be consumed or administered internally, under the supervision of a qualified physician. A specific dietary plan is designed aiming for the management of a disease or condition for which distinctive nutrition is required based on the medical evaluation as per recognized scientific principles (Rao, 2017). Example include slowly digested carbohydrates for diabetes mellitus, lactoferrin for immune enhancement, transgenic plants for oral vaccination against infectious diseases, health bars with added medications, etc.

Farmaceuticals: Farmaceuticals are products that consist of medically valuable components produced from modified agricultural crops or animals. The term farmaceutical is a combination of the words "farm" and "pharmaceuticals." Examples include ginseng tablet (standardized ginseng extract along with vitamins and minerals), silymarin syrup and capsules, and ginger capsules (Rao, 2017).

The rapid emergence of nutraceuticals has resulted from increased consumer awareness about rises in drug costs, drug insensitivity/refractoriness, and greater incidences of such drug adverse effects which complicate disease management. Their inclination toward natural products for prevention and treatment of disease have been further facilitated by poor existing alternatives for chronic diseases, impersonal exchanges with healthcare providers, desire for personalized medicines, large

Nutraceuticals. DOI: https://doi.org/10.1016/B978-0-12-821038-3.00074-4

population of aged patients, and renewed focus on preventive medicine. The overall general public perception that "natural is good" has complimented this change. Accordingly, a major percentage of the global population, particularly in developing countries like India, is looking at alternative medical therapy and nutraceuticals (including herbals/botanicals) with great expectations for the better management of disease.

Nutraceuticals accomplish desirable therapeutic outcomes with less or no side effects in comparison to pharmaceuticals (Nelson, 1999; Whitman, 2001). Some nutraceuticals have advantages over synthetic drugs and may have possible clinical uses in the prevention and management of several disorders. Nutraceuticals provide protection against and have a physiological benefit in several conditions, such as chronic inflammatory disorders, neurodegenerative diseases, diabetes mellitus, obesity, cardiovascular disorders, respiratory diseases, osteoporosis, cancer, and also as immunomodulators (Gupta et al., 2010).

The concept of nutraceutical supplements has now extended to the individualization of their use. As a result, terms like nutrigenomics—the study of the interaction between dietary components with the genome and the resulting proteonomic and metabolomic changes—and nutrigenetics—the understanding of gene-based differences in responses to dietary components and developing nutraceuticals that are most compatible with health, based on individual genetic makeup—are becoming increasingly popular. Nutrigenetics, in particular, is a nascent area that is developing rapidly and riding the wave of "personalized medicine," providing opportunities for nutraceutical product development (Palthur et al., 2010).

Global scenario for nutraceuticals

Nutraceuticals market began in Japan during the 1980s and nowadays it is being commonly used in market but, still there is no regulatory control of their marketing and production as the pharmaceutical drugs (Singh and Gujar 2013; Chintale, et al., 2013). The aim of nutraceutical regulation is largely to ensure products are safe and properly labeled. There is a perception that this lack of oversight leads to products of variable quality and with claims of questionable merit. Nutraceutical market is projected to achieve a global market size of about Rs.90 billion in 2013 at a Compound Annual Growth Rate (CAGR) of 20.24%. (Bhowmik et al., 2013). In the last twenty years the emerging nutraceuticals technology has created a global market with impressive growth rates estimated between 15% to 20% annually (Verbeke, 2005; Hilliam, 2000), with Japan, the United States (US) and the European Union (EU) as major markets (global market share 39%, 31% and 28% respectively, Nutraingredients-usa.com, 2010). It was estimated that the global market for nutraceuticals was about $ 142.1 billion in 2011 and reached about $ 250 billion by 2018 (Sector report 2013).

Globalization of the nutraceutical and functional food industries presents significant challenges to stakeholders, of which one of the important challenges is the regulatory variance between countries active in the marketplace. The global regulatory authorities are focusing on the product quality and safety worldwide. Food products are exported across the globe. Therefore maintaining safety and quality standards as per various regulatory guidelines set by the respective governments is very important. This can be a real driver for industry growth.

In the United States, for the manufacturing and marketing of nutraceuticals, legislation was passed in 1994 known as the Dietary Supplement Health and Education Act (DSHEA). The regulations under the DSHEA are different from those covering "conventional" foods and drug products. Under the 1994 DSHEA, the manufacturer of a dietary product or dietary ingredient will be responsible for ensuring product safety before it is marketed and the FDA is responsible for taking action against any unsafe dietary supplement product after it reaches the market. The federal Good Manufacturing Practice (GMP) regulation (21 CFR, Part 111) was introduced in 2007 and GMP enforcement was the responsibility of the FDA and Department of Health and Human Services. The rules specify that all domestic and foreign companies that manufacture, package, label, or hold nutraceuticals, including those involved with testing, quality control, and nutraceutical distribution in the United States, have the responsibility for assuring the safety of food products, both before and after their introduction to the market, by assuring the safety of the ingredients used in their products, evaluating the identity, purity, quality, strength, and composition of dietary supplements, preparing, packaging, and holding products in compliance with the FDA's current good manufacturing practice (cGMP) regulations. They should also submit reports of serious adverse events to the FDA.

In the EU, the European Food and Safety Authority regulates food legislation with a special focus on "food supplements." The EU defines food supplements as concentrated sources of nutrients (e.g., proteins, vitamins, and minerals) and other substances that have a beneficial nutritional effect. The EU legislation related to food supplements is Directive 2002/46/EC. The EU has prepared a list of permitted vitamin or mineral substances which may be added to food supplements for specific nutritional purposes. According to Bourne Partners, Europe also enforces stricter rules for product claims and the European market is driven on the basis of these health claims. There is variability within the EU when it comes to product claims by different countries and recommended daily allowances (RDAs) of various nutraceuticals. As a result, Europe's nutraceutical companies have formed a coalition as Food Supplements Europe to work with regulators to ensure quality and "to shape a positive regulatory environment for the future." In Japan, the nutraceutical industries are categorized into two segments. The first segment, "Foods with Nutrient Function Claims," sets the standards for the minimum and maximum daily levels of 12 vitamins and five minerals. Apart from forming these daily standards, the labels are not strictly regulated. The second segment is "Foods for Specified Health Uses" (FOSHU). These contain dietary ingredients that have reported beneficial physiological effects and promote health. Similar to other nations, disease risk reduction claims are not allowed in Japan, and FOSHU requires premarketing approval. The Consumer Affairs Agency, which is part of the Ministry of Health, Labor and Welfare, has been set up by the Japanese government. It controls the implementation of laws relating to nutrition labeling and health claims approval to maintain the standards and safety of these health foods. In China, the China Health Care Association has been appointed by the government to oversee the nutraceutical industry. There are three main entities which play an important role in policy making for this industry. The first is China's State Food and Drug Administration (SFDA), which is in charge of dietary supplements and issues the "blue hat" registration. The second is the Ministry of Health (MOH), which actually oversees the SFDA, but its main duty in the dietary supplement industry is overseeing the approval of new novel food ingredients and finally, the Administration of Quality Supervision Inspection and Quarantine, which controls all imports and exports passing across China's borders.

Nutraceuticals in India

Introduction of nutraceuticals in Indian was started by various companies in early 2000. The Indian nutraceutical industry has great prospects. Two major segments of Indian nutraceutical market are: (1) food supplements: which constitute about 60% of the market and distributed like fast moving consumer goods (FMCGs) and (2) vitamins and mineral supplements: constitutes about 40% of the market and distributed like the prescription drugs in India. Over the last decade a wide range of products have been available, giving an insight into the tremendous growth. On the one hand, there has been an increase in the incidence of diet and its related health issues, whereas, on the other hand, there is a growing awareness on the importance of nutrition and diet for long term good health.

The nutraceuticals industry in India is one of the rapid growing markets in the Asia-Pacific region. At present, the US, European Union and India are currently the world's largest nutraceuticals markets and China is expected to surpass them all by 2030 (Rao, 2017). According to a recent report, the nutraceuticals industry in India is worth about $ 2.2 billion and is likely to grow at a significant 21% Compound Annual Growth Rate (CAGR) to $ 10 billion in 2022 (Verma 2017). In Indian market, nutraceutical is a new word that has a spectacular annual growth rate of 25% which is same as Japan (Jain et al., 2018).

These have contributed to a favorable market conditions for nutraceutical industry in India.

India has a lot of advantages like qualified human resources, world class R & D facilities and varied raw material-aspects that give our country a leading edge. The growth pattern of nutraceuticals in India is very high. While the global market of nutraceuticals is growing at the rate of 7%, the Indian market has been growing by over 18-20% from many years (Majaz et al., 2012). The growth rate of functional food is about 54%, dietary supplements have 32% and the growing rate of functional beverage is 14%.

India is one of the largest producers of fruits and vegetables on the planet, however only 2% of these agricultural products is processed (Kumar et al., 2014). This may be due to poor infrastructure and inflexible laws that have been plaguing the nutraceutical system in India over the years. Moreover, regulatory agencies were equally responsible for multiple policy designs and a confusing regulatory scenario. However, for the first time, in the early 1990s, food processing laws and regulations were framed in a genuine and authentic manner for the food and nutraceutical sector in India. This framework was a combination of both state and national laws, which included the Export Act 1963, the Insecticide Act 1968, Meat Food Products Order 1973, Bureau of Indian Standards (BIS) Act 1986, and the Food Product Order 1995. In addition, the framework incorporated several other rules, acts, and orders such as the Agriculture Produce Act and the Essential Commodities Act 1995. In 1998, a subjective committee set up by the Prime Minister's office endorsed a single consolidated food regulatory authority to remove the confusion created by a multidepartment approach and all rules and acts with regard to the food sector were merged. Several public experts and also representatives of the Standing Committee of the Parliament supported this endorsement (Kumar et al., 2014). Further, in 2003, the MOH emphasized the need to devise new categories for the effective regulation of functional foods and dietary supplements which included obligatory safety inspection and evaluation of such products. It was stated that the discretionary standards should be brought under the BIS and the National Standards body. Again, in 2005, several government committees suggested a single regulatory authority for the food, governed by a single unified law, and as a result the Indian Food Safety Standard Bill, 2005 was passed. This bill was enforced as an Act in 2006 to ensure major scientific development with regards to the functional foods and dietary supplements industry.

The food processing sector laws were framed in the mid-1990s, and included state laws and various national laws. The national laws include Export (Quality Control and Inspection) Act 1963, Solvent Extracted Oil Control Order 1967, The Insecticide Act 1968, Meat Food Products Order 1973, Prevention of Food Adulteration Act (PFA) 1954 rules (Ministry of Health and Family Welfare) with the last amendments in 1986, BIS Act 1986, Environmental Protection Act 1986, Pollution Control Act 1986, Milk and Milk Products Order 1992, The Infant Milk Substitutes Feeding Bottles and Infant Food (Regulation of Production, Supply) Act 1992 and Rules 1993, Food Product Order 1995, Agriculture Produce Act, Essential Commodities Act 1995 (Ministry of Food and Consumers Affairs), Industrial license, Vegetable Oil Product Control Order 1998.

Regulatory aspects of nutraceuticals in India

Currently, India is the second most populous country in the world, likely to be the most populated country by 2027. As per the Indian definitions, nutraceuticals must have the listed ingredients along with their general properties. Nutraceuticals do not include traditional foods but contain certain important components having therapeutic activity or are formulated to fulfill a particular dietary requirement. These are notified in such a way that all the required ingredients in the foodstuffs must differ from ordinary foodstuffs. If there exists any kind of foodstuff then it must be defined as the following. These include the parts which are obtained in type of liquid extract, dry powder, or extract in the way of ethyl alcohol and hydro alcoholic, and they may be in the form of a combination or singly added from the plants/botanicals. Enzymes, vitamins, minerals, proteins, and amino acids or their components of metals should not exceed more than the limits which have been mentioned in the RDA for the Indian rules and regulations.

In India, nutraceuticals fall into two categories: functional foods (fortified food items and beverages) and dietary supplements (vitamins, minerals, protein supplements, and herbal products). Nutraceuticals are different from pharmaceuticals so they are generally not under the regulatory control of the Drugs and Cosmetic Act, 1940. Therefore, for regulation of nutraceuticals, various acts have been passed from time to time. The prevention of Food Adulteration Act was passed in 1954 and the Fruit Product Order in 1955. Similarly, there were the Meat Food Products Order, 1973 and The Edible Oils Packaging (Regulation) Order, 1998, The Solvent Extracted Oil, De-oiled Meal, and Edible Flour (Control) Order, 1967, The Milk and Milk Products Order, 1990, Essential Commodities Act, 1955 relating to food (Ahmad et al., 2011).To bring the entire Act under one roof and to unify licensing procedures, common application forms and procedures for registration of food businesses, central licensing, safety, sanitary, and hygienic conditions, the Prime Minister's council on trade and industry appointed a single food regulatory authority in 1998 which recommends a unified rule on food and agriculture industries.

In 2002, a national nonprofit association was constituted and the main objectives of this association were that every food manufacturer would provide scientific proof for the efficacy and safety of their products for consumers. In 2003, a MOH expert group under food law reported that there should be mandatory safety testing for these products. In 2005, standing committees of parliament on agriculture and other committees submitted their 12th reports, which demanded a single regulatory body and integrated law. The Indian Food Safety and Standards Bill was passed in 2005 and signed by the President on August 23, 2006. Finally, the government of India implemented the Food Safety and Standards Authority of India (FSSAI) in 2006. The Food Safety and Standards Act (FSSA) aims to integrate the food safety laws in the country in order to systematically and scientifically develop the food processing industry and paradigm shift from a regulatory regime to self-compliance. With the FSSAI Act being constituted in 2006, various rules and regulations related to nutraceuticals have been framed and came into existence in 2008. The Act came with two main objectives: (1) to introduce a single statute relating to food, and (2) to provide for scientific development of the food processing industry. The Food Safety and Standards Authority of India (FSSAI), 2006 Act consists of 12 chapters with 101 sections and two schedules. Chapter IV article 22 of the Act addresses nutraceutical, functional food, and dietary supplements and the need to regulate these products such that anyone can manufacture, sell, distribute, or import these products. FSSAI is the regulatory body for food and nutrition and is responsible for framing rules to govern the nutraceuticals market in India. The FSSAI defines "foods for special dietary uses or functional foods or nutraceuticals or health supplements." According to FSSA 2006, nutraceuticals are: (1) foods processed or formulated to satisfy particular dietary requirements for a physiological condition or specific disorder and may contain one or more of the following ingredients, namely: (a) plants or botanicals or their parts in the form of powder, concentrate or extract in water, ethyl alcohol or hydroalcoholic extract, single or in combination; (b) minerals, vitamins, proteins, metals, their compounds, or amino acids (in amounts not exceeding the recommended daily allowance for Indians) or enzymes (within permissible limits); (c) substances from animal origin; and (d) a dietary substance for use by human beings to supplement the diet by increasing the total dietary intake. Such products may be formulated in the form of powders, granules, tablets, capsules, liquids, jelly, and other dosage forms but not parenteral, and are meant for oral administration. This does not include a drug as defined in Ayurvedic, Sidha, and Unani systems in the Drugs and Cosmetics Act, 1940. This also does not include a narcotic drug or a psychotropic substance as defined in the Schedule of the Narcotic Drugs and Psychotropic Substances Act, 1985 and rules made there under and substances listed in Schedules E and EI of the Drugs and Cosmetics Rules, 1945. According to FSSA 2006 (implemented) registration and licensing for manufacturing is required.

Registration and licensing requirements

- To obtain registration and license in accordance with the procedure laid down in FSSAI (licensing and registration of food business) regulation 2011, is mandatory for every food business operator in the country (Verma 2017).
- A manufacturer can start business only after the registration or having a valid license.
- Petty food manufacturers of annual turnover less than Rs. 12 lacs need to register with the commissioner and turnover more than Rs. 12 lacs must get a food license from the FSSAI office.
- An application for the grant of a license shall be made in form B of schedule 2. An application should be filled for a license to be granted. This license shall be issued within 60 days from the date of issue of the application ID number (Jain et al., 2018).

After the issue of an application ID number, it is the responsibility of the licensing authority to inform the food safety officer to inspect the premises in the manner prescribed by the FSSAI in accordance with these regulations.

The licensing authority shall issue a license in format C under schedule 2 of these regulations. Registrations or licenses granted under these regulations shall be valid and subsisting, unless otherwise specified, for a period of 1–5 years.

For import, it is necessary to obtain an NOC from FSSAI and IEC. According to FSS regulations 2010 and 2011, registration of a manufacturing site is required and a manufacturing license in India has to be obtained from the state licensing authority except for those specified under schedule I. For import, according to draft consultation FSS regulations, 2011, an import license is needed from the central licensing authority. Several documents including the registration application, declaration form, application for manufacturing license, blueprint of processing unit, list of officials, list of equipment/machinery, photo ID, list of food category to be manufactured, analysis report, food safety management plan, pesticide residue report, recall plan, etc., need to be submitted. Temporary stall holders are exempted from the license but need to get their businesses registered with the local municipality or *Panchayat*. The FSSA incorporates the salient provisions of the Prevention of Food Adulteration Act 1954 and is based on international legislations, instrumentalities, and the Codex Alimentarius Commission (Palthur et al., 2010). In total, eight laws under the Food Safety and Standards Act (FSSA), 2006, govern the food sector and establish the Food Safety and Standards Authority of India (FSSAI), which regulates the sector and other allied committees.

FSSAI: Constitution and Responsibilities

Manufacture, storage, distribution, sale and import of nutraceuticals in India are regulated under the Food Safety and Standards Act, 2006. This Act consolidated the laws relating to food and established the Food Safety and Standards Authority of India for laying down science based standards for articles of food. However, no separate regulations for nutraceuticals exist in India. The Food Safety and Standard Rules, 2011 are effective from 5th May, 2011. The Food Safety and Standard Authority have issued regulations and under the new rule each state will have a food safety commissioner who would be the implementing agency.

A chairperson and total of 22 members constitute the FSSAI. The chairperson of the FSSAI can be either a civil servant not below the rank of Secretary, or an eminent food scientist. There should be seven ex officio members from various ministries and they should not be below the post of Joint Secretary. Every 3 years five members should be appointed from the States and Union Territories by rotation. The Authority contains two representatives each from the consumer organizations and food industry, three food technologists, two members from a farmer's organization, and one from a retail organization. The FSSAI works along with several scientific panels and a Central Advisory Committee to lay down standards for food safety. The FSSA and State Food Safety Authorities have powers to monitor and regulate the food business operators. The Commissioner of Food Safety of each state appoints a Designated Officer (DO), at the level of Sub-Divisional Officer. The DO of a specific district has the responsibility to issue or cancel licenses, prohibit sale of food articles that violate specified standards, and receive reports and samples of food articles from Food Safety Officers and get those analyzed. The Act provides for a graded penalty structure where the punishment depends on the severity of the violation, such as a fine is imposed on offenses such as manufacturing, selling, storing, or importing substandard or misbranded food. More severe offenses such as manufacturing, distributing, selling, or importing unsafe food, which results in injury, could incur a prison sentence that could be extended to life imprisonment in the case that the violation causes death. Even hawkers, vendors, temporary stall holders, and minor manufacturers who make their own food, could be fined up to INR 25,000 if they violate the specified standards.

The stringent regulations in quality and safety of nutraceutical products will benefit the industry on the whole. Regulatory requirements of these products throughout their life cycle are to be known for effective marketing strategies such as a fast and cost-effective approval process. The regulatory requirement for an industry not only involves product registration, rather it involves regulatory requirements right from their development to post-marketing surveillance such as site registration, approval process, timelines for approval, fees, import requirements, clinical trial applications, and variation applications. For effective life cycle management of a product, the regulatory department needs to have a grip on all the regulatory requirements.

As nutraceutical regulations are evolving in India, with the FSSA recently being implemented, there is a possibility that some of the content is conflicting/confusing, but for Indian industry to take shape, these need to be streamlined. In order to enter the Indian nutraceutical market, some of the very important areas to focus on include product evaluation, actual product analysis, procuring licenses, and developing India-specific health and label claims. The FSSAI has guaranteed that the licensing and registration of the nutraceuticals will be a simple process despite various parameters being involved (FSSAI, 2017).

As nutraceutical regulation is evolving in India, with the recent implementation of FSSAI, it is very important to focus on the essential areas and amend/streamline the regulations from time to time. Important aspects to consider are: product evaluation, product analysis, license procurement and development of specific health claims. Formulations are commonly used and hence appropriate care must be taken for assessing the quantity of each ingredient. The active constituent should be evaluated along with the additives with reference to factors like permissibility, dosages and standards according to the daily requirements in the Indian context. The classification of products also leaves a lot to be desired for example, food or food supplement or drug according to the FSSAI rules or the Drugs and Cosmetics Act. For product analysis, development and authentication of extracts, collection of samples, analysis of food, action plan and administrative action for incomplete procedures, etc, are of primary importance. For product registration, several licenses are needed depending on the purpose, viz. bulk drug or finished product sale, import of either, licensing for manufacturing, packaging, or marketing. Many documents are to be submitted to the government authorities for the same. Indian regulations also require development of appropriate health and label claims prior to entry in the Indian market (Keservani et al., 2019).

Health and label claims

One of the major elements to be focused on while entering the Indian market is the development of health and label claims specific to Indian regulatory guidelines. The major questions of International as well as national clients is on Indian labeling and packaging requirements, packing of consignment, need for sample material and declaration for registration, composition of consignment and approach for the same, label content, and structure–function claim and label claim. India-specific label content and claims are developed based on the results of a regulatory assessment of the product. Every package of food containing nutraceutical should have the word "NUTRACEUTICAL" written on the label with a common name of the nutraceutical. The amount of each nutraceutical ingredient in the product that either has a nutritional or physiological effect should be declared. If applicable, the quantity of nutrient shall be expressed in terms of percentage of the relevant recommended daily allowances as specified by the Indian Council of Medical Research even when the nutrient is present along with a nutraceutical as an adjunct and shall bear an advisory warning '*not to exceed the stated recommended daily usage*' (FSSAI, 2017).

Nutraceuticals from Indian Traditional System of Medicine

Herbal agents and supplements used in traditional Indian systems of medicine such as Ayurveda, Unani, or Siddha include all products intended for internal/external use for diagnosis, treatment (alleviation or prevention) of diseases in humans or animals, and manufactured exclusively in accordance with the formulae described in the authoritative books of the Ayurvedic, Siddha, and Unani systems of medicine. Some of the commonly used nutraceuticals are listed in Table 74.1. This rich traditional experience has been documented and the Government of India has given regulatory status to these preparations as

TABLE 74.1 Some nutraceuticals used in India.

Product	Health benefits
Tocopherols	Antioxidant
Ginseng	Rejuvenator, reduces lethargy, prevents aging, attenuates arthritis, and impotence
Fructo-oligosaccharides (prebiotic) *Lactobacillus* and *Bifidobacterium* (probiotic)	Improve the quality of intestinal microflora, gastrointestinal healthImprove intestinal microflora and prevent diarrhea. Effective in irritable bowel syndrome, other inflammatory bowel diseases, and gastrointestinal infections
Insoluble fibers (wheat bran)β-Glucan (oats, barley) Soluble fibers (*Psyllium*)	Reduce the risk of breast or colon cancerReduces the amounts of low-density lipoprotein (LDL) cholesterol, and risks of coronary heart disease, osteoporosis, and cancerReduce the risk of cardiovascular disease; protect against heart disease and some cancers; lower LDL and total cholesterol
Phytoestrogens	Cholesterol-lowering agent, prevent cardiovascular disorders, prostate and other cancers, cataracts and macular degeneration, boost immune function, retard the aging process, promote healing of burns, eczema, skin problems
Ashwagandha	Antistress, boosts the immune system, improves memory, reduces inflammation, antitumor
β-Carotene, lutein, lycopene	Antioxidant neutralizes free radicals, acts against cancer and cardiovascular disease, prevents night blindness and other eye problems, skin disorders, immune enhancer. Protects against toxins and infections. Reduces the risk of prostate cancer
Calcium and vitamins	Calcium supplement, joint health
Amla, ashwagandha, and pippali	Immune boosters
Predigested proteins, vitamins, minerals, and carbohydrates	Protein supplement
Long-chain omega-3 fatty acids—DHA/EPA (salmon and other fish oils) Conjugated linoleic acid (cheese, meat products)	Reduce the risk of cardiovascular disease; improve mental, visual functionsImprove body composition; decrease the risk of certain cancers
Isoflavones: daidzein, genistein	Menopause symptoms, such as hot flashes; protect against heart disease and some cancers; lower LDL and total cholesterol
Stanol ester	Lower blood cholesterol levels by inhibiting cholesterol absorption

Ayurvedic/Unani drugs. There are more than 57 books recognized by Government and they are laid down under the First Schedule of Drug and Cosmetic Act 1940. The First Schedule was inserted in a regulatory book more than 50 years ago and has been amended from time to time. The latest inclusion in 2010 was the Ayurvedic Pharmacopoeia of India.

The Indian Pharmacopoeia Commission is an autonomous body under the Ministry of Health and Family Welfare, Government of India, which publishes the Indian Pharmacopoeia, which is the legally recognized book for standards of medicines in India and specifies the standards in terms of identity, purity, and strength of medicines. The Indian Pharmacopoeia has also laid down a monograph for herbs and herbal products to maintain their quality. The quality standards of herbal agents/supplements are set as per the WHO guidelines using both conventional and modern scientific approaches (Verma, 2013). This has been developed in four national laboratories and a number of volumes on Indian medicinal plants have been published. To further intensify the initiative in regulating herbal agents, a new Ministry of Ayush has been recently set up by the government and herbal agents will now be evaluated separately. The purpose is to promote cultivation of herbs, promote research work in traditional medicines, lay down standards for quality of drugs, develop regulations to promote herbal medicines, regulate clinical research for safety and efficacy, and amend regulations from time to time for the betterment of the traditional systems of medicine in the interest of viable health care (Ravishankar and Shukla, 2007). As evidence of active progress in this direction the following phytopharmaceuticals have been approved as drugs: guglip extract and tablets—standardized Guggulu extract (*Cammiphoramukul*)—for cholesterol reduction, *Gingko biloba* tablets (*Gingko biloba* leaf extract)—for treating short-term loss of memory, ginseng tablets (standardized ginseng extract along with vitamins and minerals)—for immunity and energy, silymarin syrup and capsules—for treating liver disorders, ginger capsules—standardized ginger—for treating chemotherapy induced nausea.

The Drugs and Cosmetics Rules of 1945 and its Schedule Y regulate the various aspects dealing with research, development, and use of phytopharmaceuticals. It describes the information/data required for approval in clinical trials and/or to import or manufacture new drugs for marketing in India and is outlined along globally accepted formats such as ICH and US FDA criteria for this purpose. There are several subsections or appendices dealing with the different aspects involved in this regulatory exercise. It also comprises of a New Drug Approval Committee and a New Drug Advisory Committee. As per Gazette Notification (GSR 702) of 2013, a "phytopharmaceutical drug" includes processed or unprocessed standard material derived from plants or parts thereof or combinations of parts of plants, extracts, or fractions thereof in a dosage form for internal or external use of humans or animals and intended to be used for diagnosis, treatment, mitigation, or prevention of any disease or disorder in humans or animals, but does not include administration by the parenteral route. This notification includes basic information, human or clinical pharmacology information, pharmacognostic description, quality specification, processing of extract or phytopharmaceutical, formulation, manufacturing process, stability data, safety pharmacology, animal toxicology and, clinical trial data (phases I–III as per Schedule Y), regulatory status, marketing information, and post-marketing surveillance. In summary, phytopharmaceuticals will have to be evaluated for efficacy and safety as per Schedule Y. Recently, it has recommended that the term unprocessed standardized material should not exist. Details of contaminants should be mentioned according to daily consumption rather than per dose. Animal toxicity studies may not be needed for 100% aqueous extracts. These regulations are aimed at developing mechanisms to recognize and promote herbal agent use while assuring the quality, safety, and efficacy of such phytopharmaceuticals. Policies and regulations should be aligned with their conventional usage, and research, interactions, and training courses should be encouraged to facilitate this (The Gazette of India, 2013).

Concluding remarks and recommendations

Nutraceuticals are rapidly emerging as viable and alternative forms of treatment for a variety of prophylactic and therapeutic purposes. This has had a great impact on the cost of health care globally, and also in developing countries like India. It is therefore necessary to have stringent regulations to ensure the safe yet effective use of these agents. The currently available regulations have taken care of most of these issues, but there is always scope for amendment and updating, keeping in view the constantly changing scenario. The FSSA (2006) and the Drugs and Cosmetics Rule (1945) together ensure that most of these contentious issues are resolved. In view of the overlap existing in the various categories of nutraceutical nomenclature, some of these guidelines need to be revisited, refined, and streamlined in the interests of the consumer and the other stakeholders in the nutraceutical market.

References

Ahmad, M.F., Ashraf, S.A., Ahmad, F.A., et al., 2011. Nutraceutical market and its regulations. Am. J. Food Technol. 6, 342–347.

Bhowmik, D., Goliath, H., Kumar, B.P., et al., 2013. Nutraceutical – a bright scope and opportunity of Indian healthcare market. Pharma Innov. 1, 29.

Chaturvedi, S., Sharma, P.K., Garg, V.K., et al., 2011. Role of nutraceuticals in health promotion. Int. J. Pharm. Tech. Res. 3, 442–448.

Chintale, A.G., Kadam, V.S., Sakhare, R.S., et al., 2013. Role of nutraceuticals in various diseases: a comprehensive review. IJRPC 3, 290–299.

DeFelice, S.L., 1992. Nutraceuticals: opportunities in an emerging market. Scrip Mag. 9.

FSSAI, 2017. Key Requirements for manufacturing nutraceuticals in India.

Gupta, S.C., Kim, J.H., Prasad, S., et al., 2010. Regulation of survival, proliferation, invasion, angiogenesis, and metastasis of tumor cells through modulation of inflammatory pathways by nutraceuticals. Cancer Metastasis Rev. 29, 05–434.

Hilliam, M., 2000. Functional food: how big is the market? World Food Ingr. 12, 50–53. Available from: http://www.fssai.gov.in. The FoodSafety and Standards Regulations, 2011.

Jain, P.N., Rathod, M.H., Vineet, J.C., et al., 2018. Current regulatory requirements for registration of nutraceuticals in India and USA. Int. J. Drug Regul. Aff. 6, 22–29.

Keservani, R.K., Jarouliya, U., Sharma, A.K., 2019. Nutraceutical and functional food regulations in India. In: Bagchi, D. (Ed.), Nutraceutical and Functional Food Regulations in the United States and around the World, 3rd ed. pp. 431–444.

Kumar, A., Rajiv, S., Proudyogiki, G., 2014. Nutraceutical and functional food regulations in India. In: Bagchi, D. (Ed.), Nutraceutical and Functional Food Regulations in the United States and Around the World, 2nd ed. pp. 327–341.

L'Abbe, M.R., Dumais, L., Chao, E., Junkins, B., 2008. Health claims on foods in Canada. J Nutr 138, 1220S–1227S.

Majaz, Q., Khurshid, M.I., Nazim, S., et al., 2012. Nutraceuticals: importance and advances in medicine and health. Int. Res. J. Pharm. 3, 71–73.

Nelson, N.J., 1999. Purple carrots, margarine laced with wood pulp? Nutraceuticals move into the supermarket. J. Natl. Cancer Inst. 91, 755–757.

Palthur, M.P., Palthur, S.S.S., Chitta, S.K., 2010. Nutraceuticals: concept and regulatory scenario. Int. J. Pharm. Pharm. Sci. 2, 14–20.

Rao, K.S., 2017. Safety assessment of nutraceuticals. BEMS Rep. 3, 70–72.

Ravishankar, B., Shukla, V.J., 2007. Indian systems of medicine: a brief profile. Afr. J. Tradit. Complement. Altern. Med. 4, 319–337.

Sector report, 2013. Nutraceuticals industry. Bourne Partner 5.

Singh, M.C., Gujar, K.N., 2013. Nutraceuticals: uses, risks and regulatory scenario. Int. J. Pharm. Pharm. Sci. 5, 23–26.

The Gazette of India, 2013. GSR 702, Extra-ordinary, Part 2, Section 3, sub-section(i).

Thomsen, C., Bloch, A.S., Hasler, C.M., 1999. Position of the American dietetic association: fundamental foods. J. Am. Diet. Assoc. 99, 1278–1285.

Verbeke, W., 2005. Consumer acceptance of functional foods: socio-demographic, cognitive and attitudinal determinants. Food Qual. Prefer. 16, 45–57.

Verma, N., 2013. Herbal medicines: regulation and practice in Europe, United States and India. Int. J. Herb. Med. 1, 1–5.

Verma, A., 2017. Nutraceutical industry in India: current scenario and future trends.

Whitman, M., 2001. Understanding the perceived need for complementary and alternative nutraceuticals: lifestyle issues. Clin. J. Oncol. Nurs. 5, 190–194.

Zeisel, S.H., 1999. Regulation of "nutraceuticals. Science 285, 1853–1855.

Chapter 75

Regulatory aspects of nutraceuticals: Chinese perspective

Mildred S. Yang

Introduction

"Nutraceuticals" is a term originally proposed by the Foundation for Innovative Medicine to refer to a broad category of products including dietary supplements (such as vitamins, minerals), botanicals (such as ginseng, St. John's wort, *Gingko biloba*), functional foods (such as oat, bran, lignin, prebiotics, polyunsaturated fat products), some novel foods of transgenic origin (such as lactoferrins), oral vaccines, and some food products with added medication (DeFelice, 1995). Nutraceuticals differ from pharmaceuticals in that nutraceuticals include all substances that are extracted from food but with some health-promoting properties, while pharmaceuticals often refer to purified chemicals with a well-defined mechanism of action toward certain physiological or biochemical target for treatment of disease (DeFelice, 2002). In the United States, the term "nutraceuticals" was further clarified to be "Those diet supplements that deliver a concentrated form of a 'presumed' bioactive agent from a food, presented in a non-food matrix, and used to enhance health in dosages that exceed those that could be obtained from normal food" (Zeisel, 1999). Compared to functional food, nutraceuticals are not food, but an extract from food that can be added to food with the objective of improving health (Halsted, 2003; Shahidi, 2009; Pranaywal et al., 2013). A nutraceutical is demonstrated to have a physiological benefit or provide protection against chronic disease (Andlauer and Fürst, 2002; Hasler, 2002). In India, nutraceuticals were defined to be functional food that aids the prevention and treatment of diseases and/or disorders of diseases other than anemia (Dureja et al., 2003; Pandey et al., 2010). Health Canada distinguishes nutraceuticals from functional food based on their appearance. According to Health Canada, functional food is similar in appearance to a conventional food, consumed as part of a usual diet, and is demonstrated to have physiological benefits and/or reduce the risk of chronic disease beyond basic nutritional functions, while a nutraceutical is a product isolated or purified from foods that is generally sold in medicinal forms not usually associated with food (Hobbs, 2005). Therefore nutraceuticals could be a fortified dietary product such as vitamin D, calcium that can be added to food, such as milk, or vitamin C extracted from fruits such as orange and lemon. They, however, must be presented as a tablet, drink, or in pill form. Currently, the health food industry states that nutraceuticals include three main segments: dietary supplements, functional foods, and herbal/natural products (Rishi, 2006). Dietary supplements are vitamins and minerals. Functional foods are suitable for regulation of the functional state of the human body, and are beneficial to a specified group(s) of people. The short-term goal of functional foods and dietary supplements is to improve the quality of life and enhance health status, while its long-term goal is to increase lifespan while maintaining health (Holub, 2006).

Chinese herbal medicine, that has been in use for over 4000 years, falls into the segment of herbal/natural products. Historically, Chinese herbal medicine was synonymous with a number of different terminologies including herbal medicine, natural products, health foods, and functional foods (USFDA, 2003). Whether a natural product is deemed a medicine or a health food depends on the intention for use defined by the manufacturer. However, licensing for health food differs from medicine in several ways. For health and functional foods, the approval will be granted based on the safety, quality control, and labeling of the products. Traditional Chinese medicine (TCM), if manufactured as drugs, must also include study on efficacy, extensive clinical trial, and safety assessment (Fan et al., 2012). The following will provide a discussion on nutraceuticals (health/functional food) according to the Chinese perspective, and the evolution of regulations to manage these products.

Nutraceuticals. DOI: https://doi.org/10.1016/B978-0-12-821038-3.00075-6

From Chinese herbal medicine to functional foods

Chinese herbal medicine is probably one of the oldest arts in disease treatment. The knowledge of herbal medicine accumulated over time since the period of Emperor Huang Ti (2698–2598 BCE). The Shen Nong Ben Cao Jing (about 300 BCE–CE 200) was the first compendium of herbal medicine compiled from oral traditions. It contained records of 365 plants, illustrating their characteristic features, therapeutic properties, and methods of use. By the Tang Dynasty (CE 659), the number of medicinal herbs increased to 844 and the New Revision of Materia Medica (National Pharmacopeia) was published. They include not only plants, but also other ingredients that were found to be beneficial to health. Selected combinations of herbs would be boiled in water or soaked in alcohol to obtain a decoction for use. The combination of herbs used was based on diagnoses made by TCM philosophy with the ultimate goal of maintaining overall body homeostasis. Trading between China and neighboring countries such as Vietnam, Japan, Korea, and Malaysia helped spread the medical knowledge and, gradually, these countries incorporated Chinese herbal medicine as a major arsenal for managing the health of their people (Huang and Lapsley, 2005; Ghosh, 2013). After centuries of further development, the total number of medicinal herbs recorded in the Comprehensive Dictionary of Chinese Materia Medica (CE 1995) reached over 8000, and included plants (such as honey, lily, almond, grapes, lotus seed, *Panax ginseng*, lotus root, mung bean), animals (such as snake, ant, oyster, earthworm, etc.) as well as exotic by-products (such as musk, deer horn antler, bear bile, placenta hominis, bee pollen) (Dai and Lou, 1996). Continuous efforts are in progress to edit and revise the Chinese Materia Medica (Bensky, 2004).

As herbal medicine began to be adopted in developed countries, at least two problems arose. On the one hand, the use of herbal medicine was founded on a different therapeutic philosophy in which treatment was based on maintaining body function homeostasis. The concept of balancing Ying and Yang of different target systems (heart, liver, spleen, lung, and kidney) and their interactions represented by the five elements (i.e., water, metal, wood, fire, and soil), cannot be equated with the Western medical theory in which diseases must involve a specific target organ or biochemical molecule that can be treated by a single chemical agonist or antagonist. On the other hand, the use of natural products may result in uncertainty of quality of the raw materials used and thus, the reliability of the products produced. To be able to distinguish herbal products from common food, and to attach functional properties similar to the Western drugs suitable for disease management, Japanese scientists first announced the term "functional food" to refer to foods with "body-modulating functions," and focused their efforts on mechanistic analysis and molecular design (Arai, 2002; Shimizu, 2003). Through establishing the food for specified health use and adopting the procedures of extraction and identification of functional ingredients, studying biological efficacy and safety assessment, many isolated products, such as fucosterol from seaweed (Lee et al., 2003), auraptenes in citrus fruits, curcuminoids in ginger (Nagao et al., 2010), and a protein inhibitor found in rice (Nagata et al., 2000) were reported, each with specific functions. Further collaborative efforts between Chinese and Japanese scientists also resulted in the identification of a list of food ingredients that can be attached to some health benefits. These include monoacyglycerol, tea polyphenol, and pollen in regulating body lipids; soy bean, shark fins, orange peel in controlling obesity; honey in delaying the aging process; royal jelly, soybean oil, shark powder, spirullina, and pollen in regulating immune function; Chinese yam, poria, pig brain extract, and phospholipids in improving memory; and turtles, honey, and jujubee in improving stamina (Shimizu, 2003; Zhao, 1999).

Food that can also be used as medicine

It is a traditional belief in China that food and medicine are from the same source. Many raw herbs which are considered medicine can also be used as common foodstuff for everyday consumption. Since 1987, the Ministry of Health (MOH) of the Chinese Government promulgated the Food Sanitation Law, prohibiting the addition of medicine or drugs into food with the exception of those used as raw materials in food preparation or as seasoning. Subsequently, a list of 33 raw materials that can be used as both food and medicine was published. The MOH revised the list in 2002 and extended it to include 87 items. In October 2013, the National Health and Family Planning Commission (NHFPC), together with several governmental organizations including the National Food Safety Risk Assessment Center, the Chinese Center for Disease Control and Prevention, Chinese Academy of Medical Sciences, Institute of Medicinal Plant, etc. revised the Food Safety Law and reviewed the list with the aim of providing new information by including details such as the parts of the plants or animal body used and some specialized methods in preparation. Table 75.1A shows the list of substances that can be used as both food and medicine. In addition, Table 75.1B shows some traditional herbs that can be used in health food production, and Table 75.1C lists some substances that are forbidden to be used in health food production.

TABLE 75.1 Compilation of approved and forbidden foods/ingredients for use in health-promoting functions in China.

A. Substances that can be used as both food and medicine according to traditional practice
Clove, star anise, beans, fennel, thistle, yam, hawthorn fruit, purslanc herb, black plum, Chinese quince fruit, hemp seed, solomonseal rhizome, licorice, Dahurian angelica root, *Gingko* seed, white hyacinth bean, hyacinth flower, longan, foctid cassia seed, lily bulb, nutmeg, cinnamon bark, Indian gooseberry, finger citron fruit, almond, apricot seed (bitter, sweet), seabuckthorn fruit, Gordon euryale seed, pepper, adzuki bean, red bean, germinated barley, kelp, date, swingle, bush-cherry seed, honeysuckle flower, Chinese white olive, heartleaf houttuynia herb, ginger root, dulcis seed, wolfberry fruit, gardenia fruit, spiny amomurn fruit, boat fruited-sterculia seed, poria, citron fruit, aromatic madder, peach seed, mulberry leaf, mulberry fruit, tangerine peel, platycodon root, bitter cardamom, lotus leaf, radish seed, lotus seed, galanga rhizome, *Lophatherum* herb, fermented soybean, chrysanthemum flower, chicory grass, airpotato yam, Siberian solomonseal rhizome, perilla leaf, perilla seed, purerarìa root, black sesame seed, black pepper, pagodatree flower and bud, dandelion, grand torreya seed, wild jujube and seed, fresh imperata rhizome, fresh reed root, tangerine peel, peppermint, menthe, coix seed, longstamen onion, scallion, raspberry fruit, agastache, pogostemon, zaocys snake, oyster, donkey hide gelatin, chicken gizzard skin or lining, honey, viper
B. Some traditional herbs that can be used in health food production
Ginseng, ginseng leaves, fruit ginseng, *Panax*, tuckahoe, daji, *Ligustrum lucidum*, dogwood, cyathula, *Fritillaria* root, horse deer antler, red deer bone, salvia, wujiapi, schisandra, cohosh, aspartic, tianma, heterophylla, morinda, woody, horsetails, arctium, burdock root, psyllium, psyllium, littoralis, fritillaria, Scrophulariaceae, raw fleece-flower root, white and white peony root, cardamom, abalone, dendrobium (must provide proof can be used), digupi, angelica, zhuru, saffron, rhodiola, American ginseng, evodia, achyranthes, eucommia, eucommia leaves, complanatus, tree peony bark, aloe vera, herb, psoralen, terminalia, TPG, polygala, ophiopogon, shells, perrin, Oriental arborvitae, rhubarb, fleece-flower root, acanthopanax, thorn mei fruit, zeeland, alisma, roses, roselle, anemarrhena, apocynum, ilex, fago, laevigata, blue, magnolia, magnolia flowers, turmeric, *Citrus aurantium*, *Citrus aurantium*, baizi jen, pearls, gynostemma, fenugreek, madder, *Piper longum*, chive seed, polygonum vine, sweet, drynaria, codonopsis, *Morus alba*, mulberry, fritillaria, motherwort, gotu kola, longspur beans, dodder, wild chrysanthemum, *Ginkgo* leaf, astragalus, fritillaria, senna, gecko, cranberries, huai real, puhuang, *Tribulus terrestris*, propolis, tamarind, eclipta, cooked rhubarb, rehmannia, turtle
C. Substances that are forbidden to be used in health food production
Rhizoma of *Radix dysosma*, *Fructus rhododendrimollis*, *Semen cuphorbiaelathyridisd*, *Radix aristolochiaengchi*, *Rhizoma anisodustanguticus*, *Radix aconiticarmichaeli*, *Radix aristolochiac*, fangchi, *Folium sinica maxim*, *Semen strychni*, *Dysosmapleiantha* (Hance) Woodson, henbane seed, croton seed, mercury, perriwinkle grass, *Radix euphobiackansui*, *Rhizoma arisacrnataisconsanguincum*, *Rizoma pinelliac* ternate, *Rhizoma typhonigiganteum*, *Radix euphobiacelractealatac*, *Hydrangyrum chloraturncompositum*, *Herba lycorisradiatac*, *Caulis aristolochiae*, *Herba crotalariac*, *Folium nerium* oleander, cinnabar, poppy capsules, red mercuric oxide, *Ramulus et Folium Taxi chinesis*, *Fructus illieium hanryidiels*, mercuric oxide, *Radix strophanthi* divaricate, *Rhododendron molle* G. Don, *Pseudobulbus crernastracseu* Pleiones, lijinag, *Radix euphorbiacpekinensis*, *Triperygium hypoglaucum* Hutch, puffer fish (blowfish), *Flos rhododendrimollis*, *Lytta caraganacpallas*, *Radix* seu *herba* Derridistrifoliatac, *Folium digitalis*, *Flos daturacmetel*, *Semen pharbitidis nil*, arsenolite (white arsenic, red arsenic, arsenic), *Radix aconitikusnezoffii*, *Cortex periplocae* (periploca skin), *Herba peganiharmalae*, *Rhizoma dysosmeversipellis*, *Folium illiciilanccolati*, *Radix aconitiszechenyiani*, *Herba convallariaemajalis*, *Radix aconitibrachypodi*, *Folium thevetiac*, mylabris, sulfur, arsenic disulfide, *Radix tripterygiiwilfordii*, *Herba atropaebelladonnae*, *Rhizoma et Radix* Verairi, *Venenum bufonisgargarizans*

Source: Adapted from NHFPC, 2002. Further instruction for the management of health (functional) food. NHFPC #51, Promulgation date 03-11-2002 (in Chinese). English/Latin names are adopted from Huang, G. and Lapsley, K., 2005. Chinese health (functional) food regulation. In: Hasler and Clare (Eds.), Regulation of Functional Foods and Nutraceuticals: A Global Perspective, Wiley-Blackwell. Chapter 16.

The new resource foods (or new food ingredients)

New resource food (NRF) was a category of food source announced by the MOH of China in 1998. This includes purified ingredients or extracts of plants or animals that can be added to foods. They differ from functional foods in that NRF can be ingredients to manufacture functional foods. New resource foods include

1. Any plant, animals, and microorganisms that are not consumed as common food in China;
2. Ingredients isolated and separated from any plant, animals, and microorganisms that are not consumed as common food in China;
3. Novel microbial species used in the food processing procedure; and
4. Substances that arose from new food manufacturing techniques.

Materials that do not contain food ingredient characteristics, or are already listed under food additives (National Standard GB2760) or dietary supplements (National Standard GB14880), or previously rejected by the MOH, or have been considered illegal according to the MOH, will not be accepted as NRF. Until January 2013, the MOH had approved a list of some 72 NRFs for use. In the revamping of the government structure in 2014, NRF was renamed new food ingredients to better reflect the nature of these materials. A list of approved

TABLE 75.2 A list of selected new resource foods.

Tea blossom	Plant sterol
Suaeda salsa seed oil	Arachidonic acid oil
Sacha Inchi oil	*Gynura divaricata (L.) DC*
Sumac fruit oil	Poppyseed oil
Cordyceps guangdongensis	*Prunella vulgaris* L.
Acai	*Microcos paniculata* L.
Phylloporiaribis (Schumach:Fr.) Ryvarden	*Plumeria rubra* L. cv. Acutifolia
Chlorella pyrenoidesa	Tea camellia seed oil
Linderae aggregate leaf	Dunaliella salina (extract)
Moringaoleifera leaf	Fish oil (extract)
Sucrose polyesters	Diacylglycerol oil
Medium- and long-chain triacylglycerol oil	Earthworm protein
Wheat oligopeptides	Milk minerals
Resistant dextrin	Milk Basic Protein
Panax Ginseng C.A.Meyer	Inulin
Leuconostoc mesenteroides subsp. mesenteroides	Polyfructose
Hyriopsis cumingii polysacchride	*Lactobacillus acidophilus*
Lepidiummeyenii Walp	Xylo-oligosaccharide
Acer truncatum Bunge Seed Oil	Sodium Hyaluronate
Peony Seed Oil	Lutein Esters
Elaeagnus Mollis Diels Oil	L-Arabinose
β- hydroxyl-β-methyl butyrate (CaHMB)	*Acanthopanax sessiliflorus*
Haematococcus pluvialis	Aloe Vera Gel
Epigallocatechin Gallate(EGCG)	Gamma aminobutyric acid
stachyose	Colostrum Basic Protein
Sucrose Polyesters	Conjugated Linoleic Acid
Corn oligopeptides powder	Conjugated Linoleic Acid Glycerides
Phosphatidyl serine	*Lactobacillus plantarum*
Camellia chrysantha (Hu) Tuyama	*Eucommia ulmoides Oliv.*
Inula nervosa wall.ex DC.	Galacto-Oligosaccharides
Noni Puree	*Lactobacillusparacasei*
Yeast β-glucan	*Lactobacillus acidophilus*
Tissue culture of *Saussurea involucrata*	*Lactobacillus rhamnosus*
DHA, DHA Algal Oil	Bonepep
Raffino-oligosaccharide	Isomaltitol
Plant sterol ester	*LactobacillusPlantarum*
	Plant stanol ester
	Globin Peptide

Promulgated on January 4, 2018.
Source: Only Latin/English names extracted from https://www.sohu.com/a/216385355-652944.

NRF was revised and promulgated on January 4, which included 97 items. Table 75.2 shows a list of selected NRF approved for use in 2018.

Regulatory aspects of functional foods

Statutory authority for health (functional) food management

In 1995, the Chinese Government enacted the Food Hygiene Law (China State Council, 1995), and established that the MOH be the statutory body for implementing the Law and the National Center for Health Inspection and Supervision for approval of health foods licensing. The Law was divided into 104 articles. Many, such as food standards, inspection, imports and exports, and responses to food safety incidents, reflect topics addressed in the food laws of other nations. Articles 22 and 23 specifically discussed regulation for health foods.

food indicated to have specific health functions, the products and its description must be submitted to the administrative department of public health under the State Council for examination and approval; its hygiene standards and the measures for control of its production and marketing shall be formulated by the administrative department of public health under the State Council. (article 22); and that

The food indicated to have specific health functions may not be harmful for human health; the content of the product description shall be true, and the functions and ingredients of the product shall be identical with the information given in the product description and there shall be no false information. (article 23)

In 2003, the authority for health (functional) food management was transferred to the newly established State Food and Drug Administration (SFDA). Since then, over 4000 functional foods have been approved in which 10% were imported from other countries (Huang, 2002). Subsequently, a series of changes took place within the government structure in 2009, 2013, and 2019. In 2009, the Food Hygiene Law was replaced with the Food Safety Law (China State Council 2009; Petry and Wu, 2009). The MOH was renamed and its functions integrated into the new agency called the National Health and Family Planning Commission (NHFPC). The SFDA was transferred from the MOH to become an agency of ministerial status directly under the State Council (NPC CPPCC Annual Session 2013)—the highest authority in the Government. Together with this change, it was also renamed the China Food and Drug Administration (CFDA). The CFDA replaced a large group of overlapping regulators in charge of comprehensive supervision of the safety management of food, health food, and cosmetics, and was the competent authority of drug regulation in China (Xinhua News Agency, 2013). The latest change in the regulatory structures took place in 2018. The National Health Commission (NHC) was created to assume responsibility for all health matters on disease prevention and health promotion, allocation of public health resources, intensifying the reform of public hospitals, promoting the separation of supervision and operation, and managing the diversification of public health service providers. The Commission was to put more emphasis on improving the quality of health services and promote the separation of supervision and operation. Under the NHC, a board of experts (http://www.nhc.gov.cn/sps/s7890/202001/af8350909f2348d0-ba1215b1b707e03a.shtml) was formed with the major objectives of drafting and reviewing food safety standards, risk surveillance, and assessment. The CFDA merged with other administrative bodies to form the State Administration for Market Regulation (SAMR), responsible for matters including market entity registration, commodity prices, antimonopoly and antiunfair competition enforcement, commercial bribery, trademarks and patents infringement, food and industrial product safety, drug administration, quality inspection, certification, and accreditation. The SAMR was designated a unit under the State Council in charge of market supervision and administration. In the revised Food Safety Law of PRC, regulations concerning functional foods and special foods were detailed in Chapter 4, section 4 under the section on "special foods (http://gkml.samr.gov.cn/nsjg/fgs/201908/t20190819_306108.html)." A website (http://www.sfdachina.com/) was created in order to facilitate the search for specific types of information using an English communication medium.

Approval process for health (functional) food application

In accordance with the Food Safety Law of the People's Republic of China (2015 version), all foods claiming to have health-promotion effects must undergo an assessment and approval process in order to obtain a health food registration certificate or filing certificate. For domestic health foods produced in China, the registration shall be conducted with the SAMR (former CFDA), whereas the filing shall be carried out with the Provincial Administration for Market Regulation. For imported health foods produced in oversea factories, both the registration and filing shall be applied with SAMR. Meanwhile, oversea companies shall have a permanent Chinese representative office or appoint a Chinese agent to deal with registration or filing and obtain such certificates (http://www.sfdachina.com/info/85-1.htm). Many of the procedures remain similar those previously described (Huang and Lapsley, 2005; Yang, 2008), with minor modification. Application materials include a registration form for each functional food application, security and toxicity tests, functional tests (except for nutritional supplements), stability tests, hygiene tests, identifying tests for functional ingredients, etc. All tests should be done following the standard procedures by qualified agencies or laboratories.

Briefly, the application should first be submitted to the local provincial health administration authorities for preliminary screening. When approved, the application would be submitted to the SFDA for evaluation and approval. The detailed application dossier should be organized into five parts:

1. *Information on the enterprise and product for administrative data* which should include a health food registration application form, a copy of the legal registration certificate of enterprise or applicant, a testimony that the name of the product is unique and different from any authorized registration products of food or drugs, a guarantee document provided by applicants that there is no infringement on other patents, and a copy of a brand registration certificate.
2. *Information specific to the food and its characteristics* which should contain a report on product research and development, a description of the product's formulation

or ingredients, the analytical methods for functional or representative components, the manufacturing process description in brief and its flow chart, the specifications or standard of product and raw materials, the quality standards, and selective basis of packing materials.

3. *Laboratory testing data for the product* which should include test reports of the product and other relevant materials issued by testing agencies. Such reports would include a test application form, the requisition to the applicant by testing agencies, the toxicology test report, the functional test report, the test reports on dope, illegal drugs, etc. (for special functions only, including alleviation of physical fatigue, loss of weight, and improvement of growth functions), test report for functional components and composition, the stability test report, the hygiene test report, and other test reports to help the application (such as an identification test for raw materials or virulence of bacterial categories).
4. *Information for the consumer* containing the label and directions for use of the product with three product samples to be provided in the application.
5. *Evidence from scientific literature* which should include a summary and a literature review, published original data in scientific literature and any document from other countries' governments (that is helpful for the approval of products).

After reviewing all scientific documents by expert scientists from multiple disciplines, a certificate would be award to the approved product with an assigned serial number such as "Foodstuff/Health [year] No.##" (Huang and Lapsley, 2005). At this stage, manufacturing could begin under the supervision of the local authority to ensure that such activity would proceed according to the general hygiene regulations and good manufacturing practice. Production must also follow the approved regulatory

TABLE 75.3 List of accepted and approved functional claims for health foods.

1	Immune regulation
2	Regulate blood lipids
3	Regulate blood sugar
4	Delay aging
5	Improve memory
6	Improve vision
7	Promote lead discharge
8	Clear throat
9	Adjust blood pressure
10	Improve sleep
11	Promote lactation
12	Antimutation
13	Antifatigue
14	Resistant to hypoxia
15	Antiradiation
16	Lose weight
17	Promote growth and development
18	Improve osteoporosis
19	Improve nutritional anemia
20	Auxiliary protection for chemical liver injury
21	Beauty (remove acne/melasma/improve skin moisture and oil)
22	Improve gastrointestinal function (regulate intestinal flora/promote digestion/moisturize laxative/adjuvant protection of gastric mucosa)

Promulgated April 17, 2018, http://www.samr.gov.cn/.
Source: Also available from https://food.chemlinked.com/foodpedia/china-health-food-regulation.

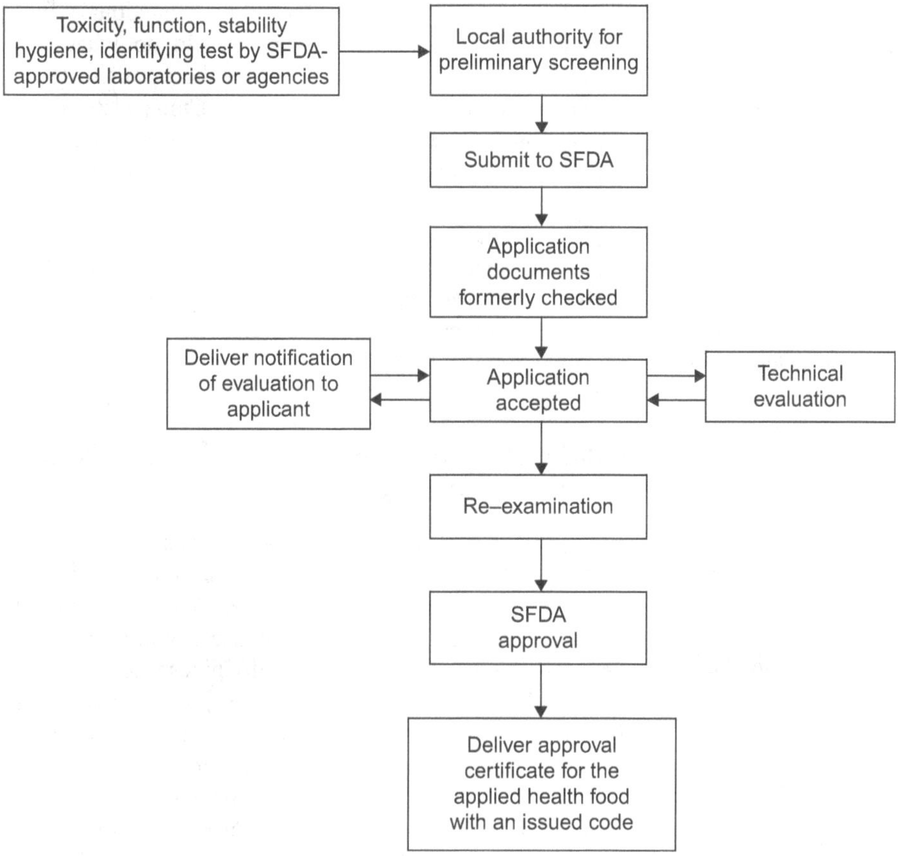

FIGURE 75.1 The "blue-hat" label for health (functional) food.

procedure to ensure proper raw materials are used, scientifically approved technical assessment procedures and methods for quality and safety controls are followed, and standard labeling and advertisement regulations are adopted. The local authority will also be responsible for testing sample products in the market to ensure compliance with quality.

Foreign manufacturers who want to sell their functional food products in China must register with the SFDA and obtain a good manufacturing practice (GMP) certificate. The following additional materials are needed for securing the certificate in China:

1. A certificate conforming to relevant local production quality control standards issued by the originating country;
2. A certificate for products having been sold for >1 year in the producing country by the certifying agency of the originating country and the Chinese embassy in the originating country;
3. Label and directions for use in the originating country;
4. Relevant product standards of the producing country or international organizations.

All the documents should be in Chinese, and the content should be consistent with the original language version.

To be in line with modern developments in scientific rigor, technical assessment procedures and methods including the standard functional assessment methods, standard toxicological assessment procedure and methods, regulation on nutrient supplements, and standard analytic methods for functional foods are continuously revised to meet the increasing demands on safety requirements. To avoid misleading consumers, the product must also be accurately labeled. The wording of health claims on labels is limited to the 27 functions listed in Table 75.3. The labeling should also include the product name, total weight, solid weight, ingredients, health function, preferred population for use, methods of consumption, date of used, storage method, operation, manufacturer's name and address, and health certificate number. A "blue-hat" symbol (Fig. 75.1) should also be added on the upper left-hand corner of the label to distinguish it from drugs.

Regulation of new resource food

To encourage research on new materials that can be added to food similar to that of nutraceuticals, the NRF was introduced. The NHC was in charge of the supervision and management of this new food source across the nation while the local health administrative authorities at or above the county level are in charge of the NRF quality within their respective administrative regions. The regulation of NRF was developed based on the provision for management of food additives (NHFPC #26, 2002). The regulation of NRF differs from functional food in that there is no need for scientific prove of efficacy as long as there is extensive literature support for its use. However, NRF must be demonstrated to be nontoxic, not damaging to health, and to not pose any acute, subacute, chronic, or other health hazards. Similar to functional food, new NRF must apply for a license before approval for production. The following information must be provided in the application:

1. Application form;
2. Scientific basis on production procedure;
3. Safety assessment report;
4. Production methods;
5. Safety standards, quality assurance, and method of determination;
6. Labeling and users' guide;
7. National and international research related to the use and safety information;
8. Other information for assessment purpose.

To reduce administrative costs, the application form for NRF does not have to be submitted to the local authority for preliminary screening. It can be submitted directly to the SFDA for processing. The main emphasis on the NRF application is on the method of extraction and identification. The method used in producing NRF must be clearly stated and consistently proven to produce the same product. Finally, unlike functional food, the approved NRF would be announced by the SFDA. No certification is issued for the approval of NRF.

Dietary supplements and their regulation

Dietary supplements include vitamins, minerals, amino acids, fungi, probiotics, nucleic acids, and extracts from wild animal and plant products. The role of dietary supplements is to supply the important nutrients to prevent nutritional deficiencies and reduce the risk of certain chronic degenerative diseases (Yang, 2005). Similar to new resource foods, dietary supplements could be consumed by either being added to food, or independently in the form of pills or beverages. China is ranked fourth in sales in the world in the dietary supplements market behind the United States, Japan, and Europe, amounting to US$15.8 billion in 2012 (USCHPA, 2014). Previously, China wished to protect the market by instituting strict regulations to control the import and sales of dietary supplements. The Food Safety Law of China stipulated that all nutrient supplements must apply for a "blue hat" certification similar to that for health (functional) food (see Fig. 75.1). The supplements should do what they claim to do, and the government has the power to recall dangerous products from the market. China also adopts international guidelines to set nutritional levels of dietary supplements (Halsted, 2003). In the 2014 reform, China intended to remove the requirements for the premarketing approval procedure and allow the import and sales of some, but not all, types of dietary supplements on a notification-based registration system by companies that have gone through a rigorous registration process with the CFDA (Grebow, 2014).

Challenges in developing herbal medicines as nutraceuticals

With the concern about the escalating costs of health care in the West, and the gradual loss of consumer confidence over the ability of conventional Western medicine in treating long-term debilitating diseases, authorities have begun to recognize that traditional Chinese herbal medicine could be a valuable resource in providing low-cost long-term health management of chronic diseases amongst the rapidly aging population (WHO, 2013). Previously, exploration of herbal medicine targeted on finding specific chemicals with therapeutic efficacies. Currently, the use of herbal extracts, or combinations of different herbs, is studied under the guidance of the treatment philosophy of traditional Chinese medical theory which aims to improve body function homeostasis, strengthen resistance to diseases, and balance the disruptive forces that brought about poor health. A survey indicated that about 70–80% of the global population rely on nonconventional medicine, mainly of herbal sources in their primary health care (WHO, 2013). However, the increase in the usage of herbal medicines has brought about concerns over the quality, efficacy, and safety of herbs available in the market (Zhang et al., 2012a,b). The major challenges to the herbal industry lie in quality assurance of the raw materials used and standardization of the herbal products (Guo et al., 2015).

Quality assurance

The quality of herbs is dictated by two important elements: the external factors and the internal characteristics (Chen and Huang, 2014). The external factors rest on the environmental conditions and poor production practice, which may result in poor internal characteristics such as contamination (e.g., toxic metals, pesticides, residues, and microbes), adulteration, and misidentification (Chen and Huang, 2014; Zhang et al., 2012a,b). The rigorous implementation of good agricultural practices (GAPs) and

GMPs would reduce the risks of external factors (Chen and Huang, 2014). Guo et al. (2015) studied the implementation process of GAP in Chinese herbal materials. The report indicated that while the Chinese government had promulgated a series of regulations for the administration of GAP through expanding the area allotted for planting herbal materials, educating farmers, and developing experts and scholars in research into GAP herbal production, problems still exist due to the inability of these workers, as well as government officials, to understand the importance of GAP implementation. There have been improvements in the training and education of workers in the industry. However, continuous use of old-fashioned cultivating techniques and disregard for the environmental effects on soil and water quality, and inadequate post-harvesting processes, have all hindered the improvement of its development.

Standardization of health (functional) food products

Characteristics such as morphology, histology, solubility, physical properties, moisture content, volatile oil, crude fiber, taste, chemical contents, DNA fingerprint, and biological characteristics are unique to each medicinal herb (Garg et al., 2012; Ernst, 2002). The variations in genetic factors produce herbal products with varying chemical compositions, and thus, biochemical efficacy, even when they are grown under the same conditions. Moreover, in most herbal products, identifying a single bioactive chemical responsible for its biological efficacy is yet to be achieved. While extensive efforts were taken to search for potential standard criteria for functional foods (Huang et al., 2012; Douillard et al., 2013), the work would require huge financial support and human efforts. One good example was the design of a standardized ginseng preparation, G115, to a product containing a fixed amount of the active ingredient, ginsenosides (Scaglione et al., 2005). Although such standardization still requires further improvement due to the lack of chemical consistency of the complex raw materials, it could be a step toward designing a health food product acceptable by the regulatory authorities (see Chapter 49 on Chinese Ginseng).

With the availability of computer technologies and advanced chemical, biochemical, and molecular biological techniques, the method of bioinformatics, a study on the "profile" of a substance could be constructed. Profiling is a method in which a set of criteria is used to determine the characteristics of a single herbal product. For example, morphological, histological, physical, and chemical profiles can be developed for a single herb for identification. The morphological profile could include color, shape, taste, smell, etc., and can be used as a first-line profile to identify a single herb while chemical profiling would be able to identify the different "active" components. Chemical profiling applied advanced analytical instrumentation to study the chemical content of a single plant (Liang et al., 2014; Chen et al. (2012), Li et al., 2014; Hou et al., 2014; Cui et al., 2012; Xu et al., 2013; Gao et al., 2013; Chen et al., 2013; Tao et al., 2013), as well as to determine the toxicology profile of herbs (Kim et al., 2014; Blacksell et al., 2014). Genetic profiling provides the fingerprint to establish the plant characteristics (Lu et al., 2015) and a profiling biomedical response may be used to study the efficacy of the product (Fu et al., 2010; Fong et al., 2013; Xiong et al., 2014; Guo et al., 2015).

Concluding remarks and future directions

In China, some 100 million people aged 60 and over had at least one chronic noncommunicable disease in 2013. It is predicted that by the year 2030, three times more people will be living with at least one such disease. There is a strong impetus for accelerating the adoption of personal health maintenance and disease-risk reduction in the healthcare system. Nutraceuticals, with their safety records, have been considered one of the attractive means for future healthcare management.

While the nutraceutical industry in China is limited to dietary supplements and traditional herbal medicine, the Chinese government has also taken careful steps to accommodate any novel chemical components which may satisfy the classification of nutraceuticals under the category of new resource food. The continuous modification of government structure and regulations were carried out through tightening their grips on certifying nutraceuticals of all categories with the intention of protecting the industry from being discredited. Continuous efforts rest in preventing unauthorized sales of poor-quality products to the public, and education of workers on the ethical issues important to the health food industry. The future of nutraceuticals relies heavily on sound scientific research in basic and medical sciences (Gu and Pei, 2017).

References

Andlauer, W., Fürst, P., 2002. Nutraceuticals: a piece of history, present status and outlook. Food Res. Int. 35, 171–176.

Arai, S., 2002. Global view on functional foods – Asian perspectives. Br. J. Nutr. 88 (Suppl. 2), S139–S143.

Bensky, D., 2004. In: Bensky, Dan, Clavey, Steven, Stoger, Erich (Eds.), Chinese Herbal Medicine: MateriaMedica, 3rd Edition September, 2004.

Blacksell, L., Byard, R.W., Musgrave, I.F., 2014. Forensic problems with the composition and content of herbal medicines. J. Forensic Leg. Med. 23, 19–21.

Chen, N.F., Huang, L.Q., 2014. Good agricultural practice (GAP) of Chinese materia medica (CMM) for ten years: achievements, problems and proposals]. Zhongguo Zhong Yao Za Zhi 39, 1143–1151.

Chen, Y., Yao, Z.H., Dai, Y., Cheng, H., We, n L.R., Zhou, G.X., et al., 2012. [Study on HPLC fingerprint of Oldenlandia diffusa]. Zhong Yao Cai 35, 886–892.

Chen, F., Li, H.L., Li, Y.H., Tan, Y.F., Zhang, J.Q., 2013. Quantitative analysis of the major constituents in Chinese medicinal preparation SuoQuan formulae by ultra fast high performance liquid chromatography/quadrupole tandem mass spectrometry. Chem. Cent. J. 7, 131.

China State Council, 1995. Food Hygiene Law of the People's Republic of China. National Law, Promulgation Date 1995-10-30. Effective Date 1995-10-30.

China State Council, 2009. Food Safety Law of the People's Republic of China. National Law #557. Promulgation Date 2009-7-8. Effective Date 2009-7-20.

Cui, P., Han, H., Wang, R., Yang, L., 2012. Identification and determination of Aconitum alkaloids in Aconitum herbs and Xiaohuoluo pill using UPLC-ESI-MS. Molecules 17, 10242–10257.

Dai, Y., Lou, X., 1996. Functional food in China. Nutr. Rev. 54, S21–S23.

DeFelice, S.L., 1995. The nutraceutical revolution: its impact on food industry R&D. Trends Food Sci. Technol. 6, 59–61.

DeFelice, S.L., 2002. FIM Rationale and Proposed Guidelines For the Nutraceutical Research & Education Act – NLEA. Presented at FIM's 10th Nutraceutical Conference November 10–11.

Douillard, F.P., Ribbera, A., Järvinen, H.M., Kant, R., Pietilä, T.E., Randazzo, C., et al., 2013. Comparative genomic and functional analysis of *Lactobacillus casei* and *Lactobacillus rhamnosus* strains marketed as probiotics. Appl. Environ. Microbiol. 79, 1923–1933.

Dureja, H., Kaushik, D., Kumar, V., 2003. Developments in nutraceuticals. Indian J. Pharmacol. 35 (6), 363–372.

Ernst, E., 2002. Toxic heavy metals and undeclared drugs in Asian herbal medicines. Trends Pharmacol. Sci. 23, 136–139.

Fan, T.P., Deal, G., Koo, H.L., Rees, D., Sun, H., Chen, S., et al., 2012. Future development of flobal regulations of Chinese herbal products. J. Ethnopharmacol. 140, 568–586.

Fong, S.Y., Liu, M., Wei, H., Löbenberg, R., Kanfer, I., Lee, V.H., et al., 2013. Establishing the pharmaceutical quality of Chinese herbal medicine: a provisional BCS classification. Mol. Pharm. 10, 1623–1643.

Fu, L., Xu, B.T., Xu, X.R., Qin, X.S., Gan, R.Y., Li, H.B., 2010. Antioxidant capacities and total phenolic contents of 56 wild fruits from South China. Molecules 15, 8602–8617.

Gao, X., Yang, X.W., Marriott, P.J., 2013. Evaluation of Coptidis Rhizoma-Euodiae Fructus couple and Zuojin products based on HPLC fingerprint chromatogram and simultaneous determination of main bioactive constituents. Pharm. Biol. 51, 1384–1392.

Garg, V., Dhar, V.J., Sharma, A., Dutt, R., 2012. Facts about standardization of herbal medicine: a review. J. Chin. Integr. Med. 10, 1077–1083.

Ghosh, D., 2013. Traditional chinese medicine—where is the evidence? Ensuring the safety, efficacy and quality of TCM requires a great deal of research and development. Available: http://www.nutraceuticalsworld.com/blog/Evidence-Based-Nutraceuticals/2013-06-10/traditional-chinese-medicine-where-is-the-evidence#sthash.NDnUY4Py.dpuf.

Grebow, J., 2014. China redrafting regulatory process for dietary supplements. Available: http://www.nutritionaloutlook.com/140707/China.

Gu, S., Pei, J., 2017. Innovating Chinese herbal medicine: from traditional health practice to scientific drug discovery. Front. Pharmacol. Available from: https://doi.org/10.3389/fphar.2017.0038116 June 2017.

Guo, D., Wu, W.Y., Ye, M., Liu, X., Cordell, G.A., 2015. A holistic approach to the quality control of traditional Chinese medicine. Science S29–S31.

Halsted, C.H., 2003. Dietary supplements and functional foods: 2 sides of a coin? Am. J. Clin. Nutr. 77, 1001S–1007S.

Hasler, C.M., 2002. Functional foods: benefits, concerns and challenges-a position paper from the american council on science and health. J. Nutr. 132, 3772–3781.

Hobbs, J.E., 2005. Evolving supply chains in the nutraceuticals and functional food industry. Can. J. Agr. Econ. 50, 559–568.

Holub, B.J., 2006. Potential for functional food ingredients to reduce health care costs. International Food and Health Innovation Conference, Malmo, Sweden, October 25–27.

Hou, J.J., Wu, W.Y., Liang, J., Yang, Z., Long, H.L., Cai, L.Y., et al., 2014. A single, multi-faceted, enhanced strategy to quantify the chromatographically diverse constituents in the roots of Euphorbia kansui. J. Pharm. Biomed. Anal. 88, 321–330.

Huang, J., 2002. Current situation and Development trend of health food, presented at the First Professional Symposium on Nutrition and Health Food, Org. by the Division of Nutrition and Health Food, Chinese Nutrition Society, Hainan.

Huang, G., Lapsley, K., 2005. Chinese health (functional) food regulation. In: Hasler, Clare (Eds.), Regulation of Functional Foods and Nutraceuticals: A Global Perspective. Wiley-Blackwell, Chapter 16.

Huang, W., Zhang, H., Liu, W., Li, C., 2012. Survey of antioxidant capacity and phenolic composition of blueberry, blackberry, and strawberry in Nanjing. J. Zhejiang Univ. Sci. B 13, 94–102.

http://www.nhc.gov.cn/sps/s7890/202001/af8350909f2348d0ba1215b1b707e03a.shtml.

http://gkml.samr.gov.cn/nsjg/fgs/201908/t20190819_306108.html.

http://www.sfdachina.com/.

http://www.sfdachina.com/info/85-1.htm.

Kim, H., Hughes, P.J., Hawes, E.M., 2014. Adverse events associated with metal contamination of traditional Chinese medicines in Korea: a clinical review. Yonsei Med. J. 55, 1177–1186.

Lee, S., Lee, Y.S., Jung, S.H., Kang, S.S., Shin, K.H., 2003. Anti-oxidant activities of fucosterol from the marine algae Pelvetia siliquosa. Arch. Pharm. Res. 26, 719–722.

Li, P.L., Liu, M.H., Hu, J.H., Su, W.W., 2014. Systematic chemical profiling of Citrus grandis 'Tomentosa' by ultra-fast liquid chromatography/diode-array detector/quadrupole time-of-flight tandem mass spectrometry. J. Pharm. Biomed. Anal. 90, 167–179.

Liang, J., Gao, H., Chen, L., Xiao, W., Wang, Z., Wang, Y., et al., 2014. Chemical profiling of an antimigraine herbal preparation, Tianshu capsule, based on the combination of HPLC, LC-DAD-MS (n), and LC-DAD-ESI-IT-TOF/MS analyses. Evid. Based Complement. Altern. Med. 2014, 580745.

Lu, Y., Lv, J., Hao, J., Niu, Y., Whent, M., Costa, J., et al., 2015. Genotype, environment, and their interactions on the phytochemical compositions and radical scavenging properties of soft winter wheat bran. LWT – Food Sci. Technol. 60, 277–283.

Nagao, K., Yamano, N., Shirouchi, B., Inoue, N., Murakami, S., Sasaki, T., et al., 2010. Effects of citrus auraptene (7-geranyloxycoumarin) on hepatic lipid metabolism in vitro and in vivo. J. Agric. Food Chem. 58, 9028–9032.

Nagata, K., Kudo, N., Abe, K., Arai, S., Tanakura, M., 2000. Three dimensional solution structure of Oryzacystatin-I, a cysteine proteinase inhibitor of the rice *Oryza sativa* L. *japonica*. Biochemistry 39, 14753–14760.

NHFPC, 2002 Further instruction for the management of health (functional) food. NHFPC #51, Promulgation date 03-11-2002 (in Chinese).

Pandey, M., Verma, R.K., Saraf, et al., 2010. Suggested that nutraceuticals: new era of medicine. Asian J. Pharm. Clin. Res. 3, 11–15.

Pranaywal, Wal, A., Nair, V.R., Nath, B., et al., 2013. Review on nutraceuticals and diet in prevention of cardiovascular diseases. Int. J. Pharm. Chem. Sci. 2, 1273–1281.

Petry, M. and Wu, B., 2009. Food Safety Law of the People's Republic of China. USDA Foreign Agricultural Service, GAIN Report CH9019.

Rishi, R.K., 2006. Nutraceutical: borderline between food and drug. Pharma. Rev. http://www.kppub.com/articles/herbal-safety-pharma-review-004/nutraceuticals-borderline-between-food-and-drugs.html. (accessed 12.02.09.).

Scaglione, F., Pannacci, M., Petrini, O., 2005. The standardised G115®*Panax ginseng* C.A. Meyer extract—a review of its properties and usage. Evid. Based Integr. Med. 2, 195. Available from: https://doi.org/10.2165/01197065-200502040-00003.

Shahidi, F., 2009. Nutraceuticals and functional foods: whole versus processed foods. Trends Food Sci. Technol. 20, 376–387.

Shimizu, T., 2003. Health claims on functional foods: the Japanese regulations and international comparison. Nutritional Res. Rev. 16, 241–252.

Tao, W., Duan, J., Zhao, R., Li, X., Yan, H., Li, J., et al., 2013. Comparison of three officinal Chinese pharmacopoeia species of Glycyrrhiza based on separation and quantification of triterpene saponins and chemometrics analysis. Food Chem. 141, 1681–1689.

USCHPA, 2014. Export-potential-report-2014-05-23.pdf.Availablein: http://uschinahpa.org/wp-content/uploads/2014/05/USFDA "Claims ThatCan Be Made for Conventional Foods and Dietary Supplements." U.S. Food and Drug Administration. September 2003. Retrieved 7 November 2010.

WHO (2013). WHO Traditional medicine strategy 2014-2023. Available: www.who.int/iris/bitstream/10665/92455/1/9789241506090-eng.pdf-422k.

Xinhua NewsAgency, 2013. China to elevate food, drug agency to general administration. Available: http://news.xinhuanet.com/english/china/2013-03/10/c-132221729.htm

Xiong, A., Yang, F., Fang, L., Yang, L., He, Y., Wan, Y.J., et al., 2014. Metabolomic and genomic evidence for compromised bile acid homeostasis by senecionine, a hepatotoxic pyrrolizidine alkaloid. Chem. Res. Toxicol. 27, 775–786.

Xu, T., Yang, M., Li, Y., Chen, X., Wang, Q., Deng, W., et al., 2013. An integrated exact mass spectrometric strategy for comprehensive and rapid characterization of phenolic compounds in licorice. Rapid Commun. Mass Spectrom. 27, 2297–2309.

Yang, Y., 2005. The implementation of nutrition labeling of foods – the challenge to nutrition theory and practice. ACTA Nutrimenta Sinica. 5, 3–6.

Yang, Y., 2008. Scientific substantiation of functional food health claims in China. J. Nutr. 138, 1199S–1205S.

Zeisel, S.H., 1999. Regulation of "nutraceuticals". Science 285, 1853–1855.

Zhang, L., Yan, J., Liu, X., Ye, Z., Yang, X., Meyboom, R., et al., 2012a. Pharmacovigilance practice and risk control of traditional Chinese medicine drugs in China: current status and future perspective. J. Ethnopharm 140, 519–525.

Zhang, J., Wider, B., Shang, H., Li, X., Ernst, E., 2012b. Quality of herbal medicines: challenges and solutions. Complement. Ther. Med. 20, 100–106.

Zhao, Y.H., 1999. Practical procedure of regulatory control in functional foods in China. Monthly Food Chemicals 10, 92–95.

Chapter 76

Uses and regulation of nutraceuticals: Australia and New Zealand

Rhian B. Cope

Introduction

Whether or not products at the human food–medicine interface are regulated as foods or medicines in Australia and New Zealand remains a somewhat complex issue. Within Australia, these products potentially fall either under the Australian Therapeutic Goods Act 1989 where they may be evaluated as a complimentary medicine or a medicine. In New Zealand, they may be regarded as a drug, or they may fall under the Dietary Supplements Regulations 1985, which in turn, falls under the Food Act 2014. However, depending on the health effects claimed, they may potentially be regarded as a "novel food" or a "food for special medical purposes" under the Food Standards Code, administered by the transnational agency Food Standards Australia New Zealand; and by a second-tier layer of legislation at the New Zealand and Australian state and territory level. In New Zealand, they could also be classified as a dietary supplement which is regulated under the Dietary Supplements Regulations 1985, which fall under the Food Act 2014. It could also be regulated as a supplemented food as per the New Zealand Food (Supplemented Food) Standard 2016.

As can already be comprehended by reading the above paragraph, the decisions on what laws, regulatory processes, and regulatory standards apply to a specific nutraceutical product in Australia and New Zealand can be very confusing. The objective of this chapter is to (hopefully) demystify some of these issues.

The Australian and New Zealand nutraceutical market

The Australian and New Zealand nutraceutical market is currently experiencing a compound annual growth rate of about 8% and has a market value of around $US10 billion dollars. The market is driven by increasing health awareness. This breaks down into three main factors: (1) the aging demography of the country, (2) the prevalence of chronic disease (notably heart health, diabetes prevention, obesity, and cholesterol regulation), and (3) obesity and other body issue concerns. Functional beverages are the largest market segment (about 36% of the market). Functional foods (about 32% of the market) and dietary supplements (about 29% of the market) are the other main market segments.

What is and is not a therapeutic good in Australia?

In Australia, a product that is ingested (i.e., swallowed) is regulated as either a food or therapeutic good (i.e., a medicine, a complementary medicine, or a medical device). The key issues of whether or not a product is a food or a medicine are the nature of the health benefit claims made by the product. Importantly, it is not necessarily the physiological effect(s) the product may have. Critically, just because a product is ingested by mouth and makes some type of health-based claim does not automatically make it a therapeutic good. Likewise, just because the physical form of the product is "medicine-like" (e.g., a capsule, a tablet, a powder, etc.) also does not automatically make it a therapeutic good, a dietary supplement, or a food.

Therapeutic goods in Australia are regulated by 26 different Acts, Regulations and Legislative Instruments! Within Australia, a therapeutic good is specifically defined in section 3 of the Therapeutic Goods Act 1989 act as the following:

> therapeutic goods means goods:
>
> (a) that are represented in any way to be, or that are, whether because of the way in which the goods are presented or for any other reason, likely to be taken to be:
>
> (i) for therapeutic use; or
>
> (ii) for use as an ingredient or component in the manufacture of therapeutic goods; or

Nutraceuticals. DOI: https://doi.org/10.1016/B978-0-12-821038-3.00076-8

(iii) for use as a container or part of a container for goods of the kind referred to in subparagraph (i) or (ii); or

(b) included in a class of goods the sole or principal use of which is, or ordinarily is, a therapeutic use or a use of a kind referred to in subparagraph (a)(ii) or (iii);

and includes biologicals, medical devices and goods declared to be therapeutic goods

The above is generally interpreted to cover any product for use in humans in connection with:

1. Preventing, diagnosing, curing, or alleviating a disease, ailment, defect, or injury;
2. Influencing, inhibiting, or modifying a physiological process;
3. Testing the susceptibility of persons to a disease or ailment;
4. Influencing, controlling, or preventing conception;
5. Testing for pregnancy;
6. Helping protect the skin from the damaging effects of UV radiation.

This also includes things that:

1. Are used as an ingredient or component in the manufacture of therapeutic goods;
2. Are used to replace or modify parts of the anatomy;
3. Make therapeutic claims on a label or in advertising (including packaging).

The following dichotomous key works through the seven fundamental questions that producers of nutraceuticals for the Australian market need to answer in order to determine where their product falls on the therapeutic good–nutraceutical interface.

1. Is the product for oral use in humans?
   - **a.** Yes: Go to 2;
   - **b.** No: the issue of the food–medicine interface does not apply and the product may be a therapeutic good.
2. Has the Secretary of the Australian Commonwealth Department of Health declared that particular goods or classes of goods are therapeutic goods (called a Section 7 Declaration under the Therapeutic Goods Act 1989)?
   - **a.** Yes: the issue of the food–medicine interface does not apply and the product is a Therapeutic Good;
   - **b.** No: Go to 3.
3. Has the Secretary of the Department of Health declared that particular goods or classes of goods are foods (called a Section 7AA declaration under the Therapeutic Goods Act 1989)?
   - **a.** Yes: The product is not a "therapeutic good." It is likely to be "food" within state/territory food regulation legislation and/or regulated under other state/territory legislation;
   - **b.** No: Go to 4.
4. Does the product fit into one of the standards in the Food Standards Code (food additive, vitamins and minerals, processing aid, novel foods etc. see http://www.foodstandards.gov.au/code/Pages/default.aspx)?
   - **a.** Yes: The product is not a "therapeutic good." It is likely to be "food" within state/territory food regulation legislation and/or regulated under other state/territory legislation;
   - **b.** No: Go to 5.
5. Does the product have a tradition of use as a food for humans in Australia and the form of the product has not been substantially altered (e.g., not purified, refined, compositionally altered, etc.)?
   - **a.** Yes: The product is not a "therapeutic good." It is likely to be "food" within state/territory food regulation legislation and/or regulated under other state/territory legislation.
   - **b.** No: Go to 6.
6. Do any of the following apply: (a) is the product represented in *any way* to be for therapeutic use (is it a therapeutic good or a therapeutic device)?; (b) likely to be *taken for therapeutic use* because of the way that it is presented?; and (c) likely to be *taken for therapeutic use* for any other reason?
   - **a.** Yes: the product is a therapeutic good and is assessed and regulated as per the Therapeutic Goods Act 1989;
   - **b.** No: Go to 7.
7. Is the product in a class of goods the sole or principal use of which is, or ordinarily is, a therapeutic use?
   - **a.** Yes: the product is a therapeutic good and is assessed and regulated as per the Therapeutic Goods Act 1989;
   - **b.** No: If it is not a biological or medical device, the product is not a "therapeutic good." It may be "food" within state/territory food regulation legislation.

If the above dichotomous key determines that the product is a therapeutic good, then some specific regulatory consequences apply:

1. The Therapeutic Goods Act 1989 enables the Therapeutic Goods Administration to regulate therapeutic goods that are imported into Australia and/or those which are shipped, transported, or sold across state or territory borders. It does not regulate products that are formulated or compounded within a state or territory that are not shipped, transported, or sold across a national, state, or territory border (e.g., compounded and sold by a practitioner only within a state or territory of Australia; also called extemporaneously

compounded). Extemporaneously compounded substances are regulated by state and territory authorities.

2. Therapeutic goods intended solely for the purpose of export are required to be listed (not registered) on the Australian Register of Therapeutic Goods (ARTG) before export is commenced.
3. The Therapeutic Goods Administration may take action against the importer, exporter, manufacturer, or supplier if the product is not included in the ARTG or is otherwise not exempt or approved under the Therapeutic Goods Act 1989.
4. If the product may be a health risk to the public (for example, it contains substances that are only available when prescribed by a health professional), the TGA can publish an alert to the public and, if necessary, order a recall of the product.
5. If a product is not a therapeutic good and it is in the ARTG, the TGA can take action under section 9F of the Therapeutic Goods Act 1989 to remove the product.

What is a complementary medicine in Australia?

Some classes of nutraceuticals, which meet the definition of a therapeutic good, may be classified as a complementary medicine in Australia. As a general rule, the premarket regulatory rigor applied to complementary medicines is lower than that applied to a medicine or medical device. The standard of evidence for demonstration of efficacy is much lower than that required for a medicine or drug. The focus of the premarket evaluation is on human safety and manufacturing quality and consistency. They may be listed on the ARTG if certain requirements are met.

A complementary medicine is defined as a therapeutic good consisting wholly or principally of one or more designated active ingredients, each of which has a clearly established identity and each of which has a traditional use. Designated active ingredients may include (but are not limited to):

1. An amino acid;
2. Charcoal;
3. A choline salt;
4. An essential oil;
5. A homeopathic preparation;
6. A microorganism, whole or extracted, except a vaccine;
7. A mineral including a mineral salt and a naturally occurring mineral;
8. A mucopolysaccharide;
9. Nonhuman animal material (or a synthetically produced substitute for material of that kind) including dried material, bone and cartilage, fats and oils, and other extracts or concentrates;
10. A lipid, including an essential fatty acid or phospholipid;
11. A substance produced by or obtained from bees, including royal jelly, bee pollen, and propolis;
12. A sugar, polysaccharide, or carbohydrate;
13. A vitamin or provitamin.

Complementary medicines are regulated as per the Australian Regulatory Guidelines for Complementary Medicines (https://www.tga.gov.au/publication/australian-regulatory-guidelines-complementary-medicines-argcm). As discussed above, extemporaneously compounded and dispensed substances may be exempt from Australian federal regulation under specific conditions; however, they may be subject to regulation by state and territory authorities.

What is and is not a therapeutic product in New Zealand and how are they regulated?

Therapeutic products and medical devices in New Zealand are potentially regulated by up to 19 different pieces of legislation! New or changed medicines and related products that are classifiable as therapeutic products require premarket evaluation to establish safety, quality, and efficacy as well as being subjected to a postmarket assessment process. The term "therapeutic product" is a comprehensive term that is applied to products that are intended to be used in or on humans for a therapeutic purpose as defined by section 4 of the Medicines Act 1981. A therapeutic product is designed to:

1. Prevent, diagnose, monitor, alleviate, treat, cure, or compensate for a disease, ailment, defect, or injury;
   or
2. Influence, inhibit, or modify a physiological process;
   or
3. Test the susceptibility of persons to a disease or ailment;
   or
4. Influence, control, or prevent conception;
   or
5. Test for pregnancy;
   or
6. Investigate, replace, or modify parts of the human anatomy.

Under the New Zealand approach, products are regarded as having a therapeutic purpose if:

1. The product contains one or more ingredient(s) that have a pharmacological action;

2. A therapeutic purpose is claimed for the product (usually on the label or in promotional material);
3. A therapeutic purpose is implied for the product (usually on the label or in promotional material);
4. The product contains a medicine listed in the First Schedule to the Medicines Regulations or a Notice in the New Zealand Gazette issued under section 106 of the Medicines Act 1981 (unless the product is in a form that cannot be administered to a human being for a therapeutic purpose).

In order for a product not to be regarded as having a therapeutic purpose, the label and promotional material must (at the very least) avoid the following:

1. A trade name that conveys an intended therapeutic purpose;
2. Words such as remedy, medicated, or therapeutic;
3. Statements that a product will/can/may prevent or treat a disease or condition or give relief from symptoms of a disease or condition;
4. Statements of traditional therapeutic use or use by ethnic groups for a therapeutic purpose;
5. Directions for use that infer a therapeutic purpose such as "dosing instructions" or instructions to "apply to the affected area";
6. Statements to the effect that the law prevents the supplier from making therapeutic claims that they consider they should be able to make about the product.

Critically, nutritional statements (i.e., statements regarding a normal biochemical or nutritional characteristic of a product) are not regarded as therapeutic claims.

Section 2 of the Medicines Act 1981 treats herbal remedies as a subcategory of therapeutic products provided that they do not contain a prescription, restricted, or pharmacy-only medical ingredient. To be defined as a herbal remedy, the product must be derived from a plant material that has been dried, crushed, subjected to aqueous or ethanolic extraction, or be a mixture of the plant-derived material with an inert substance.

Section 94 of the Medicines Act 1981 also identifies another class of substances termed "related products." These are products that are primarily foods, dentifrices, or cosmetics that also have a therapeutic purpose. Related products much not contain a prescription, restricted, or pharmacy-only medicine ingredient.

In New Zealand, dietary supplements are controlled by the Dietary Supplements Regulations 1985. These products are defined by Regulation 2 as an edible substance, in a controlled dosage form, which is intended to supplement the intake of substances normally derived from food. These substances must not be marketed or promoted for a therapeutic purpose.

What is a novel food in Australia and New Zealand?

Some nutraceuticals that are not therapeutic goods or therapeutic products may be classified as a novel food as per Australia New Zealand Food Standards Code—Standard 1.5.1. Like complementary medicines, these products require premarket approval before sale and use. Typically, a very extensive and detailed premarket evaluation is carried out by Food Standards Australia New Zealand, a transnational agency. Additional regulation may occur at the Australian state and territory level or in New Zealand.

A novel food is a substance or proposed food ingredient that has no history of traditional use. The term "*no history of traditional use*" is specifically defined as follows (https://www.legislation.gov.au/Details/F2017C00324):

> (a) a food that does not have a history of human consumption in Australia or New Zealand;
>
> or
>
> (b) a substance derived from a food, where that substance does not have a history of human consumption in Australia or New Zealand other than as a component of that food;
>
> or
>
> (c) any other substance, where that substance, or the source from which it is derived, does not have a history of human consumption as a food in Australia or New Zealand.
>
> (2) The presence of a food in a food for special medical purposes or the use of a food as a food for special medical purposes does not constitute a history of human consumption in Australia or New Zealand in relation to that food for the purposes of this section.

The premarket evaluation of novel foods in Australia typically involves an in-depth and detailed assessment of the public health and safety considerations having regard to:

1. The potential for adverse effects in humans;
   or
2. The composition or structure of the food;
   or
3. The process by which the food has been prepared;
   or
4. The source from which it is derived;
   or
5. Patterns and levels of consumption of the food;
   or
6. Any other relevant matters.

Typical categories of novel foods may include: (1) plants or animals and their components; (2) plant or animal extracts; (3) herbs, including extracts; (4) dietary macrocomponents; (5) single chemical entities; (6) microorganisms, including probiotics; and (7) foods produced from new sources, or by a process not previously applied to food.

Concluding remarks and future directions

Taking a helicopter view, the critical issue that must be addressed by a nutraceutical producer in Australia and/or New Zealand is whether or not a therapeutic claim will be made by the product. Although the Australian and New Zealand regulatory landscapes are complex, the first overriding principle in both systems is that if a product makes any sort of therapeutic claim regarding the substance in any form of media (labeling, packaging, promotional, advertising, etc.) then it will be potentially regulated as a therapeutic good or therapeutic product. The second overriding principle is that if the product contains any drug or pharmaceutical substance (including anything that is chemically related to a drug or pharmaceutical active substance), then it will potentially be regulated as a therapeutic good or therapeutic product. The third overriding principle is that, irrespective of what regulatory category a nutraceutical falls under in Australia and New Zealand, the human safety and quality characteristics of the product remain paramount. Both countries have very substantial legal and regulatory processes for the restriction and/or removal of products that have human safety concerns. If nutraceutical producers have any doubts about the legalities of their products and how they may be regulated, they should contact the Australian Therapeutic Goods Administration, New Zealand Medsafe, or Food Standards Australia New Zealand *before* producing, importing, or marketing their products.

Further reading

Commonwealth of Australia. Therapeutic Goods Act 1989. https://www.legislation.gov.au/Series/C2004A03952.

Food Standards Australia New Zealand. Advisory Committee Novel Foods. http://www.foodstandards.gov.au/industry/novel/novelcommittee/Pages/default.aspx.

Food Standards Australia New Zealand. Guidance Tool for Determining Whether a Food is Novel or Not. http://www.foodstandards.gov.au/industry/novel/documents/Guidance%20Tool%20-%20for%20website%20_2_.pdf.

Food Standards Australia New Zealand. Regulation of Novel Foods. http://www.foodstandards.gov.au/industry/novel/Pages/default.aspx.

Medsafe New Zealand. Categorisation of Products. http://www.medsafe.govt.nz/regulatory/categorisation-of-products.asp.

Medsafe New Zealand. Introductory Regulatory Guidance. http://www.medsafe.govt.nz/regulatory/regguidance.asp.

The Association of New Zealand Advertisers. Therapeutic Advertising Pre-vetting Service. https://www.anza.co.nz/Category?Action = View&Category_id = 262.

Therapeutic Goods Administration Australia. Food-Medicine Interface Guidance Tool (FMIGT). https://www.tga.gov.au/food-medicine-interface-guidance-tool-fmigt.

Therapeutic Goods Administration Australia. How therapeutic goods are regulated in Australia. https://www.tga.gov.au/how-therapeutic-goods-are-regulated-australia.

Therapeutic Goods Administration Australia. TGA Basics. https://www.tga.gov.au/tga-basics.

Chapter 77

Regulatory aspects of nutraceuticals: Japanese perspective

Keizo Umegaki

Introduction

Lifestyle-related diseases such as obesity, diabetes, hypertension, and cancer are of major concern and can be largely mitigated by improving dietary habits. Recent developments in food science and technology have revealed various physiological functionalities in foods and food components. Consumers are highly motivated to take advantage of such functionalities in an effort to improve and maintain health and, in some cases, to prevent disease. This trend can be observed worldwide, and a wide range of foods with various labels, forms, and components are distributed on the market.

In Japan, foods claiming health benefits are generally recognized as health foods, a term that is currently not clearly defined by law. This is a generic term describing foods that consumers deem to be healthful (Ohama et al., 2006; Tanaka et al., 2004). Categories of health foods recognized by consumers include nutraceuticals, foods with health claims, foods for specified health uses (FoSHU), functional foods, dietary supplements, nutritional supplements, health supplements, and natural foods.

In general, claims of health food benefits are based on in vitro and animal studies, but have not been substantiated in humans. In addition, most health foods consist of many components, and data regarding digestion, absorption, and appropriate dosages for safety and efficacy in humans have not attracted substantial consumer attention, even at the component level. With the increasing use of health foods in capsule or tablet form, consumers can occasionally mistake them for drugs. Consumers tend to think that health foods function as drugs, yet are as safe as ordinary foods, especially those made with natural ingredients. In addition, consumer purchases can be influenced by exaggerated media advertisements (Owens et al., 2014). Illegal health foods currently available that contain pharmaceutical ingredients or toxic chemicals pose a health hazard (Czepielewska et al., 2018; Hudson et al., 2018; Tucker et al., 2018). Although these illegal products are not considered to be foods and pose a serious health hazard, they continue to be distributed on the market as health foods until such time that the governmental agency responsible can demonstrate their illegality.

To manage these issues, and particularly to ensure the safety of health foods, regulatory systems are needed. Labeling on food packaging plays an important role in providing consumers with adequate and correct information and in regulating various food products marketed for health. With increasing globalization, a variety of products are distributed via the Internet, and developing measures to ensure the safety of health foods is thus an increasingly complex process (Kobayashi et al., 2018). In this chapter, the regulatory system governing health foods in Japan is described with a particular focus on foods with health claims, which consist of three types of foods: FoSHU, foods with nutrient function claims, and foods with function claims.

Historical treatment of health claims in Japan

Health claims for foods are typically regulated by laws established in individual countries. In Japan, the health claims system has evolved as a result of developments in scientific knowledge, international trends in the treatment of health claims, and the occurrence of health hazards associated with health foods. Regulatory systems are updated on an ongoing basis, and the ministerial and governmental offices responsible have responded by creating a new commission and agency. Major changes in the treatment of health claims occurred with the creation of FoSHU in 1991, foods with nutrient function claims in 2001, and foods with function claims in 2015 (Table 77.1). Notably, the regulatory system makes a rigid and consistent distinction between foods and drugs. The Pharmaceutical Affairs Act, which was revised in 2014, prohibits the use of terms

Nutraceuticals. DOI: https://doi.org/10.1016/B978-0-12-821038-3.00077-X

TABLE 77.1 Evolution of the regulation of health claims in Japan.

1952	Establishment of special nutritive foods, which consisted of two categories: "foods that can supply certain nutrients" and "foods that are suitable for a particular purpose" (currently foods for special dietary uses)
1984	Proposal of the concept of functional foods by Japanese academia
1991	Establishment of foods for specified health uses (FoSHU); currently, foods for special dietary uses is clearly defined
1996	Establishment of a nutrition labeling standards system
2001	Establishment of foods with health claims, a new regulatory system consisting of FoSHU and foods with nutrient function claims; FoSHU was incorporated into this system
2003	Establishment of the Food Safety Committee, which is in charge of safety evaluations as opposed to the Ministry of Health, Labour and Welfare (MHLW)
2005	Revision of FoSHU categories (four categories: ordinary, qualified, standardized, and claim of risk reduction)
2009	Category revision of foods for special dietary uses
	Creation of the Consumer Affairs Agency, which is in charge of the overall evaluation and approval of health claims as opposed to the MHLW
2015	Introduction of foods with function claims as a new category of foods with health claims

The MHLW was in charge of the overall evaluation and approval of health claims until the establishment of the Food Safety Committee and Consumer Affairs Agency.

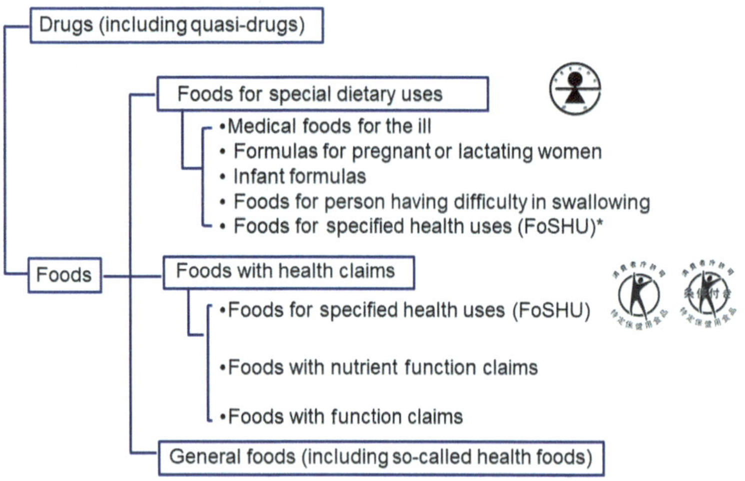

FIGURE 77.1 Current structuring of foods and drugs in Japan. **The initial FoSHU system was part of foods for special dietary uses; therefore it remains categorized within foods for special dietary uses and is also part of foods with health claims. Foods for special dietary uses and FoSHU are designated by seals, indicating that the finished product has been evaluated and granted governmental approval.*

such as disease, prevention, treatment, cure, and diagnosis for food products, even in foods with health claims.

The Food and Drug Classification by the Ministry of Health, Labour and Welfare (MHLW) of Japan is shown in Fig. 77.1. In Japan, there are only two classifications: foods and drugs (including quasi-drugs). Historically, the term "health foods" has been used without definition by law. However, foods for special dietary uses and foods with health claims are specified under the Health Promotion Law of 2002 and the Food Labeling Act of 2013. To aid in the classification of foods and drugs, the Pharmaceutical Bureau of the MHLW issued the notification "Instruction and Regulation for Disapproval Drugs" and created two lists of components: those used exclusively as drugs and those used as materials for non-drugs (no drug efficacy claims are made) (Ohama et al., 2006; Tanaka et al., 2004). These lists have been revised in response to the deregulation and opening of the market to foreign products and the appearance of novel components as potential materials for health foods.

Foods for special dietary uses

After mandatory approval, foods for special dietary uses can be claimed to be appropriate for special dietary use (Tanaka et al., 2004). Originally, the designation of foods for special dietary uses was enforced under the Japanese Nutrition Improvement Law (currently the Health

Promotion Law) of 1952. At that time, this category was part of special nutritive foods, which were divided into the following two categories: "foods that can supply certain nutrients" and "foods that are suitable for a particular purpose." Foods for special dietary uses was clearly defined in 1991, with significant revisions undertaken in 2009, resulting in the following five current categories: formulas for pregnant or lactating women, infant formulas, foods for persons with swallowing difficulties, medical foods for the ill, and FoSHU (Fig. 77.1). FoSHU was initially established in this category in 1991, so it remains categorized within foods for special dietary uses as well as foods with health claims, based on a labeling system established in 2001. Medical foods for the ill contain subcategories that consist of low-protein foods, special diets for food allergies, lactose-free foods, and high-density liquid diets. With the exception of FoSHU, foods for special dietary uses should be found in forms easily recognized as foods, that is, capsules and tablets are prohibited, and generally used following the advice of a health professional such as a physician or registered dietitian. By contrast, other health foods included in foods with health claims are designed to be used by consumers at their own discretion by referring to claims on the food labeling.

Foods for specified health uses

Differences between functional foods and foods for specified health uses

The FoSHU system was created by the MHLW in 1991 to encompass functional foods. Although functional foods have yet to be defined by law, the general concept was proposed in the 1980s by the Japanese scientific academia (Arai, 1996). According to this concept, foods typically have three functions, in the following order: (1) nutritional, (2) sensory, and (3) physiological. Functional foods focus on the third function, that is, physiological, and contain component(s) that exert a physiological function. The development of food science and technology has revealed various physiological functions of foods and/or components, including dietary fiber, oligosaccharides, lactic acid bacteria, polyphenols, and peptides. Manufacturers producing various foods take advantage of such scientific data by incorporating them into product advertisements. However, the majority of well-known functional foods are produced based solely on in vitro and animal studies, so little remains known about their safety and efficacy in humans. Whether the functionalities of raw materials can be realized in finished products as emphasized in the labeling or via claims by the manufacturer also remains unclear. Exaggerated labeling on and misleading claims about products are commonplace on the market, and these can lead to consumer confusion. To counter these issues and utilize food functionality for health promotion, the MHLW implemented the FoSHU system under the Japanese Nutrition Improvement Law [currently the Food Labeling Act enacted by the Consumer Affairs Agency (CAA)]. Functional foods and FoSHU are both designed to contribute to health maintenance and enhancement. However, FoSHU differ from common functional foods as follows: the safety and efficacy of both components as well as the finished product are evaluated using current scientific knowledge; a randomized placebo-controlled clinical trial is conducted; the necessary documents supporting the claim are confirmed by the governmental agency responsible, and the claim is legally approved and denoted by a specific seal, thereby enhancing consumer recognition (Fig. 77.1); and they aim to improve the dietary habits of users. As shown in Table 77.1, the initial FoSHU system was modified with the creation of a regulatory system for foods with health claims in 2001 and a revision of the subcategories in 2005. Initially, both the entire review process and approval of FoSHU were conducted by the MHLW. However, since 2003, safety evaluations have been conducted by the Food Safety Commission, and since 2009, overall evaluations have been conducted by the CAA and Consumer Commission.

Requirements and subcategories of foods for specified health uses

The FoSHU requirements are shown in Table 77.2. As described, foods designated as FoSHU must have proven safety and efficacy, that is, clinical studies of the final product must be conducted according to the CAA guidelines. FoSHU are foods, not drugs, and target healthy people or individuals concerned with their health, not those suffering from diseases. Therefore neither the product nor its functional components can be included in the list of ingredients (raw materials) used exclusively as drugs. The product must also be intended for use by consumers at their own discretion by referring to product claims of improved dietary habits and contributions to health maintenance and enhancement. Until 2001, FoSHU were required to be in conventional or processed form (foods for special dietary uses); however, this requirement was dropped with the introduction of foods with health claims. It is also important to confirm whether a functional ingredient is contained in the indicated amount after the approved FoSHU product is distributed on the market. Thus the National Institute of Health and Nutrition or laboratories authorized by the MHLW have been charged with performing inspections.

Currently, FoSHU consists of four groups: ordinary FoSHU, standardized FoSHU, qualified FoSHU, and FoSHU with a disease risk reduction claim. Standardized FoSHU are products that satisfy a number of criteria, such as containing ingredients at the indicated levels;

TABLE 77.2 Requirements for foods for specified health uses approval.

The product is expected to improve dietary habits and contribute to health maintenance and promotion, and satisfies the following requirements:

1. Scientific evidence for the intended health use of the product or ingredients involved must be clarified both medically and nutritionally
2. Appropriate intake levels of the product and/or related ingredients can be established both medically and nutritionally
3. The product and/or the ingredients involved are safe based on the submitted materials
4. The following matters should be clarified for the functional ingredients involved (this shall not apply when there is an appropriate reason):
   a. Physical, chemical, and biological characterization and relevant methods
   b. Methods of qualitative and quantitative analytical determination
5. The product and/or its functional ingredients do not lead to the excessive intake of sodium or sugar, or it is a beverage that does not contain alcohol
6. The functional ingredient in a foodstuff is not significantly altered from the original state
7. The product is intended to be consumed on a daily basis
8. The product and/or its functional ingredients are not included in the list of ingredients (raw materials) used exclusively as drugs (drug list)

Source: Adapted from https://www.cao.go.jp/bunken-suishin/teianbosyu/doc/tb_h30fu_05caa_124b.p

many FoSHU products with the same health claims have already been approved for years. The evaluation and approval of standardized FoSHU are simplified compared with ordinary FoSHU, that is, these products do not require a detailed review process by the Consumer Commission. Qualified FoSHU are products that have insufficient scientific evidence to meet the efficacy standards of ordinary FoSHU, but some evidence of their safety and efficacy as finished products has been presented. Unlike other types of FoSHU, qualified FoSHU must include a disclaimer on the label stating that "evidence of efficacy has not necessarily been established." Standard FoSHU and qualified FoSHU were introduced to increase the flexibility of the regulatory system. Disease risk reduction claims are approved under the following limited conditions: applicable to ordinary FoSHU (currently only approved for calcium and folic acid for the prevention of osteoporosis and neural tube defects, respectively), and the product must contain a defined amount for daily intake (300–700 mg for calcium and 0.4–1 mg for folic acid in the form of pteroylmonoglutamic acid) and include a statement of precautions and contraindications on the label. The disease risk reduction claims were introduced in reference to the Codex Committee under the WHO/FAO organization (Yamada et al., 2008).

Current foods for specified health uses claims and products

Current FoSHU claims can be classified into eight main groups (see Table 77.3). Among FoSHU products, one-third consist of claims of improved gastrointestinal conditions, followed by claims of regulating blood neutral lipid levels and body fat, blood glucose, and blood pressure. Foods contained in this category are typically beverages, yogurt, noodles, bread, cereals, cooking oil, and sausages, that is, products that can be incorporated into daily meals to improve dietary habits. Approximately 1074 FoSHU products had been approved by the end of 2019. Although the functionality of FoSHU has received considerable attention, it is also important to recognize that the safe and effective use of these products is needed to improve dietary habits, in addition to a balanced diet from staple foods, main dishes, and side dishes. Therefore such statements are indicated on the labeling of each product. More detailed information on FoSHU is available on the CAA website (https://www.caa.go.jp/policies/policy/food_labeling/).

Foods with nutrient function claims

The importance of supplying specific nutrients in certain foods has been recognized since the creation of special nutritive foods in 1952. According to nutrition labeling standards, only the content of nutrients in foods could be labeled. Health claims of functional nutrients were not allowed, even for vitamins and minerals, until the creation of the foods with health claims system in 2001. Currently, foods with health claims encompasses FoSHU, foods with nutrient function claims, and foods with function claims. Foods with nutrient function claims can be marketed not only in tablet or capsule form, but also as conventional and fresh foods. Foods with nutrient function claims can be labeled with nutrient function and freely manufactured and distributed without governmental approval under the conditions that the foods must meet established standards

TABLE 77.3 Currently approved foods for specified health uses: ingredients and health claims.

Target	Type of food (examples)	Ingredients (examples)	Health claims (examples)
Gastrointestinal conditions	Table sugar, yogurt	Oligosaccharides, lactic acid bacteria	Helps maintain favorable GI conditions
			Helps improve bowel habits
Cholesterol	Soft drinks, cooking oil	Chitosan, phytosterol	Helps lower cholesterol levels
Triacylglycerol/body fat	Cooking oil, oolong tea	Medium-chain fatty acids, polyphenols	Helps mitigate body fat gain
			Aimed at individuals concerned about body fat
Blood pressure	Instant soup, candy	Peptides, chlorogenic acid	Aimed at individuals with high blood pressure
Bone	Soft drinks	Isoflavones, vitamin K2	Promotes calcium absorption; supports bone health
Teeth	Chewing gum	Sugar alcohol	Helps maintain strong and healthy teeth
Blood glucose	Soft drinks	Indigestible dextrin, polyphenols	Aimed at individuals concerned about blood glucose level

and contain appropriate amounts of the nutrient (within the designated upper and lower daily limits) for daily consumption. The lower daily limit is based on the standard values for nutrition labeling as per the Japanese dietary reference intakes (Health Service Bureau and Ministry of Health Labour and Welfare, 2018), while the upper daily limit is generally set as the maximum amount of nutrients in quasi-drugs, which are defined as products that have minimal to moderate pharmacologic activity, but are restricted in use to specific indications.

The statements of foods with nutrient function claims are predetermined based on internationally recognized findings, that is, examples of nutrient function claims include a reference to the *Codex Alimentarius* and are easy to understand for consumers (Shimizu, 2003; Yamada et al., 2008). Precautions and contraindications should also be provided on labels for every nutrient, including information concerning excess intake. In contrast to FoSHU, foods with nutrient function claims are not subject to governmental approval; therefore these are not designated by a seal, and the claims must state that "this product has not been investigated individually by the CAA." To avoid confusion with drugs, no nutrient function claim can contain a disease name in the statement.

Currently, the nutrient functions of 13 vitamins, six minerals (calcium, magnesium, potassium, zinc, iron, and copper), and n-3 fatty acids can be stated. Concerning the risk of health hazards by overdose, making nutrient function claims for potassium is not allowed for foods with nutrient function claims that come in the form of capsules or tablets. Examples of nutrient function claims, precautions and contraindications, and upper and lower daily intake limits for foods with nutrient function claims are shown in Table 77.4. Foods with nutrient function claims are intended to supplement or complement the daily diet in cases of insufficient dietary intake due to increased age or unhealthy dietary practices, or when consumers feel that dietary supplementation is warranted (Tanaka et al., 2004). For example, selenium is an important mineral with well-recognized supplemental effectiveness. However, under foods with nutrient function claims, selenium cannot claim a nutrient function because, in general, Japanese people consume selenium at approximately twofold the recommended daily allowance (Rayman, 2005), that is, selenium supplementation is not necessary for most Japanese.

Foods with function claims

As FoSHU require governmental approval, they require not only substantial research costs to substantiate the safety and efficacy of their claims, but also abundant time to be confirmed by the responsible governmental agency. On the other hand, claims for foods with nutrient function claims are predetermined, cannot be changed, and can only be made for a limited number of nutrients. To mitigate the restrictions of these food labeling systems, foods with function claims was introduced as part of foods with health claims in 2015, in reference to the regulatory system for dietary supplements in the United States. With the introduction of foods with function claims, more products clearly labeled with certain nutritional or health functions

TABLE 77.4 Examples of nutrient function claims, precautions, and contraindications for foods with nutrient function claims in Japan.

Nutrient	Limit in food		Nutrient function claims	Precautions and contraindications
	Lower	Upper		
Vitamin D	1.65 μg	5 μg	Vitamin D promotes intestinal calcium absorption and aids in bone development	Consumption of large quantities of this product does not cure disease or promote health. Do not exceed the recommended daily intake
Vitamin B_1	0.36 mg	25 mg	Vitamin B1 aids in energy production from carbohydrates and maintains skin and mucosa in a healthy state	Consumption of large quantities of this product does not cure disease or promote health. Do not exceed the recommended daily intake
Vitamin C	30 mg	1000 mg	Vitamin C helps maintain skin and mucosa in a healthy state and has an antioxidant effect	Consumption of large quantities of this product does not cure disease or promote health. Do not exceed the recommended daily intake
Calcium	204 mg	600 mg	Calcium is necessary for the development of bones and teeth	Consumption of large quantities of this product does not cure disease or promote health. Do not exceed the recommended daily intake

Source: Adapted from https://www.caa.go.jp/policies/policy/food_labeling/health_promotion/pdf/food_labeling_cms206_20191126_11.pdf

can be made available, enabling consumers to make more informed choices. However, the government does not evaluate the safety and efficacy of foods with function claims; the onus to collect the scientific evidence and prove the health claims of the product is on the food business operators. The products include a disclaimer on the label stating that the product has not officially been endorsed by the government. In the same way as other foods with health claims, the use of medical terms for foods with function claims such as "diagnose," "prevent," "cure," and "treat" is prohibited. Food business operators are required to submitted complete documents to the secretory-general of the CAA 60 days prior to the targeted launch date according to the Guidelines on Notification of Foods with Health Claims, which are updated frequently and can be viewed on the home page of the CAA (https://www.caa.go.jp/policies/policy/food_labeling/about_-foods_with_function_claims/).

The basic approach for foods with function claims is as follows: ensuring safety, establishing requirements for scientific evidence to make function claims, providing consumers with information through proper labeling that is not misleading to consumers, and promoting voluntary and reasonable product choices by consumers. This covers all food products, that is, foods in the form of capsules or tablets, conventional foods, and fresh foods. However, the following products are not permitted as foods with function claims: those developed for people suffering from diseases, minors, pregnant women (including those planning a pregnancy), or lactating women; FoSHU; foods with nutrient function claims; alcohol-containing beverages; or those that may lead to the excessive consumption of fat, cholesterol, sugar (limited to mono- and disaccharides, excluding sugar alcohols), or sodium. In addition, foods whose functional substance(s) have not been identified as not suitable as foods with function claims.

Product safety can be substantiated by either evaluating the history of consumption by humans based on actual intake data or conducting safety tests on a finished product or functional substance(s). Evaluations of the interactions of the functional substance(s) with the products and drugs are also required. If an interaction is confirmed, the appropriateness for selling such a product must be explained scientifically. Safety in terms of sanitation and quality control during the production and manufacturing processes must also be ensured using analytical methods for the quantitative determination of functional substance(s). To help prevent health hazards, a system to collect adverse events should also be established. Product efficacy can be substantiated in two ways: the first is by clinical trial(s) with a finished product, and the second is by systematic literature review(s) on a finished product or functional substance(s).

Currently, half of the products classified as foods with function claims are in the form of capsules or tablets, while others include fresh foods, such as sprouts claiming to provide "bone health and isoflavones" and mandarin oranges cultivated in special areas claiming to provide "β-cryptoxanthin and bone health." The actual claims of foods with function claims are similar to those of FoSHU, but the products can be marketed more easily. Thus food business operators have shown substantial interest in this new food labeling system, and approximately 2621 products classified as foods with function claims were registered by the end of 2019, which is more than twice that of FoSHU, which have been implemented since 1991.

Current issues in foods with health claims

Because foods with health claims exhibit weak physiological effects, they can be chosen directly by consumers without the advice of a health professional. However, some consumers expect such food products to have effects similar to those of drugs (Chiba et al., 2014). FoSHU and foods with function claims are intended for use to improve dietary habits and lifestyle, whereas foods with nutrient function claims should be used for nutrient supplementation if habitual intake is lower than the recommended amount. Our survey of FoSHU usage revealed that only a limited number of consumers, those who appropriately referred to product labels and adjusted their lifestyle accordingly, noticed the beneficial effects of FoSHU. We also found that users of foods with nutrient function claims had high health awareness and healthy lifestyles, but did not use the appropriate amounts of foods with nutrient function claims in terms of vitamin and mineral supplementation (Sato et al., 2016). Similar findings have been reported in other countries (Dickinson, MacKay, 2014). In Japan, the inappropriate use of foods with nutrient function claims has clearly been observed in the case of folic acid. The labels on folic acid-containing FoSHU are permitted to claim a reduced risk of neural tube defects in fetuses with intake, while those on folic acid-containing foods with nutrient function claims are permitted to claim nutrient functions. However, many pregnant women initiate folic acid supplementation after discovering that they are pregnant, which is too late to reduce the risk of neural tube defects (Sato et al., 2013). Another inappropriate use of foods with nutrient function claims has been observed in preschoolers learning to eat a balanced diet in reference to the labeling on staple foods (Sato et al., 2009). These surveys revealed issues related to foods with health claims that need improvement. If consumers are able to interpret the information on labels correctly and use it appropriately, then foods with health claims are capable of promoting and maintaining health. In addition, health claims on food labels should support and be consistent with national health and nutritional policies; however, the current regulations governing health claims on food labels in Japan remain insufficient.

Efforts to ensure the safety of health foods

Manufacturing of safe products

Foods, even in the form of tablets and capsules, are freely chosen and taken by consumers, so equal priority should be given to the issues of safety and efficacy. With the increasing use of health foods, it is critical to avoid health hazards and ensure safety. Adverse events associated with health foods can occur with the intake of products of inferior quality, the excessive consumption of multiple products, consumption by individuals who are sensitive to the products/components, that is, individuals with allergies, and simultaneous intake with contraindicated drugs (Kobayashi et al., 2018). The MHLW has widely identified and disclosed illegal products, that is, those that include pharmaceutical ingredients and toxic chemicals, and has announced the development of two guidelines to ensure the safety of health foods, particularly those in capsule, tablet, or powder form: the Good Manufacturing Practice Guideline and the Guideline for the Self-investigation of Safety of Raw Materials (Ohama et al., 2006).

Providing reliable safety information

Providing consumers with access to accurate information, which helps prevent misunderstandings of the nutritional and health benefits of foods with health claims, and alerting consumers to the hazards of illegal products are important measures to ensure health food safety. Accordingly, the National Institute of Health and Nutrition has created an online database of safety and efficacy reports of health foods and components, including adverse reports, which can be disseminated to consumers and health professionals (Umegaki, 2015). In particular, adverse reports are an invaluable reference tool because they document the occurrence of similar adverse events in the past. This online database also aims to promote consumer education and communication between health professionals and consumers. An outline of the database is shown in Fig. 77.2. Two strategies have been used in disseminating information to consumers. The first

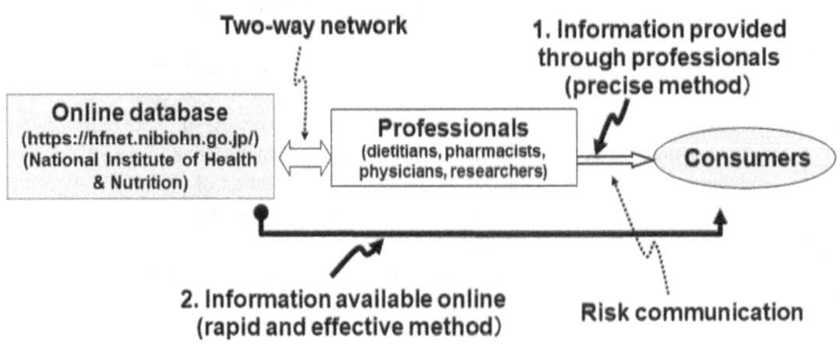

FIGURE 77.2 Outline of the strategies in place for providing information on health foods to consumers and health professionals via an online database.

provides each consumer with one-on-one attention through professionals such as dietitians, pharmacists, and physicians, whereas the second provides information through the Internet. The former can provide accurate basic knowledge and information, and the latter can provide information, such as health hazards and warning messages from the government, quickly and effectively. The detailed contents currently provided on the website (https://hfnet.nibiohn.go.jp/) include: (1) basic information regarding health and functional foods (foods with health claims evaluated by the CAA), (2) current and past health issues associated with the intake of health and functional foods, (3) information on foods and food components that have received recent attention, and (4) a database of scientifically proven information on foods and food components. This information is prepared and provided from a consumer's point of view. The government entrusts the private sector with the training or education of health food professionals, but does not license them nationally. To address this, several organizations educate, certify, and train advisory staff who educate, teach, and provide counsel about various health foods. These highly trained experts are extremely important in making the website as helpful as possible for consumers.

Collection of adverse events and causality assessment

Documenting adverse cases regarding health foods and evaluating causality are important to ensure food safety, even if the causal relationship between the health food and reported health issue is uncertain. In Japan, adverse event reports are typically collected at three sites: public health centers, the practical living information online network system operated by the National Consumer Affairs Center, and individual food product companies (Umegaki et al., 2013). Notably, the adverse event reports from the three sources are not dealt with in a consistent manner. Thus proposals have been made to investigate and characterize these reports from the viewpoint of evaluating causality using causality association rating methods, which are suitable for health foods (Ichimaru et al., 2013; Ide et al., 2015). This system is potentially useful for preventing adverse events and ensuring the safety of not only foods with health claims, which must follow government regulations, but also other health foods on the market that lack scientific evidence through rapid administrative action. As the occurrence of adverse events related to health foods is rare, and only a few events have been reported to administrative agencies, the systemic collection of adverse events is needed. To address this issue, we attempted to collect adverse events via a nationwide online survey and to clarify the prevalence and related suspected ingredients (Nishijima et al., 2018a,b). A nationwide online survey conducted on health food users may enable the safety of the ingredient(s) used in health foods to be confirmed and reevaluated, if these ingredient(s) are standardized (Nishijima et al., 2019).

Concluding remarks and future directions

The initial concept of functional foods proposed in Japan developed over time into FoSHU, followed by the creation of foods with health claims, revisions that align with international trends. These regulatory systems help to discriminate health foods that have scientific evidence confirming their safety and efficacy. In addition, a system governing foods with health claims could contribute to the improvement and maintenance of health. However, the lack of a systematic health foods naming system has led to a situation in which consumers find it difficult to recognize and understand the exact characteristics and appropriate usage of these foods. Awareness is required in regard to mistaking these foods for drugs due to aggressive health claims because adverse events may occur as the result of an overdose or drug interaction; moreover, consumers may delay the seeking of appropriate medical care. Therefore it is vital to create and maintain an environment in which consumers can clearly distinguish between foods and drugs. It is also noteworthy that foods, even those in capsule or tablet form, are not intended for the diagnosis, treatment, or cure of any disease. As long as foods with health claims are used in the intended way, that is, to improve lifestyle and dietary habits, they can be beneficial for health promotion and the prevention of lifestyle-related disease. Such circumstances can be fostered by enhancing communication between consumers and health professionals, sharing scientific information regarding health and nutrition, and encouraging cooperation between all parties concerned.

References

Arai, S., 1996. Studies on functional foods in Japan—state of the art. Biosci. Biotechnol. Biochem. 60, 9–15.

Chiba, T., Sato, Y., Nakanishi, T., et al., 2014. Inappropriate application of food for specified health uses. J. Jpn. Soc. Nutr. Food Sci. 67, 177–184 (in Japanese).

Czepielewska, E., Makarewicz-Wujec, M., Rozewski, F., et al., 2018. Drug adulteration of food supplements: a threat to public health in the European Union? Regul. Toxicol. Pharmacol. 97, 98–102.

Dickinson, A., MacKay, D., 2014. Health habits and other characteristics of dietary supplement users: a review. Nutr. J. 13, 14.

Hudson, A., Lopez, E., Almalki, A.J., et al., 2018. A review of the toxicity of compounds found in herbal dietary supplements. Planta Med. 84, 613–626.

Ichimaru, K., Ide, K., Ono, A., et al., 2013. Modification of a dendritic algorithm for evaluation of causal relationships of adverse events with health food. Jpn. J. Clin. Pharmacol. Ther. 44, 405–410 (in Japanese).

Ide, K., Yamada, H., Kitagawa, M., et al., 2015. Methods for estimating causal relationships of adverse events with dietary supplements. BMJ Open 5, e009038.

Kobayashi, E., Sato, Y., Umegaki, K., et al., 2018. Analysis of safety alerts associated with dietary supplements from Japan and overseas. Shokuhin Eiseigaku Zasshi 59, 93–98 (in Japanese).

Nishijima, C., Chiba, T., Sato, Y., Yamada, H., Umegaki, K., 2018a. Nationwide online survey method to estimate ongoing adverse events caused by supplement use: application to diarrhea. Shokuhin Eiseigaku Zasshi 59, 106–113.

Nishijima, C., Sato, Y., Chiba, T., Umegaki, K., 2018b. Nationwide online survey to complement the current voluntary reporting system for adverse events associated with dietary supplements: application to the case of skin manifestations. J. Nutr. Sci. Vitaminol. (Tokyo) 64, 277–283.

Nishijima, C., Chiba, T., Sato, Y., Umegaki, K., 2019. Nationwide online survey enables the reevaluation of the safety of *Coleus forskohlii* extract intake based on the adverse event frequencies. Nutrients 11, 866.

Ohama, H., Ikeda, H., Moriyama, H., 2006. Health foods and foods with health claims in Japan. Toxicology 221, 95–111.

Owens, C., Baergen, R., Puckett, D., 2014. Online sources of herbal product information. Am. J. Med. 127, 109–115.

Rayman, M.P., 2005. Selenium in cancer prevention: a review of the evidence and mechanism of action. Proc. Nutr. Soc. 64, 527–542.

Sato, Y., Yamagishi, A., Hashimoto, Y., Virgona, N., Hoshiyama, Y., Umegaki, K., 2009. Use of dietary supplements among preschool children in Japan. J. Nutr. Sci. Vitaminol. (Tokyo) 55, 317–325.

Sato, Y., Nakanishi, T., Chiba, T., Yokotani, K., Ishinaga, K., Takimoto, H., et al., 2013. Prevalence of inappropriate dietary supplement use among pregnant women in Japan. Asia Pac. J. Clin. Nutr. 22, 83–89.

Sato, Y., Tsubota-Utsugi, M., Chiba, T., Tsuboyama-Kasaoka, N., Takimoto, H., Nishi, N., et al., 2016. Personal behaviors including food consumption and mineral supplement use among Japanese adults: a secondary analysis from the National Health and Nutrition Survey, 2003-2010. Asia Pac. J. Clin. Nutr. 25, 385–392.

Shimizu, T., 2003. Health claims on functional foods: the Japanese regulations and an international comparison. Nutr. Res. Rev. 16, 241–252.

Tanaka, H., Kanda, F., Suguro, R., et al., 2004. Current system for regulation of health foods in Japan. Jpn. Med. Assoc. J. 47, 436–450.

Tucker, J., Fischer, T., Upjohn, L., et al., 2018. Unapproved pharmaceutical ingredients included in dietary supplements associated with US food and drug administration warnings. JAMA Netw. Open. 1, e183337.

Umegaki, K., 2015. Positive and negative aspects of food with health claims in Japan. J. Nutr. Sci. Vitaminol. (Tokyo) 61, S133–S135.

Umegaki, K., Yamada, H., Chiba, T., et al., 2013. Information sources for causality assessment of health problems related to health foods and their usefulness. Shokuhin Eiseigaku Zasshi 54, 282–289 (in Japanese).

Yamada, K., Sato-Mito, N., Nagata, J., Umegaki, K., 2008. Health claim evidence requirements in Japan. J. Nutr. 138, 1192S–1198SS.

Health Service Bureau, Ministry of Health Labour and Welfare, JAPAN 2018. Dietary reference intakes for Japanese. <https://www.mhlw.go.jp/file/06-Seisakujouhou-10900000-Kenkoukyoku/Full_DRIs2015.pdf> (Accecced 15 February 2020).

Chapter 78

Nutraceuticals: the Turkish perspective

Begum Yurdakok-Dikmen and Ayhan Filazi

Introduction

The fastest growing segments of the food industry include the discovery, development, and marketing of functional foods, nutraceuticals, and related products (Nicoletti, 2012). Several factors influence this trend, such as the recognition of the role of food in the preservation of health which is reflected through the increasing interest of consumers in their own health; the increase in life expectancy which is reflected through the increasing elderly population; increasing costs of health care by increased costs of pharmaceuticals and other healthcare costs; and the highly competitive nature of the food market with small profit margins, the advances in biotechnology and "omics," and the current consumer perception that "natural is good" (Vella et al., 2014; Wang and Li, 2015).

As part of a globalization process, modern Turkish culinary culture has changed, with ready-made and fast-food habits. On the other hand, the multicultural structural reflections including the transitions from Central Asia to the Ottoman kitchen are still present and affect the food habits and consumption of nutraceuticals. Turks in Central Asia lived mostly on meat and milk products, and were greatly affected by the climatic conditions of the region, where cereal crops were consumed in lesser quantities. As a result of the start of the modernization period, culinary culture pertaining to Turkish culture started to change in both meal customs and ingredients used in the preparation of these meals (Bilgin and Samanci, 2008). Through consumption of these ready-made foods, and fabricated ingredients, the intestinal microflora of Turkish people has changed over time. On the other hand, in Turkish culture, many types of foods and herbs have been associated with specific health benefits and as functional foods. Among the functional foods, yogurt is considered an ideal medium for delivery of beneficial functional ingredients. The word "yoğurt" is Turkish in origin and was introduced from the Turkish word "yoğurmak" meaning molding and swollen/stiffened. Historical findings attribute yogurt to the Neolithic peoples of Central Asia at around 6000 BC and people from Çatalhöyük, Konya (Anatolia), at around 5500–6000 BC.

Among the most significant thinkers and writers of the Islamic Golden Age, Ibn-i Sina (Latin form Avicenna, 980–1037), with his most famous work "el-Kânûn fi't-Tıbb: The Book of Healing" which was used as a standard text at many medieval universities, introduced yogurt in many sections including those about gastric ulcers and gout diseases. Assumed to be the first pharmacy book (Kitâbü's-Saydana Fi't-Tıb), Ebu'r-Reyhan El-Beyrunî (Latin Form Alberuni veya Aliboron, 973–1051) introduced the cheese yeast collected from the stomach chamber (the abomasum) of harvested young, unweaned calves that could be used as an antidote agonist poisonous hemlock (Yurdakök, 2013). Considered as a national drink, "ayran," a cold salted yogurt beverage, is defined as "drinkable fermented product prepared by the addition of water to yogurt culture to standardized milk" by the Turkish Food Codex (Sanli et al., 2013). Ayran was also found to be a very suitable food product to carry relevant probiotic bacteria while adding certain herbal essential oils (such as *Mentha longifolia*) (Mahmoudi, 2014). Another fermented milk product consumed in the daily lives of Turkish people is "kefir," which is made with "grains" and has its origins in the northern slopes of the Caucasus mountains. Due to the claimed health benefits of kefir, which include reduction of lactose intolerance symptoms, stimulation of the immune system, lowering cholesterol, and antimutagenic and anticarcinogenic properties, kefir has become an important functional dairy food (Guzel-Seydim et al., 2011).

According to Yilmaz et al. (2019), kefir consumption may modulate gut microbiota, and regular consumption of kefir may improve the patient's quality of life in the short term. In addition, a study reported that *Staphylococcus aureus* contamination prior to kefir fermentation is of no great importance for the public health since *S. aureus* is inhibited during the fermentation process (Kivanc and Yapici, 2019). Supernatants derived from the traditional fermented milk products kefir and ayran were found to be

Nutraceuticals. DOI: https://doi.org/10.1016/B978-0-12-821038-3.00078-1

effective in reducing the DNA damage induced by human fecal water samples in an in vitro colon cell culture model where the effect was found to be dose-dependent by ayran. This reduction in DNA damage was related to the antioxidant capacity of these fermented drinks (Grishna et al., 2011).

With its geostrategic location as a gateway to Asia, the Middle East, and Europe, Turkey plays an important role in bridging the product gap when it comes to nutritional supplements and cosmetics exportation and expansion. Therefore with the increased awareness of health trends, especially regarding nutraceuticals and functional foods, the sales of vitamins and dietary supplements among the total sales in health and wellness products has greatly increased. Vitamins and dietary supplements registered 8% current value growth in 2014 (Euromonitor, 2015). The functional foods market is growing worldwide as well as in Turkey. Among the functional dairy products light, lactose-free, omega-3 included, prebiotic, Ca, and vitamins added, prebiotic dairy products including kefir, and among functional bakery products wholemeal/rye bread, oaten/wholemeal biscuit, wholegrain crackers, light crackers/cake, and breakfast cereals are the most common products marketed and consumed in Turkey. Consumer demography in the country, such as the income and educational levels, was found to have an effect on functional food consumption. On the other hand, for some functional food types income level does not affect consumption, such as functional herbal teas which are commonly consumed, such as linden, sage, green, winter, and rosehip teas (Dogan et al., 2011).

The Turkish Association of Diabetic and Functional Food Producers Association (Diyabetik ve Fonksiyonel Gıda Üreticileri Derneği) was founded in 2008 with the vision for a healthier generation and a mission to provide the required information to the community regarding the diabetic and functional foods and to provide profit for the nation's economy through a reduction of health costs by reducing disease risks. This nongovernmental organization was established to strengthen communication, dissemination of information, and networking among interested parties from academia, industry, and government participants regarding the health benefits of functional foods (DFDG, 2012).

Herbal functional foods

Turkey is a leading producer and exporter of various herbs and other botanicals (Esiyok et al., 2004). Herbal medicines have been used in Turkey traditionally, although today modern medical science is utilized all over the country. Although traditional methods have limited use, a rich flora of medicinal plants in the country, and experienced academic people in this field creates great potential (WHO,1998). Situated in a geographical area comprising Mediterranean, continental, and oceanic climates, Turkey has rich plant diversity and the flora comprises about 9500 species, 30% of which are endemic (Özgen et al., 2012; Davis, 1988).

Herbal drugs were previously termed "folk remedies" in Turkey. The production and use of these herbal drugs are transferred between generations, as the herbalist's trade (Akbulut and Bayramoglu, 2013). Shamanic treatments by traditional folk healers have been reshaped through the old tradition according to its own principles up until now in Anatolia. Thus a Lokman healer, who is an Islamic type, has become the ancestor of contemporary folk healers from the first period of Islam until now. The Lokman healer was assumed to be the one who can speak with flowers and healing plants, not only engaged with biological diseases, but also implementing healing practices against magic, sorcery, and the evil eye (Ugurlu, 2011). Currently many stores which sell herbal remedies are named as "Lokman Hekim" in Turkey. Also, traditionally and currently, herbalists sell their products in aktars, derived from the Arabic word "akkar" meaning "the source of remedies." Aktars, which were operated under the roof of a guild in the Ottoman period, were known to make a contribution to a field of great importance in terms of health, culture, and society where the traditional health practitioners and aktars met the medicinal needs of the people. Currently aktars serve people by relying on the traditional knowledge, which includes information about how to harvest, dry, store, and use herbs, and also which herbs are used for which ailments (Bostancı-Ege and Onder-Erol, 2012). On the other hand, these herbalists' trade is not defined or regulated by any legislation. This occupation has an education infrastructure in Turkey, but there is no requirement for a certificate of competence to establish a shop. This occupation has no legal responsibilities and no competence is demanded (Ugurlu, 2011). Furthermore, pharmacies also provide herbal drugs licensed by the Ministry of Health for various ailments; however, these drugs are far less than in other European countries, mainly due to the regulation procedures (Suzgec-Selcuk and Eyisan, 2012).

Bioactive compounds including foods grown in Turkey which are used for their various potential benefits include the following: Amaranthus (*Amaranthus* spp.), caper (*Cappatis spinosa* L.), chicory (*Cichorium intybus* L), coriander (*Coriandrum sativum* L), cumin (*Carum carvi* L), dandelion (*Taraxacum officinale* L), fennel (*Foenicilum culgare* L), flaz (*Linum mucronatum*), mallow (*Malva sylvestris* L.), marjoram (*Orignaum onites* L.), milk thistle (*Sylibum marianum* L.), nettle (*Urtica urens/diotica* L.), purslane (*Portulaca oleracea*), rocket (*Eruca sativa*), rosemary (*Rosmarinus officinalis* L.), sage (*Salvia officinalis* L), savory (*Satureia hortensis*), Spanish lavender (*Lavandula oechas* L), Spanish salsify (*Scolymus hispanicusa*), sweet basil (*Ocimum basilicum* L), thyme (*Origanum vulgare* L), wild mustard (*Sinapis arvansis* L.), and wild radish (*Raphanus raphanistrum*) (Esiyok et al., 2004). Prominent worldwide among medicinal plants, nettle was the medicinal plant investigated most often; in the study by Ogur et al. (2006), 44.9% of participants among 2984, stated that nettle was used for cancer treatment in their community. Nettle was also found to be the most frequently used remedy used by patients with diabetes in Kayseri, Turkey; followed by thyme, parsley, jujuba, sour pomegranate, lettuce, and Jerusalem artichoke (Inanc et al., 2007). As an ethnopharmacological point of view, along with these plants some animals are also used for treatments such as cattle bile for eczema, wolf bones for goiter, and trout head for gastric ulcer and fractured bones (Özgen et al., 2012).

In a study by Soner et al. (2013), data collected from a questionnaire completed by both out- and inpatients admitted to the local hospital in Konya, Turkey, revealed that 48.8% of the study population used herbal medicines, with 38.3% of all respondents using or having used another prescribed medication concurrently with those herbal remedies. The most frequent reasons for using these herbal medications were as follows: maintaining health, improving weakness, weight loss, musculoskeletal pain, and cough and cold symptoms. The most frequently consumed herbs were Camellia sinensis (14.2%), R. officinalis (10.2%), and Zingiber officinale (9.1%) followed by other herbal sourced supplements including L-carnitin, coenzyme Q, chrome, linseed, horse chestnut, mangosteen and cypress, each less than 0.4%, making a total of 6%. Kav (2009) reported that patients with functional and inflammatory bowel diseases are known to use complementary and alternative medicine therapies more often than other groups for symptoms including bloating, indigestion, constipation, diarrhea, and dyspepsia. Cassia extracts and commercially prepared herbal teas, which also contain senna in significant amounts, are also used commonly and have selective effects on gastrointestinal motility. The influence of women's programs on TV and commercial advertisements on the web also have an effect on the consumption of these herbal remedies (Bostancı-Ege and Onder-Erol, 2012).

Current debates about these products are still ongoing since herbal food supplements receive their license from the Ministry of Agriculture and Forestry and not from the Ministry of Health. Some herbal remedies (such as *Fructus agni* casti natural dried extracts for menstrual cyclus modulator, *Phleum pratense* grass pollen extract for allergy) which are sold in pharmacies receive licensing from the Ministry of Health. Due to this situation, where there is no standardized production, marketing, safety, and efficacy, as well as vigilance measures ascertained, many cases of severe poisoning and even deaths have been recorded. The Turkish Pharmacists' Association applied to a judicial court for legislation regarding herbal supplements and in the coming years updated and effective measures are expected to be brought into legislation. In order to implement the national development goals, sustainable development policies are incorporated into various sectors in Turkey including medicinal and aromatic plants, which are globally recognized as a valuable source. For this purpose, several universities in Turkey have established "Medicinal and Aromatic Plants Application and Research Centers" for identification, investigation of ecological variations, cultivation, conservation, and advanced production systems of medicinal, aromatic, and ornamental plants. These centers provide scientific evidence based on their facilities for R&D on chemical, pharmacological, technological, toxicological, biotechnological, and industrial properties of medicinal/aromatic plants and create a bridge with the private sector, ensuring university–private sector cooperation. These local centers (Duzce, Dicle, Necmettin Erbakan, Artvin Coruh, Erzincan, Gumushane, Dumlupınar, Afyonkarahisar Universities, etc.) particularly provide for preservation of medicinal/aromatic/ornamental plants in provincial and surrounding provinces leading to the development of trade and industry in the region. Also, some universities offer 2-year programs on medicinal and aromatic plants (Anadolu, Ankara, Health Sciences, Nevsehir Haci Bektas Veli, Isparta Applied Sciences, Bingol, Ege Universities, etc.).

Marine nutraceuticals

Marine-derived nutraceuticals are alternative sources of synthetic ingredients that can contribute to consumers' well-being and play a vital role in human health and nutrition (Kim, 2013). As one of the most important sources of nonanimal sulfated polysaccharides, marine algae have been used for various ailments. There are studies concerning the antimicrobial and antifungal activities of marine algae collected from Turkey such as Phaeophyceae (*Dictyopteris membranaceae*, *Cystoseira barbata*, *Cystoseira compressa*, *Cystoseira mediterranea*, *Halopteris scoparia*, *Dictyota dichotoma*, *Colpomenia sinuosa*, *Ectocarpus siliculosus*, *Padina pavonica*, *Dictyota linearis*), Rhodophyceae (*Jania rubens*, *Acanthophora najadiformis*, *Laurencia papillosa*, *Hypnea musciformis*, *Gracilaria gracilis*, *Ceramium rubrum*), and Chlorophyceae (*Enteromorpha linza*, *Ulva rigida*) (Erturk and Tas, 2011; Taskin et al., 2007; Tuney et al., 2007; Haliki et al., 2005; Karabay-Yavasoglu et al., 2007). Although no suppliers for marine algae are present, fish oil and omega-3 oil are currently being marketed in Turkey. Other marine products are usually imported.

Eicosapentaenoic acid and docosahexaenoic acid are the two most beneficial omega-3 sources, found in fish oil supplements from the family of long-chain polyunsaturated fatty acids, and are essential to maintain the balanced production of hormone-like substances which help regulating many important physiological functions including blood pressure, blood clotting, nerve transmission, inflammatory and allergic responses, the functions of the kidneys and gastrointestinal tract, and the production of other hormones (Calder, 2015; American Pregnancy, 2014). Average omega-3 consumption for 19–30-year-old males and females was found to be 1.59 and 1.22 g, for 31–50 age groups 1.51 and 1.18 g; for 51–64-year-olds 1.32 and 1.08 g; for 65–74-year-olds 1.13 and 0.92 g; and for those over 75 year old 1.07 and 0.83 g in Turkey (Ministry of Health, 2014). These values were found to be around the daily intake limits set by the World Health Organization. Along with commercials and advice from dieticians/practitioners, the consumption of omega-3 has grown among Turkish customers, with a market of 25 million Euros, with more than 100 products containing omega-3 and over 4.5 million boxes sold in 2012 (Finance, 2013).

Pre- and probiotics

Among nutraceuticals and functional foods, probiotic dairy products were found to be highly encouraging due to the growing demand for health-oriented foods in accordance with the cultural background in Turkey. The average monthly yogurt consumption per household was found to be 14.96 kg in urban areas and 34.26 kg in rural areas, with 24.46 kg for the urban and rural areas combined in 2010 (Ates and Ceylan, 2010). The per capita consumption of yogurt in Turkey increased to 30 kg in 2013 which was the largest in the world. Yogurt is mostly consumed as part of meal (54%) or as cooking ingredient (38%) in 2010-2013 (Durak, 2013). Yogurt consumption per capita is higher in rural areas as rural consumers can produce yogurt themselves, mostly with milk obtained from their own animals, while the high-income earners in the urban areas purchase their yogurt (Ates and Ceylan, 2010). The probiotic fruit yogurts market is also trending (Cakmakci et al., 2006), and is defined by the Turkish Standards 4806 Milk and Dairy Products as pulped yogurt with a special technique including fruits such as apple, pear, strawberry, and banana, which was first introduced to the Turkish market in 1997.

The largest application segment for prebiotics in the country applies to bakery products, including several types of biscuits followed by dairy products, health drinks, nutrition bars, breakfast cereals, flour, jams, and beverages. Not only restricted to food and beverages, prebiotic applications are applied to dietary supplements (nutritional supplements, specialty nutrients, infant foods) and animal feed probiotics as well.

Fermented products as nutraceuticals

Fermented food products play an important role in the human diet around the world due to their health benefits, where traditional fermented foods and beverages have been an important part of people's lives. A variety of fermented milk products are produced throughout Anatolia and have been used for various ailments for centuries.

Among the fermented beverages, the national drink of Turkey "ayran" is a drinkable fermented yogurt drink traditionally produced with the addition of water to yogurt (30–50 g/100 g) and salt (0.5–1 g/100 g). With the increased demand and advancement in manufacturing technology, ayran is now produced from homogenized and pasteurized (90°C–95°C, 5–10 min) milk using ropy lactic acid cultures (*Streptococcus thermophilus* and *Lactobacillus delbrueckii* spp. *bulgaricus*), which increase the consistency and reduce serum separation (Koksoy and Kilic, 2003; Altay et al., 2013). Ayran is popular both in urban (even large fast-food companies include ayran in their standard menus) and rural areas and is offered as a "standard" drink to guests. Ayran consumption in Turkey reaches up to 750,000 tonnes annually including homemade production, and the packed ayran consumption is around 45,000 tonnes (Hurriyet, 2013).

Kefir, etymologically derived from two Turkish words "köpür" meaning "foamy" and "keyif" meaning "good feeling" (Etimolojiturkce, 2013), has health benefits for the immune system, gastrointestinal system, and cholesterol metabolism along with antitumoral, antibacterial, and antifungal properties (Altay et al., 2013). Recently antihyperlipidemic and antihyperglycemic (Alsayadi et al., 2014) effects, as well as improved efficacy and tolerability of triple therapy in eradicating *Helicobacter pylori*, have been illustrated (Bekar et al., 2011). Kefir is a viscous, self-carbonated beverage with a smooth, slightly foamy body and whitish color, and commercially produced kefir can be found in supermarkets with some being fruit flavored (strawberry, forest berries, and mix-fruit) and some light ones.

Among fermented beverages, the red-colored, cloudy, and sour soft beverage "Shalgam" (salgam) is the traditional lactic acid beverage produced from the lactic acid fermentation of black carrot, aromatic turnip (celem), rock-salt, sourdough, bulgur flour, and drinkable water served cold in large glasses. There are two main processing methods: the traditional method including two fermentation stages with sourdough fermentation (first fermentation) and carrot fermentation (second fermentation) and the direct method having only the second fermentation (Erten et al., 2008, Altay et al., 2013). Shalgam juice fermentation is carried out mainly by lactic acid bacteria (LAB). The main species isolated in the fermentations of salgam are *Lactobasillus plantarum* and *L. paracasei* spp. *paracasei*. Significant growth of *Lactobacillus brevis* and *Lactobacillus fermentum* was also observed during the fermentations depending on the production plant and *Leuconostoc mesenteroides* spp. *mesenteroides*, *Pediococcus pentosaceaceus*, and *L. delbrueckii* spp. *delbrueckii* were isolated at the beginning of fermentation, but did not grow and died off (Tanguler and Erten, 2012). Moreover, the result obtained from a study indicated that shalgam does not have cytotoxic, genotoxic, and antibacterial activity in three test systems. The authors concluded that shalgam might not be a potential risk for human cells (Tazehkand and Valipour, 2019).

Among other traditional fermented vegetable drinks, shalgam was found to be similar to the LAB fermented Korean dish "Kimchi." Lactic acid fermentation increases the shelf-life of fruits and vegetables and also enhances several beneficial properties, including nutritive value and flavors, and reduces toxicity and is assumed as a potential source of probiotics (Swain et al., 2014). Consisting of two nonacylated and three acylated cyanidin derivatives, shalgam anthocyanins have potential for high antioxidant activity (Turker et al., 2004). It is assumed to be a perfect accompanied to the Turkish meat dish "kebap" and also served with rakı, a traditional highly alcoholic beverage. It has diuretic, antiinflammatory, lactatic properties and is used for the maintenance of cholesterol, weight control, and arthritis (Kabak and Dobson, 2011). Since shalgam contains high amounts of salt, it is not recommended for hypertensive patients.

Consumed mainly for its pleasant taste and flavor in the Middle East and Balkan region, "boza" is defined by the Turkish Standards Institution (TS 9778), as "A product which is made by addition of water to cereals such as millet, maize, wheat, and rice." This yeast and lactic acid fermentation create a highly viscous and low-alcoholic liquid with a pale yellow color and sweet or sour taste (Tanguler, 2014). The origin of this ancient drink dates back to people who lived in Anatolia and Mesopotamia around 400 BC. In Turkey, while boza can actually be made at home, it is also served in some patisseries, wine shops, local restaurants, and boza houses such as "Vefa Bozacısı." This culturally popular drink has beneficial health effects such as boza-related LAB bacteriocins or bacteriocin-like peptides, which were found to possess antifungal and antibacterial activities (Holzapfel, 2015); protein hydrolysate, fractionated hydrolysates, and dialysates obtained following in vitro digestion of boza were found to contain bioactive compounds vented as a good source of angiotensin-converting enzyme-inhibitory peptides (Kancabaş and Karakaya, 2013). Recent advances in food engineering allowed boza to contain 10^6 cfu/mL viable microorganisms, which was defined as enough for functional foods using certificated probiotic starter cultures (*Lactobacillus acidophilus* LA-5, *Bifidobacterium bifidum* BB-12, and *Saccharomyces boulardii*) (Arslan et al., 2015).

Other fermented local dairy products include kurut, koumiss (kımız); cereal-based fermented products include tarhana; fermented fruits include turşu (pickle) and hardaliye; and fermented meat includes "sucuk" which is consumed widely in various parts of Turkey (Kabak and Dobson, 2011).

Sports nutritionals

The market of sports nutritionals including sports drinks and nutritional bars, and others such as protein (PRO) and weight gain powders, pre- and post-workout supplements, and energy gels is an increasing trend driven by consumer demand for healthier, more dynamic sports beverages. Isotonic, hypotonic, and hypertonic beverages are currently marketed to help athletes to replace water, electrolytes, and energy. These products and other supplements such as PRO or amino acids (AA) to carbohydrate (CHO) are defined by the Turkish Food Codex-Sports Food Regulation (2003/42). The soft drinks market in Turkey is dominated by multinational companies and mainly sold in sports nutrition retailers or in large enterprise sports markets and some of the larger supermarkets.

Energy drinks

There is no agreed definition of "energy" drinks, either in the European Union (EU) or the United States, where nonalcoholic beverages containing caffeine, taurine, and/or vitamins (often in combination with other ingredients) that are marketed for their actual or perceived effects as stimulants, energizers, and performance enhancers are attributed for these products. As per the Turkish Food Codex Energy Drinks Communiqué no 30110 (numbered 2017/4), in Turkey, an energy drink is defined as "Flavored nonalcoholic beverage containing caffeine, which may contain taurine, glucoronolactone, inositol, CHO, AA, vitamins, minerals and other food and ingredients." Energy drinks provide mental and physical stimulation, through their stimulant ingredients which are mainly caffeine, where other herbal stimulants (guarana, yerba mate, etc.), simple sugars (glucose, fructose, etc.), naturally formed glucose metabolite (glucuronolactine), AA (taurine, carnitine, creatine, etc.), herbs (*Gingko biloba*, ginseng, etc.), and vitamins could be included. The energy drink market in Turkey included 26 different brands, 15 of which are local and 11 of which are foreign in 2014

(Koc et al., 2014); which has currently reached 40 in total. Meanwhile each year many firms are imposed to administrative *fines* for *several* deficiencies from the Ministry of Food and Agriculture based on Turkish Food Codex, Directory of Energy Drinks (2017/4, published in 30 June 2017 with No:30110 in Official Gazzette); where the list of penalty firms are publicly listed.

The product properties are described in the Turkish Codex as follows: energy content no less than 45 kcal per 100 mL, caffeine up to 150 mg/L, inositol up to 100 mg/L, glucuronolactone up to 20 mg/L, and taurine up to 800 mg/L.

Regulation of nutraceuticals in Turkey

As a country at the stage of accession to the EU, Turkey signed the EU Food Legislation. After both parties (EU and Turkey) arrived at a consensus, the Turkish food legislation was changed and the implementation phase began. On this occasion, to guarantee the implementation of food safety (from farm to fork) in Turkey, the EU Food Legislation (EC 178/2002) is expected to be fully adopted.

The main target of the Turkish food and agriculture policy is to harmonize the related laws and regulations with the EU acquis communautaire. In the scope of EU harmonization, the Turkish Government issued a new Law no. 5996 on "Veterinary Services, Phytosanitary, Food and Feed" on June 13, 2010 (Turkish Official Gazette. No. 27610) with the objective of protecting and ensuring public health, food and feed safety, animal health and welfare, plant health, and consumer interests taking into consideration environmental protection. In the past, veterinary services, phytosanitary, food, and feed policies were covered by separate laws and regulations. Unlike the old bills, Law 5996 related to veterinary services, plant wealth, food and bait requirements, conditions for food safety, food codex, food hygiene, risk analysis, and traceability are some concepts that have to be applied in the food industry. For the implementation of law 5996, the Ministry of Agriculture and Rural Affairs was extensively reorganized by Decree Law No. 639 on June 8, 2011 (Turkish Official Gazette No. 27958). The Decree Law states that the Ministry has duties and responsibilities such as certification, inspection, capacity building and data collection in the field of agricultural development, agricultural land development, animal husbandry, pastoral farming, rural development, water management, plant production, fisheries, food safety and security, biodiversity conservation and environmental protection. There are five public and one semipublic institutions with relevance to food quality and safety in Turkey (the Ministry of Agriculture and Forestry, the Turkish Standards Institution, the Turkish Patent Institute, the Under-Secretary for Foreign Trade and the Turkish Accreditation Agency) (Koc et al., 2011).

The Law gives an authority for the principles and procedures related to the production, packaging, sale, import, and export of spring water, drinking water, natural mineral water, and water for medical purposes and the principles and procedures related to compliance with technical and hygienic rules, ensuring fulfillment of quality standards and monitoring and control of quality standards of potable and utility water to the Ministry of Health. The Ministry of Health is also responsible for the principles and procedures related to the production, import, export, and control of dietary foods for special medical purposes.

Currently, the Ministry of Agriculture and Forestry is the competent authority with regard to food and feed safety, veterinary, and phytosanitary issues in Turkey. Under this Ministry, the General Directorate of Food and Control (GDFC) is the most important ministerial structure for food and feed safety, veterinary and phytosanitary policies, adoption of legislation related to these policies, relevant official controls and inspections at all stages of processing, distribution and placing on the market, as well as ensuring animal health and welfare. The GDFC is the contact point for international organizations such as Codex Alimentarius Commission, European Food Safety Authority, European and Mediterranean Plant Protection Organization, World Organization for Animal Health, World Trade Organization, etc.

In Turkey, nutraceuticals are listed as "food supplements" in the Food Law 5996. This covers vitamins, minerals, other substances such as AA, essential fatty acids, various plants and herbal extracts, whether alone or in mixtures. The use and advertising of food supplements has become widespread, especially in the recent years. Under the Turkish Food Law 5996 and the Food Labeling Regulation, which also apply to food supplements, the advertisements of food and food supplements should not be misleading and should not include health claims, namely the implication of treatment and diagnosis of a disease or health disorder. Other than this, there is no specific regulation for the advertising of food supplements.

From the increase in deaths of people using some kind of food supplements—specially for weight loss or control purposes—in recent years, the Ministry of Health, Radio and Television Supreme Council (RTUK) and the Advertisement Board (within the body of the Ministry of Customs and Trade) have decided to collaborate with the Ministry of Agriculture and Forestry to combat the use of misleading information and health claims in advertisements for food supplements in Turkey. The collaboration seems to be very effectively, where the Advertisement Board and RTUK imposed some important heavy sanctions against advertisers and media channels regarding misleading food supplement advertisements.

In addition to this, "Regulation on the import, export, production, processing of food supplements and their placing on the market" was published in the Turkish Official Gazette (No. 28635) on May 2, 2013 by the Ministry. This Regulation sets forth the rules and procedures of import, export, production, processing, and approval of food supplements, their placing on the market and traceability. It also sets forth the rules concerning the responsibilities of enterprises and their registration. The Regulation further establishes a Food Supplements Commission composed of officers of the Ministry of Agriculture and Forestry and officers of the Ministry of Health. It is the first regulation specifically on food supplements in Turkey. Although it does not include any new advertising-related provisions other than those in the Food Law and the Food Labeling Regulation, it brings important provisions for the control of the food supplement market. The most noteworthy development coming within the Regulation is the introduction of the Food Supplement Tracking System which appears to be a very efficient system for monitoring food supplements, from the production stage to the point of sale, with the help of computer programs, databases, and communication infrastructures. A list of food supplements licensed up to March 1, 2020 by the Ministry of Agriculture and Forestry following the latest "Turkish Food Codex—Communiqué on Food Supplements" (Turkish Official Gazette, No. 28737, 16 August 2013) is shown in Table 78.1.

In order to decrease the export of medicinal/aromatic/ornamental plants, a directive called the National Property General Directive (Turkish Official Gazette, No. 30250, 24 November 2017) was published by the Ministry of Finance. This Directive determines the procedures and principles regarding the leasing of treasury immovable properties in order to grow medicinal, aromatic, and ornamental plants, where a list of 131 plants is included.

To conclude, the government is determined to regulate and control the food supplement market as well as to make the public aware of the correct use of food supplements. Not only does the Regulation introduce very important obligations for food supplement producers, but also the collaboration of governmental authorities seems to bring an efficient control over the producer firms and to raise awareness among the public.

The implementation of food supplements began with the "Turkish Food Codex—Communiqué on Food Supplements" (No. 2013/49), which largely reflects the existing acquis (Turkish Official Gazette, No. 28737, 16 August 2013). This communiqué set the regulations for dietary/food supplements and is completely appropriate to 2002/46/EC. It states that "food supplements means foodstuffs the purpose of which is to supplement the normal diet and which are concentrated sources of nutrients or other substances with a nutritional or physiological effect, alone or in combination, marketed in dose form, namely forms such as capsules, pastilles, tablets, pills and other similar forms, sachets of powder, ampoules of liquids, drop dispensing bottles and other similar forms of liquids and powders designed to be taken in measured small unit quantities." A number of elements from that definition are important to ascertain the products that fall under the scope. An important element is that food supplements are considered as foodstuffs. This means that Communiqué 2013/49 that cover foodstuffs is applicable to food supplements, except where specifically provided for in the food supplements legislation itself. This does not completely resolve the issue of the borderline between medicinal products and food supplements, given that specific compositional provisions are still missing, but it does indicate that when they are in line with the definition, such products should follow food law.

The Communiqué requires supplements to be demonstrated as safe, both in quantity and in quality, and also available to be used by the body. The Communiqué states that the upper safe level of vitamins and minerals established by scientific risk assessment based on generally scientific data should be taken into account, together with recommended dietary allowances of different consumer groups. Some vitamins are essential in small quantities but dangerous in large quantities, notably vitamin A. Consequently, only those supplements that have been proven to be safe may be sold without prescription. In practice, however, there appears to be little risk to supplement users experiencing adverse side effects due to excessive intake of micronutrients.

Turkish regulations concern food supplements marketed as foodstuffs and do not define them as drugs. In Turkey, the established view is that food supplements should not be labeled with drug claims: "The labeling, presentation, and advertising must not attribute to food supplements the property of preventing, treating, or curing a human disease," but can bear health properties.

In order to ensure a high level of consumer protection and to facilitate their choices, products to be placed on the market must be safe and bear adequate and appropriate labeling. The labeling shall include the following particulars:

1. The names of the categories of nutrients or substances that characterize the product or an indication of the nature of those nutrients or substances;
2. The dose of the product recommended for daily consumption;
3. A warning not to exceed the recommended daily dose;
4. A statement explaining that food supplements should not be used as a substitute for a varied diet.

The amount of the nutrients or substances with a nutritional or physiological effect present in the product shall be declared on the labeling in numerical form. In the case

TABLE 78.1 List of food supplements licensed by the Ministry of Food, Agriculture, and Livestock from January 13, 2014, to May 4, 2015.

	Number of licenced products	Product name	Main content	Main nutraceutical benefit
1	2	Shavergrass powder	*Equisetum arvense*	Silisium supplement for antiaging, antihypertensive, antiulcerative
2	2	Vitamin C	Vitamin C	Antioxidant, general health supplement
3	4	Coriander oil	*Coriandrum sativum*	Antiaging, antiinflammatory, antimicrobial, diabetes, sexual enhancement
4	3	Selenium	Selenium	Antioxidant, general health supplement, cancer prevention
5	2	Pomegranate fruit oil	*Punica granatum*	Antioxidant, anticholesterol
6	2	Black cumin oil	*Nigella sativa*	Antiaging, anticancer, gastrointestinal health, antimigraine, antidiabetic
7	2	Wheat germ oil	*Triticum vulgare*	Antioxidant
8	4	Ginger extract	*Zingiber officinale*	Antiemetic, cancer prevention, antidiabetic
9	7	Curcumin (turmeric) extract	*Curcuma longa*	Musculoskeletal disease prevention, antiinflammatory
10	3	Saw palmetto berry extract	*Serenoa repens*	Gastrointestinal health, chronic cough soother, cancer prevention
11	1	TA 65	Cycloastragenol	Antiaging
12	3	Folic acid	Folic acid	Pregnancy, nervous system
13	2	Woundwort extract	*Lavandula stoechas*	Gastrointestinal health, antinociceptive, expectorant
14	4	St. John's wort	*Hypericum perforatum*	Mental health (antidepression)
15	3	Ginseng extract	*Panax ginseng*	Mental health, cardiovascular system
16	2	Fenugreek seed extract	*Trigonell foenum-graecum*	Menopause symptomatic treatment
17	2	Broccoli extract	*Brassica oleracea* var. *italica*	Anticancer, treatment of autism
18	3	Garlic extract	*Allium sativum*	Antiinflammatory, cardiovascular system, cancer prevention
19	4	Reishi mushroom extract	*Ganoderma lucidum*	Immune system enhancing, cancer prevention, antidiabetic
20	2	Spirulina extract	*Arthrospira platensis*	Weight control, antioxidant
21	3	Olive leaf extract	*Olea europaea*	Antioxidant, antidiabetic, immune system enhancing
22	1	Vitex fruit extract	*Vitex agnus-castu*	Menopause symptomatic treatment, menstrual cycle regulation
23	2	Devil's weed extract	*Tribulus terrestris*	Male virility and general vitality
24	1	Brown seaweed	*Ascophyllum nodosum*	Antiaging, gastrointestinal health, general vitality

(*Continued*)

TABLE 78.1 (Continued)

	Number of licenced products	Product name	Main content	Main nutraceutical benefit
25	1	Large cranberry extract	*Vaccinium macrocarpon*	Urinary tract infections
26	9	Probiotic	Several probiotic microorganisms in different combinations/ concentrations	Gastrointestinal health, general vitality
27	1	*Chlorella* algae extract	*Chlorella pyrenoidosa*	Cancer prevention, stimulate immune system
28	2	Flax seed oil	*Linum usitatissimum*	Omega-3 source, gastrointestinal support, cardiovascular disease prevention, blood sugar control
29	1	Chitosan capsule	Chitosan	Weight control
30	2	Magnesium	Magnesium	Bone–muscle health, gastrointestinal health, cardiovascular health
31	7	Fish oil	Omega-3	Cardiovascular health, nervous system health, antioxidant
32	1	Fish oil and krill oil mix	Omega-3	Cardiovascular health, nervous system health, antioxidant
33	1	Red krill oil (*Euphausia superba*)	Omega-3	Cardiovascular health, nervous system health, antioxidant
34	8	Vitamin and mineral	Vitamin–mineral complex	Support immune system
35	2	Fig extract	*Ficus carica*	Gastrointestinal health (pediatric)
36	2	Choline stabilizing ortosilisic acid	Ortosilisic acid	Atherosclerosis, Alzheimer disease, immune system enhancement, collagen synthesis
37	3	Milk thistle extract	*Sylibum marianum*	Hepatoprotective
38	4	Propolis support	Bee resin	Antioxidant
39	4	Artichoke leaf extract	*Cynara scolymus*	Reduction of cholesterol levels, gastrointestinal health
40	1	Salai extract	*Boswellia serrata*	Antiinflammatory
41	4	Nettle extract	*Urtica dioica*	Urinary health, arthritis
42	2	Vegetable oil support	Mixture of several natural vegetable oils	Gastrointestinal health
43	2	Devil's claw extract	*Harpagophytum procumbens*	Cardiovascular health
44	1	Prebiotic fiber	Polydextrose	Gastrointestinal health
45	1	Narrowleaf plantain liquid extract	*Plantago lanceolata*	Wound healing

(*Continued*)

TABLE 78.1 (Continued)

	Number of licenced products	Product name	Main content	Main nutraceutical benefit
46	1	Licorice root liquid extract	*Glycyrrhiza glabra*	Gastrointestinal health, antipyretic, antitussive, mucolytic
47	1	Rosemary liquid extract	*Rosmarinus officinalis*	Analgesic, diuretic, strengthening
48	2	Ginkgo liquid extract	*Ginkgo biloba*	Antioxidant
49	1	Black grape seed liquid extract	*Vitis vinifera*	Antioxidant
50	1	White peony extract	*Paeonia officinalis*	Antihemorrhoidal and treatment of anal fissure
51	1	NemolixC3	Eggshell membrane, sunflower seed oil, bovine gelatin, frankincense, turmeric	Antiaging, antioxidant
52	1	Bee pollen	Bee pollen	Antioxidant
53	1	Pycnogenol (pine bark extract)	Pycnogenol	Antioxidant
54	1	Glucosamine sulfate	Glucosamine sulfate	Antirheumatic
55	1	Thyme extract	*Thymus vulgaris*	Respiratory health, antiinflammatory
56	1	Coenzyme Q-10	Coenzyme Q-10	Cardiovascular health, cancer treatment, antioxidant
57	2	Pumpkin seed oil	*Cucurbita pepo*	Antiinflammatory, urinary health, gastrointestinal system health
58	2	Carob tree sirup	*Ceratonia siliqua*	Gastrointestinal health, weight management

of plant/herbal products, the label must contain the following information: plant name and origin, material origin, plant seed, preparation model, activity, biological marker, toxicology, pollution, and side effects.

The regulatory framework on food supplements is fairly recent. Such products have remained under national law until 2010, subject of course to the principle of mutual recognition that proved rather ineffective to assure free movement of such products. Harmonization would only be undertaken for the elements that are essential to protect consumer health and provide correct information. The principle says that a product, lawfully manufactured in one Member State should be able to move within Turkey. The only exceptions to this principle are on grounds of public morality, public policy, or public security; the protection of health and life of humans, animals, or plants; the protection of national treasures possessing artistic, historic, or archeological value; or the protection of industrial and commercial property. Such prohibitions or restrictions must not constitute a means of arbitrary discrimination or a disguised restriction on trade between EU and Turkey.

An important element of this Communiqué is that food supplements may contain substances with a nutritional or physiological effect. Until that point, it was not legally recognized that substances possessing physiological, but nonmedicinal properties could be regulated outside medicinal law, although this was the practical reality in Turkey. Another element is the fact that in order to fall under the food supplements law the product must be a concentrated form, presented to be taken in measured small unit quantities.

The Communiqué foresees that maximum amounts of vitamins and minerals present in food supplements shall be set, taking into account the upper safe levels of vitamins and minerals established by scientific risk assessment based on generally accepted scientific data, taking into account, as appropriate, the varying degrees of sensitivity of different consumer groups; and intake of vitamins and minerals from other dietary sources. It also specifies that when the maximum levels will be set, due account should also be taken of reference intakes of vitamins and minerals for the population.

This is a good example of the application of the risk analysis process. In a first phase the upper tolerable safe intake levels (UL) of the vitamins and minerals are set by the Food Supplements Commission. The UL is the intake level that can be consumed daily over a lifetime without being likely to pose any risk to health according to available evidence. It guides the risk assessor in characterizing the risk associated with each individual vitamin and mineral. In a second phase the risk managers consider the risk assessment. They will have to determine what maximum levels will be established for use in food supplements based on the established UL and the intake of the vitamins and minerals from other food sources including water.

On the other hand, the Turkish Food Codex Regulation on nutrition and health claims made on food (Turkish Official Gazette, No. 29960, 26 January 2017) determines the rules regarding the nutritional and health declarations in foods supplied to the end consumer in order to protect the consumer at the highest level. Standards covering the rules regarding the nutrition and health declarations used in labeling, presentation, or advertisement of the food provided to the end consumer and mass consumption places are included in this directorate in detail. Therefore consumers are protected from nutritional and health claims that can be ambiguous, false, or misleading to cause suspicion of nutritional efficiency or reliability along with encouragement of excessive consumption of a certain food.

Concluding remarks and future directions

Due to its geographical location and cultural heritage, Turkish cuisine has been influenced by Europe, Asia, the Middle East, and Africa, where the fermented foods and probiotics (such as yogurt) have always been a part of daily life. As nutraceuticals and functional foods, a large sector base with a great trend of consumers toward healthier food wishes in Turkey, products with several acclaimed health benefits (energy balance, cancer prevention, cardiovascular disease, mental state and performance, bone protection/musculoskeletal disease prevention, weight management/obesity, pregnancy, child development, gastrointestinal health, oxidative stress, etc.) are currently on the market.

Turkey has a wide endemic flora, compromising many genetic resources that are potential sources of nutraceuticals and phytopharmaceuticals. In the near future, not only the growing interest in phytomedicine by multidisciplinary scientific fields, but also the potential interest of consumers would force governmental authorities to expand the research areas and adjust regulations, which would also add standardization and quality to plant and other animal source materials.

Therefore herbals should be carefully studied, evaluated, and viewed from various standpoints including studies of pharmacokinetics and bioavailability. Research and industry should work together for clarification to draw an entire picture of the chemistry, efficacy, and safety of herbs as the major constituent of nutraceuticals. Advertising through the Internet should be restricted to avoid any misguidance of consumers.

Regulations regarding the quality and safety of nutraceuticals and functional foods are expected to be formed in accordance with the local needs and not only as a reflection of EU commodities. For this purpose, the Ministry of Health and Ministry of Agriculture and Forestry should work together to create an updated list of nutraceuticals and food supplements and with transparency publish the results for public commodities. Distribution and marketing strategies should be clarified and the public should be enlightened about the benefits.

The traditional Turkish ethnopharmacology wisdom, along with the traditional probiotic habits, add value and create a good market for nutraceuticals. With the increased awareness of the public and the sustained strategies for developing nutraceuticals, such as improvement of scientific and technological capacities, Turkey is expected to be a good leader for coordinating and networking nutraceutical studies globally due to its unique strategic position.

References

Akbulut, S., Bayramoglu, M., 2013. The trade and use of some medical and aromatic herbs in Turkey. Ethno Med. 7, 67–77.

Alsayadi, M., Jawfi, Y., Belarbi, M., et al., 2014. Evaluation of anti-hyperglycemic and anti-hyperlipidemic activities of water kefir as probiotic on streptozotocin-induced diabetic Wistar rats. J. Diabetes Mellit. 4, 85–95.

Altay, F., Karbancioglu-Guler, F., Daskaya-Dikmen, C., et al., 2013. A review on traditional Turkish fermented non-alcoholic beverages: microbiota, fermentation process and quality characteristics. Int. J. Food Microbiol. 167, 44–56.

American Pregnancy, 2014. Omega-3 fish oil and pregnancy. <http://americanpregnancy.org/pregnancy-health/omega-3-fish-oil/> (Accessed 20 April 2015).

Arslan, A., Durak, A.N., Erbas, M., et al., 2015. Determination of microbiological and chemical properties of probiotic boza and its consumer acceptability. J. Am. Coll. Nutr. 34, 56–64.

Ates, H.C., Ceylan, M., 2010. Effects of socio-economic factors on the consumption of milk, yoghurt and cheese. Insights from Turkey. Br. Food J. 112, 234–250.

Bekar, O., Yilmaz, Y., Gulten, M., 2011. Kefir improves the efficacy and tolerability of triple therapy in eradicating *Helicobacter pylori*. J. Med. Food 14, 344–347.

Bilgin A., Samanci Ö., 2008. *Turkish Cuisine*. Ankara: Ministry of Culture and Tourism, 392. <http://www.muslimheritage.com/article/turkish-cuisine-book-review> (Accessed 20 April 2015).

Bostancı-Ege, G., Onder-Erol, P., 2012. The resurgence of herbal culture in Turkish society: and analysis on the social phenomenon of "Aktars". Milli Folklor 95, 315–332.

Cakmakci, S., Caglar, A., Turkoglu, H., 2006. Yoğurdun insan beslenmesindeki rolü ve önemi. Stand. ve Ekon. Derg. 384, 29–35 [article in Turkish].

Calder, P.C., 2015. Marine omega-3 fatty acids and inflammatory process: effects, mechanisms and clinical relevance. BBA- Biochim. Biophys. Acta 1851, 469–484.

Davis, P.H., 1988. Flora of Turkey and the East Aegean Islands, Vol. 10. Edinburgh University Press, Edinburgh.

Decree Law No. 639 on the establishment and functions of the Ministry of Food Agriculture and Livestock. Turkish Official Gazette No. 27958, 8 June 2011.

DFDG, 2012. Diyabetik ve Fonksiyonel Gıda Üreticileri Derneği. <http://www.dfgd.org.tr/> (Accessed 20 April 2015).

Dogan, I.S., Yildiz, O., Eyduran, E., et al., 2011. A study on determination of functional food consumption habits and awareness of consumers in Turkey. Bulg. J. Agric. Sci. 17, 246–257.

Durak, M., 2013. Current situation of dairy products and brands in Turkey. Brandtalks. <http://brandtalks.org/2013/08/turkiyede-sut-urunleri-ve-markalari-ne-durumda/> (Accessed 20 April 2015). [article in Turkish].

Erten, H., Tanguler, H., Canbas, A., 2008. A traditional Turkish lactic acid fermented beverage: shalgam (Salgam). Food Rev. Int. 24, 352–359.

Erturk, O., Tas, B., 2011. Antibacterial and antifungal effects of some marine algae. Kafkas Univ. Vet. Fak. Derg. 17 (Suppl A), S121–S124.

Esiyok, D., Otles, S., Akcicek, E., 2004. Herbs as food source in Turkey. Asian Pac. J. Cancer Prev. 5, 334–339.

Etimolojiturkce, 2013. Etimoloji Türkçe: Kefir. <http://www.etimolojiturkce.com/kelime/kefir> (Accessed 20 April 2015). [article in Turkish].

Euromonitor, 2015. Euromonitor International. Vitamins and dietary supplements in Turkey. Country report. <http://www.euromonitor.com/vitamins-and-dietary-supplements-in-turkey/report>. (Accessed 20 April 2015).

Finance, 2013. Healthy profit from fish oil. <http://finans.mynet.com/haber/iphone/foto-analiz/baligin-yagindan-cikan-saglikli-kr/90390>. (Accessed 20 April 2015). [article in Turkish].

Grishina, A., Kulikova, I., Alieva, K., et al., 2011. Antigenotoxic effect of kefir and ayran supernatants on fecal water-induced DNA damage in human colon cells. Nutr. Cancer 63, 73–79.

Guzel-Seydim, Z.B., Kok-Tas, T., Greene, A.K., et al., 2011. Review: functional properties of kefir. Crit. Rev. Food Sci. Nutr. 51, 261–268.

Haliki, A., Denizci, A.A., Cetingul, V., 2005. An investigation on antifungal activities of some marine algae (Phaeophyta, Rhodophyta). Eur. J. Fish. Aquat. Sci. 22, 13–15.

Holzapfel, W., 2015. Advances in Fermented Foods and Beverages: Improving Quality Technologies and Health Benefits. Woodhead Publishing, Cambridge.

Hurriyet, 2013. Ayran consumption increased 83 percent over 6 years. <http://www.hurriyet.com.tr/ekonomi/23170145.asp>. [Article in national gazette in Turkish].

Inanc, N., Cicek, B., Sahin, H., et al., 2007. Use of herbs by the parients with diabetes in Kayseri, Turkey. Pak. J. Nutr. 6, 310–312.

Kabak, B., Dobson, A.D., 2011. An introduction to the traditional fermented foods and beverages of Turkey. Crit. Rev. Food Sci. Nutr. 51, 248–260.

Kancabaş, A., Karakaya, S., 2013. Angiotensin-converting enzyme (ACE)-inhibitory activity of boza, a traditional fermented beverage. J. Sci. Food Agric. 93, 641–645.

Karabay-Yavasoglu, N.U., Sukatar, A., Ozdemir, G., et al., 2007. Antimicrobial activity of volatile components and various extracts of the red alga *Jania rubens*. Phytother. Res. 21, 153–156.

Kav, T., 2009. Use of complementary and alternative medicine: a survey in Turkish gastroenterology patients. BMC Complement. Altern. Med. 9, 41.

Kim, S.-K., 2013. Marine Nutraceuticals: Prospects and Perspectives. CRC Press, Taylor & Francis Group, Boca Raton, FL.

Kivanc, M., Yapici, E., 2019. Survival of *Escherichia coli* O157:H7 and *Staphylococcus aureus* during the fermentation and storage of kefir. Food Sci. Technol. 39 ((supl.1), 225–230.

Koc, A.A., Asci, S., Alpas, H., et al., 2011. Food Quality Assurance Schemes in Turkey. JRC Scientific and Technical Reports. EUR 24672 EN. Luxembourg: Publications Office of the European Union.

Koc, B., Gul, A., Akpinar, M.G., et al., 2014. Determining consumers' preferences for energy drinks consumption with conjoint analysis: a cross section study from East Mediterranean, Turkey. J. Nutr. Food Sci. 4, 324.

Koksoy, A., Kilic, M., 2003. Effects of water and salt level on rheological properties of ayran, a Turkish yoghurt drink. Int. Dairy. J. 13, 835–839.

Law No. 5996 on veterinary services, plant health, food and feed. Turkish Official Gazette. No. 27610, 13 June 2010.

Mahmoudi, E., 2014. Effect of *Mentha longifolia* L. essential oil on physicochemical properties of the bio-ayran. J. Essent. Oil Bear. Plants 17, 56–66.

Ministry of Health, 2014. Turkish Nutrition and Health Research. Report on Nutrition Status and Nutirion Habits 2010. Hacettepe University Faculty of Health Sciences, Ankara, Turkey. [Article in Turkish].

Nicoletti, M., 2012. Nutraceuticals and botanicals: overview and perspectives. Int. J. Food Sci. Nutr. 63 (Suppl 1), 2–6.

Ogur, R., Korkmaz, A., Bakır, B., 2006. Herbal treatment usage frequency, types and preferences in Turkey. Middle East J. Family Med. 4, 38–44.

Özgen, U., Kaya, Y., Houghton, P., 2012. Folk medicines in the villages of Ilıca District (Erzurum, Turkey). Turk. J. Biol. 36, 93–106.

Regulation on the import, export, production, processing of food supplements and their placing on the market. Turkish Official Gazette No. 28635, 2 May 2013.

Sanli, T., Sezgin, E., Senel, E., et al., 2013. The effect of transglutaminase on some physicochemical and sensory properties of the Turkish drinking yoghurt Ayran. Int. J. Dairy Technol. 66, 410–416.

Soner, B.C., Sahin, A.S., Sahin, T.K., 2013. A survey of Turkish hospital patients' use of herbal medicine. Eur. J. Integr. Med. 5, 547–552.

Suzgec-Selcuk, S., Eyisan, S., 2012. Türkiye'deki eczanelerde bulunan bitkisel ilaçlar. Marmara Pharm. J. 16, 164–180 [Article in Turkish].

Swain, M.R., Anandharaj, M., Ray, R.C., et al., 2014. Fermented fruits and vegetables of Asia: a potential source of probiotics. Biotechnol. Res. Int. 2014, 250424. Available from: https://doi.org/10.1155/2014/250424.

Tanguler, H., 2014. Traditional Turkish fermented cereal based products: Tarhana, Boza and Chickpea Bread. Turk. J. Agric. Food Sci. Technol. 2, 144–149.

Tanguler, H., Erten, H., 2012. Occurrence and growth of lactic acid bacteria species during the fermentation of shalgam (salgam), a traditional Turkish fermented beverage. LWT Food Sci. Technol. 46, 36–41.

Taskin, E., Ozturk, M., Taskin, E., et al., 2007. Antibacterial activities of some marine algae from the Aegean Sea (Turkey). Afr. J. Biotechnol. 6, 2746–2751.

Tazehkand, M.N., Valipour, E., 2019. Cytotoxic, antimicrobial and DNA breaking activity of Salgam. Indian J. Biochem. Biophys. 56, 69–174.

Tuney, I., Cadirci, B.H., Unal, D., et al., 2007. In antimicrobial activities of crude extracts of marıne algae from the coast of Izmir (Turkey). Fres Environ. Bull. 16, 428–434.

Turker, N., Aksay, S., Ekiz, I., 2004. Effect of stırage temperature on the stability of anthocyanins of a fermented black carrot (*Daucus carota* var. L.) beverage: shalgam. J. Agric. Food Chem. 52, 3807–3813.

Turkish Food Codex—Communiqué on Food Supplements (No: 2013/49), Turkish Official Gazette No. 28737, 16 August 2013.

Turkish Food Codex Regulation on Nutrition and Health Claims made on Food (Turkish Official Gazette, No. 29960, 26 January 2017).

Ugurlu, S., 2011. Traditional folk medicine in the Turkish folk culture. Turkish Stud. 6, 317–327.

Vella, M.N., Stratton, L.M., Sheeshka, J., et al., 2014. Functional food awareness and perceptions in relation to information sources in older adults. Nutr. J. 13, 44.

Wang, C., Li, S., 2015. Functional foods and nutraceuticals: potential role in human health. In: Ghosh, D., Bagchi, D., Konishi, R. (Eds.), Clinical Aspects of Functional Foods and Nutraceuticals. CRC Press, Taylor & Francis Group, Boca Raton, FL.

WHO, 1998. Bulletin of the World Health Organization. (document WHO/TRM/98.1). A Regulatory Situation of Herbal Medicines—A Worldwide Review. World Health Organization, Geneva.

Yilmaz, I., Dolar, M.E., Özpınar, H., 2019. Effect of administering kefir on the changes in fecal microbiota and symptoms of inflammatory bowel disease: a randomized controlled trial. Turk. J. Gastroenterol. 30 (3), 242–253.

Yurdakök, M., 2013. History of probiotics and yogurt. Çocuk Sağlığı ve Hastalıkları Derg. 56, 43–60.

Index

Note: Page numbers followed by "*f*" and "*t*" refer to figures and tables, respectively.

C

D

E

F

G

I

J

K

O

P

Q

R

S

T

U

V

W

X

Y

Z